CHEMICAL PROPERTIES HANDBOOK

CHEMICAL ENGINEERING BOOKS

CHEMICAL PROPERTIES HANDBOOK

PHYSICAL, THERMODYNAMIC, ENVIRONMENTAL, TRANSPORT, SAFETY,
AND HEALTH RELATED PROPERTIES FOR ORGANIC AND INORGANIC CHEMICALS

CARL L. YAWS

PROFESSOR OF CHEMICAL ENGINEERING
LAMAR UNIVERSITY
BEAUMONT, TEXAS

McGRAW-HILL

New York San Francisco Washington, D.C. Auckland Bogotá
Caracas Lisbon London Madrid Mexico City Milan
Montreal New Delhi San Juan Singapore
Sydney Tokyo Toronto

Library of Congress Cataloging-in-Publication Data

Yaws, Carl L.
    Chemical properties handbook : physical, thermodynamic,
environmental, transport, safety, and health related properties for
organic and inorganic chemicals / Carl L. Yaws.
        p. cm.
    Includes index.
    ISBN 0-07-073401-1 (acid-free paper)
    1. Chemicals—Handbooks, manuals, etc. 2. Chemicals—Safety
measures—Handbooks, manuals, etc. I. Title.
TP200.Y35 1999
660'.02'1—dc21   98-39275
                                                                        CIP

McGraw-Hill

A Division of The McGraw·Hill Companies

    2 3 4 5 6 7 8 9 0   KGP/KGP   9 0 3 2 1 0 9

ISBN 0-07-073401-1

*The sponsoring editor for this book was Robert Esposito, the editing
supervisor was Paul R. Sobel, and the production supervisor was
Modestine Cameron.*

Printed and bound by Quebecor/Kingsport.

McGraw-Hill books are available at special quantity discounts to use
as premiums and sales promotions, or for use in corporate training
programs. For more information, please write to the Director of
Special Sales, McGraw-Hill, 11 West 19th Street, New York, NY
10011. Or contact your local bookstore.

This book was printed on acid-free paper.

CONTENTS

CONTRIBUTORS

Li Bu — Process Engineer, Great Lakes Chemical Corporation, 3324 Chelsea Ave., P.O. Box 80035, Memphis, Tennessee 38108, U S A

Deepa R. Balundgi — Graduate Student, Chemical Engineering Department, Lamar University, P.O. Box 10053, Beaumont, Texas 77710, U S A

Daniel H. Chen — Professor, Chemical Engineering Department, Lamar University, P.O. Box 10053, Beaumont, Texas 77710, U S A

Mei Han — Process Engineer, Firestone Synthetic Rubber & Latex Company, P.O. Box 1269, Orange, Texas 77630, U S A

Jack. R. Hopper — Professor, Chemical Engineering Department, Lamar University, P. O. Box 10053, Beaumont, Texas 77710, U S A

Eric L. Jaycox — Industrial Hygienist, MS CIH, Huntsman Chemical, Port Neches, Texas 77651, U S A

Xiaoyan Lin — Chemical Process Engineer, Thermal Recovery, Inc., 5740 W. Little York #1011, Houston, Texas 77091, U S A

Sachin Nijhawan — Design Engineer (Process), Mustang Engineering Inc., 16001 Park Ten Place, Houston, Texas 77710, U S A

Ralph W. Pike — Professor, Chemical Engineering Department, Louisiana State University, Baton Rouge, Louisiana 70803-7303, U S A

Marco A. Satyro — Scientific Software Specialist, SEA++ INC., 82 Hawkwood Rd. N. W., Calgary, Alberta, Canada T3G 2J1

Sachin Sheth — Technical Support Manager, EPCON International, 16360 Park Ten Place, Houston, Texas 77084, U S A

Sivakumar Srinivasan — Graduate Student, Chemical Engineering Department, Lamar University, P.O. Box 10053, Beaumont, Texas 77710, U S A

Saumya Tripathi — Graduate Student, Chemical Engineering Department, Lamar University, P.O. Box 10053, Beaumont, Texas 77710, U S A

Xiao M. Wang — Graduate Student, Chemical Engineering Department, Lamar University, P.O. Box 10053, Beaumont, Texas 77710, U S A

Carl L. Yaws — Professor, Chemical Engineering Department, Lamar University, P. O. Box 10053, Beaumont, Texas 77710, U S A

ACKNOWLEDGMENTS

The author wishes to acknowledge special appreciation to his wife (Annette) and family (Kent, Michele, Chelsea, Brandon, Lindsay, Rebecca, Chloe, and Sarah).

Many colleagues and students have made contributions and helpful comments over the years. The author is grateful to each: Jack R. Hopper, Joe W. Miller, Jr., C. S. Fang, K. Y. Li, Keith C. Hansen, Daniel H. Chen, P. Y. Chiang, H. C. Yang, Xiang Pan, Xiaoyan Lin, Li Bu, Sachin Nijhawan, Mei Han, Sachin Sheth, Deepa R. Balundgi, Sivakumar Srinivasan, Sauma Tripathi, Xiao M. Wang, and Marco A. Satyro.

The author wishes to acknowledge that the Gulf Coast Hazardous Substance Research Center provided partial support to this work.

Carl L. Yaws, Lamar University. Beaumont, Texas

DISCLAIMER

This handbook presents a variety of data for chemical properties. It is incumbent upon the user to execute judgement in the use of the data. The author does not provide any guarantee, expressed or implied, with regard to the general or specific applicability of the data, the range of errors that may be associated with any of the data, or the appropriateness of using any of the data in any subsequent calculation, design ,or decision process. The author accepts no responsibility for damages, if any, suffered by any reader or user of this handbook as a result of decisions made or actions taken on information contained therein.

Li Bu	Process Engineer, Great Lakes Chemical Corporation, 3324 Chelsea Ave., P.O. Box 80035, Memphis, Tennessee 38108, U S A
Deepa R. Balundgi	Graduate Student, Chemical Engineering Department, Lamar University, P.O. Box 10053, Beaumont, Texas 77710, U S A
Daniel H. Chen	Professor, Chemical Engineering Department, Lamar University, P.O. Box 10053, Beaumont, Texas 77710, U S A
Mei Han	Process Engineer, Firestone Synthetic Rubber & Latex Company, P.O. Box 1269, Orange, Texas 77630, U S A
Jack R. Hopper	Professor, Chemical Engineering Department, Lamar University, P.O. Box 10053, Beaumont, Texas 77710, U S A
Eric L. Jaycox	Industrial Hygienist, MS CIH, Huntsman Chemical, Port Neches, Texas 77651, U S A
Xiaoyan Lin	Chemical Process Engineer, Thermal Recovery, Inc., 5740 W. Little York #1011, Houston, Texas 77091, U S A
Sachin Nijhawan	Design Engineer (Process), Mustang Engineering Inc., 16001 Park Ten Place, Houston, Texas 77710, U S A
Ralph W. Pike	Professor, Chemical Engineering Department, Louisiana State University, Baton Rouge, Louisiana 70803-7303, U S A
Marco A. Satyro	Scientific Software Specialist, SEA++ INC., 82 Hawkwood Rd. N. W., Calgary, Alberta, Canada T3G 2J1
Sachin Sheth	Technical Support Manager, EPCON International, 16360 Park Ten Place, Houston, Texas 77084, U S A
Sivakumar Srinivasan	Graduate Student, Chemical Engineering Department, Lamar University, P.O. Box 10053, Beaumont, Texas 77710, U S A
Saumya Tripathi	Graduate Student, Chemical Engineering Department, Lamar University, P.O. Box 10053, Beaumont, Texas 77710, U S A
Xiao M. Wang	Graduate Student, Chemical Engineering Department, Lamar University, P.O. Box 10053, Beaumont, Texas 77710, U S A
Carl L. Yaws	Professor, Chemical Engineering Department, Lamar University, P.O. Box 10053, Beaumont, Texas 77710, U S A

ACKNOWLEDGMENTS

The author wishes to acknowledge special appreciation to his wife (Annette) and family (Kent, Michele, Chelsea, Brandon, Lindsay, Rebecca, Chloe, and Sarah).

Many colleagues and students have made contributions and helpful comments over the years. The author is grateful to each : Jack R. Hopper, Joe W. Miller, Jr. C.S. Fang, K.Y. Li, Kieth C. Hansen, Daniel H. Chen, P.Y. Chiang, H.C. Yang, Xiang Pan, Xiaoyan Lin, Li Bu, Sachin Nijhawan, Mei Han, Sachin Sheth, Deepa R. Balundgi, Sivakumar Srinivasan, Sauma Tripathi, Xiao M. Wang A. Satyro.

The author wishes to acknowledge that the Gulf Coast Hazardous Substance Research Center provided partial support this work.

Carl L. Yaws, Lamar University. Beaumont, Texas

DISCLAIMER

CHEMICAL PROPERTIES HANDBOOK

Chapter 1

CRITICAL PROPERTIES AND ACENTRIC FACTOR

Carl L. Yaws, Xiaoyan Lin, Li Bu, Deepa R. Balundgi, and Saumya Tripathi
Lamar University, Beaumont, Texas

ABSTRACT

Results for critical properties and acentric factor are presented for major organic and inorganic compounds. The critical properties include critical temperature, pressure, volume, density, and compressibility factor. The chemical formula, molecular weight, freezing point, and boiling point are also given. The results are displayed in easy-to-use tabulations which are especially applicable for rapid engineering usage with the personal computer or hand calculator. The chemicals encompass hydrocarbon, oxygen, nitrogen, halogen, silicon, sulfur, and other compound types.

INTRODUCTION

Physical and thermodynamic property data for organic and inorganic chemicals are of special value to engineers in the chemical processing and petroleum refining industries. The engineering design of process equipment often requires knowledge of such properties as heat capacity, enthalpy, density, viscosity, thermal conductivity, and others.

In this article, results are presented for critical properties and acentric factor, which are usable in corresponding states correlations to determine properties such as heat capacity, enthalpy, density, viscosity, and thermal conductivity. The results are intended for initial engineering studies and are presented in an easy-to-use tabular format which is especially applicable for rapid engineering usage with the personal computer oι hand calculator.

CRITICAL PROPERTIES AND ACENTRIC FACTOR

The results for critical properties and acentric factor are shown in Tables 1-1 and 1-2 for organic and inorganic compounds. The tabulations are based on both experimental data and estimated values.

In the data collection, a literature search was conducted to identify data source publications for organics (1-44) and inorganics (1-59). Both experimental values for the property under consideration and parameter values for estimation of the property are included in the source publications. The publications were screened and copies of appropriate data were made. These data were then keyed into the computer to provide a database of critical properties for compounds for which experimental data are available. The database also served as a basis to check the accuracy of the estimation method.

Upon completion of data collection, estimation of the critical properties and acentric factor for the remaining compounds was performed. For organic compounds, the group contribution method of Joback as given by Reid, Prausnitz, and Poling (29) was primarily used for the estimation of critical temperature (T_C), pressure (P_C), and volume (V_C).

For inorganic compounds, estimates of critical temperature were based on modifications of the Guldberg-Guye rule (11), Gates-Thodos method (11) and Grosse equation (11). Estimates of other critical constants and acentric factor were primarily based on extension of the vapor pressure curve and modifications of the Benson relation (11) and Herzog proposal (11). Very limited experimental data for critical constants and acentric factor are available for inorganic compounds and elements that are solids at room temperature. Thus, the estimates for these substances should be considered rough approximations in the absence of experimental data.

Critical density (ρ_C) was determined from dividing molecular weight by critical volume:

$$\rho_C = MW / V_C \qquad (1\text{-}1)$$

where ρ_C = critical density, g/cm^3
MW = molecular weight, g/mol
V_C = critical volume, cm^3/mol

Critical compressibility factor (Z_C) was ascertained from applying the gas law at the critical point:

$$Z_C = P_C V_C / R T_C \qquad (1\text{-}2)$$

For many of the compounds, the acentric factor (ω) was estimated by the following equation which is given in Reid, Prausnitz, and Poling (29):

$$\omega = \frac{3}{7} \frac{T_B / T_C}{1 - T_B / T_C} (\log P_C) - 1 \tag{1-3}$$

where ω = acentric factor

T_B = boiling point temperature, K

T_C = critical temperature, K

P_C = critical pressure, atm

This equation for acentric factor is based on extending the vapor pressure by the Antoine type relation.

Comparisons of estimates and data for critical temperature are shown in Figs. 1-1 and 1-2 for normal alkanes and elements. Both graphs disclose favorable agreement of estimates and data.

A comparison of the estimates with experimental data was favorable for the group contribution method of Joback for organic compounds. Average absolute errors of 0.9%, 6.3%, 4.4%, and 4.6% were experienced for critical temperature (465 compounds), pressure (453 compounds), volume (345 compounds), and compressibility factor (348 compounds). Average absolute error for acentric factor (277 compounds) was about 6%.

The normal boiling (T_B) and freezing (T_F) point temperatures are also given in the table. For most compounds, data are available. For the compounds without data, the group contribution method of Joback (29) was used to estimate the boiling and freezing point temperatures for organic compounds. As discussed by Reid, Prausnitz, and Poling (29), no reliable methods are available for precise estimation of freezing point temperature. Thus, the estimates for freezing point temperature should be considered as rough approximations.

Portions of this material appeared in Hydrocarbon Processing, 68, 61 (July 1989) and are reprinted by special permission.

REFERENCES - ORGANIC COMPOUNDS

1. TRC THERMODYNAMIC TABLES - HYDROCARBONS, Vols. I - XIII, Thermodynamics Research Center, TAMU, College Station, TX (1998).
2. TRC THERMODYNAMIC TABLES - NON-HYDROCARBONS, Vols. I - XI, Thermodynamics Research Center, TAMU, College Station, TX (1997).
3. TECHNICAL DATA BOOK - PETROLEUM REFINING, Vols. I and II, American Petroleum Institute, Washington, DC (1972, 1977, 1982).
4. Daubert, T. E. and R. P. Danner, DATA COMPILATION OF PROPERTIES OF PURE COMPOUNDS, Parts 1, 2, 3, and 4, Supplements 1 and 2, DIPPR Project, AIChE, New York, NY (1985-1992).
5. Ambrose, D., VAPOUR-LIQUID CRITICAL PROPERTIES, National Physical Laboratory, Teddington, England, NPL Report Chem 107 (Feb., 1980).
6. Simmrock, K. H., R. Janowsky and A. Ohnsorge, CRITICAL DATA OF PURE SUBSTANCES, Vol. II, Parts 1 and 2, Dechema Chemistry Data Series, 6000 Frankfurt/Main, Germany (1986).
7. INTERNATIONAL CRITICAL TABLES, McGraw-Hill, New York, NY (1926).
8. Braker, W. and A. L. Mossman, MATHESON GAS DATA BOOK, 6th ed., Matheson Gas Products, Secaucaus, NJ (1980).
9. CRC HANDBOOK OF CHEMISTRY AND PHYSICS, 75th - 78th eds., CRC Press, Inc., Boca Raton, FL (1994-1997).
10. LANGE'S HANDBOOK OF CHEMISTRY, 13th and 14th eds., McGraw-Hill, New York, NY (1985, 1992).
11. PERRY'S CHEMICAL ENGINEERING HANDBOOK, 5th and 6th eds., McGraw-Hill, New York, NY (1973, 1984).
12. Landolt-Bornstein, ZAHLENWERTE UND FUNKIONEN ANS PHYSIK, CHEMEI, ASTRONOMIE UND TECHNIK, Springer-Verlag, Heidelberg, Germany (1972-1997).
13. Kaye, G. W. C. and T. H. Laby, TABLES OF PHYSICAL AND CHEMICAL CONSTANTS, Longman Group Limited, London, England (1973).
14. Raznjevic, Kuzman, HANDBOOK OF THERMODYNAMIC TABLES AND CHARTS, Hemisphere Publishing Corp., New York, NY (1976).
15. Driesbach, R. R., PHYSICAL PROPERTIES OF CHEMICAL COMPOUNDS, Vol. I (No. 15), Vol. II (No. 22), Vol. III (No. 29), Advances in Chemistry Series, American Chemical Society, Washington, DC (1955,1959,1961).
16. Vargaftik, N. B., TABLES ON THE THERMOPHYSICAL PROPERTIES OF LIQUIDS AND GASES, 2nd ed., English translation, Hemisphere Publishing Corporation, New York, NY (1975, 1983).
17. Rabinovich, V. A., editor, THERMOPHYSICAL PROPERTIES OF GASES AND LIQUIDS, translated from Russian, U. S. Dept. Commerce, Springfield, VA (1970).
18. Horvath, A. L., PHYSICAL PROPERTIES OF INORGANIC COMPOUNDS, Crane, Russak & Company, Inc., New York, NY (1975).
19. Timmermans, J., PHYSICO-CHEMICAL CONSTANTS OF PURE ORGANIC COMPOUNDS, Vols. 1 and 2, Elsevier, New York, NY (1950,1965).
20. ENCYCLOPEDIA OF CHEMICAL TECHNOLOGY, 3rd and 4th eds., John Wiley and Sons, Inc., New York, NY (1978-1997).

21. Sax, N. I. and R. J. Lewis, Jr., HAWLEY'S CONDENSED CHEMICAL DICTIONARY, 11th ed., Van Nostrand Reinhold Co., New York, NY (1987).

22. Beaton, C. F. and G. F. Hewitt, PHYSICAL PROPERTY DATA FOR THE DESIGN ENGINEER, Hemisphere Publishing Corporation, New York, NY (1989).

23. THERMOPHYSICAL PROPERTIES OF MATTER, 1st and 2nd eds., IFI/Plenum, New York, NY (1970-1976).

24. Ho, C. Y., P. E. Liley, T. Makita, and Y. Tanaka, PROPERTIES OF INORGANIC AND ORGANIC FLUIDS, Hemisphere Publishing Corporation, New York, NY (1988).

25. Verschueren, K., HANDBOOK OF ENVIRONMENTAL DATA ON ORGANIC CHEMICALS, Van Nostrand Reinhold, New York, NY (1996).

26. Lide, D. R. and H. V. Kehianian, CRC HANDBOOK OF THERMOPHYSICAL AND THERMOCHEMICAL DATA, CRC Press, Boca Raton, FL (1994).

27. Bretsznajder, S., PREDICTION OF TRANSPORT AND OTHER PHYSICAL PROPERTIES OF FLUIDS, International Series of Monographs in Chemical Engineering, Vol. 2, Pergamon Press, Oxford, England (1971).

28. Lyman, W. J., W. F. Reehl, and D. H. Rosenblatt, HANDBOOK OF CHEMICAL PROPERTY ESTIMATION METHODS, McGraw-Hill, New York, NY (1982).

29. Reid, R. C., J. M. Prausnitz, and B. E. Poling, THE PROPERTIES OF GASES AND LIQUIDS, 3rd ed. (R. C. Reid and T. K. Sherwood), 4th ed., McGraw-Hill, New York, NY (1977, 1987).

30. Baum, E. J., CHEMICAL PROPERTY ESTIMATION, Lewis Publishers, New York, NY (1998).

31. Mackay, D., W. Y. Shiu, and K. C. Ma, ILLUSTRATED HANDBOOK OF PHYSICAL-CHEMICAL PROPERTIES AND ENVIRONMENTAL FATE FOR ORGANIC CHEMICALS, Vols. 1, 2, 3, 4, and 5, Lewis Publishers, New York, NY (1992, 1992, 1993, 1995, 1997).

32. Yaws, C. L., PHYSICAL PROPERTIES, McGraw-Hill, New York, NY (1977).

33. Yaws, C. L., THERMODYNAMIC AND PHYSICAL PROPERTY DATA, Gulf Publishing Co., Houston, TX (1992).

34. Yaws, C. L. and R. W. Gallant, PHYSICAL PROPERTIES OF HYDROCARBONS, Vols. 1 (2nd ed.), 2 (3rd ed.), 3 and 4, Gulf Publishing Co., Houston, TX (1992, 1993, 1993, 1995).

35. Zwolinski, B. J. and R. C. Wilhoit, VAPOR PRESSURES AND HEATS OF VAPORIZATION OF HYDROCARBONS AND RELATED COMPOUNDS, Thermodynamic Research Center, TAMU, College Station, TX (1971).

36. Boublick, T., V. Fried, and E. Hala, THE VAPOUR PRESSURES OF PURE SUBSTANCES, 1st and 2nd eds., Elsevier, New York, NY (1975, 1984).

37. Ohe, S., COMPUTER AIDED DATA BOOK OF VAPOR PRESSURE, Data Book Publishing Company, Tokyo, Japan (1976).

38. Altunin, V. V., V. Z. Geller, E. K. Petrov, D. C. Rasskazov, and G. A. Spiridonov, THERMOPHYSICAL PROPERTIES OF FREONS, Methane Series, Parts 1 and 2, Hemisphere Publishing Corporation, New York, NY (1987).

39. Howard, P. H. and W. M. Meylan, eds., HANDBOOK OF PHYSICAL PROPERTIES OF ORGANIC CHEMICALS, CRC Press, Boca Raton, FL (1997).

40. Yaws, C. L. and others, Hydrocarbon Processing, 68, 61 (July, 1989).

41. Yaws, C. L., HANDBOOK OF VAPOR PRESSURE, Vols. 1, 2, 3, and 4, Gulf Publishing Co., Houston, TX (1994, 1994, 1994, 1995).

42. Yaws, C. L., HANDBOOK OF TRANSPORT PROPERTY DATA, Gulf Publishing Co., Houston, TX (1995).

43. Yaws, C. L., HANDBOOK OF THERMODYNAMIC DIAGRAMS, Vols. 1, 2, 3, and 4, Gulf Publishing Co., Houston, TX (1996).

44. Yaws, C. L., HANDBOOK OF CHEMICAL COMPOUND DATA FOR PROCESS SAFETY, Gulf Publishing Co., Houston, TX (1997).

REFERENCES - INORGANIC COMPOUNDS

1-29. See above REFERENCES - ORGANIC COMPOUNDS

30. Ohse, R. W., HANDBOOK OF THERMODYNAMIC AND TRANSPORT PROPERTIES OF ALKALI METALS, Blackwell Scientific Publications, London, England (1985).

31. Mellor, J. W., INORGANIC AND THEORETICAL CHEMISTRY, original volumes and supplements, Longmans, Green and Co., London, England (1956-present).

32. GMELIN'S HANDBOOK OF INORGANIC CHEMISTRY, original volumes and supplements, Weinheim Verlag Chemie (1966 - present).

33. Bailar, J. C., H. J. Emel'eus, and A. F. Trotman-Dickenson, COMPREHENSIVE INORGANIC CHEMISTRY, Pergamon Press, Elmsford, NJ (1973).

34. Samsonov, G. V., ed., HANDBOOK OF THE PHYSICO-CHEMICAL PROPERTIES OF THE ELEMENTS, Plenum, Washington, DC (1968).

35. Barin, I. and O. Knacke, THERMOCHEMICAL PROPERTIES OF INORGANIC SUBSTANCES, Springer-Verlag, New York, NY (1973).

36. Yaws, C. L. and others, Solid State Technology, 16 (1), 39 (1973).

37. Yaws, C. L. and others, Solid State Technology, 17 (1), 47 (1974).

38. Yaws, C. L. and others, Solid State Technology, 17 (11), 31 (1974).

39. Yaws, C. L. and others, Chem. Eng., 81 (12), 70 (June 10, 1974).

40. Yaws, C. L. and others, Chem. Eng., 81 (14), 85 (July 8, 1974).

41. Yaws, C. L. and others, Chem. Eng., 81 (17), 99 (August 19, 1974).

42. Yaws, C. L. and others, Chem. Eng., 81 (20), 115 (Sept. 30, 1974).

43. Yaws, C. L. and others, Chem. Eng., 81 (23), 113 (Oct. 28, 1974).

44. Yaws, C. L. and others, Chem. Eng., 81 (25), 178 (Nov. 25, 1974).

45. Yaws, C. L. and others, Chem. Eng., 81 (27), 67 (Dec. 23, 1974).

46. Yaws, C. L. and others, Chem. Eng., 82 (2), 99 (Jan. 20, 1975).

47. Yaws, C. L. and others, Chem. Eng., 82 (4), 87 (Feb. 17, 1975).

48. Yaws, C. L. and others, Solid State Technology, 21 (No.1), 43 (1978).

49. Yaws, C. L. and others, Solid State Technology, 22 (No.2), 65 (1979).

50. Yaws, C. L. and others, Solid State Technology, 24 (No.1), 87 (1981).

51. Yaws, C. L. and others, J.Ch.I.Ch.E., 12, 33 (1981).

52. Yaws, C. L. and others, J.Ch.I.Ch.E., 14, 205 (1983).

53. Yaws, C. L. and others, Ind. Eng. Chem. Process Des. Dev., 23, 48 (1984).

54. Yaws, C. L. and others, J. Chem. Eng. Data, 40 (1), 15 (1995).

55. Yaws, C. L. and others, J. Chem. Eng. Data, 40 (1), 18 (1995).

56. Yaws, C. L., PHYSICAL PROPERTIES, McGraw-Hill, New York, NY (1977).

57. Ohe, S., COMPUTER AIDED DATA BOOK OF VAPOR PRESSURE, Data Book Publishing Company, Tokyo, Japan (1976).

58. Nesmeyanov, A. N., VAPOR PRESSURE OF THE CHEMICAL ELEMENTS, Elsevier, New York, NY (1963).

59. Boublick, T., V. Fried, and E. Hala, THE VAPOUR PRESSURES OF PURE SUBSTANCES, 1st ed., 2nd ed., Elsevier, New York, NY (1975, 1984).

Figure 1-1 Critical Temperature of Normal Alkanes

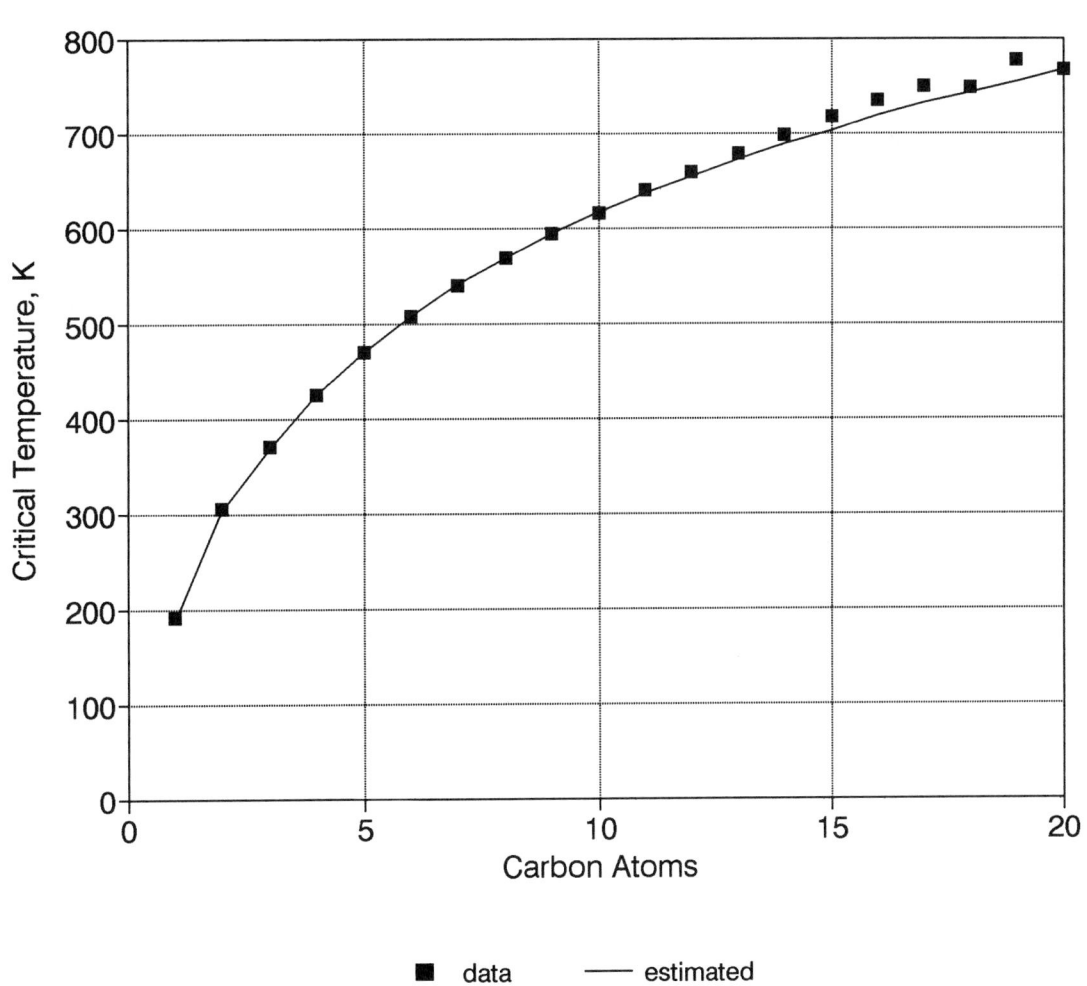

Figure 1-2 Critical Temperature of Elements

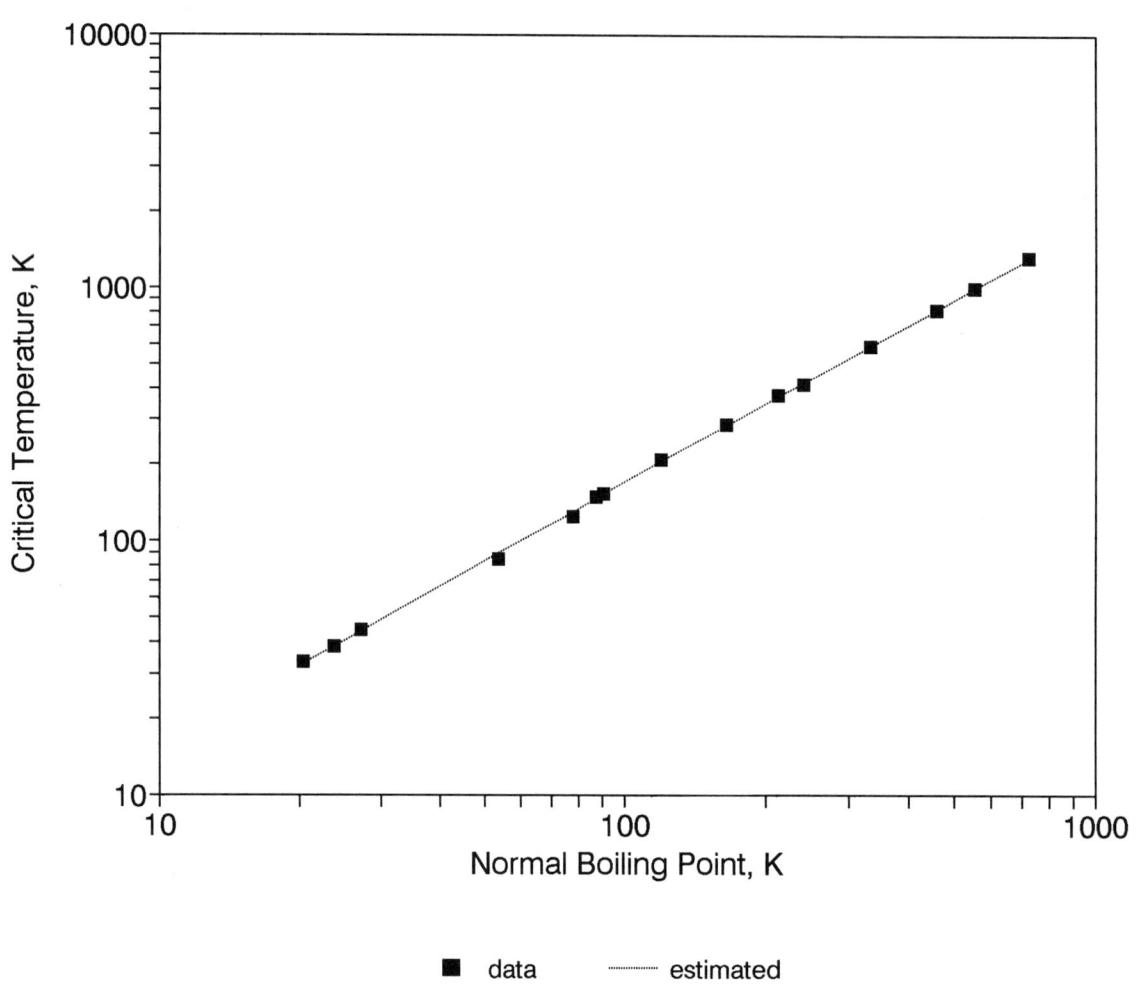

Table 1-1 CRITICAL PROPERTIES AND ACENTRIC FACTOR - ORGANIC COMPOUNDS

NO	FORMULA	NAME	MW g/mol	T_F K	T_B K	T_C K	P_C bar	V_C cm3/mol	RHO_C g/cm3	Z_C	OMEGA
1	CBrClF2	BROMOCHLORODIFLUOROMETHANE	165.365	113.65	269.14	426.15	42.54	246.0	0.6722	0.295	0.187
2	CBrCl3	BROMOTRICHLOROMETHANE	198.273	252.15	378.05	606.00	49.70	284.0	0.6981	0.280	0.192
3	CBrF3	BROMOTRIFLUOROMETHANE	148.910	105.15	215.26	340.15	39.72	200.0	0.7446	0.281	0.173
4	CBr2F2	DIBROMODIFLUOROMETHANE	209.816	163.05	295.94	478.00	53.30	249.0	0.8426	0.334	0.200
5	CClF3	CHLOROTRIFLUOROMETHANE	104.459	92.15	191.74	301.96	39.46	180.3	0.5794	0.283	0.180
6	CClN	CYANOGEN CHLORIDE	61.470	266.65	286.00	449.00	59.90	163.0	0.3771	0.262	0.320
7	CCl2F2	DICHLORODIFLUOROMETHANE	120.913	115.15	243.36	384.95	41.25	217.0	0.5572	0.280	0.180
8	CCl2O	PHOSGENE	98.916	145.37	280.71	455.00	56.74	190.2	0.5200	0.285	0.201
9	CCl3F	TRICHLOROFLUOROMETHANE	137.368	162.04	296.97	471.00	44.08	248.0	0.5539	0.279	0.184
10	CCl4	CARBON TETRACHLORIDE	153.822	250.33	349.79	556.35	45.60	276.0	0.5573	0.272	0.193
11	CF2O	CARBONYL FLUORIDE	66.007	161.89	188.58	297.00	57.60	141.0	0.4681	0.329	0.283
12	CF4	CARBON TETRAFLUORIDE	88.005	89.56	145.09	227.50	37.39	140.0	0.6286	0.277	0.186
13	CHBr3	TRIBROMOMETHANE	252.731	281.20	422.35	696.00	60.90	286.0	0.8837	0.301	0.156
14	CHClF2	CHLORODIFLUOROMETHANE	86.468	115.73	232.32	369.30	49.71	166.0	0.5209	0.269	0.219
15	CHCl2F	DICHLOROFLUOROMETHANE	102.923	138.15	282.05	451.58	51.84	196.0	0.5251	0.271	0.207
16	CHCl3	CHLOROFORM	119.377	209.63	334.33	536.40	54.72	239.0	0.4995	0.293	0.213
17	CHF3	TRIFLUOROMETHANE	70.014	117.97	190.99	298.89	48.36	133.3	0.5252	0.259	0.267
18	CHI3	TRIIODOMETHANE	393.732	396.16	491.16	794.55	53.12	349.5	1.1266	0.281	0.193
19	CHN	HYDROGEN CYANIDE	27.026	259.91	298.85	456.65	53.91	138.6	0.1950	0.197	0.410
20	CHNS	ISOTHIOCYANIC-ACID	59.086	-------	-------	-------	-------	-------	-------	-------	-------
21	CH2BrCl	BROMOCHLOROMETHANE	129.384	185.20	341.20	557.00	68.10	188.0	0.6882	0.276	0.220
22	CH2Br2	DIBROMOMETHANE	173.835	220.60	370.10	611.00	71.70	223.0	0.7795	0.315	0.210
23	CH2ClF	CHLOROFLUOROMETHANE	68.478	140.16	264.06	424.91	51.31	158.5	0.4320	0.230	0.199
24	CH2Cl2	DICHLOROMETHANE	84.932	178.01	312.90	510.00	60.80	185.0	0.4591	0.265	0.192
25	CH2F2	DIFLUOROMETHANE	52.024	137.00	221.50	351.60	58.30	121.0	0.4300	0.241	0.276
26	CH2I2	DIIODOMETHANE	267.836	279.25	455.15	747.00	54.70	272.0	0.9847	0.240	0.141
27	CH2O	FORMALDEHYDE	30.026	181.15	254.05	408.00	65.86	105.0	0.2860	0.204	0.282
28	CH2O2	FORMIC ACID	46.026	281.55	373.71	580.00	73.90	125.0	0.3682	0.192	0.473
29	CH3Br	METHYL BROMIDE	94.939	179.55	276.71	467.00	80.00	156.0	0.6086	0.321	0.192
30	CH3Cl	METHYL CHLORIDE	50.488	175.45	248.93	416.25	66.79	139.0	0.3632	0.268	0.153
31	CH3Cl3Si	METHYL TRICHLOROSILANE	149.478	195.35	339.55	517.00	35.30	340.0	0.4396	0.279	0.263
32	CH3F	METHYL FLUORIDE	34.033	131.35	194.82	317.70	58.77	113.0	0.3012	0.251	0.204
33	CH3I	METHYL IODIDE	141.939	206.70	315.58	528.00	73.70	185.0	0.7672	0.311	0.193
34	CH3NO	FORMAMIDE	45.041	275.70	493.00	771.00	78.00	163.0	0.2763	0.198	0.453
35	CH3NO2	NITROMETHANE	61.040	244.60	374.35	588.15	63.13	173.4	0.3520	0.224	0.348
36	CH3NO2	METHYL-NITRITE	61.040	256.16	261.16	-------	-------	-------	-------	-------	-------
37	CH3NO3	METHYL-NITRATE	77.040	190.86	339.16	-------	-------	-------	-------	-------	-------
38	CH4	METHANE	16.043	90.67	111.66	190.58	46.04	99.3	0.1616	0.288	0.011
39	CH4Cl2Si	METHYL DICHLOROSILANE	115.034	182.55	314.70	483.00	39.50	289.0	0.3980	0.284	0.276
40	CH4O	METHANOL	32.042	175.47	337.85	512.58	80.96	117.8	0.2720	0.224	0.566
41	CH4O3S	METHANESULFONIC ACID	96.107	292.81	561.00	-------	-------	220.0	0.4369	-------	-------
42	CH4S	METHYL MERCAPTAN	48.109	150.18	279.11	469.95	72.35	145.0	0.3318	0.268	0.146
43	CH5ClSi	METHYL CHLOROSILANE	80.589	139.05	281.85	442.00	41.70	246.0	0.3276	0.279	0.225
44	CH5N	METHYLAMINE	31.057	179.69	266.82	430.05	74.58	154.0	0.2017	0.321	0.281
45	CH6Si	METHYL SILANE	46.144	116.34	216.25	352.50	48.40	205.0	0.2251	0.339	0.139
46	CN4O8	TETRANITROMETHANE	196.033	287.05	398.85	540.00	17.40	468.0	0.4189	0.181	0.516
47	CO	CARBON MONOXIDE	28.010	68.15	81.70	132.92	34.99	93.1	0.3009	0.295	0.066
48	COS	CARBONYL SULFIDE	60.076	134.35	223.00	378.80	63.49	135.1	0.4447	0.272	0.097
49	CO2	CARBON DIOXIDE	44.010	216.58	194.67	304.19	73.82	94.0	0.4682	0.274	0.228
50	CS2	CARBON DISULFIDE	76.143	161.58	319.37	552.00	79.03	160.0	0.4759	0.276	0.108
51	C2BrF3	BROMOTRIFLUOROETHYLENE	160.921	-------	270.65	432.00	44.80	239.0	0.6733	0.298	0.175
52	C2Br2F4	1,2-DIBROMOTETRAFLUOROETHANE	259.824	162.65	320.41	487.80	33.93	341.0	0.7619	0.285	0.250
53	C2ClF3	CHLOROTRIFLUOROETHYLENE	116.470	115.00	245.30	379.15	40.53	212.0	0.5494	0.273	0.264
54	C2ClF5	CHLOROPENTAFLUOROETHANE	154.467	173.71	234.04	353.15	31.57	252.0	0.6130	0.271	0.251
55	C2Cl2F4	1,2-DICHLOROTETRAFLUOROETHANE	170.921	179.15	276.92	418.85	32.63	293.7	0.5820	0.275	0.252
56	C2Cl3F3	1,1,2-TRICHLOROTRIFLUOROETHANE	187.375	238.15	320.75	487.25	34.15	325.3	0.5760	0.274	0.255
57	C2Cl4	TETRACHLOROETHYLENE	165.833	250.80	394.40	620.00	44.90	248.0	0.6687	0.216	0.214
58	C2Cl4F2	1,1,2,2-TETRACHLORODIFLUOROETHANE	203.830	299.15	366.00	551.00	33.40	351.0	0.5807	0.264	0.291
59	C2Cl4O	TRICHLOROACETYL CHLORIDE	181.832	-------	391.15	590.00	41.00	332.0	0.5477	0.277	0.348
60	C2Cl6	HEXACHLOROETHANE	236.738	459.95	460.00	698.00	33.40	412.0	0.5746	0.237	0.221
61	C2F4	TETRAFLUOROETHYLENE	100.016	142.00	197.51	306.45	39.44	172.0	0.5815	0.266	0.226
62	C2F6	HEXAFLUOROETHANE	138.012	172.45	194.95	292.80	29.79	224.0	0.6161	0.274	0.245
63	C2HBrClF3	HALOTHANE	197.382	-------	323.35	521.00	39.20	296.0	0.6668	0.268	0.091
64	C2HClF2	2-CHLORO-1,1-DIFLUOROETHYLENE	98.479	134.65	254.55	400.55	44.58	197.0	0.4999	0.264	0.219
65	C2HCl3	TRICHLOROETHYLENE	131.388	188.40	360.10	571.00	49.10	256.0	0.5132	0.265	0.217
66	C2HCl3O	DICHLOROACETYL CHLORIDE	147.387	-------	382.15	579.00	46.10	288.0	0.5118	0.276	0.371
67	C2HCl3O	TRICHLOROACETALDEHYDE	147.387	216.00	370.85	565.00	44.10	288.0	0.5118	0.270	0.332
68	C2HCl5	PENTACHLOROETHANE	202.293	244.15	433.03	665.00	36.80	369.0	0.5482	0.246	0.246
69	C2HF3	TRIFLUOROETHENE	82.025	94.53	221.01	347.22	45.16	182.5	0.4495	0.286	0.238
70	C2HF3O2	TRIFLUOROACETIC ACID	114.024	257.90	344.95	491.25	32.58	204.0	0.5589	0.163	0.524
71	C2HF5	PENTAFLUOROETHANE	120.022	170.15	225.15	342.00	34.40	216.0	0.5557	0.261	0.259
72	C2H2	ACETYLENE	26.038	192.40	189.15	308.32	61.39	113.0	0.2305	0.271	0.187
73	C2H2Br4	1,1,2,2-TETRABROMOETHANE	345.654	273.15	516.65	824.00	46.00	401.0	0.8620	0.269	0.177
74	C2H2Cl2	1,1-DICHLOROETHYLENE	96.943	150.65	304.71	482.00	51.90	224.0	0.4328	0.290	0.272
75	C2H2Cl2	cis-1,2-DICHLOROETHYLENE	96.943	193.15	333.65	527.00	51.90	224.0	0.4328	0.265	0.264
76	C2H2Cl2	trans-1,2-DICHLOROETHYLENE	96.943	223.35	320.85	508.00	51.90	224.0	0.4328	0.275	0.264
77	C2H2Cl2O	CHLOROACETYL CHLORIDE	112.943	251.15	379.15	581.00	51.10	245.0	0.4610	0.259	0.358
78	C2H2Cl2O	DICHLOROACETALDEHYDE	112.943	223.00	362.00	555.00	49.50	245.0	0.4610	0.263	0.344

7

Table 1-1 CRITICAL PROPERTIES AND ACENTRIC FACTOR - ORGANIC COMPOUNDS (continued)

NO	FORMULA	NAME	MW g/mol	T_F K	T_B K	T_C K	P_C bar	V_C cm3/mol	RHO_C g/cm3	Z_C	OMEGA
79	C2H2Cl2O2	DICHLOROACETIC ACID	128.942	286.55	467.15	686.00	51.70	265.0	0.4866	0.240	0.555
80	C2H2Cl3F	1,1,1-TRICHLOROFLUOROETHANE	151.394	-------	366.00	565.00	39.90	294.0	0.5149	0.250	0.250
81	C2H2Cl4	1,1,1,2-TETRACHLOROETHANE	167.849	202.94	403.65	624.00	40.20	325.0	0.5165	0.252	0.242
82	C2H2Cl4	1,1,2,2-TETRACHLOROETHANE	167.849	229.35	418.25	645.00	40.90	325.0	0.5165	0.248	0.259
83	C2H2F2	1,1-DIFLUOROETHYLENE	64.035	129.15	187.50	302.80	44.58	154.0	0.4158	0.273	0.139
84	C2H2F2	cis-1,2-DIFLUOROETHENE	64.035	107.90	247.86	394.67	47.69	163.5	0.3917	0.238	0.210
85	C2H2F2	trans-1,2-DIFLUOROETHENE	64.035	107.90	247.86	394.67	47.69	163.5	0.3917	0.238	0.210
86	C2H2F4	1,1,1,2-TETRAFLUOROETHANE	102.031	172.15	247.15	380.00	36.90	203.0	0.5026	0.237	0.239
87	C2H2O	KETENE	42.037	122.00	223.34	370.00	58.10	144.0	0.2919	0.272	0.126
88	C2H2O4	OXALIC ACID	90.036	462.65	569.00	804.00	70.20	205.0	0.4392	0.215	0.918
89	C2H3Br	VINYL BROMIDE	106.950	135.35	288.95	473.00	71.80	200.0	0.5348	0.365	0.282
90	C2H3Cl	VINYL CHLORIDE	62.499	119.36	259.78	432.00	56.70	179.0	0.3492	0.283	0.101
91	C2H3ClF2	1-CHLORO-1,1-DIFLUOROETHANE	100.495	142.35	263.14	410.20	41.24	231.0	0.4350	0.279	0.237
92	C2H3ClO	ACETYL CHLORIDE	78.498	160.30	323.90	508.00	57.40	196.0	0.4005	0.266	0.334
93	C2H3ClO	CHLOROACETALDEHYDE	78.498	-------	358.00	555.00	53.70	201.0	0.3905	0.234	0.330
94	C2H3ClO2	CHLOROACETIC ACID	94.497	333.15	462.50	686.00	57.80	221.0	0.4276	0.224	0.551
95	C2H3ClO2	METHYL CHLOROFORMATE	94.497	-------	344.00	525.00	53.60	221.0	0.4276	0.271	0.393
96	C2H3Cl3	1,1,1-TRICHLOROETHANE	133.404	242.75	347.23	545.00	42.96	281.0	0.4747	0.266	0.216
97	C2H3Cl3	1,1,2-TRICHLOROETHANE	133.404	236.50	387.00	602.00	44.80	281.0	0.4747	0.252	0.260
98	C2H3F	VINYL FLUORIDE	46.044	112.65	200.95	327.80	52.39	144.0	0.3198	0.277	0.189
99	C2H3F3	1,1,1-TRIFLUOROETHANE	84.041	161.85	225.75	346.25	37.58	194.0	0.4332	0.253	0.253
100	C2H3N	ACETONITRILE	41.053	229.32	354.75	545.50	48.33	173.0	0.2373	0.184	0.338
101	C2H3NO	METHYL ISOCYANATE	57.052	256.15	312.00	505.00	51.90	190.0	0.3003	0.235	0.175
102	C2H4	ETHYLENE	28.054	104.01	169.47	282.36	50.32	129.1	0.2174	0.277	0.085
103	C2H4Br2	1,1-DIBROMOETHANE	187.862	210.15	381.15	628.00	60.30	276.0	0.6807	0.319	0.125
104	C2H4Br2	1,2-DIBROMOETHANE	187.862	282.94	404.51	650.15	54.77	261.6	0.7182	0.265	0.207
105	C2H4Cl2	1,1-DICHLOROETHANE	98.959	176.19	330.45	523.00	50.66	240.0	0.4123	0.280	0.244
106	C2H4Cl2	1,2-DICHLOROETHANE	98.959	237.49	356.59	561.00	53.70	220.0	0.4498	0.253	0.288
107	C2H4Cl2O	BIS(CHLOROMETHYL)ETHER	114.959	231.65	378.00	579.00	45.80	258.0	0.4456	0.245	0.324
108	C2H4F2	1,1-DIFLUOROETHANE	66.051	156.15	247.35	386.60	44.99	181.0	0.3649	0.253	0.263
109	C2H4F2	1,2-DIFLUOROETHANE	66.051	-------	303.65	476.00	43.40	202.0	0.3270	0.222	0.224
110	C2H4I2	1,2-DIIODOETHANE	281.863	356.16	473.16	749.91	47.30	323.5	0.8713	0.245	0.223
111	C2H4O	ACETALDEHYDE	44.053	150.15	293.55	461.00	55.50	157.0	0.2806	0.227	0.317
112	C2H4O	ETHYLENE OXIDE	44.053	161.45	283.85	469.15	71.94	140.3	0.3140	0.259	0.198
113	C2H4OS	THIOACETIC-ACID	76.113	150.16	360.16	577.34	69.21	219.5	0.3468	0.317	0.304
114	C2H4O2	ACETIC ACID	60.053	289.81	391.05	592.71	57.86	171.0	0.3512	0.201	0.462
115	C2H4O2	METHYL FORMATE	60.053	174.15	304.90	487.20	59.98	172.0	0.3491	0.255	0.254
116	C2H4S	THIACYCLOPROPANE	60.114	165.37	328.07	555.00	73.80	151.5	0.3968	0.214	0.154
117	C2H5Br	BROMOETHANE	108.966	154.55	311.50	503.80	62.32	214.9	0.5070	0.320	0.183
118	C2H5Cl	ETHYL CHLORIDE	64.514	136.75	285.42	460.35	52.69	200.0	0.3226	0.275	0.204
119	C2H5ClO	2-CHLOROETHANOL	80.514	205.65	401.75	585.00	59.20	212.0	0.3798	0.258	0.637
120	C2H5F	ETHYL FLUORIDE	48.060	129.95	235.45	375.31	50.28	164.0	0.2930	0.264	0.209
121	C2H5I	ETHYL IODIDE	155.966	162.05	345.45	561.00	59.90	238.0	0.6553	0.306	1.137
122	C2H5N	ETHYLENEIMINE	43.068	195.20	329.00	537.00	68.50	173.0	0.2489	0.265	0.089
123	C2H5NO	ACETAMIDE	59.068	354.15	494.30	761.00	66.00	215.0	0.2747	0.224	0.189
124	C2H5NO	N-METHYLFORMAMIDE	59.068	269.35	472.66	721.00	56.20	215.0	0.2747	0.202	0.192
125	C2H5NO2	NITROETHANE	75.067	183.63	387.22	593.00	51.60	236.0	0.3181	0.247	0.265
126	C2H5NO3	ETHYL-NITRATE	91.066	178.56	360.36	-------	-------	-------	-------	-------	-------
127	C2H6	ETHANE	30.070	90.35	184.55	305.42	48.80	147.9	0.2033	0.284	0.099
128	C2H6AlCl	DIMETHYLALUMINUM CHLORIDE	92.054	252.15	399.15	619.00	36.20	320.0	0.2877	0.225	0.183
129	C2H6O	DIMETHYL ETHER	46.069	131.66	248.31	400.10	53.70	170.0	0.2710	0.274	0.204
130	C2H6O	ETHANOL	46.069	159.05	351.44	516.25	63.84	166.9	0.2760	0.248	0.637
131	C2H6OS	DIMETHYL SULFOXIDE	78.135	291.67	462.15	726.00	56.50	227.0	0.3442	0.212	0.209
132	C2H6O2	ETHYLENE GLYCOL	62.068	260.15	470.45	645.00	75.30	191.0	0.3250	0.268	1.137
133	C2H6O4S	DIMETHYL SULFATE	126.133	241.35	461.95	758.00	51.60	293.0	0.4305	0.240	0.089
134	C2H6S	DIMETHYL SULFIDE	62.136	174.88	310.48	503.04	55.30	200.9	0.3093	0.266	0.189
135	C2H6S	ETHYL MERCAPTAN	62.136	125.26	308.15	499.15	54.90	207.0	0.3002	0.274	0.192
136	C2H6S2	DIMETHYL DISULFIDE	94.202	188.44	382.90	606.00	53.60	252.0	0.3738	0.268	0.265
137	C2H7N	DIMETHYLAMINE	45.084	180.96	280.03	437.65	53.09	187.0	0.2411	0.273	0.294
138	C2H7N	ETHYLAMINE	45.084	192.15	289.73	456.15	56.24	182.0	0.2477	0.270	0.285
139	C2H7NO	MONOETHANOLAMINE	61.084	283.65	444.15	638.00	68.70	225.0	0.2715	0.291	0.797
140	C2H8N2	ETHYLENEDIAMINE	60.099	284.29	390.41	593.00	62.90	264.0	0.2276	0.337	0.479
141	C2H8Si	DIMETHYL SILANE	60.171	122.93	253.55	402.00	35.60	258.0	0.2332	0.275	0.132
142	C2N2	CYANOGEN	52.036	245.25	252.00	400.15	59.78	195.0	0.2669	0.350	0.279
143	C3F6	HEXAFLUOROPROPYLENE	150.023	116.65	243.55	368.00	29.00	268.0	0.5598	0.254	0.204
144	C3F6O	HEXAFLUOROACETONE	166.023	151.15	245.88	357.14	28.37	329.0	0.5046	0.314	0.364
145	C3F8	OCTAFLUOROPROPANE	188.020	125.46	236.40	345.05	26.80	299.0	0.6288	0.279	0.326
146	C3H2N2	MALONONITRILE	66.062	304.90	491.50	715.00	40.40	248.0	0.2664	0.169	0.509
147	C3H3Cl	PROPARGYL CHLORIDE	74.510	-------	331.00	541.00	53.00	211.0	0.3531	0.249	0.152
148	C3H3N	ACRYLONITRILE	53.064	189.63	350.45	535.00	44.80	212.0	0.2503	0.214	0.350
149	C3H3NO	OXAZOLE	69.063	-------	342.65	554.00	63.20	237.0	0.2914	0.325	0.233
150	C3H4	METHYLACETYLENE	40.065	170.45	249.94	402.39	56.28	164.0	0.2443	0.276	0.216
151	C3H4	PROPADIENE	40.065	136.87	238.65	393.15	54.70	162.0	0.2473	0.271	0.160
152	C3H4Cl2	2,3-DICHLOROPROPENE	110.970	191.50	365.75	577.00	43.80	277.0	0.4006	0.253	0.206
153	C3H4O	ACROLEIN	56.064	185.45	325.84	506.00	50.00	197.0	0.2846	0.234	0.320
154	C3H4O	PROPARGYL ALCOHOL	56.064	221.35	386.75	580.00	65.30	176.0	0.3185	0.238	0.555
155	C3H4O2	ACRYLIC ACID	72.064	286.65	414.15	615.00	56.60	208.0	0.3465	0.230	0.518
156	C3H4O2	beta-PROPIOLACTONE	72.064	239.75	435.15	686.00	69.10	195.0	0.3696	0.236	0.345

8

NO	FORMULA	NAME	MW g/mol	T_F K	T_B K	T_C K	P_C bar	V_C cm3/mol	RHO_C g/cm3	Z_C	OMEGA
157	C3H4O2	VINYL FORMATE	72.064	-------	320.00	498.00	50.20	217.0	0.3321	0.263	0.285
158	C3H4O3	ETHYLENE CARBONATE	88.063	309.55	511.15	790.00	67.70	193.0	0.4563	0.199	0.416
159	C3H4O3	PYRUVIC ACID	88.063	286.75	438.15	634.52	56.50	239.0	0.3685	0.256	0.670
160	C3H5Br	3-BROMO-1-PROPENE	120.977	153.76	343.16	540.20	51.39	246.5	0.4908	0.282	0.273
161	C3H5Cl	2-CHLOROPROPENE	76.525	135.75	295.80	478.00	47.10	234.0	0.3270	0.277	0.153
162	C3H5Cl	3-CHLOROPROPENE	76.525	138.65	318.11	514.15	47.10	234.0	0.3270	0.258	0.154
163	C3H5ClO	alpha-EPICHLOROHYDRIN	92.525	215.95	389.26	610.00	49.00	233.0	0.3971	0.225	0.256
164	C3H5ClO2	METHYL CHLOROACETATE	108.524	241.03	402.97	600.00	45.00	270.0	0.4019	0.244	0.434
165	C3H5ClO2	ETHYL CHLOROFORMATE	108.524	192.00	366.00	508.15	45.00	274.0	0.3961	0.292	0.835
166	C3H5Cl3	1,2,3-TRICHLOROPROPANE	147.431	258.45	430.00	652.00	38.70	334.0	0.4414	0.238	0.306
167	C3H5I	3-IODO-1-PROPENE	167.977	173.86	375.16	595.81	45.29	272.5	0.6164	0.249	0.202
168	C3H5N	PROPIONITRILE	55.079	180.26	370.50	564.40	41.85	229.0	0.2405	0.204	0.325
169	C3H5NO	ACRYLAMIDE	71.079	357.65	465.75	710.00	57.30	260.0	0.2734	0.252	0.196
170	C3H5NO	HYDRACRYLONITRILE	71.079	227.15	494.15	690.00	48.90	243.0	0.2925	0.207	0.826
171	C3H5NO	LACTONITRILE	71.079	233.00	457.00	643.00	50.30	243.0	0.2925	0.229	0.796
172	C3H5N3O9	NITROGLYCERINE	227.088	286.15	523.00	680.00	30.00	419.0	0.5420	0.222	1.184
173	C3H6	CYCLOPROPANE	42.081	145.73	240.37	397.91	55.75	162.8	0.2585	0.274	0.134
174	C3H6	PROPYLENE	42.081	87.90	225.43	364.76	46.13	181.0	0.2325	0.275	0.142
175	C3H6Br2	1,2-DIBROMOPROPANE	201.888	217.96	413.16	634.11	54.07	321.5	0.6280	0.330	0.384
176	C3H6Cl2	1,1-DICHLOROPROPANE	112.986	-------	361.25	560.00	42.40	291.0	0.3883	0.265	0.253
177	C3H6Cl2	1,2-DICHLOROPROPANE	112.986	172.71	369.52	572.00	42.40	291.0	0.3883	0.259	0.251
178	C3H6Cl2	1,3-DICHLOROPROPANE	112.987	173.65	393.55	603.00	41.50	291.0	0.3883	0.241	0.292
179	C3H6Cl2	2,2-DICHLOROPROPANE	112.986	239.36	342.46	539.46	41.04	290.5	0.3889	0.266	0.198
180	C3H6I2	1,2-DIIODOPROPANE	295.889	253.16	500.16	780.49	42.06	373.5	0.7922	0.242	0.237
181	C3H6O	ACETONE	58.080	178.45	329.44	508.20	47.02	209.0	0.2779	0.233	0.306
182	C3H6O	ALLYL ALCOHOL	58.080	144.15	370.23	545.05	56.20	208.0	0.2792	0.258	0.572
183	C3H6O	METHYL VINYL ETHER	58.080	151.15	278.65	437.00	46.70	210.0	0.2766	0.270	0.237
184	C3H6O	n-PROPIONALDEHYDE	58.080	193.15	321.15	496.00	46.60	210.0	0.2766	0.237	0.302
185	C3H6O	1,2-PROPYLENE OXIDE	58.080	161.22	307.05	482.25	49.24	186.0	0.3123	0.228	0.271
186	C3H6O	1,3-PROPYLENE OXIDE	58.080	-------	321.00	520.00	57.50	188.0	0.3089	0.250	0.201
187	C3H6O2	ETHYL FORMATE	74.079	193.55	327.46	508.40	47.42	229.0	0.3235	0.257	0.285
188	C3H6O2	METHYL ACETATE	74.079	175.15	330.09	506.80	46.90	228.0	0.3249	0.254	0.325
189	C3H6O2	PROPIONIC ACID	74.079	252.45	414.32	604.00	45.30	230.0	0.3221	0.207	0.536
190	C3H6O2S	3-MERCAPTOPROPIONIC ACID	106.145	290.65	501.00	729.00	50.20	281.0	0.3777	0.233	0.587
191	C3H6O3	LACTIC ACID	90.079	291.15	447.00	616.00	59.65	216.9	0.4153	0.253	1.035
192	C3H6O3	METHOXYACETIC ACID	90.079	281.00	478.26	691.00	49.80	251.0	0.3589	0.218	0.630
193	C3H6O3	TRIOXANE	90.079	334.65	387.65	604.00	58.20	206.0	0.4373	0.239	0.334
194	C3H6S	THIACYCLOBUTANE	74.140	199.96	368.13	603.00	61.00	199.5	0.3716	0.228	0.195
195	C3H7Br	1-BROMOPROPANE	122.993	163.15	344.15	544.00	53.90	266.0	0.4624	0.317	0.285
196	C3H7Br	2-BROMOPROPANE	122.993	184.15	332.56	532.00	55.10	266.0	0.4624	0.331	0.243
197	C3H7Cl	ISOPROPYL CHLORIDE	78.541	155.97	308.85	489.00	45.40	247.0	0.3180	0.276	0.224
198	C3H7Cl	n-PROPYL CHLORIDE	78.541	150.35	319.67	503.15	45.80	247.0	0.3180	0.270	0.228
199	C3H7F	1-FLUOROPROPANE	62.087	114.16	269.95	422.00	41.57	221.5	0.2803	0.263	0.227
200	C3H7F	2-FLUOROPROPANE	62.087	139.80	263.81	415.68	42.00	215.5	0.2881	0.262	0.204
201	C3H7I	ISOPROPYL IODIDE	169.993	183.15	362.65	578.00	51.20	290.0	0.5862	0.309	0.238
202	C3H7I	n-PROPYL IODIDE	169.993	171.85	375.60	593.00	50.30	290.0	0.5862	0.296	0.258
203	C3H7N	ALLYLAMINE	57.095	184.95	326.45	505.00	51.70	247.0	0.2312	0.304	0.327
204	C3H7N	PROPYLENEIMINE	57.095	229.00	334.00	529.00	54.20	208.0	0.2745	0.256	0.257
205	C3H7NO	N,N-DIMETHYLFORMAMIDE	73.095	212.72	426.15	647.00	44.20	267.0	0.2738	0.219	0.376
206	C3H7NO	N-METHYLACETAMIDE	73.095	301.15	478.15	718.00	49.80	267.0	0.2738	0.223	0.435
207	C3H7NO2	1-NITROPROPANE	89.094	169.16	404.33	605.00	43.50	288.0	0.3094	0.249	0.412
208	C3H7NO2	2-NITROPROPANE	89.094	181.83	393.40	594.00	44.50	288.0	0.3094	0.260	0.376
209	C3H7NO3	PROPYL-NITRATE	105.093	173.16	383.16	-------	-------	-------	-------	-------	-------
210	C3H7NO3	ISOPROPYL-NITRATE	105.093	173.16	373.66	-------	-------	-------	-------	-------	-------
211	C3H8	PROPANE	44.096	85.46	231.11	369.82	42.49	202.9	0.2174	0.280	0.152
212	C3H8O	ISOPROPANOL	60.096	185.28	355.41	508.31	47.64	220.1	0.2730	0.248	0.669
213	C3H8O	METHYL ETHYL ETHER	60.096	160.00	280.50	437.80	43.98	221.0	0.2719	0.267	0.219
214	C3H8O	n-PROPANOL	60.096	146.95	370.35	536.71	51.70	218.5	0.2750	0.253	0.628
215	C3H8O2	2-METHOXYETHANOL	76.095	188.05	397.55	564.00	50.10	242.0	0.3144	0.259	0.731
216	C3H8O2	METHYLAL	76.095	168.35	315.00	480.60	39.52	213.0	0.3573	0.211	0.290
217	C3H8O2	1,2-PROPYLENE GLYCOL	76.095	213.15	460.75	626.00	61.00	239.0	0.3184	0.280	1.107
218	C3H8O2	1,3-PROPYLENE GLYCOL	76.095	246.45	487.55	658.00	59.20	217.0	0.3507	0.235	1.152
219	C3H8O3	GLYCEROL	92.095	291.33	563.15	723.00	40.00	264.0	0.3488	0.176	1.320
220	C3H8S	n-PROPYLMERCAPTAN	76.163	159.95	340.87	536.00	46.30	254.0	0.2999	0.264	0.235
221	C3H8S	ISOPROPYL MERCAPTAN	76.163	142.61	325.71	517.00	47.50	254.0	0.2999	0.281	0.212
222	C3H8S	ETHYL-METHYL-SULFIDE	76.156	167.20	340.15	532.80	42.50	257.5	0.2958	0.262	0.216
223	C3H9N	n-PROPYLAMINE	59.111	190.15	321.65	496.95	47.42	260.0	0.2274	0.298	0.296
224	C3H9N	ISOPROPYLAMINE	59.111	177.95	305.55	471.85	45.39	221.0	0.2675	0.256	0.279
225	C3H9N	TRIMETHYLAMINE	59.111	156.08	276.02	433.25	40.73	254.0	0.2327	0.287	0.209
226	C3H9NO	1-AMINO-2-PROPANOL	75.111	274.89	432.61	614.00	56.70	278.0	0.2702	0.309	0.794
227	C3H9NO	3-AMINO-1-PROPANOL	75.111	284.15	460.65	649.00	55.00	278.0	0.2702	0.283	0.830
228	C3H9NO	METHYLETHANOLAMINE	75.111	268.65	431.15	630.00	52.20	253.0	0.2969	0.252	0.586
229	C3H9O4P	TRIMETHYL PHOSPHATE	140.076	227.00	465.85	764.00	85.00	229.4	0.6105	0.307	-------
230	C3H10N2	1,2-PROPANEDIAMINE	74.126	236.53	392.45	587.00	52.70	316.0	0.2346	0.341	0.474
231	C3H10Si	TRIMETHYL SILANE	74.198	137.26	279.85	432.00	31.90	311.0	0.2386	0.276	0.175
232	C4Cl4S	TETRACHLOROTHIOPHENE	221.921	301.97	506.54	753.00	36.70	428.0	0.5185	0.251	0.361
233	C4Cl6	HEXACHLORO-1,3-BUTADIENE	260.760	252.15	488.15	741.00	28.40	491.0	0.5311	0.226	0.155
234	C4F8	OCTAFLUORO-2-BUTENE	200.031	138.15	270.36	392.00	23.30	347.0	0.5765	0.248	0.291

NO	FORMULA	NAME	MW g/mol	T_F K	T_B K	T_C K	P_C bar	V_C cm3/mol	RHO_C g/cm3	Z_C	OMEGA
235	C4F8	OCTAFLUOROCYCLOBUTANE	200.031	232.96	267.17	388.37	27.78	324.8	0.6159	0.279	0.356
236	C4F10	DECAFLUOROBUTANE	238.028	144.95	271.15	386.35	23.23	397.0	0.5996	0.287	0.372
237	C4H2	BUTADIYNE(BIACETYLENE)	50.060	237.16	283.46	478.02	58.63	183.5	0.2728	0.271	0.100
238	C4H2O3	MALEIC ANHYDRIDE	98.058	326.00	475.15	721.00	72.80	219.0	0.4478	0.266	0.546
239	C4H4	VINYLACETYLENE	52.076	-------	278.25	454.00	48.60	205.0	0.2540	0.264	0.118
240	C4H4N2	SUCCINONITRILE	80.089	331.30	540.15	770.00	35.40	300.0	0.2670	0.166	0.559
241	C4H4O	FURAN	68.075	187.55	304.50	490.15	55.02	218.2	0.3120	0.295	0.200
242	C4H4O2	DIKETENE	84.075	266.65	399.20	616.00	59.60	234.0	0.3593	0.272	0.382
243	C4H4O3	SUCCINIC ANHYDRIDE	100.074	393.00	536.58	811.00	67.30	223.0	0.4488	0.223	0.530
244	C4H4O4	FUMARIC ACID	116.073	560.15	563.15	771.00	49.80	297.0	0.3908	0.231	0.989
245	C4H4O4	MALEIC ACID	116.073	403.45	565.00	773.00	49.90	297.0	0.3908	0.231	0.998
246	C4H4S	THIOPHENE	84.142	234.94	357.31	579.35	56.90	219.0	0.3842	0.259	0.193
247	C4H5Cl	CHLOROPRENE	88.536	143.15	332.55	525.00	42.60	273.0	0.3243	0.266	0.193
248	C4H5N	trans-CROTONITRILE	67.090	222.00	394.38	586.00	38.80	282.0	0.2379	0.225	0.398
249	C4H5N	cis-CROTONITRILE	67.090	200.55	380.60	568.00	38.80	265.0	0.2532	0.218	0.379
250	C4H5N	METHACRYLONITRILE	67.090	237.35	363.45	554.00	38.80	265.0	0.2532	0.223	0.301
251	C4H5N	PYRROLE	67.090	249.74	403.00	639.75	62.10	230.0	0.2917	0.269	0.288
252	C4H5N	VINYLACETONITRILE	67.090	186.15	391.67	584.00	38.80	259.0	0.2590	0.207	0.378
253	C4H5NO2	METHYL CYANOACETATE	99.089	260.08	478.24	687.00	38.10	305.0	0.3249	0.203	0.549
254	C4H6	CYCLOBUTENE	54.091	153.76	275.75	446.33	52.66	195.5	0.2767	0.278	0.189
255	C4H6	1,2-BUTADIENE	54.092	136.95	284.00	444.00	45.00	219.0	0.2470	0.267	0.251
256	C4H6	1,3-BUTADIENE	54.092	164.25	268.74	425.37	43.30	220.8	0.2449	0.270	0.193
257	C4H6	DIMETHYLACETYLENE	54.092	240.91	300.13	488.15	50.80	221.0	0.2448	0.277	0.130
258	C4H6	ETHYLACETYLENE	54.092	147.43	281.22	443.20	49.50	222.0	0.2437	0.298	0.247
259	C4H6Cl2	1,3-DICHLORO-trans-2-BUTENE	124.997	-------	402.00	618.00	37.80	325.0	0.3846	0.239	0.242
260	C4H6Cl2	1,4-DICHLORO-cis-2-BUTENE	124.997	225.15	425.65	640.00	37.80	343.0	0.3644	0.244	0.331
261	C4H6Cl2	1,4-DICHLORO-trans-2-BUTENE	124.997	274.15	429.26	646.00	37.80	330.0	0.3788	0.232	0.333
262	C4H6Cl2	3,4-DICHLORO-1-BUTENE	124.997	212.00	388.00	589.00	38.50	330.0	0.3788	0.259	0.300
263	C4H6O	trans-CROTONALDEHYDE	70.091	196.65	377.25	571.00	42.50	250.0	0.2804	0.224	0.346
264	C4H6O	2,5-DIHYDROFURAN	70.091	-------	339.00	542.00	55.00	216.0	0.3245	0.264	0.229
265	C4H6O	DIVINYL ETHER	70.091	172.05	301.45	463.00	42.50	250.0	0.2804	0.276	0.291
266	C4H6O	METHACROLEIN	70.091	192.15	341.15	530.00	42.50	250.0	0.2804	0.241	0.246
267	C4H6O2	2-BUTYNE-1,4-DIOL	86.090	331.00	511.15	695.00	58.60	256.0	0.3363	0.260	1.134
268	C4H6O2	gamma-BUTYROLACTONE	86.090	229.78	477.15	739.00	59.40	265.0	0.3249	0.256	0.369
269	C4H6O2	cis-CROTONIC ACID	86.090	288.65	445.05	647.00	47.00	270.0	0.3189	0.236	0.572
270	C4H6O2	trans-CROTONIC ACID	86.090	344.55	458.15	666.00	47.00	270.0	0.3189	0.229	0.578
271	C4H6O2	METHACRYLIC ACID	86.090	288.15	434.15	643.00	47.00	270.0	0.3189	0.237	0.468
272	C4H6O2	METHYL ACRYLATE	86.090	196.32	353.35	536.00	42.50	270.0	0.3189	0.258	0.348
273	C4H6O2	VINYL ACETATE	86.090	180.35	345.65	524.00	42.50	270.0	0.3189	0.263	0.338
274	C4H6O3	ACETIC ANHYDRIDE	102.090	200.15	411.78	569.15	46.81	290.0	0.3520	0.287	0.840
275	C4H6O4	SUCCINIC ACID	118.089	461.15	591.00	806.00	47.10	300.0	0.3936	0.211	0.991
276	C4H6O5	DIGLYCOLIC ACID	134.089	421.15	610.00	820.00	44.20	331.0	0.4051	0.215	1.081
277	C4H6O5	MALIC ACID	134.089	403.15	602.00	781.00	50.70	331.0	0.4051	0.258	1.530
278	C4H6O6	TARTARIC ACID	150.088	479.15	660.00	828.00	51.80	305.0	0.4921	0.230	2.011
279	C4H7N	n-BUTYRONITRILE	69.106	161.25	390.75	582.25	37.90	278.0	0.2486	0.218	0.371
280	C4H7N	ISOBUTYRONITRILE	69.106	201.70	376.76	565.00	37.60	278.0	0.2486	0.223	0.338
281	C4H7NO	ACETONE CYANOHYDRIN	85.106	253.15	463.00	647.00	42.50	296.0	0.2875	0.234	0.733
282	C4H7NO	2-METHACRYLAMIDE	85.106	383.65	488.00	741.00	54.50	298.0	0.2856	0.264	0.421
283	C4H7NO	3-METHOXYPROPIONITRILE	85.106	210.12	439.00	638.00	36.30	324.0	0.2627	0.222	0.465
284	C4H7NO	2-PYRROLIDONE	85.106	298.15	518.15	792.00	61.70	264.0	0.3224	0.247	0.434
285	C4H8	1-BUTENE	56.107	87.80	266.90	419.59	40.20	239.9	0.2338	0.276	0.187
286	C4H8	cis-2-BUTENE	56.107	134.26	276.87	435.58	42.06	234.0	0.2398	0.272	0.203
287	C4H8	trans-2-BUTENE	56.107	167.62	274.03	428.63	41.02	238.2	0.2356	0.274	0.218
288	C4H8	CYCLOBUTANE	56.107	182.48	285.66	459.93	49.85	210.2	0.2670	0.274	0.187
289	C4H8	ISOBUTENE	56.107	132.81	266.25	417.90	39.99	238.9	0.2349	0.275	0.189
290	C4H8Br2	1,2-DIBROMOBUTANE	215.915	207.76	439.46	659.28	47.17	377.5	0.5720	0.325	0.429
291	C4H8Br2	2,3-DIBROMOBUTANE	215.915	238.66	434.16	656.96	47.69	371.5	0.5812	0.324	0.397
292	C4H8Cl2	1,4-DICHLOROBUTANE	127.013	235.85	427.05	641.00	36.10	343.0	0.3703	0.232	0.322
293	C4H8I2	1,2-DIIODOBUTANE	309.916	279.06	476.76	726.41	37.27	429.5	0.7216	0.265	0.281
294	C4H8O	n-BUTYRALDEHYDE	72.107	176.75	347.95	525.00	40.00	263.0	0.2742	0.241	0.345
295	C4H8O	ISOBUTYRALDEHYDE	72.107	208.15	337.25	507.00	41.00	263.0	0.2742	0.256	0.370
296	C4H8O	1,2-EPOXYBUTANE	72.107	123.15	336.57	526.00	43.90	258.0	0.2795	0.259	0.235
297	C4H8O	METHYL ETHYL KETONE	72.107	186.48	352.79	535.50	41.54	267.0	0.2701	0.249	0.324
298	C4H8O	ETHYL VINYL ETHER	72.107	157.35	308.70	475.15	40.73	263.0	0.2742	0.271	0.266
299	C4H8O	TETRAHYDROFURAN	72.107	164.65	338.00	540.15	51.88	223.9	0.3220	0.259	0.226
300	C4H8O2	cis-2-BUTENE-1,4-DIOL	88.106	284.15	508.15	677.88	52.00	279.0	0.3158	0.257	1.174
301	C4H8O2	trans-2-BUTENE-1,4-DIOL	88.106	300.45	510.00	681.00	52.00	279.0	0.3158	0.256	1.174
302	C4H8O2	ISOBUTYRIC ACID	88.106	227.15	427.85	609.15	40.53	292.0	0.3017	0.234	0.618
303	C4H8O2	n-BUTYRIC ACID	88.106	267.95	436.42	628.00	44.20	283.0	0.3113	0.240	0.604
304	C4H8O2	1,4-DIOXANE	88.106	284.95	374.47	587.00	52.08	238.0	0.3702	0.254	0.280
305	C4H8O2	ETHYL ACETATE	88.106	189.60	350.21	523.30	38.80	286.0	0.3081	0.255	0.366
306	C4H8O2	METHYL PROPIONATE	88.106	185.65	352.60	530.60	40.04	282.0	0.3124	0.256	0.353
307	C4H8O2	n-PROPYL FORMATE	88.106	180.25	353.97	538.00	40.63	285.0	0.3091	0.259	0.318
308	C4H8O2S	SULFOLANE	120.172	300.75	558.15	849.00	50.30	300.0	0.4006	0.214	0.382
309	C4H8S	TETRAHYDROTHIOPHENE	88.173	176.99	394.27	631.95	51.66	249.0	0.3541	0.245	0.199
310	C4H9Br	1-BROMOBUTANE	137.019	160.75	374.75	577.00	45.40	319.0	0.4295	0.302	0.323
311	C4H9Br	2-BROMOBUTANE	137.019	161.25	364.37	567.00	46.30	320.0	0.4282	0.314	0.268
312	C4H9Cl	n-BUTYL CHLORIDE	92.568	150.05	351.58	537.00	38.20	300.0	0.3086	0.257	0.274

NO	FORMULA	NAME	MW g/mol	T_F K	T_B K	T_C K	P_C bar	V_C cm3/mol	RHO_C g/cm3	Z_C	OMEGA
313	C4H9Cl	sec-BUTYL CHLORIDE	92.568	141.85	341.25	520.60	39.00	300.0	0.3086	0.270	0.291
314	C4H9Cl	tert-BUTYL CHLORIDE	92.568	247.75	323.75	507.00	39.00	300.0	0.3086	0.278	0.194
315	C4H9I	2-IODO-2-METHYLPROPANE	184.020	234.96	373.16	587.90	38.82	336.5	0.5469	0.267	0.179
316	C4H9N	PYRROLIDINE	71.122	215.31	359.72	568.55	56.13	248.7	0.2860	0.295	0.275
317	C4H9NO	N,N-DIMETHYLACETAMIDE	87.122	253.15	439.25	658.00	40.30	321.0	0.2714	0.236	0.364
318	C4H9NO	MORPHOLINE	87.122	270.05	401.15	618.00	53.40	276.0	0.3157	0.287	0.358
319	C4H9NO2	1-NITROBUTANE	103.121	191.83	426.05	624.00	38.00	341.5	0.3020	0.260	0.452
320	C4H9NO2	2-NITROBUTANE	103.121	141.16	412.85	615.00	36.00	335.5	0.3074	0.264	0.357
321	C4H10	n-BUTANE	58.123	134.86	272.65	425.18	37.97	254.9	0.2280	0.274	0.199
322	C4H10	ISOBUTANE	58.123	113.54	261.43	408.14	36.48	262.7	0.2213	0.282	0.177
323	C4H10N2	PIPERAZINE	86.137	379.15	419.15	638.00	55.30	310.0	0.2779	0.323	0.414
324	C4H10O	n-BUTANOL	74.123	183.85	390.81	562.93	44.13	274.5	0.2700	0.259	0.595
325	C4H10O	sec-BUTANOL	74.123	158.45	372.70	536.01	41.94	268.0	0.2766	0.252	0.571
326	C4H10O	tert-BUTANOL	74.123	298.97	355.57	506.20	39.72	275.0	0.2695	0.260	0.616
327	C4H10O	DIETHYL ETHER	74.123	156.85	307.58	466.70	36.38	280.0	0.2647	0.262	0.285
328	C4H10O	METHYL-PROPYL-ETHER	74.122	156.87	311.72	476.20	38.00	274.0	0.2705	0.263	0.271
329	C4H10O	METHYL ISOPROPYL ETHER	74.123	127.93	303.92	464.50	38.80	276.0	0.2686	0.277	0.279
330	C4H10O	ISOBUTANOL	74.123	165.15	380.81	547.73	42.95	272.0	0.2725	0.257	0.589
331	C4H10O2	1,3-BUTANEDIOL	90.122	196.15	480.15	643.00	50.00	292.0	0.3086	0.273	1.146
332	C4H10O2	1,4-BUTANEDIOL	90.122	293.05	501.15	667.00	48.80	297.0	0.3034	0.261	1.189
333	C4H10O2	2,3-BUTANEDIOL	90.122	280.75	453.85	611.00	51.30	267.0	0.3375	0.270	1.106
334	C4H10O2	t-BUTYL HYDROPEROXIDE	90.122	277.45	405.50	576.00	43.40	290.0	0.3108	0.263	0.668
335	C4H10O2	1,2-DIMETHOXYETHANE	90.122	215.15	357.20	536.15	38.70	270.6	0.3330	0.235	0.346
336	C4H10O2	2-ETHOXYETHANOL	90.122	-------	408.15	569.00	42.40	294.0	0.3065	0.264	0.759
337	C4H10O3	DIETHYLENE GLYCOL	106.122	262.70	518.15	744.60	46.00	312.0	0.3401	0.232	0.621
338	C4H10O4S	DIETHYL SULFATE	154.187	248.00	483.00	792.00	68.90	398.0	0.3874	0.416	0.162
339	C4H10S	n-BUTYL MERCAPTAN	90.189	157.46	371.61	569.00	39.70	307.0	0.2938	0.258	0.278
340	C4H10S	ISOBUTYL MERCAPTAN	90.189	128.31	361.64	559.00	40.60	307.0	0.2938	0.268	0.252
341	C4H10S	sec-BUTYL MERCAPTAN	90.189	133.02	358.13	554.00	40.60	307.0	0.2938	0.271	0.248
342	C4H10S	tert-BUTYL MERCAPTAN	90.189	274.26	337.37	530.00	40.60	307.0	0.2938	0.273	0.191
343	C4H10S	DIETHYL SULFIDE	90.189	169.20	365.25	557.15	39.62	318.0	0.2836	0.272	0.294
344	C4H10S	ISOPROPYL-METHYL-SULFIDE	90.183	171.65	357.90	551.00	39.00	307.5	0.2933	0.269	0.259
345	C4H10S	METHYL-PROPYL-SULFIDE	90.183	160.19	368.71	563.00	38.50	313.5	0.2877	0.266	0.285
346	C4H10S2	DIETHYL DISULFIDE	122.255	171.63	427.13	642.00	38.70	358.0	0.3415	0.260	0.346
347	C4H11N	n-BUTYLAMINE	73.138	224.05	350.55	531.90	42.00	313.0	0.2337	0.297	0.330
348	C4H11N	ISOBUTYLAMINE	73.138	188.55	340.88	513.73	42.15	312.0	0.2344	0.308	0.363
349	C4H11N	sec-BUTYLAMINE	73.138	168.65	336.15	514.30	40.00	310.0	0.2359	0.290	0.282
350	C4H11N	tert-BUTYLAMINE	73.138	206.19	317.55	483.90	38.40	293.0	0.2496	0.280	0.275
351	C4H11N	DIETHYLAMINE	73.138	223.35	328.60	496.60	37.09	301.0	0.2430	0.270	0.304
352	C4H11NO	DIMETHYLETHANOLAMINE	89.137	214.15	407.15	571.82	41.40	300.0	0.2971	0.261	0.711
353	C4H11NO2	DIETHANOLAMINE	105.137	301.15	542.04	715.00	32.70	349.0	0.3013	0.192	1.046
354	C4H11NO2	2-AMINOETHOXYETHANOL	105.137	-------	514.00	699.00	43.60	330.0	0.3186	0.248	0.969
355	C4H12N2O	N-AMINOETHYL ETHANOLAMINE	104.152	-------	517.00	698.00	44.60	387.0	0.2691	0.297	1.050
356	C4H12Si	TETRAMETHYLSILANE	88.225	174.07	299.80	450.40	28.14	357.0	0.2471	0.268	0.224
357	C4H13N3	DIETHYLENE TRIAMINE	103.167	234.15	480.25	676.00	42.20	342.0	0.3017	0.257	0.700
358	C5Cl6	HEXACHLOROCYCLOPENTADIENE	272.771	284.49	512.15	746.00	30.10	526.0	0.5186	0.255	0.369
359	C5H4O2	FURFURAL	96.086	236.65	434.85	657.00	55.12	252.0	0.3813	0.254	0.444
360	C5H5N	PYRIDINE	79.101	231.53	388.41	619.95	56.34	254.0	0.3114	0.278	0.239
361	C5H6	CYCLOPENTADIENE	66.103	188.15	314.65	507.00	51.50	225.0	0.2938	0.275	0.212
362	C5H6	2-METHYL-1-BUTENE-3-YNE	66.103	160.15	305.40	492.00	43.80	248.0	0.2665	0.266	0.137
363	C5H6	1-PENTENE-3-YNE	66.103	-------	332.40	520.00	44.00	256.0	0.2582	0.261	0.252
364	C5H6	1-PENTENE-4-YNE	66.103	-------	315.65	503.00	44.00	256.0	0.2582	0.269	0.179
365	C5H6N2	GLUTARONITRILE	94.116	244.21	559.15	782.00	31.50	352.0	0.2674	0.171	0.603
366	C5H6O2	FURFURYL ALCOHOL	98.101	258.52	443.15	632.00	53.50	263.0	0.3730	0.268	0.736
367	C5H6O3	GLUTARIC ANHYDRIDE	114.101	328.00	562.69	838.00	58.00	275.0	0.4149	0.229	0.537
368	C5H6O4	CITRACONIC ACID	130.100	356.15	607.00	829.00	42.40	340.0	0.3826	0.209	0.927
369	C5H6O4	ITACONIC ACID	130.100	438.75	601.00	821.00	42.40	340.0	0.3826	0.211	0.925
370	C5H6S	2-METHYLTHIOPHENE	98.162	209.77	385.71	610.00	48.50	275.5	0.3563	0.265	0.238
371	C5H6S	3-METHYLTHIOPHENE	98.162	204.18	388.60	615.00	49.50	275.5	0.3563	0.263	0.242
372	C5H7N	N-METHYLPYRROLE	81.117	216.91	385.89	610.00	47.70	283.0	0.2866	0.266	0.213
373	C5H7NO2	ETHYL CYANOACETATE	113.116	250.65	479.15	679.00	33.40	358.0	0.3160	0.212	0.573
374	C5H8	CYCLOPENTENE	68.118	138.13	317.38	507.00	47.90	240.0	0.2838	0.273	0.195
375	C5H8	ISOPRENE	68.118	127.27	307.21	484.00	38.50	276.0	0.2468	0.264	0.158
376	C5H8	3-METHYL-1,2-BUTADIENE	68.118	159.53	314.00	490.00	38.30	291.0	0.2341	0.274	0.187
377	C5H8	2-METHYL-1,3-BUTADIENE	68.118	127.20	307.22	483.30	37.40	266.0	0.2561	0.248	0.164
378	C5H8	1,2-PENTADIENE	68.118	135.89	318.01	500.00	38.00	276.0	0.2468	0.252	0.154
379	C5H8	cis-1,3-PENTADIENE	68.118	132.35	317.22	499.00	37.40	276.0	0.2468	0.249	0.147
380	C5H8	trans-1,3-PENTADIENE	68.118	185.71	315.17	500.00	37.40	276.0	0.2468	0.248	0.116
381	C5H8	1,4-PENTADIENE	68.118	124.86	299.11	479.00	37.40	303.0	0.2248	0.285	0.084
382	C5H8	2,3-PENTADIENE	68.118	147.50	321.40	497.00	38.00	295.0	0.2309	0.271	0.218
383	C5H8	1-PENTYNE	68.118	167.45	313.33	481.20	41.70	277.0	0.2459	0.289	0.290
384	C5H8	2-PENTYNE	68.118	163.86	329.22	521.99	42.28	277.5	0.2455	0.270	0.186
385	C5H8	3-METHYL-1-BUTYNE	68.118	183.45	302.15	463.20	42.00	275.0	0.2477	0.300	0.308
386	C5H8	SPIROPENTANE	68.118	166.11	312.19	499.74	52.13	236.5	0.2880	0.297	0.221
387	C5H8N4O12	PENTAERYTHRITOL TETRANITRATE	316.138	413.65	543.00	676.00	22.40	731.0	0.4325	0.291	1.451
388	C5H8O	CYCLOPENTANONE	84.118	221.85	403.80	626.00	58.50	258.0	0.3260	0.290	0.388
389	C5H8O	METHYL ISOPROPENYL KETONE	84.118	219.55	371.15	566.00	38.90	302.0	0.2785	0.250	0.286
390	C5H8O2	ACETYLACETONE	100.117	249.65	413.55	602.00	39.60	323.0	0.3100	0.256	0.496

11

NO	FORMULA	NAME	MW g/mol	T_F K	T_B K	T_C K	P_C bar	V_C cm3/mol	RHO_C g/cm3	Z_C	OMEGA
391	C5H8O2	ALLYL ACETATE	100.117	138.00	377.15	559.00	36.80	323.0	0.3100	0.256	0.388
392	C5H8O2	ETHYL ACRYLATE	100.117	201.95	372.65	553.00	36.80	323.0	0.3100	0.259	0.378
393	C5H8O2	METHYL METHACRYLATE	100.117	224.95	373.45	564.00	36.80	323.0	0.3100	0.253	0.317
394	C5H8O2	VINYL PROPIONATE	100.117	-------	364.35	546.00	36.80	323.0	0.3100	0.262	0.336
395	C5H8O3	2-HYDROXYETHYL ACRYLATE	116.117	213.00	484.00	662.00	39.80	359.0	0.3234	0.260	0.864
396	C5H8O3	LEVULINIC ACID	116.117	308.15	518.95	723.00	40.20	343.0	0.3385	0.229	0.787
397	C5H8O3	METHYL ACETOACETATE	116.117	193.15	444.85	642.00	37.10	343.0	0.3385	0.238	0.513
398	C5H8O4	GLUTARIC ACID	132.116	370.65	595.54	807.00	40.40	363.0	0.3640	0.219	0.959
399	C5H9N	VALERONITRILE	83.133	176.95	414.45	603.00	32.60	331.0	0.2512	0.215	0.415
400	C5H9NO	n-BUTYL ISOCYANATE	99.133	-------	388.15	568.00	34.40	360.0	0.2754	0.262	0.415
401	C5H9NO	N-METHYL-2-PYRROLIDONE	99.133	249.15	475.15	724.00	47.80	316.0	0.3137	0.251	0.358
402	C5H9NO4	L-GLUTAMIC ACID	147.131	497.15	670.00	886.00	41.34	383.3	0.3839	0.215	1.197
403	C5H10	CYCLOPENTANE	70.134	179.31	322.40	511.76	45.02	258.3	0.2715	0.273	0.194
404	C5H10	2-METHYL-1-BUTENE	70.134	135.58	304.30	465.00	34.00	292.0	0.2402	0.257	0.229
405	C5H10	2-METHYL-2-BUTENE	70.134	139.39	311.71	471.00	34.00	292.0	0.2402	0.254	0.277
406	C5H10	3-METHYL-1-BUTENE	70.134	104.66	293.21	450.37	35.16	302.1	0.2322	0.284	0.229
407	C5H10	1-PENTENE	70.134	107.93	303.11	464.78	35.29	296.0	0.2369	0.270	0.233
408	C5H10	cis-2-PENTENE	70.134	121.75	310.08	475.93	36.54	302.1	0.2322	0.279	0.241
409	C5H10	trans-2-PENTENE	70.134	132.89	309.49	475.37	36.54	302.1	0.2322	0.279	0.237
410	C5H10Br2	2,3-DIBROMO-2-METHYLBUTANE	229.942	288.00	444.01	668.37	42.55	422.5	0.5442	0.324	0.377
411	C5H10Cl2	1,5-DICHLOROPENTANE	141.040	200.35	453.15	663.00	31.90	422.0	0.3342	0.244	0.385
412	C5H10O	METHYL ISOPROPYL KETONE	86.134	181.15	367.55	553.00	38.50	310.0	0.2779	0.260	0.350
413	C5H10O	2-PENTANONE	86.134	196.29	375.46	561.08	36.94	301.0	0.2862	0.238	0.346
414	C5H10O	DIETHYL KETONE	86.134	234.18	375.14	560.95	37.39	336.0	0.2564	0.269	0.350
415	C5H10O	VALERALDEHYDE	86.134	182.00	376.15	554.00	35.00	316.0	0.2726	0.240	0.393
416	C5H10O2	n-BUTYL FORMATE	102.133	181.25	379.25	559.00	35.10	336.0	0.3040	0.254	0.384
417	C5H10O2	ETHYL PROPIONATE	102.133	199.25	372.25	546.00	33.62	345.0	0.2960	0.256	0.394
418	C5H10O2	ISOBUTYL FORMATE	102.133	177.35	371.22	551.35	38.81	352.0	0.2902	0.298	0.390
419	C5H10O2	ISOPROPYL ACETATE	102.133	199.75	361.65	538.00	35.80	336.0	0.3040	0.269	0.355
420	C5H10O2	n-PROPYL ACETATE	102.133	178.15	374.65	549.40	33.60	345.0	0.2960	0.254	0.394
421	C5H10O2	METHYL n-BUTYRATE	102.133	187.35	375.90	554.50	34.73	340.0	0.3004	0.256	0.381
422	C5H10O2	2-METHYLBUTYRIC ACID	102.133	-------	450.15	643.00	38.90	347.0	0.2943	0.252	0.589
423	C5H10O2	ISOVALERIC ACID	102.133	243.85	448.25	634.00	38.90	336.0	0.3040	0.248	0.648
424	C5H10O2	VALERIC ACID	102.133	239.15	458.65	651.00	38.10	336.0	0.3040	0.237	0.627
425	C5H10O2	TETRAHYDROFURFURYL ALCOHOL	102.133	-------	451.15	639.00	46.60	290.0	0.3522	0.254	0.703
426	C5H10O2S	3-METHYL SULFOLANE	134.199	273.65	549.15	817.00	42.40	353.0	0.3802	0.220	0.419
427	C5H10O3	DIETHYL CARBONATE	118.133	230.15	399.95	576.00	33.90	356.0	0.3318	0.252	0.485
428	C5H10O3	ETHYL LACTATE	118.133	247.15	427.65	588.00	38.60	354.0	0.3337	0.280	0.793
429	C5H10S	THIACYCLOHEXANE	102.194	292.14	414.90	657.12	46.53	295.5	0.3458	0.252	0.220
430	C5H10S	CYCLOPENTANETHIOL	102.194	155.39	405.33	629.00	42.70	310.5	0.3291	0.276	0.262
431	C5H11Br	1-BROMOPENTANE	151.046	185.26	402.74	564.76	37.68	377.5	0.4001	0.303	0.384
432	C5H11Cl	1-CHLOROPENTANE	106.595	174.15	381.54	568.00	33.50	352.0	0.3028	0.250	0.334
433	C5H11Cl	1-CHLORO-3-METHYLBUTANE	106.595	168.76	371.66	558.87	33.53	358.5	0.2973	0.259	0.293
434	C5H11Cl	2-CHLORO-2-METHYLBUTANE	106.595	199.66	358.76	548.97	33.96	353.5	0.3015	0.263	0.233
435	C5H11N	N-METHYLPYRROLIDINE	85.149	183.15	352.30	550.00	42.00	298.0	0.2857	0.274	0.227
436	C5H11N	PIPERIDINE	85.149	262.65	379.55	594.05	46.51	308.0	0.2765	0.290	0.243
437	C5H11NO	tert-BUTYLFORMAMIDE	101.148	289.15	475.15	692.00	35.60	383.0	0.2641	0.237	0.449
438	C5H12	ISOPENTANE	72.150	113.25	300.99	460.43	33.81	305.8	0.2359	0.270	0.228
439	C5H12	NEOPENTANE	72.150	256.58	282.65	433.78	31.99	303.6	0.2377	0.269	0.196
440	C5H12	n-PENTANE	72.150	143.42	309.22	469.65	33.69	312.3	0.2310	0.269	0.249
441	C5H12O	2,2-DIMETHYL-1-PROPANOL	88.150	327.15	386.25	550.00	38.80	327.0	0.2696	0.277	0.604
442	C5H12O	tert-PENTYL-ALCOHOL	88.149	327.00	386.30	549.00	39.71	323.5	0.2725	0.280	0.621
443	C5H12O	2-METHYL-1-BUTANOL	88.150	-------	401.85	565.00	38.80	327.0	0.2696	0.270	0.678
444	C5H12O	2-METHYL-2-BUTANOL	88.150	264.35	375.15	545.15	38.80	327.0	0.2696	0.280	0.483
445	C5H12O	3-METHYL-1-BUTANOL	88.150	155.95	404.35	579.45	38.80	327.0	0.2696	0.263	0.556
446	C5H12O	3-METHYL-2-BUTANOL	88.150	-------	384.65	574.00	39.60	327.0	0.2696	0.271	0.351
447	C5H12O	1-PENTANOL	88.150	195.56	410.95	586.15	38.80	326.0	0.2704	0.260	0.594
448	C5H12O	2-PENTANOL	88.150	200.00	392.15	552.00	38.80	327.0	0.2696	0.276	0.675
449	C5H12O	3-PENTANOL	88.150	204.15	388.45	547.00	38.80	327.0	0.2696	0.279	0.675
450	C5H12O	METHYL sec-BUTYL ETHER	88.150	-------	332.15	498.00	34.10	329.0	0.2679	0.271	0.306
451	C5H12O	METHYL tert-BUTYL ETHER	88.150	164.55	328.35	497.10	34.30	329.0	0.2679	0.273	0.267
452	C5H12O	METHYL ISOBUTYL ETHER	88.150	-------	331.70	497.00	34.10	329.0	0.2679	0.272	0.310
453	C5H12O	ETHYL PROPYL ETHER	88.150	145.65	337.01	500.23	33.70	339.0	0.2600	0.275	0.346
454	C5H12O2	ETHYLENE GLYCOL MONOPROPYL ETHER	104.149	183.15	424.50	582.00	36.70	347.0	0.3001	0.263	0.783
455	C5H12O2	NEOPENTYL GLYCOL	104.149	400.00	483.00	643.00	42.40	345.0	0.3019	0.274	1.143
456	C5H12O2	1,5-PENTANEDIOL	104.149	257.15	512.15	673.00	41.50	345.0	0.3019	0.256	1.220
457	C5H12O3	2-(2-METHOXYETHOXY)ETHANOL	120.148	197.15	466.75	630.00	35.40	367.0	0.3274	0.248	0.870
458	C5H12O4	PENTAERYTHRITOL	136.148	534.15	631.00	780.00	47.80	381.0	0.3573	0.281	2.120
459	C5H12S	n-PENTYL MERCAPTAN	104.216	197.45	399.79	598.00	34.70	359.0	0.2903	0.251	0.321
460	C5H12S	BUTYL-METHYL-SULFIDE	104.210	175.33	396.58	591.00	33.80	369.5	0.2820	0.265	0.332
461	C5H12S	ETHYL-PROPYL-SULFIDE	104.210	156.15	391.65	584.00	33.80	369.5	0.2820	0.268	0.329
462	C5H12S	2-METHYL-2-BUTANETHIOL	104.210	169.38	372.28	566.00	32.70	358.5	0.2907	0.288	0.243
463	C5H13N	n-PENTYLAMINE	87.165	218.15	377.65	555.00	35.80	365.0	0.2388	0.283	0.407
464	C5H13NO2	METHYL DIETHANOLAMINE	119.164	252.15	520.15	678.00	38.80	401.0	0.2972	0.276	1.302
465	C6Cl6	HEXACHLOROBENZENE	284.782	501.70	582.55	825.00	28.50	526.0	0.5414	0.219	0.497
466	C6F6	HEXAFLUOROBENZENE	186.056	278.25	353.41	516.73	32.73	335.0	0.5554	0.255	0.395
467	C6H3ClN2O4	1-CHLORO-2,4-DINITROBENZENE	202.554	326.55	588.00	813.77	34.90	478.0	0.4238	0.247	0.732
468	C6H3Cl2NO2	1,2-DICHLORO-4-NITROBENZENE	192.001	315.65	529.00	758.00	36.00	436.0	0.4404	0.249	0.539

NO	FORMULA	NAME	MW g/mol	T_F K	T_B K	T_C K	P_C bar	V_C cm3/mol	RHO_C g/cm3	Z_C	OMEGA
469	C6H3Cl3	1,2,4-TRICHLOROBENZENE	181.448	290.15	486.15	725.00	37.20	395.0	0.4594	0.244	0.358
470	C6H3N3O6	1,3,5-TRINITROBENZENE	213.106	398.40	748.00	1005.00	33.90	520.0	0.4098	0.211	0.808
471	C6H4Br2	m-DIBROMOBENZENE	235.906	266.25	491.15	761.00	46.60	372.0	0.6342	0.274	0.293
472	C6H4ClNO2	m-CHLORONITROBENZENE	157.556	317.65	508.75	742.00	39.80	432.0	0.3647	0.279	0.489
473	C6H4ClNO2	o-CHLORONITROBENZENE	157.556	306.15	519.00	757.00	39.80	432.0	0.3647	0.273	0.483
474	C6H4ClNO2	p-CHLORONITROBENZENE	157.556	356.65	515.15	751.00	39.80	432.0	0.3647	0.275	0.491
475	C6H4Cl2	m-DICHLOROBENZENE	147.003	248.39	446.23	683.95	40.70	351.0	0.4188	0.251	0.279
476	C6H4Cl2	o-DICHLOROBENZENE	147.003	256.15	453.57	705.00	40.70	351.0	0.4188	0.244	0.219
477	C6H4Cl2	p-DICHLOROBENZENE	147.003	326.14	447.21	684.75	40.70	351.0	0.4188	0.251	0.285
478	C6H4F2	m-DIFLUOROBENZENE	114.094	249.16	363.66	552.94	40.67	299.5	0.3809	0.265	0.320
479	C6H4F2	o-DIFLUOROBENZENE	114.094	239.16	364.66	554.46	40.67	299.5	0.3809	0.264	0.320
480	C6H4F2	p-DIFLUOROBENZENE	114.094	260.16	362.00	556.00	44.00	299.5	0.3809	0.266	0.299
481	C6H4N2O4	m-DINITROBENZENE	168.109	364.00	573.00	805.00	38.50	434.0	0.3873	0.250	0.682
482	C6H4N2O4	o-DINITROBENZENE	168.109	390.40	592.00	831.00	38.50	434.0	0.3873	0.242	0.687
483	C6H4N2O4	p-DINITROBENZENE	168.109	446.60	572.00	803.00	38.50	434.0	0.3873	0.250	0.686
484	C6H5Br	BROMOBENZENE	157.010	242.43	429.24	670.15	45.19	324.0	0.4846	0.263	0.251
485	C6H5Cl	MONOCHLOROBENZENE	112.558	227.95	404.87	632.35	45.19	308.0	0.3654	0.265	0.251
486	C6H5ClO	m-CHLOROPHENOL	128.558	306.00	487.00	729.00	53.20	320.0	0.4017	0.281	0.486
487	C6H5ClO	o-CHLOROPHENOL	128.558	282.00	447.53	675.00	50.00	325.0	0.3956	0.290	0.437
488	C6H5ClO	p-CHLOROPHENOL	128.558	316.00	493.11	738.00	53.20	325.0	0.3956	0.282	0.485
489	C6H5Cl2N	3,4-DICHLOROANILINE	162.018	344.65	545.00	800.00	41.10	409.0	0.3961	0.253	0.468
490	C6H5F	FLUOROBENZENE	96.104	230.94	357.88	560.09	45.51	269.0	0.3573	0.263	0.247
491	C6H5I	IODOBENZENE	204.010	241.83	461.60	721.15	45.19	351.0	0.5812	0.265	0.247
492	C6H5NO2	NITROBENZENE	123.111	278.91	483.95	719.00	44.00	349.0	0.3528	0.257	0.448
493	C6H6	BENZENE	78.114	278.68	353.24	562.16	48.98	258.9	0.3017	0.271	0.211
494	C6H6ClN	m-CHLOROANILINE	127.573	262.75	501.65	751.00	45.90	364.0	0.3505	0.268	0.420
495	C6H6ClN	o-CHLOROANILINE	127.573	481.99	481.99	722.00	45.90	364.0	0.3505	0.278	0.425
496	C6H6ClN	p-CHLOROANILINE	127.573	343.05	503.65	754.00	45.90	364.0	0.3505	0.267	0.421
497	C6H6N2	cis-DICYANO-1-BUTENE	106.127	249.00	501.00	691.00	29.50	392.0	0.2707	0.201	0.672
498	C6H6N2	trans-DICYANO-1-BUTENE	106.127	260.00	499.00	689.00	29.50	392.0	0.2707	0.202	0.664
499	C6H6N2	1,4-DICYANO-2-BUTENE	106.127	349.00	547.00	755.00	29.50	426.0	0.2491	0.200	0.667
500	C6H6N2O2	m-NITROANILINE	138.126	387.15	579.00	815.00	44.20	406.0	0.3402	0.265	0.740
501	C6H6N2O2	o-NITROANILINE	138.126	344.65	558.00	784.00	44.20	406.0	0.3402	0.275	0.741
502	C6H6N2O2	p-NITROANILINE	138.126	420.65	609.15	851.00	44.20	406.0	0.3402	0.254	0.782
503	C6H6O	PHENOL	94.113	314.06	454.99	694.25	61.30	229.0	0.4110	0.243	0.426
504	C6H6O2	1,2-BENZENEDIOL	110.112	377.60	518.65	764.00	74.90	300.0	0.3670	0.354	0.701
505	C6H6O2	1,3-BENZENEDIOL	110.112	382.00	549.65	810.00	74.90	300.0	0.3670	0.334	0.677
506	C6H6O2	p-HYDROQUINONE	110.112	444.65	558.15	822.00	74.50	300.0	0.3670	0.327	0.686
507	C6H6O3	1,2,3-BENZENETRIOL	126.112	407.00	581.85	830.00	88.10	318.0	0.3966	0.406	0.945
508	C6H6S	PHENYL MERCAPTAN	110.180	258.26	442.29	689.00	47.40	315.0	0.3498	0.261	0.263
509	C6H7N	ANILINE	93.128	267.13	457.60	699.00	53.09	270.0	0.3449	0.247	0.404
510	C6H7N	2-METHYLPYRIDINE	93.128	206.44	402.55	621.00	43.80	320.0	0.2910	0.271	0.278
511	C6H7N	3-METHYLPYRIDINE	93.128	255.01	417.29	645.00	43.80	320.0	0.2910	0.261	0.271
512	C6H7N	4-METHYLPYRIDINE	93.128	276.73	418.50	646.15	46.61	325.6	0.2860	0.283	0.302
513	C6H8	1,3-CYCLOHEXADIENE	80.130	161.00	353.49	558.00	47.30	277.0	0.2893	0.282	0.231
514	C6H8	METHYLCYCLOPENTADIENE	80.130	-------	345.93	541.00	44.30	279.0	0.2872	0.275	0.238
515	C6H8N2	ADIPONITRILE	108.143	275.64	568.15	781.00	28.30	406.0	0.2664	0.177	0.672
516	C6H8N2	METHYLGLUTARONITRILE	108.143	228.15	536.15	742.00	28.80	404.0	0.2677	0.189	0.638
517	C6H8N2	m-PHENYLENEDIAMINE	108.143	334.00	560.00	824.00	51.80	377.0	0.2869	0.285	0.543
518	C6H8N2	o-PHENYLENEDIAMINE	108.143	376.95	525.00	781.00	51.80	315.0	0.3433	0.251	0.494
519	C6H8N2	p-PHENYLENEDIAMINE	108.143	413.00	540.00	796.00	51.80	317.0	0.3411	0.248	0.539
520	C6H8N2	PHENYLHYDRAZINE	108.143	292.35	516.65	761.00	49.10	418.0	0.2587	0.324	0.535
521	C6H8N2O	BIS(CYANOETHYL)ETHER	124.142	246.85	579.00	783.00	28.30	377.0	0.3293	0.164	0.782
522	C6H8O4	DIMETHYL MALEATE	144.127	254.15	478.15	675.00	32.20	403.0	0.3576	0.231	0.562
523	C6H8O6	ASCORBIC ACID	176.126	465.15	637.00	783.00	52.90	339.0	0.5195	0.275	2.389
524	C6H8O7	CITRIC ACID	192.125	426.15	659.00	822.00	37.98	419.7	0.4578	0.233	1.857
525	C6H10	1-METHYLCYCLOPENTENE	82.145	145.96	348.95	541.99	37.90	311.2	0.2640	0.273	0.219
526	C6H10	3-METHYLCYCLOPENTENE	82.145	130.16	343.16	535.71	40.16	298.5	0.2752	0.269	0.221
527	C6H10	4-METHYLCYCLOPENTENE	82.145	112.31	348.31	543.75	40.16	298.5	0.2752	0.265	0.221
528	C6H10	CYCLOHEXENE	82.145	169.67	356.12	560.40	43.50	291.0	0.2823	0.272	0.214
529	C6H10	2,3-DIMETHYL-1,3-BUTADIENE	82.145	197.15	341.93	526.00	35.20	315.0	0.2608	0.254	0.214
530	C6H10	1,5-HEXADIENE	82.145	132.47	332.61	507.00	33.50	339.0	0.2423	0.269	0.232
531	C6H10	cis,trans-2,4-HEXADIENE	82.145	177.05	356.65	538.00	33.50	331.0	0.2482	0.248	0.275
532	C6H10	trans,trans-2,4-HEXADIENE	82.145	228.25	355.05	535.00	33.50	331.0	0.2482	0.249	0.282
533	C6H10	1-HEXYNE	82.145	141.25	344.48	516.20	36.20	322.0	0.2551	0.272	0.333
534	C6H10	2-HEXYNE	82.145	183.65	357.67	549.00	35.30	331.0	0.2482	0.256	0.221
535	C6H10	3-HEXYNE	82.145	170.05	354.35	544.00	35.30	331.0	0.2482	0.258	0.218
536	C6H10O	CYCLOHEXANONE	98.145	242.00	428.90	629.15	38.50	311.0	0.3156	0.229	0.450
537	C6H10O	MESITYL OXIDE	98.145	220.15	402.95	600.00	34.10	355.0	0.2765	0.243	0.327
538	C6H10O2	epsilon-CAPROLACTONE	114.144	271.85	514.00	771.00	46.30	352.0	0.3243	0.254	0.442
539	C6H10O2	ETHYL METHACRYLATE	114.144	-------	390.15	577.00	32.50	375.0	0.3044	0.254	0.344
540	C6H10O2	n-PROPYL ACRYLATE	114.144	-------	392.15	569.00	32.50	376.0	0.3036	0.258	0.434
541	C6H10O3	ETHYLACETOACETATE	130.144	234.15	453.95	643.00	32.70	391.0	0.3328	0.239	0.561
542	C6H10O3	PROPIONIC ANHYDRIDE	130.144	228.15	442.15	618.00	33.40	396.0	0.3286	0.257	0.618
543	C6H10O4	ADIPIC ACID	146.143	425.50	611.00	809.00	35.30	400.0	0.3654	0.210	1.054
544	C6H10O4	DIETHYL OXALATE	146.143	232.55	458.85	646.00	30.90	416.0	0.3513	0.239	0.568
545	C6H10O4	ETHYLENE GLYCOL DIACETATE	146.143	242.15	463.65	653.00	30.90	416.0	0.3513	0.237	0.560
546	C6H10O4	ETHYLIDENE DIACETATE	146.143	292.00	442.15	635.00	32.60	406.0	0.3600	0.251	0.478

13

NO	FORMULA	NAME	MW g/mol	T_F K	T_B K	T_C K	P_C bar	V_C cm3/mol	RHO_C g/cm3	Z_C	OMEGA
547	C6H11N	HEXANENITRILE	97.160	192.85	436.75	622.05	29.20	384.0	0.2530	0.217	0.474
548	C6H11NO	epsilon-CAPROLACTAM	113.159	342.36	543.15	806.00	47.70	356.0	0.3179	0.253	0.477
549	C6H11NO	CYCLOHEXANONE OXIME	113.159	363.15	481.15	715.00	46.90	369.0	0.3067	0.291	0.462
550	C6H12	CYCLOHEXANE	84.161	279.69	353.87	553.54	40.75	307.9	0.2734	0.273	0.212
551	C6H12	2,3-DIMETHYL-1-BUTENE	84.161	115.89	328.76	500.00	32.20	349.0	0.2411	0.270	0.227
552	C6H12	2,3-DIMETHYL-2-BUTENE	84.161	198.82	346.35	524.00	31.60	372.0	0.2262	0.270	0.233
553	C6H12	3,3-DIMETHYL-1-BUTENE	84.161	157.95	314.40	480.00	32.90	333.0	0.2527	0.275	0.226
554	C6H12	2-ETHYL-1-BUTENE	84.161	141.61	337.82	512.00	31.60	364.0	0.2312	0.270	0.228
555	C6H12	trans-3-METHYL-2-PENTENE	84.161	134.70	343.60	521.00	32.90	350.0	0.2405	0.270	0.207
556	C6H12	1-HEXENE	84.161	133.39	336.63	504.03	31.40	354.0	0.2377	0.265	0.280
557	C6H12	cis-2-HEXENE	84.161	132.00	342.03	513.00	31.60	359.0	0.2344	0.266	0.272
558	C6H12	trans-2-HEXENE	84.161	140.17	341.02	513.00	31.60	360.0	0.2338	0.267	0.261
559	C6H12	cis-3-HEXENE	84.161	135.33	339.60	509.00	31.70	351.0	0.2398	0.263	0.279
560	C6H12	trans-3-HEXENE	84.161	159.73	340.24	509.00	31.70	351.0	0.2398	0.263	0.285
561	C6H12	METHYLCYCLOPENTANE	84.161	130.73	344.96	532.79	37.85	318.9	0.2639	0.272	0.230
562	C6H12	2-METHYL-1-PENTENE	84.161	137.42	335.25	507.00	31.60	359.0	0.2344	0.269	0.241
563	C6H12	2-METHYL-2-PENTENE	84.161	138.07	340.45	514.00	31.60	363.0	0.2318	0.268	0.245
564	C6H12	3-METHYL-1-PENTENE	84.161	120.20	327.33	495.00	32.90	343.3	0.2452	0.274	0.264
565	C6H12	3-METHYL-cis-2-PENTENE	84.161	138.31	340.85	515.00	32.90	343.0	0.2454	0.264	0.259
566	C6H12	4-METHYL-1-PENTENE	84.161	119.51	327.01	496.00	32.20	345.0	0.2439	0.269	0.239
567	C6H12	4-METHYL-cis-2-PENTENE	84.161	138.30	329.53	499.00	32.20	346.0	0.2432	0.269	0.244
568	C6H12	4-METHYL-trans-2-PENTENE	84.161	132.35	331.75	501.00	32.20	346.0	0.2432	0.267	0.255
569	C6H12N2	TRIETHYLENEDIAMINE	112.175	434.25	447.15	655.00	39.10	382.0	0.2937	0.274	0.460
570	C6H12O	BUTYL VINYL ETHER	100.161	181.25	366.97	536.00	31.20	364.0	0.2752	0.255	0.380
571	C6H12O	CYCLOHEXANOL	100.161	296.60	434.00	625.15	37.49	322.0	0.3111	0.232	0.514
572	C6H12O	1-HEXANAL	100.161	217.15	401.45	579.00	31.10	369.0	0.2714	0.238	0.439
573	C6H12O	ETHYL ISOPROPYL KETONE	100.161	-------	386.55	567.00	33.20	369.0	0.2714	0.260	0.391
574	C6H12O	2-HEXANONE	100.161	217.35	400.85	587.05	33.24	369.0	0.2714	0.251	0.397
575	C6H12O	3-HEXANONE	100.161	217.50	396.65	582.82	33.20	364.0	0.2752	0.249	0.376
576	C6H12O	METHYL ISOBUTYL KETONE	100.161	189.15	389.65	571.40	32.73	369.0	0.2714	0.254	0.389
577	C6H12O2	n-PENTYL FORMATE	116.160	199.65	406.60	576.00	31.25	389.0	0.2986	0.254	0.528
578	C6H12O2	n-BUTYL ACETATE	116.160	199.65	399.15	579.65	31.10	389.0	0.2986	0.251	0.410
579	C6H12O2	sec-BUTYL ACETATE	116.160	174.15	385.15	561.00	31.70	389.0	0.2986	0.264	0.406
580	C6H12O2	tert-BUTYL ACETATE	116.160	-------	369.15	545.00	31.70	389.0	0.2986	0.272	0.343
581	C6H12O2	ETHYL n-BUTYRATE	116.160	175.15	394.65	571.00	30.60	421.0	0.2759	0.271	0.419
582	C6H12O2	ETHYL ISOBUTYRATE	116.160	185.00	383.00	553.15	30.40	410.0	0.2833	0.271	0.426
583	C6H12O2	ISOBUTYL ACETATE	116.160	174.30	389.80	561.00	31.60	389.0	0.2986	0.264	0.454
584	C6H12O2	n-PROPYL PROPIONATE	116.160	197.25	395.65	578.00	31.10	389.0	0.2986	0.252	0.376
585	C6H12O2	CYCLOHEXYL PEROXIDE	116.160	253.15	490.00	685.00	42.10	342.0	0.3396	0.253	0.751
586	C6H12O2	DIACETONE ALCOHOL	116.160	229.15	441.00	606.00	36.00	387.0	0.3002	0.277	0.757
587	C6H12O2	2-ETHYL BUTYRIC ACID	116.160	258.15	466.95	655.00	34.10	389.0	0.2986	0.244	0.633
588	C6H12O2	n-HEXANOIC ACID	116.160	270.15	478.85	667.00	33.50	389.0	0.2986	0.235	0.670
589	C6H12O3	2-ETHOXYETHYL ACETATE	132.159	211.45	429.45	597.00	24.62	409.0	0.3231	0.203	0.534
590	C6H12O3	HYDROXYCAPROIC ACID	132.159	334.00	576.00	758.00	36.40	402.0	0.3288	0.232	1.163
591	C6H12O3	PARALDEHYDE	132.159	285.75	397.25	579.00	35.00	365.0	0.3621	0.265	0.441
592	C6H12O3	sec-BUTYL GLYCOLATE	132.160	-------	450.65	-------	-------	-------	-------	-------	-------
593	C6H12S	THIACYCLOHEPTANE	116.221	292.14	414.90	640.07	43.74	391.5	0.2969	0.322	0.291
594	C6H13N	CYCLOHEXYLAMINE	99.176	255.45	407.65	615.00	42.00	360.0	0.2755	0.296	0.360
595	C6H13N	HEXAMETHYLENEIMINE	99.176	236.15	404.85	615.00	42.70	361.0	0.2747	0.301	0.330
596	C6H14	2,2-DIMETHYLBUTANE	86.177	174.28	322.88	488.78	30.81	358.8	0.2402	0.272	0.234
597	C6H14	2,3-DIMETHYLBUTANE	86.177	145.19	331.13	499.98	31.27	357.8	0.2409	0.269	0.248
598	C6H14	n-HEXANE	86.177	177.84	341.88	507.43	30.12	369.9	0.2330	0.264	0.305
599	C6H14	2-METHYLPENTANE	86.177	119.55	333.41	497.50	30.10	366.4	0.2352	0.267	0.278
600	C6H14	3-METHYLPENTANE	86.177	110.25	336.42	504.43	31.24	366.4	0.2352	0.273	0.274
601	C6H14N2O2	LYSINE	146.189	483.00	615.00	821.00	35.30	502.0	0.2912	0.260	1.012
602	C6H14O	2-ETHYL-1-BUTANOL	102.177	158.75	419.65	580.00	34.00	380.0	0.2689	0.268	0.714
603	C6H14O	1-HEXANOL	102.177	228.55	430.15	611.35	35.10	381.3	0.2680	0.263	0.580
604	C6H14O	2-HEXANOL	102.177	223.00	413.04	586.20	34.00	380.0	0.2689	0.265	0.566
605	C6H14O	2-METHYL-1-PENTANOL	102.177	-------	421.15	582.00	34.00	380.0	0.2689	0.267	0.726
606	C6H14O	4-METHYL-2-PENTANOL	102.177	-------	404.85	574.40	34.70	380.0	0.2689	0.276	0.572
607	C6H14O	n-BUTYL ETHYL ETHER	102.177	170.15	365.35	531.00	29.90	382.0	0.2675	0.259	0.390
608	C6H14O	DIISOPROPYL ETHER	102.177	187.65	341.45	500.05	28.78	386.0	0.2647	0.267	0.338
609	C6H14O	DI-n-PROPYL ETHER	102.177	149.95	362.79	530.60	30.28	382.0	0.2675	0.262	0.370
610	C6H14O	METHYL tert-PENTYL ETHER	102.177	-------	359.45	534.00	30.40	382.0	0.2675	0.262	0.301
611	C6H14O2	ACETAL	118.176	173.15	376.75	541.00	29.80	402.0	0.2940	0.266	0.432
612	C6H14O2	2-BUTOXYETHANOL	118.176	203.15	444.47	600.00	32.40	400.0	0.2954	0.260	0.817
613	C6H14O2	1,6-HEXANEDIOL	118.176	315.15	516.15	670.00	36.10	398.0	0.2969	0.258	1.268
614	C6H14O2	HEXYLENE GLYCOL	118.176	223.15	470.65	670.65	40.10	398.0	0.2969	0.309	1.197
615	C6H14O2S	DI-n-PROPYL SULFONE	150.242	303.00	543.00	763.00	31.10	463.0	0.3245	0.227	0.582
616	C6H14O3	DIETHYLENE GLYCOL DIMETHYL ETHER	134.175	203.15	432.91	604.00	28.60	422.0	0.3180	0.240	0.575
617	C6H14O3	DIPROPYLENE GLYCOL	134.175	233.00	504.95	654.00	35.80	415.0	0.3233	0.273	1.198
618	C6H14O3	2-(2-ETHOXYETHOXY)ETHANOL	134.175	195.15	475.15	632.00	31.40	420.0	0.3195	0.251	0.901
619	C6H14O3	TRIMETHYLOLPROPANE	134.175	331.15	562.04	709.00	39.10	416.0	0.3225	0.276	1.543
620	C6H14O4	TRIETHYLENE GLYCOL	150.175	265.79	551.00	700.00	33.20	443.0	0.3390	0.253	1.386
621	C6H14O6	SORBITOL	182.174	370.85	777.00	959.00	46.40	483.0	0.3772	0.281	2.199
622	C6H14S	n-HEXYLMERCAPTAN	118.243	192.62	425.81	623.00	30.80	412.0	0.2870	0.245	0.368
623	C6H14S	BUTYL-ETHYL-SULFIDE	118.237	178.03	417.41	609.00	30.00	425.5	0.2779	0.265	0.374
624	C6H14S	ISOPROPYL-SULFIDE	118.237	170.45	393.19	585.71	32.25	413.5	0.2859	0.274	0.316

14

NO	FORMULA	NAME	MW g/mol	T_F K	T_B K	T_C K	P_C bar	V_C cm3/mol	RHO_C g/cm3	Z_C	OMEGA
625	C6H14S	METHYL-PENTYL-SULFIDE	118.237	179.16	401.16	587.98	31.67	425.5	0.2779	0.276	0.376
626	C6H14S	PROPYL-SULFIDE	118.237	170.45	416.00	609.73	31.67	425.5	0.2779	0.266	0.376
627	C6H14S2	PROPYL-DISULFIDE	150.297	187.68	464.65	673.00	27.50	479.5	0.3134	0.277	0.370
628	C6H15Al	TRIETHYL ALUMINUN	114.167	220.65	458.15	720.15	13.58	230.0	0.4964	0.522	-------
629	C6H15Al2Cl3	ETHYL ALUMINUM SESQUICHLORIDE	247.506	253.15	482.15	791.00	-------	-------	-------	-------	-------
630	C6H15N	DIISOPROPYLAMINE	101.192	176.85	357.05	523.10	32.00	418.0	0.2421	0.308	0.388
631	C6H15N	DI-n-PROPYLAMINE	101.192	210.15	382.00	555.80	36.30	418.0	0.2421	0.328	0.465
632	C6H15N	n-HEXYLAMINE	101.192	251.85	404.65	583.00	31.80	418.0	0.2421	0.274	0.467
633	C6H15N	TRIETHYLAMINE	101.192	158.45	361.92	535.15	30.40	390.0	0.2595	0.266	0.316
634	C6H15NO	6-AMINOHEXANOL	117.191	331.00	508.00	681.00	34.40	436.0	0.2688	0.265	0.970
635	C6H15NO2	DIISOPROPANOLAMINE	133.191	318.15	521.90	672.00	36.00	454.0	0.2934	0.293	1.389
636	C6H15NO3	TRIETHANOLAMINE	149.190	294.35	613.00	787.00	24.50	472.0	0.3161	0.177	1.101
637	C6H15N3	N-AMINOETHYL PIPERAZINE	129.205	254.15	493.55	708.00	38.50	407.0	0.3175	0.266	0.555
638	C6H15O4P	TRIETHYL PHOSPHATE	182.156	216.00	484.15	794.00	10.80	1010.0	0.1804	1.650	-------
639	C6H16N2	HEXAMETHYLENEDIAMINE	116.207	313.95	475.04	663.00	32.90	475.0	0.2446	0.284	0.650
640	C6H18N3OP	HEXAMETHYL PHOSPHORAMIDE	179.202	280.15	506.15	-------	-------	-------	-------	-------	-------
641	C6H18N4	TRIETHYLENE TETRAMINE	146.236	285.15	539.65	718.00	31.70	482.0	0.3034	0.256	0.974
642	C6H18OSi2	HEXAMETHYLDISILOXANE	162.379	204.93	373.67	518.70	19.14	601.0	0.2702	0.267	0.418
643	C6H18O3Si3	HEXAMETHYLCYCLOTRISILOXANE	222.464	337.15	408.26	554.20	16.63	634.0	0.3509	0.229	0.474
644	C6H19NSi2	HEXAMETHYLDISILAZANE	161.395	-------	399.15	544.00	19.20	613.0	0.2633	0.260	0.510
645	C7H3ClF3NO2	4-CHLORO-3-NITROBENZOTRIFLUORIDE	225.554	-------	495.15	686.00	27.40	490.0	0.4603	0.235	0.607
646	C7H3Cl2F3	2,4-DICHLOROBENZOTRIFLUORIDE	215.001	247.55	450.65	646.00	28.10	443.0	0.4853	0.232	0.434
647	C7H3Cl2NO	3,4-DICHLOROPHENYL ISOCYANATE	188.012	316.15	501.00	733.00	33.30	456.0	0.4123	0.249	0.335
648	C7H4ClF3	p-CHLOROBENZOTRIFLUORIDE	180.557	237.15	412.15	601.00	30.10	399.0	0.4525	0.240	0.373
649	C7H4Cl2O	m-CHLOROBENZOYL CHLORIDE	175.014	280.00	498.00	724.00	36.80	406.0	0.4311	0.248	0.454
650	C7H4F3NO2	3-NITROBENZOTRIFLUORIDE	191.110	272.00	475.93	667.00	28.00	441.0	0.4334	0.223	0.536
651	C7H5ClO	BENZOYL CHLORIDE	140.569	272.65	470.15	697.00	40.60	367.0	0.3830	0.257	0.421
652	C7H5ClO2	o-CHLOROBENZOIC ACID	156.568	415.15	560.15	792.00	40.30	383.0	0.4088	0.234	0.664
653	C7H5Cl3	BENZOTRICHLORIDE	195.475	268.40	486.65	737.00	33.40	447.0	0.4373	0.244	0.260
654	C7H5F3	BENZOTRIFLUORIDE	146.112	244.14	375.20	565.00	33.90	356.0	0.4104	0.257	0.282
655	C7H5N	BENZONITRILE	103.123	260.40	464.15	699.35	42.15	339.0	0.3042	0.246	0.352
656	C7H5NO	PHENYL ISOCYANATE	119.123	243.15	438.75	648.00	40.60	341.0	0.3493	0.257	0.438
657	C7H5N3O6	2,4,6-TRINITROTOLUENE	227.133	354.00	573.00	795.00	30.40	480.0	0.4732	0.221	1.977
658	C7H6Cl2	BENZYL DICHLORIDE	161.030	257.00	487.00	731.00	36.50	404.0	0.3986	0.243	0.326
659	C7H6Cl2	2,4-DICHLOROTOLUENE	161.030	259.65	474.25	705.00	35.90	404.0	0.3986	0.247	0.359
660	C7H6N2O4	2,4-DINITROTOLUENE	182.136	343.00	590.00	814.00	34.00	487.0	0.3740	0.245	0.718
661	C7H6N2O4	2,5-DINITROTOLUENE	182.136	325.65	590.00	814.00	34.00	472.0	0.3859	0.237	0.740
662	C7H6N2O4	2,6-DINITROTOLUENE	182.136	339.00	558.00	770.00	36.00	487.0	0.3740	0.242	0.738
663	C7H6N2O4	3,4-DINITROTOLUENE	182.136	332.00	610.00	842.00	34.00	487.0	0.3740	0.237	0.737
664	C7H6N2O4	3,5-DINITROTOLUENE	182.136	365.65	588.00	814.00	34.00	473.0	0.3851	0.238	0.702
665	C7H6O	BENZALDEHYDE	106.124	247.15	451.90	695.00	46.50	324.0	0.3275	0.261	0.305
666	C7H6O2	BENZOIC ACID	122.123	395.52	522.40	751.00	44.70	339.1	0.3600	0.246	0.604
667	C7H6O2	p-HYDROXYBENZALDEHYDE	122.123	390.15	583.15	844.00	49.90	361.0	0.3383	0.257	0.617
668	C7H6O2	SALICYLALDEHYDE	122.123	266.15	469.65	680.00	49.90	342.0	0.3571	0.302	0.626
669	C7H6O3	SALICYLIC ACID	138.123	431.75	529.00	739.00	51.80	364.0	0.3795	0.307	0.832
670	C7H7Br	p-BROMOTOLUENE	171.037	299.95	457.50	699.00	43.70	379.0	0.4513	0.285	0.318
671	C7H7Cl	BENZYL CHLORIDE	126.585	234.15	452.55	686.00	39.10	360.0	0.3516	0.247	0.314
672	C7H7Cl	o-CHLOROTOLUENE	126.585	236.65	432.30	656.00	39.10	354.0	0.3576	0.254	0.304
673	C7H7Cl	p-CHLOROTOLUENE	126.585	280.65	435.65	660.00	39.10	360.0	0.3516	0.257	0.313
674	C7H7F	p-FLUOROTOLUENE	110.131	216.36	389.76	590.48	38.15	337.5	0.3263	0.262	0.311
675	C7H7NO	FORMANILIDE	121.137	323.15	544.15	787.00	41.10	382.0	0.3171	0.240	0.545
676	C7H7NO2	m-NITROTOLUENE	137.138	289.20	505.00	734.00	38.00	441.0	0.3110	0.275	0.490
677	C7H7NO2	o-NITROTOLUENE	137.138	269.98	495.64	720.00	38.00	441.0	0.3110	0.280	0.482
678	C7H7NO2	p-NITROTOLUENE	137.138	324.75	511.65	736.00	38.00	441.0	0.3110	0.274	0.541
679	C7H7NO3	o-NITROANISOLE	153.138	283.60	546.15	782.00	37.60	422.0	0.3629	0.244	0.561
680	C7H8	TOLUENE	92.141	178.18	383.78	591.79	41.09	315.8	0.2918	0.264	0.264
681	C7H8	1,3,5-CYCLOHEPTATRIENE	92.141	193.66	388.56	593.90	43.34	311.5	0.2958	0.266	0.324
682	C7H8O	ANISOLE	108.140	235.65	426.73	641.65	41.75	337.0	0.3209	0.264	0.369
683	C7H8O	BENZYL ALCOHOL	108.140	257.85	477.85	677.00	45.50	335.0	0.3228	0.271	0.691
684	C7H8O	m-CRESOL	108.140	285.39	475.43	705.85	45.60	312.0	0.3466	0.242	0.449
685	C7H8O	o-CRESOL	108.140	304.19	464.15	697.55	50.06	282.0	0.3835	0.243	0.434
686	C7H8O	p-CRESOL	108.140	307.93	475.13	704.65	51.50	277.0	0.3904	0.244	0.513
687	C7H8O2	GUAIACOL	124.139	304.65	478.15	697.00	47.30	353.0	0.3517	0.288	0.563
688	C7H8O2	p-METHOXYPHENOL	124.139	329.00	516.00	758.00	49.70	342.0	0.3630	0.270	0.541
689	C7H9N	BENZYLAMINE	107.155	227.15	457.65	683.50	43.20	373.0	0.2873	0.284	0.409
690	C7H9N	2,6-DIMETHYLPYRIDINE	107.155	267.00	417.20	623.75	37.80	316.0	0.3391	0.230	0.350
691	C7H9N	N-METHYLANILINE	107.155	216.15	469.02	701.55	51.98	373.0	0.2873	0.332	0.480
692	C7H9N	m-TOLUIDINE	107.155	242.75	476.55	709.15	41.54	373.0	0.2873	0.263	0.413
693	C7H9N	o-TOLUIDINE	107.155	249.47	473.55	694.15	37.49	373.0	0.2873	0.242	0.442
694	C7H9N	p-TOLUIDINE	107.155	316.90	473.40	693.15	40.00	373.0	0.2873	0.259	0.476
695	C7H10	2-NORBORNENE	94.156	319.40	368.65	583.00	39.30	337.0	0.2794	0.273	0.159
696	C7H10N2	TOLUENEDIAMINE	122.170	371.25	557.15	804.00	43.80	376.0	0.3249	0.246	0.576
697	C7H11NO	CYCLOHEXYL ISOCYANATE	125.170	-------	442.15	633.00	34.70	408.0	0.3068	0.269	0.530
698	C7H12	1-HEPTYNE	96.172	192.26	372.86	559.69	32.95	389.5	0.2469	0.276	0.293
699	C7H12O2	n-BUTYL ACRYLATE	128.171	208.55	421.00	598.00	26.30	428.0	0.2995	0.226	0.438
700	C7H12O2	ISOBUTYL ACRYLATE	128.171	212.00	405.15	580.00	29.50	428.0	0.2995	0.262	0.457
701	C7H12O2	n-PROPYL METHACRYLATE	128.171	-------	414.00	599.00	29.10	428.0	0.2995	0.250	0.401
702	C7H12O4	DIETHYL MALONATE	160.170	224.25	472.05	653.00	27.80	469.0	0.3415	0.240	0.611

NO	FORMULA	NAME	MW g/mol	T_F K	T_B K	T_C K	P_C bar	V_C cm3/mol	RHO_C g/cm3	Z_C	OMEGA
703	C7H14	CYCLOHEPTANE	98.188	265.15	391.94	604.30	38.40	359.0	0.2735	0.274	0.243
704	C7H14	1,1-DIMETYLCYCLOPENTANE	98.188	203.36	361.00	547.00	34.45	360.0	0.2727	0.273	0.272
705	C7H14	cis-1,2-DIMETHYLCYCLOPENTANE	98.188	219.26	372.68	565.15	34.45	370.0	0.2654	0.271	0.266
706	C7H14	trans-1,2-DIMETHYLCYCLOPENTANE	98.188	155.58	365.02	553.15	34.45	360.0	0.2727	0.270	0.270
707	C7H14	cis-1,3-DIMETHYLCYCLOPENTANE	98.188	139.45	363.92	551.00	34.45	360.0	0.2727	0.271	0.274
708	C7H14	trans-1,3-DIMETHYLCYCLOPENTANE	98.188	139.18	364.88	553.00	34.45	360.0	0.2727	0.270	0.270
709	C7H14	ETHYLCYCLOPENTANE	98.188	134.71	376.62	569.52	33.98	374.5	0.2622	0.269	0.272
710	C7H14	2-ETHYL-1-PENTENE	98.188	168.00	367.15	,543.00	29.50	398.0	0.2467	0.260	0.309
711	C7H14	3-ETHYL-1-PENTENE	98.188	145.67	357.26	530.00	30.30	398.0	0.2467	0.274	0.302
712	C7H14	1-HEPTENE	98.188	154.27	366.79	537.29	28.30	413.0	0.2377	0.262	0.331
713	C7H14	cis-2-HEPTENE	98.188	164.00	371.56	549.00	28.40	424.0	0.2316	0.264	0.294
714	C7H14	trans-2-HEPTENE	98.188	163.67	371.10	543.00	28.50	406.0	0.2418	0.256	0.337
715	C7H14	cis-3-HEPTENE	98.188	136.51	368.90	545.00	28.40	421.0	0.2332	0.264	0.295
716	C7H14	trans-3-HEPTENE	98.188	136.52	368.82	540.00	28.50	406.0	0.2418	0.258	0.334
717	C7H14	METHYLCYCLOHEXANE	98.188	146.58	374.08	572.19	34.71	368.0	0.2668	0.269	0.235
718	C7H14	2-METHYL-1-HEXENE	98.188	170.28	364.99	538.00	28.70	398.0	0.2467	0.255	0.309
719	C7H14	3-METHYL-1-HEXENE	98.188	145.00	357.05	528.00	29.50	398.0	0.2467	0.267	0.306
720	C7H14	4-METHYL-1-HEXENE	98.188	131.70	359.88	534.00	30.40	398.0	0.2467	0.273	0.302
721	C7H14	2,3,3-TRIMETHYL-1-BUTENE	98.188	163.30	351.04	531.00	31.40	381.0	0.2577	0.270	0.241
722	C7H14O	DIISOPROPYL KETONE	114.188	204.81	397.55	576.00	30.20	416.0	0.2745	0.262	0.405
723	C7H14O	2-HEPTANONE	114.188	238.15	424.05	611.55	29.20	421.0	0.2712	0.242	0.413
724	C7H14O	1-HEPTANAL	114.188	230.15	425.95	603.00	28.00	421.0	0.2712	0.235	0.487
725	C7H14O	1-METHYLCYCLOHEXANOL	114.188	299.15	430.15	603.00	37.90	414.0	0.2758	0.313	0.683
726	C7H14O	cis-2-METHYLCYCLOHEXANOL	114.188	280.15	438.15	614.00	37.90	414.0	0.2758	0.307	0.679
727	C7H14O	trans-2-METHYLCYCLOHEXANOL	114.188	269.15	439.65	616.00	37.90	414.0	0.2758	0.306	0.683
728	C7H14O	cis-3-METHYLCYCLOHEXANOL	114.188	267.65	441.15	618.00	37.90	414.0	0.2758	0.305	0.704
729	C7H14O	trans-3-METHYLCYCLOHEXANOL	114.188	272.65	441.15	617.00	37.90	414.0	0.2758	0.306	0.697
730	C7H14O	cis-4-METHYLCYCLOHEXANOL	114.188	-------	444.15	622.00	37.90	414.0	0.2758	0.303	0.658
731	C7H14O	trans-4-METHYLCYCLOHEXANOL	114.188	-------	444.15	622.00	37.90	414.0	0.2758	0.303	0.691
732	C7H14O	5-METHYL-2-HEXANONE	114.188	199.25	417.95	601.00	29.70	421.0	0.2712	0.250	0.434
733	C7H14O2	n-BUTYL PROPIONATE	130.187	183.63	419.75	594.00	28.00	442.0	0.2945	0.251	0.475
734	C7H14O2	ETHYL ISOVALERATE	130.187	173.85	407.45	587.95	28.40	442.0	0.2945	0.257	0.407
735	C7H14O2	ISOPENTYL ACETATE	130.187	194.65	415.25	599.00	28.40	442.0	0.2945	0.252	0.414
736	C7H14O2	n-PENTYL ACETATE	130.187	202.35	422.15	598.00	28.00	442.0	0.2945	0.249	0.490
737	C7H14O2	n-PROPYL n-BUTYRATE	130.187	177.95	416.45	594.00	28.00	442.0	0.2945	0.251	0.448
738	C7H14O2	n-HEPTANOIC ACID	130.187	265.83	496.15	680.00	29.90	442.0	0.2945	0.234	0.717
739	C7H14O3	ETHYL-3-ETHOXYPROPIONATE	146.186	-------	438.15	609.00	27.20	462.0	0.3164	0.248	0.578
740	C7H15Br	1-BROMOHEPTANE	179.100	217.05	452.05	651.00	30.80	447.0	0.4007	0.271	0.444
741	C7H15N	N-METHYLCYCLOHEXYLAMINE	113.203	264.65	422.00	622.00	34.90	393.0	0.2880	0.265	0.386
742	C7H16	2,2-DIMETHYLPENTANE	100.204	149.34	352.34	520.50	37.73	416.0	0.2409	0.267	0.288
743	C7H16	2,3-DIMETHYLPENTANE	100.204	-------	362.93	537.35	29.08	393.0	0.2550	0.256	0.292
744	C7H16	2,4-DIMETHYLPENTANE	100.204	153.91	353.64	519.79	27.37	418.0	0.2397	0.265	0.302
745	C7H16	3,3-DIMETHYLPENTANE	100.204	138.70	359.21	536.40	29.46	414.0	0.2420	0.273	0.267
746	C7H16	3-ETHYLPENTANE	100.204	154.55	366.62	540.64	28.91	416.0	0.2409	0.268	0.309
747	C7H16	n-HEPTANE	100.204	182.57	371.58	540.26	27.36	431.9	0.2320	0.263	0.351
748	C7H16	2-METHYLHEXANE	100.204	154.90	363.20	530.37	27.34	421.0	0.2380	0.261	0.328
749	C7H16	3-METHYLHEXANE	100.204	153.75	365.00	535.25	28.14	404.0	0.2480	0.255	0.322
750	C7H16	2,2,3-TRIMETHYLBUTANE	100.204	248.57	354.03	531.17	29.54	398.0	0.2518	0.266	0.250
751	C7H16O	1-HEPTANOL	116.203	239.15	449.45	631.90	31.50	435.2	0.2670	0.261	0.587
752	C7H16O	2-HEPTANOL	116.203	243.00	432.35	588.00	30.30	432.0	0.2690	0.268	0.763
753	C7H16O	5-METHYL-1-HEXANOL	116.203	-------	445.15	605.00	30.30	432.0	0.2690	0.260	0.781
754	C7H16O	ISOPROPYL-TERT-BUTYL-ETHER	116.203	177.80	378.66	558.21	28.29	428.5	0.2712	0.261	0.307
755	C7H16S	n-HEPTYL MERCAPTAN	132.270	229.92	450.09	645.00	27.70	456.0	0.2901	0.240	0.419
756	C7H16S	BUTYL-PROPYL-SULFIDE	132.263	206.66	444.16	653.50	28.51	481.5	0.2747	0.259	0.318
757	C7H16S	ETHYL-PENTYL-SULFIDE	132.263	206.66	444.16	638.37	28.51	481.5	0.2747	0.259	0.420
758	C7H16S	HEXYL-METHYL-SULFIDE	132.263	206.66	444.16	638.37	28.51	481.5	0.2747	0.259	0.420
759	C7H17N	1-AMINOHEPTANE	115.219	254.15	430.05	607.00	28.50	471.0	0.2446	0.266	0.511
760	C8H4Cl2O2	ISOPHTHALOYL CHLORIDE	203.024	317.00	549.00	768.00	33.30	471.0	0.4310	0.246	0.646
761	C8H4O3	PHTHALIC ANHYDRIDE	148.118	404.26	557.65	791.00	47.20	421.0	0.3518	0.302	0.708
762	C8H6	ETHYNYLBENZENE	102.135	242.53	418.36	655.43	44.03	337.5	0.3026	0.273	0.239
763	C8H6O4	ISOPHTHALIC ACID	166.133	619.15	753.00	1007.00	39.50	424.0	0.3918	0.200	1.062
764	C8H6O4	PHTHALIC ACID	166.133	464.15	598.00	800.00	39.50	424.0	0.3918	0.252	1.059
765	C8H6O4	TEREPHTHALIC ACID	166.133	700.15	832.00	1113.00	39.50	424.0	0.3918	0.181	1.059
766	C8H6S	BENZOTHIOPHENE	134.202	304.50	493.05	754.00	41.40	349.0	0.3845	0.230	0.296
767	C8H7N	INDOLE	117.150	273.68	526.15	790.00	43.00	431.0	0.2718	0.282	0.374
768	C8H8	STYRENE	104.152	242.54	418.31	648.00	40.00	352.0	0.2959	0.261	0.236
769	C8H8	1,3,5,7-CYCLOOCTATETRAENE	104.151	266.16	413.16	642.55	41.41	345.5	0.3015	0.268	0.244
770	C8H8O	ACETOPHENONE	120.151	293.65	475.15	701.00	38.40	376.0	0.3196	0.248	0.429
771	C8H8O	p-TOLUALDEHYDE	120.151	-------	477.15	698.00	36.70	416.0	0.2888	0.263	0.442
772	C8H8O2	METHYL BENZOATE	136.150	260.75	472.65	693.00	35.90	436.0	0.3123	0.272	0.415
773	C8H8O2	o-TOLUIC ACID	136.150	376.85	532.00	751.00	38.60	397.0	0.3429	0.245	0.657
774	C8H8O2	p-TOLUIC ACID	136.150	452.75	548.15	773.00	38.60	397.0	0.3429	0.238	0.661
775	C8H8O3	METHYL SALICYLATE	152.150	265.15	493.65	701.00	40.90	410.0	0.3711	0.288	0.632
776	C8H8O3	VANILLIN	152.150	355.00	558.00	777.00	40.10	415.0	0.3666	0.258	0.757
777	C8H9NO	ACETANILIDE	135.166	386.65	576.95	825.00	37.30	430.0	0.3143	0.234	0.564
778	C8H10	ETHYLBENZENE	106.167	178.20	409.35	617.17	36.09	373.8	0.2840	0.263	0.304
779	C8H10	m-XYLENE	106.167	225.30	412.27	617.05	35.41	375.8	0.2825	0.259	0.326
780	C8H10	o-XYLENE	106.167	247.98	417.58	630.37	37.34	369.2	0.2876	0.263	0.313

NO	FORMULA	NAME	MW g/mol	T_F K	T_B K	T_C K	P_C bar	V_C cm3/mol	RHO_C g/cm3	Z_C	OMEGA
781	C8H10	p-XYLENE	106.167	286.41	411.51	616.26	35.11	379.1	0.2801	0.260	0.326
782	C8H10O	m-ETHYLPHENOL	122.167	-------	477.66	-------	-------	-------	-------	-------	-------
783	C8H10O	p-ETHYLPHENOL	122.167	318.23	491.14	716.45	42.90	387.0	0.3157	0.279	0.524
784	C8H10O	PHENETOLE	122.167	243.63	443.15	647.15	34.25	390.0	0.3132	0.248	0.415
785	C8H10O	2-PHENYLETHANOL	122.167	247.00	492.05	684.00	39.20	387.0	0.3157	0.267	0.743
786	C8H10O	2,3-XYLENOL	122.167	345.71	490.07	722.95	49.00	360.0	0.3394	0.293	0.511
787	C8H10O	2,4-XYLENOL	122.167	297.68	484.13	707.65	44.00	390.0	0.3132	0.292	0.513
788	C8H10O	2,5-XYLENOL	122.167	347.99	484.33	707.05	49.00	350.0	0.3490	0.292	0.563
789	C8H10O	2,6-XYLENOL	122.167	318.76	474.22	701.05	43.00	390.0	0.3132	0.288	0.455
790	C8H10O	3,4-XYLENOL	122.167	338.25	500.15	729.95	50.00	350.0	0.3490	0.288	0.573
791	C8H10O	3,5-XYLENOL	122.167	336.59	494.89	715.65	36.48	480.0	0.2545	0.294	0.491
792	C8H11N	N,N-DIMETHYLANILINE	121.182	275.60	466.69	687.15	36.27	465.0	0.2606	0.295	0.403
793	C8H11N	o-ETHYLANILINE	121.182	226.55	482.65	704.00	37.40	399.0	0.3037	0.255	0.463
794	C8H11N	2,4,6-TRIMETHYLPYRIDINE	121.182	229.00	444.00	653.00	33.30	417.0	0.2906	0.256	0.376
795	C8H11NO	p-PHENETIDINE	137.181	277.00	528.00	754.00	35.70	446.0	0.3076	0.254	0.553
796	C8H12	1,5-CYCLOOCTADIENE	108.183	203.98	423.27	645.00	39.00	366.0	0.2956	0.266	0.286
797	C8H12	VINYLCYCLOHEXENE	108.183	164.00	401.00	599.00	34.30	379.0	0.2854	0.261	0.329
798	C8H12O4	1,4-CYCLOHEXANEDICARBOXYLIC ACID	172.181	585.65	669.00	889.00	34.20	464.0	0.3711	0.215	1.036
799	C8H12O4	DIETHYL MALEATE	172.181	264.35	498.15	680.00	26.10	508.0	0.3389	0.235	0.666
800	C8H14O2	n-BUTYL METHACRYLATE	142.198	-------	434.00	616.00	26.30	481.0	0.2956	0.247	0.466
801	C8H14O3	BUTYRIC ANHYDRIDE	158.197	199.85	468.15	639.00	26.90	501.0	0.3158	0.254	0.659
802	C8H14O4	DIETHYL SUCCINATE	174.197	252.35	489.65	660.00	25.30	522.0	0.3337	0.241	0.737
803	C8H16	CYCLOOCTANE	112.214	287.60	424.30	647.20	35.50	410.0	0.2737	0.271	0.236
804	C8H16	1,1-DIMETHYLCYCLOHEXANE	112.215	239.66	392.70	591.15	29.38	450.0	0.2494	0.269	0.233
805	C8H16	cis-1,2-DIMETHYLCYCLOHEXANE	112.215	223.16	402.94	606.15	29.38	460.0	0.2439	0.268	0.232
806	C8H16	trans-1,2-DIMETHYLCYCLOHEXANE	112.215	184.99	396.58	596.15	29.38	460.0	0.2439	0.273	0.238
807	C8H16	cis-1,3-DIMETHYLCYCLOHEXANE	112.215	197.58	393.24	591.15	29.38	450.0	0.2494	0.269	0.237
808	C8H16	trans-1,3-DIMETHYLCYCLOHEXANE	112.215	183.07	397.61	598.00	39.38	460.0	0.2439	0.272	0.234
809	C8H16	cis-1,4-DIMETHYLCYCLOHEXANE	112.215	185.72	397.47	598.15	29.38	460.0	0.2439	0.272	0.231
810	C8H16	trans-1,4-DIMETHYLCYCLOHEXANE	112.215	236.21	392.51	590.15	29.38	450.0	0.2494	0.269	0.237
811	C8H16	ETHYLCYCLOHEXANE	112.215	161.84	404.95	609.15	30.40	450.0	0.2494	0.270	0.246
812	C8H16	2-ETHYL-1-HEXENE	112.215	-------	393.15	574.00	30.70	399.0	0.2812	0.257	0.380
813	C8H16	1-METHYL-1-ETHYLCYCLOPENTANE	112.215	129.35	394.67	582.00	30.20	428.0	0.2622	0.267	0.330
814	C8H16	1-OCTENE	112.215	171.45	394.44	566.60	25.50	472.0	0.2377	0.256	0.375
815	C8H16	trans-2-OCTENE	112.215	185.45	398.15	577.00	25.80	484.0	0.2318	0.260	0.338
816	C8H16	trans-3-OCTENE	112.215	163.15	396.45	574.00	25.80	480.0	0.2338	0.260	0.344
817	C8H16	trans-4-OCTENE	112.215	179.37	395.41	573.00	25.80	480.0	0.2338	0.260	0.339
818	C8H16	n-PROPYLCYCLOPENTANE	112.215	155.82	404.11	603.00	30.00	425.0	0.2640	0.254	0.272
819	C8H16	2,4,4-TRIMETHYL-1-PENTENE	112.215	179.70	374.59	553.00	26.30	465.0	0.2413	0.266	0.270
820	C8H16	2,4,4-TRIMETHYL-2-PENTENE	112.215	166.84	378.06	558.00	26.30	470.0	0.2388	0.266	0.265
821	C8H16O	2-ETHYLHEXANAL	128.214	-------	433.80	607.00	25.80	474.0	0.2705	0.242	0.520
822	C8H16O	1-OCTANAL	128.214	246.00	447.15	621.00	25.50	474.0	0.2705	0.234	0.547
823	C8H16O	2-OCTANONE	128.214	252.85	445.75	624.00	26.40	469.0	0.2734	0.239	0.528
824	C8H16O2	n-BUTYL n-BUTYRATE	144.214	181.15	438.15	616.00	25.40	494.0	0.2919	0.245	0.485
825	C8H16O2	n-HEXYL ACETATE	144.214	192.25	444.65	618.00	25.40	494.0	0.2919	0.244	0.540
826	C8H16O2	ISOBUTYL ISOBUTYRATE	144.214	192.45	420.65	602.00	26.10	494.0	0.2919	0.258	0.395
827	C8H16O2	n-OCTANOIC ACID	144.214	289.65	513.05	692.00	26.90	494.0	0.2919	0.231	0.779
828	C8H16O4	DIETHYLENE GLYCOL ETHYL ETHER ACETATE	176.213	248.15	490.55	660.00	24.20	565.0	0.3119	0.249	0.715
829	C8H18	2,2-DIMETHYLHEXANE	114.231	151.97	379.99	549.80	25.30	478.0	0.2390	0.265	0.338
830	C8H18	2,3-DIMETHYLHEXANE	114.231	-------	388.76	563.40	26.30	468.2	0.2440	0.263	0.347
831	C8H18	2,4-DIMETHYLHEXANE	114.231	-------	382.58	553.50	25.60	472.0	0.2420	0.263	0.344
832	C8H18	2,5-DIMETHYLHEXANE	114.231	182.00	382.26	550.00	24.90	482.0	0.2370	0.262	0.358
833	C8H18	3,3-DIMETHYLHEXANE	114.231	147.05	385.12	562.00	26.50	442.8	0.2580	0.251	0.320
834	C8H18	3,4-DIMETHYLHEXANE	114.231	-------	390.88	568.80	26.90	466.2	0.2450	0.265	0.338
835	C8H18	3-ETHYLHEXANE	114.231	-------	391.69	565.40	26.10	455.1	0.2510	0.253	0.363
836	C8H18	3-ETHYL-2-METHYLPENTANE	114.230	158.20	388.81	567.00	27.10	445.3	0.2565	0.254	0.361
837	C8H18	3-METHYL-3-ETHYLPENTANE	114.231	182.28	391.42	576.50	28.10	455.1	0.2510	0.267	0.305
838	C8H18	2-METHYLHEPTANE	114.231	164.16	390.80	559.64	24.85	488.0	0.2341	0.261	0.377
839	C8H18	3-METHYLHEPTANE	114.231	152.60	392.08	563.67	25.46	464.0	0.2462	0.252	0.372
840	C8H18	4-METHYLHEPTANE	114.231	152.20	390.86	561.74	25.42	476.0	0.2400	0.259	0.371
841	C8H18	n-OCTANE	114.231	216.38	398.83	568.83	24.86	492.1	0.2322	0.259	0.396
842	C8H18	2,2,3-TRIMETHYLPENTANE	114.231	160.89	383.00	563.50	27.30	436.0	0.2620	0.254	0.297
843	C8H18	2,2,4-TRIMETHYLPENTANE	114.231	165.78	372.39	543.96	25.68	468.0	0.2441	0.266	0.303
844	C8H18	2,3,3-TRIMETHYLPENTANE	114.231	172.22	387.92	573.50	28.20	455.1	0.2510	0.269	0.290
845	C8H18	2,3,4-TRIMETHYLPENTANE	114.231	163.95	386.62	566.30	27.30	460.6	0.2480	0.267	0.316
846	C8H18	2,2,3,3-TETRAMETHYLBUTANE	114.230	172.47	379.60	567.80	28.70	461.8	0.2474	0.280	0.251
847	C8H18O	DI-n-BUTYL ETHER	130.230	177.95	413.44	581.00	24.60	487.0	0.2674	0.248	0.467
848	C8H18O	DI-sec-BUTYL ETHER	130.230	173.15	394.20	559.00	25.30	494.0	0.2636	0.269	0.432
849	C8H18O	DI-tert-BUTYL ETHER	130.230	195.00	380.40	550.00	25.30	487.0	0.2674	0.269	0.346
850	C8H18O	2-ETHYL-1-HEXANOL	130.230	203.15	457.75	640.25	27.30	485.0	0.2685	0.249	0.549
851	C8H18O	1-OCTANOL	130.230	257.65	468.35	652.50	28.60	490.0	0.2658	0.258	0.594
852	C8H18O	2-OCTANOL	130.230	241.55	452.95	637.15	27.30	469.0	0.2777	0.242	0.506
853	C8H18O2	DI-t-BUTYL PEROXIDE	146.230	233.15	384.15	547.00	24.80	508.0	0.2879	0.277	0.403
854	C8H18O2S	DI-n-BUTYL SULFONE	178.296	318.00	564.00	767.00	25.40	569.0	0.3133	0.227	0.688
855	C8H18O3	DIETHYLENE GLYCOL DIETHYL ETHER	162.229	228.85	462.15	624.00	23.70	558.0	0.2907	0.255	0.681
856	C8H18O3	DIETHYLENE GLYCOL MONOBUTYL EHTER	162.229	205.15	504.15	654.00	25.60	526.0	0.3084	0.248	0.937
857	C8H18O4	TRIETHYLENE GLYCOL DIMETHYL ETHER	178.229	229.35	489.15	651.00	23.10	548.0	0.3252	0.234	0.792
858	C8H18O5	TETRAETHYLENE GLYCOL	194.228	268.15	581.00	722.00	25.90	564.0	0.3444	0.243	1.578

NO	FORMULA	NAME	MW g/mol	T_F K	T_B K	T_C K	P_C bar	V_C cm3/mol	RHO_C g/cm3	Z_C	OMEGA
859	C8H18S	n-OCTYL MERCAPTAN	146.297	223.95	472.19	664.00	25.20	518.0	0.2824	0.236	0.473
860	C8H18S	tert-OCTYL MERCAPTAN	146.297	199.00	429.00	627.00	25.90	529.0	0.2766	0.263	0.307
861	C8H18S	BUTYL-SULFIDE	146.290	209.86	455.15	650.00	25.00	537.5	0.2722	0.260	0.394
862	C8H18S	ETHYL-HEXYL-SULFIDE	146.290	209.86	468.16	660.72	25.79	537.5	0.2722	0.252	0.465
863	C8H18S	HEPTYL-METHYL-SULFIDE	146.290	209.86	468.16	660.72	25.79	537.5	0.2722	0.252	0.465
864	C8H18S	PENTYL-PROPYL-SULFIDE	146.290	209.86	468.16	660.72	25.79	537.5	0.2722	0.252	0.465
865	C8H18S2	BUTYL-DISULFIDE	178.350	202.16	504.36	704.16	26.24	591.5	0.3015	0.265	0.529
866	C8H19N	DI-n-BUTYLAMINE	129.246	211.15	432.00	607.50	31.10	524.0	0.2467	0.323	0.560
867	C8H19N	DIISOBUTYLAMINE	129.246	203.15	412.25	580.00	25.70	524.0	0.2467	0.279	0.485
868	C8H19N	n-OCTYLAMINE	129.246	272.75	452.75	627.00	25.80	524.0	0.2467	0.259	0.568
869	C8H23N5	TETRAETHYLENEPENTAMINE	189.304	243.00	606.15	774.00	25.30	636.0	0.2976	0.250	1.237
870	C8H24O4Si4	OCTAMETHYLCYCLOTETRASILOXANE	296.618	290.80	448.15	586.50	13.32	970.0	0.3058	0.265	0.589
871	C9H4O5	TRIMELLITIC ANHYDRIDE	192.128	438.15	663.00	890.00	40.80	462.0	0.4159	0.255	1.050
872	C9H6N2O2	TOLUENE DIISOCYANATE	174.159	287.04	523.15	737.00	30.40	525.0	0.3317	0.260	0.553
873	C9H7N	ISOQUINOLINE	129.161	299.45	516.40	803.15	49.80	403.0	0.3205	0.301	0.289
874	C9H7N	QUINOLINE	129.161	258.25	510.75	782.15	46.60	469.0	0.2754	0.336	0.329
875	C9H7NO	8-HYDROXYQUINOLINE	145.161	346.00	540.00	788.00	43.60	414.0	0.3506	0.276	0.522
876	C9H8	INDENE	116.163	271.70	455.77	687.00	38.20	368.0	0.3157	0.246	0.335
877	C9H8O	2-METHYLBENZOFURAN	132.162	-------	470.65	698.00	36.40	405.0	0.3263	0.254	0.374
878	C9H10	INDANE	118.178	221.74	451.12	684.90	39.50	381.0	0.3102	0.264	0.309
879	C9H10	cis-PROPENYLBENZENE	118.178	211.47	443.16	664.60	34.64	411.5	0.2872	0.258	0.316
880	C9H10	trans-PROPENYLBENZENE	118.178	243.82	443.16	664.60	34.64	411.5	0.2872	0.258	0.316
881	C9H10	alpha-METHYLSTYRENE	118.178	249.95	438.65	654.00	33.60	423.0	0.2794	0.261	0.327
882	C9H10	m-METHYLSTYRENE	118.178	186.81	444.75	657.00	32.90	407.0	0.2904	0.245	0.349
883	C9H10	o-METHYLSTYRENE	118.178	204.58	442.96	659.00	34.70	407.0	0.2904	0.258	0.341
884	C9H10	p-METHYLSTYRENE	118.178	239.02	445.93	665.00	33.60	431.0	0.2742	0.262	0.318
885	C9H10O2	BENZYL ACETATE	150.177	221.65	486.65	699.00	31.80	449.0	0.3345	0.246	0.470
886	C9H10O2	ETHYL BENZOATE	150.177	238.45	486.55	698.00	31.80	489.0	0.3071	0.268	0.479
887	C9H10O3	ETHYL VANILLIN	166.177	350.65	567.00	748.00	32.70	467.0	0.3558	0.246	1.073
888	C9H11NO	p-DIMETHYLAMINOBENZALDEHYDE	149.192	348.00	588.00	832.00	30.70	471.0	0.3168	0.209	0.527
889	C9H12	CUMENE	120.194	177.14	425.56	631.15	32.09	427.7	0.2810	0.262	0.338
890	C9H12	m-ETHYLTOLUENE	120.194	177.61	434.48	637.15	28.37	490.0	0.2453	0.262	0.322
891	C9H12	o-ETHYLTOLUENE	120.194	192.35	438.33	651.15	30.40	460.0	0.2613	0.258	0.293
892	C9H12	p-ETHYLTOLUENE	120.194	210.83	435.16	640.15	29.38	470.0	0.2557	0.259	0.324
893	C9H12	MESITYLENE	120.194	228.46	437.89	637.36	31.27	433.0	0.2776	0.256	0.398
894	C9H12	n-PROPYLBENZENE	120.194	173.67	432.39	638.38	32.00	440.0	0.2732	0.265	0.346
895	C9H12	1,2,3-TRIMETHYLBENZENE	120.194	247.79	449.27	664.53	34.54	414.0	0.2903	0.259	0.366
896	C9H12	1,2,4-TRIMETHYLBENZENE	120.194	229.38	442.53	649.13	32.32	430.0	0.2795	0.258	0.379
897	C9H12O	BENZYL ETHYL ETHER	136.194	275.65	458.15	662.00	31.10	442.0	0.3081	0.250	0.433
898	C9H12O	2-PHENYL-2-PROPANOL	136.194	309.15	475.15	660.00	34.90	440.0	0.3095	0.280	0.698
899	C9H12O2	CUMENE HYDROPEROXIDE	152.193	264.26	442.70	605.00	33.40	419.0	0.3632	0.278	0.995
900	C9H14O	ISOPHORONE	138.210	265.05	488.35	715.00	33.30	456.0	0.3031	0.255	0.400
901	C9H14O6	GLYCERYL TRIACETATE	218.207	277.25	532.15	704.00	23.10	625.0	0.3491	0.247	0.839
902	C9H16	1-NONYNE	124.225	223.16	423.96	610.81	26.74	501.5	0.2477	0.264	0.382
903	C9H16O4	AZELAIC ACID	188.224	379.65	633.36	811.00	25.60	610.0	0.3086	0.232	1.172
904	C9H18	BUTYLCYCLOPENTANE	126.241	165.18	429.76	625.05	27.64	480.5	0.2627	0.256	0.354
905	C9H18	cis-1,3,5-TRIMETHYLCYCLOHEXANE	126.241	223.46	411.66	607.86	26.49	470.5	0.2683	0.247	0.274
906	C9H18	cis,trans-1,3,5-TRIMETHYLCYCLOHEXANE	126.241	188.76	413.70	602.20	26.49	470.5	0.2683	0.245	0.333
907	C9H18	ISOPROPYLCYCLOHEXANE	126.242	183.76	427.91	627.00	28.50	464.0	0.2721	0.254	0.330
908	C9H18	1-NONENE	126.242	191.78	420.02	593.25	23.30	528.0	0.2391	0.249	0.417
909	C9H18	n-PROPYLCYCLOHEXANE	126.242	178.28	429.90	639.15	28.07	477.0	0.2647	0.252	0.260
910	C9H18O	DIISOBUTYL KETONE	142.241	227.17	441.41	615.00	24.80	522.0	0.2725	0.253	0.512
911	C9H18O	1-NONANAL	142.241	255.15	468.15	640.00	23.30	527.0	0.2699	0.231	0.592
912	C9H18O2	n-BUTYL VALERATE	158.241	180.35	459.65	629.00	23.30	547.0	0.2893	0.244	0.596
913	C9H18O2	n-NONANOIC ACID	158.241	285.55	528.75	703.00	24.50	547.0	0.2893	0.229	0.831
914	C9H18O2	n-OCTYL FORMATE	158.241	234.05	471.95	645.00	23.30	547.0	0.2893	0.238	0.587
915	C9H20	3,3-DIETHYLPENTANE	128.258	240.12	419.34	610.05	26.75	473.0	0.2712	0.249	0.338
916	C9H20	2,2-DIMETHYL-3-ETHYLPENTANE	128.258	173.68	406.99	590.00	25.70	511.0	0.2510	0.268	0.335
917	C9H20	3-ETHYL-2,3-DIMETHYLPENTANE	128.257	173.67	417.86	606.80	26.85	477.0	0.2689	0.254	0.349
918	C9H20	2,4-DIMETHYL-3-ETHYLPENTANE	128.258	150.79	409.87	591.00	25.30	512.0	0.2505	0.264	0.353
919	C9H20	2,2-DIMETHYLHEPTANE	128.258	160.15	405.84	576.80	23.50	519.0	0.2471	0.254	0.390
920	C9H20	2,6-DIMETHYLHEPTANE	128.258	170.25	408.36	579.00	23.00	520.0	0.2467	0.248	0.393
921	C9H20	3-ETHYLHEPTANE	128.258	158.25	416.35	590.00	23.90	540.0	0.2429	0.257	0.408
922	C9H20	4-ETHYLHEPTANE	128.257	159.96	414.36	584.95	23.25	533.5	0.2404	0.255	0.416
923	C9H20	2,3-DIMETHYLHEPTANE	128.257	160.16	413.66	589.60	24.01	515.0	0.2490	0.252	0.385
924	C9H20	2,4-DIMETHYLHEPTANE	128.257	160.16	406.05	576.80	23.41	517.0	0.2481	0.252	0.390
925	C9H20	2,5-DIMETHYLHEPTANE	128.257	160.16	409.16	581.10	23.51	522.0	0.2457	0.254	0.393
926	C9H20	3,4-DIMETHYLHEPTANE	128.257	170.26	413.76	591.90	24.62	503.0	0.2550	0.252	0.379
927	C9H20	3,5-DIMETHYLHEPTANE	128.257	170.26	409.16	583.20	24.01	510.0	0.2515	0.253	0.385
928	C9H20	4,4-DIMETHYLHEPTANE	128.257	170.26	408.36	585.40	24.32	501.0	0.2560	0.250	0.364
929	C9H20	3-ETHYL-2-METHYLHEXANE	128.257	160.16	411.16	588.10	24.52	497.0	0.2581	0.249	0.378
930	C9H20	4-ETHYL-2-METHYLHEXANE	128.257	160.16	406.96	580.00	24.01	504.0	0.2545	0.251	0.386
931	C9H20	3-ETHYL-3-METHYLHEXANE	128.257	160.16	413.76	597.50	25.53	487.0	0.2634	0.250	0.352
932	C9H20	3-ETHYL-4-METHYLHEXANE	128.257	160.16	413.56	593.70	25.13	490.0	0.2617	0.249	0.372
933	C9H20	2,2,3-TRIMETHYLHEXANE	128.257	153.16	406.75	588.00	24.90	498.9	0.2571	0.254	0.332
934	C9H20	2,2,4-TRIMETHYLHEXANE	128.257	153.00	399.69	573.50	23.80	506.6	0.2532	0.253	0.321
935	C9H20	2,3,3-TRIMETHYLHEXANE	128.257	156.36	410.84	596.00	25.53	491.0	0.2612	0.253	0.333
936	C9H20	2,3,4-TRIMETHYLHEXANE	128.257	156.36	412.20	594.50	25.23	494.0	0.2596	0.252	0.353

18

NO	FORMULA	NAME	MW g/mol	T_F K	T_B K	T_C K	P_C bar	V_C cm3/mol	RHO_C g/cm3	Z_C	OMEGA
937	C9H20	2,3,5-TRIMETHYLHEXANE	128.257	145.36	404.50	579.20	24.01	509.0	0.2520	0.254	0.364
938	C9H20	2,4,4-TRIMETHYLHEXANE	128.257	159.78	403.81	581.50	24.32	500.0	0.2565	0.251	0.344
939	C9H20	3,3,4-TRIMETHYLHEXANE	128.257	171.96	413.62	602.30	26.24	484.0	0.2650	0.254	0.328
940	C9H20	2-METHYLOCTANE	128.258	192.78	416.43	586.75	22.90	541.0	0.2371	0.254	0.422
941	C9H20	3-METHYLOCTANE	128.258	165.55	417.38	590.15	23.41	529.0	0.2425	0.252	0.413
942	C9H20	4-METHYLOCTANE	128.258	159.95	415.59	587.65	23.41	523.0	0.2452	0.251	0.413
943	C9H20	n-NONANE	128.258	219.63	423.97	595.65	23.06	547.7	0.2342	0.255	0.438
944	C9H20	2,2,3,3-TETRAMETHYLPENTANE	128.258	263.26	413.44	610.85	27.36	478.0	0.2683	0.257	0.280
945	C9H20	2,2,3,4-TETRAMETHYLPENTANE	128.258	152.06	406.18	592.15	25.64	490.0	0.2618	0.255	0.311
946	C9H20	2,2,4,4-TETRAMETHYLPENTANE	128.258	206.95	395.44	571.35	23.61	504.0	0.2545	0.250	0.316
947	C9H20	2,3,3,4-TETRAMETHYLPENTANE	128.257	171.10	414.72	607.50	27.20	516.5	0.2483	0.246	0.313
948	C9H20	2,2,5-TRIMETHYLHEXANE	128.258	167.39	397.24	568.05	23.31	519.0	0.2471	0.256	0.357
949	C9H20O	2,6-DIMETHYL-4-HEPTANOL	144.257	208.00	451.00	603.00	25.50	538.0	0.2681	0.274	0.802
950	C9H20O	1-NONANOL	144.257	268.15	486.25	673.00	26.00	546.4	0.2640	0.254	0.594
951	C9H20O	2-NONANOL	144.257	238.15	471.65	623.00	24.80	538.0	0.2681	0.258	0.890
952	C9H20S	n-NONYL MERCAPTAN	160.324	253.05	492.95	681.00	23.10	571.0	0.2808	0.233	0.531
953	C9H20S	BUTYL-PENTYL-SULFIDE	160.317	231.16	491.16	681.56	23.45	593.5	0.2701	0.246	0.508
954	C9H20S	ETHYL-HEPTYL-SULFIDE	160.317	231.16	491.16	681.56	23.45	593.5	0.2701	0.246	0.508
955	C9H20S	HEXYL-PROPYL-SULFIDE	160.317	231.16	491.16	681.56	23.45	593.5	0.2701	0.246	0.508
956	C9H20S	METHYL-OCTYL-SULFIDE	160.317	231.16	491.16	681.56	23.45	593.5	0.2701	0.246	0.508
957	C9H21N	n-NONYLAMINE	143.272	273.15	475.35	648.00	23.60	577.0	0.2483	0.253	0.615
958	C9H21N	TRIPROPYLAMINE	143.272	179.65	429.65	577.50	22.30	576.0	0.2487	0.268	0.699
959	C10H6O8	PYROMELLITIC ACID	254.153	554.00	722.00	893.00	31.40	584.0	0.4352	0.247	1.830
960	C10H7Br	1-BROMONAPHTHALENE	207.070	279.35	554.25	824.00	37.00	453.0	0.4571	0.245	0.369
961	C10H7Cl	1-CHLORONAPHTHALENE	162.618	269.15	532.45	785.00	34.00	434.0	0.3747	0.226	0.383
962	C10H8	NAPHTHALENE	128.174	353.43	491.14	748.35	40.51	413.0	0.3103	0.269	0.302
963	C10H8	AZULENE	128.173	173.66	515.16	773.48	38.97	409.5	0.3130	0.248	0.355
964	C10H9N	QUINALDINE	143.188	272.15	519.75	773.00	29.60	490.0	0.2922	0.226	0.280
965	C10H10	m-DIVINYLBENZENE	130.189	206.25	472.65	692.00	31.20	440.0	0.2959	0.239	0.373
966	C10H10	1-METHYLINDENE	130.189	-------	471.65	703.00	34.60	440.0	0.2959	0.260	0.336
967	C10H10	2-METHYLINDENE	130.189	353.15	458.00	684.00	34.60	429.0	0.3035	0.261	0.328
968	C10H10O4	DIMETHYL PHTHALATE	194.187	272.15	556.85	766.00	27.80	530.0	0.3664	0.231	0.647
969	C10H10O4	DIMETHYL TEREPHTHALATE	194.187	413.80	561.15	772.00	27.80	529.0	0.3671	0.229	0.637
970	C10H12	DICYCLOPENTADIENE	132.205	307.00	443.00	660.00	30.60	445.0	0.2971	0.248	0.285
971	C10H12	1,2,3,4-TETRAHYDRONAPHTHALENE	132.205	237.40	480.77	720.15	36.20	441.0	0.2998	0.267	0.328
972	C10H12O	ANETHOLE	148.205	294.50	508.45	723.00	29.00	482.0	0.3075	0.233	0.485
973	C10H12O4	DIALLYL MALEATE	196.203	226.15	520.00	693.00	23.30	606.0	0.3238	0.245	0.789
974	C10H14	n-BUTYLBENZENE	134.221	185.30	456.46	660.55	28.87	497.0	0.2701	0.261	0.392
975	C10H14	sec-BUTYLBENZENE	134.221	197.72	446.48	664.54	29.51	497.0	0.2701	0.265	0.276
976	C10H14	tert-BUTYLBENZENE	134.221	215.27	442.30	660.00	29.70	492.0	0.2728	0.266	0.266
977	C10H14	1,2,3,4-TETRAMETHYLBENZENE	134.221	266.91	478.25	695.10	28.40	487.5	0.2753	0.237	0.368
978	C10H14	m-CYMENE	134.221	209.44	448.23	657.00	29.30	485.0	0.2767	0.260	0.341
979	C10H14	o-CYMENE	134.221	201.64	451.33	662.00	29.30	489.0	0.2745	0.260	0.337
980	C10H14	p-CYMENE	134.221	205.25	450.28	653.15	28.37	492.0	0.2728	0.257	0.372
981	C10H14	m-DIETHYLBENZENE	134.221	189.26	454.29	663.00	28.80	488.0	0.2750	0.255	0.350
982	C10H14	o-DIETHYLBENZENE	134.221	241.93	456.61	668.00	28.80	502.0	0.2674	0.260	0.340
983	C10H14	p-DIETHYLBENZENE	134.221	230.32	456.94	657.96	28.03	497.0	0.2701	0.255	0.404
984	C1OH14	2-ETHYL-m-XYLENE	134.221	256.89	463.19	671.00	30.20	482.0	0.2785	0.261	0.407
985	C10H14	2-ETHYL-p-XYLENE	134.221	219.52	459.98	663.00	28.80	482.0	0.2785	0.252	0.411
986	C10H14	3-ETHYL-o-XYLENE	134.221	223.64	467.11	680.00	28.80	507.0	0.2647	0.258	0.362
987	C10H14	4-ETHYL-m-XYLENE	134.221	210.27	461.59	665.00	28.80	482.0	0.2785	0.251	0.414
988	C10H14	4-ETHYL-o-XYLENE	134.221	206.22	462.93	667.00	28.80	490.0	0.2739	0.254	0.411
989	C1OH14	5-ETHYL-m-XYLENE	134.221	188.82	456.93	655.00	27.50	482.0	0.2785	0.243	0.417
990	C10H14	ISOBUTYLBENZENE	134.221	221.70	445.94	650.15	30.40	456.0	0.2943	0.256	0.381
991	C10H14	1,2,3,5-TETRAMETHYLBENZENE	134.221	249.46	471.15	679.00	29.70	482.0	0.2785	0.254	0.424
992	C10H14	1,2,4,5-TETRAMETHYLBENZENE	134.221	352.38	469.99	675.15	29.38	482.0	0.2785	0.252	0.435
993	C10H14O	p-tert-BUTYLPHENOL	150.221	371.56	512.88	734.00	33.40	493.0	0.3047	0.270	0.510
994	C10H14O2	p-tert-BUTYLCATECHOL	166.220	325.00	558.00	776.00	37.70	511.0	0.3253	0.299	0.740
995	C10H15N	N,N-DIETHYLANILINE	149.236	235.15	489.42	702.00	28.50	556.0	0.2684	0.272	0.426
996	C10H15N	2,6-DIETHYLANILINE	149.236	276.65	508.65	678.00	31.20	495.0	0.3015	0.274	0.954
997	C10H16	CAMPHENE	136.237	320.15	433.65	638.00	27.50	499.0	0.2730	0.259	0.296
998	C10H16	D-LIMONENE	136.237	199.00	449.65	660.00	27.50	524.0	0.2600	0.263	0.313
999	C10H16	alpha-PHELLANDRENE	136.237	-------	448.15	649.00	28.20	500.0	0.2725	0.261	0.381
1000	C1OH16	beta-PHELLANDRENE	136.237	-------	447.15	648.00	28.20	487.0	0.2797	0.255	0.374
1001	C10H16	alpha-PINENE	136.237	209.15	429.29	632.00	27.60	504.0	0.2703	0.265	0.286
1002	C10H16	beta-PINENE	136.237	211.61	439.19	643.00	27.60	506.0	0.2692	0.261	0.325
1003	C10H16	alpha-TERPINENE	136.237	-------	450.35	652.00	28.00	506.0	0.2692	0.261	0.376
1004	C1OH16	gamma-TERPINENE	136.237	-------	456.15	661.00	28.00	505.0	0.2698	0.257	0.375
1005	C10H16	TERPINOLENE	136.237	-------	458.15	672.00	27.70	509.0	0.2677	0.252	0.301
1006	C10H16O	CAMPHOR	152.236	453.25	480.57	709.00	29.90	460.0	0.3309	0.233	0.319
1007	C10H18	1-DECYNE	138.252	229.16	447.16	632.49	24.27	557.5	0.2480	0.257	0.426
1008	C10H18	cis-DECAHYDRONANPHTALENE	138.253	230.20	468.97	702.25	32.42	480.0	0.2880	0.267	0.294
1009	C10H18	trans-DECAHYDRONAPHTHALENE	138.253	242.79	460.46	687.05	28.37	480.0	0.2880	0.238	0.254
1010	C10H18O4	SEBACIC ACID	202.251	407.65	642.09	815.00	23.50	658.0	0.3074	0.228	1.205
1011	C10H20	n-BUTYLCYCLOHEXANE	140.269	198.42	454.13	667.00	25.70	534.0	0.2627	0.247	0.274
1012	C10H20	1-CYCLOPENTYLPENTANE	140.268	190.16	453.76	647.49	25.05	536.5	0.2615	0.250	0.398
1013	C10H20	1-DECENE	140.269	206.89	443.75	617.05	21.68	585.0	0.2398	0.247	0.465
1014	C10H20O	1-DECANAL	156.268	267.15	488.15	657.00	21.50	580.0	0.2694	0.228	0.642

NO	FORMULA	NAME	MW g/mol	T_F K	T_B K	T_C K	P_C bar	V_C cm3/mol	RHO_C g/cm3	Z_C	OMEGA
1015	C10H20O2	n-DECANOIC ACID	172.268	304.75	543.15	713.00	22.50	600.0	0.2871	0.228	0.877
1016	C10H20O2	2-ETHYLHEXYL ACETATE	172.268	180.15	471.75	639.00	21.70	600.0	0.2871	0.245	0.631
1017	C10H20O2	ISOPENTYL ISOVALERATE	172.268	215.00	467.15	637.00	22.00	600.0	0.2871	0.249	0.579
1018	C10H22	n-DECANE	142.285	243.49	447.30	618.45	21.23	603.1	0.2359	0.249	0.484
1019	C10H22	2-METHYLNONANE	142.285	198.50	440.15	610.00	21.20	583.0	0.2441	0.244	0.472
1020	C10H22	3-METHYLNONANE	142.285	188.35	440.95	613.00	21.60	583.0	0.2441	0.247	0.465
1021	C10H22	4-METHYLNONANE	142.285	174.45	438.85	610.00	21.60	583.0	0.2441	0.248	0.465
1022	C10H22	5-METHYLNONANE	142.285	185.45	438.30	610.00	21.60	583.0	0.2441	0.248	0.456
1023	C10H22	3-ETHYLOCTANE	142.284	185.46	439.66	613.60	21.89	561.0	0.2536	0.241	0.446
1024	C10H22	4-ETHYLOCTANE	142.284	185.46	436.80	609.60	21.78	552.0	0.2578	0.237	0.443
1025	C10H22	2,2-DIMETHYLOCTANE	142.285	-------	430.05	602.00	21.60	574.0	0.2479	0.248	0.429
1026	C10H22	2,3-DIMETHYLOCTANE	142.284	219.16	437.47	613.20	21.89	567.0	0.2509	0.243	0.424
1027	C10H22	2,4-DIMETHYLOCTANE	142.284	219.16	429.06	599.40	21.38	566.0	0.2514	0.243	0.430
1028	C10H22	2,5-DIMETHYLOCTANE	142.284	219.16	431.66	603.00	21.48	569.0	0.2501	0.244	0.432
1029	C10H22	2,6-DIMETHYLOCTANE	142.284	219.16	433.54	603.10	21.48	576.0	0.2470	0.247	0.453
1030	C10H22	2,7-DIMETHYLOCTANE	142.284	219.16	433.03	602.90	20.97	590.0	0.2412	0.247	0.438
1031	C10H22	3,3-DIMETHYLOCTANE	142.284	219.16	434.36	612.10	22.19	557.0	0.2554	0.243	0.404
1032	C10H22	3,4-DIMETHYLOCTANE	142.284	219.16	436.56	614.00	22.39	551.0	0.2582	0.242	0.417
1033	C10H22	3,5-DIMETHYLOCTANE	142.284	219.16	432.56	606.30	21.89	555.0	0.2564	0.241	0.424
1034	C10H22	3,6-DIMETHYLOCTANE	142.284	219.16	433.96	608.30	21.89	562.0	0.2532	0.243	0.424
1035	C10H22	4,4-DIMETHYLOCTANE	142.284	219.16	430.66	606.90	22.09	548.0	0.2596	0.240	0.402
1036	C10H22	4,5-DIMETHYLOCTANE	142.284	219.16	435.29	612.20	22.39	546.0	0.2606	0.240	0.418
1037	C10H22	4-PROPYLHEPTANE	142.284	219.16	430.66	601.00	21.78	545.0	0.2611	0.238	0.444
1038	C10H22	4-ISOPROPYLHEPTANE	142.284	219.16	432.06	607.60	22.29	537.0	0.2650	0.237	0.416
1039	C10H22	3-ETHYL-2-METHYLHEPTANE	142.284	219.16	434.36	610.90	22.29	544.0	0.2616	0.239	0.415
1040	C10H22	4-ETHYL-2-METHYLHEPTANE	142.284	219.16	429.36	601.80	21.89	545.0	0.2611	0.238	0.424
1041	C10H22	5-ETHYL-2-METHYLHEPTANE	142.284	219.16	432.86	606.70	21.89	555.0	0.2564	0.241	0.424
1042	C10H22	3-ETHYL-3-METHYLHEPTANE	142.284	219.16	436.96	620.00	23.10	532.0	0.2675	0.238	0.389
1043	C10H22	4-ETHYL-3-METHYLHEPTANE	142.284	219.16	435.36	614.30	22.80	530.0	0.2685	0.237	0.410
1044	C10H22	3-ETHYL-5-METHYLHEPTANE	142.284	219.16	431.36	606.60	22.29	541.0	0.2630	0.239	0.416
1045	C10H22	3-ETHYL-4-METHYLHEPTANE	142.284	219.16	436.16	615.50	22.80	533.0	0.2669	0.237	0.409
1046	C10H22	4-ETHYL-4-METHYLHEPTANE	142.284	219.16	433.96	615.70	23.10	525.0	0.2710	0.237	0.390
1047	C10H22	2,2,3-TRIMETHYLHEPTANE	142.284	219.16	430.76	611.70	22.70	546.0	0.2606	0.244	0.378
1048	C10H22	2,2,4-TRIMETHYLHEPTANE	142.284	219.16	421.46	594.50	21.68	552.0	0.2578	0.242	0.389
1049	C10H22	2,2,5-TRIMETHYLHEPTANE	142.284	219.16	423.96	598.00	21.68	559.0	0.2545	0.244	0.389
1050	C10H22	2,2,6-TRIMETHYLHEPTANE	142.284	219.16	422.09	593.40	21.28	573.0	0.2483	0.247	0.396
1051	C10H22	2,3,3-TRIMETHYLHEPTANE	142.284	219.16	433.36	617.50	23.20	538.0	0.2645	0.243	0.371
1052	C10H22	2,3,4-TRIMETHYLHEPTANE	142.284	219.16	433.06	613.70	22.90	538.0	0.2645	0.241	0.391
1053	C10H22	2,3,5-TRIMETHYLHEPTANE	142.284	219.16	433.86	612.80	22.39	547.0	0.2601	0.240	0.397
1054	C10H22	2,3,6-TRIMETHYLHEPTANE	142.284	219.16	429.16	604.10	21.89	560.0	0.2541	0.244	0.403
1055	C10H22	2,4,4-TRIMETHYLHEPTANE	142.284	219.16	424.16	600.30	22.19	541.0	0.2630	0.241	0.383
1056	C10H22	2,4,5-TRIMETHYLHEPTANE	142.284	219.16	429.66	606.90	22.39	544.0	0.2616	0.241	0.397
1057	C10H22	2,4,6-TRIMETHYLHEPTANE	142.284	219.16	420.76	590.30	21.48	560.0	0.2541	0.245	0.411
1058	C10H22	2,5,5-TRIMETHYLHEPTANE	142.284	219.16	425.96	602.90	22.19	550.0	0.2587	0.243	0.383
1059	C10H22	3,3,4-TRIMETHYLHEPTANE	142.284	219.16	435.06	622.10	23.71	526.0	0.2705	0.241	0.365
1060	C10H22	3,3,5-TRIMETHYLHEPTANE	142.284	219.16	428.85	609.50	23.20	578.5	0.2460	0.248	0.382
1061	C10H22	3,4,4-TRIMETHYLHEPTANE	142.284	219.16	434.26	620.90	23.71	524.0	0.2715	0.241	0.365
1062	C10H22	3,4,5-TRIMETHYLHEPTANE	142.284	219.16	435.66	612.80	22.39	547.0	0.2601	0.240	0.417
1063	C10H22	3-ISOPROPYL-2-METHYLHEXANE	142.284	219.16	439.86	623.40	22.90	529.0	0.2690	0.234	0.391
1064	C10H22	3,3-DIETHYLHEXANE	142.284	219.16	439.46	627.80	24.12	510.0	0.2790	0.236	0.377
1065	C10H22	3,4-DIETHYLHEXANE	142.284	219.165	437.06	618.80	23.30	519.0	0.2742	0.235	0.403
1066	C10H22	3-ETHYL-2,2-DIMETHYLHEXANE	142.284	219.16	429.26	611.70	23.10	526.0	0.2705	0.239	0.369
1067	C10H22	4-ETHYL-2,2-DIMETHYLHEXANE	142.284	219.16	420.16	594.60	22.19	539.0	0.2640	0.242	0.384
1068	C10H22	3-ETHYL-2,3-DIMETHYLHEXANE	142.284	219.16	436.86	626.80	24.22	516.0	0.2757	0.240	0.359
1069	C10H22	4-ETHYL-2,3-DIMETHYLHEXANE	142.284	219.16	434.06	617.30	23.41	524.0	0.2715	0.239	0.384
1070	C10H22	3-ETHYL-2,4-DIMETHYLHEXANE	142.284	219.16	433.26	616.10	23.41	522.0	0.2726	0.239	0.385
1071	C10H22	4-ETHYL-2,4-DIMETHYLHEXANE	142.284	219.16	434.26	620.90	24.71	524.0	0.2715	0.241	0.383
1072	C10H22	3-ETHYL-2,5-DIMETHYLHEXANE	142.284	219.16	427.26	603.50	22.39	537.0	0.2650	0.240	0.397
1073	C10H22	4-ETHYL-3,3-DIMETHYLHEXANE	142.284	219.16	436.06	625.70	24.22	513.0	0.2774	0.239	0.358
1074	C10H22	3-ETHYL-3,4-DIMETHYLHEXANE	142.284	219.16	435.26	624.50	24.22	511.0	0.2784	0.238	0.359
1075	C10H22	2,2,3,3-TETRAMETHYLHEXANE	142.284	219.16	433.48	623.00	25.10	573.5	0.2481	0.242	0.364
1076	C10H22	2,2,3,4-TETRAMETHYLHEXANE	142.284	219.16	431.96	620.40	23.71	525.0	0.2710	0.241	0.345
1077	C10H22	2,2,3,5-TETRAMETHYLHEXANE	142.284	219.16	421.56	601.30	22.70	540.0	0.2635	0.245	0.357
1078	C10H22	2,2,4,4-TETRAMETHYLHEXANE	142.284	219.16	426.96	610.20	22.49	535.0	0.2660	0.237	0.344
1079	C10H22	2,2,4,5-TETRAMETHYLHEXANE	142.284	219.16	421.04	598.50	22.19	544.0	0.2616	0.243	0.363
1080	C10H22	2,2,5,5-TETRAMETHYLHEXANE	142.284	260.56	410.63	581.40	21.90	573.5	0.2481	0.256	0.375
1081	C10H22	2,3,3,4-TETRAMETHYLHEXANE	142.284	260.56	437.75	633.10	24.82	514.0	0.2768	0.242	0.334
1082	C10H22	2,3,3,5-TETRAMETHYLHEXANE	142.284	260.56	426.26	610.10	23.20	531.0	0.2680	0.243	0.351
1083	C10H22	2,3,4,4-TETRAMETHYLHEXANE	142.284	260.56	434.76	626.60	24.22	518.0	0.2747	0.241	0.339
1084	C10H22	2,3,4,5-TETRAMETHYLHEXANE	142.284	260.56	429.36	613.20	23.41	530.0	0.2685	0.243	0.365
1085	C10H22	3,3,4,4-TETRAMETHYLHEXANE	142.284	260.56	443.16	646.70	25.74	506.0	0.2812	0.242	0.311
1086	C10H22	2,4-DIMETHYL-3-ISOPROPYLPENTANE	142.284	191.46	430.20	614.40	23.41	521.0	0.2731	0.239	0.365
1087	C10H22	3,3-DIETHYL-2-METHYLPENTANE	142.284	191.46	442.86	639.90	25.33	501.0	0.2840	0.239	0.346
1088	C10H22	3-ETHYL-2,2,3-TRIMETHYLPENTANE	142.284	191.46	442.66	646.00	25.74	503.0	0.2829	0.241	0.311
1089	C10H22	3-ETHYL-2,2,4-TRIMETHYLPENTANE	142.284	191.46	428.46	615.30	23.71	518.0	0.2747	0.240	0.346
1090	C10H22	3-ETHYL-2,3,4-TRIMETHYLPENTANE	142.284	191.46	442.60	642.30	25.43	506.0	0.2812	0.241	0.329
1091	C10H22	2,2,3,3,4-PENTAMETHYLPENTANE	142.284	236.71	439.21	643.80	25.84	508.0	0.2801	0.245	0.294
1092	C10H22	2,2,3,4,4-PENTAMETHYLPENTANE	142.284	234.41	432.45	627.30	24.01	521.0	0.2731	0.240	0.308

NO	FORMULA	NAME	MW g/mol	T_F K	T_B K	T_C K	P_C bar	V_C cm3/mol	RHO_C g/cm3	Z_C	OMEGA
1093	C10H22O	1-DECANOL	158.284	280.05	503.35	690.00	23.70	599.6	0.2640	0.248	0.613
1094	C10H22O	DI-n-PENTYL ETHER	158.284	203.72	459.90	622.00	20.90	593.0	0.2669	0.240	0.601
1095	C10H22O	ISODECANOL	158.284	213.15	493.00	644.00	22.80	591.0	0.2678	0.252	0.913
1096	C10H22O5	TETRAETHYLENE GLYCOL DIMETHYL ETHER	222.282	243.45	548.95	705.00	19.40	674.0	0.3298	0.223	0.965
1097	C10H22S	n-DECYL MERCAPTAN	174.351	247.56	512.35	696.00	21.30	624.0	0.2794	0.230	0.588
1098	C10H22S	BUTYL-HEXYL-SULFIDE	174.344	238.16	513.16	701.03	21.41	649.5	0.2684	0.239	0.551
1099	C10H22S	ETHYL-OCTYL-SULFIDE	174.344	238.16	513.16	701.03	21.41	649.5	0.2684	0.239	0.551
1100	C10H22S	HEPTYL-PROPYL-SULFIDE	174.344	238.16	513.16	701.03	21.41	649.5	0.2684	0.239	0.551
1101	C10H22S	METHYL-NONYL-SULFIDE	174.344	238.16	513.16	701.03	21.41	649.5	0.2684	0.239	0.551
1102	C10H22S	PENTYL-SULFIDE	174.344	238.16	513.16	701.03	21.41	649.5	0.2684	0.239	0.551
1103	C10H22S2	PENTYL-DISULFIDE	206.404	214.16	537.06	726.94	21.75	703.5	0.2934	0.253	0.614
1104	C10H23N	n-DECYLAMINE	157.299	288.85	493.65	663.00	21.80	629.0	0.2501	0.249	0.669
1105	C11H10	1-METHYLNAPHTHALENE	142.200	242.67	517.83	772.04	36.60	523.0	0.2719	0.298	0.348
1106	C11H10	2-METHYLNAPHTHALENE	142.200	307.73	514.20	761.00	32.50	507.0	0.2805	0.260	0.346
1107	C11H14O2	n-BUTYL BENZOATE	178.231	251.65	523.15	724.00	25.90	555.0	0.3211	0.239	0.575
1108	C11H16	n-PENTYLBENZENE	148.248	198.15	478.61	679.90	26.04	550.0	0.2695	0.253	0.439
1109	C11H16O	p-tert-AMYLPHENOL	164.247	366.00	535.15	751.00	29.80	546.0	0.3008	0.261	0.569
1110	C11H20	1-UNDECYNE	152.279	248.16	468.16	650.99	22.12	613.5	0.2482	0.251	0.470
1111	C11H20O2	2-ETHYLHEXYL ACRYLATE	184.279	183.15	489.15	655.00	20.70	639.0	0.2884	0.243	0.673
1112	C11H22	1-UNDECENE	154.296	223.99	465.82	638.00	20.30	642.0	0.2403	0.246	0.517
1113	C11H22	1-CYCLOPENTYLHEXANE	154.295	200.16	476.26	667.67	22.81	592.5	0.2604	0.243	0.442
1114	C11H22	PENTYLCYCLOHEXANE	154.295	215.66	476.87	674.01	23.36	584.5	0.2640	0.244	0.413
1115	C11H22O	1-UNDECANAL	170.295	273.15	506.15	672.00	20.00	632.0	0.2695	0.226	0.697
1116	C11H24	n-UNDECANE	156.312	247.57	469.08	638.76	19.66	657.0	0.2379	0.243	0.536
1117	C11H24O	1-UNDECANOL	172.311	289.05	518.15	704.00	20.80	643.0	0.2680	0.229	0.587
1118	C11H24S	UNDECYL MERCAPTAN	188.378	270.15	530.55	710.00	19.80	676.0	0.2787	0.227	0.636
1119	C11H24S	BUTYL-HEPTYL-SULFIDE	188.371	254.66	533.16	717.91	19.63	705.5	0.2670	0.232	0.592
1120	C11H24S	DECYL-METHYL-SULFIDE	188.371	254.66	533.16	717.91	19.63	705.5	0.2670	0.232	0.592
1121	C11H24S	ETHYL-NONYL-SULFIDE	188.371	254.66	533.16	717.91	19.63	705.5	0.2670	0.232	0.592
1122	C11H24S	OCTYL-PROPYL-SULFIDE	188.371	254.66	533.16	717.91	19.63	705.5	0.2670	0.232	0.592
1123	C12H8O	DIBENZOFURAN	168.195	355.65	557.86	837.80	32.00	533.0	0.3156	0.245	0.275
1124	C12H9N	DIBENZOPYRROLE	167.210	517.95	627.86	899.00	32.00	482.0	0.3469	0.210	0.494
1125	C12H10	ACENAPHTHENE	154.211	366.56	550.54	803.15	31.00	553.0	0.2789	0.257	0.381
1126	C12H10	BIPHENYL	154.211	342.37	528.15	789.26	38.47	501.6	0.3075	0.294	0.366
1127	C12H10O	DIPHENYL ETHER	170.211	300.02	531.46	763.00	31.30	502.8	0.3385	0.248	0.472
1128	C12H11N	p-AMINODIPHENYL	169.226	326.00	575.00	817.00	32.90	539.0	0.3140	0.261	0.545
1129	C12H11N	DIPHENYLAMINE	169.226	326.15	575.15	817.00	31.80	539.0	0.3140	0.252	0.530
1130	C12H11N3	p-AMINOAZOBENZENE	197.240	401.00	633.00	877.00	29.00	642.0	0.3072	0.255	0.635
1131	C12H11N3	1,3-DIPHENYLTRIAZENE	197.240	372.00	610.00	845.00	28.26	642.0	0.3072	0.258	0.623
1132	C12H12	1,2-DIMETHYLNAPHTHALENE	156.227	272.16	539.46	775.34	30.06	521.5	0.2996	0.243	0.443
1133	C12H12	1,3-DIMETHYLNAPHTHALENE	156.227	269.16	538.36	773.76	30.06	521.5	0.2996	0.244	0.443
1134	C12H12	1,4-DIMETHYLNAPHTHALENE	156.227	280.32	540.46	776.78	30.06	521.5	0.2996	0.243	0.443
1135	C12H12	1,5-DIMETHYLNAPHTHALENE	156.227	355.16	538.16	773.47	30.06	521.5	0.2996	0.244	0.443
1136	C12H12	1,6-DIMETHYLNAPHTHALENE	156.227	259.16	536.16	770.60	30.06	521.5	0.2996	0.245	0.443
1137	C12H12	1,7-DIMETHYLNAPHTHALENE	156.227	260.16	536.16	770.60	30.06	521.5	0.2996	0.245	0.443
1138	C12H12	2,3-DIMETHYLNAPHTHALENE	156.227	378.16	541.16	777.78	30.06	521.5	0.2996	0.242	0.443
1139	C12H12	2,6-DIMETHYLNAPHTHALENE	156.227	384.55	535.15	777.00	31.70	520.0	0.3004	0.255	0.418
1140	C12H12	2,7-DIMETHYLNAPHTHALENE	156.227	370.15	536.15	778.00	31.70	520.0	0.3004	0.255	0.420
1141	C12H12	1-ETHYLNAPHTHALENE	156.227	259.34	531.48	776.00	30.00	520.0	0.3004	0.242	0.363
1142	C12H12	2-ETHYLNAPHTHALENE	156.227	265.76	531.49	774.90	31.40	521.5	0.2996	0.254	OMEGA
1143	C12H12N2	p-AMINODIPHENYLAMINE	184.241	341.15	627.15	867.00	31.90	596.0	0.3091	0.264	0.694
1144	C12H12N2	HYDRAZOBENZENE	184.241	404.15	573.00	792.00	30.90	556.0	0.3314	0.261	0.680
1145	C12H14	1,2,3-TRIMETHYLINDENE	158.243	344.65	509.00	726.00	27.70	529.0	0.2991	0.243	0.441
1146	C12H14O4	DIETHYL PHTHALATE	222.241	269.15	567.15	757.00	23.30	635.0	0.3500	0.235	0.763
1147	C12H16	CYCLOHEXYLBENZENE	160.259	280.14	513.27	744.00	28.80	555.1	0.2887	0.258	0.379
1148	C12H18	m-DIISOPROPYLBENZENE	162.275	210.02	476.33	684.00	24.50	600.0	0.2705	0.258	0.359
1149	C12H18	p-DIISOPROPYLBENZENE	162.275	256.08	483.65	689.00	24.50	598.0	0.2714	0.256	0.390
1150	C12H18	n-HEXYLBENZENE	162.275	212.00	499.26	698.00	23.80	618.0	0.2626	0.253	0.478
1151	C12H18	1,2,3-TRIETHYLBENZENE	162.274	206.66	490.66	684.37	23.36	599.5	0.2707	0.246	0.479
1152	C12H18	1,2,4-TRIETHYLBENZENE	162.274	206.66	490.66	684.37	23.36	599.5	0.2707	0.246	0.479
1153	C12H18	1,3,5-TRIETHYLBENZENE	162.274	206.66	489.16	682.28	23.36	599.5	0.2707	0.247	0.479
1154	C12H18	HEXAMETHYLBENZENE	162.274	438.66	536.60	758.00	22.38	599.5	0.2707	0.216	0.396
1155	C12H20O4	DIBUTYL MALEATE	228.288	188.15	553.15	716.00	19.00	719.0	0.3175	0.229	0.899
1156	C12H22	BICYCLOHEXYL	166.307	276.78	512.19	727.00	25.60	598.0	0.2781	0.253	0.428
1157	C12H22	1-DODECYNE	166.306	254.16	488.16	668.16	20.25	669.5	0.2484	0.244	0.512
1158	C12H23N	DICYCLOHEXYLAMINE	181.321	273.05	529.00	737.00	25.20	619.0	0.2929	0.255	0.513
1159	C12H24	1-DODECENE	168.323	237.93	486.50	657.00	18.90	700.0	0.2405	0.242	0.564
1160	C12H24	1-CYCLOPENTYLHEPTANE	168.322	220.00	497.30	679.00	19.40	648.5	0.2596	0.237	0.515
1161	C12H24	1-CYCLOHEXYLHEXANE	168.322	263.60	497.86	691.81	21.33	640.5	0.2628	0.238	0.456
1162	C12H24O	1-DODECANAL	184.322	285.15	523.15	685.00	18.60	685.0	0.2691	0.224	0.754
1163	C12H24O2	n-DODECANOIC ACID	200.321	317.15	571.85	734.00	19.40	705.0	0.2841	0.224	0.967
1164	C12H26	n-DODECANE	170.338	263.57	489.47	658.20	18.24	726.8	0.2344	0.242	0.573
1165	C12H26O	DI-n-HEXYL ETHER	186.338	230.15	498.85	658.00	18.16	698.0	0.2670	0.232	0.711
1166	C12H26O	1-DODECANOL	186.338	296.95	535.00	721.00	19.30	696.0	0.2677	0.224	0.639
1167	C12H26O3	DIETHYLENE GLYCOL DI-n-BUTYL ETHER	218.337	212.95	529.15	680.00	17.60	803.0	0.2719	0.250	0.846
1168	C12H26S	n-DODECYL MERCAPTAN	202.404	265.15	547.75	724.00	18.40	729.0	0.2776	0.223	0.677
1169	C12H26S	BUTYL-OCTYL-SULFIDE	202.397	259.16	552.16	733.68	18.06	761.5	0.2658	0.225	0.631
1170	C12H26S	DECYL-ETHYL-SULFIDE	202.397	259.16	552.16	733.68	18.06	761.5	0.2658	0.225	0.631

NO	FORMULA	NAME	MW g/mol	T_F K	T_B K	T_C K	P_C bar	V_C cm3/mol	RHO_C g/cm3	Z_C	OMEGA
1171	C12H26S	HEXYL-SULFIDE	202.397	259.16	552.16	733.68	18.06	761.5	0.2658	0.225	0.631
1172	C12H26S	METHYL-UNDECYL-SULFIDE	202.397	259.16	552.16	733.68	18.06	761.5	0.2658	0.225	0.631
1173	C12H26S	NONYL-PROPYL-SULFIDE	202.397	259.16	552.16	733.68	18.06	761.5	0.2658	0.225	0.631
1174	C12H26S2	HEXYL-DISULFIDE	234.457	225.16	566.66	747.10	18.33	815.5	0.2875	0.241	0.692
1175	C12H27BO3	TRI-n-BUTYL BORATE	230.156	203.15	506.65	743.15	19.89	863.3	0.2666	0.278	0.189
1176	C12H27N	DODECYLAMINE	185.353	301.47	532.35	696.00	18.80	735.0	0.2522	0.239	0.769
1177	C12H27N	TRI-n-BUTYLAMINE	185.353	203.00	487.15	644.00	18.00	735.0	0.2522	0.247	0.694
1178	C13H10	FLUORENE	166.222	387.94	570.44	870.00	47.00	400.0	0.4156	0.260	0.349
1179	C13H10O	BENZOPHENONE	182.222	321.35	579.24	816.00	30.10	591.0	0.3083	0.262	0.545
1180	C13H12	DIPHENYLMETHANE	168.238	298.39	537.42	768.00	29.20	547.0	0.3076	0.250	0.462
1181	C13H14	1-PROPYLNAPHTHALENE	170.254	264.69	545.96	771.45	27.56	577.5	0.2948	0.248	0.488
1182	C13H14	2-PROPYLNAPHTHALENE	170.254	270.16	546.66	772.44	27.56	577.5	0.2948	0.248	0.488
1183	C13H14	2ETHYL-3-METHYLNAPHTHALENE	170.254	344.16	550.16	776.44	27.13	577.5	0.2948	0.243	0.488
1184	C13H14	2ETHYL-6-METHYLNAPHTHALENE	170.254	318.16	543.16	766.56	27.13	577.5	0.2948	0.246	0.488
1185	C13H14	2ETHYL-7-METHYLNAPHTHALENE	170.254	318.16	543.16	766.56	27.13	577.5	0.2948	0.246	0.488
1186	C13H20	n-HEPTYLBENZENE	176.302	225.15	519.25	714.00	21.80	648.0	0.2721	0.238	0.529
1187	C13H24	1-TRIDECYNE	180.333	268.16	507.16	684.11	18.61	725.5	0.2486	0.237	0.553
1188	C13H26	1-TRIDECENE	182.349	250.08	505.93	675.00	17.70	756.0	0.2412	0.238	0.612
1189	C13H26	1-CYCLOPENTYLOCTANE	182.348	229.16	516.86	702.06	19.14	704.5	0.2588	0.231	0.526
1190	C13H26	1-CYCLOHEXYLHEPTANE	182.348	242.66	518.06	708.63	19.56	696.5	0.2618	0.231	0.498
1191	C13H26O	1-TRIDECANAL	198.349	288.15	540.15	700.00	17.40	738.0	0.2688	0.221	0.785
1192	C13H26O2	n-BUTYL NONANOATE	214.348	235.15	503.00	652.00	17.40	794.0	0.2700	0.255	0.828
1193	C13H26O2	METHYL DODECANOATE	214.348	278.15	540.00	712.00	17.40	758.0	0.2828	0.223	0.690
1194	C13H28	n-TRIDECANE	184.365	267.76	508.62	675.80	17.23	770.0	0.2394	0.236	0.619
1195	C13H28O	1-TRIDECANOL	200.365	303.75	547.15	731.00	18.10	749.0	0.2675	0.223	0.627
1196	C13H28S	BUTYL-NONYL-SULFIDE	216.424	271.16	570.16	748.42	16.67	817.5	0.2647	0.219	0.667
1197	C13H28S	DECYL-PROPYL-SULFIDE	216.424	271.16	570.16	748.42	16.67	817.5	0.2647	0.219	0.667
1198	C13H28S	DODECYL-METHYL-SULFIDE	216.424	271.16	570.16	748.42	16.67	817.5	0.2647	0.219	0.667
1199	C13H28S	ETHYL-UNDECYL-SULFIDE	216.424	271.16	570.16	748.42	16.67	817.5	0.2647	0.219	0.667
1200	C13H28S	1-TRIDECANETHIOL	216.424	282.04	563.96	742.13	17.33	817.5	0.2647	0.230	0.673
1201	C14H8O2	ANTHRAQUINONE	208.216	559.15	653.05	900.00	31.50	544.0	0.3828	0.229	0.681
1202	C14H10	ANTHRACENE	178.233	489.25	615.18	873.00	29.00	554.0	0.3217	0.221	0.489
1203	C14H10	DIPHENYLACETYLENE	178.233	335.65	573.00	832.00	29.00	611.0	0.2917	0.256	0.384
1204	C14H10	PHENANTHRENE	178.233	372.38	613.45	869.25	29.00	554.0	0.3217	0.222	0.495
1205	C14H12	cis-STILBENE	180.249	268.15	535.00	757.00	27.40	584.0	0.3086	0.254	0.471
1206	C14H12	trans-STILBENE	180.249	397.35	579.65	820.00	27.40	578.0	0.3118	0.232	0.490
1207	C14H12O2	BENZYL BENZOATE	212.248	292.55	596.65	820.00	25.80	694.0	0.3058	0.263	0.623
1208	C14H14	1,1-DIPHENYLETHANE	182.265	255.20	545.78	775.00	26.80	604.0	0.3018	0.251	0.457
1209	C14H14	1,2-DIPHENYLETHANE	182.265	324.34	553.65	780.00	26.50	606.0	0.3008	0.248	0.489
1210	C14H14O	DIBENZYL ETHER	198.265	276.75	561.45	777.00	25.60	608.0	0.3261	0.241	0.591
1211	C14H16	1-n-BUTYLNAPHTHALENE	184.281	253.43	562.54	792.00	26.80	631.0	0.2920	0.257	0.495
1212	C14H16	2-BUTYLNAPHTHALENE	184.280	268.16	562.16	780.96	24.98	633.5	0.2909	0.244	0.533
1213	C14H22	n-OCTYLBENZENE	190.329	237.15	537.55	729.00	20.20	703.0	0.2707	0.234	0.572
1214	C14H22	1,2,3,4-TETRAETHYLBENZENE	190.328	284.96	524.16	708.20	19.30	711.5	0.2675	0.233	0.562
1215	C14H22	1,2,3,5-TETRAETHYLBENZENE	190.328	284.16	523.66	707.52	19.30	711.5	0.2675	0.234	0.562
1216	C14H22	1,2,4,5-TETRAETHYLBENZENE	190.328	283.16	523.16	706.85	19.30	711.5	0.2675	0.234	0.562
1217	C14H22O	p-tert-OCTYLPHENOL	206.328	358.55	563.60	765.00	22.80	704.0	0.2931	0.252	0.631
1218	C14H28	1-TETRADECENE	196.376	260.30	524.25	692.00	16.60	817.0	0.2404	0.236	0.648
1219	C14H28	1-CYCLOPENTYLNONANE	196.375	244.16	535.26	716.95	17.62	760.5	0.2582	0.225	0.566
1220	C14H28	1-CYCLOHEXYLOCTANE	196.375	253.46	536.76	723.61	18.00	752.5	0.2610	0.225	0.538
1221	C14H28O2	n-TETRADECANOIC ACID	228.375	327.55	599.35	756.00	17.00	811.0	0.2816	0.219	1.025
1222	C14H30	n-TETRADECANE	198.392	279.01	526.73	692.40	16.21	842.8	0.2354	0.237	0.662
1223	C14H30O	1-TETRADECANOL	214.392	310.65	560.15	741.00	17.00	802.0	0.2673	0.221	0.677
1224	C14H30S	BUTYL-DECYL-SULFIDE	230.451	276.16	587.16	762.23	15.44	873.5	0.2638	0.213	0.700
1225	C14H30S	DODECYL-ETHYL-SULFIDE	230.451	276.16	587.16	762.23	15.44	873.5	0.2638	0.213	0.700
1226	C14H30S	HEPTYL-SULFIDE	230.451	276.16	587.16	762.23	15.44	873.5	0.2638	0.213	0.700
1227	C14H30S	METHYL-TRIDECYL-SULFIDE	230.451	276.16	587.16	762.23	15.44	873.5	0.2638	0.213	0.700
1228	C14H30S	PROPYL-UNDECYL-SULFIDE	230.451	276.16	587.16	762.23	15.44	873.5	0.2638	0.213	0.700
1229	C14H30S	1-TETRADECANETHIOL	230.451	279.26	579.36	753.80	16.03	873.5	0.2638	0.223	0.707
1230	C14H30S2	HEPTYL-DISULFIDE	262.511	235.16	593.86	765.96	15.65	927.5	0.2830	0.228	0.758
1231	C14H31N	TETRADECYLAMINE	213.407	311.34	564.45	722.30	16.60	887.0	0.2406	0.245	0.860
1232	C15H10N2O2	DIPHENYLMETHANE-4,4'-DIISOCYANATE	250.257	311.20	609.00	802.00	22.80	712.0	0.3515	0.243	0.950
1233	C15H16O	p-CUMYLPHENOL	212.291	346.00	608.15	834.00	26.80	659.0	0.3221	0.255	0.660
1234	C15H16O2	BISPHENOL A	228.291	426.15	633.65	849.00	29.30	677.0	0.3372	0.281	0.945
1235	C15H18	1-PENTYLNAPHTHALENE	198.307	251.16	580.16	793.32	22.74	689.5	0.2876	0.238	0.575
1236	C15H18	2-PENTYLNAPHTHALENE	198.307	269.16	583.16	797.48	22.74	689.5	0.2876	0.236	0.575
1237	C15H24	n-NONYLBENZENE	204.356	249.00	555.20	741.00	18.95	790.0	0.2587	0.243	0.638
1238	C15H24O	2,6-DI-tert-BUTYL-p-CRESOL	220.355	344.00	538.00	720.00	21.10	757.0	0.2911	0.267	0.686
1239	C15H24O	NONYLPHENOL	220.355	-------	581.00	757.00	20.70	757.0	0.2911	0.249	0.900
1240	C15H28	1-PENTADECYNE	208.386	283.16	541.16	711.41	15.87	837.5	0.2488	0.225	0.628
1241	C15H30	1-PENTADECENE	210.403	269.42	541.61	708.00	15.70	875.0	0.2405	0.233	0.684
1242	C15H30	1-CYCLOPENTYLDECANE	210.402	251.03	552.54	730.64	16.29	816.5	0.2577	0.219	0.604
1243	C15H30	1-CYCLOHEXYLNONANE	210.402	262.96	554.66	737.79	16.62	808.5	0.2602	0.219	0.577
1244	C15H30O2	PENTADECANOIC ACID	242.402	326.58	612.05	766.00	16.00	864.0	0.2806	0.217	1.040
1245	C15H32	n-PENTADECANE	212.419	283.11	543.83	706.80	15.20	880.0	0.2414	0.228	0.705
1246	C15H32O	1-PENTADECANOL	228.417	317.04	578.01	722.53	15.19	894.5	0.2554	0.226	1.015
1247	C15H32S	BUTYL-UNDECYL-SULFIDE	244.478	284.16	603.16	775.15	14.34	929.5	0.2630	0.207	0.729
1248	C15H32S	DODECYL-PROPYL-SULFIDE	244.478	284.16	603.16	775.15	14.34	929.5	0.2630	0.207	0.729

NO	FORMULA	NAME	MW g/mol	T_F K	T_B K	T_C K	P_C bar	V_C cm3/mol	RHO$_C$ g/cm3	Z_C	OMEGA
1249	C15H32S	ETHYL-TRIDECYL-SULFIDE	244.478	284.16	603.16	775.15	14.34	929.5	0.2630	0.207	0.729
1250	C15H32S	METHYL-TETRADECYL-SULFIDE	244.478	284.16	603.16	775.15	14.34	929.5	0.2630	0.207	0.729
1251	C15H32S	1-PENTADECANETHIOL	244.478	290.93	593.86	764.77	14.86	929.5	0.2630	0.217	0.737
1252	C16H10	FLUORANTHENE	202.255	383.33	655.95	905.00	26.10	655.0	0.3088	0.227	0.588
1253	C16H10	PYRENE	202.255	423.81	667.95	936.00	26.10	630.0	0.3210	0.211	0.509
1254	C16H12	1-PHENYLNAPHTHALENE	204.271	318.15	607.15	849.00	26.30	656.0	0.3114	0.244	0.531
1255	C16H20	1-n-HEXYLNAPHTHALENE	212.335	255.15	595.15	813.00	22.50	741.0	0.2866	0.247	0.587
1256	C16H22O4	DIBUTYL PHTHALATE	278.348	238.15	613.15	781.00	17.50	846.0	0.3290	0.228	0.947
1257	C16H26	n-DECYLBENZENE	218.382	258.77	571.04	753.00	17.70	881.0	0.2479	0.249	0.681
1258	C16H26	PENTAETHYLBENZENE	218.381	327.66	550.16	723.64	16.22	823.5	0.2652	0.222	0.637
1259	C16H30	1-HEXADECYNE	222.413	288.16	557.16	724.26	14.72	893.5	0.2489	0.219	0.661
1260	C16H32	n-DECYLCYCLOHEXANE	224.430	271.42	570.75	751.25	16.50	858.0	0.2616	0.227	0.663
1261	C16H32	1-CYCLOPENTYLUNDECANE	224.429	263.16	568.76	743.30	15.09	872.5	0.2572	0.213	0.638
1262	C16H32	1-HEXADECENE	224.430	277.51	558.02	722.00	14.80	933.0	0.2405	0.230	0.732
1263	C16H32O2	n-HEXADECANOIC ACID	256.429	335.95	624.15	776.00	15.10	917.0	0.2796	0.215	1.083
1264	C16H34	n-HEXADECANE	226.446	291.34	560.01	720.60	14.19	930.0	0.2435	0.220	0.747
1265	C16H34O	DI-n-OCTYL ETHER	242.445	265.55	559.65	707.00	14.40	910.0	0.2664	0.223	0.934
1266	C16H34O	1-HEXADECANOL	242.445	322.35	585.15	761.00	15.10	907.0	0.2673	0.216	0.748
1267	C16H34S	BUTYL-DODECYL-SULFIDE	258.505	288.16	618.16	787.27	13.35	985.5	0.2623	0.201	0.754
1268	C16H34S	ETHYL-TETRADECYL-SULFIDE	258.505	288.16	618.16	791.68	12.88	931.5	0.2775	0.182	0.686
1269	C16H34S	METHYL-PENTADECYL-SULFIDE	258.505	288.16	618.16	787.27	13.35	985.5	0.2623	0.201	0.754
1270	C16H34S	OCTYL-SULFIDE	258.505	288.16	618.16	787.27	13.35	985.5	0.2623	0.201	0.754
1271	C16H34S	PROPYL-TRIDECYL-SULFIDE	258.505	288.16	618.16	787.27	13.35	985.5	0.2623	0.201	0.754
1272	C16H34S	1-HEXADECANETHIOL	258.505	290.93	607.16	774.68	13.82	985.5	0.2623	0.211	0.763
1273	C16H34S2	OCTYL-DISULFIDE	290.565	244.16	619.16	784.46	13.52	1039.5	0.2795	0.215	0.806
1274	C17H28	n-UNDECYLBENZENE	232.409	268.00	586.40	764.00	16.72	910.0	0.2554	0.240	0.738
1275	C17H32	1-HEPTADECYNE	236.440	295.16	572.16	736.21	13.70	949.5	0.2490	0.212	0.690
1276	C17H34	1-CYCLOPENTYLDODECANE	238.456	268.16	584.06	755.17	14.03	928.5	0.2568	0.207	0.669
1277	C17H34	1-CYCLOHEXYLUNDECANE	238.456	278.96	586.26	761.74	14.29	920.5	0.2591	0.208	0.646
1278	C17H34	1-HEPTADECENE	238.457	284.40	573.48	736.00	14.10	955.0	0.2497	0.220	0.753
1279	C17H36	n-HEPTADECANE	240.473	295.13	575.30	733.37	13.17	1005.8	0.2391	0.217	0.768
1280	C17H36O	1-HEPTADECANOL	256.472	327.05	597.15	770.00	14.30	960.0	0.2672	0.214	0.795
1281	C17H36S	BUTYL-TRIDECYL-SULFIDE	272.531	294.16	632.16	798.63	12.46	1041.5	0.2617	0.195	0.774
1282	C17H36S	ETHYL-PENTADECYL-SULFIDE	272.531	294.16	632.16	798.63	12.46	1041.5	0.2617	0.195	0.774
1283	C17H36S	HEXADECYL-METHYL-SULFIDE	272.531	294.16	632.16	798.63	12.46	1041.5	0.2617	0.195	0.774
1284	C17H36S	PROPYL-TETRADECYL-SULFIDE	272.531	294.16	632.16	798.63	12.46	1041.5	0.2617	0.195	0.774
1285	C17H36S	1-HEPTADECANETHIOL	272.531	300.37	621.16	786.01	12.88	1041.5	0.2617	0.205	0.783
1286	C18H12	CHRYSENE	228.293	531.15	714.15	979.00	23.90	690.0	0.3309	0.213	0.604
1287	C18H14	m-TERPHENYL	230.309	360.00	650.00	924.85	35.06	767.7	0.3000	0.350	0.558
1288	C18H14	o-TERPHENYL	230.309	329.35	609.00	890.95	39.01	752.6	0.3060	0.396	0.467
1289	C18H14	p-TERPHENYL	230.309	485.00	649.15	925.95	33.24	762.6	0.3020	0.329	0.528
1290	C18H15P	TRIPHENYLPHOSPHINE	262.291	354.40	650.15	1008.00	78.40	554.0	0.4734	0.518	0.452
1291	C18H15O4P	TRIPHENYL PHOSPHATE	326.288	323.15	686.65	-------	-------	-------	-------	-------	-------
1292	C18H16N2	N,N'-DIPHENYL-p-PHENYLENEDIAMINE	260.339	409.00	688.00	906.00	23.10	817.0	0.3187	0.251	0.876
1293	C18H22	2,3-DIMETHYL-2,3-DIPHENYLBUTANE	238.373	392.15	589.00	805.00	19.90	781.0	0.3052	0.232	0.521
1294	C18H22O2	DICUMYL PEROXIDE	270.371	311.15	669.00	884.00	21.80	873.0	0.3097	0.259	0.450
1295	C18H30	n-DODECYLBENZENE	246.436	275.93	600.76	774.26	15.79	1000.0	0.2464	0.245	0.786
1296	C18H30	HEXAETHYLBENZENE	246.435	401.16	571.16	734.78	13.82	935.5	0.2634	0.212	0.698
1297	C18H32O2	LINOLEIC ACID	280.451	268.15	628.00	775.00	14.10	990.0	0.2833	0.217	1.176
1298	C18H34	1-OCTADECYNE	250.467	300.16	586.16	747.33	12.77	1005.5	0.2491	0.207	0.715
1299	C18H34O2	OLEIC ACID	282.467	286.53	633.00	781.00	13.90	1000.0	0.2825	0.214	1.187
1300	C18H34O4	DIBUTYL SEBACATE	314.466	263.95	622.15	768.00	13.20	1050.0	0.2995	0.217	1.126
1301	C18H34O4	DIHEXYL ADIPATE	314.466	259.35	621.15	767.00	13.20	1030.0	0.3053	0.213	1.094
1302	C18H36	1-CYCOPENTYLTRIDECANE	252.482	278.16	598.56	766.47	13.07	984.5	0.2565	0.202	0.697
1303	C18H36	1-CYCLOHEXYLDODECANE	252.482	285.66	600.86	772.83	13.31	976.5	0.2586	0.202	0.675
1304	C18H36	1-OCTADECENE	252.484	290.76	587.97	748.00	13.40	1050.0	0.2405	0.226	0.790
1305	C18H36O2	STEARIC ACID	284.483	342.75	648.35	799.00	13.60	1020.0	0.2789	0.209	1.084
1306	C18H38	n-OCTADECANE	254.500	301.33	589.86	745.26	12.14	1070.0	0.2652	0.217	0.795
1307	C18H38O	DINONYL ETHER	270.499	-------	591.00	736.00	13.00	1020.0	0.2652	0.217	1.002
1308	C18H38O	1-OCTADECANOL	270.499	331.05	608.15	777.00	13.60	1010.0	0.2678	0.213	0.863
1309	C18H38S	BUTYL-TETRADECYL-SULFIDE	286.558	298.16	646.16	810.53	11.66	1097.5	0.2611	0.190	0.787
1310	C18H38S	ETHYL-HEXADECYL-SULFIDE	286.558	298.16	646.16	810.53	11.66	1097.5	0.2611	0.190	0.787
1311	C18H38S	HEPTADECYL-METHYL-SULFIDE	286.558	298.16	646.16	810.53	11.66	1097.5	0.2611	0.190	0.787
1312	C18H38S	NONYL-SULFIDE	286.558	298.16	646.16	810.53	11.66	1097.5	0.2611	0.190	0.787
1313	C18H38S	PENTADECYL-PROPYL-SULFIDE	286.558	298.16	646.16	810.53	11.66	1097.5	0.2611	0.190	0.787
1314	C18H38S	1-OCTADECANETHIOL	286.558	300.93	633.16	795.36	12.04	1097.5	0.2611	0.200	0.798
1315	C18H38S2	NONYL-DISULFIDE	318.618	252.16	642.16	802.30	11.79	1151.5	0.2767	0.204	0.832
1316	C19H26	1-n-NONYLNAPHTHALENE	254.415	284.15	639.00	849.00	16.80	1000.0	0.2544	0.238	0.617
1317	C19H32	n-TRIDECYLBENZENE	260.463	283.15	614.43	783.00	15.00	1060.0	0.2457	0.244	0.844
1318	C19H36	1-NONADECYNE	264.493	306.16	600.16	758.94	11.94	1061.5	0.2492	0.201	0.735
1319	C19H36O2	METHYL OLEATE	296.494	293.05	617.00	764.00	12.80	1060.0	0.2797	0.214	1.049
1320	C19H38	1-CYCLOPENTYLTETRADECANE	266.509	282.00	612.16	772.00	11.20	1040.5	0.2561	0.201	0.789
1321	C19H38	1-CYCLOHEXYLTRIDECANE	266.509	291.66	614.66	783.38	12.42	1032.5	0.2581	0.197	0.700
1322	C19H38	1-NONADECENE	266.511	296.55	602.17	760.00	12.80	1100.0	0.2423	0.223	0.841
1323	C19H38O2	NONADECANOIC ACID	298.510	341.23	659.15	810.00	13.00	1080.0	0.2764	0.208	1.070
1324	C19H40	n-NONADECANE	268.527	305.33	603.05	755.93	11.17	1130.0	0.2376	0.201	0.820
1325	C19H40O	1-NONADECANOL	284.524	334.87	631.00	775.30	11.49	1118.5	0.2544	0.199	0.976
1326	C19H40S	BUTYL-PENTADECYL-SULFIDE	300.585	303.16	659.16	821.75	10.93	1153.5	0.2606	0.185	0.794

Table 1-1 CRITICAL PROPERTIES AND ACENTRIC FACTOR - ORGANIC COMPOUNDS (continued)

NO	FORMULA	NAME	MW g/mol	T_F K	T_B K	T_C K	P_C bar	V_C cm3/mol	RHO_C g/cm3	Z_C	OMEGA
1327	C19H40S	ETHYL-HEPTADECYL-SULFIDE	300.585	303.16	659.16	821.75	10.93	1153.5	0.2606	0.185	0.794
1328	C19H40S	HEXADECYL-PROPYL-SULFIDE	300.585	303.16	659.16	821.75	10.93	1153.5	0.2606	0.185	0.794
1329	C19H40S	METHYL-OCTADECYL-SULFIDE	300.585	303.16	659.16	821.75	10.93	1153.5	0.2606	0.185	0.794
1330	C19H40S	1-NONADECANETHIOL	300.585	307.04	645.16	805.29	11.28	1153.5	0.2606	0.194	0.807
1331	C20H16	TRIPHENYLETHYLENE	256.347	342.15	669.00	908.00	21.00	860.0	0.2981	0.239	0.600
1332	C20H28	1-n-DECYLNAPHTHALENE	268.442	288.15	652.00	859.00	15.80	1070.0	0.2509	0.237	0.642
1333	C20H30O2	ABIETIC ACID	302.457	446.65	649.70	832.00	16.80	930.0	0.3252	0.226	1.129
1334	C20H31N	DEHYDROABIETYLAMINE	285.473	317.65	660.00	863.00	17.00	1020.0	0.2799	0.242	0.742
1335	C20H34	1-PHENYLTETRADECANE	274.489	289.16	627.16	792.00	14.19	1110.0	0.2473	0.240	0.869
1336	C20H38	1-EICOSYNE	278.520	309.16	613.16	769.79	11.19	1117.5	0.2492	0.195	0.750
1337	C20H40	1-CYCLOPENTYLPENTADECANE	280.536	290.00	625.00	780.00	10.20	1096.5	0.2558	0.191	0.833
1338	C20H40	1-CYCLOHEXYLTETRADECANE	280.536	297.16	627.16	792.82	11.62	1088.5	0.2577	0.192	0.719
1339	C20H40	1-EICOSENE	280.538	301.76	615.54	771.00	12.20	1160.0	0.2418	0.221	0.877
1340	C20H42	n-EICOSANE	282.553	309.59	616.93	767.04	10.40	1190.0	0.2374	0.194	0.876
1341	C20H42O	1-EICOSANOL	298.553	338.55	629.15	792.00	12.40	1120.0	0.2666	0.211	0.937
1342	C20H42S	BUTYL-HEXADECYL-SULFIDE	314.612	308.16	671.16	832.33	10.27	1209.5	0.2601	0.179	0.795
1343	C20H42S	DECYL-SULFIDE	314.612	308.16	671.16	832.33	10.27	1209.5	0.2601	0.179	0.795
1344	C20H42S	ETHYL-OCTADECYL-SULFIDE	314.612	308.16	671.16	832.33	10.27	1209.5	0.2601	0.179	0.795
1345	C20H42S	HEPTADECYL-PROPYL-SULFIDE	314.612	308.16	671.16	832.33	10.27	1209.5	0.2601	0.179	0.795
1346	C20H42S	METHYL-NONADECYL-SULFIDE	314.612	308.16	671.16	832.33	10.27	1209.5	0.2601	0.179	0.795
1347	C20H42S	1-EICOSANETHIOL	314.612	310.37	656.16	814.57	10.58	1209.5	0.2601	0.189	0.809
1348	C20H42S2	DECYL-DISULFIDE	346.672	259.16	663.16	820.08	10.38	1263.5	0.2744	0.192	0.830
1349	C21H21O4P	TRI-o-CRESYL PHOSPHATE	368.369	240.15	-------	-------	-------	-------	-------	-------	-------
1350	C21H36	1-PHENYLPENTADECANE	288.515	295.16	639.16	800.00	13.48	1140.0	0.2531	0.230	0.914
1351	C21H42	1-CYCLOPENTYLHEXADECANE	294.563	294.16	637.16	797.25	10.72	1152.5	0.2556	0.186	0.748
1352	C21H42	1-CYCLOHEXYLPENTADECANE	294.563	302.16	640.16	803.46	10.90	1144.5	0.2574	0.187	0.733
1353	C22H38	1-PHENYLHEXADECANE	302.542	300.16	651.16	808.00	12.87	1200.0	0.2521	0.230	0.964
1354	C22H44	1-CYCLOHEXYLHEXADECANE	308.590	306.76	652.16	813.42	10.24	1200.5	0.2571	0.182	0.741
1355	C22H44O2	n-BUTYL STEARATE	340.590	299.45	623.15	764.00	11.10	1230.0	0.2769	0.215	1.035
1356	C24H38O4	DIISOOCTYL PHTHALATE	390.563	-------	694.00	851.00	11.80	1420.0	0.2750	0.237	1.088
1357	C24H38O4	DIOCTYL PHTHALATE	390.563	223.15	657.15	806.00	11.80	1270.0	0.3075	0.224	1.142
1358	C24H42O	DINONYLPHENOL	346.597	-------	722.00	886.00	12.40	1220.0	0.2841	0.205	1.136
1359	C26H20	TETRAPHENYLETHYLENE	332.445	496.15	760.00	996.00	17.10	1020.0	0.3259	0.211	0.729
1360	C28H46O4	DIISODECYL PHTHALATE	446.671	227.59	723.00	887.00	10.00	1460.0	0.3059	0.198	1.076

MW - molecular weight, g/mol

T_F - freezing point temperature, K

T_B - boiling point temperature, K

T_C - critical temperature, K

P_C - critical pressure, bar

V_C - critical volume, cm3/mol

RHO_C - critical density, g/cm3

Z_C - critical compressibility factor

OMEGA - acentric factor

Table 1-2 CRITICAL PROPERTIES AND ACENTRIC FACTOR - INORGANIC COMPOUNDS

NO	FORMULA	NAME	MW g/mol	T_F K	T_B K	T_C K	P_C bar	V_C cm3/mol	RHO_C g/cm3	Z_C	OMEGA
1	Ag	SILVER	107.868	1234.00	2485.00	7480.00	5066.00	58.2	1.8534	0.474	-0.210
2	AgCl	SILVER CHLORIDE	143.321	728.15	1837.15	2992.10	100.28	744.3	0.1926	0.300	0.360
3	AgI	SILVER IODIDE	234.773	825.15	1779.15	2897.64	97.12	744.3	0.3154	0.300	0.351
4	Al	ALUMINUM	26.982	933.00	2329.15	7151.00	5458.00	39.0	0.6918	0.358	-0.230
5	AlB3H12	ALUMINUM BOROHYDRIDE	71.510	209.15	319.05	513.77	112.46	114.0	0.6275	0.300	0.436
6	AlBr3	ALUMINUN BROMIDE	266.694	390.15	529.45	763.00	28.90	310.0	0.8603	0.141	0.399
7	AlCl3	ALUMINUM CHLORIDE	133.340	465.70	453.15	629.00	26.35	261.5	0.5100	0.132	0.660
8	AlF3	ALUMINUM FLUORIDE	83.977	1313.15	1810.15	2948.13	98.81	744.3	0.1128	0.300	0.356
9	AlI3	ALUMINUM IODIDE	407.695	464.15	658.65	983.00	35.95	408.0	0.9993	0.300	0.349
10	Al2O3	ALUMINUM OXIDE	101.961	2325.00	3253.15	5335.00	1953.20	68.1	1.4965	0.300	1.200
11	Al2S3O12	ALUMINUM SULFATE	342.154	1043.20	-------	-------	------	------	------	-----	-----
12	Ar	ARGON	39.948	83.80	87.28	150.86	48.98	74.6	0.5356	0.291	0.000
13	As	ARSENIC	74.922	1090.15	885.00	1673.15	223.00	34.9	2.1468	0.056	0.121
14	AsBr3	ARSENIC TRIBROMIDE	314.634	306.15	493.15	789.01	66.40	270.7	1.1623	0.274	0.298
15	AsCl3	ARSENIC TRICHLORIDE	181.280	255.15	403.55	654.00	59.12	252.0	0.7194	0.274	0.220
16	AsF3	ARSENIC TRIFLUORIDE	131.917	267.25	329.45	530.21	87.81	137.6	0.9590	0.274	0.363
17	AsF5	ARSENIC PENTAFLUORIDE	169.914	193.35	220.35	357.73	41.13	198.1	0.8575	0.274	0.106
18	AsH3	ARSINE	77.945	156.28	210.67	373.00	64.13	132.5	0.5883	0.274	0.006
19	AsI3	ARSENIC TRIIODIDE	455.635	419.15	676.15	1101.22	99.06	277.3	1.6430	0.300	0.357
20	As2O3	ARSENIC TRIOXIDE	197.841	585.95	730.35	1189.50	161.97	183.2	1.0800	0.300	0.502
21	At	ASTATINE	210.000	575.15	607.00	1096.37	1236.30	22.1	9.4929	0.300	0.641
22	Au	GOLD	196.967	1337.33	3120.00	4398.00	6354.40	50.3	3.9158	0.300	2.973
23	B	BORON	10.811	2348.15	4133.00	7934.59	8417.50	23.5	0.4598	0.300	0.826
24	BBr3	BORON TRIBROMIDE	250.523	228.15	364.85	581.00	48.66	272.0	0.9210	0.274	0.216
25	BCl3	BORON TRICHLORIDE	117.169	166.15	285.65	451.95	38.71	266.0	0.4405	0.274	0.151
26	BF3	BORON TRIFLUORIDE	67.806	146.05	173.35	260.90	49.85	123.6	0.5485	0.284	0.430
27	BH2CO	BORINE CARBONYL	40.837	136.15	209.15	340.03	55.03	140.7	0.2902	0.274	0.188
28	BH3O3	BORIC ACID	61.833	458.15	-------	-------	------	------	------	-----	-----
29	B2D6	DEUTERODIBORANE	33.718	-------	179.87	293.74	32.17	208.0	0.1621	0.274	0.017
30	B2H5Br	DIBORANE HYDROBROMIDE	106.566	168.95	289.45	466.98	43.61	243.9	0.4369	0.274	0.142
31	B2H6	DIBORANE	27.670	107.65	180.65	289.80	40.53	173.1	0.1598	0.291	0.125
32	B3N3H6	BORINE TRIAMINE	80.501	214.95	323.75	521.20	36.34	326.7	0.2464	0.274	0.092
33	B4H10	TETRABORANE	53.323	153.25	289.25	466.66	38.84	273.7	0.1948	0.274	0.107
34	B5H9	PENTABORANE	63.126	226.35	331.55	568.45	46.41	285.1	0.2214	0.280	0.000
35	B5H11	TETRAHYDROPENTABORANE	65.142	-------	340.15	547.13	41.29	301.8	0.2158	0.274	0.134
36	B10H14	DECABORANE	122.221	372.75	486.15	791.78	59.02	334.6	0.3652	0.300	0.203
37	Ba	BARIUM	137.327	1000.15	1907.00	3572.13	381.40	233.6	0.5878	0.300	0.264
38	Be	BERYLLIUM	9.012	1560.15	2744.00	5199.80	548.80	236.3	0.0381	0.300	0.309
39	BeB2H8	BERYLLIUM BOROHYDRIDE	38.698	396.15	363.15	591.45	48.77	302.5	0.1279	0.300	0.147
40	BeBr2	BERYLLIUM BROMIDE	168.820	763.15	793.15	1216.86	100.34	302.5	0.5581	0.300	0.361
41	BeCl2	BERYLLIUM CHLORIDE	79.918	678.15	823.15	1238.03	102.09	302.5	0.2642	0.300	0.366
42	BeF2	BERYLLIUM FLUORIDE	47.009	1073.15	-------	------	------	------	------	-----	-----
43	BeI2	BERYLLIUM IODIDE	262.821	761.15	863.15	1238.03	102.09	302.5	0.8688	0.300	0.366
44	Bi	BISMUTH	208.980	544.15	1698.15	4620.00	682.59	79.4	2.6320	0.300	-0.300
45	BiBr3	BISMUTH TRIBROMIDE	448.692	491.15	734.15	1220.00	98.60	302.0	1.4857	0.300	0.288
46	BiCl3	BISMUTH TRICHLORIDE	315.338	503.15	714.15	1178.00	95.91	261.7	1.2050	0.300	0.304
47	BrF5	BROMINE PENTAFLUORIDE	174.896	211.75	313.55	470.00	57.16	187.3	0.9337	0.274	0.504
48	Br2	BROMINE	159.808	265.90	331.90	584.15	103.35	135.0	1.1838	0.287	0.119
49	C	CARBON	12.011	4247.00	4203.00	6810.00	2230.00	18.8	0.6389	0.074	1.566
50	CCl2O	PHOSGENE	98.916	145.37	280.71	455.00	56.74	190.2	0.5200	0.285	0.201
51	CF2O	CARBONYL FLUORIDE	66.007	161.89	188.58	297.00	57.60	141.0	0.4681	0.329	0.283
52	CH4N2O	UREA	60.056	405.85	465.00	705.00	90.50	218.0	0.2755	0.337	0.620
53	CH4N2S	THIOUREA	76.122	454.15	536.00	854.00	82.30	248.0	0.3069	0.287	0.359
54	CNBr	CYANOGEN BROMIDE	105.922	331.15	334.65	545.03	70.08	194.0	0.5460	0.300	0.254
55	CNCl	CYANOGEN CHLORIDE	61.470	266.65	286.00	449.00	59.90	163.0	0.3771	0.262	0.320
56	CNF	CYANOGEN FLUORIDE	45.016	-------	227.17	368.51	79.00	106.3	0.4236	0.274	0.303
57	CO	CARBON MONOXIDE	28.010	68.15	81.70	132.92	34.99	93.1	0.3009	0.295	0.066
58	COS	CARBONYL SULFIDE	60.076	134.35	223.00	378.80	63.49	135.1	0.4447	0.272	0.097
59	COSe	CARBON OXYSELENIDE	106.970	-------	251.25	406.58	86.30	107.3	0.9967	0.274	0.338
60	CO2	CARBON DIOXIDE	44.010	216.58	194.70	304.19	73.82	94.0	0.4682	0.274	0.228
61	CS2	CARBON DISULFIDE	76.143	161.58	319.37	552.00	79.03	160.0	0.4759	0.276	0.108
62	CSeS	CARBON SELENOSULFIDE	123.037	197.95	358.75	576.53	74.12	177.2	0.6944	0.274	0.316
63	C2N2	CYANOGEN	52.035	238.75	252.15	399.90	63.03	144.5	0.3601	0.274	0.312
64	C3S2	CARBON SUBSULFIDE	100.165	273.55	-------	-------	------	------	------	-----	-----
65	Ca	CALCIUM	40.078	1115.15	1762.00	3292.23	352.40	233.0	0.1720	0.300	0.254
66	CaF2	CALCIUM FLUORIDE	78.075	1691.00	2806.50	4570.85	376.91	302.5	0.2581	0.300	0.752
67	CbF5	COLUMBIUM FLUORIDE	187.898	348.65	498.15	811.32	66.90	302.5	0.6212	0.300	0.241
68	Cd	CADMIUM	112.411	594.05	1043.15	2291.00	861.70	37.9	2.9660	0.300	0.050
69	CdCl2	CADMIUM CHLORIDE	183.316	841.15	1240.15	2019.79	166.55	302.5	0.6060	0.300	0.511
70	CdF2	CADMIUM FLUORIDE	150.408	793.15	2024.15	3296.66	271.84	302.5	0.4972	0.300	0.656
71	CdI2	CADMIUM IODIDE	366.220	658.15	1069.15	1741.29	143.59	302.5	1.2106	0.300	0.467
72	CdO	CADMIUM OXIDE	128.410	-------	1832.15	2983.96	246.06	302.5	0.4245	0.300	0.626
73	ClF	CHLORINE MONOFLUORIDE	54.451	128.15	172.65	282.32	79.01	81.4	0.6690	0.274	0.276
74	ClFO3	PERCHLORYL FLUORIDE	102.449	125.41	226.49	368.40	53.70	161.0	0.6363	0.282	0.173
75	ClF3	CHLORINE TRIFLUORIDE	92.448	190.15	284.65	459.39	77.79	134.5	0.6872	0.274	0.316
76	ClF5	CHLORINE PENTAFLUORIDE	130.445	-------	260.05	415.90	52.60	230.4	0.5662	0.350	0.216
77	ClHO3S	CHLOROSULFONIC ACID	116.525	193.15	427.00	700.00	85.00	195.0	0.5976	0.285	0.301
78	ClHO4	PERCHLORIC ACID	100.458	171.95	385.00	631.00	38.60	168.0	0.5980	0.124	0.050

Table 1-2 CRITICAL PROPERTIES AND ACENTRIC FACTOR - INORGANIC COMPOUNDS (continued)

NO	FORMULA	NAME	MW g/mol	T_F K	T_B K	T_C K	P_C bar	V_C cm3/mol	RHO_C g/cm3	Z_C	OMEGA
79	ClO2	CHLORINE DIOXIDE	67.452	213.55	284.05	465.00	108.28	97.8	0.6895	0.274	0.356
80	Cl2	CHLORINE	70.905	172.12	239.12	417.15	77.11	123.8	0.5730	0.275	0.069
81	Cl2O	CHLORINE MONOXIDE	86.905	157.15	275.35	444.68	74.94	135.2	0.6430	0.274	0.303
82	Cl2O7	CHLORINE HEPTOXIDE	182.901	182.15	351.95	565.78	50.90	253.2	0.7223	0.274	0.200
83	Co	COBALT	58.933	1768.15	2528.00	7398.48	5148.70	35.8	1.6442	0.300	-0.180
84	CoCl2	COBALT CHLORIDE	129.839	1008.15	1323.15	2154.97	177.70	302.5	0.4292	0.300	0.530
85	CoNC3O4	COBALT NITROSYL TRICARBONYL	172.971	262.15	353.15	567.68	47.43	298.6	0.5793	0.300	0.178
86	Cr	CHROMIUM	51.996	2180.15	2840.00	8560.93	5784.10	36.9	1.4084	0.300	-0.200
87	CrC6O6	CHROMIUM CARBONYL	220.059	423.65	424.15	690.80	56.96	302.5	0.7275	0.300	0.193
88	CrO2Cl2	CHROMIUM OXYCHLORIDE	154.900	176.65	390.25	626.33	59.99	237.8	0.6513	0.274	0.256
89	Cs	CESIUM	132.905	301.65	963.15	2048.10	116.50	316.4	0.4201	0.216	-0.220
90	CsBr	CESIUM BROMIDE	212.809	909.15	1573.15	2562.13	211.27	302.5	0.7035	0.300	0.581
91	CsCl	CESIUM CHLORIDE	168.358	919.15	1573.15	2562.13	211.27	302.5	0.5566	0.300	0.581
92	CsF	CESIUM FLUORIDE	151.904	956.15	1524.15	2482.33	204.69	302.5	0.5022	0.300	0.572
93	CsI	CESIUM IODIDE	259.810	894.15	1553.15	2529.56	208.59	302.5	0.8589	0.300	0.577
94	Cu	COPPER	63.546	1357.77	3150.00	5123.00	6415.50	61.0	1.0417	0.300	1.601
95	CuBr	CUPROUS BROMIDE	143.450	777.15	1628.15	2651.71	218.66	302.5	0.4742	0.300	0.591
96	CuCl	CUPROUS CHLORIDE	98.999	703.00	1763.15	2435.00	236.79	256.5	0.3859	0.300	1.664
97	CuCl2	CUPRIC CHLORIDE	134.451	906.15	1266.15	2010.00	170.04	294.9	0.4560	0.300	0.623
98	CuI	COPPER IODIDE	190.450	878.15	1609.15	2620.77	216.11	302.5	0.6296	0.300	0.588
99	DCN	DEUTERIUM CYANIDE	28.034	261.15	299.35	482.63	113.56	96.8	0.2896	0.274	0.435
100	D2	DEUTERIUM	4.032	18.73	23.65	38.35	16.64	60.3	0.0669	0.314	-0.140
101	D2O	DEUTERIUM OXIDE	20.031	276.96	374.55	643.89	219.41	56.3	0.3558	0.231	0.368
102	Eu	EUROPIUM	151.965	1095.15	1742.00	5150.00	3547.90	36.2	4.1969	0.300	-0.220
103	F2	FLUORINE	37.997	53.53	84.95	144.31	52.15	66.2	0.5740	0.288	0.059
104	F2O	FLUORINE OXIDE	53.996	49.25	128.55	215.10	49.50	97.6	0.5532	0.270	0.075
105	Fe	IRON	55.847	1808.15	3000.00	9340.00	10150.00	28.0	1.9945	0.366	-0.300
106	FeC5O5	IRON PENTACARBONYL	195.899	252.15	378.15	607.20	35.24	392.5	0.4991	0.274	0.091
107	FeCl2	FERROUS CHLORIDE	126.752	945.15	1299.15	2115.88	174.48	302.5	0.4190	0.300	0.524
108	FeCl3	FERRIC CHLORIDE	162.205	577.15	592.15	964.41	79.53	302.5	0.5362	0.300	0.292
109	Fr	FRANCIUM	223.000	300.15	879.00	1606.46	1790.20	22.4	9.9626	0.300	0.682
110	Ga	GALLIUM	69.723	302.91	2517.00	7620.00	5126.30	75.3	0.9259	0.300	-0.220
111	GaCl3	GALLIUM TRICHLORIDE	176.081	350.90	474.15	694.00	38.20	263.0	0.6695	0.174	0.458
112	Gd	GADOLINIUM	157.250	1587.15	3537.15	--------	------	------	-------	-----	-----
113	Ge	GERMANIUM	72.610	1211.40	3125.00	8400.00	6364.60	32.9	2.2055	0.300	-0.040
114	GeBr4	GERMANIUM BROMIDE	392.226	299.25	462.15	740.00	44.17	381.7	1.0277	0.274	0.169
115	GeCl4	GERMANIUM CHLORIDE	214.421	223.65	357.15	574.00	39.57	330.5	0.6489	0.274	0.123
116	GeHCl3	TRICHLORO GERMANE	179.976	202.05	348.15	559.78	46.21	276.0	0.6522	0.274	0.170
117	GeH4	GERMANE	76.642	107.26	185.00	308.00	55.50	140.0	0.5474	0.303	0.151
118	Ge2H6	DIGERMANE	151.268	164.15	304.65	491.01	46.67	239.7	0.6311	0.274	0.165
119	Ge3H8	TRIGERMANE	225.894	167.55	383.95	616.37	47.05	298.4	0.7569	0.274	0.180
120	HBr	HYDROGEN BROMIDE	80.912	186.34	206.45	363.15	85.52	100.3	0.8070	0.284	0.069
121	HCN	HYDROGEN CYANIDE	27.026	259.91	298.85	456.65	53.91	138.6	0.1950	0.197	0.410
122	HCl	HYDROGEN CHLORIDE	36.461	158.97	188.15	324.65	83.09	81.0	0.4500	0.249	0.132
123	HF	HYDROGEN FLUORIDE	20.006	189.79	292.67	461.15	64.85	69.0	0.2899	0.117	0.383
124	HI	HYDROGEN IODIDE	127.912	222.38	237.55	423.85	83.10	121.9	1.0490	0.288	0.038
125	HNO3	NITRIC ACID	63.013	231.55	356.15	520.00	68.90	145.0	0.4346	0.231	0.714
126	H2	HYDROGEN	2.016	13.95	20.39	33.18	13.13	64.2	0.0314	0.305	-0.220
127	H2O	WATER	18.015	273.15	373.15	647.13	220.55	56.0	0.3220	0.229	0.345
128	H2O2	HYDROGEN PEROXIDE	34.015	272.72	423.35	730.15	216.84	77.7	0.4378	0.278	0.360
129	H2S	HYDROGEN SULFIDE	34.082	187.68	212.80	373.53	89.63	98.5	0.3460	0.284	0.083
130	H2SO4	SULFURIC ACID	98.079	283.46	610.00	925.00	64.00	177.0	0.5540	0.147	0.494
131	H2S2	HYDROGEN DISULFIDE	66.148	183.45	337.15	542.39	88.36	139.8	0.4731	0.274	0.366
132	H2Se	HYDROGEN SELENIDE	80.976	209.15	232.05	411.10	83.44	112.2	0.7215	0.274	0.064
133	H2Te	HYDROGEN TELLURIDE	129.616	224.15	271.15	438.04	71.93	138.7	0.9343	0.274	0.289
134	H3NO3S	SULFAMIC ACID	97.095	478.00	--------	-------	------	225.0	0.4315	-----	-----
135	He	HELIUM-3	3.016	1.01	3.20	3.31	1.17	72.5	0.0416	0.308	-0.470
136	He	HELIUM-4	4.003	1.76	4.22	5.20	2.28	57.3	0.0699	0.302	-0.390
137	Hf	HAFNIUM	178.490	2506.15	5960.00	21688.00	12138.00	44.6	4.0049	0.300	-0.340
138	Hg	MERCURY	200.590	234.29	629.73	1735.00	1608.00	56.4	3.5597	0.628	-0.160
139	HgBr2	MERCURIC BROMIDE	360.398	510.15	592.15	964.41	79.53	302.5	1.1914	0.300	0.292
140	HgCl2	MERCURIC CHLORIDE	271.495	550.15	577.15	939.98	77.51	302.5	0.8975	0.300	0.284
141	HgI2	MERCURIC IODIDE	454.399	532.15	627.15	1078.15	100.00	268.9	1.6897	0.300	0.189
142	IF7	IODINE HEPTAFLUORIDE	259.893	278.65	277.15	447.53	41.26	247.1	1.0518	0.274	0.122
143	I2	IODINE	253.809	386.75	458.39	819.15	116.54	155.0	1.6375	0.265	0.117
144	In	INDIUM	114.818	429.75	2323.00	6730.00	2432.00	82.6	1.3900	0.359	-0.240
145	Ir	IRIDIUM	192.220	2719.15	4450.00	15035.00	72356.00	5.2	37.0850	0.300	-0.130
146	K	POTASSIUM	39.098	336.35	1037.00	2223.00	162.12	209.0	0.1871	0.183	-0.180
147	KBr	POTASSIUM BROMIDE	119.002	1003.15	1656.15	2697.31	222.42	302.5	0.3934	0.300	0.596
148	KCl	POTASSIUM CHLORIDE	74.551	1044.00	1688.87	3470.00	180.00	625.0	0.1193	0.390	-0.120
149	KF	POTASSIUM FLUORIDE	58.097	1153.15	1775.15	2891.12	238.40	302.5	0.1921	0.300	0.617
150	KI	POTASSIUM IODIDE	166.003	996.15	1597.15	2601.22	214.50	302.5	0.5488	0.300	0.585
151	KOH	POTASSIUM HYDROXIDE	56.106	679.00	1600.00	2605.86	214.88	302.5	0.1855	0.300	0.586
152	Kr	KRYPTON	83.800	115.78	119.80	209.35	55.02	91.2	0.9189	0.288	0.000
153	La	LANTHANUM	138.906	1193.15	3643.00	9511.00	5460.00	36.5	3.8056	0.252	-0.010
154	Li	LITHIUM	6.941	453.69	1597.00	4085.00	1722.50	47.0	0.1477	0.238	-0.040
155	LiBr	LITHIUM BROMIDE	86.845	820.15	1583.15	2578.42	212.62	302.5	0.2871	0.300	0.583
156	LiCl	LITHIUM CHLORIDE	42.394	887.15	1655.15	2695.68	222.29	302.5	0.1401	0.300	0.596

NO	FORMULA	NAME	MW g/mol	T_F K	T_B K	T_C K	P_C bar	V_C cm3/mol	RHO_C g/cm3	Z_C	OMEGA
157	LiF	LITHIUM FLUORIDE	25.939	1143.15	1954.15	3182.65	262.44	302.5	0.0858	0.300	0.645
158	LiI	LITHIUM IODIDE	133.845	719.15	1444.15	2352.04	193.95	302.5	0.4425	0.300	0.556
159	Lu	LUTECIUM	174.967	1936.15	2535.00	4128.66	5162.90	20.0	8.7715	0.300	1.527
160	Mg	MAGNESIUM	24.305	923.15	1376.00	2241.04	2802.40	20.0	1.2185	0.300	1.346
161	MgCl2	MAGNESIUM CHLORIDE	95.210	985.15	1691.15	2754.32	227.12	302.5	0.3147	0.300	0.602
162	MgO	MAGNESIUM OXIDE	40.304	3105.00	3873.20	5950.00	33.91	209.5	0.1924	0.014	0.214
163	Mn	MANGANESE	54.938	1519.15	2392.00	6902.82	4871.70	35.3	1.5544	0.300	-0.160
164	MnCl2	MANGANESE CHLORIDE	125.843	923.15	1463.15	2382.98	196.50	302.5	0.4160	0.300	0.560
165	Mo	MOLYBDENUM	95.940	2895.15	5081.15	9620.00	10349.00	38.3	2.5050	0.300	0.924
166	MoF6	MOLYBDENUM FLUORIDE	209.930	290.15	309;15	498.12	50.30	225.6	0.9306	0.274	0.189
167	MoO3	MOLYBDENUM OXIDE	143.938	1068.15	1424.15	2319.46	191.26	302.5	0.4758	0.300	0.552
168	NCl3	NITROGEN TRICHLORIDE	120.365	246.15	344.15	564.00	62.10	206.9	0.5818	0.274	0.199
169	ND3	HEAVY AMMONIA	20.055	199.15	239.75	388.40	125.71	70.4	0.2850	0.274	0.447
170	NF3	NITROGEN TRIFLUORIDE	71.002	66.36	144.09	233.85	45.30	118.8	0.5979	0.277	0.126
171	NH3	AMMONIA	17.031	195.41	239.72	405.65	112.78	72.5	0.2350	0.242	0.252
172	NH3O	HYDROXYLAMINE	33.030	306.25	383.00	574.00	175.18	74.6	0.4425	0.274	0.694
173	NH4Br	AMMONIUM BROMIDE	97.943	-------	669.15	1089.82	17.94	1515.3	0.0646	0.300	-0.150
174	NH4Cl	AMMONIUM CHLORIDE	53.491	793.20	612.00	882.00	16.40	1341.5	0.0399	0.300	3.920
175	NH4I	AMMONIUM IODIDE	144.943	-------	678.05	1104.32	18.18	1515.3	0.0957	0.300	-0.150
176	NH5O	AMMONIUM HYDROXIDE	35.046	194.15	-------	-------	------	------	------	-----	-----
177	NH5S	AMMONIUM HYDROGENSULFIDE	51.112	391.15	306.45	499.10	8.22	1515.3	0.0337	0.300	-0.380
178	NO	NITRIC OXIDE	30.006	112.15	121.38	180.15	64.85	57.7	0.5200	0.250	0.585
179	NOCl	NITROSYL CHLORIDE	65.459	213.55	267.77	440.65	91.19	139.3	0.4699	0.347	0.307
180	NOF	NITROSYL FLUORIDE	49.005	139.15	217.15	352.67	112.78	71.2	0.6879	0.274	0.405
181	NO2	NITROGEN DIOXIDE	46.006	261.95	294.00	431.35	101.33	82.5	0.5577	0.233	0.849
182	N2	NITROGEN	28.013	63.15	77.35	126.10	33.94	90.1	0.3109	0.292	0.040
183	N2F4	TETRAFLUOROHYDRAZINE	104.007	111.65	198.95	309.35	37.10	213.0	0.4883	0.307	0.223
184	N2H4	HYDRAZINE	32.045	274.69	386.65	653.15	146.92	158.0	0.2028	0.427	0.314
185	N2H4C	AMMONIUM CYANIDE	44.056	309.15	304.85	491.32	109.47	102.2	0.4309	0.274	0.425
186	N2H6CO2	AMMONIUM CARBAMATE	78.071	-------	331.45	539.82	44.51	302.5	0.2581	0.300	0.120
187	N2O	NITROUS OXIDE	44.013	182.33	184.67	309.57	72.45	97.4	0.4520	0.274	0.142
188	N2O3	NITROGEN TRIOXIDE	76.012	170.00	275.15	425.00	69.90	195.0	0.3898	0.386	0.431
189	N2O4	NITROGEN TETRAOXIDE	92.011	261.90	302.22	431.15	101.33	82.5	1.1154	0.233	1.007
190	N2O5	NITROGEN PENTOXIDE	108.010	303.15	320.15	515.51	64.33	182.6	0.5917	0.274	0.266
191	Na	SODIUM	22.990	370.98	1156.00	2573.00	354.64	116.0	0.1982	0.192	-0.100
192	NaBr	SODIUM BROMIDE	102.894	1020.00	1663.82	4287.00	192.52	398.0	0.2585	0.215	-0.800
193	NaCN	SODIUM CYANIDE	49.008	836.85	1769.15	2900.00	237.60	304.5	0.1610	0.300	0.589
194	NaCl	SODIUM CHLORIDE	58.442	1073.95	1738.15	3400.00	355.00	266.0	0.2197	0.334	0.134
195	NaF	SODIUM FLUORIDE	41.988	1269.00	1982.72	5530.00	531.96	185.0	0.2270	0.214	-1.110
196	NaI	SODIUM IODIDE	149.894	924.15	1577.15	2568.65	322.13	198.9	0.7536	0.300	0.706
197	NaOH	SODIUM HYDROXIDE	39.997	596.00	1663.15	2820.00	253.31	200.0	0.2000	0.216	0.477
198	Na2SO4	SODIUM SULFATE	142.043	1157.00	-------	-------	------	------	------	-----	-----
199	Nb	NIOBIUM	92.906	2750.15	5115.00	17904.10	10418.00	42.9	2.1671	0.300	-0.310
200	Nd	NEODYMIUM	144.240	1289.15	3384.00	10665.10	6892.10	38.6	3.7368	0.300	-0.240
201	Ne	NEON	20.180	24.55	27.09	44.40	26.53	41.7	0.4839	0.300	-0.040
202	Ni	NICKEL	58.693	1728.15	2415.00	6986.15	4918.50	35.4	1.6566	0.300	-0.170
203	NiC4O4	NICKEL CARBONYL	170.735	248.15	315.65	508.40	32.39	357.5	0.4775	0.274	0.056
204	NiF2	NICKEL FLUORIDE	96.690	1723.15	2013.15	3278.75	270.37	302.5	0.3196	0.300	0.654
205	Np	NEPTUNIUM	237.000	913.15	-------	-------	------	------	------	-----	-----
206	O2	OXYGEN	31.999	54.36	90.17	154.58	50.43	73.4	0.4360	0.288	0.022
207	O3	OZONE	47.998	80.15	161.85	261.00	55.73	89.0	0.5393	0.229	0.227
208	Os	OSMIUM	190.230	3306.15	4880.00	16878.70	9938.90	42.4	4.4906	0.300	-0.300
209	OsOF5	OSMIUM OXIDE PENTAFLUORIDE	301.221	332.95	373.65	608.55	50.18	302.5	0.9958	0.300	0.155
210	OsO4	OSMIUM TETROXIDE - YELLOW	254.228	329.15	403.15	656.60	54.14	302.5	0.8404	0.300	0.178
211	OsO4	OSMIUM TETROXIDE - WHITE	254.228	315.15	403.15	656.60	54.14	302.5	0.8404	0.300	0.178
212	P	PHOSPHORUS - WHITE	30.974	317.25	553.45	993.75	83.29	297.6	0.1041	0.300	0.032
213	PBr3	PHOSPHORUS TRIBROMIDE	270.686	233.15	448.45	711.00	53.99	300.0	0.9023	0.274	0.264
214	PCl2F3	PHOSPHORUS DICHLORIDE TRIFLUORIDE	158.874	265.15	283.15	457.02	40.48	257.2	0.6178	0.274	0.118
215	PCl3	PHOSPHORUS TRICHLORIDE	137.332	181.15	349.25	563.15	56.70	260.0	0.5282	0.315	0.234
216	PCl5	PHOSPHORUS PENTACHLORIDE	208.237	433.15	433.00	646.15	58.15	277.2	0.7513	0.300	0.531
217	PH3	PHOSPHINE	33.998	139.37	185.41	324.75	65.36	113.3	0.3000	0.274	0.036
218	PH4Br	PHOSPHONIUM BROMIDE	114.910	-------	311.45	501.76	62.21	183.6	0.6260	0.274	0.254
219	PH4Cl	PHOSPHONIUM CHLORIDE	70.458	244.65	246.15	322.30	49.14	149.4	0.4716	0.274	1.640
220	PH4I	PHOSPHONIUM IODIDE	161.910	291.65	335.45	539.70	77.61	158.4	1.0221	0.274	0.326
221	POCl3	PHOSPHORUS OXYCHLORIDE	153.331	274.33	378.65	602.15	51.66	265.5	0.5774	0.274	0.240
222	PSBr3	PHOSPHORUS THIOBROMIDE	302.752	311.15	448.15	729.89	60.19	302.5	1.0008	0.300	0.209
223	PSCl3	PHOSPHORUS THIOCHLORIDE	169.398	236.95	398.15	638.82	48.57	299.6	0.5654	0.274	0.192
224	P4O6	PHOSPHORUS TRIOXIDE	219.891	295.65	446.25	714.86	52.08	312.7	0.7032	0.274	0.218
225	P4O10	PHOSPHORUS PENTOXIDE	283.889	693.15	-------	-------	------	------	------	-----	-----
226	P4S10	PHOSPHORUS PENTASULFIDE	444.555	561.15	787.15	1291.00	232.00	138.8	3.2027	0.300	0.594
227	Pb	LEAD	207.200	600.61	2024.00	5400.00	861.30	93.2	2.2232	0.179	-0.250
228	PbBr2	LEAD BROMIDE	367.008	646.15	1187.15	1933.47	159.43	302.5	1.2133	0.300	0.498
229	PbCl2	LEAD CHLORIDE	278.105	774.15	1227.15	1998.62	164.81	302.5	0.9194	0.300	0.507
230	PbF2	LEAD FLUORIDE	245.197	1128.15	1566.15	2550.73	210.33	302.5	0.8106	0.300	0.580
231	PbI2	LEAD IODIDE	461.009	675.15	1145.15	1865.07	153.79	302.5	1.5240	0.300	0.487
232	PbO	LEAD OXIDE	223.199	1163.15	1745.15	2842.26	234.37	302.5	0.7379	0.300	0.612
233	PbS	LEAD SULFIDE	239.266	1387.15	1554.15	2531.19	208.72	302.5	0.7910	0.300	0.577
234	Pd	PALLADIUM	106.420	1828.05	3385.00	10669.10	6894.10	38.6	2.7568	0.300	-0.240

27

NO	FORMULA	NAME	MW g/mol	T_F K	T_B K	T_C K	P_C bar	V_C cm3/mol	RHO_C g/cm3	Z_C	OMEGA
235	Po	POLONIUM	209.000	527.15	1235.00	3013.08	2515.30	29.9	6.9944	0.300	0.011
236	Pt	PLATINUM	195.080	2041.55	3980.00	6983.00	8105.90	759.1	0.2570	0.300	1.217
237	Ra	RADIUM	226.000	973.15	1809.00	4862.82	3684.30	32.9	6.8645	0.300	-0.100
238	Rb	RUBIDIUM	85.468	312.46	978.00	2111.10	134.00	247.0	0.3460	0.189	-0.220
239	RbBr	RUBIDIUM BROMIDE	165.372	955.15	1625.15	2646.82	218.26	302.5	0.5467	0.300	0.591
240	RbCl	RUBIDIUM CHLORIDE	120.921	988.15	1654.15	2694.06	222.15	302.5	0.3997	0.300	0.596
241	RbF	RUBIDIUM FLUORIDE	104.466	1033.15	1681.15	2738.03	225.78	302.5	0.3453	0.300	0.601
242	RbI	RUBIDIUM IODIDE	212.372	915.15	1577.15	2568.65	211.81	302.5	0.7021	0.300	0.582
243	Re	RHENIUM	186.207	3459.15	5915.00	21482.80	12047.00	32.1	5.8008	0.300	-0.340
244	Re2O7	RHENIUM HEPTOXIDE	484.410	569.15	635.55	1035.10	85.35	302.5	1.6014	0.300	0.313
245	Rh	RHODIUM	102.906	2237.15	3940.00	12906.60	8024.40	40.1	2.5649	0.300	-0.270
246	Rn	RADON	222.000	202.15	211.35	377.40	63.00	140.0	1.5857	0.281	-0.020
247	Ru	RUTHENIUM	101.070	2607.15	4500.00	15247.10	9165.00	41.5	2.4355	0.300	-0.290
248	RuF5	RUTHENIUM PENTAFLUORIDE	196.062	359.65	600.15	977.44	80.60	302.5	0.6481	0.300	0.296
249	S	SULFUR	32.066	388.36	717.82	1313.00	182.08	158.0	0.2029	0.264	0.262
250	SF4	SULFUR TETRAFLUORIDE	108.060	149.15	233.15	364.00	52.22	158.8	0.6806	0.274	0.307
251	SF6	SULFUR HEXAFLUORIDE	146.056	222.45	209.25	318.69	37.60	198.5	0.7357	0.282	0.215
252	SOBr2	THIONYL BROMIDE	207.873	220.95	412.65	661.75	64.89	232.3	0.8948	0.274	0.283
253	SOCl2	THIONYL CHLORIDE	118.971	172.00	348.75	567.00	63.63	203.0	0.5861	0.274	0.231
254	SOF2	SULFUROUS OXYFLUORIDE	86.062	162.65	228.90	371.25	59.28	142.7	0.6033	0.274	0.218
255	SO2	SULFUR DIOXIDE	64.065	200.00	263.13	430.75	78.84	122.0	0.5251	0.269	0.245
256	SO2Cl2	SULFURYL CHLORIDE	134.970	222.00	342.55	545.00	46.10	224.0	0.6025	0.228	0.176
257	SO3	SULFUR TRIOXIDE	80.064	289.95	317.90	490.85	82.07	127.1	0.6300	0.256	0.422
258	S2Cl2	SULFUR MONOCHLORIDE	135.037	193.15	411.15	659.37	62.75	239.4	0.5641	0.274	0.272
259	Sb	ANTIMONY	121.757	903.78	1898.00	5070.00	3865.60	32.7	3.7216	0.300	-0.080
260	SbBr3	ANTIMONY TRIBROMIDE	361.469	369.75	548.15	892.75	73.62	302.5	1.1949	0.300	0.269
261	SbCl3	ANTIMONY TRICHLORIDE	228.115	346.55	493.40	794.00	48.20	270.0	0.8449	0.197	0.171
262	SbCl5	ANTIMONY PENTACHLORIDE	299.021	275.95	413.15	662.54	39.42	382.9	0.7810	0.274	0.129
263	SbH3	STIBINE	124.781	185.15	255.15	440.35	73.06	157.2	0.7938	0.314	0.097
264	SbI3	ANTIMONY TRIIODIDE	502.470	440.15	674.15	1097.96	90.54	302.5	1.6611	0.300	0.330
265	Sb2O3	ANTIMONY TRIOXIDE	291.512	929.15	1698.15	2765.72	228.06	302.5	0.9637	0.300	0.604
266	Sc	SCANDIUM	44.956	1814.15	2700.00	8035.08	5499.00	36.5	1.2334	0.300	-0.190
267	Se	SELENIUM	78.960	494.15	930.00	1766.00	380.00	62.3	1.2674	0.161	0.227
268	SeCl4	SELENIUM TETRACHLORIDE	220.771	-------	464.65	743.95	61.05	277.6	0.7953	0.274	0.269
269	SeF6	SELENIUM HEXAFLUORIDE	192.950	238.45	227.35	368.80	44.75	187.7	1.0278	0.274	0.133
270	SeOCl2	SELENIUM OXYCHLORIDE	165.865	281.65	441.15	706.80	77.47	207.8	0.7981	0.274	0.340
271	SeO2	SELENIUM DIOXIDE	110.959	613.15	590.15	961.16	79.26	302.5	0.3668	0.300	0.291
272	Si	SILICON	28.086	1685.00	3513.80	5159.00	537.00	233.0	0.1205	0.292	1.494
273	SiBrCl2F	BROMODICHLOROFLUOROSILANE	197.893	160.85	308.55	497.17	38.27	295.9	0.6687	0.274	0.106
274	SiBrF3	TRIFLUOROBROMOSILANE	164.985	202.65	231.45	375.28	37.54	227.7	0.7245	0.274	0.082
275	SiBr2ClF	DIBROMOCHLOROFLUOROSILANE	242.345	173.85	332.65	535.27	39.32	310.1	0.7814	0.274	0.118
276	SiClF3	TRIFLUOROCHLOROSILANE	120.533	131.15	203.15	330.54	35.25	213.6	0.5643	0.274	0.053
277	SiCl2F2	DICHLORODIFLUOROSILANE	136.988	133.45	241.35	390.93	35.95	247.7	0.5531	0.274	0.072
278	SiCl3F	TRICHLOROFLUOROSILANE	153.442	152.35	285.35	460.49	37.24	281.7	0.5447	0.274	0.093
279	SiCl4	SILICON TETRACHLORIDE	169.896	204.30	330.00	507.00	35.93	326.0	0.5212	0.278	0.232
280	SiF4	SILICON TETRAFLUORIDE	104.079	186.35	178.35	259.00	37.19	165.0	0.6308	0.285	0.385
281	SiHBr3	TRIBROMOSILANE	268.805	199.65	384.95	610.00	47.02	350.0	0.7680	0.324	0.222
282	SiHCl3	TRICHLOROSILANE	135.452	144.95	305.00	479.00	41.70	268.0	0.5054	0.281	0.203
283	SiHF3	TRIFLUOROSILANE	86.089	141.75	178.15	291.02	39.95	165.9	0.5188	0.274	0.079
284	SiH2Br2	DIBROMOSILANE	189.909	202.95	343.65	550.00	53.00	246.0	0.7720	0.285	0.227
285	SiH2Cl2	DICHLOROSILANE	101.007	151.15	281.45	449.00	44.30	228.0	0.4430	0.271	0.177
286	SiH2F2	DIFLUOROSILANE	68.098	-------	195.35	318.21	47.59	152.3	0.4471	0.274	0.139
287	SiH2I2	DIIODOSILANE	283.910	272.15	422.65	660.00	66.88	232.0	1.2238	0.283	0.389
288	SiH3Br	MONOBROMOSILANE	111.013	179.25	275.55	454.00	56.44	177.0	0.6272	0.265	0.155
289	SiH3Cl	MONOCHLOROSILANE	66.562	155.05	242.75	396.65	48.43	174.0	0.3825	0.256	0.136
290	SiH3F	MONOFLUOROSILANE	50.108	-------	175.15	286.28	46.88	139.1	0.3602	0.274	0.125
291	SiH3I	IODOSILANE	158.014	216.15	318.55	515.00	69.41	160.0	0.9876	0.259	0.276
292	SiH4	SILANE	32.117	88.15	161.00	269.70	48.43	132.7	0.2420	0.287	0.097
293	SiO2	SILICON DIOXIDE	60.084	1883.00	2503.20	4076.87	336.18	302.5	0.1986	0.300	0.719
294	Si2Cl6	HEXACHLORODISILANE	268.887	271.95	412.15	660.96	29.11	517.3	0.5198	0.274	0.035
295	Si2F6	HEXAFLUORODISILANE	170.161	254.55	254.25	411.33	30.16	310.6	0.5478	0.274	0.022
296	Si2H5Cl	DISILANYL CHLORIDE	96.663	-------	314.70	506.89	41.58	277.7	0.3480	0.274	0.132
297	Si2H6	DISILANE	62.219	140.65	259.00	432.00	51.30	198.0	0.3142	0.283	0.102
298	Si2OCl3F3	TRICHLOROTRIFLUORODISILOXANE	235.524	-------	315.89	508.78	26.77	433.0	0.5439	0.274	0.000
299	Si2OCl6	HEXACHLORODISILOXANE	284.887	239.95	408.75	655.58	27.90	535.2	0.5323	0.274	0.022
300	Si2OH6	DISILOXANE	78.218	128.95	257.75	416.86	36.19	262.4	0.2981	0.274	0.078
301	Si3Cl8	OCTACHLOROTRISILANE	367.878	-------	484.55	775.41	24.70	715.0	0.5145	0.274	-0.010
302	Si3H8	TRISILANE	92.320	155.95	326.25	525.15	33.70	354.9	0.2601	0.274	0.070
303	Si3H9N	TRISILAZANE	107.335	167.45	321.85	518.20	31.65	372.9	0.2878	0.274	0.050
304	Si4H10	TETRASILANE	122.421	179.55	373.15	599.30	29.68	460.0	0.2661	0.274	0.037
305	Sm	SAMARIUM	150.360	1345.15	1874.00	5082.92	3816.70	33.2	4.5262	0.300	-0.100
306	Sn	TIN	118.710	505.00	2995.00	7400.00	6099.80	115.1	1.0314	0.300	0.101
307	SnBr4	STANNIC BROMIDE	438.326	304.15	477.85	764.82	43.43	401.1	1.0928	0.274	0.165
308	SnCl2	STANNOUS CHLORIDE	189.615	519.95	896.15	1459.53	120.35	302.5	0.6268	0.300	0.414
309	SnCl4	STANNIC CHLORIDE	260.521	242.95	386.15	619.85	41.24	342.4	0.7609	0.274	0.140
310	SnH4	STANNIC HYDRIDE	122.742	123.25	220.85	358.52	53.42	152.9	0.8028	0.274	0.184
311	SnI4	STANNIC IODIDE	626.328	417.65	621.15	1011.64	83.22	302.5	2.0705	0.300	0.306
312	Sr	STRONTIUM	87.620	1050.15	1630.00	4267.20	3319.80	32.1	2.7327	0.300	-0.070

Table 1-2 CRITICAL PROPERTIES AND ACENTRIC FACTOR - INORGANIC COMPOUNDS (continued)

NO	FORMULA	NAME	MW g/mol	T_F K	T_B K	T_C K	P_C bar	V_C cm3/mol	RHO_C g/cm3	Z_C	OMEGA
313	SrO	STRONTIUM OXIDE	103.619	2703.15	-------	-------	------	------	------	-----	-----
314	Ta	TANTALUM	180.948	3290.15	5565.00	19900.90	11334.00	43.8	4.1314	0.300	-0.330
315	Tc	TECNNETIUM	98.000	2430.15	5000.00	17400.80	10183.00	42.6	2.2992	0.300	-0.310
316	Te	TELLURIUM	127.600	722.66	1285.00	4840.00	2617.10	46.1	2.7660	0.300	-0.470
317	TeCl4	TELLURIUM TETRACHLORIDE	269.411	497.15	665.15	1083.31	89.33	302.5	0.8906	0.300	0.326
318	TeF6	TELLURIUM HEXAFLUORIDE	241.590	235.35	234.55	380.18	34.47	251.2	0.9616	0.274	0.057
319	Ti	TITANIUM	47.880	1941.15	3442.00	6400.00	7010.20	22.8	2.1025	0.300	0.915
320	TiCl4	TITANIUM TETRACHLORIDE	189.691	249.05	409.00	638.00	46.61	340.0	0.5579	0.299	0.284
321	Tl	THALLIUM	204.383	577.15	1745.00	4648.06	3554.00	32.6	6.2650	0.300	-0.090
322	TlBr	THALLOUS BROMIDE	284.287	733.15	1092.15	1778.75	146.68	302.5	0.9398	0.300	0.473
323	TlI	THALLOUS IODIDE	331.288	713.15	1096.15	1785.26	147.21	302.5	1.0952	0.300	0.474
324	Tm	THULIUM	168.934	1818.15	2219.15	6283.16	4519.70	34.7	4.8716	0.300	-0.150
325	U	URANIUM	238.029	1408.15	4135.00	13712.60	8421.60	40.6	5.8605	0.300	-0.270
326	UF6	URANIUM FLUORIDE	352.019	342.35	328.85	505.80	46.60	250.0	1.4081	0.277	0.318
327	V	VANADIUM	50.942	2183.15	3665.00	11787.10	7464.40	39.4	1.2933	0.300	-0.250
328	VCl4	VANADIUM TETRACHLORIDE	192.752	247.45	425.00	697.00	60.30	268.0	0.7192	0.279	0.186
329	VOCl3	VANADIUM OXYTRICHLORIDE	173.299	193.65	400.00	636.00	49.96	290.0	0.5976	0.274	0.230
330	W	TUNGSTEN	183.840	3695.15	5645.00	14756.00	11497.00	33.9	5.4230	0.300	0.077
331	WF6	TUNGSTEN FLUORIDE	297.830	272.65	290.45	468.56	46.75	228.3	1.3044	0.274	0.163
332	Xe	XENON	131.290	161.36	165.03	289.74	58.40	118.0	1.1126	0.286	0.000
333	Yb	YTTERBIUM	173.040	1097.15	1660.00	4365.92	3380.90	32.2	5.3719	0.300	-0.070
334	Yt	YTTRIUM	88.906	1799.15	3055.00	9381.32	6222.00	37.6	2.3639	0.300	-0.220
335	Zn	ZINC	65.390	692.70	1181.15	3170.00	2904.00	33.0	1.9815	0.364	0.078
336	ZnCl2	ZINC CHLORIDE	136.295	638.15	1005.15	1637.05	134.99	302.5	0.4506	0.300	0.448
337	ZnF2	ZINC FLUORIDE	103.387	1145.15	1770.15	2882.98	237.73	302.5	0.3418	0.300	0.616
338	ZnO	ZINC OXIDE	81.389	2248.20	-------	-------	------	------	------	-----	-----
339	ZnSO4	ZINC SULFATE	161.454	953.00	-------	-------	------	------	------	-----	-----
340	Zr	ZIRCONIUM	91.224	2128.15	4598.00	8802.00	9364.60	23.5	3.8909	0.300	0.859
341	ZrBr4	ZIRCONIUM BROMIDE	410.840	723.15	630.15	1026.30	84.63	302.5	1.3582	0.300	0.310
342	ZrCl4	ZIRCONIUM CHLORIDE	233.035	710.15	604.15	983.96	81.14	302.5	0.7704	0.300	0.298
343	ZrI4	ZIRCONIUM IODIDE	598.842	772.15	704.15	1146.82	94.57	302.5	1.9796	0.300	0.343

MW - molecular weight, g/mol

T_F - freezing point temperature, K

T_B - boiling point temperature, K

T_C - critical temperature, K

P_C - critical pressure, bar

V_C - critical volume, cm3/mol

RHO_C - critical density, g/cm3

Z_C - critical compressibility factor

OMEGA - acentric factor

Chapter 2

HEAT CAPACITY OF GAS

Carl L. Yaws, Xiaoyan Lin, Li Bu, Sachin Nijhawan, Deepa R. Balundgi, and Saumya Tripathi
Lamar University, Beaumont, Texas

ABSTRACT

Results for heat capacity of ideal gas as a function of temperature are presented for major organic and inorganic compounds. The results cover a wide temperature range and include hydrocarbon, oxygen, nitrogen, halogen, sulfur, silicon, and many other chemical types. The agreement between correlation and data is quite good.

INTRODUCTION

Thermodynamic properties such as heat capacity are important in the engineering design of chemical processes. In gas-phase chemical reactions, the heat capacity is required to determine the energy (heat) necessary to bring the chemical reactants up to reaction temperature. Additional uses include generalized heat exchanger and energy balance design calculations.

In this article, correlation results for heat capacity of gas are provided in an easy-to-use tabular format that is especially applicable for rapid engineering use with the personal computer or hand calculator.

HEAT CAPACITY CORRELATION

The correlation for heat capacity of the ideal gas is a series expansion in temperature:

$$C_P = A + B\,T + C\,T^2 + D\,T^3 + E\,T^4 \tag{2-1}$$

where
C_P = heat capacity of ideal gas, joule/(mol K)
A, B, C, D, and E = regression coefficients for chemical compound
T = temperature, K

The results for heat capacity of gas are given in Tables 2-1 and 2-2. The tabulations are based on regression of experimental data and estimates from an extensive literature search for organics (1-40) and inorganics (1-78). Both experimental values for the property under consideration and parameter values for estimation of the property are included in the source publications. The numerous data points were processed with a generalized least-squares computer program for minimizing the deviations.

The tabulation for organic compounds is applicable to a wide variety of substances: hydrocarbons (alkanes, olefins, acetylenes, cycloalkanes, etc.), oxygenates (alcohols, aldehydes, ketones, acids, ethers, glycols, anhydrides, etc.), halogenates (chlorinated, brominated, fluorinated, etc.), nitrogenates (nitriles, amines, cyanates, amides, etc.), sulfur compounds (mercaptans, sulfides, sulfates, etc.), silicon compounds (silanes, chlorosilanes, etc.), and many other chemical types.

The tabulation for inorganic compounds is also comprehensive: carbon oxides (carbon monoxide, carbon dioxide, etc.), nitrogen oxides (nitric oxide, nitrous oxide, etc.), sulfur oxides (sulfur dioxide, sulfur trioxide, etc.), hydrogen oxides (water, hydrogen peroxide, etc.), ammonias (ammonia, ammonium hydroxide, etc.), hydrogen halides (hydrogen chloride, hydrogen fluoride, etc.), sulfur acids (sulfuric acid, hydrogen sulfide, etc.), hydroxides (sodium hydroxide, potassium hydroxide, etc.), silicon halides (trichlorosilane, silicon tetrachloride, etc.), ureas (urea, thiourea, etc.), cyanides (hydrogen cyanide, cyanogen chloride, etc.), hydrides (silane, diborane, etc.), sodium derivatives (sodium chloride, sodium fluoride, etc.), aluminum derivatives (aluminum bromide, aluminum chloride, etc.), and many other compound types. Many elements are covered: hydrogen, nitrogen, oxygen, helium, argon, neon, chlorine, bromine, iodine, fluorine, sulfur, phosphorous, aluminum, lead, tin, mercury, sodium, magnesium, silicon, antimony, boron, iron, chromium, cobalt, titanium, tantalum, silver, gold, platinum, radon, uranium, and many others.

A comparison of correlation and actual data for heat capacity is shown in Fig. 2-1 for a representative chemical. The graph indicates good agreement of correlation and data.

EXAMPLES

The correlation results maybe used for prediction and calculation of heat capacity and other thermodynamic properties. Examples are given below.

Example 1 Estimate the heat capacity of carbon tetrachloride (CCl4) as a low-pressure gas at 500 K.

Substitution of the coefficients from the table and temperature into the equation for heat capacity yields:

HEAT CAPACITY OF GAS

$C_P = 19.816 + (3.3311E-01)(500) - (5.0511E-04)(500^2) + (3.4057E-07)(500^3) - (8.4249E-11)(500^4)$

$C_P = 97.40$ joule/(mol K)

Example 2 Calculate the energy required to heat gaseous ethyl chloride (C2H5Cl) from 300 to 600 K at low pressure.

From thermodynamics, the change in enthalpy, ΔH, at constant pressure is

$$\Delta H = \int C_P \, dT = \int (A + BT + CT^2 + DT^3 + ET^4) \, dT$$

$$\Delta H = AT + (B/2)T^2 + (C/3)T^3 + (D/4)T^4 + (E/5)T^5 \Big]_{T_1}^{T_2}$$

Substitution of the coefficients from the table and the temperature limits into the equation yields

$\Delta H = (35.946)(600 - 300) + (5.2294E-02/2)(600^2 - 300^2) + (2.0321E-04/3)(600^3 - 300^3) - (2.2795E-07/4)(600^4 - 300^4)$
    $+ (6.9123E-11/5)(600^5 - 300^5)$

$\Delta H = 24,760$ joule/mol

Portions of this material appeared in Chem. Eng., <u>95</u> (No. 7), 91 (May 9, 1988) and are reprinted by special permission.

REFERENCES – ORGANIC COMPOUNDS

1-34. See **REFERENCES - ORGANIC COMPOUNDS** in Chapter 1, CRITICAL PROPERTIES AND ACENTRIC FACTOR

35. Pedley, J. B., <u>THERMOCHEMICAL DATA AND STRUCTURES OF ORGANIC COMPOUNDS</u>, Vol. I, Thermodynamics Research Center, College Station, TX (1994).

36. Frenkel, M., K. N. Marsh, R. C. Wilhoit, G. J. Kabo, and G. N. Roganov, <u>THERMODYNAMICS OF ORGANIC COMPOUNDS IN THE GAS STATE</u>, Vols. I and II, Thermodynamics Research Center, College Station, TX (1994).

37. Suris, A. L., <u>HANDBOOK OF THERMODYNAMIC HIGH TEMPERATURE PROCESS DATA</u>, Hemisphere Publishing Corporation, New York, NY (1987).

38. Yaws, C. L., H. M. Ni, and P. Y. Chiang, Chem. Eng., <u>95</u> (7), 91 (May 9, 1988).

39. Yaws, C. L., <u>HANDBOOK OF THERMODYNAMIC DIAGRAMS</u>, Vols. 1, 2, 3, and 4, Gulf Publishing Co., Houston, TX (1996).

40. Yaws, C. L., <u>HANDBOOK OF CHEMICAL COMPOUND DATA FOR PROCESS SAFETY</u>, Gulf Publishing Co., Houston, TX (1997).

REFERENCES - INORGANIC COMPOUNDS

1-56. See **REFERENCES - INORGANIC COMPOUNDS** in Chapter 1, CRITICAL PROPERTIES AND ACENTRIC FACTOR

57. Chase, M. W. and others, <u>JANAF THERMOCHEMICAL TABLES, 1974 Supplement</u>, J. Phys. Chem. Ref. Data, <u>3(2)</u>, (1974).

58. Chase, M. W. and others, <u>JANAF THERMOCHEMICAL TABLES, 1975 Supplement</u>, J. Phys. Chem. Ref. Data, <u>4(1)</u>, (1975).

59. Chase, M. W. and others <u>JANAF THERMOCHEMICAL TABLES</u>, Parts 1 and 2, 3rd ed., J. Phys. Chem. Ref. Data, <u>4</u>, Supplement No. 1 (1985).

60. Wagman, D. D. and others, <u>THE NBS TABLES OF CHEMICAL THERMODYNAMIC PROPERTIES</u>, J. Phys. Chem. Ref. Data, <u>4</u>, Supplement No. 2 (1982).

61. Kelley, K. K., <u>CONTRIBUTIONS TO THE DATA ON THEORETICAL METALLURGY</u>, Bulletin 584, United States Government Printing Office, Washington, DC (1960).

62. Wicks, C. E. and F. E. Block, <u>THERMODYNAMIC PROPERTIES OF 65 ELEMENTS - THEIR OXIDES, HALIDES, CARBIDES AND NITRIDES</u>, Bulletin 605, United States Government Printing Office, Washington, DC (1963).

63. Wagman, D. D. and others, <u>SELECTED VALUES OF CHEMICAL THERMODYNAMIC PROPERTIES</u>, NBS Technical Note 270-3, United States Government Printing Office, Washington, DC (1968).

64. Karapet'yants, M. Kh. and M. L. Karapet'yants, <u>THERMODYNAMIC CONSTANTS OF INORGANIC AND ORGANIC COMPOUNDS</u>, Ann Arbor - Humphrey Science Publishers, Ann Arbor, MH (1970).

65. Lesieecki, M.L. and J. S. Shirk, J. Chem. Phys., <u>56</u>, 4171 (1972).

66. Sherman, R. H. and W. F. Giauque, J. Amer. Chem. Soc., <u>77</u>, 2154 (1955).

67. Frenkel, M. L. and E. A. Gusev, G. Ya. Kabo, J. Appl. Chem. of the USSR, <u>56</u> (1), 204 (1983).

68. Kunchur, N. R. and M. R. Truter, J. Chem. Soc., 2551 (1958).

69. Goodwin, R. D., J. Phys. Chem. Ref. Data, <u>14</u> (4), 849 (1985).

70. McBride, B. J. and S. Gordon, J. Chem. Phys., <u>35</u>, 2198 (1961).

71. Nagarajan, G., Bull. Soc. Chim. Belges., <u>72</u>, 524 (1963).

72. Harrison, B. and W. H. Seaton, Ind. Eng. Chem. Res., <u>27</u>, 1536 (1988).

73. Nagarajan, G. and S. B. Cotter, Z. Naturforsch. A., <u>26(11)</u>, 1800 (1971).

74. Ott, J. B. and W. F. Giauque, J. Amer. Chem. Soc., <u>82</u>, 1308 (1960).

75. Cerny, C. and E. Erdos, Chem. List., <u>47</u>, 1742 (1953).

76. Golosova, R. M., V. V. Korobov, and M. Kh. Karapet'yants, Russ. J. Phys. Chem., <u>45(5)</u>, 598 (1971).

77. Cerny, C., and E. Erdos, Collect. Czech. Chem. Commun., <u>19</u>, 646 (1954).

78. Nagarajan, G., Bull. Soc. Chim. Belg., <u>72</u>, 346 (1963).

Figure 2-1 Heat Capacity of Gas

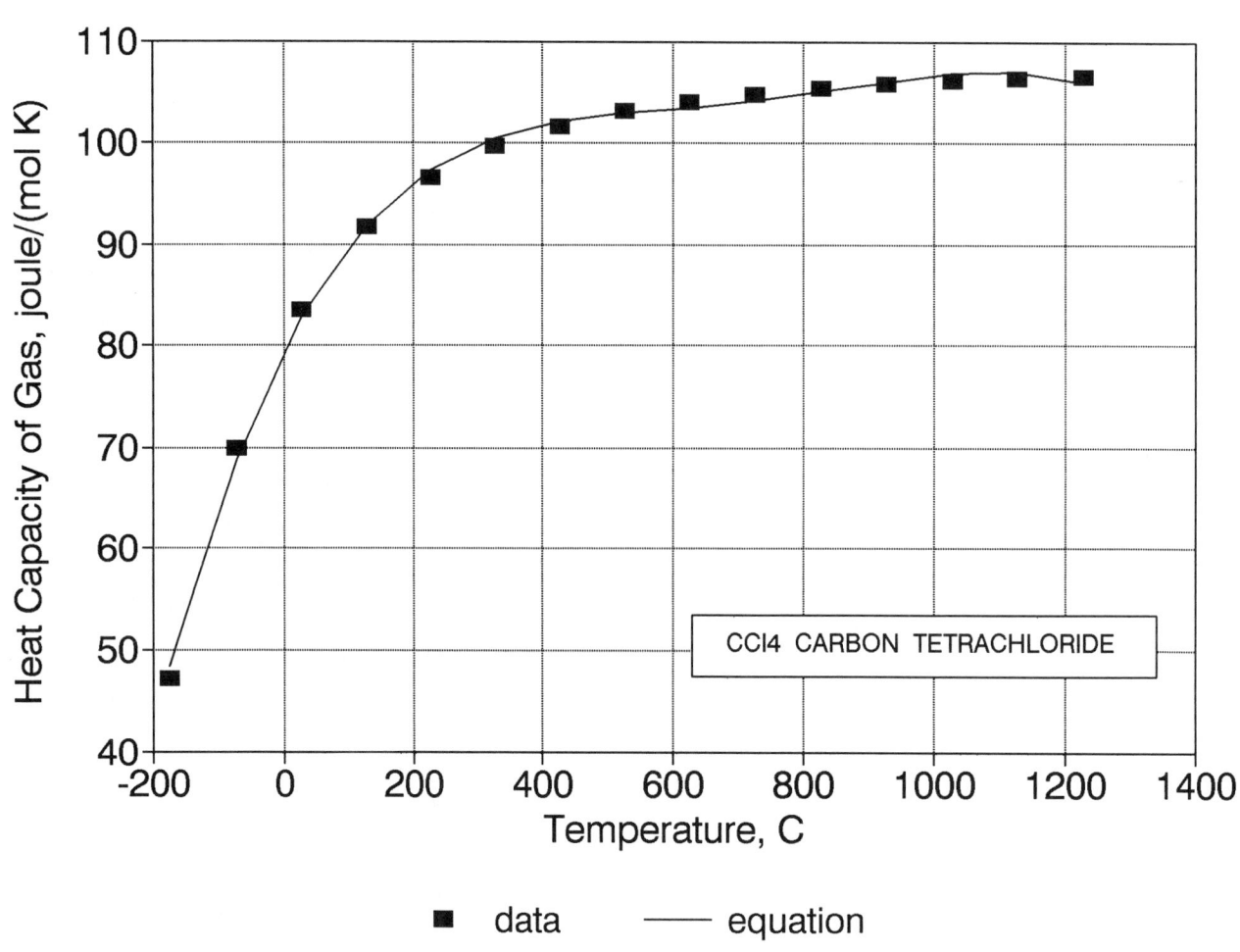

Table 2-1 HEAT CAPACITY OF GAS - ORGANIC COMPOUNDS

			$C_P = A + B\,T + C\,T^2 + D\,T^3 + E\,T^4$				$(C_P$ - joule/(mol K), T - K$)$		
NO	FORMULA	NAME	A	B	C	D	E	TMIN	TMAX
1	CBrClF2	BROMOCHLORODIFLUOROMETHANE	18.387	2.7933E-01	-3.7127E-04	2.2889E-07	-5.3229E-11	100	1500
2	CBrCl3	BROMOTRICHLOROMETHANE	24.484	3.2024E-01	-4.9096E-04	3.3359E-07	-8.2982E-11	100	1500
3	CBrF3	BROMOTRIFLUOROMETHANE	17.208	2.4770E-01	-2.9181E-04	1.6247E-07	-3.4765E-11	100	1500
4	CBr2F2	DIBROMODIFLUOROMETHANE	22.399	2.7403E-01	-3.7433E-04	2.3638E-07	-5.6067E-11	100	1500
5	CClF3	CHLOROTRIFLUOROMETHANE	13.762	2.4951E-01	-2.8194E-04	1.4962E-07	-3.0450E-11	100	1500
6	CClN	CYANOGEN CHLORIDE	21.270	1.1915E-01	-1.6822E-04	1.1457E-07	-2.9210E-11	100	1500
7	CCl2F2	DICHLORODIFLUOROMETHANE	14.877	2.8292E-01	-3.6295E-04	2.1591E-07	-4.8619E-11	100	1500
8	CCl2O	PHOSGENE	20.747	1.7972E-01	-2.3242E-04	1.4224E-07	-3.3087E-11	100	1500
9	CCl3F	TRICHLOROFLUOROMETHANE	16.636	3.1336E-01	-4.4426E-04	2.8612E-07	-6.8556E-11	89	1500
10	CCl4	CARBON TETRACHLORIDE	19.816	3.3311E-01	-5.0511E-04	3.4057E-07	-8.4249E-11	100	1500
11	CF2O	CARBONYL FLUORIDE	23.640	8.9853E-02	-2.4575E-05	-2.8140E-08	1.4023E-11	100	1500
12	CF4	CARBON TETRAFLUORIDE	15.278	1.9916E-01	-1.6369E-04	5.1686E-08	-3.1820E-12	100	1500
13	CHBr3	TRIBROMOMETHANE	33.356	1.7475E-01	-1.9516E-04	1.0725E-07	-2.3180E-11	100	1500
14	CHClF2	CHLORODIFLUOROMETHANE	20.519	1.4746E-01	-9.2440E-05	1.4379E-08	3.4356E-12	100	1500
15	CHCl2F	DICHLOROFLUOROMETHANE	33.078	1.0473E-01	-2.2510E-05	-3.8822E-08	1.8245E-11	100	1500
16	CHCl3	CHLOROFORM	22.487	1.9823E-01	-2.1676E-04	1.1636E-07	-2.4555E-11	100	1500
17	CHF3	TRIFLUOROMETHANE	23.287	9.5385E-02	2.0049E-05	-7.4432E-08	2.7428E-11	100	1500
18	CHI3	TRIIODOMETHANE	41.552	1.5326E-01	-1.5543E-04	5.8354E-08	--------	298	1000
19	CHN	HYDROGEN CYANIDE	25.766	3.7969E-02	-1.2416E-05	-3.2240E-09	2.2610E-12	100	1500
20	CHNS	ISOTHIOCYANIC-ACID	14.770	1.4484E-01	-1.4613E-04	5.5948E-08	--------	298	1000
21	CH2BrCl	BROMOCHLOROMETHANE	27.752	9.1021E-02	-8.2541E-06	-3.7449E-08	1.5156E-11	100	1500
22	CH2Br2	DIBROMOMETHANE	28.305	1.0581E-01	-4.7966E-05	-2.6711E-09	5.1497E-12	100	1500
23	CH2ClF	CHLOROFLUOROMETHANE	12.674	1.3973E-01	-8.8709E-05	2.1577E-08	--------	298	1000
24	CH2Cl2	DICHLOROMETHANE	26.694	8.3984E-02	8.9712E-06	-5.0924E-08	1.8726E-11	100	1500
25	CH2F2	DIFLUOROMETHANE	30.323	1.1176E-02	1.5809E-04	-1.6323E-07	4.7955E-11	100	1500
26	CH2I2	DIIODOMETHANE	28.918	1.1739E-01	-7.3808E-05	1.7055E-08	1.2051E-14	100	1500
27	CH2O	FORMALDEHYDE	34.428	-2.9779E-02	1.5104E-04	-1.2733E-07	3.3887E-11	50	1500
28	CH2O2	FORMIC ACID	31.745	7.4234E-03	1.8791E-04	-1.9475E-07	5.7613E-11	50	1500
29	CH3Br	METHYL BROMIDE	29.146	2.4374E-02	1.0655E-04	-1.1324E-07	3.3241E-11	100	1500
30	CH3Cl	METHYL CHLORIDE	27.385	2.6036E-02	1.0320E-04	-1.0887E-07	3.1642E-11	150	1500
31	CH3Cl3Si	METHYL TRICHLOROSILANE	56.670	2.0066E-01	-1.6721E-04	7.2533E-08	-1.2684E-11	200	1500
32	CH3F	METHYL FLUORIDE	34.077	-3.5019E-02	2.2031E-04	-1.9566E-07	5.4104E-11	100	1500
33	CH3I	METHYL IODIDE	25.635	6.6836E-02	1.2292E-05	-3.6742E-08	1.2301E-11	100	1500
34	CH3NO	FORMAMIDE	30.911	1.4363E-02	1.9281E-04	-1.9805E-07	5.8262E-11	150	1500
35	CH3NO2	NITROMETHANE	41.136	3.4367E-03	2.6380E-04	-2.6898E-07	7.9503E-11	100	1500
36	CH3NO2	METHYL-NITRITE	12.593	2.0137E-01	-1.1296E-04	2.2463E-08	--------	298	1000
37	CH3NO3	METHYL-NITRATE	17.049	2.4073E-01	-1.4966E-04	3.4904E-08	--------	298	1000
38	CH4	METHANE	34.942	-3.9957E-02	1.9184E-04	-1.5303E-07	3.9321E-11	50	1500
39	CH4Cl2Si	METHYL DICHLOROSILANE	37.250	2.3327E-01	-1.9952E-04	9.1473E-08	-1.7387E-11	200	1500
40	CH4O	METHANOL	40.046	-3.8287E-02	2.4529E-04	-2.1679E-07	5.9909E-11	100	1500
41	CH4O3S	METHANESULFONIC ACID	65.450	-1.0363E-01	6.2784E-04	-6.7194E-07	2.3114E-10	298	1000
42	CH4S	METHYL MERCAPTAN	40.307	-3.6753E-03	1.8400E-04	-1.7596E-07	5.0137E-11	100	1500
43	CH5ClSi	METHYL CHLOROSILANE	25.886	2.1064E-01	-1.3055E-04	3.6499E-08	-3.0118E-12	200	1500
44	CH5N	METHYLAMINE	40.039	-1.5108E-02	2.5012E-04	-2.3336E-07	6.5582E-11	100	1500
45	CH6Si	METHYL SILANE	25.277	1.2988E-01	5.9803E-05	-1.2080E-07	4.0036E-11	100	1500
46	CN4O8	TETRANITROMETHANE	23.733	5.5312E-01	-4.5854E-04	1.5286E-07	-1.3967E-11	298	1500
47	CO	CARBON MONOXIDE	29.556	-6.5807E-03	2.0130E-05	-1.2227E-08	2.2617E-12	60	1500
48	COS	CARBONYL SULFIDE	20.913	9.2794E-02	-9.7014E-05	5.0943E-08	-1.0615E-11	100	1500
49	CO2	CARBON DIOXIDE	27.437	4.2315E-02	-1.9555E-05	3.9968E-09	-2.9872E-13	50	5000
50	CS2	CARBON DISULFIDE	20.461	1.2299E-01	-1.6184E-04	1.0199E-07	-2.4444E-11	100	1500
51	C2BrF3	BROMOTRIFLUOROETHYLENE	33.956	2.3803E-01	-2.5448E-04	1.3108E-07	-2.6462E-11	300	1200
52	C2Br2F4	1,2-DIBROMOTETRAFLUOROETHANE	50.542	3.7777E-01	-4.8770E-04	3.1054E-07	-7.8174E-11	298	1200
53	C2ClF3	CHLOROTRIFLUOROETHYLENE	28.388	2.5871E-01	-2.9010E-04	1.5876E-07	-3.3880E-11	300	1500
54	C2ClF5	CHLOROPENTAFLUOROETHANE	24.663	3.8598E-01	-3.8927E-04	1.7751E-07	-2.9992E-11	200	1500
55	C2Cl2F4	1,2-DICHLOROTETRAFLUOROETHANE	17.183	4.8507E-01	-5.9368E-04	3.3494E-07	-7.1705E-11	273	1500
56	C2Cl3F3	1,1,2-TRICHLOROTRIFLUOROETHANE	42.456	4.0973E-01	-5.0045E-04	2.8463E-07	-6.1623E-11	200	1500
57	C2Cl4	TETRACHLOROETHYLENE	34.627	3.1065E-01	-4.5258E-04	3.2734E-07	-9.4234E-11	298	1000
58	C2Cl4F2	1,1,2,2-TETRACHLORODIFLUOROETHANE	3.788	6.9339E-01	-1.0927E-03	7.6080E-07	-1.9434E-10	298	1000
59	C2Cl4O	TRICHLOROACETYL CHLORIDE	55.547	2.5821E-01	-2.9449E-04	1.6054E-07	-3.3869E-11	298	1500
60	C2Cl6	HEXACHLOROETHANE	48.475	4.4117E-01	-5.9638E-04	3.6922E-07	-8.5631E-11	150	1500
61	C2F4	TETRAFLUOROETHYLENE	30.934	2.1587E-01	-1.9110E-04	7.5718E-08	-1.0493E-11	100	1500
62	C2F6	HEXAFLUOROETHANE	13.604	4.3503E-01	-4.8166E-04	2.4841E-07	-4.8897E-11	100	1500
63	C2HBrClF3	HALOTHANE	65.307	1.2411E-01	8.8736E-05	-1.7716E-07	6.3268E-11	100	1500
64	C2HClF2	2-CHLORO-1,1-DIFLUOROETHYLENE	19.530	2.4576E-01	-2.5701E-04	1.3429E-07	-2.7612E-11	298	1500
65	C2HCl3	TRICHLOROETHYLENE	40.879	1.6218E-01	-1.0399E-04	1.3310E-08	5.9103E-12	100	1500
66	C2HCl3O	DICHLOROACETYL CHLORIDE	38.143	2.3343E-01	-1.9070E-04	5.6320E-08	-2.5489E-12	298	1500
67	C2HCl3O	TRICHLOROACETALDEHYDE	46.176	2.3126E-01	-2.3659E-04	1.2539E-07	-2.7197E-11	298	1200
68	C2HCl5	PENTACHLOROETHANE	28.297	4.3657E-01	-5.5684E-04	3.3877E-07	-7.8834E-11	100	1500
69	C2HF3	TRIFLUOROETHENE	16.298	2.3109E-01	-1.9825E-04	6.4911E-08	--------	298	1000
70	C2HF3O2	TRIFLUOROACETIC ACID	9.274	3.9410E-01	-4.3755E-04	2.4769E-07	-5.7309E-11	298	1200
71	C2HF5	PENTAFLUOROETHANE	40.370	2.0895E-01	-4.5524E-05	-8.2827E-08	3.9062E-11	100	1500
72	C2H2	ACETYLENE	19.360	1.1519E-01	-1.2374E-04	7.2370E-08	-1.6590E-11	200	1500
73	C2H2Br4	1,1,2,2-TETRABROMOETHANE	49.111	2.5471E-01	-2.1340E-04	7.7534E-08	-9.0356E-12	150	1500
74	C2H2Cl2	1,1-DICHLOROETHYLENE	21.272	2.0394E-01	-1.8292E-04	8.2561E-08	-1.4802E-11	200	1500
75	C2H2Cl2	cis-1,2-DICHLOROETHYLENE	11.376	2.4489E-01	-2.5183E-04	1.3368E-07	-2.8416E-11	200	1500
76	C2H2Cl2	trans-1,2-DICHLOROETHYLENE	19.666	2.0908E-01	-1.9503E-04	9.4941E-08	-1.8848E-11	200	1500
77	C2H2Cl2O	CHLOROACETYL CHLORIDE	29.566	2.1377E-01	-1.8076E-04	7.9913E-08	-1.4782E-11	298	1500
78	C2H2Cl2O	DICHLOROACETALDEHYDE	39.761	1.3843E-01	9.8425E-06	-1.0396E-07	4.4962E-11	298	1200

Table 2-1 HEAT CAPACITY OF GAS - ORGANIC COMPOUNDS (continued)

$$C_P = A + BT + CT^2 + DT^3 + ET^4 \qquad (C_P \text{ - joule/(mol K)}, \; T \text{ - K})$$

NO	FORMULA	NAME	A	B	C	D	E	TMIN	TMAX
79	C2H2Cl2O2	DICHLOROACETIC ACID	48.334	1.7165E-01	-6.2134E-05	-3.6029E-08	2.3198E-11	298	1200
80	C2H2Cl3F	1,1,1-TRICHLOROFLUOROETHANE	20.183	3.8754E-01	-4.6141E-04	2.7663E-07	-6.5842E-11	298	1200
81	C2H2Cl4	1,1,1,2-TETRACHLOROETHANE	27.227	3.4932E-01	-3.7793E-04	2.0496E-07	-4.3937E-11	200	1500
82	C2H2Cl4	1,1,2,2-TETRACHLOROETHANE	20.427	3.6839E-01	-4.0365E-04	2.2067E-07	-4.7303E-11	298	1500
83	C2H2F2	1,1-DIFLUOROETHYLENE	24.354	1.2196E-01	1.1084E-05	-7.9704E-08	3.0820E-11	100	1500
84	C2H2F2	cis-1,2-DIFLUOROETHENE	7.250	2.1736E-01	-1.6905E-04	5.1484E-08	--------	298	1000
85	C2H2F2	trans-1,2-DIFLUOROETHENE	12.402	2.0202E-01	-1.5191E-04	4.4890E-08	--------	298	1000
86	C2H2F4	1,1,1,2-TETRAFLUOROETHANE	8.429	3.4966E-01	-3.3281E-04	1.5603E-07	-2.8939E-11	200	1500
87	C2H2O	KETENE	-14.704	3.1238E-01	-4.3385E-04	2.9499E-07	-7.5221E-11	200	1500
88	C2H2O4	OXALIC ACID	-5.565	1.3496E-01	1.3737E-05	-1.9105E-07	1.1311E-10	298	1000
89	C2H3Br	VINYL BROMIDE	19.032	1.4697E-01	-7.2736E-05	4.7354E-09	4.4305E-12	200	1500
90	C2H3Cl	VINYL CHLORIDE	17.193	1.4564E-01	-6.4281E-05	-3.2385E-09	6.7882E-12	200	1500
91	C2H3ClF2	1-CHLORO-1,1-DIFLUOROETHANE	20.964	2.6700E-01	-2.0774E-04	7.5759E-08	-9.7753E-12	200	1500
92	C2H3ClO	ACETYL CHLORIDE	37.484	1.0683E-01	1.3035E-05	-5.7327E-08	1.9960E-11	200	1500
93	C2H3ClO	CHLOROACETALDEHYDE	25.272	1.5004E-01	-3.4110E-05	-4.1182E-08	2.0993E-11	298	1200
94	C2H3ClO2	CHLOROACETIC ACID	9.327	2.9997E-01	-2.6947E-04	1.2616E-07	-2.4406E-11	298	1500
95	C2H3ClO2	METHYL CHLOROFORMATE	13.353	2.7827E-01	-2.0298E-04	3.2225E-08	1.9750E-11	298	900
96	C2H3Cl3	1,1,1-TRICHLOROETHANE	18.674	3.3443E-01	-3.4963E-04	1.8764E-07	-4.0744E-11	100	1500
97	C2H3Cl3	1,1,2-TRICHLOROETHANE	28.881	2.4893E-01	-1.7639E-04	5.2632E-08	-3.5668E-12	200	1500
98	C2H3F	VINYL FLUORIDE	27.617	5.4052E-02	1.3093E-04	-1.6220E-07	5.0829E-11	100	1500
99	C2H3F3	1,1,1-TRIFLUOROETHANE	33.444	1.5361E-01	3.3402E-05	-1.1974E-07	4.4424E-11	100	1500
100	C2H3N	ACETONITRILE	36.947	2.2085E-02	1.4661E-04	-1.5012E-07	4.3482E-11	100	1500
101	C2H3NO	METHYL ISOCYANATE	21.328	8.5385E-02	7.8504E-05	-1.0050E-07	2.9508E-11	298	1500
102	C2H4	ETHYLENE	32.083	-1.4831E-02	2.4774E-04	-2.3766E-07	6.8274E-11	60	1500
103	C2H4Br2	1,1-DIBROMOETHANE	21.084	2.5090E-01	-2.0060E-04	8.4960E-08	-1.5026E-11	200	1500
104	C2H4Br2	1,2-DIBROMOETHANE	47.739	1.3553E-01	1.0414E-05	-6.8462E-08	2.5192E-11	200	1500
105	C2H4Cl2	1,1-DICHLOROETHANE	15.730	2.6124E-01	-2.1489E-04	9.5761E-08	-1.8004E-11	200	1500
106	C2H4Cl2	1,2-DICHLOROETHANE	37.275	1.4362E-01	1.0378E-05	-7.8305E-08	2.8872E-11	200	1500
107	C2H4Cl2O	BIS(CHLOROMETHYL)ETHER	3.763	3.6729E-01	-3.5749E-04	1.8325E-07	-3.7910E-11	298	1500
108	C2H4F2	1,1-DIFLUOROETHANE	36.271	7.8276E-02	1.6310E-04	-2.0396E-07	6.3814E-11	100	1500
109	C2H4F2	1,2-DIFLUOROETHANE	18.309	2.0288E-01	-6.1613E-05	-3.8781E-08	2.0688E-11	100	1500
110	C2H4I2	1,2-DIIODOETHANE	25.893	2.3491E-01	-1.6837E-04	4.9162E-08	--------	298	1000
111	C2H4O	ACETALDEHYDE	34.140	4.0020E-02	1.5634E-04	-1.6445E-07	4.7248E-11	100	1500
112	C2H4O	ETHYLENE OXIDE	30.827	-7.6041E-03	3.2347E-04	-3.2747E-07	9.7271E-11	50	1500
113	C2H4OS	THIOACETIC-ACID	38.465	1.6308E-01	-7.0053E-05	4.8020E-09	--------	298	1000
114	C2H4O2	ACETIC ACID	34.850	3.7626E-02	2.8311E-04	-3.0767E-07	9.2646E-11	50	1500
115	C2H4O2	METHYL FORMATE	5.795	2.5072E-01	-1.7515E-04	6.0565E-08	-8.1015E-12	250	1500
116	C2H4S	THIACYCLOPROPANE	-11.920	2.7834E-01	-2.1566E-04	6.7312E-08	--------	298	1000
117	C2H5Br	BROMOETHANE	26.552	1.1837E-01	6.7525E-05	-1.1655E-07	3.7186E-11	100	1500
118	C2H5Cl	ETHYL CHLORIDE	35.946	5.2294E-02	2.0321E-04	-2.2795E-07	6.9123E-11	100	1500
119	C2H5ClO	2-CHLOROETHANOL	12.997	2.5587E-01	-1.6553E-04	5.2292E-08	-6.5812E-12	298	1500
120	C2H5F	ETHYL FLUORIDE	21.452	1.2080E-01	7.8409E-05	-1.2578E-07	3.9184E-11	150	1500
121	C2H5I	ETHYL IODIDE	27.759	1.1915E-01	5.9726E-05	-1.0756E-07	3.4157E-11	100	1500
122	C2H5N	ETHYLENEIMINE	12.316	1.1833E-01	1.2598E-04	-1.8322E-07	5.8831E-11	150	1500
123	C2H5NO	ACETAMIDE	17.748	1.3627E-01	1.0668E-04	-1.8647E-07	6.2842E-11	100	1500
124	C2H5NO	N-METHYLFORMAMIDE	43.449	-1.0054E-01	7.2412E-04	-8.6234E-07	3.2724E-10	298	1000
125	C2H5NO2	NITROETHANE	17.726	2.2334E-01	-2.1690E-05	-8.0889E-08	3.2223E-11	200	1500
126	C2H5NO3	ETHYL-NITRATE	3.530	3.8951E-01	-2.7423E-04	7.6538E-08	--------	298	1000
127	C2H6	ETHANE	28.146	4.3447E-02	1.8946E-04	-1.9082E-07	5.3349E-11	100	1500
128	C2H6AlCl	DIMETHYLALUMINUM CHLORIDE	13.870	3.1526E-01	-2.4008E-04	9.9159E-08	-1.7228E-11	298	TMAX
129	C2H6O	DIMETHYL ETHER	34.668	7.0293E-02	1.6530E-04	-1.7675E-07	4.9313E-11	100	1500
130	C2H6O	ETHANOL	27.091	1.1055E-01	1.0957E-04	-1.5046E-07	4.6601E-11	100	1500
131	C2H6OS	DIMETHYL SULFOXIDE	27.816	2.4839E-01	-1.3176E-04	2.3843E-08	1.6501E-12	200	1500
132	C2H6O2	ETHYLENE GLYCOL	48.218	1.9073E-01	-6.6117E-05	-1.8834E-08	1.2555E-11	200	1500
133	C2H6O4S	DIMETHYL SULFATE	23.800	3.7920E-01	-2.0385E-04	2.4893E-08	8.1594E-12	200	1500
134	C2H6S	DIMETHYL SULFIDE	35.994	1.2381E-01	5.0871E-05	-9.1708E-08	2.8274E-11	200	1500
135	C2H6S	ETHYL MERCAPTAN	47.034	4.1940E-02	2.3486E-04	-2.5035E-07	7.4049E-11	100	1500
136	C2H6S2	DIMETHYL DISULFIDE	50.010	1.4793E-01	4.2325E-05	-9.9087E-08	3.2173E-11	200	1500
137	C2H7N	DIMETHYLAMINE	30.638	1.0737E-01	1.5824E-04	-1.9418E-07	5.8509E-11	200	1500
138	C2H7N	ETHYLAMINE	30.983	1.2458E-01	1.0966E-04	-1.5256E-07	4.6640E-11	200	1500
139	C2H7NO	MONOETHANOLAMINE	-0.555	3.7003E-01	-3.1976E-04	1.5834E-07	-3.2344E-11	298	1500
140	C2H8N2	ETHYLENEDIAMINE	10.429	3.2490E-01	-1.9912E-04	6.3557E-08	-8.7124E-12	298	1500
141	C2H8Si	DIMETHYL SILANE	27.940	2.2419E-01	-4.0022E-06	-9.9567E-08	3.6936E-11	100	1500
142	C2N2	CYANOGEN	22.445	1.6837E-01	-2.3212E-04	1.5784E-07	-4.0479E-11	100	1500
143	C3F6	HEXAFLUOROPROPYLENE	-3.108	5.2270E-01	-4.6521E-04	1.9228E-07	-2.9928E-11	200	1500
144	C3F6O	HEXAFLUOROACETONE	0.451	5.2201E-01	-4.5370E-04	1.6725E-07	-1.9476E-11	200	1500
145	C3F8	OCTAFLUOROPROPANE	14.695	6.1745E-01	-6.7174E-04	3.4675E-07	-6.9257E-11	273	1500
146	C3H2N2	MALONONITRILE	27.389	1.6397E-01	-3.8642E-05	-5.4165E-08	2.7216E-11	298	1200
147	C3H3Cl	PROPARGYL CHLORIDE	20.366	2.2487E-01	-1.9722E-04	9.5424E-08	-1.9480E-11	298	1500
148	C3H3N	ACRYLONITRILE	18.425	1.8336E-01	-1.0072E-04	1.8747E-08	9.1114E-13	200	1500
149	C3H3NO	OXAZOLE	25.962	5.5802E-02	3.1816E-04	-3.7323E-07	1.1787E-10	50	1500
150	C3H4	METHYLACETYLENE	27.565	1.2037E-01	-6.0666E-06	-4.0713E-08	1.5078E-11	200	1500
151	C3H4	PROPADIENE	28.504	8.1576E-02	1.0896E-04	-1.4641E-07	4.5759E-11	50	1500
152	C3H4Cl2	2,3-DICHLOROPROPENE	29.854	2.6319E-01	-2.1492E-04	9.6796E-08	-1.8613E-11	298	1200
153	C3H4O	ACROLEIN	109.243	-5.0952E-01	1.7059E-03	-1.8068E-06	6.5983E-10	298	1000
154	C3H4O	PROPARGYL ALCOHOL	20.577	2.0827E-01	-1.1794E-04	2.3926E-08	1.9047E-13	298	1500
155	C3H4O2	ACRYLIC ACID	7.755	2.9386E-01	-2.0878E-04	7.1591E-08	-9.0960E-12	250	1500
156	C3H4O2	beta-PROPIOLACTONE	9.108	2.3694E-01	-5.7117E-05	-5.5709E-08	2.6215E-11	200	1500

Table 2-1 HEAT CAPACITY OF GAS - ORGANIC COMPOUNDS (continued)

| NO | FORMULA | NAME | $C_P = A + B\,T + C\,T^2 + D\,T^3 + E\,T^4$ | | | | $(C_P$ - joule/(mol K), T - K$)$ | | |
			A	B	C	D	E	TMIN	TMAX
157	C3H4O2	VINYL FORMATE	-19.250	4.1753E-01	-4.2109E-04	2.2583E-07	-4.7796E-11	298	1500
158	C3H4O3	ETHYLENE CARBONATE	--------	--------	--------	--------	--------	--------	--------
159	C3H4O3	PYRUVIC ACID	5.045	3.7767E-01	-3.4222E-04	1.6758E-07	-3.3836E-11	298	1500
160	C3H5Br	3-BROMO-1-PROPENE	6.651	2.9532E-01	-2.1107E-04	6.2363E-08	--------	298	1000
161	C3H5Cl	2-CHLOROPROPENE	33.157	1.7339E-01	-4.0298E-05	-3.2447E-08	1.4113E-11	298	1500
162	C3H5Cl	3-CHLOROPROPENE	24.507	1.8758E-01	-2.1732E-05	-6.0427E-08	2.4116E-11	200	1500
163	C3H5ClO	alpha-EPICHLOROHYDRIN	-27.780	5.1614E-01	-5.6307E-04	3.2430E-07	-7.2399E-11	298	1500
164	C3H5ClO2	METHYL CHLOROACETATE	-13.590	5.2003E-01	-5.7146E-04	3.3108E-07	-7.4031E-11	298	1500
165	C3H5ClO2	ETHYL CHLOROFORMATE	32.231	2.5311E-01	-6.0013E-05	-5.0872E-08	2.2667E-11	100	1500
166	C3H5Cl3	1,2,3-TRICHLOROPROPANE	45.369	2.7047E-01	-1.3637E-04	1.1641E-08	7.2782E-12	200	1500
167	C3H5I	3-IODO-1-PROPENE	10.734	3.0355E-01	-2.3002E-04	7.1450E-08	--------	298	1000
168	C3H5N	PROPIONITRILE	17.618	2.2119E-01	-1.0707E-04	1.7352E-08	1.0610E-12	200	1500
169	C3H5NO	ACRYLAMIDE	13.165	2.6213E-01	-8.5250E-05	-3.5101E-08	2.0435E-11	200	1500
170	C3H5NO	HYDRACRYLONITRILE	8.904	3.1056E-01	-2.0843E-04	6.3273E-08	-6.1415E-12	298	1500
171	C3H5NO	LACTONITRILE	30.071	2.3651E-01	-1.0389E-04	1.4435E-08	1.2163E-12	298	1500
172	C3H5N3O9	NITROGLYCERINE	--------	--------	--------	--------	--------	--------	--------
173	C3H6	CYCLOPROPANE	21.172	6.3106E-02	2.9197E-04	-3.2708E-07	9.9730E-11	100	1500
174	C3H6	PROPYLENE	31.298	7.2449E-02	1.9481E-04	-2.1582E-07	6.2974E-11	90	1500
175	C3H6Br2	1,2-DIBROMOPROPANE	13.187	3.7640E-01	-2.7601E-04	8.1906E-08	--------	298	1000
176	C3H6Cl2	1,1-DICHLOROPROPANE	36.554	2.2898E-01	-2.1999E-05	-8.4949E-08	3.4171E-11	150	1500
177	C3H6Cl2	1,2-DICHLOROPROPANE	34.575	2.4270E-01	-5.7726E-05	-4.9517E-08	2.3275E-11	200	1500
178	C3H6Cl2	1,3-DICHLOROPROPANE	37.917	2.3441E-01	-5.7287E-05	-4.5766E-08	2.1530E-11	200	1500
179	C3H6Cl2	2,2-DICHLOROPROPANE	10.720	4.0973E-01	-3.3691E-04	1.1135E-07	--------	298	1000
180	C3H6I2	1,2-DIIODOPROPANE	16.386	3.6405E-01	-2.6305E-04	7.8174E-08	--------	298	1000
181	C3H6O	ACETONE	35.918	9.3896E-02	1.8730E-04	-2.1643E-07	6.3174E-11	100	1500
182	C3H6O	ALLYL ALCOHOL	18.528	2.1287E-01	-2.8839E-05	-6.5014E-08	2.6352E-11	200	1500
183	C3H6O	METHYL VINYL ETHER	15.321	2.3596E-01	-9.9256E-05	7.6893E-09	3.3293E-12	298	1500
184	C3H6O	n-PROPIONALDEHYDE	58.911	4.8385E-03	3.3514E-04	-3.0509E-07	8.3305E-11	200	1500
185	C3H6O	1,2-PROPYLENE OXIDE	29.501	9.2545E-02	2.5626E-04	-2.9921E-07	9.0294E-11	50	1500
186	C3H6O	1,3-PROPYLENE OXIDE	-32.684	4.0929E-01	-3.6377E-04	2.0140E-07	-4.5037E-11	298	1500
187	C3H6O2	ETHYL FORMATE	36.654	1.4922E-01	1.3957E-04	-1.9500E-07	5.9055E-11	100	1500
188	C3H6O2	METHYL ACETATE	-22.287	4.8275E-01	-4.6631E-04	2.3286E-07	-4.3094E-11	298	1200
189	C3H6O2	PROPIONIC ACID	-0.970	3.8307E-01	-3.0872E-04	1.3886E-07	-2.6720E-11	298	1500
190	C3H6O2S	3-MERCAPTOPROPIONIC ACID	26.577	3.7025E-01	-2.9565E-04	1.5526E-07	-4.2314E-11	298	1200
191	C3H6O3	LACTIC ACID	4.890	4.2659E-01	-3.5416E-04	1.5688E-07	-2.9209E-11	298	1500
192	C3H6O3	METHOXYACETIC ACID	-42.782	7.8675E-01	-1.3648E-03	1.3336E-06	-5.0983E-10	298	900
193	C3H6O3	TRIOXANE	4.910	2.6407E-01	6.9003E-05	-1.9208E-07	6.6054E-11	200	1500
194	C3H6S	THIACYCLOBUTANE	-19.537	3.6008E-01	-2.2545E-04	5.4995E-08	--------	298	1000
195	C3H7Br	1-BROMOPROPANE	24.209	2.2956E-01	-2.9974E-05	-6.7459E-08	2.6960E-11	200	1500
196	C3H7Br	2-BROMOPROPANE	28.602	2.1902E-01	-4.6160E-06	-9.1820E-08	3.4572E-11	200	1500
197	C3H7Cl	ISOPROPYL CHLORIDE	45.660	9.5434E-02	2.4116E-04	-2.8322E-07	8.6109E-11	100	1500
198	C3H7Cl	n-PROPYL CHLORIDE	16.153	2.6543E-01	-9.2380E-05	-1.9679E-08	1.4135E-11	200	1500
199	C3H7F	1-FLUOROPROPANE	7.424	2.9058E-01	-1.3549E-04	1.8822E-08	--------	298	1000
200	C3H7F	2-FLUOROPROPANE	-1.643	3.3412E-01	-1.9377E-04	4.3413E-08	--------	298	1000
201	C3H7I	ISOPROPYL IODIDE	29.083	2.2383E-01	-1.7102E-05	-8.1396E-08	3.1643E-11	200	1500
202	C3H7I	n-PROPYL IODIDE	17.653	2.7777E-01	-1.1961E-04	-7.5057E-10	9.3394E-12	200	1500
203	C3H7N	ALLYLAMINE	18.274	2.6583E-01	-1.2102E-04	1.3458E-08	3.2317E-12	298	1500
204	C3H7N	PROPYLENEIMINE	-38.561	4.9894E-01	-4.5377E-04	2.2334E-07	-4.4943E-11	298	1500
205	C3H7NO	N,N-DIMETHYLFORMAMIDE	29.310	2.0837E-01	1.0912E-04	-2.1506E-07	7.2177E-11	200	1500
206	C3H7NO	N-METHYLACETAMIDE	-1.339	2.8413E-01	-7.2490E-05	-3.9834E-08	1.8676E-11	298	1500
207	C3H7NO2	1-NITROPROPANE	28.505	2.6062E-01	2.4779E-05	-1.4204E-07	5.0913E-11	200	1500
208	C3H7NO2	2-NITROPROPANE	18.837	3.1000E-01	-5.2946E-05	-9.1409E-08	3.8968E-11	200	1500
209	C3H7NO3	PROPYL-NITRATE	7.081	4.6627E-01	-3.0292E-04	7.6718E-08	--------	298	1000
210	C3H7NO3	ISOPROPYL-NITRATE	-1.985	5.0982E-01	-3.6118E-04	1.0130E-07	--------	298	1000
211	C3H8	PROPANE	28.277	1.1600E-01	1.9597E-04	-2.3271E-07	6.8669E-11	100	1500
212	C3H8O	ISOPROPANOL	25.535	2.1203E-01	5.3492E-05	-1.4727E-07	4.9406E-11	100	1500
213	C3H8O	METHYL ETHYL ETHER	35.289	1.7427E-01	9.4277E-05	-1.4826E-07	4.3296E-11	100	1500
214	C3H8O	n-PROPANOL	31.507	2.3082E-01	-7.8983E-05	6.3696E-09	8.6908E-13	100	2980
215	C3H8O2	2-METHOXYETHANOL	11.416	3.4693E-01	-1.7874E-04	3.1146E-08	1.1718E-12	298	1500
216	C3H8O2	METHYLAL	39.490	2.4609E-02	7.8528E-04	-1.0921E-06	4.5093E-10	298	1000
217	C3H8O2	1,2-PROPYLENE GLYCOL	14.404	3.2565E-01	-7.8741E-05	-1.2420E-07	7.4776E-11	298	1500
218	C3H8O2	1,3-PROPYLENE GLYCOL	7.340	3.5920E-01	-1.5544E-04	-4.8543E-08	4.6374E-11	298	1200
219	C3H8O3	GLYCEROL	9.656	4.2826E-01	-2.6797E-04	3.1794E-08	2.7745E-11	298	1200
220	C3H8S	n-PROPYLMERCAPTAN	37.035	1.9064E-01	7.2562E-05	-1.4440E-07	4.6182E-11	200	1500
221	C3H8S	ISOPROPYL MERCAPTAN	31.652	2.2715E-01	1.2837E-05	-1.0474E-07	3.6662E-11	200	1500
222	C3H8S	ETHYL-METHYL-SULFIDE	16.284	3.0620E-01	-1.4918E-04	2.7207E-08	--------	298	1000
223	C3H9N	n-PROPYLAMINE	7.637	3.3741E-01	-1.6070E-04	2.3836E-08	2.3647E-12	298	1500
224	C3H9N	ISOPROPYLAMINE	-4.758	4.0947E-01	-2.8998E-04	1.1548E-07	-2.0187E-11	298	1500
225	C3H9N	TRIMETHYLAMINE	26.377	2.1496E-01	1.0135E-04	-1.8839E-07	5.9860E-11	200	1500
226	C3H9NO	1-AMINO-2-PROPANOL	-8.396	5.1026E-01	-4.5822E-04	2.3361E-07	-4.9093E-11	298	1500
227	C3H9NO	3-AMINO-1-PROPANOL	7.276	4.0342E-01	-2.5533E-04	8.2617E-08	-1.1842E-11	298	1500
228	C3H9NO	METHYLETHANOLAMINE	-10.405	4.7860E-01	-3.7360E-04	1.6992E-07	-3.5477E-11	298	1200
229	C3H9O4P	TRIMETHYL PHOSPHATE	--------	--------	--------	--------	--------	--------	--------
230	C3H10N2	1,2-PROPANEDIAMINE	3.612	4.5467E-01	-3.0830E-04	1.1190E-07	-1.7969E-11	298	1500
231	C3H10Si	TRIMETHYL SILANE	95.333	-3.8565E-02	4.8559E-04	-4.2881E-07	1.1555E-10	298	1500
232	C4Cl4S	TETRACHLOROTHIOPHENE	11.496	5.2672E-01	-6.0693E-04	3.2882E-07	-6.8095E-11	298	1500
233	C4Cl6	HEXACHLORO-1,3-BUTADIENE	101.108	2.8291E-01	-3.0609E-04	1.5800E-07	-3.1695E-11	298	1500
234	C4F8	OCTAFLUORO-2-BUTENE	-17.951	5.6720E-01	-7.4536E-04	4.9173E-07	-1.3124E-10	298	1000

Table 2-1 HEAT CAPACITY OF GAS - ORGANIC COMPOUNDS (continued)

			$C_P = A + B\,T + C\,T^2 + D\,T^3 + E\,T^4$				(C_P - joule/(mol K), T - K)		
NO	FORMULA	NAME	A	B	C	D	E	TMIN	TMAX
235	C4F8	OCTAFLUOROCYCLOBUTANE	45.579	4.9467E-01	-4.0808E-04	1.3789E-07	-1.1769E-11	200	1500
236	C4F10	DECAFLUOROBUTANE	38.645	6.1357E-01	-3.5155E-04	-1.2037E-07	1.2661E-10	298	1000
237	C4H2	BUTADIYNE(BIACETYLENE)	24.367	2.2525E-01	-2.2609E-04	8.7977E-08	--------	298	1000
238	C4H2O3	MALEIC ANHYDRIDE	-72.015	1.0423E+00	-1.8716E-03	1.6527E-06	-5.5647E-10	298	1000
239	C4H4	VINYLACETYLENE	19.335	2.2308E-01	-1.3739E-04	3.5667E-08	-1.7996E-12	200	1500
240	C4H4N2	SUCCINONITRILE	15.172	3.2583E-01	-2.2170E-04	5.1492E-08	7.0991E-12	298	1200
241	C4H4O	FURAN	-13.779	3.3489E-01	-2.2273E-04	6.9360E-08	-8.1619E-12	200	2980
242	C4H4O2	DIKETENE	14.704	2.8851E-01	-1.3307E-04	-3.2283E-09	1.3066E-11	298	1200
243	C4H4O3	SUCCINIC ANHYDRIDE	5.950	5.3900E-01	-7.4225E-04	6.0553E-07	-2.0505E-10	298	1000
244	C4H4O4	FUMARIC ACID	15.759	5.9932E-01	-7.0854E-04	4.2980E-07	-1.0058E-10	298	1500
245	C4H4O4	MALEIC ACID	-15.115	5.8902E-01	-6.4675E-04	3.6810E-07	-8.0001E-11	298	1500
246	C4H4S	THIOPHENE	22.037	1.2481E-01	2.4505E-04	-3.3887E-07	1.1175E-10	50	1500
247	C4H5Cl	CHLOROPRENE	20.268	3.1369E-01	-2.4207E-04	9.3958E-08	-1.5001E-11	298	1500
248	C4H5N	trans-CROTONITRILE	-1.466	3.4012E-01	-2.3532E-04	7.3984E-08	-8.2896E-12	250	1500
249	C4H5N	cis-CROTONITRILE	-1.780	3.2415E-01	-2.0418E-04	5.2998E-08	-3.0376E-12	298	1500
250	C4H5N	METHACRYLONITRILE	27.067	2.4619E-01	-1.0446E-04	-2.9851E-10	7.0951E-12	298	1600
251	C4H5N	PYRROLE	-7.680	3.1201E-01	-1.1806E-04	-2.7912E-08	2.1314E-11	200	1500
252	C4H5N	VINYLACETONITRILE	21.519	2.5405E-01	-1.2044E-04	1.1212E-08	4.5239E-12	298	1500
253	C4H5NO2	METHYL CYANOACETATE	-29.108	6.1090E-01	-6.3826E-04	3.0816E-07	-4.1498E-11	298	1200
254	C4H6	CYCLOBUTENE	-27.686	3.9001E-01	-2.6413E-04	7.1350E-08	--------	298	1000
255	C4H6	1,2-BUTADIENE	30.240	1.5688E-01	6.7668E-05	-1.2597E-07	3.9234E-11	100	1500
256	C4H6	1,3-BUTADIENE	18.835	2.0473E-01	6.2485E-05	-1.7148E-07	6.0858E-11	100	1500
257	C4H6	DIMETHYLACETYLENE	41.071	1.0010E-01	1.4966E-04	-1.7838E-07	5.2575E-11	200	1500
258	C4H6	ETHYLACETYLENE	29.857	1.8734E-01	-7.0968E-06	-6.8287E-08	2.5343E-11	200	1500
259	C4H6Cl2	1,3-DICHLORO-trans-2-BUTENE	33.873	3.2586E-01	-2.1702E-04	7.3722E-08	-1.0497E-11	298	1500
260	C4H6Cl2	1,4-DICHLORO-cis-2-BUTENE	0.724	5.0324E-01	-4.5246E-04	2.3322E-07	-5.1035E-11	298	1200
261	C4H6Cl2	1,4-DICHLORO-trans-2-BUTENE	7.205	4.3947E-01	-3.8703E-04	1.8893E-07	-3.8658E-11	298	1500
262	C4H6Cl2	3,4-DICHLORO-1-BUTENE	50.561	2.2753E-01	-4.3374E-05	-1.6965E-08	-4.1852E-12	298	1000
263	C4H6O	trans-CROTONALDEHYDE	11.591	3.2301E-01	-1.3067E-04	1.0791E-08	3.1192E-12	298	1500
264	C4H6O	2,5-DIHYDROFURAN	-11.125	3.4203E-01	-1.4489E-04	-1.3031E-08	1.7017E-11	200	1500
265	C4H6O	DIVINYL ETHER	1.254	3.5057E-01	-2.4183E-04	8.7375E-08	-1.3414E-11	298	1500
266	C4H6O	METHACROLEIN	14.506	1.5922E-01	3.1118E-04	-4.2100E-07	1.4222E-10	298	1200
267	C4H6O2	2-BUTYNE-1,4-DIOL	27.892	3.0363E-01	-1.5388E-04	1.4828E-08	8.3391E-12	298	1200
268	C4H6O2	gamma-BUTYROLACTONE	22.370	1.7297E-01	2.4826E-04	-3.3530E-07	1.0555E-10	100	1500
269	C4H6O2	cis-CROTONIC ACID	0.446	4.7569E-01	-4.4446E-04	2.3234E-07	-5.0122E-11	298	1500
270	C4H6O2	trans-CROTONIC ACID	-5.229	4.5048E-01	-3.8771E-04	1.8280E-07	-3.5447E-11	298	1500
271	C4H6O2	METHACRYLIC ACID	-28.131	5.4744E-01	-5.3877E-04	2.8583E-07	-6.0864E-11	298	1200
272	C4H6O2	METHYL ACRYLATE	1.222	4.0619E-01	-2.8529E-04	9.7153E-08	-1.4073E-11	298	1200
273	C4H6O2	VINYL ACETATE	27.664	2.3366E-01	6.2106E-05	-1.6972E-07	5.7917E-11	100	1500
274	C4H6O3	ACETIC ANHYDRIDE	9.500	3.4425E-01	-8.6736E-05	-7.6769E-08	3.6721E-11	200	1500
275	C4H6O4	SUCCINIC ACID	23.417	4.7462E-01	-3.9265E-04	1.7354E-07	-3.2804E-11	298	1500
276	C4H6O5	DIGLYCOLIC ACID	0.721	6.0311E-01	-5.4772E-04	2.6192E-07	-5.2115E-11	298	1500
277	C4H6O5	MALIC ACID	13.117	5.8833E-01	-5.5236E-04	2.7180E-07	-5.5122E-11	298	1500
278	C4H6O6	TARTARIC ACID	4.526	6.8876E-01	-6.8590E-04	3.5082E-07	-7.2249E-11	298	1500
279	C4H7N	n-BUTYRONITRILE	14.849	3.4077E-01	-2.0780E-04	6.2989E-08	-7.5521E-12	200	1500
280	C4H7N	ISOBUTYRONITRILE	29.844	2.3992E-01	-5.0570E-06	-9.7993E-08	3.6314E-11	200	1500
281	C4H7NO	ACETONE CYANOHYDRIN	14.424	4.1722E-01	-3.1414E-04	1.2671E-07	-2.1994E-11	298	1500
282	C4H7NO	2-METHACRYLAMIDE	-18.333	5.8643E-01	-6.4763E-04	4.4441E-07	-1.2776E-10	298	1200
283	C4H7NO	3-METHOXYPROPIONITRILE	9.316	3.9156E-01	-2.3090E-04	5.4563E-08	-2.5331E-12	298	1500
284	C4H7NO	2-PYRROLIDONE	14.880	2.0967E-01	2.3448E-04	-3.3759E-07	1.0745E-10	100	1500
285	C4H8	1-BUTENE	24.915	2.0648E-01	5.9828E-05	-1.4166E-07	4.7053E-11	200	1500
286	C4H8	cis-2-BUTENE	29.137	1.4008E-01	1.9109E-04	-2.3717E-07	7.0962E-11	200	1500
287	C4H8	trans-2-BUTENE	40.312	1.3472E-01	1.6877E-04	-2.1140E-07	6.3263E-11	200	1500
288	C4H8	CYCLOBUTANE	22.621	8.8506E-02	3.9106E-04	-4.3201E-07	1.2970E-10	100	1500
289	C4H8	ISOBUTENE	32.918	1.8546E-01	7.7876E-05	-1.4645E-07	4.6867E-11	200	1500
290	C4H8Br2	1,2-DIBROMOBUTANE	17.413	4.5325E-01	-3.1397E-04	8.8061E-08	--------	298	1000
291	C4H8Br2	2,3-DIBROMOBUTANE	6.074	4.9421E-01	-3.5639E-04	1.0175E-07	--------	298	1000
292	C4H8Cl2	1,4-DICHLOROBUTANE	0.074	5.1333E-01	-4.3439E-04	2.0527E-07	-4.1398E-11	298	1500
293	C4H8I2	1,2-DIIODOBUTANE	23.107	4.2961E-01	-2.8503E-04	7.7103E-08	--------	298	1000
294	C4H8O	n-BUTYRALDEHYDE	64.374	6.4776E-02	3.5143E-04	-3.5371E-07	1.0082E-10	200	1500
295	C4H8O	ISOBUTYRALDEHYDE	-1.360	4.0519E-01	-2.5176E-04	6.0505E-08	6.4389E-12	298	1200
296	C4H8O	1,2-EPOXYBUTANE	6.590	3.4252E-01	-1.2004E-04	-3.3075E-08	2.1868E-11	200	1500
297	C4H8O	METHYL ETHYL KETONE	37.369	2.3045E-01	5.7387E-06	-8.8168E-08	2.9637E-11	200	1500
298	C4H8O	ETHYL VINYL ETHER	-0.946	4.0720E-01	-2.8730E-04	1.1776E-07	-2.1862E-11	298	1500
299	C4H8O	TETRAHYDROFURAN	32.887	2.4554E-02	6.0226E-04	-6.2385E-07	1.8528E-10	50	1500
300	C4H8O2	cis-2-BUTENE-1,4-DIOL	0.759	4.0598E-01	-2.1465E-04	4.8876E-08	-9.9659E-12	298	1000
301	C4H8O2	trans-2-BUTENE-1,4-DIOL	10.706	4.0048E-01	-2.5577E-04	1.1699E-07	-3.9430E-11	298	1000
302	C4H8O2	ISOBUTYRIC ACID	-32.990	5.9238E-01	-5.0629E-04	2.0791E-07	-2.4372E-11	298	1500
303	C4H8O2	n-BUTYRIC ACID	14.368	3.9591E-01	-1.8906E-04	-7.6462E-09	2.0812E-11	298	1200
304	C4H8O2	1,4-DIOXANE	-46.223	5.7263E-01	-3.8800E-04	1.1392E-07	-9.0669E-12	298	1500
305	C4H8O2	ETHYL ACETATE	69.848	8.2338E-02	3.7159E-04	-4.1129E-07	1.2369E-10	200	1500
306	C4H8O2	METHYL PROPIONATE	-131.953	1.3767E+00	-2.4790E-03	2.2378E-06	-7.6784E-10	298	1000
307	C4H8O2	n-PROPYL FORMATE	-23.921	5.9124E-01	-5.6041E-04	2.9107E-07	-6.1164E-11	298	1500
308	C4H8O2S	SULFOLANE	2.498	4.2069E-01	-9.2595E-05	-1.1878E-07	5.2922E-11	298	1500
309	C4H8S	TETRAHYDROTHIOPHENE	-6.161	3.7746E-01	-1.3544E-04	-3.9013E-08	2.5996E-11	200	1500
310	C4H9Br	1-BROMOBUTANE	29.412	2.9219E-01	-2.6578E-05	-9.7812E-08	3.7812E-11	200	1500
311	C4H9Br	2-BROMOBUTANE	-2.986	4.7747E-01	-3.5869E-04	1.4697E-07	-2.5768E-11	298	1500
312	C4H9Cl	n-BUTYL CHLORIDE	42.595	2.0225E-01	1.5618E-04	-2.4192E-07	7.7385E-11	150	1500

36

Table 2-1 HEAT CAPACITY OF GAS - ORGANIC COMPOUNDS (continued)

			$C_P = A + B T + C T^2 + D T^3 + E T^4$					$(C_P - joule/(mol K), T - K)$		
NO	FORMULA	NAME	A	B	C	D	E	TMIN	TMAX
313	C4H9Cl	sec-BUTYL CHLORIDE	41.573	2.1551E-01	1.3137E-04	-2.2399E-07	7.2509E-11	150	1500
314	C4H9Cl	tert-BUTYL CHLORIDE	18.882	3.8038E-01	-1.9282E-04	2.7361E-08	5.0124E-12	200	1500
315	C4H9I	2-IODO-2-METHYLPROPANE	-5.317	5.2116E-01	-3.9394E-04	1.2042E-07	--------	298	1000
316	C4H9N	PYRROLIDINE	-8.802	3.1151E-01	6.6087E-05	-2.0528E-07	7.1304E-11	200	1500
317	C4H9NO	N,N-DIMETHYLACETAMIDE	-21.561	5.5114E-01	-4.1728E-04	1.8505E-07	-3.6554E-11	298	1500
318	C4H9NO	MORPHOLINE	-35.984	6.7041E-01	-6.1690E-04	3.0813E-07	-6.3936E-11	298	1500
319	C4H9NO2	1-NITROBUTANE	-4.174	5.2112E-01	-3.1991E-04	7.6405E-08	--------	298	1000
320	C4H9NO2	2-NITROBUTANE	-12.062	5.5229E-01	-3.5602E-04	9.1671E-08	--------	298	1000
321	C4H10	n-BUTANE	20.056	2.8153E-01	-1.3143E-05	-9.4571E-08	3.4149E-11	200	1500
322	C4H10	ISOBUTANE	6.772	3.4147E-01	-1.0271E-04	-3.6849E-08	2.0429E-11	200	1500
323	C4H10N2	PIPERAZINE	-64.055	7.9174E-01	-6.6296E-04	2.4423E-07	-3.0828E-11	298	1500
324	C4H10O	n-BUTANOL	8.157	4.1032E-01	-2.2645E-04	6.0372E-08	-6.2802E-12	200	2980
325	C4H10O	sec-BUTANOL	22.465	3.5134E-01	-1.2858E-04	-1.1931E-08	1.2940E-11	200	1500
326	C4H10O	tert-BUTANOL	8.866	4.2394E-01	-2.4206E-04	6.1419E-08	-4.3829E-12	200	1500
327	C4H10O	DIETHYL ETHER	35.979	2.8444E-01	-1.2673E-06	-1.0128E-07	3.4529E-11	200	1500
328	C4H10O	METHYL-PROPYL-ETHER	22.326	3.3416E-01	-1.0595E-04	-5.9045E-09	--------	298	1000
329	C4H10O	METHYL ISOPROPYL ETHER	48.135	1.8178E-01	2.0299E-04	-2.6456E-07	7.9645E-11	200	1500
330	C4H10O	ISOBUTANOL	71.169	-3.9114E-01	3.1468E-03	-5.6342E-06	3.3290E-09	200	700
331	C4H10O2	1,3-BUTANEDIOL	1.100	5.2567E-01	-4.0675E-04	1.7268E-07	-3.0418E-11	298	1500
332	C4H10O2	1,4-BUTANEDIOL	-7.265	5.5344E-01	-4.5299E-04	2.0616E-07	-3.8161E-11	298	1500
333	C4H10O2	2,3-BUTANEDIOL	-1.698	5.3981E-01	-3.9103E-04	1.2195E-07	-3.1009E-12	298	1200
334	C4H10O2	t-BUTYL HYDROPEROXIDE	2.100	5.5655E-01	-4.5845E-04	2.1168E-07	-4.2051E-11	298	1500
335	C4H10O2	1,2-DIMETHOXYETHANE	6.542	4.6606E-01	-2.6242E-04	6.5670E-08	-5.4689E-12	298	1500
336	C4H10O2	2-ETHOXYETHANOL	-0.213	4.8872E-01	-3.0001E-04	8.1972E-08	-7.4878E-12	298	1500
337	C4H10O3	DIETHYLENE GLYCOL	13.906	4.8367E-01	-2.7706E-04	6.2086E-08	-1.5319E-12	200	1500
338	C4H10O4S	DIETHYL SULFATE	-16.764	6.7731E-01	-6.1231E-04	2.9140E-07	-5.6686E-11	298	1500
339	C4H10S	n-BUTYL MERCAPTAN	46.393	2.2674E-01	1.2687E-04	-1.9438E-07	5.8247E-11	200	1500
340	C4H10S	ISOBUTYL MERCAPTAN	37.177	2.8839E-01	1.8512E-05	-1.3850E-07	4.8919E-11	200	1500
341	C4H10S	sec-BUTYL MERCAPTAN	41.398	2.6634E-01	5.7961E-05	-1.6133E-07	5.3327E-11	200	1500
342	C4H10S	tert-BUTYL MERCAPTAN	30.326	3.3585E-01	-5.0433E-05	-9.9820E-08	4.0942E-11	200	1500
343	C4H10S	DIETHYL SULFIDE	49.361	2.0829E-01	1.5828E-04	-2.2475E-07	6.8109E-11	200	1500
344	C4H10S	ISOPROPYL-METHYL-SULFIDE	13.619	4.0498E-01	-2.0941E-04	5.1463E-08	--------	298	1000
345	C4H10S	METHYL-PROPYL-SULFIDE	16.011	3.9193E-01	-1.8498E-04	3.2783E-08	--------	298	1000
346	C4H10S2	DIETHYL DISULFIDE	50.958	3.3793E-01	-6.5255E-05	-8.3102E-08	3.5651E-11	200	1500
347	C4H11N	n-BUTYLAMINE	45.381	2.2649E-01	1.5750E-04	-2.4190E-07	7.5475E-11	200	1500
348	C4H11N	ISOBUTYLAMINE	-1.698	4.7082E-01	-2.7368E-04	7.1498E-08	-5.8916E-12	298	1500
349	C4H11N	sec-BUTYLAMINE	18.784	3.5654E-01	-4.1983E-05	-1.1648E-07	4.5681E-11	200	1500
350	C4H11N	tert-BUTYLAMINE	28.406	3.3042E-01	4.8961E-06	-1.5250E-07	5.6636E-11	200	1500
351	C4H11N	DIETHYLAMINE	40.851	2.3495E-01	1.6164E-04	-2.5266E-07	7.9398E-11	200	1500
352	C4H11NO	DIMETHYLETHANOLAMINE	-18.435	6.0239E-01	-4.4463E-04	1.6514E-07	-2.2558E-11	298	1200
353	C4H11NO2	DIETHANOLAMINE	-5.264	6.1929E-01	-4.9545E-04	2.1789E-07	-3.8987E-11	298	1500
354	C4H11NO2	2-AMINOETHOXYETHANOL	-5.199	6.2316E-01	-5.2386E-04	2.6796E-07	-6.4932E-11	298	1200
355	C4H12N2O	N-AMINOETHYL ETHANOLAMINE	-13.558	6.7312E-01	-5.4235E-04	2.4320E-07	-4.6505E-11	298	1500
356	C4H12Si	TETRAMETHYLSILANE	61.616	2.9614E-01	-1.0454E-05	-9.9222E-08	3.5679E-11	200	1500
357	C4H13N3	DIETHYLENE TRIAMINE	-8.147	6.7234E-01	-5.1124E-04	2.1737E-07	-4.0511E-11	298	1500
358	C5Cl6	HEXACHLOROCYCLOPENTADIENE	38.750	8.0546E-01	-1.5519E-03	1.4183E-06	-4.8251E-10	298	1000
359	C5H4O2	FURFURAL	15.470	2.9835E-01	-1.9177E-05	-1.4621E-07	5.9506E-11	100	1500
360	C5H5N	PYRIDINE	23.262	1.1251E-01	3.7351E-04	-4.5402E-07	1.4286E-10	50	1500
361	C5H6	CYCLOPENTADIENE	-25.616	4.2049E-01	-2.7180E-04	7.2995E-08	-3.9537E-12	200	1500
362	C5H6	2-METHYL-1-BUTENE-3-YNE	11.399	3.4714E-01	-2.3506E-04	7.6877E-08	-9.4000E-12	298	1500
363	C5H6	1-PENTENE-3-YNE	9.264	3.3995E-01	-2.3573E-04	8.8403E-08	-1.4533E-11	298	1500
364	C5H6	1-PENTENE-4-YNE	11.588	3.4489E-01	-2.5839E-04	1.0999E-07	-2.0576E-11	298	1500
365	C5H6N2	GLUTARONITRILE	18.975	3.9760E-01	-2.4793E-04	6.6492E-08	-6.1305E-12	298	1500
366	C5H6O2	FURFURYL ALCOHOL	-7.696	5.5491E-01	-4.9754E-04	2.3193E-07	-4.4815E-11	298	1500
367	C5H6O3	GLUTARIC ANHYDRIDE	-39.768	6.2221E-01	-3.8554E-04	1.0999E-07	-1.1564E-11	298	1500
368	C5H6O4	CITRACONIC ACID	-44.280	9.7863E-01	-1.3251E-03	9.3278E-07	-2.5668E-10	298	1200
369	C5H6O4	ITACONIC ACID	-29.874	8.3391E-01	-1.0039E-03	6.4524E-07	-1.6726E-10	298	1200
370	C5H6S	2-METHYLTHIOPHENE	-19.124	4.7660E-01	-3.3969E-04	9.6826E-08	--------	298	1000
371	C5H6S	3-METHYLTHIOPHENE	-23.147	4.9802E-01	-3.7681E-04	1.1365E-07	--------	298	1000
372	C5H7N	N-METHYLPYRROLE	-42.333	5.9879E-01	-5.0928E-04	2.4282E-07	-4.8924E-11	298	1500
373	C5H7NO2	ETHYL CYANOACETATE	-28.861	6.9064E-01	-6.7310E-04	3.4241E-07	-7.0338E-11	298	1500
374	C5H8	CYCLOPENTENE	-4.355	2.5989E-01	1.1735E-04	-2.3049E-07	7.6078E-11	200	1500
375	C5H8	ISOPRENE	-0.007	4.2828E-01	-3.0377E-04	1.1890E-07	-2.0653E-11	200	1500
376	C5H8	3-METHYL-1,2-BUTADIENE	33.913	2.2132E-01	4.9559E-05	-1.2900E-07	4.1214E-11	150	1500
377	C5H8	2-METHYL-1,3-BUTADIENE	-15.114	5.1405E-01	-4.1618E-04	1.3870E-07	--------	298	1000
378	C5H8	1,2-PENTADIENE	15.041	3.3233E-01	-1.2798E-04	-1.0733E-08	1.3140E-11	200	1500
379	C5H8	cis-1,3-PENTADIENE	11.441	3.2718E-01	-1.0552E-04	-3.1107E-08	1.8831E-11	200	1500
380	C5H8	trans-1,3-PENTADIENE	10.384	3.4613E-01	-1.4819E-04	3.3729E-09	9.3660E-12	200	1500
381	C5H8	1,4-PENTADIENE	28.759	2.2687E-01	1.0259E-04	-1.9453E-07	6.2357E-11	100	1500
382	C5H8	2,3-PENTADIENE	29.678	2.8033E-01	-7.3692E-05	-3.3295E-08	1.5621E-11	150	1500
383	C5H8	1-PENTYNE	18.298	3.5705E-01	-2.0856E-04	5.4791E-08	-4.3208E-12	200	1500
384	C5H8	2-PENTYNE	12.196	3.3446E-01	-1.5693E-04	2.5306E-08	--------	298	1000
385	C5H8	3-METHYL-1-BUTYNE	33.392	2.6136E-01	-1.6275E-05	-9.4718E-08	3.6363E-11	200	1500
386	C5H8	SPIROPENTANE	-41.488	5.3877E-01	-3.8261E-04	1.0916E-07	--------	298	1000
387	C5H8N4O12	PENTAERYTHRITOL TETRANITRATE	--------	--------	--------	--------	--------	--------	--------
388	C5H8O	CYCLOPENTANONE	-29.270	5.0540E-01	-3.4447E-04	1.3119E-07	-2.6198E-11	298	1500
389	C5H8O	METHYL ISOPROPENYL KETONE	-6.032	4.4160E-01	-2.5697E-04	5.3823E-08	2.4497E-12	298	1200
390	C5H8O2	ACETYLACETONE	-6.071	5.0469E-01	-3.9471E-04	2.1259E-07	-6.0544E-11	298	1200

Table 2-1 HEAT CAPACITY OF GAS - ORGANIC COMPOUNDS (continued)

NO	FORMULA	NAME	$C_P = A + BT + CT^2 + DT^3 + ET^4$ (C$_P$ - joule/(mol K), T - K)						
			A	B	C	D	E	TMIN	TMAX
391	C5H8O2	ALLYL ACETATE	-20.793	6.0941E-01	-5.2082E-04	2.3682E-07	-4.6892E-11	298	1500
392	C5H8O2	ETHYL ACRYLATE	-9.599	5.4466E-01	-4.1029E-04	1.6344E-07	-3.1021E-11	298	1500
393	C5H8O2	METHYL METHACRYLATE	-25.526	6.0628E-01	-5.0627E-04	2.2388E-07	-4.2286E-11	298	1500
394	C5H8O2	VINYL PROPIONATE	-1.977	4.8760E-01	-2.8547E-04	4.8284E-08	7.1078E-12	298	1500
395	C5H8O3	2-HYDROXYETHYL ACRYLATE	-36.842	8.7884E-01	-1.0766E-03	6.9101E-07	-1.7137E-10	298	1200
396	C5H8O3	LEVULINIC ACID	32.518	4.3890E-01	-2.5355E-04	6.9217E-08	-1.1605E-11	298	1500
397	C5H8O3	METHYL ACETOACETATE	-27.476	6.8761E-01	-6.6106E-04	3.5816E-07	-8.1625E-11	298	1500
398	C5H8O4	GLUTARIC ACID	2.795	7.1090E-01	-7.9710E-04	5.4496E-07	-1.6341E-10	298	1200
399	C5H9N	VALERONITRILE	20.430	3.9187E-01	-1.7706E-04	1.0768E-08	9.2411E-12	200	1500
400	C5H9NO	n-BUTYL ISOCYANATE	25.162	3.1973E-01	3.5100E-05	-1.7032E-07	5.8493E-11	298	1100
401	C5H9NO	N-METHYL-2-PYRROLIDONE	-46.964	5.6527E-01	-1.6330E-04	-1.9993E-07	1.2780E-10	298	1200
402	C5H9NO4	L-GLUTAMIC ACID	14.969	6.8061E-01	-5.5977E-04	2.4091E-07	-4.3844E-11	298	1500
403	C5H10	CYCLOPENTANE	19.735	1.1636E-01	5.1261E-04	-5.6745E-07	1.7045E-10	100	1500
404	C5H10	2-METHYL-1-BUTENE	36.231	2.5093E-01	8.9416E-05	-1.8917E-07	6.2160E-11	200	1500
405	C5H10	2-METHYL-2-BUTENE	39.489	1.9866E-01	1.8422E-04	-2.5432E-07	7.7912E-11	200	1500
406	C5H10	3-METHYL-1-BUTENE	34.280	3.1601E-01	-6.1287E-05	-6.6735E-08	2.8688E-11	200	1500
407	C5H10	1-PENTENE	37.101	2.3664E-01	1.1834E-04	-2.1139E-07	6.8054E-11	200	1500
408	C5H10	cis-2-PENTENE	24.729	2.5447E-01	1.1355E-04	-2.1629E-07	7.0502E-11	200	1500
409	C5H10	trans-2-PENTENE	38.859	1.9866E-01	1.3722E-04	-2.1992E-07	6.9261E-11	200	1500
410	C5H10Br2	2,3-DIBROMO-2-METHYLBUTANE	-5.975	6.4948E-01	-4.8539E-04	1.4253E-07	--------	298	1000
411	C5H10Cl2	1,5-DICHLOROPENTANE	4.347	5.8619E-01	-4.6054E-04	2.0428E-07	-3.9364E-11	298	1500
412	C5H10O	METHYL ISOPROPYL KETONE	-13.343	4.9441E-01	-2.5117E-04	2.5991E-08	9.7133E-12	298	1500
413	C5H10O	2-PENTANONE	42.356	2.7425E-01	6.3786E-05	-1.6870E-07	5.5342E-11	200	1500
414	C5H10O	DIETHYL KETONE	49.800	2.6897E-01	5.0669E-05	-1.5227E-07	4.9510E-11	200	1500
415	C5H10O	VALERALDEHYDE	26.913	3.5952E-01	-6.7847E-05	-7.0728E-08	2.9503E-11	273	1500
416	C5H10O2	n-BUTYL FORMATE	-23.729	6.8205E-01	-6.0993E-04	3.0105E-07	-6.1033E-11	298	1500
417	C5H10O2	ETHYL PROPIONATE	139.382	-4.8565E-01	2.3071E-03	-2.7597E-06	1.0859E-09	298	1000
418	C5H10O2	ISOBUTYL FORMATE	-26.232	6.8785E-01	-6.0138E-04	2.8357E-07	-5.3862E-11	298	1500
419	C5H10O2	ISOPROPYL ACETATE	-45.829	7.9654E-01	-7.9890E-04	4.3031E-07	-9.2988E-11	298	1500
420	C5H10O2	n-PROPYL ACETATE	21.906	4.2733E-01	-1.3858E-04	-5.4238E-08	3.0118E-11	298	1500
421	C5H10O2	METHYL n-BUTYRATE	-39.235	8.0965E-01	-9.2286E-04	6.0084E-07	-1.6118E-10	298	1200
422	C5H10O2	2-METHYLBUTYRIC ACID	-19.156	6.0607E-01	-4.1921E-04	1.3696E-07	-1.7062E-11	298	1500
423	C5H10O2	ISOVALERIC ACID	-5.803	5.9939E-01	-4.6602E-04	1.9713E-07	-3.6253E-11	298	1500
424	C5H10O2	VALERIC ACID	-12.596	6.5474E-01	-5.8609E-04	2.8904E-07	-5.8646E-11	298	1500
425	C5H10O2	TETRAHYDROFURFURYL ALCOHOL	-21.588	7.0118E-01	-6.2383E-04	3.0249E-07	-6.1395E-11	298	1500
426	C5H10O2S	3-METHYL SULFOLANE	2.742	4.4985E-01	1.4128E-04	-4.9419E-07	2.1699E-10	298	1200
427	C5H10O3	DIETHYL CARBONATE	--------	--------	--------	--------	--------	--------	--------
428	C5H10O3	ETHYL LACTATE	-30.734	8.0038E-01	-8.1025E-04	4.4743E-07	-9.9928E-11	298	1500
429	C5H10S	THIACYCLOHEXANE	-52.045	6.3212E-01	-3.3873E-04	6.1028E-08	--------	298	1000
430	C5H10S	CYCLOPENTANETHIOL	-36.386	5.8174E-01	-3.5459E-04	8.4634E-08	--------	298	1000
431	C5H11Br	1-BROMOPENTANE	1.254	5.2886E-01	-3.2580E-04	8.1718E-08	--------	298	1000
432	C5H11Cl	1-CHLOROPENTANE	22.357	4.1358E-01	-1.2716E-04	-4.7751E-08	2.7004E-11	200	1500
433	C5H11Cl	1-CHLORO-3-METHYLBUTANE	-2.501	5.5928E-01	-3.7295E-04	1.0363E-07	--------	298	1000
434	C5H11Cl	2-CHLORO-2-METHYLBUTANE	-10.851	5.8312E-01	-3.8563E-04	1.0259E-07	--------	298	1000
435	C5H11N	N-METHYLPYRROLIDINE	-63.618	6.8971E-01	-4.8249E-04	1.6504E-07	-2.1384E-11	298	1500
436	C5H11N	PIPERIDINE	-53.313	7.5541E-01	-5.6470E-04	1.6625E-07	-1.0457E-11	298	1500
437	C5H11NO	tert-BUTYLFORMAMIDE	-18.333	5.7294E-01	-2.9150E-04	-1.2192E-08	3.9907E-11	298	1500
438	C5H12	ISOPENTANE	-0.881	4.7498E-01	-2.4797E-04	6.7512E-08	-8.5343E-12	200	1500
439	C5H12	NEOPENTANE	-17.917	5.7236E-01	-4.1705E-04	2.1158E-07	-5.1006E-11	200	1500
440	C5H12	n-PENTANE	26.671	3.2324E-01	4.2820E-05	-1.6639E-07	5.6036E-11	200	1500
441	C5H12O	2,2-DIMETHYL-1-PROPANOL	-24.052	6.5427E-01	-4.7894E-04	1.6991E-07	-2.1290E-11	298	1500
442	C5H12O	tert-PENTYL-ALCOHOL	-9.454	5.6409E-01	-3.2853E-04	7.2178E-08	--------	298	1000
443	C5H12O	2-METHYL-1-BUTANOL	-19.293	6.3101E-01	-4.7958E-04	1.9533E-07	-3.1575E-11	298	1500
444	C5H12O	2-METHYL-2-BUTANOL	-16.955	6.5298E-01	-5.4260E-04	2.6069E-07	-5.4462E-11	298	1200
445	C5H12O	3-METHYL-1-BUTANOL	17.380	3.8036E-01	9.0875E-05	-3.3684E-07	1.4257E-10	298	1200
446	C5H12O	3-METHYL-2-BUTANOL	-14.366	6.2608E-01	-4.8709E-04	2.1430E-07	-4.1705E-11	298	1500
447	C5H12O	1-PENTANOL	9.175	4.7662E-01	-1.9542E-04	-1.3991E-08	2.0685E-11	200	1500
448	C5H12O	2-PENTANOL	-7.077	5.9369E-01	-4.3971E-04	1.8499E-07	-3.5235E-11	298	1400
449	C5H12O	3-PENTANOL	61.233	7.9840E-02	9.0823E-04	-1.2984E-06	5.4866E-10	298	1000
450	C5H12O	METHYL sec-BUTYL ETHER	-0.104	5.5588E-01	-3.6662E-04	1.3167E-07	-2.0764E-11	298	1500
451	C5H12O	METHYL tert-BUTYL ETHER	39.585	2.8849E-01	1.3825E-04	-2.5131E-07	8.0807E-11	200	1500
452	C5H120	METHYL ISOBUTYL ETHER	-11.287	5.9389E-01	-4.4085E-04	2.0554E-07	-4.9208E-11	298	1500
453	C5H12O	ETHYL PROPYL ETHER	14.457	4.5609E-01	-1.7751E-04	-1.1636E-08	1.6934E-11	298	1500
454	C5H12O2	ETHYLENE GLYCOL MONOPROPYL ETHER	-16.033	6.8258E-01	-5.7788E-04	2.9935E-07	-7.2869E-11	298	1200
455	C5H12O2	NEOPENTYL GLYCOL	-11.513	6.5964E-01	-4.8149E-04	1.8407E-07	-3.1786E-11	298	1500
456	C5H12O2	1,5-PENTANEDIOL	-2.425	6.1055E-01	-4.3769E-04	1.6872E-07	-2.6880E-11	298	1500
457	C5H12O3	2-(2-METHOXYETHOXY)ETHANOL	-22.744	7.9858E-01	-8.1464E-04	5.3038E-07	-1.5753E-10	298	1500
458	C5H12O4	PENTAERYTHRITOL	-4.675	7.5699E-01	-6.1689E-04	2.7195E-07	-5.3121E-11	298	1400
459	C5H12S	n-PENTYL MERCAPTAN	19.604	4.6839E-01	-2.0875E-04	4.3242E-08	-7.5836E-12	273	1500
460	C5H12S	BUTYL-METHYL-SULFIDE	18.502	4.6735E-01	-2.0221E-04	3.0437E-08	--------	298	1000
461	C5H12S	ETHYL-PROPYL-SULFIDE	15.706	4.7037E-01	-1.9889E-04	2.8258E-08	--------	298	1000
462	C5H12S	2-METHYL-2-BUTANETHIOL	3.470	5.6358E-01	-3.3987E-04	7.9358E-08	--------	298	1000
463	C5H13N	n-PENTYLAMINE	6.603	5.2073E-01	-2.5708E-04	3.9765E-08	3.5264E-12	298	1500
464	C5H13NO2	METHYL DIETHANOLAMINE	-16.817	7.4349E-01	-5.5635E-04	2.1535E-07	-3.4622E-11	298	1500
465	C6Cl6	HEXACHLOROBENZENE	40.828	6.3403E-01	-7.3328E-04	4.0581E-07	-8.6866E-11	200	1500
466	C6F6	HEXAFLUOROBENZENE	56.398	4.1193E-01	-2.7480E-04	8.4094E-08	-7.9692E-12	200	1500
467	C6H3ClN2O4	1-CHLORO-2,4-DINITROBENZENE	36.588	5.6182E-01	-3.9302E-04	1.0043E-07	-1.2252E-12	200	1500
468	C6H3Cl2NO2	1,2-DICHLORO-4-NITROBENZENE	20.678	5.6182E-01	-3.9302E-04	1.0043E-07	-1.2252E-12	200	1500

Table 2-1 HEAT CAPACITY OF GAS - ORGANIC COMPOUNDS (continued)

			$C_P = A + BT + CT^2 + DT^3 + ET^4$					$(C_P$ - joule/(mol K), T - K)	
NO	FORMULA	NAME	A	B	C	D	E	TMIN	TMAX
469	C6H3Cl3	1,2,4-TRICHLOROBENZENE	2.547	5.6873E-01	-5.5217E-04	2.6984E-07	-5.2816E-11	200	1500
470	C6H3N3O6	1,3,5-TRINITROBENZENE	52.498	5.6182E-01	-3.9302E-04	1.0043E-07	-1.2252E-12	200	1500
471	C6H4Br2	m-DIBROMOBENZENE	23.494	3.7580E-01	-1.7947E-04	-1.2488E-08	2.2028E-11	150	1500
472	C6H4ClNO2	m-CHLORONITROBENZENE	2.238	5.6182E-01	-3.9302E-04	1.0043E-07	-1.2252E-12	200	1500
473	C6H4ClNO2	o-CHLORONITROBENZENE	5.599	5.4506E-01	-3.6483E-04	8.1076E-08	3.5675E-12	200	1500
474	C6H4ClNO2	p-CHLORONITROBENZENE	2.861	5.6351E-01	-3.9804E-04	1.0486E-07	-2.4547E-12	200	1500
475	C6H4Cl2	m-DICHLOROBENZENE	-9.294	5.4269E-01	-4.8322E-04	2.1721E-07	-3.9293E-11	200	1500
476	C6H4Cl2	o-DICHLOROBENZENE	-9.539	5.4073E-01	-4.7821E-04	2.1330E-07	-3.8284E-11	200	1500
477	C6H4Cl2	p-DICHLOROBENZENE	-9.451	5.4233E-01	-4.8126E-04	2.1537E-07	-3.8775E-11	200	1500
478	C6H4F2	m-DIFLUOROBENZENE	-26.742	5.7170E-01	-4.6421E-04	1.4498E-07	---------	298	1000
479	C6H4F2	o-DIFLUOROBENZENE	-24.499	5.5919E-01	-4.4225E-04	1.3656E-07	---------	298	1000
480	C6H4F2	p-DIFLUOROBENZENE	-25.837	5.7137E-01	-4.6668E-04	1.4709E-07	---------	298	1000
481	C6H4N2O4	m-DINITROBENZENE	18.148	5.6182E-01	-3.9302E-04	1.0043E-07	-1.2252E-12	200	1500
482	C6H4N2O4	o-DINITROBENZENE	18.148	5.6182E-01	-3.9302E-04	1.0043E-07	-1.2252E-12	200	1500
483	C6H4N2O4	p-DINITROBENZENE	18.148	5.6182E-01	-3.9302E-04	1.0043E-07	-1.2252E-12	200	1500
484	C6H5Br	BROMOBENZENE	10.965	3.3493E-01	-7.7630E-05	-9.0065E-08	4.1966E-11	200	1500
485	C6H5Cl	MONOCHLOROBENZENE	27.315	2.4405E-01	9.3748E-05	-2.2377E-07	8.0155E-11	200	1500
486	C6H5ClO	m-CHLOROPHENOL	-30.256	6.6657E-01	-6.5807E-04	3.2782E-07	-6.5085E-11	298	1500
487	C6H5ClO	o-CHLOROPHENOL	-30.256	6.6657E-01	-6.5807E-04	3.2782E-07	-6.5085E-11	298	1500
488	C6H5ClO	p-CHLOROPHENOL	-30.256	6.6657E-01	-6.5807E-04	3.2782E-07	-6.5085E-11	298	1500
489	C6H5Cl2N	3,4-DICHLOROANILINE	-12.710	6.9240E-01	-7.0094E-04	3.5781E-07	-7.2645E-11	298	1500
490	C6H5F	FLUOROBENZENE	-7.810	4.1311E-01	-1.9932E-04	-6.0575E-09	2.0763E-11	200	1500
491	C6H5I	IODOBENZENE	5.014	3.8494E-01	-1.7674E-04	-1.4239E-08	2.1809E-11	200	1500
492	C6H5NO2	NITROBENZENE	-16.202	5.6182E-01	-3.9302E-04	1.0043E-07	-1.2252E-12	200	1500
493	C6H6	BENZENE	-31.368	4.7460E-01	-3.1137E-04	8.5237E-08	-5.0524E-12	200	1500
494	C6H6ClN	m-CHLOROANILINE	-24.493	6.6130E-01	-6.1856E-04	2.9376E-07	-5.6020E-11	298	1500
495	C6H6ClN	o-CHLOROANILINE	-24.493	6.6130E-01	-6.1856E-04	2.9376E-07	-5.6020E-11	298	1500
496	C6H6ClN	p-CHLOROANILINE	-24.135	6.5666E-01	-6.0355E-04	2.7595E-07	-4.8992E-11	298	1200
497	C6H6N2	cis-DICYANO-1-BUTENE	6.870	4.1827E-01	-7.9183E-05	-2.1892E-07	1.2143E-10	298	1000
498	C6H6N2	trans-DICYANO-1-BUTENE	-1.571	5.5288E-01	-4.8007E-04	2.1814E-07	-4.0133E-11	298	1500
499	C6H6N2	1,4-DICYANO-2-BUTENE	39.740	2.7876E-01	1.3446E-04	-3.5716E-07	1.5299E-10	298	1000
500	C6H6N2O2	m-NITROANILINE	12.288	5.7313E-01	-4.5651E-04	1.8410E-07	-2.9867E-11	200	1500
501	C6H6N2O2	o-NITROANILINE	12.288	5.7313E-01	-4.5651E-04	1.8410E-07	-2.9867E-11	200	1500
502	C6H6N2O2	p-NITROANILINE	12.288	5.7313E-01	-4.5651E-04	1.8410E-07	-2.9867E-11	200	1500
503	C6H6O	PHENOL	4.408	3.6338E-01	-6.0417E-05	-1.2794E-07	5.5287E-11	100	1500
504	C6H6O2	1,2-BENZENEDIOL	-24.649	6.2268E-01	-5.0070E-04	1.8696E-07	-2.5672E-11	200	1500
505	C6H6O2	1,3-BENZENEDIOL	-28.476	6.8792E-01	-6.7225E-04	3.3593E-07	-6.7411E-11	200	1500
506	C6H6O2	p-HYDROQUINONE	-23.109	6.6077E-01	-6.4518E-04	3.2817E-07	-6.7604E-11	200	1500
507	C6H6O3	1,2,3-BENZENETRIOL	-23.381	7.5537E-01	-7.6515E-04	3.8498E-07	-7.6871E-11	298	1500
508	C6H6S	PHENYL MERCAPTAN	-5.259	4.4764E-01	-2.3973E-04	2.2703E-08	1.2746E-11	200	1500
509	C6H7N	ANILINE	-22.062	5.7313E-01	-4.5651E-04	1.8410E-07	-2.9867E-11	200	1500
510	C6H7N	2-METHYLPYRIDINE	-17.819	4.7591E-01	-2.5380E-04	3.3280E-08	8.7218E-12	200	1500
511	C6H7N	3-METHYLPYRIDINE	-16.136	4.6342E-01	-2.2923E-04	1.6537E-08	1.2749E-11	200	1500
512	C6H7N	4-METHYLPYRIDINE	-18.170	4.7450E-01	-2.4996E-04	3.0282E-08	9.4921E-12	200	1500
513	C6H8	1,3-CYCLOHEXADIENE	9.959	2.5557E-01	2.0432E-04	-3.3043E-07	1.0763E-10	100	1500
514	C6H8	METHYLCYCLOPENTADIENE	-59.926	7.1827E-01	-6.8011E-04	3.3399E-07	-6.6324E-11	298	1500
515	C6H8N2	ADIPONITRILE	46.703	2.1098E-01	6.0849E-04	-1.0564E-06	4.7126E-10	298	1000
516	C6H8N2	METHYLGLUTARONITRILE	32.077	4.3754E-01	-2.3741E-04	4.6871E-08	4.8993E-13	298	1500
517	C6H8N2	m-PHENYLENEDIAMINE	-35.663	7.5211E-01	-7.0735E-04	3.3963E-07	-6.5853E-11	298	1500
518	C6H8N2	o-PHENYLENEDIAMINE	-35.663	7.5211E-01	-7.0735E-04	3.3963E-07	-6.5853E-11	298	1500
519	C6H8N2	p-PHENYLENEDIAMINE	-35.663	7.5211E-01	-7.0735E-04	3.3963E-07	-6.5853E-11	298	1500
520	C6H8N2	PHENYLHYDRAZINE	-53.505	7.8460E-01	-7.4455E-04	3.6656E-07	-7.3334E-11	298	1500
521	C6H8N2O	BIS(CYANOETHYL)ETHER	30.043	4.8287E-01	-2.4564E-04	1.9368E-08	1.2919E-11	298	1500
522	C6H8O4	DIMETHYL MALEATE	-1.849	7.6196E-01	-7.8473E-04	4.5797E-07	-1.1125E-10	298	1200
523	C6H8O6	ASCORBIC ACID	-7.357	1.0100E+00	-1.1580E-03	6.7650E-07	-1.5220E-10	298	1500
524	C6H8O7	CITRIC ACID	20.051	8.5219E-01	-8.1655E-04	4.0741E-07	-8.3420E-11	298	1500
525	C6H10	1-METHYLCYCLOPENTENE	-29.473	4.9672E-01	-2.0380E-04	4.8087E-09	---------	298	1000
526	C6H10	3-METHYLCYCLOPENTENE	-43.229	5.7170E-01	-3.2845E-04	6.9601E-08	---------	298	1000
527	C6H10	4-METHYLCYCLOPENTENE	-41.176	5.6124E-01	-3.1389E-04	6.3170E-08	---------	298	1000
528	C6H10	CYCLOHEXENE	11.332	2.5648E-01	2.9121E-04	-4.0917E-07	1.2869E-10	100	1500
529	C6H10	2,3-DIMETHYL-1,3-BUTADIENE	-14.346	6.3345E-01	-5.7820E-04	2.9473E-07	-6.2408E-11	298	1500
530	C6H10	1,5-HEXADIENE	-19.919	5.8801E-01	-4.6457E-04	2.0423E-07	-3.8563E-11	298	1500
531	C6H10	cis,trans-2,4-HEXADIENE	-21.887	5.8531E-01	-4.4663E-04	1.8361E-07	-3.2586E-11	298	1500
532	C6H10	trans,trans-2,4-HEXADIENE	-13.866	5.8123E-01	-4.6720E-04	2.1027E-07	-4.0509E-11	298	1500
533	C6H10	1-HEXYNE	19.611	4.4045E-01	-2.3708E-04	4.1265E-08	3.5316E-12	200	1500
534	C6H10	2-HEXYNE	-3.658	5.1409E-01	-3.7142E-04	1.5649E-07	-2.7818E-11	298	1500
535	C6H10	3-HEXYNE	-12.209	5.5220E-01	-4.4735E-04	2.2254E-07	-4.7781E-11	298	1500
536	C6H10O	CYCLOHEXANONE	-12.337	4.0837E-01	1.0597E-04	-2.9616E-07	1.0088E-10	200	1500
537	C6H10O	MESITYL OXIDE	16.062	5.8779E-01	-6.0312E-04	4.4456E-07	-1.4943E-10	298	1000
538	C6H10O2	epsilon-CAPROLACTONE	-59.878	7.3431E-01	-4.4185E-04	9.0409E-08	3.2030E-12	298	1500
539	C6H10O2	ETHYL METHACRYLATE	-38.063	7.5789E-01	-6.5982E-04	3.0689E-07	-5.9646E-11	298	1500
540	C6H10O2	n-PROPYL ACRYLATE	-34.027	8.8100E-01	-9.6426E-04	5.6512E-07	-1.2909E-10	298	1500
541	C6H10O3	ETHYLACETOACETATE	-38.431	8.1942E-01	-7.5317E-04	3.7707E-07	-7.8132E-11	298	1500
542	C6H10O3	PROPIONIC ANHYDRIDE	29.000	3.6844E-01	3.0361E-04	-6.2082E-07	2.4092E-10	298	1200
543	C6H10O4	ADIPIC ACID	7.418	6.7710E-01	-4.6904E-04	1.4619E-07	-1.5583E-11	298	1500
544	C6H10O4	DIETHYL OXALATE	-63.195	1.0643E+00	-1.1909E-03	6.9357E-07	-1.5592E-10	298	1500
545	C6H10O4	ETHYLENE GLYCOL DIACETATE	-63.207	1.0644E+00	-1.1911E-03	6.9374E-07	-1.5597E-10	298	1500
546	C6H10O4	ETHYLIDENE DIACETATE	-11.375	5.4896E-01	3.3144E-04	-1.0756E-06	5.5152E-10	298	1000

39

Table 2-1 HEAT CAPACITY OF GAS - ORGANIC COMPOUNDS (continued)

NO	FORMULA	NAME	$C_P = A + B T + C T^2 + D T^3 + E T^4$				(C$_P$ - joule/(mol K), T - K)		
			A	B	C	D	E	TMIN	TMAX
547	C6H11N	HEXANENITRILE	23.213	4.6598E-01	-1.9041E-04	-8.6856E-09	1.7539E-11	200	1500
548	C6H11NO	epsilon-CAPROLACTAM	-10.914	2.7864E-01	6.5610E-04	-9.5916E-07	3.6689E-10	298	1000
549	C6H11NO	CYCLOHEXANONE OXIME	-38.514	6.5385E-01	-2.3559E-04	-7.2552E-08	4.5887E-11	298	1500
550	C6H12	CYCLOHEXANE	13.783	2.0742E-01	5.3682E-04	-6.3012E-07	1.8988E-10	100	1500
551	C6H12	2,3-DIMETHYL-1-BUTENE	32.307	4.2869E-01	-1.5601E-04	-2.1440E-08	1.8781E-11	200	1500
552	C6H12	2,3-DIMETHYL-2-BUTENE	50.138	1.9488E-01	3.1558E-04	-3.7082E-07	1.0920E-10	200	1500
553	C6H12	3,3-DIMETHYL-1-BUTENE	37.579	2.8542E-01	1.8034E-04	-3.0375E-07	9.8601E-11	200	1500
554	C6H12	2-ETHYL-1-BUTENE	26.884	3.9566E-01	-5.3867E-05	-1.1252E-07	4.5467E-11	200	1500
555	C6H12	trans-3-METHYL-2-PENTENE	-16.123	5.7438E-01	-3.4767E-04	8.6822E-08	--------	298	1000
556	C6H12	1-HEXENE	32.517	3.5231E-01	2.2971E-05	-1.6592E-07	5.8510E-11	200	1500
557	C6H12	cis-2-HEXENE	35.642	2.9178E-01	1.5778E-04	-2.7413E-07	8.8092E-11	200	1500
558	C6H12	trans-2-HEXENE	49.211	2.6046E-01	1.8185E-04	-2.8038E-07	8.7781E-11	200	1500
559	C6H12	cis-3-HEXENE	58.647	1.3967E-01	4.7553E-04	-5.3528E-07	1.6155E-10	150	1500
560	C6H12	trans-3-HEXENE	66.499	1.6110E-01	3.8543E-04	-4.4785E-07	1.3530E-10	150	1500
561	C6H12	METHYLCYCLOPENTANE	-9.939	4.2528E-01	1.2521E-05	-1.8864E-07	6.4751E-11	200	1500
562	C6H12	2-METHYL-1-PENTENE	35.161	3.5515E-01	1.6698E-05	-1.6387E-07	5.7999E-11	200	1500
563	C6H12	2-METHYL-2-PENTENE	27.059	3.4858E-01	4.7035E-05	-1.9000E-07	6.5743E-11	200	1500
564	C6H12	3-METHYL-1-PENTENE	59.573	2.5875E-01	1.8945E-04	-3.0001E-07	9.6074E-11	150	1500
565	C6H12	3-METHYL-cis-2-PENTENE	58.069	1.6574E-01	4.0178E-04	-4.6769E-07	1.4145E-10	150	1500
566	C6H12	4-METHYL-1-PENTENE	26.362	3.5120E-01	3.9961E-05	-1.8537E-07	6.4392E-11	200	1500
567	C6H12	4-METHYL-cis-2-PENTENE	52.969	2.5164E-01	2.0116E-04	-3.0068E-07	9.5720E-11	200	1500
568	C6H12	4-METHYL-trans-2-PENTENE	39.751	3.7243E-01	-5.1165E-05	-9.8996E-08	3.8818E-11	200	1500
569	C6H12N2	TRIETHYLENEDIAMINE	-124.418	1.0337E+00	-8.2395E-04	3.2203E-07	-4.8974E-11	298	1500
570	C6H12O	BUTYL VINYL ETHER	-1.119	5.8419E-01	-3.7075E-04	1.2882E-07	-2.3385E-11	298	1500
571	C6H12O	CYCLOHEXANOL	17.124	3.3700E-01	2.8176E-04	-4.2713E-07	1.3215E-10	200	1500
572	C6H12O	1-HEXANAL	71.773	2.0294E-01	3.4761E-04	-4.1609E-07	1.2259E-10	200	1500
573	C6H12O	ETHYL ISOPROPYL KETONE	36.515	3.4267E-01	1.4830E-04	-3.4209E-07	1.3148E-10	298	1200
574	C6H12O	2-HEXANONE	45.050	3.5513E-01	4.8896E-05	-1.9813E-07	6.4005E-11	200	1500
575	C6H12O	3-HEXANONE	73.031	2.2148E-01	2.7912E-04	-3.6876E-07	1.1286E-10	150	1500
576	C6H12O	METHYL ISOBUTYL KETONE	2.404	5.8495E-01	-3.7647E-04	1.2418E-07	-1.7051E-11	298	1500
577	C6H12O2	n-PENTYL FORMATE	-30.624	8.1555E-01	-7.4226E-04	3.7578E-07	-7.8511E-11	298	1500
578	C6H12O2	n-BUTYL ACETATE	85.139	7.1561E-02	8.7842E-04	-1.0725E-06	3.7362E-10	298	1200
579	C6H12O2	sec-BUTYL ACETATE	-37.391	8.3030E-01	-6.9627E-04	2.6603E-07	-2.1215E-11	298	1000
580	C6H12O2	tert-BUTYL ACETATE	-66.418	1.0291E+00	-1.1512E-03	7.0828E-07	-1.7774E-10	298	1200
581	C6H12O2	ETHYL n-BUTYRATE	-16.529	7.2624E-01	-5.1890E-04	1.2305E-07	2.3936E-11	298	1200
582	C6H12O2	ETHYL ISOBUTYRATE	-62.802	9.3400E-01	-9.1286E-04	4.8405E-07	-1.0440E-10	298	1500
583	C6H12O2	ISOBUTYL ACETATE	-24.628	7.5868E-01	-5.9783E-04	2.3506E-07	-3.4297E-11	298	1500
584	C6H12O2	n-PROPYL PROPIONATE	-26.162	7.8326E-01	-6.3633E-04	2.2535E-07	-8.0808E-12	298	1200
585	C6H12O2	CYCLOHEXYL PEROXIDE	-66.681	9.6297E-01	-1.0232E-03	7.4423E-07	-2.5645E-10	298	1000
586	C6H12O2	DIACETONE ALCOHOL	13.030	5.9591E-01	-3.8411E-04	1.5153E-07	-3.8398E-11	298	1200
587	C6H12O2	2-ETHYL BUTYRIC ACID	-44.341	8.5885E-01	-8.2522E-04	4.6805E-07	-1.1743E-10	298	1200
588	C6H12O2	n-HEXANOIC ACID	-9.559	7.4029E-01	-6.6784E-04	3.7713E-07	-1.0030E-10	298	1200
589	C6H12O3	2-ETHOXYETHYL ACETATE	-34.381	8.6550E-01	-7.7015E-04	3.7720E-07	-8.0618E-11	298	1500
590	C6H12O3	HYDROXYCAPROIC ACID	-2.103	7.3513E-01	-5.5705E-04	2.2007E-07	-3.6780E-11	298	1500
591	C6H12O3	PARALDEHYDE	-129.363	1.2890E+00	-1.4032E-03	7.4424E-07	-1.4315E-10	298	1500
592	C6H12O3	sec-BUTYL GLYCOLATE	--------	--------	--------	--------	--------	--------	--------
593	C6H12S	THIACYCLOHEPTANE	-70.538	7.2789E-01	-2.4125E-04	-4.9195E-08	--------	298	1000
594	C6H13N	CYCLOHEXYLAMINE	-47.225	7.0490E-01	-3.4142E-04	1.8298E-08	1.9665E-11	298	1500
595	C6H13N	HEXAMETHYLENEIMINE	-81.362	7.9390E-01	-4.5964E-04	9.4512E-08	2.5824E-12	298	1400
596	C6H14	2,2-DIMETHYLBUTANE	-1.477	5.5644E-01	-2.4802E-04	4.1433E-08	2.4035E-13	200	1500
597	C6H14	2,3-DIMETHYLBUTANE	-25.999	6.8344E-01	-4.8517E-04	2.1262E-07	-4.3837E-11	200	1500
598	C6H14	n-HEXANE	25.924	4.1927E-01	-1.2491E-05	-1.5916E-07	5.8784E-11	200	1500
599	C6H14	2-METHYLPENTANE	-7.197	6.0097E-01	-3.4094E-04	9.5210E-08	-1.0297E-11	200	1500
600	C6H14	3-METHYLPENTANE	-7.123	5.8327E-01	-3.0338E-04	6.8016E-08	-3.9778E-12	200	1500
601	C6H14N2O2	LYSINE	3.524	8.1910E-01	-6.3474E-04	2.7688E-07	-5.6094E-11	298	1500
602	C6H14O	2-ETHYL-1-BUTANOL	-25.284	7.6092E-01	-6.1315E-04	2.7815E-07	-5.3099E-11	298	1500
603	C6H14O	1-HEXANOL	10.719	5.5767E-01	-2.1818E-04	-3.2298E-08	2.9769E-11	200	1500
604	C6H14O	2-HEXANOL	-8.756	6.7929E-01	-4.4926E-04	1.4008E-07	-1.1453E-11	298	1200
605	C6H14O	2-METHYL-1-PENTANOL	-18.096	7.1113E-01	-5.2110E-04	2.1617E-07	-4.0923E-11	298	1500
606	C6H14O	4-METHYL-2-PENTANOL	-17.282	7.3333E-01	-5.6666E-04	2.4735E-07	-4.7894E-11	298	1500
607	C6H14O	n-BUTYL ETHYL ETHER	10.663	5.7291E-01	-2.7890E-04	3.8539E-08	6.0989E-12	298	1500
608	C6H14O	DIISOPROPYL ETHER	92.068	1.1054E-01	5.8784E-04	-6.1854E-07	1.7855E-10	100	1500
609	C6H14O	DI-n-PROPYL ETHER	-26.082	7.6707E-01	-6.4734E-04	3.3945E-07	-8.3996E-11	298	1200
610	C6H14O	METHYL tert-PENTYL ETHER	-5.816	6.8044E-01	-4.6708E-04	1.7397E-07	-2.8573E-11	298	1500
611	C6H14O2	ACETAL	31.834	3.4539E-01	5.6817E-04	-1.0268E-06	4.4628E-10	298	1000
612	C6H14O2	2-BUTOXYETHANOL	-22.219	8.1143E-01	-7.1242E-04	3.9347E-07	-1.0334E-10	298	1200
613	C6H14O2	1,6-HEXANEDIOL	-2.789	7.0351E-01	-4.9185E-04	1.8516E-07	-3.0221E-11	298	1500
614	C6H14O2	HEXYLENE GLYCOL	-30.746	8.8908E-01	-8.0455E-04	4.0021E-07	-8.2866E-11	298	1500
615	C6H14O2S	DI-n-PROPYL SULFONE	16.469	6.7237E-01	-3.3664E-04	5.5367E-08	4.0249E-12	298	1500
616	C6H14O3	DIETHYLENE GLYCOL DIMETHYL ETHER	-38.307	9.9161E-01	-1.1006E-03	7.7626E-07	-2.4078E-10	298	1200
617	C6H14O3	DIPROPYLENE GLYCOL	-31.988	9.4839E-01	-8.9835E-04	4.8010E-07	-1.1001E-10	298	1200
618	C6H14O3	2-(2-ETHOXYETHOXY)ETHANOL	-15.223	8.0630E-01	-6.0939E-04	2.4852E-07	-4.4531E-11	298	1500
619	C6H14O3	TRIMETHYLOLPROPANE	-14.530	8.3299E-01	-6.6014E-04	2.7694E-07	-4.8128E-11	298	1500
620	C6H14O4	TRIETHYLENE GLYCOL	2.101	7.8650E-01	-5.6060E-04	2.0905E-07	-3.3896E-11	298	1500
621	C6H14O6	SORBITOL	-6.695	1.0042E+00	-7.9252E-04	2.0507E-07	3.0979E-11	298	1500
622	C6H14S	n-HEXYLMERCAPTAN	56.097	3.5459E-01	1.3831E-04	-2.6776E-07	8.5007E-11	200	1500
623	C6H14S	BUTYL-ETHYL-SULFIDE	14.362	5.6756E-01	-2.5704E-04	4.2116E-08	--------	298	1000
624	C6H14S	ISOPROPYL-SULFIDE	-5.059	7.2542E-01	-5.1831E-04	1.5466E-07	--------	298	1000

Table 2-1 HEAT CAPACITY OF GAS - ORGANIC COMPOUNDS (continued)

NO	FORMULA	NAME	$C_P = A + B\,T + C\,T^2 + D\,T^3 + E\,T^4$					(C$_P$ - joule/(mol K), T - K)	
			A	B	C	D	E	TMIN	TMAX
625	C6H14S	METHYL-PENTYL-SULFIDE	17.152	5.6455E-01	-2.6040E-04	4.4321E-08	--------	298	1000
626	C6H14S	PROPYL-SULFIDE	16.912	5.4572E-01	-2.1603E-04	2.5860E-08	--------	298	1000
627	C6H14S2	PROPYL-DISULFIDE	22.739	6.4496E-01	-3.5881E-04	8.0986E-08	--------	298	1000
628	C6H15Al	TRIETHYL ALUMINUM	--------	--------	--------	--------	--------	--------	--------
629	C6H15Al2Cl3	ETHYL ALUMINUM SESQUICHLORIDE	123.889	2.1683E-01	6.9390E-04	-8.2388E-07	2.5547E-10	100	1500
630	C6H15N	DIISOPROPYLAMINE	-38.690	8.3934E-01	-6.8079E-04	3.1403E-07	-6.5038E-11	298	1500
631	C6H15N	DI-n-PROPYLAMINE	-35.272	8.1873E-01	-6.6672E-04	3.1414E-07	-6.4824E-11	298	1500
632	C6H15N	n-HEXYLAMINE	62.901	3.0756E-01	2.3918E-04	-3.4644E-07	1.0365E-10	200	1500
633	C6H15N	TRIETHYLAMINE	55.793	3.3337E-01	2.2077E-04	-3.4942E-07	1.0838E-10	200	1500
634	C6H15NO	6-AMINOHEXANOL	10.049	6.6024E-01	-3.7498E-04	9.7671E-08	-1.2128E-11	298	1500
635	C6H15NO2	DIISOPROPANOLAMINE	-32.670	9.6179E-01	-8.5944E-04	4.1808E-07	-8.2801E-11	298	1200
636	C6H15NO3	TRIETHANOLAMINE	37.730	4.9155E-01	3.5952E-04	-8.2287E-07	3.5374E-10	298	1200
637	C6H15N3	N-AMINOETHYL PIPERAZINE	-18.325	7.9527E-01	-4.9153E-04	1.3652E-07	-1.2092E-11	298	1500
638	C6H15O4P	TRIETHYL PHOSPHATE	--------	--------	--------	--------	--------	--------	--------
639	C6H16N2	HEXAMETHYLENEDIAMINE	8.511	7.1350E-01	-4.7390E-04	2.1163E-07	-6.1825E-11	298	1200
640	C6H18N3OP	HEXAMETHYL PHOSPHORAMIDE	--------	--------	--------	--------	--------	--------	--------
641	C6H18N4	TRIETHYLENE TETRAMINE	-26.003	1.0161E+00	-8.1609E-04	3.6628E-07	-7.0398E-11	298	1500
642	C6H18OSi2	HEXAMETHYLDISILOXANE	96.759	5.3023E-01	-1.0701E-04	-1.0135E-07	4.3962E-11	200	1500
643	C6H18O3Si3	HEXAMETHYLCYCLOTRISILOXANE	42.372	7.1830E-01	-3.5419E-05	-3.1816E-07	1.2408E-10	100	1500
644	C6H19NSi2	HEXAMETHYLDISILAZANE	-10.219	8.5172E-01	-4.3504E-04	6.1843E-08	9.4823E-12	298	1500
645	C7H3ClF3NO2	4-CHLORO-3-NITROBENZOTRIFLUORIDE	2.462	8.0944E-01	-8.0238E-04	3.9826E-07	-7.8100E-11	298	1600
646	C7H3Cl2F3	2,4-DICHLOROBENZOTRIFLUORIDE	-18.588	8.4436E-01	-8.9784E-04	4.7892E-07	-1.0373E-10	298	1200
647	C7H3Cl2NO	3,4-DICHLOROPHENYL ISOCYANATE	-88.252	1.2181E+00	-1.8543E-03	1.3889E-06	-3.9144E-10	298	1200
648	C7H4ClF3	p-CHLOROBENZOTRIFLUORIDE	-32.088	8.0944E-01	-8.0238E-04	3.9826E-07	-7.8100E-11	298	1600
649	C7H4Cl2O	m-CHLOROBENZOYL CHLORIDE	3.438	5.7712E-01	-4.8258E-04	2.0062E-07	-3.3551E-11	298	1500
650	C7H4F3NO2	3-NITROBENZOTRIFLUORIDE	-16.178	8.0944E-01	-8.0238E-04	3.9826E-07	-7.8100E-11	298	1600
651	C7H5ClO	BENZOYL CHLORIDE	-8.923	5.4527E-01	-4.0749E-04	1.4722E-07	-2.0723E-11	298	1500
652	C7H5ClO2	o-CHLOROBENZOIC ACID	8.444	3.3784E-01	3.0050E-04	-6.8672E-07	3.0928E-10	298	1000
653	C7H5Cl3	BENZOTRICHLORIDE	-17.223	7.3674E-01	-7.4806E-04	4.2367E-07	-1.0609E-10	298	1500
654	C7H5F3	BENZOTRIFLUORIDE	-5.571	5.6388E-01	-3.3437E-04	5.2123E-08	1.1313E-11	200	1500
655	C7H5N	BENZONITRILE	-2.624	4.5843E-01	-2.6757E-04	4.3711E-08	7.3182E-12	200	1500
656	C7H5NO	PHENYL ISOCYANATE	-28.586	5.8118E-01	-4.2739E-04	1.4817E-07	-1.9292E-11	298	1500
657	C7H5N3O6	2,4,6-TRINITROTOLUENE	78.953	5.2187E-01	-2.9827E-04	6.1220E-08	1.2576E-12	200	1500
658	C7H6Cl2	BENZYL DICHLORIDE	-23.635	6.5729E-01	-5.5210E-04	2.3709E-07	-4.1699E-11	298	1500
659	C7H6Cl2	2,4-DICHLOROTOLUENE	37.983	2.1821E-01	7.2356E-04	-1.2976E-06	6.0155E-10	298	1000
660	C7H6N2O4	2,4-DINITROTOLUENE	44.603	5.2187E-01	-2.9827E-04	6.1220E-08	1.2576E-12	200	1500
661	C7H6N2O4	2,5-DINITROTOLUENE	44.603	5.2187E-01	-2.9827E-04	6.1220E-08	1.2576E-12	200	1500
662	C7H6N2O4	2,6-DINITROTOLUENE	44.603	5.2187E-01	-2.9827E-04	6.1220E-08	1.2576E-12	200	1500
663	C7H6N2O4	3,4-DINITROTOLUENE	44.603	5.2187E-01	-2.9827E-04	6.1220E-08	1.2576E-12	200	1500
664	C7H6N2O4	3,5-DINITROTOLUENE	44.603	5.2187E-01	-2.9827E-04	6.1220E-08	1.2576E-12	200	1500
665	C7H6O	BENZALDEHYDE	-0.890	4.4758E-01	-1.8566E-04	-3.6205E-08	3.1110E-11	200	1500
666	C7H6O2	BENZOIC ACID	16.158	2.8234E-01	1.7811E-04	-3.2176E-07	1.0752E-10	200	1500
667	C7H6O2	p-HYDROXYBENZALDEHYDE	-17.327	6.1293E-01	-3.6141E-04	4.2615E-08	1.8399E-11	298	1200
668	C7H6O2	SALICYLALDEHYDE	-33.117	7.0148E-01	-5.4741E-04	2.0112E-07	-2.8014E-11	298	1500
669	C7H6O3	SALICYLIC ACID	-4.230	4.2572E-01	1.4885E-04	-5.4965E-07	2.6161E-10	298	1000
670	C7H7Br	p-BROMOTOLUENE	-11.148	5.5111E-01	-3.6702E-04	1.0465E-07	-8.5052E-12	298	1500
671	C7H7Cl	BENZYL CHLORIDE	-20.027	5.0561E-01	-3.9247E-04	1.4825E-07	-2.1517E-11	200	1500
672	C7H7Cl	o-CHLOROTOLUENE	-19.597	5.4660E-01	-2.2561E-04	-1.3546E-07	1.0757E-10	298	1000
673	C7H7Cl	p-CHLOROTOLUENE	-19.597	5.4660E-01	-2.2561E-04	-1.3546E-07	1.0757E-10	298	1000
674	C7H7F	p-FLUOROTOLUENE	-33.607	6.2296E-01	-4.4225E-04	1.2414E-07	--------	298	1000
675	C7H7NO	FORMANILIDE	-58.019	7.4752E-01	-6.4061E-04	2.6844E-07	-4.4253E-11	298	1500
676	C7H7NO2	m-NITROTOLUENE	10.253	5.2187E-01	-2.9827E-04	6.1220E-08	1.2576E-12	200	1500
677	C7H7NO2	o-NITROTOLUENE	10.253	5.2187E-01	-2.9827E-04	6.1220E-08	1.2576E-12	200	1500
678	C7H7NO2	p-NITROTOLUENE	10.253	5.2187E-01	-2.9827E-04	6.1220E-08	1.2576E-12	200	1500
679	C7H7NO3	o-NITROANISOLE	-11.742	6.4927E-01	-4.3634E-04	1.1823E-07	-3.1868E-12	298	1200
680	C7H8	TOLUENE	-24.097	5.2187E-01	-2.9827E-04	6.1220E-08	1.2576E-12	200	1500
681	C7H8	1,3,5-CYCLOHEPTATRIENE	-42.702	6.8618E-01	-5.4743E-04	1.7426E-07	--------	298	1000
682	C7H8O	ANISOLE	-46.092	6.4927E-01	-4.3634E-04	1.1823E-07	-3.1868E-12	298	1200
683	C7H8O	BENZYL ALCOHOL	-38.244	5.7295E-01	-1.9618E-04	-1.7885E-07	1.2218E-10	298	1000
684	C7H8O	m-CRESOL	-28.118	6.6639E-01	-5.5684E-04	2.4989E-07	-4.7599E-11	200	1500
685	C7H8O	o-CRESOL	-10.489	5.7475E-01	-3.8585E-04	1.1736E-07	-1.1385E-11	200	1500
686	C7H8O	p-CRESOL	-21.163	6.2005E-01	-4.6259E-04	1.7288E-07	-2.5755E-11	200	1500
687	C7H8O2	GUAIACOL	-53.237	8.0557E-01	-7.1471E-04	3.1975E-07	-5.4471E-11	298	1200
688	C7H8O2	p-METHOXYPHENOL	-55.124	8.2278E-01	-7.6384E-04	3.7503E-07	-7.5815E-11	298	1500
689	C7H9N	BENZYLAMINE	-44.632	7.3792E-01	-6.4144E-04	3.0305E-07	-6.0296E-11	298	1500
690	C7H9N	2,6-DIMETHYLPYRIDINE	-12.959	5.3389E-01	-2.5211E-04	1.2256E-08	1.5481E-11	200	1500
691	C7H9N	N-METHYLANILINE	-63.167	8.2599E-01	-7.5180E-04	3.5552E-07	-6.8719E-11	298	1500
692	C7H9N	m-TOLUIDINE	-15.216	5.9535E-01	-3.8755E-04	1.1831E-07	-1.2703E-11	200	1500
693	C7H9N	o-TOLUIDINE	-15.391	6.2396E-01	-4.3867E-04	1.5449E-07	-2.1698E-11	200	1500
694	C7H9N	p-TOLUIDINE	-18.360	6.1016E-01	-4.1260E-04	1.3537E-07	-1.6747E-11	200	1500
695	C7H10	2-NORBORNENE	-75.556	7.5963E-01	-5.2049E-04	1.6449E-07	-1.7953E-11	298	1500
696	C7H10N2	TOLUENEDIAMINE	-75.447	1.0663E+00	-1.1548E-03	6.1040E-07	-1.1054E-10	298	1500
697	C7H11NO	CYCLOHEXYL ISOCYANATE	-20.869	5.1917E-01	4.6475E-05	-2.9175E-07	1.0386E-10	298	1500
698	C7H12	1-HEPTYNE	10.606	5.6170E-01	-3.2726E-04	7.6638E-08	--------	298	1000
699	C7H12O2	n-BUTYL ACRYLATE	31.423	6.2124E-01	-3.6990E-04	1.0373E-07	-1.7445E-11	298	1500
700	C7H12O2	ISOBUTYL ACRYLATE	-39.964	9.9799E-01	-1.0514E-03	5.9928E-07	-1.3516E-10	298	1500
701	C7H12O2	n-PROPYL METHACRYLATE	-43.005	8.7703E-01	-7.6204E-04	3.5635E-07	-6.9619E-11	298	1500
702	C7H12O4	DIETHYL MALONATE	-85.174	1.2290E+00	-1.3284E-03	7.4897E-07	-1.6575E-10	298	1500

41

Table 2-1 HEAT CAPACITY OF GAS - ORGANIC COMPOUNDS (continued)

NO	FORMULA	NAME	$C_P = A + B T + C T^2 + D T^3 + E T^4$					(C$_P$ - joule/(mol K), T - K)	
			A	B	C	D	E	TMIN	TMAX
703	C7H14	CYCLOHEPTANE	9.447	3.5261E-01	3.7699E-04	-5.3869E-07	1.6956E-10	100	1500
704	C7H14	1,1-DIMETYLCYCLOPENTANE	9.329	4.1173E-01	1.8706E-04	-3.7024E-07	1.2310E-10	200	1500
705	C7H14	cis-1,2-DIMETHYLCYCLOPENTANE	10.096	4.1584E-01	1.6872E-04	-3.5294E-07	1.1818E-10	200	1500
706	C7H14	trans-1,2-DIMETHYLCYCLOPENTANE	11.514	4.1203E-01	1.7044E-04	-3.5275E-07	1.1806E-10	200	1500
707	C7H14	cis-1,3-DIMETHYLCYCLOPENTANE	11.486	4.1227E-01	1.6993E-04	-3.5235E-07	1.1795E-10	200	1500
708	C7H14	trans-1,3-DIMETHYLCYCLOPENTANE	11.486	4.1227E-01	1.6993E-04	-3.5235E-07	1.1795E-10	200	1500
709	C7H14	ETHYLCYCLOPENTANE	-28.514	5.7607E-01	-9.4379E-05	-1.6445E-07	6.9435E-11	100	1500
710	C7H14	2-ETHYL-1-PENTENE	7.553	5.5924E-01	-1.9092E-04	-5.9654E-08	3.5819E-11	298	1500
711	C7H14	3-ETHYL-1-PENTENE	2.696	5.6924E-01	-1.9489E-04	-6.2627E-08	3.7450E-11	298	1500
712	C7H14	1-HEPTENE	40.754	3.9922E-01	5.3848E-05	-2.1599E-07	7.4482E-11	200	1500
713	C7H14	cis-2-HEPTENE	-7.099	5.9005E-01	-2.3765E-04	-2.1795E-08	2.4907E-11	200	1500
714	C7H14	trans-2-HEPTENE	22.872	4.6064E-01	-1.9919E-05	-1.7969E-07	6.6169E-11	150	1500
715	C7H14	cis-3-HEPTENE	-25.417	6.7669E-01	-3.7697E-04	7.3028E-08	1.3960E-12	200	1500
716	C7H14	trans-3-HEPTENE	1.557	5.6602E-01	-1.9660E-04	-5.5257E-08	3.4487E-11	150	1500
717	C7H14	METHYLCYCLOHEXANE	4.296	4.2716E-01	2.1058E-04	-3.9987E-07	1.3121E-10	200	1500
718	C7H14	2-METHYL-1-HEXENE	13.419	5.3960E-01	-1.7371E-04	-6.3050E-08	3.5189E-11	298	1500
719	C7H14	3-METHYL-1-HEXENE	2.696	5.6924E-01	-1.9489E-04	-6.2627E-08	3.7450E-11	298	1500
720	C7H14	4-METHYL-1-HEXENE	-21.774	7.0339E-01	-4.4459E-04	1.2769E-07	-1.3424E-11	150	1500
721	C7H14	2,3,3-TRIMETHYL-1-BUTENE	0.695	6.0287E-01	-2.3415E-04	-4.8958E-08	3.6689E-11	298	1500
722	C7H14O	DIISOPROPYL KETONE	-59.457	8.9163E-01	-7.3368E-04	3.3507E-07	-6.4643E-11	298	1500
723	C7H14O	2-HEPTANONE	77.352	2.5365E-01	3.8252E-04	-4.9648E-07	1.4916E-10	150	1500
724	C7H14O	1-HEPTANAL	76.088	2.6868E-01	3.5354E-04	-4.5649E-07	1.3757E-10	200	1500
725	C7H14O	1-METHYLCYCLOHEXANOL	-54.565	8.4577E-01	-5.7323E-04	2.3275E-07	-5.9702E-11	298	1200
726	C7H14O	cis-2-METHYLCYCLOHEXANOL	-47.283	7.6999E-01	-3.7628E-04	3.6819E-08	7.7738E-12	298	1200
727	C7H14O	trans-2-METHYLCYCLOHEXANOL	-47.283	7.6999E-01	-3.7628E-04	3.6819E-08	7.7738E-12	298	1200
728	C7H14O	cis-3-METHYLCYCLOHEXANOL	-47.283	7.6999E-01	-3.7628E-04	3.6819E-08	7.7738E-12	298	1200
729	C7H14O	trans-3-METHYLCYCLOHEXANOL	-47.283	7.6999E-01	-3.7628E-04	3.6819E-08	7.7738E-12	298	1200
730	C7H14O	cis-4-METHYLCYCLOHEXANOL	-47.283	7.6999E-01	-3.7628E-04	3.6819E-08	7.7738E-12	298	1200
731	C7H14O	trans-4-METHYLCYCLOHEXANOL	-47.283	7.6999E-01	-3.7628E-04	3.6819E-08	7.7738E-12	298	1200
732	C7H14O	5-METHYL-2-HEXANONE	-3.763	7.0117E-01	-4.7468E-04	1.8106E-07	-3.3882E-11	298	1500
733	C7H14O2	n-BUTYL PROPIONATE	-3.593	7.7324E-01	-5.9004E-04	2.9563E-07	-8.6992E-11	298	1500
734	C7H14O2	ETHYL ISOVALERATE	-45.699	9.9698E-01	-9.7130E-04	5.3585E-07	-1.2593E-10	298	1200
735	C7H14O2	ISOPENTYL ACETATE	-51.790	9.8897E-01	-9.0976E-04	4.6383E-07	-1.0053E-10	298	1500
736	C7H14O2	n-PENTYL ACETATE	-31.967	8.9188E-01	-7.5408E-04	3.6684E-07	-8.2457E-11	298	1200
737	C7H14O2	n-PROPYL n-BUTYRATE	-8.783	8.0270E-01	-6.4913E-04	3.4629E-07	-1.0270E-10	298	1200
738	C7H14O2	n-HEPTANOIC ACID	11.779	6.9196E-01	-4.1039E-04	1.1266E-07	-1.5295E-11	298	1500
739	C7H14O3	ETHYL-3-ETHOXYPROPIONATE	-62.701	1.1653E+00	-1.2886E-03	8.0077E-07	-2.0569E-10	298	1200
740	C7H15Br	1-BROMOHEPTANE	-11.271	7.7970E-01	-5.4848E-04	1.9542E-07	-2.5440E-11	298	1200
741	C7H15N	N-METHYLCYCLOHEXYLAMINE	-59.516	8.3575E-01	-4.3148E-04	4.6928E-08	1.6571E-11	298	1500
742	C7H16	2,2-DIMETHYLPENTANE	-19.277	7.6888E-01	-5.1678E-04	2.0476E-07	-3.7168E-11	200	1500
743	C7H16	2,3-DIMETHYLPENTANE	38.654	3.7259E-01	2.7632E-04	-4.3148E-07	1.3811E-10	100	1500
744	C7H16	2,4-DIMETHYLPENTANE	-32.996	8.7352E-01	-7.3286E-04	3.6633E-07	-7.9787E-11	200	1500
745	C7H16	3,3-DIMETHYLPENTANE	-23.909	7.8329E-01	-5.3992E-04	2.2604E-07	-4.4324E-11	200	1500
746	C7H16	3-ETHYLPENTANE	19.245	5.5072E-01	-1.4055E-04	-8.2482E-08	3.9487E-11	200	1500
747	C7H16	n-HEPTANE	26.984	5.0387E-01	-4.4748E-05	-1.6835E-07	6.5183E-11	200	1500
748	C7H16	2-METHYLHEXANE	-3.249	6.6625E-01	-3.3836E-04	6.0489E-08	2.5385E-12	200	1500
749	C7H16	3-METHYLHEXANE	-12.841	7.1358E-01	-4.2021E-04	1.1997E-07	-1.2906E-11	200	1500
750	C7H16	2,2,3-TRIMETHYLBUTANE	-21.150	7.4663E-01	-4.6699E-04	1.7923E-07	-3.4265E-11	200	1500
751	C7H16O	1-HEPTANOL	11.810	6.4236E-01	-2.4939E-04	-4.3649E-08	3.7035E-11	200	1500
752	C7H16O	2-HEPTANOL	-28.784	8.8407E-01	-7.1442E-04	3.2639E-07	-6.5218E-11	298	1500
753	C7H16O	5-METHYL-1-HEXANOL	0.274	6.8732E-01	-3.1154E-04	-1.1838E-08	3.5683E-11	298	1200
754	C7H16O	ISOPROPYL-TERT-BUTYL-ETHER	20.789	6.1547E-01	-2.6674E-04	3.0050E-08	--------	298	1000
755	C7H16S	n-HEPTYL MERCAPTAN	59.314	4.3814E-01	8.8964E-05	-2.6174E-07	8.7398E-11	200	1500
756	C7H16S	BUTYL-PROPYL-SULFIDE	15.543	6.4304E-01	-2.7440E-04	3.9835E-08	--------	298	1000
757	C7H16S	ETHYL-PENTYL-SULFIDE	13.355	6.6304E-01	-3.1221E-04	5.4413E-08		298	1000
758	C7H16S	HEXYL-METHYL-SULFIDE	16.143	6.6003E-01	-3.1560E-04	5.6630E-08	--------	298	1000
759	C7H17N	1-AMINOHEPTANE	-1.435	7.4857E-01	-4.4543E-04	1.3200E-07	-1.8684E-11	298	1500
760	C8H4Cl2O2	ISOPHTHALOYL CHLORIDE	32.762	5.2222E-01	-3.3597E-04	9.0843E-08	-9.2349E-12	298	1200
761	C8H4O3	PHTHALIC ANHYDRIDE	40.083	3.6084E-02	9.5956E-04	-1.2341E-06	4.6597E-10	298	1000
762	C8H6	ETHYNYLBENZENE	-38.345	6.5890E-01	-5.3689E-04	1.7236E-07	--------	298	1000
763	C8H6O4	ISOPHTHALIC ACID	-48.374	6.9316E-01	-4.1468E-04	7.2676E-08	8.1593E-12	298	1500
764	C8H6O4	PHTHALIC ACID	-48.374	6.9316E-01	-4.1468E-04	7.2676E-08	8.1593E-12	298	1500
765	C8H6O4	TEREPHTHALIC ACID	-48.374	6.9316E-01	-4.1468E-04	7.2676E-08	8.1593E-12	298	1500
766	C8H6S	BENZOTHIOPHENE	-47.900	8.1457E-01	-8.0881E-04	4.1421E-07	-8.5298E-11	298	1500
767	C8H7N	INDOLE	-46.384	7.0422E-01	-5.3462E-04	1.9548E-07	-2.7036E-11	200	1500
768	C8H8	STYRENE	71.201	5.4767E-02	6.4793E-04	-6.9875E-07	2.1232E-10	100	1500
769	C8H8	1,3,5,7-CYCLOOCTATETRAENE	-41.686	8.4677E-01	-4.9986E-04	1.4494E-07	--------	298	1000
770	C8H8O	ACETOPHENONE	-44.384	6.3161E-01	-3.8644E-04	9.1755E-08	-2.9424E-12	298	1500
771	C8H8O	p-TOLUALDEHYDE	-34.314	6.9270E-01	-5.3448E-04	1.9482E-07	-1.6767E-11	298	1200
772	C8H8O2	METHYL BENZOATE	-44.363	6.6062E-01	-3.0730E-04	-7.6698E-08	7.8555E-11	298	1200
773	C8H8O2	o-TOLUIC ACID	5.831	3.2470E-01	4.8726E-04	-8.7808E-07	3.7290E-10	298	1000
774	C8H8O2	p-TOLUIC ACID	5.831	3.2470E-01	4.8726E-04	-8.7808E-07	3.7290E-10	298	1000
775	C8H8O3	METHYL SALICYLATE	-64.438	9.2054E-01	-8.6046E-04	4.1813E-07	-8.1500E-11	298	1500
776	C8H8O3	VANILLIN	-24.141	7.6542E-01	-4.9618E-04	1.3823E-07	-1.9509E-11	298	1200
777	C8H9NO	ACETANILIDE	-52.405	7.8304E-01	-5.9239E-04	2.3324E-07	-4.1725E-11	298	1500
778	C8H10	ETHYLBENZENE	-20.527	5.9578E-01	-3.0849E-04	3.5621E-08	1.2409E-11	200	1500
779	C8H10	m-XYLENE	-16.725	5.6424E-01	-2.6465E-04	1.3381E-08	1.5869E-11	200	1500
780	C8H10	o-XYLENE	0.182	5.1344E-01	-2.0212E-04	-2.1615E-08	2.3212E-11	200	1500

42

Table 2-1 HEAT CAPACITY OF GAS - ORGANIC COMPOUNDS (continued)

			$C_P = A + B T + C T^2 + D T^3 + E T^4$				$(C_P$ - joule/(mol K), T - K)		
NO	FORMULA	NAME	A	B	C	D	E	TMIN	TMAX
781	C8H10	p-XYLENE	-17.360	5.6470E-01	-2.6293E-04	1.1217E-08	1.6544E-11	200	1500
782	C8H10O	m-ETHYLPHENOL	17.179	5.1321E-01	-1.5178E-04	-9.7687E-08	5.1382E-11	200	1500
783	C8H10O	p-ETHYLPHENOL	17.179	5.1321E-01	-1.5178E-04	-9.7687E-08	5.1382E-11	200	1500
784	C8H10O	PHENETOLE	-70.622	8.7227E-01	-7.3933E-04	3.3808E-07	-6.7682E-11	298	1500
785	C8H10O	2-PHENYLETHANOL	-41.581	7.7144E-01	-6.0352E-04	2.4945E-07	-4.5938E-11	298	1500
786	C8H10O	2,3-XYLENOL	5.637	6.6110E-01	-4.6084E-04	1.6135E-07	-2.2812E-11	200	1500
787	C8H10O	2,4-XYLENOL	-6.981	6.8005E-01	-4.6398E-04	1.5240E-07	-1.8683E-11	200	1500
788	C8H10O	2,5-XYLENOL	-0.097	6.4699E-01	-4.0504E-04	1.0860E-07	-7.1789E-12	200	1500
789	C8H10O	2,6-XYLENOL	-3.351	6.5837E-01	-4.2958E-04	1.3046E-07	-1.3721E-11	200	1500
790	C8H10O	3,4-XYLENOL	-5.010	7.1824E-01	-5.6191E-04	2.3462E-07	-4.1599E-11	200	1500
791	C8H10O	3,5-XYLENOL	-14.573	6.9845E-01	-4.8247E-04	1.6146E-07	-2.0501E-11	200	1500
792	C8H11N	N,N-DIMETHYLANILINE	-80.587	9.3415E-01	-7.6870E-04	3.2262E-07	-5.7595E-11	298	1500
793	C8H11N	o-ETHYLANILINE	-44.119	8.6260E-01	-7.5428E-04	3.5713E-07	-7.2513E-11	298	1500
794	C8H11N	2,4,6-TRIMETHYLPYRIDINE	-44.959	7.5625E-01	-4.7616E-04	1.2070E-07	-5.5157E-12	298	1500
795	C8H11NO	p-PHENETIDINE	-61.255	9.5389E-01	-8.2789E-04	3.7816E-07	-7.4127E-11	298	1500
796	C8H12	1,5-CYCLOOCTADIENE	4.987	4.2829E-01	1.9618E-04	-4.0302E-07	1.3486E-10	100	1500
797	C8H12	VINYLCYCLOHEXENE	-85.630	9.7156E-01	-8.6793E-04	4.1915E-07	-8.4485E-11	298	1500
798	C8H12O4	1,4-CYCLOHEXANEDICARBOXYLIC ACID	-95.482	1.1417E+00	-9.0210E-04	3.5790E-07	-6.0891E-11	298	1200
799	C8H12O4	DIETHYL MALEATE	-89.669	1.5506E+00	-2.0466E-03	1.3800E-06	-3.5435E-10	298	1200
800	C8H14O2	n-BUTYL METHACRYLATE	-51.893	1.0235E+00	-9.3080E-04	4.7360E-07	-1.0396E-10	298	1200
801	C8H14O3	BUTYRIC ANHYDRIDE	29.745	7.9376E-01	-4.1894E-04	8.7715E-08	-5.9476E-12	298	1200
802	C8H14O4	DIETHYL SUCCINATE	-50.188	1.2234E+00	-1.2888E-03	7.3303E-07	-1.6796E-10	298	1500
803	C8H16	CYCLOOCTANE	-96.998	9.5483E-01	-5.7940E-04	1.3566E-07	--------	298	1000
804	C8H16	1,1-DIMETHYLCYCLOHEXANE	12.609	4.3616E-01	3.5994E-04	-5.5592E-07	1.7723E-10	200	1500
805	C8H16	cis-1,2-DIMETHYLCYCLOHEXANE	12.356	4.6185E-01	2.8460E-04	-4.9224E-07	1.6009E-10	200	1500
806	C8H16	trans-1,2-DIMETHYLCYCLOHEXANE	5.195	5.1990E-01	1.7366E-04	-4.0995E-07	1.3877E-10	200	1500
807	C8H16	cis-1,3-DIMETHYLCYCLOHEXANE	16.162	4.4120E-01	3.2552E-04	-5.2079E-07	1.6680E-10	200	1500
808	C8H16	trans-1,3-DIMETHYLCYCLOHEXANE	17.334	4.4313E-01	3.0406E-04	-5.0089E-07	1.6161E-10	200	1500
809	C8H16	cis-1,4-DIMETHYLCYCLOHEXANE	17.334	4.4313E-01	3.0406E-04	-5.0089E-07	1.6161E-10	200	1500
810	C8H16	trans-1,4-DIMETHYLCYCLOHEXANE	7.358	4.9586E-01	2.2944E-04	-4.5522E-07	1.5104E-10	200	1500
811	C8H16	ETHYLCYCLOHEXANE	9.711	4.9505E-01	2.0606E-04	-4.2645E-07	1.4166E-10	200	1500
812	C8H16	2-ETHYL-1-HEXENE	-24.228	8.3947E-01	-6.1664E-04	2.4854E-07	-4.3134E-11	298	1500
813	C8H16	1-METHYL-1-ETHYLCYCLOPENTANE	-69.713	9.2602E-01	-6.2526E-04	2.0490E-07	-2.4889E-11	298	1500
814	C8H16	1-OCTENE	56.266	4.0665E-01	1.5805E-04	-3.2277E-07	1.0600E-10	200	1500
815	C8H16	trans-2-OCTENE	6.815	6.4056E-01	-2.3322E-04	-4.6067E-08	3.3079E-11	200	1500
816	C8H16	trans-3-OCTENE	-5.979	6.9800E-01	-3.1979E-04	9.4308E-09	1.9956E-11	200	1500
817	C8H16	trans-4-OCTENE	-5.979	6.9800E-01	-3.1979E-04	9.4308E-09	1.9956E-11	200	1500
818	C8H16	n-PROPYLCYCLOPENTANE	-51.866	7.8827E-01	-3.5255E-04	-6.8545E-09	3.1314E-11	100	1500
819	C8H16	2,4,4-TRIMETHYL-1-PENTENE	-28.020	8.6749E-01	-6.3580E-04	2.4452E-07	-3.9076E-11	298	1500
820	C8H16	2,4,4-TRIMETHYL-2-PENTENE	-23.378	8.4303E-01	-6.1037E-04	2.3775E-07	-3.9769E-11	298	1500
821	C8H16O	2-ETHYLHEXANAL	-30.667	8.9033E-01	-6.5620E-04	2.6358E-07	-4.4888E-11	298	1500
822	C8H16O	1-OCTANAL	85.407	3.0880E-01	4.0878E-04	-5.4231E-07	1.6932E-10	200	1500
823	C8H16O	2-OCTANONE	81.259	3.2148E-01	3.8818E-04	-5.4646E-07	1.7288E-10	150	1500
824	C8H16O2	n-BUTYL n-BUTYRATE	14.674	7.8762E-01	-3.9225E-04	4.2787E-08	1.5217E-11	298	1200
825	C8H16O2	n-HEXYL ACETATE	-46.949	1.0667E+00	-9.5577E-04	4.7655E-07	-9.8701E-11	298	1500
826	C8H16O2	ISOBUTYL ISOBUTYRATE	-78.721	1.1911E+00	-1.1320E-03	5.8376E-07	-1.2228E-10	298	1500
827	C8H16O2	n-OCTANOIC ACID	-20.242	9.9093E-01	-9.2370E-04	5.5027E-07	-1.5434E-10	298	1000
828	C8H16O4	DIETHYLENE GLYCOL ETHYL ETHER ACETATE	-51.637	1.1924E+00	-1.0849E-03	5.3285E-07	-1.0908E-10	298	1500
829	C8H18	2,2-DIMETHYLHEXANE	-9.102	7.9198E-01	-4.3438E-04	1.1094E-07	-9.1975E-12	200	1500
830	C8H18	2,3-DIMETHYLHEXANE	-45.246	9.7125E-01	-7.6224E-04	3.4891E-07	-7.0688E-11	200	1500
831	C8H18	2,4-DIMETHYLHEXANE	-35.844	9.7252E-01	-7.9337E-04	3.8126E-07	-8.0700E-11	200	1500
832	C8H18	2,5-DIMETHYLHEXANE	-32.363	9.0722E-01	-6.5105E-04	2.6842E-07	-4.9960E-11	200	1500
833	C8H18	3,3-DIMETHYLHEXANE	-37.306	9.6180E-01	-7.3386E-04	3.2761E-07	-6.5298E-11	200	1500
834	C8H18	3,4-DIMETHYLHEXANE	-59.720	1.0415E+00	-8.8492E-04	4.3894E-07	-9.4154E-11	200	1500
835	C8H18	3-ETHYLHEXANE	16.214	6.6412E-01	-2.2656E-04	-5.1790E-08	3.5344E-11	200	1500
836	C8H18	3-ETHYL-2-METHYLPENTANE	-7.477	7.7747E-01	-4.2844E-04	9.1763E-08	--------	298	1000
837	C8H18	3-METHYL-3-ETHYLPENTANE	12.976	6.4949E-01	-1.3948E-04	-1.2724E-07	5.6504E-11	150	1500
838	C8H18	2-METHYLHEPTANE	-3.367	7.5824E-01	-3.8216E-04	5.7358E-08	8.0178E-12	200	1500
839	C8H18	3-METHYLHEPTANE	-10.106	7.8711E-01	-4.3415E-04	9.7487E-08	-2.9023E-12	200	1500
840	C8H18	4-METHYLHEPTANE	-17.581	8.3526E-01	-5.1418E-04	1.5104E-07	-1.5855E-11	200	1500
841	C8H18	n-OCTANE	29.053	5.8016E-01	-5.7103E-05	-1.9548E-07	7.6614E-11	200	1500
842	C8H18	2,2,3-TRIMETHYLPENTANE	-48.847	9.9777E-01	-8.0693E-04	3.9894E-07	-8.7211E-11	200	1500
843	C8H18	2,2,4-TRIMETHYLPENTANE	-21.703	8.5849E-01	-5.5323E-04	2.0892E-07	-3.7285E-11	200	1500
844	C8H18	2,3,3-TRIMETHYLPENTANE	-51.678	1.0143E+00	-8.2601E-04	4.0855E-07	-8.9531E-11	200	1500
845	C8H18	2,3,4-TRIMETHYLPENTANE	-53.615	1.0646E+00	-9.6734E-04	5.2516E-07	-1.2149E-10	200	1500
846	C8H18	2,2,3,3-TETRAMETHYLBUTANE	-44.660	9.7893E-01	-6.7032E-04	1.8192E-07	--------	298	1000
847	C8H18O	DI-n-BUTYL ETHER	86.995	3.4274E-01	3.7493E-04	-5.7832E-07	1.8911E-10	200	1500
848	C8H18O	DI-sec-BUTYL ETHER	-50.577	1.1014E+00	-9.9402E-04	5.1326E-07	-1.1304E-10	298	1200
849	C8H18O	DI-tert-BUTYL ETHER	122.563	1.7324E-01	6.4796E-04	-6.9227E-07	2.0237E-10	200	1500
850	C8H18O	2-ETHYL-1-HEXANOL	-8.577	8.3851E-01	-5.0979E-04	1.4511E-07	-1.4766E-11	298	1500
851	C8H18O	1-OCTANOL	13.196	7.2493E-01	-2.7623E-04	-5.8380E-08	4.5179E-11	200	1500
852	C8H18O	2-OCTANOL	-23.930	9.4474E-01	-7.0058E-04	2.8277E-07	-4.9163E-11	298	1500
853	C8H18O2	DI-t-BUTYL PEROXIDE	-5.907	9.4614E-01	-6.7505E-04	2.4895E-07	-3.9188E-11	298	1500
854	C8H18O2S	DI-n-BUTYL SULFONE	6.590	9.1442E-01	-5.5866E-04	1.7288E-07	-1.9647E-11	298	1500
855	C8H18O3	DIETHYLENE GLYCOL DIETHYL ETHER	-31.678	1.0594E+00	-8.1318E-04	3.3723E-07	-6.1289E-11	298	1500
856	C8H18O3	DIETHYLENE GLYCOL MONOBUTYL EHTER	-15.996	9.8794E-01	-6.9921E-04	2.5671E-07	-4.0681E-11	298	1500
857	C8H18O4	TRIETHYLENE GLYCOL DIMETHYL ETHER	-15.847	1.0394E+00	-7.4905E-04	2.8115E-07	-4.5915E-11	298	1500
858	C8H18O5	TETRAETHYLENE GLYCOL	7.854	1.0016E+00	-7.0468E-04	2.6794E-07	-4.8409E-11	298	1200

Table 2-1 HEAT CAPACITY OF GAS - ORGANIC COMPOUNDS (continued)

$$C_P = A + B\,T + C\,T^2 + D\,T^3 + E\,T^4 \qquad (C_P - joule/(mol\ K),\ T - K)$$

NO	FORMULA	NAME	A	B	C	D	E	TMIN	TMAX
859	C8H18S	n-OCTYL MERCAPTAN	63.070	4.9322E-01	1.4077E-04	-3.4436E-07	1.1448E-10	200	1500
860	C8H18S	tert-OCTYL MERCAPTAN	-27.449	1.0002E+00	-7.2696E-04	2.4916E-07	-3.1819E-11	298	1500
861	C8H18S	BUTYL-SULFIDE	14.557	7.3839E-01	-3.2937E-04	5.2028E-08	--------	298	1000
862	C8H18S	ETHYL-HEXYL-SULFIDE	12.005	7.6027E-01	-3.7040E-04	6.8300E-08	--------	298	1000
863	C8H18S	HEPTYL-METHYL-SULFIDE	14.798	7.5722E-01	-3.7374E-04	7.0492E-08	--------	298	1000
864	C8H18S	PENTYL-PROPYL-SULFIDE	14.557	7.3839E-01	-3.2937E-04	5.2028E-08	--------	298	1000
865	C8H18S2	BUTYL-DISULFIDE	20.391	8.3760E-01	-4.7208E-04	1.0713E-07	--------	298	1000
866	C8H19N	DI-n-BUTYLAMINE	-36.248	1.0025E+00	-7.6232E-04	3.2620E-07	-6.0071E-11	298	1500
867	C8H19N	DIISOBUTYLAMINE	-37.761	9.8864E-01	-6.9887E-04	2.6003E-07	-4.1045E-11	298	1500
868	C8H19N	n-OCTYLAMINE	-11.736	9.0057E-01	-6.1188E-04	2.2599E-07	-3.5786E-11	298	1500
869	C8H23N5	TETRAETHYLENEPENTAMINE	-49.208	1.3828E+00	-1.1462E-03	5.2201E-07	-1.0064E-10	298	1500
870	C8H24O4Si4	OCTAMETHYLCYCLOTETRASILOXANE	-10.762	1.3722E+00	-8.6662E-04	2.2799E-07	-1.4708E-11	298	1500
871	C9H4O5	TRIMELLITIC ANHYDRIDE	-13.209	5.6342E-01	1.4947E-05	-4.4215E-07	2.1578E-10	298	1500
872	C9H6N2O2	TOLUENE DIISOCYANATE	-5.078	5.7582E-01	2.9402E-04	-9.0683E-07	4.3055E-10	298	1000
873	C9H7N	ISOQUINOLINE	-18.255	5.8698E-01	-2.5163E-04	-4.1653E-08	3.9647E-11	200	1500
874	C9H7N	QUINOLINE	-21.558	6.0838E-01	-2.9061E-04	-1.2260E-08	3.1722E-11	200	1500
875	C9H7NO	8-HYDROXYQUINOLINE	-62.510	9.7219E-01	-9.3439E-04	4.4959E-07	-8.6601E-11	298	1500
876	C9H8	INDENE	-31.858	6.2986E-01	-3.6332E-04	6.8279E-08	4.9422E-12	200	1500
877	C9H8O	2-METHYLBENZOFURAN	-59.800	8.8351E-01	-8.0230E-04	3.7939E-07	-7.2950E-11	298	1500
878	C9H10	INDANE	-132.570	1.1568E+00	-1.2134E-03	6.6759E-07	-1.4627E-10	298	1500
879	C9H10	cis-PROPENYLBENZENE	-24.332	6.9287E-01	-4.5279E-04	1.1806E-07	--------	298	1000
880	C9H10	trans-PROPENYLBENZENE	-29.349	7.2308E-01	-4.9300E-04	1.3395E-07	--------	298	1000
881	C9H1O	alpha-METHYLSTYRENE	24.923	4.3818E-01	-1.3764E-05	-1.8472E-07	7.0450E-11	200	1500
882	C9H10	m-METHYLSTYRENE	25.882	4.3319E-01	-5.1049E-06	-1.9087E-07	7.1985E-11	200	1500
883	C9H10	o-METHYLSTYRENE	25.882	4.3319E-01	-5.1049E-06	-1.9087E-07	7.1985E-11	200	1500
884	C9H10	p-METHYLSTYRENE	25.882	4.3319E-01	-5.1049E-06	-1.9087E-07	7.1985E-11	200	1500
885	C9H10O2	BENZYL ACETATE	-64.296	9.7745E-01	-8.0623E-04	2.7233E-07	-1.0633E-12	298	1200
886	C9H10O2	ETHYL BENZOATE	-91.208	1.0326E+00	-9.5250E-04	4.6492E-07	-9.0498E-11	298	1500
887	C9H10O3	ETHYL VANILLIN	-34.060	9.0152E-01	-6.1072E-04	1.8656E-07	-2.7428E-11	298	1200
888	C9H11NO	p-DIMETHYLAMINOBENZALDEHYDE	-65.931	9.6897E-01	-6.6584E-04	1.8803E-07	-1.2951E-11	298	1500
889	C9H12	CUMENE	10.149	5.1138E-01	-1.7703E-05	-2.2612E-07	8.8002E-11	200	1500
890	C9H12	m-ETHYLTOLUENE	-7.269	5.9696E-01	-2.0235E-04	-6.2005E-08	3.8692E-11	200	1500
891	C9H12	o-ETHYLTOLUENE	8.688	5.5216E-01	-1.5746E-04	-7.9142E-08	4.0458E-11	200	1500
892	C9H12	p-ETHYLTOLUENE	-13.082	6.3055E-01	-2.5570E-04	-2.7145E-08	3.0472E-11	200	1500
893	C9H12	MESITYLENE	-8.737	6.0394E-01	-2.3025E-04	-3.1411E-08	2.8748E-11	200	1500
894	C9H12	n-PROPYLBENZENE	-10.933	6.4349E-01	-2.7829E-04	-1.4431E-08	2.8143E-11	200	1500
895	C9H12	1,2,3-TRIMETHYLBENZENE	12.802	5.5862E-01	-1.9088E-04	-4.5260E-08	2.9986E-11	200	1500
896	C9H12	1,2,4-TRIMETHYLBENZENE	7.238	5.5929E-01	-1.7851E-04	-5.9043E-08	3.4337E-11	200	1500
897	C9H12O	BENZYL ETHYL ETHER	-77.310	9.9168E-01	-8.2953E-04	3.7784E-07	-7.6033E-11	298	1500
898	C9H12O	2-PHENYL-2-PROPANOL	-42.729	8.6797E-01	-6.3040E-04	2.1037E-07	-2.1293E-11	298	1200
899	C9H12O2	CUMENE HYDROPEROXIDE	-42.988	9.9727E-01	-9.1932E-04	4.9051E-07	-1.1852E-10	298	1200
900	C9H14O	ISOPHORONE	-74.479	1.1286E+00	-1.0642E-03	5.8129E-07	-1.4417E-10	298	1200
901	C9H14O6	GLYCERYL TRIACETATE	-94.112	1.5807E+00	-1.7308E-03	9.5867E-07	-2.0165E-10	298	1200
902	C9H16	1-NONYNE	7.735	7.5747E-01	-4.4627E-04	1.0602E-07	--------	298	1000
903	C9H16O4	AZELAIC ACID	-10.832	1.1553E+00	-1.1558E-03	7.2207E-07	-2.0666E-10	298	1200
904	C9H18	BUTYLCYCLOPENTANE	-116.725	1.3097E+00	-1.2439E-03	5.2915E-07	--------	298	1000
905	C9H18	cis,cis-1,3,5-TRIMETHYLCYCLOHEXANE	-59.731	9.2906E-01	-4.4221E-04	5.4375E-08	--------	298	1000
906	C9H18	cis,trans-1,3,5-TRIMETHYLCYCLOHEXANE	-61.237	9.4713E-01	-4.9400E-04	8.2170E-08	--------	298	1000
907	C9H18	ISOPROPYLCYCLOHEXANE	-72.808	1.0165E+00	-6.2550E-04	1.7248E-07	-1.4751E-11	298	1500
908	C9H18	1-NONENE	55.655	4.9879E-01	1.0927E-04	-3.1474E-07	1.0691E-10	200	1500
909	C9H18	n-PROPYLCYCLOHEXANE	25.567	5.2828E-01	2.3719E-04	-4.6758E-07	1.5413E-10	200	1500
910	C9H18O	DIISOBUTYL KETONE	-1.433	8.4406E-01	-5.7459E-04	1.9432E-07	-2.8451E-11	298	1500
911	C9H18O	1-NONANAL	84.758	3.9995E-01	3.6567E-04	-5.3736E-07	1.6751E-10	200	1500
912	C9H18O2	n-BUTYL VALERATE	-43.663	1.1626E+00	-1.0255E-03	5.0302E-07	-1.0407E-10	298	1500
913	C9H18O2	n-NONANOIC ACID	16.182	8.4306E-01	-4.3510E-04	6.1835E-08	7.4964E-12	298	1200
914	C9H18O2	n-OCTYL FORMATE	-38.335	1.1242E+00	-9.4618E-04	4.4200E-07	-8.6768E-11	298	1500
915	C9H20	3,3-DIETHYLPENTANE	37.793	6.5813E-01	-9.9686E-05	-1.5751E-07	6.2352E-11	150	1500
916	C9H20	2,2-DIMETHYL-3-ETHYLPENTANE	24.550	6.2909E-01	3.7947E-05	-2.9127E-07	1.0305E-10	150	1500
917	C9H20	3-ETHYL-2,3-DIMETHYLPENTANE	-58.911	1.0983E+00	-7.6517E-04	2.1671E-07	--------	298	1000
918	C9H20	2,4-DIMETHYL-3-ETHYLPENTANE	6.652	7.6279E-01	-2.3975E-04	-7.9564E-08	4.7011E-11	150	1500
919	C9H20	2,2-DIMETHYLHEPTANE	43.712	5.7185E-01	1.2970E-04	-3.6900E-07	1.2607E-10	150	1500
920	C9H20	2,6-DIMETHYLHEPTANE	15.557	7.2025E-01	-1.4814E-04	-1.6638E-07	7.3518E-11	150	1500
921	C9H20	3-ETHYLHEPTANE	68.581	4.4754E-01	3.1908E-04	-5.1182E-07	1.6482E-10	150	1500
922	C9H20	4-ETHYLHEPTANE	-27.273	9.4692E-01	-5.7145E-04	1.3701E-07	--------	298	1000
923	C9H20	2,3-DIMETHYLHEPTANE	-21.292	9.3010E-01	-5.5350E-04	1.3021E-07	--------	298	1000
924	C9H20	2,4-DIMETHYLHEPTANE	-35.993	9.7910E-01	-6.0785E-04	1.5039E-07	--------	298	1000
925	C9H20	2,5-DIMETHYLHEPTANE	-35.993	9.7910E-01	-6.0785E-04	1.5039E-07	--------	298	1000
926	C9H20	3,4-DIMETHYLHEPTANE	-36.744	9.8328E-01	-6.1576E-04	1.5475E-07	--------	298	1000
927	C9H20	3,5-DIMETHYLHEPTANE	-51.426	1.0322E+00	-6.6990E-04	1.7483E-07	--------	298	1000
928	C9H20	4,4-DIMETHYLHEPTANE	-34.003	1.0088E+00	-6.5873E-04	1.7450E-07	--------	298	1000
929	C9H20	3-ETHYL-2-METHYLHEXANE	-36.744	9.8328E-01	-6.1576E-04	1.5475E-07	--------	298	1000
930	C9H20	4-ETHYL-2-METHYLHEXANE	-51.426	1.0322E+00	-6.6990E-04	1.7483E-07	--------	298	1000
931	C9H20	3-ETHYL-3-METHYLHEXANE	-49.447	1.0620E+00	-7.2095E-04	1.9902E-07	--------	298	1000
932	C9H20	3-ETHYL-4-METHYLHEXANE	-52.162	1.0363E+00	-6.7772E-04	1.7914E-07	--------	298	1000
933	C9H20	2,2,3-TRIMETHYLHEXANE	-43.468	1.0452E+00	-7.0300E-04	1.9221E-07	--------	298	1000
934	C9H20	2,2,4-TRIMETHYLHEXANE	-58.158	1.0941E+00	-7.5722E-04	2.1235E-07	--------	298	1000
935	C9H20	2,3,3-TRIMETHYLHEXANE	-43.468	1.0452E+00	-7.0300E-04	1.9221E-07	--------	298	1000
936	C9H20	2,3,4-TRIMETHYLHEXANE	-46.166	1.0193E+00	-6.5965E-04	1.7226E-07	--------	298	1000

44

Table 2-1 HEAT CAPACITY OF GAS - ORGANIC COMPOUNDS (continued)

NO	FORMULA	NAME	$C_P = A + B\,T + C\,T^2 + D\,T^3 + E\,T^4$					(C$_P$ - joule/(mol K), T - K)	
			A	B	C	D	E	TMIN	TMAX
937	C9H20	2,3,5-TRIMETHYLHEXANE	-45.422	1.0152E+00	-6.5174E-04	1.6792E-07	--------	298	1000
938	C9H20	2,4,4-TRIMETHYLHEXANE	-58.158	1.0941E+00	-7.5722E-04	2.1235E-07	--------	298	1000
939	C9H20	3,3,4-TRIMETHYLHEXANE	-58.911	1.0983E+00	-7.6517E-04	2.1671E-07	--------	298	1000
940	C9H20	2-METHYLOCTANE	51.299	5.3562E-01	1.6961E-04	-4.0233E-07	1.3567E-10	150	1500
941	C9H20	3-METHYLOCTANE	43.182	5.7408E-01	1.0042E-04	-3.4988E-07	1.2161E-10	150	1500
942	C9H20	4-METHYLOCTANE	37.821	6.1030E-01	3.4470E-05	-3.0208E-07	1.0936E-10	150	1500
943	C9H20	n-NONANE	29.687	6.6821E-01	-9.6492E-05	-2.0014E-07	8.2200E-11	200	1500
944	C9H20	2,2,3,3-TETRAMETHYLPENTANE	-56.144	1.1449E+00	-9.5333E-04	4.9493E-07	-1.1283E-10	200	1500
945	C9H20	2,2,3,4-TETRAMETHYLPENTANE	-6.361	8.2329E-01	-3.2768E-04	-3.3527E-09	2.4144E-11	150	1500
946	C9H20	2,2,4,4-TETRAMETHYLPENTANE	45.985	5.7438E-01	1.1062E-04	-3.1707E-07	1.0496E-10	150	1500
947	C9H20	2,3,3,4-TETRAMETHYLPENTANE	-52.919	1.0814E+00	-7.4714E-04	2.0985E-07	--------	298	1000
948	C9H20	2,2,5-TRIMETHYLHEXANE	18.986	6.8130E-01	-5.8717E-05	-2.2116E-07	8.5439E-11	150	1500
949	C9H20O	2,6-DIMETHYL-4-HEPTANOL	-26.633	1.0358E+00	-7.3051E-04	2.6868E-07	-4.1207E-11	298	1500
950	C9H20O	1-NONANOL	13.919	8.1238E-01	-3.1365E-04	-6.4537E-08	5.1036E-11	200	1500
951	C9H20O	2-NONANOL	-15.877	9.8909E-01	-6.6014E-04	2.1150E-07	-1.9683E-11	298	1200
952	C9H20S	n-NONYL MERCAPTAN	66.762	5.6153E-01	1.4375E-04	-3.8391E-07	1.2953E-10	200	1500
953	C9H20S	BUTYL-PENTYL-SULFIDE	13.210	8.3559E-01	-3.8753E-04	6.5902E-08	--------	298	1000
954	C9H20S	ETHYL-HEPTYL-SULFIDE	10.905	8.5630E-01	-4.2668E-04	8.1270E-08	--------	298	1000
955	C9H20S	HEXYL-PROPYL-SULFIDE	13.210	8.3559E-01	-3.8753E-04	6.5902E-08	--------	298	1000
956	C9H20S	METHYL-OCTYL-SULFIDE	13.683	8.5337E-01	-4.3020E-04	8.3538E-08	--------	298	1000
957	C9H21N	n-NONYLAMINE	-12.915	9.9952E-01	-6.8214E-04	2.5663E-07	-4.1988E-11	298	1500
958	C9H21N	TRIPROPYLAMINE	-46.899	1.1357E+00	-8.4797E-04	3.3979E-07	-5.7551E-11	298	1500
959	C10H6O8	PYROMELLITIC ACID	-67.161	9.7245E-01	-6.8626E-04	1.9256E-07	-6.0900E-13	298	1500
960	C10H7Br	1-BROMONAPHTHALENE	-46.005	8.5830E-01	-7.4681E-04	3.2926E-07	-5.8743E-11	298	1500
961	C10H7Cl	1-CHLORONAPHTHALENE	-57.895	9.1225E-01	-8.3435E-04	3.8928E-07	-7.3276E-11	298	1500
962	C10H8	NAPHTHALENE	67.099	4.3239E-02	9.1740E-04	-1.0019E-06	3.0896E-10	50	1500
963	C10H8	AZULENE	-72.693	8.4458E-01	-6.2417E-04	1.7962E-07	--------	298	1000
964	C10H9N	QUINALDINE	-65.789	9.5661E-01	-8.0989E-04	3.4702E-07	-6.0324E-11	298	1500
965	C10H10	m-DIVINYLBENZENE	-35.543	8.5580E-01	-7.3218E-04	3.3133E-07	-6.2061E-11	298	1500
966	C10H10	1-METHYLINDENE	-48.066	9.3476E-01	-8.3220E-04	3.8682E-07	-7.4167E-11	298	1500
967	C10H10	2-METHYLINDENE	-49.008	9.3164E-01	-8.2626E-04	3.8402E-07	-7.3411E-11	298	1500
968	C10H10O4	DIMETHYL PHTHALATE	-60.400	9.3881E-01	-5.4837E-04	6.5470E-09	7.7122E-11	298	1200
969	C10H10O4	DIMETHYL TEREPHTHALATE	-132.654	1.3389E+00	-1.2518E-03	4.2932E-07	2.8171E-11	298	1000
970	C10H12	DICYCLOPENTADIENE	-147.089	1.3677E+00	-1.3862E-03	7.1581E-07	-1.4513E-10	298	1500
971	C10H12	1,2,3,4-TETRAHYDRONAPHTHALENE	-51.694	8.3545E-01	-5.0119E-04	1.1368E-07	-6.7703E-13	200	1500
972	C10H12O	ANETHOLE	-45.965	9.3023E-01	-7.2535E-04	2.9642E-07	-5.0285E-11	298	1500
973	C10H12O4	DIALLYL MALEATE	-47.979	1.3810E+00	-1.5324E-03	8.8301E-07	-1.9930E-10	298	1500
974	C10H14	n-BUTYLBENZENE	26.627	5.3108E-01	3.7958E-05	-2.7954E-07	1.0200E-10	200	1500
975	C10H14	sec-BUTYLBENZENE	104.794	1.1063E+00	8.1144E-04	-8.6635E-07	2.5831E-10	200	1500
976	C10H14	tert-BUTYLBENZENE	-53.936	9.6706E-01	-7.1192E-04	2.5338E-07	-3.4001E-11	275	1500
977	C10H14	1,2,3,4-TETRAMETHYLBENZENE	4.848	7.3283E-01	-4.0722E-04	8.5575E-08	--------	298	1000
978	C10H14	m-CYMENE	31.855	4.9392E-01	1.1261E-04	-3.3794E-07	1.1778E-10	200	1500
979	C10H14	o-CYMENE	-38.033	8.6610E-01	-5.5561E-04	1.6108E-07	-1.4651E-11	298	1500
980	C10H14	p-CYMENE	36.759	4.6194E-01	1.6776E-04	-3.7638E-07	1.2734E-10	200	1500
981	C10H14	m-DIETHYLBENZENE	24.047	5.5431E-01	-1.1581E-05	-2.4047E-07	9.0827E-11	200	1500
982	C10H14	o-DIETHYLBENZENE	34.267	5.3592E-01	-5.9210E-06	-2.3425E-07	8.7499E-11	200	1500
983	C10H14	p-DIETHYLBENZENE	28.958	5.2201E-01	4.4335E-05	-2.7933E-07	1.0033E-10	200	1500
984	C10H14	2-ETHYL-m-XYLENE	-32.782	9.0529E-01	-6.7917E-04	2.6936E-07	-4.5691E-11	298	1500
985	C10H14	2-ETHYL-p-XYLENE	-32.097	8.7204E-01	-6.1913E-04	2.2900E-07	-3.5721E-11	298	1500
986	C10H14	3-ETHYL-o-XYLENE	-37.474	9.3807E-01	-7.5782E-04	3.4813E-07	-7.3906E-11	298	1200
987	C10H14	4-ETHYL-m-XYLENE	-32.097	8.7204E-01	-6.1913E-04	2.2900E-07	-3.5721E-11	298	1500
988	C10H14	4-ETHYL-o-XYLENE	-32.097	8.7204E-01	-6.1913E-04	2.2900E-07	-3.5721E-11	298	1500
989	C10H14	5-ETHYL-m-XYLENE	-24.187	7.9716E-01	-4.8265E-04	1.3407E-07	-1.2432E-11	298	1500
990	C10H14	ISOBUTYLBENZENE	-70.878	1.0445E+00	-8.7778E-04	3.9154E-07	-7.1408E-11	298	1600
991	C10H14	1,2,3,5-TETRAMETHYLBENZENE	55.191	3.9958E-01	2.2251E-04	-3.8570E-07	1.2358E-10	150	1500
992	C10H14	1,2,4,5-TETRAMETHYLBENZENE	29.751	5.6692E-01	-1.1573E-04	-1.1468E-07	4.8575E-11	200	1500
993	C10H14O	p-tert-BUTYLPHENOL	-70.607	1.1546E+00	-1.0219E-03	4.6879E-07	-8.7947E-11	298	1500
994	C10H14O2	p-tert-BUTYLCATECHOL	-66.567	1.2468E+00	-1.1749E-03	5.7126E-07	-1.1278E-10	298	1500
995	C10H15N	N,N-DIETYHLANILINE	-103.246	1.2315E+00	-1.0695E-03	4.8560E-07	-9.0354E-11	298	1500
996	C10H15N	2,6-DIETHYLANILINE	-45.098	1.0631E+00	-8.9633E-04	4.0581E-07	-7.6550E-11	298	1500
997	C10H16	CAMPHENE	-154.624	1.3370E+00	-1.0808E-03	4.5075E-07	-7.7569E-11	298	1500
998	C10H16	D-LIMONENE	-75.119	1.1283E+00	-9.4007E-04	4.2286E-07	-7.9691E-11	298	1500
999	C10H16	alpha-PHELLANDRENE	-99.641	1.2604E+00	-1.1540E-03	5.6750E-07	-1.1500E-10	298	1500
1000	C10H16	beta-PHELLANDRENE	-102.519	1.2851E+00	-1.1913E-03	5.8754E-07	-1.1834E-10	298	1500
1001	C10H16	alpha-PINENE	-200.422	1.4834E+00	-1.2692E-03	5.4464E-07	-9.4071E-11	298	1500
1002	C10H16	beta-PINENE	-140.815	1.3045E+00	-1.0299E-03	4.0838E-07	-6.5221E-11	298	1500
1003	C10H16	alpha-TERPINENE	-49.869	1.1280E+00	-1.0188E-03	5.0169E-07	-1.0246E-10	298	1500
1004	C10H16	gamma-TERPINENE	-71.384	1.2266E+00	-1.1733E-03	6.0467E-07	-1.2738E-10	298	1500
1005	C10H16	TERPINOLENE	-59.866	1.0552E+00	-8.3267E-04	3.5840E-07	-6.6028E-11	298	1500
1006	C10H16O	CAMPHOR	-145.746	1.3060E+00	-9.5043E-04	3.2082E-07	-3.9497E-11	298	1500
1007	C10H18	1-DECYNE	6.678	8.5299E-01	-5.0112E-04	1.1788E-07	--------	298	1500
1008	C10H18	cis-DECAHYDRONANPHTALENE	-22.786	6.4589E-01	1.8319E-04	-4.7868E-07	1.6380E-10	200	1500
1009	C10H18	trans-DECAHYDRONAPHTHALENE	1.778	4.8130E-01	5.2802E-04	-7.6032E-07	2.4097E-10	150	1500
1010	C10H18O4	SEBACIC ACID	12.334	1.0779E+00	-7.8296E-04	2.9696E-07	-4.8452E-11	298	1500
1011	C10H20	n-BUTYLCYCLOHEXANE	90.421	2.3264E-01	9.4595E-04	-1.0571E-06	3.1928E-10	150	1500
1012	C10H20	1-CYCLOPENTYLPENTANE	-67.341	1.0922E+00	-7.0400E-04	1.9059E-07	--------	298	1000
1013	C10H20	1-DECENE	121.553	2.0974E-01	7.8760E-04	-8.6982E-07	2.6033E-10	150	1500
1014	C10H20O	1-DECANAL	210.932	-1.8370E-01	1.5621E-03	-1.4857E-06	4.2973E-10	150	1500

45

Table 2-1 HEAT CAPACITY OF GAS - ORGANIC COMPOUNDS (continued)

NO	FORMULA	NAME	$C_P = A + B T + C T^2 + D T^3 + E T^4$					$(C_P$ - joule/(mol K), T - K)	
			A	B	C	D	E	TMIN	TMAX
1015	C10H20O2	n-DECANOIC ACID	16.364	9.3012E-01	-4.7163E-04	5.8799E-08	1.1477E-11	298	1200
1016	C10H20O2	2-ETHYLHEXYL ACETATE	-18.921	1.1337E+00	-8.6008E-04	3.5303E-07	-6.1420E-11	298	1500
1017	C10H20O2	ISOPENTYL ISOVALERATE	-49.541	1.2738E+00	-1.0883E-03	5.0965E-07	-9.9751E-11	298	1500
1018	C10H22	n-DECANE	31.780	7.4489E-01	-1.0945E-04	-2.2668E-07	9.3458E-11	200	1500
1019	C10H22	2-METHYLNONANE	57.338	5.8226E-01	2.2492E-04	-4.8905E-07	1.6519E-10	150	1500
1020	C10H22	3-METHYLNONANE	49.113	6.2101E-01	1.5367E-04	-4.3414E-07	1.5034E-10	150	1500
1021	C10H22	4-METHYLNONANE	43.716	6.5491E-01	9.6751E-05	-3.9509E-07	1.4060E-10	150	1500
1022	C10H22	5-METHYLNONANE	46.011	6.3918E-01	1.2507E-04	-4.1575E-07	1.4602E-10	150	1500
1023	C10H22	3-ETHYLOCTANE	-6.595	9.6968E-01	-5.4271E-04	1.1646E-07	--------	298	1000
1024	C10H22	4-ETHYLOCTANE	-6.595	9.6968E-01	-5.4271E-04	1.1646E-07	--------	298	1000
1025	C10H22	2,2-DIMETHYLOCTANE	50.808	6.2243E-01	1.6032E-04	-4.2638E-07	1.4566E-10	150	1500
1026	C10H22	2,3-DIMETHYLOCTANE	-30.720	1.0547E+00	-6.4091E-04	1.5416E-07	--------	298	1000
1027	C10H22	2,4-DIMETHYLOCTANE	-16.035	1.0059E+00	-5.8672E-04	1.3405E-07	--------	298	1000
1028	C10H22	2,5-DIMETHYLOCTANE	-30.720	1.0547E+00	-6.4091E-04	1.5416E-07	--------	298	1000
1029	C10H22	2,6-DIMETHYLOCTANE	-30.720	1.0547E+00	-6.4091E-04	1.5416E-07	--------	298	1000
1030	C10H22	2,7-DIMETHYLOCTANE	-15.284	1.0017E+00	-5.7881E-04	1.2969E-07	--------	298	1000
1031	C10H22	3,3-DIMETHYLOCTANE	-28.772	1.0848E+00	-6.9220E-04	1.7847E-07	--------	298	1000
1032	C10H22	3,4-DIMETHYLOCTANE	-31.481	1.0590E+00	-6.4894E-04	1.5857E-07	--------	298	1000
1033	C10H22	3,5-DIMETHYLOCTANE	-46.170	1.1080E+00	-7.0316E-04	1.7870E-07	--------	298	1000
1034	C10H22	3,6-DIMETHYLOCTANE	-46.170	1.1080E+00	-7.0316E-04	1.7870E-07	--------	298	1000
1035	C10H22	4,4-DIMETHYLOCTANE	-28.772	1.0848E+00	-6.9220E-04	1.7847E-07	--------	298	1000
1036	C10H22	4,5-DIMETHYLOCTANE	-31.481	1.0590E+00	-6.4894E-04	1.5857E-07	--------	298	1000
1037	C10H22	4-PROPYLHEPTANE	-22.025	1.0228E+00	-6.0476E-04	1.4090E-07	--------	298	1000
1038	C10H22	4-ISOPROPYLHEPTANE	-31.481	1.0590E+00	-6.4894E-04	1.5857E-07	--------	298	1000
1039	C10H22	3-ETHYL-2-METHYLHEPTANE	-31.481	1.0590E+00	-6.4894E-04	1.5857E-07	--------	298	1000
1040	C10H22	4-ETHYL-2-METHYLHEPTANE	-46.170	1.1080E+00	-7.0316E-04	1.7870E-07	--------	298	1000
1041	C10H22	5-ETHYL-2-METHYLHEPTANE	-46.170	1.1080E+00	-7.0316E-04	1.7870E-07	--------	298	1000
1042	C10H22	3-ETHYL-3-METHYLHEPTANE	-44.217	1.1379E+00	-7.5438E-04	2.0299E-07	--------	298	1000
1043	C10H22	4-ETHYL-3-METHYLHEPTANE	-46.928	1.1122E+00	-7.1111E-04	1.8308E-07	--------	298	1000
1044	C10H22	3-ETHYL-5-METHYLHEPTANE	-61.605	1.1611E+00	-7.6525E-04	2.0317E-07	--------	298	1000
1045	C10H22	3-ETHYL-4-METHYLHEPTANE	-46.928	1.1122E+00	-7.1111E-04	1.8308E-07	--------	298	1000
1046	C10H22	4-ETHYL-4-METHYLHEPTANE	-44.217	1.1379E+00	-7.5438E-04	2.0299E-07	--------	298	1000
1047	C10H22	2,2,3-TRIMETHYLHEPTANE	-38.215	1.1210E+00	-7.3626E-04	1.9608E-07	--------	298	1000
1048	C10H22	2,2,4-TRIMETHYLHEPTANE	-52.894	1.1698E+00	-7.9040E-04	2.1617E-07	--------	298	1000
1049	C10H22	2,2,5-TRIMETHYLHEPTANE	-52.894	1.1698E+00	-7.9040E-04	2.1617E-07	--------	298	1000
1050	C10H22	2,2,6-TRIMETHYLHEPTANE	-37.470	1.1168E+00	-7.2839E-04	1.9175E-07	--------	298	1000
1051	C10H22	2,3,3-TRIMETHYLHEPTANE	-38.215	1.1210E+00	-7.3626E-04	1.9608E-07	--------	298	1000
1052	C10H22	2,3,4-TRIMETHYLHEPTANE	-40.930	1.0952E+00	-6.9304E-04	1.7620E-07	--------	298	1000
1053	C10H22	2,3,5-TRIMETHYLHEPTANE	-55.626	1.1442E+00	-7.4730E-04	1.9636E-07	--------	298	1000
1054	C10H22	2,3,6-TRIMETHYLHEPTANE	-40.185	1.0911E+00	-6.8517E-04	1.7187E-07	--------	298	1000
1055	C10H22	2,4,4-TRIMETHYLHEPTANE	-52.894	1.1698E+00	-7.9040E-04	2.1617E-07	--------	298	1000
1056	C10H22	2,4,5-TRIMETHYLHEPTANE	-55.626	1.1442E+00	-7.4730E-04	1.9636E-07	--------	298	1000
1057	C10H22	2,4,6-TRIMETHYLHEPTANE	-54.873	1.1400E+00	-7.3940E-04	1.9200E-07	--------	298	1000
1058	C10H22	2,5,5-TRIMETHYLHEPTANE	-52.894	1.1698E+00	-7.9040E-04	2.1617E-07	--------	298	1000
1059	C10H22	3,3,4-TRIMETHYLHEPTANE	-53.660	1.1741E+00	-7.9847E-04	2.2060E-07	--------	298	1000
1060	C10H22	3,3,5-TRIMETHYLHEPTANE	-68.028	1.2211E+00	-8.4889E-04	2.3852E-07	--------	298	1000
1061	C10H22	3,4,4-TRIMETHYLHEPTANE	-53.660	1.1741E+00	-7.9847E-04	2.2060E-07	--------	298	1000
1062	C10H22	3,4,5-TRIMETHYLHEPTANE	-56.367	1.1483E+00	-7.5517E-04	2.0067E-07	--------	298	1000
1063	C10H22	3-ISOPROPYL-2-METHYLHEXANE	-40.930	1.0952E+00	-6.9304E-04	1.7620E-07	--------	298	1000
1064	C10H22	3,3-DIETHYLHEXANE	-59.647	1.1910E+00	-8.1647E-04	2.2743E-07	--------	298	1000
1065	C10H22	3,4-DIETHYLHEXANE	-62.363	1.1653E+00	-7.7325E-04	2.0755E-07	--------	298	1000
1066	C10H22	3-ETHYL-2,2-DIMETHYLHEXANE	-53.660	1.1741E+00	-7.9847E-04	2.2060E-07	--------	298	1000
1067	C10H22	4-ETHYL-2,2-DIMETHYLHEXANE	-68.350	1.2231E+00	-8.5270E-04	2.4073E-07	--------	298	1000
1068	C10H22	3-ETHYL-2,3-DIMETHYLHEXANE	-53.660	1.1741E+00	-7.9847E-04	2.2060E-07	--------	298	1000
1069	C10H22	4-ETHYL-2,3-DIMETHYLHEXANE	-56.367	1.1483E+00	-7.5517E-04	2.0067E-07	--------	298	1000
1070	C10H22	3-ETHYL-2,4-DIMETHYLHEXANE	-56.367	1.1483E+00	-7.5517E-04	2.0067E-07	--------	298	1000
1071	C10H22	4-ETHYL-2,4-DIMETHYLHEXANE	-68.350	1.2231E+00	-8.5270E-04	2.4073E-07	--------	298	1000
1072	C10H22	3-ETHYL-2,5-DIMETHYLHEXANE	-55.739	1.1449E+00	-7.4852E-04	1.9698E-07	--------	298	1000
1073	C10H22	4-ETHYL-3,3-DIMETHYLHEXANE	-69.099	1.2273E+00	-8.6057E-04	2.4507E-07	--------	298	1000
1074	C10H22	3-ETHYL-3,4-DIMETHYLHEXANE	-69.099	1.2273E+00	-8.6057E-04	2.4507E-07	--------	298	1000
1075	C10H22	2,2,3,3-TETRAMETHYLHEXANE	-56.275	1.2191E+00	-8.6659E-04	2.5087E-07	--------	298	1000
1076	C10H22	2,2,3,4-TETRAMETHYLHEXANE	-63.111	1.2103E+00	-8.4257E-04	2.3823E-07	--------	298	1000
1077	C10H22	2,2,3,5-TETRAMETHYLHEXANE	-62.354	1.2061E+00	-8.3462E-04	2.3385E-07	--------	298	1000
1078	C10H22	2,2,4,4-TETRAMETHYLHEXANE	-75.086	1.2850E+00	-9.4002E-04	2.7825E-07	--------	298	1000
1079	C10H22	2,2,4,5-TETRAMETHYLHEXANE	-62.354	1.2061E+00	-8.3462E-04	2.3385E-07	--------	298	1000
1080	C10H22	2,2,5,5-TETRAMETHYLHEXANE	-59.655	1.2319E+00	-8.7797E-04	2.5381E-07	--------	298	1000
1081	C10H22	2,3,3,4-TETRAMETHYLHEXANE	-63.111	1.2103E+00	-8.4257E-04	2.3823E-07	--------	298	1000
1082	C10H22	2,3,3,5-TETRAMETHYLHEXANE	-62.354	1.2061E+00	-8.3462E-04	2.3385E-07	--------	298	1000
1083	C10H22	2,3,4,4-TETRAMETHYLHEXANE	-63.111	1.2103E+00	-8.4257E-04	2.3823E-07	--------	298	1000
1084	C10H22	2,3,4,5-TETRAMETHYLHEXANE	-50.367	1.1314E+00	-7.3705E-04	1.9379E-07	--------	298	1000
1085	C10H22	3,3,4,4-TETRAMETHYLHEXANE	-71.835	1.2727E+00	-9.2948E-04	2.7570E-07	--------	298	1000
1086	C10H22	2,4-DIMETHYL-3-ISOPROPYLPENTANE	-63.111	1.2103E+00	-8.4257E-04	2.3823E-07	--------	298	1000
1087	C10H22	3,3-DIETHYL-2-METHYLPENTANE	-69.099	1.2273E+00	-8.6057E-04	2.4507E-07	--------	298	1000
1088	C10H22	3-ETHYL-2,2,3-TRIMETHYLPENTANE	-71.701	1.2721E+00	-9.2864E-04	2.7531E-07	--------	298	1000
1089	C10H22	3-ETHYL-2,2,4-TRIMETHYLPENTANE	-63.111	1.2103E+00	-8.4257E-04	2.3823E-07	--------	298	1000
1090	C10H22	3-ETHYL-2,3,4-TRIMETHYLPENTANE	-63.111	1.2103E+00	-8.4257E-04	2.3823E-07	--------	298	1000
1091	C10H22	2,2,3,3,4-PENTAMETHYLPENTANE	-57.509	1.2215E+00	-8.7257E-04	2.5417E-07	--------	298	1000
1092	C10H22	2,2,3,4,4-PENTAMETHYLPENTANE	-69.860	1.2724E+00	-9.3006E-04	2.7583E-07	--------	298	1000

Table 2-1 HEAT CAPACITY OF GAS - ORGANIC COMPOUNDS (continued)

			$C_P = A + B\,T + C\,T^2 + D\,T^3 + E\,T^4$				$(C_P$ - joule/(mol K), T - K)		
NO	FORMULA	NAME	A	B	C	D	E	TMIN	TMAX
1093	C10H22O	1-DECANOL	15.733	8.9157E-01	-3.3219E-04	-8.6679E-08	6.1276E-11	200	1500
1094	C10H22O	DI-n-PENTYL ETHER	-17.080	1.0765E+00	-7.2109E-04	2.5716E-07	-3.9483E-11	298	1500
1095	C10H22O	ISODECANOL	-34.228	1.1729E+00	-8.9588E-04	3.7908E-07	-6.8013E-11	298	1500
1096	C10H22O5	TETRAETHYLENE GLYCOL DIMETHYL ETHER	-62.043	1.4737E+00	-1.2136E-03	5.2774E-07	-9.7123E-11	298	1500
1097	C10H22S	n-DECYL MERCAPTAN	70.441	6.2988E-01	1.4669E-04	-4.2347E-07	1.4459E-10	200	1500
1098	C10H22S	BUTYL-HEXYL-SULFIDE	12.094	9.3174E-01	-4.4396E-04	7.8948E-08	--------	298	1000
1099	C10H22S	ETHYL-OCTYL-SULFIDE	9.551	9.5358E-01	-4.8476E-04	9.4985E-08	--------	298	1000
1100	C10H22S	HEPTYL-PROPYL-SULFIDE	12.094	9.3174E-01	-4.4396E-04	7.8948E-08	--------	298	1000
1101	C10H22S	METHYL-NONYL-SULFIDE	12.344	9.5052E-01	-4.8811E-04	9.7178E-08	--------	298	1000
1102	C10H22S	PENTYL-SULFIDE	12.094	9.3174E-01	-4.4396E-04	7.8948E-08	--------	298	1000
1103	C10H22S2	PENTYL-DISULFIDE	17.922	1.0309E+00	-5.8676E-04	1.3407E-07	--------	298	1000
1104	C10H23N	n-DECYLAMINE	-9.825	1.0697E+00	-6.7989E-04	2.1931E-07	-2.9272E-11	298	1500
1105	C11H10	1-METHYLNAPHTHALENE	-8.325	6.5561E-01	-2.3148E-04	-8.6605E-08	5.3577E-11	200	1500
1106	C11H10	2-METHYLNAPHTHALENE	-0.899	6.1854E-01	-1.8095E-04	-1.1454E-07	5.9208E-11	200	1500
1107	C11H14O2	n-BUTYL BENZOATE	-84.783	1.1780E+00	-9.7914E-04	4.2833E-07	-7.6038E-11	298	1500
1108	C11H16	n-PENTYLBENZENE	-4.484	7.7970E-01	-2.7838E-04	-7.5740E-08	5.1215E-11	200	1500
1109	C11H16O	p-tert-AMYLPHENOL	-71.475	1.2516E+00	-1.0881E-03	4.9607E-07	-9.3224E-11	298	1500
1110	C11H20	1-UNDECYNE	5.865	9.4705E-01	-5.5333E-04	1.2835E-07		298	1000
1111	C11H20O2	2-ETHYLHEXYL ACRYLATE	-44.392	1.3753E+00	-1.2506E-03	6.2600E-07	-1.2931E-10	298	1500
1112	C11H22	1-UNDECENE	132.612	2.4011E-01	8.5062E-04	-9.4540E-07	2.8339E-10	150	1500
1113	C11H22	1-CYCLOPENTYLHEXANE	-68.262	1.1870E+00	-7.5781E-04	2.0198E-07	--------	298	1000
1114	C11H22	PENTYLCYCLOHEXANE	-65.480	1.1809E+00	-6.8379E-04	1.4580E-07	--------	298	1000
1115	C11H22O	1-UNDECANAL	2.102	1.0578E+00	-6.7455E-04	2.2287E-07	-3.1192E-11	298	1500
1116	C11H24	n-UNDECANE	125.212	3.1401E-01	7.9137E-04	-9.1410E-07	2.7568E-10	150	1500
1117	C11H24O	1-UNDECANOL	16.702	9.7755E-01	-3.6662E-04	-9.5146E-08	6.7752E-11	200	1500
1118	C11H24S	UNDECYL MERCAPTAN	74.322	6.9712E-01	1.5160E-04	-4.6448E-07	1.6008E-10	200	1500
1119	C11H24S	BUTYL-HEPTYL-SULFIDE	10.773	1.0288E+00	-5.0175E-04	9.2508E-08	--------	298	1000
1120	C11H24S	DECYL-METHYL-SULFIDE	11.252	1.0464E+00	-5.4405E-04	1.0991E-07	--------	298	1000
1121	C11H24S	ETHYL-NONYL-SULFIDE	8.459	1.0494E+00	-5.4070E-04	1.0772E-07	--------	298	1000
1122	C11H24S	OCTYL-PROPYL-SULFIDE	10.773	1.0288E+00	-5.0175E-04	9.2508E-08	--------	298	1000
1123	C12H8O	DIBENZOFURAN	-14.004	5.8793E-01	9.7079E-05	-4.4225E-07	1.6599E-10	100	1500
1124	C12H9N	DIBENZOPYRROLE	-117.362	1.3214E+00	-1.3019E-03	6.4403E-07	-1.2737E-10	298	1500
1125	C12H10	ACENAPHTHENE	-61.063	9.6388E-01	-7.2100E-04	2.6234E-07	-3.6857E-11	200	1500
1126	C12H10	BIPHENYL	-29.153	7.6716E-01	-3.4341E-04	-3.7724E-08	4.6179E-11	200	1500
1127	C12H10O	DIPHENYL ETHER	-118.442	1.2856E+00	-1.2190E-03	5.9127E-07	-1.1547E-10	298	1500
1128	C12H11N	p-AMINODIPHENYL	-95.528	1.2611E+00	-1.1955E-03	5.8242E-07	-1.1477E-10	298	1500
1129	C12H11N	DIPHENYLAMINE	-119.401	1.3060E+00	-1.2200E-03	5.8764E-07	-1.1447E-10	298	1500
1130	C12H11N3	p-AMINOAZOBENZENE	-160.867	1.7547E+00	-1.9301E-03	1.0499E-06	-2.2321E-10	298	1500
1131	C12H11N3	1,3-DIPHENYLTRIAZENE	-190.085	1.8079E+00	-1.9564E-03	1.0452E-06	-2.1809E-10	298	1500
1132	C12H12	1,2-DIMETHYLNAPHTHALENE	-59.463	1.0193E+00	-7.3258E-04	2.0709E-07	--------	298	1000
1133	C12H12	1,3-DIMETHYLNAPHTHALENE	-51.764	9.8240E-01	-6.8768E-04	1.8992E-07	--------	298	1000
1134	C12H12	1,4-DIMETHYLNAPHTHALENE	-59.463	1.0193E+00	-7.3258E-04	2.0709E-07	--------	298	1000
1135	C12H12	1,5-DIMETHYLNAPHTHALENE	-59.463	1.0193E+00	-7.3258E-04	2.0709E-07	--------	298	1000
1136	C12H12	1,6-DIMETHYLNAPHTHALENE	-51.764	9.8240E-01	-6.8768E-04	1.8992E-07	--------	298	1000
1137	C12H12	1,7-DIMETHYLNAPHTHALENE	-51.764	9.8240E-01	-6.8768E-04	1.8992E-07	--------	298	1000
1138	C12H12	2,3-DIMETHYLNAPHTHALENE	-32.046	8.8584E-01	-5.6325E-04	1.3936E-07	--------	298	1000
1139	C12H12	2,6-DIMETHYLNAPHTHALENE	-49.867	1.0029E+00	-7.9843E-04	3.3597E-07	-5.8987E-11	298	1500
1140	C12H12	2,7-DIMETHYLNAPHTHALENE	-49.867	1.0029E+00	-7.9843E-04	3.3597E-07	-5.8987E-11	298	1500
1141	C12H12	1-ETHYLNAPHTHALENE	-77.162	1.1312E+00	-9.8675E-04	4.5204E-07	-8.5024E-11	298	1500
1142	C12H12	2-ETHYLNAPHTHALENE	-56.103	1.0022E+00	-7.1651E-04	2.0322E-07	--------	298	1000
1143	C12H12N2	p-AMINODIPHENYLAMINE	-109.455	1.3879E+00	-1.3126E-03	6.3299E-07	-1.2318E-10	298	1500
1144	C12H12N2	HYDRAZOBENZENE	-116.092	1.3797E+00	-1.3149E-03	6.4282E-07	-1.2675E-10	298	1500
1145	C12H14	1,2,3-TRIMETHYLINDENE	-37.324	1.0980E+00	-9.2975E-04	4.1878E-07	-7.8898E-11	298	1500
1146	C12H14O4	DIETHYL PHTHALATE	-122.210	1.4777E+00	-1.3850E-03	6.7906E-07	-1.3210E-10	298	1500
1147	C12H16	CYCLOHEXYLBENZENE	-94.764	1.1349E+00	-6.8154E-04	1.4688E-07	2.5822E-12	298	1400
1148	C12H18	m-DIISOPROPYLBENZENE	-70.177	1.2283E+00	-9.8495E-04	4.2208E-07	-7.6005E-11	298	1500
1149	C12H18	p-DIISOPROPYLBENZENE	-70.177	1.2283E+00	-9.8495E-04	4.2208E-07	-7.6005E-11	298	1500
1150	C12H18	n-HEXYLBENZENE	39.655	6.4328E-01	6.6681E-05	-3.5564E-07	1.2779E-10	200	1500
1151	C12H18	1,2,3-TRIETHYLBENZENE	-10.544	9.5040E-01	-5.3559E-04	1.1260E-07	--------	298	1000
1152	C12H18	1,2,4-TRIETHYLBENZENE	-13.571	9.6324E-01	-5.4802E-04	1.1608E-07	--------	298	1000
1153	C12H18	1,3,5-TRIETHYLBENZENE	-25.200	1.0011E+00	-5.9132E-04	1.3307E-07	--------	298	1000
1154	C12H18	HEXAMETHYLBENZENE	3.380	1.0043E+00	-6.6300E-04	1.8064E-07	--------	298	1000
1155	C12H20O4	DIBUTYL MALEATE	68.479	9.1396E-01	-4.0061E-04	1.5325E-08	2.1770E-11	298	1500
1156	C12H22	BICYCLOHEXYL	-122.649	1.3647E+00	-7.9966E-04	1.8314E-07	-5.1919E-12	298	1500
1157	C12H22	1-DODECYNE	4.031	1.0475E+00	-6.1798E-04	1.4619E-07	--------	298	1000
1158	C12H23N	DICYCLOHEXYLAMINE	-128.032	1.4622E+00	-8.6193E-04	1.8462E-07	3.6227E-13	298	1500
1159	C12H24	1-DODECENE	83.102	6.2741E-01	2.2136E-04	-4.7938E-07	1.5890E-10	200	1500
1160	C12H24	1-CYCLOPENTYLHEPTANE	-69.308	1.2824E+00	-8.1257E-04	2.1379E-07	--------	298	1000
1161	C12H24	1-CYCLOHEXYLHEXANE	-66.659	1.2774E+00	-7.4103E-04	1.5921E-07	--------	298	1000
1162	C12H24O	1-DODECANAL	4.271	1.1290E+00	-6.7200E-04	1.8542E-07	-1.7582E-11	298	1500
1163	C12H24O2	n-DODECANOIC ACID	-4.295	1.2373E+00	-8.2209E-04	2.7680E-07	-3.8871E-11	298	1500
1164	C12H26	n-DODECANE	71.498	7.2559E-01	1.1553E-04	-4.1196E-07	1.4141E-10	200	1500
1165	C12H26O	DI-n-HEXYL ETHER	-18.114	1.2587E+00	-8.1053E-04	2.6448E-07	-3.5458E-11	298	1500
1166	C12H26O	1-DODECANOL	17.965	1.0602E+00	-3.9424E-04	-1.1142E-07	7.6380E-11	200	1500
1167	C12H26O3	DIETHYLENE GLYCOL DI-n-BUTYL ETHER	-45.437	1.5002E+00	-1.1545E-03	4.8538E-07	-8.9144E-11	298	1500
1168	C12H26S	n-DODECYL MERCAPTAN	-15.335	1.2989E+00	-9.1285E-04	3.6369E-07	-6.4587E-11	298	1500
1169	C12H26S	BUTYL-OCTYL-SULFIDE	9.659	1.1248E+00	-5.5789E-04	1.0535E-07	--------	298	1000
1170	C12H26S	DECYL-ETHYL-SULFIDE	7.109	1.1468E+00	-5.9915E-04	1.2180E-07	--------	298	1000

47

Table 2-1 HEAT CAPACITY OF GAS - ORGANIC COMPOUNDS (continued)

			$C_P = A + B T + C T^2 + D T^3 + E T^4$					$(C_P$ - joule/(mol K), T - K)	
NO	FORMULA	NAME	A	B	C	D	E	TMIN	TMAX
1171	C12H26S	HEXYL-SULFIDE	9.659	1.1248E+00	-5.5789E-04	1.0535E-07	--------	298	1000
1172	C12H26S	METHYL-UNDECYL-SULFIDE	9.886	1.1439E+00	-6.0266E-04	1.2407E-07	--------	298	1000
1173	C12H26S	NONYL-PROPYL-SULFIDE	9.659	1.1248E+00	-5.5789E-04	1.0535E-07	--------	298	1000
1174	C12H26S2	HEXYL-DISULFIDE	15.498	1.2240E+00	-7.0057E-04	1.6042E-07	--------	298	1000
1175	C12H27BO3	TRI-n-BUTYL BORATE	--------	--------	--------	--------	--------	--------	--------
1176	C12H27N	DODECYLAMINE	-18.410	1.3048E+00	-8.9181E-04	3.3488E-07	-5.5494E-11	298	1500
1177	C12H27N	TRI-n-BUTYLAMINE	-52.703	1.4429E+00	-1.0614E-03	4.2083E-07	-7.1662E-11	298	1500
1178	C13H10	FLUORENE	-74.758	1.0187E+00	-7.1721E-04	2.2140E-07	-2.0280E-11	200	1500
1179	C13H10O	BENZOPHENONE	-75.275	1.0612E+00	-7.8307E-04	2.7117E-07	-3.4873E-11	298	1500
1180	C13H12	DIPHENYLMETHANE	-100.298	1.2250E+00	-1.0670E-03	4.8040E-07	-8.7845E-11	298	1600
1181	C13H14	1-PROPYLNAPHTHALENE	-60.229	1.1155E+00	-7.8906E-04	2.1995E-07	--------	298	1000
1182	C13H14	2-PROPYLNAPHTHALENE	-53.154	1.0812E+00	-7.4705E-04	2.0346E-07	--------	298	1000
1183	C13H14	2ETHYL-3-METHYLNAPHTHALENE	-34.726	1.0000E+00	-6.4551E-04	1.6249E-07	--------	298	1000
1184	C13H14	2ETHYL-6-METHYLNAPHTHALENE	-79.040	1.2718E+00	-1.1243E-03	4.1640E-07	--------	298	1000
1185	C13H14	2ETHYL-7-METHYLNAPHTHALENE	-79.040	1.2718E+00	-1.1243E-03	4.1640E-07	--------	298	1000
1186	C13H20	n-HEPTYLBENZENE	45.886	7.0113E-01	7.7233E-05	-3.9047E-07	1.3976E-10	200	1500
1187	C13H24	1-TRIDECYNE	3.531	1.1404E+00	-6.6881E-04	1.5612E-07	--------	298	1000
1188	C13H26	1-TRIDECENE	88.663	6.8861E-01	2.2611E-04	-5.0981E-07	1.6963E-10	200	1500
1189	C13H26	1-CYCLOPENTYLOCTANE	-71.166	1.3829E+00	-8.7705E-04	2.3148E-07	--------	298	1000
1190	C13H26	1-CYCLOHEXYLHEPTANE	-68.166	1.3756E+00	-8.0040E-04	1.7339E-07	--------	298	1000
1191	C13H26O	1-TRIDECANAL	-1.839	1.2551E+00	-7.8737E-04	2.4389E-07	-2.9663E-11	298	1500
1192	C13H26O2	n-BUTYL NONANOATE	-58.657	1.6081E+00	-1.3740E-03	6.5735E-07	-1.3319E-10	298	1500
1193	C13H26O2	METHYL DODECANOATE	-28.448	1.4656E+00	-1.1419E-03	4.9828E-07	-9.3415E-11	298	1500
1194	C13H28	n-TRIDECANE	110.400	5.3321E-01	7.3984E-04	-1.0212E-06	3.2423E-10	150	1500
1195	C13H28O	1-TRIDECANOL	19.127	1.1452E+00	-4.2519E-04	-1.2106E-07	8.3188E-11	200	1500
1196	C13H28S	BUTYL-NONYL-SULFIDE	8.282	1.2223E+00	-6.1660E-04	1.1955E-07	--------	298	1000
1197	C13H28S	DECYL-PROPYL-SULFIDE	8.282	1.2223E+00	-6.1660E-04	1.1955E-07	--------	298	1000
1198	C13H28S	DODECYL-METHYL-SULFIDE	8.794	1.2397E+00	-6.5860E-04	1.3680E-07	--------	298	1000
1199	C13H28S	ETHYL-UNDECYL-SULFIDE	5.979	1.2429E+00	-6.5547E-04	1.3472E-07	--------	298	1000
1200	C13H28S	1-TRIDECANETHIOL	7.133	1.2506E+00	-6.7266E-04	1.4268E-07	--------	298	1000
1201	C14H8O2	ANTHRAQUINONE	-77.119	1.2856E+00	-1.3749E-03	9.3187E-07	-3.0596E-10	298	900
1202	C14H10	ANTHRACENE	-68.180	1.0788E+00	-8.1286E-04	2.9013E-07	-3.8196E-11	200	1500
1203	C14H10	DIPHENYLACETYLENE	-69.864	1.0800E+00	-8.1462E-04	2.8054E-07	-3.3905E-11	298	1500
1204	C14H10	PHENANTHRENE	-73.218	1.1192E+00	-8.9679E-04	3.5695E-07	-5.6466E-11	200	1500
1205	C14H12	cis-STILBENE	-96.782	1.2486E+00	-1.0880E-03	5.0166E-07	-9.5052E-11	298	1500
1206	C14H12	trans-STILBENE	-68.882	1.1174E+00	-8.3787E-04	3.0198E-07	-4.2209E-11	298	1500
1207	C14H12O2	BENZYL BENZOATE	-137.070	1.4041E+00	-1.1419E-03	4.2281E-07	-4.5129E-11	298	1200
1208	C14H14	1,1-DIPHENYLETHANE	-101.538	1.3208E+00	-1.1248E-03	4.9731E-07	-8.9846E-11	298	1600
1209	C14H14	1,2-DIPHENYLETHANE	-101.104	1.3200E+00	-1.1240E-03	4.9709E-07	-8.9901E-11	298	1600
1210	C14H14O	DIBENZYL ETHER	-133.169	1.4371E+00	-1.2020E-03	5.1612E-07	-9.1168E-11	298	1500
1211	C14H16	1-n-BUTYLNAPHTHALENE	71.101	4.0054E-01	7.3214E-04	-9.7479E-07	3.1420E-10	100	1500
1212	C14H16	2-BUTYLNAPHTHALENE	-52.467	1.1665E+00	-7.8488E-04	2.0665E-07	--------	298	1000
1213	C14H22	n-OCTYLBENZENE	51.588	7.6266E-01	8.0566E-05	-4.2003E-07	1.5047E-10	200	1500
1214	C14H22	1,2,3,4-TETRAETHYLBENZENE	-6.944	1.1986E+00	-7.5550E-04	1.9018E-07	--------	298	1000
1215	C14H22	1,2,3,5-TETRAETHYLBENZENE	-6.031	1.1701E+00	-7.0718E-04	1.6749E-07	--------	298	1000
1216	C14H22	1,2,4,5-TETRAETHYLBENZENE	20.140	1.0473E+00	-5.3568E-04	9.1090E-08	--------	298	1000
1217	C14H22O	p-tert-OCTYLPHENOL	-99.971	1.6764E+00	-1.4559E-03	6.5521E-07	-1.2088E-10	298	1500
1218	C14H28	1-TETRADECENE	167.903	3.1544E-01	1.0739E-03	-1.2004E-06	3.6046E-10	150	1500
1219	C14H28	1-CYCLOPENTYLNONANE	-71.931	1.4768E+00	-9.2922E-04	2.4200E-07	--------	298	1000
1220	C14H28	1-CYCLOHEXYLOCTANE	-66.463	1.4551E+00	-8.2822E-04	1.7154E-07	--------	298	1000
1221	C14H28O2	n-TETRADECANOIC ACID	-47.091	1.5789E+00	-1.1521E-03	4.4718E-07	-7.3810E-11	298	1500
1222	C14H30	n-TETRADECANE	115.502	6.0882E-01	6.8043E-04	-9.7091E-07	3.0756E-10	150	1500
1223	C14H30O	1-TETRADECANOL	20.531	1.2268E+00	-4.4901E-04	-1.3852E-07	9.2055E-11	200	1500
1224	C14H30S	BUTYL-DECYL-SULFIDE	7.169	1.3183E+00	-6.7275E-04	1.3239E-07	--------	298	1000
1225	C14H30S	DODECYL-ETHYL-SULFIDE	4.734	1.3396E+00	-7.1270E-04	1.4807E-07	--------	298	1000
1226	C14H30S	HEPTYL-SULFIDE	7.169	1.3183E+00	-6.7275E-04	1.3239E-07	--------	298	1000
1227	C14H30S	METHYL-TRIDECYL-SULFIDE	7.532	1.3365E+00	-7.1601E-04	1.5024E-07	--------	298	1000
1228	C14H30S	PROPYL-UNDECYL-SULFIDE	7.169	1.3183E+00	-6.7275E-04	1.3239E-07	--------	298	1000
1229	C14H30S	1-TETRADECANETHIOL	5.882	1.3473E+00	-7.2998E-04	1.5606E-07	--------	298	1000
1230	C14H30S2	HEPTYL-DISULFIDE	13.023	1.4174E+00	-8.1529E-04	1.8739E-07	--------	298	1000
1231	C14H31N	TETRADECYLAMINE	-23.409	1.5091E+00	-1.0193E-03	3.6745E-07	-5.6794E-11	298	1500
1232	C15H10N2O2	DIPHENYLMETHANE-4,4'-DIISOCYANATE	-40.814	1.1664E+00	-8.2318E-04	2.6180E-07	-2.8364E-11	298	1500
1233	C15H16O	p-CUMYLPHENOL	-105.972	1.5282E+00	-1.3205E-03	5.8321E-07	-1.0494E-10	298	1500
1234	C15H16O2	BISPHENOL A	-106.990	1.6818E+00	-1.6100E-03	7.9103E-07	-1.5526E-10	298	1600
1235	C15H18	1-PENTYLNAPHTHALENE	-64.191	1.3167E+00	-9.1609E-04	2.5301E-07	--------	298	1000
1236	C15H18	2-PENTYLNAPHTHALENE	-55.601	1.2747E+00	-8.6220E-04	2.3086E-07	--------	298	1000
1237	C15H24	n-NONYLBENZENE	0.378	1.1149E+00	-3.9485E-04	-1.2822E-07	8.2222E-11	200	1500
1238	C15H24O	2,6-DI-tert-BUTYL-p-CRESOL	-85.106	1.6932E+00	-1.3931E-03	5.9127E-07	-1.0339E-10	298	1500
1239	C15H24O	NONYLPHENOL	-51.387	1.5081E+00	-1.1306E-03	4.4257E-07	-7.2709E-11	298	1500
1240	C15H28	1-PENTADECYNE	0.246	1.3379E+00	-7.9036E-04	1.8667E-07	--------	298	1000
1241	C15H30	1-PENTADECENE	26.049	1.1942E+00	-3.8640E-04	-1.8619E-07	1.0451E-10	200	1500
1242	C15H30	1-CYCLOPENTYLDECANE	-73.011	1.5723E+00	-9.8395E-04	2.5373E-07	--------	298	1000
1243	C15H30	1-CYCLOHEXYLNONANE	-68.404	1.5558E+00	-8.9257E-04	1.8891E-07	--------	298	1000
1244	C15H30O2	PENTADECANOIC ACID	-12.682	1.5569E+00	-1.0581E-03	3.7408E-07	-5.6293E-11	298	1500
1245	C15H32	n-PENTADECANE	124.647	6.2706E-01	8.3164E-04	-1.1689E-06	3.7326E-10	150	1500
1246	C15H32O	1-PENTADECANOL	-5.044	1.4565E+00	-8.1601E-04	1.7556E-07	--------	298	1000
1247	C15H32S	BUTYL-UNDECYL-SULFIDE	5.949	1.4149E+00	-7.2973E-04	1.4562E-07	--------	298	1000
1248	C15H32S	DODECYL-PROPYL-SULFIDE	5.949	1.4149E+00	-7.2973E-04	1.4562E-07	--------	298	1000

Table 2-1 HEAT CAPACITY OF GAS - ORGANIC COMPOUNDS (continued)

$$C_P = A + B\,T + C\,T^2 + D\,T^3 + E\,T^4 \qquad (C_P - \text{joule/(mol K)},\ T - K)$$

NO	FORMULA	NAME	A	B	C	D	E	TMIN	TMAX
1249	C15H32S	ETHYL-TRIDECYL-SULFIDE	3.439	1.4367E+00	-7.7078E-04	1.6189E-07	--------	298	1000
1250	C15H32S	METHYL-TETRADECYL-SULFIDE	6.221	1.4338E+00	-7.7421E-04	1.6413E-07	--------	298	1000
1251	C15H32S	1-PENTADECANETHIOL	4.560	1.4447E+00	-7.8831E-04	1.7001E-07	--------	298	1000
1252	C16H10	FLUORANTHENE	-112.910	1.4495E+00	-1.3602E-03	6.4245E-07	-1.2194E-10	298	1500
1253	C16H10	PYRENE	27.191	5.8885E-01	2.4989E-04	-5.7314E-07	2.0114E-10	150	1500
1254	C16H12	1-PHENYLNAPHTHALENE	-114.610	1.4322E+00	-1.2669E-03	5.7235E-07	-1.0505E-10	298	1500
1255	C16H20	1-n-HEXYLNAPHTHALENE	-82.067	1.5218E+00	-1.2290E-03	5.2212E-07	-9.1550E-11	298	1500
1256	C16H22O4	DIBUTYL PHTHALATE	148.647	-3.1458E-01	4.5159E-03	-6.5996E-06	3.0370E-09	298	800
1257	C16H26	n-DECYLBENZENE	65.790	8.6789E-01	1.2244E-04	-5.0606E-07	1.7890E-10	200	1500
1258	C16H26	PENTAETHYLBENZENE	-15.092	1.4580E+00	-9.7337E-04	2.6329E-07	--------	298	1000
1259	C16H30	1-HEXADECYNE	-0.578	1.4321E+00	-8.4270E-04	1.9719E-07	--------	298	1000
1260	C16H32	n-DECYLCYCLOHEXANE	157.247	4.0618E-01	1.3486E-03	-1.5340E-06	4.6479E-10	150	1500
1261	C16H32	1-CYCLOPENTYLUNDECANE	-74.940	1.6734E+00	-1.0497E-03	2.7219E-07	--------	298	1000
1262	C16H32	1-HEXADECENE	108.689	8.5437E-01	2.7357E-04	-6.2732E-07	2.0929E-10	200	1500
1263	C16H32O2	n-HEXADECANOIC ACID	-54.487	1.8046E+00	-1.3388E-03	5.3671E-07	-9.2028E-11	298	1500
1264	C16H34	n-HEXADECANE	131.750	6.7397E-01	8.7770E-04	-1.2430E-06	3.9785E-10	150	1500
1265	C16H34O	DI-n-OCTYL ETHER	-33.322	1.7059E+00	-1.1626E-03	4.2114E-07	-6.4955E-11	298	1500
1266	C16H34O	1-HEXADECANOL	119.789	8.1674E-01	6.2364E-04	-1.0540E-06	3.4977E-10	150	1500
1267	C16H34S	BUTYL-DODECYL-SULFIDE	4.654	1.5120E+00	-7.8781E-04	1.5944E-07	--------	298	1000
1268	C16H34S	ETHYL-TETRADECYL-SULFIDE	2.086	1.5340E+00	-8.2898E-04	1.7579E-07	--------	298	1000
1269	C16H34S	METHYL-PENTADECYL-SULFIDE	4.843	1.5311E+00	-8.3266E-04	1.7816E-07	--------	298	1000
1270	C16H34S	OCTYL-SULFIDE	4.654	1.5120E+00	-7.8781E-04	1.5944E-07	--------	298	1000
1271	C16H34S	PROPYL-TRIDECYL-SULFIDE	4.654	1.5120E+00	-7.8781E-04	1.5944E-07	--------	298	1000
1272	C16H34S	1-HEXADECANETHIOL	3.223	1.5418E+00	-8.4634E-04	1.8383E-07	--------	298	1000
1273	C16H34S2	OCTYL-DISULFIDE	10.461	1.6113E+00	-9.3077E-04	2.1467E-07	--------	298	1000
1274	C17H28	n-UNDECYLBENZENE	2.711	1.2830E+00	-4.5442E-04	-1.5326E-07	9.7372E-11	200	1500
1275	C17H32	1-HEPTADECYNE	-1.487	1.5267E+00	-8.9638E-04	2.0853E-07	--------	298	1000
1276	C17H34	1-CYCLOPENTYLDODECANE	-75.781	1.7676E+00	-1.1022E-03	2.8279E-07	--------	298	1000
1277	C17H34	1-CYCLOHEXYLUNDECANE	-71.107	1.7505E+00	-1.0093E-03	2.1658E-07	--------	298	1000
1278	C17H34	1-HEPTADECENE	125.132	7.9164E-01	6.6154E-04	-1.0780E-06	3.5594E-10	150	1500
1279	C17H36	n-HEPTADECANE	111.903	9.5987E-01	2.7901E-04	-6.7520E-07	2.2545E-10	200	1500
1280	C17H36O	1-HEPTADECANOL	127.444	8.7086E-01	6.4059E-04	-1.0987E-06	3.6628E-10	150	1500
1281	C17H36S	BUTYL-TRIDECYL-SULFIDE	3.286	1.6093E+00	-8.4617E-04	1.7341E-07	--------	298	1000
1282	C17H36S	ETHYL-PENTADECYL-SULFIDE	1.074	1.6295E+00	-8.8420E-04	1.8811E-07	--------	298	1000
1283	C17H36S	HEXADECYL-METHYL-SULFIDE	3.852	1.6265E+00	-8.8768E-04	1.9038E-07	--------	298	1000
1284	C17H36S	PROPYL-TETRADECYL-SULFIDE	3.286	1.6093E+00	-8.4617E-04	1.7341E-07	--------	298	1000
1285	C17H36S	1-HEPTADECANETHIOL	2.201	1.6373E+00	-9.0169E-04	1.9621E-07	--------	298	1000
1286	C18H12	CHRYSENE	-41.414	1.1172E+00	-5.6061E-04	1.7203E-09	5.1921E-11	200	1500
1287	C18H14	m-TERPHENYL	-159.334	1.8173E+00	-1.7948E-03	9.0423E-07	-1.7969E-10	298	1600
1288	C18H14	o-TERPHENYL	-159.334	1.8173E+00	-1.7948E-03	9.0423E-07	-1.7969E-10	298	1600
1289	C18H14	p-TERPHENYL	-159.334	1.8173E+00	-1.7948E-03	9.0423E-07	-1.7969E-10	298	1600
1290	C18H15P	TRIPHENYLPHOSPHINE	-140.077	1.8099E+00	-1.7088E-03	8.3604E-07	-1.6512E-10	298	1500
1291	C18H15O4P	TRIPHENYL PHOSPHATE	--------	--------	--------	--------	--------	--------	--------
1292	C18H16N2	N,N'-DIPHENYL-p-PHENYLENEDIAMINE	-188.536	2.0474E+00	-1.9559E-03	9.5063E-07	-1.8600E-10	298	1500
1293	C18H22	2,3-DIMETHYL-2,3-DIPHENYLBUTANE	-159.612	1.9863E+00	-1.7695E-03	8.1753E-07	-1.5467E-10	298	1500
1294	C18H22O2	DICUMYL PEROXIDE	-76.744	1.6944E+00	-1.2744E-03	4.7912E-07	-7.3317E-11	298	1500
1295	C18H30	n-DODECYLBENZENE	106.695	7.7941E-01	5.8463E-04	-9.4162E-07	3.0397E-10	150	1500
1296	C18H30	HEXAETHYLBENZENE	-14.268	1.7028E+00	-1.1851E-03	3.3734E-07	--------	298	1000
1297	C18H32O2	LINOLEIC ACID	-113.518	2.1932E+00	-1.9690E-03	9.5258E-07	-1.7989E-10	298	1500
1298	C18H34	1-OCTADECYNE	-3.530	1.6283E+00	-9.6261E-04	2.2710E-07	--------	298	1000
1299	C18H34O2	OLEIC ACID	-55.663	1.9919E+00	-1.5562E-03	6.7067E-07	-1.2336E-10	298	1500
1300	C18H34O4	DIBUTYL SEBACATE	-64.966	2.2026E+00	-1.8844E-03	8.8961E-07	-1.7701E-10	298	1500
1301	C18H34O4	DIHEXYL ADIPATE	-64.614	2.0235E+00	-1.8000E-03	8.8573E-07	-1.8193E-10	298	1500
1302	C18H36	1-CYCOPENTYLTRIDECANE	-76.584	1.8616E+00	-1.1543E-03	2.9320E-07	--------	298	1000
1303	C18H36	1-CYCLOHEXYLDODECANE	-73.659	1.8553E+00	-1.0810E-03	2.3798E-07	--------	298	1000
1304	C18H36	1-OCTADECENE	216.021	4.1046E-01	1.3808E-03	-1.5473E-06	4.6513E-10	150	1500
1305	C18H36O2	STEARIC ACID	7.515	1.6996E+00	-9.2847E-04	1.7563E-07	6.6437E-12	298	1200
1306	C18H38	n-OCTADECANE	124.715	9.8653E-01	3.4273E-04	-7.4838E-07	2.4804E-10	200	1500
1307	C18H38O	DINONYL ETHER	-39.993	1.9149E+00	-1.2953E-03	4.5896E-07	-6.9698E-11	298	1500
1308	C18H38O	1-OCTADECANOL	129.533	9.2997E-01	6.6025E-04	-1.1313E-06	3.7237E-10	150	1500
1309	C18H38S	BUTYL-TETRADECYL-SULFIDE	2.285	1.7048E+00	-9.0128E-04	1.8569E-07	--------	298	1000
1310	C18H38S	ETHYL-HEXADECYL-SULFIDE	-0.274	1.7267E+00	-9.4236E-04	2.0198E-07	--------	298	1000
1311	C18H38S	HEPTADECYL-METHYL-SULFIDE	2.546	1.7235E+00	-9.4546E-04	2.0405E-07	--------	298	1000
1312	C18H38S	NONYL-SULFIDE	2.285	1.7048E+00	-9.0128E-04	1.8569E-07	--------	298	1000
1313	C18H38S	PENTADECYL-PROPYL-SULFIDE	2.285	1.7048E+00	-9.0128E-04	1.8569E-07	--------	298	1000
1314	C18H38S	1-OCTADECANETHIOL	0.874	1.7344E+00	-9.5964E-04	2.0998E-07	--------	298	1000
1315	C18H38S2	NONYL-DISULFIDE	8.113	1.8040E+00	-1.0441E-03	2.4081E-07	--------	298	1000
1316	C19H26	1-n-NONYLNAPHTHALENE	-76.861	1.7404E+00	-1.2853E-03	4.9048E-07	-7.7646E-11	298	1500
1317	C19H32	n-TRIDECYLBENZENE	86.726	1.0288E+00	1.7917E-04	-6.3099E-07	2.2067E-10	200	1500
1318	C19H36	1-NONADECYNE	-4.361	1.7226E+00	-1.0154E-03	2.3793E-07	--------	298	1000
1319	C19H36O2	METHYL OLEATE	-44.058	2.0299E+00	-1.5136E-03	6.1226E-07	-1.0656E-10	298	1500
1320	C19H38	1-CYCLOPENTYLTETRADECANE	-78.538	1.9628E+00	-1.2203E-03	3.1177E-07	--------	298	1000
1321	C19H38	1-CYCLOHEXYLTRIDECANE	-74.538	1.9494E+00	-1.1329E-03	2.4803E-07	--------	298	1000
1322	C19H38	1-NONADECENE	69.594	1.2608E+00	3.8456E-06	-6.0097E-07	2.2323E-10	200	1500
1323	C19H38O2	NONADECANOIC ACID	-23.711	1.9781E+00	-1.3539E-03	4.8185E-07	-7.2339E-11	298	1500
1324	C19H40	n-NONADECANE	132.530	1.0358E+00	3.6926E-04	-7.9581E-07	2.6362E-10	200	1500
1325	C19H40O	1-NONADECANOL	-10.671	1.8473E+00	-1.0527E-03	2.3365E-07	--------	298	1000
1326	C19H40S	BUTYL-PENTADECYL-SULFIDE	0.926	1.8020E+00	-9.5956E-04	1.9961E-07	--------	298	1000

49

Table 2-1 HEAT CAPACITY OF GAS - ORGANIC COMPOUNDS (continued)

			$C_P = A + B\,T + C\,T^2 + D\,T^3 + E\,T^4$				$(C_P$ - joule/(mol K), T - K)		
NO	FORMULA	NAME	A	B	C	D	E	TMIN	TMAX
1327	C19H40S	ETHYL-HEPTADECYL-SULFIDE	-1.379	1.8227E+00	-9.9872E-04	2.1498E-07	--------	298	1000
1328	C19H40S	HEXADECYL-PROPYL-SULFIDE	0.926	1.8020E+00	-9.5956E-04	1.9961E-07	--------	298	1000
1329	C19H40S	METHYL-OCTADECYL-SULFIDE	1.431	1.8196E+00	-1.0019E-03	2.1709E-07	--------	298	1000
1330	C19H40S	1-NONADECANETHIOL	1.139	1.8243E+00	-1.0070E-03	2.1884E-07	--------	298	1000
1331	C20H16	TRIPHENYLETHYLENE	-122.663	1.7649E+00	-1.5644E-03	7.2308E-07	-1.3678E-10	298	1500
1332	C20H28	1-n-DECYLNAPHTHALENE	-82.955	1.8664E+00	-1.4003E-03	5.4871E-07	-8.9673E-11	298	1500
1333	C20H30O2	ABIETIC ACID	-260.137	2.7296E+00	-2.3498E-03	1.0242E-06	-1.8203E-10	298	1500
1334	C20H31N	DEHYDROABIETYLAMINE	-127.428	2.3347E+00	-2.0102E-03	9.0944E-07	-1.7163E-10	298	1200
1335	C20H34	1-PHENYLTETRADECANE	-53.024	1.8455E+00	-1.1361E-03	2.7552E-07	--------	298	1000
1336	C20H38	1-EICOSYNE	-5.184	1.8167E+00	-1.0677E-03	2.4845E-07	--------	298	1000
1337	C20H40	1-CYCLOPENTYLPENTADECANE	-79.601	2.0582E+00	-1.2749E-03	3.2342E-07	--------	298	1000
1338	C20H40	1-CYCLOHEXYLTETRADECANE	-75.450	2.0437E+00	-1.1851E-03	2.5824E-07	--------	298	1000
1339	C20H40	1-EICOSENE	125.107	1.1243E+00	2.5851E-04	-7.3027E-07	2.4754E-10	200	1500
1340	C20H42	n-EICOSANE	137.730	1.0992E+00	3.6839E-04	-8.2058E-07	2.7259E-10	200	1500
1341	C20H42O	1-EICOSANOL	149.479	9.8474E-01	8.2856E-04	-1.3392E-06	4.3815E-10	150	1500
1342	C20H42S	BUTYL-HEXADECYL-SULFIDE	-0.147	1.8979E+00	-1.0156E-03	2.1245E-07	--------	298	1000
1343	C20H42S	DECYL-SULFIDE	-0.147	1.8979E+00	-1.0156E-03	2.1245E-07	--------	298	1000
1344	C20H42S	ETHYL-OCTADECYL-SULFIDE	-2.711	1.9199E+00	-1.0566E-03	2.2859E-07	--------	298	1000
1345	C20H42S	HEPTADECYL-PROPYL-SULFIDE	-0.147	1.8979E+00	-1.0156E-03	2.1245E-07	--------	298	1000
1346	C20H42S	METHYL-NONADECYL-SULFIDE	0.087	1.9168E+00	-1.0598E-03	2.3076E-07	--------	298	1000
1347	C20H42S	1-EICOSANETHIOL	-1.585	1.9278E+00	-1.0740E-03	2.3669E-07	--------	298	1000
1348	C20H42S2	DECYL-DISULFIDE	5.649	1.9974E+00	-1.1587E-03	2.6773E-07	--------	298	1000
1349	C21H21O4P	TRI-o-CRESYL PHOSPHATE	--------	--------	--------	--------	--------	--------	--------
1350	C21H36	1-PHENYLPENTADECANE	-54.969	1.9466E+00	-1.2016E-03	2.9377E-07	--------	298	1000
1351	C21H42	1-CYCLOPENTYLHEXADECANE	-80.324	2.1518E+00	-1.3266E-03	3.3374E-07	--------	298	1000
1352	C21H42	1-CYCLOHEXYLPENTADECANE	-77.977	2.1483E+00	-1.2565E-03	2.7954E-07	--------	298	1000
1353	C22H38	1-PHENYLHEXADECANE	-55.944	2.0416E+00	-1.2556E-03	3.0517E-07	--------	298	1000
1354	C22H44	1-CYCLOHEXYLHEXADECANE	-76.584	2.2301E+00	-1.2898E-03	2.8109E-07	--------	298	1000
1355	C22H44O2	n-BUTYL STEARATE	-69.046	2.5211E+00	-2.0966E-03	1.0835E-06	-2.7531E-10	298	1200
1356	C24H38O4	DIISOOCTYL PHTHALATE	-87.607	2.2843E+00	-1.3309E-03	2.5129E-07	1.4926E-11	298	1500
1357	C24H38O4	DIOCTYL PHTHALATE	-82.524	2.2568E+00	-1.2839E-03	2.1577E-07	2.8112E-11	298	1100
1358	C24H42O	DINONYLPHENOL	-21.466	2.1296E+00	-1.0229E-03	-8.3690E-08	1.5876E-10	298	1000
1359	C26H20	TETRAPHENYLETHYLENE	-149.970	2.2370E+00	-1.9635E-03	8.9829E-07	-1.6846E-10	298	1500
1360	C28H46O4	DIISODECYL PHTHALATE	-128.947	2.9258E+00	-2.1855E-03	9.4298E-07	-2.1374E-10	298	1200

C_P - heat capacity of ideal gas, joule/(mol K)

A, B, C, D, and E - regression coefficients for chemical compound

T - temperature, K

TMIN - minimum temperature, K

TMAX - maximum temperature, K

Table 2-2 HEAT CAPACITY OF GAS - INORGANIC COMPOUNDS

NO	FORMULA	NAME	$C_P = A + B T + C T^2 + D T^3 + E T^4$ (C_P - joule/(mol K), T - K)						
			A	B	C	D	E	TMIN	TMAX
1	Ag	SILVER	20.794	0.0000E+00	0.0000E+00	0.0000E+00	0.0000E+00	298	6000
2	AgCl	SILVER CHLORIDE	32.931	1.3917E-02	-1.6486E-05	8.4479E-09	-1.5693E-12	298	2000
3	AgI	SILVER IODIDE	35.710	5.3212E-03	-6.3035E-06	3.2301E-09	-6.0004E-13	298	2000
4	Al	ALUMINUM	21.636	-1.3303E-03	7.3647E-07	-1.7803E-10	1.6481E-14	298	6000
5	AlB3H12	ALUMINUM BOROHYDRIDE	-------	----------	----------	----------	----------	----	-----
6	AlBr3	ALUMINUN BROMIDE	39.535	2.0117E-01	-3.2271E-04	2.2542E-07	-5.7081E-11	100	1500
7	AlCl3	ALUMINUM CHLORIDE	34.535	2.0117E-01	-3.2271E-04	2.2542E-07	-5.7081E-11	100	1500
8	AlF3	ALUMINUM FLUORIDE	48.884	4.9935E-02	-2.4534E-05	4.8689E-09	-3.3559E-13	200	6000
9	AlI3	ALUMINUM IODIDE	69.896	2.1051E-02	-1.0811E-05	2.2023E-09	-1.5440E-13	200	6000
10	Al2O3	ALUMINUM OXIDE	-------	----------	----------	----------	----------	----	-----
11	Al2S3O12	ALUMINUM SULFATE	-------	----------	----------	----------	----------	----	-----
12	Ar	ARGON	20.786	0.0000E+00	0.0000E+00	0.0000E+00	0.0000E+00	100	1500
13	As	ARSENIC	20.720	5.2598E-05	9.7228E-08	-1.9217E-11	1.3200E-15	298	6000
14	AsBr3	ARSENIC TRIBROMIDE	68.784	5.9075E-02	-6.9453E-05	3.5616E-08	-6.6261E-12	298	2000
15	AsCl3	ARSENIC TRICHLORIDE	63.784	5.9075E-02	-6.9453E-05	3.5616E-08	-6.6261E-12	298	2000
16	AsF3	ARSENIC TRIFLUORIDE	39.648	1.2752E-01	-1.4842E-04	7.6110E-08	-1.4160E-11	298	2000
17	AsF5	ARSENIC PENTAFLUORIDE	74.312	1.2752E-01	-1.4842E-04	7.6110E-08	-1.4160E-11	298	2000
18	AsH3	ARSINE	31.578	2.2579E-04	1.2295E-04	-1.3416E-07	4.1378E-11	80	1500
19	AsI3	ARSENIC TRIIODIDE	76.357	2.1285E-02	-2.5214E-05	1.2920E-08	-2.4001E-12	298	2000
20	As2O3	ARSENIC TRIOXIDE	-------	----------	----------	----------	----------	----	-----
21	At	ASTATINE	-------	----------	----------	----------	----------	----	-----
22	Au	GOLD	20.956	-1.3652E-03	1.6529E-06	-3.2668E-10	2.2440E-14	298	6000
23	B	BORON	20.811	-4.4781E-05	3.5078E-08	-1.4447E-11	2.2709E-15	200	6000
24	BBr3	BORON TRIBROMIDE	38.762	1.4855E-01	-2.0938E-04	1.3632E-07	-3.3271E-11	298	1500
25	BCl3	BORON TRICHLORIDE	24.444	1.9076E-01	-2.6142E-04	1.6467E-07	-3.8875E-11	100	1500
26	BF3	BORON TRIFLUORIDE	22.487	1.1814E-01	-8.7099E-05	2.2344E-08	1.2182E-13	100	1500
27	BH2CO	BORINE CARBONYL	-2.568	1.7067E-01	-6.8997E-05	1.2108E-08	-7.6892E-13	298	1500
28	BH3O3	BORIC ACID	33.718	1.1026E-01	1.1706E-05	-6.4763E-08	2.3670E-11	100	1500
29	B2D6	DEUTERODIBORANE	21.184	1.7067E-01	-6.8997E-05	1.2108E-08	-7.6892E-13	100	6000
30	B2H5Br	DIBORANE HYDROBROMIDE	31.932	1.7067E-01	-6.8997E-05	1.2108E-08	-7.6892E-13	100	6000
31	B2H6	DIBORANE	19.984	1.7067E-01	-6.8997E-05	1.2108E-08	-7.6892E-13	100	6000
32	B3N3H6	BORINE TRIAMINE	-38.941	6.0750E-01	-5.9547E-04	3.0827E-07	-6.4789E-11	298	1500
33	B4H10	TETRABORANE	-66.873	5.6949E-01	-3.3162E-04	5.6690E-08	7.8563E-12	298	1500
34	B5H9	PENTABORANE	-48.121	5.6949E-01	-3.3162E-04	5.6690E-08	7.8563E-12	298	1500
35	B5H11	TETRAHYDROPENTABORANE	-47.121	5.6949E-01	-3.3162E-04	5.6690E-08	7.8563E-12	298	1500
36	B10H14	DECABORANE	-------	----------	----------	----------	----------	----	-----
37	Ba	BARIUM	22.141	-3.5147E-04	1.3612E-06	-2.6903E-10	1.8480E-14	298	6000
38	Be	BERYLLIUM	20.492	1.2418E-03	-1.1098E-06	3.0261E-10	-1.7385E-14	200	6000
39	BeB2H8	BERYLLIUM BOROHYDRIDE	-------	----------	----------	----------	----------	----	-----
40	BeBr2	BERYLLIUM BROMIDE	49.193	1.9295E-02	-9.5080E-06	1.8903E-09	-1.3044E-13	200	6000
41	BeCl2	BERYLLIUM CHLORIDE	44.579	2.5358E-02	-1.2328E-05	2.4319E-09	-1.6699E-13	200	6000
42	BeF2	BERYLLIUM FLUORIDE	36.836	3.4492E-02	-1.6319E-05	3.1695E-09	-2.1551E-13	200	6000
43	BeI2	BERYLLIUM IODIDE	51.249	1.6602E-02	-8.2631E-06	1.6524E-09	-1.1446E-13	200	6000
44	Bi	BISMUTH	20.794	0.0000E+00	0.0000E+00	0.0000E+00	0.0000E+00	298	3000
45	BiBr3	BISMUTH TRIBROMIDE	-------	----------	----------	----------	----------	----	-----
46	BiCl3	BISMUTH TRICHLORIDE	66.210	7.9478E-02	-1.5111E-04	1.3058E-07	-4.2429E-11	298	1000
47	BrF5	BROMINE PENTAFLUORIDE	27.183	3.9339E-01	-5.9604E-04	4.0800E-07	-1.0308E-10	298	1500
48	Br2	BROMINE	27.169	4.9172E-02	-8.5027E-05	6.2796E-08	-1.6556E-11	100	1500
49	C	CARBON	21.069	-7.9119E-04	5.0895E-07	-6.9132E-11	2.7021E-15	298	6000
50	CCl2O	PHOSGENE	20.747	1.7972E-01	-2.3242E-04	1.4224E-07	-3.3087E-11	100	1500
51	CF2O	CARBONYL FLUORIDE	23.640	8.9853E-02	-2.4575E-05	-2.8140E-08	1.4023E-11	100	1500
52	CH4N2O	UREA	24.856	1.4437E-01	3.8088E-05	-1.1007E-07	3.9161E-11	150	1500
53	CH4N2S	THIOUREA	21.530	2.2204E-01	-1.7193E-04	7.4203E-08	-1.3867E-11	273	1500
54	CNBr	CYANOGEN BROMIDE	31.562	7.7072E-02	-1.0251E-04	7.0456E-08	-1.8400E-11	298	1500
55	CNCl	CYANOGEN CHLORIDE	21.270	1.1915E-01	-1.6822E-04	1.1457E-07	-2.9210E-11	100	1500
56	CNF	CYANOGEN FLUORIDE	26.132	7.5002E-02	-8.3145E-05	4.9592E-08	-1.1885E-11	298	1500
57	CO	CARBON MONOXIDE	29.556	-6.5807E-03	2.0130E-05	-1.2227E-08	2.2617E-12	60	1500
58	COS	CARBONYL SULFIDE	20.913	9.2794E-02	-9.7014E-05	5.0943E-08	-1.0615E-11	100	1500
59	COSe	CARBON OXYSELENIDE	21.912	9.2794E-02	-9.7014E-05	5.0943E-08	-1.0615E-11	100	1500
60	CO2	CARBON DIOXIDE	27.437	4.2315E-02	-1.9555E-05	3.9968E-09	-2.9872E-13	50	5000
61	CS2	CARBON DISULFIDE	20.461	1.2299E-01	-1.6184E-04	1.0199E-07	-2.4444E-11	100	1500
62	CSeS	CARBON SELENOSULFIDE	21.461	1.2299E-01	-1.6184E-04	1.0199E-07	-2.4444E-11	100	1500
63	C2N2	CYANOGEN	32.265	1.1687E-01	-1.4171E-04	9.2703E-08	-2.3760E-11	298	1500
64	C3S2	CARBON SUBSULFIDE	-------	----------	----------	----------	----------	----	-----
65	Ca	CALCIUM	19.595	5.6939E-03	-6.2707E-06	2.2198E-09	-1.9289E-13	200	6000
66	CaF2	CALCIUM FLUORIDE	31.889	9.1682E-02	-1.1853E-04	6.5077E-08	-1.2762E-11	100	2000
67	CbF5	COLUMBIUM FLUORIDE	-------	----------	----------	----------	----------	----	-----
68	Cd	CADMIUM	20.794	0.0000E+00	0.0000E+00	0.0000E+00	0.0000E+00	298	6000
69	CdCl2	CADMIUM CHLORIDE	-------	----------	----------	----------	----------	----	-----
70	CdF2	CADMIUM FLUORIDE	-------	----------	----------	----------	----------	----	-----
71	CdI2	CADMIUM IODIDE	-------	----------	----------	----------	----------	----	-----
72	CdO	CADMIUM OXIDE	-------	----------	----------	----------	----------	----	-----
73	ClF	CHLORINE MONOFLUORIDE	22.567	4.7581E-02	-6.3572E-05	3.9963E-08	-9.4968E-12	298	1500
74	ClFO3	PERCHLORYL FLUORIDE	13.200	2.3797E-01	-2.5150E-04	1.2324E-07	-2.2897E-11	100	1500
75	ClF3	CHLORINE TRIFLUORIDE	21.386	2.2286E-01	-3.3105E-04	2.2357E-07	-5.5964E-11	298	1500
76	ClF5	CHLORINE PENTAFLUORIDE	15.530	4.3077E-01	-6.4695E-04	4.4026E-07	-1.1080E-10	298	1500
77	ClHO3S	CHLOROSULFONIC ACID	21.765	2.7543E-01	-3.3639E-04	2.0259E-07	-4.6684E-11	300	1500
78	ClHO4	PERCHLORIC ACID	16.002	2.7543E-01	-3.3639E-04	2.0259E-07	-4.6684E-11	298	1500

51

Table 2-2 HEAT CAPACITY OF GAS - INORGANIC COMPOUNDS (continued)

NO	FORMULA	NAME	$C_P = A + B T + C T^2 + D T^3 + E T^4$				$(C_P$ - joule/(mol K), T - K)$			
			A	B	C	D	E	TMIN	TMAX
79	ClO2	CHLORINE DIOXIDE	30.482	3.9797E-02	4.5262E-06	-3.2447E-08	1.3089E-11	50	1500
80	Cl2	CHLORINE	27.213	3.0426E-02	-3.3353E-05	1.5961E-08	-2.7021E-12	50	1500
81	Cl2O	CHLORINE MONOXIDE	25.608	1.1593E-01	-1.7038E-04	1.1417E-07	-2.8420E-11	298	1500
82	Cl2O7	CHLORINE HEPTOXIDE	110.489	3.9797E-02	4.5262E-06	-3.2447E-08	1.3089E-11	50	1500
83	Co	COBALT	20.226	1.0902E-02	-5.9195E-06	1.2340E-09	-8.3106E-14	200	6000
84	CoCl2	COBALT CHLORIDE	66.944	0.0000E+00	0.0000E+00	0.0000E+00	0.0000E+00	298	2000
85	CoNC3O4	COBALT NITROSYL TRICARBONYL	76.352	8.9853E-02	-2.4575E-05	-2.8140E-08	1.4023E-11	100	1500
86	Cr	CHROMIUM	20.646	-1.0807E-03	2.3335E-06	-4.6120E-10	3.1680E-14	298	6000
87	CrC6O6	CHROMIUM CARBONYL	-------	----------	----------	----------	----------	----	-----
88	CrO2Cl2	CHROMIUM OXYCHLORIDE	61.375	8.9853E-02	-2.4575E-05	-2.8140E-08	1.4023E-11	100	1500
89	Cs	CESIUM	17.852	1.2814E-02	-1.2546E-05	4.0771E-09	-3.4005E-13	200	6000
90	CsBr	CESIUM BROMIDE	-------	----------	----------	----------	----------	----	-----
91	CsCl	CESIUM CHLORIDE	35.901	3.2410E-03	-1.2812E-06	2.6274E-10	-1.8495E-14	200	6000
92	CsF	CESIUM FLUORIDE	33.868	6.2098E-03	-2.8937E-06	5.8810E-10	-4.1159E-14	200	6000
93	CsI	CESIUM IODIDE	36.623	2.4559E-03	-2.9093E-06	1.4908E-09	-2.7694E-13	298	2000
94	Cu	COPPER	21.474	-2.4764E-03	1.3408E-06	3.1615E-11	-2.1552E-14	200	6000
95	CuBr	CUPROUS BROMIDE	33.713	1.1461E-02	-1.3577E-05	6.9571E-09	-1.2924E-12	298	2000
96	CuCl	CUPROUS CHLORIDE	26.113	4.7580E-02	-7.3905E-05	4.9836E-08	-1.2107E-11	100	1600
97	CuCl2	CUPRIC CHLORIDE	-------	----------	----------	----------	----------	----	-----
98	CuI	COPPER IODIDE	34.537	9.0051E-03	-1.0667E-05	5.4663E-09	-1.0154E-12	298	2000
99	DCN	DEUTERIUM CYANIDE	25.967	3.7969E-02	-1.2416E-05	-3.2240E-09	2.2610E-12	100	1500
100	D2	DEUTERIUM	31.159	-1.2796E-02	2.4964E-05	-1.5015E-08	3.3248E-12	100	1500
101	D2O	DEUTERIUM OXIDE	33.308	-4.6722E-03	3.4878E-05	-2.2602E-08	4.4864E-12	100	2000
102	Eu	EUROPIUM	20.836	0.0000E+00	0.0000E+00	0.0000E+00	0.0000E+00	298	3000
103	F2	FLUORINE	27.408	1.2928E-02	7.0701E-06	-1.6302E-08	5.9789E-12	100	1500
104	F2O	FLUORINE OXIDE	16.655	1.3539E-01	-1.8807E-04	1.2034E-07	-2.8760E-11	298	1500
105	Fe	IRON	23.362	1.7756E-04	-1.0357E-06	5.7896E-10	-5.7987E-14	100	6000
106	FeC5O5	IRON PENTACARBONYL	86.030	8.9853E-02	-2.4575E-05	-2.8140E-08	1.4023E-11	100	1500
107	FeCl2	FERROUS CHLORIDE	55.454	1.8947E-02	-1.0585E-05	2.2221E-09	-1.5836E-13	200	6000
108	FeCl3	FERRIC CHLORIDE	71.720	1.8442E-02	-9.5458E-06	1.9534E-09	-1.3734E-13	200	6000
109	Fr	FRANCIUM	20.836	0.0000E+00	0.0000E+00	0.0000E+00	0.0000E+00	298	3000
110	Ga	GALLIUM	15.486	5.6271E-02	-9.3476E-05	5.6763E-08	-1.1669E-11	100	2000
111	GaCl3	GALLIUM TRICHLORIDE	53.444	1.0963E-01	-1.5367E-04	9.5770E-08	-2.2076E-11	300	1500
112	Gd	GADOLINIUM	29.589	-9.5360E-03	8.3087E-06	-3.0415E-09	3.9613E-13	298	3000
113	Ge	GERMANIUM	32.510	-5.9413E-03	-6.9056E-17	3.8797E-20	-7.3999E-24	298	2300
114	GeBr4	GERMANIUM BROMIDE	87.753	7.3780E-02	-1.0615E-04	6.8167E-08	-1.6121E-11	298	1500
115	GeCl4	GERMANIUM CHLORIDE	69.397	1.3977E-01	-2.0098E-04	1.2907E-07	-3.0523E-11	298	1500
116	GeHCl3	TRICHLORO GERMANE	24.939	2.5068E-01	-3.4090E-04	2.1707E-07	-5.2003E-11	100	1500
117	GeH4	GERMANE	-15.224	3.0554E-01	-4.0678E-04	2.6063E-07	-6.1884E-11	200	1500
118	Ge2H6	DIGERMANE	21.867	3.0554E-01	-4.0678E-04	2.6063E-07	-6.1884E-11	200	1500
119	Ge3H8	TRIGERMANE	63.478	3.0554E-01	-4.0678E-04	2.6063E-07	-6.1884E-11	200	1500
120	HBr	HYDROGEN BROMIDE	30.169	-8.0274E-03	1.6731E-05	-7.4730E-09	8.3068E-13	200	1500
121	HCN	HYDROGEN CYANIDE	25.766	3.7969E-02	-1.2416E-05	-3.2240E-09	2.2610E-12	100	1500
122	HCl	HYDROGEN CHLORIDE	29.244	-1.2615E-03	1.1210E-06	4.9676E-09	-2.4963E-12	50	1500
123	HF	HYDROGEN FLUORIDE	29.085	9.6118E-04	-4.4705E-06	6.7830E-09	-2.1975E-12	50	1500
124	HI	HYDROGEN IODIDE	29.770	-7.4945E-03	2.0687E-05	-1.1963E-08	2.1010E-12	100	1500
125	HNO3	NITRIC ACID	19.755	1.3415E-01	-6.1116E-05	-1.2343E-08	1.1106E-11	100	1500
126	H2	HYDROGEN	25.399	2.0178E-02	-3.8549E-05	3.1880E-08	-8.7585E-12	250	1500
127	H2O	WATER	33.933	-8.4186E-03	2.9906E-05	-1.7825E-08	3.6934E-12	100	1500
128	H2O2	HYDROGEN PEROXIDE	36.181	8.2657E-03	6.6420E-05	-6.9944E-08	2.0951E-11	100	1500
129	H2S	HYDROGEN SULFIDE	33.878	-1.1216E-02	5.2578E-05	-3.8397E-08	9.0281E-12	100	1500
130	H2SO4	SULFURIC ACID	9.486	3.3795E-01	-3.8078E-04	2.1308E-07	-4.6878E-11	100	1500
131	H2S2	HYDROGEN DISULFIDE	58.617	-1.1216E-02	5.2578E-05	-3.8397E-08	9.0281E-12	100	1500
132	H2Se	HYDROGEN SELENIDE	34.878	-1.1216E-02	5.2578E-05	-3.8397E-08	9.0281E-12	100	1500
133	H2Te	HYDROGEN TELLURIDE	34.878	-1.1216E-02	5.2578E-05	-3.8397E-08	9.0281E-12	100	1500
134	H3NO3S	SULFAMIC ACID	-------	----------	----------	----------	----------	----	-----
135	He	HELIUM-3	20.786	0.0000E+00	0.0000E+00	0.0000E+00	0.0000E+00	100	1500
136	He	HELIUM-4	20.786	0.0000E+00	0.0000E+00	0.0000E+00	0.0000E+00	100	1500
137	Hf	HAFNIUM	17.312	8.8863E-03	6.5082E-07	-8.0882E-10	9.4573E-14	298	6000
138	Hg	MERCURY	20.790	-1.8318E-05	2.3525E-08	-1.0144E-11	1.3685E-15	200	6000
139	HgBr2	MERCURIC BROMIDE	57.914	7.2022E-03	-3.7349E-06	7.6521E-10	-5.3849E-14	200	6000
140	HgCl2	MERCURIC CHLORIDE	53.826	1.3601E-02	-6.9949E-06	1.4260E-09	-1.0001E-13	200	6000
141	HgI2	MERCURIC IODIDE	59.646	4.4227E-03	-2.2995E-06	4.7174E-10	-3.3219E-14	200	6000
142	IF7	IODINE HEPTAFLUORIDE	38.537	5.0269E-01	-7.1344E-04	4.5817E-07	-1.0835E-10	298	1500
143	I2	IODINE	34.151	1.3930E-02	-2.0952E-05	1.4362E-08	-3.5948E-12	100	1500
144	In	INDIUM	14.644	2.1769E-02	-1.2076E-05	2.4236E-09	-1.6645E-13	298	6000
145	Ir	IRIDIUM	19.981	2.0510E-03	2.0772E-06	-7.6038E-10	9.9033E-14	298	3000
146	K	POTASSIUM	20.010	3.2000E-03	-2.8000E-06	7.2300E-10	-1.5340E-14	100	5900
147	KBr	POTASSIUM BROMIDE	35.829	3.3664E-03	-1.3392E-06	2.7463E-10	-1.9336E-14	200	6000
148	KCl	POTASSIUM CHLORIDE	28.337	4.5356E-02	-7.9274E-05	5.9311E-08	-1.5806E-11	100	1500
149	KF	POTASSIUM FLUORIDE	32.731	8.0972E-03	-3.7990E-06	7.7311E-10	-5.4167E-14	200	6000
150	KI	POTASSIUM IODIDE	36.133	3.1520E-03	-1.3495E-06	3.0109E-10	-2.2575E-14	200	6000
151	KOH	POTASSIUM HYDROXIDE	21.454	1.4891E-01	-2.5712E-04	1.9271E-07	-5.1494E-11	100	1500
152	Kr	KRYPTON	20.786	0.0000E+00	0.0000E+00	0.0000E+00	0.0000E+00	100	6200
153	La	LANTHANUM	18.456	1.7948E-02	-8.0882E-06	1.5357E-09	-1.0160E-13	298	6000
154	Li	LITHIUM	20.837	-1.1700E-04	-4.9553E-08	4.8575E-11	6.0679E-15	200	6000
155	LiBr	LITHIUM BROMIDE	30.831	1.0693E-02	-5.0365E-06	1.0064E-09	-6.9372E-14	200	6000
156	LiCl	LITHIUM CHLORIDE	29.887	1.1939E-02	-5.6863E-06	1.1407E-09	-7.9176E-14	200	6000

Table 2-2 HEAT CAPACITY OF GAS - INORGANIC COMPOUNDS (continued)

NO	FORMULA	NAME	$C_P = A + B\,T + C\,T^2 + D\,T^3 + E\,T^4$ (C_P - joule/(mol K), T - K)						
			A	B	C	D	E	TMIN	TMAX
157	LiF	LITHIUM FLUORIDE	27.741	1.4047E-02	-6.4527E-06	1.2601E-09	-8.5917E-14	200	6000
158	LiI	LITHIUM IODIDE	31.675	9.6641E-03	-4.5580E-06	9.2351E-10	-6.4516E-14	200	6000
159	Lu	LUTECIUM	15.468	2.5496E-02	-2.2434E-05	8.2121E-09	-1.0696E-12	298	3000
160	Mg	MAGNESIUM	20.549	9.9274E-04	-8.6947E-07	2.2187E-10	-8.4283E-15	200	6000
161	MgCl2	MAGNESIUM CHLORIDE	52.489	1.5354E-02	-7.8027E-06	1.5797E-09	-1.1030E-13	200	6000
162	MgO	MAGNESIUM OXIDE	27.851	1.3866E-02	-6.4322E-06	1.2585E-09	-8.5915E-14	200	6000
163	Mn	MANGANESE	20.794	0.0000E+00	0.0000E+00	0.0000E+00	0.0000E+00	298	6000
164	MnCl2	MANGANESE CHLORIDE	32.112	1.6131E-02	-1.8910E-05	9.6902E-09	-1.8001E-12	298	2000
165	Mo	MOLYBDENUM	20.075	3.7542E-03	-4.8253E-06	2.0008E-09	-1.8505E-13	200	6000
166	MoF6	MOLYBDENUM FLUORIDE	41.680	4.1131E-01	-6.0261E-04	4.0311E-07	-1.0025E-10	298	1500
167	MoO3	MOLYBDENUM OXIDE	45.345	5.4478E-02	-2.6605E-05	5.2605E-09	-3.6170E-13	200	6000
168	NCl3	NITROGEN TRICHLORIDE	30.253	1.7432E-01	-2.1734E-04	1.1642E-07	-2.2436E-11	100	2000
169	ND3	HEAVY AMMONIA	34.574	-1.2581E-02	8.8906E-05	-7.1783E-08	1.8569E-11	100	1500
170	NF3	NITROGEN TRIFLUORIDE	18.732	1.5505E-01	-1.4305E-04	5.3741E-08	-5.8443E-12	100	1500
171	NH3	AMMONIA	33.573	-1.2581E-02	8.8906E-05	-7.1783E-08	1.8569E-11	100	1500
172	NH3O·	HYDROXYLAMINE	21.935	1.0340E-01	-5.8693E-05	1.0557E-08	1.5150E-12	200	1500
173	NH4Br	AMMONIUM BROMIDE	--------	----------	----------	----------	----------	----	-----
174	NH4Cl	AMMONIUM CHLORIDE	--------	----------	----------	----------	----------	----	-----
175	NH4I	AMMONIUM IODIDE	--------	----------	----------	----------	----------	----	-----
176	NH5O	AMMONIUM HYDROXIDE	--------	----------	----------	----------	----------	----	-----
177	NH5S	AMMONIUM HYDROGENSULFIDE	--------	----------	----------	----------	----------	----	-----
178	NO	NITRIC OXIDE	33.227	-2.3626E-02	5.3156E-05	-3.7858E-08	9.1197E-12	50	1500
179	NOCl	NITROSYL CHLORIDE	28.551	7.5899E-02	-9.4410E-05	6.0476E-08	-1.5054E-11	100	1500
180	NOF	NITROSYL FLUORIDE	27.551	7.5899E-02	-9.4410E-05	6.0476E-08	-1.5054E-11	100	1500
181	NO2	NITROGEN DIOXIDE	32.791	-7.4294E-04	8.1722E-05	-8.2872E-08	2.4424E-11	50	1500
182	N2	NITROGEN	29.342	-3.5395E-03	1.0076E-05	-4.3116E-09	2.5935E-13	50	1500
183	N2F4	TETRAFLUOROHYDRAZINE	12.422	3.0609E-01	-3.1077E-04	1.3914E-07	-2.2235E-11	100	1500
184	N2H4	HYDRAZINE	23.630	9.1270E-02	2.9042E-05	-7.1858E-08	2.5093E-11	100	1500
185	N2H4C	AMMONIUM CYANIDE	52.812	9.1270E-02	2.9042E-05	-7.1858E-08	2.5093E-11	100	1500
186	N2H6CO2	AMMONIUM CARBAMATE	--------	----------	----------	----------	----------	----	-----
187	N2O	NITROUS OXIDE	23.219	6.1984E-02	-3.7989E-05	6.9671E-09	8.1421E-13	100	1500
188	N2O3	NITROGEN TRIOXIDE	28.509	1.6895E-01	-1.8161E-04	9.9662E-08	-2.1975E-11	100	1500
189	N2O4	NITROGEN TETRAOXIDE	29.587	2.2719E-01	-2.2740E-04	1.0698E-07	-1.9223E-11	50	1500
190	N2O5	NITROGEN PENTOXIDE	63.710	1.2317E-01	-5.9937E-05	1.1842E-08	-8.1522E-13	200	6000
191	Na	SODIUM	20.904	-5.7046E-04	6.8044E-07	-3.1392E-10	5.4554E-14	200	6000
192	NaBr	SODIUM BROMIDE	34.638	5.2612E-03	-2.3207E-06	4.7485E-10	-3.3382E-14	200	6000
193	NaCN	SODIUM CYANIDE	42.090	5.0830E-02	-8.5459E-05	6.2288E-08	-1.6304E-11	100	1500
194	NaCl	SODIUM CHLORIDE	26.445	5.0830E-02	-8.5459E-05	6.2288E-08	-1.6304E-11	100	1500
195	NaF	SODIUM FLUORIDE	25.450	4.4529E-02	-6.3995E-05	4.1793E-08	-1.0111E-11	100	1500
196	NaI	SODIUM IODIDE	--------	----------	----------	----------	----------	----	-----
197	NaOH	SODIUM HYDROXIDE	22.246	1.4234E-01	-2.4264E-04	1.8054E-07	-4.8026E-11	100	1500
198	Na2SO4	SODIUM SULFATE	23.349	4.0133E-01	-5.0787E-04	2.9884E-07	-6.6688E-11	100	1500
199	Nb	NIOBIUM	34.103	-1.3370E-02	5.8793E-06	-8.2040E-10	4.0084E-14	298	6000
200	Nd	NEODYMIUM	16.407	2.3782E-02	-1.4863E-05	3.4524E-09	-2.6839E-13	298	3000
201	Ne	NEON	20.786	0.0000E+00	0.0000E+00	0.0000E+00	0.0000E+00	100	1500
202	Ni	NICKEL	22.126	6.6383E-03	-5.2966E-06	1.4438E-09	-1.3443E-13	298	4500
203	NiC4O4	NICKEL CARBONYL	69.830	8.9853E-02	-2.4575E-05	-2.8140E-08	1.4023E-11	100	1500
204	NiF2	NICKEL FLUORIDE	--------	----------	----------	----------	----------	----	-----
205	Np	NEPTUNIUM	--------	----------	----------	----------	----------	----	-----
206	O2	OXYGEN	29.526	-8.8999E-03	3.8083E-05	-3.2629E-08	8.8607E-12	50	1500
207	O3	OZONE	31.467	1.4982E-02	6.7966E-05	-8.4157E-08	2.7205E-11	50	1500
208	Os	OSMIUM	22.291	-9.3555E-03	1.5950E-05	-6.4535E-09	8.4379E-13	298	3000
209	OsOF5	OSMIUM OXIDE PENTAFLUORIDE	--------	----------	----------	----------	----------	----	-----
210	OsO4	OSMIUM TETROXIDE - YELLOW	24.585	2.4934E-01	-3.3236E-04	2.1344E-07	-5.0476E-11	298	1500
211	OsO4	OSMIUM TETROXIDE - WHITE	24.585	2.4934E-01	-3.3236E-04	2.1344E-07	-5.0476E-11	298	1500
212	P	PHOSPHORUS - WHITE	20.785	-1.1821E-05	5.0830E-08	-8.1726E-11	4.3861E-14	100	1500
213	PBr3	PHOSPHORUS TRIBROMIDE	56.758	1.0427E-01	-1.6430E-04	1.1542E-07	-2.9670E-11	298	1500
214	PCl2F3	PHOSPHORUS DICHLORIDE TRIFLUORIDE	20.696	4.7099E-01	-7.8406E-04	5.6105E-07	-1.4442E-10	100	1500
215	PCl3	PHOSPHORUS TRICHLORIDE	27.213	2.4066E-01	-3.9532E-04	2.8032E-07	-7.1695E-11	100	1500
216	PCl5	PHOSPHORUS PENTACHLORIDE	25.701	4.7099E-01	-7.8406E-04	5.6105E-07	-1.4442E-10	100	1500
217	PH3	PHOSPHINE	32.964	-1.4201E-02	1.3216E-04	-1.1915E-07	3.2843E-11	100	1500
218	PH4Br	PHOSPHONIUM BROMIDE	62.034	-1.4201E-02	1.3216E-04	-1.1915E-07	3.2843E-11	100	1500
219	PH4Cl	PHOSPHONIUM CHLORIDE	61.942	-1.4201E-02	1.3216E-04	-1.1915E-07	3.2843E-11	100	1500
220	PH4I	PHOSPHONIUM IODIDE	62.037	-1.4201E-02	1.3216E-04	-1.1915E-07	3.2843E-11	100	1500
221	POCl3	PHOSPHORUS OXYCHLORIDE	23.911	3.2446E-01	-5.0571E-04	3.4836E-07	-8.7607E-11	40	1500
222	PSBr3	PHOSPHORUS THIOBROMIDE	63.322	1.6822E-01	-2.5688E-04	1.7681E-07	-4.4843E-11	298	1500
223	PSCl3	PHOSPHORUS THIOCHLORIDE	27.454	3.3554E-01	-5.4132E-04	3.7986E-07	-9.6538E-11	100	1500
224	P4O6	PHOSPHORUS TRIOXIDE	-26.248	8.7464E-01	-1.2429E-03	8.1470E-07	-1.9988E-10	298	1500
225	P4O10	PHOSPHORUS PENTOXIDE	-20.051	1.0106E+00	-1.2610E-03	7.4993E-07	-1.7240E-10	100	1500
226	P4S10	PHOSPHORUS PENTASULFIDE	90.333	6.3567E-01	-6.6145E-04	2.8692E-07	-3.7235E-11	300	1200
227	Pb	LEAD	18.543	1.0164E-02	-1.3817E-05	7.2087E-09	-1.0584E-12	298	3000
228	PbBr2	LEAD BROMIDE	--------	----------	----------	----------	----------	----	-----
229	PbCl2	LEAD CHLORIDE	53.551	7.4895E-03	-3.8716E-06	7.9170E-10	-5.5643E-14	200	6000
230	PbF2	LEAD FLUORIDE	47.569	1.6774E-02	-8.5850E-06	1.7453E-09	-1.2220E-13	200	6000
231	PbI2	LEAD IODIDE	56.891	2.1468E-03	-1.1186E-06	2.2970E-10	-1.6182E-14	200	6000
232	PbO	LEAD OXIDE	29.545	1.0301E-02	-4.4871E-06	8.4084E-10	-5.5752E-14	200	6000
233	PbS	LEAD SULFIDE	30.934	2.0057E-02	-2.3759E-05	1.2175E-08	-2.2617E-12	298	2000
234	Pd	PALLADIUM	62.467	-6.1923E-03	-1.5080E-14	2.6177E-18	-1.6819E-22	2800	5000

53

Table 2-2 HEAT CAPACITY OF GAS - INORGANIC COMPOUNDS (continued)

			$C_P = A + B T + C T^2 + D T^3 + E T^4$				$(C_P$ - joule/(mol K), T - K)		
NO	FORMULA	NAME	A	B	C	D	E	TMIN	TMAX
235	Po	POLONIUM	20.794	0.0000E+00	0.0000E+00	0.0000E+00	0.0000E+00	298	2000
236	Pt	PLATINUM	20.251	1.0042E-03	-1.8969E-17	2.6597E-21	-1.3184E-25	2000	8000
237	Ra	RADIUM	20.794	-6.4631E-14	1.0427E-16	-6.6041E-20	1.4297E-23	298	2000
238	Rb	RUBIDIUM	20.794	-6.4631E-14	1.0427E-16	-6.6041E-20	1.4297E-23	298	2000
239	RbBr	RUBIDIUM BROMIDE	-------	----------	----------	----------	----------	----	-----
240	RbCl	RUBIDIUM CHLORIDE	34.663	9.7583E-03	-1.0667E-05	5.4663E-09	-1.0154E-12	298	2000
241	RbF	RUBIDIUM FLUORIDE	-------	----------	----------	----------	----------	----	-----
242	RbI	RUBIDIUM IODIDE	-------	⊥---------	----------	----------	----------	----	-----
243	Re	RHENIUM	19.288	6.9353E-03	-8.9193E-06	3.9873E-09	-5.1245E-13	298	3000
244	Re2O7	RHENIUM HEPTOXIDE	-------	----------	----------	----------	----------	----	-----
245	Rh	RHODIUM	26.401	5.0208E-04	6.6480E-18	-1.0516E-21	5.8938E-26	1800	7000
246	Rn	RADON	20.794	-3.9576E-15	4.4438E-18	-1.9112E-21	2.7806E-25	298	3000
247	Ru	RUTHENIUM	16.548	2.1758E-02	-1.7864E-05	6.5392E-09	-8.5168E-13	298	3000
248	RuF5	RUTHENIUM PENTAFLUORIDE	-------	----------	----------	----------	----------	----	-----
249	S	SULFUR	24.624	-5.0402E-03	2.4244E-06	-4.2197E-10	2.5175E-14	200	6000
250	SF4	SULFUR TETRAFLUORIDE	15.486	3.1315E-01	-4.4453E-04	2.9083E-07	-7.1213E-11	298	1500
251	SF6	SULFUR HEXAFLUORIDE	-7.934	5.1224E-01	-6.4878E-04	3.7509E-07	-8.1524E-11	100	1500
252	SOBr2	THIONYL BROMIDE	48.491	1.0815E-01	-1.4791E-04	9.4987E-08	-2.2463E-11	298	1500
253	SOCl2	THIONYL CHLORIDE	34.838	1.4750E-01	-1.7837E-04	9.4399E-08	-1.8111E-11	100	2000
254	SOF2	SULFUROUS OXYFLUORIDE	18.639	1.9505E-01	-2.5689E-04	1.5983E-07	-3.7838E-11	298	1500
255	SO2	SULFUR DIOXIDE	29.637	3.4735E-02	9.2903E-06	-2.9885E-08	1.0937E-11	100	1500
256	SO2Cl2	SULFURYL CHLORIDE	18.553	2.9713E-01	-4.2391E-04	2.7784E-07	-6.7857E-11	100	1500
257	SO3	SULFUR TRIOXIDE	22.466	1.1981E-01	-9.0842E-05	2.5503E-08	-7.9208E-13	100	1500
258	S2Cl2	SULFUR MONOCHLORIDE	51.240	1.1549E-01	-1.6270E-04	1.0449E-07	-2.4709E-11	298	1500
259	Sb	ANTIMONY	20.794	-6.4631E-14	1.0427E-16	-6.6041E-20	1.4297E-23	298	2000
260	SbBr3	ANTIMONY TRIBROMIDE	55.926	1.2781E-01	-2.4301E-04	2.0998E-07	-6.8231E-11	298	1000
261	SbCl3	ANTIMONY TRICHLORIDE	63.593	7.1912E-02	-1.0441E-04	6.7050E-08	-1.5856E-11	298	1500
262	SbCl5	ANTIMONY PENTACHLORIDE	106.953	7.1912E-02	-1.0441E-04	6.7050E-08	-1.5856E-11	298	1500
263	SbH3	STIBINE	13.058	1.2924E-01	-1.3034E-04	6.6840E-08	-1.2435E-11	298	2000
264	SbI3	ANTIMONY TRIIODIDE	55.926	1.2781E-01	-2.4301E-04	2.0998E-07	-6.8231E-11	298	1000
265	Sb2O3	ANTIMONY TRIOXIDE	-------	----------	----------	----------	----------	----	-----
266	Sc	SCANDIUM	21.494	-3.7579E-03	2.3279E-06	-1.4837E-10	-2.5247E-15	298	7000
267	Se	SELENIUM	19.331	5.9528E-03	-2.7286E-06	5.0714E-10	-3.2857E-14	298	6000
268	SeCl4	SELENIUM TETRACHLORIDE	35.672	3.1545E-01	-5.0601E-04	3.5370E-07	-8.9635E-11	100	1500
269	SeF6	SELENIUM HEXAFLUORIDE	-6.934	5.1224E-01	-6.4878E-04	3.7509E-07	-8.1524E-11	100	1500
270	SeOCl2	SELENIUM OXYCHLORIDE	35.838	1.4750E-01	-1.7837E-04	9.4399E-08	-1.8111E-11	100	2000
271	SeO2	SELENIUM DIOXIDE	-------	----------	----------	----------	----------	----	-----
272	Si	SILICON	24.177	-6.3683E-03	4.0692E-06	-8.8857E-10	6.3891E-14	200	6000
273	SiBrCl2F	BROMODICHLOROFLUOROSILANE	46.876	2.1385E-01	-3.2354E-04	2.2226E-07	-5.6419E-11	298	1500
274	SiBrF3	TRIFLUOROBROMOSILANE	26.673	2.6781E-01	-3.7339E-04	2.4125E-07	-5.8565E-11	298	1500
275	SiBr2ClF	DIBROMOCHLOROFLUOROSILANE	57.826	2.1385E-01	-3.2354E-04	2.2226E-07	-5.6419E-11	298	1500
276	SiClF3	TRIFLUOROCHLOROSILANE	26.816	2.6781E-01	-3.7339E-04	2.4125E-07	-5.8565E-11	298	1500
277	SiCl2F2	DICHLORODIFLUOROSILANE	40.896	2.1385E-01	-3.2354E-04	2.2226E-07	-5.6419E-11	298	1500
278	SiCl3F	TRICHLOROFLUOROSILANE	49.698	2.1385E-01	-3.2354E-04	2.2226E-07	-5.6419E-11	298	1500
279	SiCl4	SILICON TETRACHLORIDE	31.672	3.1545E-01	-5.0601E-04	3.5370E-07	-8.9635E-11	100	1500
280	SiF4	SILICON TETRAFLUORIDE	18.032	2.7359E-01	-3.5566E-04	2.1540E-07	-4.9404E-11	100	1500
281	SiHBr3	TRIBROMOSILANE	29.302	2.5068E-01	-3.4090E-04	2.1707E-07	-5.2003E-11	100	1500
282	SiHCl3	TRICHLOROSILANE	24.939	2.5068E-01	-3.4090E-04	2.1707E-07	-5.2003E-11	100	1500
283	SiHF3	TRIFLUOROSILANE	13.820	2.3386E-01	-2.7143E-04	1.5566E-07	-3.4933E-11	298	1500
284	SiH2Br2	DIBROMOSILANE	25.760	1.7618E-01	-1.6179E-04	7.0860E-08	-1.1902E-11	100	1500
285	SiH2Cl2	DICHLOROSILANE	21.583	1.7618E-01	-1.6179E-04	7.0860E-08	-1.1902E-11	100	1500
286	SiH2F2	DIFLUOROSILANE	6.367	1.9374E-01	-1.6805E-04	7.1483E-08	-1.1961E-11	100	1500
287	SiH2I2	DIIODOSILANE	27.010	1.7618E-01	-1.6179E-04	7.0860E-08	-1.1902E-11	100	1500
288	SiH3Br	MONOBROMOSILANE	10.745	1.9428E-01	-1.9487E-04	1.0614E-07	-2.3863E-11	298	1500
289	SiH3Cl	MONOCHLOROSILANE	7.830	1.9428E-01	-1.9487E-04	1.0614E-07	-2.3863E-11	298	1500
290	SiH3F	MONOFLUOROSILANE	3.610	1.9169E-01	-1.7363E-04	8.3890E-08	-1.6955E-11	298	1500
291	SiH3I	IODOSILANE	9.485	1.9428E-01	-1.9487E-04	1.0614E-07	-2.3863E-11	298	1500
292	SiH4	SILANE	28.887	1.7546E-02	1.4919E-04	-1.5680E-07	4.6291E-11	100	1500
293	SiO2	SILICON DIOXIDE	29.690	4.6016E-02	-2.2333E-05	4.4152E-09	-3.0437E-13	100	6000
294	Si2Cl6	HEXACHLORODISILANE	103.559	3.1545E-01	-5.0601E-04	3.5370E-07	-8.9635E-11	100	1500
295	Si2F6	HEXAFLUORODISILANE	76.975	2.7359E-01	-3.5566E-04	2.1540E-07	-4.9404E-11	100	1500
296	Si2H5Cl	DISILANYL CHLORIDE	48.881	1.9428E-01	-1.9487E-04	1.0614E-07	-2.3863E-11	298	1500
297	Si2H6	DISILANE	27.353	1.9208E-01	-5.8767E-05	-3.4180E-08	1.8348E-11	100	1500
298	Si2OCl3F3	TRICHLOROTRIFLUORODISILOXANE	101.566	2.1385E-01	-3.2354E-04	2.2226E-07	-5.6419E-11	298	1500
299	Si2OCl6	HEXACHLORODISILOXANE	96.365	3.1545E-01	-5.0601E-04	3.5370E-07	-8.9635E-11	100	1500
300	Si2OH6	DISILOXANE	37.295	1.9208E-01	-5.8767E-05	-3.4180E-08	1.8348E-11	100	1500
301	Si3Cl8	OCTACHLOROTRISILANE	141.065	3.1545E-01	-5.0601E-04	3.5370E-07	-8.9635E-11	100	1500
302	Si3H8	TRISILANE	47.295	1.9208E-01	-5.8767E-05	-3.4180E-08	1.8348E-11	100	1500
303	Si3H9N	TRISILAZANE	81.085	1.9208E-01	-5.8767E-05	-3.4180E-08	1.8348E-11	100	1500
304	Si4H10	TETRASILANE	93.015	1.9208E-01	-5.8767E-05	-3.4180E-08	1.8348E-11	100	1500
305	Sm	SAMARIUM	28.785	8.1797E-03	-9.9705E-06	3.6498E-09	-4.7536E-13	298	3000
306	Sn	TIN	14.899	3.0942E-02	-1.9228E-05	4.4884E-09	-3.6598E-13	298	5000
307	SnBr4	STANNIC BROMIDE	46.415	3.1545E-01	-5.0601E-04	3.5370E-07	-8.9635E-11	100	1500
308	SnCl2	STANNOUS CHLORIDE	92.048	0.0000E+00	0.0000E+00	0.0000E+00	0.0000E+00	925	1500
309	SnCl4	STANNIC CHLORIDE	38.672	3.1545E-01	-5.0601E-04	3.5370E-07	-8.9635E-11	100	1500
310	SnH4	STANNIC HYDRIDE	37.102	1.7546E-02	1.4919E-04	-1.5680E-07	4.6291E-11	100	1500
311	SnI4	STANNIC IODIDE	95.285	6.0145E-02	-1.1436E-04	9.8816E-08	-3.2109E-11	298	1000
312	Sr	STRONTIUM	20.794	0.0000E+00	0.0000E+00	0.0000E+00	0.0000E+00	298	6000

54

Table 2-2 HEAT CAPACITY OF GAS - INORGANIC COMPOUNDS (continued)

NO	FORMULA	NAME	$C_P = A + B T + C T^2 + D T^3 + E T^4$ (C_P - joule/(mol K), T - K)						
			A	B	C	D	E	TMIN	TMAX
313	SrO	STRONTIUM OXIDE	26.452	3.1778E-02	-3.6851E-05	1.8883E-08	-3.5079E-12	298	2000
314	Ta	TANTALUM	19.546	2.2704E-03	1.0416E-05	-5.9989E-09	9.7275E-13	298	3000
315	Tc	TECNNETIUM	18.549	1.5699E-03	1.6967E-05	-1.1184E-08	1.9556E-12	298	3000
316	Te	TELLURIUM	21.187	-2.4502E-03	3.7389E-06	-1.3687E-09	1.7826E-13	298	3000
317	TeCl4	TELLURIUM TETRACHLORIDE	108.784	0.0000E+00	0.0000E+00	0.0000E+00	0.0000E+00	665	2000
318	TeF6	TELLURIUM HEXAFLUORIDE	-5.933	5.1224E-01	-6.4878E-04	3.7509E-07	-8.1524E-11	100	1500
319	Ti	TITANIUM	27.731	-1.3797E-02	8.5069E-06	-1.4657E-09	8.3732E-14	200	6000
320	TiCl4	TITANIUM TETRACHLORIDE	67.914	6.6801E-02	-3.5178E-05	7.2884E-09	-5.1766E-13	100	6000
321	Tl	THALLIUM	21.569	-4.0480E-03	4.9852E-06	-1.8249E-09	2.3768E-13	298	3000
322	TlBr	THALLOUS BROMIDE	35.971	4.5026E-03	-5.3337E-06	2.7331E-09	-5.0772E-13	298	2000
323	TlI	THALLOUS IODIDE	36.492	2.8653E-03	-3.3942E-06	1.7393E-09	-3.2310E-13	298	2000
324	Tm	THULIUM	--------	----------	----------	----------	----------	----	-----
325	U	URANIUM	24.151	-3.5206E-03	5.4007E-06	-1.9770E-09	2.5749E-13	298	3000
326	UF6	URANIUM FLUORIDE	-4.133	5.1224E-01	-6.4878E-04	3.7509E-07	-8.1524E-11	100	1500
327	V	VANADIUM	25.248	-6.5941E-04	4.2019E-07	-8.3346E-11	6.0274E-15	298	6000
328	VCl4	VANADIUM TETRACHLORIDE	35.481	3.4358E-01	-6.0413E-04	4.4984E-07	-1.1924E-10	50	1500
329	VOCl3	VANADIUM OXYTRICHLORIDE	29.050	3.2969E-01	-5.2847E-04	3.6323E-07	-8.9429E-11	50	1600
330	W	TUNGSTEN	10.773	4.8986E-02	-2.9559E-05	6.5548E-09	-4.7595E-13	200	6000
331	WF6	TUNGSTEN FLUORIDE	35.463	4.3695E-01	-6.4406E-04	4.3271E-07	-1.0794E-10	298	1500
332	Xe	XENON	20.786	0.0000E+00	0.0000E+00	0.0000E+00	0.0000E+00	50	1500
333	Yb	YTTERBIUM	20.794	0.0000E+00	0.0000E+00	0.0000E+00	0.0000E+00	298	3000
334	Yt	YTTRIUM	29.414	-2.0002E-02	1.2834E-05	-2.5366E-09	1.7424E-13	298	6000
335	Zn	ZINC	21.128	-1.3303E-03	7.3647E-07	-1.7803E-10	1.6481E-14	298	6000
336	ZnCl2	ZINC CHLORIDE	60.250	8.3680E-04	-2.8882E-14	1.2948E-17	-2.1389E-21	1005	2000
337	ZnF2	ZINC FLUORIDE	--------	----------	----------	----------	----------	----	-----
338	ZnO	ZINC OXIDE	24.456	3.4987E-02	-3.9761E-05	2.0374E-08	-3.7848E-12	298	2000
339	ZnSO4	ZINC SULFATE	--------	----------	----------	----------	----------	----	-----
340	Zr	ZIRCONIUM	25.546	2.0875E-04	1.0423E-06	-1.1950E-10	7.4615E-16	200	6000
341	ZrBr4	ZIRCONIUM BROMIDE	96.455	1.8772E-02	-9.7142E-06	1.9876E-09	-1.3973E-13	200	6000
342	ZrCl4	ZIRCONIUM CHLORIDE	84.606	3.5549E-02	-1.8288E-05	3.7719E-09	-2.6793E-13	200	6000
343	ZrI4	ZIRCONIUM IODIDE	101.195	1.1254E-02	-5.8578E-06	1.2028E-09	-8.4763E-14	200	6000

C_P - heat capacity of ideal gas, joule/(mol K)

A, B, C, D, and E - regression coefficients for chemical compound

T - temperature, K

TMIN - minimum temperature, K

TMAX - maximum temperature, K

Chapter 3

HEAT CAPACITY OF LIQUID

Carl L. Yaws, Xiaoyan Lin, Li Bu, Sachin Nijhawan, Deepa R. Balundgi, and Saumya Tripathi
Lamar University, Beaumont, Texas

ABSTRACT

Results for heat capacity of liquid as a function of temperature are presented for major organic and inorganic chemicals. The results cover a wide temperature range and include many compound types. The agreement between correlation and data is quite good.

INTRODUCTION

Thermodynamic properties such as liquid heat capacity are important in the engineering design of chemical processes. In liquid-phase chemical reactions, the liquid heat capacity is required to determine the energy (heat) necessary to bring the liquid chemical reactants up to reaction temperature. Additional uses include heat exchanger and energy balance design calculations.

In this article, correlation results for liquid heat capacity are provided in an easy-to-use tabular format that is especially applicable for rapid engineering use with the personal computer or hand calculator.

HEAT CAPACITY CORRELATION

The correlation for heat capacity of liquid is a series expansion in temperature:

$$C_P = A + B\,T + C\,T^2 + D\,T^3 \tag{3-1}$$

where C_P = heat capacity of liquid, joule/(mol K)
A, B, C, and D = regression coefficients for chemical compound
T = temperature, K

The results for heat capacity of liquid are given in Tables 3-1 and 3-2. In preparing the compilation, a literature search was conducted to identify data source publications for organics (1-43) and inorganics (1-104). Both experimental values for the property under consideration and parameter values for estimation of the property are included in the source publications. The publications were screened for appropriate data. The compilation resulting from the screening is based on both experimental data and estimated values.

For organic compounds, most of the estimates were based on group contribution (Cheuh-Swanson, 29), corresponding states (Lee-Kesler, 29) and boiling point methods (Yaws and co-workers). The relation of (heat capacity)(density^n)=constant was utilized to extend both experimental data and estimates. Values of n ranged from 1/2 to 1. Experimental data and estimates were then regressed to provide the same equation for all compounds.

For inorganic compounds, many of the estimates are based on the JANAF tables (57-59), Bureau of Mines bulletins (60-63), and group contribution methods. The relation of (heat capacity)(density^n)=constant was utilized to extend both experimental data and estimates.

Very limited experimental data for liquid heat capacity are available at temperatures in the region of the melting point temperature. Data in the boiling-critical point temperature interval are also very scarce. Thus, the values in the region of the melting point and in the boiling-critical point temperature interval should be considered rough approximations. The values in the intermediate region (above melting and below boiling point) are more accurate.

A comparison of correlation and actual data for liquid heat capacity is shown in Fig. 3-1 for a representative chemical. The graph discloses good agreement of correlation and data.

EXAMPLES

The correlation results maybe used for prediction and calculation of heat capacity and additional thermodynamic properties. Examples are given below.

Example 1 Estimate the liquid heat capacity of pentane (C5H12) at 298.15 K.

Substitution of the coefficients from the table and temperature into the equation for heat capacity yields

C_P = 80.641 + 6.2195E-01*298.15 − 2.2682E-03*298.15^2 + 3.7423E-06*298.15^3

C_P = 163.64 joule/(mol K)

Example 2 Calculate the energy required to heat liquid toluene (C7H8) from 300 to 500 K.

HEAT CAPACITY OF LIQUID

From thermodynamics, the change in enthalpy, ΔH, at constant pressure is

$$\Delta H = \int C_P \, dT = \int (A + B\,T + C\,T^2 + D\,T^3)\, dT$$

$$\Delta H = A\,T + (B/2)\,T^2 + (C/3)\,T^3 + (D/4)\,T^4 \ \Big]_{T_1}^{T_2}$$

Substitution of the coefficients from the table and the temperature limits into the equation yields

$\Delta H = (83.703)(500 - 300) + (5.1666\text{E-}01/2)(500^2 - 300^2) - (1.4910\text{E-}03/3)(500^3 - 300^3) + (1.9725\text{E-}06/4)(500^4 - 300^4)$

$\Delta H = 36,200$ joule/mol

Portions of this material appeared in Hydrocarbon Processing, 70, 73 (December, 1991) and Chem. Eng., 99, 130 (April, 1992). These portions are reprinted by special permission.

REFERENCES – ORGANIC COMPOUNDS

1-36. See **REFERENCES - ORGANIC COMPOUNDS** in **Chapter 2, HEAT CAPACITY OF GAS**

37. Altunin, V. V., V. Z. Geller, E. K. Petrov, D. C. Rasskazov, and G. A. Spiridonov, THERMOPHYSICAL PROPERTIES OF FREONS, Methane Series, Part 1, Hemisphere Publishing Corporation, New York, NY (1987).
38. Altunin, V. V., V. Z. Geller, E. A. Kremenevskaya, I. I. Perelshtein, and E. K. Petrov, THERMOPHYSICAL PROPERTIES OF FREONS, Methane Series, Part 2, Hemisphere Publishing Corporation, New York, NY (1987).
39. Wilhoit, R. C. and B. J. Zwolinski, PHYSICAL AND THERMODYNAMIC PROPERTIES OF ALIPHATIC ALCOHOLS, American Chemical Society, American Institute of Physics, National Bureau of Standards, New York, NW (1973).
40. Yaws, C. L. and others, Hydrocarbon Processing, 70, 73 (December, 1991).
41. Yaws, C. L. and others, Chem. Eng., 99, 130 (April, 1992).
42. Yaws, C. L., HANDBOOK OF THERMODYNAMIC DIAGRAMS, Vols. 1, 2, 3, and 4, Gulf Publishing Co., Houston, TX (1996).
43. Yaws, C. L., HANDBOOK OF CHEMICAL COMPOUND DATA FOR PROCESS SAFETY, Gulf Publishing Co., Houston, TX (1997).

REFERENCES - INORGANIC COMPOUNDS

1-64. See **REFERENCES - INORGANIC COMPOUNDS** in **Chapter 2 HEAT CAPACITY OF GAS**

65. Lyon, R. N., ed., LIQUID-METALS HANDBOOK, 2nd ed., Atomic Energy Commission, Washington, DC (1954).
66. Janz, G. J. and C. B. Allen, PHYSICAL PROPERTIES DATA COMPILATIONS RELEVANT TO ENERGY STORAGE. II. MOLTEN SALTS: DATA ON SINGLE AND MULTICOMPONENT SALT SYSTEMS, Nat. Bur.Stand., Molten Salts Data Center, Troy, NY (April 1979).
67. Fink, J. K., TABLES OF THERMODYNAMIC PROPERTIES OF SODIUM, Argonne National Lab., ANL-CEN-RSD-82-4, Chemical Engineering Division, Argonne, IL (June, 1982).
68. Mills, K. C., THERMODYNAMIC DATA FOR INORGANIC SULPHIDES, SELENIDES AND TELLURIDES, Butterworths, London, England (1974).
69. Tarakad, R. R. and R. P. Danner, AIChE J. 23(6), 944 (1977) and 23(5), 685, (1977).
70. Chueh, C. F. and A. C. Swanson, Chem. Eng. Prog. 69(7), 83 (1973).
71. Giauque, W. F. and T. M. Powell, J. Amer. Chem. Soc. 61, 1970 (1939).
72. Davis, C. M., Jr., J. Chem. Phys. 45(7), 2461 (1966).
73. Hu, J., D. White, and H. L. Johnston, J. Amer. Chem. Soc., 15, 5642 (1953).
74. Smith, T. O. and B. M. Fabuss, Wright Air Development Center, Technical Report 59-327 (1962).
75. Grosh, J. and M. S. Jhon, Proc. Nat. Acad. Sci. 54, 1004 (1965).
76. Hu, J. H. and D. White, J. Amer. Chem. Soc. 75, 1232 (1953).
77. Giauque, W. F. and R. Wieke, J. Amer. Chem. Soc., 51, 1441 (1929).
78. Forsythe, W. R. and W. F. Giauque, J. Amer. chem. Soc., 64, 48 (1948).
79. Evans, W. H. and R. Jacobson, J. Res. Nat. Bur. Stand., 55, 83 (1955).
80. Douglas, T. B. and A. F. Ball, J. Amer. Chem. Soc. 74, 2472 (1952).
81. Douglas, T. B. and L. F. Epstein, J. Amer. Chem. Soc., 77, 2144 (1955).
82. Haar, L. and J. S. Gallagher, J. Phys. Chem. Ref. Data, 7(3), 635 (1978).
83. Giauque, W. F. and J. O. Clayton, J. Amer. Chem. Soc., 55, 4875 (1933).
84. Wiebe, R. and M. J. Brevoort, J. Amer. Chem. Soc., 51, 622 (1930).
85. Scott, D. W. and G. D. Oliver, J. Amer. Chem. Soc., 71, 2293 (1949).
86. Giauque, W. F. and J. D. Kemp, J. Chem. Phys., 6, 40 (1938).
87. McCarty, R. D. and L. A. Weber, Nat. Bur. of Stand. Tech. Note 384, Washington DC (1971).
88. Giauque W. F. and H. L. Johnston, J. Amer. Chem. Soc., 51, 2300 (1929).
89. Jenkins, A. C. and F. S. Dipaolo, J. Chem. Phys., 25(2), 296 (1956).
90. Lee, B. I. and M. G. Kesler, AIChE J., 21(3), 510 (1975).
91. Ott, J. B. and W. F. Giauque, J. Amer. Chem. Soc., 82, 1308 (1960).
92. McDonald, R. A., J. Chem. Eng. Data, 12, 115 (1967).
93. Majer, V., V. Svoboda, and M. Lencka, J. Chem. Thermo., 17, 365 (1985).
94. Sherman, R. H. and W. F. Giauque, J. Amer. Chem. Soc., 77, 2154 (1955).

95. Clarke, J. T., E. B. Rifkin, and H. L. Johnston, J. Amer. Chem. Soc., 75, 781 (1953).

96. Jhon, J. S., J. Grosh, and H. Eyring, J. Phys. Chem., 71(7), 2533 (1967).

97. Pace E. L. and M. A. Reno, J. Chem. Phys., 48(3), 1231 (1968).

98. Pace, E. L. and M. A. Reno, J. Chem. Phys., 48(3), 1231 (1968).

99. Clayton, J. O. and W. F. Giauque, J. Amer. Chem. Soc., 54, 2610 (1932).

100. Kaischeu, R., Z. Phys. Chem., B40, 273 (1938).

101. Kemp, J. D. and W. F. Giauque, J. Amer. Chem. Soc., 59, 79 (1937).

102. Void, R. D., J. Amer. Chem. Soc., 59, 1515 (1937).

103. Weissler, A., J. Amer. Chem. Soc. 71, 1272 (1949).

104. Pace, E. L. and J. S. Mosser, J. Chem. Phys., 19(1), 154 (1963).

Figure 3-1 Heat Capacity of Liquid

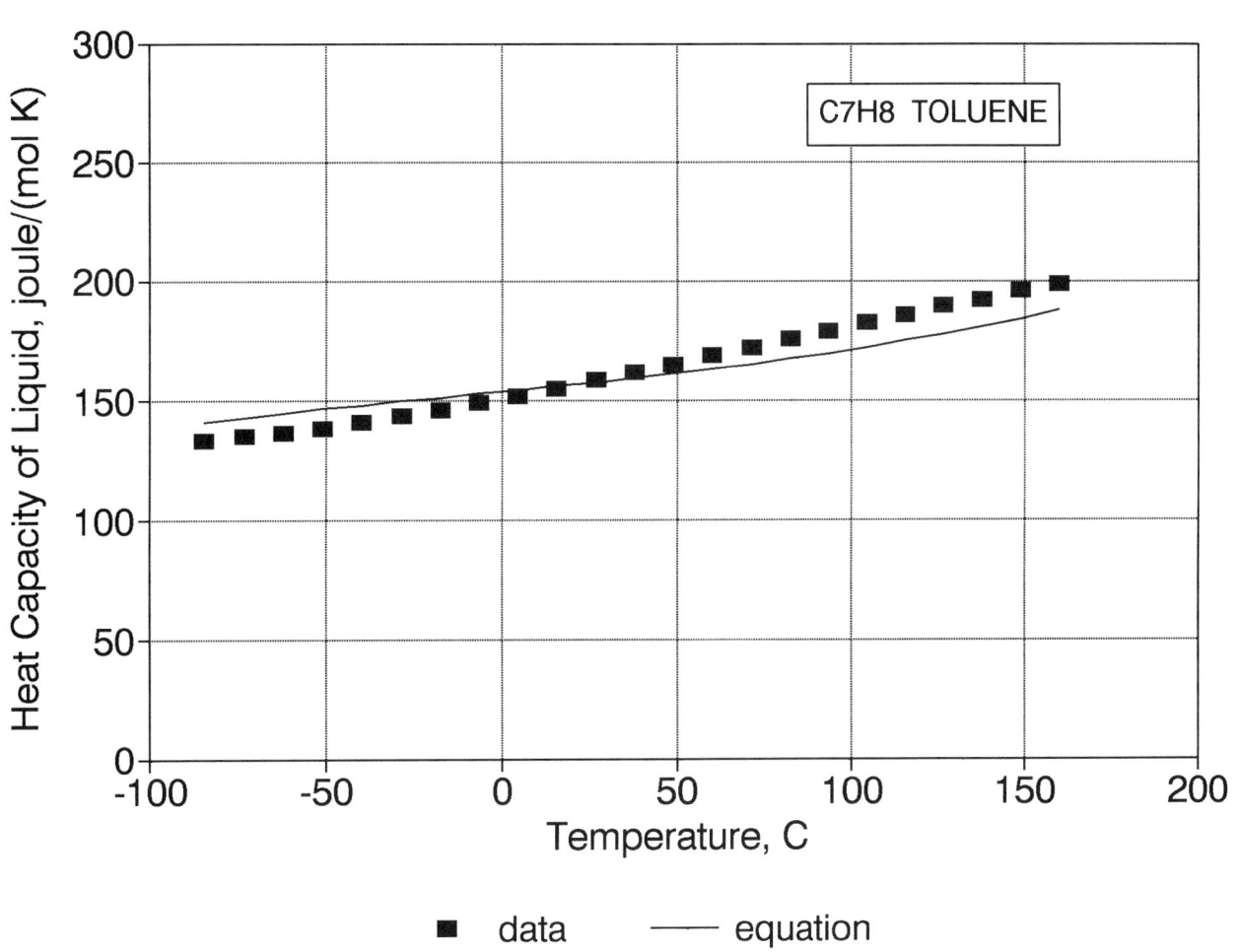

C7H8 TOLUENE

■ data   —— equation

Table 3-1 HEAT CAPACITY OF LIQUID - ORGANIC COMPOUNDS

			$C_P = A + B\,T + C\,T^2 + D\,T^3$				(C_P - joule/(mol K), T - K)		
NO	FORMULA	NAME	A	B	C	D	TMIN	TMAX	C_P @ 25 C
1	CBrClF2	BROMOCHLORODIFLUOROMETHANE	66.842	4.2713E-01	-1.7526E-03	3.3738E-06	115	384	127.81
2	CBrCl3	BROMOTRICHLOROMETHANE	32.285	7.1628E-01	-1.9205E-03	2.0950E-06	253	545	130.65
3	CBrF3	BROMOTRIFLUOROMETHANE	53.868	5.8473E-01	-2.9409E-03	6.6584E-06	106	306	143.25
4	CBr2F2	DIBROMODIFLUOROMETHANE	60.141	5.4837E-01	-1.9397E-03	2.9706E-06	164	430	129.94
5	CClF3	CHLOROTRIFLUOROMETHANE	47.972	5.5277E-01	-3.1183E-03	8.0282E-06	93	272	--------
6	CClN	CYANOGEN CHLORIDE	-130.070	2.0318E+00	-6.6150E-03	7.7883E-06	268	404	94.09
7	CCl2F2	DICHLORODIFLUOROMETHANE	53.463	4.6913E-01	-2.0770E-03	4.2398E-06	116	346	121.07
8	CCl2O	PHOSGENE	53.075	4.5299E-01	-1.6986E-03	2.8257E-06	146	410	112.03
9	CCl3F	TRICHLOROFLUOROMETHANE	29.120	3.0976E-01	-1.1066E-03	1.7185E-06	163	424	68.65
10	CCl4	CARBON TETRACHLORIDE	9.671	9.3363E-01	-2.6768E-03	3.0425E-06	251	501	130.72
11	CF2O	CARBONYL FLUORIDE	-33.419	1.9984E+00	-1.0073E-02	1.9136E-05	163	267	--------
12	CF4	CARBON TETRAFLUORIDE	25.395	9.8067E-01	-7.0731E-03	2.1219E-05	91	205	--------
13	CHBr3	TRIBROMOMETHANE	23.676	7.1569E-01	-1.6914E-03	1.6314E-06	282	626	129.94
14	CHClF2	CHLORODIFLUOROMETHANE	44.503	4.7737E-01	-2.2107E-03	4.5786E-06	117	332	111.66
15	CHCl2F	DICHLOROFLUOROMETHANE	51.597	4.1564E-01	-1.5739E-03	2.7008E-06	139	406	107.20
16	CHCl3	CHLOROFORM	28.296	6.5897E-01	-2.0353E-03	2.5901E-06	211	483	112.49
17	CHF3	TRIFLUOROMETHANE	19.142	8.4432E-01	-4.6671E-03	1.0607E-05	119	269	--------
18	CHI3	TRIIODOMETHANE	194.656	-9.1508E-02	1.8580E-04	--------	397	521	--------
19	CHN	HYDROGEN CYANIDE	-123.155	1.7769E+00	-5.8083E-03	6.9129E-06	261	411	73.52
20	CHNS	ISOTHIOCYANIC-ACID	--------	--------	--------	--------	--------	--------	--------
21	CH2BrCl	BROMOCHLOROMETHANE	48.459	3.8479E-01	-1.1676E-03	1.5606E-06	186	501	100.75
22	CH2Br2	DIBROMOMETHANE	38.738	4.8154E-01	-1.3291E-03	1.5449E-06	222	550	105.11
23	CH2ClF	CHLOROFLUOROMETHANE	86.915	-3.6025E-02	2.6561E-04	--------	141	294	99.78
24	CH2Cl2	DICHLOROMETHANE	38.941	4.9008E-01	-1.6224E-03	2.3069E-06	179	459	101.98
25	CH2F2	DIFLUOROMETHANE	21.136	7.0644E-01	-3.3337E-03	6.4757E-06	138	316	107.05
26	CH2I2	DIIODOMETHANE	49.637	5.5690E-01	-1.2500E-03	1.1654E-06	280	672	135.45
27	CH2O	FORMALDEHYDE	--------	--------	--------	--------	--------	--------	--------
28	CH2O2	FORMIC ACID	-16.110	8.7229E-01	-2.3665E-03	2.4454E-06	283	522	98.40
29	CH3Br	METHYL BROMIDE	25.042	4.9612E-01	-1.7627E-03	2.5993E-06	181	420	85.16
30	CH3Cl	METHYL CHLORIDE	11.381	6.2238E-01	-2.4353E-03	3.8333E-06	176	375	82.33
31	CH3Cl3Si	METHYL TRICHLOROSILANE	54.416	8.8694E-01	-2.8583E-03	3.8406E-06	196	465	166.56
32	CH3F	METHYL FLUORIDE	9.382	8.5352E-01	-4.4087E-03	9.1708E-06	132	286	--------
33	CH3I	METHYL IODIDE	24.222	4.6459E-01	-1.4583E-03	1.8792E-06	208	475	82.91
34	CH3NO	FORMAMIDE	36.786	5.1348E-01	-1.1422E-03	1.0480E-06	277	694	116.12
35	CH3NO2	NITROMETHANE	10.892	7.0067E-01	-1.9525E-03	2.1877E-06	246	529	104.22
36	CH3NO2	METHYL-NITRITE	--------	--------	--------	--------	--------	--------	--------
37	CH3NO3	METHYL-NITRATE	--------	--------	--------	--------	--------	--------	--------
38	CH4	METHANE	-0.018	1.1982E+00	-9.8722E-03	3.1670E-05	92	172	--------
39	CH4Cl2Si	METHYL DICHLOROSILANE	66.188	6.8839E-01	-2.3767E-03	3.3997E-06	184	435	150.26
40	CH4O	METHANOL	40.152	3.1046E-01	-1.0291E-03	1.4598E-06	176	461	79.93
41	CH4O3S	METHANESULFONIC ACID	--------	--------	--------	--------	--------	--------	--------
42	CH4S	METHYL MERCAPTAN	46.472	3.7853E-01	-1.3665E-03	2.2085E-06	151	423	96.39
43	CH5ClSi	METHYL CHLOROSILANE	68.963	6.0183E-01	-2.3367E-03	4.0231E-06	140	398	147.31
44	CH5N	METHYLAMINE	13.565	9.0836E-01	-3.4881E-03	5.2770E-06	181	387	114.19
45	CH6Si	METHYL SILANE	36.552	4.8720E-01	-2.3448E-03	4.9668E-06	117	317	105.02
46	CN4O8	TETRANITROMETHANE	-280.326	4.0096E+00	-1.1394E-02	1.1966E-05	288	486	219.46
47	CO	CARBON MONOXIDE	-19.312	2.5072E+00	-2.8970E-02	1.2745E-04	69	120	--------
48	COS	CARBONYL SULFIDE	31.903	3.9686E-01	-1.7527E-03	3.3294E-06	135	341	82.66
49	CO2	CARBON DIOXIDE	-338.956	5.2796E+00	-2.3279E-02	3.5980E-05	218	274	--------
50	CS2	CARBON DISULFIDE	39.938	2.3565E-01	-7.2098E-04	1.0443E-06	163	497	73.79
51	C2BrF3	BROMOTRIFLUOROETHYLENE	-20.750	1.4074E+00	-5.1867E-03	7.4525E-06	201	389	135.33
52	C2Br2F4	1,2-DIBROMOTETRAFLUOROETHANE	83.092	7.8843E-01	-2.7534E-03	4.1741E-06	164	439	184.03
53	C2ClF3	CHLOROTRIFLUOROETHYLENE	58.305	6.4342E-01	-2.9048E-03	5.9780E-06	116	341	150.36
54	C2ClF5	CHLOROPENTAFLUOROETHANE	-17.172	1.8274E+00	-8.0452E-03	1.3663E-05	175	318	174.61
55	C2Cl2F4	1,2-DICHLOROTETRAFLUOROETHANE	17.180	8.2104E-01	-3.1756E-03	4.9217E-06	180	377	110.13
56	C2Cl3F3	1,1,2-TRICHLOROTRIFLUOROETHANE	-25.308	1.6488E+00	-5.2824E-03	6.5167E-06	239	439	169.44
57	C2Cl4	TETRACHLOROETHYLENE	51.760	6.5780E-01	-1.6916E-03	1.8652E-06	252	558	146.95
58	C2Cl4F2	1,1,2,2-TETRACHLORODIFLUOROETHANE	-122.474	2.3932E+00	-6.5504E-03	6.6997E-06	300	496	--------
59	C2Cl4O	TRICHLOROACETYL CHLORIDE	78.666	6.5083E-01	-1.8623E-03	2.3259E-06	201	531	168.81
60	C2Cl6	HEXACHLOROETHANE	--------	--------	--------	--------	--------	--------	--------
61	C2F4	TETRAFLUOROETHYLENE	-2.266	1.5284E+00	-7.9070E-03	1.6018E-05	143	276	--------
62	C2F6	HEXAFLUOROETHANE	-152.829	3.8878E+00	-1.9409E-02	3.5499E-05	173	264	--------
63	C2HBrClF3	HALOTHANE	51.616	8.7711E-01	-2.7885E-03	3.6941E-06	201	469	163.15
64	C2HClF2	2-CHLORO-1,1-DIFLUOROETHYLENE	48.550	6.0574E-01	-2.5591E-03	4.7271E-06	136	360	126.95
65	C2HCl3	TRICHLOROETHYLENE	58.916	4.7250E-01	-1.4035E-03	1.8353E-06	189	514	123.67
66	C2HCl3O	DICHLOROACETYL CHLORIDE	68.128	6.3023E-01	-1.8345E-03	2.3116E-06	201	521	154.23
67	C2HCl3O	TRICHLOROACETALDEHYDE	51.153	7.7236E-01	-2.2695E-03	2.7747E-06	217	509	153.22
68	C2HCl5	PENTACHLOROETHANE	74.593	7.9922E-01	-2.0081E-03	2.1335E-06	245	599	190.91
69	C2HF3	TRIFLUOROETHENE	70.461	-1.7267E-02	2.4546E-04	--------	96	251	--------
70	C2HF3O2	TRIFLUOROACETIC ACID	-75.927	2.0669E+00	-6.4086E-03	7.5280E-06	259	442	170.15
71	C2HF5	PENTAFLUOROETHANE	-34.493	1.9633E+00	-8.9222E-03	1.5512E-05	171	308	168.85
72	C2H2	ACETYLENE	--------	--------	--------	--------	--------	--------	--------
73	C2H2Br4	1,1,2,2-TETRABROMOETHANE	99.752	4.2305E-01	-8.8103E-04	7.8706E-07	274	742	168.43
74	C2H2Cl2	1,1-DICHLOROETHYLENE	57.979	4.0559E-01	-1.4258E-03	2.2741E-06	152	434	112.43
75	C2H2Cl2	cis-1,2-DICHLOROETHYLENE	41.692	5.9180E-01	-1.8794E-03	2.5208E-06	194	474	117.88
76	C2H2Cl2	trans-1,2-DICHLOROETHYLENE	11.098	8.7604E-01	-2.7731E-03	3.4932E-06	224	457	118.37
77	C2H2Cl2O	CHLOROACETYL CHLORIDE	12.574	9.8386E-01	-2.7415E-03	3.0453E-06	252	523	142.92
78	C2H2Cl2O	DICHLOROACETALDEHYDE	29.386	8.0754E-01	-2.3963E-03	2.9026E-06	224	500	134.07

Table 3-1 HEAT CAPACITY OF LIQUID - ORGANIC COMPOUNDS (continued)

NO	FORMULA	NAME	$C_P = A + BT + CT^2 + DT^3$						$(C_P$ - joule/(mol K), T - K)	
			A	B	C	D	TMIN	TMAX	C_P @ 25 C	
79	C2H2Cl2O2	DICHLOROACETIC ACID	14.071	1.0731E+00	-2.5470E-03	2.4567E-06	288	617	172.72	
80	C2H2Cl3F	1,1,1-TRICHLOROFLUOROETHANE	56.323	7.1402E-01	-2.1347E-03	2.7158E-06	201	509	151.42	
81	C2H2Cl4	1,1,1,2-TETRACHLOROETHANE	74.301	5.6404E-01	-1.5412E-03	1.8552E-06	204	562	154.63	
82	C2H2Cl4	1,1,2,2-TETRACHLOROETHANE	62.631	6.7404E-01	-1.7561E-03	1.9582E-06	230	581	159.39	
83	C2H2F2	1,1-DIFLUOROETHYLENE	1.271	1.0285E+00	-5.5247E-03	1.1887E-05	130	273	--------	
84	C2H2F2	cis-1,2-DIFLUOROETHENE	79.394	-2.2240E-02	2.5396E-04	--------	109	278	--------	
85	C2H2F2	trans-1,2-DIFLUOROETHENE	79.394	-2.2240E-02	2.5396E-04	--------	109	278	--------	
86	C2H2F4	1,1,1,2-TETRAFLUOROETHANE	-14.015	1.4306E+00	-6.0220E-03	9.9834E-06	173	342	141.80	
87	C2H2O	KETENE	35.985	4.3573E-01	-1.9959E-03	4.0303E-06	123	333	95.29	
88	C2H2O4	OXALIC ACID	--------	--------	--------	--------	--------	--------	--------	
89	C2H3Br	VINYL BROMIDE	50.988	3.8234E-01	-1.4003E-03	2.3630E-06	136	426	103.13	
90	C2H3Cl	VINYL CHLORIDE	45.366	2.8792E-01	-1.1535E-03	2.1636E-06	120	389	86.01	
91	C2H3ClF2	1-CHLORO-1,1-DIFLUOROETHANE	40.999	7.3050E-01	-3.0172E-03	5.3495E-06	143	369	132.37	
92	C2H3ClO	ACETYL CHLORIDE	55.906	4.4327E-01	-1.4885E-03	2.2362E-06	161	457	115.02	
93	C2H3ClO	CHLOROACETALDEHYDE	37.389	5.9407E-01	-1.8083E-03	2.3175E-06	201	500	115.19	
94	C2H3ClO2	CHLOROACETIC ACID	-60.356	1.4722E+00	-3.3749E-03	2.9610E-06	334	617	--------	
95	C2H3ClO2	METHYL CHLOROFORMATE	47.268	7.3539E-01	-2.3261E-03	3.0667E-06	201	473	141.03	
96	C2H3Cl3	1,1,1-TRICHLOROETHANE	11.142	1.0501E+00	-3.0826E-03	3.5983E-06	244	491	145.56	
97	C2H3Cl3	1,1,2-TRICHLOROETHANE	34.934	8.5054E-01	-2.3306E-03	2.6455E-06	238	542	151.46	
98	C2H3F	VINYL FLUORIDE	37.095	5.7380E-01	-2.9484E-03	6.5829E-06	114	295	--------	
99	C2H3F3	1,1,1-TRIFLUOROETHANE	-2.885	1.3367E+00	-6.1220E-03	1.0882E-05	163	312	139.86	
100	C2H3N	ACETONITRILE	4.296	6.9400E-01	-2.0870E-03	2.4996E-06	230	491	91.94	
101	C2H3NO	METHYL ISOCYANATE	-45.350	1.1286E+00	-3.4639E-03	4.0222E-06	257	455	89.81	
102	C2H4	ETHYLENE	25.597	5.7078E-01	-3.3620E-03	8.4120E-06	105	254	--------	
103	C2H4Br2	1,1-DIBROMOETHANE	60.555	5.1074E-01	-1.3765E-03	1.6277E-06	211	565	133.61	
104	C2H4Br2	1,2-DIBROMOETHANE	16.067	8.1028E-01	-2.0095E-03	1.9894E-06	284	585	131.75	
105	C2H4Cl2	1,1-DICHLOROETHANE	57.325	5.6014E-01	-1.8136E-03	2.5617E-06	177	471	131.00	
106	C2H4Cl2	1,2-DICHLOROETHANE	26.310	7.7555E-01	-2.2271E-03	2.6109E-06	238	505	128.77	
107	C2H4Cl2O	BIS(CHLOROMETHYL)ETHER	31.958	8.0721E-01	-2.2981E-03	2.6746E-06	233	521	139.23	
108	C2H4F2	1,1-DIFLUOROETHANE	18.839	7.1043E-01	-3.0231E-03	5.2385E-06	157	348	100.76	
109	C2H4F2	1,2-DIFLUOROETHANE	2.349	9.3877E-01	-3.2379E-03	4.4459E-06	201	428	112.25	
110	C2H4I2	1,2-DIIODOETHANE	186.357	-9.3383E-02	2.1491E-04	--------	357	503	--------	
111	C2H4O	ACETALDEHYDE	45.056	4.4853E-01	-1.6607E-03	2.7000E-06	151	415	102.72	
112	C2H4O	ETHYLENE OXIDE	35.720	4.2908E-01	-1.5473E-03	2.4070E-06	162	422	89.90	
113	C2H4OS	THIOACETIC-ACID	--------	--------	--------	--------	--------	--------	--------	
114	C2H4O2	ACETIC ACID	-18.944	1.0971E+00	-2.8921E-03	2.9275E-06	291	533	128.66	
115	C2H4O2	METHYL FORMATE	42.381	5.7064E-01	-1.9727E-03	2.8945E-06	175	438	113.87	
116	C2H4S	THIACYCLOPROPANE	81.296	-8.1717E-02	3.7077E-04	--------	166	358	89.89	
117	C2H5Br	BROMOETHANE	49.401	3.8348E-01	-1.3086E-03	2.0048E-06	156	453	100.54	
118	C2H5Cl	ETHYL CHLORIDE	60.180	3.4553E-01	-1.2983E-03	2.1963E-06	138	414	106.00	
119	C2H5ClO	2-CHLOROETHANOL	58.464	4.2992E-01	-1.2482E-03	1.5263E-06	207	527	116.14	
120	C2H5F	ETHYL FLUORIDE	47.221	5.5316E-01	-2.4970E-03	4.8169E-06	131	338	117.84	
121	C2H5I	ETHYL IODIDE	58.413	3.7396E-01	-1.1485E-03	1.6316E-06	163	505	111.06	
122	C2H5N	ETHYLENEIMINE	40.410	5.2835E-01	-1.6704E-03	2.1850E-06	196	483	107.36	
123	C2H5NO	ACETAMIDE	-42.231	1.1311E+00	-2.3734E-03	1.9225E-06	355	685	--------	
124	C2H5NO	N-METHYLFORMAMIDE	35.271	6.2281E-01	-1.4573E-03	1.4059E-06	270	649	128.68	
125	C2H5NO2	NITROETHANE	70.398	5.6739E-01	-1.6579E-03	2.1408E-06	185	534	148.93	
126	C2H5NO3	ETHYL-NITRATE	--------	--------	--------	--------	--------	--------	--------	
127	C2H6	ETHANE	38.332	4.1006E-01	-2.3024E-03	5.9347E-06	91	275	--------	
128	C2H6AlCl	DIMETHYLALUMINUM CHLORIDE	14.491	9.2971E-01	-2.4828E-03	2.6678E-06	253	557	141.69	
129	C2H6O	DIMETHYL ETHER	48.074	5.6225E-01	-2.3915E-03	4.4614E-06	133	360	121.36	
130	C2H6O	ETHANOL	59.342	3.6358E-01	-1.2164E-03	1.8030E-06	160	465	107.40	
131	C2H6OS	DIMETHYL SULFOXIDE	27.879	7.8628E-01	-1.7762E-03	1.6546E-06	293	653	148.27	
132	C2H6O2	ETHYLENE GLYCOL	75.878	6.4182E-01	-1.6493E-03	1.6937E-06	261	581	165.52	
133	C2H6O4S	DIMETHYL SULFATE	85.080	6.4498E-01	-1.4269E-03	1.4557E-06	242	682	189.12	
134	C2H6S	DIMETHYL SULFIDE	50.108	5.5593E-01	-1.8618E-03	2.6910E-06	176	453	121.67	
135	C2H6S	ETHYL MERCAPTAN	72.618	3.4419E-01	-1.1990E-03	2.0330E-06	126	449	122.54	
136	C2H6S2	DIMETHYL DISULFIDE	76.236	4.9795E-01	-1.3915E-03	1.7762E-06	189	545	148.08	
137	C2H7N	DIMETHYLAMINE	36.962	9.5817E-01	-3.5846E-03	5.3990E-06	182	394	147.08	
138	C2H7N	ETHYLAMINE	15.784	8.7144E-01	-3.1108E-03	4.4673E-06	193	411	117.47	
139	C2H7NO	MONOETHANOLAMINE	23.111	1.2283E+00	-3.1218E-03	3.0714E-06	285	574	193.21	
140	C2H8N2	ETHYLENEDIAMINE	-28.801	1.5005E+00	-4.0172E-03	4.0670E-06	285	534	169.25	
141	C2H8Si	DIMETHYL SILANE	60.542	5.7509E-01	-2.4542E-03	4.7258E-06	124	362	139.10	
142	C2N2	CYANOGEN	-224.889	3.6145E+00	-1.3037E-02	1.6821E-05	246	360	139.69	
143	C3F6	HEXAFLUOROPROPYLENE	56.118	8.5816E-01	-3.9981E-03	8.2839E-06	118	331	176.12	
144	C3F6O	HEXAFLUOROACETONE	30.204	1.7703E+00	-8.1288E-03	1.4734E-05	152	321	225.95	
145	C3F8	OCTAFLUOROPROPANE	66.135	1.2101E+00	-5.8622E-03	1.2066E-05	126	311	225.61	
146	C3H2N2	MALONONITRILE	-5.793	1.0354E+00	-2.3802E-03	2.1470E-06	306	644	--------	
147	C3H3Cl	PROPARGYL CHLORIDE	32.486	5.8350E-01	-1.7987E-03	2.3489E-06	201	487	108.82	
148	C3H3N	ACRYLONITRILE	33.362	5.8644E-01	-1.8625E-03	2.4956E-06	191	482	108.79	
149	C3H3NO	OXAZOLE	57.639	4.2117E-01	-1.2531E-03	1.6223E-06	201	499	114.82	
150	C3H4	METHYLACETYLENE	15.304	7.8431E-01	-3.1665E-03	5.1375E-06	171	362	103.83	
151	C3H4	PROPADIENE	32.395	4.8533E-01	-2.0716E-03	3.8414E-06	138	354	94.76	
152	C3H4Cl2	2,3-DICHLOROPROPENE	54.917	5.5292E-01	-1.6272E-03	2.1034E-06	193	519	130.87	
153	C3H4O	ACROLEIN	48.243	5.8199E-01	-1.9335E-03	2.6860E-06	186	455	121.07	
154	C3H4O	PROPARGYL ALCOHOL	43.633	8.2639E-01	-2.3747E-03	2.8288E-06	222	522	153.90	
155	C3H4O2	ACRYLIC ACID	-18.242	1.2106E+00	-3.1160E-03	3.1409E-06	288	554	148.96	
156	C3H4O2	beta-PROPIOLACTONE	45.746	5.2593E-01	-1.3036E-03	1.3725E-06	241	617	123.05	

Table 3-1 HEAT CAPACITY OF LIQUID - ORGANIC COMPOUNDS (continued)

NO	FORMULA	NAME	$C_P = A + B T + C T^2 + D T^3$				$(C_P$ - joule/(mol K), T - K)		
			A	B	C	D	TMIN	TMAX	C_P @ 25 C
157	C3H4O2	VINYL FORMATE	27.210	9.3492E-01	-3.0910E-03	4.1784E-06	201	448	141.93
158	C3H4O3	ETHYLENE CARBONATE	36.594	7.7104E-01	-1.6229E-03	1.3975E-06	311	711	--------
159	C3H4O3	PYRUVIC ACID	5.286	9.5884E-01	-2.4314E-03	2.3929E-06	288	571	138.45
160	C3H5Br	3-BROMO-1-PROPENE	113.850	-7.9860E-02	4.1977E-04	--------	155	373	127.35
161	C3H5Cl	2-CHLOROPROPENE	64.195	4.0759E-01	-1.4675E-03	2.4639E-06	137	430	120.57
162	C3H5Cl	3-CHLOROPROPENE	62.471	3.5800E-01	-1.2093E-03	1.9291E-06	140	463	112.83
163	C3H5ClO	alpha-EPICHLOROHYDRIN	52.634	5.7412E-01	-1.5783E-03	1.8687E-06	217	549	133.04
164	C3H5ClO2	METHYL CHLOROACETATE	30.022	9.8044E-01	-2.6859E-03	3.0203E-06	242	540	163.63
165	C3H5ClO2	ETHYL CHLOROFORMATE	63.019	1.0786E+00	-3.5343E-03	4.8355E-06	193	457	198.57
166	C3H5Cl3	1,2,3-TRICHLOROPROPANE	39.473	9.3369E-01	-2.3660E-03	2.4595E-06	259	587	172.40
167	C3H5I	3-IODO-1-PROPENE	130.120	-1.0253E-01	4.5072E-04	--------	175	405	139.62
168	C3H5N	PROPIONITRILE	52.658	5.0701E-01	-1.5507E-03	2.0731E-06	181	508	120.92
169	C3H5NO	ACRYLAMIDE	-48.597	1.0637E+00	-2.3221E-03	1.9286E-06	359	639	--------
170	C3H5NO	HYDRACRYLONITRILE	104.867	6.5944E-01	-1.6508E-03	1.7649E-06	228	621	201.51
171	C3H5NO	LACTONITRILE	78.481	7.4002E-01	-1.9488E-03	2.1353E-06	234	579	182.47
172	C3H5N3O9	NITROGLYCERINE	--------	--------	--------	--------	--------	--------	--------
173	C3H6	CYCLOPROPANE	30.543	5.0198E-01	-2.1040E-03	3.7444E-06	147	358	92.42
174	C3H6	PROPYLENE	54.718	3.4512E-01	-1.6315E-03	3.8755E-06	89	328	115.30
175	C3H6Br2	1,2-DIBROMOPROPANE	157.717	-1.0292E-01	3.7858E-04	--------	219	443	160.68
176	C3H6Cl2	1,1-DICHLOROPROPANE	60.938	6.9144E-01	-2.0728E-03	2.6533E-06	201	504	153.16
177	C3H6Cl2	1,2-DICHLOROPROPANE	75.742	5.0261E-01	-1.5107E-03	2.0567E-06	174	515	145.81
178	C3H6Cl2	1,3-DICHLOROPROPANE	85.462	4.4791E-01	-1.2858E-03	1.6933E-06	175	543	149.59
179	C3H6Cl2	2,2-DICHLOROPROPANE	133.548	-2.0321E-01	6.0655E-04	--------	240	372	126.88
180	C3H6I2	1,2-DIIODOPROPANE	196.466	-1.4565E-01	4.2127E-04	--------	254	530	190.49
181	C3H6O	ACETONE	46.878	6.2652E-01	-2.0761E-03	2.9583E-06	179	457	127.53
182	C3H6O	ALLYL ALCOHOL	81.284	4.3822E-01	-1.4019E-03	2.1259E-06	145	491	143.67
183	C3H6O	METHYL VINYL ETHER	53.292	6.0476E-01	-2.3451E-03	3.8926E-06	152	393	128.31
184	C3H6O	n-PROPIONALDEHYDE	29.204	8.1621E-01	-2.7350E-03	3.7667E-06	194	446	129.26
185	C3H6O	1,2-PROPYLENE OXIDE	53.347	5.1543E-01	-1.8029E-03	2.7795E-06	162	434	120.42
186	C3H6O	1,3-PROPYLENE OXIDE	29.517	6.1685E-01	-1.9717E-03	2.6102E-06	201	468	107.34
187	C3H6O2	ETHYL FORMATE	47.479	8.1081E-01	-2.6421E-03	3.6081E-06	195	458	149.98
188	C3H6O2	METHYL ACETATE	57.308	6.3751E-01	-2.1308E-03	3.0569E-06	176	456	138.99
189	C3H6O2	PROPIONIC ACID	29.869	9.4614E-01	-2.5563E-03	2.7814E-06	253	544	158.44
190	C3H6O2S	3-MERCAPTOPROPIONIC ACID	13.125	1.2179E+00	-2.7220E-03	2.5743E-06	292	656	202.50
191	C3H6O3	LACTIC ACID	--------	--------	--------	--------	--------	--------	--------
192	C3H6O3	METHOXYACETIC ACID	42.745	1.0047E+00	-2.3923E-03	2.3077E-06	282	622	190.82
193	C3H6O3	TRIOXANE	-127.928	1.9657E+00	-4.8761E-03	4.4890E-06	336	544	--------
194	C3H6S	THIACYCLOBUTANE	109.201	-1.1182E-01	4.1243E-04	--------	201	398	112.50
195	C3H7Br	1-BROMOPROPANE	70.791	4.9718E-01	-1.5708E-03	2.2582E-06	164	490	139.24
196	C3H7Br	2-BROMOPROPANE	41.959	7.4056E-01	-2.3195E-03	3.2325E-06	185	479	142.24
197	C3H7Cl	ISOPROPYL CHLORIDE	64.547	5.2282E-01	-1.8174E-03	2.8239E-06	157	440	133.71
198	C3H7Cl	n-PROPYL CHLORIDE	67.640	4.5271E-01	-1.5406E-03	2.4002E-06	151	453	129.29
199	C3H7F	1-FLUOROPROPANE	87.495	-1.9919E-02	2.2365E-04	--------	115	300	101.44
200	C3H7F	2-FLUOROPROPANE	87.227	-3.3816E-02	2.5179E-04	--------	141	294	--------
201	C3H7I	ISOPROPYL IODIDE	66.420	5.0524E-01	-1.4866E-03	1.9647E-06	184	520	136.98
202	C3H7I	n-PROPYL IODIDE	72.223	4.3776E-01	-1.2696E-03	1.7051E-06	173	534	135.08
203	C3H7N	ALLYLAMINE	62.267	6.8530E-01	-2.3003E-03	3.1803E-06	186	455	146.40
204	C3H7N	PROPYLENEIMINE	11.421	9.1952E-01	-2.8116E-03	3.4284E-06	230	476	126.51
205	C3H7NO	N,N-DIMETHYLFORMAMIDE	63.727	6.0708E-01	-1.6163E-03	1.8560E-06	214	582	150.24
206	C3H7NO	N-METHYLACETAMIDE	16.310	8.7673E-01	-2.0013E-03	1.8233E-06	302	646	--------
207	C3H7NO2	1-NITROPROPANE	92.868	5.1750E-01	-1.4948E-03	1.9847E-06	170	545	166.88
208	C3H7NO2	2-NITROPROPANE	86.389	6.0686E-01	-1.7670E-03	2.2920E-06	183	535	171.00
209	C3H7NO3	PROPYL-NITRATE	--------	--------	--------	--------	--------	--------	--------
210	C3H7NO3	ISOPROPYL-NITRATE	--------	--------	--------	--------	--------	--------	--------
211	C3H8	PROPANE	59.642	3.2831E-01	-1.5377E-03	3.6539E-06	86	333	117.67
212	C3H8O	ISOPROPANOL	72.525	7.9553E-01	-2.6330E-03	3.6498E-06	186	457	172.39
213	C3H8O	METHYL ETHYL ETHER	53.730	7.0133E-01	-2.6597E-03	4.3166E-06	161	394	140.81
214	C3H8O	n-PROPANOL	88.080	4.0224E-01	-1.3032E-03	1.9677E-06	148	483	144.32
215	C3H8O2	2-METHOXYETHANOL	74.939	7.1255E-01	-2.1504E-03	2.8269E-06	189	508	171.15
216	C3H8O2	METHYLAL	71.690	6.9438E-01	-2.3871E-03	3.6350E-06	169	433	162.86
217	C3H8O2	1,2-PROPYLENE GLYCOL	118.614	6.7283E-01	-1.8377E-03	2.1303E-06	214	563	212.32
218	C3H8O2	1,3-PROPYLENE GLYCOL	79.062	1.1435E+00	-2.9055E-03	3.0820E-06	247	592	243.40
219	C3H8O3	GLYCEROL	132.145	8.6007E-01	-1.9745E-03	1.8068E-06	292	651	260.94
220	C3H8S	n-PROPYLMERCAPTAN	81.561	4.6417E-01	-1.4850E-03	2.1680E-06	161	482	145.40
221	C3H8S	ISOPROPYL MERCAPTAN	83.939	4.3688E-01	-1.4514E-03	2.2863E-06	144	465	145.76
222	C3H8S	ETHYL-METHYL-SULFIDE	131.553	-1.2192E-01	5.5458E-04	--------	168	370	144.50
223	C3H9N	n-PROPYLAMINE	50.389	9.7033E-01	-3.2728E-03	4.5136E-06	191	447	168.38
224	C3H9N	ISOPROPYLAMINE	57.834	8.8718E-01	-3.1071E-03	4.6044E-06	179	425	168.18
225	C3H9N	TRIMETHYLAMINE	47.960	7.3020E-01	-2.8388E-03	4.6585E-06	157	390	136.78
226	C3H9NO	1-AMINO-2-PROPANOL	11.987	1.2623E+00	-3.3222E-03	3.3900E-06	276	553	182.88
227	C3H9NO	3-AMINO-1-PROPANOL	10.165	1.3660E+00	-3.3987E-03	3.3543E-06	285	584	204.23
228	C3H9NO	METHYLETHANOLAMINE	20.704	1.1721E+00	-3.0232E-03	3.1204E-06	270	567	184.11
229	C3H9O4P	TRIMETHYL PHOSPHATE	--------	--------	--------	--------	--------	--------	--------
230	C3H10N2	1,2-PROPANEDIAMINE	26.004	1.3569E+00	-3.8503E-03	4.3652E-06	238	528	203.98
231	C3H10Si	TRIMETHYL SILANE	66.982	6.7557E-01	-2.6664E-03	4.6980E-06	138	389	155.89
232	C4Cl4S	TETRACHLOROTHIOPHENE	--------	--------	--------	--------	--------	--------	--------
233	C4Cl6	HEXACHLORO-1,3-BUTADIENE	66.730	5.3054E-01	-1.2159E-03	1.2039E-06	253	667	148.74
234	C4F8	OCTAFLUORO-2-BUTENE	49.574	8.3253E-01	-3.5878E-03	6.5923E-06	139	353	153.58

Table 3-1 HEAT CAPACITY OF LIQUID - ORGANIC COMPOUNDS (continued)

			$C_P = A + B\,T + C\,T^2 + D\,T^3$				(C_P - joule/(mol K), T - K)		
NO	FORMULA	NAME	A	B	C	D	TMIN	TMAX	C_P @ 25 C
235	C4F8	OCTAFLUOROCYCLOBUTANE	-278.444	5.0367E+00	-1.8835E-02	2.5599E-05	234	350	227.44
236	C4F10	DECAFLUOROBUTANE	74.834	1.5602E+00	-6.7549E-03	1.2165E-05	146	348	261.96
237	C4H2	BUTADIYNE(BIACETYLENE)	--------	--------	--------	--------	--------	--------	--------
238	C4H2O3	MALEIC ANHYDRIDE	-12.662	1.0564E+00	-2.3244E-03	2.0518E-06	327	649	--------
239	C4H4	VINYLACETYLENE	8.143	8.8875E-01	-3.1484E-03	4.4314E-06	201	409	110.70
240	C4H4N2	SUCCINONITRILE	-3.768	1.1557E+00	-2.4533E-03	2.0499E-06	332	693	--------
241	C4H4O	FURAN	33.281	6.5201E-01	-2.2226E-03	3.1164E-06	189	441	112.70
242	C4H4O2	DIKETENE	24.669	9.8002E-01	-2.5652E-03	2.6901E-06	268	554	160.13
243	C4H4O3	SUCCINIC ANHYDRIDE	-71.503	1.3452E+00	-2.6140E-03	1.9408E-06	394	730	--------
244	C4H4O4	FUMARIC ACID	-1816.608	1.0582E+01	-1.8229E-02	1.0941E-05	561	694	--------
245	C4H4O4	MALEIC ACID	-170.196	2.2916E+00	-4.5516E-03	3.3936E-06	404	696	--------
246	C4H4S	THIOPHENE	32.611	6.7871E-01	-1.9074E-03	2.2163E-06	236	521	124.15
247	C4H5Cl	CHLOROPRENE	78.440	4.2596E-01	-1.4073E-03	2.1860E-06	144	473	138.28
248	C4H5N	trans-CROTONITRILE	27.424	7.3182E-01	-2.0987E-03	2.4768E-06	223	527	124.70
249	C4H5N	cis-CROTONITRILE	46.235	6.6994E-01	-2.0035E-03	2.5325E-06	202	511	135.00
250	C4H5N	METHACRYLONITRILE	3.484	1.1226E+00	-3.2966E-03	3.8592E-06	238	499	147.42
251	C4H5N	PYRROLE	41.545	6.4459E-01	-1.6782E-03	1.7807E-06	251	576	131.74
252	C4H5N	VINYLACETONITRILE	57.718	5.3362E-01	-1.5729E-03	2.0388E-06	187	526	131.04
253	C4H5NO2	METHYL CYANOACETATE	34.459	9.7717E-01	-2.3802E-03	2.4111E-06	261	618	178.12
254	C4H6	CYCLOBUTENE	72.569	-2.4828E-02	1.7241E-04	--------	155	306	80.49
255	C4H6	1,2-BUTADIENE	58.969	5.0878E-01	-1.9609E-03	3.4071E-06	138	400	126.65
256	C4H6	1,3-BUTADIENE	34.680	7.3205E-01	-2.8426E-03	4.6035E-06	165	383	122.26
257	C4H6	DIMETHYLACETYLENE	-25.640	1.2894E+00	-4.1111E-03	5.0352E-06	242	439	126.80
258	C4H6	ETHYLACETYLENE	55.668	6.2486E-01	-2.3963E-03	4.0147E-06	148	399	135.36
259	C4H6Cl2	1,3-DICHLORO-trans-2-BUTENE	87.399	6.0354E-01	-1.6638E-03	2.0265E-06	201	556	173.16
260	C4H6Cl2	1,4-DICHLORO-cis-2-BUTENE	65.982	7.7047E-01	-2.0398E-03	2.2964E-06	226	576	175.23
261	C4H6Cl2	1,4-DICHLORO-trans-2-BUTENE	15.008	1.1174E+00	-2.8198E-03	2.8433E-06	275	581	172.85
262	C4H6Cl2	3,4-DICHLORO-1-BUTENE	62.876	7.5359E-01	-2.1496E-03	2.6035E-06	213	530	165.48
263	C4H6O	trans-CROTONALDEHYDE	52.946	7.2786E-01	-2.1530E-03	2.7664E-06	198	514	151.89
264	C4H6O	2,5-DIHYDROFURAN	45.389	6.3271E-01	-1.9501E-03	2.5349E-06	201	488	127.86
265	C4H6O	DIVINYL ETHER	46.544	6.8934E-01	-2.4756E-03	3.7582E-06	173	417	131.61
266	C4H6O	METHACROLEIN	42.511	6.8698E-01	-2.1841E-03	2.9238E-06	193	477	130.68
267	C4H6O2	2-BUTYNE-1,4-DIOL	-32.149	1.6409E+00	-3.7193E-03	3.2662E-06	332	626	--------
268	C4H6O2	gamma-BUTYROLACTONE	73.029	4.3498E-01	-1.0196E-03	1.0527E-06	231	665	139.99
269	C4H6O2	cis-CROTONIC ACID	-7.425	1.4354E+00	-3.5711E-03	3.4943E-06	290	582	195.72
270	C4H6O2	trans-CROTONIC ACID	-118.011	2.0711E+00	-4.7846E-03	4.1639E-06	346	599	--------
271	C4H6O2	METHACRYLIC ACID	-7.089	1.2882E+00	-3.2195E-03	3.1623E-06	289	579	174.60
272	C4H6O2	METHYL ACRYLATE	54.109	8.0399E-01	-2.5149E-03	3.3155E-06	197	482	158.13
273	C4H6O2	VINYL ACETATE	63.910	7.0656E-01	-2.2832E-03	3.1788E-06	181	472	155.86
274	C4H6O3	ACETIC ANHYDRIDE	71.831	8.8879E-01	-2.6534E-03	3.3501E-06	201	512	189.75
275	C4H6O4	SUCCINIC ACID	-416.831	3.5896E+00	-6.6188E-03	4.4587E-06	462	725	--------
276	C4H6O5	DIGLYCOLIC ACID	-207.030	2.5933E+00	-4.8935E-03	3.4671E-06	422	738	--------
277	C4H6O5	MALIC ACID	-142.811	2.5086E+00	-4.9347E-03	3.6723E-06	404	703	--------
278	C4H6O6	TARTARIC ACID	-512.873	4.4047E+00	-7.8567E-03	5.1216E-06	480	745	--------
279	C4H7N	n-BUTYRONITRILE	84.111	5.0460E-01	-1.5134E-03	2.1012E-06	162	524	155.72
280	C4H7N	ISOBUTYRONITRILE	51.909	7.8410E-01	-2.3405E-03	2.9697E-06	203	509	156.35
281	C4H7NO	ACETONE CYANOHYDRIN	55.844	1.3376E+00	-3.4362E-03	3.6169E-06	254	582	245.04
282	C4H7NO	2-METHACRYLAMIDE	-90.764	1.7110E+00	-3.5405E-03	2.7745E-06	385	667	--------
283	C4H7NO	3-METHOXYPROPIONITRILE	63.094	7.0394E-01	-1.9037E-03	2.2158E-06	211	574	162.47
284	C4H7NO	2-PYRROLIDONE	52.256	7.8791E-01	-1.6638E-03	1.4633E-06	299	713	178.05
285	C4H8	1-BUTENE	74.597	3.3434E-01	-1.3914E-03	3.0241E-06	89	378	130.74
286	C4H8	cis-2-BUTENE	58.899	5.0376E-01	-1.9765E-03	3.5035E-06	135	392	126.25
287	C4H8	trans-2-BUTENE	36.162	7.9739E-01	-3.0674E-03	4.8919E-06	169	386	130.89
288	C4H8	CYCLOBUTANE	40.687	5.3104E-01	-1.8991E-03	2.7874E-06	183	414	104.07
289	C4H8	ISOBUTENE	57.611	5.6251E-01	-2.2985E-03	4.1773E-06	134	376	131.72
290	C4H8Br2	1,2-DIBROMOBUTANE	161.110	-9.1363E-02	3.5485E-04	--------	209	469	165.41
291	C4H8Br2	2,3-DIBROMOBUTANE	164.727	-1.1861E-01	3.8568E-04	--------	240	464	163.65
292	C4H8Cl2	1,4-DICHLOROBUTANE	64.944	8.3651E-01	-2.1802E-03	2.4034E-06	237	577	184.25
293	C4H8I2	1,2-DIIODOBUTANE	194.119	-1.8047E-01	4.6690E-04	--------	280	507	181.82
294	C4H8O	n-BUTYRALDEHYDE	64.363	7.2566E-01	-2.3548E-03	3.3065E-06	178	473	159.03
295	C4H8O	ISOBUTYRALDEHYDE	31.228	1.1020E+00	-3.5601E-03	4.6662E-06	209	456	166.99
296	C4H8O	1,2-EPOXYBUTANE	87.286	3.8742E-01	-1.2905E-03	2.1365E-06	124	473	144.70
297	C4H8O	METHYL ETHYL KETONE	61.406	7.5324E-01	-2.3814E-03	3.2240E-06	187	482	159.75
298	C4H8O	ETHYL VINYL ETHER	65.133	6.4348E-01	-2.2971E-03	3.6025E-06	158	428	148.27
299	C4H8O	TETRAHYDROFURAN	63.393	4.0257E-01	-1.2686E-03	1.8275E-06	166	486	119.08
300	C4H8O2	cis-2-BUTENE-1,4-DIOL	33.870	1.5014E+00	-3.6176E-03	3.5022E-06	285	610	252.75
301	C4H8O2	trans-2-BUTENE-1,4-DIOL	4.371	1.7115E+00	-4.0527E-03	3.7939E-06	301	613	--------
302	C4H8O2	ISOBUTYRIC ACID	62.423	8.1702E-01	-2.2463E-03	2.5743E-06	228	548	174.56
303	C4H8O2	n-BUTYRIC ACID	28.210	1.1040E+00	-2.8523E-03	2.9528E-06	269	565	182.09
304	C4H8O2	1,4-DIOXANE	-20.729	1.2913E+00	-3.4298E-03	3.5408E-06	286	528	153.23
305	C4H8O2	ETHYL ACETATE	62.832	8.4097E-01	-2.6998E-03	3.6631E-06	191	471	170.66
306	C4H8O2	METHYL PROPIONATE	72.707	8.2005E-01	-2.6119E-03	3.5581E-06	187	478	179.33
307	C4H8O2	n-PROPYL FORMATE	74.311	6.9572E-01	-2.1969E-03	3.0196E-06	181	484	166.48
308	C4H8O2S	SULFOLANE	74.580	5.5933E-01	-1.1027E-03	9.3850E-07	302	764	--------
309	C4H8S	TETRAHYDROTHIOPHENE	91.345	3.1960E-01	-8.6978E-04	1.1074E-06	178	569	138.67
310	C4H9Br	1-BROMOBUTANE	86.013	4.6563E-01	-1.3952E-03	1.9565E-06	162	519	152.68
311	C4H9Br	2-BROMOBUTANE	105.227	4.5815E-01	-1.4005E-03	1.9658E-06	162	510	169.43
312	C4H9Cl	n-BUTYL CHLORIDE	86.225	5.1021E-01	-1.6387E-03	2.4698E-06	151	483	158.13

63

Table 3-1 HEAT CAPACITY OF LIQUID - ORGANIC COMPOUNDS (continued)

| NO | FORMULA | NAME | $C_P = A + B\,T + C\,T^2 + D\,T^3$ | | | | $(C_P$ - joule/(mol K), T - K$)$ | | |
			A	B	C	D	TMIN	TMAX	C_P @ 25 C
313	C4H9Cl	sec-BUTYL CHLORIDE	90.546	4.7991E-01	-1.5952E-03	2.5017E-06	143	469	158.13
314	C4H9Cl	tert-BUTYL CHLORIDE	-21.729	1.4906E+00	-4.5845E-03	5.4420E-06	249	456	159.39
315	C4H9I	2-IODO-2-METHYLPROPANE	142.728	-1.6837E-01	5.2368E-04	--------	236	403	139.08
316	C4H9N	PYRROLIDINE	56.715	7.6953E-01	-2.2572E-03	2.7490E-06	216	512	158.36
317	C4H9NO	N,N-DIMETHYLACETAMIDE	43.944	8.9037E-01	-2.2645E-03	2.3578E-06	254	592	170.59
318	C4H9NO	MORPHOLINE	22.401	1.0975E+00	-2.8725E-03	2.9705E-06	271	556	173.02
319	C4H9NO2	1-NITROBUTANE	--------	--------	--------	--------	--------	--------	--------
320	C4H9NO2	2-NITROBUTANE	--------	--------	--------	--------	--------	--------	--------
321	C4H10	n-BUTANE	62.873	5.8913E-01	-2.3588E-03	4.2257E-06	136	383	140.84
322	C4H10	ISOBUTANE	71.791	4.8472E-01	-2.0519E-03	4.0634E-06	115	367	141.61
323	C4H10N2	PIPERAZINE	-192.418	2.7211E+00	-6.1900E-03	5.1736E-06	380	574	--------
324	C4H10O	n-BUTANOL	83.877	5.6628E-01	-1.7208E-03	2.2780E-06	185	507	160.12
325	C4H10O	sec-BUTANOL	95.037	5.6593E-01	-1.8256E-03	2.6675E-06	159	482	172.18
326	C4H10O	tert-BUTANOL	-309.415	4.4863E+00	-1.2958E-02	1.3642E-05	300	456	--------
327	C4H10O	DIETHYL ETHER	75.939	7.7335E-01	-2.7936E-03	4.4383E-06	158	420	175.81
328	C4H10O	METHYL-PROPYL-ETHER	144.405	-1.6391E-01	7.7474E-04	--------	158	342	164.40
329	C4H10O	METHYL ISOPROPYL ETHER	94.789	4.6032E-01	-1.7158E-03	2.9878E-06	129	418	158.70
330	C4H10O	ISOBUTANOL	96.150	4.9462E-01	-1.5601E-03	2.2031E-06	166	493	163.33
331	C4H10O2	1,3-BUTANEDIOL	125.062	7.6284E-01	-2.0381E-03	2.4560E-06	197	579	236.42
332	C4H10O2	1,4-BUTANEDIOL	10.303	1.5972E+00	-3.8628E-03	3.7022E-06	294	600	241.24
333	C4H10O2	2,3-BUTANEDIOL	1.022	1.6059E+00	-4.1757E-03	4.2641E-06	282	550	221.65
334	C4H10O2	t-BUTYL HYDROPEROXIDE	-18.834	1.1762E+00	-3.2019E-03	3.3722E-06	278	518	136.59
335	C4H10O2	1,2-DIMETHOXYETHANE	45.096	1.1083E+00	-3.3990E-03	4.2685E-06	216	483	186.52
336	C4H10O2	2-ETHOXYETHANOL	84.952	9.3806E-01	-2.7796E-03	3.5283E-06	201	512	211.06
337	C4H10O3	DIETHYLENE GLYCOL	126.618	8.5587E-01	-1.9468E-03	1.8725E-06	264	670	258.37
338	C4H10O4S	DIETHYL SULFATE	--------	--------	--------	--------	--------	--------	--------
339	C4H10S	n-BUTYL MERCAPTAN	98.315	5.2234E-01	-1.5783E-03	2.2589E-06	158	512	173.62
340	C4H10S	ISOBUTYL MERCAPTAN	115.038	3.6933E-01	-1.1509E-03	1.8046E-06	129	503	170.67
341	C4H10S	sec-BUTYL MERCAPTAN	104.316	4.4894E-01	-1.3994E-03	2.1926E-06	134	499	171.89
342	C4H10S	tert-BUTYL MERCAPTAN	-75.206	1.9880E+00	-5.7348E-03	6.3144E-06	275	477	175.08
343	C4H10S	DIETHYL SULFIDE	90.432	5.8704E-01	-1.8123E-03	2.5217E-06	170	501	171.19
344	C4H10S	ISOPROPYL-METHYL-SULFIDE	150.789	-1.3087E-01	5.7976E-04	--------	173	388	163.31
345	C4H10S	METHYL-PROPYL-SULFIDE	159.495	-1.2092E-01	5.7992E-04	--------	161	399	174.99
346	C4H10S2	DIETHYL DISULFIDE	125.723	5.2943E-01	-1.4155E-03	1.8261E-06	173	578	206.15
347	C4H11N	n-BUTYLAMINE	32.672	1.1738E+00	-3.6137E-03	4.4206E-06	225	479	178.56
348	C4H11N	ISOBUTYLAMINE	71.394	8.9544E-01	-2.9423E-03	4.0108E-06	190	462	183.12
349	C4H11N	sec-BUTYLAMINE	81.998	6.9550E-01	-2.3225E-03	3.3532E-06	170	463	171.78
350	C4H11N	tert-BUTYLAMINE	43.328	1.2643E+00	-4.2403E-03	5.6876E-06	207	436	194.09
351	C4H11N	DIETHYLAMINE	2.800	1.4016E+00	-4.5271E-03	5.7411E-06	224	447	170.40
352	C4H11NO	DIMETHYLETHANOLAMINE	74.876	1.0965E+00	-3.1968E-03	3.9069E-06	215	515	221.16
353	C4H11NO2	DIETHANOLAMINE	76.703	1.0821E+00	-2.4860E-03	2.2497E-06	302	644	--------
354	C4H11NO2	2-AMINOETHOXYETHANOL	141.811	7.2934E-01	-1.8110E-03	2.0710E-06	201	628	253.17
355	C4H12N2O	N-AMINOETHYL ETHANOLAMINE	151.634	7.7874E-01	-1.9367E-03	2.2052E-06	201	629	270.10
356	C4H12Si	TETRAMETHYLSILANE	55.176	1.1592E+00	-4.2512E-03	6.5005E-06	175	405	195.17
357	C4H13N3	DIETHYLENE TRIAMINE	91.830	8.2839E-01	-2.0729E-03	2.2306E-06	235	608	213.67
358	C5Cl6	HEXACHLOROCYCLOPENTADIENE	37.164	5.0776E-01	-1.1331E-03	1.0504E-06	285	671	115.67
359	C5H4O2	FURFURAL	66.792	7.0755E-01	-1.8082E-03	1.9630E-06	238	591	169.03
360	C5H5N	PYRIDINE	37.150	6.9497E-01	-1.8749E-03	2.1188E-06	233	558	133.85
361	C5H6	CYCLOPENTADIENE	43.542	5.8799E-01	-1.9291E-03	2.6761E-06	189	456	118.30
362	C5H6	2-METHYL-1-BUTENE-3-YNE	61.157	5.5782E-01	-1.9126E-03	2.9378E-06	161	443	135.31
363	C5H6	1-PENTENE-3-YNE	21.792	9.3719E-01	-2.9734E-03	3.9877E-06	201	468	142.59
364	C5H6	1-PENTENE-4-YNE	36.297	8.2047E-01	-2.6840E-03	3.6127E-06	201	453	138.08
365	C5H6N2	GLUTARONITRILE	70.115	7.1706E-01	-1.5840E-03	1.5600E-06	245	704	184.45
366	C5H6O2	FURFURYL ALCOHOL	72.353	9.3499E-01	-2.4272E-03	2.5437E-06	260	569	202.77
367	C5H6O3	GLUTARIC ANHYDRIDE	41.100	9.5549E-01	-1.9004E-03	1.5421E-06	329	754	--------
368	C5H6O4	CITRACONIC ACID	-18.090	1.6511E+00	-3.2561E-03	2.5334E-06	357	746	--------
369	C5H6O4	ITACONIC ACID	-273.296	2.9306E+00	-5.4439E-03	3.7600E-06	440	739	--------
370	C5H6S	2-METHYLTHIOPHENE	141.537	-1.2849E-01	4.6171E-04	--------	211	416	144.27
371	C5H6S	3-METHYLTHIOPHENE	140.361	-1.2085E-01	4.4876E-04	--------	205	419	144.22
372	C5H7N	N-METHYLPYRROLE	41.210	7.8255E-01	-2.1567E-03	2.5583E-06	218	549	150.61
373	C5H7NO2	ETHYL CYANOACETATE	69.778	1.0987E+00	-2.7196E-03	2.8165E-06	252	611	230.24
374	C5H8	CYCLOPENTENE	64.821	3.4830E-01	-1.1840E-03	1.9109E-06	139	456	114.06
375	C5H8	ISOPRENE	80.542	4.5089E-01	-1.6166E-03	2.7723E-06	128	436	144.75
376	C5H8	3-METHYL-1,2-BUTADIENE	62.370	6.9368E-01	-2.4233E-03	3.7091E-06	161	441	152.08
377	C5H8	2-METHYL-1,3-BUTADIENE	127.014	-1.0764E-01	6.5526E-04	--------	128	337	153.17
378	C5H8	1,2-PENTADIENE	76.700	4.4815E-01	-1.5516E-03	2.5392E-06	137	450	139.68
379	C5H8	cis-1,3-PENTADIENE	79.165	4.6309E-01	-1.6164E-03	2.6748E-06	133	449	144.44
380	C5H8	trans-1,3-PENTADIENE	47.966	7.8011E-01	-2.5933E-03	3.6547E-06	187	450	146.89
381	C5H8	1,4-PENTADIENE	67.713	5.7753E-01	-2.1090E-03	3.6881E-06	126	431	150.17
382	C5H8	2,3-PENTADIENE	75.900	5.7996E-01	-2.0270E-03	3.1863E-06	149	447	153.07
383	C5H8	1-PENTYNE	42.607	9.4233E-01	-3.2983E-03	5.0254E-06	168	433	163.56
384	C5H8	2-PENTYNE	156.802	-1.4060E-01	6.6609E-04	--------	165	359	174.09
385	C5H8	3-METHYL-1-BUTYNE	35.285	1.0768E+00	-3.8240E-03	5.6180E-06	184	417	165.30
386	C5H8	SPIROPENTANE	112.077	-9.8977E-02	4.7696E-04	--------	167	342	124.97
387	C5H8N4O12	PENTAERYTHRITOL TETRANITRATE	--------	--------	--------	--------	--------	--------	--------
388	C5H8O	CYCLOPENTANONE	66.239	6.8576E-01	-1.8291E-03	2.1151E-06	223	563	164.16
389	C5H8O	METHYL ISOPROPENYL KETONE	41.822	8.4489E-01	-2.5959E-03	3.1418E-06	221	509	158.17
390	C5H8O2	ACETYLACETONE	49.793	1.1457E+00	-3.1191E-03	3.4079E-06	251	542	204.44

64

Table 3-1 HEAT CAPACITY OF LIQUID - ORGANIC COMPOUNDS (continued)

			$C_P = A + B T + C T^2 + D T^3$			(C_P - joule/(mol K), T - K)			
NO	FORMULA	NAME	A	B	C	D	TMIN	TMAX	C_P @ 25 C
391	C5H8O2	ALLYL ACETATE	114.963	5.2345E-01	-1.6371E-03	2.4946E-06	139	503	191.62
392	C5H8O2	ETHYL ACRYLATE	66.535	9.1312E-01	-2.7675E-03	3.5431E-06	203	498	186.67
393	C5H8O2	METHYL METHACRYLATE	42.365	1.0787E+00	-3.1551E-03	3.7759E-06	226	508	183.58
394	C5H8O2	VINYL PROPIONATE	64.575	8.7548E-01	-2.6832E-03	3.4751E-06	201	491	179.18
395	C5H8O3	2-HYDROXYETHYL ACRYLATE	112.540	7.7899E-01	-2.0092E-03	2.2917E-06	214	596	226.93
396	C5H8O3	LEVULINIC ACID	41.840	1.4153E+00	-3.1901E-03	2.8584E-06	309	651	--------
397	C5H8O3	METHYL ACETOACETATE	127.150	5.1044E-01	-1.3794E-03	1.6579E-06	194	578	200.66
398	C5H8O4	GLUTARIC ACID	-10.226	1.7285E+00	-3.4237E-03	2.6321E-06	372	726	--------
399	C5H9N	VALERONITRILE	100.491	6.2946E-01	-1.8110E-03	2.3635E-06	178	543	189.82
400	C5H9NO	n-BUTYL ISOCYANATE	24.141	1.1425E+00	-3.3850E-03	4.3837E-06	201	511	180.06
401	C5H9NO	N-METHYL-2-PYRROLIDONE	80.326	6.3684E-01	-1.4931E-03	1.5017E-06	250	652	177.27
402	C5H9NO4	L-GLUTAMIC ACID	--------	--------	--------	--------	--------	--------	--------
403	C5H10	CYCLOPENTANE	49.998	5.3894E-01	-1.7696E-03	2.5053E-06	180	461	119.78
404	C5H10	2-METHYL-1-BUTENE	77.452	5.7879E-01	-2.1473E-03	3.6555E-06	137	419	156.02
405	C5H10	2-METHYL-2-BUTENE	74.305	5.8384E-01	-2.1386E-03	3.5676E-06	140	424	152.82
406	C5H10	3-METHYL-1-BUTENE	90.005	3.9777E-01	-1.5462E-03	2.9909E-06	106	405	150.42
407	C5H10	1-PENTENE	88.892	4.2198E-01	-1.5803E-03	2.9790E-06	109	418	153.18
408	C5H10	cis-2-PENTENE	80.445	4.9493E-01	-1.8139E-03	3.2037E-06	123	428	151.67
409	C5H10	trans-2-PENTENE	81.528	5.2108E-01	-1.9000E-03	3.2213E-06	134	428	153.37
410	C5H10Br2	2,3-DIBROMO-2-METHYLBUTANE	180.601	-1.7099E-01	4.3932E-04	--------	289	474	168.67
411	C5H10Cl2	1,5-DICHLOROPENTANE	109.864	7.1412E-01	-1.8637E-03	2.1832E-06	201	597	214.96
412	C5H10O	METHYL ISOPROPYL KETONE	83.052	7.1419E-01	-2.1982E-03	2.9759E-06	182	498	179.46
413	C5H10O	2-PENTANONE	--------	--------	--------	--------	--------	--------	--------
414	C5H10O	DIETHYL KETONE	26.231	1.2822E+00	-3.7449E-03	4.3816E-06	235	505	191.76
415	C5H10O	VALERALDEHYDE	97.499	6.5995E-01	-2.0343E-03	2.7292E-06	183	499	185.76
416	C5H10O2	n-BUTYL FORMATE	94.303	7.3040E-01	-2.2265E-03	2.9947E-06	182	503	193.52
417	C5H10O2	ETHYL PROPIONATE	70.920	9.7543E-01	-2.9871E-03	3.8823E-06	200	491	199.10
418	C5H10O2	ISOBUTYL FORMATE	104.038	8.2157E-01	-2.5541E-03	3.4864E-06	178	496	214.35
419	C5H10O2	ISOPROPYL ACETATE	70.900	9.8618E-01	-3.0555E-03	3.9901E-06	201	484	199.07
420	C5H10O2	n-PROPYL ACETATE	91.591	7.8205E-01	-2.4341E-03	3.3267E-06	179	494	196.55
421	C5H10O2	METHYL n-BUTYRATE	90.092	8.6461E-01	-2.6506E-03	3.5164E-06	188	499	205.45
422	C5H10O2	2-METHYLBUTYRIC ACID	109.104	7.4411E-01	-1.9894E-03	2.3727E-06	201	579	217.00
423	C5H10O2	ISOVALERIC ACID	81.845	1.1114E+00	-2.9084E-03	3.1565E-06	245	571	238.34
424	C5H10O2	VALERIC ACID	79.976	8.8728E-01	-2.2799E-03	2.4715E-06	240	586	207.35
425	C5H10O2	TETRAHYDROFURFURYL ALCOHOL	115.140	5.1702E-01	-1.3875E-03	1.6517E-06	201	575	189.73
426	C5H10O2S	3-METHYL SULFOLANE	110.368	7.2683E-01	-1.4981E-03	1.3663E-06	275	735	230.11
427	C5H10O3	DIETHYL CARBONATE	56.855	1.1659E+00	-3.3274E-03	3.9006E-06	231	518	212.06
428	C5H10O3	ETHYL LACTATE	-46.239	2.1823E+00	-5.9832E-03	6.8683E-06	248	529	254.59
429	C5H10S	THIACYCLOHEXANE	184.412	-2.4478E-01	5.9183E-04	--------	293	445	164.04
430	C5H10S	CYCLOPENTANETHIOL	143.203	-7.2569E-02	3.8499E-04	--------	156	435	155.79
431	C5H11Br	1-BROMOPENTANE	143.414	-1.1807E-01	4.7786E-04	--------	186	433	150.69
432	C5H11Cl	1-CHLOROPENTANE	99.232	6.3409E-01	-1.9117E-03	2.6022E-06	175	511	187.32
433	C5H11Cl	1-CHLORO-3-METHYLBUTANE	127.911	-1.1249E-01	4.9476E-04	--------	170	402	138.35
434	C5H11Cl	2-CHLORO-2-METHYLBUTANE	129.368	-1.4760E-01	5.3859E-04	--------	201	389	133.24
435	C5H11N	N-METHYLPYRROLIDINE	73.463	6.1054E-01	-1.8779E-03	2.5390E-06	184	495	155.86
436	C5H11N	PIPERIDINE	11.977	1.2812E+00	-3.4750E-03	3.7200E-06	264	535	183.64
437	C5H11NO	tert-BUTYLFORMAMIDE	20.664	1.2624E+00	-2.9890E-03	2.8365E-06	290	623	206.52
438	C5H12	ISOPENTANE	91.474	4.4852E-01	-1.6859E-03	3.1342E-06	114	414	158.40
439	C5H12	NEOPENTANE	-186.315	3.2441E+00	-1.0910E-02	1.3428E-05	258	390	167.00
440	C5H12	n-PENTANE	80.641	6.2195E-01	-2.2682E-03	3.7423E-06	144	423	163.64
441	C5H12O	2,2-DIMETHYL-1-PROPANOL	-280.652	3.8455E+00	-1.0176E-02	9.8371E-06	328	495	--------
442	C5H12O	tert-PENTYL-ALCOHOL	239.683	-2.5343E-01	5.6873E-04	--------	328	416	--------
443	C5H12O	2-METHYL-1-BUTANOL	109.861	8.8256E-01	-2.6422E-03	3.3415E-06	201	509	226.68
444	C5H12O	2-METHYL-2-BUTANOL	-38.513	2.2919E+00	-6.5955E-03	7.2865E-06	265	491	251.62
445	C5H12O	3-METHYL-1-BUTANOL	95.590	4.8022E-01	-1.4387E-03	2.0386E-06	157	522	164.90
446	C5H12O	3-METHYL-2-BUTANOL	85.419	8.6417E-01	-2.5382E-03	3.2100E-06	201	517	202.52
447	C5H12O	1-PENTANOL	105.748	7.4623E-01	-2.1694E-03	2.7315E-06	197	528	207.78
448	C5H12O	2-PENTANOL	93.720	1.1370E+00	-3.4487E-03	4.4456E-06	201	497	243.98
449	C5H12O	3-PENTANOL	87.761	1.2602E+00	-3.8391E-03	4.9160E-06	205	492	252.51
450	C5H12O	METHYL sec-BUTYL ETHER	49.275	1.1582E+00	-3.8163E-03	5.1616E-06	201	448	192.16
451	C5H12O	METHYL tert-BUTYL ETHER	83.744	7.6602E-01	-2.6132E-03	3.9171E-06	166	447	183.66
452	C5H12O	METHYL ISOBUTYL ETHER	49.834	1.1435E+00	-3.7712E-03	5.1067E-06	201	447	190.88
453	C5H12O	ETHYL PROPYL ETHER	102.864	7.0110E-01	-2.4095E-03	3.8232E-06	147	450	199.04
454	C5H12O2	ETHYLENE GLYCOL MONOPROPYL ETHER	99.514	7.2678E-01	-2.1371E-03	2.8033E-06	184	524	200.52
455	C5H12O2	NEOPENTYL GLYCOL	-509.209	5.1325E+00	-1.1404E-02	9.1520E-06	401	579	--------
456	C5H12O2	1,5-PENTANEDIOL	67.470	1.0846E+00	-2.6842E-03	2.7543E-06	258	606	225.24
457	C5H12O3	2-(2-METHOXYETHOXY)ETHANOL	138.608	9.8731E-01	-2.6928E-03	3.2679E-06	198	567	280.22
458	C5H12O4	PENTAERYTHRITOL	-1429.747	9.5192E+00	-1.6680E-02	1.0341E-05	535	702	--------
459	C5H12S	n-PENTYL MERCAPTAN	94.716	7.5979E-01	-2.1472E-03	2.6885E-06	198	538	201.63
460	C5H12S	BUTYL-METHYL-SULFIDE	187.240	-1.4780E-01	6.3060E-04	--------	176	427	199.23
461	C5H12S	ETHYL-PROPYL-SULFIDE	180.319	-1.2105E-01	5.9709E-04	--------	157	422	197.30
462	C5H12S	2-METHYL-2-BUTANETHIOL	165.372	-1.2006E-01	5.5546E-04	--------	170	402	178.95
463	C5H13N	n-PENTYLAMINE	60.963	1.1533E+00	-3.4538E-03	4.2115E-06	219	500	209.44
464	C5H13NO2	METHYL DIETHANOLAMINE	105.151	1.3564E+00	-3.3459E-03	3.4589E-06	253	610	303.80
465	C6Cl6	HEXACHLOROBENZENE	-743.411	5.4844E+00	-9.6238E-03	6.0919E-06	503	743	--------
466	C6F6	HEXAFLUOROBENZENE	-138.486	2.9233E+00	-8.5664E-03	9.3787E-06	279	465	220.17
467	C6H3ClN2O4	1-CHLORO-2,4-DINITROBENZENE	--------	--------	--------	--------	--------	--------	--------
468	C6H3Cl2NO2	1,2-DICHLORO-4-NITROBENZENE	--------	--------	--------	--------	--------	--------	--------

Table 3-1 HEAT CAPACITY OF LIQUID - ORGANIC COMPOUNDS (continued)

			$C_P = A + B\,T + C\,T^2 + D\,T^3$				(C$_P$ - joule/(mol K), T - K)		
NO	FORMULA	NAME	A	B	C	D	TMIN	TMAX	C$_P$ @ 25 C
469	C6H3Cl3	1,2,4-TRICHLOROBENZENE	48.567	1.0127E+00	-2.3033E-03	2.1401E-06	291	653	202.48
470	C6H3N3O6	1,3,5-TRINITROBENZENE	--------	--------	--------	--------	--------	--------	--------
471	C6H4Br2	m-DIBROMOBENZENE	85.577	6.0756E-01	-1.3411E-03	1.2780E-06	267	685	181.38
472	C6H4ClNO2	m-CHLORONITROBENZENE	17.760	1.3304E+00	-2.9350E-03	2.5462E-06	319	668	--------
473	C6H4ClNO2	o-CHLORONITROBENZENE	45.550	1.1376E+00	-2.4757E-03	2.1890E-06	307	681	--------
474	C6H4ClNO2	p-CHLORONITROBENZENE	-29.177	1.6056E+00	-3.3941E-03	2.7386E-06	358	676	--------
475	C6H4Cl2	m-DICHLOROBENZENE	61.903	7.3971E-01	-1.8045E-03	1.8839E-06	249	616	171.97
476	C6H4Cl2	o-DICHLOROBENZENE	68.542	7.7568E-01	-1.8380E-03	1.8585E-06	257	635	185.68
477	C6H4Cl2	p-DICHLOROBENZENE	-32.283	1.4413E+00	-3.3088E-03	2.9587E-06	327	616	--------
478	C6H4F2	m-DIFLUOROBENZENE	146.205	-2.2760E-01	6.4167E-04	--------	250	394	135.39
479	C6H4F2	o-DIFLUOROBENZENE	143.058	-2.0858E-01	6.1680E-04	--------	240	395	135.70
480	C6H4F2	p-DIFLUOROBENZENE	149.090	-2.4236E-01	6.5284E-04	--------	261	392	134.86
481	C6H4N2O4	m-DINITROBENZENE	14.002	1.5564E+00	-3.1119E-03	2.4113E-06	365	725	--------
482	C6H4N2O4	o-DINITROBENZENE	-51.485	1.7659E+00	-3.3714E-03	2.4936E-06	391	748	--------
483	C6H4N2O4	p-DINITROBENZENE	-262.234	2.8646E+00	-5.3418E-03	3.6879E-06	448	723	--------
484	C6H5Br	BROMOBENZENE	65.399	6.4039E-01	-1.6014E-03	1.7006E-06	243	603	159.05
485	C6H5Cl	MONOCHLOROBENZENE	64.358	6.1906E-01	-1.6346E-03	1.8478E-06	229	569	152.60
486	C6H5ClO	m-CHLOROPHENOL	38.907	8.8277E-01	-1.9666E-03	1.7734E-06	307	656	--------
487	C6H5ClO	o-CHLOROPHENOL	40.234	1.0880E+00	-2.6322E-03	2.5615E-06	283	608	198.53
488	C6H5ClO	p-CHLOROPHENOL	16.057	1.0413E+00	-2.2888E-03	2.0147E-06	317	664	--------
489	C6H5Cl2N	3,4-DICHLOROANILINE	15.718	1.2502E+00	-2.5363E-03	2.0484E-06	346	720	--------
490	C6H5F	FLUOROBENZENE	35.390	8.6297E-01	-2.5132E-03	2.9781E-06	232	504	148.21
491	C6H5I	IODOBENZENE	76.484	5.4594E-01	-1.2829E-03	1.3175E-06	243	649	160.14
492	C6H5NO2	NITROBENZENE	51.773	9.1277E-01	-2.1098E-03	2.0093E-06	280	647	189.62
493	C6H6	BENZENE	-31.662	1.3043E+00	-3.6078E-03	3.8243E-06	280	506	137.87
494	C6H6ClN	m-CHLOROANILINE	67.821	9.1756E-01	-2.0690E-03	2.0013E-06	264	676	210.51
495	C6H6ClN	o-CHLOROANILINE	-744.768	5.4383E+00	-1.0437E-02	7.0862E-06	483	650	--------
496	C6H6ClN	p-CHLOROANILINE	-23.832	1.4238E+00	-3.0256E-03	2.5105E-06	344	679	--------
497	C6H6N2	cis-DICYANO-1-BUTENE	49.986	1.1770E+00	-2.9128E-03	2.9828E-06	250	622	221.04
498	C6H6N2	trans-DICYANO-1-BUTENE	34.915	1.2997E+00	-3.1934E-03	3.2026E-06	261	620	223.43
499	C6H6N2	1,4-DICYANO-2-BUTENE	-90.745	1.9916E+00	-4.2350E-03	3.4644E-06	350	680	--------
500	C6H6N2O2	m-NITROANILINE	-32.233	1.6894E+00	-3.2658E-03	2.4503E-06	388	734	--------
501	C6H6N2O2	o-NITROANILINE	36.230	1.3890E+00	-2.8446E-03	2.3237E-06	346	706	--------
502	C6H6N2O2	p-NITROANILINE	-86.794	1.8706E+00	-3.4276E-03	2.4025E-06	422	766	--------
503	C6H6O	PHENOL	38.622	1.0983E+00	-2.4897E-03	2.2802E-06	315	625	--------
504	C6H6O2	1,2-BENZENEDIOL	-52.044	1.4248E+00	-2.9108E-03	2.2679E-06	379	688	--------
505	C6H6O2	1,3-BENZENEDIOL	-7.809	1.4223E+00	-2.7824E-03	2.0976E-06	383	729	--------
506	C6H6O2	p-HYDROQUINONE	-72.810	1.6718E+00	-3.0532E-03	2.1057E-06	446	740	--------
507	C6H6O3	1,2,3-BENZENETRIOL	68.242	8.7470E-01	-1.6146E-03	1.1776E-06	408	747	--------
508	C6H6S	PHENYL MERCAPTAN	58.539	7.8070E-01	-1.8817E-03	1.9154E-06	259	620	174.80
509	C6H7N	ANILINE	63.288	9.8960E-01	-2.3583E-03	2.3296E-06	268	629	210.44
510	C6H7N	2-METHYLPYRIDINE	72.116	6.2427E-01	-1.7202E-03	2.0497E-06	207	559	159.66
511	C6H7N	3-METHYLPYRIDINE	36.448	8.8379E-01	-2.2709E-03	2.3843E-06	256	581	161.28
512	C6H7N	4-METHYLPYRIDINE	13.736	1.1295E+00	-2.8466E-03	2.8505E-06	278	582	173.01
513	C6H8	1,3-CYCLOHEXADIENE	70.721	4.7940E-01	-1.4750E-03	2.1151E-06	162	502	138.59
514	C6H8	METHYLCYCLOPENTADIENE	59.202	7.4490E-01	-2.2926E-03	2.9861E-06	201	487	156.64
515	C6H8N2	ADIPONITRILE	57.459	6.4731E-01	-1.4099E-03	1.2959E-06	277	703	159.47
516	C6H8N2	METHYLGLUTARONITRILE	85.906	8.6895E-01	-2.0198E-03	2.1182E-06	229	668	221.58
517	C6H8N2	m-PHENYLENEDIAMINE	30.904	1.1114E+00	-2.2425E-03	1.8053E-06	335	742	--------
518	C6H8N2	o-PHENYLENEDIAMINE	-56.502	1.6935E+00	-3.4068E-03	2.6393E-06	378	703	--------
519	C6H8N2	p-PHENYLENEDIAMINE	-138.577	2.1251E+00	-4.0994E-03	2.9816E-06	414	716	--------
520	C6H8N2	PHENYLHYDRAZINE	39.493	9.8286E-01	-2.1853E-03	1.9533E-06	293	685	190.05
521	C6H8N2O	BIS(CYANOETHYL)ETHER	133.852	7.5737E-01	-1.6480E-03	1.6110E-06	248	705	255.86
522	C6H8O4	DIMETHYL MALEATE	81.726	1.2265E+00	-3.0327E-03	3.1301E-06	255	608	260.78
523	C6H8O6	ASCORBIC ACID	-533.966	5.1304E+00	-9.5422E-03	6.4843E-06	466	705	--------
524	C6H8O7	CITRIC ACID	-248.283	3.4707E+00	-6.4923E-03	4.5716E-06	427	740	--------
525	C6H10	1-METHYLCYCLOPENTENE	130.320	-8.9159E-02	4.8824E-04	--------	147	379	147.14
526	C6H10	3-METHYLCYCLOPENTENE	122.081	-7.3410E-02	4.6273E-04	--------	131	373	141.33
527	C6H10	4-METHYLCYCLOPENTENE	124.412	-5.8681E-02	4.4830E-04	--------	113	378	146.77
528	C6H10	CYCLOHEXENE	75.841	4.7761E-01	-1.4586E-03	2.0271E-06	171	504	142.31
529	C6H10	2,3-DIMETHYL-1,3-BUTADIENE	63.063	9.0896E-01	-2.8708E-03	3.8191E-06	198	473	180.09
530	C6H10	1,5-HEXADIENE	100.547	4.9449E-01	-1.6992E-03	2.7799E-06	133	456	170.61
531	C6H10	cis,trans-2,4-HEXADIENE	78.343	7.2388E-01	-2.2889E-03	3.1790E-06	178	484	174.96
532	C6H10	trans,trans-2,4-HEXADIENE	28.602	1.2039E+00	-3.6471E-03	4.4353E-06	229	482	180.89
533	C6H10	1-HEXYNE	107.406	5.6639E-01	-1.8970E-03	2.9945E-06	142	465	187.01
534	C6H10	2-HEXYNE	67.189	7.6877E-01	-2.3721E-03	3.2155E-06	185	494	170.76
535	C6H10	3-HEXYNE	82.692	6.3383E-01	-1.9955E-03	2.8095E-06	171	490	168.74
536	C6H10O	CYCLOHEXANONE	68.641	8.6690E-01	-2.2835E-03	2.4978E-06	243	566	190.32
537	C6H10O	MESITYL OXIDE	74.977	1.0107E+00	-2.8245E-03	3.3225E-06	221	540	213.30
538	C6H10O2	epsilon-CAPROLACTONE	83.582	7.3015E-01	-1.5987E-03	1.4942E-06	273	694	198.77
539	C6H10O2	ETHYL METHACRYLATE	82.106	8.6863E-01	-2.5461E-03	3.2089E-06	201	519	199.81
540	C6H10O2	n-PROPYL ACRYLATE	90.971	1.0073E+00	-2.9850E-03	3.7890E-06	201	512	226.37
541	C6H10O3	ETHYLACETOACETATE	88.660	1.1455E+00	-2.9953E-03	3.2993E-06	235	579	251.36
542	C6H10O3	PROPIONIC ANHYDRIDE	83.119	1.0817E+00	-2.9281E-03	3.3352E-06	229	556	233.72
543	C6H10O4	ADIPIC ACID	-190.095	2.9078E+00	-5.4871E-03	3.8953E-06	427	728	--------
544	C6H10O4	DIETHYL OXALATE	85.853	1.2371E+00	-3.1977E-03	3.5605E-06	234	581	264.80
545	C6H10O4	ETHYLENE GLYCOL DIACETATE	85.070	1.2494E+00	-3.2095E-03	3.4466E-06	243	588	263.61
546	C6H10O4	ETHYLIDENE DIACETATE	-4.293	1.7964E+00	-4.4905E-03	4.4158E-06	293	572	249.15

Table 3-1 HEAT CAPACITY OF LIQUID - ORGANIC COMPOUNDS (continued)

$$C_P = A + B T + C T^2 + D T^3 \qquad (C_P - joule/(mol\ K),\ T - K)$$

NO	FORMULA	NAME	A	B	C	D	TMIN	TMAX	C_P @ 25 C
547	C6H11N	HEXANENITRILE	110.589	7.9476E-01	-2.1997E-03	2.7153E-06	194	560	223.97
548	C6H11NO	epsilon-CAPROLACTAM	31.964	1.2526E+00	-2.5271E-03	2.0436E-06	343	725	-------
549	C6H11NO	CYCLOHEXANONE OXIME	-54.389	1.8697E+00	-4.0224E-03	3.3103E-06	364	644	-------
550	C6H12	CYCLOHEXANE	-44.417	1.6016E+00	-4.4676E-03	4.7582E-06	281	498	162.07
551	C6H12	2,3-DIMETHYL-1-BUTENE	106.873	4.5993E-01	-1.6023E-03	2.8074E-06	117	450	175.97
552	C6H12	2,3-DIMETHYL-2-BUTENE	51.122	9.7008E-01	-3.1012E-03	4.0855E-06	200	472	172.96
553	C6H12	3,3-DIMETHYL-1-BUTENE	82.629	7.3199E-01	-2.5677E-03	4.0169E-06	159	432	179.09
554	C6H12	2-ETHYL-1-BUTENE	91.019	5.6181E-01	-1.9133E-03	3.0203E-06	143	461	168.49
555	C6H12	trans-3-METHYL-2-PENTENE	151.557	-1.0557E-01	6.1819E-04	--------	136	374	175.03
556	C6H12	1-HEXENE	100.785	5.4444E-01	-1.8726E-03	3.0757E-06	134	454	178.16
557	C6H12	cis-2-HEXENE	96.557	5.0424E-01	-1.7178E-03	2.7979E-06	133	462	168.34
558	C6H12	trans-2-HEXENE	97.745	5.6490E-01	-1.9080E-03	3.0339E-06	141	462	176.97
559	C6H12	cis-3-HEXENE	99.597	5.4675E-01	-1.8640E-03	3.0242E-06	136	458	177.06
560	C6H12	trans-3-HEXENE	86.524	7.3371E-01	-2.4616E-03	3.7047E-06	161	458	184.65
561	C6H12	METHYLCYCLOPENTANE	92.280	3.9756E-01	-1.2966E-03	2.0816E-06	132	480	150.72
562	C6H12	2-METHYL-1-PENTENE	97.368	5.5769E-01	-1.9065E-03	3.0787E-06	138	456	175.77
563	C6H12	2-METHYL-2-PENTENE	91.931	5.1877E-01	-1.7524E-03	2.8016E-06	139	463	165.07
564	C6H12	3-METHYL-1-PENTENE	116.510	5.2150E-01	-1.8268E-03	3.1737E-06	121	446	193.72
565	C6H12	3-METHYL-cis-2-PENTENE	101.713	5.4538E-01	-1.8339E-03	2.9283E-06	139	464	178.90
566	C6H12	4-METHYL-1-PENTENE	97.695	4.3999E-01	-1.5403E-03	2.6788E-06	121	446	162.96
567	C6H12	4-METHYL-cis-2-PENTENE	95.360	5.5699E-01	-1.9260E-03	3.1315E-06	139	449	173.22
568	C6H12	4-METHYL-trans-2-PENTENE	103.386	5.4752E-01	-1.8917E-03	3.1298E-06	133	451	181.42
569	C6H12N2	TRIETHYLENEDIAMINE	-608.040	5.1246E+00	-1.0863E-02	8.1963E-06	435	590	--------
570	C6H12O	BUTYL VINYL ETHER	92.734	9.1734E-01	-2.9004E-03	3.9872E-06	182	482	214.09
571	C6H12O	CYCLOHEXANOL	-47.321	1.9131E+00	-4.8388E-03	4.7281E-06	298	563	218.25
572	C6H12O	1-HEXANAL	88.621	9.1837E-01	-2.6499E-03	3.1797E-06	218	521	211.15
573	C6H12O	ETHYL ISOPROPYL KETONE	62.886	1.0637E+00	-3.1466E-03	4.0376E-06	201	510	207.32
574	C6H12O	2-HEXANONE	72.825	1.0372E+00	-2.9477E-03	3.5365E-06	218	528	213.77
575	C6H12O	3-HEXANONE	76.972	1.0374E+00	-2.9656E-03	3.5648E-06	219	525	217.14
576	C6H12O	METHYL ISOBUTYL KETONE	96.284	8.5227E-01	-2.5379E-03	3.3066E-06	190	514	212.42
577	C6H12O2	n-PENTYL FORMATE	93.856	1.0880E+00	-3.1818E-03	4.0343E-06	201	518	242.31
578	C6H12O2	n-BUTYL ACETATE	91.175	9.9902E-01	-2.9032E-03	3.6712E-06	201	522	228.25
579	C6H12O2	sec-BUTYL ACETATE	111.023	8.6334E-01	-2.6280E-03	3.6131E-06	175	505	230.57
580	C6H12O2	tert-BUTYL ACETATE	93.486	1.0976E+00	-3.3694E-03	4.3600E-06	201	491	236.76
581	C6H12O2	ETHYL n-BUTYRATE	117.263	8.2767E-01	-2.5104E-03	3.3783E-06	176	514	230.41
582	C6H12O2	ETHYL ISOBUTYRATE	93.347	8.6764E-01	-2.6632E-03	3.5669E-06	186	498	209.83
583	C6H12O2	ISOBUTYL ACETATE	118.616	8.2730E-01	-2.5236E-03	3.4564E-06	175	505	232.55
584	C6H12O2	n-PROPYL PROPIONATE	94.198	9.6107E-01	-2.8068E-03	3.5722E-06	198	520	225.91
585	C6H12O2	CYCLOHEXYL PEROXIDE	86.532	9.6569E-01	-2.3547E-03	2.4198E-06	254	617	229.27
586	C6H12O2	DIACETONE ALCOHOL	88.803	1.0612E+00	-2.9263E-03	3.3398E-06	230	545	233.58
587	C6H12O2	2-ETHYL BUTYRIC ACID	60.533	1.3454E+00	-3.3834E-03	3.5251E-06	259	590	254.32
588	C6H12O2	n-HEXANOIC ACID	72.281	1.3529E+00	-3.3288E-03	3.3399E-06	271	600	268.25
589	C6H12O3	2-ETHOXYETHYL ACETATE	78.294	1.2284E+00	-3.4999E-03	4.2007E-06	212	537	244.75
590	C6H12O3	HYDROXYCAPROIC ACID	-15.705	2.0081E+00	-4.2865E-03	3.6022E-06	335	682	--------
591	C6H12O3	PARALDEHYDE	-47.948	2.2916E+00	-6.1330E-03	6.3645E-06	287	521	258.78
592	C6H12O3	sec-BUTYL GLYCOLATE	--------	--------	--------	--------	--------	--------	--------
593	C6H12S	THIACYCLOHEPTANE	202.732	-2.1939E-01	5.5270E-04	--------	293	445	186.45
594	C6H13N	CYCLOHEXYLAMINE	42.828	1.2696E+00	-3.3783E-03	3.6184E-06	256	554	216.95
595	C6H13N	HEXAMETHYLENEIMINE	76.226	1.0918E+00	-2.9610E-03	3.3029E-06	237	554	226.08
596	C6H14	2,2-DIMETHYLBUTANE	74.689	8.4782E-01	-2.9059E-03	4.2664E-06	175	440	182.23
597	C6H14	2,3-DIMETHYLBUTANE	96.237	6.1920E-01	-2.1286E-03	3.3811E-06	146	450	181.25
598	C6H14	n-HEXANE	78.848	8.8729E-01	-2.9482E-03	4.1999E-06	179	457	192.63
599	C6H14	2-METHYLPENTANE	110.129	5.0521E-01	-1.7675E-03	3.0660E-06	121	448	184.90
600	C6H14	3-METHYLPENTANE	114.180	4.4292E-01	-1.5358E-03	2.7258E-06	111	454	181.96
601	C6H14N2O2	LYSINE	-483.631	4.3645E+00	-7.7810E-03	5.0707E-06	484	739	--------
602	C6H14O	2-ETHYL-1-BUTANOL	158.546	7.1200E-01	-2.1451E-03	2.9921E-06	160	522	259.44
603	C6H14O	1-HEXANOL	89.086	9.4770E-01	-2.5912E-03	2.9524E-06	230	550	219.55
604	C6H14O	2-HEXANOL	85.019	1.1847E+00	-3.3647E-03	3.9685E-06	224	528	244.31
605	C6H14O	2-METHYL-1-PENTANOL	110.923	1.0917E+00	-3.1733E-03	3.9840E-06	201	524	259.91
606	C6H14O	4-METHYL-2-PENTANOL	107.198	1.0434E+00	-3.0591E-03	3.8693E-06	201	517	248.90
607	C6H14O	n-BUTYL ETHYL ETHER	107.841	8.7979E-01	-2.8249E-03	4.0279E-06	171	478	225.79
608	C6H14O	DIISOPROPYL ETHER	73.085	1.1235E+00	-3.7356E-03	5.2200E-06	189	450	214.32
609	C6H14O	DI-n-PROPYL ETHER	121.584	7.0573E-01	-2.2937E-03	3.4752E-06	151	478	220.21
610	C6H14O	METHYL tert-PENTYL ETHER	72.866	1.1193E+00	-3.4929E-03	4.5718E-06	201	481	217.28
611	C6H14O2	ACETAL	116.186	8.9191E-01	-2.8062E-03	3.9321E-06	174	487	236.87
612	C6H14O2	2-BUTOXYETHANOL	141.017	1.0449E+00	-2.9601E-03	3.6163E-06	204	540	285.26
613	C6H14O2	1,6-HEXANEDIOL	-49.791	2.5975E+00	-6.1380E-03	5.6286E-06	316	603	--------
614	C6H14O2	HEXYLENE GLYCOL	148.666	1.3168E+00	-3.5373E-03	4.0785E-06	224	559	334.92
615	C6H14O2S	DI-n-PROPYL SULFONE	69.312	1.3810E+00	-2.9794E-03	2.6512E-06	304	687	--------
616	C6H14O3	DIETHYLENE GLYCOL DIMETHYL ETHER	118.254	1.1127E+00	-3.1346E-03	3.8297E-06	204	544	272.85
617	C6H14O3	DIPROPYLENE GLYCOL	144.536	1.3978E+00	-3.5903E-03	3.9399E-06	234	589	346.58
618	C6H14O3	2-(2-ETHOXYETHOXY)ETHANOL	156.817	1.0470E+00	-2.8479E-03	3.4688E-06	196	569	307.75
619	C6H14O3	TRIMETHYLOLPROPANE	111.221	1.8370E+00	-4.1267E-03	3.5496E-06	332	638	--------
620	C6H14O4	TRIETHYLENE GLYCOL	162.233	1.2720E+00	-3.0443E-03	2.9863E-06	267	630	350.02
621	C6H14O6	SORBITOL	106.032	9.4243E-01	-1.6484E-03	1.1669E-06	372	863	--------
622	C6H14S	n-HEXYLMERCAPTAN	100.678	8.7864E-01	-2.3740E-03	2.9901E-06	194	561	230.85
623	C6H14S	BUTYL-ETHYL-SULFIDE	219.696	-1.6716E-01	6.9631E-04	--------	179	447	231.75
624	C6H14S	ISOPROPYL-SULFIDE	188.855	-1.3921E-01	6.2212E-04	--------	171	423	202.65

67

Table 3-1 HEAT CAPACITY OF LIQUID - ORGANIC COMPOUNDS (continued)

			$C_P = A + B T + C T^2 + D T^3$					(C_P - joule/(mol K), T - K)	
NO	FORMULA	NAME	A	B	C	D	TMIN	TMAX	C_P @ 25 C
625	C6H14S	METHYL-PENTYL-SULFIDE	200.445	-1.5811E-01	6.6107E-04	--------	180	431	212.07
626	C6H14S	PROPYL-SULFIDE	209.800	-1.4776E-01	6.5195E-04	--------	171	446	223.70
627	C6H14S2	PROPYL-DISULFIDE	240.406	-1.4485E-01	5.8815E-04	--------	189	495	249.50
628	C6H15Al	TRIETHYL ALUMINUM	181.059	3.0401E-01	-4.3637E-04	5.8165E-07	222	648	248.32
629	C6H15Al2Cl3	ETHYL ALUMINUM SESQUICHLORIDE	--------	--------	--------	--------	--------	--------	--------
630	C6H15N	DIISOPROPYLAMINE	100.043	9.8366E-01	-3.1982E-03	4.4920E-06	178	471	228.07
631	C6H15N	DI-n-PROPYLAMINE	73.211	1.1938E+00	-3.5867E-03	4.4751E-06	211	500	228.92
632	C6H15N	n-HEXYLAMINE	40.666	1.7414E+00	-4.8489E-03	5.3464E-06	253	525	270.51
633	C6H15N	TRIETHYLAMINE	114.243	7.0577E-01	-2.2590E-03	3.3247E-06	159	482	211.97
634	C6H15NO	6-AMINOHEXANOL	-49.170	2.0330E+00	-4.6707E-03	4.1372E-06	332	613	--------
635	C6H15NO2	DIISOPROPANOLAMINE	-9.455	2.3765E+00	-5.6088E-03	5.0712E-06	319	605	--------
636	C6H15NO3	TRIETHANOLAMINE	179.047	1.2436E+00	-2.6631E-03	2.3379E-06	295	708	375.06
637	C6H15N3	N-AMINOETHYL PIPERAZINE	120.025	1.0923E+00	-2.5872E-03	2.6128E-06	255	637	284.95
638	C6H15O4P	TRIETHYL PHOSPHATE	--------	--------	--------	--------	--------	--------	--------
639	C6H16N2	HEXAMETHYLENEDIAMINE	-6.944	2.0495E+00	-4.8605E-03	4.4981E-06	315	597	--------
640	C6H18N3OP	HEXAMETHYL PHOSPHORAMIDE	--------	--------	--------	--------	--------	--------	--------
641	C6H18N4	TRIETHYLENE TETRAMINE	91.558	1.7020E+00	-3.9120E-03	3.6870E-06	286	646	348.98
642	C6H18OSi2	HEXAMETHYLDISILOXANE	85.588	1.7918E+00	-5.6915E-03	7.4549E-06	206	467	311.44
643	C6H18O3Si3	HEXAMETHYLCYCLOTRISILOXANE	-537.133	6.8511E+00	-1.7801E-02	1.6878E-05	338	499	--------
644	C6H19NSi2	HEXAMETHYLDISILAZANE	97.027	1.3771E+00	-4.2362E-03	5.4945E-06	201	490	276.68
645	C7H3ClF3NO2	4-CHLORO-3-NITROBENZOTRIFLUORIDE	--------	--------	--------	--------	--------	--------	--------
646	C7H3Cl2F3	2,4-DICHLOROBENZOTRIFLUORIDE	67.398	1.2515E+00	-3.2297E-03	3.4450E-06	249	581	244.74
647	C7H3Cl2NO	3,4-DICHLOROPHENYL ISOCYANATE	13.155	1.3501E+00	-2.9951E-03	2.6326E-06	317	660	--------
648	C7H4ClF3	p-CHLOROBENZOTRIFLUORIDE	62.232	1.1478E+00	-3.1490E-03	3.5615E-06	238	541	218.90
649	C7H4Cl2O	m-CHLOROBENZOYL CHLORIDE	57.810	9.7855E-01	-2.2472E-03	2.1291E-06	281	652	206.23
650	C7H4F3NO2	3-NITROBENZOTRIFLUORIDE	--------	--------	--------	--------	--------	--------	--------
651	C7H5ClO	BENZOYL CHLORIDE	52.817	8.6263E-01	-2.0473E-03	2.0049E-06	274	627	181.15
652	C7H5ClO2	o-CHLOROBENZOIC ACID	-195.279	2.4870E+00	-4.8165E-03	3.4930E-06	416	713	--------
653	C7H5Cl3	BENZOTRICHLORIDE	80.853	8.6290E-01	-1.9659E-03	1.8902E-06	269	663	213.47
654	C7H5F3	BENZOTRIFLUORIDE	22.859	1.2863E+00	-3.6796E-03	4.2076E-06	245	509	190.80
655	C7H5N	BENZONITRILE	60.009	8.1051E-01	-1.9418E-03	1.9429E-06	261	629	180.55
656	C7H5NO	PHENYL ISOCYANATE	57.143	7.8666E-01	-2.0256E-03	2.1812E-06	244	583	169.43
657	C7H5N3O6	2,4,6-TRINITROTOLUENE	21.181	1.7159E+00	-3.4629E-03	2.7647E-06	355	716	--------
658	C7H6Cl2	BENZYL DICHLORIDE	74.474	7.3936E-01	-1.6958E-03	1.6858E-06	258	658	188.85
659	C7H6Cl2	2,4-DICHLOROTOLUENE	46.183	5.4112E-01	-1.2859E-03	1.2850E-06	261	635	127.27
660	C7H6N2O4	2,4-DINITROTOLUENE	6.605	1.4730E+00	-2.9647E-03	2.3797E-06	344	733	--------
661	C7H6N2O4	2,5-DINITROTOLUENE	--------	--------	--------	--------	--------	--------	--------
662	C7H6N2O4	2,6-DINITROTOLUENE	--------	--------	--------	--------	--------	--------	--------
663	C7H6N2O4	3,4-DINITROTOLUENE	--------	--------	--------	--------	--------	--------	--------
664	C7H6N2O4	3,5-DINITROTOLUENE	--------	--------	--------	--------	--------	--------	--------
665	C7H6O	BENZALDEHYDE	72.865	7.0427E-01	-1.7065E-03	1.7622E-06	248	626	177.85
666	C7H6O2	BENZOIC ACID	-158.917	2.3735E+00	-4.8280E-03	3.6876E-06	397	676	--------
667	C7H6O2	p-HYDROXYBENZALDEHYDE	-6.120	1.2887E+00	-2.4382E-03	1.7829E-06	391	760	--------
668	C7H6O2	SALICYLALDEHYDE	72.299	1.1065E+00	-2.7015E-03	2.6999E-06	267	612	233.61
669	C7H6O3	SALICYLIC ACID	--------	--------	--------	--------	--------	--------	--------
670	C7H7Br	p-BROMOTOLUENE	22.692	1.1392E+00	-2.6347E-03	2.4503E-06	301	629	--------
671	C7H7Cl	BENZYL CHLORIDE	82.217	7.0948E-01	-1.7551E-03	1.8744E-06	235	617	187.41
672	C7H7Cl	o-CHLOROTOLUENE	82.902	7.9838E-01	-2.0378E-03	2.2156E-06	238	590	198.51
673	C7H7Cl	p-CHLOROTOLUENE	34.924	1.1405E+00	-2.8001E-03	2.7683E-06	282	594	199.43
674	C7H7F	p-FLUOROTOLUENE	145.658	-1.5813E-01	5.2841E-04	--------	217	420	145.48
675	C7H7NO	FORMANILIDE	31.885	1.1692E+00	-2.4418E-03	2.0593E-06	324	708	--------
676	C7H7NO2	m-NITROTOLUENE	33.670	9.9886E-01	-2.2718E-03	2.0913E-06	290	661	184.96
677	C7H7NO2	o-NITROTOLUENE	45.732	8.8396E-01	-2.0735E-03	1.9967E-06	271	648	177.89
678	C7H7NO2	p-NITROTOLUENE	-3.361	1.0676E+00	-2.3531E-03	2.0262E-06	326	662	--------
679	C7H7NO3	o-NITROANISOLE	92.024	1.0466E+00	-2.2554E-03	2.0464E-06	285	704	257.81
680	C7H8	TOLUENE	83.703	5.1666E-01	-1.4910E-03	1.9725E-06	179	533	157.49
681	C7H8	1,3,5-CYCLOHEPTATRIENE	183.184	-1.6155E-01	6.2267E-04	--------	195	419	190.37
682	C7H8O	ANISOLE	77.223	8.8434E-01	-2.3076E-03	2.5370E-06	237	577	203.00
683	C7H8O	BENZYL ALCOHOL	97.570	8.6633E-01	-2.1388E-03	2.1700E-06	259	609	223.25
684	C7H8O	m-CRESOL	78.436	1.0457E+00	-2.4251E-03	2.3031E-06	286	635	235.69
685	C7H8O	o-CRESOL	51.055	1.1941E+00	-2.7297E-03	2.5339E-06	305	628	--------
686	C7H8O	p-CRESOL	40.620	1.2687E+00	-2.8630E-03	2.6374E-06	309	634	--------
687	C7H8O2	GUAIACOL	22.222	1.2061E+00	-2.7948E-03	2.5675E-06	306	627	--------
688	C7H8O2	p-METHOXYPHENOL	30.613	1.2814E+00	-2.7284E-03	2.3197E-06	330	682	--------
689	C7H9N	BENZYLAMINE	157.089	4.3487E-01	-1.0951E-03	1.1698E-06	228	615	220.40
690	C7H9N	2,6-DIMETHYLPYRIDINE	58.073	1.0059E+00	-2.5953E-03	2.7069E-06	268	561	199.01
691	C7H9N	N-METHYLANILINE	108.285	6.1464E-01	-1.5202E-03	1.6582E-06	217	631	200.35
692	C7H9N	m-TOLUIDINE	91.578	8.1400E-01	-1.9600E-03	2.0129E-06	244	638	213.39
693	C7H9N	o-TOLUIDINE	84.427	9.0250E-01	-2.2050E-03	2.2526E-06	250	625	217.20
694	C7H9N	p-TOLUIDINE	-12.448	1.4992E+00	-3.4632E-03	3.1091E-06	318	624	--------
695	C7H10	2-NORBORNENE	-111.267	2.0246E+00	-5.2173E-03	5.0164E-06	320	525	--------
696	C7H10N2	TOLUENEDIAMINE	-32.164	1.7865E+00	-3.5394E-03	2.7332E-06	372	724	--------
697	C7H11NO	CYCLOHEXYL ISOCYANATE	105.949	8.0902E-01	-2.1979E-03	2.6405E-06	201	570	221.76
698	C7H12	1-HEPTYNE	210.215	-2.0548E-01	7.9111E-04	--------	193	403	219.27
699	C7H12O2	n-BUTYL ACRYLATE	101.239	1.1519E+00	-3.2414E-03	3.9463E-06	210	538	261.14
700	C7H12O2	ISOBUTYL ACRYLATE	95.067	1.1903E+00	-3.4348E-03	4.1928E-06	213	522	255.76
701	C7H12O2	n-PROPYL METHACRYLATE	88.969	7.9514E-01	-2.2583E-03	2.7936E-06	201	539	199.33
702	C7H12O4	DIETHYL MALONATE	116.141	1.1953E+00	-3.0852E-03	3.4713E-06	225	588	290.26

68

Table 3-1 HEAT CAPACITY OF LIQUID - ORGANIC COMPOUNDS (continued)

			$C_P = A + B\,T + C\,T^2 + D\,T^3$			(C_P - joule/(mol K), T - K)		
NO	FORMULA	NAME	A	B	C	D	TMIN	TMAX	C_P @ 25 C
703	C7H14	CYCLOHEPTANE	22.809	1.2050E+00	-3.2051E-03	3.3996E-06	266	544	187.27
704	C7H14	1,1-DIMETYLCYCLOPENTANE	67.646	9.0042E-01	-2.7291E-03	3.5161E-06	204	492	186.70
705	C7H14	cis-1,2-DIMETHYLCYCLOPENTANE	59.555	9.7636E-01	-2.8547E-03	3.4670E-06	220	509	188.78
706	C7H14	trans-1,2-DIMETHYLCYCLOPENTANE	97.923	5.4369E-01	-1.6777E-03	2.4600E-06	157	498	176.09
707	C7H14	cis-1,3-DIMETHYLCYCLOPENTANE	111.592	5.0848E-01	-1.5795E-03	2.4410E-06	140	496	187.48
708	C7H14	trans-1,3-DIMETHYLCYCLOPENTANE	108.553	4.7295E-01	-1.4677E-03	2.2606E-06	140	498	179.01
709	C7H14	ETHYLCYCLOPENTANE	109.808	4.1311E-01	-1.2645E-03	1.9260E-06	136	513	171.61
710	C7H14	2-ETHYL-1-PENTENE	117.859	7.4247E-01	-2.3420E-03	3.3135E-06	169	489	218.86
711	C7H14	3-ETHYL-1-PENTENE	135.633	5.7239E-01	-1.8716E-03	2.8576E-06	147	477	215.65
712	C7H14	1-HEPTENE	114.317	6.7244E-01	-2.1649E-03	3.2064E-06	155	484	207.34
713	C7H14	cis-2-HEPTENE	99.819	7.4221E-01	-2.3369E-03	3.3278E-06	165	494	201.57
714	C7H14	trans-2-HEPTENE	108.115	7.3356E-01	-2.3190E-03	3.3308E-06	165	489	208.96
715	C7H14	cis-3-HEPTENE	121.809	5.7886E-01	-1.8604E-03	2.8850E-06	138·	491	205.49
716	C7H14	trans-3-HEPTENE	122.104	5.6756E-01	-1.8271E-03	2.8571E-06	138	486	204.63
717	C7H14	METHYLCYCLOHEXANE	103.668	4.6217E-01	-1.3973E-03	2.0550E-06	148	515	171.72
718	C7H14	2-METHYL-1-HEXENE	106.364	8.2847E-01	-2.6235E-03	3.7203E-06	171	484	218.76
719	C7H14	3-METHYL-1-HEXENE	123.572	6.4499E-01	-2.1072E-03	3.2509E-06	146	475	214.72
720	C7H14	4-METHYL-1-HEXENE	123.548	5.3447E-01	-1.7377E-03	2.7783E-06	133	481	202.07
721	C7H14	2,3,3-TRIMETHYL-1-BUTENE	110.303	7.4985E-01	-2.4008E-03	3.5056E-06	164	478	213.37
722	C7H14O	DIISOPROPYL KETONE	85.991	1.1369E+00	-3.3568E-03	4.1558E-06	206	518	236.71
723	C7H14O	2-HEPTANONE	65.391	1.2723E+00	-3.4343E-03	3.8550E-06	239	550	241.60
724	C7H14O	1-HEPTANAL	79.243	1.2336E+00	-3.3958E-03	3.8988E-06	231	543	248.52
725	C7H14O	1-METHYLCYCLOHEXANOL	-150.105	3.0109E+00	-7.7992E-03	7.7061E-06	300	543	--------
726	C7H14O	cis-2-METHYLCYCLOHEXANOL	2.842	1.9243E+00	-5.0271E-03	5.0869E-06	281	553	264.51
727	C7H14O	trans-2-METHYLCYCLOHEXANOL	43.292	1.6245E+00	-4.2799E-03	4.4220E-06	270	554	264.39
728	C7H14O	cis-3-METHYLCYCLOHEXANOL	6.519	1.8769E+00	-4.9328E-03	5.1346E-06	269	556	263.71
729	C7H14O	trans-3-METHYLCYCLOHEXANOL	44.409	1.6304E+00	-4.2749E-03	4.3796E-06	274	555	266.59
730	C7H14O	cis-4-METHYLCYCLOHEXANOL	135.271	9.0905E-01	-2.4855E-03	3.0226E-06	201	560	265.47
731	C7H14O	trans-4-METHYLCYCLOHEXANOL	134.634	9.4209E-01	-2.6118E-03	3.1477E-06	201	560	266.77
732	C7H14O	5-METHYL-2-HEXANONE	106.371	9.1613E-01	-2.5950E-03	3.2112E-06	200	541	233.94
733	C7H14O2	n-BUTYL PROPIONATE	128.652	8.9746E-01	-2.5942E-03	3.3614E-06	185	535	254.71
734	C7H14O2	ETHYL ISOVALERATE	137.705	8.3648E-01	-2.4502E-03	3.2777E-06	175	529	256.17
735	C7H14O2	ISOPENTYL ACETATE	119.930	9.5880E-01	-2.7166E-03	3.4200E-06	196	539	254.95
736	C7H14O2	n-PENTYL ACETATE	117.364	1.0496E+00	-2.9677E-03	3.6617E-06	203	538	263.54
737	C7H14O2	n-PROPYL n-BUTYRATE	135.799	8.6257E-01	-2.4884E-03	3.2841E-06	179	535	258.81
738	C7H14O2	n-HEPTANOIC ACID	72.739	1.3618E+00	-3.3044E-03	3.3289E-06	267	612	273.24
739	C7H14O2	ETHYL-3-ETHOXYPROPIONATE	123.353	1.1571E+00	-3.2169E-03	3.9748E-06	201	548	287.73
740	C7H15Br	1-BROMOHEPTANE	118.794	8.6306E-01	-2.2453E-03	2.5620E-06	218	586	244.42
741	C7H15N	N-METHYLCYCLOHEXYLAMINE	39.040	1.4458E+00	-3.7690E-03	3.9502E-06	266	560	239.76
742	C7H16	2,2-DIMETHYLPENTANE	115.052	6.5774E-01	-2.1698E-03	3.3303E-06	150	468	206.54
743	C7H16	2,3-DIMETHYLPENTANE	85.488	9.3231E-01	-2.8826E-03	3.7624E-06	201	484	206.92
744	C7H16	2,4-DIMETHYLPENTANE	118.526	7.0061E-01	-2.3161E-03	3.4981E-06	155	468	214.24
745	C7H16	3,3-DIMETHYLPENTANE	118.692	6.0252E-01	-1.9340E-03	3.0327E-06	140	483	206.79
746	C7H16	3-ETHYLPENTANE	116.451	7.0991E-01	-2.2627E-03	3.3492E-06	156	487	215.74
747	C7H16	n-HEPTANE	101.121	9.7739E-01	-3.0712E-03	4.1844E-06	184	486	230.42
748	C7H16	2-METHYLHEXANE	118.184	7.1284E-01	-2.3129E-03	3.4493E-06	156	477	216.53
749	C7H16	3-METHYLHEXANE	126.861	6.4079E-01	-2.0615E-03	3.0647E-06	155	482	215.88
750	C7H16	2,2,3-TRIMETHYLBUTANE	0.826	1.7037E+00	-5.0650E-03	5.8856E-06	250	478	214.52
751	C7H16O	1-HEPTANOL	93.968	1.1999E+00	-3.1629E-03	3.4722E-06	240	569	262.58
752	C7H16O	2-HEPTANOL	77.603	1.6101E+00	-4.4836E-03	5.0303E-06	244	529	292.41
753	C7H16O	5-METHYL-1-HEXANOL	110.967	1.1814E+00	-3.3055E-03	4.1023E-06	201	545	278.10
754	C7H16O	ISOPROPYL-TERT-BUTYL-ETHER	218.409	-2.0906E-01	8.5669E-04	--------	179	409	232.23
755	C7H16S	n-HEPTYL MERCAPTAN	108.046	1.0802E+00	-2.7997E-03	3.1263E-06	231	581	264.09
756	C7H16S	BUTYL-PROPYL-SULFIDE	263.202	-2.1538E-01	7.5760E-04	--------	208	474	266.33
757	C7H16S	ETHYL-PENTYL-SULFIDE	264.004	-2.3234E-01	8.0486E-04	--------	208	474	266.28
758	C7H16S	HEXYL-METHYL-SULFIDE	264.004	-2.3234E-01	8.0486E-04	--------	208	474	266.28
759	C7H17N	1-AMINOHEPTANE	48.454	1.5927E+00	-4.2895E-03	4.6302E-06	255	546	264.72
760	C8H4Cl2O2	ISOPHTHALOYL CHLORIDE	44.675	1.2944E+00	-2.7615E-03	2.3807E-06	318	691	--------
761	C8H4O3	PHTHALIC ANHYDRIDE	-105.627	1.9840E+00	-3.8847E-03	2.8513E-06	405	712	--------
762	C8H6	ETHYNYLBENZENE	183.372	-1.6806E-01	5.1645E-04	--------	244	448	179.17
763	C8H6O4	ISOPHTHALIC ACID	--------	--------	--------	--------	--------	--------	--------
764	C8H6O4	PHTHALIC ACID	--------	--------	--------	--------	--------	--------	--------
765	C8H6O4	TEREPHTHALIC ACID	--------	--------	--------	--------	--------	--------	--------
766	C8H6S	BENZOTHIOPHENE	36.583	1.0240E+00	-2.2186E-03	1.9892E-06	306	679	--------
767	C8H7N	INDOLE	87.722	7.6976E-01	-1.6788E-03	1.5257E-06	275	711	208.42
768	C8H8	STYRENE	66.737	8.4051E-01	-2.1615E-03	2.3324E-06	244	583	187.00
769	C8H8	1,3,5,7-CYCLOOCTATETRAENE	236.513	-2.7092E-01	7.3503E-04	--------	267	443	221.08
770	C8H8O	ACETOPHENONE	30.911	1.1353E+00	-2.6405E-03	2.4764E-06	295	631	200.32
771	C8H8O	p-TOLUALDEHYDE	119.664	6.2791E-01	-1.5631E-03	1.7837E-06	201	628	215.20
772	C8H8O2	METHYL BENZOATE	65.554	1.0124E+00	-2.4553E-03	2.4559E-06	262	624	214.24
773	C8H8O2	o-TOLUIC ACID	-106.856	2.1576E+00	-4.4616E-03	3.5101E-06	378	676	--------
774	C8H8O2	p-TOLUIC ACID	-409.886	3.7262E+00	-7.0801E-03	4.9079E-06	454	696	--------
775	C8H8O3	METHYL SALICYLATE	97.902	1.0367E+00	-2.4663E-03	2.4373E-06	266	631	252.34
776	C8H8O3	VANILLIN	-8.950	1.7876E+00	-3.6684E-03	2.9525E-06	356	699	--------
777	C8H9NO	ACETANILIDE	-54.663	1.7295E+00	-3.3303E-03	2.4823E-06	388	743	--------
778	C8H10	ETHYLBENZENE	102.111	5.5959E-01	-1.5609E-03	2.0149E-06	179	555	183.60
779	C8H10	m-XYLENE	70.916	8.0450E-01	-2.1885E-03	2.5061E-06	226	555	182.66
780	C8H10	o-XYLENE	56.460	9.4926E-01	-2.4902E-03	2.6838E-06	249	567	189.25

Table 3-1 HEAT CAPACITY OF LIQUID - ORGANIC COMPOUNDS (continued)

			$C_P = A + B T + C T^2 + D T^3$				(C_P - joule/(mol K), T - K)		
NO	FORMULA	NAME	A	B	C	D	TMIN	TMAX	C_P @ 25 C
781	C8H10	p-XYLENE	-11.035	1.5158E+00	-3.9039E-03	3.9193E-06	287	555	197.75
782	C8H10O	m-ETHYLPHENOL	--------	--------	--------	--------	--------	--------	--------
783	C8H10O	p-ETHYLPHENOL	22.135	1.4685E+00	-3.2860E-03	2.9185E-06	319	645	--------
784	C8H10O	PHENETOLE	72.026	1.0392E+00	-2.6794E-03	2.8834E-06	245	582	220.11
785	C8H10O	2-PHENYLETHANOL	80.595	1.0256E+00	-2.5190E-03	2.6273E-06	248	616	232.08
786	C8H10O	2,3-XYLENOL	-2.686	1.6455E+00	-3.5635E-03	3.0090E-06	347	651	--------
787	C8H10O	2,4-XYLENOL	60.600	1.2899E+00	-2.9766E-03	2.7446E-06	299	637	253.34
788	C8H10O	2,5-XYLENOL	-38.995	1.8247E+00	-4.0078E-03	3.4037E-06	349	636	--------
789	C8H10O	2,6-XYLENOL	25.237	1.4346E+00	-3.2504E-03	2.9149E-06	320	631	--------
790	C8H10O	3,4-XYLENOL	15.704	1.5867E+00	-3.4378E-03	2.9258E-06	339	657	--------
791	C8H10O	3,5-XYLENOL	-53.263	2.0930E+00	-4.6697E-03	3.9694E-06	338	644	--------
792	C8H11N	N,N-DIMETHYLANILINE	60.597	1.2244E+00	-2.9639E-03	2.8755E-06	277	618	238.40
793	C8H11N	o-ETHYLANILINE	124.722	8.4471E-01	-2.0522E-03	2.1995E-06	228	634	252.44
794	C8H11N	2,4,6-TRIMETHYLPYRIDINE	96.356	9.5429E-01	-2.4678E-03	2.7316E-06	230	588	233.90
795	C8H11NO	p-PHENETIDINE	94.612	1.1123E+00	-2.4776E-03	2.3162E-06	278	679	267.40
796	C8H12	1,5-CYCLOOCTADIENE	95.884	7.9280E-01	-2.0754E-03	2.4825E-06	205	581	213.56
797	C8H12	VINYLCYCLOHEXENE	120.647	5.8357E-01	-1.6847E-03	2.2935E-06	165	539	205.67
798	C8H12O4	1,4-CYCLOHEXANEDICARBOXYLIC ACID	-1442.953	8.3711E+00	-1.3146E-02	7.3187E-06	587	800	--------
799	C8H12O4	DIETHYL MALEATE	93.009	1.3593E+00	-3.2653E-03	3.3253E-06	265	612	296.15
800	C8H14O2	n-BUTYL METHACRYLATE	135.373	1.0816E+00	-2.9999E-03	3.6584E-06	201	554	288.13
801	C8H14O3	BUTYRIC ANHYDRIDE	152.823	1.0158E+00	-2.7247E-03	3.2640E-06	201	575	299.98
802	C8H14O4	DIETHYL SUCCINATE	102.038	1.5936E+00	-4.0077E-03	4.2022E-06	253	594	332.29
803	C8H16	CYCLOOCTANE	277.940	-3.5205E-01	8.6774E-04	--------	289	454	250.11
804	C8H16	1,1-DIMETHYLCYCLOHEXANE	45.883	1.2604E+00	-3.5092E-03	3.9654E-06	241	532	214.81
805	C8H16	cis-1,2-DIMETHYLCYCLOHEXANE	73.911	1.0319E+00	-2.8669E-03	3.3195E-06	224	546	214.71
806	C8H16	trans-1,2-DIMETHYLCYCLOHEXANE	106.382	7.2688E-01	-2.1051E-03	2.7007E-06	186	537	207.55
807	C8H16	cis-1,3-DIMETHYLCYCLOHEXANE	93.200	8.4033E-01	-2.4187E-03	3.0292E-06	199	532	209.02
808	C8H16	trans-1,3-DIMETHYLCYCLOHEXANE	104.906	7.5947E-01	-2.1970E-03	2.8318E-06	184	538	211.09
809	C8H16	cis-1,4-DIMETHYLCYCLOHEXANE	103.345	7.7049E-01	-2.2240E-03	2.8466E-06	187	538	210.81
810	C8H16	trans-1,4-DIMETHYLCYCLOHEXANE	55.054	1.1928E+00	-3.3306E-03	3.7992E-06	237	531	215.30
811	C8H16	ETHYLCYCLOHEXANE	122.282	5.5767E-01	-1.6020E-03	2.1615E-06	163	548	203.43
812	C8H16	2-ETHYL-1-HEXENE	108.603	9.8631E-01	-2.8315E-03	3.6361E-06	201	517	247.34
813	C8H16	1-METHYL-1-ETHYLCYCLOPENTANE	149.266	5.2569E-01	-1.5767E-03	2.4130E-06	130	524	229.79
814	C8H16	1-OCTENE	119.984	8.3332E-01	-2.5321E-03	3.4745E-06	172	510	235.43
815	C8H16	trans-2-OCTENE	115.221	9.4317E-01	-2.7955E-03	3.6582E-06	186	519	244.88
816	C8H16	trans-3-OCTENE	132.009	7.7295E-01	-2.3321E-03	3.2525E-06	164	517	241.36
817	C8H16	trans-4-OCTENE	118.777	8.9091E-01	-2.6659E-03	3.5593E-06	180	516	241.75
818	C8H16	n-PROPYLCYCLOPENTANE	129.599	5.5399E-01	-1.5899E-03	2.2092E-06	157	543	211.99
819	C8H16	2,4,4-TRIMETHYL-1-PENTENE	105.466	8.7518E-01	-2.6953E-03	3.6641E-06	181	498	223.91
820	C8H16	2,4,4-TRIMETHYL-2-PENTENE	119.011	7.8648E-01	-2.4252E-03	3.4004E-06	168	502	228.04
821	C8H16O	2-ETHYLHEXANAL	119.983	1.0588E+00	-2.9787E-03	3.6578E-06	201	546	267.84
822	C8H16O	1-OCTANAL	98.815	1.2933E+00	-3.4387E-03	3.7394E-06	247	559	277.82
823	C8H16O	2-OCTANONE	65.104	1.5004E+00	-3.9364E-03	4.2336E-06	254	562	274.72
824	C8H16O2	n-BUTYL n-BUTYRATE	163.586	8.4755E-01	-2.3662E-03	3.0276E-06	182	554	286.18
825	C8H16O2	n-HEXYL ACETATE	151.219	1.0146E+00	-2.8015E-03	3.4906E-06	193	556	297.21
826	C8H16O2	ISOBUTYL ISOBUTYRATE	134.537	9.6926E-01	-2.7469E-03	3.4583E-06	193	542	271.00
827	C8H16O2	n-OCTANOIC ACID	70.790	1.7647E+00	-4.1521E-03	3.9451E-06	291	623	332.41
828	C8H16O4	DIETHYLENE GLYCOL ETHYL ETHER ACETATE	94.642	1.4299E+00	-3.6213E-03	3.8211E-06	249	594	300.32
829	C8H18	2,2-DIMETHYLHEXANE	134.993	7.1970E-01	-2.2691E-03	3.3443E-06	153	495	236.50
830	C8H18	2,3-DIMETHYLHEXANE	109.692	1.0911E+00	-3.2597E-03	4.1505E-06	201	507	255.25
831	C8H18	2,4-DIMETHYLHEXANE	100.402	1.1952E+00	-3.6032E-03	4.6524E-06	201	498	259.76
832	C8H18	2,5-DIMETHYLHEXANE	117.480	1.0073E+00	-3.1157E-03	4.2149E-06	183	495	252.56
833	C8H18	3,3-DIMETHYLHEXANE	162.481	6.0487E-01	-1.8604E-03	2.7473E-06	148	506	250.25
834	C8H18	3,4-DIMETHYLHEXANE	110.815	1.0643E+00	-3.1571E-03	3.9990E-06	201	512	253.47
835	C8H18	3-ETHYLHEXANE	118.557	1.0528E+00	-3.1296E-03	3.9795E-06	201	509	259.72
836	C8H18	3-ETHYL-2-METHYLPENTANE	223.040	-1.8075E-01	8.4029E-04	--------	159	419	243.85
837	C8H18	3-METHYL-3-ETHYLPENTANE	127.455	8.5726E-01	-2.5344E-03	3.3510E-06	183	519	246.57
838	C8H18	2-METHYLHEPTANE	134.965	8.1458E-01	-2.5182E-03	3.5416E-06	165	504	247.85
839	C8H18	3-METHYLHEPTANE	148.156	6.7559E-01	-2.0742E-03	3.0104E-06	154	507	244.99
840	C8H18	4-METHYLHEPTANE	143.202	7.5567E-01	-2.3377E-03	3.4039E-06	153	506	250.91
841	C8H18	n-OCTANE	82.736	1.3043E+00	-3.8254E-03	4.6459E-06	217	512	254.71
842	C8H18	2,2,3-TRIMETHYLPENTANE	145.424	6.7385E-01	-2.0472E-03	2.9087E-06	162	507	241.44
843	C8H18	2,2,4-TRIMETHYLPENTANE	122.772	7.9485E-01	-2.4977E-03	3.5652E-06	167	490	232.22
844	C8H18	2,3,3-TRIMETHYLPENTANE	137.011	7.7661E-01	-2.3224E-03	3.1604E-06	173	516	245.87
845	C8H18	2,3,4-TRIMETHYLPENTANE	133.693	7.0961E-01	-2.1657E-03	3.0284E-06	165	510	233.01
846	C8H18	2,2,3,3-TETRAMETHYLBUTANE	236.692	-1.8556E-01	8.2075E-04	--------	173	410	254.33
847	C8H18O	DI-n-BUTYL ETHER	142.477	9.7186E-01	-2.8724E-03	3.8233E-06	179	523	278.23
848	C8H18O	DI-sec-BUTYL ETHER	144.597	9.8928E-01	-3.0250E-03	4.1672E-06	174	503	281.10
849	C8H18O	DI-tert-BUTYL ETHER	111.404	1.1963E+00	-3.6509E-03	4.7788E-06	196	495	270.20
850	C8H18O	2-ETHYL-1-HEXANOL	149.560	9.7903E-01	-2.6177E-03	3.1042E-06	204	576	291.00
851	C8H18O	1-OCTANOL	93.554	1.6251E+00	-4.1118E-03	4.2771E-06	259	587	325.92
852	C8H18O	2-OCTANOL	108.193	1.3922E+00	-3.6190E-03	3.9542E-06	243	573	306.36
853	C8H18O2	DI-t-BUTYL PEROXIDE	58.496	1.8484E+00	-5.4614E-03	6.5107E-06	234	492	296.67
854	C8H18O2S	DI-n-BUTYL SULFONE	59.519	1.8350E+00	-3.9006E-03	3.3716E-06	319	690	--------
855	C8H18O3	DIETHYLENE GLYCOL DIETHYL ETHER	121.376	1.5648E+00	-4.2040E-03	4.7572E-06	230	562	340.31
856	C8H18O3	DIETHYLENE GLYCOL MONOBUTYL EHTER	178.607	1.2134E+00	-3.1845E-03	3.7197E-06	206	589	355.89
857	C8H18O4	TRIETHYLENE GLYCOL DIMETHYL ETHER	107.066	1.7928E+00	-4.5686E-03	5.1691E-06	230	586	372.47
858	C8H18O5	TETRAETHYLENE GLYCOL	148.238	1.9353E+00	-4.4926E-03	4.3610E-06	269	650	441.47

Table 3-1 HEAT CAPACITY OF LIQUID - ORGANIC COMPOUNDS (continued)

NO	FORMULA	NAME	$C_P = A + B T + C T^2 + D T^3$				(C_P - joule/(mol K), T - K)		
			A	B	C	D	TMIN	TMAX	C_P @ 25 C
859	C8H18S	n-OCTYL MERCAPTAN	139.644	1.0737E+00	-2.7255E-03	3.0400E-06	225	598	298.05
860	C8H18S	tert-OCTYL MERCAPTAN	145.080	9.9301E-01	-2.7057E-03	3.2816E-06	200	564	287.60
861	C8H18S	BUTYL-SULFIDE	280.893	-2.4129E-01	8.2276E-04	--------	211	485	282.09
862	C8H18S	ETHYL-HEXYL-SULFIDE	298.657	-2.5797E-01	8.6942E-04	--------	211	498	299.03
863	C8H18S	HEPTYL-METHYL-SULFIDE	298.657	-2.5797E-01	8.6942E-04	--------	211	498	299.03
864	C8H18S	PENTYL-PROPYL-SULFIDE	298.657	-2.5797E-01	8.6942E-04	--------	211	498	299.03
865	C8H18S2	BUTYL-DISULFIDE	302.417	-1.9723E-01	7.1586E-04	--------	203	534	307.25
866	C8H19N	DI-n-BUTYLAMINE	125.522	1.2417E+00	-3.4554E-03	4.1291E-06	212	547	298.00
867	C8H19N	DIISOBUTYLAMINE	116.945	1.1416E+00	-3.3242E-03	4.1390E-06	204	522	271.50
868	C8H19N	n-OCTYLAMINE	23.349	1.8986E+00	-4.8933E-03	5.0194E-06	274	564	287.48
869	C8H23N5	TETRAETHYLENEPENTAMINE	235.417	1.3983E+00	-3.1051E-03	3.0617E-06	244	697	457.45
870	C8H24O4Si4	OCTAMETHYLCYCLOTETRASILOXANE	-154.769	4.9615E+00	-1.3184E-02	1.3370E-05	292	528	506.88
871	C9H4O5	TRIMELLITIC ANHYDRIDE	-152.608	3.4176E+00	-5.9931E-03	4.0275E-06	439	801	--------
872	C9H6N2O2	TOLUENE DIISOCYANATE	166.184	8.4304E-01	-1.9148E-03	1.7489E-06	288	663	293.68
873	C9H7N	ISOQUINOLINE	78.662	9.0821E-01	-1.8955E-03	1.6489E-06	300	723	--------
874	C9H7N	QUINOLINE	101.271	7.4437E-01	-1.6527E-03	1.5545E-06	259	704	217.49
875	C9H7NO	8-HYDROXYQUINOLINE	35.798	1.3699E+00	-2.7904E-03	2.2710E-06	347	709	--------
876	C9H8	INDENE	43.911	1.0085E+00	-2.4187E-03	2.3969E-06	273	618	193.11
877	C9H8O	2-METHYLBENZOFURAN	127.806	6.4053E-01	-1.5902E-03	1.8163E-06	201	628	225.56
878	C9H10	INDANE	95.857	6.8757E-01	-1.7066E-03	1.8820E-06	223	616	199.03
879	C9H10	cis-PROPENYLBENZENE	208.904	-1.6810E-01	5.8261E-04	--------	212	473	210.58
880	C9H10	trans-PROPENYLBENZENE	219.352	-2.1767E-01	6.3897E-04	--------	245	473	211.25
881	C9H10	alpha-METHYLSTYRENE	56.844	1.0891E+00	-2.7746E-03	2.9320E-06	251	589	212.62
882	C9H10	m-METHYLSTYRENE	119.076	6.2029E-01	-1.6332E-03	1.9918E-06	188	591	211.62
883	C9H10	o-METHYLSTYRENE	107.320	7.0596E-01	-1.8365E-03	2.1433E-06	206	593	211.36
884	C9H10	p-METHYLSTYRENE	50.671	1.1117E+00	-2.8132E-03	3.0418E-06	240	599	212.65
885	C9H10O2	BENZYL ACETATE	115.669	9.0022E-01	-2.1868E-03	2.4005E-06	223	629	253.30
886	C9H10O2	ETHYL BENZOATE	99.390	9.7851E-01	-2.3907E-03	2.5031E-06	239	628	244.95
887	C9H10O3	ETHYL VANILLIN	-55.581	2.3898E+00	-5.0676E-03	4.1689E-06	352	673	--------
888	C9H11NO	p-DIMETHYLAMINOBENZALDEHYDE	14.062	1.4819E+00	-2.9250E-03	2.3046E-06	349	749	--------
889	C9H12	CUMENE	124.621	6.3293E-01	-1.7331E-03	2.2146E-06	178	568	217.96
890	C9H12	m-ETHYLTOLUENE	118.191	6.5993E-01	-1.8276E-03	2.2958E-06	179	573	213.33
891	C9H12	o-ETHYLTOLUENE	116.132	7.0496E-01	-1.8863E-03	2.2662E-06	193	586	218.70
892	C9H12	p-ETHYLTOLUENE	96.421	8.4441E-01	-2.2664E-03	2.6292E-06	212	576	216.40
893	C9H12	MESITYLENE	83.637	8.7859E-01	-2.3192E-03	2.5989E-06	229	574	208.31
894	C9H12	n-PROPYLBENZENE	123.471	6.1973E-01	-1.6883E-03	2.1608E-06	175	575	215.43
895	C9H12	1,2,3-TRIMETHYLBENZENE	89.589	8.7498E-01	-2.1986E-03	2.3108E-06	249	598	216.26
896	C9H12	1,2,4-TRIMETHYLBENZENE	90.157	8.5988E-01	-2.2344E-03	2.4758E-06	230	584	213.52
897	C9H12O	BENZYL ETHYL ETHER	63.263	1.2864E+00	-3.1713E-03	3.1521E-06	277	596	248.44
898	C9H12O	2-PHENYL-2-PROPANOL	-13.130	1.7631E+00	-4.2375E-03	3.9438E-06	310	594	--------
899	C9H12O2	CUMENE HYDROPEROXIDE	30.201	1.2367E+00	-3.2853E-03	3.4917E-06	265	545	199.41
900	C9H14O	ISOPHORONE	104.758	1.1287E+00	-2.6348E-03	2.5900E-06	266	644	275.71
901	C9H14O6	GLYCERYL TRIACETATE	92.681	1.9216E+00	-4.5232E-03	4.3632E-06	278	634	379.17
902	C9H16	1-NONYNE	285.120	-3.0556E-01	9.6796E-04	--------	224	454	280.06
903	C9H16O4	AZELAIC ACID	-99.273	2.9104E+00	-5.7081E-03	4.3289E-06	381	730	--------
904	C9H18	BUTYLCYCLOPENTANE	237.663	-1.5962E-01	7.2063E-04	--------	166	460	254.13
905	C9H18	cis,cis-1,3,5-TRIMETHYLCYCLOHEXANE	237.210	-2.7544E-01	8.6342E-04	--------	224	442	231.84
906	C9H18	cis,trans-1,3,5-TRIMETHYLCYCLOHEXANE	226.185	-2.1498E-01	8.0846E-04	--------	190	444	233.96
907	C9H18	ISOPROPYLCYCLOHEXANE	135.470	7.3570E-01	-2.0088E-03	2.5402E-06	185	564	243.58
908	C9H18	1-NONENE	131.207	1.0851E+00	-3.1284E-03	3.9692E-06	193	534	281.84
909	C9H18	n-PROPYLCYCLOHEXANE	140.390	6.9340E-01	-1.8670E-03	2.3693E-06	179	575	243.96
910	C9H18O	DIISOBUTYL KETONE	120.604	1.3273E+00	-3.5981E-03	4.1206E-06	228	554	305.70
911	C9H18O	1-NONANAL	109.614	1.4417E+00	-3.7140E-03	3.9055E-06	256	576	312.82
912	C9H18O2	n-BUTYL VALERATE	186.168	9.8508E-01	-2.6925E-03	3.4254E-06	181	566	331.31
913	C9H18O2	n-NONANOIC ACID	68.159	1.9333E+00	-4.4977E-03	4.2984E-06	287	633	358.69
914	C9H18O2	n-OCTYL FORMATE	121.458	1.4581E+00	-3.7740E-03	4.1758E-06	235	581	331.39
915	C9H20	3,3-DIETHYLPENTANE	100.112	1.3624E+00	-3.6579E-03	4.0967E-06	241	549	289.72
916	C9H20	2,2-DIMETHYL-3-ETHYLPENTANE	142.170	8.0821E-01	-2.3530E-03	3.1466E-06	175	531	257.37
917	C9H20	3-ETHYL-2,3-DIMETHYLPENTANE	262.700	-2.0700E-01	8.7062E-04	--------	175	448	278.37
918	C9H20	2,4-DIMETHYL-3-ETHYLPENTANE	156.947	6.7635E-01	-1.9882E-03	2.8297E-06	152	532	256.86
919	C9H20	2,2-DIMETHYLHEPTANE	162.732	8.3408E-01	-2.4872E-03	3.5029E-06	161	519	283.16
920	C9H20	2,6-DIMETHYLHEPTANE	150.823	8.6786E-01	-2.5671E-03	3.5066E-06	171	521	274.31
921	C9H20	3-ETHYLHEPTANE	163.938	7.6230E-01	-2.2423E-03	3.1228E-06	159	531	274.66
922	C9H20	4-ETHYLHEPTANE	254.613	-1.9550E-01	8.9291E-04	--------	161	446	275.70
923	C9H20	2,3-DIMETHYLHEPTANE	252.537	-1.9354E-01	8.8253E-04	--------	161	444	273.28
924	C9H20	2,4-DIMETHYLHEPTANE	243.392	-1.9716E-01	8.9422E-04	--------	161	436	264.10
925	C9H20	2,5-DIMETHYLHEPTANE	247.061	-1.9425E-01	8.8543E-04	--------	161	439	267.85
926	C9H20	3,4-DIMETHYLHEPTANE	256.472	-2.1263E-01	9.0306E-04	--------	171	444	273.35
927	C9H20	3,5-DIMETHYLHEPTANE	250.917	-2.1497E-01	9.1060E-04	--------	171	439	267.77
928	C9H20	4,4-DIMETHYLHEPTANE	250.045	-2.1516E-01	9.0978E-04	--------	171	438	266.77
929	C9H20	3-ETHYL-2-METHYLHEXANE	249.595	-1.9554E-01	8.8760E-04	--------	161	441	270.20
930	C9H20	4-ETHYL-2-METHYLHEXANE	244.519	-1.9622E-01	8.9043E-04	--------	161	437	265.17
931	C9H20	3-ETHYL-3-METHYLHEXANE	252.843	-1.8863E-01	8.6329E-04	--------	161	444	273.35
932	C9H20	3-ETHYL-4-METHYLHEXANE	252.536	-1.9298E-01	8.7872E-04	--------	161	444	273.11
933	C9H20	2,2,3-TRIMETHYLHEXANE	242.060	-1.7176E-01	8.3347E-04	--------	154	437	264.94
934	C9H20	2,2,4-TRIMETHYLHEXANE	233.575	-1.7765E-01	8.5469E-04	--------	154	430	256.58
935	C9H20	2,3,3-TRIMETHYLHEXANE	248.056	-1.7619E-01	8.3591E-04	--------	157	441	269.83
936	C9H20	2,3,4-TRIMETHYLHEXANE	249.607	-1.7997E-01	8.4982E-04	--------	157	442	271.49

71

Table 3-1 HEAT CAPACITY OF LIQUID - ORGANIC COMPOUNDS (continued)

NO	FORMULA	NAME	$C_P = A + B T + C T^2 + D T^3$				$(C_P$ - joule/(mol K), T - K)		
			A	B	C	D	TMIN	TMAX	C_P @ 25 C
937	C9H20	2,3,5-TRIMETHYLHEXANE	236.670	-1.6201E-01	8.3248E-04	--------	146	435	262.37
938	C9H20	2,4,4-TRIMETHYLHEXANE	240.788	-1.9049E-01	8.7041E-04	--------	161	434	261.37
939	C9H20	3,3,4-TRIMETHYLHEXANE	256.900	-2.0287E-01	8.6327E-04	--------	173	444	273.15
940	C9H20	2-METHYLOCTANE	129.481	1.1045E+00	-3.2083E-03	4.0849E-06	194	528	281.87
941	C9H20	3-METHYLOCTANE	157.300	7.9579E-01	-2.3303E-03	3.1813E-06	167	531	271.73
942	C9H20	4-METHYLOCTANE	154.947	7.9611E-01	-2.3367E-03	3.2605E-06	161	529	271.00
943	C9H20	n-NONANE	98.040	1.3538E+00	-3.8058E-03	4.4991E-06	221	536	282.60
944	C9H20	2,2,3,3-TETRAMETHYLPENTANE	51.432	1.6682E+00	-4.3927E-03	4.6698E-06	264	550	282.09
945	C9H20	2,2,3,4-TETRAMETHYLPENTANE	156.896	6.3296E-01	-1.8390E-03	2.6149E-06	153	533	251.44
946	C9H20	2,2,4,4-TETRAMETHYLPENTANE	110.024	1.1491E+00	-3.3554E-03	4.1855E-06	208	514	265.28
947	C9H20	2,3,3,4-TETRAMETHYLPENTANE	241.147	-1.6815E-01	8.2339E-04	--------	153	436	264.21
948	C9H20	2,2,5-TRIMETHYLHEXANE	139.690	8.6342E-01	-2.6067E-03	3.6252E-06	168	511	261.48
949	C9H20O	2,6-DIMETHYL-4-HEPTANOL	146.975	1.3141E+00	-3.6752E-03	4.4493E-06	209	543	330.01
950	C9H20O	1-NONANOL	119.340	1.7978E+00	-4.4066E-03	4.4125E-06	269	606	380.59
951	C9H20O	2-NONANOL	119.928	1.6303E+00	-4.3506E-03	4.8181E-06	239	561	346.95
952	C9H20S	n-NONYL MERCAPTAN	122.053	1.4300E+00	-3.4863E-03	3.6087E-06	254	613	334.13
953	C9H20S	BUTYL-PENTYL-SULFIDE	343.332	-3.2796E-01	9.8120E-04	--------	232	521	332.77
954	C9H20S	ETHYL-HEPTYL-SULFIDE	343.332	-3.2796E-01	9.8120E-04	--------	232	521	332.77
955	C9H20S	HEXYL-PROPYL-SULFIDE	343.332	-3.2796E-01	9.8120E-04	--------	232	521	332.77
956	C9H20S	METHYL-OCTYL-SULFIDE	343.332	-3.2796E-01	9.8120E-04	--------	232	521	332.77
957	C9H21N	n-NONYLAMINE	44.935	1.9488E+00	-4.9000E-03	4.9495E-06	274	583	321.56
958	C9H21N	TRIPROPYLAMINE	175.535	1.0070E+00	-3.0021E-03	3.9660E-06	181	520	314.03
959	C10H6O8	PYROMELLITIC ACID	-866.125	5.9659E+00	-9.5709E-03	5.5330E-06	555	804	--------
960	C10H7Br	1-BROMONAPHTHALENE	114.214	8.0622E-01	-1.6535E-03	1.4811E-06	280	742	246.86
961	C10H7Cl	1-CHLORONAPHTHALENE	108.538	8.3305E-01	-1.7833E-03	1.6721E-06	270	707	242.70
962	C10H8	NAPHTHALENE	-30.842	1.5362E+00	-3.2492E-03	2.6568E-06	354	674	--------
963	C10H8	AZULENE	233.549	-1.0264E-01	4.5809E-04	--------	175	545	244.04
964	C10H9N	QUINALDINE	86.319	1.0201E+00	-2.2474E-03	2.0895E-06	273	696	246.05
965	C10H10	m-DIVINYLBENZENE	133.444	7.6302E-01	-1.8930E-03	2.1501E-06	207	623	249.65
966	C10H10	1-METHYLINDENE	172.844	5.7594E-01	-1.4323E-03	1.6119E-06	201	633	259.96
967	C10H10	2-METHYLINDENE	-81.284	2.1991E+00	-4.9461E-03	4.1828E-06	354	616	--------
968	C10H10O4	DIMETHYL PHTHALATE	116.404	1.1694E+00	-2.5655E-03	2.4133E-06	273	689	300.96
969	C10H10O4	DIMETHYL TEREPHTHALATE	-190.453	3.1282E+00	-6.1441E-03	4.5099E-06	415	695	--------
970	C10H12	DICYCLOPENTADIENE	-32.699	1.9645E+00	-4.7301E-03	4.4311E-06	308	594	--------
971	C10H12	1,2,3,4-TETRAHYDRONAPHTHALENE	118.217	7.6524E-01	-1.8150E-03	1.8740E-06	238	648	234.70
972	C10H12O	ANETHOLE	73.791	1.8384E+00	-4.1681E-03	3.8533E-06	296	651	353.50
973	C10H12O4	DIALLYL MALEATE	169.156	1.3290E+00	-3.2856E-03	3.5482E-06	227	624	367.38
974	C10H14	n-BUTYLBENZENE	140.161	7.2011E-01	-1.8916E-03	2.3069E-06	186	594	247.85
975	C10H14	sec-BUTYLBENZENE	137.811	7.6025E-01	-1.9702E-03	2.3246E-06	199	598	250.95
976	C10H14	tert-BUTYLBENZENE	113.801	8.1538E-01	-2.1047E-03	2.3922E-06	216	594	233.22
977	C10H14	1,2,3,4-TETRAMETHYLBENZENE	257.596	-2.9458E-01	7.5542E-04	--------	268	508	236.92
978	C10H14	m-CYMENE	121.295	8.5346E-01	-2.2176E-03	2.5641E-06	210	591	246.58
979	C10H14	o-CYMENE	128.152	8.0370E-01	-2.0908E-03	2.4476E-06	203	596	246.79
980	C10H14	p-CYMENE	121.863	8.1593E-01	-2.1402E-03	2.5029E-06	206	588	241.22
981	C10H14	m-DIETHYLBENZENE	140.766	7.2215E-01	-1.8858E-03	2.2725E-06	190	597	248.67
982	C10H14	o-DIETHYLBENZENE	96.082	1.1064E+00	-2.7841E-03	2.9618E-06	243	601	256.97
983	C10H14	p-DIETHYLBENZENE	104.866	1.0430E+00	-2.6792E-03	2.9453E-06	231	592	255.72
984	C1OH14	2-ETHYL-m-XYLENE	80.529	1.3172E+00	-3.2639E-03	3.3597E-06	258	604	272.15
985	C10H14	2-ETHYL-p-XYLENE	122.523	9.8711E-01	-2.5324E-03	2.8461E-06	221	597	267.15
986	C10H14	3-ETHYL-o-XYLENE	115.338	9.8998E-01	-2.4961E-03	2.7344E-06	225	612	261.09
987	C10H14	4-ETHYL-m-XYLENE	131.880	9.1833E-01	-2.3617E-03	2.7082E-06	211	599	267.52
988	C10H14	4-ETHYL-o-XYLENE	138.038	8.7852E-01	-2.2642E-03	2.6114E-06	207	600	267.91
989	C1OH14	5-ETHYL-m-XYLENE	145.064	7.8009E-01	-2.0540E-03	2.4995E-06	190	590	261.31
990	C10H14	ISOBUTYLBENZENE	111.391	9.5474E-01	-2.4575E-03	2.7977E-06	223	585	251.74
991	C10H14	1,2,3,5-TETRAMETHYLBENZENE	101.280	1.2219E+00	-3.0164E-03	3.1337E-06	250	611	280.50
992	C10H14	1,2,4,5-TETRAMETHYLBENZENE	-157.001	2.8646E+00	-6.4974E-03	5.5536E-06	353	608	--------
993	C10H14O	p-tert-BUTYLPHENOL	-114.737	2.7001E+00	-5.6830E-03	4.5566E-06	373	661	--------
994	C10H14O2	p-tert-BUTYLCATECHOL	85.881	1.6545E+00	-3.4613E-03	2.9371E-06	326	698	--------
995	C10H15N	N,N-DIETYHLANILINE	135.108	1.0920E+00	-2.6691E-03	2.7956E-06	236	632	297.52
996	C10H15N	2,6-DIETHYLANILINE	90.510	1.6652E+00	-4.0127E-03	3.9532E-06	278	610	335.06
997	C10H16	CAMPHENE	-78.162	2.3233E+00	-5.6540E-03	5.2271E-06	321	574	--------
998	C10H16	D-LIMONENE	147.367	8.5555E-01	-2.2423E-03	2.6375E-06	200	594	273.03
999	C10H16	alpha-PHELLANDRENE	147.595	9.1044E-01	-2.4065E-03	2.8595E-06	201	584	280.91
1000	C1OH16	beta-PHELLANDRENE	141.088	9.4087E-01	-2.4715E-03	2.9581E-06	201	583	280.31
1001	C10H16	alpha-PINENE	115.716	8.4523E-01	-2.2864E-03	2.6813E-06	210	569	235.54
1002	C10H16	beta-PINENE	154.744	7.6838E-01	-2.0522E-03	2.3566E-06	213	579	263.87
1003	C10H16	alpha-TERPINENE	156.773	9.4558E-01	-2.4884E-03	2.9501E-06	201	587	295.68
1004	C1OH16	gamma-TERPINENE	157.700	9.3121E-01	-2.4245E-03	2.8522E-06	201	595	295.41
1005	C10H16	TERPINOLENE	147.614	8.5448E-01	-2.1968E-03	2.5600E-06	201	605	274.95
1006	C10H16O	CAMPHOR	-918.198	7.2532E+00	-1.4472E-02	1.0320E-05	454	638	--------
1007	C10H18	1-DECYNE	319.895	-3.4153E-01	1.0389E-03	--------	230	477	310.42
1008	C10H18	cis-DECAHYDRONANPHTALENE	140.152	9.2529E-01	-2.2387E-03	2.3877E-06	231	632	280.31
1009	C10H18	trans-DECAHYDRONAPHTHALENE	108.262	9.5784E-01	-2.3377E-03	2.4579E-06	244	618	251.18
1010	C10H18O4	SEBACIC ACID	-232.230	3.7055E+00	-7.0861E-03	5.1392E-06	409	734	--------
1011	C10H20	n-BUTYLCYCLOHEXANE	172.622	9.6521E-01	-2.4777E-03	2.9249E-06	199	600	317.67
1012	C10H20	1-CYCLOPENTYLPENTANE	277.584	-2.1950E-01	8.2818E-04	--------	191	484	285.76
1013	C10H20	1-DECENE	137.962	1.1934E+00	-3.2863E-03	3.9390E-06	208	555	306.05
1014	C10H20O	1-DECANAL	99.611	1.7386E+00	-4.3395E-03	4.4070E-06	268	591	349.03

Table 3-1 HEAT CAPACITY OF LIQUID - ORGANIC COMPOUNDS (continued)

NO	FORMULA	NAME	$C_P = A + B\,T + C\,T^2 + D\,T^3$				$(C_P$ - joule/(mol K), T - K)		
			A	B	C	D	TMIN	TMAX	C_P @ 25 C
1015	C10H20O2	n-DECANOIC ACID	82.541	2.0901E+00	-4.7509E-03	4.3243E-06	306	642	--------
1016	C10H20O2	2-ETHYLHEXYL ACETATE	196.577	1.0336E+00	-2.8064E-03	3.5277E-06	181	575	348.77
1017	C10H20O2	ISOPENTYL ISOVALERATE	196.741	1.1653E+00	-3.1036E-03	3.5764E-06	216	573	363.08
1018	C10H22	n-DECANE	79.741	1.6926E+00	-4.5287E-03	4.9769E-06	244	557	313.73
1019	C10H22	2-METHYLNONANE	148.342	1.1724E+00	-3.2619E-03	4.0293E-06	200	549	314.73
1020	C10H22	3-METHYLNONANE	163.728	1.0268E+00	-2.8681E-03	3.6165E-06	189	552	310.75
1021	C10H22	4-METHYLNONANE	170.158	9.8007E-01	-2.7635E-03	3.6333E-06	175	549	313.00
1022	C10H22	5-METHYLNONANE	163.221	1.0420E+00	-2.9260E-03	3.7311E-06	186	549	312.69
1023	C10H22	3-ETHYLOCTANE	297.286	-2.7568E-01	1.0262E-03	--------	189	471	306.31
1024	C10H22	4-ETHYLOCTANE	298.363	-2.5314E-01	1.0270E-03	--------	175	469	314.18
1025	C10H22	2,2-DIMETHYLOCTANE	150.701	1.1882E+00	-3.3380E-03	4.1326E-06	201	542	317.76
1026	C10H22	2,3-DIMETHYLOCTANE	310.889	-3.6032E-01	1.1271E-03	--------	220	467	303.65
1027	C10H22	2,4-DIMETHYLOCTANE	300.513	-3.6840E-01	1.1483E-03	--------	220	459	292.75
1028	C10H22	2,5-DIMETHYLOCTANE	303.643	-3.6525E-01	1.1403E-03	--------	220	462	296.11
1029	C10H22	2,6-DIMETHYLOCTANE	307.570	-3.6498E-01	1.1423E-03	--------	220	464	300.29
1030	C10H22	2,7-DIMETHYLOCTANE	305.167	-3.6255E-01	1.1343E-03	--------	220	463	297.91
1031	C10H22	3,3-DIMETHYLOCTANE	306.637	-3.5622E-01	1.1158E-03	--------	220	464	299.61
1032	C10H22	3,4-DIMETHYLOCTANE	309.671	-3.5905E-01	1.1231E-03	--------	220	467	302.46
1033	C10H22	3,5-DIMETHYLOCTANE	304.961	-3.6667E-01	1.1431E-03	--------	220	463	297.25
1034	C10H22	3,6-DIMETHYLOCTANE	306.433	-3.6204E-01	1.1316E-03	--------	220	464	299.08
1035	C10H22	4,4-DIMETHYLOCTANE	302.371	-3.6365E-01	1.1344E-03	--------	220	461	294.79
1036	C10H22	4,5-DIMETHYLOCTANE	308.309	-3.6336E-01	1.1340E-03	--------	220	465	300.78
1037	C10H22	4-PROPYLHEPTANE	303.307	-3.7857E-01	1.1735E-03	--------	220	461	294.75
1038	C10H22	4-ISOPROPYLHEPTANE	304.709	-3.7117E-01	1.1532E-03	--------	220	462	296.56
1039	C10H22	3-ETHYL-2-METHYLHEPTANE	307.322	-3.6599E-01	1.1403E-03	--------	220	464	299.56
1040	C10H22	4-ETHYL-2-METHYLHEPTANE	301.426	-3.7449E-01	1.1622E-03	--------	220	459	293.09
1041	C10H22	5-ETHYL-2-METHYLHEPTANE	305.343	-3.6659E-01	1.1428E-03	--------	220	463	297.63
1042	C10H22	3-ETHYL-3-METHYLHEPTANE	310.177	-3.5683E-01	1.1156E-03	--------	220	467	302.96
1043	C10H22	4-ETHYL-3-METHYLHEPTANE	308.658	-3.6576E-01	1.1388E-03	--------	220	465	300.84
1044	C10H22	3-ETHYL-5-METHYLHEPTANE	303.594	-3.6791E-01	1.1450E-03	--------	220	461	295.68
1045	C10H22	3-ETHYL-4-METHYLHEPTANE	309.662	-3.6519E-01	1.1374E-03	--------	220	466	301.89
1046	C10H22	4-ETHYL-4-METHYLHEPTANE	306.529	-3.6038E-01	1.1244E-03	--------	220	464	299.04
1047	C10H22	2,2,3-TRIMETHYLHEPTANE	301.506	-3.4804E-01	1.0938E-03	--------	220	461	294.97
1048	C10H22	2,2,4-TRIMETHYLHEPTANE	290.745	-3.6419E-01	1.1355E-03	--------	220	451	283.10
1049	C10H22	2,2,5-TRIMETHYLHEPTANE	293.497	-3.5887E-01	1.1225E-03	--------	220	454	286.28
1050	C10H22	2,2,6-TRIMETHYLHEPTANE	291.021	-3.5797E-01	1.1211E-03	--------	220	452	283.95
1051	C10H22	2,3,3-TRIMETHYLHEPTANE	304.682	-3.4485E-01	1.0852E-03	--------	220	463	298.33
1052	C10H22	2,3,4-TRIMETHYLHEPTANE	304.922	-3.5459E-01	1.1104E-03	--------	220	463	297.91
1053	C10H22	2,3,5-TRIMETHYLHEPTANE	306.245	-3.5948E-01	1.1234E-03	--------	220	464	298.93
1054	C10H22	2,3,6-TRIMETHYLHEPTANE	300.038	-3.5823E-01	1.1212E-03	--------	220	459	292.90
1055	C10H22	2,4,4-TRIMETHYLHEPTANE	293.936	-3.6068E-01	1.1262E-03	--------	220	454	286.51
1056	C10H22	2,4,5-TRIMETHYLHEPTANE	300.845	-3.5978E-01	1.1243E-03	--------	220	460	293.52
1057	C10H22	2,4,6-TRIMETHYLHEPTANE	289.813	-3.6484E-01	1.1385E-03	--------	220	451	282.25
1058	C10H22	2,5,5-TRIMETHYLHEPTANE	295.893	-3.5599E-01	1.1144E-03	--------	220	456	288.81
1059	C10H22	3,3,4-TRIMETHYLHEPTANE	306.866	-3.4385E-01	1.0821E-03	--------	220	465	300.53
1060	C10H22	3,3,5-TRIMETHYLHEPTANE	298.523	-3.4072E-01	1.0758E-03	--------	220	459	292.57
1061	C10H22	3,4,4-TRIMETHYLHEPTANE	305.850	-3.4427E-01	1.0832E-03	--------	220	464	299.50
1062	C10H22	3,4,5-TRIMETHYLHEPTANE	308.822	-3.6332E-01	1.1335E-03	--------	220	466	301.26
1063	C10H22	3-ISOPROPYL-2-METHYLHEXANE	314.525	-3.6435E-01	1.1342E-03	--------	220	470	306.71
1064	C10H22	3,3-DIETHYLHEXANE	313.202	-3.5200E-01	1.1021E-03	--------	220	469	306.22
1065	C10H22	3,4-DIETHYLHEXANE	310.903	-3.6513E-01	1.1362E-03	--------	220	467	303.04
1066	C10H22	3-ETHYL-2,2-DIMETHYLHEXANE	300.060	-3.5414E-01	1.1084E-03	--------	220	459	293.00
1067	C10H22	4-ETHYL-2,2-DIMETHYLHEXANE	288.944	-3.6124E-01	1.1276E-03	--------	220	450	281.48
1068	C10H22	3-ETHYL-2,3-DIMETHYLHEXANE	309.032	-3.4101E-01	1.0745E-03	--------	220	467	302.88
1069	C10H22	4-ETHYL-2,3-DIMETHYLHEXANE	306.225	-3.5397E-01	1.1081E-03	--------	220	464	299.19
1070	C10H22	3-ETHYL-2,4-DIMETHYLHEXANE	305.208	-3.5440E-01	1.1093E-03	--------	220	463	298.15
1071	C10H22	4-ETHYL-2,4-DIMETHYLHEXANE	305.829	-3.4416E-01	1.0831E-03	--------	220	464	299.50
1072	C10H22	3-ETHYL-2,5-DIMETHYLHEXANE	298.068	-3.6334E-01	1.1328E-03	--------	220	457	290.43
1073	C10H22	4-ETHYL-3,3-DIMETHYLHEXANE	308.177	-3.4330E-01	1.0800E-03	--------	220	466	301.83
1074	C10H22	3-ETHYL-3,4-DIMETHYLHEXANE	307.296	-3.4561E-01	1.0857E-03	--------	220	465	300.77
1075	C10H22	2,2,3,3-TETRAMETHYLHEXANE	304.361	-3.3713E-01	1.0648E-03	--------	220	463	298.50
1076	C10H22	2,2,3,4-TETRAMETHYLHEXANE	302.661	-3.4061E-01	1.0733E-03	--------	220	462	296.52
1077	C10H22	2,2,3,5-TETRAMETHYLHEXANE	289.718	-3.4622E-01	1.0888E-03	--------	220	452	283.28
1078	C10H22	2,2,4,4-TETRAMETHYLHEXANE	297.284	-3.5589E-01	1.1122E-03	--------	220	457	290.04
1079	C10H22	2,2,4,5-TETRAMETHYLHEXANE	289.594	-3.5401E-01	1.1087E-03	--------	220	451	282.60
1080	C10H22	2,2,5,5-TETRAMETHYLHEXANE	303.057	-4.8151E-01	1.2529E-03	--------	262	441	270.87
1081	C10H22	2,3,3,4-TETRAMETHYLHEXANE	335.560	-4.5519E-01	1.1905E-03	--------	262	468	305.67
1082	C10H22	2,3,3,5-TETRAMETHYLHEXANE	322.824	-4.7916E-01	1.2448E-03	--------	262	456	290.61
1083	C10H22	2,3,4,4-TETRAMETHYLHEXANE	332.559	-4.6510E-01	1.2129E-03	--------	262	465	301.70
1084	C10H22	2,3,4,5-TETRAMETHYLHEXANE	326.901	-4.7986E-01	1.2463E-03	--------	262	459	294.62
1085	C10H22	3,3,4,4-TETRAMETHYLHEXANE	341.010	-4.3712E-01	1.1503E-03	--------	262	473	312.94
1086	C10H22	2,4-DIMETHYL-3-ISOPROPYLPENTANE	286.505	-2.7595E-01	1.0077E-03	--------	192	460	293.81
1087	C10H22	3,3-DIETHYL-2-METHYLPENTANE	302.418	-2.6252E-01	9.6850E-04	--------	192	473	310.24
1088	C10H22	3-ETHYL-2,2,3-TRIMETHYLPENTANE	301.780	-2.5175E-01	9.3718E-04	--------	192	473	310.03
1089	C10H22	3-ETHYL-2,2,4-TRIMETHYLPENTANE	284.110	-2.7052E-01	9.9146E-04	--------	192	458	291.59
1090	C10H22	3-ETHYL-2,3,4-TRIMETHYLPENTANE	301.839	-2.5618E-01	9.5037E-04	--------	192	473	309.94
1091	C10H22	2,2,3,3,4-PENTAMETHYLPENTANE	319.598	-3.5534E-01	1.0459E-03	--------	238	469	306.63
1092	C10H22	2,2,3,4,4-PENTAMETHYLPENTANE	311.551	-3.7509E-01	1.1011E-03	--------	235	462	297.60

73

Table 3-1 HEAT CAPACITY OF LIQUID - ORGANIC COMPOUNDS (continued)

			$C_P = A + B T + C T^2 + D T^3$				(C_P - joule/(mol K), T - K)		
NO	FORMULA	NAME	A	B	C	D	TMIN	TMAX	C_P @ 25 C
1093	C10H22O	1-DECANOL	97.978	2.1695E+00	-5.1554E-03	5.0024E-06	281	621	419.11
1094	C10H22O	DI-n-PENTYL ETHER	172.072	1.2808E+00	-3.4843E-03	4.2101E-06	205	560	355.80
1095	C10H22O	ISODECANOL	177.290	1.4535E+00	-3.8470E-03	4.4446E-06	214	580	386.48
1096	C10H22O5	TETRAETHYLENE GLYCOL DIMETHYL ETHER	151.123	1.6274E+00	-3.8555E-03	4.0393E-06	244	635	400.67
1097	C10H22S	n-DECYL MERCAPTAN	181.298	1.2845E+00	-3.0994E-03	3.1909E-06	249	626	373.33
1098	C10H22S	BUTYL-HEXYL-SULFIDE	382.149	-3.6933E-01	1.0609E-03	--------	239	543	366.34
1099	C10H22S	ETHYL-OCTYL-SULFIDE	382.149	-3.6933E-01	1.0609E-03	--------	239	543	366.34
1100	C10H22S	HEPTYL-PROPYL-SULFIDE	382.149	-3.6933E-01	1.0609E-03	--------	239	543	366.34
1101	C10H22S	METHYL-NONYL-SULFIDE	382.149	-3.6933E-01	1.0609E-03	--------	239	543	366.34
1102	C10H22S	PENTYL-SULFIDE	382.149	-3.6933E-01	1.0609E-03	--------	239	543	366.34
1103	C10H22S2	PENTYL-DISULFIDE	358.937	-2.5192E-01	8.4135E-04	--------	215	567	358.62
1104	C10H23N	n-DECYLAMINE	22.358	2.2623E+00	-5.5160E-03	5.3443E-06	290	597	348.17
1105	C11H10	1-METHYLNAPHTHALENE	114.585	7.4739E-01	-1.6860E-03	1.6519E-06	244	695	231.33
1106	C11H10	2-METHYLNAPHTHALENE	55.527	1.1885E+00	-2.5841E-03	2.2618E-06	309	685	--------
1107	C11H14O2	n-BUTYL BENZOATE	128.190	1.0698E+00	-2.4876E-03	2.5024E-06	253	652	292.35
1108	C11H16	n-PENTYLBENZENE	163.989	8.4271E-01	-2.1430E-03	2.4924E-06	199	612	290.80
1109	C11H16O	p-tert-AMYLPHENOL	-78.043	2.5891E+00	-5.3928E-03	4.3232E-06	367	676	--------
1110	C11H20	1-UNDECYNE	362.301	-4.1921E-01	1.1557E-03	--------	249	498	340.05
1111	C11H20O2	2-ETHYLHEXYL ACRYLATE	177.773	1.3122E+00	-3.3957E-03	4.2877E-06	184	590	380.79
1112	C11H22	1-UNDECENE	131.914	1.3656E+00	-3.6108E-03	4.0918E-06	225	574	326.54
1113	C11H22	1-CYCLOPENTYLHEXANE	313.322	-2.5876E-01	9.0996E-04	--------	201	506	317.06
1114	C11H22	PENTYLCYCLOHEXANE	321.248	-2.8653E-01	9.2864E-04	--------	217	507	318.37
1115	C11H22O	1-UNDECANAL	75.031	1.9647E+00	-4.7801E-03	4.7739E-06	274	605	362.42
1116	C11H24	n-UNDECANE	94.169	1.7806E+00	-4.6303E-03	4.9675E-06	249	575	345.10
1117	C11H24O	1-UNDECANOL	89.917	2.3236E+00	-5.3887E-03	5.1024E-06	290	634	438.91
1118	C11H24S	UNDECYL MERCAPTAN	170.940	1.5240E+00	-3.5636E-03	3.4697E-06	271	639	400.50
1119	C11H24S	BUTYL-HEPTYL-SULFIDE	427.083	-4.4455E-01	1.1758E-03	--------	256	563	399.06
1120	C11H24S	DECYL-METHYL-SULFIDE	427.083	-4.4455E-01	1.1758E-03	--------	256	563	399.06
1121	C11H24S	ETHYL-NONYL-SULFIDE	427.083	-4.4455E-01	1.1758E-03	--------	256	563	399.06
1122	C11H24S	OCTYL-PROPYL-SULFIDE	427.083	-4.4455E-01	1.1758E-03	--------	256	563	399.06
1123	C12H8O	DIBENZOFURAN	3.954	1.4589E+00	-2.8450E-03	2.2112E-06	357	754	--------
1124	C12H9N	DIBENZOPYRROLE	-574.755	4.3452E+00	-7.1633E-03	4.3069E-06	519	809	--------
1125	C12H10	ACENAPHTHENE	-25.875	1.7836E+00	-3.5722E-03	2.7591E-06	368	723	--------
1126	C12H10	BIPHENYL	27.519	1.5432E+00	-3.1647E-03	2.5801E-06	343	710	--------
1127	C12H10O	DIPHENYL ETHER	109.032	1.1920E+00	-2.5769E-03	2.2960E-06	301	687	--------
1128	C12H11N	p-AMINODIPHENYL	89.305	1.3934E+00	-2.8082E-03	2.3212E-06	327	735	--------
1129	C12H11N	DIPHENYLAMINE	72.447	1.4612E+00	-2.9564E-03	2.4383E-06	327	735	--------
1130	C12H11N3	p-AMINOAZOBENZENE	-18.690	2.2198E+00	-4.0371E-03	2.8766E-06	402	789	--------
1131	C12H11N3	1,3-DIPHENYLTRIAZENE	24.880	1.9501E+00	-3.7161E-03	2.8099E-06	373	761	--------
1132	C12H12	1,2-DIMETHYLNAPHTHALENE	305.551	-2.6995E-01	6.8728E-04	--------	273	569	286.16
1133	C12H12	1,3-DIMETHYLNAPHTHALENE	302.181	-2.6258E-01	6.7797E-04	--------	270	568	284.16
1134	C12H12	1,4-DIMETHYLNAPHTHALENE	311.152	-2.8602E-01	7.0334E-04	--------	282	570	288.40
1135	C12H12	1,5-DIMETHYLNAPHTHALENE	294.400	-1.4230E-01	3.2020E-04	--------	356	568	--------
1136	C12H12	1,6-DIMETHYLNAPHTHALENE	294.192	-2.4342E-01	6.5644E-04	--------	260	566	279.97
1137	C12H12	1,7-DIMETHYLNAPHTHALENE	294.562	-2.4492E-01	6.5789E-04	--------	261	566	280.02
1138	C12H12	2,3-DIMETHYLNAPHTHALENE	302.800	-1.6514E-01	3.4091E-04	--------	379	571	--------
1139	C12H12	2,6-DIMETHYLNAPHTHALENE	-128.447	2.4458E+00	-4.9011E-03	3.7744E-06	386	699	--------
1140	C12H12	2,7-DIMETHYLNAPHTHALENE	-81.227	2.1602E+00	-4.3745E-03	3.4460E-06	371	700	--------
1141	C12H12	1-ETHYLNAPHTHALENE	143.270	9.1341E-01	-2.0099E-03	1.9109E-06	260	698	287.58
1142	C12H12	2-ETHYLNAPHTHALENE	286.838	-2.3020E-01	6.1420E-04	--------	267	561	272.80
1143	C12H12N2	p-AMINODIPHENYLAMINE	114.220	1.4456E+00	-2.7514E-03	2.1602E-06	342	780	357.90
1144	C12H12N2	HYDRAZOBENZENE	--------	--------	--------	--------	--------	--------	--------
1145	C12H14	1,2,3-TRIMETHYLINDENE	-74.461	2.5391E+00	-5.5295E-03	4.6506E-06	346	653	--------
1146	C12H14O4	DIETHYL PHTHALATE	125.381	1.5763E+00	-3.4735E-03	3.3343E-06	270	681	374.95
1147	C12H16	CYCLOHEXYLBENZENE	89.522	1.4510E+00	-3.2409E-03	3.0515E-06	281	670	314.92
1148	C12H18	m-DIISOPROPYLBENZENE	116.438	1.1828E+00	-2.9169E-03	3.3756E-06	211	616	299.26
1149	C12H18	p-DIISOPROPYLBENZENE	112.459	1.3655E+00	-3.3118E-03	3.3714E-06	257	620	314.53
1150	C12H18	n-HEXYLBENZENE	168.255	9.9045E-01	-2.4462E-03	2.7183E-06	213	628	318.15
1151	C12H18	1,2,3-TRIETHYLBENZENE	287.384	-2.3046E-01	7.8685E-04	--------	208	521	288.62
1152	C12H18	1,2,4-TRIETHYLBENZENE	287.384	-2.3046E-01	7.8685E-04	--------	208	521	288.62
1153	C12H18	1,3,5-TRIETHYLBENZENE	285.138	-2.2919E-01	7.8316E-04	--------	208	519	286.42
1154	C12H18	HEXAMETHYLBENZENE	404.309	-3.6164E-01	5.9695E-04	--------	440	567	--------
1155	C12H20O4	DIBUTYL MALEATE	261.366	1.0830E+00	-2.6152E-03	3.0453E-06	189	644	432.52
1156	C12H22	BICYCLOHEXYL	111.675	1.3662E+00	-3.1246E-03	2.9755E-06	278	654	320.12
1157	C12H22	1-DODECYNE	397.674	-4.6369E-01	1.2353E-03	--------	255	518	369.23
1158	C12H23N	DICYCLOHEXYLAMINE	142.238	1.5261E+00	-3.4477E-03	3.2968E-06	274	663	378.16
1159	C12H24	1-DODECENE	129.203	1.5842E+00	-4.0461E-03	4.3851E-06	239	591	358.08
1160	C12H24	1-CYCLOPENTYLHEPTANE	355.089	-3.3434E-01	1.0442E-03	--------	221	527	348.23
1161	C12H24	1-CYCLOHEXYLHEXANE	382.754	-4.4226E-01	1.1327E-03	--------	265	528	351.58
1162	C12H24O	1-DODECANAL	58.364	2.6887E+00	-6.4291E-03	6.1799E-06	286	617	452.28
1163	C12H24O2	n-DODECANOIC ACID	50.801	2.2580E+00	-4.9660E-03	4.3771E-06	318	661	--------
1164	C12H26	n-DODECANE	84.485	2.0358E+00	-5.0981E-03	5.2186E-06	265	592	376.59
1165	C12H26O	DI-n-HEXYL ETHER	181.379	1.6216E+00	-4.1307E-03	4.5662E-06	231	592	418.70
1166	C12H26O	1-DODECANOL	91.986	2.6300E+00	-5.9460E-03	5.4956E-06	298	649	493.22
1167	C12H26O3	DIETHYLENE GLYCOL DI-n-BUTYL ETHER	231.848	1.5569E+00	-3.9365E-03	4.4234E-06	214	612	463.35
1168	C12H26S	n-DODECYL MERCAPTAN	159.306	1.6759E+00	-3.8485E-03	3.7803E-06	266	652	417.05
1169	C12H26S	BUTYL-OCTYL-SULFIDE	464.127	-4.8601E-01	1.2540E-03	--------	260	582	430.70
1170	C12H26S	DECYL-ETHYL-SULFIDE	464.127	-4.8601E-01	1.2540E-03	--------	260	582	430.70

Table 3-1 HEAT CAPACITY OF LIQUID - ORGANIC COMPOUNDS (continued)

NO	FORMULA	NAME	$C_P = A + B T + C T^2 + D T^3$				$(C_P$ - joule/(mol K), T - K$)$		
			A	B	C	D	TMIN	TMAX	C_P @ 25 C
1171	C12H26S	HEXYL-SULFIDE	464.127	-4.8601E-01	1.2540E-03	--------	260	582	430.70
1172	C12H26S	METHYL-UNDECYL-SULFIDE	464.127	-4.8601E-01	1.2540E-03	--------	260	582	430.70
1173	C12H26S	NONYL-PROPYL-SULFIDE	464.127	-4.8601E-01	1.2540E-03	--------	260	582	430.70
1174	C12H26S2	HEXYL-DISULFIDE	414.759	-3.1117E-01	9.6959E-04	--------	226	597	408.17
1175	C12H27BO3	TRI-n-BUTYL BORATE	--------	--------	--------	--------	--------	--------	--------
1176	C12H27N	DODECYLAMINE	58.749	2.4313E+00	-5.6459E-03	5.2280E-06	302	626	--------
1177	C12H27N	TRI-n-BUTYLAMINE	222.369	1.3613E+00	-3.6182E-03	4.2792E-06	204	580	420.02
1178	C13H10	FLUORENE	12.689	1.5292E+00	-2.8199E-03	2.0561E-06	389	783	--------
1179	C13H10O	BENZOPHENONE	74.191	1.4261E+00	-2.9066E-03	2.4119E-06	322	734	--------
1180	C13H12	DIPHENYLMETHANE	81.092	1.2372E+00	-2.6723E-03	2.3822E-06	299	691	--------
1181	C13H14	1-PROPYLNAPHTHALENE	313.483	-2.6476E-01	6.9554E-04	--------	266	576	296.37
1182	C13H14	2-PROPYLNAPHTHALENE	317.168	-2.7519E-01	7.0612E-04	--------	271	577	297.89
1183	C13H14	2ETHYL-3-METHYLNAPHTHALENE	313.867	-1.4295E-01	3.3372E-04	--------	345	580	--------
1184	C13H14	2ETHYL-6-METHYLNAPHTHALENE	299.018	-1.1865E-01	3.0962E-04	--------	319	573	--------
1185	C13H14	2ETHYL-7-METHYLNAPHTHALENE	299.018	-1.1865E-01	3.0962E-04	--------	319	573	--------
1186	C13H20	n-HEPTYLBENZENE	169.683	1.1875E+00	-2.8301E-03	3.0427E-06	226	643	352.82
1187	C13H24	1-TRIDECYNE	440.407	-5.4183E-01	1.3511E-03	--------	269	537	398.96
1188	C13H26	1-TRIDECENE	133.770	1.9000E+00	-4.7012E-03	4.9057E-06	251	608	412.38
1189	C13H26	1-CYCLOPENTYLOCTANE	391.112	-3.7151E-01	1.1034E-03	--------	230	547	378.43
1190	C13H26	1-CYCLOHEXYLHEPTANE	401.460	-4.0417E-01	1.1270E-03	--------	244	548	381.14
1191	C13H26O	1-TRIDECANAL	74.377	2.4379E+00	-5.6713E-03	5.4098E-06	289	630	440.47
1192	C13H26O2	n-BUTYL NONANOATE	175.844	2.0016E+00	-5.1618E-03	5.6392E-06	236	587	463.22
1193	C13H26O2	METHYL DODECANOATE	77.645	2.5348E+00	-5.9489E-03	5.6820E-06	279	641	455.19
1194	C13H28	n-TRIDECANE	85.027	2.2008E+00	-5.3677E-03	5.4016E-06	269	608	407.21
1195	C13H28O	1-TRIDECANOL	66.215	2.8239E+00	-6.3310E-03	5.7081E-06	305	658	--------
1196	C13H28S	BUTYL-NONYL-SULFIDE	507.610	-5.5512E-01	1.3570E-03	--------	272	600	462.73
1197	C13H28S	DECYL-PROPYL-SULFIDE	507.610	-5.5512E-01	1.3570E-03	--------	272	600	462.73
1198	C13H28S	DODECYL-METHYL-SULFIDE	507.610	-5.5512E-01	1.3570E-03	--------	272	600	462.73
1199	C13H28S	ETHYL-UNDECYL-SULFIDE	507.610	-5.5512E-01	1.3570E-03	--------	272	600	462.73
1200	C13H28S	1-TRIDECANETHIOL	502.923	-5.6338E-01	1.3313E-03	--------	283	594	453.29
1201	C14H8O2	ANTHRAQUINONE	-773.219	5.1936E+00	-8.2652E-03	4.7346E-06	560	810	--------
1202	C14H10	ANTHRACENE	-269.578	3.1196E+00	-5.3371E-03	3.3589E-06	490	786	--------
1203	C14H10	DIPHENYLACETYLENE	74.380	1.3641E+00	-2.6965E-03	2.1745E-06	337	749	--------
1204	C14H10	PHENANTHRENE	56.723	1.4209E+00	-2.6607E-03	1.9757E-06	373	782	--------
1205	C14H12	cis-STILBENE	126.454	1.1465E+00	-2.5546E-03	2.4261E-06	269	681	305.50
1206	C14H12	trans-STILBENE	-127.759	2.6396E+00	-5.0660E-03	3.7231E-06	398	738	--------
1207	C14H12O2	BENZYL BENZOATE	91.066	1.4869E+00	-3.0693E-03	2.6897E-06	294	738	332.84
1208	C14H14	1,1-DIPHENYLETHANE	147.196	1.0915E+00	-2.3979E-03	2.3146E-06	256	698	320.81
1209	C14H14	1,2-DIPHENYLETHANE	49.681	1.7617E+00	-3.6945E-03	3.1244E-06	325	702	--------
1210	C14H14O	DIBENZYL ETHER	137.942	1.2915E+00	-2.8039E-03	2.5860E-06	278	699	342.31
1211	C14H16	1-n-BUTYLNAPHTHALENE	177.544	1.0600E+00	-2.2934E-03	2.1895E-06	254	713	347.73
1212	C14H16	2-BUTYLNAPHTHALENE	344.536	-2.9530E-01	7.5842E-04	--------	269	592	323.91
1213	C14H22	n-OCTYLBENZENE	175.330	1.3630E+00	-3.1673E-03	3.2877E-06	238	656	387.29
1214	C14H22	1,2,3,4-TETRAETHYLBENZENE	445.373	-5.5941E-01	1.3035E-03	--------	286	554	394.46
1215	C14H22	1,2,3,5-TETRAETHYLBENZENE	443.482	-5.5357E-01	1.2951E-03	--------	285	554	393.56
1216	C14H22	1,2,4,5-TETRAETHYLBENZENE	441.767	-5.4981E-01	1.2915E-03	--------	284	553	392.64
1217	C14H22O	p-tert-OCTYLPHENOL	-64.720	3.0804E+00	-6.3791E-03	5.1373E-06	360	689	--------
1218	C14H28	1-TETRADECENE	135.178	2.0938E+00	-5.0407E-03	5.0893E-06	261	623	446.24
1219	C14H28	1-CYCLOPENTYLNONANE	432.142	-4.4017E-01	1.2106E-03	--------	245	565	408.52
1220	C14H28	1-CYCLOHEXYLOCTANE	441.096	-4.6299E-01	1.2227E-03	--------	254	567	411.75
1221	C14H28O2	n-TETRADECANOIC ACID	80.266	2.8162E+00	-6.0100E-03	5.1299E-06	329	680	--------
1222	C14H30	n-TETRADECANE	111.814	2.2092E+00	-5.2555E-03	5.0865E-06	280	623	438.12
1223	C14H30O	1-TETRADECANOL	78.050	2.9087E+00	-6.3668E-03	5.6761E-06	312	667	--------
1224	C14H30S	BUTYL-DECYL-SULFIDE	545.147	-6.0177E-01	1.4364E-03	--------	277	617	493.42
1225	C14H30S	DODECYL-ETHYL-SULFIDE	545.147	-6.0177E-01	1.4364E-03	--------	277	617	493.42
1226	C14H30S	HEPTYL-SULFIDE	545.147	-6.0177E-01	1.4364E-03	--------	277	617	493.42
1227	C14H30S	METHYL-TRIDECYL-SULFIDE	545.147	-6.0177E-01	1.4364E-03	--------	277	617	493.42
1228	C14H30S	PROPYL-UNDECYL-SULFIDE	545.147	-6.0177E-01	1.4364E-03	--------	277	617	493.42
1229	C14H30S	1-TETRADECANETHIOL	530.238	-5.8177E-01	1.3846E-03	--------	280	609	479.86
1230	C14H30S2	HEPTYL-DISULFIDE	470.604	-3.7780E-01	1.1063E-03	--------	236	624	456.31
1231	C14H31N	TETRADECYLAMINE	30.642	2.9495E+00	-6.6279E-03	5.9265E-06	312	650	--------
1232	C15H10N2O2	DIPHENYLMETHANE-4,4'-DIISOCYANATE	92.853	1.5683E+00	-3.2530E-03	2.7758E-06	312	722	--------
1233	C15H16O	p-CUMYLPHENOL	77.060	1.9527E+00	-3.8275E-03	3.0315E-06	347	751	--------
1234	C15H16O2	BISPHENOL A	-113.816	3.2404E+00	-5.9001E-03	4.1160E-06	427	764	--------
1235	C15H18	1-PENTYLNAPHTHALENE	369.346	-2.8638E-01	7.8730E-04	--------	252	610	353.95
1236	C15H18	2-PENTYLNAPHTHALENE	384.577	-3.2852E-01	8.3280E-04	--------	270	613	360.66
1237	C15H24	n-NONYLBENZENE	184.605	1.5629E+00	-3.5850E-03	3.5802E-06	250	667	426.78
1238	C15H24O	2,6-DI-tert-BUTYL-p-CRESOL	-67.369	3.4023E+00	-7.4333E-03	6.2908E-06	345	648	--------
1239	C15H24O	NONYLPHENOL	289.541	1.1436E+00	-2.6295E-03	2.8767E-06	201	681	473.00
1240	C15H28	1-PENTADECYNE	514.760	-6.5678E-01	1.5301E-03	--------	284	571	454.96
1241	C15H30	1-PENTADECENE	132.592	2.3026E+00	-5.4038E-03	5.3048E-06	270	637	479.35
1242	C15H30	1-CYCLOPENTYLDECANE	467.727	-4.8710E-01	1.2934E-03	--------	252	583	437.47
1243	C15H30	1-CYCLOHEXYLNONANE	480.871	-5.2232E-01	1.3177E-03	--------	264	585	442.27
1244	C15H30O2	PENTADECANOIC ACID	44.353	3.0864E+00	-6.5079E-03	5.5809E-06	327	689	--------
1245	C15H32	n-PENTADECANE	94.014	2.4973E+00	-5.8025E-03	5.5554E-06	284	636	470.02
1246	C15H32O	1-PENTADECANOL	449.182	-2.3525E-01	5.8440E-04	--------	318	608	--------
1247	C15H32S	BUTYL-UNDECYL-SULFIDE	585.948	-6.6504E-01	1.5316E-03	--------	285	633	523.81
1248	C15H32S	DODECYL-PROPYL-SULFIDE	585.948	-6.6504E-01	1.5316E-03	--------	285	633	523.81

Table 3-1 HEAT CAPACITY OF LIQUID - ORGANIC COMPOUNDS (continued)

NO	FORMULA	NAME	$C_P = A + B T + C T^2 + D T^3$				$(C_P$ - joule/(mol K), T - K)		
			A	B	C	D	TMIN	TMAX	C_P @ 25 C
1249	C15H32S	ETHYL-TRIDECYL-SULFIDE	585.948	-6.6504E-01	1.5316E-03	--------	285	633	523.81
1250	C15H32S	METHYL-TETRADECYL-SULFIDE	585.948	-6.6504E-01	1.5316E-03	--------	285	633	523.81
1251	C15H32S	1-PENTADECANETHIOL	571.504	-6.5912E-01	1.4928E-03	--------	292	624	507.69
1252	C16H10	FLUORANTHENE	9.710	1.7567E+00	-3.1767E-03	2.2852E-06	384	815	--------
1253	C16H10	PYRENE	-84.718	2.2508E+00	-3.8750E-03	2.5914E-06	425	842	--------
1254	C16H12	1-PHENYLNAPHTHALENE	111.333	1.4647E+00	-2.8900E-03	2.3745E-06	319	764	--------
1255	C16H20	1-n-HEXYLNAPHTHALENE	218.069	1.2676E+00	-2.6689E-03	2.5143E-06	256	732	425.40
1256	C16H22O4	DIBUTYL PHTHALATE	230.175	1.5996E+00	-3.4574E-03	3.4963E-06	239	703	492.41
1257	C16H26	n-DECYLBENZENE	179.894	1.7867E+00	-4.0368E-03	3.9135E-06	260	678	457.48
1258	C16H26	PENTAETHYLBENZENE	456.054	-2.6062E-01	6.1397E-04	--------	329	580	--------
1259	C16H30	1-HEXADECYNE	550.396	-7.0803E-01	1.6120E-03	--------	289	587	482.60
1260	C16H32	n-DECYLCYCLOHEXANE	172.667	1.8606E+00	-4.1295E-03	3.9414E-06	272	676	464.79
1261	C16H32	1-CYCLOPENTYLUNDECANE	508.310	-5.5850E-01	1.4006E-03	--------	264	599	466.30
1262	C16H32	1-HEXADECENE	123.396	2.3682E+00	-5.4376E-03	5.2101E-06	279	650	484.19
1263	C16H32O2	n-HEXADECANOIC ACID	86.290	3.5237E+00	-7.3217E-03	6.1001E-06	337	698	--------
1264	C16H34	n-HEXADECANE	89.101	2.7062E+00	-6.1478E-03	5.7520E-06	292	649	501.90
1265	C16H34O	DI-n-OCTYL ETHER	203.551	2.3720E+00	-5.5593E-03	5.5102E-06	267	636	562.61
1266	C16H34O	1-HEXADECANOL	71.293	3.2851E+00	-6.9805E-03	6.0298E-06	323	685	--------
1267	C16H34S	BUTYL-DODECYL-SULFIDE	622.149	-7.1360E-01	1.6119E-03	--------	289	648	552.68
1268	C16H34S	ETHYL-TETRADECYL-SULFIDE	633.810	-7.9140E-01	1.7468E-03	--------	289	648	553.14
1269	C16H34S	METHYL-PENTADECYL-SULFIDE	622.149	-7.1360E-01	1.6119E-03	--------	289	648	552.68
1270	C16H34S	OCTYL-SULFIDE	622.149	-7.1360E-01	1.6119E-03	--------	289	648	552.68
1271	C16H34S	PROPYL-TRIDECYL-SULFIDE	622.149	-7.1360E-01	1.6119E-03	--------	289	648	552.68
1272	C16H34S	1-HEXADECANETHIOL	599.552	-6.8922E-01	1.5549E-03	--------	292	637	532.28
1273	C16H34S2	OCTYL-DISULFIDE	526.898	-4.5120E-01	1.2486E-03	--------	245	649	503.36
1274	C17H28	n-UNDECYLBENZENE	187.475	1.9638E+00	-4.3397E-03	4.1164E-06	269	688	496.30
1275	C17H32	1-HEPTADECYNE	588.957	-7.7667E-01	1.7148E-03	--------	296	602	509.82
1276	C17H34	1-CYCLOPENTYLDODECANE	542.579	-6.0580E-01	1.4824E-03	--------	269	614	493.73
1277	C17H34	1-CYCLOHEXYLUNDECANE	556.626	-6.3995E-01	1.5012E-03	--------	280	616	499.27
1278	C17H34	1-HEPTADECENE	137.993	2.6259E+00	-5.8785E-03	5.5338E-06	285	662	545.01
1279	C17H36	n-HEPTADECANE	113.571	2.8548E+00	-6.3960E-03	5.8757E-06	296	660	551.90
1280	C17H36O	1-HEPTADECANOL	50.508	3.5963E+00	-7.5402E-03	6.4420E-06	328	693	--------
1281	C17H36S	BUTYL-TRIDECYL-SULFIDE	660.667	-7.7595E-01	1.7050E-03	--------	295	662	580.88
1282	C17H36S	ETHYL-PENTADECYL-SULFIDE	660.667	-7.7595E-01	1.7050E-03	--------	295	662	580.88
1283	C17H36S	HEXADECYL-METHYL-SULFIDE	660.667	-7.7595E-01	1.7050E-03	--------	295	662	580.88
1284	C17H36S	PROPYL-TETRADECYL-SULFIDE	660.667	-7.7595E-01	1.7050E-03	--------	295	662	580.88
1285	C17H36S	1-HEPTADECANETHIOL	568.362	-2.2510E-01	6.2055E-04	--------	301	651	--------
1286	C18H12	CHRYSENE	-545.542	4.5719E+00	-7.0849E-03	4.0640E-06	532	881	--------
1287	C18H14	m-TERPHENYL	95.472	1.7763E+00	-3.1949E-03	2.3620E-06	361	832	--------
1288	C18H14	o-TERPHENYL	124.902	1.7076E+00	-3.2090E-03	2.5419E-06	330	802	--------
1289	C18H14	p-TERPHENYL	-341.819	3.8714E+00	-6.4058E-03	3.9822E-06	486	833	--------
1290	C18H15P	TRIPHENYLPHOSPHINE	257.405	9.1712E-01	-1.2998E-03	1.0300E-06	355	907	--------
1291	C18H15O4P	TRIPHENYL PHOSPHATE	--------	--------	--------	--------	--------	--------	--------
1292	C18H16N2	N,N'-DIPHENYL-p-PHENYLENEDIAMINE	-14.857	2.7179E+00	-4.8050E-03	3.3377E-06	410	815	--------
1293	C18H22	2,3-DIMETHYL-2,3-DIPHENYLBUTANE	-192.217	3.8745E+00	-7.5618E-03	5.6430E-06	393	725	--------
1294	C18H22O2	DICUMYL PEROXIDE	210.066	1.5142E+00	-2.8880E-03	2.3578E-06	312	796	--------
1295	C18H30	n-DODECYLBENZENE	202.922	2.0826E+00	-4.5475E-03	4.2038E-06	277	697	531.03
1296	C18H30	HEXAETHYLBENZENE	523.506	-4.5337E-01	8.1021E-04	--------	402	601	--------
1297	C18H32O2	LINOLEIC ACID	241.348	2.3065E+00	-5.0663E-03	4.7468E-06	269	698	604.48
1298	C18H34	1-OCTADECYNE	546.793	-2.5083E-01	6.7478E-04	--------	301	616	--------
1299	C18H34O2	OLEIC ACID	278.686	2.5434E+00	-5.4355E-03	4.9240E-06	288	703	684.32
1300	C18H34O4	DIBUTYL SEBACATE	235.231	2.5419E+00	-5.5054E-03	5.3382E-06	265	691	645.20
1301	C18H34O4	DIHEXYL ADIPATE	223.316	2.1857E+00	-4.7266E-03	4.6434E-06	260	690	577.90
1302	C18H36	1-CYCOPENTYLTRIDECANE	582.090	-6.7530E-01	1.5820E-03	--------	279	629	521.38
1303	C18H36	1-CYCLOHEXYLDODECANE	594.223	-7.0157E-01	1.5961E-03	--------	287	631	526.93
1304	C18H36	1-OCTADECENE	126.330	2.8997E+00	-6.4067E-03	5.8998E-06	292	673	577.72
1305	C18H36O2	STEARIC ACID	99.012	3.5874E+00	-7.2484E-03	5.9035E-06	344	719	--------
1306	C18H38	n-OCTADECANE	151.154	2.7878E+00	-6.1542E-03	5.5249E-06	302	671	--------
1307	C18H38O	DINONYL ETHER	350.435	1.5528E+00	-3.6072E-03	4.0513E-06	201	662	600.12
1308	C18H38O	1-OCTADECANOL	73.560	3.6272E+00	-7.5319E-03	6.3562E-06	332	699	--------
1309	C18H38S	BUTYL-TETRADECYL-SULFIDE	697.648	-8.2713E-01	1.7825E-03	--------	299	676	--------
1310	C18H38S	ETHYL-HEXADECYL-SULFIDE	697.648	-8.2713E-01	1.7825E-03	--------	299	676	--------
1311	C18H38S	HEPTADECYL-METHYL-SULFIDE	697.648	-8.2713E-01	1.7825E-03	--------	299	676	--------
1312	C18H38S	NONYL-SULFIDE	697.648	-8.2713E-01	1.7825E-03	--------	299	676	--------
1313	C18H38S	PENTADECYL-PROPYL-SULFIDE	697.648	-8.2713E-01	1.7825E-03	--------	299	676	--------
1314	C18H38S	1-OCTADECANETHIOL	591.688	-2.3207E-01	6.3942E-04	--------	302	663	--------
1315	C18H38S2	NONYL-DISULFIDE	581.970	-5.2612E-01	1.3841E-03	--------	253	672	548.15
1316	C19H26	1-n-NONYLNAPHTHALENE	201.833	1.8464E+00	-3.7278E-03	3.2462E-06	285	764	507.01
1317	C19H32	n-TRIDECYLBENZENE	167.086	2.4805E+00	-5.3327E-03	4.8612E-06	284	705	561.44
1318	C19H36	1-NONADECYNE	573.768	-2.6805E-01	7.0335E-04	--------	307	630	--------
1319	C19H36O2	METHYL OLEATE	183.562	2.9014E+00	-6.2576E-03	5.6990E-06	294	688	643.39
1320	C19H38	1-CYCLOPENTYLTETRADECANE	613.417	-7.1500E-01	1.6521E-03	--------	283	642	547.10
1321	C19H38	1-CYCLOHEXYLTRIDECANE	630.460	-7.5854E-01	1.6822E-03	--------	293	645	553.83
1322	C19H38	1-NONADECENE	127.073	3.0234E+00	-6.5615E-03	5.9422E-06	298	684	602.71
1323	C19H38O2	NONADECANOIC ACID	90.383	3.4899E+00	-6.9781E-03	5.6786E-06	342	729	--------
1324	C19H40	n-NONADECANE	118.433	3.2613E+00	-7.0875E-03	6.3030E-06	306	680	--------
1325	C19H40O	1-NONADECANOL	553.437	-2.9593E-01	6.8477E-04	--------	336	659	--------
1326	C19H40S	BUTYL-PENTADECYL-SULFIDE	644.561	-2.5109E-01	6.8360E-04	--------	304	689	--------

Table 3-1 HEAT CAPACITY OF LIQUID - ORGANIC COMPOUNDS (continued)

NO	FORMULA	NAME	$C_P = A + B\,T + C\,T^2 + D\,T^3$				$(C_P$ - joule/(mol K), T - K)		
			A	B	C	D	TMIN	TMAX	C_P @ 25 C
1327	C19H40S	ETHYL-HEPTADECYL-SULFIDE	644.561	-2.5109E-01	6.8360E-04	--------	304	689	--------
1328	C19H40S	HEXADECYL-PROPYL-SULFIDE	644.561	-2.5109E-01	6.8360E-04	--------	304	689	--------
1329	C19H40S	METHYL-OCTADECYL-SULFIDE	644.561	-2.5109E-01	6.8360E-04	--------	304	689	--------
1330	C19H40S	1-NONADECANETHIOL	616.944	-2.5008E-01	6.6863E-04	--------	308	675	--------
1331	C20H16	TRIPHENYLETHYLENE	141.784	1.8218E+00	-3.3616E-03	2.5723E-06	343	817	--------
1332	C20H28	1-n-DECYLNAPHTHALENE	206.361	1.9941E+00	-3.9778E-03	3.4222E-06	289	773	538.00
1333	C20H30O2	ABIETIC ACID	-670.074	7.2834E+00	-1.3314E-02	9.0556E-06	448	749	--------
1334	C20H31N	DEHYDROABIETYLAMINE	250.971	2.7890E+00	-5.4341E-03	4.4311E-06	319	777	--------
1335	C20H34	1-PHENYLTETRADECANE	573.751	-5.3762E-01	1.2674E-03	--------	290	657	526.12
1336	C20H38	1-EICOSYNE	599.040	-2.8118E-01	7.2846E-04	--------	310	643	--------
1337	C20H40	1-CYCLOPENTYLPENTADECANE	655.491	-8.1490E-01	1.8063E-03	--------	291	655	573.09
1338	C20H40	1-CYCLOHEXYLTETRADECANE	665.330	-8.1721E-01	1.7693E-03	--------	298	657	578.96
1339	C20H40	1-EICOSENE	129.465	3.2494E+00	-6.9467E-03	6.1932E-06	303	694	--------
1340	C20H42	n-EICOSANE	122.226	3.4187E+00	-7.3226E-03	6.4222E-06	311	690	--------
1341	C20H42O	1-EICOSANOL	129.055	3.8145E+00	-7.7934E-03	6.4028E-06	340	713	--------
1342	C20H42S	BUTYL-HEXADECYL-SULFIDE	670.841	-2.6868E-01	7.1249E-04	--------	309	701	--------
1343	C20H42S	DECYL-SULFIDE	670.841	-2.6868E-01	7.1249E-04	--------	309	701	--------
1344	C20H42S	ETHYL-OCTADECYL-SULFIDE	670.841	-2.6868E-01	7.1249E-04	--------	309	701	--------
1345	C20H42S	HEPTADECYL-PROPYL-SULFIDE	670.841	-2.6868E-01	7.1249E-04	--------	309	701	--------
1346	C20H42S	METHYL-NONADECYL-SULFIDE	670.841	-2.6868E-01	7.1249E-04	--------	309	701	--------
1347	C20H42S	1-EICOSANETHIOL	640.128	-2.6315E-01	6.9182E-04	--------	311	686	--------
1348	C20H42S2	DECYL-DISULFIDE	636.753	-6.0997E-01	1.5285E-03	--------	260	693	590.76
1349	C21H21O4P	TRI-o-CRESYL PHOSPHATE	--------	--------	--------	--------	--------	--------	--------
1350	C21H36	1-PHENYLPENTADECANE	607.875	-6.0106E-01	1.3706E-03	--------	296	669	550.51
1351	C21H42	1-CYCLOPENTYLHEXADECANE	686.835	-8.4813E-01	1.8437E-03	--------	295	667	597.86
1352	C21H42	1-CYCLOHEXYLPENTADECANE	614.217	-2.5150E-01	6.8235E-04	--------	303	670	--------
1353	C22H38	1-PHENYLHEXADECANE	579.507	-1.8746E-01	5.3794E-04	--------	301	681	--------
1354	C22H44	1-CYCLOHEXYLHEXADECANE	639.511	-2.6718E-01	7.0866E-04	--------	308	682	--------
1355	C22H44O2	n-BUTYL STEARATE	178.423	3.6670E+00	-7.9746E-03	7.1017E-06	300	688	--------
1356	C24H38O4	DIISOOCTYL PHTHALATE	288.702	2.7659E+00	-5.5170E-03	4.6621E-06	301	766	--------
1357	C24H38O4	DIOCTYL PHTHALATE	370.524	1.9804E+00	-4.1602E-03	4.2820E-06	224	725	704.66
1358	C24H42O	DINONYLPHENOL	268.894	3.0425E+00	-5.9340E-03	4.8935E-06	301	797	--------
1359	C26H20	TETRAPHENYLETHYLENE	-438.361	5.0591E+00	-7.9480E-03	4.7206E-06	497	896	--------
1360	C28H46O4	DIISODECYL PHTHALATE	530.837	1.9576E+00	-3.9030E-03	3.6707E-06	229	798	864.83

C_P - heat capacity of liquid, joule/(mol K)

A, B, C, and D - regression coefficients for chemical compound

T - temperature, K

TMIN - minimum temperature, K

TMAX - maximum temperature, K

Table 3-2 HEAT CAPACITY OF LIQUID - INORGANIC COMPOUNDS

NO	FORMULA	NAME	$C_P = A + B\,T + C\,T^2 + D\,T^3$						$(C_P$ - joule/(mol K), T - K)$	
			A	B	C	D	TMIN	TMAX	C_P @ 25 C	
1	Ag	SILVER	39.815	-1.1119E-02	3.5369E-06	-2.0574E-10	1234	5128	-------	
2	AgCl	SILVER CHLORIDE	61.723	1.6805E-02	-1.7007E-05	5.3955E-09	728	2394	-------	
3	AgI	SILVER IODIDE	54.778	1.3951E-02	-1.5305E-05	5.1269E-09	825	2318	-------	
4	Al	ALUMINUM	39.028	-9.9702E-03	3.4048E-06	-1.6971E-10	933	5721	-------	
5	AlB3H12	ALUMINUM BOROHYDRIDE	--------	----------	----------	----------	----	----	-------	
6	AlBr3	ALUMINUN BROMIDE	-50.050	1.3697E+00	-3.4966E-03	2.9198E-06	390	610	-------	
7	AlCl3	ALUMINUM CHLORIDE	29439.240	-1.7942E+02	3.6554E-01	-2.4786E-04	466	503	-------	
8	AlF3	ALUMINUM FLUORIDE	276.426	-2.3029E-01	1.0619E-04	-1.3480E-08	1313	2359	-------	
9	AlI3	ALUMINUM IODIDE	109.627	1.8446E-01	-5.2505E-04	4.3380E-07	464	786	-------	
10	Al2O3	ALUMINUM OXIDE	401.249	-1.7736E-01	4.5180E-05	-3.0999E-09	2325	4268	-------	
11	Al2S3O12	ALUMINUM SULFATE	--------	----------	----------	----------	----	----	-------	
12	Ar	ARGON	440.300	-1.0095E+01	8.1712E-02	-2.0216E-04	84	143	-------	
13	As	ARSENIC	--------	----------	----------	----------	----	----	-------	
14	AsBr3	ARSENIC TRIBROMIDE	178.296	-2.0895E-01	8.1465E-05	3.0548E-07	306	750	-------	
15	AsCl3	ARSENIC TRICHLORIDE	167.404	-2.8114E-01	2.4063E-04	3.8305E-07	255	621	115.12	
16	AsF3	ARSENIC TRIFLUORIDE	297.644	-1.3133E+00	2.6960E-03	-1.1127E-06	267	504	116.26	
17	AsF5	ARSENIC PENTAFLUORIDE	339.768	-2.2900E+00	6.9853E-03	-4.5162E-06	193	340	158.26	
18	AsH3	ARSINE	124.870	-7.8635E-01	2.7795E-03	-2.3029E-06	156	354	76.46	
19	AsI3	ARSENIC TRIIODIDE	--------	----------	----------	----------	----	----	-------	
20	As2O3	ARSENIC TRIOXIDE	--------	----------	----------	----------	----	----	-------	
21	At	ASTATINE	179.451	-5.9492E-01	8.2288E-04	-3.5570E-07	575	877	-------	
22	Au	GOLD	22.040	1.0159E-02	-4.9341E-06	7.7228E-10	1337	3518	-------	
23	B	BORON	44.678	-9.2975E-03	1.8846E-06	-8.6700E-11	2348	6348	-------	
24	BBr3	BORON TRIBROMIDE	151.727	-2.4545E-01	1.3117E-04	6.4708E-07	228	552	107.36	
25	BCl3	BORON TRICHLORIDE	91.710	-2.5307E-02	2.2522E-04	4.2585E-07	166	429	115.47	
26	BF3	BORON TRIFLUORIDE	229.392	-1.6881E+00	5.8510E-03	-1.3713E-06	146	248	-------	
27	BH2CO	BORINE CARBONYL	98.959	-3.7101E-01	7.8026E-04	1.1523E-06	136	323	88.24	
28	BH3O3	BORIC ACID	--------	----------	----------	----------	----	----	-------	
29	B2D6	DEUTERODIBORANE	164.803	-1.1667E+00	3.9201E-03	-1.5502E-06	170	279	-------	
30	B2H5Br	DIBORANE HYDROBROMIDE	135.027	-1.8405E-01	-9.7790E-05	1.4767E-06	169	444	110.60	
31	B2H6	DIBORANE	113.027	-5.1807E-01	1.5363E-03	1.5122E-06	108	275	-------	
32	B3N3H6	BORINE TRIAMINE	206.875	-5.3952E-01	7.7322E-04	5.7364E-07	215	495	129.95	
33	B4H10	TETRABORANE	110.148	-4.6499E-02	-4.4712E-04	1.6870E-06	153	443	101.25	
34	B5H9	PENTABORANE	198.765	-5.4259E-01	9.3825E-04	7.6148E-08	226	540	122.41	
35	B5H11	TETRAHYDROPENTABORANE	228.068	-6.1121E-01	9.0301E-04	4.0052E-07	230	520	136.72	
36	B10H14	DECABORANE	957.219	-3.6358E+00	6.5162E-03	-3.5033E-06	373	633	-------	
37	Ba	BARIUM	52.208	-1.7787E-02	7.3625E-06	-4.7512E-10	1000	2858	-------	
38	Be	BERYLLIUM	41.324	-1.2794E-02	3.8763E-06	-2.5998E-10	1560	4160	-------	
39	BeB2H8	BERYLLIUM BOROHYDRIDE	--------	----------	----------	----------	----	----	-------	
40	BeBr2	BERYLLIUM BROMIDE	1915.718	-6.2339E+00	7.0923E-03	-2.6455E-06	763	973	-------	
41	BeCl2	BERYLLIUM CHLORIDE	887.654	-2.7540E+00	3.2132E-03	-1.2045E-06	678	990	-------	
42	BeF2	BERYLLIUM FLUORIDE	-1720.570	6.2543E+00	-7.1497E-03	2.7221E-06	825	990	-------	
43	BeI2	BERYLLIUM IODIDE	1591.358	-5.0808E+00	5.7324E-03	-2.1137E-06	761	990	-------	
44	Bi	BISMUTH	34.049	-9.3007E-03	4.8099E-06	-3.8013E-10	544	3696	-------	
45	BiBr3	BISMUTH TRIBROMIDE	--------	----------	----------	----------	----	----	-------	
46	BiCl3	BISMUTH TRICHLORIDE	--------	----------	----------	----------	----	----	-------	
47	BrF5	BROMINE PENTAFLUORIDE	260.255	-5.9106E-01	3.1250E-04	2.1047E-06	212	447	167.59	
48	Br2	BROMINE	165.688	-6.5579E-01	1.3794E-03	-6.0832E-07	266	555	76.66	
49	C	CARBON	--------	----------	----------	----------	----	----	-------	
50	CCl2O	PHOSGENE	148.648	-4.2379E-01	7.8886E-04	4.8919E-07	145	432	105.39	
51	CF2O	CARBONYL FLUORIDE	483.524	-4.8061E+00	1.9191E-02	-2.2284E-05	166	282	-------	
52	CH4N2O	UREA	965.507	-5.0993E+00	1.0028E-02	-6.3799E-06	406	564	-------	
53	CH4N2S	THIOUREA	18.182	6.4960E-01	-1.1931E-03	9.1094E-07	454	811	-------	
54	CNBr	CYANOGEN BROMIDE	--------	----------	----------	----------	----	----	-------	
55	CNCl	CYANOGEN CHLORIDE	125.478	-1.0770E-02	-1.2687E-03	3.1665E-06	267	427	93.41	
56	CNF	CYANOGEN FLUORIDE	228.038	-1.4944E+00	4.4216E-03	-2.7795E-06	200	350	101.88	
57	CO	CARBON MONOXIDE	125.595	-1.7022E+00	1.0707E-02		68	126	-------	
58	COS	CARBONYL SULFIDE	86.329	-9.2189E-03	-1.1739E-03	4.1043E-06	134	378	88.01	
59	COSe	CARBON OXYSELENIDE	208.225	-1.1581E+00	3.0770E-03	-1.5229E-06	200	386	96.09	
60	CO2	CARBON DIOXIDE	-3981.020	5.2511E+01	-2.2708E-01	3.2866E-04	220	290	-------	
61	CS2	CARBON DISULFIDE	94.329	-1.5208E-01	2.1058E-04	3.2259E-07	164	540	76.25	
62	CSeS	CARBON SELENOSULFIDE	110.492	-6.3675E-02	-2.2469E-04	8.2437E-07	198	548	93.38	
63	C2N2	CYANOGEN	273.325	-1.5946E+00	4.1012E-03	-1.9851E-06	239	380	109.85	
64	C3S2	CARBON SUBSULFIDE	--------	----------	----------	----------	----	----	-------	
65	Ca	CALCIUM	57.358	-3.6594E-02	1.7589E-05	-2.1846E-09	1115	2634	-------	
66	CaF2	CALCIUM FLUORIDE	123.067	-7.3372E-03	-8.7482E-06	3.1959E-09	1691	3657	-------	
67	CbF5	COLUMBIUM FLUORIDE	--------	----------	----------	----------	----	----	-------	
68	Cd	CADMIUM	37.742	-2.2836E-02	1.8623E-05	-3.7534E-09	594	1833	-------	
69	CdCl2	CADMIUM CHLORIDE	171.964	-1.4422E-01	8.5079E-05	-9.0911E-09	841	1616	-------	
70	CdF2	CADMIUM FLUORIDE	91.946	2.2686E-02	-2.0834E-05	6.0000E-09	793	2637	-------	
71	CdI2	CADMIUM IODIDE	122.814	-4.5555E-02	6.4210E-06	1.6731E-08	658	1393	-------	
72	CdO	CADMIUM OXIDE	--------	----------	----------	----------	----	----	-------	
73	ClF	CHLORINE MONOFLUORIDE	142.655	-9.9143E-01	3.6076E-03	-1.4201E-06	128	268	-------	
74	ClFO3	PERCHLORYL FLUORIDE	78.590	1.3804E-01	-9.2420E-04	2.9980E-06	125	350	117.05	
75	ClF3	CHLORINE TRIFLUORIDE	172.729	-5.2736E-01	8.9492E-04	6.1428E-07	190	436	111.33	
76	ClF5	CHLORINE PENTAFLUORIDE	242.074	-1.0746E+00	2.7506E-03	-1.0820E-06	190	395	137.51	
77	ClHO3S	CHLOROSULFONIC ACID	140.318	1.6349E-02	-3.5109E-04	7.0485E-07	193	665	132.66	
78	ClHO4	PERCHLORIC ACID	97.789	9.2028E-02	-2.8693E-04	7.5547E-07	172	599	119.74	

Table 3-2 HEAT CAPACITY OF LIQUID - INORGANIC COMPOUNDS (continued)

			$C_P = A + B\,T + C\,T^2 + D\,T^3$				$(C_P$ - joule/(mol K), T - K)		
NO	FORMULA	NAME	A	B	C	D	TMIN	TMAX	C_P @ 25 C
79	ClO2	CHLORINE DIOXIDE	186.153	-8.0657E-01	1.8068E-03	-5.0986E-07	214	442	92.77
80	Cl2	CHLORINE	127.601	-6.0215E-01	1.5776E-03	-5.3099E-07	172	396	74.23
81	Cl2O	CHLORINE MONOXIDE	111.657	-1.3358E-01	-1.8421E-04	1.5449E-06	157	422	96.40
82	Cl2O7	CHLORINE HEPTOXIDE	176.297	-2.3104E-02	-6.0715E-04	1.6442E-06	182	537	159.01
83	Co	COBALT	50.685	-1.0565E-02	3.1419E-06	-2.0959E-10	1768	5919	-------
84	CoCl2	COBALT CHLORIDE	283.472	-3.8809E-01	2.5473E-04	-4.9830E-08	1008	1724	-------
85	CoNC3O4	COBALT NITROSYL TRICARBONYL	--------	----------	----------	----------	----	----	-------
86	Cr	CHROMIUM	48.762	-8.5202E-03	2.2133E-06	-1.2704E-10	2180	6849	-------
87	CrC6O6	CHROMIUM CARBONYL	--------	----------	----------	----------	----	----	-------
88	CrO2Cl2	CHROMIUM OXYCHLORIDE	133.278	6.7177E-02	-5.9599E-04	1.1291E-06	177	595	130.25
89	Cs	CESIUM	37.749	-1.6562E-02	1.3078E-05	9.0457E-10	302	1638	-------
90	CsBr	CESIUM BROMIDE	84.805	-3.6931E-03	-1.0394E-05	6.6543E-09	909	2050	-------
91	CsCl	CESIUM CHLORIDE	86.258	-6.5198E-03	-8.6203E-06	6.2930E-09	919	2050	-------
92	CsF	CESIUM FLUORIDE	95.415	-3.2010E-02	7.6367E-06	3.1870E-09	956	1986	-------
93	CsI	CESIUM IODIDE	76.146	-2.5229E-03	-1.0173E-05	6.3563E-09	894	2024	-------
94	Cu	COPPER	30.283	4.8707E-03	-2.8866E-06	5.3382E-10	1358	4098	-------
95	CuBr	CUPROUS BROMIDE	63.087	1.6494E-02	-2.0332E-05	7.5408E-09	777	2121	-------
96	CuCl	CUPROUS CHLORIDE	49.489	4.8240E-02	-4.2532E-05	1.2010E-08	703	1948	-------
97	CuCl2	CUPRIC CHLORIDE	193.681	-1.8981E-01	1.1338E-04	-1.4983E-08	906	1608	-------
98	CuI	COPPER IODIDE	68.360	6.4358E-03	-1.4481E-05	6.5529E-09	878	2097	-------
99	DCN	DEUTERIUM CYANIDE	255.987	-1.3002E+00	2.9514E-03	-1.4668E-06	261	458	91.82
100	D2	DEUTERIUM	189.904	-2.5841E+01	1.2676E+00	-1.9855E-02	19	23	-------
101	D2O	DEUTERIUM OXIDE	463.373	-2.6757E+00	5.3338E-03	-3.4321E-06	277	612	48.80
102	Eu	EUROPIUM	46.324	-1.2855E-02	5.6539E-06	-5.3980E-10	1095	4120	-------
103	F2	FLUORINE	83.829	-7.8518E-01	5.2305E-03	4.6617E-06	53	137	-------
104	F2O	FLUORINE OXIDE	58.551	5.6117E-02	-1.7926E-03	1.1093E-05	49	204	-------
105	Fe	IRON	63.677	-1.5748E-02	3.8125E-06	-1.6110E-10	1808	7472	-------
106	FeC5O5	IRON PENTACARBONYL	228.921	-5.1929E-01	6.4325E-04	3.8803E-07	252	577	141.56
107	FeCl2	FERROUS CHLORIDE	228.216	-2.6823E-01	1.7289E-04	-3.0882E-08	945	1693	-------
108	FeCl3	FERRIC CHLORIDE	1095.001	-4.3838E+00	6.5061E-03	-3.1125E-06	577	772	-------
109	Fr	FRANCIUM	32.919	-2.0235E-03	-3.9538E-06	7.4549E-09	300	1285	-------
110	Ga	GALLIUM	29.054	-3.9678E-03	1.4912E-06	-3.3666E-11	303	6096	-------
111	GaCl3	GALLIUM TRICHLORIDE	93.606	4.6518E-01	-1.6817E-03	1.8248E-06	351	555	-------
112	Gd	GADOLINIUM	133.256	-1.4211E-01	6.6794E-05	-9.6818E-09	1587	2646	-------
113	Ge	GERMANIUM	31.272	-3.9632E-03	1.1025E-06	-5.7618E-11	1211	6720	-------
114	GeBr4	GERMANIUM BROMIDE	209.183	-3.3541E-01	2.7356E-04	2.9745E-07	299	703	-------
115	GeCl4	GERMANIUM CHLORIDE	182.720	-3.1903E-01	2.3749E-04	7.2826E-07	224	545	128.01
116	GeHCl3	TRICHLORO GERMANE	140.720	-1.4616E-01	-1.1764E-04	9.4426E-07	202	532	111.71
117	GeH4	GERMANE	74.535	-1.9734E-01	2.7866E-04	1.8467E-06	107	293	-------
118	Ge2H6	DIGERMANE	134.135	-6.9779E-02	-4.4032E-04	1.6999E-06	164	466	119.24
119	Ge3H8	TRIGERMANE	168.171	1.0374E-01	-8.2277E-04	1.5394E-06	168	586	166.76
120	HBr	HYDROGEN BROMIDE	-99.048	2.0735E+00	-8.9095E-03	1.2673E-05	186	345	63.04
121	HCN	HYDROGEN CYANIDE	252.213	-1.4144E+00	3.0625E-03	-1.1827E-06	260	434	71.40
122	HCl	HYDROGEN CHLORIDE	73.993	-1.2946E-01	-7.8980E-05	2.6409E-06	165	308	98.37
123	HF	HYDROGEN FLUORIDE	24.415	2.0382E-01	-9.4339E-04	1.8975E-06	190	438	51.61
124	HI	HYDROGEN IODIDE	136.724	-7.2705E-01	1.9425E-03	-1.0279E-06	222	403	65.39
125	HNO3	NITRIC ACID	214.478	-7.6762E-01	1.4970E-03	-3.0208E-07	239	494	110.68
126	H2	HYDROGEN	50.607	-6.1136E+00	3.0930E-01	-4.1480E-03	14	32	-------
127	H2O	WATER	92.053	-3.9953E-02	-2.1103E-04	5.3469E-07	273	615	75.55
128	H2O2	HYDROGEN PEROXIDE	-15.248	6.7693E-01	-1.4948E-03	1.2018E-06	273	694	85.55
129	H2S	HYDROGEN SULFIDE	80.985	-1.2464E-01	-3.6053E-05	1.6942E-06	191	355	85.52
130	H2SO4	SULFURIC ACID	26.004	7.0337E-01	-1.3856E-03	1.0342E-06	298	879	139.95
131	H2S2	HYDROGEN DISULFIDE	125.844	-6.4965E-02	-3.2435E-04	1.1672E-06	183	515	108.58
132	H2Se	HYDROGEN SELENIDE	336.560	-3.0726E+00	1.1266E-02	-1.2985E-05	209	329	77.73
133	H2Te	HYDROGEN TELLURIDE	203.374	-1.1010E+00	2.7510E-03	-1.4212E-06	224	416	82.00
134	H3NO3S	SULFAMIC ACID	--------	----------	----------	----------	----	----	-------
135	He	HELIUM-3	13.021	-1.9604E+01	1.3057E+01	-2.1265E+00	1	3	-------
136	He	HELIUM-4	20.743	5.4005E+00	-7.5537E+00	1.5064E+00	2	5	-------
137	Hf	HAFNIUM	41.378	-2.2063E-03	3.6408E-07	-1.4709E-11	2506	9990	-------
138	Hg	MERCURY	30.388	-1.0980E-02	9.4412E-06	6.7418E-10	234	1648	27.97
139	HgBr2	MERCURIC BROMIDE	617.602	-2.3963E+00	3.5985E-03	-1.7236E-06	510	772	-------
140	HgCl2	MERCURIC CHLORIDE	898.706	-3.7271E+00	5.6832E-03	-2.8002E-06	550	752	-------
141	HgI2	MERCURIC IODIDE	460.181	-1.5511E+00	2.1514E-03	-9.3670E-07	532	862	-------
142	IF7	IODINE HEPTAFLUORIDE	748.653	-4.7652E+00	1.1750E-02	-7.9388E-06	279	425	161.97
143	I2	IODINE	314.003	-1.3711E+00	2.5657E-03	-1.4911E-06	387	655	-------
144	In	INDIUM	33.107	-5.9093E-03	2.3041E-06	-6.7111E-11	430	5384	-------
145	Ir	IRIDIUM	47.311	-4.7730E-03	8.6371E-07	-3.7176E-11	2719	9990	-------
146	K	POTASSIUM	35.955	-1.6870E-02	1.3313E-05	-1.8000E-09	336	2112	-------
147	KBr	POTASSIUM BROMIDE	83.136	-1.5270E-02	-1.2769E-06	3.7604E-09	1003	2158	-------
148	KCl	POTASSIUM CHLORIDE	188.929	-1.8986E-01	8.7872E-05	-8.9111E-09	1044	2776	-------
149	KF	POTASSIUM FLUORIDE	102.150	-4.2533E-02	1.4102E-05	4.2341E-10	1153	2313	-------
150	KI	POTASSIUM IODIDE	91.930	-2.7363E-02	5.2787E-06	3.0087E-09	996	2081	-------
151	KOH	POTASSIUM HYDROXIDE	71.429	4.2195E-02	-4.8017E-05	1.7182E-08	679	2085	-------
152	Kr	KRYPTON	46.229	-1.9828E-03	-1.1330E-03	8.6959E-06	116	199	-------
153	La	LANTHANUM	38.456	-4.1375E-03	1.0175E-06	-4.1321E-11	1193	7609	-------
154	Li	LITHIUM	36.127	-1.3610E-02	7.1002E-06	-2.2351E-10	454	3268	-------
155	LiBr	LITHIUM BROMIDE	63.897	1.1805E-02	-1.8001E-05	7.4021E-09	820	2063	-------
156	LiCl	LITHIUM CHLORIDE	65.139	8.1102E-03	-1.4493E-05	6.0732E-09	887	2157	-------

Table 3-2 HEAT CAPACITY OF LIQUID - INORGANIC COMPOUNDS (continued)

			$C_P = A + B T + C T^2 + D T^3$				(C_P - joule/(mol K), T - K)		
NO	FORMULA	NAME	A	B	C	D	TMIN	TMAX	C_P @ 25 C
157	LiF	LITHIUM FLUORIDE	71.665	-4.5728E-03	-4.5210E-06	2.7042E-09	1143	2546	-------
158	LiI	LITHIUM IODIDE	68.927	1.7582E-02	-2.6135E-05	1.1257E-08	719	1882	-------
159	Lu	LUTECIUM	97.002	-6.9833E-02	2.3955E-05	-2.4553E-09	1936	3303	-------
160	Mg	MAGNESIUM	71.613	-6.2494E-02	1.6395E-05	6.8222E-09	923	1793	-------
161	MgCl2	MAGNESIUM CHLORIDE	102.090	-6.3316E-03	-9.3842E-06	6.1211E-09	985	2203	-------
162	MgO	MAGNESIUM OXIDE	602.235	-3.6437E-01	7.5796E-05	-4.4385E-09	3105	4760	-------
163	Mn	MANGANESE	56.925	-1.2358E-02	3.9700E-06	-2.8287E-10	1519	5522	-------
164	MnCl2	MANGANESE CHLORIDE	123.626	-4.6248E-02	1.3019E-05	4.0716E-09	923	1906	-------
165	Mo	MOLYBDENUM	52.902	-8.8880E-03	1.4542E-06	-5.2798E-11	2895	7696	-------
166	MoF6	MOLYBDENUM FLUORIDE	403.709	-1.8394E+00	3.8023E-03	-1.3783E-06	290	473	156.76
167	MoO3	MOLYBDENUM OXIDE	332.436	-4.0139E-01	2.4240E-04	-4.2807E-08	1068	1856	-------
168	NCl3	NITROGEN TRICHLORIDE	192.643	-6.1039E-01	1.0601E-03	-6.3674E-08	246	536	103.21
169	ND3	HEAVY AMMONIA	237.684	-1.4495E+00	4.0877E-03	-2.3781E-06	199	369	105.85
170	NF3	NITROGEN TRIFLUORIDE	72.634	3.9006E-02	-1.8012E-03	1.0334E-05	66	222	-------
171	NH3	AMMONIA	-182.157	3.3618E+00	-1.4398E-02	2.0371E-05	195	385	80.16
172	NH3O	HYDROXYLAMINE	488.200	-2.1067E+00	3.8692E-03	-1.5822E-06	306	545	-------
173	NH4Br	AMMONIUM BROMIDE	--------	----------	----------	----------	----	----	-------
174	NH4Cl	AMMONIUM CHLORIDE	--------	----------	----------	----------	----	----	-------
175	NH4I	AMMONIUM IODIDE	--------	----------	----------	----------	----	----	-------
176	NH5O	AMMONIUM HYDROXIDE	--------	----------	----------	----------	----	----	-------
177	NH5S	AMMONIUM HYDROGENSULFIDE	--------	----------	----------	----------	----	----	-------
178	NO	NITRIC OXIDE	9860.145	-2.1310E+02	1.5107E+00	-3.4573E-03	113	171	-------
179	NOCl	NITROSYL CHLORIDE	181.178	-8.2714E-01	1.9451E-03	-6.3093E-07	214	419	90.75
180	NOF	NITROSYL FLUORIDE	136.591	-4.5808E-01	8.3378E-04	1.7172E-06	139	335	119.64
181	NO2	NITROGEN DIOXIDE	-916.569	9.0550E+00	-2.5996E-02	2.5257E-05	298	410	141.72
182	N2	NITROGEN	76.452	-3.5226E-01	-2.6690E-03	5.0057E-05	64	120	-------
183	N2F4	TETRAFLUOROHYDRAZINE	111.713	-2.3800E-02	-1.4518E-03	6.6483E-06	112	294	-------
184	N2H4	HYDRAZINE	71.025	2.4083E-01	-8.2965E-04	1.1247E-06	275	620	98.89
185	N2H4C	AMMONIUM CYANIDE	407.226	-1.9783E+00	4.0513E-03	-1.6224E-06	309	467	-------
186	N2H6CO2	AMMONIUM CARBAMATE	--------	----------	----------	----------	----	----	-------
187	N2O	NITROUS OXIDE	220.901	-1.6908E+00	5.6320E-03	-3.6589E-06	182	294	-------
188	N2O3	NITROGEN TRIOXIDE	149.838	-4.4366E-02	-1.5130E-03	4.8628E-06	170	404	131.00
189	N2O4	NITROGEN TETRAOXIDE	397.502	-2.0622E+00	4.3168E-03	-8.9910E-07	262	410	142.57
190	N2O5	NITROGEN PENTOXIDE	341.495	-1.4532E+00	2.8336E-03	-8.4240E-07	303	490	-------
191	Na	SODIUM	22.431	3.5211E-02	-4.4481E-05	1.6321E-08	371	2444	-------
192	NaBr	SODIUM BROMIDE	133.301	-1.1628E-01	5.3381E-05	-5.3139E-09	1020	3430	-------
193	NaCN	SODIUM CYANIDE	69.087	3.8057E-02	-4.1570E-05	1.3878E-08	837	2320	-------
194	NaCl	SODIUM CHLORIDE	95.016	-3.1081E-02	9.6789E-07	5.5116E-09	1074	3230	-------
195	NaF	SODIUM FLUORIDE	47.103	1.9241E-02	-6.2969E-06	2.8598E-09	1269	5254	-------
196	NaI	SODIUM IODIDE	72.610	-6.1075E-03	-6.7740E-06	5.1489E-09	924	2055	-------
197	NaOH	SODIUM HYDROXIDE	87.639	-4.8368E-04	-4.5423E-06	1.1863E-09	600	2700	-------
198	Na2SO4	SODIUM SULFATE	233.515	-9.5276E-03	-3.4665E-05	1.5771E-08	1157	3515	-------
199	Nb	NIOBIUM	38.868	-3.2254E-03	5.5151E-07	-2.3346E-11	2750	9990	-------
200	Nd	NEODYMIUM	53.558	-4.6994E-03	1.1701E-06	-5.1290E-11	1289	8532	-------
201	Ne	NEON	179.694	-1.2115E+01	2.9508E-01	-1.5396E-03	25	42	-------
202	Ni	NICKEL	56.355	-1.2849E-02	3.9925E-06	-2.8218E-10	1728	5589	-------
203	NiC4O4	NICKEL CARBONYL	734.180	-4.6820E+00	1.3029E-02	-1.1077E-05	248	407	202.82
204	NiF2	NICKEL FLUORIDE	666.230	-7.5541E-01	3.3374E-04	-4.6949E-08	1723	2623	-------
205	Np	NEPTUNIUM	--------	----------	----------	----------	----	----	-------
206	O2	OXYGEN	46.432	3.9506E-01	-7.0522E-03	3.9897E-05	54	147	-------
207	O3	OZONE	40.279	8.1284E-01	-4.6303E-03	1.3033E-05	90	248	-------
208	Os	OSMIUM	45.089	-5.1072E-03	8.4646E-07	-3.6497E-11	3306	9990	-------
209	OsOF5	OSMIUM OXIDE PENTAFLUORIDE	--------	----------	----------	----------	----	----	-------
210	OsO4	OSMIUM TETROXIDE - YELLOW	--------	----------	----------	----------	----	----	-------
211	OsO4	OSMIUM TETROXIDE - WHITE	--------	----------	----------	----------	----	----	-------
212	P	PHOSPHORUS - WHITE	36.970	-5.8048E-02	9.0422E-05	-3.3467E-08	317	795	-------
213	PBr3	PHOSPHORUS TRIBROMIDE	131.023	1.0737E-02	-3.5824E-04	6.7756E-07	233	675	120.34
214	PCl2F3	PHOSPHORUS DICHLORIDE TRIFLUORIDE	340.032	-1.6902E+00	3.8192E-03	-1.5226E-06	265	434	135.23
215	PCl3	PHOSPHORUS TRICHLORIDE	177.073	-4.6599E-01	8.2800E-04	1.8546E-07	240	535	116.66
216	PCl5	PHOSPHORUS PENTACHLORIDE	--------	----------	----------	----------	----	----	-------
217	PH3	PHOSPHINE	120.475	-7.7634E-01	2.8214E-03	-1.6233E-06	139	309	96.79
218	PH4Br	PHOSPHONIUM BROMIDE	109.154	6.3133E-02	-7.3509E-04	1.7678E-06	140	477	109.49
219	PH4Cl	PHOSPHONIUM CHLORIDE	2871.578	-2.7564E+01	9.1944E-02	-9.4995E-05	245	306	308.79
220	PH4I	PHOSPHONIUM IODIDE	322.737	-1.4722E+00	2.9923E-03	-1.3427E-06	292	513	114.22
221	POCl3	PHOSPHORUS OXYCHLORIDE	194.159	-4.7684E-01	1.0467E-03	-2.1262E-07	274	572	139.40
222	PSBr3	PHOSPHORUS THIOBROMIDE	--------	----------	----------	----------	----	----	-------
223	PSCl3	PHOSPHORUS THIOCHLORIDE	170.596	-1.8690E-01	-3.1267E-05	6.9512E-07	237	607	130.52
224	P4O6	PHOSPHORUS TRIOXIDE	341.331	-6.3516E-01	6.3560E-04	4.0286E-07	296	679	219.14
225	P4O10	PHOSPHORUS PENTOXIDE	--------	----------	----------	----------	----	----	-------
226	P4S10	PHOSPHORUS PENTASULFIDE	1211.094	-2.1879E+00	2.2969E-03	-6.4517E-07	561	1033	-------
227	Pb	LEAD	34.491	-7.7801E-03	3.4152E-06	-2.0984E-10	601	4320	-------
228	PbBr2	LEAD BROMIDE	114.209	1.5383E-02	-4.5212E-05	2.7510E-08	646	1547	-------
229	PbCl2	LEAD CHLORIDE	145.729	-6.4904E-02	2.1731E-05	8.1624E-09	774	1599	-------
230	PbF2	LEAD FLUORIDE	232.369	-2.1673E-01	1.1455E-04	-1.6364E-08	1128	2041	-------
231	PbI2	LEAD IODIDE	122.822	-1.7437E-02	-1.8626E-05	2.1713E-08	675	1492	-------
232	PbO	LEAD OXIDE	100.294	-5.2676E-02	2.0670E-05	-9.4681E-10	1163	2274	-------
233	PbS	LEAD SULFIDE	490.517	-7.4455E-01	4.2487E-04	-7.7903E-08	1387	2025	-------
234	Pd	PALLADIUM	39.908	-4.3823E-03	1.0119E-06	-4.6195E-11	1828	8535	-------

Table 3-2 HEAT CAPACITY OF LIQUID - INORGANIC COMPOUNDS (continued)

NO	FORMULA	NAME	$C_P = A + B\,T + C\,T^2 + D\,T^3$				$(C_P$ - joule/(mol K), T - K)		
			A	B	C	D	TMIN	TMAX	C_P @ 25 C
235	Po	POLONIUM	36.879	-1.4670E-02	1.0312E-05	-1.4413E-09	527	2410	-------
236	Pt	PLATINUM	40.386	-3.6997E-03	4.1239E-07	5.3674E-11	2042	5586	-------
237	Ra	RADIUM	--------	----------	----------	----------	----	----	-------
238	Rb	RUBIDIUM	37.152	-2.0347E-02	1.8127E-05	-3.4898E-09	313	1600	-------
239	RbBr	RUBIDIUM BROMIDE	75.323	-6.8236E-03	-6.1573E-06	4.7732E-09	955	2117	-------
240	RbCl	RUBIDIUM CHLORIDE	74.142	-1.0286E-02	-3.3799E-06	3.8837E-09	988	2155	-------
241	RbF	RUBIDIUM FLUORIDE	89.516	-2.0592E-02	1.5492E-06	3.2149E-09	1033	2190	-------
242	RbI	RUBIDIUM IODIDE	73.791	-4.0609E-03	-8.3953E-06	5.5997E-09	915	2055	-------
243	Re	RHENIUM	--------	----------	----------	----------	----	----	-------
244	Re2O7	RHENIUM HEPTOXIDE	--------	----------	----------	----------	----	----	-------
245	Rh	RHODIUM	48.042	-4.4407E-03	8.7824E-07	-3.5375E-11	2237	9990	-------
246	Rn	RADON	53.121	-1.2264E-01	1.2936E-04	8.2668E-07	202	359	49.97
247	Ru	RUTHENIUM	36.856	-3.5038E-03	6.3786E-07	-2.7334E-11	2607	9990	-------
248	RuF5	RUTHENIUM PENTAFLUORIDE	--------	----------	----------	----------	----	----	-------
249	S	SULFUR	108.050	-2.3743E-01	2.2729E-04	-5.9964E-08	443	1247	-------
250	SF4	SULFUR TETRAFLUORIDE	161.219	-4.2384E-01	2.9705E-04	2.8834E-06	149	346	137.68
251	SF6	SULFUR HEXAFLUORIDE	1614.139	-1.6346E+01	5.7400E-02	-6.3610E-05	222	303	157.07
252	SOBr2	THIONYL BROMIDE	131.854	-3.6263E-02	-2.8115E-04	7.2193E-07	221	629	115.18
253	SOCl2	THIONYL CHLORIDE	116.818	-2.0658E-02	-3.7051E-04	1.0653E-06	172	539	105.96
254	SOF2	SULFUROUS OXYFLUORIDE	163.384	-7.7319E-01	1.9849E-03	-4.3363E-08	163	353	108.15
255	SO2	SULFUR DIOXIDE	203.445	-1.0537E+00	2.6113E-03	-1.0697E-06	198	409	93.06
256	SO2Cl2	SULFURYL CHLORIDE	176.168	-2.1659E-01	-3.0189E-05	1.0404E-06	222	436	136.48
257	SO3	SULFUR TRIOXIDE	5064.851	-4.1901E+01	1.1959E-01	-1.1117E-04	290	393	256.97
258	S2Cl2	SULFUR MONOCHLORIDE	123.162	4.5179E-02	-4.6347E-04	8.6402E-07	193	626	118.33
259	Sb	ANTIMONY	38.280	-1.1202E-02	4.8901E-06	-4.3762E-10	904	4056	-------
260	SbBr3	ANTIMONY TRIBROMIDE	202.175	-3.6922E-01	4.8373E-04	-1.0419E-07	370	714	-------
261	SbCl3	ANTIMONY TRICHLORIDE	214.359	-4.4810E-01	4.3365E-04	1.5743E-07	347	754	-------
262	SbCl5	ANTIMONY PENTACHLORIDE	247.892	-5.1771E-01	5.8833E-04	3.2025E-07	276	629	154.32
263	SbH3	STIBINE	112.882	-4.4438E-01	1.1052E-03	-2.9859E-07	185	418	70.72
264	SbI3	ANTIMONY TRIIODIDE	206.905	-2.3740E-01	2.1509E-04	6.0243E-09	440	878	-------
265	Sb2O3	ANTIMONY TRIOXIDE	--------	----------	----------	----------	----	----	-------
266	Sc	SCANDIUM	41.100	-7.4985E-03	2.0970E-06	-1.2866E-10	1814	6428	-------
267	Se	SELENIUM	50.628	-4.9239E-02	4.1165E-05	-5.2086E-09	494	1413	-------
268	SeCl4	SELENIUM TETRACHLORIDE	418.234	-1.3945E+00	2.0611E-03	-6.8241E-07	400	707	-------
269	SeF6	SELENIUM HEXAFLUORIDE	521.512	-3.9988E+00	1.1778E-02	-9.4441E-06	238	350	125.94
270	SeOCl2	SELENIUM OXYCHLORIDE	176.321	-2.7503E-01	2.0506E-04	3.2930E-07	282	671	121.28
271	SeO2	SELENIUM DIOXIDE	--------	----------	----------	----------	----	----	-------
272	Si	SILICON	16.634	1.3656E-02	-5.7334E-06	7.8271E-10	1685	4127	-------
273	SiBrCl2F	BROMODICHLOROFLUOROSILANE	130.232	-3.2357E-02	-5.3260E-04	1.7311E-06	161	472	119.12
274	SiBrF3	TRIFLUOROBROMOSILANE	268.919	-1.7333E+00	5.0455E-03	-3.1338E-06	203	357	117.60
275	SiBr2ClF	DIBROMOCHLOROFLUOROSILANE	145.181	-3.2456E-02	-5.1834E-04	1.5533E-06	174	509	130.60
276	SiClF3	TRIFLUOROCHLOROSILANE	139.074	-5.1789E-01	1.0735E-03	1.9250E-06	131	314	131.11
277	SiCl2F2	DICHLORODIFLUOROSILANE	122.470	-1.2816E-01	-4.1457E-04	2.7333E-06	133	371	119.85
278	SiCl3F	TRICHLOROFLUOROSILANE	133.585	-6.6696E-02	-5.2185E-04	2.0831E-06	152	437	122.52
279	SiCl4	SILICON TETRACHLORIDE	170.175	-1.7026E-01	-4.0824E-04	1.8411E-06	204	482	131.92
280	SiF4	SILICON TETRAFLUORIDE	2140.088	-2.6918E+01	1.1633E-01	-1.6111E-04	186	246	-------
281	SiHBr3	TRIBROMOSILANE	135.537	-1.5375E-02	-3.4581E-04	8.6226E-07	200	580	123.07
282	SiHCl3	TRICHLOROSILANE	115.924	8.4832E-02	-1.0480E-03	2.5710E-06	145	455	116.20
283	SiHF3	TRIFLUOROSILANE	173.894	-1.3300E+00	4.9431E-03	-3.3017E-06	142	276	-------
284	SiH2Br2	DIBROMOSILANE	137.271	-1.6406E-01	-4.8843E-05	8.6699E-07	203	523	106.99
285	SiH2Cl2	DICHLOROSILANE	106.133	-2.9755E-02	-5.6554E-04	1.9803E-06	151	427	99.48
286	SiH2F2	DIFLUOROSILANE	151.417	-1.0050E+00	3.3413E-03	-1.6945E-06	150	302	103.89
287	SiH2I2	DIIODOSILANE	187.585	-2.8898E-01	1.6334E-04	4.9458E-07	272	627	129.05
288	SiH3Br	MONOBROMOSILANE	127.539	-3.6623E-01	5.9877E-04	6.1058E-07	179	431	87.76
289	SiH3Cl	MONOCHLOROSILANE	111.792	-3.3048E-01	4.9100E-04	1.1515E-06	155	377	87.42
290	SiH3F	MONOFLUOROSILANE	161.642	-1.3357E+00	5.1056E-03	-4.0023E-06	150	272	-------
291	SiH3I	IODOSILANE	160.802	-4.6642E-01	7.2389E-04	3.7261E-07	216	489	95.96
292	SiH4	SILANE	83.087	-3.3736E-01	8.8970E-04	2.6155E-06	88	256	-------
293	SiO2	SILICON DIOXIDE	343.910	-2.8233E-01	9.4498E-05	-9.0006E-09	1883	3261	-------
294	Si2Cl6	HEXACHLORODISILANE	303.025	-6.0011E-01	6.3865E-04	4.7265E-07	272	628	193.40
295	Si2F6	HEXAFLUORODISILANE	567.874	-3.6969E+00	9.6950E-03	-6.3974E-06	255	391	157.92
296	Si2H5Cl	DISILANYL CHLORIDE	302.034	-1.3530E+00	2.8779E-03	-1.1496E-06	250	482	123.99
297	Si2H6	DISILANE	120.754	-1.6259E-01	-8.0012E-05	1.6528E-06	141	410	108.97
298	Si2OCl3F3	TRICHLOROTRIFLUORODISILOXANE	171.753	5.7533E-02	-9.9486E-04	2.5152E-06	150	483	167.13
299	Si2OCl6	HEXACHLORODISILOXANE	234.785	-2.2292E-01	-1.1141E-04	9.5103E-07	240	623	183.62
300	Si2OH6	DISILOXANE	117.796	-5.2777E-03	-7.9248E-04	2.8120E-06	129	396	120.31
301	Si3Cl8	OCTACHLOROTRISILANE	252.739	-5.3122E-03	-5.1185E-04	9.5352E-07	250	737	230.93
302	Si3H8	TRISILANE	134.027	4.0684E-02	-7.2105E-04	1.7761E-06	156	499	129.13
303	Si3H9N	TRISILAZANE	177.472	-3.7205E-02	-6.8654E-04	2.1050E-06	167	492	161.14
304	Si4H10	TETRASILANE	186.279	5.1391E-02	-7.7967E-04	1.6714E-06	180	569	176.59
305	Sm	SAMARIUM	79.831	-2.8249E-02	1.1464E-05	-1.1147E-09	1345	4066	-------
306	Sn	TIN	32.185	-3.8132E-03	1.1297E-06	9.7579E-12	505	5920	-------
307	SnBr4	STANNIC BROMIDE	216.159	-3.0485E-01	1.9892E-04	3.3125E-07	304	727	-------
308	SnCl2	STANNOUS CHLORIDE	101.457	-9.7805E-03	-3.5809E-05	4.1992E-08	520	1168	-------
309	SnCl4	STANNIC CHLORIDE	197.452	-1.6537E-01	-1.1620E-04	8.5095E-07	243	496	160.37
310	SnH4	STANNIC HYDRIDE	98.002	-1.2797E-01	-3.1437E-04	2.7016E-06	123	341	103.60
311	SnI4	STANNIC IODIDE	267.087	-4.2069E-01	4.8074E-04	-8.4231E-08	418	809	-------
312	Sr	STRONTIUM	52.462	-2.1746E-02	1.0468E-05	-1.2018E-09	1050	3414	-------

Table 3-2 HEAT CAPACITY OF LIQUID - INORGANIC COMPOUNDS (continued)

| NO | FORMULA | NAME | $C_P = A + BT + CT^2 + DT^3$ | | | | (C_P - joule/(mol K), T - K) | | |
			A	B	C	D	TMIN	TMAX	C_P @ 25 C
313	SrO	STRONTIUM OXIDE	1393.962	-1.2874E+00	4.1182E-04	-4.3325E-08	2703	3414	-------
314	Ta	TANTALUM	50.312	-4.5380E-03	7.1533E-07	-2.9852E-11	3290	9990	-------
315	Tc	TECNNETIUM	47.318	-3.4901E-03	6.2149E-07	-2.6106E-11	2430	9990	-------
316	Te	TELLURIUM	41.053	-7.4313E-03	4.3966E-06	-4.4035E-10	723	3872	-------
317	TeCl4	TELLURIUM TETRACHLORIDE	571.987	-1.4606E+00	1.8839E-03	-7.0713E-07	497	867	-------
318	TeF6	TELLURIUM HEXAFLUORIDE	415.124	-2.9443E+00	8.3781E-03	-6.0631E-06	235	361	121.35
319	Ti	TITANIUM	64.889	-1.5128E-02	3.5781E-06	-1.7230E-10	1941	5120	-------
320	TiCl4	TITANIUM TETRACHLORIDE	167.628	-1.3344E-01	-2.2324E-04	8.7538E-07	249	606	131.20
321	Tl	THALLIUM	39.777	-1.1537E-02	5.7637E-06	-4.3596E-10	577	3718	-------
322	TlBr	THALLOUS BROMIDE	105.945	-9.3797E-02	6.0583E-05	-5.7640E-09	733	1423	-------
323	TlI	THALLOUS IODIDE	102.619	-7.0163E-02	3.8114E-05	1.4728E-09	713	1428	-------
324	Tm	THULIUM	43.955	-1.1992E-02	3.9977E-06	-3.1232E-10	1818	5027	-------
325	U	URANIUM	41.409	-2.5172E-03	5.4060E-07	-2.0495E-11	1408	9990	-------
326	UF6	URANIUM FLUORIDE	7076.942	-5.5326E+01	1.4726E-01	-1.2951E-04	342	405	-------
327	V	VANADIUM	53.424	-5.4286E-03	1.1304E-06	-4.6992E-11	2183	9430	-------
328	VCl4	VANADIUM TETRACHLORIDE	180.041	-2.2949E-01	1.8695E-04	3.1641E-07	247	662	136.62
329	VOCl3	VANADIUM OXYTRICHLORIDE	143.763	2.2899E-02	-4.9188E-04	1.0313E-06	194	604	134.20
330	W	TUNGSTEN	49.874	-7.1744E-03	1.0698E-06	-4.2705E-11	3695	9990	-------
331	WF6	TUNGSTEN FLUORIDE	383.816	-1.8493E+00	4.0558E-03	-1.5331E-06	273	445	152.35
332	Xe	XENON	48.554	-7.0584E-02	-2.3703E-04	2.6119E-06	161	275	-------
333	Yb	YTTERBIUM	49.209	-2.0206E-02	9.4875E-06	-1.0677E-09	1097	3493	-------
334	Yt	YTTRIUM	50.741	-6.9676E-03	1.7585E-06	-9.1826E-11	1799	7505	-------
335	Zn	ZINC	50.774	-3.5075E-02	1.5935E-05	-1.6078E-09	693	3012	-------
336	ZnCl2	ZINC CHLORIDE	134.015	-7.8300E-02	3.5706E-05	1.1410E-08	638	1310	-------
337	ZnF2	ZINC FLUORIDE	145.678	-5.8528E-02	1.8809E-05	9.1413E-10	1145	2306	-------
338	ZnO	ZINC OXIDE	--------	----------	----------	----------	----	----	-------
339	ZnSO4	ZINC SULFATE	--------	----------	----------	----------	----	----	-------
340	Zr	ZIRCONIUM	49.471	-5.1497E-03	8.5190E-07	-9.9895E-12	2128	7042	-------
341	ZrBr4	ZIRCONIUM BROMIDE	11832.940	-4.4975E+01	5.7514E-02	-2.4424E-05	723	821	-------
342	ZrCl4	ZIRCONIUM CHLORIDE	5534.857	-2.0485E+01	2.5785E-02	-1.0725E-05	710	787	-------
343	ZrI4	ZIRCONIUM IODIDE	5621.564	-1.9348E+01	2.2655E-02	-8.7732E-06	772	917	-------

C_P - heat capacity of liquid, joule/(mol K)

A, B, C, and D - regression coefficients for chemical compound

T - temperature, K

TMIN - minimum temperature, K

TMAX - maximum temperature, K

Chapter 4

HEAT CAPACITY OF SOLID

Carl L. Yaws, Deepa R. Balundgi, and Saumya Tripathi
Lamar University, Beaumont, Texas

ABSTRACT

Results for heat capacity of solid as a function of temperature are presented for major organic and inorganic chemicals. The results cover a wide temperature range and include many types of compounds. The agreement between correlation and data is quite good.

INTRODUCTION

Thermodynamic properties such as heat capacity are important in the engineering design of chemical processes. In unit operations involving solids at elevated temperatures, the heat capacity is required to determine the energy (heat) necessary to bring the solids up to the required processing temperature. Additional uses include heat exchanger and energy balance design calculations.

In this article, correlation results for heat capacity of solids are provided in an easy-to-use tabular format that is especially applicable for rapid engineering use with the personal computer or hand calculator.

HEAT CAPACITY CORRELATION

The correlation for heat capacity of solid is a series expansion in temperature:

$$C_P = A + B\,T + C\,T^2 \tag{4-1}$$

where C_P = heat capacity of solid, joule/(mol K)

A, B, and C = regression coefficients for chemical compound

T = temperature, K

The results for heat capacity of solid are given in Tables 4-1 and 4-2. The tabulations are applicable to a wide variety of substances.

In preparing the compilation, a literature search was conducted to identify data source publications for organics (1-38) and inorganics (1-104). Both experimental values for the property under consideration and parameter values for estimation of the property are included in the source publications. The publications were screened for appropriate data. The compilation resulting from the screening is based on both experimental data and estimated values. For organics, many of the values are based on sources from DIPPR (4). For inorganics, many of the values are based on sources from JANAF tables (57-59) and Thermophysical Properties of Matter (23). The estimates are primarily based on empirical methods of the senior author. Experimental data and estimates were then regressed to provide the same equation for all compounds.

Very limited experimental data for solid heat capacity are available at very low temperatures. Thus, the estimated values at very low temperatures should be considered rough approximations. The values for substances that are solids at room temperature are more accurate.

A comparison of correlation and actual data values for heat capacity is shown in Fig. 4-1 for a representative chemical. The graph indicates good agreement of correlation and data.

EXAMPLES

The correlation results maybe used for prediction and calculation of heat capacity and additional thermodynamic properties. Examples are given below.

Example 1 Estimate the solid heat capacity of phenol (C6H6O) at 298.15 K.

Substitution of the coefficients from the table and temperature into the equation for heat capacity yields

$C_P = 9.769 + (4.0832E-01)\,(298.15) - (1.9001E-05)\,(298.15^2)$

$C_P = 129.82$ joule/(mol K)

Example 2 Calculate the energy required to heat solid naphthalene (C10H8) from 100 to 300 K.

From thermodynamics, the change in enthalpy, ΔH, at constant pressure is

$\Delta H = \int C_P\, dT = \int (A + B\,T + C\,T^2)\, dT$

$\Delta H = A\,T + (B/2)\,T^2 + (C/3)\,T^3\ \Big]_{T_1}^{T_2}$

Substitution of the coefficients from the table and the temperature limits into the equation yields

$$\Delta H = (4.824)(300 - 100) + (5.0634E\text{-}01/2)(300^2 - 100^2) + (1.8503E\text{-}04/3)(300^3 - 100^3)$$

$$\Delta H = 22{,}820 \text{ joule/mol}$$

REFERENCES - ORGANIC COMPOUNDS

1-36. See **REFERENCES - ORGANIC COMPOUNDS** in **Chapter 2, HEAT CAPACITY OF GAS**

37. Stull, D. R., E. F. Westrum, Jr., and G. C. Sinke, THE CHEMICAL THERMODYNAMICS OF ORGANIC COMPOUNDS, John Wiley and Sons, New York, NY (1969).

38. Stull, D. R., and H. Prophet, Project Directors, JANAF THERMOCHEMICAL TABLES, 2nd edition, NSRDS-NBS 37, U.S. Government Printing Office, Washington DC (1971).

REFERENCES - INORGANIC COMPOUNDS

1-64. See **REFERENCES - INORGANIC COMPOUNDS** in **Chapter 2, HEAT CAPACITY OF GAS**

65. Lyon, R. N., ed., LIQUID-METALS HANDBOOK, 2nd ed., Atomic Energy Commission, Washington, DC (1954).

66. Fink, J. K., TABLES OF THERMODYNAMIC PROPERTIES OF SODIUM, Argonne National Lab., ANL-CEN-RSD-82-4, Chemical Engineering Division, Argonne, IL (June 1982).

67. Mills, K. C., THERMODYNAMIC DATA FOR INORGANIC SULPHIDES SELENIDES AND TELLURIDES, Butterworths, London, England (1974).

68. Booth, H. S. and D. R. Martin, BORON TRIFLUORIDE AND ITS DERIVATIVES, John Wiley and Sons, Inc., New York (1949).

69. McCarty, R. D., J. Hord, and H. M. Roder, SELECTED PROPERTIES OF HYDROGEN, Center of Chemical Engineering, National Engineering Laboratory, Nat. Bur. Stand. Monograph 168, Boulder, CO (1981).

70. Kaufmann, D. W., PHYSICAL PROPERTIES OF SODIUM CHLORIDE IN CRYSTAL, LIQUID, GAS AND ACQUEOUS SOLUTION STATES, Reinhold Pub. Corp., New York, NY (1960).

71. Roder, H. M. and L. A. Weber, eds., ASRDI OXYGEN TECHNOLOGY SURVEY, VOL. 1 : THERMAL PHYSICAL PROPERTIES, NASA-SP-3071, National Aeronautics and Space Administration, Washington, DC (1972).

72. Leadbetter, A. J., J. Phys. C : Solid State Physics, 1, 1481 (1968).

73. Long, E. A. and J. D. Kemp, J. Chem. Soc., 58(10), 1829 (1936).

74. Hu, J., D. White, and H. L. Johnston, J. Amer. Chem. Soc., 15, 5642 (1953).

75. Giauque, W. F. and R. Wiebe, J. Amer. Chem. Soc., 50, 2193 (1928).

76. Giguere, P. A. and others, Can. J. Chem., 32, 117 (1954).

77. Lindenberg, A. B., Comptes Rend. Acad. Sci. Paris, 273, 1017 (1971).

78. Damphinee, T. M. and D. L. Martin, Proc. Roy. Soc., Ser. A233, 214 (1955).

79. Krier, C. A., R.S. Craig, and W. E. Wallace, J. Phys. Chem., 61, 522 (1957).

80. Stull, D. R. and others, J. Chem. Eng. Data, 15, 52 (1970).

81. Leu, A. L. and others, Proc. Nat. Acad. Sci. USA, 72(3), 1026 (1975).

82. Brock, F. H., J. Am. Rocket Soc., 31(2), 265 (1966).

83. Overstreet, R. and W. F. Giauque, J. Amer. Chem. Soc., 59, 254 (1937).

84. Ziegler, W. T. and C. E. Messer, J. Amer. Chem. Soc., 63, 2694 (1941).

85. Sirkar, S. C. and J. Gupta, Indian J. of Phys., 12, 145 (1938).

86. McGraw, J., J. Amer. Chem. Soc., 53, 3683 (1931).

87. Brodale, G. E. and W. F. Giauque, J. Phys. Chem., 76(5), 737 (1972).

88. Paukov, I. E. and M. N. Laurent'eva, Russ. J. Phys. Chem., 43(8), (1969).

89. Shmidt, N. E., Russ. J. Inorg. Chem., 12(7), 929 (1967).

90. Brown, J. C., J. Chem. Soc. (LONDON), 987 (1903).

91. Stephenson, C. C. and others, J. Chem. Thermo., 1, 59 (1969).

92. Dewar, J. D., Proc. Roy Soc., 89A, 158 (1913).

93. Chihara, H., M. Nakamura, and K. Masukane, Bull. Chem. Soc. Japan, 46, 97 (1973).

94. Andon, R. J. L. and others, Trans. Faraday Soc., 59, 2702 (1963).

95. Clever, H. L., E. F. Westrum, and A. W. Cordes, J. Phys. Chem., 69, 1214 (1965).

96. Chao, T., Hydrocarbon Processing, 217 (Nov., 1980).

97. Eastman, E. D. and W. C. McGavock, J. Am. Chem. Soc., 59, 145 (1937).

98. Latimar, W. M., J. Amer. Chem. Soc., 44, 90 (1922).

99. Klein, M. L., J. A. Morrison, and R. D. Weir, Trans. Faraday Soc., 48, 93 (1969).

100. Anderson, C. T., J. Amer. Chem. Soc., 58, 568 (1936).

101. Wietzel, V. R., A. Anorg. Allg. Chem., 116, 71 (1921).

102. Miller, R. W., J. Amer. Chem. Soc., 50, 2653 (1928).

103. Magnus, V. A., Z. Phys., 14, 5 (1913).

104. Krestovnikov, A. N. and E. J. Feigina, Z. Obsch. Khim., 6, 1481 (1936).

Figure 4-1 Heat Capacity of Solid

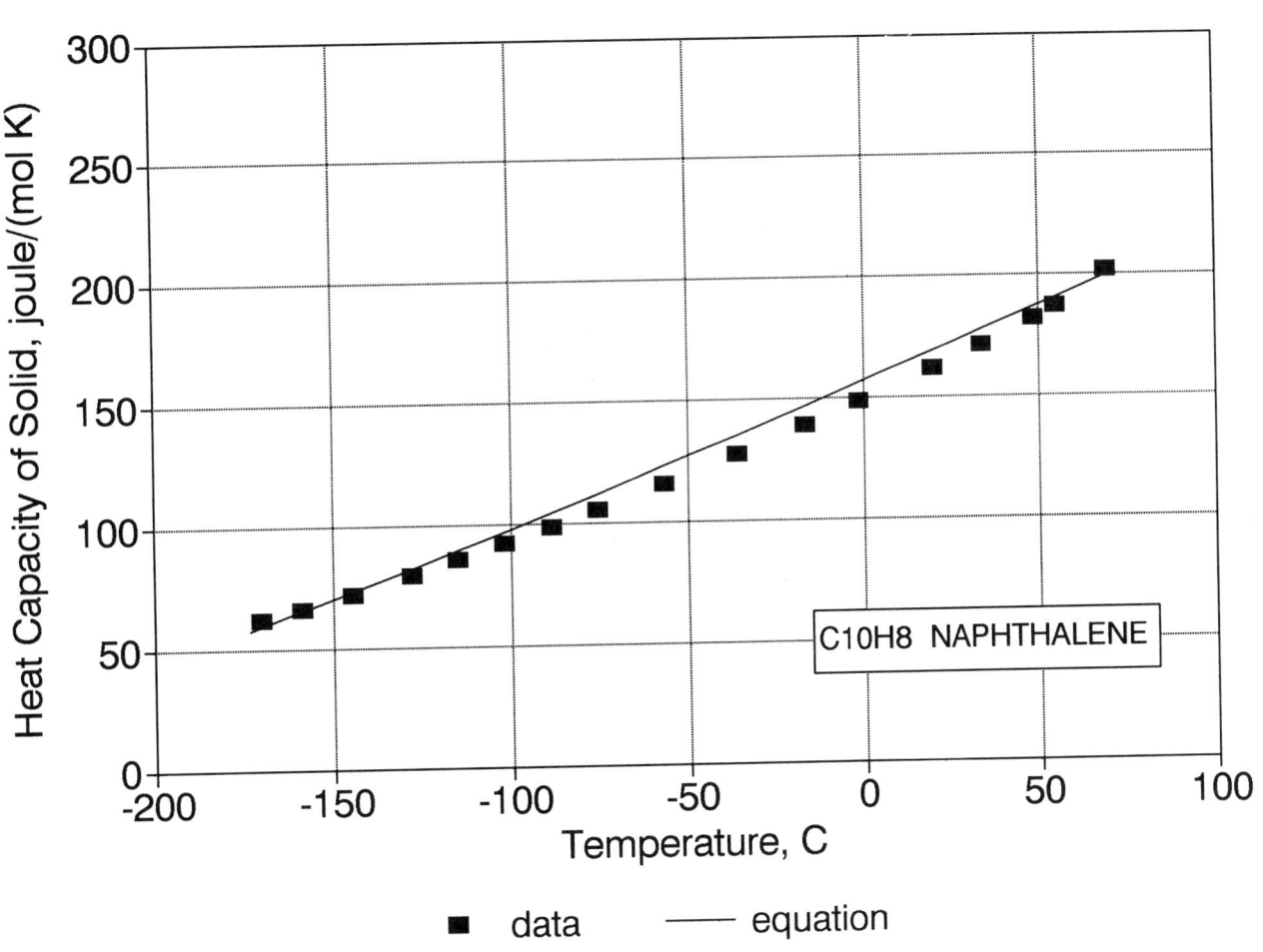

C10H8 NAPHTHALENE

■ data   —— equation

Table 4-1 HEAT CAPACITY OF SOLID - ORGANIC COMPOUNDS

NO	FORMULA	NAME	$C_P = A + B T + C T^2$					$(C_P$ - joule/(mol K), T - K$)$		
			A	B	C	TMIN	TMAX	C_P @25 C	C_P @TMIN	C_P @TMAX
1	CBrClF2	BROMOCHLORODIFLUOROMETHANE	-------	----------	----------	----	----	-------	-------	-------
2	CBrCl3	BROMOTRICHLOROMETHANE	-24.693	4.6458E-01	0.0000E+00	263	293	-------	97.563	111.500
3	CBrF3	BROMOTRIFLUOROMETHANE	-------	----------	----------	----	----	-------	-------	-------
4	CBr2F2	DIBROMODIFLUOROMETHANE	-------	----------	----------	----	----	-------	-------	-------
5	CClF3	CHLOROTRIFLUOROMETHANE	-------	----------	----------	----	----	-------	-------	-------
6	CClN	CYANOGEN CHLORIDE	-6.607	2.4792E-01	0.0000E+00	237	267	-------	52.063	59.500
7	CCl2F2	DICHLORODIFLUOROMETHANE	-------	----------	----------	----	----	-------	-------	-------
8	CCl2O	PHOSGENE	-2.144	9.6996E-01	-3.6747E-03	50	140	-------	37.167	61.654
9	CCl3F	TRICHLOROFLUOROMETHANE	-------	----------	----------	----	----	-------	-------	-------
10	CCl4	CARBON TETRACHLORIDE	36.706	2.7022E-01	3.3356E-04	90	240	-------	63.724	120.767
11	CF2O	CARBONYL FLUORIDE	139.852	-1.6982E+00	8.1497E-03	100	155	-------	51.530	72.347
12	CF4	CARBON TETRAFLUORIDE	-0.815	1.0828E+00	-3.1832E-04	15	76	-------	14.948	79.878
13	CHBr3	TRIBROMOMETHANE	-16.326	3.9625E-01	0.0000E+00	251	281	-------	83.213	95.100
14	CHClF2	CHLORODIFLUOROMETHANE	15.600	5.7600E-01	-1.5600E-03	30	116	-------	31.476	61.367
15	CHCl2F	DICHLOROFLUOROMETHANE	-------	----------	----------	----	----	-------	-------	-------
16	CHCl3	CHLOROFORM	-------	----------	----------	----	----	-------	-------	-------
17	CHF3	TRIFLUOROMETHANE	0.034	9.2330E-01	-2.9612E-03	20	118	-------	17.316	67.745
18	CHI3	TRIIODOMETHANE	-------	----------	----------	----	----	-------	-------	-------
19	CHN	HYDROGEN CYANIDE	-2.179	4.0519E-01	-6.4276E-04	100	254	-------	31.912	59.286
20	CHNS	ISOTHIOCYANIC-ACID	-------	----------	----------	----	----	-------	-------	-------
21	CH2BrCl	BROMOCHLOROMETHANE	-------	----------	----------	----	----	-------	-------	-------
22	CH2Br2	DIBROMOMETHANE	6.362	3.2792E-01	0.0000E+00	191	221	-------	68.863	78.700
23	CH2ClF	CHLOROFLUOROMETHANE	-------	----------	----------	----	----	-------	-------	-------
24	CH2Cl2	DICHLOROMETHANE	-------	----------	----------	----	----	-------	-------	-------
25	CH2F2	DIFLUOROMETHANE	-------	----------	----------	----	----	-------	-------	-------
26	CH2I2	DIIODOMETHANE	-12.871	3.2792E-01	0.0000E+00	249	279	-------	68.863	78.700
27	CH2O	FORMALDEHYDE	-------	----------	----------	----	----	-------	-------	-------
28	CH2O2	FORMIC ACID	-1.690	4.1677E-01	-6.9790E-04	100	275	-------	33.008	60.142
29	CH3Br	METHYL BROMIDE	-2.906	8.1876E-01	-2.7920E-03	15	168	-------	8.460	55.820
30	CH3Cl	METHYL CHLORIDE	-6.229	7.6944E-01	-2.0175E-03	80	175	-------	42.414	66.663
31	CH3Cl3Si	METHYL TRICHLOROSILANE	-------	----------	----------	----	----	-------	-------	-------
32	CH3F	METHYL FLUORIDE	-------	----------	----------	----	----	-------	-------	-------
33	CH3I	METHYL IODIDE	-------	----------	----------	----	----	-------	-------	-------
34	CH3NO	FORMAMIDE	25.314	6.0039E-02	6.0278E-04	85	260	-------	34.725	81.746
35	CH3NO2	NITROMETHANE	43.784	3.7261E-02	5.6099E-04	95	235	-------	52.331	83.660
36	CH3NO2	METHYL-NITRITE	-------	----------	----------	----	----	-------	-------	-------
37	CH3NO3	METHYL-NITRATE	-------	----------	----------	----	----	-------	-------	-------
38	CH4	METHANE	5.270	7.1386E-01	-3.2653E-03	21	90	-------	18.821	43.069
39	CH4Cl2Si	METHYL DICHLOROSILANE	-------	----------	----------	----	----	-------	-------	-------
40	CH4O	METHANOL	-7.783	7.1864E-01	-1.9224E-03	30	151	-------	12.046	56.899
41	CH4O3S	METHANESULFONIC ACID	-26.405	5.0000E-01	0.0000E+00	263	293	-------	105.000	120.000
42	CH4S	METHYL MERCAPTAN	-6.063	7.5863E-01	-1.8296E-03	100	150	-------	51.504	66.602
43	CH5ClSi	METHYL CHLOROSILANE	-------	----------	----------	----	----	-------	-------	-------
44	CH5N	METHYLAMINE	-8.069	6.4888E-01	-1.5032E-03	80	175	-------	34.221	59.420
45	CH6Si	METHYL SILANE	-------	----------	----------	----	----	-------	-------	-------
46	CN4O8	TETRANITROMETHANE	-48.520	1.0313E+00	0.0000E+00	257	287	-------	216.563	247.500
47	CO	CARBON MONOXIDE	-6.144	9.5427E-01	2.1432E-03	8	62	-------	1.627	61.259
48	COS	CARBONYL SULFIDE	-1.660	7.8895E-01	-3.0815E-03	80	131	-------	41.734	48.809
49	CO2	CARBON DIOXIDE	-1.630	5.4167E-01	-1.2689E-03	80	210	-------	33.583	56.162
50	CS2	CARBON DISULFIDE	0.504	7.0300E-01	-2.2463E-03	88	161	-------	44.989	55.457
51	C2BrF3	BROMOTRIFLUOROETHYLENE	-------	----------	----------	----	----	-------	-------	-------
52	C2Br2F4	1,2-DIBROMOTETRAFLUOROETHANE	-------	----------	----------	----	----	-------	-------	-------
53	C2ClF3	CHLOROTRIFLUOROETHYLENE	-0.617	1.0755E+00	-3.8793E-03	16	115	-------	15.588	71.756
54	C2ClF5	CHLOROPENTAFLUOROETHANE	102.280	-2.5508E-01	2.2733E-03	80	174	-------	96.449	126.567
55	C2Cl2F4	1,2-DICHLOROTETRAFLUOROETHANE	-------	----------	----------	----	----	-------	-------	-------
56	C2Cl3F3	1,1,2-TRICHLOROTRIFLUOROETHANE	1.446	4.4500E-01	0.0000E+00	207	237	-------	93.450	106.800
57	C2Cl4	TETRACHLOROETHYLENE	3.919	5.7213E-01	0.0000E+00	203	233	-------	120.146	137.310
58	C2Cl4F2	1,1,2,2-TETRACHLORODIFLUOROETHANE	-37.869	7.1250E-01	0.0000E+00	263	293	-------	149.625	171.000
59	C2Cl4O	TRICHLOROACETYL CHLORIDE	-------	----------	----------	----	----	-------	-------	-------
60	C2Cl6	HEXACHLOROETHANE	-41.759	7.1813E-01	0.0000E+00	268	298	172.350	150.806	172.350
61	C2F4	TETRAFLUOROETHYLENE	11.319	6.1422E-01	-6.4173E-04	30	142	-------	29.168	85.598
62	C2F6	HEXAFLUOROETHANE	144.000	-1.1141E+00	5.5049E-03	111	164	-------	88.152	109.099
63	C2HBrClF3	HALOTHANE	-------	----------	----------	----	----	-------	-------	-------
64	C2HClF2	2-CHLORO-1,1-DIFLUOROETHYLENE	40.648	3.8583E-01	0.0000E+00	105	135	-------	81.025	92.600
65	C2HCl3	TRICHLOROETHYLENE	-------	----------	----------	----	----	-------	-------	-------
66	C2HCl3O	DICHLOROACETYL CHLORIDE	-------	----------	----------	----	----	-------	-------	-------
67	C2HCl3O	TRICHLOROACETALDEHYDE	-------	----------	----------	----	----	-------	-------	-------
68	C2HCl5	PENTACHLOROETHANE	-2.673	6.4417E-01	0.0000E+00	214	244	-------	135.275	154.600
69	C2HF3	TRIFLUOROETHENE	-------	----------	----------	----	----	-------	-------	-------
70	C2HF3O2	TRIFLUOROACETIC ACID	-9.069	5.0667E-01	0.0000E+00	228	258	-------	106.400	121.600
71	C2HF5	PENTAFLUOROETHANE	-------	----------	----------	----	----	-------	-------	-------
72	C2H2	ACETYLENE	-------	----------	----------	----	----	-------	-------	-------
73	C2H2Br4	1,1,2,2-TETRABROMOETHANE	-11.906	3.5917E-01	0.0000E+00	243	273	-------	75.425	86.200
74	C2H2Cl2	1,1-DICHLOROETHYLENE	0.953	7.8296E-01	-2.0509E-03	83	151	-------	51.719	72.350
75	C2H2Cl2	cis-1,2-DICHLOROETHYLENE	-------	----------	----------	----	----	-------	-------	-------
76	C2H2Cl2	trans-1,2-DICHLOROETHYLENE	5.980	3.5917E-01	0.0000E+00	193	223	-------	75.425	86.200
77	C2H2Cl2O	CHLOROACETYL CHLORIDE	-4.795	4.3000E-01	0.0000E+00	221	251	-------	90.300	103.200
78	C2H2Cl2O	DICHLOROACETALDEHYDE	7.296	4.2917E-01	0.0000E+00	193	223	-------	90.125	103.000

86

Table 4-1 HEAT CAPACITY OF SOLID - ORGANIC COMPOUNDS (continued)

			$C_P = A + B T + C T^2$					$(C_P$ - joule/(mol K), T - K)		
NO	FORMULA	NAME	A	B	C	TMIN	TMAX	C_P @25 C	C_P @TMIN	C_P @TMAX
79	C2H2Cl2O2	DICHLOROACETIC ACID	27.888	3.5339E-01	-6.0008E-05	94	267	-------	60.437	118.005
80	C2H2Cl3F	1,1,1-TRICHLOROFLUOROETHANE	-------	----------	----------	----	----	-------	-------	-------
81	C2H2Cl4	1,1,1,2-TETRACHLOROETHANE	-------	----------	----------	----	----	-------	-------	-------
82	C2H2Cl4	1,1,2,2-TETRACHLOROETHANE	6.133	5.7583E-01	0.0000E+00	199	229	-------	120.925	138.200
83	C2H2F2	1,1-DIFLUOROETHYLENE	-------	----------	----------	----	----	-------	-------	-------
84	C2H2F2	cis-1,2-DIFLUOROETHENE	-------	----------	----------	----	----	-------	-------	-------
85	C2H2F2	trans-1,2-DIFLUOROETHENE	-------	----------	----------	----	----	-------	-------	-------
86	C2H2F4	1,1,1,2-TETRAFLUOROETHANE	-------	----------	----------	----	----	-------	-------	-------
87	C2H2O	KETENE	-------	----------	----------	----	----	-------	-------	-------
88	C2H2O4	OXALIC ACID	0.206	5.0483E-01	-4.6577E-04	40	323	109.316	19.817	114.671
89	C2H3Br	VINYL BROMIDE	-------	----------	----------	----	----	-------	-------	-------
90	C2H3Cl	VINYL CHLORIDE	-------	----------	----------	----	----	-------	-------	-------
91	C2H3ClF2	1-CHLORO-1,1-DIFLUOROETHANE	-------	----------	----------	----	----	-------	-------	-------
92	C2H3ClO	ACETYL CHLORIDE	-------	----------	----------	----	----	-------	-------	-------
93	C2H3ClO	CHLOROACETALDEHYDE	-------	----------	----------	----	----	-------	-------	-------
94	C2H3ClO2	CHLOROACETIC ACID	-55.750	5.9850E-01	0.0000E+00	303	333	-------	125.685	143.640
95	C2H3ClO2	METHYL CHLOROFORMATE	-------	----------	----------	----	----	-------	-------	-------
96	C2H3Cl3	1,1,1-TRICHLOROETHANE	7.486	6.3038E-01	-4.9575E-04	100	243	-------	65.566	131.297
97	C2H3Cl3	1,1,2-TRICHLOROETHANE	-87.513	1.6088E+00	-2.8320E-03	130	222	-------	73.774	130.129
98	C2H3F	VINYL FLUORIDE	-------	----------	----------	----	----	-------	-------	-------
99	C2H3F3	1,1,1-TRIFLUOROETHANE	2.523	8.2317E-01	-1.9302E-03	79	146	-------	55.578	81.553
100	C2H3N	ACETONITRILE	-3.723	5.6409E-01	-8.1975E-04	32	217	-------	13.273	80.062
101	C2H3NO	METHYL ISOCYANATE	-5.841	3.6167E-01	0.0000E+00	226	256	-------	75.950	86.800
102	C2H4	ETHYLENE	-0.519	4.0383E-02	-1.5314E-04	16	75	-------	0.087	1.643
103	C2H4Br2	1,1-DIBROMOETHANE	-------	----------	----------	----	----	-------	-------	-------
104	C2H4Br2	1,2-DIBROMOETHANE	14.010	5.0846E-01	-1.3967E-04	132	248	-------	78.476	131.443
105	C2H4Cl2	1,1-DICHLOROETHANE	1.286	7.8013E-01	-1.8134E-03	90	166	-------	56.790	80.772
106	C2H4Cl2	1,2-DICHLOROETHANE	11.707	3.0746E-01	2.1177E-03	80	177	-------	49.857	132.663
107	C2H4Cl2O	BIS(CHLOROMETHYL)ETHER	3.931	4.7083E-01	0.0000E+00	202	232	-------	98.875	113.000
108	C2H4F2	1,1-DIFLUOROETHANE	-------	----------	----------	----	----	-------	-------	-------
109	C2H4F2	1,2-DIFLUOROETHANE	-------	----------	----------	----	----	-------	-------	-------
110	C2H4I2	1,2-DIIODOETHANE	-------	----------	----------	----	----	-------	-------	-------
111	C2H4O	ACETALDEHYDE	-------	----------	----------	----	----	-------	-------	-------
112	C2H4O	ETHYLENE OXIDE	-6.998	7.6291E-01	-1.9333E-03	80	160	-------	41.662	65.575
113	C2H4OS	THIOACETIC-ACID	-------	----------	----------	----	----	-------	-------	-------
114	C2H4O2	ACETIC ACID	3.300	5.1151E-01	-8.2579E-04	100	290	-------	46.194	82.180
115	C2H4O2	METHYL FORMATE	-------	----------	----------	----	----	-------	-------	-------
116	C2H4S	THIACYCLOPROPANE	-------	----------	----------	----	----	-------	-------	-------
117	C2H5Br	BROMOETHANE	-------	----------	----------	----	----	-------	-------	-------
118	C2H5Cl	ETHYL CHLORIDE	-1.417	7.3946E-01	-6.7817E-04	60	135	-------	40.509	85.938
119	C2H5ClO	2-CHLOROETHANOL	-------	----------	----------	----	----	-------	-------	-------
120	C2H5F	ETHYL FLUORIDE	-------	----------	----------	----	----	-------	-------	-------
121	C2H5I	ETHYL IODIDE	-------	----------	----------	----	----	-------	-------	-------
122	C2H5N	ETHYLENEIMINE	-------	----------	----------	----	----	-------	-------	-------
123	C2H5NO	ACETAMIDE	32.750	1.2333E-01	2.2020E-04	89	345	89.095	45.490	101.591
124	C2H5NO	N-METHYLFORMAMIDE	-12.963	4.4167E-01	0.0000E+00	239	269	-------	92.750	106.000
125	C2H5NO2	NITROETHANE	20.040	3.7240E-01	0.0000E+00	80	180	-------	49.832	87.072
126	C2H5NO3	ETHYL-NITRATE	-------	----------	----------	----	----	-------	-------	-------
127	C2H6	ETHANE	-9.600	8.6284E-01	-3.0170E-04	30	80	-------	16.014	57.496
128	C2H6AlCl	DIMETHYLALUMINUM CHLORIDE	-6.328	5.2083E-01	0.0000E+00	222	252	-------	109.375	125.000
129	C2H6O	DIMETHYL ETHER	-6.999	8.7535E-01	-2.2195E-03	24	132	-------	12.930	69.770
130	C2H6O	ETHANOL	-3.189	7.0754E-01	-1.8586E-03	30	159	-------	16.364	62.264
131	C2H6OS	DIMETHYL SULFOXIDE	4.219	5.0886E-01	-3.0993E-04	20	292	-------	14.272	126.271
132	C2H6O2	ETHYLENE GLYCOL	17.783	2.4670E-01	2.7695E-04	40	240	-------	28.094	92.943
133	C2H6O4S	DIMETHYL SULFATE	-0.683	6.8333E-01	0.0000E+00	211	241	-------	143.500	164.000
134	C2H6S	DIMETHYL SULFIDE	-7.793	8.7346E-01	-2.0392E-03	15	175	-------	4.850	82.592
135	C2H6S	ETHYL MERCAPTAN	-8.486	1.0035E+00	-3.1469E-03	14	125	-------	4.946	67.836
136	C2H6S2	DIMETHYL DISULFIDE	-6.019	1.0324E+00	-2.5006E-03	18	181	-------	12.036	98.944
137	C2H7N	DIMETHYLAMINE	-8.020	8.0392E-01	-1.8406E-03	14	176	-------	3.108	76.472
138	C2H7N	ETHYLAMINE	-------	----------	----------	----	----	-------	-------	-------
139	C2H7NO	MONOETHANOLAMINE	-22.771	5.2167E-01	0.0000E+00	254	284	-------	109.550	125.200
140	C2H8N2	ETHYLENEDIAMINE	-9.813	7.3408E-01	-1.4935E-03	40	179	-------	17.081	73.749
141	C2H8Si	DIMETHYL SILANE	-------	----------	----------	----	----	-------	-------	-------
142	C2N2	CYANOGEN	7.004	4.5662E-01	-5.6992E-04	30	240	-------	20.190	83.765
143	C3F6	HEXAFLUOROPROPYLENE	-------	----------	----------	----	----	-------	-------	-------
144	C3F6O	HEXAFLUOROACETONE	4.324	1.1876E+00	-2.9129E-03	21	141	-------	27.841	113.914
145	C3F8	OCTAFLUOROPROPANE	8.252	1.2704E+00	-3.3253E-03	21	96	-------	33.408	99.566
146	C3H2N2	MALONONITRILE	-20.265	4.3792E-01	0.0000E+00	273	305	110.301	99.353	113.261
147	C3H3Cl	PROPARGYL CHLORIDE	-------	----------	----------	----	----	-------	-------	-------
148	C3H3N	ACRYLONITRILE	114.000	-8.7040E-01	3.7180E-03	164	184	-------	71.226	79.569
149	C3H3NO	OXAZOLE	-------	----------	----------	----	----	-------	-------	-------
150	C3H4	METHYLACETYLENE	-------	----------	----------	----	----	-------	-------	-------
151	C3H4	PROPADIENE	-------	----------	----------	----	----	-------	-------	-------
152	C3H4Cl2	2,3-DICHLOROPROPENE	-------	----------	----------	----	----	-------	-------	-------
153	C3H4O	ACROLEIN	-------	----------	----------	----	----	-------	-------	-------
154	C3H4O	PROPARGYL ALCOHOL	6.053	3.2458E-01	0.0000E+00	191	221	-------	68.163	77.900
155	C3H4O2	ACRYLIC ACID	-18.446	3.9542E-01	0.0000E+00	257	287	-------	83.038	94.900
156	C3H4O2	beta-PROPIOLACTONE	0.099	3.9542E-01	0.0000E+00	210	240	-------	83.038	94.900

Table 4-1 HEAT CAPACITY OF SOLID - ORGANIC COMPOUNDS (continued)

NO	FORMULA	NAME	$C_P = A + B\,T + C\,T^2$					$(C_P$ - joule/(mol K), T - K)		
			A	B	C	TMIN	TMAX	C_P @25 C	C_P @TMIN	C_P @TMAX
157	C3H4O2	VINYL FORMATE	-------	----------	----------	----	----	----	-------	-------
158	C3H4O3	ETHYLENE CARBONATE	-28.554	4.9104E-01	0.0000E+00	268	298	117.850	103.119	117.850
159	C3H4O3	PYRUVIC ACID	-21.817	4.6667E-01	0.0000E+00	257	287	-------	98.000	112.000
160	C3H5Br	3-BROMO-1-PROPENE	-------	----------	----------	----	----	-------	-------	-------
161	C3H5Cl	2-CHLOROPROPENE	-------	----------	----------	----	----	-------	-------	-------
162	C3H5Cl	3-CHLOROPROPENE	-------	----------	----------	----	----	-------	-------	-------
163	C3H5ClO	alpha-EPICHLOROHYDRIN	-------	----------	----------	----	----	-------	-------	-------
164	C3H5ClO2	METHYL CHLOROACETATE	-0.560	5.4375E-01	0.0000E+00	211	241	-------	114.188	130.500
165	C3H5ClO2	ETHYL CHLOROFORMATE	-------	----------	----------	----	----	-------	-------	-------
166	C3H5Cl3	1,2,3-TRICHLOROPROPANE	-7.726	4.1875E-01	0.0000E+00	228	258	-------	87.938	100.500
167	C3H5I	3-IODO-1-PROPENE	-------	----------	----------	----	----	-------	-------	-------
168	C3H5N	PROPIONITRILE	-5.426	8.2851E-01	-1.8182E-03	80	180	-------	49.219	84.841
169	C3H5NO	ACRYLAMIDE	-25.136	4.7292E-01	0.0000E+00	263	293	-------	99.313	113.500
170	C3H5NO	HYDRACRYLONITRILE	6.077	4.7292E-01	0.0000E+00	197	227	-------	99.313	113.500
171	C3H5NO	LACTONITRILE	3.325	4.7500E-01	0.0000E+00	203	233	-------	99.750	114.000
172	C3H5N3O9	NITROGLYCERINE	-58.072	1.2583E+00	0.0000E+00	256	286	-------	264.250	302.000
173	C3H6	CYCLOPROPANE	-9.471	8.8247E-01	-2.5520E-03	25	146	-------	10.996	64.916
174	C3H6	PROPYLENE	-7.390	9.2926E-01	-1.2937E-03	13	87	-------	4.472	63.664
175	C3H6Br2	1,2-DIBROMOPROPANE	-------	----------	----------	----	----	-------	-------	-------
176	C3H6Cl2	1,1-DICHLOROPROPANE	-------	----------	----------	----	----	-------	-------	-------
177	C3H6Cl2	1,2-DICHLOROPROPANE	-------	----------	----------	----	----	-------	-------	-------
178	C3H6Cl2	1,3-DICHLOROPROPANE	-------	----------	----------	----	----	-------	-------	-------
179	C3H6Cl2	2,2-DICHLOROPROPANE	-------	----------	----------	----	----	-------	-------	-------
180	C3H6I2	1,2-DIIODOPROPANE	-------	----------	----------	----	----	-------	-------	-------
181	C3H6O	ACETONE	-3.893	9.3523E-01	-2.1879E-03	30	178	-------	22.195	93.325
182	C3H6O	ALLYL ALCOHOL	-------	----------	----------	----	----	-------	-------	-------
183	C3H6O	METHYL VINYL ETHER	-------	----------	----------	----	----	-------	-------	-------
184	C3H6O	n-PROPIONALDEHYDE	-------	----------	----------	----	----	-------	-------	-------
185	C3H6O	1,2-PROPYLENE OXIDE	-5.434	8.7481E-01	-2.6508E-03	88	161	-------	51.021	66.698
186	C3H6O	1,3-PROPYLENE OXIDE	-------	----------	----------	----	----	-------	-------	-------
187	C3H6O2	ETHYL FORMATE	-------	----------	----------	----	----	-------	-------	-------
188	C3H6O2	METHYL ACETATE	-------	----------	----------	----	----	-------	-------	-------
189	C3H6O2	PROPIONIC ACID	7.275	5.2452E-01	-4.2631E-04	40	250	-------	27.574	111.760
190	C3H6O2S	3-MERCAPTOPROPIONIC ACID	19.898	4.3245E-01	1.9893E-04	85	245	-------	58.233	137.894
191	C3H6O3	LACTIC ACID	24.322	3.5249E-01	0.0000E+00	90	280	-------	56.046	123.019
192	C3H6O3	METHOXYACETIC ACID	-22.379	5.4583E-01	0.0000E+00	251	281	-------	114.625	131.000
193	C3H6O3	TRIOXANE	6.231	4.6169E-01	-3.5488E-04	100	335	112.337	48.851	120.992
194	C3H6S	THIACYCLOBUTANE	-------	----------	----------	----	----	-------	-------	-------
195	C3H7Br	1-BROMOPROPANE	-------	----------	----------	----	----	-------	-------	-------
196	C3H7Br	2-BROMOPROPANE	127.900	-1.0795E+00	5.3240E-03	117	180	-------	74.504	105.795
197	C3H7Cl	ISOPROPYL CHLORIDE	-------	----------	----------	----	----	-------	-------	-------
198	C3H7Cl	n-PROPYL CHLORIDE	-------	----------	----------	----	----	-------	-------	-------
199	C3H7F	1-FLUOROPROPANE	-------	----------	----------	----	----	-------	-------	-------
200	C3H7F	2-FLUOROPROPANE	-------	----------	----------	----	----	-------	-------	-------
201	C3H7I	ISOPROPYL IODIDE	-------	----------	----------	----	----	-------	-------	-------
202	C3H7I	n-PROPYL IODIDE	-------	----------	----------	----	----	-------	-------	-------
203	C3H7N	ALLYLAMINE	-------	----------	----------	----	----	-------	-------	-------
204	C3H7N	PROPYLENEIMINE	5.317	4.8333E-01	0.0000E+00	199	229	-------	101.500	116.000
205	C3H7NO	N,N-DIMETHYLFORMAMIDE	23.451	2.2590E-02	0.0000E+00	83	208	-------	25.326	28.150
206	C3H7NO	N-METHYLACETAMIDE	-29.454	5.5417E-01	0.0000E+00	263	293	-------	116.375	133.000
207	C3H7NO2	1-NITROPROPANE	-------	----------	----------	----	----	-------	-------	-------
208	C3H7NO2	2-NITROPROPANE	-------	----------	----------	----	----	-------	-------	-------
209	C3H7NO3	PROPYL-NITRATE	-------	----------	----------	----	----	-------	-------	-------
210	C3H7NO3	ISOPROPYL-NITRATE	-------	----------	----------	----	----	-------	-------	-------
211	C3H8	PROPANE	-11.230	1.0590E+00	-3.6000E-03	30	84	-------	17.300	52.324
212	C3H8O	ISOPROPANOL	-4.203	7.0986E-01	-1.0630E-03	80	177	-------	45.782	88.025
213	C3H8O	METHYL ETHYL ETHER	-------	----------	----------	----	----	-------	-------	-------
214	C3H8O	n-PROPANOL	-7.637	8.6899E-01	-2.0922E-03	12	147	-------	2.489	74.882
215	C3H8O2	2-METHOXYETHANOL	-------	----------	----------	----	----	-------	-------	-------
216	C3H8O2	METHYLAL	-13.696	1.2512E+00	-3.6199E-03	70	168	-------	56.151	94.351
217	C3H8O2	1,2-PROPYLENE GLYCOL	-------	----------	----------	----	----	-------	-------	-------
218	C3H8O2	1,3-PROPYLENE GLYCOL	-------	----------	----------	----	----	-------	-------	-------
219	C3H8O3	GLYCEROL	-1.052	5.8027E-01	-3.8973E-04	158	291	-------	81.034	134.921
220	C3H8S	n-PROPYLMERCAPTAN	-9.552	1.0230E+00	-2.7800E-03	22	137	-------	11.410	78.340
221	C3H8S	ISOPROPYL MERCAPTAN	-6.738	1.0470E+00	-3.0891E-03	15	110	-------	7.909	71.155
222	C3H8S	ETHYL-METHYL-SULFIDE	-------	----------	----------	----	----	-------	-------	-------
223	C3H9N	n-PROPYLAMINE	-5.566	7.9362E-01	-1.4690E-03	15	188	-------	6.008	91.799
224	C3H9N	ISOPROPYLAMINE	4.002	5.6979E-01	-3.4429E-04	25	178	-------	18.031	94.493
225	C3H9N	TRIMETHYLAMINE	-6.147	8.6589E-01	-1.7406E-03	70	150	-------	45.936	84.572
226	C3H9NO	1-AMINO-2-PROPANOL	-22.051	6.3292E-01	0.0000E+00	245	275	-------	132.913	151.900
227	C3H9NO	3-AMINO-1-PROPANOL	-29.268	6.6292E-01	0.0000E+00	254	284	-------	139.213	159.100
228	C3H9NO	METHYLETHANOLAMINE	-18.145	6.3333E-01	0.0000E+00	239	269	-------	133.000	152.000
229	C3H9O4P	TRIMETHYL PHOSPHATE	10.833	8.3333E-01	0.0000E+00	197	227	-------	175.000	200.000
230	C3H10N2	1,2-PROPANEDIAMINE	-1662.77	7.9209E+00	0.0000E+00	225	231	-------	121.965	169.490
231	C3H10Si	TRIMETHYL SILANE	-------	----------	----------	----	----	-------	-------	-------
232	C4Cl4S	TETRACHLOROTHIOPHENE	-34.671	6.5417E-01	0.0000E+00	263	293	-------	137.375	157.000
233	C4Cl6	HEXACHLORO-1,3-BUTADIENE	-9.416	7.7500E-01	0.0000E+00	222	252	-------	162.750	186.000
234	C4F8	OCTAFLUORO-2-BUTENE	-------	----------	----------	----	----	-------	-------	-------

Table 4-1 HEAT CAPACITY OF SOLID - ORGANIC COMPOUNDS (continued)

NO	FORMULA	NAME	$C_P = A + B\,T + C\,T^2$					$(C_P$ - joule/(mol K), T - K)		
			A	B	C	TMIN	TMAX	C_P @25 C	C_P @TMIN	C_P @TMAX
235	C4F8	OCTAFLUOROCYCLOBUTANE	6.311	8.6552E-01	7.5199E-04	17	134	-------	21.643	136.295
236	C4F10	DECAFLUOROBUTANE	-------	----------	----------	----	----	-------	-------	-------
237	C4H2	BUTADIYNE(BIACETYLENE)	-------	----------	----------	----	----	-------	-------	-------
238	C4H2O3	MALEIC ANHYDRIDE	32.500	2.0960E-01	2.7330E-04	89	320	119.287	53.193	127.673
239	C4H4	VINYLACETYLENE	-------	----------	----------	----	----	-------	-------	-------
240	C4H4N2	SUCCINONITRILE	177.700	-4.5776E-01	1.1744E-03	240	331	145.615	135.483	154.901
241	C4H4O	FURAN	7.911	5.9455E-01	-1.7149E-03	31	150	-------	24.845	58.509
242	C4H4O2	DIKETENE	-15.102	5.6667E-01	0.0000E+00	237	267	-------	119.000	136.000
243	C4H4O3	SUCCINIC ANHYDRIDE	-26.442	4.9750E-01	0.0000E+00	263	293	-------	104.475	119.400
244	C4H4O4	FUMARIC ACID	23.580	4.4635E-01	-1.6605E-04	90	300	141.898	62.406	142.541
245	C4H4O4	MALEIC ACID	27.560	3.6440E-01	0.0000E+00	60	350	136.206	49.424	155.100
246	C4H4S	THIOPHENE	359.016	-2.9004E+00	7.5097E-03	179	231	-------	80.471	89.759
247	C4H5Cl	CHLOROPRENE	-------	----------	----------	----	----	-------	-------	-------
248	C4H5N	trans-CROTONITRILE	7.800	4.3333E-01	0.0000E+00	192	222	-------	91.000	104.000
249	C4H5N	cis-CROTONITRILE	17.095	4.3333E-01	0.0000E+00	171	201	-------	91.000	104.000
250	C4H5N	METHACRYLONITRILE	1.148	4.3333E-01	0.0000E+00	207	237	-------	91.000	104.000
251	C4H5N	PYRROLE	5.685	4.3742E-01	-5.2808E-04	80	250	-------	37.299	81.989
252	C4H5N	VINYLACETONITRILE	-------	----------	----------	----	----	-------	-------	-------
253	C4H5NO2	METHYL CYANOACETATE	-11.546	5.7500E-01	0.0000E+00	230	260	-------	120.750	138.000
254	C4H6	CYCLOBUTENE	-------	----------	----------	----	----	-------	-------	-------
255	C4H6	1,2-BUTADIENE	-11.329	1.0030E+00	-2.8938E-03	15	135	-------	3.065	71.339
256	C4H6	1,3-BUTADIENE	-2.811	7.2243E-01	-1.1845E-03	15	162	-------	7.759	83.137
257	C4H6	DIMETHYLACETYLENE	-14.400	1.1267E+00	-3.1590E-03	15	140	-------	1.790	81.422
258	C4H6	ETHYLACETYLENE	-9.594	1.0018E+00	-2.9908E-03	77	140	-------	49.810	72.035
259	C4H6Cl2	1,3-DICHLORO-trans-2-BUTENE	8.662	5.8333E-01	0.0000E+00	195	225	-------	122.500	140.000
260	C4H6Cl2	1,4-DICHLORO-cis-2-BUTENE	-19.864	5.8167E-01	0.0000E+00	244	274	-------	122.150	139.600
261	C4H6Cl2	1,4-DICHLORO-trans-2-BUTENE	-------	----------	----------	----	----	-------	-------	-------
262	C4H6Cl2	3,4-DICHLORO-1-BUTENE	-------	----------	----------	----	----	-------	-------	-------
263	C4H6O	trans-CROTONALDEHYDE	-------	----------	----------	----	----	-------	-------	-------
264	C4H6O	2,5-DIHYDROFURAN	-------	----------	----------	----	----	-------	-------	-------
265	C4H6O	DIVINYL ETHER	-------	----------	----------	----	----	-------	-------	-------
266	C4H6O	METHACROLEIN	-------	----------	----------	----	----	-------	-------	-------
267	C4H6O2	2-BUTYNE-1,4-DIOL	-26.929	5.0667E-01	0.0000E+00	263	293	-------	106.400	121.600
268	C4H6O2	gamma-BUTYROLACTONE	12.744	4.0153E-01	2.3346E-04	30	230	-------	25.000	117.334
269	C4H6O2	cis-CROTONIC ACID	-24.649	5.0667E-01	0.0000E+00	259	289	-------	106.400	121.600
270	C4H6O2	trans-CROTONIC ACID	-63.187	1.0866E+00	0.0000E+00	268	298	260.790	228.191	260.790
271	C4H6O2	METHACRYLIC ACID	-20.985	4.3583E-01	0.0000E+00	258	288	-------	91.525	104.600
272	C4H6O2	METHYL ACRYLATE	10.570	6.8740E-01	-1.0590E-03	60	196	-------	48.002	104.675
273	C4H6O2	VINYL ACETATE	-------	----------	----------	----	----	-------	-------	-------
274	C4H6O3	ACETIC ANHYDRIDE	-------	----------	----------	----	----	-------	-------	-------
275	C4H6O4	SUCCINIC ACID	26.200	4.2520E-01	0.0000E+00	90	323	152.973	64.468	163.540
276	C4H6O5	DIGLYCOLIC ACID	-38.224	7.1917E-01	0.0000E+00	263	293	-------	151.025	172.600
277	C4H6O5	MALIC ACID	-38.204	7.2083E-01	0.0000E+00	263	293	-------	151.375	173.000
278	C4H6O6	TARTARIC ACID	82.119	3.1697E-01	0.0000E+00	309	323	-------	180.110	184.500
279	C4H7N	n-BUTYRONITRILE	-------	----------	----------	----	----	-------	-------	-------
280	C4H7N	ISOBUTYRONITRILE	-------	----------	----------	----	----	-------	-------	-------
281	C4H7NO	ACETONE CYANOHYDRIN	-7.098	5.8417E-01	0.0000E+00	222	252	-------	122.675	140.200
282	C4H7NO	2-METHACRYLAMIDE	0.445	6.8296E-01	-8.7103E-04	60	300	126.641	38.287	126.941
283	C4H7NO	3-METHOXYPROPIONITRILE	-------	----------	----------	----	----	-------	-------	-------
284	C4H7NO	2-PYRROLIDONE	16.334	3.4866E-01	1.3970E-04	60	285	-------	37.757	127.050
285	C4H8	1-BUTENE	-11.985	1.1487E+00	-3.5829E-03	40	88	-------	28.231	61.263
286	C4H8	cis-2-BUTENE	-13.014	1.0572E+00	-2.6455E-03	30	134	-------	16.320	81.235
287	C4H8	trans-2-BUTENE	-8.042	7.9284E-01	-7.4103E-04	87	160	-------	55.644	100.063
288	C4H8	CYCLOBUTANE	-------	----------	----------	----	----	-------	-------	-------
289	C4H8	ISOBUTENE	34.263	6.8540E-02	2.0724E-03	80	130	-------	53.010	78.197
290	C4H8Br2	1,2-DIBROMOBUTANE	-------	----------	----------	----	----	-------	-------	-------
291	C4H8Br2	2,3-DIBROMOBUTANE	-------	----------	----------	----	----	-------	-------	-------
292	C4H8Cl2	1,4-DICHLOROBUTANE	2.647	6.6167E-01	0.0000E+00	206	236	-------	138.950	158.800
293	C4H8I2	1,2-DIIODOBUTANE	-------	----------	----------	----	----	-------	-------	-------
294	C4H8O	n-BUTYRALDEHYDE	18.160	4.7245E-01	8.9900E-05	50	165	-------	42.007	98.562
295	C4H8O	ISOBUTYRALDEHYDE	-------	----------	----------	----	----	-------	-------	-------
296	C4H8O	1,2-EPOXYBUTANE	-6.231	9.8128E-01	-2.6769E-03	20	144	-------	12.324	79.537
297	C4H8O	METHYL ETHYL KETONE	-3.725	1.0075E+00	-2.3752E-03	98	186	-------	72.325	101.550
298	C4H8O	ETHYL VINYL ETHER	-------	----------	----------	----	----	-------	-------	-------
299	C4H8O	TETRAHYDROFURAN	-7.297	8.9887E-01	-2.2471E-03	20	165	-------	9.781	79.783
300	C4H8O2	cis-2-BUTENE-1,4-DIOL	-25.901	5.8667E-01	0.0000E+00	254	284	-------	123.200	140.800
301	C4H8O2	trans-2-BUTENE-1,4-DIOL	-35.464	5.8667E-01	0.0000E+00	270	300	139.451	123.200	140.800
302	C4H8O2	ISOBUTYRIC ACID	7.597	5.8667E-01	0.0000E+00	197	227	-------	123.200	140.800
303	C4H8O2	n-BUTYRIC ACID	7.504	4.8110E-01	8.8969E-04	121	221	-------	78.394	157.280
304	C4H8O2	1,4-DIOXANE	29.970	1.7640E-01	5.3800E-04	70	260	-------	44.954	112.203
305	C4H8O2	ETHYL ACETATE	51.063	1.4108E-01	1.1674E-03	90	190	-------	73.216	119.778
306	C4H8O2	METHYL PROPIONATE	-------	----------	----------	----	----	-------	-------	-------
307	C4H8O2	n-PROPYL FORMATE	-------	----------	----------	----	----	-------	-------	-------
308	C4H8O2S	SULFOLANE	-34.060	6.4083E-01	0.0000E+00	263	293	-------	134.575	153.800
309	C4H8S	TETRAHYDROTHIOPHENE	11.372	5.3482E-01	-5.8653E-04	33	174	-------	28.134	86.659
310	C4H9Br	1-BROMOBUTANE	-------	----------	----------	----	----	-------	-------	-------
311	C4H9Br	2-BROMOBUTANE	-------	----------	----------	----	----	-------	-------	-------
312	C4H9Cl	n-BUTYL CHLORIDE	-------	----------	----------	----	----	-------	-------	-------

Table 4-1 HEAT CAPACITY OF SOLID - ORGANIC COMPOUNDS (continued)

			$C_P = A + B T + C T^2$					$(C_P$ - joule/(mol K), T - K)		
NO	FORMULA	NAME	A	B	C	TMIN	TMAX	C_P @25 C	C_P @TMIN	C_P @TMAX
313	C4H9Cl	sec-BUTYL CHLORIDE	-------	----------	----------	----	----	-------	-------	-------
314	C4H9Cl	tert-BUTYL CHLORIDE	208.150	-2.2670E+00	1.0005E-02	122	180	-------	80.517	123.786
315	C4H9I	2-IODO-2-METHYLPROPANE	-------	----------	----------	----	----	-------	-------	-------
316	C4H9N	PYRROLIDINE	1.682	6.7136E-01	-1.2135E-03	80	215	-------	47.625	89.978
317	C4H9NO	N,N-DIMETHYLACETAMIDE	-9.049	6.8817E-01	0.0000E+00	223	253	-------	144.515	165.160
318	C4H9NO	MORPHOLINE	-19.958	6.6417E-01	0.0000E+00	240	270	-------	139.475	159.400
319	C4H9NO2	1-NITROBUTANE	-------	----------	----------	----	----	-------	-------	-------
320	C4H9NO2	2-NITROBUTANE	-------	----------	----------	----	----	-------	-------	-------
321	C4H10	n-BUTANE	-13.596	1.1691E+00	-3.5716E-03	20	108	-------	8.358	71.011
322	C4H10	ISOBUTANE	110.211	-1.8703E+00	1.4353E-02	77	107	-------	51.295	74.414
323	C4H10N2	PIPERAZINE	-39.420	7.4167E-01	0.0000E+00	263	293	-------	155.750	178.000
324	C4H10O	n-BUTANOL	-8.483	9.2644E-01	-1.8804E-03	11	179	-------	1.666	97.002
325	C4H10O	sec-BUTANOL	-5.751	9.2342E-01	-2.0567E-03	84	158	-------	57.434	88.930
326	C4H10O	tert-BUTANOL	5.123	4.8054E-01	4.4981E-04	148	280	-------	85.788	174.938
327	C4H10O	DIETHYL ETHER	-4.998	1.0375E+00	-2.3253E-03	30	157	-------	24.034	100.522
328	C4H10O	METHYL-PROPYL-ETHER	-------	----------	----------	----	----	-------	-------	-------
329	C4H10O	METHYL ISOPROPYL ETHER	-8.716	1.1263E+00	-2.9283E-03	15	128	-------	7.520	87.453
330	C4H10O	ISOBUTANOL	-6.531	8.9849E-01	-1.9318E-03	15	160	-------	6.512	87.773
331	C4H10O2	1,3-BUTANEDIOL	-------	----------	----------	----	----	-------	-------	-------
332	C4H10O2	1,4-BUTANEDIOL	-35.367	6.6667E-01	0.0000E+00	263	293	-------	140.000	160.000
333	C4H10O2	2,3-BUTANEDIOL	-27.167	6.6667E-01	0.0000E+00	251	281	-------	140.000	160.000
334	C4H10O2	t-BUTYL HYDROPEROXIDE	-24.967	6.6667E-01	0.0000E+00	247	277	-------	140.000	160.000
335	C4H10O2	1,2-DIMETHOXYETHANE	-------	----------	----------	----	----	-------	-------	-------
336	C4H10O2	2-ETHOXYETHANOL	-------	----------	----------	----	----	-------	-------	-------
337	C4H10O3	DIETHYLENE GLYCOL	-16.741	7.3750E-01	0.0000E+00	233	263	-------	154.875	177.000
338	C4H10O4S	DIETHYL SULFATE	-7.233	9.0417E-01	0.0000E+00	218	248	-------	189.875	217.000
339	C4H10S	n-BUTYL MERCAPTAN	-7.433	1.0831E+00	-2.5104E-03	34	150	-------	26.172	98.627
340	C4H10S	ISOBUTYL MERCAPTAN	-5.583	1.0695E+00	-2.2723E-03	16	128	-------	11.426	94.241
341	C4H10S	sec-BUTYL MERCAPTAN	-5.532	1.0767E+00	-2.7406E-03	16	133	-------	11.231	89.197
342	C4H10S	tert-BUTYL MERCAPTAN	-0.347	7.7192E-01	-3.5317E-04	70	150	-------	51.957	107.555
343	C4H10S	DIETHYL SULFIDE	-11.070	1.0877E+00	-2.4400E-03	16	159	-------	5.678	100.313
344	C4H10S	ISOPROPYL-METHYL-SULFIDE	-------	----------	----------	----	----	-------	-------	-------
345	C4H10S	METHYL-PROPYL-SULFIDE	-------	----------	----------	----	----	-------	-------	-------
346	C4H10S2	DIETHYL DISULFIDE	-------	----------	----------	----	----	-------	-------	-------
347	C4H11N	n-BUTYLAMINE	10.607	6.6500E-01	0.0000E+00	194	224	-------	139.650	159.600
348	C4H11N	ISOBUTYLAMINE	-------	----------	----------	----	----	-------	-------	-------
349	C4H11N	sec-BUTYLAMINE	-------	----------	----------	----	----	-------	-------	-------
350	C4H11N	tert-BUTYLAMINE	41.600	4.9000E-02	1.7400E-03	91	202	-------	60.578	122.700
351	C4H11N	DIETHYLAMINE	11.211	6.7333E-01	0.0000E+00	193	223	-------	141.400	161.600
352	C4H11NO	DIMETHYLETHANOLAMINE	19.280	7.4583E-01	0.0000E+00	184	214	-------	156.625	179.000
353	C4H11NO2	DIETHANOLAMINE	-49.328	8.0667E-01	0.0000E+00	271	301	191.180	169.400	193.600
354	C4H11NO2	2-AMINOETHOXYETHANOL	-------	----------	----------	----	----	-------	-------	-------
355	C4H12N2O	N-AMINOETHYL ETHANOLAMINE	-------	----------	----------	----	----	-------	-------	-------
356	C4H12Si	TETRAMETHYLSILANE	-7.100	1.1840E+00	-1.9600E-03	26	171	-------	22.381	137.928
357	C4H13N3	DIETHYLENE TRIAMINE	5.675	9.7000E-01	0.0000E+00	204	234	-------	203.700	232.800
358	C5Cl6	HEXACHLOROCYCLOPENTADIENE	-35.870	8.0625E-01	0.0000E+00	254	284	-------	169.313	193.500
359	C5H4O2	FURFURAL	1.534	4.5792E-01	0.0000E+00	207	237	-------	96.163	109.900
360	C5H5N	PYRIDINE	9.128	4.6179E-01	-4.9275E-04	20	232	-------	18.167	89.627
361	C5H6	CYCLOPENTADIENE	-------	----------	----------	----	----	-------	-------	-------
362	C5H6	2-METHYL-1-BUTENE-3-YNE	-------	----------	----------	----	----	-------	-------	-------
363	C5H6	1-PENTENE-3-YNE	-------	----------	----------	----	----	-------	-------	-------
364	C5H6	1-PENTENE-4-YNE	-------	----------	----------	----	----	-------	-------	-------
365	C5H6N2	GLUTARONITRILE	-3.094	9.6812E-01	-2.0584E-03	25	200	-------	19.822	108.196
366	C5H6O2	FURFURYL ALCOHOL	-9.962	5.3792E-01	0.0000E+00	229	259	-------	112.963	129.100
367	C5H6O3	GLUTARIC ANHYDRIDE	-32.355	6.0875E-01	0.0000E+00	263	293	-------	127.838	146.100
368	C5H6O4	CITRACONIC ACID	-36.098	6.7917E-01	0.0000E+00	263	293	-------	142.625	163.000
369	C5H6O4	ITACONIC ACID	-36.098	6.7917E-01	0.0000E+00	263	293	-------	142.625	163.000
370	C5H6S	2-METHYLTHIOPHENE	-------	----------	----------	----	----	-------	-------	-------
371	C5H6S	3-METHYLTHIOPHENE	-------	----------	----------	----	----	-------	-------	-------
372	C5H7N	N-METHYLPYRROLE	5.602	7.0980E-01	-1.4984E-03	30	202	-------	25.541	87.795
373	C5H7NO2	ETHYL CYANOACETATE	-6.310	5.9250E-01	0.0000E+00	221	251	-------	124.425	142.200
374	C5H8	CYCLOPENTENE	129.410	-1.0182E+00	4.9210E-03	92	138	-------	77.388	82.658
375	C5H8	ISOPRENE	-6.640	1.0492E+00	-3.2118E-03	20	115	-------	13.418	71.655
376	C5H8	3-METHYL-1,2-BUTADIENE	-9.903	1.0260E+00	-2.3665E-03	16	160	-------	5.708	93.551
377	C5H8	2-METHYL-1,3-BUTADIENE	-------	----------	----------	----	----	-------	-------	-------
378	C5H8	1,2-PENTADIENE	-------	----------	----------	----	----	-------	-------	-------
379	C5H8	cis-1,3-PENTADIENE	-9.200	1.1070E+00	-3.3800E-03	12	132	-------	3.864	78.106
380	C5H8	trans-1,3-PENTADIENE	5.285	4.6220E-01	2.285	102	186	-------	76.366	171.696
381	C5H8	1,4-PENTADIENE	-2.295	9.8833E-01	-2.2196E-03	29	120	-------	24.835	84.337
382	C5H8	2,3-PENTADIENE	-11.387	1.1476E+00	-3.0138E-03	12	148	-------	1.671	92.318
383	C5H8	1-PENTYNE	-------	----------	----------	----	----	-------	-------	-------
384	C5H8	2-PENTYNE	-------	----------	----------	----	----	-------	-------	-------
385	C5H8	3-METHYL-1-BUTYNE	-------	----------	----------	----	----	-------	-------	-------
386	C5H8	SPIROPENTANE	-------	----------	----------	----	----	-------	-------	-------
387	C5H8N4O12	PENTAERYTHRITOL TETRANITRATE	-93.192	1.7583E+00	0.0000E+00	263	293	-------	369.250	422.000
388	C5H8O	CYCLOPENTANONE	9.930	5.4708E-01	0.0000E+00	192	222	-------	114.888	131.300
389	C5H8O	METHYL ISOPROPENYL KETONE	11.162	5.4583E-01	0.0000E+00	190	220	-------	114.625	131.000
390	C5H8O2	ACETYLACETONE	42.620	3.6789E-01	0.0000E+00	90	255	-------	75.730	136.358

90

Table 4-1 HEAT CAPACITY OF SOLID - ORGANIC COMPOUNDS (continued)

NO	FORMULA	NAME	$C_P = A + B T + C T^2$					$(C_P$ - joule/(mol K), T - K$)$		
			A	B	C	TMIN	TMAX	C_P @25 C	C_P @TMIN	C_P @TMAX
391	C5H8O2	ALLYL ACETATE	-------	----------	----------	----	----	-------	-------	-------
392	C5H8O2	ETHYL ACRYLATE	-------	----------	----------	----	----	-------	-------	-------
393	C5H8O2	METHYL METHACRYLATE	-1.294	8.0787E-01	-4.9741E-04	80	215	-------	60.152	149.406
394	C5H8O2	VINYL PROPIONATE	-------	----------	----------	----	----	-------	-------	-------
395	C5H8O3	2-HYDROXYETHYL ACRYLATE	18.563	6.8750E-01	0.0000E+00	183	213	-------	144.375	165.000
396	C5H8O3	LEVULINIC ACID	-36.607	6.8875E-01	0.0000E+00	263	293	-------	144.638	165.300
397	C5H8O3	METHYL ACETOACETATE	-------	----------	----------	----	----	-------	-------	-------
398	C5H8O4	GLUTARIC ACID	-40.312	7.5846E-01	0.0000E+00	263	293	-------	159.276	182.030
399	C5H9N	VALERONITRILE	-------	----------	----------	----	----	-------	-------	-------
400	C5H9NO	n-BUTYL ISOCYANATE	-------	----------	----------	----	----	-------	-------	-------
401	C5H9NO	N-METHYL-2-PYRROLIDONE	-6.363	6.9542E-01	0.0000E+00	219	249	-------	146.038	166.900
402	C5H9NO4	L-GLUTAMIC ACID	-13.217	1.0968E+00	-1.9620E-03	29	205	-------	16.469	129.082
403	C5H10	CYCLOPENTANE	-12.035	1.1423E+00	-4.0626E-03	26	119	-------	15.067	66.406
404	C5H10	2-METHYL-1-BUTENE	-9.323	1.1079E+00	-3.1194E-03	12	136	-------	3.523	83.549
405	C5H10	2-METHYL-2-BUTENE	-12.202	1.1851E+00	-2.9367E-03	12	139	-------	1.596	95.928
406	C5H10	3-METHYL-1-BUTENE	-9.689	1.1240E+00	-3.5000E-03	15	105	-------	6.384	69.611
407	C5H10	1-PENTENE	-14.419	1.3985E+00	-5.2926E-03	12	100	-------	1.601	72.508
408	C5H10	cis-2-PENTENE	-13.188	1.2147E+00	-3.2365E-03	32	117	-------	22.178	84.724
409	C5H10	trans-2-PENTENE	-13.769	1.2076E+00	-3.3744E-03	13	122	-------	1.359	83.260
410	C5H10Br2	2,3-DIBROMO-2-METHYLBUTANE	-------	----------	----------	----	----	-------	-------	-------
411	C5H10Cl2	1,5-DICHLOROPENTANE	-------	----------	----------	----	----	-------	-------	-------
412	C5H10O	METHYL ISOPROPYL KETONE	-8.349	1.0740E+00	-1.5680E-03	11	169	-------	3.670	128.336
413	C5H10O	2-PENTANONE	3.699	8.6733E-01	-9.2988E-04	30	196	-------	28.882	138.150
414	C5H10O	DIETHYL KETONE	-2.605	1.0108E+00	-1.6337E-03	80	225	-------	67.800	142.067
415	C5H10O	VALERALDEHYDE	-------	----------	----------	----	----	-------	-------	-------
416	C5H10O2	n-BUTYL FORMATE	-------	----------	----------	----	----	-------	-------	-------
417	C5H10O2	ETHYL PROPIONATE	-------	----------	----------	----	----	-------	-------	-------
418	C5H10O2	ISOBUTYL FORMATE	-------	----------	----------	----	----	-------	-------	-------
419	C5H10O2	ISOPROPYL ACETATE	-------	----------	----------	----	----	-------	-------	-------
420	C5H10O2	n-PROPYL ACETATE	-------	----------	----------	----	----	-------	-------	-------
421	C5H10O2	METHYL n-BUTYRATE	-------	----------	----------	----	----	-------	-------	-------
422	C5H10O2	2-METHYLBUTYRIC ACID	-------	----------	----------	----	----	-------	-------	-------
423	C5H10O2	ISOVALERIC ACID	-2.687	6.9792E-01	0.0000E+00	214	244	-------	146.563	167.500
424	C5H10O2	VALERIC ACID	-6.460	9.6724E-01	-1.4786E-03	80	239	-------	61.456	140.288
425	C5H10O2	TETRAHYDROFURFURYL ALCOHOL	-------	----------	----------	----	----	-------	-------	-------
426	C5H10O2S	3-METHYL SULFOLANE	-24.326	7.2292E-01	0.0000E+00	244	274	-------	151.813	173.500
427	C5H10O3	DIETHYL CARBONATE	7.572	7.6875E-01	0.0000E+00	200	230	-------	161.438	184.500
428	C5H10O3	ETHYL LACTATE	-6.265	7.6875E-01	0.0000E+00	218	248	-------	161.438	184.500
429	C5H10S	THIACYCLOHEXANE	-------	----------	----------	----	----	-------	-------	-------
430	C5H10S	CYCLOPENTANETHIOL	-------	----------	----------	----	----	-------	-------	-------
431	C5H11Br	1-BROMOPENTANE	-------	----------	----------	----	----	-------	-------	-------
432	C5H11Cl	1-CHLOROPENTANE	-------	----------	----------	----	----	-------	-------	-------
433	C5H11Cl	1-CHLORO-3-METHYLBUTANE	-------	----------	----------	----	----	-------	-------	-------
434	C5H11Cl	2-CHLORO-2-METHYLBUTANE	-------	----------	----------	----	----	-------	-------	-------
435	C5H11N	N-METHYLPYRROLIDINE	-------	----------	----------	----	----	-------	-------	-------
436	C5H11N	PIPERIDINE	-15.959	7.0458E-01	0.0000E+00	233	263	-------	147.963	169.100
437	C5H11NO	tert-BUTYLFORMAMIDE	-38.091	7.7500E-01	0.0000E+00	259	289	-------	162.750	186.000
438	C5H12	ISOPENTANE	-10.547	1.2547E+00	-3.3494E-03	9	113	-------	0.474	88.590
439	C5H12	NEOPENTANE	105.567	-2.6562E-01	1.6489E-03	140	255	-------	100.699	145.056
440	C5H12	n-PENTANE	-11.568	1.2081E+00	-3.2359E-03	12	135	-------	2.791	92.424
441	C5H12O	2,2-DIMETHYL-1-PROPANOL	-37.581	7.0708E-01	0.0000E+00	263	293	-------	148.488	169.700
442	C5H12O	tert-PENTYL-ALCOHOL	-------	----------	----------	----	----	-------	-------	-------
443	C5H12O	2-METHYL-1-BUTANOL	-------	----------	----------	----	----	-------	-------	-------
444	C5H12O	2-METHYL-2-BUTANOL	-35.700	1.2380E+00	-1.7300E-03	92	264	-------	63.921	170.671
445	C5H12O	3-METHYL-1-BUTANOL	-------	----------	----------	----	----	-------	-------	-------
446	C5H12O	3-METHYL-2-BUTANOL	-------	----------	----------	----	----	-------	-------	-------
447	C5H12O	1-PENTANOL	-1.283	9.4108E-01	-1.7549E-03	32	178	-------	26.670	110.646
448	C5H12O	2-PENTANOL	-------	----------	----------	----	----	-------	-------	-------
449	C5H12O	3-PENTANOL	-------	----------	----------	----	----	-------	-------	-------
450	C5H12O	METHYL sec-BUTYL ETHER	-------	----------	----------	----	----	-------	-------	-------
451	C5H12O	METHYL tert-BUTYL ETHER	-1.362	8.1138E-01	5.9830E-05	70	165	-------	55.727	133.770
452	C5H12O	METHYL ISOBUTYL ETHER	-------	----------	----------	----	----	-------	-------	-------
453	C5H12O	ETHYL PROPYL ETHER	-12.093	1.3107E+00	-3.5016E-03	20	146	-------	12.719	104.521
454	C5H12O2	ETHYLENE GLYCOL MONOPROPYL ETHER	-------	----------	----------	----	----	-------	-------	-------
455	C5H12O2	NEOPENTYL GLYCOL	-41.346	7.7792E-01	0.0000E+00	263	293	-------	163.363	186.700
456	C5H12O2	1,5-PENTANEDIOL	-13.341	7.7792E-01	0.0000E+00	227	257	-------	163.363	186.700
457	C5H12O3	2-(2-METHOXYETHOXY)ETHANOL	-------	----------	----------	----	----	-------	-------	-------
458	C5H12O4	PENTAERYTHRITOL	694.780	-3.7257E+00	6.8221E-03	298	463	-------	190.402	432.617
459	C5H12S	n-PENTYL MERCAPTAN	-8.913	1.1364E+00	-2.1159E-03	20	187	-------	13.472	129.478
460	C5H12S	BUTYL-METHYL-SULFIDE	-------	----------	----------	----	----	-------	-------	-------
461	C5H12S	ETHYL-PROPYL-SULFIDE	-------	----------	----------	----	----	-------	-------	-------
462	C5H12S	2-METHYL-2-BUTANETHIOL	-------	----------	----------	----	----	-------	-------	-------
463	C5H13N	n-PENTYLAMINE	17.143	7.8458E-01	0.0000E+00	188	218	-------	164.763	188.300
464	C5H13NO2	METHYL DIETHANOLAMINE	-11.254	9.2625E-01	0.0000E+00	222	252	-------	194.513	222.300
465	C6Cl6	HEXACHLOROBENZENE	9.857	1.0438E+00	-1.3221E-03	45	320	203.547	54.151	208.498
466	C6F6	HEXAFLUOROBENZENE	-7.700	1.0685E+00	-1.2370E-03	90	278	-------	78.471	193.857
467	C6H3ClN2O4	1-CHLORO-2,4-DINITROBENZENE	-48.583	9.1667E-01	0.0000E+00	263	293	-------	192.500	220.000
468	C6H3Cl2NO2	1,2-DICHLORO-4-NITROBENZENE	-41.075	7.7500E-01	0.0000E+00	263	293	-------	162.750	186.000

Table 4-1 HEAT CAPACITY OF SOLID - ORGANIC COMPOUNDS (continued)

| | | | \multicolumn{3}{c}{$C_P = A + B T + C T^2$} | | | \multicolumn{3}{c}{(C_P - joule/(mol K), T - K)} | | |
NO	FORMULA	NAME	A	B	C	TMIN	TMAX	C_P @25 C	C_P @TMIN	C_P @TMAX
469	C6H3Cl3	1,2,4-TRICHLOROBENZENE	-31.720	6.3250E-01	0.0000E+00	260	290	-------	132.825	151.800
470	C6H3N3O6	1,3,5-TRINITROBENZENE	-56.092	1.0583E+00	0.0000E+00	263	293	-------	222.250	254.000
471	C6H4Br2	m-DIBROMOBENZENE	-14.766	5.6250E-01	0.0000E+00	236	266	-------	118.125	135.000
472	C6H4ClNO2	m-CHLORONITROBENZENE	-37.515	7.0583E-01	0.0000E+00	263	293	-------	148.225	169.400
473	C6H4ClNO2	o-CHLORONITROBENZENE	-37.515	7.0583E-01	0.0000E+00	263	293	-------	148.225	169.400
474	C6H4ClNO2	p-CHLORONITROBENZENE	-37.515	7.0583E-01	0.0000E+00	263	293	-------	148.225	169.400
475	C6H4Cl2	m-DICHLOROBENZENE	8.983	4.7654E-01	0.0000E+00	191	221	-------	100.074	114.370
476	C6H4Cl2	o-DICHLOROBENZENE	7.277	4.7408E-01	0.0000E+00	195	225	-------	99.558	113.780
477	C6H4Cl2	p-DICHLOROBENZENE	14.717	5.2608E-01	-2.9842E-04	50	270	-------	40.275	135.005
478	C6H4F2	m-DIFLUOROBENZENE	-------	----------	----------	----	----	-------	-------	-------
479	C6H4F2	o-DIFLUOROBENZENE	-------	----------	----------	----	----	-------	-------	-------
480	C6H4F2	p-DIFLUOROBENZENE	-------	----------	----------	----	----	-------	-------	-------
481	C6H4N2O4	m-DINITROBENZENE	59.300	2.1400E-01	8.1300E-04	110	363	195.374	92.795	244.295
482	C6H4N2O4	o-DINITROBENZENE	41.200	3.7800E-01	4.2000E-04	110	390	191.236	88.003	252.558
483	C6H4N2O4	p-DINITROBENZENE	65.510	4.4015E-01	3.6401E-05	293	447	199.977	197.668	269.365
484	C6H5Br	BROMOBENZENE	37.300	1.0900E-01	1.0500E-03	90	240	-------	55.612	123.934
485	C6H5Cl	MONOCHLOROBENZENE	-3.361	6.4419E-01	-6.0310E-04	80	226	-------	44.314	111.421
486	C6H5ClO	m-CHLOROPHENOL	-30.033	5.6667E-01	0.0000E+00	263	293	-------	119.000	136.000
487	C6H5ClO	o-CHLOROPHENOL	-30.033	5.6667E-01	0.0000E+00	263	293	-------	119.000	136.000
488	C6H5ClO	p-CHLOROPHENOL	-30.033	5.6667E-01	0.0000E+00	263	293	-------	119.000	136.000
489	C6H5Cl2N	3,4-DICHLOROANILINE	-37.763	7.1250E-01	0.0000E+00	263	293	-------	149.625	171.000
490	C6H5F	FLUOROBENZENE	12.597	3.3976E-01	4.7480E-04	90	230	-------	47.017	115.853
491	C6H5I	IODOBENZENE	25.900	2.8260E-01	4.2700E-04	90	242	-------	54.793	119.213
492	C6H5NO2	NITROBENZENE	38.830	2.2390E-01	5.8770E-04	70	279	-------	57.383	146.996
493	C6H6	BENZENE	9.359	3.7714E-01	1.4772E-04	100	279	-------	48.551	125.934
494	C6H6ClN	m-CHLOROANILINE	-14.655	6.4417E-01	0.0000E+00	233	263	-------	135.275	154.600
495	C6H6ClN	o-CHLOROANILINE	-20.001	6.4417E-01	0.0000E+00	241	271	-------	135.275	154.600
496	C6H6ClN	p-CHLOROANILINE	-39.894	6.1375E-01	0.0000E+00	275	305	143.096	128.888	147.300
497	C6H6N2	cis-DICYANO-1-BUTENE	-5.813	6.4583E-01	0.0000E+00	219	249	-------	135.625	155.000
498	C6H6N2	trans-DICYANO-1-BUTENE	-12.917	6.4583E-01	0.0000E+00	230	260	-------	135.625	155.000
499	C6H6N2	1,4-DICYANO-2-BUTENE	-70.396	6.4583E-01	0.0000E+00	319	349	-------	135.625	155.000
500	C6H6N2O2	m-NITROANILINE	32.100	3.3000E-01	4.5000E-04	110	344	170.492	73.974	198.679
501	C6H6N2O2	o-NITROANILINE	31.200	3.1860E-01	4.9900E-04	110	321	170.548	72.412	185.016
502	C6H6N2O2	p-NITROANILINE	27.205	3.5671E-01	4.5044E-04	110	421	173.599	72.030	256.959
503	C6H6O	PHENOL	9.769	4.0832E-01	-1.9001E-05	20	314	129.821	17.928	136.128
504	C6H6O2	1,2-BENZENEDIOL	5.070	4.5140E-01	0.0000E+00	110	298	139.655	54.859	139.655
505	C6H6O2	1,3-BENZENEDIOL	16.900	2.9660E-01	3.1700E-04	110	344	133.511	53.472	156.289
506	C6H6O2	p-HYDROQUINONE	7.700	4.2400E-01	0.0000E+00	44	445	134.116	26.398	196.571
507	C6H6O3	1,2,3-BENZENETRIOL	-81.751	7.2992E-01	0.0000E+00	322	352	-------	153.283	175.180
508	C6H6S	PHENYL MERCAPTAN	74.730	-1.0600E-01	1.3320E-03	130	258	-------	83.461	136.202
509	C6H7N	ANILINE	12.511	3.6035E-01	3.4393E-04	40	267	-------	27.475	133.314
510	C6H7N	2-METHYLPYRIDINE	1.256	7.7916E-01	-1.5805E-03	34	189	-------	25.678	91.981
511	C6H7N	3-METHYLPYRIDINE	11.864	5.8950E-01	-9.5040E-04	37	207	-------	32.504	93.202
512	C6H7N	4-METHYLPYRIDINE	-21.150	5.7583E-01	0.0000E+00	247	277	-------	120.925	138.200
513	C6H8	1,3-CYCLOHEXADIENE	1524.900	-2.4904E+01	1.0747E-01	115	160	-------	82.231	291.492
514	C6H8	METHYLCYCLOPENTADIENE	-------	----------	----------	----	----	-------	-------	-------
515	C6H8N2	ADIPONITRILE	-25.809	7.2417E-01	0.0000E+00	246	276	-------	152.075	173.800
516	C6H8N2	METHYLGLUTARONITRILE	8.581	7.2417E-01	0.0000E+00	198	228	-------	152.075	173.800
517	C6H8N2	m-PHENYLENEDIAMINE	-38.489	7.2417E-01	0.0000E+00	263	293	-------	152.075	173.800
518	C6H8N2	o-PHENYLENEDIAMINE	-38.489	7.2417E-01	0.0000E+00	263	293	-------	152.075	173.800
519	C6H8N2	p-PHENYLENEDIAMINE	-38.489	7.2417E-01	0.0000E+00	263	293	-------	152.075	173.800
520	C6H8N2	PHENYLHYDRAZINE	-37.910	7.2417E-01	0.0000E+00	262	292	-------	152.075	173.800
521	C6H8N2O	BIS(CYANOETHYL)ETHER	-5.451	7.9583E-01	0.0000E+00	217	247	-------	167.125	191.000
522	C6H8O4	DIMETHYL MALEATE	-11.202	7.9167E-01	0.0000E+00	224	254	-------	166.250	190.000
523	C6H8O6	ASCORBIC ACID	-49.562	9.3250E-01	0.0000E+00	263	293	-------	195.825	223.800
524	C6H8O7	CITRIC ACID	-53.327	1.0033E+00	0.0000E+00	263	293	-------	210.700	240.800
525	C6H10	1-METHYLCYCLOPENTENE	-------	----------	----------	----	----	-------	-------	-------
526	C6H10	3-METHYLCYCLOPENTENE	-------	----------	----------	----	----	-------	-------	-------
527	C6H10	4-METHYLCYCLOPENTENE	-------	----------	----------	----	----	-------	-------	-------
528	C6H10	CYCLOHEXENE	165.370	-1.3084E+00	5.7570E-03	139	170	-------	94.819	109.106
529	C6H10	2,3-DIMETHYL-1,3-BUTADIENE	-------	----------	----------	----	----	-------	-------	-------
530	C6H10	1,5-HEXADIENE	-------	----------	----------	----	----	-------	-------	-------
531	C6H10	cis,trans-2,4-HEXADIENE	-------	----------	----------	----	----	-------	-------	-------
532	C6H10	trans,trans-2,4-HEXADIENE	6.903	5.8750E-01	0.0000E+00	198	228	-------	123.375	141.000
533	C6H10	1-HEXYNE	-------	----------	----------	----	----	-------	-------	-------
534	C6H10	2-HEXYNE	-------	----------	----------	----	----	-------	-------	-------
535	C6H10	3-HEXYNE	-------	----------	----------	----	----	-------	-------	-------
536	C6H10O	CYCLOHEXANONE	-1.342	6.7083E-01	0.0000E+00	212	242	-------	140.875	161.000
537	C6H10O	MESITYL OXIDE	13.316	6.7083E-01	0.0000E+00	190	220	-------	140.875	161.000
538	C6H10O2	epsilon-CAPROLACTONE	-11.236	7.4167E-01	0.0000E+00	225	255	-------	155.750	178.000
539	C6H10O2	ETHYL METHACRYLATE	12.396	7.2917E-01	0.0000E+00	193	223	-------	153.125	175.000
540	C6H10O2	n-PROPYL ACRYLATE	-------	----------	----------	----	----	-------	-------	-------
541	C6H10O3	ETHYLACETOACETATE	4.680	8.0000E-01	0.0000E+00	204	234	-------	168.000	192.000
542	C6H10O3	PROPIONIC ANHYDRIDE	9.628	8.1250E-01	0.0000E+00	198	228	-------	170.625	195.000
543	C6H10O4	ADIPIC ACID	-56.301	9.6821E-01	0.0000E+00	268	298	232.370	203.324	232.370
544	C6H10O4	DIETHYL OXALATE	6.488	8.7083E-01	0.0000E+00	203	233	-------	182.875	209.000
545	C6H10O4	ETHYLENE GLYCOL DIACETATE	-1.872	8.7083E-01	0.0000E+00	212	242	-------	182.875	209.000
546	C6H10O4	ETHYLIDENE DIACETATE	-45.283	8.7083E-01	0.0000E+00	262	292	-------	182.875	209.000

Table 4-1 HEAT CAPACITY OF SOLID - ORGANIC COMPOUNDS (continued)

NO	FORMULA	NAME	$C_P = A + B\,T + C\,T^2$					C_P - joule/(mol K), T - K)		
			A	B	C	TMIN	TMAX	C_P @25 C	C_P @TMIN	C_P @TMAX
547	C6H11N	HEXANENITRILE	-------	----------	----------	----	----	-------	-------	-------
548	C6H11NO	epsilon-CAPROLACTAM	29.000	2.8000E-01	5.4300E-04	35	342	160.751	39.465	188.506
549	C6H11NO	CYCLOHEXANONE OXIME	-42.842	8.0833E-01	0.0000E+00	263	293	-------	169.750	194.000
550	C6H12	CYCLOHEXANE	-4.329	7.9566E-01	-1.4603E-03	80	181	-------	49.978	91.798
551	C6H12	2,3-DIMETHYL-1-BUTENE	-------	--------	----------	----	----	-------	-------	-------
552	C6H12	2,3-DIMETHYL-2-BUTENE	-14.873	1.3454E+00	-2.8136E-03	20	185	-------	10.909	137.871
553	C6H12	3,3-DIMETHYL-1-BUTENE	87.730	2.9270E-01	0.0000E+00	130	151	-------	125.635	132.016
554	C6H12	2-ETHYL-1-BUTENE	-------	----------	----------	----	----	-------	-------	-------
555	C6H12	trans-3-METHYL-2-PENTENE	-------	----------	----------	----	----	-------	-------	-------
556	C6H12	1-HEXENE	-13.019	1.3140E+00	-3.6567E-03	11	132	-------	0.993	96.722
557	C6H12	cis-2-HEXENE	-------	----------	----------	----	----	-------	-------	-------
558	C6H12	trans-2-HEXENE	-------	----------	----------	----	----	-------	-------	-------
559	C6H12	cis-3-HEXENE	-------	----------	----------	----	----	-------	-------	-------
560	C6H12	trans-3-HEXENE	-------	----------	----------	----	----	-------	-------	-------
561	C6H12	METHYLCYCLOPENTANE	-5.380	1.0181E+00	-2.5105E-03	25	131	-------	18.503	84.803
562	C6H12	2-METHYL-1-PENTENE	-------	----------	----------	----	----	-------	-------	-------
563	C6H12	2-METHYL-2-PENTENE	-------	----------	----------	----	----	-------	-------	-------
564	C6H12	3-METHYL-1-PENTENE	-------	----------	----------	----	----	-------	-------	-------
565	C6H12	3-METHYL-cis-2-PENTENE	-------	----------	----------	----	----	-------	-------	-------
566	C6H12	4-METHYL-1-PENTENE	-------	----------	----------	----	----	-------	-------	-------
567	C6H12	4-METHYL-cis-2-PENTENE	-------	----------	----------	----	----	-------	-------	-------
568	C6H12	4-METHYL-trans-2-PENTENE	-------	----------	----------	----	----	-------	-------	-------
569	C6H12N2	TRIETHYLENEDIAMINE	10.073	4.0599E-01	3.3682E-04	100	350	161.060	54.040	193.391
570	C6H12O	BUTYL VINYL ETHER	-------	----------	----------	----	----	-------	-------	-------
571	C6H12O	CYCLOHEXANOL	1.537	6.0387E-01	-3.5509E-04	80	265	-------	47.574	136.627
572	C6H12O	1-HEXANAL	-------	----------	----------	----	----	-------	-------	-------
573	C6H12O	ETHYL ISOPROPYL KETONE	-------	----------	----------	----	----	-------	-------	-------
574	C6H12O	2-HEXANONE	19.961	6.5857E-01	1.3675E-04	40	208	-------	46.362	162.545
575	C6H12O	3-HEXANONE	-10.072	1.2452E+00	-2.0767E-03	13	140	-------	5.193	123.645
576	C6H12O	METHYL ISOBUTYL KETONE	-------	----------	----------	----	----	-------	-------	-------
577	C6H12O2	n-PENTYL FORMATE	-------	----------	----------	----	----	-------	-------	-------
578	C6H12O2	n-BUTYL ACETATE	-------	----------	----------	----	----	-------	-------	-------
579	C6H12O2	sec-BUTYL ACETATE	-------	----------	----------	----	----	-------	-------	-------
580	C6H12O2	tert-BUTYL ACETATE	-34.758	8.0833E-01	0.0000E+00	253	283	-------	169.750	194.000
581	C6H12O2	ETHYL n-BUTYRATE	-------	----------	----------	----	----	-------	-------	-------
582	C6H12O2	ETHYL ISOBUTYRATE	-------	----------	----------	----	----	-------	-------	-------
583	C6H12O2	ISOBUTYL ACETATE	-------	----------	----------	----	----	-------	-------	-------
584	C6H12O2	n-PROPYL PROPIONATE	-------	----------	----------	----	----	-------	-------	-------
585	C6H12O2	CYCLOHEXYL PEROXIDE	-10.641	8.0917E-01	0.0000E+00	223	253	-------	169.925	194.200
586	C6H12O2	DIACETONE ALCOHOL	8.327	7.6750E-01	0.0000E+00	199	229	-------	161.175	184.200
587	C6H12O2	2-ETHYL BUTYRIC ACID	-14.686	8.0917E-01	0.0000E+00	228	258	-------	169.925	194.200
588	C6H12O2	n-HEXANOIC ACID	-23.140	7.6750E-01	0.0000E+00	240	270	-------	161.175	184.200
589	C6H12O3	2-ETHOXYETHYL ACETATE	-------	----------	----------	----	----	-------	-------	-------
590	C6H12O3	HYDROXYCAPROIC ACID	-46.596	8.7917E-01	0.0000E+00	263	293	-------	184.625	211.000
591	C6H12O3	PARALDEHYDE	-1.330	9.8367E-01	-9.2315E-04	30	286	-------	27.349	204.375
592	C6H12O3	sec-BUTYL GLYCOLATE	-------	----------	----------	----	----	-------	-------	-------
593	C6H12S	THIACYCLOHEPTANE	-------	----------	----------	----	----	-------	-------	-------
594	C6H13N	CYCLOHEXYLAMINE	-12.605	8.1583E-01	0.0000E+00	225	255	-------	171.325	195.800
595	C6H13N	HEXAMETHYLENEIMINE	3.141	8.1583E-01	0.0000E+00	206	236	-------	171.325	195.800
596	C6H14	2,2-DIMETHYLBUTANE	21.500	1.1730E+00	-2.8330E-03	128	174	-------	125.228	139.882
597	C6H14	2,3-DIMETHYLBUTANE	-9.794	1.1721E+00	-2.4229E-03	75	136	-------	64.405	104.828
598	C6H14	n-HEXANE	-7.908	1.1255E+00	-2.0973E-03	15	178	-------	8.502	125.910
599	C6H14	2-METHYLPENTANE	-12.332	1.3504E+00	-3.8871E-03	10	120	-------	0.784	93.555
600	C6H14	3-METHYLPENTANE	-13.327	1.4237E+00	-4.3470E-03	16	110	-------	7.863	90.795
601	C6H14N2O2	LYSINE	-58.521	1.1042E+00	0.0000E+00	263	293	-------	231.875	265.000
602	C6H14O	2-ETHYL-1-BUTANOL	-------	----------	----------	----	----	-------	-------	-------
603	C6H14O	1-HEXANOL	-13.896	1.2083E+00	-2.3797E-03	108	200	-------	88.500	132.583
604	C6H14O	2-HEXANOL	13.883	8.1667E-01	0.0000E+00	193	223	-------	171.500	196.000
605	C6H14O	2-METHYL-1-PENTANOL	13.912	8.1833E-01	0.0000E+00	193	223	-------	171.850	196.400
606	C6H14O	4-METHYL-2-PENTANOL	-------	----------	----------	----	----	-------	-------	-------
607	C6H14O	n-BUTYL ETHYL ETHER	-------	----------	----------	----	----	-------	-------	-------
608	C6H14O	DIISOPROPYL ETHER	-7.016	1.1576E+00	-1.8875E-03	10	188	-------	4.371	143.738
609	C6H14O	DI-n-PROPYL ETHER	-11.700	1.3740E+00	-3.5800E-03	25	150	-------	20.413	113.835
610	C6H14O	METHYL tert-PENTYL ETHER	-------	----------	----------	----	----	-------	-------	-------
611	C6H14O2	ACETAL	-------	----------	----------	----	----	-------	-------	-------
612	C6H14O2	2-BUTOXYETHANOL	-------	----------	----------	----	----	-------	-------	-------
613	C6H14O2	1,6-HEXANEDIOL	-66.821	8.8917E-01	0.0000E+00	285	315	198.284	186.725	213.400
614	C6H14O2	HEXYLENE GLYCOL	14.982	8.8917E-01	0.0000E+00	193	223	-------	186.725	213.400
615	C6H14O2S	DI-n-PROPYL SULFONE	-52.117	9.8333E-01	0.0000E+00	263	293	-------	206.500	236.000
616	C6H14O3	DIETHYLENE GLYCOL DIMETHYL ETHER	-------	----------	----------	----	----	-------	-------	-------
617	C6H14O3	DIPROPYLENE GLYCOL	6.720	9.6000E-01	0.0000E+00	203	233	-------	201.600	230.400
618	C6H14O3	2-(2-ETHOXYETHOXY)ETHANOL	-------	----------	----------	----	----	-------	-------	-------
619	C6H14O3	TRIMETHYLOLPROPANE	-87.047	1.3590E+00	0.0000E+00	274	304	318.152	285.399	326.170
620	C6H14O4	TRIETHYLENE GLYCOL	-26.585	1.0308E+00	0.0000E+00	236	266	-------	216.475	247.400
621	C6H14O6	SORBITOL	-80.022	1.2672E+00	0.0000E+00	273	303	297.784	266.105	304.120
622	C6H14S	n-HEXYLMERCAPTAN	-12.256	1.3081E+00	-2.5943E-03	15	186	-------	7.102	141.417
623	C6H14S	BUTYL-ETHYL-SULFIDE	-------	----------	----------	----	----	-------	-------	-------
624	C6H14S	ISOPROPYL-SULFIDE	-------	----------	----------	----	----	-------	-------	-------

Table 4-1 HEAT CAPACITY OF SOLID - ORGANIC COMPOUNDS (continued)

NO	FORMULA	NAME	$C_P = A + B\,T + C\,T^2$					$(C_P$ - joule/(mol K), T - K)		
			A	B	C	TMIN	TMAX	C_P @25 C	C_P @TMIN	C_P @TMAX
625	C6H14S	METHYL-PENTYL-SULFIDE	-------	----------	----------	----	----	-------	-------	-------
626	C6H14S	PROPYL-SULFIDE	-------	----------	----------	----	----	-------	-------	-------
627	C6H14S2	PROPYL-DISULFIDE	-------	----------	----------	----	----	-------	-------	-------
628	C6H15Al	TRIETHYL ALUMINUN	17.334	8.9583E-01	0.0000E+00	191	221	-------	188.125	215.000
629	C6H15Al2Cl3	ETHYL ALUMINUM SESQUICHLORIDE	-70.446	1.3292E+00	0.0000E+00	263	293	-------	279.125	319.000
630	C6H15N	DIISOPROPYLAMINE	-------	----------	----------	----	----	-------	-------	-------
631	C6H15N	DI-n-PROPYLAMINE	-------	----------	----------	----	----	-------	-------	-------
632	C6H15N	n-HEXYLAMINE	-10.616	8.9583E-01	0.0000E+00	222	252	-------	188.125	215.000
633	C6H15N	TRIETHYLAMINE	-------	----------	----------	----	----	-------	-------	-------
634	C6H15NO	6-AMINOHEXANOL	-58.000	9.6667E-01	0.0000E+00	270	300	230.212	203.000	232.000
635	C6H15NO2	DIISOPROPANOLAMINE	-55.143	1.0375E+00	0.0000E+00	263	293	-------	217.875	249.000
636	C6H15NO3	TRIETHANOLAMINE	-58.908	1.1083E+00	0.0000E+00	263	293	-------	232.750	266.000
637	C6H15N3	N-AMINOETHYL PIPERAZINE	-15.742	1.1125E+00	0.0000E+00	224	254	-------	233.625	267.000
638	C6H15O4P	TRIETHYL PHOSPHATE	-------	----------	----------	----	----	-------	-------	-------
639	C6H16N2	HEXAMETHYLENEDIAMINE	-55.497	1.0442E+00	0.0000E+00	263	293	-------	219.275	250.600
640	C6H18N3OP	HEXAMETHYL PHOSPHORAMIDE	-56.177	1.3992E+00	0.0000E+00	250	280	-------	293.825	335.800
641	C6H18N4	TRIETHYLENE TETRAMINE	-71.108	1.3417E+00	0.0000E+00	263	293	-------	281.750	322.000
642	C6H18OSi2	HEXAMETHYLDISILOXANE	-14.300	1.9135E+00	-3.1400E-03	12	205	-------	7.419	245.965
643	C6H18O3Si3	HEXAMETHYLCYCLOTRISILOXANE	7.671	1.4742E+00	-8.0035E-04	100	335	376.056	147.087	411.895
644	C6H19NSi2	HEXAMETHYLDISILAZANE	-------	----------	----------	----	----	-------	-------	-------
645	C7H3ClF3NO2	4-CHLORO-3-NITROBENZOTRIFLUORIDE	-------	----------	----------	----	----	-------	-------	-------
646	C7H3Cl2F3	2,4-DICHLOROBENZOTRIFLUORIDE	-6.175	8.1792E-01	0.0000E+00	218	248	-------	171.763	196.300
647	C7H3Cl2NO	3,4-DICHLOROPHENYL ISOCYANATE	-39.043	7.3458E-01	0.0000E+00	263	293	-------	154.263	176.300
648	C7H4ClF3	p-CHLOROBENZOTRIFLUORIDE	2.136	7.4958E-01	0.0000E+00	207	237	-------	157.413	179.900
649	C7H4Cl2O	m-CHLOROBENZOYL CHLORIDE	-26.650	6.6625E-01	0.0000E+00	250	280	-------	139.913	159.900
650	C7H4F3NO2	3-NITROBENZOTRIFLUORIDE	-28.533	8.9167E-01	0.0000E+00	242	272	-------	187.250	214.000
651	C7H5ClO	BENZOYL CHLORIDE	-19.522	5.9792E-01	0.0000E+00	243	273	-------	125.563	143.500
652	C7H5ClO2	o-CHLOROBENZOIC ACID	-45.000	7.5000E-01	0.0000E+00	270	300	178.613	157.500	180.000
653	C7H5Cl3	BENZOTRICHLORIDE	-21.123	7.4375E-01	0.0000E+00	238	268	-------	156.188	178.500
654	C7H5F3	BENZOTRIFLUORIDE	14.405	6.1336E-01	2.6021E-05	10	244	-------	20.542	165.702
655	C7H5N	BENZONITRILE	-10.753	5.2708E-01	0.0000E+00	230	260	-------	110.688	126.500
656	C7H5NO	PHENYL ISOCYANATE	-1.883	5.9792E-01	0.0000E+00	213	243	-------	125.563	143.500
657	C7H5N3O6	2,4,6-TRINITROTOLUENE	-237.700	2.8490E+00	-3.1570E-03	173	353	331.093	160.955	374.699
658	C7H6Cl2	BENZYL DICHLORIDE	-------	----------	----------	----	----	-------	-------	-------
659	C7H6Cl2	2,4-DICHLOROTOLUENE	-13.272	6.7542E-01	0.0000E+00	230	260	-------	141.838	162.100
660	C7H6N2O4	2,4-DINITROTOLUENE	-51.982	1.0796E+00	0.0000E+00	258	288	-------	226.713	259.100
661	C7H6N2O4	2,5-DINITROTOLUENE	-50.792	9.5833E-01	0.0000E+00	263	293	-------	201.250	230.000
662	C7H6N2O4	2,6-DINITROTOLUENE	-31.625	9.5833E-01	0.0000E+00	243	273	-------	201.250	230.000
663	C7H6N2O4	3,4-DINITROTOLUENE	-31.625	9.5833E-01	0.0000E+00	243	273	-------	201.250	230.000
664	C7H6N2O4	3,5-DINITROTOLUENE	-50.792	9.5833E-01	0.0000E+00	263	293	-------	201.250	230.000
665	C7H6O	BENZALDEHYDE	8.544	6.0220E-01	-6.0833E-04	29	213	-------	25.440	109.289
666	C7H6O2	BENZOIC ACID	5.773	5.5931E-01	-2.2217E-04	100	396	152.782	59.483	192.236
667	C7H6O2	p-HYDROXYBENZALDEHYDE	-31.890	6.0000E-01	0.0000E+00	263	293	-------	126.000	144.000
668	C7H6O2	SALICYLALDEHYDE	-15.701	6.0042E-01	0.0000E+00	236	266	-------	126.088	144.100
669	C7H6O3	SALICYLIC ACID	36.780	3.1990E-01	3.7930E-04	93	295	-------	69.768	164.208
670	C7H7Br	p-BROMOTOLUENE	-32.266	6.0708E-01	0.0000E+00	263	293	-------	127.488	145.700
671	C7H7Cl	BENZYL CHLORIDE	4.831	8.2583E-01	0.0000E+00	204	234	-------	173.425	198.200
672	C7H7Cl	o-CHLOROTOLUENE	2.034	6.0708E-01	0.0000E+00	207	237	-------	127.488	145.700
673	C7H7Cl	p-CHLOROTOLUENE	-24.678	6.0708E-01	0.0000E+00	251	281	-------	127.488	145.700
674	C7H7F	p-FLUOROTOLUENE	-------	----------	----------	----	----	-------	-------	-------
675	C7H7NO	FORMANILIDE	-32.709	6.1542E-01	0.0000E+00	263	293	-------	129.238	147.700
676	C7H7NO2	m-NITROTOLUENE	-36.839	7.4875E-01	0.0000E+00	259	289	-------	157.238	179.700
677	C7H7NO2	o-NITROTOLUENE	-22.448	7.4875E-01	0.0000E+00	240	270	-------	157.238	179.700
678	C7H7NO2	p-NITROTOLUENE	-41.766	7.1825E-01	0.0000E+00	268	298	172.380	150.833	172.380
679	C7H7NO3	o-NITROANISOLE	-35.537	8.1958E-01	0.0000E+00	253	283	-------	172.113	196.700
680	C7H8	TOLUENE	-1.330	9.0564E-01	-2.3442E-03	70	178	-------	50.578	85.607
681	C7H8	1,3,5-CYCLOHEPTATRIENE	-------	----------	----------	----	----	-------	-------	-------
682	C7H8O	ANISOLE	2.682	6.0958E-01	0.0000E+00	206	236	-------	128.013	146.300
683	C7H8O	BENZYL ALCOHOL	137.900	-1.6260E+00	8.6600E-03	90	180	-------	61.706	125.804
684	C7H8O	m-CRESOL	16.761	4.4184E-01	4.5137E-05	30	285	-------	30.057	146.535
685	C7H8O	o-CRESOL	1.283	6.2049E-01	-3.5595E-04	100	304	154.641	59.772	157.094
686	C7H8O	p-CRESOL	10.062	5.3796E-01	-2.1915E-04	30	308	150.975	26.004	154.938
687	C7H8O2	GUAIACOL	-36.098	6.7917E-01	0.0000E+00	263	293	-------	142.625	163.000
688	C7H8O2	p-METHOXYPHENOL	-36.164	6.8042E-01	0.0000E+00	263	293	-------	142.888	163.300
689	C7H9N	BENZYLAMINE	8.829	6.8708E-01	0.0000E+00	197	227	-------	144.288	164.900
690	C7H9N	2,6-DIMETHYLPYRIDINE	-18.551	6.8708E-01	0.0000E+00	237	267	-------	144.288	164.900
691	C7H9N	N-METHYLANILINE	-------	----------	----------	----	----	-------	-------	-------
692	C7H9N	m-TOLUIDINE	-1.889	6.8708E-01	0.0000E+00	213	243	-------	144.288	164.900
693	C7H9N	o-TOLUIDINE	-6.507	6.8708E-01	0.0000E+00	219	249	-------	144.288	164.900
694	C7H9N	p-TOLUIDINE	-163.690	1.1517E+00	0.0000E+00	250	317	179.689	124.235	201.284
695	C7H10	2-NORBORNENE	-31.474	5.4125E-01	0.0000E+00	268	298	129.900	113.663	129.900
696	C7H10N2	TOLUENEDIAMINE	-44.402	8.3542E-01	0.0000E+00	263	293	-------	175.438	200.500
697	C7H11NO	CYCLOHEXYL ISOCYANATE	-------	----------	----------	----	----	-------	-------	-------
698	C7H12	1-HEPTYNE	-------	----------	----------	----	----	-------	-------	-------
699	C7H12O2	n-BUTYL ACRYLATE	-------	----------	----------	----	----	-------	-------	-------
700	C7H12O2	ISOBUTYL ACRYLATE	-------	----------	----------	----	----	-------	-------	-------
701	C7H12O2	n-PROPYL METHACRYLATE	-------	----------	----------	----	----	-------	-------	-------
702	C7H12O4	DIETHYL MALONATE	15.468	9.8208E-01	0.0000E+00	194	224	-------	206.238	235.700

94

Table 4-1 HEAT CAPACITY OF SOLID - ORGANIC COMPOUNDS (continued)

			$C_P = A + B T + C T^2$					$(C_P$ - joule/(mol K), T - K$)$		
NO	FORMULA	NAME	A	B	C	TMIN	TMAX	C_P @25 C	C_P @TMIN	C_P @TMAX
703	C7H14	CYCLOHEPTANE	141.780	-3.6760E-01	1.5054E-03	216	260	--------	132.530	148.006
704	C7H14	1,1-DIMETYLCYCLOPENTANE	--------	----------	----------	----	----	--------	--------	-------
705	C7H14	cis-1,2-DIMETHYLCYCLOPENTANE	146.740	-2.7824E-01	1.3409E-03	147	213	--------	134.797	148.337
706	C7H14	trans-1,2-DIMETHYLCYCLOPENTANE	--------	----------	----------	----	----	--------	--------	-------
707	C7H14	cis-1,3-DIMETHYLCYCLOPENTANE	--------	----------	----------	----	----	--------	--------	-------
708	C7H14	trans-1,3-DIMETHYLCYCLOPENTANE	-7.145	1.1262E+00	-2.8470E-03	21	133	--------	14.745	92.342
709	C7H14	ETHYLCYCLOPENTANE	-10.177	1.3201E+00	-4.0288E-03	12	127	--------	5.255	92.449
710	C7H14	2-ETHYL-1-PENTENE	--------	----------	----------	----	----	--------	--------	-------
711	C7H14	3-ETHYL-1-PENTENE	--------	----------	----------	----	----	--------	--------	-------
712	C7H14	1-HEPTENE	8.206	9.6369E-01	-1.6212E-03	70	150	--------	67.721	116.283
713	C7H14	cis-2-HEPTENE	--------	----------	----------	----	----	--------	--------	-------
714	C7H14	trans-2-HEPTENE	--------	----------	----------	----	----	--------	--------	-------
715	C7H14	cis-3-HEPTENE	--------	----------	----------	----	----	--------	--------	-------
716	C7H14	trans-3-HEPTENE	--------	----------	----------	----	----	--------	--------	-------
717	C7H14	METHYLCYCLOHEXANE	-7.266	9.8159E-01	-2.1347E-03	79	147	--------	57.144	90.750
718	C7H14	2-METHYL-1-HEXENE	--------	----------	----------	----	----	--------	--------	-------
719	C7H14	3-METHYL-1-HEXENE	--------	----------	----------	----	----	--------	--------	-------
720	C7H14	4-METHYL-1-HEXENE	--------	----------	----------	----	----	--------	--------	-------
721	C7H14	2,3,3-TRIMETHYL-1-BUTENE	--------	----------	----------	----	----	--------	--------	-------
722	C7H14O	DIISOPROPYL KETONE	1.774	1.0355E+00	-8.0423E-04	28	193	--------	29.789	171.653
723	C7H14O	2-HEPTANONE	1.572	8.4958E-01	0.0000E+00	208	238	--------	178.413	203.900
724	C7H14O	1-HEPTANAL	25.961	6.5498E-01	1.1489E-04	80	210	--------	79.095	168.574
725	C7H14O	1-METHYLCYCLOHEXANOL	-45.155	8.4958E-01	0.0000E+00	263	293	--------	178.413	203.900
726	C7H14O	cis-2-METHYLCYCLOHEXANOL	-34.111	8.4958E-01	0.0000E+00	250	280	--------	178.413	203.900
727	C7H14O	trans-2-METHYLCYCLOHEXANOL	-24.765	8.4958E-01	0.0000E+00	239	269	--------	178.413	203.900
728	C7H14O	cis-3-METHYLCYCLOHEXANOL	-23.433	8.4750E-01	0.0000E+00	238	268	--------	177.975	203.400
729	C7H14O	trans-3-METHYLCYCLOHEXANOL	-27.739	8.4958E-01	0.0000E+00	243	273	--------	178.413	203.900
730	C7H14O	cis-4-METHYLCYCLOHEXANOL	--------	----------	----------	----	----	--------	--------	-------
731	C7H14O	trans-4-METHYLCYCLOHEXANOL	--------	----------	----------	----	----	--------	--------	-------
732	C7H14O	5-METHYL-2-HEXANONE	--------	----------	----------	----	----	--------	--------	-------
733	C7H14O2	n-BUTYL PROPIONATE	--------	----------	----------	----	----	--------	--------	-------
734	C7H14O2	ETHYL ISOVALERATE	--------	----------	----------	----	----	--------	--------	-------
735	C7H14O2	ISOPENTYL ACETATE	--------	----------	----------	----	----	--------	--------	-------
736	C7H14O2	n-PENTYL ACETATE	--------	----------	----------	----	----	--------	--------	-------
737	C7H14O2	n-PROPYL n-BUTYRATE	--------	----------	----------	----	----	--------	--------	-------
738	C7H14O2	n-HEPTANOIC ACID	164.213	-1.1939E+00	6.1499E-03	105	255	--------	106.654	259.660
739	C7H14O3	ETHYL-3-ETHOXYPROPIONATE	16.858	9.9167E-01	0.0000E+00	193	223	--------	208.250	238.000
740	C7H15Br	1-BROMOHEPTANE	50.730	4.0160E-01	5.7550E-04	113	203	--------	103.273	156.066
741	C7H15N	N-METHYLCYCLOHEXYLAMINE	-22.801	9.2500E-01	0.0000E+00	235	265	--------	194.250	222.000
742	C7H16	2,2-DIMETHYLPENTANE	82.800	-5.1460E-01	5.8700E-03	89	149	--------	83.402	136.976
743	C7H16	2,3-DIMETHYLPENTANE	-6.500	1.3230E+00	-3.2738E-03	10	80	--------	6.403	78.388
744	C7H16	2,4-DIMETHYLPENTANE	-4.100	1.1193E+00	-1.4041E-03	82	154	--------	78.178	134.902
745	C7H16	3,3-DIMETHYLPENTANE	-8.543	1.2245E+00	-2.4904E-03	20	133	--------	14.951	110.094
746	C7H16	3-ETHYLPENTANE	--------	----------	----------	----	----	--------	--------	-------
747	C7H16	n-HEPTANE	-10.556	1.2945E+00	-2.4627E-03	20	183	--------	14.349	143.695
748	C7H16	2-METHYLHEXANE	-1.601	1.1826E+00	-2.0698E-03	30	155	--------	32.015	131.926
749	C7H16	3-METHYLHEXANE	166.450	-8.0900E-02	8.6370E-04	93	154	--------	166.381	174.429
750	C7H16	2,2,3-TRIMETHYLBUTANE	97.860	2.1930E-01	6.3380E-04	121	249	--------	133.824	191.537
751	C7H16O	1-HEPTANOL	35.532	5.7472E-01	1.4924E-05	80	239	--------	81.605	173.829
752	C7H16O	2-HEPTANOL	-2.789	9.2958E-01	0.0000E+00	213	243	--------	195.213	223.100
753	C7H16O	5-METHYL-1-HEXANOL	--------	----------	----------	----	----	--------	--------	-------
754	C7H16O	ISOPROPYL-TERT-BUTYL-ETHER	--------	----------	----------	----	----	--------	--------	-------
755	C7H16S	n-HEPTYL MERCAPTAN	-4.796	1.1861E+00	-1.4446E-03	26	230	--------	24.711	191.556
756	C7H16S	BUTYL-PROPYL-SULFIDE	--------	----------	----------	----	----	--------	--------	-------
757	C7H16S	ETHYL-PENTYL-SULFIDE	--------	----------	----------	----	----	--------	--------	-------
758	C7H16S	HEXYL-METHYL-SULFIDE	--------	----------	----------	----	----	--------	--------	-------
759	C7H17N	1-AMINOHEPTANE	-14.250	1.0071E+00	0.0000E+00	224	254	--------	211.488	241.700
760	C8H4Cl2O2	ISOPHTHALOYL CHLORIDE	-40.633	7.6667E-01	0.0000E+00	263	293	--------	161.000	184.000
761	C8H4O3	PHTHALIC ANHYDRIDE	26.320	3.9060E-01	2.1260E-04	50	350	161.676	46.382	189.074
762	C8H6	ETHYNYLBENZENE	--------	----------	----------	----	----	--------	--------	-------
763	C8H6O4	ISOPHTHALIC ACID	-69.755	8.4042E-01	0.0000E+00	293	323	180.816	176.488	201.700
764	C8H6O4	PHTHALIC ACID	22.000	5.5600E-01	0.0000E+00	90	350	187.771	72.040	216.600
765	C8H6O4	TEREPHTHALIC ACID	-26.000	6.3400E-01	0.0000E+00	318	398	--------	175.612	226.332
766	C8H6S	BENZOTHIOPHENE	5.607	5.0478E-01	2.1384E-04	100	252	--------	58.224	146.275
767	C8H7N	INDOLE	-21.499	6.3833E-01	0.0000E+00	244	274	--------	134.050	153.200
768	C8H8	STYRENE	7.101	4.8630E-01	9.0330E-04	100	240	--------	64.764	175.871
769	C8H8	1,3,5,7-CYCLOOCTATETRAENE	--------	----------	----------	----	----	--------	--------	-------
770	C8H8O	ACETOPHENONE	-33.836	6.4083E-01	0.0000E+00	263	293	--------	134.575	153.800
771	C8H8O	p-TOLUALDEHYDE	--------	----------	----------	----	----	--------	--------	-------
772	C8H8O2	METHYL BENZOATE	-14.767	7.1167E-01	0.0000E+00	231	261	--------	149.450	170.800
773	C8H8O2	o-TOLUIC ACID	-42.265	7.2871E-01	0.0000E+00	268	298	174.999	153.029	174.890
774	C8H8O2	p-TOLUIC ACID	-40.849	7.0429E-01	0.0000E+00	268	298	169.136	147.901	169.030
775	C8H8O3	METHYL SALICYLATE	-19.680	7.8250E-01	0.0000E+00	235	265	--------	164.325	187.800
776	C8H8O3	VANILLIN	-41.517	7.8333E-01	0.0000E+00	263	293	--------	164.500	188.000
777	C8H9NO	ACETANILIDE	-115.731	7.8917E-01	0.0000E+00	357	387	--------	165.725	189.400
778	C8H10	ETHYLBENZENE	-4.874	1.0061E+00	-2.2358E-03	20	170	--------	14.353	101.543
779	C8H10	m-XYLENE	3.913	8.4357E-01	-1.4305E-03	80	225	--------	62.244	121.357
780	C8H10	o-XYLENE	0.302	7.8782E-01	-8.1442E-04	20	247	--------	15.732	145.206

Table 4-1 HEAT CAPACITY OF SOLID - ORGANIC COMPOUNDS (continued)

NO	FORMULA	NAME	$C_P = A + B\,T + C\,T^2$					$(C_P$ - joule/(mol K), T - K)		
			A	B	C	TMIN	TMAX	C_P @25 C	C_P @TMIN	C_P @TMAX
781	C8H10	p-XYLENE	0.872	8.0786E-01	-9.5350E-04	153	286	-------	102.154	153.928
782	C8H10O	m-ETHYLPHENOL	-------	----------	----------	----	----	-------	-------	-------
783	C8H10O	p-ETHYLPHENOL	-50.130	8.6208E-01	0.0000E+00	268	298	206.900	181.038	206.900
784	C8H10O	PHENETOLE	68.013	1.0151E+00	0.0000E+00	143	173	-------	213.176	243.630
785	C8H10O	2-PHENYLETHANOL	-5.046	7.2083E-01	0.0000E+00	217	247	-------	151.375	173.000
786	C8H10O	2,3-XYLENOL	-38.312	7.2083E-01	0.0000E+00	263	293	-------	151.375	173.000
787	C8H10O	2,4-XYLENOL	-38.312	7.2083E-01	0.0000E+00	263	293	-------	151.375	173.000
788	C8H10O	2,5-XYLENOL	-38.312	7.2083E-01	0.0000E+00	263	293	-------	151.375	173.000
789	C8H10O	2,6-XYLENOL	-38.312	7.2083E-01	0.0000E+00	263	293	-------	151.375	173.000
790	C8H10O	3,4-XYLENOL	-38.312	7.2083E-01	0.0000E+00	263	293	-------	151.375	173.000
791	C8H10O	3,5-XYLENOL	-38.312	7.2083E-01	0.0000E+00	263	293	-------	151.375	173.000
792	C8H11N	N,N-DIMETHYLANILINE	-28.421	7.9833E-01	0.0000E+00	246	276	-------	167.650	191.600
793	C8H11N	o-ETHYLANILINE	10.738	7.9833E-01	0.0000E+00	197	227	-------	167.650	191.600
794	C8H11N	2,4,6-TRIMETHYLPYRIDINE	8.800	8.0000E-01	0.0000E+00	199	229	-------	168.000	192.000
795	C8H11NO	p-PHENETIDINE	-32.221	8.7083E-01	0.0000E+00	247	277	-------	182.875	209.000
796	C8H12	1,5-CYCLOOCTADIENE	-------	----------	----------	----	----	-------	-------	-------
797	C8H12	VINYLCYCLOHEXENE	-------	----------	----------	----	----	-------	-------	-------
798	C8H12O4	1,4-CYCLOHEXANEDICARBOXYLIC ACID	-53.814	1.0125E+00	0.0000E+00	263	293	-------	212.625	243.000
799	C8H12O4	DIETHYL MALEATE	-24.654	1.0125E+00	0.0000E+00	234	264	-------	212.625	243.000
800	C8H14O2	n-BUTYL METHACRYLATE	-------	----------	----------	----	----	-------	-------	-------
801	C8H14O3	BUTYRIC ANHYDRIDE	-------	----------	----------	----	----	-------	-------	-------
802	C8H14O4	DIETHYL SUCCINATE	-13.503	1.0933E+00	0.0000E+00	222	252	-------	229.600	262.400
803	C8H16	CYCLOOCTANE	-------	----------	----------	----	----	-------	-------	-------
804	C8H16	1,1-DIMETHYLCYCLOHEXANE	-1.562	9.3396E-01	-1.5232E-03	23	142	-------	18.880	100.151
805	C8H16	cis-1,2-DIMETHYLCYCLOHEXANE	-3.820	9.8512E-01	-1.6844E-03	22	168	-------	16.746	114.084
806	C8H16	trans-1,2-DIMETHYLCYCLOHEXANE	-7.038	1.0419E+00	-1.8746E-03	12	172	-------	5.185	116.891
807	C8H16	cis-1,3-DIMETHYLCYCLOHEXANE	-4.891	9.9783E-01	-1.5718E-03	97	182	-------	77.072	124.711
808	C8H16	trans-1,3-DIMETHYLCYCLOHEXANE	-3.795	9.9148E-01	-1.6578E-03	19	167	-------	14.872	115.645
809	C8H16	cis-1,4-DIMETHYLCYCLOHEXANE	-4.339	1.0090E+00	-1.7773E-03	19	174	-------	14.350	117.261
810	C8H16	trans-1,4-DIMETHYLCYCLOHEXANE	-6.722	1.0218E+00	-1.6474E-03	12	176	-------	5.568	121.982
811	C8H16	ETHYLCYCLOHEXANE	-5.813	1.0768E+00	-2.1102E-03	84	156	-------	69.749	110.815
812	C8H16	2-ETHYL-1-HEXENE	-------	----------	----------	----	----	-------	-------	-------
813	C8H16	1-METHYL-1-ETHYLCYCLOPENTANE	-------	----------	----------	----	----	-------	-------	-------
814	C8H16	1-OCTENE	-13.439	1.4959E+00	-3.5955E-03	11	151	-------	2.949	130.642
815	C8H16	trans-2-OCTENE	-------	----------	----------	----	----	-------	-------	-------
816	C8H16	trans-3-OCTENE	-------	----------	----------	----	----	-------	-------	-------
817	C8H16	trans-4-OCTENE	-------	----------	----------	----	----	-------	-------	-------
818	C8H16	n-PROPYLCYCLOPENTANE	-9.173	1.3018E+00	-3.0974E-03	25	156	-------	21.437	118.467
819	C8H16	2,4,4-TRIMETHYL-1-PENTENE	-4.998	1.1609E+00	-1.4462E-03	81	166	-------	79.456	147.728
820	C8H16	2,4,4-TRIMETHYL-2-PENTENE	45.300	1.3670E-01	4.0970E-03	81	144	-------	83.333	149.677
821	C8H16O	2-ETHYLHEXANAL	-------	----------	----------	----	----	-------	-------	-------
822	C8H16O	1-OCTANAL	-5.775	9.6250E-01	0.0000E+00	216	246	-------	202.125	231.000
823	C8H16O	2-OCTANONE	-4.352	1.2540E+00	-1.6558E-03	80	253	-------	85.373	206.868
824	C8H16O2	n-BUTYL n-BUTYRATE	-------	----------	----------	----	----	-------	-------	-------
825	C8H16O2	n-HEXYL ACETATE	-------	----------	----------	----	----	-------	-------	-------
826	C8H16O2	ISOBUTYL ISOBUTYRATE	-------	----------	----------	----	----	-------	-------	-------
827	C8H16O2	n-OCTANOIC ACID	241.472	-1.6054E+00	6.4938E-03	155	280	-------	148.652	301.081
828	C8H16O4	DIETHYLENE GLYCOL ETHYL ETHER ACETATE	-9.563	1.1733E+00	0.0000E+00	218	248	-------	246.400	281.600
829	C8H18	2,2-DIMETHYLHEXANE	-------	----------	----------	----	----	-------	-------	-------
830	C8H18	2,3-DIMETHYLHEXANE	-------	----------	----------	----	----	-------	-------	-------
831	C8H18	2,4-DIMETHYLHEXANE	-------	----------	----------	----	----	-------	-------	-------
832	C8H18	2,5-DIMETHYLHEXANE	-------	----------	----------	----	----	-------	-------	-------
833	C8H18	3,3-DIMETHYLHEXANE	-------	----------	----------	----	----	-------	-------	-------
834	C8H18	3,4-DIMETHYLHEXANE	-------	----------	----------	----	----	-------	-------	-------
835	C8H18	3-ETHYLHEXANE	-------	----------	----------	----	----	-------	-------	-------
836	C8H18	3-ETHYL-2-METHYLPENTANE	-------	----------	----------	----	----	-------	-------	-------
837	C8H18	3-METHYL-3-ETHYLPENTANE	-------	----------	----------	----	----	-------	-------	-------
838	C8H18	2-METHYLHEPTANE	-12.372	1.4704E+00	-3.1167E-03	20	164	-------	15.789	145.015
839	C8H18	3-METHYLHEPTANE	-14.209	1.5211E+00	-3.3521E-03	30	153	-------	28.408	139.854
840	C8H18	4-METHYLHEPTANE	-------	----------	----------	----	----	-------	-------	-------
841	C8H18	n-OCTANE	-8.951	1.3011E+00	-2.1009E-03	80	216	-------	81.690	174.207
842	C8H18	2,2,3-TRIMETHYLPENTANE	-------	----------	----------	----	----	-------	-------	-------
843	C8H18	2,2,4-TRIMETHYLPENTANE	4.600	9.6370E-01	-3.6600E-04	10	165	-------	14.200	153.646
844	C8H18	2,3,3-TRIMETHYLPENTANE	-------	----------	----------	----	----	-------	-------	-------
845	C8H18	2,3,4-TRIMETHYLPENTANE	44.466	5.8678E-01	0.0000E+00	160	164	-------	138.351	140.481
846	C8H18	2,2,3,3-TETRAMETHYLBUTANE	-------	----------	----------	----	----	-------	-------	-------
847	C8H18O	DI-n-BUTYL ETHER	-------	----------	----------	----	----	-------	-------	-------
848	C8H18O	DI-sec-BUTYL ETHER	-------	----------	----------	----	----	-------	-------	-------
849	C8H18O	DI-tert-BUTYL ETHER	-------	----------	----------	----	----	-------	-------	-------
850	C8H18O	2-ETHYL-1-HEXANOL	-------	----------	----------	----	----	-------	-------	-------
851	C8H18O	1-OCTANOL	41.740	5.7900E-01	2.7800E-04	102	217	-------	103.627	180.404
852	C8H18O	2-OCTANOL	78.530	-9.7000E-03	2.7190E-03	102	201	-------	105.775	186.539
853	C8H18O2	DI-t-BUTYL PEROXIDE	6.336	9.2500E-01	0.0000E+00	203	233	-------	194.250	222.000
854	C8H18O2S	DI-n-BUTYL SULFONE	-64.042	1.2083E+00	0.0000E+00	263	293	-------	253.750	290.000
855	C8H18O3	DIETHYLENE GLYCOL DIETHYL ETHER	13.194	1.1833E+00	0.0000E+00	199	229	-------	248.500	284.000
856	C8H18O3	DIETHYLENE GLYCOL MONOBUTYL EHTER	-------	----------	----------	----	----	-------	-------	-------
857	C8H18O4	TRIETHYLENE GLYCOL DIMETHYL ETHER	33.880	1.1049E+00	-9.1900E-04	90	229	-------	125.877	238.709
858	C8H18O5	TETRAETHYLENE GLYCOL	-37.275	1.3242E+00	0.0000E+00	238	268	-------	278.075	317.800

Table 4-1 HEAT CAPACITY OF SOLID - ORGANIC COMPOUNDS (continued)

| NO | FORMULA | NAME | $C_P = A + B T + C T^2$ | | | | | $(C_P$ - joule/(mol K), T - K$)$ | | |
			A	B	C	TMIN	TMAX	C_P @25 C	C_P @TMIN	C_P @TMAX
859	C8H18S	n-OCTYL MERCAPTAN	17.120	1.0667E+00	0.0000E+00	194	224	-------	224.000	256.000
860	C8H18S	tert-OCTYL MERCAPTAN	-------	----------	----------	----	----	-------	-------	-------
861	C8H18S	BUTYL-SULFIDE	-------	----------	----------	----	----	-------	-------	-------
862	C8H18S	ETHYL-HEXYL-SULFIDE	-------	----------	----------	----	----	-------	-------	-------
863	C8H18S	HEPTYL-METHYL-SULFIDE	-------	----------	----------	----	----	-------	-------	-------
864	C8H18S	PENTYL-PROPYL-SULFIDE	-------	----------	----------	----	----	-------	-------	-------
865	C8H18S2	BUTYL-DISULFIDE	-------	----------	----------	----	----	-------	-------	-------
866	C8H19N	DI-n-BUTYLAMINE	-------	----------	----------	----	----	-------	-------	-------
867	C8H19N	DIISOBUTYLAMINE	-------	----------	----------	----	----	-------	-------	-------
868	C8H19N	n-OCTYLAMINE	-36.066	1.1183E+00	0.0000E+00	242	272	-------	234.850	268.400
869	C8H23N5	TETRAETHYLENEPENTAMINE	-5.135	1.7117E+00	0.0000E+00	213	243	-------	359.450	410.800
870	C8H24O4Si4	OCTAMETHYLCYCLOTETRASILOXANE	147.930	9.4850E-01	0.0000E+00	175	250	-------	313.918	385.055
871	C9H4O5	TRIMELLITIC ANHYDRIDE	-608.740	2.8777E+00	-2.5946E-06	298	380	-------	249.016	484.411
872	C9H6N2O2	TOLUENE DIISOCYANATE	-41.376	8.7958E-01	0.0000E+00	257	287	-------	184.713	211.100
873	C9H7N	ISOQUINOLINE	-35.588	6.6958E-01	0.0000E+00	263	293	-------	140.613	160.700
874	C9H7N	QUINOLINE	69.492	-3.5624E-01	2.8466E-03	90	258	-------	60.488	167.339
875	C9H7NO	8-HYDROXYQUINOLINE	-39.308	7.4167E-01	0.0000E+00	263	293	-------	155.750	178.000
876	C9H8	INDENE	10.429	5.0740E-01	2.5934E-05	15	272	-------	18.046	150.204
877	C9H8O	2-METHYLBENZOFURAN	-------	----------	----------	----	----	-------	-------	-------
878	C9H10	INDANE	18.792	5.0875E-01	-2.4497E-04	40	222	-------	38.750	119.557
879	C9H10	cis-PROPENYLBENZENE	-------	----------	----------	----	----	-------	-------	-------
880	C9H10	trans-PROPENYLBENZENE	-------	----------	----------	----	----	-------	-------	-------
881	C9H1O	alpha-METHYLSTYRENE	-6.778	6.8125E-01	0.0000E+00	220	250	-------	143.063	163.500
882	C9H10	m-METHYLSTYRENE	-------	----------	----------	----	----	-------	-------	-------
883	C9H10	o-METHYLSTYRENE	-------	----------	----------	----	----	-------	-------	-------
884	C9H10	p-METHYLSTYRENE	0.668	6.8125E-01	0.0000E+00	209	239	-------	143.063	163.500
885	C9H10O2	BENZYL ACETATE	15.101	8.2292E-01	0.0000E+00	192	222	-------	172.813	197.500
886	C9H10O2	ETHYL BENZOATE	1.276	8.2292E-01	0.0000E+00	208	238	-------	172.813	197.500
887	C9H10O3	ETHYL VANILLIN	-47.503	8.9375E-01	0.0000E+00	263	293	-------	187.688	214.500
888	C9H11NO	p-DIMETHYLAMINOBENZALDEHYDE	-47.700	9.0000E-01	0.0000E+00	263	293	-------	189.000	216.000
889	C9H12	CUMENE	-0.034	9.3772E-01	-1.2691E-03	15	172	-------	13.332	123.507
890	C9H12	m-ETHYLTOLUENE	-------	----------	----------	----	----	-------	-------	-------
891	C9H12	o-ETHYLTOLUENE	-------	----------	----------	----	----	-------	-------	-------
892	C9H12	p-ETHYLTOLUENE	-------	----------	----------	----	----	-------	-------	-------
893	C9H12	MESITYLENE	19.437	7.5383E-01	-8.4533E-04	45	228	-------	51.641	147.536
894	C9H12	n-PROPYLBENZENE	-7.091	1.1479E+00	-2.4662E-03	70	174	-------	61.175	117.838
895	C9H12	1,2,3-TRIMETHYLBENZENE	21.629	5.0652E-01	9.3681E-04	50	219	-------	49.291	177.213
896	C9H12	1,2,4-TRIMETHYLBENZENE	40.490	4.8928E-01	0.0000E+00	180	229	-------	128.560	152.721
897	C9H12O	BENZYL ETHYL ETHER	-29.664	8.3208E-01	0.0000E+00	246	276	-------	174.738	199.700
898	C9H12O	2-PHENYL-2-PROPANOL	-44.292	8.3333E-01	0.0000E+00	263	293	-------	175.000	200.000
899	C9H12O2	CUMENE HYDROPEROXIDE	-21.905	9.0292E-01	0.0000E+00	234	264	-------	189.613	216.700
900	C9H14O	ISOPHORONE	-22.848	9.1208E-01	0.0000E+00	235	265	-------	191.538	218.900
901	C9H14O6	GLYCERYL TRIACETATE	-47.168	1.2663E+00	0.0000E+00	247	277	-------	265.913	303.900
902	C9H16	1-NONYNE	-------	----------	----------	----	----	-------	-------	-------
903	C9H16O4	AZELAIC ACID	-3.794	1.2046E+00	0.0000E+00	213	243	-------	252.963	289.100
904	C9H18	BUTYLCYCLOPENTANE	-------	----------	----------	----	----	-------	-------	-------
905	C9H18	cis,cis-1,3,5-TRIMETHYLCYCLOHEXANE	-------	----------	----------	----	----	-------	-------	-------
906	C9H18	cis,trans-1,3,5-TRIMETHYLCYCLOHEXANE	-------	----------	----------	----	----	-------	-------	-------
907	C9H18	ISOPROPYLCYCLOHEXANE	-------	----------	----------	----	----	-------	-------	-------
908	C9H18	1-NONENE	-------	----------	----------	----	----	-------	-------	-------
909	C9H18	n-PROPYLCYCLOHEXANE	-5.997	1.1335E+00	-1.6869E-03	70	178	-------	65.083	142.453
910	C9H18O	DIISOBUTYL KETONE	13.739	1.0708E+00	0.0000E+00	197	227	-------	224.875	257.000
911	C9H18O	1-NONANAL	-16.242	1.0721E+00	0.0000E+00	225	255	-------	225.138	257.300
912	C9H18O2	n-BUTYL VALERATE	-------	----------	----------	----	----	-------	-------	-------
913	C9H18O2	n-NONANOIC ACID	-52.174	1.1429E+00	0.0000E+00	256	286	-------	240.013	274.300
914	C9H18O2	n-OCTYL FORMATE	6.850	1.1417E+00	0.0000E+00	204	234	-------	239.750	274.000
915	C9H20	3,3-DIETHYLPENTANE	-1.137	1.2371E+00	-1.6237E-03	30	208	-------	34.515	186.077
916	C9H20	2,2-DIMETHYL-3-ETHYLPENTANE	-------	----------	----------	----	----	-------	-------	-------
917	C9H20	3-ETHYL-2,3-DIMETHYLPENTANE	-------	----------	----------	----	----	-------	-------	-------
918	C9H20	2,4-DIMETHYL-3-ETHYLPENTANE	-------	----------	----------	----	----	-------	-------	-------
919	C9H20	2,2-DIMETHYLHEPTANE	-------	----------	----------	----	----	-------	-------	-------
920	C9H20	2,6-DIMETHYLHEPTANE	-------	----------	----------	----	----	-------	-------	-------
921	C9H20	3-ETHYLHEPTANE	-------	----------	----------	----	----	-------	-------	-------
922	C9H20	4-ETHYLHEPTANE	-------	----------	----------	----	----	-------	-------	-------
923	C9H20	2,3-DIMETHYLHEPTANE	-------	----------	----------	----	----	-------	-------	-------
924	C9H20	2,4-DIMETHYLHEPTANE	-------	----------	----------	----	----	-------	-------	-------
925	C9H20	2,5-DIMETHYLHEPTANE	-------	----------	----------	----	----	-------	-------	-------
926	C9H20	3,4-DIMETHYLHEPTANE	-------	----------	----------	----	----	-------	-------	-------
927	C9H20	3,5-DIMETHYLHEPTANE	-------	----------	----------	----	----	-------	-------	-------
928	C9H20	4,4-DIMETHYLHEPTANE	-------	----------	----------	----	----	-------	-------	-------
929	C9H20	3-ETHYL-2-METHYLHEXANE	-------	----------	----------	----	----	-------	-------	-------
930	C9H20	4-ETHYL-2-METHYLHEXANE	-------	----------	----------	----	----	-------	-------	-------
931	C9H20	3-ETHYL-3-METHYLHEXANE	-------	----------	----------	----	----	-------	-------	-------
932	C9H20	3-ETHYL-4-METHYLHEXANE	-------	----------	----------	----	----	-------	-------	-------
933	C9H20	2,2,3-TRIMETHYLHEXANE	-------	----------	----------	----	----	-------	-------	-------
934	C9H20	2,2,4-TRIMETHYLHEXANE	-------	----------	----------	----	----	-------	-------	-------
935	C9H20	2,3,3-TRIMETHYLHEXANE	-------	----------	----------	----	----	-------	-------	-------
936	C9H20	2,3,4-TRIMETHYLHEXANE	-------	----------	----------	----	----	-------	-------	-------

97

Table 4-1 HEAT CAPACITY OF SOLID - ORGANIC COMPOUNDS (continued)

NO	FORMULA	NAME	$C_P = A + B T + C T^2$					$(C_P$ - joule/(mol K), T - K)$		
			A	B	C	TMIN	TMAX	C_P @25 C	C_P @TMIN	C_P @TMAX
937	C9H20	2,3,5-TRIMETHYLHEXANE	-------	----------	----------	----	----	-------	-------	-------
938	C9H20	2,4,4-TRIMETHYLHEXANE	-------	----------	----------	----	----	-------	-------	-------
939	C9H20	3,3,4-TRIMETHYLHEXANE	-------	----------	----------	----	----	-------	-------	-------
940	C9H20	2-METHYLOCTANE	-------	----------	----------	----	----	-------	-------	-------
941	C9H20	3-METHYLOCTANE	-------	----------	----------	----	----	-------	-------	-------
942	C9H20	4-METHYLOCTANE	-------	----------	----------	----	----	-------	-------	-------
943	C9H20	n-NONANE	-8.091	1.3659E+00	-1.8786E-03	116	217	-------	125.168	199.955
944	C9H20	2,2,3,3-TETRAMETHYLPENTANE	-9.850	1.2867E+00	-1.5480E-03	20	174	-------	15.265	167.505
945	C9H20	2,2,3,4-TETRAMETHYLPENTANE	-------	----------	----------	----	----	-------	-------	-------
946	C9H20	2,2,4,4-TETRAMETHYLPENTANE	5.438	8.4939E-01	9.7511E-04	20	207	-------	22.816	222.555
947	C9H20	2,3,3,4-TETRAMETHYLPENTANE	-------	----------	----------	----	----	-------	-------	-------
948	C9H20	2,2,5-TRIMETHYLHEXANE	-------	----------	----------	----	----	-------	-------	-------
949	C9H20O	2,6-DIMETHYL-4-HEPTANOL	-------	----------	----------	----	----	-------	-------	-------
950	C9H20O	1-NONANOL	-32.431	1.1521E+00	0.0000E+00	238	268	-------	241.938	276.500
951	C9H20O	2-NONANOL	2.135	1.1542E+00	0.0000E+00	208	238	-------	242.375	277.000
952	C9H20S	n-NONYL MERCAPTAN	-15.334	1.1750E+00	0.0000E+00	223	253	-------	246.750	282.000
953	C9H20S	BUTYL-PENTYL-SULFIDE	-------	----------	----------	----	----	-------	-------	-------
954	C9H20S	ETHYL-HEPTYL-SULFIDE	-------	----------	----------	----	----	-------	-------	-------
955	C9H20S	HEXYL-PROPYL-SULFIDE	-------	----------	----------	----	----	-------	-------	-------
956	C9H20S	METHYL-OCTYL-SULFIDE	-------	----------	----------	----	----	-------	-------	-------
957	C9H21N	n-NONYLAMINE	-40.761	1.2296E+00	0.0000E+00	243	273	-------	258.213	295.100
958	C9H21N	TRIPROPYLAMINE	-------	----------	----------	----	----	-------	-------	-------
959	C10H6O8	PYROMELLITIC ACID	-59.404	1.1208E+00	0.0000E+00	263	293	-------	235.375	269.000
960	C10H7Br	1-BROMONAPHTHALENE	-27.578	7.0083E-01	0.0000E+00	249	279	-------	147.175	168.200
961	C10H7Cl	1-CHLORONAPHTHALENE	-20.429	7.0083E-01	0.0000E+00	239	269	-------	147.175	168.200
962	C10H8	NAPHTHALENE	4.824	5.0634E-01	1.8503E-04	100	353	172.238	57.308	206.893
963	C10H8	AZULENE	-------	----------	----------	----	----	-------	-------	-------
964	C10H9N	QUINALDINE	-25.104	7.8083E-01	0.0000E+00	242	272	-------	163.975	187.400
965	C10H10	m-DIVINYLBENZENE	-------	----------	----------	----	----	-------	-------	-------
966	C10H10	1-METHYLINDENE	-37.869	7.1250E-01	0.0000E+00	263	293	-------	149.625	171.000
967	C10H10	2-METHYLINDENE	-37.869	7.1250E-01	0.0000E+00	263	293	-------	149.625	171.000
968	C10H10O4	DIMETHYL PHTHALATE	-32.016	9.9583E-01	0.0000E+00	242	272	-------	209.125	239.000
969	C10H10O4	DIMETHYL TEREPHTHALATE	137.330	2.6200E-01	5.0000E-04	317	404	-------	270.831	324.573
970	C10H12	DICYCLOPENTADIENE	-53.098	7.9250E-01	0.0000E+00	277	307	183.186	166.425	190.200
971	C10H12	1,2,3,4-TETRAHYDRONAPHTHALENE	5.380	6.5363E-01	-2.3607E-04	124	237	-------	82.616	147.236
972	C10H12O	ANETHOLE	-45.713	8.6250E-01	0.0000E+00	263	293	-------	181.125	207.000
973	C10H12O4	DIALLYL MALEATE	14.900	1.0758E+00	0.0000E+00	196	226	-------	225.925	258.200
974	C10H14	n-BUTYLBENZENE	-4.751	1.1654E+00	-1.8847E-03	80	185	-------	76.414	146.474
975	C10H14	sec-BUTYLBENZENE	-------	----------	----------	----	----	-------	-------	-------
976	C10H14	tert-BUTYLBENZENE	32.500	4.0870E-01	1.1130E-03	92	215	-------	79.644	172.059
977	C10H14	1,2,3,4-TETRAMETHYLBENZENE	-------	----------	----------	----	----	-------	-------	-------
978	C10H14	m-CYMENE	-------	----------	----------	----	----	-------	-------	-------
979	C10H14	o-CYMENE	-------	----------	----------	----	----	-------	-------	-------
980	C10H14	p-CYMENE	49.440	2.8030E-01	1.5500E-03	92	180	-------	88.460	149.946
981	C10H14	m-DIETHYLBENZENE	-------	----------	----------	----	----	-------	-------	-------
982	C10H14	o-DIETHYLBENZENE	-1.658	8.7250E-01	0.0000E+00	212	242	-------	183.225	209.400
983	C10H14	p-DIETHYLBENZENE	8.463	8.7250E-01	0.0000E+00	200	230	-------	183.225	209.400
984	C1OH14	2-ETHYL-m-XYLENE	-14.708	8.7083E-01	0.0000E+00	227	257	-------	182.875	209.000
985	C10H14	2-ETHYL-p-XYLENE	17.835	8.7083E-01	0.0000E+00	190	220	-------	182.875	209.000
986	C10H14	3-ETHYL-o-XYLENE	14.274	8.7250E-01	0.0000E+00	194	224	-------	183.225	209.400
987	C10H14	4-ETHYL-m-XYLENE	-------	----------	----------	----	----	-------	-------	-------
988	C10H14	4-ETHYL-o-XYLENE	-------	----------	----------	----	----	-------	-------	-------
989	C1OH14	5-ETHYL-m-XYLENE	-------	----------	----------	----	----	-------	-------	-------
990	C10H14	ISOBUTYLBENZENE	15.967	8.7250E-01	0.0000E+00	192	222	-------	183.225	209.400
991	C10H14	1,2,3,5-TETRAMETHYLBENZENE	20.900	8.0800E-01	-5.1400E-04	92	249	-------	91.171	190.477
992	C10H14	1,2,4,5-TETRAMETHYLBENZENE	22.840	7.5380E-01	-3.7700E-04	92	352	214.073	89.136	241.651
993	C10H14O	p-tert-BUTYLPHENOL	-50.138	9.4333E-01	0.0000E+00	263	293	-------	198.100	226.400
994	C10H14O2	p-tert-BUTYLCATECHOL	-53.903	1.0142E+00	0.0000E+00	263	293	-------	212.975	243.400
995	C10H15N	N,N-DIETHYLANILINE	4.951	1.0208E+00	0.0000E+00	205	235	-------	214.375	245.000
996	C10H15N	2,6-DIETHYLANILINE	-37.414	1.0208E+00	0.0000E+00	247	277	-------	214.375	245.000
997	C10H16	CAMPHENE	-53.295	8.8825E-01	0.0000E+00	270	300	211.537	186.533	213.180
998	C10H16	D-LIMONENE	-------	----------	----------	----	----	-------	-------	-------
999	C10H16	alpha-PHELLANDRENE	-50.571	9.5417E-01	0.0000E+00	263	293	-------	200.375	229.000
1000	C1OH16	beta-PHELLANDRENE	-50.571	9.5417E-01	0.0000E+00	263	293	-------	200.375	229.000
1001	C10H16	alpha-PINENE	-------	----------	----------	----	----	-------	-------	-------
1002	C10H16	beta-PINENE	-------	----------	----------	----	----	-------	-------	-------
1003	C10H16	alpha-TERPINENE	-------	----------	----------	----	----	-------	-------	-------
1004	C1OH16	gamma-TERPINENE	-------	----------	----------	----	----	-------	-------	-------
1005	C10H16	TERPINOLENE	-------	----------	----------	----	----	-------	-------	-------
1006	C10H16O	CAMPHOR	230.050	1.3811E-01	0.0000E+00	293	452	271.227	270.537	292.414
1007	C10H18	1-DECYNE	-------	----------	----------	----	----	-------	-------	-------
1008	C10H18	cis-DECAHYDRONANPHTALENE	3.789	7.6501E-01	-5.6368E-04	25	216	-------	22.561	142.793
1009	C10H18	trans-DECAHYDRONAPHTHALENE	-2.881	1.0325E+00	0.0000E+00	213	243	-------	216.825	247.800
1010	C10H18O4	SEBACIC ACID	-69.937	1.3158E+00	0.0000E+00	263	293	-------	276.325	315.800
1011	C10H20	n-BUTYLCYCLOHEXANE	-8.420	1.3246E+00	-2.4352E-03	20	183	-------	16.717	152.328
1012	C10H20	1-CYCLOPENTYLPENTANE	-------	----------	----------	----	----	-------	-------	-------
1013	C10H20	1-DECENE	-11.788	1.5467E+00	-2.5491E-03	21	196	-------	19.050	193.309
1014	C10H20O	1-DECANAL	-20.294	1.1833E+00	0.0000E+00	227	257	-------	248.500	284.000

98

Table 4-1 HEAT CAPACITY OF SOLID - ORGANIC COMPOUNDS (continued)

NO	FORMULA	NAME	$C_P = A + B\,T + C\,T^2$					$(C_P$ - joule/(mol K), T - K)		
			A	B	C	TMIN	TMAX	C_P @25 C	C_P @TMIN	C_P @TMAX
1015	C10H20O2	n-DECANOIC ACID	244.282	-1.9766E+00	8.9547E-03	90	305	450.971	138.921	473.558
1016	C10H20O2	2-ETHYLHEXYL ACETATE	-------	----------	----------	----	----	-------	-------	-------
1017	C10H20O2	ISOPENTYL ISOVALERATE	-------	----------	----------	----	----	-------	-------	-------
1018	C10H22	n-DECANE	-7.702	1.4074E+00	-1.7457E-03	80	243	-------	93.718	231.492
1019	C10H22	2-METHYLNONANE	299.608	-3.7540E+00	1.9209E-02	80	199	-------	122.227	311.324
1020	C10H22	3-METHYLNONANE	92.871	-2.3410E-01	4.7645E-03	80	188	-------	104.635	217.802
1021	C10H22	4-METHYLNONANE	249.416	-3.1673E+00	1.8336E-02	80	174	-------	113.377	254.877
1022	C10H22	5-METHYLNONANE	112.162	-5.2081E-01	5.9968E-03	80	185	-------	108.877	221.819
1023	C10H22	3-ETHYLOCTANE	-------	----------	----------	----	----	-------	-------	-------
1024	C10H22	4-ETHYLOCTANE	-------	----------	----------	----	----	-------	-------	-------
1025	C10H22	2,2-DIMETHYLOCTANE	-------	----------	----------	----	----	-------	-------	-------
1026	C10H22	2,3-DIMETHYLOCTANE	-------	----------	----------	----	----	-------	-------	-------
1027	C10H22	2,4-DIMETHYLOCTANE	-------	----------	----------	----	----	-------	-------	-------
1028	C10H22	2,5-DIMETHYLOCTANE	-------	----------	----------	----	----	-------	-------	-------
1029	C10H22	2,6-DIMETHYLOCTANE	-------	----------	----------	----	----	-------	-------	-------
1030	C10H22	2,7-DIMETHYLOCTANE	-------	----------	----------	----	----	-------	-------	-------
1031	C10H22	3,3-DIMETHYLOCTANE	-------	----------	----------	----	----	-------	-------	-------
1032	C10H22	3,4-DIMETHYLOCTANE	-------	----------	----------	----	----	-------	-------	-------
1033	C10H22	3,5-DIMETHYLOCTANE	-------	----------	----------	----	----	-------	-------	-------
1034	C10H22	3,6-DIMETHYLOCTANE	-------	----------	----------	----	----	-------	-------	-------
1035	C10H22	4,4-DIMETHYLOCTANE	-------	----------	----------	----	----	-------	-------	-------
1036	C10H22	4,5-DIMETHYLOCTANE	-------	----------	----------	----	----	-------	-------	-------
1037	C10H22	4-PROPYLHEPTANE	-------	----------	----------	----	----	-------	-------	-------
1038	C10H22	4-ISOPROPYLHEPTANE	-------	----------	----------	----	----	-------	-------	-------
1039	C10H22	3-ETHYL-2-METHYLHEPTANE	-------	----------	----------	----	----	-------	-------	-------
1040	C10H22	4-ETHYL-2-METHYLHEPTANE	-------	----------	----------	----	----	-------	-------	-------
1041	C10H22	5-ETHYL-2-METHYLHEPTANE	-------	----------	----------	----	----	-------	-------	-------
1042	C10H22	3-ETHYL-3-METHYLHEPTANE	-------	----------	----------	----	----	-------	-------	-------
1043	C10H22	4-ETHYL-3-METHYLHEPTANE	-------	----------	----------	----	----	-------	-------	-------
1044	C10H22	3-ETHYL-5-METHYLHEPTANE	-------	----------	----------	----	----	-------	-------	-------
1045	C10H22	3-ETHYL-4-METHYLHEPTANE	-------	----------	----------	----	----	-------	-------	-------
1046	C10H22	4-ETHYL-4-METHYLHEPTANE	-------	----------	----------	----	----	-------	-------	-------
1047	C10H22	2,2,3-TRIMETHYLHEPTANE	-------	----------	----------	----	----	-------	-------	-------
1048	C10H22	2,2,4-TRIMETHYLHEPTANE	-------	----------	----------	----	----	-------	-------	-------
1049	C10H22	2,2,5-TRIMETHYLHEPTANE	-------	----------	----------	----	----	-------	-------	-------
1050	C10H22	2,2,6-TRIMETHYLHEPTANE	-------	----------	----------	----	----	-------	-------	-------
1051	C10H22	2,3,3-TRIMETHYLHEPTANE	-------	----------	----------	----	----	-------	-------	-------
1052	C10H22	2,3,4-TRIMETHYLHEPTANE	-------	----------	----------	----	----	-------	-------	-------
1053	C10H22	2,3,5-TRIMETHYLHEPTANE	-------	----------	----------	----	----	-------	-------	-------
1054	C10H22	2,3,6-TRIMETHYLHEPTANE	-------	----------	----------	----	----	-------	-------	-------
1055	C10H22	2,4,4-TRIMETHYLHEPTANE	-------	----------	----------	----	----	-------	-------	-------
1056	C10H22	2,4,5-TRIMETHYLHEPTANE	-------	----------	----------	----	----	-------	-------	-------
1057	C10H22	2,4,6-TRIMETHYLHEPTANE	-------	----------	----------	----	----	-------	-------	-------
1058	C10H22	2,5,5-TRIMETHYLHEPTANE	-------	----------	----------	----	----	-------	-------	-------
1059	C10H22	3,3,4-TRIMETHYLHEPTANE	-------	----------	----------	----	----	-------	-------	-------
1060	C10H22	3,3,5-TRIMETHYLHEPTANE	-------	----------	----------	----	----	-------	-------	-------
1061	C10H22	3,4,4-TRIMETHYLHEPTANE	-------	----------	----------	----	----	-------	-------	-------
1062	C10H22	3,4,5-TRIMETHYLHEPTANE	-------	----------	----------	----	----	-------	-------	-------
1063	C10H22	3-ISOPROPYL-2-METHYLHEXANE	-------	----------	----------	----	----	-------	-------	-------
1064	C10H22	3,3-DIETHYLHEXANE	-------	----------	----------	----	----	-------	-------	-------
1065	C10H22	3,4-DIETHYLHEXANE	-------	----------	----------	----	----	-------	-------	-------
1066	C10H22	3-ETHYL-2,2-DIMETHYLHEXANE	-------	----------	----------	----	----	-------	-------	-------
1067	C10H22	4-ETHYL-2,2-DIMETHYLHEXANE	-------	----------	----------	----	----	-------	-------	-------
1068	C10H22	3-ETHYL-2,3-DIMETHYLHEXANE	-------	----------	----------	----	----	-------	-------	-------
1069	C10H22	4-ETHYL-2,3-DIMETHYLHEXANE	-------	----------	----------	----	----	-------	-------	-------
1070	C10H22	3-ETHYL-2,4-DIMETHYLHEXANE	-------	----------	----------	----	----	-------	-------	-------
1071	C10H22	4-ETHYL-2,4-DIMETHYLHEXANE	-------	----------	----------	----	----	-------	-------	-------
1072	C10H22	3-ETHYL-2,5-DIMETHYLHEXANE	-------	----------	----------	----	----	-------	-------	-------
1073	C10H22	4-ETHYL-3,3-DIMETHYLHEXANE	-------	----------	----------	----	----	-------	-------	-------
1074	C10H22	3-ETHYL-3,4-DIMETHYLHEXANE	-------	----------	----------	----	----	-------	-------	-------
1075	C10H22	2,2,3,3-TETRAMETHYLHEXANE	-------	----------	----------	----	----	-------	-------	-------
1076	C10H22	2,2,3,4-TETRAMETHYLHEXANE	-------	----------	----------	----	----	-------	-------	-------
1077	C10H22	2,2,3,5-TETRAMETHYLHEXANE	-------	----------	----------	----	----	-------	-------	-------
1078	C10H22	2,2,4,4-TETRAMETHYLHEXANE	-------	----------	----------	----	----	-------	-------	-------
1079	C10H22	2,2,4,5-TETRAMETHYLHEXANE	-------	----------	----------	----	----	-------	-------	-------
1080	C10H22	2,2,5,5-TETRAMETHYLHEXANE	-------	----------	----------	----	----	-------	-------	-------
1081	C10H22	2,3,3,4-TETRAMETHYLHEXANE	-------	----------	----------	----	----	-------	-------	-------
1082	C10H22	2,3,3,5-TETRAMETHYLHEXANE	-------	----------	----------	----	----	-------	-------	-------
1083	C10H22	2,3,4,4-TETRAMETHYLHEXANE	-------	----------	----------	----	----	-------	-------	-------
1084	C10H22	2,3,4,5-TETRAMETHYLHEXANE	-------	----------	----------	----	----	-------	-------	-------
1085	C10H22	3,3,4,4-TETRAMETHYLHEXANE	-------	----------	----------	----	----	-------	-------	-------
1086	C10H22	2,4-DIMETHYL-3-ISOPROPYLPENTANE	-------	----------	----------	----	----	-------	-------	-------
1087	C10H22	3,3-DIETHYL-2-METHYLPENTANE	-------	----------	----------	----	----	-------	-------	-------
1088	C10H22	3-ETHYL-2,2,3-TRIMETHYLPENTANE	-------	----------	----------	----	----	-------	-------	-------
1089	C10H22	3-ETHYL-2,2,4-TRIMETHYLPENTANE	-------	----------	----------	----	----	-------	-------	-------
1090	C10H22	3-ETHYL-2,3,4-TRIMETHYLPENTANE	-------	----------	----------	----	----	-------	-------	-------
1091	C10H22	2,2,3,3,4-PENTAMETHYLPENTANE	-------	----------	----------	----	----	-------	-------	-------
1092	C10H22	2,2,3,4,4-PENTAMETHYLPENTANE	-------	----------	----------	----	----	-------	-------	-------

Table 4-1 HEAT CAPACITY OF SOLID - ORGANIC COMPOUNDS (continued)

NO	FORMULA	NAME	$C_P = A + BT + CT^2$					$(C_P$ - joule/(mol K), T - K)		
			A	B	C	TMIN	TMAX	C_P @25 C	C_P @TMIN	C_P @TMAX
1093	C10H22O	1-DECANOL	-50.596	1.2633E+00	0.0000E+00	250	280	-------	265.300	303.200
1094	C10H22O	DI-n-PENTYL ETHER	-------	----------	----------	----	----	-------	-------	-------
1095	C10H22O	ISODECANOL	-------	----------	----------	----	----	-------	-------	-------
1096	C10H22O5	TETRAETHYLENE GLYCOL DIMETHYL ETHER	-5.333	1.5458E+00	0.0000E+00	213	243	-------	324.625	371.000
1097	C10H22S	n-DECYL MERCAPTAN	24.926	9.7074E-01	2.4630E-04	51	248	-------	74.626	280.337
1098	C10H22S	BUTYL-HEXYL-SULFIDE	-------	----------	----------	----	----	-------	-------	-------
1099	C10H22S	ETHYL-OCTYL-SULFIDE	-------	----------	----------	----	----	-------	-------	-------
1100	C10H22S	HEPTYL-PROPYL-SULFIDE	-------	----------	----------	----	----	-------	-------	-------
1101	C10H22S	METHYL-NONYL-SULFIDE	-------	----------	----------	----	----	-------	-------	-------
1102	C10H22S	PENTYL-SULFIDE	-------	----------	----------	----	----	-------	-------	-------
1103	C10H22S2	PENTYL-DISULFIDE	-------	----------	----------	----	----	-------	-------	-------
1104	C10H23N	n-DECYLAMINE	-65.500	1.3408E+00	0.0000E+00	259	289	-------	281.575	321.800
1105	C11H10	1-METHYLNAPHTHALENE	0.729	7.7009E-01	-5.7996E-04	15	243	-------	12.150	153.453
1106	C11H10	2-METHYLNAPHTHALENE	5.423	6.6908E-01	-1.1430E-04	100	308	194.747	71.187	200.494
1107	C11H14O2	n-BUTYL BENZOATE	-12.184	1.0458E+00	0.0000E+00	222	252	-------	219.625	251.000
1108	C11H16	n-PENTYLBENZENE	-------	----------	----------	----	----	-------	-------	-------
1109	C11H16O	p-tert-AMYLPHENOL	-56.051	1.0546E+00	0.0000E+00	263	293	-------	221.463	253.100
1110	C11H20	1-UNDECYNE	-------	----------	----------	----	----	-------	-------	-------
1111	C11H20O2	2-ETHYLHEXYL ACRYLATE	-------	----------	----------	----	----	-------	-------	-------
1112	C11H22	1-UNDECENE	-10.181	1.5773E+00	-2.2155E-03	118	217	-------	145.594	227.946
1113	C11H22	1-CYCLOPENTYLHEXANE	-------	----------	----------	----	----	-------	-------	-------
1114	C11H22	PENTYLCYCLOHEXANE	-------	----------	----------	----	----	-------	-------	-------
1115	C11H22O	1-UNDECANAL	-42.957	1.2958E+00	0.0000E+00	243	273	-------	272.125	311.000
1116	C11H24	n-UNDECANE	-4.481	1.5629E+00	-2.3378E-03	40	214	-------	54.060	222.954
1117	C11H24O	1-UNDECANOL	-67.423	1.3746E+00	0.0000E+00	259	289	-------	288.663	329.900
1118	C11H24S	UNDECYL MERCAPTAN	-42.210	1.4000E+00	0.0000E+00	240	270	-------	294.000	336.000
1119	C11H24S	BUTYL-HEPTYL-SULFIDE	-------	----------	----------	----	----	-------	-------	-------
1120	C11H24S	DECYL-METHYL-SULFIDE	-------	----------	----------	----	----	-------	-------	-------
1121	C11H24S	ETHYL-NONYL-SULFIDE	-------	----------	----------	----	----	-------	-------	-------
1122	C11H24S	OCTYL-PROPYL-SULFIDE	-------	----------	----------	----	----	-------	-------	-------
1123	C12H8O	DIBENZOFURAN	-40.633	7.6667E-01	0.0000E+00	263	293	-------	161.000	184.000
1124	C12H9N	DIBENZOPYRROLE	-44.823	8.4333E-01	0.0000E+00	263	293	-------	177.100	202.400
1125	C12H10	ACENAPHTHENE	5.620	5.4669E-01	3.6356E-04	100	367	200.934	63.924	254.865
1126	C12H10	BIPHENYL	36.860	3.4260E-01	6.3960E-04	185	342	195.862	122.131	228.839
1127	C12H10O	DIPHENYL ETHER	16.787	6.5093E-01	7.0817E-05	30	300	217.156	36.378	218.439
1128	C12H11N	p-AMINODIPHENYL	-49.025	9.2500E-01	0.0000E+00	263	293	-------	194.250	222.000
1129	C12H11N	DIPHENYLAMINE	-58.807	9.9421E-01	0.0000E+00	269	299	237.616	208.784	238.610
1130	C12H11N3	p-AMINOAZOBENZENE	-69.035	1.1506E+00	0.0000E+00	270	300	274.011	241.623	276.140
1131	C12H11N3	1,3-DIPHENYLTRIAZENE	-68.500	1.1417E+00	0.0000E+00	270	300	271.888	239.750	274.000
1132	C12H12	1,2-DIMETHYLNAPHTHALENE	-------	----------	----------	----	----	-------	-------	-------
1133	C12H12	1,3-DIMETHYLNAPHTHALENE	-------	----------	----------	----	----	-------	-------	-------
1134	C12H12	1,4-DIMETHYLNAPHTHALENE	-------	----------	----------	----	----	-------	-------	-------
1135	C12H12	1,5-DIMETHYLNAPHTHALENE	-------	----------	----------	----	----	-------	-------	-------
1136	C12H12	1,6-DIMETHYLNAPHTHALENE	-------	----------	----------	----	----	-------	-------	-------
1137	C12H12	1,7-DIMETHYLNAPHTHALENE	-------	----------	----------	----	----	-------	-------	-------
1138	C12H12	2,3-DIMETHYLNAPHTHALENE	-------	----------	----------	----	----	-------	-------	-------
1139	C12H12	2,6-DIMETHYLNAPHTHALENE	11.075	7.3623E-01	-2.5796E-04	22	337	207.650	26.937	229.770
1140	C12H12	2,7-DIMETHYLNAPHTHALENE	19.711	6.2158E-01	5.0747E-05	38	361	209.547	43.223	250.979
1141	C12H12	1-ETHYLNAPHTHALENE	-16.520	8.5417E-01	0.0000E+00	229	259	-------	179.375	205.000
1142	C12H12	2-ETHYLNAPHTHALENE	-------	----------	----------	----	----	-------	-------	-------
1143	C12H12N2	p-AMINODIPHENYLAMINE	-56.959	1.0717E+00	0.0000E+00	263	293	-------	225.050	257.200
1144	C12H12N2	HYDRAZOBENZENE	-56.959	1.0717E+00	0.0000E+00	263	293	-------	225.050	257.200
1145	C12H14	1,2,3-TRIMETHYLINDENE	-49.695	9.3500E-01	0.0000E+00	263	293	-------	196.350	224.400
1146	C12H14O4	DIETHYL PHTHALATE	10.992	1.3565E+00	-1.3824E-03	30	269	-------	50.442	275.944
1147	C12H16	CYCLOHEXYLBENZENE	-29.182	8.8023E-01	0.0000E+00	220	280	-------	164.469	217.406
1148	C12H18	m-DIISOPROPYLBENZENE	-------	----------	----------	----	----	-------	-------	-------
1149	C12H18	p-DIISOPROPYLBENZENE	-17.608	1.0950E+00	0.0000E+00	226	256	-------	229.950	262.800
1150	C12H18	n-HEXYLBENZENE	-------	----------	----------	----	----	-------	-------	-------
1151	C12H18	1,2,3-TRIETHYLBENZENE	-------	----------	----------	----	----	-------	-------	-------
1152	C12H18	1,2,4-TRIETHYLBENZENE	-------	----------	----------	----	----	-------	-------	-------
1153	C12H18	1,3,5-TRIETHYLBENZENE	-------	----------	----------	----	----	-------	-------	-------
1154	C12H18	HEXAMETHYLBENZENE	-------	----------	----------	----	----	-------	-------	-------
1155	C12H20O4	DIBUTYL MALEATE	-------	----------	----------	----	----	-------	-------	-------
1156	C12H22	BICYCLOHEXYL	-46.159	1.2550E+00	0.0000E+00	247	277	-------	263.550	301.200
1157	C12H22	1-DODECYNE	-------	----------	----------	----	----	-------	-------	-------
1158	C12H23N	DICYCLOHEXYLAMINE	-46.338	1.4042E+00	0.0000E+00	243	273	-------	294.875	337.000
1159	C12H24	1-DODECENE	-6.158	1.5234E+00	-9.9773E-04	80	213	-------	109.328	272.947
1160	C12H24	1-CYCLOPENTYLHEPTANE	-------	----------	----------	----	----	-------	-------	-------
1161	C12H24	1-CYCLOHEXYLHEXANE	-------	----------	----------	----	----	-------	-------	-------
1162	C12H24O	1-DODECANAL	-63.473	1.4058E+00	0.0000E+00	255	285	-------	295.225	337.400
1163	C12H24O2	n-DODECANOIC ACID	202.564	-1.1243E+00	6.4399E-03	90	310	439.816	153.540	472.903
1164	C12H26	n-DODECANE	-6.075	1.5507E+00	-1.7008E-03	100	260	-------	131.986	282.128
1165	C12H26O	DI-n-HEXYL ETHER	14.635	1.4858E+00	0.0000E+00	200	230	-------	312.025	356.600
1166	C12H26O	1-DODECANOL	-78.972	1.4858E+00	0.0000E+00	263	293	-------	312.025	356.600
1167	C12H26O3	DIETHYLENE GLYCOL DI-n-BUTYL ETHER	13.194	1.1833E+00	0.0000E+00	199	229	-------	248.500	284.000
1168	C12H26S	n-DODECYL MERCAPTAN	-38.312	1.5233E+00	0.0000E+00	235	265	-------	319.900	365.600
1169	C12H26S	BUTYL-OCTYL-SULFIDE	-------	----------	----------	----	----	-------	-------	-------
1170	C12H26S	DECYL-ETHYL-SULFIDE	-------	----------	----------	----	----	-------	-------	-------

Table 4-1 HEAT CAPACITY OF SOLID - ORGANIC COMPOUNDS (continued)

NO	FORMULA	NAME	$C_P = A + B\,T + C\,T^2$					$(C_P$ - joule/(mol K), T - K)		
			A	B	C	TMIN	TMAX	C_P @25 C	C_P @TMIN	C_P @TMAX
1171	C12H26S	HEXYL-SULFIDE	-------	----------	----------	----	----	-------	--------	-------
1172	C12H26S	METHYL-UNDECYL-SULFIDE	-------	----------	----------	----	----	-------	--------	-------
1173	C12H26S	NONYL-PROPYL-SULFIDE	-------	----------	----------	----	----	-------	--------	-------
1174	C12H26S2	HEXYL-DISULFIDE	-------	----------	----------	----	----	-------	--------	-------
1175	C12H27BO3	TRI-n-BUTYL BORATE	-------	----------	----------	----	----	-------	--------	-------
1176	C12H27N	DODECYLAMINE	-83.091	1.5633E+00	0.0000E+00	263	293	-------	328.300	375.200
1177	C12H27N	TRI-n-BUTYLAMINE	-------	----------	----------	----	----	-------	--------	-------
1178	C13H10	FLUORENE	58.400	8.3600E-02	1.3470E-03	288	388	203.065	194.202	293.552
1179	C13H10O	BENZOPHENONE	-10.965	8.0530E-01	6.1852E-05	123	293	-------	89.146	230.425
1180	C13H12	DIPHENYLMETHANE	44.860	3.0550E-01	1.0830E-03	75	298	232.216	73.864	232.445
1181	C13H14	1-PROPYLNAPHTHALENE	-------	----------	----------	----	----	-------	--------	-------
1182	C13H14	2-PROPYLNAPHTHALENE	-------	----------	----------	----	----	-------	--------	-------
1183	C13H14	2ETHYL-3-METHYLNAPHTHALENE	-------	----------	----------	----	----	-------	--------	-------
1184	C13H14	2ETHYL-6-METHYLNAPHTHALENE	-------	----------	----------	----	----	-------	--------	-------
1185	C13H14	2ETHYL-7-METHYLNAPHTHALENE	-------	----------	----------	----	----	-------	--------	-------
1186	C13H20	n-HEPTYLBENZENE	17.944	1.2083E+00	0.0000E+00	195	225	-------	253.750	290.000
1187	C13H24	1-TRIDECYNE	-------	----------	----------	----	----	-------	--------	-------
1188	C13H26	1-TRIDECENE	-14.536	1.4421E+00	0.0000E+00	220	250	-------	302.838	346.100
1189	C13H26	1-CYCLOPENTYLOCTANE	-------	----------	----------	----	----	-------	--------	-------
1190	C13H26	1-CYCLOHEXYLHEPTANE	-------	----------	----------	----	----	-------	--------	-------
1191	C13H26O	1-TRIDECANAL	-73.027	1.5167E+00	0.0000E+00	258	288	-------	318.500	364.000
1192	C13H26O2	n-BUTYL NONANOATE	7.699	1.5875E+00	0.0000E+00	205	235	-------	333.375	381.000
1193	C13H26O2	METHYL DODECANOATE	-60.579	1.5879E+00	0.0000E+00	248	278	-------	333.463	381.100
1194	C13H28	n-TRIDECANE	-3.445	1.5775E+00	-1.2807E-03	138	255	-------	189.254	315.553
1195	C13H28O	1-TRIDECANOL	-84.885	1.5971E+00	0.0000E+00	263	293	-------	335.388	383.300
1196	C13H28S	BUTYL-NONYL-SULFIDE	-------	----------	----------	----	----	-------	--------	-------
1197	C13H28S	DECYL-PROPYL-SULFIDE	-------	----------	----------	----	----	-------	--------	-------
1198	C13H28S	DODECYL-METHYL-SULFIDE	-------	----------	----------	----	----	-------	--------	-------
1199	C13H28S	ETHYL-UNDECYL-SULFIDE	-------	----------	----------	----	----	-------	--------	-------
1200	C13H28S	1-TRIDECANETHIOL	-------	----------	----------	----	----	-------	--------	-------
1201	C14H8O2	ANTHRAQUINONE	59.031	6.0737E-01	0.0000E+00	280	520	240.118	229.095	374.863
1202	C14H10	ANTHRACENE	11.100	5.8160E-01	2.7900E-05	30	489	186.984	28.573	302.131
1203	C14H10	DIPHENYLACETYLENE	31.240	5.3250E-01	4.0100E-04	100	336	225.651	88.500	255.151
1204	C14H10	PHENANTHRENE	7.296	6.7861E-01	-2.5026E-06	20	270	-------	20.867	190.340
1205	C14H12	cis-STILBENE	-25.828	9.1750E-01	0.0000E+00	238	268	-------	192.675	220.200
1206	C14H12	trans-STILBENE	28.200	5.9100E-01	2.8800E-04	92	397	230.008	85.196	308.505
1207	C14H12O2	BENZYL BENZOATE	-55.659	1.0592E+00	0.0000E+00	263	293	-------	222.425	254.200
1208	C14H14	1,1-DIPHENYLETHANE	54.240	2.8750E-01	1.2360E-03	102	232	-------	96.370	187.208
1209	C14H14	1,2-DIPHENYLETHANE	60.350	2.5600E-01	1.3120E-03	75	324	253.305	86.930	281.399
1210	C14H14O	DIBENZYL ETHER	-39.200	1.0667E+00	0.0000E+00	247	277	-------	224.000	256.000
1211	C14H16	1-n-BUTYLNAPHTHALENE	-14.493	1.0792E+00	0.0000E+00	223	253	-------	226.625	259.000
1212	C14H16	2-BUTYLNAPHTHALENE	-------	----------	----------	----	----	-------	--------	-------
1213	C14H22	n-OCTYLBENZENE	3.753	1.3167E+00	0.0000E+00	207	237	-------	276.500	316.000
1214	C14H22	1,2,3,4-TETRAETHYLBENZENE	-------	----------	----------	----	----	-------	--------	-------
1215	C14H22	1,2,3,5-TETRAETHYLBENZENE	-------	----------	----------	----	----	-------	--------	-------
1216	C14H22	1,2,4,5-TETRAETHYLBENZENE	-------	----------	----------	----	----	-------	--------	-------
1217	C14H22O	p-tert-OCTYLPHENOL	-73.538	1.3875E+00	0.0000E+00	263	293	-------	291.375	333.000
1218	C14H28	1-TETRADECENE	-31.533	1.5533E+00	0.0000E+00	230	260	-------	326.200	372.800
1219	C14H28	1-CYCLOPENTYLNONANE	-------	----------	----------	----	----	-------	--------	-------
1220	C14H28	1-CYCLOHEXYLOCTANE	-------	----------	----------	----	----	-------	--------	-------
1221	C14H28O2	n-TETRADECANOIC ACID	184.313	-6.8332E-01	5.4508E-03	90	320	465.123	166.966	523.814
1222	C14H30	n-TETRADECANE	-6.108	1.6991E+00	-1.5681E-03	80	279	-------	119.783	345.880
1223	C14H30O	1-TETRADECANOL	-90.798	1.7083E+00	0.0000E+00	263	293	-------	358.750	410.000
1224	C14H30S	BUTYL-DECYL-SULFIDE	-------	----------	----------	----	----	-------	--------	-------
1225	C14H30S	DODECYL-ETHYL-SULFIDE	-------	----------	----------	----	----	-------	--------	-------
1226	C14H30S	HEPTYL-SULFIDE	-------	----------	----------	----	----	-------	--------	-------
1227	C14H30S	METHYL-TRIDECYL-SULFIDE	-------	----------	----------	----	----	-------	--------	-------
1228	C14H30S	PROPYL-UNDECYL-SULFIDE	-------	----------	----------	----	----	-------	--------	-------
1229	C14H30S	1-TETRADECANETHIOL	-------	----------	----------	----	----	-------	--------	-------
1230	C14H30S2	HEPTYL-DISULFIDE	-------	----------	----------	----	----	-------	--------	-------
1231	C14H31N	TETRADECYLAMINE	-94.917	1.7858E+00	0.0000E+00	263	293	-------	375.025	428.600
1232	C15H10N2O2	DIPHENYLMETHANE-4,4'-DIISOCYANATE	-64.925	1.2250E+00	0.0000E+00	263	293	-------	257.250	294.000
1233	C15H16O	p-CUMYLPHENOL	-62.695	1.1796E+00	0.0000E+00	263	293	-------	247.713	283.100
1234	C15H16O2	BISPHENOL A	-66.460	1.2504E+00	0.0000E+00	263	293	-------	262.588	300.100
1235	C15H18	1-PENTYLNAPHTHALENE	-------	----------	----------	----	----	-------	--------	-------
1236	C15H18	2-PENTYLNAPHTHALENE	-------	----------	----------	----	----	-------	--------	-------
1237	C15H24	n-NONYLBENZENE	-19.500	2.1667E+00	0.0000E+00	219	249	-------	455.000	520.000
1238	C15H24O	2,6-DI-tert-BUTYL-p-CRESOL	-79.703	1.4996E+00	0.0000E+00	263	293	-------	314.913	359.900
1239	C15H24O	NONYLPHENOL	-57.209	1.4996E+00	0.0000E+00	248	278	-------	314.913	359.900
1240	C15H28	1-PENTADECYNE	-------	----------	----------	----	----	-------	--------	-------
1241	C15H30	1-PENTADECENE	-49.033	1.6667E+00	0.0000E+00	239	269	-------	350.000	400.000
1242	C15H30	1-CYCLOPENTYLDECANE	-------	----------	----------	----	----	-------	--------	-------
1243	C15H30	1-CYCLOHEXYLNONANE	-------	----------	----------	----	----	-------	--------	-------
1244	C15H30O2	PENTADECANOIC ACID	164.349	-2.8297E-01	4.3694E-03	90	315	468.396	174.274	508.771
1245	C15H32	n-PENTADECANE	1.031	1.6367E+00	-8.4319E-04	80	271	-------	126.572	382.537
1246	C15H32O	1-PENTADECANOL	-------	----------	----------	----	----	-------	--------	-------
1247	C15H32S	BUTYL-UNDECYL-SULFIDE	-------	----------	----------	----	----	-------	--------	-------
1248	C15H32S	DODECYL-PROPYL-SULFIDE	-------	----------	----------	----	----	-------	--------	-------

Table 4-1 HEAT CAPACITY OF SOLID - ORGANIC COMPOUNDS (continued)

| NO | FORMULA | NAME | $C_P = A + B T + C T^2$ | | | | | $(C_P$ - joule/(mol K), T - K) | | |
			A	B	C	TMIN	TMAX	C_P @25 C	C_P @TMIN	C_P @TMAX
1249	C15H32S	ETHYL-TRIDECYL-SULFIDE	-------	----------	----------	----	----	-------	-------	-------
1250	C15H32S	METHYL-TETRADECYL-SULFIDE	-------	----------	----------	----	----	-------	-------	-------
1251	C15H32S	1-PENTADECANETHIOL	-------	----------	----------	----	----	-------	-------	-------
1252	C16H10	FLUORANTHENE	8.223	6.8669E-01	2.5114E-04	50	383	235.285	43.186	308.356
1253	C16H10	PYRENE	40.400	3.0650E-01	1.0940E-03	130	424	229.032	98.734	366.796
1254	C16H12	1-PHENYLNAPHTHALENE	-52.087	9.8000E-01	0.0000E+00	263	293	-------	205.800	235.200
1255	C16H20	1-n-HEXYLNAPHTHALENE	-19.695	1.3000E+00	0.0000E+00	225	255	-------	273.000	312.000
1256	C16H22O4	DIBUTYL PHTHALATE	3.076	1.6625E+00	0.0000E+00	208	238	-------	349.125	399.000
1257	C16H26	n-DECYLBENZENE	-28.906	1.5400E+00	0.0000E+00	229	259	-------	323.400	369.600
1258	C16H26	PENTAETHYLBENZENE	-------	----------	----------	----	----	-------	-------	-------
1259	C16H30	1-HEXADECYNE	-------	----------	----------	----	----	-------	-------	-------
1260	C16H32	n-DECYLCYCLOHEXANE	-2.919	1.6482E+00	-1.5098E-03	100	271	-------	146.803	333.212
1261	C16H32	1-CYCLOPENTYLUNDECANE	-------	----------	----------	----	----	-------	-------	-------
1262	C16H32	1-HEXADECENE	61.212	-1.2084E-01	8.6087E-03	100	278	-------	135.214	690.646
1263	C16H32O2	n-HEXADECANOIC ACID	13.037	1.5351E+00	1.6129E-05	100	320	472.161	166.708	505.921
1264	C16H34	n-HEXADECANE	-1.938	1.7934E+00	-1.3253E-03	100	291	-------	164.146	408.030
1265	C16H34O	DI-n-OCTYL ETHER	-49.290	1.9292E+00	0.0000E+00	236	266	-------	405.125	463.000
1266	C16H34O	1-HEXADECANOL	96.620	5.8352E-01	2.0967E-03	80	322	456.982	156.721	502.587
1267	C16H34S	BUTYL-DODECYL-SULFIDE	-------	----------	----------	----	----	-------	-------	-------
1268	C16H34S	ETHYL-TETRADECYL-SULFIDE	-------	----------	----------	----	----	-------	-------	-------
1269	C16H34S	METHYL-PENTADECYL-SULFIDE	-------	----------	----------	----	----	-------	-------	-------
1270	C16H34S	OCTYL-SULFIDE	-------	----------	----------	----	----	-------	-------	-------
1271	C16H34S	PROPYL-TRIDECYL-SULFIDE	-------	----------	----------	----	----	-------	-------	-------
1272	C16H34S	1-HEXADECANETHIOL	-------	----------	----------	----	----	-------	-------	-------
1273	C16H34S2	OCTYL-DISULFIDE	-------	----------	----------	----	----	-------	-------	-------
1274	C17H28	n-UNDECYLBENZENE	-46.200	1.6500E+00	0.0000E+00	238	268	-------	346.500	396.000
1275	C17H32	1-HEPTADECYNE	-83.990	1.8917E+00	0.0000E+00	254	284	-------	397.250	454.000
1276	C17H34	1-CYCLOPENTYLDODECANE	-------	----------	----------	----	----	-------	-------	-------
1277	C17H34	1-CYCLOHEXYLUNDECANE	-------	----------	----------	----	----	-------	-------	-------
1278	C17H34	1-HEPTADECENE	-------	----------	----------	----	----	-------	-------	-------
1279	C17H36	n-HEPTADECANE	1.710	1.9073E+00	-1.5240E-03	30	271	-------	58.392	406.334
1280	C17H36O	1-HEPTADECANOL	-108.537	2.0421E+00	0.0000E+00	263	293	-------	428.838	490.100
1281	C17H36S	BUTYL-TRIDECYL-SULFIDE	-------	----------	----------	----	----	-------	-------	-------
1282	C17H36S	ETHYL-PENTADECYL-SULFIDE	-------	----------	----------	----	----	-------	-------	-------
1283	C17H36S	HEXADECYL-METHYL-SULFIDE	-------	----------	----------	----	----	-------	-------	-------
1284	C17H36S	PROPYL-TETRADECYL-SULFIDE	-------	----------	----------	----	----	-------	-------	-------
1285	C17H36S	1-HEPTADECANETHIOL	-------	----------	----------	----	----	-------	-------	-------
1286	C18H12	CHRYSENE	-55.409	1.0425E+00	0.0000E+00	263	293	-------	218.925	250.200
1287	C18H14	m-TERPHENYL	-59.661	1.1225E+00	0.0000E+00	263	293	-------	235.725	269.400
1288	C18H14	o-TERPHENYL	39.450	5.8620E-01	6.6700E-04	90	325	273.517	97.611	300.488
1289	C18H14	p-TERPHENYL	32.370	8.1230E-01	8.6890E-05	297	349	282.281	281.201	326.280
1290	C18H15P	TRIPHENYLPHOSPHINE	63.600	4.1600E-01	1.3970E-03	102	299	311.815	120.496	312.252
1291	C18H15O4P	TRIPHENYL PHOSPHATE	-130.035	2.1673E+00	0.0000E+00	270	300	516.131	455.123	520.140
1292	C18H16N2	N,N'-DIPHENYL-p-PHENYLENEDIAMINE	-75.304	1.4208E+00	0.0000E+00	263	293	-------	298.375	341.000
1293	C18H22	2,3-DIMETHYL-2,3-DIPHENYLBUTANE	-76.625	1.4417E+00	0.0000E+00	263	293	-------	302.750	346.000
1294	C18H22O2	DICUMYL PEROXIDE	-87.682	1.6497E+00	0.0000E+00	263	293	-------	346.439	395.930
1295	C18H30	n-DODECYLBENZENE	-64.824	1.8042E+00	0.0000E+00	246	276	-------	378.875	433.000
1296	C18H30	HEXAETHYLBENZENE	-------	----------	----------	----	----	-------	-------	-------
1297	C18H32O2	LINOLEIC ACID	-57.027	2.0258E+00	0.0000E+00	238	268	-------	425.425	486.200
1298	C18H34	1-OCTADECYNE	-------	----------	----------	----	----	-------	-------	-------
1299	C18H34O2	OLEIC ACID	-77.434	1.6642E+00	0.0000E+00	257	287	-------	349.475	399.400
1300	C18H34O4	DIBUTYL SEBACATE	-75.329	2.2058E+00	0.0000E+00	244	274	-------	463.225	529.400
1301	C18H34O4	DIHEXYL ADIPATE	-42.683	2.2058E+00	0.0000E+00	229	259	-------	463.225	529.400
1302	C18H36	1-CYCOPENTYLTRIDECANE	-------	----------	----------	----	----	-------	-------	-------
1303	C18H36	1-CYCLOHEXYLDODECANE	-------	----------	----------	----	----	-------	-------	-------
1304	C18H36	1-OCTADECENE	-78.847	1.5533E+00	0.0000E+00	261	291	-------	326.200	372.800
1305	C18H36O2	STEARIC ACID	-118.770	1.9795E+00	0.0000E+00	270	300	471.418	415.695	475.080
1306	C18H38	n-OCTADECANE	34.884	1.5273E+00	-1.7550E-04	40	301	474.643	95.695	479.165
1307	C18H38O	DINONYL ETHER	-------	----------	----------	----	----	-------	-------	-------
1308	C18H38O	1-OCTADECANOL	-114.450	2.1533E+00	0.0000E+00	263	293	-------	452.200	516.800
1309	C18H38S	BUTYL-TETRADECYL-SULFIDE	-------	----------	----------	----	----	-------	-------	-------
1310	C18H38S	ETHYL-HEXADECYL-SULFIDE	-------	----------	----------	----	----	-------	-------	-------
1311	C18H38S	HEPTADECYL-METHYL-SULFIDE	-------	----------	----------	----	----	-------	-------	-------
1312	C18H38S	NONYL-SULFIDE	-------	----------	----------	----	----	-------	-------	-------
1313	C18H38S	PENTADECYL-PROPYL-SULFIDE	-------	----------	----------	----	----	-------	-------	-------
1314	C18H38S	1-OCTADECANETHIOL	-------	----------	----------	----	----	-------	-------	-------
1315	C18H38S2	NONYL-DISULFIDE	-------	----------	----------	----	----	-------	-------	-------
1316	C19H26	1-n-NONYLNAPHTHALENE	-72.130	1.6338E+00	0.0000E+00	254	284	-------	343.088	392.100
1317	C19H32	n-TRIDECYLBENZENE	-80.852	1.8738E+00	0.0000E+00	253	283	-------	393.488	449.700
1318	C19H36	1-NONADECYNE	-------	----------	----------	----	----	-------	-------	-------
1319	C19H36O2	METHYL OLEATE	-115.406	2.1754E+00	0.0000E+00	263	293	-------	456.838	522.100
1320	C19H38	1-CYCLOPENTYLTETRADECANE	-------	----------	----------	----	----	-------	-------	-------
1321	C19H38	1-CYCLOHEXYLTRIDECANE	-------	----------	----------	----	----	-------	-------	-------
1322	C19H38	1-NONADECENE	-119.462	2.1125E+00	0.0000E+00	267	297	-------	443.625	507.000
1323	C19H38O2	NONADECANOIC ACID	190.219	-2.0129E-01	4.8165E-03	90	341	558.362	211.116	682.359
1324	C19H40	n-NONADECANE	-116.598	2.1938E+00	0.0000E+00	263	293	-------	460.688	526.500
1325	C19H40O	1-NONADECANOL	-------	----------	----------	----	----	-------	-------	-------
1326	C19H40S	BUTYL-PENTADECYL-SULFIDE	-------	----------	----------	----	----	-------	-------	-------

Table 4-1 HEAT CAPACITY OF SOLID - ORGANIC COMPOUNDS (continued)

NO	FORMULA	NAME	$C_P = A + B T + C T^2$					$(C_P$ - joule/(mol K), T - K)		
			A	B	C	TMIN	TMAX	C_P@25 C	C_P@TMIN	C_P@TMAX
1327	C19H40S	ETHYL-HEPTADECYL-SULFIDE	-------	----------	----------	----	----	-------	-------	-------
1328	C19H40S	HEXADECYL-PROPYL-SULFIDE	-------	----------	----------	----	----	-------	-------	-------
1329	C19H40S	METHYL-OCTADECYL-SULFIDE	-------	----------	----------	----	----	-------	-------	-------
1330	C19H40S	1-NONADECANETHIOL	-------	----------	----------	----	----	-------	-------	-------
1331	C20H16	TRIPHENYLETHYLENE	45.450	6.7620E-01	7.0200E-04	102	322	309.462	121.644	336.311
1332	C20H28	1-n-DECYLNAPHTHALENE	-84.022	1.7450E+00	0.0000E+00	258	288	-------	366.450	418.800
1333	C20H30O2	ABIETIC ACID	-104.528	1.9667E+00	0.0000E+00	263	293	-------	413.000	472.000
1334	C20H31N	DEHYDROABIETYLAMINE	-104.675	1.9750E+00	0.0000E+00	263	293	-------	414.750	474.000
1335	C20H34	1-PHENYLTETRADECANE	-------	----------	----------	----	----	-------	-------	-------
1336	C20H38	1-EICOSYNE	-137.416	2.2250E+00	0.0000E+00	272	302	525.968	467.250	534.000
1337	C20H40	1-CYCLOPENTYLPENTADECANE	-------	----------	----------	----	----	-------	-------	-------
1338	C20H40	1-CYCLOHEXYLTETRADECANE	-------	----------	----------	----	----	-------	-------	-------
1339	C20H40	1-EICOSENE	-------	----------	----------	----	----	-------	-------	-------
1340	C20H42	n-EICOSANE	135.800	5.0800E-01	3.1100E-03	94	263	-------	210.595	483.877
1341	C20H42O	1-EICOSANOL	-126.276	2.3758E+00	0.0000E+00	263	293	-------	498.925	570.200
1342	C20H42S	BUTYL-HEXADECYL-SULFIDE	-------	----------	----------	----	----	-------	-------	-------
1343	C20H42S	DECYL-SULFIDE	-------	----------	----------	----	----	-------	-------	-------
1344	C20H42S	ETHYL-OCTADECYL-SULFIDE	-------	----------	----------	----	----	-------	-------	-------
1345	C20H42S	HEPTADECYL-PROPYL-SULFIDE	-------	----------	----------	----	----	-------	-------	-------
1346	C20H42S	METHYL-NONADECYL-SULFIDE	-------	----------	----------	----	----	-------	-------	-------
1347	C20H42S	1-EICOSANETHIOL	-------	----------	----------	----	----	-------	-------	-------
1348	C20H42S2	DECYL-DISULFIDE	-------	----------	----------	----	----	-------	-------	-------
1349	C21H21O4P	TRI-o-CRESYL PHOSPHATE	-0.281	1.8750E+00	0.0000E+00	210	240	-------	393.750	450.000
1350	C21H36	1-PHENYLPENTADECANE	-------	----------	----------	----	----	-------	-------	-------
1351	C21H42	1-CYCLOPENTYLHEXADECANE	-------	----------	----------	----	----	-------	-------	-------
1352	C21H42	1-CYCLOHEXYLPENTADECANE	-------	----------	----------	----	----	-------	-------	-------
1353	C22H38	1-PHENYLHEXADECANE	-------	----------	----------	----	----	-------	-------	-------
1354	C22H44	1-CYCLOHEXYLHEXADECANE	-------	----------	----------	----	----	-------	-------	-------
1355	C22H44O2	n-BUTYL STEARATE	-137.614	2.5892E+00	0.0000E+00	263	293	-------	543.725	621.400
1356	C24H38O4	DIISOOCTYL PHTHALATE	43.421	2.5542E+00	0.0000E+00	193	223	-------	536.375	613.000
1357	C24H38O4	DIOCTYL PHTHALATE	-25.113	3.5386E+00	-6.5413E-03	10	170	-------	9.619	387.407
1358	C24H42O	DINONYLPHENOL	-------	----------	----------	----	----	-------	-------	-------
1359	C26H20	TETRAPHENYLETHYLENE	62.000	8.2640E-01	8.5900E-04	102	346	384.751	155.130	450.344
1360	C28H46O4	DIISODECYL PHTHALATE	33.238	2.6783E+00	0.0000E+00	198	228	-------	562.450	642.800

C_P - heat capacity of solid, joule/(mol K)

A, B, and C - regression coefficients for chemical compound

T - temperature, K

TMIN - minimum temperature, K

TMAX - maximum temperature, K

Table 4-2 HEAT CAPACITY OF SOLID - INORGANIC COMPOUNDS

NO	FORMULA	NAME	$C_P = A + B\,T + C\,T^2$					$(C_P$ - joule/(mol K), T - K)		
			A	B	C	TMIN	TMAX	C_P @25 C	C_P @TMIN	CP @TMAX
1	Ag	SILVER	23.724	4.9940E-03	2.7977E-07	203	925	25.237	24.751	28.584
2	AgCl	SILVER CHLORIDE	50.156	5.1457E-01	-9.4711E-04	133	268	-------	101.734	120.030
3	AgI	SILVER IODIDE	24.520	9.9943E-02	1.1671E-06	298	420	54.422	54.407	66.702
4	Al	ALUMINUM	19.080	1.5558E-02	-5.1316E-07	205	873	23.673	22.247	32.272
5	AlB3H12	ALUMINUM BOROHYDRIDE	-------	----------	----------	----	----	-------	-------	-------
6	AlBr3	ALUMINUN BROMIDE	43.303	3.1699E-01	-4.1823E-04	100	300	100.636	70.820	100.760
7	AlCl3	ALUMINUM CHLORIDE	12.954	4.1933E-01	-5.1004E-04	100	466	92.638	49.787	97.617
8	AlF3	ALUMINUM FLUORIDE	-16.325	4.8626E-01	-6.0348E-04	156	298	75.006	44.770	75.006
9	AlI3	ALUMINUM IODIDE	83.941	1.7290E-02	1.1048E-04	298	400	98.916	98.916	108.533
10	Al2O3	ALUMINUM OXIDE	-8.121	3.8687E-01	-3.1623E-04	290	540	79.112	77.475	108.573
11	Al2S3O12	ALUMINUM SULFATE	108.779	5.8215E-01	-2.8719E-04	250	900	256.818	236.367	400.087
12	Ar	ARGON	29.664	2.8264E-02	8.3620E-04	34	84	-------	31.575	37.905
13	As	ARSENIC	16.186	3.9867E-02	-3.7447E-05	105	291	-------	19.946	24.616
14	AsBr3	ARSENIC TRIBROMIDE	30.148	1.0594E-01	4.1357E-04	256	306	98.495	84.418	101.342
15	AsCl3	ARSENIC TRICHLORIDE	37.480	9.4009E-02	4.5825E-04	205	255	-------	76.052	91.299
16	AsF3	ARSENIC TRIFLUORIDE	35.397	9.5384E-02	4.3905E-04	217	267	-------	76.842	92.247
17	AsF5	ARSENIC PENTAFLUORIDE	53.868	9.8211E-02	6.8511E-04	143	193	-------	82.025	98.470
18	AsH3	ARSINE	32.315	4.0056E-01	-2.1426E-03	32	106	-------	43.005	50.724
19	AsI3	ARSENIC TRIIODIDE	-------	----------	----------	----	----	-------	-------	-------
20	As2O3	ARSENIC TRIOXIDE	35.125	2.0292E-01	4.2593E-07	298	582	95.662	95.631	153.366
21	At	ASTATINE	-9.840	3.5352E-02	6.7318E-05	525	575	-------	27.290	32.761
22	Au	GOLD	21.992	1.3011E-02	-5.2303E-06	280	946	25.406	25.227	29.619
23	B	BORON	3.491	3.5220E-02	-1.2922E-05	298	1200	12.844	12.839	27.148
24	BBr3	BORON TRIBROMIDE	40.186	8.7001E-02	4.8836E-04	178	228	-------	71.184	85.456
25	BCl3	BORON TRICHLORIDE	45.153	7.2447E-02	6.2373E-04	116	166	-------	61.983	74.409
26	BF3	BORON TRIFLUORIDE	40.048	5.2447E-02	4.9958E-04	80	140	-------	47.441	57.183
27	BH2CO	BORINE CARBONYL	34.342	4.7654E-02	5.5315E-04	86	136	-------	42.553	51.084
28	BH3O3	BORIC ACID	75.571	8.6396E-02	-1.1746E-04	283	293	-------	90.613	90.803
29	B2D6	DEUTERODIBORANE	33.891	5.5367E-02	4.6139E-04	120	170	-------	47.179	56.638
30	B2H5Br	DIBORANE HYDROBROMIDE	50.405	8.1941E-02	6.8887E-04	119	169	-------	69.899	83.912
31	B2H6	DIBORANE	-13.463	9.8057E-01	-3.0124E-03	18	105	-------	3.420	56.131
32	B3N3H6	BORINE TRIAMINE	51.292	1.0385E-01	6.2959E-04	165	215	-------	85.552	102.704
33	B4H10	TETRABORANE	48.530	7.3256E-02	7.0950E-04	103	153	-------	63.658	76.420
34	B5H9	PENTABORANE	46.140	9.8967E-02	5.6119E-04	176	226	-------	81.045	97.293
35	B5H11	TETRAHYDROPENTABORANE	50.719	1.1087E-01	6.1592E-04	180	230	-------	90.631	108.800
36	B10H14	DECABORANE	16.411	2.4844E-01	1.4323E-03	100	300	217.805	55.579	219.850
37	Ba	BARIUM	-39.805	3.4949E-02	3.6782E-05	950	1000	-------	26.607	31.942
38	Be	BERYLLIUM	-57.482	2.5391E-02	1.6814E-05	1510	1560	-------	19.207	23.058
39	BeB2H8	BERYLLIUM BOROHYDRIDE	-------	----------	----------	----	----	-------	-------	-------
40	BeBr2	BERYLLIUM BROMIDE	46.717	7.7704E-02	-4.3644E-05	298	700	66.004	66.004	79.724
41	BeCl2	BERYLLIUM CHLORIDE	13.579	2.3209E-01	-2.0441E-04	100	600	64.606	34.744	79.247
42	BeF2	BERYLLIUM FLUORIDE	5.119	1.9804E-01	-1.3888E-04	243	300	51.820	45.056	52.037
43	BeI2	BERYLLIUM IODIDE	34.146	1.4975E-01	-1.0956E-04	298	700	69.055	69.055	85.286
44	Bi	BISMUTH	33.909	-6.3884E-02	1.2271E-04	256	523	25.771	25.597	34.064
45	BiBr3	BISMUTH TRIBROMIDE	-------	----------	----------	----	----	-------	-------	-------
46	BiCl3	BISMUTH TRICHLORIDE	-------	----------	----------	----	----	-------	-------	-------
47	BrF5	BROMINE PENTAFLUORIDE	66.326	1.3218E-01	8.1716E-04	162	212	-------	109.085	130.954
48	Br2	BROMINE	35.762	6.6370E-02	1.1649E-04	100	266	-------	43.564	61.646
49	C	CARBON	-0.832	3.4846E-02	-1.3233E-05	200	1100	8.381	5.608	21.487
50	CCl2O	PHOSGENE	44.529	1.1558E-01	-1.2710E-04	100	140	-------	54.816	58.180
51	CF2O	CARBONYL FLUORIDE	139.852	-1.6982E+00	8.1497E-03	100	155	-------	51.530	72.347
52	CH4N2O	UREA	13.513	2.6827E-01	2.1285E-04	80	400	95.389	35.111	124.226
53	CH4N2S	THIOUREA	41.342	1.9298E-01	1.7569E-04	183	273	-------	82.580	107.163
54	CNBr	CYANOGEN BROMIDE	-------	----------	----------	----	----	-------	0.000	0.000
55	CNCl	CYANOGEN CHLORIDE	51.224	5.0607E-02	-7.3357E-05	256	267	-------	59.372	59.502
56	CNF	CYANOGEN FLUORIDE	35.009	6.5889E-02	4.3926E-04	150	200	-------	54.776	65.758
57	CO	CARBON MONOXIDE	21.830	-4.7101E-02	6.4073E-03	30	62	-------	26.184	43.540
58	COS	CARBONYL SULFIDE	41.728	-9.9603E-03	2.7100E-04	100	131	-------	43.442	45.078
59	COSe	CARBON OXYSELENIDE	36.396	6.8499E-02	4.5666E-04	150	200	-------	56.946	68.362
60	CO2	CARBON DIOXIDE	41.195	3.1470E-02	6.4118E-05	150	210	-------	47.359	50.632
61	CS2	CARBON DISULFIDE	27.036	2.0040E-01	-2.9552E-04	80	161	-------	41.177	51.652
62	CSeS	CARBON SELENOSULFIDE	39.729	7.4040E-02	5.0044E-04	148	198	-------	61.638	73.995
63	C2N2	CYANOGEN	34.853	7.9797E-02	4.2277E-04	189	239	-------	64.976	78.003
64	C3S2	CARBON SUBSULFIDE	-------	----------	----------	----	----	-------	-------	-------
65	Ca	CALCIUM	8.654	1.4203E-01	-3.0955E-04	112	201	-------	20.641	24.691
66	CaF2	CALCIUM FLUORIDE	59.264	3.2222E-02	-2.4645E-07	298	1400	68.849	68.844	103.892
67	CbF5	COLUMBIUM FLUORIDE	-------	----------	----------	----	----	-------	0.000	0.000
68	Cd	CADMIUM	27.714	-1.5910E-02	3.5412E-05	295	543	26.118	26.102	29.519
69	CdCl2	CADMIUM CHLORIDE	44.568	1.6722E-01	-2.2377E-04	170	300	74.533	66.528	74.595
70	CdF2	CADMIUM FLUORIDE	-61.146	8.4770E-02	1.1407E-04	743	793	-------	64.848	77.848
71	CdI2	CADMIUM IODIDE	63.343	8.6748E-02	-1.2988E-04	170	300	77.661	74.336	77.678
72	CdO	CADMIUM OXIDE	26.599	2.0312E-03	7.8868E-07	152	290	-------	26.926	27.256
73	ClF	CHLORINE MONOFLUORIDE	38.617	5.1389E-02	6.5757E-04	78	128	-------	46.650	56.002
74	ClFO3	PERCHLORYL FLUORIDE	48.105	6.3068E-02	8.3634E-04	75	125	-------	57.617	69.168
75	ClF3	CHLORINE TRIFLUORIDE	46.901	8.4221E-02	6.0093E-04	140	190	-------	70.508	84.643
76	ClF5	CHLORINE PENTAFLUORIDE	56.016	1.0052E-01	7.1799E-04	140	190	-------	84.161	101.034
77	ClHO3S	CHLOROSULFONIC ACID	57.540	1.0481E-01	7.3214E-04	143	193	-------	87.546	105.098
78	ClHO4	PERCHLORIC ACID	27.955	1.8747E-01	5.8079E-04	90	171	-------	49.532	77.045

Table 4-2 HEAT CAPACITY OF SOLID - INORGANIC COMPOUNDS (continued)

NO	FORMULA	NAME	$C_P = A + B\,T + C\,T^2$ A	B	C	TMIN	TMAX	C_P @25 C	C_P @TMIN	CP @TMAX
								(C_P - joule/(mol K), T - K)		
79	ClO2	CHLORINE DIOXIDE	35.791	7.1963E-02	4.4000E-04	164	214	-------	59.330	71.224
80	Cl2	CHLORINE	37.173	7.3016E-02	-1.6956E-05	120	165	-------	45.691	48.759
81	Cl2O	CHLORINE MONOXIDE	44.729	6.8783E-02	6.4194E-04	107	157	-------	59.469	71.391
82	Cl2O7	CHLORINE HEPTOXIDE	71.740	1.2406E-01	9.3878E-04	132	182	-------	104.529	125.485
83	Co	COBALT	20.505	1.4439E-02	1.1737E-06	273	973	24.915	24.537	35.666
84	CoCl2	COBALT CHLORIDE	36.930	2.3639E-01	-3.2545E-04	189	304	78.480	69.955	78.700
85	CoNC3O4	COBALT NITROSYL TRICARBONYL	-------	----------	----------	----	----	-------	-------	-------
86	Cr	CHROMIUM	20.903	1.0088E-02	4.4764E-07	298	1073	23.951	23.950	32.243
87	CrC6O6	CHROMIUM CARBONYL	72.739	1.1182E-01'	-3.6525E-05	298	1600	102.832	102.819	158.148
88	CrO2Cl2	CHROMIUM OXYCHLORIDE	60.149	1.0137E-01	8.0039E-04	127	177	-------	85.826	103.033
89	Cs	CESIUM	54.569	-2.7723E-01	7.0339E-04	180	280	-------	27.458	32.091
90	CsBr	CESIUM BROMIDE	-63.045	6.5975E-02	7.6791E-05	859	909	-------	50.320	60.409
91	CsCl	CESIUM CHLORIDE	42.677	4.5084E-02	-4.1502E-05	168	299	52.429	49.082	52.454
92	CsF	CESIUM FLUORIDE	46.583	1.8144E-02	-3.4115E-07	298	900	51.963	51.963	62.636
93	CsI	CESIUM IODIDE	47.506	2.3242E-02	-1.8658E-05	182	298	52.777	51.113	52.780
94	Cu	COPPER	22.041	8.8444E-03	-1.0071E-06	373	1273	-------	25.200	31.668
95	CuBr	CUPROUS BROMIDE	32.876	1.1299E-01	-1.3280E-04	141	300	54.759	46.198	54.821
96	CuCl	CUPROUS CHLORIDE	36.581	4.1723E-02	-5.6685E-06	298	703	48.517	48.511	63.111
97	CuCl2	CUPRIC CHLORIDE	62.816	3.6636E-02	-1.7248E-05	295	906	72.206	72.123	81.852
98	CuI	COPPER IODIDE	50.826	1.1961E-02	8.8842E-10	298	861	54.195	54.194	60.928
99	DCN	DEUTERIUM CYANIDE	28.541	7.4140E-02	3.5113E-04	211	261	-------	59.851	71.850
100	D2	DEUTERIUM	7.399	-3.7404E-02	1.1962E-03	18	19	-------	7.114	7.121
101	D2O	DEUTERIUM OXIDE	10.981	1.0162E-01	-3.9516E-05	150	271	-------	25.335	35.580
102	Eu	EUROPIUM	27.276	-4.4175E-03	1.5046E-05	298	1050	27.296	27.296	39.225
103	F2	FLUORINE	17.167	5.1968E-01	-6.0949E-04	40	53	-------	36.979	43.216
104	F2O	FLUORINE OXIDE	37.708	-2.3750E-03	3.1666E-03	40	49	-------	37.708	45.272
105	Fe	IRON	26.748	-1.5339E-02	3.8405E-05	295	973	25.589	25.566	48.183
106	FeC5O5	IRON PENTACARBONYL	46.979	1.1583E-01	5.7298E-04	202	252	-------	93.809	112.615
107	FeCl2	FERROUS CHLORIDE	26.313	3.3533E-01	-5.6735E-04	125	300	75.857	59.403	75.850
108	FeCl3	FERRIC CHLORIDE	35.813	2.6813E-01	-2.0735E-04	100	500	97.325	60.553	118.042
109	Fr	FRANCIUM	7.709	2.5889E-02	1.0350E-04	250	300	24.628	20.661	24.803
110	Ga	GALLIUM	-2.609	3.0069E-01	-8.4389E-04	30	200	-------	5.652	23.773
111	GaCl3	GALLIUM TRICHLORIDE	106.769	4.3788E-02	-2.9899E-05	298	350	117.166	117.162	118.432
112	Gd	GADOLINIUM	28.620	-1.2462E-03	5.1377E-06	295	1300	28.706	28.700	35.683
113	Ge	GERMANIUM	22.813	2.7105E-03	1.7652E-06	298	1210	23.778	23.778	28.678
114	GeBr4	GERMANIUM BROMIDE	34.276	1.1435E-01	4.5879E-04	249	299	109.155	91.282	109.582
115	GeCl4	GERMANIUM CHLORIDE	48.924	1.0350E-01	5.9601E-04	174	224	-------	84.869	101.883
116	GeHCl3	TRICHLORO GERMANE	46.711	8.8784E-02	5.8391E-04	152	202	-------	73.710	88.487
117	GeH4	GERMANE	64.699	-1.0214E-01	5.3995E-04	121	165	-------	60.248	62.577
118	Ge2H6	DIGERMANE	56.058	8.9101E-02	7.8056E-04	114	164	-------	76.400	91.716
119	Ge3H8	TRIGERMANE	79.537	1.2845E-01	1.0928E-03	118	168	-------	109.736	131.736
120	HBr	HYDROGEN BROMIDE	61.418	-2.8724E-01	1.2784E-03	125	185	-------	45.488	52.060
121	HCN	HYDROGEN CYANIDE	37.296	1.6112E-02	1.2395E-04	150	254	-------	42.501	49.400
122	HCl	HYDROGEN CHLORIDE	10.738	3.4011E-01	-5.3026E-04	100	158	-------	39.446	51.238
123	HF	HYDROGEN FLUORIDE	22.830	-4.4537E-02	3.2996E-04	120	190	-------	22.237	26.245
124	HI	HYDROGEN IODIDE	51.979	-1.0285E-01	3.9534E-04	150	218	-------	45.447	48.345
125	HNO3	NITRIC ACID	45.780	2.1034E-02	1.2484E-04	150	230	-------	51.744	57.222
126	H2	HYDROGEN	5.707	2.1779E-03	-3.8386E-05	12	14	-------	5.728	5.730
127	H2O	WATER	9.695	7.4955E-02	-1.5584E-05	150	273	-------	20.588	29.006
128	H2O2	HYDROGEN PEROXIDE	6.280	2.0680E-01	-9.4030E-05	100	260	-------	26.020	53.692
129	H2S	HYDROGEN SULFIDE	-4.533	5.0824E-01	-6.7042E-04	20	104	-------	5.364	40.903
130	H2SO4	SULFURIC ACID	-34.353	7.0211E-01	-6.1154E-04	240	280	-------	98.928	114.292
131	H2S2	HYDROGEN DISULFIDE	48.591	8.4543E-02	6.3352E-04	133	183	-------	71.155	85.420
132	H2Se	HYDROGEN SELENIDE	27.219	5.3554E-02	3.3650E-04	159	209	-------	44.265	53.139
133	H2Te	HYDROGEN TELLURIDE	29.549	6.2670E-02	3.5986E-04	174	224	-------	51.377	61.677
134	H3NO3S	SULFAMIC ACID	107.167	1.2416E-01	-1.6938E-04	283	293	-------	128.739	129.005
135	He	HELIUM-3	3.458	5.8448E-03	-1.1931E-04	0.5	1	-------	3.461	3.464
136	He	HELIUM-4	4.332	1.8473E-02	-3.8295E-04	1	1.7	-------	4.350	4.363
137	Hf	HAFNIUM	23.443	7.6784E-03	-4.4021E-08	298	1200	25.728	25.728	32.593
138	Hg	MERCURY	21.905	2.3673E-02	1.7107E-05	111	228	-------	24.739	28.196
139	HgBr2	MERCURIC BROMIDE	66.088	3.1968E-02	-3.4642E-06	295	500	75.311	75.217	81.206
140	HgCl2	MERCURIC CHLORIDE	60.263	5.7350E-02	-3.8857E-05	295	500	73.908	73.800	79.224
141	HgI2	MERCURIC IODIDE	61.657	6.2969E-02	-3.0243E-05	298	400	77.743	77.743	82.006
142	IF7	IODINE HEPTAFLUORIDE	45.380	1.3137E-01	5.7454E-04	229	279	-------	105.456	126.597
143	I2	IODINE	45.389	9.6004E-03	8.7002E-05	100	387	55.985	47.219	62.115
144	In	INDIUM	37.324	-8.1142E-02	1.5100E-04	250	423	26.555	26.475	30.040
145	Ir	IRIDIUM	23.071	5.9876E-03	-2.8121E-08	273	1273	24.854	24.704	30.649
146	K	POTASSIUM	66.981	-3.0185E-01	5.9102E-04	269	323	29.522	28.555	31.156
147	KBr	POTASSIUM BROMIDE	40.786	6.0279E-02	-7.3900E-05	175	300	52.189	49.072	52.219
148	KCl	POTASSIUM CHLORIDE	46.432	1.2844E-02	7.0364E-06	200	1000	50.887	49.282	66.312
149	KF	POTASSIUM FLUORIDE	21.885	1.6529E-01	-2.4806E-04	157	325	49.113	41.756	49.399
150	KI	POTASSIUM IODIDE	47.867	1.0075E-02	7.0779E-06	199	773	51.500	50.147	59.884
151	KOH	POTASSIUM HYDROXIDE	50.276	4.4209E-02	5.3533E-07	250	522	63.504	61.362	73.499
152	Kr	KRYPTON	24.794	3.0753E-02	6.4752E-04	66	116	-------	28.840	34.621
153	La	LANTHANUM	22.848	1.1320E-02	-6.0436E-07	298	550	26.169	26.169	28.891
154	Li	LITHIUM	6.609	8.8913E-02	-8.8029E-05	100	454	25.293	14.620	28.828
155	LiBr	LITHIUM BROMIDE	47.461	-3.9333E-03	3.1178E-05	298	800	49.059	49.059	64.268
156	LiCl	LITHIUM CHLORIDE	-50.341	5.5246E-02	6.5993E-05	837	887	-------	42.159	50.610

Table 4-2 HEAT CAPACITY OF SOLID - INORGANIC COMPOUNDS (continued)

| NO | FORMULA | NAME | $C_P = A + B T + C T^2$ | | | | | $(C_P$ - joule/(mol K), T - K) | | |
			A	B	C	TMIN	TMAX	C_P @25 C	C_P @TMIN	CP @TMAX
157	LiF	LITHIUM FLUORIDE	8.651	1.5700E-01	-1.5495E-04	237	425	41.686	37.190	47.388
158	LiI	LITHIUM IODIDE	42.366	2.4847E-02	4.7563E-06	298	700	50.197	50.197	62.090
159	Lu	LUTECIUM	27.895	-6.7080E-03	8.6883E-06	298	1200	26.667	26.667	32.357
160	Mg	MAGNESIUM	20.435	1.7050E-02	-7.2350E-06	295	543	24.875	24.835	27.561
161	MgCl2	MAGNESIUM CHLORIDE	21.836	1.4404E-02	1.3651E-06	273	1073	26.252	25.872	38.863
162	MgO	MAGNESIUM OXIDE	10.461	1.1153E-01	-8.0123E-05	200	800	36.590	29.561	48.403
163	Mn	MANGANESE	21.836	1.4404E-02	1.3651E-06	273	1073	26.252	25.872	38.863
164	MnCl2	MANGANESE CHLORIDE	40.128	1.8663E-01	-2.5519E-04	182	300	73.086	65.616	73.151
165	Mo	MOLYBDENUM	22.481	5.1793E-03	9.2710E-07	273	1073	24.107	23.965	29.106
166	MoF6	MOLYBDENUM FLUORIDE	-0.504	4.0934E-01	-6.1477E-04	166	290	-------	50.485	66.502
167	MoO3	MOLYBDENUM OXIDE	3.999	3.5817E-01	-4.0826E-04	187	300	74.494	56.736	74.705
168	NCl3	NITROGEN TRICHLORIDE	109.267	1.0374E-01	-1.6170E-04	236	246	-------	124.743	125.004
169	ND3	HEAVY AMMONIA	38.628	7.2403E-02	4.8544E-04	149	199	-------	60.226	72.300
170	NF3	NITROGEN TRIFLUORIDE	44.332	3.0001E-02	1.8338E-03	16	66	-------	45.314	54.398
171	NH3	AMMONIA	15.849	1.4204E-01	-9.4648E-05	120	190	-------	31.531	39.420
172	NH3O	HYDROXYLAMINE	63.697	4.1483E-02	-4.7138E-05	273	293	-------	71.509	71.805
173	NH4Br	AMMONIUM BROMIDE	-------	----------	----------	----	----	-------	-------	-------
174	NH4Cl	AMMONIUM CHLORIDE	23.118	1.5000E-01	-3.0267E-05	458	700	-------	85.486	113.285
175	NH4I	AMMONIUM IODIDE	60.317	7.1680E-02	6.5275E-08	298	824	81.694	81.683	119.425
176	NH5O	AMMONIUM HYDROXIDE	-------	----------	----------	----	----	-------	-------	-------
177	NH5S	AMMONIUM HYDROGENSULFIDE	-------	----------	----------	----	----	-------	-------	-------
178	NO	NITRIC OXIDE	38.549	4.6751E-02	7.5224E-04	62	112	-------	44.360	53.253
179	NOCl	NITROSYL CHLORIDE	34.245	6.8855E-02	4.2100E-04	164	214	-------	56.768	68.148
180	NOF	NITROSYL FLUORIDE	48.362	6.8137E-02	7.6429E-04	89	139	-------	60.511	72.642
181	NO2	NITROGEN DIOXIDE	57.668	5.4614E-02	-7.9741E-05	251	262	-------	66.352	66.503
182	N2	NITROGEN	24.334	2.8866E-01	1.1577E-03	37	63	-------	36.599	47.114
183	N2F4	TETRAFLUOROHYDRAZINE	56.290	6.8051E-02	1.1038E-03	62	112	-------	64.681	77.648
184	N2H4	HYDRAZINE	13.455	1.8799E-01	-1.1636E-04	100	275	-------	31.091	56.313
185	N2H4C	AMMONIUM CYANIDE	30.761	1.1064E-01	4.2694E-04	259	309	101.700	88.106	105.769
186	N2H6CO2	AMMONIUM CARBAMATE	-------	----------	----------	----	----	-------	-------	-------
187	N2O	NITROUS OXIDE	39.037	4.6160E-03	2.7450E-04	100	182	-------	42.244	49.001
188	N2O3	NITROGEN TRIOXIDE	56.675	9.2588E-02	7.7157E-04	120	170	-------	78.896	94.714
189	N2O4	NITROGEN TETRAOXIDE	6.214	6.1115E-01	-8.3225E-04	30	258	-------	23.901	108.539
190	N2O5	NITROGEN PENTOXIDE	-7.816	8.2387E-01	-1.0947E-03	40	298	-------	23.387	140.486
191	Na	SODIUM	21.907	1.9663E-02	-7.9455E-06	150	300	27.063	24.677	27.090
192	NaBr	SODIUM BROMIDE	44.540	2.4857E-02	-8.4931E-06	200	1000	51.196	49.172	60.904
193	NaCN	SODIUM CYANIDE	68.363	9.1158E-04	-9.5302E-08	298	835	68.626	68.626	69.057
194	NaCl	SODIUM CHLORIDE	41.293	3.3607E-02	-1.3927E-05	200	1074	50.075	47.458	61.322
195	NaF	SODIUM FLUORIDE	37.684	2.9026E-02	-6.1728E-06	200	1200	45.790	43.243	63.627
196	NaI	SODIUM IODIDE	42.869	4.2504E-02	-3.8911E-05	221	301	52.083	50.356	52.140
197	NaOH	SODIUM HYDROXIDE	51.234	1.3088E-02	2.3359E-05	250	500	57.213	55.966	63.618
198	Na2SO4	SODIUM SULFATE	12.202	5.8138E-01	-6.0649E-04	59	458	131.629	44.434	151.256
199	Nb	NIOBIUM	24.242	2.5083E-03	9.0985E-07	273	1273	25.071	24.995	28.910
200	Nd	NEODYMIUM	25.372	3.1587E-03	1.3371E-05	298	800	27.502	27.502	36.456
201	Ne	NEON	27.742	8.9744E-02	-3.5263E-03	20	25	-------	28.126	27.820
202	Ni	NICKEL	25.667	-1.3406E-02	4.7730E-05	275	624	25.913	25.588	35.906
203	NiC4O4	NICKEL CARBONYL	67.850	1.6356E-01	8.2543E-04	198	248	-------	132.668	159.265
204	NiF2	NICKEL FLUORIDE	16.804	2.2128E-01	-2.0923E-04	198	298	64.181	52.427	64.191
205	Np	NEPTUNIUM	40.611	-1.2992E-01	3.4926E-04	240	320	32.922	29.547	34.800
206	O2	OXYGEN	-16.683	1.5925E+00	-2.9859E-03	13	44	-------	4.211	47.314
207	O3	OZONE	52.933	4.8055E-02	1.5939E-03	30	80	-------	55.830	67.023
208	Os	OSMIUM	23.617	3.8552E-03	-1.0204E-07	273	1273	24.758	24.663	28.360
209	OsOF5	OSMIUM OXIDE PENTAFLUORIDE	-------	----------	----------	----	----	-------	-------	-------
210	OsO4	OSMIUM TETROXIDE - YELLOW	-------	----------	----------	----	----	-------	-------	-------
211	OsO4	OSMIUM TETROXIDE - WHITE	-------	----------	----------	----	----	-------	-------	-------
212	P	PHOSPHORUS - WHITE	20.081	1.3495E-02	-6.0878E-06	250	870	23.563	23.074	27.214
213	PBr3	PHOSPHORUS TRIBROMIDE	43.654	9.7012E-02	5.2968E-04	183	233	-------	79.190	95.066
214	PCl2F3	PHOSPHORUS DICHLORIDE TRIFLUORIDE	40.254	1.0710E-01	4.9778E-04	215	265	-------	86.338	103.647
215	PCl3	PHOSPHORUS TRICHLORIDE	14.056	6.4389E-01	-1.3640E-03	30	181	-------	32.145	85.936
216	PCl5	PHOSPHORUS PENTACHLORIDE	-18.289	2.5859E+00	-7.1709E-03	55	150	-------	102.368	208.167
217	PH3	PHOSPHINE	32.564	4.5929E-02	5.1393E-04	89	139	-------	40.773	48.948
218	PH4Br	PHOSPHONIUM BROMIDE	55.446	1.4306E-01	6.8122E-04	210	260	-------	115.529	138.691
219	PH4Cl	PHOSPHONIUM CHLORIDE	81.338	1.9230E-01	9.8794E-04	195	245	-------	156.201	187.517
220	PH4I	PHOSPHONIUM IODIDE	29.623	9.3594E-02	3.8731E-04	242	292	-------	74.857	89.865
221	POCl3	PHOSPHORUS OXYCHLORIDE	15.788	5.7784E-01	-7.8868E-04	30	274	-------	32.413	114.953
222	PSBr3	PHOSPHORUS THIOBROMIDE	-------	----------	----------	----	----	-------	-------	-------
223	PSCl3	PHOSPHORUS THIOCHLORIDE	109.985	9.4402E-02	-1.4869E-04	226	237	-------	123.725	124.005
224	P4O6	PHOSPHORUS TRIOXIDE	54.638	1.7759E-01	-7.2293E-04	246	296	-------	141.886	170.331
225	P4O10	PHOSPHORUS PENTOXIDE	77.355	4.3770E-01	-2.9083E-04	200	325	182.001	153.261	188.763
226	P4S10	PHOSPHORUS PENTASULFIDE	178.412	4.0088E-01	-1.9305E-04	200	560	280.773	250.866	342.362
227	Pb	LEAD	23.167	1.1576E-02	-2.0490E-06	295	600	26.436	26.404	29.375
228	PbBr2	LEAD BROMIDE	74.103	9.4205E-03	2.4482E-05	200	600	79.088	76.967	88.569
229	PbCl2	LEAD CHLORIDE	68.272	2.9836E-02	-7.9575E-07	298	700	77.097	77.097	88.768
230	PbF2	LEAD FLUORIDE	60.786	3.9651E-02	-3.6260E-06	298	500	72.285	72.285	79.705
231	PbI2	LEAD IODIDE	71.526	1.7372E-02	9.8152E-06	199	291	-------	75.380	77.413
232	PbO	LEAD OXIDE	26.341	8.1702E-02	-5.5025E-05	200	700	45.809	40.480	56.570
233	PbS	LEAD SULFIDE	45.695	1.2221E-02	-1.6617E-06	200	1200	49.191	48.073	57.967
234	Pd	PALLADIUM	20.384	1.3990E-02	-3.5278E-06	273	1273	24.241	23.942	32.476

106

Table 4-2 HEAT CAPACITY OF SOLID - INORGANIC COMPOUNDS (continued)

			$C_P = A + B T + C T^2$					$(C_P$ - joule/(mol K), T - K)		
NO	FORMULA	NAME	A	B	C	TMIN	TMAX	C_P @25 C	C_P @TMIN	CP @TMAX
235	Po	POLONIUM	-4.766	2.6485E-02	5.5506E-05	477	527	-------	20.508	24.620
236	Pt	PLATINUM	24.141	5.8138E-03	-3.6149E-07	273	1273	25.842	25.701	30.956
237	Ra	RADIUM	-------			----	----	----	-------	-------
238	Rb	RUBIDIUM	34.171	-7.7866E-02	2.2344E-04	220	298	30.818	27.855	30.818
239	RbBr	RUBIDIUM BROMIDE	28.806	-3.0893E-02	8.2449E-05	206	249	-------	25.942	26.228
240	RbCl	RUBIDIUM CHLORIDE	-60.936	5.4660E-02	5.8264E-05	938	988	-------	41.624	49.968
241	RbF	RUBIDIUM FLUORIDE	20.097	5.0212E-02	6.9574E-06	575	950	-------	51.269	74.077
242	RbI	RUBIDIUM IODIDE	52.207	-1.5192E-02	5.9005E-05	208	277	-------	51.603	52.524
243	Re	RHENIUM	24.622	4.4294E-03	-4.0178E-07	278	1200	25.907	25.821	29.359
244	Re2O7	RHENIUM HEPTOXIDE	-------			----	----	----	-------	-------
245	Rh	RHODIUM	22.703	8.6231E-03	5.5752E-07	273	1273	25.324	25.100	34.586
246	Rn	RADON	16.748	3.1849E-02	2.0932E-04	152	202	-------	26.439	31.740
247	Ru	RUTHENIUM	21.440	6.7856E-03	2.4643E-08	273	1273	23.466	23.295	30.118
248	RuF5	RUTHENIUM PENTAFLUORIDE	-------			----	----	----	-------	-------
249	S	SULFUR	2.003	1.2002E-01	-1.6202E-04	30	383	23.383	5.457	24.203
250	SF4	SULFUR TETRAFLUORIDE	56.979	8.4329E-02	8.5052E-04	99	149	-------	73.701	88.477
251	SF6	SULFUR HEXAFLUORIDE	132.989	1.1479E-01	-1.9275E-04	213	223	-------	148.694	149.005
252	SOBr2	THIONYL BROMIDE	44.426	9.2699E-02	5.4226E-04	171	221	-------	76.120	91.380
253	SOCl2	THIONYL CHLORIDE	49.569	8.1741E-02	6.7001E-04	122	172	-------	69.514	83.451
254	SOF2	SULFUROUS OXYFLUORIDE	42.989	6.7849E-02	6.2293E-04	113	163	-------	58.276	69.959
255	SO2	SULFUR DIOXIDE	40.783	8.4215E-02	4.9247E-05	120	198	-------	51.598	59.354
256	SO2Cl2	SULFURYL CHLORIDE	88.854	7.7261E-02	-1.3072E-04	212	222	-------	99.358	99.564
257	SO3	SULFUR TRIOXIDE	67.647	2.1122E-01	8.8025E-04	240	290	-------	169.010	202.893
258	S2Cl2	SULFUR MONOCHLORIDE	51.315	9.3468E-02	6.5294E-04	143	193	-------	78.075	93.728
259	Sb	ANTIMONY	23.548	3.9315E-03	3.2951E-06	253	823	25.013	24.753	29.016
260	SbBr3	ANTIMONY TRIBROMIDE	15.325	1.0375E-01	3.2449E-04	320	370	-------	81.676	98.050
261	SbCl3	ANTIMONY TRICHLORIDE	11.046	4.0794E-01	-2.9186E-04	298	346	106.729	106.693	117.253
262	SbCl5	ANTIMONY PENTACHLORIDE	44.283	1.2599E-01	5.5758E-04	226	276	-------	101.216	121.508
263	SbH3	STIBINE	29.381	5.1531E-02	3.8129E-04	135	185	-------	43.310	51.993
264	SbI3	ANTIMONY TRIIODIDE	-0.016	1.1966E-01	3.0670E-04	390	440	-------	93.354	112.069
265	Sb2O3	ANTIMONY TRIOXIDE	-------	----------	----------	----	----	----	-------	-------
266	Sc	SCANDIUM	25.032	8.7561E-04	5.2965E-06	298	1300	25.763	25.763	35.121
267	Se	SELENIUM	10.199	8.3618E-02	-1.0910E-04	228	300	25.432	23.583	25.466
268	SeCl4	SELENIUM TETRACHLORIDE	-40.036	1.4073E-01	2.6646E-04	528	578	-------	108.619	130.396
269	SeF6	SELENIUM HEXAFLUORIDE	38.482	8.7963E-02	4.6677E-04	188	238	-------	71.635	85.996
270	SeOCl2	SELENIUM OXYCHLORIDE	33.417	9.8652E-02	4.2587E-04	232	282	-------	79.123	94.985
271	SeO2	SELENIUM DIOXIDE	-------	----------	----------	----	----	----	-------	-------
272	Si	SILICON	16.149	1.6356E-02	-6.4577E-06	273	1273	20.451	20.133	26.505
273	SiBrCl2F	BROMODICHLOROFLUOROSILANE	56.783	8.8862E-02	8.0165E-04	111	161	-------	76.484	91.817
274	SiBrF3	TRIFLUOROBROMOSILANE	40.814	7.7802E-02	5.0967E-04	153	203	-------	64.567	77.512
275	SiBr2ClF	DIBROMOCHLOROFLUOROSILANE	60.376	1.0043E-01	8.1088E-04	124	174	-------	85.252	102.344
276	SiClF3	TRIFLUOROCHLOROSILANE	49.810	6.7350E-02	8.2995E-04	81	131	-------	60.741	72.919
277	SiCl2F2	DICHLORODIFLUOROSILANE	54.914	7.5148E-02	9.0052E-04	83	133	-------	67.456	80.979
278	SiCl3F	TRICHLOROFLUOROSILANE	58.596	8.8069E-02	8.6047E-04	102	152	-------	76.624	91.985
279	SiCl4	SILICON TETRACHLORIDE	64.049	-2.0639E-01	1.3933E-03	100	200	-------	57.343	78.504
280	SiF4	SILICON TETRAFLUORIDE	45.328	-1.2605E-01	2.3604E-03	88	182	-------	52.576	100.757
281	SiHBr3	TRIBROMOSILANE	51.925	9.7560E-02	6.5192E-04	150	200	-------	81.124	97.388
282	SiHCl3	TRICHLOROSILANE	57.611	8.3538E-02	8.7981E-04	95	145	-------	73.474	88.205
283	SiHF3	TRIFLUOROSILANE	38.843	5.5440E-02	6.0426E-04	92	142	-------	49.016	58.843
284	SiH2Br2	DIBROMOSILANE	44.518	8.4985E-02	5.5564E-04	153	203	-------	70.515	84.652
285	SiH2Cl2	DICHLOROSILANE	47.365	7.0781E-02	6.9976E-04	101	151	-------	61.684	74.051
286	SiH2F2	DIFLUOROSILANE	35.139	5.2220E-02	5.2220E-04	100	150	-------	45.583	54.722
287	SiH2I2	DIIODOSILANE	37.861	1.0516E-01	4.7337E-04	222	272	-------	84.583	101.541
288	SiH3Br	MONOBROMOSILANE	37.996	6.4819E-02	5.0150E-04	129	179	-------	54.751	65.728
289	SiH3Cl	MONOCHLOROSILANE	37.531	5.7142E-02	5.4395E-04	105	155	-------	49.537	59.468
290	SiH3F	MONOFLUOROSILANE	31.550	4.6886E-02	4.6886E-04	100	150	-------	40.927	49.132
291	SiH3I	IODOSILANE	37.637	7.6660E-02	4.6139E-04	166	216	-------	63.111	75.763
292	SiH4	SILANE	-7.317	8.1259E-01	1.2337E-03	15	56	-------	5.234	42.180
293	SiO2	SILICON DIOXIDE	2.478	1.6522E-01	-9.6769E-05	100	848	43.137	18.033	72.999
294	Si2Cl6	HEXACHLORODISILANE	56.937	1.5794E-01	7.1162E-04	222	272	-------	127.048	152.518
295	Si2F6	HEXAFLUORODISILANE	48.325	1.2079E-01	5.9052E-04	205	255	-------	97.740	117.335
296	Si2H5Cl	DISILANYL CHLORIDE	41.418	1.0088E-01	5.0442E-04	200	250	-------	81.772	98.166
297	Si2H6	DISILANE	51.815	7.3552E-02	8.1139E-04	91	141	-------	65.150	78.212
298	Si2OCl3F3	TRICHLOROTRIFLUORODISILOXANE	82.922	1.2323E-01	1.2323E-03	100	150	-------	107.568	129.133
299	Si2OCl6	HEXACHLORODISILOXANE	64.722	1.4914E-01	7.8518E-04	190	240	-------	121.382	145.717
300	Si2OH6	DISILOXANE	58.646	7.8377E-02	9.9274E-04	79	129	-------	71.021	85.260
301	Si3Cl8	OCTACHLOROTRISILANE	76.638	1.8667E-01	9.3335E-04	200	250	-------	151.306	181.640
302	Si3H8	TRISILANE	63.227	9.6677E-02	9.1247E-04	106	156	-------	83.712	100.495
303	Si3H9N	TRISILAZANE	75.786	1.2234E-01	1.0416E-03	117	167	-------	104.523	125.478
304	Si4H10	TETRASILANE	80.629	1.3774E-01	1.0632E-03	130	180	-------	116.318	139.637
305	Sm	SAMARIUM	13.079	6.2040E-02	-2.8947E-05	298	1100	29.003	29.003	46.297
306	Sn	TIN	-3.365	2.5009E-02	5.4956E-05	455	505	-------	19.397	23.286
307	SnBr4	STANNIC BROMIDE	35.390	1.2247E-01	4.8188E-04	254	304	114.740	97.641	117.216
308	SnCl2	STANNOUS CHLORIDE	-12.757	7.7242E-02	1.6436E-04	470	520	-------	59.842	71.839
309	SnCl4	STANNIC CHLORIDE	55.154	1.2919E-01	6.6954E-04	193	243	-------	105.007	126.059
310	SnH4	STANNIC HYDRIDE	44.826	5.8069E-02	7.9275E-04	73	123	-------	53.333	64.025
311	SnI4	STANNIC IODIDE	6.535	1.3957E-01	3.7964E-04	368	418	-------	109.164	131.050
312	Sr	STRONTIUM	22.354	1.5847E-02	-5.1449E-06	100	800	26.622	23.888	31.740

Table 4-2 HEAT CAPACITY OF SOLID - INORGANIC COMPOUNDS (continued)

| NO | FORMULA | NAME | $C_P = A + B T + C T^2$ | | | | | $(C_P$ - Joule/(mol K), T - K) | | |
			A	B	C	TMIN	TMAX	C_P @25 C	C_P @TMIN	CP @TMAX
313	SrO	STRONTIUM OXIDE	0.722	2.9389E-01	-4.9060E-04	111	298	44.736	27.336	44.736
314	Ta	TANTALUM	23.477	5.5650E-03	-9.4258E-07	295	1700	25.053	25.037	30.214
315	Tc	TECNNETIUM	-145.809	3.6390E-02	1.5289E-05	2380	2430	-------	27.418	32.914
316	Te	TELLURIUM	19.623	3.4265E-02	-4.6177E-05	148	301	25.734	23.680	25.754
317	TeCl4	TELLURIUM TETRACHLORIDE	-22.094	1.8691E-01	4.1799E-04	447	497	-------	145.056	174.137
318	TeF6	TELLURIUM HEXAFLUORIDE	38.227	8.5945E-02	4.6369E-04	185	235	-------	70.087	84.138
319	Ti	TITANIUM	27.700	-1.8215E-02	3.1069E-05	311	1033	-------	25.041	42.045
320	TiCl4	TITANIUM TETRACHLORIDE	51.686	3.4258E-01	-1.8317E-04	100	200	-------	84.112	112.875
321	Tl	THALLIUM	25.359	-1.5755E-03	2.4836E-05	237	493	27.097	26.376	30.622
322	TlBr	THALLOUS BROMIDE	-34.087	5.6827E-02	8.3184E-05	683	733	-------	43.556	52.288
323	TlI	THALLOUS IODIDE	-34.132	6.1028E-02	9.2027E-05	663	713	-------	46.809	56.193
324	Tm	THULIUM	22.604	9.7180E-03	-8.6499E-07	295	1300	25.424	25.395	33.775
325	U	URANIUM	26.137	-7.6442E-03	3.1495E-05	373	933	-------	27.670	46.428
326	UF6	URANIUM FLUORIDE	115.827	5.1120E-03	5.6052E-04	230	337	167.178	146.654	181.288
327	V	VANADIUM	24.724	1.8308E-03	4.0263E-06	273	1273	25.628	25.525	33.581
328	VCl4	VANADIUM TETRACHLORIDE	113.793	1.0593E-01	-1.6334E-04	237	247	-------	129.725	130.005
329	VOCl3	VANADIUM OXYTRICHLORIDE	58.092	1.0606E-01	7.3834E-04	144	194	-------	88.564	106.319
330	W	TUNGSTEN	23.246	4.1342E-03	1.8110E-07	273	1200	24.495	24.389	28.468
331	WF6	TUNGSTEN FLUORIDE	43.797	1.2203E-01	5.4806E-04	223	273	-------	98.135	117.809
332	Xe	XENON	20.284	3.1819E-02	2.8573E-04	111	161	-------	27.371	32.858
333	Yb	YTTERBIUM	11.298	6.4252E-02	-5.2673E-05	298	550	25.772	25.772	30.703
334	Yt	YTTRIUM	24.359	6.8860E-03	2.0884E-07	260	1200	26.430	26.163	32.923
335	Zn	ZINC	24.331	-8.1713E-04	1.3464E-05	253	673	25.285	24.986	29.880
336	ZnCl2	ZINC CHLORIDE	48.329	1.1147E-01	-1.1501E-04	241	346	71.339	68.516	73.141
337	ZnF2	ZINC FLUORIDE	25.281	1.9848E-01	-2.1104E-04	234	299	65.696	60.123	65.788
338	ZnO	ZINC OXIDE	11.190	1.1882E-01	-1.0002E-04	150	473	37.725	26.762	45.018
339	ZnSO4	ZINC SULFATE	67.372	1.0066E-01	-8.7483E-06	295	953	96.608	96.307	155.360
340	Zr	ZIRCONIUM	27.648	-5.8836E-03	9.3058E-06	333	1013	-------	26.721	31.239
341	ZrBr4	ZIRCONIUM BROMIDE	98.211	1.1209E-01	-8.7052E-05	200	700	123.891	117.146	134.015
342	ZrCl4	ZIRCONIUM CHLORIDE	69.798	2.4786E-01	-2.6840E-04	217	300	119.839	110.893	120.001
343	ZrI4	ZIRCONIUM IODIDE	64.049	-2.0639E-01	1.3933E-03	100	200	-------	57.343	78.504

C_P - heat capacity of solid, joule/(mol K)

A, B, and C - regression coefficients for chemical compound

T - temperature, K

TMIN - minimum temperature, K

TMAX - maximum temperature, K

Chapter 5

ENTHALPY OF VAPORIZATION

Carl L. Yaws, Xiaoyan Lin, Li Bu, Sachin Nijhawan, Deepa R. Balundgi, and Saumya Tripathi
Lamar University, Beaumont, Texas

ABSTRACT

Results for enthalpy of vaporization are presented for major organic and inorganic compounds. The complete temperature range for the liquid is covered from freezing to the critical point for most of the compounds. The results are displayed in easy-to-use tabulations that are especially applicable for rapid engineering usage with the personal computer or hand calculator.

INTRODUCTION

Physical and thermodynamic property data such as enthalpy of vaporization are of special value to engineers in the chemical processing and petroleum refining industries. As an example, knowledge of the enthalpy of vaporization is required in the design of heat exchangers for vaporizing liquids. Other examples of usage include reboilers and overhead condensers in distillation. In this article, results for enthalpy of vaporization as a function of temperature are presented for a wide variety of compounds.

ENTHALPY OF VAPORIZATION CORRELATION

A modified Watson equation was selected for enthalpy of vaporization as a function of temperature:

$$\Delta H_{vap} = A \left(1 - T/T_C\right)^n \tag{5-1}$$

where
ΔH_{vap} = enthalpy of vaporization, kjoule/mol
A, T_C, and n = regression coefficients for chemical compound
T = temperature, K

The results for enthalpy of vaporization are given in Tables 5-1 and 5-2. In preparing the tabulations, a literature search was conducted to identify data source publications for organics (1-41) and inorganics (1-93). Both experimental values for the property under consideration and parameter values for estimation of the property are included in the source publications. The publications were screened for appropriate data. The compilation resulting from the screening is based on both experimental data and estimated values. In the absence of experimental data, estimates were primarily based on the Riedel equation (29). Experimental data and estimates were then regressed to provide the same equation for all compounds.

The tabulation discloses the temperature range for which the equation may be used. The respective minimum and maximum temperatures are denoted by TMIN and TMAX. The temperature T_B is the normal boiling point (temperature at which the vapor pressure is 1 atm). Results for enthalpy of vaporization at the normal boiling point are provided in the last column.

A comparison of calculated and experimental data values for enthalpy of vaporization is shown in Fig. 5-1 for a representative chemical. The graph indicates good agreement of correlation and data.

EXAMPLES

The correlation results may be used for prediction and calculation of enthalpy of vaporization. Examples are given below.

Example 1 Estimate the enthalpy of vaporization of carbon tetrafluoride (CF4) at 183.15 K.

Substitution of the regression coefficients from the table and temperature into the equation for enthalpy of vaporization yields

$\Delta H_{vap} = (16.6594) \left(1 - 183.15/227.5\right)^{0.349}$

$\Delta H_{vap} = 9.415$ kjoule/mol

Example 2 Estimate the enthalpy of vaporization of ethane (C2H6) at 200 K.

Substitution of the regression coefficients from the table and temperature into the equation for enthalpy of vaporization yields

$\Delta H_{vap} = (21.342) \left(1 - 200/305.42\right)^{0.403}$

$\Delta H_{vap} = 13.90$ kjoule/mol

Portions of this material appeared in Hydrocarbon Processing, 69, 87 (June, 1990) and are reprinted by special permission.

REFERENCES – ORGANIC COMPOUNDS

1-34. See **REFERENCES - ORGANIC COMPOUNDS** in **Chapter 1, CRITICAL PROPERTIES AND ACENTRIC FACTOR**

35. Edmister, W. C., APPLIED HYDROCARBON THERMODYNAMICS, Vols. 1 and 2, Vol 2 (2nd ed.), Gulf Publishing Co., Houston, TX (1961, 1974, 1984).

36. Zwolinski, B. J. and R. C. Wilhoit, VAPOR PRESSURES AND HEATS OF VAPORIZATION OF HYDROCARBONS AND RELATED COMPOUNDS, Thermodynamic Research Center, TAMU, College Station, TX (1971).

37. Wilhoit, R. C. and B. J. Zwolinski, PHYSICAL AND THERMODYNAMIC PROPERTIES OF ALIPHATIC ALCOHOLS, American Chemical Society, American Institute of Physics, National Bureau of Standards, New York, NY (1973).

38. Boublick, T., V. Fried and E. Hala, THE VAPOUR PRESSURES OF PURE SUBSTANCES, 1st ed., 2nd ed., Elsevier, New York, NY (1975, 1984).

39. Ohe, S., COMPUTER AIDED DATA BOOK OF VAPOR PRESSURE, Data Book Publishing Company, Tokyo, Japan (1976).

40. Howard, P. H. and W. M. Meylan, eds., HANDBOOK OF PHYSICAL PROPERTIES OF ORGANIC CHEMICALS, CRC Press, Boca Raton, FL (1997).

41. Yaws, C. L. and others, Hydrocarbon Processing, 69, 87 (June, 1990).

REFERENCES - INORGANIC COMPOUNDS

1-56. See **REFERENCES - INORGANIC COMPOUNDS** in **Chapter 1, CRITICAL PROPERTIES AND ACENTRIC FACTOR**

57. Lyon, R. N., ed., LIQUID-METALS HANDBOOK, 2nd ed., Atomic Energy Commission, Washington, DC (1954).

58. Karapet'yants, M. Kh. and M. L. Karapet'yants, THERMODYNAMIC CONSTANTS OF INORGANIC AND ORGANIC COMPOUNDS, Ann Arbor - Humphrey Science Publishers, Ann Arbor, MH (1970).

59. Fink, J. K., TABLES OF THERMODYNAMIC PROPERTIES OF SODIUM, Argonne National Lab., ANL-CEN-RSD-82-4, Chemical Engineering Division, Argonne, IL (June 1982).

60. Mills, K. C., THERMODYNAMIC DATA FOR INORGANIC SULPHIDES SELENIDES AND TELLURIDES, Butterworths, London, England (1974).

61. Cox, J. D.and G. Pilcher, THERMOCHEMISTRY OF ORGANIC AND ORGANOMETALLIC COMPOUNDS, Academic Press, New York, NY (1970).

62. Kazavchinskii, Ya.Z., and others, HEAVY WATER THERMOPHYSICAL PROPERTIES, U. S. Department of Commerce, NBS, Washington, DC (1971).

63. Whalley, E., THE THERMODYNAMIC AND TRANSPORT PROPERTIES OF HEAVY WATER, Proc. Conf. Thermodyn. Trans. Prop. of Fluids, pp. 15-26, London, England (July, 1957).

64. Moore, G. A. and T. R. Shives, IRON (99.9 + %), Metals Handbook, 8th ed., 1206 (1961).

65. McCarty, R. D., HYDROGEN TECHNOLOGICAL SURVEY - THERMOPHYSICAL PROPERTIES, Prepared for Aerospace Research and Data Institute, NASA Lewis Research Center (1975).

66. McCarty, R. D., J. Hord, and H. M. Roder, SELECTED PROPERTIES OF HYDROGEN (ENGINEERING DESIGN DATA), Center for Chemical Engineering, National Engineering Laboratory, Nat. Bur. Stand. Monograph, 168, Boulder, CO (1981).

67. Haar, L., J. S. Gallagher, and G. S. Kell, NBS/NRC STEAM TABLES, THERMODYNAMIC AND TRANSPORT PROPERTIES AND COMPUTER PROGRAMS FOR VAPOR AND LIQUID STATES OF WATER IN SI UNITS, Hemisphere Publishing Corporation, Washington, DC (1984).

68. Keenan, J. H. and others, STEAM TABLES. THERMODYNAMIC PROPERTIES OF WATER INCLUDING VAPOR, LIQUID AND SOLID PHASES, John Wiley & Sons, Inc., New York, NY (1969).

69. Kaufmann, D. W., ed., SODIUM CHLORIDE, Reinhold Pub. Corp., New York, NY (1960).

70. Parish, M. B., J. Chem. Eng. Data, 6(4), 592 (1961).

71. Leider, H. R., O. H. Krikorian, and D. A. Young, Carbon, 11, 555 (1973).

72. Frank, A. and K. Clausius, Z. Phys. Chem., 42, 395 (1939).

73. Mascherpa, G., Rev. Chim. Miner., 2, 379 (1965).

74. Grubitsch, H. and E. Suppan, Monatsch. Chem., 93, 246 (1962).

75. Watson, K. M.,Ind. Eng. Chem. 23, 360 (1931).

76. Fischer, W. and O. Juberman, Z. Anorg. Chem., 227, 227 (1936).

77. Clausius, K. and G. Faber, Z. Phys. Chem., 51, 352 (1942).

78. Duisman, J. A. and S. A. Stern, J. Chem. Eng. Data, 14(4), 457 (1969).

79. Stern, S. A., J. L. Mullhaupt and W. B. Kay, Chem. Rev., 60, 185 (1960).

80. Foley, W. T. and P. A. Giguere, Can. J. Chem., 29, 895 (1951).

81. Abrahams, B. M., D. W. Osborne and B. Weinstock, Phys. Rev., 80(3), 336 (1980).

82. Gibbons, R. M. and C. McKinley, Adv. Cryog. Eng., 13, 375 (1968).

83. Wagman, D. and others, J. Phys. Chem. Ref. Data, Suppl. No. 2 (1982).

84. Lindenberg, A. B., Comptes Rend. Acad. Sci. Paris, 273, 1017 (1971).

85. Johnston, H. L. and W. F. Giauque, J. Amer. Chem. Soc., 51, 94 (1929).

86. Wanger, W., Cryogenics, 12(3), 214 (1972).

87. Couch, E. J., K. A. Kobe and L. J. Herth, J. Chem. Eng. Data, 6(2), 229 (1961).

88. Daniels, F. and A. C. Bright, J. Chem. Soc., 42, 1131 (1920).

89. Ladenburg, R. and R. Minkowski, Z. Phys., 6, 153 (1921).

90. West, W. A. and A. W. C. Menzies, J. Phys. Chem., 33(2), 1880 (1929).

91. Awbery, J. H., Proc. Phys. Soc. London, 39, 417 (1927).

92. Rigby, W., Chem. Ind., 1508 (1969).

93. Hyne, R. A. and P. F. Telly, J. Chem. Soc. 2348 (1961).

Figure 5-1 Enthalpy of Vaporization

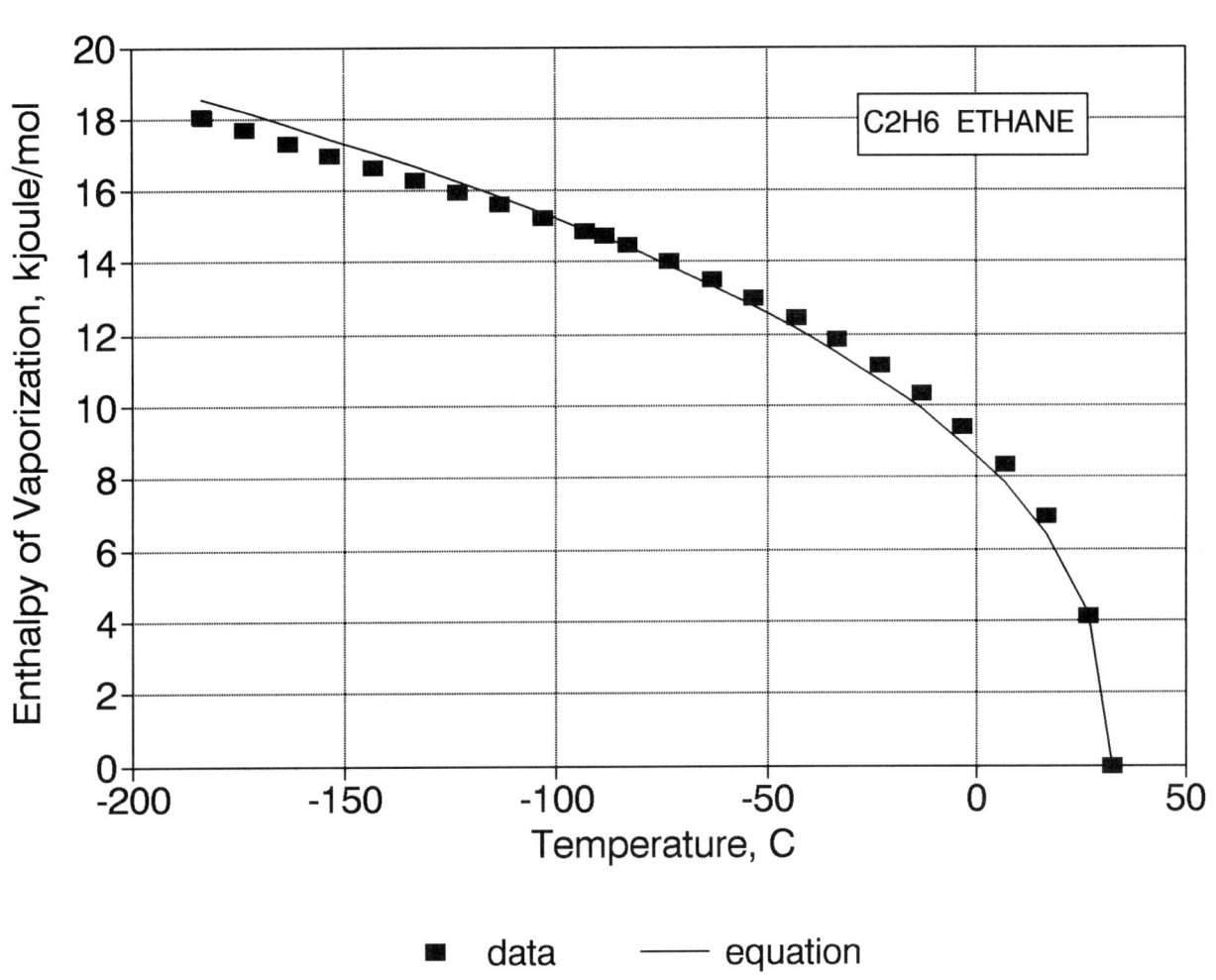

Table 5-1 ENTHALPY OF VAPORIZATION - ORGANIC COMPOUNDS

NO	FORMULA	NAME	HVAP = A (1 - T/T$_c$)^n			(HVAP - kjoule/mol, T - K)			
			A	T$_c$	n	TMIN	TMAX	T$_B$	HVAP @ T$_B$
1	CBrClF2	BROMOCHLORODIFLUOROMETHANE	32.403	426.15	0.371	113.65	426.15	269.14	22.37
2	CBrCl3	BROMOTRICHLOROMETHANE	45.857	606.00	0.363	252.15	606.00	378.05	32.16
3	CBrF3	BROMOTRIFLUOROMETHANE	25.070	340.15	0.357	105.15	340.15	215.26	17.53
4	CBr2F2	DIBROMODIFLUOROMETHANE	39.175	478.00	0.399	163.05	478.00	295.94	26.65
5	CClF3	CHLOROTRIFLUOROMETHANE	18.614	301.96	0.175	92.15	301.96	191.74	15.60
6	CClN	CYANOGEN CHLORIDE	38.096	449.00	0.345	266.65	449.00	286.00	26.86
7	CCl2F2	DICHLORODIFLUOROMETHANE	30.930	384.95	0.406	115.15	384.95	243.36	20.61
8	CCl2O	PHOSGENE	35.610	455.00	0.378	145.37	455.00	280.71	24.78
9	CCl3F	TRICHLOROFLUOROMETHANE	37.119	471.20	0.391	162.04	471.20	296.97	25.16
10	CCl4	CARBON TETRACHLORIDE	37.890	556.35	0.241	250.33	556.35	349.79	29.84
11	CF2O	CARBONYL FLUORIDE	29.048	297.00	0.466	161.89	297.00	188.58	18.16
12	CF4	CARBON TETRAFLUORIDE	16.659	227.50	0.349	89.56	227.50	145.09	11.69
13	CHBr3	TRIBROMOMETHANE	55.620	696.00	0.421	281.20	696.00	422.35	37.54
14	CHClF2	CHLORODIFLUOROMETHANE	29.612	369.30	0.386	115.73	369.30	232.32	20.19
15	CHCl2F	DICHLOROFLUOROMETHANE	29.763	451.58	0.193	138.15	451.58	282.05	24.64
16	CHCl3	CHLOROFORM	42.953	536.40	0.375	209.63	536.40	334.33	29.79
17	CHF3	TRIFLUOROMETHANE	23.858	298.89	0.355	117.97	298.89	190.99	16.62
18	CHI3	TRIIODOMETHANE	61.071	794.55	0.380	396.16	794.55	491.16	42.36
19	CHN	HYDROGEN CYANIDE	42.384	456.65	0.428	259.83	456.65	298.85	26.90
20	CHNS	ISOTHIOCYANIC-ACID	--------	--------	------	--------	--------	--------	------
21	CH2BrCl	BROMOCHLOROMETHANE	44.363	557.00	0.374	185.20	557.00	341.20	31.12
22	CH2Br2	DIBROMOMETHANE	48.196	611.00	0.377	220.60	611.00	370.10	33.93
23	CH2ClF	CHLOROFLUOROMETHANE	32.899	424.91	0.380	140.16	424.91	264.06	22.74
24	CH2Cl2	DICHLOROMETHANE	41.910	510.00	0.410	178.01	510.00	312.90	28.38
25	CH2F2	DIFLUOROMETHANE	28.364	351.60	0.341	137.00	351.60	221.50	20.21
26	CH2I2	DIIODOMETHANE	61.958	747.00	0.446	279.25	747.00	455.15	40.74
27	CH2O	FORMALDEHYDE	30.940	408.00	0.297	181.15	408.00	254.05	23.16
28	CH2O2	FORMIC ACID	5.514	580.00	-1.338	281.45	373.71	373.71	21.99
29	CH3Br	METHYL BROMIDE	32.009	467.00	0.308	179.47	467.00	276.71	24.28
30	CH3Cl	METHYL CHLORIDE	32.534	416.25	0.452	175.43	416.25	248.93	21.55
31	CH3Cl3Si	METHYL TRICHLOROSILANE	38.200	517.00	0.287	195.35	517.00	339.55	28.10
32	CH3F	METHYL FLUORIDE	25.088	317.70	0.369	131.35	317.70	194.82	17.67
33	CH3I	METHYL IODIDE	35.967	528.00	0.302	206.70	528.00	315.58	27.32
34	CH3NO	FORMAMIDE	57.390	771.00	0.170	450.00	771.00	493.00	48.25
35	CH3NO2	NITROMETHANE	47.330	588.15	0.297	244.60	588.15	374.35	35.04
36	CH3NO2	METHYL-NITRITE	--------	--------	------	--------	--------	--------	------
37	CH3NO3	METHYL-NITRATE	--------	--------	------	--------	--------	--------	------
38	CH4	METHANE	10.312	190.58	0.265	90.67	190.58	111.66	8.16
39	CH4Cl2Si	METHYL DICHLOROSILANE	36.769	483.00	0.313	182.55	483.00	314.70	26.43
40	CH4O	METHANOL	52.723	512.58	0.377	175.47	512.58	337.85	35.14
41	CH4O3S	METHANESULFONIC ACID	--------	--------	------	--------	--------	--------	------
42	CH4S	METHYL MERCAPTAN	34.579	469.95	0.379	150.18	469.95	279.11	24.57
43	CH5ClSi	METHYL CHLOROSILANE	32.830	442.00	0.331	139.05	442.00	281.85	23.46
44	CH5N	METHYLAMINE	38.656	430.05	0.405	179.69	430.05	266.82	26.11
45	CH6Si	METHYL SILANE	23.222	352.50	0.322	116.34	352.50	216.25	17.10
46	CN4O8	TETRANITROMETHANE	60.691	540.00	0.400	287.05	540.00	398.85	35.48
47	CO	CARBON MONOXIDE	8.003	132.92	0.318	68.15	132.92	81.70	5.91
48	COS	CARBONYL SULFIDE	25.695	378.80	0.363	134.30	378.80	223.00	18.61
49	CO2	CARBON DIOXIDE	15.326	304.19	0.227	216.58	304.19	194.67	------
50	CS2	CARBON DISULFIDE	34.997	552.00	0.299	161.11	552.00	270.65	28.61
51	C2BrF3	BROMOTRIFLUOROETHYLENE	32.370	432.00	0.369	173.00	432.00	320.41	19.64
52	C2Br2F4	1,2-DIBROMOTETRAFLUOROETHANE	39.518	487.80	0.364	162.65	487.80	245.30	30.64
53	C2ClF3	CHLOROTRIFLUOROETHYLENE	31.800	379.15	0.382	115.00	379.15	319.37	15.70
54	C2ClF5	CHLOROPENTAFLUOROETHANE	28.993	353.15	0.373	173.71	353.15	234.04	19.33
55	C2Cl2F4	1,2-DICHLOROTETRAFLUOROETHANE	34.508	418.85	0.369	179.15	418.85	276.92	23.15
56	C2Cl3F3	1,1,2-TRICHLOROTRIFLUOROETHANE	41.060	487.25	0.383	236.75	487.25	320.75	27.22
57	C2Cl4	TETRACHLOROETHYLENE	50.840	620.00	0.388	250.80	620.00	394.40	34.34
58	C2Cl4F2	1,1,2,2-TETRACHLORODIFLUOROETHANE	48.875	551.00	0.393	299.15	551.00	366.00	31.83
59	C2Cl4O	TRICHLOROACETYL CHLORIDE	43.249	590.00	0.170	273.15	590.00	391.15	35.95
60	C2Cl6	HEXACHLOROETHANE	73.750	698.00	0.380	460.00	698.00	460.00	49.00
61	C2F4	TETRAFLUOROETHYLENE	25.129	306.45	0.386	142.00	306.45	197.51	16.86
62	C2F6	HEXAFLUOROETHANE	24.933	292.80	0.398	173.10	292.80	194.95	16.12
63	C2HBrClF3	HALOTHANE	44.527	521.00	0.505	223.15	521.00	323.35	27.29
64	C2HClF2	2-CHLORO-1,1-DIFLUOROETHYLENE	32.425	400.55	0.383	134.65	400.55	254.55	22.03
65	C2HCl3	TRICHLOROETHYLENE	46.915	571.00	0.396	188.40	571.00	360.10	31.62
66	C2HCl3O	DICHLOROACETYL CHLORIDE	47.311	579.00	0.248	298.15	579.00	382.15	36.20
67	C2HCl3O	TRICHLOROACETALDEHYDE	51.328	565.00	0.382	216.00	565.00	370.85	34.13
68	C2HCl5	PENTACHLOROETHANE	60.787	665.00	0.469	244.15	665.00	433.03	37.09
69	C2HF3	TRIFLUOROETHENE	28.119	347.22	0.380	94.53	347.22	221.01	19.14
70	C2HF3O2	TRIFLUOROACETIC ACID	--------	--------	------	--------	--------	--------	------
71	C2HF5	PENTAFLUOROETHANE	28.858	342.00	0.382	170.15	342.00	225.15	19.15
72	C2H2	ACETYLENE	16.740	308.32	-0.056	192.40	200.00	189.15	------
73	C2H2Br4	1,1,2,2-TETRABROMOETHANE	85.450	824.00	0.554	273.15	824.00	516.65	49.48
74	C2H2Cl2	1,1-DICHLOROETHYLENE	40.660	482.00	0.424	150.59	482.00	304.71	26.61
75	C2H2Cl2	cis-1,2-DICHLOROETHYLENE	42.213	527.00	0.349	193.15	527.00	333.65	29.75
76	C2H2Cl2	trans-1,2-DICHLOROETHYLENE	38.924	508.00	0.322	223.35	508.00	320.85	28.22
77	C2H2Cl2O	CHLOROACETYL CHLORIDE	41.512	581.00	0.129	251.15	581.00	379.15	36.22
78	C2H2Cl2O	DICHLOROACETALDEHYDE	51.064	555.00	0.381	223.00	555.00	362.00	34.15

112

Table 5-1 ENTHALPY OF VAPORIZATION - ORGANIC COMPOUNDS (continued)

			HVAP = A (1 - T/T$_c$)^n					(HVAP - kjoule/mol, T - K)	
NO	FORMULA	NAME	A	T$_c$	n	TMIN	TMAX	T$_B$	HVAP @ T$_B$
79	C2H2Cl2O2	DICHLOROACETIC ACID	77.385	686.00	0.391	286.54	686.00	467.15	49.51
80	C2H2Cl3F	1,1,1-TRICHLOROFLUOROETHANE	46.195	565.00	0.366	173.00	565.00	366.00	31.53
81	C2H2Cl4	1,1,1,2-TETRACHLOROETHANE	54.371	624.00	0.412	202.94	624.00	403.65	35.41
82	C2H2Cl4	1,1,2,2-TETRACHLOROETHANE	62.450	645.00	0.463	229.35	645.00	418.25	38.49
83	C2H2F2	1,1-DIFLUOROETHYLENE	21.713	302.80	0.366	129.15	302.80	187.50	15.25
84	C2H2F2	cis-1,2-DIFLUOROETHENE	30.968	394.67	0.380	107.90	394.67	247.86	21.27
85	C2H2F2	trans-1,2-DIFLUOROETHENE	30.968	394.67	0.380	107.90	394.67	247.86	21.27
86	C2H2F4	1,1,1,2-TETRAFLUOROETHANE	30.842	380.00	0.367	172.15	380.00	247.15	20.97
87	C2H2O	KETENE	21.373	370.00	0.105	200.00	370.00	223.34	19.39
88	C2H2O4	OXALIC ACID	114.676	804.00	0.372	462.65	804.00	569.00	72.57
89	C2H3Br	VINYL BROMIDE	33.400	473.00	0.319	135.35	473.00	288.95	24.72
90	C2H3Cl	VINYL CHLORIDE	35.586	432.00	0.483	119.36	432.00	259.78	22.82
91	C2H3ClF2	1-CHLORO-1,1-DIFLUOROETHANE	33.504	410.20	0.390	142.35	410.20	263.14	22.46
92	C2H3ClO	ACETYL CHLORIDE	43.400	508.00	0.362	160.30	508.00	323.90	30.05
93	C2H3ClO	CHLOROACETALDEHYDE	49.876	555.00	0.367	293.00	555.00	358.00	34.10
94	C2H3ClO2	CHLOROACETIC ACID	--------	--------	------	--------	--------	--------	------
95	C2H3ClO2	METHYL CHLOROFORMATE	50.751	525.00	0.385	192.00	525.00	344.00	33.68
96	C2H3Cl3	1,1,1-TRICHLOROETHANE	44.083	545.00	0.385	242.75	545.00	347.23	29.84
97	C2H3Cl3	1,1,2-TRICHLOROETHANE	46.865	602.00	0.294	236.50	602.00	387.00	34.62
98	C2H3F	VINYL FLUORIDE	25.960	327.80	0.367	112.65	327.80	200.95	18.32
99	C2H3F3	1,1,1-TRIFLUOROETHANE	28.185	346.25	0.351	161.82	346.25	225.75	19.46
100	C2H3N	ACETONITRILE	43.082	545.50	0.335	229.32	545.50	354.75	30.30
101	C2H3NO	METHYL ISOCYANATE	37.963	505.00	0.363	256.15	505.00	312.00	26.77
102	C2H4	ETHYLENE	19.986	282.36	0.431	103.97	282.36	169.47	13.46
103	C2H4Br2	1,1-DIBROMOETHANE	57.147	628.00	0.526	210.15	628.00	381.15	34.97
104	C2H4Br2	1,2-DIBROMOETHANE	42.935	650.15	0.187	282.85	650.15	404.51	35.79
105	C2H4Cl2	1,1-DICHLOROETHANE	42.180	523.00	0.367	176.19	523.00	330.45	29.23
106	C2H4Cl2	1,2-DICHLOROETHANE	45.426	561.00	0.342	237.49	561.00	356.59	32.16
107	C2H4Cl2O	BIS(CHLOROMETHYL)ETHER	52.061	579.00	0.379	231.65	579.00	378.00	34.86
108	C2H4F2	1,1-DIFLUOROETHANE	31.420	386.60	0.361	156.15	386.60	247.35	21.73
109	C2H4F2	1,2-DIFLUOROETHANE	37.739	476.00	0.360	215.00	476.00	303.65	26.18
110	C2H4I2	1,2-DIIODOETHANE	59.709	749.91	0.380	356.16	749.91	473.16	40.88
111	C2H4O	ACETALDEHYDE	44.950	461.00	0.594	150.15	461.00	293.55	24.63
112	C2H4O	ETHYLENE OXIDE	36.474	469.15	0.377	160.71	469.15	283.85	25.70
113	C2H4OS	THIOACETIC-ACID	49.971	577.34	0.380	150.16	577.34	360.16	34.46
114	C2H4O2	ACETIC ACID	11.575	592.71	-0.650	289.81	391.05	391.05	23.33
115	C2H4O2	METHYL FORMATE	43.997	487.20	0.453	174.15	487.20	304.90	28.19
116	C2H4S	THIACYCLOPROPANE	41.024	555.00	0.380	165.37	555.00	328.07	29.20
117	C2H5Br	BROMOETHANE	37.570	503.80	0.323	154.55	503.80	311.50	27.53
118	C2H5Cl	ETHYL CHLORIDE	35.233	460.35	0.365	134.80	460.35	285.42	24.75
119	C2H5ClO	2-CHLOROETHANOL	53.406	585.00	0.214	205.65	585.00	401.75	41.66
120	C2H5F	ETHYL FLUORIDE	27.155	375.31	0.312	129.95	375.31	235.45	19.96
121	C2H5I	ETHYL IODIDE	41.018	561.00	0.326	162.05	561.00	345.45	30.03
122	C2H5N	ETHYLENEIMINE	49.400	537.00	0.466	195.20	537.00	329.00	31.75
123	C2H5NO	ACETAMIDE	67.370	761.00	0.183	354.15	761.00	494.30	55.61
124	C2H5NO	N-METHYLFORMAMIDE	66.276	721.00	0.318	269.35	721.00	472.66	47.22
125	C2H5NO2	NITROETHANE	50.820	593.00	0.323	183.63	593.00	387.22	36.10
126	C2H5NO3	ETHYL-NITRATE	--------	--------	------	--------	--------	--------	------
127	C2H6	ETHANE	21.342	305.42	0.403	90.35	305.42	184.55	14.69
128	C2H6AlCl	DIMETHYLALUMINUM CHLORIDE	53.692	619.00	0.436	252.15	619.00	399.15	34.19
129	C2H6O	DIMETHYL ETHER	27.769	400.10	0.261	131.65	400.10	248.31	21.56
130	C2H6O	ETHANOL	43.122	516.25	0.079	300.00	516.25	351.44	39.40
131	C2H6OS	DIMETHYL SULFOXIDE	64.400	726.00	0.406	291.67	726.00	462.15	42.70
132	C2H6O2	ETHYLENE GLYCOL	88.200	645.00	0.397	260.15	645.00	470.45	52.49
133	C2H6O4S	DIMETHYL SULFATE	70.370	758.00	0.556	241.35	758.00	461.95	41.72
134	C2H6S	DIMETHYL SULFIDE	38.350	503.04	0.369	174.88	503.04	310.48	26.91
135	C2H6S	ETHYL MERCAPTAN	38.449	499.15	0.373	125.26	499.15	308.15	26.87
136	C2H6S2	DIMETHYL DISULFIDE	49.210	606.00	0.360	188.44	606.00	382.90	34.34
137	C2H7N	DIMETHYLAMINE	41.070	437.65	0.424	180.96	437.65	280.03	26.64
138	C2H7N	ETHYLAMINE	42.923	456.15	0.447	192.15	456.15	289.73	27.35
139	C2H7NO	MONOETHANOLAMINE	74.024	638.00	0.304	283.65	638.00	444.15	51.53
140	C2H8N2	ETHYLENEDIAMINE	57.520	593.00	0.338	284.29	593.00	390.41	40.01
141	C2H8Si	DIMETHYL SILANE	29.574	402.00	0.389	122.93	402.00	253.55	20.07
142	C2N2	CYANOGEN	33.850	400.15	0.371	245.25	400.15	252.00	23.41
143	C3F6	HEXAFLUOROPROPYLENE	29.323	368.00	0.381	116.65	368.00	243.55	19.40
144	C3F6O	HEXAFLUOROACETONE	33.791	357.14	0.378	147.70	357.14	245.88	21.74
145	C3F8	OCTAFLUOROPROPANE	30.670	345.05	0.383	125.46	345.05	236.40	19.70
146	C3H2N2	MALONONITRILE	60.998	715.00	0.240	304.99	715.00	491.50	46.14
147	C3H3Cl	PROPARGYL CHLORIDE	39.888	541.00	0.371	293.00	541.00	331.00	28.08
148	C3H3N	ACRYLONITRILE	41.537	535.00	0.273	189.63	535.00	350.50	31.06
149	C3H3NO	OXAZOLE	45.149	554.00	0.388	189.15	554.00	342.65	31.06
150	C3H4	METHYLACETYLENE	32.800	402.39	0.400	170.45	402.39	249.94	22.25
151	C3H4	PROPADIENE	29.430	393.15	0.382	136.87	393.15	238.65	20.60
152	C3H4Cl2	2,3-DICHLOROPROPENE	44.618	577.00	0.362	191.50	577.00	365.75	31.01
153	C3H4O	ACROLEIN	39.414	506.00	0.296	185.45	506.00	325.84	29.03
154	C3H40	PROPARGYL ALCOHOL	66.498	580.00	0.411	221.35	580.00	386.75	42.33
155	C3H4O2	ACRYLIC ACID	62.977	615.00	0.351	286.65	615.00	414.15	42.52
156	C3H4O2	beta-PROPIOLACTONE	62.600	686.00	0.372	239.75	686.00	435.15	43.06

113

Table 5-1 ENTHALPY OF VAPORIZATION - ORGANIC COMPOUNDS (continued)

			$HVAP = A (1 - T/T_c)^n$				(HVAP - kjoule/mol, T - K)		
NO	FORMULA	NAME	A	T_c	n	TMIN	TMAX	T_B	HVAP @ T_B
157	C3H4O2	VINYL FORMATE	49.568	498.00	0.450	200.00	498.00	320.00	31.20
158	C3H4O3	ETHYLENE CARBONATE	73.400	790.00	0.337	309.55	790.00	511.15	51.68
159	C3H4O3	PYRUVIC ACID	60.322	634.52	0.236	286.75	634.52	438.15	45.74
160	C3H5Br	3-BROMO-1-PROPENE	45.415	540.20	0.380	153.76	540.20	343.16	30.96
161	C3H5Cl	2-CHLOROPROPENE	34.273	478.00	0.355	135.75	478.00	295.80	24.34
162	C3H5Cl	3-CHLOROPROPENE	36.739	514.15	0.349	138.65	514.15	318.11	26.24
163	C3H5ClO	alpha-EPICHLOROHYDRIN	41.461	610.00	0.106	320.00	610.00	389.26	37.23
164	C3H5ClO2	METHYL CHLOROACETATE	60.967	600.00	0.391	241.03	600.00	402.97	39.45
165	C3H5ClO2	ETHYL CHLOROFORMATE	71.110	508.15	0.399	192.00	508.15	366.00	42.77
166	C3H5Cl3	1,2,3-TRICHLOROPROPANE	57.600	652.00	0.378	258.45	652.00	430.00	38.33
167	C3H5I	3-IODO-1-PROPENE	39.116	595.81	0.380	138.70	595.81	375.16	26.82
168	C3H5N	PROPIONITRILE	46.240	564.40	0.320	180.26	564.40	370.50	32.85
169	C3H5NO	ACRYLAMIDE	87.147	710.00	0.117	357.65	710.00	465.75	76.92
170	C3H5NO	HYDRACRYLONITRILE	60.635	690.00	0.078	227.15	690.00	494.15	54.96
171	C3H5NO	LACTONITRILE	87.164	643.00	0.391	233.00	643.00	457.00	53.67
172	C3H5N3O9	NITROGLYCERINE	114.973	680.00	0.383	286.15	680.00	523.00	65.58
173	C3H6	CYCLOPROPANE	28.060	397.91	0.361	145.59	397.91	240.37	20.08
174	C3H6	PROPYLENE	26.098	364.76	0.358	87.90	364.76	225.43	18.49
175	C3H6Br2	1,2-DIBROMOPROPANE	61.221	634.11	0.380	217.96	634.11	413.16	41.01
176	C3H6Cl2	1,1-DICHLOROPROPANE	47.738	560.00	0.392	200.00	560.00	361.25	31.81
177	C3H6Cl2	1,2-DICHLOROPROPANE	42.132	572.00	0.264	172.71	572.00	369.52	32.03
178	C3H6Cl2	1,3-DICHLOROPROPANE	51.452	603.00	0.367	173.65	603.00	393.55	34.90
179	C3H6Cl2	2,2-DICHLOROPROPANE	42.947	539.46	0.380	239.36	539.46	342.46	29.29
180	C3H6I2	1,2-DIIODOPROPANE	63.223	780.49	0.380	253.16	780.49	500.16	42.84
181	C3H6O	ACETONE	49.244	508.20	0.481	260.00	508.20	329.44	29.79
182	C3H6O	ALLYL ALCOHOL	49.291	545.05	0.188	240.00	545.05	370.23	39.80
183	C3H6O	METHYL VINYL ETHER	38.646	437.00	0.427	151.15	437.00	278.65	25.05
184	C3H6O	n-PROPIONALDEHYDE	40.655	496.00	0.336	193.15	496.00	321.15	28.64
185	C3H6O	1,2-PROPYLENE OXIDE	40.176	482.25	0.366	161.22	482.25	307.05	27.73
186	C3H6O	1,3-PROPYLENE OXIDE	40.994	520.00	0.377	255.00	520.00	321.00	28.54
187	C3H6O2	ETHYL FORMATE	43.925	508.40	0.376	193.55	508.40	327.46	29.79
188	C3H6O2	METHYL ACETATE	44.968	506.80	0.365	175.15	506.80	330.09	30.61
189	C3H6O2	PROPIONIC ACID	41.271	604.00	0.241	390.00	604.00	414.32	31.22
190	C3H6O2S	3-MERCAPTOPROPIONIC ACID	71.600	729.00	0.261	290.65	729.00	501.00	52.86
191	C3H6O3	LACTIC ACID	95.565	616.00	0.384	291.15	616.00	447.00	58.16
192	C3H6O3	METHOXYACETIC ACID	86.180	691.00	0.414	281.00	691.00	478.26	52.92
193	C3H6O3	TRIOXANE	55.823	604.00	0.388	334.65	604.00	387.65	37.48
194	C3H6S	THIACYCLOBUTANE	46.099	603.00	0.380	199.96	603.00	368.13	32.22
195	C3H7Br	1-BROMOPROPANE	41.810	544.00	0.322	163.15	544.00	344.15	30.29
196	C3H7Br	2-BROMOPROPANE	40.098	532.00	0.336	184.15	532.00	332.56	28.84
197	C3H7Cl	ISOPROPYL CHLORIDE	40.560	489.00	0.419	155.97	489.00	308.85	26.69
198	C3H7Cl	n-PROPYL CHLORIDE	41.205	503.15	0.387	150.35	503.15	319.67	27.89
199	C3H7F	1-FLUOROPROPANE	33.807	422.00	0.380	114.16	422.00	269.95	22.94
200	C3H7F	2-FLUOROPROPANE	32.419	415.68	0.380	139.80	415.68	263.81	22.11
201	C3H7I	ISOPROPYL IODIDE	43.643	578.00	0.336	183.15	578.00	362.65	31.32
202	C3H7I	n-PROPYL IODIDE	43.767	593.00	0.307	171.85	593.00	375.60	32.16
203	C3H7N	ALLYLAMINE	45.369	505.00	0.377	184.95	505.00	326.45	30.66
204	C3H7N	PROPYLENEIMINE	44.050	529.00	0.374	229.00	529.00	334.00	30.33
205	C3H7NO	N,N-DIMETHYLFORMAMIDE	59.355	647.00	0.381	212.72	647.00	426.15	39.41
206	C3H7NO	N-METHYLACETAMIDE	73.423	718.00	0.390	301.15	718.00	478.15	47.88
207	C3H7NO2	1-NITROPROPANE	53.462	605.00	0.320	169.16	605.00	404.33	37.56
208	C3H7NO2	2-NITROPROPANE	50.400	594.00	0.315	181.83	594.00	393.40	35.80
209	C3H7NO3	PROPYL-NITRATE	--------	--------	------	--------	--------	--------	-------
210	C3H7NO3	ISOPROPYL-NITRATE	--------	--------	------	--------	--------	--------	-------
211	C3H8	PROPANE	26.890	369.82	0.365	85.44	369.82	231.11	18.80
212	C3H8O	ISOPROPANOL	58.982	508.31	0.326	350.00	508.31	355.41	39.87
213	C3H8O	METHYL ETHYL ETHER	46.070	437.80	0.535	160.00	437.80	280.50	26.64
214	C3H8O	n-PROPANOL	70.179	536.71	0.451	146.95	536.71	370.35	41.38
215	C3H8O2	2-METHOXYETHANOL	56.090	564.00	0.288	225.00	564.00	397.55	39.47
216	C3H8O2	METHYLAL	41.313	480.60	0.372	168.35	480.60	315.00	27.79
217	C3H8O2	1,2-PROPYLENE GLYCOL	80.700	626.00	0.295	213.15	626.00	460.75	54.48
218	C3H8O2	1,3-PROPYLENE GLYCOL	87.601	658.00	0.307	246.45	658.00	487.55	57.86
219	C3H8O3	GLYCEROL	104.153	723.00	0.301	291.33	723.00	563.15	66.13
220	C3H8S	n-PROPYLMERCAPTAN	43.296	536.00	0.370	159.95	536.00	340.87	29.79
221	C3H8S	ISOPROPYL MERCAPTAN	41.364	517.00	0.385	142.61	517.00	325.71	28.21
222	C3H8S	ETHYL-METHYL-SULFIDE	43.417	532.80	0.380	167.20	532.80	340.15	29.50
223	C3H9N	n-PROPYLAMINE	42.535	496.95	0.350	190.15	496.95	321.65	29.54
224	C3H9N	ISOPROPYLAMINE	44.664	471.85	0.440	177.95	471.85	305.55	28.23
225	C3H9N	TRIMETHYLAMINE	34.600	433.25	0.384	156.08	433.25	276.02	23.44
226	C3H9NO	1-AMINO-2-PROPANOL	69.000	614.00	0.317	274.89	614.00	432.61	46.88
227	C3H9NO	3-AMINO-1-PROPANOL	89.299	649.00	0.388	274.89	649.00	460.65	55.26
228	C3H9NO	METHYLETHANOLAMINE	72.608	630.00	0.389	268.65	630.00	431.15	46.36
229	C3H9O4P	TRIMETHYL PHOSPHATE	--------	--------	------	--------	--------	--------	-------
230	C3H10N2	1,2-PROPANEDIAMINE	60.230	587.00	0.378	236.53	587.00	392.45	39.68
231	C3H10Si	TRIMETHYL SILANE	32.324	432.00	0.366	137.26	432.00	279.85	22.06
232	C4Cl4S	TETRACHLOROTHIOPHENE	75.497	753.00	0.422	301.97	753.00	506.54	47.12
233	C4Cl6	HEXACHLORO-1,3-BUTADIENE	78.705	741.00	0.588	252.15	741.00	488.15	41.82
234	C4F8	OCTAFLUORO-2-BUTENE	33.532	392.00	0.370	138.15	392.00	270.36	21.75

Table 5-1 ENTHALPY OF VAPORIZATION - ORGANIC COMPOUNDS (continued)

			HVAP = A $(1 - T/T_c)^n$			(HVAP - kjoule/mol, T - K)			
NO	FORMULA	NAME	A	T_C	n	TMIN	TMAX	T_B	HVAP @ T_B
235	C4F8	OCTAFLUOROCYCLOBUTANE	36.820	388.37	0.396	232.96	388.37	267.17	23.22
236	C4F10	DECAFLUOROBUTANE	40.215	386.35	0.455	144.95	386.35	271.15	23.19
237	C4H2	BUTADIYNE(BIACETYLENE)	45.694	478.02	0.380	237.16	478.02	283.46	32.47
238	C4H2O3	MALEIC ANHYDRIDE	59.574	721.00	0.216	325.72	721.00	475.15	47.22
239	C4H4	VINYLACETYLENE	36.157	454.00	0.441	179.00	454.00	278.25	23.79
240	C4H4N2	SUCCINONITRILE	82.935	770.00	0.350	331.16	770.00	540.15	54.32
241	C4H4O	FURAN	40.816	490.15	0.419	187.55	490.15	304.50	27.17
242	C4H4O2	DIKETENE	51.516	616.00	0.271	266.65	616.00	399.20	38.82
243	C4H4O3	SUCCINIC ANHYDRIDE	71.005	811.00	0.250	393.00	811.00	536.58	54.15
244	C4H4O4	FUMARIC ACID	112.151	771.00	0.359	560.15	771.00	563.15	70.05
245	C4H4O4	MALEIC ACID	116.339	773.00	0.376	403.45	773.00	565.00	71.02
246	C4H4S	THIOPHENE	46.602	579.35	0.397	234.94	579.35	357.31	31.85
247	C4H5Cl	CHLOROPRENE	38.616	525.00	0.324	143.15	525.00	332.55	27.90
248	C4H5N	trans-CROTONITRILE	49.577	586.00	0.297	222.00	586.00	394.38	35.57
249	C4H5N	cis-CROTONITRILE	52.053	568.00	0.391	200.55	568.00	380.60	33.74
250	C4H5N	METHACRYLONITRILE	41.374	554.00	0.281	237.35	554.00	363.45	30.65
251	C4H5N	PYRROLE	59.350	639.75	0.420	249.74	639.75	403.00	39.09
252	C4H5N	VINYLACETONITRILE	50.368	584.00	0.319	186.15	584.00	391.67	35.34
253	C4H5NO2	METHYL CYANOACETATE	76.410	687.00	0.382	260.08	687.00	478.24	48.48
254	C4H6	CYCLOBUTENE	37.194	446.33	0.380	153.76	446.33	275.75	25.81
255	C4H6	1,2-BUTADIENE	30.344	444.00	0.221	200.00	444.00	284.00	24.22
256	C4H6	1,3-BUTADIENE	35.170	425.37	0.448	200.00	425.37	268.74	22.48
257	C4H6	DIMETHYLACETYLENE	28.554	488.15	0.081	280.00	488.15	300.13	26.43
258	C4H6	ETHYLACETYLENE	39.079	443.20	0.463	147.43	443.20	281.22	24.52
259	C4H6Cl2	1,3-DICHLORO-trans-2-BUTENE	56.846	618.00	0.440	276.00	618.00	402.00	35.80
260	C4H6Cl2	1,4-DICHLORO-cis-2-BUTENE	57.961	640.00	0.378	225.15	640.00	425.65	38.33
261	C4H6Cl2	1,4-DICHLORO-trans-2-BUTENE	58.490	646.00	0.376	274.15	646.00	429.26	38.79
262	C4H6Cl2	3,4-DICHLORO-1-BUTENE	51.419	589.00	0.376	212.00	589.00	388.00	34.32
263	C4H6O	trans-CROTONALDEHYDE	52.200	571.00	0.375	196.65	571.00	377.25	34.81
264	C4H6O	2,5-DIHYDROFURAN	44.139	542.00	0.380	273.00	542.00	339.00	30.39
265	C4H6O	DIVINYL ETHER	39.673	463.00	0.374	172.05	463.00	301.45	26.76
266	C4H6O	METHACROLEIN	37.248	530.00	0.218	250.00	530.00	341.15	29.74
267	C4H6O2	2-BUTYNE-1,4-DIOL	114.180	695.00	0.383	331.00	695.00	511.15	68.61
268	C4H6O2	gamma-BUTYROLACTONE	64.573	739.00	0.397	229.78	739.00	477.15	42.77
269	C4H6O2	cis-CROTONIC ACID	71.825	647.00	0.331	288.65	647.00	445.05	48.85
270	C4H6O2	trans-CROTONIC ACID	77.785	666.00	0.399	344.55	666.00	458.15	48.88
271	C4H6O2	METHACRYLIC ACID	41.968	643.00	0.208	288.15	643.00	434.15	33.22
272	C4H6O2	METHYL ACRYLATE	46.800	536.00	0.347	196.32	536.00	353.35	32.21
273	C4H6O2	VINYL ACETATE	45.805	524.00	0.353	180.35	524.00	345.65	31.31
274	C4H6O3	ACETIC ANHYDRIDE	58.520	569.15	0.280	200.15	569.15	411.78	40.83
275	C4H6O4	SUCCINIC ACID	119.929	806.00	0.370	461.15	806.00	591.00	73.55
276	C4H6O5	DIGLYCOLIC ACID	128.964	820.00	0.372	421.15	820.00	610.00	77.70
277	C4H6O5	MALIC ACID	155.429	781.00	0.366	403.15	781.00	602.00	90.65
278	C4H6O6	TARTARIC ACID	188.440	828.00	0.337	479.15	828.00	660.00	110.08
279	C4H7N	n-BUTYRONITRILE	49.578	582.25	0.325	161.25	582.25	390.75	34.54
280	C4H7N	ISOBUTYRONITRILE	50.997	565.00	0.371	201.70	565.00	376.76	33.92
281	C4H7NO	ACETONE CYANOHYDRIN	57.428	647.00	0.291	252.15	647.00	463.00	39.83
282	C4H7NO	2-METHACRYLAMIDE	74.164	741.00	0.390	383.65	741.00	488.00	48.77
283	C4H7NO	3-METHOXYPROPIONITRILE	59.865	638.00	0.322	210.12	638.00	439.00	41.14
284	C4H7NO	2-PYRROLIDONE	78.223	792.00	0.366	298.15	792.00	518.15	53.03
285	C4H8	1-BUTENE	33.390	419.59	0.393	87.80	419.59	266.90	22.44
286	C4H8	cis-2-BUTENE	34.490	435.58	0.383	134.26	435.58	276.87	23.43
287	C4H8	trans-2-BUTENE	33.200	428.63	0.364	167.60	428.63	274.03	22.91
288	C4H8	CYCLOBUTANE	33.357	459.93	0.340	182.48	459.93	285.66	23.98
289	C4H8	ISOBUTENE	32.950	417.90	0.389	132.81	417.90	266.25	22.21
290	C4H8Br2	1,2-DIBROMOBUTANE	64.667	659.28	0.380	207.76	659.28	439.46	42.60
291	C4H8Br2	2,3-DIBROMOBUTANE	63.961	656.96	0.380	238.66	656.96	434.16	42.41
292	C4H8Cl2	1,4-DICHLOROBUTANE	57.353	641.00	0.376	235.85	641.00	427.05	37.96
293	C4H8I2	1,2-DIIODOBUTANE	61.882	726.41	0.380	279.06	726.41	476.76	41.24
294	C4H8O	n-BUTYRALDEHYDE	47.940	525.00	0.378	176.75	525.00	347.95	31.79
295	C4H8O	ISOBUTYRALDEHYDE	44.990	507.00	0.347	208.15	507.00	337.25	30.78
296	C4H8O	1,2-EPOXYBUTANE	42.433	526.00	0.367	123.15	526.00	336.57	29.17
297	C4H8O	METHYL ETHYL KETONE	50.652	535.50	0.450	300.00	535.50	352.79	31.22
298	C4H8O	ETHYL VINYL ETHER	39.830	475.15	0.375	157.35	475.15	308.70	26.88
299	C4H8O	TETRAHYDROFURAN	44.439	540.15	0.391	164.65	540.15	338.00	30.26
300	C4H8O2	cis-2-BUTENE-1,4-DIOL	83.100	677.88	0.203	284.15	677.88	508.15	62.74
301	C4H8O2	trans-2-BUTENE-1,4-DIOL	80.400	681.00	0.177	300.45	681.00	510.00	62.95
302	C4H8O2	ISOBUTYRIC ACID	65.379	609.15	0.329	227.15	609.15	427.85	43.88
303	C4H8O2	n-BUTYRIC ACID	18.437	628.00	-0.757	320.00	436.42	436.42	45.29
304	C4H8O2	1,4-DIOXANE	36.047	587.00	0.049	340.00	587.00	374.47	34.30
305	C4H8O2	ETHYL ACETATE	49.346	523.30	0.385	189.60	523.30	350.21	32.23
306	C4H8O2	METHYL PROPIONATE	50.320	530.60	0.402	185.65	530.60	352.60	32.44
307	C4H8O2	n-PROPYL FORMATE	45.553	538.00	0.329	280.00	538.00	353.97	32.01
308	C4H8O2S	SULFOLANE	80.596	849.00	0.375	300.75	849.00	558.15	53.93
309	C4H8S	TETRAHYDROTHIOPHENE	51.360	631.95	0.399	176.98	631.95	394.27	34.77
310	C4H9Br	1-BROMOBUTANE	51.840	577.00	0.387	160.75	577.00	374.75	34.55
311	C4H9Br	2-BROMOBUTANE	47.810	567.00	0.379	161.25	567.00	364.37	32.37
312	C4H9Cl	n-BUTYL CHLORIDE	46.720	537.00	0.394	150.05	537.00	351.58	30.73

Table 5-1 ENTHALPY OF VAPORIZATION - ORGANIC COMPOUNDS (continued)

NO	FORMULA	NAME	$HVAP = A(1 - T/T_c)^n$					(HVAP - kjoule/mol, T - K)	
			A	T_C	n	TMIN	TMAX	T_B	HVAP @ T_B
313	C4H9Cl	sec-BUTYL CHLORIDE	41.570	520.60	0.331	141.85	520.60	341.25	29.21
314	C4H9Cl	tert-BUTYL CHLORIDE	39.946	507.00	0.385	247.75	507.00	323.75	27.00
315	C4H9I	2-IODO-2-METHYLPROPANE	44.551	587.90	0.380	234.96	587.90	373.16	30.38
316	C4H9N	PYRROLIDINE	51.909	568.55	0.422	215.31	568.55	359.72	34.02
317	C4H9NO	N,N-DIMETHYLACETAMIDE	61.356	658.00	0.374	253.15	658.00	439.25	40.64
318	C4H9NO	MORPHOLINE	56.358	618.00	0.375	270.05	618.00	401.15	38.05
319	C4H9NO2	1-NITROBUTANE	60.194	624.00	0.380	191.83	624.00	426.05	38.91
320	C4H9NO2	2-NITROBUTANE	56.194	615.00	0.380	141.16	615.00	412.85	36.82
321	C4H10	n-BUTANE	33.020	425.18	0.377	134.86	425.18	272.65	22.44
322	C4H10	ISOBUTANE	31.954	408.14	0.392	113.54	408.14	261.43	21.40
323	C4H10N2	PIPERAZINE	65.337	638.00	0.416	379.15	638.00	419.15	41.87
324	C4H10O	n-BUTANOL	63.024	562.93	0.318	220.00	562.93	390.81	43.24
325	C4H10O	sec-BUTANOL	75.278	536.01	0.512	158.45	536.01	372.70	40.96
326	C4H10O	tert-BUTANOL	107.467	506.20	0.813	340.00	506.20	355.57	40.11
327	C4H10O	DIETHYL ETHER	41.972	466.70	0.407	156.85	466.70	307.58	27.09
328	C4H10O	METHYL-PROPYL-ETHER	46.435	476.20	0.380	156.87	476.20	311.72	31.00
329	C4H10O	METHYL ISOPROPYL ETHER	39.301	464.50	0.371	127.93	464.50	303.92	26.50
330	C4H10O	ISOBUTANOL	86.900	547.73	0.583	165.15	547.73	380.81	43.47
331	C4H10O2	1,3-BUTANEDIOL	79.000	643.00	0.243	196.15	643.00	480.15	56.58
332	C4H10O2	1,4-BUTANEDIOL	80.089	667.00	0.179	293.05	667.00	501.15	62.43
333	C4H10O2	2,3-BUTANEDIOL	72.800	611.00	0.204	280.75	611.00	453.85	55.19
334	C4H10O2	t-BUTYL HYDROPEROXIDE	--------	--------	------	--------	--------	--------	------
335	C4H10O2	1,2-DIMETHOXYETHANE	46.790	536.15	0.349	215.15	536.15	357.20	31.90
336	C4H10O2	2-ETHOXYETHANOL	60.038	569.00	0.310	203.00	569.00	408.15	40.58
337	C4H10O3	DIETHYLENE GLYCOL	94.746	744.60	0.431	262.70	744.60	518.15	56.72
338	C4H10O4S	DIETHYL SULFATE	--------	--------	------	--------	--------	--------	------
339	C4H10S	n-BUTYL MERCAPTAN	49.751	569.00	0.394	157.46	569.00	371.61	32.78
340	C4H10S	ISOBUTYL MERCAPTAN	46.682	559.00	0.382	128.31	559.00	361.64	31.36
341	C4H10S	sec-BUTYL MERCAPTAN	46.111	554.00	0.384	133.02	554.00	358.13	30.93
342	C4H10S	tert-BUTYL MERCAPTAN	43.145	530.00	0.408	274.26	530.00	337.37	28.55
343	C4H10S	DIETHYL SULFIDE	47.167	557.15	0.364	169.20	557.15	365.25	32.00
344	C4H10S	ISOPROPYL-METHYL-SULFIDE	45.731	551.00	0.380	171.65	551.00	357.90	30.70
345	C4H10S	METHYL-PROPYL-SULFIDE	48.062	563.00	0.380	160.19	563.00	368.71	32.08
346	C4H10S2	DIETHYL DISULFIDE	57.892	642.00	0.396	171.63	642.00	427.13	37.53
347	C4H11N	n-BUTYLAMINE	48.716	531.90	0.386	224.05	531.90	350.55	32.16
348	C4H11N	ISOBUTYLAMINE	48.814	513.73	0.425	188.55	513.73	340.88	30.73
349	C4H11N	sec-BUTYLAMINE	49.179	514.30	0.466	168.65	514.30	336.15	30.01
350	C4H11N	tert-BUTYLAMINE	41.511	483.90	0.382	206.19	483.90	317.55	27.61
351	C4H11N	DIETHYLAMINE	43.370	496.60	0.375	223.35	496.60	328.60	28.89
352	C4H11NO	DIMETHYLETHANOLAMINE	73.988	571.82	0.403	214.15	571.82	407.15	44.80
353	C4H11NO2	DIETHANOLAMINE	103.400	715.00	0.333	301.15	715.00	542.04	64.46
354	C4H11NO2	2-AMINOETHOXYETHANOL	104.887	699.00	0.386	293.15	699.00	517.00	62.39
355	C4H12N2O	N-AMINOETHYL ETHANOLAMINE	111.030	698.00	0.393	273.15	698.00	514.00	65.75
356	C4H12Si	TETRAMETHYLSILANE	37.309	450.40	0.393	174.07	450.40	299.80	24.26
357	C4H13N3	DIETHYLENE TRIAMINE	63.500	676.00	0.222	234.15	676.00	480.25	48.23
358	C5Cl6	HEXACHLOROCYCLOPENTADIENE	76.590	746.00	0.432	284.49	746.00	512.15	46.40
359	C5H4O2	FURFURAL	59.186	657.00	0.313	236.65	657.00	434.85	42.15
360	C5H5N	PYRIDINE	53.461	619.95	0.408	231.51	619.95	388.41	35.77
361	C5H6	CYCLOPENTADIENE	33.400	507.00	0.268	188.15	507.00	314.65	25.76
362	C5H6	2-METHYL-1-BUTENE-3-YNE	36.469	492.00	0.386	160.15	492.00	305.40	25.08
363	C5H6	1-PENTENE-3-YNE	38.010	520.00	0.304	150.00	520.00	332.40	27.88
364	C5H6	1-PENTENE-4-YNE	37.619	503.00	0.363	150.00	503.00	315.65	26.29
365	C5H6N2	GLUTARONITRILE	74.337	782.00	0.265	244.21	782.00	559.15	53.30
366	C5H6O2	FURFURYL ALCOHOL	68.853	632.00	0.288	258.52	632.00	443.15	48.62
367	C5H6O3	GLUTARIC ANHYDRIDE	79.362	838.00	0.294	328.00	838.00	562.69	57.21
368	C5H6O4	CITRACONIC ACID	121.190	829.00	0.384	356.15	829.00	607.00	73.07
369	C5H6O4	ITACONIC ACID	117.944	821.00	0.374	438.75	821.00	601.00	72.07
370	C5H6S	2-METHYLTHIOPHENE	49.586	610.00	0.380	209.77	610.00	385.71	33.90
371	C5H6S	3-METHYLTHIOPHENE	50.070	615.00	0.380	204.18	615.00	388.60	34.25
372	C5H7N	N-METHYLPYRROLE	55.305	610.00	0.455	216.91	610.00	385.89	35.07
373	C5H7NO2	ETHYL CYANOACETATE	102.063	679.00	0.380	250.65	679.00	479.15	64.12
374	C5H8	CYCLOPENTENE	34.747	507.00	0.234	290.00	507.00	317.38	27.60
375	C5H8	ISOPRENE	39.310	484.00	0.425	127.27	484.00	307.21	25.62
376	C5H8	3-METHYL-1,2-BUTADIENE	41.233	490.00	0.426	159.53	490.00	314.00	26.66
377	C5H8	2-METHYL-1,3-BUTADIENE	38.257	483.30	0.380	127.20	483.30	307.22	26.07
378	C5H8	1,2-PENTADIENE	42.497	500.00	0.453	135.89	500.00	318.01	26.89
379	C5H8	cis-1,3-PENTADIENE	42.393	499.00	0.458	132.35	499.00	317.22	26.70
380	C5H8	trans-1,3-PENTADIENE	35.992	500.00	0.305	250.00	500.00	315.17	26.57
381	C5H8	1,4-PENTADIENE	39.614	479.00	0.491	124.86	479.00	299.11	24.49
382	C5H8	2,3-PENTADIENE	43.062	497.00	0.422	147.50	497.00	321.40	27.76
383	C5H8	1-PENTYNE	39.532	481.20	0.351	167.45	481.20	313.33	27.32
384	C5H8	2-PENTYNE	42.704	521.99	0.380	163.86	521.99	329.22	29.25
385	C5H8	3-METHYL-1-BUTYNE	37.940	463.20	0.357	183.45	463.20	302.15	26.02
386	C5H8	SPIROPENTANE	38.800	499.74	0.380	166.11	499.74	312.19	26.74
387	C5H8N4O12	PENTAERYTHRITOL TETRANITRATE	125.724	676.00	0.360	413.65	676.00	543.00	70.02
388	C5H8O	CYCLOPENTANONE	50.900	626.00	0.292	221.85	626.00	403.80	37.62
389	C5H8O	METHYL ISOPROPENYL KETONE	48.711	566.00	0.375	219.55	566.00	371.15	32.66
390	C5H8O2	ACETYLACETONE	63.700	602.00	0.383	249.95	602.00	413.55	40.83

116

Table 5-1 ENTHALPY OF VAPORIZATION - ORGANIC COMPOUNDS (continued)

			HVAP = A $(1 - T/T_c)^n$				(HVAP - kjoule/mol, T - K)		
NO	FORMULA	NAME	A	T_c	n	TMIN	TMAX	T_B	HVAP @ T_B
391	C5H8O2	ALLYL ACETATE	47.280	559.00	0.306	138.00	559.00	377.15	33.53
392	C5H8O2	ETHYL ACRYLATE	53.630	553.00	0.414	201.95	553.00	372.65	33.72
393	C5H8O2	METHYL METHACRYLATE	54.170	564.00	0.421	224.95	564.00	373.45	34.30
394	C5H8O2	VINYL PROPIONATE	41.639	546.00	0.218	364.35	546.00	364.35	32.76
395	C5H8O3	2-HYDROXYETHYL ACRYLATE	80.132	662.00	0.302	213.00	662.00	484.00	53.89
396	C5H8O3	LEVULINIC ACID	--------	--------	------	--------	--------	--------	------
397	C5H8O3	METHYL ACETOACETATE	67.861	642.00	0.384	193.15	642.00	444.85	43.12
398	C5H8O4	GLUTARIC ACID	126.410	807.00	0.412	370.65	807.00	595.54	72.80
399	C5H9N	VALERONITRILE	56.540	603.00	0.358	176.95	603.00	414.45	37.29
400	C5H9NO	n-BUTYL ISOCYANATE	56.246	568.00	0.388	193.00	568.00	388.15	36.00
401	C5H9NO	N-METHYL-2-PYRROLIDONE	63.749	724.00	0.332	249.15	724.00	475.15	44.72
402	C5H9NO4	L-GLUTAMIC ACID	146.263	886.00	0.363	497.15	886.00	670.00	87.62
403	C5H10	CYCLOPENTANE	43.625	511.76	0.473	280.00	511.76	322.40	27.26
404	C5H10	2-METHYL-1-BUTENE	43.291	465.00	0.495	300.00	465.00	304.30	25.58
405	C5H10	2-METHYL-2-BUTENE	39.380	471.00	0.366	139.39	471.00	311.71	26.48
406	C5H10	3-METHYL-1-BUTENE	36.382	450.37	0.380	104.66	450.37	293.21	24.39
407	C5H10	1-PENTENE	39.508	464.78	0.407	107.93	464.78	303.11	25.71
408	C5H10	cis-2-PENTENE	40.630	475.93	0.404	121.75	475.93	310.08	26.54
409	C5H10	trans-2-PENTENE	39.540	475.37	0.389	132.89	475.37	309.49	26.25
410	C5H10Br2	2,3-DIBROMO-2-METHYLBUTANE	62.943	668.37	0.380	288.00	668.37	444.01	41.57
411	C5H10Cl2	1,5-DICHLOROPENTANE	63.367	663.00	0.379	200.35	663.00	453.15	40.97
412	C5H10O	METHYL ISOPROPYL KETONE	51.411	553.00	0.386	250.00	553.00	367.55	33.72
413	C5H10O	2-PENTANONE	51.693	561.08	0.393	196.29	561.08	375.46	33.47
414	C5H10O	DIETHYL KETONE	49.640	560.95	0.354	234.18	560.95	375.14	33.57
415	C5H10O	VALERALDEHYDE	52.738	554.00	0.374	182.00	554.00	376.15	34.48
416	C5H10O2	n-BUTYL FORMATE	48.700	559.00	0.267	181.25	559.00	379.25	35.97
417	C5H10O2	ETHYL PROPIONATE	52.708	546.00	0.397	199.25	546.00	372.25	33.45
418	C5H10O2	ISOBUTYL FORMATE	56.025	551.35	0.447	177.35	551.35	371.22	33.98
419	C5H10O2	ISOPROPYL ACETATE	52.355	538.00	0.435	199.75	538.00	361.65	32.23
420	C5H10O2	n-PROPYL ACETATE	55.070	549.40	0.409	178.15	549.40	374.65	34.47
421	C5H10O2	METHYL n-BUTYRATE	51.915	554.50	0.392	187.35	554.50	375.90	33.30
422	C5H10O2	2-METHYLBUTYRIC ACID	74.773	643.00	0.393	193.00	643.00	450.15	46.58
423	C5H10O2	ISOVALERIC ACID	51.204	634.00	0.138	243.85	634.00	448.25	43.22
424	C5H10O2	VALERIC ACID	81.000	651.00	0.420	239.15	651.00	458.65	48.54
425	C5H10O2	TETRAHYDROFURFURYL ALCOHOL	83.749	639.00	0.420	193.00	639.00	451.15	50.08
426	C5H10O2S	3-METHYL SULFOLANE	80.832	817.00	0.381	273.65	817.00	549.15	52.85
427	C5H10O3	DIETHYL CARBONATE	59.300	576.00	0.376	230.15	576.00	399.95	37.97
428	C5H10O3	ETHYL LACTATE	80.229	588.00	0.409	247.15	588.00	427.65	47.15
429	C5H10S	THIACYCLOHEXANE	52.571	657.12	0.380	292.14	657.12	414.90	35.98
430	C5H10S	CYCLOPENTANETHIOL	52.328	629.00	0.380	155.39	629.00	405.33	35.33
431	C5H11Br	1-BROMOPENTANE	55.410	564.76	0.380	185.26	564.76	402.74	34.48
432	C5H11Cl	1-CHLOROPENTANE	54.239	568.00	0.457	320.00	568.00	381.54	32.60
433	C5H11Cl	1-CHLORO-3-METHYLBUTANE	48.272	558.87	0.380	168.76	558.87	371.66	31.86
434	C5H11Cl	2-CHLORO-2-METHYLBUTANE	44.309	548.97	0.380	199.66	548.97	358.76	29.62
435	C5H11N	N-METHYLPYRROLIDINE	43.966	550.00	0.371	183.15	550.00	352.30	30.08
436	C5H11N	PIPERIDINE	50.120	594.05	0.394	262.65	594.05	379.55	33.55
437	C5H11NO	tert-BUTYLFORMAMIDE	70.745	692.00	0.384	289.15	692.00	475.15	45.31
438	C5H12	ISOPENTANE	37.692	460.43	0.395	113.25	460.43	300.99	24.79
439	C5H12	NEOPENTANE	34.060	433.78	0.383	256.58	433.78	282.65	22.74
440	C5H12	n-PENTANE	39.854	469.65	0.398	143.42	469.65	309.22	25.99
441	C5H12O	2,2-DIMETHYL-1-PROPANOL	72.285	550.00	0.461	327.15	550.00	386.25	41.35
442	C5H12O	tert-PENTYL-ALCOHOL	68.413	549.00	0.380	327.00	549.00	386.30	43.10
443	C5H12O	2-METHYL-1-BUTANOL	79.530	565.00	0.463	203.00	565.00	401.85	44.75
444	C5H12O	2-METHYL-2-BUTANOL	78.205	545.15	0.573	264.35	545.15	375.15	40.11
445	C5H12O	3-METHYL-1-BUTANOL	83.370	579.45	0.537	155.95	579.45	404.35	43.84
446	C5H12O	3-METHYL-2-BUTANOL	87.180	574.00	0.678	188.00	574.00	384.65	41.10
447	C5H12O	1-PENTANOL	83.100	586.15	0.511	195.56	586.15	410.95	44.83
448	C5H12O	2-PENTANOL	77.440	552.00	0.467	200.00	552.00	392.15	43.41
449	C5H12O	3-PENTANOL	63.390	547.00	0.326	300.00	547.00	388.45	42.33
450	C5H12O	METHYL sec-BUTYL ETHER	43.770	498.00	0.378	150.00	498.00	332.15	28.89
451	C5H12O	METHYL tert-BUTYL ETHER	45.028	497.10	0.434	300.00	497.10	328.35	28.17
452	C5H12O	METHYL ISOBUTYL ETHER	42.680	497.00	0.380	150.00	497.00	331.70	28.09
453	C5H12O	ETHYL PROPYL ETHER	54.365	500.23	0.606	145.65	500.23	337.01	27.58
454	C5H12O2	ETHYLENE GLYCOL MONOPROPYL ETHER	57.800	582.00	0.245	183.15	582.00	424.50	41.96
455	C5H12O2	NEOPENTYL GLYCOL	104.118	643.00	0.371	400.00	643.00	483.00	62.15
456	C5H12O2	1,5-PENTANEDIOL	144.530	673.00	0.555	500.00	673.00	512.15	65.31
457	C5H12O3	2-(2-METHOXYETHOXY)ETHANOL	85.746	630.00	0.455	197.15	630.00	466.75	46.38
458	C5H12O4	PENTAERYTHRITOL	137.028	780.00	0.162	534.15	780.00	631.00	104.80
459	C5H12S	n-PENTYL MERCAPTAN	53.578	598.00	0.381	197.45	598.00	399.79	35.18
460	C5H12S	BUTYL-METHYL-SULFIDE	52.347	591.00	0.380	175.33	591.00	396.58	34.31
461	C5H12S	ETHYL-PROPYL-SULFIDE	52.960	584.00	0.380	156.15	584.00	391.65	34.73
462	C5H12S	2-METHYL-2-BUTANETHIOL	47.143	566.00	0.380	169.38	566.00	372.28	31.37
463	C5H13N	n-PENTYLAMINE	54.993	555.00	0.397	218.15	555.00	377.65	34.96
464	C5H13NO2	METHYL DIETHANOLAMINE	124.400	678.00	0.396	252.15	678.00	520.15	69.85
465	C6Cl6	HEXACHLOROBENZENE	130.295	825.00	0.622	501.70	825.00	582.55	60.83
466	C6F6	HEXAFLUOROBENZENE	50.350	516.73	0.402	278.25	516.73	353.41	31.69
467	C6H3ClN2O4	1-CHLORO-2,4-DINITROBENZENE	104.880	813.77	0.388	326.55	813.77	588.00	63.77
468	C6H3Cl2NO2	1,2-DICHLORO-4-NITROBENZENE	83.778	758.00	0.386	315.65	758.00	529.00	52.78

Table 5-1 ENTHALPY OF VAPORIZATION - ORGANIC COMPOUNDS (continued)

NO	FORMULA	NAME	A	T_c	n	TMIN	TMAX	T_B	HVAP @ T_B
								HVAP = A $(1 - T/T_c)^n$	(HVAP - kjoule/mol, T - K)
469	C6H3Cl3	1,2,4-TRICHLOROBENZENE	69.001	725.00	0.403	290.15	725.00	486.15	44.11
470	C6H3N3O6	1,3,5-TRINITROBENZENE	82.753	1005.00	0.221	398.40	1005.00	748.00	61.22
471	C6H4Br2	m-DIBROMOBENZENE	63.216	761.00	0.352	266.25	761.00	491.15	43.89
472	C6H4ClNO2	m-CHLORONITROBENZENE	78.485	742.00	0.385	317.65	742.00	508.75	50.27
473	C6H4ClNO2	o-CHLORONITROBENZENE	80.864	757.00	0.394	306.15	757.00	519.00	51.26
474	C6H4ClNO2	p-CHLORONITROBENZENE	79.523	751.00	0.384	356.65	751.00	515.15	50.97
475	C6H4Cl2	m-DICHLOROBENZENE	56.920	683.95	0.358	248.39	683.95	446.23	38.99
476	C6H4Cl2	o-DICHLOROBENZENE	62.088	705.00	0.428	256.15	705.00	453.57	39.94
477	C6H4Cl2	p-DICHLOROBENZENE	59.791	684.75	0.386	326.14	684.75	447.21	39.73
478	C6H4F2	m-DIFLUOROBENZENE	49.103	552.94	0.380	249.16	552.94	363.66	32.67
479	C6H4F2	o-DIFLUOROBENZENE	48.412	554.46	0.380	239.16	554.46	364.66	32.21
480	C6H4F2	p-DIFLUOROBENZENE	48.760	556.00	0.380	260.16	556.00	362.00	32.68
481	C6H4N2O4	m-DINITROBENZENE	99.691	805.00	0.386	364.00	805.00	573.00	61.67
482	C6H4N2O4	o-DINITROBENZENE	103.483	831.00	0.388	390.08	831.00	592.00	63.81
483	C6H4N2O4	p-DINITROBENZENE	99.697	803.00	0.387	446.60	803.00	572.00	61.56
484	C6H5Br	BROMOBENZENE	55.523	670.15	0.377	242.43	670.15	429.24	37.75
485	C6H5Cl	MONOCHLOROBENZENE	46.062	632.35	0.246	370.00	632.35	404.87	35.82
486	C6H5ClO	m-CHLOROPHENOL	59.620	729.00	0.238	306.00	729.00	487.00	45.86
487	C6H5ClO	o-CHLOROPHENOL	62.673	675.00	0.410	282.00	675.00	447.53	40.12
488	C6H5ClO	p-CHLOROPHENOL	62.870	738.00	0.263	316.00	738.00	493.11	47.04
489	C6H5Cl2N	3,4-DICHLOROANILINE	83.140	800.00	0.385	344.65	800.00	545.00	53.54
490	C6H5F	FLUOROBENZENE	45.834	560.09	0.372	230.94	560.09	357.88	31.38
491	C6H5I	IODOBENZENE	60.150	721.15	0.387	241.83	721.15	461.60	40.50
492	C6H5NO2	NITROBENZENE	67.414	719.00	0.380	278.91	719.00	483.95	44.08
493	C6H6	BENZENE	49.888	562.16	0.489	278.68	562.16	353.24	30.75
494	C6H6ClN	m-CHLOROANILINE	77.215	751.00	0.467	262.75	751.00	501.65	46.14
495	C6H6ClN	o-CHLOROANILINE	69.808	722.00	0.369	271.05	722.00	481.99	46.50
496	C6H6ClN	p-CHLOROANILINE	73.401	754.00	0.372	343.05	754.00	503.65	48.71
497	C6H6N2	cis-DICYANO-1-BUTENE	85.030	691.00	0.384	249.00	691.00	501.00	51.79
498	C6H6N2	trans-DICYANO-1-BUTENE	84.171	689.00	0.383	260.00	689.00	499.00	51.39
499	C6H6N2	1,4-DICYANO-2-BUTENE	91.569	755.00	0.377	349.00	755.00	547.00	56.32
500	C6H6N2O2	m-NITROANILINE	111.880	815.00	0.467	387.15	815.00	579.00	62.72
501	C6H6N2O2	o-NITROANILINE	109.486	784.00	0.476	344.65	784.00	558.00	60.56
502	C6H6N2O2	p-NITROANILINE	121.086	851.00	0.466	420.65	851.00	609.15	67.37
503	C6H6O	PHENOL	77.397	694.25	0.462	314.06	694.25	454.99	47.31
504	C6H6O2	1,2-BENZENEDIOL	67.750	764.00	0.190	377.60	764.00	518.65	54.60
505	C6H6O2	1,3-BENZENEDIOL	100.840	810.00	0.397	382.00	810.00	549.65	64.26
506	C6H6O2	p-HYDROQUINONE	96.593	822.00	0.362	444.65	822.00	558.15	64.02
507	C6H6O3	1,2,3-BENZENETRIOL	86.221	830.00	0.192	407.00	830.00	581.85	68.38
508	C6H6S	PHENYL MERCAPTAN	62.237	689.00	0.441	258.27	689.00	442.29	39.57
509	C6H7N	ANILINE	72.038	699.00	0.459	267.13	699.00	457.60	44.22
510	C6H7N	2-METHYLPYRIDINE	55.512	621.00	0.411	206.44	621.00	402.55	36.13
511	C6H7N	3-METHYLPYRIDINE	57.482	645.00	0.415	255.01	645.00	417.29	37.31
512	C6H7N	4-METHYLPYRIDINE	55.527	646.15	0.367	276.73	646.15	418.50	37.86
513	C6H8	1,3-CYCLOHEXADIENE	44.171	558.00	0.366	161.00	558.00	353.49	30.59
514	C6H8	METHYLCYCLOPENTADIENE	44.043	541.00	0.375	150.00	541.00	345.93	30.04
515	C6H8N2	ADIPONITRILE	98.047	781.00	0.410	275.64	781.00	568.15	57.54
516	C6H8N2	METHYLGLUTARONITRILE	89.543	742.00	0.389	228.15	742.00	536.15	54.38
517	C6H8N2	m-PHENYLENEDIAMINE	77.618	824.00	0.289	334.00	824.00	560.00	55.86
518	C6H8N2	o-PHENYLENEDIAMINE	83.165	781.00	0.387	376.95	781.00	525.00	54.01
519	C6H8N2	p-PHENYLENEDIAMINE	87.980	796.00	0.386	413.00	796.00	540.00	56.78
520	C6H8N2	PHENYLHYDRAZINE	68.169	761.00	0.228	292.35	761.00	516.65	52.61
521	C6H8N2O	BIS(CYANOETHYL)ETHER	104.992	783.00	0.391	246.85	783.00	579.00	62.05
522	C6H8O4	DIMETHYL MALEATE	74.673	675.00	0.374	254.15	675.00	478.15	47.10
523	C6H8O6	ASCORBIC ACID	199.700	783.00	0.327	465.15	783.00	637.00	115.31
524	C6H8O7	CITRIC ACID	181.850	822.00	0.356	426.15	822.00	659.00	102.23
525	C6H10	1-METHYLCYCLOPENTENE	43.011	541.99	0.380	145.96	541.99	348.95	29.05
526	C6H10	3-METHYLCYCLOPENTENE	42.597	535.71	0.380	130.16	535.71	343.16	28.87
527	C6H10	4-METHYLCYCLOPENTENE	43.232	543.75	0.380	112.31	543.75	348.31	29.31
528	C6H10	CYCLOHEXENE	36.523	560.40	0.159	300.00	560.40	356.12	31.11
529	C6H10	2,3-DIMETHYL-1,3-BUTADIENE	44.559	526.00	0.412	197.15	526.00	341.93	28.91
530	C6H10	1,5-HEXADIENE	40.959	507.00	0.372	132.47	507.00	332.61	27.54
531	C6H10	cis,trans-2,4-HEXADIENE	46.017	538.00	0.378	177.05	538.00	356.65	30.51
532	C6H10	trans,trans-2,4-HEXADIENE	46.665	535.00	0.385	228.25	535.00	355.05	30.68
533	C6H10	1-HEXYNE	45.750	516.20	0.370	141.25	516.20	344.48	30.45
534	C6H10	2-HEXYNE	49.100	549.00	0.439	183.65	549.00	357.67	30.91
535	C6H10	3-HEXYNE	48.080	544.00	0.436	170.05	544.00	354.35	30.37
536	C6H10O	CYCLOHEXANONE	53.720	629.15	0.285	242.00	629.15	428.90	38.76
537	C6H10O	MESITYL OXIDE	56.711	600.00	0.406	220.15	600.00	402.95	36.09
538	C6H10O2	epsilon-CAPROLACTONE	58.118	771.00	0.249	271.85	771.00	514.00	44.21
539	C6H10O2	ETHYL METHACRYLATE	53.906	577.00	0.399	223.15	577.00	390.15	34.38
540	C6H10O2	n-PROPYL ACRYLATE	57.395	569.00	0.387	273.15	569.00	392.15	36.51
541	C6H10O3	ETHYLACETOACETATE	63.814	643.00	0.311	234.15	643.00	453.95	43.61
542	C6H10O3	PROPIONIC ANHYDRIDE	63.740	618.00	0.317	228.15	618.00	442.15	42.79
543	C6H10O4	ADIPIC ACID	110.800	809.00	0.297	425.50	809.00	611.00	72.94
544	C6H10O4	DIETHYL OXALATE	74.100	646.00	0.400	232.55	646.00	458.85	45.14
545	C6H10O4	ETHYLENE GLYCOL DIACETATE	74.718	653.00	0.403	242.15	653.00	463.65	45.37
546	C6H10O4	ETHYLIDENE DIACETATE	73.701	635.00	0.380	292.00	635.00	442.15	46.86

118

Table 5-1 ENTHALPY OF VAPORIZATION - ORGANIC COMPOUNDS (continued)

NO	FORMULA	NAME	$HVAP = A\left(1 - T/T_c\right)^n$				(HVAP - kjoule/mol, T - K)		
			A	T_c	n	TMIN	TMAX	T_B	HVAP @ T_B
547	C6H11N	HEXANENITRILE	63.091	622.05	0.418	192.85	622.05	436.75	38.03
548	C6H11NO	epsilon-CAPROLACTAM	83.529	806.00	0.378	342.36	806.00	543.15	54.69
549	C6H11NO	CYCLOHEXANONE OXIME	74.919	715.00	0.399	363.15	715.00	481.15	47.97
550	C6H12	CYCLOHEXANE	49.060	553.54	0.486	279.69	553.54	353.87	29.89
551	C6H12	2,3-DIMETHYL-1-BUTENE	42.628	500.00	0.410	115.89	500.00	328.76	27.47
552	C6H12	2,3-DIMETHYL-2-BUTENE	46.772	524.00	0.429	198.92	524.00	346.35	29.41
553	C6H12	3,3-DIMETHYL-1-BUTENE	38.888	480.00	0.381	157.95	480.00	314.40	25.93
554	C6H12	2-ETHYL-1-BUTENE	44.441	512.00	0.417	141.61	512.00	337.82	28.35
555	C6H12	trans-3-METHYL-2-PENTENE	44.105	521.00	0.380	134.70	521.00	343.60	29.29
556	C6H12	1-HEXENE	43.880	504.03	0.387	133.39	504.03	336.63	28.64
557	C6H12	cis-2-HEXENE	43.620	513.00	0.377	132.00	513.00	342.03	28.83
558	C6H12	trans-2-HEXENE	45.745	513.00	0.414	140.17	513.00	341.02	29.10
559	C6H12	cis-3-HEXENE	44.215	509.00	0.387	135.33	509.00	339.60	28.88
560	C6H12	trans-3-HEXENE	44.765	509.00	0.390	159.73	509.00	340.24	29.10
561	C6H12	METHYLCYCLOPENTANE	44.379	532.79	0.397	130.73	532.79	344.96	29.34
562	C6H12	2-METHYL-1-PENTENE	44.724	507.00	0.421	137.42	507.00	335.25	28.35
563	C6H12	2-METHYL-2-PENTENE	45.650	514.00	0.421	138.07	514.00	340.45	28.90
564	C6H12	3-METHYL-1-PENTENE	40.443	495.00	0.363	120.20	495.00	327.33	27.30
565	C6H12	3-METHYL-cis-2-PENTENE	44.280	515.00	0.393	138.31	515.00	340.85	28.92
566	C6H12	4-METHYL-1-PENTENE	42.628	496.00	0.410	119.51	496.00	327.01	27.41
567	C6H12	4-METHYL-cis-2-PENTENE	47.360	499.00	0.503	138.30	499.00	329.53	27.51
568	C6H12	4-METHYL-trans-2-PENTENE	47.551	501.00	0.493	132.35	501.00	331.75	27.85
569	C6H12N2	TRIETHYLENEDIAMINE	68.298	655.00	0.394	434.25	655.00	447.15	43.45
570	C6H12O	BUTYL VINYL ETHER	51.418	536.00	0.387	181.15	536.00	366.97	32.90
571	C6H12O	CYCLOHEXANOL	85.741	625.15	0.527	296.60	625.15	434.00	45.92
572	C6H12O	1-HEXANAL	55.267	579.00	0.350	217.15	579.00	401.45	36.54
573	C6H12O	ETHYL ISOPROPYL KETONE	53.870	567.00	0.406	200.00	567.00	386.55	33.84
574	C6H12O	2-HEXANONE	56.790	587.05	0.382	217.35	587.05	400.85	36.62
575	C6H12O	3-HEXANONE	56.580	582.82	0.413	217.50	582.82	396.65	35.32
576	C6H12O	METHYL ISOBUTYL KETONE	57.680	571.40	0.416	189.15	571.40	389.65	35.82
577	C6H12O2	n-PENTYL FORMATE	64.029	576.00	0.396	199.65	576.00	406.60	39.44
578	C6H12O2	n-BUTYL ACETATE	57.750	579.65	0.393	199.65	579.65	399.15	36.51
579	C6H12O2	sec-BUTYL ACETATE	56.523	561.00	0.421	174.15	561.00	385.15	34.68
580	C6H12O2	tert-BUTYL ACETATE	50.584	545.00	0.389	283.15	545.00	369.15	32.58
581	C6H12O2	ETHYL n-BUTYRATE	51.982	571.00	0.334	175.15	571.00	394.65	35.11
582	C6H12O2	ETHYL ISOBUTYRATE	50.580	553.15	0.331	185.00	553.15	383.00	34.24
583	C6H12O2	ISOBUTYL ACETATE	55.003	561.00	0.360	174.30	561.00	389.80	35.88
584	C6H12O2	n-PROPYL PROPIONATE	56.304	578.00	0.396	197.25	578.00	395.65	35.66
585	C6H12O2	CYCLOHEXYL PEROXIDE	--------	--------	------	---------	---------	--------	------
586	C6H12O2	DIACETONE ALCOHOL	64.998	606.00	0.290	380.00	606.00	441.00	44.57
587	C6H12O2	2-ETHYL BUTYRIC ACID	78.936	655.00	0.394	258.15	655.00	466.95	48.28
588	C6H12O2	n-HEXANOIC ACID	95.200	667.00	0.476	270.15	667.00	478.85	52.12
589	C6H12O3	2-ETHOXYETHYL ACETATE	65.950	597.00	0.386	211.45	597.00	429.45	40.38
590	C6H12O3	HYDROXYCAPROIC ACID	126.916	758.00	0.383	334.00	758.00	576.00	73.49
591	C6H12O3	PARALDEHYDE	48.502	579.00	0.217	320.00	579.00	397.25	37.72
592	C6H12O3	sec-BUTYL GLYCOLATE	--------	--------	------	---------	---------	--------	------
593	C6H12S	THIACYCLOHEPTANE	55.024	640.07	0.380	292.14	640.07	414.90	37.00
594	C6H13N	CYCLOHEXYLAMINE	57.160	615.00	0.421	255.45	615.00	407.65	36.17
595	C6H13N	HEXAMETHYLENEIMINE	61.600	615.00	0.443	236.15	615.00	404.85	38.28
596	C6H14	2,2-DIMETHYLBUTANE	39.699	488.78	0.377	174.28	488.78	322.88	26.42
597	C6H14	2,3-DIMETHYLBUTANE	42.022	499.98	0.391	145.19	499.98	331.13	27.49
598	C6H14	n-HEXANE	45.610	507.43	0.401	177.84	507.43	341.88	29.11
599	C6H14	2-METHYLPENTANE	42.780	497.50	0.384	119.55	497.50	333.41	27.94
600	C6H14	3-METHYLPENTANE	43.250	504.43	0.387	110.25	504.43	336.42	28.26
601	C6H14N2O2	LYSINE	124.093	821.00	0.374	498.00	821.00	615.00	73.99
602	C6H14O	2-ETHYL-1-BUTANOL	79.061	580.00	0.470	250.00	580.00	419.65	43.20
603	C6H14O	1-HEXANOL	84.151	611.35	0.498	228.55	611.35	430.15	45.93
604	C6H14O	2-HEXANOL	91.216	586.20	0.621	223.00	586.20	413.04	42.77
605	C6H14O	2-METHYL-1-PENTANOL	91.454	582.00	0.539	320.00	582.00	421.15	45.73
606	C6H14O	4-METHYL-2-PENTANOL	87.567	574.40	0.613	320.00	574.40	404.85	41.45
607	C6H14O	n-BUTYL ETHYL ETHER	50.817	531.00	0.408	170.15	531.00	365.35	31.59
608	C6H14O	DIISOPROPYL ETHER	37.048	500.05	0.196	300.00	500.05	341.45	29.58
609	C6H14O	DI-n-PROPYL ETHER	51.040	530.60	0.400	149.95	530.60	362.79	32.21
610	C6H14O	METHYL tert-PENTYL ETHER	46.558	534.00	0.375	160.00	534.00	359.45	30.61
611	C6H14O2	ACETAL	59.142	541.00	0.437	173.15	541.00	376.75	35.13
612	C6H14O2	2-BUTOXYETHANOL	69.762	600.00	0.306	203.15	600.00	444.47	46.15
613	C6H14O2	1,6-HEXANEDIOL	102.370	670.00	0.296	315.15	670.00	516.15	66.23
614	C6H14O2	HEXYLENE GLYCOL	108.571	621.00	0.402	223.15	621.00	470.65	61.39
615	C6H14O2S	DI-n-PROPYL SULFONE	87.322	763.00	0.385	303.00	763.00	543.00	54.10
616	C6H14O3	DIETHYLENE GLYCOL DIMETHYL ETHER	73.382	604.00	0.457	203.15	604.00	432.91	41.23
617	C6H14O3	DIPROPYLENE GLYCOL	117.495	654.00	0.430	233.00	654.00	504.95	62.21
618	C6H14O3	2-(2-ETHOXYETHOXY)ETHANOL	85.200	632.00	0.431	195.15	632.00	475.15	46.73
619	C6H14O3	TRIMETHYLOLPROPANE	93.600	709.00	0.138	331.15	709.00	562.04	75.33
620	C6H14O4	TRIETHYLENE GLYCOL	130.880	700.00	0.460	265.79	700.00	551.00	64.24
621	C6H14O6	SORBITOL	254.100	959.00	0.389	370.85	959.00	777.00	133.12
622	C6H14S	n-HEXYLMERCAPTAN	59.221	623.00	0.390	192.62	623.00	425.81	37.81
623	C6H14S	BUTYL-ETHYL-SULFIDE	57.009	609.00	0.380	178.03	609.00	417.41	36.74
624	C6H14S	ISOPROPYL-SULFIDE	53.767	585.71	0.380	170.45	585.71	393.19	35.23

Table 5-1 ENTHALPY OF VAPORIZATION - ORGANIC COMPOUNDS (continued)

			HVAP = A (1 - T/T$_c$)^n				(HVAP - kjoule/mol, T - K)		
NO	FORMULA	NAME	A	T$_C$	n	TMIN	TMAX	T$_B$	HVAP @ T$_B$
625	C6H14S	METHYL-PENTYL-SULFIDE	55.558	587.98	0.380	179.16	587.98	401.16	35.94
626	C6H14S	PROPYL-SULFIDE	56.665	609.73	0.380	170.45	609.73	416.00	36.65
627	C6H14S2	PROPYL-DISULFIDE	65.459	673.00	0.380	187.68	673.00	464.65	41.92
628	C6H15Al	TRIETHYL ALUMINUM	89.103	720.15	0.380	233.15	720.15	458.15	60.68
629	C6H15Al2Cl3	ETHYL ALUMINUM SESQUICHLORIDE	--------	--------	------	--------	---------	--------	------
630	C6H15N	DIISOPROPYLAMINE	50.059	523.10	0.436	176.85	523.10	357.05	30.35
631	C6H15N	DI-n-PROPYLAMINE	54.470	555.80	0.377	210.15	555.80	382.00	35.14
632	C6H15N	n-HEXYLAMINE	56.273	583.00	0.342	251.85	583.00	404.65	37.53
633	C6H15N	TRIETHYLAMINE	46.726	535.15	0.367	158.45	535.15	361.92	30.89
634	C6H15NO	6-AMINOHEXANOL	83.954	681.00	0.235	331.00	681.00	508.00	60.84
635	C6H15NO2	DIISOPROPANOLAMINE	124.856	672.00	0.376	318.15	672.00	521.90	71.06
636	C6H15NO3	TRIETHANOLAMINE	120.600	787.00	0.354	294.35	787.00	613.00	70.68
637	C6H15N3	N-AMINOETHYL PIPERAZINE	80.423	708.00	0.397	254.15	708.00	493.55	50.06
638	C6H15O4P	TRIETHYL PHOSPHATE	--------	--------	------	--------	---------	--------	------
639	C6H16N2	HEXAMETHYLENEDIAMINE	80.490	663.00	0.393	313.95	663.00	475.04	49.04
640	C6H18N3OP	HEXAMETHYL PHOSPHORAMIDE	--------	--------	------	--------	---------	--------	------
641	C6H18N4	TRIETHYLENE TETRAMINE	110.130	718.00	0.399	285.15	718.00	539.65	63.18
642	C6H18OSi2	HEXAMETHYLDISILOXANE	49.465	518.70	0.363	204.93	518.70	373.67	31.15
643	C6H18O3Si3	HEXAMETHYLCYCLOTRISILOXANE	61.930	554.20	0.426	337.15	554.20	408.26	35.08
644	C6H19NSi2	HEXAMETHYLDISILAZANE	54.050	544.00	0.338	293.15	544.00	399.15	34.56
645	C7H3ClF3NO2	4-CHLORO-3-NITROBENZOTRIFLUORIDE	80.152	686.00	0.386	293.15	686.00	495.15	48.91
646	C7H3Cl2F3	2,4-DICHLOROBENZOTRIFLUORIDE	65.073	646.00	0.384	247.55	646.00	450.65	41.11
647	C7H3Cl2NO	3,4-DICHLOROPHENYL ISOCYANATE	68.765	733.00	0.415	316.15	733.00	501.00	42.66
648	C7H4ClF3	p-CHLOROBENZOTRIFLUORIDE	52.688	601.00	0.313	237.15	601.00	412.15	36.67
649	C7H4Cl2O	m-CHLOROBENZOYL CHLORIDE	70.300	724.00	0.352	280.00	724.00	498.00	46.66
650	C7H4F3NO2	3-NITROBENZOTRIFLUORIDE	72.175	667.00	0.373	272.00	667.00	475.93	45.28
651	C7H5ClO	BENZOYL CHLORIDE	54.739	697.00	0.202	272.65	697.00	470.15	43.63
652	C7H5ClO2	o-CHLOROBENZOIC ACID	--------	--------	------	--------	---------	--------	------
653	C7H5Cl3	BENZOTRICHLORIDE	61.290	737.00	0.371	268.40	737.00	486.65	41.06
654	C7H5F3	BENZOTRIFLUORIDE	50.606	565.00	0.395	244.14	565.00	375.20	32.89
655	C7H5N	BENZONITRILE	71.138	699.35	0.472	260.40	699.35	464.15	42.53
656	C7H5NO	PHENYL ISOCYANATE	55.068	648.00	0.282	243.15	648.00	438.75	40.04
657	C7H5N3O6	2,4,6-TRINITROTOLUENE	81.549	795.00	0.380	354.00	795.00	573.00	50.22
658	C7H6Cl2	BENZYL DICHLORIDE	58.526	731.00	0.309	257.00	731.00	487.00	41.70
659	C7H6Cl2	2,4-DICHLOROTOLUENE	76.405	705.00	0.511	259.65	705.00	474.25	43.18
660	C7H6N2O4	2,4-DINITROTOLUENE	109.587	814.00	0.380	343.00	814.00	590.00	67.11
661	C7H6N2O4	2,5-DINITROTOLUENE	105.621	814.00	0.389	325.65	814.00	590.00	63.94
662	C7H6N2O4	2,6-DINITROTOLUENE	99.553	770.00	0.388	339.00	770.00	558.00	60.36
663	C7H6N2O4	3,4-DINITROTOLUENE	109.018	842.00	0.389	332.00	842.00	610.00	66.03
664	C7H6N2O4	3,5-DINITROTOLUENE	73.070	814.00	0.165	365.65	814.00	588.00	59.14
665	C7H6O	BENZALDEHYDE	55.180	695.00	0.257	247.15	695.00	451.90	42.13
666	C7H6O2	BENZOIC ACID	79.554	751.00	0.380	395.52	751.00	522.40	50.63
667	C7H6O2	p-HYDROXYBENZALDEHYDE	82.298	844.00	0.250	390.15	844.00	583.15	61.36
668	C7H6O2	SALICYLALDEHYDE	82.129	680.00	0.406	266.15	680.00	469.65	51.00
669	C7H6O3	SALICYLIC ACID	--------	--------	------	--------	---------	--------	------
670	C7H7Br	p-BROMOTOLUENE	58.531	699.00	0.344	299.95	699.00	457.50	40.61
671	C7H7Cl	BENZYL CHLORIDE	61.120	686.00	0.380	234.15	686.00	452.55	40.58
672	C7H7Cl	o-CHLOROTOLUENE	57.594	656.00	0.378	236.65	656.00	432.30	38.35
673	C7H7Cl	p-CHLOROTOLUENE	58.941	660.00	0.384	280.65	660.00	435.65	38.95
674	C7H7F	p-FLUOROTOLUENE	58.331	590.48	0.380	216.36	590.48	389.76	38.71
675	C7H7NO	FORMANILIDE	87.362	787.00	0.385	323.15	787.00	544.15	55.56
676	C7H7NO2	m-NITROTOLUENE	58.527	734.00	0.216	289.20	734.00	505.00	45.51
677	C7H7NO2	o-NITROTOLUENE	64.370	720.00	0.288	269.98	720.00	495.64	46.01
678	C7H7NO2	p-NITROTOLUENE	58.100	736.00	0.185	324.75	736.00	511.65	46.64
679	C7H7NO3	o-NITROANISOLE	88.335	782.00	0.388	283.60	782.00	546.15	55.48
680	C7H8	TOLUENE	50.139	591.79	0.383	178.18	591.79	383.78	33.59
681	C7H8	1,3,5-CYCLOHEPTATRIENE	52.884	593.90	0.380	193.66	593.90	388.65	35.32
682	C7H8O	ANISOLE	58.530	641.65	0.364	235.65	641.65	426.73	39.31
683	C7H8O	BENZYL ALCOHOL	81.440	677.00	0.372	257.85	677.00	477.85	51.66
684	C7H8O	m-CRESOL	81.650	705.85	0.462	285.39	705.85	475.43	48.68
685	C7H8O	o-CRESOL	79.206	697.55	0.503	304.19	697.55	464.15	45.67
686	C7H8O	p-CRESOL	78.700	704.65	0.411	307.93	704.65	475.13	49.63
687	C7H8O2	GUAIACOL	62.300	697.00	0.260	304.65	697.00	478.15	46.10
688	C7H8O2	p-METHOXYPHENOL	84.140	758.00	0.390	329.00	758.00	516.00	53.90
689	C7H9N	BENZYLAMINE	67.120	683.50	0.384	227.15	683.50	457.65	43.87
690	C7H9N	2,6-DIMETHYLPYRIDINE	59.470	623.75	0.397	267.00	623.75	417.20	38.35
691	C7H9N	N-METHYLANILINE	63.832	701.55	0.310	216.15	701.55	469.02	45.33
692	C7H9N	m-TOLUIDINE	67.855	709.15	0.365	242.75	709.15	476.55	45.17
693	C7H9N	o-TOLUIDINE	71.212	694.15	0.397	249.47	694.15	473.55	45.18
694	C7H9N	p-TOLUIDINE	70.248	693.15	0.400	316.90	693.15	473.40	44.37
695	C7H10	2-NORBORNENE	44.528	583.00	0.390	319.40	583.00	368.65	30.14
696	C7H10N2	TOLUENEDIAMINE	115.327	804.00	0.579	420.00	804.00	557.15	58.21
697	C7H11NO	CYCLOHEXYL ISOCYANATE	69.513	633.00	0.387	193.00	633.00	442.15	43.71
698	C7H12	1-HEPTYNE	48.368	559.69	0.380	192.26	559.69	372.86	31.88
699	C7H12O2	n-BUTYL ACRYLATE	55.784	598.00	0.327	208.55	598.00	421.00	37.46
700	C7H12O2	ISOBUTYL ACRYLATE	59.814	580.00	0.388	212.00	580.00	405.15	37.56
701	C7H12O2	n-PROPYL METHACRYLATE	57.034	599.00	0.354	223.00	599.00	414.00	37.63
702	C7H12O4	DIETHYL MALONATE	73.150	653.00	0.360	224.25	653.00	472.05	46.09

120

Table 5-1 ENTHALPY OF VAPORIZATION - ORGANIC COMPOUNDS (continued)

NO	FORMULA	NAME	A	T_c	n	TMIN	TMAX	T_B	HVAP @ T_B
					$HVAP = A (1 - T/T_c)^n$		(HVAP - kjoule/mol, T - K)		
703	C7H14	CYCLOHEPTANE	50.021	604.30	0.382	265.12	604.30	391.94	33.55
704	C7H14	1,1-DIMETYLCYCLOPENTANE	53.692	547.00	0.520	300.00	547.00	361.00	30.64
705	C7H14	cis-1,2-DIMETHYLCYCLOPENTANE	47.597	565.15	0.377	219.26	565.15	372.68	31.71
706	C7H14	trans-1,2-DIMETHYLCYCLOPENTANE	47.560	553.15	0.391	155.58	553.15	365.02	31.20
707	C7H14	cis-1,3-DIMETHYLCYCLOPENTANE	45.520	551.00	0.363	139.45	551.00	363.92	30.75
708	C7H14	trans-1,3-DIMETHYLCYCLOPENTANE	48.600	553.00	0.406	139.18	553.00	364.88	31.37
709	C7H14	ETHYLCYCLOPENTANE	49.239	569.52	0.391	134.71	569.52	376.62	32.25
710	C7H14	2-ETHYL-1-PENTENE	49.236	543.00	0.401	168.00	543.00	367.15	31.33
711	C7H14	3-ETHYL-1-PENTENE	47.925	530.00	0.398	145.67	530.00	357.26	30.68
712	C7H14	1-HEPTENE	49.717	537.29	0.391	154.27	537.29	366.79	31.74
713	C7H14	cis-2-HEPTENE	47.591	549.00	0.375	164.00	549.00	371.56	31.16
714	C7H14	trans-2-HEPTENE	48.443	543.00	0.368	163.67	543.00	371.10	31.73
715	C7H14	cis-3-HEPTENE	50.987	545.00	0.424	136.51	545.00	368.90	31.58
716	C7H14	trans-3-HEPTENE	48.165	540.00	0.370	136.52	540.00	368.82	31.49
717	C7H14	METHYLCYCLOHEXANE	49.420	572.19	0.415	146.58	572.19	374.08	31.82
718	C7H14	2-METHYL-1-HEXENE	48.162	538.00	0.386	170.28	538.00	364.99	31.08
719	C7H14	3-METHYL-1-HEXENE	48.631	528.00	0.407	145.00	528.00	357.05	30.73
720	C7H14	4-METHYL-1-HEXENE	46.828	534.00	0.379	131.70	534.00	359.88	30.62
721	C7H14	2,3,3-TRIMETHYL-1-BUTENE	45.351	531.00	0.404	163.30	531.00	351.04	29.29
722	C7H14O	DIISOPROPYL KETONE	55.954	576.00	0.377	204.81	576.00	397.55	35.97
723	C7H14O	2-HEPTANONE	62.190	611.55	0.403	238.15	611.55	424.05	38.62
724	C7H14O	1-HEPTANAL	61.813	603.00	0.380	230.15	603.00	425.95	38.80
725	C7H14O	1-METHYLCYCLOHEXANOL	75.367	603.00	0.396	299.15	603.00	430.15	45.95
726	C7H14O	cis-2-METHYLCYCLOHEXANOL	76.130	614.00	0.400	280.15	614.00	438.15	46.17
727	C7H14O	trans-2-METHYLCYCLOHEXANOL	78.500	616.00	0.417	269.15	616.00	439.65	46.60
728	C7H14O	cis-3-METHYLCYCLOHEXANOL	80.680	618.00	0.425	267.65	618.00	441.15	47.41
729	C7H14O	trans-3-METHYLCYCLOHEXANOL	79.960	617.00	0.423	272.65	617.00	441.15	47.02
730	C7H14O	cis-4-METHYLCYCLOHEXANOL	85.048	622.00	0.447	298.15	622.00	444.15	48.60
731	C7H14O	trans-4-METHYLCYCLOHEXANOL	83.070	622.00	0.431	293.00	622.00	444.15	48.43
732	C7H14O	5-METHYL-2-HEXANONE	61.610	601.00	0.397	199.25	601.00	417.95	38.43
733	C7H14O2	n-BUTYL PROPIONATE	62.528	594.00	0.392	183.63	594.00	419.75	38.66
734	C7H14O2	ETHYL ISOVALERATE	57.350	587.95	0.380	173.85	587.95	407.45	36.61
735	C7H14O2	ISOPENTYL ACETATE	59.546	599.00	0.389	194.65	599.00	415.25	37.60
736	C7H14O2	n-PENTYL ACETATE	66.120	598.00	0.441	202.35	598.00	422.15	38.54
737	C7H14O2	n-PROPYL n-BUTYRATE	57.692	594.00	0.357	177.95	594.00	416.45	37.49
738	C7H14O2	n-HEPTANOIC ACID	91.500	680.00	0.436	265.83	680.00	496.15	51.73
739	C7H14O3	ETHYL-3-ETHOXYPROPIONATE	70.230	609.00	0.394	223.00	609.00	438.15	42.56
740	C7H15Br	1-BROMOHEPTANE	61.887	651.00	0.348	217.05	651.00	452.05	40.97
741	C7H15N	N-METHYLCYCLOHEXYLAMINE	59.857	622.00	0.386	264.65	622.00	422.00	38.63
742	C7H16	2,2-DIMETHYLPENTANE	44.802	520.50	0.377	149.34	520.50	352.34	29.26
743	C7H16	2,3-DIMETHYLPENTANE	46.866	537.35	0.382	160.00	537.35	362.93	30.49
744	C7H16	2,4-DIMETHYLPENTANE	46.860	519.79	0.397	153.91	519.79	353.64	29.80
745	C7H16	3,3-DIMETHYLPENTANE	45.350	536.40	0.381	138.70	536.40	359.21	29.74
746	C7H16	3-ETHYLPENTANE	47.325	540.64	0.373	154.55	540.64	366.62	31.01
747	C7H16	n-HEPTANE	49.730	540.26	0.386	182.56	540.26	371.58	31.73
748	C7H16	2-METHYLHEXANE	49.917	530.37	0.408	154.90	530.37	363.20	31.17
749	C7H16	3-METHYLHEXANE	48.924	535.25	0.393	153.75	535.25	365.00	31.19
750	C7H16	2,2,3-TRIMETHYLBUTANE	43.600	531.17	0.372	248.57	531.17	354.03	28.98
751	C7H16O	1-HEPTANOL	97.650	631.90	0.564	239.15	631.90	449.45	48.46
752	C7H16O	2-HEPTANOL	71.122	588.00	0.345	243.00	588.00	432.35	44.96
753	C7H16O	5-METHYL-1-HEXANOL	81.640	605.00	0.398	293.15	605.00	445.15	48.07
754	C7H16O	ISOPROPYL-TERT-BUTYL-ETHER	49.017	558.21	0.380	177.80	558.21	378.66	31.85
755	C7H16S	n-HEPTYL MERCAPTAN	64.179	645.00	0.387	229.92	645.00	450.09	40.39
756	C7H16S	BUTYL-PROPYL-SULFIDE	58.075	653.50	0.380	206.66	653.50	444.16	37.68
757	C7H16S	ETHYL-PENTYL-SULFIDE	63.302	638.37	0.380	206.66	638.37	444.16	40.28
758	C7H16S	HEXYL-METHYL-SULFIDE	63.302	638.37	0.380	206.66	638.37	444.16	40.28
759	C7H17N	1-AMINOHEPTANE	62.415	607.00	0.358	254.15	607.00	430.05	40.15
760	C8H4Cl2O2	ISOPHTHALOYL CHLORIDE	92.327	768.00	0.386	317.00	768.00	549.00	56.88
761	C8H4O3	PHTHALIC ANHYDRIDE	55.495	791.00	0.038	404.15	791.00	557.65	52.98
762	C8H6	ETHYNYLBENZENE	53.166	655.43	0.380	242.53	655.43	418.36	36.13
763	C8H6O4	ISOPHTHALIC ACID	--------	--------	------	--------	--------	--------	------
764	C8H6O4	PHTHALIC ACID	124.140	800.00	0.374	464.15	800.00	598.00	74.19
765	C8H6O4	TEREPHTHALIC ACID	--------	--------	------	--------	--------	--------	------
766	C8H6S	BENZOTHIOPHENE	65.524	754.00	0.375	304.50	754.00	493.05	44.01
767	C8H7N	INDOLE	75.800	790.00	0.405	273.68	790.00	526.15	48.62
768	C8H8	STYRENE	65.327	648.00	0.558	350.00	648.00	418.31	36.62
769	C8H8	1,3,5,7-CYCLOOCTATETRAENE	53.839	642.55	0.380	266.16	642.55	413.16	36.40
770	C8H8O	ACETOPHENONE	64.600	701.00	0.334	292.81	701.00	475.15	44.25
771	C8H8O	p-TOLUALDEHYDE	61.165	698.00	0.255	350.00	698.00	477.15	45.61
772	C8H8O2	METHYL BENZOATE	68.432	693.00	0.385	260.75	693.00	472.65	44.02
773	C8H8O2	o-TOLUIC ACID	91.444	751.00	0.388	376.85	751.00	532.00	56.69
774	C8H8O2	p-TOLUIC ACID	93.577	773.00	0.382	452.75	773.00	548.15	58.39
775	C8H8O3	METHYL SALICYLATE	83.860	701.00	0.392	265.15	701.00	493.65	52.02
776	C8H8O3	VANILLIN	101.285	777.00	0.386	355.00	777.00	558.00	62.12
777	C8H9NO	ACETANILIDE	91.958	825.00	0.378	386.65	825.00	576.95	58.39
778	C8H10	ETHYLBENZENE	54.788	617.17	0.388	178.15	617.17	409.35	35.91
779	C8H10	m-XYLENE	60.216	617.05	0.458	225.30	617.05	412.27	36.33
780	C8H10	o-XYLENE	55.606	630.37	0.375	247.98	630.37	417.58	37.00

Table 5-1 ENTHALPY OF VAPORIZATION - ORGANIC COMPOUNDS (continued)

NO	FORMULA	NAME	HVAP = A (1 - T/T$_c$)^n					(HVAP - kjoule/mol, T - K)	
			A	T$_C$	n	TMIN	TMAX	T$_B$	HVAP @ T$_B$
781	C8H10	p-XYLENE	52.910	616.26	0.354	286.41	616.26	411.51	35.82
782	C8H10O	m-ETHYLPHENOL	--------	--------	------	--------	--------	--------	------
783	C8H10O	p-ETHYLPHENOL	78.406	716.45	0.391	318.23	716.45	491.14	49.88
784	C8H10O	PHENETOLE	63.538	647.15	0.379	243.63	647.15	443.15	41.02
785	C8H10O	2-PHENYLETHANOL	69.907	684.00	0.249	247.00	684.00	492.05	50.95
786	C8H10O	2,3-XYLENOL	74.260	722.95	0.360	345.71	722.95	490.07	49.39
787	C8H10O	2,4-XYLENOL	75.830	707.65	0.383	345.71	707.65	484.13	48.77
788	C8H10O	2,5-XYLENOL	72.328	707.05	0.333	347.99	707.05	484.33	49.23
789	C8H10O	2,6-XYLENOL	68.539	701.05	0.359	318.76	701.05	474.22	45.71
790	C8H10O	3,4-XYLENOL	80.619	729.95	0.372	338.25	729.95	500.15	52.45
791	C8H10O	3,5-XYLENOL	85.430	715.65	0.456	336.59	715.65	494.89	49.97
792	C8H11N	N,N-DIMETHYLANILINE	67.877	687.15	0.405	275.60	687.15	466.69	42.83
793	C8H11N	o-ETHYLANILINE	72.457	704.00	0.378	226.55	704.00	482.65	46.79
794	C8H11N	2,4,6-TRIMETHYLPYRIDINE	66.274	653.00	0.454	229.00	653.00	444.00	39.51
795	C8H11NO	p-PHENETIDINE	84.681	754.00	0.389	277.00	754.00	528.00	53.00
796	C8H12	1,5-CYCLOOCTADIENE	57.411	645.00	0.400	203.98	645.00	423.27	37.45
797	C8H12	VINYLCYCLOHEXENE	53.750	599.00	0.382	164.00	599.00	401.00	35.21
798	C8H12O4	1,4-CYCLOHEXANEDICARBOXYLIC ACID	134.081	889.00	0.363	585.65	889.00	669.00	80.76
799	C8H12O4	DIETHYL MALEATE	67.227	680.00	0.250	264.35	680.00	498.15	48.34
800	C8H14O2	n-BUTYL METHACRYLATE	53.144	616.00	0.229	350.00	616.00	434.00	40.20
801	C8H14O3	BUTYRIC ANHYDRIDE	80.900	639.00	0.416	199.85	639.00	468.15	46.73
802	C8H14O4	DIETHYL SUCCINATE	87.639	660.00	0.406	252.35	660.00	489.65	50.57
803	C8H16	CYCLOOCTANE	53.858	647.20	0.380	287.60	647.20	424.30	35.92
804	C8H16	1,1-DIMETHYLCYCLOHEXANE	50.424	591.15	0.404	239.66	591.15	392.70	32.44
805	C8H16	cis-1,2-DIMETHYLCYCLOHEXANE	52.848	606.15	0.416	223.16	606.15	402.94	33.54
806	C8H16	trans-1,2-DIMETHYLCYCLOHEXANE	51.194	596.15	0.405	184.99	596.15	396.58	32.87
807	C8H16	cis-1,3-DIMETHYLCYCLOHEXANE	51.530	591.15	0.415	197.58	591.15	393.24	32.72
808	C8H16	trans-1,3-DIMETHYLCYCLOHEXANE	52.970	598.00	0.426	183.05	598.00	397.61	33.25
809	C8H16	cis-1,4-DIMETHYLCYCLOHEXANE	52.600	598.15	0.423	185.72	598.15	397.47	33.14
810	C8H16	trans-1,4-DIMETHYLCYCLOHEXANE	50.471	590.15	0.403	236.21	590.15	392.51	32.48
811	C8H16	ETHYLCYCLOHEXANE	54.124	609.15	0.421	161.84	609.15	404.95	34.16
812	C8H16	2-ETHYL-1-HEXENE	50.637	574.00	0.342	173.00	574.00	393.15	34.11
813	C8H16	1-METHYL-1-ETHYLCYCLOPENTANE	50.481	582.00	0.369	129.35	582.00	394.67	33.23
814	C8H16	1-OCTENE	55.443	566.60	0.401	171.45	566.60	394.44	34.39
815	C8H16	trans-2-OCTENE	52.114	577.00	0.373	185.45	577.00	398.15	33.67
816	C8H16	trans-3-OCTENE	52.470	574.00	0.378	163.15	574.00	396.45	33.67
817	C8H16	trans-4-OCTENE	53.364	573.00	0.392	179.37	573.00	395.41	33.72
818	C8H16	n-PROPYLCYCLOPENTANE	55.488	603.00	0.426	155.81	603.00	404.11	34.59
819	C8H16	2,4,4-TRIMETHYL-1-PENTENE	48.938	553.00	0.404	179.70	553.00	374.59	30.99
820	C8H16	2,4,4-TRIMETHYL-2-PENTENE	52.770	558.00	0.444	166.84	558.00	378.06	31.93
821	C8H16O	2-ETHYLHEXANAL	65.967	607.00	0.386	200.00	607.00	433.80	40.65
822	C8H16O	1-OCTANAL	66.984	621.00	0.368	246.00	621.00	447.15	41.93
823	C8H16O	2-OCTANONE	83.766	624.00	0.508	252.85	624.00	445.75	44.32
824	C8H16O2	n-BUTYL n-BUTYRATE	65.316	616.00	0.389	181.15	616.00	438.15	40.29
825	C8H16O2	n-HEXYL ACETATE	65.535	618.00	0.363	192.25	618.00	444.65	41.31
826	C8H16O2	ISOBUTYL ISOBUTYRATE	63.200	602.00	0.431	192.45	602.00	420.65	37.68
827	C8H16O2	n-OCTANOIC ACID	107.340	692.00	0.477	289.65	692.00	513.05	56.31
828	C8H16O4	DIETHYLENE GLYCOL ETHYL ETHER ACETATE	67.328	660.00	0.258	248.15	660.00	490.55	47.41
829	C8H18	2,2-DIMETHYLHEXANE	52.211	549.80	0.407	151.97	549.80	379.99	32.37
830	C8H18	2,3-DIMETHYLHEXANE	51.600	563.40	0.377	272.04	563.40	388.76	33.18
831	C8H18	2,4-DIMETHYLHEXANE	50.874	553.50	0.382	272.04	553.50	382.58	32.48
832	C8H18	2,5-DIMETHYLHEXANE	52.445	550.00	0.397	182.00	550.00	382.26	32.73
833	C8H18	3,3-DIMETHYLHEXANE	51.708	562.00	0.398	147.05	562.00	385.12	32.64
834	C8H18	3,4-DIMETHYLHEXANE	51.624	568.80	0.377	272.04	568.80	390.88	33.31
835	C8H18	3-ETHYLHEXANE	52.754	565.40	0.380	272.04	565.40	391.69	33.69
836	C8H18	3-ETHYL-2-METHYLPENTANE	51.179	567.00	0.380	158.20	567.00	388.81	32.97
837	C8H18	3-METHYL-3-ETHYLPENTANE	50.350	576.50	0.377	182.28	576.50	391.42	32.81
838	C8H18	2-METHYLHEPTANE	59.503	559.64	0.481	164.16	559.64	390.80	33.44
839	C8H18	3-METHYLHEPTANE	55.167	563.67	0.406	152.60	563.67	392.08	34.04
840	C8H18	4-METHYLHEPTANE	54.857	561.74	0.405	152.20	561.74	390.86	33.88
841	C8H18	n-OCTANE	59.077	568.83	0.439	216.38	568.83	398.83	34.77
842	C8H18	2,2,3-TRIMETHYLPENTANE	50.473	563.50	0.396	160.89	563.50	383.00	32.16
843	C8H18	2,2,4-TRIMETHYLPENTANE	42.901	543.96	0.281	165.78	543.96	372.39	31.02
844	C8H18	2,3,3-TRIMETHYLPENTANE	49.910	573.50	0.383	172.22	573.50	387.92	32.40
845	C8H18	2,3,4-TRIMETHYLPENTANE	50.752	566.30	0.385	163.95	566.30	386.62	32.62
846	C8H18	2,2,3,3-TETRAMETHYLBUTANE	47.805	567.80	0.380	172.47	567.80	379.60	31.42
847	C8H18O	DI-n-BUTYL ETHER	61.030	581.00	0.395	177.95	581.00	413.44	37.35
848	C8H18O	DI-sec-BUTYL ETHER	54.555	559.00	0.380	173.15	559.00	394.20	34.30
849	C8H18O	DI-tert-BUTYL ETHER	50.332	550.00	0.380	195.00	550.00	380.40	32.19
850	C8H18O	2-ETHYL-1-HEXANOL	111.600	640.25	0.661	203.15	640.25	457.75	48.68
851	C8H18O	1-OCTANOL	98.380	652.50	0.546	257.65	652.50	468.35	49.31
852	C8H18O	2-OCTANOL	97.600	637.15	0.598	241.55	637.15	452.95	46.47
853	C8H18O2	DI-t-BUTYL PEROXIDE	54.251	547.00	0.396	233.15	547.00	384.15	33.58
854	C8H18O2S	DI-n-BUTYL SULFONE	95.599	767.00	0.388	318.00	767.00	564.00	57.08
855	C8H18O3	DIETHYLENE GLYCOL DIETHYL ETHER	74.753	624.00	0.380	228.85	624.00	462.15	44.76
856	C8H18O3	DIETHYLENE GLYCOL MONOBUTYL EHTER	100.656	654.00	0.422	205.15	654.00	504.15	54.05
857	C8H18O4	TRIETHYLENE GLYCOL DIMETHYL ETHER	88.451	651.00	0.397	229.35	651.00	489.15	50.90
858	C8H18O5	TETRAETHYLENE GLYCOL	150.027	722.00	0.398	268.15	722.00	581.00	78.32

Table 5-1 ENTHALPY OF VAPORIZATION - ORGANIC COMPOUNDS (continued)

NO	FORMULA	NAME	$HVAP = A(1 - T/T_c)^n$			(HVAP - kjoule/mol, T - K)			
			A	T_c	n	TMIN	TMAX	T_B	HVAP @ T_B
859	C8H18S	n-OCTYL MERCAPTAN	68.700	664.00	0.386	223.95	664.00	472.19	42.54
860	C8H18S	tert-OCTYL MERCAPTAN	55.125	627.00	0.376	199.00	627.00	429.00	35.74
861	C8H18S	BUTYL-SULFIDE	62.759	650.00	0.380	209.86	650.00	455.15	39.71
862	C8H18S	ETHYL-HEXYL-SULFIDE	68.656	660.72	0.380	209.86	660.72	468.16	42.97
863	C8H18S	HEPTYL-METHYL-SULFIDE	68.656	660.72	0.380	209.86	660.72	468.16	42.97
864	C8H18S	PENTYL-PROPYL-SULFIDE	68.656	660.72	0.380	209.86	660.72	468.16	42.97
865	C8H18S2	BUTYL-DISULFIDE	75.631	704.16	0.380	202.16	704.16	504.36	46.86
866	C8H19N	DI-n-BUTYLAMINE	62.330	607.50	0.345	211.15	607.50	432.00	40.61
867	C8H19N	DIISOBUTYLAMINE	54.267	580.00	0.315	203.15	580.00	412.25	36.71
868	C8H19N	n-OCTYLAMINE	63.814	627.00	0.325	272.75	627.00	452.75	42.09
869	C8H23N5	TETRAETHYLENEPENTAMINE	139.175	774.00	0.405	243.00	774.00	606.15	74.94
870	C8H24O4Si4	OCTAMETHYLCYCLOTETRASILOXANE	75.515	586.50	0.455	290.80	586.50	448.15.	39.14
871	C9H4O5	TRIMELLITIC ANHYDRIDE	181.596	890.00	0.580	550.00	890.00	663.00	82.22
872	C9H6N2O2	TOLUENE DIISOCYANATE	81.541	737.00	0.385	287.04	737.00	523.15	50.64
873	C9H7N	ISOQUINOLINE	73.600	803.15	0.438	299.45	803.15	516.40	46.88
874	C9H7N	QUINOLINE	72.060	782.15	0.398	258.25	782.15	510.75	47.29
875	C9H7NO	8-HYDROXYQUINOLINE	86.509	788.00	0.394	348.00	788.00	540.00	54.86
876	C9H8	INDENE	66.030	687.00	0.439	271.70	687.00	455.77	40.94
877	C9H8O	2-METHYLBENZOFURAN	65.990	698.00	0.379	290.00	698.00	470.65	43.14
878	C9H10	INDANE	62.107	684.90	0.420	221.74	684.90	451.12	39.54
879	C9H10	cis-PROPENYLBENZENE	58.821	664.60	0.380	211.47	664.60	443.16	38.74
880	C9H10	trans-PROPENYLBENZENE	58.821	664.60	0.380	243.82	664.60	443.16	38.74
881	C9H1O	alpha-METHYLSTYRENE	58.201	654.00	0.371	249.95	654.00	438.65	38.54
882	C9H10	m-METHYLSTYRENE	60.430	657.00	0.378	186.81	657.00	444.75	39.42
883	C9H10	o-METHYLSTYRENE	62.997	659.00	0.406	204.58	659.00	442.96	40.06
884	C9H10	p-METHYLSTYRENE	61.939	665.00	0.406	239.02	665.00	445.93	39.46
885	C9H10O2	BENZYL ACETATE	76.940	699.00	0.458	221.65	699.00	486.65	44.58
886	C9H10O2	ETHYL BENZOATE	65.108	698.00	0.316	238.45	698.00	486.55	44.64
887	C9H10O3	ETHYL VANILLIN	118.900	748.00	0.383	350.65	748.00	567.00	69.05
888	C9H11NO	p-DIMETHYLAMINOBENZALDEHYDE	72.300	832.00	0.246	348.00	832.00	588.00	53.47
889	C9H12	CUMENE	57.201	631.15	0.363	177.14	631.15	425.56	38.07
890	C9H12	m-ETHYLTOLUENE	61.228	637.15	0.418	177.61	637.15	434.48	37.93
891	C9H12	o-ETHYLTOLUENE	60.553	651.15	0.415	192.35	651.15	438.33	38.07
892	C9H12	p-ETHYLTOLUENE	61.086	640.15	0.413	210.83	640.15	435.16	38.17
893	C9H12	MESITYLENE	60.070	637.36	0.365	228.46	637.36	437.89	39.31
894	C9H12	n-PROPYLBENZENE	60.107	638.38	0.397	173.55	638.38	432.39	38.36
895	C9H12	1,2,3-TRIMETHYLBENZENE	58.787	664.53	0.345	247.79	664.53	449.27	39.85
896	C9H12	1,2,4-TRIMETHYLBENZENE	57.823	649.13	0.342	229.38	649.13	442.53	39.09
897	C9H12O	BENZYL ETHYL ETHER	62.273	662.00	0.341	275.65	662.00	458.15	41.67
898	C9H12O	2-PHENYL-2-PROPANOL	66.767	660.00	0.248	309.15	660.00	475.15	48.70
899	C9H12O2	CUMENE HYDROPEROXIDE	--------	--------	------	--------	--------	--------	------
900	C9H14O	ISOPHORONE	58.780	715.00	0.285	265.05	715.00	488.35	42.37
901	C9H14O6	GLYCERYL TRIACETATE	98.262	704.00	0.393	277.25	704.00	532.15	56.45
902	C9H16	1-NONYNE	58.222	610.81	0.380	223.16	610.81	423.96	37.12
903	C9H16O4	AZELAIC ACID	135.082	811.00	0.377	379.65	811.00	633.36	76.20
904	C9H18	BUTYLCYCLOPENTANE	56.572	625.05	0.380	165.18	625.05	429.76	36.36
905	C9H18	cis,cis-1,3,5-TRIMETHYLCYCLOHEXANE	51.485	607.86	0.380	223.46	607.86	411.66	33.50
906	C9H18	cis,trans-1,3,5-TRIMETHYLCYCLOHEXANE	54.444	602.20	0.380	188.76	602.20	413.70	35.02
907	C9H18	ISOPROPYLCYCLOHEXANE	56.960	627.00	0.384	183.76	627.00	427.91	36.66
908	C9H18	1-NONENE	61.407	593.25	0.411	191.78	593.25	420.02	37.02
909	C9H18	n-PROPYLCYCLOHEXANE	59.920	639.15	0.444	178.25	639.15	429.90	36.50
910	C9H18O	DIISOBUTYL KETONE	66.134	615.00	0.406	227.17	615.00	441.41	39.57
911	C9H18O	1-NONANAL	72.548	640.00	0.376	255.15	640.00	468.15	44.25
912	C9H18O2	n-BUTYL VALERATE	66.330	629.00	0.321	180.35	629.00	459.65	43.53
913	C9H18O2	n-NONANOIC ACID	110.862	703.00	0.468	285.65	703.00	528.75	57.71
914	C9H18O2	n-OCTYL FORMATE	77.340	645.00	0.425	234.05	645.00	471.95	44.21
915	C9H20	3,3-DIETHYLPENTANE	53.622	610.05	0.360	240.12	610.05	419.34	35.28
916	C9H20	2,2-DIMETHYL-3-ETHYLPENTANE	53.749	590.00	0.382	173.68	590.00	406.99	34.37
917	C9H20	3-ETHYL-2,3-DIMETHYLPENTANE	55.016	606.80	0.380	173.67	606.80	417.86	35.31
918	C9H20	2,4-DIMETHYL-3-ETHYLPENTANE	54.246	591.00	0.377	150.79	591.00	409.87	34.73
919	C9H20	2,2-DIMETHYLHEPTANE	56.720	576.80	0.396	160.15	576.80	405.84	35.04
920	C9H20	2,6-DIMETHYLHEPTANE	58.107	579.00	0.406	170.25	579.00	408.36	35.38
921	C9H20	3-ETHYLHEPTANE	57.608	590.00	0.382	158.25	590.00	416.35	36.11
922	C9H20	4-ETHYLHEPTANE	58.541	584.95	0.380	159.96	584.95	414.36	36.65
923	C9H20	2,3-DIMETHYLHEPTANE	57.171	589.60	0.380	160.16	589.60	413.66	36.11
924	C9H20	2,4-DIMETHYLHEPTANE	56.149	576.80	0.380	160.16	576.80	406.05	35.36
925	C9H20	2,5-DIMETHYLHEPTANE	56.558	581.10	0.380	160.16	581.10	409.16	35.61
926	C9H20	3,4-DIMETHYLHEPTANE	57.382	591.90	0.380	170.26	591.90	413.76	36.36
927	C9H20	3,5-DIMETHYLHEPTANE	56.442	583.20	0.380	170.26	583.20	409.16	35.65
928	C9H20	4,4-DIMETHYLHEPTANE	55.695	585.40	0.380	170.26	585.40	408.36	35.36
929	C9H20	3-ETHYL-2-METHYLHEXANE	56.794	588.10	0.380	160.16	588.10	411.16	35.98
930	C9H20	4-ETHYL-2-METHYLHEXANE	56.447	580.00	0.380	160.16	580.00	406.96	35.65
931	C9H20	3-ETHYL-3-METHYLHEXANE	55.932	597.50	0.380	160.16	597.50	413.76	35.73
932	C9H20	3-ETHYL-4-METHYLHEXANE	57.271	593.70	0.380	160.16	593.70	413.56	36.40
933	C9H20	2,2,3-TRIMETHYLHEXANE	54.376	588.00	0.380	153.16	588.00	406.75	34.77
934	C9H20	2,2,4-TRIMETHYLHEXANE	53.542	573.50	0.380	153.00	573.50	399.69	34.02
935	C9H20	2,3,3-TRIMETHYLHEXANE	54.541	596.00	0.380	156.36	596.00	410.84	34.98
936	C9H20	2,3,4-TRIMETHYLHEXANE	55.928	594.50	0.380	156.36	594.50	412.20	35.69

Table 5-1 ENTHALPY OF VAPORIZATION - ORGANIC COMPOUNDS (continued)

			HVAP = A $(1 - T/T_c)^n$			(HVAP - kjoule/mol, T - K)		
NO	FORMULA	NAME	A	T_C	n	TMIN	TMAX	T_B	HVAP @ T_B
937	C9H20	2,3,5-TRIMETHYLHEXANE	54.893	579.20	0.380	145.36	579.20	404.50	34.81
938	C9H20	2,4,4-TRIMETHYLHEXANE	53.835	581.50	0.380	159.78	581.50	403.81	34.31
939	C9H20	3,3,4-TRIMETHYLHEXANE	54.630	602.30	0.380	171.96	602.30	413.62	35.15
940	C9H20	2-METHYLOCTANE	59.521	586.75	0.397	192.78	586.75	416.43	36.43
941	C9H20	3-METHYLOCTANE	59.280	590.15	0.397	165.55	590.15	417.38	36.40
942	C9H20	4-METHYLOCTANE	58.567	587.65	0.392	159.95	587.65	415.59	36.19
943	C9H20	n-NONANE	59.378	595.65	0.377	219.63	595.65	423.97	37.15
944	C9H20	2,2,3,3-TETRAMETHYLPENTANE	53.924	610.85	0.396	263.26	610.85	413.44	34.48
945	C9H20	2,2,3,4-TETRAMETHYLPENTANE	55.257	592.15	0.414	152.06	592.15	406.18	34.21
946	C9H20	2,2,4,4-TETRAMETHYLPENTANE	50.787	571.35	0.378	206.95	571.35	395.44	32.54
947	C9H20	2,3,3,4-TETRAMETHYLPENTANE	54.038	607.50	0.380	171.10	607.50	414.72	34.94
948	C9H20	2,2,5-TRIMETHYLHEXANE	54.046	568.05	0.396	167.39	568.05	397.24	33.58
949	C9H20O	2,6-DIMETHYL-4-HEPTANOL	77.187	603.00	0.381	208.00	603.00	451.00	45.66
950	C9H20O	1-NONANOL	107.160	673.00	0.574	268.15	673.00	486.25	51.34
951	C9H20O	2-NONANOL	90.580	623.00	0.398	238.15	623.00	471.65	51.58
952	C9H20S	n-NONYL MERCAPTAN	75.340	681.00	0.403	253.05	681.00	492.95	44.85
953	C9H20S	BUTYL-PENTYL-SULFIDE	74.130	681.56	0.380	231.16	681.56	491.16	45.66
954	C9H20S	ETHYL-HEPTYL-SULFIDE	74.130	681.56	0.380	231.16	681.56	491.16	45.66
955	C9H20S	HEXYL-PROPYL-SULFIDE	74.130	681.56	0.380	231.16	681.56	491.16	45.66
956	C9H20S	METHYL-OCTYL-SULFIDE	74.130	681.56	0.380	231.16	681.56	491.16	45.66
957	C9H21N	n-NONYLAMINE	70.700	648.00	0.343	273.15	648.00	475.35	44.92
958	C9H21N	TRIPROPYLAMINE	60.752	577.50	0.380	179.65	577.50	429.65	36.20
959	C10H6O8	PYROMELLITIC ACID	188.507	893.00	0.335	554.00	893.00	722.00	108.35
960	C10H7Br	1-BROMONAPHTHALENE	77.917	824.00	0.383	279.35	824.00	554.25	50.80
961	C10H7Cl	1-CHLORONAPHTHALENE	78.664	785.00	0.411	269.15	785.00	532.45	49.36
962	C10H8	NAPHTHALENE	76.150	748.35	0.526	353.43	748.35	491.14	43.42
963	C10H8	AZULENE	84.164	773.48	0.380	173.66	773.48	515.16	55.48
964	C10H9N	QUINALDINE	66.934	773.00	0.387	272.15	773.00	519.75	43.46
965	C10H10	m-DIVINYLBENZENE	63.731	692.00	0.364	206.25	692.00	472.65	41.95
966	C10H10	1-METHYLINDENE	64.449	703.00	0.384	350.00	703.00	471.65	42.06
967	C10H10	2-METHYLINDENE	62.265	684.00	0.385	353.15	684.00	458.00	40.65
968	C10H10O4	DIMETHYL PHTHALATE	85.511	766.00	0.340	272.15	766.00	556.85	55.00
969	C10H10O4	DIMETHYL TEREPHTHALATE	72.322	772.00	0.242	413.80	772.00	561.15	52.83
970	C10H12	DICYCLOPENTADIENE	56.501	660.00	0.373	307.00	660.00	443.00	37.31
971	C10H12	1,2,3,4-TETRAHYDRONAPHTHALENE	59.455	720.15	0.308	350.00	720.15	480.77	42.35
972	C10H12O	ANETHOLE	76.102	723.00	0.384	294.50	723.00	508.45	47.73
973	C10H12O4	DIALLYL MALEATE	93.915	693.00	0.396	226.15	693.00	520.00	54.21
974	C10H14	n-BUTYLBENZENE	61.981	660.55	0.365	185.30	660.55	456.46	40.37
975	C10H14	sec-BUTYLBENZENE	66.188	664.54	0.472	197.72	664.54	446.48	39.12
976	C10H14	tert-BUTYLBENZENE	64.314	660.00	0.464	215.27	660.00	442.30	38.44
977	C10H14	1,2,3,4-TETRAMETHYLBENZENE	70.088	695.10	0.380	266.91	695.10	478.25	45.02
978	C10H14	m-CYMENE	61.494	657.00	0.395	209.44	657.00	448.23	39.10
979	C10H14	o-CYMENE	62.180	662.00	0.399	201.64	662.00	451.33	39.38
980	C10H14	p-CYMENE	64.270	653.15	0.411	205.25	653.15	450.28	39.75
981	C10H14	m-DIETHYLBENZENE	66.070	663.00	0.427	189.26	663.00	454.29	40.33
982	C10H14	o-DIETHYLBENZENE	65.100	668.00	0.418	241.93	668.00	456.61	40.24
983	C10H14	p-DIETHYLBENZENE	63.534	657.96	0.378	230.32	657.96	456.94	40.58
984	C1OH14	2-ETHYL-m-XYLENE	65.477	671.00	0.383	256.89	671.00	463.19	41.79
985	C10H14	2-ETHYL-p-XYLENE	61.878	663.00	0.352	219.52	663.00	459.98	40.80
986	C10H14	3-ETHYL-o-XYLENE	67.497	680.00	0.416	223.64	680.00	467.11	41.64
987	C10H14	4-ETHYL-m-XYLENE	62.138	665.00	0.352	210.27	665.00	461.59	40.95
988	C10H14	4-ETHYL-o-XYLENE	65.793	667.00	0.385	206.22	667.00	462.93	41.70
989	C1OH14	5-ETHYL-m-XYLENE	63.322	655.00	0.369	188.82	655.00	456.93	40.73
990	C10H14	ISOBUTYLBENZENE	67.524	650.15	0.480	320.00	650.15	445.94	38.73
991	C10H14	1,2,3,5-TETRAMETHYLBENZENE	66.027	679.00	0.370	249.46	679.00	471.15	42.61
992	C10H14	1,2,4,5-TETRAMETHYLBENZENE	65.063	675.15	0.358	352.38	675.15	469.99	42.48
993	C10H14O	p-tert-BUTYLPHENOL	84.460	734.00	0.427	371.56	734.00	512.88	50.60
994	C10H14O2	p-tert-BUTYLCATECHOL	100.839	776.00	0.394	325.00	776.00	558.00	61.15
995	C10H15N	N,N-DIETYHLANILINE	75.892	702.00	0.430	235.15	702.00	489.42	45.41
996	C10H15N	2,6-DIETHYLANILINE	102.077	678.00	0.396	276.65	678.00	508.65	58.93
997	C10H16	CAMPHENE	56.105	638.00	0.382	320.15	638.00	433.65	36.32
998	C10H16	D-LIMONENE	60.442	660.00	0.380	199.00	660.00	449.65	39.14
999	C10H16	alpha-PHELLANDRENE	61.809	649.00	0.373	220.00	649.00	448.15	39.91
1000	C1OH16	beta-PHELLANDRENE	62.962	648.00	0.398	220.00	648.00	447.15	39.50
1001	C10H16	alpha-PINENE	55.440	632.00	0.386	209.15	632.00	429.29	35.74
1002	C10H16	beta-PINENE	60.035	643.00	0.402	211.61	643.00	439.19	37.83
1003	C10H16	alpha-TERPINENE	62.165	652.00	0.385	220.00	652.00	450.35	39.57
1004	C1OH16	gamma-TERPINENE	63.865	661.00	0.393	220.00	661.00	456.15	40.30
1005	C10H16	TERPINOLENE	66.655	672.00	0.353	200.00	672.00	458.15	44.49
1006	C10H16O	CAMPHOR	64.250	709.00	0.383	453.25	709.00	480.57	41.64
1007	C10H18	1-DECYNE	63.219	632.49	0.380	229.16	632.49	447.16	39.65
1008	C10H18	cis-DECAHYDRONANPHTALENE	64.211	702.25	0.434	230.20	702.25	468.97	39.80
1009	C10H18	trans-DECAHYDRONAPHTHALENE	52.979	687.05	0.285	350.00	687.05	460.46	38.62
1010	C10H18O4	SEBACIC ACID	130.559	815.00	0.342	407.65	815.00	642.09	76.83
1011	C10H20	n-BUTYLCYCLOHEXANE	65.519	667.00	0.460	198.42	667.00	454.13	38.74
1012	C10H20	1-CYCLOPENTYLPENTANE	61.613	647.49	0.380	190.16	647.49	453.76	38.95
1013	C10H20	1-DECENE	65.577	617.05	0.402	206.89	617.05	443.75	39.36
1014	C10H20O	1-DECANAL	80.027	657.00	0.396	267.15	657.00	488.15	46.73

Table 5-1 ENTHALPY OF VAPORIZATION - ORGANIC COMPOUNDS (continued)

			HVAP = A $(1 - T/T_c)^n$				(HVAP - kjoule/mol, T - K)		
NO	FORMULA	NAME	A	T_c	n	TMIN	TMAX	T_B	HVAP @ T_B
1015	C10H20O2	n-DECANOIC ACID	148.500	713.00	0.615	304.75	713.00	543.15	61.46
1016	C10H20O2	2-ETHYLHEXYL ACETATE	82.325	639.00	0.474	180.15	639.00	471.75	43.61
1017	C10H20O2	ISOPENTYL ISOVALERATE	74.100	637.00	0.409	215.00	637.00	467.15	43.15
1018	C10H22	n-DECANE	71.428	618.45	0.451	243.49	618.45	447.30	40.02
1019	C10H22	2-METHYLNONANE	65.666	610.00	0.425	198.50	610.00	440.15	38.14
1020	C10H22	3-METHYLNONANE	66.226	613.00	0.435	188.35	613.00	440.95	38.11
1021	C10H22	4-METHYLNONANE	62.016	610.00	0.375	174.45	610.00	438.85	38.51
1022	C10H22	5-METHYLNONANE	65.682	610.00	0.432	185.45	610.00	438.30	37.98
1023	C10H22	3-ETHYLOCTANE	63.361	613.60	0.380	188.36	613.60	439.66	39.24
1024	C10H22	4-ETHYLOCTANE	62.818	609.60	0.380	174.46	609.60	436.80	38.91
1025	C10H22	2,2-DIMETHYLOCTANE	59.810	602.00	0.380	225.00	602.00	430.05	37.15
1026	C10H22	2,3-DIMETHYLOCTANE	61.420	613.20	0.380	219.16	613.20	437.47	38.20
1027	C10H22	2,4-DIMETHYLOCTANE	61.008	599.40	0.380	219.16	599.40	429.06	37.82
1028	C10H22	2,5-DIMETHYLOCTANE	61.011	603.00	0.380	219.16	603.00	431.66	37.82
1029	C10H22	2,6-DIMETHYLOCTANE	61.868	603.10	0.380	219.16	603.10	433.54	38.20
1030	C10H22	2,7-DIMETHYLOCTANE	61.886	602.90	0.380	219.16	602.90	433.03	38.24
1031	C10H22	3,3-DIMETHYLOCTANE	60.176	612.10	0.380	219.16	612.10	434.36	37.61
1032	C10H22	3,4-DIMETHYLOCTANE	60.957	614.00	0.380	219.16	614.00	436.56	38.03
1033	C10H22	3,5-DIMETHYLOCTANE	60.682	606.30	0.380	219.16	606.30	432.56	37.74
1034	C10H22	3,6-DIMETHYLOCTANE	61.216	608.30	0.380	219.16	608.30	433.96	38.07
1035	C10H22	4,4-DIMETHYLOCTANE	59.573	606.90	0.380	219.16	606.90	430.66	37.24
1036	C10H22	4,5-DIMETHYLOCTANE	60.689	612.20	0.380	219.16	612.20	435.29	37.87
1037	C10H22	4-PROPYLHEPTANE	61.543	601.00	0.380	219.16	601.00	430.66	38.12
1038	C10H22	4-ISOPROPYLHEPTANE	60.025	607.60	0.380	219.16	607.60	432.06	37.45
1039	C10H22	3-ETHYL-2-METHYLHEPTANE	60.488	610.90	0.380	219.16	610.90	434.36	37.74
1040	C10H22	4-ETHYL-2-METHYLHEPTANE	60.077	601.80	0.380	219.16	601.80	429.36	37.36
1041	C10H22	5-ETHYL-2-METHYLHEPTANE	60.684	606.70	0.380	219.16	606.70	432.86	37.74
1042	C10H22	3-ETHYL-3-METHYLHEPTANE	59.665	620.00	0.380	219.16	620.00	436.96	37.53
1043	C10H22	4-ETHYL-3-METHYLHEPTANE	60.171	614.30	0.380	219.16	614.30	435.36	37.66
1044	C10H22	3-ETHYL-5-METHYLHEPTANE	60.428	606.60	0.380	219.16	606.60	431.36	37.70
1045	C10H22	3-ETHYL-4-METHYLHEPTANE	60.299	615.50	0.380	219.16	615.50	436.16	37.74
1046	C10H22	4-ETHYL-4-METHYLHEPTANE	59.337	615.70	0.380	219.16	615.70	433.96	37.32
1047	C10H22	2,2,3-TRIMETHYLHEPTANE	58.957	611.70	0.380	219.16	611.70	430.76	37.11
1048	C10H22	2,2,4-TRIMETHYLHEPTANE	58.183	594.50	0.380	219.16	594.50	421.46	36.40
1049	C10H22	2,2,5-TRIMETHYLHEPTANE	58.587	598.00	0.380	219.16	598.00	423.96	36.65
1050	C10H22	2,2,6-TRIMETHYLHEPTANE	58.902	593.40	0.380	219.16	593.40	422.09	36.74
1051	C10H22	2,3,3-TRIMETHYLHEPTANE	58.909	617.50	0.380	219.16	617.50	433.36	37.20
1052	C10H22	2,3,4-TRIMETHYLHEPTANE	59.400	613.70	0.380	219.16	613.70	433.06	37.32
1053	C10H22	2,3,5-TRIMETHYLHEPTANE	59.314	612.80	0.380	219.16	612.80	433.86	37.15
1054	C10H22	2,3,6-TRIMETHYLHEPTANE	59.769	604.10	0.380	219.16	604.10	429.16	37.32
1055	C10H22	2,4,4-TRIMETHYLHEPTANE	57.938	600.30	0.380	219.16	600.30	424.16	36.36
1056	C10H22	2,4,5-TRIMETHYLHEPTANE	59.244	606.90	0.380	219.16	606.90	429.66	37.11
1057	C10H22	2,4,6-TRIMETHYLHEPTANE	59.285	590.30	0.380	219.16	590.30	420.76	36.90
1058	C10H22	2,5,5-TRIMETHYLHEPTANE	58.600	602.90	0.380	219.16	602.90	425.96	36.78
1059	C10H22	3,3,4-TRIMETHYLHEPTANE	58.726	622.10	0.380	219.16	622.10	435.06	37.20
1060	C10H22	3,3,5-TRIMETHYLHEPTANE	58.182	609.50	0.380	219.16	609.50	428.85	36.65
1061	C10H22	3,4,4-TRIMETHYLHEPTANE	58.531	620.90	0.380	219.16	620.90	434.26	37.07
1062	C10H22	3,4,5-TRIMETHYLHEPTANE	60.012	612.80	0.380	219.16	612.80	435.66	37.45
1063	C10H22	3-ISOPROPYL-2-METHYLHEXANE	58.263	623.40	0.380	219.16	623.40	439.86	36.61
1064	C10H22	3,3-DIETHYLHEXANE	59.435	627.80	0.380	219.16	627.80	439.46	37.61
1065	C10H22	3,4-DIETHYLHEXANE	59.984	618.80	0.380	219.16	618.80	437.06	37.66
1066	C10H22	3-ETHYL-2,2-DIMETHYLHEXANE	58.110	611.70	0.380	219.16	611.70	429.26	36.69
1067	C10H22	4-ETHYL-2,2-DIMETHYLHEXANE	57.741	594.60	0.380	219.16	594.60	420.16	36.23
1068	C10H22	3-ETHYL-2,3-DIMETHYLHEXANE	58.551	626.80	0.380	219.16	626.80	436.86	37.20
1069	C10H22	4-ETHYL-2,3-DIMETHYLHEXANE	59.078	617.30	0.380	219.16	617.30	434.06	37.24
1070	C10H22	3-ETHYL-2,4-DIMETHYLHEXANE	58.950	616.10	0.380	219.16	616.10	433.26	37.15
1071	C10H22	4-ETHYL-2,4-DIMETHYLHEXANE	58.531	620.90	0.380	219.16	620.90	434.26	37.07
1072	C10H22	3-ETHYL-2,5-DIMETHYLHEXANE	58.777	603.50	0.380	219.16	603.50	427.26	36.82
1073	C10H22	4-ETHYL-3,3-DIMETHYLHEXANE	58.414	625.70	0.380	219.16	625.70	436.06	37.11
1074	C10H22	3-ETHYL-3,4-DIMETHYLHEXANE	58.222	624.50	0.380	219.16	624.50	435.26	36.99
1075	C10H22	2,2,3,3-TETRAMETHYLHEXANE	57.149	623.00	0.380	219.16	623.00	433.48	36.36
1076	C10H22	2,2,3,4-TETRAMETHYLHEXANE	57.248	620.40	0.380	219.16	620.40	431.96	36.40
1077	C10H22	2,2,3,5-TETRAMETHYLHEXANE	57.332	601.30	0.380	219.16	601.30	421.56	36.23
1078	C10H22	2,2,4,4-TETRAMETHYLHEXANE	55.382	610.20	0.380	219.16	610.20	426.96	35.06
1079	C10H22	2,2,4,5-TETRAMETHYLHEXANE	56.845	598.50	0.380	219.16	598.50	421.04	35.82
1080	C10H22	2,2,5,5-TETRAMETHYLHEXANE	56.183	581.40	0.380	260.56	581.40	410.63	35.27
1081	C10H22	2,3,3,4-TETRAMETHYLHEXANE	57.823	633.10	0.380	260.56	633.10	437.75	36.99
1082	C10H22	2,3,3,5-TETRAMETHYLHEXANE	57.223	610.10	0.380	260.56	610.10	426.26	36.28
1083	C10H22	2,3,4,4-TETRAMETHYLHEXANE	57.470	626.60	0.380	260.56	626.60	434.76	36.65
1084	C10H22	2,3,4,5-TETRAMETHYLHEXANE	58.260	613.20	0.380	260.56	613.20	429.36	36.86
1085	C10H22	3,3,4,4-TETRAMETHYLHEXANE	54.337	646.70	0.380	260.56	646.70	443.16	35.02
1086	C10H22	2,4-DIMETHYL-3-ISOPROPYLPENTANE	57.466	614.40	0.380	191.46	614.40	430.20	36.36
1087	C10H22	3,3-DIETHYL-2-METHYLPENTANE	58.718	639.90	0.380	191.46	639.90	442.86	37.53
1088	C10H22	3-ETHYL-2,2,3-TRIMETHYLPENTANE	56.607	646.00	0.380	191.46	646.00	442.66	36.48
1089	C10H22	3-ETHYL-2,2,4-TRIMETHYLPENTANE	56.860	615.30	0.380	191.46	615.30	428.46	36.15
1090	C10H22	3-ETHYL-2,3,4-TRIMETHYLPENTANE	57.720	642.30	0.380	191.46	642.30	442.60	37.03
1091	C10H22	2,2,3,3,4-PENTAMETHYLPENTANE	56.208	643.80	0.380	236.71	643.80	439.21	36.36
1092	C10H22	2,2,3,4,4-PENTAMETHYLPENTANE	55.067	627.30	0.380	234.41	627.30	432.45	35.31

Table 5-1 ENTHALPY OF VAPORIZATION - ORGANIC COMPOUNDS (continued)

NO	FORMULA	NAME	$HVAP = A(1 - T/T_c)^n$				(HVAP - kjoule/mol, T - K)		
			A	T_c	n	TMIN	TMAX	T_B	HVAP @ T_B
1093	C10H22O	1-DECANOL	108.960	690.00	0.562	280.05	690.00	503.35	52.26
1094	C10H22O	DI-n-PENTYL ETHER	61.955	622.00	0.295	203.72	622.00	459.90	41.67
1095	C10H22O	ISODECANOL	95.035	644.00	0.399	213.15	644.00	493.00	53.28
1096	C10H22O5	TETRAETHYLENE GLYCOL DIMETHYL ETHER	86.406	705.00	0.266	243.45	705.00	548.95	57.85
1097	C10H22S	n-DECYL MERCAPTAN	82.370	696.00	0.417	247.56	696.00	512.35	47.26
1098	C10H22S	BUTYL-HEXYL-SULFIDE	79.664	701.03	0.380	238.16	701.03	513.16	48.30
1099	C10H22S	ETHYL-OCTYL-SULFIDE	79.664	701.03	0.380	238.16	701.03	513.16	48.30
1100	C10H22S	HEPTYL-PROPYL-SULFIDE	79.664	701.03	0.380	238.16	701.03	513.16	48.30
1101	C10H22S	METHYL-NONYL-SULFIDE	79.664	701.03	0.380	238.16	701.03	513.16	48.30
1102	C10H22S	PENTYL-SULFIDE	79.664	701.03	0.380	238.16	701.03	513.16	48.30
1103	C10H22S2	PENTYL-DISULFIDE	85.016	726.94	0.380	214.16	726.94	537.06	51.05
1104	C10H23N	n-DECYLAMINE	73.090	663.00	0.326	288.85	663.00	493.65	46.84
1105	C11H10	1-METHYLNAPHTHALENE	72.738	772.04	0.416	242.67	772.04	517.83	45.82
1106	C11H10	2-METHYLNAPHTHALENE	72.937	761.00	0.404	307.73	761.00	514.20	46.28
1107	C11H14O2	n-BUTYL BENZOATE	73.031	724.00	0.307	251.65	724.00	523.15	49.27
1108	C11H16	n-PENTYLBENZENE	64.400	679.90	0.346	198.15	679.90	478.61	42.27
1109	C11H16O	p-tert-AMYLPHENOL	85.282	751.00	0.389	366.00	751.00	535.15	52.51
1110	C11H20	1-UNDECYNE	68.110	650.99	0.380	248.16	650.99	468.16	42.04
1111	C11H20O2	2-ETHYLHEXYL ACRYLATE	67.100	655.00	0.281	183.15	655.00	489.15	45.61
1112	C11H22	1-UNDECENE	73.575	638.00	0.433	223.99	638.00	465.82	41.73
1113	C11H22	1-CYCLOPENTYLHEXANE	66.188	667.67	0.380	200.16	667.67	476.26	41.17
1114	C11H22	PENTYLCYCLOHEXANE	65.151	674.01	0.380	215.66	674.01	476.87	40.84
1115	C11H22O	1-UNDECANAL	78.172	672.00	0.346	273.15	672.00	506.15	48.17
1116	C11H24	n-UNDECANE	73.511	638.76	0.413	247.57	638.76	469.08	42.52
1117	C11H24O	1-UNDECANOL	116.070	704.00	0.600	289.05	704.00	518.15	52.20
1118	C11H24S	UNDECYL MERCAPTAN	86.800	710.00	0.410	270.15	710.00	530.55	49.39
1119	C11H24S	BUTYL-HEPTYL-SULFIDE	85.071	717.91	0.380	254.66	717.91	533.16	50.79
1120	C11H24S	DECYL-METHYL-SULFIDE	85.071	717.91	0.380	254.66	717.91	533.16	50.79
1121	C11H24S	ETHYL-NONYL-SULFIDE	85.071	717.91	0.380	254.66	717.91	533.16	50.79
1122	C11H24S	OCTYL-PROPYL-SULFIDE	85.071	717.91	0.380	254.66	717.91	533.16	50.79
1123	C12H8O	DIBENZOFURAN	63.346	837.80	0.302	355.65	837.80	557.86	45.49
1124	C12H9N	DIBENZOPYRROLE	87.440	899.00	0.330	517.95	899.00	627.86	58.87
1125	C12H10	ACENAPHTHENE	71.842	803.15	0.298	366.56	803.15	550.54	50.90
1126	C12H10	BIPHENYL	77.536	789.26	0.414	342.37	789.26	528.15	49.05
1127	C12H10O	DIPHENYL ETHER	69.970	763.00	0.336	300.10	763.00	531.46	46.87
1128	C12H11N	p-AMINODIPHENYL	91.122	817.00	0.390	326.00	817.00	575.00	56.70
1129	C12H11N	DIPHENYLAMINE	83.711	817.00	0.346	326.15	817.00	575.15	54.94
1130	C12H11N3	p-AMINOAZOBENZENE	104.940	877.00	0.389	401.00	877.00	633.00	63.80
1131	C12H11N3	1,3-DIPHENYLTRIAZENE	100.312	845.00	0.390	372.00	845.00	610.00	60.90
1132	C12H12	1,2-DIMETHYLNAPHTHALENE	78.915	775.34	0.380	272.16	775.34	539.46	50.21
1133	C12H12	1,3-DIMETHYLNAPHTHALENE	78.460	773.76	0.380	269.16	773.76	538.36	49.92
1134	C12H12	1,4-DIMETHYLNAPHTHALENE	78.770	776.78	0.380	280.82	776.78	540.46	50.12
1135	C12H12	1,5-DIMETHYLNAPHTHALENE	78.435	773.47	0.380	355.16	773.47	538.16	49.90
1136	C12H12	1,6-DIMETHYLNAPHTHALENE	78.138	770.60	0.380	259.16	770.60	536.16	49.71
1137	C12H12	1,7-DIMETHYLNAPHTHALENE	78.138	770.60	0.380	260.16	770.60	536.16	49.71
1138	C12H12	2,3-DIMETHYLNAPHTHALENE	78.869	777.78	0.380	378.16	777.78	541.16	50.18
1139	C12H12	2,6-DIMETHYLNAPHTHALENE	73.779	777.00	0.355	383.32	777.00	535.15	48.75
1140	C12H12	2,7-DIMETHYLNAPHTHALENE	73.396	778.00	0.349	368.81	778.00	536.15	48.82
1141	C12H12	1-ETHYLNAPHTHALENE	78.348	776.00	0.424	259.34	776.00	531.48	48.01
1142	C12H12	2-ETHYLNAPHTHALENE	74.713	774.90	0.380	265.76	774.90	531.49	48.12
1143	C12H12N2	p-AMINODIPHENYLAMINE	109.456	867.00	0.395	341.15	867.00	627.15	65.89
1144	C12H12N2	HYDRAZOBENZENE	98.343	792.00	0.391	404.15	792.00	573.00	59.49
1145	C12H14	1,2,3-TRIMETHYLINDENE	64.200	726.00	0.281	344.65	726.00	509.00	45.73
1146	C12H14O4	DIETHYL PHTHALATE	104.231	757.00	0.431	269.15	757.00	567.15	57.43
1147	C12H16	CYCLOHEXYLBENZENE	81.460	744.00	0.467	280.14	744.00	513.27	47.15
1148	C12H18	m-DIISOPROPYLBENZENE	70.653	684.00	0.444	210.02	684.00	476.33	41.62
1149	C12H18	p-DIISOPROPYLBENZENE	73.206	689.00	0.441	256.08	689.00	483.65	42.92
1150	C12H18	n-HEXYLBENZENE	73.268	698.00	0.389	212.00	698.00	499.26	44.95
1151	C12H18	1,2,3-TRIETHYLBENZENE	72.282	684.37	0.380	206.66	684.37	490.66	44.74
1152	C12H18	1,2,4-TRIETHYLBENZENE	72.282	684.37	0.380	206.66	684.37	490.66	44.74
1153	C12H18	1,3,5-TRIETHYLBENZENE	72.058	682.28	0.380	206.66	682.28	489.16	44.61
1154	C12H18	HEXAMETHYLBENZENE	73.427	758.00	0.380	438.66	758.00	536.60	46.00
1155	C12H20O4	DIBUTYL MALEATE	104.061	716.00	0.393	275.00	716.00	553.15	58.15
1156	C12H22	BICYCLOHEXYL	68.541	727.00	0.354	276.78	727.00	512.19	44.52
1157	C12H22	1-DODECYNE	73.060	668.16	0.380	254.16	668.16	488.16	44.38
1158	C12H23N	DICYCLOHEXYLAMINE	73.509	737.00	0.347	273.05	737.00	529.00	47.39
1159	C12H24	1-DODECENE	78.802	657.00	0.437	237.93	657.00	486.50	43.70
1160	C12H24	1-CYCLOPENTYLHEPTANE	71.533	679.00	0.380	220.00	679.00	497.30	43.35
1161	C12H24	1-CYCLOHEXYLHEXANE	69.465	691.81	0.380	263.60	691.81	497.86	42.84
1162	C12H24O	1-DODECANAL	83.180	685.00	0.348	285.15	685.00	523.15	50.35
1163	C12H24O2	n-DODECANOIC ACID	130.530	734.00	0.491	316.98	734.00	571.85	62.19
1164	C12H26	n-DODECANE	77.166	658.20	0.407	263.57	658.20	489.47	44.34
1165	C12H26O	DI-n-HEXYL ETHER	84.417	658.00	0.396	230.15	658.00	498.85	48.12
1166	C12H26O	1-DODECANOL	119.420	721.00	0.579	296.95	721.00	535.00	54.50
1167	C12H26O3	DIETHYLENE GLYCOL DI-n-BUTYL ETHER	68.969	680.00	0.197	212.95	680.00	529.15	51.27
1168	C12H26S	n-DODECYL MERCAPTAN	81.424	724.00	0.334	265.15	724.00	547.75	50.79
1169	C12H26S	BUTYL-OCTYL-SULFIDE	90.429	733.68	0.380	259.16	733.68	552.16	53.19
1170	C12H26S	DECYL-ETHYL-SULFIDE	90.429	733.68	0.380	259.16	733.68	552.16	53.19

Table 5-1 ENTHALPY OF VAPORIZATION - ORGANIC COMPOUNDS (continued)

			$HVAP = A(1 - T/T_c)^n$				(HVAP - kjoule/mol, T - K)	
NO	FORMULA	NAME	A	T_C	n	TMIN	TMAX	T_B	HVAP @ T_B
1171	C12H26S	HEXYL-SULFIDE	90.429	733.68	0.380	259.16	733.68	552.16	53.19
1172	C12H26S	METHYL-UNDECYL-SULFIDE	90.429	733.68	0.380	259.16	733.68	552.16	53.19
1173	C12H26S	NONYL-PROPYL-SULFIDE	90.429	733.68	0.380	259.16	733.68	552.16	53.19
1174	C12H26S2	HEXYL-DISULFIDE	94.045	747.10	0.380	225.16	747.10	566.66	54.81
1175	C12H27BO3	TRI-n-BUTYL BORATE	56.900	743.15	0.373	203.15	743.15	506.65	37.12
1176	C12H27N	DODECYLAMINE	85.100	696.00	0.343	301.47	696.00	532.35	51.79
1177	C12H27N	TRI-n-BUTYLAMINE	81.663	644.00	0.397	203.00	644.00	487.15	46.61
1178	C13H10	FLUORENE	78.720	870.00	0.369	387.94	870.00	570.44	53.12
1179	C13H10O	BENZOPHENONE	78.700	816.00	0.304	321.35	816.00	579.24	54.03
1180	C13H12	DIPHENYLMETHANE	79.558	768.00	0.387	298.39	768.00	537.42	49.94
1181	C13H14	1-PROPYLNAPHTHALENE	82.034	771.45	0.380	264.69	771.45	545.96	51.41
1182	C13H14	2-PROPYLNAPHTHALENE	82.141	772.44	0.380	270.16	772.44	546.66	51.47
1183	C13H14	2ETHYL-3-METHYLNAPHTHALENE	82.513	776.44	0.380	344.16	776.44	550.16	51.65
1184	C13H14	2ETHYL-6-METHYLNAPHTHALENE	81.463	766.56	0.380	318.16	766.56	543.16	50.99
1185	C13H14	2ETHYL-7-METHYLNAPHTHALENE	81.463	766.56	0.380	318.16	766.56	543.16	50.99
1186	C13H20	n-HEPTYLBENZENE	81.340	714.00	0.410	225.15	714.00	519.25	47.75
1187	C13H24	1-TRIDECYNE	78.024	684.11	0.380	268.16	684.11	507.16	46.67
1188	C13H26	1-TRIDECENE	77.146	675.00	0.372	250.08	675.00	505.93	46.10
1189	C13H26	1-CYCLOPENTYLOCTANE	75.325	702.06	0.380	229.16	702.06	516.86	45.40
1190	C13H26	1-CYCLOHEXYLHEPTANE	73.948	708.63	0.380	242.66	708.63	518.06	44.89
1191	C13H26O	1-TRIDECANAL	95.624	700.00	0.414	288.15	700.00	540.15	51.88
1192	C13H26O2	n-BUTYL NONANOATE	91.050	652.00	0.399	235.15	652.00	503.00	50.52
1193	C13H26O2	METHYL DODECANOATE	95.360	712.00	0.428	278.15	712.00	540.00	51.92
1194	C13H28	n-TRIDECANE	81.877	675.80	0.416	267.76	675.80	508.62	45.79
1195	C13H28O	1-TRIDECANOL	121.920	731.00	0.588	303.75	731.00	547.15	54.15
1196	C13H28S	BUTYL-NONYL-SULFIDE	95.677	748.42	0.380	271.16	748.42	570.16	55.47
1197	C13H28S	DECYL-PROPYL-SULFIDE	95.677	748.42	0.380	271.16	748.42	570.16	55.47
1198	C13H28S	DODECYL-METHYL-SULFIDE	95.677	748.42	0.380	271.16	748.42	570.16	55.47
1199	C13H28S	ETHYL-UNDECYL-SULFIDE	95.677	748.42	0.380	271.16	748.42	570.16	55.47
1200	C13H28S	1-TRIDECANETHIOL	92.101	742.13	0.380	282.04	742.13	563.96	53.56
1201	C14H8O2	ANTHRAQUINONE	79.700	900.00	0.194	559.15	900.00	653.05	62.02
1202	C14H10	ANTHRACENE	79.861	873.00	0.307	488.93	873.00	615.18	54.92
1203	C14H10	DIPHENYLACETYLENE	79.562	832.00	0.382	335.65	832.00	573.00	50.94
1204	C14H10	PHENANTHRENE	87.928	869.25	0.398	372.38	869.25	613.45	54.04
1205	C14H12	cis-STILBENE	83.470	757.00	0.451	268.15	757.00	535.00	48.00
1206	C14H12	trans-STILBENE	90.530	820.00	0.408	397.35	820.00	579.65	54.87
1207	C14H12O2	BENZYL BENZOATE	98.129	820.00	0.395	292.55	820.00	596.65	58.71
1208	C14H14	1,1-DIPHENYLETHANE	82.250	775.00	0.402	255.20	775.00	545.78	50.40
1209	C14H14	1,2-DIPHENYLETHANE	80.610	780.00	0.371	324.34	780.00	553.65	50.94
1210	C14H14O	DIBENZYL ETHER	93.930	777.00	0.436	276.75	777.00	561.45	53.70
1211	C14H16	1-n-BUTYLNAPHTHALENE	86.593	792.00	0.403	253.43	792.00	562.54	52.56
1212	C14H16	2-BUTYLNAPHTHALENE	86.920	780.96	0.380	268.16	780.96	562.16	53.60
1213	C14H22	n-OCTYLBENZENE	86.707	729.00	0.428	237.15	729.00	537.55	48.92
1214	C14H22	1,2,3,4-TETRAETHYLBENZENE	81.509	708.20	0.380	284.96	708.20	524.16	48.84
1215	C14H22	1,2,3,5-TETRAETHYLBENZENE	81.433	707.52	0.380	284.16	707.52	523.66	48.80
1216	C14H22	1,2,4,5-TETRAETHYLBENZENE	81.355	706.85	0.380	283.16	706.85	523.16	48.75
1217	C14H22O	p-tert-OCTYLPHENOL	91.851	765.00	0.391	358.55	765.00	563.60	54.51
1218	C14H28	1-TETRADECENE	87.015	692.00	0.418	260.30	692.00	524.25	48.12
1219	C14H28	1-CYCLOPENTYLNONANE	79.583	716.95	0.380	244.16	716.95	535.26	47.24
1220	C14H28	1-CYCLOHEXYLOCTANE	78.319	723.61	0.380	253.46	723.61	536.76	46.82
1221	C14H28O2	n-TETRADECANOIC ACID	127.600	756.00	0.439	327.55	756.00	599.35	63.94
1222	C14H30	n-TETRADECANE	86.885	692.40	0.418	279.01	692.40	526.73	47.79
1223	C14H30O	1-TETRADECANOL	129.100	741.00	0.590	310.65	741.00	560.15	56.18
1224	C14H30S	BUTYL-DECYL-SULFIDE	100.740	762.23	0.380	276.16	762.23	587.16	57.60
1225	C14H30S	DODECYL-ETHYL-SULFIDE	100.740	762.23	0.380	276.16	762.23	587.16	57.60
1226	C14H30S	HEPTYL-SULFIDE	100.740	762.23	0.380	276.16	762.23	587.16	57.60
1227	C14H30S	METHYL-TRIDECYL-SULFIDE	100.740	762.23	0.380	276.16	762.23	587.16	57.60
1228	C14H30S	PROPYL-UNDECYL-SULFIDE	100.740	762.23	0.380	276.16	762.23	587.16	57.60
1229	C14H30S	1-TETRADECANETHIOL	96.316	753.80	0.380	279.26	753.80	579.36	55.23
1230	C14H30S2	HEPTYL-DISULFIDE	103.305	765.96	0.380	235.16	765.96	593.86	58.58
1231	C14H31N	TETRADECYLAMINE	92.500	722.30	0.334	311.34	722.30	564.45	55.66
1232	C15H10N2O2	DIPHENYLMETHANE-4,4'-DIISOCYANATE	99.640	802.00	0.182	311.20	802.00	609.00	76.89
1233	C15H16O	p-CUMYLPHENOL	101.680	834.00	0.389	346.00	834.00	608.15	61.17
1234	C15H16O2	BISPHENOL A	128.771	849.00	0.422	426.15	849.00	633.65	72.18
1235	C15H18	1-PENTYLNAPHTHALENE	92.290	793.32	0.380	251.16	793.32	580.16	56.01
1236	C15H18	2-PENTYLNAPHTHALENE	92.738	797.48	0.380	269.16	797.48	583.16	56.29
1237	C15H24	n-NONYLBENZENE	88.087	741.00	0.387	249.00	741.00	555.20	51.57
1238	C15H24O	2,6-DI-tert-BUTYL-p-CRESOL	92.023	720.00	0.405	344.00	720.00	538.00	52.72
1239	C15H24O	NONYLPHENOL	110.048	757.00	0.394	279.15	757.00	581.00	61.94
1240	C15H28	1-PENTADECYNE	87.588	711.41	0.380	283.16	711.41	541.16	50.87
1241	C15H30	1-PENTADECENE	91.426	708.00	0.426	269.42	708.00	541.61	49.34
1242	C15H30	1-CYCLOPENTYLDECANE	83.774	730.64	0.380	251.03	730.64	552.54	49.00
1243	C15H30	1-CYCLOHEXYLNONANE	82.559	737.79	0.380	262.96	737.79	554.66	48.62
1244	C15H30O2	PENTADECANOIC ACID	123.149	766.00	0.416	325.68	766.00	612.05	63.17
1245	C15H32	n-PENTADECANE	91.667	706.80	0.419	283.10	706.80	543.83	49.57
1246	C15H32O	1-PENTADECANOL	127.154	722.53	0.380	317.04	722.53	578.01	68.98
1247	C15H32S	BUTYL-UNDECYL-SULFIDE	105.513	775.15	0.380	284.16	775.15	603.16	59.54
1248	C15H32S	DODECYL-PROPYL-SULFIDE	105.513	775.15	0.380	284.16	775.15	603.16	59.54

127

Table 5-1 ENTHALPY OF VAPORIZATION - ORGANIC COMPOUNDS (continued)

NO	FORMULA	NAME	HVAP = A (1 - T/T$_c$)^n				(HVAP - kjoule/mol, T - K)		
			A	T$_c$	n	TMIN	TMAX	T$_B$	HVAP @ T$_B$
1249	C15H32S	ETHYL-TRIDECYL-SULFIDE	105.513	775.15	0.380	284.16	775.15	603.16	59.54
1250	C15H32S	METHYL-TETRADECYL-SULFIDE	105.513	775.15	0.380	284.16	775.15	603.16	59.54
1251	C15H32S	1-PENTADECANETHIOL	104.736	764.77	0.380	290.93	764.77	593.86	59.27
1252	C16H10	FLUORANTHENE	84.656	905.00	0.299	383.33	905.00	655.95	57.56
1253	C16H10	PYRENE	103.700	936.00	0.464	423.81	936.00	667.95	58.05
1254	C16H12	1-PHENYLNAPHTHALENE	93.445	849.00	0.387	318.15	849.00	607.15	57.48
1255	C16H20	1-n-HEXYLNAPHTHALENE	98.061	813.00	0.418	255.15	813.00	595.15	56.55
1256	C16H22O4	DIBUTYL PHTHALATE	121.495	781.00	0.433	238.15	781.00	613.15	62.44
1257	C16H26	n-DECYLBENZENE	94.351	753.00	0.399	258.77	753.00	571.04	53.53
1258	C16H26	PENTAETHYLBENZENE	89.870	723.64	0.380	327.66	723.64	550.16	52.23
1259	C16H30	1-HEXADECYNE	92.189	724.26	0.380	288.16	724.26	557.16	52.80
1260	C16H32	n-DECYLCYCLOHEXANE	101.312	751.25	0.490	271.42	751.25	570.75	50.37
1261	C16H32	1-CYCLOPENTYLUNDECANE	88.236	743.30	0.380	263.16	743.30	568.76	50.88
1262	C16H32	1-HEXADECENE	90.707	722.00	0.375	277.51	722.00	558.02	52.03
1263	C16H32O2	n-HEXADECANOIC ACID	92.115	776.00	0.208	335.66	776.00	624.15	65.61
1264	C16H34	n-HEXADECANE	96.680	720.60	0.422	291.32	720.60	560.01	51.31
1265	C16H34O	DI-n-OCTYL ETHER	105.054	707.00	0.393	265.55	707.00	559.65	56.72
1266	C16H34O	1-HEXADECANOL	136.960	761.00	0.574	322.35	761.00	585.15	59.07
1267	C16H34S	BUTYL-DODECYL-SULFIDE	109.852	787.27	0.380	288.16	787.27	618.16	61.23
1268	C16H34S	ETHYL-TETRADECYL-SULFIDE	103.411	791.68	0.380	288.16	791.68	618.16	58.09
1269	C16H34S	METHYL-PENTADECYL-SULFIDE	109.852	787.27	0.380	288.16	787.27	618.16	61.23
1270	C16H34S	OCTYL-SULFIDE	109.852	787.27	0.380	288.16	787.27	618.16	61.23
1271	C16H34S	PROPYL-TRIDECYL-SULFIDE	109.852	787.27	0.380	288.16	787.27	618.16	61.23
1272	C16H34S	1-HEXADECANETHIOL	108.899	774.68	0.380	290.93	774.68	607.16	60.86
1273	C16H34S2	OCTYL-DISULFIDE	114.967	784.46	0.380	244.16	784.46	619.16	63.62
1274	C17H28	n-UNDECYLBENZENE	105.900	764.00	0.436	268.00	764.00	586.40	56.06
1275	C17H32	1-HEPTADECYNE	96.561	736.21	0.380	295.16	736.21	572.16	54.58
1276	C17H34	1-CYCLOPENTYLDODECANE	92.457	755.17	0.380	268.16	755.17	584.06	52.59
1277	C17H34	1-CYCLOHEXYLUNDECANE	90.927	761.74	0.380	278.96	761.74	586.26	52.05
1278	C17H34	1-HEPTADECENE	96.960	736.00	0.393	284.40	736.00	573.48	53.55
1279	C17H36	n-HEPTADECANE	102.000	733.37	0.433	295.13	733.37	575.30	52.48
1280	C17H36O	1-HEPTADECANOL	146.578	770.00	0.595	327.05	770.00	597.15	60.26
1281	C17H36S	BUTYL-TRIDECYL-SULFIDE	113.656	798.63	0.380	294.16	798.63	632.16	62.63
1282	C17H36S	ETHYL-PENTADECYL-SULFIDE	113.656	798.63	0.380	294.16	798.63	632.16	62.63
1283	C17H36S	HEXADECYL-METHYL-SULFIDE	113.656	798.63	0.380	294.16	798.63	632.16	62.63
1284	C17H36S	PROPYL-TETRADECYL-SULFIDE	113.656	798.63	0.380	294.16	798.63	632.16	62.63
1285	C17H36S	1-HEPTADECANETHIOL	112.815	786.01	0.380	300.37	786.01	621.16	62.32
1286	C18H12	CHRYSENE	100.270	979.00	0.287	531.15	979.00	714.15	68.90
1287	C18H14	m-TERPHENYL	101.267	924.85	0.377	360.00	924.85	650.00	64.09
1288	C18H14	o-TERPHENYL	93.136	890.95	0.403	329.35	890.95	609.00	58.58
1289	C18H14	p-TERPHENYL	102.942	925.95	0.401	485.00	925.95	649.15	63.43
1290	C18H15P	TRIPHENYLPHOSPHINE	106.482	1008.00	0.434	354.40	1008.00	650.15	67.93
1291	C18H15O4P	TRIPHENYL PHOSPHATE	--------	--------	------	--------	--------	--------	------
1292	C18H16N2	N,N'-DIPHENYL-p-PHENYLENEDIAMINE	128.571	906.00	0.388	409.00	906.00	688.00	73.98
1293	C18H22	2,3-DIMETHYL-2,3-DIPHENYLBUTANE	87.820	805.00	0.384	392.15	805.00	589.00	52.99
1294	C18H22O2	DICUMYL PEROXIDE	94.157	884.00	0.430	311.15	884.00	669.00	51.27
1295	C18H30	n-DODECYLBENZENE	107.892	774.26	0.429	275.93	774.26	600.76	56.80
1296	C18H30	HEXAETHYLBENZENE	97.039	734.78	0.380	401.16	734.78	571.16	54.84
1297	C18H32O2	LINOLEIC ACID	133.937	775.00	0.399	268.15	775.00	628.00	69.00
1298	C18H34	1-OCTADECYNE	100.483	747.33	0.380	300.16	747.33	586.16	56.10
1299	C18H34O2	OLEIC ACID	134.633	781.00	0.394	286.53	781.00	633.00	69.91
1300	C18H34O4	DIBUTYL SEBACATE	121.040	768.00	0.361	263.95	768.00	622.15	66.45
1301	C18H34O4	DIHEXYL ADIPATE	129.407	767.00	0.414	259.35	767.00	621.15	65.09
1302	C18H36	1-CYCOPENTYLTRIDECANE	96.704	766.47	0.380	278.16	766.47	598.56	54.31
1303	C18H36	1-CYCLOHEXYLDODECANE	94.798	772.83	0.380	285.66	772.83	600.86	53.56
1304	C18H36	1-OCTADECENE	107.900	748.00	0.430	290.76	748.00	587.97	55.60
1305	C18H36O2	STEARIC ACID	131.712	799.00	0.417	342.75	799.00	648.35	65.69
1306	C18H38	n-OCTADECANE	106.950	745.26	0.451	301.33	745.26	589.86	52.74
1307	C18H38O	DINONYL ETHER	--------	--------	------	--------	--------	--------	------
1308	C18H38O	1-OCTADECANOL	147.500	777.00	0.566	331.05	777.00	608.15	62.17
1309	C18H38S	BUTYL-TETRADECYL-SULFIDE	117.006	810.53	0.380	298.16	810.53	646.16	63.81
1310	C18H38S	ETHYL-HEXADECYL-SULFIDE	117.006	810.53	0.380	298.16	810.53	646.16	63.81
1311	C18H38S	HEPTADECYL-METHYL-SULFIDE	117.006	810.53	0.380	298.16	810.53	646.16	63.81
1312	C18H38S	NONYL-SULFIDE	117.006	810.53	0.380	298.16	810.53	646.16	63.81
1313	C18H38S	PENTADECYL-PROPYL-SULFIDE	117.006	810.53	0.380	298.16	810.53	646.16	63.81
1314	C18H38S	1-OCTADECANETHIOL	115.954	795.36	0.380	300.93	795.36	633.16	63.37
1315	C18H38S2	NONYL-DISULFIDE	120.785	802.30	0.380	252.16	802.30	642.16	65.48
1316	C19H26	1-n-NONYLNAPHTHALENE	100.781	849.00	0.391	284.15	849.00	639.00	58.37
1317	C19H32	n-TRIDECYLBENZENE	111.300	783.00	0.431	283.15	783.00	614.43	57.42
1318	C19H36	1-NONADECYNE	104.127	758.94	0.380	306.16	758.94	600.16	57.46
1319	C19H36O2	METHYL OLEATE	122.000	764.00	0.395	293.05	764.00	617.00	63.63
1320	C19H38	1-CYCLOPENTYLTETRADECANE	101.846	772.00	0.380	282.00	772.00	612.16	55.98
1321	C19H38	1-CYCLOHEXYLTRIDECANE	98.756	783.38	0.380	291.66	783.38	614.66	55.10
1322	C19H38	1-NONADECENE	114.172	760.00	0.436	296.55	760.00	602.17	57.54
1323	C19H38O2	NONADECANOIC ACID	132.575	810.00	0.416	341.23	810.00	659.15	65.89
1324	C19H40	n-NONADECANE	114.743	755.93	0.448	305.33	755.93	603.05	56.07
1325	C19H40O	1-NONADECANOL	133.621	775.30	0.380	334.87	775.30	631.00	70.53
1326	C19H40S	BUTYL-PENTADECYL-SULFIDE	119.524	821.75	0.380	303.16	821.75	659.16	64.58

128

Table 5-1 ENTHALPY OF VAPORIZATION - ORGANIC COMPOUNDS (continued)

NO	FORMULA	NAME	$HVAP = A(1 - T/T_c)^n$					(HVAP - kjoule/mol, T - K)	
			A	T_c	n	TMIN	TMAX	T_B	HVAP @ T_B
1327	C19H40S	ETHYL-HEPTADECYL-SULFIDE	119.524	821.75	0.380	303.16	821.75	659.16	64.58
1328	C19H40S	HEXADECYL-PROPYL-SULFIDE	119.524	821.75	0.380	303.16	821.75	659.16	64.58
1329	C19H40S	METHYL-OCTADECYL-SULFIDE	119.524	821.75	0.380	303.16	821.75	659.16	64.58
1330	C19H40S	1-NONADECANETHIOL	118.517	805.29	0.380	307.04	805.29	645.16	64.15
1331	C20H16	TRIPHENYLETHYLENE	105.867	908.00	0.388	342.15	908.00	669.00	63.07
1332	C20H28	1-n-DECYLNAPHTHALENE	103.972	859.00	0.390	288.15	859.00	652.00	59.69
1333	C20H30O2	ABIETIC ACID	144.830	832.00	0.456	446.65	832.00	649.70	72.48
1334	C20H31N	DEHYDROABIETYLAMINE	112.506	863.00	0.390	317.65	863.00	660.00	63.98
1335	C20H34	1-PHENYLTETRADECANE	104.834	792.00	0.380	289.16	792.00	627.16	57.74
1336	C20H38	1-EICOSYNE	107.181	769.79	0.380	309.16	769.79	613.16	58.53
1337	C20H40	1-CYCLOPENTYLPENTADECANE	106.540	780.00	0.380	290.00	780.00	625.00	57.66
1338	C20H40	1-CYCLOHEXYLTETRADECANE	102.631	792.82	0.380	297.16	792.82	627.16	56.61
1339	C20H40	1-EICOSENE	121.918	771.00	0.469	301.76	771.00	615.54	57.53
1340	C20H42	n-EICOSANE	115.000	767.04	0.409	309.59	767.04	616.93	59.01
1341	C20H42O	1-EICOSANOL	165.548	792.00	0.593	338.55	792.00	629.15	64.80
1342	C20H42S	BUTYL-HEXADECYL-SULFIDE	121.163	832.33	0.380	308.16	832.33	671.16	64.93
1343	C20H42S	DECYL-SULFIDE	121.163	832.33	0.380	308.16	832.33	671.16	64.93
1344	C20H42S	ETHYL-OCTADECYL-SULFIDE	121.163	832.33	0.380	308.16	832.33	671.16	64.93
1345	C20H42S	HEPTADECYL-PROPYL-SULFIDE	121.163	832.33	0.380	308.16	832.33	671.16	64.93
1346	C20H42S	METHYL-NONADECYL-SULFIDE	121.163	832.33	0.380	308.16	832.33	671.16	64.93
1347	C20H42S	1-EICOSANETHIOL	120.123	814.57	0.380	310.37	814.57	656.16	64.48
1348	C20H42S2	DECYL-DISULFIDE	123.517	820.08	0.380	259.16	820.08	663.16	65.89
1349	C21H21O4P	TRI-o-CRESYL PHOSPHATE	--------	--------	-----	--------	--------	--------	-----
1350	C21H36	1-PHENYLPENTADECANE	109.302	800.00	0.380	295.16	800.00	639.16	59.41
1351	C21H42	1-CYCLOPENTYLHEXADECANE	109.198	797.25	0.380	294.16	797.25	637.16	59.33
1352	C21H42	1-CYCLOHEXYLPENTADECANE	106.244	803.46	0.380	302.16	803.46	640.16	57.99
1353	C22H38	1-PHENYLHEXADECANE	113.890	808.00	0.380	300.16	808.00	651.16	61.09
1354	C22H44	1-CYCLOHEXYLHEXADECANE	109.730	813.42	0.380	306.76	813.42	652.16	59.33
1355	C22H44O2	n-BUTYL STEARATE	129.614	764.00	0.472	299.45	764.00	623.15	58.35
1356	C24H38O4	DIISOOCTYL PHTHALATE	144.123	851.00	0.418	223.00	851.00	694.00	71.11
1357	C24H38O4	DIOCTYL PHTHALATE	145.786	806.00	0.468	298.00	806.00	657.15	66.13
1358	C24H42O	DINONYLPHENOL	153.288	886.00	0.432	350.00	886.00	722.00	73.97
1359	C26H20	TETRAPHENYLETHYLENE	126.414	996.00	0.378	496.15	996.00	760.00	73.35
1360	C28H46O4	DIISODECYL PHTHALATE	142.256	887.00	0.410	350.00	887.00	723.00	71.20

HVAP - enthalpy of vaporization, kjoule/mol

A, T_c, and n - regression coefficients of chemical compound

T - temperature, K

TMAX - maximum temperature, K

TMIN - minimun temperature, K

T_B - boiling temperature, K

Table 5-2 ENTHALPY OF VAPORIZATION - INORGANIC COMPOUNDS

NO	FORMULA	NAME	HVAP = A (1 - T/T$_c$)^n			(HVAP - kjoule/mol, T - K)			
			A	T$_c$	n	TMIN	TMAX	T$_B$	HVAP @ T$_B$
1	Ag	SILVER	313.400	7480.00	0.420	1234.00	6410.00	2485.00	264.51
2	AgCl	SILVER CHLORIDE	272.448	2992.10	0.380	728.15	2992.10	1837.15	189.75
3	AgI	SILVER IODIDE	261.492	2897.64	0.380	825.15	2897.64	1779.15	182.12
4	Al	ALUMINUM	352.914	7151.00	0.380	933.00	7151.00	2329.15	303.83
5	AlB3H12	ALUMINUM BOROHYDRIDE	50.277	513.77	0.380	209.15	513.77	319.05	34.77
6	AlBr3	ALUMINUN BROMIDE	75.053	763.00	0.380	390.15	763.00	529.45	47.86
7	AlCl3	ALUMINUM CHLORIDE	39.790	629.00	0.380	465.75	629.00	-------	------
8	AlF3	ALUMINUM FLUORIDE	267.338	2948.13	0.380	1313.15	2948.13	1810.15	186.19
9	AlI3	ALUMINUM IODIDE	90.067	983.00	0.380	464.15	983.00	658.65	59.10
10	Al2O3	ALUMINUM OXIDE	865.697	5335.00	0.380	2325.00	5335.00	3253.15	605.43
11	Al2S3O12	ALUMINUM SULFATE	---------	--------	------	--------	--------	--------	------
12	Ar	ARGON	8.729	150.86	0.352	83.78	150.86	87.28	6.44
13	As	ARSENIC	42.546	1673.15	0.380	1090.15	1673.15	885.00	31.96
14	AsBr3	ARSENIC TRIBROMIDE	67.829	789.01	0.380	306.15	789.01	493.15	46.72
15	AsCl3	ARSENIC TRICHLORIDE	51.702	654.00	0.380	255.15	654.00	403.55	35.90
16	AsF3	ARSENIC TRIFLUORIDE	48.527	530.21	0.380	267.25	530.21	329.45	33.55
17	AsF5	ARSENIC PENTAFLUORIDE	24.778	357.73	0.380	193.35	357.73	220.35	17.22
18	AsH3	ARSINE	22.170	373.00	0.356	156.23	373.00	210.67	16.49
19	AsI3	ARSENIC TRIIODIDE	99.929	1101.22	0.380	419.15	1101.22	676.15	69.60
20	As2O3	ARSENIC TRIOXIDE	122.755	1189.50	0.380	585.95	1189.50	730.35	85.50
21	At	ASTATINE	121.497	1096.37	0.380	575.15	1096.37	607.00	89.42
22	Au	GOLD	512.802	4398.00	0.380	1337.33	4398.00	3120.00	320.62
23	B	BORON	698.871	7934.59	0.380	2348.15	7934.59	4133.00	528.40
24	BBr3	BORON TRIBROMIDE	45.860	581.00	0.380	228.15	581.00	364.85	31.50
25	BCl3	BORON TRICHLORIDE	36.526	451.95	0.423	166.15	451.95	285.65	23.93
26	BF3	BORON TRIFLUORIDE	26.220	260.90	0.384	146.05	260.90	173.35	17.24
27	BH2CO	BORINE CARBONYL	25.958	340.03	0.380	136.15	340.03	209.15	18.06
28	BH3O3	BORIC ACID	---------	--------	------	--------	--------	--------	------
29	B2D6	DEUTERODIBORANE	18.114	293.74	0.380	150.00	293.74	179.87	12.64
30	B2H5Br	DIBORANE HYDROBROMIDE	33.799	466.98	0.380	168.95	466.98	289.45	23.40
31	B2H6	DIBORANE	20.782	289.80	0.383	107.65	289.80	180.65	14.30
32	B3N3H6	BORINE TRIAMINE	35.506	521.20	0.380	214.95	521.20	323.75	24.55
33	B4H10	TETRABORANE	32.359	466.66	0.380	153.25	466.66	289.25	22.41
34	B5H9	PENTABORANE	34.195	568.45	0.380	226.35	568.45	331.55	24.52
35	B5H11	TETRAHYDROPENTABORANE	39.242	547.13	0.380	225.00	547.13	340.15	27.12
36	B10H14	DECABORANE	61.465	791.78	0.380	372.75	791.78	486.15	42.81
37	Ba	BARIUM	201.813	3572.13	0.380	1000.15	3572.13	1907.00	151.00
38	Be	BERYLLIUM	447.677	5199.80	0.380	1560.15	5199.80	2744.00	336.64
39	BeB2H8	BERYLLIUM BOROHYDRIDE	43.056	591.45	0.380	396.15	591.45	363.15	29.99
40	BeBr2	BERYLLIUM BROMIDE	138.967	1216.86	0.380	763.15	1216.86	793.15	93.07
41	BeCl2	BERYLLIUM CHLORIDE	154.297	1238.03	0.380	678.15	1238.03	823.15	101.84
42	BeF2	BERYLLIUM FLUORIDE	---------	--------	------	--------	--------	--------	------
43	BeI2	BERYLLIUM IODIDE	191.486	1238.03	0.380	761.15	1238.03	863.15	121.61
44	Bi	BISMUTH	217.024	4620.00	0.380	544.15	4620.00	1698.15	182.34
45	BiBr3	BISMUTH TRIBROMIDE	103.091	1220.00	0.380	491.15	1220.00	734.15	72.66
46	BiCl3	BISMUTH TRICHLORIDE	101.319	1178.00	0.380	503.15	1178.00	714.15	71.10
47	BrF5	BROMINE PENTAFLUORIDE	49.886	470.00	0.380	211.75	470.00	313.55	32.84
48	Br2	BROMINE	41.279	584.15	0.380	265.90	584.15	331.90	30.00
49	C	CARBON	1201.893	6810.00	0.380	4203.00	6810.00	4203.00	834.46
50	CCl2O	PHOSGENE	35.600	455.00	0.378	145.37	455.00	280.71	24.78
51	CF2O	CARBONYL FLUORIDE	29.060	297.00	0.466	161.89	297.00	188.58	18.16
52	CH4N2O	UREA	87.864	705.00	0.380	298.10	705.00	465.00	58.34
53	CH4N2S	THIOUREA	81.130	854.00	0.397	454.15	854.00	536.00	54.82
54	CNBr	CYANOGEN BROMIDE	44.682	545.03	0.380	331.15	545.03	334.65	31.12
55	CNCl	CYANOGEN CHLORIDE	38.090	449.00	0.345	266.65	449.00	286.00	26.86
56	CNF	CYANOGEN FLUORIDE	31.776	368.51	0.380	217.00	368.51	227.17	22.08
57	CO	CARBON MONOXIDE	8.003	132.92	0.318	68.15	132.92	81.70	5.91
58	COS	CARBONYL SULFIDE	25.700	378.80	0.363	134.30	378.80	223.00	18.61
59	COSe	CARBON OXYSELENIDE	36.298	406.58	0.380	160.00	406.58	251.25	25.18
60	CO2	CARBON DIOXIDE	18.260	304.19	0.240	216.58	304.19	-------	------
61	CS2	CARBON DISULFIDE	34.980	552.00	0.299	161.11	552.00	319.37	27.03
62	CSeS	CARBON SELENOSULFIDE	50.449	576.53	0.380	197.95	576.53	358.75	34.85
63	C2N2	CYANOGEN	34.938	399.90	0.380	238.75	399.90	252.15	23.93
64	C3S2	CARBON SUBSULFIDE	---------	--------	------	--------	--------	--------	------
65	Ca	CALCIUM	200.376	3292.23	0.380	1115.15	3292.23	1762.00	149.76
66	CaF2	CALCIUM FLUORIDE	308.700	4570.85	0.380	2806.50	4570.85	2806.50	215.00
67	CbF5	COLUMBIUM FLUORIDE	65.558	811.32	0.380	348.65	811.32	498.15	45.66
68	Cd	CADMIUM	125.610	2291.00	0.380	594.05	2291.00	1043.15	99.71
69	CdCl2	CADMIUM CHLORIDE	209.868	2019.79	0.380	841.15	2019.79	1240.15	146.17
70	CdF2	CADMIUM FLUORIDE	383.450	3296.66	0.380	793.15	3296.66	2024.15	267.06
71	CdI2	CADMIUM IODIDE	174.386	1741.29	0.380	658.15	1741.29	1069.15	121.45
72	CdO	CADMIUM OXIDE	339.546	2983.96	0.380	1732.00	2983.96	1832.15	236.48
73	ClF	CHLORINE MONOFLUORIDE	23.664	282.32	0.380	128.15	282.32	172.65	16.52
74	ClFO3	PERCHLORYL FLUORIDE	27.743	368.40	0.380	125.41	368.40	226.49	19.32
75	ClF3	CHLORINE TRIFLUORIDE	40.167	459.39	0.380	190.15	459.39	284.65	27.82
76	ClF5	CHLORINE PENTAFLUORIDE	33.185	415.90	0.380	200.00	415.90	260.05	22.85
77	ClHO3S	CHLOROSULFONIC ACID	61.560	700.00	0.443	193.15	700.00	427.00	40.55
78	ClHO4	PERCHLORIC ACID	38.207	631.00	0.380	273.15	631.00	385.00	26.71

130

Table 5-2 ENTHALPY OF VAPORIZATION - INORGANIC COMPOUNDS (continued)

NO	FORMULA	NAME	HVAP = A (1 - T/T$_c$)^n			(HVAP - kjoule/mol, T - K)			
			A	T$_C$	n	TMIN	TMAX	T$_B$	HVAP @ T$_B$
79	ClO2	CHLORINE DIOXIDE	37.810	465.00	0.383	213.55	465.00	284.05	26.34
80	Cl2	CHLORINE	28.560	417.15	0.401	172.12	417.15	239.12	20.29
81	Cl2O	CHLORINE MONOXIDE	38.352	444.68	0.380	157.15	444.68	275.35	26.57
82	Cl2O7	CHLORINE HEPTOXIDE	43.805	565.78	0.380	182.15	565.78	351.95	30.27
83	Co	COBALT	471.837	7398.48	0.380	1768.15	7398.48	2528.00	402.53
84	CoCl2	COBALT CHLORIDE	227.449	2154.97	0.380	1008.15	2154.97	1323.15	158.41
85	CoNC3O4	COBALT NITROSYL TRICARBONYL	42.898	567.68	0.380	262.15	567.68	353.15	29.64
86	Cr	CHROMIUM	409.383	8560.93	0.380	2180.15	8560.93	2840.00	351.24
87	CrC6O6	CHROMIUM CARBONYL	53.005	690.80	0.380	423.65	690.80	424.15	36.92
88	CrO2Cl2	CHROMIUM OXYCHLORIDE	51.533	626.33	0.380	176.65	626.33	390.25	35.57
89	Cs	CESIUM	83.560	2048.10	0.380	301.65	2048.10	963.15	65.64
90	CsBr	CESIUM BROMIDE	281.655	2562.13	0.380	909.15	2562.13	1573.15	196.16
91	CsCl	CESIUM CHLORIDE	281.655	2562.13	0.380	919.15	2562.13	1573.15	196.16
92	CsF	CESIUM FLUORIDE	270.893	2482.33	0.380	956.15	2482.33	1524.15	188.67
93	CsI	CESIUM IODIDE	277.255	2529.56	0.380	894.15	2529.56	1553.15	193.10
94	Cu	COPPER	414.111	5123.00	0.380	1357.77	5123.00	3150.00	288.17
95	CuBr	CUPROUS BROMIDE	293.811	2651.71	0.380	777.15	2651.71	1628.15	204.63
96	CuCl	CUPROUS CHLORIDE	104.543	2435.00	0.380	1079.50	2435.00	1763.15	64.09
97	CuCl2	CUPRIC CHLORIDE	230.440	2010.00	0.380	906.15	2010.00	1266.15	157.95
98	CuI	COPPER IODIDE	289.603	2620.77	0.380	878.15	2620.77	1609.15	201.70
99	DCN	DEUTERIUM CYANIDE	47.148	482.63	0.380	261.15	482.63	299.35	32.63
100	D2	DEUTERIUM	1.657	38.35	0.352	18.73	38.35	23.65	1.18
101	D2O	DEUTERIUM OXIDE	59.672	643.89	0.405	276.97	643.89	374.55	41.91
102	Eu	EUROPIUM	208.558	5150.00	0.380	1095.15	5150.00	1742.00	178.27
103	F2	FLUORINE	10.549	144.31	0.546	53.48	144.31	84.95	6.50
104	F2O	FLUORINE OXIDE	14.337	215.10	0.380	49.25	215.10	128.55	10.14
105	Fe	IRON	408.070	9340.00	0.380	1808.10	9340.00	3000.00	352.26
106	FeC5O5	IRON PENTACARBONYL	41.259	607.20	0.380	252.15	607.20	378.15	28.49
107	FeCl2	FERROUS CHLORIDE	222.343	2115.88	0.380	945.15	2115.88	1299.15	154.86
108	FeCl3	FERRIC CHLORIDE	82.151	964.41	0.380	577.15	964.41	592.15	57.22
109	Fr	FRANCIUM	182.446	1606.46	0.380	300.15	1606.46	879.00	135.02
110	Ga	GALLIUM	301.858	7620.00	0.380	302.91	7620.00	2517.00	259.20
111	GaCl3	GALLIUM TRICHLORIDE	62.570	694.00	0.307	350.90	694.00	474.15	43.96
112	Gd	GADOLINIUM	---------	--------	------	--------	--------	--------	------
113	Ge	GERMANIUM	398.319	8400.00	0.380	1211.40	8400.00	3125.00	333.77
114	GeBr4	GERMANIUM BROMIDE	55.305	740.00	0.380	299.25	740.00	462.15	38.12
115	GeCl4	GERMANIUM CHLORIDE	40.642	574.00	0.380	223.65	574.00	357.15	28.08
116	GeHCl3	TRICHLORO GERMANE	41.877	559.78	0.380	202.05	559.78	348.15	28.94
117	GeH4	GERMANE	18.100	308.00	0.246	107.26	308.00	185.00	14.45
118	Ge2H6	DIGERMANE	36.542	491.01	0.380	164.15	491.01	304.65	25.29
119	Ge3H8	TRIGERMANE	46.671	616.37	0.380	167.55	616.37	383.95	32.22
120	HBr	HYDROGEN BROMIDE	25.300	363.15	0.401	185.15	363.15	206.45	18.06
121	HCN	HYDROGEN CYANIDE	41.447	456.65	0.387	259.83	456.65	298.85	27.48
122	HCl	HYDROGEN CHLORIDE	30.540	324.65	0.647	158.97	324.65	188.15	17.43
123	HF	HYDROGEN FLUORIDE	41.668	461.15	0.380	189.79	461.15	292.67	28.42
124	HI	HYDROGEN IODIDE	45.080	423.85	0.810	222.38	423.85	237.55	23.16
125	HNO3	NITRIC ACID	70.600	520.00	0.693	293.15	520.00	356.15	31.71
126	H2	HYDROGEN	0.659	33.18	0.380	13.95	33.18	20.39	0.46
127	H2O	WATER	52.053	647.13	0.321	273.16	647.13	373.15	39.50
128	H2O2	HYDROGEN PEROXIDE	61.900	730.15	0.325	272.74	730.15	423.35	46.72
129	H2S	HYDROGEN SULFIDE	21.230	373.53	0.219	187.68	373.53	212.80	17.65
130	H2SO4	SULFURIC ACID	50.115	925.00	0.380	599.10	925.00	610.00	33.28
131	H2S2	HYDROGEN DISULFIDE	49.804	542.39	0.380	183.45	542.39	337.15	34.43
132	H2Se	HYDROGEN SELENIDE	26.961	411.10	0.380	209.15	411.10	232.05	19.66
133	H2Te	HYDROGEN TELLURIDE	37.265	438.04	0.380	224.15	438.04	271.15	25.83
134	H3NO3S	SULFAMIC ACID	---------	--------	------	--------	--------	--------	------
135	He	HELIUM-3	---------	--------	------	--------	--------	--------	------
136	He	HELIUM-4	0.154	5.20	0.380	1.76	5.20	4.22	0.08
137	Hf	HAFNIUM	738.298	21688.00	0.380	2506.15	21688.00	5960.00	653.44
138	Hg	MERCURY	69.200	1735.00	0.355	234.31	1735.00	629.73	58.96
139	HgBr2	MERCURIC BROMIDE	82.151	964.41	0.380	510.15	964.41	592.15	57.22
140	HgCl2	MERCURIC CHLORIDE	79.459	939.98	0.380	550.15	939.98	577.15	55.34
141	HgI2	MERCURIC IODIDE	81.785	1078.10	0.380	532.15	1078.10	627.15	58.73
142	IF7	IODINE HEPTAFLUORIDE	31.639	447.53	0.380	278.65	447.53	277.15	21.92
143	I2	IODINE	55.400	819.15	0.351	386.75	819.15	458.39	41.54
144	In	INDIUM	266.609	6730.00	0.380	429.75	6730.00	2323.00	226.99
145	Ir	IRIDIUM	642.920	15035.00	0.380	2719.15	15035.00	4450.00	562.65
146	K	POTASSIUM	98.753	2223.00	0.380	336.35	2223.00	1037.00	77.78
147	KBr	POTASSIUM BROMIDE	300.028	2697.31	0.380	1003.15	2697.31	1656.15	208.96
148	KCl	POTASSIUM CHLORIDE	186.282	3470.00	0.380	1044.00	3470.00	1688.87	144.58
149	KF	POTASSIUM FLUORIDE	326.668	2891.12	0.380	1153.15	2891.12	1775.15	227.51
150	KI	POTASSIUM IODIDE	286.950	2601.22	0.380	996.15	2601.22	1597.15	199.85
151	KOH	POTASSIUM HYDROXIDE	142.670	2605.86	0.380	1596.00	2605.86	1600.00	99.37
152	Kr	KRYPTON	12.553	209.35	0.380	115.78	209.35	119.80	9.09
153	La	LANTHANUM	482.826	9511.00	0.380	1193.15	9511.00	3643.00	401.87
154	Li	LITHIUM	163.204	4085.00	0.380	453.69	4085.00	1597.00	135.18
155	LiBr	LITHIUM BROMIDE	283.860	2578.42	0.380	820.15	2578.42	1583.15	197.70
156	LiCl	LITHIUM CHLORIDE	299.806	2695.68	0.380	887.15	2695.68	1655.15	208.81

Table 5-2 ENTHALPY OF VAPORIZATION - INORGANIC COMPOUNDS (continued)

			HVAP = A (1 - T/T$_c$)^n				(HVAP - kjoule/mol, T - K)		
NO	FORMULA	NAME	A	T$_c$	n	TMIN	TMAX	T$_B$	HVAP @ T$_B$
157	LiF	LITHIUM FLUORIDE	367.352	3182.65	0.380	1143.15	3182.65	1954.15	255.85
158	LiI	LITHIUM IODIDE	253.462	2352.04	0.380	719.15	2352.04	1444.15	176.53
159	Lu	LUTECIUM	984.839	4128.66	0.380	1936.15	4128.66	2535.00	685.91
160	Mg	MAGNESIUM	183.744	2241.04	0.380	923.15	2241.04	1376.00	127.97
161	MgCl2	MAGNESIUM CHLORIDE	307.828	2754.32	0.380	985.15	2754.32	1691.15	214.39
162	MgO	MAGNESIUM OXIDE	449.150	5950.00	0.317	3105.00	5950.00	3873.20	321.62
163	Mn	MANGANESE	254.916	6902.82	0.380	1519.15	6902.82	2392.00	216.86
164	MnCl2	MANGANESE CHLORIDE	257.586	2382.98	0.380	923.15	2382.98	1463.15	179.40
165	Mo	MOLYBDENUM	777.758	9620.00	0.380	2895.15	9620.00	5081.15	584.62
166	MoF6	MOLYBDENUM FLUORIDE	38.092	498.12	0.380	290.15	498.12	309.15	26.36
167	MoO3	MOLYBDENUM OXIDE	249.133	2319.46	0.380	1068.15	2319.46	1424.15	173.51
168	NCl3	NITROGEN TRICHLORIDE	43.544	564.00	0.380	246.15	564.00	344.15	30.44
169	ND3	HEAVY AMMONIA	38.309	388.40	0.380	199.15	388.40	239.75	26.60
170	NF3	NITROGEN TRIFLUORIDE	15.920	233.85	0.344	66.36	233.85	144.09	11.45
171	NH3	AMMONIA	31.523	405.65	0.364	195.41	405.65	239.72	22.77
172	NH3O	HYDROXYLAMINE	83.487	574.00	0.380	306.25	574.00	383.00	54.96
173	NH4Br	AMMONIUM BROMIDE	51.730	1089.82	0.380	793.00	1089.82	-------	------
174	NH4Cl	AMMONIUM CHLORIDE	65.837	882.00	0.380	793.20	882.00	-------	------
175	NH4I	AMMONIUM IODIDE	52.787	1104.32	0.380	793.00	1104.32	-------	------
176	NH5O	AMMONIUM HYDROXIDE	--------	--------	------	--------	--------	-------	------
177	NH5S	AMMONIUM HYDROGENSULFIDE	13.818	499.10	0.380	391.15	499.10	306.45	9.62
178	NO	NITRIC OXIDE	22.740	180.15	0.456	109.50	180.15	121.38	13.64
179	NOCl	NITROSYL CHLORIDE	36.720	440.65	0.382	213.55	440.65	267.77	25.69
180	NOF	NITROSYL FLUORIDE	33.495	352.67	0.380	139.15	352.67	217.15	23.29
181	NO2	NITROGEN DIOXIDE	38.116	431.35	0.380	294.30	431.35	294.00	24.67
182	N2	NITROGEN	9.430	126.10	0.533	63.15	126.10	77.35	5.68
183	N2F4	TETRAFLUOROHYDRAZINE	17.720	309.35	0.176	111.65	309.35	198.95	14.78
184	N2H4	HYDRAZINE	59.794	653.15	0.471	274.68	653.15	386.65	39.20
185	N2H4C	AMMONIUM CYANIDE	47.584	491.32	0.380	309.15	491.32	304.85	32.93
186	N2H6CO2	AMMONIUM CARBAMATE	38.049	539.82	0.380	400.00	539.82	-------	------
187	N2O	NITROUS OXIDE	26.860	309.57	0.452	182.30	309.57	184.67	17.82
188	N2O3	NITROGEN TRIOXIDE	58.576	425.00	0.380	246.00	425.00	275.15	39.42
189	N2O4	NITROGEN TETRAOXIDE	28.911	431.15	0.380	294.25	431.15	302.22	18.27
190	N2O5	NITROGEN PENTOXIDE	87.000	515.51	0.380	263.00	515.51	320.15	60.17
191	Na	SODIUM	111.698	2573.00	0.380	370.98	2573.00	1156.00	89.04
192	NaBr	SODIUM BROMIDE	160.750	4287.00	0.380	1720.00	4287.00	1663.82	133.38
193	NaCN	SODIUM CYANIDE	154.810	2900.00	0.380	1773.10	2900.00	1769.15	108.24
194	NaCl	SODIUM CHLORIDE	170.710	3400.00	0.380	1738.00	3400.00	1738.15	130.05
195	NaF	SODIUM FLUORIDE	176.150	5530.00	0.380	2060.00	5530.00	1982.72	148.80
196	NaI	SODIUM IODIDE	309.814	2568.65	0.380	924.15	2568.65	1577.15	215.78
197	NaOH	SODIUM HYDROXIDE	281.536	2820.00	0.380	596.00	2820.00	1663.15	200.67
198	Na2SO4	SODIUM SULFATE	--------	--------	------	--------	--------	-------	------
199	Nb	NIOBIUM	789.261	17904.10	0.380	2750.15	17904.10	5115.00	694.54
200	Nd	NEODYMIUM	327.250	10665.10	0.380	1289.15	10665.10	3384.00	283.07
201	Ne	NEON	2.484	44.40	0.380	24.55	44.40	27.09	1.74
202	Ni	NICKEL	460.380	6986.15	0.380	1728.15	6986.15	2415.00	391.85
203	NiC4O4	NICKEL CARBONYL	33.034	508.40	0.380	248.15	508.40	315.65	22.85
204	NiF2	NICKEL FLUORIDE	380.914	3278.75	0.380	1723.15	3278.75	2013.15	265.30
205	Np	NEPTUNIUM	--------	--------	------	--------	--------	-------	------
206	O2	OXYGEN	8.040	154.58	0.201	54.35	154.58	90.17	6.74
207	O3	OZONE	16.020	261.00	0.203	80.15	261.00	161.85	13.17
208	Os	OSMIUM	724.238	16878.70	0.380	3306.15	16878.70	4880.00	636.16
209	OsOF5	OSMIUM OXIDE PENTAFLUORIDE	44.741	608.55	0.380	332.95	608.55	373.65	31.16
210	OsO4	OSMIUM TETROXIDE - YELLOW	49.537	656.60	0.380	329.15	656.60	403.15	34.50
211	OsO4	OSMIUM TETROXIDE - WHITE	49.537	656.60	0.380	315.15	656.60	403.15	34.50
212	P	PHOSPHORUS - WHITE	70.687	993.75	0.380	317.25	993.75	553.45	51.88
213	PBr3	PHOSPHORUS TRIBROMIDE	59.114	711.00	0.380	233.15	711.00	448.45	40.48
214	PCl2F3	PHOSPHORUS DICHLORIDE TRIFLUORIDE	32.135	457.02	0.380	265.15	457.02	283.15	22.26
215	PCl3	PHOSPHORUS TRICHLORIDE	44.020	563.15	0.417	221.15	563.15	349.25	29.41
216	PCl5	PHOSPHORUS PENTACHLORIDE	70.321	646.15	0.380	433.15	646.15	433.00	46.14
217	PH3	PHOSPHINE	17.680	324.75	0.275	139.37	324.75	185.41	14.01
218	PH4Br	PHOSPHONIUM BROMIDE	41.209	501.76	0.380	300.00	501.76	311.45	28.51
219	PH4Cl	PHOSPHONIUM CHLORIDE	67.013	322.30	0.380	244.65	322.30	246.15	38.73
220	PH4I	PHOSPHONIUM IODIDE	47.689	539.70	0.380	291.65	539.70	335.45	32.97
221	POCl3	PHOSPHORUS OXYCHLORIDE	38.576	602.15	0.380	298.15	602.15	378.65	26.47
222	PSBr3	PHOSPHORUS THIOBROMIDE	57.022	729.89	0.380	311.15	729.89	448.15	39.71
223	PSCl3	PHOSPHORUS THIOCHLORIDE	49.012	638.82	0.380	236.95	638.82	398.15	33.82
224	P4O6	PHOSPHORUS TRIOXIDE	56.504	714.86	0.380	295.65	714.86	446.25	38.95
225	P4O10	PHOSPHORUS PENTOXIDE	--------	--------	------	--------	--------	-------	------
226	P4S10	PHOSPHORUS PENTASULFIDE	141.452	1291.00	0.380	561.15	1291.00	787.15	98.93
227	Pb	LEAD	214.192	5400.00	0.380	600.61	5400.00	2024.00	179.18
228	PbBr2	LEAD BROMIDE	198.760	1933.47	0.380	646.15	1933.47	1187.15	138.43
229	PbCl2	LEAD CHLORIDE	207.134	1998.62	0.380	774.15	1998.62	1227.15	144.26
230	PbF2	LEAD FLUORIDE	280.114	2550.73	0.380	1128.15	2550.73	1566.15	195.09
231	PbI2	LEAD IODIDE	190.026	1865.07	0.380	675.15	1865.07	1145.15	132.35
232	PbO	LEAD OXIDE	319.920	2842.26	0.380	1163.15	2842.26	1745.15	222.82
233	PbS	LEAD SULFIDE	277.475	2531.19	0.380	1387.15	2531.19	1554.15	193.25
234	Pd	PALLADIUM	455.350	10669.10	0.380	1828.05	10669.10	3385.00	393.88

132

Table 5-2 ENTHALPY OF VAPORIZATION - INORGANIC COMPOUNDS (continued)

			HVAP = A $(1 - T/T_c)^n$			(HVAP - kjoule/mol, T - K)		
NO	FORMULA	NAME	A	T_c	n	TMIN	TMAX	T_B	HVAP @ T_B
235	Po	POLONIUM	123.170	3013.08	0.380	527.15	3013.08	1235.00	100.80
236	Pt	PLATINUM	714.481	6983.00	0.380	2041.55	6983.00	3980.00	518.47
237	Ra	RADIUM	155.862	4862.82	0.380	973.15	4862.82	1809.00	130.61
238	Rb	RUBIDIUM	87.165	2111.10	0.380	312.46	2111.10	978.00	68.81
239	RbBr	RUBIDIUM BROMIDE	293.146	2646.82	0.380	955.15	2646.82	1625.15	204.17
240	RbCl	RUBIDIUM CHLORIDE	299.584	2694.06	0.380	988.15	2694.06	1654.15	208.65
241	RbF	RUBIDIUM FLUORIDE	305.596	2738.03	0.380	1033.15	2738.03	1681.15	212.84
242	RbI	RUBIDIUM IODIDE	282.537	2568.65	0.380	915.15	2568.65	1577.15	196.78
243	Re	RHENIUM	798.859	21482.80	0.380	3459.15	21482.80	5915.00	706.84
244	Re2O7	RHENIUM HEPTOXIDE	90.026	1035.10	0.380	569.15	1035.10	635.55	62.70
245	Rh	RHODIUM	570.393	12906.60	0.380	2237.15	12906.60	3940.00	496.67
246	Rn	RADON	24.730	377.40	0.380	202.15	377.40	211.35	18.10
247	Ru	RUTHENIUM	641.164	15247.10	0.380	2607.15	15247.10	4500.00	561.37
248	RuF5	RUTHENIUM PENTAFLUORIDE	83.593	977.44	0.380	359.65	977.44	600.15	58.22
249	S	SULFUR	12.994	1313.00	0.380	388.36	1313.00	717.82	9.62
250	SF4	SULFUR TETRAFLUORIDE	31.741	364.00	0.380	149.15	364.00	233.15	21.52
251	SF6	SULFUR HEXAFLUORIDE	24.170	318.69	0.359	233.15	318.69	209.25	16.47
252	SOBr2	THIONYL BROMIDE	55.998	661.75	0.380	220.95	661.75	412.65	38.63
253	SOCl2	THIONYL CHLORIDE	41.170	567.00	0.381	172.00	567.00	348.75	28.63
254	SOF2	SULFUROUS OXYFLUORIDE	29.294	371.25	0.380	162.65	371.25	228.90	20.35
255	SO2	SULFUR DIOXIDE	46.900	430.75	0.636	197.67	430.75	263.13	25.72
256	SO2Cl2	SULFURYL CHLORIDE	40.750	545.00	0.380	222.00	545.00	342.55	27.96
257	SO3	SULFUR TRIOXIDE	73.370	490.85	0.565	289.95	490.85	317.90	40.71
258	S2Cl2	SULFUR MONOCHLORIDE	55.195	659.37	0.380	193.15	659.37	411.15	38.08
259	Sb	ANTIMONY	81.255	5070.00	0.380	903.78	5070.00	1898.00	67.99
260	SbBr3	ANTIMONY TRIBROMIDE	74.301	892.75	0.380	369.75	892.75	548.15	51.75
261	SbCl3	ANTIMONY TRICHLORIDE	90.323	794.00	0.698	346.55	794.00	493.40	45.87
262	SbCl5	ANTIMONY PENTACHLORIDE	47.225	662.54	0.380	275.95	662.54	413.15	32.58
263	SbH3	STIBINE	30.105	440.35	0.380	185.15	440.35	255.15	21.66
264	SbI3	ANTIMONY TRIIODIDE	97.134	1097.96	0.380	440.15	1097.96	674.15	67.65
265	Sb2O3	ANTIMONY TRIOXIDE	309.391	2765.72	0.380	929.15	2765.72	1698.15	215.48
266	Sc	SCANDIUM	366.939	8035.08	0.380	1814.15	8035.08	2700.00	314.06
267	Se	SELENIUM	35.428	1766.00	0.380	494.15	1766.00	930.00	26.66
268	SeCl4	SELENIUM TETRACHLORIDE	62.100	743.95	0.380	450.00	743.95	464.65	42.80
269	SeF6	SELENIUM HEXAFLUORIDE	26.418	368.80	0.380	238.45	368.80	227.35	18.35
270	SeOCl2	SELENIUM OXYCHLORIDE	63.378	706.80	0.380	281.65	706.80	441.15	43.70
271	SeO2	SELENIUM DIOXIDE	81.791	961.16	0.380	613.15	961.16	590.15	56.97
272	Si	SILICON	502.415	5159.00	0.380	1685.00	5159.00	3513.80	325.42
273	SiBrCl2F	BROMODICHLOROFLUOROSILANE	34.438	497.17	0.380	160.85	497.17	308.55	23.83
274	SiBrF3	TRIFLUOROBROMOSILANE	25.228	375.28	0.380	202.65	375.28	231.45	17.52
275	SiBr2ClF	DIBROMOCHLOROFLUOROSILANE	37.643	535.27	0.380	173.85	535.27	332.65	26.02
276	SiClF3	TRIFLUOROCHLOROSILANE	21.418	330.54	0.380	131.15	330.54	203.15	14.91
277	SiCl2F2	DICHLORODIFLUOROSILANE	25.942	390.93	0.380	133.45	390.93	241.35	18.01
278	SiCl3F	TRICHLOROFLUOROSILANE	31.390	460.49	0.380	152.35	460.49	285.35	21.74
279	SiCl4	SILICON TETRACHLORIDE	41.700	507.00	0.383	204.30	507.00	330.00	27.86
280	SiF4	SILICON TETRAFLUORIDE	14.874	259.00	0.380	186.35	259.00	178.35	9.55
281	SiHBr3	TRIBROMOSILANE	48.458	610.00	0.380	199.65	610.00	384.95	33.17
282	SiHCl3	TRICHLOROSILANE	36.130	479.00	0.351	144.95	479.00	305.00	25.33
283	SiHF3	TRIFLUOROSILANE	19.507	291.02	0.380	141.75	291.02	178.15	13.61
284	SiH2Br2	DIBROMOSILANE	43.880	550.00	0.380	202.95	550.00	343.65	30.23
285	SiH2Cl2	DICHLOROSILANE	36.242	449.00	0.369	151.15	449.00	281.45	25.20
286	SiH2F2	DIFLUOROSILANE	22.955	318.21	0.380	151.00	318.21	195.35	15.99
287	SiH2I2	DIIODOSILANE	62.336	660.00	0.380	272.15	660.00	422.65	42.26
288	SiH3Br	MONOBROMOSILANE	33.350	454.00	0.380	179.25	454.00	275.55	23.39
289	SiH3Cl	MONOCHLOROSILANE	28.475	396.65	0.380	155.05	396.65	242.75	19.87
290	SiH3F	MONOFLUOROSILANE	20.297	286.28	0.380	155.00	286.28	175.15	14.17
291	SiH3I	IODOSILANE	43.219	515.00	0.380	216.15	515.00	318.55	29.97
292	SiH4	SILANE	16.660	269.70	0.320	88.48	269.70	161.00	12.46
293	SiO2	SILICON DIOXIDE	496.135	4076.87	0.380	1883.00	4076.87	2503.20	345.54
294	Si2Cl6	HEXACHLORODISILANE	41.738	660.96	0.380	271.95	660.96	412.15	28.79
295	Si2F6	HEXAFLUORODISILANE	25.541	411.33	0.380	254.55	411.33	254.25	17.72
296	Si2H5Cl	DISILANYL CHLORIDE	36.265	506.89	0.380	310.00	506.89	314.70	25.09
297	Si2H6	DISILANE	29.270	432.00	0.351	143.85	432.00	259.00	21.23
298	Si2OCl3F3	TRICHLOROTRIFLUORODISILOXANE	30.502	508.78	0.380	310.00	508.78	315.89	21.10
299	Si2OCl6	HEXACHLORODISILOXANE	40.639	655.58	0.380	239.95	655.58	408.75	28.04
300	Si2OH6	DISILOXANE	27.884	416.86	0.380	128.95	416.86	257.75	19.34
301	Si3Cl8	OCTACHLOROTRISILANE	45.917	775.41	0.380	450.00	775.41	484.55	31.63
302	Si3H8	TRISILANE	34.749	525.15	0.380	155.95	525.15	326.25	24.03
303	Si3H9N	TRISILAZANE	33.400	518.20	0.380	167.45	518.20	321.85	23.10
304	Si4H10	TETRASILANE	37.950	599.30	0.380	179.55	599.30	373.15	26.20
305	Sm	SAMARIUM	233.547	5082.92	0.380	1345.15	5082.92	1874.00	196.10
306	Sn	TIN	340.788	7400.00	0.380	505.08	7400.00	2995.00	279.82
307	SnBr4	STANNIC BROMIDE	56.902	764.82	0.380	304.15	764.82	477.85	39.21
308	SnCl2	STANNOUS CHLORIDE	139.643	1459.53	0.380	519.95	1459.53	896.15	97.26
309	SnCl4	STANNIC CHLORIDE	44.770	619.85	0.380	242.95	619.85	386.15	30.90
310	SnH4	STANNIC HYDRIDE	27.240	358.52	0.380	123.25	358.52	220.85	18.93
311	SnI4	STANNIC IODIDE	87.399	1011.64	0.380	417.65	1011.64	621.15	60.87
312	Sr	STRONTIUM	167.437	4267.20	0.380	1050.15	4267.20	1630.00	139.45

133

Table 5-2 ENTHALPY OF VAPORIZATION - INORGANIC COMPOUNDS (continued)

| NO | FORMULA | NAME | $HVAP = A (1 - T/T_c)^n$ | | | | (HVAP - kjoule/mol, T - K) | | |
			A	T_c	n	TMIN	TMAX	T_B	HVAP @ T_B
313	SrO	STRONTIUM OXIDE	---------	--------	------	--------	-------	-------	------
314	Ta	TANTALUM	856.093	19900.90	0.380	3290.15	19900.90	5565.00	755.77
315	Tc	TECNNETIUM	668.703	17400.80	0.380	2430.15	17400.80	5000.00	587.93
316	Te	TELLURIUM	56.795	4840.00	0.380	722.66	4840.00	1285.00	50.51
317	TeCl4	TELLURIUM TETRACHLORIDE	95.468	1083.31	0.380	497.15	1083.31	665.15	66.49
318	TeF6	TELLURIUM HEXAFLUORIDE	24.753	380.18	0.380	235.35	380.18	234.55	17.19
319	Ti	TITANIUM	584.041	6400.00	0.380	1941.15	6400.00	3442.00	435.59
320	TiCl4	TITANIUM TETRACHLORIDE	46.800	638.00	0.287	249.00	638.00	409.00	34.88
321	Tl	THALLIUM	193.469	4648.06	0.380	577.15	4648.06	1745.00	161.78
322	TlBr	THALLOUS BROMIDE	179.096	1778.75	0.380	733.15	1778.75	1092.15	124.74
323	TlI	THALLOUS IODIDE	179.918	1785.26	0.380	713.15	1785.26	1096.15	125.31
324	Tm	THULIUM	291.499	6283.16	0.380	1818.15	6283.16	2219.15	247.02
325	U	URANIUM	482.118	13712.60	0.380	1408.15	13712.60	4135.00	420.65
326	UF6	URANIUM FLUORIDE	44.980	505.80	0.380	342.35	505.80	328.85	30.18
327	V	VANADIUM	527.947	11787.10	0.380	2183.15	11787.10	3665.00	458.28
328	VCl4	VANADIUM TETRACHLORIDE	45.149	697.00	0.245	247.45	697.00	425.00	35.86
329	VOCl3	VANADIUM OXYTRICHLORIDE	24.727	636.00	0.380	400.00	636.00	400.00	16.97
330	W	TUNGSTEN	971.447	14756.00	0.380	3695.15	14756.00	5645.00	808.81
331	WF6	TUNGSTEN FLUORIDE	34.775	468.56	0.380	272.65	468.56	290.45	24.08
332	Xe	XENON	17.484	289.74	0.380	161.36	289.74	165.03	12.69
333	Yb	YTTERBIUM	185.746	4365.92	0.380	1097.15	4365.92	1660.00	154.87
334	Yt	YTTRIUM	411.674	9381.32	0.380	1799.15	9381.32	3055.00	354.43
335	Zn	ZINC	137.616	3170.00	0.380	692.70	3170.00	1181.15	115.27
336	ZnCl2	ZINC CHLORIDE	161.388	1637.05	0.380	638.15	1637.05	1005.15	112.40
337	ZnF2	ZINC FLUORIDE	325.542	2882.98	0.380	1145.15	2882.98	1770.15	226.73
338	ZnO	ZINC OXIDE	539.740	8000.00	0.380	2223.00	8000.00	-------	------
339	ZnSO4	ZINC SULFATE	---------	--------	------	--------	-------	-------	------
340	Zr	ZIRCONIUM	773.798	8802.00	0.380	2128.15	8802.00	4598.00	584.36
341	ZrBr4	ZIRCONIUM BROMIDE	89.040	1026.30	0.380	723.15	1026.30	630.15	62.01
342	ZrCl4	ZIRCONIUM CHLORIDE	84.316	983.96	0.380	710.15	983.96	604.15	58.72
343	ZrI4	ZIRCONIUM IODIDE	102.721	1146.82	0.380	772.15	1146.82	704.15	71.54

HVAP - enthalpy of vaporization, kjoule/mol

A, T_c, and n - regression coefficients of chemical compound

T - temperature, K

TMAX - maximum temperature, K

TMIN - minimun temperature, K

T_B - boiling temperature, K

Chapter 6

ENTHALPY OF FUSION

Carl L. Yaws, Deepa R. Balundgi, and Saumya Tripathi
Lamar University, Beaumont, Texas

ABSTRACT

Results for enthalpy of fusion are presented for major organic and inorganic chemicals. The results are displayed in easy-to-use tabulations that are especially applicable for rapid engineering usage with the personal computer or hand calculator.

INTRODUCTION

Physical and thermodynamic property data such as enthalpy of fusion are of special value to engineers in the chemical processing and petroleum refining industries. As an example, knowledge of the enthalpy of fusion is required in the design of heat exchangers for melting solids. In this article, results for enthalpy of fusion are presented for a wide variety of compounds.

ENTHALPY OF FUSION

The results for enthalpy of fusion are given in Tables 6-1 and 6-2. The tabulations are applicable to a wide variety of substances including hydrocarbons, oxygenates, halogenates, nitrogenates, sulfur compounds, silicon compounds, and many other chemical types. The compilation is based on screening appropriate publications for organics (1-35) and inorganics (1-78). The tabulation is arranged by chemical formula to provide ease of use in quickly locating data.

EXAMPLES

The tabulation of results maybe used for engineering problems involving enthalpy of fusion. Examples are given below.

Example 1 Determine the enthalpy of fusion of phenol (C_6H_6O) at the freezing point.

Inspection of the table discloses the following for the enthalpy of fusion at the freezing point:

ΔH_{fus} = 11.514 kjoule/mol

Example 2 Determine energy required to melt 10 mols of naphthalene ($C_{10}H_8$) at the freezing point.

Substitution of data from the table into the following equation yields

Q = (mass) (ΔH_{fus}) = (10 mols) (18.979 kjoule/mol)

Q = 189.79 kjoule

REFERENCES – ORGANIC COMPOUNDS

1-34. See **REFERENCES - ORGANIC COMPOUNDS** in Chapter 1, CRITICAL PROPERTIES AND ACENTRIC FACTOR
35. Howard, P. H. and W. M. Meylan, eds., HANDBOOK OF PHYSICAL PROPERTIES OF ORGANIC CHEMICALS, CRC Press, Boca Raton, FL (1997).

REFERENCES - INORGANIC COMPOUNDS

1-64. See **REFERENCES - INORGANIC COMPOUNDS** in Chapter 2, HEAT CAPACITY OF GAS
65. Fink, J. K., TABLES OF THERMODYNAMIC PROPERTIES OF SODIUM, Argonne National Lab., ANL-CEN-RSD-82-4, Chemical Engineering Division, Argonne, IL (June 1982).
66. Touloukian, Y. S. and C. Y. Ho, eds., PROPERTIES OF NONMETALLIC FLUID ELEMENTS, McGraw-Hill Book Co., New York, NY (1981).
67. Pace, E. L. and M. A. Reno, J. Chem. Phys., 48(3), 1231 (1968).
68. Della Gatta, G. and D. Ferro, Thermochim. Acta, 122, 143 (1987).
69. Koehler, J. K. and W. F. Giauque, J. Amer. Chem. Soc., 80, 2659 (1958).
70. Mascherpa, G., Rev. Chim. Miner., 2, 379 (1965).
71. Stern, S. A., J.L. Mullhaupt and W. B. Kay, Chem. Rev., 60, 185 (1960).
72. Forsythe, W. R. and W. F. Giauque, J. Amer. Chem. Soc., 64, 48 (1948).
73. Giguere, P. A. and others, Can. J. Chem., 32, 117 (1954).
74. Douglas, T. B. and others, J. Amer. Chem. Soc., 77, 2144 (1955).
75. Chang, D. R. and R. A. Howald, High Temperature Science, 15, 209 (1982).
76. Pace, E. L. and J. S. Mosser, J. Chem. Phys., 19(1), 154 (1963).
77. Donaldson, R. E. and O. R. Quayle, J. Amer. Chem. Soc., 72, 35 (1950).
78. Whittaker, A. G. and D. M. Yost, J. Chem. Phys., 17(2), 188 (1949).

Table 6-1 ENTHALPY OF FUSION - ORGANIC COMPOUNDS

NO	FORMULA	NAME	MW	T_F	HFUS @ T_F		
			g/mol	K	kjoule/mol	kjoule/kg	BTU/lb
1	CBrClF2	BROMOCHLORODIFLUOROMETHANE	165.365	113.65	----------	----------	----------
2	CBrCl3	BROMOTRICHLOROMETHANE	198.273	252.15	----------	----------	----------
3	CBrF3	BROMOTRIFLUOROMETHANE	148.910	105.15	----------	----------	----------
4	CBr2F2	DIBROMODIFLUOROMETHANE	209.816	163.05	----------	----------	----------
5	CClF3	CHLOROTRIFLUOROMETHANE	104.459	92.15	----------	----------	----------
6	CClN	CYANOGEN CHLORIDE	61.470	266.65	----------	----------	----------
7	CCl2F2	DICHLORODIFLUOROMETHANE	120.913	115.15	----------	----------	----------
8	CCl2O	PHOSGENE	98.916	145.37	5.738	58.009	24.944
9	CCl3F	TRICHLOROFLUOROMETHANE	137.368	162.04	6.897	50.208	21.590
10	CCl4	CARBON TETRACHLORIDE	153.822	250.33	2.535	16.480	7.086
11	CF2O	CARBONYL FLUORIDE	66.007	161.89	6.708	101.626	43.699
12	CF4	CARBON TETRAFLUORIDE	88.005	89.56	0.712	8.090	3.479
13	CHBr3	TRIBROMOMETHANE	252.731	281.20	11.600	45.899	`19.736
14	CHClF2	CHLORODIFLUOROMETHANE	86.468	115.73	4.123	47.682	20.503
15	CHCl2F	DICHLOROFLUOROMETHANE	102.923	138.15	----------	----------	----------
16	CHCl3	CHLOROFORM	119.377	209.63	9.540	79.915	34.363
17	CHF3	TRIFLUOROMETHANE	70.014	117.97	4.059	57.974	24.929
18	CHI3	TRIIODOMETHANE	393.732	396.16	----------	----------	----------
19	CHN	HYDROGEN CYANIDE	27.026	259.91	8.406	311.034	133.745
20	CHNS	ISOTHIOCYANIC-ACID	59.086		--------	----------	----------
21	CH2BrCl	BROMOCHLOROMETHANE	129.384	185.20	----------	----------	----------
22	CH2Br2	DIBROMOMETHANE	173.835	220.60	----------	----------	----------
23	CH2ClF	CHLOROFLUOROMETHANE	68.478	140.16	----------	----------	----------
24	CH2Cl2	DICHLOROMETHANE	84.932	178.01	4.602	54.185	23.299
25	CH2F2	DIFLUOROMETHANE	52.024	137.00	----------	----------	----------
26	CH2I2	DIIODOMETHANE	267.836	279.25	12.050	44.990	19.346
27	CH2O	FORMALDEHYDE	30.026	181.15	7.050	234.797	100.962
28	CH2O2	FORMIC ACID	46.026	281.55	12.698	275.888	118.632
29	CH3Br	METHYL BROMIDE	94.939	179.55	5.979	62.977	27.080
30	CH3Cl	METHYL CHLORIDE	50.488	175.45	6.548	129.694	55.768
31	CH3Cl3Si	METHYL TRICHLOROSILANE	149.478	195.35	8.945	59.842	25.732
32	CH3F	METHYL FLUORIDE	34.033	131.35	----------	----------	----------
33	CH3I	METHYL IODIDE	141.939	206.70	----------	----------	----------
34	CH3NO	FORMAMIDE	45.041	275.70	7.980	177.172	76.184
35	CH3NO2	NITROMETHANE	61.040	244.60	9.703	158.961	68.353
36	CH3NO2	METHYL-NITRITE	61.040	256.16	----------	----------	----------
37	CH3NO3	METHYL-NITRATE	77.040	190.86	----------	----------	----------
38	CH4	METHANE	16.043	90.67	0.941	58.655	25.222
39	CH4Cl2Si	METHYL DICHLOROSILANE	115.034	182.55	9.870	85.801	36.894
40	CH4O	METHANOL	32.042	175.47	3.205	100.025	43.011
41	CH4O3S	METHANESULFONIC ACID	96.107	292.81	----------	----------	----------
42	CH4S	METHYL MERCAPTAN	48.109	150.18	5.904	122.721	52.770
43	CH5ClSi	METHYL CHLOROSILANE	80.589	139.05	----------	----------	----------
44	CH5N	METHYLAMINE	31.057	179.69	6.134	197.508	84.928
45	CH6Si	METHYL SILANE	46.144	116.34	----------	----------	----------
46	CN4O8	TETRANITROMETHANE	196.033	287.05	----------	----------	----------
47	CO	CARBON MONOXIDE	28.010	68.15	0.841	30.025	12.911
48	COS	CARBONYL SULFIDE	60.076	134.35	4.728	78.700	33.841
49	CO2	CARBON DIOXIDE	44.010	216.58	9.019	204.931	88.120
50	CS2	CARBON DISULFIDE	76.143	161.58	4.393	57.694	24.808
51	C2BrF3	BROMOTRIFLUOROETHYLENE	160.921	--------	----------	----------	----------
52	C2Br2F4	1,2-DIBROMOTETRAFLUOROETHANE	259.824	162.65	----------	----------	----------
53	C2ClF3	CHLOROTRIFLUOROETHYLENE	116.470	115.00	5.553	47.678	20.501
54	C2ClF5	CHLOROPENTAFLUOROETHANE	154.467	173.71	1.878	12.158	5.228
55	C2Cl2F4	1,2-DICHLOROTETRAFLUOROETHANE	170.921	179.15	----------	----------	----------
56	C2Cl3F3	1,1,2-TRICHLOROTRIFLUOROETHANE	187.375	238.15	----------	----------	----------
57	C2Cl4	TETRACHLOROETHYLENE	165.833	250.80	10.460	63.076	27.122
58	C2Cl4F2	1,1,2,2-TETRACHLORODIFLUOROETHANE	203.830	299.15	----------	----------	----------
59	C2Cl4O	TRICHLOROACETYL CHLORIDE	181.832	--------	----------	----------	----------
60	C2Cl6	HEXACHLOROETHANE	236.738	459.95	9.749	41.181	17.708
61	C2F4	TETRAFLUOROETHYLENE	100.016	142.00	7.715	77.138	33.169
62	C2F6	HEXAFLUOROETHANE	138.012	172.45	2.686	19.462	8.369
63	C2HBrClF3	HALOTHANE	197.382	--------	----------	----------	----------
64	C2HClF2	2-CHLORO-1,1-DIFLUOROETHYLENE	98.479	134.65	----------	----------	----------
65	C2HCl3	TRICHLOROETHYLENE	131.388	188.40	----------	----------	----------
66	C2HCl3O	DICHLOROACETYL CHLORIDE	147.387	--------	----------	----------	----------
67	C2HCl3O	TRICHLOROACETALDEHYDE	147.387	216.00	----------	----------	----------
68	C2HCl5	PENTACHLOROETHANE	202.293	244.15	11.340	56.057	24.105
69	C2HF3	TRIFLUOROETHENE	82.025	94.53	----------	----------	----------
70	C2HF3O2	TRIFLUOROACETIC ACID	114.024	257.90	----------	----------	----------
71	C2HF5	PENTAFLUOROETHANE	120.022	170.15	----------	----------	----------
72	C2H2	ACETYLENE	26.038	192.40	3.766	144.635	62.193
73	C2H2Br4	1,1,2,2-TETRABROMOETHANE	345.654	273.15	----------	----------	----------
74	C2H2Cl2	1,1-DICHLOROETHYLENE	96.943	150.65	6.515	67.204	28.898
75	C2H2Cl2	cis-1,2-DICHLOROETHYLENE	96.943	193.15	7.205	74.322	31.958
76	C2H2Cl2	trans-1,2-DICHLOROETHYLENE	96.943	223.35	11.983	123.609	53.152
77	C2H2Cl2O	CHLOROACETYL CHLORIDE	112.943	251.15	----------	----------	----------
78	C2H2Cl2O	DICHLOROACETALDEHYDE	112.943	223.00	----------	----------	----------

136

Table 6-1 ENTHALPY OF FUSION - ORGANIC COMPOUNDS (continued)

NO	FORMULA	NAME	MW g/mol	T$_F$ K	HFUS @ T$_F$ kjoule/mol	kjoule/kg	BTU/lb
79	C2H2Cl2O2	DICHLOROACETIC ACID	128.942	286.55	---------	---------	---------
80	C2H2Cl3F	1,1,1-TRICHLOROFLUOROETHANE	151.394	--------	---------	---------	---------
81	C2H2Cl4	1,1,1,2-TETRACHLOROETHANE	167.849	202.94	---------	---------	---------
82	C2H2Cl4	1,1,2,2-TETRACHLOROETHANE	167.849	229.35	9.172	54.644	23.497
83	C2H2F2	1,1-DIFLUOROETHYLENE	64.035	129.15	---------	---------	---------
84	C2H2F2	cis-1,2-DIFLUOROETHENE	64.035	107.90	---------	---------	---------
85	C2H2F2	trans-1,2-DIFLUOROETHENE	64.035	107.90	---------	---------	---------
86	C2H2F4	1,1,1,2-TETRAFLUOROETHANE	102.031	172.15	---------	---------	---------
87	C2H2O	KETENE	42.037,	122.00	---------	---------	---------
88	C2H2O4	OXALIC ACID	90.036	462.65	---------	---------	---------
89	C2H3Br	VINYL BROMIDE	106.950	135.35	5.504	51.463	22.129
90	C2H3Cl	VINYL CHLORIDE	62.499	119.36	4.744	75.905	32.639
91	C2H3ClF2	1-CHLORO-1,1-DIFLUOROETHANE	100.495	142.35	---------	---------	---------
92	C2H3ClO	ACETYL CHLORIDE	78.498	160.30	275.310	3507.223	1508.106
93	C2H3ClO	CHLOROACETALDEHYDE	78.498	--------	---------	---------	---------
94	C2H3ClO2	CHLOROACETIC ACID	94.497	333.15	12.285	130.004	55.902
95	C2H3ClO2	METHYL CHLOROFORMATE	94.497	--------	---------	---------	---------
96	C2H3Cl3	1,1,1-TRICHLOROETHANE	133.404	242.75	2.350	17.616	7.575
97	C2H3Cl3	1,1,2-TRICHLOROETHANE	133.404	236.50	11.297	84.683	36.414
98	C2H3F	VINYL FLUORIDE	46.044	112.65	---------	---------	---------
99	C2H3F3	1,1,1-TRIFLUOROETHANE	84.041	161.85	6.192	73.678	31.682
100	C2H3N	ACETONITRILE	41.053	229.32	8.912	217.085	93.347
101	C2H3NO	METHYL ISOCYANATE	57.052	256.15	---------	---------	---------
102	C2H4	ETHYLENE	28.054	104.01	3.351	119.448	51.363
103	C2H4Br2	1,1-DIBROMOETHANE	187.862	210.15	10.630	56.584	24.331
104	C2H4Br2	1,2-DIBROMOETHANE	187.862	282.94	10.962	58.351	25.091
105	C2H4Cl2	1,1-DICHLOROETHANE	98.959	176.19	7.870	79.528	34.197
106	C2H4Cl2	1,2-DICHLOROETHANE	98.959	237.49	8.828	89.209	38.360
107	C2H4Cl2O	BIS(CHLOROMETHYL)ETHER	114.959	231.65	---------	---------	---------
108	C2H4F2	1,1-DIFLUOROETHANE	66.051	156.15	---------	---------	---------
109	C2H4F2	1,2-DIFLUOROETHANE	66.051	--------	---------	---------	---------
110	C2H4I2	1,2-DIIODOETHANE	281.863	356.16	---------	---------	---------
111	C2H4O	ACETALDEHYDE	44.053	150.15	3.222	73.139	31.450
112	C2H4O	ETHYLENE OXIDE	44.053	161.45	5.171	117.381	50.474
113	C2H4OS	THIOACETIC-ACID	76.113	150.16	---------	---------	---------
114	C2H4O2	ACETIC ACID	60.053	289.81	11.715	195.078	83.883
115	C2H4O2	METHYL FORMATE	60.053	174.15	7.531	125.406	53.925
116	C2H4S	THIACYCLOPROPANE	60.114	165.37	---------	---------	---------
117	C2H5Br	BROMOETHANE	108.966	154.55	5.858	53.760	23.117
118	C2H5Cl	ETHYL CHLORIDE	64.514	136.75	4.452	69.008	29.674
119	C2H5ClO	2-CHLOROETHANOL	80.514	205.65	---------	---------	---------
120	C2H5F	ETHYL FLUORIDE	48.060	129.95	---------	---------	---------
121	C2H5I	ETHYL IODIDE	155.966	162.05	---------	---------	---------
122	C2H5N	ETHYLENEIMINE	43.068	195.20	5.870	136.296	58.607
123	C2H5NO	ACETAMIDE	59.068	354.15	15.707	265.914	114.343
124	C2H5NO	N-METHYLFORMAMIDE	59.068	269.35	14.300	242.094	104.100
125	C2H5NO2	NITROETHANE	75.067	183.63	9.853	131.256	56.440
126	C2H5NO3	ETHYL-NITRATE	91.066	178.56	---------	---------	---------
127	C2H6	ETHANE	30.070	90.35	2.859	95.078	40.884
128	C2H6AlCl	DIMETHYLALUMINUM CHLORIDE	92.054	252.15	---------	---------	---------
129	C2H6O	DIMETHYL ETHER	46.069	131.66	4.937	107.165	46.081
130	C2H6O	ETHANOL	46.069	159.05	4.931	107.035	46.025
131	C2H6OS	DIMETHYL SULFOXIDE	78.135	291.67	13.933	178.320	76.677
132	C2H6O2	ETHYLENE GLYCOL	62.068	260.15	9.958	160.437	68.988
133	C2H6O4S	DIMETHYL SULFATE	126.133	241.35	---------	---------	---------
134	C2H6S	DIMETHYL SULFIDE	62.136	174.88	7.985	128.508	55.259
135	C2H6S	ETHYL MERCAPTAN	62.136	125.26	4.975	80.066	34.429
136	C2H6S2	DIMETHYL DISULFIDE	94.202	188.44	9.192	97.578	41.958
137	C2H7N	DIMETHYLAMINE	45.084	180.96	5.941	131.776	56.664
138	C2H7N	ETHYLAMINE	45.084	192.15	9.350	207.391	89.178
139	C2H7NO	MONOETHANOLAMINE	61.084	283.65	20.496	335.538	144.281
140	C2H8N2	ETHYLENEDIAMINE	60.099	284.29	22.583	375.763	161.578
141	C2H8Si	DIMETHYL SILANE	60.171	122.93	---------	---------	---------
142	C2N2	CYANOGEN	52.036	245.25	8.109	155.834	67.009
143	C3F6	HEXAFLUOROPROPYLENE	150.023	116.65	---------	---------	---------
144	C3F6O	HEXAFLUOROACETONE	166.023	151.15	8.380	50.475	21.704
145	C3F8	OCTAFLUOROPROPANE	188.020	125.46	0.477	2.537	1.091
146	C3H2N2	MALONONITRILE	66.062	304.90	10.790	163.331	70.233
147	C3H3Cl	PROPARGYL CHLORIDE	74.510	--------	---------	---------	---------
148	C3H3N	ACRYLONITRILE	53.064	189.63	6.230	117.405	50.484
149	C3H3NO	OXAZOLE	69.063	--------	---------	---------	---------
150	C3H4	METHYLACETYLENE	40.065	170.45	---------	---------	---------
151	C3H4	PROPADIENE	40.065	136.87	---------	---------	---------
152	C3H4Cl2	2,3-DICHLOROPROPENE	110.970	191.50	---------	---------	---------
153	C3H4O	ACROLEIN	56.064	185.45	1.020	18.193	7.823
154	C3H4O	PROPARGYL ALCOHOL	56.064	221.35	---------	---------	---------
155	C3H4O2	ACRYLIC ACID	72.064	286.65	11.129	154.432	66.406
156	C3H4O2	beta-PROPIOLACTONE	72.064	239.75	9.410	130.578	56.149

137

Table 6-1 ENTHALPY OF FUSION - ORGANIC COMPOUNDS (continued)

NO	FORMULA	NAME	MW g/mol	T_F K	HFUS @ T_F kjoule/mol	kjoule/kg	BTU/lb
157	C3H4O2	VINYL FORMATE	72.064	--------	---------	---------	---------
158	C3H4O3	ETHYLENE CARBONATE	88.063	309.55	10.042	114.032	49.034
159	C3H4O3	PYRUVIC ACID	88.063	286.75	---------	---------	---------
160	C3H5Br	3-BROMO-1-PROPENE	120.977	153.76	---------	---------	---------
161	C3H5Cl	2-CHLOROPROPENE	76.525	135.75	---------	---------	---------
162	C3H5Cl	3-CHLOROPROPENE	76.525	138.65	---------	---------	---------
163	C3H5ClO	alpha-EPICHLOROHYDRIN	92.525	215.95	10.460	113.051	48.612
164	C3H5ClO2	METHYL CHLOROACETATE	108.524	241.03	11.288	104.014	44.726
165	C3H5ClO2	ETHYL CHLOROFORMATE	108.524	192.00	---------	---------	---------
166	C3H5Cl3	1,2,3-TRICHLOROPROPANE	147.431	258.45	---------	---------	---------
167	C3H5I	3-IODO-1-PROPENE	167.977	173.86	---------	---------	---------
168	C3H5N	PROPIONITRILE	55.079	180.26	5.029	91.305	39.261
169	C3H5NO	ACRYLAMIDE	71.079	357.65	---------	---------	---------
170	C3H5NO	HYDRACRYLONITRILE	71.079	227.15	7.550	106.220	45.675
171	C3H5NO	LACTONITRILE	71.079	233.00	12.800	180.081	77.435
172	C3H5N3O9	NITROGLYCERINE	227.088	286.15	26.600	117.135	50.368
173	C3H6	CYCLOPROPANE	42.081	145.73	5.439	129.251	55.578
174	C3H6	PROPYLENE	42.081	87.90	3.003	71.362	30.686
175	C3H6Br2	1,2-DIBROMOPROPANE	201.888	217.96	---------	---------	---------
176	C3H6Cl2	1,1-DICHLOROPROPANE	112.986	--------	---------	---------	---------
177	C3H6Cl2	1,2-DICHLOROPROPANE	112.986	172.71	---------	---------	---------
178	C3H6Cl2	1,3-DICHLOROPROPANE	112.987	173.65	---------	---------	---------
179	C3H6Cl2	2,2-DICHLOROPROPANE	112.986	239.36	---------	---------	---------
180	C3H6I2	1,2-DIIODOPROPANE	295.889	253.16	---------	---------	---------
181	C3H6O	ACETONE	58.080	178.45	5.691	97.986	42.134
182	C3H6O	ALLYL ALCOHOL	58.080	144.15	5.120	88.154	37.906
183	C3H6O	METHYL VINYL ETHER	58.080	151.15	5.990	103.134	44.347
184	C3H6O	n-PROPIONALDEHYDE	58.080	193.15	9.750	167.872	72.185
185	C3H6O	1,2-PROPYLENE OXIDE	58.080	161.22	6.531	112.448	48.353
186	C3H6O	1,3-PROPYLENE OXIDE	58.080	--------	---------	---------	---------
187	C3H6O2	ETHYL FORMATE	74.079	193.55	9.205	124.259	53.431
188	C3H6O2	METHYL ACETATE	74.079	175.15	7.970	107.588	46.263
189	C3H6O2	PROPIONIC ACID	74.079	252.45	10.660	143.900	61.877
190	C3H6O2S	3-MERCAPTOPROPIONIC ACID	106.145	290.65	16.970	159.876	68.747
191	C3H6O3	LACTIC ACID	90.079	291.15	11.350	126.001	54.180
192	C3H6O3	METHOXYACETIC ACID	90.079	281.00	10.500	116.564	50.123
193	C3H6O3	TRIOXANE	90.079	334.65	14.623	162.335	69.804
194	C3H6S	THIACYCLOBUTANE	74.140	199.96	---------	---------	---------
195	C3H7Br	1-BROMOPROPANE	122.993	163.15	6.527	53.068	22.819
196	C3H7Br	2-BROMOPROPANE	122.993	184.15	6.527	53.068	22.819
197	C3H7Cl	ISOPROPYL CHLORIDE	78.541	155.97	---------	---------	---------
198	C3H7Cl	n-PROPYL CHLORIDE	78.541	150.35	5.544	70.587	30.353
199	C3H7F	1-FLUOROPROPANE	62.087	114.16	---------	---------	---------
200	C3H7F	2-FLUOROPROPANE	62.087	139.80	---------	---------	---------
201	C3H7I	ISOPROPYL IODIDE	169.993	183.15	---------	---------	---------
202	C3H7I	n-PROPYL IODIDE	169.993	171.85	---------	---------	---------
203	C3H7N	ALLYLAMINE	57.095	184.95	9.490	166.214	71.472
204	C3H7N	PROPYLENEIMINE	57.095	229.00	---------	---------	---------
205	C3H7NO	N,N-DIMETHYLFORMAMIDE	73.095	212.72	16.154	221.000	95.030
206	C3H7NO	N-METHYLACETAMIDE	73.095	301.15	16.300	222.997	95.889
207	C3H7NO2	1-NITROPROPANE	89.094	169.16	---------	---------	---------
208	C3H7NO2	2-NITROPROPANE	89.094	181.83	---------	---------	---------
209	C3H7NO3	PROPYL-NITRATE	105.093	173.16	---------	---------	---------
210	C3H7NO3	ISOPROPYL-NITRATE	105.093	173.16	---------	---------	---------
211	C3H8	PROPANE	44.096	85.46	3.524	79.917	34.364
212	C3H8O	ISOPROPANOL	60.096	185.28	5.410	90.023	38.710
213	C3H8O	METHYL ETHYL ETHER	60.096	160.00	7.980	132.788	57.099
214	C3H8O	n-PROPANOL	60.096	146.95	5.372	89.390	38.438
215	C3H8O2	2-METHOXYETHANOL	76.095	188.05	8.330	109.468	47.071
216	C3H8O2	METHYLAL	76.095	168.35	8.330	109.468	47.071
217	C3H8O2	1,2-PROPYLENE GLYCOL	76.095	213.15	7.570	99.481	42.777
218	C3H8O2	1,3-PROPYLENE GLYCOL	76.095	246.45	12.000	157.698	67.810
219	C3H8O3	GLYCEROL	92.095	291.33	18.284	198.534	85.370
220	C3H8S	n-PROPYLMERCAPTAN	76.163	159.95	5.477	71.912	30.922
221	C3H8S	ISOPROPYL MERCAPTAN	76.163	142.61	5.736	75.312	32.384
222	C3H8S	ETHYL-METHYL-SULFIDE	76.156	167.20	---------	---------	---------
223	C3H9N	n-PROPYLAMINE	59.111	190.15	11.700	197.933	85.111
224	C3H9N	ISOPROPYLAMINE	59.111	177.95	7.325	123.919	53.285
225	C3H9N	TRIMETHYLAMINE	59.111	156.08	6.544	110.707	47.604
226	C3H9NO	1-AMINO-2-PROPANOL	75.111	274.89	13.557	180.493	77.612
227	C3H9NO	3-AMINO-1-PROPANOL	75.111	284.15	15.900	211.687	91.025
228	C3H9NO	METHYLETHANOLAMINE	75.111	268.65	8.760	116.627	50.150
229	C3H9O4P	TRIMETHYL PHOSPHATE	140.076	227.00	---------	---------	---------
230	C3H10N2	1,2-PROPANEDIAMINE	74.126	236.53	18.423	248.536	106.871
231	C3H10Si	TRIMETHYL SILANE	74.198	137.26	---------	---------	---------
232	C4Cl4S	TETRACHLOROTHIOPHENE	221.921	301.97	---------	---------	---------
233	C4Cl6	HEXACHLORO-1,3-BUTADIENE	260.760	252.15	---------	---------	---------
234	C4F8	OCTAFLUORO-2-BUTENE	200.031	138.15	---------	---------	---------

Table 6-1 ENTHALPY OF FUSION - ORGANIC COMPOUNDS (continued)

NO	FORMULA	NAME	MW g/mol	T_F K	HFUS @ T_F kjoule/mol	kjoule/kg	BTU/lb
235	C4F8	OCTAFLUOROCYCLOBUTANE	200.031	232.96	2.770	13.848	5.955
236	C4F10	DECAFLUOROBUTANE	238.028	144.95	----------	----------	----------
237	C4H2	BUTADIYNE(BIACETYLENE)	50.060	237.16	----------	----------	----------
238	C4H2O3	MALEIC ANHYDRIDE	98.058	326.00	13.550	138.184	59.419
239	C4H4	VINYLACETYLENE	52.076	--------	----------	----------	----------
240	C4H4N2	SUCCINONITRILE	80.089	331.30	3.703	46.236	19.882
241	C4H4O	FURAN	68.075	187.55	3.803	55.865	24.022
242	C4H4O2	DIKETENE	84.075	266.65	14.430	171.632	73.802
243	C4H4O3	SUCCINIC ANHYDRIDE	100.074	393.00	20.415	203.999	87.720
244	C4H4O4	FUMARIC ACID	116.073	560.15	----------	----------	----------
245	C4H4O4	MALEIC ACID	116.073	403.45	----------	----------	----------
246	C4H4S	THIOPHENE	84.142	234.94	5.086	60.445	25.992
247	C4H5Cl	CHLOROPRENE	88.536	143.15	----------	----------	----------
248	C4H5N	trans-CROTONITRILE	67.090	222.00	9.820	146.371	62.939
249	C4H5N	cis-CROTONITRILE	67.090	200.55	8.540	127.292	54.735
250	C4H5N	METHACRYLONITRILE	67.090	237.35	----------	----------	----------
251	C4H5N	PYRROLE	67.090	249.74	7.908	117.872	50.685
252	C4H5N	VINYLACETONITRILE	67.090	186.15	10.100	150.544	64.734
253	C4H5NO2	METHYL CYANOACETATE	99.089	260.08	----------	----------	----------
254	C4H6	CYCLOBUTENE	54.091	153.76	----------	----------	----------
255	C4H6	1,2-BUTADIENE	54.092	136.95	6.962	128.707	55.344
256	C4H6	1,3-BUTADIENE	54.092	164.25	7.985	147.619	63.476
257	C4H6	DIMETHYLACETYLENE	54.092	240.91	9.234	170.709	73.405
258	C4H6	ETHYLACETYLENE	54.092	147.43	6.029	111.458	47.927
259	C4H6Cl2	1,3-DICHLORO-trans-2-BUTENE	124.997	--------	----------	----------	----------
260	C4H6Cl2	1,4-DICHLORO-cis-2-BUTENE	124.997	225.15	----------	----------	----------
261	C4H6Cl2	1,4-DICHLORO-trans-2-BUTENE	124.997	274.15	----------	----------	----------
262	C4H6Cl2	3,4-DICHLORO-1-BUTENE	124.997	212.00	----------	----------	----------
263	C4H6O	trans-CROTONALDEHYDE	70.091	196.65	8.860	126.407	54.355
264	C4H6O	2,5-DIHYDROFURAN	70.091	--------	----------	----------	----------
265	C4H6O	DIVINYL ETHER	70.091	172.05	7.950	113.424	48.772
266	C4H6O	METHACROLEIN	70.091	192.15	----------	----------	----------
267	C4H6O2	2-BUTYNE-1,4-DIOL	86.090	331.00	----------	----------	----------
268	C4H6O2	gamma-BUTYROLACTONE	86.090	229.78	9.570	111.163	47.800
269	C4H6O2	cis-CROTONIC ACID	86.090	288.65	----------	----------	----------
270	C4H6O2	trans-CROTONIC ACID	86.090	344.55	12.979	150.761	64.827
271	C4H6O2	METHACRYLIC ACID	86.090	288.15	----------	----------	----------
272	C4H6O2	METHYL ACRYLATE	86.090	196.32	11.129	129.272	55.587
273	C4H6O2	VINYL ACETATE	86.090	180.35	5.370	62.377	26.822
274	C4H6O3	ACETIC ANHYDRIDE	102.090	200.15	10.502	102.870	44.234
275	C4H6O4	SUCCINIC ACID	118.089	461.15	----------	----------	----------
276	C4H6O5	DIGLYCOLIC ACID	134.089	421.15	----------	----------	----------
277	C4H6O5	MALIC ACID	134.089	403.15	----------	----------	----------
278	C4H6O6	TARTARIC ACID	150.088	479.15	----------	----------	----------
279	C4H7N	n-BUTYRONITRILE	69.106	161.25	5.021	72.656	31.242
280	C4H7N	ISOBUTYRONITRILE	69.106	201.70	10.300	149.046	64.090
281	C4H7NO	ACETONE CYANOHYDRIN	85.106	253.15	----------	----------	----------
282	C4H7NO	2-METHACRYLAMIDE	85.106	383.65	----------	----------	----------
283	C4H7NO	3-METHOXYPROPIONITRILE	85.106	210.12	15.900	186.826	80.335
284	C4H7NO	2-PYRROLIDONE	85.106	298.15	----------	----------	----------
285	C4H8	1-BUTENE	56.107	87.80	3.848	68.583	29.491
286	C4H8	cis-2-BUTENE	56.107	134.26	7.310	130.287	56.023
287	C4H8	trans-2-BUTENE	56.107	167.62	9.758	173.918	74.785
288	C4H8	CYCLOBUTANE	56.107	182.48	1.088	19.392	8.338
289	C4H8	ISOBUTENE	56.107	132.81	5.931	105.709	45.455
290	C4H8Br2	1,2-DIBROMOBUTANE	215.915	207.76	----------	----------	----------
291	C4H8Br2	2,3-DIBROMOBUTANE	215.915	238.66	----------	----------	----------
292	C4H8Cl2	1,4-DICHLOROBUTANE	127.013	235.85	----------	----------	----------
293	C4H8I2	1,2-DIIODOBUTANE	309.916	279.06	----------	----------	----------
294	C4H8O	n-BUTYRALDEHYDE	72.107	176.75	11.104	153.993	66.217
295	C4H8O	ISOBUTYRALDEHYDE	72.107	208.15	12.000	166.419	71.560
296	C4H8O	1,2-EPOXYBUTANE	72.107	123.15	----------	----------	----------
297	C4H8O	METHYL ETHYL KETONE	72.107	186.48	8.439	117.034	50.325
298	C4H8O	ETHYL VINYL ETHER	72.107	157.35	1.850	25.656	11.032
299	C4H8O	TETRAHYDROFURAN	72.107	164.65	8.540	118.435	50.927
300	C4H8O2	cis-2-BUTENE-1,4-DIOL	88.106	284.15	14.500	164.574	70.767
301	C4H8O2	trans-2-BUTENE-1,4-DIOL	88.106	300.45	14.500	164.574	70.767
302	C4H8O2	ISOBUTYRIC ACID	88.106	227.15	5.021	56.988	24.505
303	C4H8O2	n-BUTYRIC ACID	88.106	267.95	11.590	131.546	56.565
304	C4H8O2	1,4-DIOXANE	88.106	284.95	12.845	145.790	62.690
305	C4H8O2	ETHYL ACETATE	88.106	189.60	10.480	118.948	51.147
306	C4H8O2	METHYL PROPIONATE	88.106	185.65	10.100	114.635	49.293
307	C4H8O2	n-PROPYL FORMATE	88.106	180.25	13.200	149.820	64.422
308	C4H8O2S	SULFOLANE	120.172	300.75	1.373	11.425	4.913
309	C4H8S	TETRAHYDROTHIOPHENE	88.173	176.99	7.351	83.370	35.849
310	C4H9Br	1-BROMOBUTANE	137.019	160.75	6.694	48.855	21.007
311	C4H9Br	2-BROMOBUTANE	137.019	161.25	----------	----------	----------
312	C4H9Cl	n-BUTYL CHLORIDE	92.568	150.05	----------	----------	----------

Table 6-1 ENTHALPY OF FUSION - ORGANIC COMPOUNDS (continued)

NO	FORMULA	NAME	MW	T$_F$	HFUS @ T$_F$		
			g/mol	K	kjoule/mol	kjoule/kg	BTU/lb
313	C4H9Cl	sec-BUTYL CHLORIDE	92.568	141.85	---------	---------	---------
314	C4H9Cl	tert-BUTYL CHLORIDE	92.568	247.75	2.092	22.600	9.718
315	C4H9I	2-IODO-2-METHYLPROPANE	184.020	234.96	---------	---------	---------
316	C4H9N	PYRROLIDINE	71.122	215.31	8.577	120.596	51.856
317	C4H9NO	N,N-DIMETHYLACETAMIDE	87.122	253.15	10.418	119.579	51.419
318	C4H9NO	MORPHOLINE	87.122	270.05	9.200	105.599	45.408
319	C4H9NO2	1-NITROBUTANE	103.121	191.83	---------	---------	---------
320	C4H9NO2	2-NITROBUTANE	103.121	141.16	---------	---------	---------
321	C4H10	n-BUTANE	58.123	134.86	4.661	80.192	34.483
322	C4H10	ISOBUTANE	58.123	113.54	4.540	78.110	33.587
323	C4H10N2	PIPERAZINE	86.137	379.15	11.700	135.830	58.407
324	C4H10O	n-BUTANOL	74.123	183.85	9.372	126.438	54.369
325	C4H10O	sec-BUTANOL	74.123	158.45	5.971	80.555	34.639
326	C4H10O	tert-BUTANOL	74.123	298.97	6.694	90.309	38.833
327	C4H10O	DIETHYL ETHER	74.123	156.85	7.190	97.001	41.710
328	C4H10O	METHYL-PROPYL-ETHER	74.122	156.87	---------	---------	---------
329	C4H10O	METHYL ISOPROPYL ETHER	74.123	127.93	5.850	78.923	33.937
330	C4H10O	ISOBUTANOL	74.123	165.15	6.322	85.291	36.675
331	C4H10O2	1,3-BUTANEDIOL	90.122	196.15	8.320	92.319	39.697
332	C4H10O2	1,4-BUTANEDIOL	90.122	293.05	16.300	180.866	77.772
333	C4H10O2	2,3-BUTANEDIOL	90.122	280.75	11.300	125.386	53.916
334	C4H10O2	t-BUTYL HYDROPEROXIDE	90.122	277.45	---------	---------	---------
335	C4H10O2	1,2-DIMETHOXYETHANE	90.122	215.15	12.556	139.322	59.909
336	C4H10O2	2-ETHOXYETHANOL	90.122	--------			
337	C4H10O3	DIETHYLENE GLYCOL	106.122	262.70	16.400	154.539	66.452
338	C4H10O4S	DIETHYL SULFATE	154.187	248.00	---------	---------	---------
339	C4H10S	n-BUTYL MERCAPTAN	90.189	157.46	10.460	115.979	49.871
340	C4H10S	ISOBUTYL MERCAPTAN	90.189	128.31	4.982	55.240	23.753
341	C4H10S	sec-BUTYL MERCAPTAN	90.189	133.02	6.477	71.816	30.881
342	C4H10S	tert-BUTYL MERCAPTAN	90.189	274.26	2.482	27.520	11.834
343	C4H10S	DIETHYL SULFIDE	90.189	169.20	11.904	131.989	56.755
344	C4H10S	ISOPROPYL-METHYL-SULFIDE	90.183	171.65	---------	---------	---------
345	C4H10S	METHYL-PROPYL-SULFIDE	90.183	160.19	---------	---------	---------
346	C4H10S2	DIETHYL DISULFIDE	122.255	171.63	---------	---------	---------
347	C4H11N	n-BUTYLAMINE	73.138	224.05	14.800	202.357	87.014
348	C4H11N	ISOBUTYLAMINE	73.138	188.55	9.990	136.591	58.734
349	C4H11N	sec-BUTYLAMINE	73.138	168.65	8.930	122.098	52.502
350	C4H11N	tert-BUTYLAMINE	73.138	206.19	0.882	12.059	5.186
351	C4H11N	DIETHYLAMINE	73.138	223.35	11.400	155.870	67.024
352	C4H11NO	DIMETHYLETHANOLAMINE	89.137	214.15	---------	---------	---------
353	C4H11NO2	DIETHANOLAMINE	105.137	301.15	25.104	238.774	102.673
354	C4H11NO2	2-AMINOETHOXYETHANOL	105.137	--------			
355	C4H12N2O	N-AMINOETHYL ETHANOLAMINE	104.152	--------	---------	---------	---------
356	C4H12Si	TETRAMETHYLSILANE	88.225	174.07	6.874	77.914	33.503
357	C4H13N3	DIETHYLENE TRIAMINE	103.167	234.15	19.700	190.953	82.110
358	C5Cl6	HEXACHLOROCYCLOPENTADIENE	272.771	284.49	11.402	41.801	17.974
359	C5H4O2	FURFURAL	96.086	236.65	14.351	149.356	64.223
360	C5H5N	PYRIDINE	79.101	231.53	8.280	104.676	45.011
361	C5H6	CYCLOPENTADIENE	66.103	188.15	---------	---------	---------
362	C5H6	2-METHYL-1-BUTENE-3-YNE	66.103	160.15	---------	---------	---------
363	C5H6	1-PENTENE-3-YNE	66.103	--------	---------	---------	---------
364	C5H6	1-PENTENE-4-YNE	66.103	--------	---------	---------	---------
365	C5H6N2	GLUTARONITRILE	94.116	244.21	12.580	133.665	57.476
366	C5H6O2	FURFURYL ALCOHOL	98.101	258.52	13.131	133.852	57.556
367	C5H6O3	GLUTARIC ANHYDRIDE	114.101	328.00	---------	---------	---------
368	C5H6O4	CITRACONIC ACID	130.100	356.15	---------	---------	---------
369	C5H6O4	ITACONIC ACID	130.100	438.75	---------	---------	---------
370	C5H6S	2-METHYLTHIOPHENE	98.162	209.77	---------	---------	---------
371	C5H6S	3-METHYLTHIOPHENE	98.162	204.18	---------	---------	---------
372	C5H7N	N-METHYLPYRROLE	81.117	216.91	0.941	11.601	4.988
373	C5H7NO2	ETHYL CYANOACETATE	113.116	250.65	---------	---------	---------
374	C5H8	CYCLOPENTENE	68.118	138.13	3.364	49.385	21.236
375	C5H8	ISOPRENE	68.118	127.27	4.925	72.301	31.089
376	C5H8	3-METHYL-1,2-BUTADIENE	68.118	159.53	7.956	116.797	50.223
377	C5H8	2-METHYL-1,3-BUTADIENE	68.118	127.20	---------	---------	---------
378	C5H8	1,2-PENTADIENE	68.118	135.89	---------	---------	---------
379	C5H8	cis-1,3-PENTADIENE	68.118	132.35	5.639	82.783	35.597
380	C5H8	trans-1,3-PENTADIENE	68.118	185.71	7.144	104.877	45.097
381	C5H8	1,4-PENTADIENE	68.118	124.86	6.073	89.154	38.336
382	C5H8	2,3-PENTADIENE	68.118	147.50	---------	---------	---------
383	C5H8	1-PENTYNE	68.118	167.45	---------	---------	---------
384	C5H8	2-PENTYNE	68.118	163.86	---------	---------	---------
385	C5H8	3-METHYL-1-BUTYNE	68.118	183.45	---------	---------	---------
386	C5H8	SPIROPENTANE	68.118	166.11	---------	---------	---------
387	C5H8N4O12	PENTAERYTHRITOL TETRANITRATE	316.138	413.65	97.900	309.675	133.160
388	C5H8O	CYCLOPENTANONE	84.118	221.85	---------	---------	---------
389	C5H8O	METHYL ISOPROPENYL KETONE	84.118	219.55	---------	---------	---------
390	C5H8O2	ACETYLACETONE	100.117	249.65	14.497	144.801	62.264

140

Table 6-1 ENTHALPY OF FUSION - ORGANIC COMPOUNDS (continued)

NO	FORMULA	NAME	MW g/mol	T_F K	HFUS @ T_F		
					kjoule/mol	kjoule/kg	BTU/lb
391	C5H8O2	ALLYL ACETATE	100.117	138.00	----------	----------	----------
392	C5H8O2	ETHYL ACRYLATE	100.117	201.95	----------	----------	----------
393	C5H8O2	METHYL METHACRYLATE	100.117	224.95	14.435	144.181	61.998
394	C5H8O2	VINYL PROPIONATE	100.117	--------	----------	----------	----------
395	C5H8O3	2-HYDROXYETHYL ACRYLATE	116.117	213.00	----------	----------	----------
396	C5H8O3	LEVULINIC ACID	116.117	308.15	9.220	79.403	34.143
397	C5H8O3	METHYL ACETOACETATE	116.117	193.15	----------	----------	----------
398	C5H8O4	GLUTARIC ACID	132.116	370.65	20.676	156.499	67.294
399	C5H9N	VALERONITRILE	83.133	176.95	4.728	56.873	24.455
400	C5H9NO	n-BUTYL ISOCYANATE	99.133	--------	----------	----------	----------
401	C5H9NO	N-METHYL-2-PYRROLIDONE	99.133	249.15	----------	----------	----------
402	C5H9NO4	L-GLUTAMIC ACID	147.131	497.15	----------	----------	----------
403	C5H10	CYCLOPENTANE	70.134	179.31	0.609	8.683	3.734
404	C5H10	2-METHYL-1-BUTENE	70.134	135.58	7.911	112.798	48.503
405	C5H10	2-METHYL-2-BUTENE	70.134	139.39	7.598	108.335	46.584
406	C5H10	3-METHYL-1-BUTENE	70.134	104.66	5.360	76.425	32.863
407	C5H10	1-PENTENE	70.134	107.93	5.807	82.799	35.603
408	C5H10	cis-2-PENTENE	70.134	121.75	7.112	101.406	43.605
409	C5H10	trans-2-PENTENE	70.134	132.89	8.352	119.086	51.207
410	C5H10Br2	2,3-DIBROMO-2-METHYLBUTANE	229.942	288.00	----------	----------	----------
411	C5H10Cl2	1,5-DICHLOROPENTANE	141.040	200.35	----------	----------	----------
412	C5H10O	METHYL ISOPROPYL KETONE	86.134	181.15	9.343	108.471	46.642
413	C5H10O	2-PENTANONE	86.134	196.29	10.623	123.331	53.032
414	C5H10O	DIETHYL KETONE	86.134	234.18	11.594	134.604	57.880
415	C5H10O	VALERALDEHYDE	86.134	182.00	15.000	174.147	74.883
416	C5H10O2	n-BUTYL FORMATE	102.133	181.25	13.900	136.097	58.522
417	C5H10O2	ETHYL PROPIONATE	102.133	199.25	12.300	120.431	51.785
418	C5H10O2	ISOBUTYL FORMATE	102.133	177.35	11.200	109.661	47.154
419	C5H10O2	ISOPROPYL ACETATE	102.133	199.75	8.880	86.945	37.387
420	C5H10O2	n-PROPYL ACETATE	102.133	178.15	11.200	109.661	47.154
421	C5H10O2	METHYL n-BUTYRATE	102.133	187.35	11.500	112.598	48.417
422	C5H10O2	2-METHYLBUTYRIC ACID	102.133	--------	----------	----------	----------
423	C5H10O2	ISOVALERIC ACID	102.133	243.85	7.322	71.691	30.827
424	C5H10O2	VALERIC ACID	102.133	239.15	14.163	138.672	59.629
425	C5H10O2	TETRAHYDROFURFURYL ALCOHOL	102.133	--------	----------	----------	----------
426	C5H10O2S	3-METHYL SULFOLANE	134.199	273.65	----------	----------	----------
427	C5H10O3	DIETHYL CARBONATE	118.133	230.15	----------	----------	----------
428	C5H10O3	ETHYL LACTATE	118.133	247.15	----------	----------	----------
429	C5H10S	THIACYCLOHEXANE	102.194	292.14	----------	----------	----------
430	C5H10S	CYCLOPENTANETHIOL	102.194	155.39	----------	----------	----------
431	C5H11Br	1-BROMOPENTANE	151.046	185.26	----------	----------	----------
432	C5H11Cl	1-CHLOROPENTANE	106.595	174.15	----------	----------	----------
433	C5H11Cl	1-CHLORO-3-METHYLBUTANE	106.595	168.76	----------	----------	----------
434	C5H11Cl	2-CHLORO-2-METHYLBUTANE	106.595	199.66	----------	----------	----------
435	C5H11N	N-METHYLPYRROLIDINE	85.149	183.15	----------	----------	----------
436	C5H11N	PIPERIDINE	85.149	262.65	11.422	134.141	57.681
437	C5H11NO	tert-BUTYLFORMAMIDE	101.148	289.15	----------	----------	----------
438	C5H12	ISOPENTANE	72.150	113.25	5.151	71.393	30.699
439	C5H12	NEOPENTANE	72.150	256.58	3.146	43.604	18.750
440	C5H12	n-PENTANE	72.150	143.42	8.393	116.327	50.021
441	C5H12O	2,2-DIMETHYL-1-PROPANOL	88.150	327.15	12.500	141.804	60.976
442	C5H12O	tert-PENTYL-ALCOHOL	88.149	327.00	----------	----------	----------
443	C5H12O	2-METHYL-1-BUTANOL	88.150	--------	----------	----------	----------
444	C5H12O	2-METHYL-2-BUTANOL	88.150	264.35	4.456	50.550	21.737
445	C5H12O	3-METHYL-1-BUTANOL	88.150	155.95	6.610	74.986	32.244
446	C5H12O	3-METHYL-2-BUTANOL	88.150	--------	----------	----------	----------
447	C5H12O	1-PENTANOL	88.150	195.56	9.791	111.072	47.761
448	C5H12O	2-PENTANOL	88.150	200.00	8.480	96.200	41.366
449	C5H12O	3-PENTANOL	88.150	204.15	9.080	103.006	44.293
450	C5H12O	METHYL sec-BUTYL ETHER	88.150	--------	----------	----------	----------
451	C5H12O	METHYL tert-BUTYL ETHER	88.150	164.55	7.600	86.217	37.073
452	C5H120	METHYL ISOBUTYL ETHER	88.150	--------	----------	----------	----------
453	C5H120	ETHYL PROPYL ETHER	88.150	145.65	8.395	95.235	40.951
454	C5H12O2	ETHYLENE GLYCOL MONOPROPYL ETHER	104.149	183.15	11.400	109.459	47.067
455	C5H12O2	NEOPENTYL GLYCOL	104.149	400.00	21.770	209.027	89.882
456	C5H12O2	1,5-PENTANEDIOL	104.149	257.15	15.728	151.014	64.936
457	C5H12O3	2-(2-METHOXYETHOXY)ETHANOL	120.148	197.15	17.700	147.318	63.347
458	C5H12O4	PENTAERYTHRITOL	136.148	534.15	20.200	148.368	63.798
459	C5H12S	n-PENTYL MERCAPTAN	104.216	197.45	17.531	168.218	72.334
460	C5H12S	BUTYL-METHYL-SULFIDE	104.210	175.33	----------	----------	----------
461	C5H12S	ETHYL-PROPYL-SULFIDE	104.210	156.15	----------	----------	----------
462	C5H12S	2-METHYL-2-BUTANETHIOL	104.210	169.38	----------	----------	----------
463	C5H13N	n-PENTYLAMINE	87.165	218.15	17.600	201.916	86.824
464	C5H13NO2	METHYL DIETHANOLAMINE	119.164	252.15	----------	----------	----------
465	C6Cl6	HEXACHLOROBENZENE	284.782	501.70	25.520	89.612	38.533
466	C6F6	HEXAFLUOROBENZENE	186.056	278.25	11.590	62.293	26.786
467	C6H3ClN2O4	1-CHLORO-2,4-DINITROBENZENE	202.554	326.55	----------	----------	----------
468	C6H3Cl2NO2	1,2-DICHLORO-4-NITROBENZENE	192.001	315.65	----------	----------	----------

Table 6-1 ENTHALPY OF FUSION - ORGANIC COMPOUNDS (continued)

NO	FORMULA	NAME	MW g/mol	T_F K	HFUS @ T_F kjoule/mol	kjoule/kg	BTU/lb
469	C6H3Cl3	1,2,4-TRICHLOROBENZENE	181.448	290.15	15.565	85.782	36.886
470	C6H3N3O6	1,3,5-TRINITROBENZENE	213.106	398.40	15.000	70.388	30.267
471	C6H4Br2	m-DIBROMOBENZENE	235.906	266.25	13.220	56.039	24.097
472	C6H4ClNO2	m-CHLORONITROBENZENE	157.556	317.65	20.782	131.902	56.718
473	C6H4ClNO2	o-CHLORONITROBENZENE	157.556	306.15	---------	---------	---------
474	C6H4ClNO2	p-CHLORONITROBENZENE	157.556	356.65	14.101	89.498	38.484
475	C6H4Cl2	m-DICHLOROBENZENE	147.003	248.39	12.590	85.645	36.827
476	C6H4Cl2	o-DICHLOROBENZENE	147.003	256.15	12.660	86.121	37.032
477	C6H4Cl2	p-DICHLOROBENZENE	147.003	326.14	18.200	123.807	53.237
478	C6H4F2	m-DIFLUOROBENZENE	114.094	249.16	---------	---------	---------
479	C6H4F2	o-DIFLUOROBENZENE	114.094	239.16	---------	---------	---------
480	C6H4F2	p-DIFLUOROBENZENE	114.094	260.16	---------	---------	---------
481	C6H4N2O4	m-DINITROBENZENE	168.109	364.00	17.360	103.266	44.405
482	C6H4N2O4	o-DINITROBENZENE	168.109	390.08	22.840	135.864	58.422
483	C6H4N2O4	p-DINITROBENZENE	168.109	446.60	28.120	167.272	71.927
484	C6H5Br	BROMOBENZENE	157.010	242.43	10.627	67.684	29.104
485	C6H5Cl	MONOCHLOROBENZENE	112.558	227.95	9.556	84.898	36.506
486	C6H5ClO	m-CHLOROPHENOL	128.558	306.00	13.600	105.789	45.489
487	C6H5ClO	o-CHLOROPHENOL	128.558	282.00	13.502	105.027	45.161
488	C6H5ClO	p-CHLOROPHENOL	128.558	316.00	13.600	105.789	45.489
489	C6H5Cl2N	3,4-DICHLOROANILINE	162.018	344.65	---------	---------	---------
490	C6H5F	FLUOROBENZENE	96.104	230.94	11.305	117.633	50.582
491	C6H5I	IODOBENZENE	204.010	241.83	9.749	47.787	20.548
492	C6H5NO2	NITROBENZENE	123.111	278.91	11.632	94.484	40.628
493	C6H6	BENZENE	78.114	278.68	9.866	126.303	54.310
494	C6H6ClN	m-CHLOROANILINE	127.573	262.75	10.152	79.578	34.219
495	C6H6ClN	o-CHLOROANILINE	127.573	481.99	11.883	93.147	40.053
496	C6H6ClN	p-CHLOROANILINE	127.573	343.05	19.840	155.519	66.873
497	C6H6N2	cis-DICYANO-1-BUTENE	106.127	249.00	---------	---------	---------
498	C6H6N2	trans-DICYANO-1-BUTENE	106.127	260.00	---------	---------	---------
499	C6H6N2	1,4-DICYANO-2-BUTENE	106.127	349.00	---------	---------	---------
500	C6H6N2O2	m-NITROANILINE	138.126	387.15	23.681	171.445	73.721
501	C6H6N2O2	o-NITROANILINE	138.126	344.65	16.108	116.618	50.146
502	C6H6N2O2	p-NITROANILINE	138.126	420.65	21.087	152.665	65.646
503	C6H6O	PHENOL	94.113	314.06	11.514	122.342	52.607
504	C6H6O2	1,2-BENZENEDIOL	110.112	377.60	22.760	206.699	88.880
505	C6H6O2	1,3-BENZENEDIOL	110.112	382.00	21.000	190.715	82.007
506	C6H6O2	p-HYDROQUINONE	110.112	444.65	27.108	246.186	105.860
507	C6H6O3	1,2,3-BENZENETRIOL	126.112	407.00	9.370	74.299	31.949
508	C6H6S	PHENYL MERCAPTAN	110.180	258.26	11.450	103.921	44.686
509	C6H7N	ANILINE	93.128	267.13	10.540	113.178	48.666
510	C6H7N	2-METHYLPYRIDINE	93.128	206.44	9.724	104.415	44.899
511	C6H7N	3-METHYLPYRIDINE	93.128	255.01	14.180	152.264	65.473
512	C6H7N	4-METHYLPYRIDINE	93.128	276.73	11.573	124.270	53.436
513	C6H8	1,3-CYCLOHEXADIENE	80.130	161.00	4.205	52.477	22.565
514	C6H8	METHYLCYCLOPENTADIENE	80.130	---------	---------	---------	---------
515	C6H8N2	ADIPONITRILE	108.143	275.64	23.076	213.384	91.755
516	C6H8N2	METHYLGLUTARONITRILE	108.143	228.15	---------	---------	---------
517	C6H8N2	m-PHENYLENEDIAMINE	108.143	334.00	18.800	173.844	74.753
518	C6H8N2	o-PHENYLENEDIAMINE	108.143	376.95	22.600	208.983	89.862
519	C6H8N2	p-PHENYLENEDIAMINE	108.143	413.00	26.800	247.820	106.563
520	C6H8N2	PHENYLHYDRAZINE	108.143	292.35	16.429	151.919	65.325
521	C6H8N2O	BIS(CYANOETHYL)ETHER	124.142	246.85	17.200	138.551	59.577
522	C6H8O4	DIMETHYL MALEATE	144.127	254.15	14.640	101.577	43.678
523	C6H8O6	ASCORBIC ACID	176.126	465.15	---------	---------	---------
524	C6H8O7	CITRIC ACID	192.125	426.15	---------	---------	---------
525	C6H10	1-METHYLCYCLOPENTENE	82.145	145.96	---------	---------	---------
526	C6H10	3-METHYLCYCLOPENTENE	82.145	130.16	---------	---------	---------
527	C6H10	4-METHYLCYCLOPENTENE	82.145	112.31	---------	---------	---------
528	C6H10	CYCLOHEXENE	82.145	169.67	3.293	40.088	17.238
529	C6H10	2,3-DIMETHYL-1,3-BUTADIENE	82.145	197.15	---------	---------	---------
530	C6H10	1,5-HEXADIENE	82.145	132.47	---------	---------	---------
531	C6H10	cis,trans-2,4-HEXADIENE	82.145	177.05	---------	---------	---------
532	C6H10	trans,trans-2,4-HEXADIENE	82.145	228.25	---------	---------	---------
533	C6H10	1-HEXYNE	82.145	141.25	---------	---------	---------
534	C6H10	2-HEXYNE	82.145	183.65	---------	---------	---------
535	C6H10	3-HEXYNE	82.145	170.05	---------	---------	---------
536	C6H10O	CYCLOHEXANONE	98.145	242.00	1.190	12.125	5.214
537	C6H10O	MESITYL OXIDE	98.145	220.15	---------	---------	---------
538	C6H10O2	epsilon-CAPROLACTONE	114.144	271.85	---------	---------	---------
539	C6H10O2	ETHYL METHACRYLATE	114.144	---------	---------	---------	---------
540	C6H10O2	n-PROPYL ACRYLATE	114.144	---------	---------	---------	---------
541	C6H10O3	ETHYLACETOACETATE	130.144	234.15	---------	---------	---------
542	C6H10O3	PROPIONIC ANHYDRIDE	130.144	228.15	---------	---------	---------
543	C6H10O4	ADIPIC ACID	146.143	425.50	16.700	114.272	49.137
544	C6H10O4	DIETHYL OXALATE	146.143	232.55	---------	---------	---------
545	C6H10O4	ETHYLENE GLYCOL DIACETATE	146.143	242.15	---------	---------	---------
546	C6H10O4	ETHYLIDENE DIACETATE	146.143	292.00	---------	---------	---------

Table 6-1 ENTHALPY OF FUSION - ORGANIC COMPOUNDS (continued)

NO	FORMULA	NAME	MW g/mol	T_F K	HFUS @ T_F kjoule/mol	kjoule/kg	BTU/lb
547	C6H11N	HEXANENITRILE	97.160	192.85	16.620	171.058	73.555
548	C6H11NO	epsilon-CAPROLACTAM	113.159	342.36	16.134	142.578	61.309
549	C6H11NO	CYCLOHEXANONE OXIME	113.159	363.15	---------	---------	---------
550	C6H12	CYCLOHEXANE	84.161	279.69	2.740	32.557	13.999
551	C6H12	2,3-DIMETHYL-1-BUTENE	84.161	115.89	1.071	12.726	5.472
552	C6H12	2,3-DIMETHYL-2-BUTENE	84.161	198.82	6.452	76.663	32.965
553	C6H12	3,3-DIMETHYL-1-BUTENE	84.161	157.95	1.070	12.714	5.467
554	C6H12	2-ETHYL-1-BUTENE	84.161	141.61	7.573	89.982	38.692
555	C6H12	trans-3-METHYL-2-PENTENE	84.161	134.70	---------	---------	---------
556	C6H12	1-HEXENE	84.161	133.39	9.348	111.073	47.761
557	C6H12	cis-2-HEXENE	84.161	132.00	8.858	105.251	45.258
558	C6H12	trans-2-HEXENE	84.161	140.17	8.255	98.086	42.177
559	C6H12	cis-3-HEXENE	84.161	135.33	8.247	97.991	42.136
560	C6H12	trans-3-HEXENE	84.161	159.73	11.080	131.652	56.611
561	C6H12	METHYLCYCLOPENTANE	84.161	130.73	6.929	82.330	35.402
562	C6H12	2-METHYL-1-PENTENE	84.161	137.42	7.029	83.518	35.913
563	C6H12	2-METHYL-2-PENTENE	84.161	138.07	8.033	95.448	41.043
564	C6H12	3-METHYL-1-PENTENE	84.161	120.20	3.600	42.775	18.393
565	C6H12	3-METHYL-cis-2-PENTENE	84.161	138.31	5.920	70.341	30.247
566	C6H12	4-METHYL-1-PENTENE	84.161	119.51	3.556	42.252	18.169
567	C6H12	4-METHYL-cis-2-PENTENE	84.161	138.30	7.364	87.499	37.625
568	C6H12	4-METHYL-trans-2-PENTENE	84.161	132.35	7.155	85.016	36.557
569	C6H12N2	TRIETHYLENEDIAMINE	112.175	434.25	6.067	54.085	23.257
570	C6H12O	BUTYL VINYL ETHER	100.161	181.25	13.800	137.778	59.245
571	C6H12O	CYCLOHEXANOL	100.161	296.60	1.783	17.801	7.655
572	C6H12O	1-HEXANAL	100.161	217.15	18.900	188.696	81.139
573	C6H12O	ETHYL ISOPROPYL KETONE	100.161	--------	---------	---------	---------
574	C6H12O	2-HEXANONE	100.161	217.35	---------	---------	---------
575	C6H12O	3-HEXANONE	100.161	217.50	13.490	134.683	57.914
576	C6H12O	METHYL ISOBUTYL KETONE	100.161	189.15	---------	---------	---------
577	C6H12O2	n-PENTYL FORMATE	116.160	199.65	---------	---------	---------
578	C6H12O2	n-BUTYL ACETATE	116.160	199.65	14.400	123.967	53.306
579	C6H12O2	sec-BUTYL ACETATE	116.160	174.15	---------	---------	---------
580	C6H12O2	tert-BUTYL ACETATE	116.160	--------	---------	---------	---------
581	C6H12O2	ETHYL n-BUTYRATE	116.160	175.15	12.500	107.610	46.272
582	C6H12O2	ETHYL ISOBUTYRATE	116.160	185.00	---------	---------	---------
583	C6H12O2	ISOBUTYL ACETATE	116.160	174.30	12.400	106.749	45.902
584	C6H12O2	n-PROPYL PROPIONATE	116.160	197.25	15.000	129.132	55.527
585	C6H12O2	CYCLOHEXYL PEROXIDE	116.160	253.15	---------	---------	---------
586	C6H12O2	DIACETONE ALCOHOL	116.160	229.15	6.040	51.997	22.359
587	C6H12O2	2-ETHYL BUTYRIC ACID	116.160	258.15	---------	---------	---------
588	C6H12O2	n-HEXANOIC ACID	116.160	270.15	15.062	129.666	55.756
589	C6H12O3	2-ETHOXYETHYL ACETATE	132.159	211.45	17.200	130.146	55.963
590	C6H12O3	HYDROXYCAPROIC ACID	132.159	334.00	---------	---------	---------
591	C6H12O3	PARALDEHYDE	132.159	285.75	13.844	104.753	45.044
592	C6H12O3	sec-BUTYL GLYCOLATE	132.160	--------	---------	---------	---------
593	C6H12S	THIACYCLOHEPTANE	116.221	292.14	---------	---------	---------
594	C6H13N	CYCLOHEXYLAMINE	99.176	255.45	13.700	138.138	59.399
595	C6H13N	HEXAMETHYLENEIMINE	99.176	236.15	---------	---------	---------
596	C6H14	2,2-DIMETHYLBUTANE	86.177	174.28	0.579	6.719	2.889
597	C6H14	2,3-DIMETHYLBUTANE	86.177	145.19	0.799	9.272	3.987
598	C6H14	n-HEXANE	86.177	177.84	13.079	151.769	65.261
599	C6H14	2-METHYLPENTANE	86.177	119.55	6.268	72.734	31.276
600	C6H14	3-METHYLPENTANE	86.177	110.25	5.303	61.536	26.461
601	C6H14N2O2	LYSINE	146.189	483.00	---------	---------	---------
602	C6H14O	2-ETHYL-1-BUTANOL	102.177	158.75	8.220	80.449	34.593
603	C6H14O	1-HEXANOL	102.177	228.55	15.397	150.689	64.796
604	C6H14O	2-HEXANOL	102.177	223.00	---------	---------	---------
605	C6H14O	2-METHYL-1-PENTANOL	102.177	--------	---------	---------	---------
606	C6H14O	4-METHYL-2-PENTANOL	102.177	--------	---------	---------	---------
607	C6H14O	n-BUTYL ETHYL ETHER	102.177	170.15	12.500	122.337	52.605
608	C6H14O	DIISOPROPYL ETHER	102.177	187.65	11.046	108.107	46.486
609	C6H14O	DI-n-PROPYL ETHER	102.177	149.95	11.000	107.656	46.292
610	C6H14O	METHYL tert-PENTYL ETHER	102.177	--------	---------	---------	---------
611	C6H14O2	ACETAL	118.176	173.15	---------	---------	---------
612	C6H14O2	2-BUTOXYETHANOL	118.176	203.15	19.100	161.623	69.498
613	C6H14O2	1,6-HEXANEDIOL	118.176	315.15	26.800	226.780	97.516
614	C6H14O2	HEXYLENE GLYCOL	118.176	223.15	14.800	125.237	53.852
615	C6H14O2S	DI-n-PROPYL SULFONE	150.242	303.00	---------	---------	---------
616	C6H14O3	DIETHYLENE GLYCOL DIMETHYL ETHER	134.175	203.15	13.586	101.256	43.540
617	C6H14O3	DIPROPYLENE GLYCOL	134.175	233.00	22.700	169.182	72.748
618	C6H14O3	2-(2-ETHOXYETHOXY)ETHANOL	134.175	195.15	16.300	121.483	52.238
619	C6H14O3	TRIMETHYLOLPROPANE	134.175	331.15	24.606	183.387	78.857
620	C6H14O4	TRIETHYLENE GLYCOL	150.175	265.79	31.200	207.758	89.336
621	C6H14O6	SORBITOL	182.174	370.85	32.013	175.728	75.563
622	C6H14S	n-HEXYLMERCAPTAN	118.243	192.62	18.010	152.313	65.495
623	C6H14S	BUTYL-ETHYL-SULFIDE	118.237	178.03	---------	---------	---------
624	C6H14S	ISOPROPYL-SULFIDE	118.237	170.45	---------	---------	---------

143

Table 6-1 ENTHALPY OF FUSION - ORGANIC COMPOUNDS (continued)

NO	FORMULA	NAME	MW g/mol	T_F K	HFUS @ T_F kjoule/mol	kjoule/kg	BTU/lb
625	C6H14S	METHYL-PENTYL-SULFIDE	118.237	179.16	---------	---------	---------
626	C6H14S	PROPYL-SULFIDE	118.237	170.45	---------	---------	---------
627	C6H14S2	PROPYL-DISULFIDE	150.297	187.68	---------	---------	---------
628	C6H15Al	TRIETHYL ALUMINUN	114.167	220.65	---------	---------	---------
629	C6H15Al2Cl3	ETHYL ALUMINUM SESQUICHLORIDE	247.506	253.15	---------	---------	---------
630	C6H15N	DIISOPROPYLAMINE	101.192	176.85	6.530	64.531	27.748
631	C6H15N	DI-n-PROPYLAMINE	101.192	210.15	14.500	143.292	61.616
632	C6H15N	n-HEXYLAMINE	101.192	251.85	21.200	209.503	90.086
633	C6H15N	TRIETHYLAMINE	101.192	158.45	---------	---------	---------
634	C6H15NO	6-AMINOHEXANOL	117.191	331.00	34.900	297.804	128.056
635	C6H15NO2	DIISOPROPANOLAMINE	133.191	318.15	18.700	140.400	60.372
636	C6H15NO3	TRIETHANOLAMINE	149.190	294.35	27.191	182.258	78.371
637	C6H15N3	N-AMINOETHYL PIPERAZINE	129.205	254.15	---------	---------	---------
638	C6H15O4P	TRIETHYL PHOSPHATE	182.156	216.00	---------	---------	---------
639	C6H16N2	HEXAMETHYLENEDIAMINE	116.207	313.95	40.376	347.449	149.403
640	C6H18N3OP	HEXAMETHYL PHOSPHORAMIDE	179.202	280.15	16.945	94.558	40.660
641	C6H18N4	TRIETHYLENE TETRAMINE	146.236	285.15	33.400	228.398	98.211
642	C6H18OSi2	HEXAMETHYLDISILOXANE	162.379	204.93	11.920	73.409	31.566
643	C6H18O3Si3	HEXAMETHYLCYCLOTRISILOXANE	222.464	337.15	15.500	69.674	29.960
644	C6H19NSi2	HEXAMETHYLDISILAZANE	161.395	---------	---------	---------	---------
645	C7H3ClF3NO2	4-CHLORO-3-NITROBENZOTRIFLUORIDE	225.554	---------	---------	---------	---------
646	C7H3Cl2F3	2,4-DICHLOROBENZOTRIFLUORIDE	215.001	247.55	---------	---------	---------
647	C7H3Cl2NO	3,4-DICHLOROPHENYL ISOCYANATE	188.012	316.15	---------	---------	---------
648	C7H4ClF3	p-CHLOROBENZOTRIFLUORIDE	180.557	237.15	---------	---------	---------
649	C7H4Cl2O	m-CHLOROBENZOYL CHLORIDE	175.014	280.00	---------	---------	---------
650	C7H4F3NO2	3-NITROBENZOTRIFLUORIDE	191.110	272.00	---------	---------	---------
651	C7H5ClO	BENZOYL CHLORIDE	140.569	272.65	---------	---------	---------
652	C7H5ClO2	o-CHLOROBENZOIC ACID	156.568	415.15	25.756	164.504	70.737
653	C7H5Cl3	BENZOTRICHLORIDE	195.475	268.40	---------	---------	---------
654	C7H5F3	BENZOTRIFLUORIDE	146.112	244.14	13.782	94.325	40.560
655	C7H5N	BENZONITRILE	103.123	260.40	10.878	105.486	45.359
656	C7H5NO	PHENYL ISOCYANATE	119.123	243.15	---------	---------	---------
657	C7H5N3O6	2,4,6-TRINITROTOLUENE	227.133	354.00	21.200	93.337	40.135
658	C7H6Cl2	BENZYL DICHLORIDE	161.030	257.00	---------	---------	---------
659	C7H6Cl2	2,4-DICHLOROTOLUENE	161.030	259.65	---------	---------	---------
660	C7H6N2O4	2,4-DINITROTOLUENE	182.136	343.00	20.126	110.500	47.515
661	C7H6N2O4	2,5-DINITROTOLUENE	182.136	325.65	---------	---------	---------
662	C7H6N2O4	2,6-DINITROTOLUENE	182.136	339.00	---------	---------	---------
663	C7H6N2O4	3,4-DINITROTOLUENE	182.136	332.00	---------	---------	---------
664	C7H6N2O4	3,5-DINITROTOLUENE	182.136	365.65	---------	---------	---------
665	C7H6O	BENZALDEHYDE	106.124	247.15	9.322	87.841	37.771
666	C7H6O2	BENZOIC ACID	122.123	395.52	18.075	148.007	63.643
667	C7H6O2	p-HYDROXYBENZALDEHYDE	122.123	390.15	24.000	196.523	84.505
668	C7H6O2	SALICYLALDEHYDE	122.123	266.15	---------	---------	---------
669	C7H6O3	SALICYLIC ACID	138.123	431.75	19.585	141.794	60.971
670	C7H7Br	p-BROMOTOLUENE	171.037	299.95	14.932	87.303	37.540
671	C7H7Cl	BENZYL CHLORIDE	126.585	234.15	---------	---------	---------
672	C7H7Cl	o-CHLOROTOLUENE	126.585	236.65	8.370	66.122	28.432
673	C7H7Cl	p-CHLOROTOLUENE	126.585	280.65	12.980	102.540	44.092
674	C7H7F	p-FLUOROTOLUENE	110.131	216.36	---------	---------	---------
675	C7H7NO	FORMANILIDE	121.139	323.15	15.600	128.778	55.374
676	C7H7NO2	m-NITROTOLUENE	137.138	289.20	14.058	102.510	44.079
677	C7H7NO2	o-NITROTOLUENE	137.138	269.98	11.476	83.682	35.983
678	C7H7NO2	p-NITROTOLUENE	137.138	324.75	16.676	121.600	52.288
679	C7H7NO3	o-NITROANISOLE	153.138	283.60	---------	---------	---------
680	C7H8	TOLUENE	92.141	178.18	6.636	72.020	30.969
681	C7H8	1,3,5-CYCLOHEPTATRIENE	92.140	193.66	---------	---------	---------
682	C7H8O	ANISOLE	108.140	235.65	11.200	103.569	44.535
683	C7H8O	BENZYL ALCOHOL	108.140	257.85	8.972	82.967	35.676
684	C7H8O	m-CRESOL	108.140	285.39	10.707	99.011	42.575
685	C7H8O	o-CRESOL	108.140	304.19	15.820	146.292	62.905
686	C7H8O	p-CRESOL	108.140	307.93	12.707	117.505	50.527
687	C7H8O2	GUAIACOL	124.139	304.65	12.500	100.694	43.298
688	C7H8O2	p-METHOXYPHENOL	124.139	329.00	16.200	130.499	56.115
689	C7H9N	BENZYLAMINE	107.155	227.15	13.400	125.052	53.773
690	C7H9N	2,6-DIMETHYLPYRIDINE	107.155	267.00	10.043	93.724	40.301
691	C7H9N	N-METHYLANILINE	107.155	216.15	9.520	88.843	38.203
692	C7H9N	m-TOLUIDINE	107.155	242.75	3.891	36.312	15.614
693	C7H9N	o-TOLUIDINE	107.155	249.47	7.530	70.272	30.217
694	C7H9N	p-TOLUIDINE	107.155	316.90	18.912	176.492	75.892
695	C7H10	2-NORBORNENE	94.156	319.40	---------	---------	---------
696	C7H10N2	TOLUENEDIAMINE	122.170	371.25	19.874	162.675	69.950
697	C7H11NO	CYCLOHEXYL ISOCYANATE	125.170	---------	---------	---------	---------
698	C7H12	1-HEPTYNE	96.172	192.26	---------	---------	---------
699	C7H12O2	n-BUTYL ACRYLATE	128.171	208.55	---------	---------	---------
700	C7H12O2	ISOBUTYL ACRYLATE	128.171	212.00	---------	---------	---------
701	C7H12O2	n-PROPYL METHACRYLATE	128.171	---------	---------	---------	---------
702	C7H12O4	DIETHYL MALONATE	160.170	224.25	---------	---------	---------

144

Table 6-1 ENTHALPY OF FUSION - ORGANIC COMPOUNDS (continued)

NO	FORMULA	NAME	MW g/mol	T$_F$ K	HFUS @ T$_F$ kjoule/mol	kjoule/kg	BTU/lb
703	C7H14	CYCLOHEPTANE	98.188	265.15	1.883	19.177	8.246
704	C7H14	1,1-DIMETYLCYCLOPENTANE	98.188	203.36	1.080	10.999	4.730
705	C7H14	cis-1,2-DIMETHYLCYCLOPENTANE	98.188	219.26	1.657	16.876	7.257
706	C7H14	trans-1,2-DIMETHYLCYCLOPENTANE	98.188	155.58	7.167	72.993	31.387
707	C7H14	cis-1,3-DIMETHYLCYCLOPENTANE	98.188	139.45	7.397	75.335	32.394
708	C7H14	trans-1,3-DIMETHYLCYCLOPENTANE	98.188	139.18	0.007	0.071	0.031
709	C7H14	ETHYLCYCLOPENTANE	98.188	134.71	6.870	69.968	30.086
710	C7H14	2-ETHYL-1-PENTENE	98.188	168.00	---------	---------	---------
711	C7H14	3-ETHYL-1-PENTENE	98.188	145.67	5.020	51.126	21.984
712	C7H14	1-HEPTENE	98.188	154.27	12.401	126.299	54.308
713	C7H14	cis-2-HEPTENE	98.188	164.00	---------	---------	---------
714	C7H14	trans-2-HEPTENE	98.188	163.67	12.000	122.215	52.552
715	C7H14	cis-3-HEPTENE	98.188	136.51	---------	---------	---------
716	C7H14	trans-3-HEPTENE	98.188	136.52	10.500	106.938	45.983
717	C7H14	METHYLCYCLOHEXANE	98.188	146.58	6.751	68.756	29.565
718	C7H14	2-METHYL-1-HEXENE	98.188	170.28	13.000	132.399	56.932
719	C7H14	3-METHYL-1-HEXENE	98.188	145.00	---------	---------	---------
720	C7H14	4-METHYL-1-HEXENE	98.188	131.70	7.500	76.384	32.845
721	C7H14	2,3,3-TRIMETHYL-1-BUTENE	98.188	163.30	0.790	8.046	3.460
722	C7H14O	DIISOPROPYL KETONE	114.188	204.81	11.180	97.909	42.101
723	C7H14O	2-HEPTANONE	114.188	238.15	---------	---------	---------
724	C7H14O	1-HEPTANAL	114.188	230.15	23.585	206.545	88.814
725	C7H14O	1-METHYLCYCLOHEXANOL	114.188	299.15	7.060	61.828	26.586
726	C7H14O	cis-2-METHYLCYCLOHEXANOL	114.188	280.15	9.640	84.422	36.302
727	C7H14O	trans-2-METHYLCYCLOHEXANOL	114.188	269.15	12.500	109.469	47.071
728	C7H14O	cis-3-METHYLCYCLOHEXANOL	114.188	267.65	12.000	105.090	45.189
729	C7H14O	trans-3-METHYLCYCLOHEXANOL	114.188	272.65	12.000	105.090	45.189
730	C7H14O	cis-4-METHYLCYCLOHEXANOL	114.188	---------	---------	---------	---------
731	C7H14O	trans-4-METHYLCYCLOHEXANOL	114.188	---------	---------	---------	---------
732	C7H14O	5-METHYL-2-HEXANONE	114.188	199.25	---------	---------	---------
733	C7H14O2	n-BUTYL PROPIONATE	130.187	183.63	17.200	132.118	56.811
734	C7H14O2	ETHYL ISOVALERATE	130.187	173.85	---------	---------	---------
735	C7H14O2	ISOPENTYL ACETATE	130.187	194.65	---------	---------	---------
736	C7H14O2	n-PENTYL ACETATE	130.187	202.35	16.500	126.741	54.499
737	C7H14O2	n-PROPYL n-BUTYRATE	130.187	177.95	17.800	136.726	58.792
738	C7H14O2	n-HEPTANOIC ACID	130.187	265.83	15.437	118.576	50.988
739	C7H14O3	ETHYL-3-ETHOXYPROPIONATE	146.186	---------	---------	---------	---------
740	C7H15Br	1-BROMOHEPTANE	179.100	217.05	12.100	67.560	29.051
741	C7H15N	N-METHYLCYCLOHEXYLAMINE	113.203	264.65	11.600	102.471	44.062
742	C7H16	2,2-DIMETHYLPENTANE	100.204	149.34	5.824	58.121	24.992
743	C7H16	2,3-DIMETHYLPENTANE	100.204	---------	---------	---------	---------
744	C7H16	2,4-DIMETHYLPENTANE	100.204	153.91	6.845	68.311	29.374
745	C7H16	3,3-DIMETHYLPENTANE	100.204	138.70	7.067	70.526	30.326
746	C7H16	3-ETHYLPENTANE	100.204	154.55	9.548	95.286	40.973
747	C7H16	n-HEPTANE	100.204	182.57	14.054	140.254	60.309
748	C7H16	2-METHYLHEXANE	100.204	154.90	9.184	91.653	39.411
749	C7H16	3-METHYLHEXANE	100.204	153.75	9.460	94.407	40.595
750	C7H16	2,2,3-TRIMETHYLBUTANE	100.204	248.57	2.261	22.564	9.703
751	C7H16O	1-HEPTANOL	116.203	239.15	18.175	156.407	67.255
752	C7H16O	2-HEPTANOL	116.203	243.00	---------	---------	---------
753	C7H16O	5-METHYL-1-HEXANOL	116.203	---------	---------	---------	---------
754	C7H16O	ISOPROPYL-TERT-BUTYL-ETHER	116.203	177.80	---------	---------	---------
755	C7H16S	n-HEPTYL MERCAPTAN	132.270	229.92	25.380	191.880	82.509
756	C7H16S	BUTYL-PROPYL-SULFIDE	132.263	206.66	---------	---------	---------
757	C7H16S	ETHYL-PENTYL-SULFIDE	132.263	206.66	---------	---------	---------
758	C7H16S	HEXYL-METHYL-SULFIDE	132.263	206.66	---------	---------	---------
759	C7H17N	1-AMINOHEPTANE	115.219	254.15	29.400	255.166	109.721
760	C8H4Cl2O2	ISOPHTHALOYL CHLORIDE	203.024	317.00	---------	---------	---------
761	C8H4O3	PHTHALIC ANHYDRIDE	148.118	404.26	23.430	158.185	68.019
762	C8H6	ETHYNYLBENZENE	102.135	242.53	---------	---------	---------
763	C8H6O4	ISOPHTHALIC ACID	166.133	619.15	---------	---------	---------
764	C8H6O4	PHTHALIC ACID	166.133	464.15	---------	---------	---------
765	C8H6O4	TEREPHTHALIC ACID	166.133	700.15	---------	---------	---------
766	C8H6S	BENZOTHIOPHENE	134.202	304.50	11.827	88.128	37.895
767	C8H7N	INDOLE	117.150	273.68	9.000	76.825	33.035
768	C8H8	STYRENE	104.152	242.54	10.950	105.135	45.208
769	C8H8	1,3,5,7-CYCLOOCTATETRAENE	104.151	266.16	---------	---------	---------
770	C8H8O	ACETOPHENONE	120.151	293.65	---------	---------	---------
771	C8H8O	p-TOLUALDEHYDE	120.151	---------	---------	---------	---------
772	C8H8O2	METHYL BENZOATE	136.150	260.75	9.736	71.509	30.749
773	C8H8O2	o-TOLUIC ACID	136.150	376.85	20.167	148.123	63.693
774	C8H8O2	p-TOLUIC ACID	136.150	452.75	22.719	166.867	71.753
775	C8H8O3	METHYL SALICYLATE	152.150	265.15	---------	---------	---------
776	C8H8O3	VANILLIN	152.150	355.00	---------	---------	---------
777	C8H9NO	ACETANILIDE	135.166	386.65	21.653	160.196	68.884
778	C8H10	ETHYLBENZENE	106.167	178.20	9.184	86.505	37.197
779	C8H10	m-XYLENE	106.167	225.30	11.569	108.970	46.857
780	C8H10	o-XYLENE	106.167	247.98	13.598	128.081	55.075

Table 6-1 ENTHALPY OF FUSION - ORGANIC COMPOUNDS (continued)

NO	FORMULA	NAME	MW g/mol	T_F K	HFUS @ T_F kjoule/mol	kjoule/kg	BTU/lb
781	C8H10	p-XYLENE	106.167	286.41	17.113	161.189	69.311
782	C8H10O	m-ETHYLPHENOL	122.167	--------	---------	---------	---------
783	C8H10O	p-ETHYLPHENOL	122.167	318.23	17.000	139.154	59.836
784	C8H10O	PHENETOLE	122.167	243.63	11.600	94.952	40.829
785	C8H10O	2-PHENYLETHANOL	122.167	247.00	---------	---------	---------
786	C8H10O	2,3-XYLENOL	122.167	345.71	19.870	162.646	69.938
787	C8H10O	2,4-XYLENOL	122.167	297.68	12.840	105.102	45.194
788	C8H10O	2,5-XYLENOL	122.167	347.99	13.900	113.779	48.925
789	C8H10O	2,6-XYLENOL	122.167	318.76	15.200	124.420	53.501
790	C8H10O	3,4-XYLENOL	122.167	338.25	13.140	107.558	46.250
791	C8H10O	3,5-XYLENOL	122.167	336.59	17.410	142.510	61.279
792	C8H11N	N,N-DIMETHYLANILINE	121.182	275.60	---------	---------	---------
793	C8H11N	o-ETHYLANILINE	121.182	226.55	12.200	100.675	43.290
794	C8H11N	2,4,6-TRIMETHYLPYRIDINE	121.182	229.00	9.535	78.683	33.834
795	C8H11NO	p-PHENETIDINE	137.181	277.00	---------	---------	---------
796	C8H12	1,5-CYCLOOCTADIENE	108.183	203.98	9.828	90.846	39.064
797	C8H12	VINYLCYCLOHEXENE	108.183	164.00	---------	---------	---------
798	C8H12O4	1,4-CYCLOHEXANEDICARBOXYLIC ACID	172.181	585.65	---------	---------	---------
799	C8H12O4	DIETHYL MALEATE	172.181	264.35	---------	---------	---------
800	C8H14O2	n-BUTYL METHACRYLATE	142.198	--------	---------	---------	---------
801	C8H14O3	BUTYRIC ANHYDRIDE	158.197	199.85	---------	---------	---------
802	C8H14O4	DIETHYL SUCCINATE	174.197	252.35	---------	---------	---------
803	C8H16	CYCLOOCTANE	112.214	287.60	---------	---------	---------
804	C8H16	1,1-DIMETHYLCYCLOHEXANE	112.214	239.66	2.021	18.010	7.744
805	C8H16	cis-1,2-DIMETHYLCYCLOHEXANE	112.215	223.16	1.645	14.659	6.304
806	C8H16	trans-1,2-DIMETHYLCYCLOHEXANE	112.215	184.99	10.422	92.875	39.936
807	C8H16	cis-1,3-DIMETHYLCYCLOHEXANE	112.215	197.58	10.821	96.431	41.465
808	C8H16	trans-1,3-DIMETHYLCYCLOHEXANE	112.215	183.07	9.866	87.921	37.806
809	C8H16	cis-1,4-DIMETHYLCYCLOHEXANE	112.215	185.72	9.308	82.948	35.668
810	C8H16	trans-1,4-DIMETHYLCYCLOHEXANE	112.215	236.21	12.332	109.896	47.255
811	C8H16	ETHYLCYCLOHEXANE	112.215	161.84	8.334	74.268	31.935
812	C8H16	2-ETHYL-1-HEXENE	112.215	--------	---------	---------	---------
813	C8H16	1-METHYL-1-ETHYLCYCLOPENTANE	112.215	129.35	---------	---------	---------
814	C8H16	1-OCTENE	112.215	171.45	15.313	136.461	58.678
815	C8H16	trans-2-OCTENE	112.215	185.45	---------	---------	---------
816	C8H16	trans-3-OCTENE	112.215	163.15	---------	---------	---------
817	C8H16	trans-4-OCTENE	112.215	179.37	---------	---------	---------
818	C8H16	n-PROPYLCYCLOPENTANE	112.215	155.82	10.033	89.409	38.446
819	C8H16	2,4,4-TRIMETHYL-1-PENTENE	112.215	179.70	8.766	78.118	33.591
820	C8H16	2,4,4-TRIMETHYL-2-PENTENE	112.215	166.84	6.794	60.544	26.034
821	C8H16O	2-ETHYLHEXANAL	128.214	--------	---------	---------	---------
822	C8H16O	1-OCTANAL	128.214	246.00	26.130	203.800	87.634
823	C8H16O	2-OCTANONE	128.214	252.85	24.419	190.455	81.896
824	C8H16O2	n-BUTYL n-BUTYRATE	144.214	181.15	---------	---------	---------
825	C8H16O2	n-HEXYL ACETATE	144.214	192.25	16.900	117.187	50.390
826	C8H16O2	ISOBUTYL ISOBUTYRATE	144.214	192.45	---------	---------	---------
827	C8H16O2	n-OCTANOIC ACID	144.214	289.65	21.380	148.252	63.748
828	C8H16O4	DIETHYLENE GLYCOL ETHYL ETHER ACETATE	176.213	248.15	25.400	144.144	61.982
829	C8H18	2,2-DIMETHYLHEXANE	114.231	151.97	6.778	59.336	25.514
830	C8H18	2,3-DIMETHYLHEXANE	114.231	--------	---------	---------	---------
831	C8H18	2,4-DIMETHYLHEXANE	114.231	--------	---------	---------	---------
832	C8H18	2,5-DIMETHYLHEXANE	114.231	182.00	12.954	113.402	48.763
833	C8H18	3,3-DIMETHYLHEXANE	114.231	147.05	7.113	62.269	26.775
834	C8H18	3,4-DIMETHYLHEXANE	114.231	--------	---------	---------	---------
835	C8H18	3-ETHYLHEXANE	114.231	--------	---------	---------	---------
836	C8H18	3-ETHYL-2-METHYLPENTANE	114.230	158.20	---------	---------	---------
837	C8H18	3-METHYL-3-ETHYLPENTANE	114.231	182.28	10.800	94.545	40.654
838	C8H18	2-METHYLHEPTANE	114.231	164.16	11.878	103.982	44.712
839	C8H18	3-METHYLHEPTANE	114.231	152.60	11.627	101.785	43.768
840	C8H18	4-METHYLHEPTANE	114.231	152.20	10.837	94.869	40.794
841	C8H18	n-OCTANE	114.231	216.38	20.740	181.562	78.072
842	C8H18	2,2,3-TRIMETHYLPENTANE	114.231	160.89	8.619	75.452	32.445
843	C8H18	2,2,4-TRIMETHYLPENTANE	114.231	165.78	9.196	80.504	34.617
844	C8H18	2,3,3-TRIMETHYLPENTANE	114.231	172.22	0.858	7.511	3.230
845	C8H18	2,3,4-TRIMETHYLPENTANE	114.231	163.95	9.268	81.134	34.888
846	C8H18	2,2,3,3-TETRAMETHYLBUTANE	114.231	172.47	---------	---------	---------
847	C8H18O	DI-n-BUTYL ETHER	130.230	177.95	16.600	127.467	54.811
848	C8H18O	DI-sec-BUTYL ETHER	130.230	173.15	---------	---------	---------
849	C8H18O	DI-tert-BUTYL ETHER	130.230	195.00	8.380	64.348	27.670
850	C8H18O	2-ETHYL-1-HEXANOL	130.230	203.15	---------	---------	---------
851	C8H18O	1-OCTANOL	130.230	257.65	42.258	324.487	139.530
852	C8H18O	2-OCTANOL	130.230	241.55	19.700	151.271	65.046
853	C8H18O2	DI-t-BUTYL PEROXIDE	146.230	233.15	---------	---------	---------
854	C8H18O2S	DI-n-BUTYL SULFONE	178.296	318.00	---------	---------	---------
855	C8H18O3	DIETHYLENE GLYCOL DIETHYL ETHER	162.229	228.85	27.100	167.048	71.831
856	C8H18O3	DIETHYLENE GLYCOL MONOBUTYL EHTER	162.229	205.15	20.300	125.132	53.807
857	C8H18O4	TRIETHYLENE GLYCOL DIMETHYL ETHER	178.229	229.35	23.710	133.031	57.203
858	C8H18O5	TETRAETHYLENE GLYCOL	194.228	268.15	36.600	188.438	81.028

146

Table 6-1 ENTHALPY OF FUSION - ORGANIC COMPOUNDS (continued)

NO	FORMULA	NAME	MW g/mol	T_F K	HFUS @ T_F		
					kjoule/mol	kjoule/kg	BTU/lb
859	C8H18S	n-OCTYL MERCAPTAN	146.297	223.95	24.000	164.050	70.541
860	C8H18S	tert-OCTYL MERCAPTAN	146.297	199.00	---------	---------	---------
861	C8H18S	BUTYL-SULFIDE	146.290	209.86	---------	---------	---------
862	C8H18S	ETHYL-HEXYL-SULFIDE	146.290	209.86	---------	---------	---------
863	C8H18S	HEPTYL-METHYL-SULFIDE	146.290	209.86	---------	---------	---------
864	C8H18S	PENTYL-PROPYL-SULFIDE	146.290	209.86	---------	---------	---------
865	C8H18S2	BUTYL-DISULFIDE	178.350	202.16	---------	---------	---------
866	C8H19N	DI-n-BUTYLAMINE	129.246	211.15	19.000	147.006	63.213
867	C8H19N	DIISOBUTYLAMINE	129.246	203.15	14.000	108.321	46.578
868	C8H19N	n-OCTYLAMINE	129.246	272.75	28.600	221.283	95.152
869	C8H23N5	TETRAETHYLENEPENTAMINE	189.304	243.00	33.200	175.379	75.413
870	C8H24O4Si4	OCTAMETHYLCYCLOTETRASILOXANE	296.618	290.80	18.000	60.684	26.094
871	C9H4O5	TRIMELLITIC ANHYDRIDE	192.128	438.15	29.037	151.134	64.987
872	C9H6N2O2	TOLUENE DIISOCYANATE	174.159	287.04	---------	---------	---------
873	C9H7N	ISOQUINOLINE	129.161	299.45	13.180	102.043	43.879
874	C9H7N	QUINOLINE	129.161	258.25	10.800	83.617	35.955
875	C9H7NO	8-HYDROXYQUINOLINE	145.161	346.00	---------	---------	---------
876	C9H8	INDENE	116.163	271.70	10.200	87.808	37.757
877	C9H8O	2-METHYLBENZOFURAN	132.162	--------	---------	---------	---------
878	C9H10	INDANE	118.178	221.74	8.598	72.755	31.285
879	C9H10	cis-PROPENYLBENZENE	118.178	211.47	---------	---------	---------
880	C9H10	trans-PROPENYLBENZENE	118.178	243.82	---------	---------	---------
881	C9H1O	alpha-METHYLSTYRENE	118.178	249.95	11.924	100.899	43.386
882	C9H10	m-METHYLSTYRENE	118.178	186.81	---------	---------	---------
883	C9H10	o-METHYLSTYRENE	118.178	204.58	---------	---------	---------
884	C9H10	p-METHYLSTYRENE	118.178	239.02	---------	---------	---------
885	C9H10O2	BENZYL ACETATE	150.177	221.65	---------	---------	---------
886	C9H10O2	ETHYL BENZOATE	150.177	238.45	---------	---------	---------
887	C9H10O3	ETHYL VANILLIN	166.177	350.65	---------	---------	---------
888	C9H11NO	p-DIMETHYLAMINOBENZALDEHYDE	149.192	348.00	---------	---------	---------
889	C9H12	CUMENE	120.194	177.14	7.326	60.951	26.209
890	C9H12	m-ETHYLTOLUENE	120.194	177.61	7.615	63.356	27.243
891	C9H12	o-ETHYLTOLUENE	120.194	192.35	9.962	82.883	35.640
892	C9H12	p-ETHYLTOLUENE	120.194	210.83	13.355	111.112	47.778
893	C9H12	MESITYLENE	120.194	228.46	9.514	79.155	34.037
894	C9H12	n-PROPYLBENZENE	120.194	173.67	9.268	77.109	33.157
895	C9H12	1,2,3-TRIMETHYLBENZENE	120.194	247.79	8.180	68.057	29.264
896	C9H12	1,2,4-TRIMETHYLBENZENE	120.194	229.38	13.192	109.756	47.195
897	C9H12O	BENZYL ETHYL ETHER	136.194	275.65	15.400	113.074	48.622
898	C9H12O	2-PHENYL-2-PROPANOL	136.194	309.15	---------	---------	---------
899	C9H12O2	CUMENE HYDROPEROXIDE	152.193	264.26	---------	---------	---------
900	C9H14O	ISOPHORONE	138.210	265.05	---------	---------	---------
901	C9H14O6	GLYCERYL TRIACETATE	218.207	277.25	---------	---------	---------
902	C9H16	1-NONYNE	124.225	223.16	---------	---------	---------
903	C9H16O4	AZELAIC ACID	188.224	379.65	---------	---------	---------
904	C9H18	BUTYLCYCLOPENTANE	126.241	165.18	---------	---------	---------
905	C9H18	cis,cis-1,3,5-TRIMETHYLCYCLOHEXANE	126.241	223.46	---------	---------	---------
906	C9H18	cis,trans-1,3,5-TRIMETHYLCYCLOHEXANE	126.241	188.76	---------	---------	---------
907	C9H18	ISOPROPYLCYCLOHEXANE	126.242	183.76	---------	---------	---------
908	C9H18	1-NONENE	126.242	191.78	17.991	142.512	61.280
909	C9H18	n-PROPYLCYCLOHEXANE	126.242	178.28	10.372	82.160	35.329
910	C9H18O	DIISOBUTYL KETONE	142.241	227.17	---------	---------	---------
911	C9H18O	1-NONANAL	142.241	255.15	34.500	242.546	104.295
912	C9H18O2	n-BUTYL VALERATE	158.241	180.35	---------	---------	---------
913	C9H18O2	n-NONANOIC ACID	158.241	285.55	20.300	128.285	55.163
914	C9H18O2	n-OCTYL FORMATE	158.241	234.05	---------	---------	---------
915	C9H20	3,3-DIETHYLPENTANE	128.258	240.12	10.090	78.670	33.828
916	C9H20	2,2-DIMETHYL-3-ETHYLPENTANE	128.258	173.68	10.200	79.527	34.197
917	C9H20	3-ETHYL-2,3-DIMETHYLPENTANE	128.257	173.67	---------	---------	---------
918	C9H20	2,4-DIMETHYL-3-ETHYLPENTANE	128.258	150.79	7.200	56.137	24.139
919	C9H20	2,2-DIMETHYLHEPTANE	128.258	160.15	8.950	69.781	30.006
920	C9H20	2,6-DIMETHYLHEPTANE	128.258	170.25	13.000	101.358	43.584
921	C9H20	3-ETHYLHEPTANE	128.258	158.25	16.000	124.749	53.642
922	C9H20	4-ETHYLHEPTANE	128.257	159.96	---------	---------	---------
923	C9H20	2,3-DIMETHYLHEPTANE	128.257	160.16	---------	---------	---------
924	C9H20	2,4-DIMETHYLHEPTANE	128.257	160.16	---------	---------	---------
925	C9H20	2,5-DIMETHYLHEPTANE	128.257	160.16	---------	---------	---------
926	C9H20	3,4-DIMETHYLHEPTANE	128.257	170.26	---------	---------	---------
927	C9H20	3,5-DIMETHYLHEPTANE	128.257	170.26	---------	---------	---------
928	C9H20	4,4-DIMETHYLHEPTANE	128.257	170.26	---------	---------	---------
929	C9H20	3-ETHYL-2-METHYLHEXANE	128.257	160.16	---------	---------	---------
930	C9H20	4-ETHYL-2-METHYLHEXANE	128.257	160.16	---------	---------	---------
931	C9H20	3-ETHYL-3-METHYLHEXANE	128.257	160.16	---------	---------	---------
932	C9H20	3-ETHYL-4-METHYLHEXANE	128.257	160.16	---------	---------	---------
933	C9H20	2,2,3-TRIMETHYLHEXANE	128.257	153.16	---------	---------	---------
934	C9H20	2,2,4-TRIMETHYLHEXANE	128.257	153.00	---------	---------	---------
935	C9H20	2,3,3-TRIMETHYLHEXANE	128.257	156.36	---------	---------	---------
936	C9H20	2,3,4-TRIMETHYLHEXANE	128.257	156.36	---------	---------	---------

Table 6-1 ENTHALPY OF FUSION - ORGANIC COMPOUNDS (continued)

NO	FORMULA	NAME	MW	T_F	HFUS @ T_F		
			g/mol	K	kjoule/mol	kjoule/kg	BTU/lb
937	C9H20	2,3,5-TRIMETHYLHEXANE	128.257	145.36	----------	----------	--------
938	C9H20	2,4,4-TRIMETHYLHEXANE	128.257	159.78	----------	----------	--------
939	C9H20	3,3,4-TRIMETHYLHEXANE	128.257	171.96	----------	----------	--------
940	C9H20	2-METHYLOCTANE	128.258	192.78	18.000	140.342	60.347
941	C9H20	3-METHYLOCTANE	128.258	165.55	17.000	132.545	56.994
942	C9H20	4-METHYLOCTANE	128.258	159.95	16.000	124.749	53.642
943	C9H20	n-NONANE	128.258	219.63	15.468	120.601	51.858
944	C9H20	2,2,3,3-TETRAMETHYLPENTANE	128.258	263.26	2.300	17.933	7.711
945	C9H20	2,2,3,4-TETRAMETHYLPENTANE	128.258	152.06	0.500	3.898	1.676
946	C9H20	2,2,4,4-TETRAMETHYLPENTANE	128.258	206.95	9.700	75.629	32.520
947	C9H20	2,3,3,4-TETRAMETHYLPENTANE	128.257	171.10	----------	----------	--------
948	C9H20	2,2,5-TRIMETHYLHEXANE	128.258	167.39	6.200	48.340	20.786
949	C9H20O	2,6-DIMETHYL-4-HEPTANOL	144.257	208.00			
950	C9H20O	1-NONANOL	144.257	268.15	28.800	199.644	85.847
951	C9H20O	2-NONANOL	144.257	238.15	14.300	99.129	42.625
952	C9H20S	n-NONYL MERCAPTAN	160.324	253.05	33.000	205.833	88.508
953	C9H20S	BUTYL-PENTYL-SULFIDE	160.317	231.16	----------	----------	--------
954	C9H20S	ETHYL-HEPTYL-SULFIDE	160.317	231.16	----------	----------	--------
955	C9H20S	HEXYL-PROPYL-SULFIDE	160.317	231.16	----------	----------	--------
956	C9H20S	METHYL-OCTYL-SULFIDE	160.317	231.16	----------	----------	--------
957	C9H21N	n-NONYLAMINE	143.272	273.15	34.300	239.405	102.944
958	C9H21N	TRIPROPYLAMINE	143.272	179.65	----------	----------	--------
959	C10H6O8	PYROMELLITIC ACID	254.153	554.00	----------	----------	--------
960	C10H7Br	1-BROMONAPHTHALENE	207.070	279.35	----------	----------	--------
961	C10H7Cl	1-CHLORONAPHTHALENE	162.618	269.15	----------	----------	--------
962	C10H8	NAPHTHALENE	128.174	353.43	18.979	148.072	63.671
963	C10H8	AZULENE	128.173	173.66	----------	----------	--------
964	C10H9N	QUINALDINE	143.188	272.15	----------	----------	--------
965	C10H10	m-DIVINYLBENZENE	130.189	206.25	----------	----------	--------
966	C10H10	1-METHYLINDENE	130.189	--------	----------	----------	--------
967	C10H10	2-METHYLINDENE	130.189	353.15	----------	----------	--------
968	C10H10O4	DIMETHYL PHTHALATE	194.187	272.15	----------	----------	--------
969	C10H10O4	DIMETHYL TEREPHTHALATE	194.187	413.80	31.630	162.884	70.040
970	C10H12	DICYCLOPENTADIENE	132.205	307.00	0.500	3.782	1.626
971	C10H12	1,2,3,4-TETRAHYDRONAPHTHALENE	132.205	237.40	12.450	94.172	40.494
972	C10H12O	ANETHOLE	148.205	294.50	16.010	108.026	46.451
973	C10H12O4	DIALLYL MALEATE	196.203	226.15	----------	----------	--------
974	C10H14	n-BUTYLBENZENE	134.221	185.30	11.222	83.608	35.952
975	C10H14	sec-BUTYLBENZENE	134.221	197.72	9.832	73.252	31.498
976	C10H14	tert-BUTYLBENZENE	134.221	215.27	8.393	62.531	26.888
977	C10H14	1,2,3,4-TETRAMETHYLBENZENE	134.221	266.91	----------	----------	--------
978	C10H14	m-CYMENE	134.221	209.44	13.682	101.936	43.833
979	C10H14	o-CYMENE	134.221	201.64	10.000	74.504	32.037
980	C10H14	p-CYMENE	134.221	205.25	9.661	71.978	30.951
981	C10H14	m-DIETHYLBENZENE	134.221	189.26	10.966	81.701	35.131
982	C10H14	o-DIETHYLBENZENE	134.221	241.93	16.778	125.003	53.751
983	C10H14	p-DIETHYLBENZENE	134.221	230.32	10.586	78.870	33.914
984	C10H14	2-ETHYL-m-XYLENE	134.221	256.89	14.710	109.595	47.126
985	C10H14	2-ETHYL-p-XYLENE	134.221	219.52	15.190	113.172	48.664
986	C10H14	3-ETHYL-o-XYLENE	134.221	223.64	13.640	101.623	43.698
987	C10H14	4-ETHYL-m-XYLENE	134.221	210.27	12.930	96.334	41.423
988	C10H14	4-ETHYL-o-XYLENE	134.221	206.22	12.050	89.777	38.604
989	C1OH14	5-ETHYL-m-XYLENE	134.221	188.82	8.954	66.711	28.686
990	C10H14	ISOBUTYLBENZENE	134.221	221.70	12.510	93.204	40.078
991	C10H14	1,2,3,5-TETRAMETHYLBENZENE	134.221	249.46	10.720	79.868	34.343
992	C10H14	1,2,4,5-TETRAMETHYLBENZENE	134.221	352.38	21.004	156.488	67.290
993	C10H14O	p-tert-BUTYLPHENOL	150.221	371.56	----------	----------	--------
994	C10H14O2	p-tert-BUTYLCATECHOL	166.220	325.00	----------	----------	--------
995	C10H15N	N,N-DIETHYLANILINE	149.236	235.15	8.490	56.890	24.463
996	C10H15N	2,6-DIETHYLANILINE	149.236	276.65	----------	----------	--------
997	C10H16	CAMPHENE	136.237	320.15	31.800	233.417	100.369
998	C10H16	D-LIMONENE	136.237	199.00	----------	----------	--------
999	C10H16	alpha-PHELLANDRENE	136.237	--------	----------	----------	--------
1000	C1OH16	beta-PHELLANDRENE	136.237	--------	----------	----------	--------
1001	C10H16	alpha-PINENE	136.237	209.15	----------	----------	--------
1002	C10H16	beta-PINENE	136.237	211.61	----------	----------	--------
1003	C10H16	alpha-TERPINENE	136.237	--------	----------	----------	--------
1004	C1OH16	gamma-TERPINENE	136.237	--------	----------	----------	--------
1005	C10H16	TERPINOLENE	136.237	--------	----------	----------	--------
1006	C10H16O	CAMPHOR	152.236	453.25	6.820	44.799	19.264
1007	C10H18	1-DECYNE	138.252	229.16	----------	----------	--------
1008	C10H18	cis-DECAHYDRONANPHTALENE	138.253	230.20	9.489	68.635	29.513
1009	C10H18	trans-DECAHYDRONAPHTHALENE	138.253	242.79	14.410	104.229	44.819
1010	C10H18O4	SEBACIC ACID	202.251	407.65	----------	----------	--------
1011	C10H20	n-BUTYLCYCLOHEXANE	140.268	198.42	14.160	100.949	43.408
1012	C10H20	1-CYCLOPENTYLPENTANE	140.268	190.16	----------	----------	--------
1013	C10H20	1-DECENE	140.269	206.89	13.807	98.432	42.326
1014	C10H20O	1-DECANAL	156.268	267.15	34.500	220.775	94.933

Table 6-1 ENTHALPY OF FUSION - ORGANIC COMPOUNDS (continued)

NO	FORMULA	NAME	MW g/mol	T$_F$ K	HFUS @ T$_F$ kjoule/mol	kjoule/kg	BTU/lb
1015	C10H20O2	n-DECANOIC ACID	172.268	304.75	27.798	161.365	69.387
1016	C10H20O2	2-ETHYLHEXYL ACETATE	172.268	180.15	---------	---------	---------
1017	C10H20O2	ISOPENTYL ISOVALERATE	172.268	215.00	---------	---------	---------
1018	C10H22	n-DECANE	142.285	243.49	28.715	201.813	86.780
1019	C10H22	2-METHYLNONANE	142.285	198.50	17.500	122.993	52.887
1020	C10H22	3-METHYLNONANE	142.285	188.35	18.800	132.129	56.816
1021	C10H22	4-METHYLNONANE	142.285	174.45	15.200	106.828	45.936
1022	C10H22	5-METHYLNONANE	142.285	185.45	16.600	116.667	50.167
1023	C10H22	3-ETHYLOCTANE	142.284	185.46	---------	---------	---------
1024	C10H22	4-ETHYLOCTANE	142.284	185.46	---------	---------	---------
1025	C10H22	2,2-DIMETHYLOCTANE	142.285	--------	---------	---------	---------
1026	C10H22	2,3-DIMETHYLOCTANE	142.284	219.16	---------	---------	---------
1027	C10H22	2,4-DIMETHYLOCTANE	142.284	219.16	---------	---------	---------
1028	C10H22	2,5-DIMETHYLOCTANE	142.284	219.16	---------	---------	---------
1029	C10H22	2,6-DIMETHYLOCTANE	142.284	219.16	---------	---------	---------
1030	C10H22	2,7-DIMETHYLOCTANE	142.284	219.16	---------	---------	---------
1031	C10H22	3,3-DIMETHYLOCTANE	142.284	219.16	---------	---------	---------
1032	C10H22	3,4-DIMETHYLOCTANE	142.284	219.16	---------	---------	---------
1033	C10H22	3,5-DIMETHYLOCTANE	142.284	219.16	---------	---------	---------
1034	C10H22	3,6-DIMETHYLOCTANE	142.284	219.16	---------	---------	---------
1035	C10H22	4,4-DIMETHYLOCTANE	142.284	219.16	---------	---------	---------
1036	C10H22	4,5-DIMETHYLOCTANE	142.284	219.16	---------	---------	---------
1037	C10H22	4-PROPYLHEPTANE	142.284	219.16	---------	---------	---------
1038	C10H22	4-ISOPROPYLHEPTANE	142.284	219.16	---------	---------	---------
1039	C10H22	3-ETHYL-2-METHYLHEPTANE	142.284	219.16	---------	---------	---------
1040	C10H22	4-ETHYL-2-METHYLHEPTANE	142.284	219.16	---------	---------	---------
1041	C10H22	5-ETHYL-2-METHYLHEPTANE	142.284	219.16	---------	---------	---------
1042	C10H22	3-ETHYL-3-METHYLHEPTANE	142.284	219.16	---------	---------	---------
1043	C10H22	4-ETHYL-3-METHYLHEPTANE	142.284	219.16	---------	---------	---------
1044	C10H22	3-ETHYL-5-METHYLHEPTANE	142.284	219.16	---------	---------	---------
1045	C10H22	3-ETHYL-4-METHYLHEPTANE	142.284	219.16	---------	---------	---------
1046	C10H22	4-ETHYL-4-METHYLHEPTANE	142.284	219.16	---------	---------	---------
1047	C10H22	2,2,3-TRIMETHYLHEPTANE	142.284	219.16	---------	---------	---------
1048	C10H22	2,2,4-TRIMETHYLHEPTANE	142.284	219.16	---------	---------	---------
1049	C10H22	2,2,5-TRIMETHYLHEPTANE	142.284	219.16	---------	---------	---------
1050	C10H22	2,2,6-TRIMETHYLHEPTANE	142.284	219.16	---------	---------	---------
1051	C10H22	2,3,3-TRIMETHYLHEPTANE	142.284	219.16	---------	---------	---------
1052	C10H22	2,3,4-TRIMETHYLHEPTANE	142.284	219.16	---------	---------	---------
1053	C10H22	2,3,5-TRIMETHYLHEPTANE	142.284	219.16	---------	---------	---------
1054	C10H22	2,3,6-TRIMETHYLHEPTANE	142.284	219.16	---------	---------	---------
1055	C10H22	2,4,4-TRIMETHYLHEPTANE	142.284	219.16	---------	---------	---------
1056	C10H22	2,4,5-TRIMETHYLHEPTANE	142.284	219.16	---------	---------	---------
1057	C10H22	2,4,6-TRIMETHYLHEPTANE	142.284	219.16	---------	---------	---------
1058	C10H22	2,5,5-TRIMETHYLHEPTANE	142.284	219.16	---------	---------	---------
1059	C10H22	3,3,4-TRIMETHYLHEPTANE	142.284	219.16	---------	---------	---------
1060	C10H22	3,3,5-TRIMETHYLHEPTANE	142.284	219.16	---------	---------	---------
1061	C10H22	3,4,4-TRIMETHYLHEPTANE	142.284	219.16	---------	---------	---------
1062	C10H22	3,4,5-TRIMETHYLHEPTANE	142.284	219.16	---------	---------	---------
1063	C10H22	3-ISOPROPYL-2-METHYLHEXANE	142.284	219.16	---------	---------	---------
1064	C10H22	3,3-DIETHYLHEXANE	142.284	219.16	---------	---------	---------
1065	C10H22	3,4-DIETHYLHEXANE	142.284	219.16	---------	---------	---------
1066	C10H22	3-ETHYL-2,2-DIMETHYLHEXANE	142.284	219.16	---------	---------	---------
1067	C10H22	4-ETHYL-2,2-DIMETHYLHEXANE	142.284	219.16	---------	---------	---------
1068	C10H22	3-ETHYL-2,3-DIMETHYLHEXANE	142.284	219.16	---------	---------	---------
1069	C10H22	4-ETHYL-2,3-DIMETHYLHEXANE	142.284	219.16	---------	---------	---------
1070	C10H22	3-ETHYL-2,4-DIMETHYLHEXANE	142.284	219.16	---------	---------	---------
1071	C10H22	4-ETHYL-2,4-DIMETHYLHEXANE	142.284	219.16	---------	---------	---------
1072	C10H22	3-ETHYL-2,5-DIMETHYLHEXANE	142.284	219.16	---------	---------	---------
1073	C10H22	4-ETHYL-3,3-DIMETHYLHEXANE	142.284	219.16	---------	---------	---------
1074	C10H22	3-ETHYL-3,4-DIMETHYLHEXANE	142.284	219.16	---------	---------	---------
1075	C10H22	2,2,3,3-TETRAMETHYLHEXANE	142.284	219.16	---------	---------	---------
1076	C10H22	2,2,3,4-TETRAMETHYLHEXANE	142.284	219.16	---------	---------	---------
1077	C10H22	2,2,3,5-TETRAMETHYLHEXANE	142.284	219.16	---------	---------	---------
1078	C10H22	2,2,4,4-TETRAMETHYLHEXANE	142.284	219.16	---------	---------	---------
1079	C10H22	2,2,4,5-TETRAMETHYLHEXANE	142.284	219.16	---------	---------	---------
1080	C10H22	2,2,5,5-TETRAMETHYLHEXANE	142.284	260.56	---------	---------	---------
1081	C10H22	2,3,3,4-TETRAMETHYLHEXANE	142.284	260.56	---------	---------	---------
1082	C10H22	2,3,3,5-TETRAMETHYLHEXANE	142.284	260.56	---------	---------	---------
1083	C10H22	2,3,4,4-TETRAMETHYLHEXANE	142.284	260.56	---------	---------	---------
1084	C10H22	2,3,4,5-TETRAMETHYLHEXANE	142.284	260.56	---------	---------	---------
1085	C10H22	3,3,4,4-TETRAMETHYLHEXANE	142.284	260.56	---------	---------	---------
1086	C10H22	2,4-DIMETHYL-3-ISOPROPYLPENTANE	142.284	191.46	---------	---------	---------
1087	C10H22	3,3-DIETHYL-2-METHYLPENTANE	142.284	191.46	---------	---------	---------
1088	C10H22	3-ETHYL-2,2,3-TRIMETHYLPENTANE	142.284	191.46	---------	---------	---------
1089	C10H22	3-ETHYL-2,2,4-TRIMETHYLPENTANE	142.284	191.46	---------	---------	---------
1090	C10H22	3-ETHYL-2,3,4-TRIMETHYLPENTANE	142.284	191.46	---------	---------	---------
1091	C10H22	2,2,3,3,4-PENTAMETHYLPENTANE	142.284	236.71	---------	---------	---------
1092	C10H22	2,2,3,4,4-PENTAMETHYLPENTANE	142.284	234.41	---------	---------	---------

Table 6-1 ENTHALPY OF FUSION - ORGANIC COMPOUNDS (continued)

NO	FORMULA	NAME	MW g/mol	T_F K	HFUS @ T_F kjoule/mol	kjoule/kg	BTU/lb
1093	C10H22O	1-DECANOL	158.284	280.05	37.656	237.901	102.298
1094	C10H22O	DI-n-PENTYL ETHER	158.284	203.72	25.800	162.998	70.089
1095	C10H22O	ISODECANOL	158.284	213.15	17.300	109.297	46.998
1096	C10H22O5	TETRAETHYLENE GLYCOL DIMETHYL ETHER	222.282	243.45	25.400	114.269	49.136
1097	C10H22S	n-DECYL MERCAPTAN	174.351	247.56	31.000	177.802	76.455
1098	C10H22S	BUTYL-HEXYL-SULFIDE	174.344	238.16	---------	---------	---------
1099	C10H22S	ETHYL-OCTYL-SULFIDE	174.344	238.16	---------	---------	---------
1100	C10H22S	HEPTYL-PROPYL-SULFIDE	174.344	238.16	---------	---------	---------
1101	C10H22S	METHYL-NONYL-SULFIDE	174.344	238.16	---------	---------	---------
1102	C10H22S	PENTYL-SULFIDE	174.344	238.16	---------	---------	---------
1103	C10H22S2	PENTYL-DISULFIDE	206.404	214.16	---------	---------	---------
1104	C10H23N	n-DECYLAMINE	157.299	288.85	33.800	214.877	92.397
1105	C11H10	1-METHYLNAPHTHALENE	142.200	242.67	6.945	48.840	21.001
1106	C11H10	2-METHYLNAPHTHALENE	142.200	307.73	12.125	85.267	36.665
1107	C11H14O2	n-BUTYL BENZOATE	178.231	251.65	---------	---------	---------
1108	C11H16	n-PENTYLBENZENE	148.248	198.15	15.240	102.801	44.204
1109	C11H16O	p-tert-AMYLPHENOL	164.247	366.00	---------	---------	---------
1110	C11H20	1-UNDECYNE	152.279	248.16	---------	---------	---------
1111	C11H20O2	2-ETHYLHEXYL ACRYLATE	184.279	183.15	---------	---------	---------
1112	C11H22	1-UNDECENE	154.296	223.99	16.991	110.120	47.351
1113	C11H22	1-CYCLOPENTYLHEXANE	154.295	200.16	---------	---------	---------
1114	C11H22	PENTYLCYCLOHEXANE	154.295	215.66	---------	---------	---------
1115	C11H22O	1-UNDECANAL	170.295	273.15	38.330	225.080	96.784
1116	C11H24	n-UNDECANE	156.312	247.57	22.179	141.889	61.012
1117	C11H24O	1-UNDECANOL	172.311	289.05	37.400	217.049	93.331
1118	C11H24S	UNDECYL MERCAPTAN	188.378	270.15	39.900	211.808	91.078
1119	C11H24S	BUTYL-HEPTYL-SULFIDE	188.371	254.66	---------	---------	---------
1120	C11H24S	DECYL-METHYL-SULFIDE	188.371	254.66	---------	---------	---------
1121	C11H24S	ETHYL-NONYL-SULFIDE	188.371	254.66	---------	---------	---------
1122	C11H24S	OCTYL-PROPYL-SULFIDE	188.371	254.66	---------	---------	---------
1123	C12H8O	DIBENZOFURAN	168.195	355.65	22.700	134.962	58.034
1124	C12H9N	DIBENZOPYRROLE	167.210	517.95	29.429	176.000	75.680
1125	C12H10	ACENAPHTHENE	154.211	366.56	21.462	139.173	59.844
1126	C12H10	BIPHENYL	154.211	342.37	18.577	120.465	51.800
1127	C12H10O	DIPHENYL ETHER	170.211	300.02	17.216	101.145	43.492
1128	C12H11N	p-AMINODIPHENYL	169.226	326.00	---------	---------	---------
1129	C12H11N	DIPHENYLAMINE	169.226	326.15	18.702	110.515	47.521
1130	C12H11N3	p-AMINOAZOBENZENE	197.240	401.00	---------	---------	---------
1131	C12H11N3	1,3-DIPHENYLTRIAZENE	197.240	372.00	---------	---------	---------
1132	C12H12	1,2-DIMETHYLNAPHTHALENE	156.227	272.16	---------	---------	---------
1133	C12H12	1,3-DIMETHYLNAPHTHALENE	156.227	269.16	---------	---------	---------
1134	C12H12	1,4-DIMETHYLNAPHTHALENE	156.227	280.82	---------	---------	---------
1135	C12H12	1,5-DIMETHYLNAPHTHALENE	156.227	355.16	---------	---------	---------
1136	C12H12	1,6-DIMETHYLNAPHTHALENE	156.227	259.16	---------	---------	---------
1137	C12H12	1,7-DIMETHYLNAPHTHALENE	156.227	260.16	---------	---------	---------
1138	C12H12	2,3-DIMETHYLNAPHTHALENE	156.227	378.16	---------	---------	---------
1139	C12H12	2,6-DIMETHYLNAPHTHALENE	156.227	384.55	22.300	142.741	61.379
1140	C12H12	2,7-DIMETHYLNAPHTHALENE	156.227	370.15	23.351	149.468	64.271
1141	C12H12	1-ETHYLNAPHTHALENE	156.227	259.34	16.300	104.335	44.864
1142	C12H12	2-ETHYLNAPHTHALENE	156.227	265.76	---------	---------	---------
1143	C12H12N2	p-AMINODIPHENYLAMINE	184.241	341.15	---------	---------	---------
1144	C12H12N2	HYDRAZOBENZENE	184.241	404.15	17.630	95.690	41.147
1145	C12H14	1,2,3-TRIMETHYLINDENE	158.243	344.65	---------	---------	---------
1146	C12H14O4	DIETHYL PHTHALATE	222.241	269.15	17.984	80.921	34.796
1147	C12H16	CYCLOHEXYLBENZENE	160.259	280.14	15.270	95.283	40.972
1148	C12H18	m-DIISOPROPYLBENZENE	162.275	210.02	---------	---------	---------
1149	C12H18	p-DIISOPROPYLBENZENE	162.275	256.08	---------	---------	---------
1150	C12H18	n-HEXYLBENZENE	162.275	212.00	18.410	113.449	48.783
1151	C12H18	1,2,3-TRIETHYLBENZENE	162.274	206.66	---------	---------	---------
1152	C12H18	1,2,4-TRIETHYLBENZENE	162.274	206.66	---------	---------	---------
1153	C12H18	1,3,5-TRIETHYLBENZENE	162.274	206.66	---------	---------	---------
1154	C12H18	HEXAMETHYLBENZENE	162.274	438.66	---------	---------	---------
1155	C12H20O4	DIBUTYL MALEATE	228.288	188.15	---------	---------	---------
1156	C12H22	BICYCLOHEXYL	166.307	276.78	---------	---------	---------
1157	C12H22	1-DODECYNE	166.306	254.16	---------	---------	---------
1158	C12H23N	DICYCLOHEXYLAMINE	181.321	273.05	---------	---------	---------
1159	C12H24	1-DODECENE	168.323	237.93	19.908	118.273	50.857
1160	C12H24	1-CYCLOPENTYLHEPTANE	168.322	220.00	---------	---------	---------
1161	C12H24	1-CYCLOHEXYLHEXANE	168.322	263.60	---------	---------	---------
1162	C12H24O	1-DODECANAL	184.322	285.15	42.210	229.001	98.471
1163	C12H24O2	n-DODECANOIC ACID	200.321	317.15	36.295	181.184	77.909
1164	C12H26	n-DODECANE	170.338	263.57	36.836	216.252	92.989
1165	C12H26O	DI-n-HEXYL ETHER	186.338	230.15	30.500	163.681	70.383
1166	C12H26O	1-DODECANOL	186.338	296.95	31.380	168.404	72.414
1167	C12H26O3	DIETHYLENE GLYCOL DI-n-BUTYL ETHER	218.337	212.95	30.600	140.150	60.265
1168	C12H26S	n-DODECYL MERCAPTAN	202.404	265.15	39.000	192.684	82.854
1169	C12H26S	BUTYL-OCTYL-SULFIDE	202.397	259.16	---------	---------	---------
1170	C12H26S	DECYL-ETHYL-SULFIDE	202.397	259.16	---------	---------	---------

150

Table 6-1 ENTHALPY OF FUSION - ORGANIC COMPOUNDS (continued)

NO	FORMULA	NAME	MW g/mol	T_F K	HFUS @ T_F kjoule/mol	kjoule/kg	BTU/lb
1171	C12H26S	HEXYL-SULFIDE	202.397	259.16	---------	---------	---------
1172	C12H26S	METHYL-UNDECYL-SULFIDE	202.397	259.16	---------	---------	---------
1173	C12H26S	NONYL-PROPYL-SULFIDE	202.397	259.16	---------	---------	---------
1174	C12H26S2	HEXYL-DISULFIDE	234.457	225.16	---------	---------	---------
1175	C12H27BO3	TRI-n-BUTYL BORATE	230.156	203.15	---------	---------	---------
1176	C12H27N	DODECYLAMINE	185.353	301.47	43.500	234.687	100.916
1177	C12H27N	TRI-n-BUTYLAMINE	185.353	203.00	---------	---------	---------
1178	C13H10	FLUORENE	166.222	387.94	19.578	117.782	50.646
1179	C13H10O	BENZOPHENONE	182.222	321.35	16.903	92.760	39.887
1180	C13H12	DIPHENYLMETHANE	168.238	298.39	18.242	108.430	46.625
1181	C13H14	1-PROPYLNAPHTHALENE	170.254	264.69	---------	---------	---------
1182	C13H14	2-PROPYLNAPHTHALENE	170.254	270.16	---------	---------	---------
1183	C13H14	2ETHYL-3-METHYLNAPHTHALENE	170.254	344.16	---------	---------	---------
1184	C13H14	2ETHYL-6-METHYLNAPHTHALENE	170.254	318.16	---------	---------	---------
1185	C13H14	2ETHYL-7-METHYLNAPHTHALENE	170.254	318.16	---------	---------	---------
1186	C13H20	n-HEPTYLBENZENE	176.302	225.15	19.300	109.471	47.073
1187	C13H24	1-TRIDECYNE	180.333	268.16	---------	---------	---------
1188	C13H26	1-TRIDECENE	182.349	250.08	22.830	125.199	53.836
1189	C13H26	1-CYCLOPENTYLOCTANE	182.348	229.16	---------	---------	---------
1190	C13H26	1-CYCLOHEXYLHEPTANE	182.348	242.66	---------	---------	---------
1191	C13H26O	1-TRIDECANAL	198.349	288.15	46.100	232.419	99.940
1192	C13H26O2	n-BUTYL NONANOATE	214.348	235.15	---------	---------	---------
1193	C13H26O2	METHYL DODECANOATE	214.348	278.15	---------	---------	---------
1194	C13H28	n-TRIDECANE	184.365	267.76	28.501	154.590	66.474
1195	C13H28O	1-TRIDECANOL	200.365	303.75	45.120	225.189	96.831
1196	C13H28S	BUTYL-NONYL-SULFIDE	216.424	271.16	---------	---------	---------
1197	C13H28S	DECYL-PROPYL-SULFIDE	216.424	271.16	---------	---------	---------
1198	C13H28S	DODECYL-METHYL-SULFIDE	216.424	271.16	---------	---------	---------
1199	C13H28S	ETHYL-UNDECYL-SULFIDE	216.424	271.16	---------	---------	---------
1200	C13H28S	1-TRIDECANETHIOL	216.424	282.04	---------	---------	---------
1201	C14H8O2	ANTHRAQUINONE	208.216	559.15	32.552	156.338	67.225
1202	C14H10	ANTHRACENE	178.233	489.25	29.370	164.784	70.857
1203	C14H10	DIPHENYLACETYLENE	178.233	335.65	21.402	120.079	51.634
1204	C14H10	PHENANTHRENE	178.233	372.38	16.463	92.368	39.718
1205	C14H12	cis-STILBENE	180.249	268.15	---------	---------	---------
1206	C14H12	trans-STILBENE	180.249	397.35	27.828	154.386	66.386
1207	C14H12O2	BENZYL BENZOATE	212.248	292.55	---------	---------	---------
1208	C14H14	1,1-DIPHENYLETHANE	182.265	255.20	17.573	96.415	41.458
1209	C14H14	1,2-DIPHENYLETHANE	182.265	324.34	30.543	167.575	72.057
1210	C14H14O	DIBENZYL ETHER	198.265	276.75	20.209	101.929	43.830
1211	C14H16	1-n-BUTYLNAPHTHALENE	184.281	253.43	25.100	136.205	58.568
1212	C14H16	2-BUTYLNAPHTHALENE	184.280	268.16	---------	---------	---------
1213	C14H22	n-OCTYLBENZENE	190.329	237.15	22.500	118.216	50.833
1214	C14H22	1,2,3,4-TETRAETHYLBENZENE	190.328	284.96	---------	---------	---------
1215	C14H22	1,2,3,5-TETRAETHYLBENZENE	190.328	284.16	---------	---------	---------
1216	C14H22	1,2,4,5-TETRAETHYLBENZENE	190.328	283.16	---------	---------	---------
1217	C14H22O	p-tert-OCTYLPHENOL	206.328	358.55	---------	---------	---------
1218	C14H28	1-TETRADECENE	196.376	260.30	25.470	129.700	55.771
1219	C14H28	1-CYCLOPENTYLNONANE	196.375	244.16	---------	---------	---------
1220	C14H28	1-CYCLOHEXYLOCTANE	196.375	253.46	---------	---------	---------
1221	C14H28O2	n-TETRADECANOIC ACID	228.375	327.55	45.100	197.482	84.917
1222	C14H30	n-TETRADECANE	198.392	279.01	45.070	227.176	97.686
1223	C14H30O	1-TETRADECANOL	214.392	310.65	49.510	230.932	99.301
1224	C14H30S	BUTYL-DECYL-SULFIDE	230.451	276.16	---------	---------	---------
1225	C14H30S	DODECYL-ETHYL-SULFIDE	230.451	276.16	---------	---------	---------
1226	C14H30S	HEPTYL-SULFIDE	230.451	276.16	---------	---------	---------
1227	C14H30S	METHYL-TRIDECYL-SULFIDE	230.451	276.16	---------	---------	---------
1228	C14H30S	PROPYL-UNDECYL-SULFIDE	230.451	276.16	---------	---------	---------
1229	C14H30S	1-TETRADECANETHIOL	230.451	279.26	---------	---------	---------
1230	C14H30S2	HEPTYL-DISULFIDE	262.511	235.16	---------	---------	---------
1231	C14H31N	TETRADECYLAMINE	213.407	311.34	51.000	238.980	102.761
1232	C15H10N2O2	DIPHENYLMETHANE-4,4'-DIISOCYANATE	250.257	311.20	---------	---------	---------
1233	C15H16O	p-CUMYLPHENOL	212.291	346.00	21.677	102.110	43.907
1234	C15H16O2	BISPHENOL A	228.291	426.15	29.286	128.284	55.162
1235	C15H18	1-PENTYLNAPHTHALENE	198.307	251.16	---------	---------	---------
1236	C15H18	2-PENTYLNAPHTHALENE	198.307	269.16	---------	---------	---------
1237	C15H24	n-NONYLBENZENE	204.356	249.00	28.870	141.273	60.747
1238	C15H24O	2,6-DI-tert-BUTYL-p-CRESOL	220.355	344.00	---------	---------	---------
1239	C15H24O	NONYLPHENOL	220.355	--------	---------	---------	---------
1240	C15H28	1-PENTADECYNE	208.386	283.16	---------	---------	---------
1241	C15H30	1-PENTADECENE	210.403	269.42	29.570	140.540	60.432
1242	C15H30	1-CYCLOPENTYLDECANE	210.402	251.03	---------	---------	---------
1243	C15H30	1-CYCLOHEXYLNONANE	210.402	262.96	---------	---------	---------
1244	C15H30O2	PENTADECANOIC ACID	242.402	325.68	41.526	171.310	73.664
1245	C15H32	n-PENTADECANE	212.419	283.11	34.593	162.853	70.027
1246	C15H32O	1-PENTADECANOL	228.417	317.04	---------	---------	---------
1247	C15H32S	BUTYL-UNDECYL-SULFIDE	244.478	284.16	---------	---------	---------
1248	C15H32S	DODECYL-PROPYL-SULFIDE	244.478	284.16	---------	---------	---------

151

Table 6-1 ENTHALPY OF FUSION - ORGANIC COMPOUNDS (continued)

NO	FORMULA	NAME	MW g/mol	T_F K	HFUS @ T_F		
					kjoule/mol	kjoule/kg	BTU/lb
1249	C15H32S	ETHYL-TRIDECYL-SULFIDE	244.478	284.16	---------	---------	---------
1250	C15H32S	METHYL-TETRADECYL-SULFIDE	244.478	284.16	---------	---------	---------
1251	C15H32S	1-PENTADECANETHIOL	244.478	290.93	---------	---------	---------
1252	C16H10	FLUORANTHENE	202.255	383.33	18.730	92.606	39.821
1253	C16H10	PYRENE	202.255	423.81	17.360	85.832	36.908
1254	C16H12	1-PHENYLNAPHTHALENE	204.271	318.15	---------	---------	---------
1255	C16H20	1-n-HEXYLNAPHTHALENE	212.335	255.15	---------	---------	---------
1256	C16H22O4	DIBUTYL PHTHALATE	278.348	238.15	---------	---------	---------
1257	C16H26	n-DECYLBENZENE	218.382	258.77	32.635	149.440	64.259
1258	C16H26	PENTAETHYLBENZENE	218.381	327.66	---------	---------	---------
1259	C16H30	1-HEXADECYNE	222.413	288.16	---------	---------	---------
1260	C16H32	n-DECYLCYCLOHEXANE	224.430	271.42	38.600	171.991	· 73.956
1261	C16H32	1-CYCLOPENTYLUNDECANE	224.429	263.16	---------	---------	---------
1262	C16H32	1-HEXADECENE	224.430	277.51	30.192	134.527	57.847
1263	C16H32O2	n-HEXADECANOIC ACID	256.429	335.95	53.711	209.458	90.067
1264	C16H34	n-HEXADECANE	226.446	291.34	53.358	235.632	101.322
1265	C16H34O	DI-n-OCTYL ETHER	242.445	265.55	47.600	196.333	84.423
1266	C16H34O	1-HEXADECANOL	242.445	322.35	58.380	240.797	103.543
1267	C16H34S	BUTYL-DODECYL-SULFIDE	258.505	288.16	---------	---------	---------
1268	C16H34S	ETHYL-TETRADECYL-SULFIDE	258.505	288.16	---------	---------	---------
1269	C16H34S	METHYL-PENTADECYL-SULFIDE	258.505	288.16	---------	---------	---------
1270	C16H34S	OCTYL-SULFIDE	258.505	288.16	---------	---------	---------
1271	C16H34S	PROPYL-TRIDECYL-SULFIDE	258.505	288.16	---------	---------	---------
1272	C16H34S	1-HEXADECANETHIOL	258.505	290.93	---------	---------	---------
1273	C16H34S2	OCTYL-DISULFIDE	290.565	244.16	---------	---------	---------
1274	C17H28	n-UNDECYLBENZENE	232.409	268.00	36.000	154.899	66.607
1275	C17H32	1-HEPTADECYNE	236.440	295.16	31.400	132.803	57.105
1276	C17H34	1-CYCLOPENTYLDODECANE	238.456	268.16	---------	---------	---------
1277	C17H34	1-CYCLOHEXYLUNDECANE	238.456	278.96	---------	---------	---------
1278	C17H34	1-HEPTADECENE	238.457	284.40	---------	---------	---------
1279	C17H36	n-HEPTADECANE	240.473	295.13	40.459	168.248	72.346
1280	C17H36O	1-HEPTADECANOL	256.472	327.05	64.000	249.540	107.302
1281	C17H36S	BUTYL-TRIDECYL-SULFIDE	272.531	294.16	---------	---------	---------
1282	C17H36S	ETHYL-PENTADECYL-SULFIDE	272.531	294.16	---------	---------	---------
1283	C17H36S	HEXADECYL-METHYL-SULFIDE	272.531	294.16	---------	---------	---------
1284	C17H36S	PROPYL-TETRADECYL-SULFIDE	272.531	294.16	---------	---------	---------
1285	C17H36S	1-HEPTADECANETHIOL	272.531	300.37	---------	---------	---------
1286	C18H12	CHRYSENE	228.293	531.15	26.200	114.765	49.349
1287	C18H14	m-TERPHENYL	230.309	360.00	24.090	104.599	44.977
1288	C18H14	o-TERPHENYL	230.309	329.35	17.191	74.643	32.097
1289	C18H14	p-TERPHENYL	230.309	485.00	33.726	146.438	62.968
1290	C18H15P	TRIPHENYLPHOSPHINE	262.291	354.40	19.690	75.069	32.280
1291	C18H15O4P	TRIPHENYL PHOSPHATE	326.288	323.15	23.970	73.463	31.589
1292	C18H16N2	N,N'-DIPHENYL-p-PHENYLENEDIAMINE	260.339	409.00	---------	---------	---------
1293	C18H22	2,3-DIMETHYL-2,3-DIPHENYLBUTANE	238.373	392.15	---------	---------	---------
1294	C18H22O2	DICUMYL PEROXIDE	270.371	311.15	20.400	75.452	32.444
1295	C18H30	n-DODECYLBENZENE	246.436	275.93	40.166	162.988	70.085
1296	C18H30	HEXAETHYLBENZENE	246.435	401.16	---------	---------	---------
1297	C18H32O2	LINOLEIC ACID	280.451	268.15	---------	---------	---------
1298	C18H34	1-OCTADECYNE	250.467	300.16	---------	---------	---------
1299	C18H34O2	OLEIC ACID	282.467	286.53	---------	---------	---------
1300	C18H34O4	DIBUTYL SEBACATE	314.466	263.95	---------	---------	---------
1301	C18H34O4	DIHEXYL ADIPATE	314.466	259.35	---------	---------	---------
1302	C18H36	1-CYCOPENTYLTRIDECANE	252.482	278.16	---------	---------	---------
1303	C18H36	1-CYCLOHEXYLDODECANE	252.482	285.66	---------	---------	---------
1304	C18H36	1-OCTADECENE	252.484	290.76	32.600	129.117	55.520
1305	C18H36O2	STEARIC ACID	284.483	342.75	61.209	215.159	92.518
1306	C18H38	n-OCTADECANE	254.500	301.33	61.965	243.477	104.695
1307	C18H38O	DINONYL ETHER	270.499	--------	---------	---------	---------
1308	C18H38O	1-OCTADECANOL	270.499	331.05	74.057	273.779	117.725
1309	C18H38S	BUTYL-TETRADECYL-SULFIDE	286.558	298.16	---------	---------	---------
1310	C18H38S	ETHYL-HEXADECYL-SULFIDE	286.558	298.16	---------	---------	---------
1311	C18H38S	HEPTADECYL-METHYL-SULFIDE	286.558	298.16	---------	---------	---------
1312	C18H38S	NONYL-SULFIDE	286.558	298.16	---------	---------	---------
1313	C18H38S	PENTADECYL-PROPYL-SULFIDE	286.558	298.16	---------	---------	---------
1314	C18H38S	1-OCTADECANETHIOL	286.558	300.93	---------	---------	---------
1315	C18H38S2	NONYL-DISULFIDE	318.618	252.16	---------	---------	---------
1316	C19H26	1-n-NONYLNAPHTHALENE	254.415	284.15	---------	---------	---------
1317	C19H32	n-TRIDECYLBENZENE	260.463	283.15	43.932	168.669	72.528
1318	C19H36	1-NONADECYNE	264.493	306.16	---------	---------	---------
1319	C19H36O2	METHYL OLEATE	296.494	293.05·	---------	---------	---------
1320	C19H38	1-CYCLOPENTYLTETRADECANE	266.509	282.00	---------	---------	---------
1321	C19H38	1-CYCLOHEXYLTRIDECANE	266.509	291.66	---------	---------	---------
1322	C19H38	1-NONADECENE	266.511	296.55	33.500	125.698	54.050
1323	C19H38O2	NONADECANOIC ACID	298.510	341.23	57.618	193.019	82.998
1324	C19H40	n-NONADECANE	268.527	305.33	45.815	170.616	73.365
1325	C19H40O	1-NONADECANOL	284.524	334.87	---------	---------	---------
1326	C19H40S	BUTYL-PENTADECYL-SULFIDE	300.585	303.16	---------	---------	---------

Table 6-1 ENTHALPY OF FUSION - ORGANIC COMPOUNDS (continued)

NO	FORMULA	NAME	MW g/mol	T$_F$ K	HFUS @ T$_F$ kjoule/mol	kjoule/kg	BTU/lb
1327	C19H40S	ETHYL-HEPTADECYL-SULFIDE	300.585	303.16	----------	----------	----------
1328	C19H40S	HEXADECYL-PROPYL-SULFIDE	300.585	303.16	----------	----------	----------
1329	C19H40S	METHYL-OCTADECYL-SULFIDE	300.585	303.16	----------	----------	----------
1330	C19H40S	1-NONADECANETHIOL	300.585	307.04	----------	----------	----------
1331	C20H16	TRIPHENYLETHYLENE	256.347	342.15	----------	----------	----------
1332	C20H28	1-n-DECYLNAPHTHALENE	268.442	288.15	----------	----------	----------
1333	C20H30O2	ABIETIC ACID	302.457	446.65	----------	----------	----------
1334	C20H31N	DEHYDROABIETYLAMINE	285.473	317.65	----------	----------	----------
1335	C20H34	1-PHENYLTETRADECANE	274.489	289.16	----------	----------	----------
1336	C20H38	1-EICOSYNE	278.520	309.16	34.300	123.151	52.955
1337	C20H40	1-CYCLOPENTYLPENTADECANE	280.536	290.00	----------	----------	----------
1338	C20H40	1-CYCLOHEXYLTETRADECANE	280.536	297.16	----------	----------	----------
1339	C20H40	1-EICOSENE	280.538	301.76	----------	----------	----------
1340	C20H42	n-EICOSANE	282.553	309.59	69.873	247.292	106.335
1341	C20H42O	1-EICOSANOL	298.553	338.55	99.161	332.139	142.820
1342	C20H42S	BUTYL-HEXADECYL-SULFIDE	314.612	308.16	----------	----------	----------
1343	C20H42S	DECYL-SULFIDE	314.612	308.16	----------	----------	----------
1344	C20H42S	ETHYL-OCTADECYL-SULFIDE	314.612	308.16	----------	----------	----------
1345	C20H42S	HEPTADECYL-PROPYL-SULFIDE	314.612	308.16	----------	----------	----------
1346	C20H42S	METHYL-NONADECYL-SULFIDE	314.612	308.16	----------	----------	----------
1347	C20H42S	1-EICOSANETHIOL	314.612	310.37	----------	----------	----------
1348	C20H42S2	DECYL-DISULFIDE	346.672	259.16	----------	----------	----------
1349	C21H21O4P	TRI-o-CRESYL PHOSPHATE	368.369	240.15	----------	----------	----------
1350	C21H36	1-PHENYLPENTADECANE	288.515	295.16	----------	----------	----------
1351	C21H42	1-CYCLOPENTYLHEXADECANE	294.563	294.16	----------	----------	----------
1352	C21H42	1-CYCLOHEXYLPENTADECANE	294.563	302.16	----------	----------	----------
1353	C22H38	1-PHENYLHEXADECANE	302.542	300.16	----------	----------	----------
1354	C22H44	1-CYCLOHEXYLHEXADECANE	308.590	306.76	----------	----------	----------
1355	C22H44O2	n-BUTYL STEARATE	340.590	299.45	56.902	167.069	71.840
1356	C24H38O4	DIISOOCTYL PHTHALATE	390.563	--------	----------	----------	----------
1357	C24H38O4	DIOCTYL PHTHALATE	390.563	223.15	----------	----------	----------
1358	C24H42O	DINONYLPHENOL	346.597	--------	----------	----------	----------
1359	C26H20	TETRAPHENYLETHYLENE	332.445	496.15	----------	----------	----------
1360	C28H46O4	DIISODECYL PHTHALATE	446.671	227.59	----------	----------	----------

MW - molecular weight, g/mol

T$_F$ - freezing point temperature, K

HFUS - enthalpy of fusion

153

Table 6-2 ENTHALPY OF FUSION - INORGANIC COMPOUNDS

NO	FORMULA	NAME	MW g/mol	T_F K	HFUS @ T_F kjoule/mol	HFUS @ T_F kjoule/kg	HFUS @ T_F BTU/lb
1	Ag	SILVER	107.868	1234.00	11.940	110.691	47.597
2	AgCl	SILVER CHLORIDE	143.321	728.15	13.200	92.101	39.603
3	AgI	SILVER IODIDE	234.773	825.15	9.410	40.081	17.235
4	Al	ALUMINUM	26.982	933.00	10.711	396.968	170.696
5	AlB3H12	ALUMINUM BOROHYDRIDE	71.510	209.15	---------	---------	---------
6	AlBr3	ALUMINUN BROMIDE	266.694	390.15	11.250	42.183	18.139
7	AlCl3	ALUMINUM CHLORIDE	133.340	465.70	35.355	265.149	114.014
8	AlF3	ALUMINUM FLUORIDE	83.977	2325.00	107.000	1274.158	547.888
9	AlI3	ALUMINUM IODIDE	407.695	464.15	15.900	39.000	16.770
10	Al2O3	ALUMINUM OXIDE	101.961	2325.00	111.100	1089.632	468.542
11	Al2S3O12	ALUMINUM SULFATE	342.154	1043.20	---------	---------	---------
12	Ar	ARGON	39.948	83.80	1.182	29.588	12.723
13	As	ARSENIC	74.922	1090.00	31.900	425.776	183.084
14	AsBr3	ARSENIC TRIBROMIDE	314.634	306.15	11.700	37.186	15.990
15	AsCl3	ARSENIC TRICHLORIDE	181.280	255.15	10.100	55.715	23.957
16	AsF3	ARSENIC TRIFLUORIDE	131.917	267.25	10.400	78.837	33.900
17	AsF5	ARSENIC PENTAFLUORIDE	169.914	193.35	---------	---------	---------
18	AsH3	ARSINE	77.945	156.28	1.195	15.331	6.592
19	AsI3	ARSENIC TRIIODIDE	455.635	419.15	---------	---------	---------
20	As2O3	ARSENIC TRIOXIDE	197.841	585.95	---------	---------	---------
21	At	ASTATINE	210.000	575.00	23.800	113.333	48.733
22	Au	GOLD	196.967	1337.58	12.700	64.478	27.725
23	B	BORON	10.811	2573.00	50.200	4643.419	1996.670
24	BBr3	BORON TRIBROMIDE	250.523	228.15	---------	---------	---------
25	BCl3	BORON TRICHLORIDE	117.169	166.15	2.109	18.000	7.740
26	BF3	BORON TRIFLUORIDE	67.806	146.05	4.184	61.705	26.533
27	BH2CO	BORINE CARBONYL	40.837	136.15	---------	---------	---------
28	BH3O3	BORIC ACID	61.833	458.15	---------	---------	---------
29	B2D6	DEUTERODIBORANE	33.718	---------	---------	---------	---------
30	B2H5Br	DIBORANE HYDROBROMIDE	106.566	168.95	---------	---------	---------
31	B2H6	DIBORANE	27.670	107.65	4.473	161.655	69.512
32	B3N3H6	BORINE TRIAMINE	80.501	214.95	---------	---------	---------
33	B4H10	TETRABORANE	53.323	153.25	---------	---------	---------
34	B5H9	PENTABORANE	63.126	226.35	---------	---------	---------
35	B5H11	TETRAHYDROPENTABORANE	65.142	---------	---------	---------	---------
36	B10H14	DECABORANE	122.221	372.75	---------	---------	---------
37	Ba	BARIUM	137.327	1002.00	7.660	55.779	23.985
38	Be	BERYLLIUM	9.012	1551.00	9.800	1087.439	467.599
39	BeB2H8	BERYLLIUM BOROHYDRIDE	38.698	396.15	---------	---------	---------
40	BeBr2	BERYLLIUM BROMIDE	168.820	763.15	9.800	58.050	24.961
41	BeCl2	BERYLLIUM CHLORIDE	79.918	678.15	8.660	108.361	46.595
42	BeF2	BERYLLIUM FLUORIDE	47.009	1073.15	4.760	101.257	43.541
43	BeI2	BERYLLIUM IODIDE	262.821	761.15	21.000	79.902	34.358
44	Bi	BISMUTH	208.980	544.50	10.480	50.148	21.564
45	BiBr3	BISMUTH TRIBROMIDE	448.692	491.15	---------	---------	---------
46	BiCl3	BISMUTH TRICHLORIDE	315.338	503.15	10.900	34.566	14.863
47	BrF5	BROMINE PENTAFLUORIDE	174.896	211.75	5.670	32.419	13.940
48	Br2	BROMINE	159.808	265.90	10.573	66.161	28.449
49	C	CARBON	12.011	4247.00	104.600	8708.684	3744.734
50	CCl2O	PHOSGENE	98.916	145.37	5.738	58.009	24.944
51	CF2O	CARBONYL FLUORIDE	66.007	161.89	6.708	101.626	43.699
52	CH4N2O	UREA	60.056	405.85	14.790	246.270	105.896
53	CH4N2S	THIOUREA	76.122	454.15	---------	---------	---------
54	CNBr	CYANOGEN BROMIDE	105.922	331.15	---------	---------	---------
55	CNCl	CYANOGEN CHLORIDE	61.470	266.65	---------	---------	---------
56	CNF	CYANOGEN FLUORIDE	45.016	---------	---------	---------	---------
57	CO	CARBON MONOXIDE	28.010	68.15	0.841	30.025	12.911
58	COS	CARBONYL SULFIDE	60.076	134.35	4.728	78.700	33.841
59	COSe	CARBON OXYSELENIDE	106.970	---------	---------	---------	---------
60	CO2	CARBON DIOXIDE	44.010	216.58	9.019	204.931	88.120
61	CS2	CARBON DISULFIDE	76.143	161.58	4.393	57.694	24.808
62	CSeS	CARBON SELENOSULFIDE	123.037	197.95	---------	---------	---------
63	C2N2	CYANOGEN	52.035	238.75	---------	---------	---------
64	C3S2	CARBON SUBSULFIDE	100.165	273.55	---------	---------	---------
65	Ca	CALCIUM	40.078	1112.00	9.330	232.796	100.102
66	CaF2	CALCIUM FLUORIDE	78.075	1691.00	29.706	380.480	163.607
67	CbF5	COLUMBIUM FLUORIDE	187.898	348.65	---------	---------	---------
68	Cd	CADMIUM	112.411	594.10	6.110	54.354	23.372
69	CdCl2	CADMIUM CHLORIDE	183.316	841.15	---------	---------	---------
70	CdF2	CADMIUM FLUORIDE	150.408	793.15	---------	---------	---------
71	CdI2	CADMIUM IODIDE	366.220	658.15	---------	---------	---------
72	CdO	CADMIUM OXIDE	128.410	---------	---------	---------	---------
73	ClF	CHLORINE MONOFLUORIDE	54.451	128.15	---------	---------	---------
74	ClFO3	PERCHLORYL FLUORIDE	102.449	125.41	3.834	37.423	16.092
75	ClF3	CHLORINE TRIFLUORIDE	92.448	190.15	---------	---------	---------
76	ClF5	CHLORINE PENTAFLUORIDE	130.445	---------	---------	---------	---------
77	ClHO3S	CHLOROSULFONIC ACID	116.525	193.15	---------	---------	---------
78	ClHO4	PERCHLORIC ACID	100.458	171.95	6.933	69.014	29.676

154

Table 6-2 ENTHALPY OF FUSION - INORGANIC COMPOUNDS (continued)

NO	FORMULA	NAME	MW g/mol	T$_F$ K	HFUS @ T$_F$ kjoule/mol	kjoule/kg	BTU/lb
79	ClO2	CHLORINE DIOXIDE	67.452	213.55	---------	---------	---------
80	Cl2	CHLORINE	70.905	172.12	6.406	90.346	38.849
81	Cl2O	CHLORINE MONOXIDE	86.905	157.15	---------	---------	---------
82	Cl2O7	CHLORINE HEPTOXIDE	182.901	182.15	---------	---------	---------
83	Co	COBALT	58.933	1768.00	15.200	257.920	110.906
84	CoCl2	COBALT CHLORIDE	129.839	1008.15	45.000	346.583	149.031
85	CoNC3O4	COBALT NITROSYL TRICARBONYL	172.971	262.15	---------	---------	---------
86	Cr	CHROMIUM	51.996	2130.00	21.000	403.877	173.667
87	CrC6O6	CHROMIUM CARBONYL	220.061	423.65	---------	---------	---------
88	CrO2Cl2	CHROMIUM OXYCHLORIDE	154.900	176.65	---------	---------	---------
89	Cs	CESIUM	132.905	301.55	2.090	15.726	6.762
90	CsBr	CESIUM BROMIDE	212.809	909.15	---------	---------	---------
91	CsCl	CESIUM CHLORIDE	168.358	919.15	15.900	94.442	40.610
92	CsF	CESIUM FLUORIDE	151.904	956.15	21.700	142.853	61.427
93	CsI	CESIUM IODIDE	259.810	894.15	---------	---------	---------
94	Cu	COPPER	63.546	1356.60	13.000	204.576	87.968
95	CuBr	CUPROUS BROMIDE	143.450	777.15	---------	---------	---------
96	CuCl	CUPROUS CHLORIDE	98.999	703.00	10.960	110.708	47.605
97	CuCl2	CUPRIC CHLORIDE	134.451	906.15	20.400	151.728	65.243
98	CuI	COPPER IODIDE	190.450	878.15	---------	---------	---------
99	DCN	DEUTERIUM CYANIDE	28.034	261.15	---------	---------	---------
100	D2	DEUTERIUM	4.032	18.73	0.199	49.355	21.223
101	D2O	DEUTERIUM OXIDE	20.031	276.96	6.339	316.459	136.078
102	Eu	EUROPIUM	151.965	1095.00	10.500	69.095	29.711
103	F2	FLUORINE	37.997	53.53	0.510	13.422	5.772
104	F2O	FLUORINE OXIDE	53.996	49.25	---------	---------	---------
105	Fe	IRON	55.847	1808.15	15.355	274.948	118.227
106	FeC5O5	IRON PENTACARBONYL	195.899	252.15	---------	---------	---------
107	FeCl2	FERROUS CHLORIDE	126.752	945.15	43.010	339.324	145.909
108	FeCl3	FERRIC CHLORIDE	162.205	577.15	43.100	265.713	114.257
109	Fr	FRANCIUM	223.000	300.00	---------	---------	---------
110	Ga	GALLIUM	69.723	302.93	5.590	80.174	34.475
111	GaCl3	GALLIUM TRICHLORIDE	176.081	350.90	21.800	123.807	53.237
112	Gd	GADOLINIUM	157.250	1586.00	15.500	98.569	42.385
113	Ge	GERMANIUM	72.610	1210.60	34.700	477.896	205.495
114	GeBr4	GERMANIUM BROMIDE	392.226	299.25	---------	---------	---------
115	GeCl4	GERMANIUM CHLORIDE	214.421	223.65	---------	---------	---------
116	GeHCl3	TRICHLORO GERMANE	179.976	202.05	---------	---------	---------
117	GeH4	GERMANE	76.642	107.26	0.836	10.908	4.690
118	Ge2H6	DIGERMANE	151.268	164.15	---------	---------	---------
119	Ge3H8	TRIGERMANE	225.894	167.55	---------	---------	---------
120	HBr	HYDROGEN BROMIDE	80.912	186.34	2.406	29.736	12.786
121	HCN	HYDROGEN CYANIDE	27.026	259.91	8.406	311.034	133.745
122	HCl	HYDROGEN CHLORIDE	36.461	158.97	1.998	54.798	23.563
123	HF	HYDROGEN FLUORIDE	20.006	189.79	4.580	228.931	98.440
124	HI	HYDROGEN IODIDE	127.912	222.38	2.872	22.453	9.655
125	HNO3	NITRIC ACID	63.013	231.55	10.473	166.204	71.468
126	H2	HYDROGEN	2.016	13.95	0.117	58.036	24.955
127	H2O	WATER	18.015	273.15	6.002	333.167	143.262
128	H2O2	HYDROGEN PEROXIDE	34.015	272.72	12.498	367.426	157.993
129	H2S	HYDROGEN SULFIDE	34.082	187.68	2.377	69.744	29.990
130	H2SO4	SULFURIC ACID	98.079	283.46	10.711	109.208	46.959
131	H2S2	HYDROGEN DISULFIDE	66.148	183.45	---------	---------	---------
132	H2Se	HYDROGEN SELENIDE	80.976	209.15	---------	---------	---------
133	H2Te	HYDROGEN TELLURIDE	129.616	224.15	---------	---------	---------
134	H3NO3S	SULFAMIC ACID	97.095	478.00	---------	---------	---------
135	He	HELIUM-3	3.016	1.01	---------	---------	---------
136	He	HELIUM-4	4.003	1.76	0.050	12.491	5.371
137	Hf	HAFNIUM	178.490	2503.00	25.500	142.865	61.432
138	Hg	MERCURY	200.590	234.29	2.295	11.441	4.920
139	HgBr2	MERCURIC BROMIDE	360.398	510.15	17.900	49.667	21.357
140	HgCl2	MERCURIC CHLORIDE	271.495	550.15	19.410	71.493	30.742
141	HgI2	MERCURIC IODIDE	454.399	532.15	18.900	41.593	17.885
142	IF7	IODINE HEPTAFLUORIDE	259.893	278.65	---------	---------	---------
143	I2	IODINE	253.809	386.75	15.517	61.137	26.289
144	In	INDIUM	114.818	429.32	3.270	28.480	12.246
145	Ir	IRIDIUM	192.220	2683.00	26.400	137.343	59.057
146	K	POTASSIUM	39.098	336.35	2.335	59.722	25.680
147	KBr	POTASSIUM BROMIDE	119.002	1003.15	25.500	214.282	92.141
148	KCl	POTASSIUM CHLORIDE	74.551	1044.00	26.284	352.564	151.603
149	KF	POTASSIUM FLUORIDE	58.097	1153.15	27.200	468.183	201.318
150	KI	POTASSIUM IODIDE	166.003	996.15	24.000	144.576	62.168
151	KOH	POTASSIUM HYDROXIDE	56.106	679.00	8.619	153.620	66.057
152	Kr	KRYPTON	83.800	115.78	1.640	19.570	8.415
153	La	LANTHANUM	138.906	1194.00	10.040	72.279	31.080
154	Li	LITHIUM	6.941	453.69	3.000	432.214	185.852
155	LiBr	LITHIUM BROMIDE	86.845	820.15	---------	---------	---------
156	LiCl	LITHIUM CHLORIDE	42.394	887.15	19.900	469.406	201.845

Table 6-2 ENTHALPY OF FUSION - INORGANIC COMPOUNDS (continued)

NO	FORMULA	NAME	MW g/mol	T_F K	HFUS @ T_F kjoule/mol	kjoule/kg	BTU/lb
157	LiF	LITHIUM FLUORIDE	25.939	1143.15	27.090	1044.373	449.081
158	LiI	LITHIUM IODIDE	133.845	719.15	14.600	109.081	46.905
159	Lu	LUTECIUM	174.967	1936.00	19.200	109.735	47.186
160	Mg	MAGNESIUM	24.305	922.00	9.040	371.940	159.934
161	MgCl2	MAGNESIUM CHLORIDE	95.210	985.15	43.100	452.684	194.654
162	MgO	MAGNESIUM OXIDE	40.304	3105.00	57.650	1430.379	615.063
163	Mn	MANGANESE	54.938	1517.00	14.400	262.114	112.709
164	MnCl2	MANGANESE CHLORIDE	125.843	923.15	30.700	243.955	104.901
165	Mo	MOLYBDENUM	95.940	2890.00	27.600	287.680	123.702
166	MoF6	MOLYBDENUM FLUORIDE	209.930	290.15	4.330	20.626	8.869
167	MoO3	MOLYBDENUM OXIDE	143.938	1068.15	48.000	333.477	143.395
168	NCl3	NITROGEN TRICHLORIDE	120.365	246.15	---------	---------	---------
169	ND3	HEAVY AMMONIA	20.055	199.15	---------	---------	---------
170	NF3	NITROGEN TRIFLUORIDE	71.002	66.36	0.398	5.605	2.410
171	NH3	AMMONIA	17.031	195.41	5.657	332.159	142.828
172	NH3O	HYDROXYLAMINE	33.030	306.25	16.500	499.546	214.805
173	NH4Br	AMMONIUM BROMIDE	97.943	---------	---------	---------	---------
174	NH4Cl	AMMONIUM CHLORIDE	53.491	793.20	---------	---------	---------
175	NH4I	AMMONIUM IODIDE	144.943	---------	21.000	144.885	62.300
176	NH5O	AMMONIUM HYDROXIDE	35.046	194.15	---------	---------	---------
177	NH5S	AMMONIUM HYDROGENSULFIDE	51.112	391.15	---------	---------	---------
178	NO	NITRIC OXIDE	30.006	112.15	2.301	76.685	32.974
179	NOCl	NITROSYL CHLORIDE	65.459	213.55	5.983	91.401	39.302
180	NOF	NITROSYL FLUORIDE	49.005	139.15	---------	---------	---------
181	NO2	NITROGEN DIOXIDE	46.006	261.95	---------	---------	---------
182	N2	NITROGEN	28.013	63.15	0.720	25.702	11.052
183	N2F4	TETRAFLUOROHYDRAZINE	104.007	111.65	---------	---------	---------
184	N2H4	HYDRAZINE	32.045	274.69	12.657	394.976	169.840
185	N2H4C	AMMONIUM CYANIDE	44.056	309.15	---------	---------	---------
186	N2H6CO2	AMMONIUM CARBAMATE	78.071	---------	---------	---------	---------
187	N2O	NITROUS OXIDE	44.013	182.33	6.540	148.592	63.895
188	N2O3	NITROGEN TRIOXIDE	76.012	170.00	---------	---------	---------
189	N2O4	NITROGEN TETRAOXIDE	92.011	261.90	14.652	159.242	68.474
190	N2O5	NITROGEN PENTOXIDE	108.010	303.15	34.640	320.711	137.906
191	Na	SODIUM	22.990	370.98	2.603	113.223	48.686
192	NaBr	SODIUM BROMIDE	102.894	1020.00	26.108	253.737	109.107
193	NaCN	SODIUM CYANIDE	49.008	836.85	15.481	315.887	135.831
194	NaCl	SODIUM CHLORIDE	58.442	1073.95	28.158	481.811	207.179
195	NaF	SODIUM FLUORIDE	41.988	1269.00	33.346	794.179	341.497
196	NaI	SODIUM IODIDE	149.894	924.15	23.600	157.445	67.701
197	NaOH	SODIUM HYDROXIDE	39.997	596.00	6.611	165.287	71.074
198	Na2SO4	SODIUM SULFATE	142.043	1157.00	23.849	167.900	72.197
199	Nb	NIOBIUM	92.906	2741.00	27.200	292.769	125.891
200	Nd	NEODYMIUM	144.240	1294.00	7.113	49.314	21.205
201	Ne	NEON	20.180	24.55	0.328	16.254	6.989
202	Ni	NICKEL	58.693	1726.00	17.600	299.865	128.942
203	NiC4O4	NICKEL CARBONYL	170.735	248.15	---------	---------	---------
204	NiF2	NICKEL FLUORIDE	96.690	1723.15	---------	---------	---------
205	Np	NEPTUNIUM	237.000	913.00	9.460	39.916	17.164
206	O2	OXYGEN	31.999	54.36	0.444	13.875	5.966
207	O3	OZONE	47.998	80.15	2.092	43.585	18.742
208	Os	OSMIUM	190.230	3327.00	29.300	154.024	66.230
209	OsOF5	OSMIUM OXIDE PENTAFLUORIDE	301.221	332.95	---------	---------	---------
210	OsO4	OSMIUM TETROXIDE - YELLOW	254.228	329.15	9.800	38.548	16.576
211	OsO4	OSMIUM TETROXIDE - WHITE	254.228	315.15	9.800	38.548	16.576
212	P	PHOSPHORUS - WHITE	30.974	870.00	18.828	607.865	261.382
213	PBr3	PHOSPHORUS TRIBROMIDE	270.686	233.15	---------	---------	---------
214	PCl2F3	PHOSPHORUS DICHLORIDE TRIFLUORIDE	158.874	265.15	---------	---------	---------
215	PCl3	PHOSPHORUS TRICHLORIDE	137.332	181.15	7.080	51.554	22.168
216	PCl5	PHOSPHORUS PENTACHLORIDE	208.237	433.15	---------	---------	---------
217	PH3	PHOSPHINE	33.998	139.37	1.130	33.237	14.292
218	PH4Br	PHOSPHONIUM BROMIDE	114.910	---------	---------	---------	---------
219	PH4Cl	PHOSPHONIUM CHLORIDE	70.458	244.65	---------	---------	---------
220	PH4I	PHOSPHONIUM IODIDE	161.910	291.65	---------	---------	---------
221	POCl3	PHOSPHORUS OXYCHLORIDE	153.331	274.33	13.104	85.462	36.749
222	PSBr3	PHOSPHORUS THIOBROMIDE	302.752	311.15	---------	---------	---------
223	PSCl3	PHOSPHORUS THIOCHLORIDE	169.398	236.95	---------	---------	---------
224	P4O6	PHOSPHORUS TRIOXIDE	219.891	295.65	---------	---------	---------
225	P4O10	PHOSPHORUS PENTOXIDE	283.889	693.15	27.196	95.798	41.193
226	P4S10	PHOSPHORUS PENTASULFIDE	444.555	561.15	41.100	92.452	39.754
227	Pb	LEAD	207.200	600.65	5.121	24.715	10.628
228	PbBr2	LEAD BROMIDE	367.008	646.15	16.440	44.795	19.262
229	PbCl2	LEAD CHLORIDE	278.105	774.15	21.900	78.747	33.861
230	PbF2	LEAD FLUORIDE	245.197	1128.15	14.700	59.952	25.779
231	PbI2	LEAD IODIDE	461.009	675.15	23.400	50.758	21.826
232	PbO	LEAD OXIDE	223.199	1163.15	---------	---------	---------
233	PbS	LEAD SULFIDE	239.266	1387.15	19.000	79.410	34.146
234	Pd	PALLADIUM	106.420	1825.00	17.200	161.624	69.498

Table 6-2 ENTHALPY OF FUSION - INORGANIC COMPOUNDS (continued)

NO	FORMULA	NAME	MW g/mol	T_F K	HFUS @ T_F		
					kjoule/mol	kjoule/kg	BTU/lb
235	Po	POLONIUM	209.000	527.00	10.000	47.847	20.574
236	Pt	PLATINUM	195.080	2045.00	19.700	100.984	43.423
237	Ra	RADIUM	226.000	973.00	7.150	31.637	13.604
238	Rb	RUBIDIUM	85.468	312.20	2.200	25.741	11.068
239	RbBr	RUBIDIUM BROMIDE	165.372	955.15	15.500	93.728	40.303
240	RbCl	RUBIDIUM CHLORIDE	120.921	988.15	18.400	152.165	65.431
241	RbF	RUBIDIUM FLUORIDE	104.466	1033.15	17.300	165.604	71.210
242	RbI	RUBIDIUM IODIDE	212.372	915.15	12.500	58.859	25.309
243	Re	RHENIUM	186.207	3453.00	33.100	177.759	76.436
244	Re2O7	RHENIUM HEPTOXIDE	484.410	569.15	---------	---------	---------
245	Rh	RHODIUM	102.906	2239.00	21.550	209.414	90.048
246	Rn	RADON	222.000	202.00	2.700	12.162	5.230
247	Ru	RUTHENIUM	101.070	2583.00	23.700	234.491	100.831
248	RuF5	RUTHENIUM PENTAFLUORIDE	196.062	359.65	---------	---------	---------
249	S	SULFUR	32.066	388.36	1.727	53.858	23.159
250	SF4	SULFUR TETRAFLUORIDE	108.060	149.15	---------	---------	---------
251	SF6	SULFUR HEXAFLUORIDE	146.056	222.45	5.024	34.398	14.791
252	SOBr2	THIONYL BROMIDE	207.873	220.95	---------	---------	---------
253	SOCl2	THIONYL CHLORIDE	118.971	172.00	---------	---------	---------
254	SOF2	SULFUROUS OXYFLUORIDE	86.062	162.65	---------	---------	---------
255	SO2	SULFUR DIOXIDE	64.065	200.00	7.402	115.539	49.682
256	SO2Cl2	SULFURYL CHLORIDE	134.970	222.00	---------	---------	---------
257	SO3	SULFUR TRIOXIDE	80.064	289.95	7.532	94.075	40.452
258	S2Cl2	SULFUR MONOCHLORIDE	135.037	193.15	---------	---------	---------
259	Sb	ANTIMONY	121.757	903.89	20.900	171.653	73.811
260	SbBr3	ANTIMONY TRIBROMIDE	361.469	369.75	---------	---------	---------
261	SbCl3	ANTIMONY TRICHLORIDE	228.115	346.55	12.693	55.643	23.926
262	SbCl5	ANTIMONY PENTACHLORIDE	299.021	275.95	---------	---------	---------
263	SbH3	STIBINE	124.781	185.15	---------	---------	---------
264	SbI3	ANTIMONY TRIIODIDE	502.470	440.15	---------	---------	---------
265	Sb2O3	ANTIMONY TRIOXIDE	291.512	929.15	---------	---------	---------
266	Sc	SCANDIUM	44.956	1814.00	15.900	353.679	152.082
267	Se	SELENIUM	78.960	490.00	5.100	64.590	27.774
268	SeCl4	SELENIUM TETRACHLORIDE	220.771	---------	---------	---------	---------
269	SeF6	SELENIUM HEXAFLUORIDE	192.950	238.45	---------	---------	---------
270	SeOCl2	SELENIUM OXYCHLORIDE	165.865	281.65	---------	---------	---------
271	SeO2	SELENIUM DIOXIDE	110.959	613.15	---------	---------	---------
272	Si	SILICON	28.086	1685.00	50.208	1787.652	768.690
273	SiBrCl2F	BROMODICHLOROFLUOROSILANE	197.893	160.85	---------	---------	---------
274	SiBrF3	TRIFLUOROBROMOSILANE	164.985	202.65	---------	---------	---------
275	SiBr2ClF	DIBROMOCHLOROFLUOROSILANE	242.345	173.85	---------	---------	---------
276	SiClF3	TRIFLUOROCHLOROSILANE	120.533	131.15	---------	---------	---------
277	SiCl2F2	DICHLORODIFLUOROSILANE	136.988	133.45	---------	---------	---------
278	SiCl3F	TRICHLOROFLUOROSILANE	153.442	152.35	---------	---------	---------
279	SiCl4	SILICON TETRACHLORIDE	169.896	204.30	7.657	45.069	19.380
280	SiF4	SILICON TETRAFLUORIDE	104.079	186.35	9.381	90.133	38.757
281	SiHBr3	TRIBROMOSILANE	268.805	199.65	---------	---------	---------
282	SiHCl3	TRICHLOROSILANE	135.452	144.95	---------	---------	---------
283	SiHF3	TRIFLUOROSILANE	86.089	141.75	---------	---------	---------
284	SiH2Br2	DIBROMOSILANE	189.909	202.95	---------	---------	---------
285	SiH2Cl2	DICHLOROSILANE	101.007	151.15	---------	---------	---------
286	SiH2F2	DIFLUOROSILANE	68.098	---------	---------	---------	---------
287	SiH2I2	DIIODOSILANE	283.910	272.15	---------	---------	---------
288	SiH3Br	MONOBROMOSILANE	111.013	179.25	---------	---------	---------
289	SiH3Cl	MONOCHLOROSILANE	66.562	155.05	---------	---------	---------
290	SiH3F	MONOFLUOROSILANE	50.108	---------	---------	---------	---------
291	SiH3I	IODOSILANE	158.014	216.15	---------	---------	---------
292	SiH4	SILANE	32.117	88.15	0.667	20.768	8.930
293	SiO2	SILICON DIOXIDE	60.084	1883.00	8.514	141.702	60.932
294	Si2Cl6	HEXACHLORODISILANE	268.887	271.95	---------	---------	---------
295	Si2F6	HEXAFLUORODISILANE	170.161	254.55	---------	---------	---------
296	Si2H5Cl	DISILANYL CHLORIDE	96.663	---------	---------	---------	---------
297	Si2H6	DISILANE	62.219	140.65	---------	---------	---------
298	Si2OCl3F3	TRICHLOROTRIFLUORODISILOXANE	235.524	---------	---------	---------	---------
299	Si2OCl6	HEXACHLORODISILOXANE	284.887	239.95	---------	---------	---------
300	Si2OH6	DISILOXANE	78.218	128.95	---------	---------	---------
301	Si3Cl8	OCTACHLOROTRISILANE	367.878	---------	---------	---------	---------
302	Si3H8	TRISILANE	92.320	155.95	---------	---------	---------
303	Si3H9N	TRISILAZANE	107.335	167.45	---------	---------	---------
304	Si4H10	TETRASILANE	122.421	179.55	---------	---------	---------
305	Sm	SAMARIUM	150.360	1350.00	10.900	72.493	31.172
306	Sn	TIN	118.710	505.12	7.200	60.652	26.080
307	SnBr4	STANNIC BROMIDE	438.326	304.15	12.000	27.377	11.772
308	SnCl2	STANNOUS CHLORIDE	189.615	519.95	12.800	67.505	29.027
309	SnCl4	STANNIC CHLORIDE	260.521	242.95	9.200	35.314	15.185
310	SnH4	STANNIC HYDRIDE	122.742	123.25	---------	---------	---------
311	SnI4	STANNIC IODIDE	626.328	417.65	---------	---------	---------
312	Sr	STRONTIUM	87.620	1042.00	9.160	104.542	44.953

Table 6-2 ENTHALPY OF FUSION - INORGANIC COMPOUNDS (continued)

NO	FORMULA	NAME	MW g/mol	T_F K	HFUS @ T_F		
					kjoule/mol	kjoule/kg	BTU/lb
313	SrO	STRONTIUM OXIDE	103.619	2703.15	75.000	723.805	311.236
314	Ta	TANTALUM	180.948	3269.00	31.400	173.531	74.618
315	Tc	TECHNETIUM	98.000	2445.00	23.810	242.959	104.472
316	Te	TELLURIUM	127.600	722.70	13.500	105.799	45.494
317	TeCl4	TELLURIUM TETRACHLORIDE	269.411	497.15	---------	---------	---------
318	TeF6	TELLURIUM HEXAFLUORIDE	241.590	235.35	---------	---------	---------
319	Ti	TITANIUM	47.880	1933.00	20.900	436.508	187.698
320	TiCl4	TITANIUM TETRACHLORIDE	189.691	249.05	9.966	52.538	22.591
321	Tl	THALLIUM	204.383	576.70	4.310	21.088	9.068
322	TlBr	THALLOUS BROMIDE	284.287	733.15	25.100	88.291	37.965
323	TlI	THALLOUS IODIDE	331.288	713.15	13.100	39.543	17.003
324	Tm	THULIUM	168.934	1818.00	18.400	108.918	46.835
325	U	URANIUM	238.029	1405.50	15.500	65.118	28.001
326	UF6	URANIUM FLUORIDE	352.019	342.35	---------	---------	---------
327	V	VANADIUM	50.942	2160.00	17.600	345.491	148.561
328	VCl4	VANADIUM TETRACHLORIDE	192.752	247.45	2.301	11.938	5.133
329	VOCl3	VANADIUM OXYTRICHLORIDE	173.299	193.65	---------	---------	---------
330	W	TUNGSTEN	183.840	3680.00	35.200	191.471	82.332
331	WF6	TUNGSTEN FLUORIDE	297.830	272.65	4.100	13.766	5.919
332	Xe	XENON	131.290	161.36	2.295	17.480	7.517
333	Yb	YTTERBIUM	173.040	1097.00	9.200	53.167	22.862
334	Yt	YTTRIUM	88.906	1799.15	---------	---------	---------
335	Zn	ZINC	65.390	692.70	7.385	112.938	48.563
336	ZnCl2	ZINC CHLORIDE	136.295	638.15	---------	---------	---------
337	ZnF2	ZINC FLUORIDE	103.387	1145.15	---------	---------	---------
338	ZnO	ZINC OXIDE	81.389	2248.00	---------	---------	---------
339	ZnSO4	ZINC SULFATE	161.454	953.00	---------	---------	---------
340	Zr	ZIRCONIUM	91.224	2125.00	23.000	252.127	108.414
341	ZrBr4	ZIRCONIUM BROMIDE	410.840	723.15	---------	---------	---------
342	ZrCl4	ZIRCONIUM CHLORIDE	233.035	710.15	50.000	214.560	92.261
343	ZrI4	ZIRCONIUM IODIDE	598.842	772.15	---------	---------	---------

MW - molecular weight, g/mol

T_F - freezing point temperature, K

HFUS - enthalpy of fusion

Chapter 7

VAPOR PRESSURE

Carl L. Yaws, Xiaoyan Lin, Li Bu, Deepa R. Balundgi, and Saumya Tripathi
Lamar University, Beaumont, Texas

ABSTRACT

Results for vapor pressure as a function of temperature are presented for major organic and inorganic chemicals. The coefficients in the equation for vapor pressure are displayed in easy-to-use tabulations that are especially applicable for rapid engineering usage with the personal computer or hand calculator. The chemicals encompass many compound types.

INTRODUCTION

Physical and thermodynamic property data such as vapor pressure are of special value to engineers in the chemical processing and petroleum refining industries. As an example, knowledge of the vapor pressure of the compound is required in the design of a storage vessel to contain the compound. In hazard analysis and vent system technology, vapor pressure at the specified temperature is important. In vapor-liquid operations, such as distillation, knowledge of vapor pressure (and activity coefficients) is required for determining K-values. In this article, results for vapor pressure as a function of temperature are presented.

VAPOR PRESSURE CORRELATION

The Antoine-type equation with extended terms was selected for correlation of vapor pressure as a function of temperature:

$$\log_{10} P = A + B/T + C \log_{10} T + D\,T + E\,T^2 \tag{7-1}$$

where
P = vapor pressure, mm Hg
A, B, C, D, and E = regression coefficients for chemical compound
T = temperature, K

The results for vapor pressure are given in Tables 7-1 and 7-2. The temperature range for which the equation may be used to predict vapor pressure is denoted by the respective minimum and maximum temperatures (TMIN and TMAX).

The tabulation for organic compounds is applicable to a wide variety of substances: hydrocarbons (alkanes, olefins, acetylenes, cycloalkanes, aromatics, etc.), oxygenates (alcohols, aldehydes, ketones, acids, ethers, glycols, anhydrides, etc.), halogenates (chlorinated, brominated, fluorinated, etc.), nitrogenates (nitriles, amines, cyanates, amides, etc.), sulfur compounds (mercaptans, sulfides, sulfates, etc.), silicon compounds (silanes, chlorosilanes, etc.), and many other types.

The tabulation for inorganic compounds provides coverage for a wide range of substances: carbon oxides (carbon monoxide, carbon dioxide, etc.), nitrogen oxides (nitric oxide, nitrous oxide, etc.), sulfur oxides (sulfur dioxide, sulfur trioxide, etc.), hydrogen oxides (water, hydrogen peroxide, etc.), ammonias (ammonia, ammonium hydroxide, etc.), hydrogen halides (hydrogen chloride, hydrogen fluoride, etc.), sulfur acids (sulfuric acid, hydrogen sulfide, etc.), hydroxides (sodium hydroxide, potassium hydroxide, etc.), silicon halides (trichlorosilane, silicon tetrachloride, etc.), ureas (urea, thiourea, etc.), cyanides (hydrogen cyanide, cyanogen chloride, etc.), hydrides (silane, diborane, etc.), sodium derivatives (sodium chloride, sodium fluoride, etc.), aluminum derivatives (aluminum borohydride, aluminum fluoride, etc.), and many other compound types. Many elements (total = 82) are covered: hydrogen, nitrogen, oxygen, helium, argon, neon, chlorine, bromine, iodine, fluorine, sulfur, phosphorous, aluminum, lead, tin, mercury, sodium, magnesium, silicon, antimony, boron, iron, chromium, cobalt, titanium, tantalum, silver, gold, platinum, radon, uranium, and many others.

In preparing the compilation, a literature search was conducted to identify data source publications for organics (1-41) and inorganics (1-61). Both experimental values for the property under consideration and parameter values for estimation of the property are included in the source publications. The publications were screened for appropriate data. The compilation resulting from the screening is based on both experimental data and estimated values. In the absence of experimental data, estimates were primarily based on Riedel equation (29) and on adjusting the A value in the equation to match the boiling point temperature of the compound. The estimates of the other coefficients for the compound were based on the same values of the compound's brother (closest member of same chemical family). Experimental data and estimates were then regressed to provide the same equation for all compounds.

A comparison of calculated values and experimental data for vapor pressure is shown in Fig. 7-1 for

VAPOR PRESSURE

EXAMPLES

The tabulated values may be used for prediction and calculation of vapor pressure. Examples are given below.

Example 1 Estimate the vapor pressure of methanol (CH4O) at a temperature of 25.13 C (298.28 K).

Substitution of the coefficients from the table and temperature into the equation for vapor pressure yields

$$\log_{10} P = 45.6171 - 3.2447E+03/298.28 - (1.3988E+01)\,(\log_{10}(298.28)) + (6.6365E-03)\,(298.28) - (1.0507E-13)\,(298.28^2)$$
$$= 2.1034$$

$$P = 10^{2.1034}$$

$$P = 126.88 \text{ mm Hg}$$

The calculated and data values compare favorably (126.88 vs 127.90, deviation = 0.80%).

Example 2 Estimate the vapor pressure of acetone (C3H6O) at a temperature of 47.35 C (320.50 K).

Substitution of the coefficients from the table and temperature into the equation for vapor pressure yields

$$\log_{10} P = 28.5884 - 2.4690E+03/320.50 - (7.3510E+00)\,(\log_{10}(320.50)) + (2.8025E-10)\,(320.50) + (2.7361E-06)\,(320.50^2)$$
$$= 2.7456$$

$$P = 10^{2.7456}$$

$$P = 556.71 \text{ mm Hg}$$

The calculated and data values compare favorably (556.71 vs 558.40, deviation = 0.30%).

REFERENCES – ORGANIC COMPOUNDS

1-34. See **REFERENCES - ORGANIC COMPOUNDS** in Chapter 1, CRITICAL PROPERTIES AND ACENTRIC FACTOR

35. Zwolinski, B. J. and R. C. Wilhoit, VAPOR PRESSURES AND HEATS OF VAPORIZATION OF HYDROCARBONS AND RELATED COMPOUNDS, Thermodynamic Research Center, TAMU, College Station, TX (1971).

36. Wilhoit, R. C. and B. J. Zwolinski, PHYSICAL AND THERMODYNAMIC PROPERTIES OF ALIPHATIC ALCOHOLS, American Chemical Society, American Institute of Physics, National Bureau of Standards, New York, NY (1973).

37. Boublick, T., V. Fried and E. Hala, THE VAPOUR PRESSURES OF PURE SUBSTANCES, 1st and 2nd eds., Elsevier, New York, NY (1975, 1984).

38. Ohe, S., COMPUTER AIDED DATA BOOK OF VAPOR PRESSURE, Data Book Publishing Company, Tokyo, Japan (1976).

39. Howard, P. H. and W. M. Meylan, eds., HANDBOOK OF PHYSICAL PROPERTIES OF ORGANIC CHEMICALS, CRC Press, Boca Raton, FL (1997).

40. Yaws, C. L., HANDBOOK OF VAPOR PRESSURE, Vols. 1, 2, 3, and 4, Gulf Publishing Co., Houston, TX (1994,1994,1994,1995).

41. Yaws, C. L., HANDBOOK OF CHEMICAL COMPOUND DATA FOR PROCESS SAFETY, Gulf Publishing Co., Houston, TX (1997).

REFERENCES - INORGANIC COMPOUNDS

1-56. See **REFERENCES - INORGANIC COMPOUNDS** in Chapter 1, CRITICAL PROPERTIES AND ACENTRIC FACTOR

57. Daubert, T. E. and R. P. Danner, DATA COMPILATION OF PROPERTIES OF PURE COMPOUNDS, Parts 1, 2, 3, and 4, Supplements 1 and 2, DIPPR Project, AIChE, New York, NY (1985-1992).

58. Nesmeyanov, A. N., VAPOR PRESSURE OF THE CHEMICAL ELEMENTS, Elsevier, New York, NY (1963).

59. Boublick, T., V. Fried, and E. Hala, THE VAPOUR PRESSURES OF PURE SUBSTANCES, 1st and 2nd eds., Elsevier, New York, NY (1975, 1984).

60. Ohe, S., COMPUTER AIDED DATA BOOK OF VAPOR PRESSURE, Data Book Publishing Company, Tokyo, Japan (1976).

61. Hultgren, R., P. D. Desai, D. T. Hawkins, M. Gleiser, K. K. Kelley, and D. D. Wagman, SELECTED VALUES OF THE THERMODYNAMIC PROPERTIES OF THE ELEMENTS, American Society for Metals, Metals Park, OH (1973).

Figure 7-1 Vapor Pressure

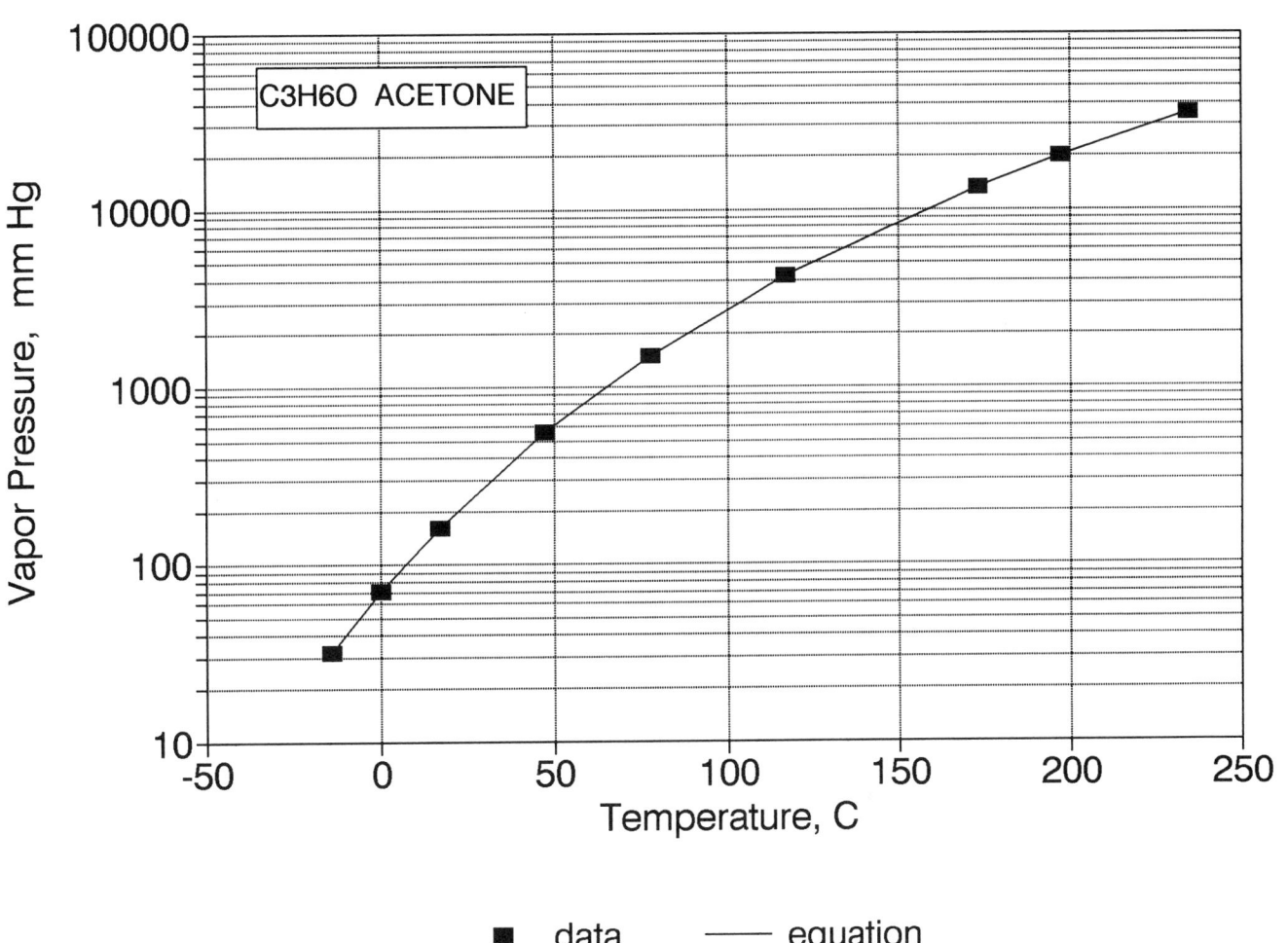

Table 7-1 VAPOR PRESSURE - ORGANIC COMPOUNDS

$$\log_{10} P = A + B/T + C \log_{10} T + D\,T + E\,T^2 \qquad (P - mm\ Hg,\ T - K)$$

NO	FORMULA	NAME	A	B	C	D	E	TMIN	TMAX
1	CBrClF2	BROMOCHLORODIFLUOROMETHANE	27.6647	-1.9063E+03	-7.4195E+00	-3.7276E-11	4.5327E-06	114.00	426.15
2	CBrCl3	BROMOTRICHLOROMETHANE	-4.3224	-1.8417E+03	6.1642E+00	-1.2539E-02	6.4762E-06	252.15	606.00
3	CBrF3	BROMOTRIFLUOROMETHANE	25.1401	-1.4666E+03	-6.7461E+00	-1.6045E-10	6.2569E-06	105.15	340.15
4	CBr2F2	DIBROMODIFLUOROMETHANE	-3.5618	-1.4303E+03	6.1979E+00	-1.7310E-02	1.2349E-05	163.05	478.00
5	CClF3	CHLOROTRIFLUOROMETHANE	6.4418	-1.0028E+03	1.6554E+00	-1.4218E-02	1.6777E-05	92.15	301.96
6	CClN	CYANOGEN CHLORIDE	-9.0499	-1.2832E+03	8.2432E+00	-1.6132E-02	9.6127E-06	266.65	449.00
7	CCl2F2	DICHLORODIFLUOROMETHANE	52.5701	-2.2537E+03	-1.8265E+01	1.2980E-02	2.0286E-13	115.15	384.95
8	CCl2O	PHOSGENE	46.6551	-2.4657E+03	-1.5351E+01	9.2288E-03	2.1645E-13	145.37	455.00
9	CCl3F	TRICHLOROFLUOROMETHANE	44.0884	-2.5022E+03	-1.4193E+01	7.8086E-03	1.3769E-13	162.04	471.20
10	CCl4	CARBON TETRACHLORIDE	31.9407	-2.6614E+03	-8.5763E+00	-6.7136E-10	2.9732E-06	250.33	556.35
11	CF2O	CARBONYL FLUORIDE	-71.0204	-3.7123E+02	4.1621E+01	-1.2131E-01	1.1355E-04	161.89	297.00
12	CF4	CARBON TETRAFLUORIDE	-19.0860	-5.7612E+02	1.6125E+01	-7.7262E-02	1.0889E-04	89.56	227.50
13	CHBr3	TRIBROMOMETHANE	-10.2943	-2.1700E+03	9.1193E+00	-1.6495E-02	7.4917E-06	281.20	696.00
14	CHClF2	CHLORODIFLUOROMETHANE	40.3847	-2.0731E+03	-1.2317E+01	-2.5116E-10	1.0498E-05	115.73	369.30
15	CHCl2F	DICHLOROFLUOROMETHANE	6.8004	-1.5999E+03	1.6648E+00	-1.0603E-02	8.3356E-06	138.15	451.58
16	CHCl3	CHLOROFORM	56.6178	-3.2462E+03	-1.8700E+01	9.5150E-03	1.1553E-12	209.63	536.40
17	CHF3	TRIFLUOROMETHANE	34.4731	-1.5956E+03	-1.0394E+01	2.1739E-09	1.2925E-05	117.97	298.89
18	CHI3	TRIIODOMETHANE	257.5138	-1.0203E+04	-9.8604E+01	7.4485E-02	-2.1057E-05	410.15	794.55
19	CHN	HYDROGEN CYANIDE	-57.5717	-3.5182E+02	2.9640E+01	-4.7820E-02	2.8550E-05	259.91	456.65
20	CHNS	ISOTHIOCYANIC-ACID	---------	---------	---------	---------	---------	---------	---------
21	CH2BrCl	BROMOCHLOROMETHANE	2.7704	-2.0139E+03	3.7817E+00	-1.3241E-02	8.1979E-06	185.20	557.00
22	CH2Br2	DIBROMOMETHANE	35.3525	-3.0445E+03	-9.5972E+00	5.8258E-10	2.9443E-06	220.60	611.00
23	CH2ClF	CHLOROFLUOROMETHANE	455.6813	-9.4306E+03	-1.9947E+02	2.9158E-01	-1.5799E-04	218.15	424.91
24	CH2Cl2	DICHLOROMETHANE	32.5609	-2.5166E+03	-8.8015E+00	1.2934E-10	3.3194E-06	178.01	510.00
25	CH2F2	DIFLUOROMETHANE	48.7353	-2.0703E+03	-1.6775E+01	1.2809E-02	3.3704E-13	137.00	351.60
26	CH2I2	DIIODOMETHANE	42.4554	-4.0527E+03	-1.1711E+01	-2.8020E-10	2.2058E-06	279.25	747.00
27	CH2O	FORMALDEHYDE	41.9603	-2.1355E+03	-1.3765E+01	9.5680E-03	-5.1101E-12	181.15	408.00
28	CH2O2	FORMIC ACID	27.9278	-2.5976E+03	-7.2489E+00	6.4110E-10	3.9421E-06	281.55	580.00
29	CH3Br	METHYL BROMIDE	29.3988	-2.0406E+03	-7.9966E+00	-4.1899E-10	5.0174E-06	179.47	467.00
30	CH3Cl	METHYL CHLORIDE	25.7244	-1.7503E+03	-6.7151E+00	-1.2956E-09	4.4341E-06	175.43	416.25
31	CH3Cl3Si	METHYL TRICHLOROSILANE	20.1131	-2.0730E+03	-4.4776E+00	3.1120E-10	1.7615E-06	195.35	517.00
32	CH3F	METHYL FLUORIDE	55.2801	-1.8879E+03	-2.0268E+01	1.8980E-02	-3.1692E-12	131.35	317.70
33	CH3I	METHYL IODIDE	-20.3718	-1.2536E+03	1.3645E+01	-2.6955E-02	1.6389E-05	206.70	528.00
34	CH3NO	FORMAMIDE	12.0765	-3.6151E+03	-3.2887E-01	9.1384E-10	-4.0029E-06	275.70	493.00
35	CH3NO2	NITROMETHANE	35.8372	-3.0979E+03	-9.7786E+00	-4.3921E-10	3.4336E-06	244.60	588.15
36	CH3NO2	METHYL-NITRITE	---------	---------	---------	---------	---------	---------	---------
37	CH3NO3	METHYL-NITRATE	---------	---------	---------	---------	---------	---------	---------
38	CH4	METHANE	14.6667	-5.7097E+02	-3.3373E+00	2.1999E-09	1.3096E-05	90.67	190.58
39	CH4Cl2Si	METHYL DICHLOROSILANE	32.5266	-2.3539E+03	-9.0702E+00	7.2608E-10	4.9896E-06	182.55	483.00
40	CH4O	METHANOL	45.6171	-3.2447E+03	-1.3988E+01	6.6365E-03	-1.0507E-13	175.47	512.58
41	CH4O3S	METHANESULFONIC ACID	-58.6166	-3.3580E+03	3.2320E+01	-5.3174E-02	2.6886E-05	292.81	561.00
42	CH4S	METHYL MERCAPTAN	-1.7459	-1.7090E+03	6.6886E+00	-2.5397E-02	1.8914E-05	150.18	469.95
43	CH5ClSi	METHYL CHLOROSILANE	39.5604	-2.3459E+03	-1.1829E+01	3.0032E-10	7.8573E-06	139.05	442.00
44	CH5N	METHYLAMINE	30.5366	-2.2074E+03	-8.0919E+00	-2.7828E-11	3.5234E-06	179.69	430.05
45	CH6Si	METHYL SILANE	-1.8743	-9.7410E+02	5.4592E+00	-2.0420E-02	2.0048E-05	116.34	352.50
46	CN4O8	TETRANITROMETHANE	-2.1747	-2.5151E+03	6.4929E+00	-1.7444E-02	9.0071E-06	287.05	540.00
47	CO	CARBON MONOXIDE	51.8145	-7.8824E+02	-2.2734E+01	5.1225E-02	6.1896E-11	68.15	132.92
48	COS	CARBONYL SULFIDE	36.8556	-1.7187E+03	-1.2036E+01	8.9612E-03	-1.1127E-13	134.35	378.80
49	CO2	CARBON DIOXIDE	35.0169	-1.5119E+03	-1.1334E+01	9.3368E-03	1.7136E-09	216.58	304.19
50	CS2	CARBON DISULFIDE	25.1475	-2.0439E+03	-6.7794E+00	3.4828E-03	-1.0105E-14	161.58	552.00
51	C2BrF3	BROMOTRIFLUOROETHYLENE	-3.4016	-1.3242E+03	6.2543E+00	-1.8654E-02	1.3759E-05	113.00	432.00
52	C2Br2F4	1,2-DIBROMOTETRAFLUOROETHANE	6.4970	-1.8310E+03	1.9094E+00	-1.0876E-02	7.6921E-06	162.65	487.70
53	C2ClF3	CHLOROTRIFLUOROETHYLENE	81.1728	-2.9152E+03	-3.0175E+01	2.3253E-02	1.1177E-13	115.00	379.15
54	C2ClF5	CHLOROPENTAFLUOROETHANE	-12.0581	-1.0154E+03	1.0548E+01	-3.0256E-02	2.4967E-05	173.71	353.15
55	C2Cl2F4	1,2-DICHLOROTETRAFLUOROETHANE	59.2670	-2.7486E+03	-2.0474E+01	1.2814E-02	1.5578E-12	179.15	418.85
56	C2Cl3F3	1,1,2-TRICHLOROTRIFLUOROETHANE	33.7192	-2.5323E+03	-9.3175E+00	1.4550E-08	3.9414E-06	238.15	487.25
57	C2Cl4	TETRACHLOROETHYLENE	23.3960	-2.6888E+03	-5.3312E+00	-1.1436E-10	9.2370E-07	250.80	620.00
58	C2Cl4F2	1,1,2,2-TETRACHLORODIFLUOROETHANE	-38.1282	-1.2604E+03	2.1347E+01	-3.4436E-02	1.7407E-05	299.15	551.00
59	C2Cl4O	TRICHLOROACETYL CHLORIDE	-13.2301	-2.0316E+03	1.0702E+01	-2.0537E-02	1.0411E-05	273.15	590.00
60	C2Cl6	HEXACHLOROETHANE	-366.2243	6.6789E+03	1.5563E+02	-1.5563E-01	5.5600E-05	306.15	698.00
61	C2F4	TETRAFLUOROETHYLENE	30.4038	-1.5265E+03	-8.7830E+00	-2.6364E-09	9.4381E-06	142.00	306.45
62	C2F6	HEXAFLUOROETHANE	21.4351	-1.2293E+03	-5.6610E+00	3.6729E-03	4.8788E-11	172.45	292.80
63	C2HBrClF3	HALOTHANE	-26.6804	-1.5205E+03	1.7746E+01	-3.9345E-02	2.3399E-05	223.15	521.00
64	C2HClF2	2-CHLORO-1,1-DIFLUOROETHYLENE	4.0143	-1.4420E+03	3.2509E+00	-1.6543E-02	1.4174E-05	134.65	400.55
65	C2HCl3	TRICHLOROETHYLENE	23.6735	-2.3763E+03	-5.8275E+00	1.9586E-03	2.8882E-14	188.40	571.00
66	C2HCl3O	DICHLOROACETYL CHLORIDE	-25.0907	-1.7624E+03	1.5817E+01	-2.6632E-02	1.3107E-05	298.15	579.00
67	C2HCl3O	TRICHLOROACETALDEHYDE	-0.9737	-2.2553E+03	5.7694E+00	-1.6755E-02	9.6236E-06	216.00	565.00
68	C2HCl5	PENTACHLOROETHANE	72.1134	-4.6522E+03	-2.3777E+01	9.6908E-03	-3.3886E-13	244.15	665.00
69	C2HF3	TRIFLUOROETHENE	117.5502	-2.8369E+03	-4.9719E+01	7.7625E-02	-4.9716E-05	180.15	244.15
70	C2HF3O2	TRIFLUOROACETIC ACID	63.4449	-3.6769E+03	-2.1130E+01	1.0777E-02	4.8480E-12	257.90	491.25
71	C2HF5	PENTAFLUOROETHANE	-19.0457	-9.3948E+02	1.4257E+01	-4.0845E-02	3.4587E-05	170.15	342.00
72	C2H2	ACETYLENE	72.6005	-2.3098E+03	-2.7223E+01	2.3721E-02	-3.8593E-10	189.00	308.32
73	C2H2Br4	1,1,2,2-TETRABROMOETHANE	77.0296	-6.6208E+03	-2.2976E+01	-2.9801E-10	3.7484E-06	273.15	824.00
74	C2H2Cl2	1,1-DICHLOROETHYLENE	-16.5419	-1.6655E+03	1.3923E+01	-4.0958E-02	2.9995E-05	150.65	482.00
75	C2H2Cl2	cis-1,2-DICHLOROETHYLENE	55.9403	-3.1677E+03	-1.8572E+01	9.8828E-03	5.7644E-14	193.15	527.00
76	C2H2Cl2	trans-1,2-DICHLOROETHYLENE	48.4574	-3.0496E+03	-1.4694E+01	-2.1262E-09	7.3465E-06	223.35	508.00
77	C2H2Cl2O	CHLOROACETYL CHLORIDE	65.2863	-4.1288E+03	-2.1342E+01	9.2900E-03	-9.9800E-13	251.15	581.00
78	C2H2Cl2O	DICHLOROACETALDEHYDE	-10.6633	-2.1012E+03	1.0310E+01	-2.4503E-02	1.4040E-05	223.00	555.00

162

Table 7-1 VAPOR PRESSURE - ORGANIC COMPOUNDS (continued)

$$\log_{10} P = A + B/T + C \log_{10} T + D\,T + E\,T^2 \qquad (P - mm\ Hg,\ T - K)$$

NO	FORMULA	NAME	A	B	C	D	E	TMIN	TMAX
79	C2H2Cl2O2	DICHLOROACETIC ACID	-7.2806	-3.3706E+03	9.3771E+00	-2.0832E-02	9.5091E-06	286.55	686.00
80	C2H2Cl3F	1,1,1-TRICHLOROFLUOROETHANE	7.3234	-2.1968E+03	1.7372E+00	-1.0298E-02	6.5023E-06	173.00	565.00
81	C2H2Cl4	1,1,1,2-TETRACHLOROETHANE	-1.0712	-2.5074E+03	6.1536E+00	-1.8763E-02	1.0462E-05	202.94	624.00
82	C2H2Cl4	1,1,2,2-TETRACHLOROETHANE	56.2356	-4.4615E+03	-1.6556E+01	-3.5724E-10	4.0425E-06	229.35	645.00
83	C2H2F2	1,1-DIFLUOROETHYLENE	-5.4320	-8.3177E+02	7.4635E+00	-2.7851E-02	2.8548E-05	129.15	302.80
84	C2H2F2	cis-1,2-DIFLUOROETHENE	86.1060	-2.3731E+03	-3.5088E+01	4.8677E-02	-2.7786E-05	195.15	277.75
85	C2H2F2	trans-1,2-DIFLUOROETHENE	86.1060	-2.3731E+03	-3.5088E+01	4.8677E-02	-2.7786E-05	195.15	277.75
86	C2H2F4	1,1,1,2-TETRAFLUOROETHANE	-10.4006	-1.1582E+03	9.8998E+00	-2.8823E-02	2.2877E-05	172.15	380.00
87	C2H2O	KETENE	15.4553	-1.3110E+03	-2.8700E+00	-2.0015E-10	6.8688E-07	122.00	370.00
88	C2H2O4	OXALIC ACID	-79.0432	-3.0545E+03	3.9089E+01	-4.4389E-02	1.4994E-05	462.65	804.00
89	C2H3Br	VINYL BROMIDE	-10.9281	-1.1619E+03	9.3115E+00	-2.2655E-02	1.7522E-05	135.35	473.00
90	C2H3Cl	VINYL CHLORIDE	52.9654	-2.5016E+03	-1.7914E+01	1.0821E-02	-4.5310E-14	119.36	432.00
91	C2H3ClF2	1-CHLORO-1,1-DIFLUOROETHANE	12.3975	-1.3494E+03	-1.8528E+00	-3.2595E-11	1.3316E-06	142.35	410.20
92	C2H3ClO	ACETYL CHLORIDE	79.4009	-3.7788E+03	-2.7926E+01	1.6209E-02	-9.9562E-14	160.30	508.00
93	C2H3ClO	CHLOROACETALDEHYDE	-31.3274	-1.4938E+03	1.8610E+01	-3.1242E-02	1.5886E-05	293.00	555.00
94	C2H3ClO2	CHLOROACETIC ACID	42.6726	-4.5970E+03	-1.1348E+01	-2.8515E-10	1.7995E-06	333.15	686.00
95	C2H3ClO2	METHYL CHLOROFORMATE	2.4278	-2.3168E+03	4.7351E+00	-1.7920E-02	1.1303E-05	192.00	525.00
96	C2H3Cl3	1,1,1-TRICHLOROETHANE	36.5468	-2.8421E+03	-1.0205E+01	-2.6369E-09	3.7075E-06	242.75	545.00
97	C2H3Cl3	1,1,2-TRICHLOROETHANE	25.0845	-2.7368E+03	-5.9182E+00	2.5155E-10	1.1831E-06	236.50	602.00
98	C2H3F	VINYL FLUORIDE	35.4702	-1.4160E+03	-1.2180E+01	1.2462E-02	1.5601E-13	112.65	327.80
99	C2H3F3	1,1,1-TRIFLUOROETHANE	27.0988	-1.6340E+03	-7.3449E+00	5.3839E-09	6.0250E-06	161.85	346.25
100	C2H3N	ACETONITRILE	23.1953	-2.3389E+03	-5.4954E+00	7.9894E-10	2.3293E-06	229.32	545.50
101	C2H3NO	METHYL ISOCYANATE	-20.1597	-1.1878E+03	1.3274E+01	-2.4414E-02	1.3907E-05	256.15	505.00
102	C2H4	ETHYLENE	18.7964	-9.9962E+02	-4.5788E+00	9.9746E-11	6.7880E-06	104.01	282.36
103	C2H4Br2	1,1-DIBROMOETHANE	25.1101	-2.8244E+03	-5.7669E+00	3.5499E-11	4.5284E-07	210.15	628.00
104	C2H4Br2	1,2-DIBROMOETHANE	16.8759	-2.4267E+03	-3.0891E+00	-6.0088E-10	3.5901E-07	282.94	650.15
105	C2H4Cl2	1,1-DICHLOROETHANE	33.3800	-2.6102E+03	-9.1336E+00	-2.8388E-11	3.7323E-06	176.19	523.00
106	C2H4Cl2	1,2-DICHLOROETHANE	48.4226	-3.1803E+03	-1.5370E+01	7.2935E-03	2.6844E-14	237.49	561.00
107	C2H4Cl2O	BIS(CHLOROMETHYL)ETHER	-3.4945	-2.2305E+03	6.7740E+00	-1.7332E-02	9.5511E-06	231.65	579.00
108	C2H4F2	1,1-DIFLUOROETHANE	43.6454	-2.1089E+03	-1.4530E+01	1.0181E-02	4.1529E-14	156.15	386.60
109	C2H4F2	1,2-DIFLUOROETHANE	-10.9499	-1.4207E+03	9.6589E+00	-2.2317E-02	1.4152E-05	215.00	476.00
110	C2H4I2	1,2-DIIODOETHANE	-12.1695	-2.5085E+03	9.9834E+00	-1.6693E-02	6.9027E-06	371.15	749.91
111	C2H4O	ACETALDEHYDE	87.3702	-3.6822E+03	-3.1548E+01	2.0114E-02	5.5341E-13	150.15	461.00
112	C2H4O	ETHYLENE OXIDE	39.9235	-2.3595E+03	-1.2517E+01	6.9835E-03	-1.1101E-13	160.71	469.15
113	C2H4OS	THIOACETIC-ACID	139.8046	-6.9268E+03	-4.8917E+01	1.9908E-02	1.4987E-06	313.15	577.34
114	C2H4O2	ACETIC ACID	28.3756	-2.9734E+03	-7.0320E+00	-1.5051E-09	2.1806E-06	289.81	592.71
115	C2H4O2	METHYL FORMATE	28.9576	-2.3582E+03	-7.4848E+00	7.4384E-10	2.7013E-06	174.15	487.20
116	C2H4S	THIACYCLOPROPANE	53.8292	-2.9773E+03	-1.8176E+01	1.2693E-02	-2.8763E-06	238.15	555.00
117	C2H5Br	BROMOETHANE	36.1816	-2.5170E+03	-1.0329E+01	-2.3368E-10	5.4956E-06	154.55	503.80
118	C2H5Cl	ETHYL CHLORIDE	28.3448	-2.0788E+03	-7.5387E+00	-1.6384E-11	4.0550E-06	136.75	460.35
119	C2H5ClO	2-CHLOROETHANOL	8.5478	-2.5196E+03	7.0198E-01	-3.9592E-03	2.2786E-06	205.65	585.00
120	C2H5F	ETHYL FLUORIDE	1.3504	-1.1935E+03	4.0341E+00	-1.6032E-02	1.4458E-05	129.95	375.31
121	C2H5I	ETHYL IODIDE	31.4422	-2.5719E+03	-8.4867E+00	-9.0736E-11	3.5710E-06	162.05	561.00
122	C2H5N	ETHYLENEIMINE	8.7006	-2.2935E+03	1.8063E+00	-1.2965E-02	8.0129E-06	195.20	537.00
123	C2H5NO	ACETAMIDE	-413.1683	8.1328E+03	1.7290E+02	-1.6059E-01	5.3892E-05	354.15	761.00
124	C2H5NO	N-METHYLFORMAMIDE	0.9176	-3.2184E+03	5.1705E+00	-1.3637E-02	6.2034E-06	269.35	721.00
125	C2H5NO2	NITROETHANE	2.2734	-2.3901E+03	4.2953E+00	-1.4564E-02	8.7219E-06	183.63	593.00
126	C2H5NO3	ETHYL-NITRATE	123.5101	-4.7884E+03	-4.8020E+01	5.0887E-02	-2.2419E-05	273.15	335.15
127	C2H6	ETHANE	20.6973	-1.1341E+03	-5.2514E+00	-9.8774E-11	6.7329E-06	90.35	305.42
128	C2H6AlCl	DIMETHYLALUMINUM CHLORIDE	15.2933	-2.5478E+03	-1.9137E+00	-3.3108E-03	1.6919E-06	252.15	619.00
129	C2H6O	DIMETHYL ETHER	20.2699	-1.5914E+03	-4.6530E+00	-1.3178E-10	2.5623E-06	131.65	400.10
130	C2H6O	ETHANOL	23.8442	-2.8642E+03	-5.0474E+00	3.7448E-11	2.7361E-07	159.05	516.25
131	C2H6OS	DIMETHYL SULFOXIDE	45.4653	-4.0439E+03	-1.3210E+01	1.0981E-07	6.4145E-06	291.67	465.27
132	C2H6O2	ETHYLENE GLYCOL	82.4062	-6.3472E+03	-2.5433E+01	-2.3732E-09	8.7467E-06	260.15	645.00
133	C2H6O4S	DIMETHYL SULFATE	33.9406	-3.8530E+03	-8.5921E+00	-1.1705E-10	8.2260E-07	241.35	758.00
134	C2H6S	DIMETHYL SULFIDE	37.2604	-2.4251E+03	-1.1384E+01	5.8122E-03	8.5893E-14	174.88	503.04
135	C2H6S	ETHYL MERCAPTAN	29.2763	-2.2725E+03	-7.7769E+00	-3.8954E-11	3.5170E-06	125.26	499.15
136	C2H6S2	DIMETHYL DISULFIDE	36.2320	-3.1241E+03	-9.9328E+00	2.2831E-11	3.1730E-06	188.44	606.00
137	C2H7N	DIMETHYLAMINE	36.9182	-2.4965E+03	-1.0417E+01	-1.6287E-09	4.6496E-06	180.96	437.65
138	C2H7N	ETHYLAMINE	33.2962	-2.4307E+03	-9.0779E+00	-1.3848E-09	3.8183E-06	192.15	456.15
139	C2H7NO	MONOETHANOLAMINE	72.9125	-5.8595E+03	-2.1914E+01	-7.1511E-10	5.9841E-06	283.65	638.00
140	C2H8N2	ETHYLENEDIAMINE	94.0887	-5.2914E+03	-3.2204E+01	1.4860E-02	1.8928E-13	284.29	593.00
141	C2H8Si	DIMETHYL SILANE	40.1984	-2.1333E+03	-1.2269E+01	3.7308E-10	9.1918E-06	122.93	402.00
142	C2N2	CYANOGEN	36.3490	-2.1975E+03	-1.0483E+01	9.3862E-08	6.6894E-06	245.25	400.15
143	C3F6	HEXAFLUOROPROPYLENE	22.6203	-1.5575E+03	-5.8544E+00	2.6148E-03	-9.7789E-14	116.65	368.00
144	C3F6O	HEXAFLUOROACETONE	-8.0192	-1.4689E+03	1.0266E+01	-3.9760E-02	3.4907E-05	147.70	357.14
145	C3F8	OCTAFLUOROPROPANE	9.4776	-1.2635E+03	-1.4699E-01	-4.9570E-03	4.7635E-06	125.46	345.05
146	C3H2N2	MALONONITRILE	29.9195	-3.6823E+03	-7.4177E+00	7.4993E-10	1.7299E-06	304.99	715.00
147	C3H3Cl	PROPARGYL CHLORIDE	-26.7697	-1.0574E+03	1.5749E+01	-2.4927E-02	1.2863E-05	293.00	541.00
148	C3H3N	ACRYLONITRILE	35.9210	-2.7763E+03	-1.0101E+01	-3.1547E-10	4.7299E-06	189.63	535.00
149	C3H3NO	OXAZOLE	3.6339	-2.0244E+03	3.3305E+00	-1.2169E-02	7.4895E-06	189.15	554.00
150	C3H4	METHYLACETYLENE	49.7385	-2.3298E+03	-1.6810E+01	1.1084E-02	-9.6739E-13	170.45	402.39
151	C3H4	PROPADIENE	47.2240	-2.2683E+03	-1.4957E+01	1.0812E-10	1.2699E-05	136.87	393.15
152	C3H4Cl2	2,3-DICHLOROPROPENE	3.6589	-2.0331E+03	3.0769E+00	-1.0877E-02	6.4885E-06	191.50	577.00
153	C3H4O	ACROLEIN	57.9815	-3.0933E+03	-1.9638E+01	1.1486E-02	-2.3854E-14	185.45	506.00
154	C3H4O	PROPARGYL ALCOHOL	46.9547	-3.5359E+03	-1.4526E+01	6.8506E-03	2.0410E-13	221.35	580.00
155	C3H4O2	ACRYLIC ACID	23.0607	-3.1347E+03	-4.8813E+00	4.3690E-04	-4.9161E-13	286.65	615.00
156	C3H4O2	beta-PROPIOLACTONE	3.8203	-2.9265E+03	3.7646E+00	-1.2141E-02	5.9783E-06	239.75	686.00

163

Table 7-1 VAPOR PRESSURE - ORGANIC COMPOUNDS (continued)

			$\log_{10} P = A + B/T + C \log_{10} T + D T + E T^2$					(P - mm Hg, T - K)	
NO	FORMULA	NAME	A	B	C	D	E	TMIN	TMAX
157	C3H4O2	VINYL FORMATE	11.5695	-2.1425E+03	5.8720E-02	-8.4252E-03	5.3903E-06	200.00	498.00
158	C3H4O3	ETHYLENE CARBONATE	-7.8597	-3.6466E+03	9.4034E+00	-1.8774E-02	7.6320E-06	309.55	790.00
159	C3H4O3	PYRUVIC ACID	-21.8344	-2.2634E+03	1.4247E+01	-2.2375E-02	1.0674E-05	286.75	634.52
160	C3H5Br	3-BROMO-1-PROPENE	127.6280	-4.7573E+03	-4.9438E+01	4.8325E-02	-1.7990E-05	290.15	540.20
161	C3H5Cl	2-CHLOROPROPENE	7.2269	-1.6159E+03	1.3121E+00	-9.3181E-03	7.1679E-06	135.75	478.00
162	C3H5Cl	3-CHLOROPROPENE	8.6085	-1.7512E+03	6.0792E-01	-7.1675E-03	5.2228E-06	138.65	514.15
163	C3H5ClO	alpha-EPICHLOROHYDRIN	24.7640	-2.8846E+03	-5.6252E+00	-1.1011E-10	5.3331E-07	215.95	610.00
164	C3H5ClO2	METHYL CHLOROACETATE	-12.3383	-2.5698E+03	1.1516E+01	-2.6540E-02	1.4106E-05	241.03	600.00
165	C3H5ClO2	ETHYL CHLOROFORMATE	5.0713	-3.4293E+03	6.0125E+00	-2.9459E-02	1.8989E-05	192.00	503.15
166	C3H5Cl3	1,2,3-TRICHLOROPROPANE	-3.9501	-2.4501E+03	6.6887E+00	-1.4991E-02	7.3402E-06	258.45	652.00
167	C3H5I	3-IODO-1-PROPENE	-16.9304	-1.7651E+03	1.2181E+01	-2.2472E-02	1.1296E-05	293.15	595.81
168	C3H5N	PROPIONITRILE	33.7908	-2.9113E+03	-9.1506E+00	2.1173E-11	3.2756E-06	180.26	564.40
169	C3H5NO	ACRYLAMIDE	17.0034	-4.4434E+03	-1.7158E+00	2.0063E-06	-8.0394E-10	357.65	476.50
170	C3H5NO	HYDRACRYLONITRILE	14.6476	-3.2574E+03	-1.9631E+00	1.1058E-10	4.4389E-07	227.15	690.00
171	C3H5NO	LACTONITRILE	7.8257	-4.2877E+03	4.3978E+00	-2.0821E-02	1.0779E-05	233.00	643.00
172	C3H5N3O9	NITROGLYCERINE	42.9395	-6.2087E+03	-1.0088E+01	-1.9266E-03	8.8162E-07	286.15	680.00
173	C3H6	CYCLOPROPANE	37.8180	-1.8661E+03	-1.2278E+01	8.5712E-03	-2.9652E-13	145.59	397.91
174	C3H6	PROPYLENE	24.5390	-1.5072E+03	-6.4800E+00	-4.2845E-11	5.4982E-06	89.00	364.76
175	C3H6Br2	1,2-DIBROMOPROPANE	89.3805	-4.6368E+03	-3.2139E+01	2.4424E-02	-7.6841E-06	323.15	525.15
176	C3H6Cl2	1,1-DICHLOROPROPANE	34.1365	-2.8930E+03	-9.2386E+00	-6.6629E-06	2.9381E-06	200.00	560.00
177	C3H6Cl2	1,2-DICHLOROPROPANE	5.4819	-2.1918E+03	2.6014E+00	-1.1751E-02	7.3435E-06	172.71	572.00
178	C3H6Cl2	1,3-DICHLOROPROPANE	41.0896	-3.4298E+03	-1.1573E+01	1.0184E-10	3.4843E-06	173.65	603.00
179	C3H6Cl2	2,2-DICHLOROPROPANE	-35.0912	-1.1632E+03	2.0045E+01	-3.3493E-02	1.7325E-05	267.15	539.46
180	C3H6I2	1,2-DIIODOPROPANE	953.9088	-3.0672E+04	-3.7702E+02	2.9971E-01	-8.7922E-05	398.15	780.49
181	C3H6O	ACETONE	28.5884	-2.4690E+03	-7.3510E+00	2.8025E-10	2.7361E-06	178.45	508.20
182	C3H6O	ALLYL ALCOHOL	21.3978	-2.9525E+03	-3.8137E+00	-2.7145E-03	1.8811E-06	144.15	545.05
183	C3H6O	METHYL VINYL ETHER	9.2628	-1.8259E+03	1.3189E+00	-1.3985E-02	1.0826E-05	151.15	437.00
184	C3H6O	n-PROPIONALDEHYDE	26.1637	-2.3059E+03	-6.5289E+00	-2.3065E-10	2.5454E-06	193.15	496.00
185	C3H6O	1,2-PROPYLENE OXIDE	38.5381	-2.6310E+03	-1.1104E+01	4.2178E-10	5.5025E-06	161.22	482.25
186	C3H6O	1,3-PROPYLENE OXIDE	-8.8497	-1.5158E+03	8.4393E+00	-1.8036E-02	1.0561E-05	230.00	520.00
187	C3H6O2	ETHYL FORMATE	29.9404	-2.5263E+03	-7.8090E+00	-1.0111E-09	2.7447E-06	193.55	508.40
188	C3H6O2	METHYL ACETATE	33.7235	-2.7204E+03	-9.1182E+00	-9.4316E-11	3.3102E-06	175.15	506.80
189	C3H6O2	PROPIONIC ACID	20.2835	-3.1165E+03	-3.6015E+00	-1.3892E-03	7.1801E-07	252.45	604.00
190	C3H6O2S	3-MERCAPTOPROPIONIC ACID	20.6694	-3.7287E+03	-3.8113E+00	-1.3575E-04	5.9461E-08	290.65	729.00
191	C3H6O3	LACTIC ACID	-27.0836	-3.9661E+03	2.0233E+01	-4.2176E-02	2.0310E-05	291.15	616.00
192	C3H6O3	METHOXYACETIC ACID	89.9021	-7.0625E+03	-2.7494E+01	-2.3273E-09	6.2004E-06	281.00	691.00
193	C3H6O3	TRIOXANE	-41.2762	-1.3985E+03	2.2474E+01	-3.2609E-02	1.4871E-05	334.65	604.40
194	C3H6S	THIACYCLOBUTANE	55.0098	-3.3957E+03	-1.8134E+01	1.0468E-02	-1.6820E-06	268.15	603.00
195	C3H7Br	1-BROMOPROPANE	-9.0284	-1.8916E+03	9.6910E+00	-2.7013E-02	1.7877E-05	163.15	544.00
196	C3H7Br	2-BROMOPROPANE	31.3032	-2.4924E+03	-8.4645E+00	1.6459E-10	3.7859E-06	184.15	532.00
197	C3H7Cl	ISOPROPYL CHLORIDE	115.1259	-4.6278E+03	-4.2197E+01	2.5260E-02	-9.8906E-13	155.97	489.00
198	C3H7Cl	n-PROPYL CHLORIDE	32.3325	-2.4850E+03	-8.8052E+00	8.9119E-11	3.6803E-06	150.35	503.15
199	C3H7F	1-FLUOROPROPANE	78.2889	-2.8266E+03	-3.0340E+01	3.8006E-02	-1.9813E-05	203.15	305.15
200	C3H7F	2-FLUOROPROPANE	-7.4089	-1.4919E+03	9.3510E+00	-3.1910E-02	2.4740E-05	224.15	415.68
201	C3H7I	ISOPROPYL IODIDE	33.2023	-2.7569E+03	-9.0585E+00	-1.2099E-10	3.5044E-06	183.15	578.00
202	C3H7I	n-PROPYL IODIDE	1.2733	-2.0214E+03	4.1138E+00	-1.2477E-02	7.6780E-06	171.85	593.00
203	C3H7N	ALLYLAMINE	2.0234	-2.0506E+03	4.5707E+00	-1.6976E-02	1.1139E-05	184.95	505.00
204	C3H7N	PROPYLENEIMINE	-8.2010	-1.7425E+03	8.5486E+00	-1.9631E-02	1.1456E-05	229.00	529.00
205	C3H7NO	N,N-DIMETHYLFORMAMIDE	-47.9857	-2.3850E+03	2.8800E+01	-5.8596E-02	3.1386E-05	212.72	647.00
206	C3H7NO	N-METHYLACETAMIDE	-13.3059	-3.0464E+03	1.1431E+01	-2.1324E-02	9.2790E-06	301.15	718.00
207	C3H7NO2	1-NITROPROPANE	2.0216	-2.5960E+03	4.8178E+00	-1.7338E-02	1.0562E-05	169.16	605.00
208	C3H7NO2	2-NITROPROPANE	1.2047	-2.3533E+03	4.6729E+00	-1.4843E-02	8.8798E-06	181.83	594.00
209	C3H7NO3	PROPYL-NITRATE	63.7382	-3.5281E+03	-2.2829E+01	2.2850E-02	-9.6496E-06	273.15	345.15
210	C3H7NO3	ISOPROPYL-NITRATE	373.3215	-1.0882E+04	-1.5319E+02	1.6946E-01	-7.6001E-05	273.15	345.15
211	C3H8	PROPANE	21.4469	-1.4627E+03	-5.2610E+00	3.2820E-11	3.7349E-06	85.46	369.82
212	C3H8O	ISOPROPANOL	38.2363	-3.5513E+03	-1.0031E+01	-3.4740E-10	1.7367E-06	185.28	508.31
213	C3H8O	METHYL ETHYL ETHER	149.5481	-5.6024E+03	-5.5552E+01	3.3138E-02	2.3724E-12	160.00	437.80
214	C3H8O	n-PROPANOL	31.5155	-3.4570E+03	-7.5235E+00	-4.2870E-11	1.3029E-07	146.95	536.71
215	C3H8O2	2-METHOXYETHANOL	153.3786	-7.1181E+03	-5.5450E+01	2.9037E-02	-4.8680E-13	188.05	564.00
216	C3H8O2	METHYLAL	3.1192	-1.8848E+03	3.8791E+00	-1.6092E-02	1.1272E-05	168.35	480.60
217	C3H8O2	1,2-PROPYLENE GLYCOL	90.2930	-6.6968E+03	-2.8109E+01	-1.3326E-10	9.3651E-06	213.15	626.00
218	C3H8O2	1,3-PROPYLENE GLYCOL	27.4723	-4.0200E+03	-6.2839E+00	-6.7098E-10	2.2952E-06	246.45	658.00
219	C3H8O3	GLYCEROL	-62.7929	-3.6585E+03	3.4249E+01	-5.1940E-02	2.2830E-05	291.15	723.00
220	C3H8S	n-PROPYLMERCAPTAN	35.1293	-2.7533E+03	-9.7127E+00	-3.1666E-11	3.6742E-06	159.95	536.00
221	C3H8S	ISOPROPYL MERCAPTAN	35.0477	-2.6208E+03	-9.7700E+00	2.6817E-11	4.0579E-06	142.61	517.00
222	C3H8S	ETHYL-METHYL-SULFIDE	26.9924	-2.5860E+03	-6.1825E+00	-4.3595E-03	5.4459E-06	247.15	532.80
223	C3H9N	n-PROPYLAMINE	24.6402	-2.3152E+03	-5.8711E+00	-4.6258E-11	1.5820E-06	190.15	496.95
224	C3H9N	ISOPROPYLAMINE	54.7199	-3.0557E+03	-1.7947E+01	9.0585E-03	8.8279E-13	177.95	471.85
225	C3H9N	TRIMETHYLAMINE	58.6807	-2.6860E+03	-2.0360E+01	1.3131E-02	-6.5630E-13	156.08	433.25
226	C3H9NO	1-AMINO-2-PROPANOL	-15.5527	-3.5262E+03	1.4251E+01	-3.2309E-02	1.5978E-05	274.89	614.00
227	C3H9NO	3-AMINO-1-PROPANOL	-12.9378	-3.9097E+03	1.3186E+01	-3.0020E-02	1.4189E-05	284.15	649.00
228	C3H9NO	METHYLETHANOLAMINE	-11.0444	-3.1081E+03	1.1331E+01	-2.5709E-02	1.2719E-05	268.65	630.00
229	C3H9O4P	TRIMETHYL PHOSPHATE	72.2176	-4.5518E+03	-2.4613E+01	1.3118E-02	-4.7970E-14	227.00	764.00
230	C3H10N2	1,2-PROPANEDIAMINE	56.4745	-4.0665E+03	-1.7745E+01	7.1207E-03	-5.6869E-13	236.53	587.00
231	C3H10Si	TRIMETHYL SILANE	4.9825	-1.4831E+03	2.4501E+00	-1.2940E-02	1.0503E-05	137.26	432.00
232	C4Cl4S	TETRACHLOROTHIOPHENE	22.5714	-3.6797E+03	-4.5565E+00	-2.6662E-04	1.1279E-07	301.97	753.00
233	C4Cl6	HEXACHLORO-1,3-BUTADIENE	35.5910	-4.1380E+03	-9.0606E+00	2.9025E-10	5.9012E-07	252.15	741.00
234	C4F8	OCTAFLUORO-2-BUTENE	33.3539	-1.9951E+03	-1.0130E+01	5.7197E-03	4.0766E-13	138.15	392.00

Table 7-1 VAPOR PRESSURE - ORGANIC COMPOUNDS (continued)

			$\log_{10} P = A + B/T + C \log_{10} T + D T + E T^2$					(P - mm Hg, T - K)	
NO	FORMULA	NAME	A	B	C	D	E	TMIN	TMAX
235	C4F8	OCTAFLUOROCYCLOBUTANE	-41.2314	-8.0603E+02	2.4009E+01	-5.1039E-02	3.5019E-05	232.96	388.37
236	C4F10	DECAFLUOROBUTANE	0.8538	-1.6238E+03	5.4162E+00	-2.4729E-02	2.0982E-05	144.95	386.35
237	C4H2	BUTADIYNE(BIACETYLENE)	1450.9484	-2.7756E+04	-6.4784E+02	1.0221E+00	-6.3551E-04	195.15	275.15
238	C4H2O3	MALEIC ANHYDRIDE	-42.9778	-1.6928E+03	2.2430E+01	-2.7916E-02	1.1707E-05	326.00	721.00
239	C4H4	VINYLACETYLENE	36.4556	-2.3569E+03	-1.0425E+01	-8.3758E-10	4.8441E-06	179.95	454.00
240	C4H4N2	SUCCINONITRILE	-10.5172	-3.6943E+03	1.0379E+01	-1.9214E-02	7.7119E-06	331.30	770.00
241	C4H4O	FURAN	24.9555	-2.1624E+03	-6.1066E+00	-2.4185E-10	2.0858E-06	187.55	490.15
242	C4H4O2	DIKETENE	17.2147	-2.6139E+03	-3.0030E+00	-3.6848E-10	1.4806E-07	266.65	616.00
243	C4H8O3	SUCCINIC ANHYDRIDE	44.8614	-5.0087E+03	-1.2203E+01	-1.1319E-01	2.3049E-06	393.00	811.00
244	C4H4O4	FUMARIC ACID	-179.4520	6.1572E+02	7.8000E+01	-7.1951E-02	2.2729E-05	560.15	771.00
245	C4H4O4	MALEIC ACID	-52.2664	-4.1120E+03	2.9193E+01	-4.0009E-02	1.4680E-05	403.45	773.00
246	C4H4S	THIOPHENE	36.6016	-2.9794E+03	-1.0104E+01	1.1445E-09	3.2472E-06	234.94	579.35
247	C4H5Cl	CHLOROPRENE	8.9353	-1.9176E+03	7.3836E-01	-8.4743E-03	6.0224E-06	143.15	525.00
248	C4H5N	trans-CROTONITRILE	43.4500	-3.2699E+03	-1.3368E+01	6.1392E-03	5.5131E-13	222.00	586.00
249	C4H5N	cis-CROTONITRILE	-4.2755	-2.0100E+03	6.6525E+00	-1.6048E-02	9.5110E-06	200.55	568.00
250	C4H5N	METHACRYLONITRILE	29.2563	-2.5669E+03	-7.7127E+00	1.0699E-09	3.2622E-06	237.35	554.00
251	C4H5N	PYRROLE	54.1597	-4.2745E+03	-1.5873E+01	-4.5171E-10	4.2238E-06	249.74	639.75
252	C4H5N	VINYLACETONITRILE	2.8218	-2.3780E+03	3.9704E+00	-1.3908E-02	8.3551E-06	186.15	584.00
253	C4H5NO2	METHYL CYANOACETATE	1.1272	-3.5604E+03	5.7936E+00	-1.7133E-02	8.1441E-06	260.08	687.00
254	C4H6	CYCLOBUTENE	68.1140	-2.6364E+03	-2.5728E+01	2.9287E-02	-1.2602E-05	196.15	446.33
255	C4H6	1,2-BUTADIENE	21.1068	-1.7990E+03	-4.9387E+00	-4.7821E-11	2.7787E-06	136.95	444.00
256	C4H6	1,3-BUTADIENE	30.0572	-1.9891E+03	-8.2922E+00	2.5664E-10	5.1334E-06	164.25	425.37
257	C4H6	DIMETHYLACETYLENE	21.8577	-2.0010E+03	-5.1054E+00	1.1247E-03	-3.2651E-12	240.91	488.15
258	C4H6	ETHYLACETYLENE	43.8278	-2.4255E+03	-1.4141E+01	8.2138E-03	7.4889E-14	147.43	443.20
259	C4H6Cl2	1,3-DICHLORO-trans-2-BUTENE	55.5016	-4.1951E+03	-1.6481E+01	-1.7816E-11	4.5523E-06	276.00	618.00
260	C4H6Cl2	1,4-DICHLORO-cis-2-BUTENE	4.7707	-2.6927E+03	3.1645E+00	-1.1767E-02	6.1930E-06	225.15	640.00
261	C4H6Cl2	1,4-DICHLORO-trans-2-BUTENE	-10.2934	-2.3720E+03	9.5309E+00	-1.8798E-02	9.0511E-06	274.15	646.00
262	C4H6Cl2	3,4-DICHLORO-1-BUTENE	2.0344	-2.3243E+03	4.2431E+00	-1.3712E-02	7.7644E-06	212.00	589.00
263	C4H6O	trans-CROTONALDEHYDE	34.3167	-3.1138E+03	-9.1374E+00	-4.2364E-10	2.5322E-06	196.65	571.00
264	C4H6O	2,5-DIHYDROFURAN	-21.9588	-1.3933E+03	1.4178E+01	-2.4916E-02	1.3232E-05	273.00	542.00
265	C4H6O	DIVINYL ETHER	-6.4142	-1.6906E+03	8.4954E+00	-2.6007E-02	1.8488E-05	172.05	463.00
266	C4H6O	METHACROLEIN	1.0846	-1.9184E+03	4.4532E+00	-1.4392E-02	9.0180E-06	192.15	530.00
267	C4H6O2	2-BUTYNE-1,4-DIOL	-32.5031	-4.7285E+03	2.2569E+01	-4.1243E-02	1.7537E-05	331.00	695.00
268	C4H6O2	gamma-BUTYROLACTONE	10.8996	-2.8661E+03	-3.8645E-01	-2.6501E-03	1.2711E-06	229.78	739.00
269	C4H6O2	cis-CROTONIC ACID	-13.8773	-3.0350E+03	1.2204E+01	-2.4859E-02	1.1689E-05	288.65	647.00
270	C4H6O2	trans-CROTONIC ACID	45.9341	-4.7347E+03	-1.2459E+01	3.1515E-09	2.0192E-06	344.55	666.00
271	C4H6O2	METHACRYLIC ACID	17.5903	-3.2940E+03	-2.5930E+00	6.4226E-12	-1.5429E-06	288.15	643.00
272	C4H6O2	METHYL ACRYLATE	47.0416	-3.1218E+03	-1.4860E+01	7.1646E-03	3.4547E-14	196.32	536.00
273	C4H6O2	VINYL ACETATE	12.7220	-2.1770E+03	-9.1458E-01	-4.5688E-03	2.9673E-06	180.35	524.00
274	C4H6O3	ACETIC ANHYDRIDE	43.5021	-3.8643E+03	-1.2162E+01	-2.1843E-09	3.3250E-06	200.15	448.15
275	C4H6O4	SUCCINIC ACID	-81.7799	-3.2780E+03	4.0443E+01	-4.6257E-02	1.5606E-05	461.15	806.00
276	C4H6O5	DIGLYCOLIC ACID	-49.4206	-4.8216E+03	2.8168E+01	-3.8017E-02	1.3265E-05	421.15	820.00
277	C4H6O5	MALIC ACID	-63.1758	-5.9090E+03	3.6133E+01	-5.2316E-02	1.9119E-05	403.15	781.00
278	C4H6O6	TARTARIC ACID	-137.3095	-5.7193E+03	6.6921E+01	-7.6973E-02	2.5181E-05	479.15	828.00
279	C4H7N	n-BUTYRONITRILE	4.8780	-2.5505E+03	3.6306E+00	-1.6630E-02	1.0604E-05	161.25	582.25
280	C4H7N	ISOBUTYRONITRILE	2.9839	-2.3552E+03	4.0400E+00	-1.4788E-02	8.7671E-06	201.70	565.00
281	C4H7NO	ACETONE CYANOHYDRIN	-85.1224	-8.9206E+02	4.1877E+01	-6.0910E-02	3.0327E-05	253.15	647.00
282	C4H7NO	2-METHACRYLAMIDE	-32.1945	-2.3705E+03	1.8470E+01	-2.4536E-02	9.4360E-06	383.65	741.00
283	C4H7NO	3-METHOXYPROPIONITRILE	35.8541	-3.5204E+03	-1.0010E+01	3.4139E-03	-7.9103E-14	210.12	638.00
284	C4H7NO	2-PYRROLIDONE	0.4237	-3.6221E+03	5.4272E+00	-1.2988E-02	5.3697E-06	298.15	792.00
285	C4H8	1-BUTENE	27.3116	-1.9235E+03	-7.2064E+00	7.4852E-12	3.6481E-06	87.80	419.59
286	C4H8	cis-2-BUTENE	31.5551	-2.1054E+03	-8.7864E+00	-1.0602E-10	5.0886E-06	134.26	435.58
287	C4H8	trans-2-BUTENE	43.0938	-2.2458E+03	-1.4152E+01	9.0594E-03	-1.9908E-13	167.62	428.63
288	C4H8	CYCLOBUTANE	26.5442	-1.9783E+03	-6.9448E+00	2.5342E-10	3.8845E-06	182.48	459.93
289	C4H8	ISOBUTENE	39.2295	-2.1094E+03	-1.2567E+01	7.7304E-03	-1.3659E-13	132.81	417.90
290	C4H8Br2	1,2-DIBROMOBUTANE	138.6495	-5.8082E+03	-5.3138E+01	4.7750E-02	-1.6041E-05	281.15	659.28
291	C4H8Br2	2,3-DIBROMOBUTANE	123.7728	-5.4664E+03	-4.6795E+01	4.0611E-02	-1.3278E-05	278.15	656.96
292	C4H8Cl2	1,4-DICHLOROBUTANE	1.1907	-2.5846E+03	4.6334E+00	-1.3364E-02	6.8967E-06	235.85	641.00
293	C4H8I2	1,2-DIIODOBUTANE	975.5011	-3.0154E+04	-3.8894E+02	3.2576E-01	-1.0110E-04	383.15	726.41
294	C4H8O	n-BUTYRALDEHYDE	66.8411	-3.6784E+03	-2.2609E+01	1.1697E-02	2.9647E-13	176.75	525.00
295	C4H8O	ISOBUTYRALDEHYDE	89.6241	-4.2317E+03	-3.1724E+01	1.7799E-02	2.8054E-12	208.15	507.20
296	C4H8O	1,2-EPOXYBUTANE	13.3628	-2.1142E+03	-1.0214E+00	-6.4578E-03	4.8665E-06	123.15	526.00
297	C4H8O	METHYL ETHYL KETONE	47.7060	-3.0965E+03	-1.5184E+01	7.4846E-03	-1.7084E-13	186.48	535.50
298	C4H8O	ETHYL VINYL ETHER	5.5570	-1.8563E+03	2.6934E+00	-1.4082E-02	1.0248E-05	157.35	475.15
299	C4H8O	TETRAHYDROFURAN	34.8700	-2.7523E+03	-9.5958E+00	1.9889E-10	3.5465E-06	164.65	540.15
300	C4H8O2	cis-2-BUTENE-1,4-DIOL	92.8466	-7.4351E+03	-2.8620E+01	3.2422E-09	8.1439E-06	284.15	677.88
301	C4H8O2	trans-2-BUTENE-1,4-DIOL	69.3513	-6.9220E+03	-2.1957E+01	9.0107E-03	-1.2354E-12	300.45	681.00
302	C4H8O2	ISOBUTYRIC ACID	11.3037	-3.1625E+03	7.2630E-01	-8.9331E-03	4.8215E-06	227.15	609.15
303	C4H8O2	n-BUTYRIC ACID	8.0847	-3.3219E+03	2.4312E+00	-1.1734E-02	5.7992E-06	267.95	628.00
304	C4H8O2	1,4-DIOXANE	20.5761	-2.4658E+03	-4.3645E+00	-2.7053E-10	8.5235E-07	284.95	587.00
305	C4H8O2	ETHYL ACETATE	0.6955	-2.2498E+03	5.4643E+00	-1.9451E-02	1.2362E-05	189.60	523.30
306	C4H8O2	METHYL PROPIONATE	35.4850	-2.9788E+03	-9.6340E+00	-1.5818E-11	3.0270E-06	185.65	530.60
307	C4H8O2	n-PROPYL FORMATE	28.6983	-2.6926E+03	-7.2435E+00	-8.7226E-11	1.9456E-06	180.25	538.00
308	C4H8O2S	SULFOLANE	-4.4873	-3.6491E+03	7.3819E+00	-1.4643E-02	5.7933E-06	300.75	849.00
309	C4H8S	TETRAHYDROTHIOPHENE	31.3909	-3.0225E+03	-8.1508E+00	1.5440E-10	2.0231E-06	176.99	631.95
310	C4H9Br	1-BROMOBUTANE	74.7061	-4.0663E+03	-2.5610E+01	1.3166E-02	1.4210E-13	160.75	577.00
311	C4H9Br	2-BROMOBUTANE	10.0875	-2.3152E+03	6.6773E-01	-9.2242E-03	5.9862E-06	161.25	567.00
312	C4H9Cl	n-BUTYL CHLORIDE	35.7808	-2.8632E+03	-9.8957E+00	5.1598E-11	3.5488E-06	150.05	537.00

Table 7-1 VAPOR PRESSURE - ORGANIC COMPOUNDS (continued)

$$\log_{10} P = A + B/T + C \log_{10} T + D T + E T^2 \qquad \text{(P - mm Hg, T - K)}$$

NO	FORMULA	NAME	A	B	C	D	E	TMIN	TMAX
313	C4H9Cl	sec-BUTYL CHLORIDE	28.2992	-2.4353E+03	-7.3590E+00	-1.3048E-11	3.0834E-06	141.85	520.60
314	C4H9Cl	tert-BUTYL CHLORIDE	-15.9627	-1.3296E+03	1.1483E+01	-2.2322E-02	1.2859E-05	247.75	507.00
315	C4H9I	2-IODO-2-METHYLPROPANE	-98.3670	2.8353E+02	4.6127E+01	-5.8084E-02	2.5345E-05	315.15	587.90
316	C4H9N	PYRROLIDINE	50.2467	-3.6404E+03	-1.4826E+01	6.5779E-10	5.0196E-06	215.31	568.55
317	C4H9NO	N,N-DIMETHYLACETAMIDE	27.8419	-3.3379E+03	-6.6412E+00	2.4260E-10	9.8824E-07	253.15	658.00
318	C4H9NO	MORPHOLINE	23.4107	-2.8912E+03	-5.1633E+00	-1.9719E-09	7.3526E-07	270.05	618.00
319	C4H9NO2	1-NITROBUTANE	671.7377	-2.0216E+04	-2.7069E+02	2.4868E-01	-8.5859E-05	359.15	624.00
320	C4H9NO2	2-NITROBUTANE	105.2369	-4.8267E+03	-3.9220E+01	3.3350E-02	-1.0807E-05	348.15	615.00
321	C4H10	n-BUTANE	27.0441	-1.9049E+03	-7.1805E+00	-6.6845E-11	4.2190E-06	134.86	425.18
322	C4H10	ISOBUTANE	31.2541	-1.9532E+03	-8.8060E+00	8.9246E-11	5.7501E-06	113.54	408.14
323	C4H10N2	PIPERAZINE	-55.9690	-1.3030E+03	2.8525E+01	-3.7190E-02	1.5599E-05	379.15	638.00
324	C4H10O	n-BUTANOL	39.6673	-4.0017E+03	-1.0295E+01	-3.2572E-10	8.6672E-07	183.85	562.93
325	C4H10O	sec-BUTANOL	49.4476	-4.2487E+03	-1.3793E+01	6.2736E-11	2.1988E-06	158.45	536.01
326	C4H10O	tert-BUTANOL	71.8181	-4.9966E+03	-2.1805E+01	1.9238E-08	5.8247E-06	298.97	506.20
327	C4H10O	DIETHYL ETHER	41.7519	-2.7410E+03	-1.2270E+01	-3.1948E-10	5.9802E-06	156.85	466.70
328	C4H10O	METHYL-PROPYL-ETHER	448.6238	-1.1762E+04	-1.8629E+02	2.0930E-01	-8.9491E-05	273.15	476.20
329	C4H10O	METHYL ISOPROPYL ETHER	10.1637	-1.9189E+03	6.8923E-01	-1.1682E-02	9.3524E-06	127.93	464.50
330	C4H10O	ISOBUTANOL	109.2803	-6.3060E+03	-3.6947E+01	1.4462E-02	-3.9480E-13	165.15	547.73
331	C4H10O2	1,3-BUTANEDIOL	109.4540	-7.4377E+03	-3.6627E+01	1.4845E-02	-7.6634E-14	196.15	643.00
332	C4H10O2	1,4-BUTANEDIOL	22.4534	-4.2023E+03	-4.2015E+00	-7.4539E-10	6.1761E-07	293.05	667.00
333	C4H10O2	2,3-BUTANEDIOL	46.6247	-4.7864E+03	-1.2792E+01	-8.5522E-10	3.8460E-06	280.75	611.00
334	C4H10O2	t-BUTYL HYDROPEROXIDE	11.5999	-2.7658E+03	-2.1182E-01	-4.1964E-03	2.1416E-06	277.45	576.00
335	C4H10O2	1,2-DIMETHOXYETHANE	34.8262	-2.7674E+03	-1.0083E+01	4.3212E-03	4.0618E-13	215.15	536.15
336	C4H10O2	2-ETHOXYETHANOL	115.8686	-6.0128E+03	-4.0900E+01	2.0888E-02	1.6481E-15	183.00	569.00
337	C4H10O3	DIETHYLENE GLYCOL	6.5069	-4.6273E+03	4.6273E+00	-1.8361E-02	8.2923E-06	262.70	744.60
338	C4H10O4S	DIETHYL SULFATE	37.5380	-4.0299E+03	-1.0340E+01	2.9825E-03	-1.4666E-11	248.00	483.00
339	C4H10S	n-BUTYL MERCAPTAN	36.2672	-3.0452E+03	-9.9743E+00	-9.1432E-11	3.2087E-06	157.46	569.00
340	C4H10S	ISOBUTYL MERCAPTAN	35.2778	-2.9105E+03	-9.6829E+00	3.9960E-11	3.2308E-06	128.31	559.00
341	C4H10S	sec-BUTYL MERCAPTAN	10.0200	-2.3218E+03	1.1050E+00	-1.3014E-02	9.2202E-06	133.02	554.00
342	C4H10S	tert-BUTYL MERCAPTAN	32.9088	-2.6250E+03	-8.9422E+00	-1.0585E-10	3.1646E-06	274.26	530.00
343	C4H10S	DIETHYL SULFIDE	2.8992	-2.2313E+03	4.1589E+00	-1.6341E-02	1.0499E-05	169.20	557.15
344	C4H10S	ISOPROPYL-METHYL-SULFIDE	60.9138	-3.4687E+03	-2.0565E+01	1.2370E-02	-1.9629E-06	260.15	551.00
345	C4H10S	METHYL-PROPYL-SULFIDE	53.4630	-3.4390E+03	-1.7201E+01	7.8569E-03	-1.1672E-08	269.15	563.00
346	C4H10S2	DIETHYL DISULFIDE	4.9562	-2.8307E+03	3.6929E+00	-1.6131E-02	9.4578E-06	171.63	642.00
347	C4H11N	n-BUTYLAMINE	25.0711	-2.5701E+03	-5.8985E+00	7.9399E-10	1.1920E-06	224.05	531.90
348	C4H11N	ISOBUTYLAMINE	-6.8794	-2.0381E+03	8.8812E+00	-2.5376E-02	1.6317E-05	188.55	513.73
349	C4H11N	sec-BUTYLAMINE	6.5307	-2.0596E+03	2.2551E+00	-1.2393E-02	8.3513E-06	168.65	514.30
350	C4H11N	tert-BUTYLAMINE	24.1625	-2.1505E+03	-6.0687E+00	2.1369E-03	-9.2692E-14	206.19	483.90
351	C4H11N	DIETHYLAMINE	32.6260	-2.4918E+03	-9.3285E+00	3.9900E-03	1.1732E-12	223.35	496.60
352	C4H11NO	DIMETHYLETHANOLAMINE	1.6173	-3.5071E+03	6.9312E+00	-2.6316E-02	1.5111E-05	214.15	571.82
353	C4H11NO2	DIETHANOLAMINE	122.0877	-8.8422E+03	-4.0422E+01	1.4062E-02	1.1986E-12	301.15	542.04
354	C4H11NO2	2-AMINOETHOXYETHANOL	0.8861	-5.3757E+03	8.4953E+00	-2.7084E-02	1.2483E-05	273.15	698.00
355	C4H12N2O	N-AMINOETHYL ETHANOLAMINE	-20.3645	-4.6961E+03	1.7262E+01	-3.6406E-02	1.6261E-05	293.15	699.00
356	C4H12Si	TETRAMETHYLSILANE	27.2755	-2.1081E+03	-7.1240E+00	-4.0697E-11	3.1382E-06	174.07	450.40
357	C4H13N3	DIETHYLENE TRIAMINE	-23.5969	-2.6670E+03	1.5808E+01	-2.8417E-02	1.4246E-05	234.15	676.00
358	C5Cl6	HEXACHLOROCYCLOPENTADIENE	-9.7942	-3.3161E+03	1.0171E+01	-2.1115E-02	9.2045E-06	284.49	746.00
359	C5H4O2	FURFURAL	32.0337	-3.4930E+03	-8.1424E+00	2.2074E-10	1.9582E-06	236.65	657.00
360	C5H5N	PYRIDINE	33.5541	-3.1318E+03	-8.8646E+00	7.1293E-12	2.2813E-06	231.51	619.95
361	C5H6	CYCLOPENTADIENE	-22.1335	-1.1705E+03	1.4556E+01	-3.0457E-02	1.9817E-05	188.15	507.00
362	C5H6	2-METHYL-1-BUTENE-3-YNE	39.3298	-2.3664E+03	-1.2384E+01	6.7937E-03	8.2122E-14	160.15	492.00
363	C5H6	1-PENTENE-3-YNE	41.9647	-2.8411E+03	-1.2372E+01	-1.0010E-10	5.9824E-06	150.00	520.00
364	C5H6	1-PENTENE-4-YNE	47.2587	-2.6717E+03	-1.5453E+01	8.5851E-03	-1.2301E-13	150.00	503.00
365	C5H6N2	GLUTARONITRILE	60.1486	-5.3093E+03	-1.8776E+01	6.8232E-03	-1.3443E-13	244.21	782.00
366	C5H6O2	FURFURYL ALCOHOL	31.4348	-3.7851E+03	-7.9113E+00	2.0812E-03	1.2964E-13	258.52	632.00
367	C5H6O3	GLUTARIC ANHYDRIDE	42.8029	-5.1785E+03	-1.1378E+01	-4.9820E-11	1.8154E-06	328.00	838.00
368	C5H6O4	CITRACONIC ACID	-11.6730	-5.4940E+03	1.2578E+01	-2.4297E-02	9.0728E-06	356.15	829.00
369	C5H6O4	ITACONIC ACID	-58.5861	-3.8726E+03	3.0953E+01	-3.7898E-02	1.2924E-05	438.75	821.00
370	C5H6S	2-METHYLTHIOPHENE	62.0938	-3.8769E+03	-2.0540E+01	1.0500E-02	-1.1254E-06	282.15	610.00
371	C5H6S	3-METHYLTHIOPHENE	55.1117	-3.6703E+03	-1.7715E+01	8.0917E-03	-3.7915E-07	284.15	615.00
372	C5H7N	N-METHYLPYRROLE	36.2741	-3.2734E+03	-9.7612E+00	-2.8918E-10	2.2419E-06	216.91	610.00
373	C5H7NO2	ETHYL CYANOACETATE	5.0305	-3.1246E+03	1.6321E+00	1.6027E-11	-6.0406E-15	250.65	679.00
374	C5H8	CYCLOPENTENE	30.1132	-2.3537E+03	-8.0609E+00	-5.7786E-11	3.4591E-06	138.13	507.00
375	C5H8	ISOPRENE	32.4693	-2.2755E+03	-9.4314E+00	4.1627E-03	-3.0643E-14	127.27	484.00
376	C5H8	3-METHYL-1,2-BUTADIENE	26.7883	-2.2642E+03	-6.7693E+00	-1.6862E-10	2.0892E-06	159.53	490.00
377	C5H8	2-METHYL-1,3-BUTADIENE	-1.9750	-1.7013E+03	6.1902E+00	-2.0627E-02	1.4120E-05	255.15	483.30
378	C5H8	1,2-PENTADIENE	26.6297	-2.2985E+03	-6.6707E+00	-2.5834E-11	1.7060E-06	135.89	500.00
379	C5H8	cis-1,3-PENTADIENE	31.3245	-2.3715E+03	-8.7812E+00	3.1436E-03	-2.2535E-14	132.35	499.00
380	C5H8	trans-1,3-PENTADIENE	27.3730	-2.2348E+03	-7.2358E+00	2.1441E-03	-6.6289E-14	132.00	500.00
381	C5H8	1,4-PENTADIENE	23.7408	-2.0505E+03	-5.6970E+00	-5.9671E-11	1.1242E-06	124.86	479.00
382	C5H8	2,3-PENTADIENE	29.2390	-2.4236E+03	-7.6062E+00	-1.1983E-10	2.4346E-06	147.50	497.00
383	C5H8	1-PENTYNE	33.8369	-2.4684E+03	-9.4301E+00	6.1345E-10	4.6760E-06	167.45	481.20
384	C5H8	2-PENTYNE	-52.3434	-8.4024E+02	2.7976E+01	-4.6584E-02	2.4750E-05	240.15	521.99
385	C5H8	3-METHYL-1-BUTYNE	-39.2632	-1.1773E+03	2.3847E+01	-5.4802E-02	3.7819E-05	183.45	463.20
386	C5H8	SPIROPENTANE	140.1393	-4.7087E+03	-5.5343E+01	5.8108E-02	-2.3249E-05	276.15	499.74
387	C5H8N4O12	PENTAERYTHRITOL TETRANITRATE	-130.9886	-2.7510E+03	6.3602E+01	-8.1824E-02	3.1971E-05	413.65	676.00
388	C5H8O	CYCLOPENTANONE	47.8103	-3.7609E+03	-1.3972E+01	-5.8657E-10	4.8771E-06	221.85	626.00
389	C5H8O	METHYL ISOPROPENYL KETONE	-0.1399	-2.1308E+03	5.0651E+00	-1.4555E-02	8.3111E-06	219.55	566.00
390	C5H8O2	ACETYLACETONE	-6.0090	-2.7795E+03	8.6491E+00	-2.1654E-02	1.1292E-05	249.95	602.00

166

Table 7-1 VAPOR PRESSURE - ORGANIC COMPOUNDS (continued)

			$\log_{10} P = A + B/T + C \log_{10} T + D T + E T^2$					(P - mm Hg, T - K)	
NO	FORMULA	NAME	A	B	C	D	E	TMIN	TMAX
391	C5H8O2	ALLYL ACETATE	26.8017	-2.6874E+03	-6.6456E+00	3.4187E-11	2.2725E-06	138.00	559.00
392	C5H8O2	ETHYL ACRYLATE	55.0109	-3.5904E+03	-1.7694E+01	8.0510E-03	-4.8864E-13	201.95	553.00
393	C5H8O2	METHYL METHACRYLATE	106.8960	-5.2741E+03	-3.7654E+01	1.8620E-02	-3.6507E-13	224.95	564.00
394	C5H8O2	VINYL PROPIONATE	-67.2714	-4.7030E+02	3.3190E+01	-4.4868E-02	2.0880E-05	362.00	546.00
395	C5H8O3	2-HYDROXYETHYL ACRYLATE	9.6617	-3.9805E+03	2.3231E+00	-1.3294E-02	7.0087E-06	213.00	662.00
396	C5H8O3	LEVULINIC ACID	-12.7087	-4.1736E+03	1.2081E+01	-2.3812E-02	1.1829E-05	308.15	556.15
397	C5H8O3	METHYL ACETOACETATE	12.9412	-3.4913E+03	7.2725E-01	-1.2394E-02	6.9235E-06	193.15	642.00
398	C5H8O4	GLUTARIC ACID	109.3324	-9.6083E+03	-3.4494E+01	9.0733E-03	1.4829E-12	370.65	807.00
399	C5H9N	VALERONITRILE	43.7757	-3.7040E+03	-1.2460E+01	-3.4293E-11	3.8460E-06	176.95	603.00
400	C5H9NO	n-BUTYL ISOCYANATE	6.0766	-2.6575E+03	3.1413E+00	-1.5075E-02	9.0562E-06	193.00	568.00
401	C5H9NO	N-METHYL-2-PYRROLIDONE	-0.8097	-3.0788E+03	5.8572E+00	-1.4916E-02	6.9974E-06	249.15	724.00
402	C5H9NO4	L-GLUTAMIC ACID	-84.7291	-4.4621E+03	4.1913E+01	-4.5554E-02	1.4115E-05	497.15	886.00
403	C5H10	CYCLOPENTANE	29.1547	-2.3512E+03	-7.6965E+00	-1.6212E-10	3.1250E-06	179.28	511.76
404	C5H10	2-METHYL-1-BUTENE	30.2418	-2.2723E+03	-8.1482E+00	5.2331E-11	3.6802E-06	135.58	465.00
405	C5H10	2-METHYL-2-BUTENE	33.7539	-2.4260E+03	-9.4429E+00	9.8488E-11	4.7156E-06	139.39	471.00
406	C5H10	3-METHYL-1-BUTENE	31.1486	-2.1764E+03	-8.6146E+00	5.9672E-11	4.7555E-06	104.66	450.37
407	C5H10	1-PENTENE	36.2741	-2.4452E+03	-1.0405E+01	-7.4629E-11	5.4070E-06	110.00	464.78
408	C5H10	cis-2-PENTENE	34.0427	-2.4524E+03	-9.5014E+00	-5.0816E-11	4.3638E-06	121.75	475.93
409	C5H10	trans-2-PENTENE	30.6231	-2.3239E+03	-8.2648E+00	-1.7049E-11	3.6666E-06	132.89	475.37
410	C5H10Br2	2,3-DIBROMO-2-METHYLBUTANE	-166.9586	1.6951E+03	7.3382E+01	-7.5898E-02	2.7650E-05	354.15	668.37
411	C5H10Cl2	1,5-DICHLOROPENTANE	10.1987	-3.1090E+03	1.1611E+00	-1.0363E-02	5.5991E-06	200.35	663.00
412	C5H10O	METHYL ISOPROPYL KETONE	131.9592	-5.9468E+03	-4.7557E+01	2.4756E-02	-2.0397E-13	181.15	553.00
413	C5H10O	2-PENTANONE	18.3056	-2.3477E+03	-3.6667E+00	7.1502E-04	1.0912E-13	196.29	561.08
414	C5H10O	DIETHYL KETONE	32.2560	-2.9431E+03	-8.5068E+00	-4.5720E-10	2.6177E-06	234.18	560.95
415	C5H10O	VALERALDEHYDE	45.7561	-3.5036E+03	-1.3292E+01	-4.3157E-10	4.7373E-06	182.00	554.00
416	C5H10O2	n-BUTYL FORMATE	20.4567	-2.5589E+03	-4.2259E+00	7.1544E-11	4.8667E-07	181.25	559.00
417	C5H10O2	ETHYL PROPIONATE	43.7540	-3.4774E+03	-1.2477E+01	1.6521E-11	3.9087E-06	199.25	546.00
418	C5H10O2	ISOBUTYL FORMATE	16.5919	-2.3290E+03	-2.9026E+00	2.7455E-11	1.6584E-07	177.35	551.35
419	C5H10O2	ISOPROPYL ACETATE	22.2064	-2.4989E+03	-4.8975E+00	-2.7852E-10	8.3385E-07	199.75	538.00
420	C5H10O2	n-PROPYL ACETATE	43.0548	-3.4692E+03	-1.2217E+01	2.4748E-10	3.7508E-06	178.15	549.40
421	C5H10O2	METHYL n-BUTYRATE	44.5661	-3.5234E+03	-1.2770E+01	-1.4010E-10	4.0354E-06	187.35	554.50
422	C5H10O2	2-METHYLBUTYRIC ACID	15.1348	-3.7848E+03	1.7558E-01	-1.2814E-02	7.1577E-06	193.00	643.00
423	C5H10O2	ISOVALERIC ACID	3.7904	-3.6011E+03	5.0920E+00	-1.8480E-02	9.4559E-06	243.85	634.00
424	C5H10O2	VALERIC ACID	15.3454	-3.9024E+03	-2.4353E-02	-1.1099E-02	5.6315E-06	239.15	651.00
425	C5H10O2	TETRAHYDROFURFURYL ALCOHOL	-11.5234	-2.9418E+03	1.1327E+01	-2.7120E-02	1.5221E-05	193.00	630.00
426	C5H10O2S	3-METHYL SULFOLANE	6.6099	-3.8689E+03	2.7869E+00	-1.0238E-02	4.3052E-06	273.65	817.00
427	C5H10O3	DIETHYL CARBONATE	75.6312	-4.9957E+03	-2.3680E+01	1.8618E-09	8.4774E-06	230.15	576.00
428	C5H10O3	ETHYL LACTATE	32.0863	-2.9164E+03	-9.5666E+00	6.5114E-03	4.5645E-13	247.15	588.00
429	C5H10S	THIACYCLOHEXANE	56.9128	-3.9263E+03	-1.8271E+01	8.0697E-03	-5.0144E-07	302.15	657.12
430	C5H10S	CYCLOPENTANETHIOL	114.1137	-5.4064E+03	-4.1699E+01	3.0074E-02	-8.1552E-06	354.15	629.00
431	C5H11Br	1-BROMOPENTANE	955.2944	-2.6563E+04	-3.9056E+02	3.8087E-01	-1.3843E-04	314.15	564.76
432	C5H11Cl	1-CHLOROPENTANE	-4.4886	-2.2604E+03	7.8088E+00	-2.3675E-02	1.4884E-05	174.15	568.00
433	C5H11Cl	1-CHLORO-3-METHYLBUTANE	223.0545	-7.3403E+03	-8.8795E+01	8.6724E-02	-3.2149E-05	312.15	558.87
434	C5H11Cl	2-CHLORO-2-METHYLBUTANE	-77.5603	-2.7352E+02	3.7920E+01	-5.2500E-02	2.4568E-05	280.15	548.97
435	C5H11N	N-METHYLPYRROLIDINE	4.2451	-2.0090E+03	2.9553E+00	-1.1618E-02	7.2587E-06	183.15	550.00
436	C5H11N	PIPERIDINE	2.3169	-2.1085E+03	3.5103E+00	-9.6044E-03	4.9394E-06	262.65	594.05
437	C5H11NO	tert-BUTYLFORMAMIDE	-7.8846	-3.0206E+03	9.0045E+00	-1.8723E-02	8.4726E-06	289.15	692.00
438	C5H12	ISOPENTANE	29.2963	-2.1762E+03	-7.8830E+00	-4.6512E-11	3.8997E-06	113.25	460.43
439	C5H12	NEOPENTANE	26.6662	-1.9307E+03	-7.0448E+00	7.4104E-09	3.9463E-06	256.58	433.78
440	C5H12	n-PENTANE	33.3239	-2.4227E+03	-9.2354E+00	9.0199E-11	4.1050E-06	143.42	469.65
441	C5H12O	2,2-DIMETHYL-1-PROPANOL	-130.9341	-4.9954E+02	6.3932E+01	-9.6510E-02	4.6921E-05	327.15	550.00
442	C5H12O	tert-PENTYL-ALCOHOL	-419.8340	6.9337E+03	1.8329E+02	-2.1490E-01	9.1359E-05	328.15	549.00
443	C5H12O	2-METHYL-1-BUTANOL	176.1269	-8.7997E+03	-6.2366E+01	2.7514E-02	-7.2309E-13	203.00	565.00
444	C5H12O	2-METHYL-2-BUTANOL	-17.2086	-3.0396E+03	1.5691E+01	-4.0746E-02	2.1944E-05	264.35	545.15
445	C5H12O	3-METHYL-1-BUTANOL	35.8184	-4.3519E+03	-7.6449E+00	-7.4737E-03	4.7915E-06	161.00	579.45
446	C5H12O	3-METHYL-2-BUTANOL	85.7112	-5.9542E+03	-2.6392E+01	1.4969E-11	5.9121E-06	188.00	574.00
447	C5H12O	1-PENTANOL	71.2535	-5.4977E+03	-2.1366E+01	3.8108E-10	5.0339E-06	195.56	586.15
448	C5H12O	2-PENTANOL	13.3731	-3.8492E+03	2.4579E+00	-2.3532E-02	1.4141E-05	200.00	552.00
449	C5H12O	3-PENTANOL	20.2685	-3.4913E+03	-2.0432E+00	-1.0453E-02	6.2649E-06	204.15	547.00
450	C5H12O	METHYL sec-BUTYL ETHER	9.1153	-2.1112E+03	1.2696E+00	-1.2214E-02	8.8206E-06	150.00	498.00
451	C5H12O	METHYL tert-BUTYL ETHER	4.7409	-1.9493E+03	3.0770E+00	-1.4463E-02	1.0039E-05	164.55	497.10
452	C5H12O	METHYL ISOBUTYL ETHER	6.2559	-2.0615E+03	2.5628E+00	-1.4245E-02	9.9920E-06	160.00	497.00
453	C5H12O	ETHYL PROPYL ETHER	60.0270	-3.6280E+03	-1.8751E+01	-2.4191E-11	8.9552E-06	145.65	500.23
454	C5H12O2	ETHYLENE GLYCOL MONOPROPYL ETHER	-19.0128	-2.5324E+03	1.4629E+01	-3.3547E-02	2.0289E-05	183.15	582.00
455	C5H12O2	NEOPENTYL GLYCOL	103.2376	-8.4730E+03	-3.1418E+01	-5.8384E-07	6.4629E-06	400.00	643.00
456	C5H12O2	1,5-PENTANEDIOL	42.1731	-5.3023E+03	-1.0917E+01	-7.7321E-12	2.4225E-06	257.15	673.00
457	C5H12O3	2-(2-METHOXYETHOXY)ETHANOL	186.1982	-9.4372E+03	-6.6416E+01	3.0358E-02	9.1914E-13	250.00	630.00
458	C5H12O4	PENTAERYTHRITOL	81.2393	-9.9990E+03	-2.3002E+01	9.7212E-04	3.1478E-06	534.15	633.35
459	C5H12S	n-PENTYL MERCAPTAN	-1.1525	-2.4630E+03	6.1419E+00	-1.8809E-02	1.0844E-05	197.45	598.00
460	C5H12S	BUTYL-METHYL-SULFIDE	66.7879	-4.0859E+03	-2.2320E+01	1.1587E-02	-1.3035E-06	290.15	591.00
461	C5H12S	ETHYL-PROPYL-SULFIDE	70.2978	-4.1217E+03	-2.3837E+01	1.3303E-02	-1.9415E-06	287.15	584.00
462	C5H12S	2-METHYL-2-BUTANETHIOL	34.6541	-2.9752E+03	-9.3212E+00	-7.9016E-04	3.4339E-06	270.15	566.00
463	C5H13N	n-PENTYLAMINE	-4.5117	-2.2580E+03	7.4447E+00	-1.9689E-02	1.1339E-05	218.15	555.00
464	C5H13NO2	METHYL DIETHANOLAMINE	8.3233	-6.2875E+03	6.7885E+00	-3.0317E-02	1.4678E-05	252.15	678.00
465	C6Cl6	HEXACHLOROBENZENE	-134.3625	-1.5459E+03	6.1748E+01	-6.5123E-02	2.0872E-05	501.70	825.00
466	C6F6	HEXAFLUOROBENZENE	-38.8085	-1.3422E+03	2.2204E+01	-3.8813E-02	2.1000E-05	278.25	516.73
467	C6H3ClN2O4	1-CHLORO-2,4-DINITROBENZENE	-3.2922	-4.8517E+03	8.2836E+00	-1.8820E-02	7.3627E-06	326.55	813.77
468	C6H3Cl2NO2	1,2-DICHLORO-4-NITROBENZENE	-7.2030	-3.6532E+03	8.9786E+00	-1.8070E-02	7.4738E-06	315.65	758.00

Table 7-1 VAPOR PRESSURE - ORGANIC COMPOUNDS (continued)

			$\log_{10} P = A + B/T + C \log_{10} T + D T + E T^2$					(P - mm Hg, T - K)		
NO	FORMULA	NAME	A	B	C	D	E	TMIN	TMAX
469	C6H3Cl3	1,2,4-TRICHLOROBENZENE	15.5947	-2.8920E+03	-2.5549E+00	2.0384E-04	-7.0601E-14	290.15	725.00
470	C6H3N3O6	1,3,5-TRINITROBENZENE	19.2854	-4.4976E+03	-3.6158E+00	-5.2495E-09	1.5596E-12	398.40	748.00
471	C6H4Br2	m-DIBROMOBENZENE	59.2974	-4.6960E+03	-1.8644E+01	6.7598E-03	-2.5567E-13	266.25	761.00
472	C6H4ClNO2	m-CHLORONITROBENZENE	-10.5500	-3.2885E+03	1.0124E+01	-1.8725E-02	7.7944E-06	317.65	742.00
473	C6H4ClNO2	o-CHLORONITROBENZENE	-3.5744	-3.5871E+03	7.3361E+00	-1.6133E-02	6.7639E-06	306.15	757.00
474	C6H4ClNO2	p-CHLORONITROBENZENE	-21.4396	-3.0103E+03	1.4461E+01	-2.2062E-02	8.7001E-06	356.65	751.00
475	C6H4Cl2	m-DICHLOROBENZENE	3.2904	-2.5839E+03	3.2893E+00	-9.5398E-03	4.6397E-06	248.39	683.95
476	C6H4Cl2	o-DICHLOROBENZENE	31.3614	-3.5226E+03	-7.8886E+00	-2.2250E-10	1.1842E-06	256.15	705.00
477	C6H4Cl2	p-DICHLOROBENZENE	36.2276	-3.6756E+03	-9.6308E+00	-1.3372E-09	1.9905E-06	326.14	684.75
478	C6H4F2	m-DIFLUOROBENZENE	93.7777	-4.4188E+03	-3.4029E+01	2.5614E-02	-6.9770E-06	309.15	552.94
479	C6H4F2	o-DIFLUOROBENZENE	88.5722	-4.1066E+03	-3.2664E+01	2.9267E-02	-1.0923E-05	304.15	405.15
480	C6H4F2	p-DIFLUOROBENZENE	87.9772	-4.2909E+03	-3.1542E+01	2.2689E-02	-5.7306E-06	307.15	556.00
481	C6H4N2O4	m-DINITROBENZENE	-19.7265	-4.1175E+03	1.4670E+01	-2.3736E-02	8.9178E-06	364.00	803.00
482	C6H4N2O4	o-DINITROBENZENE	-24.2463	-4.1140E+03	1.6344E+01	-2.4085E-02	8.6190E-06	390.08	831.00
483	C6H4N2O4	p-DINITROBENZENE	-57.4971	-2.8352E+03	2.9400E+01	-3.4235E-02	1.1758E-05	446.60	803.00
484	C6H5Br	BROMOBENZENE	-9.4583	-2.3551E+03	9.2584E+00	-1.9386E-02	9.6324E-06	242.43	670.15
485	C6H5Cl	MONOCHLOROBENZENE	19.4343	-2.5801E+03	-3.9391E+00	-4.4005E-11	4.9583E-07	227.95	632.35
486	C6H5ClO	m-CHLOROPHENOL	21.6353	-3.2353E+03	-4.6016E+00	-1.9099E-10	1.0770E-06	306.00	729.00
487	C6H5ClO	o-CHLOROPHENOL	18.2631	-2.6520E+03	-3.6728E+00	8.3047E-10	1.3841E-06	282.00	675.00
488	C6H5ClO	p-CHLOROPHENOL	-9.8143	-2.5177E+03	8.3886E+00	-1.2250E-02	5.1440E-06	316.00	738.00
489	C6H5Cl2N	3,4-DICHLOROANILINE	-15.2685	-3.3857E+03	1.1926E+01	-1.9227E-02	7.4179E-06	344.65	800.00
490	C6H5F	FLUOROBENZENE	-5.4849	-1.8597E+03	7.1515E+00	-1.6467E-02	9.2622E-06	230.94	560.09
491	C6H5I	IODOBENZENE	32.7342	-3.5824E+03	-8.4197E+00	2.0073E-10	1.5910E-06	241.83	721.15
492	C6H5NO2	NITROBENZENE	-54.4937	-2.1123E+03	2.9321E+01	-4.4839E-02	2.0162E-05	278.91	719.00
493	C6H6	BENZENE	31.7718	-2.7254E+03	-8.4443E+00	-5.3534E-09	2.7187E-06	278.68	562.16
494	C6H6ClN	m-CHLOROANILINE	65.6033	-5.3779E+03	-2.0518E+01	6.7867E-03	2.1167E-13	262.75	751.00
495	C6H6ClN	o-CHLOROANILINE	90.6491	-6.0410E+03	-3.0118E+01	1.1564E-02	4.8388E-13	271.05	722.00
496	C6H6ClN	p-CHLOROANILINE	-15.3259	-2.8592E+03	1.1527E+01	-1.8071E-02	7.2359E-06	343.05	754.00
497	C6H6N2	cis-DICYANO-1-BUTENE	5.8017	-4.1358E+03	4.4579E+00	-1.7654E-02	8.5171E-06	249.00	691.00
498	C6H6N2	trans-DICYANO-1-BUTENE	2.4678	-4.0104E+03	5.7900E+00	-1.8851E-02	8.9537E-06	260.00	689.00
499	C6H6N2	1,4-DICYANO-2-BUTENE	-20.1232	-3.7427E+03	1.4764E+01	-2.4708E-02	9.8055E-06	349.00	755.00
500	C6H6N2O2	m-NITROANILINE	125.1147	-8.0609E+03	-4.3204E+01	1.9083E-02	1.9535E-12	387.15	579.00
501	C6H6N2O2	o-NITROANILINE	-112.5774	-1.5945E+03	5.4577E+01	-7.6775E-02	3.6152E-05	344.65	558.00
502	C6H6N2O2	p-NITROANILINE	56.1642	-5.3655E+03	-1.7958E+01	9.0920E-03	7.0305E-10	420.65	609.15
503	C6H6O	PHENOL	23.5332	-3.4961E+03	-4.8990E+00	1.2160E-04	9.6537E-13	314.06	694.25
504	C6H6O2	1,2-BENZENEDIOL	-75.0168	-1.1958E+03	3.5449E+01	-3.8535E-02	1.4681E-05	377.60	764.00
505	C6H6O2	1,3-BENZENEDIOL	-40.1526	-3.6125E+03	2.3198E+01	-3.1795E-02	1.1659E-05	382.00	810.00
506	C6H6O2	p-HYDROQUINONE	46.0254	-5.5833E+03	-1.2677E+01	3.0019E-03	-8.0308E-12	444.65	822.00
507	C6H6O3	1,2,3-BENZENETRIOL	13.8033	-4.0143E+03	-1.4894E+00	1.7902E-09	2.9165E-07	407.00	830.00
508	C6H6S	PHENYL MERCAPTAN	-5.4919	-2.8549E+03	8.1770E+00	-1.9494E-02	9.2817E-06	258.26	689.00
509	C6H7N	ANILINE	124.3764	-7.1676E+03	-4.2763E+01	1.7336E-02	5.7138E-15	267.13	699.00
510	C6H7N	2-METHYLPYRIDINE	34.3728	-3.2825E+03	-9.0927E+00	-3.6324E-10	2.1425E-06	206.44	621.00
511	C6H7N	3-METHYLPYRIDINE	35.2679	-3.4346E+03	-9.3555E+00	-1.3286E-10	2.0641E-06	255.01	645.00
512	C6H7N	4-METHYLPYRIDINE	-18.9075	-2.0520E+03	1.3080E+01	-2.2709E-02	1.0877E-05	276.73	646.15
513	C6H8	1,3-CYCLOHEXADIENE	32.7055	-2.7281E+03	-8.8297E+00	4.2150E-11	3.1600E-06	161.00	558.00
514	C6H8	METHYLCYCLOPENTADIENE	9.7525	-2.1337E+03	6.9389E-01	-9.3574E-03	6.4104E-06	150.00	541.00
515	C6H8N2	ADIPONITRILE	7.1898	-4.7244E+03	3.7992E+00	-1.5064E-02	6.4906E-06	275.64	781.00
516	C6H8N2	METHYLGLUTARONITRILE	14.6669	-4.5558E+03	6.0200E-01	-1.2402E-02	5.9533E-06	228.15	742.00
517	C6H8N2	m-PHENYLENEDIAMINE	-2.1314	-3.4126E+03	5.5842E+00	-9.6378E-03	3.7115E-06	334.00	824.00
518	C6H8N2	o-PHENYLENEDIAMINE	-23.8872	-3.0704E+03	1.5395E+01	-2.1989E-02	8.2796E-06	376.95	781.00
519	C6H8N2	p-PHENYLENEDIAMINE	-36.2709	-2.9092E+03	2.0434E+01	-2.5933E-02	9.2817E-06	413.00	796.00
520	C6H8N2	PHENYLHYDRAZINE	71.3143	-6.1509E+03	-2.1292E+01	7.2424E-11	4.6365E-06	292.35	761.00
521	C6H8N2O	BIS(CYANOETHYL)ETHER	57.3170	-6.5835E+03	-1.5829E+01	2.9371E-10	1.9717E-06	246.85	783.00
522	C6H8O4	DIMETHYL MALEATE	-1.9173	-3.4310E+03	7.0795E+00	-1.9039E-02	9.2324E-06	254.15	675.00
523	C6H8O6	ASCORBIC ACID	-175.2051	-5.5764E+03	8.4273E+01	-9.9268E-02	3.3909E-05	465.15	783.00
524	C6H8O7	CITRIC ACID	-75.5828	-7.1822E+03	4.2237E+01	-5.8401E-02	2.0221E-05	426.15	822.00
525	C6H10	1-METHYLCYCLOPENTENE	30.8840	-2.6219E+03	-8.2001E+00	1.5229E-04	2.5331E-06	268.15	541.99
526	C6H10	3-METHYLCYCLOPENTENE	84.2905	-3.6414E+03	-3.1463E+01	3.0590E-02	-1.2356E-05	263.15	394.15
527	C6H10	4-METHYLCYCLOPENTENE	52.1568	-3.1062E+03	-1.7205E+01	1.0261E-02	-1.6256E-06	271.15	543.75
528	C6H10	CYCLOHEXENE	52.1749	-3.2380E+03	-1.6878E+01	8.0388E-03	1.3259E-13	169.67	560.40
529	C6H10	2,3-DIMETHYL-1,3-BUTADIENE	29.9755	-2.5577E+03	-7.8544E+00	2.2361E-10	2.4591E-06	197.15	526.00
530	C6H10	1,5-HEXADIENE	10.5886	-2.0106E+03	2.8813E-01	-9.5620E-03	7.1640E-06	132.47	507.00
531	C6H10	cis,trans-2,4-HEXADIENE	5.6098	-2.1512E+03	2.6753E+00	-1.2840E-02	8.2559E-06	177.00	538.00
532	C6H10	trans,trans-2,4-HEXADIENE	40.7122	-3.0869E+03	-1.1634E+01	2.3357E-09	4.1939E-06	228.25	535.00
533	C6H10	1-HEXYNE	55.7231	-3.2541E+03	-1.8405E+01	9.5814E-03	9.2278E-14	141.25	516.20
534	C6H10	2-HEXYNE	51.6017	-3.3176E+03	-1.6451E+01	7.1637E-03	1.6871E-13	183.65	549.00
535	C6H10	3-HEXYNE	18.3265	-2.2166E+03	-3.6371E+00	2.2419E-04	-1.9213E-14	170.05	544.00
536	C6H10O	CYCLOHEXANONE	70.5022	-4.4120E+03	-2.3605E+01	1.1205E-02	-1.5648E-13	242.00	629.15
537	C6H10O	MESITYL OXIDE	34.0136	-3.2840E+03	-8.9451E+00	-1.0706E-10	1.9953E-06	220.15	600.00
538	C6H10O2	epsilon-CAPROLACTONE	-35.1025	-1.9390E+03	1.9302E+01	-2.6524E-02	1.1577E-05	271.85	771.00
539	C6H10O2	ETHYL METHACRYLATE	27.9574	-2.8696E+03	-6.9383E+00	-2.3481E-10	1.6779E-06	223.15	577.00
540	C6H10O2	n-PROPYL ACRYLATE	-14.2111	-2.1969E+03	1.1608E+01	-2.3737E-02	1.2300E-05	273.15	569.00
541	C6H10O3	ETHYLACETOACETATE	13.2623	-3.0925E+03	-7.3852E-01	-4.6405E-03	2.3962E-06	234.15	643.00
542	C6H10O3	PROPIONIC ANHYDRIDE	-3.9636	-2.9323E+03	7.6716E+00	-2.0145E-02	1.0758E-05	228.15	618.00
543	C6H10O4	ADIPIC ACID	59.7715	-7.3614E+03	-1.6370E+01	1.7452E-06	2.0667E-06	425.50	610.65
544	C6H10O4	DIETHYL OXALATE	5.0526	-3.5060E+03	4.2854E+00	-1.6974E-02	8.7698E-06	232.55	646.00
545	C6H10O4	ETHYLENE GLYCOL DIACETATE	167.0849	-1.0545E+04	-5.4181E+01	1.2204E-11	1.3938E-05	242.15	653.00
546	C6H10O4	ETHYLIDENE DIACETATE	150.2951	-9.1567E+03	-4.8925E+01	1.0422E-09	1.3981E-05	292.00	635.00

Table 7-1 VAPOR PRESSURE - ORGANIC COMPOUNDS (continued)

$$\log_{10} P = A + B/T + C \log_{10} T + D T + E T^2 \qquad \text{(P - mm Hg, T - K)}$$

NO	FORMULA	NAME	A	B	C	D	E	TMIN	TMAX
547	C6H11N	HEXANENITRILE	11.0057	-2.6129E+03	-3.5500E-01	-3.6747E-03	2.0975E-06	192.85	622.05
548	C6H11NO	epsilon-CAPROLACTAM	-6.5636	-3.5773E+03	8.1932E+00	-1.4866E-02	5.7323E-06	342.36	806.00
549	C6H11NO	CYCLOHEXANONE OXIME	-30.2431	-2.4875E+03	1.7987E+01	-2.5651E-02	1.0303E-05	363.15	715.00
550	C6H12	CYCLOHEXANE	48.5529	-3.0874E+03	-1.5521E+01	7.3830E-03	6.3563E-12	279.69	553.54
551	C6H12	2,3-DIMETHYL-1-BUTENE	30.3612	-2.4492E+03	-8.0866E+00	5.7386E-11	2.9830E-06	115.89	500.00
552	C6H12	2,3-DIMETHYL-2-BUTENE	36.2976	-2.7371E+03	-1.0600E+01	4.0594E-03	5.2266E-14	198.92	524.00
553	C6H12	3,3-DIMETHYL-1-BUTENE	53.9034	-2.8800E+03	-1.8024E+01	1.0035E-02	-2.8642E-13	157.95	480.00
554	C6H12	2-ETHYL-1-BUTENE	30.7807	-2.4852E+03	-8.5324E+00	3.0560E-03	1.6108E-14	141.61	512.00
555	C6H12	trans-3-METHYL-2-PENTENE	-2.3218	-1.9687E+03	6.3933E+00	-1.9584E-02	1.2261E-05	250.15	521.00
556	C6H12	1-HEXENE	33.4486	-2.6221E+03	-9.1784E+00	3.0930E-12	3.6780E-06	133.39	504.03
557	C6H12	cis-2-HEXENE	30.8810	-2.4822E+03	-8.6590E+00	3.5070E-03	3.9648E-14	132.00	513.00
558	C6H12	trans-2-HEXENE	33.4508	-2.6886E+03	-9.1025E+00	-8.3643E-11	3.1688E-06	140.17	513.00
559	C6H12	cis-3-HEXENE	32.6260	-2.6286E+03	-8.8452E+00	2.1251E-11	3.3095E-06	135.33	509.00
560	C6H12	trans-3-HEXENE	32.6960	-2.6457E+03	-8.8555E+00	-2.9450E-11	3.2926E-06	159.73	509.00
561	C6H12	METHYLCYCLOPENTANE	32.4766	-2.6434E+03	-8.7933E+00	2.0749E-11	3.2158E-06	130.73	532.79
562	C6H12	2-METHYL-1-PENTENE	32.9509	-2.6171E+03	-8.9572E+00	-8.7635E-11	3.1710E-06	137.42	507.00
563	C6H12	2-METHYL-2-PENTENE	30.0876	-2.5736E+03	-7.8673E+00	-2.1219E-11	2.3548E-06	138.07	514.00
564	C6H12	3-METHYL-1-PENTENE	35.0173	-2.5652E+03	-9.8547E+00	2.2120E-11	4.5149E-06	120.20	495.00
565	C6H12	3-METHYL-cis-2-PENTENE	29.4561	-2.5243E+03	-7.6890E+00	-4.0644E-11	2.6159E-06	138.31	515.00
566	C6H12	4-METHYL-1-PENTENE	44.7746	-2.7364E+03	-1.4283E+01	7.3100E-03	4.8402E-14	119.51	496.00
567	C6H12	4-METHYL-cis-2-PENTENE	33.8373	-2.5887E+03	-9.3331E+00	-9.8033E-11	3.6761E-06	138.30	499.00
568	C6H12	4-METHYL-trans-2-PENTENE	33.4412	-2.5983E+03	-9.1711E+00	-1.1438E-10	3.5389E-06	132.35	501.00
569	C6H12N2	TRIETHYLENEDIAMINE	57.6476	-4.9631E+03	-1.6755E+01	1.7547E-07	3.6887E-06	434.25	655.00
570	C6H12O	BUTYL VINYL ETHER	3.5099	-2.4111E+03	4.2976E+00	-1.8090E-02	1.1571E-05	181.15	536.00
571	C6H12O	CYCLOHEXANOL	49.9123	-4.8446E+03	-1.3711E+01	3.5451E-09	1.5932E-06	296.60	625.15
572	C6H12O	1-HEXANAL	-10.8651	-2.3852E+03	1.0619E+01	-2.5662E-02	1.4529E-05	217.15	579.00
573	C6H12O	ETHYL ISOPROPYL KETONE	87.6742	-5.4447E+03	-2.7894E+01	-8.4222E-10	9.7551E-06	200.00	567.00
574	C6H12O	2-HEXANONE	4.0508	-2.6276E+03	3.7783E+00	-1.4342E-02	8.0592E-06	217.35	587.05
575	C6H12O	3-HEXANONE	41.5000	-3.5485E+03	-1.1617E+01	2.3758E-10	3.2773E-06	217.50	582.82
576	C6H12O	METHYL ISOBUTYL KETONE	64.1919	-4.3587E+03	-1.9766E+01	-3.9997E-10	7.1020E-06	189.15	571.40
577	C6H12O2	n-PENTYL FORMATE	8.6264	-3.0941E+03	2.6316E+00	-1.6152E-02	9.4969E-06	199.65	576.00
578	C6H12O2	n-BUTYL ACETATE	4.3830	-2.7134E+03	3.9835E+00	-1.6575E-02	9.7246E-06	199.65	579.15
579	C6H12O2	sec-BUTYL ACETATE	11.3615	-2.5445E+03	2.5232E-01	-8.6981E-03	5.5066E-06	174.15	561.00
580	C6H12O2	tert-BUTYL ACETATE	-28.0818	-1.5744E+03	1.7253E+01	-3.0433E-02	1.5894E-05	283.15	545.00
581	C6H12O2	ETHYL n-BUTYRATE	1.4298	-2.4683E+03	5.0326E+00	-1.8024E-02	1.1247E-05	175.15	571.00
582	C6H12O2	ETHYL ISOBUTYRATE	6.6661	-2.3847E+03	2.1818E+00	-1.0957E-02	6.8003E-06	185.00	553.15
583	C6H12O2	ISOBUTYL ACETATE	35.1224	-3.2426E+03	-9.3893E+00	1.7877E-11	2.6707E-06	174.30	561.00
584	C6H12O2	n-PROPYL PROPIONATE	28.0816	-3.0218E+03	-6.8307E+00	-3.3989E-10	1.1397E-06	197.25	578.00
585	C6H12O2	CYCLOHEXYL PEROXIDE	16.8158	-4.5516E+03	-4.3507E-02	-1.2089E-02	5.8187E-06	253.15	685.00
586	C6H12O2	DIACETONE ALCOHOL	-10.1327	-2.2331E+03	8.6771E+00	-1.4450E-02	7.7882E-06	229.15	606.00
587	C6H12O2	2-ETHYL BUTYRIC ACID	0.0986	-3.6483E+03	6.6758E+00	-2.0070E-02	9.8395E-06	258.15	655.00
588	C6H12O2	n-HEXANOIC ACID	55.7058	-5.6602E+03	-1.5458E+01	1.0823E-09	1.8718E-06	270.15	667.00
589	C6H12O3	2-ETHOXYETHYL ACETATE	1.9276	-3.1451E+03	5.7407E+00	-2.1017E-02	1.1834E-05	211.45	597.00
590	C6H12O3	HYDROXYCAPROIC ACID	-16.5030	-5.7732E+03	1.5880E+01	-3.2654E-02	1.3186E-05	334.00	758.00
591	C6H12O3	PARALDEHYDE	-23.0555	-2.0812E+03	1.5478E+01	-2.8496E-02	1.4336E-05	285.75	579.00
592	C6H12O3	sec-BUTYL GLYCOLATE	83.6124	-4.8819E+03	-2.9906E+01	2.4934E-02	-8.7092E-06	301.45	450.65
593	C6H12S	THIACYCLOHEPTANE	176.8865	-7.0818E+03	-6.7575E+01	5.5282E-02	-1.7234E-05	351.15	640.07
594	C6H13N	CYCLOHEXYLAMINE	62.6197	-4.0549E+03	-2.0524E+01	9.2844E-03	-1.5497E-12	255.45	615.00
595	C6H13N	HEXAMETHYLENEIMINE	50.4247	-4.1023E+03	-1.4587E+01	-7.6705E-10	3.7877E-06	236.15	615.00
596	C6H14	2,2-DIMETHYLBUTANE	33.1285	-2.4527E+03	-9.2016E+00	-4.7077E-10	4.1755E-06	174.28	488.78
597	C6H14	2,3-DIMETHYLBUTANE	33.6319	-2.5524E+03	-9.3142E+00	1.4759E-10	3.9140E-06	145.19	499.98
598	C6H14	n-HEXANE	69.7378	-3.6278E+03	-2.3927E+01	1.2810E-02	-1.6844E-13	177.84	507.43
599	C6H14	2-METHYLPENTANE	30.7477	-2.4888E+03	-8.2295E+00	-2.3723E-11	3.2402E-06	119.55	497.50
600	C6H14	3-METHYLPENTANE	35.2848	-2.6773E+03	-9.8546E+00	2.2352E-11	4.0277E-06	110.25	504.43
601	C6H14N2O2	LYSINE	146.3052	-1.1586E+04	-4.7671E+01	1.3593E-02	-8.6096E-11	483.00	821.00
602	C6H14O	2-ETHYL-1-BUTANOL	217.1721	-1.0215E+04	-7.8194E+01	3.6089E-02	4.3427E-12	250.00	580.00
603	C6H14O	1-HEXANOL	53.9686	-4.9501E+03	-1.5199E+01	-6.6922E-10	2.3647E-06	228.55	611.35
604	C6H14O	2-HEXANOL	53.6472	-4.7399E+03	-1.5189E+01	1.9796E-10	2.5900E-06	223.00	586.20
605	C6H14O	2-METHYL-1-PENTANOL	26.1511	-3.5143E+03	-5.6950E+00	1.9370E-10	1.1685E-07	223.00	582.00
606	C6H14O	4-METHYL-2-PENTANOL	43.2285	-4.0171E+03	-1.1821E+01	-1.3977E-10	2.4177E-06	183.00	574.40
607	C6H14O	n-BUTYL ETHYL ETHER	8.5224	-2.4667E+03	1.9513E+00	-1.4047E-02	9.2664E-06	170.15	531.00
608	C6H14O	DIISOPROPYL ETHER	15.9552	-2.0276E+03	-2.8551E+00	2.7662E-04	-9.9111E-14	187.65	500.05
609	C6H14O	DI-n-PROPYL ETHER	44.0232	-3.2820E+03	-1.2792E+01	1.2682E-10	4.8776E-06	149.95	530.60
610	C6H14O	METHYL tert-PENTYL ETHER	9.0032	-2.2495E+03	1.2518E+00	-1.1267E-02	7.5982E-06	160.00	534.00
611	C6H14O2	ACETAL	32.9053	-3.1266E+03	-8.7033E+00	1.8687E-03	-2.0390E-13	173.15	541.00
612	C6H14O2	2-BUTOXYETHANOL	-39.3735	-3.0058E+03	2.5696E+01	-5.7339E-02	3.2713E-05	203.15	600.00
613	C6H14O2	1,6-HEXANEDIOL	98.5706	-7.8238E+03	-3.1753E+01	1.0862E-02	-9.3985E-15	315.15	670.00
614	C6H14O2	HEXYLENE GLYCOL	12.4830	-5.5212E+03	4.5945E+00	-2.8890E-02	1.5535E-05	223.15	621.00
615	C6H14O2S	DI-n-PROPYL SULFONE	45.5346	-5.3305E+03	-1.2180E+01	-7.2205E-10	1.5917E-06	303.00	763.00
616	C6H14O3	DIETHYLENE GLYCOL DIMETHYL ETHER	-25.6309	-2.4739E+03	1.7797E+01	-3.8906E-02	2.2088E-05	203.15	604.00
617	C6H14O3	DIPROPYLENE GLYCOL	-34.4044	-2.5539E+03	1.9845E+01	-3.0216E-02	1.5480E-05	233.00	654.00
618	C6H14O3	2-(2-ETHOXYETHOXY)ETHANOL	108.9937	-7.4542E+03	-3.4699E+01	-7.3427E-11	1.0904E-05	195.15	632.00
619	C6H14O3	TRIMETHYLOLPROPANE	-98.9783	-2.5253E+03	4.8303E+01	-6.1680E-02	2.5945E-05	331.15	709.00
620	C6H14O4	TRIETHYLENE GLYCOL	13.3551	-3.8387E+03	-1.3933E+00	1.4247E-10	9.8759E-07	265.79	700.00
621	C6H14O6	SORBITOL	8.3436	-1.3304E+04	1.0812E+01	-3.4183E-02	1.1537E-05	370.85	959.00
622	C6H14S	n-HEXYLMERCAPTAN	3.6922	-2.8348E+03	4.2831E+00	-1.6786E-02	9.5563E-06	192.62	623.00
623	C6H14S	BUTYL-ETHYL-SULFIDE	77.6473	-4.6358E+03	-2.6411E+01	1.4144E-02	-2.0253E-06	306.15	609.00
624	C6H14S	ISOPROPYL-SULFIDE	81.9806	-4.3714E+03	-2.8802E+01	1.8793E-02	-4.1562E-06	286.15	585.71

Table 7-1 VAPOR PRESSURE - ORGANIC COMPOUNDS (continued)

$$\log_{10} P = A + B/T + C \log_{10} T + D\,T + E\,T^2 \qquad (P\text{ - mm Hg, }T\text{ - K})$$

NO	FORMULA	NAME	A	B	C	D	E	TMIN	TMAX
625	C6H14S	METHYL-PENTYL-SULFIDE	-77.9132	-8.5910E+02	3.8748E+01	-5.4706E-02	2.4907E-05	343.15	587.98
626	C6H14S	PROPYL-SULFIDE	101.1564	-5.2143E+03	-3.6171E+01	2.3976E-02	-5.6621E-06	305.15	609.73
627	C6H14S2	PROPYL-DISULFIDE	149.9561	-7.0370E+03	-5.5867E+01	4.2982E-02	-1.3632E-05	346.15	501.15
628	C6H15Al	TRIETHYL ALUMINUM	42.4467	-5.4613E+03	-1.0406E+01	9.1793E-05	-1.8982E-14	220.65	720.15
629	C6H15Al2Cl3	ETHYL ALUMINUM SESQUICHLORIDE	1009.6078	-3.7457E+04	-3.8125E+02	1.9481E-01	-5.3609E-11	253.15	500.00
630	C6H15N	DIISOPROPYLAMINE	198.8840	-7.9159E+03	-7.3734E+01	4.0300E-02	-3.8643E-13	176.85	523.10
631	C6H15N	DI-n-PROPYLAMINE	2.0630	-2.6611E+03	5.1601E+00	-1.8720E-02	1.1000E-05	210.15	555.80
632	C6H15N	n-HEXYLAMINE	57.8555	-3.9804E+03	-1.8579E+01	8.1204E-03	1.0791E-12	251.85	583.00
633	C6H15N	TRIETHYLAMINE	8.8604	-2.2482E+03	1.2692E+00	-1.1021E-02	7.4404E-06	158.45	535.15
634	C6H15NO	6-AMINOHEXANOL	-215.9646	1.2568E+03	9.6830E+01	-1.1493E-01	4.9388E-05	331.00	681.00
635	C6H15NO2	DIISOPROPANOLAMINE	-36.5701	-5.3225E+03	2.5621E+01	-4.9745E-02	2.1952E-05	318.15	672.00
636	C6H15NO3	TRIETHANOLAMINE	135.3206	-1.0312E+04	-4.4637E+01	1.4368E-02	-1.7552E-13	294.35	787.00
637	C6H15N3	N-AMINOETHYL PIPERAZINE	-2.4787	-3.7348E+03	7.7097E+00	-2.0701E-02	9.7725E-06	254.15	708.00
638	C6H15O4P	TRIETHYL PHOSPHATE	93.8194	-5.3692E+03	-3.2942E+01	1.7761E-02	-4.3029E-14	216.00	794.00
639	C6H16N2	HEXAMETHYLENEDIAMINE	-22.2161	-3.1489E+03	1.5888E+01	-2.8891E-02	1.2928E-05	313.95	663.00
640	C6H18N3OP	HEXAMETHYL PHOSPHORAMIDE	79.7221	-6.0554E+03	-2.4895E+01	-7.1377E-08	9.5549E-06	280.15	506.15
641	C6H18N4	TRIETHYLENE TETRAMINE	-6.5589	-5.1778E+03	1.1362E+01	-2.9298E-02	1.3055E-05	285.15	718.00
642	C6H18OSi2	HEXAMETHYLDISILOXANE	13.1786	-2.2417E+03	-1.3026E+00	-3.3036E-03	2.0403E-06	204.93	518.70
643	C6H18O3Si3	HEXAMETHYLCYCLOTRISILOXANE	37.4172	-3.4814E+03	-1.0103E+01	1.0086E-08	2.2130E-06	337.15	554.20
644	C6H19NSi2	HEXAMETHYLDISILAZANE	90.5970	-4.9110E+03	-3.1339E+01	1.5295E-02	1.3413E-12	293.15	544.00
645	C7H3ClF3NO2	4-CHLORO-3-NITROBENZOTRIFLUORIDE	-9.2380	-3.4778E+03	1.0265E+01	-2.2184E-02	1.0033E-05	293.15	686.00
646	C7H3Cl2F3	2,4-DICHLOROBENZOTRIFLUORIDE	0.9827	-2.9598E+03	5.2153E+00	-1.5452E-02	7.7787E-06	247.55	646.00
647	C7H3Cl2NO	3,4-DICHLOROPHENYL ISOCYANATE	-656.2031	9.6595E+03	2.9354E+02	-4.0817E-01	2.0633E-04	316.15	526.95
648	C7H4ClF3	p-CHLOROBENZOTRIFLUORIDE	-2.8084	-2.4667E+03	6.5704E+00	-1.7128E-02	9.1387E-06	237.15	601.00
649	C7H4Cl2O	m-CHLOROBENZOYL CHLORIDE	33.2875	-3.8982E+03	-8.7948E+00	2.3275E-03	-1.9629E-13	280.00	724.00
650	C7H4F3NO2	3-NITROBENZOTRIFLUORIDE	92.4123	-6.0515E+03	-3.0845E+01	1.2137E-02	-1.1882E-13	272.00	667.00
651	C7H5ClO	BENZOYL CHLORIDE	1.1537	-2.6751E+03	4.0043E+00	-8.9221E-03	4.1180E-06	272.65	697.00
652	C7H5ClO2	o-CHLOROBENZOIC ACID	-42.9847	-3.1867E+03	2.3694E+01	-3.0284E-02	1.0828E-05	415.15	792.00
653	C7H5Cl3	BENZOTRICHLORIDE	0.4912	-2.7285E+03	4.4706E+00	-1.0580E-02	4.7621E-06	268.40	737.00
654	C7H5F3	BENZOTRIFLUORIDE	28.3157	-2.6906E+03	-7.4330E+00	2.3153E-03	3.9138E-13	244.14	565.00
655	C7H5N	BENZONITRILE	5.6061	-2.8639E+03	2.4465E+00	-8.4626E-03	3.9790E-06	260.40	699.35
656	C7H5NO	PHENYL ISOCYANATE	26.0365	-3.1145E+03	-6.1914E+00	1.2247E-11	1.5742E-06	243.15	648.00
657	C7H5N3O6	2,4,6-TRINITROTOLUENE	6.3156	-2.6756E+03	-4.6215E-02	6.1747E-09	-2.3743E-12	354.00	518.30
658	C7H6Cl2	BENZYL DICHLORIDE	2.5175	-2.6069E+03	3.1664E+00	-7.3890E-03	3.4017E-06	257.00	731.00
659	C7H6Cl2	2,4-DICHLOROTOLUENE	31.9325	-3.7438E+03	-8.0123E+00	-7.5077E-11	1.2500E-06	259.65	705.00
660	C7H6N2O4	2,4-DINITROTOLUENE	11.5966	-3.0079E+03	-1.6468E+00	1.5949E-03	-1.8722E-14	343.00	814.00
661	C7H6N2O4	2,5-DINITROTOLUENE	-3.2242	-4.8907E+03	8.2992E+00	-1.8972E-02	7.4327E-06	325.65	814.00
662	C7H6N2O4	2,6-DINITROTOLUENE	-14.5673	-4.2746E+03	1.2904E+01	-2.3800E-02	9.4513E-06	339.00	770.00
663	C7H6N2O4	3,4-DINITROTOLUENE	-1.5478	-5.1010E+03	7.5705E+00	-1.7734E-02	6.7654E-06	332.00	842.00
664	C7H6N2O4	3,5-DINITROTOLUENE	-5.7421	-3.2067E+03	6.4644E+00	-8.3308E-03	3.1010E-06	365.65	814.00
665	C7H6O	BENZALDEHYDE	28.4711	-3.4489E+03	-6.8363E+00	-2.8173E-10	9.5236E-07	247.15	695.00
666	C7H6O2	BENZOIC ACID	-140.0388	8.0479E+01	6.2611E+01	-6.5321E-02	2.4596E-05	395.52	524.00
667	C7H6O2	p-HYDROXYBENZALDEHYDE	11.2084	-4.0163E+03	6.0941E-02	-3.4771E-03	1.2348E-06	390.15	844.00
668	C7H6O2	SALICYLALDEHYDE	31.2437	-3.5351E+03	-8.0244E+00	7.8869E-10	2.7203E-06	266.15	680.00
669	C7H6O3	SALICYLIC ACID	177.3858	-1.2871E+04	-5.6301E+01	-1.6667E-07	1.1353E-05	431.75	739.00
670	C7H7Br	p-BROMOTOLUENE	12.8209	-2.6568E+03	-1.4314E+00	-8.9677E-04	3.9733E-07	299.95	699.00
671	C7H7Cl	BENZYL CHLORIDE	12.1503	-2.9139E+03	-3.7120E-01	-5.2889E-03	2.6296E-06	234.15	686.00
672	C7H7Cl	o-CHLOROTOLUENE	33.2792	-3.4099E+03	-8.6743E+00	6.1874E-10	1.8987E-06	236.65	656.00
673	C7H7Cl	p-CHLOROTOLUENE	61.8901	-4.3760E+03	-1.9840E+01	7.7991E-03	1.0781E-13	280.65	660.00
674	C7H7F	p-FLUOROTOLUENE	107.7550	-5.0398E+03	-3.9443E+01	2.9500E-02	-8.2554E-06	341.15	590.48
675	C7H7NO	FORMANILIDE	-5.9746	-3.8576E+03	8.4845E+00	-1.7089E-02	6.8480E-06	323.15	787.00
676	C7H7NO2	m-NITROTOLUENE	12.1169	-2.7684E+03	-1.4768E+00	4.6539E-04	-1.8362E-14	289.20	734.00
677	C7H7NO2	o-NITROTOLUENE	7.8266	-2.9908E+03	1.1064E+00	-4.9168E-03	2.2375E-06	269.98	720.00
678	C7H7NO2	p-NITROTOLUENE	9.9641	-2.6549E+03	-8.0182E-01	5.3926E-04	-4.1090E-14	324.75	736.00
679	C7H7NO3	o-NITROANISOLE	3.7708	-4.1871E+03	4.6593E+00	-1.4263E-02	6.0583E-06	283.60	782.00
680	C7H8	TOLUENE	34.0775	-3.0379E+03	-9.1635E+00	1.0289E-11	2.7035E-06	178.18	591.79
681	C7H8	1,3,5-CYCLOHEPTATRIENE	142.6727	-5.4014E+03	-5.5950E+01	5.8830E-02	-2.5652E-05	273.15	340.15
682	C7H8O	ANISOLE	-8.1053	-2.5386E+03	9.0289E+00	-2.0462E-02	1.0536E-05	235.65	641.65
683	C7H8O	BENZYL ALCOHOL	-36.2189	-3.3475E+03	2.3337E+01	-4.4600E-02	2.1443E-05	257.85	677.00
684	C7H8O	m-CRESOL	105.5280	-6.9748E+03	-3.5083E+01	1.2508E-02	-2.4317E-12	285.39	705.85
685	C7H8O	o-CRESOL	89.4591	-6.0489E+03	-2.9483E+01	1.0936E-02	1.9933E-12	304.19	697.55
686	C7H8O	p-CRESOL	122.8998	-7.6175E+03	-4.1637E+01	1.5709E-02	-8.9199E-13	307.93	704.65
687	C7H8O2	GUAIACOL	62.5937	-5.2602E+03	-1.8589E+01	4.2117E-09	4.8237E-06	304.65	697.00
688	C7H8O2	p-METHOXYPHENOL	-11.8605	-3.5151E+03	1.0904E+01	-1.9681E-02	7.9826E-06	329.00	758.00
689	C7H9N	BENZYLAMINE	7.0445	-3.2078E+03	2.6513E+00	-1.1970E-02	6.0428E-06	227.15	683.50
690	C7H9N	2,6-DIMETHYLPYRIDINE	46.2468	-3.9493E+03	-1.3162E+01	-1.1004E-09	3.4080E-06	267.00	623.75
691	C7H9N	N-METHYLANILINE	-7.2448	-2.9648E+03	8.9163E+00	-2.0616E-02	1.0451E-05	216.15	701.55
692	C7H9N	m-TOLUIDINE	7.0317	-3.2034E+03	2.3006E+00	-9.7791E-03	4.6824E-06	242.75	709.15
693	C7H9N	o-TOLUIDINE	96.5685	-6.2643E+03	-3.2263E+01	1.2361E-02	6.2915E-13	249.47	694.15
694	C7H9N	p-TOLUIDINE	-13.9927	-2.5795E+03	1.0832E+01	-1.7705E-02	7.6741E-06	316.90	693.15
695	C7H10	2-NORBORNENE	-32.8921	-1.0869E+03	1.8252E+01	-2.6719E-02	1.2726E-05	319.40	583.00
696	C7H10N2	TOLUENEDIAMINE	43.8539	-5.4930E+03	-1.1472E+01	-3.9637E-10	1.2598E-06	371.25	804.00
697	C7H11NO	CYCLOHEXYL ISOCYANATE	12.0255	-3.4717E+03	1.2043E+00	-1.3523E-02	7.6331E-06	193.00	633.00
698	C7H12	1-HEPTYNE	257.7236	-8.1481E+03	-1.0333E+02	1.0182E-01	-3.7742E-05	287.15	559.69
699	C7H12O2	n-BUTYL ACRYLATE	37.5709	-3.3554E+03	-1.0814E+01	3.9510E-03	2.3590E-14	208.55	598.00
700	C7H12O2	ISOBUTYL ACRYLATE	2.7565	-2.7679E+03	4.7435E+00	-1.7391E-02	9.9204E-06	212.00	580.00
701	C7H12O2	n-PROPYL METHACRYLATE	0.3871	-2.6502E+03	5.4688E+00	-1.6946E-02	9.2970E-06	223.00	599.00
702	C7H12O4	DIETHYL MALONATE	64.4785	-5.0283E+03	-2.0379E+01	7.5224E-03	-4.3640E-14	224.25	653.00

Table 7-1 VAPOR PRESSURE - ORGANIC COMPOUNDS (continued)

$$\log_{10} P = A + B/T + C \log_{10} T + D T + E T^2 \qquad (P - mm\ Hg,\ T - K)$$

NO	FORMULA	NAME	A	B	C	D	E	TMIN	TMAX
703	C7H14	CYCLOHEPTANE	54.0858	-3.6109E+03	-1.7331E+01	7.5272E-03	1.7553E-12	265.12	604.30
704	C7H14	1,1-DIMETYLCYCLOPENTANE	58.1943	-3.4151E+03	-1.9294E+01	9.6704E-03	-2.4361E-15	203.36	547.00
705	C7H14	cis-1,2-DIMETHYLCYCLOPENTANE	36.3623	-3.0025E+03	-1.0070E+01	-1.0435E-09	3.3726E-06	219.26	565.15
706	C7H14	trans-1,2-DIMETHYLCYCLOPENTANE	36.8109	-2.9536E+03	-1.0275E+01	-4.6212E-12	3.6730E-06	155.58	553.15
707	C7H14	cis-1,3-DIMETHYLCYCLOPENTANE	35.4255	-2.7286E+03	-1.0444E+01	4.6608E-03	1.7565E-14	139.45	551.00
708	C7H14	trans-1,3-DIMETHYLCYCLOPENTANE	53.1912	-3.3121E+03	-1.7277E+01	8.3107E-03	5.0896E-14	139.18	553.00
709	C7H14	ETHYLCYCLOPENTANE	36.3631	-3.0448E+03	-1.0038E+01	3.5007E-11	3.2347E-06	134.71	569.52
710	C7H14	2-ETHYL-1-PENTENE	45.6909	-3.3443E+03	-1.3400E+01	9.4764E-11	4.9466E-06	168.00	543.00
711	C7H14	3-ETHYL-1-PENTENE	32.1696	-2.7728E+03	-8.5671E+00	-4.3108E-13	2.6975E-06	145.67	530.00
712	C7H14	1-HEPTENE	38.1255	-3.0640E+03	-1.0679E+01	1.2244E-10	3.6680E-06	154.27	537.29
713	C7H14	cis-2-HEPTENE	9.1082	-2.3022E+03	1.1634E+00	-1.0763E-02	7.0739E-06	164.00	549.00
714	C7H14	trans-2-HEPTENE	35.9810	-2.9861E+03	-9.9356E+00	-3.7562E-11	3.4663E-06	163.67	543.00
715	C7H14	cis-3-HEPTENE	51.1587	-3.3361E+03	-1.6342E+01	7.3561E-03	-4.5258E-14	136.51	545.00
716	C7H14	trans-3-HEPTENE	11.4048	-2.4198E+03	4.8920E-01	-1.1842E-02	8.4421E-06	136.52	540.00
717	C7H14	METHYLCYCLOHEXANE	38.0955	-3.0738E+03	-1.0684E+01	-5.1766E-11	3.5282E-06	146.58	572.19
718	C7H14	2-METHYL-1-HEXENE	31.6484	-2.7938E+03	-8.3748E+00	-3.2642E-11	2.5818E-06	170.28	538.00
719	C7H14	3-METHYL-1-HEXENE	38.3500	-3.0067E+03	-1.0784E+01	-1.4078E-10	3.7671E-06	145.00	528.00
720	C7H14	4-METHYL-1-HEXENE	13.7716	-2.3436E+03	-8.5755E-01	-8.2352E-03	5.9905E-06	131.70	534.00
721	C7H14	2,3,3-TRIMETHYL-1-BUTENE	34.3198	-2.7439E+03	-9.4453E+00	4.8805E-11	3.4012E-06	163.30	531.00
722	C7H14O	DIISOPROPYL KETONE	39.9665	-3.4805E+03	-1.1093E+01	-3.3888E-10	3.1900E-06	204.81	576.00
723	C7H14O	2-HEPTANONE	-13.0256	-2.6425E+03	1.1879E+01	-2.7571E-02	1.4560E-05	238.15	611.55
724	C7H14O	1-HEPTANAL	82.0370	-5.0309E+03	-2.7566E+01	1.2058E-02	6.7371E-13	230.15	603.00
725	C7H14O	1-METHYLCYCLOHEXANOL	-47.6960	-2.4407E+03	2.7425E+01	-4.6818E-02	2.2498E-05	299.15	603.00
726	C7H14O	cis-2-METHYLCYCLOHEXANOL	-28.3266	-2.2272E+03	1.7515E+01	-2.9010E-02	1.4232E-05	280.15	614.00
727	C7H14O	trans-2-METHYLCYCLOHEXANOL	33.9333	-3.7421E+03	-8.6881E+00	1.0473E-09	2.1842E-06	269.15	616.00
728	C7H14O	cis-3-METHYLCYCLOHEXANOL	14.1810	-3.7184E+03	2.7148E-01	-1.0466E-02	5.2141E-06	267.65	618.00
729	C7H14O	trans-3-METHYLCYCLOHEXANOL	6.7678	-3.5847E+03	3.5943E+00	-1.5274E-02	7.5766E-06	272.65	617.00
730	C7H14O	cis-4-METHYLCYCLOHEXANOL	-111.8683	-1.5989E+03	5.5870E+01	-8.3457E-02	3.8048E-05	325.15	622.00
731	C7H14O	trans-4-METHYLCYCLOHEXANOL	-77.5317	-2.0679E+03	4.0418E+01	-6.1911E-02	2.8176E-05	327.15	622.00
732	C7H14O	5-METHYL-2-HEXANONE	48.0969	-4.0180E+03	-1.3841E+01	-4.8110E-10	3.8749E-06	199.25	601.00
733	C7H14O2	n-BUTYL PROPIONATE	34.9074	-3.5538E+03	-9.0906E+00	9.7933E-11	1.6987E-06	183.63	594.00
734	C7H14O2	ETHYL ISOVALERATE	13.3878	-2.8353E+03	-3.0743E-01	-9.0271E-03	5.5521E-06	173.85	587.95
735	C7H14O2	ISOPENTYL ACETATE	10.0856	-2.8848E+03	1.2945E+00	-1.1623E-02	6.7499E-06	194.65	599.00
736	C7H14O2	n-PENTYL ACETATE	7.8848	-3.0696E+03	2.7085E+00	-1.5165E-02	8.7135E-06	202.35	598.00
737	C7H14O2	n-PROPYL n-BUTYRATE	9.3498	-2.8326E+03	1.5675E+00	-1.2107E-02	7.3132E-06	177.95	594.00
738	C7H14O2	n-HEPTANOIC ACID	202.0065	-1.1589E+04	-6.9941E+01	2.5732E-02	4.2361E-13	265.83	680.00
739	C7H14O3	ETHYL-3-ETHOXYPROPIONATE	-5.1616	-3.2227E+03	9.1650E+00	-2.6421E-02	1.4383E-05	223.00	609.00
740	C7H15Br	1-BROMOHEPTANE	43.3327	-4.0389E+03	-1.2105E+01	-1.5959E-10	3.0522E-06	217.05	651.00
741	C7H15N	N-METHYLCYCLOHEXYLAMINE	-13.2174	-2.3986E+03	1.1171E+01	-2.2656E-02	1.1306E-05	264.65	622.00
742	C7H16	2,2-DIMETHYLPENTANE	6.2875	-2.1682E+03	2.6936E+00	-1.5525E-02	1.0917E-05	149.34	520.50
743	C7H16	2,3-DIMETHYLPENTANE	39.7737	-2.9050E+03	-1.2012E+01	5.1334E-03	-2.3807E-14	160.00	537.35
744	C7H16	2,4-DIMETHYLPENTANE	35.9436	-2.8460E+03	-9.9938E+00	8.0693E-11	3.6419E-06	153.91	519.79
745	C7H16	3,3-DIMETHYLPENTANE	30.2570	-2.6313E+03	-7.9839E+00	4.6848E-13	2.7170E-06	138.70	536.40
746	C7H16	3-ETHYLPENTANE	8.5463	-2.2979E+03	1.5503E+00	-1.2233E-02	8.2670E-06	154.55	540.64
747	C7H16	n-HEPTANE	65.0257	-3.8188E+03	-2.1684E+01	1.0387E-02	1.0206E-11	182.56	540.26
748	C7H16	2-METHYLHEXANE	54.1075	-3.3785E+03	-1.7547E+01	8.2594E-03	-3.4967E-14	154.90	530.37
749	C7H16	3-METHYLHEXANE	35.2535	-2.9310E+03	-9.6667E+00	-5.2026E-11	3.2107E-06	153.75	535.25
750	C7H16	2,2,3-TRIMETHYLBUTANE	32.3633	-2.6614E+03	-8.7743E+00	-7.6870E-10	3.2006E-06	248.57	531.17
751	C7H16O	1-HEPTANOL	-19.9205	-4.3239E+03	1.8794E+01	-5.0553E-02	2.6161E-05	239.15	631.90
752	C7H16O	2-HEPTANOL	53.5603	-4.6821E+03	-1.5411E+01	7.1219E-10	4.1355E-06	243.00	588.00
753	C7H16O	5-METHYL-1-HEXANOL	-22.3023	-3.2580E+03	1.6815E+01	-3.4479E-02	1.6736E-05	293.15	605.00
754	C7H16O	ISOPROPYL-TERT-BUTYL-ETHER	480.2536	-1.3423E+04	-1.9683E+02	2.0283E-01	-7.8469E-05	314.15	558.21
755	C7H16S	n-HEPTYL MERCAPTAN	-7.0418	-2.8798E+03	9.1280E+00	-2.2899E-02	1.1894E-05	229.92	645.00
756	C7H16S	BUTYL-PROPYL-SULFIDE	362.7259	-1.2483E+04	-1.4264E+02	1.2046E-01	-3.8508E-05	372.15	653.50
757	C7H16S	ETHYL-PENTYL-SULFIDE	757.8673	-2.3337E+04	-3.0359E+02	2.6757E-01	-8.8823E-05	372.15	638.37
758	C7H16S	HEXYL-METHYL-SULFIDE	757.8673	-2.3337E+04	-3.0359E+02	2.6757E-01	-8.8823E-05	372.15	638.37
759	C7H17N	1-AMINOHEPTANE	55.9576	-4.4793E+03	-1.6540E+01	-9.1120E-10	4.8463E-06	254.15	607.00
760	C8H4Cl2O2	ISOPHTHALOYL CHLORIDE	48.7180	-5.6827E+03	-1.3141E+01	-5.4832E-10	1.6803E-06	317.00	768.00
761	C8H4O3	PHTHALIC ANHYDRIDE	30.6331	-3.8783E+03	-7.8671E+00	1.1148E-09	2.5885E-06	404.15	791.00
762	C8H6	ETHYNYLBENZENE	77.3201	-4.7127E+03	-2.6032E+01	1.2744E-02	-1.4952E-06	355.15	655.43
763	C8H6O4	ISOPHTHALIC ACID	101.8930	-1.2470E+04	-2.9146E+01	1.2145E-08	2.4529E-06	619.15	1007.00
764	C8H6O4	PHTHALIC ACID	-90.3221	-3.2214E+03	4.4109E+01	-5.0056E-02	1.6895E-05	464.15	800.00
765	C8H6O4	TEREPHTHALIC ACID	105.8916	-1.4001E+04	-3.0009E+01	-2.1837E-07	2.0825E-06	700.15	1113.00
766	C8H6S	BENZOTHIOPHENE	-9.5352	-2.6947E+03	8.8858E+00	-1.5478E-02	6.5159E-06	304.50	754.00
767	C8H7N	INDOLE	94.1625	-6.9431E+03	-3.0613E+01	9.9280E-03	1.7461E-13	273.68	790.00
768	C8H8	STYRENE	55.8621	-4.0240E+03	-1.7609E+01	6.6842E-03	1.9438E-13	242.54	648.00
769	C8H8	1,3,5,7-CYCLOOCTATETRAENE	174.8742	-6.4461E+03	-6.9148E+01	7.2135E-02	-3.1103E-05	273.15	350.15
770	C8H8O	ACETOPHENONE	55.5798	-4.5101E+03	-1.7284E+01	6.4184E-03	6.5557E-13	292.80	701.00
771	C8H8O	p-TOLUALDEHYDE	-6.7169	-3.0255E+03	8.4112E+00	-1.7602E-02	7.9137E-06	289.85	698.00
772	C8H8O2	METHYL BENZOATE	-13.6342	-2.9133E+03	1.1773E+01	-2.3979E-02	1.1324E-05	260.75	693.00
773	C8H8O2	o-TOLUIC ACID	-35.8816	-3.2354E+03	2.1133E+01	-3.0165E-02	1.1587E-05	376.85	751.00
774	C8H8O2	p-TOLUIC ACID	-67.6587	-2.2339E+03	3.3347E+01	-3.7709E-02	1.3130E-05	452.75	773.00
775	C8H8O3	METHYL SALICYLATE	202.6840	-1.2160E+04	-6.6670E+01	-1.8009E-09	1.8060E-05	265.15	701.00
776	C8H8O3	VANILLIN	-25.5830	-4.0860E+03	1.7515E+01	-2.8177E-02	1.0912E-05	355.00	777.00
777	C8H9NO	ACETANILIDE	-29.5448	-3.4172E+03	1.7932E+01	-2.4444E-02	8.8193E-06	386.65	825.00
778	C8H10	ETHYLBENZENE	36.1998	-3.3402E+03	-9.7970E+00	-1.1467E-11	2.5758E-06	178.15	617.17
779	C8H10	m-XYLENE	34.6803	-3.2981E+03	-9.2570E+00	-4.3563E-10	2.4103E-06	225.30	617.05
780	C8H10	o-XYLENE	37.2413	-3.4573E+03	-1.0126E+01	9.0676E-11	2.6123E-06	247.98	630.37

171

Table 7-1 VAPOR PRESSURE - ORGANIC COMPOUNDS (continued)

			$\log_{10} P = A + B/T + C \log_{10} T + D T + E T^2$				(P - mm Hg, T - K)		
NO	FORMULA	NAME	A	B	C	D	E	TMIN	TMAX
781	C8H10	p-XYLENE	60.0531	-4.0159E+03	-1.9441E+01	8.2881E-03	-2.3647E-12	286.41	616.26
782	C8H10O	m-ETHYLPHENOL	211.0890	-9.5120E+03	-7.9105E+01	5.7745E-02	-1.7291E-05	423.57	503.30
783	C8H10O	p-ETHYLPHENOL	16.9092	-3.7255E+03	-1.7886E+00	-4.2275E-03	1.8002E-06	318.23	716.45
784	C8H10O	PHENETOLE	-8.3543	-2.7728E+03	9.4482E+00	-2.1842E-02	1.1038E-05	243.63	647.15
785	C8H10O	2-PHENYLETHANOL	-9.2064	-3.1412E+03	9.5151E+00	-1.9088E-02	9.2863E-06	247.00	684.00
786	C8H10O	2,3-XYLENOL	82.9273	-6.0367E+03	-2.6948E+01	9.7390E-03	2.5196E-12	345.71	722.95
787	C8H10O	2,4-XYLENOL	53.3866	-5.1516E+03	-1.5095E+01	-1.3196E-09	2.8455E-06	345.71	707.65
788	C8H10O	2,5-XYLENOL	47.5888	-4.8102E+03	-1.3186E+01	-1.0208E-09	2.7045E-06	347.99	707.05
789	C8H10O	2,6-XYLENOL	87.1964	-5.8721E+03	-2.8853E+01	1.1130E-02	2.2316E-12	318.76	701.05
790	C8H10O	3,4-XYLENOL	68.6521	-6.150E+03	-2.0184E+01	-1.1259E-10	4.0266E-06	338.25	729.95
791	C8H10O	3,5-XYLENOL	-44.9150	-2.8912E+03	2.5704E+01	-3.9714E-02	1.6464E-05	336.59	715.65
792	C8H11N	N,N-DIMETHYLANILINE	20.1770	-3.1095E+03	-4.0127E+00	5.8538E-10	3.5387E-07	275.60	687.15
793	C8H11N	o-ETHYLANILINE	8.6419	-3.5422E+03	2.2380E+00	-1.2061E-02	5.9601E-06	226.55	704.00
794	C8H11N	2,4,6-TRIMETHYLPYRIDINE	38.3952	-3.5792E+03	-1.0994E+01	3.7210E-03	-8.0514E-14	229.00	653.00
795	C8H11NO	p-PHENETIDINE	3.1746	-3.9881E+03	4.8784E+00	-1.4857E-02	6.5179E-06	277.00	754.00
796	C8H12	1,5-CYCLOOCTADIENE	4.4346	-2.7129E+03	3.5539E+00	-1.3761E-02	7.5139E-06	203.98	645.00
797	C8H12	VINYLCYCLOHEXENE	72.8256	-4.2313E+03	-2.4638E+01	1.1831E-02	5.7377E-14	164.00	599.00
798	C8H12O4	1,4-CYCLOHEXANEDICARBOXYLIC ACID	-154.1402	-1.1010E+04	6.7478E+01	-5.9160E-02	1.6947E-05	585.65	889.00
799	C8H12O4	DIETHYL MALEATE	26.9138	-3.7706E+03	-6.2019E+00	1.5946E-03	1.0585E-06	264.35	680.00
800	C8H14O2	n-BUTYL METHACRYLATE	5.9469	-3.0212E+03	3.2591E+00	-1.4190E-02	7.6392E-06	223.00	616.00
801	C8H14O3	BUTYRIC ANHYDRIDE	111.9208	-6.7589E+03	-3.8201E+01	1.5853E-02	3.1976E-13	199.85	639.00
802	C8H14O4	DIETHYL SUCCINATE	-9.0047	-4.0194E+03	1.1644E+01	-3.0216E-02	1.4883E-05	252.35	660.00
803	C8H16	CYCLOOCTANE	73.2250	-4.4541E+03	-2.4825E+01	1.3715E-02	-2.4083E-06	367.15	647.20
804	C8H16	1,1-DIMETHYLCYCLOHEXANE	33.1329	-3.0084E+03	-8.8498E+00	-4.3621E-10	2.3704E-06	239.66	591.15
805	C8H16	cis-1,2-DIMETHYLCYCLOHEXANE	32.1635	-3.0728E+03	-8.4344E+00	6.8943E-10	1.9558E-06	223.16	606.15
806	C8H16	trans-1,2-DIMETHYLCYCLOHEXANE	31.9364	-2.9889E+03	-8.4129E+00	-6.9874E-11	2.1641E-06	184.99	596.15
807	C8H16	cis-1,3-DIMETHYLCYCLOHEXANE	32.4775	-3.0067E+03	-8.5896E+00	7.0258E-11	2.1739E-06	197.58	591.15
808	C8H16	trans-1,3-DIMETHYLCYCLOHEXANE	32.4384	-3.0550E+03	-8.5372E+00	2.2892E-10	2.0099E-06	183.05	598.00
809	C8H16	cis-1,4-DIMETHYLCYCLOHEXANE	31.9151	-3.0253E+03	-8.3613E+00	5.7055E-12	1.9673E-06	185.72	598.15
810	C8H16	trans-1,4-DIMETHYLCYCLOHEXANE	32.5731	-2.9872E+03	-8.6494E+00	-2.1355E-09	2.2946E-06	236.21	590.15
811	C8H16	ETHYLCYCLOHEXANE	32.7090	-3.1283E+03	-8.6023E+00	-3.9268E-11	1.9935E-06	161.84	609.15
812	C8H16	2-ETHYL-1-HEXENE	-0.9882	-2.3855E+03	6.1309E+00	-2.0135E-02	1.2580E-05	173.00	574.00
813	C8H16	1-METHYL-1-ETHYLCYCLOPENTANE	11.1029	-2.5757E+03	6.9091E-01	-1.2137E-02	8.3378E-06	135.00	582.00
814	C8H16	1-OCTENE	56.1183	-3.7657E+03	-1.8006E+01	7.7387E-03	-1.3036E-13	171.45	566.60
815	C8H16	trans-2-OCTENE	6.0888	-2.4722E+03	2.6638E+00	-1.2993E-02	7.8755E-06	185.45	577.00
816	C8H16	trans-3-OCTENE	9.0877	-2.5796E+03	1.5749E+00	-1.2824E-02	8.2120E-06	163.15	574.00
817	C8H16	trans-4-OCTENE	3.8148	-2.5476E+03	4.1541E+00	-1.7663E-02	1.0898E-05	179.37	573.00
818	C8H16	n-PROPYLCYCLOPENTANE	33.9220	-3.2097E+03	-8.9914E+00	-3.2992E-11	2.0684E-06	155.81	603.00
819	C8H16	2,4,4-TRIMETHYL-1-PENTENE	2.4008	-2.2801E+03	4.5221E+00	-1.7723E-02	1.1158E-05	179.70	553.00
820	C8H16	2,4,4-TRIMETHYL-2-PENTENE	33.6284	-3.0090E+03	-8.9606E+00	2.3421E-10	2.1539E-06	166.84	558.00
821	C8H16O	2-ETHYLHEXANAL	8.1124	-3.2254E+03	2.8038E+00	-1.5918E-02	9.0787E-06	200.00	607.00
822	C8H16O	1-OCTANAL	64.3916	-5.0161E+03	-1.9394E+01	6.8258E-10	5.5416E-06	246.00	621.00
823	C8H16O	2-OCTANONE	-8.6901	-3.6873E+03	1.1757E+01	-3.2793E-02	1.6676E-05	252.85	624.00
824	C8H16O2	n-BUTYL n-BUTYRATE	13.3448	-3.2742E+03	3.3579E-01	-1.1943E-02	7.0298E-06	181.15	616.00
825	C8H16O2	n-HEXYL ACETATE	8.8519	-3.2317E+03	2.2858E+00	-1.4351E-02	8.2306E-06	192.25	618.00
826	C8H16O2	ISOBUTYL ISOBUTYRATE	19.4898	-3.1402E+03	-2.8082E+00	-5.5690E-03	3.2425E-06	192.45	602.00
827	C8H16O2	n-OCTANOIC ACID	110.8655	-8.1565E+03	-3.6088E+01	1.1152E-02	9.9900E-13	289.65	692.00
828	C8H16O4	DIETHYLENE GLYCOL ETHYL ETHER ACETATE	45.9906	-4.3020E+03	-1.3729E+01	5.2997E-03	-2.5976E-13	248.15	660.00
829	C8H18	2,2-DIMETHYLHEXANE	38.7670	-3.1841E+03	-1.0857E+01	1.9275E-12	3.4797E-06	151.97	549.80
830	C8H18	2,3-DIMETHYLHEXANE	57.3778	-3.7143E+03	-1.8599E+01	8.2907E-03	-2.8441E-12	272.04	563.40
831	C8H18	2,4-DIMETHYLHEXANE	56.2877	-3.6225E+03	-1.8225E+01	8.1864E-03	8.7232E-12	272.04	553.50
832	C8H18	2,5-DIMETHYLHEXANE	40.0260	-3.2647E+03	-1.1282E+01	-6.5408E-10	3.6200E-06	182.00	550.00
833	C8H18	3,3-DIMETHYLHEXANE	38.0712	-3.1736E+03	-1.0617E+01	6.3090E-11	3.3817E-06	147.05	562.00
834	C8H18	3,4-DIMETHYLHEXANE	38.6119	-3.2685E+03	-1.0752E+01	3.6386E-09	3.2771E-06	272.04	568.80
835	C8H18	3-ETHYLHEXANE	40.2079	-3.3651E+03	-1.1285E+01	-5.4180E-09	3.4199E-06	272.04	565.40
836	C8H18	3-ETHYL-2-METHYLPENTANE	70.5211	-4.0062E+03	-2.4167E+01	1.4484E-02	-2.5304E-06	282.15	567.00
837	C8H18	3-METHYL-3-ETHYLPENTANE	35.2518	-3.0871E+03	-9.6172E+00	-2.3414E-11	2.9375E-06	182.28	576.50
838	C8H18	2-METHYLHEPTANE	37.6930	-3.2611E+03	-1.0391E+01	-1.0524E-12	3.0560E-06	164.16	559.64
839	C8H18	3-METHYLHEPTANE	52.8828	-3.6231E+03	-1.6804E+01	7.1828E-03	7.4077E-14	152.60	563.67
840	C8H18	4-METHYLHEPTANE	40.2080	-3.3661E+03	-1.1279E+01	-8.7855E-11	3.4055E-06	152.20	561.74
841	C8H18	n-OCTANE	29.0948	-3.0114E+03	-7.2653E+00	-2.2696E-11	1.4680E-06	216.38	568.83
842	C8H18	2,2,3-TRIMETHYLPENTANE	35.9540	-3.0569E+03	-9.8896E+00	-7.2916E-11	3.1060E-06	160.89	563.50
843	C8H18	2,2,4-TRIMETHYLPENTANE	50.3422	-3.2789E+03	-1.6111E+01	7.4260E-03	-9.1804E-14	165.78	543.96
844	C8H18	2,3,3-TRIMETHYLPENTANE	33.9671	-2.9982E+03	-9.1858E+00	-2.1839E-10	2.8100E-06	172.22	573.50
845	C8H18	2,3,4-TRIMETHYLPENTANE	34.1565	-3.0232E+03	-9.2267E+00	2.7691E-11	2.7828E-06	163.95	566.30
846	C8H18	2,2,3,3-TETRAMETHYLBUTANE	403.3685	-1.2917E+04	-1.6038E+02	1.4286E-01	-4.8604E-05	376.15	567.80
847	C8H18O	DI-n-BUTYL ETHER	12.9321	-3.0416E+03	4.2929E-01	-1.2370E-02	7.5943E-06	177.95	581.00
848	C8H18O	DI-sec-BUTYL ETHER	4.7651	-2.7465E+03	4.2606E+00	-2.0225E-02	1.2846E-05	173.15	559.00
849	C8H18O	DI-tert-BUTYL ETHER	-3.1154	-2.2501E+03	7.0062E+00	-2.1131E-02	1.2902E-05	195.00	550.00
850	C8H18O	2-ETHYL-1-HEXANOL	182.5024	-9.9679E+03	-6.3556E+01	2.4581E-02	3.4324E-13	203.15	640.25
851	C8H18O	1-OCTANOL	-26.3876	-4.2263E+03	2.1093E+01	-5.0048E-02	2.4611E-05	257.65	652.50
852	C8H18O	2-OCTANOL	-16.9480	-4.2841E+03	1.7277E+01	-4.7705E-02	2.4444E-05	241.55	637.00
853	C8H18O2	DI-t-BUTYL PEROXIDE	-17.3901	-2.1221E+03	1.3506E+01	-3.0330E-02	1.7215E-05	233.15	547.00
854	C8H18O2S	DI-n-BUTYL SULFONE	-6.7614	-4.2955E+03	9.4799E+00	-2.0356E-02	8.3388E-06	318.00	767.00
855	C8H18O3	DIETHYLENE GLYCOL DIETHYL ETHER	31.0938	-3.7273E+03	-7.6847E+00	-2.9675E-10	1.5639E-06	228.85	624.00
856	C8H18O3	DIETHYLENE GLYCOL MONOBUTYL EHTER	-46.2629	-4.4530E+03	3.1671E+01	-7.5265E-02	4.0651E-05	205.15	654.00
857	C8H18O4	TRIETHYLENE GLYCOL DIMETHYL ETHER	9.9777	-4.3785E+03	3.2930E+00	-1.9174E-02	9.9170E-06	229.35	651.00
858	C8H18O5	TETRAETHYLENE GLYCOL	104.9113	-1.0373E+04	-3.1015E+01	-1.6045E-09	4.6066E-06	268.15	722.00

172

Table 7-1 VAPOR PRESSURE - ORGANIC COMPOUNDS (continued)

NO	FORMULA	NAME	$\log_{10} P = A + B/T + C \log_{10} T + D T + E T^2$					(P - mm Hg, T - K)	
			A	B	C	D	E	TMIN	TMAX
859	C8H18S	n-OCTYL MERCAPTAN	50.9589	-4.6326E+03	-1.4583E+01	3.0340E-10	3.2768E-06	223.95	664.00
860	C8H18S	tert-OCTYL MERCAPTAN	7.2993	-2.6253E+03	1.9354E+00	-1.0430E-02	5.8477E-06	199.00	627.00
861	C8H18S	BUTYL-SULFIDE	796.0429	-2.4760E+04	-3.1803E+02	2.7497E-01	-8.9514E-05	379.15	650.00
862	C8H18S	ETHYL-HEXYL-SULFIDE	1400.4298	-4.2566E+04	-5.6059E+02	4.7845E-01	-1.5340E-04	388.15	660.72
863	C8H18S	HEPTYL-METHYL-SULFIDE	1400.4298	-4.2566E+04	-5.6059E+02	4.7845E-01	-1.5340E-04	·388.15	660.72
864	C8H18S	PENTYL-PROPYL-SULFIDE	1400.4298	-4.2566E+04	-5.6059E+02	4.7845E-01	-1.5340E-04	388.15	660.72
865	C8H18S2	BUTYL-DISULFIDE	230.2473	-1.0332E+04	-8.6093E+01	5.8851E-02	-1.5236E-05	374.15	704.16
866	C8H19N	DI-n-BUTYLAMINE	103.4087	-6.4571E+03	-3.3284E+01	1.5763E-10	1.1476E-05	211.15	607.50
867	C8H19N	DIISOBUTYLAMINE	22.0141	-2.8234E+03	-4.7498E+00	-7.4792E-11	7.8720E-07	203.15	580.00
868	C8H19N	n-OCTYLAMINE	-28.0873	-2.3939E+03	1.7839E+01	-3.1586E-02	1.5502E-05	272.75	627.00
869	C8H23N5	TETRAETHYLENEPENTAMINE	24.9007	-7.4002E+03	-4.9209E-01	-1.9264E-02	8.7883E-06	243.00	774.00
870	C8H24O4Si4	OCTAMETHYLCYCLOTETRASILOXANE	-40.6427	-2.5644E+03	2.4617E+01	-4.5950E-02	2.2737E-05	290.80	586.50
871	C9H4O5	TRIMELLITIC ANHYDRIDE	-39.6819	-5.4887E+03	2.3878E+01	-3.1856E-02	1.0428E-05	438.15	890.00
872	C9H6N2O2	TOLUENE DIISOCYANATE	-18.2929	-3.6537E+03	1.4544E+01	-2.8216E-02	1.2344E-05	287.04	737.00
873	C9H7N	ISOQUINOLINE	45.5737	-4.4715E+03	-1.3308E+01	4.0186E-03	-6.4589E-14	299.45	803.15
874	C9H7N	QUINOLINE	76.5432	-5.7748E+03	-2.4619E+01	8.4666E-03	3.5586E-13	258.25	782.15
875	C9H7NO	8-HYDROXYQUINOLINE	-12.7774	-3.5666E+03	1.1091E+01	-1.8860E-02	7.3318E-06	346.00	788.00
876	C9H8	INDENE	94.3667	-5.7427E+03	-3.1916E+01	1.3095E-02	-1.4035E-12	271.70	687.00
877	C9H8O	2-METHYLBENZOFURAN	59.0455	-4.6509E+03	-1.8481E+01	6.6104E-03	-3.8444E-13	290.00	698.00
878	C9H10	INDANE	37.3577	-3.7337E+03	-1.0040E+01	6.3179E-11	2.2062E-06	221.74	684.90
879	C9H10	cis-PROPENYLBENZENE	64.8029	-4.0396E+03	-2.2673E+01	1.8969E-02	-6.6567E-06	291.15	454.15
880	C9H10	trans-PROPENYLBENZENE	-253.5058	4.3311E+03	1.0817E+02	-1.0629E-01	3.7879E-05	378.15	664.60
881	C9H1O	alpha-METHYLSTYRENE	-0.8626	-2.5638E+03	5.3807E+00	-1.3516E-02	6.7181E-06	249.95	654.00
882	C9H10	m-METHYLSTYRENE	11.6959	-2.9912E+03	3.3334E+01	-8.8935E-03	4.9793E-06	186.81	657.00
883	C9H10	o-METHYLSTYRENE	36.8413	-3.7269E+03	-9.7997E+00	1.4115E-10	1.9658E-06	204.58	659.00
884	C9H10	p-METHYLSTYRENE	50.6506	-4.0628E+03	-1.5524E+01	5.5381E-03	-1.1313E-13	239.02	665.00
885	C9H10O2	BENZYL ACETATE	46.1904	-4.6053E+03	-1.2820E+01	1.6574E-10	2.5462E-06	221.65	699.00
886	C9H10O2	ETHYL BENZOATE	40.8047	-3.9985E+03	-1.1793E+01	4.0697E-03	-1.2372E-13	238.45	698.00
887	C9H10O3	ETHYL VANILLIN	-27.5098	-5.0251E+03	1.9836E+01	-3.5011E-02	1.3941E-05	350.65	748.00
888	C9H11NO	p-DIMETHYLAMINOBENZALDEHYDE	31.1247	-4.1807E+03	-8.1760E+00	2.5709E-03	3.2343E-13	348.00	832.00
889	C9H12	CUMENE	-0.9234	-2.9558E+03	7.1685E+00	-2.5369E-02	1.4858E-05	177.14	631.15
890	C9H12	m-ETHYLTOLUENE	39.8909	-3.6042E+03	-1.1466E+01	3.5274E-03	7.3492E-14	177.61	637.15
891	C9H12	o-ETHYLTOLUENE	15.1142	-2.9821E+03	-1.2619E+00	-6.3248E-03	3.5155E-06	192.35	651.15
892	C9H12	p-ETHYLTOLUENE	46.9026	-3.8382E+03	-1.4154E+01	4.9305E-03	-1.3901E-13	210.83	640.15
893	C9H12	MESITYLENE	37.6361	-3.6753E+03	-1.0156E+01	-1.0068E-10	2.4232E-06	228.46	637.36
894	C9H12	n-PROPYLBENZENE	39.8219	-3.6978E+03	-1.0962E+01	8.7429E-11	2.6959E-06	173.67	638.38
895	C9H12	1,2,3-TRIMETHYLBENZENE	2.7492	-2.6428E+03	3.6120E+00	-1.0213E-02	5.0553E-06	247.79	664.53
896	C9H12	1,2,4-TRIMETHYLBENZENE	2.1667	-2.6318E+03	4.0350E+00	-1.1776E-02	6.0956E-06	229.38	649.13
897	C9H12O	BENZYL ETHYL ETHER	27.6421	-3.4249E+03	-6.5804E+00	9.3417E-10	1.0547E-06	275.65	662.00
898	C9H12O	2-PHENYL-2-PROPANOL	10.6041	-3.1597E+03	1.4321E-01	-3.8988E-03	1.7578E-06	309.15	660.00
899	C9H12O2	CUMENE HYDROPEROXIDE	65.7553	-7.2353E+03	-1.7947E+01	-2.8383E-08	4.8915E-06	264.26	442.70
900	C9H14O	ISOPHORONE	33.9350	-3.5233E+03	-9.5117E+00	3.5518E-03	-7.9256E-14	265.05	715.00
901	C9H14O6	GLYCERYL TRIACETATE	0.5931	-4.6647E+03	7.3556E+00	-2.2339E-02	1.0189E-05	277.25	704.00
902	C9H16	1-NONYNE	168.0426	-7.0162E+03	-6.3306E+01	4.6976E-02	-1.2272E-05	323.15	610.81
903	C9H16O4	AZELAIC ACID	129.6139	-1.1868E+04	-3.9405E+01	-2.0707E-08	5.9824E-06	379.65	811.00
904	C9H18	BUTYLCYCLOPENTANE	63.5929	-4.3226E+03	-2.0646E+01	8.7543E-03	-2.7299E-07	314.15	625.05
905	C9H18	cis,cis-1,3,5-TRIMETHYLCYCLOHEXANE	-323.4899	5.4756E+03	1.3990E+02	-1.5273E-01	5.9984E-05	351.15	607.86
906	C9H18	cis,trans-1,3,5-TRIMETHYLCYCLOHEXANE	-211.1731	2.6774E+03	9.3218E+01	-1.0535E-01	4.2323E-05	353.15	602.20
907	C9H18	ISOPROPYLCYCLOHEXANE	38.6401	-3.5537E+03	-1.0621E+01	8.7017E-11	2.7087E-06	183.76	627.00
908	C9H18	1-NONENE	60.6089	-4.2023E+03	-1.9446E+01	7.8308E-03	1.5910E-13	191.78	593.25
909	C9H18	n-PROPYLCYCLOHEXANE	33.6844	-3.4070E+03	-8.8018E+00	-7.6792E-11	1.6258E-06	178.28	639.15
910	C9H18O	DIISOBUTYL KETONE	-12.9061	-2.8810E+03	1.2212E+01	-2.9636E-02	1.5909E-05	227.17	615.00
911	C9H18O	1-NONANAL	67.9442	-5.4374E+03	-2.0455E+01	2.7722E-10	5.3570E-06	255.15	640.00
912	C9H18O2	n-BUTYL VALERATE	16.6975	-3.4064E+03	-1.5457E+00	-6.8086E-03	3.9686E-06	180.35	629.00
913	C9H18O2	n-NONANOIC ACID	100.7681	-8.4475E+03	-3.0662E+01	8.9301E-10	5.6788E-06	285.55	703.00
914	C9H18O2	n-OCTYL FORMATE	-14.4563	-2.5677E+03	1.1553E+01	-2.2727E-02	1.1727E-05	234.05	645.00
915	C9H20	3,3-DIETHYLPENTANE	-12.5469	-2.1828E+03	1.0630E+01	-2.2169E-02	1.1665E-05	240.12	610.05
916	C9H20	2,2-DIMETHYL-3-ETHYLPENTANE	10.1040	-2.6142E+03	9.0709E-01	-1.0373E-02	6.3652E-06	173.68	590.00
917	C9H20	3-ETHYL-2,3-DIMETHYLPENTANE	113.9533	-5.3218E+03	-4.1997E+01	3.1893E-02	-9.1036E-06	303.15	606.80
918	C9H20	2,4-DIMETHYL-3-ETHYLPENTANE	12.5531	-2.7306E+03	3.5965E-02	-1.0320E-02	6.6977E-06	150.79	591.00
919	C9H20	2,2-DIMETHYLHEPTANE	45.1437	-3.7011E+03	-1.2947E+01	7.5743E-11	3.8061E-06	160.15	576.80
920	C9H20	2,6-DIMETHYLHEPTANE	49.6777	-3.7128E+03	-1.5362E+01	5.8912E-03	1.0266E-13	170.25	579.00
921	C9H20	3-ETHYLHEPTANE	12.6472	-2.8943E+03	2.7706E-01	-1.1564E-02	7.3625E-06	158.25	590.00
922	C9H20	4-ETHYLHEPTANE	93.5549	-4.9603E+03	-3.3114E+01	2.1111E-02	-4.5301E-06	303.15	584.95
923	C9H20	2,3-DIMETHYLHEPTANE	72.9382	-4.4209E+03	-2.4596E+01	1.2655E-02	-1.4424E-06	302.15	589.60
924	C9H20	2,4-DIMETHYLHEPTANE	43.3671	-3.6628E+03	-1.2245E+01	-3.1999E-04	3.6684E-06	297.15	576.80
925	C9H20	2,5-DIMETHYLHEPTANE	55.7744	-3.9934E+03	-1.7395E+01	4.9547E-03	1.6265E-06	300.15	581.10
926	C9H20	3,4-DIMETHYLHEPTANE	73.3252	-4.4089E+03	-2.4822E+01	1.3200E-02	-1.7419E-06	302.15	591.90
927	C9H20	3,5-DIMETHYLHEPTANE	51.8497	-3.8750E+03	-1.5826E+01	3.6661E-03	2.0126E-06	299.15	583.20
928	C9H20	4,4-DIMETHYLHEPTANE	64.8742	-4.1098E+03	-2.1453E+01	1.0302E-02	-7.4066E-07	297.15	585.40
929	C9H20	3-ETHYL-2-METHYLHEXANE	69.4666	-4.2845E+03	-2.3255E+01	1.1739E-02	-1.2011E-06	300.15	588.10
930	C9H20	4-ETHYL-2-METHYLHEXANE	54.6479	-3.9134E+03	-1.7035E+01	5.0564E-03	1.4792E-06	298.15	580.00
931	C9H20	3-ETHYL-3-METHYLHEXANE	65.4400	-4.1356E+03	-2.1749E+01	1.0999E-02	-1.1938E-06	301.15	597.50
932	C9H20	3-ETHYL-4-METHYLHEXANE	66.9215	-4.2282E+03	-2.2222E+01	1.0861E-02	-9.5641E-07	302.15	593.70
933	C9H20	2,2,3-TRIMETHYLHEXANE	39.8016	-3.4508E+03	-1.1093E+01	4.0116E-06	3.0574E-06	296.15	588.00
934	C9H20	2,2,4-TRIMETHYLHEXANE	50.2219	-3.6674E+03	-1.5411E+01	4.1515E-03	1.6925E-06	291.15	573.50
935	C9H20	2,3,3-TRIMETHYLHEXANE	56.0669	-3.8539E+03	-1.7923E+01	7.3435E-03	1.3032E-07	298.15	596.00
936	C9H20	2,3,4-TRIMETHYLHEXANE	74.6496	-4.3667E+03	-2.5532E+01	1.4575E-02	-2.4380E-06	300.15	594.50

Table 7-1 VAPOR PRESSURE - ORGANIC COMPOUNDS (continued)

$$\log_{10} P = A + B/T + C \log_{10} T + D T + E T^2 \qquad (P - mm\ Hg,\ T - K)$$

NO	FORMULA	NAME	A	B	C	D	E	TMIN	TMAX
937	C9H20	2,3,5-TRIMETHYLHEXANE	48.3587	-3.7091E+03	-1.4491E+01	2.6885E-03	2.3246E-06	295.15	579.20
938	C9H20	2,4,4-TRIMETHYLHEXANE	63.6395	-3.9565E+03	-2.1186E+01	1.1037E-02	-1.2484E-06	293.15	581.50
939	C9H20	3,3,4-TRIMETHYLHEXANE	59.8798	-3.9554E+03	-1.9537E+01	9.1750E-03	-6.4602E-07	300.15	602.30
940	C9H20	2-METHYLOCTANE	6.0191	-2.8579E+03	3.4068E+00	-1.6572E-02	9.8047E-06	192.78	586.75
941	C9H20	3-METHYLOCTANE	9.8147	-2.9609E+03	1.9061E+00	-1.5675E-02	9.7961E-06	165.55	590.10
942	C9H20	4-METHYLOCTANE	11.2012	-2.9467E+03	1.2133E+00	-1.4423E-02	9.1770E-06	159.95	587.65
943	C9H20	n-NONANE	8.8817	-2.8042E+03	1.5262E+00	-1.0464E-02	5.7972E-06	219.63	595.65
944	C9H20	2,2,3,3-TETRAMETHYLPENTANE	35.4216	-3.2760E+03	-9.5678E+00	9.0298E-10	2.4355E-06	263.26	610.85
945	C9H20	2,2,3,4-TETRAMETHYLPENTANE	52.3443	-3.6477E+03	-1.6608E+01	6.9991E-03	-5.1010E-14	152.06	592.15
946	C9H20	2,2,4,4-TETRAMETHYLPENTANE	-3.8184	-2.2442E+03	7.0671E+00	-1.9644E-02	1.1435E-05	206.95	571.35
947	C9H20	2,3,3,4-TETRAMETHYLPENTANE	65.9876	-4.0672E+03	-2.2199E+01	1.2445E-02	-2.0351E-06	300.15	607.50
948	C9H20	2,2,5-TRIMETHYLHEXANE	7.8816	-2.6422E+03	2.3902E+00	-1.5376E-02	9.7931E-06	167.39	568.05
949	C9H20O	2,6-DIMETHYL-4-HEPTANOL	10.9218	-4.1891E+03	3.2601E+00	-2.1994E-02	1.2379E-05	208.00	603.00
950	C9H20O	1-NONANOL	111.7949	-8.3502E+03	-3.4786E+01	3.3682E-10	7.2697E-06	268.15	673.00
951	C9H20O	2-NONANOL	-1.1889	-4.3218E+03	8.8539E+00	-2.9377E-02	1.5343E-05	238.15	623.00
952	C9H20S	n-NONYL MERCAPTAN	-12.6269	-3.2744E+03	1.1911E+01	-2.6433E-02	1.2780E-05	253.05	681.00
953	C9H20S	BUTYL-PENTYL-SULFIDE	77.5993	-4.4994E+03	-2.7270E+01	1.8520E-02	-5.2382E-06	402.15	533.15
954	C9H20S	ETHYL-HEPTYL-SULFIDE	77.5993	-4.4994E+03	-2.7270E+01	1.8520E-02	-5.2382E-06	402.15	533.15
955	C9H20S	HEXYL-PROPYL-SULFIDE	77.5993	-4.4994E+03	-2.7270E+01	1.8520E-02	-5.2382E-06	402.15	533.15
956	C9H20S	METHYL-OCTYL-SULFIDE	77.5993	-4.4994E+03	-2.7270E+01	1.8520E-02	-5.2382E-06	402.15	533.15
957	C9H21N	n-NONYLAMINE	87.6004	-5.8022E+03	-2.9187E+01	1.1826E-02	-2.1798E-12	273.15	648.00
958	C9H21N	TRIPROPYLAMINE	45.7821	-4.3495E+03	-1.2623E+01	-6.2525E-11	2.4778E-06	179.65	577.50
959	C10H6O8	PYROMELLITIC ACID	-175.6445	-3.9334E+03	7.9624E+01	-7.6697E-02	2.2512E-05	554.00	893.00
960	C10H7Br	1-BROMONAPHTHALENE	5.4815	-3.6780E+03	2.9858E+00	-9.7457E-03	4.0451E-06	279.35	824.00
961	C10H7Cl	1-CHLORONAPHTHALENE	93.7760	-6.6521E+03	-3.0859E+01	1.0748E-02	-3.2318E-14	269.15	785.00
962	C10H8	NAPHTHALENE	34.9161	-3.9357E+03	-9.0648E+00	-2.0672E-09	1.5550E-06	353.43	748.35
963	C10H8	AZULENE	448.1038	-1.6769E+04	-1.7310E+02	1.2871E-01	-3.5996E-05	433.15	773.48
964	C10H9N	QUINALDINE	-14.3595	-4.4014E+03	1.4558E+01	-3.4414E-02	1.4987E-05	272.15	773.00
965	C10H10	m-DIVINYLBENZENE	38.5125	-3.7960E+03	-1.0930E+01	3.4461E-03	-3.2401E-16	206.25	692.00
966	C10H10	1-METHYLINDENE	-25.1496	-2.1593E+03	1.5324E+01	-2.2019E-02	9.0801E-06	350.00	703.00
967	C10H10	2-METHYLINDENE	-28.2567	-1.9586E+03	1.6553E+01	-2.3305E-02	9.7151E-06	353.15	684.00
968	C10H10O4	DIMETHYL PHTHALATE	12.6974	-4.1989E+03	3.4630E-01	-7.6524E-03	3.3490E-06	272.15	766.00
969	C10H10O4	DIMETHYL TEREPHTHALATE	-23.1247	-2.4471E+03	1.3496E+01	-1.5109E-02	5.5490E-06	403.80	772.00
970	C10H12	DICYCLOPENTADIENE	26.8270	-3.1207E+03	-6.4697E+00	-3.7793E-07	1.1384E-06	307.00	443.00
971	C10H12	1,2,3,4-TETRAHYDRONAPHTHALENE	39.9174	-4.1320E+03	-1.0780E+01	1.9691E-10	2.0405E-06	237.40	720.15
972	C10H12O	ANETHOLE	-6.3301	-3.3205E+03	8.4249E+00	-1.7871E-02	7.8280E-06	294.50	723.00
973	C10H12O4	DIALLYL MALEATE	15.4205	-4.7760E+03	9.5106E-01	-1.5469E-02	7.7640E-06	226.15	693.00
974	C10H14	n-BUTYLBENZENE	49.9687	-4.3981E+03	-1.4352E+01	4.2054E-11	3.4379E-06	185.30	660.55
975	C10H14	sec-BUTYLBENZENE	61.5904	-4.5093E+03	-1.9522E+01	6.9865E-03	7.8205E-14	197.72	664.54
976	C10H14	tert-BUTYLBENZENE	41.4522	-3.9027E+03	-1.1410E+01	2.4230E-10	2.2517E-06	215.27	660.00
977	C10H14	1,2,3,4-TETRAMETHYLBENZENE	-32.4220	-2.1756E+03	1.8715E+01	-2.6796E-02	1.1010E-05	353.15	695.10
978	C10H14	m-CYMENE	4.9207	-2.9235E+03	3.4809E+00	-1.3932E-02	7.4450E-06	209.44	657.00
979	C10H14	o-CYMENE	7.1102	-3.0029E+03	2.5743E+00	-1.2918E-02	6.9705E-06	201.64	662.00
980	C10H14	p-CYMENE	-5.5137	-3.0256E+03	8.9840E+00	-2.5597E-02	1.3823E-05	205.25	653.15
981	C10H14	m-DIETHYLBENZENE	37.1857	-3.8380E+03	-9.8677E+00	6.3157E-11	1.7741E-06	189.26	663.00
982	C10H14	o-DIETHYLBENZENE	3.4308	-2.9876E+03	4.1014E+00	-1.3931E-02	6.9359E-06	241.93	668.00
983	C10H14	p-DIETHYLBENZENE	-2.4793	-2.8942E+03	6.7988E+00	-1.8269E-02	9.3732E-06	230.32	657.96
984	C1OH14	2-ETHYL-m-XYLENE	62.1333	-4.6665E+03	-1.9724E+01	7.3435E-03	-1.1916E-13	256.89	671.00
985	C10H14	2-ETHYL-p-XYLENE	11.0144	-2.9801E+03	2.7007E-01	-6.7946E-03	3.5409E-06	219.52	663.00
986	C10H14	3-ETHYL-o-XYLENE	36.9976	-3.9360E+03	-9.7603E+00	7.2747E-11	1.6679E-06	223.64	680.00
987	C10H14	4-ETHYL-m-XYLENE	10.6755	-3.0166E+03	5.3834E-01	-7.7304E-03	4.0929E-06	210.27	665.00
988	C10H14	4-ETHYL-o-XYLENE	60.0226	-4.5814E+03	-1.8945E+01	7.0278E-03	1.4101E-14	206.22	667.00
989	C1OH14	5-ETHYL-m-XYLENE	63.6620	-4.6313E+03	-2.0395E+01	7.8833E-03	4.4259E-14	188.82	655.00
990	C10H14	ISOBUTYLBENZENE	-7.0438	-2.6892E+03	8.7843E+00	-2.1426E-02	1.1248E-05	221.70	650.15
991	C10H14	1,2,3,5-TETRAMETHYLBENZENE	-3.9778	-2.9600E+03	7.3226E+00	-1.7725E-02	8.6365E-06	249.46	679.00
992	C10H14	1,2,4,5-TETRAMETHYLBENZENE	-51.3593	-1.6523E+03	2.6656E+01	-3.5721E-02	1.5018E-05	352.38	675.15
993	C10H14O	p-tert-BUTYLPHENOL	-54.7404	-2.4727E+03	2.8991E+01	-3.9356E-02	1.5430E-05	371.56	734.00
994	C10H14O2	p-tert-BUTYLCATECHOL	-7.8630	-4.5064E+03	1.0224E+01	-2.1429E-02	8.6391E-06	325.00	776.00
995	C10H15N	N,N-DIETHYLANILINE	34.5048	-4.1694E+03	-8.6686E+00	-7.2562E-11	8.8939E-07	235.15	702.00
996	C10H15N	2,6-DIETHYLANILINE	-4.0230	-4.7696E+03	9.9926E+00	-2.7756E-02	1.2952E-05	276.65	678.00
997	C10H16	CAMPHENE	32.9148	-3.3183E+03	-8.6235E+00	-1.4100E-09	1.9327E-06	320.15	638.00
998	C10H16	D-LIMONENE	9.3771	-2.8246E+03	1.0584E+00	-8.9107E-03	4.8462E-06	199.00	660.00
999	C10H16	alpha-PHELLANDRENE	12.3991	-2.9959E+03	-1.6574E-01	-7.0035E-03	3.6922E-06	220.00	649.00
1000	C1OH16	beta-PHELLANDRENE	10.4546	-3.0154E+03	8.9414E-01	-9.3665E-03	4.9425E-06	220.00	648.00
1001	C10H16	alpha-PINENE	21.4735	-2.7156E+03	-5.0076E+00	2.8146E-03	-1.5389E-06	209.15	632.00
1002	C10H16	beta-PINENE	46.3728	-3.9789E+03	-1.3284E+01	-1.3113E-10	3.4783E-06	211.61	643.00
1003	C10H16	alpha-TERPINENE	43.9494	-4.0418E+03	-1.2314E+01	4.3012E-10	2.8828E-06	220.00	652.00
1004	C1OH16	gamma-TERPINENE	4.2428	-3.0193E+03	3.8560E+00	-1.4383E-02	7.5045E-06	220.00	661.00
1005	C10H16	TERPINOLENE	29.6203	-3.6749E+03	-7.1907E+00	9.0533E-04	-1.1109E-15	200.00	672.00
1006	C10H16O	CAMPHOR	115.6738	-7.1537E+03	-3.9077E+01	1.4335E-02	-6.1661E-11	453.25	709.00
1007	C10H18	1-DECYNE	-16.5555	-2.2598E+03	1.1813E+01	-1.8849E-02	8.0521E-06	351.15	632.49
1008	C10H18	cis-DECAHYDRONAPHTALENE	45.6345	-4.2100E+03	-1.2881E+01	-7.8083E-11	2.8637E-06	230.20	702.25
1009	C10H18	trans-DECAHYDRONAPHTHALENE	76.1002	-5.5300E+03	-2.5814E+01	9.7608E-03	-2.5814E-13	242.79	687.05
1010	C10H18O4	SEBACIC ACID	165.2175	-1.2613E+04	-5.4591E+01	1.6475E-02	-3.3040E-12	407.65	815.00
1011	C10H20	n-BUTYLCYCLOHEXANE	33.6340	-3.6479E+03	-8.6428E+00	-9.6941E-11	1.1897E-06	198.42	667.00
1012	C10H20	1-CYCLOPENTYLPENTANE	44.3550	-4.1544E+03	-1.2314E+01	-5.2585E-04	3.0850E-06	333.15	647.49
1013	C10H20	1-DECENE	2.2678	-3.1244E+03	5.4320E+00	-2.0137E-02	1.1221E-05	206.89	617.05
1014	C10H20O	1-DECANAL	82.5061	-6.3937E+03	-2.5303E+01	-1.9398E-09	6.3146E-06	267.15	657.00

Table 7-1 VAPOR PRESSURE - ORGANIC COMPOUNDS (continued)

$$\log_{10} P = A + B/T + C \log_{10} T + D T + E T^2 \qquad (P - mm Hg, T - K)$$

NO	FORMULA	NAME	A	B	C	D	E	TMIN	TMAX
1015	C10H20O2	n-DECANOIC ACID	196.1175	-1.3955E+04	-6.2586E+01	7.4527E-09	1.2291E-05	304.75	713.00
1016	C10H20O2	2-ETHYLHEXYL ACETATE	18.0496	-3.9772E+03	-9.0811E-01	-1.2571E-02	7.2550E-06	180.15	639.00
1017	C10H20O2	ISOPENTYL ISOVALERATE	22.2542	-2.9750E+03	-5.2266E+00	2.0253E-03	-1.8744E-14	215.00	637.00
1018	C10H22	n-DECANE	26.5125	-3.3584E+03	-6.1174E+00	-3.3225E-10	4.8554E-07	243.49	618.45
1019	C10H22	2-METHYLNONANE	56.8132	-4.6022E+03	-1.6769E+01	-1.5533E-10	4.4072E-06	198.50	610.00
1020	C10H22	3-METHYLNONANE	64.6522	-4.5644E+03	-2.0802E+01	8.1382E-03	-3.4025E-14	188.35	613.00
1021	C10H22	4-METHYLNONANE	12.4486	-3.0918E+03	4.2102E-01	-1.1259E-02	6.7772E-06	174.45	610.00
1022	C10H22	5-METHYLNONANE	3.7845	-3.2217E+03	5.2149E+00	-2.2500E-02	1.3173E-05	185.45	610.00
1023	C10H22	3-ETHYLOCTANE	81.3943	-5.0355E+03	-2.7491E+01	1.3341E-02	-1.3742E-06	323.15	613.60
1024	C10H22	4-ETHYLOCTANE	87.2735	-5.1369E+03	-2.9989E+01	1.6016E-02	-2.3588E-06	321.15	609.60
1025	C10H22	2,2-DIMETHYLOCTANE	3.7585	-2.7486E+03	3.8574E+00	-1.4111E-02	7.6942E-06	225.00	602.00
1026	C10H22	2,3-DIMETHYLOCTANE	66.6393	-4.6014E+03	-2.1496E+01	7.8420E-03	5.1865E-07	321.15	613.20
1027	C10H22	2,4-DIMETHYLOCTANE	35.4998	-3.7636E+03	-8.6134E+00	-5.0412E-03	5.3779E-06	316.15	599.40
1028	C10H22	2,5-DIMETHYLOCTANE	60.2445	-4.4047E+03	-1.8847E+01	5.0920E-03	1.6422E-06	317.15	603.00
1029	C10H22	2,6-DIMETHYLOCTANE	156.5719	-6.7262E+03	-5.9349E+01	4.9770E-02	-1.7181E-05	318.15	462.15
1030	C10H22	2,7-DIMETHYLOCTANE	38.7548	-3.9152E+03	-9.8475E+00	-4.2320E-03	5.1351E-06	319.15	602.90
1031	C10H22	3,3-DIMETHYLOCTANE	54.9080	-4.2103E+03	-1.6830E+01	3.9908E-03	1.7435E-06	318.15	612.10
1032	C10H22	3,4-DIMETHYLOCTANE	68.1212	-4.6019E+03	-2.2192E+01	8.8808E-03	5.5663E-08	320.15	614.00
1033	C10H22	3,5-DIMETHYLOCTANE	67.0778	-4.5538E+03	-2.1748E+01	8.2992E-03	3.6514E-07	317.15	606.30
1034	C10H22	3,6-DIMETHYLOCTANE	49.3284	-4.1335E+03	-1.4367E+01	9.0414E-04	3.0570E-06	319.15	608.30
1035	C10H22	4,4-DIMETHYLOCTANE	52.7481	-4.1112E+03	-1.5989E+01	3.3029E-03	2.0078E-06	315.15	606.90
1036	C10H22	4,5-DIMETHYLOCTANE	76.0376	-4.7828E+03	-2.5490E+01	1.2197E-02	-1.1443E-06	319.15	612.20
1037	C10H22	4-PROPYLHEPTANE	51.7167	-4.1866E+03	-1.5336E+01	1.7503E-03	2.8445E-06	317.15	601.00
1038	C10H22	4-ISOPROPYLHEPTANE	80.0469	-4.8377E+03	-2.7213E+01	1.4110E-02	-1.8540E-06	316.15	607.60
1039	C10H22	3-ETHYL-2-METHYLHEPTANE	71.0698	-4.6406E+03	-2.3462E+01	1.0298E-02	-4.7207E-07	318.15	610.90
1040	C10H22	4-ETHYL-2-METHYLHEPTANE	76.3971	-4.7458E+03	-2.5661E+01	1.2347E-02	-1.1103E-06	315.15	601.80
1041	C10H22	5-ETHYL-2-METHYLHEPTANE	70.5597	-4.6436E+03	-2.3186E+01	9.7061E-03	-1.4400E-07	317.15	606.70
1042	C10H22	3-ETHYL-3-METHYLHEPTANE	67.2602	-4.4892E+03	-2.2061E+01	9.7119E-03	-5.1833E-07	319.15	620.00
1043	C10H22	4-ETHYL-3-METHYLHEPTANE	83.5934	-4.9398E+03	-2.8698E+01	1.5712E-02	-2.5128E-06	318.15	614.30
1044	C10H22	3-ETHYL-5-METHYLHEPTANE	50.7351	-4.1090E+03	-1.5062E+01	2.0425E-03	2.5513E-06	317.15	606.60
1045	C10H22	3-ETHYL-4-METHYLHEPTANE	80.9936	-4.8859E+03	-2.7607E+01	1.4585E-02	-2.1028E-06	319.15	615.50
1046	C10H22	4-ETHYL-4-METHYLHEPTANE	68.5490	-4.4902E+03	-2.2630E+01	1.0382E-02	-7.5401E-07	317.15	615.70
1047	C10H22	2,2,3-TRIMETHYLHEPTANE	42.3175	-3.8075E+03	-1.1774E+01	-4.1100E-04	3.2063E-06	315.15	611.70
1048	C10H22	2,2,4-TRIMETHYLHEPTANE	34.4057	-3.5644E+03	-8.4449E+00	-4.1996E-03	4.8827E-06	308.15	594.50
1049	C10H22	2,2,5-TRIMETHYLHEPTANE	27.8150	-3.4273E+03	-5.6827E+00	-7.0320E-03	5.9070E-06	310.15	598.00
1050	C10H22	2,2,6-TRIMETHYLHEPTANE	8.6749	-2.9752E+03	2.3524E+00	-1.5554E-02	9.2247E-06	310.15	593.40
1051	C10H22	2,3,3-TRIMETHYLHEPTANE	49.0603	-3.9786E+03	-1.4603E+01	2.5999E-03	2.0186E-06	316.15	617.50
1052	C10H22	2,3,4-TRIMETHYLHEPTANE	65.4081	-4.4359E+03	-2.1249E+01	8.6379E-03	-1.4051E-08	317.15	613.70
1053	C10H22	2,3,5-TRIMETHYLHEPTANE	92.7669	-5.1409E+03	-3.2519E+01	1.9489E-02	-3.8700E-06	317.15	612.80
1054	C10H22	2,3,6-TRIMETHYLHEPTANE	41.1335	-3.8435E+03	-1.1089E+01	-1.9948E-03	4.0805E-06	315.15	604.10
1055	C10H22	2,4,4-TRIMETHYLHEPTANE	54.0724	-4.0427E+03	-1.6677E+01	4.4287E-03	1.5529E-06	310.15	600.30
1056	C10H22	2,4,5-TRIMETHYLHEPTANE	58.3551	-4.2448E+03	-1.8307E+01	5.5537E-03	1.2052E-06	314.15	606.90
1057	C10H22	2,4,6-TRIMETHYLHEPTANE	15.9590	-3.1803E+03	-5.9365E-01	-1.2952E-02	8.3992E-06	309.15	590.30
1058	C10H22	2,5,5-TRIMETHYLHEPTANE	29.7678	-3.4725E+03	-6.5430E+00	-5.8798E-03	5.3640E-06	312.15	602.90
1059	C10H22	3,3,4-TRIMETHYLHEPTANE	55.3693	-4.1259E+03	-1.7269E+01	5.5026E-03	8.6995E-07	317.15	622.10
1060	C10H22	3,3,5-TRIMETHYLHEPTANE	79.7191	-4.7188E+03	-2.7294E+01	1.5052E-02	-2.4141E-06	313.15	609.50
1061	C10H22	3,4,4-TRIMETHYLHEPTANE	57.3601	-4.1652E+03	-1.8105E+01	6.3641E-03	5.6007E-07	316.15	620.90
1062	C10H22	3,4,5-TRIMETHYLHEPTANE	108.3906	-5.5548E+03	-3.8975E+01	2.5878E-02	-6.1851E-06	318.15	612.80
1063	C10H22	3-ISOPROPYL-2-METHYLHEXANE	165.0874	-6.9768E+03	-6.2423E+01	4.8743E-02	-1.4382E-05	318.15	623.40
1064	C10H22	3,3-DIETHYLHEXANE	72.4818	-4.6008E+03	-2.4312E+01	1.2372E-02	-1.6299E-06	320.15	627.80
1065	C10H22	3,4-DIETHYLHEXANE	89.6991	-5.0846E+03	-3.1270E+01	1.8468E-02	-3.5895E-06	319.15	618.80
1066	C10H22	3-ETHYL-2,2-DIMETHYLHEXANE	51.3900	-3.9790E+03	-1.5644E+01	3.8686E-03	1.5493E-06	313.15	611.70
1067	C10H22	4-ETHYL-2,2-DIMETHYLHEXANE	19.5418	-3.1584E+03	-2.3512E+00	-9.9985E-03	6.9575E-06	307.15	594.60
1068	C10H22	3-ETHYL-2,3-DIMETHYLHEXANE	69.7644	-4.4848E+03	-2.3261E+01	1.1587E-02	-1.3978E-06	318.15	626.80
1069	C10H22	4-ETHYL-2,3-DIMETHYLHEXANE	75.0688	-4.6609E+03	-2.5306E+01	1.2913E-02	-1.6500E-06	317.15	617.30
1070	C10H22	3-ETHYL-2,4-DIMETHYLHEXANE	77.0647	-4.6991E+03	-2.6148E+01	1.3800E-02	-1.9752E-06	316.15	616.10
1071	C10H22	4-ETHYL-2,4-DIMETHYLHEXANE	99.8986	-5.2673E+03	-3.5632E+01	2.3335E-02	-5.4868E-06	316.15	620.90
1072	C10H22	3-ETHYL-2,5-DIMETHYLHEXANE	64.4906	-4.3655E+03	-2.0893E+01	8.2671E-03	2.0753E-07	312.15	603.50
1073	C10H22	4-ETHYL-3,3-DIMETHYLHEXANE	64.6460	-4.3415E+03	-2.1165E+01	9.6098E-03	-7.0000E-07	317.15	625.70
1074	C10H22	3-ETHYL-3,4-DIMETHYLHEXANE	66.2861	-4.3712E+03	-2.1859E+01	1.0342E-02	-9.6623E-07	317.15	624.50
1075	C10H22	2,2,3,3-TETRAMETHYLHEXANE	110.5241	-5.4530E+03	-4.0210E+01	2.8578E-02	-7.5620E-06	314.15	623.00
1076	C10H22	2,2,3,4-TETRAMETHYLHEXANE	64.6414	-4.2837E+03	-2.1222E+01	9.7824E-03	-7.4964E-07	314.15	620.40
1077	C10H22	2,2,3,5-TETRAMETHYLHEXANE	11.0588	-2.9272E+03	1.0937E+00	-1.3086E-02	7.9386E-06	308.15	601.30
1078	C10H22	2,2,4,4-TETRAMETHYLHEXANE	120.0491	-5.5764E+03	-4.4295E+01	3.2983E-02	-9.2000E-06	308.15	610.20
1079	C10H22	2,2,4,5-TETRAMETHYLHEXANE	44.3313	-3.7422E+03	-1.2716E+01	7.0194E-04	2.8949E-06	307.15	598.50
1080	C10H22	2,2,5,5-TETRAMETHYLHEXANE	15.1639	-2.9837E+03	-5.2179E-01	-1.2202E-02	8.0498E-06	300.15	581.40
1081	C10H22	2,3,3,4-TETRAMETHYLHEXANE	56.2570	-4.0994E+03	-1.7787E+01	6.6874E-03	2.1253E-07	318.15	633.10
1082	C10H22	2,3,3,5-TETRAMETHYLHEXANE	39.5641	-3.6386E+03	-1.0800E+01	-7.8227E-04	3.2261E-06	310.15	610.10
1083	C10H22	2,3,4,4-TETRAMETHYLHEXANE	62.6524	-4.2449E+03	-2.0417E+01	9.1248E-03	-5.8440E-07	315.15	626.60
1084	C10H22	2,3,4,5-TETRAMETHYLHEXANE	40.4650	-3.7289E+03	-1.1065E+01	-8.8788E-04	3.3247E-06	313.15	613.20
1085	C10H22	3,3,4,4-TETRAMETHYLHEXANE	70.9826	-4.4512E+03	-2.3980E+01	1.3216E-02	-2.2933E-06	320.15	646.70
1086	C10H22	2,4-DIMETHYL-3-ISOPROPYLPENTANE	86.9811	-4.8692E+03	-3.0383E+01	1.8437E-02	-3.7507E-06	313.15	614.40
1087	C10H22	3,3-DIETHYL-2-METHYLPENTANE	72.7173	-4.5749E+03	-2.4526E+01	1.3099E-02	-2.0730E-06	322.15	639.90
1088	C10H22	3-ETHYL-2,2,3-TRIMETHYLPENTANE	74.5660	-4.5329E+03	-2.5472E+01	1.4700E-02	-3.8218E-06	320.15	646.00
1089	C10H22	3-ETHYL-2,2,4-TRIMETHYLPENTANE	61.1872	-4.1607E+03	-1.9352E+01	8.5615E-03	-3.1058E-07	311.15	615.30
1090	C10H22	3-ETHYL-2,3,4-TRIMETHYLPENTANE	83.5354	-4.8126E+03	-2.9052E+01	1.7675E-02	-3.7440E-06	320.15	642.30
1091	C10H22	2,2,3,3,4-PENTAMETHYLPENTANE	60.0195	-4.1084E+03	-1.9550E+01	9.2480E-03	-9.4842E-07	317.15	643.80
1092	C10H22	2,2,3,4,4-PENTAMETHYLPENTANE	96.0096	-4.9637E+03	-3.4432E+01	2.3655E-02	-5.9877E-06	311.15	627.30

175

Table 7-1 VAPOR PRESSURE - ORGANIC COMPOUNDS (continued)

			$\log_{10} P = A + B/T + C \log_{10} T + D T + E T^2$					(P - mm Hg, T - K)		
NO	FORMULA	NAME	A	B	C	D	E	TMIN	TMAX
1093	C10H22O	1-DECANOL	103.0308	-8.1526E+03	-3.1641E+01	-7.2300E-10	6.0332E-06	280.05	690.00
1094	C10H22O	DI-n-PENTYL ETHER	31.5633	-3.5807E+03	-7.9975E+00	-2.3601E-10	1.8944E-06	203.72	622.00
1095	C10H22O	ISODECANOL	13.5831	-4.8371E+03	2.5431E+00	-2.1303E-02	1.1413E-05	213.15	644.00
1096	C10H22O5	TETRAETHYLENE GLYCOL DIMETHYL ETHER	14.0128	-4.3341E+03	-1.2383E-01	-7.1725E-03	3.4491E-06	243.45	705.00
1097	C10H22S	n-DECYL MERCAPTAN	67.3579	-5.9003E+03	-1.9956E+01	4.9800E-10	4.2317E-06	247.56	696.00
1098	C10H22S	BUTYL-HEXYL-SULFIDE	69.9661	-4.2921E+03	-2.4230E+01	1.5724E-02	-4.2537E-06	416.15	560.15
1099	C10H22S	ETHYL-OCTYL-SULFIDE	69.9661	-4.2921E+03	-2.4230E+01	1.5724E-02	-4.2537E-06	416.15	560.15
1100	C10H22S	HEPTYL-PROPYL-SULFIDE	69.9661	-4.2921E+03	-2.4230E+01	1.5724E-02	-4.2537E-06	416.15	560.15
1101	C10H22S	METHYL-NONYL-SULFIDE	69.9661	-4.2921E+03	-2.4230E+01	1.5724E-02	-4.2537E-06	416.15	560.15
1102	C10H22S	PENTYL-SULFIDE	69.9661	-4.2921E+03	-2.4230E+01	1.5724E-02	-4.2537E-06	416.15	560.15
1103	C10H22S2	PENTYL-DISULFIDE	281.8811	-1.2799E+04	-1.0509E+02	6.7955E-02	-1.6575E-05	401.15	726.94
1104	C10H23N	n-DECYLAMINE	-33.9072	-2.7517E+03	2.0663E+01	-3.4930E-02	1.6215E-05	288.85	663.00
1105	C11H10	1-METHYLNAPHTHALENE	29.8895	-3.9535E+03	-7.2253E+00	2.1109E-11	8.9552E-07	242.67	772.04
1106	C11H10	2-METHYLNAPHTHALENE	56.2052	-5.2563E+03	-1.6195E+01	8.1583E-11	3.0253E-06	307.73	761.00
1107	C11H14O2	n-BUTYL BENZOATE	7.6313	-3.5509E+03	2.0943E+00	-9.2518E-03	4.3287E-06	251.65	724.00
1108	C11H16	n-PENTYLBENZENE	34.2755	-3.6829E+03	-9.3387E+00	2.7727E-03	-8.8315E-15	198.15	679.90
1109	C11H16O	p-tert-AMYLPHENOL	-29.9224	-3.0806E+03	1.8272E+01	-2.6676E-02	1.0399E-05	366.00	751.00
1110	C11H20	1-UNDECYNE	25.5196	-3.6568E+03	-5.0668E+00	-4.2478E-03	3.1571E-06	353.15	650.99
1111	C11H20O2	2-ETHYLHEXYL ACRYLATE	34.3532	-3.8584E+03	-9.3203E+00	3.0189E-03	-2.7271E-14	183.15	655.00
1112	C11H22	1-UNDECENE	60.2094	-5.0417E+03	-1.7776E+01	9.0476E-10	4.2624E-06	223.99	638.00
1113	C11H22	1-CYCLOPENTYLHEXANE	48.3003	-4.5708E+03	-1.3563E+01	-4.1718E-04	3.0684E-06	351.15	667.67
1114	C11H22	PENTYLCYCLOHEXANE	32.6784	-4.0505E+03	-7.4298E+00	-5.0121E-03	4.3369E-06	351.15	674.01
1115	C11H22O	1-UNDECANAL	-31.8129	-3.1374E+03	2.0433E+01	-3.7256E-02	1.7537E-05	273.15	672.00
1116	C11H24	n-UNDECANE	82.9230	-5.6085E+03	-2.7327E+01	1.0469E-02	7.0870E-13	247.57	638.76
1117	C11H24O	1-UNDECANOL	188.0700	-1.1188E+04	-6.4590E+01	2.2644E-02	3.3487E-12	289.05	704.00
1118	C11H24S	UNDECYL MERCAPTAN	191.0623	-1.0913E+04	-6.6394E+01	2.5057E-02	-5.1267E-13	270.15	710.00
1119	C11H24S	BUTYL-HEPTYL-SULFIDE	64.2099	-4.1311E+03	-2.1958E+01	1.3714E-02	-3.5731E-06	428.15	584.15
1120	C11H24S	DECYL-METHYL-SULFIDE	64.2099	-4.1311E+03	-2.1958E+01	1.3714E-02	-3.5731E-06	428.15	584.15
1121	C11H24S	ETHYL-NONYL-SULFIDE	64.2099	-4.1311E+03	-2.1958E+01	1.3714E-02	-3.5731E-06	428.15	584.15
1122	C11H24S	OCTYL-PROPYL-SULFIDE	64.2099	-4.1311E+03	-2.1958E+01	1.3714E-02	-3.5731E-06	428.15	584.15
1123	C12H8O	DIBENZOFURAN	17.6646	-3.1989E+03	-3.3464E+00	6.0686E-10	4.4676E-07	355.65	837.80
1124	C12H9N	DIBENZOPYRROLE	-119.8570	-3.2537E+02	5.2568E+01	-4.6797E-02	1.4113E-05	517.95	899.00
1125	C12H10	ACENAPHTHENE	28.8173	-4.1623E+03	-6.7750E+00	-1.0872E-09	6.3928E-07	366.56	803.15
1126	C12H10	BIPHENYL	53.0479	-5.3509E+03	-1.4955E+01	2.1039E-09	2.4345E-06	342.37	789.26
1127	C12H10O	DIPHENYL ETHER	-26.9635	-2.5909E+03	1.6420E+01	-2.4334E-02	1.0244E-05	300.02	763.00
1128	C12H11N	p-AMINODIPHENYL	-3.8805	-4.0795E+03	7.5647E+00	-1.5752E-02	6.1473E-06	326.00	817.00
1129	C12H11N	DIPHENYLAMINE	9.7736	-3.9008E+03	9.1207E-01	-5.8980E-03	2.3012E-06	326.15	817.00
1130	C12H11N3	p-AMINOAZOBENZENE	-19.0955	-4.2712E+03	1.3826E+01	-2.0210E-02	6.9384E-06	401.00	877.00
1131	C12H11N3	1,3-DIPHENYLTRIAZENE	-13.8496	-4.2319E+03	1.1797E+01	-1.9348E-02	7.0094E-06	372.00	845.00
1132	C12H12	1,2-DIMETHYLNAPHTHALENE	80.9531	-6.5659E+03	-2.5417E+01	5.8637E-03	1.2737E-06	402.15	775.34
1133	C12H12	1,3-DIMETHYLNAPHTHALENE	84.8587	-5.6887E+03	-2.9242E+01	1.8312E-02	-4.7895E-06	421.15	585.15
1134	C12H12	1,4-DIMETHYLNAPHTHALENE	-60.7398	-1.2148E+03	2.8693E+01	-2.8126E-02	9.2388E-06	421.15	776.78
1135	C12H12	1,5-DIMETHYLNAPHTHALENE	71.6143	-5.9394E+03	-2.2479E+01	6.7276E-03	2.3898E-07	423.15	773.47
1136	C12H12	1,6-DIMETHYLNAPHTHALENE	31.1131	-4.1403E+03	-7.5477E+00	-6.4435E-04	1.5135E-06	421.15	770.60
1137	C12H12	1,7-DIMETHYLNAPHTHALENE	150.1682	-8.2324E+03	-5.4206E+01	3.4631E-02	-9.1270E-06	423.15	595.15
1138	C12H12	2,3-DIMETHYLNAPHTHALENE	71.1090	-5.9471E+03	-2.2272E+01	6.5823E-03	2.5956E-07	428.15	777.78
1139	C12H12	2,6-DIMETHYLNAPHTHALENE	-6.9795	-2.9488E+03	7.4483E+00	-1.1581E-02	4.3391E-06	383.32	777.00
1140	C12H12	2,7-DIMETHYLNAPHTHALENE	-1.7726	-3.1042E+03	5.3361E+00	-9.6632E-03	3.6806E-06	368.81	778.00
1141	C12H12	1-ETHYLNAPHTHALENE	7.5650	-3.7597E+03	2.6035E+00	-1.1581E-02	5.1365E-06	259.34	776.00
1142	C12H12	2-ETHYLNAPHTHALENE	138.9424	-7.4909E+03	-5.0409E+01	3.4108E-02	-9.5362E-06	391.15	566.15
1143	C12H12N2	p-AMINODIPHENYLAMINE	-1.7258	-5.0658E+03	7.3819E+00	-1.6555E-02	6.1361E-06	341.15	867.00
1144	C12H12N2	HYDRAZOBENZENE	16.8982	-5.0039E+03	-3.5846E-01	-9.9629E-03	4.2938E-06	404.15	573.00
1145	C12H14	1,2,3-TRIMETHYLINDENE	17.5434	-3.2091E+03	-3.0901E+00	1.9203E-10	7.8034E-08	344.65	726.00
1146	C12H14O4	DIETHYL PHTHALATE	72.1438	-7.0747E+03	-2.1029E+01	-3.2404E-10	3.4691E-06	269.15	757.00
1147	C12H16	CYCLOHEXYLBENZENE	-14.0614	-3.5138E+03	1.2631E+01	-2.6285E-02	1.1568E-05	280.14	744.00
1148	C12H18	m-DIISOPROPYLBENZENE	31.8443	-3.8099E+03	-7.8983E+00	8.3393E-11	8.1613E-07	210.02	684.00
1149	C12H18	p-DIISOPROPYLBENZENE	9.3996	-3.4567E+03	1.8257E+00	-1.1471E-02	5.4813E-06	256.08	689.00
1150	C12H18	n-HEXYLBENZENE	6.7694	-3.6050E+03	3.3416E+00	-1.5306E-02	7.8479E-06	212.00	698.00
1151	C12H18	1,2,3-TRIETHYLBENZENE	-358.0208	7.9022E+03	1.4783E+02	-1.2821E-01	4.1279E-05	419.15	684.37
1152	C12H18	1,2,4-TRIETHYLBENZENE	67.3455	-4.5199E+03	-2.3280E+01	1.7857E-02	-5.7468E-06	319.15	493.15
1153	C12H18	1,3,5-TRIETHYLBENZENE	-396.9101	9.0363E+03	1.6348E+02	-1.4157E-01	4.5564E-05	418.15	682.28
1154	C12H18	HEXAMETHYLBENZENE	148.1217	-7.3688E+03	-5.4569E+01	3.7882E-02	-1.0004E-05	451.15	758.00
1155	C12H20O4	DIBUTYL MALEATE	2.4103	-5.0321E+03	6.9147E+00	-2.2687E-02	1.0287E-05	275.00	716.00
1156	C12H22	BICYCLOHEXYL	43.7149	-4.4737E+03	-1.2082E+01	1.0564E-10	2.4453E-06	276.78	727.00
1157	C12H22	1-DODECYNE	-7.8023	-2.8587E+03	8.3798E+00	-1.5468E-02	6.5514E-06	386.15	668.16
1158	C12H23N	DICYCLOHEXYLAMINE	-50.5512	-2.7053E+03	2.8273E+01	-4.5702E-02	2.0443E-05	273.05	737.00
1159	C12H24	1-DODECENE	-8.5899	-3.5241E+03	1.0806E+01	-2.8161E-02	1.4267E-05	237.93	657.00
1160	C12H24	1-CYCLOPENTYLHEPTANE	48.0744	-4.8876E+03	-1.3081E+01	-2.0206E-03	3.6908E-06	368.15	679.00
1161	C12H24	1-CYCLOHEXYLHEXANE	44.2276	-4.6610E+03	-1.1795E+01	-2.0286E-03	3.3751E-06	367.15	691.81
1162	C12H24O	1-DODECANAL	-47.6290	-3.1249E+03	2.7611E+01	-4.6717E-02	2.1392E-05	285.15	685.00
1163	C12H24O2	n-DODECANOIC ACID	-83.0980	-4.7634E+03	4.6211E+01	-7.6239E-02	3.2073E-05	317.15	734.00
1164	C12H26	n-DODECANE	-5.6532	-3.4698E+03	9.0272E+00	-2.3185E-02	1.1235E-05	263.57	658.20
1165	C12H26O	DI-n-HEXYL ETHER	10.0096	-4.1641E+03	2.8039E+00	-1.7099E-02	8.7760E-06	230.15	658.00
1166	C12H26O	1-DODECANOL	114.5440	-9.2060E+03	-3.5275E+01	-1.3222E-09	6.1548E-06	296.95	721.00
1167	C12H26O3	DIETHYLENE GLYCOL DI-n-BUTYL ETHER	61.7424	-5.1004E+03	-1.9671E+01	8.2329E-03	1.0941E-13	212.95	680.00
1168	C12H26S	n-DODECYL MERCAPTAN	90.3277	-6.5839E+03	-2.9741E+01	1.0992E-02	-1.5813E-13	265.15	724.00
1169	C12H26S	BUTYL-OCTYL-SULFIDE	488.0936	-1.9912E+04	-1.8598E+02	1.3095E-01	-3.7088E-05	407.15	582.15
1170	C12H26S	DECYL-ETHYL-SULFIDE	59.3379	-3.9911E+03	-2.0048E+01	1.2080E-02	-3.0384E-06	439.15	608.15

176

Table 7-1 VAPOR PRESSURE - ORGANIC COMPOUNDS (continued)

$$\log_{10} P = A + B/T + C \log_{10} T + D\,T + E\,T^2 \qquad (P - \text{mm Hg}, T - \text{K})$$

NO	FORMULA	NAME	A	B	C	D	E	TMIN	TMAX
1171	C12H26S	HEXYL-SULFIDE	59.3379	-3.9911E+03	-2.0048E+01	1.2080E-02	-3.0384E-06	439.15	608.15
1172	C12H26S	METHYL-UNDECYL-SULFIDE	59.3379	-3.9911E+03	-2.0048E+01	1.2080E-02	-3.0384E-06	439.15	608.15
1173	C12H26S	NONYL-PROPYL-SULFIDE	59.3379	-3.9911E+03	-2.0048E+01	1.2080E-02	-3.0384E-06	439.15	608.15
1174	C12H26S2	HEXYL-DISULFIDE	317.1013	-1.4847E+04	-1.1741E+02	7.1405E-02	-1.6270E-05	426.15	747.10
1175	C12H27BO3	TRI-n-BUTYL BORATE	8.3425	-2.6942E+03	7.8090E-01	-5.9864E-03	3.0070E-06	203.15	743.15
1176	C12H27N	DODECYLAMINE	120.8152	-7.8738E+03	-4.0876E+01	1.5576E-02	6.7451E-13	301.47	696.00
1177	C12H27N	TRI-n-BUTYLAMINE	14.3754	-4.1372E+03	9.5698E-01	-1.5617E-02	8.5491E-06	203.00	644.00
1178	C13H10	FLUORENE	53.9382	-5.3622E+03	-1.6059E+01	4.5696E-03	8.1430E-13	387.94	870.00
1179	C13H10O	BENZOPHENONE	16.4144	-3.8064E+03	-2.3984E+00	-7.4544E-04	2.9345E-07	321.35	816.00
1180	C13H12	DIPHENYLMETHANE	50.8894	-5.2749E+03	-1.4246E+01	-4.2994E-10	2.4197E-06	298.39	768.00
1181	C13H14	1-PROPYLNAPHTHALENE	131.8028	-7.9574E+03	-4.6101E+01	2.4187E-02	-4.5747E-06	428.15	771.45
1182	C13H14	2-PROPYLNAPHTHALENE	144.8838	-8.3997E+03	-5.1219E+01	2.7910E-02	-5.5943E-06	433.15	772.44
1183	C13H14	2ETHYL-3-METHYLNAPHTHALENE	256.5800	-1.2145E+04	-9.4901E+01	5.9570E-02	-1.4277E-05	469.15	776.44
1184	C13H14	2ETHYL-6-METHYLNAPHTHALENE	117.2705	-7.6092E+03	-4.0107E+01	1.8664E-02	-2.8127E-06	465.15	766.56
1185	C13H14	2ETHYL-7-METHYLNAPHTHALENE	117.2705	-7.6092E+03	-4.0107E+01	1.8664E-02	-2.8127E-06	465.15	766.56
1186	C13H20	n-HEPTYLBENZENE	89.2811	-6.4093E+03	-2.9248E+01	1.0328E-02	6.2451E-14	225.15	714.00
1187	C13H24	1-TRIDECYNE	33.4188	-4.3334E+03	-7.7611E+00	-3.6928E-03	3.3986E-06	393.15	684.11
1188	C13H26	1-TRIDECENE	8.1909	-3.7115E+03	2.5607E+00	-1.2858E-02	6.2798E-06	250.08	675.00
1189	C13H26	1-CYCLOPENTYLOCTANE	43.3958	-5.0519E+03	-1.0882E+01	-4.6584E-03	4.4687E-06	385.15	702.06
1190	C13H26	1-CYCLOHEXYLHEPTANE	41.7798	-4.8887E+03	-1.0490E+01	-3.9757E-03	3.9933E-06	384.15	708.63
1191	C13H26O	1-TRIDECANAL	161.5042	-9.7660E+03	-5.5591E+01	2.1036E-02	5.5498E-13	288.15	700.00
1192	C13H26O2	n-BUTYL NONANOATE	6.5223	-4.4759E+03	5.0441E+00	-2.2415E-02	1.1463E-05	235.15	652.00
1193	C13H26O2	METHYL DODECANOATE	68.5001	-6.4740E+03	-1.9996E+01	4.9208E-10	3.4462E-06	278.15	712.00
1194	C13H28	n-TRIDECANE	49.2391	-4.9649E+03	-1.3769E+01	-2.1146E-09	2.5902E-06	267.76	675.80
1195	C13H28O	1-TRIDECANOL	119.2170	-9.5810E+03	-3.6776E+01	1.4106E-09	6.2604E-06	303.75	731.00
1196	C13H28S	BUTYL-NONYL-SULFIDE	55.2758	-3.8714E+03	-1.8466E+01	1.0771E-02	-2.6240E-06	450.15	630.15
1197	C13H28S	DECYL-PROPYL-SULFIDE	55.2758	-3.8714E+03	-1.8466E+01	1.0771E-02	-2.6240E-06	450.15	630.15
1198	C13H28S	DODECYL-METHYL-SULFIDE	55.2758	-3.8714E+03	-1.8466E+01	1.0771E-02	-2.6240E-06	450.15	630.15
1199	C13H28S	ETHYL-UNDECYL-SULFIDE	55.2758	-3.8714E+03	-1.8466E+01	1.0771E-02	-2.6240E-06	450.15	630.15
1200	C13H28S	1-TRIDECANETHIOL	177.1486	-1.0227E+04	-6.2559E+01	3.1360E-02	-5.3595E-06	424.15	742.13
1201	C14H8O2	ANTHRAQUINONE	144.9121	-1.2120E+04	-4.5070E+01	1.1051E-03	6.2679E-06	559.15	653.05
1202	C14H10	ANTHRACENE	-120.0992	4.4780E+00	5.2574E+01	-4.7696E-02	1.5020E-05	488.93	873.00
1203	C14H10	DIPHENYLACETYLENE	-5.1555	-3.4335E+03	7.2102E+00	-1.3104E-02	5.0073E-06	335.65	832.00
1204	C14H10	PHENANTHRENE	50.2858	-5.7409E+03	-1.3935E+01	-8.8520E-10	2.1343E-06	372.38	869.25
1205	C14H12	cis-STILBENE	33.3864	-3.9885E+03	-1.0330E+01	1.3550E-02	-7.3213E-06	268.15	535.00
1206	C14H12	trans-STILBENE	68.6303	-6.3776E+03	-2.1015E+01	5.7183E-03	1.8334E-12	397.35	820.00
1207	C14H12O2	BENZYL BENZOATE	-1.5640	-4.6284E+03	7.3630E+00	-1.8259E-02	7.4580E-06	292.55	820.00
1208	C14H14	1,1-DIPHENYLETHANE	7.7920	-3.8630E+03	2.4667E+00	-1.1120E-02	4.9686E-06	255.20	775.00
1209	C14H14	1,2-DIPHENYLETHANE	48.5573	-5.2841E+03	-1.3410E+01	-1.0073E-09	2.1338E-06	324.34	780.00
1210	C14H14O	DIBENZYL ETHER	1.7972	-3.8773E+03	3.9185E+00	-4.9549E-03	1.4918E-14	276.75	777.00
1211	C14H16	1-n-BUTYLNAPHTHALENE	49.7689	-5.5181E+03	-1.3708E+01	1.2616E-10	1.9679E-06	253.43	792.00
1212	C14H16	2-BUTYLNAPHTHALENE	1878.6219	-6.3688E+04	-7.3714E+02	5.6673E-01	-1.7095E-04	412.15	567.15
1213	C14H22	n-OCTYLBENZENE	1.8919	-4.1324E+03	6.1473E+00	-2.0294E-02	9.6879E-06	237.15	729.00
1214	C14H22	1,2,3,4-TETRAETHYLBENZENE	62.7022	-4.6397E+03	-2.1207E+01	1.5260E-02	-4.6147E-06	339.15	523.15
1215	C14H22	1,2,3,5-TETRAETHYLBENZENE	-298.2878	6.4585E+03	1.2247E+02	-1.0039E-01	3.0644E-05	447.15	707.52
1216	C14H22	1,2,4,5-TETRAETHYLBENZENE	-316.0617	7.0146E+03	1.2952E+02	-1.0593E-01	3.2289E-05	447.15	706.85
1217	C14H22O	p-tert-OCTYLPHENOL	112.8024	-8.1574E+03	-3.7219E+01	1.2314E-02	6.6452E-12	358.55	765.00
1218	C14H28	1-TETRADECENE	100.1124	-7.0403E+03	-3.3050E+01	1.1591E-02	-1.3156E-14	260.30	692.00
1219	C14H28	1-CYCLOPENTYLNONANE	51.7890	-5.6110E+03	-1.3889E+01	-3.1493E-03	4.0422E-06	399.15	716.95
1220	C14H28	1-CYCLOHEXYLOCTANE	44.7900	-5.2699E+03	-1.1386E+01	-4.0422E-03	4.0233E-06	399.15	723.61
1221	C14H28O2	n-TETRADECANOIC ACID	-79.3115	-4.6936E+03	4.3573E+01	-6.8469E-02	2.7919E-05	327.55	756.00
1222	C14H30	n-TETRADECANE	106.1056	-7.3461E+03	-3.5195E+01	1.2356E-02	-8.3950E-13	279.01	692.40
1223	C14H30O	1-TETRADECANOL	130.2611	-1.0422E+04	-4.0360E+01	1.7910E-09	6.8167E-06	310.65	741.00
1224	C14H30S	BUTYL-DECYL-SULFIDE	51.8592	-3.7684E+03	-1.7144E+01	9.7076E-03	-2.2970E-06	460.15	651.15
1225	C14H30S	DODECYL-ETHYL-SULFIDE	51.8592	-3.7684E+03	-1.7144E+01	9.7076E-03	-2.2970E-06	460.15	651.15
1226	C14H30S	HEPTYL-SULFIDE	51.8592	-3.7684E+03	-1.7144E+01	9.7076E-03	-2.2970E-06	460.15	651.15
1227	C14H30S	METHYL-TRIDECYL-SULFIDE	51.8592	-3.7684E+03	-1.7144E+01	9.7076E-03	-2.2970E-06	460.15	651.15
1228	C14H30S	PROPYL-UNDECYL-SULFIDE	51.8592	-3.7684E+03	-1.7144E+01	9.7076E-03	-2.2970E-06	460.15	651.15
1229	C14H30S	1-TETRADECANETHIOL	174.4855	-1.0525E+04	-6.0972E+01	2.8387E-02	-4.2369E-06	437.15	753.80
1230	C14H30S2	HEPTYL-DISULFIDE	313.8133	-1.5713E+04	-1.1450E+02	6.3386E-02	-1.2849E-05	448.15	765.96
1231	C14H31N	TETRADECYLAMINE	127.2644	-8.5930E+03	-4.2895E+01	1.5716E-02	4.4317E-13	311.34	722.30
1232	C15H10N2O2	DIPHENYLMETHANE-4,4'-DIISOCYANATE	-15.5268	-4.7726E+03	1.3371E+01	-2.5298E-02	1.1904E-05	311.20	609.00
1233	C15H16O	p-CUMYLPHENOL	-6.5190	-4.5451E+03	9.0675E+00	-1.7864E-02	6.7285E-06	346.00	834.00
1234	C15H16O2	BISPHENOL A	172.1183	-1.4667E+04	-5.4854E+01	1.1985E-02	9.7628E-12	426.15	849.00
1235	C15H18	1-PENTYLNAPHTHALENE	213.6784	-1.1340E+04	-7.7250E+01	4.3708E-02	-9.2840E-06	458.15	793.32
1236	C15H18	2-PENTYLNAPHTHALENE	244.7128	-1.2467E+04	-8.9196E+01	5.1526E-02	-1.1188E-05	463.15	797.48
1237	C15H24	n-NONYLBENZENE	-0.9235	-4.2232E+03	7.3073E+00	-2.0964E-02	9.7152E-06	249.00	741.00
1238	C15H24O	2,6-DI-tert-BUTYL-p-CRESOL	69.7769	-6.5661E+03	-2.0402E+01	4.0687E-09	3.5418E-06	344.00	720.00
1239	C15H24O	NONYLPHENOL	6.1076	-5.4241E+03	5.3661E+00	-2.0127E-02	8.7825E-06	279.15	757.00
1240	C15H28	1-PENTADECYNE	16.5889	-4.2684E+03	-5.9774E-01	-1.0847E-02	5.7477E-06	413.15	711.41
1241	C15H30	1-PENTADECENE	113.3410	-7.8282E+03	-3.7725E+01	1.3147E-02	4.8168E-13	269.42	708.00
1242	C15H30	1-CYCLOPENTYLDECANE	56.1979	-6.0504E+03	-1.5301E+01	-2.9629E-03	4.0349E-06	414.15	730.64
1243	C15H30	1-CYCLOHEXYLNONANE	47.2365	-5.6428E+03	-1.2038E+01	-4.3786E-03	4.1572E-06	414.15	737.79
1244	C15H30O2	PENTADECANOIC ACID	245.4056	-1.7856E+04	-7.8444E+01	-4.1719E-06	1.4029E-05	325.68	766.40
1245	C15H32	n-PENTADECANE	116.5157	-8.0410E+03	-3.8799E+01	1.3398E-02	-4.4444E-13	283.10	706.80
1246	C15H32O	1-PENTADECANOL	4883.7775	-1.7356E+05	-1.8909E+03	1.3107E+00	-3.4597E-04	503.15	722.53
1247	C15H32S	BUTYL-UNDECYL-SULFIDE	48.9050	-3.6775E+03	-1.6007E+01	8.8160E-03	-2.0298E-06	469.15	672.15
1248	C15H32S	DODECYL-PROPYL-SULFIDE	48.9050	-3.6775E+03	-1.6007E+01	8.8160E-03	-2.0298E-06	469.15	672.15

Table 7-1 VAPOR PRESSURE - ORGANIC COMPOUNDS (continued)

			$\log_{10} P = A + B/T + C \log_{10} T + D T + E T^2$					(P - mm Hg, T - K)	
NO	FORMULA	NAME	A	B	C	D	E	TMIN	TMAX
1249	C15H32S	ETHYL-TRIDECYL-SULFIDE	48.9050	-3.6775E+03	-1.6007E+01	8.8160E-03	-2.0298E-06	469.15	672.15
1250	C15H32S	METHYL-TETRADECYL-SULFIDE	48.9050	-3.6775E+03	-1.6007E+01	8.8160E-03	-2.0298E-06	469.15	672.15
1251	C15H32S	1-PENTADECANETHIOL	159.2799	-1.0391E+04	-5.4523E+01	2.2127E-02	-2.3039E-06	449.15	764.77
1252	C16H10	FLUORANTHENE	70.6802	-6.4840E+03	-2.2241E+01	7.2184E-03	-6.3035E-13	383.33	905.00
1253	C16H10	PYRENE	70.7671	-6.9413E+03	-2.1790E+01	6.0727E-03	1.5767E-12	423.81	936.00
1254	C16H12	1-PHENYLNAPHTHALENE	0.1631	-4.3345E+03	5.8605E+00	-1.3899E-02	5.3712E-06	318.15	849.00
1255	C16H20	1-n-HEXYLNAPHTHALENE	52.8437	-6.1027E+03	-1.4546E+01	4.4942E-11	1.8419E-06	255.15	813.00
1256	C16H22O4	DIBUTYL PHTHALATE	152.6750	-1.0754E+04	-5.1170E+01	1.6933E-02	2.4948E-14	238.15	781.00
1257	C16H26	n-DECYLBENZENE	-4.4754	-4.4669E+03	9.1965E+00	-2.4010E-02	1.0848E-05	258.77	753.00
1258	C16H26	PENTAETHYLBENZENE	-176.2373	2.4449E+03	7.3939E+01	-6.0885E-02	1.8275E-05	359.15	723.64
1259	C16H30	1-HEXADECYNE	136.7323	-7.6251E+03	-5.0137E+01	3.8522E-02	-1.2734E-05	445.15	724.26
1260	C16H32	n-DECYLCYCLOHEXANE	108.4769	-7.9333E+03	-3.5723E+01	1.1860E-02	-3.0571E-13	271.42	751.25
1261	C16H32	1-CYCLOPENTYLUNDECANE	36.3759	-5.7482E+03	-7.1096E+00	-1.0164E-02	6.1164E-06	427.15	743.30
1262	C16H32	1-HEXADECENE	10.5925	-4.4210E+03	1.9605E+00	-1.2372E-02	5.5619E-06	277.51	722.00
1263	C16H32O2	n-HEXADECANOIC ACID	34.6559	-5.2645E+03	-8.8645E+00	2.3028E-03	-3.6120E-13	335.66	776.00
1264	C16H34	n-HEXADECANE	99.1091	-7.5333E+03	-3.2251E+01	1.0453E-02	1.2328E-12	291.32	720.60
1265	C16H34O	DI-n-OCTYL ETHER	4.7440	-5.1545E+03	6.1040E+00	-2.2754E-02	1.0549E-05	265.55	707.00
1266	C16H34O	1-HEXADECANOL	218.9440	-1.3794E+04	-7.4743E+01	2.4485E-02	2.1992E-12	322.35	761.00
1267	C16H34S	BUTYL-DODECYL-SULFIDE	46.4778	-3.6016E+03	-1.5078E+01	8.1055E-03	-1.8220E-06	477.15	691.15
1268	C16H34S	ETHYL-TETRADECYL-SULFIDE	46.4778	-3.6016E+03	-1.5078E+01	8.1055E-03	-1.8220E-06	477.15	691.15
1269	C16H34S	METHYL-PENTADECYL-SULFIDE	46.4778	-3.6016E+03	-1.5078E+01	8.1055E-03	-1.8220E-06	477.15	691.15
1270	C16H34S	OCTYL-SULFIDE	46.4778	-3.6016E+03	-1.5078E+01	8.1055E-03	-1.8220E-06	477.15	691.15
1271	C16H34S	PROPYL-TRIDECYL-SULFIDE	46.4778	-3.6016E+03	-1.5078E+01	8.1055E-03	-1.8220E-06	477.15	691.15
1272	C16H34S	1-HEXADECANETHIOL	146.3286	-1.0322E+04	-4.8992E+01	1.6719E-02	-6.5786E-07	461.15	774.68
1273	C16H34S2	OCTYL-DISULFIDE	281.8096	-1.5461E+04	-1.0076E+02	4.9300E-02	-8.3216E-06	469.15	784.46
1274	C17H28	n-UNDECYLBENZENE	124.1549	-8.8970E+03	-4.1223E+01	1.3662E-02	-8.1321E-14	268.00	764.00
1275	C17H32	1-HEPTADECYNE	226.5937	-1.0745E+04	-8.5256E+01	6.6064E-02	-1.9726E-05	457.15	736.21
1276	C17H34	1-CYCLOPENTYLDODECANE	18.2871	-5.4528E+03	2.8681E-01	-1.6297E-02	7.7805E-06	441.15	755.17
1277	C17H34	1-CYCLOHEXYLUNDECANE	30.3012	-5.6687E+03	-4.7892E+00	-1.1229E-02	6.0766E-06	440.15	761.74
1278	C17H34	1-HEPTADECENE	87.0835	-7.7982E+03	-2.6138E+01	1.9006E-09	4.5584E-06	284.40	736.00
1279	C17H36	n-HEPTADECANE	173.4039	-1.0943E+04	-5.9212E+01	2.0705E-02	-1.3433E-12	295.13	733.37
1280	C17H36O	1-HEPTADECANOL	-2.1471	-6.6524E+03	1.1311E+01	-3.3612E-02	1.3573E-05	327.05	770.00
1281	C17H36S	BUTYL-TRIDECYL-SULFIDE	44.3249	-3.5329E+03	-1.4258E+01	7.4918E-03	-1.6466E-06	485.15	709.15
1282	C17H36S	ETHYL-PENTADECYL-SULFIDE	44.3249	-3.5329E+03	-1.4258E+01	7.4918E-03	-1.6466E-06	485.15	709.15
1283	C17H36S	HEXADECYL-METHYL-SULFIDE	44.3249	-3.5329E+03	-1.4258E+01	7.4918E-03	-1.6466E-06	485.15	709.15
1284	C17H36S	PROPYL-TETRADECYL-SULFIDE	44.3249	-3.5329E+03	-1.4258E+01	7.4918E-03	-1.6466E-06	485.15	709.15
1285	C17H36S	1-HEPTADECANETHIOL	98.9718	-8.9675E+03	-3.0316E+01	3.1006E-03	2.8533E-06	471.15	786.01
1286	C18H12	CHRYSENE	-50.1566	-3.4381E+03	2.5178E+01	-2.4620E-02	7.0144E-06	531.15	979.00
1287	C18H14	m-TERPHENYL	-14.7175	-4.3577E+03	1.1935E+01	-1.8441E-02	6.4370E-06	360.00	924.85
1288	C18H14	o-TERPHENYL	-8.0641	-4.0928E+03	9.1076E+00	-1.6326E-02	6.0467E-06	329.35	890.95
1289	C18H14	p-TERPHENYL	-39.6342	-3.2661E+03	2.1080E+01	-2.2574E-02	6.9092E-06	485.00	925.95
1290	C18H15P	TRIPHENYLPHOSPHINE	-2.5257	-4.7269E+03	6.7467E+00	-1.2384E-02	4.1401E-06	354.40	1008.00
1291	C18H15O4P	TRIPHENYL PHOSPHATE	28.0972	-5.6684E+03	-5.9768E+00	-3.1567E-09	1.0751E-12	323.15	686.65
1292	C18H16N2	N,N'-DIPHENYL-p-PHENYLENEDIAMINE	-19.7400	-5.5157E+03	1.5157E+01	-2.3357E-02	7.8058E-06	409.00	906.00
1293	C18H22	2,3-DIMETHYL-2,3-DIPHENYLBUTANE	90.1236	-7.2210E+03	-2.8976E+01	8.9691E-03	-4.1974E-13	392.15	805.00
1294	C18H22O2	DICUMYL PEROXIDE	20.0849	-1.6206E+03	-5.7800E+00	2.3158E-03	1.2360E-12	311.15	669.00
1295	C18H30	n-DODECYLBENZENE	145.6916	-1.0165E+04	-4.8761E+01	1.5985E-02	4.8810E-13	275.93	774.26
1296	C18H30	HEXAETHYLBENZENE	10.9941	-3.9288E+03	7.7647E-01	-8.2286E-03	4.0556E-06	407.15	734.78
1297	C18H32O2	LINOLEIC ACID	40.6453	-7.5442E+03	-7.5552E+00	-1.0656E-02	5.2640E-06	268.15	628.00
1298	C18H34	1-OCTADECYNE	329.5375	-1.4335E+04	-1.2543E+02	9.2940E-02	-2.7568E-05	469.15	747.33
1299	C18H34O2	OLEIC ACID	78.6973	-8.8227E+03	-2.2472E+01	4.8353E-11	2.6578E-06	286.53	633.15
1300	C18H34O4	DIBUTYL SEBACATE	25.1840	-6.3289E+03	-2.7197E+00	-1.0001E-02	4.3607E-06	274.15	768.00
1301	C18H34O4	DIHEXYL ADIPATE	12.9950	-6.6542E+03	3.9639E+00	-2.3327E-02	1.0417E-05	259.35	767.00
1302	C18H36	1-CYCLOPENTYLTRIDECANE	-12.7007	-4.7034E+03	1.2652E+01	-2.5643E-02	1.0193E-05	453.15	766.47
1303	C18H36	1-CYCLOHEXYLDODECANE	20.2835	-5.6053E+03	-6.1610E-01	-1.4798E-02	7.0062E-06	452.15	772.83
1304	C18H36	1-OCTADECENE	125.2363	-1.0086E+04	-3.8883E+01	-1.1233E-09	7.1580E-06	290.76	748.00
1305	C18H36O2	STEARIC ACID	-40.3638	-4.7724E+03	2.4502E+01	-3.7665E-02	1.4595E-05	342.75	799.00
1306	C18H38	n-OCTADECANE	-15.0772	-4.8702E+03	1.4501E+01	-3.1625E-02	1.3478E-05	301.33	745.26
1307	C18H38O	DINONYL ETHER	32.8943	-6.3222E+03	-5.3308E+00	-1.0980E-02	5.5688E-06	273.00	591.00
1308	C18H38O	1-OCTADECANOL	30.1086	-7.2605E+03	-2.9558E+00	-1.5338E-02	6.1296E-06	331.05	777.00
1309	C18H38S	BUTYL-TETRADECYL-SULFIDE	42.3262	-3.4681E+03	-1.3499E+01	6.9364E-03	-1.4913E-06	493.15	727.15
1310	C18H38S	ETHYL-HEXADECYL-SULFIDE	42.3262	-3.4681E+03	-1.3499E+01	6.9364E-03	-1.4913E-06	493.15	727.15
1311	C18H38S	HEPTADECYL-METHYL-SULFIDE	42.3262	-3.4681E+03	-1.3499E+01	6.9364E-03	-1.4913E-06	493.15	727.15
1312	C18H38S	NONYL-SULFIDE	42.3262	-3.4681E+03	-1.3499E+01	6.9364E-03	-1.4913E-06	493.15	727.15
1313	C18H38S	PENTADECYL-PROPYL-SULFIDE	42.3262	-3.4681E+03	-1.3499E+01	6.9364E-03	-1.4913E-06	493.15	727.15
1314	C18H38S	1-OCTADECANETHIOL	58.7548	-7.8553E+03	-1.4451E+01	-8.3488E-03	5.7427E-06	482.15	795.36
1315	C18H38S2	NONYL-DISULFIDE	193.1989	-1.3103E+04	-6.5470E+01	2.2376E-02	-1.1247E-06	488.15	802.30
1316	C19H26	1-n-NONYLNAPHTHALENE	9.5431	-4.9776E+03	2.4344E+00	-1.2052E-02	4.8826E-06	284.15	849.00
1317	C19H32	n-TRIDECYLBENZENE	160.3924	-1.1093E+04	-5.3875E+01	1.7532E-02	3.7270E-13	283.15	783.00
1318	C19H36	1-NONADECYNE	448.8162	-1.8597E+04	-1.7170E+02	1.2507E-01	-3.6011E-05	481.15	758.94
1319	C19H36O2	METHYL OLEATE	79.3010	-8.3180E+03	-2.2989E+01	1.4073E-09	3.1603E-06	293.05	764.00
1320	C19H38	1-CYCLOPENTYLTETRADECANE	-110.6422	-1.5625E+03	5.0810E+01	-5.1913E-02	1.6687E-05	465.15	772.00
1321	C19H38	1-CYCLOHEXYLTRIDECANE	-5.4644	-4.9857E+03	9.6230E+00	-2.2358E-02	8.9036E-06	464.15	783.38
1322	C19H38	1-NONADECENE	95.5175	-8.7380E+03	-2.8710E+01	-7.3561E-10	4.6274E-06	296.55	760.00
1323	C19H38O2	NONADECANOIC ACID	-165.8469	-3.5582E+03	8.5521E+01	-1.2752E-01	4.9131E-05	341.23	809.90
1324	C19H40	n-NONADECANE	76.7647	-7.7205E+03	-2.2376E+01	6.5102E-11	3.1141E-06	305.33	755.93
1325	C19H40O	1-NONADECANOL	3144.6124	-1.2162E+05	-1.1981E+03	7.5841E-01	-1.8279E-04	547.15	775.30
1326	C19H40S	BUTYL-PENTADECYL-SULFIDE	40.6881	-3.4142E+03	-1.2881E+01	6.4926E-03	-1.3696E-06	500.15	743.15

Table 7-1 VAPOR PRESSURE - ORGANIC COMPOUNDS (continued)

$$\log_{10} P = A + B/T + C \log_{10} T + D T + E T^2 \qquad (P - mm\ Hg, T - K)$$

NO	FORMULA	NAME	A	B	C	D	E	TMIN	TMAX
1327	C19H40S	ETHYL-HEPTADECYL-SULFIDE	40.6881	-3.4142E+03	-1.2881E+01	6.4926E-03	-1.3696E-06	500.15	743.15
1328	C19H40S	HEXADECYL-PROPYL-SULFIDE	40.6881	-3.4142E+03	-1.2881E+01	6.4926E-03	-1.3696E-06	500.15	743.15
1329	C19H40S	METHYL-OCTADECYL-SULFIDE	40.6881	-3.4142E+03	-1.2881E+01	6.4926E-03	-1.3696E-06	500.15	743.15
1330	C19H40S	1-NONADECANETHIOL	5.7906	-6.1701E+03	6.0897E+00	-2.1963E-02	8.9263E-06	492.15	805.29
1331	C20H16	TRIPHENYLETHYLENE	1.1386	-4.9715E+03	5.6613E+00	-1.3449E-02	4.8480E-06	342.15	908.00
1332	C20H28	1-n-DECYLNAPHTHALENE	9.7493	-5.1523E+03	2.4619E+00	-1.2244E-02	4.8981E-06	288.15	859.00
1333	C20H30O2	ABIETIC ACID	36.2755	-6.3461E+03	-8.7558E+00	1.5404E-06	2.3687E-06	446.65	649.70
1334	C20H31N	DEHYDROABIETYLAMINE	4.4503	-5.4572E+03	5.0824E+00	-1.5481E-02	5.9302E-06	317.65	863.00
1335	C20H34	1-PHENYLTETRADECANE	305.7678	-1.6591E+04	-1.0971E+02	5.4580E-02	-9.5588E-06	478.15	792.00
1336	C20H38	1-EICOSYNE	603.4276	-2.4137E+04	-2.3157E+02	1.6619E-01	-4.6732E-05	492.15	769.79
1337	C20H40	1-CYCLOPENTYLPENTADECANE	-161.1344	3.3660E+01	7.0326E+01	-6.4380E-02	1.9394E-05	476.15	780.00
1338	C20H40	1-CYCLOHEXYLTETRADECANE	-31.0024	-4.3362E+03	1.9690E+01	-2.9411E-02	1.0569E-05	476.15	792.82
1339	C20H40	1-EICOSENE	142.6341	-1.0430E+04	-4.7275E+01	1.4711E-02	1.0584E-12	301.76	771.00
1340	C20H42	n-EICOSANE	19.4193	-5.8699E+03	-4.4282E-01	-1.2606E-02	5.2241E-06	309.59	767.04
1341	C20H42O	1-EICOSANOL	112.0728	-1.1079E+04	-3.3201E+01	2.9672E-09	3.3864E-06	338.55	792.00
1342	C20H42S	BUTYL-HEXADECYL-SULFIDE	39.2403	-3.3660E+03	-1.2337E+01	6.1075E-03	-1.2656E-06	506.15	759.15
1343	C20H42S	DECYL-SULFIDE	39.2403	-3.3660E+03	-1.2337E+01	6.1075E-03	-1.2656E-06	506.15	759.15
1344	C20H42S	ETHYL-OCTADECYL-SULFIDE	39.2403	-3.3660E+03	-1.2337E+01	6.1075E-03	-1.2656E-06	506.15	759.15
1345	C20H42S	HEPTADECYL-PROPYL-SULFIDE	39.2403	-3.3660E+03	-1.2337E+01	6.1075E-03	-1.2656E-06	506.15	759.15
1346	C20H42S	METHYL-NONADECYL-SULFIDE	39.2403	-3.3660E+03	-1.2337E+01	6.1075E-03	-1.2656E-06	506.15	759.15
1347	C20H42S	1-EICOSANETHIOL	-43.9590	-4.5256E+03	2.5230E+01	-3.4015E-02	1.1570E-05	502.15	814.57
1348	C20H42S2	DECYL-DISULFIDE	72.7420	-9.3678E+03	-1.8552E+01	-9.3714E-03	6.4158E-06	508.15	820.08
1349	C21H21O4P	TRI-o-CRESYL PHOSPHATE	21.1624	-5.2756E+03	-3.3565E+00	8.6660E-06	-2.9202E-09	427.75	565.85
1350	C21H36	1-PHENYLPENTADECANE	318.9977	-1.7673E+04	-1.1386E+02	5.4021E-02	-8.7250E-06	489.15	800.00
1351	C21H42	1-CYCLOPENTYLHEXADECANE	-187.5470	7.5766E+02	8.0621E+01	-7.1155E-02	2.0926E-05	488.15	797.25
1352	C21H42	1-CYCLOHEXYLPENTADECANE	-89.1311	-2.4564E+03	4.2178E+01	-4.4101E-02	1.3952E-05	486.15	803.46
1353	C22H38	1-PHENYLHEXADECANE	301.1363	-1.7645E+04	-1.0606E+02	4.5880E-02	-6.1587E-06	500.15	808.00
1354	C22H44	1-CYCLOHEXYLHEXADECANE	-150.0216	-4.2008E+02	6.5603E+01	-5.8899E-02	1.7241E-05	497.15	813.42
1355	C22H44O2	n-BUTYL STEARATE	-0.7975	-5.9182E+03	8.9492E+00	-2.5746E-02	1.0847E-05	299.45	764.00
1356	C24H38O4	DIISOOCTYL PHTHALATE	24.8803	-7.5821E+03	-1.2216E+00	-1.5470E-02	6.5101E-06	260.00	851.00
1357	C24H38O4	DIOCTYL PHTHALATE	27.8473	-7.6834E+03	-2.1134E+00	-1.5234E-02	6.2365E-06	298.00	806.00
1358	C24H42O	DINONYLPHENOL	-0.7422	-7.2012E+03	8.9887E+00	-2.2697E-02	8.2181E-06	350.00	886.00
1359	C26H20	TETRAPHENYLETHYLENE	-35.5767	-4.7226E+03	2.0148E+01	-2.2596E-02	6.5760E-06	496.15	996.00
1360	C28H46O4	DIISODECYL PHTHALATE	81.7895	-7.4225E+03	-2.6916E+01	1.1502E-02	-4.3530E-14	233.00	723.00

P - vapor pressure, mm Hg

A, B, C, D, and E - regression coefficients for chemical compound

T - temperature, K

TMIN - minimum temperature, K

TMAX - maximum temperature, K

Table 7-2 VAPOR PRESSURE - INORGANIC COMPOUNDS

$$\log_{10} P = A + B/T + C \log_{10} T + D T + E T^2 \qquad (\text{P - mm Hg, T - K})$$

NO	FORMULA	NAME	A	B	C	D	E	TMIN	TMAX
1	Ag	SILVER	23.6822	-1.6026E+04	-4.5239E+00	4.0517E-04	1.0895E-16	1234.00	6410.00
2	AgCl	SILVER CHLORIDE	35.4158	-1.2320E+04	-8.7445E+00	1.7514E-03	-1.4905E-07	1185.15	1837.15
3	AgI	SILVER IODIDE	69.1111	-1.3908E+04	-1.9966E+01	4.3499E-03	-3.9755E-07	1093.15	1779.15
4	Al	ALUMINUM	9.9884	-1.3837E+04	-3.4595E-01	1.1361E-11	-1.2182E-15	933.00	2329.10
5	AlB3H12	ALUMINUM BOROHYDRIDE	65.4884	-2.8974E+03	-2.4421E+01	2.8261E-02	-1.3715E-05	220.95	319.05
6	AlBr3	ALUMINUN BROMIDE	5597.9805	-1.5335E+05	-2.2913E+03	2.1944E+00	-8.0708E-04	354.45	529.45
7	AlCl3	ALUMINUM CHLORIDE	27.8613	-6.7364E+03	-3.8080E+00	-1.3181E-05	5.2521E-09	373.15	465.75
8	AlF3	ALUMINUM FLUORIDE	4703.8748	-4.6135E+05	-1.5238E+03	3.4062E-01	-3.0079E-05	1511.15	1810.15
9	AlI3	ALUMINUM IODIDE	585.1783	-2.6198E+04	-2.1938E+02	1.3820E-01	-3.5008E-05	451.15	658.65
10	Al2O3	ALUMINUM OXIDE	14.1611	-2.8238E+04	-7.3843E-01	-3.7413E-07	2.2086E-11	2421.10	3253.10
11	Al2S3O12	ALUMINUM SULFATE	-464.7417	4.9953E+04	1.3895E+02	5.7398E-04	-1.0158E-07	845.15	1043.20
12	Ar	ARGON	14.9138	-4.5675E+02	-3.5895E+00	3.5490E-08	2.1907E-05	83.78	150.86
13	As	ARSENIC	70.7356	-9.6421E+03	-2.2451E+01	1.3810E-02	-3.8579E-06	420.00	885.00
14	AsBr3	ARSENIC TRIBROMIDE	69.8773	-4.4831E+03	-2.2441E+01	1.8936E-02	-6.1554E-06	314.95	493.15
15	AsCl3	ARSENIC TRICHLORIDE	93.9840	-4.4550E+03	-3.5080E+01	3.3505E-02	-1.3338E-05	261.75	403.55
16	AsF3	ARSENIC TRIFLUORIDE	27.7465	-2.4305E+03	-7.8250E+00	7.8140E-03	-3.3333E-06	270.65	329.45
17	AsF5	ARSENIC PENTAFLUORIDE	1944.5157	-3.0430E+04	-9.1148E+02	1.8231E+00	-1.4325E-03	155.25	220.35
18	AsH3	ARSINE	-10.3462	-7.7134E+02	9.1086E+00	-2.4654E-02	2.0673E-05	156.23	373.00
19	AsI3	ARSENIC TRIIODIDE	96.0056	-6.8120E+03	-3.3239E+01	2.0112E-02	-5.0888E-06	437.15	602.65
20	As2O3	ARSENIC TRIOXIDE	-2811.4740	7.9643E+04	1.1255E+03	-9.0385E-01	2.6735E-04	485.65	730.35
21	At	ASTATINE	-17.3579	-4.5040E+03	1.3457E+01	-2.2096E-02	9.7987E-06	279.00	607.00
22	Au	GOLD	126.1594	-2.9160E+04	-3.7975E+01	8.0922E-03	-6.6670E-07	1226.00	3120.00
23	B	BORON	38.5877	-3.6230E+04	-7.5640E+00	-1.1450E-04	5.1894E-08	1821.00	4133.00
24	BBr3	BORON TRIBROMIDE	104.0215	-4.2744E+03	-3.9717E+01	3.9970E-02	-1.6726E-05	273.15	361.05
25	BCl3	BORON TRICHLORIDE	35.3538	-2.3048E+03	-1.0108E+01	5.5834E-10	5.1455E-06	166.15	451.95
26	BF3	BORON TRIFLUORIDE	75.2534	-2.4715E+03	-2.6531E+01	8.5366E-08	4.2709E-05	144.78	260.90
27	BH2CO	BORINE CARBONYL	433.1860	-6.8163E+03	-2.0499E+02	4.5121E-01	-3.7515E-04	133.95	209.15
28	BH3O3	BORIC ACID	-81.1257	2.1815E+03	2.9986E+01	-2.2693E-05	1.0956E-08	293.15	401.15
29	B2D6	DEUTERODIBORANE	120.8038	-2.4092E+03	-5.4329E+01	1.1943E-01	-1.0820E-04	118.25	179.25
30	B2H5Br	DIBORANE HYDROBROMIDE	488.4916	-8.8892E+03	-2.2351E+02	4.1313E-01	-2.9011E-04	179.85	289.45
31	B2H6	DIBORANE	11.5597	-9.4549E+02	-1.3754E+00	-2.3575E-03	2.6782E-06	107.65	289.80
32	B3N3H6	BORINE TRIAMINE	58.7997	-2.8667E+03	-2.1466E+01	2.5002E-02	-1.2241E-05	210.15	323.75
33	B4H10	TETRABORANE	23.0730	-1.7369E+03	-6.6041E+00	8.4552E-03	-4.6056E-06	182.25	289.25
34	B5H9	PENTABORANE	-163.4159	1.4743E+03	7.5896E+01	-1.0578E-01	5.0905E-05	232.75	568.45
35	B5H11	TETRAHYDROPENTABORANE	7.6486	-1.8417E+03	2.8719E-01	-3.0007E-04	1.3472E-07	222.95	340.15
36	B10H14	DECABORANE	4813.9118	-1.2837E+05	-1.9845E+03	1.9935E+00	-7.8068E-04	333.15	436.95
37	Ba	BARIUM	-18.1369	-8.1603E+03	1.0107E+01	-5.9955E-03	9.8349E-07	638.00	1907.00
38	Be	BERYLLIUM	-4.7459	-1.6799E+04	5.4503E+00	-2.6044E-03	2.8638E-07	1097.00	2744.00
39	BeB2H8	BERYLLIUM BOROHYDRIDE	-94.5732	2.2630E+02	4.1281E+01	-2.7627E-02	8.9618E-06	274.15	363.15
40	BeBr2	BERYLLIUM BROMIDE	2669.4633	-9.8679E+04	-1.0326E+03	7.2331E-01	-1.9329E-04	562.15	747.15
41	BeCl2	BERYLLIUM CHLORIDE	-289.7317	1.0348E+04	1.0311E+02	-1.6198E-02	-9.9081E-06	564.15	760.15
42	BeF2	BERYLLIUM FLUORIDE	224.7647	-2.8762E+04	-7.0897E+01	1.7991E-02	-1.8955E-06	1145.55	1372.15
43	BeI2	BERYLLIUM IODIDE	3417.9820	-1.2770E+05	-1.3156E+03	8.8819E-01	-2.2863E-04	556.15	760.15
44	Bi	BISMUTH	1449.1712	-8.3717E+04	-5.1359E+02	2.0492E-01	-2.9858E-05	569.00	1700.00
45	BiBr3	BISMUTH TRIBROMIDE	216.7751	-1.3734E+04	-7.6683E+01	3.9665E-02	-8.4421E-06	534.15	734.15
46	BiCl3	BISMUTH TRICHLORIDE	-6.1334	-3.1936E+03	5.2620E+00	-2.4866E-03	5.0977E-07	503.65	710.55
47	BrF5	BROMINE PENTAFLUORIDE	51.1403	-2.6286E+03	-1.8290E+01	2.1840E-02	-1.0991E-05	203.85	333.55
48	Br2	BROMINE	23.7200	-2.2840E+03	-5.6145E+00	2.2602E-09	1.7888E-06	265.85	584.15
49	C	CARBON	-2.0686	-2.5987E+04	3.5504E+00	-5.3440E-04	2.9310E-08	3259.10	4399.10
50	CCl2O	PHOSGENE	46.6551	-2.4657E+03	-1.5351E+01	9.2288E-03	-4.9658E-14	145.37	455.00
51	CF2O	CARBONYL FLUORIDE	-70.1866	-3.8315E+02	4.1225E+01	-1.2049E-01	1.1291E-04	161.89	297.00
52	CH4N2O	UREA	6.7305	-4.4855E+03	1.5588E+00	-1.7855E-03	8.4006E-07	340.65	368.05
53	CH4N2S	THIOUREA	-35.4224	-2.3772E+03	1.9025E+01	-2.0819E-02	6.8609E-06	454.15	854.00
54	CNBr	CYANOGEN BROMIDE	2302.7452	-4.9937E+04	-9.9806E+02	1.3391E+00	-7.0662E-04	273.01	313.09
55	CNCl	CYANOGEN CHLORIDE	-9.0018	-1.2842E+03	8.2222E+00	-1.6105E-02	9.5986E-06	266.65	449.00
56	CNF	CYANOGEN FLUORIDE	179.8136	-4.3677E+03	-7.6561E+01	1.1639E-01	-7.2513E-05	196.75	226.35
57	CO	CARBON MONOXIDE	51.8145	-7.8824E+02	-2.2734E+01	5.1225E-02	4.6603E-11	68.15	132.92
58	COS	CARBONYL SULFIDE	36.8556	-1.7187E+03	-1.2036E+01	8.9612E-03	2.0283E-13	134.30	378.80
59	COSe	CARBON OXYSELENIDE	525.5520	-9.1072E+03	-2.4172E+02	4.5341E-01	-3.1974E-04	156.05	251.25
60	CO2	CARBON DIOXIDE	35.0187	-1.5119E+03	-1.1335E+01	9.3383E-03	7.7626E-10	216.58	304.19
61	CS2	CARBON DISULFIDE	25.1475	-2.0439E+03	-6.7794E+00	3.4828E-03	3.4373E-15	161.11	552.00
62	CSeS	CARBON SELENOSULFIDE	-101.6898	-4.5082E+02	5.2520E+01	-1.0400E-01	6.9601E-05	225.85	358.75
63	C2N2	CYANOGEN	4016.4115	-5.9887E+04	-1.9008E+03	3.9066E+00	-3.0816E-03	177.35	252.15
64	C3S2	CARBON SUBSULFIDE	-243.5223	3.1498E+03	1.0795E+02	-1.3131E-01	5.9581E-05	287.15	403.95
65	Ca	CALCIUM	-75.9862	-4.9573E+03	3.0625E+01	-1.3574E-02	1.9938E-06	625.00	1762.00
66	CaF2	CALCIUM FLUORIDE	49.0700	-2.7912E+04	-1.0510E+01	1.4288E-08	-1.0745E-12	1691.00	2806.50
67	CbF5	COLUMBIUM FLUORIDE	-42.2353	-1.1915E+03	2.0862E+01	-2.2002E-02	8.8506E-06	359.45	498.15
68	Cd	CADMIUM	-145.3985	-8.3891E+02	6.0509E+01	-4.3639E-02	1.1003E-05	393.00	1043.00
69	CdCl2	CADMIUM CHLORIDE	-4189.3800	2.2619E+05	1.4896E+03	-5.9464E-01	9.0541E-05	891.15	1240.15
70	CdF2	CADMIUM FLUORIDE	-1738.0050	1.5268E+05	5.6319E+02	-1.1301E-01	7.8517E-06	1385.15	2024.15
71	CdI2	CADMIUM IODIDE	-899.5410	3.5567E+04	3.3430E+02	-1.6865E-01	3.2257E-05	689.15	1069.15
72	CdO	CADMIUM OXIDE	42.8498	-1.5443E+04	-1.0651E+01	2.0649E-03	-1.7038E-07	1273.15	1832.15
73	ClF	CHLORINE MONOFLUORIDE	44.9144	-1.6526E+03	-1.6727E+01	3.3592E-02	-2.8601E-05	129.75	172.15
74	ClFO3	PERCHLORYL FLUORIDE	40.5028	-1.8544E+03	-1.3458E+01	9.9792E-03	3.8242E-13	125.41	368.40
75	ClF3	CHLORINE TRIFLUORIDE	25.4578	-1.6807E+03	-7.7581E+00	9.8092E-03	-5.2739E-06	192.75	284.65
76	ClF5	CHLORINE PENTAFLUORIDE	131.4931	-3.7893E+03	-5.3872E+01	7.2145E-02	-3.9712E-05	193.95	297.95
77	ClHO3S	CHLOROSULFONIC ACID	-5.6040	-2.7604E+03	8.4466E+00	-2.2029E-02	1.1699E-05	193.15	700.00
78	ClHO4	PERCHLORIC ACID	7.5448	-2.0352E+03	1.0732E+00	-7.2499E-03	4.2925E-06	171.95	631.00

180

Table 7-2 VAPOR PRESSURE - INORGANIC COMPOUNDS (continued)

$$\log_{10} P = A + B/T + C \log_{10} T + D T + E T^2 \qquad (P - mm\ Hg,\ T - K)$$

NO	FORMULA	NAME	A	B	C	D	E	TMIN	TMAX
79	ClO2	CHLORINE DIOXIDE	121.9791	-5.2893E+03	-4.1838E+01	-2.0950E-08	2.6848E-05	213.55	465.00
80	Cl2	CHLORINE	28.8659	-1.6745E+03	-8.5216E+00	5.3792E-03	-7.7867E-13	172.12	417.15
81	Cl2O	CHLORINE MONOXIDE	20.7506	-1.6377E+03	-5.5984E+00	7.4816E-03	-4.2600E-06	174.65	275.35
82	Cl2O7	CHLORINE HEPTOXIDE	5.5051	-1.7898E+03	1.0930E+00	-1.1071E-03	4.8227E-07	227.85	351.95
83	Co	COBALT	16.7750	-1.8953E+04	-1.7830E+00	2.8243E-05	-6.2846E-08	1095.00	2528.00
84	CoCl2	COBALT CHLORIDE	90.9544	-1.2963E+04	-2.7733E+01	7.3497E-03	-8.1551E-07	1043.15	1323.15
85	CoNC3O4	COBALT NITROSYL TRICARBONYL	-2295.1360	4.3067E+04	1.0148E+03	-1.4317E+00	7.7031E-04	271.85	353.15
86	Cr	CHROMIUM	-80.3456	-1.2221E+04	2.9746E+01	-6.8400E-03	5.2454E-07	1229.00	2840.00
87	CrC6O6	CHROMIUM CARBONYL	-3135.3220	6.6813E+04	1.3443E+03	-1.6151E+00	7.4196E-04	309.15	424.15
88	CrO2Cl2	CHROMIUM OXYCHLORIDE	226.3675	-7.6454E+03	-9.0472E+01	9.3809E-02	-3.9753E-05	254.75	390.25
89	Cs	CESIUM	-2.9708	-3.6286E+03	4.1641E+00	-3.6632E-03	7.9575E-07	295.00	959.00
90	CsBr	CESIUM BROMIDE	1537.6463	-1.0591E+05	-5.2949E+02	1.7992E-01	-2.3362E-05	1021.15	1573.15
91	CsCl	CESIUM CHLORIDE	438.8475	-3.4903E+04	-1.5008E+02	5.3499E-02	-7.3394E-06	1017.15	1573.15
92	CsF	CESIUM FLUORIDE	1768.1436	-1.1822E+05	-6.1137E+02	2.1264E-01	-2.8322E-05	985.15	1524.15
93	CsI	CESIUM IODIDE	276.4031	-2.6150E+04	-9.2366E+01	3.0791E-02	-4.0411E-06	1011.15	1553.15
94	Cu	COPPER	-82.6254	-1.1231E+04	3.1336E+01	-8.9074E-03	7.5663E-07	1130.00	3150.00
95	CuBr	CUPROUS BROMIDE	208.6672	-2.0301E+04	-6.8060E+01	1.9026E-02	-2.1495E-06	845.15	1628.15
96	CuCl	CUPROUS CHLORIDE	27.2556	-7.5654E+03	-6.1858E+00	1.2376E-10	-1.7508E-14	703.00	1763.10
97	CuCl2	CUPRIC CHLORIDE	-238.6809	1.7213E+04	7.5807E+01	2.5333E-05	-6.1652E-09	582.85	794.15
98	CuI	COPPER IODIDE	855.8840	-5.7746E+04	-2.9506E+02	1.0189E-01	-1.3492E-05	883.15	1609.15
99	DCN	DEUTERIUM CYANIDE	798.8279	-1.7354E+04	-3.4780E+02	4.9610E-01	-2.8192E-04	204.25	299.35
100	D2	DEUTERIUM	6.1037	-6.7085E+01	-5.7226E-01	1.6894E-02	-1.7612E-11	18.73	38.35
101	D2O	DEUTERIUM OXIDE	-12.8257	-2.1886E+03	1.0645E+01	-1.9029E-02	9.1375E-06	276.97	643.89
102	Eu	EUROPIUM	-55.5456	-6.4880E+03	2.3529E+01	-1.0970E-02	1.6487E-06	640.00	1742.00
103	F2	FLUORINE	27.1409	-5.7201E+02	-1.0015E+01	2.1078E-02	8.9567E-13	53.48	144.31
104	F2O	FLUORINE OXIDE	-24.8186	-1.9007E+02	1.5995E+01	-4.1421E-02	4.8042E-05	77.05	128.55
105	Fe	IRON	11.5549	-1.9538E+04	-6.2549E-01	-2.7182E-09	1.9086E-13	1808.15	3008.20
106	FeC5O5	IRON PENTACARBONYL	317.6618	-9.3157E+03	-1.2965E+02	1.3467E-01	-4.8175E-05	266.65	378.15
107	FeCl2	FERROUS CHLORIDE	-1350.9350	6.2772E+04	4.8764E+02	-2.0598E-01	3.2434E-05	973.15	1299.15
108	FeCl3	FERRIC CHLORIDE	64969.7170	-1.9169E+06	-2.6114E+04	2.2347E+01	-7.3076E-03	467.15	592.15
109	Fr	FRANCIUM	-38.3826	-2.2043E+03	1.8322E+01	-1.5441E-02	4.4042E-06	267.00	879.00
110	Ga	GALLIUM	-4.4968	-1.2924E+04	4.3403E+00	-9.5226E-04	2.3443E-08	954.00	2517.00
111	GaCl3	GALLIUM TRICHLORIDE	54.9805	-4.8102E+03	-1.6009E+01	-8.4664E-09	3.9331E-06	350.90	694.00
112	Gd	GADOLINIUM	9.6612	-1.0909E+04	-4.7131E-01	1.2082E-03	-3.9062E-07	728.00	1770.00
113	Ge	GERMANIUM	24.3265	-1.9756E+04	-4.9583E+00	9.5753E-04	-8.0668E-08	1230.00	3125.00
114	GeBr4	GERMANIUM BROMIDE	151.2941	-6.5181E+03	-5.7438E+01	4.8303E-02	-1.6731E-05	316.45	462.15
115	GeCl4	GERMANIUM CHLORIDE	207.3749	-6.4745E+03	-8.4232E+01	9.6332E-02	-4.5015E-05	228.15	357.15
116	GeHCl3	TRICHLORO GERMANE	175.3633	-5.9201E+03	-7.0301E+01	7.9543E-02	-3.7043E-05	231.85	348.15
117	GeH4	GERMANE	32.2743	-1.3704E+03	-9.9425E+00	5.1455E-10	1.6203E-05	107.26	308.00
118	Ge2H6	DIGERMANE	45.6425	-2.1567E+03	-1.6528E+01	2.1047E-02	-1.1265E-05	184.45	304.65
119	Ge3H8	TRIGERMANE	36.7997	-2.5324E+03	-1.2064E+01	1.1954E-02	-5.0050E-06	236.25	383.95
120	HBr	HYDROGEN BROMIDE	34.4939	-1.6379E+03	-1.0909E+01	7.5732E-03	-2.5521E-12	185.15	363.15
121	HCN	HYDROGEN CYANIDE	-57.0540	-3.6256E+02	2.9415E+01	-4.7528E-02	2.8406E-05	259.83	456.65
122	HCl	HYDROGEN CHLORIDE	43.5455	-1.6279E+03	-1.5214E+01	1.3783E-02	-1.4984E-11	158.97	324.65
123	HF	HYDROGEN FLUORIDE	23.7347	-1.7996E+03	-6.1764E+00	-5.0046E-10	6.1500E-06	189.79	461.15
124	HI	HYDROGEN IODIDE	21.4282	-1.4515E+03	-5.5756E+00	3.3878E-03	-8.9335E-12	222.38	423.85
125	HNO3	NITRIC ACID	71.7653	-4.3768E+03	-2.2769E+01	-4.5988E-07	1.1856E-05	231.55	376.10
126	H2	HYDROGEN	3.4132	-4.1316E+01	1.0947E+00	-6.6896E-10	1.4589E-04	13.95	33.18
127	H2O	WATER	29.8605	-3.1522E+03	-7.3037E+00	2.4247E-09	1.8090E-06	273.16	647.13
128	H2O2	HYDROGEN PEROXIDE	33.3222	-3.7350E+03	-8.3458E+00	-1.2351E-10	1.6917E-06	272.74	730.00
129	H2S	HYDROGEN SULFIDE	18.6383	-1.3446E+03	-4.1034E+00	3.1815E-09	2.4664E-06	187.68	373.53
130	H2SO4	SULFURIC ACID	2.0582	-4.1924E+03	3.2578E+00	-1.1224E-03	5.5371E-07	283.15	603.15
131	H2S2	HYDROGEN DISULFIDE	22.3747	-2.4043E+03	-5.5539E+00	5.8402E-03	-2.6241E-06	229.95	337.15
132	H2Se	HYDROGEN SELENIDE	573.9933	-1.0568E+04	-2.6071E+02	4.7258E-01	-3.4316E-04	157.85	232.05
133	H2Te	HYDROGEN TELLURIDE	532.0079	-1.0876E+04	-2.3540E+02	3.7306E-01	-2.3649E-04	176.75	271.15
134	H3NO3S	SULFAMIC ACID	0.1130	-6.8449E+02	5.2613E-05	-6.8934E-08	3.4587E-11	293.15	373.15
135	He	HELIUM-3	2.1239	-1.1769E+00	2.0493E+00	-1.7652E-10	9.4980E-03	1.01	3.31
136	He	HELIUM-4	2.8838	-3.9043E+00	6.7240E-01	1.1913E-01	-2.1010E-11	1.76	5.20
137	Hf	HAFNIUM	-187.3128	-4.7256E+03	5.9397E+01	-7.1673E-03	2.6648E-07	2117.00	5960.00
138	Hg	MERCURY	11.3169	-3.3515E+03	-1.1296E+00	-2.9698E-13	1.1699E-07	234.31	1735.00
139	HgBr2	MERCURIC BROMIDE	-6001.9470	1.6586E+05	2.4352E+03	-2.1646E+00	7.2767E-04	409.65	592.15
140	HgCl2	MERCURIC CHLORIDE	8091.0938	-2.3113E+05	-3.2811E+03	2.9716E+00	-1.0294E-03	409.35	577.15
141	HgI2	MERCURIC IODIDE	-9817.2740	2.9088E+05	3.9308E+03	-3.2560E+00	1.0229E-03	430.65	627.15
142	IF7	IODINE HEPTAFLUORIDE	2.2894	-1.4949E+03	2.7875E+00	-3.5044E-03	1.9005E-06	186.15	277.15
143	I2	IODINE	46.9335	-5.1763E+03	-1.1170E+01	-1.0130E-02	7.7500E-06	242.00	819.15
144	In	INDIUM	112.3721	-1.8894E+04	-3.5777E+01	1.1333E-02	-1.3444E-06	850.00	2323.00
145	Ir	IRIDIUM	-197.4829	-6.5755E+03	6.4387E+01	-9.6603E-03	5.0122E-07	1944.00	4450.00
146	K	POTASSIUM	10.8410	-4.6934E+03	-1.1916E+00	1.5875E-04	1.7454E-17	336.35	2223.00
147	KBr	POTASSIUM BROMIDE	86.7862	-1.5677E+04	-2.5680E+01	5.9097E-03	-5.6987E-07	1068.15	1656.15
148	KCl	POTASSIUM CHLORIDE	-8.0224	-9.3722E+03	6.4641E+00	-3.1639E-03	3.2745E-07	1044.00	3470.00
149	KF	POTASSIUM FLUORIDE	86.1553	-1.7030E+04	-2.5138E+01	5.3564E-03	-4.7879E-07	1158.15	1775.15
150	KI	POTASSIUM IODIDE	88.6708	-1.5026E+04	-2.6513E+01	6.3780E-03	-6.4206E-07	1018.15	1597.15
151	KOH	POTASSIUM HYDROXIDE	20.9787	-9.5262E+03	-3.8001E+00	2.9030E-11	1.2312E-08	679.00	1600.00
152	Kr	KRYPTON	-12.6883	-3.1111E+02	1.0610E+01	-3.8518E-02	5.0870E-05	115.78	209.35
153	La	LANTHANUM	190.2107	-4.1131E+04	-5.7186E+01	9.9276E-03	-6.4353E-07	1441.50	3643.00
154	Li	LITHIUM	12.1182	-8.4301E+03	-1.3510E+00	2.2909E-04	-6.2641E-18	453.69	4085.00
155	LiBr	LITHIUM BROMIDE	65.3934	-1.3245E+04	-1.8782E+01	4.4647E-03	-4.4694E-07	1021.15	1583.15
156	LiCl	LITHIUM CHLORIDE	1659.2051	-1.1514E+05	-5.7029E+02	1.9064E-01	-2.4324E-05	1056.15	1655.15

Table 7-2 VAPOR PRESSURE - INORGANIC COMPOUNDS (continued)

NO	FORMULA	NAME	$\log_{10} P = A + B/T + C \log_{10} T + D T + E T^2$ (P - mm Hg, T - K)						
			A	B	C	D	E	TMIN	TMAX
157	LiF	LITHIUM FLUORIDE	69.3367	-1.8147E+04	-1.9178E+01	3.6011E-03	-2.8544E-07	1320.15	1954.15
158	LiI	LITHIUM IODIDE	72.2997	-1.4450E+04	-2.0834E+01	5.2464E-03	-5.5773E-07	996.15	1444.15
159	Lu	LUTETIUM	-47.6746	-1.1929E+04	1.8989E+01	-4.5961E-03	3.5367E-07	1057.00	2535.00
160	Mg	MAGNESIUM	-72.6513	-4.2014E+03	3.0264E+01	-1.5558E-02	2.6429E-06	517.00	1376.00
161	MgCl2	MAGNESIUM CHLORIDE	1239.2680	-9.0236E+04	-4.2304E+02	1.3654E-01	-1.6884E-05	1051.15	1691.15
162	MgO	MAGNESIUM OXIDE	-41.2727	-1.4025E+04	1.5392E+01	-2.3614E-03	1.1265E-07	3105.00	5950.00
163	Mn	MANGANESE	-123.9176	-5.9845E+03	4.6074E+01	-1.4659E-02	1.5195E-06	924.00	2392.00
164	MnCl2	MANGANESE CHLORIDE	1275.0362	-8.8381E+04	-4.3842E+02	1.4858E-01	-1.9347E-05	1009.15	1463.15
165	Mo	MOLYBDENUM	74.9735	-4.1955E+04	-2.0072E+01	3.2166E-03	-2.2507E-07	1677.00	5100.00
166	MoF6	MOLYBDENUM FLUORIDE	300.7229	-7.9655E+03	-1.2644E+02	1.6582E-01	-8.8423E-05	207.65	309.15
167	MoO3	MOLYBDENUM OXIDE	3015.8551	-2.1859E+05	-1.0259E+03	3.1920E-01	-3.8938E-05	1007.15	1424.15
168	NCl3	NITROGEN TRICHLORIDE	88.8600	-4.2008E+03	-3.1565E+01	1.8305E-02	-6.0409E-11	246.15	367.15
169	ND3	HEAVY AMMONIA	125.5395	-3.5614E+03	-5.1766E+01	7.4954E-02	-4.4958E-05	199.15	239.75
170	NF3	NITROGEN TRIFLUORIDE	8.4514	-7.4300E+02	3.9105E-01	-1.1295E-02	1.7758E-05	66.36	233.85
171	NH3	AMMONIA	37.1575	-2.0277E+03	-1.1601E+01	7.4625E-03	-9.5811E-12	195.41	405.65
172	NH3O	HYDROXYLAMINE	2.4590	-4.0142E+03	6.3297E+00	-1.8547E-02	1.1266E-05	306.25	383.00
173	NH4Br	AMMONIUM BROMIDE	23.7261	-5.2014E+03	-5.1721E+00	2.6919E-03	-5.9955E-07	471.45	669.15
174	NH4Cl	AMMONIUM CHLORIDE	-893.5430	1.9484E+04	3.6716E+02	-3.2056E-01	1.0034E-04	553.15	793.15
175	NH4I	AMMONIUM IODIDE	0.4097	-4.4032E+03	3.5216E+00	-1.7562E-03	3.7842E-07	484.05	678.05
176	NH5O	AMMONIUM HYDROXIDE	42.3381	-2.3577E+03	-1.3597E+01	8.5486E-03	4.7298E-11	203.15	353.15
177	NH5S	AMMONIUM HYDROGENSULFIDE	178.9587	-6.0657E+03	-7.2004E+01	8.7809E-02	-4.4365E-05	222.05	306.45
178	NO	NITRIC OXIDE	61.2046	-1.5365E+03	-2.3621E+01	2.9377E-02	-1.3066E-09	109.50	180.15
179	NOCl	NITROSYL CHLORIDE	24.1469	-1.8469E+03	-6.0513E+00	9.2881E-09	4.5319E-06	213.55	440.65
180	NOF	NITROSYL FLUORIDE	478.8331	-7.7441E+03	-2.2562E+02	4.9213E-01	-4.2279E-04	141.15	217.15
181	NO2	NITROGEN DIOXIDE	32.1203	-2.2563E+03	-9.7702E+00	8.6560E-03	-5.1036E-11	261.95	431.35
182	N2	NITROGEN	23.8572	-4.7668E+02	-8.6689E+00	2.0128E-02	-2.4139E-11	63.15	126.10
183	N2F4	TETRAFLUOROHYDRAZINE	16.4473	-9.6387E+02	-4.3123E+00	5.9876E-03	2.2680E-13	111.65	309.35
184	N2H4	HYDRAZINE	31.2541	-3.1466E+03	-8.2200E+00	2.6734E-03	4.6004E-13	274.68	653.15
185	N2H4C	AMMONIUM CYANIDE	179.6608	-6.1294E+03	-7.2215E+01	8.8090E-02	-4.4545E-05	222.55	304.85
186	N2H6CO2	AMMONIUM CARBAMATE	93.9638	-4.8160E+03	-3.4502E+01	3.6858E-02	-1.6578E-05	247.05	331.45
187	N2O	NITROUS OXIDE	61.5168	-2.1016E+03	-2.2337E+01	1.8232E-02	-1.1348E-10	182.30	309.57
188	N2O3	NITROGEN TRIOXIDE	-2.5932	-1.8582E+03	7.4627E+00	-2.7402E-02	2.0570E-05	170.00	425.00
189	N2O4	NITROGEN TETRAOXIDE	-197.7926	7.7599E+02	9.9702E+01	-2.0780E-01	1.4907E-04	261.90	320.65
190	N2O5	NITROGEN PENTOXIDE	-270.4814	4.6851E+02	1.3740E+02	-2.9833E-01	2.3064E-04	236.35	305.55
191	Na	SODIUM	7.2828	-5.4249E+03	1.5358E-01	-2.0200E-04	3.7007E-08	370.97	2573.00
192	NaBr	SODIUM BROMIDE	18.4164	-1.0501E+04	-2.8637E+00	8.9249E-10	-1.1019E-13	1020.00	1720.00
193	NaCN	SODIUM CYANIDE	-2.2303	-8.2019E+03	3.8990E+00	-2.3458E-03	3.9325E-07	836.85	1769.10
194	NaCl	SODIUM CHLORIDE	22.4317	-1.1358E+04	-4.2035E+00	3.4674E-04	-3.9472E-12	1073.90	1738.10
195	NaF	SODIUM FLUORIDE	11.6744	-1.1663E+04	-9.6084E-01	1.2957E-04	-1.1232E-14	1269.00	2060.00
196	NaI	SODIUM IODIDE	69.6407	-1.4137E+04	-2.0041E+01	4.7338E-03	-4.7106E-07	1040.15	1577.15
197	NaOH	SODIUM HYDROXIDE	-48.2774	-1.9340E+03	1.7000E+01	2.9640E-11	-8.7510E-07	596.00	1830.00
198	Na2SO4	SODIUM SULFATE	-2.2687	-1.5051E+04	3.5005E+00	-1.2712E-03	1.7716E-07	1173.10	1223.10
199	Nb	NIOBIUM	-64.3485	-2.9438E+04	2.3622E+01	-3.9155E-03	2.0660E-07	2250.00	5115.00
200	Nd	NEODYMIUM	130.3312	-2.6812E+04	-3.9789E+01	8.4714E-03	-6.7800E-07	1144.00	3384.00
201	Ne	NEON	0.0175	-5.9863E+01	3.6341E+00	-8.8025E-09	-1.8115E-04	24.56	44.40
202	Ni	NICKEL	-57.4301	-1.3533E+04	2.3611E+01	-7.6670E-03	7.8143E-07	1061.00	2415.00
203	NiC4O4	NICKEL CARBONYL	24.7530	-1.9573E+03	-7.0802E+00	7.4594E-03	-3.3536E-06	250.15	315.65
204	NiF2	NICKEL FLUORIDE	634.6103	-7.2704E+04	-2.0233E+02	4.4627E-02	-3.9862E-06	1350.15	1556.15
205	Np	NEPTUNIUM	88.3172	-3.1522E+04	-2.5083E+01	4.1165E-03	-2.8679E-07	1617.99	2073.99
206	O2	OXYGEN	20.6695	-5.2697E+02	-6.7062E+00	1.2926E-02	-9.8832E-13	54.35	154.58
207	O3	OZONE	38.6910	-1.4144E+03	-1.2543E+01	-1.3045E-10	2.4393E-05	80.15	261.00
208	Os	OSMIUM	-252.0332	-6.3868E+03	8.1758E+01	-1.2735E-02	7.0592E-07	2234.00	4880.00
209	OsOF5	OSMIUM OXIDE PENTAFLUORIDE	18125.3280	-4.6581E+05	-7.4493E+03	6.8810E+00	-2.0309E-03	304.80	330.90
210	OsO4	OSMIUM TETROXIDE - YELLOW	-6976.1920	1.3882E+05	3.0478E+03	-4.0659E+00	2.0494E-03	276.35	403.15
211	OsO4	OSMIUM TETROXIDE - WHITE	-1362.6820	2.3640E+04	6.0656E+02	-8.6236E-01	4.5669E-04	267.55	403.15
212	P	PHOSPHORUS - WHITE	37.5747	-4.5200E+03	-9.8304E+00	-1.4385E-06	1.4357E-06	404.15	590.15
213	PBr3	PHOSPHORUS TRIBROMIDE	75.1763	-4.1298E+03	-2.7120E+01	2.3495E-02	-8.5043E-06	280.95	448.45
214	PCl2F3	PHOSPHORUS DICHLORIDE TRIFLUORIDE	41.7838	-2.0432E+03	-1.4841E+01	2.0345E-02	-1.1817E-05	193.35	250.35
215	PCl3	PHOSPHORUS TRICHLORIDE	56.7046	-3.2295E+03	-1.8915E+01	1.0097E-02	-7.5546E-13	181.15	374.15
216	PCl5	PHOSPHORUS PENTACHLORIDE	30.0396	-6.2579E+03	-7.4964E-01	-2.9515E-02	1.0957E-05	433.15	465.00
217	PH3	PHOSPHINE	17.6034	-1.0512E+03	-4.0706E+00	-1.5186E-09	5.1920E-06	139.37	324.75
218	PH4Br	PHOSPHONIUM BROMIDE	79.2715	-4.0989E+03	-2.8848E+01	3.2807E-02	-1.5741E-05	229.45	311.45
219	PH4Cl	PHOSPHONIUM CHLORIDE	255.5276	-6.5388E+03	-1.0929E+02	1.6761E-01	-1.0572E-04	182.15	246.15
220	PH4I	PHOSPHONIUM IODIDE	90.3833	-4.6958E+03	-3.3063E+01	3.5026E-02	-1.5625E-05	247.95	335.45
221	POCl3	PHOSPHORUS OXYCHLORIDE	89.5904	-4.4038E+03	-3.1847E+01	1.8570E-02	1.2780E-08	274.33	378.65
222	PSBr3	PHOSPHORUS THIOBROMIDE	100.8066	-6.1441E+03	-3.5921E+01	2.9018E-02	-9.8408E-06	323.15	448.15
223	PSCl3	PHOSPHORUS THIOCHLORIDE	30.9267	-2.9337E+03	-8.3323E+00	2.4967E-03	-3.2735E-11	236.95	398.15
224	P4O6	PHOSPHORUS TRIOXIDE	-9.1727	-1.7181E+03	6.7269E+00	-5.0441E-03	1.6463E-06	312.85	446.25
225	P4O10	PHOSPHORUS PENTOXIDE	-55.9316	-2.8529E+03	2.7900E+01	-2.9138E-02	9.4669E-06	693.15	758.15
226	P4S10	PHOSPHORUS PENTASULFIDE	18.3195	-5.1772E+03	-3.3346E+00	1.0110E-03	-6.7993E-14	561.15	1291.00
227	Pb	LEAD	-17.6204	-8.5777E+03	9.2106E+00	-3.9318E-03	5.4789E-07	708.00	2024.00
228	PbBr2	LEAD BROMIDE	-392.9133	9.8716E+03	1.4928E+02	-7.7340E-02	1.4423E-05	786.15	1187.15
229	PbCl2	LEAD CHLORIDE	178.2156	-1.8711E+04	-5.7943E+01	1.8346E-02	-2.4020E-06	820.15	1227.15
230	PbF2	LEAD FLUORIDE	797.7308	-6.5142E+04	-2.6816E+02	8.0837E-02	-9.4270E-06	1134.15	1566.15
231	PbI2	LEAD IODIDE	1790.0196	-9.5547E+04	-6.4509E+02	2.9522E-01	-5.2262E-05	752.15	1145.15
232	PbO	LEAD OXIDE	-357.2600	1.4278E+04	1.2467E+02	-3.7266E-02	4.2118E-06	1216.15	1745.15
233	PbS	LEAD SULFIDE	-2243.0170	1.5079E+05	7.6189E+02	-2.1739E-01	2.2835E-05	1125.15	1554.15
234	Pd	PALLADIUM	90.8138	-2.9630E+04	-2.5363E+01	3.7450E-03	-2.0394E-07	1336.00	3385.00

Table 7-2 VAPOR PRESSURE - INORGANIC COMPOUNDS (continued)

$$\log_{10} P = A + B/T + C \log_{10} T + D\,T + E\,T^2 \qquad (P - \text{mm Hg},\ T - \text{K})$$

NO	FORMULA	NAME	A	B	C	D	E	TMIN	TMAX
235	Po	POLONIUM	220.5827	-1.5775E+04	-7.8208E+01	3.8994E-02	-7.4047E-06	448.00	1235.00
236	Pt	PLATINUM	87.6383	-4.0548E+04	-2.2701E+01	1.8600E-03	-1.5887E-08	1744.00	3980.00
237	Ra	RADIUM	31.4507	-1.0221E+04	-7.3958E+00	4.1178E-04	1.3090E-07	593.00	1809.00
238	Rb	RUBIDIUM	13.7111	-4.4634E+03	-2.1307E+00	-7.2787E-05	1.8422E-07	310.00	978.00
239	RbBr	RUBIDIUM BROMIDE	78.4584	-1.4856E+04	-2.2965E+01	5.3426E-03	-5.2198E-07	1054.15	1625.15
240	RbCl	RUBIDIUM CHLORIDE	95.4464	-1.6358E+04	-2.8542E+01	6.6137E-03	-6.4090E-07	1065.15	1654.15
241	RbF	RUBIDIUM FLUORIDE	4858.3696	-3.9369E+05	-1.6193E+03	4.3432E-01	-4.5433E-05	1194.15	1681.15
242	RbI	RUBIDIUM IODIDE	87.3359	-1.5106E+04	-2.6017E+01	6.2775E-03	-6.3467E-07	1021.15	1577.15
243	Re	RHENIUM	-31.5392	-3.2254E+04	1.2215E+01	-1.2695E-03	3.7363E-08	2480.00	5915.00
244	Re2O7	RHENIUM HEPTOXIDE	-50919.0	1.6274E+06	2.0103E+04	-1.5444E+01	4.5207E-03	485.65	635.55
245	Rh	RHODIUM	-83.3270	-2.3619E+04	3.2126E+01	-8.8651E-03	7.4827E-07	1735.00	3940.00
246	Rn	RADON	168.4046	-3.3600E+03	-7.5618E+01	1.4923E-01	-1.2016E-04	128.95	211.75
247	Ru	RUTHENIUM	3.1324	-3.5567E+04	3.0770E+00	-1.1098E-03	6.9449E-08	2051.00	4500.00
248	RuF5	RUTHENIUM PENTAFLUORIDE	253.0895	-1.0396E+04	-9.8031E+01	8.5375E-02	-3.0601E-05	322.75	429.95
249	S	SULFUR	86.7925	-7.8894E+03	-2.7433E+01	7.5706E-03	5.7656E-15	388.36	1313.00
250	SF4	SULFUR TETRAFLUORIDE	340.1740	-6.8405E+03	-1.5222E+02	2.6997E-01	-1.9417E-04	160.85	223.95
251	SF6	SULFUR HEXAFLUORIDE	10.5389	-1.0352E+03	-1.1341E+00	-1.8565E-07	1.1504E-10	223.15	318.69
252	SOBr2	THIONYL BROMIDE	1.5135	-1.9716E+03	2.6530E+00	-2.2814E-03	8.4571E-07	266.45	412.65
253	SOCl2	THIONYL CHLORIDE	66.4546	-3.3385E+03	-2.3318E+01	1.5153E-02	-8.4247E-12	172.00	372.15
254	SOF2	SULFUROUS OXYFLUORIDE	238.2631	-5.2181E+03	-1.0431E+02	1.7358E-01	-1.1781E-04	174.42	229.05
255	SO2	SULFUR DIOXIDE	19.7418	-1.8132E+03	-4.1458E+00	-4.4284E-09	8.4918E-07	197.67	430.75
256	SO2Cl2	SULFURYL CHLORIDE	5.9028	-2.0407E+03	2.2796E+00	-1.0358E-02	6.0161E-06	222.00	545.00
257	SO3	SULFUR TRIOXIDE	114.0529	-6.4619E+03	-3.6784E+01	-1.7530E-07	1.1919E-06	289.95	490.85
258	S2Cl2	SULFUR MONOCHLORIDE	162.6274	-6.2944E+03	-6.3327E+01	6.1401E-02	-2.4506E-05	265.75	411.15
259	Sb	ANTIMONY	-14.9322	-1.3754E+04	1.5738E+01	-2.3835E-02	5.1947E-06	617.00	1898.00
260	SbBr3	ANTIMONY TRIBROMIDE	-68.4797	-6.2413E+02	3.0034E+01	-2.0378E-02	4.6884E-06	367.05	548.15
261	SbCl3	ANTIMONY TRICHLORIDE	68.0484	-5.1047E+03	-2.1724E+01	7.4742E-03	5.7270E-13	346.55	794.00
262	SbCl5	ANTIMONY PENTACHLORIDE	73.6650	-4.3530E+03	-2.6389E+01	2.3832E-02	-9.0652E-06	295.85	387.25
263	SbH3	STIBINE	65.7241	-2.2895E+03	-2.5586E+01	3.4766E-02	-1.7809E-05	177.87	440.35
264	SbI3	ANTIMONY TRIIODIDE	-1834.2310	5.3081E+04	7.3328E+02	-5.9001E-01	1.8014E-04	436.75	674.15
265	Sb2O3	ANTIMONY TRIOXIDE	3557.6905	-2.2934E+05	-1.2298E+03	4.1528E-01	-5.3009E-05	847.15	1698.15
266	Sc	SCANDIUM	-410.3222	1.5184E+04	1.4020E+02	-3.6544E-02	3.4523E-06	1110.00	2700.00
267	Se	SELENIUM	994.5705	-4.3994E+04	-3.7357E+02	2.2452E-01	-5.1145E-05	397.00	930.00
268	SeCl4	SELENIUM TETRACHLORIDE	62.7934	-5.6601E+03	-2.0127E+01	1.5021E-02	-4.7573E-06	347.15	464.65
269	SeF6	SELENIUM HEXAFLUORIDE	-13.7376	-9.6701E+02	1.0083E+01	-1.4935E-02	9.6551E-06	154.55	227.35
270	SeOCl2	SELENIUM OXYCHLORIDE	-31.1449	-1.5960E+03	1.5901E+01	-1.1690E-02	3.7904E-06	307.95	441.15
271	SeO2	SELENIUM DIOXIDE	-26.7091	-2.9884E+03	1.3896E+01	-7.6193E-03	1.8305E-06	430.15	590.15
272	Si	SILICON	315.0687	-7.1384E+04	-8.9680E+01	8.3445E-03	-2.5806E-09	1997.10	2560.00
273	SiBrCl2F	BROMODICHLOROFLUOROSILANE	221.2325	-5.5951E+03	-9.3501E+01	1.2431E-01	-6.0950E-05	186.65	474.30
274	SiBrF3	TRIFLUOROBROMOSILANE	-316.1829	5.0196E+03	1.4386E+02	-2.1246E-01	1.1958E-04	203.35	354.90
275	SiBr2ClF	DIBROMOCHLOROFLUOROSILANE	114.7880	-3.9867E+03	-4.5478E+01	5.1723E-02	-2.2118E-05	207.95	515.92
276	SiClF3	TRIFLUOROCHLOROSILANE	102.6712	-2.5416E+03	-4.3347E+01	7.1921E-02	-4.5231E-05	129.15	308.83
277	SiCl2F2	DICHLORODIFLUOROSILANE	65.5754	-2.0450E+03	-2.6643E+01	4.5134E-02	-2.8299E-05	148.45	367.35
278	SiCl3F	TRICHLOROFLUOROSILANE	106.0275	-3.4770E+03	-4.2166E+01	5.0626E-02	-2.3053E-05	180.55	434.85
279	SiCl4	SILICON TETRACHLORIDE	28.4503	-2.3911E+03	-7.3965E+00	-9.3193E-10	2.7569E-06	204.30	507.00
280	SiF4	SILICON TETRAFLUORIDE	250.9551	-6.9843E+03	-9.4613E+01	-5.3053E-05	1.3121E-04	186.35	259.00
281	SiHBr3	TRIBROMOSILANE	29.8264	-2.6956E+03	-8.0772E+00	2.0334E-03	1.0063E-06	242.65	617.50
282	SiHCl3	TRICHLOROSILANE	8.8008	-1.6896E+03	5.0043E-01	-6.9024E-03	5.1631E-06	144.95	479.00
283	SiHF3	TRIFLUOROSILANE	-122.7759	5.7853E+01	6.9914E+01	-2.1980E-01	2.2343E-04	121.15	276.95
284	SiH2Br2	DIBROMOSILANE	63.8405	-2.9308E+03	-2.3568E+01	2.4715E-02	-9.7683E-06	212.25	559.24
285	SiH2Cl2	DICHLOROSILANE	30.1827	-2.0844E+03	-8.2717E+00	-3.7469E-10	4.5636E-06	151.15	449.00
286	SiH2F2	DIFLUOROSILANE	23.8427	-1.5675E+03	-4.4921E+00	-1.9742E-02	3.1144E-05	126.45	310.75
287	SiH2I2	DIIODOSILANE	-132.4973	2.1632E+03	5.6398E+01	-4.9010E-02	1.6020E-05	276.95	683.81
288	SiH3Br	MONOBROMOSILANE	48.9247	-2.1416E+03	-1.7831E+01	2.1780E-02	-1.0087E-05	187.45	455.15
289	SiH3Cl	MONOCHLOROSILANE	97.0716	-3.0578E+03	-3.8232E+01	4.4033E-02	-1.8687E-05	155.35	396.79
290	SiH3F	MONOFLUOROSILANE	262.3858	-5.0922E+03	-1.1579E+02	1.8678E-01	-1.1081E-04	120.15	285.87
291	SiH3I	IODOSILANE	81.5431	-3.3034E+03	-3.0771E+01	3.1175E-02	-1.1973E-05	220.15	524.59
292	SiH4	SILANE	49.8037	-1.3946E+03	-1.8981E+01	2.2497E-02	3.0530E-13	88.48	269.70
293	SiO2	SILICON DIOXIDE	-378.5210	6.5473E+03	1.3150E+02	-3.5774E-02	3.4220E-06	1883.00	2503.20
294	Si2Cl6	HEXACHLORODISILANE	28.0093	-2.9842E+03	-7.7427E+00	6.7458E-03	-2.5044E-06	277.15	412.15
295	Si2F6	HEXAFLUORODISILANE	-3.5918	-1.9401E+03	6.6355E+00	-8.5356E-03	4.7603E-06	192.15	254.25
296	Si2H5Cl	DISILANYL CHLORIDE	50.2854	-2.4890E+03	-1.8081E+01	2.1374E-02	-1.0671E-05	226.95	291.15
297	Si2H6	DISILANE	19.4083	-1.5000E+03	-4.5432E+00	-2.0804E-10	3.3390E-06	143.85	432.00
298	Si2OCl3F3	TRICHLOROTRIFLUORODISILOXANE	-39.7219	-2.7322E+02	1.9305E+01	-1.7428E-02	7.2191E-06	235.15	316.35
299	Si2OCl6	HEXACHLORODISILOXANE	173.0145	-6.6562E+03	-6.7483E+01	6.5445E-02	-2.6121E-05	268.15	408.75
300	Si2OH6	DISILOXANE	23.4097	-1.5324E+03	-6.9541E+00	1.0056E-02	-6.1820E-06	160.65	257.75
301	Si3Cl8	OCTACHLOROTRISILANE	0.9548	-2.4388E+03	2.9120E+00	-2.1105E-03	6.5985E-07	319.45	484.55
302	Si3H8	TRISILANE	61.7030	-2.7796E+03	-2.2974E+01	2.7178E-02	-1.3469E-05	204.25	326.25
303	Si3H9N	TRISILAZANE	77.8169	-3.1701E+03	-2.9814E+01	3.5861E-02	-1.8007E-05	204.45	321.85
304	Si4H10	TETRASILANE	72.5811	-3.7126E+03	-2.6507E+01	2.6806E-02	-1.1367E-05	245.45	373.15
305	Sm	SAMARIUM	-63.3814	-6.9759E+03	2.5576E+01	-9.5945E-03	1.2128E-06	733.00	1874.00
306	Sn	TIN	-11.8452	-1.3744E+04	6.4004E+00	-9.7861E-04	-4.2795E-10	1096.00	2995.00
307	SnBr4	STANNIC BROMIDE	62.0925	-4.1212E+03	-2.1351E+01	1.6385E-02	-5.2861E-06	331.45	477.85
308	SnCl2	STANNOUS CHLORIDE	430.9683	-2.2551E+04	-1.5874E+02	9.2347E-02	-2.1208E-05	589.15	896.15
309	SnCl4	STANNIC CHLORIDE	146.4762	-5.5847E+03	-5.7244E+01	5.8634E-02	-2.4782E-05	250.45	386.15
310	SnH4	STANNIC HYDRIDE	7.8482	-9.7474E+02	-2.7435E-01	4.6026E-04	-3.3081E-07	133.15	220.85
311	SnI4	STANNIC IODIDE	-5.9850	-2.4424E+03	5.1095E+00	-2.7912E-03	6.6165E-07	429.15	621.15
312	Sr	STRONTIUM	-111.2297	-2.8962E+03	4.3785E+01	-2.0560E-02	3.2941E-06	582.00	1630.00

Table 7-2 VAPOR PRESSURE - INORGANIC COMPOUNDS (continued)

NO	FORMULA	NAME	$\log_{10} P = A + B/T + C \log_{10} T + D T + E T^2$					$(P$ - mm Hg, T - K$)$	
			A	B	C	D	E	TMIN	TMAX
313	SrO	STRONTIUM OXIDE	2848.9447	-4.0303E+05	-8.7252E+02	1.3051E-01	-7.7456E-06	2341.15	2683.15
314	Ta	TANTALUM	90.4608	-5.3107E+04	-2.4076E+01	2.9553E-03	-1.3908E-07	2511.00	5565.00
315	Tc	TECHNETIUM	-240.5191	-5.7928E+03	7.8794E+01	-1.3112E-02	7.4635E-07	1660.00	5000.00
316	Te	TELLURIUM	-130.9948	-5.2810E+03	5.7873E+01	-4.6486E-02	1.0781E-05	497.00	1285.00
317	TeCl4	TELLURIUM TETRACHLORIDE	225.5681	-1.3194E+04	-8.0899E+01	4.5316E-02	-1.0441E-05	506.15	665.15
318	TeF6	TELLURIUM HEXAFLUORIDE	103.1152	-3.1620E+03	-4.2245E+01	6.7520E-02	-4.5015E-05	161.85	234.55
319	Ti	TITANIUM	-194.8742	-8.2733E+03	6.8261E+01	-1.7329E-02	1.5517E-06	1508.00	3442.00
320	TiCl4	TITANIUM TETRACHLORIDE	65.9073	-4.0187E+03	-2.2002E+01	1.0422E-02	-8.7715E-13	249.05	638.00
321	Tl	THALLIUM	149.2845	-1.7175E+04	-4.9494E+01	1.8084E-02	-2.5186E-06	636.00	1745.00
322	TlBr	THALLOUS BROMIDE	65.7795	-9.3154E+03	-1.9913E+01	6.6371E-03	-9.3147E-07	763.15	1092.15
323	TlI	THALLOUS IODIDE	107.4718	-1.2047E+04	-3.4449E+01	1.2081E-02	-1.7669E-06	713.15	1096.15
324	Tm	THULIUM	581.9203	-4.0634E+04	-2.0847E+02	9.8996E-02	-1.8204E-05	661.00	1237.00
325	U	URANIUM	5.1916	-2.3655E+04	1.4051E+00	-6.7084E-04	6.4472E-08	1600.00	4135.00
326	UF6	URANIUM FLUORIDE	141.7284	-5.4522E+03	-5.5519E+01	6.2879E-02	-2.9623E-05	234.35	328.85
327	V	VANADIUM	52.0677	-3.1989E+04	-1.2620E+01	1.6179E-03	-1.0505E-07	1604.00	3665.00
328	VCl4	VANADIUM TETRACHLORIDE	12.7215	-2.2338E+03	-1.7693E+00	-9.8525E-11	3.4861E-07	247.45	697.00
329	VOCl3	VANADIUM OXYTRICHLORIDE	31.8135	-2.8585E+03	-8.5879E+00	3.9375E-09	3.4924E-06	193.65	400.00
330	W	TUNGSTEN	-19.5111	-3.8683E+04	8.6887E+00	-7.1355E-04	2.1195E-08	2667.00	5645.00
331	WF6	TUNGSTEN FLUORIDE	354.8192	-8.8753E+03	-1.5107E+02	2.0875E-01	-1.1715E-04	201.55	290.45
332	Xe	XENON	15.6530	-8.1035E+02	-3.9013E+00	4.7985E-03	-1.7020E-11	161.36	289.74
333	Yb	YTTERBIUM	-61.8092	-5.3349E+03	2.5566E+01	-1.1528E-02	1.7104E-06	599.00	1660.00
334	Yt	YTTRIUM	-198.7054	-2.2991E+04	6.8766E+01	-1.6442E-02	1.3848E-06	1246.00	3055.00
335	Zn	ZINC	-20.3143	-4.6362E+03	1.0073E+01	-3.8085E-03	4.8860E-07	692.70	3170.00
336	ZnCl2	ZINC CHLORIDE	256.5500	-1.7845E+04	-9.1583E+01	4.9680E-02	-1.0787E-05	701.15	1005.15
337	ZnF2	ZINC FLUORIDE	-17799.4	1.2561E+06	6.0444E+03	-1.7915E+00	2.0155E-04	1243.15	1770.15
338	ZnO	ZINC OXIDE	10.0724	-1.5790E+04	-3.0065E-02	6.5659E-06	-5.4927E-10	1773.10	2223.10
339	ZnSO4	ZINC SULFATE	8.7415	-2.2158E+04	-5.1178E-04	6.6577E-07	-3.3159E-10	293.15	378.15
340	Zr	ZIRCONIUM	95.1134	-4.3264E+04	-2.5683E+01	2.9830E-03	-1.1696E-07	1975.00	4598.00
341	ZrBr4	ZIRCONIUM BROMIDE	35.8920	-6.9028E+03	-8.7767E+00	4.6821E-03	-1.0688E-06	480.15	630.15
342	ZrCl4	ZIRCONIUM CHLORIDE	80.7853	-8.5643E+03	-2.5618E+01	1.4577E-02	-3.5146E-06	463.15	604.15
343	ZrI4	ZIRCONIUM IODIDE	95.3927	-1.0449E+04	-3.0444E+01	1.5025E-02	-3.1340E-06	537.15	704.15

P - vapor pressure, mm Hg

A, B, C, D, and E - regression coefficients for chemical compound

T - temperature, K

TMIN - minimum temperature, K

TMAX - maximum temperature, K

Chapter 8

DENSITY OF LIQUID

Carl L. Yaws, Xiaoyan Lin, Li Bu, Sachin Nijhawan, Deepa R. Balundgi, and Saumya Tripathi
Lamar University, Beaumont, Texas

ABSTRACT

Results for saturated liquid density as a function of temperature are presented for major organic and inorganic chemicals. The results are displayed in easy-to-use tabulations that are especially applicable for rapid engineering usage with the personal computer or hand calculator.

INTRODUCTION

Liquid density data are important in process engineering design, such as sizing of storage vessels that contain the basic raw materials and products for a plant. In distillation, stripping and absorption, liquid density data are required in the determination of flooding and calculation of column diameter. Additional usage is encountered in various heat, mass, and momentum transfer operations. In this article, results for liquid density as a function of temperature are presented for a wide variety of chemicals.

LIQUID DENSITY CORRELATION

The modified form of the Rackett equation was selected for correlation of saturated liquid density as a function of temperature:

$$\text{density} = A \, B^{-(1-T/T_c)^n} \tag{8-1}$$

where
density = saturated liquid density, g/ml
A, B, and n = regression coefficients for chemical compound
T = temperature, K
T_c = critical temperature, K

The results for saturated liquid density are given in Tables 8-1 and 8-2. The compilation is based on screening appropriate data source publications for organics (1-40) and inorganics (1-120). Both experimental values for the property under consideration and parameter values for estimation of the property are included in the source publications. The tabulation is arranged by chemical formula to provide ease of use in quickly locating data. The temperature range for which the equation may be used is denoted by TMIN and TMAX. For many of the compounds, the temperature range corresponds to the freezing point and critical point. Values for liquid density at 25 C are provided in the last column.

A comparison of calculations and data for saturated liquid density is shown in Fig. 8-1 for a representative chemical. The graph indicates good agreement of calculations and data.

EXAMPLES

The correlation results may be used for prediction and calculation of liquid density. Examples are given below.

Example 1 Estimate the liquid density of methanol (CH4O) at -20 C (253.15 K).

Substitution of the coefficients from the table and the temperature into the equation for liquid density yields

$$\text{density} = (0.27197) (0.27192^{-(1-253.15/512.58)^{0.2331}})$$

density = 0.826 g/ml

The calculated and data values compare favorably (0.826 vs 0.829, deviation = 0.4%).

Example 2 Estimate the liquid density of methyl chloride (CH3Cl) at 100 C (373.15 K).

Substitution of the coefficients from the table and the temperature into the equation for liquid density yields

$$\text{density} = (0.35821) (0.26109^{-(1-373.15/416.25)^{0.2869}})$$

density = 0.722 g/ml

The calculated and data values compare favorably (0.722 vs 0.725, deviation = 0.4%).

Portions of this material appeared in Hydrocarbon Processing, 70, 103 (January, 1991) and are reprinted by special permission.

REFERENCES – ORGANIC COMPOUNDS

1-34. See **REFERENCES - ORGANIC COMPOUNDS** in **Chapter 1, CRITICAL PROPERTIES AND ACENTRIC FACTOR**

35. Wilhoit, R. C. and B. J. Zwolinski, <u>PHYSICAL AND THERMODYNAMIC PROPERTIES OF ALIPHATIC ALCOHOLS</u>, American Chemical Society, American Institute of Physics, National Bureau of Standards, New York, NW (1973).

36. McKetta, J. J. and C. C. Hsu, J. Chem. Eng. Data, $\underline{9}$(1), 45 (1964)

37. Howard, P. H. and W. M. Meylan, eds., <u>HANDBOOK OF PHYSICAL PROPERTIES OF ORGANIC CHEMICALS</u>, CRC Press, Boca Raton, FL (1997).

38. Yaws, C. L. and others, Hydrocarbon Processing, $\underline{70}$, 103 (January, 1991).

39. Yaws, C. L., <u>HANDBOOK OF THERMODYNAMIC DIAGRAMS</u>, Vols. 1, 2, 3, and 4, Gulf Publishing Co., Houston, TX (1996).

40. Yaws, C. L., <u>HANDBOOK OF CHEMICAL COMPOUND DATA FOR PROCESS SAFETY</u>, Gulf Publishing Co., Houston, TX (1997).

REFERENCES – INORGANIC COMPOUNDS

1-56. See **REFERENCES - INORGANIC COMPOUNDS** in **Chapter 1, CRITICAL PROPERTIES AND ACENTRIC FACTOR**

57. Touloukian, T. S. and C. Y. Ho, eds., <u>PROPERTIES OF NONMETALLIC FLUID ELEMENTS</u>, McGraw-Hill Book Co., New York, NY(1981).

58. Lyon, R. N., ed., <u>LIQUID-METALS HANDBOOK</u>, Atomic Energy Commission and Dept. of Navy, Washington, DC (1954).

59. Emsley, J., <u>THE ELEMENTS</u>, 2nd ed., Clarendon Press, Oxford University Press, New York, NY (1991).

60. Perry, D. L. and S. L. Phillips, <u>HANDBOOK OF INORGANIC COMPOUNDS</u>, CRC Press, New York, NY (1995).

61. Oshe, R. E., ed., <u>HANDBOOK OF THERMODYNAMIC AND TRANSPORT PROPERTIES OF ALKALI METALS</u>, Blackwell Scientific Publishers, Oxford, England (1985).

62. Van Horn, K. R., ed., <u>ALUMINUM</u>, Vol. 1, American Society for Metals, Metals Park, Ohio (1967).

63. Janz, G. L., C. B. Bansal, N. P. Bansal, R. M. Murphy, and R. P. T. Tompkins, <u>PHYSICAL PROPERTIES DATA COMPILATIONS RELEVANT TO ENERGY STORAGE. II. MOLTEN SALTS: DATA ON SINGLE AND MULTICOMPONENT SALT SYSTEMS</u>, Nat. Bur. Stand., Molten Salts Data Center, Troy, NY (April, 1979).

64. Janz, G. J., A. T. Ward, and R. D. Reeves, <u>MOLTEN SALT DATA</u>, Technical Bulletin Series, Rensselaer Polytechnic Institute, Troy, NY (1964).

65. Murgulescu, I. G. and M. Serban, Rev. Roum. Chim., $\underline{19}$, 1417(1974).

66. Waseda, Y. and K. Suzuki, Phys. Status Solidi B: Basic Research, $\underline{57}$, 351 (1973).

67. Nisel'son, L. A. and T. D. Sokolova, Russ. J. Inorg. Chem., $\underline{10}$, 827 (1965).

68. Forster, S., Cryogenics, $\underline{3}$, 176 (1963).

69. Saji, Y. and S. Kobayashi, Cryogenics, $\underline{4}$, 136 (1964).

70. Maitland, G. C. and E. B. Smith, J. Chem. Eng. Data, $\underline{17}$ (2), 150 (1972).

71. Kestin, J., E. A. Knierim, B. Mason, S. T. Majafi, S. T. Ro, and M. Wadman, J. Phys. Chem. Ref. Data, $\underline{13}$ (1), 229 (1984).

72. Runovskaya, I. V., A. D. Zorin, and G. G. Devyatykh, Russ. J. Inorg. Chem., $\underline{15}$, 1338 (1970).

73. Rankine, A. O. and C. J. Smith, Phil. Mag. $\underline{42}$, 601 (1921).

74. Luchinskii, G. P., Zh. Obsch. Khim., $\underline{7}$, 2116 (1937).

75. Reichenberg, D., AIChE J., $\underline{19}$, 854 (1973).

76. Reichenberg, D., AIChE J., $\underline{21}$, 181 (1975).

77. Stiel, L. T. and G. Thodos, AIChE J., $\underline{10}$, 266 (1964).

78. Boon, J. P., J. C. Legros, and G. Thomaes, Physica, $\underline{33}$, 547 (1967).

79. Rao, R. V. G. and K. N. Swamy, Z. Phys. Chem. Leipzig, $\underline{255}$ (2), 300 (1974).

80. Rudenko, N. S. and L. W. Schubrukow, Phys. Zeit. der Sowjetunion $\underline{6}$, 470 (1934).

81. Herreman, W., W. Grevendork, and A. DeBock, J. Chem. Phys., $\underline{53}$ (1), 185 (1970).

82. Titani, T., Bull. Chem. Soc. Japan, $\underline{8}$, 255 (1933).

83. Lewis, J. R., J. Amer. Chem. Soc., $\underline{47}$, 626 (1925).

84. Kuliffeev, V. K., V. I. Panchishnyi, and G. P. Standevich, Isv. Vyssh. Ucheb. Zaved. Tsvet. Met., $\underline{11(2)}$, 116 (1968).

85. Simkin, J. and R. L. Jarry, J. Phys. Chem., $\underline{61}$, 503 (1957).

86. Usanovich, M., T. Sumarokova, and V. Udovenko, Acta Physicochim. USSR, $\underline{11}$, 505 (1939).

87. Kestin, J., J. V. Sengers, B. Kamgar-Parsi and J. M. H. Levelt Sengers, J. Phys. Chem. Ref. Data, $\underline{13}$ (2), 601 (1984).

88. Matsunaga, N. and A. Nagashima, J. Phys. Chem. Ref. Data, $\underline{12}$ (4), 933 (1983).

89. Hanley, H. J. M. and R. Prydz, J. Phys. Chem. Ref. Data, $\underline{1}$ (4), 1101 (1972).

90. Greenwood, N. N. and K. Wade, J. Inorg. Nucl. Chem., $\underline{3}$, 349 (1957).

91. McIntosh, D. and B. Steele, Z. Phys. Chem., $\underline{55}$, 129 (1906).

92. Schuil, A. E., Phil. Mag., $\underline{28}$, 679 (1939).

93. Simons, J. H. and R. D. Dresdner, J. Amer. Chem. Soc., $\underline{66}$, 1070 (1944).

94. Baker, C. E., J. Chem. Phys., $\underline{46}$, 2846 (1967).

95. Stern, S. A., J. L. Mullhaupt and W. B. Kay, Chem. Rev., $\underline{60}$, 185 (1960).

96. Mason, D. M., I. Petker, and S. P. Vango, J. Phys. Chem., $\underline{59}$, 511 (1955).

97. Bingham, E. C. and S. B. Stone, J. Phys. Chem., $\underline{27}$, 701 (1923).

98. Miskidzh'yan, S. P. and N. A. Trifonov, Zh. Obshch. Khim., $\underline{17}$, 1033 (1947).

99. Naumova, A. S., Zh. Obshch. Khim., $\underline{19}$, 1228 (1949).

100. Briner, E., B. Susz, and P. Favarger, Helv. Chim. Acta, $\underline{18}$, 375 (1935).

101. Taylor, E. G., L. M. Lynne, and A. G. Follous, Can. J. Chem., $\underline{29}$, 439 (1951).

102. Friend, J. N. and W. D. Hargreaves, Phil. Mag., $\underline{34}$, 810 (1943).

103. Maass. O. and W. H. Hatcher, J. Amer. Chem. Soc., $\underline{42}$, 2548 (1920).

104. Bright, N. F. H., H. Hutchison, and D. Smith, J. Soc. Chem. Ind., $\underline{65}$, 385 (1946).

105. Morozov, I. R., J. Appl. Chem. (USSR), $\underline{24}$, 975 (1951).

106. Andrade, E. N. Da C. and E. R. Dobbs, Proc. Roy. Soc. London, 211A, $\underline{12}$ (1952).

107. Leu, A-L. and H. Eyring, Proc. Nat. Acad. Sci. USA, $\underline{72}$ (3) 1026 (1975).

108. Krynicki, K. and J. W. Hennel, Acta Physica Polonica, $\underline{24}$ (8), 269 (1963).

109. Mason, D. M. and O. W. Wilcox, B. H. Sage, J. Phys. Chem., $\underline{56}$, 1008 (1952).

110. Janz, G. J., J. Phys. Chem. Ref. Data, $\underline{9}$ (4), 791 (1980).

111. Janz, G. J. and G. L. Gardner, U. Krebs, and R. P. T. Tomkins, J. Phys. Chem. Ref. Data, $\underline{3}$ (1), 1 (1974).

112. Gossink, R. G. and J. M. Stevels, Inorg. Chem., $\underline{11}$ (9), 2180 (1972).

113. Jenkins, A. C. and F. S. Dipaolo, J. Chem. Phys., $\underline{25}$ (2), 296 (1956).

114. Lutschinsky, G. P., Zeit. Fur. Anorganische und Allgemeine Chemie, $\underline{223}$, 210 (1935).

115. Gutmann, V., Monatshofte Fur Chemie, $\underline{83}$, 164 (1952).

116. Bacon, R. F. and R. Fanelli, J. Amer. Chem. Soc., $\underline{15}$, 639 (1943).

117. Veda, K. and K. Kigoshi, J. Inorg. Nucl. Chem., $\underline{36}$, 989 (1974).

118. Nisel'son, L. A. and P. P. Pugachevich, T. D. Sokolova and, R. A. Bederdinov, Russ. J. Inorg. Chem., $\underline{10}$ (6), 705 (1965).

119. Runovskaya, I. V. and A. D. Zorin, G. G. Devyatykh, Russ. J. Inorg. Chem., $\underline{15}$ (9), (1970).

120. Bacon, J. F. and A. A. Hasapis, J. Appl. Phys., $\underline{30}$ (9), 1470 (1959).

Figure 8-1 Density of Liquid

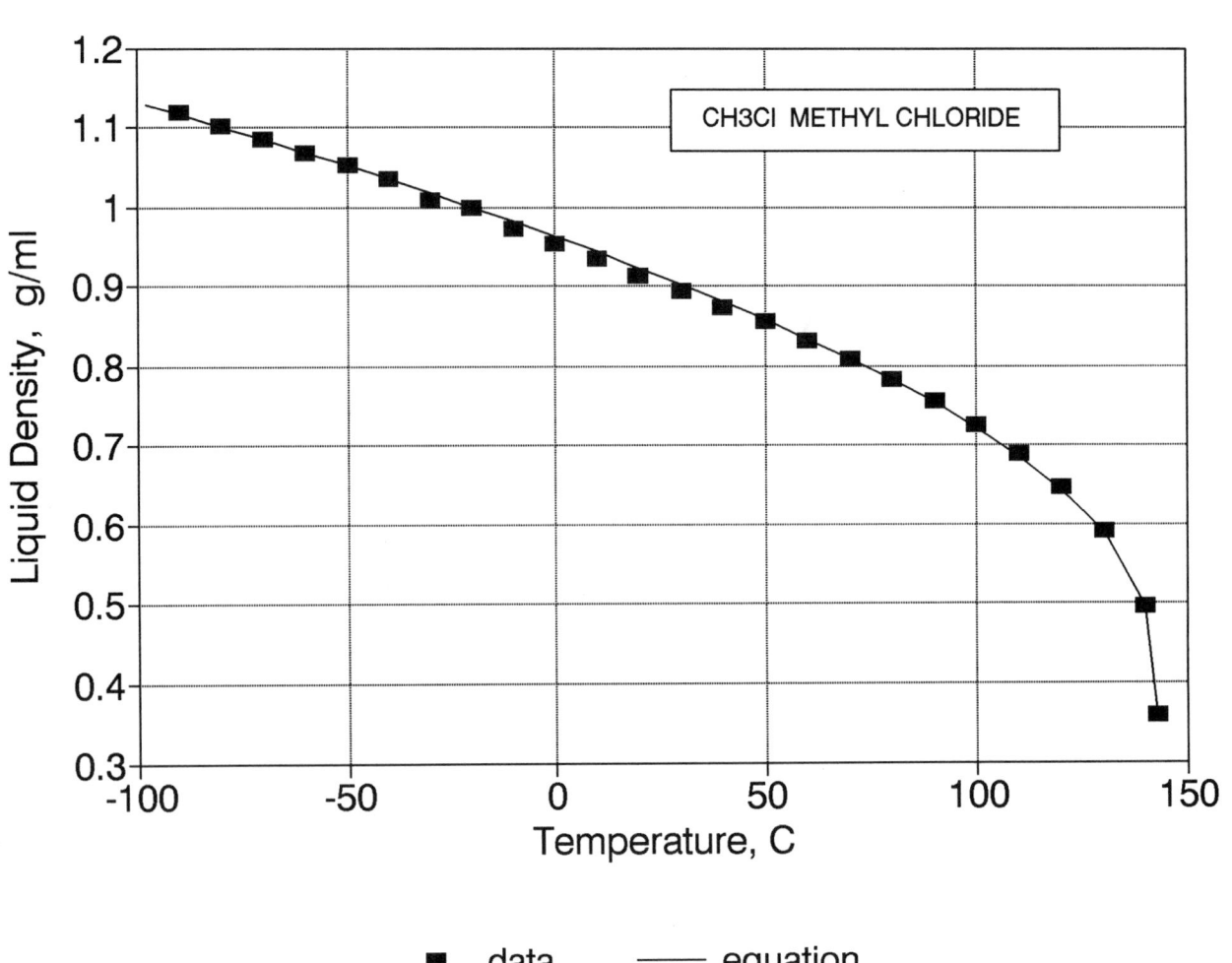

Table 8-1 DENSITY OF LIQUID - ORGANIC COMPOUNDS

			density = A B $^{-(1-T/T_c)^n}$					(density - g/ml, T - K)	
NO	FORMULA	NAME	A	B	n	T_c	TMIN	TMAX	density @ 25 C
1	CBrClF2	BROMOCHLORODIFLUOROMETHANE	0.67201	0.25624	0.26440	426.15	113.65	426.15	1.810
2	CBrCl3	BROMOTRICHLOROMETHANE	0.69849	0.28000	0.28570	606.00	252.15	606.00	1.994
3	CBrF3	BROMOTRIFLUOROMETHANE	0.74110	0.26995	0.28006	340.15	105.15	340.15	1.536
4	CBr2F2	DIBROMODIFLUOROMETHANE	0.84264	0.27721	0.26256	478.00	163.05	478.00	2.274
5	CClF3	CHLOROTRIFLUOROMETHANE	0.58753	0.27896	0.29070	301.96	92.15	301.96	0.841
6	CClN	CYANOGEN CHLORIDE	0.37713	0.22974	0.23860	449.00	266.65	449.00	1.172
7	CCl2F2	DICHLORODIFLUOROMETHANE	0.57494	0.27880	0.29650	384.95	115.15	384.95	1.307
8	CCl2O	PHOSGENE	0.51316	0.27119	0.27201	455.00	145.37	455.00	1.363
9	CCl3F	TRICHLOROFLUOROMETHANE	0.56082	0.27556	0.28571	471.20	162.04	471.20	1.477
10	CCl4	CARBON TETRACHLORIDE	0.56607	0.27663	0.29000	556.35	250.33	556.35	1.583
11	CF2O	CARBONYL FLUORIDE	0.46780	0.32913	0.28570	297.00	161.89	297.00	--------
12	CF4	CARBON TETRAFLUORIDE	0.62949	0.28390	0.29095	227.50	89.56	227.50	--------
13	CHBr3	TRIBROMOMETHANE	0.88375	0.24757	0.30060	696.00	281.20	696.00	2.876
14	CHClF2	CHLORODIFLUOROMETHANE	0.51858	0.26605	0.28123	369.30	115.73	369.30	1.193
15	CHCl2F	DICHLOROFLUOROMETHANE	0.52414	0.27110	0.28571	451.58	138.15	451.58	1.367
16	CHCl3	CHLOROFORM	0.49807	0.25274	0.28766	536.40	209.63	536.40	1.480
17	CHF3	TRIFLUOROMETHANE	0.52435	0.25490	0.28910	298.89	117.97	298.89	0.667
18	CHI3	TRIIODOMETHANE	1.12660	0.28100	0.28571	794.55	396.16	794.55	--------
19	CHN	HYDROGEN CYANIDE	0.19501	0.18589	0.28206	456.65	259.91	456.65	0.680
20	CHNS	ISOTHIOCYANIC-ACID	--------	--------	--------	--------	--------	--------	--------
21	CH2BrCl	BROMOCHLOROMETHANE	0.68686	0.27700	0.28570	557.00	185.20	557.00	1.926
22	CH2Br2	DIBROMOMETHANE	0.77951	0.24834	0.27583	611.00	220.60	611.00	2.482
23	CH2ClF	CHLOROFLUOROMETHANE	0.44380	0.23000	0.28571	424.91	140.16	424.91	1.256
24	CH2Cl2	DICHLOROMETHANE	0.45965	0.25678	0.29020	510.00	178.01	510.00	1.318
25	CH2F2	DIFLUOROMETHANE	0.42519	0.25220	0.28100	351.60	137.00	351.60	0.957
26	CH2I2	DIIODOMETHANE	0.98475	0.24998	0.26540	747.00	279.25	747.00	3.306
27	CH2O	FORMALDEHYDE	0.26192	0.22240	0.28570	408.00	181.15	408.00	0.736
28	CH2O2	FORMIC ACID	0.36821	0.24296	0.23663	580.00	281.55	580.00	1.214
29	CH3Br	METHYL BROMIDE	0.60859	0.26292	0.28030	467.00	179.55	467.00	1.662
30	CH3Cl	METHYL CHLORIDE	0.35821	0.26109	0.28690	416.25	175.45	416.25	0.913
31	CH3Cl3Si	METHYL TRICHLOROSILANE	0.43982	0.26111	0.27788	517.00	195.35	517.00	1.266
32	CH3F	METHYL FLUORIDE	0.29788	0.24153	0.28540	317.70	131.35	317.70	0.566
33	CH3I	METHYL IODIDE	0.76723	0.25854	0.26790	528.00	206.70	528.00	2.265
34	CH3NO	FORMAMIDE	0.27633	0.20352	0.25178	771.00	275.70	771.00	1.129
35	CH3NO2	NITROMETHANE	0.35200	0.23918	0.29030	588.15	244.60	588.15	1.129
36	CH3NO2	METHYL-NITRITE	--------	--------	--------	--------	--------	--------	--------
37	CH3NO3	METHYL-NITRATE	--------	--------	--------	--------	--------	--------	--------
38	CH4	METHANE	0.15998	0.28810	0.27700	190.58	90.67	190.58	--------
39	CH4Cl2Si	METHYL DICHLOROSILANE	0.39804	0.28209	0.22529	483.00	182.55	483.00	1.103
40	CH4O	METHANOL	0.27197	0.27192	0.23310	512.58	175.47	512.58	0.787
41	CH4O3S	METHANESULFONIC ACID	--------	--------	--------	--------	--------	--------	1.477
42	CH4S	METHYL MERCAPTAN	0.33179	0.28018	0.28523	469.95	150.18	469.95	0.862
43	CH5ClSi	METHYL CHLOROSILANE	0.32761	0.26257	0.26569	442.00	139.05	442.00	0.884
44	CH5N	METHYLAMINE	0.20168	0.21405	0.22750	430.05	179.69	430.05	0.655
45	CH6Si	METHYL SILANE	0.22509	0.26757	0.28799	352.50	116.34	352.50	0.486
46	CN4O8	TETRANITROMETHANE	0.41867	0.18144	0.28570	540.00	287.05	540.00	1.626
47	CO	CARBON MONOXIDE	0.29818	0.27655	0.29053	132.92	68.15	132.92	--------
48	COS	CARBONYL SULFIDE	0.44482	0.28943	0.27140	378.80	134.35	378.80	1.005
49	CO2	CARBON DIOXIDE	0.46382	0.26160	0.29030	304.19	216.58	304.19	0.713
50	CS2	CARBON DISULFIDE	0.47589	0.28749	0.32260	552.00	161.58	552.00	1.256
51	C2BrF3	BROMOTRIFLUOROETHYLENE	0.67050	0.24000	0.30000	432.00	173.00	432.00	1.830
52	C2Br2F4	1,2-DIBROMOTETRAFLUOROETHANE	0.76194	0.26172	0.26547	487.80	162.65	487.80	2.162
53	C2ClF3	CHLOROTRIFLUOROETHYLENE	0.55068	0.25930	0.30740	379.15	115.00	379.15	1.275
54	C2ClF5	CHLOROPENTAFLUOROETHANE	0.61452	0.28454	0.28530	353.15	173.71	353.15	1.287
55	C2Cl2F4	1,2-DICHLOROTETRAFLUOROETHANE	0.57270	0.26920	0.27450	418.85	179.15	418.85	1.455
56	C2Cl3F3	1,1,2-TRICHLOROTRIFLUOROETHANE	0.57596	0.27178	0.28040	487.25	238.15	487.25	1.564
57	C2Cl4	TETRACHLOROETHYLENE	0.66671	0.32758	0.35630	620.00	250.80	620.00	1.613
58	C2Cl4F2	1,1,2,2-TETRACHLORODIFLUOROETHANE	0.57239	0.26740	0.28570	551.00	299.15	551.00	--------
59	C2Cl4O	TRICHLOROACETYL CHLORIDE	0.55730	0.27270	0.28570	590.00	273.15	590.00	1.613
60	C2Cl6	HEXACHLOROETHANE	0.57460	0.23700	0.28571	698.00	459.95	698.00	--------
61	C2F4	TETRAFLUOROETHYLENE	0.57587	0.26880	0.28571	306.45	142.00	306.45	0.920
62	C2F6	HEXAFLUOROETHANE	0.61625	0.26693	0.29794	292.80	172.45	292.80	--------
63	C2HBrClF3	HALOTHANE	0.66571	0.26831	0.28571	521.00	223.15	521.00	1.869
64	C2HClF2	2-CHLORO-1,1-DIFLUOROETHYLENE	0.50014	0.26765	0.28820	400.55	134.65	400.55	1.217
65	C2HCl3	TRICHLOROETHYLENE	0.50416	0.26952	0.28571	571.00	188.40	571.00	1.458
66	C2HCl3O	DICHLOROACETYL CHLORIDE	0.52331	0.26970	0.28570	579.00	298.15	579.00	1.519
67	C2HCl3O	TRICHLOROACETALDEHYDE	0.51178	0.26596	0.27840	565.00	216.00	565.00	1.499
68	C2HCl5	PENTACHLOROETHANE	0.54822	0.26628	0.28491	665.00	244.15	665.00	1.675
69	C2HF3	TRIFLUOROETHENE	0.44950	0.28600	0.28571	347.22	94.53	347.22	0.919
70	C2HF3O2	TRIFLUOROACETIC ACID	0.55894	0.27475	0.30290	491.25	257.90	491.25	1.480
71	C2HF5	PENTAFLUOROETHANE	0.55633	0.26100	0.28571	342.00	170.15	342.00	1.174
72	C2H2	ACETYLENE	0.23017	0.27091	0.29171	308.32	192.40	308.32	0.377
73	C2H2Br4	1,1,2,2-TETRABROMOETHANE	0.86181	0.26030	0.21400	824.00	273.15	824.00	2.927
74	C2H2Cl2	1,1-DICHLOROETHYLENE	0.43624	0.29000	0.28571	482.00	150.65	482.00	1.117
75	C2H2Cl2	cis-1,2-DICHLOROETHYLENE	0.43930	0.26128	0.28570	527.00	193.15	527.00	1.265
76	C2H2Cl2	trans-1,2-DICHLOROETHYLENE	0.44616	0.26700	0.28570	508.00	223.35	508.00	1.244
77	C2H2Cl2O	CHLOROACETYL CHLORIDE	0.47008	0.25420	0.28570	581.00	251.15	581.00	1.434
78	C2H2Cl2O	DICHLOROACETALDEHYDE	0.47708	0.25395	0.28571	555.00	223.00	555.00	1.433

189

Table 8-1 DENSITY OF LIQUID - ORGANIC COMPOUNDS (continued)

| NO | FORMULA | NAME | density = A B $^{-(1-T/T_c)^n}$ | | | | (density - g/ml, T - K) | | |
			A	B	n	T_c	TMIN	TMAX	density @ 25 C
79	C2H2Cl2O2	DICHLOROACETIC ACID	0.48656	0.24720	0.32550	686.00	286.55	686.00	1.553
80	C2H2Cl3F	1,1,1-TRICHLOROFLUOROETHANE	0.51435	0.25000	0.28571	565.00	173.00	565.00	1.575
81	C2H2Cl4	1,1,1,2-TETRACHLOROETHANE	0.50079	0.25970	0.28570	624.00	202.94	624.00	1.535
82	C2H2Cl4	1,1,2,2-TETRACHLOROETHANE	0.51649	0.25953	0.29593	645.00	229.35	645.00	1.587
83	C2H2F2	1,1-DIFLUOROETHYLENE	0.40029	0.23820	0.30940	302.80	129.15	302.80	0.594
84	C2H2F2	cis-1,2-DIFLUOROETHENE	0.39170	0.23800	0.28571	394.67	107.90	394.67	1.023
85	C2H2F2	trans-1,2-DIFLUOROETHENE	0.39170	0.23800	0.28571	394.67	107.90	394.67	1.023
86	C2H2F4	1,1,1,2-TETRAFLUOROETHANE	0.49834	0.24141	0.31372	380.00	172.15	380.00	1.199
87	C2H2O	KETENE	0.29188	0.27200	0.28571	370.00	122.00	370.00	0.660
88	C2H2O4	OXALIC ACID	0.43975	0.21500	0.28571	804.00	462.65	804.00	--------
89	C2H3Br	VINYL BROMIDE	0.53515	0.25321	0.28900	473.00	135.35	473.00	1.499
90	C2H3Cl	VINYL CHLORIDE	0.34897	0.27070	0.27160	432.00	119.36	432.00	0.903
91	C2H3ClF2	1-CHLORO-1,1-DIFLUOROETHANE	0.42636	0.24537	0.29800	410.20	142.35	410.20	1.107
92	C2H3ClO	ACETYL CHLORIDE	0.39703	0.26869	0.28570	508.00	160.30	508.00	1.102
93	C2H3ClO	CHLOROACETALDEHYDE	0.38125	0.23960	0.28571	555.00	293.00	555.00	1.200
94	C2H3ClO2	CHLOROACETIC ACID	0.41192	0.23248	0.28571	686.00	333.15	686.00	--------
95	C2H3ClO2	METHYL CHLOROFORMATE	0.42760	0.26789	0.27814	525.00	192.00	525.00	1.213
96	C2H3Cl3	1,1,1-TRICHLOROETHANE	0.47476	0.27258	0.29333	545.00	242.75	545.00	1.330
97	C2H3Cl3	1,1,2-TRICHLOROETHANE	0.47455	0.25475	0.31000	602.00	236.50	602.00	1.435
98	C2H3F	VINYL FLUORIDE	0.32311	0.27390	0.28571	327.80	112.65	327.80	0.620
99	C2H3F3	1,1,1-TRIFLUOROETHANE	0.42934	0.26210	0.26250	346.25	161.85	346.25	0.953
100	C2H3N	ACETONITRILE	0.23730	0.22642	0.28128	545.50	229.32	545.50	0.779
101	C2H3NO	METHYL ISOCYANATE	0.30084	0.23442	0.28571	505.00	256.15	505.00	0.926
102	C2H4	ETHYLENE	0.21428	0.28061	0.28571	282.36	104.01	282.36	--------
103	C2H4Br2	1,1-DIBROMOETHANE	0.68067	0.26364	0.29825	628.00	210.15	628.00	2.045
104	C2H4Br2	1,2-DIBROMOETHANE	0.71466	0.26634	0.28571	650.15	282.94	650.15	2.169
105	C2H4Cl2	1,1-DICHLOROETHANE	0.41231	0.26533	0.28700	523.00	176.19	523.00	1.168
106	C2H4Cl2	1,2-DICHLOROETHANE	0.46501	0.28742	0.31041	561.00	237.49	561.00	1.246
107	C2H4Cl2O	BIS(CHLOROMETHYL)ETHER	0.43043	0.25410	0.28570	579.00	231.65	579.00	1.312
108	C2H4F2	1,1-DIFLUOROETHANE	0.36462	0.25640	0.27970	386.60	156.15	386.60	0.898
109	C2H4F2	1,2-DIFLUOROETHANE	0.32624	0.22210	0.28570	476.00	215.00	476.00	1.016
110	C2H4I2	1,2-DIIODOETHANE	0.87130	0.24500	0.28571	749.91	356.16	749.91	--------
111	C2H4O	ACETALDEHYDE	0.28207	0.26004	0.27760	461.00	150.15	461.00	0.774
112	C2H4O	ETHYLENE OXIDE	0.31402	0.26089	0.28253	469.15	161.45	469.15	0.862
113	C2H4OS	THIOACETIC-ACID	0.41630	0.31700	0.28571	577.34	150.16	577.34	1.059
114	C2H4O2	ACETIC ACID	0.35182	0.26954	0.26843	592.71	289.81	592.71	1.043
115	C2H4O2	METHYL FORMATE	0.34143	0.25838	0.27680	487.20	174.15	487.20	0.967
116	C2H4S	THIACYCLOPROPANE	0.29230	0.21400	0.28571	555.00	165.37	555.00	1.007
117	C2H5Br	BROMOETHANE	0.50699	0.25918	0.27980	503.80	154.55	503.80	1.450
118	C2H5Cl	ETHYL CHLORIDE	0.32259	0.27464	0.23140	460.35	136.75	460.35	0.890
119	C2H5ClO	2-CHLOROETHANOL	0.37979	0.26160	0.21893	585.00	205.65	585.00	1.196
120	C2H5F	ETHYL FLUORIDE	0.29307	0.27099	0.24420	375.31	129.95	375.31	0.712
121	C2H5I	ETHYL IODIDE	0.65415	0.25912	0.29853	561.00	162.05	561.00	1.920
122	C2H5N	ETHYLENEIMINE	0.24895	0.23289	0.23357	537.00	195.20	537.00	0.831
123	C2H5NO	ACETAMIDE	0.28126	0.21906	0.28570	761.00	354.15	761.00	--------
124	C2H5NO	N-METHYLFORMAMIDE	0.27473	0.22427	0.27470	721.00	269.35	721.00	0.999
125	C2H5NO2	NITROETHANE	0.31810	0.23655	0.27800	593.00	183.63	593.00	1.043
126	C2H5NO3	ETHYL-NITRATE	--------	--------	--------	--------	--------	--------	--------
127	C2H6	ETHANE	0.20087	0.27330	0.28330	305.42	90.35	305.42	0.315
128	C2H6AlCl	DIMETHYLALUMINUM CHLORIDE	0.28732	0.22536	0.28571	619.00	252.15	619.00	0.988
129	C2H6O	DIMETHYL ETHER	0.26390	0.26325	0.28060	400.10	131.66	400.10	0.655
130	C2H6O	ETHANOL	0.26570	0.26395	0.23670	516.25	159.05	516.25	0.787
131	C2H6OS	DIMETHYL SULFOXIDE	0.34418	0.25344	0.32197	726.00	291.67	726.00	1.095
132	C2H6O2	ETHYLENE GLYCOL	0.32503	0.25499	0.17200	645.00	260.15	645.00	1.110
133	C2H6O4S	DIMETHYL SULFATE	0.43048	0.25926	0.37020	758.00	241.35	758.00	1.322
134	C2H6S	DIMETHYL SULFIDE	0.30676	0.26780	0.28571	503.04	174.88	503.04	0.850
135	C2H6S	ETHYL MERCAPTAN	0.30092	0.26940	0.27866	499.15	125.26	499.15	0.833
136	C2H6S2	DIMETHYL DISULFIDE	0.37382	0.27705	0.31143	606.00	188.44	606.00	1.057
137	C2H7N	DIMETHYLAMINE	0.24110	0.26785	0.24800	437.65	180.96	437.65	0.650
138	C2H7N	ETHYLAMINE	0.24773	0.25651	0.28589	456.15	192.15	456.15	0.677
139	C2H7NO	MONOETHANOLAMINE	0.27149	0.22411	0.20150	638.00	283.65	638.00	1.014
140	C2H8N2	ETHYLENEDIAMINE	0.22765	0.20726	0.20173	593.00	284.29	593.00	0.893
141	C2H8Si	DIMETHYL SILANE	0.23323	0.26351	0.28421	402.00	122.93	402.00	0.578
142	C2N2	CYANOGEN	0.26685	0.20984	0.20635	400.15	245.25	400.15	0.866
143	C3F6	HEXAFLUOROPROPYLENE	0.55973	0.24195	0.31130	368.00	116.65	368.00	1.304
144	C3F6O	HEXAFLUOROACETONE	0.50475	0.23393	0.22900	357.14	151.15	357.14	1.321
145	C3F8	OCTAFLUOROPROPANE	0.62735	0.27240	0.28170	345.05	125.46	345.05	1.317
146	C3H2N2	MALONONITRILE	0.26638	0.20459	0.26090	715.00	304.90	715.00	--------
147	C3H3Cl	PROPARGYL CHLORIDE	0.35313	0.25459	0.31320	541.00	293.00	541.00	1.024
148	C3H3N	ACRYLONITRILE	0.25030	0.22930	0.28939	535.00	189.63	535.00	0.801
149	C3H3NO	OXAZOLE	0.29157	0.32500	0.28571	554.00	189.15	554.00	0.718
150	C3H4	METHYLACETYLENE	0.24368	0.26448	0.27900	402.39	170.45	402.39	0.607
151	C3H4	PROPADIENE	0.24731	0.27017	0.30300	393.15	136.87	393.15	0.579
152	C3H4Cl2	2,3-DICHLOROPROPENE	0.40062	0.25367	0.30630	577.00	191.50	577.00	1.201
153	C3H4O	ACROLEIN	0.28459	0.26124	0.24890	506.00	185.45	506.00	0.834
154	C3H4O	PROPARGYL ALCOHOL	0.30488	0.24900	0.28570	580.00	221.35	580.00	0.945
155	C3H4O2	ACRYLIC ACID	0.34645	0.25822	0.30701	615.00	286.65	615.00	1.046
156	C3H4O2	beta-PROPIOLACTONE	0.36994	0.23600	0.28571	686.00	239.75	686.00	1.262

Table 8-1 DENSITY OF LIQUID - ORGANIC COMPOUNDS (continued)

NO	FORMULA	NAME	density = A B $^{-(1-T/Tc)^n}$						(density - g/ml, T - K)
			A	B	n	T_c	TMIN	TMAX	density @ 25 C
157	C3H4O2	VINYL FORMATE	0.33220	0.25151	0.29455	498.00	200.00	498.00	0.954
158	C3H4O3	ETHYLENE CARBONATE	0.38298	0.23700	0.28600	790.00	309.55	790.00	--------
159	C3H4O3	PYRUVIC ACID	0.36842	0.23910	0.23400	634.52	286.75	634.52	1.265
160	C3H5Br	3-BROMO-1-PROPENE	0.50790	0.28200	0.28571	540.20	153.76	540.20	1.389
161	C3H5Cl	2-CHLOROPROPENE	0.33385	0.27165	0.28571	478.00	135.75	478.00	0.895
162	C3H5Cl	3-CHLOROPROPENE	0.32483	0.25957	0.28570	514.15	138.65	514.15	0.931
163	C3H5ClO	alpha-EPICHLOROHYDRIN	0.39710	0.26476	0.30308	610.00	215.95	610.00	1.174
164	C3H5ClO2	METHYL CHLOROACETATE	0.40193	0.25108	0.30950	600.00	241.03	600.00	1.229
165	C3H5ClO2	ETHYL CHLOROFORMATE	0.39599	0.26090	0.28300	508.15	192.00	508.15	1.127
166	C3H5Cl3	1,2,3-TRICHLOROPROPANE	0.42698	0.24650	0.28570	652.00	258.45	652.00	1.384
167	C3H5I	3-IODO-1-PROPENE	0.58800	0.24900	0.28571	595.81	173.86	595.81	1.839
168	C3H5N	PROPIONITRILE	0.24012	0.23452	0.28040	564.40	180.26	564.40	0.777
169	C3H5NO	ACRYLAMIDE	0.27378	0.25200	0.28571	710.00	357.65	710.00	--------
170	C3H5NO	HYDRACRYLONITRILE	0.29250	0.23676	0.22451	690.00	227.15	690.00	1.040
171	C3H5NO	LACTONITRILE	0.29250	0.24607	0.23340	643.00	233.00	643.00	0.983
172	C3H5N3O9	NITROGLYCERINE	0.54190	0.27886	0.29995	680.00	286.15	680.00	1.586
173	C3H6	CYCLOPROPANE	0.25880	0.27400	0.28571	397.91	145.73	397.91	0.619
174	C3H6	PROPYLENE	0.23314	0.27517	0.30246	364.76	87.90	364.76	0.504
175	C3H6Br2	1,2-DIBROMOPROPANE	0.76340	0.33000	0.28571	634.11	217.96	634.11	1.925
176	C3H6Cl2	1,1-DICHLOROPROPANE	0.38737	0.26561	0.28571	560.00	200.00	560.00	1.126
177	C3H6Cl2	1,2-DICHLOROPROPANE	0.38661	0.26055	0.28571	572.00	172.71	572.00	1.150
178	C3H6Cl2	1,3-DICHLOROPROPANE	0.38827	0.26224	0.27100	603.00	173.65	603.00	1.181
179	C3H6Cl2	2,2-DICHLOROPROPANE	0.38630	0.26600	0.28571	539.46	239.36	539.46	1.106
180	C3H6I2	1,2-DIIODOPROPANE	0.74510	0.24200	0.28571	780.49	253.16	780.49	2.566
181	C3H6O	ACETONE	0.27728	0.25760	0.29903	508.20	178.45	508.20	0.786
182	C3H6O	ALLYL ALCOHOL	0.28347	0.25408	0.28571	545.05	144.15	545.05	0.845
183	C3H6O	METHYL VINYL ETHER	0.27658	0.26436	0.25805	437.00	151.15	437.00	0.744
184	C3H6O	n-PROPIONALDEHYDE	0.26909	0.24390	0.28600	496.00	193.15	496.00	0.796
185	C3H6O	1,2-PROPYLENE OXIDE	0.31226	0.27634	0.29353	482.25	161.22	482.25	0.823
186	C3H6O	1,3-PROPYLENE OXIDE	0.30473	0.25347	0.28571	520.00	255.00	520.00	0.894
187	C3H6O2	ETHYL FORMATE	0.33311	0.26940	0.29354	508.40	193.55	508.40	0.917
188	C3H6O2	METHYL ACETATE	0.32119	0.25855	0.27450	506.80	175.15	506.80	0.927
189	C3H6O2	PROPIONIC ACID	0.32283	0.25916	0.27644	604.00	252.45	604.00	0.988
190	C3H6O2S	3-MERCAPTOPROPIONIC ACID	0.37774	0.23621	0.40466	729.00	290.65	729.00	1.213
191	C3H6O3	LACTIC ACID	0.39816	0.26350	0.28570	616.00	291.15	616.00	1.201
192	C3H6O3	METHOXYACETIC ACID	0.35888	0.25150	0.27447	691.00	281.00	691.00	1.170
193	C3H6O3	TRIOXANE	0.40372	0.25860	0.28570	604.00	334.65	604.00	--------
194	C3H6S	THIACYCLOBUTANE	0.30040	0.22800	0.28571	603.00	199.96	603.00	1.014
195	C3H7Br	1-BROMOPROPANE	0.46239	0.26021	0.29165	544.00	163.15	544.00	1.345
196	C3H7Br	2-BROMOPROPANE	0.46222	0.24800	0.38000	532.00	184.15	532.00	1.282
197	C3H7Cl	ISOPROPYL CHLORIDE	0.31887	0.27503	0.28571	489.00	155.97	489.00	0.855
198	C3H7Cl	n-PROPYL CHLORIDE	0.31422	0.27365	0.28570	503.15	150.35	503.15	0.856
199	C3H7F	1-FLUOROPROPANE	0.30710	0.26300	0.28571	422.00	114.16	422.00	0.787
200	C3H7F	2-FLUOROPROPANE	0.28810	0.26200	0.28571	415.68	139.80	415.68	0.733
201	C3H7I	ISOPROPYL IODIDE	0.58619	0.26661	0.30232	578.00	183.15	578.00	1.695
202	C3H7I	n-PROPYL IODIDE	0.58619	0.26130	0.30121	593.00	171.85	593.00	1.739
203	C3H7N	ALLYLAMINE	0.23116	0.23966	0.20760	505.00	184.95	505.00	0.757
204	C3H7N	PROPYLENEIMINE	0.27419	0.25660	0.28600	529.00	229.00	529.00	0.802
205	C3H7NO	N,N-DIMETHYLFORMAMIDE	0.27376	0.23013	0.27630	647.00	212.72	647.00	0.945
206	C3H7NO	N-METHYLACETAMIDE	0.27376	0.23568	0.27379	718.00	301.15	718.00	--------
207	C3H7NO2	1-NITROPROPANE	0.30936	0.24470	0.27365	605.00	169.16	605.00	0.996
208	C3H7NO2	2-NITROPROPANE	0.30936	0.24668	0.27368	594.00	181.83	594.00	0.983
209	C3H7NO3	PROPYL-NITRATE	--------	--------	--------	--------	--------	--------	--------
210	C3H7NO3	ISOPROPYL-NITRATE	--------	--------	--------	--------	--------	--------	--------
211	C3H8	PROPANE	0.22151	0.27744	0.28700	369.82	85.46	369.82	0.493
212	C3H8O	ISOPROPANOL	0.26785	0.26475	0.24300	508.31	185.28	508.31	0.783
213	C3H8O	METHYL ETHYL ETHER	0.28400	0.28920	0.28950	437.80	160.00	437.80	0.692
214	C3H8O	n-PROPANOL	0.27684	0.27200	0.24940	536.71	146.95	536.71	0.802
215	C3H8O2	2-METHOXYETHANOL	0.31877	0.25504	0.28570	564.00	188.05	564.00	0.960
216	C3H8O2	METHYLAL	0.35726	0.30576	0.31755	480.60	168.35	480.60	0.854
217	C3H8O2	1,2-PROPYLENE GLYCOL	0.31839	0.26106	0.20459	626.00	213.15	626.00	1.033
218	C3H8O2	1,3-PROPYLENE GLYCOL	0.32832	0.25080	0.28570	658.00	246.45	658.00	1.052
219	C3H8O3	GLYCEROL	0.34908	0.24902	0.15410	723.00	291.33	723.00	1.257
220	C3H8S	n-PROPYLMERCAPTAN	0.29987	0.27903	0.26920	536.00	159.95	536.00	0.836
221	C3H8S	ISOPROPYL MERCAPTAN	0.29957	0.28094	0.28571	517.00	142.61	517.00	0.809
222	C3H8S	ETHYL-METHYL-SULFIDE	0.28820	0.26200	0.28571	532.80	167.20	532.80	0.832
223	C3H9N	n-PROPYLAMINE	0.22763	0.23878	0.24610	496.95	190.15	496.95	0.714
224	C3H9N	ISOPROPYLAMINE	0.26757	0.28280	0.29720	471.85	177.95	471.85	0.684
225	C3H9N	TRIMETHYLAMINE	0.23283	0.25703	0.26870	433.25	156.08	433.25	0.629
226	C3H9NO	1-AMINO-2-PROPANOL	0.27018	0.23107	0.22134	614.00	274.89	614.00	0.957
227	C3H9NO	3-AMINO-1-PROPANOL	0.30490	0.25110	0.28570	649.00	284.15	649.00	0.972
228	C3H9NO	METHYLETHANOLAMINE	0.29664	0.25232	0.28570	630.00	268.65	630.00	0.934
229	C3H9O4P	TRIMETHYL PHOSPHATE	--------	--------	--------	--------	--------	--------	1.202
230	C3H10N2	1,2-PROPANEDIAMINE	0.23458	0.21184	0.25620	587.00	236.53	587.00	0.856
231	C3H10Si	TRIMETHYL SILANE	0.23859	0.26428	0.29230	432.00	137.26	432.00	0.614
232	C4Cl4S	TETRACHLOROTHIOPHENE	0.51829	0.25100	0.28570	753.00	301.97	753.00	--------
233	C4Cl6	HEXACHLORO-1,3-BUTADIENE	0.47514	0.25300	0.28570	741.00	252.15	741.00	1.556
234	C4F8	OCTAFLUORO-2-BUTENE	0.57656	0.25410	0.28126	392.00	138.15	392.00	1.442

Table 8-1 DENSITY OF LIQUID - ORGANIC COMPOUNDS (continued)

| NO | FORMULA | NAME | density = A B $^{-(1-T/Tc)^n}$ | | | | (density - g/ml, T - K) | | |
			A	B	n	T_c	TMIN	TMAX	density @ 25 C
235	C4F8	OCTAFLUOROCYCLOBUTANE	0.60936	0.26015	0.27770	388.37	232.96	388.37	1.495
236	C4F10	DECAFLUOROBUTANE	0.59730	0.25600	0.26700	386.35	144.95	386.35	1.497
237	C4H2	BUTADIYNE(BIACETYLENE)	0.26400	0.27100	0.28571	478.02	237.16	478.02	0.709
238	C4H2O3	MALEIC ANHYDRIDE	0.44777	0.26141	0.35584	721.00	326.00	721.00	--------
239	C4H4	VINYLACETYLENE	0.25447	0.26348	0.28571	454.00	179.95	454.00	0.680
240	C4H4N2	SUCCINONITRILE	0.26696	0.21793	0.27023	770.00	331.30	770.00	--------
241	C4H4O	FURAN	0.31281	0.24724	0.26050	490.15	187.55	490.15	0.935
242	C4H4O2	DIKETENE	0.35838	0.27300	0.28571	616.00	266.65	616.00	1.050
243	C4H8O3	SUCCINIC ANHYDRIDE	0.44790	0.22300	0.28571	811.00	393.00	811.00	--------
244	C4H4O4	FUMARIC ACID	0.39036	0.23100	0.28571	771.00	560.15	771.00	--------
245	C4H4O4	MALEIC ACID	0.39013	0.23100	0.29000	773.00	403.45	773.00	--------
246	C4H4S	THIOPHENE	0.38423	0.28195	0.30770	579.35	234.94	579.35	1.059
247	C4H5Cl	CHLOROPRENE	0.33147	0.26427	0.27860	525.00	143.15	525.00	0.950
248	C4H5N	trans-CROTONITRILE	0.23818	0.22430	0.28570	586.00	222.00	586.00	0.807
249	C4H5N	cis-CROTONITRILE	0.25317	0.23495	0.28223	568.00	200.55	568.00	0.819
250	C4H5N	METHACRYLONITRILE	0.25317	0.23730	0.29632	554.00	237.35	554.00	0.795
251	C4H5N	PYRROLE	0.29168	0.24703	0.24793	639.75	249.74	639.75	0.965
252	C4H5N	VINYLACETONITRILE	0.25904	0.23915	0.28991	584.00	186.15	584.00	0.829
253	C4H5NO2	METHYL CYANOACETATE	0.32490	0.22700	0.31950	687.00	260.08	687.00	1.119
254	C4H6	CYCLOBUTENE	0.27670	0.27800	0.28571	446.33	153.76	446.33	0.704
255	C4H6	1,2-BUTADIENE	0.24700	0.26696	0.28570	444.00	136.95	444.00	0.646
256	C4H6	1,3-BUTADIENE	0.24597	0.27227	0.29074	425.37	164.25	425.37	0.615
257	C4H6	DIMETHYLACETYLENE	0.25168	0.26924	0.28571	488.15	240.91	488.15	0.686
258	C4H6	ETHYLACETYLENE	0.24366	0.26147	0.28310	443.20	147.43	443.20	0.648
259	C4H6Cl2	1,3-DICHLORO-trans-2-BUTENE	0.38461	0.26734	0.27883	618.00	276.00	618.00	1.153
260	C4H6Cl2	1,4-DICHLORO-cis-2-BUTENE	0.36468	0.24348	0.28571	640.00	225.15	640.00	1.188
261	C4H6Cl2	1,4-DICHLORO-trans-2-BUTENE	0.36248	0.24270	0.28570	646.00	274.15	646.00	1.187
262	C4H6Cl2	3,4-DICHLORO-1-BUTENE	0.37997	0.25861	0.28570	589.00	212.00	589.00	1.148
263	C4H6O	trans-CROTONALDEHYDE	0.28036	0.24575	0.32348	571.00	196.65	571.00	0.847
264	C4H6O	2,5-DIHYDROFURAN	0.32469	0.26347	0.28571	542.00	273.00	542.00	0.939
265	C4H6O	DIVINYL ETHER	0.28036	0.27600	0.28571	463.00	172.05	463.00	0.731
266	C4H6O	METHACROLEIN	0.27638	0.24460	0.28570	530.00	192.15	530.00	0.840
267	C4H6O2	2-BUTYNE-1,4-DIOL	0.33578	0.26000	0.28571	695.00	331.00	695.00	--------
268	C4H6O2	gamma-BUTYROLACTONE	0.32486	0.24071	0.26500	739.00	229.78	739.00	1.125
269	C4H6O2	cis-CROTONIC ACID	0.31113	0.24175	0.28570	647.00	288.65	647.00	1.023
270	C4H6O2	trans-CROTONIC ACID	0.30460	0.23990	0.28570	666.00	344.55	666.00	--------
271	C4H6O2	METHACRYLIC ACID	0.31044	0.24380	0.28570	643.00	288.15	643.00	1.012
272	C4H6O2	METHYL ACRYLATE	0.32153	0.25534	0.28571	536.00	196.32	536.00	0.949
273	C4H6O2	VINYL ACETATE	0.31843	0.25803	0.28270	524.00	180.35	524.00	0.926
274	C4H6O3	ACETIC ANHYDRIDE	0.33578	0.24080	0.26990	569.15	200.15	569.15	1.077
275	C4H6O4	SUCCINIC ACID	0.39352	0.21091	0.28571	806.00	461.15	806.00	--------
276	C4H6O5	DIGLYCOLIC ACID	0.40433	0.21500	0.28571	820.00	421.15	820.00	--------
277	C4H6O5	MALIC ACID	0.40579	0.25800	0.28571	781.00	403.15	781.00	--------
278	C4H6O6	TARTARIC ACID	0.49068	0.23000	0.28571	828.00	479.15	828.00	--------
279	C4H7N	n-BUTYRONITRILE	0.24862	0.24331	0.28586	582.25	161.25	582.25	0.786
280	C4H7N	ISOBUTYRONITRILE	0.24858	0.24449	0.29985	565.00	201.70	565.00	0.766
281	C4H7NO	ACETONE CYANOHYDRIN	0.28033	0.23985	0.28570	647.00	253.15	647.00	0.928
282	C4H7NO	2-METHACRYLAMIDE	0.28517	0.26400	0.29000	741.00	383.65	741.00	--------
283	C4H7NO	3-METHOXYPROPIONITRILE	0.26257	0.22180	0.28570	638.00	210.12	638.00	0.924
284	C4H7NO	2-PYRROLIDONE	0.32236	0.24173	0.29639	792.00	298.15	792.00	1.108
285	C4H8	1-BUTENE	0.23224	0.26630	0.28530	419.59	87.80	419.59	0.588
286	C4H8	cis-2-BUTENE	0.24085	0.27053	0.28571	435.58	134.26	435.58	0.617
287	C4H8	trans-2-BUTENE	0.23730	0.27217	0.28571	428.63	167.62	428.63	0.599
288	C4H8	CYCLOBUTANE	0.26698	0.29634	0.23814	459.93	182.48	459.93	0.689
289	C4H8	ISOBUTENE	0.23181	0.26660	0.27964	417.90	132.81	417.90	0.589
290	C4H8Br2	1,2-DIBROMOBUTANE	0.69270	0.32500	0.28571	659.28	207.76	659.28	1.785
291	C4H8Br2	2,3-DIBROMOBUTANE	0.68750	0.32400	0.28571	656.96	238.66	656.96	1.774
292	C4H8Cl2	1,4-DICHLOROBUTANE	0.37030	0.25940	0.29729	641.00	235.85	641.00	1.135
293	C4H8I2	1,2-DIIODOBUTANE	0.72770	0.26500	0.28571	726.41	279.06	726.41	2.280
294	C4H8O	n-BUTYRALDEHYDE	0.26623	0.24820	0.28570	525.00	176.75	525.00	0.797
295	C4H8O	ISOBUTYRALDEHYDE	0.27294	0.25695	0.28571	507.00	208.15	507.00	0.784
296	C4H8O	1,2-EPOXYBUTANE	0.28219	0.25650	0.28570	526.00	123.15	526.00	0.824
297	C4H8O	METHYL ETHYL KETONE	0.26760	0.25140	0.28570	535.50	186.48	535.50	0.799
298	C4H8O	ETHYL VINYL ETHER	0.27750	0.26793	0.28571	475.15	157.35	475.15	0.749
299	C4H8O	TETRAHYDROFURAN	0.32205	0.28084	0.29120	540.15	164.65	540.15	0.880
300	C4H8O2	cis-2-BUTENE-1,4-DIOL	0.32803	0.24780	0.28570	677.88	284.15	677.88	1.070
301	C4H8O2	trans-2-BUTENE-1,4-DIOL	0.32674	0.24765	0.28570	681.00	300.45	681.00	--------
302	C4H8O2	ISOBUTYRIC ACID	0.30227	0.25490	0.26860	609.15	227.15	609.15	0.946
303	C4H8O2	n-BUTYRIC ACID	0.31132	0.26192	0.27997	628.00	267.95	628.00	0.953
304	C4H8O2	1,4-DIOXANE	0.37018	0.28130	0.30477	587.00	284.95	587.00	1.029
305	C4H8O2	ETHYL ACETATE	0.30654	0.25856	0.27800	523.30	189.60	523.30	0.894
306	C4H8O2	METHYL PROPIONATE	0.30991	0.25865	0.27700	530.60	185.65	530.60	0.909
307	C4H8O2	n-PROPYL FORMATE	0.30848	0.26134	0.28000	538.00	180.25	538.00	0.900
308	C4H8O2S	SULFOLANE	0.40060	0.26983	0.30400	849.00	300.75	849.00	--------
309	C4H8S	TETRAHYDROTHIOPHENE	0.35416	0.29726	0.24882	631.95	176.99	631.95	0.997
310	C4H9Br	1-BROMOBUTANE	0.42953	0.26278	0.28909	577.00	160.75	577.00	1.269
311	C4H9Br	2-BROMOBUTANE	0.42818	0.27695	0.24000	567.00	161.25	567.00	1.253
312	C4H9Cl	n-BUTYL CHLORIDE	0.30871	0.26429	0.29540	537.00	150.05	537.00	0.880

Table 8-1 DENSITY OF LIQUID - ORGANIC COMPOUNDS (continued)

NO	FORMULA	NAME	density = A B $^{-(1-T/Tc)^n}$				(density - g/ml, T - K)		
			A	B	n	T_C	TMIN	TMAX	density @ 25 C
313	C4H9Cl	sec-BUTYL CHLORIDE	0.30859	0.26946	0.27913	520.60	141.85	520.60	0.868
314	C4H9Cl	tert-BUTYL CHLORIDE	0.30888	0.27727	0.28571	507.00	247.75	507.00	0.836
315	C4H9I	2-IODO-2-METHYLPROPANE	0.52230	0.26700	0.28571	587.90	234.96	587.90	1.536
316	C4H9N	PYRROLIDINE	0.28603	0.26233	0.26330	568.55	215.31	568.55	0.860
317	C4H9NO	N,N-DIMETHYLACETAMIDE	0.27141	0.23272	0.27021	658.00	253.15	658.00	0.937
318	C4H9NO	MORPHOLINE	0.31566	0.25657	0.25640	618.00	270.05	618.00	0.996
319	C4H9NO2	1-NITROBUTANE	0.31630	0.26000	0.28571	624.00	191.83	624.00	0.968
320	C4H9NO2	2-NITROBUTANE	0.32480	0.26400	0.28571	615.00	141.16	615.00	0.978
321	C4H10	n-BUTANE	0.22827	0.27240	0.28630	425.18	134.86	425.18	0.573
322	C4H10	ISOBUTANE	0.22281	0.27294	0.27301	408.14	113.54	408.14	0.552
323	C4H10N2	PIPERAZINE	0.27801	0.32300	0.28571	638.00	379.15	638.00	--------
324	C4H10O	n-BUTANOL	0.26891	0.26674	0.24570	562.93	183.85	562.93	0.806
325	C4H10O	sec-BUTANOL	0.27343	0.26350	0.26040	536.01	158.45	536.01	0.805
326	C4H10O	tert-BUTANOL	0.26921	0.25650	0.27370	506.20	298.97	506.20	--------
327	C4H10O	DIETHYL ETHER	0.27267	0.27608	0.29358	466.70	156.85	466.70	0.708
328	C4H10O	METHYL-PROPYL-ETHER	0.26370	0.26300	0.28571	476.20	156.87	476.20	0.723
329	C4H10O	METHYL ISOPROPYL ETHER	0.26856	0.28478	0.24440	464.50	127.93	464.50	0.714
330	C4H10O	ISOBUTANOL	0.26984	0.27206	0.23435	547.73	165.15	547.73	0.797
331	C4H10O2	1,3-BUTANEDIOL	0.32441	0.25980	0.28570	643.00	196.15	643.00	1.002
332	C4H10O2	1,4-BUTANEDIOL	0.31558	0.25131	0.28570	667.00	293.05	667.00	1.013
333	C4H10O2	2,3-BUTANEDIOL	0.33718	0.26990	0.28600	611.00	280.75	611.00	0.994
334	C4H10O2	t-BUTYL HYDROPEROXIDE	0.30445	0.26825	0.28570	576.00	277.45	576.00	0.886
335	C4H10O2	1,2-DIMETHOXYETHANE	0.29886	0.26183	0.28571	536.15	215.15	536.15	0.865
336	C4H10O2	2-ETHOXYETHANOL	0.31086	0.25983	0.28570	569.00	183.00	569.00	0.925
337	C4H10O3	DIETHYLENE GLYCOL	0.34013	0.26112	0.24220	744.60	262.70	744.60	1.114
338	C4H10O4S	DIETHYL SULFATE	0.38743	0.26131	0.40840	792.00	248.00	792.00	1.172
339	C4H10S	n-BUTYL MERCAPTAN	0.29378	0.27053	0.29944	569.00	157.46	569.00	0.837
340	C4H10S	ISOBUTYL MERCAPTAN	0.29378	0.28222	0.25951	559.00	128.31	559.00	0.830
341	C4H10S	sec-BUTYL MERCAPTAN	0.29378	0.27024	0.30682	554.00	133.02	554.00	0.825
342	C4H10S	tert-BUTYL MERCAPTAN	0.29377	0.27351	0.32002	530.00	274.26	530.00	0.795
343	C4H10S	DIETHYL SULFIDE	0.28226	0.26333	0.27445	557.15	169.20	557.15	0.832
344	C4H10S	ISOPROPYL-METHYL-SULFIDE	0.28830	0.26900	0.28571	551.00	171.65	551.00	0.825
345	C4H10S	METHYL-PROPYL-SULFIDE	0.28790	0.26600	0.28571	563.00	160.19	563.00	0.837
346	C4H10S2	DIETHYL DISULFIDE	0.34157	0.27764	0.30060	642.00	171.63	642.00	0.988
347	C4H11N	n-BUTYLAMINE	0.23379	0.24470	0.24278	531.90	224.05	531.90	0.741
348	C4H11N	ISOBUTYLAMINE	0.23429	0.24789	0.23650	513.73	188.55	513.73	0.730
349	C4H11N	sec-BUTYLAMINE	0.23593	0.25100	0.24711	514.30	168.65	514.30	0.720
350	C4H11N	tert-BUTYLAMINE	0.24960	0.27438	0.25400	483.90	206.19	483.90	0.688
351	C4H11N	DIETHYLAMINE	0.24111	0.25362	0.27280	496.60	223.35	496.60	0.702
352	C4H11NO	DIMETHYLETHANOLAMINE	0.29716	0.26120	0.28571	571.82	214.15	571.82	0.882
353	C4H11NO2	DIETHANOLAMINE	0.30126	0.23968	0.18920	715.00	301.15	715.00	--------
354	C4H11NO2	2-AMINOETHOXYETHANOL	0.31903	0.24723	0.28571	699.00	293.15	699.00	1.051
355	C4H12N2O	N-AMINOETHYL ETHANOLAMINE	0.31637	0.25300	0.28570	698.00	273.15	698.00	1.022
356	C4H12Si	TETRAMETHYLSILANE	0.24713	0.26978	0.29380	450.40	174.07	450.40	0.641
357	C4H13N3	DIETHYLENE TRIAMINE	0.30174	0.25671	0.28571	676.00	234.15	676.00	0.954
358	C5Cl6	HEXACHLOROCYCLOPENTADIENE	0.52106	0.25404	0.28570	746.00	284.49	746.00	1.703
359	C5H4O2	FURFURAL	0.37235	0.26030	0.28570	657.00	236.65	657.00	1.155
360	C5H5N	PYRIDINE	0.30752	0.24333	0.30450	619.95	231.53	619.95	0.979
361	C5H6	CYCLOPENTADIENE	0.29310	0.27553	0.28571	507.00	188.15	507.00	0.797
362	C5H6	2-METHYL-1-BUTENE-3-YNE	0.26653	0.27671	0.30821	492.00	160.15	492.00	0.699
363	C5H6	1-PENTENE-3-YNE	0.25821	0.24054	0.36452	520.00	150.00	520.00	0.734
364	C5H6	1-PENTENE-4-YNE	0.26065	0.26680	0.28570	503.00	150.00	503.00	0.724
365	C5H6N2	GLUTARONITRILE	0.26738	0.21579	0.34480	782.00	244.21	782.00	0.981
366	C5H6O2	FURFURYL ALCOHOL	0.37299	0.27753	0.23180	632.00	258.52	632.00	1.127
367	C5H6O3	GLUTARIC ANHYDRIDE	0.41477	0.22900	0.28571	838.00	328.00	838.00	--------
368	C5H6O4	CITRACONIC ACID	0.37251	0.21484	0.28570	829.00	356.15	829.00	--------
369	C5H6O4	ITACONIC ACID	0.37298	0.21666	0.28570	821.00	438.75	821.00	--------
370	C5H6S	2-METHYLTHIOPHENE	0.33860	0.26500	0.28571	610.00	209.77	610.00	1.014
371	C5H6S	3-METHYLTHIOPHENE	0.33660	0.26300	0.28571	615.00	204.18	615.00	1.016
372	C5H7N	N-METHYLPYRROLE	0.28663	0.23565	0.34379	610.00	216.91	610.00	0.903
373	C5H7NO2	ETHYL CYANOACETATE	0.31595	0.23894	0.29300	679.00	250.65	679.00	1.058
374	C5H8	CYCLOPENTENE	0.28148	0.27498	0.28571	507.00	138.13	507.00	0.767
375	C5H8	ISOPRENE	0.24604	0.26488	0.28571	484.00	127.27	484.00	0.676
376	C5H8	3-METHYL-1,2-BUTADIENE	0.23409	0.24625	0.29041	490.00	159.53	490.00	0.681
377	C5H8	2-METHYL-1,3-BUTADIENE	0.23400	0.24800	0.28571	483.30	127.20	483.30	0.675
378	C5H8	1,2-PENTADIENE	0.24681	0.26440	0.28786	500.00	135.89	500.00	0.688
379	C5H8	cis-1,3-PENTADIENE	0.23992	0.25594	0.28570	499.00	132.35	499.00	0.686
380	C5H8	trans-1,3-PENTADIENE	0.24681	0.26773	0.30401	500.00	185.71	500.00	0.671
381	C5H8	1,4-PENTADIENE	0.22482	0.22721	0.33730	479.00	124.86	479.00	0.653
382	C5H8	2,3-PENTADIENE	0.23091	0.24613	0.27012	497.00	147.50	497.00	0.690
383	C5H8	1-PENTYNE	0.24591	0.23520	0.35300	481.20	167.45	481.20	0.688
384	C5H8	2-PENTYNE	0.25230	0.27000	0.28571	521.99	163.86	521.99	0.705
385	C5H8	3-METHYL-1-BUTYNE	0.24770	0.26008	0.30807	463.20	183.45	463.20	0.660
386	C5H8	SPIROPENTANE	0.28800	0.29700	0.28571	499.74	166.11	499.74	0.735
387	C5H8N4O12	PENTAERYTHRITOL TETRANITRATE	0.43297	0.29100	0.28571	676.00	413.65	676.00	--------
388	C5H8O	CYCLOPENTANONE	0.32605	0.27182	0.31334	626.00	221.85	626.00	0.945
389	C5H8O	METHYL ISOPROPENYL KETONE	0.27709	0.25094	0.28570	566.00	219.55	566.00	0.846
390	C5H8O2	ACETYLACETONE	0.30995	0.25792	0.25054	602.00	249.65	602.00	0.971

Table 8-1 DENSITY OF LIQUID - ORGANIC COMPOUNDS (continued)

NO	FORMULA	NAME	density = A B $^{-(1-T/T_c)^n}$						(density - g/ml, T - K)
			A	B	n	T_c	TMIN	TMAX	density @ 25 C
391	C5H8O2	ALLYL ACETATE	0.30889	0.25663	0.28571	559.00	138.00	559.00	0.922
392	C5H8O2	ETHYL ACRYLATE	0.31019	0.25833	0.28571	553.00	201.95	553.00	0.918
393	C5H8O2	METHYL METHACRYLATE	0.30985	0.25357	0.28571	564.00	224.95	564.00	0.937
394	C5H8O2	VINYL PROPIONATE	0.30976	0.26200	0.28571	546.00	364.35	546.00	--------
395	C5H8O3	2-HYDROXYETHYL ACRYLATE	0.32342	0.25961	0.28570	662.00	213.00	662.00	1.008
396	C5H8O3	LEVULINIC ACID	0.33854	0.24979	0.25720	723.00	308.15	723.00	--------
397	C5H8O3	METHYL ACETOACETATE	0.33853	0.26807	0.21300	642.00	193.15	642.00	1.072
398	C5H8O4	GLUTARIC ACID	0.36394	0.24540	0.25522	807.00	370.65	807.00	--------
399	C5H9N	VALERONITRILE	0.25121	0.24826	0.27970	603.00	176.95	603.00	0.794
400	C5H9NO	n-BUTYL ISOCYANATE	0.27538	0.20710	0.41200	568.00	193.00	568.00	0.877
401	C5H9NO	N-METHYL-2-PYRROLIDONE	0.31372	0.25432	0.27360	724.00	249.15	724.00	1.025
402	C5H9NO4	L-GLUTAMIC ACID	0.38403	0.21500	0.28571	886.00	497.15	886.00	--------
403	C5H10	CYCLOPENTANE	0.27236	0.27247	0.28571	511.76	179.31	511.76	0.750
404	C5H10	2-METHYL-1-BUTENE	0.23663	0.26065	0.28570	465.00	135.58	465.00	0.645
405	C5H10	2-METHYL-2-BUTENE	0.23693	0.25700	0.28571	471.00	139.39	471.00	0.657
406	C5H10	3-METHYL-1-BUTENE	0.23215	0.26924	0.26388	450.37	104.66	450.37	0.622
407	C5H10	1-PENTENE	0.23787	0.26648	0.29050	464.78	107.93	464.78	0.635
408	C5H10	cis-2-PENTENE	0.23217	0.25138	0.29862	475.93	121.75	475.93	0.650
409	C5H10	trans-2-PENTENE	0.23215	0.26198	0.27760	475.37	132.89	475.37	0.643
410	C5H10Br2	2,3-DIBROMO-2-METHYLBUTANE	0.54420	0.32400	0.28571	668.37	288.00	668.37	1.410
411	C5H10Cl2	1,5-DICHLOROPENTANE	0.33398	0.24438	0.28571	663.00	200.35	663.00	1.096
412	C5H10O	METHYL ISOPROPYL KETONE	0.27526	0.26204	0.28570	553.00	181.15	553.00	0.805
413	C5H10O	2-PENTANONE	0.28617	0.26662	0.32850	561.08	196.29	561.08	0.802
414	C5H10O	DIETHYL KETONE	0.25635	0.24291	0.27364	560.95	234.18	560.95	0.810
415	C5H10O	VALERALDEHYDE	0.27258	0.27025	0.24456	554.00	182.00	554.00	0.805
416	C5H10O2	n-BUTYL FORMATE	0.30397	0.26606	0.27850	559.00	181.25	559.00	0.887
417	C5H10O2	ETHYL PROPIONATE	0.30663	0.26280	0.29420	546.00	199.25	546.00	0.884
418	C5H10O2	ISOBUTYL FORMATE	0.29013	0.25670	0.26830	551.35	177.35	551.35	0.875
419	C5H10O2	ISOPROPYL ACETATE	0.30574	0.26734	0.28600	538.00	199.75	538.00	0.871
420	C5H10O2	n-PROPYL ACETATE	0.29499	0.25600	0.27830	549.40	178.15	549.40	0.883
421	C5H10O2	METHYL n-BUTYRATE	0.29809	0.25700	0.27720	554.50	187.35	554.50	0.893
422	C5H10O2	2-METHYLBUTYRIC ACID	0.29438	0.25244	0.28571	643.00	193.00	643.00	0.932
423	C5H10O2	ISOVALERIC ACID	0.30463	0.26610	0.27430	634.00	243.85	634.00	0.926
424	C5H10O2	VALERIC ACID	0.30423	0.26363	0.28151	651.00	239.15	651.00	0.934
425	C5H10O2	TETRAHYDROFURFURYL ALCOHOL	0.35218	0.28281	0.23330	639.00	193.00	639.00	1.048
426	C5H10O2S	3-METHYL SULFOLANE	0.38016	0.26878	0.31440	817.00	273.65	817.00	1.188
427	C5H10O3	DIETHYL CARBONATE	0.33184	0.26670	0.28570	576.00	230.15	576.00	0.970
428	C5H10O3	ETHYL LACTATE	0.33372	0.21190	0.45530	588.00	247.15	588.00	1.027
429	C5H10S	THIACYCLOHEXANE	0.30760	0.25200	0.28571	657.12	292.14	657.12	0.981
430	C5H10S	CYCLOPENTANETHIOL	0.32910	0.27600	0.28571	629.00	155.39	629.00	0.961
431	C5H11Br	1-BROMOPENTANE	0.46230	0.30300	0.28571	564.76	185.26	564.76	1.212
432	C5H11Cl	1-CHLOROPENTANE	0.30284	0.26950	0.28040	568.00	174.15	568.00	0.878
433	C5H11Cl	1-CHLORO-3-METHYLBUTANE	0.29190	0.25900	0.28571	558.87	168.76	558.87	0.865
434	C5H11Cl	2-CHLORO-2-METHYLBUTANE	0.29560	0.26300	0.28571	548.97	199.66	548.97	0.860
435	C5H11N	N-METHYLPYRROLIDINE	0.28582	0.27360	0.28600	550.00	183.15	550.00	0.806
436	C5H11N	PIPERIDINE	0.27645	0.25460	0.27140	594.05	262.65	594.05	0.858
437	C5H11NO	tert-BUTYLFORMAMIDE	0.26401	0.23707	0.28571	692.00	289.15	692.00	0.899
438	C5H12	ISOPENTANE	0.23725	0.27610	0.28673	460.43	113.25	460.43	0.616
439	C5H12	NEOPENTANE	0.23295	0.27380	0.29195	433.78	256.58	433.78	0.586
440	C5H12	n-PENTANE	0.23143	0.26923	0.28215	469.65	143.42	469.65	0.621
441	C5H12O	2,2-DIMETHYL-1-PROPANOL	0.27001	0.27700	0.28571	550.00	327.15	550.00	--------
442	C5H12O	tert-PENTYL-ALCOHOL	0.29300	0.28000	0.28571	549.00	327.00	549.00	--------
443	C5H12O	2-METHYL-1-BUTANOL	0.26957	0.26829	0.23220	565.00	203.00	565.00	0.814
444	C5H12O	2-METHYL-2-BUTANOL	0.26959	0.25962	0.26470	545.15	264.35	545.15	0.805
445	C5H12O	3-METHYL-1-BUTANOL	0.27219	0.26080	0.28570	579.45	155.95	579.45	0.812
446	C5H12O	3-METHYL-2-BUTANOL	0.27665	0.26440	0.28500	574.00	188.00	574.00	0.814
447	C5H12O	1-PENTANOL	0.26923	0.26730	0.25060	586.15	195.56	586.15	0.812
448	C5H12O	2-PENTANOL	0.27942	0.26670	0.28570	552.00	200.00	552.00	0.805
449	C5H12O	3-PENTANOL	0.28353	0.26520	0.28570	547.00	204.15	547.00	0.818
450	C5H12O	METHYL sec-BUTYL ETHER	0.26892	0.26996	0.28570	498.00	150.00	498.00	0.737
451	C5H12O	METHYL tert-BUTYL ETHER	0.26791	0.27032	0.28290	497.10	164.55	497.10	0.735
452	C5H12O	METHYL ISOBUTYL ETHER	0.26663	0.27282	0.28570	497.00	150.00	497.00	0.725
453	C5H12O	ETHYL PROPYL ETHER	0.26206	0.26600	0.29200	500.23	145.65	500.23	0.724
454	C5H12O2	ETHYLENE GLYCOL MONOPROPYL ETHER	0.30303	0.26066	0.28570	582.00	183.15	582.00	0.906
455	C5H12O2	NEOPENTYL GLYCOL	0.30146	0.27400	0.28571	643.00	400.00	643.00	--------
456	C5H12O2	1,5-PENTANEDIOL	0.30820	0.25064	0.28570	673.00	257.15	673.00	0.994
457	C5H12O3	2-(2-METHOXYETHOXY)ETHANOL	0.32247	0.25180	0.28570	630.00	197.15	630.00	1.017
458	C5H12O4	PENTAERYTHRITOL	0.35711	0.28100	0.29000	780.00	534.15	780.00	--------
459	C5H12S	n-PENTYL MERCAPTAN	0.29030	0.27123	0.30131	598.00	197.45	598.00	0.838
460	C5H12S	BUTYL-METHYL-SULFIDE	0.28260	0.26500	0.28571	591.00	175.33	591.00	0.838
461	C5H12S	ETHYL-PROPYL-SULFIDE	0.28440	0.26800	0.28571	584.00	156.15	584.00	0.832
462	C5H12S	2-METHYL-2-BUTANETHIOL	0.30050	0.28800	0.28571	566.00	169.38	566.00	0.821
463	C5H13N	n-PENTYLAMINE	0.23881	0.24876	0.25190	555.00	218.15	555.00	0.751
464	C5H13NO2	METHYL DIETHANOLAMINE	0.32253	0.25430	0.28570	678.00	252.15	678.00	1.029
465	C6Cl6	HEXACHLOROBENZENE	0.54029	0.21900	0.28571	825.00	501.70	825.00	--------
466	C6F6	HEXAFLUOROBENZENE	0.55556	0.25947	0.27853	516.73	278.25	516.73	1.606
467	C6H3ClN2O4	1-CHLORO-2,4-DINITROBENZENE	0.42377	0.23584	0.24220	813.77	326.55	813.77	--------
468	C6H3Cl2NO2	1,2-DICHLORO-4-NITROBENZENE	0.44041	0.24309	0.27433	758.00	315.65	758.00	--------

Table 8-1 DENSITY OF LIQUID - ORGANIC COMPOUNDS (continued)

NO	FORMULA	NAME	density = A B$^{-(1-T/Tc)^n}$				(density - g/ml, T - K)		
			A	B	n	T_C	TMIN	TMAX	density @ 25 C
469	C6H3Cl3	1,2,4-TRICHLOROBENZENE	0.44365	0.25240	0.28570	725.00	290.15	725.00	1.449
470	C6H3N3O6	1,3,5-TRINITROBENZENE	0.40975	0.21100	0.28571	1005.00	398.40	1005.00	--------
471	C6H4Br2	m-DIBROMOBENZENE	0.63364	0.27420	0.28571	761.00	266.25	761.00	1.947
472	C6H4ClNO2	m-CHLORONITROBENZENE	0.36472	0.22340	0.24499	742.00	317.65	742.00	--------
473	C6H4ClNO2	o-CHLORONITROBENZENE	0.40415	0.24650	0.28570	757.00	306.15	757.00	--------
474	C6H4ClNO2	p-CHLORONITROBENZENE	0.36473	0.23023	0.22680	751.00	356.65	751.00	--------
475	C6H4Cl2	m-DICHLOROBENZENE	0.41882	0.26147	0.31526	683.95	248.39	683.95	1.283
476	C6H4Cl2	o-DICHLOROBENZENE	0.41887	0.26112	0.30815	705.00	256.15	705.00	1.301
477	C6H4Cl2	p-DICHLOROBENZENE	0.41880	0.26276	0.30788	684.75	326.14	684.75	--------
478	C6H4F2	m-DIFLUOROBENZENE	0.40070	0.26500	0.28571	552.94	249.16	552.94	1.162
479	C6H4F2	o-DIFLUOROBENZENE	0.39520	0.26400	0.28571	554.46	239.16	554.46	1.150
480	C6H4F2	p-DIFLUOROBENZENE	0.40120	0.26600	0.28571	556.00	260.16	556.00	1.162
481	C6H4N2O4	m-DINITROBENZENE	0.38739	0.23724	0.22716	805.00	364.00	805.00	--------
482	C6H4N2O4	o-DINITROBENZENE	0.38708	0.24200	0.28571	831.00	390.08	831.00	--------
483	C6H4N2O4	p-DINITROBENZENE	0.38776	0.25000	0.28571	803.00	446.60	803.00	--------
484	C6H5Br	BROMOBENZENE	0.48497	0.26632	0.28210	670.15	242.43	670.15	1.487
485	C6H5Cl	MONOCHLOROBENZENE	0.37818	0.27648	0.29036	632.35	227.95	632.35	1.101
486	C6H5ClO	m-CHLOROPHENOL	0.41332	0.27300	0.28570	729.00	306.00	729.00	--------
487	C6H5ClO	o-CHLOROPHENOL	0.39564	0.25865	0.27140	675.00	282.00	675.00	1.255
488	C6H5ClO	p-CHLOROPHENOL	0.39568	0.25180	0.29540	738.00	316.00	738.00	--------
489	C6H5Cl2N	3,4-DICHLOROANILINE	0.40679	0.24610	0.28570	800.00	344.65	800.00	--------
490	C6H5F	FLUOROBENZENE	0.35747	0.27277	0.28291	560.09	230.94	560.09	1.019
491	C6H5I	IODOBENZENE	0.58162	0.26381	0.28950	721.15	241.83	721.15	1.822
492	C6H5NO2	NITROBENZENE	0.36140	0.24731	0.28571	719.00	278.91	719.00	1.199
493	C6H6	BENZENE	0.30090	0.26770	0.28180	562.16	278.68	562.16	0.873
494	C6H6ClN	m-CHLOROANILINE	0.35047	0.23202	0.32410	751.00	262.75	751.00	1.211
495	C6H6ClN	o-CHLOROANILINE	0.35054	0.24214	0.26040	722.00	481.99	722.00	--------
496	C6H6ClN	p-CHLOROANILINE	0.35048	0.23739	0.28500	754.00	343.05	754.00	--------
497	C6H6N2	cis-DICYANO-1-BUTENE	0.27111	0.20100	0.28571	691.00	249.00	691.00	1.062
498	C6H6N2	trans-DICYANO-1-BUTENE	0.27055	0.20200	0.28571	689.00	260.00	689.00	1.054
499	C6H6N2	1,4-DICYANO-2-BUTENE	0.24885	0.20040	0.28571	755.00	349.00	755.00	--------
500	C6H6N2O2	m-NITROANILINE	0.34128	0.26400	0.28571	815.00	387.15	815.00	--------
501	C6H6N2O2	o-NITROANILINE	0.34058	0.27500	0.28571	784.00	344.65	784.00	--------
502	C6H6N2O2	p-NITROANILINE	0.34240	0.25200	0.28571	851.00	420.65	851.00	--------
503	C6H6O	PHENOL	0.41476	0.32162	0.32120	694.25	314.06	694.25	--------
504	C6H6O2	1,2-BENZENEDIOL	0.36704	0.25257	0.26690	764.00	377.60	764.00	--------
505	C6H6O2	1,3-BENZENEDIOL	0.36705	0.25652	0.24327	810.00	382.00	810.00	--------
506	C6H6O2	p-HYDROQUINONE	0.36707	0.32700	0.28571	822.00	444.65	822.00	--------
507	C6H6O3	1,2,3-BENZENETRIOL	0.39752	0.40500	0.28571	830.00	407.00	830.00	--------
508	C6H6S	PHENYL MERCAPTAN	0.34977	0.26326	0.30798	689.00	258.26	689.00	1.073
509	C6H7N	ANILINE	0.31190	0.25000	0.28571	699.00	267.13	699.00	1.018
510	C6H7N	2-METHYLPYRIDINE	0.29103	0.24646	0.27179	621.00	206.44	621.00	0.940
511	C6H7N	3-METHYLPYRIDINE	0.29102	0.24342	0.28343	645.00	255.01	645.00	0.952
512	C6H7N	4-METHYLPYRIDINE	0.28600	0.24104	0.27524	646.15	276.73	646.15	0.950
513	C6H8	1,3-CYCLOHEXADIENE	0.28927	0.25874	0.31490	558.00	161.00	558.00	0.837
514	C6H8	METHYLCYCLOPENTADIENE	0.28769	0.27431	0.28571	541.00	150.00	541.00	0.805
515	C6H8N2	ADIPONITRILE	0.26638	0.23008	0.28379	781.00	275.64	781.00	0.960
516	C6H8N2	METHYLGLUTARONITRILE	0.26768	0.21875	0.35460	742.00	228.15	742.00	0.950
517	C6H8N2	m-PHENYLENEDIAMINE	0.28684	0.21430	0.25364	824.00	334.00	824.00	--------
518	C6H8N2	o-PHENYLENEDIAMINE	0.34379	0.25093	0.28571	781.00	376.95	781.00	--------
519	C6H8N2	p-PHENYLENEDIAMINE	0.34134	0.24797	0.28571	796.00	413.00	796.00	--------
520	C6H8N2	PHENYLHYDRAZINE	0.25872	0.19684	0.24048	761.00	292.35	761.00	1.094
521	C6H8N2O	BIS(CYANOETHYL)ETHER	0.32929	0.26359	0.30190	783.00	246.85	783.00	1.044
522	C6H8O4	DIMETHYL MALEATE	0.35765	0.25098	0.29120	675.00	254.15	675.00	1.148
523	C6H8O6	ASCORBIC ACID	0.52025	0.27509	0.28571	783.00	465.15	783.00	--------
524	C6H8O7	CITRIC ACID	0.44643	0.23915	0.28570	822.00	426.15	822.00	--------
525	C6H10	1-METHYLCYCLOPENTENE	0.27600	0.27300	0.28571	541.99	145.96	541.99	0.776
526	C6H10	3-METHYLCYCLOPENTENE	0.26790	0.26900	0.28571	535.71	130.16	535.71	0.759
527	C6H10	4-METHYLCYCLOPENTENE	0.26490	0.26500	0.28571	543.75	112.31	543.75	0.763
528	C6H10	CYCLOHEXENE	0.28236	0.27161	0.28571	560.40	169.67	560.40	0.806
529	C6H10	2,3-DIMETHYL-1,3-BUTADIENE	0.25998	0.27194	0.28900	526.00	197.15	526.00	0.723
530	C6H10	1,5-HEXADIENE	0.24231	0.26619	0.26770	507.00	132.47	507.00	0.688
531	C6H10	cis,trans-2,4-HEXADIENE	0.24751	0.26020	0.28910	538.00	177.05	538.00	0.719
532	C6H10	trans,trans-2,4-HEXADIENE	0.24817	0.26626	0.28209	535.00	228.25	535.00	0.710
533	C6H10	1-HEXYNE	0.25511	0.27185	0.27710	516.20	141.25	516.20	0.712
534	C6H10	2-HEXYNE	0.24817	0.25248	0.31611	549.00	183.65	549.00	0.727
535	C6H10	3-HEXYNE	0.24596	0.26065	0.28571	544.00	170.05	544.00	0.718
536	C6H10O	CYCLOHEXANONE	0.31558	0.27183	0.27167	629.15	242.00	629.15	0.942
537	C6H10O	MESITYL OXIDE	0.27648	0.25438	0.28484	600.00	220.15	600.00	0.852
538	C6H10O2	epsilon-CAPROLACTONE	0.32432	0.25420	0.28570	771.00	271.85	771.00	1.067
539	C6H10O2	ETHYL METHACRYLATE	0.30100	0.25690	0.28571	577.00	223.15	577.00	0.908
540	C6H10O2	n-PROPYL ACRYLATE	0.30211	0.25955	0.28571	569.00	273.15	569.00	0.900
541	C6H10O3	ETHYLACETOACETATE	0.31956	0.24910	0.28570	643.00	234.15	643.00	1.023
542	C6H10O3	PROPIONIC ANHYDRIDE	0.32780	0.25807	0.28571	618.00	228.15	618.00	1.007
543	C6H10O4	ADIPIC ACID	0.36537	0.25915	0.28200	809.00	425.50	809.00	--------
544	C6H10O4	DIETHYL OXALATE	0.35132	0.25332	0.33419	646.00	232.55	646.00	1.073
545	C6H10O4	ETHYLENE GLYCOL DIACETATE	0.33859	0.24565	0.28570	653.00	242.15	653.00	1.101
546	C6H10O4	ETHYLIDENE DIACETATE	0.35995	0.26895	0.29542	635.00	292.00	635.00	1.069

Table 8-1 DENSITY OF LIQUID - ORGANIC COMPOUNDS (continued)

| NO | FORMULA | NAME | density = A B$^{-(1-T/T_c)^{\wedge}n}$ | | | | (density - g/ml, T - K) | | |
			A	B	n	T_c	TMIN	TMAX	density @ 25 C
547	C6H11N	HEXANENITRILE	0.25303	0.24921	0.28637	622.05	192.85	622.05	0.801
548	C6H11NO	epsilon-CAPROLACTAM	0.31765	0.25357	0.28571	806.00	342.36	806.00	-------
549	C6H11NO	CYCLOHEXANONE OXIME	0.30678	0.29100	0.28571	715.00	363.15	715.00	-------
550	C6H12	CYCLOHEXANE	0.27376	0.27408	0.28511	553.54	279.69	553.54	0.773
551	C6H12	2,3-DIMETHYL-1-BUTENE	0.24347	0.26774	0.28571	500.00	115.89	500.00	0.673
552	C6H12	2,3-DIMETHYL-2-BUTENE	0.22624	0.24252	0.26403	524.00	198.82	524.00	0.703
553	C6H12	3,3-DIMETHYL-1-BUTENE	0.25273	0.28406	0.29910	480.00	157.95	480.00	0.648
554	C6H12	2-ETHYL-1-BUTENE	0.23121	0.25140	0.27523	512.00	141.61	512.00	0.685
555	C6H12	trans-3-METHYL-2-PENTENE	0.24820	0.27000	0.28571	521.00	134.70	521.00	0.693
556	C6H12	1-HEXENE	0.23941	0.26605	0.28571	504.03	133.39	504.03	0.667
557	C6H12	cis-2-HEXENE	0.23443	0.25626	0.27838	513.00	132.00	513.00	0.683
558	C6H12	trans-2-HEXENE	0.23733	0.26272	0.28571	513.00	140.17	513.00	0.673
559	C6H12	cis-3-HEXENE	0.23934	0.26339	0.28571	509.00	135.33	509.00	0.675
560	C6H12	trans-3-HEXENE	0.23978	0.26234	0.29552	509.00	159.73	509.00	0.673
561	C6H12	METHYLCYCLOPENTANE	0.26391	0.27042	0.28276	532.79	130.73	532.79	0.745
562	C6H12	2-METHYL-1-PENTENE	0.23956	0.26336	0.28570	507.00	137.42	507.00	0.675
563	C6H12	2-METHYL-2-PENTENE	0.23865	0.26076	0.28570	514.00	138.07	514.00	0.681
564	C6H12	3-METHYL-1-PENTENE	0.24515	0.27450	0.28442	495.00	120.20	495.00	0.663
565	C6H12	3-METHYL-cis-2-PENTENE	0.24536	0.26738	0.28391	515.00	138.31	515.00	0.689
566	C6H12	4-METHYL-1-PENTENE	0.24191	0.27164	0.28571	496.00	119.51	496.00	0.659
567	C6H12	4-METHYL-cis-2-PENTENE	0.24204	0.26986	0.28571	499.00	138.30	499.00	0.665
568	C6H12	4-METHYL-trans-2-PENTENE	0.24129	0.26962	0.28571	501.00	132.35	501.00	0.664
569	C6H12N2	TRIETHYLENEDIAMINE	0.29394	0.27400	0.28571	655.00	434.25	655.00	-------
570	C6H12O	BUTYL VINYL ETHER	0.26764	0.26200	0.28570	536.00	181.25	536.00	0.774
571	C6H12O	CYCLOHEXANOL	0.29681	0.24340	0.28570	625.15	296.60	625.15	0.960
572	C6H12O	1-HEXANAL	0.27144	0.27131	0.24447	579.00	217.15	579.00	0.810
573	C6H12O	ETHYL ISOPROPYL KETONE	0.27144	0.24527	0.34305	567.00	200.00	567.00	0.806
574	C6H12O	2-HEXANONE	0.27144	0.26073	0.29630	587.05	217.35	587.05	0.807
575	C6H12O	3-HEXANONE	0.27517	0.26557	0.28660	582.82	217.50	582.82	0.810
576	C6H12O	METHYL ISOBUTYL KETONE	0.26654	0.25887	0.28571	571.40	189.15	571.40	0.796
577	C6H12O2	n-PENTYL FORMATE	0.29848	0.25775	0.30860	576.00	199.65	576.00	0.881
578	C6H12O2	n-BUTYL ACETATE	0.29857	0.26028	0.30900	579.65	199.65	579.65	0.876
579	C6H12O2	sec-BUTYL ACETATE	0.29861	0.26088	0.30420	561.00	174.15	561.00	0.868
580	C6H12O2	tert-BUTYL ACETATE	0.29861	0.26927	0.26996	545.00	283.15	545.00	0.861
581	C6H12O2	ETHYL n-BUTYRATE	0.27594	0.24632	0.26390	571.00	175.15	571.00	0.874
582	C6H12O2	ETHYL ISOBUTYRATE	0.29410	0.26107	0.28571	553.15	185.00	553.15	0.863
583	C6H12O2	ISOBUTYL ACETATE	0.29861	0.26575	0.28366	561.00	174.30	561.00	0.869
584	C6H12O2	n-PROPYL PROPIONATE	0.29860	0.26150	0.30232	578.00	197.25	578.00	0.877
585	C6H12O2	CYCLOHEXYL PEROXIDE	0.32637	0.26309	0.28571	685.00	253.15	685.00	1.015
586	C6H12O2	DIACETONE ALCOHOL	0.30017	0.26031	0.25110	606.00	229.15	606.00	0.934
587	C6H12O2	2-ETHYL BUTYRIC ACID	0.29861	0.25707	0.31103	655.00	258.15	655.00	0.919
588	C6H12O2	n-HEXANOIC ACID	0.29910	0.26580	0.27650	667.00	270.15	667.00	0.921
589	C6H12O3	2-ETHOXYETHYL ACETATE	0.28801	0.22760	0.28570	597.00	211.45	597.00	0.970
590	C6H12O3	HYDROXYCAPROIC ACID	0.32901	0.23200	0.28571	758.00	334.00	758.00	-------
591	C6H12O3	PARALDEHYDE	0.36208	0.28486	0.31341	579.00	285.75	579.00	0.985
592	C6H12O3	sec-BUTYL GLYCOLATE	-------	-------	-------	-------	-------	-------	-------
593	C6H12S	THIACYCLOHEPTANE	0.29690	0.32200	0.28571	640.07	292.14	640.07	0.766
594	C6H13N	CYCLOHEXYLAMINE	0.27551	0.25517	0.27070	615.00	255.45	615.00	0.863
595	C6H13N	HEXAMETHYLENEIMINE	0.27475	0.25340	0.25620	615.00	236.15	615.00	0.875
596	C6H14	2,2-DIMETHYLBUTANE	0.24016	0.27670	0.27990	488.78	174.28	488.78	0.644
597	C6H14	2,3-DIMETHYLBUTANE	0.23970	0.27042	0.28571	499.98	145.19	499.98	0.658
598	C6H14	n-HEXANE	0.23242	0.26500	0.27810	507.43	177.84	507.43	0.656
599	C6H14	2-METHYLPENTANE	0.23466	0.26724	0.28570	497.50	119.55	497.50	0.648
600	C6H14	3-METHYLPENTANE	0.23520	0.26598	0.27917	504.43	110.25	504.43	0.660
601	C6H14N2O2	LYSINE	0.29076	0.26000	0.28571	821.00	483.00	821.00	-------
602	C6H14O	2-ETHYL-1-BUTANOL	0.26890	0.26035	0.24730	580.00	158.75	580.00	0.829
603	C6H14O	1-HEXANOL	0.26801	0.26725	0.25400	611.35	228.55	611.35	0.816
604	C6H14O	2-HEXANOL	0.26879	0.26275	0.26945	586.20	223.00	586.20	0.810
605	C6H14O	2-METHYL-1-PENTANOL	0.27621	0.26006	0.28570	582.00	223.00	582.00	0.827
606	C6H14O	4-METHYL-2-PENTANOL	0.27674	0.26827	0.28571	574.40	183.00	574.40	0.805
607	C6H14O	n-BUTYL ETHYL ETHER	0.26107	0.26506	0.28570	531.00	170.15	531.00	0.745
608	C6H14O	DIISOPROPYL ETHER	0.26218	0.26974	0.28571	500.05	187.65	500.05	0.721
609	C6H14O	DI-n-PROPYL ETHER	0.26185	0.26780	0.28570	530.60	149.95	530.60	0.741
610	C6H14O	METHYL tert-PENTYL ETHER	0.26600	0.26301	0.28571	534.00	160.00	534.00	0.766
611	C6H14O2	ACETAL	0.28990	0.27006	0.28571	541.00	173.15	541.00	0.821
612	C6H14O2	2-BUTOXYETHANOL	0.29544	0.26661	0.25455	600.00	203.15	600.00	0.896
613	C6H14O2	1,6-HEXANEDIOL	0.30487	0.25120	0.28570	670.00	315.15	670.00	-------
614	C6H14O2	HEXYLENE GLYCOL	0.32315	0.28400	0.28570	621.00	223.15	621.00	0.918
615	C6H14O2S	DI-n-PROPYL SULFONE	0.32450	0.25643	0.30567	763.00	303.00	763.00	-------
616	C6H14O3	DIETHYLENE GLYCOL DIMETHYL ETHER	0.30306	0.25214	0.28570	604.00	203.15	604.00	0.942
617	C6H14O3	DIPROPYLENE GLYCOL	0.33335	0.26500	0.28570	654.00	233.00	654.00	1.018
618	C6H14O3	2-(2-ETHOXYETHOXY)ETHANOL	0.31485	0.25465	0.28570	632.00	195.15	632.00	0.984
619	C6H14O3	TRIMETHYLOLPROPANE	0.32253	0.25893	0.15756	709.00	331.15	709.00	-------
620	C6H14O4	TRIETHYLENE GLYCOL	0.33903	0.26071	0.20960	700.00	265.79	700.00	1.122
621	C6H14O6	SORBITOL	0.37718	0.23255	0.21731	959.00	370.85	959.00	-------
622	C6H14S	n-HEXYLMERCAPTAN	0.28700	0.25553	0.37222	623.00	192.62	623.00	0.837
623	C6H14S	BUTYL-ETHYL-SULFIDE	0.27850	0.26500	0.28571	609.00	178.03	609.00	0.833
624	C6H14S	ISOPROPYL-SULFIDE	0.28590	0.27400	0.28571	585.71	170.45	585.71	0.822

Table 8-1 DENSITY OF LIQUID - ORGANIC COMPOUNDS (continued)

| NO | FORMULA | NAME | density = A B $^{-(1-T/Tc)^n}$ | | | | | | (density - g/ml, T - K) |
			A	B	n	T_C	TMIN	TMAX	density @ 25 C
625	C6H14S	METHYL-PENTYL-SULFIDE	0.29300	0.27600	0.28571	587.98	179.16	587.98	0.839
626	C6H14S	PROPYL-SULFIDE	0.27920	0.26600	0.28571	609.73	170.45	609.73	0.833
627	C6H14S2	PROPYL-DISULFIDE	0.32230	0.27700	0.28571	673.00	187.68	673.00	0.955
628	C6H15Al	TRIETHYL ALUMINUM	0.49638	0.47987	0.65550	720.15	220.65	720.15	0.833
629	C6H15Al2Cl3	ETHYL ALUMINUM SESQUICHLORIDE	--------	--------	--------	--------	--------	--------	1.092
630	C6H15N	DIISOPROPYLAMINE	0.24256	0.25786	0.27100	523.10	176.85	523.10	0.713
631	C6H15N	DI-n-PROPYLAMINE	0.24260	0.25356	0.27430	555.80	210.15	555.80	0.737
632	C6H15N	n-HEXYLAMINE	0.24243	0.25403	0.25235	583.00	251.85	583.00	0.761
633	C6H15N	TRIETHYLAMINE	0.25995	0.27386	0.28720	535.15	158.45	535.15	0.724
634	C6H15NO	6-AMINOHEXANOL	0.26868	0.26500	0.28571	681.00	331.00	681.00	--------
635	C6H15NO2	DIISOPROPANOLAMINE	0.29337	0.24629	0.22229	672.00	318.15	672.00	--------
636	C6H15NO3	TRIETHANOLAMINE	0.31608	0.24801	0.20350	787.00	294.35	787.00	1.120
637	C6H15N3	N-AMINOETHYL PIPERAZINE	0.31714	0.26644	0.28570	708.00	254.15	708.00	0.983
638	C6H15O4P	TRIETHYL PHOSPHATE	0.18062	0.13488	0.25660	794.00	216.00	794.00	1.066
639	C6H16N2	HEXAMETHYLENEDIAMINE	0.24507	0.28300	0.28571	663.00	313.95	663.00	--------
640	C6H18N3OP	HEXAMETHYL PHOSPHORAMIDE	--------	--------	--------	--------	--------	--------	1.020
641	C6H18N4	TRIETHYLENE TETRAMINE	0.30364	0.25574	0.28571	718.00	285.15	718.00	0.978
642	C6H18OSi2	HEXAMETHYLDISILOXANE	0.27010	0.26680	0.28570	518.70	204.93	518.70	0.760
643	C6H18O3Si3	HEXAMETHYLCYCLOTRISILOXANE	0.35076	0.28820	0.30930	554.20	337.15	554.20	--------
644	C6H19NSi2	HEXAMETHYLDISILAZANE	0.26371	0.25980	0.28570	544.00	293.15	544.00	0.772
645	C7H3ClF3NO2	4-CHLORO-3-NITROBENZOTRIFLUORIDE	0.44961	0.24100	0.28570	686.00	293.15	686.00	1.506
646	C7H3Cl2F3	2,4-DICHLOROBENZOTRIFLUORIDE	0.45914	0.24500	0.28570	646.00	247.55	646.00	1.492
647	C7H3Cl2NO	3,4-DICHLOROPHENYL ISOCYANATE	0.41231	0.23920	0.27449	733.00	316.15	733.00	--------
648	C7H4ClF3	p-CHLOROBENZOTRIFLUORIDE	0.41091	0.26470	0.28570	601.00	237.15	601.00	1.226
649	C7H4Cl2O	m-CHLOROBENZOYL CHLORIDE	0.43142	0.24800	0.28571	724.00	280.00	724.00	1.430
650	C7H4F3NO2	3-NITROBENZOTRIFLUORIDE	0.41564	0.23215	0.28570	667.00	272.00	667.00	1.426
651	C7H5ClO	BENZOYL CHLORIDE	0.38080	0.25862	0.28600	697.00	272.65	697.00	1.206
652	C7H5ClO2	o-CHLOROBENZOIC ACID	0.42130	0.22744	0.28570	792.00	415.15	792.00	--------
653	C7H5Cl3	BENZOTRICHLORIDE	0.41969	0.25387	0.28570	737.00	268.40	737.00	1.369
654	C7H5F3	BENZOTRIFLUORIDE	0.40096	0.26310	0.28570	565.00	244.14	565.00	1.178
655	C7H5N	BENZONITRILE	0.30420	0.24793	0.28410	699.35	260.40	699.35	1.001
656	C7H5NO	PHENYL ISOCYANATE	0.34958	0.25680	0.28570	648.00	243.15	648.00	1.093
657	C7H5N3O6	2,4,6-TRINITROTOLUENE	0.47319	0.26352	0.28170	795.00	354.00	795.00	--------
658	C7H6Cl2	BENZYL DICHLORIDE	0.39858	0.25970	0.31940	731.00	257.00	731.00	1.247
659	C7H6Cl2	2,4-DICHLOROTOLUENE	0.38727	0.25466	0.28570	705.00	259.65	705.00	1.247
660	C7H6N2O4	2,4-DINITROTOLUENE	0.37400	0.22633	0.29149	814.00	343.00	814.00	--------
661	C7H6N2O4	2,5-DINITROTOLUENE	0.38619	0.23693	0.28571	814.00	325.65	814.00	--------
662	C7H6N2O4	2,6-DINITROTOLUENE	0.40005	0.24180	0.28571	770.00	339.00	770.00	--------
663	C7H6N2O4	3,4-DINITROTOLUENE	0.37454	0.23617	0.28571	842.00	332.00	842.00	--------
664	C7H6N2O4	3,5-DINITROTOLUENE	0.38541	0.23741	0.28571	814.00	365.65	814.00	--------
665	C7H6O	BENZALDEHYDE	0.32759	0.25780	0.28500	695.00	247.15	695.00	1.040
666	C7H6O2	BENZOIC ACID	0.35235	0.24812	0.28570	751.00	395.52	751.00	--------
667	C7H6O2	p-HYDROXYBENZALDEHYDE	0.33830	0.24380	0.24234	844.00	390.15	844.00	--------
668	C7H6O2	SALICYLALDEHYDE	0.35733	0.25345	0.26250	680.00	266.15	680.00	1.162
669	C7H6O3	SALICYLIC ACID	--------	--------	--------	--------	--------	--------	--------
670	C7H7Br	p-BROMOTOLUENE	0.45128	0.26245	0.30082	699.00	299.95	699.00	--------
671	C7H7Cl	BENZYL CHLORIDE	0.34198	0.25374	0.28570	686.00	234.15	686.00	1.097
672	C7H7Cl	o-CHLOROTOLUENE	0.35762	0.26985	0.28396	656.00	236.65	656.00	1.077
673	C7H7Cl	p-CHLOROTOLUENE	0.35170	0.26860	0.28754	660.00	280.65	660.00	1.063
674	C7H7F	p-FLUOROTOLUENE	0.33130	0.26200	0.28571	590.48	216.36	590.48	0.991
675	C7H7NO	FORMANILIDE	0.31704	0.24000	0.29000	787.00	323.15	787.00	--------
676	C7H7NO2	m-NITROTOLUENE	0.31097	0.22103	0.27206	734.00	289.20	734.00	1.152
677	C7H7NO2	o-NITROTOLUENE	0.31098	0.21877	0.27087	720.00	269.98	720.00	1.158
678	C7H7NO2	p-NITROTOLUENE	0.31098	0.22476	0.25628	736.00	324.75	736.00	--------
679	C7H7NO3	o-NITROANISOLE	0.36287	0.24287	0.28830	782.00	283.60	782.00	1.244
680	C7H8	TOLUENE	0.29999	0.27108	0.29889	591.79	178.18	591.79	0.865
681	C7H8	1,3,5-CYCLOHEPTATRIENE	0.29800	0.26600	0.28571	593.90	193.66	593.90	0.882
682	C7H8O	ANISOLE	0.32089	0.26119	0.28010	641.65	235.65	641.65	0.990
683	C7H8O	BENZYL ALCOHOL	0.32318	0.26404	0.22400	677.00	257.85	677.00	1.041
684	C7H8O	m-CRESOL	0.34663	0.28268	0.27070	705.85	285.39	705.85	1.030
685	C7H8O	o-CRESOL	0.38355	0.30518	0.30990	697.55	304.19	697.55	--------
686	C7H8O	p-CRESOL	0.38059	0.29920	0.33410	704.65	307.93	704.65	--------
687	C7H8O2	GUAIACOL	0.35171	0.25826	0.27125	697.00	304.65	697.00	--------
688	C7H8O2	p-METHOXYPHENOL	0.36301	0.26523	0.28450	758.00	329.00	758.00	--------
689	C7H9N	BENZYLAMINE	0.28727	0.26525	0.13446	683.50	227.15	683.50	0.981
690	C7H9N	2,6-DIMETHYLPYRIDINE	0.33911	0.30490	0.27170	623.75	267.00	623.75	0.918
691	C7H9N	N-METHYLANILINE	0.28754	0.24324	0.25374	701.55	216.15	701.55	0.982
692	C7H9N	m-TOLUIDINE	0.28725	0.23990	0.26905	709.15	242.75	709.15	0.985
693	C7H9N	o-TOLUIDINE	0.28729	0.23820	0.25730	694.15	249.47	694.15	0.994
694	C7H9N	p-TOLUIDINE	0.28727	0.24079	0.26158	693.15	316.90	693.15	--------
695	C7H10	2-NORBORNENE	0.27963	0.27300	0.28571	583.00	319.40	583.00	--------
696	C7H10N2	TOLUENEDIAMINE	0.32454	0.24665	0.28571	804.00	371.25	804.00	--------
697	C7H11NO	CYCLOHEXYL ISOCYANATE	0.33492	0.24640	0.28570	633.00	193.00	633.00	1.077
698	C7H12	1-HEPTYNE	0.25840	0.27600	0.28571	559.69	192.26	559.69	0.728
699	C7H12O2	n-BUTYL ACRYLATE	0.29947	0.25838	0.30843	598.00	208.55	598.00	0.894
700	C7H12O2	ISOBUTYL ACRYLATE	0.29833	0.26282	0.28571	580.00	212.00	580.00	0.885
701	C7H12O2	n-PROPYL METHACRYLATE	0.29277	0.25580	0.28571	599.00	223.00	599.00	0.897
702	C7H12O4	DIETHYL MALONATE	0.34151	0.25619	0.31612	653.00	224.25	653.00	1.050

Table 8-1 DENSITY OF LIQUID - ORGANIC COMPOUNDS (continued)

NO	FORMULA	NAME	density = A B $^{-(1-T/Tc)^n}$				(density - g/ml, T - K)		
			A	B	n	T_C	TMIN	TMAX	density @ 25 C
703	C7H14	CYCLOHEPTANE	0.27591	0.27199	0.28571	604.30	265.15	604.30	0.806
704	C7H14	1,1-DIMETYLCYCLOPENTANE	0.27275	0.27729	0.30163	547.00	203.36	547.00	0.750
705	C7H14	cis-1,2-DIMETHYLCYCLOPENTANE	0.26682	0.26980	0.28571	565.15	219.26	565.15	0.768
706	C7H14	trans-1,2-DIMETHYLCYCLOPENTANE	0.27274	0.27793	0.30976	553.15	155.58	553.15	0.747
707	C7H14	cis-1,3-DIMETHYLCYCLOPENTANE	0.27274	0.27941	0.31401	551.00	139.45	551.00	0.740
708	C7H14	trans-1,3-DIMETHYLCYCLOPENTANE	0.27274	0.28050	0.30466	553.00	139.18	553.00	0.744
709	C7H14	ETHYLCYCLOPENTANE	0.26220	0.26936	0.27770	569.52	134.71	569.52	0.763
710	C7H14	2-ETHYL-1-PENTENE	0.24670	0.27353	0.26632	543.00	168.00	543.00	0.704
711	C7H14	3-ETHYL-1-PENTENE	0.24670	0.28384	0.24201	530.00	145.67	530.00	0.692
712	C7H14	1-HEPTENE	0.23776	0.26305	0.27420	537.29	154.27	537.29	0.693
713	C7H14	cis-2-HEPTENE	0.23158	0.24867	0.28858	549.00	164.00	549.00	0.703
714	C7H14	trans-2-HEPTENE	0.24185	0.26451	0.28672	543.00	163.67	543.00	0.697
715	C7H14	cis-3-HEPTENE	0.23323	0.25446	0.27941	545.00	136.51	545.00	0.698
716	C7H14	trans-3-HEPTENE	0.24184	0.26560	0.28571	540.00	136.52	540.00	0.694
717	C7H14	METHYLCYCLOHEXANE	0.26688	0.27041	0.29270	572.19	146.58	572.19	0.766
718	C7H14	2-METHYL-1-HEXENE	0.24670	0.26774	0.29214	538.00	170.28	538.00	0.698
719	C7H14	3-METHYL-1-HEXENE	0.24670	0.27383	0.28221	528.00	145.00	528.00	0.687
720	C7H14	4-METHYL-1-HEXENE	0.24670	0.27130	0.28372	534.00	131.70	534.00	0.694
721	C7H14	2,3,3-TRIMETHYL-1-BUTENE	0.25771	0.27903	0.29600	531.00	163.30	531.00	0.701
722	C7H14O	DIISOPROPYL KETONE	0.27449	0.23385	0.26180	576.00	204.81	576.00	0.912
723	C7H14O	2-HEPTANONE	0.27123	0.26114	0.30447	611.55	238.15	611.55	0.811
724	C7H14O	1-HEPTANAL	0.27123	0.26249	0.28953	603.00	230.15	603.00	0.813
725	C7H14O	1-METHYLCYCLOHEXANOL	0.27581	0.22362	0.31329	603.00	299.15	603.00	--------
726	C7H14O	cis-2-METHYLCYCLOHEXANOL	0.27581	0.24052	0.23687	614.00	280.15	614.00	0.932
727	C7H14O	trans-2-METHYLCYCLOHEXANOL	0.27582	0.24655	0.22625	616.00	269.15	616.00	0.921
728	C7H14O	cis-3-METHYLCYCLOHEXANOL	0.27582	0.23722	0.28185	618.00	267.65	618.00	0.911
729	C7H14O	trans-3-METHYLCYCLOHEXANOL	0.27582	0.24921	0.21950	617.00	272.65	617.00	0.918
730	C7H14O	cis-4-METHYLCYCLOHEXANOL	0.30881	0.27100	0.28570	622.00	294.85	622.00	0.913
731	C7H14O	trans-4-METHYLCYCLOHEXANOL	0.27581	0.24578	0.25051	622.00	293.00	622.00	0.908
732	C7H14O	5-METHYL-2-HEXANONE	0.26429	0.25680	0.28571	601.00	199.25	601.00	0.808
733	C7H14O2	n-BUTYL PROPIONATE	0.28675	0.25740	0.28571	594.00	183.63	594.00	0.872
734	C7H14O2	ETHYL ISOVALERATE	0.28914	0.26160	0.28570	587.95	173.85	587.95	0.865
735	C7H14O2	ISOPENTYL ACETATE	0.29449	0.26505	0.30090	599.00	194.65	599.00	0.867
736	C7H14O2	n-PENTYL ACETATE	0.29459	0.26480	0.29320	598.00	202.35	598.00	0.872
737	C7H14O2	n-PROPYL n-BUTYRATE	0.29457	0.26444	0.29790	594.00	177.95	594.00	0.868
738	C7H14O2	n-HEPTANOIC ACID	0.29454	0.25920	0.30590	680.00	265.83	680.00	0.913
739	C7H14O3	ETHYL-3-ETHOXYPROPIONATE	0.31642	0.25721	0.32154	609.00	223.00	609.00	0.945
740	C7H15Br	1-BROMOHEPTANE	0.37782	0.26974	0.28571	651.00	217.05	651.00	1.135
741	C7H15N	N-METHYLCYCLOHEXYLAMINE	0.28781	0.26542	0.28570	622.00	264.65	622.00	0.865
742	C7H16	2,2-DIMETHYLPENTANE	0.24513	0.27670	0.28300	520.50	149.34	520.50	0.673
743	C7H16	2,3-DIMETHYLPENTANE	0.25315	0.28614	0.27130	537.35	160.00	537.35	0.691
744	C7H16	2,4-DIMETHYLPENTANE	0.23961	0.27530	0.26950	519.79	153.91	519.79	0.668
745	C7H16	3,3-DIMETHYLPENTANE	0.24793	0.27010	0.30840	536.40	138.70	536.40	0.687
746	C7H16	3-ETHYLPENTANE	0.24259	0.26444	0.29140	540.64	154.55	540.64	0.695
747	C7H16	n-HEPTANE	0.23237	0.26020	0.27910	540.26	182.57	540.26	0.682
748	C7H16	2-METHYLHEXANE	0.23793	0.26970	0.27900	530.37	154.90	530.37	0.674
749	C7H16	3-METHYLHEXANE	0.24512	0.28060	0.26250	535.25	153.75	535.25	0.684
750	C7H16	2,2,3-TRIMETHYLBUTANE	0.24943	0.27920	0.28000	531.17	248.57	531.17	0.687
751	C7H16O	1-HEPTANOL	0.26702	0.26320	0.27300	631.90	239.15	631.90	0.820
752	C7H16O	2-HEPTANOL	0.26895	0.26582	0.25340	588.00	243.00	588.00	0.814
753	C7H16O	5-METHYL-1-HEXANOL	0.26900	0.25072	0.33120	605.00	293.15	605.00	0.812
754	C7H16O	ISOPROPYL-TERT-BUTYL-ETHER	0.25480	0.26100	0.28571	558.21	177.80	558.21	0.750
755	C7H16S	n-HEPTYL MERCAPTAN	0.28445	0.27097	0.30330	645.00	229.92	645.00	0.839
756	C7H16S	BUTYL-PROPYL-SULFIDE	0.26970	0.25900	0.28571	653.50	206.66	653.50	0.839
757	C7H16S	ETHYL-PENTYL-SULFIDE	0.27150	0.25900	0.28571	638.37	206.66	638.37	0.839
758	C7H16S	HEXYL-METHYL-SULFIDE	0.27150	0.25900	0.28571	638.37	206.66	638.37	0.839
759	C7H17N	1-AMINOHEPTANE	0.24462	0.25289	0.26588	607.00	254.15	607.00	0.772
760	C8H4Cl2O2	ISOPHTHALOYL CHLORIDE	0.42325	0.25015	0.28570	768.00	317.00	768.00	--------
761	C8H4O3	PHTHALIC ANHYDRIDE	0.35183	0.23220	0.23460	791.00	404.26	791.00	--------
762	C8H6	ETHYNYLBENZENE	0.30250	0.27300	0.28571	655.43	242.53	655.43	0.901
763	C8H6O4	ISOPHTHALIC ACID	0.39189	0.20000	0.28571	1007.00	619.15	1007.00	--------
764	C8H6O4	PHTHALIC ACID	0.39150	0.25200	0.28571	800.00	464.15	800.00	--------
765	C8H6O4	TEREPHTHALIC ACID	0.39179	0.18100	0.28571	1113.00	700.15	1113.00	--------
766	C8H6S	BENZOTHIOPHENE	0.38453	0.25807	0.33304	754.00	304.50	754.00	--------
767	C8H7N	INDOLE	0.27526	0.21250	0.23240	790.00	273.68	790.00	1.102
768	C8H8	STYRENE	0.29383	0.26315	0.28570	648.00	242.54	648.00	0.900
769	C8H8	1,3,5,7-CYCLOOCTATETRAENE	0.30150	0.26800	0.28571	642.55	266.16	642.55	0.907
770	C8H8O	ACETOPHENONE	0.31955	0.25470	0.29062	701.00	293.65	701.00	1.024
771	C8H8O	p-TOLUALDEHYDE	0.30688	0.24760	0.28570	698.00	298.85	698.00	--------
772	C8H8O2	METHYL BENZOATE	0.31228	0.23519	0.26760	693.00	260.75	693.00	1.085
773	C8H8O2	o-TOLUIC ACID	0.34096	0.24685	0.28571	751.00	376.85	751.00	--------
774	C8H8O2	p-TOLUIC ACID	0.34357	0.23800	0.28571	773.00	452.75	773.00	--------
775	C8H8O3	METHYL SALICYLATE	0.37125	0.26414	0.26120	701.00	265.15	701.00	1.175
776	C8H8O3	VANILLIN	0.36605	0.25800	0.28571	777.00	355.00	777.00	--------
777	C8H9NO	ACETANILIDE	0.31014	0.23700	0.28570	825.00	386.65	825.00	--------
778	C8H10	ETHYLBENZENE	0.28889	0.26438	0.29213	617.17	178.20	617.17	0.865
779	C8H10	m-XYLENE	0.27866	0.25925	0.27243	617.05	225.30	617.05	0.861
780	C8H10	o-XYLENE	0.28381	0.26083	0.27410	630.37	247.98	630.37	0.876

198

Table 8-1 DENSITY OF LIQUID - ORGANIC COMPOUNDS (continued)

			density = A B$^{-(1-T/T_c)^n}$				(density - g/ml, T - K)		
NO	FORMULA	NAME	A	B	n	T_C	TMIN	TMAX	density @ 25 C
781	C8H10	p-XYLENE	0.27984	0.26003	0.27900	616.26	286.41	616.26	0.858
782	C8H10O	m-ETHYLPHENOL	--------	--------	--------	--------	--------	--------	--------
783	C8H10O	p-ETHYLPHENOL	0.32672	0.26930	0.28570	716.45	318.23	716.45	--------
784	C8H10O	PHENETOLE	0.30533	0.25467	0.28571	647.15	243.63	647.15	0.961
785	C8H10O	2-PHENYLETHANOL	0.31567	0.24898	0.30277	684.00	247.00	684.00	1.016
786	C8H10O	2,3-XYLENOL	0.33990	0.29300	0.28571	722.95	345.71	722.95	--------
787	C8H10O	2,4-XYLENOL	0.31327	0.25992	0.24870	707.65	297.68	707.65	1.015
788	C8H10O	2,5-XYLENOL	0.34916	0.28495	0.29700	707.05	347.99	707.05	--------
789	C8H10O	2,6-XYLENOL	0.31293	0.28800	0.28571	701.05	318.76	701.05	--------
790	C8H10O	3,4-XYLENOL	0.34912	0.28645	0.28070	729.95	338.25	729.95	--------
791	C8H10O	3,5-XYLENOL	0.25458	0.20987	0.22630	715.65	336.59	715.65	--------
792	C8H11N	N,N-DIMETHYLANILINE	0.26088	0.22868	0.23350	687.15	275.60	687.15	0.949
793	C8H11N	o-ETHYLANILINE	0.30381	0.25486	0.28570	704.00	226.55	704.00	0.977
794	C8H11N	2,4,6-TRIMETHYLPYRIDINE	0.29043	0.25592	0.28571	653.00	229.00	653.00	0.913
795	C8H11NO	p-PHENETIDINE	0.31545	0.24764	0.28570	754.00	277.00	754.00	1.057
796	C8H12	1,5-CYCLOOCTADIENE	0.29559	0.25915	0.34755	645.00	203.98	645.00	0.878
797	C8H12	VINYLCYCLOHEXENE	0.27918	0.26687	0.28570	599.00	164.00	599.00	0.826
798	C8H12O4	1,4-CYCLOHEXANEDICARBOXYLIC ACID	0.37054	0.21500	0.28571	889.00	585.65	889.00	--------
799	C8H12O4	DIETHYL MALEATE	0.33894	0.28278	0.33357	680.00	264.35	680.00	0.961
800	C8H14O2	n-BUTYL METHACRYLATE	0.28691	0.25450	0.28570	616.00	223.00	616.00	0.891
801	C8H14O3	BUTYRIC ANHYDRIDE	0.30988	0.25847	0.28570	639.00	199.85	639.00	0.960
802	C8H14O4	DIETHYL SUCCINATE	0.33370	0.25864	0.29522	660.00	252.35	660.00	1.036
803	C8H16	CYCLOOCTANE	0.27790	0.27100	0.28571	647.20	287.60	647.20	0.830
804	C8H16	1,1-DIMETHYLCYCLOHEXANE	0.24937	0.25143	0.27758	591.15	239.66	591.15	0.777
805	C8H16	cis-1,2-DIMETHYLCYCLOHEXANE	0.24395	0.24358	0.26809	606.15	223.16	606.15	0.792
806	C8H16	trans-1,2-DIMETHYLCYCLOHEXANE	0.24395	0.25026	0.26580	596.15	184.99	596.15	0.772
807	C8H16	cis-1,3-DIMETHYLCYCLOHEXANE	0.24936	0.25583	0.28393	591.15	197.58	591.15	0.762
808	C8H16	trans-1,3-DIMETHYLCYCLOHEXANE	0.24395	0.24438	0.27796	598.00	183.07	598.00	0.781
809	C8H16	cis-1,4-DIMETHYLCYCLOHEXANE	0.24395	0.24531	0.27709	598.15	185.72	598.15	0.779
810	C8H16	trans-1,4-DIMETHYLCYCLOHEXANE	0.24937	0.25855	0.27750	590.15	236.21	590.15	0.759
811	C8H16	ETHYLCYCLOHEXANE	0.24942	0.25384	0.26750	609.15	161.84	609.15	0.784
812	C8H16	2-ETHYL-1-HEXENE	0.28124	0.29741	0.34186	574.00	173.00	574.00	0.723
813	C8H16	1-METHYL-1-ETHYLCYCLOPENTANE	0.26372	0.26556	0.28571	582.00	129.35	582.00	0.777
814	C8H16	1-OCTENE	0.23682	0.25649	0.28571	566.60	171.45	566.60	0.711
815	C8H16	trans-2-OCTENE	0.23618	0.25552	0.28570	577.00	185.45	577.00	0.716
816	C8H16	trans-3-OCTENE	0.23614	0.25690	0.28570	574.00	163.15	574.00	0.711
817	C8H16	trans-4-OCTENE	0.23618	0.25730	0.28570	573.00	179.37	573.00	0.710
818	C8H16	n-PROPYLCYCLOPENTANE	0.26406	0.27006	0.29032	603.00	155.82	603.00	0.773
819	C8H16	2,4,4-TRIMETHYL-1-PENTENE	0.24383	0.26325	0.28570	553.00	179.70	553.00	0.711
820	C8H16	2,4,4-TRIMETHYL-2-PENTENE	0.24369	0.26104	0.28570	558.00	166.84	558.00	0.717
821	C8H16O	2-ETHYLHEXANAL	0.26165	0.25050	0.28570	607.00	200.00	607.00	0.819
822	C8H16O	1-OCTANAL	0.27049	0.27235	0.24964	621.00	246.00	621.00	0.816
823	C8H16O	2-OCTANONE	0.27337	0.26616	0.29628	624.00	252.85	624.00	0.815
824	C8H16O2	n-BUTYL n-BUTYRATE	0.29193	0.27108	0.27650	616.00	181.15	616.00	0.866
825	C8H16O2	n-HEXYL ACETATE	0.29194	0.26544	0.29810	618.00	192.25	618.00	0.868
826	C8H16O2	ISOBUTYL ISOBUTYRATE	0.28483	0.26402	0.28570	602.00	192.45	602.00	0.852
827	C8H16O2	n-OCTANOIC ACID	0.29231	0.26676	0.28020	692.00	289.65	692.00	0.903
828	C8H16O4	DIETHYLENE GLYCOL ETHYL ETHER ACETATE	0.31206	0.24902	0.28570	660.00	248.15	660.00	1.006
829	C8H18	2,2-DIMETHYLHEXANE	0.23673	0.26386	0.27774	549.80	151.97	549.80	0.692
830	C8H18	2,3-DIMETHYLHEXANE	0.24189	0.26817	0.26950	563.40	272.04	563.40	0.708
831	C8H18	2,4-DIMETHYLHEXANE	0.24718	0.27280	0.29850	553.50	272.04	553.50	0.693
832	C8H18	2,5-DIMETHYLHEXANE	0.23576	0.26329	0.27820	550.00	182.00	550.00	0.690
833	C8H18	3,3-DIMETHYLHEXANE	0.25513	0.29013	0.25680	562.00	147.05	562.00	0.707
834	C8H18	3,4-DIMETHYLHEXANE	0.24176	0.26579	0.26761	568.80	272.04	568.80	0.716
835	C8H18	3-ETHYLHEXANE	0.24802	0.27726	0.26570	565.40	272.04	565.40	0.710
836	C8H18	3-ETHYL-2-METHYLPENTANE	0.23480	0.25400	0.28571	567.00	158.20	567.00	0.711
837	C8H18	3-METHYL-3-ETHYLPENTANE	0.25100	0.27254	0.28169	576.50	182.28	576.50	0.724
838	C8H18	2-METHYLHEPTANE	0.23177	0.25900	0.27130	559.64	164.16	559.64	0.696
839	C8H18	3-METHYLHEPTANE	0.24461	0.27441	0.27130	563.67	152.60	563.67	0.702
840	C8H18	4-METHYLHEPTANE	0.23935	0.25882	0.28324	561.74	152.20	561.74	0.713
841	C8H18	n-OCTANE	0.22807	0.25476	0.26940	568.83	216.38	568.83	0.699
842	C8H18	2,2,3-TRIMETHYLPENTANE	0.26064	0.29071	0.27340	563.50	160.89	563.50	0.712
843	C8H18	2,2,4-TRIMETHYLPENTANE	0.24563	0.27373	0.28460	543.96	165.78	543.96	0.690
844	C8H18	2,3,3-TRIMETHYLPENTANE	0.25089	0.27446	0.27410	573.50	172.22	573.50	0.722
845	C8H18	2,3,4-TRIMETHYLPENTANE	0.24307	0.26794	0.26474	566.30	163.95	566.30	0.716
846	C8H18	2,2,3,3-TETRAMETHYLBUTANE	0.24740	0.28000	0.28571	567.80	172.47	567.80	0.692
847	C8H18O	DI-n-BUTYL ETHER	0.25521	0.25986	0.28570	581.00	177.95	581.00	0.764
848	C8H18O	DI-sec-BUTYL ETHER	0.26379	0.26874	0.28571	559.00	173.15	559.00	0.759
849	C8H18O	DI-tert-BUTYL ETHER	0.26676	0.27010	0.28571	550.00	195.00	550.00	0.760
850	C8H18O	2-ETHYL-1-HEXANOL	0.26851	0.26127	0.27730	640.25	203.15	640.25	0.830
851	C8H18O	1-OCTANOL	0.26568	0.26126	0.28090	652.50	257.65	652.50	0.823
852	C8H18O	2-OCTANOL	0.27765	0.27264	0.29430	637.15	241.55	637.15	0.817
853	C8H18O2	DI-t-BUTYL PEROXIDE	0.28537	0.27942	0.28570	547.00	233.15	547.00	0.790
854	C8H18O2S	DI-n-BUTYL SULFONE	0.31335	0.25806	0.30522	767.00	318.00	767.00	--------
855	C8H18O3	DIETHYLENE GLYCOL DIETHYL ETHER	0.29057	0.25504	0.28570	624.00	228.85	624.00	0.904
856	C8H18O3	DIETHYLENE GLYCOL MONOBUTYL EHTER	0.30082	0.25390	0.28570	654.00	205.15	654.00	0.952
857	C8H18O4	TRIETHYLENE GLYCOL DIMETHYL ETHER	0.32523	0.24622	0.39180	651.00	229.35	651.00	0.980
858	C8H18O5	TETRAETHYLENE GLYCOL	0.33834	0.24768	0.28570	722.00	268.15	722.00	1.122

Table 8-1 DENSITY OF LIQUID - ORGANIC COMPOUNDS (continued)

| NO | FORMULA | NAME | density = A B $^{-(1-T/Tc)^n}$ | | | | (density - g/ml, T - K) | | |
			A	B	n	T_C	TMIN	TMAX	density @ 25 C
859	C8H18S	n-OCTYL MERCAPTAN	0.28243	0.27171	0.29980	664.00	223.95	664.00	0.840
860	C8H18S	tert-OCTYL MERCAPTAN	0.27675	0.26263	0.28571	627.00	199.00	627.00	0.841
861	C8H18S	BUTYL-SULFIDE	0.27110	0.26000	0.28571	650.00	209.86	650.00	0.840
862	C8H18S	ETHYL-HEXYL-SULFIDE	0.26290	0.25200	0.28571	660.72	209.86	660.72	0.840
863	C8H18S	HEPTYL-METHYL-SULFIDE	0.26290	0.25200	0.28571	660.72	209.86	660.72	0.840
864	C8H18S	PENTYL-PROPYL-SULFIDE	0.26290	0.25200	0.28571	660.72	209.86	660.72	0.840
865	C8H18S2	BUTYL-DISULFIDE	0.30030	0.26500	0.28571	704.16	202.16	704.16	0.934
866	C8H19N	DI-n-BUTYLAMINE	0.24734	0.25824	0.28270	607.50	211.15	607.50	0.757
867	C8H19N	DIISOBUTYLAMINE	0.24737	0.26249	0.27200	580.00	203.15	580.00	0.743
868	C8H19N	n-OCTYLAMINE	0.25035	0.25550	0.28571	627.00	272.75	627.00	0.779
869	C8H23N5	TETRAETHYLENEPENTAMINE	0.29764	0.25004	0.28571	774.00	243.00	774.00	0.994
870	C8H24O4Si4	OCTAMETHYLCYCLOTETRASILOXANE	0.30628	0.25230	0.27690	586.50	290.80	586.50	0.949
871	C9H4O5	TRIMELLITIC ANHYDRIDE	0.41542	0.25500	0.28571	890.00	438.15	890.00	--------
872	C9H6N2O2	TOLUENE DIISOCYANATE	0.33173	0.24582	0.14390	737.00	287.04	737.00	1.220
873	C9H7N	ISOQUINOLINE	0.32054	0.24749	0.27890	803.15	299.45	803.15	--------
874	C9H7N	QUINOLINE	0.27550	0.21440	0.23616	782.15	258.25	782.15	1.090
875	C9H7NO	8-HYDROXYQUINOLINE	0.35000	0.27600	0.29000	788.00	346.00	788.00	--------
876	C9H8	INDENE	0.31515	0.25400	0.31000	687.00	271.70	687.00	0.994
877	C9H8O	2-METHYLBENZOFURAN	0.32648	0.25390	0.28570	698.00	290.00	698.00	1.051
878	C9H10	INDANE	0.31019	0.26117	0.30223	684.90	221.74	684.90	0.960
879	C9H10	cis-PROPENYLBENZENE	0.28840	0.25800	0.28571	664.60	211.47	664.60	0.904
880	C9H10	trans-PROPENYLBENZENE	0.28770	0.25800	0.28571	664.60	243.82	664.60	0.902
881	C9H1O	alpha-METHYLSTYRENE	0.27938	0.24510	0.29512	654.00	249.95	654.00	0.905
882	C9H10	m-METHYLSTYRENE	0.29036	0.25698	0.29075	657.00	186.81	657.00	0.908
883	C9H10	o-METHYLSTYRENE	0.29037	0.25672	0.29306	659.00	204.58	659.00	0.908
884	C9H10	p-METHYLSTYRENE	0.27420	0.22722	0.34660	665.00	239.02	665.00	0.916
885	C9H10O2	BENZYL ACETATE	0.33445	0.25370	0.33360	699.00	221.65	699.00	1.045
886	C9H10O2	ETHYL BENZOATE	0.30711	0.23878	0.28487	698.00	238.45	698.00	1.042
887	C9H10O3	ETHYL VANILLIN	0.35518	0.24600	0.28571	748.00	350.65	748.00	--------
888	C9H11NO	p-DIMETHYLAMINOBENZALDEHYDE	0.29323	0.22580	0.28570	832.00	348.00	832.00	--------
889	C9H12	CUMENE	0.28240	0.26180	0.29000	631.15	177.14	631.15	0.860
890	C9H12	m-ETHYLTOLUENE	0.24529	0.22803	0.25957	637.15	177.61	637.15	0.860
891	C9H12	o-ETHYLTOLUENE	0.26129	0.23963	0.27075	651.15	192.35	651.15	0.877
892	C9H12	p-ETHYLTOLUENE	0.25573	0.23831	0.27196	640.15	210.83	640.15	0.857
893	C9H12	MESITYLENE	0.27770	0.25909	0.27982	637.36	228.46	637.36	0.861
894	C9H12	n-PROPYLBENZENE	0.27349	0.25270	0.29130	638.38	173.67	638.38	0.860
895	C9H12	1,2,3-TRIMETHYLBENZENE	0.29063	0.27026	0.26080	664.53	247.79	664.53	0.891
896	C9H12	1,2,4-TRIMETHYLBENZENE	0.27966	0.25948	0.27724	649.13	229.38	649.13	0.872
897	C9H12O	BENZYL ETHYL ETHER	0.30813	0.26925	0.26320	662.00	275.65	662.00	0.945
898	C9H12O	2-PHENYL-2-PROPANOL	0.30953	0.25658	0.26019	660.00	309.15	660.00	--------
899	C9H12O2	CUMENE HYDROPEROXIDE	0.36351	0.27800	0.28600	605.00	264.26	605.00	1.043
900	C9H14O	ISOPHORONE	0.29347	0.26380	0.28570	715.00	265.05	715.00	0.920
901	C9H14O6	GLYCERYL TRIACETATE	0.34964	0.24630	0.28570	704.00	277.25	704.00	1.158
902	C9H16	1-NONYNE	0.25050	0.26400	0.28571	610.81	223.16	610.81	0.752
903	C9H16O4	AZELAIC ACID	0.30855	0.23160	0.28571	811.00	379.65	811.00	--------
904	C9H18	BUTYLCYCLOPENTANE	0.25160	0.25600	0.28571	625.05	165.18	625.05	0.781
905	C9H18	cis,cis-1,3,5-TRIMETHYLCYCLOHEXANE	0.24190	0.24700	0.28571	607.86	223.46	607.86	0.767
906	C9H18	cis,trans-1,3,5-TRIMETHYLCYCLOHEXANE	0.22580	0.24500	0.28571	602.20	188.76	602.20	0.718
907	C9H18	ISOPROPYLCYCLOHEXANE	0.27207	0.27126	0.29805	627.00	183.76	627.00	0.798
908	C9H18	1-NONENE	0.23551	0.25321	0.28571	593.25	191.78	593.25	0.725
909	C9H18	n-PROPYLCYCLOHEXANE	0.26469	0.26694	0.30020	639.15	178.28	639.15	0.790
910	C9H18O	DIISOBUTYL KETONE	0.27249	0.27240	0.28130	615.00	227.17	615.00	0.802
911	C9H18O	1-NONANAL	0.26990	0.27291	0.25011	640.00	255.15	640.00	0.819
912	C9H18O2	n-BUTYL VALERATE	0.28929	0.26646	0.29650	629.00	180.35	629.00	0.863
913	C9H18O2	n-NONANOIC ACID	0.28928	0.25819	0.31610	703.00	285.55	703.00	0.902
914	C9H18O2	n-OCTYL FORMATE	0.28929	0.26348	0.30888	645.00	234.05	645.00	0.870
915	C9H20	3,3-DIETHYLPENTANE	0.27116	0.28977	0.29339	610.05	240.12	610.05	0.750
916	C9H20	2,2-DIMETHYL-3-ETHYLPENTANE	0.24980	0.26899	0.28571	590.00	173.68	590.00	0.731
917	C9H20	3-ETHYL-2,3-DIMETHYLPENTANE	0.24260	0.25400	0.28571	606.80	173.67	606.80	0.751
918	C9H20	2,4-DIMETHYL-3-ETHYLPENTANE	0.24837	0.26588	0.28571	591.00	150.79	591.00	0.734
919	C9H20	2,2-DIMETHYLHEPTANE	0.24712	0.27229	0.29376	576.80	160.15	576.80	0.707
920	C9H20	2,6-DIMETHYLHEPTANE	0.24665	0.27283	0.29230	579.00	170.25	579.00	0.706
921	C9H20	3-ETHYLHEPTANE	0.24292	0.26538	0.27904	590.00	158.25	590.00	0.723
922	C9H20	4-ETHYLHEPTANE	0.23750	0.25500	0.28571	584.95	159.96	584.95	0.724
923	C9H20	2,3-DIMETHYLHEPTANE	0.23390	0.25200	0.28571	589.60	160.16	589.60	0.722
924	C9H20	2,4-DIMETHYLHEPTANE	0.23220	0.25200	0.28571	576.80	160.16	576.80	0.711
925	C9H20	2,5-DIMETHYLHEPTANE	0.23350	0.25400	0.28571	581.10	160.16	581.10	0.713
926	C9H20	3,4-DIMETHYLHEPTANE	0.23540	0.25200	0.28571	591.90	170.26	591.90	0.727
927	C9H20	3,5-DIMETHYLHEPTANE	0.23440	0.25300	0.28571	583.20	170.26	583.20	0.719
928	C9H20	4,4-DIMETHYLHEPTANE	0.23280	0.25000	0.28571	585.40	170.26	585.40	0.721
929	C9H20	3-ETHYL-2-METHYLHEXANE	0.23420	0.24900	0.28571	588.10	160.16	588.10	0.729
930	C9H20	4-ETHYL-2-METHYLHEXANE	0.23360	0.25100	0.28571	580.00	160.16	580.00	0.719
931	C9H20	3-ETHYL-3-METHYLHEXANE	0.23620	0.25000	0.28571	597.50	160.16	597.50	0.737
932	C9H20	3-ETHYL-4-METHYLHEXANE	0.23570	0.24900	0.28571	593.70	160.16	593.70	0.736
933	C9H20	2,2,3-TRIMETHYLHEXANE	0.23660	0.25400	0.28571	588.00	153.16	588.00	0.725
934	C9H20	2,2,4-TRIMETHYLHEXANE	0.23380	0.25300	0.28571	573.50	153.00	573.50	0.713
935	C9H20	2,3,3-TRIMETHYLHEXANE	0.23790	0.25300	0.28571	596.00	156.36	596.00	0.734
936	C9H20	2,3,4-TRIMETHYLHEXANE	0.23760	0.25200	0.28571	594.50	156.36	594.50	0.735

Table 8-1 DENSITY OF LIQUID - ORGANIC COMPOUNDS (continued)

NO	FORMULA	NAME	density = A B$^{-(1-T/Tc)^n}$				(density - g/ml, T - K)		
			A	B	n	T_C	TMIN	TMAX	density @ 25 C
937	C9H20	2,3,5-TRIMETHYLHEXANE	0.23550	0.25400	0.28571	579.20	145.36	579.20	0.718
938	C9H20	2,4,4-TRIMETHYLHEXANE	0.23360	0.25100	0.28571	581.50	159.78	581.50	0.720
939	C9H20	3,3,4-TRIMETHYLHEXANE	0.24010	0.25400	0.28571	602.30	171.96	602.30	0.741
940	C9H20	2-METHYLOCTANE	0.23414	0.25713	0.28570	586.75	192.78	586.75	0.710
941	C9H20	3-METHYLOCTANE	0.24245	0.26711	0.28007	590.15	165.55	590.15	0.717
942	C9H20	4-METHYLOCTANE	0.24524	0.26571	0.29984	587.65	159.95	587.65	0.716
943	C9H20	n-NONANE	0.23364	0.25556	0.28571	595.65	219.63	595.65	0.715
944	C9H20	2,2,3,3-TETRAMETHYLPENTANE	0.26833	0.28394	0.29712	610.85	263.26	610.85	0.753
945	C9H20	2,2,3,4-TETRAMETHYLPENTANE	0.26175	0.28166	0.29201	592.15	152.06	592.15	0.735
946	C9H20	2,2,4,4-TETRAMETHYLPENTANE	0.25448	0.27885	0.28631	571.35	206.95	571.35	0.716
947	C9H20	2,3,3,4-TETRAMETHYLPENTANE	0.23690	0.25100	0.28571	592.60	152.00	592.60	0.735
948	C9H20	2,2,5-TRIMETHYLHEXANE	0.24536	0.26753	0.29540	568.05	167.39	568.05	0.707
949	C9H20O	2,6-DIMETHYL-4-HEPTANOL	0.27341	0.26836	0.28570	603.00	208.00	603.00	0.807
950	C9H20O	1-NONANOL	0.26400	0.26493	0.26313	673.00	268.15	673.00	0.824
951	C9H20O	2-NONANOL	0.26812	0.26341	0.27220	623.00	238.15	623.00	0.820
952	C9H20S	n-NONYL MERCAPTAN	0.28078	0.27027	0.30570	681.00	253.05	681.00	0.841
953	C9H20S	BUTYL-PENTYL-SULFIDE	0.25560	0.24600	0.28571	681.56	231.16	681.56	0.840
954	C9H20S	ETHYL-HEPTYL-SULFIDE	0.25560	0.24600	0.28571	681.56	231.16	681.56	0.840
955	C9H20S	HEXYL-PROPYL-SULFIDE	0.25560	0.24600	0.28571	681.56	231.16	681.56	0.840
956	C9H20S	METHYL-OCTYL-SULFIDE	0.25560	0.24600	0.28571	681.56	231.16	681.56	0.840
957	C9H21N	n-NONYLAMINE	0.24805	0.25300	0.28571	648.00	273.15	648.00	0.785
958	C9H21N	TRIPROPYLAMINE	0.24950	0.26668	0.24500	577.50	179.65	577.50	0.754
959	C10H6O8	PYROMELLITIC ACID	0.43510	0.24700	0.28571	893.00	554.00	893.00	--------
960	C10H7Br	1-BROMONAPHTHALENE	0.45710	0.26057	0.30299	824.00	279.35	824.00	1.478
961	C10H7Cl	1-CHLORONAPHTHALENE	0.37470	0.26597	0.31459	785.00	269.15	785.00	1.171
962	C10H8	NAPHTHALENE	0.30619	0.25037	0.27300	748.35	353.43	748.35	--------
963	C10H8	AZULENE	0.31300	0.24800	0.28571	773.48	173.66	773.48	1.053
964	C10H9N	QUINALDINE	0.29044	0.22705	0.28571	773.00	272.15	773.00	1.055
965	C10H10	m-DIVINYLBENZENE	0.29589	0.25859	0.30295	692.00	206.25	692.00	0.925
966	C10H10	1-METHYLINDENE	0.29588	0.26316	0.21234	703.00	350.00	703.00	--------
967	C10H10	2-METHYLINDENE	0.30348	0.26254	0.23950	684.00	353.15	684.00	--------
968	C10H10O4	DIMETHYL PHTHALATE	0.36641	0.25428	0.30720	766.00	272.15	766.00	1.189
969	C10H10O4	DIMETHYL TEREPHTHALATE	0.36710	0.26885	0.26120	772.00	413.80	772.00	--------
970	C10H12	DICYCLOPENTADIENE	0.29725	0.24800	0.28570	660.00	307.00	660.00	--------
971	C10H12	1,2,3,4-TETRAHYDRONAPHTHALENE	0.29840	0.25750	0.26770	720.15	237.40	720.15	0.967
972	C10H12O	ANETHOLE	0.30747	0.25656	0.29525	723.00	294.50	723.00	0.984
973	C10H12O4	DIALLYL MALEATE	0.32384	0.24500	0.28571	693.00	226.15	693.00	1.073
974	C10H14	n-BUTYLBENZENE	0.27043	0.25325	0.28960	660.55	185.30	660.55	0.858
975	C10H14	sec-BUTYLBENZENE	0.27558	0.26013	0.28570	664.54	197.72	664.54	0.858
976	C10H14	tert-BUTYLBENZENE	0.27832	0.26103	0.28570	660.00	215.27	660.00	0.863
977	C10H14	1,2,3,4-TETRAMETHYLBENZENE	0.26430	0.23700	0.28571	695.10	266.91	695.10	0.901
978	C10H14	m-CYMENE	0.27674	0.25906	0.29420	657.00	209.44	657.00	0.857
979	C10H14	o-CYMENE	0.27447	0.25298	0.28800	662.00	201.64	662.00	0.873
980	C10H14	p-CYMENE	0.27296	0.25742	0.28750	653.15	205.25	653.15	0.852
981	C10H14	m-DIETHYLBENZENE	0.27326	0.25662	0.28571	663.00	189.26	663.00	0.860
982	C10H14	o-DIETHYLBENZENE	0.27466	0.25340	0.28570	668.00	241.93	668.00	0.876
983	C10H14	p-DIETHYLBENZENE	0.27075	0.25397	0.28571	657.96	230.32	657.96	0.858
984	C1OH14	2-ETHYL-m-XYLENE	0.27847	0.25308	0.29133	671.00	256.89	671.00	0.886
985	C10H14	2-ETHYL-p-XYLENE	0.27846	0.25625	0.29320	663.00	219.52	663.00	0.873
986	C10H14	3-ETHYL-o-XYLENE	0.26474	0.24012	0.28485	680.00	223.64	680.00	0.888
987	C10H14	4-ETHYL-m-XYLENE	0.27847	0.25636	0.29537	665.00	210.27	665.00	0.872
988	C10H14	4-ETHYL-o-XYLENE	0.27405	0.25435	0.28571	667.00	206.22	667.00	0.871
989	C1OH14	5-ETHYL-m-XYLENE	0.27847	0.25932	0.29461	655.00	188.82	655.00	0.861
990	C10H14	ISOBUTYLBENZENE	0.29434	0.27495	0.32244	650.15	221.70	650.15	0.849
991	C10H14	1,2,3,5-TETRAMETHYLBENZENE	0.27771	0.25427	0.28571	679.00	249.46	679.00	0.887
992	C10H14	1,2,4,5-TETRAMETHYLBENZENE	0.27679	0.25384	0.28571	675.15	352.38	675.15	--------
993	C10H14O	p-tert-BUTYLPHENOL	0.31037	0.26490	0.28570	734.00	371.56	734.00	--------
994	C10H14O2	p-tert-BUTYLCATECHOL	0.35024	0.27730	0.28570	776.00	325.00	776.00	--------
995	C10H15N	N,N-DIETYHLANILINE	0.26888	0.23810	0.26203	702.00	235.15	702.00	0.931
996	C10H15N	2,6-DIETHYLANILINE	0.30132	0.27412	0.28571	678.00	276.65	678.00	0.902
997	C10H16	CAMPHENE	0.27302	0.25606	0.26469	638.00	320.15	638.00	--------
998	C10H16	D-LIMONENE	0.26000	0.25157	0.27210	660.00	199.00	660.00	0.839
999	C10H16	alpha-PHELLANDRENE	0.27306	0.26074	0.28570	649.00	220.00	649.00	0.843
1000	C1OH16	beta-PHELLANDRENE	0.27974	0.26440	0.31500	648.00	220.00	648.00	0.837
1001	C10H16	alpha-PINENE	0.27032	0.25601	0.26040	632.00	209.15	632.00	0.857
1002	C10H16	beta-PINENE	0.26924	0.26424	0.20685	643.00	211.61	643.00	0.867
1003	C10H16	alpha-TERPINENE	0.26915	0.26144	0.28571	652.00	220.00	652.00	0.830
1004	C1OH16	gamma-TERPINENE	0.26951	0.25754	0.28570	661.00	220.00	661.00	0.845
1005	C10H16	TERPINOLENE	0.26777	0.25225	0.28571	672.00	200.00	672.00	0.858
1006	C10H16O	CAMPHOR	0.33140	0.23300	0.28571	709.00	453.25	709.00	--------
1007	C10H18	1-DECYNE	0.24540	0.25700	0.28571	632.49	229.16	632.49	0.762
1008	C10H18	cis-DECAHYDRONANPHTALENE	0.28899	0.26613	0.28720	702.25	230.20	702.25	0.894
1009	C10H18	trans-DECAHYDRONAPHTHALENE	0.28679	0.26991	0.29520	687.05	242.79	687.05	0.868
1010	C10H18O4	SEBACIC ACID	0.30716	0.22835	0.28571	815.00	407.65	815.00	--------
1011	C10H20	n-BUTYLCYCLOHEXANE	0.26269	0.26560	0.30240	667.00	198.42	667.00	0.796
1012	C10H20	1-CYCLOPENTYLPENTANE	0.24630	0.25000	0.28571	647.49	190.16	647.49	0.787
1013	C10H20	1-DECENE	0.23981	0.25776	0.28562	617.05	206.89	617.05	0.737
1014	C10H20O	1-DECANAL	0.26943	0.26956	0.26840	657.00	267.15	657.00	0.821

Table 8-1 DENSITY OF LIQUID - ORGANIC COMPOUNDS (continued)

| NO | FORMULA | NAME | density = A B $^{-(1-T/Tc)^n}$ | | | | (density - g/ml, T - K) | | |
			A	B	n	T_C	TMIN	TMAX	density @ 25 C
1015	C10H20O2	n-DECANOIC ACID	0.28725	0.27072	0.26860	713.00	304.75	713.00	--------
1016	C10H20O2	2-ETHYLHEXYL ACETATE	0.27669	0.25430	0.28570	639.00	180.15	639.00	0.869
1017	C10H20O2	ISOPENTYL ISOVALERATE	0.28711	0.28038	0.24390	637.00	215.00	637.00	0.854
1018	C10H22	n-DECANE	0.23276	0.25240	0.28570	618.45	243.49	618.45	0.728
1019	C10H22	2-METHYLNONANE	0.24406	0.26475	0.30083	610.00	198.50	610.00	0.723
1020	C10H22	3-METHYLNONANE	0.24406	0.26569	0.28692	613.00	188.35	613.00	0.729
1021	C10H22	4-METHYLNONANE	0.24406	0.26162	0.30463	610.00	174.45	610.00	0.728
1022	C10H22	5-METHYLNONANE	0.24406	0.26379	0.29452	610.00	185.45	610.00	0.729
1023	C10H22	3-ETHYLOCTANE	0.22690	0.24100	0.28571	613.60	185.46	613.60	0.736
1024	C10H22	4-ETHYLOCTANE	0.22370	0.23700	0.28571	609.60	185.46	609.60	0.734
1025	C10H22	2,2-DIMETHYLOCTANE	0.24788	0.27137	0.29314	602.00	225.00	602.00	0.721
1026	C10H22	2,3-DIMETHYLOCTANE	0.22790	0.24300	0.28571	613.20	219.16	613.20	0.734
1027	C10H22	2,4-DIMETHYLOCTANE	0.22600	0.24300	0.28571	599.40	219.16	599.40	0.723
1028	C10H22	2,5-DIMETHYLOCTANE	0.22750	0.24400	0.28571	603.00	219.16	603.00	0.726
1029	C10H22	2,6-DIMETHYLOCTANE	0.22890	0.24700	0.28571	603.10	219.16	603.10	0.723
1030	C10H22	2,7-DIMETHYLOCTANE	0.22780	0.24700	0.28571	602.90	219.16	602.90	0.720
1031	C10H22	3,3-DIMETHYLOCTANE	0.22830	0.24300	0.28571	612.10	219.16	612.10	0.735
1032	C10H22	3,4-DIMETHYLOCTANE	0.22920	0.24200	0.28571	614.00	219.16	614.00	0.741
1033	C10H22	3,5-DIMETHYLOCTANE	0.22680	0.24100	0.28571	606.30	219.16	606.30	0.733
1034	C10H22	3,6-DIMETHYLOCTANE	0.22800	0.24300	0.28571	608.30	219.16	608.30	0.732
1035	C10H22	4,4-DIMETHYLOCTANE	0.22540	0.24000	0.28571	606.90	219.16	606.90	0.731
1036	C10H22	4,5-DIMETHYLOCTANE	0.22850	0.24000	0.28571	612.20	219.16	612.20	0.743
1037	C10H22	4-PROPYLHEPTANE	0.22490	0.23800	0.28571	601.00	219.16	601.00	0.732
1038	C10H22	4-ISOPROPYLHEPTANE	0.22430	0.23700	0.28571	607.60	219.16	607.60	0.735
1039	C10H22	3-ETHYL-2-METHYLHEPTANE	0.22680	0.23900	0.28571	610.90	219.16	610.90	0.740
1040	C10H22	4-ETHYL-2-METHYLHEPTANE	0.22480	0.23800	0.28571	601.80	219.16	601.80	0.732
1041	C10H22	5-ETHYL-2-METHYLHEPTANE	0.22640	0.24100	0.28571	606.70	219.16	606.70	0.732
1042	C10H22	3-ETHYL-3-METHYLHEPTANE	0.22690	0.23800	0.28571	620.00	219.16	620.00	0.746
1043	C10H22	4-ETHYL-3-METHYLHEPTANE	0.22690	0.23700	0.28571	614.30	219.16	614.30	0.746
1044	C10H22	3-ETHYL-5-METHYLHEPTANE	0.22640	0.23900	0.28571	606.60	219.16	606.60	0.737
1045	C10H22	3-ETHYL-4-METHYLHEPTANE	0.22680	0.23700	0.28571	615.50	219.16	615.50	0.747
1046	C10H22	4-ETHYL-4-METHYLHEPTANE	0.22690	0.23700	0.28571	615.70	219.16	615.70	0.747
1047	C10H22	2,2,3-TRIMETHYLHEPTANE	0.23020	0.24400	0.28571	611.70	219.16	611.70	0.738
1048	C10H22	2,2,4-TRIMETHYLHEPTANE	0.22620	0.24200	0.28571	594.50	219.16	594.50	0.724
1049	C10H22	2,2,5-TRIMETHYLHEPTANE	0.22750	0.24400	0.28571	598.00	219.16	598.00	0.724
1050	C10H22	2,2,6-TRIMETHYLHEPTANE	0.22900	0.24700	0.28571	593.40	219.16	593.40	0.720
1051	C10H22	2,3,3-TRIMETHYLHEPTANE	0.23080	0.24300	0.28571	617.50	219.16	617.50	0.745
1052	C10H22	2,3,4-TRIMETHYLHEPTANE	0.22960	0.24100	0.28571	613.70	219.16	613.70	0.745
1053	C10H22	2,3,5-TRIMETHYLHEPTANE	0.22780	0.24000	0.28571	612.80	219.16	612.80	0.741
1054	C10H22	2,3,6-TRIMETHYLHEPTANE	0.22880	0.24400	0.28571	604.10	219.16	604.10	0.731
1055	C10H22	2,4,4-TRIMETHYLHEPTANE	0.22690	0.24100	0.28571	600.30	219.16	600.30	0.731
1056	C10H22	2,4,5-TRIMETHYLHEPTANE	0.22810	0.24100	0.28571	606.90	219.16	606.90	0.737
1057	C10H22	2,4,6-TRIMETHYLHEPTANE	0.22750	0.24500	0.28571	590.30	219.16	590.30	0.719
1058	C10H22	2,5,5-TRIMETHYLHEPTANE	0.22980	0.24300	0.28571	602.90	219.16	602.90	0.736
1059	C10H22	3,3,4-TRIMETHYLHEPTANE	0.23100	0.24100	0.28571	622.10	219.16	622.10	0.752
1060	C10H22	3,3,5-TRIMETHYLHEPTANE	0.23380	0.24800	0.28571	609.50	219.16	609.50	0.739
1061	C10H22	3,4,4-TRIMETHYLHEPTANE	0.23140	0.24100	0.28571	620.90	219.16	620.90	0.753
1062	C10H22	3,4,5-TRIMETHYLHEPTANE	0.23110	0.24000	0.28571	612.80	219.16	612.80	0.752
1063	C10H22	3-ISOPROPYL-2-METHYLHEXANE	0.22260	0.23400	0.28571	623.40	219.16	623.40	0.744
1064	C10H22	3,3-DIETHYLHEXANE	0.22780	0.23600	0.28571	627.80	219.16	627.80	0.757
1065	C10H22	3,4-DIETHYLHEXANE	0.22500	0.23500	0.28571	618.80	219.16	618.80	0.747
1066	C10H22	3-ETHYL-2,2-DIMETHYLHEXANE	0.22820	0.23900	0.28571	611.70	219.16	611.70	0.745
1067	C10H22	4-ETHYL-2,2-DIMETHYLHEXANE	0.22820	0.24200	0.28571	594.60	219.16	594.60	0.730
1068	C10H22	3-ETHYL-2,3-DIMETHYLHEXANE	0.23190	0.24000	0.28571	626.80	219.16	626.80	0.760
1069	C10H22	4-ETHYL-2,3-DIMETHYLHEXANE	0.22970	0.23900	0.28571	617.30	219.16	617.30	0.752
1070	C10H22	3-ETHYL-2,4-DIMETHYLHEXANE	0.22970	0.23900	0.28571	616.10	219.16	616.10	0.751
1071	C10H22	4-ETHYL-2,4-DIMETHYLHEXANE	0.23110	0.24100	0.28571	620.90	219.16	620.90	0.752
1072	C10H22	3-ETHYL-2,5-DIMETHYLHEXANE	0.22760	0.24000	0.28571	603.50	219.16	603.50	0.737
1073	C10H22	4-ETHYL-3,3-DIMETHYLHEXANE	0.23120	0.23900	0.28571	625.70	219.16	625.70	0.760
1074	C10H22	3-ETHYL-3,4-DIMETHYLHEXANE	0.23050	0.23800	0.28571	624.50	219.16	624.50	0.760
1075	C10H22	2,2,3,3-TETRAMETHYLHEXANE	0.23420	0.24200	0.28571	623.00	219.16	623.00	0.761
1076	C10H22	2,2,3,4-TETRAMETHYLHEXANE	0.23080	0.24100	0.28571	620.40	219.16	620.40	0.751
1077	C10H22	2,2,3,5-TETRAMETHYLHEXANE	0.23070	0.24500	0.28571	601.30	219.16	601.30	0.733
1078	C10H22	2,2,4,4-TETRAMETHYLHEXANE	0.22610	0.23700	0.28571	610.20	219.16	610.20	0.742
1079	C10H22	2,2,4,5-TETRAMETHYLHEXANE	0.22890	0.24300	0.28571	598.50	219.16	598.50	0.731
1080	C10H22	2,2,5,5-TETRAMETHYLHEXANE	0.23560	0.25600	0.28571	581.40	260.56	581.40	0.715
1081	C10H22	2,3,3,4-TETRAMETHYLHEXANE	0.23450	0.24200	0.28571	633.10	260.56	633.10	0.765
1082	C10H22	2,3,3,5-TETRAMETHYLHEXANE	0.23160	0.24300	0.28571	610.10	260.56	610.10	0.745
1083	C10H22	2,3,4,4-TETRAMETHYLHEXANE	0.23850	0.24100	0.28571	626.60	260.56	626.60	0.779
1084	C10H22	2,3,4,5-TETRAMETHYLHEXANE	0.23150	0.24300	0.28571	613.20	260.56	613.20	0.746
1085	C10H22	3,3,4,4-TETRAMETHYLHEXANE	0.23710	0.24200	0.28571	646.70	260.56	646.70	0.779
1086	C10H22	2,4-DIMETHYL-3-ISOPROPYLPENTANE	0.23090	0.23900	0.28571	614.40	191.46	614.40	0.754
1087	C10H22	3,3-DIETHYL-2-METHYLPENTANE	0.23440	0.23900	0.28571	639.90	191.46	639.90	0.775
1088	C10H22	3-ETHYL-2,2,3-TRIMETHYLPENTANE	0.23610	0.24100	0.28571	646.00	191.46	646.00	0.778
1089	C10H22	3-ETHYL-2,2,4-TRIMETHYLPENTANE	0.23120	0.24000	0.28571	615.30	191.46	615.30	0.753
1090	C10H22	3-ETHYL-2,3,4-TRIMETHYLPENTANE	0.23510	0.24100	0.28571	642.30	191.46	642.30	0.773
1091	C10H22	2,2,3,3,4-PENTAMETHYLPENTANE	0.23920	0.24500	0.28571	643.80	236.71	643.80	0.777
1092	C10H22	2,2,3,4,4-PENTAMETHYLPENTANE	0.23300	0.24000	0.28571	627.30	234.41	627.30	0.764

Table 8-1 DENSITY OF LIQUID - ORGANIC COMPOUNDS (continued)

			density = A B$^{-(1-T/Tc)^n}$				(density - g/ml, T - K)		
NO	FORMULA	NAME	A	B	n	T_C	TMIN	TMAX	density @ 25 C
1093	C10H22O	1-DECANOL	0.26416	0.26113	0.29040	690.00	280.05	690.00	0.825
1094	C10H22O	DI-n-PENTYL ETHER	0.26692	0.27155	0.29940	622.00	203.72	622.00	0.780
1095	C10H22O	ISODECANOL	0.26547	0.25389	0.28571	644.00	213.15	644.00	0.837
1096	C10H22O5	TETRAETHYLENE GLYCOL DIMETHYL ETHER	0.32980	0.25837	0.35270	705.00	243.45	705.00	1.006
1097	C10H22S	n-DECYL MERCAPTAN	0.27941	0.27937	0.26122	696.00	247.56	696.00	0.841
1098	C10H22S	BUTYL-HEXYL-SULFIDE	0.24780	0.23900	0.28571	701.03	238.16	701.03	0.841
1099	C10H22S	ETHYL-OCTYL-SULFIDE	0.24780	0.23900	0.28571	701.03	238.16	701.03	0.841
1100	C10H22S	HEPTYL-PROPYL-SULFIDE	0.24780	0.23900	0.28571	701.03	238.16	701.03	0.841
1101	C10H22S	METHYL-NONYL-SULFIDE	0.24780	0.23900	0.28571	701.03	238.16	701.03	0.841
1102	C10H22S	PENTYL-SULFIDE	0.24780	0.23900	0.28571	701.03	238.16	701.03	0.841
1103	C10H22S2	PENTYL-DISULFIDE	0.28160	0.25300	0.28571	726.94	214.16	726.94	0.918
1104	C10H23N	n-DECYLAMINE	0.24715	0.25170	0.28570	663.00	288.85	663.00	0.791
1105	C11H10	1-METHYLNAPHTHALENE	0.27189	0.22408	0.25709	772.04	242.67	772.04	1.017
1106	C11H10	2-METHYLNAPHTHALENE	0.28252	0.23843	0.25590	761.00	307.73	761.00	--------
1107	C11H14O2	n-BUTYL BENZOATE	0.32114	0.26431	0.29710	724.00	251.65	724.00	1.001
1108	C11H16	n-PENTYLBENZENE	0.26967	0.25746	0.28130	679.90	198.15	679.90	0.855
1109	C11H16O	p-tert-AMYLPHENOL	0.30033	0.26100	0.28571	751.00	366.00	751.00	--------
1110	C11H20	1-UNDECYNE	0.24090	0.25100	0.28571	650.99	248.16	650.99	0.769
1111	C11H20O2	2-ETHYLHEXYL ACRYLATE	0.28838	0.24214	0.39520	655.00	183.15	655.00	0.880
1112	C11H22	1-UNDECENE	0.24035	0.25645	0.29033	638.00	223.99	638.00	0.747
1113	C11H22	1-CYCLOPENTYLHEXANE	0.24000	0.24300	0.28571	667.67	200.16	667.67	0.793
1114	C11H22	PENTYLCYCLOHEXANE	0.24250	0.24400	0.28571	674.01	215.66	674.01	0.800
1115	C11H22O	1-UNDECANAL	0.26946	0.26210	0.30963	672.00	273.15	672.00	0.823
1116	C11H24	n-UNDECANE	0.23143	0.24999	0.28571	638.76	247.57	638.76	0.737
1117	C11H24O	1-UNDECANOL	0.26798	0.26292	0.30200	704.00	289.05	704.00	0.831
1118	C11H24S	UNDECYL MERCAPTAN	0.27867	0.28063	0.25705	710.00	270.15	710.00	0.841
1119	C11H24S	BUTYL-HEPTYL-SULFIDE	0.24020	0.23200	0.28571	717.91	254.66	717.91	0.841
1120	C11H24S	DECYL-METHYL-SULFIDE	0.24020	0.23200	0.28571	717.91	254.66	717.91	0.841
1121	C11H24S	ETHYL-NONYL-SULFIDE	0.24020	0.23200	0.28571	717.91	254.66	717.91	0.841
1122	C11H24S	OCTYL-PROPYL-SULFIDE	0.24020	0.23200	0.28571	717.91	254.66	717.91	0.841
1123	C12H8O	DIBENZOFURAN	0.31555	0.22920	0.29450	837.80	355.65	837.80	--------
1124	C12H9N	DIBENZOPYRROLE	0.34727	0.21000	0.28571	899.00	517.95	899.00	--------
1125	C12H10	ACENAPHTHENE	0.27888	0.22047	0.24020	803.15	366.56	803.15	--------
1126	C12H10	BIPHENYL	0.30766	0.25375	0.27892	789.26	342.37	789.26	--------
1127	C12H10O	DIPHENYL ETHER	0.34314	0.27600	0.26661	763.00	300.02	763.00	--------
1128	C12H11N	p-AMINODIPHENYL	0.31403	0.26100	0.28571	817.00	326.00	817.00	--------
1129	C12H11N	DIPHENYLAMINE	0.31410	0.24546	0.28330	817.00	326.15	817.00	--------
1130	C12H11N3	p-AMINOAZOBENZENE	0.30716	0.25500	0.28571	877.00	401.00	877.00	--------
1131	C12H11N3	1,3-DIPHENYLTRIAZENE	0.30751	0.25800	0.28571	845.00	372.00	845.00	--------
1132	C12H12	1,2-DIMETHYLNAPHTHALENE	0.29600	0.24300	0.28571	775.34	272.16	775.34	1.014
1133	C12H12	1,3-DIMETHYLNAPHTHALENE	0.29380	0.24400	0.28571	773.76	269.16	773.76	1.003
1134	C12H12	1,4-DIMETHYLNAPHTHALENE	0.29560	0.24300	0.28571	776.78	280.82	776.78	1.013
1135	C12H12	1,5-DIMETHYLNAPHTHALENE	0.29960	0.24400	0.28571	773.47	355.16	773.47	--------
1136	C12H12	1,6-DIMETHYLNAPHTHALENE	0.29400	0.24500	0.28571	770.60	259.16	770.60	0.999
1137	C12H12	1,7-DIMETHYLNAPHTHALENE	0.29400	0.24500	0.28571	770.60	260.16	770.60	0.999
1138	C12H12	2,3-DIMETHYLNAPHTHALENE	0.29960	0.24200	0.28571	777.78	378.16	777.78	--------
1139	C12H12	2,6-DIMETHYLNAPHTHALENE	0.30044	0.24605	0.31910	777.00	384.55	777.00	--------
1140	C12H12	2,7-DIMETHYLNAPHTHALENE	0.30044	0.24608	0.31993	778.00	370.15	778.00	--------
1141	C12H12	1-ETHYLNAPHTHALENE	0.30045	0.25375	0.26400	776.00	259.34	776.00	1.004
1142	C12H12	2-ETHYLNAPHTHALENE	0.29970	0.25400	0.28571	774.90	265.76	774.90	0.988
1143	C12H12N2	p-AMINODIPHENYLAMINE	0.30883	0.26400	0.28571	867.00	341.15	867.00	--------
1144	C12H12N2	HYDRAZOBENZENE	0.33125	0.26100	0.28571	792.00	404.15	792.00	--------
1145	C12H14	1,2,3-TRIMETHYLINDENE	0.29914	0.24275	0.28571	726.00	344.65	726.00	--------
1146	C12H14O4	DIETHYL PHTHALATE	0.34999	0.25154	0.35250	757.00	269.15	757.00	1.113
1147	C12H16	CYCLOHEXYLBENZENE	0.29085	0.25100	0.32273	744.00	280.14	744.00	0.939
1148	C12H18	m-DIISOPROPYLBENZENE	0.27046	0.23535	0.40400	684.00	210.02	684.00	0.852
1149	C12H18	p-DIISOPROPYLBENZENE	0.26927	0.25774	0.28570	689.00	256.08	689.00	0.853
1150	C12H18	n-HEXYLBENZENE	0.26391	0.25217	0.28570	698.00	212.00	698.00	0.855
1151	C12H18	1,2,3-TRIETHYLBENZENE	0.26440	0.24600	0.28571	684.37	206.66	684.37	0.870
1152	C12H18	1,2,4-TRIETHYLBENZENE	0.26440	0.24600	0.28571	684.37	206.66	684.37	0.870
1153	C12H18	1,3,5-TRIETHYLBENZENE	0.27070	0.24700	0.28571	682.28	206.66	682.28	0.887
1154	C12H18	HEXAMETHYLBENZENE	0.27070	0.21600	0.28571	758.00	438.66	758.00	--------
1155	C12H20O4	DIBUTYL MALEATE	0.31751	0.26117	0.30740	716.00	188.15	716.00	0.991
1156	C12H22	BICYCLOHEXYL	0.27812	0.26185	0.28130	727.00	276.78	727.00	0.883
1157	C12H22	1-DODECYNE	0.23540	0.24400	0.28571	668.16	254.16	668.16	0.775
1158	C12H23N	DICYCLOHEXYLAMINE	0.29292	0.26661	0.29800	737.00	273.05	737.00	0.909
1159	C12H24	1-DODECENE	0.24047	0.25418	0.29662	657.00	237.93	657.00	0.756
1160	C12H24	1-CYCLOPENTYLHEPTANE	0.23790	0.23700	0.28571	679.00	220.00	679.00	0.806
1161	C12H24	1-CYCLOHEXYLHEXANE	0.26280	0.23800	0.28571	691.81	263.60	691.81	0.892
1162	C12H24O	1-DODECANAL	0.24814	0.24260	0.28570	685.00	285.15	685.00	0.826
1163	C12H24O2	n-DODECANOIC ACID	0.28418	0.26550	0.29330	734.00	317.15	734.00	--------
1164	C12H26	n-DODECANE	0.23403	0.25183	0.28960	658.20	263.57	658.20	0.745
1165	C12H26O	DI-n-HEXYL ETHER	0.26696	0.27208	0.30250	658.00	230.15	658.00	0.790
1166	C12H26O	1-DODECANOL	0.26716	0.26218	0.31100	721.00	296.95	721.00	0.830
1167	C12H26O3	DIETHYLENE GLYCOL DI-n-BUTYL ETHER	0.27194	0.24994	0.28571	680.00	212.95	680.00	0.881
1168	C12H26S	n-DODECYL MERCAPTAN	0.27765	0.27018	0.31186	724.00	265.15	724.00	0.842
1169	C12H26S	BUTYL-OCTYL-SULFIDE	0.23280	0.22500	0.28571	733.68	259.16	733.68	0.842
1170	C12H26S	DECYL-ETHYL-SULFIDE	0.23280	0.22500	0.28571	733.68	259.16	733.68	0.842

Table 8-1 DENSITY OF LIQUID - ORGANIC COMPOUNDS (continued)

NO	FORMULA	NAME	density = A B $^{-(1-T/Tc)^n}$				(density - g/ml, T - K)		
			A	B	n	T_c	TMIN	TMAX	density @ 25 C
1171	C12H26S	HEXYL-SULFIDE	0.23280	0.22500	0.28571	733.68	259.16	733.68	0.842
1172	C12H26S	METHYL-UNDECYL-SULFIDE	0.23280	0.22500	0.28571	733.68	259.16	733.68	0.842
1173	C12H26S	NONYL-PROPYL-SULFIDE	0.23280	0.22500	0.28571	733.68	259.16	733.68	0.842
1174	C12H26S2	HEXYL-DISULFIDE	0.26530	0.24100	0.28571	747.10	225.16	747.10	0.908
1175	C12H27BO3	TRI-n-BUTYL BORATE	0.26660	0.24121	0.39033	743.15	203.15	743.15	0.854
1176	C12H27N	DODECYLAMINE	0.25256	0.26384	0.27740	696.00	301.47	696.00	--------
1177	C12H27N	TRI-n-BUTYLAMINE	0.25305	0.26523	0.27430	644.00	203.00	644.00	0.775
1178	C13H10	FLUORENE	0.41540	0.26000	0.28571	870.00	387.94	870.00	--------
1179	C13H10O	BENZOPHENONE	0.30836	0.23634	0.27310	816.00	321.35	816.00	--------
1180	C13H12	DIPHENYLMETHANE	0.30387	0.25318	0.28570	768.00	298.39	768.00	--------
1181	C13H14	1-PROPYLNAPHTHALENE	0.29340	0.24800	0.28571	771.45	264.69	771.45	0.987
1182	C13H14	2-PROPYLNAPHTHALENE	0.28930	0.24800	0.28571	772.44	270.16	772.44	0.973
1183	C13H14	2ETHYL-3-METHYLNAPHTHALENE	0.29480	0.24300	0.28571	776.44	344.16	776.44	--------
1184	C13H14	2ETHYL-6-METHYLNAPHTHALENE	0.29480	0.24600	0.28571	766.56	318.16	766.56	--------
1185	C13H14	2ETHYL-7-METHYLNAPHTHALENE	0.29480	0.24600	0.28571	766.56	318.16	766.56	--------
1186	C13H20	n-HEPTYLBENZENE	0.27207	0.25684	0.32095	714.00	225.15	714.00	0.853
1187	C13H24	1-TRIDECYNE	0.22990	0.23700	0.28571	684.11	268.16	684.11	0.781
1188	C13H26	1-TRIDECENE	0.24121	0.25311	0.30402	675.00	250.08	675.00	0.762
1189	C13H26	1-CYCLOPENTYLOCTANE	0.22920	0.23100	0.28571	702.06	229.16	702.06	0.801
1190	C13H26	1-CYCLOHEXYLHEPTANE	0.23050	0.23100	0.28571	708.63	242.66	708.63	0.807
1191	C13H26O	1-TRIDECANAL	0.26877	0.26107	0.32170	700.00	288.15	700.00	0.827
1192	C13H26O2	n-BUTYL NONANOATE	0.27000	0.25481	0.28570	652.00	235.15	652.00	0.851
1193	C13H26O2	METHYL DODECANOATE	0.28508	0.22100	0.28571	712.00	278.15	712.00	1.039
1194	C13H28	n-TRIDECANE	0.23767	0.25040	0.31200	675.80	267.76	675.80	0.754
1195	C13H28O	1-TRIDECANOL	0.24656	0.24200	0.28570	731.00	303.75	731.00	--------
1196	C13H28S	BUTYL-NONYL-SULFIDE	0.22650	0.21900	0.28571	748.42	271.16	748.42	0.842
1197	C13H28S	DECYL-PROPYL-SULFIDE	0.22650	0.21900	0.28571	748.42	271.16	748.42	0.842
1198	C13H28S	DODECYL-METHYL-SULFIDE	0.22650	0.21900	0.28571	748.42	271.16	748.42	0.842
1199	C13H28S	ETHYL-UNDECYL-SULFIDE	0.22650	0.21900	0.28571	748.42	271.16	748.42	0.842
1200	C13H28S	1-TRIDECANETHIOL	0.23670	0.23000	0.28571	742.13	282.04	742.13	0.842
1201	C14H8O2	ANTHRAQUINONE	0.36491	0.24020	0.28570	900.00	559.15	900.00	--------
1202	C14H10	ANTHRACENE	0.32172	0.26069	0.23530	873.00	489.25	873.00	--------
1203	C14H10	DIPHENYLACETYLENE	0.29187	0.25600	0.28571	832.00	335.65	832.00	--------
1204	C14H10	PHENANTHRENE	0.32181	0.25228	0.24860	869.25	372.38	869.25	--------
1205	C14H12	cis-STILBENE	0.30849	0.25436	0.28571	757.00	268.15	757.00	1.011
1206	C14H12	trans-STILBENE	0.31224	0.23200	0.28571	820.00	397.35	820.00	--------
1207	C14H12O2	BENZYL BENZOATE	0.30582	0.22320	0.32700	820.00	292.55	820.00	1.115
1208	C14H14	1,1-DIPHENYLETHANE	0.30176	0.25065	0.30352	775.00	255.20	775.00	0.996
1209	C14H14	1,2-DIPHENYLETHANE	0.30031	0.24800	0.29000	780.00	324.34	780.00	--------
1210	C14H14O	DIBENZYL ETHER	0.31276	0.25120	0.28570	777.00	276.75	777.00	1.042
1211	C14H16	1-n-BUTYLNAPHTHALENE	0.29206	0.25258	0.28340	792.00	253.43	792.00	0.973
1212	C14H16	2-BUTYLNAPHTHALENE	0.28120	0.24400	0.28571	780.96	268.16	780.96	0.962
1213	C14H22	n-OCTYLBENZENE	0.27074	0.25617	0.32630	729.00	237.15	729.00	0.853
1214	C14H22	1,2,3,4-TETRAETHYLBENZENE	0.25410	0.23300	0.28571	708.20	284.96	708.20	0.883
1215	C14H22	1,2,3,5-TETRAETHYLBENZENE	0.25350	0.23400	0.28571	707.52	284.16	707.52	0.878
1216	C14H22	1,2,4,5-TETRAETHYLBENZENE	0.25270	0.23400	0.28571	706.85	283.16	706.85	0.875
1217	C14H22O	p-tert-OCTYLPHENOL	0.29349	0.25200	0.29000	765.00	358.55	765.00	--------
1218	C14H28	1-TETRADECENE	0.24037	0.25196	0.30449	692.00	260.30	692.00	0.768
1219	C14H28	1-CYCLOPENTYLNONANE	0.22380	0.22500	0.28571	716.95	244.16	716.95	0.804
1220	C14H28	1-CYCLOHEXYLOCTANE	0.22490	0.22500	0.28571	723.61	253.46	723.61	0.810
1221	C14H28O2	n-TETRADECANOIC ACID	0.28169	0.26965	0.28360	756.00	327.55	756.00	--------
1222	C14H30	n-TETRADECANE	0.23556	0.25588	0.27348	692.40	279.01	692.40	0.758
1223	C14H30O	1-TETRADECANOL	0.26759	0.26293	0.31824	741.00	310.65	741.00	--------
1224	C14H30S	BUTYL-DECYL-SULFIDE	0.22020	0.21300	0.28571	762.23	276.16	762.23	0.843
1225	C14H30S	DODECYL-ETHYL-SULFIDE	0.22020	0.21300	0.28571	762.23	276.16	762.23	0.843
1226	C14H30S	HEPTYL-SULFIDE	0.22020	0.21300	0.28571	762.23	276.16	762.23	0.843
1227	C14H30S	METHYL-TRIDECYL-SULFIDE	0.22020	0.21300	0.28571	762.23	276.16	762.23	0.843
1228	C14H30S	PROPYL-UNDECYL-SULFIDE	0.22020	0.21300	0.28571	762.23	276.16	762.23	0.843
1229	C14H30S	1-TETRADECANETHIOL	0.22970	0.22300	0.28571	753.80	279.26	753.80	0.842
1230	C14H30S2	HEPTYL-DISULFIDE	0.24920	0.22800	0.28571	765.96	235.16	765.96	0.900
1231	C14H31N	TETRADECYLAMINE	0.24069	0.24508	0.28571	722.30	311.34	722.30	--------
1232	C15H10N2O2	DIPHENYLMETHANE-4,4'-DIISOCYANATE	0.35154	0.24341	0.28571	802.00	311.20	802.00	--------
1233	C15H16O	p-CUMYLPHENOL	0.32176	0.25500	0.28571	834.00	346.00	834.00	--------
1234	C15H16O2	BISPHENOL A	0.33722	0.28100	0.28571	849.00	426.15	849.00	--------
1235	C15H18	1-PENTYLNAPHTHALENE	0.27440	0.23800	0.28571	793.32	251.16	793.32	0.962
1236	C15H18	2-PENTYLNAPHTHALENE	0.26940	0.23600	0.28571	797.48	269.16	797.48	0.953
1237	C15H24	n-NONYLBENZENE	0.25876	0.24856	0.30140	741.00	249.00	741.00	0.852
1238	C15H24O	2,6-DI-tert-BUTYL-p-CRESOL	0.29532	0.26300	0.28571	720.00	344.00	720.00	--------
1239	C15H24O	NONYLPHENOL	0.29109	0.25460	0.29244	757.00	279.15	757.00	0.949
1240	C15H28	1-PENTADECYNE	0.22000	0.22500	0.28571	711.41	283.16	711.41	0.789
1241	C15H30	1-PENTADECENE	0.24047	0.25056	0.31110	708.00	269.42	708.00	0.773
1242	C15H30	1-CYCLOPENTYLDECANE	0.21840	0.21900	0.28571	730.64	251.03	730.64	0.807
1243	C15H30	1-CYCLOHEXYLNONANE	0.21930	0.21900	0.28571	737.79	262.96	737.79	0.813
1244	C15H30O2	PENTADECANOIC ACID	0.28056	0.25872	0.33432	766.00	325.68	766.00	--------
1245	C15H32	n-PENTADECANE	0.24137	0.25375	0.31579	706.80	283.11	706.80	0.765
1246	C15H32O	1-PENTADECANOL	0.25540	0.22600	0.28571	722.53	317.04	722.53	--------
1247	C15H32S	BUTYL-UNDECYL-SULFIDE	0.21400	0.20700	0.28571	775.15	284.16	775.15	0.843
1248	C15H32S	DODECYL-PROPYL-SULFIDE	0.21400	0.20700	0.28571	775.15	284.16	775.15	0.843

Table 8-1 DENSITY OF LIQUID - ORGANIC COMPOUNDS (continued)

NO	FORMULA	NAME	density = A B $^{-(1-T/Tc)^n}$				(density - g/ml, T - K)		
			A	B	n	T_C	TMIN	TMAX	density @ 25 C
1249	C15H32S	ETHYL-TRIDECYL-SULFIDE	0.21400	0.20700	0.28571	775.15	284.16	775.15	0.843
1250	C15H32S	METHYL-TETRADECYL-SULFIDE	0.21400	0.20700	0.28571	775.15	284.16	775.15	0.843
1251	C15H32S	1-PENTADECANETHIOL	0.22360	0.21700	0.28571	764.77	290.93	764.77	0.843
1252	C16H10	FLUORANTHENE	0.30876	0.22722	0.28571	905.00	383.33	905.00	--------
1253	C16H10	PYRENE	0.31757	0.21360	0.28571	936.00	423.81	936.00	--------
1254	C16H12	1-PHENYLNAPHTHALENE	0.31139	0.24155	0.29049	849.00	318.15	849.00	--------
1255	C16H20	1-n-HEXYLNAPHTHALENE	0.28656	0.25364	0.30090	813.00	255.15	813.00	0.947
1256	C16H22O4	DIBUTYL PHTHALATE	0.32901	0.25148	0.37367	781.00	238.15	781.00	1.043
1257	C16H26	n-DECYLBENZENE	0.24794	0.23857	0.29619	753.00	258.77	753.00	0.852
1258	C16H26	PENTAETHYLBENZENE	0.26520	0.22200	0.28571	723.64	327.66	723.64	--------
1259	C16H30	1-HEXADECYNE	0.21500	0.21900	0.28571	724.26	288.16	724.26	0.793
1260	C16H32	n-DECYLCYCLOHEXANE	0.26157	0.26241	0.32209	751.25	271.42	751.25	0.815
1261	C16H32	1-CYCLOPENTYLUNDECANE	0.21300	0.21300	0.28571	743.30	263.16	743.30	0.810
1262	C16H32	1-HEXADECENE	0.24055	0.24958	0.31572	722.00	277.51	722.00	0.777
1263	C16H32O2	n-HEXADECANOIC ACID	0.27973	0.26805	0.29470	776.00	335.95	776.00	--------
1264	C16H34	n-HEXADECANE	0.24348	0.25442	0.32380	720.60	291.34	720.60	0.770
1265	C16H34O	DI-n-OCTYL ETHER	0.26642	0.27174	0.30391	707.00	265.55	707.00	0.803
1266	C16H34O	1-HEXADECANOL	0.26758	0.26300	0.32360	761.00	322.35	761.00	--------
1267	C16H34S	BUTYL-DODECYL-SULFIDE	0.20790	0.20100	0.28571	787.27	288.16	787.27	0.843
1268	C16H34S	ETHYL-TETRADECYL-SULFIDE	0.19040	0.18200	0.28571	791.68	288.16	791.68	0.844
1269	C16H34S	METHYL-PENTADECYL-SULFIDE	0.20790	0.20100	0.28571	787.27	288.16	787.27	0.843
1270	C16H34S	OCTYL-SULFIDE	0.20790	0.20100	0.28571	787.27	288.16	787.27	0.843
1271	C16H34S	PROPYL-TRIDECYL-SULFIDE	0.20790	0.20100	0.28571	787.27	288.16	787.27	0.843
1272	C16H34S	1-HEXADECANETHIOL	0.21760	0.21100	0.28571	774.68	290.93	774.68	0.843
1273	C16H34S2	OCTYL-DISULFIDE	0.23390	0.21500	0.28571	784.46	244.16	784.46	0.894
1274	C17H28	n-UNDECYLBENZENE	0.25543	0.24601	0.30840	764.00	268.00	764.00	0.851
1275	C17H32	1-HEPTADECYNE	0.20900	0.21200	0.28571	736.21	295.16	736.21	0.796
1276	C17H34	1-CYCLOPENTYLDODECANE	0.20750	0.20700	0.28571	755.17	268.16	755.17	0.812
1277	C17H34	1-CYCLOHEXYLUNDECANE	0.20920	0.20800	0.28571	761.74	278.96	761.74	0.817
1278	C17H34	1-HEPTADECENE	0.24970	0.25731	0.33430	736.00	284.40	736.00	0.782
1279	C17H36	n-HEPTADECANE	0.23953	0.25316	0.30520	733.37	295.13	733.37	0.773
1280	C17H36O	1-HEPTADECANOL	0.26717	0.25890	0.33904	770.00	327.05	770.00	--------
1281	C17H36S	BUTYL-TRIDECYL-SULFIDE	0.20180	0.19500	0.28571	798.63	294.16	798.63	0.844
1282	C17H36S	ETHYL-PENTADECYL-SULFIDE	0.20180	0.19500	0.28571	798.63	294.16	798.63	0.844
1283	C17H36S	HEXADECYL-METHYL-SULFIDE	0.20180	0.19500	0.28571	798.63	294.16	798.63	0.844
1284	C17H36S	PROPYL-TETRADECYL-SULFIDE	0.20180	0.19500	0.28571	798.63	294.16	798.63	0.844
1285	C17H36S	1-HEPTADECANETHIOL	0.26170	0.20500	0.28571	786.01	300.37	786.01	--------
1286	C18H12	CHRYSENE	0.31492	0.21285	0.28571	979.00	531.15	979.00	--------
1287	C18H14	m-TERPHENYL	0.30009	0.23686	0.29680	924.85	360.00	924.85	--------
1288	C18H14	o-TERPHENYL	0.30601	0.24035	0.31720	890.95	329.35	890.95	--------
1289	C18H14	p-TERPHENYL	0.30203	0.23661	0.29435	925.95	485.00	925.95	--------
1290	C18H15P	TRIPHENYLPHOSPHINE	0.47330	0.35320	0.55468	1008.00	354.40	1008.00	--------
1291	C18H15O4P	TRIPHENYL PHOSPHATE	--------	--------	--------	--------	--------	--------	--------
1292	C18H16N2	N,N'-DIPHENYL-p-PHENYLENEDIAMINE	0.31807	0.25100	0.28571	906.00	409.00	906.00	--------
1293	C18H22	2,3-DIMETHYL-2,3-DIPHENYLBUTANE	0.30549	0.23200	0.28571	805.00	392.15	805.00	--------
1294	C18H22O2	DICUMYL PEROXIDE	0.30963	0.25899	0.28571	884.00	311.15	884.00	--------
1295	C18H30	n-DODECYLBENZENE	0.24820	0.24353	0.28571	774.26	275.93	774.26	0.849
1296	C18H30	HEXAETHYLBENZENE	0.24150	0.21200	0.28571	734.78	401.16	734.78	--------
1297	C18H32O2	LINOLEIC ACID	0.25830	0.23756	0.28600	775.00	268.15	775.00	0.902
1298	C18H34	1-OCTADECYNE	0.24910	0.20700	0.28571	747.33	300.16	747.33	--------
1299	C18H34O2	OLEIC ACID	0.28245	0.26812	0.28970	781.00	286.53	781.00	0.888
1300	C18H34O4	DIBUTYL SEBACATE	0.29949	0.25466	0.37852	768.00	263.95	768.00	0.932
1301	C18H34O4	DIHEXYL ADIPATE	0.30531	0.25883	0.38852	767.00	259.35	767.00	0.932
1302	C18H36	1-CYCOPENTYLTRIDECANE	0.20290	0.20200	0.28571	766.47	278.16	766.47	0.814
1303	C18H36	1-CYCLOHEXYLDODECANE	0.20370	0.20200	0.28571	772.83	285.66	772.83	0.819
1304	C18H36	1-OCTADECENE	0.24046	0.24722	0.32724	748.00	290.76	748.00	0.785
1305	C18H36O2	STEARIC ACID	0.27980	0.26862	0.30250	799.00	342.75	799.00	--------
1306	C18H38	n-OCTADECANE	0.23837	0.25763	0.27400	745.26	301.33	745.26	--------
1307	C18H38O	DINONYL ETHER	0.26519	0.26640	0.33030	736.00	273.00	736.00	0.808
1308	C18H38O	1-OCTADECANOL	0.26787	0.26493	0.32717	777.00	331.05	777.00	--------
1309	C18H38S	BUTYL-TETRADECYL-SULFIDE	0.26110	0.19000	0.28571	810.53	298.16	810.53	--------
1310	C18H38S	ETHYL-HEXADECYL-SULFIDE	0.26110	0.19000	0.28571	810.53	298.16	810.53	--------
1311	C18H38S	HEPTADECYL-METHYL-SULFIDE	0.26110	0.19000	0.28571	810.53	298.16	810.53	--------
1312	C18H38S	NONYL-SULFIDE	0.26110	0.19000	0.28571	810.53	298.16	810.53	--------
1313	C18H38S	PENTADECYL-PROPYL-SULFIDE	0.26110	0.19000	0.28571	810.53	298.16	810.53	--------
1314	C18H38S	1-OCTADECANETHIOL	0.26110	0.20000	0.28571	795.36	300.93	795.36	--------
1315	C18H38S2	NONYL-DISULFIDE	0.22100	0.20400	0.28571	802.30	252.16	802.30	0.889
1316	C19H26	1-n-NONYLNAPHTHALENE	0.25442	0.22733	0.30164	849.00	284.15	849.00	0.934
1317	C19H32	n-TRIDECYLBENZENE	0.24571	0.23488	0.31981	783.00	283.15	783.00	0.851
1318	C19H36	1-NONADECYNE	0.24920	0.20100	0.28571	758.94	306.16	758.94	--------
1319	C19H36O2	METHYL OLEATE	0.27971	0.26240	0.33247	764.00	293.05	764.00	0.870
1320	C19H38	1-CYCLOPENTYLTETRADECANE	0.20220	0.20100	0.28571	772.00	282.00	772.00	0.816
1321	C19H38	1-CYCLOHEXYLTRIDECANE	0.19900	0.19700	0.28571	783.38	291.66	783.38	0.821
1322	C19H38	1-NONADECENE	0.24228	0.24836	0.33336	760.00	296.55	760.00	0.788
1323	C19H38O2	NONADECANOIC ACID	0.27638	0.26267	0.32160	810.00	341.23	810.00	--------
1324	C19H40	n-NONADECANE	0.23777	0.25012	0.30650	755.93	305.33	755.93	--------
1325	C19H40O	1-NONADECANOL	0.25440	0.19900	0.28571	775.30	334.87	775.30	--------
1326	C19H40S	BUTYL-PENTADECYL-SULFIDE	0.26060	0.18500	0.28571	821.75	303.16	821.75	--------

Table 8-1 DENSITY OF LIQUID - ORGANIC COMPOUNDS (continued)

NO	FORMULA	NAME	density = A B$^{-(1-T/T_c)^n}$						(density - g/ml, T - K)
			A	B	n	T_c	TMIN	TMAX	density @ 25 C
1327	C19H40S	ETHYL-HEPTADECYL-SULFIDE	0.26060	0.18500	0.28571	821.75	303.16	821.75	--------
1328	C19H40S	HEXADECYL-PROPYL-SULFIDE	0.26060	0.18500	0.28571	821.75	303.16	821.75	--------
1329	C19H40S	METHYL-OCTADECYL-SULFIDE	0.26060	0.18500	0.28571	821.75	303.16	821.75	--------
1330	C19H40S	1-NONADECANETHIOL	0.26060	0.19400	0.28571	805.29	307.04	805.29	--------
1331	C20H16	TRIPHENYLETHYLENE	0.29835	0.23900	0.28571	908.00	342.15	908.00	--------
1332	C20H28	1-n-DECYLNAPHTHALENE	0.25088	0.22504	0.30747	859.00	288.15	859.00	0.928
1333	C20H30O2	ABIETIC ACID	0.32502	0.22600	0.28571	832.00	446.65	832.00	--------
1334	C20H31N	DEHYDROABIETYLAMINE	0.27912	0.24231	0.28571	863.00	317.65	863.00	--------
1335	C20H34	1-PHENYLTETRADECANE	0.24460	0.24000	0.28571	792.00	289.16	792.00	0.851
1336	C20H38	1-EICOSYNE	0.24920	0.19500	0.28571	769.79	309.16	769.79	--------
1337	C20H40	1-CYCLOPENTYLPENTADECANE	0.19320	0.19100	0.28571	780.00	290.00	780.00	0.818
1338	C20H40	1-CYCLOHEXYLTETRADECANE	0.19430	0.19200	0.28571	792.82	297.16	792.82	0.822
1339	C20H40	1-EICOSENE	0.24184	0.24745	0.33591	771.00	301.76	771.00	--------
1340	C20H42	n-EICOSANE	0.23759	0.24934	0.30880	767.04	309.59	767.04	--------
1341	C20H42O	1-EICOSANOL	0.26657	0.26872	0.28870	792.00	338.55	792.00	--------
1342	C20H42S	BUTYL-HEXADECYL-SULFIDE	0.26010	0.17900	0.28571	832.33	308.16	832.33	--------
1343	C20H42S	DECYL-SULFIDE	0.26010	0.17900	0.28571	832.33	308.16	832.33	--------
1344	C20H42S	ETHYL-OCTADECYL-SULFIDE	0.26010	0.17900	0.28571	832.33	308.16	832.33	--------
1345	C20H42S	HEPTADECYL-PROPYL-SULFIDE	0.26010	0.17900	0.28571	832.33	308.16	832.33	--------
1346	C20H42S	METHYL-NONADECYL-SULFIDE	0.26010	0.17900	0.28571	832.33	308.16	832.33	--------
1347	C20H42S	1-EICOSANETHIOL	0.26010	0.18900	0.28571	814.57	310.37	814.57	--------
1348	C20H42S2	DECYL-DISULFIDE	0.20750	0.19200	0.28571	820.08	259.16	820.08	0.885
1349	C21H21O4P	TRI-o-CRESYL PHOSPHATE	--------	--------	--------	--------	--------	--------	1.165
1350	C21H36	1-PHENYLPENTADECANE	0.23510	0.23000	0.28571	800.00	295.16	800.00	0.851
1351	C21H42	1-CYCLOPENTYLHEXADECANE	0.18810	0.18600	0.28571	−797.25	294.16	797.25	0.819
1352	C21H42	1-CYCLOHEXYLPENTADECANE	0.25740	0.18700	0.28571	803.46	302.16	803.46	--------
1353	C22H38	1-PHENYLHEXADECANE	0.25210	0.23000	0.28571	808.00	300.16	808.00	--------
1354	C22H44	1-CYCLOHEXYLHEXADECANE	0.25710	0.18200	0.28571	813.42	306.76	813.42	--------
1355	C22H44O2	n-BUTYL STEARATE	0.24798	0.24000	0.28570	764.00	299.45	764.00	--------
1356	C24H38O4	DIISOOCTYL PHTHALATE	0.27512	0.23675	0.28571	851.00	254.00	851.00	0.983
1357	C24H38O4	DIOCTYL PHTHALATE	0.30753	0.25052	0.38451	806.00	223.15	806.00	0.980
1358	C24H42O	DINONYLPHENOL	0.28460	0.20500	0.28571	886.00	350.00	886.00	--------
1359	C26H20	TETRAPHENYLETHYLENE	0.32689	0.21000	0.28571	996.00	496.15	996.00	--------
1360	C28H46O4	DIISODECYL PHTHALATE	0.26334	0.23000	0.28600	887.00	227.59	887.00	0.973

density - density of liquid, g/ml

A, B, n, and T_c - regression coefficients of chemical compound

T - temperature, K

TMAX - maximum temperature, K

TMIN - minimun temperature, K

Table 8-2 DENSITY OF LIQUID - INORGANIC COMPOUNDS

NO	FORMULA	NAME	density = A B $^{-(1-T/Tc)^n}$				(density - g/ml, T - K)		
			A	B	n	T_C	TMIN	TMAX	density @ 25 C
1	Ag	SILVER	0.31819	0.03098	0.13648	6410.00	1234.00	6410.00	--------
2	AgCl	SILVER CHLORIDE	1.59790	0.30000	0.28571	2992.10	728.15	928.15	--------
3	AgI	SILVER IODIDE	1.86077	0.30000	0.28571	2897.64	825.15	1025.15	--------
4	Al	ALUMINUM	0.20122	0.07645	0.28571	7151.00	933.00	1373.15	--------
5	AlB3H12	ALUMINUM BOROHYDRIDE	--------	--------	--------	--------	--------	--------	--------
6	AlBr3	ALUMINUN BROMIDE	0.47444	0.14123	0.28571	763.00	390.15	459.80	--------
7	AlCl3	ALUMINUM CHLORIDE	0.50983	0.24349	0.31244	629.00	465.75	629.00	--------
8	AlF3	ALUMINUM FLUORIDE	0.92305	0.30000	0.28571	2948.13	1313.15	1513.15	--------
9	AlI3	ALUMINUM IODIDE	1.17262	0.30000	0.28571	983.00	464.15	561.40	--------
10	Al2O3	ALUMINUM OXIDE	1.26357	0.30000	0.28571	5335.00	2325.00	2525.00	--------
11	Al2S3O12	ALUMINUM SULFATE	--------	--------	--------	--------	--------	--------	--------
12	Ar	ARGON	0.53120	0.28600	0.29840	150.86	83.78	150.86	--------
13	As	ARSENIC	0.62605	0.05595	0.28571	1673.15	1090.15	1290.15	--------
14	AsBr3	ARSENIC TRIBROMIDE	1.06061	0.27400	0.28571	789.01	306.15	399.65	--------
15	AsCl3	ARSENIC TRICHLORIDE	0.72432	0.27400	0.28571	654.00	255.15	654.00	2.150
16	AsF3	ARSENIC TRIFLUORIDE	0.95308	0.27400	0.28571	530.21	267.25	530.21	2.649
17	AsF5	ARSENIC PENTAFLUORIDE	0.87020	0.27400	0.28571	357.73	193.35	357.73	1.890
18	AsH3	ARSINE	0.79658	0.35950	0.43880	373.00	156.23	373.00	1.321
19	AsI3	ARSENIC TRIIODIDE	1.41799	0.30000	0.28571	1101.22	419.15	619.15	--------
20	As2O3	ARSENIC TRIOXIDE	1.27968	0.30000	0.28571	1189.50	585.95	658.15	--------
21	At	ASTATINE	--------	--------	--------	--------	--------	--------	--------
22	Au	GOLD	8.35557	0.44663	0.28571	4398.00	1337.33	1537.33	--------
23	B	BORON	0.72680	0.30000	0.28571	7934.59	2348.15	2548.15	--------
24	BBr3	BORON TRIBROMIDE	0.91487	0.27400	0.28571	581.00	228.15	581.00	2.625
25	BCl3	BORON TRICHLORIDE	0.78990	0.41145	0.51112	451.95	166.15	451.95	1.318
26	BF3	BORON TRIFLUORIDE	0.54855	0.26189	0.22100	260.90	146.05	260.90	--------
27	BH2CO	BORINE CARBONYL	--------	--------	--------	--------	--------	--------	--------
28	BH3O3	BORIC ACID	--------	--------	--------	--------	--------	--------	--------
29	B2D6	DEUTERODIBORANE	--------	--------	--------	--------	--------	--------	--------
30	B2H5Br	DIBORANE HYDROBROMIDE	--------	--------	--------	--------	--------	--------	--------
31	B2H6	DIBORANE	0.15975	0.27040	0.28790	289.80	107.65	289.80	--------
32	B3N3H6	BORINE TRIAMINE	0.32656	0.27400	0.28571	521.20	214.95	521.20	0.902
33	B4H10	TETRABORANE	0.19485	0.27400	0.28571	466.66	153.25	466.66	0.513
34	B5H9	PENTABORANE	0.22962	0.27998	0.28571	568.45	226.35	568.45	0.643
35	B5H11	TETRAHYDROPENTABORANE	0.26340	0.27400	0.28571	547.13	230.00	547.13	0.741
36	B10H14	DECABORANE	0.31796	0.30000	0.28571	791.78	372.75	429.45	--------
37	Ba	BARIUM	1.11110	0.30000	0.28571	3572.13	1000.15	1200.15	--------
38	Be	BERYLLIUM	0.57566	0.30000	0.28571	5199.80	1560.15	1760.15	--------
39	BeB2H8	BERYLLIUM BOROHYDRIDE	--------	--------	--------	--------	--------	--------	--------
40	BeBr2	BERYLLIUM BROMIDE	1.28962	0.30000	0.28571	1216.86	763.15	919.72	--------
41	BeCl2	BERYLLIUM CHLORIDE	0.57918	0.30000	0.28571	1238.03	678.15	800.00	--------
42	BeF2	BERYLLIUM FLUORIDE	--------	--------	--------	--------	--------	--------	--------
43	BeI2	BERYLLIUM IODIDE	1.59610	0.30000	0.28571	1238.03	761.15	920.44	--------
44	Bi	BISMUTH	1.75617	0.16368	0.28571	4620.00	544.15	1235.15	--------
45	BiBr3	BISMUTH TRIBROMIDE	1.61236	0.30000	0.28571	1220.00	491.15	1100.00	--------
46	BiCl3	BISMUTH TRICHLORIDE	1.38859	0.30000	0.28571	1178.00	503.15	1100.00	--------
47	BrF5	BROMINE PENTAFLUORIDE	0.93371	0.27400	0.28571	470.00	211.75	470.00	2.466
48	Br2	BROMINE	1.18377	0.29527	0.32950	584.15	265.85	584.15	3.104
49	C	CARBON	0.31475	0.10000	0.28571	6810.00	4765.00	4965.00	--------
50	CCl2O	PHOSGENE	0.51316	0.27119	0.27201	455.00	145.37	455.00	1.363
51	CF2O	CARBONYL FLUORIDE	0.46780	0.32913	0.28570	297.00	161.89	297.00	--------
52	CH4N2O	UREA	0.56982	0.33700	0.28571	705.00	405.85	505.85	--------
53	CH4N2S	THIOUREA	0.30743	0.28700	0.28571	854.00	454.15	854.00	--------
54	CNBr	CYANOGEN BROMIDE	0.73999	0.30000	0.28571	545.03	331.15	332.90	--------
55	CNCl	CYANOGEN CHLORIDE	0.37713	0.22974	0.23860	449.00	266.65	449.00	1.172
56	CNF	CYANOGEN FLUORIDE	0.51072	0.27400	0.28571	368.51	200.00	368.51	1.144
57	CO	CARBON MONOXIDE	0.29818	0.27655	0.29053	132.92	68.15	132.92	--------
58	COS	CARBONYL SULFIDE	0.44482	0.28943	0.27140	378.80	134.30	378.80	1.005
59	COSe	CARBON OXYSELENIDE	0.71258	0.27400	0.28571	406.58	200.00	406.58	1.731
60	CO2	CARBON DIOXIDE	0.46382	0.26160	0.29030	304.19	216.58	304.19	0.713
61	CS2	CARBON DISULFIDE	0.47589	0.28749	0.32260	552.00	161.11	552.00	1.256
62	CSeS	CARBON SELENOSULFIDE	0.69071	0.27400	0.28571	576.53	197.95	576.53	1.977
63	C2N2	CYANOGEN	0.36145	0.27400	0.28571	399.90	238.75	399.90	0.868
64	C3S2	CARBON SUBSULFIDE	--------	--------	--------	--------	--------	--------	--------
65	Ca	CALCIUM	0.46831	0.30000	0.28571	3292.23	1115.15	1315.15	--------
66	CaF2	CALCIUM FLUORIDE	0.33727	0.10000	0.28571	4570.85	1691.00	3000.00	--------
67	CbF5	COLUMBIUM FLUORIDE	1.09002	0.30000	0.28571	811.32	348.65	423.40	--------
68	Cd	CADMIUM	3.72607	0.43396	0.28571	2291.00	594.05	873.15	--------
69	CdCl2	CADMIUM CHLORIDE	1.20842	0.30000	0.28571	2019.79	841.15	1200.00	--------
70	CdF2	CADMIUM FLUORIDE	2.01691	0.30000	0.28571	3296.66	793.15	993.15	--------
71	CdI2	CADMIUM IODIDE	1.54369	0.30000	0.28571	1741.29	658.15	1100.00	--------
72	CdO	CADMIUM OXIDE	2.20048	0.30000	0.28571	2983.96	1000.00	1200.00	--------
73	ClF	CHLORINE MONOFLUORIDE	0.60386	0.27400	0.28571	282.32	128.15	282.32	--------
74	ClFO3	PERCHLORYL FLUORIDE	0.63654	0.27200	0.29870	368.40	125.41	368.40	1.408
75	ClF3	CHLORINE TRIFLUORIDE	0.68343	0.27400	0.28571	459.39	190.15	459.39	1.785
76	ClF5	CHLORINE PENTAFLUORIDE	0.85421	0.35050	0.28571	415.90	190.00	415.90	1.774
77	ClHO3S	CHLOROSULFONIC ACID	0.59672	0.28520	0.28571	700.00	193.15	700.00	1.741
78	ClHO4	PERCHLORIC ACID	0.59920	0.26092	0.34521	631.00	171.95	631.00	1.760

207

Table 8-2 DENSITY OF LIQUID - INORGANIC COMPOUNDS (continued)

NO	FORMULA	NAME	density = A B^(-(1 - T/Tc)^n)				(density - g/ml, T - K)	
			A	B	n	T_C	TMIN	TMAX	density @ 25 C

NO	FORMULA	NAME	A	B	n	T_C	TMIN	TMAX	density @ 25 C
79	ClO2	CHLORINE DIOXIDE	0.60087	0.27400	0.28571	465.00	213.55	465.00	1.579
80	Cl2	CHLORINE	0.56600	0.27315	0.28830	417.15	172.12	417.15	1.398
81	Cl2O	CHLORINE MONOXIDE	0.60356	0.27400	0.28571	444.68	157.15	444.68	1.549
82	Cl2O7	CHLORINE HEPTOXIDE	0.65031	0.27400	0.28571	565.78	182.15	565.78	1.850
83	Co	COBALT	2.51864	0.30000	0.28571	7398.48	1768.15	1968.15	-------
84	CoCl2	COBALT CHLORIDE	1.05823	0.30000	0.28571	2154.97	1008.15	1208.15	-------
85	CoNC3O4	COBALT NITROSYL TRICARBONYL	-------	-------	-------	-------	-------	-------	-------
86	Cr	CHROMIUM	2.13536	0.30000	0.28571	8560.93	2180.15	2380.15	-------
87	CrC6O6	CHROMIUM CARBONYL	0.63518	0.30000	0.28571	690.80	423.65	690.80	-------
88	CrO2Cl2	CHROMIUM OXYCHLORIDE	0.64832	0.27400	0.28571	626.33	176.65	626.33	1.902
89	Cs	CESIUM	0.28361	0.14024	0.28571	2048.10	301.65	2030.00	-------
90	CsBr	CESIUM BROMIDE	1.05781	0.30000	0.28571	2562.13	909.15	1500.00	-------
91	CsCl	CESIUM CHLORIDE	0.93727	0.30000	0.28571	2562.13	919.15	1500.00	-------
92	CsF	CESIUM FLUORIDE	1.26025	0.30000	0.28571	2482.33	976.00	1500.00	-------
93	CsI	CESIUM IODIDE	1.07356	0.30000	0.28571	2529.56	894.15	1500.00	-------
94	Cu	COPPER	2.63621	0.30000	0.28571	5123.00	1357.77	1557.77	-------
95	CuBr	CUPROUS BROMIDE	1.48318	0.30000	0.28571	2651.71	777.15	977.15	-------
96	CuCl	CUPROUS CHLORIDE	1.94545	0.45137	0.56812	2435.00	703.00	1763.10	-------
97	CuCl2	CUPRIC CHLORIDE	1.08795	0.30000	0.28571	2010.00	906.15	1106.15	-------
98	CuI	COPPER IODIDE	1.70575	0.30000	0.28571	2620.77	878.15	1078.15	-------
99	DCN	DEUTERIUM CYANIDE	-------	-------	-------	-------	-------	-------	-------
100	D2	DEUTERIUM	0.06671	0.31500	0.28571	38.35	18.73	38.35	-------
101	D2O	DEUTERIUM OXIDE	0.35429	0.28000	0.20000	643.89	276.96	643.89	1.090
102	Eu	EUROPIUM	1.57191	0.30000	0.28571	5150.00	1095.15	1295.15	-------
103	F2	FLUORINE	0.57092	0.28518	0.29000	144.31	53.48	144.31	-------
104	F2O	FLUORINE OXIDE	0.55240	0.27016	0.28571	215.10	49.25	215.10	-------
105	Fe	IRON	0.57093	0.07000	0.28571	9340.00	1808.10	2500.00	-------
106	FeC5O5	IRON PENTACARBONYL	0.49908	0.27400	0.28571	607.20	252.15	607.20	1.451
107	FeCl2	FERROUS CHLORIDE	1.02146	0.30000	0.28571	2115.88	945.15	1145.15	-------
108	FeCl3	FERRIC CHLORIDE	1.02395	0.30000	0.28571	964.41	577.15	584.65	-------
109	Fr	FRANCIUM	-------	-------	-------	-------	-------	-------	-------
110	Ga	GALLIUM	0.47740	0.07625	0.28571	7620.00	302.91	1373.15	-------
111	GaCl3	GALLIUM TRICHLORIDE	0.66952	0.25063	0.29982	694.00	350.90	694.00	-------
112	Gd	GADOLINIUM	2.68561	0.30000	0.28571	3307.66	1587.15	1678.58	-------
113	Ge	GERMANIUM	1.73560	0.30000	0.28571	8400.00	1211.40	1411.40	-------
114	GeBr4	GERMANIUM BROMIDE	1.02768	0.27400	0.28571	740.00	299.25	380.70	-------
115	GeCl4	GERMANIUM CHLORIDE	0.64888	0.27400	0.28571	574.00	223.65	574.00	1.854
116	GeHCl3	TRICHLORO GERMANE	0.67714	0.27400	0.28571	559.78	202.05	559.78	1.919
117	GeH4	GERMANE	0.54904	0.30254	0.28571	308.00	107.26	308.00	0.859
118	Ge2H6	DIGERMANE	0.63111	0.27400	0.28571	491.01	164.15	491.01	1.701
119	Ge3H8	TRIGERMANE	0.75692	0.27400	0.28571	616.37	167.55	616.37	2.211
120	HBr	HYDROGEN BROMIDE	0.80283	0.28545	0.28571	363.15	185.15	363.15	1.728
121	HCN	HYDROGEN CYANIDE	0.19501	0.18589	0.28206	456.65	259.83	456.65	0.680
122	HCl	HYDROGEN CHLORIDE	0.44134	0.26957	0.31870	324.65	158.97	324.65	0.796
123	HF	HYDROGEN FLUORIDE	0.29041	0.17660	0.37330	461.15	189.79	461.15	0.941
124	HI	HYDROGEN IODIDE	1.04638	0.28826	0.28571	423.85	222.38	423.85	2.520
125	HNO3	NITRIC ACID	0.43471	0.23110	0.19170	520.00	231.55	373.15	1.509
126	H2	HYDROGEN	0.03125	0.34730	0.27560	33.18	13.95	33.18	-------
127	H2O	WATER	0.34710	0.27400	0.28571	647.13	273.16	647.13	1.027
128	H2O2	HYDROGEN PEROXIDE	0.43776	0.24982	0.28770	730.15	272.74	730.15	1.443
129	H2S	HYDROGEN SULFIDE	0.34832	0.28160	0.28570	373.53	187.68	373.53	0.777
130	H2SO4	SULFURIC ACID	0.42169	0.19356	0.28570	925.00	283.46	363.49	1.833
131	H2S2	HYDROGEN DISULFIDE	0.47306	0.27400	0.28571	542.39	183.45	542.39	1.326
132	H2Se	HYDROGEN SELENIDE	0.72147	0.27400	0.28571	411.10	209.15	411.10	1.766
133	H2Te	HYDROGEN TELLURIDE	0.93431	0.27400	0.28571	438.04	224.15	438.04	2.378
134	H3NO3S	SULFAMIC ACID	-------	-------	-------	-------	-------	-------	-------
135	He	HELIUM-3	0.04141	0.46660	0.22300	3.31	1.01	3.31	-------
136	He	HELIUM-4	0.06930	0.41860	0.24100	5.20	1.76	5.20	-------
137	Hf	HAFNIUM	3.75257	0.30000	0.28571	21687.96	2506.15	2706.15	-------
138	Hg	MERCURY	3.56411	0.25360	0.16155	1735.00	234.31	1735.00	13.487
139	HgBr2	MERCURIC BROMIDE	1.93930	0.30000	0.28571	964.41	510.15	800.00	-------
140	HgCl2	MERCURIC CHLORIDE	1.71179	0.30000	0.28571	939.98	550.15	800.00	-------
141	HgI2	MERCURIC IODIDE	1.93098	0.30000	0.28571	1078.10	532.15	750.00	-------
142	IF7	IODINE HEPTAFLUORIDE	1.05176	0.27400	0.28571	447.53	278.65	447.53	2.709
143	I2	IODINE	1.63746	0.33313	0.33550	819.15	386.75	819.15	-------
144	In	INDIUM	0.55791	0.07561	0.28571	6730.00	429.75	573.15	-------
145	Ir	IRIDIUM	6.41388	0.30000	0.28571	15035.00	2719.15	2919.15	-------
146	K	POTASSIUM	0.18362	0.21625	0.28571	2223.00	336.35	2220.00	-------
147	KBr	POTASSIUM BROMIDE	0.72724	0.30000	0.28571	2697.31	1003.15	1500.00	-------
148	KCl	POTASSIUM CHLORIDE	0.09536	0.04278	0.35520	3470.00	1044.00	3000.00	-------
149	KF	POTASSIUM FLUORIDE	0.66025	0.30000	0.28571	2891.12	1153.15	1500.00	-------
150	KI	POTASSIUM IODIDE	0.82666	0.30000	0.28571	2601.22	996.15	1500.00	-------
151	KOH	POTASSIUM HYDROXIDE	0.30810	0.15400	0.28571	2605.86	679.00	1500.00	-------
152	Kr	KRYPTON	0.91799	0.28840	0.29390	209.35	115.78	209.35	-------
153	La	LANTHANUM	1.50549	0.25204	0.28571	9511.00	1193.15	1393.15	-------
154	Li	LITHIUM	0.03337	0.05618	0.28571	4085.00	453.69	3670.00	-------
155	LiBr	LITHIUM BROMIDE	0.85546	0.30000	0.28571	2578.42	820.15	1500.00	-------
156	LiCl	LITHIUM CHLORIDE	0.50679	0.30000	0.28571	2695.68	887.15	1500.00	-------

Table 8-2 DENSITY OF LIQUID - INORGANIC COMPOUNDS (continued)

NO	FORMULA	NAME	density = A B$^{-(1-T/T_c)^n}$ A	B	n	T_c	TMIN	TMAX	(density - g/ml, T - K) density @ 25 C
157	LiF	LITHIUM FLUORIDE	0.62331	0.30000	0.28571	3182.65	1143.15	1700.00	-------
158	LiI	LITHIUM IODIDE	1.04484	0.30000	0.28571	2352.04	719.15	1600.00	-------
159	Lu	LUTECIUM	3.32517	0.30000	0.28571	4128.66	1936.15	2136.15	-------
160	Mg	MAGNESIUM	0.07780	0.03010	0.28571	2241.04	923.15	1023.15	-------
161	MgCl2	MAGNESIUM CHLORIDE	0.58060	0.30000	0.28571	2754.32	985.15	1600.00	-------
162	MgO	MAGNESIUM OXIDE	0.19187	0.01440	0.28571	5950.00	3105.00	5950.00	-------
163	Mn	MANGANESE	2.09497	0.30000	0.28571	6902.82	1519.15	1719.15	-------
164	MnCl2	MANGANESE CHLORIDE	0.90334	0.30000	0.28571	2382.98	923.15	1123.15	-------
165	Mo	MOLYBDENUM	3.14663	0.30000	0.28571	9620.00	2895.15	3095.15	-------
166	MoF6	MOLYBDENUM FLUORIDE	0.93061	0.27400	0.28571	498.12	290.15	498.12	2.523
167	MoO3	MOLYBDENUM OXIDE	1.56112	0.30000	0.28571	2319.46	1068.15	1268.15	-------
168	NCl3	NITROGEN TRICHLORIDE	0.57737	0.27400	0.28571	564.00	293.00	564.00	1.641
169	ND3	HEAVY AMMONIA	--------	--------	--------	--------	--------	--------	-------
170	NF3	NITROGEN TRIFLUORIDE	0.59810	0.28230	0.29140	233.85	66.36	233.85	-------
171	NH3	AMMONIA	0.23689	0.25471	0.28870	405.65	195.41	405.65	0.602
172	NH3O	HYDROXYLAMINE	0.42502	0.27400	0.28571	574.00	306.15	574.00	-------
173	NH4Br	AMMONIUM BROMIDE	0.98286	0.30000	0.28571	1089.82	800.00	1000.00	-------
174	NH4Cl	AMMONIUM CHLORIDE	0.75456	0.30000	0.28571	882.00	793.20	883.20	-------
175	NH4I	AMMONIUM IODIDE	1.00871	0.30000	0.28571	1104.32	800.00	1000.00	-------
176	NH5O	AMMONIUM HYDROXIDE	--------	--------	--------	--------	--------	--------	-------
177	NH5S	AMMONIUM HYDROGENSULFIDE	0.49634	0.30000	0.28571	499.10	391.15	455.37	-------
178	NO	NITRIC OXIDE	0.51712	0.30440	0.24200	180.15	109.50	180.15	-------
179	NOCl	NITROSYL CHLORIDE	0.46993	0.27029	0.24883	440.65	213.55	440.65	1.262
180	NOF	NITROSYL FLUORIDE	0.49514	0.27400	0.28571	352.67	139.15	352.67	1.058
181	NO2	NITROGEN DIOXIDE	0.54240	0.27210	0.24320	431.35	293.15	431.35	1.442
182	N2	NITROGEN	0.31205	0.28479	0.29250	126.10	63.15	126.10	-------
183	N2F4	TETRAFLUOROHYDRAZINE	0.54042	0.27760	0.28570	309.35	111.65	309.35	0.888
184	N2H4	HYDRAZINE	0.20285	0.16613	0.18980	653.15	274.68	653.15	1.004
185	N2H4C	AMMONIUM CYANIDE	--------	--------	--------	--------	--------	--------	-------
186	N2H6CO2	AMMONIUM CARBAMATE	--------	--------	--------	--------	--------	--------	-------
187	N2O	NITROUS OXIDE	0.44927	0.27244	0.28820	309.57	182.30	309.57	0.742
188	N2O3	NITROGEN TRIOXIDE	0.39542	0.18000	0.28571	425.00	263.15	425.00	1.331
189	N2O4	NITROGEN TETRAOXIDE	0.50649	0.23319	0.28571	431.15	261.90	431.15	1.434
190	N2O5	NITROGEN PENTOXIDE	0.55485	0.27400	0.28571	515.51	303.15	311.65	-------
191	Na	SODIUM	0.10896	0.10222	0.28571	2573.00	370.98	2500.00	-------
192	NaBr	SODIUM BROMIDE	0.25856	0.07850	0.52911	4287.00	1020.00	2500.00	-------
193	NaCN	SODIUM CYANIDE	0.17117	0.11000	0.28571	2900.00	973.15	1173.15	-------
194	NaCl	SODIUM CHLORIDE	0.22127	0.10591	0.37527	3400.00	1073.90	3400.00	-------
195	NaF	SODIUM FLUORIDE	0.22682	0.07797	0.65700	5530.00	1269.00	2500.00	-------
196	NaI	SODIUM IODIDE	0.93198	0.30000	0.28571	2568.65	924.15	1600.00	-------
197	NaOH	SODIUM HYDROXIDE	0.19975	0.09793	0.25382	2820.00	596.00	2000.00	-------
198	Na2SO4	SODIUM SULFATE	0.26141	0.10000	0.28571	3700.00	1157.00	2000.00	-------
199	Nb	NIOBIUM	2.48435	0.30000	0.28571	17904.10	2750.15	2950.15	-------
200	Nd	NEODYMIUM	2.02652	0.30000	0.28571	10665.12	1289.15	1489.15	-------
201	Ne	NEON	0.48504	0.30670	0.27860	44.40	24.56	44.40	-------
202	Ni	NICKEL	2.56376	0.30000	0.28571	6986.15	1728.15	1928.15	-------
203	NiC4O4	NICKEL CARBONYL	0.47753	0.27400	0.28571	508.40	248.15	508.40	1.306
204	NiF2	NICKEL FLUORIDE	1.52939	0.30000	0.28571	3278.75	1723.15	1923.15	-------
205	Np	NEPTUNIUM	--------	--------	--------	--------	--------	--------	-------
206	O2	OXYGEN	0.43533	0.28772	0.29240	154.58	54.35	154.58	-------
207	O3	OZONE	0.53965	0.29820	0.28800	261.00	80.15	261.00	-------
208	Os	OSMIUM	6.48474	0.30000	0.28571	16878.68	3306.15	3506.15	-------
209	OsOF5	OSMIUM OXIDE PENTAFLUORIDE	--------	--------	--------	--------	--------	--------	-------
210	OsO4	OSMIUM TETROXIDE - YELLOW	1.62701	0.30000	0.28571	656.60	329.15	366.15	-------
211	OsO4	OSMIUM TETROXIDE - WHITE	1.60780	0.30000	0.28571	656.60	315.15	359.15	-------
212	P	PHOSPHORUS - WHITE	1.02120	0.50058	0.66360	993.75	317.55	553.15	-------
213	PBr3	PHOSPHORUS TRIBROMIDE	0.93428	0.27400	0.28571	711.00	233.15	711.00	2.830
214	PCl2F3	PHOSPHORUS DICHLORIDE TRIFLUORIDE	--------	--------	--------	--------	--------	--------	-------
215	PCl3	PHOSPHORUS TRICHLORIDE	0.52825	0.25681	0.29690	563.15	200.00	349.25	1.566
216	PCl5	PHOSPHORUS PENTACHLORIDE	1.38262	0.30000	0.28571	646.15	433.15	504.20	-------
217	PH3	PHOSPHINE	0.30000	0.28918	0.36923	324.75	139.37	324.75	0.491
218	PH4Br	PHOSPHONIUM BROMIDE	--------	--------	--------	--------	--------	--------	-------
219	PH4Cl	PHOSPHONIUM CHLORIDE	--------	--------	--------	--------	--------	--------	-------
220	PH4I	PHOSPHONIUM IODIDE	1.01597	0.27400	0.28571	539.70	291.65	539.70	2.843
221	POCl3	PHOSPHORUS OXYCHLORIDE	0.76181	0.35086	0.42595	602.15	274.33	378.65	1.667
222	PSBr3	PHOSPHORUS THIOBROMIDE	0.94173	0.30000	0.28571	729.89	311.15	379.65	-------
223	PSCl3	PHOSPHORUS THIOCHLORIDE	0.55688	0.27400	0.28571	638.82	242.75	371.65	1.643
224	P4O6	PHOSPHORUS TRIOXIDE	0.70154	0.27400	0.28571	714.86	295.65	714.86	2.128
225	P4O10	PHOSPHORUS PENTOXIDE	--------	--------	--------	--------	--------	--------	-------
226	P4S10	PHOSPHORUS PENTASULFIDE	0.63096	0.30000	0.28571	1291.00	596.15	1291.00	-------
227	Pb	LEAD	1.72203	0.15282	0.28571	5400.00	600.64	1273.15	-------
228	PbBr2	LEAD BROMIDE	1.92627	0.30000	0.28571	1933.47	646.15	1500.00	-------
229	PbCl2	LEAD CHLORIDE	1.72887	0.30000	0.28571	1998.62	774.15	1500.00	-------
230	PbF2	LEAD FLUORIDE	2.79825	0.30000	0.28571	2550.73	1128.15	1328.15	-------
231	PbI2	LEAD IODIDE	2.01004	0.30000	0.28571	1865.07	675.15	875.15	-------
232	PbO	LEAD OXIDE	3.18209	0.30000	0.28571	2842.26	1163.15	1363.15	-------
233	PbS	LEAD SULFIDE	2.70282	0.30000	0.28571	2531.19	1387.15	1470.65	-------
234	Pd	PALLADIUM	3.31600	0.30000	0.28571	10669.07	1828.05	2028.05	-------

Table 8-2 DENSITY OF LIQUID - INORGANIC COMPOUNDS (continued)

NO	FORMULA	NAME	density = A B$^{-(1-T/T_c)^n}$ A	B	n	T_c	TMIN	TMAX	(density - g/ml, T - K) density @ 25 C
235	Po	POLONIUM	2.76115	0.30000	0.28571	3013.08	527.15	727.15	-------
236	Pt	PLATINUM	6.65186	0.30000	0.28571	6983.00	2041.55	2241.55	-------
237	Ra	RADIUM	1.49145	0.30000	0.28571	4862.82	973.15	1173.15	-------
238	Rb	RUBIDIUM	0.20818	0.09847	0.28571	2111.10	312.46	2030.00	-------
239	RbBr	RUBIDIUM BROMIDE	0.91773	0.30000	0.28571	2646.82	955.15	1800.00	-------
240	RbCl	RUBIDIUM CHLORIDE	0.76254	0.30000	0.28571	2694.06	988.15	1800.00	-------
241	RbF	RUBIDIUM FLUORIDE	1.19655	0.30000	0.28571	2738.03	1033.15	1233.15	-------
242	RbI	RUBIDIUM IODIDE	0.97769	0.30000	0.28571	2568.65	915.15	1800.00	-------
243	Re	RHENIUM	6.01402	0.30000	0.28571	21482.82	3459.15	3659.15	-------
244	Re2O7	RHENIUM HEPTOXIDE	2.10435	0.30000	0.28571	1035.10	569.15	602.35	-------
245	Rh	RHODIUM	3.40525	0.30000	0.28571	12906.61	2237.15	2437.15	-------
246	Rn	RADON	1.58779	0.28111	0.28571	377.40	202.15	377.40	3.578
247	Ru	RUTHENIUM	3.48198	0.30000	0.28571	15247.14	2607.15	2807.15	-------
248	RuF5	RUTHENIUM PENTAFLUORIDE	0.90813	0.30000	0.28571	977.44	359.65	559.65	-------
249	S	SULFUR	0.20468	0.10440	0.11400	1313.00	388.36	1313.00	-------
250	SF4	SULFUR TETRAFLUORIDE	0.68957	0.27400	0.28571	364.00	149.15	364.00	1.526
251	SF6	SULFUR HEXAFLUORIDE	0.73472	0.27010	0.29210	318.69	223.15	318.69	1.322
252	SOBr2	THIONYL BROMIDE	0.89295	0.27400	0.28571	661.75	220.95	661.75	2.659
253	SOCl2	THIONYL CHLORIDE	0.58606	0.27659	0.30630	567.00	172.00	567.00	1.630
254	SOF2	SULFUROUS OXYFLUORIDE	0.60330	0.27400	0.28571	371.25	162.65	371.25	1.361
255	SO2	SULFUR DIOXIDE	0.51726	0.25514	0.28930	430.75	197.67	430.75	1.366
256	SO2Cl2	SULFURYL CHLORIDE	0.60253	0.27472	0.30830	545.00	222.00	545.00	1.658
257	SO3	SULFUR TRIOXIDE	0.63003	0.19602	0.41787	490.85	289.95	490.85	1.897
258	S2Cl2	SULFUR MONOCHLORIDE	0.56169	0.27400	0.28571	659.37	193.15	659.37	1.671
259	Sb	ANTIMONY	1.69927	0.24231	0.28571	5070.00	903.78	1243.15	-------
260	SbBr3	ANTIMONY TRIBROMIDE	1.42997	0.30000	0.28571	892.75	369.75	458.95	-------
261	SbCl3	ANTIMONY TRICHLORIDE	0.77034	0.21620	0.28570	794.00	346.55	794.00	-------
262	SbCl5	ANTIMONY PENTACHLORIDE	0.78102	0.27400	0.28571	662.54	275.95	662.54	2.326
263	SbH3	STIBINE	0.90538	0.31372	0.28571	440.35	185.15	440.35	2.096
264	SbI3	ANTIMONY TRIIODIDE	1.68384	0.30000	0.28571	1097.96	440.15	640.15	-------
265	Sb2O3	ANTIMONY TRIOXIDE	1.72636	0.30000	0.28571	2765.72	929.15	1129.15	-------
266	Sc	SCANDIUM	0.90098	0.30000	0.28571	8035.08	1814.15	2014.15	-------
267	Se	SELENIUM	0.75697	0.16124	0.28571	1766.00	494.15	694.15	-------
268	SeCl4	SELENIUM TETRACHLORIDE	0.76605	0.27400	0.28571	743.95	400.00	600.00	-------
269	SeF6	SELENIUM HEXAFLUORIDE	0.85211	0.27400	0.28571	368.80	238.45	368.80	1.911
270	SeOCl2	SELENIUM OXYCHLORIDE	0.79805	0.27400	0.28571	706.80	281.65	706.80	2.414
271	SeO2	SELENIUM DIOXIDE	1.33585	0.30000	0.28571	961.16	613.15	736.82	-------
272	Si	SILICON	0.84021	0.29172	0.28571	5159.00	1685.00	1885.00	-------
273	SiBrCl2F	BROMODICHLOROFLUOROSILANE	0.66872	0.27400	0.28571	497.17	160.85	497.17	1.812
274	SiBrF3	TRIFLUOROBROMOSILANE	0.72447	0.27400	0.28571	375.28	202.65	375.28	1.651
275	SiBr2ClF	DIBROMOCHLOROFLUOROSILANE	0.81945	0.27400	0.28571	535.27	173.85	535.27	2.286
276	SiClF3	TRIFLUOROCHLOROSILANE	0.56426	0.27400	0.28571	330.54	131.15	330.54	1.099
277	SiCl2F2	DICHLORODIFLUOROSILANE	0.55307	0.27400	0.28571	390.93	133.45	390.93	1.305
278	SiCl3F	TRICHLOROFLUOROSILANE	0.54467	0.27400	0.28571	460.49	152.35	460.49	1.424
279	SiCl4	SILICON TETRACHLORIDE	0.52116	0.26416	0.28150	507.00	204.30	507.00	1.470
280	SiF4	SILICON TETRAFLUORIDE	0.67478	0.28495	0.28571	259.00	186.35	259.00	-------
281	SiHBr3	TRIBROMOSILANE	1.05904	0.32450	0.28571	610.00	199.65	610.00	2.682
282	SiHCl3	TRICHLOROSILANE	0.50539	0.25740	0.34900	479.00	144.95	479.00	1.328
283	SiHF3	TRIFLUOROSILANE	0.51882	0.27400	0.28571	291.02	141.75	291.02	-------
284	SiH2Br2	DIBROMOSILANE	0.77365	0.28512	0.28571	550.00	202.95	550.00	2.111
285	SiH2Cl2	DICHLOROSILANE	0.44591	0.26888	0.28570	449.00	151.15	449.00	1.167
286	SiH2F2	DIFLUOROSILANE	0.44709	0.27400	0.28571	318.21	150.00	318.21	0.805
287	SiH2I2	DIIODOSILANE	0.97501	0.28278	0.28571	660.00	272.15	660.00	2.825
288	SiH3Br	MONOBROMOSILANE	0.55308	0.26469	0.28571	454.00	179.25	454.00	1.473
289	SiH3Cl	MONOCHLOROSILANE	0.35290	0.25554	0.28571	396.65	155.05	396.65	0.882
290	SiH3F	MONOFLUOROSILANE	0.36018	0.27400	0.28571	286.28	150.00	286.28	-------
291	SiH3I	IODOSILANE	0.70808	0.25939	0.28571	515.00	216.15	515.00	2.031
292	SiH4	SILANE	0.24205	0.28405	0.39578	269.70	88.48	269.70	-------
293	SiO2	SILICON DIOXIDE	0.33537	0.10000	0.28571	4076.87	2000.00	2200.00	-------
294	Si2Cl6	HEXACHLORODISILANE	0.51982	0.27400	0.28571	660.96	271.95	660.96	1.547
295	Si2F6	HEXAFLUORODISILANE	0.54778	0.27400	0.28571	411.33	254.55	411.33	1.341
296	Si2H5Cl	DISILANYL CHLORIDE	0.34804	0.27400	0.28571	506.89	250.00	506.89	0.951
297	Si2H6	DISILANE	0.31422	0.27551	0.29510	432.00	143.85	432.00	0.782
298	Si2OCl3F3	TRICHLOROTRIFLUORODISILOXANE	0.54394	0.27400	0.28571	508.78	150.00	508.78	1.488
299	Si2OCl6	HEXACHLORODISILOXANE	0.53229	0.27400	0.28571	655.58	239.95	655.58	1.581
300	Si2OH6	DISILOXANE	0.31264	0.27400	0.28571	416.86	128.95	416.86	0.772
301	Si3Cl8	OCTACHLOROTRISILANE	0.51452	0.27400	0.28571	775.41	250.00	775.41	1.588
302	Si3H8	TRISILANE	0.26010	0.27400	0.28571	525.15	155.95	525.15	0.720
303	Si3H9N	TRISILAZANE	0.31120	0.27400	0.28571	518.20	167.45	518.20	0.858
304	Si4H10	TETRASILANE	0.26613	0.27400	0.28571	599.30	179.55	599.30	0.771
305	Sm	SAMARIUM	2.30412	0.30000	0.28571	5082.92	1345.15	1545.15	-------
306	Sn	TIN	0.68914	0.09454	0.28571	7400.00	505.08	977.15	-------
307	SnBr4	STANNIC BROMIDE	1.09276	0.27400	0.28571	764.82	304.15	391.00	-------
308	SnCl2	STANNOUS CHLORIDE	1.15953	0.30000	0.28571	1459.53	519.95	1200.00	-------
309	SnCl4	STANNIC CHLORIDE	0.75734	0.27400	0.28571	619.85	242.95	619.85	2.215
310	SnH4	STANNIC HYDRIDE	-------	-------	-------	-------	-------	-------	-------
311	SnI4	STANNIC IODIDE	1.51710	0.30000	0.28571	1011.64	417.65	617.65	-------
312	Sr	STRONTIUM	0.78222	0.30000	0.28571	4267.20	1050.15	1250.15	-------

Table 8-2 DENSITY OF LIQUID - INORGANIC COMPOUNDS (continued)

NO	FORMULA	NAME	density = A B$^{-(1-T/Tc)^{\wedge}n}$				(density - g/ml, T - K)		
			A	B	n	T_c	TMIN	TMAX	density @ 25 C
313	SrO	STRONTIUM OXIDE	--------	--------	--------	--------	2703.15	2703.15	-------
314	Ta	TANTALUM	4.78107	0.30000	0.28571	19900.93	3290.15	3490.15	-------
315	Tc	TECNNETIUM	3.34979	0.30000	0.28571	17400.77	2430.15	2630.15	-------
316	Te	TELLURIUM	1.83626	0.30000	0.28571	4840.00	722.66	922.66	-------
317	TeCl4	TELLURIUM TETRACHLORIDE	0.93555	0.30000	0.28571	1083.31	497.15	581.15	-------
318	TeF6	TELLURIUM HEXAFLUORIDE	1.16414	0.27400	0.28571	380.18	235.35	380.18	2.684
319	Ti	TITANIUM	1.38759	0.30000	0.28571	6400.00	1941.15	2141.15	-------
320	TiCl4	TITANIUM TETRACHLORIDE	0.55695	0.26324	0.27260	638.00	249.05	638.00	1.714
321	Tl	THALLIUM	1.79212	0.14783	0.28571	4648.06	577.15	777.00	-------
322	TlBr	THALLOUS BROMIDE	2.55912	0.30000	0.28571	1778.75	733.15	933.15	-------
323	TlI	THALLOUS IODIDE	2.38850	0.30000	0.28571	1785.26	713.15	913.15	-------
324	Tm	THULIUM	2.88674	0.30000	0.28571	6283.16	1818.15	2018.15	-------
325	U	URANIUM	5.57293	0.30000	0.28571	13712.62	1408.15	1608.15	-------
326	UF6	URANIUM FLUORIDE	1.74575	0.27705	0.28571	505.80	342.35	401.33	-------
327	V	VANADIUM	1.78294	0.30000	0.28571	11787.15	2183.15	2383.15	-------
328	VCl4	VANADIUM TETRACHLORIDE	0.71831	0.33057	0.31162	697.00	247.45	697.00	1.821
329	VOCl3	VANADIUM OXYTRICHLORIDE	0.59752	0.29177	0.31700	636.00	193.65	400.00	1.637
330	W .	TUNGSTEN	5.84023	0.30000	0.28571	14756.00	3695.15	3895.15	-------
331	WF6	TUNGSTEN FLUORIDE	1.28420	0.27400	0.28571	468.56	272.65	468.56	3.387
332	Xe	XENON	1.10648	0.28552	0.28967	289.74	161.36	289.74	-------
333	Yb	YTTERBIUM	2.12198	0.30000	0.28571	4365.92	1097.15	1297.15	-------
334	Yt	YTTRIUM	1.32860	0.30000	0.28571	9381.32	1799.15	1999.15	-------
335	Zn	ZINC	2.25387	0.29920	0.28571	3170.00	692.70	1073.15	-------
336	ZnCl2	ZINC CHLORIDE	0.84382	0.30000	0.28571	1637.05	638.15	1200.00	-------
337	ZnF2	ZINC FLUORIDE	1.61024	0.30000	0.28571	2882.98	1145.15	1345.15	-------
338	ZnO	ZINC OXIDE	--------	--------	--------	--------	--------	--------	-------
339	ZnSO4	ZINC SULFATE	--------	--------	--------	--------	--------	--------	-------
340	Zr	ZIRCONIUM	1.90680	0.30000	0.28571	8802.00	2128.15	2328.15	-------
341	ZrBr4	ZIRCONIUM BROMIDE	--------	--------	--------	--------	--------	--------	-------
342	ZrCl4	ZIRCONIUM CHLORIDE	1.08375	0.30000	0.28571	983.96	710.15	836.75	-------
343	ZrI4	ZIRCONIUM IODIDE	--------	--------	--------	--------	--------	--------	-------

density - density of liquid, g/ml

A, B, n, and T_c - regression coefficients of chemical compound

T - temperature, K

TMAX - maximum temperature, K

TMIN - minimun temperature, K

Chapter 9

SURFACE TENSION

Carl L. Yaws, Xiaoyan Lin, Li Bu, and Sachin Nijhawan
Lamar University, Beaumont, Texas

ABSTRACT

Results for surface tension are presented for major organic and inorganic chemicals. For many of the chemicals, the complete temperature range for the liquid is covered from freezing point to the critical point. The results are displayed in easy-to-use tabulations that are especially applicable for rapid engineering usage with the personal computer or hand calculator.

INTRODUCTION

Physical and thermodynamic property data such as surface tension are of special value to engineers in the chemical processing and petroleum refining industries. As an example, surface tension data are important in many chemical-process engineering applications, such as heat, mass, and momentum transfer operations that involve process equipment such as heat exchangers, distillation columns, absorption, and fluid-flow piping. In this article, results for surface tension as a function of temperature are presented for a wide variety of compounds.

SURFACE TENSION CORRELATION

A modified Othmer relation was selected for correlation of surface tension as a function of temperature:

$$\text{sigma} = A \left(1 - T/T_c\right)^n \tag{9-1}$$

where

sigma = surface tension, dynes/cm

A, T_c, and n = regression coefficients for chemical compound

T = temperature, K

The results for surface tension are given in Tables 9-1 and 9-2. The tabulations are arranged by chemical formula to provides ease of use in quickly locating data. A wide variety of substances are covered. The range for application is denoted by the respective minimum and maximum temperatures (TMIN and TMAX).

In preparing the compilation, a literature search was conducted to identify data source publications for organics (1-40) and inorganics (1-112). Both experimental values for the property under consideration and parameter values for estimation of the property are included in the source publications. The publications were screened for appropriate data. The compilation resulting from the screening is based on both experimental data and estimated values. In the absence of experimental data, estimates were primarily based on Sugden method (group contribution, 29) and Brock and Bird correlation (corresponding states, 29). Experimental data and estimates were then regressed to provide the same equation for all compounds.

A comparison of calculations and data for surface tension is shown in Fig. 9-1 for a representative chemical. The graph indicates good agreement of calculations and data.

EXAMPLES

The correlation results maybe used for prediction and calculation of surface tension. Examples are given below.

Example 1 Estimate the surface tension of carbon tetrachloride (CCl4) at 378.15 K.

Substitution of the regression coefficients from the table and temperature into the equation for surface tension yields

sigma = (66.750) (1 - 378.15/556.35)$^{1.2140}$

sigma = 16.76 dyne/cm

The calculated and data values compare favorably (16.76 vs 16.64, deviation = 0.7%).

Example 2 Estimate the surface tension of ethane (C2H6) at 133.15 K.

Substitution of the regression coefficients from the table and temperature into the equation for surface tension yields sigma = (48.984) (1 − 133.15/305.42)$^{1.2065}$

sigma = 24.55 dyne/cm

The calculated and data values compare favorably (24.55 vs 24.48, deviation = 0.3%).

Portions of this material appeared in Chem. Eng., $\underline{98}$, 140 (March, 1991) and are reprinted by special permission.

REFERENCES – ORGANIC COMPOUNDS

1-34. See **REFERENCES - ORGANIC COMPOUNDS** in **Chapter 1, CRITICAL PROPERTIES AND ACENTRIC FACTOR**

35. Maxwell, J. B., DATA BOOK ON HYDROCARBONS, D. Van Nostrand, Princeton. NJ (1958).
36. Egloff, G., PHYSICAL CONSTANTS OF HYDROCARBONS, Vols. 1-6, Reinhold Publishing Corp., New York, NY (1939-1947).
37. Jasper, J.J., J. Phys. Chem. Ref. Data, $\underline{1}$ (No.4), 841 (1972).
38. Wilhoit, R. C. and B. J. Zwolinski, PHYSICAL AND THERMODYNAMIC PROPERTIES OF ALIPHATIC ALCOHOLS, American Chemical Society, American Institute of Physics, National Bureau of Standards, New York, NY (1973).
39. Yaws, C. L. and others, Chem. Eng., $\underline{98}$, 140 (March, 1991).
40. Yaws, C. L., HANDBOOK OF CHEMICAL COMPOUND DATA FOR PROCESS SAFETY, Gulf Publishing Co., Houston, TX (1997).

REFERENCES – INORGANIC COMPOUNDS

1-56. See **REFERENCES - INORGANIC COMPOUNDS** in **Chapter 1, CRITICAL PROPERTIES AND ACENTRIC FACTOR**

57. Lyon, R. N., ed., LIQUID-METALS HANDBOOK, 2nd ed., Atomic Energy Commission, Washington, DC (1954).
58. Karapet'yants, M. Kh. and M. L. Karapet'yants, THERMODYNAMIC CONSTANTS OF INORGANIC AND ORGANIC COMPOUNDS, Ann Arbor - Humphrey Science Publishers, Ann Arbor, MH (1970).
59. Fink, J. K., TABLES OF THERMODYNAMIC PROPERTIES OF SODIUM, Argonne National Lab., ANL-CEN-RSD-82-4, Chemical Engineering Division, Argonne, IL (June 1982).
60. Mills, K. C., THERMODYNAMIC DATA FOR INORGANIC SULPHIDES SELENIDES AND TELLURIDES, Butterworths, London, England (1974).
61. Cox, J. D.and G. Pilcher, THERMOCHEMISTRY OF ORGANIC AND ORGANOMETALLIC COMPOUNDS, Academic Press, New York, NY (1970).
62. Janz, G. J. and C. B. Allen, PHYSICAL PROPERTIES DATA COMPILATIONS RELEVANT TO ENERGY STORAGE. II. MOLTEN SALTS: DATA ON SINGLE AND MULTICOMPONENT SALTSYSTEMS, Nat. Bur. Stand., Molten Salts Data Center, Troy, NY (April 1979).
63. Janz, G. J. and R. P. T. Tomkins, PHYSICAL PROPERTIES DATA COMPILATIONS RELEVANT TO ENERGY STORAGE. 4. MOLTEN SALTS: DATA ON ADDITIONAL SINGLE AND MULTI-COMPONENT SALT SYSTEMS, U.S. Government Printing Office, Washington, DC (1981).
64. Tuller, W. N., ed., THE SULPHUR DATA BOOK, McGraw-Hill, New York, NY (1954).
65. Cox, J. D. and G. Pilcher, THERMOCHEMISTRY OF ORGANIC AND ORGANOMETALLIC COMPOUNDS, Academic Press, New York, NY (1970).
66. Riddick, J. A. and W. B. Bunger, ORGANIC SOLVENTS: PHYSICAL PROPERTIES AND METHODS OF PURIFICATION, 3rd ed., Wiley Interscience, New York, NY (1970).
67. Moore, G. A. and T. R. Shives, IRON, Metals Handbook, 8th ed., 1206 (1961).
68. McCarty, R. D., J. Hord, and H. M. Roder, SELECTED PROPERTIES OF HYDROGEN (ENGINEERING DESIGN DATA), Center for Chemical Engineering, National Engineering Laboratory, Nat. Bur. Stand. Monograph 168, Boulder, CO (1981).
69. Davison, H. W., COMPILATION OF THERMOPHYSICAL PROPERTIES OF LIQUID LITHIUM, U.S. NASA Tech. Note D-4650 (1968).
70. Kaufmann, D. W., ed., SODIUM CHLORIDE, Reinhold Pub. Corp., New York, NY (1960).
71. Weiss, G., ed., HAZARDOUS CHEMICALS DATA BOOK, Noyes Data Corporation, Park Ridge, NJ (1980).
72. Neugebauer, C. A. and J. L. Margrave, J. Phys. Chem., $\underline{60}$, 1318 (1956).
73. Bernard, G. and C. H. P. Lupis, Metall. Trans., $\underline{2}$, 555 (1971).
74. Sugden, S., J. Chem. Soc. (London, Transaction), $\underline{125}$, 32 (1924).
75. Waseda, Y. and K. Suzuki, Phys. Status Solidi B: Basic Research, $\underline{57}$, 351 (1973).
76. Nisel'son, L. A. and T. D. Sokolova, Russ. J. Inorg. Chem., $\underline{10}$, 827 (1965).
77. Saji, Y. and S. Kobayashi, Cryogenics, 136 (1964).
78. Durrant, A. A., T. G. Pearson, and P. L. Robinson, J. Chem. Soc. Part 1, 730 (1934).
79. Jasper, J. J., J. Phys. Chem. Ref. Data, $\underline{1(4)}$, 841 (1972).
80. Cook, R. P. and P. L. Robinson, J. Chem. Soc. (London), 1001 (1935).
81. Jarry, R. A. and J. J. Fritz, J. Chem. Eng. Data, $\underline{3(1)}$, 34 (1958).
82. Simkin, J. and R. L. Jarry, J. Phys. Chem., $\underline{61}$, 503 (1957).
83. Usanovich, M., T. Sumarokova, and V. Udovenko, Acta Physicochim, USSR, $\underline{11}$, 505 (1939).
84. Cheesman, G. H., J. Chem. Soc., $\underline{35}$, (1930).
85. Cockett, A. H. and A. Ferguson, Phil. Mag., $\underline{28}$, 685 (1939).
86. Heicks, J. R. and others, J. Phys. Chem., $\underline{58}$, 488 (1954).
87. White D., J. H. Hu, and H. L. Johnston, J. Am. Chem. Soc., $\underline{76}$, 2584 (1954).
88. Greenwood, N. N. and K. Wade, J. Inorg. Nucl. Chem., $\underline{3}$, 349 (1957).

89. Pearson, T. G. and P. L. Robinson, J. Chem. Soc., 736 (1934).

90. Grosh, J. and others, Proc. Nat. Acad. Sci., 54, 1004 (1965).

91. Stern, S. A., J. L. Mullhapt and W. B. Kay, Chem. Rev., 60, 185 (1960).

92. Vargaftic, M. B., B. N. Volkov, and L. D. Voljak, J. Phys. Chem. Ref. Data, 12(3), 817 (1983).

93. Maass, O. and W. H. Hatcher, J. Amer. Chem. Soc., 42, 2548 (1920).

94. Sabinina, L. and L. Terpugow, Z. Phys. Chem., 173A, 237 (1935).

95. Harkins, W. D. and W. W. Ewing, J. Amer. Chem. Soc., 42, 2539 (1920).

96. Sauewold, F. and G. Drath, Z. Anorg. Alleg. Chem., 154, 79 (1926).

97. Bircumshaw, L. L., Phil. Mag. 7(8), 341 (1926).

98. Bloom, H., F. G. Davis, and D. W. James, Trans. Faraday Soc., 56, 1179 (1960).

99. Janz, G. J., J. Phys. Chem. Ref. Data 9(4), 311 (1974).

100. Sokolov, O. K., Zh. Neorg. Khim., 11(7), 1703 (1966).

101. Streng, A. G., J. Chem. Eng. Data, 6(3), 43 (1961).

102. Jenkins, A. C. and F. S. Dipaolo, J. Chem. Phys., 25(2), 296 (1956).

103. Pugachevich, P. P. and others, Russ. J. Phys. Chem., 41, 299 (1967).

104. Person, T. G. and P. L. Robinson, J. Chem. Soc., 1472 (1933).

105. Berthoud, A., Helv. Chem. Acta, 5, 513 (1922).

106. Kingery, W. D. and M. Humenik, J. Phys. Chem., 57, 359 (1953).

107. Nisel'son, L. A. and others, Russ. J. Inorg. Chem., 10(6), 705 (1965).

108. MacKenzie, C. A. and others, J. Amer. Chem. Soc., 72, 2032 (1950).

109. Lapidus, I. I., Teplofiz. Svoist. Vesh., 5, 119 (1972).

110. Pugachevich, P. P., Russ. J. Phys. Chem., 40, 1560 (1966).

111. Smith, B. L., P. R. Gardner, and E. H. Parker, J. Chem. Phys., 47(3), 1148 (1967).

112. Osipov, L. I., SURFACE CHEMISTRY, THEORY AND INDUSTRIAL APPLICATION, Reinhold Publishing Corporation, New York, NY (1964).

Figure 9-1 Surface Tension

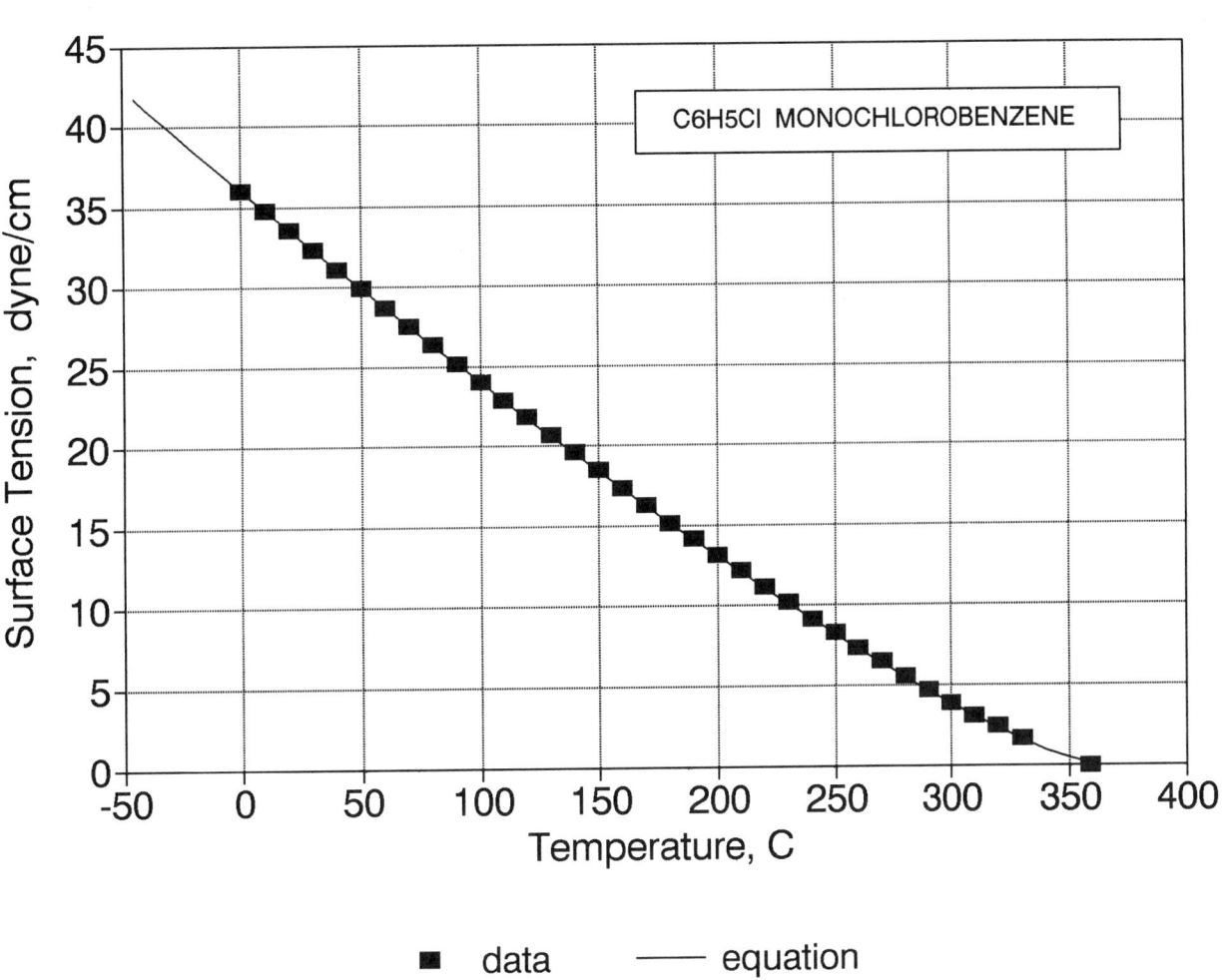

C6H5Cl MONOCHLOROBENZENE

■ data ——— equation

Table 9-1 SURFACE TENSION - ORGANIC COMPOUNDS

			sigma = A (1 - T/T$_c$)^n			(sigma - dynes/cm, T - K)		
NO	FORMULA	NAME	A	T$_c$	n	TMIN	TMAX	sigma @ 25 C
1	CBrClF2	BROMOCHLORODIFLUOROMETHANE	69.945	426.15	1.2446	113.65	426.15	15.65
2	CBrCl3	BROMOTRICHLOROMETHANE	82.862	606.00	1.2417	252.15	606.00	35.74
3	CBrF3	BROMOTRIFLUOROMETHANE	55.088	340.15	1.2553	105.15	340.15	3.99
4	CBr2F2	DIBROMODIFLUOROMETHANE	65.798	478.00	1.2074	163.05	478.00	20.21
5	CClF3	CHLOROTRIFLUOROMETHANE	63.490	301.96	1.2530	92.15	301.96	0.26
6	CClN	CYANOGEN CHLORIDE	79.260	449.00	1.1873	266.65	449.00	21.71
7	CCl2F2	DICHLORODIFLUOROMETHANE	58.000	384.95	1.2670	115.15	384.95	8.79
8	CCl2O	PHOSGENE	65.193	455.00	1.1600	145.37	455.00	18.95
9	CCl3F	TRICHLOROFLUOROMETHANE	62.722	471.20	1.2278	162.04	471.20	18.34
10	CCl4	CARBON TETRACHLORIDE	66.750	556.35	1.2140	250.33	556.35	26.29
11	CF2O	CARBONYL FLUORIDE	22.688	297.00	1.1691	161.89	297.00	------
12	CF4	CARBON TETRAFLUORIDE	56.591	227.50	1.2361	89.56	227.50	------
13	CHBr3	TRIBROMOMETHANE	86.700	696.00	1.1780	281.20	696.00	44.86
14	CHClF2	CHLORODIFLUOROMETHANE	66.600	369.30	1.2969	115.73	369.30	7.87
15	CHCl2F	DICHLOROFLUOROMETHANE	77.730	451.58	1.2500	138.15	451.58	20.16
16	CHCl3	CHLOROFORM	69.284	536.40	1.1761	209.63	536.40	26.68
17	CHF3	TRIFLUOROMETHANE	64.910	298.89	1.2743	117.97	298.89	0.03
18	CHI3	TRIIODOMETHANE	--------	--------	-------	--------	--------	------
19	CHN	HYDROGEN CYANIDE	52.256	456.65	1.0198	259.83	456.65	17.76
20	CHNS	ISOTHIOCYANIC-ACID	--------	--------	-------	--------	--------	------
21	CH2BrCl	BROMOCHLOROMETHANE	90.090	557.00	1.2354	185.20	557.00	34.96
22	CH2Br2	DIBROMOMETHANE	96.177	611.00	1.3061	220.60	611.00	40.12
23	CH2ClF	CHLOROFLUOROMETHANE	--------	--------	-------	--------	--------	------
24	CH2Cl2	DICHLOROMETHANE	88.570	510.00	1.2800	178.01	510.00	28.77
25	CH2F2	DIFLUOROMETHANE	70.182	351.60	1.2233	137.00	351.60	7.01
26	CH2I2	DIIODOMETHANE	93.610	747.00	1.2341	279.25	747.00	49.92
27	CH2O	FORMALDEHYDE	--------	--------	-------	--------	--------	------
28	CH2O2	FORMIC ACID	83.952	580.00	1.0760	281.45	580.00	38.62
29	CH3Br	METHYL BROMIDE	83.795	467.00	1.2222	179.47	467.00	24.17
30	CH3Cl	METHYL CHLORIDE	68.594	416.25	1.1966	175.43	416.25	15.19
31	CH3Cl3Si	METHYL TRICHLOROSILANE	57.880	517.00	1.2640	195.35	517.00	19.53
32	CH3F	METHYL FLUORIDE	67.673	317.70	1.2967	131.35	317.70	1.82
33	CH3I	METHYL IODIDE	86.540	528.00	1.2840	206.70	528.00	29.75
34	CH3NO	FORMAMIDE	107.989	771.00	1.2222	275.70	771.00	59.41
35	CH3NO2	NITROMETHANE	80.510	588.15	1.1275	244.60	588.15	36.27
36	CH3NO2	METHYL-NITRITE	--------	--------	-------	--------	--------	------
37	CH3NO3	METHYL-NITRATE	--------	--------	-------	--------	--------	------
38	CH4	METHANE	35.684	190.58	1.0920	90.67	190.58	------
39	CH4Cl2Si	METHYL DICHLOROSILANE	67.488	483.00	1.3756	182.55	483.00	18.01
40	CH4O	METHANOL	68.329	512.58	1.2222	273.10	512.58	23.55
41	CH4O3S	METHANESULFONIC ACID	--------	--------	-------	--------	--------	------
42	CH4S	METHYL MERCAPTAN	81.867	469.95	1.2261	150.18	469.95	23.84
43	CH5ClSi	METHYL CHLOROSILANE	47.318	442.00	1.2040	139.05	442.00	12.25
44	CH5N	METHYLAMINE	85.600	430.05	1.2556	179.69	430.05	19.41
45	CH6Si	METHYL SILANE	--------	--------	-------	--------	--------	------
46	CN4O8	TETRANITROMETHANE	68.260	540.00	1.0356	287.05	540.00	29.71
47	CO	CARBON MONOXIDE	27.959	132.92	1.1330	68.15	132.92	------
48	COS	CARBONYL SULFIDE	75.960	378.80	1.4660	134.30	378.80	7.87
49	CO2	CARBON DIOXIDE	79.970	304.19	1.2617	216.58	304.19	0.57
50	CS2	CARBON DISULFIDE	79.590	552.00	1.1909	161.11	552.00	19.72
51	C2BrF3	BROMOTRIFLUOROETHYLENE	93.237	432.00	1.3998	173.00	432.00	24.85
52	C2Br2F4	1,2-DIBROMOTETRAFLUOROETHANE	58.960	487.80	1.2557	162.65	487.80	8.49
53	C2ClF3	CHLOROTRIFLUOROETHYLENE	83.750	379.15	1.3200	115.00	379.15	30.04
54	C2ClF5	CHLOROPENTAFLUOROETHANE	62.348	353.15	1.2090	173.71	353.15	6.58
55	C2Cl2F4	1,2-DICHLOROTETRAFLUOROETHANE	51.800	418.85	1.2470	179.15	418.85	10.98
56	C2Cl3F3	1,1,2-TRICHLOROTRIFLUOROETHANE	55.380	487.25	1.2380	236.75	487.25	17.16
57	C2Cl4	TETRACHLOROETHYLENE	63.360	620.00	1.0586	250.80	620.00	31.65
58	C2Cl4F2	1,1,2,2-TETRACHLORODIFLUOROETHANE	64.160	551.00	1.2930	299.15	551.00	------
59	C2Cl4O	TRICHLOROACETYL CHLORIDE	87.290	590.00	1.2211	273.15	590.00	36.96
60	C2Cl6	HEXACHLOROETHANE	--------	--------	-------	--------	--------	------
61	C2F4	TETRAFLUOROETHYLENE	78.572	306.45	1.2500	142.00	306.45	0.86
62	C2F6	HEXAFLUOROETHANE	72.058	292.80	1.3086	173.10	292.80	------
63	C2HBrClF3	HALOTHANE	72.065	521.00	1.2785	223.15	521.00	24.33
64	C2HClF2	2-CHLORO-1,1-DIFLUOROETHYLENE	52.302	400.55	1.2559	134.65	400.55	9.43
65	C2HCl3	TRICHLOROETHYLENE	72.517	571.00	1.2537	188.40	571.00	28.73
66	C2HCl3O	DICHLOROACETYL CHLORIDE	74.324	579.00	1.2122	298.00	579.00	30.92
67	C2HCl3O	TRICHLOROACETALDEHYDE	61.250	565.00	1.2170	216.00	565.00	24.58
68	C2HCl5	PENTACHLOROETHANE	72.760	665.00	1.2610	244.15	665.00	34.37
69	C2HF3	TRIFLUOROETHENE	--------	--------	-------	--------	--------	------
70	C2HF3O2	TRIFLUOROACETIC ACID	39.066	491.25	1.1345	257.90	491.25	13.54
71	C2HF5	PENTAFLUOROETHANE	65.495	342.00	1.2525	170.15	342.00	5.00
72	C2H2	ACETYLENE	56.902	308.32	1.1346	192.40	308.32	1.19
73	C2H2Br4	1,1,2,2-TETRABROMOETHANE	101.720	824.00	1.6388	273.15	824.00	48.72
74	C2H2Cl2	1,1-DICHLOROETHYLENE	54.870	482.00	1.1743	150.59	482.00	17.69
75	C2H2Cl2	cis-1,2-DICHLOROETHYLENE	73.250	527.00	1.2550	193.15	527.00	25.71
76	C2H2Cl2	trans-1,2-DICHLOROETHYLENE	81.338	508.00	1.2264	223.35	508.00	27.50
77	C2H2Cl2O	CHLOROACETYL CHLORIDE	98.556	581.00	1.2474	251.15	581.00	40.15
78	C2H2Cl2O	DICHLOROACETALDEHYDE	105.560	555.00	1.2520	223.00	555.00	40.23

Table 9-1 SURFACE TENSION - ORGANIC COMPOUNDS (continued)

NO	FORMULA	NAME	A	T_c	n	TMIN	TMAX	sigma @ 25 C
			sigma = A $(1 - T/T_c)^n$			(sigma - dynes/cm, T - K)		
79	C2H2Cl2O2	DICHLOROACETIC ACID	62.800	686.00	1.0150	286.54	686.00	35.20
80	C2H2Cl3F	1,1,1-TRICHLOROFLUOROETHANE	103.900	565.00	1.2855	173.00	565.00	39.61
81	C2H2Cl4	1,1,1,2-TETRACHLOROETHANE	71.740	624.00	1.2584	202.94	624.00	31.67
82	C2H2Cl4	1,1,2,2-TETRACHLOROETHANE	76.200	645.00	1.2340	229.35	645.00	35.44
83	C2H2F2	1,1-DIFLUOROETHYLENE	88.229	302.80	1.4253	129.15	302.80	0.23
84	C2H2F2	cis-1,2-DIFLUOROETHENE	--------	--------	-------	--------	-------	------
85	C2H2F2	trans-1,2-DIFLUOROETHENE	--------	--------	-------	--------	-------	------
86	C2H2F4	1,1,1,2-TETRAFLUOROETHANE	90.588	380.00	1.3044	172.15	380.00	12.23
87	C2H2O	KETENE	53.573	370.00	1.2815	122.00	370.00	6.56
88	C2H2O4	OXALIC ACID	--------	--------	-------	--------	-------	------
89	C2H3Br	VINYL BROMIDE	75.152	473.00	1.3278	135.35	473.00	20.05
90	C2H3Cl	VINYL CHLORIDE	70.230	432.00	1.2710	119.36	432.00	15.84
91	C2H3ClF2	1-CHLORO-1,1-DIFLUOROETHANE	70.436	410.20	1.3563	142.35	410.20	12.12
92	C2H3ClO	ACETYL CHLORIDE	66.360	508.00	1.0890	160.30	508.00	25.34
93	C2H3ClO	CHLOROACETALDEHYDE	92.093	555.00	1.2778	293.00	555.00	34.41
94	C2H3ClO2	CHLOROACETIC ACID	75.250	686.00	1.0852	333.15	686.00	------
95	C2H3ClO2	METHYL CHLOROFORMATE	76.346	525.00	1.2047	192.00	525.00	27.78
96	C2H3Cl3	1,1,1-TRICHLOROETHANE	65.600	545.00	1.2170	242.75	545.00	25.02
97	C2H3Cl3	1,1,2-TRICHLOROETHANE	77.200	602.00	1.2100	236.50	602.00	33.75
98	C2H3F	VINYL FLUORIDE	56.613	327.80	1.2014	112.65	327.80	3.16
99	C2H3F3	1,1,1-TRIFLUOROETHANE	56.154	346.25	1.1831	161.82	346.25	5.43
100	C2H3N	ACETONITRILE	68.249	545.50	1.0970	229.32	545.50	28.66
101	C2H3NO	METHYL ISOCYANATE	84.540	505.00	1.3166	256.15	505.00	26.10
102	C2H4	ETHYLENE	52.940	282.36	1.2784	103.97	282.36	------
103	C2H4Br2	1,1-DIBROMOETHANE	94.643	628.00	1.3919	210.15	628.00	38.62
104	C2H4Br2	1,2-DIBROMOETHANE	80.970	650.15	1.2261	282.85	650.15	38.16
105	C2H4Cl2	1,1-DICHLOROETHANE	72.110	523.00	1.2530	176.19	523.00	25.04
106	C2H4Cl2	1,2-DICHLOROETHANE	80.140	561.00	1.2000	237.49	561.00	32.27
107	C2H4Cl2O	BIS(CHLOROMETHYL)ETHER	82.350	579.00	1.2564	231.65	579.00	33.18
108	C2H4F2	1,1-DIFLUOROETHANE	62.400	386.60	1.2392	156.15	386.60	10.03
109	C2H4F2	1,2-DIFLUOROETHANE	57.400	476.00	1.1716	215.00	476.00	18.11
110	C2H4I2	1,2-DIIODOETHANE	--------	--------	-------	--------	-------	------
111	C2H4O	ACETALDEHYDE	67.660	461.00	1.1940	150.15	461.00	19.53
112	C2H4O	ETHYLENE OXIDE	74.730	469.15	1.1410	160.71	469.15	23.63
113	C2H4OS	THIOACETIC-ACID	--------	--------	-------	--------	-------	------
114	C2H4O2	ACETIC ACID	57.050	592.71	1.0703	289.81	592.71	26.99
115	C2H4O2	METHYL FORMATE	80.980	487.20	1.2740	174.15	487.20	24.24
116	C2H4S	THIACYCLOPROPANE	84.165	555.00	1.2222	165.37	555.00	32.82
117	C2H5Br	BROMOETHANE	64.450	503.80	1.1256	154.55	503.80	23.51
118	C2H5Cl	ETHYL CHLORIDE	57.652	460.35	1.0880	134.80	460.35	18.53
119	C2H5ClO	2-CHLOROETHANOL	75.080	585.00	1.0100	205.65	585.00	36.55
120	C2H5F	ETHYL FLUORIDE	54.130	375.31	1.1170	129.95	375.31	9.25
121	C2H5I	ETHYL IODIDE	68.420	561.00	1.1604	162.05	561.00	28.39
122	C2H5N	ETHYLENEIMINE	88.340	537.00	1.1905	195.20	537.00	33.67
123	C2H5NO	ACETAMIDE	90.940	761.00	1.3140	354.15	761.00	------
124	C2H5NO	N-METHYLFORMAMIDE	86.510	721.00	1.5164	269.35	721.00	38.52
125	C2H5NO2	NITROETHANE	71.700	593.00	1.1700	183.63	593.00	31.66
126	C2H5NO3	ETHYL-NITRATE	--------	--------	-------	--------	-------	------
127	C2H6	ETHANE	48.984	305.42	1.2065	90.35	305.42	0.54
128	C2H6AlCl	DIMETHYLALUMINUM CHLORIDE	--------	--------	-------	--------	-------	------
129	C2H6O	DIMETHYL ETHER	60.960	400.10	1.2286	131.65	400.10	11.36
130	C2H6O	ETHANOL	67.036	516.25	1.2222	273.15	516.25	23.39
131	C2H6OS	DIMETHYL SULFOXIDE	93.260	726.00	1.4300	291.67	726.00	43.78
132	C2H6O2	ETHYLENE GLYCOL	106.491	645.00	1.2222	260.15	645.00	49.89
133	C2H6O4S	DIMETHYL SULFATE	83.933	758.00	1.5188	241.35	758.00	39.29
134	C2H6S	DIMETHYL SULFIDE	67.900	503.04	1.1500	174.88	503.04	24.17
135	C2H6S	ETHYL MERCAPTAN	60.000	499.15	1.0630	125.26	499.15	22.82
136	C2H6S2	DIMETHYL DISULFIDE	78.720	606.00	1.2804	188.44	606.00	33.07
137	C2H7N	DIMETHYLAMINE	56.215	437.65	1.0807	180.96	437.65	16.34
138	C2H7N	ETHYLAMINE	75.200	456.15	1.2690	192.15	456.15	19.58
139	C2H7NO	MONOETHANOLAMINE	108.495	638.00	1.2222	283.65	638.00	50.24
140	C2H8N2	ETHYLENEDIAMINE	91.188	593.00	1.1137	284.29	593.00	41.88
141	C2H8Si	DIMETHYL SILANE	--------	--------	-------	--------	-------	------
142	C2N2	CYANOGEN	73.193	400.15	1.2010	245.25	400.15	14.18
143	C3F6	HEXAFLUOROPROPYLENE	84.315	368.00	1.3979	116.65	368.00	8.26
144	C3F6O	HEXAFLUOROACETONE	71.624	357.14	1.1705	147.70	357.14	8.70
145	C3F8	OCTAFLUOROPROPANE	64.855	345.05	1.2195	125.46	345.05	5.69
146	C3H2N2	MALONONITRILE	99.960	715.00	1.2596	304.99	715.00	------
147	C3H3Cl	PROPARGYL CHLORIDE	86.537	541.00	1.3555	293.00	541.00	29.22
148	C3H3N	ACRYLONITRILE	63.900	535.00	1.0705	189.63	535.00	26.71
149	C3H3NO	OXAZOLE	86.200	554.00	1.1605	189.15	554.00	35.17
150	C3H4	METHYLACETYLENE	57.401	402.39	1.1898	170.45	402.39	11.51
151	C3H4	PROPADIENE	49.510	393.15	1.1620	136.87	393.15	9.50
152	C3H4Cl2	2,3-DICHLOROPROPENE	79.822	577.00	1.2883	191.50	577.00	31.28
153	C3H4O	ACROLEIN	72.351	506.00	1.2813	185.45	506.00	23.14
154	C3H4O	PROPARGYL ALCOHOL	74.570	580.00	1.0286	221.35	580.00	35.50
155	C3H4O2	ACRYLIC ACID	65.495	615.00	1.2549	286.65	615.00	28.50
156	C3H4O2	beta-PROPIOLACTONE	98.950	686.00	1.3006	239.75	686.00	47.13

Table 9-1 SURFACE TENSION - ORGANIC COMPOUNDS (continued)

NO	FORMULA	NAME	sigma = A (1 - T/T$_c$)^n			(sigma - dynes/cm, T - K)		
			A	T$_c$	n	TMIN	TMAX	sigma @ 25 C
157	C3H4O2	VINYL FORMATE	71.363	498.00	1.3095	200.00	498.00	21.59
158	C3H4O3	ETHYLENE CARBONATE	78.980	790.00	1.2816	309.55	790.00	------
159	C3H4O3	PYRUVIC ACID	91.080	634.52	1.0910	286.75	634.52	45.57
160	C3H5Br	3-BROMO-1-PROPENE	70.096	540.20	1.2222	153.76	540.20	26.28
161	C3H5Cl	2-CHLOROPROPENE	64.259	478.00	1.2620	135.75	478.00	18.72
162	C3H5Cl	3-CHLOROPROPENE	71.800	514.15	1.2900	138.65	514.15	23.46
163	C3H5ClO	alpha-EPICHLOROHYDRIN	81.950	610.00	1.2140	215.95	610.00	36.29
164	C3H5ClO2	METHYL CHLOROACETATE	84.370	600.00	1.2998	241.03	600.00	34.54
165	C3H5ClO2	ETHYL CHLOROFORMATE	74.230	508.15	1.1760	192.00	508.15	26.26
166	C3H5Cl3	1,2,3-TRICHLOROPROPANE	81.078	652.00	1.2750	258.45	652.00	37.19
167	C3H5I	3-IODO-1-PROPENE	72.658	595.81	1.2222	173.86	595.81	31.11
168	C3H5N	PROPIONITRILE	62.590	564.40	1.1364	180.26	564.40	26.65
169	C3H5NO	ACRYLAMIDE	26.700	710.00	1.2491	357.65	465.75	------
170	C3H5NO	HYDRACRYLONITRILE	82.456	690.00	1.0714	227.15	494.15	44.97
171	C3H5NO	LACTONITRILE	70.200	643.00	1.0585	233.00	643.00	36.30
172	C3H5N3O9	NITROGLYCERINE	151.320	680.00	1.9360	286.15	680.00	49.51
173	C3H6	CYCLOPROPANE	74.254	397.91	1.2599	145.59	397.91	12.99
174	C3H6	PROPYLENE	53.467	364.76	1.2058	87.90	364.76	6.88
175	C3H6Br2	1,2-DIBROMOPROPANE	73.651	634.11	1.2222	217.96	634.11	33.88
176	C3H6Cl2	1,1-DICHLOROPROPANE	66.350	560.00	1.2512	200.00	560.00	25.63
177	C3H6Cl2	1,2-DICHLOROPROPANE	75.860	572.00	1.2560	172.71	572.00	30.08
178	C3H6Cl2	1,3-DICHLOROPROPANE	75.706	603.00	1.1880	173.65	603.00	33.67
179	C3H6Cl2	2,2-DICHLOROPROPANE	--------	--------	-------	-------	-------	------
180	C3H6I2	1,2-DIIODOPROPANE	--------	--------	-------	-------	-------	------
181	C3H6O	ACETONE	62.200	508.20	1.1240	178.45	508.20	23.04
182	C3H6O	ALLYL ALCOHOL	70.232	545.05	1.2222	273.15	545.05	26.68
183	C3H6O	METHYL VINYL ETHER	64.054	437.00	1.1820	151.15	437.00	16.52
184	C3H6O	n-PROPIONALDEHYDE	73.478	496.00	1.2845	193.15	496.00	22.57
185	C3H6O	1,2-PROPYLENE OXIDE	69.465	482.25	1.2078	161.22	482.25	21.71
186	C3H6O	1,3-PROPYLENE OXIDE	83.477	520.00	1.2857	255.00	520.00	27.92
187	C3H6O2	ETHYL FORMATE	60.950	508.40	1.0996	193.55	508.40	23.08
188	C3H6O2	METHYL ACETATE	62.200	506.80	1.0480	175.15	506.80	24.54
189	C3H6O2	PROPIONIC ACID	58.456	604.00	1.1793	252.45	604.00	26.20
190	C3H6O2S	3-MERCAPTOPROPIONIC ACID	88.625	729.00	1.5402	290.65	729.00	39.43
191	C3H6O3	LACTIC ACID	84.500	616.00	1.1755	291.15	447.00	38.82
192	C3H6O3	METHOXYACETIC ACID	74.954	691.00	1.1620	281.00	691.00	38.89
193	C3H6O3	TRIOXANE	98.300	604.00	1.2421	334.65	604.00	------
194	C3H6S	THIACYCLOBUTANE	79.713	603.00	1.2222	199.96	603.00	34.63
195	C3H7Br	1-BROMOPROPANE	65.670	544.00	1.2036	163.15	544.00	25.25
196	C3H7Br	2-BROMOPROPANE	56.360	532.00	1.0980	184.15	532.00	22.86
197	C3H7Cl	ISOPROPYL CHLORIDE	58.440	489.00	1.2170	155.97	489.00	18.60
198	C3H7Cl	n-PROPYL CHLORIDE	66.870	503.15	1.2750	150.35	503.15	21.28
199	C3H7F	1-FLUOROPROPANE	--------	--------	-------	-------	-------	------
200	C3H7F	2-FLUOROPROPANE	--------	--------	-------	-------	-------	------
201	C3H7I	ISOPROPYL IODIDE	65.050	578.00	1.2320	183.15	578.00	26.62
202	C3H7I	n-PROPYL IODIDE	66.100	593.00	1.1850	171.85	593.00	28.88
203	C3H7N	ALLYLAMINE	63.900	505.00	1.0846	184.95	505.00	24.27
204	C3H7N	PROPYLENEIMINE	69.592	529.00	1.2632	229.00	529.00	24.41
205	C3H7NO	N,N-DIMETHYLFORMAMIDE	67.100	647.00	1.0800	212.72	647.00	34.43
206	C3H7NO	N-METHYLACETAMIDE	78.700	718.00	1.5650	301.15	718.00	------
207	C3H7NO2	1-NITROPROPANE	65.160	605.00	1.1500	169.16	605.00	29.85
208	C3H7NO2	2-NITROPROPANE	69.700	594.00	1.2566	181.83	594.00	29.03
209	C3H7NO3	PROPYL-NITRATE	--------	--------	-------	-------	-------	------
210	C3H7NO3	ISOPROPYL-NITRATE	--------	--------	-------	-------	-------	------
211	C3H8	PROPANE	49.624	369.82	1.1920	85.44	369.82	7.02
212	C3H8O	ISOPROPANOL	65.930	508.31	1.2222	273.15	508.31	22.40
213	C3H8O	METHYL ETHYL ETHER	61.200	437.80	1.2410	160.00	437.80	14.82
214	C3H8O	n-PROPANOL	66.660	536.71	1.2222	283.15	536.71	24.74
215	C3H8O2	2-METHOXYETHANOL	81.278	564.00	1.2222	283.15	564.00	32.41
216	C3H8O2	METHYLAL	70.967	480.60	1.2676	168.35	480.60	20.79
217	C3H8O2	1,2-PROPYLENE GLYCOL	64.270	626.00	0.9190	213.15	626.00	35.47
218	C3H8O2	1,3-PROPYLENE GLYCOL	99.179	658.00	1.2222	293.15	658.00	47.43
219	C3H8O3	GLYCEROL	124.793	723.00	1.2222	291.33	723.00	65.16
220	C3H8S	n-PROPYLMERCAPTAN	67.847	536.00	1.2692	159.95	536.00	24.19
221	C3H8S	ISOPROPYL MERCAPTAN	61.750	517.00	1.2338	142.61	517.00	21.38
222	C3H8S	ETHYL-METHYL-SULFIDE	66.542	532.80	1.2222	167.20	532.80	24.42
223	C3H9N	n-PROPYLAMINE	68.300	496.95	1.2125	190.15	496.95	22.49
224	C3H9N	ISOPROPYLAMINE	58.130	471.85	1.2070	177.95	471.85	17.40
225	C3H9N	TRIMETHYLAMINE	50.824	433.25	1.1440	156.08	433.25	13.40
226	C3H9NO	1-AMINO-2-PROPANOL	72.108	614.00	1.0321	274.89	614.00	36.31
227	C3H9NO	3-AMINO-1-PROPANOL	78.833	649.00	1.1947	284.15	649.00	37.81
228	C3H9NO	METHYLETHANOLAMINE	75.568	630.00	1.2200	268.65	630.00	34.57
229	C3H9O4P	TRIMETHYL PHOSPHATE	78.046	764.00	1.5108	227.00	764.00	36.96
230	C3H10N2	1,2-PROPANEDIAMINE	80.705	587.00	1.3223	236.53	587.00	31.60
231	C3H10Si	TRIMETHYL SILANE	--------	--------	-------	-------	-------	------
232	C4Cl4S	TETRACHLOROTHIOPHENE	76.835	753.00	1.2752	301.97	753.00	------
233	C4Cl6	HEXACHLORO-1,3-BUTADIENE	66.220	741.00	1.2820	252.15	741.00	34.23
234	C4F8	OCTAFLUORO-2-BUTENE	71.215	392.00	1.2500	138.15	392.00	11.93

218

Table 9-1 SURFACE TENSION - ORGANIC COMPOUNDS (continued)

NO	FORMULA	NAME	sigma = A (1 - T/T$_c$)^n			(sigma - dynes/cm, T - K)		
			A	T$_c$	n	TMIN	TMAX	sigma @ 25 C
235	C4F8	OCTAFLUOROCYCLOBUTANE	54.520	388.37	1.3170	232.96	388.37	7.97
236	C4F10	DECAFLUOROBUTANE	66.981	386.35	1.2175	144.95	386.35	11.09
237	C4H2	BUTADIYNE(BIACETYLENE)	64.631	478.02	1.2222	237.16	478.02	19.57
238	C4H2O3	MALEIC ANHYDRIDE	82.327	721.00	1.2311	325.72	721.00	------
239	C4H4	VINYLACETYLENE	67.539	454.00	1.2857	179.95	278.25	17.08
240	C4H4N2	SUCCINONITRILE	104.890	770.00	1.3994	331.16	770.00	------
241	C4H4O	FURAN	73.600	490.15	1.2719	187.55	490.15	22.34
242	C4H4O2	DIKETENE	54.886	616.00	1.2186	266.65	616.00	24.51
243	C4H8O3	SUCCINIC ANHYDRIDE	204.740	811.00	1.2782	393.00	811.00	------
244	C4H4O4	FUMARIC ACID	84.852	771.00	1.1920	560.15	771.00	------
245	C4H4O4	MALEIC ACID	87.562	773.00	1.2260	403.45	773.00	------
246	C4H4S	THIOPHENE	82.100	579.35	1.3280	234.94	579.35	31.44
247	C4H5Cl	CHLOROPRENE	60.742	525.00	1.1309	143.15	525.00	23.52
248	C4H5N	trans-CROTONITRILE	68.570	586.00	1.3150	222.00	586.00	26.92
249	C4H5N	cis-CROTONITRILE	73.420	568.00	1.2662	200.55	568.00	28.61
250	C4H5N	METHACRYLONITRILE	65.725	554.00	1.3016	237.35	554.00	24.04
251	C4H5N	PYRROLE	81.450	639.75	1.2870	249.74	639.75	36.32
252	C4H5N	VINYLACETONITRILE	62.584	584.00	1.0918	186.15	584.00	28.69
253	C4H5NO2	METHYL CYANOACETATE	90.070	687.00	1.3793	260.08	687.00	41.08
254	C4H6	CYCLOBUTENE	64.890	446.33	1.2222	153.76	446.33	16.86
255	C4H6	1,2-BUTADIENE	51.187	444.00	1.0571	136.95	444.00	15.78
256	C4H6	1,3-BUTADIENE	47.682	425.37	1.0507	164.25	425.37	13.41
257	C4H6	DIMETHYLACETYLENE	63.101	488.15	1.2096	240.91	488.15	20.15
258	C4H6	ETHYLACETYLENE	53.410	443.20	1.0210	147.43	443.20	17.07
259	C4H6Cl2	1,3-DICHLORO-trans-2-BUTENE	70.082	618.00	1.2050	276.00	618.00	31.69
260	C4H6Cl2	1,4-DICHLORO-cis-2-BUTENE	79.894	640.00	1.2836	225.15	640.00	35.72
261	C4H6Cl2	1,4-DICHLORO-trans-2-BUTENE	77.264	646.00	1.2771	274.15	646.00	35.05
262	C4H6Cl2	3,4-DICHLORO-1-BUTENE	77.346	589.00	1.2550	212.00	589.00	31.90
263	C4H6O	trans-CROTONALDEHYDE	67.053	571.00	1.3007	196.65	571.00	25.66
264	C4H6O	2,5-DIHYDROFURAN	75.198	542.00	1.2583	273.00	542.00	27.53
265	C4H6O	DIVINYL ETHER	50.033	463.00	1.2178	172.05	463.00	14.23
266	C4H6O	METHACROLEIN	68.197	530.00	1.2930	192.15	530.00	23.41
267	C4H6O2	2-BUTYNE-1,4-DIOL	64.685	695.00	1.1408	331.00	695.00	------
268	C4H6O2	gamma-BUTYROLACTONE	73.080	739.00	1.2384	229.78	739.00	38.54
269	C4H6O2	cis-CROTONIC ACID	58.472	647.00	1.2880	288.65	445.05	26.39
270	C4H6O2	trans-CROTONIC ACID	63.106	666.00	1.2382	344.55	666.00	------
271	C4H6O2	METHACRYLIC ACID	65.015	643.00	1.2891	288.15	643.00	29.12
272	C4H6O2	METHYL ACRYLATE	58.830	536.00	1.0920	196.32	536.00	24.23
273	C4H6O2	VINYL ACETATE	68.685	524.00	1.2500	180.35	524.00	23.99
274	C4H6O3	ACETIC ANHYDRIDE	80.370	569.15	1.2420	200.15	569.15	31.98
275	C4H6O4	SUCCINIC ACID	143.210	806.00	1.2603	461.15	806.00	------
276	C4H6O5	DIGLYCOLIC ACID	125.830	820.00	1.2557	421.15	820.00	------
277	C4H6O5	MALIC ACID	58.398	781.00	1.1086	403.15	781.00	------
278	C4H6O6	TARTARIC ACID	144.050	828.00	1.1467	479.15	828.00	------
279	C4H7N	n-BUTYRONITRILE	59.560	582.25	1.1050	161.25	582.25	26.95
280	C4H7N	ISOBUTYRONITRILE	63.927	565.00	1.3102	201.70	565.00	23.92
281	C4H7NO	ACETONE CYANOHYDRIN	71.276	647.00	1.3355	252.15	393.15	31.24
282	C4H7NO	2-METHACRYLAMIDE	42.241	741.00	1.2309	383.65	741.00	------
283	C4H7NO	3-METHOXYPROPIONITRILE	72.625	638.00	1.1470	210.12	638.00	35.26
284	C4H7NO	2-PYRROLIDONE	84.673	792.00	1.3028	298.15	792.00	45.76
285	C4H8	1-BUTENE	56.000	419.59	1.2341	87.80	419.59	12.12
286	C4H8	cis-2-BUTENE	58.810	435.58	1.2450	134.26	435.58	13.99
287	C4H8	trans-2-BUTENE	67.640	428.63	1.4050	167.60	428.63	12.72
288	C4H8	CYCLOBUTANE	61.399	459.93	1.2444	182.48	459.93	16.73
289	C4H8	ISOBUTENE	55.440	417.90	1.2453	132.81	417.90	11.69
290	C4H8Br2	1,2-DIBROMOBUTANE	--------	--------	-------	-------	-------	------
291	C4H8Br2	2,3-DIBROMOBUTANE	--------	--------	-------	-------	-------	------
292	C4H8Cl2	1,4-DICHLOROBUTANE	75.105	641.00	1.2606	235.85	641.00	34.13
293	C4H8I2	1,2-DIIODOBUTANE	--------	--------	-------	-------	-------	------
294	C4H8O	n-BUTYRALDEHYDE	70.330	525.00	1.2710	176.75	525.00	24.21
295	C4H8O	ISOBUTYRALDEHYDE	60.401	507.00	1.2288	208.15	507.00	20.31
296	C4H8O	1,2-EPOXYBUTANE	68.593	526.00	1.2820	123.15	526.00	23.47
297	C4H8O	METHYL ETHYL KETONE	59.441	535.50	1.1165	186.48	535.50	23.96
298	C4H8O	ETHYL VINYL ETHER	61.934	475.15	1.2477	157.35	475.15	18.07
299	C4H8O	TETRAHYDROFURAN	67.130	540.15	1.2310	164.65	540.15	24.98
300	C4H8O2	cis-2-BUTENE-1,4-DIOL	77.086	677.88	1.1606	284.15	677.88	39.34
301	C4H8O2	trans-2-BUTENE-1,4-DIOL	75.984	681.00	1.1626	300.45	681.00	------
302	C4H8O2	ISOBUTYRIC ACID	54.790	609.15	1.1904	227.05	609.15	24.61
303	C4H8O2	n-BUTYRIC ACID	56.230	628.00	1.1930	267.95	628.00	26.08
304	C4H8O2	1,4-DIOXANE	82.160	587.00	1.2899	284.95	587.00	32.92
305	C4H8O2	ETHYL ACETATE	59.870	523.30	1.1220	189.60	523.30	23.24
306	C4H8O2	METHYL PROPIONATE	65.570	530.60	1.2002	185.65	530.60	24.35
307	C4H8O2	n-PROPYL FORMATE	59.446	538.00	1.1256	180.25	538.00	23.94
308	C4H8O2S	SULFOLANE	85.420	849.00	1.2300	300.75	849.00	------
309	C4H8S	TETRAHYDROTHIOPHENE	74.500	631.95	1.2712	176.98	631.95	33.10
310	C4H9Br	1-BROMOBUTANE	64.350	577.00	1.2510	160.75	577.00	25.91
311	C4H9Br	2-BROMOBUTANE	61.620	567.00	1.2270	161.25	567.00	24.67
312	C4H9Cl	n-BUTYL CHLORIDE	60.040	537.00	1.1732	150.05	537.00	23.21

Table 9-1 SURFACE TENSION - ORGANIC COMPOUNDS (continued)

NO	FORMULA	NAME	sigma = A (1 - T/T_c)^n			(sigma - dynes/cm, T - K)		
			A	T_c	n	TMIN	TMAX	sigma @ 25 C
313	C4H9Cl	sec-BUTYL CHLORIDE	56.460	520.60	1.1348	141.85	520.60	21.51
314	C4H9Cl	tert-BUTYL CHLORIDE	53.800	507.00	1.1850	247.75	507.00	18.81
315	C4H9I	2-IODO-2-METHYLPROPANE	--------	--------	-------	--------	-------	------
316	C4H9N	PYRROLIDINE	76.138	568.55	1.2222	283.15	568.55	30.70
317	C4H9NO	N,N-DIMETHYLACETAMIDE	82.160	658.00	1.5070	253.15	658.00	33.09
318	C4H9NO	MORPHOLINE	81.525	618.00	1.1927	270.05	618.00	37.16
319	C4H9NO2	1-NITROBUTANE	64.652	624.00	1.2222	191.83	624.00	29.22
320	C4H9NO2	2-NITROBUTANE	65.162	615.00	1.2222	141.16	615.00	28.97
321	C4H10	n-BUTANE	52.660	425.18	1.2330	134.86	425.18	11.87
322	C4H10	ISOBUTANE	52.165	408.14	1.2723	113.54	408.14	9.84
323	C4H10N2	PIPERAZINE	24.853	638.00	1.1667	379.15	638.00	------
324	C4H10O	n-BUTANOL	64.526	562.93	1.2222	283.15	562.93	25.67
325	C4H10O	sec-BUTANOL	65.688	536.01	1.2222	273.15	536.01	24.33
326	C4H10O	tert-BUTANOL	62.590	506.20	1.2222	298.15	506.20	21.11
327	C4H10O	DIETHYL ETHER	57.356	466.70	1.2280	156.85	466.70	16.42
328	C4H10O	METHYL-PROPYL-ETHER	--------	--------	-------	--------	-------	------
329	C4H10O	METHYL ISOPROPYL ETHER	47.342	464.50	1.0967	127.93	464.50	15.35
330	C4H10O	ISOBUTANOL	62.150	547.73	1.2222	283.15	547.73	23.78
331	C4H10O2	1,3-BUTANEDIOL	77.735	643.00	1.1600	196.15	643.00	37.73
332	C4H10O2	1,4-BUTANEDIOL	76.100	667.00	1.1950	293.05	667.00	37.49
333	C4H10O2	2,3-BUTANEDIOL	71.754	611.00	1.1133	280.75	611.00	34.06
334	C4H10O2	t-BUTYL HYDROPEROXIDE	47.198	576.00	1.1120	277.45	576.00	20.98
335	C4H10O2	1,2-DIMETHOXYETHANE	55.912	536.15	1.2672	215.15	536.15	19.98
336	C4H10O2	2-ETHOXYETHANOL	73.652	569.00	1.2222	293.15	569.00	29.73
337	C4H10O3	DIETHYLENE GLYCOL	92.567	744.60	1.2222	293.15	744.60	49.54
338	C4H10O4S	DIETHYL SULFATE	66.270	792.00	1.4686	248.00	792.00	33.12
339	C4H10S	n-BUTYL MERCAPTAN	64.040	569.00	1.2554	157.46	569.00	25.22
340	C4H10S	ISOBUTYL MERCAPTAN	56.561	559.00	1.1523	128.31	559.00	23.50
341	C4H10S	sec-BUTYL MERCAPTAN	65.712	554.00	1.3000	133.02	554.00	24.07
342	C4H10S	tert-BUTYL MERCAPTAN	61.868	530.00	1.3482	274.26	530.00	20.29
343	C4H10S	DIETHYL SULFIDE	62.890	557.15	1.2223	169.20	557.15	24.66
344	C4H10S	ISOPROPYL-METHYL-SULFIDE	61.280	551.00	1.2222	171.65	551.00	23.65
345	C4H10S	METHYL-PROPYL-SULFIDE	62.741	563.00	1.2222	160.19	563.00	24.96
346	C4H10S2	DIETHYL DISULFIDE	65.500	642.00	1.2225	171.63	642.00	30.53
347	C4H11N	n-BUTYLAMINE	65.400	531.90	1.1813	224.05	531.90	24.76
348	C4H11N	ISOBUTYLAMINE	52.150	513.73	1.0097	188.55	513.73	21.70
349	C4H11N	sec-BUTYLAMINE	53.717	514.30	1.0775	168.65	514.30	21.11
350	C4H11N	tert-BUTYLAMINE	49.068	483.90	1.1155	206.19	483.90	16.86
351	C4H11N	DIETHYLAMINE	54.000	496.60	1.1010	223.35	496.60	19.67
352	C4H11NO	DIMETHYLETHANOLAMINE	74.850	571.82	1.1916	214.15	571.82	31.11
353	C4H11NO2	DIETHANOLAMINE	78.330	715.00	0.9370	301.15	542.04	------
354	C4H11NO2	2-AMINOETHOXYETHANOL	76.399	699.00	1.1927	293.15	699.00	39.31
355	C4H12N2O	N-AMINOETHYL ETHANOLAMINE	82.880	698.00	1.1780	273.15	698.00	43.05
356	C4H12Si	TETRAMETHYLSILANE	47.946	450.40	1.2573	174.07	450.40	12.26
357	C4H13N3	DIETHYLENE TRIAMINE	83.106	676.00	1.1880	234.15	676.00	41.64
358	C5Cl6	HEXACHLOROCYCLOPENTADIENE	86.513	746.00	1.2490	284.49	746.00	45.74
359	C5H4O2	FURFURAL	84.290	657.00	1.1124	236.65	657.00	43.01
360	C5H5N	PYRIDINE	81.500	619.95	1.2160	231.51	619.95	36.72
361	C5H6	CYCLOPENTADIENE	--------	--------	-------	--------	-------	------
362	C5H6	2-METHYL-1-BUTENE-3-YNE	59.343	492.00	1.3121	160.15	492.00	17.48
363	C5H6	1-PENTENE-3-YNE	68.184	520.00	1.2505	150.00	520.00	23.50
364	C5H6	1-PENTENE-4-YNE	68.000	503.00	1.2619	150.00	503.00	21.89
365	C5H6N2	GLUTARONITRILE	121.556	782.00	1.8859	244.21	470.65	49.16
366	C5H6O2	FURFURYL ALCOHOL	72.170	632.00	1.0060	258.52	632.00	37.98
367	C5H6O3	GLUTARIC ANHYDRIDE	166.840	838.00	1.2802	328.00	838.00	------
368	C5H6O4	CITRACONIC ACID	112.480	829.00	1.2873	356.15	829.00	------
369	C5H6O4	ITACONIC ACID	105.190	821.00	1.2559	438.75	821.00	------
370	C5H6S	2-METHYLTHIOPHENE	68.657	610.00	1.2222	209.77	610.00	30.24
371	C5H6S	3-METHYLTHIOPHENE	71.275	615.00	1.2222	204.18	615.00	31.69
372	C5H7N	N-METHYLPYRROLE	86.037	610.00	1.5067	216.91	610.00	31.31
373	C5H7NO2	ETHYL CYANOACETATE	70.900	679.00	1.1940	250.65	679.00	35.55
374	C5H8	CYCLOPENTENE	65.300	507.00	1.2520	138.13	507.00	21.51
375	C5H8	ISOPRENE	55.896	484.00	1.2824	127.27	484.00	16.38
376	C5H8	3-METHYL-1,2-BUTADIENE	58.397	490.00	1.3484	159.53	490.00	16.49
377	C5H8	2-METHYL-1,3-BUTADIENE	52.114	483.30	1.2222	127.20	483.30	16.13
378	C5H8	1,2-PENTADIENE	59.400	500.00	1.2871	135.89	500.00	18.48
379	C5H8	cis-1,3-PENTADIENE	60.290	499.00	1.3081	132.35	499.00	18.33
380	C5H8	trans-1,3-PENTADIENE	55.025	500.00	1.3113	185.71	500.00	16.75
381	C5H8	1,4-PENTADIENE	73.010	479.00	1.5896	124.86	479.00	15.52
382	C5H8	2,3-PENTADIENE	58.790	497.00	1.2761	147.50	497.00	18.27
383	C5H8	1-PENTYNE	56.670	481.20	1.1424	167.45	481.20	18.79
384	C5H8	2-PENTYNE	58.888	521.99	1.2222	163.86	521.99	20.92
385	C5H8	3-METHYL-1-BUTYNE	60.547	463.20	1.3093	183.45	463.20	15.68
386	C5H8	SPIROPENTANE	69.251	499.74	1.2222	166.11	499.74	22.83
387	C5H8N4O12	PENTAERYTHRITOL TETRANITRATE	37.909	676.00	1.0346	413.65	676.00	------
388	C5H8O	CYCLOPENTANONE	68.530	626.00	1.1330	221.85	626.00	32.93
389	C5H8O	METHYL ISOPROPENYL KETONE	65.500	566.00	1.2715	219.55	566.00	25.30
390	C5H8O2	ACETYLACETONE	67.380	602.00	1.1619	249.95	602.00	30.45

Table 9-1 SURFACE TENSION - ORGANIC COMPOUNDS (continued)

			sigma = A (1 - T/T$_c$)^n			(sigma - dynes/cm, T - K)		
NO	FORMULA	NAME	A	T$_c$	n	TMIN	TMAX	sigma @ 25 C
391	C5H8O2	ALLYL ACETATE	65.105	559.00	1.2162	138.00	559.00	25.76
392	C5H8O2	ETHYL ACRYLATE	58.960	553.00	1.1250	201.95	553.00	24.66
393	C5H8O2	METHYL METHACRYLATE	65.920	564.00	1.2620	224.95	564.00	25.51
394	C5H8O2	VINYL PROPIONATE	52.769	546.00	1.2402	298.00	546.00	19.81
395	C5H8O3	2-HYDROXYETHYL ACRYLATE	50.676	662.00	1.1890	213.00	662.00	24.87
396	C5H8O3	LEVULINIC ACID	78.635	723.00	1.2222	308.15	723.00	------
397	C5H8O3	METHYL ACETOACETATE	80.192	642.00	1.2337	193.15	642.00	37.12
398	C5H8O4	GLUTARIC ACID	78.398	807.00	1.1050	370.65	807.00	------
399	C5H9N	VALERONITRILE	56.210	603.00	1.0760	176.95	603.00	26.97
400	C5H9NO	n-BUTYL ISOCYANATE	102.350	568.00	1.7385	193.00	568.00	28.06
401	C5H9NO	N-METHYL-2-PYRROLIDONE	76.514	724.00	1.2190	249.15	724.00	40.07
402	C5H9NO4	L-GLUTAMIC ACID	129.130	886.00	1.2963	497.15	670.00	------
403	C5H10	CYCLOPENTANE	72.050	511.76	1.3730	179.28	511.76	21.71
404	C5H10	2-METHYL-1-BUTENE	57.002	465.00	1.2767	135.58	465.00	15.40
405	C5H10	2-METHYL-2-BUTENE	54.770	471.00	1.1636	139.39	471.00	17.06
406	C5H10	3-METHYL-1-BUTENE	49.370	450.37	1.1754	104.66	450.37	13.80
407	C5H10	1-PENTENE	50.757	464.78	1.1588	107.93	464.78	15.46
408	C5H10	cis-2-PENTENE	56.695	475.93	1.2354	121.75	475.93	16.80
409	C5H10	trans-2-PENTENE	47.275	475.37	1.0724	132.89	475.37	16.41
410	C5H10Br2	2,3-DIBROMO-2-METHYLBUTANE	--------	--------	--------	--------	--------	------
411	C5H10Cl2	1,5-DICHLOROPENTANE	73.705	663.00	1.2824	200.35	663.00	34.26
412	C5H10O	METHYL ISOPROPYL KETONE	63.985	553.00	1.2220	181.15	553.00	24.83
413	C5H10O	2-PENTANONE	61.823	561.08	1.2222	298.15	561.08	24.48
414	C5H10O	DIETHYL KETONE	54.700	560.95	1.0560	234.18	560.95	24.56
415	C5H10O	VALERALDEHYDE	57.540	554.00	1.0800	182.00	554.00	24.98
416	C5H10O2	n-BUTYL FORMATE	57.032	559.00	1.1073	181.25	559.00	24.52
417	C5H10O2	ETHYL PROPIONATE	62.910	546.00	1.2334	199.25	546.00	23.75
418	C5H10O2	ISOBUTYL FORMATE	57.780	551.35	1.1720	177.35	551.35	23.21
419	C5H10O2	ISOPROPYL ACETATE	58.080	538.00	1.2430	199.75	538.00	21.28
420	C5H10O2	n-PROPYL ACETATE	62.160	549.40	1.2240	178.15	549.40	23.86
421	C5H10O2	METHYL n-BUTYRATE	64.084	554.50	1.2418	187.35	554.50	24.58
422	C5H10O2	2-METHYLBUTYRIC ACID	58.365	643.00	1.2356	193.00	643.00	27.03
423	C5H10O2	ISOVALERIC ACID	52.660	634.00	1.1700	243.85	634.00	25.04
424	C5H10O2	VALERIC ACID	56.800	651.00	1.2257	239.15	651.00	26.81
425	C5H10O2	TETRAHYDROFURFURYL ALCOHOL	83.852	639.00	1.2222	294.75	639.00	38.90
426	C5H10O2S	3-METHYL SULFOLANE	76.790	817.00	1.2550	273.65	817.00	43.44
427	C5H10O3	DIETHYL CARBONATE	62.560	576.00	1.2150	230.15	576.00	25.80
428	C5H10O3	ETHYL LACTATE	117.360	588.00	2.0057	247.15	426.54	28.40
429	C5H10S	THIACYCLOHEXANE	70.235	657.12	1.2222	292.14	657.12	33.54
430	C5H10S	CYCLOPENTANETHIOL	68.181	629.00	1.2222	155.39	629.00	31.09
431	C5H11Br	1-BROMOPENTANE	67.065	564.76	1.2222	185.26	564.76	26.80
432	C5H11Cl	1-CHLOROPENTANE	60.310	568.00	1.1870	174.15	568.00	24.93
433	C5H11Cl	1-CHLORO-3-METHYLBUTANE	57.960	558.87	1.2222	168.76	558.87	22.82
434	C5H11Cl	2-CHLORO-2-METHYLBUTANE	56.815	548.97	1.2222	199.66	548.97	21.81
435	C5H11N	N-METHYLPYRROLIDINE	65.657	550.00	1.2438	183.15	550.00	24.85
436	C5H11N	PIPERIDINE	67.163	594.05	1.2091	262.65	594.05	28.92
437	C5H11NO	tert-BUTYLFORMAMIDE	65.928	692.00	1.2738	289.15	692.00	32.16
438	C5H12	ISOPENTANE	50.926	460.43	1.2078	113.25	460.43	14.45
439	C5H12	NEOPENTANE	47.906	433.78	1.2745	256.58	433.78	10.89
440	C5H12	n-PENTANE	52.090	469.65	1.2054	143.42	469.65	15.47
441	C5H12O	2,2-DIMETHYL-1-PROPANOL	43.025	550.00	1.1676	327.15	550.00	------
442	C5H12O	tert-PENTYL-ALCOHOL	64.154	549.00	1.2222	327.00	549.00	------
443	C5H12O	2-METHYL-1-BUTANOL	54.490	565.00	1.0358	203.00	565.00	25.05
444	C5H12O	2-METHYL-2-BUTANOL	61.930	545.15	1.2222	264.35	545.15	23.53
445	C5H12O	3-METHYL-1-BUTANOL	59.834	579.45	1.2222	273.15	579.45	24.74
446	C5H12O	3-METHYL-2-BUTANOL	61.450	574.00	1.2456	188.00	574.00	24.67
447	C5H12O	1-PENTANOL	63.074	586.15	1.2222	283.15	586.15	26.46
448	C5H12O	2-PENTANOL	55.186	552.00	1.1016	200.00	552.00	23.45
449	C5H12O	3-PENTANOL	65.540	547.00	1.1600	204.15	547.00	26.29
450	C5H12O	METHYL sec-BUTYL ETHER	56.267	498.00	1.2340	150.00	498.00	18.24
451	C5H12O	METHYL tert-BUTYL ETHER	59.353	497.10	1.2385	164.55	497.10	19.09
452	C5H12O	METHYL ISOBUTYL ETHER	52.497	497.00	1.2324	150.00	497.00	16.98
453	C5H12O	ETHYL PROPYL ETHER	52.783	500.23	1.1111	145.65	500.23	19.28
454	C5H12O2	ETHYLENE GLYCOL MONOPROPYL ETHER	66.370	582.00	1.1716	183.15	582.00	28.62
455	C5H12O2	NEOPENTYL GLYCOL	46.734	643.00	1.0953	400.00	643.00	------
456	C5H12O2	1,5-PENTANEDIOL	73.583	673.00	1.1720	257.15	673.00	37.06
457	C5H12O3	2-(2-METHOXYETHOXY)ETHANOL	76.950	630.00	1.1894	197.15	630.00	35.90
458	C5H12O4	PENTAERYTHRITOL	34.784	780.00	0.9919	534.15	780.00	------
459	C5H12S	n-PENTYL MERCAPTAN	57.960	598.00	1.1600	197.45	598.00	26.02
460	C5H12S	BUTYL-METHYL-SULFIDE	60.609	591.00	1.2222	175.33	591.00	25.69
461	C5H12S	ETHYL-PROPYL-SULFIDE	60.705	584.00	1.2222	156.15	584.00	25.35
462	C5H12S	2-METHYL-2-BUTANETHIOL	54.052	566.00	1.2222	169.38	566.00	21.66
463	C5H13N	n-PENTYLAMINE	56.256	555.00	1.0696	218.15	555.00	24.68
464	C5H13NO2	METHYL DIETHANOLAMINE	74.940	678.00	1.1620	252.15	678.00	38.22
465	C6Cl6	HEXACHLOROBENZENE	176.230	825.00	1.3046	501.70	825.00	------
466	C6F6	HEXAFLUOROBENZENE	62.277	516.73	1.2065	278.25	516.73	22.06
467	C6H3ClN2O4	1-CHLORO-2,4-DINITROBENZENE	78.340	813.77	1.0278	326.55	813.77	------
468	C6H3Cl2NO2	1,2-DICHLORO-4-NITROBENZENE	70.960	758.00	1.0240	315.65	758.00	------

Table 9-1 SURFACE TENSION - ORGANIC COMPOUNDS (continued)

NO	FORMULA	NAME	sigma = A (1 - T/T$_c$)^n			(sigma - dynes/cm, T - K)		
			A	T$_c$	n	TMIN	TMAX	sigma @ 25 C
469	C6H3Cl3	1,2,4-TRICHLOROBENZENE	87.180	725.00	1.2629	290.15	725.00	44.66
470	C6H3N3O6	1,3,5-TRINITROBENZENE	134.180	1005.00	1.3867	398.40	748.00	------
471	C6H4Br2	m-DIBROMOBENZENE	78.983	761.00	1.2239	266.25	761.00	42.98
472	C6H4ClNO2	m-CHLORONITROBENZENE	101.650	742.00	1.5110	317.65	742.00	------
473	C6H4ClNO2	o-CHLORONITROBENZENE	84.780	757.00	1.2480	306.15	757.00	------
474	C6H4ClNO2	p-CHLORONITROBENZENE	77.306	751.00	1.1370	356.65	751.00	------
475	C6H4Cl2	m-DICHLOROBENZENE	67.130	683.95	1.0940	248.39	683.95	35.88
476	C6H4Cl2	o-DICHLOROBENZENE	75.756	705.00	1.3220	256.15	705.00	36.63
477	C6H4Cl2	p-DICHLOROBENZENE	71.996	684.75	1.2374	326.14	684.75	------
478	C6H4F2	m-DIFLUOROBENZENE	65.280	552.94	1.2222	249.16	552.94	25.32
479	C6H4F2	o-DIFLUOROBENZENE	--------	--------	-------	-------	-------	------
480	C6H4F2	p-DIFLUOROBENZENE	67.461	556.00	1.2222	260.16	556.00	26.37
481	C6H4N2O4	m-DINITROBENZENE	82.520	805.00	1.0778	364.00	805.00	------
482	C6H4N2O4	o-DINITROBENZENE	87.280	831.00	1.2266	390.08	831.00	------
483	C6H4N2O4	p-DINITROBENZENE	94.900	803.00	1.2010	446.60	803.00	------
484	C6H5Br	BROMOBENZENE	74.122	670.15	1.2310	242.43	670.15	35.91
485	C6H5Cl	MONOCHLOROBENZENE	72.700	632.35	1.2420	227.95	632.35	32.93
486	C6H5ClO	m-CHLOROPHENOL	81.086	729.00	1.1857	306.00	729.00	------
487	C6H5ClO	o-CHLOROPHENOL	78.530	675.00	1.1360	282.00	675.00	40.50
488	C6H5ClO	p-CHLOROPHENOL	87.560	738.00	1.2670	316.00	738.00	------
489	C6H5Cl2N	3,4-DICHLOROANILINE	88.420	800.00	1.2489	344.65	800.00	------
490	C6H5F	FLUOROBENZENE	72.200	560.09	1.3030	230.94	560.09	26.82
491	C6H5I	IODOBENZENE	76.400	721.15	1.2600	241.83	721.15	39.01
492	C6H5NO2	NITROBENZENE	79.440	719.00	1.1362	278.91	719.00	43.23
493	C6H6	BENZENE	71.950	562.16	1.2389	278.68	562.16	28.21
494	C6H6ClN	m-CHLOROANILINE	92.363	751.00	1.4304	262.75	751.00	44.80
495	C6H6ClN	o-CHLOROANILINE	81.353	722.00	1.2222	271.05	722.00	42.43
496	C6H6ClN	p-CHLOROANILINE	79.710	754.00	1.0963	343.05	754.00	------
497	C6H6N2	cis-DICYANO-1-BUTENE	131.920	691.00	1.3606	249.00	691.00	61.18
498	C6H6N2	trans-DICYANO-1-BUTENE	127.490	689.00	1.3520	260.00	689.00	59.24
499	C6H6N2	1,4-DICYANO-2-BUTENE	90.710	755.00	1.3400	349.00	755.00	------
500	C6H6N2O2	m-NITROANILINE	97.000	815.00	1.1820	387.15	579.00	------
501	C6H6N2O2	o-NITROANILINE	83.435	784.00	1.1709	344.65	558.00	------
502	C6H6N2O2	p-NITROANILINE	109.600	851.00	1.2150	420.65	609.15	------
503	C6H6O	PHENOL	74.500	694.25	1.0767	314.06	694.25	------
504	C6H6O2	1,2-BENZENEDIOL	85.199	764.00	1.2222	373.15	764.00	------
505	C6H6O2	1,3-BENZENEDIOL	95.033	810.00	1.2222	373.15	810.00	------
506	C6H6O2	p-HYDROQUINONE	31.682	822.00	1.0926	444.65	822.00	------
507	C6H6O3	1,2,3-BENZENETRIOL	14.106	830.00	0.9188	407.00	830.00	------
508	C6H6S	PHENYL MERCAPTAN	79.270	689.00	1.2720	258.27	689.00	38.54
509	C6H7N	ANILINE	77.260	699.00	1.0800	267.13	699.00	42.38
510	C6H7N	2-METHYLPYRIDINE	75.030	621.00	1.2638	206.44	621.00	32.82
511	C6H7N	3-METHYLPYRIDINE	78.700	645.00	1.2993	255.01	645.00	35.15
512	C6H7N	4-METHYLPYRIDINE	76.843	646.15	1.2813	276.73	646.15	34.77
513	C6H8	1,3-CYCLOHEXADIENE	75.905	558.00	1.3660	161.00	558.00	26.72
514	C6H8	METHYLCYCLOPENTADIENE	63.660	541.00	1.2403	150.00	541.00	23.57
515	C6H8N2	ADIPONITRILE	--------	--------	-------	-------	-------	------
516	C6H8N2	METHYLGLUTARONITRILE	111.999	742.00	1.8954	228.15	471.55	42.29
517	C6H8N2	m-PHENYLENEDIAMINE	93.957	824.00	1.2590	334.00	824.00	------
518	C6H8N2	o-PHENYLENEDIAMINE	100.660	781.00	1.2324	376.95	781.00	------
519	C6H8N2	p-PHENYLENEDIAMINE	101.090	796.00	1.2295	413.00	796.00	------
520	C6H8N2	PHENYLHYDRAZINE	85.754	761.00	1.3041	292.35	761.00	44.84
521	C6H8N2O	BIS(CYANOETHYL)ETHER	83.375	783.00	1.1880	246.85	783.00	47.18
522	C6H8O4	DIMETHYL MALEATE	76.543	675.00	1.2342	254.15	675.00	37.28
523	C6H8O6	ASCORBIC ACID	100.400	783.00	1.0100	465.15	783.00	------
524	C6H8O7	CITRIC ACID	103.980	822.00	1.1420	426.15	822.00	------
525	C6H10	1-METHYLCYCLOPENTENE	50.547	541.99	1.2222	145.96	541.99	19.04
526	C6H10	3-METHYLCYCLOPENTENE	59.737	535.71	1.2222	130.16	535.71	22.11
527	C6H10	4-METHYLCYCLOPENTENE	59.836	543.75	1.2222	112.31	543.75	22.65
528	C6H10	CYCLOHEXENE	66.470	560.40	1.2317	169.67	560.40	26.09
529	C6H10	2,3-DIMETHYL-1,3-BUTADIENE	56.234	526.00	1.2469	197.15	526.00	19.81
530	C6H10	1,5-HEXADIENE	53.618	507.00	1.2080	132.47	507.00	18.37
531	C6H10	cis,trans-2,4-HEXADIENE	56.853	538.00	1.2600	177.05	538.00	20.54
532	C6H10	trans,trans-2,4-HEXADIENE	52.701	535.00	1.2184	228.25	535.00	19.53
533	C6H10	1-HEXYNE	57.890	516.20	1.2022	141.25	516.20	20.54
534	C6H10	2-HEXYNE	65.135	549.00	1.2793	183.65	549.00	23.91
535	C6H10	3-HEXYNE	63.003	544.00	1.2821	170.05	544.00	22.76
536	C6H10O	CYCLOHEXANONE	73.500	629.15	1.1835	242.00	629.15	34.37
537	C6H10O	MESITYL OXIDE	66.120	600.00	1.2560	220.15	600.00	27.90
538	C6H10O2	epsilon-CAPROLACTONE	74.258	771.00	1.2164	271.85	771.00	40.97
539	C6H10O2	ETHYL METHACRYLATE	60.700	577.00	1.2526	223.15	577.00	24.41
540	C6H10O2	n-PROPYL ACRYLATE	61.442	569.00	1.2250	273.15	569.00	24.75
541	C6H10O3	ETHYLACETOACETATE	65.450	643.00	1.1590	234.15	643.00	31.79
542	C6H10O3	PROPIONIC ANHYDRIDE	70.260	618.00	1.2180	228.15	618.00	31.50
543	C6H10O4	ADIPIC ACID	53.506	809.00	1.2150	425.50	611.00	------
544	C6H10O4	DIETHYL OXALATE	68.210	646.00	1.2447	232.55	646.00	31.57
545	C6H10O4	ETHYLENE GLYCOL DIACETATE	69.847	653.00	1.2290	242.15	653.00	33.01
546	C6H10O4	ETHYLIDENE DIACETATE	61.901	635.00	1.1853	292.00	635.00	29.20

222

Table 9-1 SURFACE TENSION - ORGANIC COMPOUNDS (continued)

NO	FORMULA	NAME	sigma = A (1 - T/T$_c$)^n			(sigma - dynes/cm, T - K)		
			A	T$_C$	n	TMIN	TMAX	sigma @ 25 C
547	C6H11N	HEXANENITRILE	55.631	622.05	1.0856	192.85	622.05	27.39
548	C6H11NO	epsilon-CAPROLACTAM	82.316	806.00	1.2340	342.36	806.00	------
549	C6H11NO	CYCLOHEXANONE OXIME	40.583	715.00	1.1760	363.15	715.00	------
550	C6H12	CYCLOHEXANE	65.097	553.54	1.2553	279.69	553.54	24.65
551	C6H12	2,3-DIMETHYL-1-BUTENE	53.555	500.00	1.2620	115.89	500.00	17.05
552	C6H12	2,3-DIMETHYL-2-BUTENE	58.028	524.00	1.2682	198.92	524.00	19.96
553	C6H12	3,3-DIMETHYL-1-BUTENE	46.753	480.00	1.2500	157.95	480.00	13.90
554	C6H12	2-ETHYL-1-BUTENE	56.890	512.00	1.2793	141.61	512.00	18.62
555	C6H12	trans-3-METHYL-2-PENTENE	53.423	521.00	1.2222	134.70	521.00	18.92
556	C6H12	1-HEXENE	52.174	504.03	1.1952	133.39	504.03	17.89
557	C6H12	cis-2-HEXENE	57.590	513.00	1.2678	132.00	513.00	19.10
558	C6H12	trans-2-HEXENE	54.325	513.00	1.2640	140.17	513.00	18.08
559	C6H12	cis-3-HEXENE	54.335	509.00	1.2583	135.33	509.00	17.93
560	C6H12	trans-3-HEXENE	54.660	509.00	1.2830	159.73	509.00	17.64
561	C6H12	METHYLCYCLOPENTANE	61.630	532.79	1.2757	130.73	532.79	21.65
562	C6H12	2-METHYL-1-PENTENE	53.950	507.00	1.2644	137.42	507.00	17.58
563	C6H12	2-METHYL-2-PENTENE	54.090	514.00	1.2739	138.07	514.00	17.91
564	C6H12	3-METHYL-1-PENTENE	50.560	495.00	1.2275	120.20	495.00	16.30
565	C6H12	3-METHYL-cis-2-PENTENE	54.970	515.00	1.2486	138.31	515.00	18.67
566	C6H12	4-METHYL-1-PENTENE	50.140	496.00	1.2485	119.51	496.00	15.92
567	C6H12	4-METHYL-cis-2-PENTENE	50.348	499.00	1.2528	138.30	499.00	16.10
568	C6H12	4-METHYL-trans-2-PENTENE	49.880	501.00	1.2519	132.35	501.00	16.08
569	C6H12N2	TRIETHYLENEDIAMINE	45.640	655.00	1.2098	434.25	655.00	------
570	C6H12O	BUTYL VINYL ETHER	59.214	536.00	1.2350	181.15	536.00	21.71
571	C6H12O	CYCLOHEXANOL	65.400	625.15	1.0360	296.60	625.15	33.42
572	C6H12O	1-HEXANAL	56.186	579.00	1.0665	217.15	579.00	25.97
573	C6H12O	ETHYL ISOPROPYL KETONE	67.820	567.00	1.4124	200.00	567.00	23.64
574	C6H12O	2-HEXANONE	56.500	587.05	1.1345	217.35	587.05	25.28
575	C6H12O	3-HEXANONE	57.740	582.82	1.1689	217.50	582.82	24.99
576	C6H12O	METHYL ISOBUTYL KETONE	57.130	571.40	1.2040	189.15	571.40	23.50
577	C6H12O2	n-PENTYL FORMATE	58.344	576.00	1.1332	199.65	576.00	25.54
578	C6H12O2	n-BUTYL ACETATE	59.720	579.65	1.2180	199.65	579.15	24.78
579	C6H12O2	sec-BUTYL ACETATE	58.720	561.00	1.2327	174.15	561.00	23.06
580	C6H12O2	tert-BUTYL ACETATE	55.450	545.00	1.1928	283.15	545.00	21.56
581	C6H12O2	ETHYL n-BUTYRATE	57.850	571.00	1.1950	175.15	571.00	23.94
582	C6H12O2	ETHYL ISOBUTYRATE	57.440	553.15	1.1991	185.00	553.15	22.70
583	C6H12O2	ISOBUTYL ACETATE	54.690	561.00	1.1440	174.30	561.00	22.97
584	C6H12O2	n-PROPYL PROPIONATE	59.950	578.00	1.2512	197.25	578.00	24.19
585	C6H12O2	CYCLOHEXYL PEROXIDE	65.333	685.00	1.1900	253.15	685.00	33.10
586	C6H12O2	DIACETONE ALCOHOL	62.110	606.00	1.0930	229.15	606.00	29.63
587	C6H12O2	2-ETHYL BUTYRIC ACID	92.612	655.00	1.2644	258.15	655.00	42.97
588	C6H12O2	n-HEXANOIC ACID	56.520	667.00	1.2610	270.15	667.00	26.78
589	C6H12O3	2-ETHOXYETHYL ACETATE	74.837	597.00	1.2828	211.45	597.00	30.80
590	C6H12O3	HYDROXYCAPROIC ACID	95.360	758.00	1.2183	334.00	758.00	------
591	C6H12O3	PARALDEHYDE	62.800	579.00	1.2400	285.75	579.00	25.61
592	C6H12O3	sec-BUTYL GLYCOLATE	--------	--------	--------	--------	--------	------
593	C6H12S	THIACYCLOHEPTANE	71.710	640.07	1.2222	292.14	640.07	33.32
594	C6H13N	CYCLOHEXYLAMINE	69.090	615.00	1.2045	255.45	615.00	31.08
595	C6H13N	HEXAMETHYLENEIMINE	76.609	615.00	1.2067	236.15	615.00	34.41
596	C6H14	2,2-DIMETHYLBUTANE	47.567	488.78	1.1708	174.28	488.78	15.80
597	C6H14	2,3-DIMETHYLBUTANE	50.185	499.98	1.2019	145.19	499.98	16.87
598	C6H14	n-HEXANE	56.081	507.43	1.2843	177.84	507.43	17.98
599	C6H14	2-METHYLPENTANE	50.054	497.50	1.1892	119.55	497.50	16.87
600	C6H14	3-METHYLPENTANE	53.437	504.43	1.2432	110.25	504.43	17.58
601	C6H14N2O2	LYSINE	40.873	821.00	1.1416	498.00	821.00	------
602	C6H14O	2-ETHYL-1-BUTANOL	53.603	580.00	1.0860	158.75	580.00	24.48
603	C6H14O	1-HEXANOL	50.920	611.35	1.0106	228.55	611.35	25.90
604	C6H14O	2-HEXANOL	51.290	586.20	1.0435	223.00	586.20	24.44
605	C6H14O	2-METHYL-1-PENTANOL	62.783	582.00	1.2222	273.15	582.00	26.10
606	C6H14O	4-METHYL-2-PENTANOL	47.398	574.40	1.0098	183.00	574.40	22.63
607	C6H14O	n-BUTYL ETHYL ETHER	56.280	531.00	1.2453	170.15	531.00	20.16
608	C6H14O	DIISOPROPYL ETHER	53.573	500.05	1.2488	187.65	500.05	17.26
609	C6H14O	DI-n-PROPYL ETHER	54.960	530.60	1.2254	149.95	530.60	19.99
610	C6H14O	METHYL tert-PENTYL ETHER	56.972	534.00	1.2516	160.00	534.00	20.49
611	C6H14O2	ACETAL	56.877	541.00	1.2073	173.15	541.00	21.63
612	C6H14O2	2-BUTOXYETHANOL	63.199	600.00	1.2222	283.15	600.00	27.29
613	C6H14O2	1,6-HEXANEDIOL	72.618	670.00	1.1511	315.15	670.00	------
614	C6H14O2	HEXYLENE GLYCOL	32.313	621.00	1.0970	223.15	621.00	15.77
615	C6H14O2S	DI-n-PROPYL SULFONE	70.990	763.00	1.2518	303.00	763.00	------
616	C6H14O3	DIETHYLENE GLYCOL DIMETHYL ETHER	66.547	604.00	1.2116	203.15	432.91	29.18
617	C6H14O3	DIPROPYLENE GLYCOL	67.130	654.00	1.1190	233.00	654.00	33.97
618	C6H14O3	2-(2-ETHOXYETHOXY)ETHANOL	69.240	632.00	1.1770	195.15	632.00	32.67
619	C6H14O3	TRIMETHYLOLPROPANE	63.900	709.00	0.7540	331.15	709.00	------
620	C6H14O4	TRIETHYLENE GLYCOL	91.649	700.00	1.2222	293.15	700.00	46.51
621	C6H14O6	SORBITOL	107.220	959.00	1.2999	370.85	568.15	------
622	C6H14S	n-HEXYLMERCAPTAN	69.020	623.00	1.4600	192.62	623.00	26.67
623	C6H14S	BUTYL-ETHYL-SULFIDE	59.057	609.00	1.2222	178.03	609.00	25.96
624	C6H14S	ISOPROPYL-SULFIDE	54.149	585.71	1.2222	170.45	585.71	22.70

223

Table 9-1 SURFACE TENSION - ORGANIC COMPOUNDS (continued)

NO	FORMULA	NAME	sigma = A (1 - T/T$_c$)^n			(sigma - dynes/cm, T - K)		
			A	T$_c$	n	TMIN	TMAX	sigma @ 25 C
625	C6H14S	METHYL-PENTYL-SULFIDE	60.755	587.98	1.2222	179.16	587.98	25.59
626	C6H14S	PROPYL-SULFIDE	58.576	609.73	1.2222	170.45	609.73	25.78
627	C6H14S2	PROPYL-DISULFIDE	61.425	673.00	1.2222	187.68	673.00	30.04
628	C6H15Al	TRIETHYL ALUMINUM	---------	--------	-------	-------	-------	------
629	C6H15Al2Cl3	ETHYL ALUMINUM SESQUICHLORIDE	---------	--------	-------	-------	--------	------
630	C6H15N	DIISOPROPYLAMINE	56.620	523.10	1.2856	176.85	523.10	19.13
631	C6H15N	DI-n-PROPYLAMINE	55.920	555.80	1.2005	210.15	555.80	22.22
632	C6H15N	n-HEXYLAMINE	58.440	583.00	1.1440	251.85	583.00	25.76
633	C6H15N	TRIETHYLAMINE	54.140	535.15	1.2078	158.45	535.15	20.24
634	C6H15NO	6-AMINOHEXANOL	45.730	681.00	1.1630	331.00	508.00	------
635	C6H15NO2	DIISOPROPANOLAMINE	61.437	672.00	0.9994	318.15	672.00	------
636	C6H15NO3	TRIETHANOLAMINE	70.106	787.00	0.9200	294.35	787.00	45.24
637	C6H15N3	N-AMINOETHYL PIPERAZINE	77.056	708.00	1.1990	254.15	708.00	40.01
638	C6H15O4P	TRIETHYL PHOSPHATE	64.870	794.00	1.6660	216.00	794.00	29.61
639	C6H16N2	HEXAMETHYLENEDIAMINE	33.344	663.00	1.1509	313.95	663.00	------
640	C6H18N3OP	HEXAMETHYL PHOSPHORAMIDE	---------	--------	-------	-------	--------	------
641	C6H18N4	TRIETHYLENE TETRAMINE	80.917	718.00	1.1753	285.15	718.00	43.07
642	C6H18OSi2	HEXAMETHYLDISILOXANE	45.660	518.70	1.2720	204.93	518.70	15.39
643	C6H18O3Si3	HEXAMETHYLCYCLOTRISILOXANE	---------	--------	-------	-------	--------	------
644	C6H19NSi2	HEXAMETHYLDISILAZANE	44.590	544.00	1.1280	293.15	544.00	18.20
645	C7H3ClF3NO2	4-CHLORO-3-NITROBENZOTRIFLUORIDE	65.451	686.00	1.2440	293.15	686.00	32.20
646	C7H3Cl2F3	2,4-DICHLOROBENZOTRIFLUORIDE	81.487	646.00	1.2624	247.55	646.00	37.30
647	C7H3Cl2NO	3,4-DICHLOROPHENYL ISOCYANATE	64.240	733.00	1.3556	316.15	733.00	------
648	C7H4ClF3	p-CHLOROBENZOTRIFLUORIDE	49.806	601.00	1.2300	237.15	601.00	21.44
649	C7H4Cl2O	m-CHLOROBENZOYL CHLORIDE	104.480	724.00	1.2536	280.00	724.00	53.72
650	C7H4F3NO2	3-NITROBENZOTRIFLUORIDE	84.480	667.00	1.2728	272.00	667.00	39.75
651	C7H5ClO	BENZOYL CHLORIDE	73.897	697.00	1.1577	272.65	697.00	38.72
652	C7H5ClO2	o-CHLOROBENZOIC ACID	140.720	792.00	1.2577	415.15	792.00	------
653	C7H5Cl3	BENZOTRICHLORIDE	78.070	737.00	1.2500	268.40	737.00	40.84
654	C7H5F3	BENZOTRIFLUORIDE	62.820	565.00	1.2470	244.14	565.00	24.65
655	C7H5N	BENZONITRILE	74.700	699.35	1.1910	260.40	699.35	38.54
656	C7H5NO	PHENYL ISOCYANATE	83.147	648.00	1.2330	243.15	648.00	38.88
657	C7H5N3O6	2,4,6-TRINITROTOLUENE	95.740	795.00	1.2222	353.15	795.00	------
658	C7H6Cl2	BENZYL DICHLORIDE	78.814	731.00	1.3215	257.00	731.00	39.43
659	C7H6Cl2	2,4-DICHLOROTOLUENE	75.685	705.00	1.2560	259.65	705.00	37.94
660	C7H6N2O4	2,4-DINITROTOLUENE	100.050	814.00	1.2790	343.00	743.00	------
661	C7H6N2O4	2,5-DINITROTOLUENE	97.335	814.00	1.2447	325.65	814.00	------
662	C7H6N2O4	2,6-DINITROTOLUENE	97.335	770.00	1.2447	339.00	770.00	------
663	C7H6N2O4	3,4-DINITROTOLUENE	88.010	842.00	1.2475	332.00	842.00	------
664	C7H6N2O4	3,5-DINITROTOLUENE	94.457	814.00	1.2337	365.65	814.00	------
665	C7H6O	BENZALDEHYDE	74.680	695.00	1.1930	247.15	695.00	38.27
666	C7H6O2	BENZOIC ACID	73.400	751.00	1.1060	395.52	751.00	------
667	C7H6O2	p-HYDROXYBENZALDEHYDE	---------	--------	-------	-------	--------	------
668	C7H6O2	SALICYLALDEHYDE	81.519	680.00	1.1372	266.15	680.00	42.29
669	C7H6O3	SALICYLIC ACID	---------	--------	-------	-------	--------	------
670	C7H7Br	p-BROMOTOLUENE	70.000	699.00	1.2740	299.95	699.00	------
671	C7H7Cl	BENZYL CHLORIDE	79.060	686.00	1.3380	234.15	686.00	36.86
672	C7H7Cl	o-CHLOROTOLUENE	68.880	656.00	1.2166	236.65	656.00	32.95
673	C7H7Cl	p-CHLOROTOLUENE	61.700	660.00	1.1330	280.65	660.00	31.23
674	C7H7F	p-FLUOROTOLUENE	65.276	590.48	1.2222	216.36	590.48	27.64
675	C7H7NO	FORMANILIDE	69.190	787.00	1.0310	323.15	787.00	------
676	C7H7NO2	m-NITROTOLUENE	76.680	734.00	1.2224	289.20	734.00	40.55
677	C7H7NO2	o-NITROTOLUENE	80.510	720.00	1.2560	269.98	720.00	41.14
678	C7H7NO2	p-NITROTOLUENE	72.160	736.00	1.1234	324.75	736.00	------
679	C7H7NO3	o-NITROANISOLE	84.800	782.00	1.2880	283.60	782.00	45.69
680	C7H8	TOLUENE	66.850	591.79	1.2456	178.18	591.79	27.93
681	C7H8	1,3,5-CYCLOHEPTATRIENE	71.678	593.90	1.2222	193.66	593.90	30.57
682	C7H8O	ANISOLE	75.870	641.65	1.2400	235.65	641.65	34.96
683	C7H8O	BENZYL ALCOHOL	109.710	677.00	1.8820	257.85	677.00	36.79
684	C7H8O	m-CRESOL	64.620	705.85	1.0779	285.39	705.85	35.76
685	C7H8O	o-CRESOL	71.010	697.55	1.1600	304.19	697.55	------
686	C7H8O	p-CRESOL	66.850	704.65	1.0960	307.93	704.65	------
687	C7H8O2	GUAIACOL	---------	--------	-------	-------	--------	------
688	C7H8O2	p-METHOXYPHENOL	88.160	758.00	1.1983	329.00	758.00	------
689	C7H9N	BENZYLAMINE	77.060	683.50	1.1834	227.15	683.50	39.11
690	C7H9N	2,6-DIMETHYLPYRIDINE	65.092	623.75	1.0759	267.00	623.75	32.34
691	C7H9N	N-METHYLANILINE	66.549	701.55	1.0658	216.15	701.55	36.90
692	C7H9N	m-TOLUIDINE	67.210	709.15	1.0550	242.75	709.15	37.80
693	C7H9N	o-TOLUIDINE	74.850	694.15	1.1176	249.47	694.15	39.97
694	C7H9N	p-TOLUIDINE	67.900	693.15	1.0570	316.90	693.15	------
695	C7H10	2-NORBORNENE	46.965	583.00	1.2663	319.40	583.00	------
696	C7H10N2	TOLUENEDIAMINE	97.340	804.00	1.2324	371.25	804.00	------
697	C7H11NO	CYCLOHEXYL ISOCYANATE	111.030	633.00	1.2560	193.00	633.00	49.90
698	C7H12	1-HEPTYNE	56.162	559.69	1.2222	192.26	559.69	22.16
699	C7H12O2	n-BUTYL ACRYLATE	62.596	598.00	1.2737	208.55	598.00	25.98
700	C7H12O2	ISOBUTYL ACRYLATE	59.924	580.00	1.2200	212.00	580.00	24.85
701	C7H12O2	n-PROPYL METHACRYLATE	58.976	599.00	1.2437	223.00	599.00	25.04
702	C7H12O4	DIETHYL MALONATE	63.900	653.00	1.1685	224.25	653.00	31.33

Table 9-1 SURFACE TENSION - ORGANIC COMPOUNDS (continued)

NO	FORMULA	NAME	A	T_c	n	TMIN	TMAX	sigma @ 25 C
						sigma = A $(1 - T/T_c)^n$		(sigma - dynes/cm, T - K)
703	C7H14	CYCLOHEPTANE	65.680	604.30	1.3080	265.12	604.30	26.99
704	C7H14	1,1-DIMETYLCYCLOPENTANE	55.282	547.00	1.2168	203.36	547.00	21.20
705	C7H14	cis-1,2-DIMETHYLCYCLOPENTANE	59.628	565.15	1.2352	219.26	565.15	23.62
706	C7H14	trans-1,2-DIMETHYLCYCLOPENTANE	57.253	553.15	1.2830	155.58	553.15	21.20
707	C7H14	cis-1,3-DIMETHYLCYCLOPENTANE	57.453	551.00	1.2963	139.45	551.00	20.93
708	C7H14	trans-1,3-DIMETHYLCYCLOPENTANE	56.916	553.00	1.2642	139.18	553.00	21.38
709	C7H14	ETHYLCYCLOPENTANE	56.983	569.52	1.2044	134.71	569.52	23.33
710	C7H14	2-ETHYL-1-PENTENE	50.971	543.00	1.1622	168.00	543.00	20.20
711	C7H14	3-ETHYL-1-PENTENE	45.895	530.00	1.0882	145.67	530.00	18.66
712	C7H14	1-HEPTENE	53.336	537.29	1.2236	154.27	537.29	19.81
713	C7H14	cis-2-HEPTENE	56.150	549.00	1.2672	164.00	549.00	20.81
714	C7H14	trans-2-HEPTENE	53.880	543.00	1.2380	163.67	543.00	20.10
715	C7H14	cis-3-HEPTENE	54.310	545.00	1.2660	136.51	545.00	19.93
716	C7H14	trans-3-HEPTENE	52.373	540.00	1.2350	136.52	540.00	19.42
717	C7H14	METHYLCYCLOHEXANE	63.100	572.19	1.3530	146.58	572.19	23.30
718	C7H14	2-METHYL-1-HEXENE	53.920	538.00	1.2505	170.28	538.00	19.63
719	C7H14	3-METHYL-1-HEXENE	50.510	528.00	1.2188	145.00	528.00	18.33
720	C7H14	4-METHYL-1-HEXENE	52.353	534.00	1.2315	131.70	534.00	19.14
721	C7H14	2,3,3-TRIMETHYL-1-BUTENE	49.530	531.00	1.2717	163.30	531.00	17.36
722	C7H14O	DIISOPROPYL KETONE	93.226	576.00	1.2430	204.81	576.00	37.67
723	C7H14O	2-HEPTANONE	61.505	611.55	1.2810	238.15	611.55	26.12
724	C7H14O	1-HEPTANAL	53.260	603.00	1.0367	230.15	603.00	26.26
725	C7H14O	1-METHYLCYCLOHEXANOL	82.975	603.00	1.3886	299.15	603.00	------
726	C7H14O	cis-2-METHYLCYCLOHEXANOL	70.542	614.00	1.1135	280.15	614.00	33.65
727	C7H14O	trans-2-METHYLCYCLOHEXANOL	65.713	616.00	1.0753	269.15	616.00	32.26
728	C7H14O	cis-3-METHYLCYCLOHEXANOL	70.400	618.00	1.2660	267.65	618.00	30.58
729	C7H14O	trans-3-METHYLCYCLOHEXANOL	59.295	617.00	1.0562	272.65	617.00	29.53
730	C7H14O	cis-4-METHYLCYCLOHEXANOL	65.674	622.00	1.1605	298.15	622.00	30.79
731	C7H14O	trans-4-METHYLCYCLOHEXANOL	48.889	622.00	1.1524	298.15	622.00	23.04
732	C7H14O	5-METHYL-2-HEXANONE	59.742	601.00	1.2380	199.25	601.00	25.57
733	C7H14O2	n-BUTYL PROPIONATE	57.230	594.00	1.1977	183.63	594.00	24.83
734	C7H14O2	ETHYL ISOVALERATE	56.630	587.95	1.2590	173.85	587.95	23.24
735	C7H14O2	ISOPENTYL ACETATE	56.986	599.00	1.2628	194.65	599.00	23.88
736	C7H14O2	n-PENTYL ACETATE	57.400	598.00	1.1980	202.35	598.00	25.10
737	C7H14O2	n-PROPYL n-BUTYRATE	58.028	594.00	1.2340	177.95	594.00	24.55
738	C7H14O2	n-HEPTANOIC ACID	54.530	680.00	1.1706	265.83	680.00	27.75
739	C7H14O3	ETHYL-3-ETHOXYPROPIONATE	65.785	609.00	1.2953	223.00	609.00	27.53
740	C7H15Br	1-BROMOHEPTANE	59.360	651.00	1.2140	217.05	651.00	28.22
741	C7H15N	N-METHYLCYCLOHEXYLAMINE	69.902	622.00	1.2228	264.65	622.00	31.47
742	C7H16	2,2-DIMETHYLPENTANE	49.260	520.50	1.2150	149.34	520.50	17.53
743	C7H16	2,3-DIMETHYLPENTANE	51.376	537.35	1.1990	160.00	537.35	19.47
744	C7H16	2,4-DIMETHYLPENTANE	49.068	519.79	1.2002	153.91	519.79	17.64
745	C7H16	3,3-DIMETHYLPENTANE	51.570	536.40	1.2257	138.70	536.40	19.07
746	C7H16	3-ETHYLPENTANE	53.470	540.64	1.2315	154.55	540.64	19.92
747	C7H16	n-HEPTANE	53.640	540.26	1.2431	182.56	540.26	19.78
748	C7H16	2-METHYLHEXANE	50.354	530.37	1.1933	154.90	530.37	18.79
749	C7H16	3-METHYLHEXANE	51.385	535.25	1.2024	153.75	535.25	19.30
750	C7H16	2,2,3-TRIMETHYLBUTANE	51.540	531.17	1.2120	248.57	531.17	18.99
751	C7H16O	1-HEPTANOL	53.770	631.90	1.0960	239.15	631.90	26.71
752	C7H16O	2-HEPTANOL	62.531	588.00	1.2222	273.15	588.00	26.34
753	C7H16O	5-METHYL-1-HEXANOL	60.830	605.00	1.3068	293.15	605.00	25.05
754	C7H16O	ISOPROPYL-TERT-BUTYL-ETHER	-------	-------	-------	-------	-------	-------
755	C7H16S	n-HEPTYL MERCAPTAN	58.274	645.00	1.2280	229.92	645.00	27.20
756	C7H16S	BUTYL-PROPYL-SULFIDE	55.675	653.50	1.2222	206.66	653.50	26.44
757	C7H16S	ETHYL-PENTYL-SULFIDE	60.555	638.37	1.2222	206.66	638.37	28.06
758	C7H16S	HEXYL-METHYL-SULFIDE	60.555	638.37	1.2222	206.66	638.37	28.06
759	C7H17N	1-AMINOHEPTANE	47.560	607.00	1.0109	254.15	607.00	24.02
760	C8H4Cl2O2	ISOPHTHALOYL CHLORIDE	87.047	768.00	1.2227	317.00	768.00	------
761	C8H4O3	PHTHALIC ANHYDRIDE	78.000	791.00	1.1560	404.15	791.00	------
762	C8H6	ETHYNYLBENZENE	68.965	655.43	1.2222	242.53	655.43	32.85
763	C8H6O4	ISOPHTHALIC ACID	--------	--------	-------	-------	-------	------
764	C8H6O4	PHTHALIC ACID	72.537	800.00	1.1528	464.15	800.00	------
765	C8H6O4	TEREPHTHALIC ACID	--------	--------	-------	-------	-------	------
766	C8H6S	BENZOTHIOPHENE	85.360	754.00	1.3545	304.50	754.00	------
767	C8H7N	INDOLE	74.095	790.00	1.2168	273.68	790.00	41.63
768	C8H8	STYRENE	68.178	648.00	1.2222	273.15	648.00	32.10
769	C8H8	1,3,5,7-CYCLOOCTATETRAENE	66.094	642.55	1.2222	266.16	642.55	30.84
770	C8H8O	ACETOPHENONE	74.203	701.00	1.1637	292.80	701.00	38.95
771	C8H8O	p-TOLUALDEHYDE	72.856	698.00	1.2512	289.85	698.00	36.28
772	C8H8O2	METHYL BENZOATE	76.550	693.00	1.2795	260.75	693.00	37.27
773	C8H8O2	o-TOLUIC ACID	69.000	751.00	1.0010	376.85	751.00	------
774	C8H8O2	p-TOLUIC ACID	92.158	773.00	1.2280	452.75	773.00	------
775	C8H8O3	METHYL SALICYLATE	77.137	701.00	1.1836	265.15	701.00	40.04
776	C8H8O3	VANILLIN	92.620	777.00	1.3364	355.00	777.00	------
777	C8H9NO	ACETANILIDE	80.202	825.00	1.2570	386.65	825.00	------
778	C8H10	ETHYLBENZENE	66.000	617.17	1.2680	178.15	617.17	28.59
779	C8H10	m-XYLENE	65.700	617.05	1.2780	225.30	617.05	28.26
780	C8H10	o-XYLENE	66.100	630.37	1.2544	247.98	630.37	29.60

Table 9-1 SURFACE TENSION - ORGANIC COMPOUNDS (continued)

NO	FORMULA	NAME	A	T_c	n	TMIN	TMAX	sigma @ 25 C
				sigma = A $(1 - T/T_c)^n$			(sigma - dynes/cm, T - K)	
781	C8H10	p-XYLENE	64.850	616.26	1.2743	286.41	616.26	27.92
782	C8H10O	m-ETHYLPHENOL	--------	--------	-------	--------	-------	------
783	C8H10O	p-ETHYLPHENOL	70.344	716.45	1.2016	318.23	716.45	------
784	C8H10O	PHENETOLE	69.090	647.15	1.2230	243.63	647.15	32.47
785	C8H10O	2-PHENYLETHANOL	78.437	684.00	1.2627	247.00	684.00	38.07
786	C8H10O	2,3-XYLENOL	60.700	722.95	1.1528	345.71	722.95	------
787	C8H10O	2,4-XYLENOL	60.545	707.65	1.1362	345.71	707.65	------
788	C8H10O	2,5-XYLENOL	60.050	707.05	1.0065	347.99	707.05	------
789	C8H10O	2,6-XYLENOL	58.386	701.05	1.1676	318.76	701.05	------
790	C8H10O	3,4-XYLENOL	64.690	729.95	1.2430	338.25	729.95	------
791	C8H10O	3,5-XYLENOL	57.270	715.65	1.0706	336.59	715.65	------
792	C8H11N	N,N-DIMETHYLANILINE	71.698	687.15	1.2132	275.60	687.15	35.95
793	C8H11N	o-ETHYLANILINE	78.624	704.00	1.2410	226.55	704.00	39.69
794	C8H11N	2,4,6-TRIMETHYLPYRIDINE	70.920	653.00	1.2522	229.00	653.00	33.04
795	C8H11NO	p-PHENETIDINE	78.663	754.00	1.2423	277.00	754.00	42.10
796	C8H12	1,5-CYCLOOCTADIENE	74.714	645.00	1.4204	203.98	645.00	30.95
797	C8H12	VINYLCYCLOHEXENE	62.623	599.00	1.2354	164.00	599.00	26.75
798	C8H12O4	1,4-CYCLOHEXANEDICARBOXYLIC ACID	129.090	889.00	1.2283	585.65	889.00	------
799	C8H12O4	DIETHYL MALEATE	66.300	680.00	1.2510	264.35	680.00	32.21
800	C8H14O2	n-BUTYL METHACRYLATE	57.505	616.00	1.2234	223.00	616.00	25.59
801	C8H14O3	BUTYRIC ANHYDRIDE	63.250	639.00	1.1880	199.85	639.00	29.98
802	C8H14O4	DIETHYL SUCCINATE	64.108	660.00	1.2083	252.35	660.00	31.01
803	C8H16	CYCLOOCTANE	62.374	647.20	1.2222	287.60	647.20	29.33
804	C8H16	1,1-DIMETHYLCYCLOHEXANE	55.500	591.15	1.2153	239.66	591.15	23.65
805	C8H16	cis-1,2-DIMETHYLCYCLOHEXANE	60.891	606.15	1.3037	223.16	606.15	25.19
806	C8H16	trans-1,2-DIMETHYLCYCLOHEXANE	54.224	596.15	1.2017	184.99	596.15	23.57
807	C8H16	cis-1,3-DIMETHYLCYCLOHEXANE	54.170	591.15	1.2460	197.58	591.15	22.59
808	C8H16	trans-1,3-DIMETHYLCYCLOHEXANE	58.130	598.00	1.2726	183.05	598.00	24.15
809	C8H16	cis-1,4-DIMETHYLCYCLOHEXANE	55.437	598.15	1.2167	185.72	598.15	23.94
810	C8H16	trans-1,4-DIMETHYLCYCLOHEXANE	55.600	590.15	1.2846	236.21	590.15	22.52
811	C8H16	ETHYLCYCLOHEXANE	60.100	609.15	1.3020	161.84	609.15	25.05
812	C8H16	2-ETHYL-1-HEXENE	64.046	574.00	1.2630	173.00	574.00	25.38
813	C8H16	1-METHYL-1-ETHYLCYCLOPENTANE	58.826	582.00	1.2380	129.35	582.00	24.18
814	C8H16	1-OCTENE	54.040	566.60	1.2468	171.45	566.60	21.29
815	C8H16	trans-2-OCTENE	54.133	577.00	1.2570	185.45	577.00	21.70
816	C8H16	trans-3-OCTENE	52.313	574.00	1.2550	163.15	574.00	20.86
817	C8H16	trans-4-OCTENE	51.805	573.00	1.2506	179.37	573.00	20.67
818	C8H16	n-PROPYLCYCLOPENTANE	56.934	603.00	1.2410	155.81	603.00	24.42
819	C8H16	2,4,4-TRIMETHYL-1-PENTENE	50.687	553.00	1.2498	179.70	553.00	19.25
820	C8H16	2,4,4-TRIMETHYL-2-PENTENE	51.920	558.00	1.2696	166.84	558.00	19.68
821	C8H16O	2-ETHYLHEXANAL	64.562	607.00	1.2440	200.00	607.00	27.86
822	C8H16O	1-OCTANAL	54.716	621.00	1.0613	246.00	621.00	27.33
823	C8H16O	2-OCTANONE	57.170	624.00	1.2180	252.85	624.00	25.91
824	C8H16O2	n-BUTYL n-BUTYRATE	55.899	616.00	1.2081	181.15	616.00	25.13
825	C8H16O2	n-HEXYL ACETATE	57.660	618.00	1.2111	192.25	618.00	25.97
826	C8H16O2	ISOBUTYL ISOBUTYRATE	53.060	602.00	1.2510	192.45	602.00	22.56
827	C8H16O2	n-OCTANOIC ACID	58.370	692.00	1.2050	289.65	692.00	29.60
828	C8H16O4	DIETHYLENE GLYCOL ETHYL ETHER ACETATE	64.713	660.00	1.2158	248.15	660.00	31.16
829	C8H18	2,2-DIMETHYLHEXANE	48.360	549.80	1.1855	151.97	549.80	19.15
830	C8H18	2,3-DIMETHYLHEXANE	50.276	563.40	1.1887	273.00	563.40	20.53
831	C8H18	2,4-DIMETHYLHEXANE	49.322	553.50	1.1933	273.00	553.50	19.59
832	C8H18	2,5-DIMETHYLHEXANE	48.650	550.00	1.1851	182.00	550.00	19.28
833	C8H18	3,3-DIMETHYLHEXANE	49.385	562.00	1.1840	147.05	562.00	20.17
834	C8H18	3,4-DIMETHYLHEXANE	52.345	568.80	1.2166	272.04	568.80	21.21
835	C8H18	3-ETHYLHEXANE	53.200	565.40	1.2355	272.04	565.40	21.08
836	C8H18	3-ETHYL-2-METHYLPENTANE	52.428	567.00	1.2222	158.20	567.00	21.06
837	C8H18	3-METHYL-3-ETHYLPENTANE	51.235	576.50	1.1910	182.28	576.50	21.53
838	C8H18	2-METHYLHEPTANE	50.283	559.64	1.2017	164.16	559.64	20.15
839	C8H18	3-METHYLHEPTANE	54.070	563.67	1.2720	152.60	563.67	20.75
840	C8H18	4-METHYLHEPTANE	50.767	561.74	1.1962	152.20	561.74	20.54
841	C8H18	n-OCTANE	52.036	568.83	1.2168	216.38	568.83	21.08
842	C8H18	2,2,3-TRIMETHYLPENTANE	49.108	563.50	1.1784	160.89	563.50	20.22
843	C8H18	2,2,4-TRIMETHYLPENTANE	47.434	543.96	1.1975	165.78	543.96	18.32
844	C8H18	2,3,3-TRIMETHYLPENTANE	50.400	573.50	1.1863	172.22	573.50	21.11
845	C8H18	2,3,4-TRIMETHYLPENTANE	50.330	566.30	1.1898	163.95	566.30	20.68
846	C8H18	2,2,3,3-TETRAMETHYLBUTANE	50.241	567.80	1.2222	172.47	567.80	20.22
847	C8H18O	DI-n-BUTYL ETHER	51.566	581.00	1.1560	177.95	581.00	22.44
848	C8H18O	DI-sec-BUTYL ETHER	51.062	559.00	1.2090	173.15	559.00	20.32
849	C8H18O	DI-tert-BUTYL ETHER	48.540	550.00	1.2181	195.00	550.00	18.75
850	C8H18O	2-ETHYL-1-HEXANOL	57.075	640.25	1.1250	203.15	640.25	28.20
851	C8H18O	1-OCTANOL	51.414	652.50	1.0488	257.65	652.50	27.10
852	C8H18O	2-OCTANOL	51.080	637.15	1.0760	241.55	637.15	25.90
853	C8H18O2	DI-t-BUTYL PEROXIDE	43.575	547.00	1.1908	233.15	547.00	17.06
854	C8H18O2S	DI-n-BUTYL SULFONE	66.820	767.00	1.2280	318.00	767.00	------
855	C8H18O3	DIETHYLENE GLYCOL DIETHYL ETHER	70.276	624.00	1.4850	228.85	624.00	26.78
856	C8H18O3	DIETHYLENE GLYCOL MONOBUTYL EHTER	69.100	654.00	1.1679	205.15	654.00	33.95
857	C8H18O4	TRIETHYLENE GLYCOL DIMETHYL ETHER	69.860	651.00	1.4684	229.35	651.00	28.42
858	C8H18O5	TETRAETHYLENE GLYCOL	82.190	722.00	1.1674	268.15	722.00	44.13

226

Table 9-1 SURFACE TENSION - ORGANIC COMPOUNDS (continued)

			$\sigma = A(1 - T/T_c)^n$			(sigma - dynes/cm, T - K)		
NO	FORMULA	NAME	A	T_c	n	TMIN	TMAX	sigma @ 25 C
859	C8H18S	n-OCTYL MERCAPTAN	57.094	664.00	1.2106	223.95	664.00	27.75
860	C8H18S	tert-OCTYL MERCAPTAN	63.233	627.00	1.2492	199.00	627.00	28.24
861	C8H18S	BUTYL-SULFIDE	56.671	650.00	1.2222	209.86	650.00	26.77
862	C8H18S	ETHYL-HEXYL-SULFIDE	59.431	660.72	1.2222	209.86	660.72	28.54
863	C8H18S	HEPTYL-METHYL-SULFIDE	59.431	660.72	1.2222	209.86	660.72	28.54
864	C8H18S	PENTYL-PROPYL-SULFIDE	59.431	660.72	1.2222	209.86	660.72	28.54
865	C8H18S2	BUTYL-DISULFIDE	58.993	704.16	1.2222	202.16	704.16	30.10
866	C8H19N	DI-n-BUTYLAMINE	55.910	607.50	1.2420	211.15	607.50	24.18
867	C8H19N	DIISOBUTYLAMINE	50.604	580.00	1.1932	203.15	580.00	21.39
868	C8H19N	n-OCTYLAMINE	59.006	627.00	1.2148	272.75	627.00	26.94
869	C8H23N5	TETRAETHYLENEPENTAMINE	77.410	774.00	1.1850	243.00	774.00	43.50
870	C8H24O4Si4	OCTAMETHYLCYCLOTETRASILOXANE	42.930	586.50	1.2097	290.80	586.50	18.19
871	C9H4O5	TRIMELLITIC ANHYDRIDE	67.588	890.00	1.1577	438.15	890.00	------
872	C9H6N2O2	TOLUENE DIISOCYANATE	95.252	737.00	1.2493	287.04	737.00	49.84
873	C9H7N	ISOQUINOLINE	85.486	803.15	1.3043	299.45	803.15	------
874	C9H7N	QUINOLINE	73.900	782.15	1.1510	258.25	782.15	42.53
875	C9H7NO	8-HYDROXYQUINOLINE	52.866	788.00	1.1953	348.00	788.00	------
876	C9H8	INDENE	41.266	687.00	1.3545	271.70	687.00	19.09
877	C9H8O	2-METHYLBENZOFURAN	69.020	698.00	1.2497	290.00	698.00	34.40
878	C9H10	INDANE	69.245	684.90	1.2434	221.74	684.90	34.02
879	C9H10	cis-PROPENYLBENZENE	63.597	664.60	1.2222	211.47	664.60	30.72
880	C9H10	trans-PROPENYLBENZENE	63.597	664.60	1.2222	243.82	664.60	30.72
881	C9H10	alpha-METHYLSTYRENE	70.239	654.00	1.3161	249.95	654.00	31.53
882	C9H10	m-METHYLSTYRENE	71.955	657.00	1.2600	186.81	657.00	33.58
883	C9H10	o-METHYLSTYRENE	72.850	659.00	1.2800	204.58	659.00	33.70
884	C9H10	p-METHYLSTYRENE	84.956	665.00	1.5170	239.02	665.00	34.46
885	C9H10O2	BENZYL ACETATE	71.267	699.00	1.2490	221.65	699.00	35.58
886	C9H10O2	ETHYL BENZOATE	67.865	698.00	1.1940	238.45	698.00	34.89
887	C9H10O3	ETHYL VANILLIN	89.511	748.00	1.1810	350.65	748.00	------
888	C9H11NO	p-DIMETHYLAMINOBENZALDEHYDE	85.915	832.00	1.2880	348.00	832.00	------
889	C9H12	CUMENE	63.800	631.15	1.3056	177.14	681.15	27.69
890	C9H12	m-ETHYLTOLUENE	67.358	637.15	1.3610	177.61	637.15	28.54
891	C9H12	o-ETHYLTOLUENE	65.340	651.15	1.2899	192.35	651.15	29.66
892	C9H12	p-ETHYLTOLUENE	65.730	640.15	1.3441	210.83	640.15	28.30
893	C9H12	MESITYLENE	60.700	637.36	1.2240	228.46	637.36	28.05
894	C9H12	n-PROPYLBENZENE	63.487	638.38	1.2764	173.55	638.38	28.43
895	C9H12	1,2,3-TRIMETHYLBENZENE	65.040	664.53	1.2583	247.79	664.53	30.75
896	C9H12	1,2,4-TRIMETHYLBENZENE	63.374	649.13	1.2605	229.38	649.13	29.19
897	C9H12O	BENZYL ETHYL ETHER	63.576	662.00	1.1220	275.65	662.00	32.48
898	C9H12O	2-PHENYL-2-PROPANOL	12.718	660.00	1.1398	309.15	660.00	------
899	C9H12O2	CUMENE HYDROPEROXIDE	53.184	442.70	1.1406	264.26	442.70	24.52
900	C9H14O	ISOPHORONE	60.490	715.00	1.2280	265.05	715.00	31.18
901	C9H14O6	GLYCERYL TRIACETATE	72.581	704.00	1.2222	293.15	704.00	37.02
902	C9H16	1-NONYNE	55.613	610.81	1.2222	223.16	610.81	24.53
903	C9H16O4	AZELAIC ACID	71.105	811.00	1.2030	379.65	811.00	------
904	C9H18	BUTYLCYCLOPENTANE	56.070	625.05	1.2222	165.18	625.05	25.39
905	C9H18	cis,cis-1,3,5-TRIMETHYLCYCLOHEXANE	52.532	607.86	1.2222	223.46	607.86	23.04
906	C9H18	cis,trans-1,3,5-TRIMETHYLCYCLOHEXANE	53.119	602.20	1.2222	188.76	602.20	23.04
907	C9H18	ISOPROPYLCYCLOHEXANE	59.500	627.00	1.3010	183.76	627.00	25.70
908	C9H18	1-NONENE	54.590	593.25	1.2652	191.78	593.25	22.56
909	C9H18	n-PROPYLCYCLOHEXANE	55.810	639.15	1.2117	178.25	639.15	26.07
910	C9H18O	DIISOBUTYL KETONE	49.400	615.00	1.1660	227.17	615.00	22.80
911	C9H18O	1-NONANAL	53.955	640.00	1.0554	255.15	640.00	27.84
912	C9H18O2	n-BUTYL VALERATE	58.106	629.00	1.2230	180.35	629.00	26.48
913	C9H18O2	n-NONANOIC ACID	57.830	703.00	1.2000	285.65	703.00	29.82
914	C9H18O2	n-OCTYL FORMATE	59.570	645.00	1.2366	234.05	645.00	27.66
915	C9H20	3,3-DIETHYLPENTANE	53.366	610.05	1.2360	240.12	610.05	23.29
916	C9H20	2,2-DIMETHYL-3-ETHYLPENTANE	53.580	590.00	1.2701	173.68	590.00	21.92
917	C9H20	3-ETHYL-2,3-DIMETHYLPENTANE	51.774	606.80	1.2222	173.67	590.40	21.92
918	C9H20	2,4-DIMETHYL-3-ETHYLPENTANE	53.703	591.00	1.2494	150.79	591.00	22.34
919	C9H20	2,2-DIMETHYLHEPTANE	51.630	576.80	1.2767	160.15	576.80	20.39
920	C9H20	2,6-DIMETHYLHEPTANE	50.170	579.00	1.2380	170.25	579.00	20.49
921	C9H20	3-ETHYLHEPTANE	54.100	590.00	1.2544	158.25	590.00	22.37
922	C9H20	4-ETHYLHEPTANE	53.434	584.95	1.2222	159.96	584.95	22.36
923	C9H20	2,3-DIMETHYLHEPTANE	51.886	589.60	1.2222	160.16	589.60	21.93
924	C9H20	2,4-DIMETHYLHEPTANE	50.832	576.80	1.2222	160.16	576.80	20.89
925	C9H20	2,5-DIMETHYLHEPTANE	50.344	581.10	1.2222	160.16	581.10	20.89
926	C9H20	3,4-DIMETHYLHEPTANE	52.694	591.90	1.2222	170.26	591.90	22.38
927	C9H20	3,5-DIMETHYLHEPTANE	51.263	583.20	1.2222	170.26	583.20	21.37
928	C9H20	4,4-DIMETHYLHEPTANE	51.567	585.40	1.2222	170.26	585.40	21.60
929	C9H20	3-ETHYL-2-METHYLHEXANE	53.025	588.10	1.2222	160.16	588.10	22.34
930	C9H20	4-ETHYL-2-METHYLHEXANE	51.507	580.00	1.2222	160.16	580.00	21.32
931	C9H20	3-ETHYL-3-METHYLHEXANE	52.996	597.50	1.2222	160.16	597.50	22.77
932	C9H20	3-ETHYL-4-METHYLHEXANE	53.482	593.70	1.2222	160.16	593.70	22.80
933	C9H20	2,2,3-TRIMETHYLHEXANE	50.852	588.00	1.2222	153.16	588.00	21.42
934	C9H20	2,2,4-TRIMETHYLHEXANE	49.256	573.50	1.2222	153.00	573.50	20.09
935	C9H20	2,3,3-TRIMETHYLHEXANE	51.244	596.00	1.2222	156.36	596.00	21.95
936	C9H20	2,3,4-TRIMETHYLHEXANE	52.293	594.50	1.2222	156.36	594.50	22.33

227

Table 9-1 SURFACE TENSION - ORGANIC COMPOUNDS (continued)

| NO | FORMULA | NAME | sigma = A (1 - T/T_c)^n | | | (sigma - dynes/cm, T - K) | | |
			A	T_c	n	TMIN	TMAX	sigma @ 25 C
937	C9H20	2,3,5-TRIMETHYLHEXANE	50.389	579.20	1.2222	145.36	579.20	20.82
938	C9H20	2,4,4-TRIMETHYLHEXANE	49.963	581.50	1.2222	159.78	581.50	20.75
939	C9H20	3,3,4-TRIMETHYLHEXANE	52.532	602.30	1.2222	171.96	602.30	22.79
940	C9H20	2-METHYLOCTANE	53.220	586.75	1.2813	192.78	416.43	21.44
941	C9H20	3-METHYLOCTANE	53.410	590.15	1.2656	165.55	590.15	21.92
942	C9H20	4-METHYLOCTANE	53.280	587.65	1.2545	159.95	587.65	21.92
943	C9H20	n-NONANE	55.400	595.65	1.3027	219.63	595.65	22.43
944	C9H20	2,2,3,3-TETRAMETHYLPENTANE	51.726	610.85	1.2143	263.26	610.85	22.94
945	C9H20	2,2,3,4-TETRAMETHYLPENTANE	49.472	592.15	1.1857	152.06	592.15	21.57
946	C9H20	2,2,4,4-TETRAMETHYLPENTANE	50.150	571.35	1.2522	206.95	571.35	19.91
947	C9H20	2,3,3,4-TETRAMETHYLPENTANE	50.666	607.50	1.2222	152.00	592.60	21.55
948	C9H20	2,2,5-TRIMETHYLHEXANE	47.943	568.05	1.2025	167.39	568.05	19.59
949	C9H20O	2,6-DIMETHYL-4-HEPTANOL	56.633	603.00	1.1700	208.00	603.00	25.50
950	C9H20O	1-NONANOL	51.820	673.00	1.0633	268.15	673.00	27.81
951	C9H20O	2-NONANOL	56.568	623.00	1.1300	238.15	623.00	27.10
952	C9H20S	n-NONYL MERCAPTAN	56.780	681.00	1.2163	253.05	681.00	28.18
953	C9H20S	BUTYL-PENTYL-SULFIDE	58.401	681.56	1.2222	231.16	681.56	28.91
954	C9H20S	ETHYL-HEPTYL-SULFIDE	58.401	681.56	1.2222	231.16	681.56	28.91
955	C9H20S	HEXYL-PROPYL-SULFIDE	58.401	681.56	1.2222	231.16	681.56	28.91
956	C9H20S	METHYL-OCTYL-SULFIDE	58.401	681.56	1.2222	231.16	681.56	28.91
957	C9H21N	n-NONYLAMINE	57.638	648.00	1.2103	273.15	648.00	27.34
958	C9H21N	TRIPROPYLAMINE	50.127	577.50	1.1095	179.65	577.50	22.39
959	C10H6O8	PYROMELLITIC ACID	72.545	893.00	1.0994	554.00	893.00	------
960	C10H7Br	1-BROMONAPHTHALENE	74.670	824.00	1.1580	279.35	824.00	44.39
961	C10H7Cl	1-CHLORONAPHTHALENE	74.843	785.00	1.2608	269.15	785.00	40.98
962	C10H8	NAPHTHALENE	83.190	748.35	1.3896	353.43	748.35	------
963	C10H8	AZULENE	75.032	773.48	1.2222	173.66	773.48	41.38
964	C10H9N	QUINALDINE	83.951	773.00	1.3260	272.15	773.00	44.00
965	C10H10	m-DIVINYLBENZENE	71.730	692.00	1.2870	206.25	692.00	34.73
966	C10H10	1-METHYLINDENE	67.876	703.00	1.2502	350.00	703.00	------
967	C10H10	2-METHYLINDENE	71.115	684.00	1.2487	353.15	684.00	------
968	C10H10O4	DIMETHYL PHTHALATE	75.056	766.00	1.2544	272.15	766.00	40.44
969	C10H10O4	DIMETHYL TEREPHTHALATE	58.131	772.00	1.0817	413.80	772.00	------
970	C10H12	DICYCLOPENTADIENE	66.436	660.00	1.2850	307.00	443.00	------
971	C10H12	1,2,3,4-TETRAHYDRONAPHTHALENE	64.027	720.15	1.2312	237.38	720.15	33.16
972	C10H12O	ANETHOLE	65.866	723.00	1.2416	281.45	723.00	34.04
973	C10H12O4	DIALLYL MALEATE	69.237	693.00	1.2860	226.15	693.00	33.59
974	C10H14	n-BUTYLBENZENE	64.900	660.55	1.3630	185.30	660.55	28.63
975	C10H14	sec-BUTYLBENZENE	61.160	664.54	1.3111	197.72	664.54	28.02
976	C10H14	tert-BUTYLBENZENE	61.234	660.00	1.3242	215.27	660.00	27.63
977	C10H14	1,2,3,4-TETRAMETHYLBENZENE	59.341	695.10	1.2222	266.91	695.10	29.92
978	C10H14	m-CYMENE	61.278	657.00	1.2440	209.44	657.00	28.88
979	C10H14	o-CYMENE	65.650	662.00	1.2530	201.64	662.00	31.01
980	C10H14	p-CYMENE	60.484	653.15	1.2430	205.25	653.15	28.35
981	C10H14	m-DIETHYLBENZENE	64.340	663.00	1.2905	189.26	663.00	29.77
982	C10H14	o-DIETHYLBENZENE	64.546	668.00	1.3093	241.93	668.00	29.76
983	C10H14	p-DIETHYLBENZENE	63.820	657.96	1.3373	230.32	657.96	28.47
984	C10OH14	2-ETHYL-m-XYLENE	70.000	671.00	1.2580	256.89	671.00	33.42
985	C10H14	2-ETHYL-p-XYLENE	66.990	663.00	1.2570	219.52	663.00	31.62
986	C10H14	3-ETHYL-o-XYLENE	71.360	680.00	1.2940	223.64	680.00	33.82
987	C10H14	4-ETHYL-m-XYLENE	66.710	665.00	1.2610	210.27	665.00	31.51
988	C10H14	4-ETHYL-o-XYLENE	65.420	667.00	1.2450	206.22	667.00	31.29
989	C10OH14	5-ETHYL-m-XYLENE	63.980	655.00	1.2500	188.82	655.00	29.95
990	C10H14	ISOBUTYLBENZENE	59.283	650.15	1.2833	221.70	650.15	26.98
991	C10H14	1,2,3,5-TETRAMETHYLBENZENE	68.250	679.00	1.2330	249.46	679.00	33.46
992	C10H14	1,2,4,5-TETRAMETHYLBENZENE	65.100	675.15	1.2100	352.38	675.15	------
993	C10H14O	p-tert-BUTYLPHENOL	63.886	734.00	1.1990	371.56	734.00	------
994	C10H14O2	p-tert-BUTYLCATECHOL	67.400	776.00	1.1508	325.00	776.00	------
995	C10H15N	N,N-DIETYHLANILINE	67.300	702.00	1.2480	235.15	702.00	33.76
996	C10H15N	2,6-DIETHYLANILINE	59.370	678.00	1.1340	276.65	678.00	30.78
997	C10H16	CAMPHENE	58.591	638.00	1.2050	320.15	638.00	------
998	C10H16	D-LIMONENE	57.560	660.00	1.2750	199.00	660.00	26.75
999	C10H16	alpha-PHELLANDRENE	63.700	649.00	1.2470	220.00	649.00	29.58
1000	C10OH16	beta-PHELLANDRENE	63.988	648.00	1.3063	220.00	648.00	28.60
1001	C10H16	alpha-PINENE	56.630	632.00	1.2276	209.15	632.00	25.87
1002	C10H16	beta-PINENE	56.758	643.00	1.2012	211.61	643.00	26.85
1003	C10H16	alpha-TERPINENE	59.415	652.00	1.2370	220.00	652.00	27.90
1004	C10OH16	gamma-TERPINENE	63.190	661.00	1.2470	220.00	661.00	29.91
1005	C10H16	TERPINOLENE	61.543	672.00	1.3040	200.00	458.15	28.65
1006	C10H16O	CAMPHOR	13.758	709.00	1.2838	453.25	709.00	------
1007	C10H18	1-DECYNE	55.505	632.49	1.2222	229.16	632.49	25.46
1008	C10H18	cis-DECAHYDRONANPHTALENE	64.760	702.25	1.2950	230.20	702.25	31.66
1009	C10H18	trans-DECAHYDRONAPHTHALENE	61.470	687.05	1.2970	242.79	687.05	29.38
1010	C10H18O4	SEBACIC ACID	77.720	815.00	1.2078	407.65	815.00	------
1011	C10H20	n-BUTYLCYCLOHEXANE	59.040	667.00	1.3513	198.42	667.00	26.51
1012	C10H20	1-CYCLOPENTYLPENTANE	55.912	647.49	1.2222	190.16	647.49	26.30
1013	C10H20	1-DECENE	54.880	617.05	1.2818	206.89	617.05	23.55
1014	C10H20O	1-DECANAL	54.777	657.00	1.1030	267.15	657.00	28.11

228

Table 9-1 SURFACE TENSION - ORGANIC COMPOUNDS (continued)

NO	FORMULA	NAME	sigma = A (1 - T/T$_c$)^n			(sigma - dynes/cm, T - K)		
			A	T$_c$	n	TMIN	TMAX	sigma @ 25 C
1015	C10H20O2	n-DECANOIC ACID	59.320	713.00	1.2056	304.75	713.00	------
1016	C10H20O2	2-ETHYLHEXYL ACETATE	40.382	639.00	1.2290	180.15	639.00	18.65
1017	C10H20O2	ISOPENTYL ISOVALERATE	51.394	637.00	1.2013	215.00	637.00	24.08
1018	C10H22	n-DECANE	55.777	618.45	1.3198	243.49	618.45	23.41
1019	C10H22	2-METHYLNONANE	51.634	610.00	1.2440	198.50	610.00	22.41
1020	C10H22	3-METHYLNONANE	52.155	613.00	1.2140	188.35	613.00	23.23
1021	C10H22	4-METHYLNONANE	54.353	610.00	1.2726	174.45	610.00	23.14
1022	C10H22	5-METHYLNONANE	52.893	610.00	1.2325	185.45	610.00	23.13
1023	C10H22	3-ETHYLOCTANE	51.170	613.60	1.2222	185.46	613.60	22.69
1024	C10H22	4-ETHYLOCTANE	50.788	609.60	1.2222	185.46	609.60	22.35
1025	C10H22	2,2-DIMETHYLOCTANE	48.840	602.00	1.2130	225.00	602.00	21.31
1026	C10H22	2,3-DIMETHYLOCTANE	50.216	613.20	1.2222	219.16	613.20	22.25
1027	C10H22	2,4-DIMETHYLOCTANE	49.337	599.40	1.2222	219.16	599.40	21.28
1028	C10H22	2,5-DIMETHYLOCTANE	49.674	603.00	1.2222	219.16	603.00	21.58
1029	C10H22	2,6-DIMETHYLOCTANE	50.539	603.10	1.2222	219.16	603.10	21.96
1030	C10H22	2,7-DIMETHYLOCTANE	50.743	602.90	1.2222	219.16	602.90	22.04
1031	C10H22	3,3-DIMETHYLOCTANE	49.823	612.10	1.2222	219.16	612.10	22.03
1032	C10H22	3,4-DIMETHYLOCTANE	50.749	614.00	1.2222	219.16	614.00	22.52
1033	C10H22	3,5-DIMETHYLOCTANE	50.039	606.30	1.2222	219.16	606.30	21.88
1034	C10H22	3,6-DIMETHYLOCTANE	50.073	608.30	1.2222	219.16	608.30	21.98
1035	C10H22	4,4-DIMETHYLOCTANE	52.493	606.90	1.2222	219.16	606.90	22.98
1036	C10H22	4,5-DIMETHYLOCTANE	50.695	612.20	1.2222	219.16	612.20	22.42
1037	C10H22	4-PROPYLHEPTANE	50.566	601.00	1.2222	219.16	601.00	21.88
1038	C10H22	4-ISOPROPYLHEPTANE	50.346	607.60	1.2222	219.16	607.60	22.07
1039	C10H22	3-ETHYL-2-METHYLHEPTANE	50.435	610.90	1.2222	219.16	610.90	22.25
1040	C10H22	4-ETHYL-2-METHYLHEPTANE	49.931	601.80	1.2222	219.16	601.80	21.64
1041	C10H22	5-ETHYL-2-METHYLHEPTANE	50.091	606.70	1.2222	219.16	606.70	21.92
1042	C10H22	3-ETHYL-3-METHYLHEPTANE	50.723	620.00	1.2222	219.16	620.00	22.76
1043	C10H22	4-ETHYL-3-METHYLHEPTANE	51.035	614.30	1.2222	219.16	614.30	22.66
1044	C10H22	3-ETHYL-5-METHYLHEPTANE	50.307	606.60	1.2222	219.16	606.60	22.01
1045	C10H22	3-ETHYL-4-METHYLHEPTANE	51.056	615.50	1.2222	219.16	615.50	22.72
1046	C10H22	4-ETHYL-4-METHYLHEPTANE	50.610	615.70	1.2222	219.16	615.70	22.53
1047	C10H22	2,2,3-TRIMETHYLHEPTANE	49.408	611.70	1.2222	219.16	611.70	21.83
1048	C10H22	2,2,4-TRIMETHYLHEPTANE	47.914	594.50	1.2222	219.16	594.50	20.46
1049	C10H22	2,2,5-TRIMETHYLHEPTANE	48.013	598.00	1.2222	219.16	598.00	20.65
1050	C10H22	2,2,6-TRIMETHYLHEPTANE	47.624	593.40	1.2222	219.16	593.40	20.29
1051	C10H22	2,3,3-TRIMETHYLHEPTANE	50.016	617.50	1.2222	219.16	617.50	22.34
1052	C10H22	2,3,4-TRIMETHYLHEPTANE	50.326	613.70	1.2222	219.16	613.70	22.32
1053	C10H22	2,3,5-TRIMETHYLHEPTANE	49.802	612.80	1.2222	219.16	612.80	22.05
1054	C10H22	2,3,6-TRIMETHYLHEPTANE	49.130	604.10	1.2222	219.16	604.10	21.39
1055	C10H22	2,4,4-TRIMETHYLHEPTANE	48.600	600.30	1.2222	219.16	600.30	21.00
1056	C10H22	2,4,5-TRIMETHYLHEPTANE	49.661	606.90	1.2222	219.16	606.90	21.74
1057	C10H22	2,4,6-TRIMETHYLHEPTANE	48.431	590.30	1.2222	219.16	590.30	20.50
1058	C10H22	2,5,5-TRIMETHYLHEPTANE	48.648	602.90	1.2222	219.16	602.90	21.13
1059	C10H22	3,3,4-TRIMETHYLHEPTANE	50.574	622.10	1.2222	219.16	622.10	22.78
1060	C10H22	3,3,5-TRIMETHYLHEPTANE	50.298	609.50	1.2222	219.16	609.50	22.13
1061	C10H22	3,4,4-TRIMETHYLHEPTANE	50.528	620.90	1.2222	219.16	620.90	22.71
1062	C10H22	3,4,5-TRIMETHYLHEPTANE	50.683	612.80	1.2222	219.16	612.80	22.44
1063	C10H22	3-ISOPROPYL-2-METHYLHEXANE	50.588	623.40	1.2222	219.16	623.40	22.84
1064	C10H22	3,3-DIETHYLHEXANE	51.842	627.80	1.2222	219.16	627.80	23.59
1065	C10H22	3,4-DIETHYLHEXANE	51.504	618.80	1.2222	219.16	618.80	23.06
1066	C10H22	3-ETHYL-2,2-DIMETHYLHEXANE	49.612	611.70	1.2222	219.16	611.70	21.92
1067	C10H22	4-ETHYL-2,2-DIMETHYLHEXANE	48.442	594.60	1.2222	219.16	594.60	20.69
1068	C10H22	3-ETHYL-2,3-DIMETHYLHEXANE	51.119	626.80	1.2222	219.16	626.80	23.22
1069	C10H22	4-ETHYL-2,3-DIMETHYLHEXANE	50.885	617.30	1.2222	219.16	617.30	22.72
1070	C10H22	3-ETHYL-2,4-DIMETHYLHEXANE	50.842	616.10	1.2222	219.16	616.10	22.65
1071	C10H22	4-ETHYL-2,4-DIMETHYLHEXANE	52.798	620.90	1.2222	219.16	620.90	23.73
1072	C10H22	3-ETHYL-2,5-DIMETHYLHEXANE	49.556	603.50	1.2222	219.16	603.50	21.55
1073	C10H22	4-ETHYL-3,3-DIMETHYLHEXANE	51.109	625.70	1.2222	219.16	625.70	23.17
1074	C10H22	3-ETHYL-3,4-DIMETHYLHEXANE	51.063	624.50	1.2222	219.16	624.50	23.10
1075	C10H22	2,2,3,3-TETRAMETHYLHEXANE	52.620	623.00	1.2222	219.16	623.00	23.74
1076	C10H22	2,2,3,4-TETRAMETHYLHEXANE	49.594	620.40	1.2222	219.16	620.40	22.27
1077	C10H22	2,2,3,5-TETRAMETHYLHEXANE	48.203	601.30	1.2222	219.16	601.30	20.87
1078	C10H22	2,2,4,4-TETRAMETHYLHEXANE	47.598	610.20	1.2222	219.16	610.20	20.97
1079	C10H22	2,2,4,5-TETRAMETHYLHEXANE	47.685	598.50	1.2222	219.16	598.50	20.53
1080	C10H22	2,2,5,5-TETRAMETHYLHEXANE	47.325	581.40	1.2222	260.56	581.40	19.65
1081	C10H22	2,3,3,4-TETRAMETHYLHEXANE	50.974	633.10	1.2222	260.56	633.10	23.41
1082	C10H22	2,3,3,5-TETRAMETHYLHEXANE	48.923	610.10	1.2222	260.56	610.10	21.55
1083	C10H22	2,3,4,4-TETRAMETHYLHEXANE	50.212	626.60	1.2222	260.56	626.60	22.80
1084	C10H22	2,3,4,5-TETRAMETHYLHEXANE	49.923	613.20	1.2222	260.56	613.20	22.12
1085	C10H22	3,3,4,4-TETRAMETHYLHEXANE	51.471	646.70	1.2222	260.56	646.70	24.18
1086	C10H22	2,4-DIMETHYL-3-ISOPROPYLPENTANE	49.923	614.40	1.2222	191.46	614.40	22.17
1087	C10H22	3,3-DIETHYL-2-METHYLPENTANE	52.436	639.90	1.2222	191.46	639.90	24.36
1088	C10H22	3-ETHYL-2,2,3-TRIMETHYLPENTANE	51.465	646.00	1.2222	191.46	646.00	24.15
1089	C10H22	3-ETHYL-2,2,4-TRIMETHYLPENTANE	49.524	615.30	1.2222	191.46	615.30	22.03
1090	C10H22	3-ETHYL-2,3,4-TRIMETHYLPENTANE	51.821	642.30	1.2222	191.46	642.30	24.17
1091	C10H22	2,2,3,3,4-PENTAMETHYLPENTANE	50.752	643.80	1.2222	236.71	643.80	23.73
1092	C10H22	2,2,3,4,4-PENTAMETHYLPENTANE	48.501	627.30	1.2222	234.41	627.30	22.05

Table 9-1 SURFACE TENSION - ORGANIC COMPOUNDS (continued)

NO	FORMULA	NAME	A	T_c	n	TMIN	TMAX	sigma @ 25 C
			sigma = A $(1 - T/T_c)^n$			(sigma - dynes/cm, T - K)		
1093	C10H22O	1-DECANOL	51.260	690.00	1.0455	280.05	690.00	28.37
1094	C10H22O	DI-n-PENTYL ETHER	55.370	622.00	1.2600	203.72	622.00	24.33
1095	C10H22O	ISODECANOL	56.603	644.00	1.1567	213.15	644.00	27.58
1096	C10H22O5	TETRAETHYLENE GLYCOL DIMETHYL ETHER	68.870	705.00	1.3155	243.45	705.00	33.42
1097	C10H22S	n-DECYL MERCAPTAN	52.303	696.00	1.0740	247.56	696.00	28.69
1098	C10H22S	BUTYL-HEXYL-SULFIDE	57.348	701.03	1.2222	238.16	701.03	29.14
1099	C10H22S	ETHYL-OCTYL-SULFIDE	57.348	701.03	1.2222	238.16	701.03	29.14
1100	C10H22S	HEPTYL-PROPYL-SULFIDE	57.348	701.03	1.2222	238.16	701.03	29.14
1101	C10H22S	METHYL-NONYL-SULFIDE	57.348	701.03	1.2222	238.16	701.03	29.14
1102	C10H22S	PENTYL-SULFIDE	57.348	701.03	1.2222	238.16	701.03	29.14
1103	C10H22S2	PENTYL-DISULFIDE	55.654	726.94	1.2222	214.16	726.94	29.19
1104	C10H23N	n-DECYLAMINE	56.800	663.00	1.2020	288.85	663.00	27.70
1105	C11H10	1-METHYLNAPHTHALENE	76.000	772.04	1.3300	242.67	772.04	39.71
1106	C11H10	2-METHYLNAPHTHALENE	66.442	761.00	1.2634	307.73	761.00	------
1107	C11H14O2	n-BUTYL BENZOATE	76.990	724.00	1.3722	251.65	724.00	37.17
1108	C11H16	n-PENTYLBENZENE	60.140	679.90	1.2576	198.15	679.90	29.10
1109	C11H16O	p-tert-AMYLPHENOL	14.393	751.00	1.2001	366.00	751.00	------
1110	C11H20	1-UNDECYNE	55.255	650.99	1.2222	248.16	650.99	26.14
1111	C11H20O2	2-ETHYLHEXYL ACRYLATE	65.420	655.00	1.5320	183.15	655.00	25.80
1112	C11H22	1-UNDECENE	55.086	638.00	1.2928	223.99	638.00	24.40
1113	C11H22	1-CYCLOPENTYLHEXANE	55.435	667.67	1.2222	200.16	667.67	26.90
1114	C11H22	PENTYLCYCLOHEXANE	55.129	674.01	1.2222	215.66	674.01	27.00
1115	C11H22O	1-UNDECANAL	57.440	672.00	1.2158	273.15	672.00	28.16
1116	C11H24	n-UNDECANE	55.723	638.76	1.3237	247.57	638.76	24.24
1117	C11H24O	1-UNDECANOL	53.455	704.00	1.1233	289.05	704.00	28.79
1118	C11H24S	UNDECYL MERCAPTAN	51.037	710.00	1.0460	270.15	710.00	28.87
1119	C11H24S	BUTYL-HEPTYL-SULFIDE	56.267	717.91	1.2222	254.66	717.91	29.20
1120	C11H24S	DECYL-METHYL-SULFIDE	56.267	717.91	1.2222	254.66	717.91	29.20
1121	C11H24S	ETHYL-NONYL-SULFIDE	56.267	717.91	1.2222	254.66	717.91	29.20
1122	C11H24S	OCTYL-PROPYL-SULFIDE	56.267	717.91	1.2222	254.66	717.91	29.20
1123	C12H8O	DIBENZOFURAN	69.610	837.80	1.3552	355.65	837.80	------
1124	C12H9N	DIBENZOPYRROLE	154.520	899.00	1.2993	517.95	899.00	------
1125	C12H10	ACENAPHTHENE	73.260	803.15	1.2246	366.56	803.15	------
1126	C12H10	BIPHENYL	74.700	789.26	1.3110	342.37	789.26	------
1127	C12H10O	DIPHENYL ETHER	67.020	763.00	1.1240	300.10	763.00	------
1128	C12H11N	p-AMINODIPHENYL	61.507	817.00	1.2100	326.00	817.00	------
1129	C12H11N	DIPHENYLAMINE	78.530	817.00	1.3173	326.15	817.00	------
1130	C12H11N3	p-AMINOAZOBENZENE	54.119	877.00	1.2060	401.00	877.00	------
1131	C12H11N3	1,3-DIPHENYLTRIAZENE	52.730	845.00	1.2026	372.00	845.00	------
1132	C12H12	1,2-DIMETHYLNAPHTHALENE	70.122	775.34	1.2222	272.16	775.34	38.74
1133	C12H12	1,3-DIMETHYLNAPHTHALENE	68.177	773.76	1.2222	269.16	773.76	37.61
1134	C12H12	1,4-DIMETHYLNAPHTHALENE	66.652	776.78	1.2222	280.82	776.78	36.88
1135	C12H12	1,5-DIMETHYLNAPHTHALENE	68.131	773.47	1.2222	355.16	773.47	------
1136	C12H12	1,6-DIMETHYLNAPHTHALENE	68.083	770.60	1.2222	259.16	770.60	37.44
1137	C12H12	1,7-DIMETHYLNAPHTHALENE	68.083	770.60	1.2222	260.16	770.60	37.44
1138	C12H12	2,3-DIMETHYLNAPHTHALENE	68.265	777.78	1.2222	378.16	777.78	------
1139	C12H12	2,6-DIMETHYLNAPHTHALENE	72.295	777.00	1.3450	383.32	777.00	------
1140	C12H12	2,7-DIMETHYLNAPHTHALENE	72.399	778.00	1.3488	368.81	778.00	------
1141	C12H12	1-ETHYLNAPHTHALENE	71.249	776.00	1.2990	259.34	776.00	37.95
1142	C12H12	2-ETHYLNAPHTHALENE	66.271	774.90	1.2222	265.76	774.90	36.60
1143	C12H12N2	p-AMINODIPHENYLAMINE	50.830	867.00	1.1860	341.15	867.00	------
1144	C12H12N2	HYDRAZOBENZENE	72.160	792.00	1.1847	404.15	792.00	------
1145	C12H14	1,2,3-TRIMETHYLINDENE	85.685	726.00	1.2625	344.65	726.00	------
1146	C12H14O4	DIETHYL PHTHALATE	69.530	757.00	1.2605	269.15	757.00	36.99
1147	C12H16	CYCLOHEXYLBENZENE	168.480	744.00	1.3540	280.14	744.00	84.22
1148	C12H18	m-DIISOPROPYLBENZENE	72.975	684.00	1.6314	210.02	684.00	28.68
1149	C12H18	p-DIISOPROPYLBENZENE	58.560	689.00	1.2503	256.08	689.00	28.82
1150	C12H18	n-HEXYLBENZENE	59.904	698.00	1.2390	212.00	698.00	30.04
1151	C12H18	1,2,3-TRIETHYLBENZENE	56.967	684.37	1.2222	206.66	684.37	28.31
1152	C12H18	1,2,4-TRIETHYLBENZENE	56.967	684.37	1.2222	206.66	684.37	28.31
1153	C12H18	1,3,5-TRIETHYLBENZENE	56.910	682.28	1.2222	206.66	682.28	28.20
1154	C12H18	HEXAMETHYLBENZENE	53.430	758.00	1.2222	438.66	758.00	------
1155	C12H20O4	DIBUTYL MALEATE	--------	--------	--------	--------	--------	------
1156	C12H22	BICYCLOHEXYL	62.810	727.00	1.2676	276.78	727.00	32.17
1157	C12H22	1-DODECYNE	55.243	668.16	1.2222	254.16	668.16	26.83
1158	C12H23N	DICYCLOHEXYLAMINE	61.670	737.00	1.2011	273.05	737.00	33.09
1159	C12H24	1-DODECENE	55.320	657.00	1.3038	237.93	657.00	25.14
1160	C12H24	1-CYCLOPENTYLHEPTANE	55.550	679.00	1.2222	220.00	679.00	27.40
1161	C12H24	1-CYCLOHEXYLHEXANE	54.781	691.81	1.2222	263.60	691.81	27.50
1162	C12H24O	1-DODECANAL	57.310	685.00	1.2122	285.15	685.00	28.67
1163	C12H24O2	n-DODECANOIC ACID	54.234	734.00	1.1399	316.98	734.00	------
1164	C12H26	n-DODECANE	55.717	658.20	1.3325	263.57	658.20	24.94
1165	C12H26O	DI-n-HEXYL ETHER	53.657	658.00	1.2424	230.15	658.00	25.35
1166	C12H26O	1-DODECANOL	52.860	721.00	1.0993	296.95	721.00	29.40
1167	C12H26O3	DIETHYLENE GLYCOL DI-n-BUTYL ETHER	53.190	680.00	1.2080	212.95	680.00	26.49
1168	C12H26S	n-DODECYL MERCAPTAN	54.756	724.00	1.2062	265.15	724.00	28.87
1169	C12H26S	BUTYL-OCTYL-SULFIDE	55.178	733.68	1.2222	259.16	733.68	29.17
1170	C12H26S	DECYL-ETHYL-SULFIDE	55.178	733.68	1.2222	259.16	733.68	29.17

Table 9-1 SURFACE TENSION - ORGANIC COMPOUNDS (continued)

NO	FORMULA	NAME	sigma = A $(1 - T/T_c)^n$			(sigma - dynes/cm, T - K)		
			A	T_c	n	TMIN	TMAX	sigma @ 25 C
1171	C12H26S	HEXYL-SULFIDE	52.996	733.68	1.2222	259.16	733.68	28.02
1172	C12H26S	METHYL-UNDECYL-SULFIDE	55.178	733.68	1.2222	259.16	733.68	29.17
1173	C12H26S	NONYL-PROPYL-SULFIDE	55.178	733.68	1.2222	259.16	733.68	29.17
1174	C12H26S2	HEXYL-DISULFIDE	58.534	747.10	1.2222	225.16	747.10	31.41
1175	C12H27BO3	TRI-n-BUTYL BORATE	58.980	743.15	1.6214	203.15	743.15	25.68
1176	C12H27N	DODECYLAMINE	51.173	696.00	1.1306	301.47	696.00	------
1177	C12H27N	TRI-n-BUTYLAMINE	51.502	644.00	1.2014	203.00	644.00	24.40
1178	C13H10	FLUORENE	179.910	870.00	1.2405	387.94	870.00	------
1179	C13H10O	BENZOPHENONE	74.300	816.00	1.2040	321.35	816.00	------
1180	C13H12	DIPHENYLMETHANE	--------	--------	-------	--------	--------	------
1181	C13H14	1-PROPYLNAPHTHALENE	65.772	771.45	1.2222	264.69	771.45	36.20
1182	C13H14	2-PROPYLNAPHTHALENE	63.965	772.44	1.2222	270.16	772.44	35.24
1183	C13H14	2ETHYL-3-METHYLNAPHTHALENE	66.120	776.44	1.2222	344.16	776.44	------
1184	C13H14	2ETHYL-6-METHYLNAPHTHALENE	65.854	766.56	1.2222	318.16	766.56	------
1185	C13H14	2ETHYL-7-METHYLNAPHTHALENE	65.854	766.56	1.2222	318.16	766.56	------
1186	C13H20	n-HEPTYLBENZENE	60.630	714.00	1.3117	225.15	714.00	29.84
1187	C13H24	1-TRIDECYNE	54.612	684.11	1.2222	268.16	684.11	27.13
1188	C13H26	1-TRIDECENE	55.800	675.00	1.3240	250.08	675.00	25.79
1189	C13H26	1-CYCLOPENTYLOCTANE	55.032	702.06	1.2222	229.16	702.06	28.00
1190	C13H26	1-CYCLOHEXYLHEPTANE	54.380	708.63	1.2222	242.66	708.63	27.90
1191	C13H26O	1-TRIDECANAL	56.536	700.00	1.2341	288.15	700.00	28.50
1192	C13H26O2	n-BUTYL NONANOATE	59.827	652.00	1.1960	235.15	652.00	28.80
1193	C13H26O2	METHYL DODECANOATE	118.420	712.00	1.2934	278.15	712.00	58.70
1194	C13H28	n-TRIDECANE	55.610	675.80	1.3363	267.76	675.80	25.55
1195	C13H28O	1-TRIDECANOL	53.740	731.00	1.1260	303.75	731.00	------
1196	C13H28S	BUTYL-NONYL-SULFIDE	54.059	748.42	1.2222	271.16	748.42	29.05
1197	C13H28S	DECYL-PROPYL-SULFIDE	54.059	748.42	1.2222	271.16	748.42	29.05
1198	C13H28S	DODECYL-METHYL-SULFIDE	54.059	748.42	1.2222	271.16	748.42	29.05
1199	C13H28S	ETHYL-UNDECYL-SULFIDE	54.059	748.42	1.2222	271.16	748.42	29.05
1200	C13H28S	1-TRIDECANETHIOL	55.538	742.13	1.2222	282.04	742.13	29.64
1201	C14H8O2	ANTHRAQUINONE	80.877	900.00	1.2056	559.15	653.05	------
1202	C14H10	ANTHRACENE	62.814	873.00	1.0356	488.93	873.00	------
1203	C14H10	DIPHENYLACETYLENE	54.940	832.00	1.2446	335.65	832.00	------
1204	C14H10	PHENANTHRENE	70.558	869.25	1.0968	372.38	869.25	------
1205	C14H12	cis-STILBENE	79.260	757.00	1.3447	268.15	535.00	40.43
1206	C14H12	trans-STILBENE	103.840	820.00	1.2763	397.35	820.00	------
1207	C14H12O2	BENZYL BENZOATE	88.200	820.00	1.4280	292.55	820.00	46.26
1208	C14H14	1,1-DIPHENYLETHANE	70.320	775.00	1.3230	255.20	775.00	36.99
1209	C14H14	1,2-DIPHENYLETHANE	65.100	780.00	1.2100	324.34	780.00	------
1210	C14H14O	DIBENZYL ETHER	73.160	777.00	1.2594	276.75	777.00	39.77
1211	C14H16	1-n-BUTYLNAPHTHALENE	64.118	792.00	1.2675	253.43	792.00	35.24
1212	C14H16	2-BUTYLNAPHTHALENE	62.820	780.96	1.2222	268.16	780.96	34.90
1213	C14H22	n-OCTYLBENZENE	60.535	729.00	1.3279	237.15	729.00	30.11
1214	C14H22	1,2,3,4-TETRAETHYLBENZENE	54.146	708.20	1.2222	284.96	708.20	27.77
1215	C14H22	1,2,3,5-TETRAETHYLBENZENE	54.157	707.52	1.2222	284.16	707.52	27.75
1216	C14H22	1,2,4,5-TETRAETHYLBENZENE	54.092	706.85	1.2222	283.16	706.85	27.69
1217	C14H22O	p-tert-OCTYLPHENOL	72.812	765.00	1.2182	358.55	765.00	------
1218	C14H28	1-TETRADECENE	56.297	692.00	1.3469	260.30	692.00	26.35
1219	C14H28	1-CYCLOPENTYLNONANE	55.175	716.95	1.2222	244.16	716.95	28.60
1220	C14H28	1-CYCLOHEXYLOCTANE	54.354	723.61	1.2222	253.46	723.61	28.40
1221	C14H28O2	n-TETRADECANOIC ACID	57.420	756.00	1.2205	327.55	756.00	------
1222	C14H30	n-TETRADECANE	56.436	692.40	1.3658	279.01	692.40	26.15
1223	C14H30O	1-TETRADECANOL	55.890	741.00	1.2455	310.65	741.00	------
1224	C14H30S	BUTYL-DECYL-SULFIDE	52.818	762.23	1.2222	276.16	762.23	28.80
1225	C14H30S	DODECYL-ETHYL-SULFIDE	52.818	762.23	1.2222	276.16	762.23	28.80
1226	C14H30S	HEPTYL-SULFIDE	53.072	762.23	1.2222	276.16	762.23	28.94
1227	C14H30S	METHYL-TRIDECYL-SULFIDE	52.818	762.23	1.2222	276.16	762.23	28.80
1228	C14H30S	PROPYL-UNDECYL-SULFIDE	52.818	762.23	1.2222	276.16	762.23	28.80
1229	C14H30S	1-TETRADECANETHIOL	54.266	753.80	1.2222	279.26	753.80	29.33
1230	C14H30S2	HEPTYL-DISULFIDE	55.577	765.96	1.2222	235.16	765.96	30.42
1231	C14H31N	TETRADECYLAMINE	56.020	722.30	1.2000	311.34	722.30	------
1232	C15H10N2O2	DIPHENYLMETHANE-4,4'-DIISOCYANATE	90.450	802.00	1.2459	311.20	802.00	------
1233	C15H16O	p-CUMYLPHENOL	74.850	834.00	1.2052	346.00	834.00	------
1234	C15H16O2	BISPHENOL A	18.628	849.00	1.1520	426.15	849.00	------
1235	C15H18	1-PENTYLNAPHTHALENE	61.734	793.32	1.2222	251.16	793.32	34.70
1236	C15H18	2-PENTYLNAPHTHALENE	60.558	797.48	1.2222	269.16	797.48	34.17
1237	C15H24	n-NONYLBENZENE	60.405	741.00	1.3371	249.00	741.00	30.35
1238	C15H24O	2,6-DI-tert-BUTYL-p-CRESOL	61.300	720.00	1.1812	344.00	720.00	------
1239	C15H24O	NONYLPHENOL	60.246	757.00	1.2046	279.15	757.00	32.96
1240	C15H28	1-PENTADECYNE	49.958	711.41	1.2222	283.16	711.41	25.72
1241	C15H30	1-PENTADECENE	56.540	708.00	1.3604	269.42	708.00	26.88
1242	C15H30	1-CYCLOPENTYLDECANE	54.839	730.64	1.2222	251.03	730.64	28.89
1243	C15H30	1-CYCLOHEXYLNONANE	54.414	737.79	1.2222	262.96	737.79	28.90
1244	C15H30O2	PENTADECANOIC ACID	56.710	766.00	1.2383	325.68	766.00	------
1245	C15H32	n-PENTADECANE	56.547	706.80	1.3710	283.10	706.80	26.68
1246	C15H32O	1-PENTADECANOL	--------	--------	-------	--------	--------	------
1247	C15H32S	BUTYL-UNDECYL-SULFIDE	51.640	775.15	1.2222	284.16	775.15	28.53
1248	C15H32S	DODECYL-PROPYL-SULFIDE	51.640	775.15	1.2222	284.16	775.15	28.53

Table 9-1 SURFACE TENSION - ORGANIC COMPOUNDS (continued)

NO	FORMULA	NAME	sigma = A (1 - T/T$_c$)^n			(sigma - dynes/cm, T - K)		
			A	T$_c$	n	TMIN	TMAX	sigma @ 25 C
1249	C15H32S	ETHYL-TRIDECYL-SULFIDE	51.640	775.15	1.2222	284.16	775.15	28.53
1250	C15H32S	METHYL-TETRADECYL-SULFIDE	51.640	775.15	1.2222	284.16	775.15	28.53
1251	C15H32S	1-PENTADECANETHIOL	52.885	764.77	1.2222	290.93	764.77	28.91
1252	C16H10	FLUORANTHENE	73.346	905.00	1.2678	383.33	905.00	------
1253	C16H10	PYRENE	10.213	936.00	1.3257	423.81	936.00	------
1254	C16H12	1-PHENYLNAPHTHALENE	75.634	849.00	1.2700	318.15	849.00	------
1255	C16H20	1-n-HEXYLNAPHTHALENE	62.260	813.00	1.2993	255.15	813.00	34.39
1256	C16H22O4	DIBUTYL PHTHALATE	59.582	781.00	1.2185	238.15	781.00	33.16
1257	C16H26	n-DECYLBENZENE	60.054	753.00	1.3430	258.77	753.00	30.52
1258	C16H26	PENTAETHYLBENZENE	51.327	723.64	1.2222	327.66	723.64	------
1259	C16H30	1-HEXADECYNE	48.986	724.26	1.2222	288.16	724.26	25.62
1260	C16H32	n-DECYLCYCLOHEXANE	58.214	751.25	1.3552	271.42	751.25	29.34
1261	C16H32	1-CYCLOPENTYLUNDECANE	49.422	743.30	1.2222	263.16	743.30	26.41
1262	C16H32	1-HEXADECENE	56.860	722.00	1.3758	277.51	722.00	27.32
1263	C16H32O2	n-HEXADECANOIC ACID	56.400	776.00	1.1865	335.66	776.00	------
1264	C16H34	n-HEXADECANE	56.990	720.60	1.3929	291.32	720.60	27.09
1265	C16H34O	DI-n-OCTYL ETHER	54.773	707.00	1.2694	265.55	707.00	27.33
1266	C16H34O	1-HEXADECANOL	56.593	761.00	1.2484	322.35	761.00	------
1267	C16H34S	BUTYL-DODECYL-SULFIDE	50.324	787.27	1.2222	288.16	787.27	28.13
1268	C16H34S	ETHYL-TETRADECYL-SULFIDE	46.992	791.68	1.2222	288.16	791.68	26.37
1269	C16H34S	METHYL-PENTADECYL-SULFIDE	50.324	787.27	1.2222	288.16	787.27	28.13
1270	C16H34S	OCTYL-SULFIDE	53.454	787.27	1.2222	288.16	787.27	29.88
1271	C16H34S	PROPYL-TRIDECYL-SULFIDE	50.324	787.27	1.2222	288.16	787.27	28.13
1272	C16H34S	1-HEXADECANETHIOL	51.474	774.68	1.2222	290.93	774.68	28.42
1273	C16H34S2	OCTYL-DISULFIDE	52.384	784.46	1.2222	244.16	784.46	29.20
1274	C17H28	n-UNDECYLBENZENE	57.520	764.00	1.2860	268.00	764.00	30.45
1275	C17H32	1-HEPTADECYNE	47.930	736.21	1.2222	295.16	736.21	25.41
1276	C17H34	1-CYCLOPENTYLDODECANE	48.368	755.17	1.2222	268.16	755.17	26.18
1277	C17H34	1-CYCLOHEXYLUNDECANE	48.294	761.74	1.2222	278.96	761.74	26.32
1278	C17H34	1-HEPTADECENE	57.230	736.00	1.3949	284.40	736.00	27.73
1279	C17H36	n-HEPTADECANE	57.080	733.37	1.3978	295.13	733.37	27.52
1280	C17H36O	1-HEPTADECANOL	58.462	770.00	1.2968	327.05	770.00	------
1281	C17H36S	BUTYL-TRIDECYL-SULFIDE	48.874	798.63	1.2222	294.16	798.63	27.61
1282	C17H36S	ETHYL-PENTADECYL-SULFIDE	48.874	798.63	1.2222	294.16	798.63	27.61
1283	C17H36S	HEXADECYL-METHYL-SULFIDE	48.874	798.63	1.2222	294.16	798.63	27.61
1284	C17H36S	PROPYL-TETRADECYL-SULFIDE	48.874	798.63	1.2222	294.16	798.63	27.61
1285	C17H36S	1-HEPTADECANETHIOL	49.992	786.01	1.2222	300.37	786.01	------
1286	C18H12	CHRYSENE	103.900	979.00	1.2983	531.15	979.00	------
1287	C18H14	m-TERPHENYL	72.743	924.85	1.3188	360.00	924.85	------
1288	C18H14	o-TERPHENYL	75.200	890.95	1.4467	329.35	890.95	------
1289	C18H14	p-TERPHENYL	69.100	925.95	1.2550	485.00	925.95	------
1290	C18H15P	TRIPHENYLPHOSPHINE	79.380	1008.00	1.7010	354.40	1008.00	------
1291	C18H15O4P	TRIPHENYL PHOSPHATE	--------	--------	-------	--------	-------	------
1292	C18H16N2	N,N'-DIPHENYL-p-PHENYLENEDIAMINE	66.628	906.00	1.1893	409.00	906.00	------
1293	C18H22	2,3-DIMETHYL-2,3-DIPHENYLBUTANE	114.000	805.00	1.2673	392.15	805.00	------
1294	C18H22O2	DICUMYL PEROXIDE	63.970	884.00	1.2267	311.15	669.00	------
1295	C18H30	n-DODECYLBENZENE	53.700	774.26	1.2256	275.93	774.26	29.59
1296	C18H30	HEXAETHYLBENZENE	48.417	734.78	1.2222	401.16	734.78	------
1297	C18H32O2	LINOLEIC ACID	59.492	775.00	1.2225	268.15	628.00	32.86
1298	C18H34	1-OCTADECYNE	46.757	747.33	1.2222	300.16	747.33	------
1299	C18H34O2	OLEIC ACID	53.290	781.00	1.0386	286.53	633.15	32.34
1300	C18H34O4	DIBUTYL SEBACATE	58.985	768.00	1.3803	263.95	768.00	29.94
1301	C18H34O4	DIHEXYL ADIPATE	77.034	767.00	1.8216	259.35	473.10	31.43
1302	C18H36	1-CYCOPENTYLTRIDECANE	47.276	766.47	1.2222	278.16	766.47	25.89
1303	C18H36	1-CYCLOHEXYLDODECANE	47.259	772.83	1.2222	285.66	772.83	26.05
1304	C18H36	1-OCTADECENE	57.541	748.00	1.4107	290.76	748.00	28.08
1305	C18H36O2	STEARIC ACID	53.650	799.00	1.1470	342.75	799.00	------
1306	C18H38	n-OCTADECANE	57.460	745.26	1.4092	301.33	745.26	------
1307	C18H38O	DINONYL ETHER	51.330	736.00	1.2393	273.00	736.00	26.97
1308	C18H38O	1-OCTADECANOL	54.600	777.00	1.2336	331.05	777.00	------
1309	C18H38S	BUTYL-TETRADECYL-SULFIDE	47.446	810.53	1.2222	298.16	810.53	------
1310	C18H38S	ETHYL-HEXADECYL-SULFIDE	47.446	810.53	1.2222	298.16	810.53	------
1311	C18H38S	HEPTADECYL-METHYL-SULFIDE	47.446	810.53	1.2222	298.16	810.53	------
1312	C18H38S	NONYL-SULFIDE	47.446	810.53	1.2222	298.16	810.53	------
1313	C18H38S	PENTADECYL-PROPYL-SULFIDE	47.446	810.53	1.2222	298.16	810.53	------
1314	C18H38S	1-OCTADECANETHIOL	48.455	795.36	1.2222	300.93	795.36	------
1315	C18H38S2	NONYL-DISULFIDE	48.948	802.30	1.2222	252.16	802.30	27.74
1316	C19H26	1-n-NONYLNAPHTHALENE	63.614	849.00	1.3443	284.15	849.00	35.56
1317	C19H32	n-TRIDECYLBENZENE	57.457	783.00	1.3335	283.15	783.00	30.32
1318	C19H36	1-NONADECYNE	45.548	758.94	1.2222	306.16	758.94	------
1319	C19H36O2	METHYL OLEATE	58.060	764.00	1.2480	293.05	764.00	31.31
1320	C19H38	1-CYCLOPENTYLTETRADECANE	43.255	772.00	1.2222	282.00	772.00	23.82
1321	C19H38	1-CYCLOHEXYLTRIDECANE	46.112	783.38	1.2222	291.66	783.38	25.68
1322	C19H38	1-NONADECENE	57.565	760.00	1.4170	296.55	760.00	28.42
1323	C19H38O2	NONADECANOIC ACID	--------	--------	-------	--------	-------	------
1324	C19H40	n-NONADECANE	56.440	755.93	1.3820	305.33	755.93	------
1325	C19H40O	1-NONADECANOL	--------	--------	-------	--------	-------	------
1326	C19H40S	BUTYL-PENTADECYL-SULFIDE	45.858	821.75	1.2222	303.16	821.75	------

Table 9-1 SURFACE TENSION - ORGANIC COMPOUNDS (continued)

NO	FORMULA	NAME	sigma = A (1 - T/T$_c$)^n			(sigma - dynes/cm, T - K)		
			A	T$_c$	n	TMIN	TMAX	sigma @ 25 C
1327	C19H40S	ETHYL-HEPTADECYL-SULFIDE	45.858	821.75	1.2222	303.16	821.75	------
1328	C19H40S	HEXADECYL-PROPYL-SULFIDE	45.858	821.75	1.2222	303.16	821.75	------
1329	C19H40S	METHYL-OCTADECYL-SULFIDE	45.858	821.75	1.2222	303.16	821.75	------
1330	C19H40S	1-NONADECANETHIOL	46.874	805.29	1.2222	307.04	805.29	------
1331	C20H16	TRIPHENYLETHYLENE	69.769	908.00	1.2540	342.15	908.00	------
1332	C20H28	1-n-DECYLNAPHTHALENE	63.397	859.00	1.3830	288.15	859.00	35.16
1333	C20H30O2	ABIETIC ACID	123.310	832.00	1.2628	446.65	649.70	------
1334	C20H31N	DEHYDROABIETYLAMINE	59.800	863.00	1.2362	317.65	863.00	------
1335	C20H34	1-PHENYLTETRADECANE	56.447	792.00	1.2222	289.16	792.00	31.69
1336	C20H38	1-EICOSYNE	44.237	769.79	1.2222	309.16	769.79	------
1337	C20H40	1-CYCLOPENTYLPENTADECANE	41.285	780.00	1.2222	290.00	780.00	22.92
1338	C20H40	1-CYCLOHEXYLTETRADECANE	44.868	792.82	1.2222	297.16	792.82	25.21
1339	C20H40	1-EICOSENE	57.948	771.00	1.4355	301.76	771.00	------
1340	C20H42	n-EICOSANE	56.480	767.04	1.3870	309.59	767.04	------
1341	C20H42O	1-EICOSANOL	54.160	792.00	1.1424	338.55	792.00	------
1342	C20H42S	BUTYL-HEXADECYL-SULFIDE	44.179	832.33	1.2222	308.16	832.33	------
1343	C20H42S	DECYL-SULFIDE	44.179	832.33	1.2222	308.16	832.33	------
1344	C20H42S	ETHYL-OCTADECYL-SULFIDE	44.179	832.33	1.2222	308.16	832.33	------
1345	C20H42S	HEPTADECYL-PROPYL-SULFIDE	44.179	832.33	1.2222	308.16	832.33	------
1346	C20H42S	METHYL-NONADECYL-SULFIDE	44.179	832.33	1.2222	308.16	832.33	------
1347	C20H42S	1-EICOSANETHIOL	45.124	814.57	1.2222	310.37	814.57	------
1348	C20H42S2	DECYL-DISULFIDE	45.290	820.08	1.2222	259.16	820.08	26.07
1349	C21H21O4P	TRI-o-CRESYL PHOSPHATE	--------	--------	--------	--------	--------	------
1350	C21H36	1-PHENYLPENTADECANE	56.257	800.00	1.2222	295.16	800.00	31.82
1351	C21H42	1-CYCLOPENTYLHEXADECANE	43.395	797.25	1.2222	294.16	797.25	24.48
1352	C21H42	1-CYCLOHEXYLPENTADECANE	43.604	803.46	1.2222	302.16	803.46	------
1353	C22H38	1-PHENYLHEXADECANE	56.434	808.00	1.2222	300.16	808.00	------
1354	C22H44	1-CYCLOHEXYLHEXADECANE	42.204	813.42	1.2222	306.76	813.42	------
1355	C22H44O2	n-BUTYL STEARATE	59.530	764.00	1.2210	299.45	764.00	------
1356	C24H38O4	DIISOOCTYL PHTHALATE	60.310	851.00	1.2500	254.00	851.00	35.18
1357	C24H38O4	DIOCTYL PHTHALATE	64.087	806.00	1.4979	223.15	806.00	32.08
1358	C24H42O	DINONYLPHENOL	15.776	886.00	1.3130	350.00	886.00	------
1359	C26H20	TETRAPHENYLETHYLENE	150.450	996.00	1.2986	496.15	996.00	------
1360	C28H46O4	DIISODECYL PHTHALATE	60.970	887.00	1.3300	227.59	723.00	35.36

sigma - surface tension, dynes/cm

A, T$_c$, and n - regression coefficients of chemical compound

T - temperature, K

TMAX - maximum temperature, K

TMIN - minimun temperature, K

Table 9-2 SURFACE TENSION - INORGANIC COMPOUNDS

$$\text{sigma} = A \left(1 - T/T_c\right)^n \qquad (\text{sigma - dynes/cm, T - K})$$

NO	FORMULA	NAME	A	T_c	n	TMIN	TMAX	T, K	sigma @ T
1	Ag	SILVER	1152.840	7480.00	1.2222	1234.00	7480.00	1234.00	924.85
2	AgCl	SILVER CHLORIDE	--------	--------	-------	--------	--------	------	------
3	AgI	SILVER IODIDE	--------	--------	-------	--------	--------	------	------
4	Al	ALUMINUM	1033.331	7151.00	1.2222	933.00	7151.00	933.00	871.03
5	AlB3H12	ALUMINUM BOROHYDRIDE	142.470	513.77	1.2222	209.15	513.77	298.15	49.30
6	AlBr3	ALUMINUN BROMIDE	--------	--------	-------	--------	--------	------	------
7	AlCl3	ALUMINUM CHLORIDE	85.633	629.00	1.5519	465.70	629.00	465.70	10.56
8	AlF3	ALUMINUM FLUORIDE	--------	--------	-------	--------	--------	------	------
9	AlI3	ALUMINUM IODIDE	--------	--------	-------	--------	--------	------	------
10	Al2O3	ALUMINUM OXIDE	1166.560	5335.00	1.2222	2325.00	5335.00	2325.00	579.56
11	Al2S3O12	ALUMINUM SULFATE	--------	--------	-------	--------	--------	------	------
12	Ar	ARGON	38.230	150.86	1.2927	83.80	150.86	83.80	13.40
13	As	ARSENIC	--------	--------	-------	--------	--------	------	------
14	AsBr3	ARSENIC TRIBROMIDE	94.144	789.01	1.2222	306.15	789.01	306.15	51.66
15	AsCl3	ARSENIC TRICHLORIDE	84.591	654.00	1.2222	255.15	654.00	298.15	40.20
16	AsF3	ARSENIC TRIFLUORIDE	114.538	530.21	1.2222	267.25	530.21	298.15	41.72
17	AsF5	ARSENIC PENTAFLUORIDE	46.608	357.73	1.2222	193.35	357.73	298.15	5.21
18	AsH3	ARSINE	66.864	373.00	1.3006	156.28	373.00	298.15	8.28
19	AsI3	ARSENIC TRIIODIDE	--------	--------	-------	--------	--------	------	------
20	As2O3	ARSENIC TRIOXIDE	--------	--------	-------	--------	--------	------	------
21	At	ASTATINE	--------	--------	-------	--------	--------	------	------
22	Au	GOLD	1783.909	4398.00	1.2222	1337.33	4398.00	1337.33	1145.38
23	B	BORON	1636.592	7934.59	1.2222	2348.15	7934.59	2348.15	1065.83
24	BBr3	BORON TRIBROMIDE	69.048	581.00	1.2222	228.15	581.00	298.15	28.65
25	BCl3	BORON TRICHLORIDE	56.810	451.95	1.1470	166.15	451.95	298.15	16.50
26	BF3	BORON TRIFLUORIDE	65.247	260.90	1.2222	146.05	260.90	146.05	23.93
27	BH2CO	BORINE CARBONYL	61.008	340.03	1.2222	136.15	340.03	298.15	4.72
28	BH3O3	BORIC ACID	--------	--------	-------	--------	--------	------	------
29	B2D6	DEUTERODIBORANE	33.206	293.74	1.2222	100.00	293.74	100.00	19.97
30	B2H5Br	DIBORANE HYDROBROMIDE	55.196	466.98	1.2222	168.95	466.98	298.15	15.92
31	B2H6	DIBORANE	46.096	289.80	1.2222	107.65	289.80	107.65	26.13
32	B3N3H6	BORINE TRIAMINE	47.906	521.20	1.2222	214.95	521.20	298.15	16.98
33	B4H10	TETRABORANE	49.076	466.66	1.2222	153.25	466.66	298.15	14.13
34	B5H9	PENTABORANE	51.430	568.45	1.2222	226.35	568.45	298.15	20.73
35	B5H11	TETRAHYDROPENTABORANE	55.631	547.13	1.2222	225.00	547.13	298.15	21.25
36	B10H14	DECABORANE	--------	--------	-------	--------	--------	------	------
37	Ba	BARIUM	410.848	3572.13	1.2222	1000.15	3572.13	1000.15	274.99
38	Be	BERYLLIUM	2497.218	5199.80	1.2222	1560.15	5199.80	1560.15	1614.73
39	BeB2H8	BERYLLIUM BOROHYDRIDE	--------	--------	-------	--------	--------	------	------
40	BeBr2	BERYLLIUM BROMIDE	--------	--------	-------	--------	--------	------	------
41	BeCl2	BERYLLIUM CHLORIDE	--------	--------	-------	--------	--------	------	------
42	BeF2	BERYLLIUM FLUORIDE	--------	--------	-------	--------	--------	------	------
43	BeI2	BERYLLIUM IODIDE	--------	--------	-------	--------	--------	------	------
44	Bi	BISMUTH	445.093	4620.00	1.2222	544.15	4620.00	544.15	381.89
45	BiBr3	BISMUTH TRIBROMIDE	131.853	1220.00	1.2222	491.15	1220.00	491.15	70.25
46	BiCl3	BISMUTH TRICHLORIDE	141.198	1178.00	1.2222	503.15	1178.00	503.15	71.47
47	BrF5	BROMINE PENTAFLUORIDE	76.954	470.00	1.2222	211.75	470.00	298.15	22.50
48	Br2	BROMINE	93.910	584.15	1.1620	265.90	584.15	298.15	40.96
49	C	CARBON	--------	--------	-------	--------	--------	------	------
50	CCl2O	PHOSGENE	65.193	455.00	1.1600	145.37	455.00	298.15	18.95
51	CF2O	CARBONYL FLUORIDE	67.776	297.00	1.2222	161.89	297.00	161.89	25.88
52	CH4N2O	UREA	--------	--------	-------	--------	--------	------	------
53	CH4N2S	THIOUREA	--------	--------	-------	--------	--------	------	------
54	CNBr	CYANOGEN BROMIDE	89.778	545.03	1.2222	331.15	545.03	331.15	28.62
55	CNCl	CYANOGEN CHLORIDE	79.260	449.00	1.1873	266.65	449.00	298.15	21.71
56	CNF	CYANOGEN FLUORIDE	89.491	368.51	1.2222	150.00	368.51	298.15	11.83
57	CO	CARBON MONOXIDE	27.959	132.92	1.1330	68.15	132.92	68.15	12.38
58	COS	CARBONYL SULFIDE	75.960	378.80	1.4660	134.35	378.80	298.15	7.87
59	COSe	CARBON OXYSELENIDE	101.334	406.58	1.2222	150.00	406.58	298.15	20.15
60	CO2	CARBON DIOXIDE	79.970	304.19	1.2617	216.58	304.19	298.15	0.57
61	CS2	CARBON DISULFIDE	79.590	552.00	1.1909	161.58	552.00	298.15	31.56
62	CSeS	CARBON SELENOSULFIDE	100.793	576.53	1.2222	197.95	576.53	298.15	41.40
63	C2N2	CYANOGEN	79.777	399.90	1.2222	238.75	399.90	298.15	14.98
64	C3S2	CARBON SUBSULFIDE	--------	--------	-------	--------	--------	------	------
65	Ca	CALCIUM	599.497	3292.23	1.2222	1115.15	3292.23	1115.15	361.62
66	CaF2	CALCIUM FLUORIDE	--------	--------	-------	--------	--------	------	------
67	CbF5	COLUMBIUM FLUORIDE	--------	--------	-------	--------	--------	------	------
68	Cd	CADMIUM	919.379	2291.00	1.2222	594.05	2291.00	594.05	637.05
69	CdCl2	CADMIUM CHLORIDE	154.548	2019.79	1.2222	841.15	2019.79	841.15	80.01
70	CdF2	CADMIUM FLUORIDE	--------	--------	-------	--------	--------	------	------
71	CdI2	CADMIUM IODIDE	--------	--------	-------	--------	--------	------	------
72	CdO	CADMIUM OXIDE	--------	--------	-------	--------	--------	------	------
73	ClF	CHLORINE MONOFLUORIDE	79.825	282.32	1.2222	128.15	282.32	128.15	38.11
74	ClFO3	PERCHLORYL FLUORIDE	66.214	368.40	1.2957	125.41	368.40	298.15	7.74
75	ClF3	CHLORINE TRIFLUORIDE	81.618	459.39	1.2222	190.15	459.39	298.15	22.70
76	ClF5	CHLORINE PENTAFLUORIDE	65.891	415.90	1.2222	190.00	415.90	298.15	14.09
77	ClHO3S	CHLOROSULFONIC ACID	87.912	700.00	1.1833	193.15	700.00	298.15	45.59
78	ClHO4	PERCHLORIC ACID	51.132	631.00	1.2222	171.95	631.00	298.15	23.40

Table 9-2 SURFACE TENSION - INORGANIC COMPOUNDS (continued)

$$\text{sigma} = A\,(1 - T/T_c)^n \qquad (\text{sigma - dynes/cm, T - K})$$

NO	FORMULA	NAME	A	T_c	n	TMIN	TMAX	T, K	sigma @ T
79	ClO2	CHLORINE DIOXIDE	126.308	465.00	1.2222	213.55	465.00	298.15	36.09
80	Cl2	CHLORINE	67.560	417.15	1.0850	172.12	417.15	298.15	17.32
81	Cl2O	CHLORINE MONOXIDE	91.930	444.68	1.2222	157.15	444.68	298.15	23.67
82	Cl2O7	CHLORINE HEPTOXIDE	69.470	565.78	1.2222	182.15	565.78	298.15	27.83
83	Co	COBALT	2629.218	7398.48	1.2222	1768.15	7398.48	1768.15	1883.04
84	CoCl2	COBALT CHLORIDE	--------	--------	-------	-------	-------	-------	-------
85	CoNC3O4	COBALT NITROSYL TRICARBONYL	64.857	567.68	1.2222	262.15	567.68	298.15	26.10
86	Cr	CHROMIUM	2330.724	8560.93	1.2222	2180.15	8560.93	2180.15	1627.34
87	CrC6O6	CHROMIUM CARBONYL	--------	--------	-------	-------	-------	-------	-------
88	CrO2Cl2	CHROMIUM OXYCHLORIDE	84.894	626.33	1.2222	176.65	626.33	298.15	38.53
89	Cs	CESIUM	86.720	2048.10	1.2222	301.65	2048.10	301.65	71.38
90	CsBr	CESIUM BROMIDE	--------	--------	-------	-------	-------	-------	-------
91	CsCl	CESIUM CHLORIDE	--------	--------	-------	-------	-------	-------	-------
92	CsF	CESIUM FLUORIDE	195.309	2482.33	1.2222	956.15	2482.33	956.15	107.78
93	CsI	CESIUM IODIDE	154.963	2529.56	1.2222	894.15	2529.56	894.15	90.93
94	Cu	COPPER	1938.012	5123.00	1.2222	1357.77	5123.00	1357.77	1330.16
95	CuBr	CUPROUS BROMIDE	--------	--------	-------	-------	-------	-------	-------
96	CuCl	CUPROUS CHLORIDE	--------	--------	-------	-------	-------	-------	-------
97	CuCl2	CUPRIC CHLORIDE	--------	--------	-------	-------	-------	-------	-------
98	CuI	COPPER IODIDE	--------	--------	-------	-------	-------	-------	-------
99	DCN	DEUTERIUM CYANIDE	140.257	482.63	1.2222	261.15	482.63	298.15	43.30
100	D2	DEUTERIUM	10.392	38.35	1.3365	18.73	38.35	18.73	4.24
101	D2O	DEUTERIUM OXIDE	153.466	643.89	1.2222	276.96	643.89	298.15	71.77
102	Eu	EUROPIUM	354.025	5150.00	1.2222	1095.15	5150.00	1095.15	264.32
103	F2	FLUORINE	40.040	144.31	1.2266	53.53	144.31	53.53	22.68
104	F2O	FLUORINE OXIDE	42.925	215.10	1.2222	49.25	215.10	49.25	31.24
105	Fe	IRON	2420.127	9340.00	1.2222	1808.15	9340.00	1808.15	1860.49
106	FeC5O5	IRON PENTACARBONYL	49.272	607.20	1.2222	252.15	607.20	298.15	21.58
107	FeCl2	FERROUS CHLORIDE	--------	--------	-------	-------	-------	-------	-------
108	FeCl3	FERRIC CHLORIDE	--------	--------	-------	-------	-------	-------	-------
109	Fr	FRANCIUM	--------	--------	-------	-------	-------	-------	-------
110	Ga	GALLIUM	747.168	7620.00	1.2222	302.91	7620.00	302.91	711.03
111	GaCl3	GALLIUM TRICHLORIDE	66.110	694.00	1.2637	350.90	694.00	350.90	27.14
112	Gd	GADOLINIUM	1514.737	3307.66	1.2222	1587.15	3307.66	1587.15	681.38
113	Ge	GERMANIUM	736.991	8400.00	1.2222	1211.40	8400.00	1211.40	609.25
114	GeBr4	GERMANIUM BROMIDE	67.626	740.00	1.2222	299.25	740.00	299.25	35.90
115	GeCl4	GERMANIUM CHLORIDE	56.194	574.00	1.2222	223.65	574.00	298.15	22.95
116	GeHCl3	TRICHLORO GERMANE	62.848	559.78	1.2222	202.05	559.78	298.15	24.81
117	GeH4	GERMANE	55.097	308.00	1.2222	107.26	308.00	298.15	0.82
118	Ge2H6	DIGERMANE	60.271	491.01	1.2222	164.15	491.01	298.15	19.23
119	Ge3H8	TRIGERMANE	66.422	616.37	1.2222	167.55	616.37	298.15	29.61
120	HBr	HYDROGEN BROMIDE	75.346	363.15	1.2222	186.34	363.15	298.15	9.20
121	HCN	HYDROGEN CYANIDE	52.256	456.65	1.0198	259.91	456.65	298.15	17.76
122	HCl	HYDROGEN CHLORIDE	85.200	324.65	1.2970	158.97	324.65	298.15	3.30
123	HF	HYDROGEN FLUORIDE	36.480	461.15	1.4122	189.79	461.15	298.15	8.40
124	HI	HYDROGEN IODIDE	63.770	423.85	1.0749	222.38	423.85	298.15	17.27
125	HNO3	NITRIC ACID	112.392	520.00	1.2222	231.55	520.00	298.15	39.68
126	H2	HYDROGEN	5.336	33.18	1.0622	13.95	33.18	13.95	2.99
127	H2O	WATER	132.674	647.13	0.9550	273.15	647.13	298.15	73.56
128	H2O2	HYDROGEN PEROXIDE	141.031	730.15	1.2222	272.72	730.15	298.15	74.26
129	H2S	HYDROGEN SULFIDE	70.920	373.53	1.2809	187.68	373.53	298.15	9.13
130	H2SO4	SULFURIC ACID	99.899	925.00	1.2222	283.46	925.00	298.15	62.09
131	H2S2	HYDROGEN DISULFIDE	116.241	542.39	1.2222	183.45	542.39	298.15	43.84
132	H2Se	HYDROGEN SELENIDE	82.166	411.10	1.2222	209.15	411.10	298.15	16.94
133	H2Te	HYDROGEN TELLURIDE	98.345	438.04	1.2222	224.15	438.04	298.15	24.37
134	H3NO3S	SULFAMIC ACID	--------	--------	-------	-------	-------	-------	-------
135	He	HELIUM-3	0.275	3.31	1.2222	1.01	3.31	1.01	0.18
136	He	HELIUM-4	0.511	5.20	1.0030	1.76	5.20	1.76	0.34
137	Hf	HAFNIUM	1892.378	21687.96	1.2222	2506.15	21688.00	2506.15	1628.65
138	Hg	MERCURY	617.571	1735.00	1.2184	234.29	1735.00	298.15	490.81
139	HgBr2	MERCURIC BROMIDE	--------	--------	-------	-------	-------	-------	-------
140	HgCl2	MERCURIC CHLORIDE	--------	--------	-------	-------	-------	-------	-------
141	HgI2	MERCURIC IODIDE	--------	--------	-------	-------	-------	-------	-------
142	IF7	IODINE HEPTAFLUORIDE	51.301	447.53	1.2222	278.65	447.53	298.15	13.42
143	I2	IODINE	234.700	819.15	2.2930	386.75	819.15	386.75	54.23
144	In	INDIUM	602.746	6730.00	1.2222	429.75	6730.00	429.75	556.04
145	Ir	IRIDIUM	2890.287	15035.00	1.2222	2719.15	15035.00	2719.15	2264.90
146	K	POTASSIUM	134.493	2223.00	1.2222	336.35	2223.00	336.35	110.06
147	KBr	POTASSIUM BROMIDE	--------	--------	-------	-------	-------	-------	-------
148	KCl	POTASSIUM CHLORIDE	200.980	3470.00	1.9505	1044.00	3470.00	1044.00	99.99
149	KF	POTASSIUM FLUORIDE	--------	--------	-------	-------	-------	-------	-------
150	KI	POTASSIUM IODIDE	136.352	2601.22	1.2222	996.15	2601.22	996.15	75.58
151	KOH	POTASSIUM HYDROXIDE	--------	--------	-------	-------	-------	-------	-------
152	Kr	KRYPTON	44.250	209.35	1.2317	115.78	209.35	115.78	16.41
153	La	LANTHANUM	857.597	9511.00	1.2222	1193.15	9511.00	1193.15	728.00
154	Li	LITHIUM	463.400	4085.00	1.2760	453.69	4085.00	453.69	398.76
155	LiBr	LITHIUM BROMIDE	--------	--------	-------	-------	-------	-------	-------
156	LiCl	LITHIUM CHLORIDE	227.108	2695.68	1.2222	887.15	2695.68	887.15	139.43

235

Table 9-2 SURFACE TENSION - INORGANIC COMPOUNDS (continued)

			sigma = A (1 - T/T$_c$)^n			(sigma - dynes/cm, T - K)			
NO	FORMULA	NAME	A	T$_c$	n	TMIN	TMAX	T, K	sigma @ T
157	LiF	LITHIUM FLUORIDE	431.521	3182.65	1.2222	1143.15	3182.65	1143.15	250.49
158	LiI	LITHIUM IODIDE	--------	--------	-------	-------	-------	------	------
159	Lu	LUTECIUM	2080.227	4128.66	1.2222	1936.15	4128.66	1936.15	959.76
160	Mg	MAGNESIUM	1104.037	2241.04	1.2222	923.15	2241.04	923.15	577.00
161	MgCl2	MAGNESIUM CHLORIDE	--------	--------	-------	-------	-------	------	------
162	MgO	MAGNESIUM OXIDE	--------	--------	-------	-------	-------	------	------
163	Mn	MANGANESE	1560.596	6902.82	1.2222	1519.15	6902.82	1519.15	1151.74
164	MnCl2	MANGANESE CHLORIDE	--------	--------	-------	-------	-------	------	------
165	Mo	MOLYBDENUM	3483.943	9620.00	1.2222	2895.15	9620.00	2895.15	2249.18
166	MoF6	MOLYBDENUM FLUORIDE	65.311	498.12	1.2222	290.15	498.12	298.15	21.41
167	MoO3	MOLYBDENUM OXIDE	--------	--------	-------	-------	-------	------	------
168	NCl3	NITROGEN TRICHLORIDE	79.176	564.00	1.2222	246.15	564.00	298.15	31.58
169	ND3	HEAVY AMMONIA	141.079	388.40	1.2222	199.15	388.40	298.15	23.70
170	NF3	NITROGEN TRIFLUORIDE	44.640	233.85	1.2222	66.36	233.85	66.36	29.69
171	NH3	AMMONIA	100.098	405.65	1.2222	195.41	405.65	298.15	19.75
172	NH3O	HYDROXYLAMINE	280.055	574.00	1.2222	306.25	574.00	306.25	110.27
173	NH4Br	AMMONIUM BROMIDE	--------	--------	-------	-------	-------	------	------
174	NH4Cl	AMMONIUM CHLORIDE	--------	--------	-------	-------	-------	------	------
175	NH4I	AMMONIUM IODIDE	--------	--------	-------	-------	-------	------	------
176	NH5O	AMMONIUM HYDROXIDE	--------	--------	-------	-------	-------	------	------
177	NH5S	AMMONIUM HYDROGENSULFIDE	--------	--------	-------	-------	-------	------	------
178	NO	NITRIC OXIDE	121.260	180.15	1.5494	112.15	180.15	112.15	26.80
179	NOCl	NITROSYL CHLORIDE	77.960	440.65	1.1810	213.55	440.65	298.15	20.55
180	NOF	NITROSYL FLUORIDE	61.495	352.67	1.2222	139.15	352.67	298.15	6.28
181	NO2	NITROGEN DIOXIDE	167.436	431.35	1.2222	261.95	431.35	298.15	39.82
182	N2	NITROGEN	28.980	126.10	1.2457	63.15	126.10	63.15	12.20
183	N2F4	TETRAFLUOROHYDRAZINE	43.167	309.35	1.2203	111.65	309.35	298.15	0.75
184	N2H4	HYDRAZINE	141.250	653.15	1.2570	274.69	653.15	298.15	65.64
185	N2H4C	AMMONIUM CYANIDE	136.540	491.32	1.2222	309.15	491.32	309.15	40.61
186	N2H6CO2	AMMONIUM CARBAMATE	57.224	539.82	1.2222	270.00	539.82	298.15	21.45
187	N2O	NITROUS OXIDE	71.370	309.57	1.2062	182.33	309.57	298.15	1.33
188	N2O3	NITROGEN TRIOXIDE	98.286	425.00	1.2222	170.00	425.00	298.15	22.42
189	N2O4	NITROGEN TETRAOXIDE	185.819	431.15	1.2222	261.90	431.15	298.15	44.14
190	N2O5	NITROGEN PENTOXIDE	84.206	515.51	1.2222	303.15	515.51	303.15	28.48
191	Na	SODIUM	231.700	2573.00	1.1986	370.98	2573.00	370.98	192.26
192	NaBr	SODIUM BROMIDE	174.770	4287.00	2.0693	1020.00	4287.00	1020.00	99.60
193	NaCN	SODIUM CYANIDE	--------	--------	-------	-------	-------	------	------
194	NaCl	SODIUM CHLORIDE	201.330	3400.00	1.4978	1073.95	3400.00	1073.95	114.02
195	NaF	SODIUM FLUORIDE	304.610	5530.00	1.9016	1269.00	5530.00	1269.00	185.55
196	NaI	SODIUM IODIDE	152.794	2568.65	1.2222	924.15	2568.65	924.15	88.59
197	NaOH	SODIUM HYDROXIDE	--------	--------	-------	-------	-------	------	------
198	Na2SO4	SODIUM SULFATE	--------	--------	-------	-------	-------	------	------
199	Nb	NIOBIUM	2327.876	17904.10	1.2222	2750.15	17904.10	2750.15	1898.62
200	Nd	NEODYMIUM	804.994	10665.12	1.2222	1289.15	10665.10	1289.15	687.72
201	Ne	NEON	15.633	44.40	1.2611	24.55	44.40	24.55	5.66
202	Ni	NICKEL	2541.848	6986.15	1.2222	1728.15	6986.15	1728.15	1796.00
203	NiC4O4	NICKEL CARBONYL	45.077	508.40	1.2222	248.15	508.40	298.15	15.32
204	NiF2	NICKEL FLUORIDE	--------	--------	-------	-------	-------	------	------
205	Np	NEPTUNIUM	--------	--------	-------	-------	-------	------	------
206	O2	OXYGEN	38.066	154.58	1.2136	54.36	154.58	54.36	22.50
207	O3	OZONE	84.280	261.00	1.8600	80.15	261.00	80.15	42.60
208	Os	OSMIUM	3268.304	16878.68	1.2222	3306.15	16878.70	3306.15	2503.83
209	OsOF5	OSMIUM OXIDE PENTAFLUORIDE	67.217	608.55	1.2222	332.95	608.55	332.95	25.53
210	OsO4	OSMIUM TETROXIDE - YELLOW	74.322	656.60	1.2222	329.15	656.60	329.15	31.76
211	OsO4	OSMIUM TETROXIDE - WHITE	74.322	656.60	1.2222	315.15	656.60	315.15	33.42
212	P	PHOSPHORUS - WHITE	110.647	993.75	1.2222	317.25	993.75	317.25	69.15
213	PBr3	PHOSPHORUS TRIBROMIDE	81.799	711.00	1.2222	233.15	711.00	298.15	42.09
214	PCl2F3	PHOSPHORUS DICHLORIDE TRIFLUORIDE	50.753	457.02	1.2222	265.15	457.02	298.15	13.95
215	PCl3	PHOSPHORUS TRICHLORIDE	70.820	563.15	1.2216	181.15	563.15	298.15	28.20
216	PCl5	PHOSPHORUS PENTACHLORIDE	--------	--------	-------	-------	-------	------	------
217	PH3	PHOSPHINE	69.131	324.75	1.4428	139.37	324.75	298.15	1.87
218	PH4Br	PHOSPHONIUM BROMIDE	80.724	501.76	1.2222	250.00	501.76	298.15	26.81
219	PH4Cl	PHOSPHONIUM CHLORIDE	123.360	322.30	1.2222	244.65	322.30	298.15	5.20
220	PH4I	PHOSPHONIUM IODIDE	102.632	539.70	1.2222	291.65	539.70	298.15	38.42
221	POCl3	PHOSPHORUS OXYCHLORIDE	74.600	602.15	1.2364	274.33	602.15	298.15	32.04
222	PSBr3	PHOSPHORUS THIOBROMIDE	85.402	729.89	1.2222	311.15	729.89	311.15	43.30
223	PSCl3	PHOSPHORUS THIOCHLORIDE	69.574	638.82	1.2222	236.95	638.82	298.15	32.26
224	P4O6	PHOSPHORUS TRIOXIDE	72.103	714.86	1.2222	295.65	714.86	298.15	37.28
225	P4O10	PHOSPHORUS PENTOXIDE	--------	--------	-------	-------	-------	------	------
226	P4S10	PHOSPHORUS PENTASULFIDE	--------	--------	-------	-------	-------	------	------
227	Pb	LEAD	527.780	5400.00	1.2222	600.61	5400.00	600.61	456.95
228	PbBr2	LEAD BROMIDE	--------	--------	-------	-------	-------	------	------
229	PbCl2	LEAD CHLORIDE	250.996	1998.62	1.2222	774.15	1998.62	774.15	137.91
230	PbF2	LEAD FLUORIDE	--------	--------	-------	-------	-------	------	------
231	PbI2	LEAD IODIDE	--------	--------	-------	-------	-------	------	------
232	PbO	LEAD OXIDE	253.003	2842.26	1.2222	1163.15	2842.26	1163.15	132.97
233	PbS	LEAD SULFIDE	--------	--------	-------	-------	-------	------	------
234	Pd	PALLADIUM	1863.962	10669.07	1.2222	1828.05	10669.10	1828.05	1481.41

Table 9-2 SURFACE TENSION - INORGANIC COMPOUNDS (continued)

$$\text{sigma} = A\left(1 - T/T_c\right)^n \qquad (\text{sigma - dynes/cm, T - K})$$

NO	FORMULA	NAME	A	T_c	n	TMIN	TMAX	T, K	sigma @ T
235	Po	POLONIUM	--------	--------	-------	-------	-------	------	------
236	Pt	PLATINUM	2668.141	6983.00	1.2222	2041.55	6983.00	2041.55	1748.42
237	Ra	RADIUM	--------	--------	-------	-------	-------	------	------
238	Rb	RUBIDIUM	108.953	2111.10	1.2222	312.46	2111.10	312.46	89.58
239	RbBr	RUBIDIUM BROMIDE	--------	--------	-------	-------	-------	------	------
240	RbCl	RUBIDIUM CHLORIDE	--------	--------	-------	-------	-------	------	------
241	RbF	RUBIDIUM FLUORIDE	--------	--------	-------	-------	-------	------	------
242	RbI	RUBIDIUM IODIDE	--------	--------	-------	-------	-------	------	------
243	Re	RHENIUM	3341.924	21482.82	1.2222	3459.15	21482.80	3459.15	2696.53
244	Re2O7	RHENIUM HEPTOXIDE	--------	--------	-------	-------	-------	------	------
245	Rh	RHODIUM	2417.180	12906.61	1.2222	2237.15	12906.60	2237.15	1915.44
246	Rn	RADON	53.696	377.40	1.2222	202.15	377.40	298.15	7.97
247	Ru	RUTHENIUM	2806.768	15247.14	1.2222	2607.15	15247.10	2607.15	2231.86
248	RuF5	RUTHENIUM PENTAFLUORIDE	--------	--------	-------	-------	-------	------	------
249	S	SULFUR	86.030	1313.00	1.0010	388.36	1313.00	388.36	60.56
250	SF4	SULFUR TETRAFLUORIDE	67.860	364.00	1.2222	149.15	364.00	298.15	8.40
251	SF6	SULFUR HEXAFLUORIDE	51.434	318.69	1.2222	222.45	318.69	298.15	1.80
252	SOBr2	THIONYL BROMIDE	93.103	661.75	1.2222	220.95	661.75	298.15	44.78
253	SOCl2	THIONYL CHLORIDE	92.457	567.00	1.4114	172.00	567.00	298.15	32.25
254	SOF2	SULFUROUS OXYFLUORIDE	68.096	371.25	1.2222	162.65	371.25	298.15	9.34
255	SO2	SULFUR DIOXIDE	87.200	430.75	1.1810	200.00	430.75	298.15	21.69
256	SO2Cl2	SULFURYL CHLORIDE	75.769	545.00	1.2222	222.00	545.00	298.15	28.78
257	SO3	SULFUR TRIOXIDE	103.340	490.85	1.2184	289.95	490.85	298.15	33.08
258	S2Cl2	SULFUR MONOCHLORIDE	88.675	659.37	1.2222	193.15	659.37	298.15	42.50
259	Sb	ANTIMONY	471.528	5070.00	1.2222	903.78	5070.00	903.78	370.93
260	SbBr3	ANTIMONY TRIBROMIDE	--------	--------	-------	-------	-------	------	------
261	SbCl3	ANTIMONY TRICHLORIDE	95.290	794.00	1.1644	346.55	794.00	346.55	48.87
262	SbCl5	ANTIMONY PENTACHLORIDE	57.152	662.54	1.2222	275.95	662.54	298.15	27.52
263	SbH3	STIBINE	72.532	440.35	1.2222	185.15	440.35	298.15	18.22
264	SbI3	ANTIMONY TRIIODIDE	--------	--------	-------	-------	-------	------	------
265	Sb2O3	ANTIMONY TRIOXIDE	--------	--------	-------	-------	-------	------	------
266	Sc	SCANDIUM	1283.547	8035.08	1.2222	1814.15	8035.08	1814.15	938.82
267	Se	SELENIUM	153.693	1766.00	1.2222	494.15	1766.00	494.15	102.90
268	SeCl4	SELENIUM TETRACHLORIDE	--------	--------	-------	-------	-------	------	------
269	SeF6	SELENIUM HEXAFLUORIDE	51.410	368.80	1.2222	238.45	368.80	298.15	6.82
270	SeOCl2	SELENIUM OXYCHLORIDE	113.623	706.80	1.2222	281.65	706.80	298.15	58.16
271	SeO2	SELENIUM DIOXIDE	--------	--------	-------	-------	-------	------	------
272	Si	SILICON	1255.794	5159.00	1.2222	1685.00	5159.00	1685.00	774.50
273	SiBrCl2F	BROMODICHLOROFLUOROSILANE	49.579	497.17	1.2222	160.85	497.17	298.15	16.19
274	SiBrF3	TRIFLUOROBROMOSILANE	43.329	375.28	1.2222	202.65	375.28	298.15	6.27
275	SiBr2ClF	DIBROMOCHLOROFLUOROSILANE	52.472	535.27	1.2222	173.85	535.27	298.15	19.40
276	SiClF3	TRIFLUOROCHLOROSILANE	38.469	330.54	1.2222	131.15	330.54	298.15	2.25
277	SiCl2F2	DICHLORODIFLUOROSILANE	42.163	390.93	1.2222	133.45	390.93	298.15	7.27
278	SiCl3F	TRICHLOROFLUOROSILANE	46.747	460.49	1.2222	152.35	460.49	298.15	13.07
279	SiCl4	SILICON TETRACHLORIDE	48.620	507.00	1.0840	204.30	507.00	298.15	18.59
280	SiF4	SILICON TETRAFLUORIDE	56.349	259.00	1.2222	186.35	259.00	186.35	11.92
281	SiHBr3	TRIBROMOSILANE	69.122	610.00	1.2222	199.65	610.00	298.15	30.44
282	SiHCl3	TRICHLOROSILANE	63.250	479.00	1.3248	144.95	479.00	298.15	17.40
283	SiHF3	TRIFLUOROSILANE	41.376	291.02	1.2222	141.75	291.02	141.75	18.30
284	SiH2Br2	DIBROMOSILANE	72.685	550.00	1.2222	202.95	550.00	298.15	27.98
285	SiH2Cl2	DICHLOROSILANE	53.418	449.00	1.1812	151.15	449.00	298.15	14.73
286	SiH2F2	DIFLUOROSILANE	51.341	318.21	1.2222	170.00	318.21	298.15	1.75
287	SiH2I2	DIIODOSILANE	105.137	660.00	1.2222	272.15	660.00	298.15	50.44
288	SiH3Br	MONOBROMOSILANE	65.933	454.00	1.2222	179.25	454.00	298.15	17.85
289	SiH3Cl	MONOCHLOROSILANE	55.653	396.65	1.2222	155.05	396.65	298.15	10.14
290	SiH3F	MONOFLUOROSILANE	48.274	286.28	1.2222	150.00	286.28	150.00	19.49
291	SiH3I	IODOSILANE	89.400	515.00	1.2222	216.15	515.00	298.15	31.06
292	SiH4	SILANE	42.633	269.70	1.2110	88.15	269.70	88.15	26.40
293	SiO2	SILICON DIOXIDE	438.690	4076.87	1.2222	1883.00	4076.87	1883.00	205.70
294	Si2Cl6	HEXACHLORODISILANE	41.695	660.96	1.2222	271.95	660.96	298.15	20.03
295	Si2F6	HEXAFLUORODISILANE	35.854	411.33	1.2222	254.55	411.33	298.15	7.41
296	Si2H5Cl	DISILANYL CHLORIDE	54.351	506.89	1.2222	250.00	506.89	298.15	18.38
297	Si2H6	DISILANE	61.937	432.00	1.2870	140.65	432.00	298.15	13.71
298	Si2OCl3F3	TRICHLOROTRIFLUORODISILOXANE	34.422	508.78	1.2222	250.00	508.78	298.15	11.71
299	Si2OCl6	HEXACHLORODISILOXANE	39.741	655.58	1.2222	239.95	655.58	298.15	18.93
300	Si2OH6	DISILOXANE	43.589	416.86	1.2222	128.95	416.86	298.15	9.39
301	Si3Cl8	OCTACHLOROTRISILANE	37.167	775.41	1.2222	250.00	775.41	298.15	20.54
302	Si3H8	TRISILANE	44.454	525.15	1.2222	155.95	525.15	298.15	15.95
303	Si3H9N	TRISILAZANE	41.420	518.20	1.2222	167.45	518.20	298.15	14.54
304	Si4H10	TETRASILANE	40.979	599.30	1.2222	179.55	599.30	298.15	17.67
305	Sm	SAMARIUM	626.089	5082.92	1.2222	1345.15	5082.92	1345.15	430.00
306	Sn	TIN	612.727	7400.00	1.2222	505.08	7400.00	505.08	562.01
307	SnBr4	STANNIC BROMIDE	66.560	764.82	1.2222	304.15	764.82	304.15	35.82
308	SnCl2	STANNOUS CHLORIDE	180.180	1459.53	1.2222	519.95	1459.53	519.95	105.18
309	SnCl4	STANNIC CHLORIDE	59.674	619.85	1.2222	242.95	619.85	298.15	26.77
310	SnH4	STANNIC HYDRIDE	60.591	358.52	1.2222	123.25	358.52	298.15	6.87
311	SnI4	STANNIC IODIDE	--------	--------	-------	-------	-------	------	------
312	Sr	STRONTIUM	426.813	4267.20	1.2222	1050.15	4267.20	1050.15	302.20

Table 9-2 SURFACE TENSION - INORGANIC COMPOUNDS (continued)

NO	FORMULA	NAME	$\sigma = A(1 - T/T_c)^n$			(sigma - dynes/cm, T - K)			
			A	T_c	n	TMIN	TMAX	T, K	sigma @ T
313	SrO	STRONTIUM OXIDE	--------	--------	-------	--------	--------	-------	------
314	Ta	TANTALUM	2668.246	19900.93	1.2222	3290.15	19900.90	3290.15	2139.45
315	Tc	TECNNETIUM	--------	--------	-------	--------	--------	-------	------
316	Te	TELLURIUM	292.137	4840.00	1.2222	722.66	4840.00	722.66	239.75
317	TeCl4	TELLURIUM TETRACHLORIDE	--------	--------	-------	--------	--------	-------	------
318	TeF6	TELLURIUM HEXAFLUORIDE	39.899	380.18	1.2222	235.35	380.18	298.15	6.12
319	Ti	TITANIUM	2574.844	6400.00	1.2222	1941.15	6400.00	1941.15	1655.44
320	TiCl4	TITANIUM TETRACHLORIDE	70.873	638.00	1.2222	249.05	638.00	298.15	32.82
321	Tl	THALLIUM	539.907	4648.06	1.2222	577.15	4648.06	577.15	459.14
322	TlBr	THALLOUS BROMIDE	--------	--------	-------	--------	--------	-------	------
323	TlI	THALLOUS IODIDE	--------	--------	-------	--------	--------	-------	------
324	Tm	THULIUM	--------	--------	-------	--------	--------	-------	------
325	U	URANIUM	1770.887	13712.62	1.2222	1408.15	13712.60	1408.15	1551.23
326	UF6	URANIUM FLUORIDE	67.457	505.80	1.2222	342.35	505.80	342.35	16.96
327	V	VANADIUM	2450.047	11787.15	1.2222	2183.15	11787.10	2183.15	1907.43
328	VCl4	VANADIUM TETRACHLORIDE	82.377	697.00	1.2222	247.45	697.00	298.15	41.64
329	VOCl3	VANADIUM OXYTRICHLORIDE	76.954	636.00	1.2222	193.65	636.00	298.15	35.52
330	W	TUNGSTEN	3535.480	14756.00	1.2222	3695.15	14756.00	3695.15	2485.71
331	WF6	TUNGSTEN FLUORIDE	59.258	468.56	1.2222	272.65	468.56	298.15	17.21
332	Xe	XENON	54.299	289.74	1.2829	161.36	289.74	161.36	19.11
333	Yb	YTTERBIUM	455.799	4365.92	1.2222	1097.15	4365.92	1097.15	320.00
334	Yt	YTTRIUM	1126.468	9381.32	1.2222	1799.15	9381.32	1799.15	868.36
335	Zn	ZINC	1066.718	3170.00	1.2222	692.70	3170.00	692.70	789.18
336	ZnCl2	ZINC CHLORIDE	--------	--------	-------	--------	--------	-------	------
337	ZnF2	ZINC FLUORIDE	--------	--------	-------	--------	--------	-------	------
338	ZnO	ZINC OXIDE	--------	--------	-------	--------	--------	-------	------
339	ZnSO4	ZINC SULFATE	--------	--------	-------	--------	--------	-------	------
340	Zr	ZIRCONIUM	2073.869	8802.00	1.2222	2128.15	8802.00	2128.15	1478.65
341	ZrBr4	ZIRCONIUM BROMIDE	--------	--------	-------	--------	--------	-------	------
342	ZrCl4	ZIRCONIUM CHLORIDE	--------	--------	-------	--------	--------	-------	------
343	ZrI4	ZIRCONIUM IODIDE	--------	--------	-------	--------	--------	-------	------

sigma - surface tension, dynes/cm

A, T_c, and n - regression coefficients of chemical compound

T - temperature, K

TMAX - maximum temperature, K

TMIN - minimun temperature, K

Chapter 10

REFRACTIVE INDEX, DIPOLE MOMENT, AND RADIUS OF GYRATION

Carl L. Yaws, Xiao M. Wang, and Marco A. Satyro*

Lamar University, Beaumont, Texas / *SEA++ INC., Calgary, Alberta

ABSTRACT

Results for refractive index, dipole moment, and radius of gyration are presented for major organic and inorganic chemicals. The chemical formula and molecular weight are also given. The results are displayed in easy-to-use tabulations which are especially applicable for rapid engineering usage with the personal computer or hand calculator. The organic chemicals encompass hydrocarbon, oxygen, nitrogen, halogen, silicon, sulfur, and other compound types.

INTRODUCTION

Physical properties such as refractive index, dipole moment, and radius of gyration are of special value to engineers in the chemical processing and petroleum refining industries. Since these properties are used in thermodynamic correlations that are involved in the design of process equipment, the results of this article are intended for initial engineering studies.

REFRACTIVE INDEX, DIPOLE MOMENT, AND RADIUS OF GYRATION

The refractive index is an indication of the manner in which a compound interacts with light. The refractive index is defined as the ratio of the speed of light in a vacuum to the speed of light in the compound. Most of the values for index of refraction are applicable to a temperature of 25 C. Exceptions to this temperature are noted in the tabulations.

The dipole moment involves the first moment of the electric charge density of the compound. Property correlations for polar compounds often require knowledge of the dipole moment. The radius of gyration is ascertained from the moment of inertia and molecular weight. This property is also used in thermodynamic correlations.

The results for refractive index, dipole moment, and radius of gyration are given in Tables 10-1 and 10-2 for organic and inorganic compounds. The tabulations are based on data source publications for organics (1-59) and inorganics (1-48). The tabulations are arranged by chemical formula to provide ease of use in quickly locating data.

REFERENCES - ORGANIC COMPOUNDS

1-34. See **REFERENCES - ORGANIC COMPOUNDS** in Chapter 1, CRITICAL PROPERTIES AND ACENTRIC FACTOR

35. Passut, C. A. and R. P. Danner, Chem. Eng. Prog. Symp. Ser., 70 (140), 30 (1974).

36. Gold, P. I. and G. J. Ogle, Chem. Eng., 76, 97 (August 11, 1969).

37. Hansch, C., A. Leo, S. H. Unger, K. H. Kum, D. Nikatiani, and E. J. Lien, J. Med. Chem., 16, 1207 (1973).

38. Kier, L. B. and L. H. Hall, MOLECULAR CONNECTIVITY IN CHEMISTRY AND DRUG RESEARCH, Academic Press, New York, NY (1976).

39. McClellan, A. L., TABLES OF EXPERIMENTAL DIPOLE MOMENTS, W. H. Freeman Publishing, San Francisco, CA (1963).

40. Lawson, D. D. and J. D. Ingham, Nature (London), 223, 614 (1969).

41. Meissner, H. P., Chem. Eng. Prog., 45, 149 (1949).

42. Skoog, D. A. and D. M. West, PRINCIPLES OF INSTRUMENTAL ANALYSIS, Holt, Rinehart and Winston, New York, NY (1971).

43. LeFevre, R. J. W., DIPOLE MOMENTS, THEIR MEASUREMENT AND APPLICATION IN CHEMISTRY, John Wiley and Sons, New York, NY (1953).

44. Moore, W. J., PHYSICAL CHEMISTRY, 4th ed., Prentice-Hall, Englewood Cliffs, NJ (1972).

45. Potenzone, R., Jr., E. Cavicchi, H. J. R. Weintraub, and A. J. Hopfinger, Comput. Chem., 1, 187 (1977).

46. Smyth, C. P., DIELECTRIC BEHAVIOR AND STRUCTURE, McGraw-Hill Book Co., New York, NY (1955).

47. Wheatley, P. J., THE DETERMINATION OF MOLECULAR STRUCTURE, Oxford University Press, London, England (1959).

48. Nelson, R. D., D. R. Lide, A. Maryott, National Bureau of Standards, NSRDS 10, Washington, DC (1967).

49. Takashi, R. and others, J. Mol. Spec., 138, 450 (1989).

50. Hayashi, M. and T. Inagusa, J. Mol. Spec., 138, 135 (1989).

51. Scappini, F., A. C. Fantoni, and W. Caminati, J. Mol. Spec., 120, 101 (1986).

52. Thompson, W. H., "A MOLECULAR ASSOCIATION FACTOR FOR USE IN THE EXTENDED THEOREM OF CORRESPONDING STATES", Ph. D. Thesis, The Pennsylvania State University, University Park, PA (1966).

53. Cygnarowicz, R.M., "THE MOLECULAR RADIUS OF GYRATION", B.S. Thesis in Engineering Science, The Pennsylvania State University, University Park, PA (1981).

54. Sutter, H. and R.H. Cole, J. Chem. Phys., 52, 132 (1970).

55. Stuper, A. J., W. E. Brugger, and P. C. Jurs, COMPUTER ASSISTED STUDIES OF CHEMICAL STRUCTURE AND BIOLOGICAL

FUNCTION, John Wiley and Sons, New York, NY (1979).

56. Bock, E. and D. Iwacha, Can. J. Chem., 46, 523 (1968).

57. Ghosh S. N. and R. Trambarulo, W. Gordy, Phys. Rev., 82, 172 (1952).

58. Shoolery, J. N. and A. H. Sharbaugh, Phys. Rev. , 82, 95 (1951).

59. Kurtz, S. S., Jr., S. Amon and A. Sankin, Ind. Eng. Chem., 42, 174 (1950).

REFERENCES - INORGANIC COMPOUNDS

1-35. See **REFERENCES - INORGANIC COMPOUNDS** in Chapter 1, **CRITICAL PROPERTIES AND ACENTRIC FACTOR**

36. Passut, C. A. and R. P. Danner, Chem. Eng. Prog. Symp. Ser., 70 (140), 30 (1974).

37. McClellan, A. L., TABLES OF EXPERIMENTAL DIPOLE MOMENTS, W. H. Freeman Publishing, San Francisco, CA (1963).

38. Stuper, A. J., W. E. Brugger, and P. C. Jurs, COMPUTER ASSISTED STUDIES OF CHEMICAL STRUCTURE AND BIOLOGICAL FUNCTION, John Wiley and Sons, New York, NY (1979).

39. Nelson, R. D., D. R. Lide, and A. Maryott, National Bureau of Standards, NSRDS 10, Washington, DC (1967).

40. Thompson, W. H., "A MOLECULAR ASSOCIATION FACTOR FOR USE IN THE EXTENDED THEOREM OF CORRESPONDING STATES", Ph. D. Thesis, The Pennsylvania State University, University Park, PA (1966).

41. Cygnarowicz, R.M., "THE MOLECULAR RADIUS OF GYRATION", B.S. Thesis in Engineering Science, The Pennsylvania State University, University Park, PA (1981).

42. Stern, S. A., J. L. Mulhaupt, and W. B. Kay, Chem. Rev., 60, 185 (1960).

43. Francis, A. W., J. Chem. Eng. Data, 5(4), 534 (1960).

44. Robinson, D. B. and N. H. Senturk, J. Chem. Therm., 11, 875 (1979).

45. Helminger, P. and F. C. DeLucia, J. Phys. Chem. Ref. Data, 2(2), 215 (1973).

46. Tolles, W. M., J. L. Kinsey, R. F. Curl, and F. H. Robert, J. Chem. Phys., 37, 927 (1962).

47. Shaulov, Y. K. and A. M. Masin, Russ. J. Phys. Chem., 47, 642 (1973).

48. Mackie, H. and P. A. G. O'Hare, Trans. Faraday Soc., 59, 309 (1963).

Table 10-1 REFRACTIVE INDEX, DIPOLE M0MENT, AND RADIUS OF GYRATION - ORGANIC COMPOUNDS

NO	FORMULA	NAME	Refractive Index		Dipole Moment		Radius of Gyration
			T, C	Value @ T	State	Debye	Angstrom
1	CBrClF2	BROMOCHLORODIFLUOROMETHANE	-------	-------	-------	-------	3.105
2	CBrCl3	BROMOTRICHLOROMETHANE	25.0	1.5060	in benzene	0.59	3.577
3	CBrF3	BROMOTRIFLUOROMETHANE	25.0	1.2380	gas	0.65	2.810
4	CBr2F2	DIBROMODIFLUOROMETHANE	14.9	1.4016	gas	0.66	3.243
5	CClF3	CHLOROTRIFLUOROMETHANE	25.0	1.1990	gas	0.51	2.894
6	CClN	CYANOGEN CHLORIDE	-------	-------	gas	2.82	1.250
7	CCl2F2	DICHLORODIFLUOROMETHANE	26.6	1.2850	gas	0.51	3.052
8	CCl2O	PHOSGENE	25.0	1.3561	gas	1.17	2.877
9	CCl3F	TRICHLOROFLUOROMETHANE	25.0	1.3740	gas	0.45	3.289
10	CCl4	CARBON TETRACHLORIDE	25.0	1.4573	gas	0.00	3.482
11	CF2O	CARBONYL FLUORIDE	-------	-------	gas	0.95	2.269
12	CF4	CARBON TETRAFLUORIDE	-73.2	1.1510	gas	0.00	2.668
13	CHBr3	TRIBROMOMETHANE	25.0	1.5956	gas	0.99	3.730
14	CHClF2	CHLORODIFLUOROMETHANE	25.0	1.2560	gas	1.42	2.689
15	CHCl2F	DICHLOROFLUOROMETHANE	25.0	1.3540	gas	1.29	3.015
16	CHCl3	CHLOROFORM	25.0	1.4431	gas	1.01	3.249
17	CHF3	TRIFLUOROMETHANE	-73.2	1.2150	gas	1.65	2.454
18	CHI3	TRIIODOMETHANE	-------	-------	in benzene	0.90	----------
19	CHN	HYDROGEN CYANIDE	25.0	1.2594	gas	2.98	0.654
20	CHNS	ISOTHIOCYANIC-ACID	-------	-------	gas	1.72	----------
21	CH2BrCl	BROMOCHLOROMETHANE	25.0	1.4808	in benzene	1.66	2.247
22	CH2Br2	DIBROMOMETHANE	25.0	1.5389	gas	1.43	2.327
23	CH2ClF	CHLOROFLUOROMETHANE	-------	-------	gas	1.82	----------
24	CH2Cl2	DICHLOROMETHANE	25.0	1.4212	gas	1.60	2.322
25	CH2F2	DIFLUOROMETHANE	25.0	1.1960	gas	1.96	1.950
26	CH2I2	DIIODOMETHANE	25.0	1.7380	in benzene	1.22	2.438
27	CH2O	FORMALDEHYDE	-------	-------	gas	2.33	1.215
28	CH2O2	FORMIC ACID	25.0	1.3693	gas	1.42	1.847
29	CH3Br	METHYL BROMIDE	25.0	1.4187	gas	1.81	1.173
30	CH3Cl	METHYL CHLORIDE	25.0	1.3362	gas	1.87	1.265
31	CH3Cl3Si	METHYL TRICHLOROSILANE	25.0	1.4085	gas	1.91	----------
32	CH3F	METHYL FLUORIDE	25.0	1.1740	gas	1.85	1.410
33	CH3I	METHYL IODIDE	25.0	1.5270	gas	1.62	1.035
34	CH3NO	FORMAMIDE	25.0	1.4468	gas	3.73	1.894
35	CH3NO2	NITROMETHANE	25.0	1.3796	gas	3.46	2.354
36	CH3NO2	METHYL-NITRITE	-------	-------	-------	-------	----------
37	CH3NO3	METHYL-NITRATE	-------	-------	in benzene	2.88	----------
38	CH4	METHANE	25.0	1.0004	gas	0.00	1.118
39	CH4Cl2Si	METHYL DICHLOROSILANE	19.9	1.3992	in benzene	1.91	3.256
40	CH4O	METHANOL	25.0	1.3265	gas	1.70	1.552
41	CH4O3S	METHANESULFONIC ACID	17.9	1.4317	-------	-------	2.817
42	CH4S	METHYL MERCAPTAN	-------	-------	gas	1.52	1.606
43	CH5ClSi	METHYL CHLOROSILANE	-------	-------	in cyclohexane	1.93	----------
44	CH5N	METHYLAMINE	25.0	1.3491	gas	1.31	1.722
45	CH6Si	METHYL SILANE	-------	-------	gas	0.73	----------
46	CN4O8	TETRANITROMETHANE	25.0	1.4358	gas	0.00	4.147
47	CO	CARBON MONOXIDE	25.0	1.0003	gas	0.11	0.558
48	COS	CARBONYL SULFIDE	-------	1.3785	gas	0.71	1.270
49	CO2	CARBON DIOXIDE	25.0	1.0004	gas	0.00	1.040
50	CS2	CARBON DISULFIDE	25.0	1.6241	gas	0.00	1.569
51	C2BrF3	BROMOTRIFLUOROETHYLENE	-------	-------	gas	0.76	3.366
52	C2Br2F4	1,2-DIBROMOTETRAFLUOROETHANE	25.0	1.3670	-------	-------	3.779
53	C2ClF3	CHLOROTRIFLUOROETHYLENE	19.9	1.0010	gas	0.40	3.293
54	C2ClF5	CHLOROPENTAFLUOROETHANE	25.0	1.2140	gas	0.52	3.587
55	C2Cl2F4	1,2-DICHLOROTETRAFLUOROETHANE	25.0	1.2880	gas	0.56	3.602
56	C2Cl3F3	1,1,2-TRICHLOROTRIFLUOROETHANE	25.0	1.3540	-------	-------	3.791
57	C2Cl4	TETRACHLOROETHYLENE	25.0	1.5055	in benzene	0.00	4.059
58	C2Cl4F2	1,1,2,2-TETRACHLORODIFLUOROETHANE	25.0	1.4130	-------	-------	3.628
59	C2Cl4O	TRICHLOROACETYL CHLORIDE	-------	-------	in benzene	1.20	4.037
60	C2Cl6	HEXACHLOROETHANE	-------	-------	gas	0.00	4.143
61	C2F4	TETRAFLUOROETHYLENE	-------	-------	-------	-------	3.232
62	C2F6	HEXAFLUOROETHANE	-73.2	1.2060	gas	0.00	3.419
63	C2HBrClF3	HALOTHANE	25.0	1.3961	-------	-------	3.735
64	C2HClF2	2-CHLORO-1,1-DIFLUOROETHYLENE	-------	-------	-------	-------	2.994
65	C2HCl3	TRICHLOROETHYLENE	25.0	1.4750	liquid	0.77	3.742
66	C2HCl3O	DICHLOROACETYL CHLORIDE	15.9	1.4638	in CCl4	1.58	3.988
67	C2HCl3O	TRICHLOROACETALDEHYDE	19.9	1.4559	gas	1.96	3.649
68	C2HCl5	PENTACHLOROETHANE	25.0	1.5005	gas	0.92	4.172
69	C2HF3	TRIFLUOROETHENE	-------	-------	-------	-------	----------
70	C2HF3O2	TRIFLUOROACETIC ACID	19.9	1.2850	gas	2.28	3.294
71	C2HF5	PENTAFLUOROETHANE	18.9	1.5012	gas	1.54	3.282
72	C2H2	ACETYLENE	-------	-------	gas	0.00	0.744
73	C2H2Br4	1,1,2,2-TETRABROMOETHANE	25.0	1.6323	in hexane	1.30	4.540
74	C2H2Cl2	1,1-DICHLOROETHYLENE	19.9	1.4247	gas	1.34	3.011
75	C2H2Cl2	cis-1,2-DICHLOROETHYLENE	19.9	1.4490	gas	1.90	2.991
76	C2H2Cl2	trans-1,2-DICHLOROETHYLENE	19.9	1.4462	in CCl4	0.00	2.634
77	C2H2Cl2O	CHLOROACETYL CHLORIDE	-------	-------	gas	2.23	3.394
78	C2H2Cl2O	DICHLOROACETALDEHYDE	-------	-------	in benzene	2.36	3.419

241

NO	FORMULA	NAME	Refractive Index		Dipole Moment		Radius of Gyration
			T, C	Value @ T	State	Debye	Angstrom
79	C2H2Cl2O2	DICHLOROACETIC ACID	19.9	1.4658	-------	-------	3.780
80	C2H2Cl3F	1,1,1-TRICHLOROFLUOROETHANE	-------	-------	-------	-------	3.476
81	C2H2Cl4	1,1,1,2-TETRACHLOROETHANE	25.0	1.4794	gas	1.29	----------
82	C2H2Cl4	1,1,2,2-TETRACHLOROETHANE	25.0	1.4914	gas	1.29	4.044
83	C2H2F2	1,1-DIFLUOROETHYLENE	-------	-------	gas	1.38	2.528
84	C2H2F2	cis-1,2-DIFLUOROETHENE	-------	-------	gas	2.42	----------
85	C2H2F2	trans-1,2-DIFLUOROETHENE	-------	-------	in benzene	0.55	----------
86	C2H2F4	1,1,1,2-TETRAFLUOROETHANE	25.0	1.0007	-------	-------	2.944
87	C2H2O	KETENE	-------	-------	gas	1.42	1.605
88	C2H2O4	OXALIC ACID	-------	-------	in dioxane	2.63	3.220
89	C2H3Br	VINYL BROMIDE	25.0	1.4350	gas	1.42	1.763
90	C2H3Cl	VINYL CHLORIDE	25.0	1.3660	gas	1.45	2.049
91	C2H3ClF2	1-CHLORO-1,1-DIFLUOROETHANE	-------	-------	gas	2.14	2.896
92	C2H3ClO	ACETYL CHLORIDE	19.9	1.3871	gas	2.72	2.712
93	C2H3ClO	CHLOROACETALDEHYDE	-------	-------	in benzene	1.99	2.668
94	C2H3ClO2	CHLOROACETIC ACID	64.9	1.4300	in benzene	2.31	3.157
95	C2H3ClO2	METHYL CHLOROFORMATE	19.9	1.3865	in benzene	2.22	2.962
96	C2H3Cl3	1,1,1-TRICHLOROETHANE	25.0	1.4313	gas	1.78	3.373
97	C2H3Cl3	1,1,2-TRICHLOROETHANE	25.0	1.4689	gas	1.25	3.717
98	C2H3F	VINYL FLUORIDE	25.0	1.3400	gas	1.43	1.934
99	C2H3F3	1,1,1-TRIFLUOROETHANE	25.0	1.2060	gas	2.32	2.767
100	C2H3N	ACETONITRILE	25.0	1.3416	gas	3.92	1.841
101	C2H3NO	METHYL ISOCYANATE	26.9	1.3630	gas	2.80	2.236
102	C2H4	ETHYLENE	25.0	1.0007	gas	0.00	1.548
103	C2H4Br2	1,1-DIBROMOETHANE	25.0	1.5101	in benzene	2.14	3.099
104	C2H4Br2	1,2-DIBROMOETHANE	25.0	1.5360	gas	1.01	2.833
105	C2H4Cl2	1,1-DICHLOROETHANE	25.0	1.4138	gas	2.06	3.095
106	C2H4Cl2	1,2-DICHLOROETHANE	25.0	1.4421	liquid	2.94	2.827
107	C2H4Cl2O	BIS(CHLOROMETHYL)ETHER	20.9	1.4350	gas	0.99	3.373
108	C2H4F2	1,1-DIFLUOROETHANE	25.0	1.2434	gas	2.27	2.514
109	C2H4F2	1,2-DIFLUOROETHANE	25.0	1.2800	gas	2.67	2.508
110	C2H4I2	1,2-DIIODOETHANE	-------	-------	in benzene	1.30	----------
111	C2H4O	ACETALDEHYDE	25.0	1.3283	gas	2.69	2.083
112	C2H4O	ETHYLENE OXIDE	6.9	1.3596	gas	1.89	1.937
113	C2H4OS	THIOACETIC-ACID	-------	-------	-------	-------	----------
114	C2H4O2	ACETIC ACID	25.0	1.3698	gas	1.74	2.610
115	C2H4O2	METHYL FORMATE	25.0	1.3415	gas	1.77	2.387
116	C2H4S	THIACYCLOPROPANE	25.0	1.4870	-------	-------	----------
117	C2H5Br	BROMOETHANE	25.0	1.4212	gas	2.03	1.987
118	C2H5Cl	ETHYL CHLORIDE	25.0	1.3652	gas	2.05	2.267
119	C2H5ClO	2-CHLOROETHANOL	19.9	1.4421	gas	1.78	2.779
120	C2H5F	ETHYL FLUORIDE	25.0	1.2621	gas	1.94	2.832
121	C2H5I	ETHYL IODIDE	25.0	1.5101	gas	1.91	1.799
122	C2H5N	ETHYLENEIMINE	25.0	1.4123	gas	1.90	1.990
123	C2H5NO	ACETAMIDE	77.9	1.4274	gas	3.76	2.621
124	C2H5NO	N-METHYLFORMAMIDE	25.0	1.4300	gas	3.83	2.428
125	C2H5NO2	NITROETHANE	25.0	1.3897	gas	3.65	2.795
126	C2H5NO3	ETHYL-NITRATE	-------	-------	in benzene	2.93	----------
127	C2H6	ETHANE	19.9	1.0047	gas	0.00	1.826
128	C2H6AlCl	DIMETHYLALUMINUM CHLORIDE	-------	-------	gas	1.63	----------
129	C2H6O	DIMETHYL ETHER	25.0	1.2984	gas	1.30	2.154
130	C2H6O	ETHANOL	25.0	1.3594	gas	1.69	2.259
131	C2H6OS	DIMETHYL SULFOXIDE	25.0	1.4773	gas	3.96	2.840
132	C2H6O2	ETHYLENE GLYCOL	25.0	1.4306	in dioxane	2.31	2.564
133	C2H6O4S	DIMETHYL SULFATE	25.0	1.3855	-------	-------	3.314
134	C2H6S	DIMETHYL SULFIDE	25.0	1.4323	gas	1.50	2.374
135	C2H6S	ETHYL MERCAPTAN	25.0	1.4278	gas	1.58	2.363
136	C2H6S2	DIMETHYL DISULFIDE	25.0	1.5230	in benzene	1.97	2.942
137	C2H7N	DIMETHYLAMINE	25.0	1.3566	gas	1.03	2.271
138	C2H7N	ETHYLAMINE	25.0	1.3627	gas	1.22	2.336
139	C2H7NO	MONOETHANOLAMINE	25.0	1.4521	in dioxane	0.78	1.826
140	C2H8N2	ETHYLENEDIAMINE	19.9	1.4568	in benzene	1.90	2.761
141	C2H8Si	DIMETHYL SILANE	-------	-------	gas	0.75	1.881
142	C2N2	CYANOGEN	-------	-------	gas	0.00	2.190
143	C3F6	HEXAFLUOROPROPYLENE	-------	-------	-------	-------	3.796
144	C3F6O	HEXAFLUOROACETONE	-------	-------	gas	0.65	3.933
145	C3F8	OCTAFLUOROPROPANE	-------	-------	-------	-------	3.736
146	C3H2N2	MALONONITRILE	33.9	1.4146	gas	3.72	3.058
147	C3H3Cl	PROPARGYL CHLORIDE	25.0	1.4317	gas	1.68	2.730
148	C3H3N	ACRYLONITRILE	25.0	1.3884	gas	3.87	2.464
149	C3H3NO	OXAZOLE	17.5	1.4285	gas	1.50	2.467
150	C3H4	METHYLACETYLENE	-40.2	1.3863	gas	0.78	1.908
151	C3H4	PROPADIENE	-34.5	1.4169	gas	0.00	1.911
152	C3H4Cl2	2,3-DICHLOROPROPENE	25.0	1.4568	gas	1.74	3.470
153	C3H4O	ACROLEIN	19.9	1.4017	gas	3.12	2.443
154	C3H4O	PROPARGYL ALCOHOL	25.0	1.4300	in benzene	1.78	2.543
155	C3H4O2	ACRYLIC ACID	25.0	1.4185	gas	1.46	2.978
156	C3H4O2	beta-PROPIOLACTONE	-------	-------	gas	4.18	2.481

NO	FORMULA	NAME	Refractive Index		Dipole Moment		Radius of Gyration
			T, C	Value @ T	State	Debye	Angstrom
157	C3H4O2	VINYL FORMATE	-------	-------	in benzene	1.66	2.680
158	C3H4O3	ETHYLENE CARBONATE	39.9	1.4199	gas	4.51	2.869
159	C3H4O3	PYRUVIC ACID	19.9	1.4280	-------	-------	3.235
160	C3H5Br	3-BROMO-1-PROPENE	25.0	1.4603	in benzene	1.81	----------
161	C3H5Cl	2-CHLOROPROPENE	25.0	1.3920	gas	1.66	2.786
162	C3H5Cl	3-CHLOROPROPENE	25.0	1.4116	gas	1.94	2.668
163	C3H5ClO	alpha-EPICHLOROHYDRIN	25.0	1.4358	in CCl4	1.80	3.043
164	C3H5ClO2	METHYL CHLOROACETATE	25.0	1.4197	-------	-------	3.539
165	C3H5ClO2	ETHYL CHLOROFORMATE	19.9	1.3952	gas	2.55	3.595
166	C3H5Cl3	1,2,3-TRICHLOROPROPANE	25.0	1.4812	-------	-------	4.263
167	C3H5I	3-IODO-1-PROPENE	25.0	1.5500	in benzene	1.62	----------
168	C3H5N	PROPIONITRILE	25.0	1.3636	gas	4.02	2.652
169	C3H5NO	ACRYLAMIDE	84.5	1.5660	-------	-------	2.950
170	C3H5NO	HYDRACRYLONITRILE	19.9	1.4256	-------	-------	3.098
171	C3H5NO	LACTONITRILE	19.9	1.4035	-------	-------	3.078
172	C3H5N3O9	NITROGLYCERINE	15.9	1.4786	in benzene	3.15	4.650
173	C3H6	CYCLOPROPANE	-------	-------	gas	0.00	2.120
174	C3H6	PROPYLENE	25.0	1.0009	gas	0.37	2.254
175	C3H6Br2	1,2-DIBROMOPROPANE	-------	-------	gas	1.24	----------
176	C3H6Cl2	1,1-DICHLOROPROPANE	25.0	1.4266	in benzene	2.08	3.495
177	C3H6Cl2	1,2-DICHLOROPROPANE	25.0	1.4368	gas	1.17	3.496
178	C3H6Cl2	1,3-DICHLOROPROPANE	25.0	1.4460	gas	2.08	3.501
179	C3H6Cl2	2,2-DICHLOROPROPANE	25.0	1.4123	gas	2.09	----------
180	C3H6I2	1,2-DIIODOPROPANE	25.0	1.6200	-------	-------	----------
181	C3H6O	ACETONE	25.0	1.3560	gas	2.88	2.746
182	C3H6O	ALLYL ALCOHOL	19.9	1.4135	gas	1.60	2.577
183	C3H6O	METHYL VINYL ETHER	25.0	1.3947	-------	-------	2.570
184	C3H6O	n-PROPIONALDEHYDE	25.0	1.3593	gas	2.52	2.643
185	C3H6O	1,2-PROPYLENE OXIDE	25.0	1.3632	gas	2.01	2.660
186	C3H6O	1,3-PROPYLENE OXIDE	25.0	1.3897	gas	1.93	2.234
187	C3H6O2	ETHYL FORMATE	25.0	1.3575	gas	1.93	2.963
188	C3H6O2	METHYL ACETATE	25.0	1.3589	gas	1.68	2.996
189	C3H6O2	PROPIONIC ACID	25.0	1.3843	gas	1.75	3.107
190	C3H6O2S	3-MERCAPTOPROPIONIC ACID	19.9	1.4940	in dioxane	2.25	3.705
191	C3H6O3	LACTIC ACID	25.0	1.4392	-------	-------	3.298
192	C3H6O3	METHOXYACETIC ACID	19.9	1.4168	gas	2.56	3.293
193	C3H6O3	TRIOXANE	-------	-------	gas	2.08	2.823
194	C3H6S	THIACYCLOBUTANE	25.0	1.5074	-------	-------	----------
195	C3H7Br	1-BROMOPROPANE	25.0	1.4317	gas	2.18	2.629
196	C3H7Br	2-BROMOPROPANE	25.0	1.4221	gas	2.21	2.615
197	C3H7Cl	ISOPROPYL CHLORIDE	25.0	1.3752	gas	2.17	2.905
198	C3H7Cl	n-PROPYL CHLORIDE	25.0	1.3858	gas	2.05	2.774
199	C3H7F	1-FLUOROPROPANE	25.0	1.3091	-------	-------	----------
200	C3H7F	2-FLUOROPROPANE	25.0	1.2992	-------	-------	----------
201	C3H7I	ISOPROPYL IODIDE	25.0	1.4961	in benzene	1.95	2.427
202	C3H7I	n-PROPYL IODIDE	25.0	1.5028	gas	2.04	2.311
203	C3H7N	ALLYLAMINE	19.9	1.4205	in benzene	1.31	2.704
204	C3H7N	PROPYLENEIMINE	25.0	1.4095	-------	-------	2.543
205	C3H7NO	N,N-DIMETHYLFORMAMIDE	25.0	1.4269	gas	3.82	3.027
206	C3H7NO	N-METHYLACETAMIDE	19.9	1.4301	gas	3.72	3.051
207	C3H7NO2	1-NITROPROPANE	25.0	1.3996	gas	3.66	3.359
208	C3H7NO2	2-NITROPROPANE	25.0	1.3924	gas	3.73	3.273
209	C3H7NO3	PROPYL-NITRATE	-------	-------	in benzene	3.01	----------
210	C3H7NO3	ISOPROPYL-NITRATE	-------	-------	-------	-------	----------
211	C3H8	PROPANE	25.0	1.2861	gas	0.00	2.431
212	C3H8O	ISOPROPANOL	25.0	1.3752	gas	1.66	2.807
213	C3H8O	METHYL ETHYL ETHER	-42.2	1.3441	gas	1.23	2.594
214	C3H8O	n-PROPANOL	25.0	1.3837	gas	1.68	2.825
215	C3H8O2	2-METHOXYETHANOL	25.0	1.4002	in benzene	2.04	2.985
216	C3H8O2	METHYLAL	25.0	1.3504	gas	0.74	2.891
217	C3H8O2	1,2-PROPYLENE GLYCOL	25.0	1.4314	liquid	3.63	3.154
218	C3H8O2	1,3-PROPYLENE GLYCOL	25.0	1.4386	in dioxane	2.52	3.169
219	C3H8O3	GLYCEROL	25.0	1.4730	liquid	4.21	3.520
220	C3H8S	n-PROPYLMERCAPTAN	25.0	1.4353	liquid	1.55	2.883
221	C3H8S	ISOPROPYL MERCAPTAN	25.0	1.4225	liquid	1.64	2.993
222	C3H8S	ETHYL-METHYL-SULFIDE	25.0	1.4374	-------	-------	----------
223	C3H9N	n-PROPYLAMINE	25.0	1.3851	gas	1.17	2.806
224	C3H9N	ISOPROPYLAMINE	25.0	1.3711	liquid	1.50	2.802
225	C3H9N	TRIMETHYLAMINE	25.0	1.3443	gas	0.61	2.766
226	C3H9NO	1-AMINO-2-PROPANOL	25.0	1.4460	-------	-------	3.108
227	C3H9NO	3-AMINO-1-PROPANOL	19.9	1.4610	ns	2.69	3.126
228	C3H9NO	METHYLETHANOLAMINE	19.9	1.4385	in benzene	2.16	3.035
229	C3H9O4P	TRIMETHYL PHOSPHATE	25.0	1.3939	in CCl4	3.03	3.640
230	C3H10N2	1,2-PROPANEDIAMINE	19.9	1.4460	-------	-------	3.170
231	C3H10Si	TRIMETHYL SILANE	-------	-------	gas	0.52	----------
232	C4Cl4S	TETRACHLOROTHIOPHENE	19.9	1.5915	in benzene	0.93	5.091
233	C4Cl6	HEXACHLORO-1,3-BUTADIENE	-------	-------	in benzene	0.20	5.299
234	C4F8	OCTAFLUORO-2-BUTENE	-------	-------	-------	-------	2.940

NO	FORMULA	NAME	Refractive Index		Dipole Moment		Radius of Gyration
			T, C	Value @ T	State	Debye	Angstrom
235	C4F8	OCTAFLUOROCYCLOBUTANE	25.0	1.2170	gas	0.00	3.889
236	C4F10	DECAFLUOROBUTANE	-------	-------	-------	-------	4.368
237	C4H2	BUTADIYNE(BIACETYLENE)			-------	-------	----------
238	C4H2O3	MALEIC ANHYDRIDE	64.4	1.4688	in dioxane	3.93	3.312
239	C4H4	VINYLACETYLENE	25.0	1.4161	gas	0.40	2.552
240	C4H4N2	SUCCINONITRILE	59.9	1.4173	in dioxane	3.83	3.361
241	C4H4O	FURAN	25.0	1.4187	gas	0.66	2.559
242	C4H4O2	DIKETENE	19.9	1.4379	gas	3.54	2.998
243	C4H4O3	SUCCINIC ANHYDRIDE	-------	-------	in benzene	3.83	3.281
244	C4H4O4	FUMARIC ACID	-------	-------	in dioxane	2.45	4.139
245	C4H4O4	MALEIC ACID	-------	-------	in dioxane	3.17	4.221
246	C4H4S	THIOPHENE	25.0	1.5257	gas	0.55	2.781
247	C4H5Cl	CHLOROPRENE	19.9	1.4583	gas	1.43	3.274
248	C4H5N	trans-CROTONITRILE	19.9	1.4225	gas	4.53	2.893
249	C4H5N	cis-CROTONITRILE	29.9	1.4134	gas	4.08	2.924
250	C4H5N	METHACRYLONITRILE	25.0	1.3989	gas	3.69	3.172
251	C4H5N	PYRROLE	25.0	1.5078	gas	1.84	2.567
252	C4H5N	VINYLACETONITRILE	25.0	1.4050	-------	-------	2.979
253	C4H5NO2	METHYL CYANOACETATE	25.0	1.4166	-------	-------	3.733
254	C4H6	CYCLOBUTENE	-------	-------	-------	-------	----------
255	C4H6	1,2-BUTADIENE	1.4	1.4205	gas	0.40	2.724
256	C4H6	1,3-BUTADIENE	-25.2	1.4293	gas	0.00	2.602
257	C4H6	DIMETHYLACETYLENE	25.0	1.3893	-------	-------	2.519
258	C4H6	ETHYLACETYLENE	-------	-------	gas	0.81	2.770
259	C4H6Cl2	1,3-DICHLORO-trans-2-BUTENE	25.0	1.4695	in benzene	2.15	4.074
260	C4H6Cl2	1,4-DICHLORO-cis-2-BUTENE	25.0	1.4887	-------	-------	3.820
261	C4H6Cl2	1,4-DICHLORO-trans-2-BUTENE	25.0	1.4863	-------	-------	3.832
262	C4H6Cl2	3,4-DICHLORO-1-BUTENE	25.0	1.4615	-------	-------	3.934
263	C4H6O	trans-CROTONALDEHYDE	19.9	1.4373	gas	3.67	3.087
264	C4H6O	2,5-DIHYDROFURAN	25.0	1.4340	in benzene	1.54	2.580
265	C4H6O	DIVINYL ETHER	-------	-------	in benzene	1.07	2.951
266	C4H6O	METHACROLEIN	19.9	1.4169	gas	2.68	3.121
267	C4H6O2	2-BUTYNE-1,4-DIOL	25.0	1.4500	in benzene	2.63	3.715
268	C4H6O2	gamma-BUTYROLACTONE	25.0	1.4348	gas	3.82	2.642
269	C4H6O2	cis-CROTONIC ACID	19.9	1.4456	-------	-------	3.342
270	C4H6O2	trans-CROTONIC ACID	79.9	1.4228	in benzene	2.13	3.493
271	C4H6O2	METHACRYLIC ACID	25.0	1.4288	in benzene	1.65	3.412
272	C4H6O2	METHYL ACRYLATE	25.0	1.4003	-------	-------	3.280
273	C4H6O2	VINYL ACETATE	25.0	1.3934	in benzene	1.79	2.969
274	C4H6O3	ACETIC ANHYDRIDE	19.9	1.3892	gas	2.80	3.579
275	C4H6O4	SUCCINIC ACID	24.9	1.3373	gas	2.20	4.159
276	C4H6O5	DIGLYCOLIC ACID	-------	-------	-------	-------	4.475
277	C4H6O5	MALIC ACID	25.0	1.3516	gas	3.12	4.209
278	C4H6O6	TARTARIC ACID	-------	-------	in dioxane	3.24	4.289
279	C4H7N	n-BUTYRONITRILE	25.0	1.3820	gas	4.07	3.088
280	C4H7N	ISOBUTYRONITRILE	25.0	1.3712	gas	4.29	3.211
281	C4H7NO	ACETONE CYANOHYDRIN	-------	-------	in benzene	3.17	3.344
282	C4H7NO	2-METHACRYLAMIDE	-------	-------	-------	-------	3.380
283	C4H7NO	3-METHOXYPROPIONITRILE	19.9	1.4032	-------	-------	3.387
284	C4H7NO	2-PYRROLIDONE	25.0	1.4860	in benzene	3.10	3.074
285	C4H8	1-BUTENE	-125.2	1.3803	gas	0.34	2.762
286	C4H8	cis-2-BUTENE	-25.2	1.3842	gas	0.30	2.833
287	C4H8	trans-2-BUTENE	-25.2	1.3932	gas	0.00	2.734
288	C4H8	CYCLOBUTANE	25.0	1.3620	gas	0.00	2.450
289	C4H8	ISOBUTENE	-25.2	1.3926	gas	0.50	2.875
290	C4H8Br2	1,2-DIBROMOBUTANE	25.0	1.5125	-------	-------	----------
291	C4H8Br2	2,3-DIBROMOBUTANE	25.0	1.5125	-------	-------	----------
292	C4H8Cl2	1,4-DICHLOROBUTANE	25.0	1.4522	gas	2.22	3.760
293	C4H8I2	1,2-DIIODOBUTANE	25.0	1.6000	-------	-------	----------
294	C4H8O	n-BUTYRALDEHYDE	25.0	1.3766	gas	2.72	3.185
295	C4H8O	ISOBUTYRALDEHYDE	25.0	1.3698	gas	2.70	3.176
296	C4H8O	1,2-EPOXYBUTANE	25.0	1.3810	in benzene	2.01	2.960
297	C4H8O	METHYL ETHYL KETONE	25.0	1.3764	benzene/dioxane	2.76	3.135
298	C4H8O	ETHYL VINYL ETHER	25.0	1.3729	gas	1.27	2.940
299	C4H8O	TETRAHYDROFURAN	25.0	1.4050	gas	1.63	2.694
300	C4H8O2	cis-2-BUTENE-1,4-DIOL	25.0	1.4716	in benzene	2.50	3.516
301	C4H8O2	trans-2-BUTENE-1,4-DIOL	-------	-------	in benzene	2.47	3.530
302	C4H8O2	ISOBUTYRIC ACID	25.0	1.3908	liquid	1.09	3.444
303	C4H8O2	n-BUTYRIC ACID	25.0	1.3958	liquid	1.23	3.610
304	C4H8O2	1,4-DIOXANE	25.0	1.4202	gas	0.00	3.017
305	C4H8O2	ETHYL ACETATE	25.0	1.3704	gas	1.78	3.468
306	C4H8O2	METHYL PROPIONATE	25.0	1.3742	in benzene	1.70	3.490
307	C4H8O2	n-PROPYL FORMATE	25.0	1.3750	in benzene	1.91	3.364
308	C4H8O2S	SULFOLANE	19.9	1.4780	in benzene	4.69	3.309
309	C4H8S	TETRAHYDROTHIOPHENE	25.0	1.5074	in benzene	1.90	2.898
310	C4H9Br	1-BROMOBUTANE	25.0	1.4378	gas	2.08	3.132
311	C4H9Br	2-BROMOBUTANE	25.0	1.4342	gas	2.23	3.123
312	C4H9Cl	n-BUTYL CHLORIDE	25.0	1.4001	gas	2.05	3.338

244

Table 10-1 REFRACTIVE INDEX, DIPOLE M0MENT, AND RADIUS OF GYRATION - ORGANIC COMPOUNDS (continued)

| NO | FORMULA | NAME | Refractive Index | | Dipole Moment | | Radius of Gyration |
			T, C	Value @ T	State	Debye	Angstrom
313	C4H9Cl	sec-BUTYL CHLORIDE	25.0	1.3941	gas	2.04	3.324
314	C4H9Cl	tert-BUTYL CHLORIDE	25.0	1.3828	gas	2.13	3.152
315	C4H9I	2-IODO-2-METHYLPROPANE	25.0	1.4890	in benzene	2.15	----------
316	C4H9N	PYRROLIDINE	25.0	1.4402	in benzene	1.58	2.700
317	C4H9NO	N,N-DIMETHYLACETAMIDE	25.0	1.4356	gas	3.81	3.368
318	C4H9NO	MORPHOLINE	25.0	1.4521	in benzene	1.54	3.042
319	C4H9NO2	1-NITROBUTANE	25.0	1.4080	in benzene	3.40	----------
320	C4H9NO2	2-NITROBUTANE	25.0	1.4019	-------	-------	----------
321	C4H10	n-BUTANE	25.0	1.3292	gas	0.00	2.886
322	C4H10	ISOBUTANE	-25.2	1.3514	gas	0.13	2.948
323	C4H10N2	PIPERAZINE	112.9	1.4460	in benzene	1.47	3.035
324	C4H10O	n-BUTANOL	25.0	1.3971	gas	1.66	3.251
325	C4H10O	sec-BUTANOL	25.0	1.3949	in benzene	1.66	3.203
326	C4H10O	tert-BUTANOL	25.0	1.3852	in benzene	1.67	3.067
327	C4H10O	DIETHYL ETHER	25.0	1.3495	gas	1.15	3.177
328	C4H10O	METHYL-PROPYL-ETHER	25.0	1.3544	in benzene	1.24	----------
329	C4H10O	METHYL ISOPROPYL ETHER	25.0	1.3576	-------	-------	3.125
330	C4H10O	ISOBUTANOL	25.0	1.3938	gas	1.64	3.332
331	C4H10O2	1,3-BUTANEDIOL	25.0	1.4390	-------	-------	3.455
332	C4H10O2	1,4-BUTANEDIOL	25.0	1.4445	liquid	3.93	3.582
333	C4H10O2	2,3-BUTANEDIOL	25.0	1.4310	in benzene	2.10	3.371
334	C4H10O2	t-BUTYL HYDROPEROXIDE	25.0	1.3983	in benzene	1.81	3.295
335	C4H10O2	1,2-DIMETHOXYETHANE	25.0	1.3781	in benzene	1.71	3.328
336	C4H10O2	2-ETHOXYETHANOL	25.0	1.4057	in benzene	2.08	3.392
337	C4H10O3	DIETHYLENE GLYCOL	25.0	1.4460	liquid	5.49	3.739
338	C4H10O4S	DIETHYL SULFATE	19.9	1.3989	-------	-------	4.239
339	C4H10S	n-BUTYL MERCAPTAN	25.0	1.4403	in benzene	1.53	3.400
340	C4H10S	ISOBUTYL MERCAPTAN	25.0	1.4360	in benzene	1.53	3.455
341	C4H10S	sec-BUTYL MERCAPTAN	25.0	1.4339	in benzene	1.55	3.362
342	C4H10S	tert-BUTYL MERCAPTAN	25.0	1.4200	in benzene	1.59	3.185
343	C4H10S	DIETHYL SULFIDE	25.0	1.4402	gas	1.54	3.212
344	C4H10S	ISOPROPYL-METHYL-SULFIDE	25.0	1.4363	-------	-------	----------
345	C4H10S	METHYL-PROPYL-SULFIDE	25.0	1.4416	-------	-------	----------
346	C4H10S2	DIETHYL DISULFIDE	25.0	1.5047	in benzene	1.96	----------
347	C4H11N	n-BUTYLAMINE	25.0	1.3987	in benzene	1.39	3.289
348	C4H11N	ISOBUTYLAMINE	25.0	1.3945	in benzene	1.27	3.230
349	C4H11N	sec-BUTYLAMINE	25.0	1.3907	in benzene	1.28	3.201
350	C4H11N	tert-BUTYLAMINE	25.0	1.3761	in benzene	1.29	3.048
351	C4H11N	DIETHYLAMINE	25.0	1.3825	gas	0.92	3.172
352	C4H11NO	DIMETHYLETHANOLAMINE	25.0	1.4277	in benzene	2.21	3.463
353	C4H11NO2	DIETHANOLAMINE	19.9	1.4747	in dioxane	0.85	3.907
354	C4H11NO2	2-AMINOETHOXYETHANOL	19.9	1.4610	-------	-------	3.737
355	C4H12N2O	N-AMINOETHYL ETHANOLAMINE	19.9	1.4861	-------	-------	3.821
356	C4H12Si	TETRAMETHYLSILANE	25.0	1.3582	liquid	0.00	----------
357	C4H13N3	DIETHYLENE TRIAMINE	25.0	1.4810	-------	-------	3.865
358	C5Cl6	HEXACHLOROCYCLOPENTADIENE	25.0	1.5626	liquid	0.72	5.310
359	C5H4O2	FURFURAL	25.0	1.5234	in benzene	3.60	3.350
360	C5H5N	PYRIDINE	25.0	1.5075	gas	2.19	2.938
361	C5H6	CYCLOPENTADIENE	19.9	1.4429	gas	0.42	2.752
362	C5H6	2-METHYL-1-BUTENE-3-YNE	25.0	1.4151	in benzene	0.55	3.224
363	C5H6	1-PENTENE-3-YNE	25.0	1.4490	gas	0.66	3.086
364	C5H6	1-PENTENE-4-YNE	25.0	1.4125	-------	-------	3.025
365	C5H6N2	GLUTARONITRILE	25.0	1.4332	-------	-------	3.972
366	C5H6O2	FURFURYL ALCOHOL	25.0	1.4831	in benzene	1.92	3.422
367	C5H6O3	GLUTARIC ANHYDRIDE	-------	-------	-------	-------	3.651
368	C5H6O4	CITRACONIC ACID	-------	-------	-------	-------	3.744
369	C5H6O4	ITACONIC ACID	-------	-------	in dioxane	2.25	4.185
370	C5H6S	2-METHYLTHIOPHENE	25.0	1.5174	in benzene	0.67	----------
371	C5H6S	3-METHYLTHIOPHENE	25.0	1.5176	in benzene	0.82	----------
372	C5H7N	N-METHYLPYRROLE	19.9	1.4875	gas	2.10	3.082
373	C5H7NO2	ETHYL CYANOACETATE	20.5	1.4179	gas	2.17	4.180
374	C5H8	CYCLOPENTENE	25.0	1.4194	gas	0.20	2.728
375	C5H8	ISOPRENE	25.0	1.4185	gas	0.25	3.213
376	C5H8	3-METHYL-1,2-BUTADIENE	25.0	1.4169	-------	-------	3.021
377	C5H8	2-METHYL-1,3-BUTADIENE	25.0	1.4185	gas	0.15	----------
378	C5H8	1,2-PENTADIENE	25.0	1.4177	-------	-------	3.262
379	C5H8	cis-1,3-PENTADIENE	25.0	1.4329	-------	-------	3.212
380	C5H8	trans-1,3-PENTADIENE	25.0	1.4267	gas	0.68	3.087
381	C5H8	1,4-PENTADIENE	25.0	1.3854	in benzene	0.38	3.049
382	C5H8	2,3-PENTADIENE	25.0	1.4251	-------	-------	3.301
383	C5H8	1-PENTYNE	25.0	1.3822	gas	0.81	3.115
384	C5H8	2-PENTYNE	25.0	1.4009	-------	-------	----------
385	C5H8	3-METHYL-1-BUTYNE	25.0	1.3695	-------	-------	3.327
386	C5H8	SPIROPENTANE	-------	-------	-------	-------	----------
387	C5H8N4O12	PENTAERYTHRITOL TETRANITRATE	-------	-------	in dioxane	2.48	6.929
388	C5H8O	CYCLOPENTANONE	19.9	1.4359	in benzene	2.93	3.165
389	C5H8O	METHYL ISOPROPENYL KETONE	25.0	1.4212	in benzene	2.74	3.480
390	C5H8O2	ACETYLACETONE	25.0	1.4465	gas	2.81	4.017

NO	FORMULA	NAME	Refractive Index		Dipole Moment		Radius of Gyration
			T, C	Value @ T	State	Debye	Angstrom
391	C5H8O2	ALLYL ACETATE	25.0	1.3985	-------	-------	3.730
392	C5H8O2	ETHYL ACRYLATE	25.0	1.4034	-------	-------	3.784
393	C5H8O2	METHYL METHACRYLATE	25.0	1.4120	in benzene	1.97	3.620
394	C5H8O2	VINYL PROPIONATE	-------	-------	-------	-------	3.867
395	C5H8O3	2-HYDROXYETHYL ACRYLATE	22.9	1.4460	-------	-------	4.188
396	C5H8O3	LEVULINIC ACID	19.9	1.4396	-------	-------	3.675
397	C5H8O3	METHYL ACETOACETATE	19.9	1.4186	-------	-------	3.934
398	C5H8O4	GLUTARIC ACID	106.9	'1.4188	in dioxane	2.64	4.591
399	C5H9N	VALERONITRILE	25.0	1.3951	in benzene	3.60	3.631
400	C5H9NO	n-BUTYL ISOCYANATE	25.0	1.4061	-------	-------	3.851
401	C5H9NO	N-METHYL-2-PYRROLIDONE	25.0	1.4690	in benzene	4.09	3.482
402	C5H9NO4	L-GLUTAMIC ACID	-------	-------	-------	-------	4.711
403	C5H10	CYCLOPENTANE	25.0	1.4036	liquid	0.00	2.850
404	C5H10	2-METHYL-1-BUTENE	25.0	1.3746	gas	0.51	3.287
405	C5H10	2-METHYL-2-BUTENE	25.0	1.3842	-------	-------	3.281
406	C5H10	3-METHYL-1-BUTENE	25.0	1.3611	gas	0.32	3.272
407	C5H10	1-PENTENE	25.0	1.3684	liquid	0.47	3.231
408	C5H10	cis-2-PENTENE	25.0	1.3798	-------	-------	3.330
409	C5H10	trans-2-PENTENE	25.0	1.3761	-------	-------	3.280
410	C5H10Br2	2,3-DIBROMO-2-METHYLBUTANE	25.0	1.5078	-------	-------	----------
411	C5H10Cl2	1,5-DICHLOROPENTANE	25.0	1.4541	in benzene	2.36	4.423
412	C5H10O	METHYL ISOPROPYL KETONE	25.0	1.3857	in benzene	2.77	3.465
413	C5H10O	2-PENTANONE	25.0	1.3880	in benzene	2.72	3.618
414	C5H10O	DIETHYL KETONE	25.0	1.3900	in CCl4	2.82	3.583
415	C5H10O	VALERALDEHYDE	25.0	1.3917	in benzene	2.57	3.516
416	C5H10O2	n-BUTYL FORMATE	25.0	1.3874	gas	1.92	3.947
417	C5H10O2	ETHYL PROPIONATE	25.0	1.3814	in benzene	1.75	3.934
418	C5H10O2	ISOBUTYL FORMATE	25.0	1.3835	in benzene	1.89	4.294
419	C5H10O2	ISOPROPYL ACETATE	25.0	1.3750	gas	1.75	3.679
420	C5H10O2	n-PROPYL ACETATE	25.0	1.3828	in benzene	1.79	3.969
421	C5H10O2	METHYL n-BUTYRATE	25.0	1.3847	in benzene	1.72	3.972
422	C5H10O2	2-METHYLBUTYRIC ACID	-------	-------	-------	-------	3.762
423	C5H10O2	ISOVALERIC ACID	25.0	1.4022	in benzene	0.63	3.762
424	C5H10O2	VALERIC ACID	25.0	1.4060	-------	-------	3.965
425	C5H10O2	TETRAHYDROFURFURYL ALCOHOL	25.0	1.4499	in dioxane	2.12	3.481
426	C5H10O2S	3-METHYL SULFOLANE	25.0	1.4756	gas	4.80	3.499
427	C5H10O3	DIETHYL CARBONATE	25.0	1.3829	gas	1.10	3.986
428	C5H10O3	ETHYL LACTATE	19.9	1.4124	in benzene	2.40	3.622
429	C5H10S	THIACYCLOHEXANE	25.0	1.5043	-------	-------	----------
430	C5H10S	CYCLOPENTANETHIOL	-------	-------	-------	-------	----------
431	C5H11Br	1-BROMOPENTANE	25.0	1.4420	liquid	1.99	----------
432	C5H11Cl	1-CHLOROPENTANE	25.0	1.4104	in benzene	1.94	3.995
433	C5H11Cl	1-CHLORO-3-METHYLBUTANE	25.0	1.4063	in benzene	1.94	----------
434	C5H11Cl	2-CHLORO-2-METHYLBUTANE	25.0	1.4023	in benzene	2.16	----------
435	C5H11N	N-METHYLPYRROLIDINE	19.9	1.4292	-------	-------	3.104
436	C5H11N	PIPERIDINE	19.9	1.4525	in benzene	1.19	3.122
437	C5H11NO	tert-BUTYLFORMAMIDE	25.0	1.4275	in dioxane	3.93	3.507
438	C5H12	ISOPENTANE	25.0	1.3509	gas	0.13	3.324
439	C5H12	NEOPENTANE	25.0	1.3390	gas	0.00	3.161
440	C5H12	n-PENTANE	25.0	1.3547	gas	0.00	3.337
441	C5H12O	2,2-DIMETHYL-1-PROPANOL	25.0	1.3915	-------	-------	3.399
442	C5H12O	tert-PENTYL-ALCOHOL	25.0	1.4024	-------	-------	----------
443	C5H12O	2-METHYL-1-BUTANOL	25.0	1.4086	-------	-------	3.612
444	C5H12O	2-METHYL-2-BUTANOL	25.0	1.4024	in benzene	1.70	3.359
445	C5H12O	3-METHYL-1-BUTANOL	25.0	1.4052	in benzene	1.80	3.684
446	C5H12O	3-METHYL-2-BUTANOL	25.0	1.4075	-------	-------	3.480
447	C5H12O	1-PENTANOL	25.0	1.4080	in CCl4	1.70	3.679
448	C5H12O	2-PENTANOL	25.0	1.4044	-------	-------	3.619
449	C5H12O	3-PENTANOL	25.0	1.4079	-------	-------	3.503
450	C5H12O	METHYL sec-BUTYL ETHER	25.0	1.3702	-------	-------	3.463
451	C5H12O	METHYL tert-BUTYL ETHER	25.0	1.3663	in benzene	1.36	3.179
452	C5H120	METHYL ISOBUTYL ETHER	25.0	1.3852	-------	-------	3.608
453	C5H12O	ETHYL PROPYL ETHER	25.0	1.3660	in benzene	1.16	3.470
454	C5H12O2	ETHYLENE GLYCOL MONOPROPYL ETHER	19.9	1.4133	-------	-------	3.822
455	C5H12O2	NEOPENTYL GLYCOL	-------	-------	-------	-------	3.660
456	C5H12O2	1,5-PENTANEDIOL	25.0	1.4487	in dioxane	2.37	4.030
457	C5H12O3	2-(2-METHOXYETHOXY)ETHANOL	25.0	1.4245	-------	-------	4.098
458	C5H12O4	PENTAERYTHRITOL	-------	-------	gas	2.00	4.159
459	C5H12S	n-PENTYL MERCAPTAN	25.0	1.4444	in benzene	1.58	3.797
460	C5H12S	BUTYL-METHYL-SULFIDE	25.0	1.4452	-------	-------	----------
461	C5H12S	ETHYL-PROPYL-SULFIDE	25.0	1.4435	-------	-------	----------
462	C5H12S	2-METHYL-2-BUTANETHIOL	25.0	1.4354	-------	-------	----------
463	C5H13N	n-PENTYLAMINE	25.0	1.4093	in benzene	1.55	3.641
464	C5H13NO2	METHYL DIETHANOLAMINE	19.9	1.4685	gas	2.86	4.366
465	C6Cl6	HEXACHLOROBENZENE	23.6	1.5691	in benzene	0.54	5.701
466	C6F6	HEXAFLUOROBENZENE	25.0	1.3761	in benzene	0.33	4.706
467	C6H3ClN2O4	1-CHLORO-2,4-DINITROBENZENE	44.9	1.5924	in benzene	3.24	5.199
468	C6H3Cl2NO2	1,2-DICHLORO-4-NITROBENZENE	24.9	1.5929	in benzene	2.17	----------

Table 10-1 REFRACTIVE INDEX, DIPOLE MOMENT, AND RADIUS OF GYRATION - ORGANIC COMPOUNDS (continued)

NO	FORMULA	NAME	Refractive Index		Dipole Moment		Radius of Gyration
			T, C	Value @ T	State	Debye	Angstrom
469	C6H3Cl3	1,2,4-TRICHLOROBENZENE	25.0	1.5693	in benzene	1.26	4.832
470	C6H3N3O6	1,3,5-TRINITROBENZENE	-------	-------	in benzene	0.41	5.769
471	C6H4Br2	m-DIBROMOBENZENE	16.9	1.6083	in benzene	1.47	4.691
472	C6H4ClNO2	m-CHLORONITROBENZENE	49.9	1.5545	gas	3.73	4.676
473	C6H4ClNO2	o-CHLORONITROBENZENE	44.9	1.5520	gas	4.64	4.367
474	C6H4ClNO2	p-CHLORONITROBENZENE	-------	-------	gas	2.83	4.494
475	C6H4Cl2	m-DICHLOROBENZENE	25.0	1.5434	gas	1.72	4.389
476	C6H4Cl2	o-DICHLOROBENZENE	25.0	1.5491	gas	2.50	4.186
477	C6H4Cl2	p-DICHLOROBENZENE	54.9	1.5285	gas	0.00	4.149
478	C6H4F2	m-DIFLUOROBENZENE	25.0	-------	gas	1.58	----------
479	C6H4F2	o-DIFLUOROBENZENE	25.0	-------	in benzene	2.40	----------
480	C6H4F2	p-DIFLUOROBENZENE	25.0	-------	-------	-------	----------
481	C6H4N2O4	m-DINITROBENZENE	-------	-------	in benzene	3.84	4.873
482	C6H4N2O4	o-DINITROBENZENE	-------	-------	in benzene	6.30	4.485
483	C6H4N2O4	p-DINITROBENZENE	-------	-------	gas	0.00	4.769
484	C6H5Br	BROMOBENZENE	25.0	1.5577	gas	1.70	3.466
485	C6H5Cl	MONOCHLOROBENZENE	19.9	1.5248	gas	1.69	3.603
486	C6H5ClO	m-CHLOROPHENOL	19.9	1.5632	in benzene	2.17	4.043
487	C6H5ClO	o-CHLOROPHENOL	25.0	1.5568	in benzene	1.33	3.837
488	C6H5ClO	p-CHLOROPHENOL	54.9	1.5419	gas	2.11	3.913
489	C6H5Cl2N	3,4-DICHLOROANILINE	-------	-------	-------	-------	4.528
490	C6H5F	FLUOROBENZENE	25.0	1.4629	gas	1.60	3.345
491	C6H5I	IODOBENZENE	19.9	1.6210	gas	1.70	3.359
492	C6H5NO2	NITROBENZENE	25.0	1.5499	gas	4.22	3.944
493	C6H6	BENZENE	25.0	1.4979	gas	0.00	3.004
494	C6H6ClN	m-CHLOROANILINE	20.7	1.5942	in benzene	2.94	4.092
495	C6H6ClN	o-CHLOROANILINE	25.0	1.5859	in benzene	1.77	3.934
496	C6H6ClN	p-CHLOROANILINE	86.9	1.5546	in benzene	2.99	3.929
497	C6H6N2	cis-DICYANO-1-BUTENE	19.9	1.4665	-------	-------	4.548
498	C6H6N2	trans-DICYANO-1-BUTENE	19.9	1.4701	-------	-------	4.061
499	C6H6N2	1,4-DICYANO-2-BUTENE	-------	-------	-------	-------	4.155
500	C6H6N2O2	m-NITROANILINE	-------	-------	in benzene	4.90	4.350
501	C6H6N2O2	o-NITROANILINE	-------	-------	in benzene	4.06	4.198
502	C6H6N2O2	p-NITROANILINE	-------	-------	in benzene	6.29	4.263
503	C6H6O	PHENOL	24.9	1.5496	gas	1.45	3.415
504	C6H6O2	1,2-BENZENEDIOL	25.0	1.6044	in benzene	2.60	3.672
505	C6H6O2	1,3-BENZENEDIOL	25.0	1.5781	in benzene	2.09	3.795
506	C6H6O2	p-HYDROQUINONE	-------	-------	in benzene	1.40	3.708
507	C6H6O3	1,2,3-BENZENETRIOL	-------	-------	-------	-------	3.960
508	C6H6S	PHENYL MERCAPTAN	25.0	1.5872	in benzene	1.23	3.608
509	C6H7N	ANILINE	25.0	1.5836	gas	1.53	3.436
510	C6H7N	2-METHYLPYRIDINE	25.0	1.4984	in benzene	1.97	3.365
511	C6H7N	3-METHYLPYRIDINE	23.9	0.1504	in benzene	2.40	3.401
512	C6H7N	4-METHYLPYRIDINE	19.9	1.5058	gas	2.75	3.409
513	C6H8	1,3-CYCLOHEXADIENE	19.9	1.4755	gas	0.44	2.835
514	C6H8	METHYLCYCLOPENTADIENE	25.0	1.4572	-------	-------	3.192
515	C6H8N2	ADIPONITRILE	25.0	1.4360	gas	3.76	3.979
516	C6H8N2	METHYLGLUTARONITRILE	25.0	1.4312	-------	-------	4.448
517	C6H8N2	m-PHENYLENEDIAMINE	-------	-------	gas	1.81	3.836
518	C6H8N2	o-PHENYLENEDIAMINE	-------	-------	gas	1.53	3.740
519	C6H8N2	p-PHENYLENEDIAMINE	-------	-------	gas	1.53	3.733
520	C6H8N2	PHENYLHYDRAZINE	25.0	1.6055	in benzene	1.67	3.776
521	C6H8N2O	BIS(CYANOETHYL)ETHER	25.0	1.4392	-------	-------	4.715
522	C6H8O4	DIMETHYL MALEATE	25.0	1.4405	in CCl4	2.48	4.952
523	C6H8O6	ASCORBIC ACID	-------	-------	in dioxane	3.96	4.891
524	C6H8O7	CITRIC ACID	19.9	1.4960	-------	-------	4.705
525	C6H10	1-METHYLCYCLOPENTENE	25.0	1.4294	-------	-------	----------
526	C6H10	3-METHYLCYCLOPENTENE	25.0	1.4184	-------	-------	----------
527	C6H10	4-METHYLCYCLOPENTENE	25.0	1.4184	-------	-------	----------
528	C6H10	CYCLOHEXENE	25.0	1.4438	gas	0.55	3.157
529	C6H10	2,3-DIMETHYL-1,3-BUTADIENE	25.0	1.4362	gas	0.51	3.570
530	C6H10	1,5-HEXADIENE	25.0	1.4010	-------	-------	3.464
531	C6H10	cis,trans-2,4-HEXADIENE	19.9	1.4560	in benzene	0.31	3.765
532	C6H10	trans,trans-2,4-HEXADIENE	19.9	1.4510	in benzene	0.31	3.525
533	C6H10	1-HEXYNE	25.0	1.3957	gas	0.83	3.691
534	C6H10	2-HEXYNE	25.0	1.4109	-------	-------	3.718
535	C6H10	3-HEXYNE	25.0	1.4088	-------	-------	3.966
536	C6H10O	CYCLOHEXANONE	25.0	1.4507	gas	3.08	3.511
537	C6H10O	MESITYL OXIDE	25.0	1.4414	in benzene	3.20	3.913
538	C6H10O2	epsilon-CAPROLACTONE	23.9	1.4481	gas	4.45	3.462
539	C6H10O2	ETHYL METHACRYLATE	25.0	1.4115	in benzene	2.15	4.197
540	C6H10O2	n-PROPYL ACRYLATE	25.0	1.4130	-------	-------	4.336
541	C6H10O3	ETHYLACETOACETATE	-------	-------	gas	2.96	4.405
542	C6H10O3	PROPIONIC ANHYDRIDE	19.9	1.4045	-------	-------	4.295
543	C6H10O4	ADIPIC ACID	25.0	1.4880	in dioxane	2.32	4.976
544	C6H10O4	DIETHYL OXALATE	19.9	1.4102	in benzene	2.49	4.668
545	C6H10O4	ETHYLENE GLYCOL DIACETATE	19.9	1.4159	in benzene	2.34	4.867
546	C6H10O4	ETHYLIDENE DIACETATE	25.0	1.3985	-------	-------	4.481

| NO | FORMULA | NAME | Refractive Index | | Dipole Moment | | Radius of Gyration |
			T, C	Value @ T	State	Debye	Angstrom
547	C6H11N	HEXANENITRILE	25.0	1.4048	-------	-------	3.971
548	C6H11NO	epsilon-CAPROLACTAM	-------	-------	in benzene	3.88	3.761
549	C6H11NO	CYCLOHEXANONE OXIME	-------	-------	in benzene	0.83	3.286
550	C6H12	CYCLOHEXANE	25.0	1.4235	gas	0.61	3.242
551	C6H12	2,3-DIMETHYL-1-BUTENE	25.0	1.3873	-------	-------	3.628
552	C6H12	2,3-DIMETHYL-2-BUTENE	25.0	1.4424	-------	-------	3.577
553	C6H12	3,3-DIMETHYL-1-BUTENE	25.0	1.3731	-------	-------	3.341
554	C6H12	2-ETHYL-1-BUTENE	25.0	1.3938	-------	-------	3.625
555	C6H12	trans-3-METHYL-2-PENTENE	-------	-------	-------	-------	----------
556	C6H12	1-HEXENE	25.0	1.3850	liquid	0.34	3.660
557	C6H12	cis-2-HEXENE	25.0	1.3947	-------	-------	3.760
558	C6H12	trans-2-HEXENE	25.0	1.3907	-------	-------	3.616
559	C6H12	cis-3-HEXENE	25.0	1.3919	in benzene	0.34	3.649
560	C6H12	trans-3-HEXENE	25.0	1.3914	gas	0.00	3.649
561	C6H12	METHYLCYCLOPENTANE	25.0	1.4070	liquid	0.00	3.299
562	C6H12	2-METHYL-1-PENTENE	25.0	1.3891	-------	-------	3.704
563	C6H12	2-METHYL-2-PENTENE	25.0	1.3974	-------	-------	3.738
564	C6H12	3-METHYL-1-PENTENE	25.0	1.3813	-------	-------	3.557
565	C6H12	3-METHYL-cis-2-PENTENE	25.0	1.3988	-------	-------	3.589
566	C6H12	4-METHYL-1-PENTENE	25.0	1.3797	-------	-------	3.679
567	C6H12	4-METHYL-cis-2-PENTENE	25.0	1.3850	-------	-------	3.621
568	C6H12	4-METHYL-trans-2-PENTENE	25.0	1.3858	-------	-------	3.766
569	C6H12N2	TRIETHYLENEDIAMINE	-------	-------	-------	-------	3.234
570	C6H12O	BUTYL VINYL ETHER	25.0	1.3997	in benzene	1.25	3.806
571	C6H12O	CYCLOHEXANOL	25.0	1.4645	in CCl4	1.86	3.601
572	C6H12O	1-HEXANAL	25.0	1.4017	-------	-------	3.924
573	C6H12O	ETHYL ISOPROPYL KETONE	25.0	1.3958	-------	-------	3.892
574	C6H12O	2-HEXANONE	25.0	1.3987	in benzene	2.68	4.093
575	C6H12O	3-HEXANONE	25.0	1.3980	-------	-------	4.009
576	C6H12O	METHYL ISOBUTYL KETONE	25.0	1.3933	liquid	2.70	3.828
577	C6H12O2	n-PENTYL FORMATE	25.0	1.3977	gas	1.90	4.159
578	C6H12O2	n-BUTYL ACETATE	25.0	1.3918	liquid	1.84	4.261
579	C6H12O2	sec-BUTYL ACETATE	25.0	1.3875	-------	-------	4.055
580	C6H12O2	tert-BUTYL ACETATE	25.0	1.3840	in benzene	1.91	3.845
581	C6H12O2	ETHYL n-BUTYRATE	25.0	1.3900	in benzene	1.76	4.298
582	C6H12O2	ETHYL ISOBUTYRATE	25.0	1.3873	in benzene	2.07	4.211
583	C6H12O2	ISOBUTYL ACETATE	25.0	1.3880	in benzene	1.87	4.276
584	C6H12O2	n-PROPYL PROPIONATE	25.0	1.3920	in benzene	1.79	4.311
585	C6H12O2	CYCLOHEXYL PEROXIDE	25.0	1.4638	-------	-------	3.861
586	C6H12O2	DIACETONE ALCOHOL	25.0	1.4219	in benzene	3.24	3.905
587	C6H12O2	2-ETHYL BUTYRIC ACID	25.0	1.4112	-------	-------	4.126
588	C6H12O2	n-HEXANOIC ACID	25.0	1.4148	liquid	1.13	4.404
589	C6H12O3	2-ETHOXYETHYL ACETATE	25.0	1.4023	in benzene	2.25	4.664
590	C6H12O3	HYDROXYCAPROIC ACID	-------	-------	-------	-------	4.727
591	C6H12O3	PARALDEHYDE	19.9	1.4050	gas	1.43	4.166
592	C6H12O3	sec-BUTYL GLYCOLATE	-------	-------	-------	-------	----------
593	C6H12S	THIACYCLOHEPTANE	-------	-------	-------	-------	----------
594	C6H13N	CYCLOHEXYLAMINE	25.0	1.4565	in hexane	1.31	3.539
595	C6H13N	HEXAMETHYLENEIMINE	23.9	1.4641	-------	-------	3.491
596	C6H14	2,2-DIMETHYLBUTANE	25.0	1.3659	gas	0.00	3.476
597	C6H14	2,3-DIMETHYLBUTANE	25.0	1.3728	gas	0.00	3.636
598	C6H14	n-HEXANE	25.0	1.3723	gas	0.00	3.769
599	C6H14	2-METHYLPENTANE	25.0	1.3687	gas	0.00	3.784
600	C6H14	3-METHYLPENTANE	25.0	1.3739	gas	0.00	3.695
601	C6H14N2O2	LYSINE	-------	-------	-------	-------	5.191
602	C6H14O	2-ETHYL-1-BUTANOL	25.0	1.4205	-------	-------	3.906
603	C6H14O	1-HEXANOL	25.0	1.4161	in benzene	1.65	4.144
604	C6H14O	2-HEXANOL	25.0	1.4128	-------	-------	3.787
605	C6H14O	2-METHYL-1-PENTANOL	25.0	1.4172	-------	-------	3.967
606	C6H14O	4-METHYL-2-PENTANOL	25.0	1.4090	-------	-------	3.843
607	C6H14O	n-BUTYL ETHYL ETHER	25.0	1.3793	in benzene	1.22	3.932
608	C6H14O	DIISOPROPYL ETHER	25.0	1.3655	gas	1.13	3.906
609	C6H14O	DI-n-PROPYL ETHER	25.0	1.3780	gas	1.21	3.914
610	C6H14O	METHYL tert-PENTYL ETHER	25.0	1.3859	-------	-------	3.465
611	C6H14O2	ACETAL	25.0	1.3682	gas	1.08	4.329
612	C6H14O2	2-BUTOXYETHANOL	25.0	1.4177	in benzene	2.08	4.192
613	C6H14O2	1,6-HEXANEDIOL	25.0	1.4485	in dioxane	2.50	4.428
614	C6H14O2	HEXYLENE GLYCOL	25.0	1.4260	in dioxane	2.90	3.918
615	C6H14O2S	DI-n-PROPYL SULFONE	29.9	1.4456	in benzene	4.47	4.706
616	C6H14O3	DIETHYLENE GLYCOL DIMETHYL ETHER	25.0	1.4043	in benzene	1.97	4.413
617	C6H14O3	DIPROPYLENE GLYCOL	19.9	1.4407	-------	-------	4.621
618	C6H14O3	2-(2-ETHOXYETHOXY)ETHANOL	25.0	1.4254	-------	-------	4.433
619	C6H14O3	TRIMETHYLOLPROPANE	-------	-------	-------	-------	4.161
620	C6H14O4	TRIETHYLENE GLYCOL	25.0	1.4550	liquid	5.58	4.710
621	C6H14O6	SORBITOL	-------	-------	-------	-------	5.008
622	C6H14S	n-HEXYLMERCAPTAN	25.0	1.4473	liquid	1.55	4.252
623	C6H14S	BUTYL-ETHYL-SULFIDE	25.0	1.4463	-------	-------	----------
624	C6H14S	ISOPROPYL-SULFIDE	25.0		-------	-------	----------

NO	FORMULA	NAME	Refractive Index		Dipole Moment		Radius of Gyration
			T, C	Value @ T	State	Debye	Angstrom
625	C6H14S	METHYL-PENTYL-SULFIDE	25.0	1.4482	-------	-------	----------
626	C6H14S	PROPYL-SULFIDE	25.0	1.4462	in benzene	1.56	----------
627	C6H14S2	PROPYL-DISULFIDE	25.0	1.4956	in benzene	1.98	----------
628	C6H15Al	TRIETHYL ALUMINUN	6.4	1.4800	in benzene	0.60	----------
629	C6H15Al2Cl3	ETHYL ALUMINUM SESQUICHLORIDE	-------	-------	-------	-------	----------
630	C6H15N	DIISOPROPYLAMINE	25.0	1.3924	-------	-------	3.806
631	C6H15N	DI-n-PROPYLAMINE	25.0	1.4018	in benzene	1.07	3.976
632	C6H15N	n-HEXYLAMINE	25.0	1.4167	in benzene	1.59	4.073
633	C6H15N	TRIETHYLAMINE	25.0	1.3980	gas	0.66	3.730
634	C6H15NO	6-AMINOHEXANOL	-------	-------	-------	-------	4.348
635	C6H15NO2	DIISOPROPANOLAMINE	29.9	1.4595	-------	-------	4.694
636	C6H15NO3	TRIETHANOLAMINE	25.0	1.4835	in dioxane	1.08	5.219
637	C6H15N3	N-AMINOETHYL PIPERAZINE	25.0	1.4983	-------	-------	4.225
638	C6H15O4P	TRIETHYL PHOSPHATE	25.0	1.4036	in benzene	3.09	4.949
639	C6H16N2	HEXAMETHYLENEDIAMINE	41.9	1.4485	-------	-------	4.483
640	C6H18N3OP	HEXAMETHYL PHOSPHORAMIDE	25.0	1.4564	in benzene	5.54	4.515
641	C6H18N4	TRIETHYLENE TETRAMINE	19.9	1.4971	-------	-------	4.861
642	C6H18OSi2	HEXAMETHYLDISILOXANE	19.9	1.3777	gas	0.66	----------
643	C6H18O3Si3	HEXAMETHYLCYCLOTRISILOXANE	-------	-------	in benzene	0.00	----------
644	C6H19NSi2	HEXAMETHYLDISILAZANE	19.9	1.4080	in benzene	0.42	----------
645	C7H3ClF3NO2	4-CHLORO-3-NITROBENZOTRIFLUORIDE	19.9	1.4893	-------	-------	5.466
646	C7H3Cl2F3	2,4-DICHLOROBENZOTRIFLUORIDE	-------	-------	-------	-------	5.233
647	C7H3Cl2NO	3,4-DICHLOROPHENYL ISOCYANATE	-------	-------	-------	-------	5.290
648	C7H4ClF3	p-CHLOROBENZOTRIFLUORIDE	25.0	1.4440	gas	1.58	4.799
649	C7H4Cl2O	m-CHLOROBENZOYL CHLORIDE	-------	-------	-------	-------	5.002
650	C7H4F3NO2	3-NITROBENZOTRIFLUORIDE	19.9	1.4715	-------	-------	5.078
651	C7H5ClO	BENZOYL CHLORIDE	19.9	1.5537	in benzene	3.16	4.267
652	C7H5ClO2	o-CHLOROBENZOIC ACID	-------	-------	in dioxane	2.45	4.537
653	C7H5Cl3	BENZOTRICHLORIDE	-------	-------	in benzene	2.04	4.579
654	C7H5F3	BENZOTRIFLUORIDE	25.0	1.4114	gas	2.86	4.161
655	C7H5N	BENZONITRILE	19.9	1.5282	gas	4.18	3.800
656	C7H5NO	PHENYL ISOCYANATE	19.9	1.5368	in CCl4	2.25	4.073
657	C7H5N3O6	2,4,6-TRINITROTOLUENE	-------	-------	in benzene	1.16	5.766
658	C7H6Cl2	BENZYL DICHLORIDE	25.0	1.5037	in benzene	2.03	4.580
659	C7H6Cl2	2,4-DICHLOROTOLUENE	21.9	1.5480	in benzene	1.95	4.668
660	C7H6N2O4	2,4-DINITROTOLUENE	25.0	1.4420	in benzene	4.32	5.070
661	C7H6N2O4	2,5-DINITROTOLUENE	-------	-------	in benzene	0.58	4.997
662	C7H6N2O4	2,6-DINITROTOLUENE	25.0	1.4790	in benzene	2.81	4.887
663	C7H6N2O4	3,4-DINITROTOLUENE	-------	-------	in benzene	6.39	4.859
664	C7H6N2O4	3,5-DINITROTOLUENE	-------	-------	in benzene	4.32	5.314
665	C7H6O	BENZALDEHYDE	25.0	1.5428	gas	2.80	3.751
666	C7H6O2	BENZOIC ACID	131.9	1.5040	in benzene	1.00	4.059
667	C7H6O2	p-HYDROXYBENZALDEHYDE	129.9	1.5105	in benzene	3.96	4.040
668	C7H6O2	SALICYLALDEHYDE	19.9	1.5718	in benzene	2.88	4.184
669	C7H6O3	SALICYLIC ACID	-------	-------	in dioxane	2.65	4.251
670	C7H7Br	p-BROMOTOLUENE	25.0	1.5486	liquid	1.90	3.891
671	C7H7Cl	BENZYL CHLORIDE	19.9	1.5391	in benzene	1.89	4.027
672	C7H7Cl	o-CHLOROTOLUENE	25.0	1.5233	gas	1.56	3.981
673	C7H7Cl	p-CHLOROTOLUENE	25.0	1.5187	gas	2.21	3.972
674	C7H7F	p-FLUOROTOLUENE	-------	-------	gas	1.68	----------
675	C7H7NO	FORMANILIDE	-------	-------	in CCl4	3.37	4.113
676	C7H7NO2	m-NITROTOLUENE	19.9	1.5468	in benzene	4.23	4.468
677	C7H7NO2	o-NITROTOLUENE	25.0	1.5474	in benzene	3.75	4.309
678	C7H7NO2	p-NITROTOLUENE	54.9	1.5312	in benzene	4.44	4.371
679	C7H7NO3	o-NITROANISOLE	25.0	1.5597	gas	4.83	4.502
680	C7H8	TOLUENE	25.0	1.4941	gas	0.36	3.472
681	C7H8	1,3,5-CYCLOHEPTATRIENE	-------	-------	-------	-------	----------
682	C7H8O	ANISOLE	25.0	1.5143	gas	1.36	3.719
683	C7H8O	BENZYL ALCOHOL	25.0	1.5384	gas	1.71	3.813
684	C7H8O	m-CRESOL	25.0	1.5396	gas	1.59	3.870
685	C7H8O	o-CRESOL	25.0	1.5442	gas	1.45	3.787
686	C7H8O	p-CRESOL	25.0	1.5391	gas	1.56	3.762
687	C7H8O2	GUAIACOL	25.0	1.5411	in benzene	2.46	4.020
688	C7H8O2	p-METHOXYPHENOL	-------	-------	-------	-------	4.040
689	C7H9N	BENZYLAMINE	19.9	1.5424	in benzene	1.38	3.837
690	C7H9N	2,6-DIMETHYLPYRIDINE	19.9	1.4976	in benzene	1.64	3.861
691	C7H9N	N-METHYLANILINE	19.9	1.5700	in benzene	1.68	3.782
692	C7H9N	m-TOLUIDINE	25.0	1.5657	in benzene	1.49	3.876
693	C7H9N	o-TOLUIDINE	25.0	1.5699	in benzene	1.60	3.774
694	C7H9N	p-TOLUIDINE	24.9	1.5540	in benzene	1.31	3.783
695	C7H10	2-NORBORNENE	-------	-------	in heptane	0.40	2.957
696	C7H10N2	TOLUENEDIAMINE	-------	-------	-------	-------	4.165
697	C7H11NO	CYCLOHEXYL ISOCYANATE	25.9	1.5341	-------	-------	4.229
698	C7H12	1-HEPTYNE	25.0	1.4060	gas	0.87	----------
699	C7H12O2	n-BUTYL ACRYLATE	25.0	1.4156	-------	-------	4.765
700	C7H12O2	ISOBUTYL ACRYLATE	-------	-------	-------	-------	4.585
701	C7H12O2	n-PROPYL METHACRYLATE	15.9	1.4200	in benzene	2.12	4.631
702	C7H12O4	DIETHYL MALONATE	19.9	1.4136	in benzene	2.54	4.710

NO	FORMULA	NAME	Refractive Index		Dipole Moment		Radius of Gyration
			T, C	Value @ T	State	Debye	Angstrom
703	C7H14	CYCLOHEPTANE	25.0	1.4424	gas	0.00	3.555
704	C7H14	1,1-DIMETYLCYCLOPENTANE	25.0	1.4109	gas	0.00	3.430
705	C7H14	cis-1,2-DIMETHYLCYCLOPENTANE	25.0	1.4196	gas	0.00	3.617
706	C7H14	trans-1,2-DIMETHYLCYCLOPENTANE	25.0	1.4094	gas	0.00	3.607
707	C7H14	cis-1,3-DIMETHYLCYCLOPENTANE	25.0	1.4063	gas	0.00	3.633
708	C7H14	trans-1,3-DIMETHYLCYCLOPENTANE	25.0	1.4081	gas	0.00	3.644
709	C7H14	ETHYLCYCLOPENTANE	25.0	1.4173	gas	0.00	3.734
710	C7H14	2-ETHYL-1-PENTENE	25.0	1.4020	-------	-------	4.181
711	C7H14	3-ETHYL-1-PENTENE	25.0	1.3955	-------	-------	3.950
712	C7H14	1-HEPTENE	25.0	1.3971	liquid	0.34	4.083
713	C7H14	cis-2-HEPTENE	25.0	1.4042	-------	-------	4.138
714	C7H14	trans-2-HEPTENE	25.0	1.4020	-------	-------	4.138
715	C7H14	cis-3-HEPTENE	25.0	1.4033	-------	-------	4.024
716	C7H14	trans-3-HEPTENE	25.0	1.4017	-------	-------	4.031
717	C7H14	METHYLCYCLOHEXANE	25.0	1.4206	gas	0.00	3.643
718	C7H14	2-METHYL-1-HEXENE	25.0	1.4008	-------	-------	4.199
719	C7H14	3-METHYL-1-HEXENE	25.0	1.3938	-------	-------	4.201
720	C7H14	4-METHYL-1-HEXENE	25.0	1.3973	-------	-------	3.972
721	C7H14	2,3,3-TRIMETHYL-1-BUTENE	25.0	1.4001	-------	-------	3.584
722	C7H14O	DIISOPROPYL KETONE	25.0	1.3976	in benzene	2.73	3.904
723	C7H14O	2-HEPTANONE	25.0	1.4066	in benzene	2.61	4.498
724	C7H14O	1-HEPTANAL	25.0	1.4094	in benzene	2.58	4.314
725	C7H14O	1-METHYLCYCLOHEXANOL	24.7	1.4587	-------	-------	3.670
726	C7H14O	cis-2-METHYLCYCLOHEXANOL	23.9	1.4633	-------	-------	3.821
727	C7H14O	trans-2-METHYLCYCLOHEXANOL	25.0	1.4597	-------	-------	3.824
728	C7H14O	cis-3-METHYLCYCLOHEXANOL	-------	-------	gas	1.77	3.931
729	C7H14O	trans-3-METHYLCYCLOHEXANOL	-------	-------	gas	1.93	3.937
730	C7H14O	cis-4-METHYLCYCLOHEXANOL	25.0	1.4584	-------	-------	3.869
731	C7H14O	trans-4-METHYLCYCLOHEXANOL	25.0	1.4544	-------	-------	3.854
732	C7H14O	5-METHYL-2-HEXANONE	25.0	1.4047	-------	-------	4.329
733	C7H14O2	n-BUTYL PROPIONATE	25.0	1.4000	in benzene	1.80	4.822
734	C7H14O2	ETHYL ISOVALERATE	25.0	1.3975	in benzene	1.97	4.481
735	C7H14O2	ISOPENTYL ACETATE	25.0	1.3981	in benzene	1.80	4.656
736	C7H14O2	n-PENTYL ACETATE	19.9	1.4028	in benzene	1.90	4.772
737	C7H14O2	n-PROPYL n-BUTYRATE	25.0	1.3976	liquid	1.75	4.832
738	C7H14O2	n-HEPTANOIC ACID	25.0	1.4210	-------	-------	4.842
739	C7H14O3	ETHYL-3-ETHOXYPROPIONATE	25.0	1.4041	-------	-------	5.086
740	C7H15Br	1-BROMOHEPTANE	25.0	1.4481	gas	2.16	4.485
741	C7H15N	N-METHYLCYCLOHEXYLAMINE	22.9	1.4530	-------	-------	3.888
742	C7H16	2,2-DIMETHYLPENTANE	25.0	1.3795	liquid	0.00	3.922
743	C7H16	2,3-DIMETHYLPENTANE	25.0	1.3895	liquid	0.00	3.917
744	C7H16	2,4-DIMETHYLPENTANE	25.0	1.3788	liquid	0.00	3.931
745	C7H16	3,3-DIMETHYLPENTANE	25.0	1.3884	liquid	0.00	3.768
746	C7H16	3-ETHYLPENTANE	25.0	1.3908	liquid	0.00	3.943
747	C7H16	n-HEPTANE	25.0	1.3851	gas	0.00	4.173
748	C7H16	2-METHYLHEXANE	25.0	1.3823	liquid	0.00	4.167
749	C7H16	3-METHYLHEXANE	25.0	1.3861	liquid	0.00	4.078
750	C7H16	2,2,3-TRIMETHYLBUTANE	25.0	1.3869	liquid	0.00	3.960
751	C7H16O	1-HEPTANOL	25.0	1.4223	in benzene	1.67	4.380
752	C7H16O	2-HEPTANOL	25.0	1.4190	in benzene	1.73	4.483
753	C7H16O	5-METHYL-1-HEXANOL	25.0	1.4220	-------	-------	4.563
754	C7H16O	ISOPROPYL-TERT-BUTYL-ETHER	-------	-------	-------	-------	----------
755	C7H16S	n-HEPTYL MERCAPTAN	25.0	1.4498	liquid	1.55	4.642
756	C7H16S	BUTYL-PROPYL-SULFIDE	-------	-------	-------	-------	----------
757	C7H16S	ETHYL-PENTYL-SULFIDE	-------	-------	-------	-------	----------
758	C7H16S	HEXYL-METHYL-SULFIDE	25.0	1.4505	-------	-------	----------
759	C7H17N	1-AMINOHEPTANE	25.0	1.4228	in benzene	1.60	4.423
760	C8H4Cl2O2	ISOPHTHALOYL CHLORIDE	25.0	1.5700	in benzene	5.16	5.520
761	C8H4O3	PHTHALIC ANHYDRIDE	-------	-------	in benzene	5.29	4.383
762	C8H6	ETHYNYLBENZENE	-------	-------	-------	-------	----------
763	C8H6O4	ISOPHTHALIC ACID	-------	-------	in dioxane	2.27	5.080
764	C8H6O4	PHTHALIC ACID	-------	-------	in dioxane	2.60	4.763
765	C8H6O4	TEREPHTHALIC ACID	-------	-------	-------	-------	4.969
766	C8H6S	BENZOTHIOPHENE	34.9	1.6332	in benzene	0.63	3.984
767	C8H7N	INDOLE	19.9	1.6300	in benzene	2.08	3.781
768	C8H8	STYRENE	25.0	1.5440	liquid	0.13	3.810
769	C8H8	1,3,5,7-CYCLOOCTATETRAENE	-------	-------	-------	-------	----------
770	C8H8O	ACETOPHENONE	19.9	1.5342	gas	3.02	4.090
771	C8H8O	p-TOLUALDEHYDE	-------	-------	in dioxane	3.30	4.133
772	C8H8O2	METHYL BENZOATE	25.0	1.5146	in benzene	2.53	4.362
773	C8H8O2	o-TOLUIC ACID	114.9	1.5120	in benzene	1.70	4.387
774	C8H8O2	p-TOLUIC ACID	-------	-------	-------	-------	4.350
775	C8H8O3	METHYL SALICYLATE	19.9	1.5239	in benzene	2.47	4.468
776	C8H8O3	VANILLIN	-------	-------	in benzene	2.87	4.704
777	C8H9NO	ACETANILIDE	-------	-------	in benzene	4.04	4.345
778	C8H10	ETHYLBENZENE	25.0	1.4932	gas	0.59	3.897
779	C8H10	m-XYLENE	25.0	1.4946	liquid	0.30	3.937
780	C8H10	o-XYLENE	25.0	1.5029	gas	0.62	3.836

NO	FORMULA	NAME	Refractive Index		Dipole Moment		Radius of Gyration
			T, C	Value @ T	State	Debye	Angstrom
781	C8H10	p-XYLENE	25.0	1.4932	gas	0.00	3.831
782	C8H10O	m-ETHYLPHENOL	-------	-------	-------	-------	----------
783	C8H10O	p-ETHYLPHENOL	25.0	1.5240	-------	-------	4.130
784	C8H10O	PHENETOLE	25.0	1.5049	gas	1.41	4.174
785	C8H10O	2-PHENYLETHANOL	19.9	1.5323	in benzene	1.65	4.279
786	C8H10O	2,3-XYLENOL	25.0	1.5420	in benzene	1.25	4.069
787	C8H10O	2,4-XYLENOL	19.9	1.5320	in benzene	1.39	4.143
788	C8H10O	2,5-XYLENOL	74.9	1.5120	in benzene	1.43	4.128
789	C8H10O	2,6-XYLENOL	19.9	1.5371	in benzene	1.41	4.068
790	C8H10O	3,4-XYLENOL	19.9	1.5442	gas	1.77	4.156
791	C8H10O	3,5-XYLENOL	-------	-------	in benzene	1.76	4.318
792	C8H11N	N,N-DIMETHYLANILINE	19.9	1.5584	gas	1.68	4.076
793	C8H11N	o-ETHYLANILINE	19.9	1.5588	-------	-------	4.157
794	C8H11N	2,4,6-TRIMETHYLPYRIDINE	25.0	1.4959	liquid	2.00	4.306
795	C8H11NO	p-PHENETIDINE	25.0	1.5528	-------	-------	4.414
796	C8H12	1,5-CYCLOOCTADIENE	25.0	1.4905	-------	-------	3.732
797	C8H12	VINYLCYCLOHEXENE	19.9	1.4641	-------	-------	3.869
798	C8H12O4	1,4-CYCLOHEXANEDICARBOXYLIC ACID	-------	-------	-------	-------	4.901
799	C8H12O4	DIETHYL MALEATE	25.0	1.4568	in benzene	2.56	5.242
800	C8H14O2	n-BUTYL METHACRYLATE	25.0	1.4215	in benzene	2.15	5.125
801	C8H14O3	BUTYRIC ANHYDRIDE	25.0	1.4105	-------	-------	4.936
802	C8H14O4	DIETHYL SUCCINATE	19.9	1.4200	in benzene	2.16	5.150
803	C8H16	CYCLOOCTANE	25.0	1.4563	-------	-------	----------
804	C8H16	1,1-DIMETHYLCYCLOHEXANE	25.0	1.4266	gas	0.00	3.668
805	C8H16	cis-1,2-DIMETHYLCYCLOHEXANE	25.0	1.4336	-------	-------	3.759
806	C8H16	trans-1,2-DIMETHYLCYCLOHEXANE	25.0	1.4247	gas	0.00	3.883
807	C8H16	cis-1,3-DIMETHYLCYCLOHEXANE	25.0	1.4206	gas	0.00	3.972
808	C8H16	trans-1,3-DIMETHYLCYCLOHEXANE	25.0	1.4284	gas	0.00	3.982
809	C8H16	cis-1,4-DIMETHYLCYCLOHEXANE	25.0	1.4273	gas	0.00	3.905
810	C8H16	trans-1,4-DIMETHYLCYCLOHEXANE	25.0	1.4185	gas	0.00	3.907
811	C8H16	ETHYLCYCLOHEXANE	25.0	1.4307	gas	0.00	3.932
812	C8H16	2-ETHYL-1-HEXENE	25.0	1.4132	-------	-------	4.579
813	C8H16	1-METHYL-1-ETHYLCYCLOPENTANE	25.0	1.4248	gas	0.00	3.773
814	C8H16	1-OCTENE	25.0	1.4062	liquid	0.34	4.457
815	C8H16	trans-2-OCTENE	25.0	1.4107	-------	-------	4.375
816	C8H16	trans-3-OCTENE	25.0	1.4102	liquid	0.00	4.415
817	C8H16	trans-4-OCTENE	25.0	1.4093	liquid	0.00	4.419
818	C8H16	n-PROPYLCYCLOPENTANE	25.0	1.4239	gas	0.00	4.135
819	C8H16	2,4,4-TRIMETHYL-1-PENTENE	25.0	1.4060	-------	-------	3.994
820	C8H16	2,4,4-TRIMETHYL-2-PENTENE	25.0	1.4135	-------	-------	4.136
821	C8H16O	2-ETHYLHEXANAL	21.9	1.4152	in benzene	2.66	4.862
822	C8H16O	1-OCTANAL	25.0	1.4156	-------	-------	4.681
823	C8H16O	2-OCTANONE	25.0	1.4133	in benzene	2.72	4.892
824	C8H16O2	n-BUTYL n-BUTYRATE	25.0	1.4029	-------	-------	5.887
825	C8H16O2	n-HEXYL ACETATE	25.0	1.4073	liquid	1.80	5.160
826	C8H16O2	ISOBUTYL ISOBUTYRATE	19.9	1.3990	-------	-------	4.768
827	C8H16O2	n-OCTANOIC ACID	25.0	1.4261	liquid	1.15	5.211
828	C8H16O4	DIETHYLENE GLYCOL ETHYL ETHER ACETATE	19.9	1.4213	-------	-------	5.692
829	C8H18	2,2-DIMETHYLHEXANE	25.0	1.3910	gas	0.00	4.392
830	C8H18	2,3-DIMETHYLHEXANE	25.0	1.3988	gas	0.00	4.363
831	C8H18	2,4-DIMETHYLHEXANE	25.0	1.3929	gas	0.00	4.267
832	C8H18	2,5-DIMETHYLHEXANE	25.0	1.3900	gas	0.00	4.452
833	C8H18	3,3-DIMETHYLHEXANE	25.0	1.3978	gas	0.00	3.717
834	C8H18	3,4-DIMETHYLHEXANE	25.0	1.4018	gas	0.00	4.230
835	C8H18	3-ETHYLHEXANE	25.0	1.3992	gas	0.00	4.571
836	C8H18	3-ETHYL-2-METHYLPENTANE	-------	-------	-------	-------	----------
837	C8H18	3-METHYL-3-ETHYLPENTANE	25.0	1.4055	-------	-------	3.928
838	C8H18	2-METHYLHEPTANE	25.0	1.3926	gas	0.00	4.570
839	C8H18	3-METHYLHEPTANE	25.0	1.3961	gas	0.00	4.490
840	C8H18	4-METHYLHEPTANE	25.0	1.3955	gas	0.00	4.520
841	C8H18	n-OCTANE	25.0	1.3951	gas	0.00	4.546
842	C8H18	2,2,3-TRIMETHYLPENTANE	25.0	1.4007	gas	0.00	3.913
843	C8H18	2,2,4-TRIMETHYLPENTANE	25.0	1.3890	gas	0.00	4.091
844	C8H18	2,3,3-TRIMETHYLPENTANE	25.0	1.4052	gas	0.00	3.960
845	C8H18	2,3,4-TRIMETHYLPENTANE	25.0	1.4020	gas	0.00	4.061
846	C8H18	2,2,3,3-TETRAMETHYLBUTANE	-------	-------	-------	-------	----------
847	C8H18O	DI-n-BUTYL ETHER	25.0	1.3968	gas	1.17	4.797
848	C8H18O	DI-sec-BUTYL ETHER	25.0	1.3930	-------	-------	4.373
849	C8H18O	DI-tert-BUTYL ETHER	19.9	1.3946	-------	-------	4.014
850	C8H18O	2-ETHYL-1-HEXANOL	25.0	1.4290	in benzene	1.76	4.809
851	C8H18O	1-OCTANOL	25.0	1.4276	in benzene	1.65	4.787
852	C8H18O	2-OCTANOL	25.0	1.4241	liquid	1.61	4.898
853	C8H18O2	DI-t-BUTYL PEROXIDE	25.0	1.3867	in benzene	0.92	4.534
854	C8H18O2S	DI-n-BUTYL SULFONE	49.9	1.4433	in benzene	4.47	5.482
855	C8H18O3	DIETHYLENE GLYCOL DIETHYL ETHER	25.0	1.4115	-------	-------	5.511
856	C8H18O3	DIETHYLENE GLYCOL MONOBUTYL EHTER	19.9	1.4306	-------	-------	5.106
857	C8H18O4	TRIETHYLENE GLYCOL DIMETHYL ETHER	25.0	1.4209	in benzene	2.24	5.414
858	C8H18O5	TETRAETHYLENE GLYCOL	25.0	1.4570	in dioxane	3.25	5.602

NO	FORMULA	NAME	Refractive Index		Dipole Moment		Radius of Gyration
			T, C	Value @ T	State	Debye	Angstrom
859	C8H18S	n-OCTYL MERCAPTAN	25.0	1.4519	in benzene	1.61	5.014
860	C8H18S	tert-OCTYL MERCAPTAN	-------	-------	-------	-------	4.264
861	C8H18S	BUTYL-SULFIDE	-------	-------	in benzene	1.57	----------
862	C8H18S	ETHYL-HEXYL-SULFIDE	-------	-------	-------	-------	----------
863	C8H18S	HEPTYL-METHYL-SULFIDE	25.0	1.4525	-------	-------	----------
864	C8H18S	PENTYL-PROPYL-SULFIDE	-------	-------	-------	-------	----------
865	C8H18S2	BUTYL-DISULFIDE	25.0	1.4903	in benzene	1.99	----------
866	C8H19N	DI-n-BUTYLAMINE	25.0	1.4152	liquid	1.06	4.723
867	C8H19N	DIISOBUTYLAMINE	25.0	1.4090	-------	-------	4.735
868	C8H19N	n-OCTYLAMINE	25.0	1.4277	in benzene	1.42	4.790
869	C8H23N5	TETRAETHYLENEPENTAMINE	19.9	1.5055	-------	-------	6.374
870	C8H24O4Si4	OCTAMETHYLCYCLOTETRASILOXANE	25.0	1.3935	in benzene	0.66	----------
871	C9H4O5	TRIMELLITIC ANHYDRIDE	-------	-------	-------	-------	5.362
872	C9H6N2O2	TOLUENE DIISOCYANATE	25.0	1.5651	gas	0.76	5.894
873	C9H7N	ISOQUINOLINE	29.9	1.6208	in benzene	0.76	4.013
874	C9H7N	QUINOLINE	25.0	1.6248	gas	2.29	4.008
875	C9H7NO	8-HYDROXYQUINOLINE	-------	-------	in benzene	2.68	4.312
876	C9H8	INDENE	25.0	1.5740	in benzene	0.67	3.839
877	C9H8O	2-METHYLBENZOFURAN	25.0	1.5460	-------	-------	1.114
878	C9H10	INDANE	25.0	1.5358	in benzene	0.54	3.850
879	C9H10	cis-PROPENYLBENZENE	-------	-------	-------	-------	----------
880	C9H10	trans-PROPENYLBENZENE	-------	-------	-------	-------	----------
881	C9H10	alpha-METHYLSTYRENE	25.0	1.5358	-------	-------	4.147
882	C9H10	m-METHYLSTYRENE	25.0	1.5385	in benzene	0.36	4.265
883	C9H10	o-METHYLSTYRENE	25.0	1.5413	in benzene	0.48	4.029
884	C9H10	p-METHYLSTYRENE	25.0	1.5395	in benzene	0.38	4.109
885	C9H10O2	BENZYL ACETATE	19.9	1.5232	in benzene	1.80	4.716
886	C9H10O2	ETHYL BENZOATE	25.0	1.5035	gas	2.00	4.765
887	C9H10O3	ETHYL VANILLIN	-------	-------	-------	-------	5.180
888	C9H11NO	p-DIMETHYLAMINOBENZALDEHYDE	99.9	1.6082	in benzene	5.58	4.667
889	C9H12	CUMENE	25.0	1.4889	liquid	0.39	4.322
890	C9H12	m-ETHYLTOLUENE	25.0	1.4941	liquid	0.33	4.265
891	C9H12	o-ETHYLTOLUENE	25.0	1.5021	liquid	0.56	4.084
892	C9H12	p-ETHYLTOLUENE	25.0	1.4924	liquid	0.00	4.189
893	C9H12	MESITYLENE	25.0	1.4968	liquid	0.00	4.375
894	C9H12	n-PROPYLBENZENE	25.0	1.4895	in benzene	0.37	4.344
895	C9H12	1,2,3-TRIMETHYLBENZENE	25.0	1.5115	liquid	0.56	4.127
896	C9H12	1,2,4-TRIMETHYLBENZENE	25.0	1.5024	liquid	0.30	4.199
897	C9H12O	BENZYL ETHYL ETHER	25.0	1.4934	-------	-------	4.581
898	C9H12O	2-PHENYL-2-PROPANOL	25.0	1.5325	-------	-------	4.285
899	C9H12O2	CUMENE HYDROPEROXIDE	19.9	1.5242	in benzene	1.79	4.532
900	C9H14O	ISOPHORONE	19.9	1.4780	in dioxane	3.99	4.276
901	C9H14O6	GLYCERYL TRIACETATE	25.0	1.4288	in benzene	2.58	6.778
902	C9H16	1-NONYNE	25.0	1.4195	-------	-------	----------
903	C9H16O4	AZELAIC ACID	110.9	1.4303	in dioxane	2.35	6.283
904	C9H18	BUTYLCYCLOPENTANE	25.0	1.4293	-------	-------	----------
905	C9H18	cis,cis-1,3,5-TRIMETHYLCYCLOHEXANE	-------	-------	-------	-------	----------
906	C9H18	cis,trans-1,3,5-TRIMETHYLCYCLOHEXANE	-------	-------	-------	-------	----------
907	C9H18	ISOPROPYLCYCLOHEXANE	25.0	1.4386	gas	0.00	4.237
908	C9H18	1-NONENE	25.0	1.4133	-------	-------	4.873
909	C9H18	n-PROPYLCYCLOHEXANE	25.0	1.4348	gas	0.00	4.367
910	C9H18O	DIISOBUTYL KETONE	19.9	1.4122	in CCl4	2.66	4.614
911	C9H18O	1-NONANAL	25.0	1.4208	-------	-------	5.038
912	C9H18O2	n-BUTYL VALERATE	19.9	1.4128	-------	-------	5.615
913	C9H18O2	n-NONANOIC ACID	25.0	1.4302	-------	-------	5.732
914	C9H18O2	n-OCTYL FORMATE	25.0	1.4141	-------	-------	5.279
915	C9H20	3,3-DIETHYLPENTANE	25.0	1.4184	in benzene	0.00	4.314
916	C9H20	2,2-DIMETHYL-3-ETHYLPENTANE	25.0	1.4010	gas	0.00	4.396
917	C9H20	3-ETHYL-2,3-DIMETHYLPENTANE	-------	-------	-------	-------	----------
918	C9H20	2,4-DIMETHYL-3-ETHYLPENTANE	25.0	1.4115	gas	0.00	4.371
919	C9H20	2,2-DIMETHYLHEPTANE	25.0	1.3993	-------	-------	4.845
920	C9H20	2,6-DIMETHYLHEPTANE	25.0	1.3983	gas	0.00	4.540
921	C9H20	3-ETHYLHEPTANE	25.0	1.4070	gas	0.00	4.953
922	C9H20	4-ETHYLHEPTANE	25.0	1.4073	-------	-------	----------
923	C9H20	2,3-DIMETHYLHEPTANE	25.0	1.4062	-------	-------	----------
924	C9H20	2,4-DIMETHYLHEPTANE	25.0	1.4011	-------	-------	----------
925	C9H20	2,5-DIMETHYLHEPTANE	25.0	1.4015	-------	-------	----------
926	C9H20	3,4-DIMETHYLHEPTANE	25.0	1.4089	-------	-------	----------
927	C9H20	3,5-DIMETHYLHEPTANE	25.0	1.4044	-------	-------	----------
928	C9H20	4,4-DIMETHYLHEPTANE	25.0	1.4053	-------	-------	----------
929	C9H20	3-ETHYL-2-METHYLHEXANE	25.0	1.4097	-------	-------	----------
930	C9H20	4-ETHYL-2-METHYLHEXANE	25.0	1.4046	-------	-------	----------
931	C9H20	3-ETHYL-3-METHYLHEXANE	25.0	1.4120	-------	-------	----------
932	C9H20	3-ETHYL-4-METHYLHEXANE	25.0	1.4128	-------	-------	----------
933	C9H20	2,2,3-TRIMETHYLHEXANE	25.0	1.4082	-------	-------	----------
934	C9H20	2,2,4-TRIMETHYLHEXANE	25.0	1.4010	-------	-------	----------
935	C9H20	2,3,3-TRIMETHYLHEXANE	25.0	1.4119	-------	-------	----------
936	C9H20	2,3,4-TRIMETHYLHEXANE	25.0	1.4120	-------	-------	----------

NO	FORMULA	NAME	Refractive Index		Dipole Moment		Radius of Gyration
			T, C	Value @ T	State	Debye	Angstrom
937	C9H20	2,3,5-TRIMETHYLHEXANE	25.0	1.4037	-------	-------	----------
938	C9H20	2,4,4-TRIMETHYLHEXANE	25.0	1.4052	-------	-------	----------
939	C9H20	3,3,4-TRIMETHYLHEXANE	25.0	1.4154	-------	-------	----------
940	C9H20	2-METHYLOCTANE	25.0	1.4008	gas	0.00	4.973
941	C9H20	3-METHYLOCTANE	25.0	1.4040	gas	0.00	5.020
942	C9H20	4-METHYLOCTANE	25.0	1.4039	gas	0.00	4.772
943	C9H20	n-NONANE	25.0	1.4031	gas	0.00	4.985
944	C9H20	2,2,3,3-TETRAMETHYLPENTANE	25.0	1.4214	gas	0.00	4.100
945	C9H20	2,2,3,4-TETRAMETHYLPENTANE	25.0	1.4125	gas	0.00	4.160
946	C9H20	2,2,4,4-TETRAMETHYLPENTANE	25.0	1.4046	gas	0.00	4.860
947	C9H20	2,3,3,4-TETRAMETHYLPENTANE	25.0	1.4200	-------	-------	----------
948	C9H20	2,2,5-TRIMETHYLHEXANE	25.0	1.3973	gas	0.00	4.658
949	C9H20O	2,6-DIMETHYL-4-HEPTANOL	25.0	1.4211	-------	-------	4.915
950	C9H20O	1-NONANOL	25.0	1.4319	in benzene	1.61	5.108
951	C9H20O	2-NONANOL	25.0	1.4290	-------	-------	5.302
952	C9H20S	n-NONYL MERCAPTAN	25.0	1.4537	-------	-------	5.344
953	C9H20S	BUTYL-PENTYL-SULFIDE	-------	-------	-------	-------	----------
954	C9H20S	ETHYL-HEPTYL-SULFIDE	-------	-------	-------	-------	----------
955	C9H20S	HEXYL-PROPYL-SULFIDE	-------	-------	-------	-------	----------
956	C9H20S	METHYL-OCTYL-SULFIDE	25.0	1.4541	-------	-------	----------
957	C9H21N	n-NONYLAMINE	25.0	1.4318	-------	-------	5.143
958	C9H21N	TRIPROPYLAMINE	25.0	1.4151	in benzene	0.74	5.270
959	C10H6O8	PYROMELLITIC ACID	-------	-------	-------	-------	5.983
960	C10H7Br	1-BROMONAPHTHALENE	19.9	1.6580	liquid	1.29	4.661
961	C10H7Cl	1-CHLORONAPHTHALENE	19.9	1.6332	liquid	1.33	4.627
962	C10H8	NAPHTHALENE	19.9	1.9320	in benzene	0.00	4.045
963	C10H8	AZULENE	-------	-------	in benzene	1.08	----------
964	C10H9N	QUINALDINE	25.0	1.6091	-------	-------	4.378
965	C10H10	m-DIVINYLBENZENE	25.0	1.5736	-------	-------	4.419
966	C10H10	1-METHYLINDENE	25.0	1.5587	-------	-------	4.230
967	C10H10	2-METHYLINDENE	25.0	1.5627	-------	-------	4.223
968	C10H10O4	DIMETHYL PHTHALATE	19.9	1.5138	in benzene	2.80	5.166
969	C10H10O4	DIMETHYL TEREPHTHALATE	-------	-------	in benzene	2.19	5.394
970	C10H12	DICYCLOPENTADIENE	34.9	1.5061	-------	-------	3.981
971	C10H12	1,2,3,4-TETRAHYDRONAPHTHALENE	25.0	1.5392	in benzene	0.22	4.162
972	C10H12O	ANETHOLE	19.9	1.5615	in benzene	1.50	4.717
973	C10H12O4	DIALLYL MALEATE	19.9	1.4699	-------	-------	6.266
974	C10H14	n-BUTYLBENZENE	25.0	1.4874	in benzene	0.37	4.849
975	C10H14	sec-BUTYLBENZENE	25.0	1.4878	in benzene	0.39	4.564
976	C10H14	tert-BUTYLBENZENE	25.0	1.4902	gas	0.70	4.318
977	C10H14	1,2,3,4-TETRAMETHYLBENZENE	25.0	1.5181	-------	-------	----------
978	C10H14	m-CYMENE	25.0	1.4905	-------	-------	4.636
979	C10H14	o-CYMENE	25.0	1.4983	-------	-------	4.439
980	C10H14	p-CYMENE	25.0	1.4885	liquid	0.00	4.557
981	C10H14	m-DIETHYLBENZENE	25.0	1.4931	liquid	0.36	4.742
982	C10H14	o-DIETHYLBENZENE	25.0	1.5011	liquid	0.59	4.398
983	C10H14	p-DIETHYLBENZENE	25.0	1.4924	liquid	0.00	4.580
984	C1OH14	2-ETHYL-m-XYLENE	25.0	1.5085	-------	-------	4.415
985	C10H14	2-ETHYL-p-XYLENE	25.0	1.5020	-------	-------	4.579
986	C10H14	3-ETHYL-o-XYLENE	25.0	1.5095	-------	-------	4.517
987	C10H14	4-ETHYL-m-XYLENE	25.0	1.5015	-------	-------	4.543
988	C10H14	4-ETHYL-o-XYLENE	25.0	1.5009	-------	-------	4.552
989	C1OH14	5-ETHYL-m-XYLENE	25.0	1.4958	gas	0.20	4.721
990	C10H14	ISOBUTYLBENZENE	25.0	1.4840	liquid	0.31	4.555
991	C10H14	1,2,3,5-TETRAMETHYLBENZENE	25.0	1.5107	-------	-------	4.516
992	C10H14	1,2,4,5-TETRAMETHYLBENZENE	25.0	1.5093	-------	-------	4.502
993	C10H14O	p-tert-BUTYLPHENOL	25.0	1.5040	in benzene	1.62	4.614
994	C10H14O2	p-tert-BUTYLCATECHOL	-------	-------	-------	-------	4.941
995	C10H15N	N,N-DIETYHLANILINE	25.0	1.5418	in benzene	1.81	4.639
996	C10H15N	2,6-DIETHYLANILINE	19.9	1.5456	-------	-------	4.730
997	C10H16	CAMPHENE	54.9	1.4562	-------	-------	3.778
998	C10H16	D-LIMONENE	25.0	1.4701	in benzene	1.57	4.508
999	C10H16	alpha-PHELLANDRENE	25.0	1.4691	-------	-------	4.458
1000	C1OH16	beta-PHELLANDRENE	25.0	1.4851	-------	-------	4.441
1001	C10H16	alpha-PINENE	25.0	1.4632	in benzene	0.36	3.418
1002	C10H16	beta-PINENE	25.0	1.4768	-------	-------	3.617
1003	C10H16	alpha-TERPINENE	25.0	1.4760	-------	-------	4.620
1004	C1OH16	gamma-TERPINENE	25.0	1.4712	-------	-------	4.615
1005	C1OH16	TERPINOLENE	19.9	1.4861	-------	-------	4.567
1006	C10H16O	CAMPHOR	-------	-------	in benzene	3.10	4.097
1007	C10H18	1-DECYNE	25.0	1.4249	-------	-------	----------
1008	C10H18	cis-DECAHYDRONANPHTALENE	25.0	1.4788	not specified	0.00	4.206
1009	C10H18	trans-DECAHYDRONAPHTHALENE	25.0	1.4674	not specified	0.00	4.314
1010	C10H18O4	SEBACIC ACID	113.2	1.4220	in dioxane	2.40	6.586
1011	C10H20	n-BUTYLCYCLOHEXANE	25.0	1.4385	gas	0.00	4.831
1012	C10H20	1-CYCLOPENTYLPENTANE	25.0	1.4336	-------	-------	----------
1013	C10H20	1-DECENE	25.0	1.4191	-------	-------	5.227
1014	C10H20O	1-DECANAL	25.0	1.4251	-------	-------	5.378

NO	FORMULA	NAME	Refractive Index		Dipole Moment		Radius of Gyration
			T, C	Value @ T	State	Debye	Angstrom
1015	C10H20O2	n-DECANOIC ACID	25.0	1.4343	-------	-------	5.932
1016	C10H20O2	2-ETHYLHEXYL ACETATE	25.0	1.4173	-------	-------	5.895
1017	C10H20O2	ISOPENTYL ISOVALERATE	25.0	1.4100	-------	-------	5.586
1018	C10H22	n-DECANE	25.0	1.4097	gas	0.00	5.184
1019	C10H22	2-METHYLNONANE	25.0	1.4075	-------	-------	5.429
1020	C10H22	3-METHYLNONANE	25.0	1.4103	-------	-------	5.286
1021	C10H22	4-METHYLNONANE	25.0	1.4095	-------	-------	5.448
1022	C10H22	5-METHYLNONANE	25.0	1.4100	-------	-------	5.318
1023	C10H22	3-ETHYLOCTANE	25.0	1.4136	-------	-------	----------
1024	C10H22	4-ETHYLOCTANE	25.0	1.4131	-------	-------	----------
1025	C10H22	2,2-DIMETHYLOCTANE	25.0	1.4060	-------	-------	5.223
1026	C10H22	2,3-DIMETHYLOCTANE	25.0	1.4127	-------	-------	----------
1027	C10H22	2,4-DIMETHYLOCTANE	25.0	1.4069	-------	-------	----------
1028	C10H22	2,5-DIMETHYLOCTANE	25.0	1.4089	-------	-------	----------
1029	C10H22	2,6-DIMETHYLOCTANE	25.0	1.4084	-------	-------	----------
1030	C10H22	2,7-DIMETHYLOCTANE	25.0	1.4062	-------	-------	----------
1031	C10H22	3,3-DIMETHYLOCTANE	25.0	1.4142	-------	-------	----------
1032	C10H22	3,4-DIMETHYLOCTANE	25.0	1.4159	-------	-------	----------
1033	C10H22	3,5-DIMETHYLOCTANE	25.0	1.4115	-------	-------	----------
1034	C10H22	3,6-DIMETHYLOCTANE	25.0	1.4115	-------	-------	----------
1035	C10H22	4,4-DIMETHYLOCTANE	25.0	1.4122	-------	-------	----------
1036	C10H22	4,5-DIMETHYLOCTANE	25.0	1.4167	-------	-------	----------
1037	C10H22	4-PROPYLHEPTANE	25.0	1.4113	-------	-------	----------
1038	C10H22	4-ISOPROPYLHEPTANE	25.0	1.4132	-------	-------	----------
1039	C10H22	3-ETHYL-2-METHYLHEPTANE	25.0	1.4151	-------	-------	----------
1040	C10H22	4-ETHYL-2-METHYLHEPTANE	25.0	1.4114	-------	-------	----------
1041	C10H22	5-ETHYL-2-METHYLHEPTANE	25.0	1.4111	-------	-------	----------
1042	C10H22	3-ETHYL-3-METHYLHEPTANE	25.0	1.4185	-------	-------	----------
1043	C10H22	4-ETHYL-3-METHYLHEPTANE	25.0	1.4183	-------	-------	----------
1044	C10H22	3-ETHYL-5-METHYLHEPTANE	25.0	1.4141	-------	-------	----------
1045	C10H22	3-ETHYL-4-METHYLHEPTANE	25.0	1.4184	-------	-------	----------
1046	C10H22	4-ETHYL-4-METHYLHEPTANE	25.0	1.4187	-------	-------	----------
1047	C10H22	2,2,3-TRIMETHYLHEPTANE	25.0	1.4145	-------	-------	----------
1048	C10H22	2,2,4-TRIMETHYLHEPTANE	25.0	1.4069	-------	-------	----------
1049	C10H22	2,2,5-TRIMETHYLHEPTANE	25.0	1.4078	-------	-------	----------
1050	C10H22	2,2,6-TRIMETHYLHEPTANE	25.0	1.4055	-------	-------	----------
1051	C10H22	2,3,3-TRIMETHYLHEPTANE	25.0	1.4179	-------	-------	----------
1052	C10H22	2,3,4-TRIMETHYLHEPTANE	25.0	1.4172	-------	-------	----------
1053	C10H22	2,3,5-TRIMETHYLHEPTANE	25.0	1.4146	-------	-------	----------
1054	C10H22	2,3,6-TRIMETHYLHEPTANE	25.0	1.4108	-------	-------	----------
1055	C10H22	2,4,4-TRIMETHYLHEPTANE	25.0	1.4120	-------	-------	----------
1056	C10H22	2,4,5-TRIMETHYLHEPTANE	25.0	1.4137	-------	-------	----------
1057	C10H22	2,4,6-TRIMETHYLHEPTANE	25.0	1.4048	-------	-------	----------
1058	C10H22	2,5,5-TRIMETHYLHEPTANE	25.0	1.4126	-------	-------	----------
1059	C10H22	3,3,4-TRIMETHYLHEPTANE	25.0	1.4213	-------	-------	----------
1060	C10H22	3,3,5-TRIMETHYLHEPTANE	25.0	1.4147	-------	-------	----------
1061	C10H22	3,4,4-TRIMETHYLHEPTANE	25.0	1.4212	-------	-------	----------
1062	C10H22	3,4,5-TRIMETHYLHEPTANE	25.0	1.4206	-------	-------	----------
1063	C10H22	3-ISOPROPYL-2-METHYLHEXANE	25.0	1.4172	-------	-------	----------
1064	C10H22	3,3-DIETHYLHEXANE	25.0	1.4235	-------	-------	----------
1065	C10H22	3,4-DIETHYLHEXANE	25.0	1.4167	-------	-------	----------
1066	C10H22	3-ETHYL-2,2-DIMETHYLHEXANE	25.0	1.4174	-------	-------	----------
1067	C10H22	4-ETHYL-2,2-DIMETHYLHEXANE	25.0	1.4107	-------	-------	----------
1068	C10H22	3-ETHYL-2,3-DIMETHYLHEXANE	25.0	1.4247	-------	-------	----------
1069	C10H22	4-ETHYL-2,3-DIMETHYLHEXANE	25.0	1.4203	-------	-------	----------
1070	C10H22	3-ETHYL-2,4-DIMETHYLHEXANE	25.0	1.4202	-------	-------	----------
1071	C10H22	4-ETHYL-2,4-DIMETHYLHEXANE	25.0	1.4212	-------	-------	----------
1072	C10H22	3-ETHYL-2,5-DIMETHYLHEXANE	25.0	1.4134	-------	-------	----------
1073	C10H22	4-ETHYL-3,3-DIMETHYLHEXANE	25.0	1.4246	-------	-------	----------
1074	C10H22	3-ETHYL-3,4-DIMETHYLHEXANE	25.0	1.4244	-------	-------	----------
1075	C10H22	2,2,3,3-TETRAMETHYLHEXANE	25.0	1.4260	-------	-------	----------
1076	C10H22	2,2,3,4-TETRAMETHYLHEXANE	25.0	1.4193	-------	-------	----------
1077	C10H22	2,2,3,5-TETRAMETHYLHEXANE	25.0	1.4119	-------	-------	----------
1078	C10H22	2,2,4,4-TETRAMETHYLHEXANE	25.0	1.4185	-------	-------	----------
1079	C10H22	2,2,4,5-TETRAMETHYLHEXANE	25.0	1.4110	-------	-------	----------
1080	C10H22	2,2,5,5-TETRAMETHYLHEXANE	25.0	1.4032	-------	-------	----------
1081	C10H22	2,3,3,4-TETRAMETHYLHEXANE	25.0	1.4275	-------	-------	----------
1082	C10H22	2,3,3,5-TETRAMETHYLHEXANE	25.0	1.4173	-------	-------	----------
1083	C10H22	2,3,4,4-TETRAMETHYLHEXANE	25.0	1.4244	-------	-------	----------
1084	C10H22	2,3,4,5-TETRAMETHYLHEXANE	25.0	1.4181	-------	-------	----------
1085	C10H22	3,3,4,4-TETRAMETHYLHEXANE	25.0	1.4346	-------	-------	----------
1086	C10H22	2,4-DIMETHYL-3-ISOPROPYLPENTANE	25.0	1.4225	-------	-------	----------
1087	C10H22	3,3-DIETHYL-2-METHYLPENTANE	25.0	1.4320	-------	-------	----------
1088	C10H22	3-ETHYL-2,2,3-TRIMETHYLPENTANE	25.0	1.4397	-------	-------	----------
1089	C10H22	3-ETHYL-2,2,4-TRIMETHYLPENTANE	25.0	1.4199	-------	-------	----------
1090	C10H22	3-ETHYL-2,3,4-TRIMETHYLPENTANE	25.0	1.4310	-------	-------	----------
1091	C10H22	2,2,3,3,4-PENTAMETHYLPENTANE	25.0	1.4341	-------	-------	----------
1092	C10H22	2,2,3,4,4-PENTAMETHYLPENTANE	25.0	1.4287	-------	-------	----------

NO	FORMULA	NAME	Refractive Index		Dipole Moment		Radius of Gyration
			T, C	Value @ T	State	Debye	Angstrom
1093	C10H22O	1-DECANOL	25.0	1.4350	in benzene	1.62	5.499
1094	C10H22O	DI-n-PENTYL ETHER	19.9	1.4119	in CCl4	1.20	5.374
1095	C10H22O	ISODECANOL	19.9	1.4352	-------	-------	5.684
1096	C10H22O5	TETRAETHYLENE GLYCOL DIMETHYL ETHER	19.9	1.4325	in benzene	2.47	6.277
1097	C10H22S	n-DECYL MERCAPTAN	25.0	1.4549	in benzene	1.59	5.712
1098	C10H22S	BUTYL-HEXYL-SULFIDE	-------	-------	-------	-------	----------
1099	C10H22S	ETHYL-OCTYL-SULFIDE	-------	-------	-------	-------	----------
1100	C10H22S	HEPTYL-PROPYL-SULFIDE	-------	-------	-------	-------	----------
1101	C10H22S	METHYL-NONYL-SULFIDE	25.0	1.4556	-------	-------	----------
1102	C10H22S	PENTYL-SULFIDE	-------	-------	-------	-------	----------
1103	C10H22S2	PENTYL-DISULFIDE	25.0	1.4867	-------	-------	----------
1104	C10H23N	n-DECYLAMINE	25.0	1.4352	in benzene	1.41	5.477
1105	C11H10	1-METHYLNAPHTHALENE	25.0	1.6151	liquid	0.51	4.435
1106	C11H10	2-METHYLNAPHTHALENE	41.9	1.6019	liquid	0.42	4.426
1107	C11H14O2	n-BUTYL BENZOATE	25.0	1.4940	-------	-------	5.505
1108	C11H16	n-PENTYLBENZENE	25.0	1.4856	-------	-------	5.168
1109	C11H16O	p-tert-AMYLPHENOL	-------	-------	-------	-------	4.810
1110	C11H20	1-UNDECYNE	25.0	1.4292	-------	-------	----------
1111	C11H20O2	2-ETHYLHEXYL ACRYLATE	19.9	1.4365	-------	-------	6.265
1112	C11H22	1-UNDECENE	25.0	1.4238	-------	-------	5.429
1113	C11H22	1-CYCLOPENTYLHEXANE	-------	-------	-------	-------	----------
1114	C11H22	PENTYLCYCLOHEXANE	25.0	1.4416	-------	-------	----------
1115	C11H22O	1-UNDECANAL	25.0	1.4288	-------	-------	5.711
1116	C11H24	n-UNDECANE	25.0	1.4151	-------	0.00	5.496
1117	C11H24O	1-UNDECANOL	25.0	1.4386	in benzene	1.67	5.808
1118	C11H24S	UNDECYL MERCAPTAN	25.0	1.4564	-------	-------	6.069
1119	C11H24S	BUTYL-HEPTYL-SULFIDE	-------	-------	-------	-------	----------
1120	C11H24S	DECYL-METHYL-SULFIDE	25.0	1.4569	-------	-------	----------
1121	C11H24S	ETHYL-NONYL-SULFIDE	-------	-------	-------	-------	----------
1122	C11H24S	OCTYL-PROPYL-SULFIDE	-------	-------	-------	-------	----------
1123	C12H8O	DIBENZOFURAN	99.0	1.6480	in benzene	0.88	4.710
1124	C12H9N	DIBENZOPYRROLE	-------	-------	in benzene	2.11	4.692
1125	C12H10	ACENAPHTHENE	19.9	1.6420	in benzene	0.25	4.468
1126	C12H10	BIPHENYL	25.0	1.5873	gas	0.00	4.834
1127	C12H10O	DIPHENYL ETHER	25.0	1.5781	in benzene	1.16	4.986
1128	C12H11N	p-AMINODIPHENYL	-------	-------	in benzene	1.76	5.027
1129	C12H11N	DIPHENYLAMINE	-------	-------	in benzene	1.08	5.082
1130	C12H11N3	p-AMINOAZOBENZENE	-------	-------	in benzene	2.48	5.663
1131	C12H11N3	1,3-DIPHENYLTRIAZENE	-------	-------	-------	-------	6.005
1132	C12H12	1,2-DIMETHYLNAPHTHALENE	25.0	1.6143	not specified	0.68	----------
1133	C12H12	1,3-DIMETHYLNAPHTHALENE	25.0	1.6080	not specified	0.36	----------
1134	C12H12	1,4-DIMETHYLNAPHTHALENE	25.0	1.6114	not specified	-------	----------
1135	C12H12	1,5-DIMETHYLNAPHTHALENE	-------	-------	not specified	0.07	----------
1136	C12H12	1,6-DIMETHYLNAPHTHALENE	25.0	1.6050	not specified	0.32	----------
1137	C12H12	1,7-DIMETHYLNAPHTHALENE	25.0	1.6054	not specified	0.54	----------
1138	C12H12	2,3-DIMETHYLNAPHTHALENE	-------	-------	not specified	0.69	----------
1139	C12H12	2,6-DIMETHYLNAPHTHALENE	-------	-------	in benzene	0.14	4.729
1140	C12H12	2,7-DIMETHYLNAPHTHALENE	-------	-------	in benzene	0.41	4.804
1141	C12H12	1-ETHYLNAPHTHALENE	25.0	1.6040	-------	-------	4.754
1142	C12H12	2-ETHYLNAPHTHALENE	25.0	1.5977	-------	-------	----------
1143	C12H12N2	p-AMINODIPHENYLAMINE	-------	-------	-------	-------	5.327
1144	C12H12N2	HYDRAZOBENZENE	-------	-------	-------	-------	5.544
1145	C12H14	1,2,3-TRIMETHYLINDENE	14.9	1.5541	-------	-------	4.780
1146	C12H14O4	DIETHYL PHTHALATE	20.9	1.5000	in benzene	2.90	5.832
1147	C12H16	CYCLOHEXYLBENZENE	25.0	1.5239	-------	-------	4.892
1148	C12H18	m-DIISOPROPYLBENZENE	25.0	1.4875	-------	-------	5.281
1149	C12H18	p-DIISOPROPYLBENZENE	25.0	1.4876	-------	-------	5.178
1150	C12H18	n-HEXYLBENZENE	25.0	1.4842	-------	-------	5.642
1151	C12H18	1,2,3-TRIETHYLBENZENE	-------	-------	-------	-------	----------
1152	C12H18	1,2,4-TRIETHYLBENZENE	-------	-------	-------	-------	----------
1153	C12H18	1,3,5-TRIETHYLBENZENE	-------	-------	in benzene	0.10	----------
1154	C12H18	HEXAMETHYLBENZENE	25.0	1.4842	in benzene	0.10	----------
1155	C12H20O4	DIBUTYL MALEATE	25.0	1.4435	-------	-------	6.463
1156	C12H22	BICYCLOHEXYL	25.0	1.4777	-------	-------	4.917
1157	C12H22	1-DODECYNE	25.0	1.4328	-------	-------	----------
1158	C12H23N	DICYCLOHEXYLAMINE	25.0	1.4823	in heptane	1.06	5.317
1159	C12H24	1-DODECENE	25.0	1.4278	-------	-------	5.747
1160	C12H24	1-CYCLOPENTYLHEPTANE	-------	-------	-------	-------	----------
1161	C12H24	1-CYCLOHEXYLHEXANE	25.0	1.4441	-------	-------	----------
1162	C12H24O	1-DODECANAL	25.0	1.4320	-------	-------	6.052
1163	C12H24O2	n-DODECANOIC ACID	25.0	1.4401	in benzene	0.76	6.601
1164	C12H26	n-DODECANE	25.0	1.4151	gas	0.00	5.914
1165	C12H26O	DI-n-HEXYL ETHER	25.0	1.4187	-------	-------	5.693
1166	C12H26O	1-DODECANOL	25.0	1.4413	in dioxane	1.70	6.119
1167	C12H26O3	DIETHYLENE GLYCOL DI-n-BUTYL ETHER	19.9	1.4235	-------	-------	6.449
1168	C12H26S	n-DODECYL MERCAPTAN	25.0	1.4576	gas	1.59	6.374
1169	C12H26S	BUTYL-OCTYL-SULFIDE	-------	-------	-------	-------	----------
1170	C12H26S	DECYL-ETHYL-SULFIDE	-------	-------	-------	-------	----------

NO	FORMULA	NAME	Refractive Index		Dipole Moment		Radius of Gyration
			T, C	Value @ T	State	Debye	Angstrom
1171	C12H26S	HEXYL-SULFIDE	-------	-------	-------	-------	----------
1172	C12H26S	METHYL-UNDECYL-SULFIDE	25.0	1.4580	-------	-------	----------
1173	C12H26S	NONYL-PROPYL-SULFIDE	-------	-------	-------	-------	----------
1174	C12H26S2	HEXYL-DISULFIDE	25.0	1.4850	-------	-------	----------
1175	C12H27BO3	TRI-n-BUTYL BORATE	25.0	1.4071	in benzene	0.78	----------
1176	C12H27N	DODECYLAMINE	25.0	1.4406	-------	-------	6.073
1177	C12H27N	TRI-n-BUTYLAMINE	25.0	1.4286	in benzene	0.78	6.690
1178	C13H10	FLUORENE	19.9	1.6470	in benzene	0.25	4.806
1179	C13H10O	BENZOPHENONE	44.9	1.5975	in benzene	2.98	5.302
1180	C13H12	DIPHENYLMETHANE	25.0	1.5752	liquid	0.26	5.141
1181	C13H14	1-PROPYLNAPHTHALENE	25.0	1.5901	-------	-------	----------
1182	C13H14	2-PROPYLNAPHTHALENE	25.0	1.5850	-------	-------	----------
1183	C13H14	2ETHYL-3-METHYLNAPHTHALENE	-------	-------	-------	-------	----------
1184	C13H14	2ETHYL-6-METHYLNAPHTHALENE	-------	-------	-------	-------	----------
1185	C13H14	2ETHYL-7-METHYLNAPHTHALENE	-------	-------	-------	-------	----------
1186	C13H20	n-HEPTYLBENZENE	25.0	1.4832	-------	-------	6.040
1187	C13H24	1-TRIDECYNE	25.0	1.4359	-------	-------	----------
1188	C13H26	1-TRIDECENE	25.0	1.4312	-------	-------	6.073
1189	C13H26	1-CYCLOPENTYLOCTANE	-------	-------	-------	-------	----------
1190	C13H26	1-CYCLOHEXYLHEPTANE	-------	-------	-------	-------	----------
1191	C13H26O	1-TRIDECANAL	25.0	1.4348	-------	-------	6.345
1192	C13H26O2	n-BUTYL NONANOATE	25.0	1.4262	-------	-------	7.273
1193	C13H26O2	METHYL DODECANOATE	19.9	1.4292	-------	-------	6.857
1194	C13H28	n-TRIDECANE	25.0	1.4235	gas	0.00	6.198
1195	C13H28O	1-TRIDECANOL	25.0	1.4433	-------	-------	6.417
1196	C13H28S	BUTYL-NONYL-SULFIDE	-------	-------	-------	-------	----------
1197	C13H28S	DECYL-PROPYL-SULFIDE	-------	-------	-------	-------	----------
1198	C13H28S	DODECYL-METHYL-SULFIDE	25.0	1.4590	-------	-------	----------
1199	C13H28S	ETHYL-UNDECYL-SULFIDE	-------	-------	-------	-------	----------
1200	C13H28S	1-TRIDECANETHIOL	25.0	1.4586	-------	-------	----------
1201	C14H8O2	ANTHRAQUINONE	-------	-------	in benzene	0.00	5.332
1202	C14H10	ANTHRACENE	19.9	1.7290	in benzene	0.00	4.980
1203	C14H10	DIPHENYLACETYLENE	-------	-------	in benzene	0.30	5.832
1204	C14H10	PHENANTHRENE	25.0	1.5480	in benzene	0.00	4.961
1205	C14H12	cis-STILBENE	-------	-------	in benzene	0.00	4.744
1206	C14H12	trans-STILBENE	-------	-------	in benzene	0.00	5.627
1207	C14H12O2	BENZYL BENZOATE	20.9	1.5681	-------	-------	6.184
1208	C14H14	1,1-DIPHENYLETHANE	25.0	1.5702	liquid	0.00	5.437
1209	C14H14	1,2-DIPHENYLETHANE	51.2	1.5704	liquid	0.00	5.610
1210	C14H14O	DIBENZYL ETHER	25.0	1.5385	in benzene	1.39	6.078
1211	C14H16	1-n-BUTYLNAPHTHALENE	25.0	1.5797	not specified	0.69	5.680
1212	C14H16	2-BUTYLNAPHTHALENE	25.0	1.5747	not specified	0.75	----------
1213	C14H22	n-OCTYLBENZENE	25.0	1.4824	-------	-------	6.498
1214	C14H22	1,2,3,4-TETRAETHYLBENZENE	-------	-------	-------	-------	----------
1215	C14H22	1,2,3,5-TETRAETHYLBENZENE	-------	-------	-------	-------	----------
1216	C14H22	1,2,4,5-TETRAETHYLBENZENE	-------	-------	-------	-------	----------
1217	C14H22O	p-tert-OCTYLPHENOL	-------	-------	-------	-------	5.601
1218	C14H28	1-TETRADECENE	25.0	1.4341	-------	-------	6.398
1219	C14H28	1-CYCLOPENTYLNONANE	-------	-------	-------	-------	----------
1220	C14H28	1-CYCLOHEXYLOCTANE	-------	-------	-------	-------	----------
1221	C14H28O2	n-TETRADECANOIC ACID	25.0	1.4445	in benzene	0.77	7.265
1222	C14H30	n-TETRADECANE	25.0	1.4269	gas	0.00	6.427
1223	C14H30O	1-TETRADECANOL	25.0	1.4454	in CCl4	1.60	6.730
1224	C14H30S	BUTYL-DECYL-SULFIDE	-------	-------	-------	-------	----------
1225	C14H30S	DODECYL-ETHYL-SULFIDE	-------	-------	-------	-------	----------
1226	C14H30S	HEPTYL-SULFIDE	-------	-------	-------	-------	----------
1227	C14H30S	METHYL-TRIDECYL-SULFIDE	25.0	1.4599	-------	-------	----------
1228	C14H30S	PROPYL-UNDECYL-SULFIDE	-------	-------	-------	-------	----------
1229	C14H30S	1-TETRADECANETHIOL	25.0	1.4595	-------	-------	----------
1230	C14H30S2	HEPTYL-DISULFIDE	25.0	1.4840	-------	-------	----------
1231	C14H31N	TETRADECYLAMINE	25.0	1.4447	-------	-------	6.704
1232	C15H10N2O2	DIPHENYLMETHANE-4,4'-DIISOCYANATE	49.9	1.5906	-------	-------	7.119
1233	C15H16O	p-CUMYLPHENOL	-------	-------	-------	-------	5.562
1234	C15H16O2	BISPHENOL A	-------	-------	-------	-------	6.815
1235	C15H18	1-PENTYLNAPHTHALENE	25.0	1.5704	-------	-------	----------
1236	C15H18	2-PENTYLNAPHTHALENE	25.0	1.5675	-------	-------	----------
1237	C15H24	n-NONYLBENZENE	25.0	1.4817	-------	-------	6.763
1238	C15H24O	2,6-DI-tert-BUTYL-p-CRESOL	-------	-------	in benzene	1.68	5.598
1239	C15H24O	NONYLPHENOL	19.9	1.5116	-------	-------	7.058
1240	C15H28	1-PENTADECYNE	25.0	1.4410	-------	-------	----------
1241	C15H30	1-PENTADECENE	25.0	1.4367	-------	-------	6.637
1242	C15H30	1-CYCLOPENTYLDECANE	-------	-------	-------	-------	----------
1243	C15H30	1-CYCLOHEXYLNONANE	-------	-------	-------	-------	----------
1244	C15H30O2	PENTADECANOIC ACID	80.0	1.4463	-------	-------	7.567
1245	C15H32	n-PENTADECANE	25.0	1.4298	gas	0.00	6.729
1246	C15H32O	1-PENTADECANOL	-------	-------	-------	-------	----------
1247	C15H32S	BUTYL-UNDECYL-SULFIDE	-------	-------	-------	-------	----------
1248	C15H32S	DODECYL-PROPYL-SULFIDE	-------	-------	-------	-------	----------

NO	FORMULA	NAME	Refractive Index		Dipole Moment		Radius of Gyration
			T, C	Value @ T	State	Debye	Angstrom
1249	C15H32S	ETHYL-TRIDECYL-SULFIDE	-------	-------	-------	-------	----------
1250	C15H32S	METHYL-TETRADECYL-SULFIDE	25.0	1.4607	-------	-------	----------
1251	C15H32S	1-PENTADECANETHIOL	25.0	1.4604	-------	-------	----------
1252	C16H10	FLUORANTHENE	-------	-------	in benzene	0.00	5.357
1253	C16H10	PYRENE	-------	-------	in benzene	0.00	5.152
1254	C16H12	1-PHENYLNAPHTHALENE	-------	-------	in benzene	0.00	5.641
1255	C16H20	1-n-HEXYLNAPHTHALENE	25.0	1.5626	-------	-------	6.518
1256	C16H22O4	DIBUTYL PHTHALATE	25.0	1.4901	in benzene	2.75	7.475
1257	C16H26	n-DECYLBENZENE	25.0	1.4811	-------	-------	7.108
1258	C16H26	PENTAETHYLBENZENE	-------	-------	-------	-------	----------
1259	C16H30	1-HEXADECYNE	25.0	1.4430	-------	-------	----------
1260	C16H32	n-DECYLCYCLOHEXANE	25.0	1.4514	-------	-------	7.225
1261	C16H32	1-CYCLOPENTYLUNDECANE	-------	-------	-------	-------	----------
1262	C16H32	1-HEXADECENE	25.0	1.4391	-------	-------	6.985
1263	C16H32O2	n-HEXADECANOIC ACID	60.0	-------	in benzene	0.72	7.725
1264	C16H34	n-HEXADECANE	25.0	1.4325	gas	0.00	7.063
1265	C16H34O	DI-n-OCTYL ETHER	25.0	1.4305	-------	-------	7.360
1266	C16H34O	1-HEXADECANOL	49.9	-------	in benzene	1.67	7.307
1267	C16H34S	BUTYL-DODECYL-SULFIDE	-------	-------	-------	-------	----------
1268	C16H34S	ETHYL-TETRADECYL-SULFIDE	-------	-------	-------	-------	----------
1269	C16H34S	METHYL-PENTADECYL-SULFIDE	25.0	1.4615	-------	-------	----------
1270	C16H34S	OCTYL-SULFIDE	-------	-------	-------	-------	----------
1271	C16H34S	PROPYL-TRIDECYL-SULFIDE	-------	-------	-------	-------	----------
1272	C16H34S	1-HEXADECANETHIOL	25.0	1.4611	-------	-------	----------
1273	C16H34S2	OCTYL-DISULFIDE	25.0	1.4820	in benzene	1.99	----------
1274	C17H28	n-UNDECYLBENZENE	25.0	1.4807	-------	-------	7.474
1275	C17H32	1-HEPTADECYNE	25.0	1.4410	-------	-------	7.325
1276	C17H34	1-CYCLOPENTYLDODECANE	-------	-------	-------	-------	----------
1277	C17H34	1-CYCLOHEXYLUNDECANE	-------	-------	-------	-------	----------
1278	C17H34	1-HEPTADECENE	25.0	1.4410	-------	-------	----------
1279	C17H36	n-HEPTADECANE	25.0	1.4348	gas	0.00	7.355
1280	C17H36O	1-HEPTADECANOL	-------	-------	-------	-------	7.590
1281	C17H36S	BUTYL-TRIDECYL-SULFIDE	-------	-------	-------	-------	----------
1282	C17H36S	ETHYL-PENTADECYL-SULFIDE	-------	-------	-------	-------	----------
1283	C17H36S	HEXADECYL-METHYL-SULFIDE	25.0	1.4621	-------	-------	----------
1284	C17H36S	PROPYL-TETRADECYL-SULFIDE	-------	-------	-------	-------	----------
1285	C17H36S	1-HEPTADECANETHIOL	-------	-------	-------	-------	----------
1286	C18H12	CHRYSENE	19.9	-------	in benzene	0.70	5.740
1287	C18H14	m-TERPHENYL	-------	-------	-------	-------	6.506
1288	C18H14	o-TERPHENYL	-------	-------	-------	-------	5.899
1289	C18H14	p-TERPHENYL	-------	-------	in benzene	0.60	6.371
1290	C18H15P	TRIPHENYLPHOSPHINE	19.9	-------	in benzene	1.39	5.601
1291	C18H15O4P	TRIPHENYL PHOSPHATE	59.9	-------	in benzene	2.81	7.103
1292	C18H16N2	N,N'-DIPHENYL-p-PHENYLENEDIAMINE	-------	-------	in benzene	1.60	6.755
1293	C18H22	2,3-DIMETHYL-2,3-DIPHENYLBUTANE	-------	-------	in benzene	0.00	5.361
1294	C18H22O2	DICUMYL PEROXIDE	25.0	-------	-------	-------	6.536
1295	C18H30	n-DODECYLBENZENE	25.0	1.4803	-------	-------	7.796
1296	C18H30	HEXAETHYLBENZENE	-------	-------	-------	-------	----------
1297	C18H32O2	LINOLEIC ACID	19.9	1.4699	in benzene	1.22	10.520
1298	C18H34	1-OCTADECYNE	-------	-------	-------	-------	----------
1299	C18H34O2	OLEIC ACID	19.9	1.4582	liquid	1.44	10.430
1300	C18H34O4	DIBUTYL SEBACATE	25.0	1.4397	not specified	2.48	10.290
1301	C18H34O4	DIHEXYL ADIPATE	25.0	1.4397	-------	-------	9.488
1302	C18H36	1-CYCOPENTYLTRIDECANE	-------	-------	-------	-------	----------
1303	C18H36	1-CYCLOHEXYLDODECANE	-------	-------	-------	-------	----------
1304	C18H36	1-OCTADECENE	25.0	1.4428	-------	-------	7.589
1305	C18H36O2	STEARIC ACID	25.0	-------	in dioxane	1.76	8.560
1306	C18H38	n-OCTADECANE	20.0	-------	gas	0.00	7.655
1307	C18H38O	DINONYL ETHER	19.9	1.4356	-------	-------	7.266
1308	C18H38O	1-OCTADECANOL	25.0	-------	in CCl4	1.70	7.930
1309	C18H38S	BUTYL-TETRADECYL-SULFIDE	-------	-------	-------	-------	----------
1310	C18H38S	ETHYL-HEXADECYL-SULFIDE	-------	-------	-------	-------	----------
1311	C18H38S	HEPTADECYL-METHYL-SULFIDE	-------	-------	-------	-------	----------
1312	C18H38S	NONYL-SULFIDE	-------	-------	-------	-------	----------
1313	C18H38S	PENTADECYL-PROPYL-SULFIDE	-------	-------	-------	-------	----------
1314	C18H38S	1-OCTADECANETHIOL	-------	-------	-------	-------	----------
1315	C18H38S2	NONYL-DISULFIDE	25.0	1.4810	-------	-------	----------
1316	C19H26	1-n-NONYLNAPHTHALENE	25.0	1.5455	gas	0.00	7.732
1317	C19H32	n-TRIDECYLBENZENE	25.0	1.4800	-------	-------	8.189
1318	C19H36	1-NONADECYNE	-------	-------	-------	-------	----------
1319	C19H36O2	METHYL OLEATE	19.9	1.4521	-------	-------	10.450
1320	C19H38	1-CYCLOPENTYLTETRADECANE	-------	-------	-------	-------	----------
1321	C19H38	1-CYCLOHEXYLTRIDECANE	-------	-------	-------	-------	----------
1322	C19H38	1-NONADECENE	25.0	1.4450	-------	-------	7.850
1323	C19H38O2	NONADECANOIC ACID	25.0	-------	-------	-------	8.686
1324	C19H40	n-NONADECANE	20.0	-------	gas	0.00	7.983
1325	C19H40O	1-NONADECANOL	-------	-------	-------	-------	----------
1326	C19H40S	BUTYL-PENTADECYL-SULFIDE	-------	-------	-------	-------	----------

NO	FORMULA	NAME	Refractive Index		Dipole Moment		Radius of Gyration
			T, C	Value @ T	State	Debye	Angstrom
1327	C19H40S	ETHYL-HEPTADECYL-SULFIDE	-------	-------	-------	-------	----------
1328	C19H40S	HEXADECYL-PROPYL-SULFIDE	-------	-------	-------	-------	----------
1329	C19H40S	METHYL-OCTADECYL-SULFIDE	-------	-------	-------	-------	----------
1330	C19H40S	1-NONADECANETHIOL	-------	-------	-------	-------	----------
1331	C20H16	TRIPHENYLETHYLENE	-------	-------	in benzene	0.60	6.512
1332	C20H28	1-n-DECYLNAPHTHALENE	25.0	1.5412	gas	0.00	8.139
1333	C20H30O2	ABIETIC ACID	-------	-------	-------	-------	6.822
1334	C20H31N	DEHYDROABIETYLAMINE	25.0	-------	-------	-------	6.103
1335	C20H34	1-PHENYLTETRADECANE	25.0	1.4797	-------	-------	----------
1336	C20H38	1-EICOSYNE	-------	---·----	-------	-------	8.106
1337	C20H40	1-CYCLOPENTYLPENTADECANE	-------	-------	-------	-------	----------
1338	C20H40	1-CYCLOHEXYLTETRADECANE	-------	-------	-------	-------	----------
1339	C20H40	1-EICOSENE	30.0	-------	-------	-------	----------
1340	C20H42	n-EICOSANE	20.0	-------	gas	0.00	8.364
1341	C20H42O	1-EICOSANOL	20.0	-------	-------	-------	8.414
1342	C20H42S	BUTYL-HEXADECYL-SULFIDE	-------	-------	-------	-------	----------
1343	C20H42S	DECYL-SULFIDE	-------	-------	-------	-------	----------
1344	C20H42S	ETHYL-OCTADECYL-SULFIDE	-------	-------	-------	-------	----------
1345	C20H42S	HEPTADECYL-PROPYL-SULFIDE	-------	-------	-------	-------	----------
1346	C20H42S	METHYL-NONADECYL-SULFIDE	-------	-------	-------	-------	----------
1347	C20H42S	1-EICOSANETHIOL	-------	-------	-------	-------	----------
1348	C20H42S2	DECYL-DISULFIDE	25.0	1.4800	-------	-------	----------
1349	C21H21O4P	TRI-o-CRESYL PHOSPHATE	25.0	1.5587	in CCl4	2.84	6.760
1350	C21H36	1-PHENYLPENTADECANE	-------	-------	-------	-------	----------
1351	C21H42	1-CYCLOPENTYLHEXADECANE	-------	-------	-------	-------	----------
1352	C21H42	1-CYCLOHEXYLPENTADECANE	-------	-------	-------	-------	----------
1353	C22H38	1-PHENYLHEXADECANE	-------	-------	-------	-------	----------
1354	C22H44	1-CYCLOHEXYLHEXADECANE	-------	-------	-------	-------	----------
1355	C22H44O2	n-BUTYL STEARATE	25.0	-------	in benzene	1.88	9.746
1356	C24H38O4	DIISOOCTYL PHTHALATE	25.0	1.4860	in CCl4	3.06	10.910
1357	C24H38O4	DIOCTYL PHTHALATE	25.0	1.4845	in benzene	2.84	11.000
1358	C24H42O	DINONYLPHENOL	-------	-------	-------	-------	11.660
1359	C26H20	TETRAPHENYLETHYLENE	-------	-------	in benzene	0.33	7.145
1360	C28H46O4	DIISODECYL PHTHALATE	25.0	1.4840	-------	-------	9.114

Refractive index applies at T.

T, C - temperature in Centigrade.

Dipole moment is given in Debye units.

The conversion factor for Debye units is 1 Debye = 3.33564E-30 C M

Radius of gyration is given in Angstrom units.

The conversion factor for Angstrom units is 1 Angstrom = 1.0E-10 meters

Table 10-2 REFRACTIVE INDEX, DIPOLE M0MENT, AND RADIUS OF GYRATION - INORGANIC COMPOUNDS

NO	FORMULA	NAME	Refractive Index		Dipole Moment		Radius of Gyration
			T, C	Value @ T	State	Debye	Angstrom
1	Ag	SILVER	-------	-------	-------	-------	----------
2	AgCl	SILVER CHLORIDE	25.0	2.0710	gas	5.70	----------
3	AgI	SILVER IODIDE	-------	-------	gas	5.10	----------
4	Al	ALUMINUM	25.0	1.4400	-------	-------	----------
5	AlB3H12	ALUMINUM BOROHYDRIDE	-------	-------	-------	-------	----------
6	AlBr3	ALUMINUN BROMIDE	-------	-------	in benzene	5.22	----------
7	AlCl3	ALUMINUM CHLORIDE	-------	-------	gas	2.03	----------
8	AlF3	ALUMINUM FLUORIDE	-------	-------	-------	-------	----------
9	AlI3	ALUMINUM IODIDE	-------	-------	in benzene	2.28	----------
10	Al2O3	ALUMINUM OXIDE	25.0	1.7600	-------	-------	----------
11	Al2S3O12	ALUMINUM SULFATE	25.0	1.4700	-------	-------	----------
12	Ar	ARGON	-188.0	1.2312	gas	0.00	1.076
13	As	ARSENIC	-------	-------	-------	-------	----------
14	AsBr3	ARSENIC TRIBROMIDE	14.0	1.6210	in dioxane	2.92	----------
15	AsCl3	ARSENIC TRICHLORIDE	16.0	1.6040	gas	1.59	----------
16	AsF3	ARSENIC TRIFLUORIDE	-------	-------	gas	2.59	----------
17	AsF5	ARSENIC PENTAFLUORIDE	-------	-------	-------	-------	----------
18	AsH3	ARSINE	-------	-------	gas	0.20	----------
19	AsI3	ARSENIC TRIIODIDE	25.0	2.2300	in dioxane	1.84	----------
20	As2O3	ARSENIC TRIOXIDE	25.0	1.7550	gas	0.13	----------
21	At	ASTATINE	-------	-------	-------	-------	----------
22	Au	GOLD	-------	-------	-------	-------	----------
23	B	BORON	-------	-------	-------	-------	----------
24	BBr3	BORON TRIBROMIDE	16.0	1.3120	liquid	0.00	3.672
25	BCl3	BORON TRICHLORIDE	5.7	1.4190	gas	0.00	3.683
26	BF3	BORON TRIFLUORIDE	25.0	1.0004	gas	0.00	3.255
27	BH2CO	BORINE CARBONYL	-------	-------	-------	-------	----------
28	BH3O3	BORIC ACID	-------	-------	-------	-------	3.948
29	B2D6	DEUTERODIBORANE	-------	-------	-------	-------	----------
30	B2H5Br	DIBORANE HYDROBROMIDE	-------	-------	-------	-------	----------
31	B2H6	DIBORANE	-------	-------	gas	0.00	----------
32	B3N3H6	BORINE TRIAMINE	-------	-------	gas	0.67	----------
33	B4H10	TETRABORANE	-------	-------	gas	0.49	----------
34	B5H9	PENTABORANE	-------	-------	gas	2.13	----------
35	B5H11	TETRAHYDROPENTABORANE	-------	-------	-------	-------	----------
36	B10H14	DECABORANE	-------	-------	in benzene	3.52	----------
37	Ba	BARIUM	-------	-------	-------	-------	----------
38	Be	BERYLLIUM	-------	-------	-------	-------	----------
39	BeB2H8	BERYLLIUM BOROHYDRIDE	-------	-------	-------	-------	----------
40	BeBr2	BERYLLIUM BROMIDE	-------	-------	-------	-------	----------
41	BeCl2	BERYLLIUM CHLORIDE	-------	-------	-------	-------	----------
42	BeF2	BERYLLIUM FLUORIDE	-------	-------	-------	-------	----------
43	BeI2	BERYLLIUM IODIDE	-------	-------	-------	-------	----------
44	Bi	BISMUTH	-------	-------	-------	-------	----------
45	BiBr3	BISMUTH TRIBROMIDE	-------	-------	-------	-------	----------
46	BiCl3	BISMUTH TRICHLORIDE	-------	-------	-------	-------	----------
47	BrF5	BROMINE PENTAFLUORIDE	-------	-------	gas	1.51	----------
48	Br2	BROMINE	25.0	1.6480	gas	0.00	0.969
49	C	CARBON	25.0	2.1500	-------	-------	----------
50	CCl2O	PHOSGENE	25.0	1.3561	gas	1.17	2.877
51	CF2O	CARBONYL FLUORIDE	-------	-------	gas	0.95	2.269
52	CH4N2O	UREA	-------	-------	in dioxane	4.56	2.600
53	CH4N2S	THIOUREA	-------	-------	in benzene	5.07	2.833
54	CNBr	CYANOGEN BROMIDE	-------	-------	-------	-------	----------
55	CNCl	CYANOGEN CHLORIDE	-------	-------	gas	2.82	1.250
56	CNF	CYANOGEN FLUORIDE	-------	-------	-------	-------	----------
57	CO	CARBON MONOXIDE	25.0	1.0003	gas	0.11	0.558
58	COS	CARBONYL SULFIDE	25.0	1.3506	gas	0.71	1.270
59	COSe	CARBON OXYSELENIDE	-------	-------	gas	0.59	----------
60	CO2	CARBON DIOXIDE	24.0	1.6630	gas	0.00	1.040
61	CS2	CARBON DISULFIDE	25.0	1.6241	gas	0.00	1.569
62	CSeS	CARBON SELENOSULFIDE	-------	-------	-------	-------	----------
63	C2N2	CYANOGEN	-------	-------	gas	0.00	1.423
64	C3S2	CARBON SUBSULFIDE	-------	-------	-------	-------	----------
65	Ca	CALCIUM	-------	-------	-------	-------	----------
66	CaF2	CALCIUM FLUORIDE	25.0	1.4338	-------	-------	----------
67	CbF5	COLUMBIUM FLUORIDE	-------	-------	-------	-------	----------
68	Cd	CADMIUM	-------	-------	-------	-------	----------
69	CdCl2	CADMIUM CHLORIDE	-------	-------	-------	-------	----------
70	CdF2	CADMIUM FLUORIDE	-------	-------	-------	-------	----------
71	CdI2	CADMIUM IODIDE	-------	-------	-------	-------	----------
72	CdO	CADMIUM OXIDE	-------	-------	-------	-------	----------
73	ClF	CHLORINE MONOFLUORIDE	-------	-------	gas	0.89	----------
74	ClFO3	PERCHLORYL FLUORIDE	-------	-------	gas	0.023	----------
75	ClF3	CHLORINE TRIFLUORIDE	-------	-------	gas	0.60	2.307
76	ClF5	CHLORINE PENTAFLUORIDE	-------	-------	-------	-------	----------
77	ClHO3S	CHLOROSULFONIC ACID	19.9	1.4331	-------	-------	----------
78	ClHO4	PERCHLORIC ACID	50.0	1.3819	-------	-------	----------

259

Table 10-2 REFRACTIVE INDEX, DIPOLE M0MENT, AND RADIUS OF GYRATION - INORGANIC COMPOUNDS (continued)

NO	FORMULA	NAME	Refractive Index		Dipole Moment		Radius of Gyration
			T, C	Value @ T	State	Debye	Angstrom
79	ClO2	CHLORINE DIOXIDE	-------	-------	gas	1.78	2.803
80	Cl2	CHLORINE	25.0	1.3786	gas	0.00	0.987
81	Cl2O	CHLORINE MONOXIDE	-------	-------	-------	-------	----------
82	Cl2O7	CHLORINE HEPTOXIDE	-------	-------	in CCl4	0.72	----------
83	Co	COBALT	-------	-------	-------	-------	----------
84	CoCl2	COBALT CHLORIDE	-------	-------	-------	-------	----------
85	CoNC3O4	COBALT NITROSYL TRICARBONYL	-------	-------	-------	-------	----------
86	Cr	CHROMIUM	-------	-------	-------	-------	----------
87	CrC6O6	CHROMIUM CARBONYL	-------	-------	-------	-------	----------
88	CrO2Cl2	CHROMIUM OXYCHLORIDE	23.0	1.5240	in CCl4	0.47	----------
89	Cs	CESIUM	-------	-------	-------	-------	----------
90	CsBr	CESIUM BROMIDE	25.0	1.6984	-------	-------	----------
91	CsCl	CESIUM CHLORIDE	25.0	1.6418	-------	-------	----------
92	CsF	CESIUM FLUORIDE	25.0	1.4718	gas	7.88	----------
93	CsI	CESIUM IODIDE	25.0	1.7876	gas	10.20	----------
94	Cu	COPPER	-------	-------	-------	-------	----------
95	CuBr	CUPROUS BROMIDE	-------	-------	-------	-------	----------
96	CuCl	CUPROUS CHLORIDE	25.0	1.9300	gas	1.82	----------
97	CuCl2	CUPRIC CHLORIDE	-------	-------	-------	-------	----------
98	CuI	COPPER IODIDE	-------	-------	-------	-------	----------
99	DCN	DEUTERIUM CYANIDE	-------	-------	-------	-------	----------
100	D2	DEUTERIUM	25.0	1.0001	-------	-------	----------
101	D2O	DEUTERIUM OXIDE	25.0	1.3280	in dioxane	1.88	----------
102	Eu	EUROPIUM	-------	-------	-------	-------	----------
103	F2	FLUORINE	-188.2	1.2000	gas	0.00	0.714
104	F2O	FLUORINE OXIDE	-------	-------	gas	0.30	----------
105	Fe	IRON	-------	-------	-------	-------	----------
106	FeC5O5	IRON PENTACARBONYL	14.0	1.5230	-------	-------	----------
107	FeCl2	FERROUS CHLORIDE	25.0	1.5670	-------	-------	----------
108	FeCl3	FERRIC CHLORIDE	-------	-------	in dioxane	1.28	----------
109	Fr	FRANCIUM	-------	-------	-------	-------	----------
110	Ga	GALLIUM	-------	-------	-------	-------	----------
111	GaCl3	GALLIUM TRICHLORIDE	-------	-------	in benzene	5.10	3.479
112	Gd	GADOLINIUM	-------	-------	-------	-------	----------
113	Ge	GERMANIUM	-------	-------	-------	-------	----------
114	GeBr4	GERMANIUM BROMIDE	26.0	1.6269	-------	-------	----------
115	GeCl4	GERMANIUM CHLORIDE	25.0	1.4614	gas	0.00	----------
116	GeHCl3	TRICHLORO GERMANE	-------	-------	-------	-------	----------
117	GeH4	GERMANE	-0.2	1.0009	-------	-------	----------
118	Ge2H6	DIGERMANE	-------	-------	-------	-------	----------
119	Ge3H8	TRIGERMANE	-------	-------	-------	-------	----------
120	HBr	HYDROGEN BROMIDE	10.0	1.3250	gas	0.79	0.157
121	HCN	HYDROGEN CYANIDE	25.0	1.2594	gas	2.98	0.654
122	HCl	HYDROGEN CHLORIDE	17.9	1.3287	gas	1.08	0.209
123	HF	HYDROGEN FLUORIDE	25.0	1.1574	gas	1.82	0.201
124	HI	HYDROGEN IODIDE	16.0	1.4660	gas	0.44	0.142
125	HNO3	NITRIC ACID	23.9	1.3970	gas	2.17	2.249
126	H2	HYDROGEN	-253.0	1.1096	gas	0.00	0.371
127	H2O	WATER	25.0	1.3325	gas	1.85	0.615
128	H2O2	HYDROGEN PEROXIDE	25.0	1.4067	gas	1.57	1.242
129	H2S	HYDROGEN SULFIDE	20.0	1.3682	gas	0.97	0.638
130	H2SO4	SULFURIC ACID	19.9	1.4183	gas	2.73	2.631
131	H2S2	HYDROGEN DISULFIDE	20.0	1.6300	in dioxane	1.18	----------
132	H2Se	HYDROGEN SELENIDE	-------	-------	solid	0.40	----------
133	H2Te	HYDROGEN TELLURIDE	-------	-------	solid	0.40	----------
134	H3NO3S	SULFAMIC ACID	25.0	1.5530	-------	-------	5.692
135	He	HELIUM-3	25.0	1.0000	gas	0.00	----------
136	He	HELIUM-4	-269.0	1.0245	gas	0.00	0.808
137	Hf	HAFNIUM	-------	-------	-------	-------	----------
138	Hg	MERCURY	25.0	1.7500	gas	0.00	----------
139	HgBr2	MERCURIC BROMIDE	-------	-------	in dioxane	1.54	----------
140	HgCl2	MERCURIC CHLORIDE	-------	-------	in dioxane	1.44	----------
141	HgI2	MERCURIC IODIDE	25.0	1.8590	in dioxane	1.68	----------
142	IF7	IODINE HEPTAFLUORIDE	-------	-------	-------	-------	----------
143	I2	IODINE	25.0	3.3400	gas	0.00	0.921
144	In	INDIUM	-------	-------	-------	-------	----------
145	Ir	IRIDIUM	-------	-------	-------	-------	----------
146	K	POTASSIUM	-------	-------	gas	0.00	----------
147	KBr	POTASSIUM BROMIDE	25.0	1.5590	in benzene	9.70	----------
148	KCl	POTASSIUM CHLORIDE	25.0	1.4897	in benzene	10.20	----------
149	KF	POTASSIUM FLUORIDE	25.0	1.3630	gas	8.59	----------
150	KI	POTASSIUM IODIDE	25.0	1.6770	gas	6.80	----------
151	KOH	POTASSIUM HYDROXIDE	-------	-------	-------	-------	----------
152	Kr	KRYPTON	-157.0	1.3032	gas	0.00	1.138
153	La	LANTHANUM	-------	-------	-------	-------	----------
154	Li	LITHIUM	-------	-------	-------	-------	----------
155	LiBr	LITHIUM BROMIDE	25.0	1.7840	gas	7.27	----------
156	LiCl	LITHIUM CHLORIDE	25.0	1.6620	gas	7.13	----------

NO	FORMULA	NAME	Refractive Index		Dipole Moment		Radius of Gyration
			T, C	Value @ T	State	Debye	Angstrom
157	LiF	LITHIUM FLUORIDE	25.0	1.3915	gas	6.33	----------
158	LiI	LITHIUM IODIDE	-------	-------	gas	7.43	----------
159	Lu	LUTECIUM	-------	-------	-------	-------	----------
160	Mg	MAGNESIUM	-------	-------	-------	-------	----------
161	MgCl2	MAGNESIUM CHLORIDE	25.0	1.5900	-------	-------	----------
162	MgO	MAGNESIUM OXIDE	25.0	1.7360	-------	-------	----------
163	Mn	MANGANESE	-------	-------	-------	-------	----------
164	MnCl2	MANGANESE CHLORIDE	-------	-------	-------	-------	----------
165	Mo	MOLYBDENUM	-------	-------	-------	-------	----------
166	MoF6	MOLYBDENUM FLUORIDE	-------	-------	-------	-------	----------
167	MoO3	MOLYBDENUM OXIDE	-------	-------	-------	-------	----------
168	NCl3	NITROGEN TRICHLORIDE	-------	-------	in benzene	0.60	3.323
169	ND3	HEAVY AMMONIA	-------	-------	gas	1.50	----------
170	NF3	NITROGEN TRIFLUORIDE	-77.0	1.3944	gas	0.24	2.385
171	NH3	AMMONIA	20.0	1.3327	gas	1.47	0.853
172	NH3O	HYDROXYLAMINE	23.4	1.4400	gas	0.59	1.339
173	NH4Br	AMMONIUM BROMIDE	25.0	1.7120	in dioxane	6.75	----------
174	NH4Cl	AMMONIUM CHLORIDE	25.0	1.6248	-------	-------	----------
175	NH4I	AMMONIUM IODIDE	25.0	1.7031	-------	-------	----------
176	NH5O	AMMONIUM HYDROXIDE	-------	-------	-------	-------	----------
177	NH5S	AMMONIUM HYDROGENSULFIDE	25.0	1.7400	-------	-------	----------
178	NO	NITRIC OXIDE	-155.2	1.3305	gas	0.15	0.549
179	NOCl	NITROSYL CHLORIDE	-------	-------	gas	1.90	1.801
180	NOF	NITROSYL FLUORIDE	-------	-------	gas	1.73	----------
181	NO2	NITROGEN DIOXIDE	19.9	1.4000	gas	0.32	1.431
182	N2	NITROGEN	-190.0	1.2053	gas	0.00	0.547
183	N2F4	TETRAFLUOROHYDRAZINE	-------	-------	gas	0.26	2.879
184	N2H4	HYDRAZINE	25.0	1.4686	gas	1.75	1.509
185	N2H4C	AMMONIUM CYANIDE	-------	-------	-------	-------	----------
186	N2H6CO2	AMMONIUM CARBAMATE	-------	-------	-------	-------	----------
187	N2O	NITROUS OXIDE	25.0	1.2380	gas	0.18	0.954
188	N2O3	NITROGEN TRIOXIDE	-------	-------	gas	2.12	2.346
189	N2O4	NITROGEN TETRAOXIDE	19.9	1.4000	gas	0.00	2.918
190	N2O5	NITROGEN PENTOXIDE	-------	-------	in CCl4	1.40	----------
191	Na	SODIUM	0.0	0.0045	gas	0.00	----------
192	NaBr	SODIUM BROMIDE	25.0	1.6141	in benzene	9.40	----------
193	NaCN	SODIUM CYANIDE	19.9	1.4520	-------	-------	----------
194	NaCl	SODIUM CHLORIDE	17.9	1.5443	gas	9.00	1.153
195	NaF	SODIUM FLUORIDE	25.0	1.3270	gas	8.16	----------
196	NaI	SODIUM IODIDE	25.0	1.7745	gas	9.24	----------
197	NaOH	SODIUM HYDROXIDE	319.9	1.4330	-------	-------	1.285
198	Na2SO4	SODIUM SULFATE	25.0	1.4640	-------	-------	----------
199	Nb	NIOBIUM	25.0	1.8000	-------	-------	----------
200	Nd	NEODYMIUM	-------	-------	-------	-------	----------
201	Ne	NEON	25.0	1.0001	gas	0.00	0.869
202	Ni	NICKEL	-------	-------	-------	-------	----------
203	NiC4O4	NICKEL CARBONYL	-------	-------	in CCl4	0.00	----------
204	NiF2	NICKEL FLUORIDE	-------	-------	-------	-------	----------
205	Np	NEPTUNIUM	-------	-------	-------	-------	----------
206	O2	OXYGEN	-181.2	1.2210	gas	0.00	0.680
207	O3	OZONE	25.0	1.2226	gas	0.53	1.544
208	Os	OSMIUM	-------	-------	-------	-------	----------
209	OsOF5	OSMIUM OXIDE PENTAFLUORIDE	-------	-------	-------	-------	----------
210	OsO4	OSMIUM TETROXIDE - YELLOW	-------	-------	-------	-------	----------
211	OsO4	OSMIUM TETROXIDE - WHITE	-------	-------	gas	0.00	----------
212	P	PHOSPHORUS - WHITE	25.0	2.1142	gas	0.00	----------
213	PBr3	PHOSPHORUS TRIBROMIDE	25.0	1.6870	in benzene	0.50	----------
214	PCl2F3	PHOSPHORUS DICHLORIDE TRIFLUORIDE	-------	-------	-------	-------	----------
215	PCl3	PHOSPHORUS TRICHLORIDE	25.0	1.5121	gas	0.78	3.436
216	PCl5	PHOSPHORUS PENTACHLORIDE	-------	-------	in CCl4	0.80	----------
217	PH3	PHOSPHINE	17.0	1.3170	gas	0.58	----------
218	PH4Br	PHOSPHONIUM BROMIDE	-------	-------	-------	-------	----------
219	PH4Cl	PHOSPHONIUM CHLORIDE	-------	-------	-------	-------	----------
220	PH4I	PHOSPHONIUM IODIDE	-------	-------	-------	-------	----------
221	POCl3	PHOSPHORUS OXYCHLORIDE	25.0	1.4838	in benzene	2.42	3.398
222	PSBr3	PHOSPHORUS THIOBROMIDE	-------	-------	-------	-------	----------
223	PSCl3	PHOSPHORUS THIOCHLORIDE	-------	-------	-------	-------	3.634
224	P4O6	PHOSPHORUS TRIOXIDE	-------	-------	-------	-------	----------
225	P4O10	PHOSPHORUS PENTOXIDE	25.0	1.4690	-------	-------	4.446
226	P4S10	PHOSPHORUS PENTASULFIDE	-------	-------	-------	-------	5.471
227	Pb	LEAD	25.0	2.0100	-------	-------	----------
228	PbBr2	LEAD BROMIDE	-------	-------	-------	-------	----------
229	PbCl2	LEAD CHLORIDE	25.0	2.1990	-------	-------	----------
230	PbF2	LEAD FLUORIDE	-------	-------	-------	-------	----------
231	PbI2	LEAD IODIDE	-------	-------	-------	-------	----------
232	PbO	LEAD OXIDE	-------	-------	gas	4.64	----------
233	PbS	LEAD SULFIDE	25.0	3.9210	gas	3.59	----------
234	Pd	PALLADIUM	-------	-------	-------	-------	----------

261

NO	FORMULA	NAME	Refractive Index		Dipole Moment		Radius of Gyration
			T, C	Value @ T	State	Debye	Angstrom
235	Po	POLONIUM	-------	-------	-------	-------	----------
236	Pt	PLATINUM	-------	-------	-------	-------	----------
237	Ra	RADIUM	-------	-------	-------	-------	----------
238	Rb	RUBIDIUM	-------	-------	-------	-------	----------
239	RbBr	RUBIDIUM BROMIDE	25.0	1.5530	gas	0.00	----------
240	RbCl	RUBIDIUM CHLORIDE	35.0	1.4930	gas	10.51	----------
241	RbF	RUBIDIUM FLUORIDE	25.0	1.3980	gas	8.55	----------
242	RbI	RUBIDIUM IODIDE	25.0	1.6474	-------	-------	----------
243	Re	RHENIUM	-------	-------	-------	-------	----------
244	Re2O7	RHENIUM HEPTOXIDE	-------	-------	-------	-------	----------
245	Rh	RHODIUM	-------	-------	-------	-------	----------
246	Rn	RADON	-------	-------	-------	-------	----------
247	Ru	RUTHENIUM	-------	-------	-------	-------	----------
248	RuF5	RUTHENIUM PENTAFLUORIDE	-------	-------	-------	-------	----------
249	S	SULFUR	125.0	1.9170	gas	0.00	----------
250	SF4	SULFUR TETRAFLUORIDE	-------	-------	gas	0.63	----------
251	SF6	SULFUR HEXAFLUORIDE	25.0	1.1670	gas	0.00	----------
252	SOBr2	THIONYL BROMIDE	-------	-------	in benzene	1.48	----------
253	SOCl2	THIONYL CHLORIDE	25.0	1.5160	gas	1.45	3.473
254	SOF2	SULFUROUS OXYFLUORIDE	-------	-------	gas	1.63	----------
255	SO2	SULFUR DIOXIDE	25.0	1.3396	gas	1.63	1.660
256	SO2Cl2	SULFURYL CHLORIDE	25.0	1.4430	gas	1.81	3.064
257	SO3	SULFUR TRIOXIDE	25.0	1.4052	gas	0.00	2.189
258	S2Cl2	SULFUR MONOCHLORIDE	14.0	1.6660	in benzene	1.07	----------
259	Sb	ANTIMONY	-------	-------	-------	-------	----------
260	SbBr3	ANTIMONY TRIBROMIDE	25.0	1.7400	in benzene	3.30	----------
261	SbCl3	ANTIMONY TRICHLORIDE	-------	-------	gas	3.93	3.031
262	SbCl5	ANTIMONY PENTACHLORIDE	22.0	1.5925	in CCl4	1.15	----------
263	SbH3	STIBINE	-------	-------	gas	0.12	----------
264	SbI3	ANTIMONY TRIIODIDE	25.0	2.3600	in benzene	0.40	----------
265	Sb2O3	ANTIMONY TRIOXIDE	25.0	2.0870	-------	-------	----------
266	Sc	SCANDIUM	-------	-------	-------	-------	----------
267	Se	SELENIUM	-------	-------	-------	-------	----------
268	SeCl4	SELENIUM TETRACHLORIDE	25.0	1.8070	-------	-------	----------
269	SeF6	SELENIUM HEXAFLUORIDE	25.0	1.8950	gas	0.00	----------
270	SeOCl2	SELENIUM OXYCHLORIDE	20.0	1.6510	in benzene	2.64	----------
271	SeO2	SELENIUM DIOXIDE	25.0	1.7600	gas	2.62	----------
272	Si	SILICON	-------	-------	-------	-------	----------
273	SiBrCl2F	BROMODICHLOROFLUOROSILANE	-------	-------	-------	-------	----------
274	SiBrF3	TRIFLUOROBROMOSILANE	-------	-------	-------	-------	----------
275	SiBr2ClF	DIBROMOCHLOROFLUOROSILANE	-------	-------	-------	-------	----------
276	SiClF3	TRIFLUOROCHLOROSILANE	-------	-------	-------	-------	----------
277	SiCl2F2	DICHLORODIFLUOROSILANE	-------	-------	-------	-------	----------
278	SiCl3F	TRICHLOROFLUOROSILANE	-------	-------	gas	0.49	----------
279	SiCl4	SILICON TETRACHLORIDE	25.0	1.4116	gas	0.00	----------
280	SiF4	SILICON TETRAFLUORIDE	-------	-------	gas	0.00	----------
281	SiHBr3	TRIBROMOSILANE	-------	-------	in heptane	0.79	----------
282	SiHCl3	TRICHLOROSILANE	25.0	1.3983	gas	0.86	3.456
283	SiHF3	TRIFLUOROSILANE	-------	-------	gas	1.27	----------
284	SiH2Br2	DIBROMOSILANE	-------	-------	-------	-------	----------
285	SiH2Cl2	DICHLOROSILANE	-122.5	1.4360	gas	1.17	2.694
286	SiH2F2	DIFLUOROSILANE	-------	-------	gas	1.55	----------
287	SiH2I2	DIIODOSILANE	-------	-------	-------	-------	----------
288	SiH3Br	MONOBROMOSILANE	-------	-------	gas	1.32	----------
289	SiH3Cl	MONOCHLOROSILANE	-------	-------	gas	1.31	----------
290	SiH3F	MONOFLUOROSILANE	-------	-------	gas	1.30	----------
291	SiH3I	IODOSILANE	-------	-------	-------	-------	----------
292	SiH4	SILANE	-------	-------	gas	0.00	----------
293	SiO2	SILICON DIOXIDE	25.0	1.5442	quartz	0.58	----------
294	Si2Cl6	HEXACHLORODISILANE	18.0	1.4748	-------	-------	----------
295	Si2F6	HEXAFLUORODISILANE	-------	-------	-------	-------	----------
296	Si2H5Cl	DISILANYL CHLORIDE	-------	-------	-------	-------	----------
297	Si2H6	DISILANE	25.0	1.5700	gas	0.00	----------
298	Si2OCl3F3	TRICHLOROTRIFLUORODISILOXANE	-------	-------	-------	-------	----------
299	Si2OCl6	HEXACHLORODISILOXANE	-------	-------	-------	-------	----------
300	Si2OH6	DISILOXANE	-------	-------	gas	0.24	----------
301	Si3Cl8	OCTACHLOROTRISILANE	-------	-------	-------	-------	----------
302	Si3H8	TRISILANE	-------	-------	-------	-------	----------
303	Si3H9N	TRISILAZANE	-------	-------	-------	-------	----------
304	Si4H10	TETRASILANE	-------	-------	-------	-------	----------
305	Sm	SAMARIUM	-------	-------	-------	-------	----------
306	Sn	TIN	-------	-------	-------	-------	----------
307	SnBr4	STANNIC BROMIDE	31.0	1.6628	in dioxane	4.13	----------
308	SnCl2	STANNOUS CHLORIDE	-------	-------	-------	-------	----------
309	SnCl4	STANNIC CHLORIDE	25.0	1.5086	gas	0.00	----------
310	SnH4	STANNIC HYDRIDE	-------	-------	-------	-------	----------
311	SnI4	STANNIC IODIDE	25.0	2.1060	in benzene	0.00	----------
312	Sr	STRONTIUM	-------	-------	-------	-------	

262

Table 10-2 REFRACTIVE INDEX, DIPOLE M0MENT, AND RADIUS OF GYRATION - INORGANIC COMPOUNDS (continued)

NO	FORMULA	NAME	Refractive Index		Dipole Moment		Radius of Gyration
			T, C	Value @ T	State	Debye	Angstrom
313	SrO	STRONTIUM OXIDE	25.0	1.8100	gas	8.90	----------
314	Ta	TANTALUM	-------	-------	-------	-------	----------
315	Tc	TECNNETIUM	-------	-------	-------	-------	----------
316	Te	TELLURIUM	25.0	1.0025	-------	-------	----------
317	TeCl4	TELLURIUM TETRACHLORIDE	-------	-------	in benzene	2.56	----------
318	TeF6	TELLURIUM HEXAFLUORIDE	-------	-------	gas	0.00	----------
319	Ti	TITANIUM	-------	-------	-------	-------	----------
320	TiCl4	TITANIUM TETRACHLORIDE	17.7	1.6120	liquid	0.00	----------
321	Tl	THALLIUM	-------	-------	-------	-------	----------
322	TlBr	THALLOUS BROMIDE	25.0	2.4000	gas	4.49	----------
323	TlI	THALLOUS IODIDE	-------	-------	gas	4.61	----------
324	Tm	THULIUM	-------	-------	-------	-------	----------
325	U	URANIUM	-------	-------	-------	-------	----------
326	UF6	URANIUM FLUORIDE	-------	-------	gas	0.00	----------
327	V	VANADIUM	25.0	3.0300	-------	-------	----------
328	VCl4	VANADIUM TETRACHLORIDE	-------	-------	in CCl4	0.00	3.564
329	VOCl3	VANADIUM OXYTRICHLORIDE	25.0	1.6300	in CCl4	0.30	3.499
330	W	TUNGSTEN	-------	-------	-------	-------	----------
331	WF6	TUNGSTEN FLUORIDE	-------	-------	-------	-------	----------
332	Xe	XENON	-112.0	1.3918	gas	0.00	1.296
333	Yb	YTTERBIUM	-------	-------	-------	-------	----------
334	Yt	YTTRIUM	-------	-------	-------	-------	----------
335	Zn	ZINC	-------	-------	-------	-------	----------
336	ZnCl2	ZINC CHLORIDE	25.0	1.6810	-------	-------	----------
337	ZnF2	ZINC FLUORIDE	-------	-------	-------	-------	----------
338	ZnO	ZINC OXIDE	25.0	2.0080	-------	-------	----------
339	ZnSO4	ZINC SULFATE	25.0	1.6580	-------	-------	----------
340	Zr	ZIRCONIUM	-------	-------	-------	-------	----------
341	ZrBr4	ZIRCONIUM BROMIDE	-------	-------	in dioxane	4.68	----------
342	ZrCl4	ZIRCONIUM CHLORIDE	-------	-------	in dioxane	5.94	----------
343	ZrI4	ZIRCONIUM IODIDE	-------	-------	in dioxane	5.36	----------

Refractive index applies at T.

T, C - temperature in Centigrade.

Dipole moment is given in Debye units.

The conversion factor for Debye units is 1 Debye = 3.33564E-30 C M

Radius of gyration is given in Angstrom units.

The conversion factor for Angstrom units is 1 Angstrom = 1.0E-10 meters

Chapter 11

ENTROPY AND ENTROPY OF FORMATION OF GAS

Carl L. Yaws

Lamar University, Beaumont, Texas

ABSTRACT

Results for entropy and entropy of formation of gas are presented for major organic and inorganic chemicals. The chemical formula and molecular weight are also given. The results are displayed in easy-to-use tabulations which are especially applicable for rapid engineering usage with the personal computer or hand calculator. The organic chemicals encompass hydrocarbon, oxygen, nitrogen, halogen, silicon, sulfur, and other compound types.

INTRODUCTION

Properties such as entropy and entropy of formation are useful in ascertaining the thermodynamics of operations encountered in the chemical processing and petroleum refining industries. As an example of such usefulness, the heat effects and equilibrium yields of chemical reactions require knowledge of the thermodynamics of the chemical reactions. Other uses include ascertaining the thermodynamics of chemical explosions.

ENTROPY AND ENTROPY OF FORMATION

The results for entropy and entropy of formation are given in Tables 11-1 and 11-2 for organic and inorganic compounds. The values apply to the ideal gas at 298.15 K. The entropy is the absolute entropy. The entropy of formation is ascertained from the appropriate thermodynamic relations for the formation of the compound from the elements. The tabulations are based on data source publications for organics (1-37) and inorganics (1-61). The tabulations are arranged by chemical formula to provide ease of use in quickly locating data.

REFERENCES – ORGANIC COMPOUNDS

1-34. See **REFERENCES - ORGANIC COMPOUNDS** in **Chapter 1, CRITICAL PROPERTIES AND ACENTRIC FACTOR**

35. Crowl, D. A. and J. F. Louvar, CHEMICAL PROCESS SAFETY, Prentice Hall, Inc., Englewood Cliffs, NJ (1990).

36. Yaws, C. L. and P. Y. Chiang, Chem. Eng., 95 (13), 81 (Sept. 26, 1988).

37. Yaws, C. L., HANDBOOK OF CHEMICAL COMPOUND DATA FOR PROCESS SAFETY, Gulf Publishing Co., Houston, TX (1997).

REFERENCES - INORGANIC COMPOUNDS

1-56. See **REFERENCES - ORGANIC COMPOUNDS** in **Chapter 1, CRITICAL PROPERTIES AND ACENTRIC FACTOR**

57. Stull, D. R., and H. Prophet, Project Directors, JANAF THERMOCHEMICAL TABLES, 2nd ed., NSRDS-NBS 37, U. S. Government Printing Office, Washington DC (1971).

58. Daubert, T. E., CHEMICAL ENGINEERING THERMODYNAMICS, McGraw-Hill, New York, NY (1985).

59. Karapet'yants, M. K. and M. L. Karapet'yants, THERMODYNAMIC CONSTANTS OF INORGANIC AND ORGANIC COMPOUNDS, translated from Russian, Ann Arbor - Humphrey Science Publishers, Ann Arbor, MI (1970).

60. Crowl, D. A. and J. F. Louvar, CHEMICAL PROCESS SAFETY, Prentice Hall, Inc., Englewood Cliffs, NJ (1990).

61. Yaws, C. L., HANDBOOK OF CHEMICAL COMPOUND DATA FOR PROCESS SAFETY, Gulf Publishing Co., Houston, TX (1997).

Table 11-1 ENTROPY AND ENTROPY OF FORMATION OF GAS - ORGANIC COMPOUNDS

NO	FORMULA	NAME	Molecular Weight g/mol	Entropy S @ 298 K, joule/(mol K)	Entropy of Formation Sf @ 298 K, joule/(mol K)
1	CBrClF2	BROMOCHLORODIFLUOROMETHANE	165.365	318.57	-78.383
2	CBrCl3	BROMOTRICHLOROMETHANE	198.273	332.92	-79.490
3	CBrF3	BROMOTRIFLUOROMETHANE	148.910	297.71	-88.144
4	CBr2F2	DIBROMODIFLUOROMETHANE	209.816	325.24	-36.559
5	CClF3	CHLOROTRIFLUOROMETHANE	104.459	285.24	-181.016
6	CClN	CYANOGEN CHLORIDE	61.470	236.22	23.310
7	CCl2F2	DICHLORODIFLUOROMETHANE	120.913	300.88	-185.242
8	CCl2O	PHOSGENE	98.916	283.74	-43.636
9	CCl3F	TRICHLOROFLUOROMETHANE	137.368	309.62	-133.322
10	CCl4	CARBON TETRACHLORIDE	153.822	309.70	-141.472
11	CF2O	CARBONYL FLUORIDE	66.007	258.79	-52.121
12	CF4	CARBON TETRAFLUORIDE	88.005	261.31	-149.589
13	CHBr3	TRIBROMOMETHANE	252.731	330.59	31.192
14	CHClF2	CHLORODIFLUOROMETHANE	86.468	280.86	-111.353
15	CHCl2F	DICHLOROFLUOROMETHANE	102.923	293.17	-107.731
16	CHCl3	CHLOROFORM	119.377	295.51	-109.743
17	CHF3	TRIFLUOROMETHANE	70.014	259.56	-115.479
18	CHI3	TRIIODOMETHANE	393.732	--------	110.414
19	CHN	HYDROGEN CYANIDE	27.026	201.67	35.083
20	CHNS	ISOTHIOCYANIC-ACID	59.086	--------	49.405
21	CH2BrCl	BROMOCHLOROMETHANE	129.384	287.00	-36.559
22	CH2Br2	DIBROMOMETHANE	173.835	292.96	4.696
23	CH2ClF	CHLOROFLUOROMETHANE	68.478	--------	-93.208
24	CH2Cl2	DICHLOROMETHANE	84.932	270.18	-88.982
25	CH2F2	DIFLUOROMETHANE	52.024	246.58	-92.336
26	CH2I2	DIIODOMETHANE	267.836	309.39	56.683
27	CH2O	FORMALDEHYDE	30.026	218.66	-20.091
28	CH2O2	FORMIC ACID	46.026	248.74	-92.604
29	CH3Br	METHYL BROMIDE	94.939	245.85	-31.863
30	CH3Cl	METHYL CHLORIDE	50.488	234.18	-78.585
31	CH3Cl3Si	METHYL TRICHLOROSILANE	149.478	351.15	-203.589
32	CH3F	METHYL FLUORIDE	34.033	222.78	-80.127
33	CH3I	METHYL IODIDE	141.939	254.01	-5.635
34	CH3NO	FORMAMIDE	45.041	248.45	-151.434
35	CH3NO2	NITROMETHANE	61.040	275.01	-227.335
36	CH3NO2	METHYL-NITRITE	61.040	--------	-218.078
37	CH3NO3	METHYL-NITRATE	77.040	--------	-302.968
38	CH4	METHANE	16.043	186.27	-80.530
39	CH4Cl2Si	METHYL DICHLOROSILANE	115.034	328.50	-180.111
40	CH4O	METHANOL	32.042	239.70	-129.666
41	CH4O3S	METHANESULFONIC ACID	96.107	189.00	--------
42	CH4S	METHYL MERCAPTAN	48.109	255.02	-43.770
43	CH5ClSi	METHYL CHLOROSILANE	80.589	298.28	-164.011
44	CH5N	METHYLAMINE	31.057	243.30	-185.376
45	CH6Si	METHYL SILANE	46.144	256.50	-159.752
46	CN4O8	TETRANITROMETHANE	196.033	--------	--------
47	CO	CARBON MONOXIDE	28.010	197.54	89.686
48	COS	CARBONYL SULFIDE	60.076	231.47	91.330
49	CO2	CARBON DIOXIDE	44.010	213.69	2.918
50	CS2	CARBON DISULFIDE	76.143	237.79	168.271
51	C2BrF3	BROMOTRIFLUOROETHYLENE	160.921	335.00	-57.018
52	C2Br2F4	1,2-DIBROMOTETRAFLUOROETHANE	259.824	387.52	-181.553
53	C2ClF3	CHLOROTRIFLUOROETHYLENE	116.470	323.00	-102.968
54	C2ClF5	CHLOROPENTAFLUOROETHANE	154.467	353.17	-280.664
55	C2Cl2F4	1,2-DICHLOROTETRAFLUOROETHANE	170.921	364.93	-275.063
56	C2Cl3F3	1,1,2-TRICHLOROTRIFLUOROETHANE	187.375	388.23	-261.010
57	C2Cl4	TETRACHLOROETHYLENE	165.833	340.85	-116.619
58	C2Cl4F2	1,1,2,2-TETRACHLORODIFLUOROETHANE	203.830	388.36	-271.742
59	C2Cl4O	TRICHLOROACETYL CHLORIDE	181.832	369.36	-190.542
60	C2Cl6	HEXACHLOROETHANE	236.738	398.74	-283.750
61	C2F4	TETRAFLUOROETHYLENE	100.016	300.01	-116.887
62	C2F6	HEXAFLUOROETHANE	138.012	332.08	-287.372
63	C2HBrClF3	HALOTHANE	197.382	360.00	-202.381
64	C2HClF2	2-CHLORO-1,1-DIFLUOROETHYLENE	98.479	304.00	-87.204
65	C2HCl3	TRICHLOROETHYLENE	131.388	325.09	-86.165
66	C2HCl3O	DICHLOROACETYL CHLORIDE	147.387	350.45	-163.274
67	C2HCl3O	TRICHLOROACETALDEHYDE	147.387	348.00	-164.347
68	C2HCl5	PENTACHLOROETHANE	202.293	381.41	-253.597
69	C2HF3	TRIFLUOROETHENE	82.025	--------	-88.244
70	C2HF3O2	TRIFLUOROACETIC ACID	114.024	333.00	-253.899
71	C2HF5	PENTAFLUOROETHANE	120.022	333.72	-249.874
72	C2H2	ACETYLENE	26.038	200.82	58.796
73	C2H2Br4	1,1,2,2-TETRABROMOETHANE	345.654	400.12	-46.353
74	C2H2Cl2	1,1-DICHLOROETHYLENE	96.943	287.87	-77.209
75	C2H2Cl2	cis-1,2-DICHLOROETHYLENE	96.943	289.61	-75.331
76	C2H2Cl2	trans-1,2-DICHLOROETHYLENE	96.943	289.85	-75.231
77	C2H2Cl2O	CHLOROACETYL CHLORIDE	112.943	325.77	-141.640
78	C2H2Cl2O	DICHLOROACETALDEHYDE	112.943	329.00	-137.515

265

Table 11-1 ENTROPY AND ENTROPY OF FORMATION OF GAS - ORGANIC COMPOUNDS (continued)

NO	FORMULA	NAME	Molecular Weight g/mol	Entropy S @ 298 K, joule/(mol K)	Entropy of Formation Sf @ 298 K, joule/(mol K)
79	C2H2Cl2O2	DICHLOROACETIC ACID	128.942	342.00	-228.073
80	C2H2Cl3F	1,1,1-TRICHLOROFLUOROETHANE	151.394	350.00	-228.073
81	C2H2Cl4	1,1,1,2-TETRACHLOROETHANE	167.849	355.85	-231.763
82	C2H2Cl4	1,1,2,2-TETRACHLOROETHANE	167.849	354.97	-225.256
83	C2H2F2	1,1-DIFLUOROETHYLENE	64.035	265.18	-79.557
84	C2H2F2	cis-1,2-DIFLUOROETHENE	64.035	--------	-76.069
85	C2H2F2	trans-1,2-DIFLUOROETHENE	64.035	--------	-76.639
86	C2H2F4	1,1,1,2-TETRAFLUOROETHANE	102.031	316.20	-231.058
87	C2H2O	KETENE	42.037	241.79	-2.683
88	C2H2O4	OXALIC ACID	90.036	343.00	-210.297
89	C2H3Br	VINYL BROMIDE	106.950	275.43	-7.983
90	C2H3Cl	VINYL CHLORIDE	62.499	273.54	-48.566
91	C2H3ClF2	1-CHLORO-1,1-DIFLUOROETHANE	100.495	306.86	-214.657
92	C2H3ClO	ACETYL CHLORIDE	78.498	294.97	-126.446
93	C2H3ClO	CHLOROACETALDEHYDE	78.498	304.00	-117.391
94	C2H3ClO2	CHLOROACETIC ACID	94.497	313.42	-210.297
95	C2H3ClO2	METHYL CHLOROFORMATE	94.497	312.00	-211.303
96	C2H3Cl3	1,1,1-TRICHLOROETHANE	133.404	320.03	-221.734
97	C2H3Cl3	1,1,2-TRICHLOROETHANE	133.404	333.42	-204.595
98	C2H3F	VINYL FLUORIDE	46.044	262.26	-46.453
99	C2H3F3	1,1,1-TRIFLUOROETHANE	84.041	287.31	-224.115
100	C2H3N	ACETONITRILE	41.053	243.47	-59.500
101	C2H3NO	METHYL ISOCYANATE	57.052	271.21	-134.429
102	C2H4	ETHYLENE	28.054	219.45	-53.061
103	C2H4Br2	1,1-DIBROMOETHANE	187.862	327.60	-97.233
104	C2H4Br2	1,2-DIBROMOETHANE	187.862	329.74	-94.986
105	C2H4Cl2	1,1-DICHLOROETHANE	98.959	305.06	-190.575
106	C2H4Cl2	1,2-DICHLOROETHANE	98.959	308.28	-187.322
107	C2H4Cl2O	BIS(CHLOROMETHYL)ETHER	114.959	360.00	-238.135
108	C2H4F2	1,1-DIFLUOROETHANE	66.051	282.42	-192.822
109	C2H4F2	1,2-DIFLUOROETHANE	66.051	288.19	-187.825
110	C2H4I2	1,2-DIIODOETHANE	281.863	--------	-40.114
111	C2H4O	ACETALDEHYDE	44.053	250.20	-110.884
112	C2H4O	ETHYLENE OXIDE	44.053	242.92	-132.584
113	C2H4OS	THIOACETIC-ACID	76.113	--------	-93.745
114	C2H4O2	ACETIC ACID	60.053	282.50	-195.036
115	C2H4O2	METHYL FORMATE	60.053	285.20	-176.388
116	C2H4S	THIACYCLOPROPANE	60.114	--------	-49.237
117	C2H5Br	BROMOETHANE	108.966	287.27	-126.446
118	C2H5Cl	ETHYL CHLORIDE	64.514	275.78	-173.436
119	C2H5ClO	2-CHLOROETHANOL	80.514	321.00	-231.427
120	C2H5F	ETHYL FLUORIDE	48.060	265.01	-174.308
121	C2H5I	ETHYL IODIDE	155.966	295.52	-99.648
122	C2H5N	ETHYLENEIMINE	43.068	250.62	-182.995
123	C2H5NO	ACETAMIDE	59.068	272.21	-263.961
124	C2H5NO	N-METHYLFORMAMIDE	59.068	280.58	-255.576
125	C2H5NO2	NITROETHANE	75.067	315.43	-323.159
126	C2H5NO3	ETHYL-NITRATE	91.066	--------	-392.789
127	C2H6	ETHANE	30.070	229.12	-173.570
128	C2H6AlCl	DIMETHYLALUMINUM CHLORIDE	92.054	--------	--------
129	C2H6O	DIMETHYL ETHER	46.069	266.69	-238.538
130	C2H6O	ETHANOL	46.069	282.59	-223.143
131	C2H6OS	DIMETHYL SULFOXIDE	78.135	306.27	-313.500
132	C2H6O2	ETHYLENE GLYCOL	62.068	323.55	-284.588
133	C2H6O4S	DIMETHYL SULFATE	126.133	335.00	-509.810
134	C2H6S	DIMETHYL SULFIDE	62.136	285.85	-149.187
135	C2H6S	ETHYL MERCAPTAN	62.136	296.06	-138.923
136	C2H6S2	DIMETHYL DISULFIDE	94.202	336.69	-130.371
137	C2H7N	DIMETHYLAMINE	45.084	272.96	-291.196
138	C2H7N	ETHYLAMINE	45.084	284.85	-279.390
139	C2H7NO	MONOETHANOLAMINE	61.084	320.24	-346.503
140	C2H8N2	ETHYLENEDIAMINE	60.099	321.83	-404.293
141	C2H8Si	DIMETHYL SILANE	60.171	300.60	-251.954
142	C2N2	CYANOGEN	52.036	241.46	39.443
143	C3F6	HEXAFLUOROPROPYLENE	150.023	483.00	-133.825
144	C3F6O	HEXAFLUOROACETONE	166.023	394.00	-335.402
145	C3F8	OCTAFLUOROPROPANE	188.020	406.00	-413.215
146	C3H2N2	MALONONITRILE	66.062	287.00	-51.987
147	C3H3Cl	PROPARGYL CHLORIDE	74.510	288.00	-36.894
148	C3H3N	ACRYLONITRILE	53.064	275.31	-34.815
149	C3H3NO	OXAZOLE	69.063	270.73	-140.600
150	C3H4	METHYLACETYLENE	40.065	248.11	-30.186
151	C3H4	PROPADIENE	40.065	243.93	-34.379
152	C3H4Cl2	2,3-DICHLOROPROPENE	110.970	336.00	-165.319
153	C3H4O	ACROLEIN	56.064	296.94	-83.917
154	C3H4O	PROPARGYL ALCOHOL	56.064	293.00	-87.875
155	C3H4O2	ACRYLIC ACID	72.064	315.01	-168.271
156	C3H4O2	beta-PROPIOLACTONE	72.064	286.14	-197.250

Table 11-1 ENTROPY AND ENTROPY OF FORMATION OF GAS - ORGANIC COMPOUNDS (continued)

NO	FORMULA	NAME	Molecular Weight g/mol	Entropy S @ 298 K, joule/(mol K)	Entropy of Formation Sf @ 298 K, joule/(mol K)
157	C3H4O2	VINYL FORMATE	72.064	305.00	-177.763
158	C3H4O3	ETHYLENE CARBONATE	88.063	260.00	-325.004
159	C3H4O3	PYRUVIC ACID	88.063	318.00	-268.321
160	C3H5Br	3-BROMO-1-PROPENE	120.977	--------	-102.599
161	C3H5Cl	2-CHLOROPROPENE	76.525	302.00	-153.279
162	C3H5Cl	3-CHLOROPROPENE	76.525	307.00	-148.348
163	C3H5ClO	alpha-EPICHLOROHYDRIN	92.525	319.32	-238.336
164	C3H5ClO2	METHYL CHLOROACETATE	108.524	373.00	-288.445
165	C3H5ClO2	ETHYL CHLOROFORMATE	108.524	346.00	-315.278
166	C3H5Cl3	1,2,3-TRICHLOROPROPANE	147.431	382.92	-295.120
167	C3H5I	3-IODO-1-PROPENE	167.977^	--------	-81.670
168	C3H5N	PROPIONITRILE	55.079	286.60	-152.675
169	C3H5NO	ACRYLAMIDE	71.079	301.08	-241.825
170	C3H5NO	HYDRACRYLONITRILE	71.079	331.00	-210.968
171	C3H5NO	LACTONITRILE	71.079	408.00	-133.825
172	C3H5N3O9	NITROGLYCERINE	227.088	--------	--------
173	C3H6	CYCLOPROPANE	42.081	237.65	-171.357
174	C3H6	PROPYLENE	42.081	266.60	-141.875
175	C3H6Br2	1,2-DIBROMOPROPANE	201.888	--------	-184.940
176	C3H6Cl2	1,1-DICHLOROPROPANE	112.986	344.80	-287.104
177	C3H6Cl2	1,2-DICHLOROPROPANE	112.986	354.80	-277.042
178	C3H6Cl2	1,3-DICHLOROPROPANE	112.987	367.60	-264.665
179	C3H6Cl2	2,2-DICHLOROPROPANE	112.986	--------	-305.786
180	C3H6I2	1,2-DIIODOPROPANE	295.889	--------	-129.264
181	C3H6O	ACETONE	58.080	295.35	-216.401
182	C3H6O	ALLYL ALCOHOL	58.080	307.57	-203.790
183	C3H6O	METHYL VINYL ETHER	58.080	308.00	-203.589
184	C3H6O	n-PROPIONALDEHYDE	58.080	304.40	-206.574
185	C3H6O	1,2-PROPYLENE OXIDE	58.080	286.73	-224.686
186	C3H6O	1,3-PROPYLENE OXIDE	58.080	265.26	-246.185
187	C3H6O2	ETHYL FORMATE	74.079	328.20	-285.762
188	C3H6O2	METHYL ACETATE	74.079	319.80	-296.830
189	C3H6O2	PROPIONIC ACID	74.079	323.00	-291.129
190	C3H6O2S	3-MERCAPTOPROPIONIC ACID	106.145	438.00	-208.083
191	C3H6O3	LACTIC ACID	90.079	364.00	-352.172
192	C3H6O3	METHOXYACETIC ACID	90.079	360.00	-355.526
193	C3H6O3	TRIOXANE	90.079	284.89	-431.595
194	C3H6S	THIACYCLOBUTANE	74.140	--------	-155.492
195	C3H7Br	1-BROMOPROPANE	122.993	330.87	-219.319
196	C3H7Br	2-BROMOPROPANE	122.993	316.02	-234.211
197	C3H7Cl	ISOPROPYL CHLORIDE	78.541	305.94	-281.503
198	C3H7Cl	n-PROPYL CHLORIDE	78.541	315.47	-266.477
199	C3H7F	1-FLUOROPROPANE	62.087	--------	-271.239
200	C3H7F	2-FLUOROPROPANE	62.087	--------	-283.347
201	C3H7I	ISOPROPYL IODIDE	169.993	324.47	-207.681
202	C3H7I	n-PROPYL IODIDE	169.993	336.06	-196.176
203	C3H7N	ALLYLAMINE	57.095	312.00	-257.588
204	C3H7N	PROPYLENEIMINE	57.095	286.00	-282.408
205	C3H7NO	N,N-DIMETHYLFORMAMIDE	73.095	326.03	-346.436
206	C3H7NO	N-METHYLACETAMIDE	73.095	320.00	-352.172
207	C3H7NO2	1-NITROPROPANE	89.094	355.64	-419.286
208	C3H7NO2	2-NITROPROPANE	89.094	347.69	-427.168
209	C3H7NO3	PROPYL-NITRATE	105.093	--------	-492.135
210	C3H7NO3	ISOPROPYL-NITRATE	105.093	--------	-504.209
211	C3H8	PROPANE	44.096	270.20	-269.596
212	C3H8O	ISOPROPANOL	60.096	309.91	-332.048
213	C3H8O	METHYL ETHYL ETHER	60.096	310.62	-331.343
214	C3H8O	n-PROPANOL	60.096	324.72	-317.156
215	C3H8O2	2-METHOXYETHANOL	76.095	358.00	-385.712
216	C3H8O2	METHYLAL	76.095	335.72	-408.821
217	C3H8O2	1,2-PROPYLENE GLYCOL	76.095	351.87	-392.487
218	C3H8O2	1,3-PROPYLENE GLYCOL	76.095	358.90	-385.444
219	C3H8O3	GLYCEROL	92.095	396.39	-450.478
220	C3H8S	n-PROPYLMERCAPTAN	76.163	336.50	-234.915
221	C3H8S	ISOPROPYL MERCAPTAN	76.163	324.30	-247.124
222	C3H8S	ETHYL-METHYL-SULFIDE	76.156	--------	-238.269
223	C3H9N	n-PROPYLAMINE	59.111	324.18	-376.220
224	C3H9N	ISOPROPYLAMINE	59.111	312.38	-388.160
225	C3H9N	TRIMETHYLAMINE	59.111	288.78	-411.739
226	C3H9NO	1-AMINO-2-PROPANOL	75.111	365.00	-439.376
227	C3H9NO	3-AMINO-1-PROPANOL	75.111	367.00	-436.022
228	C3H9NO	METHYLETHANOLAMINE	75.111	344.00	-459.165
229	C3H9O4P	TRIMETHYL PHOSPHATE	140.076	--------	--------
230	C3H10N2	1,2-PROPANEDIAMINE	74.126	360.10	-501.459
231	C3H10Si	TRIMETHYL SILANE	74.198	331.02	-357.874
232	C4Cl4S	TETRACHLOROTHIOPHENE	221.921	--------	--------
233	C4Cl6	HEXACHLORO-1,3-BUTADIENE	260.760	458.82	-233.071
234	C4F8	OCTAFLUORO-2-BUTENE	200.031	435.00	-402.482

Table 11-1 ENTROPY AND ENTROPY OF FORMATION OF GAS - ORGANIC COMPOUNDS (continued)

NO	FORMULA	NAME	Molecular Weight g/mol	Entropy S @ 298 K, joule/(mol K)	Entropy of Formation Sf @ 298 K, joule/(mol K)
235	C4F8	OCTAFLUOROCYCLOBUTANE	200.031	400.37	-433.205
236	C4F10	DECAFLUOROBUTANE	238.028	481.00	-570.183
237	C4H2	BUTADIYNE(BIACETYLENE)	50.060	--------	96.696
238	C4H2O3	MALEIC ANHYDRIDE	98.058	317.00	-145.229
239	C4H4	VINYLACETYLENE	52.076	279.37	-4.629
240	C4H4N2	SUCCINONITRILE	80.089	326.98	-148.616
241	C4H4O	FURAN	68.075	267.14	-119.302
242	C4H4O2	DIKETENE	84.075	307.50	-181.788
243	C4H4O3	SUCCINIC ANHYDRIDE	100.074	335.00	-255.241
244	C4H4O4	FUMARIC ACID	116.073	393.00	-301.157
245	C4H4O4	MALEIC ACID	116.073	393.30	-300.889
246	C4H4S	THIOPHENE	84.142	278.65	-37.062
247	C4H5Cl	CHLOROPRENE	88.536	320.58	-140.299
248	C4H5N	trans-CROTONITRILE	67.090	298.36	-146.772
249	C4H5N	cis-CROTONITRILE	67.090	304.00	-141.204
250	C4H5N	METHACRYLONITRILE	67.090	223.80	-221.264
251	C4H5N	PYRROLE	67.090	270.54	-174.442
252	C4H5N	VINYLACETONITRILE	67.090	302.00	-141.539
253	C4H5NO2	METHYL CYANOACETATE	99.089	381.00	-268.321
254	C4H6	CYCLOBUTENE	54.091	--------	-150.998
255	C4H6	1,2-BUTADIENE	54.092	293.01	-121.550
256	C4H6	1,3-BUTADIENE	54.092	278.74	-135.871
257	C4H6	DIMETHYLACETYLENE	54.092	283.30	-131.209
258	C4H6	ETHYLACETYLENE	54.092	290.83	-123.797
259	C4H6Cl2	1,3-DICHLORO-trans-2-BUTENE	124.997	353.00	-284.622
260	C4H6Cl2	1,4-DICHLORO-cis-2-BUTENE	124.997	536.00	-101.627
261	C4H6Cl2	1,4-DICHLORO-trans-2-BUTENE	124.997	383.00	-254.637
262	C4H6Cl2	3,4-DICHLORO-1-BUTENE	124.997	371.00	-266.644
263	C4H6O	trans-CROTONALDEHYDE	70.091	333.00	-184.136
264	C4H6O	2,5-DIHYDROFURAN	70.091	284.84	-232.366
265	C4H6O	DIVINYL ETHER	70.091	334.00	-183.196
266	C4H6O	METHACROLEIN	70.091	332.00	-182.458
267	C4H6O2	2-BUTYNE-1,4-DIOL	86.090	358.00	-261.613
268	C4H6O2	gamma-BUTYROLACTONE	86.090	305.40	-314.305
269	C4H6O2	cis-CROTONIC ACID	86.090	351.00	-268.321
270	C4H6O2	trans-CROTONIC ACID	86.090	356.69	-263.022
271	C4H6O2	METHACRYLIC ACID	86.090	350.41	-269.059
272	C4H6O2	METHYL ACRYLATE	86.090	365.64	-253.832
273	C4H6O2	VINYL ACETATE	86.090	328.00	-291.799
274	C4H6O3	ACETIC ANHYDRIDE	102.090	389.95	-332.182
275	C4H6O4	SUCCINIC ACID	118.089	403.00	-422.271
276	C4H6O5	DIGLYCOLIC ACID	134.089	441.00	-486.332
277	C4H6O5	MALIC ACID	134.089	450.00	-476.270
278	C4H6O6	TARTARIC ACID	150.088	486.00	-543.351
279	C4H7N	n-BUTYRONITRILE	69.106	325.43	-250.210
280	C4H7N	ISOBUTYRONITRILE	69.106	313.30	-262.284
281	C4H7NO	ACETONE CYANOHYDRIN	85.106	336.00	-342.210
282	C4H7NO	2-METHACRYLAMIDE	85.106	361.00	-317.290
283	C4H7NO	3-METHOXYPROPIONITRILE	85.106	364.00	-314.271
284	C4H7NO	2-PYRROLIDONE	85.106	301.17	-377.059
285	C4H8	1-BUTENE	56.107	307.83	-239.577
286	C4H8	cis-2-BUTENE	56.107	300.83	-244.340
287	C4H8	trans-2-BUTENE	56.107	296.48	-248.667
288	C4H8	CYCLOBUTANE	56.107	265.39	-279.691
289	C4H8	ISOBUTENE	56.107	293.59	-251.451
290	C4H8Br2	1,2-DIBROMOBUTANE	215.915	--------	-288.512
291	C4H8Br2	2,3-DIBROMOBUTANE	215.915	--------	-302.432
292	C4H8Cl2	1,4-DICHLOROBUTANE	127.013	389.00	-379.339
293	C4H8I2	1,2-DIIODOBUTANE	309.916	--------	-235.351
294	C4H8O	n-BUTYRALDEHYDE	72.107	343.65	-302.700
295	C4H8O	ISOBUTYRALDEHYDE	72.107	313.97	-333.557
296	C4H8O	1,2-EPOXYBUTANE	72.107	322.00	-325.742
297	C4H8O	METHYL ETHYL KETONE	72.107	338.11	-309.576
298	C4H8O	ETHYL VINYL ETHER	72.107	349.00	-298.742
299	C4H8O	TETRAHYDROFURAN	72.107	297.29	-350.495
300	C4H8O2	cis-2-BUTENE-1,4-DIOL	88.106	387.00	-363.274
301	C4H8O2	trans-2-BUTENE-1,4-DIOL	88.106	387.00	-363.274
302	C4H8O2	ISOBUTYRIC ACID	88.106	341.25	-409.022
303	C4H8O2	n-BUTYRIC ACID	88.106	362.00	-386.651
304	C4H8O2	1,4-DIOXANE	88.106	299.78	-450.344
305	C4H8O2	ETHYL ACETATE	88.106	359.70	-387.456
306	C4H8O2	METHYL PROPIONATE	88.106	359.60	-390.743
307	C4H8O2	n-PROPYL FORMATE	88.106	372.54	-377.293
308	C4H8O2S	SULFOLANE	120.172	344.00	-439.376
309	C4H8S	TETRAHYDROTHIOPHENE	88.173	309.36	-267.751
310	C4H9Br	1-BROMOBUTANE	137.019	369.82	-316.720
311	C4H9Br	2-BROMOBUTANE	137.019	361.00	-316.317
312	C4H9Cl	n-BUTYL CHLORIDE	92.568	355.17	-363.877

Table 11-1 ENTROPY AND ENTROPY OF FORMATION OF GAS - ORGANIC COMPOUNDS (continued)

NO	FORMULA	NAME	Molecular Weight g/mol	Entropy S @ 298 K, joule/(mol K)	Entropy of Formation Sf @ 298 K, joule/(mol K)
313	C4H9Cl	sec-BUTYL CHLORIDE	92.568	351.39	-362.334
314	C4H9Cl	tert-BUTYL CHLORIDE	92.568	317.53	-399.665
315	C4H9I	2-IODO-2-METHYLPROPANE	184.020	--------	-326.279
316	C4H9N	PYRROLIDINE	71.122	309.49	-396.713
317	C4H9NO	N,N-DIMETHYLACETAMIDE	87.122	351.00	-457.823
318	C4H9NO	MORPHOLINE	87.122	232.00	-576.891
319	C4H9NO2	1-NITROBUTANE	103.121	--------	-516.720
320	C4H9NO2	2-NITROBUTANE	103.121	--------	-527.788
321	C4H10	n-BUTANE	58.123	309.91	-365.588
322	C4H10	ISOBUTANE	58.123	295.39	-381.150
323	C4H10N2	PIPERAZINE	86.137	321.00	-545.698
324	C4H10O	n-BUTANOL	74.123	362.75	-415.093
325	C4H10O	sec-BUTANOL	74.123	359.03	-419.151
326	C4H10O	tert-BUTANOL	74.123	326.27	-452.021
327	C4H10O	DIETHYL ETHER	74.123	341.00	-435.586
328	C4H10O	METHYL-PROPYL-ETHER	74.122	--------	-428.710
329	C4H10O	METHYL ISOPROPYL ETHER	74.123	341.60	-439.913
330	C4H10O	ISOBUTANOL	74.123	348.34	-499.681
331	C4H10O2	1,3-BUTANEDIOL	90.122	401.52	-430.018
332	C4H10O2	1,4-BUTANEDIOL	90.122	380.91	-479.322
333	C4H10O2	2,3-BUTANEDIOL	90.122	399.00	-480.631
334	C4H10O2	t-BUTYL HYDROPEROXIDE	90.122	364.00	-516.854
335	C4H10O2	1,2-DIMETHOXYETHANE	90.122	385.00	-495.858
336	C4H10O2	2-ETHOXYETHANOL	90.122	399.00	-482.978
337	C4H10O3	DIETHYLENE GLYCOL	106.122	440.00	-544.021
338	C4H10O4S	DIETHYL SULFATE	154.187	417.00	-701.996
339	C4H10S	n-BUTYL MERCAPTAN	90.189	375.20	-332.450
340	C4H10S	ISOBUTYL MERCAPTAN	90.189	362.84	-344.793
341	C4H10S	sec-BUTYL MERCAPTAN	90.189	366.73	-340.869
342	C4H10S	tert-BUTYL MERCAPTAN	90.189	337.86	-369.646
343	C4H10S	DIETHYL SULFIDE	90.189	368.00	-339.594
344	C4H10S	ISOPROPYL-METHYL-SULFIDE	90.183	--------	-348.315
345	C4H10S	METHYL-PROPYL-SULFIDE	90.183	--------	-335.972
346	C4H10S2	DIETHYL DISULFIDE	122.255	--------	-325.004
347	C4H11N	n-BUTYLAMINE	73.138	363.00	-473.755
348	C4H11N	ISOBUTYLAMINE	73.138	351.00	-485.997
349	C4H11N	sec-BUTYLAMINE	73.138	351.04	-485.695
350	C4H11N	tert-BUTYLAMINE	73.138	327.73	-498.876
351	C4H11N	DIETHYLAMINE	73.138	352.21	-484.555
352	C4H11NO	DIMETHYLETHANOLAMINE	89.137	384.00	-555.425
353	C4H11NO2	DIETHANOLAMINE	105.137	428.78	-613.114
354	C4H11NO2	2-AMINOETHOXYETHANOL	105.137	446.00	-597.015
355	C4H12N2O	N-AMINOETHYL ETHANOLAMINE	104.152	447.00	-653.396
356	C4H12Si	TETRAMETHYLSILANE	88.225	358.90	-466.275
357	C4H13N3	DIETHYLENE TRIAMINE	103.167	444.00	-714.909
358	C5Cl6	HEXACHLOROCYCLOPENTADIENE	272.771	456.00	-241.489
359	C5H4O2	FURFURAL	96.086	333.30	-161.563
360	C5H5N	PYRIDINE	79.101	282.80	-167.835
361	C5H6	CYCLOPENTADIENE	66.103	280.33	-140.097
362	C5H6	2-METHYL-1-BUTENE-3-YNE	66.103	278.00	-140.869
363	C5H6	1-PENTENE-3-YNE	66.103	314.00	-107.329
364	C5H6	1-PENTENE-4-YNE	66.103	307.00	-114.037
365	C5H6N2	GLUTARONITRILE	94.116	368.60	-244.843
366	C5H6O2	FURFURYL ALCOHOL	98.101	408.00	-217.676
367	C5H6O3	GLUTARIC ANHYDRIDE	114.101	353.00	-375.650
368	C5H6O4	CITRACONIC ACID	130.100	434.00	-395.774
369	C5H6O4	ITACONIC ACID	130.100	438.00	-392.420
370	C5H6S	2-METHYLTHIOPHENE	98.162	--------	-131.645
371	C5H6S	3-METHYLTHIOPHENE	98.162	--------	-130.941
372	C5H7N	N-METHYLPYRROLE	81.117	307.50	-274.023
373	C5H7NO2	ETHYL CYANOACETATE	113.116	412.92	-373.537
374	C5H8	CYCLOPENTENE	68.118	289.66	-261.144
375	C5H8	ISOPRENE	68.118	315.64	-235.351
376	C5H8	3-METHYL-1,2-BUTADIENE	68.118	321.51	-231.125
377	C5H8	2-METHYL-1,3-BUTADIENE	68.118	--------	-235.184
378	C5H8	1,2-PENTADIENE	68.118	334.80	-217.374
379	C5H8	cis-1,3-PENTADIENE	68.118	322.80	-226.497
380	C5H8	trans-1,3-PENTADIENE	68.118	315.60	-231.125
381	C5H8	1,4-PENTADIENE	68.118	334.00	-217.374
382	C5H8	2,3-PENTADIENE	68.118	329.10	-226.061
383	C5H8	1-PENTYNE	68.118	329.80	-221.030
384	C5H8	2-PENTYNE	68.118	--------	-219.051
385	C5H8	3-METHYL-1-BUTYNE	68.118	318.90	-231.830
386	C5H8	SPIROPENTANE	68.118	--------	-268.590
387	C5H8N4O12	PENTAERYTHRITOL TETRANITRATE	316.138	--------	--------
388	C5H8O	CYCLOPENTANONE	84.118	325.00	-328.492
389	C5H8O	METHYL ISOPROPENYL KETONE	84.118	360.00	-293.476
390	C5H8O2	ACETYLACETONE	100.117	403.00	-352.172

Table 11-1 ENTROPY AND ENTROPY OF FORMATION OF GAS - ORGANIC COMPOUNDS (continued)

NO	FORMULA	NAME	Molecular Weight g/mol	Entropy S @ 298 K, joule/(mol K)	Entropy of Formation Sf @ 298 K, joule/(mol K)
391	C5H8O2	ALLYL ACETATE	100.117	405.00	-352.172
392	C5H8O2	ETHYL ACRYLATE	100.117	406.64	-349.086
393	C5H8O2	METHYL METHACRYLATE	100.117	401.00	-354.754
394	C5H8O2	VINYL PROPIONATE	100.117	385.07	-370.954
395	C5H8O3	2-HYDROXYETHYL ACRYLATE	116.117	451.00	-409.190
396	C5H8O3	LEVULINIC ACID	116.117	426.00	-432.668
397	C5H8O3	METHYL ACETOACETATE	116.117	443.00	-415.898
398	C5H8O4	GLUTARIC ACID	132.116	443.00	-516.519
399	C5H9N	VALERONITRILE	83.133	366.14	-346.403
400	C5H9NO	n-BUTYL ISOCYANATE	99.133	382.00	-432.534
401	C5H9NO	N-METHYL-2-PYRROLIDONE	99.133	338.36	-476.170
402	C5H9NO4	L-GLUTAMIC ACID	147.131	490.00	-630.555
403	C5H10	CYCLOPENTANE	70.134	292.88	-388.596
404	C5H10	2-METHYL-1-BUTENE	70.134	341.96	-341.875
405	C5H10	2-METHYL-2-BUTENE	70.134	338.57	-342.814
406	C5H10	3-METHYL-1-BUTENE	70.134	333.47	-347.879
407	C5H10	1-PENTENE	70.134	345.81	-335.536
408	C5H10	cis-2-PENTENE	70.134	346.27	-335.100
409	C5H10	trans-2-PENTENE	70.134	340.41	-341.003
410	C5H10Br2	2,3-DIBROMO-2-METHYLBUTANE	229.942	--------	-421.130
411	C5H10Cl2	1,5-DICHLOROPENTANE	141.040	429.00	-475.600
412	C5H10O	METHYL ISOPROPYL KETONE	86.134	369.90	-413.886
413	C5H10O	2-PENTANONE	86.134	378.61	-407.781
414	C5H10O	DIETHYL KETONE	86.134	370.10	-416.368
415	C5H10O	VALERALDEHYDE	86.134	382.89	-400.939
416	C5H10O2	n-BUTYL FORMATE	102.133	411.96	-474.325
417	C5H10O2	ETHYL PROPIONATE	102.133	402.50	-483.985
418	C5H10O2	ISOBUTYL FORMATE	102.133	406.60	-479.960
419	C5H10O2	ISOPROPYL ACETATE	102.133	390.20	-496.394
420	C5H10O2	n-PROPYL ACETATE	102.133	402.30	-484.320
421	C5H10O2	METHYL n-BUTYRATE	102.133	398.80	-487.674
422	C5H10O2	2-METHYLBUTYRIC ACID	102.133	390.00	-489.686
423	C5H10O2	ISOVALERIC ACID	102.133	390.00	-495.388
424	C5H10O2	VALERIC ACID	102.133	402.00	-484.186
425	C5H10O2	TETRAHYDROFURFURYL ALCOHOL	102.133	374.00	-513.835
426	C5H10O2S	3-METHYL SULFOLANE	134.199	372.00	-546.638
427	C5H10O3	DIETHYL CARBONATE	118.133	404.00	-585.108
428	C5H10O3	ETHYL LACTATE	118.133	456.00	-533.289
429	C5H10S	THIACYCLOHEXANE	102.194	--------	-390.106
430	C5H10S	CYCLOPENTANETHIOL	102.194	--------	-351.970
431	C5H11Br	1-BROMOPENTANE	151.046	--------	-413.986
432	C5H11Cl	1-CHLOROPENTANE	106.595	394.85	-461.144
433	C5H11Cl	1-CHLORO-3-METHYLBUTANE	106.595	--------	-458.326
434	C5H11Cl	2-CHLORO-2-METHYLBUTANE	106.595	--------	-489.787
435	C5H11N	N-METHYLPYRROLIDINE	85.149	329.00	-513.567
436	C5H11N	PIPERIDINE	85.149	337.00	-506.121
437	C5H11NO	tert-BUTYLFORMAMIDE	101.148	374.00	-570.183
438	C5H12	ISOPENTANE	72.150	343.63	-468.422
439	C5H12	NEOPENTANE	72.150	305.89	-505.618
440	C5H12	n-PENTANE	72.150	349.45	-463.089
441	C5H12O	2,2-DIMETHYL-1-PROPANOL	88.150	365.60	-549.019
442	C5H12O	tert-PENTYL-ALCOHOL	88.149	--------	-547.711
443	C5H12O	2-METHYL-1-BUTANOL	88.150	393.51	-521.147
444	C5H12O	2-METHYL-2-BUTANOL	88.150	362.75	-551.501
445	C5H12O	3-METHYL-1-BUTANOL	88.150	388.00	-526.882
446	C5H12O	3-METHYL-2-BUTANOL	88.150	385.17	-529.465
447	C5H12O	1-PENTANOL	88.150	402.50	-511.924
448	C5H12O	2-PENTANOL	88.150	396.40	-518.229
449	C5H12O	3-PENTANOL	88.150	382.00	-531.846
450	C5H12O	METHYL sec-BUTYL ETHER	88.150	384.00	-530.605
451	C5H12O	METHYL tert-BUTYL ETHER	88.150	352.96	-561.597
452	C5H12O	METHYL ISOBUTYL ETHER	88.150	381.00	-533.289
453	C5H12O	ETHYL PROPYL ETHER	88.150	388.10	-526.581
454	C5H12O2	ETHYLENE GLYCOL MONOPROPYL ETHER	104.149	438.00	-580.245
455	C5H12O2	NEOPENTYL GLYCOL	104.149	403.00	-613.785
456	C5H12O2	1,5-PENTANEDIOL	104.149	433.25	-583.599
457	C5H12O3	2-(2-METHOXYETHOXY)ETHANOL	120.148	476.00	-546.705
458	C5H12O4	PENTAERYTHRITOL	136.148	437.19	-785.008
459	C5H12S	n-PENTYL MERCAPTAN	104.216	415.39	-428.576
460	C5H12S	BUTYL-METHYL-SULFIDE	104.210	--------	-432.064
461	C5H12S	ETHYL-PROPYL-SULFIDE	104.210	--------	-429.851
462	C5H12S	2-METHYL-2-BUTANETHIOL	104.210	--------	-456.918
463	C5H13N	n-PENTYLAMINE	87.165	393.30	-579.876
464	C5H13NO2	METHYL DIETHANOLAMINE	119.164	469.00	-707.697
465	C6Cl6	HEXACHLOROBENZENE	284.782	436.62	-261.848
466	C6F6	HEXAFLUOROBENZENE	186.056	382.20	-259.064
467	C6H3ClN2O4	1-CHLORO-2,4-DINITROBENZENE	202.554	436.00	-506.121
468	C6H3Cl2NO2	1,2-DICHLORO-4-NITROBENZENE	192.001	405.00	-350.159

NO	FORMULA	NAME	Molecular Weight g/mol	Entropy S @ 298 K, joule/(mol K)	Entropy of Formation Sf @ 298 K, joule/(mol K)
469	C6H3Cl3	1,2,4-TRICHLOROBENZENE	181.448	380.91	-183.834
470	C6H3N3O6	1,3,5-TRINITROBENZENE	213.106	443.00	--------
471	C6H4Br2	m-DIBROMOBENZENE	235.906	366.90	-80.765
472	C6H4ClNO2	m-CHLORONITROBENZENE	157.556	373.00	78.819
473	C6H4ClNO2	o-CHLORONITROBENZENE	157.556	370.83	-338.085
474	C6H4ClNO2	p-CHLORONITROBENZENE	157.556	367.00	-341.439
475	C6H4Cl2	m-DICHLOROBENZENE	147.003	343.53	-174.878
476	C6H4Cl2	o-DICHLOROBENZENE	147.003	341.85	-176.824
477	C6H4Cl2	p-DICHLOROBENZENE	147.003	336.74	-181.586
478	C6H4F2	m-DIFLUOROBENZENE	114.094	--------	-177.662
479	C6H4F2	o-DIFLUOROBENZENE	114.094	--------	-176.119
480	C6H4F2	p-DIFLUOROBENZENE	114.094	--------	-182.425
481	C6H4N2O4	m-DINITROBENZENE	168.109	406.00	--------
482	C6H4N2O4	o-DINITROBENZENE	168.109	406.00	--------
483	C6H4N2O4	p-DINITROBENZENE	168.109	394.00	--------
484	C6H5Br	BROMOBENZENE	157.010	324.39	-112.393
485	C6H5Cl	MONOCHLOROBENZENE	112.558	313.47	-158.712
486	C6H5ClO	m-CHLOROPHENOL	128.558	351.00	-223.713
487	C6H5ClO	o-CHLOROPHENOL	128.558	351.00	-223.713
488	C6H5ClO	p-CHLOROPHENOL	128.558	345.00	-229.750
489	C6H5Cl2N	3,4-DICHLOROANILINE	162.018	377.00	-303.203
490	C6H5F	FLUOROBENZENE	96.104	302.63	-159.416
491	C6H5I	IODOBENZENE	204.010	334.05	-84.622
492	C6H5NO2	NITROBENZENE	123.111	359.00	-303.203
493	C6H6	BENZENE	78.114	269.20	-156.733
494	C6H6ClN	m-CHLOROANILINE	127.573	356.00	-277.377
495	C6H6ClN	o-CHLOROANILINE	127.573	356.00	-277.377
496	C6H6ClN	p-CHLOROANILINE	127.573	356.00	-277.377
497	C6H6N2	cis-DICYANO-1-BUTENE	106.127	380.00	-238.135
498	C6H6N2	trans-DICYANO-1-BUTENE	106.127	374.00	-244.843
499	C6H6N2	1,4-DICYANO-2-BUTENE	106.127	394.00	-224.719
500	C6H6N2O2	m-NITROANILINE	138.126	383.00	-441.053
501	C6H6N2O2	o-NITROANILINE	138.126	383.00	-440.047
502	C6H6N2O2	p-NITROANILINE	138.126	376.81	-444.407
503	C6H6O	PHENOL	94.113	314.81	-212.879
504	C6H6O2	1,2-BENZENEDIOL	110.112	345.00	-285.091
505	C6H6O2	1,3-BENZENEDIOL	110.112	318.00	-313.265
506	C6H6O2	p-HYDROQUINONE	110.112	344.17	-287.037
507	C6H6O3	1,2,3-BENZENETRIOL	126.112	382.00	-352.172
508	C6H6S	PHENYL MERCAPTAN	110.180	336.90	-120.946
509	C6H7N	ANILINE	93.128	319.16	-267.751
510	C6H7N	2-METHYLPYRIDINE	93.128	325.01	-262.016
511	C6H7N	3-METHYLPYRIDINE	93.128	324.97	-261.982
512	C6H7N	4-METHYLPYRIDINE	93.128	319.47	-268.154
513	C6H8	1,3-CYCLOHEXADIENE	80.130	303.40	-253.228
514	C6H8	METHYLCYCLOPENTADIENE	80.130	317.00	-238.135
515	C6H8N2	ADIPONITRILE	108.143	400.03	-348.180
516	C6H8N2	METHYLGLUTARONITRILE	108.143	390.00	-358.880
517	C6H8N2	m-PHENYLENEDIAMINE	108.143	360.00	-388.395
518	C6H8N2	o-PHENYLENEDIAMINE	108.143	360.00	-388.395
519	C6H8N2	p-PHENYLENEDIAMINE	108.143	349.00	-398.457
520	C6H8N2	PHENYLHYDRAZINE	108.143	351.00	-397.216
521	C6H8N2O	BIS(CYANOETHYL)ETHER	124.142	419.00	-431.327
522	C6H8O4	DIMETHYL MALEATE	144.127	499.00	-467.550
523	C6H8O6	ASCORBIC ACID	176.126	434.00	-737.884
524	C6H8O7	CITRIC ACID	192.125	555.00	-717.760
525	C6H10	1-METHYLCYCLOPENTENE	82.145	--------	-360.792
526	C6H10	3-METHYLCYCLOPENTENE	82.145	--------	-356.599
527	C6H10	4-METHYLCYCLOPENTENE	82.145	--------	-358.276
528	C6H10	CYCLOHEXENE	82.145	310.75	-376.388
529	C6H10	2,3-DIMETHYL-1,3-BUTADIENE	82.145	355.00	-331.712
530	C6H10	1,5-HEXADIENE	82.145	374.00	-314.942
531	C6H10	cis,trans-2,4-HEXADIENE	82.145	356.00	-330.706
532	C6H10	trans,trans-2,4-HEXADIENE	82.145	356.00	-331.041
533	C6H10	1-HEXYNE	82.145	369.40	-318.397
534	C6H10	2-HEXYNE	82.145	372.00	-315.278
535	C6H10	3-HEXYNE	82.145	376.00	-311.924
536	C6H10O	CYCLOHEXANONE	98.145	322.17	-467.449
537	C6H10O	MESITYL OXIDE	98.145	398.11	-391.682
538	C6H10O2	epsilon-CAPROLACTONE	114.144	362.08	-530.236
539	C6H10O2	ETHYL METHACRYLATE	114.144	442.00	-449.438
540	C6H10O2	n-PROPYL ACRYLATE	114.144	446.00	-446.084
541	C6H10O3	ETHYLACETOACETATE	130.144	484.00	-509.810
542	C6H10O3	PROPIONIC ANHYDRIDE	130.144	470.00	-524.904
543	C6H10O4	ADIPIC ACID	146.143	498.06	-598.927
544	C6H10O4	DIETHYL OXALATE	146.143	527.00	-569.512
545	C6H10O4	ETHYLENE GLYCOL DIACETATE	146.143	507.00	-590.307
546	C6H10O4	ETHYLIDENE DIACETATE	146.143	494.00	-603.723

Table 11-1 ENTROPY AND ENTROPY OF FORMATION OF GAS - ORGANIC COMPOUNDS (continued)

NO	FORMULA	NAME	Molecular Weight g/mol	Entropy S @ 298 K, joule/(mol K)	Entropy of Formation Sf @ 298 K, joule/(mol K)
547	C6H11N	HEXANENITRILE	97.160	405.83	-443.032
548	C6H11NO	epsilon-CAPROLACTAM	113.159	363.59	-587.255
549	C6H11NO	CYCLOHEXANONE OXIME	113.159	377.00	-573.872
550	C6H12	CYCLOHEXANE	84.161	298.24	-519.537
551	C6H12	2,3-DIMETHYL-1-BUTENE	84.161	360.00	-452.021
552	C6H12	2,3-DIMETHYL-2-BUTENE	84.161	354.00	-452.993
553	C6H12	3,3-DIMETHYL-1-BUTENE	84.161	343.80	-473.923
554	C6H12	2-ETHYL-1-BUTENE	84.161	368.00	-441.087
555	C6H12	trans-3-METHYL-2-PENTENE	84.161	--------	-435.888
556	C6H12	1-HEXENE	84.161	384.64	-433.071
557	C6H12	cis-2-HEXENE	84.161	368.00	-431.226
558	C6H12	trans-2-HEXENE	84.161	369.00	-437.129
559	C6H12	cis-3-HEXENE	84.161	379.61	-438.102
560	C6H12	trans-3-HEXENE	84.161	374.84	-442.864
561	C6H12	METHYLCYCLOPENTANE	84.161	339.91	-477.813
562	C6H12	2-METHYL-1-PENTENE	84.161	374.00	-435.586
563	C6H12	2-METHYL-2-PENTENE	84.161	364.00	-439.242
564	C6H12	3-METHYL-1-PENTENE	84.161	376.81	-440.919
565	C6H12	3-METHYL-cis-2-PENTENE	84.161	378.44	-439.242
566	C6H12	4-METHYL-1-PENTENE	84.161	368.00	-449.908
567	C6H12	4-METHYL-cis-2-PENTENE	84.161	373.34	-444.273
568	C6H12	4-METHYL-trans-2-PENTENE	84.161	368.28	-449.338
569	C6H12N2	TRIETHYLENEDIAMINE	112.175	358.00	-651.685
570	C6H12O	BUTYL VINYL ETHER	100.161	428.00	-492.370
571	C6H12O	CYCLOHEXANOL	100.161	327.69	-592.453
572	C6H12O	1-HEXANAL	100.161	422.14	-496.495
573	C6H12O	ETHYL ISOPROPYL KETONE	100.161	406.90	-513.500
574	C6H12O	2-HEXANONE	100.161	417.86	-502.264
575	C6H12O	3-HEXANONE	100.161	409.20	-511.152
576	C6H12O	METHYL ISOBUTYL KETONE	100.161	406.77	-513.601
577	C6H12O2	n-PENTYL FORMATE	116.160	447.30	-577.562
578	C6H12O2	n-BUTYL ACETATE	116.160	442.50	-581.251
579	C6H12O2	sec-BUTYL ACETATE	116.160	439.10	-583.599
580	C6H12O2	tert-BUTYL ACETATE	116.160	406.20	-617.139
581	C6H12O2	ETHYL n-BUTYRATE	116.160	441.70	-581.251
582	C6H12O2	ETHYL ISOBUTYRATE	116.160	435.00	-588.965
583	C6H12O2	ISOBUTYL ACETATE	116.160	441.00	-578.568
584	C6H12O2	n-PROPYL PROPIONATE	116.160	442.00	-580.916
585	C6H12O2	CYCLOHEXYL PEROXIDE	116.160	380.00	-642.965
586	C6H12O2	DIACETONE ALCOHOL	116.160	438.00	-586.282
587	C6H12O2	2-ETHYL BUTYRIC ACID	116.160	423.00	-600.369
588	C6H12O2	n-HEXANOIC ACID	116.160	441.00	-582.257
589	C6H12O3	2-ETHOXYETHYL ACETATE	132.159	493.00	-633.909
590	C6H12O3	HYDROXYCAPROIC ACID	132.159	467.00	-657.387
591	C6H12O3	PARALDEHYDE	132.159	337.48	-787.926
592	C6H12O3	sec-BUTYL GLYCOLATE	132.160	--------	--------
593	C6H12S	THIACYCLOHEPTANE	116.221	--------	-487.674
594	C6H13N	CYCLOHEXYLAMINE	99.176	351.00	-627.905
595	C6H13N	HEXAMETHYLENEIMINE	99.176	356.52	-622.371
596	C6H14	2,2-DIMETHYLBUTANE	86.177	358.34	-590.106
597	C6H14	2,3-DIMETHYLBUTANE	86.177	365.92	-582.526
598	C6H14	n-HEXANE	86.177	388.74	-559.920
599	C6H14	2-METHYLPENTANE	86.177	380.89	-567.801
600	C6H14	3-METHYLPENTANE	86.177	383.00	-568.506
601	C6H14N2O2	LYSINE	146.189	529.00	-815.026
602	C6H14O	2-ETHYL-1-BUTANOL	102.177	427.15	-623.814
603	C6H14O	1-HEXANOL	102.177	441.50	-609.324
604	C6H14O	2-HEXANOL	102.177	436.00	-614.791
605	C6H14O	2-METHYL-1-PENTANOL	102.177	433.00	-617.642
606	C6H14O	4-METHYL-2-PENTANOL	102.177	424.58	-626.363
607	C6H14O	n-BUTYL ETHYL ETHER	102.177	429.00	-621.835
608	C6H14O	DIISOPROPYL ETHER	102.177	381.58	-660.540
609	C6H14O	DI-n-PROPYL ETHER	102.177	422.50	-628.274
610	C6H14O	METHYL tert-PENTYL ETHER	102.177	408.00	-643.971
611	C6H14O2	ACETAL	118.176	454.00	-699.312
612	C6H14O2	2-BUTOXYETHANOL	118.176	478.00	-674.157
613	C6H14O2	1,6-HEXANEDIOL	118.176	471.12	-681.972
614	C6H14O2	HEXYLENE GLYCOL	118.176	456.00	-696.830
615	C6H14O2S	DI-n-PROPYL SULFONE	150.242	470.00	-715.412
616	C6H14O3	DIETHYLENE GLYCOL DIMETHYL ETHER	134.175	503.00	-752.977
617	C6H14O3	DIPROPYLENE GLYCOL	134.175	511.00	-744.592
618	C6H14O3	2-(2-ETHOXYETHOXY)ETHANOL	134.175	517.00	-737.884
619	C6H14O3	TRIMETHYLOLPROPANE	134.175	488.32	-767.667
620	C6H14O4	TRIETHYLENE GLYCOL	150.175	557.98	-800.168
621	C6H14O6	SORBITOL	182.174	645.00	-919.001
622	C6H14S	n-HEXYLMERCAPTAN	118.243	454.70	-525.943
623	C6H14S	BUTYL-ETHYL-SULFIDE	118.237	--------	-527.251
624	C6H14S	ISOPROPYL-SULFIDE	118.237	--------	-564.682

Table 11-1 ENTROPY AND ENTROPY OF FORMATION OF GAS - ORGANIC COMPOUNDS (continued)

NO	FORMULA	NAME	Molecular Weight g/mol	Entropy S @ 298 K, joule/(mol K)	Entropy of Formation Sf @ 298 K, joule/(mol K)
625	C6H14S	METHYL-PENTYL-SULFIDE	118.237	--------	-529.465
626	C6H14S	PROPYL-SULFIDE	118.237	--------	-531.846
627	C6H14S2	PROPYL-DISULFIDE	150.297	--------	-517.122
628	C6H15Al	TRIETHYL ALUMINUN	114.167	--------	--------
629	C6H15Al2Cl3	ETHYL ALUMINUM SESQUICHLORIDE	247.506	--------	--------
630	C6H15N	DIISOPROPYLAMINE	101.192	412.00	-697.300
631	C6H15N	DI-n-PROPYLAMINE	101.192	429.00	-680.195
632	C6H15N	n-HEXYLAMINE	101.192	441.00	-668.455
633	C6H15N	TRIETHYLAMINE	101.192	405.43	-703.907
634	C6H15NO	6-AMINOHEXANOL	117.191	486.00	-726.145
635	C6H15NO2	DIISOPROPANOLAMINE	133.191	513.00	-801.610
636	C6H15NO3	TRIETHANOLAMINE	149.190	537.77	-879.255
637	C6H15N3	N-AMINOETHYL PIPERAZINE	129.205	451.00	-851.585
638	C6H15O4P	TRIETHYL PHOSPHATE	182.156	--------	--------
639	C6H16N2	HEXAMETHYLENEDIAMINE	116.207	474.13	-796.042
640	C6H18N3OP	HEXAMETHYL PHOSPHORAMIDE	179.202	--------	--------
641	C6H18N4	TRIETHYLENE TETRAMINE	146.236	564.00	-1028.509
642	C6H18OSi2	HEXAMETHYLDISILOXANE	162.379	534.90	-814.758
643	C6H18O3Si3	HEXAMETHYLCYCLOTRISILOXANE	222.464	412.10	--------
644	C6H19NSi2	HEXAMETHYLDISILAZANE	161.395	--------	--------
645	C7H3ClF3NO2	4-CHLORO-3-NITROBENZOTRIFLUORIDE	225.554	464.05	-488.278
646	C7H3Cl2F3	2,4-DICHLOROBENZOTRIFLUORIDE	215.001	439.57	-323.461
647	C7H3Cl2NO	3,4-DICHLOROPHENYL ISOCYANATE	188.012	406.00	-251.115
648	C7H4ClF3	p-CHLOROBENZOTRIFLUORIDE	180.557	403.00	-313.835
649	C7H4Cl2O	m-CHLOROBENZOYL CHLORIDE	175.014	400.37	-226.430
650	C7H4F3NO2	3-NITROBENZOTRIFLUORIDE	191.110	433.00	-472.916
651	C7H5ClO	BENZOYL CHLORIDE	140.569	368.19	-212.410
652	C7H5ClO2	o-CHLOROBENZOIC ACID	156.568	399.82	-283.314
653	C7H5Cl3	BENZOTRICHLORIDE	195.475	401.46	-299.581
654	C7H5F3	BENZOTRIFLUORIDE	146.112	372.59	-297.803
655	C7H5N	BENZONITRILE	103.123	321.04	-141.036
656	C7H5NO	PHENYL ISOCYANATE	119.123	348.00	-216.669
657	C7H5N3O6	2,4,6-TRINITROTOLUENE	227.133	461.00	--------
658	C7H6Cl2	BENZYL DICHLORIDE	161.030	386.00	-268.992
659	C7H6Cl2	2,4-DICHLOROTOLUENE	161.030	378.95	-275.901
660	C7H6N2O4	2,4-DINITROTOLUENE	182.136	414.00	--------
661	C7H6N2O4	2,5-DINITROTOLUENE	182.136	--------	--------
662	C7H6N2O4	2,6-DINITROTOLUENE	182.136	408.00	--------
663	C7H6N2O4	3,4-DINITROTOLUENE	182.136	414.00	--------
664	C7H6N2O4	3,5-DINITROTOLUENE	182.136	--------	--------
665	C7H6O	BENZALDEHYDE	106.124	335.90	-198.558
666	C7H6O2	BENZOIC ACID	122.123	368.99	-267.617
667	C7H6O2	p-HYDROXYBENZALDEHYDE	122.123	388.00	-248.197
668	C7H6O2	SALICYLALDEHYDE	122.123	388.00	-248.935
669	C7H6O3	SALICYLIC ACID	138.123	399.82	-339.225
670	C7H7Br	p-BROMOTOLUENE	171.037	364.00	-209.961
671	C7H7Cl	BENZYL CHLORIDE	126.585	361.00	-247.191
672	C7H7Cl	o-CHLOROTOLUENE	126.585	348.00	-260.607
673	C7H7Cl	p-CHLOROTOLUENE	126.585	342.00	-266.644
674	C7H7F	p-FLUOROTOLUENE	110.131	--------	-258.762
675	C7H7NO	FORMANILIDE	121.139	329.00	-366.594
676	C7H7NO2	m-NITROTOLUENE	137.138	387.00	-409.861
677	C7H7NO2	o-NITROTOLUENE	137.138	387.00	-410.867
678	C7H7NO2	p-NITROTOLUENE	137.138	381.00	-415.898
679	C7H7NO3	o-NITROANISOLE	153.138	401.00	-499.413
680	C7H8	TOLUENE	92.141	319.74	-241.523
681	C7H8	1,3,5-CYCLOHEPTATRIENE	92.140	--------	-246.554
682	C7H8O	ANISOLE	108.140	361.00	-303.975
683	C7H8O	BENZYL ALCOHOL	108.140	365.47	-299.514
684	C7H8O	m-CRESOL	108.140	356.04	-307.899
685	C7H8O	o-CRESOL	108.140	352.59	-307.060
686	C7H8O	p-CRESOL	108.140	350.75	-316.988
687	C7H8O2	GUAIACOL	124.139	397.00	-368.942
688	C7H8O2	p-METHOXYPHENOL	124.139	397.00	-368.942
689	C7H9N	BENZYLAMINE	107.155	358.00	-364.816
690	C7H9N	2,6-DIMETHYLPYRIDINE	107.155	356.09	-367.936
691	C7H9N	N-METHYLANILINE	107.155	341.00	-382.358
692	C7H9N	m-TOLUIDINE	107.155	354.30	-368.942
693	C7H9N	o-TOLUIDINE	107.155	349.30	-375.650
694	C7H9N	p-TOLUIDINE	107.155	353.40	-368.942
695	C7H10	2-NORBORNENE	94.156	314.00	-379.004
696	C7H10N2	TOLUENEDIAMINE	122.170	397.64	-486.903
697	C7H11NO	CYCLOHEXYL ISOCYANATE	125.170	377.00	-579.574
698	C7H12	1-HEPTYNE	96.172	--------	-415.663
699	C7H12O2	n-BUTYL ACRYLATE	128.171	485.12	-543.518
700	C7H12O2	ISOBUTYL ACRYLATE	128.171	474.00	-553.413
701	C7H12O2	n-PROPYL METHACRYLATE	128.171	481.00	-546.705
702	C7H12O4	DIETHYL MALONATE	160.170	554.00	-680.865

NO	FORMULA	NAME	Molecular Weight g/mol	Entropy S @ 298 K, joule/(mol K)	Entropy of Formation Sf @ 298 K, joule/(mol K)
703	C7H14	CYCLOHEPTANE	98.188	342.34	-611.571
704	C7H14	1,1-DIMETYLCYCLOPENTANE	98.188	359.28	-594.734
705	C7H14	cis-1,2-DIMETHYLCYCLOPENTANE	98.188	366.14	-587.858
706	C7H14	trans-1,2-DIMETHYLCYCLOPENTANE	98.188	366.81	-587.154
707	C7H14	cis-1,3-DIMETHYLCYCLOPENTANE	98.188	366.81	-587.121
708	C7H14	trans-1,3-DIMETHYLCYCLOPENTANE	98.188	366.81	-587.154
709	C7H14	ETHYLCYCLOPENTANE	98.188	378.32	-575.650
710	C7H14	2-ETHYL-1-PENTENE	98.188	420.00	-534.161
711	C7H14	3-ETHYL-1-PENTENE	98.188	415.00	-539.158
712	C7H14	1-HEPTENE	98.188	423.59	-530.304
713	C7H14	cis-2-HEPTENE	98.188	423.00	-531.276
714	C7H14	trans-2-HEPTENE	98.188	423.00	-531.276
715	C7H14	cis-3-HEPTENE	98.188	424.00	-530.270
716	C7H14	trans-3-HEPTENE	98.188	424.00	-530.270
717	C7H14	METHYLCYCLOHEXANE	98.188	343.34	-610.599
718	C7H14	2-METHYL-1-HEXENE	98.188	418.40	-535.771
719	C7H14	3-METHYL-1-HEXENE	98.188	420.80	-533.356
720	C7H14	4-METHYL-1-HEXENE	98.188	421.00	-533.289
721	C7H14	2,3,3-TRIMETHYL-1-BUTENE	98.188	380.40	-573.772
722	C7H14O	DIISOPROPYL KETONE	114.188	457.00	-601.711
723	C7H14O	2-HEPTANONE	114.188	456.89	-599.195
724	C7H14O	1-HEPTANAL	114.188	461.38	-594.868
725	C7H14O	1-METHYLCYCLOHEXANOL	114.188	375.00	-681.670
726	C7H14O	cis-2-METHYLCYCLOHEXANOL	114.188	391.00	-665.672
727	C7H14O	trans-2-METHYLCYCLOHEXANOL	114.188	391.00	-665.672
728	C7H14O	cis-3-METHYLCYCLOHEXANOL	114.188	391.00	-665.672
729	C7H14O	trans-3-METHYLCYCLOHEXANOL	114.188	391.00	-665.672
730	C7H14O	cis-4-METHYLCYCLOHEXANOL	114.188	380.00	-676.673
731	C7H14O	trans-4-METHYLCYCLOHEXANOL	114.188	380.00	-676.673
732	C7H14O	5-METHYL-2-HEXANONE	114.188	446.00	-610.431
733	C7H14O2	n-BUTYL PROPIONATE	130.187	481.30	-677.847
734	C7H14O2	ETHYL ISOVALERATE	130.187	482.00	-677.511
735	C7H14O2	ISOPENTYL ACETATE	130.187	476.30	-682.878
736	C7H14O2	n-PENTYL ACETATE	130.187	481.80	-677.511
737	C7H14O2	n-PROPYL n-BUTYRATE	130.187	481.30	-677.847
738	C7H14O2	n-HEPTANOIC ACID	130.187	480.00	-679.524
739	C7H14O3	ETHYL-3-ETHOXYPROPIONATE	146.186	533.00	-727.822
740	C7H15Br	1-BROMOHEPTANE	179.100	486.00	-609.425
741	C7H15N	N-METHYLCYCLOHEXYLAMINE	113.203	377.00	-737.884
742	C7H16	2,2-DIMETHYLPENTANE	100.204	392.58	-691.699
743	C7H16	2,3-DIMETHYLPENTANE	100.204	414.55	-670.501
744	C7H16	2,4-DIMETHYLPENTANE	100.204	396.89	-687.909
745	C7H16	3,3-DIMETHYLPENTANE	100.204	398.15	-684.823
746	C7H16	3-ETHYLPENTANE	100.204	411.54	-673.017
747	C7H16	n-HEPTANE	100.204	427.98	-656.616
748	C7H16	2-METHYLHEXANE	100.204	420.41	-664.598
749	C7H16	3-METHYLHEXANE	100.204	425.93	-660.406
750	C7H16	2,2,3-TRIMETHYLBUTANE	100.204	383.34	-701.258
751	C7H16O	1-HEPTANOL	116.203	480.45	-706.591
752	C7H16O	2-HEPTANOL	116.203	475.00	-711.051
753	C7H16O	5-METHYL-1-HEXANOL	116.203	467.00	-721.114
754	C7H16O	ISOPROPYL-TERT-BUTYL-ETHER	116.203	--------	-769.143
755	C7H16S	n-HEPTYL MERCAPTAN	132.270	493.89	-623.210
756	C7H16S	BUTYL-PROPYL-SULFIDE	132.263	--------	-623.210
757	C7H16S	ETHYL-PENTYL-SULFIDE	132.263	--------	-624.484
758	C7H16S	HEXYL-METHYL-SULFIDE	132.263	--------	-626.732
759	C7H17N	1-AMINOHEPTANE	115.219	480.00	-765.789
760	C8H4Cl2O2	ISOPHTHALOYL CHLORIDE	203.024	464.00	-271.675
761	C8H4O3	PHTHALIC ANHYDRIDE	148.118	399.53	-215.093
762	C8H6	ETHYNYLBENZENE	102.135	--------	-115.646
763	C8H6O4	ISOPHTHALIC ACID	166.133	442.37	-404.863
764	C8H6O4	PHTHALIC ACID	166.133	442.37	-405.333
765	C8H6O4	TEREPHTHALIC ACID	166.133	448.00	-398.793
766	C8H6S	BENZOTHIOPHENE	134.202	211.00	-257.253
767	C8H7N	INDOLE	117.150	328.00	-270.669
768	C8H8	STYRENE	104.152	345.10	-222.841
769	C8H8	1,3,5,7-CYCLOOCTATETRAENE	104.151	--------	-241.087
770	C8H8O	ACETOPHENONE	120.151	384.51	-286.198
771	C8H8O	p-TOLUALDEHYDE	120.151	389.00	-281.737
772	C8H8O2	METHYL BENZOATE	136.150	414.00	-358.544
773	C8H8O2	o-TOLUIC ACID	136.150	406.14	-367.097
774	C8H8O2	p-TOLUIC ACID	136.150	406.14	-367.097
775	C8H8O3	METHYL SALICYLATE	152.150	456.00	-420.258
776	C8H8O3	VANILLIN	152.150	465.00	-409.190
777	C8H9NO	ACETANILIDE	135.166	369.00	-462.754
778	C8H10	ETHYLBENZENE	106.167	360.45	-338.051
779	C8H10	m-XYLENE	106.167	357.69	-340.869
780	C8H10	o-XYLENE	106.167	352.75	-345.766

Table 11-1 ENTROPY AND ENTROPY OF FORMATION OF GAS - ORGANIC COMPOUNDS (continued)

NO	FORMULA	NAME	Molecular Weight g/mol	Entropy S @ 298 K, joule/(mol K)	Entropy of Formation Sf @ 298 K, joule/(mol K)
781	C8H10	p-XYLENE	106.167	352.10	-346.067
782	C8H10O	m-ETHYLPHENOL	122.167	--------	--------
783	C8H10O	p-ETHYLPHENOL	122.167	390.54	-410.766
784	C8H10O	PHENETOLE	122.167	402.00	-399.296
785	C8H10O	2-PHENYLETHANOL	122.167	405.00	-396.277
786	C8H10O	2,3-XYLENOL	122.167	390.08	-415.361
787	C8H10O	2,4-XYLENOL	122.167	397.72	-403.555
788	C8H10O	2,5-XYLENOL	122.167	395.76	-409.056
789	C8H10O	2,6-XYLENOL	122.167	389.72	-411.571
790	C8H10O	3,4-XYLENOL	122.167	391.16	-410.129
791	C8H10O	3,5-XYLENOL	122.167	391.69	-409.626
792	C8H11N	N,N-DIMETHYLANILINE	121.182	366.00	-493.812
793	C8H11N	o-ETHYLANILINE	121.182	402.00	-458.494
794	C8H11N	2,4,6-TRIMETHYLPYRIDINE	121.182	477.00	-384.035
795	C8H11NO	p-PHENETIDINE	137.181	437.00	-525.239
796	C8H12	1,5-CYCLOOCTADIENE	108.183	350.51	-479.624
797	C8H12	VINYLCYCLOHEXENE	108.183	378.00	-452.457
798	C8H12O4	1,4-CYCLOHEXANEDICARBOXYLIC ACID	172.181	453.00	-784.840
799	C8H12O4	DIETHYL MALEATE	172.181	582.00	-657.387
800	C8H14O2	n-BUTYL METHACRYLATE	142.198	521.00	-640.617
801	C8H14O3	BUTYRIC ANHYDRIDE	158.197	549.00	-719.940
802	C8H14O4	DIETHYL SUCCINATE	174.197	587.00	-781.486
803	C8H16	CYCLOOCTANE	112.214	--------	-723.394
804	C8H16	1,1-DIMETHYLCYCLOHEXANE	112.215	365.01	-725.239
805	C8H16	cis-1,2-DIMETHYLCYCLOHEXANE	112.215	374.51	-715.680
806	C8H16	trans-1,2-DIMETHYLCYCLOHEXANE	112.215	370.91	-719.369
807	C8H16	cis-1,3-DIMETHYLCYCLOHEXANE	112.215	370.45	-719.772
808	C8H16	trans-1,3-DIMETHYLCYCLOHEXANE	112.215	376.23	-714.003
809	C8H16	cis-1,4-DIMETHYLCYCLOHEXANE	112.215	370.45	-719.772
810	C8H16	trans-1,4-DIMETHYLCYCLOHEXANE	112.215	364.80	-725.507
811	C8H16	ETHYLCYCLOHEXANE	112.215	382.59	-707.697
812	C8H16	2-ETHYL-1-HEXENE	112.215	459.00	-631.494
813	C8H16	1-METHYL-1-ETHYLCYCLOPENTANE	112.215	401.00	-689.485
814	C8H16	1-OCTENE	112.215	462.54	-627.704
815	C8H16	trans-2-OCTENE	112.215	462.00	-628.543
816	C8H16	trans-3-OCTENE	112.215	464.00	-626.530
817	C8H16	trans-4-OCTENE	112.215	458.00	-632.567
818	C8H16	n-PROPYLCYCLOPENTANE	112.215	417.27	-673.017
819	C8H16	2,4,4-TRIMETHYL-1-PENTENE	112.215	429.00	-661.412
820	C8H16	2,4,4-TRIMETHYL-2-PENTENE	112.215	425.00	-665.437
821	C8H16O	2-ETHYLHEXANAL	128.214	495.00	-698.004
822	C8H16O	1-OCTANAL	128.214	500.63	-692.135
823	C8H16O	2-OCTANONE	128.214	496.20	-696.965
824	C8H16O2	n-BUTYL n-BUTYRATE	144.214	521.50	-774.107
825	C8H16O2	n-HEXYL ACETATE	144.214	520.90	-774.778
826	C8H16O2	ISOBUTYL ISOBUTYRATE	144.214	510.00	-784.840
827	C8H16O2	n-OCTANOIC ACID	144.214	520.00	-774.778
828	C8H16O4	DIETHYLENE GLYCOL ETHYL ETHER ACETATE	176.213	611.00	-889.552
829	C8H18	2,2-DIMETHYLHEXANE	114.231	432.71	-789.636
830	C8H18	2,3-DIMETHYLHEXANE	114.231	451.96	-776.891
831	C8H18	2,4-DIMETHYLHEXANE	114.231	447.60	-775.214
832	C8H18	2,5-DIMETHYLHEXANE	114.231	442.33	-781.788
833	C8H18	3,3-DIMETHYLHEXANE	114.231	438.23	-782.760
834	C8H18	3,4-DIMETHYLHEXANE	114.231	451.58	-772.531
835	C8H18	3-ETHYLHEXANE	114.231	457.86	-762.703
836	C8H18	3-ETHYL-2-METHYLPENTANE	114.230	--------	-779.675
837	C8H18	3-METHYL-3-ETHYLPENTANE	114.231	424.72	-787.825
838	C8H18	2-METHYLHEPTANE	114.231	459.57	-765.521
839	C8H18	3-METHYLHEPTANE	114.231	465.51	-759.182
840	C8H18	4-METHYLHEPTANE	114.231	457.39	-767.500
841	C8H18	n-OCTANE	114.231	467.23	-754.151
842	C8H18	2,2,3-TRIMETHYLPENTANE	114.231	423.88	-795.673
843	C8H18	2,2,4-TRIMETHYLPENTANE	114.231	422.96	-797.652
844	C8H18	2,3,3-TRIMETHYLPENTANE	114.231	427.02	-789.368
845	C8H18	2,3,4-TRIMETHYLPENTANE	114.231	428.40	-792.722
846	C8H18	2,2,3,3-TETRAMETHYLBUTANE	114.230	--------	-831.461
847	C8H18O	DI-n-BUTYL ETHER	130.230	501.40	-822.908
848	C8H18O	DI-sec-BUTYL ETHER	130.230	501.00	-860.641
849	C8H18O	DI-tert-BUTYL ETHER	130.230	425.00	-895.992
850	C8H18O	2-ETHYL-1-HEXANOL	130.230	496.60	-826.497
851	C8H18O	1-OCTANOL	130.230	519.23	-803.958
852	C8H18O	2-OCTANOL	130.230	515.00	-808.653
853	C8H18O2	DI-t-BUTYL PEROXIDE	146.230	495.00	-931.075
854	C8H18O2S	DI-n-BUTYL SULFONE	178.296	548.00	-911.622
855	C8H18O3	DIETHYLENE GLYCOL DIETHYL ETHER	162.229	585.00	-942.479
856	C8H18O3	DIETHYLENE GLYCOL MONOBUTYL EHTER	162.229	596.00	-932.417
857	C8H18O4	TRIETHYLENE GLYCOL DIMETHYL ETHER	178.229	622.00	-1009.559
858	C8H18O5	TETRAETHYLENE GLYCOL	194.228	674.00	-1083.045

NO	FORMULA	NAME	Molecular Weight g/mol	Entropy S @ 298 K, joule/(mol K)	Entropy of Formation Sf @ 298 K, joule/(mol K)
859	C8H18S	n-OCTYL MERCAPTAN	146.297	533.14	-720.610
860	C8H18S	tert-OCTYL MERCAPTAN	146.297	470.00	-591.984
861	C8H18S	BUTYL-SULFIDE	146.290	--------	-726.212
862	C8H18S	ETHYL-HEXYL-SULFIDE	146.290	--------	-721.885
863	C8H18S	HEPTYL-METHYL-SULFIDE	146.290	--------	-724.099
864	C8H18S	PENTYL-PROPYL-SULFIDE	146.290	--------	-720.476
865	C8H18S2	BUTYL-DISULFIDE	178.350	--------	-711.924
866	C8H19N	DI-n-BUTYLAMINE	129.246	508.00	-874.057
867	C8H19N	DIISOBUTYLAMINE	129.246	485.00	-896.864
868	C8H19N	n-OCTYLAMINE	129.246	520.00	-862.083
869	C8H23N5	TETRAETHYLENEPENTAMINE	189.304	683.00	-1343.217
870	C8H24O4Si4	OCTAMETHYLCYCLOTETRASILOXANE	296.618	--------	--------
871	C9H4O5	TRIMELLITIC ANHYDRIDE	192.128	349.00	-476.270
872	C9H6N2O2	TOLUENE DIISOCYANATE	174.159	402.00	-437.934
873	C9H7N	ISOQUINOLINE	129.161	365.20	-239.141
874	C9H7N	QUINOLINE	129.161	365.60	-238.806
875	C9H7NO	8-HYDROXYQUINOLINE	145.161	396.00	-310.247
876	C9H8	INDENE	116.163	336.87	-237.095
877	C9H8O	2-METHYLBENZOFURAN	132.162	302.00	-373.134
878	C9H10	INDANE	118.178	234.00	-470.569
879	C9H10	cis-PROPENYLBENZENE	118.178	--------	-320.510
880	C9H10	trans-PROPENYLBENZENE	118.178	--------	-323.897
881	C9H1O	alpha-METHYLSTYRENE	118.178	383.67	-320.510
882	C9H10	m-METHYLSTYRENE	118.178	389.50	-314.607
883	C9H10	o-METHYLSTYRENE	118.178	383.70	-320.510
884	C9H10	p-METHYLSTYRENE	118.178	383.67	-320.510
885	C9H10O2	BENZYL ACETATE	150.177	458.99	-450.579
886	C9H10O2	ETHYL BENZOATE	150.177	455.00	-456.146
887	C9H10O3	ETHYL VANILLIN	166.177	506.00	-506.456
888	C9H11NO	p-DIMETHYLAMINOBENZALDEHYDE	149.192	434.00	-532.953
889	C9H12	CUMENE	120.194	388.57	-446.252
890	C9H12	m-ETHYLTOLUENE	120.194	404.17	-430.522
891	C9H12	o-ETHYLTOLUENE	120.194	399.20	-435.586
892	C9H12	p-ETHYLTOLUENE	120.194	398.90	-435.854
893	C9H12	MESITYLENE	120.194	385.60	-449.505
894	C9H12	n-PROPYLBENZENE	120.194	400.66	-434.077
895	C9H12	1,2,3-TRIMETHYLBENZENE	120.194	380.50	-449.908
896	C9H12	1,2,4-TRIMETHYLBENZENE	120.194	396.10	-438.940
897	C9H12O	BENZYL ETHYL ETHER	136.194	439.00	-498.742
898	C9H12O	2-PHENYL-2-PROPANOL	136.194	688.00	-248.197
899	C9H12O2	CUMENE HYDROPEROXIDE	152.193	455.00	-584.940
900	C9H14O	ISOPHORONE	138.210	422.00	-646.319
901	C9H14O6	GLYCERYL TRIACETATE	218.207	676.00	-904.578
902	C9H16	1-NONYNE	124.225	--------	-610.297
903	C9H16O4	AZELAIC ACID	188.224	600.00	-905.584
904	C9H18	BUTYLCYCLOPENTANE	126.241	--------	-770.283
905	C9H18	cis,cis-1,3,5-TRIMETHYLCYCLOHEXANE	126.241	--------	-836.089
906	C9H18	cis,trans-1,3,5-TRIMETHYLCYCLOHEXANE	126.241	--------	-826.530
907	C9H18	ISOPROPYLCYCLOHEXANE	126.242	405.00	-821.734
908	C9H18	1-NONENE	126.242	501.49	-725.105
909	C9H18	n-PROPYLCYCLOHEXANE	126.242	419.53	-807.043
910	C9H18O	DIISOBUTYL KETONE	142.241	509.00	-820.392
911	C9H18O	1-NONANAL	142.241	539.88	-789.368
912	C9H18O2	n-BUTYL VALERATE	158.241	572.00	-859.970
913	C9H18O2	n-NONANOIC ACID	158.241	559.00	-873.386
914	C9H18O2	n-OCTYL FORMATE	158.241	564.70	-867.013
915	C9H20	3,3-DIETHYLPENTANE	128.258	436.50	-895.589
916	C9H20	2,2-DIMETHYL-3-ETHYLPENTANE	128.258	461.00	-896.998
917	C9H20	3-ETHYL-2,3-DIMETHYLPENTANE	128.257	--------	-887.875
918	C9H20	2,4-DIMETHYL-3-ETHYLPENTANE	128.258	471.10	-887.339
919	C9H20	2,2-DIMETHYLHEPTANE	128.258	471.80	-884.119
920	C9H20	2,6-DIMETHYLHEPTANE	128.258	481.50	-880.027
921	C9H20	3-ETHYLHEPTANE	128.258	497.40	-861.211
922	C9H20	4-ETHYLHEPTANE	128.257	--------	-861.211
923	C9H20	2,3-DIMETHYLHEPTANE	128.257	--------	-868.489
924	C9H20	2,4-DIMETHYLHEPTANE	128.257	--------	-868.523
925	C9H20	2,5-DIMETHYLHEPTANE	128.257	--------	-868.523
926	C9H20	3,4-DIMETHYLHEPTANE	128.257	--------	-865.571
927	C9H20	3,5-DIMETHYLHEPTANE	128.257	--------	-871.340
928	C9H20	4,4-DIMETHYLHEPTANE	128.257	--------	-880.698
929	C9H20	3-ETHYL-2-METHYLHEXANE	128.257	--------	-868.523
930	C9H20	4-ETHYL-2-METHYLHEXANE	128.257	--------	-874.258
931	C9H20	3-ETHYL-3-METHYLHEXANE	128.257	--------	-874.962
932	C9H20	3-ETHYL-4-METHYLHEXANE	128.257	--------	-868.523
933	C9H20	2,2,3-TRIMETHYLHEXANE	128.257	--------	-891.263
934	C9H20	2,2,4-TRIMETHYLHEXANE	128.257	--------	-891.263
935	C9H20	2,3,3-TRIMETHYLHEXANE	128.257	--------	-887.875
936	C9H20	2,3,4-TRIMETHYLHEXANE	128.257	--------	-878.618

276

NO	FORMULA	NAME	Molecular Weight g/mol	Entropy S @ 298 K, joule/(mol K)	Entropy of Formation Sf @ 298 K, joule/(mol K)
937	C9H20	2,3,5-TRIMETHYLHEXANE	128.257	--------	-887.339
938	C9H20	2,4,4-TRIMETHYLHEXANE	128.257	--------	-887.875
939	C9H20	3,3,4-TRIMETHYLHEXANE	128.257	--------	-882.140
940	C9H20	2-METHYLOCTANE	128.258	498.80	-861.244
941	C9H20	3-METHYLOCTANE	128.258	504.70	-855.442
942	C9H20	4-METHYLOCTANE	128.258	502.70	-855.442
943	C9H20	n-NONANE	128.258	506.40	-851.384
944	C9H20	2,2,3,3-TETRAMETHYLPENTANE	128.258	436.60	-910.750
945	C9H20	2,2,3,4-TETRAMETHYLPENTANE	128.258	451.20	-904.310
946	C9H20	2,2,4,4-TETRAMETHYLPENTANE	128.258	430.60	-925.641
947	C9H20	2,3,3,4-TETRAMETHYLPENTANE	128.257	--------	-906.691
948	C9H20	2,2,5-TRIMETHYLHEXANE	128.258	460.07	-896.998
949	C9H20O	2,6-DIMETHYL-4-HEPTANOL	144.257	520.00	-939.125
950	C9H20O	1-NONANOL	144.257	558.56	-901.191
951	C9H20O	2-NONANOL	144.257	548.00	-912.292
952	C9H20S	n-NONYL MERCAPTAN	160.324	572.39	-817.843
953	C9H20S	BUTYL-PENTYL-SULFIDE	160.317	--------	-817.877
954	C9H20S	ETHYL-HEPTYL-SULFIDE	160.317	--------	-819.118
955	C9H20S	HEXYL-PROPYL-SULFIDE	160.317	--------	-817.843
956	C9H20S	METHYL-OCTYL-SULFIDE	160.317	--------	-821.499
957	C9H21N	n-NONYLAMINE	143.272	559.00	-959.383
958	C9H21N	TRIPROPYLAMINE	143.272	530.00	-988.395
959	C10H6O8	PYROMELLITIC ACID	254.153	619.00	-637.263
960	C10H7Br	1-BROMONAPHTHALENE	207.070	386.00	-204.494
961	C10H7Cl	1-CHLORONAPHTHALENE	162.618	375.00	-252.222
962	C10H8	NAPHTHALENE	128.174	333.15	-243.602
963	C10H8	AZULENE	128.173	--------	-241.355
964	C10H9N	QUINALDINE	143.188	350.00	-392.420
965	C10H10	m-DIVINYLBENZENE	130.189	412.00	-298.239
966	C10H10	1-METHYLINDENE	130.189	354.00	-355.526
967	C10H10	2-METHYLINDENE	130.189	361.00	-348.818
968	C10H10O4	DIMETHYL PHTHALATE	194.187	660.00	-459.500
969	C10H10O4	DIMETHYL TEREPHTHALATE	194.187	550.00	-546.705
970	C10H12	DICYCLOPENTADIENE	132.205	240.00	-600.034
971	C10H12	1,2,3,4-TETRAHYDRONAPHTHALENE	132.205	369.64	-471.206
972	C10H12O	ANETHOLE	148.205	464.00	-477.947
973	C10H12O4	DIALLYL MALEATE	196.203	639.00	-610.431
974	C10H14	n-BUTYLBENZENE	134.221	439.49	-531.578
975	C10H14	sec-BUTYLBENZENE	134.221	427.60	-543.686
976	C10H14	tert-BUTYLBENZENE	134.221	395.90	-575.314
977	C10H14	1,2,3,4-TETRAMETHYLBENZENE	134.221	--------	-554.587
978	C10H14	m-CYMENE	134.221	431.50	-539.997
979	C10H14	o-CYMENE	134.221	426.70	-544.692
980	C10H14	p-CYMENE	134.221	426.30	-545.028
981	C10H14	m-DIETHYLBENZENE	134.221	439.50	-531.712
982	C10H14	o-DIETHYLBENZENE	134.221	434.80	-536.743
983	C10H14	p-DIETHYLBENZENE	134.221	433.10	-537.045
984	C1OH14	2-ETHYL-m-XYLENE	134.221	421.00	-550.495
985	C10H14	2-ETHYL-p-XYLENE	134.221	436.60	-534.899
986	C10H14	3-ETHYL-o-XYLENE	134.221	421.00	-550.226
987	C10H14	4-ETHYL-m-XYLENE	134.221	436.60	-534.630
988	C10H14	4-ETHYL-o-XYLENE	134.221	436.60	-534.932
989	C1OH14	5-ETHYL-m-XYLENE	134.221	426.10	-545.363
990	C10H14	ISOBUTYLBENZENE	134.221	433.70	-537.783
991	C10H14	1,2,3,5-TETRAMETHYLBENZENE	134.221	422.50	-548.549
992	C10H14	1,2,4,5-TETRAMETHYLBENZENE	134.221	418.50	-552.474
993	C10H14O	p-tert-BUTYLPHENOL	150.221	447.00	-626.866
994	C10H14O2	p-tert-BUTYLCATECHOL	166.220	478.00	-697.635
995	C10H15N	N,N-DIETYHLANILINE	149.236	386.60	-745.262
996	C10H15N	2,6-DIETHYLANILINE	149.236	473.00	-659.064
997	C10H16	CAMPHENE	136.237	365.00	--------
998	C10H16	D-LIMONENE	136.237	--------	--------
999	C10H16	alpha-PHELLANDRENE	136.237	429.00	-671.809
1000	C1OH16	beta-PHELLANDRENE	136.237	422.00	-679.725
1001	C10H16	alpha-PINENE	136.237	473.00	-629.549
1002	C10H16	beta-PINENE	136.237	404.00	-698.642
1003	C10H16	alpha-TERPINENE	136.237	434.50	-666.108
1004	C1OH16	gamma-TERPINENE	136.237	457.00	-646.319
1005	C10H16	TERPINOLENE	136.237	--------	--------
1006	C10H16O	CAMPHOR	152.236	391.00	-813.349
1007	C10H18	1-DECYNE	138.252	--------	-707.697
1008	C10H18	cis-DECAHYDRONANPHTALENE	138.253	377.73	-854.469
1009	C10H18	trans-DECAHYDRONAPHTHALENE	138.253	374.55	-857.723
1010	C10H18O4	SEBACIC ACID	202.251	640.00	-1002.516
1011	C10H20	n-BUTYLCYCLOHEXANE	140.269	458.48	-904.276
1012	C10H20	1-CYCLOPENTYLPENTANE	140.268	--------	-867.684
1013	C10H20	1-DECENE	140.269	542.93	-822.338
1014	C10H20O	1-DECANAL	156.268	579.12	-886.735

277

Table 11-1 ENTROPY AND ENTROPY OF FORMATION OF GAS - ORGANIC COMPOUNDS (continued)

NO	FORMULA	NAME	Molecular Weight g/mol	Entropy S @ 298 K, joule/(mol K)	Entropy of Formation Sf @ 298 K, joule/(mol K)
1015	C10H20O2	n-DECANOIC ACID	172.268	599.00	-970.317
1016	C10H20O2	2-ETHYLHEXYL ACETATE	172.268	605.00	-962.603
1017	C10H20O2	ISOPENTYL ISOVALERATE	172.268	589.00	-979.138
1018	C10H22	n-DECANE	142.285	545.70	-948.784
1019	C10H22	2-METHYLNONANE	142.285	538.02	-958.880
1020	C10H22	3-METHYLNONANE	142.285	543.90	-953.144
1021	C10H22	4-METHYLNONANE	142.285	541.90	-953.144
1022	C10H22	5-METHYLNONANE	142.285	536.50	-958.880
1023	C10H22	3-ETHYLOCTANE	142.284	--------	-958.880
1024	C10H22	4-ETHYLOCTANE	142.284	--------	-953.144
1025	C10H22	2,2-DIMETHYLOCTANE	142.285	511.00	-981.754
1026	C10H22	2,3-DIMETHYLOCTANE	142.284	--------	-966.158
1027	C10H22	2,4-DIMETHYLOCTANE	142.284	--------	-966.192
1028	C10H22	2,5-DIMETHYLOCTANE	142.284	--------	-966.158
1029	C10H22	2,6-DIMETHYLOCTANE	142.284	--------	-966.158
1030	C10H22	2,7-DIMETHYLOCTANE	142.284	--------	-977.696
1031	C10H22	3,3-DIMETHYLOCTANE	142.284	--------	-972.631
1032	C10H22	3,4-DIMETHYLOCTANE	142.284	--------	-963.240
1033	C10H22	3,5-DIMETHYLOCTANE	142.284	--------	-963.240
1034	C10H22	3,6-DIMETHYLOCTANE	142.284	--------	-969.009
1035	C10H22	4,4-DIMETHYLOCTANE	142.284	--------	-972.631
1036	C10H22	4,5-DIMETHYLOCTANE	142.284	--------	-968.975
1037	C10H22	4-PROPYLHEPTANE	142.284	--------	-968.003
1038	C10H22	4-ISOPROPYLHEPTANE	142.284	--------	-971.927
1039	C10H22	3-ETHYL-2-METHYLHEPTANE	142.284	--------	-966.192
1040	C10H22	4-ETHYL-2-METHYLHEPTANE	142.284	--------	-966.192
1041	C10H22	5-ETHYL-2-METHYLHEPTANE	142.284	--------	-971.960
1042	C10H22	3-ETHYL-3-METHYLHEPTANE	142.284	--------	-972.631
1043	C10H22	4-ETHYL-3-METHYLHEPTANE	142.284	--------	-963.240
1044	C10H22	3-ETHYL-5-METHYLHEPTANE	142.284	--------	-966.192
1045	C10H22	3-ETHYL-4-METHYLHEPTANE	142.284	--------	-966.192
1046	C10H22	4-ETHYL-4-METHYLHEPTANE	142.284	--------	-972.631
1047	C10H22	2,2,3-TRIMETHYLHEPTANE	142.284	--------	-988.898
1048	C10H22	2,2,4-TRIMETHYLHEPTANE	142.284	--------	-988.898
1049	C10H22	2,2,5-TRIMETHYLHEPTANE	142.284	--------	-988.932
1050	C10H22	2,2,6-TRIMETHYLHEPTANE	142.284	--------	-994.835
1051	C10H22	2,3,3-TRIMETHYLHEPTANE	142.284	--------	-985.678
1052	C10H22	2,3,4-TRIMETHYLHEPTANE	142.284	--------	-976.287
1053	C10H22	2,3,5-TRIMETHYLHEPTANE	142.284	--------	-976.287
1054	C10H22	2,3,6-TRIMETHYLHEPTANE	142.284	--------	-979.205
1055	C10H22	2,4,4-TRIMETHYLHEPTANE	142.284	--------	-985.712
1056	C10H22	2,4,5-TRIMETHYLHEPTANE	142.284	--------	-976.287
1057	C10H22	2,4,6-TRIMETHYLHEPTANE	142.284	--------	-985.008
1058	C10H22	2,5,5-TRIMETHYLHEPTANE	142.284	--------	-985.712
1059	C10H22	3,3,4-TRIMETHYLHEPTANE	142.284	--------	-979.775
1060	C10H22	3,3,5-TRIMETHYLHEPTANE	142.284	--------	-979.809
1061	C10H22	3,4,4-TRIMETHYLHEPTANE	142.284	--------	-979.775
1062	C10H22	3,4,5-TRIMETHYLHEPTANE	142.284	--------	-976.287
1063	C10H22	3-ISOPROPYL-2-METHYLHEXANE	142.284	--------	-984.974
1064	C10H22	3,3-DIETHYLHEXANE	142.284	--------	-981.788
1065	C10H22	3,4-DIETHYLHEXANE	142.284	--------	-977.696
1066	C10H22	3-ETHYL-2,2-DIMETHYLHEXANE	142.284	--------	-988.932
1067	C10H22	4-ETHYL-2,2-DIMETHYLHEXANE	142.284	--------	-994.835
1068	C10H22	3-ETHYL-2,3-DIMETHYLHEXANE	142.284	--------	-979.809
1069	C10H22	4-ETHYL-2,3-DIMETHYLHEXANE	142.284	--------	-979.239
1070	C10H22	3-ETHYL-2,4-DIMETHYLHEXANE	142.284	--------	-976.287
1071	C10H22	4-ETHYL-2,4-DIMETHYLHEXANE	142.284	--------	-985.712
1072	C10H22	3-ETHYL-2,5-DIMETHYLHEXANE	142.284	--------	-979.239
1073	C10H22	4-ETHYL-3,3-DIMETHYLHEXANE	142.284	--------	-985.712
1074	C10H22	3-ETHYL-3,4-DIMETHYLHEXANE	142.284	--------	-979.809
1075	C10H22	2,2,3,3-TETRAMETHYLHEXANE	142.284	--------	-1008.419
1076	C10H22	2,2,3,4-TETRAMETHYLHEXANE	142.284	--------	-999.161
1077	C10H22	2,2,3,5-TETRAMETHYLHEXANE	142.284	--------	-1001.979
1078	C10H22	2,2,4,4-TETRAMETHYLHEXANE	142.284	--------	-1008.452
1079	C10H22	2,2,4,5-TETRAMETHYLHEXANE	142.284	--------	-1001.945
1080	C10H22	2,2,5,5-TETRAMETHYLHEXANE	142.284	--------	-1023.310
1081	C10H22	2,3,3,4-TETRAMETHYLHEXANE	142.284	--------	-992.856
1082	C10H22	2,3,3,5-TETRAMETHYLHEXANE	142.284	--------	-998.625
1083	C10H22	2,3,4,4-TETRAMETHYLHEXANE	142.284	--------	-992.856
1084	C10H22	2,3,4,5-TETRAMETHYLHEXANE	142.284	--------	-995.103
1085	C10H22	3,3,4,4-TETRAMETHYLHEXANE	142.284	--------	-1005.065
1086	C10H22	2,4-DIMETHYL-3-ISOPROPYLPENTANE	142.284	--------	-1007.178
1087	C10H22	3,3-DIETHYL-2-METHYLPENTANE	142.284	--------	-994.835
1088	C10H22	3-ETHYL-2,2,3-TRIMETHYLPENTANE	142.284	--------	-1002.683
1089	C10H22	3-ETHYL-2,2,4-TRIMETHYLPENTANE	142.284	--------	-1008.419
1090	C10H22	3-ETHYL-2,3,4-TRIMETHYLPENTANE	142.284	--------	-998.625
1091	C10H22	2,2,3,3,4-PENTAMETHYLPENTANE	142.284	--------	-1021.466
1092	C10H22	2,2,3,4,4-PENTAMETHYLPENTANE	142.284	--------	-1030.589

Table 11-1 ENTROPY AND ENTROPY OF FORMATION OF GAS - ORGANIC COMPOUNDS (continued)

NO	FORMULA	NAME	Molecular Weight g/mol	Entropy S @ 298 K, joule/(mol K)	Entropy of Formation Sf @ 298 K, joule/(mol K)
1093	C10H22O	1-DECANOL	158.284	597.48	-998.591
1094	C10H22O	DI-n-PENTYL ETHER	158.284	585.00	-1011.236
1095	C10H22O	ISODECANOL	158.284	585.00	-1009.559
1096	C10H22O5	TETRAETHYLENE GLYCOL DIMETHYL ETHER	222.282	730.00	-1277.880
1097	C10H22S	n-DECYL MERCAPTAN	174.351	611.64	-915.244
1098	C10H22S	BUTYL-HEXYL-SULFIDE	174.344	--------	-915.110
1099	C10H22S	ETHYL-OCTYL-SULFIDE	174.344	--------	-916.519
1100	C10H22S	HEPTYL-PROPYL-SULFIDE	174.344	--------	-915.110
1101	C10H22S	METHYL-NONYL-SULFIDE	174.344	--------	-918.766
1102	C10H22S	PENTYL-SULFIDE	174.344	--------	-921.013
1103	C10H22S2	PENTYL-DISULFIDE	206.404	--------	-906.557
1104	C10H23N	n-DECYLAMINE	157.299	599.00	-1055.710
1105	C11H10	1-METHYLNAPHTHALENE	142.200	377.44	-338.185
1106	C11H10	2-METHYLNAPHTHALENE	142.200	380.03	-335.536
1107	C11H14O2	n-BUTYL BENZOATE	178.231	539.00	-643.971
1108	C11H16	n-PENTYLBENZENE	148.248	479.18	-628.409
1109	C11H16O	p-tert-AMYLPHENOL	164.247	481.00	-729.230
1110	C11H20	1-UNDECYNE	152.279	--------	-805.065
1111	C11H20O2	2-ETHYLHEXYL ACRYLATE	184.279	643.00	-932.417
1112	C11H22	1-UNDECENE	154.296	581.93	-919.738
1113	C11H22	1-CYCLOPENTYLHEXANE	154.295	--------	-964.917
1114	C11H22	PENTYLCYCLOHEXANE	154.295	--------	-1001.677
1115	C11H22O	1-UNDECANAL	170.295	620.00	-982.056
1116	C11H24	n-UNDECANE	156.312	585.00	-1046.051
1117	C11H24O	1-UNDECANOL	172.311	636.26	-1095.992
1118	C11H24S	UNDECYL MERCAPTAN	188.378	650.90	-1012.477
1119	C11H24S	BUTYL-HEPTYL-SULFIDE	188.371	--------	-1012.510
1120	C11H24S	DECYL-METHYL-SULFIDE	188.371	--------	-1016.133
1121	C11H24S	ETHYL-NONYL-SULFIDE	188.371	--------	-1013.919
1122	C11H24S	OCTYL-PROPYL-SULFIDE	188.371	--------	-1012.510
1123	C12H8O	DIBENZOFURAN	168.195	305.60	-387.724
1124	C12H9N	DIBENZOPYRROLE	167.210	245.00	-507.798
1125	C12H10	ACENAPHTHENE	154.211	368.90	-352.843
1126	C12H10	BIPHENYL	154.211	392.67	-328.660
1127	C12H10O	DIPHENYL ETHER	170.210	419.00	-404.830
1128	C12H11N	p-AMINODIPHENYL	169.226	434.00	-449.438
1129	C12H11N	DIPHENYLAMINE	169.226	402.00	-479.624
1130	C12H11N3	p-AMINOAZOBENZENE	197.240	503.00	-570.183
1131	C12H11N3	1,3-DIPHENYLTRIAZENE	197.240	495.00	--------
1132	C12H12	1,2-DIMETHYLNAPHTHALENE	156.227	--------	-445.011
1133	C12H12	1,3-DIMETHYLNAPHTHALENE	156.227	--------	-442.462
1134	C12H12	1,4-DIMETHYLNAPHTHALENE	156.227	--------	-450.746
1135	C12H12	1,5-DIMETHYLNAPHTHALENE	156.227	--------	-450.746
1136	C12H12	1,6-DIMETHYLNAPHTHALENE	156.227	--------	-442.462
1137	C12H12	1,7-DIMETHYLNAPHTHALENE	156.227	--------	-442.462
1138	C12H12	2,3-DIMETHYLNAPHTHALENE	156.227	--------	-440.953
1139	C12H12	2,6-DIMETHYLNAPHTHALENE	156.227	413.70	-443.166
1140	C12H12	2,7-DIMETHYLNAPHTHALENE	156.227	413.70	-443.166
1141	C12H12	1-ETHYLNAPHTHALENE	156.227	420.90	-433.775
1142	C12H12	2-ETHYLNAPHTHALENE	156.227	--------	-431.092
1143	C12H12N2	p-AMINODIPHENYLAMINE	184.241	449.00	-593.661
1144	C12H12N2	HYDRAZOBENZENE	184.241	452.00	-590.307
1145	C12H14	1,2,3-TRIMETHYLINDENE	158.243	416.00	-567.835
1146	C12H14O4	DIETHYL PHTHALATE	222.241	742.00	-651.685
1147	C12H16	CYCLOHEXYLBENZENE	160.259	428.48	-684.957
1148	C12H18	m-DIISOPROPYLBENZENE	162.275	499.00	-745.028
1149	C12H18	p-DIISOPROPYLBENZENE	162.275	488.00	-755.995
1150	C12H18	n-HEXYLBENZENE	162.275	517.90	-725.675
1151	C12H18	1,2,3-TRIETHYLBENZENE	162.274	--------	-736.307
1152	C12H18	1,2,4-TRIETHYLBENZENE	162.274	--------	-725.541
1153	C12H18	1,3,5-TRIETHYLBENZENE	162.274	--------	-735.905
1154	C12H18	HEXAMETHYLBENZENE	162.274	--------	-791.212·
1155	C12H20O4	DIBUTYL MALEATE	228.288	740.00	-1043.099
1156	C12H22	BICYCLOHEXYL	166.307	450.00	-1055.174
1157	C12H22	1-DODECYNE	166.306	--------	-902.331
1158	C12H23N	DICYCLOHEXYLAMINE	181.321	474.00	-1192.185
1159	C12H24	1-DODECENE	168.323	621.11	-1017.105
1160	C12H24	1-CYCLOPENTYLHEPTANE	168.322	--------	-1062.318
1161	C12H24	1-CYCLOHEXYLHEXANE	168.322	--------	-1098.943
1162	C12H24O	1-DODECANAL	184.322	659.00	-1079.121
1163	C12H24O2	n-DODECANOIC ACID	200.321	683.00	-1157.136
1164	C12H26	n-DODECANE	170.338	624.20	-1143.418
1165	C12H26O	DI-n-HEXYL ETHER	186.338	658.60	-1210.129
1166	C12H26O	1-DODECANOL	186.338	675.21	-1193.225
1167	C12H26O3	DIETHYLENE GLYCOL DI-n-BUTYL ETHER	218.337	743.00	-1331.545
1168	C12H26S	n-DODECYL MERCAPTAN	202.404	690.14	-1109.878
1169	C12H26S	BUTYL-OCTYL-SULFIDE	202.397	--------	-1109.878
1170	C12H26S	DECYL-ETHYL-SULFIDE	202.397	--------	-1111.152

Table 11-1 ENTROPY AND ENTROPY OF FORMATION OF GAS - ORGANIC COMPOUNDS (continued)

NO	FORMULA	NAME	Molecular Weight g/mol	Entropy S @ 298 K, joule/(mol K)	Entropy of Formation Sf @ 298 K, joule/(mol K)
1171	C12H26S	HEXYL-SULFIDE	202.397	--------	-1115.646
1172	C12H26S	METHYL-UNDECYL-SULFIDE	202.397	--------	-1113.399
1173	C12H26S	NONYL-PROPYL-SULFIDE	202.397	--------	-1109.878
1174	C12H26S2	HEXYL-DISULFIDE	234.457	--------	-1101.191
1175	C12H27BO3	TRI-n-BUTYL BORATE	230.156	--------	--------
1176	C12H27N	DODECYLAMINE	185.353	678.00	-1249.338
1177	C12H27N	TRI-n-BUTYLAMINE	185.353	657.00	-1271.172
1178	C13H10	FLUORENE	166.222	381.50	-346.134
1179	C13H10O	BENZOPHENONE	182.222	440.00	-389.066
1180	C13H12	DIPHENYLMETHANE	168.238	440.00	-419.252
1181	C13H14	1-PROPYLNAPHTHALENE	170.254	--------	-529.767
1182	C13H14	2-PROPYLNAPHTHALENE	170.254	--------	-528.358
1183	C13H14	2ETHYL-3-METHYLNAPHTHALENE	170.254	--------	-530.740
1184	C13H14	2ETHYL-6-METHYLNAPHTHALENE	170.254	--------	-532.953
1185	C13H14	2ETHYL-7-METHYLNAPHTHALENE	170.254	--------	-534.228
1186	C13H20	n-HEPTYLBENZENE	176.302	556.80	-823.042
1187	C13H24	1-TRIDECYNE	180.333	--------	-999.732
1188	C13H26	1-TRIDECENE	182.349	660.31	-1114.372
1189	C13H26	1-CYCLOPENTYLOCTANE	182.348	--------	-1159.718
1190	C13H26	1-CYCLOHEXYLHEPTANE	182.348	--------	-1196.311
1191	C13H26O	1-TRIDECANAL	198.349	698.00	-1176.589
1192	C13H26O2	n-BUTYL NONANOATE	214.348	717.80	-1259.098
1193	C13H26O2	METHYL DODECANOATE	214.348	728.00	-1248.700
1194	C13H28	n-TRIDECANE	184.365	663.43	-1240.818
1195	C13H28O	1-TRIDECANOL	200.365	712.83	-1291.900
1196	C13H28S	BUTYL-NONYL-SULFIDE	216.424	--------	-1207.144
1197	C13H28S	DECYL-PROPYL-SULFIDE	216.424	--------	-1207.144
1198	C13H28S	DODECYL-METHYL-SULFIDE	216.424	--------	-1210.766
1199	C13H28S	ETHYL-UNDECYL-SULFIDE	216.424	--------	-1208.553
1200	C13H28S	1-TRIDECANETHIOL	216.424	--------	-1207.144
1201	C14H8O2	ANTHRAQUINONE	208.216	--------	--------
1202	C14H10	ANTHRACENE	178.233	392.60	-340.768
1203	C14H10	DIPHENYLACETYLENE	178.233	466.01	-267.181
1204	C14H10	PHENANTHRENE	178.233	394.50	-338.756
1205	C14H12	cis-STILBENE	180.249	453.25	-410.532
1206	C14H12	trans-STILBENE	180.249	459.00	-405.836
1207	C14H12O2	BENZYL BENZOATE	212.248	545.00	-523.897
1208	C14H14	1,1-DIPHENYLETHANE	182.265	479.00	-516.519
1209	C14H14	1,2-DIPHENYLETHANE	182.265	479.00	-516.519
1210	C14H14O	DIBENZYL ETHER	198.265	524.00	-572.531
1211	C14H16	1-n-BUTYLNAPHTHALENE	184.281	497.20	-627.302
1212	C14H16	2-BUTYLNAPHTHALENE	184.280	--------	-624.618
1213	C14H22	n-OCTYLBENZENE	190.329	596.00	-920.443
1214	C14H22	1,2,3,4-TETRAETHYLBENZENE	190.328	--------	-936.441
1215	C14H22	1,2,3,5-TETRAETHYLBENZENE	190.328	--------	-930.404
1216	C14H22	1,2,4,5-TETRAETHYLBENZENE	190.328	--------	-934.463
1217	C14H22O	p-tert-OCTYLPHENOL	206.328	567.00	-1052.155
1218	C14H28	1-TETRADECENE	196.376	699.52	-1211.773
1219	C14H28	1-CYCLOPENTYLNONANE	196.375	--------	-1256.951
1220	C14H28	1-CYCLOHEXYLOCTANE	196.375	--------	-1293.577
1221	C14H28O2	n-TETRADECANOIC ACID	228.375	754.00	-1361.060
1222	C14H30	n-TETRADECANE	198.392	702.68	-1338.085
1223	C14H30O	1-TETRADECANOL	214.392	751.78	-1389.301
1224	C14H30S	BUTYL-DECYL-SULFIDE	230.451	--------	-1304.545
1225	C14H30S	DODECYL-ETHYL-SULFIDE	230.451	--------	-1305.786
1226	C14H30S	HEPTYL-SULFIDE	230.451	--------	-1310.280
1227	C14H30S	METHYL-TRIDECYL-SULFIDE	230.451	--------	-1308.033
1228	C14H30S	PROPYL-UNDECYL-SULFIDE	230.451	--------	-1304.511
1229	C14H30S	1-TETRADECANETHIOL	230.451	--------	-1304.511
1230	C14H30S2	HEPTYL-DISULFIDE	262.511	--------	-1295.824
1231	C14H31N	TETRADECYLAMINE	213.407	756.00	-1445.581
1232	C15H10N2O2	DIPHENYLMETHANE-4,4'-DIISOCYANATE	250.257	--------	--------
1233	C15H16O	p-CUMYLPHENOL	212.291	532.00	-699.648
1234	C15H16O2	BISPHENOL A	228.291	543.54	-792.152
1235	C15H18	1-PENTYLNAPHTHALENE	198.307	--------	-724.099
1236	C15H18	2-PENTYLNAPHTHALENE	198.307	--------	-721.449
1237	C15H24	n-NONYLBENZENE	204.356	636.16	-1017.676
1238	C15H24O	2,6-DI-tert-BUTYL-p-CRESOL	220.355	606.00	-707.697
1239	C15H24O	NONYLPHENOL	220.355	673.00	-1082.341
1240	C15H28	1-PENTADECYNE	208.386	--------	-1194.365
1241	C15H30	1-PENTADECENE	210.403	742.00	-1309.006
1242	C15H30	1-CYCLOPENTYLDECANE	210.402	--------	-1354.352
1243	C15H30	1-CYCLOHEXYLNONANE	210.402	--------	-1390.978
1244	C15H30O2	PENTADECANOIC ACID	242.402	796.00	-1452.289
1245	C15H32	n-PENTADECANE	212.419	741.93	-1435.452
1246	C15H32O	1-PENTADECANOL	228.417	--------	-1486.534
1247	C15H32S	BUTYL-UNDECYL-SULFIDE	244.478	--------	-1401.778
1248	C15H32S	DODECYL-PROPYL-SULFIDE	244.478	--------	-1401.778

NO	FORMULA	NAME	Molecular Weight g/mol	Entropy S @ 298 K, joule/(mol K)	Entropy of Formation Sf @ 298 K, joule/(mol K)
1249	C15H32S	ETHYL-TRIDECYL-SULFIDE	244.478	--------	-1403.153
1250	C15H32S	METHYL-TETRADECYL-SULFIDE	244.478	--------	-1405.434
1251	C15H32S	1-PENTADECANETHIOL	244.478	--------	-1401.912
1252	C16H10	FLUORANTHENE	202.255	418.50	-326.346
1253	C16H10	PYRENE	202.255	402.00	-342.110
1254	C16H12	1-PHENYLNAPHTHALENE	204.271	459.00	-416.267
1255	C16H20	1-n-HEXYLNAPHTHALENE	212.335	578.50	-819.051
1256	C16H22O4	DIBUTYL PHTHALATE	278.348	900.00	-1038.068
1257	C16H26	n-DECYLBENZENE	218.382	673.70	-1115.076
1258	C16H26	PENTAETHYLBENZENE	218.381	--------	-1140.902
1259	C16H30	1-HEXADECYNE	222.413	--------	-1291.766
1260	C16H32	n-DECYLCYCLOHEXANE	224.430	694.42	-1488.345
1261	C16H32	1-CYCLOPENTYLUNDECANE	224.429	--------	-1451.585
1262	C16H32	1-HEXADECENE	224.430	777.96	-1406.406
1263	C16H32O2	n-HEXADECANOIC ACID	256.429	833.00	-1552.910
1264	C16H34	n-HEXADECANE	226.446	778.31	-1532.853
1265	C16H34O	DI-n-OCTYL ETHER	242.445	822.00	-1592.152
1266	C16H34O	1-HEXADECANOL	242.445	832.40	-1583.934
1267	C16H34S	BUTYL-DODECYL-SULFIDE	258.505	--------	-1499.178
1268	C16H34S	ETHYL-TETRADECYL-SULFIDE	258.505	--------	-1500.587
1269	C16H34S	METHYL-PENTADECYL-SULFIDE	258.505	--------	-1502.801
1270	C16H34S	OCTYL-SULFIDE	258.505	--------	-1504.914
1271	C16H34S	PROPYL-TRIDECYL-SULFIDE	258.505	--------	-1499.178
1272	C16H34S	1-HEXADECANETHIOL	258.505	--------	-1499.178
1273	C16H34S2	OCTYL-DISULFIDE	290.565	--------	-1490.491
1274	C17H28	n-UNDECYLBENZENE	232.409	712.60	-1212.477
1275	C17H32	1-HEPTADECYNE	236.440	818.00	-1388.999
1276	C17H34	1-CYCLOPENTYLDODECANE	238.456	--------	-1548.985
1277	C17H34	1-CYCLOHEXYLUNDECANE	238.456	--------	-1585.611
1278	C17H34	1-HEPTADECENE	238.457	--------	-1503.639
1279	C17H36	n-HEPTADECANE	240.473	820.23	-1630.086
1280	C17H36O	1-HEPTADECANOL	256.472	868.64	-1681.201
1281	C17H36S	BUTYL-TRIDECYL-SULFIDE	272.531	--------	-1596.545
1282	C17H36S	ETHYL-PENTADECYL-SULFIDE	272.531	--------	-1597.820
1283	C17H36S	HEXADECYL-METHYL-SULFIDE	272.531	--------	-1600.067
1284	C17H36S	PROPYL-TETRADECYL-SULFIDE	272.531	--------	-1596.545
1285	C17H36S	1-HEPTADECANETHIOL	272.531	--------	-1596.545
1286	C18H12	CHRYSENE	228.293	423.30	-463.458
1287	C18H14	m-TERPHENYL	230.309	526.00	-489.686
1288	C18H14	o-TERPHENYL	230.309	526.00	-489.686
1289	C18H14	p-TERPHENYL	230.309	523.00	-493.040
1290	C18H15P	TRIPHENYLPHOSPHINE	262.291	--------	--------
1291	C18H15O4P	TRIPHENYL PHOSPHATE	326.288	--------	--------
1292	C18H16N2	N,N'-DIPHENYL-p-PHENYLENEDIAMINE	260.339	526.00	-815.026
1293	C18H22	2,3-DIMETHYL-2,3-DIPHENYLBUTANE	238.373	573.00	-965.286
1294	C18H22O2	DICUMYL PEROXIDE	270.371	665.00	-1077.981
1295	C18H30	n-DODECYLBENZENE	246.436	751.57	-1309.710
1296	C18H30	HEXAETHYLBENZENE	246.435	--------	-1364.146
1297	C18H32O2	LINOLEIC ACID	280.451	902.00	-1494.885
1298	C18H34	1-OCTADECYNE	250.467	--------	-1486.399
1299	C18H34O2	OLEIC ACID	282.467	910.06	-1616.938
1300	C18H34O4	DIBUTYL SEBACATE	314.466	981.00	-1750.797
1301	C18H34O4	DIHEXYL ADIPATE	314.466	902.00	-1831.293
1302	C18H36	1-CYCOPENTYLTRIDECANE	252.482	--------	-1645.950
1303	C18H36	1-CYCLOHEXYLDODECANE	252.482	--------	-1683.012
1304	C18H36	1-OCTADECENE	252.484	856.42	-1601.040
1305	C18H36O2	STEARIC ACID	284.483	904.75	-1752.809
1306	C18H38	n-OCTADECANE	254.500	859.68	-1727.486
1307	C18H38O	DINONYL ETHER	270.499	901.00	-1785.678
1308	C18H38O	1-OCTADECANOL	270.499	907.59	-1778.568
1309	C18H38S	BUTYL-TETRADECYL-SULFIDE	286.558	--------	-1693.812
1310	C18H38S	ETHYL-HEXADECYL-SULFIDE	286.558	--------	-1695.221
1311	C18H38S	HEPTADECYL-METHYL-SULFIDE	286.558	--------	-1697.434
1312	C18H38S	NONYL-SULFIDE	286.558	--------	-1699.547
1313	C18H38S	PENTADECYL-PROPYL-SULFIDE	286.558	--------	-1693.812
1314	C18H38S	1-OCTADECANETHIOL	286.558	--------	-1693.946
1315	C18H38S2	NONYL-DISULFIDE	318.618	--------	-1685.259
1316	C19H26	1-n-NONYLNAPHTHALENE	254.415	697.00	-1109.475
1317	C19H32	n-TRIDECYLBENZENE	260.463	790.50	-1407.111
1318	C19H36	1-NONADECYNE	264.493	--------	-1583.666
1319	C19H36O2	METHYL OLEATE	296.494	956.00	-1707.194
1320	C19H38	1-CYCLOPENTYLTETRADECANE	266.509	--------	-1743.619
1321	C19H38	1-CYCLOHEXYLTRIDECANE	266.509	--------	-1780.379
1322	C19H38	1-NONADECENE	266.511	899.00	-1698.306
1323	C19H38O2	NONADECANOIC ACID	298.510	953.00	-1842.361
1324	C19H40	n-NONADECANE	268.527	898.93	-1824.753
1325	C19H40O	1-NONADECANOL	284.524	--------	-1875.968
1326	C19H40S	BUTYL-PENTADECYL-SULFIDE	300.585	--------	-1791.212

Table 11-1 ENTROPY AND ENTROPY OF FORMATION OF GAS - ORGANIC COMPOUNDS (continued)

NO	FORMULA	NAME	Molecular Weight g/mol	Entropy S @ 298 K, joule/(mol K)	Entropy of Formation Sf @ 298 K, joule/(mol K)
1327	C19H40S	ETHYL-HEPTADECYL-SULFIDE	300.585	--------	-1792.453
1328	C19H40S	HEXADECYL-PROPYL-SULFIDE	300.585	--------	-1791.179
1329	C19H40S	METHYL-OCTADECYL-SULFIDE	300.585	--------	-1794.701
1330	C19H40S	1-NONADECANETHIOL	300.585	--------	-1791.212
1331	C20H16	TRIPHENYLETHYLENE	256.347	582.00	-576.891
1332	C20H28	1-n-DECYLNAPHTHALENE	268.442	736.00	-1206.775
1333	C20H30O2	ABIETIC ACID	302.457	648.00	-1630.387
1334	C20H31N	DEHYDROABIETYLAMINE	285.473	525.00	-1710.548
1335	C20H34	1-PHENYLTETRADECANE	274.489	--------	-1504.343
1336	C20H38	1-EICOSYNE	278.520	934.89	-1681.033
1337	C20H40	1-CYCLOPENTYLPENTADECANE	280.536	--------	-1841.020
1338	C20H40	1-CYCLOHEXYLTETRADECANE	280.536	--------	-1877.645
1339	C20H40	1-EICOSENE	280.538	--------	-1795.673
1340	C20H42	n-EICOSANE	282.553	938.18	-1922.120
1341	C20H42O	1-EICOSANOL	298.553	985.50	-1973.235
1342	C20H42S	BUTYL-HEXADECYL-SULFIDE	314.612	--------	-1888.445
1343	C20H42S	DECYL-SULFIDE	314.612	--------	-1894.315
1344	C20H42S	ETHYL-OCTADECYL-SULFIDE	314.612	--------	-1889.854
1345	C20H42S	HEPTADECYL-PROPYL-SULFIDE	314.612	--------	-1888.445
1346	C20H42S	METHYL-NONADECYL-SULFIDE	314.612	--------	-1892.101
1347	C20H42S	1-EICOSANETHIOL	314.612	--------	-1888.580
1348	C20H42S2	DECYL-DISULFIDE	346.672	--------	-1879.893
1349	C21H21O4P	TRI-o-CRESYL PHOSPHATE	368.369	--------	--------
1350	C21H36	1-PHENYLPENTADECANE	288.515	--------	-1601.744
1351	C21H42	1-CYCLOPENTYLHEXADECANE	294.563	--------	-1938.286
1352	C21H42	1-CYCLOHEXYLPENTADECANE	294.563	--------	-1975.046
1353	C22H38	1-PHENYLHEXADECANE	302.542	--------	-1698.977
1354	C22H44	1-CYCLOHEXYLHEXADECANE	308.590	--------	-2072.279
1355	C22H44O2	n-BUTYL STEARATE	340.590	1084.50	-2119.369
1356	C24H38O4	DIISOOCTYL PHTHALATE	390.563	1030.00	-1998.994
1357	C24H38O4	DIOCTYL PHTHALATE	390.563	1149.00	-1879.658
1358	C24H42O	DINONYLPHENOL	346.597	1120.00	-1862.150
1359	C26H20	TETRAPHENYLETHYLENE	332.445	693.00	-761.362
1360	C28H46O4	DIISODECYL PHTHALATE	446.671	1221.60	-2352.138

Entropy is the absolute entropy of the ideal gas at ambient conditions (298.15 K, 1 atm).

Entropy of formation applies to the ideal gas at ambient conditions (298.15 K, 1 atm).

Table 11-2 ENTROPY AND ENTROPY OF FORMATION OF GAS - INORGANIC COMPOUNDS

NO	FORMULA	NAME	Molecular Weight g/mol	Entropy S @ 298 K, joule/(mol K)	Entropy of Formation Sf @ 298 K, joule/(mol K)
1	Ag	SILVER	107.868	173.00	130.470
2	AgCl	SILVER CHLORIDE	143.321	-------	-------
3	AgI	SILVER IODIDE	234.773	-------	-------
4	Al	ALUMINUM	26.982	164.54	136.170
5	AlB3H12	ALUMINUM BOROHYDRIDE	71.510	-------	-449.440
6	AlBr3	ALUMINUN BROMIDE	266.694	-------	89.880
7	AlCl3	ALUMINUM CHLORIDE	133.340	314.29	-44.090
8	AlF3	ALUMINUM FLUORIDE	83.977	-------	-55.010
9	AlI3	ALUMINUM IODIDE	407.695	-------	152.210
10	Al2O3	ALUMINUM OXIDE	101.961	50.92	-------
11	Al2S3O12	ALUMINUM SULFATE	342.154	239.30	-------
12	Ar	ARGON	39.948	154.73	0.000
13	As	ARSENIC	74.922	-------	138.930
14	AsBr3	ARSENIC TRIBROMIDE	314.634	-------	97.270
15	AsCl3	ARSENIC TRICHLORIDE	181.280	-------	-42.260
16	AsF3	ARSENIC TRIFLUORIDE	131.917	-------	-50.310
17	AsF5	ARSENIC PENTAFLUORIDE	169.914	-------	345.220
18	AsH3	ARSINE	77.945	222.70	-8.390
19	AsI3	ARSENIC TRIIODIDE	455.635	-------	-------
20	As2O3	ARSENIC TRIOXIDE	197.841	-------	-------
21	At	ASTATINE	210.000	-------	-------
22	Au	GOLD	196.967	-------	133.050
23	B	BORON	10.811	-------	147.580
24	BBr3	BORON TRIBROMIDE	250.523	-------	90.220
25	BCl3	BORON TRICHLORIDE	117.169	290.07	-50.650
26	BF3	BORON TRIFLUORIDE	67.806	254.26	-55.680
27	BH2CO	BORINE CARBONYL	40.837	-------	-------
28	BH3O3	BORIC ACID	61.833	295.14	-------
29	B2D6	DEUTERODIBORANE	33.718	-------	-------
30	B2H5Br	DIBORANE HYDROBROMIDE	106.566	-------	-------
31	B2H6	DIBORANE	27.670	232.11	-171.390
32	B3N3H6	BORINE TRIAMINE	80.501	-------	-407.960
33	B4H10	TETRABORANE	53.323	-------	-------
34	B5H9	PENTABORANE	63.126	-------	-341.440
35	B5H11	TETRAHYDROPENTABORANE	65.142	-------	-------
36	B10H14	DECABORANE	122.221	-------	-620.650
37	Ba	BARIUM	137.327	-------	114.040
38	Be	BERYLLIUM	9.012	-------	125.440
39	BeB2H8	BERYLLIUM BOROHYDRIDE	38.698	-------	-------
40	BeBr2	BERYLLIUM BROMIDE	168.820	-------	116.040
41	BeCl2	BERYLLIUM CHLORIDE	79.918	-------	9.810
42	BeF2	BERYLLIUM FLUORIDE	47.009	-------	16.570
43	BeI2	BERYLLIUM IODIDE	262.821	-------	169.040
44	Bi	BISMUTH	208.980	-------	130.470
45	BiBr3	BISMUTH TRIBROMIDE	448.692	-------	-32.530
46	BiCl3	BISMUTH TRICHLORIDE	315.338	-------	-32.280
47	BrF5	BROMINE PENTAFLUORIDE	174.896	-------	-262.620
48	Br2	BROMINE	159.808	245.37	93.240
49	C	CARBON	12.011	157.99	152.270
50	CCl2O	PHOSGENE	98.916	283.74	-47.630
51	CF2O	CARBONYL FLUORIDE	66.007	258.79	-55.150
52	CH4N2O	UREA	60.056	267.00	-------
53	CH4N2S	THIOUREA	76.122	303.10	-------
54	CNBr	CYANOGEN BROMIDE	105.922	-------	70.170
55	CNCl	CYANOGEN CHLORIDE	61.470	236.22	23.480
56	CNF	CYANOGEN FLUORIDE	45.016	-------	22.760
57	CO	CARBON MONOXIDE	28.010	197.54	89.550
58	COS	CARBONYL SULFIDE	60.076	231.47	91.230
59	COSe	CARBON OXYSELENIDE	106.970	-------	-------
60	CO2	CARBON DIOXIDE	44.010	213.69	3.020
61	CS2	CARBON DISULFIDE	76.143	237.79	166.020
62	CSeS	CARBON SELENOSULFIDE	123.037	-------	-------
63	C2N2	CYANOGEN	52.035	-------	30.680
64	C3S2	CARBON SUBSULFIDE	100.165	-------	-------
65	Ca	CALCIUM	40.078	-------	113.370
66	CaF2	CALCIUM FLUORIDE	78.075	274.37	-------
67	CbF5	COLUMBIUM FLUORIDE	187.898	-------	-------
68	Cd	CADMIUM	112.411	-------	115.490
69	CdCl2	CADMIUM CHLORIDE	183.316	-------	-------
70	CdF2	CADMIUM FLUORIDE	150.408	-------	-------
71	CdI2	CADMIUM IODIDE	366.220	-------	-------
72	CdO	CADMIUM OXIDE	128.410	-------	-------
73	ClF	CHLORINE MONOFLUORIDE	54.451	-------	5.030
74	ClFO3	PERCHLORYL FLUORIDE	102.449	278.86	-241.490
75	ClF3	CHLORINE TRIFLUORIDE	92.448	-------	-134.830
76	ClF5	CHLORINE PENTAFLUORIDE	130.445	-------	-------
77	ClHO3S	CHLOROSULFONIC ACID	116.525	308.00	-------
78	ClHO4	PERCHLORIC ACID	100.458	181.89	-290.600

Table 11-2 ENTROPY AND ENTROPY OF FORMATION OF GAS - INORGANIC COMPOUNDS (continued)

NO	FORMULA	NAME	Molecular Weight g/mol	Entropy S @ 298 K, joule/(mol K)	Entropy of Formation Sf @ 298 K, joule/(mol K)
79	ClO2	CHLORINE DIOXIDE	67.452	256.84	-60.370
80	Cl2	CHLORINE	70.905	222.97	0.000
81	Cl2O	CHLORINE MONOXIDE	86.905	-------	-59.030
82	Cl2O7	CHLORINE HEPTOXIDE	182.901	-------	-376.370
83	Co	COBALT	58.933	-------	148.920
84	CoCl2	COBALT CHLORIDE	129.839	-------	-------
85	CoNC3O4	COBALT NITROSYL TRICARBONYL	172.971	-------	-------
86	Cr	CHROMIUM	51.996	-------	150.260
87	CrC6O6	CHROMIUM CARBONYL	220.059	-------	-------
88	CrO2Cl2	CHROMIUM OXYCHLORIDE	154.900	-------	-122.420
89	Cs	CESIUM	132.905	-------	90.220
90	CsBr	CESIUM BROMIDE	212.809	-------	-------
91	CsCl	CESIUM CHLORIDE	168.358	-------	-------
92	CsF	CESIUM FLUORIDE	151.904	-------	-------
93	CsI	CESIUM IODIDE	259.810	-------	-------
94	Cu	COPPER	63.546	-------	133.150
95	CuBr	CUPROUS BROMIDE	143.450	-------	-------
96	CuCl	CUPROUS CHLORIDE	98.999	237.10	-------
97	CuCl2	CUPRIC CHLORIDE	134.451	108.07	-------
98	CuI	COPPER IODIDE	190.450	-------	-------
99	DCN	DEUTERIUM CYANIDE	28.034	-------	-------
100	D2	DEUTERIUM	4.032	144.86	0.000
101	D2O	DEUTERIUM OXIDE	20.031	198.34	-49.130
102	Eu	EUROPIUM	151.965	-------	111.020
103	F2	FLUORINE	37.997	202.69	0.000
104	F2O	FLUORINE OXIDE	53.996	-------	-57.690
105	Fe	IRON	55.847	180.49	152.940
106	FeC5O5	IRON PENTACARBONYL	195.899	-------	-145.950
107	FeCl2	FERROUS CHLORIDE	126.752	-------	-234.350
108	FeCl3	FERRIC CHLORIDE	162.205	-------	2.810
109	Fr	FRANCIUM	223.000	-------	87.780
110	Ga	GALLIUM	69.723	-------	127.790
111	GaCl3	GALLIUM TRICHLORIDE	176.081	334.40	-------
112	Gd	GADOLINIUM	157.250	-------	126.450
113	Ge	GERMANIUM	72.610	-------	136.840
114	GeBr4	GERMANIUM BROMIDE	392.226	-------	60.370
115	GeCl4	GERMANIUM CHLORIDE	214.421	-------	-129.130
116	GeHCl3	TRICHLORO GERMANE	179.976	-------	-------
117	GeH4	GERMANE	76.642	217.12	-75.800
118	Ge2H6	DIGERMANE	151.268	-------	-------
119	Ge3H8	TRIGERMANE	225.894	-------	-------
120	HBr	HYDROGEN BROMIDE	80.912	198.59	57.350
121	HCN	HYDROGEN CYANIDE	27.026	201.67	34.880
122	HCl	HYDROGEN CHLORIDE	36.461	186.80	10.060
123	HF	HYDROGEN FLUORIDE	20.006	173.67	7.040
124	HI	HYDROGEN IODIDE	127.912	206.49	83.180
125	HNO3	NITRIC ACID	63.013	266.37	-202.580
126	H2	HYDROGEN	2.016	130.57	0.000
127	H2O	WATER	18.015	188.72	-44.270
128	H2O2	HYDROGEN PEROXIDE	34.015	232.88	-102.970
129	H2S	HYDROGEN SULFIDE	34.082	205.59	42.930
130	H2SO4	SULFURIC ACID	98.079	298.70	-------
131	H2S2	HYDROGEN DISULFIDE	66.148	-------	-------
132	H2Se	HYDROGEN SELENIDE	80.976	-------	46.290
133	H2Te	HYDROGEN TELLURIDE	129.616	-------	-------
134	H3NO3S	SULFAMIC ACID	97.095	-------	-------
135	He	HELIUM-3	3.016	-------	-------
136	He	HELIUM-4	4.003	126.04	0.000
137	Hf	HAFNIUM	178.490	-------	143.220
138	Hg	MERCURY	200.590	174.96	99.280
139	HgBr2	MERCURIC BROMIDE	360.398	-------	91.860
140	HgCl2	MERCURIC CHLORIDE	271.495	-------	-4.420
141	HgI2	MERCURIC IODIDE	454.399	-------	143.940
142	IF7	IODINE HEPTAFLUORIDE	259.893	-------	-417.800
143	I2	IODINE	253.809	260.58	144.560
144	In	INDIUM	114.818	-------	116.050
145	Ir	IRIDIUM	192.220	-------	158.980
146	K	POTASSIUM	39.098	160.34	95.590
147	KBr	POTASSIUM BROMIDE	119.002	-------	109.850
148	KCl	POTASSIUM CHLORIDE	74.551	238.97	62.780
149	KF	POTASSIUM FLUORIDE	58.097	-------	60.460
150	KI	POTASSIUM IODIDE	166.003	-------	-296.100
151	KOH	POTASSIUM HYDROXIDE	56.106	236.27	13.750
152	Kr	KRYPTON	83.800	163.97	0.000
153	La	LANTHANUM	138.906	-------	125.440
154	Li	LITHIUM	6.941	138.77	109.680
155	LiBr	LITHIUM BROMIDE	86.845	-------	119.030
156	LiCl	LITHIUM CHLORIDE	42.394	-------	72.240

NO	FORMULA	NAME	Molecular Weight g/mol	Entropy S @ 298 K, joule/(mol K)	Entropy of Formation Sf @ 298 K, joule/(mol K)
157	LiF	LITHIUM FLUORIDE	25.939	-------	69.730
158	LiI	LITHIUM IODIDE	133.845	-------	144.960
159	Lu	LUTECIUM	174.967	-------	133.490
160	Mg	MAGNESIUM	24.305	-------	116.050
161	MgCl2	MAGNESIUM CHLORIDE	95.210	-------	1.600
162	MgO	MAGNESIUM OXIDE	40.304	213.16	77.900
163	Mn	MANGANESE	54.938	-------	141.540
164	MnCl2	MANGANESE CHLORIDE	125.843	-------	42.100
165	Mo	MOLYBDENUM	95.940	-------	152.940
166	MoF6	MOLYBDENUM FLUORIDE	209.930	-------	-301.710
167	MoO3	MOLYBDENUM OXIDE	143.938	-------	-------
168	NCl3	NITROGEN TRICHLORIDE	120.365	297.37	-------
169	ND3	HEAVY AMMONIA	20.055	-------	-------
170	NF3	NITROGEN TRIFLUORIDE	71.002	260.73	-139.190
171	NH3	AMMONIA	17.031	192.67	-98.940
172	NH3O	HYDROXYLAMINE	33.030	235.70	-------
173	NH4Br	AMMONIUM BROMIDE	97.943	-------	-------
174	NH4Cl	AMMONIUM CHLORIDE	53.491	94.60	-------
175	NH4I	AMMONIUM IODIDE	144.943	-------	-------
176	NH5O	AMMONIUM HYDROXIDE	35.046	165.56	-------
177	NH5S	AMMONIUM HYDROGENSULFIDE	51.112	-------	-------
178	NO	NITRIC OXIDE	30.006	210.60	12.350
179	NOCl	NITROSYL CHLORIDE	65.459	261.63	-47.990
180	NOF	NITROSYL FLUORIDE	49.005	-------	-51.920
181	NO2	NITROGEN DIOXIDE	46.006	239.92	-60.710
182	N2	NITROGEN	28.013	191.50	0.000
183	N2F4	TETRAFLUOROHYDRAZINE	104.007	301.00	-296.160
184	N2H4	HYDRAZINE	32.045	238.61	-214.660
185	N2H4C	AMMONIUM CYANIDE	44.056	-------	-------
186	N2H6CO2	AMMONIUM CARBAMATE	78.071	-------	-------
187	N2O	NITROUS OXIDE	44.013	219.85	-74.120
188	N2O3	NITROGEN TRIOXIDE	76.012	308.45	-187.150
189	N2O4	NITROGEN TETRAOXIDE	92.011	304.28	-297.500
190	N2O5	NITROGEN PENTOXIDE	108.010	346.00	-348.150
191	Na	SODIUM	22.990	153.71	102.300
192	NaBr	SODIUM BROMIDE	102.894	241.12	113.540
193	NaCN	SODIUM CYANIDE	49.008	249.43	-------
194	NaCl	SODIUM CHLORIDE	58.442	229.70	66.760
195	NaF	SODIUM FLUORIDE	41.988	51.98	64.690
196	NaI	SODIUM IODIDE	149.894	-------	-------
197	NaOH	SODIUM HYDROXIDE	39.997	228.33	-------
198	Na2SO4	SODIUM SULFATE	142.043	346.75	-------
199	Nb	NIOBIUM	92.906	-------	150.260
200	Nd	NEODYMIUM	144.240	-------	118.060
201	Ne	NEON	20.180	146.21	0.000
202	Ni	NICKEL	58.693	-------	151.600
203	NiC4O4	NICKEL CARBONYL	170.735	-------	-------
204	NiF2	NICKEL FLUORIDE	96.690	-------	-------
205	Np	NEPTUNIUM	237.000	-------	-------
206	O2	OXYGEN	31.999	205.04	0.000
207	O3	OZONE	47.998	238.82	-68.760
208	Os	OSMIUM	190.230	-------	154.280
209	OsOF5	OSMIUM OXIDE PENTAFLUORIDE	301.221	-------	-------
210	OsO4	OSMIUM TETROXIDE - YELLOW	254.228	-------	-148.920
211	OsO4	OSMIUM TETROXIDE - WHITE	254.228	-------	-148.920
212	P	PHOSPHORUS - WHITE	30.974	163.09	122.090
213	PBr3	PHOSPHORUS TRIBROMIDE	270.686	-------	78.820
214	PCl2F3	PHOSPHORUS DICHLORIDE TRIFLUORIDE	158.874	-------	-------
215	PCl3	PHOSPHORUS TRICHLORIDE	137.332	311.58	-64.400
216	PCl5	PHOSPHORUS PENTACHLORIDE	208.237	364.58	-234.450
217	PH3	PHOSPHINE	33.998	210.23	-26.830
218	PH4Br	PHOSPHONIUM BROMIDE	114.910	-------	-------
219	PH4Cl	PHOSPHONIUM CHLORIDE	70.458	-------	-------
220	PH4I	PHOSPHONIUM IODIDE	161.910	-------	-------
221	POCl3	PHOSPHORUS OXYCHLORIDE	153.331	325.38	-152.940
222	PSBr3	PHOSPHORUS THIOBROMIDE	302.752	-------	89.660
223	PSCl3	PHOSPHORUS THIOCHLORIDE	169.398	337.23	-51.920
224	P4O6	PHOSPHORUS TRIOXIDE	219.891	-------	-360.780
225	P4O10	PHOSPHORUS PENTOXIDE	283.889	403.76	-729.600
226	P4S10	PHOSPHORUS PENTASULFIDE	444.555	381.70	-------
227	Pb	LEAD	207.200	-------	110.680
228	PbBr2	LEAD BROMIDE	367.008	-------	127.860
229	PbCl2	LEAD CHLORIDE	278.105	-------	32.880
230	PbF2	LEAD FLUORIDE	245.197	-------	-------
231	PbI2	LEAD IODIDE	461.009	-------	182.910
232	PbO	LEAD OXIDE	223.199	-------	71.570
233	PbS	LEAD SULFIDE	239.266	-------	-------
234	Pd	PALLADIUM	106.420	-------	129.130

NO	FORMULA	NAME	Molecular Weight g/mol	Entropy S @ 298 K, joule/(mol K)	Entropy of Formation Sf @ 298 K, joule/(mol K)
235	Po	POLONIUM	209.000	-------	126.060
236	Pt	PLATINUM	195.080	-------	150.260
237	Ra	RADIUM	226.000	-------	97.270
238	Rb	RUBIDIUM	85.468	-------	93.240
239	RbBr	RUBIDIUM BROMIDE	165.372	-------	-------
240	RbCl	RUBIDIUM CHLORIDE	120.921	-------	-------
241	RbF	RUBIDIUM FLUORIDE	104.466	-------	-------
242	RbI	RUBIDIUM IODIDE	212.372	-------	-------
243	Re	RHENIUM	186.207	-------	151.940
244	Re2O7	RHENIUM HEPTOXIDE	484.410	-------	-355.530
245	Rh	RHODIUM	102.906	-------	154.620
246	Rn	RADON	222.000	-------	-------
247	Ru	RUTHENIUM	101.070	-------	157.300
248	RuF5	RUTHENIUM PENTAFLUORIDE	196.062	-------	-------
249	S	SULFUR	32.066	167.71	135.840
250	SF4	SULFUR TETRAFLUORIDE	108.060	-------	-138.190
251	SF6	SULFUR HEXAFLUORIDE	146.056	291.63	-348.820
252	SOBr2	THIONYL BROMIDE	207.873	-------	-------
253	SOCl2	THIONYL CHLORIDE	118.971	309.77	-47.630
254	SOF2	SULFUROUS OXYFLUORIDE	86.062	-------	-58.590
255	SO2	SULFUR DIOXIDE	64.065	248.11	11.070
256	SO2Cl2	SULFURYL CHLORIDE	134.970	311.83	-147.580
257	SO3	SULFUR TRIOXIDE	80.064	256.51	-82.510
258	S2Cl2	SULFUR MONOCHLORIDE	135.037	-------	134.830
259	Sb	ANTIMONY	121.757	-------	134.720
260	SbBr3	ANTIMONY TRIBROMIDE	361.469	-------	-1403.660
261	SbCl3	ANTIMONY TRICHLORIDE	228.115	337.80	-42.100
262	SbCl5	ANTIMONY PENTACHLORIDE	299.021	-------	-201.240
263	SbH3	STIBINE	124.781	-------	-9.060
264	SbI3	ANTIMONY TRIIODIDE	502.470	-------	-------
265	Sb2O3	ANTIMONY TRIOXIDE	291.512	-------	-------
266	Sc	SCANDIUM	44.956	-------	140.200
267	Se	SELENIUM	78.960	-------	134.500
268	SeCl4	SELENIUM TETRACHLORIDE	220.771	-------	-------
269	SeF6	SELENIUM HEXAFLUORIDE	192.950	-------	-335.400
270	SeOCl2	SELENIUM OXYCHLORIDE	165.865	-------	-------
271	SeO2	SELENIUM DIOXIDE	110.959	-------	-122.660
272	Si	SILICON	28.086	167.97	149.250
273	SiBrCl2F	BROMODICHLOROFLUOROSILANE	197.893	-------	-------
274	SiBrF3	TRIFLUOROBROMOSILANE	164.985	-------	-------
275	SiBr2ClF	DIBROMOCHLOROFLUOROSILANE	242.345	-------	-------
276	SiClF3	TRIFLUOROCHLOROSILANE	120.533	-------	-125.500
277	SiCl2F2	DICHLORODIFLUOROSILANE	136.988	-------	-------
278	SiCl3F	TRICHLOROFLUOROSILANE	153.442	-------	-------
279	SiCl4	SILICON TETRACHLORIDE	169.896	330.83	-134.160
280	SiF4	SILICON TETRAFLUORIDE	104.079	282.49	-141.540
281	SiHBr3	TRIBROMOSILANE	268.805	-------	36.560
282	SiHCl3	TRICHLOROSILANE	135.452	313.59	-103.970
283	SiHF3	TRIFLUOROSILANE	86.089	-------	-110.400
284	SiH2Br2	DIBROMOSILANE	189.909	-------	-------
285	SiH2Cl2	DICHLOROSILANE	101.007	286.62	-87.010
286	SiH2F2	DIFLUOROSILANE	68.098	-------	-91.220
287	SiH2I2	DIIODOSILANE	283.910	-------	-------
288	SiH3Br	MONOBROMOSILANE	111.013	-------	-------
289	SiH3Cl	MONOCHLOROSILANE	66.562	-------	-75.780
290	SiH3F	MONOFLUOROSILANE	50.108	-------	-77.110
291	SiH3I	IODOSILANE	158.014	-------	-------
292	SiH4	SILANE	32.117	204.51	-75.800
293	SiO2	SILICON DIOXIDE	60.084	228.86	4.880
294	Si2Cl6	HEXACHLORODISILANE	268.887	-------	-------
295	Si2F6	HEXAFLUORODISILANE	170.161	-------	-------
296	Si2H5Cl	DISILANYL CHLORIDE	96.663	-------	-------
297	Si2H6	DISILANE	62.219	271.50	-157.640
298	Si2OCl3F3	TRICHLOROTRIFLUORODISILOXANE	235.524	-------	-------
299	Si2OCl6	HEXACHLORODISILOXANE	284.887	-------	-------
300	Si2OH6	DISILOXANE	78.218	-------	-------
301	Si3Cl8	OCTACHLOROTRISILANE	367.878	-------	-------
302	Si3H8	TRISILANE	92.320	-------	-------
303	Si3H9N	TRISILAZANE	107.335	-------	-------
304	Si4H10	TETRASILANE	122.421	-------	-------
305	Sm	SAMARIUM	150.360	-------	113.700
306	Sn	TIN	118.710	-------	117.390
307	SnBr4	STANNIC BROMIDE	438.326	-------	-------
308	SnCl2	STANNOUS CHLORIDE	189.615	-------	-------
309	SnCl4	STANNIC CHLORIDE	260.521	-------	-131.810
310	SnH4	STANNIC HYDRIDE	122.742	-------	-85.530
311	SnI4	STANNIC IODIDE	626.328	-------	-------
312	Sr	STRONTIUM	87.620	-------	112.360

NO	FORMULA	NAME	Molecular Weight g/mol	Entropy S @ 298 K, joule/(mol K)	Entropy of Formation Sf @ 298 K, joule/(mol K)
313	SrO	STRONTIUM OXIDE	103.619	-------	-1674.160
314	Ta	TANTALUM	180.948	-------	143.220
315	Tc	TECNNETIUM	98.000	-------	246.370
316	Te	TELLURIUM	127.600	-------	132.820
317	TeCl4	TELLURIUM TETRACHLORIDE	269.411	-------	-------
318	TeF6	TELLURIUM HEXAFLUORIDE	241.590	-------	-------
319	Ti	TITANIUM	47.880	-------	149.590
320	TiCl4	TITANIUM TETRACHLORIDE	189.691	354.79	-123.760
321	Tl	THALLIUM	204.383	-------	116.720
322	TlBr	THALLOUS BROMIDE	284.287	-------	-------
323	TlI	THALLOUS IODIDE	331.288	-------	-------
324	Tm	THULIUM	168.934	-------	116.380
325	U	URANIUM	238.029	-------	149.590
326	UF6	URANIUM FLUORIDE	352.019	-------	-280.730
327	V	VANADIUM	50.942	-------	-805.630
328	VCl4	VANADIUM TETRACHLORIDE	192.752	362.29	-112.360
329	VOCl3	VANADIUM OXYTRICHLORIDE	173.299	344.29	-------
330	W	TUNGSTEN	183.840	-------	141.870
331	WF6	TUNGSTEN FLUORIDE	297.830	-------	-300.520
332	Xe	XENON	131.290	169.57	-------
333	Yb	YTTERBIUM	173.040	-------	113.700
334	Yt	YTTRIUM	88.906	-------	134.830
335	Zn	ZINC	65.390	160.98	119.400
336	ZnCl2	ZINC CHLORIDE	136.295	-------	-------
337	ZnF2	ZINC FLUORIDE	103.387	-------	-------
338	ZnO	ZINC OXIDE	81.389	43.64	-------
339	ZnSO4	ZINC SULFATE	161.454	110.50	-------
340	Zr	ZIRCONIUM	91.224	-------	141.870
341	ZrBr4	ZIRCONIUM BROMIDE	410.840	-------	71.530
342	ZrCl4	ZIRCONIUM CHLORIDE	233.035	-------	-115.560
343	ZrI4	ZIRCONIUM IODIDE	598.842	-------	169.900

Entropy is the absolute entropy of the ideal gas at ambient conditions (298.15 K, 1 atm).

Entropy of formation applies to the ideal gas at ambient conditions (298.15 K, 1 atm).

Chapter 12

ENTHALPY OF FORMATION

Carl L. Yaws, Li Bu, Sachin Nijhawan, Jack R. Hopper, and Ralph W. Pike*

Lamar University, Beaumont, Texas / *Louisiana State University, Baton Rouge, Louisiana

ABSTRACT

Results for enthalpy of formation are presented for major organic and inorganic chemicals. The major chemicals include many compound types. The results are provided in easy-to-use tabulations which are especially applicable for rapid engineering usage with the personal computer or hand calculator. The agreement of correlation and data is quite good.

INTRODUCTION

Enthalpy of formation for individual compounds in chemical reactions is required to determine the heat of reaction, $\Delta H_{reaction}$ and associated heating and cooling requirements:

$$\Delta H_{reaction} = \Sigma \ (n\Delta H_f)_{products} \ - \ \Sigma \ (n\Delta H_f)_{reactants} \tag{12-1}$$

If $\Delta H_{reaction} < 0$, then the chemical reaction is exothermic and cooling is needed to maintain the reaction temperature. If $\Delta H_{reaction} > 0$, the reaction is endothermic and heating is required to conduct the chemical reaction. Since the heat effects of a reaction maybe determined from the enthalpy of formation for individual compounds, results for enthalpy of formation are presented in this article for major chemicals.

CORRELATION FOR ENTHALPY OF FORMATION

The correlation for enthalpy of formation of the ideal gas is a series expansion in temperature:

$$\Delta H_f = A + B \ T + C \ T^2 \tag{12-2}$$

where ΔH_f = enthalpy of formation of ideal gas, kjoule/mol
A, B, and C = regression coefficient for chemical compound
T = temperature, K

The results for enthalpy of formation are given in Tables 12-1 and 12-2. The tabulations are arranged by chemical formula to provides ease of use in quickly locating data. A wide variety of substances are covered.

In preparing the compilation, a literature search was conducted to identify data source publications for organics (1-40) and inorganics (1-60). The publications were screened for appropriate data. The compilation resulted from the screening.

For organics, the range for application is denoted by the respective minimum and maximum temperatures (TMIN and TMAX). In the absence of data, values at 298.15 K were extended to higher temperatures by integration of the appropriate thermodynamic equations involving heat capacities. The numerous data points were processed with a generalized least-square computer program for minimizing the deviation. The spot values at room temperature (298.15 K) are the actual data. The spot values at elevated temperature (500 K) are calculated from the correlation. Since water and hydrogen chloride are of major industrial importance, regression coefficients are provided for these compounds at the end of the table.

For inorganics, results are also given for internal energy of formation and entropy at room temperature (298.15 K). These thermodynamic properties are useful in computing the energy of a chemical explosion. As given by Crowl and Louvar (20), the equations for a chemical explosion involve the change in Helmholtz free energy which may be calculated from data for internal energy of formation and entropy:

$$\text{Explosion energy}_{limit} = - \Delta A_{reaction} \tag{12-3}$$

$$\Delta A_{reaction} = \Delta U_{reaction} - T \ \Delta S_{reaction} \tag{12-4}$$

$$\Delta U_{reaction} = \Sigma \ (n\Delta U_f)_{products} \ - \ \Sigma \ (n\Delta U_f)_{reactants} \tag{12-5}$$

$$\Delta S_{reaction} = \Sigma \ (nS)_{products} \ - \ \Sigma \ (nS)_{reactants} \ - R \ n \ \Sigma \ (x_i \ \ln \ x_i)_{products} \tag{12-6}$$

The last term $(R \ n \ \Sigma \ x_i \ \ln \ x_i)$ in the above equation represents the entropy of mixing. The R is the universal gas constant.

The actual energy release in a chemical explosion will be less than the limiting value given by the change in Helmholtz energy because of thermal effects and irreversibility. In an example in Crowl and Louvar

(20), 12% of the limiting value is suggested for a vapor cloud explosion in a partially confined area (ethylene explosion in a ditch). For vapor cloud explosion in an unconfined area, 2% of the limiting value is suggested by the same authors.

A comparison of correlation and data is shown in Fig. 12-1 for a representative chemical. The graph discloses favorable agreement of correlation and data.

EXAMPLES

The correlation results can be used to make calculations for enthalpy of formation and heat of reaction. Examples are given below.

Example 1 Let us estimate the enthalpy of formation of methane (CH4) as a low-pressure gas at 500 K.

We obtain correlation constants (A, B, C) for methane from the table and substitute them into the equation at a temperature of 500 K to get

ΔH_f = -63.425 + (-4.3355E-01) (500) + (1.7220E-05) (500^2)

ΔH_f = -80.80 kjoule/mol

Example 2   Calculate the heat of the reaction for the dehydrogenation of 1-butene (C4H8) to 1,3 butadiene (C4H6) at a reaction temperature of 900 K:

C4H8(g) → C4H6(g) + H2(g)

The heat of the reaction may be determined from enthalpy of formation at 900 K for the products and reactants:

$\Delta H_{reaction} = \Delta H_{f,C4H6} + \Delta H_{f,H2} - \Delta H_{f,C4H8}$

Using coefficients for 1-butene and 1,3-butadiene from the table and the equation for enthalpy of formation, we obtain

$\Delta H_{f,C4H8}$ = 21.822 + (-8.5458E-02) (900) + (3.8902E-05) (900^2) = -23.58
$\Delta H_{f,H2}$ = 0
$\Delta H_{f,C4H6}$ = 123.286 + (-5.1225E-02) (900) + (2.3192E-05) (900^2) = 95.97

Substitution of ΔH_f values at 900 K into the equation for the heat of the reaction yields

$\Delta H_{reaction}$ = 95.97 + 0 - (-23.58)

$\Delta H_{reaction}$ = 119.55 kjoule/mol

Since $\Delta H_{reaction}$ > 0, the reaction is endothermic and would require heating to maintain the reaction temperature.

Portions of this material appeared in Chem. Eng., 95 (No. 13), 81 (Sept. 26, 1988) and are reprinted by special permission.

REFERENCES – ORGANIC COMPOUNDS

1-34. See **REFERENCES - ORGANIC COMPOUNDS** in Chapter 1, **CRITICAL PROPERTIES AND ACENTRIC FACTOR**
35. Smith, J. M. and H. C. Van Ness, INTRODUCTION TO CHEMICAL ENGINEERING THERMODYNAMICS, 4th ed., McGraw-Hill, New York, NY (1987).
36. Pedley, J. B., THERMOCHEMICAL DATA AND STRUCTURES OF ORGANIC COMPOUNDS, Vol. I, Thermodynamics Research Center, College Station, TX (1994).
37. Frenkel, M., K. N. Marsh, R. C. Wilhoit, G. J. Kabo, and G. N. Roganov, THERMODYNAMICS OF ORGANIC COMPOUNDS IN THE GAS STATE, Vols. I and II, Thermodynamics Research Center, College Station, TX (1994).
38. Crowl, D. A. and J. F. Louvar, CHEMICAL PROCESS SAFETY, Prentice Hall, Inc., Englewood Cliffs, NJ (1990).
39. Yaws, C. L. and P. Y. Chiang, Chem. Eng., 95 (13), 81 (Sept. 26, 1988).
40. Yaws, C. L., HANDBOOK OF CHEMICAL COMPOUND DATA FOR PROCESS SAFETY, Gulf Publishing Co., Houston, TX (1997).

REFERENCES - INORGANIC COMPOUNDS

1-56. See **REFERENCES - INORGANIC COMPOUNDS** in Chapter 1, **CRITICAL PROPERTIES AND ACENTRIC FACTOR**
57. Daubert, T. E., CHEMICAL ENGINEERING THERMODYNAMICS, McGraw-Hill, New York, NY 1985).
58. Karapet'yants, M. Kh. and M. L. Karapet'yants, THERMODYNAMIC CONSTANTS OF INORGANIC AND ORGANIC COMPOUNDS, translated from Russian, Ann Arbor - Humphrey Science Publishers, Ann Arbor, MI (1970).
59. Crowl, D. A. and J. F. Louvar, CHEMICAL PROCESS SAFETY, Prentice Hall, Inc., Englewood Cliffs, NJ (1990).
60. Yaws, C. L., HANDBOOK OF CHEMICAL COMPOUND DATA FOR PROCESS SAFETY, Gulf Publishing Co., Houston, TX (1997).

Figure 12-1 Enthalpy of Formation

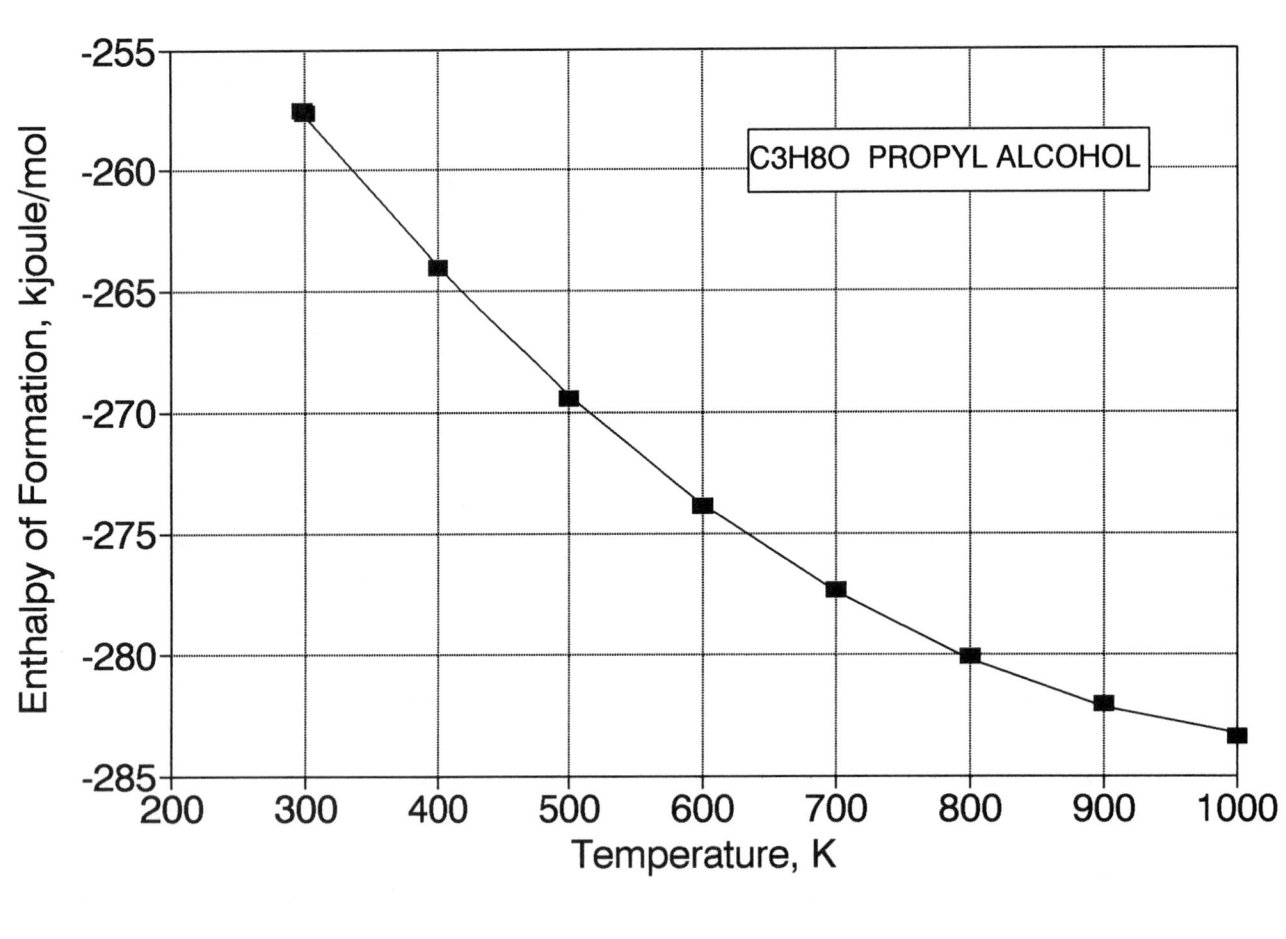

Table 12-1 ENTHALPY OF FORMATION OF GAS - ORGANIC COMPOUNDS

			$Hf = A + B T + C T^2$					(Hf - kjoule/mol, T - K)	
NO	FORMULA	NAME	A	B	C	TMIN	TMAX	Hf @ 298 K	Hf @ 500 K
1	CBrClF2	BROMOCHLORODIFLUOROMETHANE	-447.716	1.9664E-03	2.5405E-06	333	1000	-431.37	-446.10
2	CBrCl3	BROMOTRICHLOROMETHANE	-57.889	1.2183E-02	-1.8701E-06	333	1000	-38.90	-52.26
3	CBrF3	BROMOTRIFLUOROMETHANE	-663.863	-3.4821E-03	4.7946E-06	333	1000	-648.94	-664.41
4	CBr2F2	DIBROMODIFLUOROMETHANE	-424.249	2.1691E-02	2.1549E-06	333	1000	-386.60	-412.86
5	CClF3	CHLOROTRIFLUOROMETHANE	-706.510	-7.0337E-03	7.0145E-06	298	1000	-707.93	-708.27
6	CClN	CYANOGEN CHLORIDE	136.095	7.2566E-03	-3.5187E-06	298	1000	137.95	138.84
7	CCl2F2	DICHLORODIFLUOROMETHANE	-493.461	-7.5944E-04	3.9602E-06	298	1000	-493.29	-492.85
8	CCl2O	PHOSGENE	-219.628	2.3616E-03	-6.7978E-07	298	1000	-218.95	-218.62
9	CCl3F	TRICHLOROFLUOROMETHANE	-286.592	4.9380E-03	1.4901E-06	298	1000	-284.93	-283.75
10	CCl4	CARBON TETRACHLORIDE	-103.244	9.0605E-03	9.5332E-07	298	1000	-100.42	-98.48
11	CF2O	CARBONYL FLUORIDE	-636.813	-8.2130E-03	3.7389E-06	298	1000	-638.90	-639.98
12	CF4	CARBON TETRAFLUORIDE	-930.396	-1.1697E-02	8.9680E-06	298	1000	-933.03	-934.00
13	CHBr3	TRIBROMOMETHANE	-27.833	-8.3580E-03	6.9128E-06	333	1000	16.70	-30.28
14	CHClF2	CHLORODIFLUOROMETHANE	-478.189	-2.1958E-02	1.2380E-05	298	1000	-483.67	-486.07
15	CHCl2F	DICHLOROFLUOROMETHANE	-281.303	-1.4763E-02	8.7128E-06	298	1000	-284.93	-286.51
16	CHCl3	CHLOROFORM	-98.906	-1.0149E-02	7.1513E-06	298	1000	-101.25	-102.19
17	CHF3	TRIFLUOROMETHANE	-691.657	-2.3233E-02	1.1683E-05	298	1000	-697.51	-700.35
18	CHI3	TRIIODOMETHANE	116.960	-1.1825E-03	2.7648E-06	459	1000	210.87	117.06
19	CHN	HYDROGEN CYANIDE	135.847	-2.1831E-03	-4.3739E-07	298	1000	135.14	134.65
20	CHNS	ISOTHIOCYANIC ACID	140.808	-5.2396E-02	2.7098E-05	389	717	127.61	121.38
21	CH2BrCl	BROMOCHLOROMETHANE	-51.279	-2.7332E-02	1.4131E-05	333	1000	-42.70	-61.41
22	CH2Br2	DIBROMOMETHANE	-39.382	-2.5218E-02	1.3158E-05	333	1000	-14.80	-48.70
23	CH2ClF	CHLOROFLUOROMETHANE	-256.086	-3.2741E-02	1.6116E-05	298	1000	-264.43	-268.43
24	CH2Cl2	DICHLOROMETHANE	-88.943	-2.5399E-02	1.2304E-05	298	1000	-95.40	-98.57
25	CH2F2	DIFLUOROMETHANE	-443.881	-3.4874E-02	1.5618E-05	298	1000	-452.88	-457.41
26	CH2I2	DIIODOMETHANE	59.496	-1.7356E-02	8.7416E-06	459	1000	117.99	53.00
27	CH2O	FORMALDEHYDE	-109.671	-2.3151E-02	7.8697E-06	298	1000	-115.90	-119.28
28	CH2O2	FORMIC ACID	-371.635	-2.7450E-02	1.3101E-05	298	1000	-378.61	-382.09
29	CH3Br	BROMOMETHANE	-44.104	-3.5166E-02	1.5043E-05	333	1000	-37.66	-57.93
30	CH3Cl	CHLOROMETHANE	-76.576	-3.7541E-02	1.6128E-05	298	1000	-86.32	-91.31
31	CH3Cl3Si	METHYL TRICHLOROSILANE	-564.422	-2.9614E-02	1.6391E-05	298	1000	-571.80	-575.13
32	CH3F	FLUOROMETHANE	-223.112	-4.1235E-02	1.7070E-05	298	1000	-233.89	-239.46
33	CH3I	IODOMETHANE	-9.253	-3.1909E-02	1.3598E-05	459	1000	13.97	-21.81
34	CH3NO	FORMAMIDE	-173.892	-4.8701E-02	2.4811E-05	298	1000	-186.19	-192.04
35	CH3NO2	NITROMETHANE	-62.208	-5.0455E-02	2.7142E-05	298	1000	-74.73	-80.65
36	CH3NO2	METHYL NITRITE	-53.279	-4.4434E-02	2.7099E-05	298	1000	-64.02	-68.72
37	CH3NO3	METHYL NITRATE	-109.621	-4.5622E-02	2.9048E-05	298	1000	-120.50	-125.17
38	CH4	METHANE	-63.425	-4.3355E-02	1.7220E-05	298	1000	-74.85	-80.80
39	CH4Cl2Si	METHYL DICHLOROSILANE	-391.563	-4.1755E-02	2.2435E-05	298	1000	-402.00	-406.83
40	CH4O	METHANOL	-188.188	-4.9823E-02	2.0791E-05	298	1000	-201.17	-207.90
41	CH4O3S	METHANESULFONIC ACID	--------	--------	--------	--------	--------	--------	--------
42	CH4S	METHANETHIOL	-0.434	-8.8588E-02	4.3430E-05	389	717	-22.97	-33.87
43	CH5ClSi	METHYL CHLOROSILANE	-200.674	-5.6902E-02	2.9462E-05	298	1000	-215.00	-221.76
44	CH5N	METHYLAMINE	-7.489	-6.0538E-02	2.7800E-05	298	1000	-23.01	-30.81
45	CH6Si	METHYL SILANE	-11.276	-7.0690E-02	3.6336E-05	298	1000	-29.10	-37.54
46	CN4O8	TETRANITROMETHANE	89.859	-3.9117E-02	4.1602E-05	298	1000	82.30	80.70
47	CO	CARBON MONOXIDE	-112.190	8.1182E-03	-8.0425E-06	298	1000	-110.54	-110.14
48	COS	CARBONYL SULFIDE	-128.369	-3.9573E-02	1.9738E-05	389	717	-138.41	-143.22
49	CO2	CARBON DIOXIDE	-393.422	1.5913E-04	-1.3945E-06	298	1000	-393.51	-393.69
50	CS2	CARBON DISULFIDE	137.658	-8.1634E-02	4.2246E-05	389	717	117.07	107.40
51	C2BrF3	BROMOTRIFLUOROETHYLENE	-472.149	5.7751E-03	-7.5542E-07	333	1000	-455.00	-469.45
52	C2Br2F4	1,2-DIBROMOTETRAFLUOROETHANE	-958.447	1.4430E-02	1.8995E-06	333	1000	-922.99	-950.76
53	C2ClF3	CHLOROTRIFLUOROETHYLENE	-574.131	4.8078E-03	-4.4688E-07	298	1000	-572.70	-571.84
54	C2ClF5	CHLOROPENTAFLUOROETHANE	-1112.316	1.1868E-02	5.7455E-08	298	1000	-1108.76	-1106.37
55	C2Cl2F4	1,2-DICHLOROTETRAFLUOROETHANE	-890.313	7.4617E-03	6.0509E-06	298	1000	-887.43	-885.07
56	C2Cl3F3	1,1,2-TRICHLOROTRIFLUOROETHANE	-699.565	1.4133E-02	3.8610E-06	298	1000	-694.96	-691.53
57	C2Cl4	TETRACHLOROETHENE	-15.596	1.2256E-02	-2.2132E-06	298	1000	-12.13	-10.02
58	C2Cl4F2	1,1,2,2-TETRACHLORODIFLUOROETHANE	-195.746	-2.1446E-02	1.2383E-05	298	1000	-699.57	-203.37
59	C2Cl4O	TRICHLOROACETYL CHLORIDE	-309.803	1.3286E-02	-1.3557E-06	298	1000	-305.93	-303.50
60	C2Cl6	HEXACHLOROETHANE	-147.858	2.1266E-02	3.5458E-07	298	1000	-141.42	-137.14
61	C2F4	TETRAFLUOROETHENE	-658.980	9.5324E-04	1.2640E-06	298	1000	-658.56	-658.19
62	C2F6	HEXAFLUOROETHANE	-1342.771	-4.2689E-03	9.7542E-06	298	1000	-1343.06	-1342.47
63	C2HBrClF3	HALOTHANE	-717.147	-1.6006E-02	1.5880E-05	333	1000	-705.00	-721.18
64	C2HClF2	2-CHLORO-1,1-DIFLUOROETHYLENE	-327.215	-7.3896E-03	4.3109E-06	298	1000	-329.00	-329.83
65	C2HCl3	TRICHLOROETHENE	-9.155	-2.6745E-03	3.3639E-06	298	1000	-9.62	-9.65
66	C2HCl3O	DICHLOROACETYL CHLORIDE	-279.829	-5.1697E-03	5.3044E-06	298	1000	-280.83	-281.09
67	C2HCl3O	TRICHLOROACETALDEHYDE	-196.945	-1.6143E-03	4.5770E-06	298	1000	-197.00	-196.61
68	C2HCl5	PENTACHLOROETHANE	-143.235	4.9308E-04	8.3387E-06	298	1000	-142.26	-140.90
69	C2HF3	TRIFLUOROETHENE	-487.938	-1.1444E-02	6.0417E-06	298	1000	-490.78	-492.15
70	C2HF3O2	TRIFLUOROACETIC ACID	-1018.266	-1.5776E-02	1.2917E-05	298	1000	-1021.70	-1022.93
71	C2HF5	PENTAFLUOROETHANE	-1100.197	-1.9818E-02	1.5969E-05	298	1000	-1104.60	-1106.11
72	C2H2	ACETYLENE(ETHYNE)	227.216	-3.5467E-04	-3.9611E-06	298	1000	226.73	226.05
73	C2H2Br4	1,1,2,2-TETRABROMOETHANE	-47.627	-1.5752E-02	1.4681E-05	333	1000	10.88	-51.83
74	C2H2Cl2	1,1-DICHLOROETHENE	6.367	-1.5909E-02	8.0362E-06	298	1000	2.38	0.42
75	C2H2Cl2	cis-1,2-DICHLOROETHENE	1.808	-1.8410E-02	9.2985E-06	298	1000	-2.80	-5.07
76	C2H2Cl2	trans-1,2-DICHLOROETHENE	3.897	-1.7136E-02	8.5157E-06	298	1000	-0.42	-2.54
77	C2H2Cl2O	CHLOROACETYL CHLORIDE	-240.374	-2.0797E-02	1.0711E-05	298	1000	-245.60	-248.09
78	C2H2Cl2O	DICHLOROACETALDEHYDE	--------	--------	--------	--------	--------	-201.00	--------

Table 12-1 ENTHALPY OF FORMATION OF GAS - ORGANIC COMPOUNDS (continued)

NO	FORMULA	NAME	A	B	C	TMIN	TMAX	Hf @ 298 K	Hf @ 500 K
				Hf = A + B T + C T²				(Hf - kjoule/mol, T - K)	
79	C2H2Cl2O2	DICHLOROACETIC ACID	-430.539	-2.1863E-02	1.1519E-05	298	1000	-436.00	-438.59
80	C2H2Cl3F	1,1,1-TRICHLOROFLUOROETHANE	-298.881	-1.4639E-02	1.3075E-05	298	1000	-302.00	-302.93
81	C2H2Cl4	1,1,1,2-TETRACHLOROETHANE	-145.904	-1.6562E-02	1.5380E-05	298	1000	-149.40	-150.34
82	C2H2Cl4	1,1,2,2-TETRACHLOROETHANE	-148.704	-1.8399E-02	1.5508E-05	298	1000	-152.72	-154.03
83	C2H2F2	1,1-DIFLUOROETHENE	-331.214	-2.1989E-02	1.0211E-05	298	1000	-336.81	-339.66
84	C2H2F2	cis-1,2-DIFLUOROETHENE	-316.026	-2.3975E-02	1.0843E-05	298	1000	-322.17	-325.30
85	C2H2F2	trans-1,2-DIFLUOROETHENE	-316.566	-2.1847E-02	9.7968E-06	298	1000	-322.17	-325.04
86	C2H2F4	1,1,1,2-TETRAFLUOROETHANE	-889.229	-2.7639E-02	1.7727E-05	298	1000	-895.79	-898.62
87	C2H2O	KETENE	-58.145	-1.0334E-02	1.7951E-06	298	1000	-61.09	-62.86
88	C2H2O4	OXALIC ACID	-703.242	-6.7002E-02	-5.9332E-06	298	1000	-723.70	-738.23
89	C2H3Br	BROMOETHYLENE	69.965	-2.7833E-02	1.1854E-05	333	1000	78.37	59.01
90	C2H3Cl	CHLOROETHENE	36.302	-3.0397E-02	1.3219E-05	298	1000	28.45	24.41
91	C2H3ClF2	1-CHLORO-1,1-DIFLUOROETHANE	-478.677	-3.3859E-02	1.9281E-05	298	1000	-487.00	-490.79
92	C2H3ClO	ACETYL CHLORIDE	-235.517	-3.2544E-02	1.4337E-05	298	1000	-243.93	-248.20
93	C2H3ClO	CHLOROACETALDEHYDE	-181.747	-3.6186E-02	1.7141E-05	298	1000	-191.00	-195.55
94	C2H3ClO2	CHLOROACETIC ACID	-427.013	-3.7047E-02	1.9987E-05	298	1000	-436.20	-440.54
95	C2H3ClO2	METHYL CHLOROFORMATE	-415.492	-3.4122E-02	1.7768E-05	298	1000	-424.00	-428.11
96	C2H3Cl3	1,1,1-TRICHLOROETHANE	-136.019	-2.6621E-02	1.7816E-05	298	1000	-142.30	-144.88
97	C2H3Cl3	1,1,2-TRICHLOROETHANE	-131.466	-2.9213E-02	1.7876E-05	298	1000	-138.49	-141.60
98	C2H3F	FLUOROETHENE	-130.390	-3.2813E-02	1.3880E-05	298	1000	-138.91	-143.33
99	C2H3F3	1,1,1-TRIFLUOROETHANE	-736.384	-3.7184E-02	2.0074E-05	298	1000	-745.59	-749.96
100	C2H3N	ACETONITRILE	95.935	-3.0757E-02	1.2350E-05	298	1000	87.86	83.64
101	C2H3NO	METHYL ISOCYANATE	-43.761	-5.3946E-02	2.3456E-05	298	1000	-57.78	-64.87
102	C2H4	ETHYLENE	63.053	-4.1076E-02	1.6598E-05	298	1000	52.30	46.66
103	C2H4Br2	1,1-DIBROMOETHANE	-61.112	-4.2545E-02	2.2938E-05	333	1000	-40.80	-76.65
104	C2H4Br2	1,2-DIBROMOETHANE	-61.722	-3.5121E-02	2.2663E-05	333	1000	-38.91	-73.62
105	C2H4Cl2	1,1-DICHLOROETHANE	-119.156	-4.3078E-02	2.2469E-05	298	1000	-129.91	-135.08
106	C2H4Cl2	1,2-DICHLOROETHANE	-119.407	-4.0489E-02	1.9654E-05	298	1000	-129.70	-134.74
107	C2H4Cl2O	BIS(CHLOROMETHYL)ETHER	-228.847	-4.9132E-02	2.6997E-05	298	1000	-241.00	-246.66
108	C2H4F2	1,1-DIFLUOROETHANE	-480.867	-5.0597E-02	2.4519E-05	298	1000	-493.71	-500.04
109	C2H4F2	1,2-DIFLUOROETHANE	-419.096	-4.7903E-02	2.6192E-05	298	1000	-431.00	-436.50
110	C2H4I2	1,2-DIIODOETHANE	12.032	-3.4388E-02	1.7856E-05	459	1000	66.53	-0.70
111	C2H4O	ACETALDEHYDE	-154.122	-4.7166E-02	2.0279E-05	298	1000	-166.36	-172.64
112	C2H4O	ETHYLENE OXIDE	-38.880	-5.4041E-02	2.5601E-05	298	1000	-52.63	-59.50
113	C2H4OS	THIOACETIC ACID	-162.021	-7.8747E-02	3.9862E-05	389	717	-181.96	-191.43
114	C2H4O2	ACETIC ACID	-422.584	-4.8354E-02	2.3337E-05	298	1000	-434.84	-440.93
115	C2H4O2	METHYL FORMATE	-337.586	-4.8154E-02	2.3236E-05	298	1000	-349.78	-355.85
116	C2H4S	THIACYCLOPROPANE	106.583	-9.7328E-02	5.2124E-05	389	717	82.22	70.95
117	C2H5Br	BROMOETHANE	-66.061	-5.3856E-02	2.5352E-05	333	1000	-64.02	-86.65
118	C2H5Cl	CHLOROETHANE	-96.985	-5.7861E-02	2.7636E-05	298	1000	-111.71	-119.01
119	C2H5ClO	2-CHLOROETHANOL	-246.553	-6.0775E-02	2.9503E-05	298	1000	-262.00	-269.56
120	C2H5F	FLUOROETHANE	-245.672	-6.1827E-02	2.8661E-05	298	1000	-261.50	-269.42
121	C2H5I	IODOETHANE	-27.250	-5.0631E-02	2.3385E-05	459	1000	-8.37	-46.72
122	C2H5N	ETHYLENIMINE	140.152	-6.6739E-02	3.4285E-05	298	1000	123.43	115.35
123	C2H5NO	ACETAMIDE	-219.755	-7.1816E-02	3.1395E-05	298	1000	-238.30	-247.81
124	C2H5NO	N-METHYLFORMAMIDE	-163.973	-7.8343E-02	3.1686E-05	298	1000	-184.43	-195.22
125	C2H5NO2	NITROETHANE	-83.513	-7.1714E-02	3.8952E-05	298	1000	-101.25	-109.63
126	C2H5NO3	ETHYL NITRATE	-137.808	-6.7032E-02	4.0944E-05	298	1000	-153.97	-161.09
127	C2H6	ETHANE	-66.735	-6.9337E-02	3.0379E-05	298	1000	-84.68	-93.81
128	C2H6AlCl	DIMETHYLALUMINUM CHLORIDE	--------	--------	--------	--------	--------	--------	--------
129	C2H6O	METHYL ETHER	-165.519	-7.1915E-02	3.2382E-05	298	1000	-184.05	-193.38
130	C2H6O	ETHYL ALCOHOL	-216.961	-6.9572E-02	3.1744E-05	298	1000	-234.81	-243.81
131	C2H6OS	DIMETHYL SULFOXIDE	--------	--------	--------	--------	--------	-209.16	--------
132	C2H6O2	ETHYLENE GLYCOL	-377.811	-4.5844E-02	2.3144E-05	298	1000	-389.32	-394.95
133	C2H6O4S	DIMETHYL SULFATE	--------	--------	--------	--------	--------	-687.00	--------
134	C2H6S	METHYL SULFIDE	-9.822	-1.0937E-01	5.5128E-05	389	717	-37.53	-50.73
135	C2H6S	ETHANETHIOL	-17.746	-1.1264E-01	5.8727E-05	389	717	-46.11	-59.39
136	C2H6S2	METHYL DISULFIDE	13.666	-1.5127E-01	8.1751E-05	389	717	-24.14	-41.53
137	C2H7N	DIMETHYLAMINE	2.793	-8.5680E-02	4.2790E-05	298	1000	-18.83	-29.35
138	C2H7N	ETHYLAMINE	-25.881	-7.9492E-02	3.8730E-05	298	1000	-46.02	-55.94
139	C2H7NO	MONOETHANOLAMINE	-189.419	-8.2034E-02	4.0429E-05	298	1000	-210.19	-220.33
140	C2H8N2	ETHYLENEDIAMINE	6.903	-9.6113E-02	4.8830E-05	298	1000	-17.32	-28.95
141	C2H8Si	DIMETHYL SILANE	-72.731	-8.7220E-02	4.4999E-05	298	1000	-94.70	-105.09
142	C2N2	CYANOGEN	305.353	1.4469E-02	-7.8065E-06	298	1000	308.95	310.64
143	C3F6	HEXAFLUOROPROPYLENE	-1079.177	-1.2017E-02	2.8826E-05	298	1000	-1080.00	-1077.98
144	C3F6O	HEXAFLUOROACETONE	-1456.635	-1.9133E-02	2.3874E-05	298	1000	-1460.00	-1460.23
145	C3F8	OCTAFLUOROPROPANE	-1703.918	-2.9918E-03	1.5769E-05	298	1000	-1703.20	-1701.47
146	C3H2N2	MALONONITRILE	269.564	-1.6055E-02	7.8461E-06	298	1000	265.50	263.50
147	C3H3Cl	PROPARGYL CHLORIDE	163.904	-1.8401E-02	6.7291E-06	298	1000	159.00	156.39
148	C3H3N	ACRYLONITRILE	191.577	-2.5353E-02	1.0174E-05	298	1000	184.93	181.44
149	C3H3NO	OXAZOLE	-2.191	-5.3814E-02	2.9915E-05	298	1000	-15.52	-21.62
150	C3H4	PROPYNE(METHYLACETYLENE)	193.472	-2.9925E-02	1.0181E-05	298	1000	185.43	181.05
151	C3H4	ALLENE(PROPADIENE)	200.497	-3.1684E-02	1.2071E-05	298	1000	192.13	187.67
152	C3H4Cl2	2,3-DICHLOROPROPENE	-38.803	-3.4907E-02	1.4800E-05	298	1000	-47.90	-52.56
153	C3H4O	ACROLEIN	-66.702	-5.9478E-02	3.8716E-05	298	1000	-81.00	-86.76
154	C3H4O	PROPARGYL ALCOHOL	51.038	-3.3751E-02	1.3763E-05	298	1000	42.20	37.60
155	C3H4O2	ACRYLIC ACID	-325.038	-4.4058E-02	2.0926E-05	298	1000	-336.23	-341.84
156	C3H4O2	beta-PROPIOLACTONE	-283.247	-5.5203E-02	2.9171E-05	298	1000	-297.06	-303.56

Table 12-1 ENTHALPY OF FORMATION OF GAS - ORGANIC COMPOUNDS (continued)

NO	FORMULA	NAME	Hf = A + B T + C T²					(Hf - kjoule/mol, T - K)	
			A	B	C	TMIN	TMAX	Hf @ 298 K	Hf @ 500 K
157	C3H4O2	VINYL FORMATE	-249.599	-4.8960E-02	2.3674E-05	298	1000	-262.00	-268.16
158	C3H4O3	ETHYLENE CARBONATE	--------	--------	--------	--------	--------	-506.90	--------
159	C3H4O3	PYRUVIC ACID	-613.439	-4.5879E-02	2.2898E-05	298	1000	-625.00	-630.65
160	C3H5Br	3-BROMO-1-PROPENE	49.836	-5.8078E-02	2.8292E-05	333	1000	49.37	27.87
161	C3H5Cl	2-CHLOROPROPENE	-8.486	-4.7533E-02	1.8967E-05	298	1000	-21.00	-27.51
162	C3H5Cl	3-CHLORO-1-PROPENE	12.752	-5.2020E-02	2.3468E-05	298	1000	-0.63	-7.39
163	C3H5ClO	alpha-EPICHLOROHYDRIN	-93.261	-5.8631E-02	3.1808E-05	298	1000	-107.80	-114.62
164	C3H5ClO2	METHYL CHLOROACETATE	-434.678	-5.7256E-02	2.9657E-05	298	1000	-449.00	-455.89
165	C3H5ClO2	ETHYL CHLOROFORMATE	-442.363	-6.3078E-02	3.5206E-05	298	1000	-458.00	-465.10
166	C3H5Cl3	1,2,3-TRICHLOROPROPANE	-173.439	-4.9723E-02	2.7132E-05	298	1000	-185.77	-191.52
167	C3H5I	3-IODO-1-PROPENE	74.094	-3.9425E-02	1.8529E-05	459	1000	95.81	59.01
168	C3H5N	PROPIONITRILE	64.375	-5.3392E-02	2.4006E-05	298	1000	50.63	43.68
169	C3H5NO	ACRYLAMIDE	-154.424	-6.2668E-02	3.4291E-05	298	1000	-170.00	-177.19
170	C3H5NO	HYDRACRYLONITRILE	-84.095	-5.5973E-02	2.7163E-05	298	1000	-98.30	-105.29
171	C3H5NO	LACTONITRILE	-50.598	-5.3550E-02	2.9943E-05	298	1000	-63.90	-69.89
172	C3H5N3O9	NITROGLYCERINE	--------	--------	--------	--------	--------	-270.90	--------
173	C3H6	CYCLOPROPANE	71.797	-7.2889E-02	3.4947E-05	298	1000	53.30	44.09
174	C3H6	PROPENE	37.334	-6.5191E-02	2.8085E-05	298	1000	20.42	11.76
175	C3H6Br2	1,2-DIBROMOPROPANE	-90.387	-5.6153E-02	3.0781E-05	333	1000	-72.80	-110.77
176	C3H6Cl2	1,1-DICHLOROPROPANE	-134.091	-6.6697E-02	3.5505E-05	298	1000	-150.80	-158.56
177	C3H6Cl2	1,2-DICHLOROPROPANE	-149.373	-6.5321E-02	3.4556E-05	298	1000	-165.69	-173.39
178	C3H6Cl2	1,3-DICHLOROPROPANE	-145.607	-6.3116E-02	3.2029E-05	298	1000	-161.50	-169.16
179	C3H6Cl2	2,2-DICHLOROPROPANE	-162.666	-5.3061E-02	2.9584E-05	298	1000	-175.73	-181.80
180	C3H6I2	1,2-DIIODOPROPANE	-14.124	-5.3752E-02	2.8988E-05	459	1000	35.98	-33.75
181	C3H6O	ACETONE	-199.175	-7.1484E-02	3.2534E-05	298	1000	-217.57	-226.78
182	C3H6O	ALLYL ALCOHOL	-115.102	-6.5898E-02	3.0046E-05	298	1000	-132.01	-140.54
183	C3H6O	METHYL VINYL ETHER	-90.188	-6.8926E-02	3.0825E-05	298	1000	-108.00	-116.94
184	C3H6O	PROPIONALDEHYDE	-175.519	-6.4417E-02	2.9498E-05	298	1000	-192.05	-200.35
185	C3H6O	PROPYLENE OXIDE	-74.450	-7.2182E-02	3.4979E-05	298	1000	-92.76	-101.80
186	C3H6O	1,3-PROPYLENE OXIDE	-52.443	-8.9155E-02	4.4114E-05	298	1000	-75.06	-85.99
187	C3H6O2	ETHYL FORMATE	-368.553	-7.8446E-02	4.0910E-05	298	1000	-388.30	-397.55
188	C3H6O2	METHYL ACETATE	-392.226	-7.0347E-02	3.4601E-05	298	1000	-410.00	-418.75
189	C3H6O2	PROPIONIC ACID	-435.858	-6.9795E-02	3.4769E-05	298	1000	-453.50	-462.06
190	C3H6O2S	3-MERCAPTOPROPIONIC ACID	--------	--------	--------	--------	--------	-405.94	--------
191	C3H6O3	LACTIC ACID	-604.002	-6.7963E-02	3.5454E-05	298	1000	-621.00	-629.12
192	C3H6O3	METHOXYACETIC ACID	-544.610	-7.3188E-02	3.7871E-05	298	1000	-563.00	-571.74
193	C3H6O3	TRIOXANE	-440.640	-1.0212E-01	5.7145E-05	298	1000	-465.90	-477.41
194	C3H6S	THIACYCLOBUTANE	93.860	-1.3110E-01	7.1249E-05	389	717	61.13	46.12
195	C3H7Br	1-BROMOPROPANE	-84.923	-7.3994E-02	3.5315E-05	333	1000	-87.86	-113.09
196	C3H7Br	2-BROMOPROPANE	-95.424	-6.9781E-02	3.4367E-05	333	1000	-97.07	-121.72
197	C3H7Cl	2-CHLOROPROPANE	-127.432	-7.5446E-02	3.7838E-05	298	1000	-146.44	-155.70
198	C3H7Cl	1-CHLOROPROPANE	-110.186	-7.8496E-02	3.7907E-05	298	1000	-130.12	-139.96
199	C3H7F	1-FLUOROPROPANE	-260.425	-8.1437E-02	3.8952E-05	298	1000	-281.16	-291.41
200	C3H7F	2-FLUOROPROPANE	-268.207	-8.0772E-02	3.9205E-05	298	1000	-288.70	-298.79
201	C3H7I	2-IODOPROPANE	-57.697	-6.4852E-02	3.1455E-05	459	1000	-41.84	-82.26
202	C3H7I	1-IODOPROPANE	-45.451	-6.7567E-02	3.2127E-05	459	1000	-30.54	-71.20
203	C3H7N	ALLYLAMINE	77.446	-7.5418E-02	3.6210E-05	298	1000	58.20	48.79
204	C3H7N	PROPYLENEIMINE	110.468	-8.6958E-02	4.6267E-05	298	1000	88.80	78.56
205	C3H7NO	N,N-DIMETHYLFORMAMIDE	-170.316	-8.5913E-02	4.6677E-05	298	1000	-191.70	-201.60
206	C3H7NO	N-METHYLACETAMIDE	-212.192	-1.0811E-01	4.9242E-05	298	TMAX	-240.00	-253.94
207	C3H7NO2	1-NITROPROPANE	-102.019	-9.1341E-02	4.9417E-05	298	1000	-124.68	-135.33
208	C3H7NO2	2-NITROPROPANE	-117.759	-9.0617E-02	4.9614E-05	298	1000	-140.16	-150.66
209	C3H7NO3	PROPYL NITRATE	-153.072	-8.6320E-02	5.1133E-05	298	1000	-174.05	-183.45
210	C3H7NO3	ISOPROPYL NITRATE	-170.180	-8.6036E-02	5.1718E-05	298	1000	-191.00	-200.27
211	C3H8	PROPANE	-80.697	-9.0500E-02	4.2104E-05	298	1000	-103.85	-115.42
212	C3H8O	ISOPROPYL ALCOHOL	-250.362	-8.7902E-02	4.3171E-05	298	1000	-272.59	-283.52
213	C3H8O	ETHYL METHYL ETHER	-193.083	-9.1203E-02	4.2572E-05	298	1000	-216.44	-228.04
214	C3H8O	PROPYL ALCOHOL	-233.953	-9.2123E-02	4.2848E-05	298	1000	-257.53	-269.30
215	C3H8O2	2-METHOXYETHANOL	-409.537	-9.6306E-02	4.7007E-05	298	1000	-434.00	-445.94
216	C3H8O2	METHYLAL	-321.847	-1.0370E-01	4.9483E-05	298	1000	-348.20	-361.33
217	C3H8O2	1,2-PROPYLENE GLYCOL	-398.229	-9.1676E-02	4.4602E-05	298	1000	-421.50	-432.92
218	C3H8O2	1,3-PROPYLENE GLYCOL	-368.415	-9.2906E-02	4.3969E-05	298	1000	-392.10	-403.88
219	C3H8O3	GLYCEROL	-559.438	-9.2185E-02	4.5003E-05	298	1000	-582.80	-594.28
220	C3H8S	1-PROPANETHIOL	-33.528	-1.3716E-01	7.3710E-05	389	717	-67.86	-83.68
221	C3H8S	2-PROPANETHIOL	-42.790	-1.3363E-01	7.1781E-05	389	717	-76.23	-91.66
222	C3H8S	ETHYL METHYL SULFIDE	-25.630	-1.3522E-01	7.1049E-05	389	717	-59.62	-75.48
223	C3H9N	PROPYLAMINE	-47.070	-1.0008E-01	4.9480E-05	298	1000	-72.38	-84.74
224	C3H9N	ISOPROPYLAMINE	-57.723	-1.0310E-01	5.1500E-05	298	1000	-83.80	-96.40
225	C3H9N	TRIMETHYLAMINE	2.507	-1.0526E-01	5.4501E-05	298	1000	-23.85	-36.50
226	C3H9NO	1-AMINO-2-PROPANOL	-213.563	-1.0091E-01	5.0937E-05	298	1000	-239.00	-251.28
227	C3H9NO	3-AMINO-1-PROPANOL	-194.322	-1.0541E-01	5.2327E-05	298	1000	-221.00	-233.94
228	C3H9NO	METHYLETHANOLAMINE	-170.349	-1.0995E-01	5.6329E-05	298	1000	-198.00	-211.24
229	C3H9O4P	TRIMETHYL PHOSPHATE	--------	--------	--------	--------	--------	-1080.00	--------
230	C3H10N2	1,2-PROPANEDIAMINE	-24.728	-1.1540E-01	5.9976E-05	298	1000	-53.68	-67.44
231	C3H10Si	TRIMETHYL SILANE	-126.965	-1.1449E-01	5.2311E-05	298	1000	-156.61	-171.13
232	C4Cl4S	TETRACHLOROTHIOPHENE	--------	--------	--------	--------	--------	--------	--------
233	C4Cl6	HEXACHLORO-1,3-BUTADIENE	-55.274	3.3435E-02	-1.5510E-05	298	1000	-46.74	-42.43
234	C4F8	OCTAFLUORO-2-BUTENE	-1633.398	-5.3022E-02	-9.6538E-06	298	1000	-1650.00	-1662.32

293

Table 12-1 ENTHALPY OF FORMATION OF GAS - ORGANIC COMPOUNDS (continued)

NO	FORMULA	NAME	Hf = A + B T + C T^2					(Hf - kjoule/mol, T - K)	
			A	B	C	TMIN	TMAX	Hf @ 298 K	Hf @ 500 K
235	C4F8	OCTAFLUOROCYCLOBUTANE	-1528.290	-4.1673E-03	1.5385E-05	298	1000	-1528.00	-1526.53
236	C4F10	DECAFLUOROBUTANE	-2141.526	-2.4704E-03	2.1936E-05	298	1000	-2140.00	-2137.28
237	C4H2	BUTADIYNE(BIACETYLENE)	469.152	1.6064E-02	-1.2284E-05	298	1000	472.79	474.11
238	C4H2O3	MALEIC ANHYDRIDE	-401.302	6.6610E-03	1.0283E-05	298	1000	-398.30	-395.40
239	C4H4	1-BUTEN-3-YNE(VINYLACETYLENE)	310.867	-2.3259E-02	7.4567E-06	298	1000	304.60	301.10
240	C4H4N2	SUCCINONITRILE	218.772	-3.6015E-02	1.8239E-05	298	1000	209.70	205.32
241	C4H4O	FURAN	-21.889	-5.0643E-02	2.4459E-05	298	1000	-34.69	-41.10
242	C4H4O2	DIKETENE	-178.774	-4.5275E-02	2.2964E-05	298	1000	-190.20	-195.67
243	C4H4O3	SUCCINIC ANHYDRIDE	-517.026	-2.9042E-02	1.7456E-05	298	1000	-524.10	-527.18
244	C4H4O4	FUMARIC ACID	-669.983	-1.2032E-02	1.6860E-05	298	1000	-671.95	-671.78
245	C4H4O4	MALEIC ACID	-664.332	-4.6007E-02	2.3844E-05	298	1000	-675.80	-681.37
246	C4H4S	THIOPHENE	139.428	-9.5701E-02	5.4074E-05	389	717	115.73	105.10
247	C4H5Cl	CHLOROPRENE	82.850	-3.7410E-02	1.4699E-05	298	1000	73.01	67.82
248	C4H5N	trans-CROTONITRILE	154.054	-5.1864E-02	2.3129E-05	298	1000	140.70	133.90
249	C4H5N	cis-CROTONITRILE	157.898	-5.6133E-02	2.4696E-05	298	1000	143.40	136.00
250	C4H5N	METHACRYLONITRILE	109.101	-4.3431E-02	2.1234E-05	298	1000	98.03	92.69
251	C4H5N	PYRROLE	124.871	-6.6624E-02	3.6192E-05	298	1000	108.28	100.61
252	C4H5N	VINYLACETONITRILE	170.086	-4.7387E-02	2.0818E-05	298	1000	157.80	151.60
253	C4H5NO2	METHYL CYANOACETATE	-223.292	-5.4738E-02	2.7677E-05	298	1000	-237.00	-243.74
254	C4H6	CYCLOBUTENE	147.570	-7.0270E-02	3.3478E-05	298	1000	129.70	120.80
255	C4H6	1,2-BUTADIENE	176.565	-5.5166E-02	2.3429E-05	298	1000	162.21	154.84
256	C4H6	1,3-BUTADIENE	123.286	-5.1225E-02	2.3192E-05	298	1000	110.16	103.47
257	C4H6	2-BUTYNE(DIMETHYLACETYLENE)	161.942	-5.9413E-02	2.3730E-05	298	1000	146.31	138.17
258	C4H6	1-BUTYNE(ETHYLACETYLENE)	178.841	-5.2246E-02	2.1448E-05	298	1000	165.18	158.08
259	C4H6Cl2	1,3-DICHLORO-trans-2-BUTENE	-64.256	-5.6920E-02	2.5291E-05	298	1000	-79.00	-86.42
260	C4H6Cl2	1,4-DICHLORO-cis-2-BUTENE	92.360	-6.0200E-02	4.1816E-05	298	1000	78.20	72.71
261	C4H6Cl2	1,4-DICHLORO-trans-2-BUTENE	-349.980	-7.0973E-02	3.4678E-05	298	1000	-66.20	-376.80
262	C4H6Cl2	3,4-DICHLORO-1-BUTENE	-48.316	-6.3449E-02	3.2432E-05	298	1000	-64.40	-71.93
263	C4H6O	trans-CROTONALDEHYDE	-87.770	-6.7073E-02	4.6761E-05	298	1000	-103.60	-109.62
264	C4H6O	2,5-DIHYDROFURAN	-88.640	-7.9747E-02	4.0185E-05	298	1000	-108.78	-118.47
265	C4H6O	DIVINYL ETHER	2.876	-6.5314E-02	2.8892E-05	298	1000	-14.00	-22.56
266	C4H6O	METHACROLEIN	-90.882	-8.6810E-02	5.2829E-05	298	1000	-112.00	-121.08
267	C4H6O2	2-BUTYNE-1,4-DIOL	-139.359	-6.0231E-02	2.5876E-05	298	1000	-155.00	-163.01
268	C4H6O2	gamma-BUTYROLACTONE	-355.977	-9.2433E-02	5.0472E-05	298	1000	-379.00	-389.57
269	C4H6O2	cis-CROTONIC ACID	-344.928	-5.6717E-02	3.1101E-05	298	1000	-359.00	-365.51
270	C4H6O2	trans-CROTONIC ACID	-351.683	-6.6601E-02	3.1687E-05	298	1000	-368.65	-377.06
271	C4H6O2	METHACRYLIC ACID	--------	--------	--------	--------	--------	-367.94	--------
272	C4H6O2	METHYL ACRYLATE	-316.034	-6.6931E-02	3.2668E-05	298	1000	-333.00	-341.33
273	C4H6O2	VINYL ACETATE	-296.867	-7.5158E-02	3.9850E-05	298	1000	-315.70	-324.48
274	C4H6O3	ACETIC ANHYDRIDE	-554.715	-8.4124E-02	4.3618E-05	298	1000	-575.72	-585.87
275	C4H6O4	SUCCINIC ACID	-808.058	-5.9654E-02	3.1941E-05	298	1000	-822.90	-829.90
276	C4H6O5	DIGLYCOLIC ACID	-927.422	-7.1564E-02	4.0228E-05	298	1000	-945.00	-953.15
277	C4H6O5	MALIC ACID	-975.099	-6.0811E-02	3.4441E-05	298	1000	-990.00	-996.89
278	C4H6O6	TARTARIC ACID	-1144.745	-6.3181E-02	3.7749E-05	298	1000	-1160.00	-1166.90
279	C4H7N	BUTYRONITRILE	52.413	-7.2015E-02	3.4304E-05	298	1000	34.06	24.98
280	C4H7N	ISOBUTYRONITRILE	43.551	-7.1546E-02	3.4701E-05	298	1000	25.40	16.45
281	C4H7NO	ACETONE CYANOHYDRIN	-115.775	-6.8364E-02	3.4742E-05	298	1000	-133.00	-141.27
282	C4H7NO	2-METHACRYLAMIDE	-165.561	-7.9472E-02	4.7169E-05	298	1000	-185.00	-193.50
283	C4H7NO	3-METHOXYPROPIONITRILE	-59.166	-7.8811E-02	4.0336E-05	298	1000	-79.00	-88.49
284	C4H7NO	2-PYRROLIDONE	-169.984	-1.0937E-01	6.0365E-05	298	1000	-197.15	-209.58
285	C4H8	1-BUTENE	21.822	-8.5458E-02	3.8902E-05	298	1000	-0.13	-11.18
286	C4H8	cis-2-BUTENE	17.621	-9.5445E-02	4.2622E-05	298	1000	-6.99	-19.45
287	C4H8	trans-2-BUTENE	10.652	-8.4475E-02	3.7370E-05	298	1000	-11.17	-22.24
288	C4H8	CYCLOBUTANE	52.777	-1.0357E-01	5.1744E-05	298	1000	26.65	13.93
289	C4H8	2-METHYLPROPENE	4.126	-8.1626E-02	3.6619E-05	298	1000	-16.90	-27.53
290	C4H8Br2	1,2-DIBROMOBUTANE	-112.190	-7.4032E-02	3.8801E-05	333	1000	-99.16	-139.51
291	C4H8Br2	2,3-DIBROMOBUTANE	-115.211	-7.4776E-02	3.9997E-05	333	1000	-102.09	-142.60
292	C4H8Cl2	1,4-DICHLOROBUTANE	-156.919	-8.8065E-02	4.6056E-05	298	1000	-179.00	-189.44
293	C4H8I2	1,2-DIIODOBUTANE	-33.628	-7.1588E-02	3.6983E-05	459	1000	11.92	-60.18
294	C4H8O	BUTYRALDEHYDE	-183.623	-8.3856E-02	3.9792E-05	298	1000	-205.02	-215.60
295	C4H8O	ISOBUTYRALDEHYDE	-192.599	-9.0525E-02	4.4390E-05	298	1000	-215.60	-226.76
296	C4H8O	1,2-EPOXYBUTANE	-86.351	-9.3268E-02	4.6178E-05	298	1000	-110.00	-121.44
297	C4H8O	2-BUTANONE	-216.593	-8.4826E-02	3.9084E-05	298	1000	-238.36	-249.24
298	C4H8O	ETHYL VINYL ETHER	-117.703	-9.0593E-02	4.2490E-05	298	1000	-140.90	-152.38
299	C4H8O	TETRAHYDROFURAN	-151.894	-1.2820E-01	6.6590E-05	298	1000	-184.18	-199.35
300	C4H8O2	cis-2-BUTENE-1,4-DIOL	-269.243	-1.0127E-01	4.8880E-05	298	1000	-295.00	-307.66
301	C4H8O2	trans-2-BUTENE-1,4-DIOL	-275.953	-9.4568E-02	4.5982E-05	298	1000	-300.00	-311.74
302	C4H8O2	ISOBUTYRIC ACID	-459.626	-9.7491E-02	4.9550E-05	298	1000	-484.13	-495.98
303	C4H8O2	n-BUTYRIC ACID	-448.775	-8.5107E-02	4.2361E-05	298	1000	-470.28	-480.74
304	C4H8O2	p-DIOXANE	-286.843	-1.1418E-01	6.2626E-05	298	1000	-315.06	-328.28
305	C4H8O2	ETHYL ACETATE	-420.199	-8.9885E-02	4.4497E-05	298	1000	-442.92	-454.02
306	C4H8O2	METHYL PROPIONATE	-406.792	-8.1945E-02	4.0486E-05	298	1000	-427.50	-437.64
307	C4H8O2	n-PROPYL FORMATE	-381.610	-9.0464E-02	4.5787E-05	298	1000	-404.38	-415.40
308	C4H8O2S	SULFOLANE	--------	--------	--------	--------	--------	-390.00	--------
309	C4H8S	THIACYCLOPENTANE	4.529	-1.5419E-01	8.5487E-05	389	717	-33.81	-51.19
310	C4H9Br	1-BROMOBUTANE	-99.663	-9.3148E-02	4.5175E-05	333	1000	-107.32	-134.94
311	C4H9Br	2-BROMOBUTANE	-113.161	-9.1316E-02	4.6175E-05	333	1000	-120.08	-147.27
312	C4H9Cl	1-CHLOROBUTANE	-122.282	-9.8696E-02	4.8547E-05	298	1000	-147.28	-159.49

Table 12-1 ENTHALPY OF FORMATION OF GAS - ORGANIC COMPOUNDS (continued)

NO	FORMULA	NAME	$Hf = A + B T + C T^2$					(Hf - kjoule/mol, T - K)	
			A	B	C	TMIN	TMAX	Hf @ 298 K	Hf @ 500 K
313	C4H9Cl	2-CHLOROBUTANE	-136.658	-9.8613E-02	4.9869E-05	298	1000	-161.50	-173.50
314	C4H9Cl	2-CHLORO-2-METHYLPROPANE	-160.666	-8.9981E-02	4.6275E-05	298	1000	-183.26	-194.09
315	C4H9I	2-IODO-2-METHYLPROPANE	-88.738	-7.1496E-02	3.7656E-05	459	1000	-73.64	-115.07
316	C4H9N	PYRROLIDINE	28.931	-1.2990E-01	6.7199E-05	298	1000	-3.60	-19.22
317	C4H9NO	N,N-DIMETHYLACETAMIDE	-196.111	-1.1651E-01	6.4425E-05	298	1000	-225.00	-238.26
318	C4H9NO	MORPHOLINE	-130.717	-1.0313E-01	5.9259E-05	298	1000	-156.00	-167.47
319	C4H9NO2	1-NITROBUTANE	-116.278	-1.1129E-01	5.9827E-05	298	1000	-143.93	-156.97
320	C4H9NO2	2-NITROBUTANE	-135.562	-1.1335E-01	6.2241E-05	298	1000	-163.59	-176.68
321	C4H10	BUTANE	-98.186	-1.0974E-01	5.2254E-05	298	1000	-126.15	-139.99
322	C4H10	2-METHYLPROPANE(ISOBUTANE)	-106.746	-1.0929E-01	5.2693E-05	298	1000	-134.52	-148.22
323	C4H10N2	PIPERAZINE	49.662	-1.1387E-01	6.9865E-05	298	1000	22.30	10.19
324	C4H10O	BUTYL ALCOHOL	-245.806	-1.1235E-01	5.3505E-05	298	1000	-274.43	-288.61
325	C4H10O	sec- BUTYL ALCOHOL	-265.140	-1.0695E-01	5.1693E-05	298	1000	-292.29	-305.69
326	C4H10O	tert-BUTYL ALCOHOL	-299.190	-1.0543E-01	5.2333E-05	298	1000	-325.81	-338.82
327	C4H10O	ETHYL ETHER	-223.739	-1.1173E-01	5.3459E-05	298	1000	-252.21	-266.24
328	C4H10O	METHYL PROPYL ETHER	-209.263	-1.1173E-01	5.3459E-05	298	1000	-237.73	-251.76
329	C4H10O	METHYL ISOPROPYL ETHER	-223.267	-1.1345E-01	5.5605E-05	298	1000	-252.04	-266.09
330	C4H10O2	1,4-BUTANEDIOL	-397.935	-1.1311E-01	5.4556E-05	298	1000	-426.70	-440.85
331	C4H10O	ISOBUTANOL	-230.711	-2.1771E-01	1.5801E-04	298	1000	-283.22	-300.06
332	C4H10O2	1,3-BUTANEDIOL	-406.820	-1.0942E-01	5.3306E-05	298	1000	-434.60	-448.20
333	C4H10O2	2,3-BUTANEDIOL	-455.344	-1.0701E-01	5.4215E-05	298	1000	-482.30	-495.30
334	C4H10O2	t-BUTYL HYDROPEROXIDE	-217.370	-1.0241E-01	5.3817E-05	298	1000	-243.00	-255.12
335	C4H10O2	1,2-DIMETHOXYETHANE	-316.235	-1.1770E-01	5.9109E-05	298	1000	-346.00	-360.33
336	C4H10O2	2-ETHOXYETHANOL	-370.303	-1.1741E-01	5.8390E-05	298	1000	-400.00	-414.41
337	C4H10O3	DIETHYLENE GLYCOL	-541.246	-1.1850E-01	5.9181E-05	298	1000	-571.20	-585.70
338	C4H10O4S	DIETHYL SULFATE	--------	--------	--------	--------	--------	-756.30	--------
339	C4H10S	1-BUTANETHIOL	-48.091	-1.6072E-01	8.9090E-05	389	717	-88.07	-106.18
340	C4H10S	2-METHYL-1-PROPANETHIOL	-58.538	-1.5467E-01	8.3048E-05	389	717	-97.24	-115.11
341	C4H10S	2-BUTANETHIOL	-57.467	-1.5522E-01	8.4186E-05	389	717	-96.23	-114.03
342	C4H10S	2-METHYL-2-PROPANETHIOL	-71.036	-1.5550E-01	8.8471E-05	389	717	-109.50	-126.67
343	C4H10S	ETHYLSULFIDE	-43.401	-1.6052E-01	8.7366E-05	389	717	-83.47	-101.82
344	C4H10S	ISOPROPYL METHYL SULFIDE	-50.442	-1.5998E-01	8.6672E-05	389	717	-90.42	-108.76
345	C4H10S	METHYL PROPYL SULFIDE	-41.907	-1.5918E-01	8.5462E-05	389	717	-81.76	-100.13
346	C4H10S2	ETHYL DISULFIDE	-26.606	-1.9330E-01	1.0753E-04	389	717	-74.64	-96.37
347	C4H11N	BUTYLAMINE	-61.681	-1.2030E-01	6.0112E-05	298	1000	-92.05	-106.80
348	C4H11N	ISOBUTYLAMINE	-67.021	-1.2579E-01	6.3331E-05	298	1000	-98.80	-114.08
349	C4H11N	sec-BUTYLAMINE	-73.525	-1.2192E-01	6.2124E-05	298	1000	-104.18	-118.96
350	C4H11N	tert-BUTYLAMINE	-90.768	-1.1679E-01	6.1840E-05	298	1000	-119.87	-133.70
351	C4H11N	DIETHYLAMINE	-40.873	-1.2518E-01	6.3559E-05	298	1000	-72.38	-87.57
352	C4H11NO	DIMETHYLETHANOLAMINE	-169.518	-1.2980E-01	6.8012E-05	298	1000	-202.00	-217.42
353	C4H11NO2	DIETHANOLAMINE	-364.502	-1.2878E-01	6.5960E-05	298	1000	-396.88	-412.40
354	C4H11NO2	2-AMINOETHOXYETHANOL	-332.434	-1.2951E-01	6.6326E-05	298	1000	-365.00	-380.61
355	C4H12N2O	N-AMINOETHYL ETHANOLAMINE	-155.270	-1.4295E-01	7.5358E-05	298	1000	-191.00	-207.91
356	C4H12Si	TETRAMETHYLSILANE	-202.738	-1.1908E-01	5.7340E-05	298	1000	-233.20	-247.94
357	C4H13N3	DIETHYLENE TRIAMINE	32.596	-1.5425E-01	8.2523E-05	298	1000	-5.86	-23.90
358	C5Cl6	HEXACHLOROCYCLOPENTADIENE	-111.938	3.9599E-02	-1.9629E-05	298	1000	-102.00	-97.05
359	C5H4O2	FURFURAL	-140.171	-4.4791E-02	2.7121E-05	298	1000	-151.04	-155.79
360	C5H5N	PYRIDINE	156.938	-6.7082E-02	3.4853E-05	298	1000	140.16	132.11
361	C5H6	CYCLOPENTADIENE	148.605	-7.0020E-02	3.3874E-05	298	1000	130.80	122.06
362	C5H6	2-METHYL-1-BUTENE-3-YNE	271.939	-4.5969E-02	1.9940E-05	298	1000	260.00	253.94
363	C5H6	1-PENTENE-3-YNE	262.414	-5.0951E-02	2.0259E-05	298	1000	249.00	242.00
364	C5H6	1-PENTENE-4-YNE	281.625	-4.7767E-02	1.8528E-05	298	1000	269.00	262.37
365	C5H6N2	GLUTARONITRILE	184.345	-5.7088E-02	2.9493E-05	298	1000	170.00	163.17
366	C5H6O2	FURFURYL ALCOHOL	-206.139	-5.2240E-02	3.0325E-05	298	1000	-218.90	-224.68
367	C5H6O3	GLUTARIC ANHYDRIDE	-327.190	-9.6097E-02	7.4816E-05	298	1000	-349.00	-356.53
368	C5H6O4	CITRACONIC ACID	-729.829	-4.4550E-02	3.2902E-05	298	1000	-740.00	-743.88
369	C5H6O4	ITACONIC ACID	-716.230	-5.3713E-02	3.4432E-05	298	1000	-729.00	-734.48
370	C5H6S	2-METHYLTHIOPHENE	112.972	-1.1734E-01	6.3660E-05	389	717	83.68	70.22
371	C5H6S	3-METHYLTHIOPHENE	112.002	-1.1660E-01	6.2199E-05	389	717	82.80	69.25
372	C5H7N	N-METHYLPYRROLE	124.327	-8.7147E-02	5.2134E-05	298	1000	103.10	93.79
373	C5H7NO2	ETHYL CYANOACETATE	-296.787	-7.7557E-02	4.0360E-05	298	1000	-316.14	-325.48
374	C5H8	CYCLOPENTENE	61.367	-1.1168E-01	5.3120E-05	298	1000	32.93	18.81
375	C5H8	ISOPRENE	95.122	-7.6059E-02	3.6732E-05	298	1000	75.73	66.28
376	C5H8	3-METHYL-1,2-BUTADIENE	148.239	-7.1898E-02	3.2194E-05	298	1000	129.70	120.34
377	C5H8	2-METHYL-1,3-BUTADIENE	93.061	-6.8053E-02	3.2019E-05	298	1000	75.73	67.04
378	C5H8	1,2-PENTADIENE	163.645	-7.0396E-02	3.2479E-05	298	1000	145.60	136.57
379	C5H8	cis-1,3-PENTADIENE	99.111	-8.1697E-02	3.7863E-05	298	1000	78.24	67.73
380	C5H8	trans-1,3-PENTADIENE	95.901	-7.0697E-02	3.2654E-05	298	1000	77.82	68.72
381	C5H8	1,4-PENTADIENE	123.470	-7.0304E-02	3.2185E-05	298	1000	105.44	96.36
382	C5H8	2,3-PENTADIENE	158.716	-7.8153E-02	3.4391E-05	298	1000	138.49	128.24
383	C5H8	1-PENTYNE	162.427	-6.9831E-02	3.0589E-05	298	1000	144.35	135.16
384	C5H8	2-PENTYNE	150.017	-8.1341E-02	3.4696E-05	298	1000	128.87	118.02
385	C5H8	3-METHYL-1-BUTYNE	154.737	-7.1337E-02	3.2282E-05	298	1000	136.40	127.14
386	C5H8	SPIROPENTANE	207.936	-9.2182E-02	4.7288E-05	298	1000	185.23	173.67
387	C5H8N4O12	PENTAERYTHRITOL TETRANITRATE	--------	--------	--------	--------	--------	-386.70	--------
388	C5H8O	CYCLOPENTANONE	-167.028	-1.0662E-01	5.1669E-05	298	1000	-194.14	-207.42
389	C5H8O	METHYL ISOPROPENYL KETONE	-156.293	-9.2569E-02	4.3139E-05	298	1000	-180.00	-191.79
390	C5H8O2	ACETYLACETONE	-354.342	-1.0076E-01	4.8832E-05	298	1000	-380.00	-392.52

295

Table 12-1 ENTHALPY OF FORMATION OF GAS - ORGANIC COMPOUNDS (continued)

NO	FORMULA	NAME	Hf = A + B T + C T²					(Hf - kjoule/mol, T - K)	
			A	B	C	TMIN	TMAX	Hf @ 298 K	Hf @ 500 K
391	C5H8O2	ALLYL ACETATE	-312.068	-8.7297E-02	4.4419E-05	298	1000	-334.00	-344.61
392	C5H8O2	ETHYL ACRYLATE	-327.044	-8.9230E-02	4.4970E-05	298	1000	-349.53	-360.42
393	C5H8O2	METHYL METHACRYLATE	-323.788	-9.3438E-02	4.6603E-05	298	1000	-347.36	-358.86
394	C5H8O2	VINYL PROPIONATE	-324.393	-9.0835E-02	4.4998E-05	298	1000	-347.36	-358.56
395	C5H8O3	2-HYDROXYETHYL ACRYLATE	-475.771	-7.0121E-02	3.9480E-05	298	1000	-493.00	-500.96
396	C5H8O3	LEVULINIC ACID	-586.125	-8.2480E-02	4.1140E-05	298	1000	-607.00	-617.08
397	C5H8O3	METHYL ACETOACETATE	-564.498	-9.7275E-02	4.9034E-05	298	1000	-589.00	-600.88
398	C5H8O4	GLUTARIC ACID	-824.047	-7.9998E-02	4.2638E-05	298	1000	-844.00	-853.39
399	C5H9N	VALERONITRILE	35.032	-9.2977E-02	4.6419E-05	298	1000	11.46	0.15
400	C5H9NO	n-BUTYL ISOCYANATE	-97.348	-1.1650E-01	5.6772E-05	298	1000	-127.00	-141.40
401	C5H9NO	N-METHYL-2-PYRROLIDONE	-161.941	-1.3543E-01	7.5490E-05	298	1000	-195.39	-210.79
402	C5H9NO4	L-GLUTAMIC ACID	-800.490	-9.5572E-02	5.3904E-05	298	1000	-824.00	-834.80
403	C5H10	CYCLOPENTANE	-42.028	-1.3962E-01	6.9856E-05	298	1000	-77.24	-94.37
404	C5H10	2-METHYL-1-BUTENE	-9.632	-1.0462E-01	4.9551E-05	298	1000	-36.32	-49.56
405	C5H10	2-METHYL-2-BUTENE	-13.966	-1.1128E-01	5.0823E-05	298	1000	-42.55	-56.90
406	C5H10	3-METHYL-1-BUTENE	-5.470	-9.1956E-02	4.3447E-05	298	1000	-28.95	-40.59
407	C5H10	1-PENTENE	5.866	-1.0485E-01	4.9179E-05	298	1000	-20.92	-34.26
408	C5H10	cis-2-PENTENE	1.075	-1.1401E-01	5.3124E-05	298	1000	-28.07	-42.65
409	C5H10	trans-2-PENTENE	-4.418	-1.0655E-01	4.8936E-05	298	1000	-31.76	-45.46
410	C5H10Br2	2,3-DIBROMO-2-METHYLBUTANE	-148.909	-8.9584E-02	4.9760E-05	333	1000	-138.91	-181.26
411	C5H10Cl2	1,5-DICHLOROPENTANE	-172.715	-1.0872E-01	5.6848E-05	298	1000	-200.00	-212.86
412	C5H10O	METHYL ISOPROPYL KETONE	-231.643	-1.2126E-01	5.9502E-05	298	1000	-262.40	-277.40
413	C5H10O	2-PENTANONE	-231.170	-1.0792E-01	5.1409E-05	298	1000	-258.65	-272.28
414	C5H10O	DIETHYL KETONE	-230.936	-1.1046E-01	5.2823E-05	298	1000	-259.20	-272.96
415	C5H10O	VALERALDEHYDE	-201.340	-1.0410E-01	5.0476E-05	298	1000	-227.82	-240.81
416	C5H10O2	n-BUTYL FORMATE	-397.195	-1.1090E-01	5.6535E-05	298	1000	-425.09	-438.51
417	C5H10O2	ETHYL PROPIONATE	-435.295	-1.1241E-01	5.6559E-05	298	1000	-463.60	-477.36
418	C5H10O2	ISOBUTYL FORMATE	-408.333	-1.1160E-01	5.7951E-05	298	1000	-436.30	-449.65
419	C5H10O2	ISOPROPYL ACETATE	-454.193	-1.0992E-01	5.7018E-05	298	1000	-481.70	-494.90
420	C5H10O2	n-PROPYL ACETATE	-436.416	-1.1228E-01	5.6405E-05	298	1000	-464.80	-478.46
421	C5H10O2	METHYL n-BUTYRATE	-423.818	-1.0645E-01	5.3139E-05	298	1000	-450.70	-463.76
422	C5H10O2	2-METHYLBUTYRIC ACID	-468.168	-1.1883E-01	6.1155E-05	298	1000	-498.00	-512.29
423	C5H10O2	ISOVALERIC ACID	-487.519	-1.0801E-01	5.5144E-05	298	1000	-514.70	-527.74
424	C5H10O2	VALERIC ACID	-464.094	-1.0397E-01	5.1824E-05	298	1000	-490.36	-503.12
425	C5H10O2	TETRAHYDROFURFURYL ALCOHOL	-343.497	-1.0404E-01	5.7904E-05	298	1000	-369.20	-381.04
426	C5H10O2S	3-METHYL SULFOLANE	--------	--------	--------	--------	--------	-420.00	--------
427	C5H10O3	DIETHYL CARBONATE	--------	--------	--------	--------	--------	-639.10	--------
428	C5H10O3	ETHYL LACTATE	-607.726	-1.0892E-01	5.6411E-05	298	1000	-635.00	-648.09
429	C5H10S	THIACYCLOHEXANE	-14.743	-2.0139E-01	1.2916E-04	389	717	-63.26	-83.15
430	C5H10S	CYCLOPENTANETHIOL	-1.680	-1.8602E-01	1.0342E-04	389	717	-47.91	-68.83
431	C5H11Br	1-BROMOPENTANE	-116.918	-1.1199E-01	5.4890E-05	333	1000	-129.16	-159.19
432	C5H11Cl	1-CHLOROPENTANE	-144.900	-1.1865E-01	5.8902E-05	298	1000	-174.89	-189.49
433	C5H11Cl	1-CHLORO-3-METHYLBUTANE	-151.716	-1.1363E-01	5.7367E-05	298	1000	-180.33	-194.19
434	C5H11Cl	2-CHLORO-2-METHYLBUTANE	-173.272	-1.1676E-01	6.0522E-05	298	1000	-202.51	-216.52
435	C5H11N	N-METHYLPYRROLIDINE	30.001	-1.5266E-01	8.0839E-05	298	1000	-8.12	-26.12
436	C5H11N	PIPERIDINE	-20.220	-1.1768E-01	6.8458E-05	298	1000	-48.90	-61.95
437	C5H11NO	tert-BUTYLFORMAMIDE	-237.838	-1.3552E-01	6.7413E-05	298	1000	-284.00	-288.74
438	C5H12	2-METHYLBUTANE(ISOPENTANE)	-121.118	-1.3184E-01	6.5174E-05	298	1000	-154.47	-170.74
439	C5H12	2,2-DIMETHYPROPANE	-134.244	-1.2632E-01	6.4480E-05	298	1000	-165.98	-181.29
440	C5H12	PENTANE	-113.399	-1.3001E-01	6.2902E-05	298	1000	-146.44	-162.68
441	C5H12O	2,2-DIMETHYL-1-PROPANOL	-286.047	-1.3153E-01	6.7762E-05	298	1000	-319.07	-334.87
442	C5H12O	tert-PENTYL ALCOHOL	-295.763	-1.3221E-01	6.6283E-05	298	1000	-329.07	-345.30
443	C5H12O	2-METHYL-1-BUTANOL	-268.146	-1.3391E-01	6.5934E-05	298	1000	-302.09	-318.62
444	C5H12O	2-METHYL-2-BUTANOL	-297.235	-1.2864E-01	6.4967E-05	298	1000	-329.70	-345.31
445	C5H12O	3-METHYL-1-BUTANOL	-267.530	-1.3638E-01	6.7266E-05	298	1000	-302.09	-318.90
446	C5H12O	3-METHYL-2-BUTANOL	-281.226	-1.3036E-01	6.4788E-05	298	1000	-314.22	-330.21
447	C5H12O	PENTYL ALCOHOL	-268.776	-1.3225E-01	6.3906E-05	298	1000	-302.38	-318.93
448	C5H12O	2-PENTANOL	-280.934	-1.2960E-01	6.3885E-05	298	1000	-313.80	-329.76
449	C5H12O	3-PENTANOL	-283.523	-1.3111E-01	6.4913E-05	298	1000	-316.73	-332.85
450	C5H12O	METHYL sec-BUTYL ETHER	-241.893	-1.3100E-01	6.6071E-05	298	1000	-275.00	-290.88
451	C5H12O	METHYL-tert-BUTYL-ETHER	-259.730	-1.3157E-01	6.6224E-05	298	1000	-292.88	-308.96
452	C5H12O	METHYL ISOBUTYL ETHER	-231.485	-1.3635E-01	6.8059E-05	298	1000	-266.00	-282.65
453	C5H12O	ETHYL PROPYL ETHER	--------	--------	--------	--------	--------	-272.20	--------
454	C5H12O2	ETHYLENE GLYCOL MONOPROPYL ETHER	-386.235	-1.3757E-01	6.8870E-05	298	1000	-421.00	-437.80
455	C5H12O2	NEOPENTYL GLYCOL	-413.681	-1.3303E-01	6.9471E-05	298	1000	-447.00	-462.83
456	C5H12O2	1,5-PENTANEDIOL	-414.627	-1.3535E-01	6.6034E-05	298	1000	-449.00	-465.79
457	C5H12O3	2-(2-METHOXYETHOXY)ETHANOL	-494.733	-1.4002E-01	7.1214E-05	298	1000	-530.00	-546.94
458	C5H12O4	PENTAERYTHRITOL	-743.005	-1.3494E-01	7.1106E-05	298	1000	-776.70	-792.70
459	C5H12S	1-PENTANETHIOL	-63.015	-1.8268E-01	1.0174E-04	389	717	-108.41	-128.92
460	C5H12S	BUTYL-METHYL SULFIDE	-56.626	-1.8296E-01	1.0103E-04	389	717	-102.17	-122.85
461	C5H12S	ETHYL PROPYL SULFIDE	-58.396	-1.8581E-01	1.0316E-04	389	717	-104.60	-125.51
462	C5H12S	2-METHYL-2-BUTANETHIOL	-83.019	-1.7726E-01	9.9261E-05	389	717	-127.03	-146.83
463	C5H13N	n-PENTYLAMINE	-75.659	-1.4696E-01	7.4613E-05	298	1000	-112.76	-130.49
464	C5H13NO2	METHYL DIETHANOLAMINE	-341.890	-1.5257E-01	8.0564E-05	298	1000	-380.00	-398.04
465	C6Cl6	HEXACHLOROBENZENE	-41.593	2.6920E-02	-3.9867E-06	298	1000	-33.89	-29.13
466	C6F6	HEXAFLUOROBENZENE	-960.479	1.2850E-02	-5.2530E-08	298	1000	-956.63	-954.07
467	C6H3ClN2O4	1-CHLORO-2,4-DINITROBENZENE	--------	--------	--------	--------	--------	35.10	--------
468	C6H3Cl2NO2	1,2-DICHLORO-4-NITROBENZENE	--------	--------	--------	--------	--------	14.60	--------

296

Table 12-1 ENTHALPY OF FORMATION OF GAS - ORGANIC COMPOUNDS (continued)

NO	FORMULA	NAME	$Hf = A + B T + C T^2$					(Hf - kjoule/mol, T - K)	
			A	B	C	TMIN	TMAX	Hf @ 298 K	Hf @ 500 K
469	C6H3Cl3	1,2,4-TRICHLOROBENZENE	-6.889	-2.0662E-02	1.3754E-05	298	1000	-11.76	-13.78
470	C6H3N3O6	1,3,5-TRINITROBENZENE	--------	--------	--------	--------	--------	-37.40	--------
471	C6H4Br2	m-DIBROMOBENZENE	104.015	-3.7890E-02	2.0963E-05	333	1000	125.52	90.31
472	C6H4ClNO2	m-CHLORONITROBENZENE	47.552	-4.4311E-02	3.0414E-05	298	1000	37.20	33.00
473	C6H4ClNO2	o-CHLORONITROBENZENE	47.573	-4.4371E-02	3.0470E-05	298	1000	37.20	33.01
474	C6H4ClNO2	p-CHLORONITROBENZENE	47.296	-4.3374E-02	3.0159E-05	298	1000	37.20	33.15
475	C6H4Cl2	m-DICHLOROBENZENE	35.708	-3.7400E-02	2.0212E-05	298	1000	26.44	22.06
476	C6H4Cl2	o-DICHLOROBENZENE	39.320	-3.7757E-02	2.0384E-05	298	1000	29.96	25.54
477	C6H4Cl2	p-DICHLOROBENZENE	32.210	-3.7170E-02	2.0225E-05	298	1000	23.01	18.68
478	C6H4F2	m-DIFLUOROBENZENE	-299.323	-4.2807E-02	2.2316E-05	298	1000	-309.99	-315.15
479	C6H4F2	o-DIFLUOROBENZENE	-283.516	-4.4576E-02	2.4650E-05	298	1000	-294.51	-299.64
480	C6H4F2	p-DIFLUOROBENZENE	-296.729	-4.2191E-02	2.2098E-05	298	1000	-307.23	-312.30
481	C6H4N2O4	m-DINITROBENZENE	--------	--------	--------	--------	--------	-27.60	--------
482	C6H4N2O4	o-DINITROBENZENE	--------	--------	--------	--------	--------	-1.80	--------
483	C6H4N2O4	p-DINITROBENZENE	--------	--------	--------	--------	--------	-38.60	--------
484	C6H5Br	BROMOBENZENE	102.370	-5.2308E-02	2.5004E-05	333	1000	105.02	82.47
485	C6H5Cl	CHLOROBENZENE	65.513	-5.4195E-02	2.6773E-05	298	1000	51.84	45.11
486	C6H5ClO	m-CHLOROPHENOL	-142.122	-4.5368E-02	2.4868E-05	298	1000	-153.30	-158.59
487	C6H5ClO	o-CHLOROPHENOL	-111.822	-4.5368E-02	2.4868E-05	298	1000	-123.00	-128.29
488	C6H5ClO	p-CHLOROPHENOL	-134.622	-4.5368E-02	2.4868E-05	298	1000	-145.80	-151.09
489	C6H5Cl2N	3,4-DICHLOROANILINE	42.089	-3.9705E-02	2.4825E-05	298	1000	32.60	28.44
490	C6H5F	FLUOROBENZENE	-102.157	-5.7304E-02	2.8787E-05	298	1000	-116.57	-123.61
491	C6H5I	IODOBENZENE	142.156	-4.5903E-02	2.1666E-05	459	1000	162.55	124.62
492	C6H5NO2	NITROBENZENE	--------	--------	--------	--------	--------	67.60	--------
493	C6H6	BENZENE	101.403	-7.2136E-02	3.2877E-05	298	1000	82.93	73.55
494	C6H6ClN	m-CHLOROANILINE	71.321	-5.6953E-02	3.1707E-05	298	1000	57.30	50.77
495	C6H6ClN	o-CHLOROANILINE	71.321	-5.6953E-02	3.1707E-05	298	1000	57.30	50.77
496	C6H6ClN	p-CHLOROANILINE	71.328	-5.6995E-02	3.1736E-05	298	1000	57.30	50.76
497	C6H6N2	cis-DICYANO-1-BUTENE	286.356	-6.1282E-02	3.1334E-05	298	1000	271.00	263.55
498	C6H6N2	trans-DICYANO-1-BUTENE	280.444	-5.3301E-02	2.6640E-05	298	1000	267.00	260.45
499	C6H6N2	1,4-DICYANO-2-BUTENE	275.161	-5.6076E-02	2.8091E-05	298	1000	261.00	254.15
500	C6H6N2O2	m-NITROANILINE	--------	--------	--------	--------	--------	58.50	--------
501	C6H6N2O2	o-NITROANILINE	--------	--------	--------	--------	--------	63.80	--------
502	C6H6N2O2	p-NITROANILINE	--------	--------	--------	--------	--------	59.50	--------
503	C6H6O	PHENOL	-80.956	-6.1053E-02	3.0058E-05	298	1000	-96.36	-103.97
504	C6H6O2	1,2-BENZENEDIOL	-257.336	-6.0268E-02	3.5352E-05	298	1000	-272.00	-278.63
505	C6H6O2	1,3-BENZENEDIOL	-261.832	-5.2375E-02	2.9199E-05	298	1000	-274.70	-280.72
506	C6H6O2	p-HYDROQUINONE	-248.442	-5.3218E-02	2.7820E-05	298	1000	-261.71	-268.10
507	C6H6O3	1,2,3-BENZENETRIOL	-435.109	-4.5346E-02	2.7301E-05	298	1000	-446.00	-450.96
508	C6H6S	BENZENETHIOL	140.377	-1.1600E-01	6.4321E-05	389	717	111.55	98.46
509	C6H7N	ANILINE	105.261	-7.3513E-02	3.7553E-05	298	1000	86.86	77.89
510	C6H7N	2-PICOLINE	121.445	-8.9575E-02	4.5898E-05	298	1000	98.95	88.13
511	C6H7N	3-PICOLINE	128.775	-9.0086E-02	4.6041E-05	298	1000	106.15	95.24
512	C6H7N	4-METHYLPYRIDINE	124.765	-8.9452E-02	4.4619E-05	298	1000	102.13	91.19
513	C6H8	1,3-CYCLOHEXADIENE	132.441	-1.0401E-01	5.3455E-05	298	1000	106.20	93.80
514	C6H8	METHYLCYCLOPENTADIENE	120.775	-8.2997E-02	4.2855E-05	298	1000	100.00	89.99
515	C6H8N2	ADIPONITRILE	167.739	-7.3266E-02	3.8533E-05	298	1000	149.50	140.74
516	C6H8N2	METHYLGLUTARONITRILE	161.196	-7.5795E-02	3.8040E-05	298	1000	142.00	132.81
517	C6H8N2	m-PHENYLENEDIAMINE	110.563	-7.8612E-02	4.3576E-05	298	1000	91.20	82.15
518	C6H8N2	o-PHENYLENEDIAMINE	110.563	-7.8612E-02	4.3576E-05	298	1000	91.20	82.15
519	C6H8N2	p-PHENYLENEDIAMINE	110.563	-7.8612E-02	4.3576E-05	298	1000	91.20	82.15
520	C6H8N2	PHENYLHYDRAZINE	226.381	-9.2031E-02	4.8886E-05	298	1000	203.50	192.59
521	C6H8N2O	BIS(CYANOETHYL)ETHER	40.592	-8.0835E-02	4.3050E-05	298	1000	20.40	10.94
522	C6H8O4	DIMETHYL MALEATE	-601.447	-7.7872E-02	4.2154E-05	298	1000	-620.80	-629.84
523	C6H8O6	ASCORBIC ACID	-937.527	-5.7814E-02	3.9487E-05	298	1000	-951.00	-956.56
524	C6H8O7	CITRIC ACID	-1372.319	-7.3425E-02	4.4682E-05	298	1000	-1390.00	-1397.86
525	C6H10	1-METHYLCYCLOPENTENE	27.217	-1.2878E-01	6.2739E-05	298	1000	-5.44	-21.49
526	C6H10	3-METHYLCYCLOPENTENE	41.144	-1.2810E-01	6.2342E-05	298	1000	8.66	-7.32
527	C6H10	4-METHYLCYCLOPENTENE	47.355	-1.2842E-01	6.2375E-05	298	1000	14.77	-1.26
528	C6H10	CYCLOHEXENE	24.718	-1.2061E-01	6.3727E-05	298	1000	-5.36	-19.66
529	C6H10	2,3-DIMETHYL-1,3-BUTADIENE	67.363	-8.7578E-02	4.2542E-05	298	1000	45.10	34.21
530	C6H10	1,5-HEXADIENE	110.446	-1.0303E-01	4.8485E-05	298	1000	84.10	71.05
531	C6H10	cis,trans-2,4-HEXADIENE	74.144	-1.0453E-01	4.8911E-05	298	1000	47.40	34.11
532	C6H10	trans,trans-2,4-HEXADIENE	68.364	-9.8002E-02	4.6147E-05	298	1000	43.30	30.90
533	C6H10	1-HEXYNE	147.172	-9.1743E-02	4.2233E-05	298	1000	123.64	111.86
534	C6H10	2-HEXYNE	131.718	-1.0356E-01	4.6816E-05	298	1000	105.00	91.64
535	C6H10	3-HEXYNE	133.406	-1.0619E-01	4.7874E-05	298	1000	106.00	92.28
536	C6H10O	CYCLOHEXANONE	-194.765	-1.4365E-01	8.1559E-05	298	1000	-230.12	-246.20
537	C6H10O	MESITYL OXIDE	-172.547	-9.0710E-02	4.5697E-05	298	1000	-195.56	-206.48
538	C6H10O2	epsilon-CAPROLACTONE	-390.955	-1.4306E-01	8.3591E-05	298	1000	-425.93	-441.59
539	C6H10O2	ETHYL METHACRYLATE	-352.056	-1.1489E-01	5.7771E-05	298	1000	-381.00	-395.06
540	C6H10O2	n-PROPYL ACRYLATE	-341.886	-8.9415E-02	4.9362E-05	298	1000	-364.00	-374.25
541	C6H10O3	ETHYLACETOACETATE	-530.103	-1.1905E-01	6.0833E-05	298	1000	-560.00	-574.42
542	C6H10O3	PROPIONIC ANHYDRIDE	-597.459	-1.1615E-01	5.9986E-05	298	1000	-626.50	-640.54
543	C6H10O4	ADIPIC ACID	-837.415	-1.1055E-01	5.8036E-05	298	1000	-865.04	-878.18
544	C6H10O4	DIETHYL OXALATE	-712.339	-1.1043E-01	5.8458E-05	298	1000	-739.80	-752.94
545	C6H10O4	ETHYLENE GLYCOL DIACETATE	-779.540	-1.1042E-01	5.8448E-05	298	1000	-807.00	-820.14
546	C6H10O4	ETHYLIDENE DIACETATE	-783.363	-1.1546E-01	6.0496E-05	298	1000	-812.00	-825.97

297

Table 12-1 ENTHALPY OF FORMATION OF GAS - ORGANIC COMPOUNDS (continued)

NO	FORMULA	NAME	Hf = A + B T + C T²					(Hf - kjoule/mol, T - K)	
			A	B	C	TMIN	TMAX	Hf @ 298 K	Hf @ 500 K
547	C6H11N	HEXANENITRILE	19.548	-1.1395E-01	5.7881E-05	298	1000	-9.25	-22.96
548	C6H11NO	epsilon-CAPROLACTAM	-199.773	-1.8658E-01	1.0123E-04	298	1000	-246.20	-267.76
549	C6H11NO	CYCLOHEXANONE OXIME	-156.673	-1.5478E-01	9.6604E-05	298	1000	-194.00	-209.91
550	C6H12	CYCLOHEXANE	-81.822	-1.6705E-01	9.2830E-05	298	1000	-123.14	-142.14
551	C6H12	2,3-DIMETHYL-1-BUTENE	-27.857	-1.0946E-01	5.2559E-05	298	1000	-55.73	-69.45
552	C6H12	2,3-DIMETHYL-2-BUTENE	-23.094	-1.4116E-01	6.6450E-05	298	1000	-59.20	-77.06
553	C6H12	3,3-DIMETHYL-1-BUTENE	-9.850	-1.3070E-01	6.1961E-05	298	1000	-43.14	-59.71
554	C6H12	2-ETHYL-1-BUTENE	-20.694	-1.2141E-01	5.8622E-05	298	1000	-51.55	-66.74
555	C6H12	trans-3-METHYL-2-PENTENE	-25.264	-1.3092E-01	6.1831E-05	298	1000	-58.66	-75.27
556	C6H12	1-HEXENE	-9.811	-1.2505E-01	5.9735E-05	298	1000	-41.67	-57.40
557	C6H12	cis-2-HEXENE	-18.319	-1.3353E-01	6.3492E-05	298	1000	-52.34	-69.21
558	C6H12	trans-2-HEXENE	-21.685	-1.2596E-01	5.9153E-05	298	1000	-53.89	-69.87
559	C6H12	cis-3-HEXENE	-13.416	-1.3447E-01	6.4337E-05	298	1000	-47.61	-64.57
560	C6H12	trans-3-HEXENE	-23.002	-1.2317E-01	5.8204E-05	298	1000	-54.43	-70.04
561	C6H12	METHYLCYCLOPENTANE	-67.643	-1.5503E-01	7.8379E-05	298	1000	-106.69	-125.56
562	C6H12	2-METHYL-1-PENTENE	-21.529	-1.2075E-01	5.8082E-05	298	1000	-52.26	-67.38
563	C6H12	2-METHYL-2-PENTENE	-26.352	-1.3092E-01	6.1831E-05	298	1000	-59.75	-76.35
564	C6H12	3-METHYL-1-PENTENE	-17.019	-1.1029E-01	5.3597E-05	298	1000	-45.02	-58.76
565	C6H12	cis-3-METHYL-2-PENTENE	-24.343	-1.3092E-01	6.1831E-05	298	1000	-57.74	-74.34
566	C6H12	4-METHYL-1-PENTENE	-10.701	-1.3063E-01	6.0835E-05	298	1000	-44.10	-60.81
567	C6H12	cis-4-METHYL-2-PENTENE	-18.946	-1.2321E-01	5.8643E-05	298	1000	-50.33	-65.89
568	C6H12	trans-4-METHYL-2-PENTENE	-25.457	-1.1314E-01	5.3589E-05	298	1000	-54.35	-68.63
569	C6H12N2	TRIETHYLENEDIAMINE	96.971	-1.8213E-01	1.0824E-04	298	1000	52.70	32.97
570	C6H12O	BUTYL VINYL ETHER	-149.334	-1.3246E-01	6.4787E-05	298	1000	-183.00	-199.37
571	C6H12O	CYCLOHEXANOL	-255.672	-1.5846E-01	9.1123E-05	298	1000	-294.55	-312.12
572	C6H12O	HEXANAL	-216.535	-1.2536E-01	6.1752E-05	298	1000	-248.15	-263.78
573	C6H12O	ETHYL ISOPROPYL KETONE	-251.649	-1.3526E-01	6.5710E-05	298	1000	-286.10	-302.85
574	C6H12O	2-HEXANONE	-247.631	-1.2609E-01	6.0435E-05	298	1000	-279.83	-295.57
575	C6H12O	3-HEXANONE	-245.795	-1.2484E-01	6.0281E-05	298	1000	-277.70	-293.14
576	C6H12O	METHYL ISOBUTYL KETONE	-256.000	-1.2779E-01	6.2353E-05	298	1000	-288.49	-304.30
577	C6H12O2	n-PENTYL FORMATE	-415.397	-1.3049E-01	6.6815E-05	298	1000	-448.20	-463.94
578	C6H12O2	n-BUTYL ACETATE	-451.564	-1.3543E-01	6.9868E-05	298	1000	-485.60	-501.81
579	C6H12O2	sec-BUTYL ACETATE	-471.366	-1.2958E-01	6.7433E-05	298	1000	-503.80	-519.30
580	C6H12O2	tert-BUTYL ACETATE	-491.273	-1.2703E-01	6.6626E-05	298	1000	-523.00	-538.13
581	C6H12O2	ETHYL n-BUTYRATE	-453.335	-1.2748E-01	6.3881E-05	298	1000	-485.50	-501.10
582	C6H12O2	ETHYL ISOBUTYRATE	-464.668	-1.3914E-01	7.1130E-05	298	1000	-499.60	-516.46
583	C6H12O2	ISOBUTYL ACETATE	-461.621	-1.3144E-01	6.6631E-05	298	1000	-494.70	-510.68
584	C6H12O2	n-PROPYL PROPIONATE	-450.965	-1.2743E-01	6.3956E-05	298	1000	-483.10	-498.69
585	C6H12O2	CYCLOHEXYL PEROXIDE	-182.945	-1.5208E-01	9.0857E-05	298	1000	-220.00	-236.27
586	C6H12O2	DIACETONE ALCOHOL	-512.891	-1.3019E-01	6.5526E-05	298	1000	-545.80	-561.60
587	C6H12O2	2-ETHYL BUTYRIC ACID	-482.132	-1.3882E-01	7.1394E-05	298	1000	-517.00	-533.69
588	C6H12O2	n-HEXANOIC ACID	-481.672	-1.2667E-01	6.4445E-05	298	1000	-513.60	-528.90
589	C6H12O3	2-ETHOXYETHYL ACETATE	-576.812	-1.3672E-01	7.1421E-05	298	1000	-611.00	-627.32
590	C6H12O3	HYDROXYCAPROIC ACID	-650.762	-1.2886E-01	6.7624E-05	298	1000	-683.00	-698.29
591	C6H12O3	PARALDEHYDE	-609.484	-1.4468E-01	7.7388E-05	298	1000	-645.30	-662.48
592	C6H12O3	sec-BUTYL GLYCOLATE	--------	--------	--------	--------	--------	--------	--------
593	C6H12S	THIACYCLOHEPTANE	-1.714	-2.5258E-01	1.7597E-04	389	717	-61.34	-84.01
594	C6H13N	CYCLOHEXYLAMINE	-62.824	-1.7145E-01	9.9225E-05	298	1000	-104.90	-123.74
595	C6H13N	HEXAMETHYLENEIMINE	12.799	-1.9035E-01	1.0697E-04	298	1000	-34.18	-55.63
596	C6H14	2,2-DIMETHYLBUTANE	-147.725	-1.5036E-01	7.6099E-05	298	1000	-185.56	-203.88
597	C6H14	2,3-DIMETHYLBUTANE	-139.269	-1.5235E-01	7.5663E-05	298	1000	-177.78	-196.53
598	C6H14	HEXANE	-129.114	-1.5013E-01	7.3458E-05	298	1000	-167.19	-185.82
599	C6H14	2-METHYLPENTANE	-137.114	-1.4707E-01	7.2785E-05	298	1000	-174.31	-192.45
600	C6H14	3-METHYLPENTANE	-133.549	-1.5013E-01	7.3458E-05	298	1000	-171.63	-190.25
601	C6H14N2O2	LYSINE	-422.367	-1.5537E-01	8.4191E-05	298	1000	-461.00	-479.00
602	C6H14O	2-ETHYL-1-BUTANOL	-287.088	-1.5407E-01	7.6404E-05	298	1000	-326.10	-345.02
603	C6H14O	HEXYL ALCOHOL	-280.972	-1.5237E-01	7.4446E-05	298	1000	-319.62	-338.55
604	C6H14O	2-HEXANOL	-295.147	-1.5183E-01	7.6472E-05	298	1000	-333.50	-351.95
605	C6H14O	2-METHYL-1-PENTANOL	-287.423	-1.5682E-01	7.7723E-05	298	1000	-327.15	-346.40
606	C6H14O	4-METHYL-2-PENTANOL	-304.295	-1.5085E-01	7.5600E-05	298	1000	-342.42	-360.82
607	C6H14O	n-BUTYL ETHYL ETHER	-253.180	-1.5482E-01	7.7303E-05	298	1000	-292.40	-311.26
608	C6H14O	ISOPROPYL ETHER	-280.303	-1.5182E-01	7.4433E-05	298	1000	-318.82	-337.61
609	C6H14O	PROPYL ETHER	-254.362	-1.5182E-01	7.4433E-05	298	1000	-292.88	-311.67
610	C6H14O	METHYL tert-PENTYL ETHER	-274.188	-1.5019E-01	7.7160E-05	298	1000	-312.00	-330.00
611	C6H14O2	ACETAL	--------	--------	--------	--------	--------	-453.50	--------
612	C6H14O2	2-BUTOXYETHANOL	-400.999	-1.5852E-01	8.0053E-05	298	1000	-441.00	-460.24
613	C6H14O2	1,6-HEXANEDIOL	-430.272	-1.5600E-01	7.6941E-05	298	1000	-469.82	-489.04
614	C6H14O2	HEXYLENE GLYCOL	-498.113	-1.4649E-01	7.6707E-05	298	1000	-534.76	-552.18
615	C6H14O2S	DI-n-PROPYL SULFONE	--------	--------	--------	--------	--------	-468.30	--------
616	C6H14O3	DIETHYLENE GLYCOL DIMETHYL ETHER	-470.359	-1.6178E-01	8.3667E-05	298	1000	-511.00	-530.33
617	C6H14O3	DIPROPYLENE GLYCOL	-590.070	-1.5166E-01	7.9376E-05	298	1000	-628.00	-646.05
618	C6H14O3	2-(2-ETHOXYETHOXY)ETHANOL	-524.341	-1.6185E-01	8.3258E-05	298	1000	-565.00	-584.45
619	C6H14O3	TRIMETHYLOLPROPANE	-601.499	-1.5442E-01	8.0313E-05	298	1000	-640.19	-658.63
620	C6H14O4	TRIETHYLENE GLYCOL	-684.367	-1.6209E-01	8.3384E-05	298	1000	-725.09	-744.57
621	C6H14O6	SORBITOL	-1104.680	-1.4247E-01	7.6390E-05	298	1000	-1140.00	-1156.82
622	C6H14S	1-HEXANETHIOL	-78.241	-2.0412E-01	1.1350E-04	389	717	-128.99	-151.92
623	C6H14S	BUTYL ETHYL SULFIDE	-73.408	-2.0828E-01	1.1606E-04	389	717	-125.19	-148.54
624	C6H14S	ISOPROPYL SULFIDE	-93.596	-1.9179E-01	1.0673E-04	389	717	-141.25	-162.81

Table 12-1 ENTHALPY OF FORMATION OF GAS - ORGANIC COMPOUNDS (continued)

NO	FORMULA	NAME	Hf = A + B T + C T²					(Hf - kjoule/mol, T - K)	
			A	B	C	TMIN	TMAX	Hf @ 298 K	Hf @ 500 K
625	C6H14S	METHYL PENTYL SULFIDE	-71.760	-2.0494E-01	1.1349E-04	389	717	-122.76	-145.86
626	C6H14S	PROPYL SULFIDE	-73.078	-2.1068E-01	1.1845E-04	389	717	-125.35	-148.80
627	C6H14S2	PROPYL DISULFIDE	-56.986	-2.4337E-01	1.3868E-04	389	717	-117.19	-144.00
628	C6H15Al	TRIETHYL ALUMINUM	--------	--------	--------	--------	--------	-163.60	--------
629	C6H15Al2Cl3	ETHYL ALUMINUM SESQUICHLORIDE	--------	--------	--------	--------	--------	--------	--------
630	C6H15N	DIISOPROPYLAMINE	-107.861	-1.6851E-01	8.8963E-05	298	1000	-150.00	-169.88
631	C6H15N	DI-n-PROPYLAMINE	-73.116	-1.7059E-01	8.7884E-05	298	1000	-116.00	-136.44
632	C6H15N	n-HEXYLAMINE	-89.912	-1.7046E-01	8.7156E-05	298	1000	-133.00	-153.35
633	C6H15N	TRIETHYLAMINE	-57.919	-1.6610E-01	8.5747E-05	298	1000	-99.58	-119.53
634	C6H15NO	6-AMINOHEXANOL	-240.749	-1.6755E-01	8.5206E-05	298	1000	-283.00	-303.22
635	C6H15NO2	DIISOPROPANOLAMINE	-412.902	-1.6496E-01	8.8109E-05	298	1000	-454.00	-473.35
636	C6H15NO3	TRIETHANOLAMINE	-519.041	-1.7335E-01	9.3476E-05	298	1000	-562.08	-582.35
637	C6H15N3	N-AMINOETHYL PIPERAZINE	70.946	-1.8516E-01	1.0269E-04	298	1000	25.10	4.04
638	C6H15O4P	TRIETHYL PHOSPHATE	--------	--------	--------	--------	--------	-1244.70	--------
639	C6H16N2	HEXAMETHYLENEDIAMINE	-54.225	-1.7772E-01	9.2755E-05	298	1000	-98.83	-119.90
640	C6H18N3OP	HEXAMETHYL PHOSPHORAMIDE	--------	--------	--------	--------	--------	-467.00	--------
641	C6H18N4	TRIETHYLENE TETRAMINE	56.138	-2.1294E-01	1.1702E-04	298	1000	3.35	-21.08
642	C6H18Osi2	HEXAMETHYLDISILOXANE	-737.844	-1.6203E-01	9.4960E-05	298	1000	-777.72	-795.12
643	C6H18O3Si3	HEXAMETHYLCYCLOTRISILOXANE	-1718.857	-2.1334E-01	1.3724E-04	298	1000	-1770.00	-1791.22
644	C6H19NSi2	HEXAMETHYLDISILAZANE	-421.688	-2.2398E-01	1.3101E-04	298	1000	-476.60	-500.92
645	C7H3ClF3NO2	4-CHLORO-3-NITROBENZOTRIFLUORIDE	--------	--------	--------	--------	--------	-634.25	--------
646	C7H3Cl2F3	2,4-DICHLOROBENZOTRIFLUORIDE	-646.766	-2.2256E-02	1.9946E-05	298	1000	-651.45	-652.91
647	C7H3Cl2NO	3,4-DICHLOROPHENYL ISOCYANATE	1.850	-1.9982E-02	1.1492E-05	298	1000	-2.97	-5.27
648	C7H4ClF3	p-CHLOROBENZOTRIFLUORIDE	-609.625	-4.2542E-02	2.8008E-05	298	1000	-619.65	-623.89
649	C7H4Cl2O	m-CHLOROBENZOYL CHLORIDE	-155.890	-3.7445E-02	1.8965E-05	298	1000	-165.31	-169.87
650	C7H4F3NO2	3-NITROBENZOTRIFLUORIDE	--------	--------	--------	--------	--------	-602.00	--------
651	C7H5ClO	BENZOYL CHLORIDE	-90.945	-5.5881E-02	2.5384E-05	298	1000	-105.30	-112.54
652	C7H5ClO2	o-CHLOROBENZOIC ACID	-306.303	-7.3276E-02	3.3699E-05	298	1000	-325.00	-334.52
653	C7H5Cl3	BENZOTRICHLORIDE	-0.177	-5.0073E-02	3.0015E-05	298	1000	-12.34	-17.71
654	C7H5F3	A,A,A-TRIFLUOROTOLUENE	-584.547	-6.3517E-02	3.6434E-05	298	1000	-600.07	-607.20
655	C7H5N	BENZONITRILE	230.953	-4.7911E-02	2.3229E-05	298	1000	218.82	212.80
656	C7H5NO	PHENYL ISOCYANATE	68.140	-6.5958E-02	3.1826E-05	298	1000	51.40	43.12
657	C7H5N3O6	2,4,6-TRINITROTOLUENE	--------	--------	--------	--------	--------	-67.10	--------
658	C7H6Cl2	BENZYL DICHLORIDE	30.148	-6.8312E-02	3.5170E-05	298	1000	13.00	4.78
659	C7H6Cl2	2,4-DICHLOROTOLUENE	-0.121	-5.4754E-02	3.0207E-05	298	1000	-13.60	-19.95
660	C7H6N2O4	2,4-DINITROTOLUENE	--------	--------	--------	--------	--------	-64.20	--------
661	C7H6N2O4	2,5-DINITROTOLUENE	--------	--------	--------	--------	--------	-34.00	--------
662	C7H6N2O4	2,6-DINITROTOLUENE	--------	--------	--------	--------	--------	-51.10	--------
663	C7H6N2O4	3,4-DINITROTOLUENE	--------	--------	--------	--------	--------	-14.00	--------
664	C7H6N2O4	3,5-DINITROTOLUENE	--------	--------	--------	--------	--------	-43.00	--------
665	C7H6O	BENZALDEHYDE	-20.526	-6.4381E-02	3.2376E-05	298	1000	-36.80	-44.62
666	C7H6O2	BENZOIC ACID	-266.140	-9.3579E-02	4.1667E-05	298	1000	-290.20	-302.51
667	C7H6O2	p-HYDROXYBENZALDEHYDE	-198.898	-6.0884E-02	4.3667E-05	298	1000	-213.00	-218.42
668	C7H6O2	SALICYLALDEHYDE	-200.534	-6.1831E-02	4.3238E-05	298	1000	-214.95	-220.64
669	C7H6O3	SALICYLIC ACID	-445.007	-8.3877E-02	3.9181E-05	298	1000	-466.35	-477.15
670	C7H7Br	p-BROMOTOLUENE	84.138	-7.5530E-02	3.6700E-05	333	1000	80.40	55.55
671	C7H7Cl	BENZYL CHLORIDE	43.607	-8.8733E-02	1.7349E-05	298	1000	18.70	3.58
672	C7H7Cl	o-CHLOROTOLUENE	37.384	-7.6510E-02	3.9523E-05	298	1000	18.20	9.01
673	C7H7Cl	p-CHLOROTOLUENE	37.384	-7.6510E-02	3.9523E-05	298	1000	18.20	9.01
674	C7H7F	p-FLUOROTOLUENE	-127.733	-8.0517E-02	4.0422E-05	298	1000	-148.03	-157.89
675	C7H7NO	FORMANILIDE	-31.691	-9.1706E-02	4.0994E-05	298	1000	-55.20	-67.30
676	C7H7NO2	m-NITROTOLUENE	--------	--------	--------	--------	--------	20.80	--------
677	C7H7NO2	o-NITROTOLUENE	--------	--------	--------	--------	--------	45.50	--------
678	C7H7NO2	p-NITROTOLUENE	--------	--------	--------	--------	--------	31.00	--------
679	C7H7NO3	o-NITROANISOLE	--------	--------	--------	--------	--------	-84.50	--------
680	C7H8	TOLUENE	74.320	-9.5998E-02	4.7011E-05	298	1000	50.00	38.07
681	C7H8	1,3,5-CYCLOHEPTATRIENE	200.418	-7.4291E-02	3.8682E-05	298	1000	181.88	172.97
682	C7H8O	ANISOLE	-41.700	-1.0405E-01	5.1683E-05	298	1000	-68.00	-80.81
683	C7H8O	BENZYL ALCOHOL	-74.280	-1.0339E-01	5.1231E-05	298	1000	-100.40	-113.17
684	C7H8O	m-CRESOL	-110.617	-8.6082E-02	4.2890E-05	298	1000	-132.34	-142.93
685	C7H8O	o-CRESOL	-108.508	-7.9613E-02	3.9667E-05	298	1000	-128.62	-138.40
686	C7H8O	p-CRESOL	-103.583	-8.6224E-02	4.2560E-05	298	1000	-125.39	-136.06
687	C7H8O2	GUAIACOL	-225.490	-9.4005E-02	4.8734E-05	298	1000	-249.00	-260.31
688	C7H8O2	p-METHOXYPHENOL	-225.503	-9.3929E-02	4.8689E-05	298	1000	-249.00	-260.30
689	C7H9N	BENZYLAMINE	88.329	-1.0296E-01	5.0315E-05	298	1000	62.23	49.43
690	C7H9N	2,6-DIMETHYLPYRIDINE	87.027	-1.1173E-01	5.5421E-05	298	1000	58.70	45.02
691	C7H9N	N-METHYLANILINE	113.492	-1.0182E-01	5.2516E-05	298	1000	88.00	75.71
692	C7H9N	m-TOLUIDINE	89.638	-1.0142E-01	5.0941E-05	298	1000	64.00	51.66
693	C7H9N	o-TOLUIDINE	81.987	-9.5780E-02	5.0468E-05	298	1000	58.00	46.71
694	C7H9N	p-TOLUIDINE	70.680	-1.0157E-01	5.0898E-05	298	1000	45.00	32.62
695	C7H10	2-NORBORNENE	122.779	-1.3321E-01	7.5914E-05	298	1000	90.00	75.15
696	C7H10N2	TOLUENEDIAMINE	82.212	-9.6583E-02	5.2805E-05	298	1000	58.37	47.12
697	C7H11NO	CYCLOHEXYL ISOCYANATE	-98.238	-1.6860E-01	9.3875E-05	298	1000	-140.00	-159.07
698	C7H12	1-HEPTYNE	131.620	-1.1201E-01	5.2898E-05	298	1000	103.01	88.84
699	C7H12O2	n-BUTYL ACRYLATE	-367.962	-1.0897E-01	5.9732E-05	298	1000	-395.05	-407.51
700	C7H12O2	ISOBUTYL ACRYLATE	-366.608	-1.1066E-01	6.0959E-05	298	1000	-394.00	-406.70
701	C7H12O2	n-PROPYL METHACRYLATE	-367.983	-1.3513E-01	6.8357E-05	298	1000	-402.00	-418.46
702	C7H12O4	DIETHYL MALONATE	-746.869	-1.3740E-01	7.3131E-05	298	1000	-781.00	-797.29

Table 12-1 ENTHALPY OF FORMATION OF GAS - ORGANIC COMPOUNDS (continued)

NO	FORMULA	NAME	A	B	C	TMIN	TMAX	Hf @ 298 K	Hf @ 500 K
				Hf = A + B T + C T²				(Hf - kjoule/mol, T - K)	
703	C7H14	CYCLOHEPTANE	-71.722	-1.9175E-01	1.0424E-04	298	1000	-119.33	-141.54
704	C7H14	1,1-DIMETHYLCYCLOPENTANE	-94.650	-1.7452E-01	9.1458E-05	298	1000	-138.28	-159.05
705	C7H14	cis-1,2-DIMETHYLCYCLOPENTANE	-86.257	-1.7289E-01	8.9927E-05	298	1000	-129.54	-150.22
706	C7H14	trans-1,2-DIMETHYLCYCLOPENTANE	-93.558	-1.7216E-01	8.9182E-05	298	1000	-136.69	-157.34
707	C7H14	cis-1,3-DIMETHYLCYCLOPENTANE	-92.722	-1.7216E-01	8.9182E-05	298	1000	-135.85	-156.51
708	C7H14	trans-1,3-DIMETHYLCYCLOPENTANE	-90.462	-1.7216E-01	8.9182E-05	298	1000	-133.60	-154.25
709	C7H14	ETHYLCYCLOPENTANE	-83.199	-1.7422E-01	8.8672E-05	298	1000	-127.07	-148.14
710	C7H14	2-ETHYL-1-PENTENE	-36.950	-1.4875E-01	7.4076E-05	298	1000	-74.64	-92.81
711	C7H14	3-ETHYL-1-PENTENE	-25.746	-1.5147E-01	7.5593E-05	298	1000	-64.10	-82.58
712	C7H14	1-HEPTENE	-25.409	-1.4519E-01	7.0333E-05	298	1000	-62.30	-80.42
713	C7H14	cis-2-HEPTENE	-28.735	-1.5908E-01	7.7429E-05	298	1000	-69.20	-88.92
714	C7H14	trans-2-HEPTENE	-35.526	-1.5353E-01	7.5023E-05	298	1000	-74.60	-93.53
715	C7H14	cis-3-HEPTENE	-28.057	-1.6017E-01	7.8630E-05	298	1000	-68.70	-88.48
716	C7H14	trans-3-HEPTENE	-34.772	-1.5411E-01	7.5862E-05	298	1000	-73.90	-92.86
717	C7H14	METHYLCYCLOHEXANE	-110.905	-1.7788E-01	1.0003E-04	298	1000	-154.77	-174.84
718	C7H14	2-METHYL-1-HEXENE	-39.818	-1.4734E-01	7.2697E-05	298	1000	-77.23	-95.31
719	C7H14	3-METHYL-1-HEXENE	-28.336	-1.5147E-01	7.5593E-05	298	1000	-66.69	-85.17
720	C7H14	4-METHYL-1-HEXENE	-28.560	-1.5072E-01	7.5055E-05	298	1000	-66.70	-85.16
721	C7H14	2,3,3-TRIMETHYL-1-BUTENE	-49.075	-1.4513E-01	7.5653E-05	298	1000	-85.50	-102.73
722	C7H14O	DIISOPROPYL KETONE	-268.409	-1.7016E-01	8.5053E-05	298	1000	-311.40	-332.22
723	C7H14O	2-HEPTANONE	-262.601	-1.4856E-01	7.2453E-05	298	1000	-300.45	-318.77
724	C7H14O	HEPTANAL	-227.526	-1.4418E-01	7.1425E-05	298	1000	-264.01	-281.76
725	C7H14O	1-METHYLCYCLOHEXANOL	-288.526	-1.7683E-01	1.0158E-04	298	1000	-332.00	-351.55
726	C7H14O	cis-2-METHYLCYCLOHEXANOL	-282.569	-1.8080E-01	1.0399E-04	298	1000	-327.00	-346.97
727	C7H14O	trans-2-METHYLCYCLOHEXANOL	-308.169	-1.8080E-01	1.0399E-04	298	1000	-352.60	-372.57
728	C7H14O	cis-3-METHYLCYCLOHEXANOL	-497.253	-1.4835E-01	7.6111E-05	298	1000	-350.90	-552.40
729	C7H14O	trans-3-METHYLCYCLOHEXANOL	-284.669	-1.8080E-01	1.0399E-04	298	1000	-329.10	-349.07
730	C7H14O	cis-4-METHYLCYCLOHEXANOL	-303.069	-1.8080E-01	1.0399E-04	298	1000	-347.50	-367.47
731	C7H14O	trans-4-METHYLCYCLOHEXANOL	-322.769	-1.8080E-01	1.0399E-04	298	1000	-367.20	-387.17
732	C7H14O	5-METHYL-2-HEXANONE	-265.737	-1.5087E-01	7.4674E-05	298	1000	-304.00	-322.50
733	C7H14O2	n-BUTYL PROPIONATE	-464.759	-1.5023E-01	7.6556E-05	298	1000	-502.60	-520.73
734	C7H14O2	ETHYL ISOVALERATE	-489.570	-1.4893E-01	7.6158E-05	298	1000	-527.00	-545.00
735	C7H14O2	ISOPENTYL ACETATE	-472.950	-1.5537E-01	8.0371E-05	298	1000	-511.90	-530.54
736	C7H14O2	n-PENTYL ACETATE	-466.567	-1.5498E-01	7.9717E-05	298	1000	-505.50	-524.13
737	C7H14O2	n-PROPYL n-BUTYRATE	-467.443	-1.5034E-01	7.6690E-05	298	1000	-505.30	-523.44
738	C7H14O2	n-HEPTANOIC ACID	--------	--------	--------	--------	--------	-534.60	--------
739	C7H14O3	ETHYL-3-ETHOXYPROPIONATE	-595.163	-1.5097E-01	7.8025E-05	298	1000	-633.00	-651.14
740	C7H15Br	1-BROMOHEPTANE	-142.781	-1.6151E-01	8.2700E-05	333	1000	-168.00	-202.86
741	C7H15N	N-METHYLCYCLOHEXYLAMINE	-54.746	-1.9675E-01	1.1414E-04	298	1000	-103.00	-124.59
742	C7H16	2,2-DIMETHYLPENTANE	-163.038	-1.7028E-01	8.4061E-05	298	1000	-206.15	-227.16
743	C7H16	2,3-DIMETHYLPENTANE	-156.134	-1.7028E-01	8.4061E-05	298	1000	-199.24	-220.26
744	C7H16	2,4-DIMETHYLPENTANE	-158.896	-1.7028E-01	8.4057E-05	298	1000	-202.00	-223.02
745	C7H16	3,3-DIMETHYLPENTANE	-158.436	-1.7028E-01	8.4057E-05	298	1000	-201.54	-222.56
746	C7H16	3-ETHYLPENTANE	-146.553	-1.7028E-01	8.4057E-05	298	1000	-189.66	-210.68
747	C7H16	HEPTANE	-144.670	-1.7028E-01	8.4057E-05	298	1000	-187.78	-208.79
748	C7H16	2-METHYLHEXANE	-151.825	-1.7028E-01	8.4061E-05	298	1000	-194.93	-215.95
749	C7H16	3-METHYLHEXANE	-149.189	-1.7028E-01	8.4057E-05	298	1000	-192.30	-213.31
750	C7H16	2,2,3-TRIMETHYLBUTANE	-162.134	-1.7004E-01	8.7257E-05	298	1000	-204.81	-225.34
751	C7H16O	HEPTYL-ALCOHOL	-291.165	-1.7250E-01	8.5002E-05	298	1000	-334.85	-356.17
752	C7H16O	2-HEPTANOL	-311.820	-1.7147E-01	8.7496E-05	298	TMAX	-355.00	-375.68
753	C7H16O	5-METHYL-1-HEXANOL	-297.211	-1.7734E-01	8.9484E-05	298	1000	-342.00	-363.51
754	C7H16O	ISOPROPYL-TERT-BUTYL-ETHER	-314.687	-1.7156E-01	8.4726E-05	298	1000	-358.15	-379.28
755	C7H16S	1-HEPTANETHIOL	-93.450	-2.2603E-01	1.2590E-04	389	717	-149.62	-174.99
756	C7H16S	BUTYL PROPYL SULFIDE	-88.119	-2.3316E-01	1.3142E-04	389	717	-145.94	-171.85
757	C7H16S	ETHYL PENTYL SULFIDE	-88.684	-2.2992E-01	1.2820E-04	389	717	-145.81	-171.59
758	C7H16S	HEXYL METHYL SULFIDE	-87.031	-2.2651E-01	1.2551E-04	389	717	-143.39	-168.91
759	C7H17N	1-AMINOHEPTANE	-105.803	-1.8708E-01	9.5141E-05	298	1000	-153.00	-175.56
760	C8H4Cl2O2	ISOPHTHALOYL CHLORIDE	-291.881	-3.9238E-02	1.7637E-05	298	1000	-302.00	-307.09
761	C8H4O3	PHTHALIC ANHYDRIDE	-371.308	-8.4063E-02	3.5447E-05	298	1000	-393.13	-404.48
762	C8H6	ETHYNYLBENZENE	340.017	-4.9660E-02	2.2255E-05	298	1000	327.27	320.75
763	C8H6O4	ISOPHTHALIC ACID	-667.093	-1.1323E-01	4.9298E-05	298	1000	-696.30	-711.38
764	C8H6O4	PHTHALIC ACID	-634.123	-1.1323E-01	4.9298E-05	298	1000	-663.33	-678.41
765	C8H6O4	TEREPHTHALIC ACID	-688.693	-1.1323E-01	4.9298E-05	298	1000	-717.90	-732.98
766	C8H6S	BENZOTHIOPHENE	--------	--------	--------	--------	--------	166.30	--------
767	C8H7N	INDOLE	177.626	-8.3827E-02	4.3237E-05	298	1000	156.60	146.52
768	C8H8	STYRENE	167.879	-8.0354E-02	3.7418E-05	298	1000	147.36	137.06
769	C8H8	1,3,5,7-CYCLOOCTATETRAENE	318.704	-8.1831E-02	4.0449E-05	298	1000	298.03	287.90
770	C8H8O	ACETOPHENONE	-57.361	-1.1280E-01	4.8462E-05	298	1000	-86.60	-101.64
771	C8H8O	p-TOLUALDEHYDE	-47.261	-9.1511E-02	4.2560E-05	298	1000	-70.70	-82.38
772	C8H8O2	METHYL BENZOATE	-258.629	-1.1424E-01	5.2031E-05	298	1000	-287.90	-302.74
773	C8H8O2	o-TOLUIC ACID	-293.177	-1.1564E-01	5.3062E-05	298	1000	-322.80	-337.73
774	C8H8O2	p-TOLUIC ACID	-293.177	-1.1564E-01	5.3062E-05	298	1000	-322.80	-337.73
775	C8H8O3	METHYL SALICYLATE	-437.897	-1.0374E-01	4.8667E-05	298	1000	-464.30	-477.60
776	C8H8O3	VANILLIN	-347.013	-9.2602E-02	6.0820E-05	298	1000	-369.00	-378.11
777	C8H9NO	ACETANILIDE	-97.485	-1.2211E-01	5.8889E-05	298	1000	-128.50	-143.82
778	C8H10	ETHYLBENZENE	58.099	-1.1129E-01	5.3183E-05	298	1000	29.79	15.75
779	C8H10	m-XYLENE	46.618	-1.1480E-01	5.3371E-05	298	1000	17.24	2.56
780	C8H10	o-XYLENE	46.756	-1.0824E-01	4.9928E-05	298	1000	19.00	5.12

300

Table 12-1 ENTHALPY OF FORMATION OF GAS - ORGANIC COMPOUNDS (continued)

NO	FORMULA	NAME	$Hf = A + B T + C T^2$					(Hf - kjoule/mol, T - K)	
			A	B	C	TMIN	TMAX	Hf @ 298 K	Hf @ 500 K
781	C8H10	p-XYLENE	47.806	-1.1644E-01	5.3672E-05	298	1000	17.95	3.01
782	C8H10O	m-ETHYLPHENOL	--------	--------	--------	--------	--------	--------	--------
783	C8H10O	p-ETHYLPHENOL	-117.978	-1.0243E-01	4.9778E-05	298	1000	-144.05	-156.75
784	C8H10O	PHENETOLE	-70.148	-1.2514E-01	6.2772E-05	298	1000	-101.70	-117.03
785	C8H10O	2-PHENYLETHANOL	-91.986	-1.1420E-01	5.5180E-05	298	1000	-121.00	-135.29
786	C8H10O	2,3-XYLENOL	-135.112	-8.8671E-02	4.8415E-05	298	1000	-157.19	-167.34
787	C8H10O	2,4-XYLENOL	-138.482	-9.7766E-02	5.2586E-05	298	1000	-162.88	-174.22
788	C8H10O	2,5-XYLENOL	-137.396	-9.7134E-02	5.2436E-05	298	1000	-161.63	-172.85
789	C8H10O	2,6-XYLENOL	-137.065	-9.8691E-02	5.2592E-05	298	1000	-161.75	-173.26
790	C8H10O	3,4-XYLENOL	-134.642	-8.8163E-02	4.8271E-05	298	1000	-156.56	-166.66
791	C8H10O	3,5-XYLENOL	-136.086	-1.0198E-01	5.4706E-05	298	1000	-161.54	-173.40
792	C8H11N	N,N-DIMETHYLANILINE	134.526	-1.3635E-01	7.1737E-05	298	1000	100.50	84.28
793	C8H11N	o-ETHYLANILINE	62.188	-1.1486E-01	5.8487E-05	298	1000	33.30	19.38
794	C8H11N	2,4,6-TRIMETHYLPYRIDINE	57.845	-1.3692E-01	7.1065E-05	298	1000	23.50	7.15
795	C8H11NO	p-PHENETIDINE	-71.265	-1.2736E-01	6.7486E-05	298	1000	-103.00	-118.08
796	C8H12	1,5-CYCLOOCTADIENE	139.715	-1.5214E-01	8.5174E-05	298	1000	102.00	84.94
797	C8H12	VINYLCYCLOHEXENE	99.895	-1.3956E-01	7.4335E-05	298	1000	65.10	48.70
798	C8H12O4	1,4-CYCLOHEXANEDICARBOXYLIC ACID	-835.405	-1.7174E-01	1.0365E-04	298	1000	-877.00	-895.36
799	C8H12O4	DIETHYL MALEATE	-717.456	-9.4041E-02	5.8562E-05	298	1000	-740.00	-749.84
800	C8H14O2	n-BUTYL METHACRYLATE	-382.881	-1.5547E-01	7.9035E-05	298	1000	-422.00	-440.86
801	C8H14O3	BUTYRIC ANHYDRIDE	-628.571	-1.3777E-01	9.9590E-05	298	1000	-660.65	-672.56
802	C8H14O4	DIETHYL SUCCINATE	-815.438	-1.4241E-01	7.4550E-05	298	1000	-851.00	-868.00
803	C8H16	CYCLOOCTANE	-71.827	-2.1661E-01	1.1578E-04	298	1000	-125.77	-151.18
804	C8H16	1,1-DIMETHYLCYCLOHEXANE	-130.382	-2.0582E-01	1.1772E-04	298	1000	-181.00	-203.86
805	C8H16	C-1,2-DIMETHYLCYCLOHEXANE	-122.905	-1.9999E-01	1.1263E-04	298	1000	-172.17	-194.74
806	C8H16	T-1,2-DIMETHYLCYCLOHEXANE	-131.913	-1.9551E-01	1.1093E-04	298	1000	-180.00	-201.94
807	C8H16	C-1,3-DIMETHYLCYCLOHEXANE	-135.419	-2.0059E-01	1.1394E-04	298	1000	-184.77	-207.23
808	C8H16	T-1,3-DIMETHYLCYCLOHEXANE	-127.549	-1.9858E-01	1.1087E-04	298	1000	-176.56	-199.12
809	C8H16	C-1,4-DIMETHYLCYCLOHEXANE	-127.633	-1.9858E-01	1.1087E-04	298	1000	-176.65	-199.20
810	C8H16	T-1,4-DIMETHYLCYCLOHEXANE	-135.913	-1.9809E-01	1.1280E-04	298	1000	-184.60	-206.76
811	C8H16	ETHYLCYCLOHEXANE	-123.344	-1.9633E-01	1.1022E-04	298	1000	-171.75	-193.95
812	C8H16	2-ETHYL-1-HEXENE	-54.553	-1.6767E-01	8.3622E-05	298	1000	-97.00	-117.48
813	C8H16	1-METHYL-1-ETHYLCYCLOPENTANE	-105.480	-1.9752E-01	1.0419E-04	298	1000	-154.90	-178.19
814	C8H16	1-OCTENE	-41.002	-1.6529E-01	8.0839E-05	298	1000	-82.93	-103.44
815	C8H16	trans-2-OCTENE	-51.330	-1.7314E-01	8.5291E-05	298	1000	-95.30	-116.58
816	C8H16	trans-3-OCTENE	-50.429	-1.7429E-01	8.6520E-05	298	1000	-94.60	-115.94
817	C8H16	trans-4-OCTENE	-50.429	-1.7429E-01	8.6520E-05	298	1000	-94.60	-115.94
818	C8H16	PROPYLCYCLOPENTANE	-99.215	-1.9418E-01	9.9081E-05	298	1000	-148.07	-171.53
819	C8H16	2,4,4-TRIMETHYL-1-PENTENE	-69.125	-1.6434E-01	8.5220E-05	298	1000	-110.40	-129.99
820	C8H16	2,4,4-TRIMETHYL-2-PENTENE	-63.181	-1.6552E-01	8.4472E-05	298	1000	-104.90	-124.82
821	C8H16O	2-ETHYLHEXANAL	-254.325	-1.7924E-01	9.0252E-05	298	1000	-299.60	-321.38
822	C8H16O	OCTANAL	-248.111	-1.6447E-01	8.2132E-05	298	1000	-289.66	-309.81
823	C8H16O	2-OCTANONE	-278.758	-1.6835E-01	8.2572E-05	298	1000	-321.60	-342.29
824	C8H16O2	n-BUTYL n-BUTYRATE	-483.962	-1.6743E-01	9.9385E-05	298	1000	-524.90	-542.83
825	C8H16O2	n-HEXYL ACETATE	-482.375	-1.7477E-01	8.9658E-05	298	1000	-526.30	-547.34
826	C8H16O2	ISOBUTYL ISOBUTYRATE	-501.511	-1.8156E-01	9.4041E-05	298	1000	-547.00	-568.78
827	C8H16O2	n-OCTANOIC ACID	-513.734	-1.6791E-01	8.6174E-05	298	1000	-556.00	-576.15
828	C8H16O4	DIETHYLENE GLYCOL ETHYL ETHER ACETATE	-731.025	-1.8041E-01	9.5628E-05	298	1000	-776.00	-797.32
829	C8H18	2,2-DIMETHYLHEXANE	-176.615	-1.9025E-01	9.4496E-05	298	1000	-224.72	-248.12
830	C8H18	2,3-DIMETHYLHEXANE	-165.820	-1.9025E-01	9.4496E-05	298	1000	-213.93	-237.32
831	C8H18	2,4-DIMETHYLHEXANE	-171.301	-1.9025E-01	9.4496E-05	298	1000	-219.41	-242.80
832	C8H18	2,5-DIMETHYLHEXANE	-174.523	-1.9025E-01	9.4496E-05	298	1000	-222.63	-246.02
833	C8H18	3,3-DIMETHYLHEXANE	-172.008	-1.9025E-01	9.4496E-05	298	1000	-220.12	-243.51
834	C8H18	3,4-DIMETHYLHEXANE	-164.900	-1.9025E-01	9.4496E-05	298	1000	-213.01	-236.40
835	C8H18	3-ETHYLHEXANE	-162.766	-1.9025E-01	9.4496E-05	298	1000	-210.87	-234.27
836	C8H18	3-ETHYL-2-METHYLPENTANE	-163.101	-1.9025E-01	9.4496E-05	298	1000	-211.21	-234.60
837	C8H18	3-ETHYL-3-METHYLPENTANE	-166.866	-1.9025E-01	9.4496E-05	298	1000	-214.97	-238.37
838	C8H18	2-METHYLHEPTANE	-167.368	-1.9025E-01	9.4496E-05	298	1000	-215.48	-238.87
839	C8H18	3-METHYLHEPTANE	-164.523	-1.9025E-01	9.4496E-05	298	1000	-212.63	-236.03
840	C8H18	4-METHYLHEPTANE	-163.979	-1.9025E-01	9.4496E-05	298	1000	-212.09	-235.48
841	C8H18	OCTANE	-160.339	-1.9025E-01	9.4491E-05	298	1000	-208.45	-231.84
842	C8H18	2,2,3-TRIMETHYLPENTANE	-172.008	-1.9025E-01	9.4496E-05	298	1000	-220.12	-243.51
843	C8H18	2,2,4-TRIMETHYLPENTANE	-176.029	-1.9025E-01	9.4496E-05	298	1000	-224.14	-247.53
844	C8H18	2,3,3-TRIMETHYLPENTANE	-168.331	-1.9025E-01	9.4496E-05	298	1000	-216.44	-239.83
845	C8H18	2,3,4-TRIMETHYLPENTANE	-169.335	-1.9025E-01	9.4496E-05	298	1000	-217.44	-240.84
846	C8H18	2,2,3,3-TETRAMETHYLBUTANE	-180.895	-1.8156E-01	9.8684E-05	298	1000	-225.89	-247.00
847	C8H18O	BUTYL ETHER	-285.366	-1.9183E-01	9.5366E-05	298	1000	-333.88	-357.44
848	C8H18O	sec-BUTYL ETHER	-312.143	-1.9183E-01	9.5366E-05	298	1000	-360.66	-384.22
849	C8H18O	tert-BUTYL ETHER	-316.327	-1.9183E-01	9.5366E-05	298	1000	-364.84	-388.40
850	C8H18O	2-ETHYL-1-HEXANOL	-315.247	-1.9802E-01	9.9572E-05	298	1000	-365.30	-389.36
851	C8H18O	OCTYL ALCOHOL	-308.352	-1.9264E-01	9.5600E-05	298	1000	-357.06	-380.77
852	C8H18O	2-OCTANOL	-327.637	-1.9255E-01	9.8723E-05	298	1000	-376.10	-399.23
853	C8H18O2	DI-t-BUTYL PEROXIDE	-294.494	-1.8581E-01	9.7806E-05	298	1000	-341.00	-362.95
854	C8H18O2S	DI-n-BUTYL SULFONE	--------	--------	--------	--------	--------	-509.80	--------
855	C8H18O3	DIETHYLENE GLYCOL DIETHYL ETHER	-528.498	-2.0566E-01	1.0756E-04	298	1000	-580.00	-604.44
856	C8H18O3	DIETHYLENE GLYCOL MONOBUTYL EHTER	-554.961	-2.0331E-01	1.0522E-04	298	1000	-606.00	-630.31
857	C8H18O4	TRIETHYLENE GLYCOL DIMETHYL ETHER	-624.184	-2.0732E-01	1.0952E-04	298	1000	-676.00	-700.46
858	C8H18O5	TETRAETHYLENE GLYCOL	-830.684	-2.0581E-01	1.0684E-04	298	1000	-882.32	-906.88

Table 12-1 ENTHALPY OF FORMATION OF GAS - ORGANIC COMPOUNDS (continued)

NO	FORMULA	NAME	Hf = A + B T + C T²					(Hf - kjoule/mol, T - K)	
			A	B	C	TMIN	TMAX	Hf @ 298 K	Hf @ 500 K
859	C8H18S	1-OCTANETHIOL	-108.679	-2.4760E-01	1.3792E-04	389	717	-170.21	-198.00
860	C8H18S	tert-OCTYL MERCAPTAN	--------	--------	--------	--------	--------	-205.00	--------
861	C8H18S	BUTYL SULFIDE	-104.211	-2.5444E-01	1.4318E-04	389	717	-167.32	-195.63
862	C8H18S	ETHYL HEXYL SULFIDE	-104.006	-2.5100E-01	1.3965E-04	389	717	-166.40	-194.59
863	C8H18S	HEPTYL METHYL SULFIDE	-102.290	-2.4794E-01	1.3734E-04	389	717	-163.97	-191.93
864	C8H18S	PENTYL PROPYL SULFIDE	-103.458	-2.5444E-01	1.4318E-04	389	717	-166.57	-194.88
865	C8H18S2	BUTY DISULFIDE	-87.425	-2.8685E-01	1.6310E-04	389	717	-158.41	-190.07
866	C8H19N	DI-n-BUTYLAMINE	-103.230	-2.1250E-01	1.1026E-04	298	1000	-156.60	-181.91
867	C8H19N	DIISOBUTYLAMINE	-125.433	-2.1465E-01	1.1268E-04	298	1000	-179.20	-204.59
868	C8H19N	n-OCTYLAMINE	-579.066	-2.5038E-01	1.2965E-04	298	1000	-174.00	-671.84
869	C8H23N5	TETRAETHYLENEPENTAMINE	79.459	-2.7070E-01	1.5100E-04	298	1000	12.60	-18.14
870	C8H24O4Si4	OCTAMETHYLCYCLOTETRASILOXANE	-2031.337	-3.1622E-01	1.7175E-04	298	1000	-2110.00	-2146.51
871	C9H4O5	TRIMELLITIC ANHYDRIDE	-758.199	-7.9367E-02	4.1184E-05	298	1000	-778.00	-787.59
872	C9H6N2O2	TOLUENE DIISOCYANATE	-210.290	-7.2077E-02	5.9419E-05	298	1000	-226.15	-231.47
873	C9H7N	ISOQUINOLINE	229.225	-8.3071E-02	4.3508E-05	298	1000	208.40	198.57
874	C9H7N	QUINOLINE	242.888	-8.2284E-02	4.3452E-05	298	1000	222.30	212.61
875	C9H7NO	8-HYDROXYQUINOLINE	36.935	-6.3881E-02	3.8039E-05	298	1000	21.50	14.50
876	C9H8	INDENE	186.907	-9.2749E-02	4.4777E-05	298	1000	163.28	151.73
877	C9H8O	2-METHYLBENZOFURAN	12.889	-8.4040E-02	4.2422E-05	298	1000	-8.25	-18.53
878	C9H10	INDANE	92.139	-1.2411E-01	5.9831E-05	298	1000	60.70	45.04
879	C9H10	cis-PROPENYLBENZENE	147.511	-1.0219E-01	4.7304E-05	298	1000	121.34	108.24
880	C9H10	trans-PROPENYLBENZENE	142.704	-1.0009E-01	4.6991E-05	298	1000	117.15	104.41
881	C9H10	alpha-METHYLSTYRENE	139.143	-1.0219E-01	4.7304E-05	298	1000	112.97	99.87
882	C9H10	m-METHYLSTYRENE	141.654	-1.0219E-01	4.7304E-05	298	1000	115.48	102.38
883	C9H10	o-METHYLSTYRENE	144.582	-1.0219E-01	4.7304E-05	298	1000	118.41	105.31
884	C9H10	p-METHYLSTYRENE	140.817	-1.0219E-01	4.7304E-05	298	1000	114.64	101.55
885	C9H10O2	BENZYL ACETATE	-280.642	-1.1392E-01	5.9043E-05	298	1000	-309.16	-322.84
886	C9H10O2	ETHYL BENZOATE	-249.456	-1.3539E-01	6.3216E-05	298	1000	-284.00	-301.35
887	C9H10O3	ETHYL VANILLIN	-376.807	-1.1368E-01	7.2509E-05	298	1000	-404.00	-415.52
888	C9H11NO	p-DIMETHYLAMINOBENZALDEHYDE	4.490	-1.3452E-01	8.3538E-05	298	1000	-27.90	-41.89
889	C9H12	CUMENE	36.884	-1.3023E-01	6.3911E-05	298	1000	3.93	-12.25
890	C9H12	m-ETHYLTOLUENE	31.613	-1.3174E-01	6.2890E-05	298	1000	-1.92	-18.54
891	C9H12	o-ETHYLTOLUENE	33.100	-1.2499E-01	5.9266E-05	298	1000	1.21	-14.58
892	C9H12	p-ETHYLTOLUENE	30.716	-1.3322E-01	6.3086E-05	298	1000	-3.26	-20.12
893	C9H12	MESITYLENE	19.092	-1.3719E-01	6.3609E-05	298	1000	-16.07	-33.60
894	C9H12	PROPYLBENZENE	40.971	-1.3067E-01	6.3463E-05	298	1000	7.82	-8.50
895	C9H12	1,2,3-TRIMETHYLBENZENE	24.674	-1.3318E-01	6.0664E-05	298	1000	-9.58	-26.75
896	C9H12	1,2,4-TRIMETHYLBENZENE	20.256	-1.3321E-01	6.1321E-05	298	1000	-13.93	-31.02
897	C9H12O	BENZYL ETHYL ETHER	-77.933	-1.4763E-01	7.5901E-05	298	1000	-115.00	-132.77
898	C9H12O	2-PHENYL-2-PROPANOL	-271.489	-1.3274E-01	6.6424E-05	298	1000	-305.00	-321.25
899	C9H12O2	CUMENE HYDROPEROXIDE	-46.529	-1.2717E-01	6.6177E-05	298	1000	-78.40	-93.57
900	C9H14O	ISOPHORONE	-213.362	-1.5212E-01	8.4199E-05	298	1000	-251.00	-268.37
901	C9H14O6	GLYCERYL TRIACETATE	-1216.601	-1.4708E-01	8.2198E-05	298	1000	-1252.70	-1269.59
902	C9H16	1-NONYNE	100.487	-1.5226E-01	7.3994E-05	298	1000	61.80	42.85
903	C9H16O4	AZELAIC ACID	-886.623	-1.6150E-01	8.5525E-05	298	1000	-927.00	-945.99
904	C9H18	BUTYLCYCLOPENTANE	-113.730	-2.1699E-01	1.1208E-04	298	1000	-168.28	-194.21
905	C9H18	cis-cis-1,3,5-TRIMETHYLCYCLOHEXANE	-160.440	-2.2370E-01	1.2813E-04	298	1000	-215.39	-240.26
906	C9H18	cis-trans-1,3,5-TRIMETHYLCYCLOHEXANE	-152.252	-2.1975E-01	1.2208E-04	298	1000	-206.56	-231.61
907	C9H18	ISOPROPYLCYCLOHEXANE	-140.627	-2.2375E-01	1.2496E-04	298	1000	-196.00	-221.26
908	C9H18	1-NONENE	-56.530	-1.8549E-01	9.1479E-05	298	1000	-103.51	-126.40
909	C9H18	PROPYLCYCLOHEXANE	-141.001	-2.1184E-01	1.1810E-04	298	1000	-193.30	-217.40
910	C9H18O	DIISOBUTYL KETONE	137.020	-6.8394E-02	3.4220E-05	298	1000	-357.60	111.38
911	C9H18O	NONANAL	-263.630	-1.8483E-01	9.2835E-05	298	1000	-310.29	-332.84
912	C9H18O2	n-BUTYL VALERATE	-512.437	-1.9003E-01	9.7486E-05	298	1000	-560.20	-583.08
913	C9H18O2	n-NONANOIC ACID	-527.747	-1.8948E-01	9.7805E-05	298	1000	-575.40	-598.04
914	C9H18O2	n-OCTYL FORMATE	-462.189	-1.9288E-01	9.9940E-05	298	1000	-510.60	-533.64
915	C9H20	3,3-DIETHYLPENTANE	-178.502	-2.1399E-01	1.1177E-04	298	1000	-231.96	-257.55
916	C9H20	3-ETHYL-2,2-DIMETHYLPENTANE	-185.590	-2.1071E-01	1.0945E-04	298	1000	-238.32	-263.59
917	C9H20	3-ETHYL-2,3-DIMETHYLPENTANE	-180.699	-2.1148E-01	1.1051E-04	298	1000	-233.55	-258.81
918	C9H20	3-ETHYL-2,4-DIMETHYLPENTANE	-180.770	-2.1589E-01	1.0965E-04	298	1000	-235.06	-261.30
919	C9H20	2,2-DIMETHYLHEPTANE	-196.451	-2.0103E-01	1.0391E-04	298	1000	-246.86	-270.99
920	C9H20	2,6-DIMETHYLHEPTANE	-190.460	-2.1002E-01	1.0597E-04	298	1000	-243.38	-268.98
921	C9H20	3-ETHYLHEPTANE	-176.874	-2.1260E-01	1.0711E-04	298	1000	-230.45	-256.40
922	C9H20	4-ETHYLHEPTANE	-176.874	-2.1260E-01	1.0711E-04	298	1000	-230.45	-256.40
923	C9H20	2,3-DIMETHYLHEPTANE	-182.699	-2.1004E-01	1.0584E-04	298	1000	-235.64	-261.26
924	C9H20	2,4-DIMETHYLHEPTANE	-186.586	-2.1416E-01	1.0841E-04	298	1000	-240.50	-266.56
925	C9H20	2,5-DIMETHYLHEPTANE	-186.586	-2.1416E-01	1.0841E-04	298	1000	-240.50	-266.56
926	C9H20	3,4-DIMETHYLHEPTANE	-178.787	-2.1437E-01	1.0847E-04	298	1000	-232.76	-258.85
927	C9H20	3,5-DIMETHYLHEPTANE	-182.653	-2.1864E-01	1.1118E-04	298	1000	-237.61	-264.18
928	C9H20	4,4-DIMETHYLHEPTANE	-190.192	-2.0519E-01	1.0638E-04	298	1000	-241.58	-266.19
929	C9H20	3-ETHYL-2-METHYLHEXANE	-178.787	-2.1437E-01	1.0847E-04	298	1000	-232.76	-258.85
930	C9H20	4-ETHYL-2-METHYLHEXANE	-182.653	-2.1864E-01	1.1118E-04	298	1000	-237.61	-264.18
931	C9H20	3-ETHYL-3-METHYLHEXANE	-183.828	-2.0981E-01	1.0923E-04	298	1000	-236.31	-261.42
932	C9H20	3-ETHYL-4-METHYLHEXANE	-174.887	-2.1870E-01	1.1112E-04	298	1000	-229.87	-256.46
933	C9H20	2,2,3-TRIMETHYLHEXANE	-189.389	-2.0710E-01	1.0784E-04	298	1000	-241.21	-265.98
934	C9H20	2,2,4-TRIMETHYLHEXANE	-190.364	-2.1154E-01	1.1070E-04	298	1000	-243.22	-268.46
935	C9H20	2,3,3-TRIMETHYLHEXANE	-187.004	-2.0710E-01	1.0784E-04	298	1000	-238.82	-263.59
936	C9H20	2,3,4-TRIMETHYLHEXANE	-180.770	-2.1589E-01	1.0965E-04	298	1000	-235.06	-261.30

Table 12-1 ENTHALPY OF FORMATION OF GAS - ORGANIC COMPOUNDS (continued)

NO	FORMULA	NAME	Hf = A + B T + C T²					(Hf - kjoule/mol, T - K)	
			A	B	C	TMIN	TMAX	Hf @ 298 K	Hf @ 500 K
937	C9H20	2,3,5-TRIMETHYLHEXANE	-188.535	-2.1583E-01	1.0972E-04	298	1000	-242.80	-269.02
938	C9H20	2,4,4-TRIMETHYLHEXANE	-187.979	-2.1154E-01	1.1070E-04	298	1000	-240.83	-266.08
939	C9H20	3,3,4-TRIMETHYLHEXANE	-183.084	-2.1148E-01	1.1051E-04	298	1000	-235.94	-261.20
940	C9H20	2-METHYLOCTANE	-184.627	-2.0407E-01	1.0198E-04	298	1000	-236.19	-261.17
941	C9H20	3-METHYLOCTANE	-180.770	-2.0838E-01	1.0459E-04	298	1000	-233.34	-258.81
942	C9H20	4-METHYLOCTANE	-180.770	-2.0838E-01	1.0459E-04	298	1000	-233.34	-258.81
943	C9H20	NONANE	-175.883	-2.1036E-01	1.0501E-04	298	1000	-229.03	-254.81
944	C9H20	2,2,3,3-TETRAMETHYLPENTANE	-186.615	-2.0361E-01	1.0897E-04	298	1000	-237.23	-261.18
945	C9H20	2,2,3,4-TETRAMETHYLPENTANE	-184.803	-2.0877E-01	1.0913E-04	298	1000	-236.98	-261.91
946	C9H20	2,2,4,4-TETRAMETHYLPENTANE	-191.263	-2.0428E-01	1.1009E-04	298	1000	-241.96	-265.88
947	C9H20	2,3,3,4-TETRAMETHYLPENTANE	-184.050	-2.0877E-01	1.0913E-04	298	1000	-236.23	-261.15
948	C9H20	2,2,5-TRIMETHYLHEXANE	-202.213	-2.0707E-01	1.0797E-04	298	1000	-254.01	-278.76
949	C9H20O	2,6-DIMETHYL-4-HEPTANOL	-355.907	-2.1527E-01	1.1131E-04	298	1000	-410.00	-435.72
950	C9H20O	NONYL ALCOHOL	-333.130	-2.1276E-01	1.0614E-04	298	1000	-386.89	-412.98
951	C9H20O	2-NONANOL	-344.161	-2.1365E-01	1.0909E-04	298	1000	-398.00	-423.71
952	C9H20S	1-NONANETHIOL	-123.788	-2.6986E-01	1.5062E-04	389	717	-190.83	-221.06
953	C9H20S	BUTYL-PENTYL-SULFIDE	-119.324	-2.7670E-01	1.5589E-04	389	717	-187.95	-218.70
954	C9H20S	ETHYL-HEPTYL-SULFIDE	-119.119	-2.7327E-01	1.5236E-04	389	717	-187.02	-217.66
955	C9H20S	HEXYL-PROPYL-SULFIDE	-118.529	-2.7671E-01	1.5589E-04	389	717	-187.15	-217.91
956	C9H20S	METHYL-OCTYL-SULFIDE	-117.466	-2.6986E-01	1.4966E-04	389	717	-184.60	-214.98
957	C9H21N	n-NONYLAMINE	-137.483	-2.2815E-01	1.1659E-04	298	1000	-195.00	-222.41
958	C9H21N	TRIPROPYLAMINE	-102.637	-2.3321E-01	1.2285E-04	298	1000	-161.00	-188.53
959	C10H6O8	PYROMELLITIC ACID	-1441.255	-1.4865E-01	6.0126E-05	298	1000	-1480.00	-1500.55
960	C10H7Br	1-BROMONAPHTHALENE	183.741	-6.8169E-02	3.3917E-05	333	1000	182.00	158.14
961	C10H7Cl	1-CHLORONAPHTHALENE	--------	--------	--------	--------	--------	119.80	--------
962	C10H8	NAPHTHALENE	173.657	-8.9278E-02	4.2685E-05	298	1000	150.96	139.69
963	C10H8	AZULENE	303.616	-9.3416E-02	4.4765E-05	298	1000	279.91	268.10
964	C10H9N	QUINALDINE	201.143	-9.7147E-02	5.2183E-05	298	1000	177.00	165.62
965	C10H10	m-DIVINYLBENZENE	234.637	-8.8595E-02	4.1629E-05	298	1000	212.00	200.75
966	C10H10	1-METHYLINDENE	143.132	-8.4924E-02	4.5615E-05	298	1000	122.00	112.07
967	C10H10	2-METHYLINDENE	136.706	-8.7096E-02	4.6507E-05	298	1000	115.00	104.78
968	C10H10O4	DIMETHYL PHTHALATE	-625.546	-1.4636E-01	6.6683E-05	298	1000	-663.00	-682.05
969	C10H10O4	DIMETHYL TEREPHTHALATE	-600.845	-1.4036E-01	6.0329E-05	298	1000	-637.00	-655.94
970	C10H12	DICYCLOPENTADIENE	228.396	-1.3248E-01	7.6971E-05	298	1000	196.10	181.40
971	C10H12	1,2,3,4-TETRAHYDRONAPHTHALENE	62.714	-1.4405E-01	7.5619E-05	298	1000	26.61	9.60
972	C10H12O	ANETHOLE	-7.789	-1.3302E-01	6.5494E-05	298	1000	-41.50	-57.93
973	C10H12O4	DIALLYL MALEATE	-496.232	-8.7546E-02	5.6831E-05	298	1000	-517.00	-525.80
974	C10H14	BUTYLBENZENE	24.380	-1.5078E-01	7.3969E-05	298	1000	-13.81	-32.52
975	C10H14	sec-BUTYLBENZENE	22.746	-1.5537E-01	7.6610E-05	298	1000	-16.90	-35.79
976	C10H14	tert-BUTYLBENZENE	15.797	-1.4893E-01	7.6523E-05	298	1000	-21.63	-39.54
977	C10H14	1,2,3,4-TETRAMETHYLBENZENE	-7.189	-1.3616E-01	6.4978E-05	298	1000	-41.92	-59.02
978	C10H14	m-CYMENE	8.917	-1.5657E-01	7.7412E-05	298	1000	-30.90	-50.02
979	C10H14	o-CYMENE	12.981	-1.5575E-01	7.8155E-05	298	1000	-26.40	-45.36
980	C10H14	p-CYMENE	11.284	-1.5817E-01	7.7663E-05	298	1000	-29.00	-48.38
981	C10H14	m-DIETHYLBENZENE	15.810	-1.4842E-01	7.2266E-05	298	1000	-21.84	-40.34
982	C10H14	o-DIETHYLBENZENE	17.102	-1.4198E-01	6.8910E-05	298	1000	-18.95	-36.66
983	C10H14	p-DIETHYLBENZENE	15.922	-1.5033E-01	7.2797E-05	298	1000	-22.26	-41.04
984	C10H14	2-ETHYL-m-XYLENE	9.805	-1.4203E-01	6.9861E-05	298	1000	-26.23	-43.75
985	C10H14	2-ETHYL-p-XYLENE	5.777	-1.4894E-01	7.1609E-05	298	1000	-32.18	-50.79
986	C10H14	3-ETHYL-o-XYLENE	10.335	-1.4182E-01	6.9723E-05	298	1000	-25.65	-43.15
987	C10H14	4-ETHYL-m-XYLENE	7.157	-1.4894E-01	7.1609E-05	298	1000	-30.80	-49.41
988	C10H14	4-ETHYL-o-XYLENE	5.867	-1.4894E-01	7.1609E-05	298	1000	-32.09	-50.70
989	C10H14	5-ETHYL-m-XYLENE	4.762	-1.5701E-01	7.4273E-05	298	1000	-35.40	-55.18
990	C10H14	ISOBUTYLBENZENE	18.411	-1.5314E-01	7.5762E-05	298	1000	-20.34	-39.22
991	C10H14	1,2,3,5-TETRAMETHYLBENZENE	-8.491	-1.4201E-01	6.6911E-05	298	1000	-44.81	-62.77
992	C10H14	1,2,4,5-TETRAMETHYLBENZENE	-8.663	-1.4271E-01	6.6325E-05	298	1000	-45.27	-63.44
993	C10H14O	p-tert-BUTYLPHENOL	-164.682	-1.4208E-01	7.6348E-05	298	1000	-200.00	-216.64
994	C10H14O2	p-tert-BUTYLCATECHOL	-343.423	-1.3220E-01	7.3527E-05	298	1000	-376.00	-391.14
995	C10H15N	N,N-DIETYHLANILINE	84.795	-1.7874E-01	9.4438E-05	298	1000	40.20	19.03
996	C10H15N	2,6-DIETHYLANILINE	18.033	-1.5307E-01	7.7884E-05	298	1000	-20.50	-39.03
997	C10H16	CAMPHENE	24.226	-2.1618E-01	1.2640E-04	298	1000	-28.60	-52.26
998	C10H16	D-LIMONENE	37.210	-1.7465E-01	9.2824E-05	298	1000	-6.40	-26.91
999	C10H16	alpha-PHELLANDRENE	33.541	-1.7265E-01	9.4055E-05	298	1000	-9.30	-29.27
1000	C10H16	beta-PHELLANDRENE	36.364	-1.6980E-01	9.3481E-05	298	1000	-5.66	-25.17
1001	C10H16	alpha-PINENE	84.670	-2.2942E-01	1.2968E-04	298	1000	28.30	2.38
1002	C10H16	beta-PINENE	89.040	-2.0677E-01	1.2285E-04	298	1000	38.70	16.37
1003	C10H16	alpha-TERPINENE	16.094	-1.4739E-01	7.9478E-05	298	1000	-20.60	-37.73
1004	C10H16	gamma-TERPINENE	25.352	-1.4932E-01	8.1379E-05	298	1000	-11.70	-28.96
1005	C10H16	TERPINOLENE	--------	--------	--------	--------	--------	--------	--------
1006	C10H16O	CAMPHOR	-202.152	-2.2454E-01	1.3127E-04	298	1000	-257.00	-281.61
1007	C10H18	1-DECYNE	84.906	-1.7226E-01	8.4458E-05	298	1000	41.21	19.89
1008	C10H18	cis-DECAHYDRONAPHTHALENE	-108.219	-2.4430E-01	1.3164E-04	298	1000	-168.95	-197.46
1009	C10H18	trans-DECAHYDRONAPHTHALENE	-121.424	-2.4475E-01	1.3190E-04	298	1000	-182.30	-210.82
1010	C10H18O4	SEBACIC ACID	-876.274	-1.8264E-01	9.7076E-05	298	1000	-921.90	-943.32
1011	C10H20	BUTYLCYCLOHEXANE	-155.733	-2.3243E-01	1.2900E-04	298	1000	-213.17	-239.70
1012	C10H20	1-CYCLOPENTYLPENTANE	-129.963	-2.3445E-01	1.2021E-04	298	1000	-188.91	-217.13
1013	C10H20	1-DECENE	-72.195	-2.0530E-01	1.0175E-04	298	1000	-124.14	-149.40
1014	C10H20O	DECANAL	-279.320	-2.0457E-01	1.0311E-04	298	1000	-330.91	-355.83

303

Table 12-1 ENTHALPY OF FORMATION OF GAS - ORGANIC COMPOUNDS (continued)

NO	FORMULA	NAME	Hf = A + B T + C T² A	B	C	TMIN	TMAX	Hf @ 298 K	Hf @ 500 K
1015	C10H20O2	n-DECANOIC ACID	-541.442	-2.1030E-01	1.0891E-04	298	1000	-594.30	-619.36
1016	C10H20O2	2-ETHYLHEXYL ACETATE	-517.576	-2.0874E-01	1.0815E-04	298	1000	-570.00	-594.91
1017	C10H20O2	ISOPENTYL ISOVALERATE	-529.874	-2.1190E-01	1.1017E-04	298	1000	-583.00	-608.28
1018	C10H22	DECANE	-191.468	-2.3050E-01	1.1559E-04	298	1000	-249.66	-277.82
1019	C10H22	2-METHYLNONANE	-200.845	-2.2168E-01	1.1107E-04	298	1000	-256.81	-283.92
1020	C10H22	3-METHYLNONANE	-196.933	-2.2616E-01	1.1381E-04	298	1000	-253.97	-281.56
1021	C10H22	4-METHYLNONANE	-196.933	-2.2616E-01	1.1381E-04	298	1000	-253.97	-281.56
1022	C10H22	5-METHYLNONANE	-196.933	-2.2616E-01	1.1381E-04	298	1000	-253.97	-281.56
1023	C10H22	3-ETHYLOCTANE	-193.016	-2.3049E-01	1.1645E-04	298	1000	-251.08	-279.15
1024	C10H22	4-ETHYLOCTANE	-193.016	-2.3049E-01	1.1645E-04	298	1000	-251.08	-279.15
1025	C10H22	2,2-DIMETHYLOCTANE	-212.602	-2.1889E-01	1.1318E-04	298	1000	-267.48	-293.75
1026	C10H22	2,3-DIMETHYLOCTANE	-202.669	-2.3229E-01	1.1794E-04	298	1000	-261.12	-289.33
1027	C10H22	2,4-DIMETHYLOCTANE	-198.903	-2.2768E-01	1.1498E-04	298	1000	-256.27	-284.00
1028	C10H22	2,5-DIMETHYLOCTANE	-202.669	-2.3229E-01	1.1794E-04	298	1000	-261.12	-289.33
1029	C10H22	2,6-DIMETHYLOCTANE	-202.669	-2.3229E-01	1.1794E-04	298	1000	-261.12	-289.33
1030	C10H22	2,7-DIMETHYLOCTANE	-206.623	-2.2781E-01	1.1519E-04	298	1000	-264.01	-291.73
1031	C10H22	3,3-DIMETHYLOCTANE	-206.313	-2.2322E-01	1.1583E-04	298	1000	-262.21	-288.96
1032	C10H22	3,4-DIMETHYLOCTANE	-194.966	-2.3212E-01	1.1768E-04	298	1000	-253.38	-281.61
1033	C10H22	3,5-DIMETHYLOCTANE	-198.765	-2.3661E-01	1.2056E-04	298	1000	-258.24	-286.93
1034	C10H22	3,6-DIMETHYLOCTANE	-198.765	-2.3661E-01	1.2056E-04	298	1000	-258.24	-286.93
1035	C10H22	4,4-DIMETHYLOCTANE	-206.313	-2.2322E-01	1.1583E-04	298	1000	-262.21	-288.96
1036	C10H22	4,5-DIMETHYLOCTANE	-194.966	-2.3212E-01	1.1768E-04	298	1000	-253.38	-281.61
1037	C10H22	4-PROPYLHEPTANE	-193.016	-2.3049E-01	1.1645E-04	298	1000	-251.08	-279.15
1038	C10H22	4-ISOPROPYLHEPTANE	-192.707	-2.3212E-01	1.1768E-04	298	1000	-251.12	-279.35
1039	C10H22	3-ETHYL-2-METHYLHEPTANE	-194.966	-2.3212E-01	1.1768E-04	298	1000	-253.38	-281.61
1040	C10H22	4-ETHYL-2-METHYLHEPTANE	-198.765	-2.3661E-01	1.2056E-04	298	1000	-258.24	-286.93
1041	C10H22	5-ETHYL-2-METHYLHEPTANE	-198.765	-2.3661E-01	1.2056E-04	298	1000	-258.24	-286.93
1042	C10H22	3-ETHYL-3-METHYLHEPTANE	-200.012	-2.2753E-01	1.1843E-04	298	1000	-256.94	-284.17
1043	C10H22	4-ETHYL-3-METHYLHEPTANE	-191.050	-2.3645E-01	1.2031E-04	298	1000	-250.50	-279.20
1044	C10H22	3-ETHYL-5-METHYLHEPTANE	-194.949	-2.4067E-01	1.2295E-04	298	1000	-255.39	-284.55
1045	C10H22	3-ETHYL-4-METHYLHEPTANE	-191.050	-2.3645E-01	1.2031E-04	298	1000	-250.50	-279.20
1046	C10H22	4-ETHYL-4-METHYLHEPTANE	-200.012	-2.2753E-01	1.1843E-04	298	1000	-256.94	-284.17
1047	C10H22	2,2,3-TRIMETHYLHEPTANE	-205.564	-2.2489E-01	1.1711E-04	298	1000	-261.83	-288.73
1048	C10H22	2,2,4-TRIMETHYLHEPTANE	-206.543	-2.2925E-01	1.1986E-04	298	1000	-263.84	-291.20
1049	C10H22	2,2,5-TRIMETHYLHEPTANE	-214.451	-2.2925E-01	1.1986E-04	298	1000	-271.75	-299.11
1050	C10H22	2,2,6-TRIMETHYLHEPTANE	-218.384	-2.2487E-01	1.1720E-04	298	1000	-274.64	-301.52
1051	C10H22	2,3,3-TRIMETHYLHEPTANE	-203.179	-2.2489E-01	1.1711E-04	298	1000	-259.45	-286.35
1052	C10H22	2,3,4-TRIMETHYLHEPTANE	-196.870	-2.3389E-01	1.1901E-04	298	1000	-255.68	-284.06
1053	C10H22	2,3,5-TRIMETHYLHEPTANE	-200.702	-2.3826E-01	1.2180E-04	298	1000	-260.54	-289.38
1054	C10H22	2,3,6-TRIMETHYLHEPTANE	-204.602	-2.3395E-01	1.1919E-04	298	1000	-263.42	-291.78
1055	C10H22	2,4,4-TRIMETHYLHEPTANE	-204.158	-2.2925E-01	1.1986E-04	298	1000	-261.46	-288.82
1056	C10H22	2,4,5-TRIMETHYLHEPTANE	-200.702	-2.3826E-01	1.2180E-04	298	1000	-260.54	-289.38
1057	C10H22	2,4,6-TRIMETHYLHEPTANE	-193.401	-2.3822E-01	1.2190E-04	298	1000	-253.22	-282.04
1058	C10H22	2,5,5-TRIMETHYLHEPTANE	-212.066	-2.2925E-01	1.1986E-04	298	1000	-269.37	-296.73
1059	C10H22	3,3,4-TRIMETHYLHEPTANE	-199.305	-2.2904E-01	1.1955E-04	298	1000	-256.56	-283.93
1060	C10H22	3,3,5-TRIMETHYLHEPTANE	-200.209	-2.3371E-01	1.2258E-04	298	1000	-258.57	-286.42
1061	C10H22	3,4,4-TRIMETHYLHEPTANE	-199.305	-2.2904E-01	1.1955E-04	298	1000	-256.56	-283.93
1062	C10H22	3,4,5-TRIMETHYLHEPTANE	-192.995	-2.3822E-01	1.2165E-04	298	1000	-252.84	-281.69
1063	C10H22	3-ISOPROPYL-2-METHYLHEXANE	-196.870	-2.3389E-01	1.1901E-04	298	1000	-255.68	-284.06
1064	C10H22	3,3-DIETHYLHEXANE	-193.694	-2.3191E-01	1.2109E-04	298	1000	-251.67	-279.38
1065	C10H22	3,4-DIETHYLHEXANE	-187.054	-2.4094E-01	1.2306E-04	298	1000	-247.57	-276.76
1066	C10H22	3-ETHYL-2,2-DIMETHYLHEXANE	-201.690	-2.2903E-01	1.1955E-04	298	1000	-258.95	-286.32
1067	C10H22	4-ETHYL-2,2-DIMETHYLHEXANE	-202.594	-2.3372E-01	1.2258E-04	298	1000	-260.96	-288.81
1068	C10H22	3-ETHYL-2,3-DIMETHYLHEXANE	-196.920	-2.2904E-01	1.1955E-04	298	1000	-254.18	-281.55
1069	C10H22	4-ETHYL-2,3-DIMETHYLHEXANE	-192.995	-2.3822E-01	1.2165E-04	298	1000	-252.84	-281.69
1070	C10H22	3-ETHYL-2,4-DIMETHYLHEXANE	-192.995	-2.3822E-01	1.2165E-04	298	1000	-252.84	-281.69
1071	C10H22	4-ETHYL-2,4-DIMETHYLHEXANE	-197.824	-2.3371E-01	1.2258E-04	298	1000	-256.19	-284.04
1072	C10H22	3-ETHYL-2,5-DIMETHYLHEXANE	-200.702	-2.3826E-01	1.2180E-04	298	1000	-260.54	-289.38
1073	C10H22	4-ETHYL-3,3-DIMETHYLHEXANE	-195.318	-2.3366E-01	1.2242E-04	298	1000	-253.68	-281.54
1074	C10H22	3-ETHYL-3,4-DIMETHYLHEXANE	-192.933	-2.3366E-01	1.2242E-04	298	1000	-251.29	-279.16
1075	C10H22	2,2,3,3-TETRAMETHYLHEXANE	-202.786	-2.2141E-01	1.1819E-04	298	1000	-257.86	-283.94
1076	C10H22	2,2,3,4-TETRAMETHYLHEXANE	-195.677	-2.3086E-01	1.2096E-04	298	1000	-253.34	-280.87
1077	C10H22	2,2,3,5-TETRAMETHYLHEXANE	-211.300	-2.3101E-01	1.2120E-04	298	1000	-268.99	-296.51
1078	C10H22	2,2,4,4-TETRAMETHYLHEXANE	-201.150	-2.2630E-01	1.2184E-04	298	1000	-257.32	-283.84
1079	C10H22	2,2,4,5-TETRAMETHYLHEXANE	-208.455	-2.3101E-01	1.2120E-04	298	1000	-266.14	-293.66
1080	C10H22	2,2,5,5-TETRAMETHYLHEXANE	-230.158	-2.2183E-01	1.1911E-04	298	1000	-285.27	-311.30
1081	C10H22	2,3,3,4-TETRAMETHYLHEXANE	-196.138	-2.3086E-01	1.2095E-04	298	1000	-253.80	-281.33
1082	C10H22	2,3,3,5-TETRAMETHYLHEXANE	-201.008	-2.3101E-01	1.2120E-04	298	1000	-258.70	-286.21
1083	C10H22	2,3,4,4-TETRAMETHYLHEXANE	-193.292	-2.3086E-01	1.2096E-04	298	1000	-250.96	-278.48
1084	C10H22	2,3,4,5-TETRAMETHYLHEXANE	-198.861	-2.3551E-01	1.2027E-04	298	1000	-258.03	-286.55
1085	C10H22	3,3,4,4-TETRAMETHYLHEXANE	-196.468	-2.2579E-01	1.2086E-04	298	1000	-252.59	-279.15
1086	C10H22	2,4-DIMETHYL-3-ISOPROPYLPENTANE	-200.363	-2.3086E-01	1.2095E-04	298	1000	-258.03	-285.55
1087	C10H22	3,3-DIETHYL-2-METHYLPENTANE	-190.548	-2.3366E-01	1.2242E-04	298	1000	-248.91	-276.77
1088	C10H22	3-ETHYL-2,2,3-TRIMETHYLPENTANE	-196.468	-2.2579E-01	1.2086E-04	298	1000	-252.59	-279.15
1089	C10H22	3-ETHYL-2,2,4-TRIMETHYLPENTANE	-195.677	-2.3086E-01	1.2096E-04	298	1000	-253.34	-280.87
1090	C10H22	3-ETHYL-2,3,4-TRIMETHYLPENTANE	-193.753	-2.3086E-01	1.2095E-04	298	1000	-251.42	-278.94
1091	C10H22	2,2,3,3,4-PENTAMETHYLPENTANE	-191.857	-2.2208E-01	1.1773E-04	298	1000	-247.19	-273.47
1092	C10H22	2,2,3,4,4-PENTAMETHYLPENTANE	-191.560	-2.2343E-01	1.2021E-04	298	1000	-247.02	-273.22

Table 12-1 ENTHALPY OF FORMATION OF GAS - ORGANIC COMPOUNDS (continued)

NO	FORMULA	NAME	$Hf = A + B T + C T^2$						(Hf - kjoule/mol, T - K)	
			A	B	C	TMIN	TMAX	Hf @ 298 K	Hf @ 500 K
1093	C10H22O	DECYL ALCOHOL	-345.109	-2.2961E-01	1.1389E-04	298	1000	-403.25	-431.44
1094	C10H22O	DI-n-PENTYL ETHER	-313.845	-2.3861E-01	1.2181E-04	298	1000	-374.00	-402.70
1095	C10H22O	ISODECANOL	-349.420	-2.3615E-01	1.1964E-04	298	1000	-409.00	-437.58
1096	C10H22O5	TETRAETHYLENE GLYCOL DIMETHYL ETHER	-819.876	-2.5722E-01	1.3658E-04	298	1000	-884.00	-914.34
1097	C10H22S	1-DECANETHIOL	-139.126	-2.9116E-01	1.6240E-04	389	717	-211.46	-244.11
1098	C10H22S	BUTYL-HEXYL-SULFIDE	-134.553	-2.9827E-01	1.6791E-04	389	717	-208.53	-241.71
1099	C10H22S	ETHYL-OCTYL-SULFIDE	-134.390	-2.9484E-01	1.6438E-04	389	717	-207.65	-240.71
1100	C10H22S	HEPTYL-PROPYL-SULFIDE	-133.800	-2.9827E-01	1.6791E-04	389	717	-207.78	-240.96
1101	C10H22S	METHYL-NONYL-SULFIDE	-132.675	-2.9178E-01	1.6206E-04	389	717	-205.23	-238.05
1102	C10H22S	PENTYL-SULFIDE	-134.553	-2.9827E-01	1.6791E-04	389	717	-208.53	-241.71
1103	C10H22S2	PENTYL-DISULFIDE	-117.859	-3.3021E-01	1.8725E-04	389	717	-199.62	-236.15
1104	C10H23N	n-DECYLAMINE	-152.522	-2.4803E-01	1.2718E-04	298	1000	-215.00	-244.74
1105	C11H10	1-METHYLNAPHTHALENE	143.101	-1.0387E-01	5.1388E-05	298	1000	116.86	104.01
1106	C11H10	2-METHYLNAPHTHALENE	142.612	-1.0442E-01	5.0551E-05	298	1000	116.11	103.04
1107	C11H14O2	n-BUTYL BENZOATE	-322.139	-1.7654E-01	8.4988E-05	298	1000	-367.00	-389.16
1108	C11H16	PENTYLBENZENE	8.791	-1.7092E-01	8.4554E-05	298	1000	-34.43	-55.53
1109	C11H16O	p-tert-AMYLPHENOL	-171.396	-1.6313E-01	8.7416E-05	298	1000	-212.00	-231.11
1110	C11H20	1-UNDECYNE	69.266	-1.9221E-01	9.4868E-05	298	1000	20.59	-3.12
1111	C11H20O2	2-ETHYLHEXYL ACRYLATE	-421.941	-1.9332E-01	1.0493E-04	298	1000	-470.00	-492.37
1112	C11H22	1-UNDECENE	-87.793	-2.2544E-01	1.1235E-04	298	1000	-144.77	-172.42
1113	C11H22	1-CYCLOPENTYLHEXANE	-145.553	-2.5440E-01	1.3061E-04	298	1000	-209.49	-240.10
1114	C11H22	PENTYLCYCLOHEXANE	-171.494	-2.5186E-01	1.3907E-04	298	1000	-233.80	-262.66
1115	C11H22O	1-UNDECANAL	-293.144	-2.2893E-01	1.1576E-04	298	1000	-351.00	-378.67
1116	C11H24	UNDECANE	-207.108	-2.5045E-01	1.2600E-04	298	1000	-270.29	-300.84
1117	C11H24O	UNDECYL ALCOHOL	-358.443	-2.5278E-01	1.2710E-04	298	1000	-422.21	-453.06
1118	C11H24S	1-UNDECANETHIOL	-154.289	-3.1307E-01	1.7480E-04	389	717	-232.04	-267.13
1119	C11H24S	BUTYL-HEPTYL-SULFIDE	-149.758	-3.2020E-01	1.8031E-04	389	717	-229.16	-264.78
1120	C11H24S	DECYL-METHYL-SULFIDE	-147.904	-3.1334E-01	1.7408E-04	389	717	-225.81	-261.06
1121	C11H24S	ETHYL-NONYL-SULFIDE	-149.553	-3.1675E-01	1.7678E-04	389	717	-228.24	-263.74
1122	C11H24S	OCTYL-PROPYL-SULFIDE	-148.879	-3.2019E-01	1.8031E-04	389	717	-228.28	-263.90
1123	C12H8O	DIBENZOFURAN	109.415	-1.0434E-01	5.5816E-05	298	1000	83.40	71.20
1124	C12H9N	DIBENZOPYRROLE	230.548	-8.6914E-02	5.2341E-05	298	1000	209.60	200.18
1125	C12H10	ACENAPHTHENE	181.778	-1.0652E-01	5.4770E-05	298	1000	155.00	142.21
1126	C12H10	BIPHENYL	209.418	-1.0832E-01	5.3509E-05	298	1000	182.09	168.64
1127	C12H10O	DIPHENYL ETHER	72.575	-1.1356E-01	5.9724E-05	298	1000	44.30	30.73
1128	C12H11N	p-AMINODIPHENYL	211.465	-1.1025E-01	5.7965E-05	298	1000	184.00	170.83
1129	C12H11N	DIPHENYLAMINE	233.708	-1.2766E-01	6.8309E-05	298	1000	202.00	186.95
1130	C12H11N3	p-AMINOAZOBENZENE	354.296	-1.0364E-01	6.8346E-05	298	1000	330.00	319.56
1131	C12H11N3	1,3-DIPHENYLTRIAZENE	360.821	-1.2326E-01	7.8204E-05	298	1000	331.60	318.74
1132	C12H12	1,2-DIMETHYLNAPHTHALENE	113.951	-1.2076E-01	6.1003E-05	298	1000	83.55	68.82
1133	C12H12	1,3-DIMETHYLNAPHTHALENE	112.458	-1.2131E-01	6.0166E-05	298	1000	81.80	66.85
1134	C12H12	1,4-DIMETHYLNAPHTHALENE	112.905	-1.2076E-01	6.1003E-05	298	1000	82.51	67.78
1135	C12H12	1,5-DIMETHYLNAPHTHALENE	112.194	-1.2076E-01	6.1003E-05	298	1000	81.80	67.07
1136	C12H12	1,6-DIMETHYLNAPHTHALENE	113.169	-1.2131E-01	6.0166E-05	298	1000	82.51	67.56
1137	C12H12	1,7-DIMETHYLNAPHTHALENE	112.458	-1.2131E-01	6.0166E-05	298	1000	81.80	66.85
1138	C12H12	2,3-DIMETHYLNAPHTHALENE	114.947	-1.2315E-01	5.8881E-05	298	1000	83.55	68.09
1139	C12H12	2,6-DIMETHYLNAPHTHALENE	113.031	-1.2015E-01	5.8279E-05	298	1000	82.51	67.52
1140	C12H12	2,7-DIMETHYLNAPHTHALENE	113.048	-1.2017E-01	5.8304E-05	298	1000	82.51	67.54
1141	C12H12	1-ETHYLNAPHTHALENE	127.097	-1.2098E-01	6.1103E-05	298	1000	96.65	81.88
1142	C12H12	2-ETHYLNAPHTHALENE	126.574	-1.2137E-01	6.0120E-05	298	1000	95.90	80.92
1143	C12H12N2	p-AMINODIPHENYLAMINE	237.736	-1.2924E-01	7.2619E-05	298	1000	206.00	191.27
1144	C12H12N2	HYDRAZOBENZENE	353.510	-1.3808E-01	7.1214E-05	298	1000	319.00	302.28
1145	C12H14	1,2,3-TRIMETHYLINDENE	81.516	-1.1941E-01	6.3623E-05	298	1000	51.70	37.72
1146	C12H14O4	DIETHYL PHTHALATE	-640.195	-1.8918E-01	8.9602E-05	298	1000	-688.30	-712.39
1147	C12H16	CYCLOHEXYLBENZENE	30.061	-2.0590E-01	1.1431E-04	298	1000	-20.92	-44.31
1148	C12H18	m-DIISOPROPYLBENZENE	-28.759	-1.9338E-01	9.6994E-05	298	1000	-77.60	-101.20
1149	C12H18	p-DIISOPROPYLBENZENE	-28.759	-1.9338E-01	9.6994E-05	298	1000	-77.60	-101.20
1150	C12H18	HEXYLBENZENE	-6.766	-1.9105E-01	9.5123E-05	298	1000	-55.02	-78.51
1151	C12H18	1,2,3-TRIETHYLBENZENE	-21.216	-1.8403E-01	8.9299E-05	298	1000	-67.99	-90.91
1152	C12H18	1,2,4-TRIETHYLBENZENE	-24.329	-1.8428E-01	9.0148E-05	298	1000	-71.09	-93.93
1153	C12H18	1,3,5-TRIETHYLBENZENE	-27.084	-1.8800E-01	9.2232E-05	298	1000	-74.73	-98.02
1154	C12H18	HEXAMETHYLBENZENE	-66.220	-1.5633E-01	7.8760E-05	298	1000	-105.69	-124.70
1155	C12H20O4	DIBUTYL MALEATE	-635.928	-2.0505E-01	1.0149E-04	298	1000	-688.00	-713.08
1156	C12H22	BICYCLOHEXYL	-200.724	-2.9151E-01	1.7188E-04	298	1000	-272.00	-303.51
1157	C12H22	1-DODECYNE	53.731	-2.1250E-01	1.0555E-04	298	1000	-0.04	-26.13
1158	C12H23N	DICYCLOHEXYLAMINE	-110.761	-3.0528E-01	1.8365E-04	298	1000	-185.00	-217.49
1159	C12H24	1-DODECENE	-103.270	-2.4581E-01	1.2309E-04	298	1000	-165.35	-195.40
1160	C12H24	1-CYCLOPENTYLHEPTANE	-161.088	-2.7469E-01	1.4130E-04	298	1000	-230.12	-263.11
1161	C12H24	1-CYCLOHEXYLHEXANE	-187.016	-2.7204E-01	1.4959E-04	298	1000	-254.39	-285.64
1162	C12H24O	1-DODECANAL	-311.190	-2.5025E-01	1.2753E-04	298	1000	-374.34	-404.43
1163	C12H24O2	n-DODECANOIC ACID	--------	--------	--------	--------	--------	-642.00	--------
1164	C12H26	DODECANE	-222.627	-2.7066E-01	1.3662E-04	298	1000	-290.87	-323.80
1165	C12H26O	DI-n-HEXYL ETHER	-344.953	-2.8013E-01	1.4382E-04	298	1000	-415.50	-449.06
1166	C12H26O	DODECYL ALCOHOL	-373.949	-2.7321E-01	1.3788E-04	298	1000	-442.83	-476.08
1167	C12H26O3	DIETHYLENE GLYCOL DI-n-BUTYL ETHER	-590.003	-2.8740E-01	1.5043E-04	298	1000	-662.00	-696.09
1168	C12H26S	1-DODECANETHIOL	-169.465	-3.3499E-01	1.8712E-04	389	717	-252.67	-290.18
1169	C12H26S	BUTYL-OCTYL-SULFIDE	-164.891	-3.4210E-01	1.9263E-04	389	717	-249.74	-287.79
1170	C12H26S	DECYL-ETHYL-SULFIDE	-166.302	-3.3839E-01	1.8885E-04	389	717	-250.37	-288.28

305

Table 12-1 ENTHALPY OF FORMATION OF GAS - ORGANIC COMPOUNDS (continued)

NO	FORMULA	NAME	Hf = A + B T + C T²					(Hf - kjoule/mol, T - K)	
			A	B	C	TMIN	TMAX	Hf @ 298 K	Hf @ 500 K
1171	C12H26S	HEXYL-SULFIDE	-164.891	-3.4210E-01	1.9263E-04	389	717	-249.74	-287.79
1172	C12H26S	METHYL-UNDECYL-SULFIDE	-163.080	-3.3533E-01	1.8654E-04	389	717	-246.44	-284.11
1173	C12H26S	NONYL-PROPYL-SULFIDE	-164.138	-3.4211E-01	1.9263E-04	389	717	-248.99	-287.03
1174	C12H26S2	HEXYL-DISULFIDE	-148.143	-3.7432E-01	2.1230E-04	389	717	-240.83	-282.23
1175	C12H27BO3	TRI-n-BUTYL BORATE	--------	--------	--------	--------	--------	-1147.00	--------
1176	C12H27N	DODECYLAMINE	-183.122	-2.8945E-01	1.4891E-04	298	1000	-256.00	-290.62
1177	C12H27N	TRI-n-BUTYLAMINE	-159.282	-2.9448E-01	1.5515E-04	298	1000	-233.00	-267.73
1178	C13H10	FLUORENE	215.700	-1.1460E-01	5.8808E-05	298	1000	186.90	173.10
1179	C13H10O	BENZOPHENONE	89.679	-1.2105E-01	5.9124E-05	298	1000	59.00	43.94
1180	C13H12	DIPHENYLMETHANE	187.533	-1.3233E-01	6.4432E-05	298	1000	154.00	137.48
1181	C13H14	1-PROPYLNAPHTHALENE	109.997	-1.4045E-01	7.1467E-05	298	1000	74.68	57.64
1182	C13H14	2-PROPYLNAPHTHALENE	108.935	-1.3961E-01	6.9705E-05	298	1000	73.47	56.56
1183	C13H14	2-ETHYL-3-METHYLNAPHTHALENE	101.307	-1.4001E-01	6.8350E-05	298	1000	65.77	48.39
1184	C13H14	2-ETHYL-6-METHYLNAPHTHALENE	95.165	-1.3307E-01	6.3325E-05	298	1000	61.30	44.46
1185	C13H14	2-ETHYL-7-METHYLNAPHTHALENE	94.676	-1.3168E-01	6.2396E-05	298	1000	60.92	44.44
1186	C13H20	1-PHENYLHEPTANE	-22.304	-2.1132E-01	1.0576E-04	298	1000	-75.65	-101.52
1187	C13H24	1-TRIDECYNE	38.150	-2.3254E-01	1.1606E-04	298	1000	-20.63	-49.10
1188	C13H26	1-TRIDECENE	-118.905	-2.6579E-01	1.3354E-04	298	1000	-185.98	-218.42
1189	C13H26	1-CYCLOPENTYLOCTANE	-176.669	-2.9473E-01	1.5180E-04	298	1000	-250.71	-286.08
1190	C13H26	1-CYCLOHEXYLHEPTANE	-202.665	-2.9196E-01	1.6004E-04	298	1000	-275.01	-308.63
1191	C13H26O	1-TRIDECANAL	-323.856	-2.7012E-01	1.3773E-04	298	1000	-392.00	-424.48
1192	C13H26O2	n-BUTYL NONANOATE	-558.963	-2.7259E-01	1.4115E-04	298	1000	-627.40	-659.97
1193	C13H26O2	METHYL DODECANOATE	-544.190	-2.7084E-01	1.3970E-04	298	1000	-612.30	-644.68
1194	C13H28	TRIDECANE	-238.304	-2.9046E-01	1.4691E-04	298	1000	-311.50	-346.81
1195	C13H28O	1-TRIDECANOL	-389.581	-2.9316E-01	1.4827E-04	298	1000	-463.46	-499.09
1196	C13H28S	BUTYL-NONYL-SULFIDE	-167.184	-4.3363E-01	2.8067E-04	389	717	-270.37	-313.83
1197	C13H28S	DECYL-PROPYL-SULFIDE	-179.439	-3.6355E-01	2.0446E-04	389	717	-269.62	-310.10
1198	C13H28S	DODECYL-METHYL-SULFIDE	-178.310	-3.5690E-01	1.9856E-04	389	717	-267.02	-307.12
1199	C13H28S	ETHYL-UNDECYL-SULFIDE	-179.958	-3.6031E-01	2.0125E-04	389	717	-269.45	-309.80
1200	C13H28S	1-TRIDECANETHIOL	-184.632	-3.5690E-01	1.9952E-04	389	717	-273.26	-313.20
1201	C14H8O2	ANTHRAQUINONE	-75.986	-7.8031E-02	4.3832E-05	298	1000	-95.20	-104.04
1202	C14H10	ANTHRACENE	255.757	-1.0192E-01	5.1871E-05	298	1000	230.10	217.76
1203	C14H10	DIPHENYLACETYLENE	455.661	-1.0037E-01	4.7752E-05	298	1000	430.12	417.41
1204	C14H10	PHENANTHRENE	232.090	-9.9304E-02	5.0512E-05	298	1000	207.10	195.07
1205	C14H12	cis-STILBENE	278.989	-1.3263E-01	6.2865E-05	298	1000	245.18	228.39
1206	C14H12	trans-STILBENE	273.199	-1.2642E-01	6.0279E-05	298	1000	241.00	225.06
1207	C14H12O2	BENZYL BENZOATE	-149.022	-1.6604E-01	8.1128E-05	298	1000	-191.00	-211.76
1208	C14H14	1,1-DIPHENYLETHANE	154.821	-1.5343E-01	7.5596E-05	298	1000	116.00	97.01
1209	C14H14	1,2-DIPHENYLETHANE	181.751	-1.5317E-01	7.5515E-05	298	1000	143.00	124.05
1210	C14H14O	DIBENZYL ETHER	63.789	-1.7811E-01	9.3537E-05	298	1000	19.30	-1.88
1211	C14H16	1-BUTYLNAPHTHALENE	93.412	-1.6063E-01	8.1965E-05	298	1000	53.05	33.59
1212	C14H16	2-BUTYLNAPHTHALENE	92.960	-1.6135E-01	8.1333E-05	298	1000	52.30	32.62
1213	C14H22	1-PHENYLOCTANE	-37.842	-2.3155E-01	1.1642E-04	298	1000	-96.23	-124.51
1214	C14H22	1,2,3,4-TETRAETHYLBENZENE	-71.827	-2.0406E-01	1.0327E-04	298	1000	-123.26	-148.04
1215	C14H22	1,2,3,5-TETRAETHYLBENZENE	-69.810	-2.0996E-01	1.0524E-04	298	1000	-122.84	-148.48
1216	C14H22	1,2,4,5-TETRAETHYLBENZENE	-69.986	-2.1047E-01	1.0450E-04	298	1000	-123.26	-149.09
1217	C14H22O	p-tert-OCTYLPHENOL	252.536	-1.0875E-01	5.5373E-05	298	1000	-291.00	212.01
1218	C14H28	1-TETRADECENE	-134.411	-2.8591E-01	1.4409E-04	298	1000	-206.52	-241.35
1219	C14H28	1-CYCLOPENTYLNONANE	-192.313	-3.1465E-01	1.6220E-04	298	1000	-271.33	-309.09
1220	C14H28	1-CYCLOHEXYLOCTANE	-218.154	-3.1231E-01	1.7073E-04	298	1000	-295.60	-331.62
1221	C14H28O2	n-TETRADECANOIC ACID	-617.472	-3.0443E-01	1.5915E-04	298	1000	-693.80	-729.90
1222	C14H30	TETRADECANE	-253.835	-3.1085E-01	1.5769E-04	298	1000	-332.13	-369.84
1223	C14H30O	1-TETRADECANOL	-405.220	-3.1313E-01	1.5868E-04	298	1000	-484.09	-522.11
1224	C14H30S	BUTYL-DECYL-SULFIDE	-195.531	-3.8483E-01	2.1623E-04	389	717	-291.00	-333.89
1225	C14H30S	DODECYL-ETHYL-SULFIDE	-195.230	-3.8187E-01	2.1328E-04	389	717	-290.08	-332.85
1226	C14H30S	HEPTYL-SULFIDE	-195.489	-3.8483E-01	2.1624E-04	389	717	-290.96	-333.85
1227	C14H30S	METHYL-TRIDECYL-SULFIDE	-193.514	-3.7881E-01	2.1095E-04	389	717	-287.65	-330.18
1228	C14H30S	PROPYL-UNDECYL-SULFIDE	-194.736	-3.8483E-01	2.1624E-04	389	717	-290.20	-333.09
1229	C14H30S	1-TETRADECANETHIOL	-219.212	-3.0867E-01	1.5044E-04	389	717	-293.88	-335.94
1230	C14H30S2	HEPTYL-DISULFIDE	-178.582	-4.1781E-01	2.3672E-04	389	717	-282.04	-328.31
1231	C14H31N	TETRADECYLAMINE	-214.866	-3.3050E-01	1.7077E-04	298	1000	-298.00	-337.42
1232	C15H10N2O2	DIPHENYLMETHANE-4,4'-DIISOCYANATE	-25.466	-1.2031E-01	6.0428E-05	298	1000	-55.80	-70.52
1233	C15H16O	p-CUMYLPHENOL	-24.023	-1.6693E-01	8.8788E-05	298	1000	-65.60	-85.29
1234	C15H16O2	BISPHENOL A	-208.442	-1.5026E-01	8.2010E-05	298	1000	-245.60	-263.07
1235	C15H18	1-PENTYLNAPHTHALENE	77.760	-1.8050E-01	9.2458E-05	298	1000	32.43	10.63
1236	C15H18	2-PENTYLNAPHTHALENE	77.308	-1.8119E-01	9.1734E-05	298	1000	31.67	9.65
1237	C15H24	1-PHENYLNONANE	-53.505	-2.5137E-01	1.2672E-04	298	1000	-116.86	-147.51
1238	C15H24O	2,6-DI-tert-BUTYL-p-CRESOL	-281.074	-2.4111E-01	1.3041E-04	298	1000	-341.00	-369.02
1239	C15H24O	NONYLPHENOL	-228.782	-2.4773E-01	1.2828E-04	298	1000	-291.00	-320.58
1240	C15H28	1-PENTADECYNE	7.001	-2.7273E-01	1.3712E-04	298	1000	-61.84	-95.09
1241	C15H30	1-PENTADECENE	-150.172	-3.0572E-01	1.5438E-04	298	1000	-227.23	-264.44
1242	C15H30	1-CYCLOPENTYLDECANE	-207.916	-3.3477E-01	1.7277E-04	298	1000	-291.96	-332.11
1243	C15H30	1-CYCLOHEXYLNONANE	-233.718	-3.3252E-01	1.8133E-04	298	1000	-316.23	-354.64
1244	C15H30O2	PENTADECANOIC ACID	-620.587	-3.1215E-01	1.6223E-04	298	1000	-699.00	-736.11
1245	C15H32	PENTADECANE	-269.383	-3.3113E-01	1.6836E-04	298	1000	-352.75	-392.86
1246	C15H32O	1-PENTADECANOL	-420.743	-3.3326E-01	1.6925E-04	298	1000	-504.67	-545.06
1247	C15H32S	BUTYL-UNDECYL-SULFIDE	-210.698	-4.0674E-01	2.2863E-04	389	717	-311.58	-356.91
1248	C15H32S	DODECYL-PROPYL-SULFIDE	-209.945	-4.0675E-01	2.2863E-04	389	717	-310.83	-356.16

Table 12-1 ENTHALPY OF FORMATION OF GAS - ORGANIC COMPOUNDS (continued)

NO	FORMULA	NAME	$Hf = A + BT + CT^2$					(Hf - kjoule/mol, T - K)	
			A	B	C	TMIN	TMAX	Hf @ 298 K	Hf @ 500 K
1249	C15H32S	ETHYL-TRIDECYL-SULFIDE	-210.342	-4.0414E-01	2.2598E-04	389	717	-310.70	-355.92
1250	C15H32S	METHYL-TETRADECYL-SULFIDE	-206.577	-4.0986E-01	2.3128E-04	389	717	-308.28	-353.69
1251	C15H32S	1-PENTADECANETHIOL	-215.079	-4.0045E-01	2.2399E-04	389	717	-314.51	-359.31
1252	C16H10	FLUORANTHENE	309.398	-8.3025E-02	4.5244E-05	298	1000	288.90	279.20
1253	C16H10	PYRENE	--------	--------	--------	--------	--------	225.00	--------
1254	C16H12	1-PHENYLNAPHTHALENE	278.006	-1.2292E-01	6.1052E-05	298	1000	247.00	231.81
1255	C16H20	1-n-HEXYLNAPHTHALENE	65.665	-2.0404E-01	1.0524E-04	298	1000	14.40	-10.05
1256	C16H22O4	DIBUTYL PHTHALATE	-678.046	-2.8955E-01	1.4976E-04	298	1000	-750.90	-785.38
1257	C16H26	1-PHENYLDECANE	-69.053	-2.7164E-01	1.3739E-04	298	1000	-137.49	-170.53
1258	C16H26	PENTAETHYLBENZENE	-116.420	-2.3422E-01	1.2113E-04	298	1000	-175.18	-203.25
1259	C16H30	1-HEXADECYNE	-8.587	-2.9288E-01	1.4771E-04	298	1000	-82.47	-118.10
1260	C16H32	1-CYCLOHEXYLDECANE	-249.484	-3.5196E-01	1.9143E-04	298	1000	-336.85	-377.61
1261	C16H32	1-CYCLOPENTYLUNDECANE	-223.413	-3.5503E-01	1.8337E-04	298	1000	-312.54	-355.08
1262	C16H32	1-HEXADECENE	-165.661	-3.2601E-01	1.6506E-04	298	1000	-247.82	-287.40
1263	C16H32O2	n-HEXADECANOIC ACID	-650.399	-3.4531E-01	1.8040E-04	298	1000	-737.00	-777.95
1264	C16H34	HEXADECANE	-284.968	-3.5109E-01	1.7876E-04	298	1000	-373.34	-415.82
1265	C16H34O	DI-n-OCTYL ETHER	-406.781	-3.6268E-01	1.8749E-04	298	1000	-498.00	-541.25
1266	C16H34O	1-HEXADECANOL	-436.308	-3.5339E-01	1.7982E-04	298	1000	-525.26	-568.05
1267	C16H34S	BUTYL-DODECYL-SULFIDE	-225.873	-4.2865E-01	2.4096E-04	389	717	-332.21	-379.96
1268	C16H34S	ETHYL-TETRADECYL-SULFIDE	-225.572	-4.2572E-01	2.3800E-04	389	717	-331.29	-378.93
1269	C16H34S	METHYL-PENTADECYL-SULFIDE	-223.735	-4.2321E-01	2.3625E-04	389	717	-328.86	-376.28
1270	C16H34S	OCTYL-SULFIDE	-225.827	-4.2865E-01	2.4096E-04	389	717	-332.17	-379.91
1271	C16H34S	PROPYL-TRIDECYL-SULFIDE	-225.120	-4.2865E-01	2.4096E-04	389	717	-331.46	-379.21
1272	C16H34S	1-HEXADECANETHIOL	-230.283	-4.2208E-01	2.3601E-04	389	717	-335.10	-382.32
1273	C16H34S2	OCTYL-DISULFIDE	-209.029	-4.6137E-01	2.6119E-04	389	717	-323.30	-374.42
1274	C17H28	1-PHENYLUNDECANE	-84.626	-2.9169E-01	1.4792E-04	298	1000	-158.07	-193.49
1275	C17H32	1-HEPTADECYNE	-24.190	-3.1298E-01	1.5823E-04	298	1000	-103.09	-141.12
1276	C17H34	1-CYCLOPENTYLDODECANE	-241.731	-3.7517E-01	1.9397E-04	298	1000	-335.89	-380.82
1277	C17H34	1-CYCLOHEXYLUNDECANE	-264.939	-3.7246E-01	2.0220E-04	298	1000	-357.44	-400.62
1278	C17H34	1-HEPTADECENE	-181.247	-3.4605E-01	1.7557E-04	298	1000	-268.40	-310.38
1279	C17H36	HEPTADECANE	-300.524	-3.7123E-01	1.8936E-04	298	1000	-393.92	-438.80
1280	C17H36O	1-HEPTADECANOL	-451.914	-3.7345E-01	1.9030E-04	298	1000	-545.89	-591.06
1281	C17H36S	BUTYL-TRIDECYL-SULFIDE	-241.036	-4.5058E-01	2.5335E-04	389	717	-352.79	-402.99
1282	C17H36S	ETHYL-PENTADECYL-SULFIDE	-240.844	-4.4735E-01	2.5015E-04	389	717	-351.92	-401.98
1283	C17H36S	HEXADECYL-METHYL-SULFIDE	-238.358	-4.4786E-01	2.5141E-04	389	717	-349.49	-399.43
1284	C17H36S	PROPYL-TETRADECYL-SULFIDE	-240.283	-4.5058E-01	2.5336E-04	389	717	-352.04	-402.23
1285	C17H36S	1-HEPTADECANETHIOL	-245.492	-4.4401E-01	2.4840E-04	389	717	-355.72	-405.39
1286	C18H12	CHRYSENE	293.383	-1.2134E-01	6.3145E-05	298	1000	262.90	248.50
1287	C18H14	m-TERPHENYL	311.129	-1.3525E-01	6.6219E-05	298	1000	277.00	260.06
1288	C18H14	o-TERPHENYL	311.129	-1.3525E-01	6.6219E-05	298	1000	277.00	260.06
1289	C18H14	p-TERPHENYL	311.129	-1.3525E-01	6.6219E-05	298	1000	277.00	260.06
1290	C18H15P	TRIPHENYLPHOSPHINE	--------	--------	--------	--------	--------	320.20	--------
1291	C18H15O4P	TRIPHENYL PHOSPHATE	--------	--------	--------	--------	--------	-757.30	--------
1292	C18H16N2	N,N'-DIPHENYL-p-PHENYLENEDIAMINE	365.182	-1.8010E-01	1.0146E-04	298	1000	321.00	300.50
1293	C18H22	2,3-DIMETHYL-2,3-DIPHENYLBUTANE	116.841	-2.3213E-01	1.2540E-04	298	1000	59.20	32.13
1294	C18H22O2	DICUMYL PEROXIDE	-20.392	-2.3544E-01	1.2264E-04	298	1000	-79.40	-107.45
1295	C18H30	1-PHENYLDODECANE	-100.265	-3.1165E-01	1.5832E-04	298	1000	-178.70	-216.51
1296	C18H30	HEXAETHYLBENZENE	-159.766	-2.5818E-01	1.3624E-04	298	1000	-224.26	-254.80
1297	C18H32O2	LINOLEIC ACID	-454.177	-3.3929E-01	1.6828E-04	298	1000	-540.00	-581.75
1298	C18H34	1-OCTADECYNE	-39.684	-3.3326E-01	1.6888E-04	298	TMAX	-123.68	-164.09
1299	C18H34O2	OLEIC ACID	-582.748	-3.5441E-01	1.8363E-04	298	1000	-671.78	-714.05
1300	C18H34O4	DIBUTYL SEBACATE	-972.794	-3.4865E-01	1.8368E-04	298	1000	-1060.00	-1101.20
1301	C18H34O4	DIHEXYL ADIPATE	-920.964	-3.7565E-01	1.4273E-04	298	1000	-1020.00	-1073.11
1302	C18H36	1-CYCOPENTYLTRIDECANE	-254.596	-3.9514E-01	2.0440E-04	298	1000	-353.76	-401.07
1303	C18H36	1-CYCLOHEXYLDODECANE	-280.554	-3.9243E-01	2.1267E-04	298	1000	-378.07	-423.60
1304	C18H36	1-OCTADECENE	-196.811	-3.6625E-01	1.8617E-04	298	1000	-289.03	-333.40
1305	C18H36O2	STEARIC ACID	-672.844	-3.7517E-01	1.9606E-04	298	1000	-767.00	-811.41
1306	C18H38	OCTADECANE	-316.097	-3.9135E-01	1.9989E-04	298	1000	-414.55	-461.80
1307	C18H38O	DINONYL ETHER	-437.406	-4.0424E-01	2.0964E-04	298	1000	-539.00	-587.12
1308	C18H38O	1-OCTADECANOL	-467.437	-3.9371E-01	2.0095E-04	298	1000	-566.47	-614.06
1309	C18H38S	BUTYL-TETRADECYL-SULFIDE	-256.308	-4.7216E-01	2.6538E-04	389	717	-373.42	-426.04
1310	C18H38S	ETHYL-HEXADECYL-SULFIDE	-256.052	-4.6898E-01	2.6216E-04	389	717	-372.50	-425.01
1311	C18H38S	HEPTADECYL-METHYL-SULFIDE	-254.237	-4.6627E-01	2.6016E-04	389	717	-370.07	-422.33
1312	C18H38S	NONYL-SULFIDE	-256.308	-4.7216E-01	2.6538E-04	389	717	-373.42	-426.04
1313	C18H38S	PENTADECYL-PROPYL-SULFIDE	-255.555	-4.7216E-01	2.6538E-04	389	717	-372.67	-425.29
1314	C18H38S	1-OCTADECANETHIOL	-260.626	-4.6593E-01	2.6072E-04	389	717	-376.31	-428.41
1315	C18H38S2	NONYL-DISULFIDE	-239.467	-5.0484E-01	2.8561E-04	389	717	-364.51	-420.49
1316	C19H26	1-n-NONYLNAPHTHALENE	21.002	-2.7166E-01	1.3803E-04	298	1000	-47.50	-80.32
1317	C19H32	1-PHENYLTRIDECANE	-115.809	-3.3179E-01	1.6891E-04	298	1000	-199.28	-239.48
1318	C19H36	1-NONADECYNE	-55.321	-3.5323E-01	1.7932E-04	298	1000	-144.31	-187.10
1319	C19H36O2	METHYL OLEATE	-532.381	-3.7255E-01	1.9287E-04	298	1000	-626.00	-670.44
1320	C19H38	1-CYCLOPENTYLTETRADECANE	-270.144	-4.1542E-01	2.1507E-04	298	1000	-374.38	-424.09
1321	C19H38	1-CYCLOHEXYLTRIDECANE	-296.114	-4.1254E-01	2.2319E-04	298	1000	-398.65	-446.59
1322	C19H38	1-NONADECENE	-212.342	-3.8644E-01	1.9678E-04	298	1000	-309.62	-356.37
1323	C19H38O2	NONADECANOIC ACID	-686.322	-3.9424E-01	2.0546E-04	298	1000	-785.30	-832.08
1324	C19H40	NONADECANE	-331.691	-4.1132E-01	2.1033E-04	298	1000	-435.14	-484.77
1325	C19H40O	1-NONADECANOL	-483.043	-4.1369E-01	2.1139E-04	298	1000	-587.10	-637.04
1326	C19H40S	BUTYL-PENTADECYL-SULFIDE	-283.244	-4.2986E-01	1.9400E-04	389	717	-394.05	-449.68

Table 12-1 ENTHALPY OF FORMATION OF GAS - ORGANIC COMPOUNDS (continued)

| NO | FORMULA | NAME | Hf = A + B T + C T² | | | | | (Hf - kjoule/mol, T - K) | |
			A	B	C	TMIN	TMAX	Hf @ 298 K	Hf @ 500 K
1327	C19H40S	ETHYL-HEPTADECYL-SULFIDE	-271.161	-4.9124E-01	2.7486E-04	389	717	-393.13	-448.07
1328	C19H40S	HEXADECYL-PROPYL-SULFIDE	-270.571	-4.9467E-01	2.7840E-04	389	717	-393.25	-448.31
1329	C19H40S	METHYL-OCTADECYL-SULFIDE	-269.508	-4.8785E-01	2.7217E-04	389	717	-390.70	-445.39
1330	C19H40S	1-NONADECANETHIOL	-275.834	-4.8785E-01	2.7312E-04	389	717	-396.94	-451.48
1331	C20H16	TRIPHENYLETHYLENE	379.398	-1.5599E-01	7.7536E-05	298	1000	340.00	320.79
1332	C20H28	1-n-DECYLNAPHTHALENE	5.297	-2.9153E-01	1.4824E-04	298	1000	-68.20	-103.41
1333	C20H30O2	ABIETIC ACID	-450.790	-3.5730E-01	2.1909E-04	298	1000	-537.00	-574.67
1334	C20H31N	DEHYDROABIETYLAMINE	-48.979	-3.0188E-01	1.7398E-04	298	1000	-123.00	-156.43
1335	C20H34	1-PHENYLTETRADECANE	-131.432	-3.5174E-01	1.7930E-04	298	1000	-219.91	-262.48
1336	C20H38	1-EICOSYNE	-70.869	-3.7336E-01	1.8988E-04	298	1000	-164.89	-210.08
1337	C20H40	1-CYCLOPENTYLPENTADECANE	-285.767	-4.3539E-01	2.2550E-04	298	1000	-395.01	-447.09
1338	C20H40	1-CYCLOHEXYLTETRADECANE	-311.733	-4.3263E-01	2.3371E-04	298	1000	-419.28	-469.62
1339	C20H40	1-EICOSENE	-227.969	-4.0640E-01	2.0717E-04	298	1000	-330.24	-379.38
1340	C20H42	EICOSANE	-347.285	-4.3145E-01	2.2089E-04	298	1000	-455.76	-507.79
1341	C20H42O	1-EICOSANOL	-498.691	-4.3367E-01	2.2187E-04	298	1000	-607.73	-660.06
1342	C20H42S	BUTYL-HEXADECYL-SULFIDE	-286.596	-5.1626E-01	2.9042E-04	389	717	-414.63	-472.12
1343	C20H42S	DECYL-SULFIDE	-286.596	-5.1626E-01	2.9042E-04	389	717	-414.63	-472.12
1344	C20H42S	ETHYL-OCTADECYL-SULFIDE	-286.433	-5.1283E-01	2.8689E-04	389	717	-413.76	-471.13
1345	C20H42S	HEPTADECYL-PROPYL-SULFIDE	-285.843	-5.1626E-01	2.9042E-04	389	717	-413.88	-471.37
1346	C20H42S	METHYL-NONADECYL-SULFIDE	-284.717	-5.0978E-01	2.8457E-04	389	717	-411.33	-468.46
1347	C20H42S	1-EICOSANETHIOL	-291.169	-5.0915E-01	2.8491E-04	389	717	-417.56	-474.52
1348	C20H42S2	DECYL-DISULFIDE	-269.902	-5.4819E-01	3.0977E-04	389	717	-405.72	-466.55
1349	C21H21O4P	TRI-o-CRESYL PHOSPHATE	--------	--------	--------	--------	--------	-851.00	--------
1350	C21H36	1-PHENYLPENTADECANE	-147.022	-3.7187E-01	1.8986E-04	298	1000	-240.54	-285.49
1351	C21H42	1-CYCLOPENTYLHEXADECANE	-301.365	-4.5534E-01	2.3594E-04	298	1000	-415.60	-470.05
1352	C21H42	1-CYCLOHEXYLPENTADECANE	-327.310	-4.5275E-01	2.4430E-04	298	1000	-439.91	-492.61
1353	C22H38	1-PHENYLHEXADECANE	-162.578	-3.9202E-01	2.0045E-04	298	1000	-261.12	-308.47
1354	C22H44	1-CYCLOHEXYLHEXADECANE	-342.720	-4.7342E-01	2.5522E-04	298	1000	-460.49	-515.63
1355	C22H44O2	n-BUTYL STEARATE	-704.384	-4.5783E-01	2.3897E-04	298	1000	-819.27	-873.56
1356	C24H38O4	DIISOOCTYL PHTHALATE	-854.131	-4.4764E-01	2.2602E-04	298	1000	-967.00	-1021.44
1357	C24H38O4	DIOCTYL PHTHALATE	-853.426	-4.4938E-01	2.2737E-04	298	1000	-966.72	-1021.27
1358	C24H42O	DINONYLPHENOL	-380.940	-4.2350E-01	2.2265E-04	298	1000	-487.00	-537.03
1359	C26H20	TETRAPHENYLETHYLENE	486.457	-1.9196E-01	9.5847E-05	298	1000	438.00	414.44
1360	C28H46O4	DIISODECYL PHTHALATE	-926.335	-5.3104E-01	2.7036E-04	298	1000	-1060.10	-1124.27

Hf - enthalpy of formation of ideal gas, kjoule/mol

A, B, and C - regression coefficients for chemical compound

T - temperature, K

TMIN - minimum temperature, K

TMAX - maximum temperature, K

Additional compounds are:

H2O(g) A = -238.41,   B = -1.2256E-02,   C = 2.7656E-06

HCl(g) A = -91.264,   B = -3.5626E-03,   C = 4.0022E-07

Table 12-2 ENTHALPY AND INTERNAL ENERGY OF FORMATION - INORGANIC COMPOUNDS

NO	FORMULA	NAME	Molecular Weight g/mol	State	Hf @ 298 K kjoule/mol	Uf @ 298 K kjoule/mol	S @ 298 K joule/(mol K)
1	Ag	SILVER	107.868	solid	0.00	0.00	42.60
2	AgCl	SILVER CHLORIDE	143.321	solid	-127.00	-125.76	96.30
3	AgI	SILVER IODIDE	234.773	solid	-61.80	-61.80	115.50
4	Al	ALUMINUM	26.982	solid	0.00	0.00	28.30
5	AlB3H12	ALUMINUM BOROHYDRIDE	71.510	gas	13.00	25.39	379.20
6	AlBr3	ALUMINUN BROMIDE	266.694	solid	-527.18	-527.19	-------
7	AlCl3	ALUMINUM CHLORIDE	133.340	solid	-704.20	-700.48	110.70
8	AlF3	ALUMINUM FLUORIDE	83.977	solid	-1510.40	-1506.68	66.50
9	AlI3	ALUMINUM IODIDE	407.695	solid	-313.80	-313.80	159.00
10	Al2O3	ALUMINUM OXIDE	101.961	solid	-1675.70	-1671.98	50.90
11	Al2S3O12	ALUMINUM SULFATE	342.154	solid	-3440.80	-3425.93	239.30
12	Ar	ARGON	39.948	gas	0.00	0.00	154.80
13	As	ARSENIC	74.922	solid	0.00	0.00	35.10
14	AsBr3	ARSENIC TRIBROMIDE	314.634	gas	-130.00	-132.48	-------
15	AsCl3	ARSENIC TRICHLORIDE	181.280	gas	-261.50	-260.26	327.20
16	AsF3	ARSENIC TRIFLUORIDE	131.917	gas	-785.80	-784.56	289.10
17	AsF5	ARSENIC PENTAFLUORIDE	169.914	gas	-1069.43	-1065.71	326.35
18	AsH3	ARSINE	77.945	gas	66.40	67.64	222.80
19	AsI3	ARSENIC TRIIODIDE	455.635	solid	-58.20	-58.20	213.10
20	As2O3	ARSENIC TRIOXIDE	197.841	solid	-639.73	-636.02	-------
21	At	ASTATINE	210.000	solid	0.00	0.00	-------
22	Au	GOLD	196.967	solid	0.00	0.00	47.40
23	B	BORON	10.811	solid	0.00	0.00	5.90
24	BBr3	BORON TRIBROMIDE	250.523	gas	-205.60	-208.08	324.20
25	BCl3	BORON TRICHLORIDE	117.169	gas	-403.80	-402.56	290.10
26	BF3	BORON TRIFLUORIDE	67.806	gas	-1136.00	-1134.76	254.40
27	BH2CO	BORINE CARBONYL	40.837	gas	-------	-------	-------
28	BH3O3	BORIC ACID	61.833	solid	-1094.30	-1086.86	88.80
29	B2D6	DEUTERODIBORANE	33.718	gas	12.72	17.68	-------
30	B2H5Br	DIBORANE HYDROBROMIDE	106.566	gas	-------	-------	-------
31	B2H6	DIBORANE	27.670	gas	35.60	40.56	232.10
32	B3N3H6	BORINE TRIAMINE	80.501	liquid	-541.00	-529.85	199.60
33	B4H10	TETRABORANE	53.323	gas	66.10	88.41	-------
34	B5H9	PENTABORANE	63.126	gas	73.20	81.88	275.90
35	B5H11	TETRAHYDROPENTABORANE	65.142	gas	103.30	114.45	-------
36	B10H14	DECABORANE	122.221	gas	-------	-------	-------
37	Ba	BARIUM	137.327	solid	0.00	0.00	62.80
38	Be	BERYLLIUM	9.012	solid	0.00	0.00	9.50
39	BeB2H8	BERYLLIUM BOROHYDRIDE	38.698	gas	-------	-------	-------
40	BeBr2	BERYLLIUM BROMIDE	168.820	solid	-353.50	-353.50	-------
41	BeCl2	BERYLLIUM CHLORIDE	79.918	solid	-490.40	-487.92	82.70
42	BeF2	BERYLLIUM FLUORIDE	47.009	solid	-1026.80	-1024.32	53.40
43	BeI2	BERYLLIUM IODIDE	262.821	solid	-192.50	-192.50	-------
44	Bi	BISMUTH	208.980	solid	0.00	0.00	56.70
45	BiBr3	BISMUTH TRIBROMIDE	448.692	solid	-271.96	-271.96	-------
46	BiCl3	BISMUTH TRICHLORIDE	315.338	solid	-379.10	-375.38	177.00
47	BrF5	BROMINE PENTAFLUORIDE	174.896	gas	-428.90	-425.18	320.20
48	Br2	BROMINE	159.808	gas	30.90	28.42	245.50
49	C	CARBON	12.011	solid	0.00	0.00	5.70
50	CCl2O	PHOSGENE	98.916	gas	-219.10	-217.86	283.50
51	CF2O	CARBONYL FLUORIDE	66.007	gas	-638.90	-637.66	-------
52	CH4N2O	UREA	60.056	solid	-333.60	-324.92	-------
53	CH4N2S	THIOUREA	76.122	gas	-93.10	-88.14	303.10
54	CNBr	CYANOGEN BROMIDE	105.922	gas	186.20	184.96	248.30
55	CNCl	CYANOGEN CHLORIDE	61.470	gas	138.00	138.00	236.20
56	CNF	CYANOGEN FLUORIDE	45.016	gas	-12.55	-12.55	224.70
57	CO	CARBON MONOXIDE	28.010	gas	-110.50	-111.74	197.70
58	COS	CARBONYL SULFIDE	60.076	gas	-142.00	-143.24	231.60
59	COSe	CARBON OXYSELENIDE	106.970	gas	-------	-------	-------
60	CO2	CARBON DIOXIDE	44.010	gas	-393.50	-393.50	213.80
61	CS2	CARBON DISULFIDE	76.143	gas	116.60	114.12	237.80
62	CSeS	CARBON SELENOSULFIDE	123.037	gas	-------	-------	256.23
63	C2N2	CYANOGEN	52.035	gas	306.70	306.70	241.90
64	C3S2	CARBON SUBSULFIDE	100.165	gas	-------	-------	-------
65	Ca	CALCIUM	40.078	solid	0.00	0.00	41.60
66	CaF2	CALCIUM FLUORIDE	78.075	solid	-1228.00	-1225.52	68.50
67	CbF5	COLUMBIUM FLUORIDE	187.898	solid	-1813.76	-1807.57	160.25
68	Cd	CADMIUM	112.411	solid	0.00	0.00	51.80
69	CdCl2	CADMIUM CHLORIDE	183.316	solid	-391.50	-389.02	115.30
70	CdF2	CADMIUM FLUORIDE	150.408	solid	-700.40	-697.92	77.40
71	CdI2	CADMIUM IODIDE	366.220	solid	-203.30	-203.30	161.10
72	CdO	CADMIUM OXIDE	128.410	solid	-258.40	-257.16	54.80
73	ClF	CHLORINE MONOFLUORIDE	54.451	gas	-50.30	-50.30	217.90
74	ClFO3	PERCHLORYL FLUORIDE	102.449	gas	-23.80	-20.08	279.00
75	ClF3	CHLORINE TRIFLUORIDE	92.448	gas	-163.20	-160.72	281.60
76	ClF5	CHLORINE PENTAFLUORIDE	130.445	gas	-------	-------	-------
77	ClHO3S	CHLOROSULFONIC ACID	116.525	gas	-------	-------	-------
78	ClHO4	PERCHLORIC ACID	100.458	gas	5.23	10.19	-------

Table 12-2 ENTHALPY AND INTERNAL ENERGY OF FORMATION - INORGANIC COMPOUNDS (continued)

NO	FORMULA	NAME	Molecular Weight g/mol	State	Hf @ 298 K kjoule/mol	Uf @ 298 K kjoule/mol	S @ 298 K joule/(mol K)
79	ClO2	CHLORINE DIOXIDE	67.452	gas	102.50	103.74	256.80
80	Cl2	CHLORINE	70.905	gas	0.00	0.00	223.10
81	Cl2O	CHLORINE MONOXIDE	86.905	gas	80.30	81.54	266.20
82	Cl2O7	CHLORINE HEPTOXIDE	182.901	gas	286.94	295.61	-------
83	Co	COBALT	58.933	solid	0.00	0.00	30.00
84	CoCl2	COBALT CHLORIDE	129.839	solid	-312.50	-310.02	109.20
85	CoNC3O4	COBALT NITROSYL TRICARBONYL	172.971	gas	-------	-------	-------
86	Cr	CHROMIUM	51.996	solid	0.00	0.00	23.80
87	CrC6O6	CHROMIUM CARBONYL	220.059	solid	-1075.62	-1068.19	-------
88	CrO2Cl2	CHROMIUM OXYCHLORIDE	154.900	gas	-538.10	-535.62	329.80
89	Cs	CESIUM	132.905	solid	0.00	0.00	85.20
90	CsBr	CESIUM BROMIDE	212.809	solid	-405.80	-405.79	113.10
91	CsCl	CESIUM CHLORIDE	168.358	solid	-443.00	-441.76	101.20
92	CsF	CESIUM FLUORIDE	151.904	solid	-553.50	-552.26	92.80
93	CsI	CESIUM IODIDE	259.810	solid	-346.60	-346.60	123.10
94	Cu	COPPER	63.546	solid	0.00	0.00	33.20
95	CuBr	CUPROUS BROMIDE	143.450	solid	-104.60	-104.60	96.10
96	CuCl	CUPROUS CHLORIDE	98.999	solid	-137.20	-135.96	86.20
97	CuCl2	CUPRIC CHLORIDE	134.451	solid	-220.10	-217.62	108.10
98	CuI	COPPER IODIDE	190.450	solid	-67.80	-67.80	96.70
99	DCN	DEUTERIUM CYANIDE	28.034	gas	-------	-------	-------
100	D2	DEUTERIUM	4.032	gas	0.00	0.00	144.86
101	D2O	DEUTERIUM OXIDE	20.031	gas	-249.20	-247.96	198.34
102	Eu	EUROPIUM	151.965	solid	0.00	0.00	77.80
103	F2	FLUORINE	37.997	gas	0.00	0.00	202.80
104	F2O	FLUORINE OXIDE	53.996	gas	24.70	25.94	247.40
105	Fe	IRON	55.847	solid	0.00	0.00	27.30
106	FeC5O5	IRON PENTACARBONYL	195.899	gas	-747.68	-743.96	444.34
107	FeCl2	FERROUS CHLORIDE	126.752	solid	-341.80	-339.32	118.00
108	FeCl3	FERRIC CHLORIDE	162.205	solid	-399.50	-395.78	142.30
109	Fr	FRANCIUM	223.000	solid	0.00	0.00	95.40
110	Ga	GALLIUM	69.723	solid	0.00	0.00	40.90
111	GaCl3	GALLIUM TRICHLORIDE	176.081	solid	-524.70	-520.98	142.00
112	Gd	GADOLINIUM	157.250	solid	0.00	0.00	68.10
113	Ge	GERMANIUM	72.610	solid	0.00	0.00	31.10
114	GeBr4	GERMANIUM BROMIDE	392.226	gas	-300.00	-302.48	396.20
115	GeCl4	GERMANIUM CHLORIDE	214.421	gas	-495.80	-493.32	347.70
116	GeHCl3	TRICHLORO GERMANE	179.976	gas	-------	-------	-------
117	GeH4	GERMANE	76.642	gas	90.80	93.28	217.10
118	Ge2H6	DIGERMANE	151.268	gas	162.30	167.26	-------
119	Ge3H8	TRIGERMANE	225.894	gas	226.80	234.24	-------
120	HBr	HYDROGEN BROMIDE	80.912	gas	-36.30	-37.54	198.70
121	HCN	HYDROGEN CYANIDE	27.026	gas	135.10	135.10	201.80
122	HCl	HYDROGEN CHLORIDE	36.461	gas	-92.30	-92.30	186.90
123	HF	HYDROGEN FLUORIDE	20.006	gas	-273.30	-273.30	173.80
124	HI	HYDROGEN IODIDE	127.912	gas	26.50	25.26	206.60
125	HNO3	NITRIC ACID	63.013	gas	-135.10	-131.38	266.40
126	H2	HYDROGEN	2.016	gas	0.00	0.00	130.70
127	H2O	WATER	18.015	gas	-241.80	-240.56	188.80
128	H2O2	HYDROGEN PEROXIDE	34.015	gas	-136.30	-133.82	232.70
129	H2S	HYDROGEN SULFIDE	34.082	gas	-20.60	-20.60	205.80
130	H2SO4	SULFURIC ACID	98.079	gas	-735.13	-730.17	298.70
131	H2S2	HYDROGEN DISULFIDE	66.148	gas	15.50	15.50	-------
132	H2Se	HYDROGEN SELENIDE	80.976	gas	29.70	29.70	219.00
133	H2Te	HYDROGEN TELLURIDE	129.616	gas	99.60	99.60	-------
134	H3NO3S	SULFAMIC ACID	97.095	solid	-674.90	-666.22	-------
135	He	HELIUM-3	3.016	gas	0.00	0.00	-------
136	He	HELIUM-4	4.003	gas	0.00	0.00	126.04
137	Hf	HAFNIUM	178.490	solid	0.00	0.00	43.60
138	Hg	MERCURY	200.590	liquid	0.00	0.00	75.90
139	HgBr2	MERCURIC BROMIDE	360.398	solid	-170.70	-170.70	172.00
140	HgCl2	MERCURIC CHLORIDE	271.495	solid	-224.30	-221.82	146.00
141	HgI2	MERCURIC IODIDE	454.399	solid	-105.40	-105.40	180.00
142	IF7	IODINE HEPTAFLUORIDE	259.893	gas	-956.04	-949.85	349.78
143	I2	IODINE	253.809	solid	0.00	0.00	116.10
144	In	INDIUM	114.818	solid	0.00	0.00	57.80
145	Ir	IRIDIUM	192.220	solid	0.00	0.00	35.50
146	K	POTASSIUM	39.098	solid	0.00	0.00	64.70
147	KBr	POTASSIUM BROMIDE	119.002	solid	-393.80	-393.80	95.90
148	KCl	POTASSIUM CHLORIDE	74.551	solid	-436.50	-435.26	82.60
149	KF	POTASSIUM FLUORIDE	58.097	solid	-567.30	-566.06	66.60
150	KI	POTASSIUM IODIDE	166.003	solid	-327.90	-327.90	106.30
151	KOH	POTASSIUM HYDROXIDE	56.106	gas	-232.63	-232.63	236.27
152	Kr	KRYPTON	83.800	gas	0.00	0.00	164.10
153	La	LANTHANUM	138.906	solid	0.00	0.00	56.90
154	Li	LITHIUM	6.941	solid	0.00	0.00	29.10
155	LiBr	LITHIUM BROMIDE	86.845	solid	-351.20	-351.20	74.30
156	LiCl	LITHIUM CHLORIDE	42.394	solid	-408.60	-407.36	59.30

NO	FORMULA	NAME	Molecular Weight g/mol	State	Hf @ 298 K kjoule/mol	Uf @ 298 K kjoule/mol	S @ 298 K joule/(mol K)
157	LiF	LITHIUM FLUORIDE	25.939	solid	-616.00	-614.76	35.70
158	LiI	LITHIUM IODIDE	133.845	solid	-270.40	-270.40	86.80
159	Lu	LUTECIUM	174.967	solid	0.00	0.00	51.00
160	Mg	MAGNESIUM	24.305	solid	0.00	0.00	32.70
161	MgCl2	MAGNESIUM CHLORIDE	95.210	solid	-641.30	-638.82	89.60
162	MgO	MAGNESIUM OXIDE	40.304	solid	-601.60	-600.36	27.00
163	Mn	MANGANESE	54.938	solid	0.00	0.00	32.00
164	MnCl2	MANGANESE CHLORIDE	125.843	solid	-481.30	-478.82	118.20
165	Mo	MOLYBDENUM	95.940	solid	0.00	0.00	28.70
166	MoF6	MOLYBDENUM FLUORIDE	209.930	gas	-1557.70	-1552.74	350.50
167	MoO3	MOLYBDENUM OXIDE	143.938	solid	-745.10	-741.38	77.70
168	NCl3	NITROGEN TRICHLORIDE	120.365	liquid	230.00	234.96	-------
169	ND3	HEAVY AMMONIA	20.055	gas	-------	-------	203.89
170	NF3	NITROGEN TRIFLUORIDE	71.002	gas	-132.10	-129.62	260.80
171	NH3	AMMONIA	17.031	gas	-45.90	-43.42	192.80
172	NH3O	HYDROXYLAMINE	33.030	solid	-114.20	-108.00	-------
173	NH4Br	AMMONIUM BROMIDE	97.943	solid	-270.80	-264.60	113.00
174	NH4Cl	AMMONIUM CHLORIDE	53.491	solid	-314.40	-306.96	94.60
175	NH4I	AMMONIUM IODIDE	144.943	solid	-201.40	-195.20	117.00
176	NH5O	AMMONIUM HYDROXIDE	35.046	liquid	-361.20	-352.52	165.60
177	NH5S	AMMONIUM HYDROGENSULFIDE	51.112	gas	-------	-------	-------
178	NO	NITRIC OXIDE	30.006	gas	90.30	90.30	210.80
179	NOCl	NITROSYL CHLORIDE	65.459	gas	51.70	52.94	261.70
180	NOF	NITROSYL FLUORIDE	49.005	gas	-66.50	-65.26	248.10
181	NO2	NITROGEN DIOXIDE	46.006	gas	33.20	34.44	240.10
182	N2	NITROGEN	28.013	gas	0.00	0.00	191.60
183	N2F4	TETRAFLUOROHYDRAZINE	104.007	gas	-8.40	-3.44	301.20
184	N2H4	HYDRAZINE	32.045	gas	95.40	100.36	238.50
185	N2H4C	AMMONIUM CYANIDE	44.056	solid	0.40	7.84	-------
186	N2H6CO2	AMMONIUM CARBAMATE	78.071	solid	-645.05	-632.65	133.47
187	N2O	NITROUS OXIDE	44.013	gas	82.10	83.34	219.90
188	N2O3	NITROGEN TRIOXIDE	76.012	gas	83.70	87.42	312.30
189	N2O4	NITROGEN TETRAOXIDE	92.011	gas	9.20	14.16	304.30
190	N2O5	NITROGEN PENTOXIDE	108.010	gas	11.30	17.50	355.70
191	Na	SODIUM	22.990	solid	0.00	0.00	51.30
192	NaBr	SODIUM BROMIDE	102.894	solid	-361.10	-361.10	86.80
193	NaCN	SODIUM CYANIDE	49.008	solid	-87.50	-86.26	115.60
194	NaCl	SODIUM CHLORIDE	58.442	solid	-411.20	-409.96	72.10
195	NaF	SODIUM FLUORIDE	41.988	solid	-576.60	-575.36	51.10
196	NaI	SODIUM IODIDE	149.894	solid	-287.80	-287.80	98.50
197	NaOH	SODIUM HYDROXIDE	39.997	solid	-425.60	-423.12	64.50
198	Na2SO4	SODIUM SULFATE	142.043	solid	-1387.10	-1382.14	149.60
199	Nb	NIOBIUM	92.906	solid	0.00	0.00	36.40
200	Nd	NEODYMIUM	144.240	solid	0.00	0.00	71.50
201	Ne	NEON	20.180	gas	0.00	0.00	146.30
202	Ni	NICKEL	58.693	solid	0.00	0.00	29.90
203	NiC4O4	NICKEL CARBONYL	170.735	gas	-220.08	-217.60	399.15
204	NiF2	NICKEL FLUORIDE	96.690	solid	-651.40	-648.92	73.60
205	Np	NEPTUNIUM	237.000	solid	0.00	0.00	50.63
206	O2	OXYGEN	31.999	gas	0.00	0.00	205.20
207	O3	OZONE	47.998	gas	142.70	143.94	238.90
208	Os	OSMIUM	190.230	solid	0.00	0.00	32.60
209	OsOF5	OSMIUM OXIDE PENTAFLUORIDE	301.221	gas	-------	-------	-------
210	OsO4	OSMIUM TETROXIDE - YELLOW	254.228	gas	-337.20	-334.72	293.80
211	OsO4	OSMIUM TETROXIDE - WHITE	254.228	gas	-337.20	-334.72	293.80
212	P	PHOSPHORUS - WHITE	30.974	solid	0.00	0.00	41.10
213	PBr3	PHOSPHORUS TRIBROMIDE	270.686	gas	-139.30	-141.78	348.10
214	PCl2F3	PHOSPHORUS DICHLORIDE TRIFLUORIDE	158.874	gas	-------	-------	-------
215	PCl3	PHOSPHORUS TRICHLORIDE	137.332	gas	-287.00	-285.76	311.80
216	PCl5	PHOSPHORUS PENTACHLORIDE	208.237	gas	-374.90	-371.18	364.60
217	PH3	PHOSPHINE	33.998	gas	5.40	6.64	210.20
218	PH4Br	PHOSPHONIUM BROMIDE	114.910	solid	-127.61	-122.65	110.04
219	PH4Cl	PHOSPHONIUM CHLORIDE	70.458	solid	-145.18	-138.99	-------
220	PH4I	PHOSPHONIUM IODIDE	161.910	solid	-69.87	-64.92	123.01
221	POCl3	PHOSPHORUS OXYCHLORIDE	153.331	gas	-558.50	-556.02	325.50
222	PSBr3	PHOSPHORUS THIOBROMIDE	302.752	gas	-263.59	-266.07	-------
223	PSCl3	PHOSPHORUS THIOCHLORIDE	169.398	gas	-363.17	-361.93	337.23
224	P4O6	PHOSPHORUS TRIOXIDE	219.891	gas	-2144.47	-2139.51	346.73
225	P4O10	PHOSPHORUS PENTOXIDE	283.889	gas	-2834.00	-2824.08	403.76
226	P4S10	PHOSPHORUS PENTASULFIDE	444.555	solid	-309.00	-309.00	381.70
227	Pb	LEAD	207.200	solid	0.00	0.00	64.80
228	PbBr2	LEAD BROMIDE	367.008	solid	-278.70	-278.70	161.50
229	PbCl2	LEAD CHLORIDE	278.105	solid	-359.40	-356.92	136.00
230	PbF2	LEAD FLUORIDE	245.197	solid	-664.00	-661.52	110.50
231	PbI2	LEAD IODIDE	461.009	solid	-175.50	-175.50	174.90
232	PbO	LEAD OXIDE	223.199	solid	-217.30	-216.06	68.70
233	PbS	LEAD SULFIDE	239.266	solid	-100.40	-100.40	91.20
234	Pd	PALLADIUM	106.420	solid	0.00	0.00	37.60

NO	FORMULA	NAME	Molecular Weight g/mol	State	Hf @ 298 K kjoule/mol	Uf @ 298 K kjoule/mol	S @ 298 K joule/(mol K)
235	Po	POLONIUM	209.000	solid	0.00	0.00	-------
236	Pt	PLATINUM	195.080	solid	0.00	0.00	41.60
237	Ra	RADIUM	226.000	solid	0.00	0.00	71.00
238	Rb	RUBIDIUM	85.468	solid	0.00	0.00	76.80
239	RbBr	RUBIDIUM BROMIDE	165.372	solid	-394.60	-394.60	110.00
240	RbCl	RUBIDIUM CHLORIDE	120.921	solid	-435.40	-434.16	95.90
241	RbF	RUBIDIUM FLUORIDE	104.466	solid	-557.70	-556.46	-------
242	RbI	RUBIDIUM IODIDE	212.372	solid	-333.80	-333.80	118.40
243	Re	RHENIUM	186.207	solid	0.00	0.00	36.90
244	Re2O7	RHENIUM HEPTOXIDE	484.410	solid	-1240.10	-1231.42	207.10
245	Rh	RHODIUM	102.906	solid	0.00	0.00	31.50
246	Rn	RADON	222.000	gas	0.00	0.00	176.20
247	Ru	RUTHENIUM	101.070	soild	0.00	0.00	28.50
248	RuF5	RUTHENIUM PENTAFLUORIDE	196.062	solid	-892.87	-886.67	-------
249	S	SULFUR	32.066	solid	0.00	0.00	32.10
250	SF4	SULFUR TETRAFLUORIDE	108.060	gas	-763.20	-760.72	299.60
251	SF6	SULFUR HEXAFLUORIDE	146.056	gas	-1220.50	-1215.54	291.50
252	SOBr2	THIONYL BROMIDE	207.873	gas	-74.06	-75.30	332.80
253	SOCl2	THIONYL CHLORIDE	118.971	gas	-212.50	-211.26	309.80
254	SOF2	SULFUROUS OXYFLUORIDE	86.062	gas	-713.51	-712.27	278.70
255	SO2	SULFUR DIOXIDE	64.065	gas	-296.80	-296.80	248.20
256	SO2Cl2	SULFURYL CHLORIDE	134.970	gas	-364.00	-361.52	311.90
257	SO3	SULFUR TRIOXIDE	80.064	gas	-395.70	-394.46	256.80
258	S2Cl2	SULFUR MONOCHLORIDE	135.037	liquid	-59.40	-56.92	-------
259	Sb	ANTIMONY	121.757	solid	0.00	0.00	45.70
260	SbBr3	ANTIMONY TRIBROMIDE	361.469	solid	-259.40	-259.40	207.10
261	SbCl3	ANTIMONY TRICHLORIDE	228.115	solid	-382.20	-378.48	184.10
262	SbCl5	ANTIMONY PENTACHLORIDE	299.021	gas	-394.34	-390.62	401.83
263	SbH3	STIBINE	124.781	gas	145.10	146.34	232.80
264	SbI3	ANTIMONY TRIIODIDE	502.470	solid	-100.40	-100.40	-------
265	Sb2O3	ANTIMONY TRIOXIDE	291.512	solid	-698.73	-695.01	-------
266	Sc	SCANDIUM	44.956	solid	0.00	0.00	34.60
267	Se	SELENIUM	78.960	solid	0.00	0.00	42.40
268	SeCl4	SELENIUM TETRACHLORIDE	220.771	solid	-189.54	-184.58	184.10
269	SeF6	SELENIUM HEXAFLUORIDE	192.950	gas	-1117.00	-1112.04	313.90
270	SeOCl2	SELENIUM OXYCHLORIDE	165.865	gas	-25.10	-23.86	-------
271	SeO2	SELENIUM DIOXIDE	110.959	solid	-225.40	-222.92	-------
272	Si	SILICON	28.086	soild	0.00	0.00	18.80
273	SiBrCl2F	BROMODICHLOROFLUOROSILANE	197.893	gas	-------	-------	-------
274	SiBrF3	TRIFLUOROBROMOSILANE	164.985	gas	-------	-------	-------
275	SiBr2ClF	DIBROMOCHLOROFLUOROSILANE	242.345	gas	-------	-------	-------
276	SiClF3	TRIFLUOROCHLOROSILANE	120.533	gas	-1317.96	-1315.48	307.82
277	SiCl2F2	DICHLORODIFLUOROSILANE	136.988	gas	-------	-------	-------
278	SiCl3F	TRICHLOROFLUOROSILANE	153.442	gas	-840.98	-838.51	335.95
279	SiCl4	SILICON TETRACHLORIDE	169.896	gas	-657.00	-654.52	330.70
280	SiF4	SILICON TETRAFLUORIDE	104.079	gas	-1615.00	-1612.52	282.80
281	SiHBr3	TRIBROMOSILANE	268.805	gas	-317.60	-318.84	348.60
282	SiHCl3	TRICHLOROSILANE	135.452	gas	-513.00	-510.52	313.90
283	SiHF3	TRIFLUOROSILANE	86.089	gas	-1184.07	-1181.59	271.90
284	SiH2Br2	DIBROMOSILANE	189.909	gas	-------	-------	309.70
285	SiH2Cl2	DICHLOROSILANE	101.007	gas	-320.49	-318.01	285.70
286	SiH2F2	DIFLUOROSILANE	68.098	gas	-811.70	-809.22	262.00
287	SiH2I2	DIIODOSILANE	283.910	gas	-------	-------	-------
288	SiH3Br	MONOBROMOSILANE	111.013	gas	-------	-------	262.40
289	SiH3Cl	MONOCHLOROSILANE	66.562	gas	-171.54	-169.07	250.70
290	SiH3F	MONOFLUOROSILANE	50.108	gas	-439.32	-436.84	238.40
291	SiH3I	IODOSILANE	158.014	gas	-------	-------	270.90
292	SiH4	SILANE	32.117	gas	34.30	36.78	204.60
293	SiO2	SILICON DIOXIDE	60.084	soild	-910.70	-908.22	41.50
294	Si2Cl6	HEXACHLORODISILANE	268.887	gas	-------	-------	-------
295	Si2F6	HEXAFLUORODISILANE	170.161	gas	-------	-------	-------
296	Si2H5Cl	DISILANYL CHLORIDE	96.663	gas	-------	-------	-------
297	Si2H6	DISILANE	62.219	gas	80.30	85.26	272.70
298	Si2OCl3F3	TRICHLOROTRIFLUORODISILOXANE	235.524	gas	-------	-------	-------
299	Si2OCl6	HEXACHLORODISILOXANE	284.887	gas	-------	-------	-------
300	Si2OH6	DISILOXANE	78.218	gas	-------	-------	-------
301	Si3Cl8	OCTACHLOROTRISILANE	367.878	gas	-------	-------	-------
302	Si3H8	TRISILANE	92.320	gas	120.90	128.34	-------
303	Si3H9N	TRISILAZANE	107.335	gas	-------	-------	-------
304	Si4H10	TETRASILANE	122.421	liquid	-294.55	-282.16	-------
305	Sm	SAMARIUM	150.360	soild	0.00	0.00	69.60
306	Sn	TIN	118.710	soild	0.00	0.00	51.20
307	SnBr4	STANNIC BROMIDE	438.326	solid	-377.40	-377.40	264.40
308	SnCl2	STANNOUS CHLORIDE	189.615	soild	-325.10	-322.62	-------
309	SnCl4	STANNIC CHLORIDE	260.521	gas	-471.50	-469.02	365.80
310	SnH4	STANNIC HYDRIDE	122.742	gas	162.80	165.28	227.70
311	SnI4	STANNIC IODIDE	626.328	gas	-------	-------	446.10
312	Sr	STRONTIUM	87.620	solid	0.00	0.00	52.30

312

Table 12-2 ENTHALPY AND INTERNAL ENERGY OF FORMATION - INORGANIC COMPOUNDS (continued)

NO	FORMULA	NAME	Molecular Weight g/mol	State	Hf @ 298 K kjoule/mol	Uf @ 298 K kjoule/mol	S @ 298 K joule/(mol K)
313	SrO	STRONTIUM OXIDE	103.619	solid	-592.00	-590.76	54.40
314	Ta	TANTALUM	180.948	solid	0.00	0.00	41.50
315	Tc	TECNNETIUM	98.000	solid	0.00	0.00	-------
316	Te	TELLURIUM	127.600	solid	0.00	0.00	49.70
317	TeCl4	TELLURIUM TETRACHLORIDE	269.411	solid	-326.40	-321.44	-------
318	TeF6	TELLURIUM HEXAFLUORIDE	241.590	gas	-1318.00	-1313.04	-------
319	Ti	TITANIUM	47.880	solid	0.00	0.00	30.70
320	TiCl4	TITANIUM TETRACHLORIDE	189.691	gas	-763.20	-760.72	353.20
321	Tl	THALLIUM	204.383	solid	0.00	0.00	64.20
322	TlBr	THALLOUS BROMIDE	284.287	solid	-173.20	-173.20	120.50
323	TlI	THALLOUS IODIDE	331.288	solid	-123.80	-123.80	127.60
324	Tm	THULIUM	168.934	solid	0.00	0.00	74.00
325	U	URANIUM	238.029	solid	0.00	0.00	50.20
326	UF6	URANIUM FLUORIDE	352.019	gas	-2147.40	-2142.44	377.90
327	V	VANADIUM	50.942	solid	0.00	0.00	28.90
328	VCl4	VANADIUM TETRACHLORIDE	192.752	gas	-525.50	-523.02	362.40
329	VOCl3	VANADIUM OXYTRICHLORIDE	173.299	gas	-695.59	-693.11	344.29
330	W	TUNGSTEN	183.840	solid	0.00	0.00	32.60
331	WF6	TUNGSTEN FLUORIDE	297.830	gas	-1721.70	-1716.74	341.10
332	Xe	XENON	131.290	gas	0.00	0.00	169.70
333	Yb	YTTERBIUM	173.040	solid	0.00	0.00	59.90
334	Yt	YTTRIUM	88.906	solid	0.00	0.00	-------
335	Zn	ZINC	65.390	solid	0.00	0.00	41.60
336	ZnCl2	ZINC CHLORIDE	136.295	solid	-415.10	-412.62	111.50
337	ZnF2	ZINC FLUORIDE	103.387	solid	-764.40	-761.92	73.70
338	ZnO	ZINC OXIDE	81.389	solid	-350.50	-349.26	43.70
339	ZnSO4	ZINC SULFATE	161.454	solid	-982.80	-977.84	110.50
340	Zr	ZIRCONIUM	91.224	solid	0.00	0.00	39.00
341	ZrBr4	ZIRCONIUM BROMIDE	410.840	solid	-760.70	-760.70	-------
342	ZrCl4	ZIRCONIUM CHLORIDE	233.035	solid	-980.50	-975.54	181.60
343	ZrI4	ZIRCONIUM IODIDE	598.842	solid	-481.60	-481.60	-------

Hf - enthalpy of formation, kjoule/mol

Uf - internal energy of formation, kjoule/mol

S - entropy, joule/(mol K)

Chapter 13

GIBBS ENERGY OF FORMATION

Carl L. Yaws, Li Bu, Sachin Nijhawan, Jack R. Hopper, and Raph W. Pike*

Lamar University, Beaumont, Texas / *Louisiana State University, Baton Rouge, Louisiana

ABSTRACT

Results for Gibbs energy of formation are presented for major organic and inorganic chemicals. The major chemicals include many compound types. The results are provided in easy-to-use tabulations that are especially applicable for rapid engineering usage with the personal computer or hand calculator. The agreement of correlation and data is quite good.

INTRODUCTION

Gibbs energy of formation is important in the analysis of chemical reactions. Values for individual compounds (reactants and products) are required to determine the change in Gibbs energy for the reaction. This change is significant because of the associated chemical equilibrium for the reaction. If the change in Gibbs energy is negative, the thermodynamics for the reaction are favorable. On the other hand, if the change in Gibbs energy is highly positive, the thermodynamics for the reaction are not favorable.

The chemical equilibrium for a reaction is associated with the change in Gibbs free energy (ΔG_r) for the reaction:

$$\Delta G_{reaction} = \Sigma \, (n\Delta G_f)_{products} - \Sigma \, (n\Delta G_f)_{reactants} \qquad (13\text{-}1)$$

The changes in Gibbs energy for a reaction may be used in preliminary work to determine if a reaction is thermodynamically favorable at a given temperature. For thermodynamic equilibrium, the following rough criteria is useful for quick screening of chemical reactions:

$\Delta G_{reaction} < 0$ kjoule/mol   [reaction favorable]

$0 < \Delta G_{reaction} < 50$ kjoule/mol   [reaction possibly favorable]

$\Delta G_{reaction} > 50$ kjoule/mol   [reaction not favorable]

CORRELATION FOR GIBBS ENERGY OF FORMATION

The correlation for Gibbs energy of formation is a series expansion in temperature:

$$\Delta G_f = A + B \, T + C \, T^2 \qquad (13\text{-}2)$$

where ΔG_f = Gibbs energy of formation of ideal gas, kjoule/mol

A, B, and C = regression coefficients for chemical compound

T = temperature, K

The results for Gibbs energy of formation are given in Tables 13-1 and 13-2. The tabulations are arranged by chemical formula to provide ease of use in quickly locating data. A wide variety of substances are covered.

In preparing the compilation, a literature search was conducted to identify data source publications for organics (1-37) and inorganics (1-61). The publications were screened for appropriate data. The compilation resulted from the screening.

For organics, the range for application is denoted by the respective minimum and maximum temperatures (TMIN and TMAX). In the absence of data, values at 298.15 K were extended to higher temperatures by integration of the appropriate thermodynamic equations involving heat capacities. The numerous data points were processed with a generalized least-square computer program for minimizing the deviation. The spot values at room temperature (298.15 K) are the actual data. The spot values at elevated temperature (500 K) are calculated from the correlation. Since water and hydrogen chloride are of major industrial importance, regression coefficients are provided for these compounds at the end of the table.

For inorganics, results are also given for Helmholtz and entropy of formation at room temperature (298.15 K). These thermodynamic properties are useful in computing the energy of a chemical explosion.

The thermodynamic equations for computing the energy of a chemical explosion are given by Crowl and Louvar (35). The energy of a chemical explosion involves work of expansion (dW = PdV) resulting from the explosion. As the expansion occurs, this energy is transferred from the explosion. At constant temperature, the change in Helmholtz energy (dA = - PdV) is related to such expansion

314

work. Thus, it is convenient to utilize the change in Helmholtz energy to ascertain the energy of an explosion:

$$\text{Explosion energy}_{limit} = -\Delta A_{reaction} \tag{13-3}$$

Since thermal effects and irreversibility are involved in an explosion, this equation represents a limiting or maximum value for the explosion energy.

For an explosion reaction, the change in Helmholtz energy may be determined from Helmholtz energy of formation for the products and reactants according to equation given below:

$$\Delta A_{reaction} = \Sigma\,(n\Delta A_f)_{products} - \Sigma\,(n\Delta A_f)_{reactants} \tag{13-4}$$

The actual energy release in a chemical explosion will be less than the limiting value given by the change in Helmholtz energy because of thermal effects and irreversibility. In an example in Crowl and Louvar (35), 12% of the limiting value is suggested for a vapor cloud explosion in a partially confined area (ethylene explosion in a ditch). For vapor cloud explosion in an unconfined area, 2% of the limiting value is suggested by the same authors.

A comparison of calculated and data values is shown in Fig. 13-1 for a representative chemical. The graph discloses favorable agreement of correlation and data.

EXAMPLES

The results can be used to make calculations for Gibbs energy of formation and the change in Gibbs energy for reaction. Examples are given below.

Example 1 Estimate the Gibbs free energy of formation of methane (CH_4) as a low-pressure gas at 500 K.

Correlation constants (A, B, C) for methane from the table are substituted into the equation at a temperature of 500 K:

$\Delta G_f = -75.262 + (7.5925E\text{-}02)\,(500) + (1.8700E\text{-}05)\,(500^2)$

$\Delta G_f = -32.63$ kjoule/mol

Example 2 Calculate the change in Gibbs free energy for the reaction of methanol and oxygen to produce formaldehyde and water at reaction temperature of 600 K:

$$CH4O(g) + 0.5\,O2(g) \rightarrow CH2O(g) + H2O(g)$$

The change in Gibbs free energy of reaction may be determined from Gibbs free energy of formation for the products and reactants:

$\Delta G_{reaction} = \Delta G_{f,CH2O} + \Delta G_{f,H2O} - \Delta G_{f,CH4O} - 0.5\,\Delta G_{f,O2}$

Using correlation constants from the table at temperature of 600 K, we obtain

$\Delta G_{f,CH2O} = -115.972 + (1.6630E\text{-}02)\,(600) + (1.1381E\text{-}05)\,(600^2) = -101.90$
$\Delta G_{f,H2O}\ = -241.740 + (4.1740E\text{-}02)\,(600) + (7.4281E\text{-}06)\,(600^2) = -214.02$
$\Delta G_{f,CH4O} = -201.860 + (1.2542E\text{-}01)\,(600) + (2.0345E\text{-}05)\,(600^2) = -119.28$
$\Delta G_{f,O2}\ = 0$

Substitution of ΔG_f values into the equation for Gibbs free energy of the reaction yields

$\Delta G_{reaction} = -101.9 + (-214.02) - (-119.28) - 0$

$\Delta G_{reaction} = -196.64$ kjoule/mol

Since the change in Gibbs free energy for the reaction is negative, the thermodynamics for the reaction are favorable.

Portions of this material appeared in Hydrocarbon Processing, 67, 81 (Nov., 1988) and are reprinted by special permission.

REFERENCES – ORGANIC COMPOUNDS

1-37. See **REFERENCES - ORGANIC COMPOUNDS** in Chapter 11, **ENTROPY AND ENTROPY OF FORMATION OF GAS**

REFERENCES - INORGANIC COMPOUNDS

1-61. See **REFERENCES - INORGANIC COMPOUNDS** in Chapter 11, **ENTROPY AND ENTROPY OF FORMATION OF GAS**

Figure 13-1 Gibbs Energy of Formation

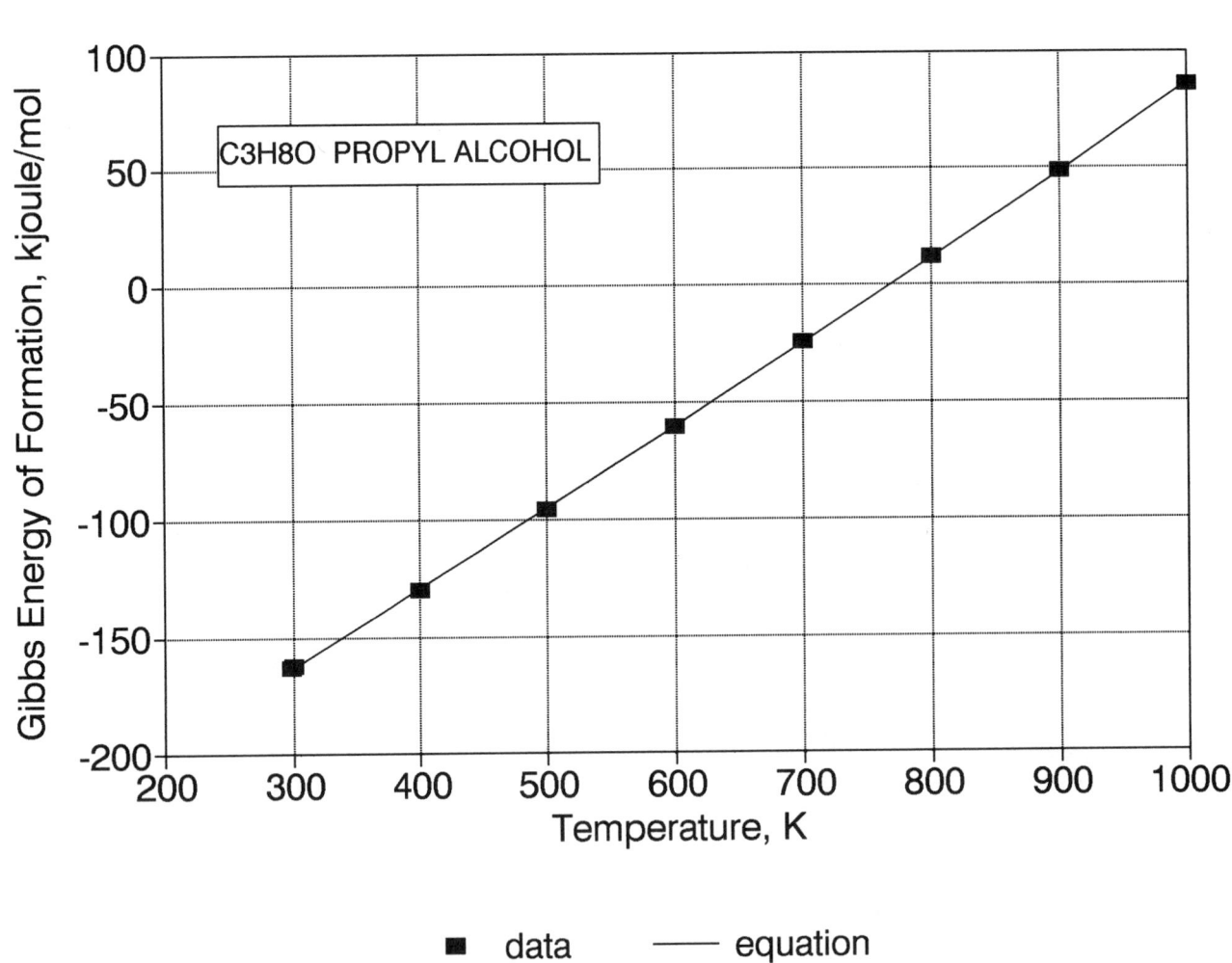

Table 13-1 GIBBS ENERGY OF FORMATION OF GAS - ORGANIC COMPOUNDS

NO	FORMULA	NAME	Gf = A + B T + C T²					(Gf - kjoule/mol, T - K)	
			A	B	C	TMIN	TMAX	Gf @ 298 K	Gf @ 500 K
1	CBrClF2	BROMOCHLORODIFLUOROMETHANE	-447.311	1.2794E-01	-4.4565E-06	333	1000	-408.00	-384.46
2	CBrCl3	BROMOTRICHLOROMETHANE	-54.665	1.2976E-01	-8.4521E-06	333	1000	-15.20	8.10
3	CBrF3	BROMOTRIFLUOROMETHANE	-664.953	1.3721E-01	-2.2103E-06	333	1000	-622.66	-596.90
4	CBr2F2	DIBROMODIFLUOROMETHANE	-418.443	1.3915E-01	-2.0391E-05	333	1000	-375.70	-353.96
5	CClF3	CHLOROTRIFLUOROMETHANE	-694.327	1.3545E-01	6.8525E-07	298	1000	-653.96	-626.43
6	CClN	CYANOGEN CHLORIDE	138.077	-2.2913E-02	-2.4942E-06	298	1000	131.00	126.00
7	CCl2F2	DICHLORODIFLUOROMETHANE	-481.827	1.4482E-01	4.3398E-06	298	1000	-438.06	-408.33
8	CCl2O	PHOSGENE	-219.038	4.4372E-02	-1.3968E-06	298	1000	-205.94	-197.20
9	CCl3F	TRICHLOROFLUOROMETHANE	-284.764	1.3487E-01	-5.3707E-06	298	1000	-245.18	-218.67
10	CCl4	CARBON TETRACHLORIDE	-100.838	1.4561E-01	-8.6766E-06	298	1000	-58.24	-30.20
11	CF2O	CARBONYL FLUORIDE	-639.112	5.1862E-02	2.9462E-06	298	1000	-623.36	-612.45
12	CF4	CARBON TETRAFLUORIDE	-933.816	1.5206E-01	6.7334E-08	298	1000	-888.43	-857.77
13	CHBr3	TRIBROMOMETHANE	-30.208	1.1033E-01	-1.8500E-07	333	1000	7.40	24.91
14	CHClF2	CHLORODIFLUOROMETHANE	-482.201	1.0493E-01	4.2461E-06	298	1000	.-450.47	-428.67
15	CHCl2F	DICHLOROFLUOROMETHANE	-283.827	1.0295E-01	3.0690E-06	298	1000	-252.81	-231.58
16	CHCl3	CHLOROFORM	-101.846	1.1137E-01	8.6302E-07	298	1000	-68.53	-45.95
17	CHF3	TRIFLUOROMETHANE	-698.170	1.1528E-01	7.1054E-06	298	1000	-663.08	-638.76
18	CHI3	TRIIODOMETHANE	116.349	1.0932E-01	-2.3735E-06	459	1000	177.95	170.42
19	CHN	HYDROGEN CYANIDE	135.278	-3.6253E-02	2.3053E-06	298	1000	124.68	117.73
20	CHNS	ISOTHIOCYANIC ACID	129.895	-6.6463E-02	3.1215E-05	389	717	112.88	104.47
21	CH2BrCl	BROMOCHLOROMETHANE	-58.800	8.2499E-02	8.3991E-06	333	1000	-31.80	-15.45
22	CH2Br2	DIBROMOMETHANE	-46.334	8.8015E-02	7.6027E-06	333	1000	-16.20	-0.43
23	CH2ClF	CHLOROFLUOROMETHANE	-262.549	8.3336E-02	1.0838E-05	298	1000	-236.64	-218.17
24	CH2Cl2	DICHLOROMETHANE	-95.965	8.8080E-02	8.5438E-06	298	1000	-68.87	-49.79
25	CH2F2	DIFLUOROMETHANE	-453.610	9.0621E-02	1.2721E-05	298	1000	-425.35	-405.12
26	CH2I2	DIIODOMETHANE	53.345	9.2134E-02	3.1467E-06	459	1000	101.09	100.20
27	CH2O	FORMALDEHYDE	-115.972	1.6630E-02	1.1381E-05	298	1000	-109.91	-104.81
28	CH2O2	FORMIC ACID	-379.192	9.1431E-02	9.5249E-06	298	1000	-351.00	-331.10
29	CH3Br	BROMOMETHANE	-55.241	8.0199E-02	1.1207E-05	333	1000	-28.16	-12.34
30	CH3Cl	CHLOROMETHANE	-86.903	7.5722E-02	1.4823E-05	298	1000	-62.89	-45.34
31	CH3Cl3Si	METHYL TRICHLOROSILANE	-572.529	2.0329E-01	8.1114E-06	298	1000	-511.10	-468.86
32	CH3F	FLUOROMETHANE	-234.475	7.6613E-02	1.6807E-05	298	1000	-210.00	-191.97
33	CH3I	IODOMETHANE	-20.627	8.3254E-02	8.0701E-06	459	1000	15.65	23.02
34	CH3NO	FORMAMIDE	-187.267	1.4992E-01	1.5392E-05	298	1000	-141.04	-108.46
35	CH3NO2	NITROMETHANE	-76.173	2.2736E-01	1.4355E-05	298	1000	-6.95	41.10
36	CH3NO2	METHYL NITRITE	-65.640	2.2023E-01	9.2945E-06	298	1000	1.00	46.80
37	CH3NO3	METHYL NITRATE	-122.358	3.0626E-01	8.2197E-06	298	1000	-30.17	32.83
38	CH4	METHANE	-75.262	7.5925E-02	1.8700E-05	298	1000	-50.84	-32.63
39	CH4Cl2Si	METHYL DICHLOROSILANE	-403.035	1.7954E-01	1.2022E-05	298	1000	-348.30	-310.26
40	CH4O	METHANOL	-201.860	1.2542E-01	2.0345E-05	298	1000	-162.51	-134.06
41	CH4O3S	METHANESULFONIC ACID	--------	--------	--------	--------	--------	-459.00	--------
42	CH4S	METHANETHIOL	-19.483	1.6198E-02	5.2738E-05	389	717	-9.92	1.80
43	CH5ClSi	METHYL CHLOROSILANE	-216.302	1.6253E-01	1.7508E-05	298	1000	-166.10	-130.66
44	CH5N	METHYLAMINE	-24.115	1.8179E-01	2.2182E-05	298	1000	32.26	72.32
45	CH6Si	METHYL SILANE	-30.687	1.5773E-01	2.2023E-05	298	1000	18.53	53.68
46	CN4O8	TETRANITROMETHANE	--------	--------	--------	--------	--------	--------	--------
47	CO	CARBON MONOXIDE	-109.885	-9.2218E-02	1.4547E-06	298	1000	-137.28	-155.63
48	COS	CARBONYL SULFIDE	-136.376	-1.0575E-01	2.5162E-05	389	717	-165.64	-182.96
49	CO2	CARBON DIOXIDE	-393.360	-3.8212E-03	1.3322E-06	298	1000	-394.38	-394.94
50	CS2	CARBON DISULFIDE	121.242	-1.9750E-01	5.0587E-05	389	717	66.90	35.14
51	C2BrF3	BROMOTRIFLUOROETHYLENE	-470.635	1.0551E-01	-4.1674E-06	333	1000	-438.00	-418.92
52	C2Br2F4	1,2-DIBROMOTETRAFLUOROETHANE	-954.624	2.8150E-01	-1.4115E-05	333	1000	-868.86	-817.40
53	C2ClF3	CHLOROTRIFLUOROETHYLENE	-572.876	1.0470E-01	-3.6626E-06	298	1000	-542.00	-521.44
54	C2ClF5	CHLOROPENTAFLUOROETHANE	-1108.395	2.8235E-01	-6.2604E-06	298	1000	-1025.08	-968.79
55	C2Cl2F4	1,2-DICHLOROTETRAFLUOROETHANE	-888.420	2.8167E-01	-1.2046E-05	298	1000	-805.42	-750.60
56	C2Cl3F3	1,1,2-TRICHLOROTRIFLUOROETHANE	-696.127	2.7082E-01	-1.6970E-05	298	1000	-617.14	-564.96
57	C2Cl4	TETRACHLOROETHENE	-12.176	1.1919E-01	-7.6720E-06	298	1000	22.64	45.50
58	C2Cl4F2	1,1,2,2-TETRACHLORODIFLUOROETHANE	-201.688	1.3803E-01	5.2053E-06	298	1000	-618.55	-131.37
59	C2Cl4O	TRICHLOROACETYL CHLORIDE	-306.217	1.9455E-01	-9.7473E-06	298	1000	-249.12	-211.38
60	C2Cl6	HEXACHLOROETHANE	-142.168	2.9199E-01	-1.8349E-05	298	1000	-56.82	-0.76
61	C2F4	TETRAFLUOROETHENE	-658.653	1.1776E-01	-1.8295E-06	298	1000	-623.71	-600.23
62	C2F6	HEXAFLUOROETHANE	-1344.410	2.9411E-01	-7.6769E-06	298	1000	-1257.38	-1199.28
63	C2HBrClF3	HALOTHANE	-721.624	2.5354E-01	-2.8445E-06	333	1000	-644.66	-595.57
64	C2HClF2	2-CHLORO-1,1-DIFLUOROETHYLENE	-329.299	8.7627E-02	1.6721E-06	298	1000	-303.00	-285.07
65	C2HCl3	TRICHLOROETHENE	-9.886	8.7331E-02	-1.0392E-06	298	1000	16.07	33.52
66	C2HCl3O	DICHLOROACETYL CHLORIDE	-281.362	1.6537E-01	-1.2833E-06	298	1000	-232.15	-199.00
67	C2HCl3O	TRICHLOROACETALDEHYDE	-197.420	1.6673E-01	-3.3136E-06	298	1000	-148.00	-114.89
68	C2HCl5	PENTACHLOROETHANE	-143.224	2.5953E-01	-9.0227E-06	298	1000	-66.65	-15.73
69	C2HF3	TRIFLUOROETHENE	-491.035	8.7927E-02	3.6243E-06	298	1000	-464.47	-446.17
70	C2HF3O2	TRIFLUOROACETIC ACID	-1022.794	2.5748E-01	-3.1555E-07	298	1000	-946.00	-894.13
71	C2HF5	PENTAFLUOROETHANE	-1105.771	2.5354E-01	1.0248E-07	298	1000	-1030.10	-978.98
72	C2H2	ACETYLENE(ETHYNE)	227.187	-6.1675E-02	4.3960E-06	298	1000	209.20	197.45
73	C2H2Br4	1,1,2,2-TETRABROMOETHANE	-52.080	2.3690E-01	-1.9570E-06	333	1000	24.70	65.88
74	C2H2Cl2	1,1-DICHLOROETHENE	1.994	7.6745E-02	5.2089E-06	298	1000	25.40	41.67
75	C2H2Cl2	cis-1,2-DICHLOROETHENE	-3.264	7.5010E-02	5.9732E-06	298	1000	19.66	35.73
76	C2H2Cl2	trans-1,2-DICHLOROETHENE	-0.772	7.4474E-02	5.8258E-06	298	1000	22.01	37.92
77	C2H2Cl2O	CHLOROACETYL CHLORIDE	-246.112	1.4122E-01	6.4023E-06	298	1000	-203.37	-173.90
78	C2H2Cl2O	DICHLOROACETALDEHYDE	--------	--------	--------	--------	--------	-160.00	--------

317

Table 13-1 GIBBS ENERGY OF FORMATION OF GAS - ORGANIC COMPOUNDS (continued)

NO	FORMULA	NAME	A	B	C	TMIN	TMAX	Gf @ 298 K	Gf @ 500 K
				Gf = A + B T + C T^2				(Gf - kjoule/mol, T - K)	
79	C2H2Cl2O2	DICHLOROACETIC ACID	-436.586	2.2788E-01	6.4353E-06	298	1000	-368.00	-321.04
80	C2H2Cl3F	1,1,1-TRICHLOROFLUOROETHANE	-303.032	2.3175E-01	-1.2733E-06	298	1000	-234.00	-187.48
81	C2H2Cl4	1,1,1,2-TETRACHLOROETHANE	-150.569	2.3608E-01	-1.9593E-06	298	1000	-80.30	-33.02
82	C2H2Cl4	1,1,2,2-TETRACHLOROETHANE	-153.859	2.2902E-01	-5.6905E-07	298	1000	-85.56	-39.49
83	C2H2F2	1,1-DIFLUOROETHENE	-337.271	7.8515E-02	7.9741E-06	298	1000	-313.09	-296.02
84	C2H2F2	cis-1,2-DIFLUOROETHENE	-322.591	7.4587E-02	9.0983E-06	298	1000	-299.49	-283.02
85	C2H2F2	trans-1,2-DIFLUOROETHENE	-322.543	7.5116E-02	8.3671E-06	298	1000	-299.32	-282.89
86	C2H2F4	1,1,1,2-TETRAFLUOROETHANE	-896.977	2.3331E-01	4.7491E-06	298	1000	-826.90	-779.13
87	C2H2O	KETENE	-60.986	2.0204E-04	6.7660E-06	298	1000	-60.29	-59.19
88	C2H2O4	OXALIC ACID	-721.681	1.8457E-01	6.1174E-05	298	1000	-661.00	-614.10
89	C2H3Br	BROMOETHYLENE	61.115	5.6200E-02	8.8699E-06	333	1000	80.75	91.43
90	C2H3Cl	CHLOROETHENE	31.557	3.2864E-02	1.8788E-05	298	1000	42.93	52.69
91	C2H3ClF2	1-CHLORO-1,1-DIFLUOROETHANE	-488.054	2.1528E-01	8.5003E-06	298	1000	-423.00	-378.29
92	C2H3ClO	ACETYL CHLORIDE	-244.471	1.2418E-01	1.2494E-05	298	1000	-206.23	-179.26
93	C2H3ClO	CHLOROACETALDEHYDE	-191.684	1.1549E-01	1.2734E-05	298	1000	-156.00	-130.75
94	C2H3ClO2	CHLOROACETIC ACID	-437.303	2.1050E-01	1.0346E-05	298	1000	-373.50	-329.47
95	C2H3ClO2	METHYL CHLOROFORMATE	-424.987	2.1121E-01	1.0134E-05	298	1000	-361.00	-316.85
96	C2H3Cl3	1,1,1-TRICHLOROETHANE	-143.436	2.2407E-01	3.9301E-06	298	1000	-76.19	-30.42
97	C2H3Cl3	1,1,2-TRICHLOROETHANE	-139.616	2.0630E-01	5.9155E-06	298	1000	-77.49	-34.99
98	C2H3F	FLUOROETHENE	-139.376	4.3674E-02	1.3376E-05	298	1000	-125.06	-114.20
99	C2H3F3	1,1,1-TRIFLUOROETHANE	-746.790	2.2467E-01	9.9950E-06	298	1000	-678.77	-631.96
100	C2H3N	ACETONITRILE	87.546	5.6339E-02	1.3233E-05	298	1000	105.60	119.02
101	C2H3NO	METHYL ISOCYANATE	-58.510	1.2997E-01	2.1224E-05	298	1000	-17.70	11.78
102	C2H4	ETHYLENE	51.752	4.9338E-02	1.7284E-05	298	1000	68.12	80.74
103	C2H4Br2	1,1-DIBROMOETHANE	-72.857	1.9018E-01	1.2053E-05	333	1000	-11.81	25.25
104	C2H4Br2	1,2-DIBROMOETHANE	-72.860	1.9515E-01	3.5902E-06	333	1000	-10.59	25.61
105	C2H4Cl2	1,1-DICHLOROETHANE	-131.060	1.9006E-01	1.2939E-05	298	1000	-73.09	-32.79
106	C2H4Cl2	1,2-DICHLOROETHANE	-130.549	1.8562E-01	1.3775E-05	298	1000	-73.85	-34.29
107	C2H4Cl2O	BIS(CHLOROMETHYL)ETHER	-242.473	2.3857E-01	1.3276E-05	298	1000	-170.00	-119.87
108	C2H4F2	1,1-DIFLUOROETHANE	-494.858	1.9101E-01	1.6948E-05	298	1000	-436.22	-395.12
109	C2H4F2	1,2-DIFLUOROETHANE	-432.308	1.8774E-01	1.3231E-05	298	1000	-375.00	-335.13
110	C2H4I2	1,2-DIIODOETHANE	1.065	1.9267E-01	5.9773E-06	459	1000	78.49	98.89
111	C2H4O	ACETALDEHYDE	-167.052	1.0714E-01	1.8665E-05	298	1000	-133.30	-108.82
112	C2H4O	ETHYLENE OXIDE	-53.838	1.3048E-01	1.8741E-05	298	1000	-13.10	16.09
113	C2H4OS	THIOACETIC ACID	-178.876	6.9466E-02	4.6167E-05	389	717	-154.01	-132.60
114	C2H4O2	ACETIC ACID	-435.963	1.9346E-01	1.6362E-05	298	1000	-376.69	-335.14
115	C2H4O2	METHYL FORMATE	-350.914	1.7474E-01	1.6322E-05	298	1000	-297.19	-259.46
116	C2H4S	THIACYCLOPROPANE	85.451	2.2403E-02	5.2946E-05	389	717	96.90	109.89
117	C2H5Br	BROMOETHANE	-83.274	1.7808E-01	1.4544E-05	333	1000	-26.32	9.40
118	C2H5Cl	CHLOROETHANE	-112.949	1.7104E-01	1.9992E-05	298	1000	-60.00	-22.43
119	C2H5ClO	2-CHLOROETHANOL	-263.294	2.2897E-01	2.0560E-05	298	1000	-193.00	-143.67
120	C2H5F	FLUOROETHANE	-262.745	1.7117E-01	2.2082E-05	298	1000	-209.53	-171.64
121	C2H5I	IODOETHANE	-45.122	1.8148E-01	1.1291E-05	459	1000	21.34	48.44
122	C2H5N	ETHYLENIMINE	121.662	1.8205E-01	2.0521E-05	298	1000	177.99	217.82
123	C2H5NO	ACETAMIDE	-239.564	2.5915E-01	2.7701E-05	298	1000	-159.60	-103.07
124	C2H5NO	N-METHYLFORMAMIDE	-185.581	2.4879E-01	3.2786E-05	298	1000	-108.23	-52.99
125	C2H5NO2	NITROETHANE	-103.424	3.2377E-01	1.9797E-05	298	1000	-4.90	63.41
126	C2H5NO3	ETHYL NITRATE	-156.474	3.9637E-01	1.3852E-05	298	1000	-36.86	45.18
127	C2H6	ETHANE	-85.787	1.6858E-01	2.6853E-05	298	1000	-32.93	5.21
128	C2H6AlCl	DIMETHYLALUMINUM CHLORIDE	--------	--------	--------	--------	--------	--------	--------
129	C2H6O	METHYL ETHER	-185.257	2.3378E-01	2.7075E-05	298	1000	-112.93	-61.60
130	C2H6O	ETHYL ALCOHOL	-236.103	2.1904E-01	2.5659E-05	298	1000	-168.28	-120.17
131	C2H6OS	DIMETHYL SULFOXIDE	--------	--------	--------	--------	--------	-115.69	--------
132	C2H6O2	ETHYLENE GLYCOL	-390.502	2.8378E-01	1.4492E-05	298	1000	-304.47	-244.99
133	C2H6O4S	DIMETHYL SULFATE	--------	--------	--------	--------	--------	-535.00	--------
134	C2H6S	METHYL SULFIDE	-33.596	1.1704E-01	6.2933E-05	389	717	6.95	40.66
135	C2H6S	ETHANETHIOL	-42.415	1.0784E-01	6.1979E-05	389	717	-4.69	27.00
136	C2H6S2	METHYL DISULFIDE	-18.556	8.6388E-02	8.3731E-05	389	717	14.73	45.57
137	C2H7N	DIMETHYLAMINE	-20.837	2.8875E-01	2.7736E-05	298	1000	67.99	130.47
138	C2H7N	ETHYLAMINE	-47.732	2.7622E-01	2.6932E-05	298	1000	37.28	97.11
139	C2H7NO	MONOETHANOLAMINE	-212.059	3.4381E-01	2.7055E-05	298	1000	-106.88	-33.39
140	C2H8N2	ETHYLENEDIAMINE	-19.589	4.0181E-01	3.0306E-05	298	1000	103.22	188.89
141	C2H8Si	DIMETHYL SILANE	-96.693	2.4963E-01	2.6980E-05	298	1000	-19.58	34.87
142	C2N2	CYANOGEN	309.375	-3.9443E-02	-4.0226E-06	298	1000	297.19	288.65
143	C3F6	HEXAFLUOROPROPYLENE	-1082.801	1.4891E-01	-1.9609E-05	298	1000	-1040.10	-1013.25
144	C3F6O	HEXAFLUOROACETONE	-1462.242	3.4532E-01	-8.8454E-06	298	1000	-1360.00	-1291.79
145	C3F8	OCTAFLUOROPROPANE	-1705.086	4.2367E-01	-1.4050E-05	298	1000	-1580.00	-1496.76
146	C3H2N2	MALONONITRILE	265.122	5.1483E-02	5.3360E-06	298	1000	281.00	292.20
147	C3H3Cl	PROPARGYL CHLORIDE	158.887	3.4531E-02	8.5425E-06	298	1000	170.00	178.29
148	C3H3N	ACRYLONITRILE	184.585	3.2496E-02	1.0680E-05	298	1000	195.31	203.50
149	C3H3NO	OXAZOLE	-17.028	1.4077E-01	1.4380E-05	298	1000	26.40	56.95
150	C3H4	PROPYNE(METHYLACETYLENE)	185.351	2.5730E-02	1.4771E-05	298	1000	194.43	201.91
151	C3H4	ALLENE(PROPADIENE)	191.839	3.0752E-02	1.4194E-05	298	1000	202.38	210.76
152	C3H4Cl2	2,3-DICHLOROPROPENE	-48.360	1.6229E-01	1.4081E-05	298	1000	1.39	36.31
153	C3H4O	ACROLEIN	-82.981	8.6786E-02	1.0501E-05	298	1000	-55.98	-36.96
154	C3H4O	PROPARGYL ALCOHOL	41.785	8.4688E-02	1.4135E-05	298	1000	68.40	87.66
155	C3H4O2	ACRYLIC ACID	-337.271	1.6672E-01	1.5209E-05	298	1000	-286.06	-250.11
156	C3H4O2	beta-PROPIOLACTONE	-298.465	1.9650E-01	1.6276E-05	298	1000	-238.25	-196.14

318

Table 13-1 GIBBS ENERGY OF FORMATION OF GAS - ORGANIC COMPOUNDS (continued)

NO	FORMULA	NAME	Gf = A + B T + C T²					(Gf - kjoule/mol, T - K)	
			A	B	C	TMIN	TMAX	Gf @ 298 K	Gf @ 500 K
157	C3H4O2	VINYL FORMATE	-263.172	1.7625E-01	1.6472E-05	298	1000	-209.00	-170.93
158	C3H4O3	ETHYLENE CARBONATE	--------	--------	--------	--------	--------	-410.00	--------
159	C3H4O3	PYRUVIC ACID	-626.151	2.6728E-01	1.4737E-05	298	1000	-545.00	-488.83
160	C3H5Br	3-BROMO-1-PROPENE	30.623	1.5305E-01	1.3603E-05	333	1000	79.96	110.55
161	C3H5Cl	2-CHLOROPROPENE	-21.467	1.4824E-01	2.0434E-05	298	1000	24.70	57.76
162	C3H5Cl	3-CHLORO-1-PROPENE	-1.565	1.4516E-01	1.9444E-05	298	1000	43.60	75.88
163	C3H5ClO	alpha-EPICHLOROHYDRIN	-109.522	2.3861E-01	1.6254E-05	298	1000	-36.74	13.85
164	C3H5ClO2	METHYL CHLOROACETATE	-450.559	2.8789E-01	1.7275E-05	298	1000	-363.00	-302.29
165	C3H5ClO2	ETHYL CHLOROFORMATE	-459.719	3.1534E-01	1.6798E-05	298	1000	-364.00	-297.85
166	C3H5Cl3	1,2,3-TRICHLOROPROPANE	-187.124	2.9501E-01	1.3833E-05	298	1000	-97.78	-36.16
167	C3H5I	3-IODO-1-PROPENE	60.299	1.6149E-01	8.7329E-06	459	1000	120.16	143.23
168	C3H5N	PROPIONITRILE	49.759	1.4902E-01	2.0199E-05	298	1000	96.15	129.32
169	C3H5NO	ACRYLAMIDE	-171.699	2.4168E-01	1.7303E-05	298	1000	-97.90	-46.54
170	C3H5NO	HYDRACRYLONITRILE	-99.550	2.0892E-01	1.8864E-05	298	1000	-35.40	9.63
171	C3H5NO	LACTONITRILE	-65.277	1.3359E-01	1.4325E-05	298	1000	-24.00	5.10
172	C3H5N3O9	NITROGLYCERINE	--------	--------	--------	--------	--------	--------	--------
173	C3H6	CYCLOPROPANE	51.643	1.6873E-01	2.4956E-05	298	1000	104.39	142.25
174	C3H6	PROPENE	19.412	1.3685E-01	2.5749E-05	298	1000	62.72	94.27
175	C3H6Br2	1,2-DIBROMOPROPANE	-108.382	2.8677E-01	1.0766E-05	333	1000	-17.66	37.72
176	C3H6Cl2	1,1-DICHLOROPROPANE	-152.405	2.8592E-01	1.9562E-05	298	1000	-65.20	-4.55
177	C3H6Cl2	1,2-DICHLOROPROPANE	-167.410	2.7636E-01	1.9161E-05	298	1000	-83.09	-24.44
178	C3H6Cl2	1,3-DICHLOROPROPANE	-162.984	2.6293E-01	2.0036E-05	298	1000	-82.59	-26.51
179	C3H6Cl2	2,2-DICHLOROPROPANE	-177.370	3.0654E-01	1.3973E-05	298	1000	-84.56	-20.61
180	C3H6I2	1,2-DIIODOPROPANE	-33.335	2.8806E-01	7.3960E-06	459	1000	74.52	112.55
181	C3H6O	ACETONE	-218.777	2.1177E-01	2.6619E-05	298	1000	-153.05	-106.24
182	C3H6O	ALLYL ALCOHOL	-133.300	2.0021E-01	2.4167E-05	298	1000	-71.25	-27.15
183	C3H6O	METHYL VINYL ETHER	-109.075	1.9865E-01	2.6170E-05	298	1000	-47.30	-3.21
184	C3H6O	PROPIONALDEHYDE	-193.326	2.0318E-01	2.3446E-05	298	1000	-130.46	-85.87
185	C3H6O	PROPYLENE OXIDE	-94.384	2.2200E-01	2.4413E-05	298	1000	-25.77	22.72
186	C3H6O	1,3-PROPYLENE OXIDE	-76.948	2.4277E-01	2.9450E-05	298	1000	-1.66	51.80
187	C3H6O2	ETHYL FORMATE	-390.061	2.8368E-01	2.3926E-05	298	1000	-303.10	-242.24
188	C3H6O2	METHYL ACETATE	-411.714	2.9492E-01	2.3110E-05	298	1000	-321.50	-258.48
189	C3H6O2	PROPIONIC ACID	-455.116	2.8903E-01	2.2657E-05	298	1000	-366.70	-304.94
190	C3H6O2S	3-MERCAPTOPROPIONIC ACID	--------	--------	--------	--------	--------	-343.90	--------
191	C3H6O3	LACTIC ACID	-622.820	3.5146E-01	2.0323E-05	298	1000	-516.00	-442.01
192	C3H6O3	METHOXYACETIC ACID	-564.785	3.5404E-01	2.2385E-05	298	1000	-457.00	-382.17
193	C3H6O3	TRIOXANE	-468.806	4.3220E-01	2.6895E-05	298	1000	-337.22	-245.98
194	C3H6S	THIACYCLOBUTANE	64.908	1.2220E-01	6.8322E-05	389	717	107.49	143.09
195	C3H7Br	1-BROMOPROPANE	-108.651	2.7327E-01	1.9373E-05	333	1000	-22.47	32.83
196	C3H7Br	2-BROMOPROPANE	-117.884	2.8906E-01	1.7035E-05	333	1000	-27.24	30.90
197	C3H7Cl	2-CHLOROPROPANE	-148.229	2.7943E-01	2.4248E-05	298	1000	-62.51	-2.45
198	C3H7Cl	1-CHLOROPROPANE	-131.804	2.6332E-01	2.6786E-05	298	1000	-50.67	6.55
199	C3H7F	1-FLUOROPROPANE	-282.862	2.6769E-01	2.8109E-05	298	1000	-200.29	-141.99
200	C3H7F	2-FLUOROPROPANE	-290.498	2.8041E-01	2.7211E-05	298	1000	-204.22	-143.49
201	C3H7I	2-IODOPROPANE	-80.618	2.9402E-01	1.3004E-05	459	1000	20.08	69.64
202	C3H7I	1-IODOPROPANE	-69.399	2.8216E-01	1.4073E-05	459	1000	27.95	75.20
203	C3H7N	ALLYLAMINE	56.742	2.5389E-01	2.6068E-05	298	1000	135.00	190.20
204	C3H7N	PROPYLENEIMINE	86.393	2.8204E-01	2.5101E-05	298	1000	173.00	233.69
205	C3H7NO	N,N-DIMETHYLFORMAMIDE	-193.997	3.4603E-01	2.4054E-05	298	1000	-88.41	-14.97
206	C3H7NO	N-METHYLACETAMIDE	-241.902	3.4546E-01	3.9969E-05	298	1000	-135.00	-59.18
207	C3H7NO2	1-NITROPROPANE	-127.329	4.1957E-01	2.5482E-05	298	1000	0.33	88.83
208	C3H7NO2	2-NITROPROPANE	-142.923	4.2813E-01	2.4633E-05	298	1000	-12.80	77.30
209	C3H7NO3	PROPYL NITRATE	-177.035	4.9531E-01	1.9599E-05	298	1000	-27.32	75.52
210	C3H7NO3	ISOPROPYL NITRATE	-194.076	5.0799E-01	1.8793E-05	298	1000	-40.67	64.62
211	C3H8	PROPANE	-105.603	2.6475E-01	3.2500E-05	298	1000	-23.47	34.90
212	C3H8O	ISOPROPYL ALCOHOL	-274.540	3.2915E-01	2.9243E-05	298	1000	-173.59	-102.73
213	C3H8O	ETHYL METHYL ETHER	-218.142	3.2628E-01	3.2734E-05	298	1000	-117.65	-46.82
214	C3H8O	PROPYL ALCOHOL	-259.317	3.1232E-01	3.3063E-05	298	1000	-162.97	-94.89
215	C3H8O2	2-METHOXYETHANOL	-436.051	3.8189E-01	3.2356E-05	298	1000	-319.00	-237.02
216	C3H8O2	METHYLAL	-350.538	4.0487E-01	3.5664E-05	298	1000	-226.31	-139.19
217	C3H8O2	1,2-PROPYLENE GLYCOL	-423.516	3.8905E-01	3.0842E-05	298	1000	-304.48	-221.28
218	C3H8O2	1,3-PROPYLENE GLYCOL	-394.053	3.8130E-01	3.2459E-05	298	1000	-277.18	-195.29
219	C3H8O3	GLYCEROL	-584.906	4.4735E-01	3.0773E-05	298	1000	-448.49	-353.54
220	C3H8S	1-PROPANETHIOL	-63.596	1.9853E-01	7.3301E-05	389	717	2.18	53.99
221	C3H8S	2-PROPANETHIOL	-72.286	2.1257E-01	7.0456E-05	389	717	-2.55	51.61
222	C3H8S	ETHYL METHYL SULFIDE	-55.315	2.0160E-01	7.3615E-05	389	717	11.42	63.89
223	C3H9N	PROPYLAMINE	-74.693	3.7313E-01	3.2876E-05	298	1000	39.79	120.09
224	C3H9N	ISOPROPYLAMINE	-86.121	3.8485E-01	3.3435E-05	298	1000	31.93	114.66
225	C3H9N	TRIMETHYLAMINE	-26.507	4.0989E-01	3.2215E-05	298	1000	98.91	186.49
226	C3H9NO	1-AMINO-2-PROPANOL	-241.423	4.3684E-01	3.2054E-05	298	1000	-108.00	-14.99
227	C3H9NO	3-AMINO-1-PROPANOL	-223.371	4.3254E-01	3.4468E-05	298	1000	-91.00	1.52
228	C3H9NO	METHYLETHANOLAMINE	-200.690	4.5681E-01	3.4127E-05	298	1000	-61.10	36.24
229	C3H9O4P	TRIMETHYL PHOSPHATE	--------	--------	--------	--------	--------	--------	--------
230	C3H10N2	1,2-PROPANEDIAMINE	-56.563	4.9943E-01	3.4985E-05	298	1000	95.83	201.90
231	C3H10Si	TRIMETHYL SILANE	-158.072	3.4875E-01	4.2926E-05	298	1000	-49.91	27.03
232	C4Cl4S	TETRACHLOROTHIOPHENE	--------	--------	--------	--------	--------	--------	--------
233	C4Cl6	HEXACHLORO-1,3-BUTADIENE	-46.026	2.3459E-01	-1.1949E-05	298	1000	22.75	68.28
234	C4F8	OCTAFLUORO-2-BUTENE	-1648.042	3.7947E-01	5.3261E-05	298	1000	-1530.00	-1444.99

Table 13-1 GIBBS ENERGY OF FORMATION OF GAS - ORGANIC COMPOUNDS (continued)

NO	FORMULA	NAME	Gf = A + B T + C T²					(Gf - kjoule/mol, T - K)	
			A	B	C	TMIN	TMAX	Gf @ 298 K	Gf @ 500 K
235	C4F8	OCTAFLUOROCYCLOBUTANE	-1530.065	4.4414E-01	-1.3877E-05	298	1000	-1398.84	-1311.47
236	C4F10	DECAFLUOROBUTANE	-2142.722	5.8551E-01	-2.1023E-05	298	1000	-1970.00	-1855.22
237	C4H2	BUTADIYNE(BIACETYLENE)	473.689	-9.9298E-02	-6.6000E-07	298	1000	443.96	423.88
238	C4H2O3	MALEIC ANHYDRIDE	-399.655	1.5467E-01	-1.6170E-05	298	1000	-355.00	-326.36
239	C4H4	1-BUTEN-3-YNE(VINYLACETYLENE)	304.509	1.1570E-03	1.1754E-05	298	1000	305.98	308.03
240	C4H4N2	SUCCINONITRILE	208.827	1.4776E-01	1.1376E-05	298	1000	254.01	285.55
241	C4H4O	FURAN	-35.958	1.1791E-01	1.7007E-05	298	1000	0.88	27.25
242	C4H4O2	DIKETENE	-191.236	1.8049E-01	1.4353E-05	298	1000	-136.00	-97.40
243	C4H4O3	SUCCINIC ANHYDRIDE	-525.044	2.5618E-01	6.4319E-06	298	1000	-448.00	-395.35
244	C4H4O4	FUMARIC ACID	-673.485	3.0836E-01	-7.3523E-06	298	1000	-582.16	-521.15
245	C4H4O4	MALEIC ACID	-677.158	3.0084E-01	1.3729E-05	298	1000	-586.09	-523.30
246	C4H4S	THIOPHENE	118.623	1.2478E-02	4.9100E-05	389	717	126.78	137.14
247	C4H5Cl	CHLOROPRENE	72.585	1.3649E-01	1.6203E-05	298	1000	114.84	144.88
248	C4H5N	trans-CROTONITRILE	139.750	1.4356E-01	1.9556E-05	298	1000	184.46	216.42
249	C4H5N	cis-CROTONITRILE	142.441	1.3738E-01	2.1557E-05	298	1000	185.50	216.52
250	C4H5N	METHACRYLONITRILE	97.215	2.1914E-01	1.4710E-05	298	1000	164.00	210.46
251	C4H5N	PYRROLE	106.519	1.7405E-01	1.8688E-05	298	1000	160.29	198.21
252	C4H5N	VINYLACETONITRILE	157.113	1.3785E-01	1.8391E-05	298	1000	200.00	230.64
253	C4H5NO2	METHYL CYANOACETATE	-238.548	2.6783E-01	1.7034E-05	298	1000	-157.00	-100.37
254	C4H6	CYCLOBUTENE	128.171	1.4815E-01	2.4315E-05	298	1000	174.72	208.33
255	C4H6	1,2-BUTADIENE	161.456	1.1686E-01	2.2211E-05	298	1000	198.45	225.44
256	C4H6	1,3-BUTADIENE	109.172	1.3296E-01	1.9003E-05	298	1000	150.67	180.40
257	C4H6	2-BUTYNE(DIMETHYLACETYLENE)	145.727	1.2497E-01	2.5536E-05	298	1000	185.43	214.60
258	C4H6	1-BUTYNE(ETHYLACETYLENE)	164.525	1.1892E-01	2.1726E-05	298	1000	202.09	229.41
259	C4H6Cl2	1,3-DICHLORO-trans-2-BUTENE	-79.860	2.8038E-01	2.1834E-05	298	1000	5.86	65.79
260	C4H6Cl2	1,4-DICHLORO-cis-2-BUTENE	75.742	1.0692E-01	7.6969E-06	298	1000	108.50	131.12
261	C4H6Cl2	1,4-DICHLORO-trans-2-BUTENE	-369.628	2.6691E-01	2.3576E-05	298	1000	9.72	-230.28
262	C4H6Cl2	3,4-DICHLORO-1-BUTENE	-65.623	2.6404E-01	2.0203E-05	298	1000	15.10	71.45
263	C4H6O	trans-CROTONALDEHYDE	-106.171	1.8945E-01	8.6539E-06	298	1000	-48.70	-9.28
264	C4H6O	2,5-DIHYDROFURAN	-110.604	2.3000E-01	2.5513E-05	298	1000	-39.50	10.78
265	C4H6O	DIVINYL ETHER	-15.074	1.7863E-01	2.5003E-05	298	1000	40.62	80.49
266	C4H6O	METHACROLEIN	-114.785	1.8527E-01	1.8695E-05	298	1000	-57.60	-17.48
267	C4H6O2	2-BUTYNE-1,4-DIOL	-155.894	2.5684E-01	2.3863E-05	298	1000	-77.00	-21.51
268	C4H6O2	gamma-BUTYROLACTONE	-381.390	3.1362E-01	2.5769E-05	298	1000	-285.29	-218.14
269	C4H6O2	cis-CROTONIC ACID	-360.595	2.6842E-01	1.5526E-05	298	1000	-279.00	-222.50
270	C4H6O2	trans-CROTONIC ACID	-370.060	2.6013E-01	2.3112E-05	298	1000	-290.23	-234.22
271	C4H6O2	METHACRYLIC ACID	--------	--------	--------	--------	--------	-287.72	--------
272	C4H6O2	METHYL ACRYLATE	-334.519	2.5153E-01	2.2359E-05	298	1000	-257.32	-203.17
273	C4H6O2	VINYL ACETATE	-317.521	2.9048E-01	2.2166E-05	298	1000	-228.70	-166.74
274	C4H6O3	ACETIC ANHYDRIDE	-578.076	3.3162E-01	2.5188E-05	298	1000	-476.68	-405.97
275	C4H6O4	SUCCINIC ACID	-824.582	4.2219E-01	1.7001E-05	298	1000	-697.00	-609.24
276	C4H6O5	DIGLYCOLIC ACID	-947.340	4.8793E-01	1.8278E-05	298	1000	-800.00	-698.80
277	C4H6O5	MALIC ACID	-992.048	4.7792E-01	1.5225E-05	298	1000	-848.00	-749.28
278	C4H6O6	TARTARIC ACID	-1162.439	5.4675E-01	1.3666E-05	298	1000	-998.00	-885.65
279	C4H7N	BUTYRONITRILE	32.613	2.4673E-01	2.5125E-05	298	1000	108.66	162.26
280	C4H7N	ISOBUTYRONITRILE	23.812	2.5960E-01	2.4170E-05	298	1000	103.60	159.65
281	C4H7NO	ACETONE CYANOHYDRIN	-134.626	3.4051E-01	2.1529E-05	298	1000	-30.97	41.01
282	C4H7NO	2-METHACRYLAMIDE	-187.448	3.1918E-01	1.8311E-05	298	1000	-90.40	-23.28
283	C4H7NO	3-METHOXYPROPIONITRILE	-80.895	3.1245E-01	2.4540E-05	298	1000	14.70	81.46
284	C4H7NO	2-PYRROLIDONE	-200.086	3.7682E-01	2.9777E-05	298	TMAX	-84.73	-4.23
285	C4H8	1-BUTENE	-1.692	2.3442E-01	3.1582E-05	298	1000	71.30	123.42
286	C4H8	cis-2-BUTENE	-8.619	2.3793E-01	3.6144E-05	298	1000	65.86	119.38
287	C4H8	trans-2-BUTENE	-12.497	2.4252E-01	3.2484E-05	298	1000	62.97	116.89
288	C4H8	CYCLOBUTANE	24.216	2.7677E-01	3.3570E-05	298	1000	110.04	170.99
289	C4H8	2-METHYLPROPENE	-18.295	2.4609E-01	3.0860E-05	298	1000	58.07	112.46
290	C4H8Br2	1,2-DIBROMOBUTANE	-135.879	3.9161E-01	1.6038E-05	333	1000	-13.14	63.94
291	C4H8Br2	2,3-DIBROMOBUTANE	-139.240	4.0686E-01	1.5194E-05	333	1000	-11.92	67.99
292	C4H8Cl2	1,4-DICHLOROBUTANE	-181.192	3.7784E-01	2.6458E-05	298	1000	-65.90	14.34
293	C4H8I2	1,2-DIIODOBUTANE	-58.978	3.9654E-01	1.1910E-05	459	1000	82.09	142.27
294	C4H8O	BUTYRALDEHYDE	-206.702	2.9866E-01	2.9382E-05	298	1000	-114.77	-50.03
295	C4H8O	ISOBUTYRALDEHYDE	-217.480	3.2985E-01	3.0302E-05	298	1000	-116.15	-44.98
296	C4H8O	1,2-EPOXYBUTANE	-112.002	3.2229E-01	3.0738E-05	298	1000	-12.88	56.83
297	C4H8O	2-BUTANONE	-239.887	3.0451E-01	3.0949E-05	298	1000	-146.06	-79.90
298	C4H8O	ETHYL VINYL ETHER	-142.593	2.9383E-01	3.2279E-05	298	1000	-51.83	12.39
299	C4H8O	TETRAHYDROFURAN	-187.060	3.4702E-01	3.9333E-05	298	1000	-79.68	-3.72
300	C4H8O2	cis-2-BUTENE-1,4-DIOL	-297.153	3.5909E-01	3.4512E-05	298	1000	-186.69	-108.98
301	C4H8O2	trans-2-BUTENE-1,4-DIOL	-301.973	3.5932E-01	3.1980E-05	298	1000	-191.69	-114.32
302	C4H8O2	ISOBUTYRIC ACID	-486.608	4.0717E-01	3.0484E-05	298	1000	-362.18	-275.40
303	C4H8O2	n-BUTYRIC ACID	-472.274	3.8416E-01	2.7623E-05	298	1000	-355.00	-273.29
304	C4H8O2	p-DIOXANE	-318.550	4.5161E-01	3.0870E-05	298	1000	-180.79	-85.03
305	C4H8O2	ETHYL ACETATE	-444.940	3.8444E-01	2.9614E-05	298	1000	-327.40	-245.32
306	C4H8O2	METHYL PROPIONATE	-429.469	3.8848E-01	2.6801E-05	298	1000	-311.00	-228.53
307	C4H8O2	n-PROPYL FORMATE	-406.624	3.7532E-01	2.8529E-05	298	1000	-291.89	-211.83
308	C4H8O2S	SULFOLANE	--------	--------	--------	--------	--------	-259.00	--------
309	C4H8S	THIACYCLOPENTANE	-29.540	2.2963E-01	7.8684E-05	389	717	46.02	104.95
310	C4H9Br	1-BROMOBUTANE	-129.464	3.7268E-01	2.3816E-05	333	1000	-12.89	62.83
311	C4H9Br	2-BROMOBUTANE	-142.489	3.7386E-01	2.1226E-05	333	1000	-25.77	49.75
312	C4H9Cl	1-CHLOROBUTANE	-149.464	3.6039E-01	3.2740E-05	298	1000	-38.79	38.92

320

NO	FORMULA	NAME	$Gf = A + B T + C T^2$					$(Gf - kjoule/mol, T - K)$	
			A	B	C	TMIN	TMAX	Gf @ 298 K	Gf @ 500 K
313	C4H9Cl	2-CHLOROBUTANE	-163.844	3.5979E-01	3.1321E-05	298	1000	-53.47	23.88
314	C4H9Cl	2-CHLORO-2-METHYLPROPANE	-185.492	3.9793E-01	2.7814E-05	298	1000	-64.10	20.43
315	C4H9I	2-IODO-2-METHYLPROPANE	-113.913	4.1708E-01	1.1495E-05	459	1000	23.64	97.50
316	C4H9N	PYRROLIDINE	-6.977	3.9482E-01	3.9541E-05	298	1000	114.68	200.32
317	C4H9NO	N,N-DIMETHYLACETAMIDE	-228.242	4.5804E-01	3.1465E-05	298	1000	-88.50	8.64
318	C4H9NO	MORPHOLINE	-159.313	5.7932E-01	2.5288E-05	298	1000	16.00	136.67
319	C4H9NO2	1-NITROBUTANE	-147.096	5.1673E-01	3.1532E-05	298	1000	10.13	119.15
320	C4H9NO2	2-NITROBUTANE	-166.954	5.2869E-01	3.0809E-05	298	1000	-6.23	105.09
321	C4H10	BUTANE	-128.375	3.6047E-01	3.8256E-05	298	1000	-17.15	61.43
322	C4H10	2-METHYLPROPANE(ISOBUTANE)	-136.801	3.7641E-01	3.7497E-05	298	1000	-20.88	60.78
323	C4H10N2	PIPERAZINE	17.821	5.5259E-01	2.2901E-05	298	1000	185.00	299.84
324	C4H10O	BUTYL ALCOHOL	-276.720	4.0989E-01	3.9100E-05	298	1000	-150.67	-62.00
325	C4H10O	sec- BUTYL ALCOHOL	-294.630	4.1501E-01	3.6367E-05	298	1000	-167.32	-78.03
326	C4H10O	tert-BUTYL ALCOHOL	-328.304	4.4891E-01	3.4424E-05	298	1000	-191.04	-95.25
327	C4H10O	ETHYL ETHER	-254.382	4.3005E-01	3.8901E-05	298	1000	-122.34	-29.63
328	C4H10O	METHYL PROPYL ETHER	-239.994	4.2359E-01	3.8666E-05	298	1000	-109.91	-18.53
329	C4H10O	METHYL ISOPROPYL ETHER	-254.443	4.3538E-01	3.8072E-05	298	1000	-120.88	-27.24
330	C4H10O2	1,4-BUTANEDIOL	-429.118	4.9506E-01	3.8566E-05	298	1000	-277.72	-171.95
331	C4H10O	ISOBUTANOL	-287.586	4.3450E-01	2.8373E-05	298	1000	-155.01	-63.24
332	C4H10O2	1,3-BUTANEDIOL	-436.981	4.7514E-01	3.6780E-05	298	1000	-291.69	-190.21
333	C4H10O2	2,3-BUTANEDIOL	-484.893	4.7808E-01	3.3787E-05	298	1000	-339.00	-237.41
334	C4H10O2	t-BUTYL HYDROPEROXIDE	-245.636	5.1550E-01	3.0419E-05	298	1000	-88.90	19.72
335	C4H10O2	1,2-DIMETHOXYETHANE	-348.667	4.9222E-01	3.7866E-05	298	1000	-198.16	-93.09
336	C4H10O2	2-ETHOXYETHANOL	-402.679	4.7927E-01	3.8251E-05	298	1000	-256.00	-153.48
337	C4H10O3	DIETHYLENE GLYCOL	-573.918	5.4040E-01	3.8370E-05	298	1000	-409.00	-294.13
338	C4H10O4S	DIETHYL SULFATE	--------	--------	--------	--------	--------	-547.00	--------
339	C4H10S	1-BUTANETHIOL	-83.578	2.9249E-01	8.2267E-05	389	717	11.05	83.24
340	C4H10S	2-METHYL-1-PROPANETHIOL	-92.734	3.0510E-01	8.1489E-05	389	717	5.56	80.19
341	C4H10S	2-BUTANETHIOL	-91.727	3.0122E-01	8.1363E-05	389	717	5.40	79.22
342	C4H10S	2-METHYL-2-PROPANETHIOL	-105.410	3.3263E-01	7.7086E-05	389	717	0.71	80.17
343	C4H10S	ETHYLSULFIDE	-78.870	2.9895E-01	8.3578E-05	389	717	17.78	91.50
344	C4H10S	ISOPROPYL METHYL SULFIDE	-85.853	3.0788E-01	8.3375E-05	389	717	13.43	88.93
345	C4H10S	METHYL PROPYL SULFIDE	-77.076	2.9483E-01	8.4110E-05	389	717	18.41	91.37
346	C4H10S2	ETHYL DISULFIDE	-68.436	2.7360E-01	1.0156E-04	389	717	22.26	93.75
347	C4H11N	BUTYLAMINE	-94.752	4.7000E-01	3.9129E-05	298	1000	49.20	149.99
348	C4H11N	ISOBUTYLAMINE	-101.656	4.8218E-01	4.0321E-05	298	1000	46.10	149.51
349	C4H11N	sec-BUTYLAMINE	-107.194	4.8307E-01	3.8180E-05	298	1000	40.63	143.89
350	C4H11N	tert-BUTYLAMINE	-123.074	4.9817E-01	3.4076E-05	298	1000	28.87	134.53
351	C4H11N	DIETHYLAMINE	-75.362	4.8136E-01	3.9617E-05	298	1000	72.09	175.22
352	C4H11NO	DIMETHYLETHANOLAMINE	-205.382	5.5381E-01	3.8678E-05	298	1000	-36.40	81.19
353	C4H11NO2	DIETHANOLAMINE	-400.057	6.1044E-01	3.9952E-05	298	1000	-214.08	-84.85
354	C4H11NO2	2-AMINOETHOXYETHANOL	-368.179	5.9427E-01	4.0200E-05	298	1000	-187.00	-60.99
355	C4H12N2O	N-AMINOETHYL ETHANOLAMINE	-194.770	6.5190E-01	4.2133E-05	298	1000	3.81	141.72
356	C4H12Si	TETRAMETHYLSILANE	-235.280	4.5965E-01	4.1311E-05	298	1000	-94.18	4.87
357	C4H13N3	DIETHYLENE TRIAMINE	-10.007	7.1391E-01	4.4288E-05	298	1000	207.29	358.02
358	C5Cl6	HEXACHLOROCYCLOPENTADIENE	-100.885	2.4199E-01	-1.2665E-05	298	1000	-30.00	16.95
359	C5H4O2	FURFURAL	-152.572	1.6333E-01	9.6362E-06	298	1000	-102.87	-68.50
360	C5H5N	PYRIDINE	138.330	1.6730E-01	2.0124E-05	298	1000	190.20	227.01
361	C5H6	CYCLOPENTADIENE	129.314	1.3722E-01	2.3799E-05	298	1000	172.57	203.87
362	C5H6	2-METHYL-1-BUTENE-3-YNE	259.349	1.3716E-01	1.8086E-05	298	1000	302.00	332.45
363	C5H6	1-PENTENE-3-YNE	248.489	1.0195E-01	2.1953E-05	298	1000	281.00	304.95
364	C5H6	1-PENTENE-4-YNE	268.583	1.0864E-01	2.1074E-05	298	1000	303.00	328.17
365	C5H6N2	GLUTARONITRILE	168.609	2.4366E-01	1.7509E-05	298	1000	243.00	294.82
366	C5H6O2	FURFURYL ALCOHOL	-220.654	2.1927E-01	1.2436E-05	298	1000	-154.00	-107.91
367	C5H6O3	GLUTARIC ANHYDRIDE	-353.847	3.8966E-01	3.9400E-06	298	1000	-237.00	-158.03
368	C5H6O4	CITRACONIC ACID	-742.355	4.0221E-01	3.2537E-06	298	1000	-622.00	-540.44
369	C5H6O4	ITACONIC ACID	-731.261	3.9663E-01	9.3100E-06	298	1000	-612.00	-530.62
370	C5H6S	2-METHYLTHIOPHENE	87.227	1.0101E-01	6.1872E-05	389	717	122.93	153.20
371	C5H6S	3-METHYLTHIOPHENE	86.422	9.9868E-02	6.2512E-05	389	717	121.84	151.98
372	C5H7N	N-METHYLPYRROLE	100.239	2.7685E-01	1.9477E-05	298	1000	184.80	243.54
373	C5H7NO2	ETHYL CYANOACETATE	-318.344	3.7320E-01	2.3099E-05	298	1000	-204.77	-125.98
374	C5H8	CYCLOPENTENE	30.597	2.5615E-01	3.8896E-05	298	1000	110.79	168.40
375	C5H8	ISOPRENE	74.244	2.3174E-01	2.6082E-05	298	1000	145.90	196.63
376	C5H8	3-METHYL-1,2-BUTADIENE	128.505	2.2625E-01	2.7245E-05	298	1000	198.61	248.44
377	C5H8	2-METHYL-1,3-BUTADIENE	74.305	2.3206E-01	2.4048E-05	298	1000	145.85	196.35
378	C5H8	1,2-PENTADIENE	144.351	2.1311E-01	2.5804E-05	298	1000	210.41	257.36
379	C5H8	cis-1,3-PENTADIENE	76.492	2.2279E-01	2.9198E-05	298	1000	145.77	195.19
380	C5H8	trans-1,3-PENTADIENE	76.388	2.2756E-01	2.5564E-05	298	1000	146.73	196.56
381	C5H8	1,4-PENTADIENE	104.148	2.1318E-01	2.5915E-05	298	1000	170.25	217.22
382	C5H8	2,3-PENTADIENE	137.273	2.2032E-01	3.0252E-05	298	1000	205.89	255.00
383	C5H8	1-PENTYNE	143.177	2.1581E-01	2.7158E-05	298	1000	210.25	257.97
384	C5H8	2-PENTYNE	127.758	2.1215E-01	3.2628E-05	298	1000	194.18	241.99
385	C5H8	3-METHYL-1-BUTYNE	135.083	2.2757E-01	2.6485E-05	298	1000	205.52	255.49
386	C5H8	SPIROPENTANE	182.924	2.6664E-01	2.9124E-05	298	1000	265.31	323.53
387	C5H8N4O12	PENTAERYTHRITOL TETRANITRATE	--------	--------	--------	--------	--------	--------	--------
388	C5H8O	CYCLOPENTANONE	-196.392	3.2411E-01	3.6176E-05	298	1000	-96.20	-25.29
389	C5H8O	METHYL ISOPROPENYL KETONE	-181.761	2.8848E-01	3.3185E-05	298	1000	-92.50	-29.23
390	C5H8O2	ACETYLACETONE	-382.032	3.4766E-01	3.4317E-05	298	1000	-275.00	-199.62

NO	FORMULA	NAME	Gf = A + B T + C T^2					(Gf - kjoule/mol, T - K)	
			A	B	C	TMIN	TMAX	Gf @ 298 K	Gf @ 500 K
391	C5H8O2	ALLYL ACETATE	-336.239	3.5060E-01	2.7223E-05	298	1000	-229.00	-154.13
392	C5H8O2	ETHYL ACRYLATE	-351.701	3.4693E-01	2.8364E-05	298	1000	-245.45	-171.15
393	C5H8O2	METHYL METHACRYLATE	-349.636	3.5237E-01	3.0129E-05	298	1000	-241.59	-165.92
394	C5H8O2	VINYL PROPIONATE	-349.481	3.6822E-01	2.9682E-05	298	1000	-236.76	-157.95
395	C5H8O3	2-HYDROXYETHYL ACRYLATE	-495.271	4.1070E-01	1.7889E-05	298	1000	-371.00	-285.45
396	C5H8O3	LEVULINIC ACID	-608.829	4.2990E-01	2.6837E-05	298	1000	-478.00	-387.17
397	C5H8O3	METHYL ACETOACETATE	-591.395	4.1366E-01	3.0880E-05	298	1000	-465.00	-376.85
398	C5H8O4	GLUTARIC ACID	-846.155	5.1598E-01	2.3107E-05	298	1000	-690.00	-582.39
399	C5H9N	VALERONITRILE	9.512	3.4287E-01	3.0370E-05	298	1000	114.74	188.54
400	C5H9NO	n-BUTYL ISOCYANATE	-129.333	4.2734E-01	3.9417E-05	298	1000	1.96	94.19
401	C5H9NO	N-METHYL-2-PYRROLIDONE	-199.425	4.7756E-01	3.5689E-05	298	1000	-53.42	48.28
402	C5H9NO4	L-GLUTAMIC ACID	-827.006	6.3230E-01	2.4411E-05	298	1000	-636.00	-504.76
403	C5H10	CYCLOPENTANE	-80.495	3.8452E-01	4.5171E-05	298	1000	38.62	123.06
404	C5H10	2-METHYL-1-BUTENE	-38.406	3.3676E-01	3.6799E-05	298	1000	65.61	139.17
405	C5H10	2-METHYL-2-BUTENE	-44.552	3.3614E-01	4.0962E-05	298	1000	59.66	133.76
406	C5H10	3-METHYL-1-BUTENE	-30.771	3.4337E-01	3.2419E-05	298	1000	74.77	149.02
407	C5H10	1-PENTENE	-22.938	3.3006E-01	3.7449E-05	298	1000	79.12	151.45
408	C5H10	cis-2-PENTENE	-30.293	3.2915E-01	4.0930E-05	298	1000	71.84	144.51
409	C5H10	trans-2-PENTENE	-33.667	3.3463E-01	3.9095E-05	298	1000	69.91	143.42
410	C5H10Br2	2,3-DIBROMO-2-METHYLBUTANE	-177.701	5.2964E-01	1.6349E-05	333	1000	-13.35	91.21
411	C5H10Cl2	1,5-DICHLOROPENTANE	-202.639	4.7350E-01	3.2761E-05	298	1000	-58.20	42.30
412	C5H10O	METHYL ISOPROPYL KETONE	-265.054	4.0942E-01	4.0362E-05	298	1000	-139.00	-50.25
413	C5H10O	2-PENTANONE	-260.842	4.0269E-01	3.7677E-05	298	1000	-137.07	-50.08
414	C5H10O	DIETHYL KETONE	-261.163	4.1025E-01	3.8597E-05	298	1000	-135.06	-46.39
415	C5H10O	VALERALDEHYDE	-229.987	3.9652E-01	3.5518E-05	298	1000	-108.28	-22.85
416	C5H10O2	n-BUTYL FORMATE	-427.826	4.7196E-01	3.4635E-05	298	1000	-283.67	-183.19
417	C5H10O2	ETHYL PROPIONATE	-466.392	4.8142E-01	3.5746E-05	298	1000	-319.30	-216.74
418	C5H10O2	ISOBUTYL FORMATE	-439.182	4.7834E-01	3.3754E-05	298	1000	-293.20	-191.58
419	C5H10O2	ISOPROPYL ACETATE	-484.647	4.9518E-01	3.3148E-05	298	1000	-333.70	-228.77
420	C5H10O2	n-PROPYL ACETATE	-467.320	4.8076E-01	3.6139E-05	298	1000	-320.40	-217.90
421	C5H10O2	METHYL n-BUTYRATE	-453.220	4.8471E-01	3.4390E-05	298	1000	-305.30	-202.27
422	C5H10O2	2-METHYLBUTYRIC ACID	-501.004	4.8757E-01	3.6506E-05	298	1000	-352.00	-248.09
423	C5H10O2	ISOVALERIC ACID	-517.323	4.9295E-01	3.3719E-05	298	1000	-367.00	-262.42
424	C5H10O2	VALERIC ACID	-492.802	4.8120E-01	3.3679E-05	298	1000	-346.00	-243.78
425	C5H10O2	TETRAHYDROFURFURYL ALCOHOL	-372.291	5.1486E-01	2.7500E-05	298	1000	-216.00	-107.99
426	C5H10O2S	3-METHYL SULFOLANE	--------	--------	--------	--------	--------	-257.02	--------
427	C5H10O3	DIETHYL CARBONATE	--------	--------	--------	--------	--------	-464.65	--------
428	C5H10O3	ETHYL LACTATE	-637.892	5.3196E-01	3.2968E-05	298	1000	-476.00	-363.67
429	C5H10S	THIACYCLOHEXANE	-59.519	3.5205E-01	8.4273E-05	389	717	53.05	137.57
430	C5H10S	CYCLOPENTANETHIOL	-43.019	3.0720E-01	9.3808E-05	389	717	57.03	134.04
431	C5H11Br	1-BROMOPENTANE	-152.750	4.7242E-01	2.7996E-05	333	1000	-5.73	90.46
432	C5H11Cl	1-CHLOROPENTANE	-177.625	4.5745E-01	3.8765E-05	298	1000	-37.40	60.79
433	C5H11Cl	1-CHLORO-3-METHYLBUTANE	-183.095	4.5557E-01	3.6144E-05	298	1000	-43.68	53.73
434	C5H11Cl	2-CHLORO-2-METHYLBUTANE	-205.548	4.8810E-01	3.5442E-05	298	1000	-56.48	47.36
435	C5H11N	N-METHYLPYRROLIDINE	-12.182	5.1221E-01	4.4621E-05	298	1000	145.00	255.08
436	C5H11N	PIPERIDINE	-53.009	5.1032E-01	2.7666E-05	298	1000	102.00	209.07
437	C5H11NO	tert-BUTYLFORMAMIDE	-275.061	5.2145E-01	4.4449E-05	298	1000	-114.00	-3.23
438	C5H12	2-METHYLBUTANE(ISOPENTANE)	-157.445	4.6399E-01	4.3411E-05	298	1000	-14.81	85.40
439	C5H12	2,2-DIMETHYPROPANE	-169.088	5.0284E-01	3.9605E-05	298	1000	-15.23	92.23
440	C5H12	PENTANE	-149.141	4.5748E-01	4.4417E-05	298	1000	-8.37	90.70
441	C5H12O	2,2-DIMETHYL-1-PROPANOL	-322.375	5.4662E-01	4.0369E-05	298	1000	-155.38	-38.97
442	C5H12O	tert-PENTYL ALCOHOL	-332.312	5.4443E-01	4.2371E-05	298	1000	-165.77	-49.51
443	C5H12O	2-METHYL-1-BUTANOL	-305.051	5.1639E-01	4.4343E-05	298	1000	-146.71	-35.77
444	C5H12O	2-METHYL-2-BUTANOL	-332.682	5.4788E-01	4.0976E-05	298	1000	-165.27	-48.50
445	C5H12O	3-METHYL-1-BUTANOL	-305.111	5.2208E-01	4.5046E-05	298	1000	-145.00	-32.81
446	C5H12O	3-METHYL-2-BUTANOL	-317.145	5.2515E-01	4.2579E-05	298	1000	-156.36	-43.92
447	C5H12O	PENTYL ALCOHOL	-305.155	5.0637E-01	4.5232E-05	298	1000	-149.75	-40.66
448	C5H12O	2-PENTANOL	-316.615	5.1346E-01	4.2919E-05	298	1000	-159.29	-49.16
449	C5H12O	3-PENTANOL	-319.635	5.2730E-01	4.3096E-05	298	1000	-158.16	-45.21
450	C5H12O	METHYL sec-BUTYL ETHER	-277.926	5.2648E-01	4.1955E-05	298	1000	-116.80	-4.20
451	C5H12O	METHYL-tert-BUTYL-ETHER	-296.013	5.5806E-01	4.2094E-05	298	1000	-125.44	-6.46
452	C5H120	METHYL ISOBUTYL ETHER	-269.004	5.2865E-01	4.4353E-05	298	1000	-107.00	6.41
453	C5H12O	ETHYL PROPYL ETHER	--------	--------	--------	--------	--------	-115.20	--------
454	C5H12O2	ETHYLENE GLYCOL MONOPROPYL ETHER	-424.154	5.7608E-01	4.4399E-05	298	1000	-248.00	-125.01
455	C5H12O2	NEOPENTYL GLYCOL	-450.426	6.1191E-01	3.9900E-05	298	1000	-264.00	-134.50
456	C5H12O2	1,5-PENTANEDIOL	-451.911	5.7833E-01	4.5466E-05	298	1000	-275.00	-151.38
457	C5H12O3	2-(2-METHOXYETHOXY)ETHANOL	-533.362	5.4333E-01	4.4004E-05	298	1000	-367.00	-250.70
458	C5H12O4	PENTAERYTHRITOL	-780.351	7.8393E-01	3.9672E-05	298	1000	-542.65	-378.47
459	C5H12S	1-PENTANETHIOL	-103.448	3.8392E-01	9.2649E-05	389	717	19.37	111.67
460	C5H12S	BUTYL-METHYL SULFIDE	-97.062	3.8661E-01	9.3888E-05	389	717	26.65	119.71
461	C5H12S	ETHYL PROPYL SULFIDE	-99.575	3.8453E-01	9.4240E-05	389	717	23.56	116.25
462	C5H12S	2-METHYL-2-BUTANETHIOL	-122.516	4.1515E-01	8.8234E-05	389	717	9.20	107.12
463	C5H13N	n-PENTYLAMINE	-116.075	5.7548E-01	4.6588E-05	298	1000	60.13	183.31
464	C5H13NO2	METHYL DIETHANOLAMINE	-384.069	7.0631E-01	4.4787E-05	298	1000	-169.00	-19.72
465	C6Cl6	HEXACHLOROBENZENE	-34.385	2.6934E-01	-1.8686E-05	298	1000	44.18	95.61
466	C6F6	HEXAFLUOROBENZENE	-957.239	2.6465E-01	-1.1629E-05	298	1000	-879.39	-827.82
467	C6H3ClN2O4	1-CHLORO-2,4-DINITROBENZENE	--------	--------	--------	--------	--------	186.00	--------
468	C6H3Cl2NO2	1,2-DICHLORO-4-NITROBENZENE	--------	--------	--------	--------	--------	119.00	--------

Table 13-1 GIBBS ENERGY OF FORMATION OF GAS - ORGANIC COMPOUNDS (continued)

			Gf = A + B T + C T^2				(Gf - kjoule/mol, T - K)		
NO	FORMULA	NAME	A	B	C	TMIN	TMAX	Gf @ 298 K	Gf @ 500 K
469	C6H3Cl3	1,2,4-TRICHLOROBENZENE	-12.668	1.8574E-01	3.0771E-06	298	1000	43.05	80.97
470	C6H3N3O6	1,3,5-TRINITROBENZENE	--------	--------	--------	--------	--------	--------	--------
471	C6H4Br2	m-DIBROMOBENZENE	93.578	1.7392E-01	1.0248E-05	333	1000	149.60	183.10
472	C6H4ClNO2	m-CHLORONITROBENZENE	35.149	-7.4130E-02	5.6551E-06	298	1000	13.70	-0.50
473	C6H4ClNO2	o-CHLORONITROBENZENE	35.167	3.4271E-01	5.6786E-06	298	1000	138.00	207.94
474	C6H4ClNO2	p-CHLORONITROBENZENE	35.150	3.4629E-01	5.1382E-06	298	1000	139.00	209.58
475	C6H4Cl2	m-DICHLOROBENZENE	25.238	1.7541E-01	1.0191E-05	298	1000	78.58	115.49
476	C6H4Cl2	o-DICHLOROBENZENE	28.788	1.7721E-01	1.0481E-05	298	1000	82.68	120.02
477	C6H4Cl2	p-DICHLOROBENZENE	21.845	1.8209E-01	1.0115E-05	298	1000	77.15	115.42
478	C6H4F2	m-DIFLUOROBENZENE	-311.294	1.7786E-01	1.2420E-05	298	1000	-257.02	-219.26
479	C6H4F2	o-DIFLUOROBENZENE	-295.933	1.7694E-01	1.1637E-05	298	1000	-242.00	-204.55
480	C6H4F2	p-DIFLUOROBENZENE	-308.558	1.8283E-01	1.2013E-05	298	1000	-252.84	-214.14
481	C6H4N2O4	m-DINITROBENZENE	--------	--------	--------	--------	--------	--------	--------
482	C6H4N2O4	o-DINITROBENZENE	--------	--------	--------	--------	--------	--------	--------
483	C6H4N2O4	p-DINITROBENZENE	--------	--------	--------	--------	--------	--------	--------
484	C6H5Br	BROMOBENZENE	85.584	1.6480E-01	1.3600E-05	333	1000	138.53	171.39
485	C6H5Cl	CHLOROBENZENE	50.479	1.5737E-01	1.7618E-05	298	1000	99.16	133.57
486	C6H5ClO	m-CHLOROPHENOL	-154.786	2.2457E-01	1.2143E-05	298	1000	-86.60	-39.47
487	C6H5ClO	o-CHLOROPHENOL	-124.486	2.2457E-01	1.2143E-05	298	1000	-56.30	-9.17
488	C6H5ClO	p-CHLOROPHENOL	-147.286	2.3061E-01	1.2143E-05	298	1000	-77.30	-28.95
489	C6H5Cl2N	3,4-DICHLOROANILINE	30.968	3.0600E-01	7.4844E-06	298	1000	123.00	185.84
490	C6H5F	FLUOROBENZENE	-118.073	1.5842E-01	1.8022E-05	298	1000	-69.04	-34.36
491	C6H5I	IODOBENZENE	125.947	1.6680E-01	9.7930E-06	459	1000	187.78	211.80
492	C6H5NO2	NITROBENZENE	--------	--------	--------	--------	--------	158.00	--------
493	C6H6	BENZENE	81.512	1.5282E-01	2.6522E-05	298	1000	129.66	164.55
494	C6H6ClN	m-CHLOROANILINE	55.476	2.7843E-01	1.4868E-05	298	1000	140.00	198.41
495	C6H6ClN	o-CHLOROANILINE	55.476	2.7843E-01	1.4868E-05	298	1000	140.00	198.41
496	C6H6ClN	p-CHLOROANILINE	55.468	2.7845E-01	1.4865E-05	298	1000	140.00	198.41
497	C6H6N2	cis-DICYANO-1-BUTENE	269.346	2.3736E-01	1.8865E-05	298	1000	342.00	392.75
498	C6H6N2	trans-DICYANO-1-BUTENE	265.703	2.4350E-01	1.7141E-05	298	1000	340.00	391.74
499	C6H6N2	1,4-DICYANO-2-BUTENE	259.694	2.2309E-01	1.8054E-05	298	1000	328.00	375.76
500	C6H6N2O2	m-NITROANILINE	--------	--------	--------	--------	--------	190.00	--------
501	C6H6N2O2	o-NITROANILINE	--------	--------	--------	--------	--------	195.00	--------
502	C6H6N2O2	p-NITROANILINE	--------	--------	--------	--------	--------	192.00	--------
503	C6H6O	PHENOL	-97.896	2.1140E-01	1.9980E-05	298	1000	-32.89	12.80
504	C6H6O2	1,2-BENZENEDIOL	-274.125	2.8740E-01	1.3885E-05	298	1000	-187.00	-126.95
505	C6H6O2	1,3-BENZENEDIOL	-276.440	3.1447E-01	1.3554E-05	298	1000	-181.30	-115.81
506	C6H6O2	p-HYDROQUINONE	-263.237	2.8688E-01	1.5725E-05	298	1000	-176.13	-115.87
507	C6H6O3	1,2,3-BENZENETRIOL	-447.876	3.5513E-01	9.4551E-06	298	1000	-341.00	-267.95
508	C6H6S	BENZENETHIOL	114.894	9.1284E-02	6.0204E-05	389	717	147.61	175.67
509	C6H7N	ANILINE	84.822	2.6707E-01	2.2598E-05	298	1000	166.69	224.01
510	C6H7N	2-PICOLINE	96.687	2.6032E-01	2.7720E-05	298	1000	177.07	233.78
511	C6H7N	3-PICOLINE	103.829	2.6047E-01	2.7853E-05	298	1000	184.26	241.03
512	C6H7N	4-METHYLPYRIDINE	100.133	2.6520E-01	2.9087E-05	298	1000	182.08	240.01
513	C6H8	1,3-CYCLOHEXADIENE	103.907	2.5010E-01	3.2471E-05	298	1000	181.70	237.08
514	C6H8	METHYLCYCLOPENTADIENE	97.759	2.3722E-01	2.5182E-05	298	1000	171.00	222.67
515	C6H8N2	ADIPONITRILE	147.367	3.4813E-01	2.1406E-05	298	1000	253.31	326.78
516	C6H8N2	METHYLGLUTARONITRILE	140.389	3.5614E-01	2.4552E-05	298	1000	249.00	324.60
517	C6H8N2	m-PHENYLENEDIAMINE	88.681	3.8980E-01	2.0686E-05	298	1000	207.00	288.75
518	C6H8N2	o-PHENYLENEDIAMINE	88.681	3.8980E-01	2.0686E-05	298	1000	207.00	288.75
519	C6H8N2	p-PHENYLENEDIAMINE	88.681	3.9986E-01	2.0686E-05	298	1000	210.00	293.78
520	C6H8N2	PHENYLHYDRAZINE	200.802	3.9736E-01	2.6431E-05	298	1000	321.93	406.09
521	C6H8N2O	BIS(CYANOETHYL)ETHER	18.301	4.3047E-01	2.3483E-05	298	1000	149.00	239.41
522	C6H8O4	DIMETHYL MALEATE	-622.986	4.6753E-01	2.1806E-05	298	1000	-481.40	-383.77
523	C6H8O6	ASCORBIC ACID	-953.797	7.4441E-01	7.3932E-06	298	1000	-731.00	-579.74
524	C6H8O7	CITRIC ACID	-1392.842	7.2195E-01	1.5158E-05	298	1000	-1176.00	-1028.08
525	C6H10	1-METHYLCYCLOPENTENE	-8.313	3.5608E-01	4.3218E-05	298	1000	102.13	180.53
526	C6H10	3-METHYLCYCLOPENTENE	5.787	3.5193E-01	4.3095E-05	298	1000	114.98	192.53
527	C6H10	4-METHYLCYCLOPENTENE	11.922	3.5346E-01	4.3307E-05	298	1000	121.59	199.48
528	C6H10	CYCLOHEXENE	-8.701	3.7574E-01	3.5208E-05	298	1000	106.86	187.97
529	C6H10	2,3-DIMETHYL-1,3-BUTADIENE	43.249	3.2814E-01	2.9625E-05	298	1000	144.00	214.72
530	C6H10	1,5-HEXADIENE	82.108	3.0963E-01	3.6474E-05	298	1000	178.00	246.04
531	C6H10	cis,trans-2,4-HEXADIENE	45.372	3.2526E-01	3.7244E-05	298	1000	146.00	217.32
532	C6H10	trans,trans-2,4-HEXADIENE	41.417	3.2595E-01	3.4688E-05	298	1000	142.00	213.06
533	C6H10	1-HEXYNE	121.936	3.1308E-01	3.3426E-05	298	1000	218.57	286.84
534	C6H10	2-HEXYNE	103.341	3.0815E-01	3.8820E-05	298	1000	199.00	267.12
535	C6H10	3-HEXYNE	104.303	3.0457E-01	3.9930E-05	298	1000	199.00	266.57
536	C6H10O	CYCLOHEXANONE	-234.483	4.6963E-01	3.6411E-05	298	1000	-90.75	9.43
537	C6H10O	MESITYL OXIDE	-197.360	3.8799E-01	2.9406E-05	298	1000	-78.78	3.99
538	C6H10O2	epsilon-CAPROLACTONE	-430.575	5.3416E-01	3.3787E-05	298	1000	-267.84	-155.05
539	C6H10O2	ETHYL METHACRYLATE	-383.839	4.4679E-01	3.6581E-05	298	1000	-247.00	-151.30
540	C6H10O2	n-PROPYL ACRYLATE	-366.655	4.4685E-01	2.3999E-05	298	1000	-231.00	-137.23
541	C6H10O3	ETHYLACETOACETATE	-563.049	5.0772E-01	3.6907E-05	298	1000	-408.00	-299.96
542	C6H10O3	PROPIONIC ANHYDRIDE	-629.720	5.2392E-01	3.5106E-05	298	1000	-470.00	-358.98
543	C6H10O4	ADIPIC ACID	-868.011	5.9791E-01	3.2721E-05	298	1000	-686.47	-560.88
544	C6H10O4	DIETHYL OXALATE	-743.049	5.6968E-01	3.1879E-05	298	1000	-570.00	-450.24
545	C6H10O4	ETHYLENE GLYCOL DIACETATE	-810.248	5.9048E-01	3.1883E-05	298	1000	-631.00	-507.04
546	C6H10O4	ETHYLIDENE DIACETATE	-815.684	6.0479E-01	3.3501E-05	298	1000	-632.00	-504.92

Table 13-1 GIBBS ENERGY OF FORMATION OF GAS - ORGANIC COMPOUNDS (continued)

NO	FORMULA	NAME	$Gf = A + B T + C T^2$					$(Gf$ - kjoule/mol, T - K$)$	
			A	B	C	TMIN	TMAX	Gf @ 298 K	Gf @ 500 K
547	C6H11N	HEXANENITRILE	-11.733	4.3932E-01	3.6216E-05	298	1000	122.84	216.98
548	C6H11NO	epsilon-CAPROLACTAM	-251.243	5.8651E-01	5.2302E-05	298	1000	-71.11	55.09
549	C6H11NO	CYCLOHEXANONE OXIME	-199.483	5.8145E-01	3.0513E-05	298	1000	-22.90	98.87
550	C6H12	CYCLOHEXANE	-127.917	5.2032E-01	4.4706E-05	298	1000	31.76	143.42
551	C6H12	2,3-DIMETHYL-1-BUTENE	-57.956	4.4708E-01	3.7750E-05	298	1000	79.04	175.02
552	C6H12	2,3-DIMETHYL-2-BUTENE	-61.872	4.4549E-01	5.0110E-05	298	1000	75.86	173.40
553	C6H12	3,3-DIMETHYL-1-BUTENE	-45.896	4.6814E-01	4.5620E-05	298	1000	98.16	199.58
554	C6H12	2-ETHYL-1-BUTENE	-54.167	4.3620E-01	4.1384E-05	298	1000	79.96	174.28
555	C6H12	trans-3-METHYL-2-PENTENE	-61.309	4.2960E-01	4.6053E-05	298	1000	71.30	165.01
556	C6H12	1-HEXENE	-44.217	4.2735E-01	4.3352E-05	298	1000	87.45	180.29
557	C6H12	cis-2-HEXENE	-55.001	4.2475E-01	4.6828E-05	298	1000	76.23	169.08
558	C6H12	trans-2-HEXENE	-56.244	4.3026E-01	4.4958E-05	298	1000	76.44	170.12
559	C6H12	cis-3-HEXENE	-50.434	4.3223E-01	4.6533E-05	298	1000	83.01	177.31
560	C6H12	trans-3-HEXENE	-56.854	4.3669E-01	4.3504E-05	298	1000	77.61	172.37
561	C6H12	METHYLCYCLOPENTANE	-110.437	4.7401E-01	4.9123E-05	298	1000	35.77	138.85
562	C6H12	2-METHYL-1-PENTENE	-54.752	4.3022E-01	4.1468E-05	298	1000	77.61	170.73
563	C6H12	2-METHYL-2-PENTENE	-62.375	4.3289E-01	4.6135E-05	298	1000	71.21	165.60
564	C6H12	3-METHYL-1-PENTENE	-47.336	4.3630E-01	3.7470E-05	298	1000	86.44	180.18
565	C6H12	cis-3-METHYL-2-PENTENE	-60.369	4.3290E-01	4.6127E-05	298	1000	73.22	167.61
566	C6H12	4-METHYL-1-PENTENE	-46.656	4.4313E-01	4.6863E-05	298	1000	90.04	186.63
567	C6H12	cis-4-METHYL-2-PENTENE	-52.910	4.3887E-01	4.2824E-05	298	1000	82.13	177.23
568	C6H12	trans-4-METHYL-2-PENTENE	-56.547	4.4368E-01	3.9806E-05	298	1000	79.62	175.25
569	C6H12N2	TRIETHYLENEDIAMINE	46.372	6.5869E-01	4.0856E-05	298	1000	247.00	385.93
570	C6H12O	BUTYL VINYL ETHER	-185.752	4.8691E-01	4.4479E-05	298	1000	-36.20	68.82
571	C6H12O	CYCLOHEXANOL	-299.464	5.9552E-01	3.9211E-05	298	1000	-117.91	8.10
572	C6H12O	HEXANAL	-251.064	4.9255E-01	4.1576E-05	298	1000	-100.12	5.61
573	C6H12O	ETHYL ISOPROPYL KETONE	-288.777	5.0729E-01	4.5992E-05	298	1000	-133.00	-23.64
574	C6H12O	2-HEXANONE	-282.224	4.9588E-01	4.3727E-05	298	1000	-130.08	-23.35
575	C6H12O	3-HEXANONE	-279.937	5.0445E-01	4.3085E-05	298	1000	-125.30	-16.94
576	C6H12O	METHYL ISOBUTYL KETONE	-291.134	5.0824E-01	4.3056E-05	298	1000	-135.36	-26.25
577	C6H12O2	n-PENTYL FORMATE	-451.432	5.7490E-01	4.0480E-05	298	1000	-276.00	-153.86
578	C6H12O2	n-BUTYL ACETATE	-488.898	5.7839E-01	4.1620E-05	298	1000	-312.30	-189.30
579	C6H12O2	sec-BUTYL ACETATE	-507.225	5.8204E-01	3.8957E-05	298	1000	-329.80	-206.46
580	C6H12O2	tert-BUTYL ACETATE	-526.466	6.1616E-01	3.7586E-05	298	1000	-339.00	-208.99
581	C6H12O2	ETHYL n-BUTYRATE	-488.545	5.7786E-01	4.0927E-05	298	1000	-312.20	-189.38
582	C6H12O2	ETHYL ISOBUTYRATE	-503.187	5.8662E-01	4.3075E-05	298	1000	-324.00	-199.11
583	C6H12O2	ISOBUTYL ACETATE	-497.955	5.7570E-01	4.1364E-05	298	1000	-322.20	-199.76
584	C6H12O2	n-PROPYL PROPIONATE	-486.177	5.7768E-01	4.0778E-05	298	1000	-309.90	-187.14
585	C6H12O2	CYCLOHEXYL PEROXIDE	-224.984	6.4784E-01	3.4101E-05	298	1000	-28.30	107.46
586	C6H12O2	DIACETONE ALCOHOL	-548.710	5.8215E-01	4.1811E-05	298	1000	-371.00	-247.18
587	C6H12O2	2-ETHYL BUTYRIC ACID	-520.467	5.9773E-01	4.2747E-05	298	1000	-338.00	-210.92
588	C6H12O2	n-HEXANOIC ACID	-516.574	5.7896E-01	3.9883E-05	298	1000	-340.00	-217.13
589	C6H12O3	2-ETHOXYETHYL ACETATE	-614.661	6.3251E-01	4.0795E-05	298	1000	-422.00	-288.21
590	C6H12O3	HYDROXYCAPROIC ACID	-686.362	6.5583E-01	3.8297E-05	298	1000	-487.00	-348.88
591	C6H12O3	PARALDEHYDE	-649.874	7.8954E-01	4.0634E-05	298	1000	-410.38	-244.95
592	C6H12S	sec-BUTYL GLYCOLATE	--------	--------	--------	--------	--------	--------	--------
593	C6H12S	THIACYCLOHEPTANE	-58.172	4.4956E-01	9.0710E-05	389	717	84.06	189.29
594	C6H13N	CYCLOHEXYLAMINE	-110.185	6.3129E-01	4.1717E-05	298	1000	82.31	215.89
595	C6H13N	HEXAMETHYLENEIMINE	-39.803	6.2438E-01	4.9453E-05	298	1000	151.38	284.75
596	C6H14	2,2-DIMETHYLBUTANE	-189.225	5.8649E-01	4.7623E-05	298	1000	-9.62	115.93
597	C6H14	2,3-DIMETHYLBUTANE	-181.310	5.7783E-01	4.9722E-05	298	1000	-4.10	120.04
598	C6H14	HEXANE	-170.447	5.5417E-01	5.0303E-05	298	1000	-0.25	119.21
599	C6H14	2-METHYLPENTANE	-177.675	5.6303E-01	4.8313E-05	298	1000	-5.02	115.92
600	C6H14	3-METHYLPENTANE	-174.861	5.6271E-01	5.0351E-05	298	1000	-2.13	119.08
601	C6H14N2O2	LYSINE	-465.293	8.1474E-01	4.3515E-05	298	1000	-218.00	-47.05
602	C6H14O	2-ETHYL-1-BUTANOL	-329.538	6.1860E-01	5.0499E-05	298	1000	-140.11	-7.61
603	C6H14O	HEXYL ALCOHOL	-322.918	6.0346E-01	5.1175E-05	298	1000	-137.95	-8.39
604	C6H14O	2-HEXANOL	-336.956	6.1022E-01	4.8633E-05	298	1000	-150.20	-19.69
605	C6H14O	2-METHYL-1-PENTANOL	-330.604	6.1216E-01	5.1497E-05	298	1000	-143.00	-11.65
606	C6H14O	4-METHYL-2-PENTANOL	-345.854	6.2173E-01	4.8645E-05	298	1000	-155.67	-22.83
607	C6H14O	n-BUTYL ETHYL ETHER	-295.723	6.1625E-01	5.0452E-05	298	1000	-107.00	25.02
608	C6H14O	ISOPROPYL ETHER	-322.051	6.5458E-01	5.0837E-05	298	1000	-121.88	17.95
609	C6H14O	PROPYL ETHER	-296.085	6.2222E-01	5.0914E-05	298	1000	-105.56	27.75
610	C6H14O	METHYL tert-PENTYL ETHER	-315.534	6.4028E-01	4.6621E-05	298	1000	-120.00	16.26
611	C6H14O2	ACETAL	--------	--------	--------	--------	--------	-245.00	--------
612	C6H14O2	2-BUTOXYETHANOL	-444.684	6.6973E-01	5.0480E-05	298	1000	-240.00	-97.20
613	C6H14O2	1,6-HEXANEDIOL	-473.236	6.7634E-01	5.1586E-05	298	1000	-266.49	-122.17
614	C6H14O2	HEXYLENE GLYCOL	-538.609	6.9511E-01	4.3647E-05	298	1000	-327.00	-180.14
615	C6H14O2S	DI-n-PROPYL SULFONE	--------	--------	--------	--------	--------	-255.00	--------
616	C6H14O3	DIETHYLENE GLYCOL DIMETHYL ETHER	-514.963	7.4974E-01	4.9517E-05	298	1000	-286.50	-127.71
617	C6H14O3	DIPROPYLENE GLYCOL	-632.018	7.4293E-01	4.5175E-05	298	1000	-406.00	-249.26
618	C6H14O3	2-(2-ETHOXYETHOXY)ETHANOL	-569.028	7.3475E-01	4.9846E-05	298	1000	-345.00	-189.19
619	C6H14O3	TRIMETHYLOLPROPANE	-644.172	7.6543E-01	4.6596E-05	298	1000	-411.31	-249.81
620	C6H14O4	TRIETHYLENE GLYCOL	-729.114	7.9699E-01	4.9930E-05	298	1000	-486.52	-318.13
621	C6H14O6	SORBITOL	-1144.342	9.2002E-01	4.0072E-05	298	1000	-866.00	-674.31
622	C6H14S	1-HEXANETHIOL	-123.627	4.7685E-01	1.0281E-04	389	717	27.82	140.50
623	C6H14S	BUTYL ETHYL SULFIDE	-119.839	4.7788E-01	1.0404E-04	389	717	32.01	145.11
624	C6H14S	ISOPROPYL SULFIDE	-136.423	5.1958E-01	9.5795E-05	389	717	27.11	147.32

Table 13-1 GIBBS ENERGY OF FORMATION OF GAS - ORGANIC COMPOUNDS (continued)

| NO | FORMULA | NAME | $G_f = A + B T + C T^2$ | | | | | (Gf - kjoule/mol, T - K) | |
			A	B	C	TMIN	TMAX	Gf @ 298 K	Gf @ 500 K
625	C6H14S	METHYL PENTYL SULFIDE	-117.395	4.8025E-01	1.0343E-04	389	717	35.10	148.59
626	C6H14S	PROPYL SULFIDE	-119.954	4.8210E-01	1.0475E-04	389	717	33.22	147.28
627	C6H14S2	PROPYL DISULFIDE	-110.285	4.5705E-01	1.2224E-04	389	717	36.99	148.80
628	C6H15Al	TRIETHYL ALUMINUM	--------	--------	--------	--------	--------	--------	--------
629	C6H15Al2Cl3	ETHYL ALUMINUM SESQUICHLORIDE	--------	--------	--------	--------	--------	--------	--------
630	C6H15N	DIISOPROPYLAMINE	-154.372	6.9531E-01	4.9650E-05	298	1000	57.90	205.70
631	C6H15N	DI-n-PROPYLAMINE	-120.143	6.7655E-01	5.2559E-05	298	1000	86.80	231.27
632	C6H15N	n-HEXYLAMINE	-136.609	6.6266E-01	5.3819E-05	298	1000	66.30	208.18
633	C6H15N	TRIETHYLAMINE	-103.700	7.0072E-01	5.1024E-05	298	1000	110.29	259.42
634	C6H15NO	6-AMINOHEXANOL	-286.883	7.2157E-01	5.2849E-05	298	1000	-66.50	87.12
635	C6H15NO2	DIISOPROPANOLAMINE	-458.534	8.0088E-01	4.7361E-05	298	1000	-215.00	-46.26
636	C6H15NO3	TRIETHANOLAMINE	-567.104	8.7966E-01	4.8631E-05	298	1000	-299.93	-115.12
637	C6H15N3	N-AMINOETHYL PIPERAZINE	19.811	8.5251E-01	4.9551E-05	298	1000	279.00	458.45
638	C6H15O4P	TRIETHYL PHOSPHATE	--------	--------	--------	--------	--------	--------	--------
639	C6H16N2	HEXAMETHYLENEDIAMINE	-103.169	7.9265E-01	5.3661E-05	298	1000	138.51	306.57
640	C6H18N3OP	HEXAMETHYL PHOSPHORAMIDE	--------	--------	--------	--------	--------	--------	--------
641	C6H18N4	TRIETHYLENE TETRAMINE	-2.718	1.0292E+00	5.7955E-05	298	1000	310.00	526.39
642	C6H18Osi2	HEXAMETHYLDISILOXANE	-782.238	8.1651E-01	3.9027E-05	298	1000	-534.80	-364.23
643	C6H18O3Si3	HEXAMETHYLCYCLOTRISILOXANE	--------	--------	--------	--------	--------	--------	--------
644	C6H19NSi2	HEXAMETHYLDISILAZANE	--------	--------	--------	--------	--------	--------	--------
645	C7H3ClF3NO2	4-CHLORO-3-NITROBENZOTRIFLUORIDE	--------	--------	--------	--------	--------	-488.67	--------
646	C7H3Cl2F3	2,4-DICHLOROBENZOTRIFLUORIDE	-653.165	3.2961E-01	-2.1938E-06	298	1000	-555.01	-488.91
647	C7H3Cl2NO	3,4-DICHLOROPHENYL ISOCYANATE	-3.825	2.5241E-01	4.6157E-06	298	1000	71.90	123.53
648	C7H4ClF3	p-CHLOROBENZOTRIFLUORIDE	-621.567	3.1783E-01	6.5531E-06	298	1000	-526.08	-461.01
649	C7H4Cl2O	m-CHLOROBENZOYL CHLORIDE	-166.250	2.2567E-01	1.1784E-05	298	1000	-97.80	-50.47
650	C7H4F3NO2	3-NITROBENZOTRIFLUORIDE	--------	--------	--------	--------	--------	-461.00	--------
651	C7H5ClO	BENZOYL CHLORIDE	-106.343	2.0915E-01	2.0639E-05	298	1000	-41.97	3.39
652	C7H5ClO2	o-CHLOROBENZOIC ACID	-326.639	2.8014E-01	2.6334E-05	298	1000	-240.53	-179.99
653	C7H5Cl3	BENZOTRICHLORIDE	-14.061	3.0152E-01	1.1038E-05	298	1000	76.98	139.46
654	C7H5F3	A,A,A-TRIFLUOROTOLUENE	-602.334	3.0018E-01	1.5014E-05	298	1000	-511.28	-448.49
655	C7H5N	BENZONITRILE	217.638	1.3974E-01	1.5912E-05	298	1000	260.87	291.49
656	C7H5NO	PHENYL ISOCYANATE	49.903	2.1430E-01	2.2360E-05	298	1000	116.00	162.64
657	C7H5N3O6	2,4,6-TRINITROTOLUENE	--------	--------	--------	--------	--------	--------	--------
658	C7H6Cl2	BENZYL DICHLORIDE	11.271	2.6779E-01	2.0973E-05	298	1000	93.20	150.41
659	C7H6Cl2	2,4-DICHLOROTOLUENE	-15.390	2.7698E-01	1.4499E-05	298	1000	68.66	126.73
660	C7H6N2O4	2,4-DINITROTOLUENE	--------	--------	--------	--------	--------	--------	--------
661	C7H6N2O4	2,5-DINITROTOLUENE	--------	--------	--------	--------	--------	--------	--------
662	C7H6N2O4	2,6-DINITROTOLUENE	--------	--------	--------	--------	--------	--------	--------
663	C7H6N2O4	3,4-DINITROTOLUENE	--------	--------	--------	--------	--------	--------	--------
664	C7H6N2O4	3,5-DINITROTOLUENE	--------	--------	--------	--------	--------	--------	--------
665	C7H6O	BENZALDEHYDE	-38.243	1.9652E-01	2.0695E-05	298	1000	22.40	65.19
666	C7H6O2	BENZOIC ACID	-292.081	2.6249E-01	3.5028E-05	298	1000	-210.41	-152.08
667	C7H6O2	p-HYDROXYBENZALDEHYDE	-215.867	2.5532E-01	6.0535E-06	298	1000	-139.00	-86.69
668	C7H6O2	SALICYLALDEHYDE	-217.773	2.5555E-01	7.2447E-06	298	1000	-140.73	-88.19
669	C7H6O3	SALICYLIC ACID	-468.296	3.3602E-01	2.9513E-05	298	1000	-365.21	-292.91
670	C7H7Br	p-BROMOTOLUENE	63.326	2.5879E-01	2.5507E-05	333	1000	143.00	199.10
671	C7H7Cl	BENZYL CHLORIDE	19.277	2.2761E-01	5.5993E-06	298	1000	92.40	147.08
672	C7H7Cl	o-CHLOROTOLUENE	16.226	2.5943E-01	2.3325E-05	298	1000	95.90	151.77
673	C7H7Cl	p-CHLOROTOLUENE	16.226	2.6547E-01	2.3325E-05	298	1000	97.70	154.79
674	C7H7F	p-FLUOROTOLUENE	-150.014	2.5689E-01	2.5593E-05	298	1000	-70.88	-15.17
675	C7H7NO	FORMANILIDE	-57.142	3.6191E-01	3.4139E-05	298	1000	54.10	132.35
676	C7H7NO2	m-NITROTOLUENE	--------	--------	--------	--------	--------	143.00	--------
677	C7H7NO2	o-NITROTOLUENE	--------	--------	--------	--------	--------	168.00	--------
678	C7H7NO2	p-NITROTOLUENE	--------	--------	--------	--------	--------	155.00	--------
679	C7H7NO3	o-NITROANISOLE	--------	--------	--------	--------	--------	64.40	--------
680	C7H8	TOLUENE	47.813	2.3831E-01	3.1916E-05	298	1000	122.01	174.95
681	C7H8	1,3,5-CYCLOHEPTATRIENE	179.884	2.4584E-01	2.2285E-05	298	1000	255.39	308.38
682	C7H8O	ANISOLE	-70.433	3.0089E-01	3.3879E-05	298	1000	22.63	88.48
683	C7H8O	BENZYL ALCOHOL	-102.870	2.9660E-01	3.3705E-05	298	1000	-11.10	53.86
684	C7H8O	m-CRESOL	-134.426	3.0573E-01	2.7772E-05	298	1000	-40.54	25.38
685	C7H8O	o-CRESOL	-130.472	3.0478E-01	2.5844E-05	298	1000	-37.07	28.38
686	C7H8O	p-CRESOL	-127.446	3.1458E-01	2.8223E-05	298	1000	-30.88	36.90
687	C7H8O2	GUAIACOL	-251.563	3.6806E-01	2.8318E-05	298	1000	-139.00	-60.46
688	C7H8O2	p-METHOXYPHENOL	-251.548	3.6801E-01	2.8318E-05	298	1000	-139.00	-60.46
689	C7H9N	BENZYLAMINE	59.902	3.6126E-01	3.4354E-05	298	1000	171.00	249.12
690	C7H9N	2,6-DIMETHYLPYRIDINE	56.305	3.6380E-01	3.6740E-05	298	1000	168.40	247.39
691	C7H9N	N-METHYLANILINE	85.261	3.8119E-01	3.0953E-05	298	1000	202.00	283.59
692	C7H9N	m-TOLUIDINE	61.721	3.6575E-01	3.2648E-05	298	1000	174.00	252.76
693	C7H9N	o-TOLUIDINE	55.596	3.7419E-01	2.8416E-05	298	1000	170.00	249.80
694	C7H9N	p-TOLUIDINE	42.711	3.6573E-01	3.2794E-05	298	1000	155.00	233.78
695	C7H10	2-NORBORNENE	85.953	3.8111E-01	3.3533E-05	298	1000	203.00	284.89
696	C7H10N2	TOLUENEDIAMINE	55.313	4.8830E-01	2.6122E-05	298	1000	203.54	305.99
697	C7H11NO	CYCLOHEXYL ISOCYANATE	-144.715	5.8013E-01	4.4930E-05	298	1000	32.80	156.58
698	C7H12	1-HEPTYNE	100.831	4.0998E-01	3.9550E-05	298	1000	226.94	315.71
699	C7H12O2	n-BUTYL ACRYLATE	-397.980	5.4320E-01	3.0024E-05	298	1000	-233.00	-118.88
700	C7H12O2	ISOBUTYL ACRYLATE	-397.245	5.5418E-01	2.9872E-05	298	1000	-229.00	-112.69
701	C7H12O2	n-PROPYL METHACRYLATE	-405.340	5.4370E-01	4.2659E-05	298	1000	-239.00	-122.82
702	C7H12O4	DIETHYL MALONATE	-785.088	6.8135E-01	3.9259E-05	298	1000	-578.00	-434.60

Table 13-1 GIBBS ENERGY OF FORMATION OF GAS - ORGANIC COMPOUNDS (continued)

NO	FORMULA	NAME	A	B	C	TMIN	TMAX	Gf @ 298 K	Gf @ 500 K
				Gf = A + B T + C T²			(Gf - kjoule/mol, T - K)		
703	C7H14	CYCLOHEPTANE	-124.727	6.1171E-01	5.3325E-05	298	1000	63.01	194.46
704	C7H14	1,1-DIMETHYLCYCLOPENTANE	-142.804	5.9241E-01	5.2208E-05	298	1000	39.04	166.45
705	C7H14	cis-1,2-DIMETHYLCYCLOPENTANE	-133.964	5.8514E-01	5.2430E-05	298	1000	45.73	171.71
706	C7H14	trans-1,2-DIMETHYLCYCLOPENTANE	-141.076	5.8436E-01	5.2455E-05	298	1000	38.37	164.22
707	C7H14	cis-1,3-DIMETHYLCYCLOPENTANE	-140.241	5.8437E-01	5.2448E-05	298	1000	39.20	165.06
708	C7H14	trans-1,3-DIMETHYLCYCLOPENTANE	-137.981	5.8437E-01	5.2451E-05	298	1000	41.46	167.32
709	C7H14	ETHYLCYCLOPENTANE	-131.223	5.7136E-01	5.4772E-05	298	1000	44.56	168.15
710	C7H14	2-ETHYL-1-PENTENE	-77.843	5.2878E-01	4.8632E-05	298	1000	84.62	198.70
711	C7H14	3-ETHYL-1-PENTENE	-67.404	5.3388E-01	4.9318E-05	298	1000	96.65	211.86
712	C7H14	1-HEPTENE	-65.354	5.2430E-01	4.9439E-05	298	1000	95.81	209.16
713	C7H14	cis-2-HEPTENE	-72.467	5.2446E-01	5.3797E-05	298	1000	89.20	203.21
714	C7H14	trans-2-HEPTENE	-77.654	5.2440E-01	5.1795E-05	298	1000	83.80	197.50
715	C7H14	cis-3-HEPTENE	-72.159	5.2421E-01	5.3348E-05	298	1000	89.40	203.28
716	C7H14	trans-3-HEPTENE	-77.137	5.2415E-01	5.1274E-05	298	1000	84.20	197.76
717	C7H14	METHYLCYCLOHEXANE	-160.038	6.1255E-01	4.6303E-05	298	1000	27.28	157.81
718	C7H14	2-METHYL-1-HEXENE	-80.291	5.2984E-01	4.8919E-05	298	1000	82.51	196.86
719	C7H14	3-METHYL-1-HEXENE	-69.994	5.2807E-01	4.9318E-05	298	1000	92.33	206.37
720	C7H14	4-METHYL-1-HEXENE	-70.078	5.2833E-01	4.9099E-05	298	1000	92.30	206.36
721	C7H14	2,3,3-TRIMETHYL-1-BUTENE	-89.051	5.7100E-01	4.3897E-05	298	1000	85.57	207.42
722	C7H14O	DIISOPROPYL KETONE	-315.348	5.9670E-01	5.4973E-05	298	1000	-132.00	-3.26
723	C7H14O	2-HEPTANONE	-303.317	5.9217E-01	5.0354E-05	298	1000	-121.80	5.36
724	C7H14O	HEPTANAL	-267.231	5.8999E-01	4.7332E-05	298	1000	-86.65	39.60
725	C7H14O	1-METHYLCYCLOHEXANOL	-337.357	6.8462E-01	4.3819E-05	298	1000	-128.76	15.91
726	C7H14O	cis-2-METHYLCYCLOHEXANOL	-332.505	6.6882E-01	4.4647E-05	298	1000	-128.53	13.07
727	C7H14O	trans-2-METHYLCYCLOHEXANOL	-358.105	6.6882E-01	4.4647E-05	298	1000	-154.13	-12.53
728	C7H14O	cis-3-METHYLCYCLOHEXANOL	-538.112	6.7593E-01	4.6109E-05	298	1000	-152.43	-188.62
729	C7H14O	trans-3-METHYLCYCLOHEXANOL	-334.605	6.6882E-01	4.4647E-05	298	1000	-130.63	10.97
730	C7H14O	cis-4-METHYLCYCLOHEXANOL	-353.005	6.7982E-01	4.4647E-05	298	1000	-145.75	-1.93
731	C7H14O	trans-4-METHYLCYCLOHEXANOL	-372.705	6.7982E-01	4.4647E-05	298	1000	-165.45	-21.63
732	C7H14O	5-METHYL-2-HEXANONE	-307.224	6.0476E-01	4.9767E-05	298	1000	-122.00	7.60
733	C7H14O2	n-BUTYL PROPIONATE	-506.173	6.7413E-01	4.7118E-05	298	1000	-300.50	-157.33
734	C7H14O2	ETHYL ISOVALERATE	-530.731	6.7460E-01	4.6234E-05	298	1000	-325.00	-181.87
735	C7H14O2	ISOPENTYL ACETATE	-515.914	6.8054E-01	4.7254E-05	298	1000	-308.30	-163.83
736	C7H14O2	n-PENTYL ACETATE	-509.358	6.7452E-01	4.7732E-05	298	1000	-303.50	-160.17
737	C7H14O2	n-PROPYL n-BUTYRATE	-508.898	6.7423E-01	4.7059E-05	298	1000	-303.20	-160.02
738	C7H14O2	n-HEPTANOIC ACID	--------	--------	--------	--------	--------	-332.00	--------
739	C7H14O3	ETHYL-3-ETHOXYPROPIONATE	-636.938	7.2568E-01	4.5929E-05	298	1000	-416.00	-262.62
740	C7H15Br	1-BROMOHEPTANE	-187.264	6.5200E-01	5.0360E-05	333	1000	13.70	151.33
741	C7H15N	N-METHYLCYCLOHEXYLAMINE	-109.107	7.4200E-01	4.7578E-05	298	1000	117.00	273.79
742	C7H16	2,2-DIMETHYLPENTANE	-209.894	6.8563E-01	5.6352E-05	298	1000	0.08	147.01
743	C7H16	2,3-DIMETHYLPENTANE	-203.029	6.6456E-01	5.6286E-05	298	1000	0.67	143.32
744	C7H16	2,4-DIMETHYLPENTANE	-205.762	6.8189E-01	5.6332E-05	298	1000	3.10	149.27
745	C7H16	3,3-DIMETHYLPENTANE	-205.327	6.7887E-01	5.6327E-05	298	1000	2.64	148.19
746	C7H16	3-ETHYLPENTANE	-193.374	6.6687E-01	5.6450E-05	298	1000	11.00	154.17
747	C7H16	HEPTANE	-191.520	6.5052E-01	5.6444E-05	298	1000	7.99	147.85
748	C7H16	2-METHYLHEXANE	-198.645	6.5837E-01	5.6475E-05	298	1000	3.22	144.66
749	C7H16	3-METHYLHEXANE	-196.033	6.5427E-01	5.6454E-05	298	1000	4.60	145.22
750	C7H16	2,2,3-TRIMETHYLBUTANE	-209.101	6.9811E-01	5.2621E-05	298	1000	4.27	153.11
751	C7H16O	HEPTYL-ALCOHOL	-338.701	7.0060E-01	5.7147E-05	298	1000	-124.18	25.89
752	C7H16O	2-HEPTANOL	-359.090	7.0689E-01	5.3670E-05	298	1000	-143.00	7.77
753	C7H16O	5-METHYL-1-HEXANOL	-346.030	7.1579E-01	5.6661E-05	298	1000	-127.00	26.03
754	C7H16O	ISOPROPYL-TERT-BUTYL-ETHER	-361.844	7.6268E-01	5.6918E-05	298	1000	-128.83	33.73
755	C7H16S	1-HEPTANETHIOL	-143.858	5.6969E-01	1.1320E-04	389	717	36.19	169.29
756	C7H16S	BUTYL PROPYL SULFIDE	-140.231	5.6947E-01	1.1450E-04	389	717	39.87	173.13
757	C7H16S	ETHYL PENTYL SULFIDE	-140.038	5.7058E-01	1.1447E-04	389	717	40.38	173.87
758	C7H16S	HEXYL METHYL SULFIDE	-137.588	5.7293E-01	1.1389E-04	389	717	43.47	177.35
759	C7H17N	1-AMINOHEPTANE	-157.280	7.6047E-01	5.9082E-05	298	1000	75.32	237.73
760	C8H4Cl2O2	ISOPHTHALOYL CHLORIDE	-302.656	2.6905E-01	1.4760E-05	298	1000	-221.00	-164.44
761	C8H4O3	PHTHALIC ANHYDRIDE	-394.491	2.0866E-01	3.3736E-05	298	1000	-329.00	-281.72
762	C8H6	ETHYNYLBENZENE	326.263	1.1297E-01	1.8500E-05	298	1000	361.75	387.37
763	C8H6O4	ISOPHTHALIC ACID	-698.412	3.9767E-01	4.3701E-05	298	1000	-575.59	-488.65
764	C8H6O4	PHTHALIC ACID	-665.442	3.9814E-01	4.3701E-05	298	1000	-542.48	-455.45
765	C8H6O4	TEREPHTHALIC ACID	-720.012	3.9160E-01	4.3701E-05	298	1000	-599.00	-513.29
766	C8H6S	BENZOTHIOPHENE	--------	--------	--------	--------	--------	243.00	--------
767	C8H7N	INDOLE	154.446	2.6933E-01	2.5623E-05	298	1000	237.30	295.52
768	C8H8	STYRENE	145.657	2.1917E-01	2.8490E-05	298	1000	213.80	262.36
769	C8H8	1,3,5,7-CYCLOOCTATETRAENE	296.084	2.3877E-01	2.6787E-05	298	1000	369.91	422.17
770	C8H8O	ACETOPHENONE	-88.415	2.7780E-01	4.4493E-05	298	1000	-1.27	61.61
771	C8H8O	p-TOLUALDEHYDE	-72.444	2.7680E-01	3.2882E-05	298	1000	13.30	74.17
772	C8H8O2	METHYL BENZOATE	-290.211	3.5257E-01	4.1827E-05	298	1000	-181.00	-103.47
773	C8H8O2	o-TOLUIC ACID	-325.099	3.6099E-01	4.2055E-05	298	1000	-213.35	-134.09
774	C8H8O2	p-TOLUIC ACID	-325.099	3.6099E-01	4.2055E-05	298	1000	-213.35	-134.09
775	C8H8O3	METHYL SALICYLATE	-466.664	4.1620E-01	3.6378E-05	298	1000	-339.00	-249.47
776	C8H8O3	VANILLIN	-372.752	4.1627E-01	1.4954E-05	298	1000	-247.00	-160.88
777	C8H9NO	ACETANILIDE	-131.217	4.5815E-01	4.1496E-05	298	1000	9.47	108.23
778	C8H10	ETHYLBENZENE	27.421	3.3327E-01	3.8542E-05	298	1000	130.58	203.69
779	C8H10	m-XYLENE	15.063	3.3452E-01	4.1387E-05	298	1000	118.87	192.67
780	C8H10	o-XYLENE	17.048	3.3940E-01	3.9428E-05	298	1000	122.09	196.61

Table 13-1 GIBBS ENERGY OF FORMATION OF GAS - ORGANIC COMPOUNDS (continued)

NO	FORMULA	NAME	Gf = A + B T + C T²					(Gf - kjoule/mol, T - K)	
			A	B	C	TMIN	TMAX	Gf @ 298 K	Gf @ 500 K
781	C8H10	p-XYLENE	15.763	3.3952E-01	4.2301E-05	298	1000	121.13	196.10
782	C8H10O	m-ETHYLPHENOL	--------	--------	--------	--------	--------	--------	--------
783	C8H10O	p-ETHYLPHENOL	-146.124	4.0624E-01	3.4752E-05	298	1000	-21.58	65.69
784	C8H10O	PHENETOLE	-104.745	3.9619E-01	4.0042E-05	298	1000	17.35	103.36
785	C8H10O	2-PHENYLETHANOL	-123.501	3.9185E-01	3.8768E-05	298	1000	-2.85	82.12
786	C8H10O	2,3-XYLENOL	-159.508	4.1480E-01	2.4694E-05	298	1000	-33.35	54.07
787	C8H10O	2,4-XYLENOL	-165.405	4.0263E-01	2.7967E-05	298	1000	-42.56	42.90
788	C8H10O	2,5-XYLENOL	-164.128	4.0813E-01	2.7633E-05	298	1000	-39.67	46.85
789	C8H10O	2,6-XYLENOL	-164.222	4.1021E-01	2.8768E-05	298	1000	-39.04	48.07
790	C8H10O	3,4-XYLENOL	-158.934	4.0987E-01	2.4340E-05	298	1000	-34.28	52.09
791	C8H10O	3,5-XYLENOL	-164.187	4.0866E-01	2.9286E-05	298	1000	-39.41	47.46
792	C8H11N	N,N-DIMETHYLANILINE	96.739	4.9296E-01	4.0087E-05	298	1000	247.73	353.24
793	C8H11N	o-ETHYLANILINE	30.442	4.5611E-01	3.5902E-05	298	1000	170.00	267.47
794	C8H11N	2,4,6-TRIMETHYLPYRIDINE	20.053	3.8170E-01	4.1554E-05	298	1000	138.00	221.29
795	C8H11NO	p-PHENETIDINE	-106.572	5.2479E-01	3.6945E-05	298	1000	53.60	165.06
796	C8H12	1,5-CYCLOOCTADIENE	97.898	4.7968E-01	4.0339E-05	298	1000	245.00	347.82
797	C8H12	VINYLCYCLOHEXENE	61.300	4.5165E-01	4.0299E-05	298	1000	200.00	297.20
798	C8H12O4	1,4-CYCLOHEXANEDICARBOXYLIC ACID	-883.136	7.9250E-01	3.6900E-05	298	1000	-643.00	-477.66
799	C8H12O4	DIETHYL MALEATE	-743.715	6.6340E-01	1.8151E-05	298	1000	-544.00	-407.48
800	C8H14O2	n-BUTYL METHACRYLATE	-425.843	6.3727E-01	4.8745E-05	298	1000	-231.00	-95.02
801	C8H14O3	BUTYRIC ANHYDRIDE	-666.571	7.3417E-01	1.3786E-05	298	1000	-446.00	-296.04
802	C8H14O4	DIETHYL SUCCINATE	-854.920	7.8048E-01	4.2213E-05	298	1000	-618.00	-454.13
803	C8H16	CYCLOOCTANE	-131.710	7.2244E-01	6.2236E-05	298	1000	89.91	245.07
804	C8H16	1,1-DIMETHYLCYCLOHEXANE	-187.269	7.2867E-01	5.1482E-05	298	1000	35.23	189.94
805	C8H16	C-1,2-DIMETHYLCYCLOHEXANE	-178.140	7.1809E-01	5.1867E-05	298	1000	41.21	193.87
806	C8H16	T-1,2-DIMETHYLCYCLOHEXANE	-185.997	7.2249E-01	4.9658E-05	298	1000	34.48	187.66
807	C8H16	C-1,3-DIMETHYLCYCLOHEXANE	-190.817	7.2267E-01	5.1077E-05	298	1000	29.83	183.29
808	C8H16	T-1,3-DIMETHYLCYCLOHEXANE	-182.416	7.1581E-01	5.2435E-05	298	1000	36.32	188.60
809	C8H16	C-1,4-DIMETHYLCYCLOHEXANE	-182.489	7.2157E-01	5.2435E-05	298	1000	37.95	191.40
810	C8H16	T-1,4-DIMETHYLCYCLOHEXANE	-190.646	7.2867E-01	5.0049E-05	298	1000	31.71	186.20
811	C8H16	ETHYLCYCLOHEXANE	-177.580	7.0980E-01	5.1198E-05	298	1000	39.25	190.12
812	C8H16	2-ETHYL-1-HEXENE	-100.692	6.2575E-01	5.4603E-05	298	1000	91.28	225.84
813	C8H16	1-METHYL-1-ETHYLCYCLOPENTANE	-159.963	6.8690E-01	5.8357E-05	298	1000	50.67	198.08
814	C8H16	1-OCTENE	-86.500	6.2135E-01	5.5457E-05	298	1000	104.22	238.04
815	C8H16	trans-2-OCTENE	-98.900	6.2155E-01	5.7599E-05	298	1000	92.10	226.28
816	C8H16	trans-3-OCTENE	-98.367	6.2020E-01	5.7204E-05	298	1000	92.20	226.04
817	C8H16	trans-4-OCTENE	-98.367	6.2624E-01	5.7204E-05	298	1000	94.00	229.05
818	C8H16	PROPYLCYCLOPENTANE	-152.696	6.6824E-01	6.0934E-05	298	1000	52.59	196.66
819	C8H16	2,4,4-TRIMETHYL-1-PENTENE	-114.410	6.5812E-01	5.0117E-05	298	1000	86.80	227.18
820	C8H16	2,4,4-TRIMETHYL-2-PENTENE	-108.751	6.6106E-01	5.1933E-05	298	1000	93.50	234.76
821	C8H16O	2-ETHYLHEXANAL	-303.689	6.9263E-01	5.7423E-05	298	1000	-91.49	56.98
822	C8H16O	OCTANAL	-293.423	6.8704E-01	5.3338E-05	298	1000	-83.30	63.43
823	C8H16O	2-OCTANONE	-324.915	6.8937E-01	5.6566E-05	298	1000	-113.80	33.91
824	C8H16O2	n-BUTYL n-BUTYRATE	-530.093	7.7820E-01	3.8507E-05	298	1000	-294.10	-131.37
825	C8H16O2	n-HEXYL ACETATE	-530.628	7.7126E-01	5.4069E-05	298	1000	-295.30	-131.48
826	C8H16O2	ISOBUTYL ISOBUTYRATE	-551.746	7.8235E-01	5.5043E-05	298	1000	-313.00	-146.81
827	C8H16O2	n-OCTANOIC ACID	-559.983	7.7075E-01	5.2154E-05	298	1000	-325.00	-161.57
828	C8H16O4	DIETHYLENE GLYCOL ETHYL ETHER ACETATE	-780.997	8.8870E-01	5.2391E-05	298	1000	-510.78	-323.55
829	C8H18	2,2-DIMETHYLHEXANE	-229.039	7.8345E-01	6.2253E-05	298	1000	10.71	178.25
830	C8H18	2,3-DIMETHYLHEXANE	-218.217	7.7062E-01	6.2282E-05	298	1000	17.70	182.67
831	C8H18	2,4-DIMETHYLHEXANE	-223.697	7.6894E-01	6.2289E-05	298	1000	11.72	176.35
832	C8H18	2,5-DIMETHYLHEXANE	-226.952	7.7568E-01	6.2197E-05	298	1000	10.46	176.44
833	C8H18	3,3-DIMETHYLHEXANE	-224.411	7.7653E-01	6.2269E-05	298	1000	13.26	179.42
834	C8H18	3,4-DIMETHYLHEXANE	-217.260	7.6617E-01	6.2351E-05	298	1000	17.32	181.41
835	C8H18	3-ETHYLHEXANE	-215.139	7.5633E-01	6.2330E-05	298	1000	16.53	178.61
836	C8H18	3-ETHYL-2-METHYLPENTANE	-215.532	7.7359E-01	6.2197E-05	298	1000	21.25	186.81
837	C8H18	3-ETHYL-3-METHYLPENTANE	-219.251	7.8158E-01	6.2315E-05	298	1000	19.92	187.12
838	C8H18	2-METHYLHEPTANE	-219.792	7.5940E-01	6.2231E-05	298	1000	12.76	175.47
839	C8H18	3-METHYLHEPTANE	-216.883	7.5284E-01	6.2428E-05	298	1000	13.72	175.15
840	C8H18	4-METHYLHEPTANE	-216.353	7.6119E-01	6.2325E-05	298	1000	16.74	179.82
841	C8H18	OCTANE	-212.692	7.4774E-01	6.2361E-05	298	1000	16.40	176.77
842	C8H18	2,2,3-TRIMETHYLPENTANE	-224.335	7.8913E-01	6.2502E-05	298	1000	17.11	185.86
843	C8H18	2,2,4-TRIMETHYLPENTANE	-228.383	7.9125E-01	6.2364E-05	298	1000	13.68	182.84
844	C8H18	2,3,3-TRIMETHYLPENTANE	-220.671	7.8284E-01	6.2446E-05	298	1000	18.91	186.36
845	C8H18	2,3,4-TRIMETHYLPENTANE	-221.742	7.8655E-01	6.2266E-05	298	1000	18.91	187.10
846	C8H18	2,2,3,3-TETRAMETHYLBUTANE	-231.148	8.3203E-01	5.0497E-05	298	1000	22.01	197.49
847	C8H18O	BUTYL ETHER	-337.796	8.1490E-01	6.4263E-05	298	1000	-88.53	85.72
848	C8H18O	sec-BUTYL ETHER	-364.965	8.5433E-01	6.2752E-05	298	1000	-104.06	77.89
849	C8H18O	tert-BUTYL ETHER	-369.106	8.8957E-01	6.2818E-05	298	1000	-97.70	91.38
850	C8H18O	2-ETHYL-1-HEXANOL	-369.740	8.2025E-01	6.3656E-05	298	1000	-118.88	56.30
851	C8H18O	OCTYL ALCOHOL	-361.387	7.9753E-01	6.3199E-05	298	1000	-117.36	53.18
852	C8H18O	2-OCTANOL	-380.697	8.0412E-01	5.9853E-05	298	1000	-135.00	36.33
853	C8H18O2	DI-t-BUTYL PEROXIDE	-345.750	9.2854E-01	5.5093E-05	298	1000	-63.40	132.29
854	C8H18O2S	DI-n-BUTYL SULFONE	--------	--------	--------	--------	--------	-238.00	--------
855	C8H18O3	DIETHYLENE GLYCOL DIETHYL ETHER	-585.290	9.3960E-01	6.1553E-05	298	1000	-299.00	-100.10
856	C8H18O3	DIETHYLENE GLYCOL MONOBUTYL EHTER	-611.060	9.2865E-01	6.2055E-05	298	1000	-328.00	-131.22
857	C8H18O4	TRIETHYLENE GLYCOL DIMETHYL ETHER	-681.439	1.0073E+00	6.0939E-05	298	1000	-375.00	-162.53
858	C8H18O5	TETRAETHYLENE GLYCOL	-887.475	1.0794E+00	6.2486E-05	298	1000	-559.41	-332.13

327

Table 13-1 GIBBS ENERGY OF FORMATION OF GAS - ORGANIC COMPOUNDS (continued)

NO	FORMULA	NAME	Gf = A + B T + C T²					(Gf - kjoule/mol, T - K)	
			A	B	C	TMIN	TMAX	Gf @ 298 K	Gf @ 500 K
859	C8H18S	1-OCTANETHIOL	-163.961	6.6230E-01	1.2366E-04	389	717	44.64	198.10
860	C8H18S	tert-OCTYL MERCAPTAN	--------	--------	--------	--------	--------	-28.50	--------
861	C8H18S	BUTYL SULFIDE	-161.106	6.6757E-01	1.2527E-04	389	717	49.20	204.00
862	C8H18S	ETHYL HEXYL SULFIDE	-160.100	6.6283E-01	1.2547E-04	389	717	48.83	202.68
863	C8H18S	HEPTYL METHYL SULFIDE	-157.568	6.6496E-01	1.2493E-04	389	717	51.92	206.14
864	C8H18S	PENTYL PROPYL SULFIDE	-160.310	6.6161E-01	1.2550E-04	389	717	48.24	201.87
865	C8H18S2	BUTYL DISULFIDE	-150.651	6.4275E-01	1.4278E-04	389	717	53.85	206.42
866	C8H19N	DI-n-BUTYLAMINE	-161.781	8.6980E-01	6.4754E-05	298	1000	104.00	289.31
867	C8H19N	DIISOBUTYLAMINE	-184.622	8.9361E-01	6.4010E-05	298	1000	88.20	278.18
868	C8H19N	n-OCTYLAMINE	-648.009	1.1517E+00	7.6661E-05	298	1000	83.03	-53.00
869	C8H23N5	TETRAETHYLENEPENTAMINE	4.557	1.3460E+00	7.1259E-05	298	1000	413.08	695.35
870	C8H24O4Si4	OCTAMETHYLCYCLOTETRASILOXANE	--------	--------	--------	--------	--------	--------	--------
871	C9H4O5	TRIMELLITIC ANHYDRIDE	-780.283	4.7597E-01	2.3712E-05	298	1000	-636.00	-536.37
872	C9H6N2O2	TOLUENE DIISOCYANATE	-230.617	4.5239E-01	-1.0771E-06	298	1000	-95.58	-4.69
873	C9H7N	ISOQUINOLINE	206.332	2.3774E-01	2.4899E-05	298	1000	279.70	331.43
874	C9H7N	QUINOLINE	220.198	2.3771E-01	2.4278E-05	298	1000	293.50	345.12
875	C9H7NO	8-HYDROXYQUINOLINE	19.044	3.1361E-01	1.3944E-05	298	1000	114.00	179.33
876	C9H8	INDENE	161.411	2.3289E-01	3.1741E-05	298	1000	233.97	285.79
877	C9H8O	2-METHYLBENZOFURAN	-10.388	3.7147E-01	2.6532E-05	298	1000	103.00	181.98
878	C9H10	INDANE	57.715	4.6673E-01	4.1892E-05	298	1000	201.00	301.55
879	C9H10	cis-PROPENYLBENZENE	119.443	3.1471E-01	3.7129E-05	298	1000	216.90	286.08
880	C9H10	trans-PROPENYLBENZENE	115.154	3.1893E-01	3.5506E-05	298	1000	213.72	283.50
881	C9H10	alpha-METHYLSTYRENE	111.072	3.1472E-01	3.7121E-05	298	1000	208.53	277.71
882	C9H10	m-METHYLSTYRENE	113.573	3.0888E-01	3.7107E-05	298	1000	209.28	277.29
883	C9H10	o-METHYLSTYRENE	116.520	3.1469E-01	3.7144E-05	298	1000	213.97	283.15
884	C9H10	p-METHYLSTYRENE	112.745	3.1472E-01	3.7119E-05	298	1000	210.20	279.38
885	C9H10O2	BENZYL ACETATE	-312.192	4.4923E-01	3.4434E-05	298	1000	-174.82	-78.97
886	C9H10O2	ETHYL BENZOATE	-286.904	4.5010E-01	4.7989E-05	298	1000	-148.00	-49.86
887	C9H10O3	ETHYL VANILLIN	-408.385	5.1376E-01	2.0549E-05	298	1000	-253.00	-146.37
888	C9H11NO	p-DIMETHYLAMINOBENZALDEHYDE	-32.865	5.4016E-01	2.6619E-05	298	1000	131.00	243.87
889	C9H12	CUMENE	0.983	4.4175E-01	4.3406E-05	298	1000	136.98	232.71
890	C9H12	m-ETHYLTOLUENE	-4.705	4.2485E-01	4.5613E-05	298	1000	126.44	219.13
891	C9H12	o-ETHYLTOLUENE	-1.301	4.2958E-01	4.3769E-05	298	1000	131.08	224.43
892	C9H12	p-ETHYLTOLUENE	-5.947	4.2950E-01	4.6751E-05	298	1000	126.69	220.49
893	C9H12	MESITYLENE	-18.595	4.4166E-01	4.9679E-05	298	1000	117.95	214.65
894	C9H12	PROPYLBENZENE	4.889	4.2937E-01	4.4012E-05	298	1000	137.24	230.58
895	C9H12	1,2,3-TRIMETHYLBENZENE	-11.858	4.4140E-01	4.9389E-05	298	1000	124.56	221.19
896	C9H12	1,2,4-TRIMETHYLBENZENE	-16.358	4.3119E-01	4.8564E-05	298	1000	116.94	211.38
897	C9H12O	BENZYL ETHYL ETHER	-118.731	4.9609E-01	4.5423E-05	298	1000	33.70	140.67
898	C9H12O	2-PHENYL-2-PROPANOL	-308.136	2.4451E-01	4.2741E-05	298	1000	-231.00	-175.19
899	C9H12O2	CUMENE HYDROPEROXIDE	-81.655	5.8302E-01	3.8376E-05	298	1000	96.00	219.45
900	C9H14O	ISOPHORONE	-255.442	6.4740E-01	4.0725E-05	298	1000	-58.30	78.44
901	C9H14O6	GLYCERYL TRIACETATE	-1257.655	9.0829E-01	3.7792E-05	298	1000	-983.00	-794.06
902	C9H16	1-NONYNE	58.588	6.0406E-01	5.1578E-05	298	1000	243.76	373.52
903	C9H16O4	AZELAIC ACID	-931.177	9.0370E-01	4.7372E-05	298	1000	-657.00	-467.48
904	C9H18	BUTYLCYCLOPENTANE	-173.390	7.6514E-01	6.7065E-05	298	1000	61.38	225.95
905	C9H18	cis-cis-1,3,5-TRIMETHYLCYCLOHEXANE	-222.259	8.4009E-01	5.5762E-05	298	1000	33.89	211.73
906	C9H18	cis-trans-1,3,5-TRIMETHYLCYCLOHEXANE	-212.955	8.2810E-01	5.8630E-05	298	1000	39.87	215.75
907	C9H18	ISOPROPYLCYCLOHEXANE	-202.336	8.2288E-01	5.9190E-05	298	1000	49.00	223.90
908	C9H18	1-NONENE	-107.582	7.1838E-01	6.1441E-05	298	1000	112.68	266.97
909	C9H18	PROPYLCYCLOHEXANE	-199.501	8.0874E-01	5.6176E-05	298	1000	47.32	218.91
910	C9H18O	DIISOBUTYL KETONE	118.056	2.5080E-01	2.1856E-05	298	1000	-113.00	248.92
911	C9H18O	NONANAL	-314.560	7.8403E-01	5.9438E-05	298	1000	-74.94	92.31
912	C9H18O2	n-BUTYL VALERATE	-564.897	8.5611E-01	5.8800E-05	298	1000	-303.80	-122.14
913	C9H18O2	n-NONANOIC ACID	-579.923	8.6909E-01	5.8317E-05	298	1000	-315.00	-130.80
914	C9H18O2	n-OCTYL FORMATE	-515.407	8.6350E-01	5.8758E-05	298	1000	-252.10	-68.97
915	C9H20	3,3-DIETHYLPENTANE	-237.693	8.9338E-01	6.4026E-05	298	1000	35.06	225.00
916	C9H20	3-ETHYL-2,2-DIMETHYLPENTANE	-243.872	8.9430E-01	6.3760E-05	298	1000	29.12	219.22
917	C9H20	3-ETHYL-2,3-DIMETHYLPENTANE	-239.241	8.8583E-01	6.3120E-05	298	1000	31.17	219.46
918	C9H20	3-ETHYL-2,4-DIMETHYLPENTANE	-240.323	8.8226E-01	6.8132E-05	298	1000	29.50	217.84
919	C9H20	2,2-DIMETHYLHEPTANE	-251.912	8.8046E-01	6.1573E-05	298	1000	16.74	203.71
920	C9H20	2,6-DIMETHYLHEPTANE	-248.402	8.7460E-01	6.6904E-05	298	1000	19.00	205.62
921	C9H20	3-ETHYLHEPTANE	-235.494	8.5556E-01	6.7981E-05	298	1000	26.32	209.28
922	C9H20	4-ETHYLHEPTANE	-235.494	8.5556E-01	6.7981E-05	298	1000	26.32	209.28
923	C9H20	2,3-DIMETHYLHEPTANE	-240.580	8.6274E-01	6.7176E-05	298	1000	23.30	207.59
924	C9H20	2,4-DIMETHYLHEPTANE	-245.667	8.6328E-01	6.7859E-05	298	1000	18.45	202.94
925	C9H20	2,5-DIMETHYLHEPTANE	-245.667	8.6328E-01	6.7859E-05	298	1000	18.45	202.94
926	C9H20	3,4-DIMETHYLHEPTANE	-237.951	8.6040E-01	6.7955E-05	298	1000	25.31	209.24
927	C9H20	3,5-DIMETHYLHEPTANE	-243.061	8.6678E-01	6.8579E-05	298	1000	22.18	207.47
928	C9H20	4,4-DIMETHYLHEPTANE	-246.853	8.7750E-01	6.2463E-05	298	1000	21.00	207.51
929	C9H20	3-ETHYL-2-METHYLHEXANE	-237.903	8.6316E-01	6.8037E-05	298	1000	26.19	210.68
930	C9H20	4-ETHYL-2-METHYLHEXANE	-243.012	8.6950E-01	6.8697E-05	298	1000	23.05	208.91
931	C9H20	3-ETHYL-3-METHYLHEXANE	-241.791	8.7215E-01	6.3284E-05	298	1000	24.56	210.11
932	C9H20	3-ETHYL-4-METHYLHEXANE	-235.257	8.6369E-01	6.8766E-05	298	1000	29.08	213.78
933	C9H20	2,2,3-TRIMETHYLHEXANE	-246.598	8.8845E-01	6.2515E-05	298	1000	24.52	213.26
934	C9H20	2,2,4-TRIMETHYLHEXANE	-248.840	8.8895E-01	6.3241E-05	298	1000	22.51	211.44
935	C9H20	2,3,3-TRIMETHYLHEXANE	-244.225	8.8514E-01	6.2489E-05	298	1000	25.90	213.97
936	C9H20	2,3,4-TRIMETHYLHEXANE	-240.368	8.7376E-01	6.8017E-05	298	1000	26.90	213.52

328

NO	FORMULA	NAME	$Gf = A + B T + C T^2$					(Gf - kjoule/mol, T - K)	
			A	B	C	TMIN	TMAX	Gf @ 298 K	Gf @ 500 K
937	C9H20	2,3,5-TRIMETHYLHEXANE	-248.145	8.8262E-01	6.7777E-05	298	1000	21.76	210.11
938	C9H20	2,4,4-TRIMETHYLHEXANE	-246.460	8.8561E-01	6.3235E-05	298	1000	23.89	212.16
939	C9H20	3,3,4-TRIMETHYLHEXANE	-241.552	8.7981E-01	6.3300E-05	298	1000	27.07	214.18
940	C9H20	2-METHYLOCTANE	-240.805	8.5474E-01	6.6289E-05	298	1000	20.59	203.14
941	C9H20	3-METHYLOCTANE	-238.219	8.4955E-01	6.6996E-05	298	1000	21.71	203.30
942	C9H20	4-METHYLOCTANE	-238.219	8.4955E-01	6.6996E-05	298	1000	21.71	203.30
943	C9H20	NONANE	-233.826	8.4477E-01	6.8451E-05	298	1000	24.81	205.67
944	C9H20	2,2,3,3-TETRAMETHYLPENTANE	-242.926	9.1022E-01	5.8406E-05	298	1000	34.31	226.79
945	C9H20	2,2,3,4-TETRAMETHYLPENTANE	-242.527	9.0198E-01	6.2433E-05	298	1000	32.64	224.07
946	C9H20	2,2,4,4-TETRAMETHYLPENTANE	-247.825	9.2587E-01	5.7635E-05	298	1000	34.02	229.52
947	C9H20	2,3,3,4-TETRAMETHYLPENTANE	-241.799	9.0451E-01	6.2351E-05	298	1000	34.10	226.05
948	C9H20	2,2,5-TRIMETHYLHEXANE	-259.495	8.9459E-01	6.2167E-05	298	1000	13.43	203.34
949	C9H20O	2,6-DIMETHYL-4-HEPTANOL	-415.232	9.3465E-01	6.5963E-05	298	1000	-130.00	68.58
950	C9H20O	NONYL ALCOHOL	-391.755	8.9463E-01	6.9190E-05	298	1000	-118.20	72.86
951	C9H20O	2-NONANOL	-402.984	9.0671E-01	6.6958E-05	298	1000	-126.00	67.11
952	C9H20S	1-NONANETHIOL	-184.234	7.5533E-01	1.3380E-04	389	717	53.01	226.88
953	C9H20S	BUTYL-PENTYL-SULFIDE	-181.275	7.5451E-01	1.3571E-04	389	717	55.90	229.91
954	C9H20S	ETHYL-HEPTYL-SULFIDE	-180.361	7.5587E-01	1.3553E-04	389	717	57.20	231.46
955	C9H20S	HEXYL-PROPYL-SULFIDE	-180.489	7.5455E-01	1.3567E-04	389	717	56.69	230.70
956	C9H20S	METHYL-OCTYL-SULFIDE	-177.894	7.5828E-01	1.3475E-04	389	717	60.33	234.93
957	C9H21N	n-NONYLAMINE	-200.252	9.5320E-01	7.1523E-05	298	1000	91.04	294.23
958	C9H21N	TRIPROPYLAMINE	-166.963	9.8524E-01	6.9062E-05	298	1000	133.69	342.92
959	C10H6O8	PYROMELLITIC ACID	-1482.378	6.2515E-01	6.1948E-05	298	1000	-1290.00	-1154.32
960	C10H7Br	1-BROMONAPHTHALENE	164.865	2.4939E-01	2.2028E-05	333	1000	242.97	295.07
961	C10H7Cl	1-CHLORONAPHTHALENE	--------	--------	--------	--------	--------	195.00	--------
962	C10H8	NAPHTHALENE	148.988	2.4014E-01	3.0705E-05	298	1000	223.59	276.74
963	C10H8	AZULENE	277.715	2.3826E-01	3.1796E-05	298	1000	351.87	404.80
964	C10H9N	QUINALDINE	174.214	3.9250E-01	2.7475E-05	298	1000	294.00	377.33
965	C10H10	m-DIVINYLBENZENE	210.225	2.9388E-01	3.1340E-05	298	1000	300.92	365.00
966	C10H10	1-METHYLINDENE	119.635	3.5533E-01	2.4116E-05	298	1000	228.00	303.33
967	C10H10	2-METHYLINDENE	112.622	3.4837E-01	2.5040E-05	298	1000	219.00	293.07
968	C10H10O4	DIMETHYL PHTHALATE	-666.083	4.5231E-01	5.3404E-05	298	1000	-526.00	-426.58
969	C10H10O4	DIMETHYL TEREPHTHALATE	-639.864	5.3852E-01	5.4542E-05	298	1000	-474.00	-356.97
970	C10H12	DICYCLOPENTADIENE	191.482	6.0473E-01	3.1253E-05	298	1000	375.00	501.66
971	C10H12	1,2,3,4-TETRAHYDRONAPHTHALENE	23.033	4.6879E-01	4.3036E-05	298	1000	167.10	268.19
972	C10H12O	ANETHOLE	-44.460	4.7330E-01	4.4024E-05	298	1000	101.00	203.19
973	C10H12O4	DIALLYL MALEATE	-520.699	6.1753E-01	1.4529E-05	298	1000	-335.00	-208.30
974	C10H14	BUTYLBENZENE	-17.190	5.2633E-01	5.0190E-05	298	1000	144.68	258.52
975	C10H14	sec-BUTYLBENZENE	-19.614	5.3551E-01	5.2319E-05	298	1000	145.20	261.22
976	C10H14	tert-BUTYLBENZENE	-25.313	5.7232E-01	4.5975E-05	298	1000	149.90	272.34
977	C10H14	1,2,3,4-TETRAMETHYLBENZENE	-44.515	5.4762E-01	4.7618E-05	298	1000	123.43	241.20
978	C10H14	m-CYMENE	-33.966	5.3304E-01	5.2097E-05	298	1000	130.10	245.58
979	C10H14	o-CYMENE	-29.887	5.3971E-01	5.0222E-05	298	1000	136.00	252.53
980	C10H14	p-CYMENE	-32.012	5.3755E-01	5.3214E-05	298	1000	133.50	250.07
981	C10H14	m-DIETHYLBENZENE	-25.143	5.2637E-01	4.9852E-05	298	1000	136.69	250.51
982	C10H14	o-DIETHYLBENZENE	-21.957	5.3091E-01	4.8193E-05	298	1000	141.08	255.55
983	C10H14	p-DIETHYLBENZENE	-25.535	5.3121E-01	5.0922E-05	298	1000	137.86	252.80
984	C1OH14	2-ETHYL-m-XYLENE	-29.292	5.4514E-01	4.7201E-05	298	1000	137.90	255.08
985	C10H14	2-ETHYL-p-XYLENE	-35.180	5.2807E-01	5.1229E-05	298	1000	127.30	241.66
986	C10H14	3-ETHYL-o-XYLENE	-28.705	5.4486E-01	4.7164E-05	298	1000	138.40	255.52
987	C10H14	4-ETHYL-m-XYLENE	-33.800	5.2780E-01	5.1229E-05	298	1000	128.60	242.91
988	C10H14	4-ETHYL-o-XYLENE	-35.090	5.2810E-01	5.1229E-05	298	1000	127.40	241.77
989	C1OH14	5-ETHYL-m-XYLENE	-38.352	5.3705E-01	5.5365E-05	298	1000	127.20	244.02
990	C10H14	ISOBUTYLBENZENE	-23.849	5.3290E-01	5.0226E-05	298	1000	140.00	255.16
991	C10H14	1,2,3,5-TETRAMETHYLBENZENE	-47.443	5.4079E-01	5.0474E-05	298	1000	118.74	235.57
992	C10H14	1,2,4,5-TETRAMETHYLBENZENE	-47.801	5.4403E-01	5.1676E-05	298	1000	119.45	237.14
993	C10H14O	p-tert-BUTYLPHENOL	-204.052	6.2690E-01	4.0190E-05	298	1000	-13.10	119.45
994	C10H14O2	p-tert-BUTYLCATECHOL	-380.162	6.9979E-01	3.4676E-05	298	1000	-168.00	-21.60
995	C10H15N	N,N-DIETYHLANILINE	35.299	7.4415E-01	5.2244E-05	298	1000	262.40	420.44
996	C10H15N	2,6-DIETHYLANILINE	-24.217	6.5554E-01	4.8016E-05	298	1000	176.00	315.56
997	C10H16	CAMPHENE	--------	--------	--------	--------	--------	--------	--------
998	C10H16	D-LIMONENE	--------	--------	--------	--------	--------	--------	--------
999	C10H16	alpha-PHELLANDRENE	-14.238	6.7225E-01	4.7693E-05	298	1000	191.00	333.81
1000	C1OH16	beta-PHELLANDRENE	-10.662	6.8096E-01	4.5846E-05	298	1000	197.00	341.28
1001	C10H16	alpha-PINENE	20.950	6.3430E-01	5.8165E-05	298	1000	216.00	352.64
1002	C10H16	beta-PINENE	31.728	7.0581E-01	4.6695E-05	298	1000	247.00	396.31
1003	C10H16	alpha-TERPINENE	-24.606	6.6549E-01	4.1712E-05	298	1000	178.00	318.57
1004	C1OH16	gamma-TERPINENE	-15.959	6.4667E-01	4.1231E-05	298	1000	181.00	317.68
1005	C10H16	TERPINOLENE	--------	--------	--------	--------	--------	--------	--------
1006	C10H16O	CAMPHOR	-264.412	8.1997E-01	5.2810E-05	298	1000	-14.50	158.77
1007	C10H18	1-DECYNE	37.488	7.0119E-01	5.7516E-05	298	1000	252.21	402.46
1008	C10H18	cis-DECAHYDRONAPHTHALENE	-175.788	8.5412E-01	6.9065E-05	298	1000	85.81	268.54
1009	C10H18	trans-DECAHYDRONAPHTHALENE	-189.001	8.5677E-01	6.9466E-05	298	1000	73.43	256.75
1010	C10H18O4	SEBACIC ACID	-926.660	1.0006E+00	5.3206E-05	298	1000	-623.00	-413.05
1011	C10H20	BUTYLCYCLOHEXANE	-219.884	9.0572E-01	6.2305E-05	298	1000	56.44	248.55
1012	C10H20	1-CYCLOPENTYLPENTANE	-194.532	8.6222E-01	7.3010E-05	298	1000	69.79	254.83
1013	C10H20	1-DECENE	-128.709	8.1530E-01	6.7574E-05	298	1000	121.04	295.84
1014	C10H20O	DECANAL	-335.675	8.8103E-01	6.5416E-05	298	1000	-66.53	121.20

Table 13-1 GIBBS ENERGY OF FORMATION OF GAS - ORGANIC COMPOUNDS (continued)

NO	FORMULA	NAME	Gf = A + B T + C T²					(Gf - kjoule/mol, T - K)	
			A	B	C	TMIN	TMAX	Gf @ 298 K	Gf @ 500 K
1015	C10H20O2	n-DECANOIC ACID	-599.347	9.6574E-01	6.4372E-05	298	1000	-305.00	-100.38
1016	C10H20O2	2-ETHYLHEXYL ACETATE	-575.113	9.5847E-01	6.3713E-05	298	1000	-283.00	-79.95
1017	C10H20O2	ISOPENTYL ISOVALERATE	-588.383	9.7576E-01	6.4083E-05	298	1000	-291.07	-84.48
1018	C10H22	DECANE	-255.000	9.4201E-01	7.4254E-05	298	1000	33.22	234.57
1019	C10H22	2-METHYLNONANE	-261.869	9.5206E-01	7.1758E-05	298	1000	29.08	232.10
1020	C10H22	3-METHYLNONANE	-259.232	9.4674E-01	7.2527E-05	298	1000	30.21	232.27
1021	C10H22	4-METHYLNONANE	-259.232	9.4674E-01	7.2527E-05	298	1000	30.21	232.27
1022	C10H22	5-METHYLNONANE	-259.270	9.5261E-01	7.2471E-05	298	1000	31.92	235.15
1023	C10H22	3-ETHYLOCTANE	-256.591	9.5302E-01	7.3335E-05	298	1000	34.81	238.25
1024	C10H22	4-ETHYLOCTANE	-256.563	9.4713E-01	7.3410E-05	298	1000	33.10	235.36
1025	C10H22	2,2-DIMETHYLOCTANE	-272.907	9.7753E-01	6.7222E-05	298	1000	25.23	232.66
1026	C10H22	2,3-DIMETHYLOCTANE	-266.771	9.6074E-01	7.3210E-05	298	1000	26.94	231.90
1027	C10H22	2,4-DIMETHYLOCTANE	-261.653	9.6011E-01	7.2592E-05	298	1000	31.80	236.55
1028	C10H22	2,5-DIMETHYLOCTANE	-266.771	9.6074E-01	7.3210E-05	298	1000	26.94	231.90
1029	C10H22	2,6-DIMETHYLOCTANE	-266.771	9.6074E-01	7.3210E-05	298	1000	26.94	231.90
1030	C10H22	2,7-DIMETHYLOCTANE	-269.396	9.7167E-01	7.2527E-05	298	1000	27.49	234.57
1031	C10H22	3,3-DIMETHYLOCTANE	-267.928	9.6911E-01	6.7859E-05	298	1000	27.78	233.59
1032	C10H22	3,4-DIMETHYLOCTANE	-259.027	9.5779E-01	7.3332E-05	298	1000	33.81	238.20
1033	C10H22	3,5-DIMETHYLOCTANE	-264.140	9.5839E-01	7.3992E-05	298	1000	28.95	233.55
1034	C10H22	3,6-DIMETHYLOCTANE	-264.094	9.6400E-01	7.4100E-05	298	1000	30.67	236.43
1035	C10H22	4,4-DIMETHYLOCTANE	-267.928	9.6911E-01	6.7859E-05	298	1000	27.78	233.59
1036	C10H22	4,5-DIMETHYLOCTANE	-258.993	9.6344E-01	7.3443E-05	298	1000	35.52	241.09
1037	C10H22	4-PROPYLHEPTANE	-256.533	9.6193E-01	7.3483E-05	298	1000	37.53	242.80
1038	C10H22	4-ISOPROPYLHEPTANE	-256.743	9.6635E-01	7.3427E-05	298	1000	38.66	244.79
1039	C10H22	3-ETHYL-2-METHYLHEPTANE	-258.970	9.6045E-01	7.3515E-05	298	1000	34.69	239.63
1040	C10H22	4-ETHYL-2-METHYLHEPTANE	-264.077	9.6109E-01	7.4107E-05	298	1000	29.83	235.00
1041	C10H22	5-ETHYL-2-METHYLHEPTANE	-264.106	9.6694E-01	7.4061E-05	298	1000	31.55	237.88
1042	C10H22	3-ETHYL-3-METHYLHEPTANE	-262.888	9.6960E-01	6.8641E-05	298	1000	33.05	239.07
1043	C10H22	4-ETHYL-3-METHYLHEPTANE	-256.383	9.5829E-01	7.4126E-05	298	1000	36.69	241.29
1044	C10H22	3-ETHYL-5-METHYLHEPTANE	-261.414	9.6136E-01	7.5079E-05	298	1000	32.68	238.04
1045	C10H22	3-ETHYL-4-METHYLHEPTANE	-256.320	9.6100E-01	7.4238E-05	298	1000	37.57	242.74
1046	C10H22	4-ETHYL-4-METHYLHEPTANE	-262.888	9.6960E-01	6.8641E-05	298	1000	33.05	239.07
1047	C10H22	2,2,3-TRIMETHYLHEPTANE	-267.615	9.8561E-01	6.8083E-05	298	1000	33.01	242.21
1048	C10H22	2,2,4-TRIMETHYLHEPTANE	-269.875	9.8622E-01	6.8723E-05	298	1000	31.00	240.42
1049	C10H22	2,2,5-TRIMETHYLHEPTANE	-277.782	9.8622E-01	6.8723E-05	298	1000	23.10	232.51
1050	C10H22	2,2,6-TRIMETHYLHEPTANE	-280.476	9.9171E-01	6.7751E-05	298	1000	21.97	232.32
1051	C10H22	2,3,3-TRIMETHYLHEPTANE	-265.181	9.8212E-01	6.8178E-05	298	1000	34.43	242.93
1052	C10H22	2,3,4-TRIMETHYLHEPTANE	-261.391	9.7099E-01	7.3509E-05	298	1000	35.40	242.48
1053	C10H22	2,3,5-TRIMETHYLHEPTANE	-266.466	9.7144E-01	7.4281E-05	298	1000	30.54	237.83
1054	C10H22	2,3,6-TRIMETHYLHEPTANE	-269.149	9.7393E-01	7.3404E-05	298	1000	28.53	236.17
1055	C10H22	2,4,4-TRIMETHYLHEPTANE	-267.448	9.8275E-01	6.8799E-05	298	1000	32.43	241.13
1056	C10H22	2,4,5-TRIMETHYLHEPTANE	-266.466	9.7144E-01	7.4281E-05	298	1000	30.54	237.83
1057	C10H22	2,4,6-TRIMETHYLHEPTANE	-259.166	9.8025E-01	7.4100E-05	298	1000	40.46	249.48
1058	C10H22	2,5,5-TRIMETHYLHEPTANE	-275.356	9.8275E-01	6.8799E-05	298	1000	24.52	233.22
1059	C10H22	3,3,4-TRIMETHYLHEPTANE	-262.647	9.7725E-01	6.8661E-05	298	1000	35.56	243.15
1060	C10H22	3,3,5-TRIMETHYLHEPTANE	-264.902	9.7782E-01	6.9318E-05	298	1000	33.56	241.34
1061	C10H22	3,4,4-TRIMETHYLHEPTANE	-262.647	9.7725E-01	6.8661E-05	298	1000	35.56	243.15
1062	C10H22	3,4,5-TRIMETHYLHEPTANE	-259.802	9.7141E-01	7.5693E-05	298	1000	38.24	244.82
1063	C10H22	3-ISOPROPYL-2-METHYLHEXANE	-261.387	9.7967E-01	7.3499E-05	298	1000	37.99	246.82
1064	C10H22	3,3-DIETHYLHEXANE	-257.797	9.7908E-01	6.9505E-05	298	1000	41.05	249.12
1065	C10H22	3,4-DIETHYLHEXANE	-253.621	9.7301E-01	7.5062E-05	298	1000	43.93	251.65
1066	C10H22	3-ETHYL-2,2-DIMETHYLHEXANE	-264.982	9.8620E-01	6.8786E-05	298	1000	35.90	245.32
1067	C10H22	4-ETHYL-2,2-DIMETHYLHEXANE	-267.225	9.9254E-01	6.9429E-05	298	1000	35.65	246.40
1068	C10H22	3-ETHYL-2,3-DIMETHYLHEXANE	-260.267	9.7727E-01	6.8648E-05	298	1000	37.95	245.53
1069	C10H22	4-ETHYL-2,3-DIMETHYLHEXANE	-258.811	9.7449E-01	7.4202E-05	298	1000	39.12	246.99
1070	C10H22	3-ETHYL-2,4-DIMETHYLHEXANE	-258.787	9.7149E-01	7.4304E-05	298	1000	38.24	245.53
1071	C10H22	4-ETHYL-2,4-DIMETHYLHEXANE	-262.419	9.8330E-01	6.9521E-05	298	1000	37.70	246.61
1072	C10H22	3-ETHYL-2,5-DIMETHYLHEXANE	-266.527	9.7459E-01	7.4067E-05	298	1000	31.42	239.29
1073	C10H22	4-ETHYL-3,3-DIMETHYLHEXANE	-259.929	9.8334E-01	6.9567E-05	298	1000	40.21	249.13
1074	C10H22	3-ETHYL-3,4-DIMETHYLHEXANE	-257.579	9.7763E-01	6.9541E-05	298	1000	40.84	248.62
1075	C10H22	2,2,3,3-TETRAMETHYLHEXANE	-264.021	1.0077E+00	6.3741E-05	298	1000	42.80	255.75
1076	C10H22	2,2,3,4-TETRAMETHYLHEXANE	-259.481	9.9664E-01	6.8878E-05	298	1000	44.56	256.06
1077	C10H22	2,2,3,5-TETRAMETHYLHEXANE	-275.197	9.9984E-01	6.8575E-05	298	1000	29.75	241.87
1078	C10H22	2,2,4,4-TETRAMETHYLHEXANE	-263.859	1.0089E+00	6.3813E-05	298	1000	43.35	256.52
1079	C10H22	2,2,4,5-TETRAMETHYLHEXANE	-272.355	9.9985E-01	6.8569E-05	298	1000	32.59	244.71
1080	C10H22	2,2,5,5-TETRAMETHYLHEXANE	-291.593	1.0233E+00	6.2952E-05	298	1000	19.83	235.81
1081	C10H22	2,3,3,4-TETRAMETHYLHEXANE	-260.022	9.9071E-01	6.8648E-05	298	1000	42.22	252.50
1082	C10H22	2,3,3,5-TETRAMETHYLHEXANE	-264.912	9.9652E-01	6.8562E-05	298	1000	39.04	250.49
1083	C10H22	2,3,4,4-TETRAMETHYLHEXANE	-257.185	9.9074E-01	6.8625E-05	298	1000	45.06	255.34
1084	C10H22	2,3,4,5-TETRAMETHYLHEXANE	-263.908	9.9031E-01	7.3463E-05	298	1000	38.66	249.61
1085	C10H22	3,3,4,4-TETRAMETHYLHEXANE	-258.982	1.0048E+00	6.4558E-05	298	1000	47.07	259.55
1086	C10H22	2,4-DIMETHYL-3-ISOPROPYLPENTANE	-264.215	1.0048E+00	6.8707E-05	298	1000	42.26	255.39
1087	C10H22	3,3-DIETHYL-2-METHYLPENTANE	-255.118	9.9226E-01	6.9725E-05	298	1000	47.70	258.45
1088	C10H22	3-ETHYL-2,2,3-TRIMETHYLPENTANE	-258.971	1.0024E+00	6.4555E-05	298	1000	46.36	258.35
1089	C10H22	3-ETHYL-2,2,4-TRIMETHYLPENTANE	-259.533	1.0062E+00	6.8727E-05	298	1000	47.32	260.73
1090	C10H22	3-ETHYL-2,3,4-TRIMETHYLPENTANE	-257.589	9.9632E-01	6.8759E-05	298	1000	46.32	257.76
1091	C10H22	2,2,3,3,4-PENTAMETHYLPENTANE	-253.281	1.0201E+00	6.4778E-05	298	1000	57.36	272.97
1092	C10H22	2,2,3,4,4-PENTAMETHYLPENTANE	-253.379	1.0306E+00	6.3360E-05	298	1000	60.25	277.75

Table 13-1 GIBBS ENERGY OF FORMATION OF GAS - ORGANIC COMPOUNDS (continued)

NO	FORMULA	NAME	$Gf = A + BT + CT^2$ A	B	C	TMIN	TMAX	(Gf - kjoule/mol, T - K) Gf @ 298 K	Gf @ 500 K
1093	C10H22O	DECYL ALCOHOL	-408.594	9.9155E-01	7.5283E-05	298	1000	-105.52	106.00
1094	C10H22O	DI-n-PENTYL ETHER	-379.502	1.0048E+00	7.4895E-05	298	1000	-72.50	141.60
1095	C10H22O	ISODECANOL	-414.464	1.0030E+00	7.4900E-05	298	1000	-108.00	105.75
1096	C10H22O5	TETRAETHYLENE GLYCOL DIMETHYL ETHER	-891.081	1.2766E+00	7.4538E-05	298	1000	-503.00	-234.17
1097	C10H22S	1-DECANETHIOL	-204.390	8.4798E-01	1.4421E-04	389	717	61.42	255.65
1098	C10H22S	BUTYL-HEXYL-SULFIDE	-201.497	8.4742E-01	1.4603E-04	389	717	64.31	258.72
1099	C10H22S	ETHYL-OCTYL-SULFIDE	-200.446	8.4814E-01	1.4635E-04	389	717	65.61	260.21
1100	C10H22S	HEPTYL-PROPYL-SULFIDE	-200.752	8.4746E-01	1.4599E-04	389	717	65.06	259.48
1101	C10H22S	METHYL-NONYL-SULFIDE	-198.094	8.5098E-01	1.4518E-04	389	717	68.70	263.69
1102	C10H22S	PENTYL-SULFIDE	-201.402	8.5291E-01	1.4621E-04	389	717	66.07	261.61
1103	C10H22S2	PENTYL-DISULFIDE	-190.970	8.2801E-01	1.6386E-04	389	717	70.67	264.00
1104	C10H23N	n-DECYLAMINE	-220.776	1.0493E+00	7.7276E-05	298	1000	99.76	323.21
1105	C11H10	1-METHYLNAPHTHALENE	114.458	3.3490E-01	3.4180E-05	298	1000	217.69	290.45
1106	C11H10	2-METHYLNAPHTHALENE	113.834	3.3151E-01	3.5442E-05	298	1000	216.15	288.45
1107	C11H14O2	n-BUTYL BENZOATE	-370.896	6.3716E-01	6.0175E-05	298	1000	-175.00	-37.27
1108	C11H16	PENTYLBENZENE	-38.343	6.2290E-01	5.6186E-05	298	1000	152.93	287.15
1109	C11H16O	p-tert-AMYLPHENOL	-216.543	7.2880E-01	4.6504E-05	298	1000	5.42	159.48
1110	C11H20	1-UNDECYNE	16.390	7.9808E-01	6.3658E-05	298	1000	260.62	431.34
1111	C11H20O2	2-ETHYLHEXYL ACRYLATE	-475.348	9.3213E-01	5.3986E-05	298	1000	-192.00	4.22
1112	C11H22	1-UNDECENE	-149.838	9.1236E-01	7.3538E-05	298	1000	129.45	324.73
1113	C11H22	1-CYCLOPENTYLHEXANE	-215.718	9.5953E-01	7.8827E-05	298	1000	78.20	283.76
1114	C11H22	PENTYLCYCLOHEXANE	-241.092	1.0031E+00	6.7958E-05	298	1000	64.85	277.43
1115	C11H22O	1-UNDECANAL	-356.067	9.7476E-01	7.3121E-05	298	1000	-58.20	149.59
1116	C11H24	UNDECANE	-276.105	1.0389E+00	8.0407E-05	298	1000	41.59	263.44
1117	C11H24O	UNDECYL ALCOHOL	-428.070	1.0886E+00	8.1234E-05	298	1000	-95.44	136.55
1118	C11H24S	1-UNDECANETHIOL	-224.542	9.4068E-01	1.5465E-04	389	717	69.83	284.46
1119	C11H24S	BUTYL-HEPTYL-SULFIDE	-221.656	9.4009E-01	1.5642E-04	389	717	72.72	287.50
1120	C11H24S	DECYL-METHYL-SULFIDE	-218.271	9.4391E-01	1.5535E-04	389	717	77.15	292.52
1121	C11H24S	ETHYL-NONYL-SULFIDE	-220.622	9.4107E-01	1.5652E-04	389	717	74.06	289.04
1122	C11H24S	OCTYL-PROPYL-SULFIDE	-220.786	9.4013E-01	1.5638E-04	389	717	73.60	288.37
1123	C12H8O	DIBENZOFURAN	80.597	3.8703E-01	2.9962E-05	298	1000	199.00	281.60
1124	C12H9N	DIBENZOPYRROLE	206.206	5.1272E-01	1.8386E-05	298	1000	361.00	467.16
1125	C12H10	ACENAPHTHENE	152.395	3.5061E-01	3.2893E-05	298	1000	260.20	335.92
1126	C12H10	BIPHENYL	179.415	3.2590E-01	3.5313E-05	298	1000	280.08	351.19
1127	C12H10O	DIPHENYL ETHER	40.997	4.0476E-01	3.3176E-05	298	1000	165.00	251.67
1128	C12H11N	p-AMINODIPHENYL	180.827	4.4924E-01	3.2270E-05	298	1000	318.00	413.51
1129	C12H11N	DIPHENYLAMINE	198.246	4.8000E-01	3.6216E-05	298	1000	345.00	447.30
1130	C12H11N3	p-AMINOAZOBENZENE	325.007	5.8114E-01	1.5422E-05	298	1000	500.00	619.43
1131	C12H11N3	1,3-DIPHENYLTRIAZENE	--------	--------	--------	--------	--------	--------	--------
1132	C12H12	1,2-DIMETHYLNAPHTHALENE	80.611	4.4220E-01	3.8339E-05	298	1000	216.23	311.29
1133	C12H12	1,3-DIMETHYLNAPHTHALENE	78.976	4.3881E-01	3.9571E-05	298	1000	213.72	308.27
1134	C12H12	1,4-DIMETHYLNAPHTHALENE	79.522	4.4807E-01	3.8242E-05	298	1000	216.90	313.12
1135	C12H12	1,5-DIMETHYLNAPHTHALENE	78.821	4.4803E-01	3.8272E-05	298	1000	216.19	312.40
1136	C12H12	1,6-DIMETHYLNAPHTHALENE	79.690	4.3879E-01	3.9579E-05	298	1000	214.43	308.98
1137	C12H12	1,7-DIMETHYLNAPHTHALENE	78.976	4.3881E-01	3.9571E-05	298	1000	213.72	308.27
1138	C12H12	2,3-DIMETHYLNAPHTHALENE	81.060	4.3524E-01	4.2654E-05	298	1000	215.02	309.34
1139	C12H12	2,6-DIMETHYLNAPHTHALENE	79.922	4.3839E-01	4.0694E-05	298	1000	214.64	309.29
1140	C12H12	2,7-DIMETHYLNAPHTHALENE	79.918	4.3842E-01	4.0642E-05	298	1000	214.64	309.29
1141	C12H12	1-ETHYLNAPHTHALENE	93.648	4.3103E-01	3.8318E-05	298	1000	225.98	318.74
1142	C12H12	2-ETHYLNAPHTHALENE	93.053	4.2755E-01	3.9627E-05	298	1000	224.43	316.73
1143	C12H12N2	p-AMINODIPHENYLAMINE	201.742	5.9666E-01	3.2995E-05	298	1000	383.00	508.32
1144	C12H12N2	HYDRAZOBENZENE	315.121	5.8933E-01	4.1760E-05	298	1000	495.00	620.23
1145	C12H14	1,2,3-TRIMETHYLINDENE	48.576	5.6668E-01	3.4633E-05	298	1000	221.00	340.57
1146	C12H14O4	DIETHYL PHTHALATE	-692.605	6.4449E-01	6.5600E-05	298	1000	-494.00	-353.96
1147	C12H16	CYCLOHEXYLBENZENE	-26.783	6.8595E-01	5.5026E-05	298	1000	183.30	329.95
1148	C12H18	m-DIISOPROPYLBENZENE	-82.083	7.3941E-01	6.2186E-05	298	1000	144.53	303.17
1149	C12H18	p-DIISOPROPYLBENZENE	-82.083	7.5037E-01	6.2186E-05	298	1000	147.80	308.65
1150	C12H18	HEXYLBENZENE	-59.463	7.1997E-01	6.2172E-05	298	1000	161.34	316.07
1151	C12H18	1,2,3-TRIETHYLBENZENE	-71.837	7.2867E-01	6.2504E-05	298	1000	151.54	308.13
1152	C12H18	1,2,4-TRIETHYLBENZENE	-72.938	7.1000E-01	6.8084E-05	298	1000	145.23	299.08
1153	C12H18	1,3,5-TRIETHYLBENZENE	-78.848	7.2900E-01	6.2716E-05	298	1000	144.68	301.33
1154	C12H18	HEXAMETHYLBENZENE	-109.218	7.8642E-01	5.0247E-05	298	1000	130.21	296.56
1155	C12H20O4	DIBUTYL MALEATE	-692.195	1.0347E+00	6.7886E-05	298	1000	-377.00	-157.87
1156	C12H22	BICYCLOHEXYL	-281.216	1.0626E+00	6.7826E-05	298	1000	42.60	267.06
1157	C12H22	1-DODECYNE	-4.724	8.9500E-01	6.9758E-05	298	1000	268.99	460.22
1158	C12H23N	DICYCLOHEXYLAMINE	-195.186	1.2030E+00	6.7090E-05	298	1000	170.45	423.07
1159	C12H24	1-DODECENE	-170.907	1.0093E+00	7.9654E-05	298	1000	137.90	353.67
1160	C12H24	1-CYCLOPENTYLHEPTANE	-236.801	1.0564E+00	8.5011E-05	298	1000	86.61	312.65
1161	C12H24	1-CYCLOHEXYLHEXANE	-262.182	1.1000E+00	7.4071E-05	298	1000	73.26	306.36
1162	C12H24O	1-DODECANAL	-379.989	1.0718E+00	7.8912E-05	298	1000	-52.60	175.64
1163	C12H24O2	n-DODECANOIC ACID	--------	--------	--------	--------	--------	-297.00	--------
1164	C12H26	DODECANE	-297.182	1.1359E+00	8.6457E-05	298	1000	50.04	292.38
1165	C12H26O	DI-n-HEXYL ETHER	-422.041	1.2030E+00	8.7102E-05	298	1000	-54.70	201.25
1166	C12H26O	DODECYL ALCOHOL	-449.239	1.1858E+00	8.7192E-05	298	1000	-87.07	165.44
1167	C12H26O3	DIETHYLENE GLYCOL DI-n-BUTYL ETHER	-669.302	1.3272E+00	8.6028E-05	298	1000	-265.00	15.82
1168	C12H26S	1-DODECANETHIOL	-244.797	1.0337E+00	1.6475E-04	389	717	78.24	313.24
1169	C12H26S	BUTYL-OCTYL-SULFIDE	-241.769	1.0327E+00	1.6697E-04	389	717	81.17	316.31
1170	C12H26S	DECYL-ETHYL-SULFIDE	-242.383	1.0341E+00	1.6662E-04	389	717	80.92	316.31

Table 13-1 GIBBS ENERGY OF FORMATION OF GAS - ORGANIC COMPOUNDS (continued)

NO	FORMULA	NAME	$Gf = A + B T + C T^2$					(Gf - kjoule/mol, T - K)	
			A	B	C	TMIN	TMAX	Gf @ 298 K	Gf @ 500 K
1171	C12H26S	HEXYL-SULFIDE	-241.877	1.0389E+00	1.6649E-04	389	717	82.89	319.20
1172	C12H26S	METHYL-UNDECYL-SULFIDE	-238.536	1.0370E+00	1.6539E-04	389	717	85.52	321.30
1173	C12H26S	NONYL-PROPYL-SULFIDE	-241.027	1.0327E+00	1.6692E-04	389	717	81.92	317.06
1174	C12H26S2	HEXYL-DISULFIDE	-231.339	1.0136E+00	1.8450E-04	389	717	87.49	321.59
1175	C12H27BO3	TRI-n-BUTYL BORATE	--------	--------	--------	--------	--------	--------	--------
1176	C12H27N	DODECYLAMINE	-262.759	1.2421E+00	8.9727E-05	298	1000	116.49	380.72
1177	C12H27N	TRI-n-BUTYLAMINE	-240.469	1.2670E+00	8.7268E-05	298	1000	146.00	414.84
1178	C13H10	FLUORENE	184.059	3.4384E-01	3.5434E-05	298	1000	290.10	364.84
1179	C13H10O	BENZOPHENONE	56.242	3.8495E-01	4.0392E-05	298	1000	175.00	258.81
1180	C13H12	DIPHENYLMETHANE	150.944	4.1485E-01	4.4277E-05	298	1000	279.00	369.44
1181	C13H14	1-PROPYLNAPHTHALENE	71.181	5.2690E-01	4.3941E-05	298	1000	232.63	345.62
1182	C13H14	2-PROPYLNAPHTHALENE	70.476	5.2339E-01	4.5280E-05	298	1000	231.00	343.49
1183	C13H14	2-ETHYL-3-METHYLNAPHTHALENE	62.759	5.2530E-01	4.7025E-05	298	1000	224.01	337.17
1184	C13H14	2-ETHYL-6-METHYLNAPHTHALENE	58.499	5.2712E-01	4.6332E-05	298	1000	220.20	333.64
1185	C13H14	2-ETHYL-7-METHYLNAPHTHALENE	58.499	5.2712E-01	4.6332E-05	298	1000	220.20	333.64
1186	C13H20	1-PHENYLHEPTANE	-80.556	8.1688E-01	6.8286E-05	298	1000	169.74	344.95
1187	C13H24	1-TRIDECYNE	-25.829	9.9206E-01	7.5746E-05	298	1000	277.44	489.14
1188	C13H26	1-TRIDECENE	-192.059	1.1064E+00	8.5626E-05	298	1000	146.27	382.55
1189	C13H26	1-CYCLOPENTYLOCTANE	-257.926	1.1536E+00	9.0881E-05	298	1000	95.06	341.59
1190	C13H26	1-CYCLOHEXYLHEPTANE	-283.301	1.1970E+00	8.0042E-05	298	1000	81.67	335.23
1191	C13H26O	1-TRIDECANAL	-398.134	1.1689E+00	8.5060E-05	298	1000	-41.20	207.56
1192	C13H26O2	n-BUTYL NONANOATE	-634.156	1.2540E+00	8.3174E-05	298	1000	-252.00	13.62
1193	C13H26O2	METHYL DODECANOATE	-618.787	1.2426E+00	8.3419E-05	298	1000	-240.00	23.38
1194	C13H28	TRIDECANE	-318.273	1.2328E+00	9.2592E-05	298	1000	58.45	321.26
1195	C13H28O	1-TRIDECANOL	-470.339	1.2840E+00	9.3318E-05	298	1000	-78.28	194.99
1196	C13H28S	BUTYL-NONYL-SULFIDE	-261.912	1.1252E+00	1.7759E-04	389	717	89.54	345.07
1197	C13H28S	DECYL-PROPYL-SULFIDE	-261.150	1.1251E+00	1.7763E-04	389	717	90.29	345.82
1198	C13H28S	DODECYL-METHYL-SULFIDE	-258.629	1.1295E+00	1.7609E-04	389	717	93.97	350.13
1199	C13H28S	ETHYL-UNDECYL-SULFIDE	-260.976	1.1265E+00	1.7738E-04	389	717	90.88	346.61
1200	C13H28S	1-TRIDECANETHIOL	-264.946	1.1264E+00	1.7521E-04	389	717	86.65	342.04
1201	C14H8O2	ANTHRAQUINONE	--------	--------	--------	--------	--------	--------	--------
1202	C14H10	ANTHRACENE	227.622	3.3843E-01	3.1962E-05	298	1000	331.70	404.83
1203	C14H10	DIPHENYLACETYLENE	427.918	2.6311E-01	3.4725E-05	298	1000	509.78	568.15
1204	C14H10	PHENANTHRENE	204.659	3.3658E-01	3.1129E-05	298	1000	308.10	380.73
1205	C14H12	cis-STILBENE	242.391	4.0465E-01	4.6262E-05	298	1000	367.58	456.28
1206	C14H12	trans-STILBENE	238.326	4.0038E-01	4.3756E-05	298	1000	362.00	449.45
1207	C14H12O2	BENZYL BENZOATE	-195.059	5.1928E-01	5.5009E-05	298	1000	-34.80	78.33
1208	C14H14	1,1-DIPHENYLETHANE	112.429	5.1175E-01	5.0518E-05	298	1000	270.00	380.94
1209	C14H14	1,2-DIPHENYLETHANE	139.432	5.1179E-01	5.0385E-05	298	1000	297.00	407.92
1210	C14H14O	DIBENZYL ETHER	14.474	5.7106E-01	5.2635E-05	298	1000	190.00	313.16
1211	C14H16	1-BUTYLNAPHTHALENE	49.101	6.2380E-01	5.0256E-05	298	1000	240.08	373.57
1212	C14H16	2-BUTYLNAPHTHALENE	48.475	6.2035E-01	5.1496E-05	298	1000	238.53	371.52
1213	C14H22	1-PHENYLOCTANE	-101.612	9.1381E-01	7.4333E-05	298	1000	178.20	373.88
1214	C14H22	1,2,3,4-TETRAETHYLBENZENE	-128.011	9.3083E-01	6.4969E-05	298	1000	155.94	353.65
1215	C14H22	1,2,3,5-TETRAETHYLBENZENE	-127.656	9.2414E-01	6.7695E-05	298	1000	154.56	351.34
1216	C14H22	1,2,4,5-TETRAETHYLBENZENE	-127.872	9.2707E-01	6.9124E-05	298	1000	155.35	352.94
1217	C14H22O	p-tert-OCTYLPHENOL	222.792	3.3800E-01	3.4662E-05	298	1000	22.70	400.46
1218	C14H28	1-TETRADECENE	-213.140	1.2035E+00	9.1571E-05	298	1000	154.77	411.52
1219	C14H28	1-CYCLOPENTYLNONANE	-279.030	1.2505E+00	9.7036E-05	298	1000	103.43	370.46
1220	C14H28	1-CYCLOHEXYLOCTANE	-304.390	1.2940E+00	8.6187E-05	298	1000	90.08	364.16
1221	C14H28O2	n-TETRADECANOIC ACID	-701.390	1.3559E+00	9.1493E-05	298	1000	-288.00	-0.57
1222	C14H30	TETRADECANE	-339.495	1.3301E+00	9.8399E-05	298	1000	66.82	350.15
1223	C14H30O	1-TETRADECANOL	-491.525	1.3812E+00	9.9188E-05	298	1000	-69.87	223.88
1224	C14H30S	BUTYL-DECYL-SULFIDE	-282.126	1.2181E+00	1.8779E-04	389	717	97.95	373.85
1225	C14H30S	DODECYL-ETHYL-SULFIDE	-281.237	1.2195E+00	1.8745E-04	389	717	99.24	375.39
1226	C14H30S	HEPTYL-SULFIDE	-282.103	1.2238E+00	1.8786E-04	389	717	99.70	376.78
1227	C14H30S	METHYL-TRIDECYL-SULFIDE	-278.793	1.2219E+00	1.8683E-04	389	717	102.34	378.87
1228	C14H30S	PROPYL-UNDECYL-SULFIDE	-281.344	1.2181E+00	1.8773E-04	389	717	98.74	374.65
1229	C14H30S	1-TETRADECANETHIOL	-285.169	1.2193E+00	1.8537E-04	389	717	95.06	370.83
1230	C14H30S2	HEPTYL-DISULFIDE	-271.597	1.1987E+00	2.0566E-04	389	717	104.31	379.16
1231	C14H31N	TETRADECYLAMINE	-305.818	1.4379E+00	1.0167E-04	298	1000	133.00	438.54
1232	C15H10N2O2	DIPHENYLMETHANE-4,4'-DIISOCYANATE	--------	--------	--------	--------	--------	--------	--------
1233	C15H16O	p-CUMYLPHENOL	-70.279	6.9915E-01	4.8134E-05	298	1000	143.00	291.33
1234	C15H16O2	BISPHENOL A	-250.212	7.9374E-01	4.0946E-05	298	1000	-9.42	156.90
1235	C15H18	1-PENTYLNAPHTHALENE	27.930	7.2052E-01	5.6005E-05	298	1000	248.32	402.19
1236	C15H18	2-PENTYLNAPHTHALENE	27.301	7.1700E-01	5.7364E-05	298	1000	246.77	400.14
1237	C15H24	1-PHENYLNONANE	-122.814	1.0110E+00	8.0315E-05	298	1000	186.56	402.75
1238	C15H24O	2,6-DI-tert-BUTYL-p-CRESOL	-347.767	7.0758E-01	6.7598E-05	298	1000	-130.00	22.92
1239	C15H24O	NONYLPHENOL	-297.079	1.0775E+00	7.5664E-05	298	1000	31.70	260.57
1240	C15H28	1-PENTADECYNE	-68.040	1.1860E+00	8.7866E-05	298	1000	294.26	546.93
1241	C15H30	1-PENTADECENE	-234.403	1.3006E+00	9.7536E-05	298	1000	163.05	440.31
1242	C15H30	1-CYCLOPENTYLDECANE	-300.184	1.3476E+00	1.0295E-04	298	1000	111.84	399.35
1243	C15H30	1-CYCLOHEXYLNONANE	-325.560	1.3912E+00	9.2165E-05	298	1000	98.49	393.07
1244	C15H30O2	PENTADECANOIC ACID	-706.530	1.4458E+00	9.5003E-05	298	1000	-266.00	40.12
1245	C15H32	PENTADECANE	-360.562	1.4269E+00	1.0458E-04	298	1000	75.23	379.03
1246	C15H32O	1-PENTADECANOL	-512.580	1.4780E+00	1.0540E-04	298	1000	-61.46	252.78
1247	C15H32S	BUTYL-UNDECYL-SULFIDE	-302.378	1.3111E+00	1.9792E-04	389	717	106.36	402.66
1248	C15H32S	DODECYL-PROPYL-SULFIDE	-301.620	1.3111E+00	1.9794E-04	389	717	107.11	403.41

NO	FORMULA	NAME	Gf = A + B T + C T²					(Gf - kjoule/mol, T - K)	
			A	B	C	TMIN	TMAX	Gf @ 298 K	Gf @ 500 K
1249	C15H32S	ETHYL-TRIDECYL-SULFIDE	-301.464	1.3125E+00	1.9758E-04	389	717	107.65	404.18
1250	C15H32S	METHYL-TETRADECYL-SULFIDE	-298.940	1.3145E+00	1.9728E-04	389	717	110.75	407.64
1251	C15H32S	1-PENTADECANETHIOL	-305.323	1.3120E+00	1.9578E-04	389	717	103.47	399.60
1252	C16H10	FLUORANTHENE	286.251	3.2759E-01	2.2551E-05	298	1000	386.20	455.68
1253	C16H10	PYRENE	--------	--------	--------	--------	--------	327.00	
1254	C16H12	1-PHENYLNAPHTHALENE	243.959	4.1325E-01	3.9796E-05	298	1000	371.11	460.53
1255	C16H20	1-n-HEXYLNAPHTHALENE	9.385	8.1495E-01	6.2681E-05	298	1000	258.60	432.53
1256	C16H22O4	DIBUTYL PHTHALATE	-757.626	1.0310E+00	8.9229E-05	298	1000	-441.40	-219.84
1257	C16H26	1-PHENYLDECANE	-143.980	1.1082E+00	8.6141E-05	298	1000	194.97	431.65
1258	C16H26	PENTAETHYLBENZENE	-181.006	1.1365E+00	7.1696E-05	298	1000	164.98	405.18
1259	C16H30	1-HEXADECYNE	-89.177	1.2831E+00	9.3850E-05	298	1000	302.67	575.83
1260	C16H32	1-CYCLOHEXYLDECANE	-346.683	1.4883E+00	9.8015E-05	298	1000	106.90	421.95
1261	C16H32	1-CYCLOPENTYLUNDECANE	-321.252	1.4445E+00	1.0913E-04	298	1000	120.25	428.26
1262	C16H32	1-HEXADECENE	-255.460	1.3976E+00	1.0363E-04	298	1000	171.50	469.23
1263	C16H32O2	n-HEXADECANOIC ACID	-745.566	1.5469E+00	1.0395E-04	298	1000	-274.00	53.85
1264	C16H34	HEXADECANE	-381.697	1.5241E+00	1.1046E-04	298	1000	83.68	407.96
1265	C16H34O	DI-n-OCTYL ETHER	-506.598	1.5838E+00	1.1146E-04	298	1000	-23.30	313.17
1266	C16H34O	1-HEXADECANOL	-533.675	1.5751E+00	1.1135E-04	298	1000	-53.01	281.71
1267	C16H34S	BUTYL-DODECYL-SULFIDE	-322.605	1.4041E+00	2.0806E-04	389	717	114.77	431.44
1268	C16H34S	ETHYL-TETRADECYL-SULFIDE	-321.539	1.4049E+00	2.0836E-04	389	717	116.11	433.00
1269	C16H34S	METHYL-PENTADECYL-SULFIDE	-319.146	1.4075E+00	2.0743E-04	389	717	119.20	436.49
1270	C16H34S	OCTYL-SULFIDE	-322.409	1.4092E+00	2.0873E-04	389	717	116.52	434.36
1271	C16H34S	PROPYL-TRIDECYL-SULFIDE	-321.856	1.4041E+00	2.0804E-04	389	717	115.52	432.20
1272	C16H34S	1-HEXADECANETHIOL	-325.520	1.4048E+00	2.0614E-04	389	717	111.88	428.43
1273	C16H34S2	OCTYL-DISULFIDE	-312.192	1.3851E+00	2.2550E-04	389	717	121.09	436.73
1274	C17H28	1-PHENYLUNDECANE	-165.020	1.2050E+00	9.2372E-05	298	1000	203.43	460.58
1275	C17H32	1-HEPTADECYNE	-110.335	1.3802E+00	9.9805E-05	298	1000	311.04	604.71
1276	C17H34	1-CYCLOPENTYLDODECANE	-345.061	1.5413E+00	1.1522E-04	298	1000	125.94	454.42
1277	C17H34	1-CYCLOHEXYLUNDECANE	-367.817	1.5854E+00	1.0407E-04	298	1000	115.31	450.88
1278	C17H34	1-HEPTADECENE	-276.569	1.4947E+00	1.0958E-04	298	1000	179.91	498.16
1279	C17H36	HEPTADECANE	-402.745	1.6209E+00	1.1669E-04	298	1000	92.09	436.87
1280	C17H36O	1-HEPTADECANOL	-554.820	1.6721E+00	1.1740E-04	298	1000	-44.64	310.59
1281	C17H36S	BUTYL-TRIDECYL-SULFIDE	-342.679	1.4965E+00	2.1883E-04	389	717	123.22	460.27
1282	C17H36S	ETHYL-PENTADECYL-SULFIDE	-341.808	1.4980E+00	2.1838E-04	389	717	124.47	461.78
1283	C17H36S	HEXADECYL-METHYL-SULFIDE	-339.362	1.5003E+00	2.1778E-04	389	717	127.57	465.26
1284	C17H36S	PROPYL-TETRADECYL-SULFIDE	-341.939	1.4965E+00	2.1877E-04	389	717	123.97	461.02
1285	C17H36S	1-HEPTADECANETHIOL	-345.747	1.4978E+00	2.1627E-04	389	717	120.29	457.22
1286	C18H12	CHRYSENE	259.993	4.6089E-01	3.6880E-05	298	1000	401.08	499.66
1287	C18H14	m-TERPHENYL	273.543	4.8653E-01	4.4484E-05	298	1000	423.00	527.93
1288	C18H14	o-TERPHENYL	273.543	4.8653E-01	4.4484E-05	298	1000	423.00	527.93
1289	C18H14	p-TERPHENYL	273.543	4.8988E-01	4.4484E-05	298	1000	424.00	529.61
1290	C18H15P	TRIPHENYLPHOSPHINE	--------	--------	--------	--------	--------	--------	--------
1291	C18H15O4P	TRIPHENYL PHOSPHATE	--------	--------	--------	--------	--------	--------	--------
1292	C18H16N2	N,N'-DIPHENYL-p-PHENYLENEDIAMINE	314.991	8.1956E-01	4.5644E-05	298	1000	564.00	736.18
1293	C18H22	2,3-DIMETHYL-2,3-DIPHENYLBUTANE	52.517	9.6576E-01	6.4991E-05	298	1000	347.00	551.64
1294	C18H22O2	DICUMYL PEROXIDE	-85.401	1.0744E+00	7.0976E-05	298	1000	242.00	469.52
1295	C18H30	1-PHENYLDODECANE	-186.163	1.3021E+00	9.8370E-05	298	1000	211.79	489.46
1296	C18H30	HEXAETHYLBENZENE	-231.018	1.3613E+00	7.6183E-05	298	1000	182.46	468.68
1297	C18H32O2	LINOLEIC ACID	-547.796	1.4843E+00	1.1087E-04	298	1000	-94.30	222.06
1298	C18H34	1-OCTADECYNE	-131.478	1.4774E+00	1.0574E-04	298	1000	319.49	633.63
1299	C18H34O2	OLEIC ACID	-680.402	1.6097E+00	1.0826E-04	298	1000	-189.69	151.51
1300	C18H34O4	DIBUTYL SEBACATE	-1069.044	1.7466E+00	1.0307E-04	298	1000	-538.00	-169.99
1301	C18H34O4	DIHEXYL ADIPATE	-1024.382	1.7922E+00	1.6685E-04	298	1000	-474.00	-86.59
1302	C18H36	1-CYCOPENTYLTRIDECANE	-363.548	1.6385E+00	1.2095E-04	298	1000	136.98	485.93
1303	C18H36	1-CYCLOHEXYLDODECANE	-388.881	1.6821E+00	1.1029E-04	298	1000	123.72	479.73
1304	C18H36	1-OCTADECENE	-297.655	1.5915E+00	1.1579E-04	298	1000	188.32	527.03
1305	C18H36O2	STEARIC ACID	-776.121	1.7456E+00	1.1312E-04	298	1000	-244.40	124.94
1306	C18H38	OCTADECANE	-423.911	1.7180E+00	1.2257E-04	298	1000	100.50	465.76
1307	C18H38O	DINONYL ETHER	-548.693	1.7769E+00	1.2350E-04	298	1000	-6.60	370.66
1308	C18H38O	1-OCTADECANOL	-575.862	1.7690E+00	1.2355E-04	298	1000	-36.19	339.52
1309	C18H38S	BUTYL-TETRADECYL-SULFIDE	-362.931	1.5895E+00	2.2894E-04	389	717	131.59	489.04
1310	C18H38S	ETHYL-HEXADECYL-SULFIDE	-361.840	1.5902E+00	2.2925E-04	389	717	132.93	490.58
1311	C18H38S	HEPTADECYL-METHYL-SULFIDE	-359.398	1.5926E+00	2.2863E-04	389	717	136.02	494.06
1312	C18H38S	NONYL-SULFIDE	-362.874	1.5950E+00	2.2922E-04	389	717	133.30	491.93
1313	C18H38S	PENTADECYL-PROPYL-SULFIDE	-362.194	1.5896E+00	2.2887E-04	389	717	132.34	489.80
1314	C18H38S	1-OCTADECANETHIOL	-365.792	1.5899E+00	2.2735E-04	389	717	128.74	486.02
1315	C18H38S2	NONYL-DISULFIDE	-352.424	1.5702E+00	2.4668E-04	389	717	137.95	494.33
1316	C19H26	1-n-NONYLNAPHTHALENE	-53.826	1.1022E+00	8.5761E-05	298	1000	283.29	518.69
1317	C19H32	1-PHENYLTRIDECANE	-207.235	1.3990E+00	1.0443E-04	298	1000	220.25	518.38
1318	C19H36	1-NONADECYNE	-152.616	1.5744E+00	1.1175E-04	298	1000	327.86	662.51
1319	C19H36O2	METHYL OLEATE	-635.006	1.6993E+00	1.1401E-04	298	1000	-117.00	243.16
1320	C19H38	1-CYCLOPENTYLTETRADECANE	-384.618	1.7355E+00	1.2721E-04	298	1000	145.48	514.94
1321	C19H38	1-CYCLOHEXYLTRIDECANE	-409.956	1.7791E+00	1.1633E-04	298	1000	132.17	508.86
1322	C19H38	1-NONADECENE	-318.805	1.6887E+00	1.2163E-04	298	1000	196.73	555.96
1323	C19H38O2	NONADECANOIC ACID	-794.872	1.8346E+00	1.1940E-04	298	1000	-236.00	152.25
1324	C19H40	NONADECANE	-444.994	1.8150E+00	1.2867E-04	298	1000	108.91	494.67
1325	C19H40O	1-NONADECANOL	-597.017	1.8661E+00	1.2948E-04	298	1000	-27.78	368.40
1326	C19H40S	BUTYL-PENTADECYL-SULFIDE	-380.286	1.6678E+00	2.5645E-04	389	717	140.00	517.73

Table 13-1 GIBBS ENERGY OF FORMATION OF GAS - ORGANIC COMPOUNDS (continued)

NO	FORMULA	NAME	Gf = A + B T + C T²					(Gf - kjoule/mol, T - K)	
			A	B	C	TMIN	TMAX	Gf @ 298 K	Gf @ 500 K
1327	C19H40S	ETHYL-HEPTADECYL-SULFIDE	-382.174	1.6836E+00	2.3903E-04	389	717	141.29	519.37
1328	C19H40S	HEXADECYL-PROPYL-SULFIDE	-382.299	1.6822E+00	2.3931E-04	389	717	140.79	518.61
1329	C19H40S	METHYL-OCTADECYL-SULFIDE	-379.714	1.6859E+00	2.3849E-04	389	717	144.39	522.84
1330	C19H40S	1-NONADECANETHIOL	-386.117	1.6833E+00	2.3717E-04	389	717	137.11	514.81
1331	C20H16	TRIPHENYLETHYLENE	336.279	5.7257E-01	5.0645E-05	298	1000	512.00	635.22
1332	C20H28	1-n-DECYLNAPHTHALENE	-75.011	1.1990E+00	9.1904E-05	298	1000	291.60	547.48
1333	C20H30O2	ABIETIC ACID	-550.120	1.6486E+00	7.3252E-05	298	1000	-50.90	292.47
1334	C20H31N	DEHYDROABIETYLAMINE	-132.586	1.7174E+00	7.3715E-05	298	1000	387.00	744.53
1335	C20H34	1-PHENYLTETRADECANE	-228.416	1.4962E+00	1.1037E-04	298	1000	228.61	547.26
1336	C20H38	1-EICOSYNE	-173.613	1.6710E+00	1.1807E-04	298	1000	336.31	691.42
1337	C20H40	1-CYCLOPENTYLPENTADECANE	-405.771	1.8327E+00	1.3312E-04	298	1000	153.89	543.84
1338	C20H40	1-CYCLOHEXYLTETRADECANE	-431.135	1.8762E+00	1.2227E-04	298	1000	140.54	537.55
1339	C20H40	1-EICOSENE	-339.913	1.7856E+00	1.2774E-04	298	1000	205.14	584.83
1340	C20H42	EICOSANE	-466.086	1.9119E+00	1.3473E-04	298	1000	117.32	523.55
1341	C20H42O	1-EICOSANOL	-618.212	1.9633E+00	1.3535E-04	298	1000	-19.41	397.28
1342	C20H42S	BUTYL-HEXADECYL-SULFIDE	-403.297	1.7751E+00	2.4959E-04	389	717	148.41	546.63
1343	C20H42S	DECYL-SULFIDE	-403.294	1.7809E+00	2.4954E-04	389	717	150.16	549.53
1344	C20H42S	ETHYL-OCTADECYL-SULFIDE	-402.324	1.7761E+00	2.4974E-04	389	717	149.70	548.14
1345	C20H42S	HEPTADECYL-PROPYL-SULFIDE	-402.579	1.7752E+00	2.4943E-04	389	717	149.16	547.39
1346	C20H42S	METHYL-NONADECYL-SULFIDE	-399.878	1.7786E+00	2.4886E-04	389	717	152.80	551.62
1347	C20H42S	1-EICOSANETHIOL	-406.155	1.7754E+00	2.4814E-04	389	717	145.52	543.56
1348	C20H42S2	DECYL-DISULFIDE	-392.688	1.7552E+00	2.6793E-04	389	717	154.77	551.91
1349	C21H21O4P	TRI-o-CRESYL PHOSPHATE	--------	--------	--------	--------	--------	--------	--------
1350	C21H36	1-PHENYLPENTADECANE	-249.518	1.5931E+00	1.1646E-04	298	1000	237.02	576.15
1351	C21H42	1-CYCLOPENTYLHEXADECANE	-426.839	1.9296E+00	1.3923E-04	298	1000	162.30	572.75
1352	C21H42	1-CYCLOHEXYLPENTADECANE	-452.259	1.9733E+00	1.2822E-04	298	1000	148.95	566.43
1353	C22H38	1-PHENYLHEXADECANE	-270.675	1.6903E+00	1.2232E-04	298	1000	245.43	605.07
1354	C22H44	1-CYCLOHEXYLHEXADECANE	-473.340	2.0702E+00	1.3439E-04	298	1000	157.36	595.35
1355	C22H44O2	n-BUTYL STEARATE	-830.487	2.1108E+00	1.3819E-04	298	1000	-187.38	259.45
1356	C24H38O4	DIISOOCTYL PHTHALATE	-977.652	1.9874E+00	1.4225E-04	298	1000	-371.00	51.60
1357	C24H38O4	DIOCTYL PHTHALATE	-977.379	1.8680E+00	1.4243E-04	298	1000	-406.30	-7.77
1358	C24H42O	DINONYLPHENOL	-497.701	1.8558E+00	1.2599E-04	298	1000	68.20	461.71
1359	C26H20	TETRAPHENYLETHYLENE	433.415	7.5618E-01	6.1938E-05	298	1000	665.00	826.99
1360	C28H46O4	DIISODECYL PHTHALATE	-1072.772	2.3391E+00	1.6675E-04	298	1000	-358.81	138.46

Gf - Gibbs energy of formation of ideal gas, kjoule/mol

A, B, and C - regression coefficients for chemical compound

T - temperature, K

TMIN - minimum temperature, K

TMAX - maximum temperature, K

Additional compounds are:

H2O(g) A = -241.74, B = 4.1740E-02, C = 7.4281E-06

HCl(g) A = -92.209, B = -1.1226E-02, C = 2.6966E-06

Table 13-2 GIBBS AND HELMHOLTZ ENERGY OF FORMATION - INORGANIC COMPOUNDS

NO	FORMULA	NAME	Molecular Weight g/mol	State	Gf @ 298 K kjoule/mol	Af @ 298 K kjoule/mol	Sf @ 298 K joule/(mol K)
1	Ag	SILVER	107.868	solid	0.00	0.00	0.0000
2	AgCl	SILVER CHLORIDE	143.321	solid	-109.80	-108.56	-0.0577
3	AgI	SILVER IODIDE	234.773	solid	-66.20	-66.20	0.0148
4	Al	ALUMINUM	26.982	solid	0.00	0.00	0.0000
5	AlB3H12	ALUMINUM BOROHYDRIDE	71.510	gas	147.00	159.39	-0.4494
6	AlBr3	ALUMINUN BROMIDE	266.694	solid	-504.39	-504.40	-0.0764
7	AlCl3	ALUMINUM CHLORIDE	133.340	solid	-628.80	-625.08	-0.2529
8	AlF3	ALUMINUM FLUORIDE	83.977	solid	-1431.10	-1427.38	-0.2660
9	AlI3	ALUMINUM IODIDE	407.695	solid	-300.80	-300.80	-0.0436
10	Al2O3	ALUMINUM OXIDE	101.961	solid	-1582.30	-1578.58	-0.3133
11	Al2S3O12	ALUMINUM SULFATE	342.154	solid	-3099.90	-3085.03	-1.1434
12	Ar	ARGON	39.948	gas	0.00	0.00	0.0000
13	As	ARSENIC	74.922	solid	0.00	0.00	0.0000
14	AsBr3	ARSENIC TRIBROMIDE	314.634	gas	-159.00	-161.48	0.0973
15	AsCl3	ARSENIC TRICHLORIDE	181.280	gas	-248.90	-247.66	-0.0423
16	AsF3	ARSENIC TRIFLUORIDE	131.917	gas	-770.80	-769.56	-0.0503
17	AsF5	ARSENIC PENTAFLUORIDE	169.914	gas	-1172.36	-1168.64	0.3452
18	AsH3	ARSINE	77.945	gas	68.90	70.14	-0.0084
19	AsI3	ARSENIC TRIIODIDE	455.635	solid	-59.40	-59.40	0.0040
20	As2O3	ARSENIC TRIOXIDE	197.841	solid	-569.02	-565.31	-0.2372
21	At	ASTATINE	210.000	solid	0.00	0.00	0.0000
22	Au	GOLD	196.967	solid	0.00	0.00	0.0000
23	B	BORON	10.811	solid	0.00	0.00	0.0000
24	BBr3	BORON TRIBROMIDE	250.523	gas	-232.50	-234.98	0.0902
25	BCl3	BORON TRICHLORIDE	117.169	gas	-388.70	-387.46	-0.0507
26	BF3	BORON TRIFLUORIDE	67.806	gas	-1119.40	-1118.16	-0.0557
27	BH2CO	BORINE CARBONYL	40.837	gas	--------	--------	--------
28	BH3O3	BORIC ACID	61.833	solid	-968.90	-961.46	-0.4206
29	B2D6	DEUTERODIBORANE	33.718	gas	--------	--------	--------
30	B2H5Br	DIBORANE HYDROBROMIDE	106.566	gas	--------	--------	--------
31	B2H6	DIBORANE	27.670	gas	86.70	91.66	-0.1714
32	B3N3H6	BORINE TRIAMINE	80.501	liquid	-392.70	-381.55	-0.4974
33	B4H10	TETRABORANE	53.323	gas	--------	--------	--------
34	B5H9	PENTABORANE	63.126	gas	175.00	183.68	-0.3414
35	B5H11	TETRAHYDROPENTABORANE	65.142	gas	--------	--------	--------
36	B10H14	DECABORANE	122.221	gas	--------	--------	--------
37	Ba	BARIUM	137.327	solid	0.00	0.00	0.0000
38	Be	BERYLLIUM	9.012	solid	0.00	0.00	0.0000
39	BeB2H8	BERYLLIUM BOROHYDRIDE	38.698	gas	--------	--------	--------
40	BeBr2	BERYLLIUM BROMIDE	168.820	solid	--------	-353.50	--------
41	BeCl2	BERYLLIUM CHLORIDE	79.918	solid	-445.60	-443.12	-0.1503
42	BeF2	BERYLLIUM FLUORIDE	47.009	solid	-979.40	-976.92	-0.1590
43	BeI2	BERYLLIUM IODIDE	262.821	solid	--------	--------	--------
44	Bi	BISMUTH	208.980	solid	0.00	0.00	0.0000
45	BiBr3	BISMUTH TRIBROMIDE	448.692	solid	--------	--------	--------
46	BiCl3	BISMUTH TRICHLORIDE	315.338	solid	-315.00	-311.28	-0.2150
47	BrF5	BROMINE PENTAFLUORIDE	174.896	gas	-350.60	-346.88	-0.2626
48	Br2	BROMINE	159.808	gas	3.10	0.62	0.0932
49	C	CARBON	12.011	solid	0.00	0.00	0.0000
50	CCl2O	PHOSGENE	98.916	gas	-204.90	-203.66	-0.0476
51	CF2O	CARBONYL FLUORIDE	66.007	gas	-623.36	-622.12	-0.0521
52	CH4N2O	UREA	60.056	solid	--------	--------	--------
53	CH4N2S	THIOUREA	76.122	gas	--------	--------	--------
54	CNBr	CYANOGEN BROMIDE	105.922	gas	165.30	164.06	0.0701
55	CNCl	CYANOGEN CHLORIDE	61.470	gas	131.00	131.00	0.0235
56	CNF	CYANOGEN FLUORIDE	45.016	gas	-19.33	-19.33	0.0227
57	CO	CARBON MONOXIDE	28.010	gas	-137.20	-138.44	0.0896
58	COS	CARBONYL SULFIDE	60.076	gas	-169.20	-170.44	0.0912
59	COSe	CARBON OXYSELENIDE	106.970	gas	--------	--------	--------
60	CO2	CARBON DIOXIDE	44.010	gas	-394.40	-394.40	0.0030
61	CS2	CARBON DISULFIDE	76.143	gas	67.10	64.62	0.1660
62	CSeS	CARBON SELENOSULFIDE	123.037	gas	--------	--------	--------
63	C2N2	CYANOGEN	52.035	gas	--------	--------	--------
64	C3S2	CARBON SUBSULFIDE	100.165	gas	--------	--------	--------
65	Ca	CALCIUM	40.078	solid	0.00	0.00	0.0000
66	CaF2	CALCIUM FLUORIDE	78.075	solid	-1175.60	-1173.12	-0.1758
67	CbF5	COLUMBIUM FLUORIDE	187.898	solid	-1699.62	-1693.43	-0.3828
68	Cd	CADMIUM	112.411	solid	0.00	0.00	0.0000
69	CdCl2	CADMIUM CHLORIDE	183.316	solid	-343.90	-341.42	-0.1597
70	CdF2	CADMIUM FLUORIDE	150.408	solid	-647.70	-645.22	-0.1768
71	CdI2	CADMIUM IODIDE	366.220	solid	-201.40	-201.40	-0.0064
72	CdO	CADMIUM OXIDE	128.410	solid	-228.70	-227.46	-0.0996
73	ClF	CHLORINE MONOFLUORIDE	54.451	gas	-51.80	-51.80	0.0050
74	ClFO3	PERCHLORYL FLUORIDE	102.449	gas	48.20	51.92	-0.2415
75	ClF3	CHLORINE TRIFLUORIDE	92.448	gas	-123.00	-120.52	-0.1348
76	ClF5	CHLORINE PENTAFLUORIDE	130.445	gas	--------	--------	--------
77	ClHO3S	CHLOROSULFONIC ACID	116.525	gas	--------	--------	--------
78	ClHO4	PERCHLORIC ACID	100.458	gas	91.87	96.83	-0.2906

Table 13-2 GIBBS AND HELMHOLTZ ENERGY OF FORMATION - INORGANIC COMPOUNDS (continued)

NO	FORMULA	NAME	Molecular Weight g/mol	State	Gf @ 298 K kjoule/mol	Af @ 298 K kjoule/mol	Sf @ 298 K joule/(mol K)
79	ClO2	CHLORINE DIOXIDE	67.452	gas	120.50	121.74	-0.0604
80	Cl2	CHLORINE	70.905	gas	0.00	0.00	0.0000
81	Cl2O	CHLORINE MONOXIDE	86.905	gas	97.90	99.14	-0.0590
82	Cl2O7	CHLORINE HEPTOXIDE	182.901	gas	399.15	407.83	-0.3764
83	Co	COBALT	58.933	solid	0.00	0.00	0.0000
84	CoCl2	COBALT CHLORIDE	129.839	solid	-269.80	-267.32	-0.1432
85	CoNC3O4	COBALT NITROSYL TRICARBONYL	172.971	gas	-------	-------	-------
86	Cr	CHROMIUM	51.996	solid	0.00	0.00	0.0000
87	CrC6O6	CHROMIUM CARBONYL	220.059	solid	-981.98	-974.55	-0.3141
88	CrO2Cl2	CHROMIUM OXYCHLORIDE	154.900	gas	-501.60	-499.12	-0.1224
89	Cs	CESIUM	132.905	solid	0.00	0.00	0.0000
90	CsBr	CESIUM BROMIDE	212.809	solid	-391.40	-391.39	-0.0483
91	CsCl	CESIUM CHLORIDE	168.358	solid	-414.50	-413.26	-0.0956
92	CsF	CESIUM FLUORIDE	151.904	solid	-525.50	-524.26	-0.0939
93	CsI	CESIUM IODIDE	259.810	solid	-340.60	-340.60	-0.0201
94	Cu	COPPER	63.546	solid	0.00	0.00	0.0000
95	CuBr	CUPROUS BROMIDE	143.450	solid	-100.80	-100.80	-0.0128
96	CuCl	CUPROUS CHLORIDE	98.999	solid	-119.90	-118.66	-0.0580
97	CuCl2	CUPRIC CHLORIDE	134.451	solid	-175.70	-173.22	-0.1489
98	CuI	COPPER IODIDE	190.450	solid	-69.50	-69.50	0.0057
99	DCN	DEUTERIUM CYANIDE	28.034	gas	-------	-------	-------
100	D2	DEUTERIUM	4.032	gas	0.00	0.00	0.0000
101	D2O	DEUTERIUM OXIDE	20.031	gas	-234.59	-233.35	-0.0490
102	Eu	EUROPIUM	151.965	solid	0.00	0.00	0.0000
103	F2	FLUORINE	37.997	gas	0.00	0.00	0.0000
104	F2O	FLUORINE OXIDE	53.996	gas	41.90	43.14	-0.0577
105	Fe	IRON	55.847	solid	0.00	0.00	0.0000
106	FeC5O5	IRON PENTACARBONYL	195.899	gas	-704.17	-700.45	-0.1460
107	FeCl2	FERROUS CHLORIDE	126.752	solid	-302.30	-299.82	-0.1325
108	FeCl3	FERRIC CHLORIDE	162.205	solid	-334.00	-330.28	-0.2197
109	Fr	FRANCIUM	223.000	solid	0.00	0.00	0.0000
110	Ga	GALLIUM	69.723	solid	0.00	0.00	0.0000
111	GaCl3	GALLIUM TRICHLORIDE	176.081	solid	-454.80	-451.08	-0.2345
112	Gd	GADOLINIUM	157.250	solid	0.00	0.00	0.0000
113	Ge	GERMANIUM	72.610	solid	0.00	0.00	0.0000
114	GeBr4	GERMANIUM BROMIDE	392.226	gas	-318.00	-320.48	0.0604
115	GeCl4	GERMANIUM CHLORIDE	214.421	gas	-457.30	-454.82	-0.1291
116	GeHCl3	TRICHLORO GERMANE	179.976	gas	-------	-------	-------
117	GeH4	GERMANE	76.642	gas	113.40	115.88	-0.0758
118	Ge2H6	DIGERMANE	151.268	gas	-------	-------	-------
119	Ge3H8	TRIGERMANE	225.894	gas	-------	-------	-------
120	HBr	HYDROGEN BROMIDE	80.912	gas	-53.40	-54.64	0.0574
121	HCN	HYDROGEN CYANIDE	27.026	gas	124.70	124.70	0.0349
122	HCl	HYDROGEN CHLORIDE	36.461	gas	-95.30	-95.30	0.0101
123	HF	HYDROGEN FLUORIDE	20.006	gas	-275.40	-275.40	0.0070
124	HI	HYDROGEN IODIDE	127.912	gas	1.70	0.46	0.0832
125	HNO3	NITRIC ACID	63.013	gas	-74.70	-70.98	-0.2026
126	H2	HYDROGEN	2.016	gas	0.00	0.00	0.0000
127	H2O	WATER	18.015	gas	-228.60	-227.36	-0.0443
128	H2O2	HYDROGEN PEROXIDE	34.015	gas	-105.60	-103.12	-0.1030
129	H2S	HYDROGEN SULFIDE	34.082	gas	-33.40	-33.40	0.0429
130	H2SO4	SULFURIC ACID	98.079	gas	-653.47	-648.51	-0.2739
131	H2S2	HYDROGEN DISULFIDE	66.148	gas	-------	-------	-------
132	H2Se	HYDROGEN SELENIDE	80.976	gas	15.90	15.90	0.0463
133	H2Te	HYDROGEN TELLURIDE	129.616	gas	-------	-------	-------
134	H3NO3S	SULFAMIC ACID	97.095	solid	-------	-------	-------
135	He	HELIUM-3	3.016	gas	0.00	0.00	0.0000
136	He	HELIUM-4	4.003	gas	0.00	0.00	0.0000
137	Hf	HAFNIUM	178.490	solid	0.00	0.00	0.0000
138	Hg	MERCURY	200.590	liquid	0.00	0.00	0.0000
139	HgBr2	MERCURIC BROMIDE	360.398	solid	-153.10	-153.10	-0.0590
140	HgCl2	MERCURIC CHLORIDE	271.495	solid	-178.60	-176.12	-0.1533
141	HgI2	MERCURIC IODIDE	454.399	solid	-101.70	-101.70	-0.0124
142	IF7	IODINE HEPTAFLUORIDE	259.893	gas	-831.48	-825.28	-0.4178
143	I2	IODINE	253.809	solid	0.00	0.00	0.0000
144	In	INDIUM	114.818	solid	0.00	0.00	0.0000
145	Ir	IRIDIUM	192.220	solid	0.00	0.00	0.0000
146	K	POTASSIUM	39.098	solid	0.00	0.00	0.0000
147	KBr	POTASSIUM BROMIDE	119.002	solid	-380.70	-380.70	-0.0439
148	KCl	POTASSIUM CHLORIDE	74.551	solid	-408.53	-407.26	-0.0939
149	KF	POTASSIUM FLUORIDE	58.097	solid	-537.80	-536.56	-0.0989
150	KI	POTASSIUM IODIDE	166.003	solid	-324.90	-324.90	-0.0101
151	KOH	POTASSIUM HYDROXIDE	56.106	gas	-233.76	-233.76	0.0038
152	Kr	KRYPTON	83.800	gas	0.00	0.00	0.0000
153	La	LANTHANUM	138.906	solid	0.00	0.00	0.0000
154	Li	LITHIUM	6.941	solid	0.00	0.00	0.0000
155	LiBr	LITHIUM BROMIDE	86.845	solid	-342.00	-342.00	-0.0309
156	LiCl	LITHIUM CHLORIDE	42.394	solid	-384.40	-383.16	-0.0812

336

Table 13-2 GIBBS AND HELMHOLTZ ENERGY OF FORMATION - INORGANIC COMPOUNDS (continued)

NO	FORMULA	NAME	Molecular Weight g/mol	State	Gf @ 298 K kjoule/mol	Af @ 298 K kjoule/mol	Sf @ 298 K joule/(mol K)
157	LiF	LITHIUM FLUORIDE	25.939	solid	-587.70	-586.46	-0.0949
158	LiI	LITHIUM IODIDE	133.845	solid	-270.30	-270.30	-0.0003
159	Lu	LUTECIUM	174.967	solid	0.00	0.00	0.0000
160	Mg	MAGNESIUM	24.305	solid	0.00	0.00	0.0000
161	MgCl2	MAGNESIUM CHLORIDE	95.210	solid	-591.80	-589.32	-0.1660
162	MgO	MAGNESIUM OXIDE	40.304	solid	-569.30	-568.06	-0.1083
163	Mn	MANGANESE	54.938	solid	0.00	0.00	0.0000
164	MnCl2	MANGANESE CHLORIDE	125.843	solid	-440.50	-438.02	-0.1368
165	Mo	MOLYBDENUM	95.940	solid	0.00	0.00	0.0000
166	MoF6	MOLYBDENUM FLUORIDE	209.930	gas	-1472.20	-1467.24	-0.2868
167	MoO3	MOLYBDENUM OXIDE	143.938	solid	-668.00	-664.28	-0.2586
168	NCl3	NITROGEN TRICHLORIDE	120.365	liquid	--------	--------	--------
169	ND3	HEAVY AMMONIA	20.055	gas	--------	--------	--------
170	NF3	NITROGEN TRIFLUORIDE	71.002	gas	-90.60	-88.12	-0.1392
171	NH3	AMMONIA	17.031	gas	-16.40	-13.92	-0.0989
172	NH3O	HYDROXYLAMINE	33.030	solid	--------	--------	--------
173	NH4Br	AMMONIUM BROMIDE	97.943	solid	-175.20	-169.00	-0.3206
174	NH4Cl	AMMONIUM CHLORIDE	53.491	solid	-202.90	-195.46	-0.3740
175	NH4I	AMMONIUM IODIDE	144.943	solid	-112.50	-106.30	-0.2982
176	NH5O	AMMONIUM HYDROXIDE	35.046	liquid	-254.00	-245.32	-0.3596
177	NH5S	AMMONIUM HYDROGENSULFIDE	51.112	gas	--------	--------	--------
178	NO	NITRIC OXIDE	30.006	gas	86.60	86.60	0.0124
179	NOCl	NITROSYL CHLORIDE	65.459	gas	66.10	67.34	-0.0483
180	NOF	NITROSYL FLUORIDE	49.005	gas	-51.00	-49.76	-0.0520
181	NO2	NITROGEN DIOXIDE	46.006	gas	51.30	52.54	-0.0607
182	N2	NITROGEN	28.013	gas	0.00	0.00	0.0000
183	N2F4	TETRAFLUOROHYDRAZINE	104.007	gas	79.90	84.86	-0.2962
184	N2H4	HYDRAZINE	32.045	gas	159.40	164.36	-0.2147
185	N2H4C	AMMONIUM CYANIDE	44.056	solid	--------	--------	--------
186	N2H6CO2	AMMONIUM CARBAMATE	78.071	solid	-448.06	-435.67	-0.6607
187	N2O	NITROUS OXIDE	44.013	gas	104.20	105.44	-0.0741
188	N2O3	NITROGEN TRIOXIDE	76.012	gas	139.50	143.22	-0.1872
189	N2O4	NITROGEN TETRAOXIDE	92.011	gas	97.90	102.86	-0.2975
190	N2O5	NITROGEN PENTOXIDE	108.010	gas	115.10	121.30	-0.3482
191	Na	SODIUM	22.990	solid	0.00	0.00	0.0000
192	NaBr	SODIUM BROMIDE	102.894	solid	-349.00	-349.00	-0.0406
193	NaCN	SODIUM CYANIDE	49.008	solid	-76.40	-75.16	-0.0372
194	NaCl	SODIUM CHLORIDE	58.442	solid	-384.10	-382.86	-0.0909
195	NaF	SODIUM FLUORIDE	41.988	solid	-546.30	-545.06	-0.1016
196	NaI	SODIUM IODIDE	149.894	solid	-286.10	-286.10	-0.0057
197	NaOH	SODIUM HYDROXIDE	39.997	solid	-379.50	-377.02	-0.1546
198	Na2SO4	SODIUM SULFATE	142.043	solid	-1270.20	-1265.24	-0.3921
199	Nb	NIOBIUM	92.906	solid	0.00	0.00	0.0000
200	Nd	NEODYMIUM	144.240	solid	0.00	0.00	0.0000
201	Ne	NEON	20.180	gas	0.00	0.00	0.0000
202	Ni	NICKEL	58.693	solid	0.00	0.00	0.0000
203	NiC4O4	NICKEL CARBONYL	170.735	gas	-567.35	-564.87	1.1648
204	NiF2	NICKEL FLUORIDE	96.690	solid	-604.10	-601.62	-0.1586
205	Np	NEPTUNIUM	237.000	solid	0.00	0.00	0.0000
206	O2	OXYGEN	31.999	gas	0.00	0.00	0.0000
207	O3	OZONE	47.998	gas	163.20	164.44	-0.0688
208	Os	OSMIUM	190.230	solid	0.00	0.00	0.0000
209	OsOF5	OSMIUM OXIDE PENTAFLUORIDE	301.221	gas	--------	--------	--------
210	OsO4	OSMIUM TETROXIDE - YELLOW	254.228	gas	-292.80	-290.32	-0.1489
211	OsO4	OSMIUM TETROXIDE - WHITE	254.228	gas	-292.80	-290.32	-0.1489
212	P	PHOSPHORUS - WHITE	30.974	solid	0.00	0.00	0.0000
213	PBr3	PHOSPHORUS TRIBROMIDE	270.686	gas	-162.80	-165.28	0.0788
214	PCl2F3	PHOSPHORUS DICHLORIDE TRIFLUORIDE	158.874	gas	--------	--------	--------
215	PCl3	PHOSPHORUS TRICHLORIDE	137.332	gas	-267.80	-266.56	-0.0644
216	PCl5	PHOSPHORUS PENTACHLORIDE	208.237	gas	-305.00	-301.28	-0.2345
217	PH3	PHOSPHINE	33.998	gas	13.40	14.64	-0.0268
218	PH4Br	PHOSPHONIUM BROMIDE	114.910	solid	-47.70	-42.74	-0.2680
219	PH4Cl	PHOSPHONIUM CHLORIDE	70.458	solid	-70.71	-64.51	-0.2498
220	PH4I	PHOSPHONIUM IODIDE	161.910	solid	0.84	5.79	-0.2372
221	POCl3	PHOSPHORUS OXYCHLORIDE	153.331	gas	-512.90	-510.42	-0.1529
222	PSBr3	PHOSPHORUS THIOBROMIDE	302.752	gas	-290.32	-292.80	0.0897
223	PSCl3	PHOSPHORUS THIOCHLORIDE	169.398	gas	-347.69	-346.45	-0.0519
224	P4O6	PHOSPHORUS TRIOXIDE	219.891	gas	-2036.90	-2031.94	-0.3608
225	P4O10	PHOSPHORUS PENTOXIDE	283.889	gas	-2621.80	-2611.88	-0.7117
226	P4S10	PHOSPHORUS PENTASULFIDE	444.555	solid	-278.00	-278.00	-0.1040
227	Pb	LEAD	207.200	solid	0.00	0.00	0.0000
228	PbBr2	LEAD BROMIDE	367.008	solid	-261.90	-261.90	-0.0564
229	PbCl2	LEAD CHLORIDE	278.105	solid	-314.10	-311.62	-0.1519
230	PbF2	LEAD FLUORIDE	245.197	solid	-617.10	-614.62	-0.1573
231	PbI2	LEAD IODIDE	461.009	solid	-173.60	-173.60	-0.0064
232	PbO	LEAD OXIDE	223.199	solid	-187.90	-186.66	-0.0986
233	PbS	LEAD SULFIDE	239.266	solid	-98.70	-98.70	-0.0057
234	Pd	PALLADIUM	106.420	solid	0.00	0.00	0.0000

337

NO	FORMULA	NAME	Molecular Weight g/mol	State	Gf @ 298 K kjoule/mol	Af @ 298 K kjoule/mol	Sf @ 298 K joule/(mol K)
235	Po	POLONIUM	209.000	solid	0.00	0.00	0.0000
236	Pt	PLATINUM	195.080	solid	0.00	0.00	0.0000
237	Ra	RADIUM	226.000	solid	0.00	0.00	0.0000
238	Rb	RUBIDIUM	85.468	solid	0.00	0.00	0.0000
239	RbBr	RUBIDIUM BROMIDE	165.372	solid	-381.80	-381.80	-0.0429
240	RbCl	RUBIDIUM CHLORIDE	120.921	solid	-407.80	-406.56	-0.0926
241	RbF	RUBIDIUM FLUORIDE	104.466	solid	-------	-------	-------
242	RbI	RUBIDIUM IODIDE	212.372	solid	-328.90	-328.90	-0.0164
243	Re	RHENIUM	186.207	solid	0.00	0.00	0.0000
244	Re2O7	RHENIUM HEPTOXIDE	484.410	solid	-1066.00	-1057.32	-0.5839
245	Rh	RHODIUM	102.906	solid	0.00	0.00	0.0000
246	Rn	RADON	222.000	gas	0.00	0.00	0.0000
247	Ru	RUTHENIUM	101.070	soild	0.00	0.00	0.0000
248	RuF5	RUTHENIUM PENTAFLUORIDE	196.062	solid	-------	-------	-------
249	S	SULFUR	32.066	solid	0.00	0.00	0.0000
250	SF4	SULFUR TETRAFLUORIDE	108.060	gas	-722.00	-719.52	-0.1382
251	SF6	SULFUR HEXAFLUORIDE	146.056	gas	-1116.50	-1111.54	-0.3488
252	SOBr2	THIONYL BROMIDE	207.873	gas	-------	-------	-------
253	SOCl2	THIONYL CHLORIDE	118.971	gas	-198.30	-197.06	-0.0476
254	SOF2	SULFUROUS OXYFLUORIDE	86.062	gas	-696.05	-694.81	-0.0586
255	SO2	SULFUR DIOXIDE	64.065	gas	-300.10	-300.10	0.0111
256	SO2Cl2	SULFURYL CHLORIDE	134.970	gas	-320.00	-317.52	-0.1476
257	SO3	SULFUR TRIOXIDE	80.064	gas	-371.10	-369.86	-0.0825
258	S2Cl2	SULFUR MONOCHLORIDE	135.037	liquid	-------	-------	-------
259	Sb	ANTIMONY	121.757	solid	0.00	0.00	0.0000
260	SbBr3	ANTIMONY TRIBROMIDE	361.469	solid	-239.30	-239.30	-0.0674
261	SbCl3	ANTIMONY TRICHLORIDE	228.115	solid	-323.70	-319.98	-0.1962
262	SbCl5	ANTIMONY PENTACHLORIDE	299.021	gas	-334.34	-330.63	-0.2012
263	SbH3	STIBINE	124.781	gas	147.80	149.04	-0.0091
264	SbI3	ANTIMONY TRIIODIDE	502.470	solid	-------	-------	-------
265	Sb2O3	ANTIMONY TRIOXIDE	291.512	solid	-------	-------	-------
266	Sc	SCANDIUM	44.956	solid	0.00	0.00	0.0000
267	Se	SELENIUM	78.960	solid	0.00	0.00	0.0000
268	SeCl4	SELENIUM TETRACHLORIDE	220.771	solid	-97.49	-92.53	-0.3087
269	SeF6	SELENIUM HEXAFLUORIDE	192.950	gas	-1017.00	-1012.04	-0.3354
270	SeOCl2	SELENIUM OXYCHLORIDE	165.865	gas	-------	-------	-------
271	SeO2	SELENIUM DIOXIDE	110.959	solid	-------	-------	-------
272	Si	SILICON	28.086	soild	0.00	0.00	0.0000
273	SiBrCl2F	BROMODICHLOROFLUOROSILANE	197.893	gas	-------	-------	-------
274	SiBrF3	TRIFLUOROBROMOSILANE	164.985	gas	-------	-------	-------
275	SiBr2ClF	DIBROMOCHLOROFLUOROSILANE	242.345	gas	-------	-------	-------
276	SiClF3	TRIFLUOROCHLOROSILANE	120.533	gas	-1280.54	-1278.06	-0.1255
277	SiCl2F2	DICHLORODIFLUOROSILANE	136.988	gas	-------	-------	-------
278	SiCl3F	TRICHLOROFLUOROSILANE	153.442	gas	-805.69	-803.21	-0.1184
279	SiCl4	SILICON TETRACHLORIDE	169.896	gas	-617.00	-614.52	-0.1342
280	SiF4	SILICON TETRAFLUORIDE	104.079	gas	-1572.80	-1570.32	-0.1415
281	SiHBr3	TRIBROMOSILANE	268.805	gas	-328.50	-329.74	0.0366
282	SiHCl3	TRICHLOROSILANE	135.452	gas	-482.00	-479.52	-0.1040
283	SiHF3	TRIFLUOROSILANE	86.089	gas	-1151.16	-1148.68	-0.1104
284	SiH2Br2	DIBROMOSILANE	189.909	gas	-------	-------	-------
285	SiH2Cl2	DICHLOROSILANE	101.007	gas	-294.93	-292.45	-0.0857
286	SiH2F2	DIFLUOROSILANE	68.098	gas	-784.50	-782.02	-0.0912
287	SiH2I2	DIIODOSILANE	283.910	gas	-------	-------	-------
288	SiH3Br	MONOBROMOSILANE	111.013	gas	-------	-------	-------
289	SiH3Cl	MONOCHLOROSILANE	66.562	gas	-148.95	-146.47	-0.0758
290	SiH3F	MONOFLUOROSILANE	50.108	gas	-416.33	-413.85	-0.0771
291	SiH3I	IODOSILANE	158.014	gas	-------	-------	-------
292	SiH4	SILANE	32.117	gas	56.90	59.38	-0.0758
293	SiO2	SILICON DIOXIDE	60.084	soild	-856.30	-853.82	-0.1825
294	Si2Cl6	HEXACHLORODISILANE	268.887	gas	-------	-------	-------
295	Si2F6	HEXAFLUORODISILANE	170.161	gas	-------	-------	-------
296	Si2H5Cl	DISILANYL CHLORIDE	96.663	gas	-------	-------	-------
297	Si2H6	DISILANE	62.219	gas	127.30	132.26	-0.1576
298	Si2OCl3F3	TRICHLOROTRIFLUORODISILOXANE	235.524	gas	-------	-------	-------
299	Si2OCl6	HEXACHLORODISILOXANE	284.887	gas	-------	-------	-------
300	Si2OH6	DISILOXANE	78.218	gas	-------	-------	-------
301	Si3Cl8	OCTACHLOROTRISILANE	367.878	gas	-------	-------	-------
302	Si3H8	TRISILANE	92.320	gas	-------	128.34	-------
303	Si3H9N	TRISILAZANE	107.335	gas	-------	-------	-------
304	Si4H10	TETRASILANE	122.421	liquid	-------	-------	-------
305	Sm	SAMARIUM	150.360	soild	0.00	0.00	0.0000
306	Sn	TIN	118.710	soild	0.00	0.00	0.0000
307	SnBr4	STANNIC BROMIDE	438.326	solid	-350.20	-350.20	-0.0912
308	SnCl2	STANNOUS CHLORIDE	189.615	soild	-------	-------	-------
309	SnCl4	STANNIC CHLORIDE	260.521	gas	-432.20	-429.72	-0.1318
310	SnH4	STANNIC HYDRIDE	122.742	gas	188.30	190.78	-0.0855
311	SnI4	STANNIC IODIDE	626.328	gas	-------	-------	-------
312	Sr	STRONTIUM	87.620	solid	0.00	0.00	0.0000

Table 13-2 GIBBS AND HELMHOLTZ ENERGY OF FORMATION - INORGANIC COMPOUNDS (continued)

NO	FORMULA	NAME	Molecular Weight g/mol	State	Gf @ 298 K kjoule/mol	Af @ 298 K kjoule/mol	Sf @ 298 K joule/(mol K)
313	SrO	STRONTIUM OXIDE	103.619	solid	-561.90	-560.66	-0.1010
314	Ta	TANTALUM	180.948	solid	0.00	0.00	0.0000
315	Tc	TECNNETIUM	98.000	solid	0.00	0.00	0.0000
316	Te	TELLURIUM	127.600	solid	0.00	0.00	0.0000
317	TeCl4	TELLURIUM TETRACHLORIDE	269.411	solid	-------	-------	-------
318	TeF6	TELLURIUM HEXAFLUORIDE	241.590	gas	-------	-------	-------
319	Ti	TITANIUM	47.880	solid	0.00	0.00	0.0000
320	TiCl4	TITANIUM TETRACHLORIDE	189.691	gas	-726.30	-723.82	-0.1238
321	Tl	THALLIUM	204.383	solid	0.00	0.00	0.0000
322	TlBr	THALLOUS BROMIDE	284.287	solid	-167.40	-167.40	-0.0195
323	TlI	THALLOUS IODIDE	331.288	solid	-125.40	-125.40	0.0054
324	Tm	THULIUM	168.934	solid	0.00	0.00	0.0000
325	U	URANIUM	238.029	solid	0.00	0.00	0.0000
326	UF6	URANIUM FLUORIDE	352.019	gas	-2063.70	-2058.74	-0.2807
327	V	VANADIUM	50.942	solid	0.00	0.00	0.0000
328	VCl4	VANADIUM TETRACHLORIDE	192.752	gas	-492.00	-489.52	-0.1124
329	VOCl3	VANADIUM OXYTRICHLORIDE	173.299	gas	-659.28	-656.80	-0.1218
330	W	TUNGSTEN	183.840	solid	0.00	0.00	0.0000
331	WF6	TUNGSTEN FLUORIDE	297.830	gas	-1632.10	-1627.14	-0.3005
332	Xe	XENON	131.290	gas	0.00	0.00	0.0000
333	Yb	YTTERBIUM	173.040	solid	0.00	0.00	0.0000
334	Yt	YTTRIUM	88.906	solid	0.00	0.00	0.0000
335	Zn	ZINC	65.390	solid	0.00	0.00	0.0000
336	ZnCl2	ZINC CHLORIDE	136.295	solid	-369.40	-366.92	-0.1533
337	ZnF2	ZINC FLUORIDE	103.387	solid	-713.30	-710.82	-0.1714
338	ZnO	ZINC OXIDE	81.389	solid	-320.50	-319.26	-0.1006
339	ZnSO4	ZINC SULFATE	161.454	solid	-871.50	-866.54	-0.3733
340	Zr	ZIRCONIUM	91.224	solid	0.00	0.00	0.0000
341	ZrBr4	ZIRCONIUM BROMIDE	410.840	solid	-------	-------	-------
342	ZrCl4	ZIRCONIUM CHLORIDE	233.035	solid	-889.90	-884.94	-0.3039
343	ZrI4	ZIRCONIUM IODIDE	598.842	solid	-------	-------	-------

Gf - Gibbs energy of formation, kjoule/mol

Af - Helmholtz energy of formation, kjoule/mol

Sf - entropy of formation, joule/(mol K)

Chapter 14

SOLUBILITY PARAMETER, LIQUID VOLUME, AND VAN DER WAALS AREA AND VOLUME

Carl L. Yaws, Xiao M. Wang, and Marco A. Satyro*

Lamar University, Beaumont, Texas / *SEA++ INC., Calgary, Alberta

ABSTRACT

Results for solubility parameter, liquid volume, and Van der Waals area and volume are presented for major organic and inorganic chemicals. The chemical formula and molecular weight are also given. The results are displayed in easy-to-use tabulations that are especially applicable for rapid engineering usage with the personal computer or hand calculator.

INTRODUCTION

Properties such as solubility parameter and liquid volume are useful in the modeling of phase equilibrium in the chemical processing and petroleum refining industries. As an example of such usefulness in vapor-liquid operations, calculation of activity coefficients using regular solution methods for phase equilibrium in distillation, stripping, and absorption requires knowledge of solubility parameter and liquid volume for the species in the mixture.

SOLUBILITY PARAMETER AND LIQUID VOLUME

The results for solubility parameter and liquid volume are given in Tables 14-1 and 14-2 for organic and inorganic compounds. The values for solubility parameter are ascertained from data for the heat of vaporization and liquid volume (molecular weight/density): $[(Hvap - RT)/v]^{0.5}$. For compounds that are liquids at ambient conditions, the values apply at 25 C. For compounds that are gases at ambient conditions, the values apply at the normal boiling point temperature. For compounds that are solids at ambient conditions, the values apply at the melting point temperature. The tabulations are based on data source publications for organics (1-40) and inorganics (1-120). The tabulations are arranged by chemical formula to provide ease of use in quickly locating data.

VAN DER WAALS AREA AND VOLUME

The results for Van der Waals area and volume are also given in Tables 14-1 and 14-2 for organic and inorganic compounds. The tabulations are based on data source publications for organics (1-5) and inorganics (1-5). Van der Waals area and volume involve the surface area and volume of an atom and may be ascertained from bond distances, contact distances, and shapes of atoms. Methods of calculation are described by Bondi (2).

REFERENCES - SOLUBILITY PARAMETER AND LIQUID VOLUME - ORGANIC COMPOUNDS

1-34. See **REFERENCES - ORGANIC COMPOUNDS** in Chapter 5, **ENTHALPY OF VAPORIZATION**
35-40. See **REFERENCES - ORGANIC COMPOUNDS** in Chapter 8, **DENSITY OF LIQUID**

REFERENCES - SOLUBILITY PARAMETER AND LIQUID VOLUME - INORGANIC COMPOUNDS

1-56. See **REFERENCES - INORGANIC COMPOUNDS** in Chapter 5, **ENTHALPY OF VAPORIZATION**
57-120. See **REFERENCES - INORGANIC COMPOUNDS** in Chapter 8, **DENSITY OF LIQUID**

REFERENCES - VAN DER WAALS AREA AND VOLUME - ORGANIC COMPOUNDS

1. Daubert, T. E. and R. P. Danner, DATA COMPILATION OF PROPERTIES OF PURE COMPOUNDS, Parts 1, 2, 3, and 4, Supplements 1 and 2, DIPPR Project, AIChE, New York, NY (1985-1992).
2. Bondi, A., PHYSICAL PROPERTIES OF MOLECULAR CRYSTALS, LIQUIDS AND GLASSES, John Wiley and Sons, Inc., New York, NY (1968).
3. Bondi, A., J. Phys. Chem., 68, 441 (1964).
4. Edward, J. T., J. Chem. Educ., 47, 261 (1970).
5. Vera, J. H., S. G. Sayegh and G. A. Ratcliff, Fluid Phase Equilibria, 1, 113 (1977).

REFERENCES - VAN DER WAALS AREA AND VOLUME - INORGANIC COMPOUNDS

1-5. See above **REFERENCES - VAN DER WAALS AREA AND VOLUME - ORGANIC COMPOUNDS**

Table 14-1 SOLUBILITY PARAMETER, LIQUID VOLUME, AND VAN DER WAALS AREA AND VOLUME - ORGANIC COMPOUNDS

NO	FORMULA	NAME	Molecular Weight, g/mol	Solubility Parameter (joule/cm³)^0.5	Liquid Volume cm³/mol	Van Der Waals Area, cm²/mol	Van Der Waals Volume, cm³/mol
1	CBrClF2	BROMOCHLORODIFLUOROMETHANE	165.365	15.258	86.482	6.200E+09	42.170
2	CBrCl3	BROMOTRICHLOROMETHANE	198.273	18.324	99.431	7.550E+09	54.650
3	CBrF3	BROMOTRIFLUOROMETHANE	148.910	14.514	74.727	7.590E+09	35.330
4	CBr2F2	DIBROMODIFLUOROMETHANE	209.816	16.217	91.991	6.360E+09	43.570
5	CClF3	CHLOROTRIFLUOROMETHANE	104.459	14.292	68.591	5.270E+09	33.570
6	CClN	CYANOGEN CHLORIDE	61.470	21.833	51.356	3.990E+09	26.320
7	CCl2F2	DICHLORODIFLUOROMETHANE	120.913	15.112	81.375	5.940E+09	39.810
8	CCl2O	PHOSGENE	98.916	17.836	70.547	5.200E+09	34.900
9	CCl3F	TRICHLOROFLUOROMETHANE	137.368	15.631	92.851	6.610E+09	46.050
10	CCl4	CARBON TETRACHLORIDE	153.822	17.577	97.149	7.280E+09	52.300
11	CF2O	CARBONYL FLUORIDE	66.007	16.450	61.324	3.800E+09	23.140
12	CF4	CARBON TETRAFLUORIDE	88.005	13.834	54.773	4.600E+09	27.330
13	CHBr3	TRIBROMOMETHANE	252.731	21.726	87.872	6.810E+09	49.980
14	CHClF2	CHLORODIFLUOROMETHANE	86.468	17.270	61.230	4.690E+09	31.020
15	CHCl2F	DICHLOROFLUOROMETHANE	102.923	17.449	73.214	5.270E+09	35.740
16	CHCl3	CHLOROFORM	119.377	19.028	80.662	7.660E+09	43.500
17	CHF3	TRIFLUOROMETHANE	70.014	17.651	48.238	3.870E+09	23.940
18	CHI3	TRIIODOMETHANE	393.732	18.826	123.260	8.010E+09	64.320
19	CHN	HYDROGEN CYANIDE	27.026	24.788	39.770	3.025E+09	18.130
20	CHNS	ISOTHIOCYANIC-ACID	59.086	-------	--------	---------	-------
21	CH2BrCl	BROMOCHLOROMETHANE	129.384	21.424	67.161	5.260E+09	37.070
22	CH2Br2	DIBROMOMETHANE	173.835	22.344	70.041	5.530E+09	39.430
23	CH2ClF	CHLOROFLUOROMETHANE	68.478	20.137	50.674	4.250E+09	27.570
24	CH2Cl2	DICHLOROMETHANE	84.932	20.378	64.428	4.990E+09	34.710
25	CH2F2	DIFLUOROMETHANE	52.024	20.626	43.171	3.550E+09	21.670
26	CH2I2	DIIODOMETHANE	267.836	24.055	81.018	6.430E+09	50.930
27	CH2O	FORMALDEHYDE	30.026	23.937	36.740	3.090E+09	18.600
28	CH2O2	FORMIC ACID	46.026	21.442	37.923	3.634E+09	22.740
29	CH3Br	METHYL BROMIDE	94.939	19.951	55.211	4.200E+09	28.070
30	CH3Cl	METHYL CHLORIDE	50.488	19.714	50.123	3.920E+09	25.290
31	CH3Cl3Si	METHYL TRICHLOROSILANE	149.478	15.227	118.043	8.090E+09	61.680
32	CH3F	METHYL FLUORIDE	34.033	20.373	38.668	3.220E+09	19.390
33	CH3I	METHYL IODIDE	141.939	20.172	62.667	4.600E+09	32.850
34	CH3NO	FORMAMIDE	45.041	39.276	39.891	4.110E+09	25.460
35	CH3NO2	NITROMETHANE	61.040	25.758	54.086	4.690E+09	30.470
36	CH3NO2	METHYL-NITRITE	61.040	-------	--------	---------	-------
37	CH3NO3	METHYL-NITRATE	77.040	-------	--------	---------	-------
38	CH4	METHANE	16.043	11.618	37.832	2.880E+09	17.050
39	CH4Cl2Si	METHYL DICHLOROSILANE	115.034	15.403	104.287	7.080E+09	52.800
40	CH4O	METHANOL	32.042	29.523	40.702	3.580E+09	21.710
41	CH4O3S	METHANESULFONIC ACID	96.107	-------	--------	6.180E+09	42.010
42	CH4S	METHYL MERCAPTAN	48.109	20.262	54.206	4.390E+09	28.480
43	CH5ClSi	METHYL CHLOROSILANE	80.589	15.438	88.606	6.060E+09	43.920
44	CH5N	METHYLAMINE	31.057	23.116	44.713	3.860E+09	24.210
45	CH6Si	METHYL SILANE	46.144	14.263	75.217	5.050E+09	35.040
46	CN4O8	TETRANITROMETHANE	196.033	18.559	120.559	1.020E+10	70.530
47	CO	CARBON MONOXIDE	28.010	6.402	35.456	2.780E+09	16.200
48	COS	CARBONYL SULFIDE	60.076	18.130	50.984	4.053E+09	25.900
49	CO2	CARBON DIOXIDE	44.010	14.564	37.278	3.230E+09	19.700
50	CS2	CARBON DISULFIDE	76.143	20.411	60.639	4.754E+09	31.200
51	C2BrF3	BROMOTRIFLUOROETHYLENE	160.921	15.624	82.979	6.600E+09	41.820
52	C2Br2F4	1,2-DIBROMOTETRAFLUOROETHANE	259.824	14.580	120.152	8.560E+09	58.340
53	C2ClF3	CHLOROTRIFLUOROETHYLENE	116.470	15.604	79.373	6.490E+09	40.260
54	C2ClF5	CHLOROPENTAFLUOROETHANE	154.467	13.185	99.993	7.570E+09	40.260
55	C2Cl2F4	1,2-DICHLOROTETRAFLUOROETHANE	170.921	13.608	112.568	8.240E+09	55.140
56	C2Cl3F3	1,1,2-TRICHLOROTRIFLUOROETHANE	187.375	14.760	119.791	9.580E+09	61.380
57	C2Cl4	TETRACHLOROETHYLENE	165.833	18.956	102.806	8.500E+09	58.980
58	C2Cl4F2	1,1,2,2-TETRACHLORODIFLUOROETHANE	203.830	16.420	124.037	8.910E+09	67.620
59	C2Cl4O	TRICHLOROACETYL CHLORIDE	181.832	18.502	112.738	8.860E+09	63.370
60	C2Cl6	HEXACHLOROETHANE	236.738	17.780	142.922	1.092E+10	80.100
61	C2F4	TETRAFLUOROETHYLENE	100.016	15.261	65.335	5.820E+09	34.020
62	C2F6	HEXAFLUOROETHANE	138.012	12.956	86.369	6.900E+09	42.660
63	C2HBrClF3	HALOTHANE	197.382	15.845	105.623	7.900E+09	54.130
64	C2HClF2	2-CHLORO-1,1-DIFLUOROETHYLENE	98.479	16.460	73.500	5.790E+09	37.100
65	C2HCl3	TRICHLOROETHYLENE	131.388	18.760	90.134	7.130E+09	49.580
66	C2HCl3O	DICHLOROACETYL CHLORIDE	147.387	19.942	97.018	7.610E+09	54.580
67	C2HCl3O	TRICHLOROACETALDEHYDE	147.387	19.153	98.311	7.830E+09	55.190
68	C2HCl5	PENTACHLOROETHANE	202.293	18.983	120.767	9.670E+09	71.310
69	C2HF3	TRIFLUOROETHENE	82.025	15.561	71.464	4.990E+09	30.640
70	C2HF3O2	TRIFLUOROACETIC ACID	114.024	21.621	77.053	6.510E+09	41.070
71	C2HF5	PENTAFLUOROETHANE	120.022	14.668	80.298	6.320E+09	40.110
72	C2H2	ACETYLENE	26.038	18.813	41.795	3.480E+09	23.100
73	C2H2Br4	1,1,2,2-TETRABROMOETHANE	345.654	23.305	118.106	9.460E+09	71.160
74	C2H2Cl2	1,1-DICHLOROETHYLENE	96.943	16.813	86.815	6.110E+09	41.430
75	C2H2Cl2	cis-1,2-DICHLOROETHYLENE	96.943	19.476	76.641	5.760E+09	40.180
76	C2H2Cl2	trans-1,2-DICHLOROETHYLENE	96.943	18.549	77.902	5.760E+09	40.180
77	C2H2Cl2O	CHLOROACETYL CHLORIDE	112.943	22.097	78.784	6.550E+09	45.170
78	C2H2Cl2O	DICHLOROACETALDEHYDE	112.943	21.251	78.819	6.580E+09	46.400

341

NO	FORMULA	NAME	Molecular Weight, g/mol	Solubility Parameter (joule/cm³)⁰·⁵	Liquid Volume cm³/mol	Van Der Waals Area, cm²/mol	Van Der Waals Volume, cm³/mol
79	C2H2Cl2O2	DICHLOROACETIC ACID	128.942	26.761	83.010	7.270E+09	51.000
80	C2H2Cl3F	1,1,1-TRICHLOROFLUOROETHANE	151.394	18.421	96.146	7.960E+09	56.280
81	C2H2Cl4	1,1,1,2-TETRACHLOROETHANE	167.849	18.912	109.378	8.550E+09	61.900
82	C2H2Cl4	1,1,2,2-TETRACHLOROETHANE	167.849	20.485	105.754	8.420E+09	62.520
83	C2H2F2	1,1-DIFLUOROETHYLENE	64.035	15.749	55.194	4.670E+09	28.390
84	C2H2F2	cis-1,2-DIFLUOROETHENE	64.035	18.620	55.397	4.670E+09	28.390
85	C2H2F2	trans-1,2-DIFLUOROETHENE	64.035	18.620	55.397	4.670E+09	28.390
86	C2H2F4	1,1,1,2-TETRAFLUOROETHANE	102.031	16.024	73.675	5.950E+09	37.560
87	C2H2O	KETENE	42.037	18.189	53.011	4.027E+09	25.710
88	C2H2O4	OXALIC ACID	90.036	35.977	61.458	6.120E+09	39.480
89	C2H3Br	VINYL BROMIDE	106.950	17.823	70.242	5.070E+09	35.530
90	C2H3Cl	VINYL CHLORIDE	62.499	17.869	64.715	4.740E+09	32.030
91	C2H3ClF2	1-CHLORO-1,1-DIFLUOROETHANE	100.495	15.558	83.734	6.240E+09	41.240
92	C2H3ClO	ACETYL CHLORIDE	78.498	20.189	71.233	5.520E+09	36.990
93	C2H3ClO	CHLOROACETALDEHYDE	78.498	23.166	65.424	5.520E+09	36.990
94	C2H3ClO2	CHLOROACETIC ACID	94.497	33.113	68.644	6.210E+09	41.370
95	C2H3ClO2	METHYL CHLOROFORMATE	94.497	20.975	77.878	6.120E+09	40.490
96	C2H3Cl3	1,1,1-TRICHLOROETHANE	133.404	17.301	100.280	7.580E+09	53.720
97	C2H3Cl3	1,1,2-TRICHLOROETHANE	133.404	19.819	92.990	7.360E+09	53.110
98	C2H3F	VINYL FLUORIDE	46.044	17.710	53.091	4.040E+09	26.130
99	C2H3F3	1,1,1-TRIFLUOROETHANE	84.041	15.743	70.939	5.570E+09	35.000
100	C2H3N	ACETONITRILE	41.053	24.094	52.677	4.310E+09	28.370
101	C2H3NO	METHYL ISOCYANATE	57.052	20.133	61.624	4.930E+09	32.530
102	C2H4	ETHYLENE	28.054	12.437	49.238	3.720E+09	23.880
103	C2H4Br2	1,1-DIBROMOETHANE	187.862	20.405	91.850	6.870E+09	49.650
104	C2H4Br2	1,2-DIBROMOETHANE	187.862	20.656	86.610	6.860E+09	49.260
105	C2H4Cl2	1,1-DICHLOROETHANE	98.959	18.330	84.721	6.330E+09	44.930
106	C2H4Cl2	1,2-DICHLOROETHANE	98.959	20.250	79.442	6.300E+09	43.700
107	C2H4Cl2O	BIS(CHLOROMETHYL)ETHER	114.959	20.573	87.650	6.900E+09	47.400
108	C2H4F2	1,1-DIFLUOROETHANE	66.051	17.381	65.134	4.890E+09	31.890
109	C2H4F2	1,2-DIFLUOROETHANE	66.051	19.211	65.028	4.900E+09	31.900
110	C2H4I2	1,2-DIIODOETHANE	281.863	20.883	100.399	7.660E+09	58.820
111	C2H4O	ACETALDEHYDE	44.053	19.819	56.497	4.490E+09	28.810
112	C2H4O	ETHYLENE OXIDE	44.053	21.624	49.909	3.300E+09	24.160
113	C2H4OS	THIOACETIC-ACID	76.113	22.203	71.885	---------	-------
114	C2H4O2	ACETIC ACID	60.053	18.356	57.580	5.180E+09	33.300
115	C2H4O2	METHYL FORMATE	60.053	20.503	62.084	5.090E+09	32.510
116	C2H4S	THIACYCLOPROPANE	60.114	21.711	59.685	4.000E+09	31.260
117	C2H5Br	BROMOETHANE	108.966	18.475	75.149	5.550E+09	38.300
118	C2H5Cl	ETHYL CHLORIDE	64.514	17.731	71.179	5.270E+09	35.520
119	C2H5ClO	2-CHLOROETHANOL	80.514	25.384	67.311	5.960E+09	40.120
120	C2H5F	ETHYL FLUORIDE	48.060	17.499	58.780	4.570E+09	29.620
121	C2H5I	ETHYL IODIDE	155.966	19.078	81.216	5.950E+09	43.080
122	C2H5N	ETHYLENEIMINE	43.068	24.617	51.796	3.690E+09	28.540
123	C2H5NO	ACETAMIDE	59.068	31.119	58.996	5.460E+09	35.690
124	C2H5NO	N-METHYLFORMAMIDE	59.068	30.520	59.120	5.480E+09	36.450
125	C2H5NO2	NITROETHANE	75.067	22.996	72.001	6.020E+09	40.700
126	C2H5NO3	ETHYL-NITRATE	91.066	-------	--------	---------	-------
127	C2H6	ETHANE	30.070	12.375	55.203	4.240E+09	27.340
128	C2H6AlCl	DIMETHYLALUMINUM CHLORIDE	92.054	20.152	93.180	---------	-------
129	C2H6O	DIMETHYL ETHER	46.069	17.572	63.147	4.840E+09	31.040
130	C2H6O	ETHANOL	46.069	26.421	58.515	4.930E+09	31.940
131	C2H6OS	DIMETHYL SULFOXIDE	78.135	26.338	71.327	6.180E+09	42.880
132	C2H6O2	ETHYLENE GLYCOL	62.068	34.478	55.915	5.620E+09	36.540
133	C2H6O4S	DIMETHYL SULFATE	126.133	23.078	95.419	7.640E+09	62.440
134	C2H6S	DIMETHYL SULFIDE	62.136	18.513	73.091	5.540E+09	38.140
135	C2H6S	ETHYL MERCAPTAN	62.136	18.270	74.613	5.750E+09	38.710
136	C2H6S2	DIMETHYL DISULFIDE	94.202	20.123	89.118	7.250E+09	50.040
137	C2H7N	DIMETHYLAMINE	45.084	19.011	67.255	5.230E+09	35.420
138	C2H7N	ETHYLAMINE	45.084	19.493	65.634	5.210E+09	34.440
139	C2H7NO	MONOETHANOLAMINE	61.084	31.191	60.264	5.740E+09	39.040
140	C2H8N2	ETHYLENEDIAMINE	60.099	25.261	67.293	9.660E+09	41.540
141	C2H8Si	DIMETHYL SILANE	60.171	13.791	94.453	6.320E+09	44.780
142	C2N2	CYANOGEN	52.036	19.750	54.655	4.380E+09	29.400
143	C3F6	HEXAFLUOROPROPYLENE	150.023	13.358	97.373	8.120E+09	49.350
144	C3F6O	HEXAFLUOROACETONE	166.023	13.496	108.157	8.500E+09	54.360
145	C3F8	OCTAFLUOROPROPANE	188.020	12.300	117.172	9.200E+09	57.990
146	C3H2N2	MALONONITRILE	66.062	28.435	62.859	5.730E+09	39.630
147	C3H3Cl	PROPARGYL CHLORIDE	74.510	19.317	72.769	5.870E+09	41.450
148	C3H3N	ACRYLONITRILE	53.064	21.555	66.237	5.130E+09	35.100
149	C3H3NO	OXAZOLE	69.063	17.947	96.175	4.070E+09	34.330
150	C3H4	METHYLACETYLENE	40.065	18.393	59.616	4.840E+09	33.270
151	C3H4	PROPADIENE	40.065	17.551	60.429	4.706E+09	30.840
152	C3H4Cl2	2,3-DICHLOROPROPENE	110.970	18.554	92.403	7.440E+09	51.040
153	C3H4O	ACROLEIN	56.064	20.345	67.190	6.390E+09	44.020
154	C3H4O	PROPARGYL ALCOHOL	56.064	28.134	59.326	5.530E+09	37.870
155	C3H4O2	ACRYLIC ACID	72.064	26.229	68.927	6.000E+09	39.930
156	C3H4O2	beta-PROPIOLACTONE	72.064	29.036	57.119	4.900E+09	35.660

342

NO	FORMULA	NAME	Molecular Weight, g/mol	Solubility Parameter (joule/cm^3)$^{0.5}$	Liquid Volume cm^3/mol	Van Der Waals Area, cm^2/mol	Van Der Waals Volume, cm^3/mol
157	C3H4O2	VINYL FORMATE	72.064	20.055	75.549	5.900E+09	39.250
158	C3H4O3	ETHYLENE CARBONATE	88.063	30.035	65.957	7.530E+09	38.220
159	C3H4O3	PYRUVIC ACID	88.063	26.653	69.629	6.780E+09	45.110
160	C3H5Br	3-BROMO-1-PROPENE	120.977	18.868	87.067	6.370E+09	45.040
161	C3H5Cl	2-CHLOROPROPENE	76.525	16.021	85.229	6.410E+09	42.860
162	C3H5Cl	3-CHLOROPROPENE	76.525	17.321	82.216	6.090E+09	42.260
163	C3H5ClO	alpha-EPICHLOROHYDRIN	92.525	22.442	78.779	5.670E+09	42.560
164	C3H5ClO2	METHYL CHLOROACETATE	108.524	22.348	88.338	7.470E+09	50.720
165	C3H5ClO2	ETHYL CHLOROFORMATE	108.524	22.215	96.255	7.470E+09	50.720
166	C3H5Cl3	1,2,3-TRICHLOROPROPANE	147.431	20.148	106.522	8.690E+09	62.720
167	C3H5I	3-IODO-1-PROPENE	167.977	16.756	91.342	6.770E+09	49.820
168	C3H5N	PROPIONITRILE	55.079	21.866	70.857	5.660E+09	38.600
169	C3H5NO	ACRYLAMIDE	71.079	30.333	84.011	6.280E+09	42.430
170	C3H5NO	HYDRACRYLONITRILE	71.079	28.512	68.323	6.350E+09	43.200
171	C3H5NO	LACTONITRILE	71.079	30.174	72.292	6.340E+09	43.190
172	C3H5N3O9	NITROGLYCERINE	227.088	25.034	143.138	1.272E+10	88.740
173	C3H6	CYCLOPROPANE	42.081	17.331	60.207	4.050E+09	30.690
174	C3H6	PROPYLENE	42.081	13.152	68.802	5.060E+09	34.080
175	C3H6Br2	1,2-DIBROMOPROPANE	201.888	20.852	104.903	8.200E+09	60.100
176	C3H6Cl2	1,1-DICHLOROPROPANE	112.986	18.122	100.356	7.680E+09	55.160
177	C3H6Cl2	1,2-DICHLOROPROPANE	112.986	18.395	98.287	7.640E+09	54.540
178	C3H6Cl2	1,3-DICHLOROPROPANE	112.987	19.819	95.653	7.650E+09	53.930
179	C3H6Cl2	2,2-DICHLOROPROPANE	112.986	16.898	102.113	7.680E+09	55.160
180	C3H6I2	1,2-DIIODOPROPANE	295.889	20.860	115.317	9.000E+09	69.660
181	C3H6O	ACETONE	58.08	19.774	73.931	5.840E+09	39.040
182	C3H6O	ALLYL ALCOHOL	58.08	24.795	68.702	5.750E+09	38.680
183	C3H6O	METHYL VINYL ETHER	58.08	17.361	75.432	5.660E+09	37.780
184	C3H6O	n-PROPIONALDEHYDE	58.08	19.372	72.944	5.840E+09	39.040
185	C3H6O	1,2-PROPYLENE OXIDE	58.08	19.110	70.549	4.640E+09	34.400
186	C3H6O	1,3-PROPYLENE OXIDE	58.08	20.480	64.983	4.650E+09	34.390
187	C3H6O2	ETHYL FORMATE	74.079	18.939	80.828	6.440E+09	42.740
188	C3H6O2	METHYL ACETATE	74.079	19.435	79.890	6.440E+09	42.540
189	C3H6O2	PROPIONIC ACID	74.079	19.459	74.963	6.530E+09	43.420
190	C3H6O2S	3-MERCAPTOPROPIONIC ACID	106.145	26.169	87.527	7.900E+09	55.010
191	C3H6O3	LACTIC ACID	90.079	30.903	75.009	7.200E+09	48.010
192	C3H6O3	METHOXYACETIC ACID	90.079	29.227	76.963	7.130E+09	47.120
193	C3H6O3	TRIOXANE	90.079	22.331	76.242	5.850E+09	41.790
194	C3H6S	THIACYCLOBUTANE	74.14	21.276	73.110	5.350E+09	38.700
195	C3H7Br	1-BROMOPROPANE	122.993	18.083	91.430	6.900E+09	48.530
196	C3H7Br	2-BROMOPROPANE	122.993	17.068	95.927	6.890E+09	48.520
197	C3H7Cl	ISOPROPYL CHLORIDE	78.541	16.455	91.835	6.610E+09	45.740
198	C3H7Cl	n-PROPYL CHLORIDE	78.541	17.041	91.706	6.620E+09	45.750
199	C3H7F	1-FLUOROPROPANE	62.087	16.661	74.544	5.920E+09	39.850
200	C3H7F	2-FLUOROPROPANE	62.087	15.887	78.919	5.910E+09	39.840
201	C3H7I	ISOPROPYL IODIDE	169.993	17.784	100.303	7.290E+09	53.300
202	C3H7I	n-PROPYL IODIDE	169.993	18.329	97.758	7.300E+09	53.310
203	C3H7N	ALLYLAMINE	57.095	19.927	75.374	6.030E+09	41.180
204	C3H7N	PROPYLENEIMINE	57.095	20.466	71.207	5.030E+09	38.760
205	C3H7NO	N,N-DIMETHYLFORMAMIDE	73.095	23.965	77.374	6.840E+09	46.810
206	C3H7NO	N-METHYLACETAMIDE	73.095	27.205	76.853	6.830E+09	47.120
207	C3H7NO2	1-NITROPROPANE	89.094	21.289	89.468	7.370E+09	50.930
208	C3H7NO2	2-NITROPROPANE	89.094	20.477	90.592	7.360E+09	50.920
209	C3H7NO3	PROPYL-NITRATE	105.093	--------	--------	--------	-------
210	C3H7NO3	ISOPROPYL-NITRATE	105.093	--------	--------	--------	-------
211	C3H8	PROPANE	44.096	13.091	75.642	5.590E+09	37.570
212	C3H8O	ISOPROPANOL	60.096	23.575	76.784	6.270E+09	42.160
213	C3H8O	METHYL ETHYL ETHER	60.096	17.000	84.121	6.190E+09	41.270
214	C3H8O	n-PROPANOL	60.096	24.557	74.939	6.280E+09	42.170
215	C3H8O2	2-METHOXYETHANOL	76.095	23.204	79.291	6.880E+09	45.870
216	C3H8O2	METHYLAL	76.095	17.189	89.127	6.790E+09	44.970
217	C3H8O2	1,2-PROPYLENE GLYCOL	76.095	29.516	73.694	6.960E+09	46.760
218	C3H8O2	1,3-PROPYLENE GLYCOL	76.095	31.173	72.364	6.970E+09	46.770
219	C3H8O3	GLYCEROL	92.095	34.315	73.289	7.650E+09	51.360
220	C3H8S	n-PROPYLMERCAPTAN	76.163	18.021	91.068	7.100E+09	48.940
221	C3H8S	ISOPROPYL MERCAPTAN	76.163	17.004	94.177	7.100E+09	48.930
222	C3H8S	ETHYL-METHYL-SULFIDE	76.156	17.891	91.583	6.890E+09	48.370
223	C3H9N	n-PROPYLAMINE	59.111	18.513	82.795	6.560E+09	44.670
224	C3H9N	ISOPROPYLAMINE	59.111	17.442	86.429	6.550E+09	44.660
225	C3H9N	TRIMETHYLAMINE	59.111	15.311	90.216	6.590E+09	45.340
226	C3H9NO	1-AMINO-2-PROPANOL	75.111	26.085	78.494	7.240E+09	49.300
227	C3H9NO	3-AMINO-1-PROPANOL	75.111	29.635	77.287	7.250E+09	49.270
228	C3H9NO	METHYLETHANOLAMINE	75.111	25.935	80.447	7.270E+09	50.250
229	C3H9O4P	TRIMETHYL PHOSPHATE	140.076	-------	116.523	9.460E+09	66.810
230	C3H10N2	1,2-PROPANEDIAMINE	74.126	22.432	86.627	7.520E+09	51.760
231	C3H10Si	TRIMETHYL SILANE	74.198	13.009	116.614	7.590E+09	54.530
232	C4Cl4S	TETRACHLOROTHIOPHENE	221.921	21.197	129.740	1.102E+10	79.800
233	C4Cl6	HEXACHLORO-1,3-BUTADIENE	260.76	18.226	167.561	1.336E+10	93.480
234	C4F8	OCTAFLUORO-2-BUTENE	200.031	12.273	129.458	1.032E+10	64.120

343

Table 14-1 SOLUBILITY PARAMETER, LIQUID VOLUME, AND VAN DER WAALS AREA AND VOLUME - ORGANIC COMPOUNDS (continued)

NO	FORMULA	NAME	Molecular Weight, g/mol	Solubility Parameter (joule/cm³)⁰·⁵	Liquid Volume cm³/mol	Van Der Waals Area, cm²/mol	Van Der Waals Volume, cm³/mol
235	C4F8	OCTAFLUOROCYCLOBUTANE	200.031	13.018	123.887	9.200E+09	61.320
236	C4F10	DECAFLUOROBUTANE	238.028	11.869	148.612	1.150E+10	73.320
237	C4H2	BUTADIYNE(BIACETYLENE)	50.06	20.881	69.070	3.840E+09	31.280
238	C4H2O3	MALEIC ANHYDRIDE	98.058	25.874	74.142	5.960E+09	44.040
239	C4H4	VINYLACETYLENE	52.076	17.035	74.015	5.660E+09	40.010
240	C4H4N2	SUCCINONITRILE	80.089	28.750	81.052	7.080E+09	49.860
241	C4H4O	FURAN	68.075	18.541	72.826	4.920E+09	37.580
242	C4H4O2	DIKETENE	84.075	22.508	80.095	6.020E+09	42.380
243	C4H4O3	SUCCINIC ANHYDRIDE	100.074	29.691	64.547	5.900E+09	43.900
244	C4H4O4	FUMARIC ACID	116.073	24.662	108.115	8.280E+09	56.420
245	C4H4O4	MALEIC ACID	116.073	30.504	91.147	8.280E+09	56.420
246	C4H4S	THIOPHENE	84.142	20.220	79.477	5.620E+09	44.680
247	C4H5Cl	CHLOROPRENE	88.536	17.046	93.155	7.230E+09	49.600
248	C4H5N	trans-CROTONITRILE	67.09	21.280	83.157	6.470E+09	45.310
249	C4H5N	cis-CROTONITRILE	67.09	21.088	81.921	6.470E+09	45.310
250	C4H5N	METHACRYLONITRILE	67.09	19.094	84.406	6.780E+09	45.320
251	C4H5N	PYRROLE	67.09	24.909	69.499	5.310E+09	41.960
252	C4H5N	VINYLACETONITRILE	67.09	21.561	80.947	6.480E+09	45.340
253	C4H5NO2	METHYL CYANOACETATE	99.089	25.809	88.588	7.860E+09	53.800
254	C4H6	CYCLOBUTENE	54.091	17.836	73.918	4.000E+09	32.240
255	C4H6	1,2-BUTADIENE	54.092	16.361	81.651	6.060E+09	41.040
256	C4H6	1,3-BUTADIENE	54.092	15.607	83.114	5.880E+09	40.820
257	C4H6	DIMETHYLACETYLENE	54.092	17.464	78.904	6.200E+09	43.440
258	C4H6	ETHYLACETYLENE	54.092	16.554	80.951	6.190E+09	43.500
259	C4H6Cl2	1,3-DICHLORO-trans-2-BUTENE	124.997	19.226	108.409	8.780E+09	61.240
260	C4H6Cl2	1,4-DICHLORO-cis-2-BUTENE	124.997	20.275	105.218	8.460E+09	60.240
261	C4H6Cl2	1,4-DICHLORO-trans-2-BUTENE	124.997	20.412	105.286	8.460E+09	60.640
262	C4H6Cl2	3,4-DICHLORO-1-BUTENE	124.997	18.420	108.919	8.480E+09	61.280
263	C4H6O	trans-CROTONALDEHYDE	70.091	21.168	82.784	6.650E+09	45.750
264	C4H6O	2,5-DIHYDROFURAN	70.091	20.079	74.665	5.460E+09	41.100
265	C4H6O	DIVINYL ETHER	70.091	15.981	95.874	6.480E+09	44.520
266	C4H6O	METHACROLEIN	70.091	18.825	83.420	6.960E+09	45.760
267	C4H6O2	2-BUTYNE-1,4-DIOL	86.09	32.130	83.670	7.580E+09	52.640
268	C4H6O2	gamma-BUTYROLACTONE	86.09	25.593	76.538	6.250E+09	45.890
269	C4H6O2	cis-CROTONIC ACID	86.09	25.807	84.165	7.340E+09	50.130
270	C4H6O2	trans-CROTONIC ACID	86.09	24.975	88.663	7.340E+09	50.350
271	C4H6O2	METHACRYLIC ACID	86.09	20.102	85.107	7.650E+09	50.140
272	C4H6O2	METHYL ACRYLATE	86.09	19.022	90.713	7.260E+09	49.280
273	C4H6O2	VINYL ACETATE	86.09	18.425	92.936	7.260E+09	49.280
274	C4H6O3	ACETIC ANHYDRIDE	102.09	21.802	94.801	8.640E+09	54.400
275	C4H6O4	SUCCINIC ACID	118.089	30.763	88.495	8.820E+09	59.500
276	C4H6O5	DIGLYCOLIC ACID	134.089	31.661	94.915	9.420E+09	63.200
277	C4H6O5	MALIC ACID	134.089	32.459	109.887	1.550E+10	64.530
278	C4H6O6	TARTARIC ACID	150.088	37.552	97.039	1.018E+10	68.680
279	C4H7N	n-BUTYRONITRILE	69.106	20.697	87.900	7.010E+09	48.830
280	C4H7N	ISOBUTYRONITRILE	69.106	20.005	90.267	7.000E+09	48.820
281	C4H7NO	ACETONE CYANOHYDRIN	85.106	22.273	91.736	7.890E+09	53.410
282	C4H7NO	2-METHACRYLAMIDE	85.106	22.762	101.559	8.400E+09	56.320
283	C4H7NO	3-METHOXYPROPIONITRILE	85.106	22.438	92.127	7.610E+09	52.530
284	C4H7NO	2-PYRROLIDONE	85.106	28.792	76.827	6.640E+09	50.470
285	C4H8	1-BUTENE	56.107	13.748	89.621	6.410E+09	44.310
286	C4H8	cis-2-BUTENE	56.107	15.543	87.450	6.400E+09	44.280
287	C4H8	trans-2-BUTENE	56.107	15.188	89.415	6.400E+09	44.280
288	C4H8	CYCLOBUTANE	56.107	16.430	80.043	5.400E+09	40.920
289	C4H8	ISOBUTENE	56.107	14.955	89.423	6.710E+09	44.290
290	C4H8Br2	1,2-DIBROMOBUTANE	215.915	20.118	120.988	9.560E+09	69.720
291	C4H8Br2	2,3-DIBROMOBUTANE	215.915	19.933	121.684	9.550E+09	69.710
292	C4H8Cl2	1,4-DICHLOROBUTANE	127.013	19.570	111.877	9.000E+09	64.160
293	C4H8I2	1,2-DIIODOBUTANE	309.916	18.819	135.942	1.048E+10	79.280
294	C4H8O	n-BUTYRALDEHYDE	72.107	18.933	90.474	7.190E+09	49.270
295	C4H8O	ISOBUTYRALDEHYDE	72.107	18.235	92.014	7.190E+09	49.270
296	C4H8O	1,2-EPOXYBUTANE	72.107	18.120	87.530	5.990E+09	44.610
297	C4H8O	METHYL ETHYL KETONE	72.107	18.796	90.204	7.190E+09	49.270
298	C4H8O	ETHYL VINYL ETHER	72.107	16.125	96.239	7.010E+09	48.010
299	C4H8O	TETRAHYDROFURAN	72.107	19.129	81.942	6.000E+09	44.620
300	C4H8O2	cis-2-BUTENE-1,4-DIOL	88.106	29.446	82.346	7.780E+09	53.480
301	C4H8O2	trans-2-BUTENE-1,4-DIOL	88.106	29.101	82.698	7.780E+09	53.480
302	C4H8O2	ISOBUTYRIC ACID	88.106	23.157	93.122	7.870E+09	53.860
303	C4H8O2	n-BUTYRIC ACID	88.106	20.263	92.457	7.880E+09	53.870
304	C4H8O2	1,4-DIOXANE	88.106	20.163	85.663	5.900E+09	46.620
305	C4H8O2	ETHYL ACETATE	88.106	18.346	98.594	7.790E+09	52.770
306	C4H8O2	METHYL PROPIONATE	88.106	18.627	96.940	7.790E+09	52.770
307	C4H8O2	n-PROPYL FORMATE	88.106	18.340	97.942	7.790E+09	52.970
308	C4H8O2S	SULFOLANE	120.172	26.304	95.280	8.000E+09	61.220
309	C4H8S	TETRAHYDROTHIOPHENE	88.173	20.545	88.437	6.700E+09	51.700
310	C4H9Br	1-BROMOBUTANE	137.019	18.420	108.000	8.250E+09	58.760
311	C4H9Br	2-BROMOBUTANE	137.019	17.514	109.392	8.250E+09	58.900
312	C4H9Cl	n-BUTYL CHLORIDE	92.568	17.297	105.194	7.970E+09	55.980

344

NO	FORMULA	NAME	Molecular Weight, g/mol	Solubility Parameter (joule/cm^3)$^{0.5}$	Liquid Volume cm^3/mol	Van Der Waals Area, cm^2/mol	Van Der Waals Volume, cm^3/mol
313	C4H9Cl	sec-BUTYL CHLORIDE	92.568	16.461	106.633	7.960E+09	55.970
314	C4H9Cl	tert-BUTYL CHLORIDE	92.568	15.297	110.734	8.180E+09	56.580
315	C4H9I	2-IODO-2-METHYLPROPANE	184.02	16.234	119.790	8.290E+09	63.530
316	C4H9N	PYRROLIDINE	71.122	20.702	82.743	6.390E+09	49.000
317	C4H9NO	N,N-DIMETHYLACETAMIDE	87.122	22.353	93.029	8.190E+09	57.040
318	C4H9NO	MORPHOLINE	87.122	22.079	87.481	6.990E+09	52.700
319	C4H9NO2	1-NITROBUTANE	103.121	20.452	106.497	8.720E+09	61.160
320	C4H9NO2	2-NITROBUTANE	103.121	19.763	105.481	8.710E+09	61.150
321	C4H10	n-BUTANE	58.123	14.453	96.553	6.940E+09	47.800
322	C4H10	ISOBUTANE	58.123	14.027	97.704	6.930E+09	47.790
323	C4H10N2	PIPERAZINE	86.137	17.962	129.371	7.380E+09	57.080
324	C4H10O	n-BUTANOL	74.123	23.289	91.943	7.620E+09	52.400
325	C4H10O	sec-BUTANOL	74.123	22.630	92.118	7.620E+09	52.390
326	C4H10O	tert-BUTANOL	74.123	21.492	94.861	7.620E+09	52.380
327	C4H10O	DIETHYL ETHER	74.123	15.532	104.666	7.540E+09	51.500
328	C4H10O	METHYL-PROPYL-ETHER	74.122	16.953	102.547	7.540E+09	51.500
329	C4H10O	METHYL ISOPROPYL ETHER	74.123	15.318	103.871	7.530E+09	51.490
330	C4H10O	ISOBUTANOL	74.123	23.751	93.028	7.620E+09	52.390
331	C4H10O2	1,3-BUTANEDIOL	90.122	26.974	89.913	8.310E+09	56.990
332	C4H10O2	1,4-BUTANEDIOL	90.122	27.958	88.987	8.320E+09	57.000
333	C4H10O2	2,3-BUTANEDIOL	90.122	25.950	90.630	8.300E+09	56.980
334	C4H10O2	t-BUTYL HYDROPEROXIDE	90.122	20.822	101.696	8.420E+09	56.080
335	C4H10O2	1,2-DIMETHOXYETHANE	90.122	17.732	104.207	8.140E+09	55.200
336	C4H10O2	2-ETHOXYETHANOL	90.122	21.541	97.456	8.230E+09	56.100
337	C4H10O3	DIETHYLENE GLYCOL	106.122	27.775	95.268	8.920E+09	60.700
338	C4H10O4S	DIETHYL SULFATE	154.187	20.335	131.602	1.034E+10	82.900
339	C4H10S	n-BUTYL MERCAPTAN	90.189	17.931	107.774	8.454E+09	59.170
340	C4H10S	ISOBUTYL MERCAPTAN	90.189	17.266	108.724	8.453E+09	59.160
341	C4H10S	sec-BUTYL MERCAPTAN	90.189	17.053	109.347	8.450E+09	59.160
342	C4H10S	tert-BUTYL MERCAPTAN	90.189	15.794	113.503	8.450E+09	59.150
343	C4H10S	DIETHYL SULFIDE	90.189	17.507	108.363	8.240E+09	58.600
344	C4H10S	ISOPROPYL-METHYL-SULFIDE	90.183	16.982	109.348	8.230E+09	58.590
345	C4H10S	METHYL-PROPYL-SULFIDE	90.183	17.665	107.706	8.240E+09	58.600
346	C4H10S2	DIETHYL DISULFIDE	122.255	18.583	123.739	9.530E+09	70.490
347	C4H11N	n-BUTYLAMINE	73.138	18.276	98.758	7.910E+09	54.900
348	C4H11N	ISOBUTYLAMINE	73.138	17.662	100.254	7.900E+09	54.890
349	C4H11N	sec-BUTYLAMINE	73.138	17.288	101.575	7.900E+09	54.890
350	C4H11N	tert-BUTYLAMINE	73.138	15.735	106.291	8.100E+09	54.880
351	C4H11N	DIETHYLAMINE	73.138	16.468	104.234	7.930E+09	55.880
352	C4H11NO	DIMETHYLETHANOLAMINE	89.137	22.786	101.094	8.630E+09	60.170
353	C4H11NO2	DIETHANOLAMINE	105.137	29.486	96.253	9.310E+09	65.080
354	C4H11NO2	2-AMINOETHOXYETHANOL	105.137	28.651	100.040	9.200E+09	63.200
355	C4H12N2O	N-AMINOETHYL ETHANOLAMINE	104.152	29.161	101.960	9.590E+09	67.580
356	C4H12Si	TETRAMETHYLSILANE	88.225	12.606	137.700	8.867E+09	64.270
357	C4H13N3	DIETHYLENE TRIAMINE	103.167	22.212	108.088	9.870E+09	70.080
358	C5Cl6	HEXACHLOROCYCLOPENTADIENE	272.771	19.184	160.155	1.386E+10	96.810
359	C5H4O2	FURFURAL	96.086	23.644	83.165	6.120E+09	47.260
360	C5H5N	PYRIDINE	79.101	21.804	80.834	6.650E+09	45.500
361	C5H6	CYCLOPENTADIENE	66.103	16.961	82.925	5.670E+09	43.860
362	C5H6	2-METHYL-1-BUTENE-3-YNE	66.103	15.588	94.567	7.310E+09	50.220
363	C5H6	1-PENTENE-3-YNE	66.103	17.268	90.076	7.020E+09	50.180
364	C5H6	1-PENTENE-4-YNE	66.103	16.443	91.252	7.010E+09	50.240
365	C5H6N2	GLUTARONITRILE	94.116	25.612	95.976	8.430E+09	60.090
366	C5H6O2	FURFURYL ALCOHOL	98.101	25.089	87.064	6.650E+09	51.190
367	C5H6O3	GLUTARIC ANHYDRIDE	114.101	29.331	76.559	7.250E+09	54.100
368	C5H6O4	CITRACONIC ACID	130.1	31.698	94.239	9.930E+09	66.630
369	C5H6O4	ITACONIC ACID	130.1	28.861	102.020	9.940E+09	66.660
370	C5H6S	2-METHYLTHIOPHENE	98.162	19.265	96.853	7.740E+09	58.350
371	C5H6S	3-METHYLTHIOPHENE	98.162	19.423	96.585	7.740E+09	58.350
372	C5H7N	N-METHYLPYRROLE	81.117	20.643	89.818	6.670E+09	51.880
373	C5H7NO2	ETHYL CYANOACETATE	113.116	27.259	106.925	9.210E+09	63.810
374	C5H8	CYCLOPENTENE	68.118	17.156	88.841	6.050E+09	46.810
375	C5H8	ISOPRENE	68.118	15.333	100.775	7.530E+09	51.030
376	C5H8	3-METHYL-1,2-BUTADIENE	68.118	15.860	100.081	7.410E+09	51.250
377	C5H8	2-METHYL-1,3-BUTADIENE	68.118	15.455	100.853	7.410E+09	51.250
378	C5H8	1,2-PENTADIENE	68.118	16.106	99.067	7.410E+09	51.270
379	C5H8	cis-1,3-PENTADIENE	68.118	16.016	99.275	7.220E+09	51.020
380	C5H8	trans-1,3-PENTADIENE	68.118	15.746	101.516	7.220E+09	51.020
381	C5H8	1,4-PENTADIENE	68.118	14.552	104.248	7.230E+09	51.050
382	C5H8	2,3-PENTADIENE	68.118	16.469	98.728	7.400E+09	51.240
383	C5H8	1-PENTYNE	68.118	16.106	98.995	7.540E+09	53.730
384	C5H8	2-PENTYNE	68.118	17.171	96.582	7.530E+09	53.720
385	C5H8	3-METHYL-1-BUTYNE	68.118	15.176	103.209	7.530E+09	53.720
386	C5H8	SPIROPENTANE	68.118	16.422	92.702	--------	-------
387	C5H8N4O12	PENTAERYTHRITOL TETRANITRATE	316.138	17.382	284.671	1.800E+10	126.250
388	C5H8O	CYCLOPENTANONE	84.118	21.103	89.057	7.000E+09	52.620
389	C5H8O	METHYL ISOPROPENYL KETONE	84.118	18.579	99.403	8.310E+09	55.990
390	C5H8O2	ACETYLACETONE	100.117	21.264	103.121	8.790E+09	60.970

NO	FORMULA	NAME	Molecular Weight, g/mol	Solubility Parameter (joule/cm³)⁰·⁵	Liquid Volume cm³/mol	Van Der Waals Area, cm²/mol	Van Der Waals Volume, cm³/mol
391	C5H8O2	ALLYL ACETATE	100.117	17.948	108.543	7.810E+09	59.510
392	C5H8O2	ETHYL ACRYLATE	100.117	18.276	109.087	8.610E+09	59.510
393	C5H8O2	METHYL METHACRYLATE	100.117	18.608	106.829	8.910E+09	59.490
394	C5H8O2	VINYL PROPIONATE	100.117	15.639	110.990	8.610E+09	59.510
395	C5H8O3	2-HYDROXYETHYL ACRYLATE	116.117	23.640	115.213	9.300E+09	64.110
396	C5H8O3	LEVULINIC ACID	116.117	29.493	109.248	9.480E+09	65.600
397	C5H8O3	METHYL ACETOACETATE	116.117	21.680	108.329	9.390E+09	64.470
398	C5H8O4	GLUTARIC ACID	132.116	29.493	109.248	1.017E+10	69.730
399	C5H9N	VALERONITRILE	83.133	19.988	104.649	8.360E+09	59.060
400	C5H9NO	n-BUTYL ISOCYANATE	99.133	18.735	112.992	8.730E+09	61.260
401	C5H9NO	N-METHYL-2-PYRROLIDONE	99.133	23.353	96.699	8.000E+09	60.390
402	C5H9NO4	L-GLUTAMIC ACID	147.131	30.289	113.693	1.113E+10	76.820
403	C5H10	CYCLOPENTANE	70.134	16.682	93.509	6.750E+09	49.690
404	C5H10	2-METHYL-1-BUTENE	70.134	14.695	108.681	8.060E+09	54.520
405	C5H10	2-METHYL-2-BUTENE	70.134	15.247	106.707	8.050E+09	54.490
406	C5H10	3-METHYL-1-BUTENE	70.134	14.010	111.820	7.750E+09	54.530
407	C5H10	1-PENTENE	70.134	14.599	110.474	7.760E+09	54.540
408	C5H10	cis-2-PENTENE	70.134	15.162	107.951	7.750E+09	54.510
409	C5H10	trans-2-PENTENE	70.134	14.973	109.095	7.750E+09	54.510
410	C5H10Br2	2,3-DIBROMO-2-METHYLBUTANE	229.942	17.121	163.088	8.830E+09	79.950
411	C5H10Cl2	1,5-DICHLOROPENTANE	141.04	19.322	128.733	1.035E+10	74.390
412	C5H10O	METHYL ISOPROPYL KETONE	86.134	18.253	106.975	8.530E+09	59.490
413	C5H10O	2-PENTANONE	86.134	18.284	107.397	8.540E+09	59.500
414	C5H10O	DIETHYL KETONE	86.134	18.260	106.401	8.540E+09	59.500
415	C5H10O	VALERALDEHYDE	86.134	18.603	106.975	8.540E+09	59.500
416	C5H10O2	n-BUTYL FORMATE	102.133	17.986	115.158	9.140E+09	63.200
417	C5H10O2	ETHYL PROPIONATE	102.133	17.667	115.480	9.140E+09	63.000
418	C5H10O2	ISOBUTYL FORMATE	102.133	17.824	116.759	9.130E+09	63.190
419	C5H10O2	ISOPROPYL ACETATE	102.133	17.120	117.238	9.130E+09	62.990
420	C5H10O2	n-PROPYL ACETATE	102.133	18.005	115.712	9.140E+09	63.000
421	C5H10O2	METHYL n-BUTYRATE	102.133	17.713	114.383	9.140E+09	63.000
422	C5H10O2	2-METHYLBUTYRIC ACID	102.133	22.615	109.622	9.220E+09	63.870
423	C5H10O2	ISOVALERIC ACID	102.133	20.073	110.254	9.220E+09	63.870
424	C5H10O2	VALERIC ACID	102.133	23.458	109.308	9.220E+09	63.880
425	C5H10O2	TETRAHYDROFURFURYL ALCOHOL	102.133	25.197	97.432	8.030E+09	59.440
426	C5H10O2S	3-METHYL SULFOLANE	134.199	24.080	113.002	9.340E+09	71.440
427	C5H10O3	DIETHYL CARBONATE	118.133	18.708	121.727	9.740E+09	66.700
428	C5H10O3	ETHYL LACTATE	118.133	22.382	114.993	9.820E+09	67.590
429	C5H10S	THIACYCLOHEXANE	102.194	19.422	104.186	8.050E+09	58.150
430	C5H10S	CYCLOPENTANETHIOL	102.194	19.029	106.357	8.050E+09	61.310
431	C5H11Br	1-BROMOPENTANE	151.046	17.729	124.658	9.600E+09	68.990
432	C5H11Cl	1-CHLOROPENTANE	106.595	16.952	121.433	9.320E+09	66.210
433	C5H11Cl	1-CHLORO-3-METHYLBUTANE	106.595	16.526	123.211	9.320E+09	66.210
434	C5H11Cl	2-CHLORO-2-METHYLBUTANE	106.595	15.666	123.966	9.320E+09	66.210
435	C5H11N	N-METHYLPYRROLIDINE	85.149	16.970	105.655	7.543E+09	58.920
436	C5H11N	PIPERIDINE	85.149	18.938	99.269	7.740E+09	59.230
437	C5H11NO	tert-BUTYLFORMAMIDE	101.148	22.008	112.511	5.720E+09	67.560
438	C5H12	ISOPENTANE	72.15	13.858	117.098	8.280E+09	58.020
439	C5H12	NEOPENTANE	72.15	13.062	119.529	8.480E+09	58.010
440	C5H12	n-PENTANE	72.15	14.439	116.126	8.290E+09	58.030
441	C5H12O	2,2-DIMETHYL-1-PROPANOL	88.15	19.265	121.103	9.170E+09	62.610
442	C5H12O	tert-PENTYL-ALCOHOL	88.149	20.164	112.596	8.960E+09	62.610
443	C5H12O	2-METHYL-1-BUTANOL	88.15	22.274	108.266	8.970E+09	62.620
444	C5H12O	2-METHYL-2-BUTANOL	88.15	20.759	109.541	9.170E+09	62.610
445	C5H12O	3-METHYL-1-BUTANOL	88.15	22.322	108.528	8.970E+09	62.620
446	C5H12O	3-METHYL-2-BUTANOL	88.15	21.607	108.313	8.960E+09	62.610
447	C5H12O	1-PENTANOL	88.15	22.576	108.534	8.980E+09	62.630
448	C5H12O	2-PENTANOL	88.15	21.671	109.452	8.970E+09	62.620
449	C5H12O	3-PENTANOL	88.15	21.150	107.732	8.970E+09	62.620
450	C5H12O	METHYL sec-BUTYL ETHER	88.15	15.445	119.531	8.880E+09	61.720
451	C5H12O	METHYL tert-BUTYL ETHER	88.15	15.117	119.887	9.080E+09	61.710
452	C5H120	METHYL ISOBUTYL ETHER	88.15	15.078	121.645	8.880E+09	61.720
453	C5H120	ETHYL PROPYL ETHER	88.15	15.411	121.741	8.890E+09	61.730
454	C5H12O2	ETHYLENE GLYCOL MONOPROPYL ETHER	104.149	20.006	114.958	9.580E+09	66.320
455	C5H12O2	NEOPENTYL GLYCOL	104.149	23.111	129.612	9.860E+09	67.210
456	C5H12O2	1,5-PENTANEDIOL	104.149	26.711	104.809	9.670E+09	67.230
457	C5H12O3	2-(2-METHOXYETHOXY)ETHANOL	120.148	22.823	118.172	1.018E+10	70.030
458	C5H12O4	PENTAERYTHRITOL	136.148	26.651	153.735	1.124E+10	76.410
459	C5H12S	n-PENTYL MERCAPTAN	104.216	17.640	124.405	9.808E+09	69.400
460	C5H12S	BUTYL-METHYL-SULFIDE	104.21	17.387	124.400	9.590E+09	68.830
461	C5H12S	ETHYL-PROPYL-SULFIDE	104.21	17.394	125.228	9.590E+09	68.830
462	C5H12S	2-METHYL-2-BUTANETHIOL	104.21	16.125	126.912	9.590E+09	72.840
463	C5H13N	n-PENTYLAMINE	87.165	17.890	116.054	9.260E+09	65.150
464	C5H13NO2	METHYL DIETHANOLAMINE	119.164	28.858	115.779	1.067E+10	75.000
465	C6Cl6	HEXACHLOROBENZENE	284.782	20.395	164.896	1.272E+10	106.700
466	C6F6	HEXAFLUOROBENZENE	186.056	16.917	115.838	7.860E+09	63.240
467	C6H3ClN2O4	1-CHLORO-2,4-DINITROBENZENE	202.554	24.977	133.452	1.081E+10	86.400
468	C6H3Cl2NO2	1,2-DICHLORO-4-NITROBENZENE	192.001	22.546	128.700	1.007E+10	81.600

NO	FORMULA	NAME	Molecular Weight, g/mol	Solubility Parameter (joule/cm³)^0.5	Liquid Volume cm³/mol	Van Der Waals Area, cm²/mol	Van Der Waals Volume, cm³/mol
469	C6H3Cl3	1,2,4-TRICHLOROBENZENE	181.448	20.618	125.250	1.233E+10	100.980
470	C6H3N3O6	1,3,5-TRINITROBENZENE	213.106	22.862	135.247	1.155E+10	91.200
471	C6H4Br2	m-DIBROMOBENZENE	235.906	20.431	121.166	8.600E+09	70.920
472	C6H4ClNO2	m-CHLORONITROBENZENE	157.556	22.774	116.906	8.780E+09	70.520
473	C6H4ClNO2	o-CHLORONITROBENZENE	157.556	23.326	116.522	8.780E+09	70.520
474	C6H4ClNO2	p-CHLORONITROBENZENE	157.556	22.067	121.431	8.780E+09	70.520
475	C6H4Cl2	m-DICHLOROBENZENE	147.003	19.574	114.533	8.220E+09	67.320
476	C6H4Cl2	o-DICHLOROBENZENE	147.003	20.311	112.969	8.220E+09	67.300
477	C6H4Cl2	p-DICHLOROBENZENE	147.003	19.328	117.406	8.220E+09	67.320
478	C6H4F2	m-DIFLUOROBENZENE	114.094	18.632	98.225	6.820E+09	55.520
479	C6H4F2	o-DIFLUOROBENZENE	114.094	18.413	99.194	6.820E+09	55.520
480	C6H4F2	p-DIFLUOROBENZENE	114.094	18.588	98.208	6.820E+09	55.520
481	C6H4N2O4	m-DINITROBENZENE	168.109	24.782	123.726	9.700E+09	76.920
482	C6H4N2O4	o-DINITROBENZENE	168.109	24.171	132.941	9.700E+09	76.920
483	C6H4N2O4	p-DINITROBENZENE	168.109	21.870	144.433	9.700E+09	76.920
484	C6H5Br	BROMOBENZENE	157.01	19.945	105.567	7.340E+09	60.160
485	C6H5Cl	MONOCHLOROBENZENE	112.558	19.264	102.264	7.140E+09	57.840
486	C6H5ClO	m-CHLOROPHENOL	128.558	22.061	102.372	7.870E+09	63.360
487	C6H5ClO	o-CHLOROPHENOL	128.558	21.391	102.433	7.870E+09	63.360
488	C6H5ClO	p-CHLOROPHENOL	128.558	22.623	100.918	7.870E+09	63.360
489	C6H5Cl2N	3,4-DICHLOROANILINE	162.018	23.031	120.741	9.260E+09	75.340
490	C6H5F	FLUOROBENZENE	96.104	18.441	94.287	6.310E+09	50.840
491	C6H5I	IODOBENZENE	204.01	20.367	111.976	7.720E+09	64.680
492	C6H5NO2	NITROBENZENE	123.111	22.612	102.717	7.760E+09	61.840
493	C6H6	BENZENE	78.114	18.706	89.485	6.000E+09	48.400
494	C6H6ClN	m-CHLOROANILINE	127.573	23.566	105.335	9.090E+09	66.440
495	C6H6ClN	o-CHLOROANILINE	127.573	22.763	125.503	9.090E+09	66.440
496	C6H6ClN	p-CHLOROANILINE	127.573	22.651	108.582	8.160E+09	66.100
497	C6H6N2	cis-DICYANO-1-BUTENE	106.127	25.694	99.931	9.240E+09	66.800
498	C6H6N2	trans-DICYANO-1-BUTENE	106.127	25.464	100.648	9.240E+09	66.800
499	C6H6N2	1,4-DICYANO-2-BUTENE	106.127	25.043	110.963	9.240E+10	66.800
500	C6H6N2O2	m-NITROANILINE	138.126	24.401	133.671	8.710E+09	69.060
501	C6H6N2O2	o-NITROANILINE	138.126	24.309	135.797	8.710E+09	69.060
502	C6H6N2O2	p-NITROANILINE	138.126	25.536	129.745	8.710E+09	69.060
503	C6H6O	PHENOL	94.113	24.102	89.091	6.790E+09	53.800
504	C6H6O2	1,2-BENZENEDIOL	110.112	24.328	95.262	7.520E+09	59.400
505	C6H6O2	1,3-BENZENEDIOL	110.112	28.330	93.575	7.520E+09	59.400
506	C6H6O2	p-HYDROQUINONE	110.112	23.753	122.588	8.460E+09	59.980
507	C6H6O3	1,2,3-BENZENETRIOL	126.112	21.925	150.529	8.010E+09	62.500
508	C6H6S	PHENYL MERCAPTAN	110.18	21.163	102.698	8.650E+09	60.650
509	C6H7N	ANILINE	93.128	24.126	91.505	7.040E+09	56.380
510	C6H7N	2-METHYLPYRIDINE	93.128	20.080	99.073	8.120E+09	56.670
511	C6H7N	3-METHYLPYRIDINE	93.128	20.710	97.826	8.120E+09	56.670
512	C6H7N	4-METHYLPYRIDINE	93.128	20.638	98.079	8.120E+09	56.670
513	C6H8	1,3-CYCLOHEXADIENE	80.13	17.973	95.767	7.020E+09	54.090
514	C6H8	METHYLCYCLOPENTADIENE	80.13	17.400	99.544	7.320E+09	54.320
515	C6H8N2	ADIPONITRILE	108.143	26.317	112.663	9.780E+09	70.320
516	C6H8N2	METHYLGLUTARONITRILE	108.143	24.944	113.838	9.770E+09	70.310
517	C6H8N2	m-PHENYLENEDIAMINE	108.143	25.592	97.724	9.020E+09	64.980
518	C6H8N2	o-PHENYLENEDIAMINE	108.143	24.750	100.072	9.020E+09	64.980
519	C6H8N2	p-PHENYLENEDIAMINE	108.143	24.811	102.198	9.020E+09	64.980
520	C6H8N2	PHENYLHYDRAZINE	108.143	24.303	98.820	8.740E+09	65.980
521	C6H8N2O	BIS(CYANOETHYL)ETHER	124.142	26.663	118.927	1.038E+10	74.020
522	C6H8O4	DIMETHYL MALEATE	144.127	21.417	125.502	1.080E+10	74.680
523	C6H8O6	ASCORBIC ACID	176.126	34.062	124.846	1.175E+10	80.950
524	C6H8O7	CITRIC ACID	192.125	31.847	134.762	1.202E+10	90.390
525	C6H10	1-METHYLCYCLOPENTENE	82.145	16.626	105.897	7.660E+09	58.770
526	C6H10	3-METHYLCYCLOPENTENE	82.145	16.307	108.289	7.660E+09	58.770
527	C6H10	4-METHYLCYCLOPENTENE	82.145	16.552	107.624	7.660E+09	58.770
528	C6H10	CYCLOHEXENE	82.145	17.611	101.890	7.400E+09	57.040
529	C6H10	2,3-DIMETHYL-1,3-BUTADIENE	82.145	15.998	113.650	9.180E+09	61.240
530	C6H10	1,5-HEXADIENE	82.145	15.031	119.365	8.580E+09	61.280
531	C6H10	cis,trans-2,4-HEXADIENE	82.145	16.581	114.309	8.560E+09	61.220
532	C6H10	trans,trans-2,4-HEXADIENE	82.145	16.534	115.653	8.560E+09	61.220
533	C6H10	1-HEXYNE	82.145	16.329	115.436	8.890E+09	63.960
534	C6H10	2-HEXYNE	82.145	16.915	113.022	8.900E+09	63.900
535	C6H10	3-HEXYNE	82.145	16.603	114.372	8.900E+09	63.900
536	C6H10O	CYCLOHEXANONE	98.145	20.143	104.142	8.350E+09	62.850
537	C6H10O	MESITYL OXIDE	98.145	18.737	115.173	9.650E+09	66.190
538	C6H10O2	epsilon-CAPROLACTONE	114.144	21.400	106.955	8.950E+09	66.550
539	C6H10O2	ETHYL METHACRYLATE	114.144	17.354	125.712	1.026E+10	69.720
540	C6H10O2	n-PROPYL ACRYLATE	114.144	17.883	126.897	9.960E+09	69.740
541	C6H10O3	ETHYLACETOACETATE	130.144	19.842	127.254	1.074E+10	74.700
542	C6H10O3	PROPIONIC ANHYDRIDE	130.144	19.520	129.257	1.074E+10	74.900
543	C6H10O4	ADIPIC ACID	146.143	25.226	133.942	1.150E+10	80.400
544	C6H10O4	DIETHYL OXALATE	146.143	20.162	136.204	1.134E+10	78.200
545	C6H10O4	ETHYLENE GLYCOL DIACETATE	146.143	20.532	132.713	1.134E+10	78.200
546	C6H10O4	ETHYLIDENE DIACETATE	146.143	20.143	136.652	1.133E+10	78.190

NO	FORMULA	NAME	Molecular Weight, g/mol	Solubility Parameter (joule/cm³)^0.5	Liquid Volume cm³/mol	Van Der Waals Area, cm²/mol	Van Der Waals Volume, cm³/mol
547	C6H11N	HEXANENITRILE	97.16	19.382	121.268	9.710E+09	69.290
548	C6H11NO	epsilon-CAPROLACTAM	113.159	24.251	110.390	9.340E+09	70.930
549	C6H11NO	CYCLOHEXANONE OXIME	113.159	19.923	134.607	9.808E+09	69.400
550	C6H12	CYCLOHEXANE	84.161	16.700	108.860	8.100E+09	61.400
551	C6H12	2,3-DIMETHYL-1-BUTENE	84.161	14.670	125.033	9.400E+09	64.740
552	C6H12	2,3-DIMETHYL-2-BUTENE	84.161	15.867	119.643	9.700E+09	64.700
553	C6H12	3,3-DIMETHYL-1-BUTENE	84.161	13.703	129.891	9.300E+09	64.750
554	C6H12	2-ETHYL-1-BUTENE	84.161	15.201	122.899	9.410E+09	64.750
555	C6H12	trans-3-METHYL-2-PENTENE	84.161	15.579	121.391	9.410E+09	64.750
556	C6H12	1-HEXENE	84.161	15.047	126.103	9.110E+09	64.770
557	C6H12	cis-2-HEXENE	84.161	15.320	123.312	9.100E+09	64.740
558	C6H12	trans-2-HEXENE	84.161	15.341	125.041	9.100E+09	64.740
559	C6H12	cis-3-HEXENE	84.161	15.242	124.641	9.100E+09	64.740
560	C6H12	trans-3-HEXENE	84.161	15.291	125.146	9.100E+09	64.740
561	C6H12	METHYLCYCLOPENTANE	84.161	16.173	113.042	8.090E+09	60.170
562	C6H12	2-METHYL-1-PENTENE	84.161	15.065	124.725	9.410E+09	64.750
563	C6H12	2-METHYL-2-PENTENE	84.161	15.375	123.525	9.400E+09	64.720
564	C6H12	3-METHYL-1-PENTENE	84.161	14.436	126.982	9.100E+09	64.760
565	C6H12	3-METHYL-cis-2-PENTENE	84.161	15.414	122.227	9.400E+09	64.720
566	C6H12	4-METHYL-1-PENTENE	84.161	14.478	127.690	9.100E+09	64.760
567	C6H12	4-METHYL-cis-2-PENTENE	84.161	14.610	126.648	9.090E+09	64.730
568	C6H12	4-METHYL-trans-2-PENTENE	84.161	14.734	126.739	9.090E+09	64.730
569	C6H12N2	TRIETHYLENEDIAMINE	112.175	16.634	147.764	7.420E+09	67.760
570	C6H12O	BUTYL VINYL ETHER	100.161	16.462	129.403	9.710E+09	68.470
571	C6H12O	CYCLOHEXANOL	100.161	23.672	104.294	8.780E+09	64.840
572	C6H12O	1-HEXANAL	100.161	18.079	123.691	9.890E+09	69.700
573	C6H12O	ETHYL ISOPROPYL KETONE	100.161	17.325	124.313	9.880E+09	69.720
574	C6H12O	2-HEXANONE	100.161	18.137	124.119	9.890E+09	69.730
575	C6H12O	3-HEXANONE	100.161	17.898	123.644	8.540E+09	69.730
576	C6H12O	METHYL ISOBUTYL KETONE	100.161	17.825	125.762	9.880E+09	69.720
577	C6H12O2	n-PENTYL FORMATE	116.16	18.580	131.815	1.049E+10	73.430
578	C6H12O2	n-BUTYL ACETATE	116.16	17.582	132.553	1.049E+10	73.230
579	C6H12O2	sec-BUTYL ACETATE	116.16	16.982	133.841	1.048E+10	73.220
580	C6H12O2	tert-BUTYL ACETATE	116.16	16.040	134.844	1.070E+10	73.200
581	C6H12O2	ETHYL n-BUTYRATE	116.16	16.943	132.890	1.049E+10	73.430
582	C6H12O2	ETHYL ISOBUTYRATE	116.16	16.504	134.609	1.050E+10	73.200
583	C6H12O2	ISOBUTYL ACETATE	116.16	17.168	133.595	1.048E+10	73.220
584	C6H12O2	n-PROPYL PROPIONATE	116.16	17.326	132.479	1.049E+10	73.230
585	C6H12O2	CYCLOHEXYL PEROXIDE	116.16	-------	114.496	8.810E+09	68.530
586	C6H12O2	DIACETONE ALCOHOL	116.16	19.018	124.336	1.077E+10	74.310
587	C6H12O2	2-ETHYL BUTYRIC ACID	116.16	21.734	126.341	1.057E+10	74.320
588	C6H12O2	n-HEXANOIC ACID	116.16	23.447	126.108	1.058E+10	74.330
589	C6H12O3	2-ETHOXYETHYL ACETATE	132.159	18.775	136.199	1.090E+10	76.930
590	C6H12O3	HYDROXYCAPROIC ACID	132.159	29.125	116.525	1.126E+10	78.920
591	C6H12O3	PARALDEHYDE	132.159	17.644	134.142	9.870E+09	72.450
592	C6H12O3	sec-BUTYL GLYCOLATE	132.16	-------	--------	---------	-------
593	C6H12S	THIACYCLOHEPTANE	116.221	16.411	151.792	9.400E+09	67.620
594	C6H13N	CYCLOHEXYLAMINE	99.176	18.828	114.968	8.490E+09	66.200
595	C6H13N	HEXAMETHYLENEIMINE	99.176	19.576	113.354	9.090E+09	69.460
596	C6H14	2,2-DIMETHYLBUTANE	86.177	13.772	133.712	9.830E+09	68.240
597	C6H14	2,3-DIMETHYLBUTANE	86.177	14.353	131.050	9.620E+09	68.240
598	C6H14	n-HEXANE	86.177	14.988	131.306	9.640E+09	68.260
599	C6H14	2-METHYLPENTANE	86.177	14.418	132.931	9.630E+09	68.250
600	C6H14	3-METHYLPENTANE	86.177	14.674	130.575	9.630E+09	68.250
601	C6H14N2O2	LYSINE	146.189	21.577	176.756	1.251E+10	88.520
602	C6H14O	2-ETHYL-1-BUTANOL	102.177	20.896	123.267	1.032E+10	72.850
603	C6H14O	1-HEXANOL	102.177	21.488	125.214	1.033E+10	72.860
604	C6H14O	2-HEXANOL	102.177	21.115	126.073	1.032E+10	72.850
605	C6H14O	2-METHYL-1-PENTANOL	102.177	20.273	123.503	1.032E+10	72.850
606	C6H14O	4-METHYL-2-PENTANOL	102.177	19.365	126.970	1.031E+10	72.840
607	C6H14O	n-BUTYL ETHYL ETHER	102.177	15.707	137.073	1.024E+10	71.960
608	C6H14O	DIISOPROPYL ETHER	102.177	14.354	141.775	1.022E+10	71.940
609	C6H14O	DI-n-PROPYL ETHER	102.177	15.755	137.816	1.022E+10	71.960
610	C6H14O	METHYL tert-PENTYL ETHER	102.177	15.437	133.420	1.043E+10	71.940
611	C6H14O2	ACETAL	118.176	16.506	143.890	1.083E+10	75.650
612	C6H14O2	2-BUTOXYETHANOL	118.176	20.252	131.839	1.093E+10	76.560
613	C6H14O2	1,6-HEXANEDIOL	118.176	25.907	122.480	1.102E+10	77.460
614	C6H14O2	HEXYLENE GLYCOL	118.176	25.082	128.715	9.740E+09	69.390
615	C6H14O2S	DI-n-PROPYL SULFONE	150.242	21.921	144.293	1.224E+10	88.560
616	C6H14O3	DIETHYLENE GLYCOL DIMETHYL ETHER	134.175	18.980	142.400	1.144E+10	79.360
617	C6H14O3	DIPROPYLENE GLYCOL	134.175	25.832	131.845	1.160E+10	81.140
618	C6H14O3	2-(2-ETHOXYETHOXY)ETHANOL	134.175	21.367	136.308	1.153E+10	80.260
619	C6H14O3	TRIMETHYLOLPROPANE	134.175	26.053	122.371	1.190E+10	82.030
620	C6H14O4	TRIETHYLENE GLYCOL	150.175	27.184	133.855	1.220E+10	84.860
621	C6H14O6	SORBITOL	182.174	39.888	130.109	1.374E+10	95.820
622	C6H14S	n-HEXYLMERCAPTAN	118.243	17.545	141.216	1.120E+10	79.630
623	C6H14S	BUTYL-ETHYL-SULFIDE	118.237	17.137	141.905	1.094E+10	79.060
624	C6H14S	ISOPROPYL-SULFIDE	118.237	16.375	143.782	1.092E+10	79.040

NO	FORMULA	NAME	Molecular Weight, g/mol	Solubility Parameter (joule/cm³)⁰·⁵	Liquid Volume cm³/mol	Van Der Waals Area, cm²/mol	Van Der Waals Volume, cm³/mol
625	C6H14S	METHYL-PENTYL-SULFIDE	118.237	16.842	140.964	1.094E+10	79.060
626	C6H14S	PROPYL-SULFIDE	118.237	17.084	141.939	1.092E+10	79.060
627	C6H14S2	PROPYL-DISULFIDE	150.297	17.810	157.403	1.224E+10	90.960
628	C6H15Al	TRIETHYL ALUMINUN	114.167	22.637	137.117	---------	-------
629	C6H15Al2Cl3	ETHYL ALUMINUM SESQUICHLORIDE	247.506	-------	--------	---------	-------
630	C6H15N	DIISOPROPYLAMINE	101.192	15.056	141.919	1.061E+10	76.320
631	C6H15N	DI-n-PROPYLAMINE	101.192	16.699	137.288	1.063E+10	76.340
632	C6H15N	n-HEXYLAMINE	101.192	17.681	132.996	1.061E+10	75.360
633	C6H15N	TRIETHYLAMINE	101.192	15.176	139.672	1.064E+10	76.030
634	C6H15NO	6-AMINOHEXANOL	117.191	21.789	145.478	1.130E+10	79.960
635	C6H15NO2	DIISOPROPANOLAMINE	133.191	26.611	134.763	1.199E+10	85.520
636	C6H15NO3	TRIETHANOLAMINE	149.19	27.325	133.149	1.271E+10	89.830
637	C6H15N3	N-AMINOETHYL PIPERAZINE	129.205	21.767	131.424	1.106E+10	84.330
638	C6H15O4P	TRIETHYL PHOSPHATE	182.156	-------	170.861	1.350E+10	97.500
639	C6H16N2	HEXAMETHYLENEDIAMINE	116.207	19.017	165.782	1.580E+10	82.460
640	C6H18N3OP	HEXAMETHYL PHOSPHORAMIDE	179.202	17.554	175.700	1.351E+10	97.400
641	C6H18N4	TRIETHYLENE TETRAMINE	146.236	24.043	149.508	1.356E+10	98.620
642	C6H18OSi2	HEXAMETHYLDISILOXANE	162.379	12.578	213.585	1.417E+10	105.700
643	C6H18O3Si3	HEXAMETHYLCYCLOTRISILOXANE	222.464	12.448	249.999	1.535E+10	123.500
644	C6H19NSi2	HEXAMETHYLDISILAZANE	161.395	13.632	209.042	1.475E+10	110.730
645	C7H3ClF3NO2	4-CHLORO-3-NITROBENZOTRIFLUORIDE	225.554	20.323	149.741	1.144E+10	88.530
646	C7H3Cl2F3	2,4-DICHLOROBENZOTRIFLUORIDE	215.001	18.406	144.105	1.190E+10	84.540
647	C7H3Cl2NO	3,4-DICHLOROPHENYL ISOCYANATE	188.012	19.665	133.931	1.150E+10	80.320
648	C7H4ClF3	p-CHLOROBENZOTRIFLUORIDE	180.557	16.714	147.324	9.680E+09	75.050
649	C7H4Cl2O	m-CHLOROBENZOYL CHLORIDE	175.014	21.358	122.412	9.630E+09	77.040
650	C7H4F3NO2	3-NITROBENZOTRIFLUORIDE	191.11	20.328	133.993	1.060E+10	81.450
651	C7H5ClO	BENZOYL CHLORIDE	140.569	20.387	116.552	9.010E+09	68.630
652	C7H5ClO2	o-CHLOROBENZOIC ACID	156.568	27.495	112.187	9.080E+09	68.500
653	C7H5Cl3	BENZOTRICHLORIDE	195.475	18.349	142.805	1.079E+10	85.890
654	C7H5F3	BENZOTRIFLUORIDE	146.112	16.654	124.041	8.780E+09	67.170
655	C7H5N	BENZONITRILE	103.123	21.810	103.034	7.520E+09	60.540
656	C7H5NO	PHENYL ISOCYANATE	119.123	20.049	108.988	8.821E+09	61.940
657	C7H5N3O6	2,4,6-TRINITROTOLUENE	227.133	20.032	155.112	1.217E+10	102.350
658	C7H6Cl2	BENZYL DICHLORIDE	161.03	19.134	129.154	9.510E+09	77.100
659	C7H6Cl2	2,4-DICHLOROTOLUENE	161.03	19.794	129.180	1.081E+10	78.110
660	C7H6N2O4	2,4-DINITROTOLUENE	182.136	25.060	137.209	1.112E+10	88.070
661	C7H6N2O4	2,5-DINITROTOLUENE	182.136	24.846	135.881	1.112E+10	88.070
662	C7H6N2O4	2,6-DINITROTOLUENE	182.136	23.676	136.750	1.112E+10	88.070
663	C7H6N2O4	3,4-DINITROTOLUENE	182.136	24.989	139.241	1.112E+10	88.070
664	C7H6N2O4	3,5-DINITROTOLUENE	182.136	21.203	140.544	1.112E+10	88.070
665	C7H6O	BENZALDEHYDE	106.124	21.610	102.013	7.700E+09	60.980
666	C7H6O2	BENZOIC ACID	122.123	22.432	112.449	9.274E+09	65.370
667	C7H6O2	p-HYDROXYBENZALDEHYDE	122.123	25.048	107.175	8.430E+09	66.500
668	C7H6O2	SALICYLALDEHYDE	122.123	24.388	105.053	8.670E+09	64.900
669	C7H6O3	SALICYLIC ACID	138.123	-------	--------	1.079E+10	75.820
670	C7H7Br	p-BROMOTOLUENE	171.037	19.334	122.416	9.740E+09	71.970
671	C7H7Cl	BENZYL CHLORIDE	126.585	20.123	115.429	8.450E+09	67.690
672	C7H7Cl	o-CHLOROTOLUENE	126.585	19.200	117.495	8.530E+09	68.990
673	C7H7Cl	p-CHLOROTOLUENE	126.585	19.288	119.104	8.530E+09	68.990
674	C7H7F	p-FLUOROTOLUENE	110.131	19.481	111.134	7.830E+09	63.090
675	C7H7NO	FORMANILIDE	121.139	24.713	112.327	9.370E+09	70.580
676	C7H7NO2	m-NITROTOLUENE	137.138	20.457	119.003	9.180E+09	72.990
677	C7H7NO2	o-NITROTOLUENE	137.138	21.100	118.411	9.180E+09	72.990
678	C7H7NO2	p-NITROTOLUENE	137.138	20.145	121.894	9.180E+09	72.990
679	C7H7NO3	o-NITROANISOLE	153.138	23.992	123.068	9.810E+09	76.990
680	C7H8	TOLUENE	92.141	18.346	106.556	7.420E+09	59.510
681	C7H8	1,3,5-CYCLOHEPTATRIENE	92.14	19.096	104.469	7.000E+09	56.420
682	C7H8O	ANISOLE	108.14	20.106	109.195	7.930E+09	62.410
683	C7H8O	BENZYL ALCOHOL	108.14	24.649	103.927	8.110E+09	64.110
684	C7H8O	m-CRESOL	108.14	24.080	104.996	8.180E+09	65.030
685	C7H8O	o-CRESOL	108.14	23.398	104.368	8.180E+09	65.030
686	C7H8O	p-CRESOL	108.14	23.827	104.957	8.180E+09	65.030
687	C7H8O2	GUAIACOL	124.139	21.477	110.825	8.780E+09	85.350
688	C7H8O2	p-METHOXYPHENOL	124.139	24.181	110.607	8.000E+09	63.400
689	C7H9N	BENZYLAMINE	107.155	21.694	109.175	8.400E+09	66.600
690	C7H9N	2,6-DIMETHYLPYRIDINE	107.155	19.292	116.775	8.860E+09	69.100
691	C7H9N	N-METHYLANILINE	107.155	21.686	109.090	8.320E+09	66.790
692	C7H9N	m-TOLUIDINE	107.155	22.104	108.743	8.460E+09	67.530
693	C7H9N	o-TOLUIDINE	107.155	22.493	107.749	8.460E+09	67.530
694	C7H9N	p-TOLUIDINE	107.155	21.739	110.838	8.460E+09	67.530
695	C7H10	2-NORBORNENE	94.156	15.840	119.630	6.780E+09	60.050
696	C7H10N2	TOLUENEDIAMINE	122.17	24.340	116.515	9.020E+09	68.410
697	C7H11NO	CYCLOHEXYL ISOCYANATE	125.17	21.119	116.250	8.580E+09	72.820
698	C7H12	1-HEPTYNE	96.172	15.983	132.097	1.101E+10	77.630
699	C7H12O2	n-BUTYL ACRYLATE	128.171	17.121	143.354	1.131E+10	80.170
700	C7H12O2	ISOBUTYL ACRYLATE	128.171	17.177	144.839	1.080E+10	78.300
701	C7H12O2	n-PROPYL METHACRYLATE	128.171	17.190	142.863	1.161E+10	79.950
702	C7H12O4	DIETHYL MALONATE	160.17	19.202	152.563	1.269E+10	88.430

NO	FORMULA	NAME	Molecular Weight, g/mol	Solubility Parameter (joule/cm³)^0.5	Liquid Volume cm³/mol	Van Der Waals Area, cm²/mol	Van Der Waals Volume, cm³/mol
703	C7H14	CYCLOHEPTANE	98.188	17.214	121.809	9.450E+09	71.610
704	C7H14	1,1-DIMETYLCYCLOPENTANE	98.188	15.457	130.926	9.070E+09	70.450
705	C7H14	cis-1,2-DIMETHYLCYCLOPENTANE	98.188	16.164	127.822	9.430E+09	71.590
706	C7H14	trans-1,2-DIMETHYLCYCLOPENTANE	98.188	15.760	131.474	9.430E+09	71.590
707	C7H14	cis-1,3-DIMETHYLCYCLOPENTANE	98.188	15.491	132.647	8.860E+09	704.500
708	C7H14	trans-1,3-DIMETHYLCYCLOPENTANE	98.188	15.818	131.919	8.860E+09	70.450
709	C7H14	ETHYLCYCLOPENTANE	98.188	16.339	128.749	9.440E+09	70.400
710	C7H14	2-ETHYL-1-PENTENE	98.188	15.449	139.476	1.076E+10	74.980
711	C7H14	3-ETHYL-1-PENTENE	98.188	15.016	141.950	1.045E+10	74.990
712	C7H14	1-HEPTENE	98.188	15.433	141.713	1.046E+10	75.000
713	C7H14	cis-2-HEPTENE	98.188	15.368	139.719	1.045E+10	74.970
714	C7H14	trans-2-HEPTENE	98.188	15.457	140.890	1.045E+10	74.970
715	C7H14	cis-3-HEPTENE	98.188	15.545	140.571	1.045E+10	74.970
716	C7H14	trans-3-HEPTENE	98.188	15.340	141.522	1.045E+10	74.970
717	C7H14	METHYLCYCLOHEXANE	98.188	16.270	128.192	8.870E+09	70.460
718	C7H14	2-METHYL-1-HEXENE	98.188	15.270	140.574	1.076E+10	74.980
719	C7H14	3-METHYL-1-HEXENE	98.188	15.008	142.904	1.045E+10	74.990
720	C7H14	4-METHYL-1-HEXENE	98.188	15.012	141.440	1.045E+10	74.990
721	C7H14	2,3,3-TRIMETHYL-1-BUTENE	98.188	14.636	140.154	1.095E+10	74.960
722	C7H14O	DIISOPROPYL KETONE	114.188	17.881	125.223	1.122E+10	79.940
723	C7H14O	2-HEPTANONE	114.188	17.884	140.783	1.124E+10	79.960
724	C7H14O	1-HEPTANAL	114.188	17.944	140.440	1.124E+10	80.000
725	C7H14O	1-METHYLCYCLOHEXANOL	114.188	21.080	123.655	1.012E+10	76.190
726	C7H14O	cis-2-METHYLCYCLOHEXANOL	114.188	21.353	122.550	1.012E+10	76.190
727	C7H14O	trans-2-METHYLCYCLOHEXANOL	114.188	21.457	124.009	1.012E+10	76.190
728	C7H14O	cis-3-METHYLCYCLOHEXANOL	114.188	21.606	125.318	1.012E+10	76.190
729	C7H14O	trans-3-METHYLCYCLOHEXANOL	114.188	21.592	124.439	1.012E+10	76.190
730	C7H14O	cis-4-METHYLCYCLOHEXANOL	114.188	22.090	125.127	1.012E+10	76.190
731	C7H14O	trans-4-METHYLCYCLOHEXANOL	114.188	21.885	125.742	1.012E+10	76.190
732	C7H14O	5-METHYL-2-HEXANONE	114.188	17.738	141.298	1.123E+10	79.950
733	C7H14O2	n-BUTYL PROPIONATE	130.187	17.377	149.316	1.184E+10	83.460
734	C7H14O2	ETHYL ISOVALERATE	130.187	16.572	150.547	1.183E+10	83.450
735	C7H14O2	ISOPENTYL ACETATE	130.187	16.932	150.230	1.178E+10	83.450
736	C7H14O2	n-PENTYL ACETATE	130.187	17.609	149.281	1.184E+10	83.460
737	C7H14O2	n-PROPYL n-BUTYRATE	130.187	16.837	149.976	1.184E+10	83.660
738	C7H14O2	n-HEPTANOIC ACID	130.187	21.949	142.541	1.193E+10	84.340
739	C7H14O3	ETHYL-3-ETHOXYPROPIONATE	146.186	18.228	154.738	1.244E+10	87.160
740	C7H15Br	1-BROMOHEPTANE	179.1	17.355	157.800	1.230E+10	89.450
741	C7H15N	N-METHYLCYCLOHEXYLAMINE	113.203	18.347	130.824	1.121E+10	78.550
742	C7H16	2,2-DIMETHYLPENTANE	100.204	14.203	148.889	1.118E+10	78.470
743	C7H16	2,3-DIMETHYLPENTANE	100.204	14.840	144.947	1.097E+10	78.470
744	C7H16	2,4-DIMETHYLPENTANE	100.204	14.358	150.024	1.097E+10	78.470
745	C7H16	3,3-DIMETHYLPENTANE	100.204	14.533	145.865	1.118E+10	78.470
746	C7H16	3-ETHYLPENTANE	100.204	15.044	144.112	1.098E+10	78.480
747	C7H16	n-HEPTANE	100.204	15.208	147.014	1.099E+10	78.490
748	C7H16	2-METHYLHEXANE	100.204	14.932	148.743	1.098E+10	78.480
749	C7H16	3-METHYLHEXANE	100.204	15.020	146.491	1.098E+10	78.480
750	C7H16	2,2,3-TRIMETHYLBUTANE	100.204	14.247	145.883	1.117E+10	78.460
751	C7H16O	1-HEPTANOL	116.203	21.517	141.796	1.168E+10	83.090
752	C7H16O	2-HEPTANOL	116.203	20.600	142.743	1.167E+10	83.080
753	C7H16O	5-METHYL-1-HEXANOL	116.203	20.448	143.096	1.167E+10	83.080
754	C7H16O	ISOPROPYL-TERT-BUTYL-ETHER	116.203	14.857	154.894	1.167E+10	83.080
755	C7H16S	n-HEPTYL MERCAPTAN	132.27	17.453	157.629	1.254E+10	89.860
756	C7H16S	BUTYL-PROPYL-SULFIDE	132.263	16.631	157.612	1.229E+10	89.290
757	C7H16S	ETHYL-PENTYL-SULFIDE	132.263	17.336	157.587	1.229E+10	89.290
758	C7H16S	HEXYL-METHYL-SULFIDE	132.263	17.336	157.587	1.229E+10	89.290
759	C7H17N	1-AMINOHEPTANE	115.219	17.652	149.328	1.196E+10	85.590
760	C8H4Cl2O2	ISOPHTHALOYL CHLORIDE	203.024	22.300	145.899	1.140E+10	89.960
761	C8H4O3	PHTHALIC ANHYDRIDE	148.118	20.333	122.498	8.220E+09	68.820
762	C8H6	ETHYNYLBENZENE	102.135	18.725	113.333	9.720E+09	72.280
763	C8H6O4	ISOPHTHALIC ACID	166.133	32.469	124.479	1.072E+10	82.800
764	C8H6O4	PHTHALIC ACID	166.133	24.357	144.741	1.072E+10	82.800
765	C8H6O4	TEREPHTHALIC ACID	166.133	-------	117.016	1.062E+10	82.800
766	C8H6S	BENZOTHIOPHENE	134.202	21.468	111.598	7.720E+09	68.640
767	C8H7N	INDOLE	117.15	23.776	106.291	7.410E+09	65.920
768	C8H8	STYRENE	104.152	19.127	115.714	8.270E+09	66.250
769	C8H8	1,3,5,7-CYCLOOCTATETRAENE	104.151	18.669	114.775	8.000E+09	64.480
770	C8H8O	ACETOPHENONE	120.151	20.888	117.366	8.930E+09	70.410
771	C8H8O	p-TOLUALDEHYDE	120.151	21.354	119.047	9.000E+09	71.330
772	C8H8O2	METHYL BENZOATE	136.15	20.476	125.527	9.530E+09	73.910
773	C8H8O2	o-TOLUIC ACID	136.15	22.922	126.888	9.690E+09	75.710
774	C8H8O2	p-TOLUIC ACID	136.15	22.042	129.817	9.690E+09	75.710
775	C8H8O3	METHYL SALICYLATE	152.15	22.404	129.518	1.020E+10	78.630
776	C8H8O3	VANILLIN	152.15	24.058	133.204	1.045E+10	81.350
777	C8H9NO	ACETANILIDE	135.166	22.976	131.040	1.004E+10	79.290
778	C8H10	ETHYLBENZENE	106.167	18.043	122.681	8.800E+09	69.740
779	C8H10	m-XYLENE	106.167	18.090	123.345	8.840E+09	70.660
780	C8H10	o-XYLENE	106.167	18.453	121.142	8.840E+09	70.660

NO	FORMULA	NAME	Molecular Weight, g/mol	Solubility Parameter (joule/cm³)^0.5	Liquid Volume cm³/mol	Van Der Waals Area, cm²/mol	Van Der Waals Volume, cm³/mol
781	C8H10	p-XYLENE	106.167	17.838	123.782	8.840E+09	70.660
782	C8H10O	m-ETHYLPHENOL	122.167	19.393	127.126	9.430E+09	74.460
783	C8H10O	p-ETHYLPHENOL	122.167	21.995	123.319	9.430E+09	74.460
784	C8H10O	PHENETOLE	122.167	19.393	127.126	9.280E+09	72.640
785	C8H10O	2-PHENYLETHANOL	122.167	21.989	120.221	9.370E+09	73.540
786	C8H10O	2,3-XYLENOL	122.167	20.758	129.686	9.330E+09	73.780
787	C8H10O	2,4-XYLENOL	122.167	21.153	120.312	9.330E+09	73.780
788	C8H10O	2,5-XYLENOL	122.167	20.914	125.340	9.330E+09	73.780
789	C8H10O	2,6-XYLENOL	122.167	19.565	137.056	9.330E+09	73.780
790	C8H10O	3,4-XYLENOL	122.167	22.866	122.481	9.330E+09	73.780
791	C8H10O	3,5-XYLENOL	122.167	22.195	124.136	9.330E+09	73.780
792	C8H11N	N,N-DIMETHYLANILINE	121.182	20.076	127.627	9.680E+09	76.710
793	C8H11N	o-ETHYLANILINE	121.182	21.316	124.046	9.810E+09	77.760
794	C8H11N	2,4,6-TRIMETHYLPYRIDINE	121.182	18.964	132.788	1.107E+10	78.950
795	C8H11NO	p-PHENETIDINE	137.181	22.738	129.824	1.041E+10	81.460
796	C8H12	1,5-CYCLOOCTADIENE	108.183	18.528	123.244	9.720E+09	74.800
797	C8H12	VINYLCYCLOHEXENE	108.183	17.223	130.928	9.720E+09	74.820
798	C8H12O4	1,4-CYCLOHEXANEDICARBOXYLIC ACID	172.181	23.927	150.022	1.209E+10	92.820
799	C8H12O4	DIETHYL MALEATE	172.181	17.630	179.211	1.350E+10	95.140
800	C8H14O2	n-BUTYL METHACRYLATE	142.198	17.141	159.662	1.296E+10	90.180
801	C8H14O3	BUTYRIC ANHYDRIDE	158.197	19.050	164.811	1.344E+10	95.360
802	C8H14O4	DIETHYL SUCCINATE	174.197	19.832	168.213	1.404E+10	98.200
803	C8H16	CYCLOOCTANE	112.214	17.228	135.155	1.080E+10	75.760
804	C8H16	1,1-DIMETHYLCYCLOHEXANE	112.215	15.673	144.469	1.099E+10	81.820
805	C8H16	cis-1,2-DIMETHYLCYCLOHEXANE	112.215	16.249	141.645	1.078E+10	81.820
806	C8H16	trans-1,2-DIMETHYLCYCLOHEXANE	112.215	15.777	145.347	1.078E+10	81.820
807	C8H16	cis-1,3-DIMETHYLCYCLOHEXANE	112.215	15.642	147.282	1.078E+10	81.820
808	C8H16	trans-1,3-DIMETHYLCYCLOHEXANE	112.215	16.041	143.768	1.078E+10	81.820
809	C8H16	cis-1,4-DIMETHYLCYCLOHEXANE	112.215	15.981	144.106	1.078E+10	81.820
810	C8H16	trans-1,4-DIMETHYLCYCLOHEXANE	112.215	15.501	147.895	1.078E+10	81.820
811	C8H16	ETHYLCYCLOHEXANE	112.215	16.360	143.117	1.022E+10	80.690
812	C8H16	2-ETHYL-1-HEXENE	112.215	15.425	155.247	1.211E+10	85.210
813	C8H16	1-METHYL-1-ETHYLCYCLOPENTANE	112.215	15.839	144.502	1.099E+10	81.820
814	C8H16	1-OCTENE	112.215	15.638	157.862	1.181E+10	85.230
815	C8H16	trans-2-OCTENE	112.215	15.412	156.816	1.180E+10	85.200
816	C8H16	trans-3-OCTENE	112.215	15.373	157.813	1.180E+10	85.200
817	C8H16	trans-4-OCTENE	112.215	15.409	158.077	1.180E+10	85.200
818	C8H16	n-PROPYLCYCLOPENTANE	112.215	16.394	145.193	1.078E+10	80.620
819	C8H16	2,4,4-TRIMETHYL-1-PENTENE	112.215	14.522	157.915	1.230E+10	85.190
820	C8H16	2,4,4-TRIMETHYL-2-PENTENE	112.215	14.980	156.437	1.229E+10	85.160
821	C8H16O	2-ETHYLHEXANAL	128.214	17.577	156.518	1.258E+10	90.180
822	C8H16O	1-OCTANAL	128.214	17.872	157.040	1.259E+10	90.190
823	C8H16O	2-OCTANONE	128.214	19.152	157.391	1.259E+10	90.190
824	C8H16O2	n-BUTYL n-BUTYRATE	144.214	16.977	166.575	1.319E+10	93.890
825	C8H16O2	n-HEXYL ACETATE	144.214	17.197	166.101	1.319E+10	93.690
826	C8H16O2	ISOBUTYL ISOBUTYRATE	144.214	16.228	169.312	1.317E+10	93.670
827	C8H16O2	n-OCTANOIC ACID	144.214	22.324	159.632	1.328E+10	94.790
828	C8H16O4	DIETHYLENE GLYCOL ETHYL ETHER ACETATE	176.213	17.750	175.101	1.439E+10	101.090
829	C8H18	2,2-DIMETHYLHEXANE	114.231	14.662	165.124	1.253E+10	88.700
830	C8H18	2,3-DIMETHYLHEXANE	114.231	15.015	161.285	1.232E+10	88.700
831	C8H18	2,4-DIMETHYLHEXANE	114.231	14.651	164.796	1.232E+10	88.700
832	C8H18	2,5-DIMETHYLHEXANE	114.231	14.741	165.556	1.232E+10	88.700
833	C8H18	3,3-DIMETHYLHEXANE	114.231	14.883	161.605	1.253E+10	88.700
834	C8H18	3,4-DIMETHYLHEXANE	114.231	15.136	159.465	1.232E+10	88.700
835	C8H18	3-ETHYLHEXANE	114.231	15.202	160.974	1.233E+10	88.710
836	C8H18	3-ETHYL-2-METHYLPENTANE	114.23	14.978	160.765	1.232E+10	88.700
837	C8H18	3-METHYL-3-ETHYLPENTANE	114.231	15.057	157.845	1.253E+10	88.700
838	C8H18	2-METHYLHEPTANE	114.231	15.081	164.232	1.233E+10	88.710
839	C8H18	3-METHYLHEPTANE	114.231	15.314	162.727	1.233E+10	88.710
840	C8H18	4-METHYLHEPTANE	114.231	15.376	160.321	1.233E+10	88.700
841	C8H18	n-OCTANE	114.231	15.360	163.507	1.234E+10	88.720
842	C8H18	2,2,3-TRIMETHYLPENTANE	114.231	14.769	160.344	1.252E+10	88.690
843	C8H18	2,2,4-TRIMETHYLPENTANE	114.231	14.051	165.452	1.252E+10	88.690
844	C8H18	2,3,3-TRIMETHYLPENTANE	114.231	14.920	158.155	1.252E+10	88.690
845	C8H18	2,3,4-TRIMETHYLPENTANE	114.231	14.934	159.513	1.231E+10	88.690
846	C8H18	2,2,3,3-TETRAMETHYLBUTANE	114.23	14.258	165.002	1.231E+10	88.690
847	C8H18O	DI-n-BUTYL ETHER	130.23	15.969	170.349	1.288E+10	91.920
848	C8H18O	DI-sec-BUTYL ETHER	130.23	14.953	171.577	1.292E+10	92.400
849	C8H18O	DI-tert-BUTYL ETHER	130.23	14.277	171.326	1.332E+10	92.380
850	C8H18O	2-ETHYL-1-HEXANOL	130.23	21.308	156.974	1.302E+10	93.310
851	C8H18O	1-OCTANOL	130.23	20.733	158.229	1.303E+10	93.320
852	C8H18O	2-OCTANOL	130.23	20.108	159.389	1.302E+10	93.310
853	C8H18O2	DI-t-BUTYL PEROXIDE	146.23	14.182	185.122	1.392E+10	96.080
854	C8H18O2S	DI-n-BUTYL SULFONE	178.296	20.413	180.109	1.494E+10	109.020
855	C8H18O3	DIETHYLENE GLYCOL DIETHYL ETHER	162.229	17.651	179.481	1.414E+10	99.820
856	C8H18O3	DIETHYLENE GLYCOL MONOBUTYL EHTER	162.229	21.034	170.414	1.420E+10	101.000
857	C8H18O4	TRIETHYLENE GLYCOL DIMETHYL ETHER	178.229	19.175	181.957	1.474E+10	103.520
858	C8H18O5	TETRAETHYLENE GLYCOL	194.228	26.208	173.146	1.552E+10	109.020

351

NO	FORMULA	NAME	Molecular Weight, g/mol	Solubility Parameter (joule/cm³)⁰·⁵	Liquid Volume cm³/mol	Van Der Waals Area, cm²/mol	Van Der Waals Volume, cm³/mol
859	C8H18S	n-OCTYL MERCAPTAN	146.297	17.294	174.195	1.390E+10	100.090
860	C8H18S	tert-OCTYL MERCAPTAN	146.297	15.311	173.885	1.386E+10	100.050
861	C8H18S	BUTYL-SULFIDE	146.29	16.463	174.244	1.364E+10	99.520
862	C8H18S	ETHYL-HEXYL-SULFIDE	146.29	17.305	174.238	1.364E+10	99.520
863	C8H18S	HEPTYL-METHYL-SULFIDE	146.29	17.305	174.238	1.364E+10	99.520
864	C8H18S	PENTYL-PROPYL-SULFIDE	146.29	17.305	174.238	1.364E+10	99.520
865	C8H18S2	BUTYL-DISULFIDE	178.35	17.559	190.951	1.494E+10	111.420
866	C8H19N	DI-n-BUTYLAMINE	129.246	16.576	170.715	1.333E+10	96.800
867	C8H19N	DIISOBUTYLAMINE	129.246	15.302	174.062	1.327E+10	96.780
868	C8H19N	n-OCTYLAMINE	129.246	17.229	165.977	1.330E+10	95.800
869	C8H23N5	TETRAETHYLENEPENTAMINE	189.304	24.237	190.365	2.341E+10	154.680
870	C8H24O4Si4	OCTAMETHYLCYCLOTETRASILOXANE	296.618	12.922	312.420	2.122E+10	166.200
871	C9H4O5	TRIMELLITIC ANHYDRIDE	192.128	26.481	150.016	1.076E+10	87.640
872	C9H6N2O2	TOLUENE DIISOCYANATE	174.159	21.851	142.752	9.040E+09	86.670
873	C9H7N	ISOQUINOLINE	129.161	22.053	118.250	7.650E+09	71.100
874	C9H7N	QUINOLINE	129.161	21.940	118.548	7.640E+09	71.100
875	C9H7NO	8-HYDROXYQUINOLINE	145.161	21.701	139.636	1.066E+10	75.820
876	C9H8	INDENE	116.163	20.467	116.856	7.930E+09	68.890
877	C9H8O	2-METHYLBENZOFURAN	132.162	20.131	125.754	8.830E+09	72.570
878	C9H10	INDANE	118.178	19.406	123.121	7.900E+09	71.270
879	C9H10	cis-PROPENYLBENZENE	118.178	18.440	130.675	1.055E+10	80.790
880	C9H10	trans-PROPENYLBENZENE	118.178	18.418	130.993	1.055E+10	80.790
881	C9H10	alpha-METHYLSTYRENE	118.178	18.345	130.641	9.920E+09	76.460
882	C9H10	m-METHYLSTYRENE	118.178	18.715	130.213	9.660E+09	77.400
883	C9H10	o-METHYLSTYRENE	118.178	18.970	130.196	9.660E+09	77.400
884	C9H10	p-METHYLSTYRENE	118.178	18.915	129.067	9.660E+09	77.400
885	C9H10O2	BENZYL ACETATE	150.177	19.945	143.697	1.088E+10	84.150
886	C9H10O2	ETHYL BENZOATE	150.177	19.019	144.076	1.088E+10	84.140
887	C9H10O3	ETHYL VANILLIN	166.177	24.958	145.132	1.023E+10	85.480
888	C9H11NO	p-DIMETHYLAMINOBENZALDEHYDE	149.192	20.606	142.211	1.144E+10	90.130
889	C9H12	CUMENE	120.194	17.512	139.798	1.014E+10	79.960
890	C9H12	m-ETHYLTOLUENE	120.194	17.857	139.698	1.019E+10	80.890
891	C9H12	o-ETHYLTOLUENE	120.194	18.013	137.112	1.019E+10	80.890
892	C9H12	p-ETHYLTOLUENE	120.194	17.849	140.241	1.019E+10	80.890
893	C9H12	MESITYLENE	120.194	18.007	139.524	1.026E+10	81.830
894	C9H12	n-PROPYLBENZENE	120.194	17.805	139.831	1.012E+10	79.970
895	C9H12	1,2,3-TRIMETHYLBENZENE	120.194	18.344	134.910	1.026E+10	81.810
896	C9H12	1,2,4-TRIMETHYLBENZENE	120.194	17.945	137.779	1.026E+10	81.810
897	C9H12O	BENZYL ETHYL ETHER	136.194	18.309	144.088	1.073E+10	83.670
898	C9H12O	2-PHENYL-2-PROPANOL	136.194	19.820	138.754	1.103E+10	84.550
899	C9H12O2	CUMENE HYDROPEROXIDE	152.193	21.497	145.892	1.162E+10	88.250
900	C9H14O	ISOPHORONE	138.21	17.857	150.287	1.235E+10	89.980
901	C9H14O6	GLYCERYL TRIACETATE	218.207	20.163	188.505	1.620E+10	114.000
902	C9H16	1-NONYNE	124.225	16.075	165.093	1.304E+10	95.070
903	C9H16O4	AZELAIC ACID	188.224	23.962	179.864	1.557E+10	111.060
904	C9H18	BUTYLCYCLOPENTANE	126.241	16.066	161.723	1.213E+10	90.850
905	C9H18	cis,cis-1,3,5-TRIMETHYLCYCLOHEXANE	126.241	15.063	164.696	1.311E+10	97.800
906	C9H18	cis,trans-1,3,5-TRIMETHYLCYCLOHEXANE	126.241	14.993	175.788	1.311E+10	97.800
907	C9H18	ISOPROPYLCYCLOHEXANE	126.242	16.291	158.146	1.156E+10	90.910
908	C9H18	1-NONENE	126.242	15.832	174.009	1.316E+10	95.460
909	C9H18	n-PROPYLCYCLOHEXANE	126.242	16.377	159.758	1.157E+10	90.920
910	C9H18O	DIISOBUTYL KETONE	142.241	16.454	177.418	1.392E+10	100.400
911	C9H18O	1-NONANAL	142.241	17.766	173.661	1.394E+10	100.420
912	C9H18O2	n-BUTYL VALERATE	158.241	16.759	183.333	1.454E+10	103.920
913	C9H18O2	n-NONANOIC ACID	158.241	21.768	175.415	1.463E+10	104.800
914	C9H18O2	n-OCTYL FORMATE	158.241	17.696	181.863	1.454E+10	104.120
915	C9H20	3,3-DIETHYLPENTANE	128.258	15.224	171.012	1.388E+10	98.930
916	C9H20	2,2-DIMETHYL-3-ETHYLPENTANE	128.258	14.832	175.433	1.387E+10	98.920
917	C9H20	3-ETHYL-2,3-DIMETHYLPENTANE	128.257	15.316	170.827	1.387E+10	98.920
918	C9H20	2,4-DIMETHYL-3-ETHYLPENTANE	128.258	14.971	174.682	1.366E+10	98.920
919	C9H20	2,2-DIMETHYLHEPTANE	128.258	14.853	181.517	1.388E+10	98.930
920	C9H20	2,6-DIMETHYLHEPTANE	128.258	14.991	181.725	1.367E+10	98.930
921	C9H20	3-ETHYLHEPTANE	128.258	15.298	177.515	1.368E+10	98.940
922	C9H20	4-ETHYLHEPTANE	128.257	15.430	177.133	1.368E+10	98.940
923	C9H20	2,3-DIMETHYLHEPTANE	128.257	15.240	177.664	1.388E+10	98.930
924	C9H20	2,4-DIMETHYLHEPTANE	128.257	14.915	180.289	1.388E+10	98.930
925	C9H20	2,5-DIMETHYLHEPTANE	128.257	15.009	179.987	1.388E+10	98.930
926	C9H20	3,4-DIMETHYLHEPTANE	128.257	15.341	176.306	1.388E+10	98.930
927	C9H20	3,5-DIMETHYLHEPTANE	128.257	15.066	178.505	1.388E+10	98.930
928	C9H20	4,4-DIMETHYLHEPTANE	128.257	15.003	177.767	1.388E+10	98.930
929	C9H20	3-ETHYL-2-METHYLHEXANE	128.257	15.256	175.856	1.388E+10	98.930
930	C9H20	4-ETHYL-2-METHYLHEXANE	128.257	15.059	178.293	1.388E+10	98.930
931	C9H20	3-ETHYL-3-METHYLHEXANE	128.257	15.261	174.031	1.388E+10	98.930
932	C9H20	3-ETHYL-4-METHYLHEXANE	128.257	15.427	174.190	1.388E+10	98.930
933	C9H20	2,2,3-TRIMETHYLHEXANE	128.257	14.862	176.933	1.387E+10	98.920
934	C9H20	2,2,4-TRIMETHYLHEXANE	128.257	14.537	179.986	1.387E+10	98.920
935	C9H20	2,3,3-TRIMETHYLHEXANE	128.257	15.025	174.629	1.387E+10	98.920
936	C9H20	2,3,4-TRIMETHYLHEXANE	128.257	15.228	174.425	1.387E+10	98.920

NO	FORMULA	NAME	Molecular Weight, g/mol	Solubility Parameter (joule/cm³)^0.5	Liquid Volume cm³/mol	Van Der Waals Area, cm²/mol	Van Der Waals Volume, cm³/mol
937	C9H20	2,3,5-TRIMETHYLHEXANE	128.257	14.818	178.656	1.387E+10	98.920
938	C9H20	2,4,4-TRIMETHYLHEXANE	128.257	14.699	178.137	1.387E+10	98.920
939	C9H20	3,3,4-TRIMETHYLHEXANE	128.257	15.140	173.009	1.387E+10	98.920
940	C9H20	2-METHYLOCTANE	128.258	15.321	180.716	1.368E+10	98.940
941	C9H20	3-METHYLOCTANE	128.258	15.385	178.934	1.368E+10	98.940
942	C9H20	4-METHYLOCTANE	128.258	15.296	179.060	1.368E+10	98.940
943	C9H20	n-NONANE	128.258	15.527	179.321	1.369E+10	98.950
944	C9H20	2,2,3,3-TETRAMETHYLPENTANE	128.258	15.108	170.329	1.407E+10	98.910
945	C9H20	2,2,3,4-TETRAMETHYLPENTANE	128.258	14.928	174.450	1.386E+10	98.910
946	C9H20	2,2,4,4-TETRAMETHYLPENTANE	128.258	14.163	179.232	1.407E+10	98.910
947	C9H20	2,3,3,4-TETRAMETHYLPENTANE	128.257	15.012	174.553	1.407E+10	98.910
948	C9H20	2,2,5-TRIMETHYLHEXANE	128.258	14.430	181.404	1.387E+10	98.920
949	C9H20O	2,6-DIMETHYL-4-HEPTANOL	144.257	17.868	178.729	1.435E+10	103.500
950	C9H20O	1-NONANOL	144.257	20.582	174.982	1.438E+10	103.550
951	C9H20O	2-NONANOL	144.257	19.573	176.011	1.437E+10	103.540
952	C9H20S	n-NONYL MERCAPTAN	160.324	17.332	190.615	1.520E+10	110.320
953	C9H20S	BUTYL-PENTYL-SULFIDE	160.317	17.297	190.840	1.499E+10	109.750
954	C9H20S	ETHYL-HEPTYL-SULFIDE	160.317	17.297	190.840	1.499E+10	109.750
955	C9H20S	HEXYL-PROPYL-SULFIDE	160.317	17.297	190.840	1.499E+10	109.750
956	C9H20S	METHYL-OCTYL-SULFIDE	160.317	17.297	190.840	1.499E+10	109.750
957	C9H21N	n-NONYLAMINE	143.272	17.323	182.439	1.466E+10	106.050
958	C9H21N	TRIPROPYLAMINE	143.272	15.154	189.955	1.469E+10	106.720
959	C10H6O8	PYROMELLITIC ACID	254.153	25.516	202.282	1.544E+10	117.240
960	C10H7Br	1-BROMONAPHTHALENE	207.07	21.231	140.070	9.890E+09	86.570
961	C10H7Cl	1-CHLORONAPHTHALENE	162.618	21.161	138.860	9.440E+09	82.640
962	C10H8	NAPHTHALENE	128.174	19.188	130.825	8.420E+09	74.640
963	C10H8	AZULENE	128.173	23.544	121.715	8.420E+09	74.640
964	C10H9N	QUINALDINE	143.188	19.752	135.721	1.151E+10	82.250
965	C10H10	m-DIVINYLBENZENE	130.189	18.747	140.689	1.030E+10	82.540
966	C10H10	1-METHYLINDENE	130.189	18.312	134.209	9.260E+09	79.110
967	C10H10	2-METHYLINDENE	130.189	17.790	139.437	9.580E+09	79.100
968	C10H10O4	DIMETHYL PHTHALATE	194.187	20.675	163.360	1.306E+10	99.460
969	C10H10O4	DIMETHYL TEREPHTHALATE	194.187	17.710	180.564	1.324E+10	101.060
970	C10H12	DICYCLOPENTADIENE	132.205	17.445	138.585	9.300E+09	81.460
971	C10H12	1,2,3,4-TETRAHYDRONAPHTHALENE	132.205	19.524	136.703	9.120E+09	80.980
972	C10H12O	ANETHOLE	148.205	19.880	150.687	1.160E+10	91.050
973	C10H12O4	DIALLYL MALEATE	196.203	19.930	182.908	1.514E+10	108.620
974	C10H14	n-BUTYLBENZENE	134.221	17.448	156.494	1.147E+10	90.200
975	C10H14	sec-BUTYLBENZENE	134.221	17.428	156.400	1.146E+10	90.190
976	C10H14	tert-BUTYLBENZENE	134.221	17.228	155.593	1.166E+10	90.180
977	C10H14	1,2,3,4-TETRAMETHYLBENZENE	134.221	19.072	148.921	1.448E+10	103.080
978	C10H14	m-CYMENE	134.221	17.129	156.591	1.153E+10	91.110
979	C10H14	o-CYMENE	134.221	17.386	153.786	1.153E+10	91.110
980	C10H14	p-CYMENE	134.221	17.376	157.444	1.153E+10	91.110
981	C10H14	m-DIETHYLBENZENE	134.221	17.670	156.030	1.154E+10	91.120
982	C10H14	o-DIETHYLBENZENE	134.221	17.764	153.280	1.154E+10	91.120
983	C10H14	p-DIETHYLBENZENE	134.221	17.534	156.430	1.154E+10	91.120
984	C10H14	2-ETHYL-m-XYLENE	134.221	18.137	151.424	1.161E+10	92.040
985	C10H14	2-ETHYL-p-XYLENE	134.221	17.611	153.715	1.161E+10	92.040
986	C10H14	3-ETHYL-o-XYLENE	134.221	18.302	151.126	1.134E+10	89.640
987	C10H14	4-ETHYL-m-XYLENE	134.221	17.650	153.864	1.161E+10	92.040
988	C10H14	4-ETHYL-o-XYLENE	134.221	17.988	154.172	1.161E+10	92.040
989	C10H14	5-ETHYL-m-XYLENE	134.221	17.569	155.924	1.161E+10	92.040
990	C10H14	ISOBUTYLBENZENE	134.221	16.990	158.078	1.137E+10	89.390
991	C10H14	1,2,3,5-TETRAMETHYLBENZENE	134.221	18.325	151.385	1.168E+10	92.960
992	C10H14	1,2,4,5-TETRAMETHYLBENZENE	134.221	17.156	159.747	1.132E+10	89.760
993	C10H14O	p-tert-BUTYLPHENOL	150.221	19.064	163.408	1.336E+10	96.280
994	C10H14O2	p-tert-BUTYLCATECHOL	166.22	22.309	158.223	1.290E+10	98.800
995	C10H15N	N,N-DIETYHLANILINE	149.236	18.913	160.371	1.247E+10	97.980
996	C10H15N	2,6-DIETHYLANILINE	149.236	21.806	165.399	1.260E+10	99.100
997	C10H16	CAMPHENE	136.237	15.839	160.729	1.019E+10	88.450
998	C10H16	D-LIMONENE	136.237	16.764	162.332	1.302E+10	95.240
999	C10H16	alpha-PHELLANDRENE	136.237	16.992	161.557	1.213E+10	94.080
1000	C10H16	beta-PHELLANDRENE	136.237	16.949	162.839	1.214E+10	94.110
1001	C10H16	alpha-PINENE	136.237	16.032	158.957	1.189E+10	91.840
1002	C10H16	beta-PINENE	136.237	16.784	157.054	1.190E+10	91.870
1003	C10H16	alpha-TERPINENE	136.237	16.865	164.065	1.301E+10	95.210
1004	C10H16	gamma-TERPINENE	136.237	17.253	161.191	1.301E+10	95.210
1005	C10H16	TERPINOLENE	136.237	18.051	158.722	1.293E+10	95.320
1006	C10H16O	CAMPHOR	152.236	16.023	154.669	1.258E+10	96.860
1007	C10H18	1-DECYNE	138.252	16.114	181.546	1.439E+10	105.300
1008	C10H18	cis-DECAHYDRONANPHTALENE	138.253	17.625	154.615	1.137E+10	94.260
1009	C10H18	trans-DECAHYDRONAPHTHALENE	138.253	17.043	159.325	1.137E+10	94.260
1010	C10H18O4	SEBACIC ACID	202.251	22.543	196.071	1.692E+10	120.880
1011	C10H20	n-BUTYLCYCLOHEXANE	140.269	16.400	176.272	1.292E+10	101.150
1012	C10H20	1-CYCLOPENTYLPENTANE	140.268	16.114	178.135	1.348E+10	101.080
1013	C10H20	1-DECENE	140.269	15.852	190.320	1.451E+10	105.690
1014	C10H20O	1-DECANAL	156.268	17.834	190.278	1.529E+10	110.650

NO	FORMULA	NAME	Molecular Weight, g/mol	Solubility Parameter (joule/cm³)^0.5	Liquid Volume cm³/mol	Van Der Waals Area, cm²/mol	Van Der Waals Volume, cm³/mol
1015	C10H20O2	n-DECANOIC ACID	172.268	22.984	194.712	1.598E+10	115.250
1016	C10H20O2	2-ETHYLHEXYL ACETATE	172.268	17.193	198.289	1.588E+10	114.140
1017	C10H20O2	ISOPENTYL ISOVALERATE	172.268	16.478	201.701	1.587E+10	114.130
1018	C10H22	n-DECANE	142.285	15.538	195.342	1.504E+10	109.180
1019	C10H22	2-METHYLNONANE	142.285	15.440	196.780	1.503E+10	109.170
1020	C10H22	3-METHYLNONANE	142.285	15.538	195.076	1.503E+10	109.170
1021	C10H22	4-METHYLNONANE	142.285	15.298	195.426	1.503E+10	109.170
1022	C10H22	5-METHYLNONANE	142.285	15.460	195.298	1.503E+10	109.170
1023	C10H22	3-ETHYLOCTANE	142.284	15.545	193.341	1.503E+10	109.170
1024	C10H22	4-ETHYLOCTANE	142.284	15.442	193.821	1.503E+10	110.170
1025	C10H22	2,2-DIMETHYLOCTANE	142.285	14.868	197.405	1.523E+10	109.160
1026	C10H22	2,3-DIMETHYLOCTANE	142.284	15.271	193.853	1.523E+10	109.160
1027	C10H22	2,4-DIMETHYLOCTANE	142.284	15.032	196.923	1.523E+10	109.160
1028	C10H22	2,5-DIMETHYLOCTANE	142.284	15.089	195.902	1.523E+10	109.160
1029	C10H22	2,6-DIMETHYLOCTANE	142.284	15.171	196.662	1.523E+10	109.160
1030	C10H22	2,7-DIMETHYLOCTANE	142.284	15.135	197.632	1.523E+10	109.160
1031	C10H22	3,3-DIMETHYLOCTANE	142.284	15.111	193.623	1.523E+10	109.160
1032	C10H22	3,4-DIMETHYLOCTANE	142.284	15.287	192.018	1.523E+10	109.160
1033	C10H22	3,5-DIMETHYLOCTANE	142.284	15.129	194.171	1.523E+10	109.160
1034	C10H22	3,6-DIMETHYLOCTANE	142.284	15.205	194.264	1.523E+10	109.160
1035	C10H22	4,4-DIMETHYLOCTANE	142.284	14.967	194.646	1.523E+10	109.160
1036	C10H22	4,5-DIMETHYLOCTANE	142.284	15.265	191.468	1.523E+10	109.160
1037	C10H22	4-PROPYLHEPTANE	142.284	15.208	194.362	1.503E+10	109.170
1038	C10H22	4-ISOPROPYLHEPTANE	142.284	15.074	193.510	1.503E+10	109.170
1039	C10H22	3-ETHYL-2-METHYLHEPTANE	142.284	15.196	192.370	1.523E+10	109.160
1040	C10H22	4-ETHYL-2-METHYLHEPTANE	142.284	15.020	194.363	1.523E+10	109.160
1041	C10H22	5-ETHYL-2-METHYLHEPTANE	142.284	15.119	194.473	1.523E+10	109.160
1042	C10H22	3-ETHYL-3-METHYLHEPTANE	142.284	15.194	190.719	1.523E+10	109.160
1043	C10H22	4-ETHYL-3-METHYLHEPTANE	142.284	15.240	190.614	1.523E+10	109.160
1044	C10H22	3-ETHYL-5-METHYLHEPTANE	142.284	15.137	193.152	1.523E+10	109.160
1045	C10H22	3-ETHYL-4-METHYLHEPTANE	142.284	15.264	190.579	1.523E+10	109.160
1046	C10H22	4-ETHYL-4-METHYLHEPTANE	142.284	15.140	190.475	1.523E+10	109.160
1047	C10H22	2,2,3-TRIMETHYLHEPTANE	142.284	14.982	192.718	1.522E+10	109.150
1048	C10H22	2,2,4-TRIMETHYLHEPTANE	142.284	14.647	196.618	1.522E+10	110.150
1049	C10H22	2,2,5-TRIMETHYLHEPTANE	142.284	14.725	196.436	1.522E+10	111.150
1050	C10H22	2,2,6-TRIMETHYLHEPTANE	142.284	14.699	197.617	1.522E+10	112.150
1051	C10H22	2,3,3-TRIMETHYLHEPTANE	142.284	15.069	190.997	1.522E+10	113.150
1052	C10H22	2,3,4-TRIMETHYLHEPTANE	142.284	15.116	191.058	1.522E+10	114.150
1053	C10H22	2,3,5-TRIMETHYLHEPTANE	142.284	15.063	191.997	1.522E+10	115.150
1054	C10H22	2,3,6-TRIMETHYLHEPTANE	142.284	14.978	194.674	1.522E+10	116.150
1055	C10H22	2,4,4-TRIMETHYLHEPTANE	142.284	14.714	194.718	1.522E+10	117.150
1056	C10H22	2,4,5-TRIMETHYLHEPTANE	142.284	14.986	193.003	1.522E+10	118.150
1057	C10H22	2,4,6-TRIMETHYLHEPTANE	142.284	14.722	197.946	1.522E+10	119.150
1058	C10H22	2,5,5-TRIMETHYLHEPTANE	142.284	14.869	193.298	1.522E+10	120.150
1059	C10H22	3,3,4-TRIMETHYLHEPTANE	142.284	15.141	189.091	1.522E+10	121.150
1060	C10H22	3,3,5-TRIMETHYLHEPTANE	142.284	14.874	192.535	1.522E+10	122.150
1061	C10H22	3,4,4-TRIMETHYLHEPTANE	142.284	15.118	188.877	1.522E+10	123.150
1062	C10H22	3,4,5-TRIMETHYLHEPTANE	142.284	15.266	189.255	1.522E+10	124.150
1063	C10H22	3-ISOPROPYL-2-METHYLHEXANE	142.284	14.994	191.358	1.523E+10	109.160
1064	C10H22	3,3-DIETHYLHEXANE	142.284	15.311	187.903	1.523E+10	109.160
1065	C10H22	3,4-DIETHYLHEXANE	142.284	15.243	190.435	1.523E+10	109.160
1066	C10H22	3-ETHYL-2,2-DIMETHYLHEXANE	142.284	14.930	191.110	1.522E+10	124.150
1067	C10H22	4-ETHYL-2,2-DIMETHYLHEXANE	142.284	14.653	194.884	1.522E+10	124.150
1068	C10H22	3-ETHYL-2,3-DIMETHYLHEXANE	142.284	15.212	187.271	1.522E+10	124.150
1069	C10H22	4-ETHYL-2,3-DIMETHYLHEXANE	142.284	15.158	189.311	1.522E+10	124.150
1070	C10H22	3-ETHYL-2,4-DIMETHYLHEXANE	142.284	15.131	189.428	1.522E+10	124.150
1071	C10H22	4-ETHYL-2,4-DIMETHYLHEXANE	142.284	15.108	189.123	1.522E+10	124.150
1072	C10H22	3-ETHYL-2,5-DIMETHYLHEXANE	142.284	14.903	193.118	1.522E+10	124.150
1073	C10H22	4-ETHYL-3,3-DIMETHYLHEXANE	142.284	15.187	187.289	1.522E+10	124.150
1074	C10H22	3-ETHYL-3,4-DIMETHYLHEXANE	142.284	15.154	187.316	1.522E+10	124.150
1075	C10H22	2,2,3,3-TETRAMETHYLHEXANE	142.284	15.009	187.066	1.521E+10	125.140
1076	C10H22	2,2,3,4-TETRAMETHYLHEXANE	142.284	14.918	189.416	1.521E+10	126.140
1077	C10H22	2,2,3,5-TETRAMETHYLHEXANE	142.284	14.663	194.014	1.521E+10	127.140
1078	C10H22	2,2,4,4-TETRAMETHYLHEXANE	142.284	14.525	191.703	1.521E+10	128.140
1079	C10H22	2,2,4,5-TETRAMETHYLHEXANE	142.284	14.565	194.524	1.521E+10	129.140
1080	C10H22	2,2,5,5-TETRAMETHYLHEXANE	142.284	14.221	199.125	1.521E+10	130.140
1081	C10H22	2,3,3,4-TETRAMETHYLHEXANE	142.284	15.194	185.913	1.521E+10	131.140
1082	C10H22	2,3,3,5-TETRAMETHYLHEXANE	142.284	14.803	191.063	1.521E+10	132.140
1083	C10H22	2,3,4,4-TETRAMETHYLHEXANE	142.284	15.247	182.738	1.521E+10	133.140
1084	C10H22	2,3,4,5-TETRAMETHYLHEXANE	142.284	14.968	190.838	1.521E+10	134.140
1085	C10H22	3,3,4,4-TETRAMETHYLHEXANE	142.284	14.885	182.722	1.521E+10	135.140
1086	C10H22	2,4-DIMETHYL-3-ISOPROPYLPENTANE	142.284	14.953	188.609	1.522E+10	124.150
1087	C10H22	3,3-DIETHYL-2-METHYLPENTANE	142.284	15.448	183.477	1.522E+10	124.150
1088	C10H22	3-ETHYL-2,2,3-TRIMETHYLPENTANE	142.284	15.200	182.918	1.521E+10	135.140
1089	C10H22	3-ETHYL-2,2,4-TRIMETHYLPENTANE	142.284	14.861	188.928	1.521E+10	135.140
1090	C10H22	3-ETHYL-2,3,4-TRIMETHYLPENTANE	142.284	15.297	184.006	1.521E+10	135.140
1091	C10H22	2,2,3,3,4-PENTAMETHYLPENTANE	142.284	15.121	183.236	1.520E+10	136.130
1092	C10H22	2,2,3,4,4-PENTAMETHYLPENTANE	142.284	14.764	186.342	1.520E+10	137.130

NO	FORMULA	NAME	Molecular Weight, g/mol	Solubility Parameter (joule/cm^3)$^{0.5}$	Liquid Volume cm^3/mol	Van Der Waals Area, cm^2/mol	Van Der Waals Volume, cm^3/mol
1093	C10H22O	1-DECANOL	158.284	20.012	191.772	1.573E+10	113.780
1094	C10H22O	DI-n-PENTYL ETHER	158.284	15.480	202.956	1.565E+10	112.890
1095	C10H22O	ISODECANOL	158.284	19.277	189.219	1.572E+10	113.770
1096	C10H22O5	TETRAETHYLENE GLYCOL DIMETHYL ETHER	222.282	18.067	221.053	1.804E+10	127.680
1097	C10H22S	n-DECYL MERCAPTAN	174.351	17.398	207.319	1.658E+10	120.550
1098	C10H22S	BUTYL-HEXYL-SULFIDE	174.344	17.301	207.343	1.634E+10	119.980
1099	C10H22S	ETHYL-OCTYL-SULFIDE	174.344	17.301	207.343	1.634E+10	119.980
1100	C10H22S	HEPTYL-PROPYL-SULFIDE	174.344	17.301	207.343	1.634E+10	119.980
1101	C10H22S	METHYL-NONYL-SULFIDE	174.344	17.301	207.343	1.634E+10	119.980
1102	C10H22S	PENTYL-SULFIDE	174.344	17.301	207.343	1.634E+10	119.980
1103	C10H22S2	PENTYL-DISULFIDE	206.404	17.275	224.786	1.764E+10	131.880
1104	C10H23N	n-DECYLAMINE	157.299	17.031	198.903	1.600E+10	116.000
1105	C11H10	1-METHYLNAPHTHALENE	142.2	20.176	139.800	9.840E+09	85.110
1106	C11H10	2-METHYLNAPHTHALENE	142.2	19.870	143.396	9.840E+09	85.110
1107	C11H14O2	n-BUTYL BENZOATE	178.231	18.289	178.114	1.370E+10	105.400
1108	C11H16	n-PENTYLBENZENE	148.248	17.023	173.453	1.285E+10	100.430
1109	C11H16O	p-tert-AMYLPHENOL	164.247	18.651	180.266	1.783E+10	131.670
1110	C11H20	1-UNDECYNE	152.279	16.122	198.085	1.678E+10	120.730
1111	C11H20O2	2-ETHYLHEXYL ACRYLATE	184.279	16.072	209.413	1.670E+10	120.880
1112	C11H22	1-UNDECENE	154.296	16.093	206.674	1.586E+10	115.920
1113	C11H22	1-CYCLOPENTYLHEXANE	154.295	16.088	194.667	1.483E+10	111.310
1114	C11H22	PENTYLCYCLOHEXANE	154.295	16.055	192.832	1.427E+10	111.380
1115	C11H22O	1-UNDECANAL	170.295	17.217	206.888	1.664E+10	120.880
1116	C11H24	n-UNDECANE	156.312	15.991	212.084	1.639E+10	119.410
1117	C11H24O	1-UNDECANOL	172.311	19.751	207.461	1.708E+10	124.010
1118	C11H24S	UNDECYL MERCAPTAN	188.378	17.290	223.958	1.790E+10	130.780
1119	C11H24S	BUTYL-HEPTYL-SULFIDE	188.371	17.284	223.938	1.769E+10	130.210
1120	C11H24S	DECYL-METHYL-SULFIDE	188.371	17.284	223.938	1.769E+10	130.210
1121	C11H24S	ETHYL-NONYL-SULFIDE	188.371	17.284	223.938	1.769E+10	130.210
1122	C11H24S	OCTYL-PROPYL-SULFIDE	188.371	17.284	223.938	1.769E+10	130.210
1123	C12H8O	DIBENZOFURAN	168.195	18.231	152.418	9.380E+09	86.640
1124	C12H9N	DIBENZOPYRROLE	167.21	20.824	141.977	1.207E+10	95.880
1125	C12H10	ACENAPHTHENE	154.211	19.481	149.792	9.900E+09	90.980
1126	C12H10	BIPHENYL	154.211	19.383	155.536	1.066E+10	91.680
1127	C12H10O	DIPHENYL ETHER	170.211	18.774	160.750	1.126E+10	95.380
1128	C12H11N	p-AMINODIPHENYL	169.226	20.660	168.703	1.164E+10	99.700
1129	C12H11N	DIPHENYLAMINE	169.226	20.550	159.724	1.301E+10	102.800
1130	C12H11N3	p-AMINOAZOBENZENE	197.24	19.752	203.818	---------	-------
1131	C12H11N3	1,3-DIPHENYLTRIAZENE	197.24	19.438	203.535	1.205E+10	108.420
1132	C12H12	1,2-DIMETHYLNAPHTHALENE	156.227	20.246	154.039	1.126E+10	96.260
1133	C12H12	1,3-DIMETHYLNAPHTHALENE	156.227	20.065	155.818	1.126E+10	96.260
1134	C12H12	1,4-DIMETHYLNAPHTHALENE	156.227	20.222	154.185	1.126E+10	96.260
1135	C12H12	1,5-DIMETHYLNAPHTHALENE	156.227	19.245	159.688	1.126E+10	96.260
1136	C12H12	1,6-DIMETHYLNAPHTHALENE	156.227	19.974	156.408	1.126E+10	96.260
1137	C12H12	1,7-DIMETHYLNAPHTHALENE	156.227	19.974	156.408	1.126E+10	96.260
1138	C12H12	2,3-DIMETHYLNAPHTHALENE	156.227	18.974	161.358	1.126E+10	96.260
1139	C12H12	2,6-DIMETHYLNAPHTHALENE	156.227	18.045	168.376	1.126E+10	96.260
1140	C12H12	2,7-DIMETHYLNAPHTHALENE	156.227	18.299	166.242	1.126E+10	96.260
1141	C12H12	1-ETHYLNAPHTHALENE	156.227	19.853	155.579	1.123E+10	95.350
1142	C12H12	2-ETHYLNAPHTHALENE	156.227	19.420	158.135	1.123E+10	95.350
1143	C12H12N2	p-AMINODIPHENYLAMINE	184.241	21.506	188.046	1.428E+10	109.880
1144	C12H12N2	HYDRAZOBENZENE	184.241	19.539	186.003	1.400E+10	110.680
1145	C12H14	1,2,3-TRIMETHYLINDENE	158.243	17.644	162.898	1.257E+10	99.530
1146	C12H14O4	DIETHYL PHTHALATE	222.241	20.207	199.686	1.576E+10	119.920
1147	C12H16	CYCLOHEXYLBENZENE	160.259	19.004	170.712	1.410E+10	101.840
1148	C12H18	m-DIISOPROPYLBENZENE	162.275	16.578	190.372	1.404E+10	101.000
1149	C12H18	p-DIISOPROPYLBENZENE	162.275	16.930	190.236	1.404E+10	101.000
1150	C12H18	n-HEXYLBENZENE	162.275	17.249	189.906	1.408E+10	109.860
1151	C12H18	1,2,3-TRIETHYLBENZENE	162.274	17.277	186.538	1.483E+10	104.460
1152	C12H18	1,2,4-TRIETHYLBENZENE	162.274	17.277	186.538	1.483E+10	104.460
1153	C12H18	1,3,5-TRIETHYLBENZENE	162.274	17.408	182.972	1.483E+10	104.460
1154	C12H18	HEXAMETHYLBENZENE	162.274	16.487	181.078	1.714E+10	114.780
1155	C12H20O4	DIBUTYL MALEATE	228.288	18.830	230.471	1.890E+10	136.060
1156	C12H22	BICYCLOHEXYL	166.307	16.991	188.376	1.464E+10	115.860
1157	C12H22	1-DODECYNE	166.306	16.137	214.621	1.709E+10	125.760
1158	C12H23N	DICYCLOHEXYLAMINE	181.321	17.189	199.418	1.642E+10	119.400
1159	C12H24	1-DODECENE	168.323	16.137	222.796	1.721E+10	126.150
1160	C12H24	1-CYCLOPENTYLHEPTANE	168.322	16.222	208.789	1.618E+10	121.540
1161	C12H24	1-CYCLOHEXYLHEXANE	168.322	16.851	188.734	1.562E+10	121.610
1162	C12H24O	1-DODECANAL	184.322	17.163	223.054	1.799E+10	131.110
1163	C12H24O2	n-DODECANOIC ACID	200.321	20.493	229.221	1.868E+10	135.710
1164	C12H26	n-DODECANE	170.338	15.911	228.633	1.774E+10	129.640
1165	C12H26O	DI-n-HEXYL ETHER	186.338	16.466	235.984	1.834E+10	133.340
1166	C12H26O	1-DODECANOL	186.338	19.485	224.405	1.843E+10	134.240
1167	C12H26O3	DIETHYLENE GLYCOL DI-n-BUTYL ETHER	218.337	15.444	247.754	1.954E+10	140.740
1168	C12H26S	n-DODECYL MERCAPTAN	202.404	16.534	240.475	1.928E+10	141.010
1169	C12H26S	BUTYL-OCTYL-SULFIDE	202.397	17.266	240.482	1.904E+10	140.440
1170	C12H26S	DECYL-ETHYL-SULFIDE	202.397	17.266	240.482	1.904E+10	140.440

NO	FORMULA	NAME	Molecular Weight, g/mol	Solubility Parameter (joule/cm³)^0.5	Liquid Volume cm³/mol	Van Der Waals Area, cm²/mol	Van Der Waals Volume, cm³/mol
1171	C12H26S	HEXYL-SULFIDE	202.397	17.266	240.482	1.904E+10	140.440
1172	C12H26S	METHYL-UNDECYL-SULFIDE	202.397	17.266	240.482	1.904E+10	140.440
1173	C12H26S	NONYL-PROPYL-SULFIDE	202.397	17.266	240.482	1.904E+10	140.440
1174	C12H26S2	HEXYL-DISULFIDE	234.457	17.044	258.244	2.034E+10	152.340
1175	C12H27BO3	TRI-n-BUTYL BORATE	230.156	12.851	269.528	---------	-------
1176	C12H27N	DODECYLAMINE	185.353	16.948	235.120	1.870E+10	137.000
1177	C12H27N	TRI-n-BUTYLAMINE	185.353	16.010	239.212	1.874E+10	137.410
1178	C13H10	FLUORENE	166.222	21.646	128.237	1.019E+10	93.670
1179	C13H10O	BENZOPHENONE	182.222	19.662	167.928	1.362E+10	106.420
1180	C13H12	DIPHENYLMETHANE	168.238	19.415	167.834	1.042E+10	93.410
1181	C13H14	1-PROPYLNAPHTHALENE	170.254	19.505	172.576	1.258E+10	105.580
1182	C13H14	2-PROPYLNAPHTHALENE	170.254	19.387	174.972	1.258E+10	105.580
1183	C13H14	2ETHYL-3-METHYLNAPHTHALENE	170.254	19.028	174.518	1.335E+10	109.020
1184	C13H14	2ETHYL-6-METHYLNAPHTHALENE	170.254	19.182	173.389	1.335E+10	109.020
1185	C13H14	2ETHYL-7-METHYLNAPHTHALENE	170.254	19.182	173.389	1.335E+10	109.020
1186	C13H20	n-HEPTYLBENZENE	176.302	17.417	206.668	1.555E+10	120.890
1187	C13H24	1-TRIDECYNE	180.333	16.156	231.001	1.844E+10	135.990
1188	C13H26	1-TRIDECENE	182.349	15.791	239.176	1.856E+10	136.380
1189	C13H26	1-CYCLOPENTYLOCTANE	182.348	16.040	227.656	1.753E+10	131.770
1190	C13H26	1-CYCLOHEXYLHEPTANE	182.348	15.973	225.821	1.697E+10	131.840
1191	C13H26O	1-TRIDECANAL	198.349	17.506	239.979	1.934E+10	141.340
1192	C13H26O2	n-BUTYL NONANOATE	214.348	16.539	251.827	1.994E+10	144.840
1193	C13H26O2	METHYL DODECANOATE	214.348	18.867	206.389	1.994E+10	144.840
1194	C13H28	n-TRIDECANE	184.365	15.901	244.445	1.909E+10	139.850
1195	C13H28O	1-TRIDECANOL	200.365	18.893	240.632	1.978E+10	144.470
1196	C13H28S	BUTYL-NONYL-SULFIDE	216.424	17.244	256.926	2.039E+10	150.670
1197	C13H28S	DECYL-PROPYL-SULFIDE	216.424	17.244	256.926	2.039E+10	150.670
1198	C13H28S	DODECYL-METHYL-SULFIDE	216.424	17.244	256.926	2.039E+10	150.670
1199	C13H28S	ETHYL-UNDECYL-SULFIDE	216.424	17.244	256.926	2.039E+10	150.670
1200	C13H28S	1-TRIDECANETHIOL	216.424	16.886	257.023	2.039E+10	154.680
1201	C14H8O2	ANTHRAQUINONE	208.216	17.803	193.625	1.204E+10	106.840
1202	C14H10	ANTHRACENE	178.233	17.809	182.940	1.084E+10	99.560
1203	C14H10	DIPHENYLACETYLENE	178.233	18.213	188.466	1.238E+10	105.640
1204	C14H10	PHENANTHRENE	178.233	20.067	167.078	1.084E+10	99.560
1205	C14H12	cis-STILBENE	180.249	18.961	178.369	1.276E+10	108.620
1206	C14H12	trans-STILBENE	180.249	19.538	172.317	1.276E+10	108.620
1207	C14H12O2	BENZYL BENZOATE	212.248	20.453	190.348	1.397E+10	115.510
1208	C14H14	1,1-DIPHENYLETHANE	182.265	18.873	183.008	1.311E+10	110.500
1209	C14H14	1,2-DIPHENYLETHANE	182.265	18.548	184.074	1.312E+10	110.540
1210	C14H14O	DIBENZYL ETHER	198.265	19.661	190.344	1.396E+10	115.840
1211	C14H16	1-n-BUTYLNAPHTHALENE	184.281	19.101	189.358	1.393E+10	115.810
1212	C14H16	2-BUTYLNAPHTHALENE	184.28	19.101	191.646	1.393E+10	115.810
1213	C14H22	n-OCTYLBENZENE	190.329	17.294	223.227	1.690E+10	131.120
1214	C14H22	1,2,3,4-TETRAETHYLBENZENE	190.328	17.202	215.428	1.988E+10	144.000
1215	C14H22	1,2,3,5-TETRAETHYLBENZENE	190.328	17.137	216.785	1.988E+10	144.000
1216	C14H22	1,2,4,5-TETRAETHYLBENZENE	190.328	17.097	217.524	1.988E+10	144.000
1217	C14H22O	p-tert-OCTYLPHENOL	206.328	17.553	223.185	1.801E+10	125.520
1218	C14H28	1-TETRADECENE	196.376	16.094	255.825	1.991E+10	146.610
1219	C14H28	1-CYCLOPENTYLNONANE	196.375	15.987	244.146	1.888E+10	142.000
1220	C14H28	1-CYCLOHEXYLOCTANE	196.375	15.933	242.372	1.832E+10	142.070
1221	C14H28O2	n-TETRADECANOIC ACID	228.375	19.081	265.672	2.138E+10	155.950
1222	C14H30	n-TETRADECANE	198.392	15.901	261.791	2.044E+10	150.100
1223	C14H30O	1-TETRADECANOL	214.392	18.703	260.444	2.113E+10	154.700
1224	C14H30S	BUTYL-DECYL-SULFIDE	230.451	17.205	273.473	2.174E+10	160.900
1225	C14H30S	DODECYL-ETHYL-SULFIDE	230.451	17.205	273.473	2.174E+10	160.900
1226	C14H30S	HEPTYL-SULFIDE	230.451	17.205	273.473	2.174E+10	160.900
1227	C14H30S	METHYL-TRIDECYL-SULFIDE	230.451	17.205	273.473	2.174E+10	160.900
1228	C14H30S	PROPYL-UNDECYL-SULFIDE	230.451	17.205	273.473	2.174E+10	160.900
1229	C14H30S	1-TETRADECANETHIOL	230.451	16.785	273.543	2.174E+10	164.910
1230	C14H30S2	HEPTYL-DISULFIDE	262.511	16.887	291.678	2.304E+10	172.800
1231	C14H31N	TETRADECYLAMINE	213.407	16.624	267.881	2.140E+10	157.200
1232	C15H10N2O2	DIPHENYLMETHANE-4,4'-DIISOCYANATE	250.257	20.607	208.490	1.790E+10	130.700
1233	C15H16O	p-CUMYLPHENOL	212.291	19.749	204.266	1.542E+10	126.300
1234	C15H16O2	BISPHENOL A	228.291	19.651	239.238	1.824E+10	134.550
1235	C15H18	1-PENTYLNAPHTHALENE	198.307	19.035	206.099	1.528E+10	124.040
1236	C15H18	2-PENTYLNAPHTHALENE	198.307	19.000	208.146	1.528E+10	124.040
1237	C15H24	n-NONYLBENZENE	204.356	17.047	239.778	1.822E+10	141.350
1238	C15H24O	2,6-DI-tert-BUTYL-p-CRESOL	220.355	16.605	246.067	1.910E+10	145.000
1239	C15H24O	NONYLPHENOL	220.355	19.455	232.204	1.890E+10	146.100
1240	C15H28	1-PENTADECYNE	208.386	16.138	264.090	2.114E+10	156.450
1241	C15H30	1-PENTADECENE	210.403	16.027	272.207	2.126E+10	156.840
1242	C15H30	1-CYCLOPENTYLDECANE	210.402	15.933	260.621	2.023E+10	152.230
1243	C15H30	1-CYCLOHEXYLNONANE	210.402	15.886	258.906	1.967E+10	152.300
1244	C15H30O2	PENTADECANOIC ACID	242.402	18.403	280.904	2.273E+10	166.180
1245	C15H32	n-PENTADECANE	212.419	15.920	277.673	2.179E+10	160.330
1246	C15H32O	1-PENTADECANOL	228.417	19.810	253.446	2.248E+10	164.930
1247	C15H32S	BUTYL-UNDECYL-SULFIDE	244.478	17.146	290.002	2.309E+10	171.130
1248	C15H32S	DODECYL-PROPYL-SULFIDE	244.478	17.146	290.002	2.309E+10	171.130

NO	FORMULA	NAME	Molecular Weight, g/mol	Solubility Parameter (joule/cm³)^0.5	Liquid Volume cm³/mol	Van Der Waals Area, cm²/mol	Van Der Waals Volume, cm³/mol
1249	C15H32S	ETHYL-TRIDECYL-SULFIDE	244.478	17.146	290.002	2.309E+10	171.130
1250	C15H32S	METHYL-TETRADECYL-SULFIDE	244.478	17.146	290.002	2.309E+10	171.130
1251	C15H32S	1-PENTADECANETHIOL	244.478	17.049	290.123	2.309E+10	175.140
1252	C16H10	FLUORANTHENE	202.255	19.277	184.693	1.126E+10	109.040
1253	C16H10	PYRENE	202.255	20.763	173.680	1.126E+10	109.040
1254	C16H12	1-PHENYLNAPHTHALENE	204.271	19.912	189.925	1.296E+10	116.500
1255	C16H20	1-n-HEXYLNAPHTHALENE	212.335	18.720	224.156	1.663E+10	136.270
1256	C16H22O4	DIBUTYL PHTHALATE	278.348	18.984	266.983	2.116E+10	160.840
1257	C16H26	n-DECYLBENZENE	218.382	17.066	256.336	1.950E+10	150.800
1258	C16H26	PENTAETHYLBENZENE	218.381	17.215	231.969	2.335E+10	167.900
1259	C16H30	1-HEXADECYNE	222.413	16.119	280.492	2.249E+10	166.680
1260	C16H32	n-DECYLCYCLOHEXANE	224.43	16.677	275.287	2.102E+10	162.530
1261	C16H32	1-CYCLOPENTYLUNDECANE	224.429	15.910	277.072	2.158E+10	162.460
1262	C16H32	1-HEXADECENE	224.43	15.773	288.658	2.261E+10	167.070
1263	C16H32O2	n-HEXADECANOIC ACID	256.429	16.212	300.929	2.408E+10	176.410
1264	C16H34	n-HEXADECANE	226.446	15.938	294.064	2.314E+10	170.560
1265	C16H34O	DI-n-OCTYL ETHER	242.445	16.505	301.971	2.374E+10	174.260
1266	C16H34O	1-HEXADECANOL	242.445	18.105	296.365	2.383E+10	175.160
1267	C16H34S	BUTYL-DODECYL-SULFIDE	258.505	17.060	306.484	2.444E+10	181.360
1268	C16H34S	ETHYL-TETRADECYL-SULFIDE	258.505	16.550	306.428	2.444E+10	181.360
1269	C16H34S	METHYL-PENTADECYL-SULFIDE	258.505	17.060	306.484	2.444E+10	181.360
1270	C16H34S	OCTYL-SULFIDE	258.505	17.060	306.484	2.444E+10	181.360
1271	C16H34S	PROPYL-TRIDECYL-SULFIDE	258.505	17.060	306.484	2.444E+10	181.360
1272	C16H34S	1-HEXADECANETHIOL	258.505	16.945	306.679	2.444E+10	185.370
1273	C16H34S2	OCTYL-DISULFIDE	290.565	16.951	325.008	2.574E+10	193.260
1274	C17H28	n-UNDECYLBENZENE	232.409	17.424	272.967	2.092E+10	161.740
1275	C17H32	1-HEPTADECYNE	236.44	15.844	297.015	2.396E+10	177.300
1276	C17H34	1-CYCLOPENTYLDODECANE	238.456	15.866	293.626	2.293E+10	172.690
1277	C17H34	1-CYCLOHEXYLUNDECANE	238.456	15.796	291.821	2.237E+10	172.760
1278	C17H34	1-HEPTADECENE	238.457	15.843	305.081	2.423E+10	179.060
1279	C17H36	n-HEPTADECANE	240.473	15.924	311.120	2.449E+10	180.790
1280	C17H36O	1-HEPTADECANOL	256.472	18.113	313.115	2.518E+10	185.390
1281	C17H36S	BUTYL-TRIDECYL-SULFIDE	272.531	16.938	323.050	2.579E+10	191.590
1282	C17H36S	ETHYL-PENTADECYL-SULFIDE	272.531	16.938	323.050	2.579E+10	191.590
1283	C17H36S	HEXADECYL-METHYL-SULFIDE	272.531	16.938	323.050	2.579E+10	191.590
1284	C17H36S	PROPYL-TETRADECYL-SULFIDE	272.531	16.938	323.050	2.579E+10	191.590
1285	C17H36S	1-HEPTADECANETHIOL	272.531	18.693	261.713	2.579E+10	195.600
1286	C18H12	CHRYSENE	228.293	19.487	210.334	1.326E+10	125.160
1287	C18H14	m-TERPHENYL	230.309	19.147	221.157	1.508E+10	133.400
1288	C18H14	o-TERPHENYL	230.309	18.427	219.652	1.508E+10	133.400
1289	C18H14	p-TERPHENYL	230.309	17.392	239.386	1.508E+10	133.400
1290	C18H15P	TRIPHENYLPHOSPHINE	262.291	18.677	244.466	1.779E+10	147.960
1291	C18H15O4P	TRIPHENYL PHOSPHATE	326.288	16.652	233.324	2.223E+10	163.320
1292	C18H16N2	N,N'-DIPHENYL-p-PHENYLENEDIAMINE	260.339	19.632	255.464	1.729E+10	151.160
1293	C18H22	2,3-DIMETHYL-2,3-DIPHENYLBUTANE	238.373	16.652	233.324	1.914E+10	153.020
1294	C18H22O2	DICUMYL PEROXIDE	270.371	16.889	264.723	2.028E+10	160.420
1295	C18H30	n-DODECYLBENZENE	246.436	17.120	290.423	2.218E+10	171.240
1296	C18H30	HEXAETHYLBENZENE	246.435	15.220	295.918	2.676E+10	191.800
1297	C18H32O2	LINOLEIC ACID	280.451	18.627	310.779	2.358E+10	189.830
1298	C18H34	1-OCTADECYNE	250.467	17.626	258.048	2.519E+10	187.140
1299	C18H34O2	OLEIC ACID	282.467	18.504	318.201	2.624E+10	193.350
1300	C18H34O4	DIBUTYL SEBACATE	314.466	17.126	337.270	2.754E+10	200.960
1301	C18H34O4	DIHEXYL ADIPATE	314.466	17.483	337.306	2.750E+10	201.000
1302	C18H36	1-CYCOPENTYLTRIDECANE	252.482	15.831	310.103	2.428E+10	182.920
1303	C18H36	1-CYCLOHEXYLDODECANE	252.482	15.732	308.244	2.372E+10	182.990
1304	C18H36	1-OCTADECENE	252.484	16.184	321.590	2.531E+10	187.530
1305	C18H36O2	STEARIC ACID	284.483	17.397	335.233	2.678E+10	196.870
1306	C18H38	n-OCTADECANE	254.5	15.799	329.141	2.584E+10	191.020
1307	C18H38O	DINONYL ETHER	270.499	-------	334.735	2.644E+10	194.720
1308	C18H38O	1-OCTADECANOL	270.499	17.739	333.572	2.653E+10	195.620
1309	C18H38S	BUTYL-TETRADECYL-SULFIDE	286.558	19.357	255.706	2.714E+10	201.820
1310	C18H38S	ETHYL-HEXADECYL-SULFIDE	286.558	19.357	255.706	2.714E+10	201.820
1311	C18H38S	HEPTADECYL-METHYL-SULFIDE	286.558	19.357	255.706	2.714E+10	201.820
1312	C18H38S	NONYL-SULFIDE	286.558	19.357	255.706	2.714E+10	201.820
1313	C18H38S	PENTADECYL-PROPYL-SULFIDE	286.558	19.357	255.706	2.714E+10	201.820
1314	C18H38S	1-OCTADECANETHIOL	286.558	18.712	269.283	2.714E+10	205.830
1315	C18H38S2	NONYL-DISULFIDE	318.618	16.601	358.368	2.844E+10	213.720
1316	C19H26	1-n-NONYLNAPHTHALENE	254.415	17.409	272.495	2.060E+10	166.150
1317	C19H32	n-TRIDECYLBENZENE	260.463	16.966	305.895	2.362E+10	182.270
1318	C19H36	1-NONADECYNE	264.493	17.671	265.868	2.654E+10	197.370
1319	C19H36O2	METHYL OLEATE	296.494	16.948	340.710	2.750E+10	202.700
1320	C19H38	1-CYCLOPENTYLTETRADECANE	266.509	15.861	326.459	2.563E+10	193.150
1321	C19H38	1-CYCLOHEXYLTRIDECANE	266.509	15.680	324.762	2.507E+10	193.220
1322	C19H38	1-NONADECENE	266.511	16.258	338.082	2.666E+10	197.760
1323	C19H38O2	NONADECANOIC ACID	298.51	17.089	351.966	2.813E+10	207.100
1324	C19H40	n-NONADECANE	268.527	15.984	346.128	2.719E+10	201.250
1325	C19H40O	1-NONADECANOL	284.524	19.256	283.179	2.788E+10	205.850
1326	C19H40S	BUTYL-PENTADECYL-SULFIDE	300.585	19.297	262.711	2.849E+10	212.050

357

NO	FORMULA	NAME	Molecular Weight, g/mol	Solubility Parameter (joule/cm³)^0.5	Liquid Volume cm³/mol	Van Der Waals Area, cm²/mol	Van Der Waals Volume, cm³/mol
1327	C19H40S	ETHYL-HEPTADECYL-SULFIDE	300.585	19.297	262.711	2.849E+10	212.050
1328	C19H40S	HEXADECYL-PROPYL-SULFIDE	300.585	19.297	262.711	2.849E+10	212.050
1329	C19H40S	METHYL-OCTADECYL-SULFIDE	300.585	19.297	262.711	2.849E+10	212.050
1330	C19H40S	1-NONADECANETHIOL	300.585	18.666	276.111	2.849E+10	216.060
1331	C20H16	TRIPHENYLETHYLENE	256.347	18.615	246.072	1.732E+10	148.550
1332	C20H28	1-n-DECYLNAPHTHALENE	268.442	17.202	289.211	2.210E+10	176.400
1333	C20H30O2	ABIETIC ACID	302.457	18.663	282.074	2.393E+10	185.700
1334	C20H31N	DEHYDROABIETYLAMINE	285.473	17.604	294.987	2.204E+10	183.740
1335	C20H34	1-PHENYLTETRADECANE	274.489	16.247	322.495	2.567E+10	195.060
1336	C20H38	1-EICOSYNE	278.52	16.263	272.413	2.801E+10	207.990
1337	C20H40	1-CYCLOPENTYLPENTADECANE	280.536	15.854	343.124	2.698E+10	203.380
1338	C20H40	1-CYCLOHEXYLTETRADECANE	280.536	15.623	341.336	2.642E+10	203.450
1339	C20H40	1-EICOSENE	280.538	16.262	355.733	2.828E+10	209.750
1340	C20H42	n-EICOSANE	282.553	15.770	363.962	2.854E+10	211.480
1341	C20H42O	1-EICOSANOL	298.553	17.816	365.918	2.923E+10	216.080
1342	C20H42S	BUTYL-HEXADECYL-SULFIDE	314.612	19.231	267.890	2.984E+10	222.280
1343	C20H42S	DECYL-SULFIDE	314.612	19.231	267.890	2.984E+10	222.280
1344	C20H42S	ETHYL-OCTADECYL-SULFIDE	314.612	19.231	267.890	2.984E+10	222.280
1345	C20H42S	HEPTADECYL-PROPYL-SULFIDE	314.612	19.231	267.890	2.984E+10	222.280
1346	C20H42S	METHYL-NONADECYL-SULFIDE	314.612	19.231	267.890	2.984E+10	222.280
1347	C20H42S	1-EICOSANETHIOL	314.612	18.564	282.986	2.984E+10	226.290
1348	C20H42S2	DECYL-DISULFIDE	346.672	16.100	391.749	3.114E+10	234.180
1349	C21H21O4P	TRI-o-CRESYL PHOSPHATE	368.369	13.782	316.216	2.666E+10	196.770
1350	C21H36	1-PHENYLPENTADECANE	288.515	16.209	339.047	2.702E+10	205.290
1351	C21H42	1-CYCLOPENTYLHEXADECANE	294.563	15.725	359.581	2.833E+10	213.610
1352	C21H42	1-CYCLOHEXYLPENTADECANE	294.563	18.067	264.378	2.777E+10	213.680
1353	C22H38	1-PHENYLHEXADECANE	302.542	16.751	331.323	2.837E+10	215.520
1354	C22H44	1-CYCLOHEXYLHEXADECANE	308.59	18.134	270.994	2.912E+10	223.910
1355	C22H44O2	n-BUTYL STEARATE	340.59	15.849	398.245	3.209E+10	236.910
1356	C24H38O4	DIISOOCTYL PHTHALATE	390.563	17.223	397.200	3.212E+10	244.260
1357	C24H38O4	DIOCTYL PHTHALATE	390.563	16.980	398.538	3.212E+10	244.260
1358	C24H42O	DINONYLPHENOL	346.597	19.759	297.487	3.111E+10	239.100
1359	C26H20	TETRAPHENYLETHYLENE	332.445	18.180	282.307	2.206E+10	190.180
1360	C28H46O4	DIISODECYL PHTHALATE	446.671	16.676	458.950	3.736E+10	283.600

Table 14-2 SOLUBILITY PARAMETER, LIQUID VOLUME, AND VAN DER WAALS AREA AND VOLUME - INORGANIC COMPOUNDS

NO	FORMULA	NAME	Molecular Weight, g/mol	Solubility Parameter (joule/cm³)^0.5	Liquid Volume cm³/mol	Van Der Waals Area, cm²/mol	Van Der Waals Volume, cm³/mol
1	Ag	SILVER	107.868	154.280	11.604	----------	-------
2	AgCl	SILVER CHLORIDE	143.321	-------	29.507	----------	-------
3	AgI	SILVER IODIDE	234.773	-------	42.250	----------	-------
4	Al	ALUMINUM	26.982	128.950	11.337	----------	-------
5	AlB3H12	ALUMINUM BOROHYDRIDE	71.510	-------	--------	----------	-------
6	AlBr3	ALUMINUN BROMIDE	266.694	-------	114.039	----------	-------
7	AlCl3	ALUMINUM CHLORIDE	133.340	18.628	103.514	----------	-------
8	AlF3	ALUMINUM FLUORIDE	83.977	-------	32.893	----------	-------
9	AlI3	ALUMINUM IODIDE	407.695	-------	127.513	----------	-------
10	Al2O3	ALUMINUM OXIDE	101.961	-------	29.029	----------	-------
11	Al2S3O12	ALUMINUM SULFATE	342.154	-------	--------	----------	-------
12	Ar	ARGON	39.948	14.138	28.586	2.670E+09	16.800
13	As	ARSENIC	74.922	-------	14.174	----------	-------
14	AsBr3	ARSENIC TRIBROMIDE	314.634	-------	96.294	----------	-------
15	AsCl3	ARSENIC TRICHLORIDE	181.280	-------	84.316	----------	-------
16	AsF3	ARSENIC TRIFLUORIDE	131.917	-------	49.791	----------	-------
17	AsF5	ARSENIC PENTAFLUORIDE	169.914	-------	72.924	----------	-------
18	AsH3	ARSINE	77.945	17.504	48.100	----------	-------
19	AsI3	ARSENIC TRIIODIDE	455.635	-------	112.448	----------	-------
20	As2O3	ARSENIC TRIOXIDE	197.841	-------	57.342	----------	-------
21	At	ASTATINE	210.000	-------	--------	----------	-------
22	Au	GOLD	196.967	-------	11.397	----------	-------
23	B	BORON	10.811	-------	5.006	----------	-------
24	BBr3	BORON TRIBROMIDE	250.523	-------	95.447	----------	-------
25	BCl3	BORON TRICHLORIDE	117.169	15.736	87.071	8.320E+09	58.120
26	BF3	BORON TRIFLUORIDE	67.806	19.136	43.145	5.730E+09	38.560
27	BH2CO	BORINE CARBONYL	40.837	-------	--------	----------	-------
28	BH3O3	BORIC ACID	61.833	-------	--------	6.650E+09	45.520
29	B2D6	DEUTERODIBORANE	33.718	-------	--------	----------	-------
30	B2H5Br	DIBORANE HYDROBROMIDE	106.566	-------	--------	----------	-------
31	B2H6	DIBORANE	27.670	14.083	64.533	----------	-------
32	B3N3H6	BORINE TRIAMINE	80.501	-------	89.262	----------	-------
33	B4H10	TETRABORANE	53.323	-------	102.498	----------	-------
34	B5H9	PENTABORANE	63.126	-------	98.200	----------	-------
35	B5H11	TETRAHYDROPENTABORANE	65.142	-------	87.956	----------	-------
36	B10H14	DECABORANE	122.221	-------	140.869	----------	-------
37	Ba	BARIUM	137.327	-------	41.301	----------	-------
38	Be	BERYLLIUM	9.012	-------	5.278	----------	-------
39	BeB2H8	BERYLLIUM BOROHYDRIDE	38.698	-------	--------	----------	-------
40	BeBr2	BERYLLIUM BROMIDE	168.820	-------	52.786	----------	-------
41	BeCl2	BERYLLIUM CHLORIDE	79.918	-------	52.848	----------	-------
42	BeF2	BERYLLIUM FLUORIDE	47.009	-------	--------	----------	-------
43	BeI2	BERYLLIUM IODIDE	262.821	-------	65.837	----------	-------
44	Bi	BISMUTH	208.980	-------	20.758	----------	-------
45	BiBr3	BISMUTH TRIBROMIDE	448.692	-------	98.440	----------	-------
46	BiCl3	BISMUTH TRICHLORIDE	315.338	-------	81.332	----------	-------
47	BrF5	BROMINE PENTAFLUORIDE	174.896	-------	70.923	----------	-------
48	Br2	BROMINE	159.808	23.660	51.480	4.180E+09	28.800
49	C	CARBON	12.011	76.877	7.455	----------	-------
50	CCl2O	PHOSGENE	98.916	17.836	70.547	5.200E+09	34.900
51	CF2O	CARBONYL FLUORIDE	66.007	16.450	61.324	3.800E+09	23.140
52	CH4N2O	UREA	60.056	-------	44.984	4.640E+09	32.340
53	CH4N2S	THIOUREA	76.122	24.914	90.638	5.390E+09	36.890
54	CNBr	CYANOGEN BROMIDE	105.922	-------	56.952	----------	-------
55	CNCl	CYANOGEN CHLORIDE	61.470	21.833	51.356	3.990E+09	26.320
56	CNF	CYANOGEN FLUORIDE	45.016	-------	32.931	----------	-------
57	CO	CARBON MONOXIDE	28.010	6.402	35.456	2.780E+09	16.200
58	COS	CARBONYL SULFIDE	60.076	18.130	50.984	4.050E+09	25.900
59	COSe	CARBON OXYSELENIDE	106.970	-------	56.147	----------	-------
60	CO2	CARBON DIOXIDE	44.010	14.564	37.272	3.230E+09	19.700
61	CS2	CARBON DISULFIDE	76.143	20.411	60.639	4.750E+09	31.200
62	CSeS	CARBON SELENOSULFIDE	123.037	-------	62.241	----------	-------
63	C2N2	CYANOGEN	52.035	-------	54.351	----------	-------
64	C3S2	CARBON SUBSULFIDE	100.165	-------	--------	----------	-------
65	Ca	CALCIUM	40.078	-------	29.361	----------	-------
66	CaF2	CALCIUM FLUORIDE	78.075	-------	30.774	----------	-------
67	CbF5	COLUMBIUM FLUORIDE	187.898	-------	61.820	----------	-------
68	Cd	CADMIUM	112.411	-------	14.022	----------	-------
69	CdCl2	CADMIUM CHLORIDE	183.316	-------	54.036	----------	-------
70	CdF2	CADMIUM FLUORIDE	150.408	-------	24.504	----------	-------
71	CdI2	CADMIUM IODIDE	366.220	-------	82.914	----------	-------
72	CdO	CADMIUM OXIDE	128.410	-------	--------	----------	-------
73	ClF	CHLORINE MONOFLUORIDE	54.451	-------	33.568	----------	-------
74	ClFO3	PERCHLORYL FLUORIDE	102.449	16.981	60.457	----------	-------
75	ClF3	CHLORINE TRIFLUORIDE	92.448	-------	50.656	----------	-------
76	ClF5	CHLORINE PENTAFLUORIDE	130.446	-------	69.166	----------	-------
77	ClHO3S	CHLOROSULFONIC ACID	116.525	26.116	66.941	5.880E+09	40.580
78	ClHO4	PERCHLORIC ACID	100.458	22.255	57.085	----------	-------

NO	FORMULA	NAME	Molecular Weight, g/mol	Solubility Parameter (joule/cm^3)$^{0.5}$	Liquid Volume cm^3/mol	Van Der Waals Area, cm^2/mol	Van Der Waals Volume, cm^3/mol
79	ClO2	CHLORINE DIOXIDE	67.452	23.936	41.769	---------	-------
80	Cl2	CHLORINE	70.905	20.148	45.389	3.810E+09	24.100
81	Cl2O	CHLORINE MONOXIDE	86.905	-------	53.904	---------	-------
82	Cl2O7	CHLORINE HEPTOXIDE	182.901	-------	98.881	---------	-------
83	Co	COBALT	58.933	-------	7.684	---------	-------
84	CoCl2	COBALT CHLORIDE	129.839	-------	44.893	---------	-------
85	CoNC3O4	COBALT NITROSYL TRICARBONYL	172.971	-------	--------	---------	-------
86	Cr	CHROMIUM	51.996	-------	8.049	---------	-------
87	CrC6O6	CHROMIUM CARBONYL	220.059	-------	138.376	---------	-------
88	CrO2Cl2	CHROMIUM OXYCHLORIDE	154.900	-------	81.436	---------	-------
89	Cs	CESIUM	132.905	-------	71.721	---------	-------
90	CsBr	CESIUM BROMIDE	212.809	-------	69.542	---------	-------
91	CsCl	CESIUM CHLORIDE	168.358	-------	62.205	---------	-------
92	CsF	CESIUM FLUORIDE	151.904	-------	--------	---------	-------
93	CsI	CESIUM IODIDE	259.810	-------	83.601	---------	-------
94	Cu	COPPER	63.546	-------	8.003	---------	-------
95	CuBr	CUPROUS BROMIDE	143.450	-------	32.506	---------	-------
96	CuCl	CUPROUS CHLORIDE	98.999	45.989	26.420	---------	-------
97	CuCl2	CUPRIC CHLORIDE	134.451	-------	44.809	---------	-------
98	CuI	COPPER IODIDE	190.450	-------	38.241	---------	-------
99	DCN	DEUTERIUM CYANIDE	28.034	-------	--------	---------	-------
100	D2	DEUTERIUM	4.032	6.264	25.112	1.380E+09	6.210
101	D2O	DEUTERIUM OXIDE	20.031	49.210	18.372	---------	-------
102	Eu	EUROPIUM	151.965	-------	31.402	---------	-------
103	F2	FLUORINE	37.997	15.184	25.237	2.500E+09	14.200
104	F2O	FLUORINE OXIDE	53.996	-------	35.637	---------	-------
105	Fe	IRON	55.847	212.120	8.024	---------	-------
106	FeC5O5	IRON PENTACARBONYL	195.899	-------	134.983	---------	-------
107	FeCl2	FERROUS CHLORIDE	126.752	-------	44.895	---------	-------
108	FeCl3	FERRIC CHLORIDE	162.205	-------	62.647	---------	-------
109	Fr	FRANCIUM	223.000	-------	--------	---------	-------
110	Ga	GALLIUM	69.723	-------	--------	---------	-------
111	GaCl3	GALLIUM TRICHLORIDE	176.081	23.527	85.783	---------	-------
112	Gd	GADOLINIUM	157.250	-------	21.565	---------	-------
113	Ge	GERMANIUM	72.610	-------	13.226	---------	-------
114	GeBr4	GERMANIUM BROMIDE	392.226	-------	124.969	---------	-------
115	GeCl4	GERMANIUM CHLORIDE	214.421	-------	115.627	---------	-------
116	GeHCl3	TRICHLORO GERMANE	179.976	-------	93.780	---------	-------
117	GeH4	GERMANE	76.642	15.230	55.645	---------	-------
118	Ge2H6	DIGERMANE	151.268	-------	88.949	---------	-------
119	Ge3H8	TRIGERMANE	225.894	-------	102.183	---------	-------
120	HBr	HYDROGEN BROMIDE	80.912	20.850	37.597	2.990E+09	17.890
121	HCN	HYDROGEN CYANIDE	27.026	24.788	39.770	3.030E+09	18.130
122	HCl	HYDROGEN CHLORIDE	36.461	21.894	30.556	2.750E+09	16.020
123	HF	HYDROGEN FLUORIDE	20.006	-------	20.945	1.830E+09	9.090
124	HI	HYDROGEN IODIDE	127.912	19.867	45.718	3.980E+09	25.370
125	HNO3	NITRIC ACID	63.013	29.619	41.771	3.910E+09	24.860
126	H2	HYDROGEN	2.016	6.648	28.604	1.380E+09	6.210
127	H2O	WATER	18.015	47.813	18.069	2.260E+09	12.370
128	H2O2	HYDROGEN PEROXIDE	34.015	45.926	23.577	2.920E+09	16.080
129	H2S	HYDROGEN SULFIDE	34.082	18.000	36.142	3.100E+09	18.720
130	H2SO4	SULFURIC ACID	98.079	25.624	53.510	6.420E+09	43.800
131	H2S2	HYDROGEN DISULFIDE	66.148	-------	49.884	---------	-------
132	H2Se	HYDROGEN SELENIDE	80.976	-------	40.433	---------	-------
133	H2Te	HYDROGEN TELLURIDE	129.616	-------	51.928	---------	-------
134	H3NO3S	SULFAMIC ACID	97.095	-------	--------	5.800E+09	38.880
135	He	HELIUM-3	3.016	-------	50.972	---------	-------
136	He	HELIUM-4	4.003	1.222	32.257	1.480E+09	6.920
137	Hf	HAFNIUM	178.490	-------	14.874	---------	-------
138	Hg	MERCURY	200.590	64.691	14.872	---------	-------
139	HgBr2	MERCURIC BROMIDE	360.398	-------	70.381	---------	-------
140	HgCl2	MERCURIC CHLORIDE	271.495	-------	62.186	---------	-------
141	HgI2	MERCURIC IODIDE	454.399	-------	87.330	---------	-------
142	IF7	IODINE HEPTAFLUORIDE	259.893	-------	--------	---------	-------
143	I2	IODINE	253.809	25.359	63.835	3.250E+09	19.800
144	In	INDIUM	114.818	-------	16.329	---------	-------
145	Ir	IRIDIUM	192.220	-------	9.611	---------	-------
146	K	POTASSIUM	39.098	41.783	49.392	---------	-------
147	KBr	POTASSIUM BROMIDE	119.002	-------	57.024	---------	-------
148	KCl	POTASSIUM CHLORIDE	74.551	40.906	48.721	---------	-------
149	KF	POTASSIUM FLUORIDE	58.097	-------	31.069	---------	-------
150	KI	POTASSIUM IODIDE	166.003	-------	70.353	---------	-------
151	KOH	POTASSIUM HYDROXIDE	56.106	-------	32.732	---------	-------
152	Kr	KRYPTON	83.800	15.276	34.648	---------	-------
153	La	LANTHANUM	138.906	-------	24.490	---------	-------
154	Li	LITHIUM	6.941	111.140	12.851	---------	-------
155	LiBr	LITHIUM BROMIDE	86.845	-------	34.502	---------	-------
156	LiCl	LITHIUM CHLORIDE	42.394	-------	28.572	---------	-------

NO	FORMULA	NAME	Molecular Weight, g/mol	Solubility Parameter (joule/cm³)⁰·⁵	Liquid Volume cm³/mol	Van Der Waals Area, cm²/mol	Van Der Waals Volume, cm³/mol
157	LiF	LITHIUM FLUORIDE	25.939	-------	14.415	---------	-------
158	LiI	LITHIUM IODIDE	133.845	-------	43.296	---------	-------
159	Lu	LUTECIUM	174.967	-------	19.265	---------	-------
160	Mg	MAGNESIUM	24.305	-------	15.396	---------	-------
161	MgCl2	MAGNESIUM CHLORIDE	95.21	-------	56.761	---------	-------
162	MgO	MAGNESIUM OXIDE	40.304	220.610	6.772	---------	-------
163	Mn	MANGANESE	54.938	-------	8.544	---------	-------
164	MnCl2	MANGANESE CHLORIDE	125.843	-------	48.912	---------	-------
165	Mo	MOLYBDENUM	95.94	-------	10.283	---------	-------
166	MoF6	MOLYBDENUM FLUORIDE	209.93	-------	83.199	---------	-------
167	MoO3	MOLYBDENUM OXIDE	143.938	-------	33.604	---------	-------
168	NCl3	NITROGEN TRICHLORIDE	120.365	-------	73.370	5.690E+09	41.050
169	ND3	HEAVY AMMONIA	20.055	-------	--------	---------	-------
170	NF3	NITROGEN TRIFLUORIDE	71.002	14.996	45.599	---------	-------
171	NH3	AMMONIA	17.031	29.217	24.993	2.450E+09	13.800
172	NH3O	HYDROXYLAMINE	33.03	-------	--------	3.200E+09	18.580
173	NH4Br	AMMONIUM BROMIDE	97.943	-------	--------	---------	-------
174	NH4Cl	AMMONIUM CHLORIDE	53.491	-------	37.953	---------	-------
175	NH4I	AMMONIUM IODIDE	144.943	-------	--------	---------	-------
176	NH5O	AMMONIUM HYDROXIDE	35.046	-------	42.123	---------	-------
177	NH5S	AMMONIUM HYDROGENSULFIDE	51.112	-------	47.330	---------	-------
178	NO	NITRIC OXIDE	30.006	23.138	23.427	2.470E+09	13.900
179	NOCl	NITROSYL CHLORIDE	65.459	21.794	49.406	---------	-------
180	NOF	NITROSYL FLUORIDE	49.005	-------	36.957	---------	-------
181	NO2	NITROGEN DIOXIDE	46.006	33.555	31.663	2.850E+09	16.800
182	N2	NITROGEN	28.013	9.082	34.677	2.720E+09	15.800
183	N2F4	TETRAFLUOROHYDRAZINE	104.007	13.310	74.077	5.060E+09	32.660
184	N2H4	HYDRAZINE	32.045	36.225	31.932	3.480E+09	21.080
185	N2H4C	AMMONIUM CYANIDE	44.056	-------	--------	---------	-------
186	N2H6CO2	AMMONIUM CARBAMATE	78.071	-------	--------	---------	-------
187	N2O	NITROUS OXIDE	44.013	20.863	36.002	3.130E+09	18.900
188	N2O3	NITROGEN TRIOXIDE	76.012	32.805	53.817	---------	-------
189	N2O4	NITROGEN TETRAOXIDE	92.011	-------	64.186	---------	-------
190	N2O5	NITROGEN PENTOXIDE	108.01	-------	71.267	---------	-------
191	Na	SODIUM	22.99	63.921	23.818	---------	-------
192	NaBr	SODIUM BROMIDE	102.894	50.600	43.922	---------	-------
193	NaCN	SODIUM CYANIDE	49.008	-------	40.170	---------	-------
194	NaCl	SODIUM CHLORIDE	58.442	57.354	37.712	---------	-------
195	NaF	SODIUM FLUORIDE	41.988	74.999	21.566	---------	-------
196	NaI	SODIUM IODIDE	149.894	-------	55.725	---------	-------
197	NaOH	SODIUM HYDROXIDE	39.997	-------	22.464	---------	-------
198	Na2SO4	SODIUM SULFATE	142.043	-------	68.659	---------	-------
199	Nb	NIOBIUM	92.906	-------	11.865	---------	-------
200	Nd	NEODYMIUM	144.24	-------	22.302	---------	-------
201	Ne	NEON	20.18	9.417	16.762	---------	-------
202	Ni	NICKEL	58.693	-------	7.544	---------	-------
203	NiC4O4	NICKEL CARBONYL	170.735	-------	130.748	---------	-------
204	NiF2	NICKEL FLUORIDE	96.69	-------	23.895	---------	-------
205	Np	NEPTUNIUM	237	-------	--------	---------	-------
206	O2	OXYGEN	31.999	8.182	28.020	2.350E+09	13.000
207	O3	OZONE	47.998	18.954	35.601	---------	-------
208	Os	OSMIUM	190.23	-------	9.464	---------	-------
209	OsOF5	OSMIUM OXIDE PENTAFLUORIDE	301.221	-------	--------	---------	-------
210	OsO4	OSMIUM TETROXIDE - YELLOW	254.228	-------	58.239	---------	-------
211	OsO4	OSMIUM TETROXIDE - WHITE	254.228	-------	58.239	---------	-------
212	P	PHOSPHORUS - WHITE	30.974	56.678	17.747	2.010E+09	10.440
213	PBr3	PHOSPHORUS TRIBROMIDE	270.686	-------	95.635	---------	-------
214	PCl2F3	PHOSPHORUS DICHLORIDE TRIFLUORIDE	158.874	-------	--------	---------	-------
215	PCl3	PHOSPHORUS TRICHLORIDE	137.332	18.398	87.687	---------	-------
216	PCl5	PHOSPHORUS PENTACHLORIDE	208.237	-------	62.669	9.690E+09	68.540
217	PH3	PHOSPHINE	33.998	16.517	45.716	3.230E+09	19.700
218	PH4Br	PHOSPHONIUM BROMIDE	114.91	-------	--------	---------	-------
219	PH4Cl	PHOSPHONIUM CHLORIDE	70.458	-------	--------	---------	-------
220	PH4I	PHOSPHONIUM IODIDE	161.91	-------	56.955	---------	-------
221	POCl3	PHOSPHORUS OXYCHLORIDE	153.331	19.808	92.005	7.520E+09	52.100
222	PSBr3	PHOSPHORUS THIOBROMIDE	302.752	-------	115.091	---------	-------
223	PSCl3	PHOSPHORUS THIOCHLORIDE	169.398	-------	103.120	---------	-------
224	P4O6	PHOSPHORUS TRIOXIDE	219.891	-------	103.336	---------	-------
225	P4O10	PHOSPHORUS PENTOXIDE	283.889	-------	--------	---------	-------
226	P4S10	PHOSPHORUS PENTASULFIDE	444.555	-------	--------	---------	-------
227	Pb	LEAD	207.2	-------	--------	---------	-------
228	PbBr2	LEAD BROMIDE	367.008	-------	65.230	---------	-------
229	PbCl2	LEAD CHLORIDE	278.105	-------	56.477	---------	-------
230	PbF2	LEAD FLUORIDE	245.197	-------	31.630	---------	-------
231	PbI2	LEAD IODIDE	461.009	-------	79.549	---------	-------
232	PbO	LEAD OXIDE	223.199	-------	24.895	---------	-------
233	PbS	LEAD SULFIDE	239.266	-------	33.910	---------	-------
234	Pd	PALLADIUM	106.42	-------	10.253	---------	-------

NO	FORMULA	NAME	Molecular Weight, g/mol	Solubility Parameter (Joule/cm³)⁰·⁵	Liquid Volume cm³/mol	Van Der Waals Area, cm²/mol	Van Der Waals Volume, cm³/mol
235	Po	POLONIUM	209	-------	24.218	---------	-------
236	Pt	PLATINUM	195.08	-------	9.853	---------	-------
237	Ra	RADIUM	226	-------	48.971	---------	-------
238	Rb	RUBIDIUM	85.468	-------	44.844	---------	-------
239	RbBr	RUBIDIUM BROMIDE	165.372	-------	62.466	---------	-------
240	RbCl	RUBIDIUM CHLORIDE	120.921	-------	55.126	---------	-------
241	RbF	RUBIDIUM FLUORIDE	104.466	-------	30.504	---------	-------
242	RbI	RUBIDIUM IODIDE	212.372	-------	75.137	---------	-------
243	Re	RHENIUM	186.207	-------	9.852	---------	-------
244	Re2O7	RHENIUM HEPTOXIDE	484.41	-------	88.276	---------	-------
245	Rh	RHODIUM	102.906	-------	9.662	---------	-------
246	Rn	RADON	222	-------	51.247	---------	-------
247	Ru	RUTHENIUM	101.07	-------	9.272	---------	-------
248	RuF5	RUTHENIUM PENTAFLUORIDE	196.062	-------	75.094	---------	-------
249	S	SULFUR	32.066	20.245	17.870	2.060E+09	10.800
250	SF4	SULFUR TETRAFLUORIDE	108.06	-------	59.614	---------	-------
251	SF6	SULFUR HEXAFLUORIDE	146.056	12.694	79.170	8.200E+09	46.800
252	SOBr2	THIONYL BROMIDE	207.873	-------	78.188		
253	SOCl2	THIONYL CHLORIDE	118.971	19.762	73.010	5.540E+09	37.740
254	SOF2	SULFUROUS OXYFLUORIDE	86.062	-------	53.300		-------
255	SO2	SULFUR DIOXIDE	64.065	12.273	43.796	4.030E+09	25.730
256	SO2Cl2	SULFURYL CHLORIDE	134.97	18.436	81.417	7.500E+09	54.340
257	SO3	SULFUR TRIOXIDE	80.064	31.093	42.195	4.930E+09	32.540
258	S2Cl2	SULFUR MONOCHLORIDE	135.037	-------	80.821	---------	-------
259	Sb	ANTIMONY	121.757	-------	18.758	---------	-------
260	SbBr3	ANTIMONY TRIBROMIDE	361.469	-------	89.939	---------	-------
261	SbCl3	ANTIMONY TRICHLORIDE	228.115	24.725	80.697	---------	-------
262	SbCl5	ANTIMONY PENTACHLORIDE	299.021	-------	128.552	---------	-------
263	SbH3	STIBINE	124.781	-------	55.748	---------	-------
264	SbI3	ANTIMONY TRIIODIDE	502.47	-------	105.469	---------	-------
265	Sb2O3	ANTIMONY TRIOXIDE	291.512	-------	57.859	---------	-------
266	Sc	SCANDIUM	44.956	-------	16.295	---------	-------
267	Se	SELENIUM	78.96	-------	19.804	---------	-------
268	SeCl4	SELENIUM TETRACHLORIDE	220.771	-------	--------	---------	-------
269	SeF6	SELENIUM HEXAFLUORIDE	192.95	-------	--------	---------	-------
270	SeOCl2	SELENIUM OXYCHLORIDE	165.865	-------	68.698	---------	-------
271	SeO2	SELENIUM DIOXIDE	110.959	-------	33.748	---------	-------
272	Si	SILICON	28.086	-------	11.123	---------	-------
273	SiBrCl2F	BROMODICHLOROFLUOROSILANE	197.893	-------	109.231	---------	-------
274	SiBrF3	TRIFLUOROBROMOSILANE	164.985	-------	85.100	---------	-------
275	SiBr2ClF	DIBROMOCHLOROFLUOROSILANE	242.345	-------	106.013	---------	-------
276	SiClF3	TRIFLUOROCHLOROSILANE	120.533	-------	79.700	---------	-------
277	SiCl2F2	DICHLORODIFLUOROSILANE	136.988	-------	92.600	---------	-------
278	SiCl3F	TRICHLOROFLUOROSILANE	153.442	-------	105.500	---------	-------
279	SiCl4	SILICON TETRACHLORIDE	169.896	15.344	115.560	7.830E+09	60.820
280	SiF4	SILICON TETRAFLUORIDE	104.079	-------	--------	---------	-------
281	SiHBr3	TRIBROMOSILANE	268.805	-------	100.232	---------	-------
282	SiHCl3	TRICHLOROSILANE	135.452	15.081	102.004	6.820E+09	51.940
283	SiHF3	TRIFLUOROSILANE	86.089	-------	61.800	---------	-------
284	SiH2Br2	DIBROMOSILANE	189.909	-------	89.957	---------	-------
285	SiH2Cl2	DICHLOROSILANE	101.007	16.489	84.077	5.810E+09	43.060
286	SiH2F2	DIFLUOROSILANE	68.098	-------	56.800	---------	-------
287	SiH2I2	DIIODOSILANE	283.91	-------	100.501	---------	-------
288	SiH3Br	MONOBROMOSILANE	111.013	-------	72.526	---------	-------
289	SiH3Cl	MONOCHLOROSILANE	66.562	-------	66.599	---------	-------
290	SiH3F	MONOFLUOROSILANE	50.108	-------	51.800	---------	-------
291	SiH3I	IODOSILANE	158.014	-------	77.784	---------	-------
292	SiH4	SILANE	32.117	14.201	55.126	3.780E+09	25.300
293	SiO2	SILICON DIOXIDE	60.084	-------	26.823	5.030E+09	33.580
294	Si2Cl6	HEXACHLORODISILANE	268.887	-------	173.787	---------	-------
295	Si2F6	HEXAFLUORODISILANE	170.161	-------	--------	---------	-------
296	Si2H5Cl	DISILANYL CHLORIDE	96.663	-------	101.689	---------	-------
297	Si2H6	DISILANE	62.219	16.055	74.015	5.930E+09	43.640
298	Si2OCl3F3	TRICHLOROTRIFLUORODISILOXANE	235.524	-------	158.296	---------	-------
299	Si2OCl6	HEXACHLORODISILOXANE	284.887	-------	180.193	---------	-------
300	Si2OH6	DISILOXANE	78.218	-------	93.600	---------	-------
301	Si3Cl8	OCTACHLOROTRISILANE	367.878	-------	231.661	---------	-------
302	Si3H8	TRISILANE	92.32	-------	128.148	---------	-------
303	Si3H9N	TRISILAZANE	107.335	-------	125.169	---------	-------
304	Si4H10	TETRASILANE	122.421	-------	158.809	---------	-------
305	Sm	SAMARIUM	150.36	-------	21.663	---------	-------
306	Sn	TIN	118.71	-------	17.072	---------	-------
307	SnBr4	STANNIC BROMIDE	438.326	-------	130.870	---------	-------
308	SnCl2	STANNOUS CHLORIDE	189.615	-------	56.564	---------	-------
309	SnCl4	STANNIC CHLORIDE	260.521	-------	117.592	---------	-------
310	SnH4	STANNIC HYDRIDE	122.742	-------	--------	---------	-------
311	SnI4	STANNIC IODIDE	626.328	-------	146.791	---------	-------
312	Sr	STRONTIUM	87.62	-------	36.893	---------	-------

NO	FORMULA	NAME	Molecular Weight, g/mol	Solubility Parameter (joule/cm^3)$^{0.5}$	Liquid Volume cm^3/mol	Van Der Waals Area, cm^2/mol	Van Der Waals Volume, cm^3/mol
313	SrO	STRONTIUM OXIDE	103.619	-------	---------	---------	-------
314	Ta	TANTALUM	180.948	-------	12.063	---------	-------
315	Tc	TECNNETIUM	98	-------	9.233	---------	-------
316	Te	TELLURIUM	127.6	-------	22.011	---------	-------
317	TeCl4	TELLURIUM TETRACHLORIDE	269.411	-------	104.864	---------	-------
318	TeF6	TELLURIUM HEXAFLUORIDE	241.59	-------	---------	---------	-------
319	Ti	TITANIUM	47.88	-------	11.650	---------	-------
320	TiCl4	TITANIUM TETRACHLORIDE	189.691	18.181	110.669	---------	-------
321	Tl	THALLIUM	204.383	-------	18.101	---------	-------
322	TlBr	THALLOUS BROMIDE	284.287	-------	39.485	---------	-------
323	TlI	THALLOUS IODIDE	331.288	-------	48.988	---------	-------
324	Tm	THULIUM	168.934	-------	19.636	---------	-------
325	U	URANIUM	238.029	-------	13.293	---------	-------
326	UF6	URANIUM FLUORIDE	352.019	-------	79.599	---------	-------
327	V	VANADIUM	50.942	-------	9.179	---------	-------
328	VCl4	VANADIUM TETRACHLORIDE	192.752	18.671	105.853	---------	-------
329	VOCl3	VANADIUM OXYTRICHLORIDE	173.299	16.316	105.849	---------	-------
330	W	TUNGSTEN	183.84	-------	10.386	---------	-------
331	WF6	TUNGSTEN FLUORIDE	297.83	-------	86.866	---------	-------
332	Xe	XENON	131.29	15.909	44.449	---------	-------
333	Yb	YTTERBIUM	173.04	-------	26.917	---------	-------
334	Yt	YTTRIUM	88.906	-------	21.554	---------	-------
335	Zn	ZINC	65.39	109.450	9.423	---------	-------
336	ZnCl2	ZINC CHLORIDE	136.295	-------	56.778	---------	-------
337	ZnF2	ZINC FLUORIDE	103.387	-------	22.652	---------	-------
338	ZnO	ZINC OXIDE	81.389	-------	---------	---------	-------
339	ZnSO4	ZINC SULFATE	161.454	-------	---------	---------	-------
340	Zr	ZIRCONIUM	91.224	-------	15.728	---------	-------
341	ZrBr4	ZIRCONIUM BROMIDE	410.84	-------	---------	---------	-------
342	ZrCl4	ZIRCONIUM CHLORIDE	233.035	-------	93.257	---------	-------
343	ZrI4	ZIRCONIUM IODIDE	598.842	-------	---------	---------	-------

Chapter 15

SOLUBILITY IN WATER AND OCTANOL-WATER PARTITION COEFFICIENT

Carl L. Yaws, Sachin Nijhawan, Li Bu, Deepa R. Balundgi, and Saumya Tripathi
Lamar University, Beaumont, Texas

ABSTRACT

Results for water solubility and octanol-water partition coefficient are presented for major organic chemicals. The results are provided in an easy-to-use table that is especially applicable for rapid engineering usage with the personal computer or hand calculator. Typical water solubility values are 786 ppm (wt) for carbon tetrachloride (CCl_4) and 0.002 ppm (wt) for tetradecane ($C_{14}H_{30}$).

INTRODUCTION

Physical and thermodynamic property data for organic chemicals are of special value to engineers in the chemical processing and petroleum refining industries. In this article results are presented for water solubility and octanol-water partition coefficient. The compilation of data is intended for use in engineering and environmental studies. As an example of such usage, solubility data are useful in determining the assessment and distribution of a chemical spill upon its contact with water.

SOLUBILITY IN WATER

The results for solubility of organic compounds in water are given in Table 15-1. The presented values are applicable to a wide variety of substances including alkanes, olefins, diolefins, alkynes, cycloalkanes, aromatics, fluorocarbons, chlorocarbons, bromocarbons, iodocarbons, alcohols, acids, ketones, aldehydes, ethers, esters, amines, nitriles, sulfides, and thiols. The tabulation also gives the molecular weight, freezing point, and normal boiling point. The last two columns provide the data for solubility in water on a weight and mole basis. The compilation is based on both experimental and estimated values.

The tabulation is arranged by carbon number (C, C2, C3, ...,C28) to provide ease of use in quickly locating solubility data using the chemical formula.

For the tabulation, the solubility data for liquids (compounds with boiling point greater than 25 C) apply for the aqueous liquid phase in contact with a vapor which contains the compound, water, and air. The solubility data for gases (compounds with boiling point less than 25 C) apply for the aqueous liquid phase in contact with a vapor which contains only the compound and water.

In preparing the compilation, a literature search was conducted to identify data source publications (1-190). Both experimental values for the property under consideration and parameter values for estimation of the property are included in the source publications. The publications were screened and copies of appropriate data were made. These data were then keyed-in to the computer to provide a database of solubility values for organic compounds for which experimental data are available. The database also served as a basis to check the accuracy of the estimation method.

Upon completion of data collection, estimation of solubility for compounds was performed using the boiling point correlation developed by Yaws and co-workers (185-190):

$$\log_{10} S = A + BT_B + CT_B^2 + DT_B^3 \qquad (15-1)$$

where
S = solubility in water at 25 C, parts per million by weight, ppm (wt)
T_B = boiling point temperature of compound, K
A, B, C, and D = regression coefficients for chemical family

Results from the correlation are in favorable agreement with experimental data for solubility in water.

Estimation of solubility for hydrocarbons and organic oxygen compounds was accomplished using the following regression coefficients:

alkanes (paraffins)

A = -17.652 (normal and isomers)
B = 177.811E-03
C = -500.907E-06
D = 411.124E-09

cycloalkanes (naphthenes)

A = -16.7 for cyclohexanes
A = -16.9 for cyclopentanes
B = 177.811E-03
C = -500.907E-06
D = 411.124E-09

benzenes (aromatics)

    A = -24.008 (no and single substitutions)
    A = -23.650 (double and triple substitutions)
    B = 221.196E-03
    C = -555.632E-06
    D = 418.830E-09

ketones

    A = 45.223 (normal and isomers)
    B = -2.3859E-01
    C = 4.8739E-04
    D = -3.7160E-07

aldehydes

    A = 20.4898 (normal and isomers)
    B = -9.0310E-02
    C = 1.9223E-04
    D = -1.7856E-07

alcohols

    A = 45.6398 (normal and isomers)
    B = -2.3859E-01
    C = 4.8739E-04
    D = -3.7160E-07

ethers

    A = 7.510 (normal and isomers)
    B = 3.2057E-03
    C = -4.0887E-05
    D = 4.7284E-09

This correlation is applicable to substances that are liquids at ambient conditions (25 C, 1 atm). For hydrocarbons (alkanes, cycloalkanes, and benzenes), the range for boiling point temperatures is 298 to 561 K. For organic oxygen compounds (alcohols, ketones, ethers, and aldehydes), the range for boiling point temperatures is 298 to 625 K. The correlation is not applicable to solids since a different solubility curve is obtained for solids.

    Estimation of solubility for the remaining compounds (olefins, diolefins, alkynes, sulfides, thiols, and esters) was also accomplished using a modified boiling point method. Due to the small database, the estimates for these compounds should be considered rough approximations. The estimates for hydrocarbons and organic oxygen compounds are more accurate.

    A comparison of experimental and estimated data values for solubility in water is shown in Fig. 15-1 for paraffins, naphthenes, and aromatics. The graph discloses favorable agreement of experimental and estimated values.

OCTANOL-WATER PARTITION COEFFICIENT

    The octanol-water partition coefficient is the ratio of a chemicals concentration in an octanol phase to its concentration in an aqueous phase:

    Kow = (concentration in octanol phase) / (concentration in aqueous phase)   (15-2)

    The results for octanol-water partition coefficient are also given in Table 15-1. Both experimental and estimated values are provided in the compilation that is based on data source publications for organic compounds (1-60). Many of the estimates are based on the atom-fragment contribution method as described by Meylan and Howard (57). The tabulation is arranged by chemical formula to provide ease of use in quickly locating data.

    Properties such as octanol-water partition coefficient (Kow) are useful in studies involving the environmental fate of chemicals. As an example of such usefulness in environmental applications, Lyman, Reehl, and Rosenblatt (42) discuss how water solubility, soil-sediment coefficient, and biological concentrations for aquatic life can be related to Kow. Chemicals with low Kow values tend to have high solubilities in water, low soil-sediment coefficients, and small biological concentrations for aquatic life.

EXAMPLES

    The tabulation and correlation maybe used for determining solubility of organic compounds in water. Examples are shown below.

Example 1 A chemical spill of carbon tetrachloride (CCl4) occurs into a body of water at ambient conditions (25 C, 1 atm). Determine the concentration of carbon tetrachloride in the water.

    Using the chemical formula (CCl4) for carbon tetrachloride, inspection of the table discloses that the solubility in water is

S = 785.7 ppm (wt)

Example 2 A chemical spill of ethylbenzene (C8H10) occurs into a body of water at ambient conditions (25 C, 1 atm). Determine the concentration of ethylbenzene in the water.

Using the chemical formula (C8H10) for ethylbenzene, inspection of the table discloses that the solubility in water is

S = 165.1 ppm (wt)

Example 3 Estimate the solubility of pentane (C5H12) in water at a temperature of 298.15 K (25 C).

Substitution of the regression coefficients (A, B, C, and D) and boiling temperature (309.22 K) into the equation for solubility of paraffins (alkanes) in water yields

$$\log_{10} S = -17.652 + (177.811E\text{-}03)\ (309.22) - (500.907E\text{-}06)\ (309.22^2) + (411.124E\text{-}09)\ (309.22^3) = 1.59106$$

$$S = 10^{1.59106}$$

S = 39.00 ppm (wt)

The calculated and data values compare favorably (39.00 vs 38.50, deviation = 0.50/38.50 = 1.3%).

Portions of this material appeared in Chem. Eng., 97, 177 (April,1990), Chem. Eng., 97, 115 (July, 90), Pollution Engineering, 22, 70 (Oct., 1990), Oil & Gas Journal, 89, 79 (April 8, 1991), Oil & Gas Journal, 89, 86 (Sept. 16, 1991) and Oil & Gas Journal, 95 (35), 80 (Aug. 29, 1994). These portions are reprinted by special permission.

REFERENCES - WATER SOLUBILITY - ORGANIC COMPOUNDS

1-34. See **REFERENCES - ORGANIC COMPOUNDS** in Chapter 1, **CRITICAL PROPERTIES AND ACENTRIC FACTOR**

35. Sørensen, J. M. and W. Arlt, LIQUID-LIQUID EQUILIBRIUM DATA COLLECTION, Vol. V, part 1, Dechema Chemistry Data Series, 6000 Frankfurt/Main, Germany (1979).

36. Macedo, E. A. and P. Rasmussen, LIQUID-LIQUID EQUILIBRIUM DATA COLLECTION, Supplement 1, Vol. V, Part 4, Dechema Chemistry Data Series, 6000 Frankfurt/Main, Germany (1987).

37. Stephen, H. and T. Stephen, SOLUBILITIES OF INORGANIC AND ORGANIC COMPOUNDS, Vol. 1, Part 1, Macmillan Company, New York, NY (1963).

38. Horvath, A. L., HALOGENATED HYDROCARBONS, Marcel Dekker, Inc., New York, NY (1982).

39. Yalkowsky, S. H. and S. Banerjee, AQUEOUS SOLUBILITY, Marcel Dekker, Inc., New York, NY (1992).

40. Freier, R. K., AQUEOUS SOLUTIONS - DATA FOR INORGANIC AND ORGANIC COMPOUNDS, Vols. 1 and 2, Walter de Gruyter, Berlin, Germany - New York, NY (1976).

41. Curme, G. O., Jr., ed., GLYCOLS, American Chemical Society Series, Reinhold Publishing Corporation, New York, NY (1953).

42. Kertes, A. S., ed., SOLUBILITY DATA SERIES, IUPAC, Vol. 15, Alcohols with Water, Pergamon Press, Oxford, England (1984).

43. Markley, K. S., FATTY ACIDS, Interscience Publishers, Inc., New York, NY (1947).

44. HANDBOOK OF CHEMISTRY, 10th (revised) - 13th eds., McGraw-Hill, New York, NY (1969-1985).

45. Hildebrand, J. L. and R. L. Scott, THE SOLUBILITY OF NONELECTROLYTES, 3rd ed., Reinhold Publishing Corporation, New York, NY (1950).

46. Acree, W. E., Jr., THERMODYNAMIC PROPERTIES OF NONELECTROLYTE SOLUTIONS, Academic Press, New York, NY (1984).

47. Kertes, A. S., O. Levy, and G. Y. Markovits, EXPERIMENTAL THERMODYNAMICS OF NONPOLAR FLUIDS, Dack, M. R. J., ed., Vol. II, Chapter 15, Solubility, 725, Butterworth, London (1975).

48. Miller, M. M., S. P. Waslk, G. L. Huang, W. Y. Shiu, and D. Mackay, Environ. Sci. Technol., 19(6), 522 (1985).

49. Price, L. C., Bull. Am. Assoc. Petrol. Geologists, 60(2), 213 (1976).

50. Mackay, D. and W. Y. Shiu, J. Phys. Chem. Ref. Data, 10(4), 1175 (1981).

51. McAuliffe, C., J. Phy. Chem., 70, 1267 (1966).

52. Baker, E. G., FUNDAMENTAL ASPECTS OF PETROLEUM GEOCHEMISTRY, Nagy, B. and U. Colombo, eds., Chapter 7, 299, Elsevier Publishing Co., New York, NY (1967).

53. Sutton, C. and J. A. Calder, Environ. Sci. Techn., 8(7), 654 (1974).

54. Yalkowsky, S. H., S. Banerjee, and R. S. Pearlman, J. Phys. Chem. Ref. Data, 13(2), 555 (1984).

55. Irmann, F., Chem. Ing. Tech. 37(8), 789 (1965).

56. Inga, R. F. and J. J. McKetta, Petrol. Refiner, 40(3),191 (1961).

57. Peake, E. and G. W. Hodgson, J. Amer. Oil Chem. Soc., 43, 215 (1966).

58. Peake, E. and G. W. Hodgson, J. Amer. Oil Chem. Soc., 44, 696 (1967).

59. McAuliffe, C., Science, 158, 478 (1969).

60. Franks, F., Nature, 210, 87 (1966).

61. Wen, W. Y. and J. H. Hung, J. Phys. Chem., 74, 170 (1970).

62. Speece, R. E. and N. N. Nirmalakhandan, Environ. Sci. Technol., 22(3), 328 (1988).

63. Polak, J. and B. C. Y. Lu, Can. J. Chem., 51, 4018 (1973).

64. Mackay, D., W. Y. Shiu, and R. P. Sutherland, Environ. Sci. Technol., 13(3), 333 (1979).

65. McAuliffe, C., Nature (London), 200, 1092 (1963).

66. Hansch, C., J. E. Quinlan, and G. L. Lawrence, J. Org. Chem., 33(1), 347 (1968).

67. Wilhelm, E., CRC Crit. Rev. Anal. Chem., 16(2), 129 (1985).
68. Culberson, O. L. and J. J. McKetta, Trans. AIME, 192, 233 (1951).
69. Davis, J. E. and J. J. McKetta, Petrol. Refiner, 39[3], 205 (1960).
70. Price, L. C., Bull. Am. Assoc. Petrol. Geologists, 63, 1572 (1979).
71. Culberson, O. L., A. B. Horn, and J. J. McKetta, Trans. AIME, 189, 1 (1950).
72. Culberson, O. L. and J. J. McKetta, Trans. AIME, 189, 319 (1950).
73..Azarnoosh, A. and J. J. McKetta, Petrol. Refiner, 37[11], 275 (1958).
74. Kobayashi, R. and D. Katz, Ind. Eng. Chem., 45, 440 (1953).
75. Brooks, W. B., G. B. Gibbs, and J. J. McKetta, Petrol. Refiner, 30[10], 118 (1951).
76. LeBreton, J. G. and J. J. McKetta, Hydrocarbon Processing, 43[6], 136 (1964).
77. Reamer, H. H., B. H. Sage, and W. N. Lacey, Ind. Eng. Chem., 44, 609 (1952).
78. Reed, C. D. and J. J. McKetta, Petrol. Refiner, 38[4], 159 (1959).
79. Kudchadker, A. P. and J. J. McKetta, Petrol. Refiner, 40[9], 231 (1961).
80. Leinonen, P. J. and D. Mackay, Can. J. Chem. Eng., 51[2], 230 (1973).
81. Baker, E. G., Science, 129, 871 (1959).
82. Rebert, C. J. and K. E. Hayworth, AICHE J., 13[1], 118 (1967).
83. Schatzberg, P., J. Phys. Chem., 67, 776 (April, 1963).
84. Kudchadker, A. P. and J. J. McKetta, AICHE J., 7, 707 (1961).
85. Bradbury, E. J., D. McNulty, R. L. Savage and E. E. McSweeney, Ind. Eng. Chem., 44, 211 (1952).
86. Davis, J. E. and J. J. McKetta, J. Chem. Eng. Data, 5, 374 (1960).
87. Li, C. C. and J. J. McKetta, J. Chem. Eng. Data, 8[2], 271 (1963).
88. Azarnoosh, A. and J. J. McKetta, J. Chem. Eng. Data, 4, 211 (1959).
89. Sanchez, M. and H. Lentz, High Temperatures-High Pressures, 5, 689 (1973).
90. Brooks, W. B., J. E. Haughn, and J. J. McKetta, Petrol. Refiner, 34[8], 129 (1955).
91. Brooks, W. B. and J. J. McKetta, Petrol. Refiner, 34[2], 143 (1955).
92. Leland, T. W., J. J. McKetta and K. A. Kobe, Ind, Eng. Chem., 47, 1265 (1955).
93. Schwarz, F. P., Anal. Chem., 52, 10 (1980).
94. Natarajan, G. S. and K. A. Venkatachalam, J. Chem. Eng. Data, 17[3], 328 (1972).
95. Reed, C. D. and J. J. McKetta, J. Chem. Eng. Data, 4, 294 (1959).
96. Inga, R. F. and J. J. McKetta, J. Chem. Eng. Data, 6, 337 (1961).
97. Simpson, L. B. and F. P. Lovell, J. Chem. Eng. Data, 7, 498 (1962).
98. Alexander, D. M., J. Phys. Chem., 63, 1021 (1959).
99. Arnold, D. S., C. A. Plank, and E. E. Erickson, Chem. Eng. Data. Ser., 3, 253 (1958).
100. Bohon, R. L. and W. F. Claussen, J. Am. Chem. Soc., 73, 1571 (1951).
101. Franks, F., M. Gent and H. H. Johnson, J. Chem. Soc., 2716 (1963).
102. Klevens, H. B., J. Phys. Colloid Chem., 54, 283 (1950).
103. Kudchadlker, A. P. and J. J. McKetta, Petrol. Refiner, 41[3], 191 (1962).
104. Rebert, C. J. and W. B. Kay, AIChE J., 5[3], 285 (1959).
105. Anthony, R. G. and J. J. McKetta, J. Chem. Eng. Data, 12, 17 (1967).
106. Anthony, R. G. and J. J. McKetta, J. Chem. Eng. Data, 12, 21 (1967).
107. Li, C. C. and J. J. McKetta, Petrol. Refiner, 42[3], 135 (1963); 42[6], 107 (1963).
108. Mackay, D. and W. Y. Shiu, J. Chem. Eng. Data, 22, 399 (1977).
109. Schwarz, F. P., J. Chem. Eng. Data, 22, 273 (1977).
110. Wauchope, R. D. and F. W. Getzen, J. Chem. Eng. Data, 17[1], 38 (1972).
111. Kabadi, V. N. and R. P. Danner, Hydrocarbon Processing, 58(5), 245 (1979).
112. Baker, R. J., B. J. Donelan, L. J. Peterson, W. E. Acree, Jr. and C. C. Tsai, Phys. Chem. Liq., 16, 279 (1987).
113. Herz, W., and F.Hiebenthal, Zeit. fuer Anorg. Chem. 177, 363 (1928).
114. Ledbury, W. and C. W. Frost, J. Soc. Chem. Ind., Lond., 46, 120 (1927).
115. Fuhner, H., Berichte der Deutschen Chem. Gesell., 57, 510 (1924).
116. Belfort, G., SELECTIVE ADSORPTION OF ORGANIC HOMOLOGS ONTO ACTIVATED CARBON FROM DILUTE AQUEOUS SOLUTIONS - SOLVOPHOBIC INTERACTION APPROACH: DEVELOPMENT AND TEST OF THEORY, CARBON: THEORY AND APPLICATION, 2, 207, Ann Arbor Science Publishers, Inc. (1981).
117. Simpson, L. B. and F. P. Lovell, J. Chem. & Eng. Data, 7, 498 (1962).
118. Krzyzanowska, T. and J. Szeliga, Nafta (Katowice, Poland), 34, 413 (1978).
119. Allott, P. R., A. Steward, V. Flook and W. W. Mapleson, Brit. J. Anaesth., 45, 294 (1973).
120. Marshall H. and D. Bain, J. Chem. Soc. (Lond.), 97, 1074 (1910).
121. Findlay, A. and A. N. Campbell, J. Chem. Soc., (Lond.), 1768 (1928).
122. Amoore, J. E. and R. G. Buttery, Chem. Senses and Flavor, 3, 57 (1978).
123. Clarke, G. A., T. R. Williams, and R. W. Taft, J. Am. Chem. Soc., 84, 2292 (1962).
124. Copley, M. J., E. Ginsberg, G. F. Zellhoefer, and C. S. Marvel, J. Am. Chem. Soc., 63, 254 (1941).
125. Doolittle, A. K., Ind. & Eng. Chem., 27, 1169 (1935).
126. Conway, J. B. and J. J. Norton, Ind. & Eng. Chem., 43, 1433 (1951).
127. Merrill, E. J., J. Pharm. Sc., 54, 1670 (1965).
128. Lamouroux, F., Comptes Rendus Hebdoa. des Seances de l'Acad. des Sc., 128, 998 (1899).
129. Ginnings, P. M., D. Plonk, and E. Carter, J. Am. Chem. Soc., 62, 1923 (1940).
130. Bidner, M. S. and M. de Santiago, Chem. Eng. Sc., 26, 1484 (1971).
131. McBain, J. W. and P. H. Richards, Ind. & Eng. Chem., 38, 642 (1946).
132. Mullin, J. M., CRYSTALLISATION, 425, CRC Press, Cleveland, Ohio (1972).
133. Desvergnes, L., Chim. E. Indust. (Milan), 25, 3 (1931).

134. Eckert, J. W., Phytopath., 52, 642 (1962).

135. Hafkenscheid, T. L. and E. Tomlinson, J. Chromatogr., 218, 409 (1981).

136. Iwamoto, E., Y. Hiyama, and Y. Yamamoto, J. Sol. Chem., 6, 371 (1977).

137. Nathan, M. F., Chem. Eng. (New York), 85, 93 (1978).

138. Banerjee, S., S. H. Yalkowsky, and S. C. Valvani, Env. Sc. & Tech., 14, 1227 (1980).

139. Chiou, C. T. and D. W. Schmedding, Env. Sc. & Tech., 16, 4 (1982).

140. Knox, J. and M. B. Richards, J. Chem. Soc. (Lond.), 115, 508 (1919).

141. McAuliffe, C., J. de Pharm. et de Chim., 70, 1267 (1966).

142. Othmer, D. F., R. E. White and E. Trueger, Ind. & Eng. Chem., 33, 1240 (1941).

143. Lozano, F. J., J. Chem. & Eng. Data, 26, 131 (1981).

144. Hobson, R. W., R. J. Hartman and E. W. Kanning, J. Am. Chem. Soc., 63, 2094 (1941).

145. Vesala, A., Acta Chem. Scand., Series A:Phy. & Inorg. Chem., 28, 839 (1974).

146. Desvergnes, L., Mon. Sc., 14, 121 (1924).

147. Mitchell, A. G., L. S. C. Wan and S. G. Bjaastad, J. Pharm. & Pharmaco., 16, 632 (1964).

148. Andrews, L. J. and R. M. Keefer, J. Am. Chem. Soc., 72, 3113 (1950).

149. Banerjee, S., Env. Sc. & Tech., 18, 587 (1984).

150. Evans, B. K., K. C. James, and D. K. Luscombe, J. Pharm. Sc., 67, 277 (1978).

151. Bergen, Jr., R. L., and F. A. Long, J. Phy. Chem., 60, 1131 (1956).

152. Cox, J. D., J. Chem. Soc. (Lond.), 3183 (1954).

153. Lloyd, B. A., S. O. Thompson, and J. B. Ferguson, Can. J. Res., Sec. B: Chem. Sc., 15, 98 (1937).

154. Dixon, M. R., C. E. Rehberg, and C. H. Fisher, J. Am. Chem. Soc., 70, 3733 (1948).

155. Mackay, D. and others, Env. Sc. & Tech., 19, 522 (1985).

156. Deno, N. C. and H. E. Berkheimer, J. Chem. & Eng. Data, 5, 1 (1960).

157. Mitchell, A. G. and L. S. C. Wan, J. Pharm. Sc., 53, 1467 (1964).

158. Lindstrom, R. E. and C. H. Lee, J. Pharm. & Pharmaco., 32, 245 (1980).

159. Ikeda, Y. and others, Yakugaku Zasshi (Tokyo), 102, 83 (1982).

160. Mahieu, J., Bull. des Soc. Chim. Belges, 45, 667 (1936).

161. Marcus, Y., L. E. Asher, J. Hormadaly and E. Pross, Hydrometallurgy, 7, 27 (1981).

162. Azaz, E. and M. Donbrow, J. Colloid & Interface Sc., 57, 11 (1976).

163. Massaldi, H. A. and C. J. King, J. Chem. & Eng. Data, 18, 393 (1973).

164. Albert, A., Chem. & Ind. (London), 252, (1956).

165. Klevens, H. B., Chem. Rev., 47, 1 (1950).

166. Mackay, D. and W. Y. Shiu, J. Chem. & Eng. Data, 22, 399 (1977).

167. Merckel, J. H. C., Recueil des Travaux Chim. des Pays-Bas, 56, 811 (1937).

168. Almgren, M., F. Grieser, and J. K. Thomas, J. Am. Chem. Soc., 101, 279 (1979).

169. Leyder, F. and P. Boulanger, Bull. Env. Contamin. & Toxic., 30, 152 (1983).

170. Burris, D. R. and W. G. MacIntyre, Env. Sc. & Tech., 20, 296 (1986).

171. Lu, P-Y. and R. L. Metcalf, Env. Health Perspect., 10, 269 (1975).

172. Price, L., "The Solubility of Hydrocarbons and Petroleum in Water as Applied to the Primary Migration of Petroleum", PHD Thesis (1973).

173. Martin, H., PESTICIDE MANUAL, British Crop Protection Council (1977).

174. Lu, P-Y., R. L. Metcalf and E. M. Carlson, Env. Health Perspect., 24, 201 (1978).

175. Hashimoto, Y., K. Tokura, K. Ozaki, and W. M. J. Strachan, Chemosphere, 11, 991 (1982).

176. Eganhouse, R. P. and J. A. Calder, Geochim. et Cosmochim. Acta, 40, 555 (1976).

177. Bowman, M. C., J. R. King, and C. L. Holder, Internat. J. Env. Anal. Chem., 4, 205 (1976).

178. Fendler, J. H., and others, J. Am. Chem. Soc., 97, 89 (1975).

179. Eik-Nes, K., J. A. Schellman, R. Lumry, and L. T. Samuels, J. Bio. Chem., 206, 411 (1954)

180. Andrews, L. J. and R. M. Keefer, J. Am. Chem. Soc., 72, 5034 (1950).

181. Akiyoshi, M., T. Deguchi, and I. Sanemasa, Bull. Chem. Soc. Jap. (Nippon Kagakukai Bull.), 60, 3935 (1987).

182. Hollifield, J. C., Bull. Env. Contamin. & Toxic., 23, 579 (1979).

183. Back, E. and B. Steenberg, Acta Chem. Scand., 4, 810 (1950).

184. Howard, P. H. and W. M. Meylan, eds., HANDBOOK OF PHYSICAL PROPERTIES OF ORGANIC CHEMICALS, CRC Press, Boca Raton, FL (1997).

185. Yaws, C. L., H. C. Yang, J. R. Hopper, and K. C. Hansen, Chem. Eng., 97(4), 177 (April, 1990).

186. Yaws, C. L., H. C. Yang, J. R. Hopper, and K. C. Hansen, Chem. Eng., 97(7), 115 (July, 1990).

187. Yaws, C. L. and others, Pollution Engineering, 22, 70 (Oct., 1990).

188. Yaws, C. L. and Xiang Pan, Oil & Gas Journal, 89, 79 (April 8, 1991).

189. Yaws, C. L., Xiang Pan, and D. G. Piper, Jr., Oil & Gas Journal, 89, 86 (Sept. 16, 1991).

190. Yaws, C. L., Xiang Pan, and D. G. Piper, Jr., Oil & Gas Journal, 95, 80 (Aug. 29, 1994).

REFERENCES – OCTANOL-WATER PARTITION COEFFICIENT - ORGANIC COMPOUNDS

1. Hansch, C. and A. J. Leo, SUBSTITUENT CONSTANTS FOR CORRELATION ANALYSIS IN CHEMISTRY AND BIOLOGY, John Wiley and Sons, New York, NY (1979).

2. Hansch, C. and others, J. Med. Chem., 16, 1207 (1973).

3. Hansch, C., J. E. Quinlan, and G. L. Lawrence, J Org. Chem., 33, 347 (1968).

4. Hansch, C. and A. Leo, J. Org. Chem., 36, 1539 (1971).

5. Hansch, C., A. Leo, and D. Elkins, Chem. Rev., 71, 525 (1971).

6. Rekker, R. F., THE HYDROPHOBIC FRAGMENT CONSTANT, Elsevier Scientific Publishing Co., New York, NY (1977).

7. Rekker, R. F. and G.G. Nys, Chim. Ther., 8, 521 (1973).
8. Rekker, R. F. and G.G. Nys, Eur. J. Med. Chem. – Chim. Ther., 9, 361 (1974).
9. Kenaga, E.E. and C.A.I. Goring, "RELATIONSHIP BETWEEN WATER SOLUBILITY, SOIL SORPTION, OCTANOL-WATER PARTITIONING, AND BIOCONCENTRATION OF CHEMICALS IN BIOTA", American Society for Testing and Materials, Third Aquatic toxicology Symposium, New Orleans, LA (October 17-18, 1978).
10. Rao, P.S.C. and J.M. Davidson, ENVIRONMENTAL IMPACT OF NONPOINT SOURCE POLLUTION, M.R. Overcash and J.M. Davidson, eds., Ann Arbor Science Publishers, Inc., Ann Arbor, MI (1980).
11. Karickhoff, S.W., D.S. Brown, and T.A. Scott, Water Res., 13, 241 (1979).
12. Holmes, H.L., STRUCTURE-ACTIVITY RELATIONSHIPS FOR SOME CONJUGATED HETEROENOID COMPOUNDS, CATECHOL MONOETHERS AND MORPHINE ALKALOIDS, Vols. 1 and 2, Defence
Research Establishment, Suffield, Ralston, Alberta, Canada (1975).
13. Holmes, H.L. and C.E. Lough, (NTIS AD A030683), Suffield Technical Note No. 365, Defence Research Establishment, Suffield, Ralston, Alberta (July, 1976).
14. Carlson, R., R. Carlson and H.Kopperman, J.Chromatogr., 107, 210 (1975).
15. Chiou, C.T., V.H. Freed, D.W. Schmedding, and R.L. Kohnert, Environ. Sci. Technol., 11, 475 (1977).
16. Chou, J.T. and P.C. Jurs, J. Chem. Inf. Comput. Sci., 19, 172 (1979).
17. Currie, D.J., C.E. Lough, R.F. Silver, and H.L. Holmes, Can. J. Chem., 44, 1035 (1966).
18. Dyott T.M., A.J. Stuper, and G.S. Zander, J.Chem. Inf. Comput. Sci., 20, 28-35 (1980).
19. Fujita, T., J. Iwasa, and C. Hansch, J. Am. Chem. Soc., 86, 5175 (1964).
20. Hall, L.H. and L.B. Kier, J. Pharm. Sci., 66, 642 (1977).
21. Hopfinger. A.J. and R.D. Battershell, J. Med. Chem., 19, 569 (1976).
22. Karickhoff, S.W. and D.S. Brown, Report No. EPA-600/4-79-032, U.S. Environmental Protection Agency, Athens, GA (1979).
23. Saeger, V.W., O. Hicks, R.G. Kaley, P.R. Michael, J.P. Mieure and E.S. Tucker, Environ. Sci. Technol., 13, 840 (1979).
24. Seiler, P., Eur. J. Med. Chem. – Chim. Ther., 9, 473 (1974).
25. Unger, S.H., J. Cook, and J. Hollenberg, J. Pharm. Sci., 67, 1374 (1978).
26. Veith, G.D., N..M. Austin, and R.T. Morris, Water Res., 13, 43 (1979).
27. Veith, G.D. and R.T. Morris, Report EPA-600/3-78-049, U.S. Environmental Protection Agency, Duluth, MN (1978).
28. Yalkowsky, S.H., R.J. Orr, and S.C. Valvani, Ind. Eng. Chem. Fundam., 18, 351 (1979).
29. Yalkowsky, S.H. and S.C.Valvani, J. Chem. Eng. Data, 24, 127 (1979).
30. Yamana, T., J. Pharm. Sci., 66, 747 (1977).
31. Funasaki N., S. Hada, and S. Neya, and K. Machida, J. Phys. Chem., 88(24), 5786 (1984).
32. Chem. Inspect. Test Inst., Japan Chemical Industry Ecology – Toxicol. and Inform. Center, ISBN 4 - 89074-101-1 (1992).
33. Martiska, A. and V. Bekarek, ACTA Univ. Palacki olomuc., Fac. Rerum Nat. 97 (Chem. 29) 63-7 (1990).
34. Jaber, H. M. and others, EPA-600/6-84/009, Menlo Park, CA (1984).
35. Neely, W. B. and G. E. Blau, ENVIRONMENTAL EXPOSURE FROM CHEMICALS, Vol. 1, CRC Press, Baca Raton, FL (1985).
36. Tanii, H. and K. Hashimoto, Arch. Toxicol., 57(2), 88 (1985).
37. Deneer, J. W. and others, Aquatic Toxicol., 13(3), 195 (1988).
38. Tanii, H and K. Hashimoto, Arch. Toxicol., 55(1), 47 (1984).
39. Sasaki, H. and others, J. Med. Chem., 34(2), 628 (1991).
40. Catz, P. and D. R. Friend, Int. J. Pharm., 55(1), 17 (1989).
41. Sasaki, H., et al., Int. J. Pharm. 44, 15-24 (1988).
42. Lyman, W. J., W. F. Reehl, and D. A. Rosenblatt, HANDBOOK OF CHEMICAL PROPERTY ESTIMATION METHODS, McGraw Hill Book Company, New York, NY (1982).
43. Suzuki, T., J. Computer Aided Molecular Design, 5, 149 (1991).
44. Valvani, S. C. and others, J. Pharm. Sci., 70, 502 (1981).
45. Debruijn, J. and others, Environ. Toxicol. Chem., 8, 499 (1989).
46. Hansch, C. and others , Med. Chem., 34(2), 786 (1991).
47. Hansch, C., A.J. Leo, Medchem Project, Claremont, CA, Pomona College. Issue # 26, 1985.
48. Hansch C., A. Leo, and A.J. Pomona, Pomona College Medicinal Project, Claremont, CA 91711, Log P Database (July, 1987).
49. Hanch, C., A. J. Leo, and D. Hoekman, ACS Professional Reference Book, "Exploring QSAR. Hydrophobic, Electronic and Steric Factors", American Chemical Society, Washington, D C (1995).
50. Dunn, W.J. and others, Quant Struct-Act. Relat., 2, 156 (1983).
51. Tanii, H., H. Tsuji, and K. Hashimoto, Toxicol. Lett., 30(1), 13 (1986).
52. Nakagawa, Y., K. Izumi, N. Oikawa, T. Sotomatsu, M. Shigemura, and T. Fujita, Environ. Toxicol. Chem., 11, 901 (1992).
53. Debnath, A.K., G.Debnath, A.J. Shusterman, and C.Hansch, Environ. Mol. Mutagen., 19(1), 37 (1992).
54. Funasaki , N., S. Hada, and S. Neya, J. Chromatogr., 361, 33 (1986).
55. Miller, M.M., Wasik, S.P., Huang, G., Shiu, W., and Mackay, D., Environ. Sci. Technol., 19, 522 (1985).
56. Sangster, J., LogKow Databank, Sangster Research Laboratories, Montreal, Quebec, Canada, (1994).
57. Meylan, W. M. and P. H. Howard, J. Pham. Sci. 84, 83 (1995).
58. Meylan, W.M. and P. H.Howard, Environ. Toxicol. Chem., 10, 1283 (1991).
59. Meylan, W.M. and P.H. Howard, J. Pharm. Sci., 84, 83 (1995).
60. Meylan, W.M. and others, Environ. Toxicol. Chem., 15, 100 (1996).

Figure 15-1 Solubility in Water

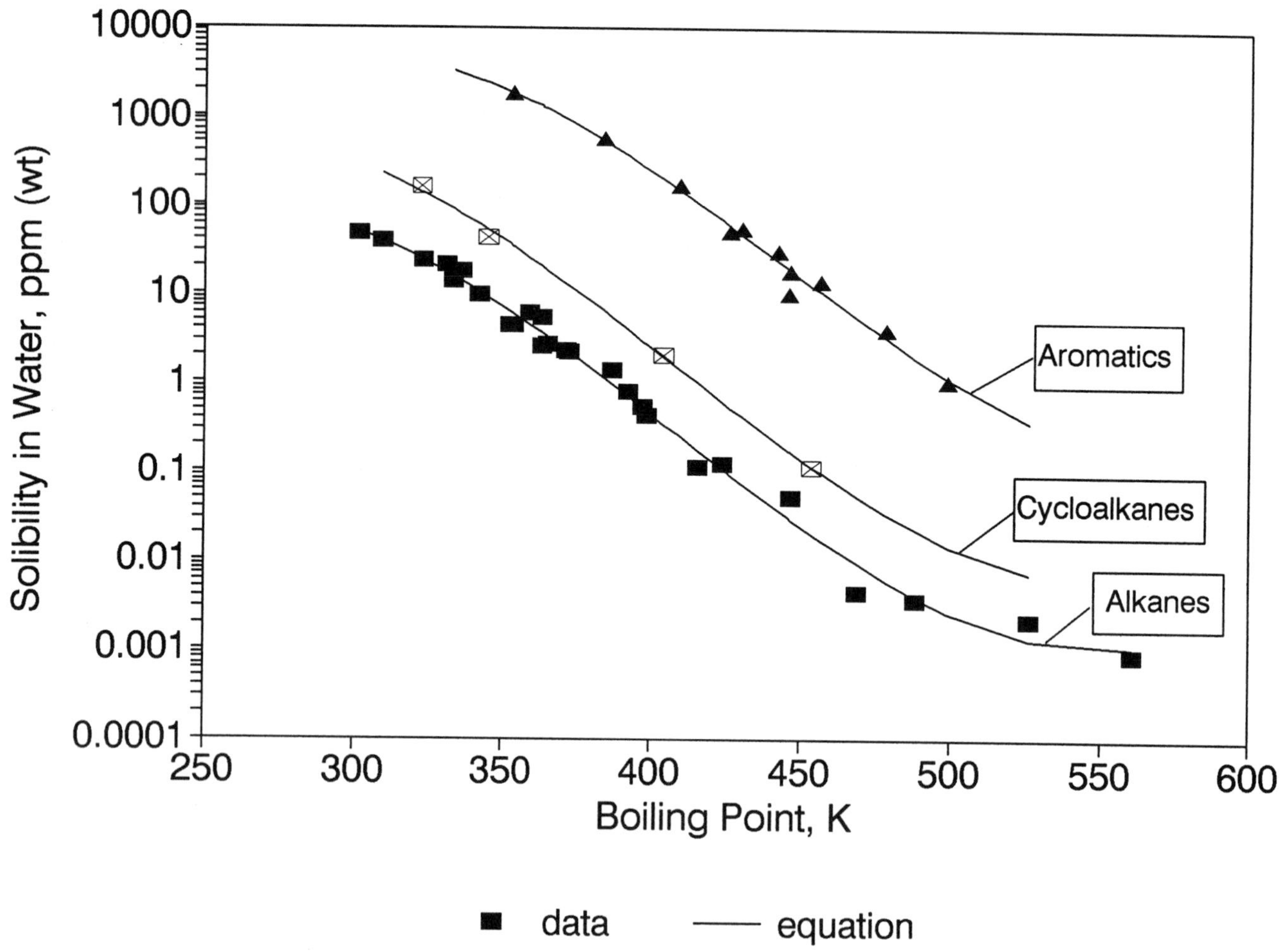

Table 15-1 SOLUBILITY IN WATER AND OCTANOL-WATER PARTITION COEFFICIENT- ORGANIC COMPOUNDS

NO	FORMULA	NAME	Solubility in Water, S (ppm - parts per million)				Octanol-Water Partition Coefficient	
			Code	T, C	S @ T, ppm (wt)	S @ T, ppm (mol)	Code	log₁₀ Kow
1	CBrClF2	BROMOCHLORODIFLUOROMETHANE	1	2.0	9.9100E+02	1.0806E+02	2	1.9
2	CBrCl3	BROMOTRICHLOROMETHANE	----	------	----------	----------	2	2.53
3	CBrF3	BROMOTRIFLUOROMETHANE	1	25.0	3.0000E+02	3.6303E+01	1	1.86
4	CBr2F2	DIBROMODIFLUOROMETHANE	----	------	----------	----------	2	1.99
5	CClF3	CHLOROTRIFLUOROMETHANE	1	25.0	9.0000E+01	1.5523E+01	1	1.65
6	CClN	CYANOGEN CHLORIDE	1	25.0	3.0000E+04	8.9826E+03	----	--------
7	CCl2F2	DICHLORODIFLUOROMETHANE	1	25.0	3.0000E+02	4.4709E+01	1	2.16
8	CCl2O	PHOSGENE	1	25.0	6.8255E+03	1.2501E+03	2	-0.71
9	CCl3F	TRICHLOROFLUOROMETHANE	1	30.0	1.0800E+03	1.4177E+02	1	2.53
10	CCl4	CARBON TETRACHLORIDE	1	25.0	7.8570E+02	9.2082E+01	1	2.83
11	CF2O	CARBONYL FLUORIDE	----	------	----------	----------	2	-1.34
12	CF4	CARBON TETRAFLUORIDE	1	25.0	1.6000E+01	3.2753E+00	1	1.18
13	CHBr3	TRIBROMOMETHANE	1	25.0	3.1100E+03	2.2233E+02	1	2.4
14	CHClF2	CHLORODIFLUOROMETHANE	1	25.0	2.7700E+03	5.7838E+02	1	1.08
15	CHCl2F	DICHLOROFLUOROMETHANE	1	25.0	1.8800E+04	3.3425E+03	1	1.55
16	CHCl3	CHLOROFORM	1	25.0	7.5000E+03	1.1391E+03	1	1.97
17	CHF3	TRIFLUOROMETHANE	1	25.0	9.0000E+02	2.3173E+02	1	0.64
18	CHI3	TRIIODOMETHANE	1	25.0	1.0000E+02	4.5759E+00	2	3.03
19	CHN	HYDROGEN CYANIDE	1	25.0	1.0000E+06	1.0000E+06	----	--------
20	CHNS	ISOTHIOCYANIC-ACID	----	------			----	--------
21	CH2BrCl	BROMOCHLOROMETHANE	1	25.0	1.5000E+04	2.1159E+03	1	1.41
22	CH2Br2	DIBROMOMETHANE	1	25.0	1.1000E+04	1.1513E+03	1	1.7
23	CH2ClF	CHLOROFLUOROMETHANE	1	25.0	1.0520E+04	2.7892E+03	1	0.51
24	CH2Cl2	DICHLOROMETHANE	1	25.0	1.9380E+04	4.1744E+03	1	1.25
25	CH2F2	DIFLUOROMETHANE	1	25.0	4.3880E+03	1.5239E+03	1	0.2
26	CH2I2	DIIODOMETHANE	1	25.0	1.2400E+03	8.3501E+01	1	2.3
27	CH2O	FORMALDEHYDE	1	25.0	1.0000E+06	1.0000E+06	1	0.35
28	CH2O2	FORMIC ACID	1	25.0	1.0000E+06	1.0000E+06	1	-0.54
29	CH3Br	METHYL BROMIDE	1	25.0	1.3410E+04	2.5725E+03	1	1.19
30	CH3Cl	METHYL CHLORIDE	1	25.0	5.9000E+03	2.1132E+03	1	0.91
31	CH3Cl3Si	METHYL TRICHLOROSILANE	----	------	----------	----------	2	2.01
32	CH3F	METHYL FLUORIDE	1	15.0	2.3900E+03	1.2665E+03	1	0.51
33	CH3I	METHYL IODIDE	1	25.0	2.6070E+04	3.3859E+03	1	1.51
34	CH3NO	FORMAMIDE	1	20.0	1.0000E+06	1.0000E+06	1	-1.51
35	CH3NO2	NITROMETHANE	1	25.0	1.0000E+04	2.9723E+03	1	-0.35
36	CH3NO2	METHYL-NITRITE	----	------	----------	----------	2	0.88
37	CH3NO3	METHYL-NITRATE	----	------	----------	----------	2	0.76
38	CH4	METHANE	1	25.0	2.4400E+01	2.7399E+01	1	1.09
39	CH4Cl2Si	METHYL DICHLOROSILANE	----	------	----------	----------	2	1.7
40	CH4O	METHANOL	1	25.0	1.0000E+06	1.0000E+06	1	-0.77
41	CH4O3S	METHANESULFONIC ACID	----	------	----------	----------	2	-2.38
42	CH4S	METHYL MERCAPTAN	1	15.0	2.4000E+04	9.1241E+03	2	0.78
43	CH5ClSi	METHYL CHLOROSILANE	----	------	----------	----------	----	--------
44	CH5N	METHYLAMINE	1	25.0	5.1920E+05	3.8514E+05	1	-0.57
45	CH6Si	METHYL SILANE	----	------	----------	----------	----	--------
46	CN4O8	TETRANITROMETHANE	----	------	----------	----------	2	-2.05
47	CO	CARBON MONOXIDE	1	25.0	2.3770E+01	1.5288E+01	----	--------
48	COS	CARBONYL SULFIDE	1	25.0	1.1500E+03	3.4513E+02	2	-1.33
49	CO2	CARBON DIOXIDE	1	25.0	1.9500E+03	7.9913E+02	2	-1.33
50	CS2	CARBON DISULFIDE	1	25.0	1.8790E+03	4.4520E+02	1	2.14
51	C2BrF3	BROMOTRIFLUOROETHYLENE	----	------	----------	----------	2	1.55
52	C2Br2F4	1,2-DIBROMOTETRAFLUOROETHANE	----	------	----------	----------	2	2.96
53	C2ClF3	CHLOROTRIFLUOROETHYLENE	----	------	----------	----------	2	1.65
54	C2ClF5	CHLOROPENTAFLUOROETHANE	1	25.0	5.8000E+01	6.7647E+00	2	2.47
55	C2Cl2F4	1,2-DICHLOROTETRAFLUOROETHANE	1	25.0	1.3700E+02	1.4442E+01	1	2.82
56	C2Cl3F3	1,1,2-TRICHLOROTRIFLUOROETHANE	1	25.0	1.7000E+02	1.6347E+01	1	3.16
57	C2Cl4	TETRACHLOROETHYLENE	1	25.0	1.5000E+02	1.6297E+01	1	3.4
58	C2Cl4F2	1,1,2,2-TETRACHLORODIFLUOROETHANE	1	25.0	1.2000E+02	1.0607E+01	2	3.41
59	C2Cl4O	TRICHLOROACETYL CHLORIDE	----	------	----------	----------	2	0.88
60	C2Cl6	HEXACHLOROETHANE	1	25.0	8.0000E+00	6.0878E-01	1	3.91
61	C2F4	TETRAFLUOROETHYLENE	1	25.0	1.5840E+02	2.8535E+01	2	1.21
62	C2F6	HEXAFLUOROETHANE	1	25.0	7.8970E+00	1.0308E+00	1	2
63	C2HBrClF3	HALOTHANE	1	24.0	3.7200E+03	3.4068E+02	1	2.3
64	C2HClF2	2-CHLORO-1,1-DIFLUOROETHYLENE	----	------	----------	----------	2	1.6
65	C2HCl3	TRICHLOROETHYLENE	1	25.0	1.1000E+03	1.5097E+02	1	2.42
66	C2HCl3O	DICHLOROACETYL CHLORIDE	----	------	----------	----------	2	-0.04
67	C2HCl3O	TRICHLOROACETALDEHYDE	----	------	----------	----------	1	0.99
68	C2HCl5	PENTACHLOROETHANE	1	25.0	5.0000E+02	4.4547E+01	1	3.22
69	C2HF3	TRIFLUOROETHENE	----	------	----------	----------	----	--------
70	C2HF3O2	TRIFLUOROACETIC ACID	----	------	----------	----------	2	0.5
71	C2HF5	PENTAFLUOROETHANE	----	------	----------	----------	2	1.55
72	C2H2	ACETYLENE	1	25.0	1.0000E+03	6.9209E+02	1	0.37
73	C2H2Br4	1,1,2,2-TETRABROMOETHANE	1	25.0	6.7537E+02	3.5222E+01	2	2.55
74	C2H2Cl2	1,1-DICHLOROETHYLENE	1	25.0	3.3450E+03	6.2330E+02	1	2.13
75	C2H2Cl2	cis-1,2-DICHLOROETHYLENE	1	25.0	3.5000E+03	6.5227E+02	1	1.86
76	C2H2Cl2	trans-1,2-DICHLOROETHYLENE	1	25.0	6.3000E+03	1.1768E+03	1	2.09
77	C2H2Cl2O	CHLOROACETYL CHLORIDE	----	------	----------	----------	2	-0.22
78	C2H2Cl2O	DICHLOROACETALDEHYDE	----	------	----------	----------	----	--------

371

NO	FORMULA	NAME	Solubility in Water, S (ppm - parts per million)				Octanol-Water Partition Coefficient	
			Code	T, C	S @ T, ppm (wt)	S @ T, ppm (mol)	Code	log₁₀ Kow
79	C2H2Cl2O2	DICHLOROACETIC ACID	1	20.0	1.0000E+06	1.0000E+06	1	0.92
80	C2H2Cl3F	1,1,1-TRICHLOROFLUOROETHANE	----	------	----------	----------	----	-------
81	C2H2Cl4	1,1,1,2-TETRACHLOROETHANE	1	25.0	1.1000E+03	1.1818E+02	2	2.93
82	C2H2Cl4	1,1,2,2-TETRACHLOROETHANE	1	25.0	2.9000E+03	3.1206E+02	1	2.39
83	C2H2F2	1,1-DIFLUOROETHYLENE	1	25.0	1.6490E+02	4.6397E+01	1	1.24
84	C2H2F2	cis-1,2-DIFLUOROETHENE	----	------	----------	----------	----	-------
85	C2H2F2	trans-1,2-DIFLUOROETHENE	----	------	----------	----------	----	-------
86	C2H2F4	1,1,1,2-TETRAFLUOROETHANE	----	------	----------	----------	2	1.68
87	C2H2O	KETENE	----	------	----------	----------	2	-0.52
88	C2H2O4	OXALIC ACID	1	20.0	8.3000E+04	1.7788E+04	2	-2.22
89	C2H3Br	VINYL BROMIDE	----	------	----------	----------	1	1.57
90	C2H3Cl	VINYL CHLORIDE	1	25.0	2.6970E+03	7.7889E+02	2	1.62
91	C2H3ClF2	1-CHLORO-1,1-DIFLUOROETHANE	----	------	----------	----------	2	2.05
92	C2H3ClO	ACETYL CHLORIDE	----	------	----------	----------	2	-0.47
93	C2H3ClO	CHLOROACETALDEHYDE	----	------	----------	----------	2	0.09
94	C2H3ClO2	CHLOROACETIC ACID	----	------	----------	----------	1	0.22
95	C2H3ClO2	METHYL CHLOROFORMATE	----	------	----------	----------	2	0.14
96	C2H3Cl3	1,1,1-TRICHLOROETHANE	1	25.0	1.0000E+03	1.3516E+02	1	2.49
97	C2H3Cl3	1,1,2-TRICHLOROETHANE	1	25.0	4.3930E+03	5.9550E+02	1	1.89
98	C2H3F	VINYL FLUORIDE	----	------	----------	----------	2	1.19
99	C2H3F3	1,1,1-TRIFLUOROETHANE	----	------	----------	----------	2	1.74
100	C2H3N	ACETONITRILE	1	25.0	1.0000E+06	1.0000E+06	1	-0.34
101	C2H3NO	METHYL ISOCYANATE	----	------	----------	----------	2	0.79
102	C2H4	ETHYLENE	1	25.0	1.3100E+02	8.4126E+01	1	1.13
103	C2H4Br2	1,1-DIBROMOETHANE	----	------	----------	----------	2	1.94
104	C2H4Br2	1,2-DIBROMOETHANE	1	25.0	4.1700E+03	4.0139E+02	1	1.96
105	C2H4Cl2	1,1-DICHLOROETHANE	1	25.0	5.0320E+03	9.1984E+02	1	1.79
106	C2H4Cl2	1,2-DICHLOROETHANE	1	25.0	8.6790E+03	1.5913E+03	1	1.48
107	C2H4Cl2O	BIS(CHLOROMETHYL)ETHER	----	------	----------	----------	2	0.57
108	C2H4F2	1,1-DIFLUOROETHANE	1	27.5	2.5000E+03	6.8310E+02	1	0.75
109	C2H4F2	1,2-DIFLUOROETHANE	----	------	----------	----------	2	1.21
110	C2H4I2	1,2-DIIODOETHANE	----	------	----------	----------	1	2.71
111	C2H4O	ACETALDEHYDE	1	25.0	1.0000E+06	1.0000E+06	1	-0.34
112	C2H4O	ETHYLENE OXIDE	1	25.0	1.0000E+06	1.0000E+06	1	-0.3
113	C2H4OS	THIOACETIC-ACID	----	------	----------	----------	----	-------
114	C2H4O2	ACETIC ACID	1	25.0	1.0000E+06	1.0000E+06	1	-0.17
115	C2H4O2	METHYL FORMATE	1	20.0	2.3800E+04	7.2606E+03	1	0.03
116	C2H4S	THIACYCLOPROPANE	----	------	----------	----------	----	-------
117	C2H5Br	BROMOETHANE	1	25.0	9.0000E+03	1.4992E+03	1	1.61
118	C2H5Cl	ETHYL CHLORIDE	1	20.0	9.0510E+03	2.5440E+03	1	1.43
119	C2H5CLO	2-CHLOROETHANOL	----	------	----------	----------	1	0.03
120	C2H5F	ETHYL FLUORIDE	1	25.0	2.1580E+03	8.1001E+02	2	1.26
121	C2H5I	ETHYL IODIDE	1	20.0	3.9000E+03	4.5203E+02	1	2
122	C2H5N	ETHYLENEIMINE	1	25.0	1.0000E+06	1.0000E+06	2	-0.28
123	C2H5NO	ACETAMIDE	1	24.5	7.0500E+05	4.2159E+05	1	-1.26
124	C2H5NO	N-METHYLFORMAMIDE	----	------	----------	----------	1	-0.97
125	C2H5NO2	NITROETHANE	1	25.0	5.0140E+04	1.2510E+04	1	0.18
126	C2H5NO3	ETHYL-NITRATE	----	------	----------	----------	2	1.25
127	C2H6	ETHANE	1	25.0	6.0400E+01	3.6187E+01	1	1.81
128	C2H6AlCl	DIMETHYLALUMINUM CHLORIDE	----	------	----------	----------	----	-------
129	C2H6O	DIMETHYL ETHER	1	18.0	6.5180E+04	2.6542E+04	1	0.1
130	C2H6O	ETHANOL	1	25.0	1.0000E+06	1.0000E+06	1	-0.31
131	C2H6OS	DIMETHYL SULFOXIDE	----	------	----------	----------	1	-1.35
132	C2H6O2	ETHYLENE GLYCOL	1	25.0	1.0000E+06	1.0000E+06	1	-1.36
133	C2H6O4S	DIMETHYL SULFATE	1	18.0	2.8000E+04	4.0975E+03	1	1.16
134	C2H6S	DIMETHYL SULFIDE	1	25.0	1.9610E+04	5.7658E+03	2	0.92
135	C2H6S	ETHYL MERCAPTAN	1	25.0	1.4780E+04	4.3306E+03	2	1.27
136	C2H6S2	DIMETHYL DISULFIDE	----	------	----------	----------	1	1.77
137	C2H7N	DIMETHYLAMINE	1	40.0	6.1980E+05	3.9446E+05	1	-0.38
138	C2H7N	ETHYLAMINE	1	20.0	1.0000E+06	1.0000E+06	1	-0.13
139	C2H7NO	MONOETHANOLAMINE	1	25.0	1.0000E+06	1.0000E+06	1	-1.31
140	C2H8N2	ETHYLENEDIAMINE	1	20.0	1.0000E+06	1.0000E+06	1	-2.04
141	C2H8Si	DIMETHYL SILANE	----	------	----------	----------	----	-------
142	C2N2	CYANOGEN	1	20.0	9.4870E+03	3.3049E+03	1	0.07
143	C3F6	HEXAFLUOROPROPYLENE	----	------	----------	----------	2	2.12
144	C3F6O	HEXAFLUOROACETONE	----	------	----------	----------	1	1.46
145	C3F8	OCTAFLUOROPROPANE	1	15.0	5.8450E+00	5.6004E-01	2	3.12
146	C3H2N2	MALONONITRILE	1	20.0	1.1760E+05	3.5069E+04	1	-0.6
147	C3H3Cl	PROPARGYL CHLORIDE	----	------	----------	----------	----	-------
148	C3H3N	ACRYLONITRILE	1	20.0	7.3500E+04	2.6226E+04	1	0.25
149	C3H3NO	OXAZOLE	----	------	----------	----------	1	0.12
150	C3H4	METHYLACETYLENE	1	25.0	3.6400E+03	1.6400E+03	1	0.94
151	C3H4	PROPADIENE	----	------	----------	----------	1	1.45
152	C3H4Cl2	2,3-DICHLOROPROPENE	1	25.0	2.1500E+03	3.4966E+02	2	2.42
153	C3H4O	ACROLEIN	1	20.0	2.1100E+05	7.9132E+04	1	-0.01
154	C3H4O	PROPARGYL ALCOHOL	1	20.0	1.0000E+06	1.0000E+06	1	-0.38
155	C3H4O2	ACRYLIC ACID	1	25.0	1.0000E+06	1.0000E+06	1	0.35
156	C3H4O2	beta-PROPIOLACTONE	----	------	----------	----------	2	-0.8

NO	FORMULA	NAME	Solubility in Water, S (ppm - parts per million)				Octanol-Water Partition Coefficient	
			Code	T, C	S @ T, ppm (wt)	S @ T, ppm (mol)	Code	log₁₀ Kow
157	C3H4O2	VINYL FORMATE	----	------	----------	----------	2	0.18
158	C3H4O3	ETHYLENE CARBONATE	----	------	----------	----------	2	0.12
159	C3H4O3	PYRUVIC ACID	1	20.0	1.0000E+06	1.0000E+06	2	-1.24
160	C3H5Br	3-BROMO-1-PROPENE	1	25.0	3.8220E+03	5.7100E+02	1	1.79
161	C3H5Cl	2-CHLOROPROPENE	----	------	----------	----------	1	2
162	C3H5Cl	3-CHLOROPROPENE	1	25.0	4.0000E+03	9.4454E+02	2	1.93
163	C3H5ClO	alpha-EPICHLOROHYDRIN	----	------	----------	----------	1	0.45
164	C3H5ClO2	METHYL CHLOROACETATE	----	------	----------	----------	2	0.63
165	C3H5ClO2	ETHYL CHLOROFORMATE	----	------	----------	----------	2	0.63
166	C3H5Cl3	1,2,3-TRICHLOROPROPANE	1	25.0	1.9000E+03	2.3255E+02	1	2.27
167	C3H5I	3-IODO-1-PROPENE	----	------	----------	----------	----	-------
168	C3H5N	PROPIONITRILE	1	25.0	9.3382E+04	3.2591E+04	1	0.16
169	C3H5NO	ACRYLAMIDE	1	30.0	6.8300E+05	3.5320E+05	1	-0.67
170	C3H5NO	HYDRACRYLONITRILE	----	------	----------	----------	2	-1.12
171	C3H5NO	LACTONITRILE	----	------	----------	----------	1	-0.94
172	C3H5N3O9	NITROGLYCERINE	1	20.0	1.3800E+03	1.0962E+02	1	1.62
173	C3H6	CYCLOPROPANE	1	21.1	5.3770E+02	2.3026E+02	1	1.72
174	C3H6	PROPYLENE	1	25.0	2.0000E+02	8.5630E+01	1	1.77
175	C3H6Br2	1,2-DIBROMOPROPANE	1	25.0	1.4280E+03	1.2759E+02	2	2.43
176	C3H6Cl2	1,1-DICHLOROPROPANE	----	------	----------	----------	----	-------
177	C3H6Cl2	1,2-DICHLOROPROPANE	1	25.0	2.7500E+03	4.3949E+02	2	2.25
178	C3H6Cl2	1,3-DICHLOROPROPANE	1	25.0	2.7300E+03	4.3628E+02	1	2
179	C3H6Cl2	2,2-DICHLOROPROPANE	----	------	----------	----------	----	-------
180	C3H6I2	1,2-DIIODOPROPANE	----	------	----------	----------	----	-------
181	C3H6O	ACETONE	1	25.0	1.0000E+06	1.0000E+06	1	-0.24
182	C3H6O	ALLYL ALCOHOL	1	25.0	1.0000E+06	1.0000E+06	1	0.17
183	C3H6O	METHYL VINYL ETHER	----	------	----------	----------	2	0.42
184	C3H6O	n-PROPIONALDEHYDE	1	25.0	4.0450E+05	1.7402E+05	1	0.59
185	C3H6O	1,2-PROPYLENE OXIDE	1	20.0	4.0500E+05	1.7432E+05	1	0.03
186	C3H6O	1,3-PROPYLENE OXIDE	----	------	----------	----------	1	-0.14
187	C3H6O2	ETHYL FORMATE	1	18.0	1.0000E+05	2.6310E+04	1	0.23
188	C3H6O2	METHYL ACETATE	1	20.0	2.4360E+05	7.2630E+04	1	0.18
189	C3H6O2	PROPIONIC ACID	1	25.0	1.0000E+06	1.0000E+06	1	0.33
190	C3H6O2S	3-MERCAPTOPROPIONIC ACID	----	------	----------	----------	1	0.43
191	C3H6O3	LACTIC ACID	----	------	----------	----------	1	-0.72
192	C3H6O3	METHOXYACETIC ACID	1	20.0	1.0000E+06	1.0000E+06	2	-0.68
193	C3H6O3	TRIOXANE	1	25.0	1.7500E+05	4.0696E+04	1	-0.43
194	C3H6S	THIACYCLOBUTANE	----	------	----------	----------	2	1.3
195	C3H7Br	1-BROMOPROPANE	1	20.0	2.4460E+03	3.5902E+02	1	2.1
196	C3H7Br	2-BROMOPROPANE	1	20.0	2.8730E+03	4.2185E+02	1	2.14
197	C3H7Cl	ISOPROPYL CHLORIDE	1	20.0	3.1000E+03	7.1275E+02	1	1.9
198	C3H7Cl	n-PROPYL CHLORIDE	1	20.0	2.7000E+03	6.2059E+02	1	2.04
199	C3H7F	1-FLUOROPROPANE	1	14.0	3.8560E+03	1.1219E+03	2	1.76
200	C3H7F	2-FLUOROPROPANE	1	15.0	3.6630E+03	1.0656E+03	----	-------
201	C3H7I	ISOPROPYL IODIDE	1	20.0	1.4000E+03	1.4855E+02	1	2.89
202	C3H7I	n-PROPYL IODIDE	1	23.5	1.0700E+03	1.1350E+02	2	2.57
203	C3H7N	ALLYAMINE	1	20.0	1.0000E+06	1.0000E+06	1	0.03
204	C3H7N	PROPYLENEIMINE	----	------	----------	----------	2	0.13
205	C3H7NO	N,N-DIMETHYLFORMAMIDE	----	------	----------	----------	1	-1.01
206	C3H7NO	N-METHYLACETAMIDE	----	------	----------	----------	1	-1.05
207	C3H7NO2	1-NITROPROPANE	1	20.0	1.3810E+04	2.8235E+03	----	-------
208	C3H7NO2	2-NITROPROPANE	1	20.0	1.6720E+04	3.4265E+03	1	0.93
209	C3H7NO3	PROPYL-NITRATE	----	------	----------	----------	2	1.74
210	C3H7NO3	ISOPROPYL-NITRATE	----	------	----------	----------	----	-------
211	C3H8	PROPANE	1	25.0	6.2400E+01	2.5494E+01	1	2.36
212	C3H8O	ISOPROPANOL	1	25.0	1.0000E+06	1.0000E+06	1	0.05
213	C3H8O	METHYL ETHYL ETHER	2	25.0	9.7258E+04	3.1286E+04	2	0.56
214	C3H8O	n-PROPANOL	1	25.0	1.0000E+06	1.0000E+06	1	0.25
215	C3H8O2	2-METHOXYETHANOL	----	------	----------	----------	1	-0.77
216	C3H8O2	METHYLAL	1	20.0	2.3000E+05	6.6045E+04	----	0
217	C3H8O2	1,2-PROPYLENE GLYCOL	1	25.0	1.0000E+06	1.0000E+06	1	-0.92
218	C3H8O2	1,3-PROPYLENE GLYCOL	----	------	----------	----------	1	-1.04
219	C3H8O3	GLYCEROL	1	25.0	1.0000E+06	1.0000E+06	1	-1.76
220	C3H8S	n-PROPYLMERCAPTAN	2	25.0	3.4045E+03	8.0736E+02	1	1.81
221	C3H8S	ISOPROPYL MERCAPTAN	2	25.0	6.6208E+03	1.5740E+03	2	1.68
222	C3H8S	ETHYL-METHYL-SULFIDE	2	25.0	8.0663E+03	1.9199E+03	1	1.54
223	C3H9N	n-PROPYLAMINE	1	25.0	1.0000E+06	1.0000E+06	1	0.48
224	C3H9N	ISOPROPYLAMINE	1	20.0	1.0000E+06	1.0000E+06	1	0.26
225	C3H9N	TRIMETHYLAMINE	1	30.0	4.7090E+05	2.1337E+05	1	0.16
226	C3H9NO	1-AMINO-2-PROPANOL	1	25.0	1.0000E+06	1.0000E+06	1	-0.96
227	C3H9NO	3-AMINO-1-PROPANOL	----	------	----------	----------	1	-1.12
228	C3H9NO	METHYLETHANOLAMINE	----	------	----------	----------	1	-0.94
229	C3H9O4P	TRIMETHYL PHOSPHATE	1	25.0	5.0000E+05	1.1395E+05	1	-0.65
230	C3H10N2	1,2-PROPANEDIAMINE	----	------	----------	----------	2	-1.2
231	C3H10Si	TRIMETHYL SILANE	----	------	----------	----------	----	-------
232	C4Cl4S	TETRACHLOROTHIOPHENE	----	------	----------	----------	----	-------
233	C4Cl6	HEXACHLORO-1,3-BUTADIENE	1	25.0	3.2300E+00	2.2315E-01	1	4.78
234	C4F8	OCTAFLUORO-2-BUTENE	----	------	----------	----------	2	3.03

NO	FORMULA	NAME	Code	T, C	Solubility in Water, S (ppm - parts per million)		Octanol-Water Partition Coefficient	
					S @ T, ppm (wt)	S @ T, ppm (mol)	Code	log₁₀ Kow
235	C4F8	OCTAFLUOROCYCLOBUTANE	1	26.0	5.0000E+01	4.5033E+00	2	1.29
236	C4F10	DECAFLUOROBUTANE	----	------	----------	----------	2	4.09
237	C4H2	BUTADIYNE(BIACETYLENE)	1	25.0	9.4510E+03	3.4218E+03	2	1.3
238	C4H2O3	MALEIC ANHYDRIDE	----	------	----------	----------	2	1.62
239	C4H4	VINYLACETYLENE	1	30.0	1.7860E+03	6.1857E+02	2	1.4
240	C4H4N2	SUCCINONITRILE	1	20.0	1.2690E+05	3.1658E+04	1	-0.99
241	C4H4O	FURAN	1	25.0	9.9010E+03	2.6394E+03	1	1.34
242	C4H4O2	DIKETENE	----	------	----------	----------	2	-0.39
243	C4H8O3	SUCCINIC ANHYDRIDE	----	------	----------	----------	2	0.81
244	C4H4O4	FUMARIC ACID	1	25.0	7.0000E+03	1.0929E+03	1	0.46
245	C4H4O4	MALEIC ACID	1	25.0	4.4070E+05	1.0897E+05	1	-0.48
246	C4H4S	THIOPHENE	1	25.0	3.0150E+03	6.4705E+02	1	1.81
247	C4H5Cl	CHLOROPRENE	----	------	----------	----------	2	2.53
248	C4H5N	trans-CROTONITRILE	----	------	----------	----------	----	-------
249	C4H5N	cis-CROTONITRILE	----	------	----------	----------	----	-------
250	C4H5N	METHACRYLONITRILE	----	------	----------	----------	1	0.68
251	C4H5N	PYRROLE	----	------	----------	----------	1	0.75
252	C4H5N	VINYLACETONITRILE	----	------	----------	----------	1	0.4
253	C4H5NO2	METHYL CYANOACETATE	----	------	----------	----------	1	-0.47
254	C4H6	CYCLOBUTENE	----	------	----------	----------	----	-------
255	C4H6	1,2-BUTADIENE	2	25.0	6.5238E+02	2.1737E+02	2	2.2
256	C4H6	1,3-BUTADIENE	1	25.0	7.3500E+02	2.4491E+02	1	1.99
257	C4H6	DIMETHYLACETYLENE	2	25.0	2.3621E+03	7.8791E+02	1	1.46
258	C4H6	ETHYLACETYLENE	1	25.0	2.8700E+03	9.5767E+02	2	1.53
259	C4H6Cl2	1,3-DICHLORO-trans-2-BUTENE	----	------	----------	----------	----	-------
260	C4H6Cl2	1,4-DICHLORO-cis-2-BUTENE	----	------	----------	----------	2	2.6
261	C4H6Cl2	1,4-DICHLORO-trans-2-BUTENE	----	------	----------	----------	2	2.6
262	C4H6Cl2	3,4-DICHLORO-1-BUTENE	----	------	----------	----------	2	2.6
263	C4H6O	trans-CROTONALDEHYDE	1	20.0	1.5000E+05	4.3389E+04	2	0.6
264	C4H6O	2,5-DIHYDROFURAN	----	------	----------	----------	1	0.46
265	C4H6O	DIVINYL ETHER	1	37.0	3.8410E+03	9.9005E+02	2	0.78
266	C4H6O	METHACROLEIN	1	25.0	6.4000E+04	1.7271E+04	2	0.74
267	C4H6O2	2-BUTYNE-1,4-DIOL	----	------	----------	----------	2	-0.93
268	C4H6O2	gamma-BUTYROLACTONE	1	20.0	1.0000E+06	1.0000E+06	1	-0.64
269	C4H6O2	cis-CROTONIC ACID	1	20.0	1.0000E+06	1.0000E+06	2	0.85
270	C4H6O2	trans-CROTONIC ACID	----	------	----------	----------	2	0.85
271	C4H6O2	METHACRYLIC ACID	----	------	----------	----------	1	0.93
272	C4H6O2	METHYL ACRYLATE	1	23.0	5.0000E+04	1.0894E+04	1	0.8
273	C4H6O2	VINYL ACETATE	1	25.0	2.5530E+04	5.4524E+03	1	0.73
274	C4H6O3	ACETIC ANHYDRIDE	1	25.0	1.5100E+05	3.0430E+04	2	-0.58
275	C4H6O4	SUCCINIC ACID	1	25.0	8.3230E+04	1.3661E+04	1	-0.59
276	C4H6O5	DIGLYCOLIC ACID	----	------	----------	----------	----	-------
277	C4H6O5	MALIC ACID	1	26.0	5.9200E+05	1.6314E+05	----	-------
278	C4H6O6	TARTARIC ACID	1	20.0	1.7100E+05	2.4161E+04	----	-------
279	C4H7N	n-BUTYRONITRILE	1	25.0	3.1950E+04	8.5304E+03	1	0.53
280	C4H7N	ISOBUTYRONITRILE	----	------	----------	----------	1	0.46
281	C4H7NO	ACETONE CYANOHYDRIN	----	------	----------	----------	2	-0.03
282	C4H7NO	2-METHACRYLAMIDE	----	------	----------	----------	----	-------
283	C4H7NO	3-METHOXYPROPIONITRILE	----	------	----------	----------	----	-------
284	C4H7NO	2-PYRROLIDONE	1	20.0	1.0000E+06	1.0000E+06	1	-0.85
285	C4H8	1-BUTENE	1	25.0	2.2200E+02	7.1291E+01	1	2.4
286	C4H8	cis-2-BUTENE	2	25.0	2.2814E+02	7.3265E+01	1	2.33
287	C4H8	trans-2-BUTENE	2	25.0	2.3505E+02	7.5482E+01	1	2.31
288	C4H8	CYCLOBUTANE	----	------	----------	----------	2	2.19
289	C4H8	ISOBUTENE	1	25.0	2.6300E+02	8.4460E+01	1	2.34
290	C4H8Br2	1,2-DIBROMOBUTANE	----	------	----------	----------	2	2.92
291	C4H8Br2	2,3-DIBROMOBUTANE	----	------	----------	----------	2	2.85
292	C4H8Cl2	1,4-DICHLOROBUTANE	----	------	----------	----------	2	2.81
293	C4H8I2	1,2-DIIODOBUTANE	----	------	----------	----------	----	-------
294	C4H8O	n-BUTYRALDEHYDE	1	25.0	8.3720E+04	2.2318E+04	1	0.88
295	C4H8O	ISOBUTYRALDEHYDE	1	25.0	8.9000E+04	2.3826E+04	2	0.74
296	C4H8O	1,2-EPOXYBUTANE	----	------	----------	----------	2	0.86
297	C4H8O	METHYL ETHYL KETONE	1	25.0	2.4830E+05	7.6234E+04	1	0.29
298	C4H8O	ETHYL VINYL ETHER	----	------	----------	----------	1	1.04
299	C4H8O	TETRAHYDROFURAN	1	20.0	1.0000E+06	1.0000E+06	1	0.46
300	C4H8O2	cis-2-BUTENE-1,4-DIOL	----	------	----------	----------	2	-0.43
301	C4H8O2	trans-2-BUTENE-1,4-DIOL	----	------	----------	----------	2	-0.43
302	C4H8O2	ISOBUTYRIC ACID	1	20.0	2.0000E+05	4.8631E+04	1	0.94
303	C4H8O2	n-BUTYRIC ACID	1	20.0	1.0000E+06	1.0000E+06	1	0.79
304	C4H8O2	1,4-DIOXANE	1	25.0	1.0000E+06	1.0000E+06	1	-0.27
305	C4H8O2	ETHYL ACETATE	1	25.0	7.3720E+04	1.6013E+04	1	0.73
306	C4H8O2	METHYL PROPIONATE	2	25.0	4.9284E+04	1.0488E+04	1	0.84
307	C4H8O2	n-PROPYL FORMATE	1	22.0	2.2000E+04	4.5785E+03	1	0.83
308	C4H8O2S	SULFOLANE	----	------	----------	----------	1	-0.77
309	C4H8S	TETRAHYDROTHIOPHENE	----	------	----------	----------	2	1.79
310	C4H9Br	1-BROMOBUTANE	1	25.0	6.1630E+02	8.1073E+01	1	2.75
311	C4H9Br	2-BROMOBUTANE	----	------	----------	----------	2	2.58
312	C4H9Cl	n-BUTYL CHLORIDE	1	25.0	7.4000E+02	1.4410E+02	1	2.64

Table 15-1 SOLUBILITY IN WATER AND OCTANOL-WATER PARTITION COEFFICIENT- ORGANIC COMPOUNDS (continued)

| NO | FORMULA | NAME | Solubility in Water, S (ppm - parts per million) | | | | Octanol-Water Partition Coefficient | |
			Code	T, C	S @ T, ppm (wt)	S @ T, ppm (mol)	Code	log₁₀ Kow
313	C4H9Cl	sec-BUTYL CHLORIDE	1	25.0	1.0000E+03	1.9477E+02	1	2.33
314	C4H9Cl	tert-BUTYL CHLORIDE	1	14.9	2.8790E+03	5.6159E+02	2	2.45
315	C4H9I	2-IODO-2-METHYLPROPANE	----	------	----------	----------	----	-------
316	C4H9N	PYRROLIDINE	1	25.0	1.0000E+06	1.0000E+06	1	0.46
317	C4H9NO	N,N-DIMETHYLACETAMIDE	1	25.0	1.0000E+06	1.0000E+06	1	-0.77
318	C4H9NO	MORPHOLINE	1	20.0	1.0000E+06	1.0000E+06	1	-0.86
319	C4H9NO2	1-NITROBUTANE	----	------	----------	----------	1	1.47
320	C4H9NO2	2-NITROBUTANE	----	------	----------	----------	----	-------
321	C4H10	n-BUTANE	1	25.0	6.1400E+01	1.9031E+01	1	2.89
322	C4H10	ISOBUTANE	1	25.0	4.8900E+01	1.5157E+01	----	-------
323	C4H10N2	PIPERAZINE	----	------	----------	----------	1	-1.5
324	C4H10O	n-BUTANOL	1	25.0	7.4600E+04	1.9216E+04	1	0.88
325	C4H10O	sec-BUTANOL	1	25.0	1.8350E+05	5.1792E+04	1	0.61
326	C4H10O	tert-BUTANOL	1	25.0	1.0000E+06	1.0000E+06	1	0.35
327	C4H10O	DIETHYL ETHER	1	25.0	6.0880E+04	1.5511E+04	1	0.89
328	C4H10O	METHYL-PROPYL-ETHER	1	25.0	3.0520E+04	7.5932E+03	1	1.21
329	C4H10O	METHYL ISOPROPYL ETHER	1	25.0	6.5070E+04	1.6634E+04	2	0.98
330	C4H10O	ISOBUTANOL	1	25.0	8.1000E+04	2.0972E+04	1	0.76
331	C4H10O2	1,3-BUTANEDIOL	----	------	----------	----------	2	-0.29
332	C4H10O2	1,4-BUTANEDIOL	1	20.0	1.0000E+06	1.0000E+06	1	-0.83
333	C4H10O2	2,3-BUTANEDIOL	----	------	----------	----------	2	-0.36
334	C4H10O2	t-BUTYL HYDROPEROXIDE	----	------	----------	----------	2	0.94
335	C4H10O2	1,2-DIMETHOXYETHANE	----	------	----------	----------	1	-0.21
336	C4H10O2	2-ETHOXYETHANOL	1	20.0	1.0000E+06	1.0000E+06	1	-0.32
337	C4H10O3	DIETHYLENE GLYCOL	1	25.0	1.0000E+06	1.0000E+06	2	-1.47
338	C4H10O4S	DIETHYL SULFATE	----	------	----------	----------	1	1.14
339	C4H10S	n-BUTYL MERCAPTAN	1	25.0	6.0000E+02	1.1991E+02	1	2.28
340	C4H10S	ISOBUTYL MERCAPTAN	2	25.0	1.1966E+03	2.3925E+02	2	2.18
341	C4H10S	sec-BUTYL MERCAPTAN	2	25.0	1.4408E+03	2.8814E+02	2	2.18
342	C4H10S	tert-BUTYL MERCAPTAN	2	25.0	4.0025E+03	8.0207E+02	2	2.14
343	C4H10S	DIETHYL SULFIDE	1	20.0	3.1200E+03	6.2477E+02	1	1.95
344	C4H10S	ISOPROPYL-METHYL-SULFIDE	2	25.0	3.3408E+03	6.6915E+02	2	1.83
345	C4H10S	METHYL-PROPYL-SULFIDE	2	25.0	1.8684E+03	3.7379E+02	2	1.9
346	C4H10S2	DIETHYL DISULFIDE	----	------	----------	----------	2	2.86
347	C4H11N	n-BUTYLAMINE	1	25.0	1.0000E+06	1.0000E+06	1	0.97
348	C4H11N	ISOBUTYLAMINE	1	20.0	1.0000E+06	1.0000E+06	1	0.73
349	C4H11N	sec-BUTYLAMINE	1	25.0	1.0000E+06	1.0000E+06	2	0.76
350	C4H11N	tert-BUTYLAMINE	----	------	----------	----------	1	0.4
351	C4H11N	DIETHYLAMINE	1	25.0	1.0000E+06	1.0000E+06	1	0.58
352	C4H11NO	DIMETHYLETHANOLAMINE	1	20.0	1.0000E+06	1.0000E+06	2	-0.94
353	C4H11NO2	DIETHANOLAMINE	1	20.0	1.0000E+06	1.0000E+06	1	-1.43
354	C4H12N2O	N-AMINOETHYL ETHANOLAMINE	----	------	----------	----------	----	-------
355	C4H11NO2	2-AMINOETHOXYETHANOL	----	------	----------	----------	2	-2.13
356	C4H12Si	TETRAMETHYLSILANE	----	------	----------	----------	2	2.72
357	C4H13N3	DIETHYLENE TRIAMINE	1	20.0	1.0000E+06	1.0000E+06	2	-2.13
358	C5Cl6	HEXACHLOROCYCLOPENTADIENE	----	------	----------	----------	1	5.04
359	C5H4O2	FURFURAL	1	25.0	7.9000E+04	1.5828E+04	1	0.41
360	C5H5N	PYRIDINE	1	25.0	1.0000E+06	1.0000E+06	1	0.65
361	C5H6	CYCLOPENTADIENE	----	------	----------	----------	2	2.25
362	C5H6	2-METHYL-1-BUTENE-3-YNE	----	------	----------	----------	----	-------
363	C5H6	1-PENTENE-3-YNE	----	------	----------	----------	----	-------
364	C5H6	1-PENTENE-4-YNE	----	------	----------	----------	----	-------
365	C5H6N2	GLUTARONITRILE	----	------	----------	----------	1	-0.72
366	C5H6O2	FURFURYL ALCOHOL	1	20.0	1.0000E+06	1.0000E+06	1	0.28
367	C5H6O3	GLUTARIC ANHYDRIDE	----	------	----------	----------	2	1.3
368	C5H6O4	CITRACONIC ACID	1	25.0	7.8300E+05	3.3317E+05	----	-------
369	C5H6O4	ITACONIC ACID	1	20.0	7.6800E+04	1.1388E+04	----	-------
370	C5H6S	2-METHYLTHIOPHENE	----	------	----------	----------	1	2.33
371	C5H6S	3-METHYLTHIOPHENE	----	------	----------	----------	1	2.34
372	C5H7N	N-METHYLPYRROLE	----	------	----------	----------	1	1.21
373	C5H7NO2	ETHYL CYANOACETATE	1	25.0	2.0000E+04	3.2397E+03	2	1.01
374	C5H8	CYCLOPENTENE	1	25.0	5.3500E+02	1.4155E+02	2	2.47
375	C5H8	ISOPRENE	2	25.0	5.3112E+02	1.4052E+02	1	2.42
376	C5H8	3-METHYL-1,2-BUTADIENE	2	25.0	4.2430E+02	1.1225E+02	2	2.61
377	C5H8	2-METHYL-1,3-BUTADIENE	1	25.0	6.4200E+02	1.6987E+02	1	2.42
378	C5H8	1,2-PENTADIENE	2	25.0	3.6738E+02	9.7185E+01	2	2.55
379	C5H8	cis-1,3-PENTADIENE	2	25.0	3.7820E+02	1.0005E+02	1	2.4
380	C5H8	trans-1,3-PENTADIENE	2	25.0	4.0718E+02	1.0772E+02	1	2.44
381	C5H8	1,4-PENTADIENE	1	25.0	5.5800E+02	1.4763E+02	1	2.48
382	C5H8	2,3-PENTADIENE	2	25.0	3.2321E+02	8.5499E+01	----	-------
383	C5H8	1-PENTYNE	2	25.0	1.5710E+03	4.1596E+02	1	1.98
384	C5H8	2-PENTYNE	2	25.0	8.5226E+02	2.2554E+02	2	2.08
385	C5H8	3-METHYL-1-BUTYNE	2	25.0	2.2336E+03	5.9168E+02	----	-------
386	C5H8	SPIROPENTANE	----	------	----------	----------	2	2.46
387	C5H8N4O12	PENTAERYTHRITOL TETRANITRATE	1	25.0	2.0000E+00	1.1397E-01	2	2.38
388	C5H8O	CYCLOPENTANONE	----	------	----------	----------	1	0.63
389	C5H8O	METHYL ISOPROPENYL KETONE	1	25.0	4.7000E+04	1.0452E+04	----	-------
390	C5H8O2	ACETYLACETONE	1	20.0	1.0900E+05	2.1539E+04	1	0.4

Table 15-1 SOLUBILITY IN WATER AND OCTANOL-WATER PARTITION COEFFICIENT- ORGANIC COMPOUNDS (continued)

NO	FORMULA	NAME	Code	T, C	S @ T, ppm (wt)	S @ T, ppm (mol)	Code	log₁₀ Kow
391	C5H8O2	ALLYL ACETATE	----	------	-----------	-----------	1	0.97
392	C5H8O2	ETHYL ACRYLATE	1	25.0	2.0000E+04	3.6588E+03	1	1.32
393	C5H8O2	METHYL METHACRYLATE	----	------	-----------	-----------	1	1.38
394	C5H8O2	VINYL PROPIONATE	----	------	-----------	-----------	2	1.22
395	C5H8O3	2-HYDROXYETHYL ACRYLATE	----	------	-----------	-----------	1	-0.21
396	C5H8O3	LEVULINIC ACID	----	------	-----------	-----------	1	-0.49
397	C5H8O3	METHYL ACETOACETATE	----	------	-----------	-----------	2	-0.69
398	C5H8O4	GLUTARIC ACID	1	65.0	5.3270E+05	1.3453E+05	1	-0.29
399	C5H9N	VALERONITRILE	----	------	-----------	-----------	1	1.12
400	C5H9NO	n-BUTYL ISOCYANATE	----	------	-----------	-----------	2	2.26
401	C5H9NO	N-METHYL-2-PYRROLIDONE	----	------	-----------	-----------	1	-0.38
402	C5H9NO4	L-GLUTAMIC ACID	1	25.0	8.6000E+03	1.0610E+03	1	-3.69
403	C5H10	CYCLOPENTANE	1	25.0	1.5600E+02	4.0076E+01	1	3
404	C5H10	2-METHYL-1-BUTENE	2	25.0	1.9346E+02	4.9700E+01	2	2.72
405	C5H10	2-METHYL-2-BUTENE	2	25.0	1.5303E+02	3.9312E+01	2	2.64
406	C5H10	3-METHYL-1-BUTENE	1	25.0	1.3000E+02	3.3396E+01	2	2.59
407	C5H10	1-PENTENE	1	25.0	1.4800E+02	3.8020E+01	2	2.66
408	C5H10	cis-2-PENTENE	1	25.0	2.0300E+02	5.2152E+01	2	2.58
409	C5H10	trans-2-PENTENE	1	25.0	2.0300E+02	5.2152E+01	2	2.58
410	C5H10Br2	2,3-DIBROMO-2-METHYLBUTANE	----	------	-----------	-----------	----	-------
411	C5H10Cl2	1,5-DICHLOROPENTANE	----	------	-----------	-----------	2	3.3
412	C5H10O	METHYL ISOPROPYL KETONE	1	25.0	6.0800E+04	1.3359E+04	1	0.84
413	C5H10O	2-PENTANONE	1	25.0	5.5360E+04	1.2109E+04	1	0.91
414	C5H10O	DIETHYL KETONE	1	25.0	3.4000E+04	7.3076E+03	1	0.99
415	C5H10O	VALERALDEHYDE	1	25.0	1.1700E+04	2.4699E+03	2	1.31
416	C5H10O2	n-BUTYL FORMATE	1	27.0	7.5580E+03	1.3415E+03	2	1.3
417	C5H10O2	ETHYL PROPIONATE	1	25.0	2.2000E+04	3.9521E+03	1	1.21
418	C5H10O2	ISOBUTYL FORMATE	2	25.0	1.8248E+04	3.2678E+03	2	1.23
419	C5H10O2	ISOPROPYL ACETATE	1	25.0	2.9000E+04	5.2404E+03	1	1.03
420	C5H10O2	n-PROPYL ACETATE	2	25.0	1.5066E+04	2.6908E+03	1	1.24
421	C5H10O2	METHYL n-BUTYRATE	1	25.0	1.5000E+04	2.6789E+03	1	1.29
422	C5H10O2	2-METHYLBUTYRIC ACID	----	------	-----------	-----------	----	-------
423	C5H10O2	ISOVALERIC ACID	1	20.0	4.0700E+04	7.4280E+03	1	1.16
424	C5H10O2	VALERIC ACID	1	25.0	2.4000E+04	4.3187E+03	1	1.39
425	C5H10O2	TETRAHYDROFURFURYL ALCOHOL	----	------	-----------	-----------	2	-0.11
426	C5H10O2S	3-METHYL SULFOLANE	----	------	-----------	-----------	2	0.18
427	C5H10O3	DIETHYL CARBONATE	1	20.0	1.8800E+04	2.9134E+03	1	1.21
428	C5H10O3	ETHYL LACTATE	1	20.0	1.0000E+06	1.0000E+06	2	-0.18
429	C5H10S	THIACYCLOHEXANE	----	------	-----------	-----------	2	2.28
430	C5H10S	CYCLOPENTANETHIOL	----	------	-----------	-----------	2	2.55
431	C5H11Br	1-BROMOPENTANE	1	25.0	1.2660E+02	1.5101E+01	1	3.37
432	C5H11Cl	1-CHLOROPENTANE	1	25.0	8.9330E+01	1.5098E+01	1	2.73
433	C5H11Cl	1-CHLORO-3-METHYLBUTANE	----	------	-----------	-----------	----	-------
434	C5H11Cl	2-CHLORO-2-METHYLBUTANE	1	25.0	3.3220E+03	5.6299E+02	1	2.52
435	C5H11N	N-METHYLPYRROLIDINE	----	------	-----------	-----------	1	0.92
436	C5H11N	PIPERIDINE	----	------	-----------	-----------	1	0.84
437	C5H11NO	tert-BUTYLFORMAMIDE	----	------	-----------	-----------	----	-------
438	C5H12	ISOPENTANE	1	25.0	4.7800E+01	1.1936E+01	1	2.3
439	C5H12	NEOPENTANE	1	25.0	3.3200E+01	8.2899E+00	1	3.11
440	C5H12	n-PENTANE	1	25.0	3.8500E+01	9.6133E+00	1	3.39
441	C5H12O	2,2-DIMETHYL-1-PROPANOL	----	------	-----------	-----------	1	1.31
442	C5H12O	tert-PENTYL-ALCOHOL	1	25.0	1.1000E+05	2.4637E+04	1	0.89
443	C5H12O	2-METHYL-1-BUTANOL	1	25.0	3.0000E+04	6.2809E+03	1	1.29
444	C5H12O	2-METHYL-2-BUTANOL	1	25.0	1.1000E+04	2.2679E+03	1	0.89
445	C5H12O	3-METHYL-1-BUTANOL	1	25.0	2.7000E+04	5.6391E+03	1	1.16
446	C5H12O	3-METHYL-2-BUTANOL	1	25.0	5.6000E+04	1.1978E+04	1	1.28
447	C5H12O	1-PENTANOL	1	25.0	2.2000E+04	4.5762E+03	1	1.51
448	C5H12O	2-PENTANOL	1	25.0	4.5000E+04	9.5380E+03	1	1.19
449	C5H12O	3-PENTANOL	1	25.0	5.2000E+04	1.1086E+04	1	1.21
450	C5H12O	METHYL sec-BUTYL ETHER	2	25.0	2.0572E+04	4.2742E+03	2	1.47
451	C5H12O	METHYL tert-BUTYL ETHER	1	25.0	5.1260E+04	1.0921E+04	1	0.94
452	C5H12O	METHYL ISOBUTYL ETHER	2	25.0	2.0977E+04	4.3598E+03	2	1.47
453	C5H12O	ETHYL PROPYL ETHER	2	25.0	1.6575E+04	3.4327E+03	2	1.54
454	C5H12O2	ETHYLENE GLYCOL MONOPROPYL ETHER	----	------	-----------	-----------	2	0.08
455	C5H12O2	NEOPENTYL GLYCOL	----	------	-----------	-----------	----	-------
456	C5H12O2	1,5-PENTANEDIOL	1	20.0	1.0000E+06	1.0000E+06	2	0.27
457	C5H12O3	2-(2-METHOXYETHOXY)ETHANOL	1	20.0	1.0000E+06	1.0000E+06	2	-1.18
458	C5H12O4	PENTAERYTHRITOL	1	20.0	6.0000E+04	8.3752E+03	1	-1.69
459	C5H12S	n-PENTYL MERCAPTAN	2	25.0	1.3696E+02	2.3677E+01	2	2.74
460	C5H12S	BUTYL-METHYL-SULFIDE	2	25.0	3.7895E+02	6.5530E+01	2	2.39
461	C5H12S	ETHYL-PROPYL-SULFIDE	2	25.0	5.0583E+02	8.7480E+01	2	2.39
462	C5H12S	2-METHYL-2-BUTANETHIOL	2	25.0	6.6926E+02	1.1576E+02	2	2.63
463	C5H13N	n-PENTYLAMINE	----	------	-----------	-----------	1	1.49
464	C5H13NO2	METHYL DIETHANOLAMINE	----	------	-----------	-----------	2	-1.5
465	C6Cl6	HEXACHLOROBENZENE	1	25.0	4.7000E-03	2.9732E-04	1	5.73
466	C6F6	HEXAFLUOROBENZENE	----	------	-----------	-----------	1	2.55
467	C6H3ClN2O4	1-CHLORO-2,4-DINITROBENZENE	1	15.0	8.0000E+00	7.1152E-01	1	2.17
468	C3H3Cl2NO2	1,2-DICHLORO-4-NITROBENZENE	----	------	-----------	-----------	1	3.12

Table 15-1 SOLUBILITY IN WATER AND OCTANOL-WATER PARTITION COEFFICIENT- ORGANIC COMPOUNDS (continued)

			Solubility in Water, S (ppm - parts per million)				Octanol-Water Partition Coefficient	
NO	FORMULA	NAME	Code	T, C	S @ T, ppm (wt)	S @ T, ppm (mol)	Code	log_{10} Kow
469	C6H3Cl3	1,2,4-TRICHLOROBENZENE	1	25.0	3.4570E+01	3.4324E+00	1	4.02
470	C6H3N3O6	1,3,5-TRINITROBENZENE	1	15.0	2.7800E+02	2.3507E+01	1	1.18
471	C6H4Br2	m-DIBROMOBENZENE	1	25.0	9.8350E+01	7.5112E+00	1	3.75
472	C6H4ClNO2	m-CHLORONITROBENZENE	1	20.0	2.7290E+02	3.1211E+01	1	2.46
473	C6H4ClNO2	o-CHLORONITROBENZENE	1	20.0	4.4120E+02	5.0467E+01	1	2.24
474	C6H4ClNO2	p-CHLORONITROBENZENE	1	20.0	2.2480E+02	2.5709E+01	1	2.39
475	C6H4Cl2	m-DICHLOROBENZENE	1	25.0	1.2300E+02	1.5075E+01	1	3.53
476	C6H4Cl2	o-DICHLOROBENZENE	1	25.0	9.2320E+01	1.1315E+01	1	3.43
477	C6H4Cl2	p-DICHLOROBENZENE	1	25.0	8.0000E+01	9.8046E+00	1	3.44
478	C6H4F2	m-DIFLUOROBENZENE	1	25.0	1.1400E+02	1.8002E+01	1	2.21
479	C6H4F2	o-DIFLUOROBENZENE	1	25.0	1.1400E+03	1.8017E+02	1	2.37
480	C6H4F2	p-DIFLUOROBENZENE	1	25.0	1.2230E+03	1.9331E+02	1	2.13
481	C6H4N2O4	m-DINITROBENZENE	1	25.0	5.7490E+02	6.1639E+01	1	1.49
482	C6H4N2O4	o-DINITROBENZENE	1	25.0	1.3300E+02	1.4254E+01	1	1.69
483	C6H4N2O4	p-DINITROBENZENE	1	25.0	6.8760E+01	7.3690E+00	1	1.46
484	C6H5Br	BROMOBENZENE	1	25.0	4.1000E+02	4.7060E+01	1	2.99
485	C6H5Cl	MONOCHLOROBENZENE	1	25.0	3.9070E+02	6.2552E+01	1	2.84
486	C6H5ClO	m-CHLOROPHENOL	1	20.0	2.5300E+04	3.6242E+03	1	2.5
487	C6H5ClO	o-CHLOROPHENOL	1	25.0	1.1350E+04	1.6062E+03	1	2.15
488	C6H5ClO	p-CHLOROPHENOL	1	20.0	2.6300E+04	3.7707E+03	1	2.39
489	C6H5Cl2N	3,4-DICHLOROANILINE	----	------	----------	----------	1	2.69
490	C6H5F	FLUOROBENZENE	1	25.0	1.5500E+03	2.9092E+02	1	2.27
491	C6H5I	IODOBENZENE	1	25.0	2.0070E+02	1.7726E+01	1	3.25
492	C6H5NO2	NITROBENZENE	1	25.0	1.9360E+03	2.8377E+02	1	1.85
493	C6H6	BENZENE	1	25.0	1.7550E+03	4.0529E+02	1	2.13
494	C6H6ClN	m-CHLOROANILINE	1	20.0	5.4420E+03	7.7209E+02	1	1.88
495	C6H6ClN	o-CHLOROANILINE	1	20.0	3.7650E+03	5.3339E+02	1	1.9
496	C6H6ClN	p-CHLOROANILINE	1	20.0	2.7550E+03	3.8997E+02	1	1.83
497	C6H6N2	cis-DICYANO-1-BUTENE	----	------	----------	----------	----	-------
498	C6H6N2	trans-DICYANO-1-BUTENE	----	------	----------	----------	----	-------
499	C6H6N2	1,4-DICYANO-2-BUTENE	----	------	----------	----------	----	-------
500	C6H6N2O2	m-NITROANILINE	1	25.0	9.0000E+02	1.1747E+02	1	1.37
501	C6H6N2O2	o-NITROANILINE	1	25.0	1.2600E+03	1.6451E+02	1	1.85
502	C6H6N2O2	p-NITROANILINE	1	25.0	6.0000E+02	7.8295E+01	1	1.39
503	C6H6O	PHENOL	1	25.0	8.0190E+04	1.6414E+04	1	1.46
504	C6H6O2	1,2-BENZENEDIOL	----	------	----------	----------	1	0.88
505	C6H6O2	1,3-BENZENEDIOL	1	30.0	6.9700E+05	2.7344E+05	1	0.8
506	C6H6O2	p-HYDROQUINONE	1	25.0	7.0000E+04	1.2165E+04	1	0.59
507	C6H6O3	1,2,3-BENZENETRIOL	1	25.0	3.8000E+05	8.0504E+04	----	-------
508	C6H6S	PHENYL MERCAPTAN	----	------	----------	----------	1	2.52
509	C6H7N	ANILINE	1	25.0	3.4160E+04	6.7953E+03	1	0.9
510	C6H7N	2-METHYLPYRIDINE	1	25.0	1.0000E+06	1.0000E+06	1	1.11
511	C6H7N	3-METHYLPYRIDINE	1	25.0	1.0000E+06	1.0000E+06	1	1.2
512	C6H7N	4-METHYLPYRIDINE	1	20.0	1.0000E+06	1.0000E+06	1	1.22
513	C6H8	1,3-CYCLOHEXADIENE	1	25.0	7.0000E+02	1.5746E+02	1	2.47
514	C6H8	METHYLCYCLOPENTADIENE	----	------	----------	----------	----	-------
515	C6H8N2	ADIPONITRILE	----	------	----------	----------	1	-0.32
516	C6H8N2	METHYLGLUTARONITRILE	----	------	----------	----------	2	0.28
517	C6H8N2	m-PHENYLENEDIAMINE	1	25.0	3.5100E+05	8.2648E+04	1	-0.33
518	C6H8N2	o-PHENYLENEDIAMINE	1	35.0	4.1500E+04	7.1610E+03	1	0.15
519	C6H8N2	p-PHENYLENEDIAMINE	1	24.0	3.8000E+04	6.5373E+03	1	-0.3
520	C6H8N2	PHENYLHYDRAZINE	----	------	----------	----------	1	1.25
521	C6H8N2O	BIS(CYANOETHYL)ETHER	----	------	----------	----------	----	-------
522	C6H8O4	DIMETHYL MALEATE	----	------	----------	----------	----	-------
523	C6H8O6	ASCORBIC ACID	----	------	----------	----------	1	-1.64
524	C6H8O7	CITRIC ACID	1	20.0	5.4100E+05	9.9520E+04	1	-1.72
525	C6H10	1-METHYLCYCLOPENTENE	----	------	----------	----------	----	-------
526	C6H10	3-METHYLCYCLOPENTENE	----	------	----------	----------	----	-------
527	C6H10	4-METHYLCYCLOPENTENE	----	------	----------	----------	----	-------
528	C6H10	CYCLOHEXENE	1	25.0	2.1310E+02	4.6742E+01	1	2.86
529	C6H10	2,3-DIMETHYL-1,3-BUTADIENE	2	25.0	1.3328E+02	2.9233E+01	2	3.13
530	C6H10	1,5-HEXADIENE	2	25.0	2.0369E+02	4.4678E+01	1	2.87
531	C6H10	cis,trans-2,4-HEXADIENE	2	25.0	6.4062E+01	1.4050E+01	1	2.8
532	C6H10	trans,trans-2,4-HEXADIENE	2	25.0	6.9599E+01	1.5264E+01	1	3.01
533	C6H10	1-HEXYNE	1	25.0	3.6000E+02	7.8973E+01	1	2.73
534	C6H10	2-HEXYNE	2	25.0	2.1973E+02	4.8196E+01	----	-------
535	C6H10	3-HEXYNE	2	25.0	2.6101E+02	5.7253E+01	----	-------
536	C6H10O	CYCLOHEXANONE	1	25.0	9.3190E+04	1.8514E+04	1	0.81
537	C6H10O	MESITYL OXIDE	1	20.0	2.8000E+04	5.2598E+03	2	1.37
538	C6H10O2	epsilon-CAPROLACTONE	----	------	----------	----------	2	0.68
539	C6H10O2	ETHYL METHACRYLATE	----	------	----------	----------	1	1.94
540	C6H10O2	n-PROPYL ACRYLATE	----	------	----------	----------	2	1.71
541	C6H10O3	ETHYLACETOACETATE	1	16.5	1.1100E+05	1.6990E+04	1	0.25
542	C6H10O3	PROPIONIC ANHYDRIDE	----	------	----------	----------	2	0.4
543	C6H10O4	ADIPIC ACID	1	34.1	2.9880E+04	3.7824E+03	1	0.08
544	C6H10O4	DIETHYL OXALATE	----	------	----------	----------	1	0.56
545	C6H10O4	ETHYLENE GLYCOL DIACETATE	1	22.0	1.2460E+05	1.7243E+04	2	0.4
546	C6H10O4	ETHYLIDENE DIACETATE	----	------	----------	----------	----	-------

NO	FORMULA	NAME	Solubility in Water, S (ppm - parts per million)				Octanol-Water Partition Coefficient	
			Code	T, C	S @ T, ppm (wt)	S @ T, ppm (mol)	Code	log₁₀ Kow
547	C6H11N	HEXANENITRILE	----	------	----------	----------	1	1.66
548	C6H11NO	epsilon-CAPROLACTAM	----	------	----------	----------	1	-0.19
549	C6H11NO	CYCLOHEXANONE OXIME	----	------	----------	----------	1	0.84
550	C6H12	CYCLOHEXANE	1	25.0	5.6100E+01	1.2009E+01	1	3.44
551	C6H12	2,3-DIMETHYL-1-BUTENE	2	25.0	8.0091E+01	1.7145E+01	2	3.13
552	C6H12	2,3-DIMETHYL-2-BUTENE	2	25.0	3.5932E+01	7.6915E+00	2	3.19
553	C6H12	3,3-DIMETHYL-1-BUTENE	2	25.0	1.3951E+02	2.9865E+01	2	3.04
554	C6H12	2-ETHYL-1-BUTENE	2	25.0	5.3804E+01	1.1517E+01	2	3.21
555	C6H12	trans-3-METHYL-2-PENTENE	2	25.0	4.1044E+01	8.7859E+00	----	-------
556	C6H12	1-HEXENE	1	25.0	6.9690E+01	1.4918E+01	1	3.39
557	C6H12	cis-2-HEXENE	2	25.0	4.4230E+01	9.4679E+00	2	3.07
558	C6H12	trans-2-HEXENE	2	25.0	4.6386E+01	9.9296E+00	2	3.07
559	C6H12	cis-3-HEXENE	2	25.0	4.9566E+01	1.0610E+01	----	-------
560	C6H12	trans-3-HEXENE	2	25.0	4.8112E+01	1.0299E+01	----	-------
561	C6H12	METHYLCYCLOPENTANE	1	25.0	4.2640E+01	9.1276E+00	1	3.37
562	C6H12	2-METHYL-1-PENTENE	1	25.0	7.8000E+01	1.6697E+01	2	3.21
563	C6H12	2-METHYL-2-PENTENE	2	25.0	4.7642E+01	1.0198E+01	2	3.13
564	C6H12	3-METHYL-1-PENTENE	2	25.0	8.5010E+01	1.8198E+01	2	3.08
565	C6H12	3-METHYL-cis-2-PENTENE	2	25.0	4.6758E+01	1.0009E+01	----	-------
566	C6H12	4-METHYL-1-PENTENE	1	25.0	4.8020E+01	1.0279E+01	2	3.08
567	C6H12	4-METHYL-cis-2-PENTENE	2	25.0	7.7532E+01	1.6597E+01	2	3
568	C6H12	4-METHYL-trans-2-PENTENE	2	25.0	7.0503E+01	1.5092E+01	2	3
569	C6H12N2	TRIETHYLENEDIAMINE	----	------	----------	----------	2	-0.83
570	C6H12O	BUTYL VINYL ETHER	----	------	----------	----------	2	1.89
571	C6H12O	CYCLOHEXANOL	1	25.0	3.8160E+04	7.0852E+03	1	1.23
572	C6H12O	1-HEXANAL	1	30.0	5.6440E+03	1.0199E+03	1	1.78
573	C6H12O	ETHYL ISOPROPYL KETONE	----	------	----------	----------	2	1.16
574	C6H12O	2-HEXANONE	1	25.0	1.6400E+04	2.9899E+03	1	1.38
575	C6H12O	3-HEXANONE	1	25.0	1.4700E+04	2.6762E+03	2	1.24
576	C6H12O	METHYL ISOBUTYL KETONE	1	25.0	1.9000E+04	3.4714E+03	1	1.31
577	C6H12O2	n-PENTYL FORMATE	2	25.0	2.3562E+03	3.6615E+02	----	-------
578	C6H12O2	n-BUTYL ACETATE	1	25.0	6.8000E+03	1.0607E+03	1	1.78
579	C6H12O2	sec-BUTYL ACETATE	1	20.0	6.2000E+03	9.6661E+02	1	1.72
580	C6H12O2	tert-BUTYL ACETATE	2	25.0	2.0462E+04	3.2292E+03	1	1.76
581	C6H12O2	ETHYL n-BUTYRATE	1	22.0	6.1680E+03	9.6159E+02	2	1.85
582	C6H12O2	ETHYL ISOBUTYRATE	2	25.0	9.3710E+03	1.4649E+03	2	1.77
583	C6H12O2	ISOBUTYL ACETATE	1	25.0	6.3000E+03	9.8228E+02	1	1.78
584	C6H12O2	n-PROPYL PROPIONATE	2	25.0	4.4905E+03	6.9907E+02	2	1.85
585	C6H12O2	CYCLOHEXYL PEROXIDE	----	------	----------	----------	----	-------
586	C6H12O2	DIACETONE ALCOHOL	----	------	----------	----------	2	-0.34
587	C6H12O2	2-ETHYL BUTYRIC ACID	----	------	----------	----------	1	1.68
588	C6H12O2	n-HEXANOIC ACID	1	20.0	9.6800E+03	1.5136E+03	1	1.92
589	C6H12O3	2-ETHOXYETHYL ACETATE	1	20.0	2.3000E+05	3.9124E+04	2	0.59
590	C6H12O3	HYDROXYCAPROIC ACID	----	------	----------	----------	1	0.81
591	C6H12O3	PARALDEHYDE	1	30.0	1.1200E+05	1.6902E+04	1	0.67
592	C6H12O3	sec-BUTYL GLYCOLATE	----	------	----------	----------	----	-------
593	C6H12S	THIACYCLOHEPTANE	----	------	----------	----------	----	-------
594	C6H13N	CYCLOHEXYLAMINE	----	------	----------	----------	1	1.49
595	C6H13N	HEXAMETHYLENEIMINE	----	------	----------	----------	2	1.68
596	C6H14	2,2-DIMETHYLBUTANE	1	25.0	2.3820E+01	4.9796E+00	1	3.82
597	C6H14	2,3-DIMETHYLBUTANE	1	25.0	2.0540E+01	4.2939E+00	1	3.42
598	C6H14	n-HEXANE	1	25.0	1.3310E+01	2.7824E+00	1	3.9
599	C6H14	2-METHYLPENTANE	1	25.0	1.3800E+01	2.8849E+00	2	3.21
600	C6H14	3-METHYLPENTANE	1	25.0	1.7910E+01	3.7441E+00	1	3.6
601	C6H14N2O2	LYSINE	----	------	----------	----------	1	-3.05
602	C6H14O	2-ETHYL-1-BUTANOL	1	25.0	1.0000E+04	1.7778E+03	2	1.75
603	C6H14O	1-HEXANOL	1	25.0	5.8750E+03	1.0409E+03	1	2.03
604	C6H14O	2-HEXANOL	1	25.0	1.4000E+04	2.4972E+03	1	1.76
605	C6H14O	2-METHYL-1-PENTANOL	1	25.0	8.1000E+03	1.4377E+03	2	1.75
606	C6H14O	4-METHYL-2-PENTANOL	1	27.0	1.5000E+04	2.6778E+03	2	1.43
607	C6H14O	n-BUTYL ETHYL ETHER	----	------	----------	----------	1	2.03
608	C6H14O	DIISOPROPYL ETHER	1	25.0	1.1250E+04	2.0021E+03	1	1.52
609	C6H14O	DI-n-PROPYL ETHER	1	25.0	3.8200E+03	6.7564E+02	1	2.03
610	C6H14O	METHYL tert-PENTYL ETHER	2	25.0	5.4758E+03	9.6982E+02	2	1.92
611	C6H14O2	ACETAL	1	25.0	4.4000E+04	6.9673E+03	1	0.84
612	C6H14O2	2-BUTOXYETHANOL	1	20.0	1.0000E+06	1.0000E+06	1	0.83
613	C6H14O2	1,6-HEXANEDIOL	----	------	----------	----------	2	0.76
614	C6H14O2	HEXYLENE GLYCOL	----	------	----------	----------	2	0.58
615	C6H14O2S	DI-n-PROPYL SULFONE	----	------	----------	----------	1	0.39
616	C6H14O3	DIETHYLENE GLYCOL DIMETHYL ETHER	----	------	----------	----------	1	-0.36
617	C6H14O3	DIPROPYLENE GLYCOL	1	25.0	1.0000E+06	1.0000E+06	----	-------
618	C6H14O3	2-(2-ETHOXYETHOXY)ETHANOL	1	20.0	1.0000E+06	1.0000E+06	1	-0.54
619	C6H14O3	TRIMETHYLOLPROPANE	----	------	----------	----------	1	-1.48
620	C6H14O4	TRIETHYLENE GLYCOL	1	20.0	1.0000E+06	1.0000E+06	2	-1.98
621	C6H14O6	SORBITOL	1	25.0	1.0000E+06	1.0000E+06	1	-2.2
622	C6H14S	n-HEXYLMERCAPTAN	2	25.0	2.9727E+01	4.5293E+00	2	3.23
623	C6H14S	BUTYL-ETHYL-SULFIDE	2	25.0	1.1113E+02	1.6934E+01	2	2.88
624	C6H14S	ISOPROPYL-SULFIDE	2	25.0	4.6228E+02	7.0462E+01	----	-------

378

NO	FORMULA	NAME	Solubility in Water, S (ppm - parts per million)				Octanol-Water Partition Coefficient	
			Code	T, C	S @ T, ppm (wt)	S @ T, ppm (mol)	Code	$\log_{10}$ Kow
625	C6H14S	METHYL-PENTYL-SULFIDE	2	25.0	2.8943E+02	4.4109E+01	2	2.88
626	C6H14S	PROPYL-SULFIDE	2	25.0	1.2072E+02	1.8395E+01	2	2.88
627	C6H14S2	PROPYL-DISULFIDE	----	------	----------	----------	2	3.84
628	C6H15Al	TRIETHYL ALUMINUN	----	------	----------	----------	----	-------
629	C6H15Al2Cl3	ETHYL ALUMINUM SESQUICHLORIDE	----	------	----------	----------	----	-------
630	C6H15N	DIISOPROPYLAMINE	----	------	----------	----------	1	1.4
631	C6H15N	DI-n-PROPYLAMINE	1	36.1	2.9100E+04	5.3076E+03	1	1.67
632	C6H15N	n-HEXYLAMINE	1	25.0	1.2000E+04	2.1576E+03	1	2.06
633	C6H15N	TRIETHYLAMINE	1	25.0	7.3670E+04	1.3961E+04	1	1.45
634	C6H15NO	6-AMINOHEXANOL	----	------	----------	----------	----	-------
635	C6H15NO2	DIISOPROPANOLAMINE	1	25.0	1.0000E+06	1.0000E+06	1	-0.82
636	C6H15NO3	TRIETHANOLAMINE	1	25.0	1.0000E+06	1.0000E+06	1	-1
637	C6H15N3	N-AMINOETHYL PIPERAZINE	----	------	----------	----------	----	-------
638	C6H15O4P	TRIETHYL PHOSPHATE	1	25.0	5.0000E+05	8.9998E+04	1	0.8
639	C6H16N2	HEXAMETHYLENEDIAMINE	1	25.0	1.0000E+06	1.0000E+06	2	0.35
640	C6H18N3OP	HEXAMETHYL PHOSPHORAMIDE	----	------	----------	----------	1	0.28
641	C6H18N4	TRIETHYLENE TETRAMINE	1	25.0	1.0000E+06	1.0000E+06	2	-2.65
642	C6H18OSi2	HEXAMETHYLDISILOXANE	----	------	----------	----------	1	4.2
643	C6H18O3Si3	HEXAMETHYLCYCLOTRISILOXANE	----	------	----------	----------	2	4.47
644	C6H19NSi2	HEXAMETHYLDISILAZANE	----	------	----------	----------	2	2.62
645	C7H3ClF3NO2	4-CHLORO-3-NITROBENZOTRIFLUORIDE	----	------	----------	----------	----	-------
646	C7H3Cl2F3	2,4-DICHLOROBENZOTRIFLUORIDE	----	------	----------	----------	2	4.24
647	C7H3Cl2NO	3,4-DICHLOROPHENYL ISOCYANATE	----	------	----------	----------	2	3.88
648	C7H4ClF3	p-CHLOROBENZOTRIFLUORIDE	----	------	----------	----------	2	3.6
649	C7H4Cl2O	m-CHLOROBENZOYL CHLORIDE	----	------	----------	----------	2	2.08
650	C7H4F3NO2	3-NITROBENZOTRIFLUORIDE	----	------	----------	----------	1	2.62
651	C7H5ClO	BENZOYL CHLORIDE	----	------	----------	----------	2	1.44
652	C7H5ClO2	o-CHLOROBENZOIC ACID	1	25.0	2.1000E+03	2.4208E+02	1	2.05
653	C7H5Cl3	BENZOTRICHLORIDE	----	------	----------	----------	2	3.9
654	C7H5F3	BENZOTRIFLUORIDE	1	25.0	4.5000E+02	5.5505E+01	1	3.01
655	C7H5N	BENZONITRILE	1	100.0	1.0000E+04	1.7615E+03	1	1.56
656	C7H5NO	PHENYL ISOCYANATE	1	20.0	7.3499E-01	1.1115E-01	2	2.59
657	C7H5N3O6	2,4,6-TRINITROTOLUENE	1	20.0	1.2000E+02	9.5188E+00	1	1.6
658	C7H6Cl2	BENZYL DICHLORIDE	----	------	----------	----------	2	2.97
659	C7H6Cl2	2,4-DICHLOROTOLUENE	----	------	----------	----------	1	4.24
660	C7H6N2O4	2,4-DINITROTOLUENE	1	22.0	2.7000E+02	2.6712E+01	1	1.98
661	C7H6N2O4	2,5-DINITROTOLUENE	----	------	----------	----------	2	2.18
662	C7H6N2O4	2,6-DINITROTOLUENE	----	------	----------	----------	1	2.1
663	C7H6N2O4	3,4-DINITROTOLUENE	----	------	----------	----------	1	2.08
664	C7H6N2O4	3,5-DINITROTOLUENE	----	------	----------	----------	----	-------
665	C7H6O	BENZALDEHYDE	1	25.0	6.5700E+03	1.1214E+03	1	1.48
666	C7H6O2	BENZOIC ACID	1	25.1	3.3900E+03	5.0153E+02	1	1.87
667	C7H6O2	p-HYDROXYBENZALDEHYDE	1	30.0	1.2900E+04	1.9241E+03	1	1.35
668	C7H6O2	SALICYLALDEHYDE	1	86.0	1.7000E+04	2.5446E+03	1	1.81
669	C7H6O3	SALICYLIC ACID	1	25.0	2.2280E+03	2.9116E+02	1	2.26
670	C7H7Br	p-BROMOTOLUENE	----	------	----------	----------	1	3.42
671	C7H7Cl	BENZYL CHLORIDE	----	------	----------	----------	1	2.3
672	C7H7Cl	o-CHLOROTOLUENE	----	------	----------	----------	1	3.42
673	C7H7Cl	p-CHLOROTOLUENE	1	20.0	1.0630E+02	1.5130E+01	1	3.33
674	C7H7F	p-FLUOROTOLUENE	----	------	----------	----------	1	2.58
675	C7H7NO	FORMANILIDE	----	------	----------	----------	1	1.15
676	C7H7NO2	m-NITROTOLUENE	1	30.0	5.0000E+02	6.5711E+01	1	2.45
677	C7H7NO2	o-NITROTOLUENE	----	------	----------	----------	1	2.3
678	C7H7NO2	p-NITROTOLUENE	1	14.5	4.0000E+01	5.2547E+00	1	2.37
679	C7H7NO3	o-NITROANISOLE	----	------	----------	----------	1	1.73
680	C7H8	TOLUENE	1	25.0	5.4240E+02	1.0609E+02	1	2.73
681	C7H8	1,3,5-CYCLOHEPTATRIENE	1	25.0	6.2010E+02	1.2130E+02	----	-------
682	C7H8O	ANISOLE	1	25.0	1.4000E+02	2.3325E+01	1	2.11
683	C7H8O	BENZYL ALCOHOL	1	25.0	4.2900E+04	7.4117E+03	1	1.1
684	C7H8O	m-CRESOL	1	20.0	2.1800E+04	3.6989E+03	1	1.96
685	C7H8O	o-CRESOL	1	20.0	2.4500E+04	4.1665E+03	1	1.95
686	C7H8O	p-CRESOL	1	20.0	1.9400E+04	3.2849E+03	1	1.94
687	C7H8O2	GUAIACOL	1	37.0	1.3160E+03	1.9119E+02	1	1.32
688	C7H8O2	p-METHOXYPHENOL	1	37.0	4.9660E+02	7.2097E+01	1	1.34
689	C7H9N	BENZYLAMINE	1	25.0	3.1400E+03	5.2928E+02	1	1.09
690	C7H9N	2,6-DIMETHYLPYRIDINE	----	------	----------	----------	1	1.68
691	C7H9N	N-METHYLANILINE	1	25.0	5.6240E+03	9.4996E+02	1	1.66
692	C7H9N	m-TOLUIDINE	1	20.0	1.5030E+04	2.5589E+03	1	1.4
693	C7H9N	o-TOLUIDINE	1	25.0	1.5000E+04	2.5537E+03	1	1.32
694	C7H9N	p-TOLUIDINE	1	15.0	6.5000E+03	1.0987E+03	1	1.39
695	C7H10	2-NORBORNENE	----	------	----------	----------	2	2.85
696	C7H10N2	TOLUENEDIAMINE	----	------	----------	----------	1	0.14
697	C7H11NO	CYCLOHEXYL ISOCYANATE	----	------	----------	----------	2	3.06
698	C7H12	1-HEPTYNE	1	25.0	9.4000E+01	1.7609E+01	2	3.01
699	C7H12O2	n-BUTYL ACRYLATE	1	20.0	1.6000E+03	2.2520E+02	1	2.36
700	C7H12O2	ISOBUTYL ACRYLATE	1	23.0	2.0000E+03	2.8159E+02	1	2.22
701	C7H12O2	n-PROPYL METHACRYLATE	----	------	----------	----------	2	2.26
702	C7H12O4	DIETHYL MALONATE	1	37.0	2.3220E+04	2.6666E+03	1	0.96

NO	FORMULA	NAME	Solubility in Water, S (ppm - parts per million)				Octanol-Water Partition Coefficient	
			Code	T, C	S @ T, ppm (wt)	S @ T, ppm (mol)	Code	log₁₀ Kow
703	C7H14	CYCLOHEPTANE	1	25.0	3.0000E+01	5.5044E+00	1	4
704	C7H14	1,1-DIMETYLCYCLOPENTANE	2	25.0	2.2530E+01	4.1338E+00	2	3.56
705	C7H14	cis-1,2-DIMETHYLCYCLOPENTANE	2	25.0	1.1910E+01	2.1852E+00	2	3.52
706	C7H14	trans-1,2-DIMETHYLCYCLOPENTANE	2	25.0	1.8155E+01	3.3310E+00	2	3.52
707	C7H14	cis-1,3-DIMETHYLCYCLOPENTANE	2	25.0	1.9267E+01	3.5351E+00	2	3.52
708	C7H14	trans-1,3-DIMETHYLCYCLOPENTANE	2	25.0	1.8293E+01	3.3563E+00	2	3.52
709	C7H14	ETHYLCYCLOPENTANE	2	25.0	9.5450E+00	1.7513E+00	2	3.59
710	C7H14	2-ETHYL-1-PENTENE	2	25.0	1.2191E+01	2.2367E+00	----	-------
711	C7H14	3-ETHYL-1-PENTENE	2	25.0	2.0691E+01	3.7964E+00	2	3.57
712	C7H14	1-HEPTENE	1	25.0	1.8160E+01	3.3319E+00	1	3.99
713	C7H14	cis-2-HEPTENE	2	25.0	9.5587E+00	1.7538E+00	2	3.56
714	C7H14	trans-2-HEPTENE	2	25.0	9.8062E+00	1.7992E+00	2	3.56
715	C7H14	cis-3-HEPTENE	2	25.0	1.1074E+01	2.0319E+00	2	3.56
716	C7H14	trans-3-HEPTENE	2	25.0	1.1123E+01	2.0409E+00	2	3.56
717	C7H14	METHYLCYCLOHEXANE	1	25.0	1.4020E+01	2.5723E+00	1	3.61
718	C7H14	2-METHYL-1-HEXENE	2	25.0	1.3712E+01	2.5158E+00	2	3.7
719	C7H14	3-METHYL-1-HEXENE	2	25.0	2.0919E+01	3.8382E+00	2	3.57
720	C7H14	4-METHYL-1-HEXENE	2	25.0	1.8029E+01	3.3079E+00	2	3.57
721	C7H14	2,3,3-TRIMETHYL-1-BUTENE	2	25.0	2.8471E+01	5.2238E+00	2	3.59
722	C7H14O	DIISOPROPYL KETONE	1	25.0	5.7000E+03	9.0360E+02	1	1.86
723	C7H14O	2-HEPTANONE	1	25.0	4.3000E+03	6.8086E+02	1	1.98
724	C7H14O	1-HEPTANAL	1	30.0	1.5160E+03	2.3948E+02	2	2.29
725	C7H14O	1-METHYLCYCLOHEXANOL	----	------	----------	----------	2	2.09
726	C7H14O	cis-2-METHYLCYCLOHEXANOL	----	------	----------	----------	1	1.84
727	C7H14O	trans-2-METHYLCYCLOHEXANOL	----	------	----------	----------	1	1.82
728	C7H14O	cis-3-METHYLCYCLOHEXANOL	----	------	----------	----------	2	2.05
729	C7H14O	trans-3-METHYLCYCLOHEXANOL	----	------	----------	----------	2	2.05
730	C7H14O	cis-4-METHYLCYCLOHEXANOL	----	------	----------	----------	2	2.05
731	C7H14O	trans-4-METHYLCYCLOHEXANOL	1	20.0	3.5000E+04	5.6895E+03	2	2.05
732	C7H14O	5-METHYL-2-HEXANONE	----	------	----------	----------	1	1.88
733	C7H14O2	n-BUTYL PROPIONATE	2	25.0	1.0872E+03	1.5059E+02	2	2.34
734	C7H14O2	ETHYL ISOVALERATE	2	25.0	2.2409E+03	3.1069E+02	2	2.26
735	C7H14O2	ISOPENTYL ACETATE	1	25.0	2.0000E+03	2.7723E+02	2	2.26
736	C7H14O2	n-PENTYL ACETATE	1	25.0	1.7300E+03	2.3975E+02	2	2.34
737	C7H14O2	n-PROPYL n-BUTYRATE	1	17.0	1.6200E+03	2.2449E+02	2	2.34
738	C7H14O2	n-HEPTANOIC ACID	1	20.0	2.4400E+03	3.3835E+02	1	2.42
739	C7H14O3	ETHYL-3-ETHOXYPROPIONATE	1	25.0	3.7970E+02	4.6807E+01	----	-------
740	C7H15Br	1-BROMOHEPTANE	1	25.0	6.6450E+00	6.6840E-01	1	4.36
741	C7H15N	N-METHYLCYCLOHEXYLAMINE	----	------	----------	----------	----	-------
742	C7H16	2,2-DIMETHYLPENTANE	1	25.0	4.4000E+00	7.9105E-01	2	3.67
743	C7H16	2,3-DIMETHYLPENTANE	1	25.0	5.2500E+00	9.4387E-01	2	3.63
744	C7H16	2,4-DIMETHYLPENTANE	1	25.0	4.4100E+00	7.9285E-01	2	3.63
745	C7H16	3,3-DIMETHYLPENTANE	1	25.0	5.9400E+00	1.0679E+00	2	3.67
746	C7H16	3-ETHYLPENTANE	2	25.0	2.9458E+00	5.2960E-01	2	3.71
747	C7H16	n-HEPTANE	1	25.0	2.2400E+00	4.0272E-01	1	4.66
748	C7H16	2-METHYLHEXANE	1	25.0	2.5400E+00	4.5665E-01	2	3.71
749	C7H16	3-METHYLHEXANE	1	25.0	2.6400E+00	4.7463E-01	2	3.71
750	C7H16	2,2,3-TRIMETHYLBUTANE	2	25.0	5.7410E+00	1.0321E+00	2	3.59
751	C7H16O	1-HEPTANOL	1	25.0	1.7400E+03	2.7015E+02	1	2.62
752	C7H16O	2-HEPTANOL	1	30.0	3.3000E+03	5.1303E+02	1	2.31
753	C7H16O	5-METHYL-1-HEXANOL	----	------	----------	----------	----	-------
754	C7H16O	ISOPROPYL-TERT-BUTYL-ETHER	1	25.0	5.0000E+02	7.7548E+01	----	-------
755	C7H16S	n-HEPTYL MERCAPTAN	2	25.0	7.6360E+00	1.0400E+00	2	3.72
756	C7H16S	BUTYL-PROPYL-SULFIDE	2	25.0	2.4126E+01	3.2862E+00	----	-------
757	C7H16S	ETHYL-PENTYL-SULFIDE	2	25.0	2.4126E+01	3.2862E+00	----	-------
758	C7H16S	HEXYL-METHYL-SULFIDE	2	25.0	2.4126E+01	3.2862E+00	----	-------
759	C7H17N	1-AMINOHEPTANE	----	------	----------	----------	1	2.57
760	C8H4Cl2O2	ISOPHTHALOYL CHLORIDE	----	------	----------	----------	----	-------
761	C8H4O3	PHTHALIC ANHYDRIDE	1	25.0	6.0000E+03	7.3362E+02	1	1.6
762	C8H6	ETHYNYLBENZENE	1	25.0	4.5620E+02	8.0497E+01	1	2.53
763	C8H6O4	ISOPHTHALIC ACID	1	25.0	1.3000E+02	1.4098E+01	1	1.66
764	C8H6O4	PHTHALIC ACID	1	25.0	1.4150E+03	1.5363E+02	1	0.73
765	C8H6O4	TEREPHTHALIC ACID	1	20.0	1.5000E+01	1.6266E+00	1	2
766	C8H6S	BENZOTHIOPHENE	1	20.0	1.3018E+02	1.7477E+01	1	3.12
767	C8H7N	INDOLE	1	20.0	1.8744E+03	2.8870E+02	1	2.14
768	C8H8	STYRENE	1	25.0	3.2160E+02	5.5641E+01	1	2.95
769	C8H8	1,3,5,7-CYCLOOCTATETRAENE	----	------	----------	----------	1	3.08
770	C8H8O	ACETOPHENONE	1	25.0	6.8420E+03	1.0319E+03	1	1.58
771	C8H8O	p-TOLUALDEHYDE	1	25.0	2.2710E+03	3.4116E+02	2	2.26
772	C8H8O2	METHYL BENZOATE	1	25.0	1.9720E+03	2.6138E+02	1	2.12
773	C8H8O2	o-TOLUIC ACID	1	25.0	1.1850E+03	1.5696E+02	1	2.46
774	C8H8O2	p-TOLUIC ACID	1	25.0	4.0000E+02	5.2945E+01	1	2.27
775	C8H8O3	METHYL SALICYLATE	1	30.0	7.0000E+02	8.2933E+01	1	2.55
776	C8H8O3	VANILLIN	1	25.0	1.1020E+04	1.3176E+03	1	1.21
777	C8H9NO	ACETANILIDE	1	25.0	6.1000E+03	8.1733E+02	1	1.16
778	C8H10	ETHYLBENZENE	1	25.0	1.6510E+02	2.8019E+01	1	3.15
779	C8H10	m-XYLENE	1	25.0	1.7400E+02	2.9530E+01	1	3.2
780	C8H10	o-XYLENE	1	25.0	2.2080E+02	3.7473E+01	1	3.12

NO	FORMULA	NAME	Solubility in Water, S (ppm - parts per million)				Octanol-Water Partition Coefficient	
			Code	T, C	S @ T, ppm (wt)	S @ T, ppm (mol)	Code	log$_{10}$ Kow
781	C8H10	p-XYLENE	1	25.0	2.0170E+02	3.4231E+01	1	3.15
782	C8H10O	m-ETHYLPHENOL	----	------	----------	----------	1	2.4
783	C8H10O	p-ETHYLPHENOL	----	------	----------	----------	1	2.58
784	C8H10O	PHENETOLE	----	------	----------	----------	1	2.51
785	C8H10O	2-PHENYLETHANOL	1	25.0	4.9000E+03	7.2560E+02	1	1.36
786	C8H10O	2,3-XYLENOL	1	25.0	4.5690E+03	6.7639E+02	2	2.61
787	C8H10O	2,4-XYLENOL	1	25.0	7.8680E+03	1.1681E+03	1	2.3
788	C8H10O	2,5-XYLENOL	1	25.0	3.5430E+03	5.2404E+02	1	2.33
789	C8H10O	2,6-XYLENOL	1	25.0	6.0470E+03	8.9632E+02	1	2.36
790	C8H10O	3,4-XYLENOL	1	25.0	4.7640E+03	7.0537E+02	1	2.23
791	C8H10O	3,5-XYLENOL	1	25.0	4.8870E+03	7.2366E+02	1	2.35
792	C8H11N	N,N-DIMETHYLANILINE	1	25.0	1.1050E+03	1.6442E+02	1	2.31
793	C8H11N	o-ETHYLANILINE	----	------	----------	----------	1	1.74
794	C8H11N	2,4,6-TRIMETHYLPYRIDINE	----	------	----------	----------	1	1.88
795	C8H11NO	p-PHENETIDINE	----	------	----------	----------	----	-------
796	C8H12	1,5-CYCLOOCTADIENE	----	------	----------	----------	1	3.16
797	C8H12	VINYLCYCLOHEXENE	1	25.0	5.0000E+01	8.3265E+00	1	3.93
798	C8H12O4	1,4-CYCLOHEXANEDICARBOXYLIC ACID	1	17.0	8.0000E+02	8.3763E+01	----	-------
799	C8H12O4	DIETHYL MALEATE	----	------	----------	----------	2	2.2
800	C8H14O2	n-BUTYL METHACRYLATE	----	------	----------	----------	1	2.88
801	C8H14O3	BUTYRIC ANHYDRIDE	----	------	----------	----------	2	1.39
802	C8H14O4	DIETHYL SUCCINATE	----	------	----------	----------	1	1.2
803	C8H16	CYCLOOCTANE	1	25.0	7.9000E+00	1.2683E+00	1	4.45
804	C8H16	1,1-DIMETHYLCYCLOHEXANE	2	25.0	5.9890E+00	9.6147E-01	2	4.05
805	C8H16	cis-1,2-DIMETHYLCYCLOHEXANE	1	25.0	6.0000E+00	9.6325E-01	2	4.01
806	C8H16	trans-1,2-DIMETHYLCYCLOHEXANE	2	25.0	4.7707E+00	7.6588E-01	2	4.01
807	C8H16	cis-1,3-DIMETHYLCYCLOHEXANE	2	25.0	5.8027E+00	9.3157E-01	2	4.01
808	C8H16	trans-1,3-DIMETHYLCYCLOHEXANE	2	25.0	4.4905E+00	7.2090E-01	2	4.01
809	C8H16	cis-1,4-DIMETHYLCYCLOHEXANE	2	25.0	4.5276E+00	7.2686E-01	2	4.01
810	C8H16	trans-1,4-DIMETHYLCYCLOHEXANE	1	25.0	3.8400E+00	6.1648E-01	2	4.01
811	C8H16	ETHYLCYCLOHEXANE	2	25.0	2.9140E+00	4.6781E-01	2	4.08
812	C8H16	2-ETHYL-1-HEXENE	2	25.0	2.7754E+00	4.4556E-01	2	4.19
813	C8H16	1-METHYL-1-ETHYLCYCLOPENTANE	2	25.0	3.3671E+00	5.4055E-01	----	-------
814	C8H16	1-OCTENE	1	25.0	4.0960E+00	6.5757E-01	1	4.57
815	C8H16	trans-2-OCTENE	2	25.0	2.0697E+00	3.3227E-01	2	4.06
816	C8H16	trans-3-OCTENE	2	25.0	2.2871E+00	3.6718E-01	----	-------
817	C8H16	trans-4-OCTENE	2	25.0	2.4311E+00	3.9029E-01	2	4.06
818	C8H16	n-PROPYLCYCLOPENTANE	1	25.0	2.0400E+00	3.2750E-01	2	4.08
819	C8H16	2,4,4-TRIMETHYL-1-PENTENE	2	25.0	8.0696E+00	1.2955E+00	1	4.55
820	C8H16	2,4,4-TRIMETHYL-2-PENTENE	2	25.0	6.6337E+00	1.0650E+00	2	4
821	C8H16O	2-ETHYLHEXANAL	----	------	----------	----------	2	2.71
822	C8H16O	1-OCTANAL	1	30.0	3.7030E+02	5.2046E+01	2	2.78
823	C8H16O	2-OCTANONE	1	20.0	9.0000E+02	1.2655E+02	1	2.37
824	C8H16O2	n-BUTYL n-BUTYRATE	2	25.0	3.7788E+02	4.7220E+01	2	2.83
825	C8H16O2	n-HEXYL ACETATE	1	25.0	5.1050E+02	6.3799E+01	2	2.83
826	C8H16O2	ISOBUTYL ISOBUTYRATE	1	25.0	5.7000E+02	7.1239E+01	2	2.68
827	C8H16O2	n-OCTANOIC ACID	1	20.0	6.8000E+02	8.4995E+01	1	3.05
828	C8H16O4	DIETHYLENE GLYCOL ETHYL ETHER ACETATE	1	20.0	1.0000E+06	1.0000E+06	2	0.32
829	C8H18	2,2-DIMETHYLHEXANE	2	25.0	1.3953E+00	2.2004E-01	2	4.16
830	C8H18	2,3-DIMETHYLHEXANE	2	25.0	8.4167E-01	1.3274E-01	2	4.12
831	C8H18	2,4-DIMETHYLHEXANE	2	25.0	1.2030E+00	1.8972E-01	2	1.9
832	C8H18	2,5-DIMETHYLHEXANE	2	25.0	1.2253E+00	1.9324E-01	2	4.12
833	C8H18	3,3-DIMETHYLHEXANE	2	25.0	1.0393E+00	1.6391E-01	2	4.16
834	C8H18	3,4-DIMETHYLHEXANE	2	25.0	7.4391E-01	1.1732E-01	2	4.12
835	C8H18	3-ETHYLHEXANE	2	25.0	7.0956E-01	1.1190E-01	2	4.2
836	C8H18	3-ETHYL-2-METHYLPENTANE	2	25.0	8.3923E-01	1.3235E-01	2	4.12
837	C8H18	3-METHYL-3-ETHYLPENTANE	2	25.0	7.2083E-01	1.1368E-01	2	4.16
838	C8H18	2-METHYLHEPTANE	2	25.0	7.4739E-01	1.1787E-01	2	4.2
839	C8H18	3-METHYLHEPTANE	1	25.0	7.9200E-01	1.2490E-01	2	4.2
840	C8H18	4-METHYLHEPTANE	2	25.0	7.4478E-01	1.1746E-01	2	4.2
841	C8H18	n-OCTANE	1	25.0	4.3100E-01	6.7972E-02	1	5.18
842	C8H18	2,2,3-TRIMETHYLPENTANE	2	25.0	1.1743E+00	1.8520E-01	2	4.09
843	C8H18	2,2,4-TRIMETHYLPENTANE	2	25.0	2.1425E+00	3.3789E-01	2	4.09
844	C8H18	2,3,3-TRIMETHYLPENTANE	2	25.0	8.8376E-01	1.3938E-01	2	4.09
845	C8H18	2,3,4-TRIMETHYLPENTANE	1	25.0	2.3020E+00	3.6304E-01	2	4.05
846	C8H18	2,2,3,3-TETRAMETHYLBUTANE	2	25.0	1.4266E+00	2.2499E-01	2	4.05
847	C8H18O	DI-n-BUTYL ETHER	1	24.8	1.8500E+03	2.5632E+02	1	3.21
848	C8H18O	DI-sec-BUTYL ETHER	2	25.0	7.7485E+02	1.0726E+02	2	2.87
849	C8H18O	DI-tert-BUTYL ETHER	2	25.0	1.7238E+03	2.3881E+02	2	2.79
850	C8H18O	2-ETHYL-1-HEXANOL	1	25.0	1.0000E+02	1.3834E+01	2	2.73
851	C8H18O	1-OCTANOL	1	25.0	5.4000E+02	7.4734E+01	1	3
852	C8H18O	2-OCTANOL	1	25.0	4.0000E+03	5.5524E+02	1	2.9
853	C8H18O2	DI-t-BUTYL PEROXIDE	----	------	----------	----------	----	-------
854	C8H18O2S	DI-n-BUTYL SULFONE	----	------	----------	----------	----	-------
855	C8H18O3	DIETHYLENE GLYCOL DIETHYL ETHER	1	20.0	1.0000E+06	1.0000E+06	1	0.39
856	C8H18O3	DIETHYLENE GLYCOL MONOBUTYL EHTER	1	20.0	1.0000E+06	1.0000E+06	1	0.56
857	C8H18O4	TRIETHYLENE GLYCOL DIMETHYL ETHER	----	------	----------	----------	2	-0.76
858	C8H18O5	TETRAETHYLENE GLYCOL	1	20.0	1.0000E+06	1.0000E+06	2	-2.02

NO	FORMULA	NAME	Solubility in Water, S (ppm - parts per million)				Octanol-Water Partition Coefficient	
			Code	T, C	S @ T, ppm (wt)	S @ T, ppm (mol)	Code	log$_{10}$ Kow
859	C8H18S	n-OCTYL MERCAPTAN	2	25.0	2.5132E+00	3.0948E-01	2	4.21
860	C8H18S	tert-OCTYL MERCAPTAN	2	25.0	2.4729E+01	3.0452E+00	2	3.99
861	C8H18S	BUTYL-SULFIDE	2	25.0	1.3388E+01	1.6487E+00	2	3.87
862	C8H18S	ETHYL-HEXYL-SULFIDE	2	25.0	6.9677E+00	8.5805E-01	2	3.87
863	C8H18S	HEPTYL-METHYL-SULFIDE	2	25.0	6.9677E+00	8.5805E-01	2	3.87
864	C8H18S	PENTYL-PROPYL-SULFIDE	2	25.0	6.9677E+00	8.5805E-01	2	3.87
865	C8H18S2	BUTYL-DISULFIDE	----	------	----------	----------	2	4.82
866	C8H19N	DI-n-BUTYLAMINE	1	25.0	3.1000E+03	4.3325E+02	1	2.83
867	C8H19N	DIISOBUTYLAMINE	----	------	----------	----------	2	2.63
868	C8H19N	n-OCTYLAMINE	1	25.0	2.0000E+02	2.7882E+01	1	2.9
869	C8H23N5	TETRAETHYLENEPENTAMINE	1	20.0	1.0000E+06	1.0000E+06	2	-3.16
870	C8H24O4Si4	OCTAMETHYLCYCLOTETRASILOXANE	----	------	----------	----------	1	5.1
871	C9H4O5	TRIMELLITIC ANHYDRIDE	----	------	----------	----------	2	1.95
872	C9H6N2O2	TOLUENE DIISOCYANATE	----	------	----------	----------	----	-------
873	C9H7N	ISOQUINOLINE	1	20.0	4.5206E+03	6.3299E+02	1	2.08
874	C9H7N	QUINOLINE	1	20.0	6.7164E+03	9.4223E+02	1	2.03
875	C9H7NO	8-HYDROXYQUINOLINE	1	20.0	5.5560E+02	6.8986E+01	1	2.02
876	C9H8	INDENE	----	------	----------	----------	1	2.92
877	C9H8O	2-METHYLBENZOFURAN	----	------	----------	----------	1	3.22
878	C9H10	INDANE	1	25.0	1.0926E+02	1.6656E+01	1	3.18
879	C9H10	cis-PROPENYLBENZENE	----	------	----------	----------	----	-------
880	C9H10	trans-PROPENYLBENZENE	----	------	----------	----------	2	3.31
881	C9H1O	alpha-METHYLSTYRENE	----	------	----------	----------	1	3.48
882	C9H10	m-METHYLSTYRENE	1	25.0	8.9000E+01	1.3568E+01	2	3.44
883	C9H10	o-METHYLSTYRENE	----	------	----------	----------	2	3.44
884	C9H10	p-METHYLSTYRENE	1	25.0	8.9000E+01	1.3568E+01	2	3.35
885	C9H10O2	BENZYL ACETATE	----	------	----------	----------	1	1.96
886	C9H10O2	ETHYL BENZOATE	1	25.0	7.2000E+02	8.6425E+01	1	2.64
887	C9H10O3	ETHYL VANILLIN	----	------	----------	----------	1	1.58
888	C9H11NO	p-DIMETHYLAMINOBENZALDEHYDE	----	------	----------	----------	1	1.81
889	C9H12	CUMENE	1	25.0	5.0000E+01	7.4945E+00	1	3.66
890	C9H12	m-ETHYLTOLUENE	2	25.0	8.2910E+01	1.2427E+01	1	3.98
891	C9H12	o-ETHYLTOLUENE	1	25.0	9.3050E+01	1.3948E+01	1	3.53
892	C9H12	p-ETHYLTOLUENE	1	25.0	9.4850E+01	1.4218E+01	1	3.63
893	C9H12	MESITYLENE	1	25.0	4.8200E+01	7.2246E+00	1	3.42
894	C9H12	n-PROPYLBENZENE	2	25.0	9.3245E+01	1.3977E+01	1	3.69
895	C9H12	1,2,3-TRIMETHYLBENZENE	2	25.0	3.6012E+01	5.3976E+00	1	3.66
896	C9H12	1,2,4-TRIMETHYLBENZENE	2	25.0	5.2662E+01	7.8931E+00	1	3.78
897	C9H12O	BENZYL ETHYL ETHER	----	------	----------	----------	1	2.16
898	C9H12O	2-PHENYL-2-PROPANOL	----	------	----------	----------	2	1.95
899	C9H12O2	CUMENE HYDROPEROXIDE	----	------	----------	----------	2	1.55
900	C9H14O	ISOPHORONE	1	25.0	1.2000E+04	1.5806E+03	1	1.7
901	C9H14O6	GLYCERYL TRIACETATE	----	------	----------	----------	1	0.25
902	C9H16	1-NONYNE	1	25.0	7.2000E+00	1.0441E+00	2	3.99
903	C9H16O4	AZELAIC ACID	1	20.0	2.4470E+03	2.3472E+02	1	1.57
904	C9H18	BUTYLCYCLOPENTANE	2	25.0	4.3077E-01	6.1472E-02	----	-------
905	C9H18	cis,cis-1,3,5-TRIMETHYLCYCLOHEXANE	2	25.0	1.9617E+00	2.7994E-01	----	-------
906	C9H18	cis,trans-1,3,5-TRIMETHYLCYCLOHEXANE	2	25.0	1.7397E+00	2.4826E-01	----	-------
907	C9H18	ISOPROPYLCYCLOHEXANE	2	25.0	7.5939E-01	1.0837E-01	2	4.5
908	C9H18	1-NONENE	1	25.0	1.1170E+00	1.5940E-01	1	5.15
909	C9H18	n-PROPYLCYCLOHEXANE	2	25.0	6.7726E-01	9.6646E-02	2	4.58
910	C9H18O	DIISOBUTYL KETONE	1	23.5	2.6400E+03	3.3513E+02	2	2.56
911	C9H18O	1-NONANAL	1	30.0	1.0510E+02	1.3312E+01	2	3.27
912	C9H18O2	n-BUTYL VALERATE	2	25.0	1.9130E+02	2.1782E+01	----	-------
913	C9H18O2	n-NONANOIC ACID	1	20.0	2.6000E+02	2.9607E+01	1	3.42
914	C9H18O2	n-OCTYL FORMATE	2	25.0	6.5327E+01	7.4376E+00	----	-------
915	C9H20	3,3-DIETHYLPENTANE	2	25.0	1.3955E-01	1.9602E-02	2	4.65
916	C9H20	2,2-DIMETHYL-3-ETHYLPENTANE	2	25.0	2.8855E-01	4.0529E-02	2	4.58
917	C9H20	3-ETHYL-2,3-DIMETHYLPENTANE	2	25.0	1.5219E-01	2.1376E-02	----	-------
918	C9H20	2,4-DIMETHYL-3-ETHYLPENTANE	2	25.0	2.4347E-01	3.4198E-02	2	4.54
919	C9H20	2,2-DIMETHYLHEPTANE	2	25.0	3.0880E-01	4.3374E-02	2	4.65
920	C9H20	2,6-DIMETHYLHEPTANE	2	25.0	2.6615E-01	3.7383E-02	2	4.65
921	C9H20	3-ETHYLHEPTANE	2	25.0	1.6628E-01	2.3356E-02	2	4.69
922	C9H20	4-ETHYLHEPTANE	2	25.0	1.8690E-01	2.6252E-02	2	4.69
923	C9H20	2,3-DIMETHYLHEPTANE	2	25.0	1.9476E-01	2.7356E-02	2	4.65
924	C9H20	2,4-DIMETHYLHEPTANE	2	25.0	3.0500E-01	4.2840E-02	2	4.65
925	C9H20	2,5-DIMETHYLHEPTANE	2	25.0	2.5388E-01	3.5660E-02	2	4.65
926	C9H20	3,4-DIMETHYLHEPTANE	2	25.0	1.9362E-01	2.7195E-02	2	4.65
927	C9H20	3,5-DIMETHYLHEPTANE	2	25.0	2.5388E-01	3.5660E-02	2	4.65
928	C9H20	4,4-DIMETHYLHEPTANE	2	25.0	2.6615E-01	3.7383E-02	2	4.65
929	C9H20	3-ETHYL-2-METHYLHEXANE	2	25.0	2.2565E-01	3.1694E-02	2	4.65
930	C9H20	4-ETHYL-2-METHYLHEXANE	2	25.0	2.8906E-01	4.0601E-02	2	4.65
931	C9H20	3-ETHYL-3-METHYLHEXANE	2	25.0	1.9362E-01	2.7195E-02	2	4.65
932	C9H20	3-ETHYL-4-METHYLHEXANE	2	25.0	1.9591E-01	2.7517E-02	2	4.65
933	C9H20	2,2,3-TRIMETHYLHEXANE	2	25.0	2.9266E-01	4.1107E-02	2	4.58
934	C9H20	2,2,4-TRIMETHYLHEXANE	2	25.0	4.4376E-01	6.2330E-02	2	4.58
935	C9H20	2,3,3-TRIMETHYLHEXANE	2	25.0	2.2994E-01	3.2298E-02	2	4.58
936	C9H20	2,3,4-TRIMETHYLHEXANE	2	25.0	2.1224E-01	2.9811E-02	2	4.54

NO	FORMULA	NAME	Solubility in Water, S (ppm - parts per million)				Octanol-Water Partition Coefficient	
			Code	T, C	S @ T, ppm (wt)	S @ T, ppm (mol)	Code	$\log_{10}$ Kow
937	C9H20	2,3,5-TRIMETHYLHEXANE	2	25.0	3.3420E-01	4.6942E-02	2	4.54
938	C9H20	2,4,4-TRIMETHYLHEXANE	2	25.0	3.4809E-01	4.8892E-02	2	4.58
939	C9H20	3,3,4-TRIMETHYLHEXANE	2	25.0	1.9522E-01	2.7420E-02	2	4.58
940	C9H20	2-METHYLOCTANE	2	25.0	1.6550E-01	2.3246E-02	2	4.69
941	C9H20	3-METHYLOCTANE	2	25.0	1.5653E-01	2.1986E-02	2	4.69
942	C9H20	4-METHYLOCTANE	1	25.0	1.1500E-01	1.6153E-02	2	4.69
943	C9H20	n-NONANE	1	25.0	1.2200E-01	1.7136E-02	2	4.76
944	C9H20	2,2,3,3-TETRAMETHYLPENTANE	2	25.0	1.9730E-01	2.7712E-02	2	4.54
945	C9H20	2,2,3,4-TETRAMETHYLPENTANE	2	25.0	3.0267E-01	4.2513E-02	2	4.5
946	C9H20	2,2,4,4-TETRAMETHYLPENTANE	2	25.0	5.6969E-01	8.0019E-02	2	4.54
947	C9H20	2,3,3,4-TETRAMETHYLPENTANE	2	25.0	1.8299E-01	2.5703E-02	2	4.5
948	C9H20	2,2,5-TRIMETHYLHEXANE	2	25.0	5.1255E-01	7.1992E-02	2	4.58
949	C9H20O	2,6-DIMETHYL-4-HEPTANOL	1	25.0	1.0000E+03	1.2499E+02	1	3.08
950	C9H20O	1-NONANOL	1	25.0	1.2980E+02	1.6211E+01	1	3.77
951	C9H20O	2-NONANOL	1	15.0	2.5966E+02	3.2434E+01	----	-------
952	C9H20S	n-NONYL MERCAPTAN	2	25.0	1.0444E+00	1.1736E-01	2	4.7
953	C9H20S	BUTYL-PENTYL-SULFIDE	2	25.0	2.5611E+00	2.8780E-01	2	4.37
954	C9H20S	ETHYL-HEPTYL-SULFIDE	2	25.0	2.5611E+00	2.8780E-01	2	4.37
955	C9H20S	HEXYL-PROPYL-SULFIDE	2	25.0	2.5611E+00	2.8780E-01	2	4.37
956	C9H20S	METHYL-OCTYL-SULFIDE	2	25.0	2.5611E+00	2.8780E-01	2	4.37
957	C9H21N	n-NONYLAMINE	-----	------	----------	----------	2	3.29
958	C9H21N	TRIPROPYLAMINE	1	25.0	7.4790E+02	9.4102E+01	1	2.79
959	C10H6O8	PYROMELLITIC ACID	1	16.0	1.4000E+04	1.0054E+03	2	0.15
960	C10H7Br	1-BROMONAPHTHALENE	1	21.0	9.3180E+00	8.1067E-01	2	4.06
961	C10H7Cl	1-CHLORONAPHTHALENE	1	25.0	2.2400E+01	2.4815E+00	1	3.9
962	C10H8	NAPHTHALENE	1	25.0	3.2050E+01	4.5048E+00	1	3.3
963	C10H8	AZULENE	----	------	----------	----------	1	3.2
964	C10H9N	QUINALDINE	----	------	----------	----------	1	2.59
965	C10H10	m-DIVINYLBENZENE	----	------	----------	----------	2	3.8
966	C10H10	1-METHYLINDENE	----	------	----------	----------	----	-------
967	C10H10	2-METHYLINDENE	----	------	----------	----------	----	-------
968	C10H10O4	DIMETHYL PHTHALATE	1	20.0	4.2920E+03	3.9973E+02	1	1.56
969	C10H10O4	DIMETHYL TEREPHTHALATE	----	------	----------	----------	1	2.25
970	C10H12	DICYCLOPENTADIENE	----	------	----------	----------	2	3.51
971	C10H12	1,2,3,4-TETRAHYDRONAPHTHALENE	1	28.0	4.6700E+01	6.3639E+00	1	3.49
972	C10H12O	ANETHOLE	1	25.0	1.1100E+02	1.3494E+01	2	3.39
973	C10H12O4	DIALLYL MALEATE	----	------	----------	----------	2	2.9
974	C10H14	n-BUTYLBENZENE	1	25.0	1.3820E+01	1.8549E+00	1	4.38
975	C10H14	sec-BUTYLBENZENE	2	25.0	4.2142E+01	5.6564E+00	1	4.57
976	C10H14	tert-BUTYLBENZENE	2	25.0	5.3350E+01	7.1609E+00	1	4.11
977	C10H14	1,2,3,4-TETRAMETHYLBENZENE	2	25.0	7.3385E+00	9.8498E-01	1	4
978	C10H14	m-CYMENE	2	25.0	3.8184E+01	5.1250E+00	1	4.5
979	C10H14	o-CYMENE	2	25.0	3.2072E+01	4.3047E+00	1	4.38
980	C10H14	p-CYMENE	2	25.0	3.4022E+01	4.5664E+00	1	4.1
981	C10H14	m-DIETHYLBENZENE	2	25.0	2.7164E+01	3.6460E+00	1	4.57
982	C10H14	o-DIETHYLBENZENE	2	25.0	2.3859E+01	3.2023E+00	1	4.42
983	C10H14	p-DIETHYLBENZENE	2	25.0	2.3423E+01	3.1438E+00	1	4.58
984	C10H14	2-ETHYL-m-XYLENE	2	25.0	1.6557E+01	2.2223E+00	1	4.28
985	C10H14	2-ETHYL-p-XYLENE	2	25.0	1.9776E+01	2.6544E+00	1	4.43
986	C10H14	3-ETHYL-o-XYLENE	2	25.0	1.3351E+01	1.7919E+00	1	4.34
987	C10H14	4-ETHYL-m-XYLENE	2	25.0	1.8088E+01	2.4277E+00	1	4.47
988	C10H14	4-ETHYL-o-XYLENE	2	25.0	1.6797E+01	2.2544E+00	1	4.5
989	C1OH14	5-ETHYL-m-XYLENE	2	25.0	2.3436E+01	3.1456E+00	1	4.55
990	C10H14	ISOBUTYLBENZENE	2	25.0	4.3445E+01	5.8313E+00	1	4.68
991	C10H14	1,2,3,5-TETRAMETHYLBENZENE	2	25.0	1.0720E+01	1.4388E+00	1	4.1
992	C10H14	1,2,4,5-TETRAMETHYLBENZENE	1	25.0	3.4800E+00	4.6708E-01	1	4
993	C10H14O	p-tert-BUTYLPHENOL	1	25.0	5.8000E+02	6.9591E+01	1	3.31
994	C10H14O2	p-tert-BUTYLCATECHOL	1	25.0	2.0000E-07	2.1676E-08	----	-------
995	C10H15N	N,N-DIETYHLANILINE	1	12.0	1.4400E+04	1.7606E+03	1	3.31
996	C10H15N	2,6-DIETHYLANILINE	1	26.7	6.7000E+02	8.0927E+01	2	3.15
997	C10H16	CAMPHENE	----	------	----------	----------	2	4.35
998	C10H16	D-LIMONENE	1	25.0	1.3800E+01	1.8248E+00	2	4.83
999	C10H16	alpha-PHELLANDRENE	----	------	----------	----------	2	4.62
1000	C1OH16	beta-PHELLANDRENE	----	------	----------	----------	2	4.7
1001	C10H16	alpha-PINENE	----	------	----------	----------	1	4.83
1002	C10H16	beta-PINENE	----	------	----------	----------	2	4.35
1003	C10H16	alpha-TERPINENE	----	------	----------	----------	----	-------
1004	C1OH16	gamma-TERPINENE	----	------	----------	----------	1	4.5
1005	C10H16	TERPINOLENE	----	------	----------	----------	1	4.47
1006	C10H16O	CAMPHOR	1	25.0	1.5680E+03	1.8581E+02	2	2.34
1007	C10H18	1-DECYNE	2	25.0	1.3311E+00	1.7345E-01	----	-------
1008	C10H18	cis-DECAHYDRONANPHTALENE	1	25.0	8.8900E-01	1.1584E-01	2	4.2
1009	C10H18	trans-DECAHYDRONAPHTHALENE	1	25.0	8.8900E-01	1.1584E-01	2	4.2
1010	C10H18O4	SEBACIC ACID	1	20.0	1.0000E+03	8.9154E+01	2	2.19
1011	C10H20	n-BUTYLCYCLOHEXANE	2	25.0	1.7779E-01	2.2833E-02	2	5.07
1012	C10H20	1-CYCLOPENTYLPENTANE	1	25.0	1.1500E-01	1.4770E-02	----	-------
1013	C10H20	1-DECENE	2	25.0	1.4780E-01	1.8982E-02	2	5.12
1014	C10H20O	1-DECANAL	----	------	----------	----------	2	3.76

NO	FORMULA	NAME	Solubility in Water, S (ppm - parts per million)				Octanol-Water Partition Coefficient	
			Code	T, C	S @ T, ppm (wt)	S @ T, ppm (mol)	Code	log₁₀ Kow
1015	C10H20O2	n-DECANOIC ACID	1	20.0	1.5000E+02	1.5688E+01	1	4.09
1016	C10H20O2	2-ETHYLHEXYL ACETATE	2	25.0	6.5940E+01	6.8961E+00	2	3.74
1017	C10H20O2	ISOPENTYL ISOVALERATE	2	25.0	8.2083E+01	8.5845E+00	2	3.66
1018	C10H22	n-DECANE	1	25.0	5.2000E-02	6.5838E-03	1	5.01
1019	C10H22	2-METHYLNONANE	2	25.0	4.2345E-02	5.3614E-03	2	5.18
1020	C10H22	3-METHYLNONANE	2	25.0	4.0502E-02	5.1280E-03	2	5.18
1021	C10H22	4-METHYLNONANE	2	25.0	4.5533E-02	5.7650E-03	2	5.18
1022	C10H22	5-METHYLNONANE	2	25.0	4.6957E-02	5.9453E-03	2	5.18
1023	C10H22	3-ETHYLOCTANE	2	25.0	4.3518E-02	5.5099E-03	2	5.18
1024	C10H22	4-ETHYLOCTANE	2	25.0	5.1087E-02	6.4682E-03	2	5.18
1025	C10H22	2,2-DIMETHYLOCTANE	2	25.0	7.4992E-02	9.4949E-03	2	5.11
1026	C10H22	2,3-DIMETHYLOCTANE	2	25.0	4.9197E-02	6.2289E-03	2	5.11
1027	C10H22	2,4-DIMETHYLOCTANE	2	25.0	7.9380E-02	1.0051E-02	2	5.11
1028	C10H22	2,5-DIMETHYLOCTANE	2	25.0	6.8388E-02	8.6588E-03	2	5.11
1029	C10H22	2,6-DIMETHYLOCTANE	2	25.0	6.1438E-02	7.7788E-03	2	5.11
1030	C10H22	2,7-DIMETHYLOCTANE	2	25.0	6.3247E-02	8.0079E-03	2	5.11
1031	C10H22	3,3-DIMETHYLOCTANE	2	25.0	5.8642E-02	7.4249E-03	2	5.11
1032	C10H22	3,4-DIMETHYLOCTANE	2	25.0	5.1782E-02	6.5563E-03	2	5.11
1033	C10H22	3,5-DIMETHYLOCTANE	2	25.0	6.4964E-02	8.2252E-03	2	5.11
1034	C10H22	3,6-DIMETHYLOCTANE	2	25.0	5.9989E-02	7.5954E-03	2	5.11
1035	C10H22	4,4-DIMETHYLOCTANE	2	25.0	7.2415E-02	9.1687E-03	2	5.11
1036	C10H22	4,5-DIMETHYLOCTANE	2	25.0	5.5632E-02	7.0438E-03	2	5.11
1037	C10H22	4-PROPYLHEPTANE	2	25.0	7.2415E-02	9.1687E-03	2	5.18
1038	C10H22	4-ISOPROPYLHEPTANE	2	25.0	6.6843E-02	8.4632E-03	2	5.11
1039	C10H22	3-ETHYL-2-METHYLHEPTANE	2	25.0	5.8642E-02	7.4249E-03	2	5.11
1040	C10H22	4-ETHYL-2-METHYLHEPTANE	2	25.0	7.8023E-02	9.8787E-03	2	5.11
1041	C10H22	5-ETHYL-2-METHYLHEPTANE	2	25.0	6.3862E-02	8.0858E-03	2	5.11
1042	C10H22	3-ETHYL-3-METHYLHEPTANE	2	25.0	5.0628E-02	6.4102E-03	2	5.11
1043	C10H22	4-ETHYL-3-METHYLHEPTANE	2	25.0	5.5412E-02	7.0159E-03	2	5.11
1044	C10H22	3-ETHYL-5-METHYLHEPTANE	2	25.0	6.9571E-02	8.8086E-03	2	5.11
1045	C10H22	3-ETHYL-4-METHYLHEPTANE	2	25.0	5.2964E-02	6.7059E-03	2	5.11
1046	C10H22	4-ETHYL-4-METHYLHEPTANE	2	25.0	5.9989E-02	7.5954E-03	2	5.11
1047	C10H22	2,2,3-TRIMETHYLHEPTANE	2	25.0	7.2001E-02	9.1163E-03	2	5.07
1048	C10H22	2,2,4-TRIMETHYLHEPTANE	2	25.0	1.2330E-01	1.5611E-02	2	5.07
1049	C10H22	2,2,5-TRIMETHYLHEPTANE	2	25.0	1.0660E-01	1.3497E-02	2	5.07
1050	C10H22	2,2,6-TRIMETHYLHEPTANE	2	25.0	1.1885E-01	1.5048E-02	2	5.07
1051	C10H22	2,3,3-TRIMETHYLHEPTANE	2	25.0	6.2070E-02	7.8589E-03	2	5.07
1052	C10H22	2,3,4-TRIMETHYLHEPTANE	2	25.0	6.3139E-02	7.9942E-03	2	5.07
1053	C10H22	2,3,5-TRIMETHYLHEPTANE	2	25.0	6.0331E-02	7.6386E-03	2	5.07
1054	C10H22	2,3,6-TRIMETHYLHEPTANE	2	25.0	7.8925E-02	9.9929E-03	2	5.07
1055	C10H22	2,4,4-TRIMETHYLHEPTANE	2	25.0	1.0537E-01	1.3341E-02	2	5.07
1056	C10H22	2,4,5-TRIMETHYLHEPTANE	2	25.0	7.6690E-02	9.7099E-03	2	5.07
1057	C10H22	2,4,6-TRIMETHYLHEPTANE	2	25.0	1.2844E-01	1.6262E-02	2	5.07
1058	C10H22	2,5,5-TRIMETHYLHEPTANE	2	25.0	9.4926E-02	1.2019E-02	2	5.07
1059	C10H22	3,3,4-TRIMETHYLHEPTANE	2	25.0	5.6361E-02	7.1361E-03	2	5.07
1060	C10H22	3,3,5-TRIMETHYLHEPTANE	2	25.0	8.0345E-02	1.0173E-02	2	5.07
1061	C10H22	3,4,4-TRIMETHYLHEPTANE	2	25.0	5.8976E-02	7.4671E-03	2	5.07
1062	C10H22	3,4,5-TRIMETHYLHEPTANE	2	25.0	5.4480E-02	6.8979E-03	2	5.07
1063	C10H22	3-ISOPROPYL-2-METHYLHEXANE	2	25.0	4.3035E-02	5.4488E-03	2	5.11
1064	C10H22	3,3-DIETHYLHEXANE	2	25.0	4.4006E-02	5.5718E-03	2	5.11
1065	C10H22	3,4-DIETHYLHEXANE	2	25.0	5.0344E-02	6.3742E-03	2	5.11
1066	C10H22	3-ETHYL-2,2-DIMETHYLHEXANE	2	25.0	7.8472E-02	9.9356E-03	2	5.08
1067	C10H22	4-ETHYL-2,2-DIMETHYLHEXANE	2	25.0	1.3302E-01	1.6842E-02	2	5.08
1068	C10H22	3-ETHYL-2,3-DIMETHYLHEXANE	2	25.0	5.0914E-02	6.4464E-03	2	5.08
1069	C10H22	4-ETHYL-2,3-DIMETHYLHEXANE	2	25.0	5.9649E-02	7.5524E-03	2	5.08
1070	C10H22	3-ETHYL-2,4-DIMETHYLHEXANE	2	25.0	6.2424E-02	7.9037E-03	2	5.08
1071	C10H22	4-ETHYL-2,4-DIMETHYLHEXANE	2	25.0	5.8976E-02	7.4671E-03	2	5.08
1072	C10H22	3-ETHYL-2,5-DIMETHYLHEXANE	2	25.0	8.8054E-02	1.1149E-02	2	5.08
1073	C10H22	4-ETHYL-3,3-DIMETHYLHEXANE	2	25.0	5.3263E-02	6.7438E-03	2	5.08
1074	C10H22	3-ETHYL-3,4-DIMETHYLHEXANE	2	25.0	5.5727E-02	7.0557E-03	2	5.08
1075	C10H22	2,2,3,3-TETRAMETHYLHEXANE	2	25.0	6.1648E-02	7.8054E-03	1	5.03
1076	C10H22	2,2,3,4-TETRAMETHYLHEXANE	2	25.0	6.7226E-02	8.5117E-03	2	5.03
1077	C10H22	2,2,3,5-TETRAMETHYLHEXANE	2	25.0	1.2258E-01	1.5520E-02	2	5.03
1078	C10H22	2,2,4,4-TETRAMETHYLHEXANE	2	25.0	8.9593E-02	1.1344E-02	2	5.03
1079	C10H22	2,2,4,5-TETRAMETHYLHEXANE	2	25.0	1.2636E-01	1.5998E-02	2	4.99
1080	C10H22	2,2,5,5-TETRAMETHYLHEXANE	2	25.0	2.3281E-01	2.9476E-02	2	5.03
1081	C10H22	2,3,3,4-TETRAMETHYLHEXANE	2	25.0	4.8429E-02	6.1317E-03	2	5.03
1082	C10H22	2,3,3,5-TETRAMETHYLHEXANE	2	25.0	9.3293E-02	1.1812E-02	2	5.03
1083	C10H22	2,3,4,4-TETRAMETHYLHEXANE	2	25.0	5.7327E-02	7.2584E-03	2	5.03
1084	C10H22	2,3,4,5-TETRAMETHYLHEXANE	2	25.0	7.8023E-02	9.8787E-03	2	5.03
1085	C10H22	3,3,4,4-TETRAMETHYLHEXANE	2	25.0	3.5840E-02	4.5378E-03	2	5.03
1086	C10H22	2,4-DIMETHYL-3-ISOPROPYLPENTANE	2	25.0	7.4350E-02	9.4136E-03	2	5.07
1087	C10H22	3,3-DIETHYL-2-METHYLPENTANE	2	25.0	3.6438E-02	4.6135E-03	2	5.07
1088	C10H22	3-ETHYL-2,2,3-TRIMETHYLPENTANE	2	25.0	3.6842E-02	4.6647E-03	2	5.03
1089	C10H22	3-ETHYL-2,2,4-TRIMETHYLPENTANE	2	25.0	8.2168E-02	1.0404E-02	2	5.03
1090	C10H22	3-ETHYL-2,3,4-TRIMETHYLPENTANE	2	25.0	3.6964E-02	4.6802E-03	2	5.03
1091	C10H22	2,2,3,3,4-PENTAMETHYLPENTANE	2	25.0	4.4625E-02	5.6501E-03	2	4.97
1092	C10H22	2,2,3,4,4-PENTAMETHYLPENTANE	2	25.0	6.5372E-02	8.2770E-03	2	4.97

NO	FORMULA	NAME	Solubility in Water, S (ppm - parts per million)			Octanol-Water Partition Coefficient	
			Code	T, C	S @ T, ppm (wt)	S @ T, ppm (mol)	Code	$\log_{10}$ Kow
1093	C10H22O	1-DECANOL	1	25.0	3.7020E+01	4.2135E+00	1	4.57
1094	C10H22O	DI-n-PENTYL ETHER	2	25.0	1.8643E+01	2.1219E+00	2	4
1095	C10H22O	ISODECANOL	----	------	----------	----------	2	3.71
1096	C10H22O5	TETRAETHYLENE GLYCOL DIMETHYL ETHER	1	20.0	1.0000E+06	1.0000E+06	----	-------
1097	C10H22S	n-DECYL MERCAPTAN	2	25.0	5.5652E-01	5.7503E-02	2	5.2
1098	C10H22S	BUTYL-HEXYL-SULFIDE	2	25.0	1.2476E+00	1.2892E-01	2	4.86
1099	C10H22S	ETHYL-OCTYL-SULFIDE	2	25.0	1.2476E+00	1.2892E-01	2	4.86
1100	C10H22S	HEPTYL-PROPYL-SULFIDE	2	25.0	1.2476E+00	1.2892E-01	2	4.86
1101	C10H22S	METHYL-NONYL-SULFIDE	2	25.0	1.2476E+00	1.2892E-01	2	4.86
1102	C10H22S	PENTYL-SULFIDE	2	25.0	1.2476E+00	1.2892E-01	2	4.86
1103	C10H22S2	PENTYL-DISULFIDE	----	------	----------	----------	2	5.8
1104	C10H23N	n-DECYLAMINE	----	------	----------	----------	2	3.78
1105	C11H10	1-METHYLNAPHTHALENE	1	25.0	2.8040E+01	3.5524E+00	1	3.87
1106	C11H10	2-METHYLNAPHTHALENE	1	25.0	2.5400E+01	3.2179E+00	1	3.86
1107	C11H14O2	n-BUTYL BENZOATE	----	------	----------	----------	1	3.84
1108	C11H16	n-PENTYLBENZENE	1	25.0	3.8400E+00	4.6664E-01	1	4.9
1109	C11H16O	p-tert-AMYLPHENOL	1	25.0	1.6800E+02	1.8429E+01	2	3.91
1110	C11H20	1-UNDECYNE	2	25.0	4.5278E-01	5.3565E-02	----	-------
1111	C11H20O2	2-ETHYLHEXYL ACRYLATE	1	23.0	1.0000E+02	9.7768E+00	2	4.09
1112	C11H22	1-UNDECENE	2	25.0	4.6747E-02	5.4580E-03	2	5.61
1113	C11H22	1-CYCLOPENTYLHEXANE	2	25.0	3.7952E-02	4.4312E-03	----	-------
1114	C11H22	PENTYLCYCLOHEXANE	2	25.0	5.8525E-02	6.8332E-03	2	5.56
1115	C11H22O	1-UNDECANAL	----	------	----------	----------	----	-------
1116	C11H24	n-UNDECANE	1	25.0	4.4000E-02	5.0710E-03	2	5.74
1117	C11H24O	1-UNDECANOL	2	25.0	2.0510E+01	2.1443E+00	1	4.72
1118	C11H24S	UNDECYL MERCAPTAN	2	25.0	3.7873E-01	3.6218E-02	2	5.69
1119	C11H24S	BUTYL-HEPTYL-SULFIDE	2	25.0	8.3626E-01	7.9976E-02	2	5.36
1120	C11H24S	DECYL-METHYL-SULFIDE	2	25.0	8.3626E-01	7.9976E-02	2	5.36
1121	C11H24S	ETHYL-NONYL-SULFIDE	2	25.0	8.3626E-01	7.9976E-02	2	5.36
1122	C11H24S	OCTYL-PROPYL-SULFIDE	2	25.0	8.3626E-01	7.9976E-02	2	5.36
1123	C12H8O	DIBENZOFURAN	1	25.0	1.0024E+01	1.0737E+00	1	4.12
1124	C12H9N	DIBENZOPYRROLE	1	25.0	9.0750E-01	9.7773E-02	1	3.72
1125	C12H10	ACENAPHTHENE	1	25.0	3.8861E+00	4.5398E-01	1	3.92
1126	C12H10	BIPHENYL	1	25.0	7.0000E+00	8.1775E-01	----	-------
1127	C12H10O	DIPHENYL ETHER	1	25.0	1.8042E+01	1.9096E+00	1	4.21
1128	C12H11N	p-AMINODIPHENYL	----	------	----------	----------	1	2.86
1129	C12H11N	DIPHENYLAMINE	1	20.0	5.3000E+01	5.6424E+00	1	3.5
1130	C12H11N3	p-AMINOAZOBENZENE	1	18.0	1.3000E+02	1.1875E+01	1	3.41
1131	C12H11N3	1,3-DIPHENYLTRIAZENE	----	------	----------	----------	2	3.99
1132	C12H12	1,2-DIMETHYLNAPHTHALENE	1	25.0	7.9990E+00	9.2240E-01	1	4.31
1133	C12H12	1,3-DIMETHYLNAPHTHALENE	1	25.0	7.9990E+00	9.2240E-01	1	4.42
1134	C12H12	1,4-DIMETHYLNAPHTHALENE	1	25.0	1.1400E+01	1.3146E+00	1	4.37
1135	C12H12	1,5-DIMETHYLNAPHTHALENE	1	25.0	3.3750E+00	3.8918E-01	1	4.38
1136	C12H12	1,6-DIMETHYLNAPHTHALENE	----	------	----------	----------	2	4.26
1137	C12H12	1,7-DIMETHYLNAPHTHALENE	----	------	----------	----------	1	4.44
1138	C12H12	2,3-DIMETHYLNAPHTHALENE	1	25.0	3.0000E+00	3.4594E-01	1	4.4
1139	C12H12	2,6-DIMETHYLNAPHTHALENE	1	25.0	2.0000E+00	2.3063E-01	1	4.31
1140	C12H12	2,7-DIMETHYLNAPHTHALENE	----	------	----------	----------	2	4.26
1141	C12H12	1-ETHYLNAPHTHALENE	1	25.0	1.0700E+01	1.2339E+00	1	4.4
1142	C12H12	2-ETHYLNAPHTHALENE	1	25.0	8.0000E+00	9.2251E-01	1	4.38
1143	C12H12N2	p-AMINODIPHENYLAMINE	1	25.0	5.2000E+02	5.0869E+01	2	1.82
1144	C12H12N2	HYDRAZOBENZENE	----	------	----------	----------	1	2.94
1145	C12H14	1,2,3-TRIMETHYLINDENE	----	------	----------	----------	----	-------
1146	C12H14O4	DIETHYL PHTHALATE	1	25.0	1.0000E+03	8.1135E+01	1	2.47
1147	C12H16	CYCLOHEXYLBENZENE	----	------	----------	----------	2	4.81
1148	C12H18	m-DIISOPROPYLBENZENE	2	25.0	8.1227E+00	9.0174E-01	2	4.9
1149	C12H18	p-DIISOPROPYLBENZENE	2	25.0	5.5393E+00	6.1495E-01	2	4.9
1150	C12H18	n-HEXYLBENZENE	1	25.0	1.0180E+00	1.1301E-01	1	5.52
1151	C12H18	1,2,3-TRIETHYLBENZENE	2	25.0	3.8857E+00	4.3138E-01	----	-------
1152	C12H18	1,2,4-TRIETHYLBENZENE	2	25.0	3.8857E+00	4.3138E-01	----	-------
1153	C12H18	1,3,5-TRIETHYLBENZENE	2	25.0	4.1874E+00	4.6487E-01	2	5.11
1154	C12H18	HEXAMETHYLBENZENE	----	------	----------	----------	1	5.11
1155	C12H20O4	DIBUTYL MALEATE	----	------	----------	----------	2	4.16
1156	C12H22	BICYCLOHEXYL	----	------	----------	----------	2	5.86
1157	C12H22	1-DODECYNE	2	25.0	1.8719E-01	2.0277E-02	----	-------
1158	C12H23N	DICYCLOHEXYLAMINE	----	------	----------	----------	----	-------
1159	C12H24	1-DODECENE	2	25.0	1.8446E-02	1.9742E-03	2	6.1
1160	C12H24	1-CYCLOPENTYLHEPTANE	2	25.0	1.6215E-02	1.7355E-03	----	-------
1161	C12H24	1-CYCLOHEXYLHEXANE	2	25.0	2.5197E-02	2.6968E-03	2	6.05
1162	C12H24O	1-DODECANAL	----	------	----------	----------	2	4.75
1163	C12H24O2	n-DODECANOIC ACID	1	20.0	5.5000E+01	4.9464E+00	1	4.6
1164	C12H26	n-DODECANE	1	25.0	3.7000E-03	3.9131E-04	1	6.1
1165	C12H26O	DI-n-HEXYL ETHER	2	25.0	3.4385E+00	3.3243E-01	2	4.98
1166	C12H26O	1-DODECANOL	1	25.0	4.0000E+00	3.8672E-01	1	5.13
1167	C12H26O3	DIETHYLENE GLYCOL DI-n-BUTYL ETHER	1	20.0	3.0000E+03	2.4821E+02	1	1.92
1168	C12H26S	n-DODECYL MERCAPTAN	2	25.0	3.2624E-01	2.9037E-02	2	6.18
1169	C12H26S	BUTYL-OCTYL-SULFIDE	2	25.0	7.4657E-01	6.6451E-02	2	5.85
1170	C12H26S	DECYL-ETHYL-SULFIDE	2	25.0	7.4657E-01	6.6451E-02	2	5.85

NO	FORMULA	NAME	Solubility in Water, S (ppm - parts per million)				Octanol-Water Partition Coefficient	
			Code	T, C	S @ T, ppm (wt)	S @ T, ppm (mol)	Code	log₁₀ Kow
1171	C12H26S	HEXYL-SULFIDE	2	25.0	7.4657E-01	6.6451E-02	2	5.85
1172	C12H26S	METHYL-UNDECYL-SULFIDE	2	25.0	7.4657E-01	6.6451E-02	2	5.85
1173	C12H26S	NONYL-PROPYL-SULFIDE	2	25.0	7.4657E-01	6.6451E-02	2	5.85
1174	C12H26S2	HEXYL-DISULFIDE	----	------	----------	----------	2	6.78
1175	C12H27BO3	TRI-n-BUTYL BORATE	----	------	----------	----------	----	-------
1176	C12H27N	DODECYLAMINE	----	------	----------	----------	----	-------
1177	C12H27N	TRI-n-BUTYLAMINE	1	25.0	1.4180E+02	1.3784E+01	2	4.46
1178	C13H10	FLUORENE	1	25.0	1.8949E+00	2.0537E-01	1	4.18
1179	C13H10O	BENZOPHENONE	1	25.0	1.3670E+02	1.3516E+01	1	3.18
1180	C13H12	DIPHENYLMETHANE	1	25.0	1.4115E+01	1.5115E+00	1	4.14
1181	C13H14	1-PROPYLNAPHTHALENE	----	------	----------	----------	----	-------
1182	C13H14	2-PROPYLNAPHTHALENE	----	------	----------	----------	----	-------
1183	C13H14	2ETHYL-3-METHYLNAPHTHALENE	----	------	----------	----------	----	-------
1184	C13H14	2ETHYL-6-METHYLNAPHTHALENE	----	------	----------	----------	----	-------
1185	C13H14	2ETHYL-7-METHYLNAPHTHALENE	----	------	----------	----------	----	-------
1186	C13H20	n-HEPTYLBENZENE	2	25.0	1.0780E+00	1.1015E-01	2	5.49
1187	C13H24	1-TRIDECYNE	2	25.0	9.6052E-02	9.5954E-03	----	-------
1188	C13H26	1-TRIDECENE	2	25.0	9.1901E-03	9.0793E-04	2	6.59
1189	C13H26	1-CYCLOPENTYLOCTANE	2	25.0	9.0239E-03	8.9152E-04	----	-------
1190	C13H26	1-CYCLOHEXYLHEPTANE	2	25.0	1.3899E-02	1.3732E-03	2	6.54
1191	C13H26O	1-TRIDECANAL	----	------	----------	----------	----	-------
1192	C13H26O2	n-BUTYL NONANOATE	----	------	----------	----------	2	5.28
1193	C13H26O2	METHYL DODECANOATE	----	------	----------	----------	2	5.28
1194	C13H28	n-TRIDECANE	2	25.0	1.9904E-03	1.9449E-04	2	6.73
1195	C13H28O	1-TRIDECANOL	2	25.0	2.1500E+00	1.9331E-01	1	5.82
1196	C13H28S	BUTYL-NONYL-SULFIDE	2	25.0	8.8324E-01	7.3520E-02	2	6.35
1197	C13H28S	DECYL-PROPYL-SULFIDE	2	25.0	8.8324E-01	7.3520E-02	2	6.35
1198	C13H28S	DODECYL-METHYL-SULFIDE	2	25.0	8.8324E-01	7.3520E-02	2	6.35
1199	C13H28S	ETHYL-UNDECYL-SULFIDE	2	25.0	8.8324E-01	7.3520E-02	2	6.35
1200	C13H28S	1-TRIDECANETHIOL	2	25.0	3.5226E-01	2.9322E-02	----	-------
1201	C14H8O2	ANTHRAQUINONE	1	25.0	1.3530E+00	1.1706E-01	1	3.39
1202	C14H10	ANTHRACENE	1	22.0	5.3470E-02	5.4045E-03	1	4.45
1203	C14H10	DIPHENYLACETYLENE	----	------	----------	----------	1	4.78
1204	C14H10	PHENANTHRENE	1	25.0	1.1817E+00	1.1944E-01	1	4.46
1205	C14H12	cis-STILBENE	----	------	----------	----------	2	4.52
1206	C14H12	trans-STILBENE	1	25.0	2.9000E-01	2.8984E-02	1	4.81
1207	C14H12O2	BENZYL BENZOATE	----	------	----------	----------	1	3.97
1208	C14H14	1,1-DIPHENYLETHANE	----	------	----------	----------	2	4.15
1209	C14H14	1,2-DIPHENYLETHANE	----	------	----------	----------	1	4.79
1210	C14H14O	DIBENZYL ETHER	----	------	----------	----------	1	3.31
1211	C14H16	1-n-BUTYLNAPHTHALENE	----	------	----------	----------	----	-------
1212	C14H16	2-BUTYLNAPHTHALENE	----	------	----------	----------	----	-------
1213	C14H22	n-OCTYLBENZENE	2	25.0	5.6956E-01	5.3910E-02	1	6.3
1214	C14H22	1,2,3,4-TETRAETHYLBENZENE	2	25.0	8.9385E-01	8.4606E-02	----	-------
1215	C14H22	1,2,3,5-TETRAETHYLBENZENE	2	25.0	9.1060E-01	8.6191E-02	----	-------
1216	C14H22	1,2,4,5-TETRAETHYLBENZENE	2	25.0	9.2777E-01	8.7816E-02	----	-------
1217	C14H22O	p-tert-OCTYLPHENOL	----	------	----------	----------	2	5.28
1218	C14H28	1-TETRADECENE	2	25.0	5.7921E-03	5.3135E-04	2	7.08
1219	C14H28	1-CYCLOPENTYLNONANE	2	25.0	6.4718E-03	5.9371E-04	----	-------
1220	C14H28	1-CYCLOHEXYLOCTANE	2	25.0	1.0088E-02	9.2545E-04	2	7.03
1221	C14H28O2	n-TETRADECANOIC ACID	1	20.0	2.0000E+01	1.5777E+00	1	6.11
1222	C14H30	n-TETRADECANE	1	25.0	2.2000E-03	1.9977E-04	1	7.2
1223	C14H30O	1-TETRADECANOL	1	25.0	3.0000E-01	2.5209E-02	1	6.03
1224	C14H30S	BUTYL-DECYL-SULFIDE	----	------	----------	----------	2	6.84
1225	C14H30S	DODECYL-ETHYL-SULFIDE	----	------	----------	----------	2	6.84
1226	C14H30S	HEPTYL-SULFIDE	----	------	----------	----------	2	6.84
1227	C14H30S	METHYL-TRIDECYL-SULFIDE	----	------	----------	----------	2	6.84
1228	C14H30S	PROPYL-UNDECYL-SULFIDE	----	------	----------	----------	2	6.84
1229	C14H30S	1-TETRADECANETHIOL	----	------	----------	----------	----	-------
1230	C14H30S2	HEPTYL-DISULFIDE	----	------	----------	----------	2	7.76
1231	C14H31N	TETRADECYLAMINE	----	------	----------	----------	2	5.75
1232	C15H10N2O2	DIPHENYLMETHANE-4,4'-DIISOCYANATE	----	------	----------	----------	2	5.22
1233	C15H16O	p-CUMYLPHENOL	----	------	----------	----------	----	-------
1234	C15H16O2	BISPHENOL A	1	20.0	3.5000E+02	2.7628E+01	1	3.32
1235	C15H18	1-PENTYLNAPHTHALENE	----	------	----------	----------	----	-------
1236	C15H18	2-PENTYLNAPHTHALENE	----	------	----------	----------	----	-------
1237	C15H24	n-NONYLBENZENE	2	25.0	3.6656E-01	3.2314E-02	1	7.11
1238	C15H24O	2,6-DI-tert-BUTYL-p-CRESOL	----	------	----------	----------	1	5.1
1239	C15H24O	NONYLPHENOL	1	25.0	7.0000E+00	5.7228E-01	----	-------
1240	C15H28	1-PENTADECYNE	2	25.0	5.0076E-02	4.3291E-03	----	-------
1241	C15H30	1-PENTADECENE	2	25.0	4.6010E-03	3.9395E-04	2	7.57
1242	C15H30	1-CYCLOPENTYLDECANE	2	25.0	5.9341E-03	5.0809E-04	----	-------
1243	C15H30	1-CYCLOHEXYLNONANE	2	25.0	9.4603E-03	8.1001E-04	2	7.52
1244	C15H30O2	PENTADECANOIC ACID	1	20.0	1.2000E+01	8.9183E-01	2	6.47
1245	C15H32	n-PENTADECANE	2	25.0	1.0658E-03	9.0386E-05	2	7.71
1246	C15H32O	1-PENTADECANOL	1	25.0	8.9000E-02	7.0193E-03	2	6.24
1247	C15H32S	BUTYL-UNDECYL-SULFIDE	----	------	----------	----------	2	7.34
1248	C15H32S	DODECYL-PROPYL-SULFIDE	----	------	----------	----------	2	7.34

NO	FORMULA	NAME	Solubility in Water, S (ppm - parts per million)				Octanol-Water Partition Coefficient	
			Code	T, C	S @ T, ppm (wt)	S @ T, ppm (mol)	Code	log₁₀ Kow
1249	C15H32S	ETHYL-TRIDECYL-SULFIDE	----	------	----------	----------	2	7.34
1250	C15H32S	METHYL-TETRADECYL-SULFIDE	----	------	----------	----------	2	7.34
1251	C15H32S	1-PENTADECANETHIOL	----	------	----------	----------	----	-------
1252	C16H10	FLUORANTHENE	1	27.0	2.4068E-01	2.1438E-02	1	5.16
1253	C16H10	PYRENE	1	25.0	1.2944E-01	1.1529E-02	1	4.88
1254	C16H12	1-PHENYLNAPHTHALENE	----	------	----------	----------	----	-------
1255	C16H20	1-n-HEXYLNAPHTHALENE	----	------	----------	----------	----	-------
1256	C16H22O4	DIBUTYL PHTHALATE	1	20.0	1.0800E+01	6.9900E-01	1	4.72
1257	C16H26	n-DECYLBENZENE	2	25.0	2.9329E-01	2.4194E-02	1	7.35
1258	C16H26	PENTAETHYLBENZENE	----	------	----------	----------	----	-------
1259	C16H30	1-HEXADECYNE	2	25.0	4.9412E-02	4.0022E-03	----	-------
1260	C16H32	n-DECYLCYCLOHEXANE	2	25.0	1.1236E-02	9.0191E-04	2	8.01
1261	C16H32	1-CYCLOPENTYLUNDECANE	2	25.0	6.8530E-03	5.5010E-04	----	-------
1262	C16H32	1-HEXADECENE	2	25.0	4.5786E-03	3.6752E-04	2	8.06
1263	C16H32O2	n-HEXADECANOIC ACID	1	20.0	7.2000E+00	5.0583E-01	1	7.17
1264	C16H34	n-HEXADECANE	1	25.0	9.0000E-04	7.1600E-05	2	8.25
1265	C16H34O	DI-n-OCTYL ETHER	2	25.0	1.3755E+00	1.0221E-01	----	-------
1266	C16H34O	1-HEXADECANOL	1	25.0	3.5000E-02	2.6007E-03	2	6.73
1267	C16H34S	BUTYL-DODECYL-SULFIDE	----	------	----------	----------	2	7.83
1268	C16H34S	ETHYL-TETRADECYL-SULFIDE	----	------	----------	----------	2	7.83
1269	C16H34S	METHYL-PENTADECYL-SULFIDE	----	------	----------	----------	2	7.83
1270	C16H34S	OCTYL-SULFIDE	----	------	----------	----------	2	7.83
1271	C16H34S	PROPYL-TRIDECYL-SULFIDE	----	------	----------	----------	2	7.83
1272	C16H34S	1-HEXADECANETHIOL	----	------	----------	----------	2	8.14
1273	C16H34S2	OCTYL-DISULFIDE	----	------	----------	----------	2	8.74
1274	C17H28	n-UNDECYLBENZENE	----	------	----------	----------	1	8.14
1275	C17H32	1-HEPTADECYNE	----	------	----------	----------	2	8.55
1276	C17H34	1-CYCLOPENTYLDODECANE	----	------	----------	----------	----	-------
1277	C17H34	1-CYCLOHEXYLUNDECANE	----	------	----------	----------	----	-------
1278	C17H34	1-HEPTADECENE	----	------	----------	----------	2	8.55
1279	C17H36	n-HEPTADECANE	2	25.0	1.5000E-03	1.1237E-04	2	8.69
1280	C17H36O	1-HEPTADECANOL	1	25.0	8.0000E-03	5.6193E-04	2	7.23
1281	C17H36S	BUTYL-TRIDECYL-SULFIDE	----	------	----------	----------	2	8.33
1282	C17H36S	ETHYL-PENTADECYL-SULFIDE	----	------	----------	----------	2	8.33
1283	C17H36S	HEXADECYL-METHYL-SULFIDE	----	------	----------	----------	2	8.33
1284	C17H36S	PROPYL-TETRADECYL-SULFIDE	----	------	----------	----------	2	8.33
1285	C17H36S	1-HEPTADECANETHIOL	----	------	----------	----------	----	-------
1286	C18H12	CHRYSENE	1	24.0	1.7122E-02	1.3511E-03	1	5.5
1287	C18H14	m-TERPHENYL	1	25.0	1.5110E+00	1.1819E-01	2	5.52
1288	C18H14	o-TERPHENYL	1	25.0	1.2390E+00	9.6916E-02	2	5.52
1289	C18H14	p-TERPHENYL	1	25.0	1.7960E-02	1.4048E-03	1	6.03
1290	C18H15P	TRIPHENYLPHOSPHINE	----	------	----------	----------	1	5.69
1291	C18H15O4P	TRIPHENYL PHOSPHATE	1	24.0	7.3000E-01	4.0305E-02	1	4.59
1292	C18H16N2	N,N'-DIPHENYL-p-PHENYLENEDIAMINE	----	------	----------	----------	2	4.04
1293	C18H22	2,3-DIMETHYL-2,3-DIPHENYLBUTANE	----	------	----------	----------	----	-------
1294	C18H22O2	DICUMYL PEROXIDE	----	------	----------	----------	1	5.5
1295	C18H30	n-DODECYLBENZENE	----	------	----------	----------	1	8.65
1296	C18H30	HEXAETHYLBENZENE	----	------	----------	----------	----	-------
1297	C18H32O2	LINOLEIC ACID	----	------	----------	----------	1	7.05
1298	C18H34	1-OCTADECYNE	----	------	----------	----------	----	-------
1299	C18H34O2	OLEIC ACID	----	------	----------	----------	2	7.73
1300	C18H34O4	DIBUTYL SEBACATE	----	------	----------	----------	2	6.3
1301	C18H34O4	DIHEXYL ADIPATE	----	------	----------	----------	2	6.3
1302	C18H36	1-CYCOPENTYLTRIDECANE	----	------	----------	----------	----	-------
1303	C18H36	1-CYCLOHEXYLDODECANE	----	------	----------	----------	2	9
1304	C18H36	1-OCTADECENE	----	------	----------	----------	2	9.04
1305	C18H36O2	STEARIC ACID	1	20.0	2.9000E+00	1.8364E-01	1	8.23
1306	C18H38	n-OCTADECANE	1	25.0	2.1000E-03	1.4865E-04	2	9.18
1307	C18H38O	DINONYL ETHER	----	------	----------	----------	2	7.92
1308	C18H38O	1-OCTADECANOL	1	34.0	1.1000E-03	7.3259E-05	2	7.7
1309	C18H38S	BUTYL-TETRADECYL-SULFIDE	----	------	----------	----------	2	8.82
1310	C18H38S	ETHYL-HEXADECYL-SULFIDE	----	------	----------	----------	2	8.82
1311	C18H38S	HEPTADECYL-METHYL-SULFIDE	----	------	----------	----------	2	8.82
1312	C18H38S	NONYL-SULFIDE	----	------	----------	----------	2	8.82
1313	C18H38S	PENTADECYL-PROPYL-SULFIDE	----	------	----------	----------	2	8.82
1314	C18H38S	1-OCTADECANETHIOL	----	------	----------	----------	2	9.12
1315	C18H38S2	NONYL-DISULFIDE	----	------	----------	----------	2	9.72
1316	C19H26	1-n-NONYLNAPHTHALENE	----	------	----------	----------	----	-------
1317	C19H32	n-TRIDECYLBENZENE	----	------	----------	----------	1	9.36
1318	C19H36	1-NONADECYNE	----	------	----------	----------	----	-------
1319	C19H36O2	METHYL OLEATE	----	------	----------	----------	2	8.02
1320	C19H38	1-CYCLOPENTYLTETRADECANE	----	------	----------	----------	----	-------
1321	C19H38	1-CYCLOHEXYLTRIDECANE	----	------	----------	----------	2	9.49
1322	C19H38	1-NONADECENE	----	------	----------	----------	2	9.54
1323	C19H38O2	NONADECANOIC ACID	----	------	----------	----------	----	-------
1324	C19H40	n-NONADECANE	2	25.0	2.0000E-03	1.3418E-04	2	9.67
1325	C19H40O	1-NONADECANOL	----	------	----------	----------	----	-------
1326	C19H40S	BUTYL-PENTADECYL-SULFIDE	----	------	----------	----------	2	9.32

Table 15-1 SOLUBILITY IN WATER AND OCTANOL-WATER PARTITION COEFFICIENT- ORGANIC COMPOUNDS (continued)

NO	FORMULA	NAME	Solubility in Water, S (ppm - parts per million)				Octanol-Water Partition Coefficient	
			Code	T, C	S @ T, ppm (wt)	S @ T, ppm (mol)	Code	log$_{10}$ Kow
1327	C19H40S	ETHYL-HEPTADECYL-SULFIDE	----	------	----------	----------	2	9.32
1328	C19H40S	HEXADECYL-PROPYL-SULFIDE	----	------	----------	----------	2	9.32
1329	C19H40S	METHYL-OCTADECYL-SULFIDE	----	------	----------	----------	2	9.32
1330	C19H40S	1-NONADECANETHIOL	----	------	----------	----------	----	-------
1331	C20H16	TRIPHENYLETHYLENE	----	------	----------	----------	----	-------
1332	C20H28	1-n-DECYLNAPHTHALENE	----	------	----------	----------	----	-------
1333	C20H30O2	ABIETIC ACID	1	20.0	4.8390E+01	2.8823E+00	----	-------
1334	C20H31N	DEHYDROABIETYLAMINE	----	------	----------	----------	----	-------
1335	C20H34	1-PHENYLTETRADECANE	----	------	----------	----------	----	-------
1336	C20H38	1-EICOSYNE	----	------	----------	----------	----	-------
1337	C20H40	1-CYCLOPENTYLPENTADECANE	----	------	----------	----------	----	-------
1338	C20H40	1-CYCLOHEXYLTETRADECANE	----	------	----------	----------	2	9.98
1339	C20H40	1-EICOSENE	----	------	----------	----------	2	10.03
1340	C20H42	n-EICOSANE	1	25.0	1.9000E-03	1.2114E-04	2	10.16
1341	C20H42O	1-EICOSANOL	----	------	----------	----------	2	8.7
1342	C20H42S	BUTYL-HEXADECYL-SULFIDE	----	------	----------	----------	2	9.81
1343	C20H42S	DECYL-SULFIDE	----	------	----------	----------	2	9.81
1344	C20H42S	ETHYL-OCTADECYL-SULFIDE	----	------	----------	----------	2	9.81
1345	C20H42S	HEPTADECYL-PROPYL-SULFIDE	----	------	----------	----------	2	9.81
1346	C20H42S	METHYL-NONADECYL-SULFIDE	----	------	----------	----------	2	9.81
1347	C20H42S	1-EICOSANETHIOL	----	------	----------	----------	----	-------
1348	C20H42S2	DECYL-DISULFIDE	----	------	----------	----------	2	10.7
1349	C21H21O4P	TRI-o-CRESYL PHOSPHATE	1	24.0	7.4000E-02	3.6190E-03	----	-------
1350	C21H36	1-PHENYLPENTADECANE	----	------	----------	----------	2	9.42
1351	C21H42	1-CYCLOPENTYLHEXADECANE	----	------	----------	----------	----	-------
1352	C21H42	1-CYCLOHEXYLPENTADECANE	----	------	----------	----------	----	-------
1353	C22H38	1-PHENYLHEXADECANE	----	------	----------	----------	2	9.91
1354	C22H44	1-CYCLOHEXYLHEXADECANE	----	------	----------	----------	----	-------
1355	C22H44O2	n-BUTYL STEARATE	----	------	----------	----------	2	9.7
1356	C24H38O4	DIISOOCTYL PHTHALATE	----	------	----------	----------	2	8.39
1357	C24H38O4	DIOCTYL PHTHALATE	1	24.0	2.8500E-01	1.3146E-02	1	7.6
1358	C24H42O	DINONYLPHENOL	----	------	----------	----------	----	-------
1359	C26H20	TETRAPHENYLETHYLENE	----	------	----------	----------	----	-------
1360	C28H46O4	DIISODECYL PHTHALATE	1	24.0	2.8000E-01	1.1293E-02	2	10.36

Code: 1 - experimental,   2 - estimate

S - solubility in water, ppm

ppm = parts per million

T - temperature, C

Kow - octanol-water partition coefficient

Chapter 16

SOLUBILITY IN WATER CONTAINING SALT

Carl L. Yaws
Lamar University, Beaumont, Texas

ABSTRACT

Results for variation of water solubility with salt concentration are presented for 217 hydrocarbons. The results for solubility in salt water are applicable for the complete range of salt concentrations including water without salt (X=0) to water saturated with salt [X=358,700 ppm (wt)]. Correlation and experimental results are in favorable agreement. The results are provided in an easy-to-use table that is especially applicable for rapid engineering usage with the personal computer or hand calculator.

INTRODUCTION

Physical and thermodynamic property data are required for the design and operation of industrial processes. In particular, the water solubility of substances is becoming increasingly important because of more and more stringent regulations regarding health, safety, and environment.

In this article, results are presented for water solubility of hydrocarbons. The results are applicable for the complete range of salt concentrations, including water without salt to water saturated with salt. The results are intended for use in initial engineering and environmental applications. As an example of such usage, solubility values issuing from the correlation are useful in determining the distribution of a hydrocarbon spill upon its contact with seawater. Solubility values at other salt concentrations may also be ascertained.

SALT WATER SOLUBILITY CORRELATION

The correlation for solubility of hydrocarbons in water containing salt is based on a series expansion in salt concentration:

$$\log_{10} S = A + B X + C X^2 \tag{16-1}$$

where
S = solubility in salt water at 25 C, parts per million by weight, ppm (wt)
X = concentration of salt (NaCl) in water, parts per million by weight, ppm (wt)
A, B, and C = correlation constants

The correlation constants (A, B, and C) are given in Table 16-1. The correlation constants in the table were determined from regression of data for water solubility. Both experimental values for the property under consideration and parameter values for estimation of the property are included in the source publications (1-193). The presented values are applicable to a wide variety of hydrocarbons (alkanes, naphthenes and aromatics with no, single, and multiple substitutions). The tabulation is arranged by carbon number (C5, C6, C7,etc.) for ease of use in quickly locating data using the chemical formula.

The tabulated values for solubility of hydrocarbons in water apply to conditions of saturation in which the hydrocarbon is in equilibrium with water. For saturation, the system pressure is approximately equal to the sum of vapor pressures of hydrocarbon and water.

A comparison of correlation and actual experimental data values for water solubility is shown in Figs. 16-1, 16-2, and 16-3 for representative hydrocarbons (pentane, methylcyclopentane, and benzene). In the figures, solubility values are plotted at salt concentrations ranging from water without salt to water saturated with salt. The graphs disclose favorable agreement of correlation and experimental data.

Portions of this material appeared in Pollution Engineering, 24, 46 (Sept. 15, 1992), Pollution Engineering, 26 (1), 70 (1994), and Pollution Engineering, 27 (6), 78 (June, 1995). These portions are reprinted by special permission.

REFERENCES – ORGANIC COMPOUNDS

1-190. See REFERENCES – WATER SOLUBILITY - ORGANIC COMPOUNDS in Chapter 15, SOLUBILITY IN WATER AND OCTANOL-WATER PARTITION COEFFICIENT
191. Yaws, C. L., Pollution Engineering, 14, 46 (Sept. 15, 1992).
192. Yaws, C. L. and X. Lin, Pollution Engineering, 26, 70 (January, 1994).
193. Yaws, C. L. and X. Lin, Pollution Engineering, 27, 78 (June, 1995).

Figure 16-1 Solubility of Pentane in Salt Water

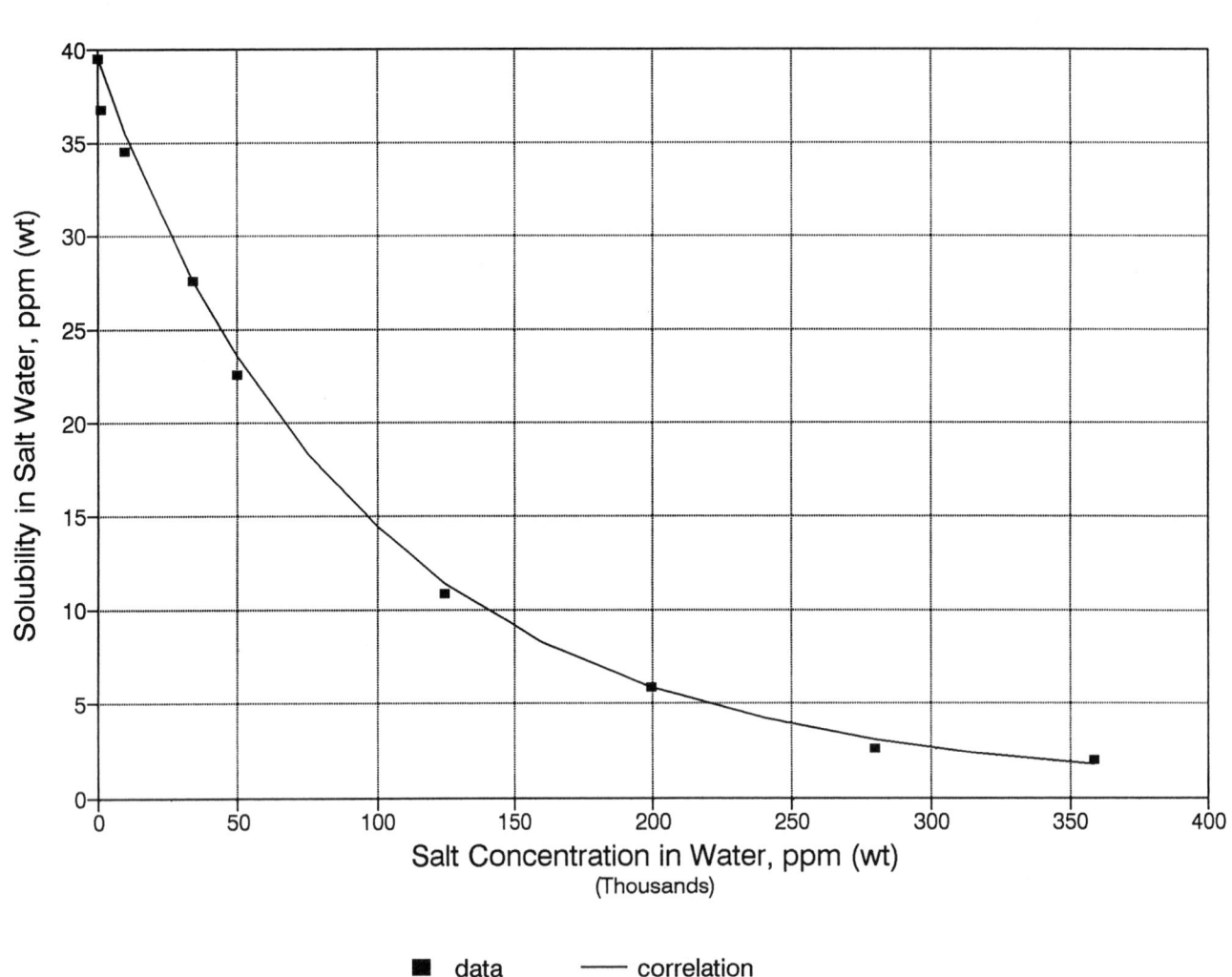

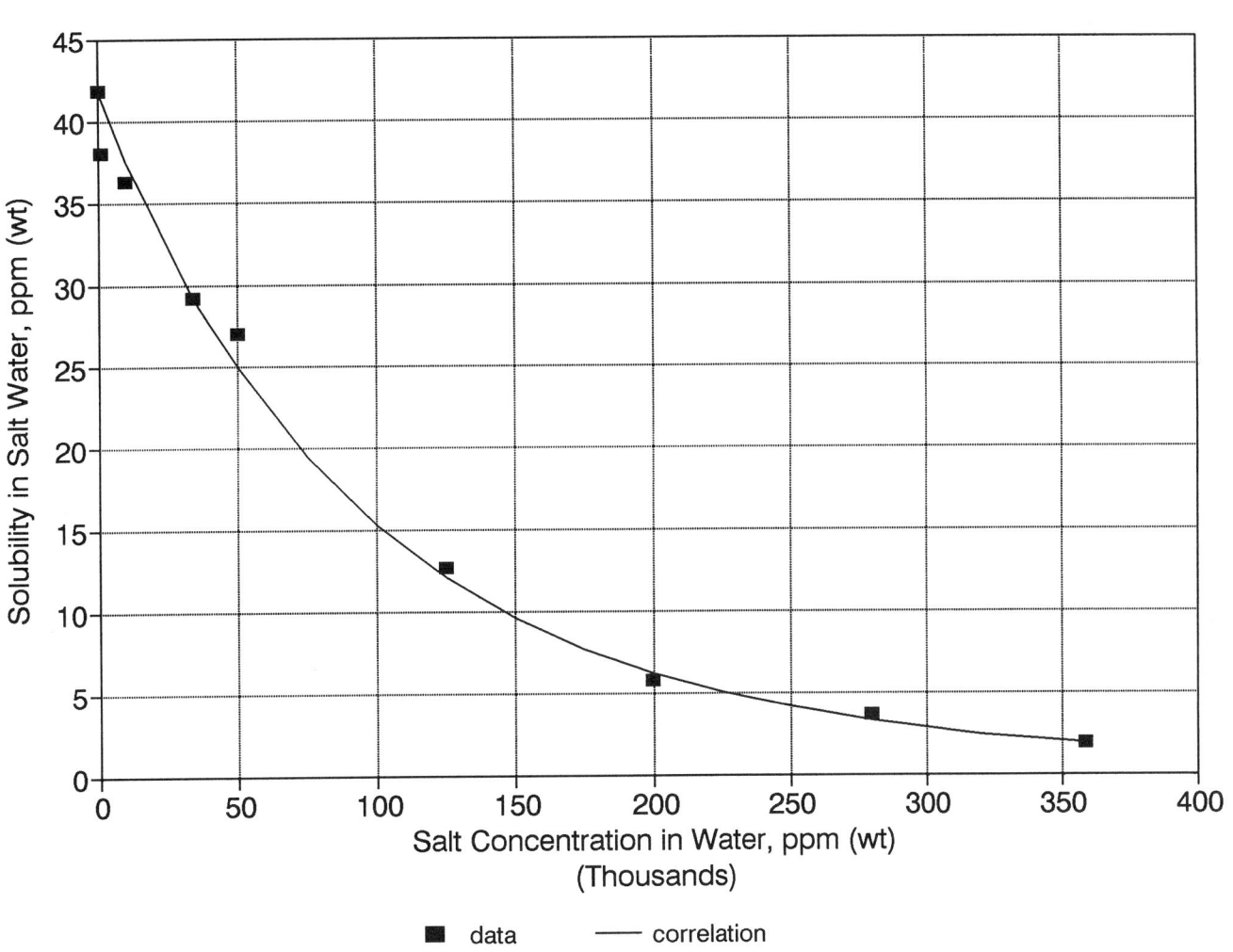

Figure 16-2 Solubility of Methylcyclopentane in Salt Water

Figure 16-3 Solubility of Benzene in Salt Water

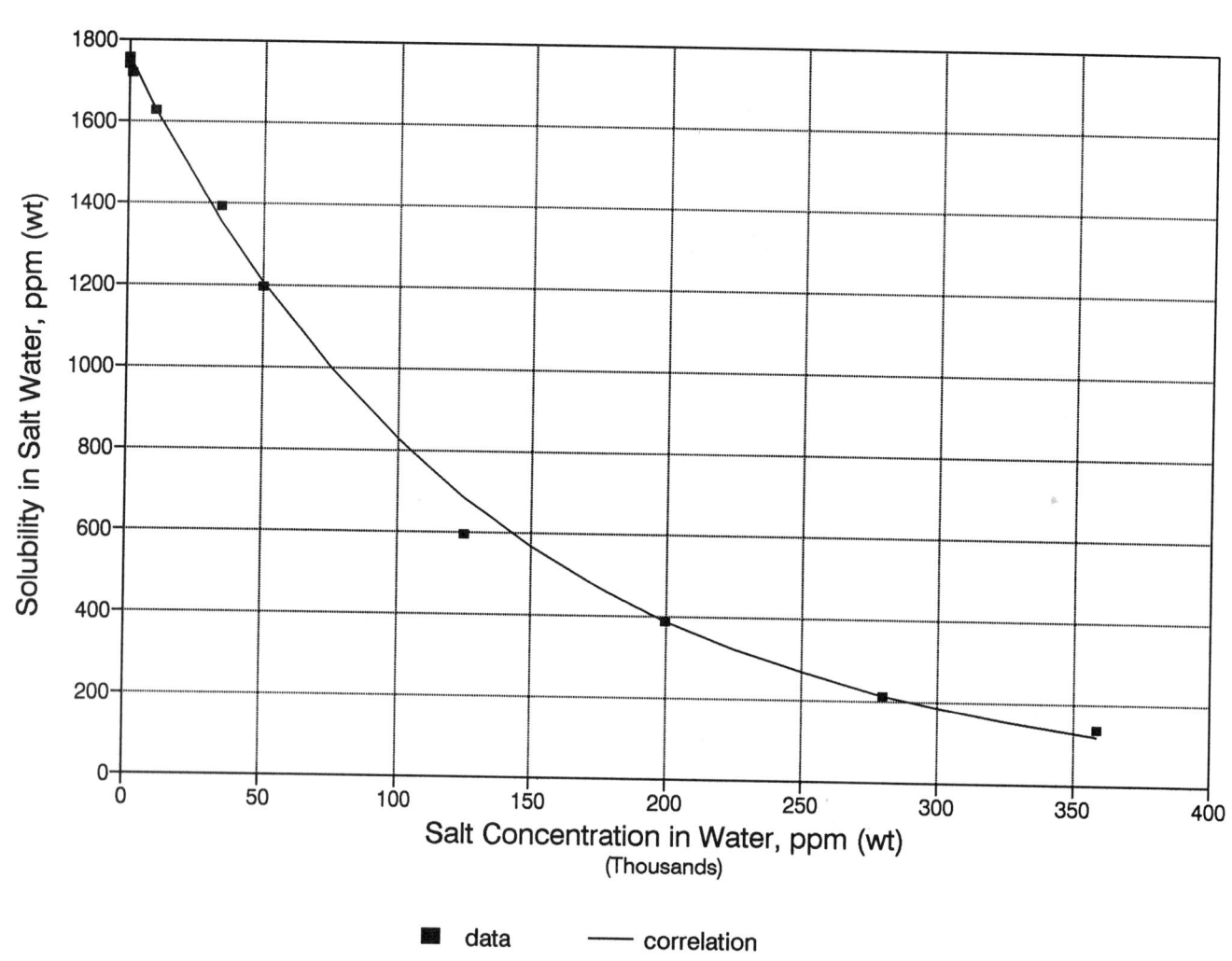

Table 16-1 SOLUBILITY IN SALT WATER - ORGANIC COMPOUNDS

NO	FORMULA	NAME	MW, g/mol	T_B, K	A	B	C	S @ X=0	S @ X=34,472
					$log_{10} S = A + B X + C X^2$			S - solubility, ppm(wt), X(salt) - ppm (wt)	
1	C5H10	CYCLOPENTANE	70.134	322.41	2.2041	-4.5987E-06	2.2993E-12	1.6000E+02	1.1180E+02
2	C5H12	PENTANE	72.150	309.22	1.5966	-4.5956E-06	2.2978E-12	3.9500E+01	2.7600E+01
3	C5H12	2-METHYLBUTANE(ISOPENTANE)	72.150	301.15	1.6794	-3.8600E-06	1.9300E-12	4.7800E+01	3.5370E+01
4	C6H6	BENZENE	78.113	353.31	3.2443	-3.2714E-06	1.6357E-16	1.7550E+03	1.3540E+03
5	C6H12	CYCLOHEXANE	84.161	353.90	1.7490	-4.5987E-06	2.2993E-12	5.6100E+01	3.9190E+01
6	C6H12	METHYLCYCLOPENTANE	84.161	344.97	1.6212	-4.5987E-06	2.2993E-12	4.1800E+01	2.9200E+01
7	C6H14	HEXANE	86.177	341.93	0.9763	-3.8600E-06	1.9300E-12	9.4700E+00	7.0070E+00
8	C6H14	2-METHYLPENTANE	86.177	333.40	1.1139	-3.8600E-06	1.9300E-12	1.3000E+01	9.6190E+00
9	C6H14	3-METHYLPENTANE	86.177	336.35	1.2530	-3.8600E-06	1.9300E-12	1.7910E+01	1.3250E+01
10	C6H14	2,2-DIMETHYLBUTANE	86.177	322.87	1.3769	-3.8600E-06	1.9300E-12	2.3820E+01	1.7630E+01
11	C6H14	2,3-DIMETHYLBUTANE	86.177	331.16	1.2810	-3.8600E-06	1.9300E-12	1.9100E+01	1.4130E+01
12	C7H8	TOLUENE	92.140	383.73	2.7343	-3.8393E-06	1.9196E-12	5.4240E+02	4.0200E+02
13	C7H14	ETHYLCYCLOPENTANE	98.188	376.60	0.9803	-4.5987E-06	2.2994E-12	9.5560E+00	6.6760E+00
14	C7H14	1,1-DIMETHYLCYCLOPENTANE	98.188	361.00	1.3528	-4.5987E-06	2.2993E-12	2.2530E+01	1.5740E+01
15	C7H14	C-1,2-DIMETHYLCYCLOPENTANE	98.188	372.68	1.0759	-4.5987E-06	2.2994E-12	1.1910E+01	8.3200E+00
16	C7H14	T-1,2-DIMETHYLCYCLOPENTANE	98.188	365.01	1.2592	-4.5987E-06	2.2993E-12	1.8160E+01	1.2690E+01
17	C7H14	C-1,3-DIMETHYLCYCLOPENTANE	98.188	363.92	1.2848	-4.5987E-06	2.2994E-12	1.9270E+01	1.3460E+01
18	C7H14	T-1,3-DIMETHYLCYCLOPENTANE	98.188	364.87	1.2625	-4.5987E-06	2.2993E-12	1.8300E+01	1.2790E+01
19	C7H14	METHYLCYCLOHEXANE	98.188	374.08	1.2041	-4.5987E-06	2.2994E-12	1.6000E+01	1.1180E+01
20	C7H16	HEPTANE	100.203	371.56	0.3502	-3.8600E-06	1.9300E-12	2.2400E+00	1.6570E+00
21	C7H16	2-METHYLHEXANE	100.203	363.19	0.4048	-3.8600E-06	1.9300E-12	2.5400E+00	1.8790E+00
22	C7H16	3-METHYLHEXANE	100.203	364.98	0.4216	-3.8600E-06	1.9300E-12	2.6400E+00	1.9540E+00
23	C7H16	3-ETHYLPENTANE	100.203	366.57	0.4704	-3.8600E-06	1.9300E-12	2.9540E+00	2.1860E+00
24	C7H16	2,2-DIMETHYLPENTANE	100.203	352.35	0.6435	-3.8600E-06	1.9300E-12	4.4000E+00	3.2560E+00
25	C7H16	2,3-DIMETHYLPENTANE	100.203	363.05	0.7202	-3.8600E-06	1.9300E-12	5.2500E+00	3.8850E+00
26	C7H16	2,4-DIMETHYLPENTANE	100.203	353.77	0.6444	-3.8600E-06	1.9300E-12	4.4100E+00	3.2630E+00
27	C7H16	3,3-DIMETHYLPENTANE	100.203	359.22	0.7723	-3.8600E-06	1.9300E-12	5.9200E+00	4.3810E+00
28	C7H16	2,2,3-TRIMETHYLBUTANE	100.203	354.03	0.7590	-3.8600E-06	1.9300E-12	5.7410E+00	4.2480E+00
29	C8H10	ETHYLBENZENE	106.167	409.17	2.2178	-5.0916E-06	2.5458E-12	1.6510E+02	1.1100E+02
30	C8H10	M-XYLENE	106.167	412.22	2.2405	-6.3514E-06	3.1757E-12	1.7400E+02	1.0600E+02
31	C8H10	O-XYLENE	106.167	417.46	2.3441	-6.8318E-06	3.4159E-12	2.2080E+02	1.2960E+02
32	C8H10	P-XYLENE	106.167	411.44	2.3047	-7.6669E-06	3.8335E-12	2.0170E+02	1.1090E+02
33	C8H16	PROPYLCYCLOPENTANE	112.214	404.10	0.3096	-4.5987E-06	2.2994E-12	2.0400E+00	1.4250E+00
34	C8H16	ETHYLCYCLOHEXANE	112.214	404.94	0.4647	-4.5987E-06	2.2994E-12	2.9160E+00	2.0370E+00
35	C8H16	1,1-DIMETHYLCYCLOHEXANE	112.214	392.70	0.7774	-4.5987E-06	2.2994E-12	5.9890E+00	4.1840E+00
36	C8H16	C-1,2-DIMETHYLCYCLOHEXANE	112.214	402.94	0.7782	-4.5987E-06	2.2993E-12	6.0000E+00	4.1920E+00
37	C8H16	T-1,2-DIMETHYLCYCLOHEXANE	112.214	396.71	0.6753	-4.5987E-06	2.2994E-12	4.7340E+00	3.3080E+00
38	C8H16	C-1,3-DIMETHYLCYCLOHEXANE	112.214	393.25	0.7634	-4.5987E-06	2.2993E-12	5.7990E+00	4.0510E+00
39	C8H16	T-1,3-DIMETHYLCYCLOHEXANE	112.214	397.60	0.6525	-4.5987E-06	2.2993E-12	4.4930E+00	3.1380E+00
40	C8H16	C-1,4-DIMETHYLCYCLOHEXANE	112.214	397.56	0.6536	-4.5987E-06	2.2993E-12	4.5040E+00	3.1460E+00
41	C8H16	T-1,4-DIMETHYLCYCLOHEXANE	112.214	392.50	0.5843	-4.5987E-06	2.2993E-12	3.8400E+00	2.6820E+00
42	C8H18	OCTANE	114.230	398.77	-0.3655	-3.8600E-06	1.9300E-12	4.3100E-01	3.1900E-01
43	C8H18	2-METHYLHEPTANE	114.230	390.80	-0.1264	-3.8599E-06	1.9300E-12	7.4740E-01	5.5310E-01
44	C8H18	3-METHYLHEPTANE	114.230	392.07	-0.1013	-3.8600E-06	1.9300E-12	7.9200E-01	5.8600E-01
45	C8H18	4-METHYLHEPTANE	114.230	390.92	-0.1295	-3.8600E-06	1.9300E-12	7.4220E-01	5.4920E-01
46	C8H18	3-ETHYLHEXANE	114.230	391.70	-0.1493	-3.8601E-06	1.9300E-12	7.0910E-01	5.2470E-01
47	C8H18	2,2-DIMETHYLHEXANE	114.230	380.01	0.1442	-3.8600E-06	1.9300E-12	1.3940E+00	1.0310E+00
48	C8H18	2,3-DIMETHYLHEXANE	114.230	388.78	-0.0754	-3.8600E-06	1.9300E-12	8.4070E-01	6.2210E-01
49	C8H18	2,4-DIMETHYLHEXANE	114.230	382.82	0.0743	-3.8600E-06	1.9300E-12	1.1870E+00	8.7810E-01
50	C8H18	2,5-DIMETHYLHEXANE	114.230	381.76	0.1007	-3.8599E-06	1.9300E-12	1.2610E+00	9.3310E-01
51	C8H18	3,3-DIMETHYLHEXANE	114.230	385.12	0.0167	-3.8600E-06	1.9300E-12	1.0390E+00	7.6900E-01
52	C8H18	3,4-DIMETHYLHEXANE	114.230	391.85	-0.1530	-3.8600E-06	1.9300E-12	7.0300E-01	5.2030E-01
53	C8H18	3-ETHYL-2-METHYLPENTANE	114.230	388.81	-0.0761	-3.8600E-06	1.9300E-12	8.3920E-01	6.2110E-01
54	C8H18	3-ETHYL-3-METHYLPENTANE	114.230	391.42	-0.1422	-3.8600E-06	1.9300E-12	7.2080E-01	5.3340E-01
55	C8H18	2,2,3-TRIMETHYLPENTANE	114.230	383.25	0.0636	-3.8601E-06	1.9300E-12	1.1580E+00	8.5670E-01
56	C8H18	2,2,4-TRIMETHYLPENTANE	114.230	372.38	0.3466	-3.8600E-06	1.9300E-12	2.2210E+00	1.6440E+00
57	C8H18	2,3,3-TRIMETHYLPENTANE	114.230	387.89	-0.0529	-3.8600E-06	1.9300E-12	8.8530E-01	6.5510E-01
58	C8H18	2,3,4-TRIMETHYLPENTANE	114.230	386.58	0.1335	-3.8600E-06	1.9300E-12	1.3600E+00	1.0060E+00
59	C8H18	2,2,3,3-TETRAMETHYLBUTANE	114.230	379.60	0.1543	-3.8600E-06	1.9300E-12	1.4270E+00	1.0560E+00
60	C9H12	PROPYLBENZENE	120.194	430.17	1.7174	-3.3289E-06	1.6645E-12	5.2160E+01	4.0240E+01
61	C9H12	CUMENE	120.194	426.31	1.6990	-2.0834E-06	1.0417E-12	5.0000E+01	4.2500E+01
62	C9H12	M-ETHYLTOLUENE	120.194	434.50	1.9181	-4.5175E-06	2.2588E-12	8.2820E+01	5.8220E+01
63	C9H12	O-ETHYLTOLUENE	120.194	438.30	1.9687	-4.5176E-06	2.2588E-12	9.3050E+01	6.5410E+01
64	C9H12	P-ETHYLTOLUENE	120.194	435.20	1.9770	-4.5176E-06	2.2588E-12	9.4850E+01	6.6670E+01
65	C9H12	1,2,3-TRIMETHYLBENZENE	120.194	449.27	1.8163	-3.8268E-06	1.9134E-12	6.5510E+01	4.8600E+01
66	C9H12	1,2,4-TRIMETHYLBENZENE	120.194	442.26	1.7562	-4.6787E-06	2.3394E-12	5.7040E+01	3.9600E+01
67	C9H12	MESITYLENE	120.194	437.89	1.6830	-5.5347E-06	2.7673E-12	4.8200E+01	3.1300E+01
68	C9H18	BUTYLCYCLOPENTANE	126.241	429.76	-0.3657	-4.5988E-06	2.2994E-12	4.3080E-01	3.0100E-01
69	C9H18	PROPYLCYCLOHEXANE	126.241	429.90	-0.1692	-4.5986E-06	2.2993E-12	6.7730E-01	4.7320E-01
70	C9H18	C-C-135TRIMETHYLCYCLOHEXANE	126.241	411.66	0.2926	-4.5987E-06	2.2994E-12	1.9620E+00	1.3700E+00
71	C9H18	C-T-135TRIMETHYLCYCLOHEXANE	126.241	413.70	0.2405	-4.5987E-06	2.2994E-12	1.7400E+00	1.2150E+00
72	C9H20	NONANE	128.257	423.81	-0.9136	-3.8600E-06	1.9300E-12	1.2200E-01	9.0290E-02
73	C9H20	2-METHYLOCTANE	128.257	416.19	-0.7752	-3.8602E-06	1.9301E-12	1.6780E-01	1.2420E-01
74	C9H20	3-METHYLOCTANE	128.257	417.37	-0.8052	-3.8604E-06	1.9302E-12	1.5660E-01	1.1590E-01
75	C9H20	4-METHYLOCTANE	128.257	415.58	-0.9393	-3.8600E-06	1.9300E-12	1.1500E-01	8.5100E-02
76	C9H20	3-ETHYLHEPTANE	128.257	416.16	-0.7744	-3.8604E-06	1.9302E-12	1.6810E-01	1.2440E-01
77	C9H20	4-ETHYLHEPTANE	128.257	414.36	-0.7284	-3.8596E-06	1.9298E-12	1.8690E-01	1.3830E-01
78	C9H20	2,2-DIMETHYLHEPTANE	128.257	405.85	-0.5106	-3.8602E-06	1.9301E-12	3.0860E-01	2.2840E-01

Table 16-1 SOLUBILITY IN SALT WATER - ORGANIC COMPOUNDS (continued)

$$\log_{10} S = A + BX + CX^2 \qquad S \text{ - solubility, ppm(wt),} \quad X(salt) \text{ - ppm (wt)}$$

NO	FORMULA	NAME	MW, g/mol	T_B, K	A	B	C	S @ X=0	S @ X=34,472
79	C9H20	2,3-DIMETHYLHEPTANE	128.257	413.66	-0.7104	-3.8602E-06	1.9301E-12	1.9480E-01	1.4420E-01
80	C9H20	2,4-DIMETHYLHEPTANE	128.257	406.05	-0.5157	-3.8600E-06	1.9300E-12	3.0500E-01	2.2570E-01
81	C9H20	2,5-DIMETHYLHEPTANE	128.257	409.16	-0.5953	-3.8597E-06	1.9299E-12	2.5390E-01	1.8790E-01
82	C9H20	2,6-DIMETHYLHEPTANE	128.257	408.37	-0.5751	-3.8600E-06	1.9300E-12	2.6600E-01	1.9680E-01
83	C9H20	3,3-DIMETHYLHEPTANE	128.257	410.17	-0.6212	-3.8599E-06	1.9299E-12	2.3920E-01	1.7700E-01
84	C9H20	3,4-DIMETHYLHEPTANE	128.257	413.76	-0.7131	-3.8604E-06	1.9302E-12	1.9360E-01	1.4330E-01
85	C9H20	3,5-DIMETHYLHEPTANE	128.257	409.16	-0.5953	-3.8597E-06	1.9299E-12	2.5390E-01	1.8790E-01
86	C9H20	4,4-DIMETHYLHEPTANE	128.257	408.36	-0.5750	-3.8603E-06	1.9301E-12	2.6610E-01	1.9690E-01
87	C9H20	3-ETHYL-2-METHYLHEXANE	128.257	411.16	-0.6467	-3.8603E-06	1.9302E-12	2.2560E-01	1.6690E-01
88	C9H20	4-ETHYL-2-METHYLHEXANE	128.257	406.96	-0.5390	-3.8602E-06	1.9301E-12	2.8910E-01	2.1390E-01
89	C9H20	3-ETHYL-3-METHYLHEXANE	128.257	413.76	-0.7131	-3.8604E-06	1.9302E-12	1.9360E-01	1.4330E-01
90	C9H20	3-ETHYL-4-METHYLHEXANE	128.257	413.56	-0.7080	-3.8596E-06	1.9298E-12	1.9590E-01	1.4500E-01
91	C9H20	2,2,3-TRIMETHYLHEXANE	128.257	406.75	-0.5336	-3.8599E-06	1.9299E-12	2.9270E-01	2.1660E-01
92	C9H20	2,2,4-TRIMETHYLHEXANE	128.257	399.69	-0.3528	-3.8601E-06	1.9300E-12	4.4380E-01	3.2840E-01
93	C9H20	2,2,5-TRIMETHYLHEXANE	128.257	397.24	-0.2675	-3.8601E-06	1.9301E-12	5.4010E-01	3.9970E-01
94	C9H20	2,3,3-TRIMETHYLHEXANE	128.257	410.84	-0.6385	-3.8597E-06	1.9299E-12	2.2990E-01	1.7010E-01
95	C9H20	2,3,4-TRIMETHYLHEXANE	128.257	412.20	-0.6733	-3.8598E-06	1.9299E-12	2.1220E-01	1.5700E-01
96	C9H20	2,3,5-TRIMETHYLHEXANE	128.257	404.50	-0.4760	-3.8599E-06	1.9300E-12	3.3420E-01	2.4730E-01
97	C9H20	2,4,4-TRIMETHYLHEXANE	128.257	403.81	-0.4583	-3.8602E-06	1.9301E-12	3.4810E-01	2.5760E-01
98	C9H20	3,3,4-TRIMETHYLHEXANE	128.257	413.62	-0.7095	-3.8598E-06	1.9299E-12	1.9520E-01	1.4450E-01
99	C9H20	3,3-DIETHYLPENTANE	128.257	419.30	-0.8542	-3.8595E-06	1.9298E-12	1.3990E-01	1.0350E-01
100	C9H20	3-ETHYL-2,2-DIMETHYLPENTANE	128.257	406.99	-0.5399	-3.8600E-06	1.9300E-12	2.8850E-01	2.1350E-01
101	C9H20	3-ETHYL-2,3-DIMETHYLPENTANE	128.257	417.86	-0.8176	-3.8598E-06	1.9299E-12	1.5220E-01	1.1260E-01
102	C9H20	3-ETHYL-2,4-DIMETHYLPENTANE	128.257	409.85	-0.6130	-3.8601E-06	1.9301E-12	2.4380E-01	1.8040E-01
103	C9H20	2,2,3,3-TETRAMETHYLPENTANE	128.257	413.44	-0.7049	-3.8602E-06	1.9301E-12	1.9730E-01	1.4600E-01
104	C9H20	2,2,3,4-TETRAMETHYLPENTANE	128.257	406.18	-0.5190	-3.8599E-06	1.9299E-12	3.0270E-01	2.2400E-01
105	C9H20	2,2,4,4-TETRAMETHYLPENTANE	128.257	395.44	-0.2444	-3.8599E-06	1.9300E-12	5.6970E-01	4.2150E-01
106	C9H20	2,3,3,4-TETRAMETHYLPENTANE	128.257	414.72	-0.7375	-3.8600E-06	1.9300E-12	1.8300E-01	1.3540E-01
107	C10H14	BUTYLBENZENE	134.221	456.43	1.1407	-8.5605E-06	4.2803E-12	1.3820E+01	7.0910E+00
108	C10H14	M-DIETHYLBENZENE	134.221	454.25	1.4350	-4.5175E-06	2.2588E-12	2.7230E+01	1.9140E+01
109	C10H14	O-DIETHYLBENZENE	134.221	457.02	1.3677	-4.5175E-06	2.2588E-12	2.3320E+01	1.6390E+01
110	C10H14	P-DIETHYLBENZENE	134.221	456.94	1.3696	-4.5175E-06	2.2588E-12	2.3420E+01	1.6460E+01
111	C10H14	1,2,3,4-TETRAMETHYLBENZENE	134.221	478.25	0.8656	-4.5175E-06	2.2588E-12	7.3390E+00	5.1590E+00
112	C10H14	1,2,3,5-TETRAMETHYLBENZENE	134.221	471.25	1.0278	-4.5175E-06	2.2588E-12	1.0660E+01	7.4950E+00
113	C10H20	1-CYCLOPENTYLPENTANE	140.268	453.76	-0.9393	-4.5995E-06	2.2997E-12	1.1500E-01	8.0330E-02
114	C10H20	BUTYLCYCLOHEXANE	140.268	454.10	-0.7493	-4.5991E-06	2.2996E-12	1.7810E-01	1.2440E-01
115	C10H22	DECANE	142.284	446.86	-1.2840	-3.8600E-06	1.9300E-12	5.2000E-02	3.8480E-02
116	C10H22	2-METHYLNONANE	142.284	440.16	-1.3737	-3.8608E-06	1.9304E-12	4.2300E-02	3.1300E-02
117	C10H22	3-METHYLNONANE	142.284	440.96	-1.3925	-3.8600E-06	1.9300E-12	4.0500E-02	2.9970E-02
118	C10H22	4-METHYLNONANE	142.284	438.86	-1.3420	-3.8600E-06	1.9300E-12	4.5500E-02	3.3670E-02
119	C10H22	5-METHYLNONANE	142.284	438.26	-1.3270	-3.8615E-06	1.9307E-12	4.7100E-02	3.4850E-02
120	C10H22	3-ETHYLOCTANE	142.284	439.66	-1.3615	-3.8600E-06	1.9300E-12	4.3500E-02	3.2190E-02
121	C10H22	4-ETHYLOCTANE	142.284	436.80	-1.2916	-3.8614E-06	1.9307E-12	5.1100E-02	3.7810E-02
122	C10H22	2,2-DIMETHYLOCTANE	142.284	430.06	-1.1255	-3.8591E-06	1.9295E-12	7.4900E-02	5.5430E-02
123	C10H22	2,3-DIMETHYLOCTANE	142.284	437.47	-1.3080	-3.8593E-06	1.9297E-12	4.9200E-02	3.6410E-02
124	C10H22	2,4-DIMETHYLOCTANE	142.284	429.06	-1.1002	-3.8591E-06	1.9296E-12	7.9400E-02	5.8760E-02
125	C10H22	2,5-DIMETHYLOCTANE	142.284	431.66	-1.1649	-3.8590E-06	1.9295E-12	6.8400E-02	5.0630E-02
126	C10H22	2,6-DIMETHYLOCTANE	142.284	433.54	-1.2118	-3.8589E-06	1.9294E-12	6.1400E-02	4.5440E-02
127	C10H22	2,7-DIMETHYLOCTANE	142.284	433.03	-1.1993	-3.8595E-06	1.9297E-12	6.3200E-02	4.6770E-02
128	C10H22	3,3-DIMETHYLOCTANE	142.284	434.36	-1.2321	-3.8612E-06	1.9306E-12	5.8600E-02	4.3360E-02
129	C10H22	3,4-DIMETHYLOCTANE	142.284	436.56	-1.2857	-3.8607E-06	1.9303E-12	5.1800E-02	3.8330E-02
130	C10H22	3,5-DIMETHYLOCTANE	142.284	432.56	-1.1871	-3.8600E-06	1.9300E-12	6.5000E-02	4.8100E-02
131	C10H22	3,6-DIMETHYLOCTANE	142.284	433.96	-1.2218	-3.8600E-06	1.9300E-12	6.0000E-02	4.4400E-02
132	C10H22	4,4-DIMETHYLOCTANE	142.284	430.66	-1.1403	-3.8590E-06	1.9295E-12	7.2400E-02	5.3580E-02
133	C10H22	4,5-DIMETHYLOCTANE	142.284	435.29	-1.2549	-3.8612E-06	1.9306E-12	5.5600E-02	4.1140E-02
134	C10H22	4-PROPYLHEPTANE	142.284	430.66	-1.1403	-3.8590E-06	1.9295E-12	7.2400E-02	5.3580E-02
135	C10H22	4-ISOPROPYLHEPTANE	142.284	432.06	-1.1752	-3.8605E-06	1.9303E-12	6.6800E-02	4.9430E-02
136	C10H22	3-ETHYL-2-METHYLHEPTANE	142.284	434.36	-1.2321	-3.8612E-06	1.9306E-12	5.8600E-02	4.3360E-02
137	C10H22	4-ETHYL-2-METHYLHEPTANE	142.284	429.36	-1.1079	-3.8600E-06	1.9300E-12	7.8000E-02	5.7720E-02
138	C10H22	5-ETHYL-2-METHYLHEPTANE	142.284	432.86	-1.1945	-3.8589E-06	1.9295E-12	6.3900E-02	4.7290E-02
139	C10H22	3-ETHYL-3-METHYLHEPTANE	142.284	436.96	-1.2958	-3.8614E-06	1.9307E-12	5.0600E-02	3.7440E-02
140	C10H22	4-ETHYL-3-METHYLHEPTANE	142.284	435.36	-1.2565	-3.8587E-06	1.9294E-12	5.5400E-02	4.1000E-02
141	C10H22	3-ETHYL-5-METHYLHEPTANE	142.284	431.36	-1.1574	-3.8610E-06	1.9305E-12	6.9600E-02	5.1500E-02
142	C10H22	3-ETHYL-4-METHYLHEPTANE	142.284	436.16	-1.2757	-3.8600E-06	1.9300E-12	5.3000E-02	3.9220E-02
143	C10H22	4-ETHYL-4-METHYLHEPTANE	142.284	433.96	-1.2218	-3.8600E-06	1.9300E-12	6.0000E-02	4.4400E-02
144	C10H22	2,2,3-TRIMETHYLHEPTANE	142.284	430.76	-1.1427	-3.8600E-06	1.9300E-12	7.2000E-02	5.3280E-02
145	C10H22	2,2,4-TRIMETHYLHEPTANE	142.284	421.46	-0.9090	-3.8603E-06	1.9301E-12	1.2330E-01	9.1250E-02
146	C10H22	2,2,5-TRIMETHYLHEPTANE	142.284	423.96	-0.9722	-3.8606E-06	1.9303E-12	1.0660E-01	7.8890E-02
147	C10H22	2,2,6-TRIMETHYLHEPTANE	142.284	422.09	-0.9248	-3.8594E-06	1.9297E-12	1.1890E-01	8.7990E-02
148	C10H22	2,3,3-TRIMETHYLHEPTANE	142.284	433.36	-1.2069	-3.8611E-06	1.9306E-12	6.2100E-02	4.5950E-02
149	C10H22	2,3,4-TRIMETHYLHEPTANE	142.284	433.06	-1.2000	-3.8611E-06	1.9306E-12	6.3100E-02	4.6690E-02
150	C10H22	2,3,5-TRIMETHYLHEPTANE	142.284	433.86	-1.2197	-3.8606E-06	1.9303E-12	6.0300E-02	4.4620E-02
151	C10H22	2,3,6-TRIMETHYLHEPTANE	142.284	429.16	-1.1029	-3.8591E-06	1.9296E-12	7.8900E-02	5.8390E-02
152	C10H22	2,4,4-TRIMETHYLHEPTANE	142.284	424.16	-0.9772	-3.8593E-06	1.9297E-12	1.0540E-01	7.7990E-02
153	C10H22	2,4,5-TRIMETHYLHEPTANE	142.284	429.66	-1.1152	-3.8595E-06	1.9298E-12	7.6700E-02	5.6760E-02
154	C10H22	2,4,6-TRIMETHYLHEPTANE	142.284	420.76	-0.8914	-3.8595E-06	1.9297E-12	1.2840E-01	9.5030E-02
155	C10H22	2,5,5-TRIMETHYLHEPTANE	142.284	425.96	-1.0227	-3.8593E-06	1.9296E-12	9.4900E-02	7.0240E-02
156	C10H22	3,3,4-TRIMETHYLHEPTANE	142.284	435.06	-1.2487	-3.8588E-06	1.9294E-12	5.6400E-02	4.1740E-02

Table 16-1 SOLUBILITY IN SALT WATER - ORGANIC COMPOUNDS (continued)

NO	FORMULA	NAME	MW, g/mol	T_B, K	A	B	C	S @ X=0	S @ X=34,472
					$\log_{10} S = A + B X + C X^2$			S - solubility, ppm(wt), X(salt) - ppm (wt)	
157	C10H22	3,3,5-TRIMETHYLHEPTANE	142.284	428.85	-1.0953	-3.8604E-06	1.9302E-12	8.0300E-02	5.9420E-02
158	C10H22	3,4,4-TRIMETHYLHEPTANE	142.284	434.26	-1.2291	-3.8600E-06	1.9300E-12	5.9000E-02	4.3660E-02
159	C10H22	3,4,5-TRIMETHYLHEPTANE	142.284	435.66	-1.2636	-3.8600E-06	1.9300E-12	5.4500E-02	4.0330E-02
160	C10H22	3-ISOPROPYL-2-METHYLHEXANE	142.284	439.86	-1.3665	-3.8600E-06	1.9300E-12	4.3000E-02	3.1820E-02
161	C10H22	3,3-DIETHYLHEXANE	142.284	439.46	-1.3565	-3.8600E-06	1.9300E-12	4.4000E-02	3.2560E-02
162	C10H22	3,4-DIETHYLHEXANE	142.284	437.06	-1.2984	-3.8607E-06	1.9303E-12	5.0300E-02	3.7220E-02
163	C10H22	3-ETHYL-2,2-DIMETHYLHEXANE	142.284	429.26	-1.1051	-3.8600E-06	1.9300E-12	7.8500E-02	5.8090E-02
164	C10H22	4-ETHYL-2,2-DIMETHYLHEXANE	142.284	420.16	-0.8761	-3.8600E-06	1.9300E-12	1.3300E-01	9.8430E-02
165	C10H22	3-ETHYL-2,3-DIMETHYLHEXANE	142.284	436.86	-1.2933	-3.8586E-06	1.9293E-12	5.0900E-02	3.7670E-02
166	C10H22	4-ETHYL-2,3-DIMETHYLHEXANE	142.284	434.06	-1.2248	-3.8612E-06	1.9306E-12	5.9600E-02	4.4100E-02
167	C10H22	3-ETHYL-2,4-DIMETHYLHEXANE	142.284	433.26	-1.2048	-3.8589E-06	1.9294E-12	6.2400E-02	4.6180E-02
168	C10H22	4-ETHYL-2,4-DIMETHYLHEXANE	142.284	434.26	-1.2291	-3.8600E-06	1.9300E-12	5.9000E-02	4.3660E-02
169	C10H22	3-ETHYL-2,5-DIMETHYLHEXANE	142.284	427.26	-1.0550	-3.8608E-06	1.9304E-12	8.8100E-02	6.5190E-02
170	C10H22	4-ETHYL-3,3-DIMETHYLHEXANE	142.284	436.06	-1.2733	-3.8606E-06	1.9303E-12	5.3300E-02	3.9440E-02
171	C10H22	3-ETHYL-3,4-DIMETHYLHEXANE	142.284	435.26	-1.2541	-3.8594E-06	1.9297E-12	5.5700E-02	4.1220E-02
172	C10H22	2,2,3,3-TETRAMETHYLHEXANE	142.284	433.48	-1.2104	-3.8611E-06	1.9306E-12	6.1600E-02	4.5580E-02
173	C10H22	2,2,3,4-TETRAMETHYLHEXANE	142.284	431.96	-1.1726	-3.8595E-06	1.9297E-12	6.7200E-02	4.9730E-02
174	C10H22	2,2,3,5-TETRAMETHYLHEXANE	142.284	421.56	-0.9115	-3.8606E-06	1.9303E-12	1.2260E-01	9.0720E-02
175	C10H22	2,2,4,4-TETRAMETHYLHEXANE	142.284	426.96	-1.0477	-3.8608E-06	1.9304E-12	8.9600E-02	6.6300E-02
176	C10H22	2,2,4,5-TETRAMETHYLHEXANE	142.284	421.04	-0.8983	-3.8595E-06	1.9297E-12	1.2640E-01	9.3530E-02
177	C10H22	2,2,5,5-TETRAMETHYLHEXANE	142.284	410.63	-0.6330	-3.8601E-06	1.9301E-12	2.3280E-01	1.7230E-01
178	C10H22	2,3,3,4-TETRAMETHYLHEXANE	142.284	437.75	-1.3152	-3.8586E-06	1.9293E-12	4.8400E-02	3.5820E-02
179	C10H22	2,3,3,5-TETRAMETHYLHEXANE	142.284	426.26	-1.0301	-3.8604E-06	1.9302E-12	9.3300E-02	6.9040E-02
180	C10H22	2,3,4,4-TETRAMETHYLHEXANE	142.284	434.76	-1.2418	-3.8606E-06	1.9303E-12	5.7300E-02	4.2400E-02
181	C10H22	2,3,4,5-TETRAMETHYLHEXANE	142.284	429.36	-1.1079	-3.8600E-06	1.9300E-12	7.8000E-02	5.7720E-02
182	C10H22	3,3,4,4-TETRAMETHYLHEXANE	142.284	443.16	-1.4461	-3.8610E-06	1.9305E-12	3.5800E-02	2.6490E-02
183	C10H22	24DIMETHYL3ISOPROPYLPENTANE	142.284	430.20	-1.1290	-3.8605E-06	1.9302E-12	7.4300E-02	5.4980E-02
184	C10H22	33-DIETHYL-2-METHYLPENTANE	142.284	442.86	-1.4389	-3.8581E-06	1.9290E-12	3.6400E-02	2.6940E-02
185	C10H22	3ETHYL-223TRIMETHYLPENTANE	142.284	442.66	-1.4342	-3.8609E-06	1.9305E-12	3.6800E-02	2.7230E-02
186	C10H22	3ETHYL-224TRIMETHYLPENTANE	142.284	428.46	-1.0851	-3.8596E-06	1.9298E-12	8.2200E-02	6.0830E-02
187	C10H22	3ETHYL-234TRIMETHYLPENTANE	142.284	442.60	-1.4318	-3.8600E-06	1.9300E-12	3.7000E-02	2.7380E-02
188	C10H22	22334-PENTAMETHYLPENTANE	142.284	439.21	-1.3507	-3.8616E-06	1.9308E-12	4.4600E-02	3.3000E-02
189	C10H22	22344-PENTAMETHYLPENTANE	142.284	432.45	-1.1844	-3.8589E-06	1.9295E-12	6.5400E-02	4.8400E-02
190	C11H16	PENTYLBENZENE	148.247	478.62	0.5843	-3.3289E-06	1.6645E-12	3.8400E+00	2.9620E+00
191	C11H22	1-CYCLOPENTYLHEXANE	154.295	476.26	-1.4202	-4.5965E-06	2.2983E-12	3.8000E-02	2.6550E-02
192	C11H22	PENTYLCYCLOHEXANE	154.295	476.87	-1.2328	-4.5974E-06	2.2987E-12	5.8500E-02	4.0870E-02
193	C11H24	UNDECANE	156.311	468.70	-2.3565	-3.8443E-06	1.9221E-12	4.4000E-03	3.2600E-03
194	C12H18	HEXYLBENZENE	162.274	499.26	0.0075	-3.3289E-06	1.6644E-12	1.0180E+00	7.8470E-01
195	C12H18	1,2,3-TRIETHYLBENZENE	162.274	490.66	0.5895	-4.5175E-06	2.2588E-12	3.8860E+00	2.7320E+00
196	C12H18	1,2,4-TRIETHYLBENZENE	162.274	490.66	0.5895	-4.5175E-06	2.2588E-12	3.8860E+00	2.7320E+00
197	C12H18	1,3,5-TRIETHYLBENZENE	162.274	489.16	0.6219	-4.5176E-06	2.2588E-12	4.1870E+00	2.9430E+00
198	C12H24	1-CYCLOPENTYLHEPTANE	168.322	497.30	-1.7905	-4.5950E-06	2.2975E-12	1.6200E-02	1.1320E-02
199	C12H24	1-CYCLOHEXYLHEXANE	168.322	497.86	-1.5986	-4.6015E-06	2.3007E-12	2.5200E-02	1.7600E-02
200	C12H26	DODECANE	170.337	488.61	-2.4318	-3.1231E-06	1.5616E-12	3.7000E-03	2.9000E-03
201	C13H20	1-PHENYLHEPTANE	176.301	519.16	-0.3239	-3.3291E-06	1.6646E-12	4.7440E-01	3.6590E-01
202	C13H26	1-CYCLOPENTYLOCTANE	182.348	516.86	-2.0458	-4.5927E-06	2.2964E-12	9.0000E-03	6.2890E-03
203	C13H26	1-CYCLOHEXYLHEPTANE	182.348	518.06	-1.8570	-4.5988E-06	2.2994E-12	1.3900E-02	9.7100E-03
204	C13H28	TRIDECANE	184.364	507.77	-2.6990	-3.8600E-06	1.9300E-12	2.0000E-03	1.4800E-03
205	C14H22	1-PHENYLOCTANE	190.328	537.56	-0.6026	-3.3292E-06	1.6646E-12	2.4970E-01	1.9260E-01
206	C14H22	1,2,3,4-TETRAETHYLBENZENE	190.328	524.16	-0.0487	-4.5176E-06	2.2588E-12	8.9390E-01	6.2840E-01
207	C14H22	1,2,3,5-TETRAETHYLBENZENE	190.328	523.66	-0.0407	-4.5176E-06	2.2588E-12	9.1060E-01	6.4010E-01
208	C14H22	1,2,4,5-TETRAETHYLBENZENE	190.328	523.16	-0.0325	-4.5176E-06	2.2588E-12	9.2780E-01	6.5230E-01
209	C14H28	1-CYCLOPENTYLNONANE	196.375	535.26	-2.1871	-4.6006E-06	2.3003E-12	6.5000E-03	4.5400E-03
210	C14H28	1-CYCLOHEXYLOCTANE	196.375	536.76	-1.9957	-4.5905E-06	2.2953E-12	1.0100E-02	7.0600E-03
211	C14H30	TETRADECANE	198.391	526.14	-2.6576	-3.3052E-06	1.6526E-12	2.2000E-03	1.7000E-03
212	C15H24	1-PHENYLNONANE	204.355	555.16	-0.7934	-3.3292E-06	1.6646E-12	1.6090E-01	1.2410E-01
213	C15H30	1-CYCLOPENTYLDECANE	210.402	552.54	-2.2291	-4.6035E-06	2.3017E-12	5.9000E-03	4.1200E-03
214	C15H30	1-CYCLOHEXYLNONANE	210.402	554.66	-2.0223	-4.5917E-06	2.2958E-12	9.5000E-03	6.6400E-03
215	C15H32	PENTADECANE	212.418	543.59	-2.9586	-3.9231E-06	1.9616E-12	1.1000E-03	8.1000E-04
216	C16H34	HEXADECANE	226.445	560.50	-3.0458	-1.0396E-05	5.1978E-12	9.0000E-04	4.0000E-04
217	C17H36	HEPTADECANE	240.471	576.00	-2.8539	-3.8106E-06	1.9053E-12	1.4000E-03	1.0400E-03

S - solubility in salt water at 25 C, parts per million by weight, ppm (wt)

X - concentration of salt (NaCl) in water, parts per million by weight, ppm (wt)

A, B, and C - correlation constants for compound

MW - molecular weight of compound, g/mol

T_B - boiling point temperature of compound, K

X = 0 is water without salt.

X = 34,472 is sea water.

Chapter 17

SOLUBILITY IN WATER AS A FUNCTION OF TEMPERATURE

Carl L. Yaws and Daniel H. Chen
Lamar University, Beaumont, Texas

ABSTRACT

Results for variation of water solubility with temperature are presented for 217 hydrocarbons in an easy-to-use tabular format that is especially applicable for rapid engineering usage with the personal computer or hand calculator. The results cover a range of 25-121 C (77-250 F), which includes temperatures encountered in air and steam stripping operations. Correlation and experimental results are in favorable agreement.

INTRODUCTION

Thermodynamic and physical property data are necessary for the design and operation of industrial processes. In particular, water solubility is becoming increasingly important in view of more and more stringent regulations regarding health, safety, and environment.

In this article, results are presented for water solubility of hydrocarbons as a function of temperature. Solubility values issuing from the correlation are applicable at ambient and elevated temperatures, such as those experienced in air and steam stripping operations.

WATER SOLUBILITY CORRELATION

The correlation for water solubility of hydrocarbons as a function of temperature is based on a series expansion in reciprocal temperature:

$$\log_{10} S = A + B/T + C/T^2 \tag{17-1}$$

where

S = solubility in water, parts per million by weight, ppm(wt)
T = temperature, K
A, B, and C = correlation constants

The correlation constants (A, B, and C) are given in Table 17-1. The correlation constants in the table were determined from regression of the data from sources for water solubility. Both experimental values for the property under consideration and parameter values for estimation of the property are included in the source publications (1-194). The presented values are applicable to a wide variety of hydrocarbons (alkanes, naphthenes, and aromatics with no, single, and multiple substitutions). The tabulation is arranged by carbon number (C5, C6, C7, etc.) for ease of use in quickly locating data using the chemical formula.

The tabulated values for solubility of hydrocarbons in water apply to conditions of saturation in which the hydrocarbon is in equilibrium with water. For saturation, the system pressure is approximately equal to the sum of vapor pressures of hydrocarbon and water.

A comparison of correlation and actual experimental data values for water solubility is shown in Figs. 17-1, 17-2, and 17-3 for representative hydrocarbons (hexane, cyclopentane and benzene). In the figures, solubility values are plotted at temperatures ranging from 25 C to 121 C. This range covers temperatures encountered in air and steam stripping operations. The graphs disclose favorable agreement of correlation and experimental data.

Portions of this material appeared in Chem. Eng., 100, 108 (Feb., 1993), Chem. Eng., 100, 122 (Oct., 1993), and Chem. Eng., 102, 113 (Feb., 1995). These portions are reprinted by special permission.

REFERENCES – ORGANIC COMPOUNDS

1-190. See REFERENCES – WATER SOLUBILITY - ORGANIC COMPOUNDS in Chapter 15, SOLUBILITY IN WATER AND OCTANOL-WATER PARTITION COEFFICIENT
191. McDevit, W. F. and F. A. Long, J. Am. Chem. Soc., 74, 1773 (1952).
192. Yaws, C. L., X. Pan, and X. Lin, Chem. Eng., 100, 108 (Feb., 1993).
193. Yaws, C. L., X. Lin, and Li Bu, Chem. Eng., 100, 122 (Oct., 1993).
194. Yaws, C. L., Li Bu, and S. Nijhawan, Chem. Eng., 102, 113 (Feb., 1995).

Figure 17-1 Solubility of Hexane in Water

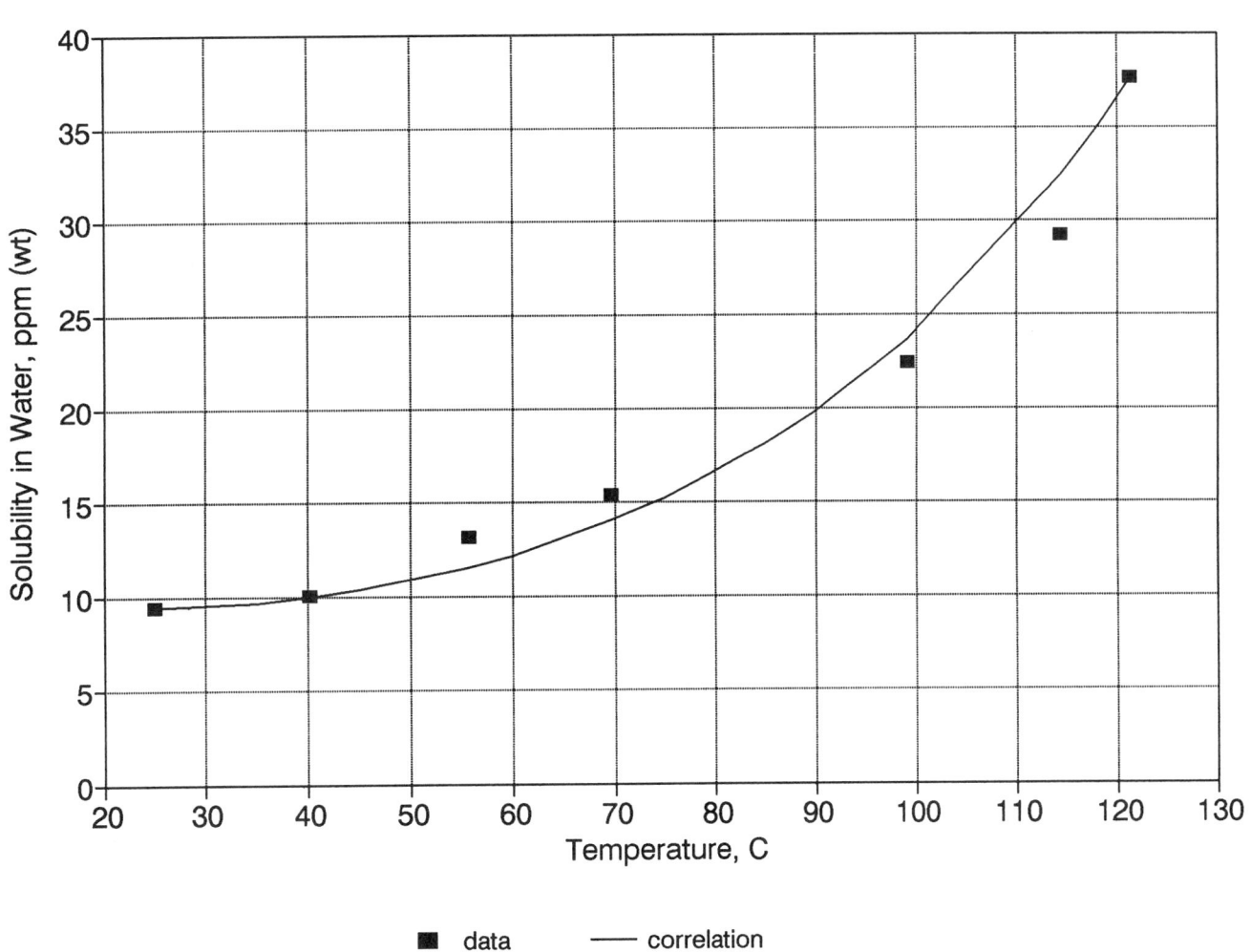

Figure 17-2 Solubility of Cyclopentane in Water

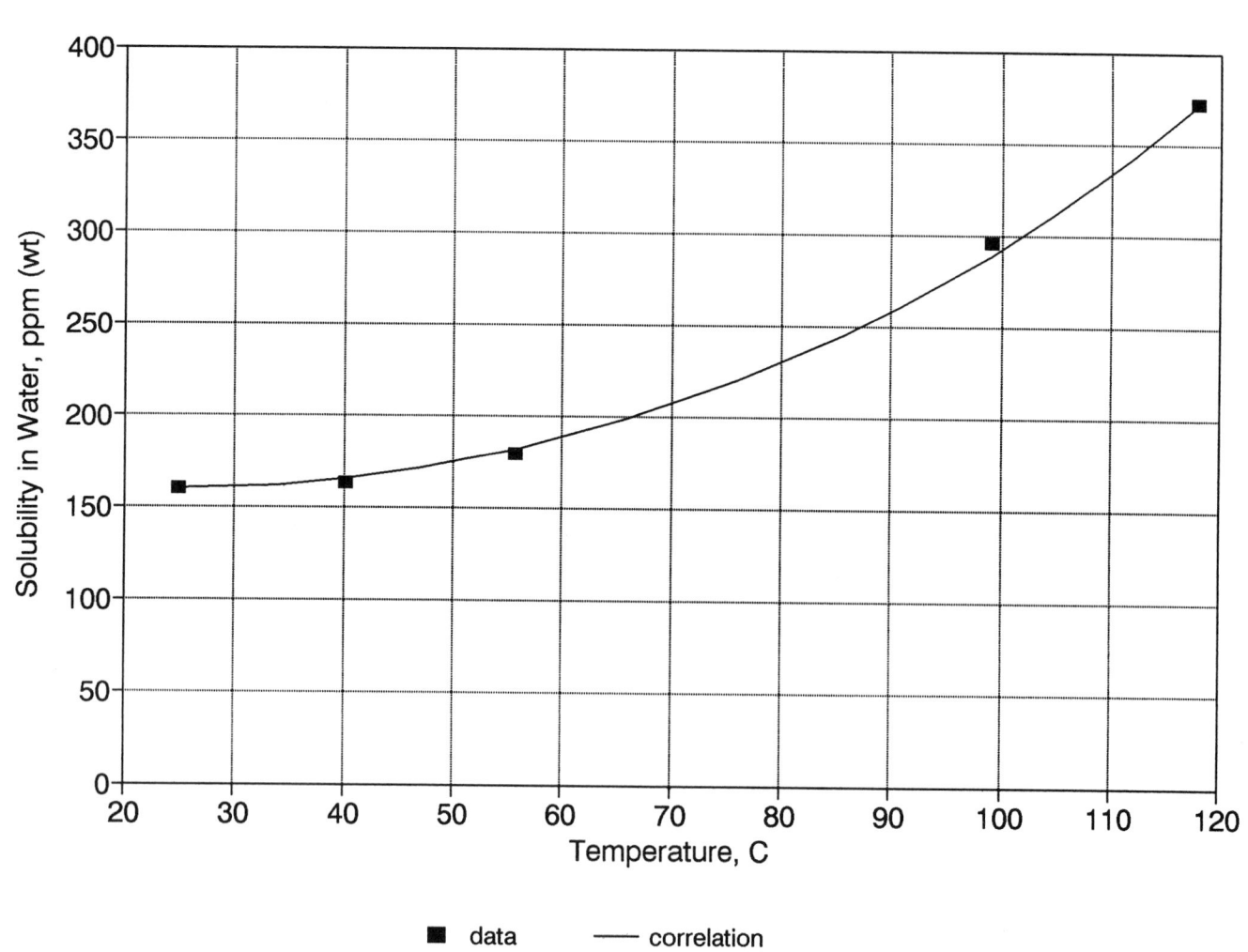

Figure 17-3 Solubility of Benzene in Water

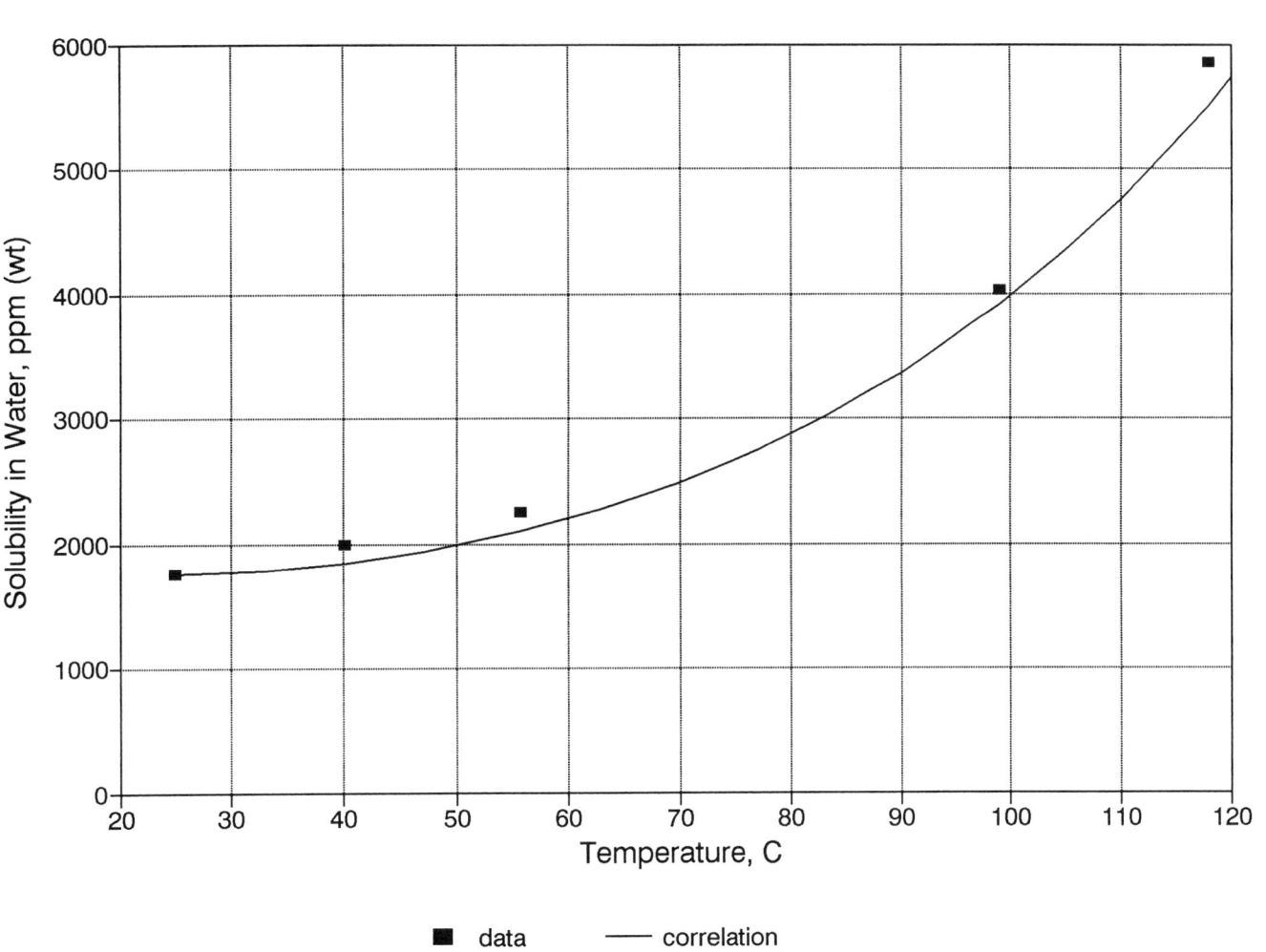

Table 17-1 SOLUBILITY IN WATER - ORGANIC COMPOUNDS

					$\log_{10} S = A + B/T + C/T^2$			S - solubility, ppm (wt), T - K	
NO	FORMULA	NAME	MW, g/mol	T_B, K	A	B	C	S @ 25 C	S @ 100 C
1	C5H10	CYCLOPENTANE	70.134	322.41	8.665	-3850.808	5.73770E+05	1.6000E+02	2.9240E+02
2	C5H12	PENTANE	72.150	309.22	9.036	-4433.922	6.60654E+05	3.9500E+01	7.9120E+01
3	C5H12	2-METHYLBUTANE(ISOPENTANE)	72.150	301.15	9.473	-4645.254	6.92143E+05	4.7800E+01	9.8870E+01
4	C6H6	BENZENE	78.113	353.31	11.994	-5214.537	7.76966E+05	1.7550E+03	3.9780E+03
5	C6H12	CYCLOHEXANE	84.161	353.90	6.403	-2773.806	4.13297E+05	5.6100E+01	8.6640E+01
6	C6H12	METHYLCYCLOPENTANE	84.161	344.97	12.430	-6436.630	9.59058E+05	4.2640E+01	1.1700E+02
7	C6H14	HEXANE	86.177	341.93	10.992	-5969.484	8.89453E+05	9.4700E+00	2.4120E+01
8	C6H14	2-METHYLPENTANE	86.177	333.40	10.606	-5657.127	8.42912E+05	1.3000E+01	3.1560E+01
9	C6H14	3-METHYLPENTANE	86.177	336.35	8.198	-4138.952	6.16704E+05	1.7910E+01	3.4290E+01
10	C6H14	2,2-DIMETHYLBUTANE	86.177	322.87	9.473	-4825.311	7.18971E+05	2.3820E+01	5.0720E+01
11	C6H14	2,3-DIMETHYLBUTANE	86.177	331.16	9.198	-4718.533	7.03061E+05	1.9100E+01	4.0000E+01
12	C7H8	TOLUENE	92.140	383.73	15.471	-7591.270	1.13110E+06	5.4240E+02	1.7810E+03
13	C7H14	ETHYLCYCLOPENTANE	98.188	376.60	14.360	-7974.533	1.18821E+06	9.5560E+00	3.3310E+01
14	C7H14	1,1-DIMETHYLCYCLOPENTANE	98.188	361.00	12.634	-6723.665	1.00183E+06	2.2530E+01	6.4600E+01
15	C7H14	C-1,2-DIMETHYLCYCLOPENTANE	98.188	372.68	13.907	-7647.583	1.13949E+06	1.1910E+01	3.9440E+01
16	C7H14	T-1,2-DIMETHYLCYCLOPENTANE	98.188	365.01	13.057	-7031.220	1.04765E+06	1.8160E+01	5.4720E+01
17	C7H14	C-1,3-DIMETHYLCYCLOPENTANE	98.188	363.92	12.940	-6946.544	1.03504E+06	1.9270E+01	5.7210E+01
18	C7H14	T-1,3-DIMETHYLCYCLOPENTANE	98.188	364.87	13.042	-7020.299	1.04602E+06	1.8300E+01	5.5050E+01
19	C7H14	METHYLCYCLOHEXANE	98.188	374.08	13.091	-7084.522	1.05559E+06	1.6000E+01	4.8570E+01
20	C7H16	HEPTANE	100.203	371.56	12.811	-7426.389	1.10653E+06	2.2400E+00	7.1780E+00
21	C7H16	2-METHYLHEXANE	100.203	363.19	14.617	-8470.165	1.26205E+06	2.5400E+00	9.5880E+00
22	C7H16	3-METHYLHEXANE	100.203	364.98	13.872	-8016.202	1.19441E+06	2.6400E+00	9.2790E+00
23	C7H16	3-ETHYLPENTANE	100.203	366.57	12.675	-7274.170	1.08385E+06	2.9540E+00	9.2270E+00
24	C7H16	2,2-DIMETHYLPENTANE	100.203	352.35	13.729	-7799.080	1.16206E+06	4.4000E+00	1.4930E+01
25	C7H16	2,3-DIMETHYLPENTANE	100.203	363.05	9.695	-5349.230	7.97035E+05	5.2500E+00	1.2130E+01
26	C7H16	2,4-DIMETHYLPENTANE	100.203	353.77	13.343	-7568.576	1.12772E+06	4.4100E+00	1.4420E+01
27	C7H16	3,3-DIMETHYLPENTANE	100.203	359.22	12.478	-6976.429	1.03949E+06	5.9200E+00	1.7670E+01
28	C7H16	2,2,3-TRIMETHYLBUTANE	100.203	354.03	11.474	-6386.198	9.51544E+05	5.7410E+00	1.5610E+01
29	C8H10	ETHYLBENZENE	106.167	409.17	13.365	-6643.672	9.89907E+05	1.6510E+02	4.6780E+02
30	C8H10	M-XYLENE	106.167	412.22	14.743	-7451.440	1.11026E+06	1.7400E+02	5.5930E+02
31	C8H10	O-XYLENE	106.167	417.46	13.112	-6417.607	9.56223E+05	2.2080E+02	6.0390E+02
32	C8H10	P-XYLENE	106.167	411.44	13.731	-6810.190	1.01472E+06	2.0170E+02	5.8610E+02
33	C8H16	PROPYLCYCLOPENTANE	112.214	404.10	17.327	-10142.371	1.51121E+06	2.0400E+00	9.9960E+00
34	C8H16	ETHYLCYCLOHEXANE	112.214	404.94	19.501	-11345.871	1.69053E+06	2.9160E+00	1.7240E+01
35	C8H16	1,1-DIMETHYLCYCLOHEXANE	112.214	392.70	14.606	-8241.599	1.22800E+06	5.9890E+00	2.1810E+01
36	C8H16	C-1,2-DIMETHYLCYCLOHEXANE	112.214	402.94	11.610	-6455.995	9.61943E+05	6.0000E+00	1.6490E+01
37	C8H16	T-1,2-DIMETHYLCYCLOHEXANE	112.214	396.71	15.082	-8586.692	1.27942E+06	4.7340E+00	1.8160E+01
38	C8H16	C-1,3-DIMETHYLCYCLOHEXANE	112.214	393.25	14.671	-8288.887	1.23504E+06	5.7990E+00	2.1260E+01
39	C8H16	T-1,3-DIMETHYLCYCLOHEXANE	112.214	397.60	15.189	-8663.467	1.29086E+06	4.4930E+00	1.7480E+01
40	C8H16	C-1,4-DIMETHYLCYCLOHEXANE	112.214	397.56	15.184	-8660.000	1.29034E+06	4.5040E+00	1.7500E+01
41	C8H16	T-1,4-DIMETHYLCYCLOHEXANE	112.214	392.50	17.766	-10240.173	1.52579E+06	3.8400E+00	1.9110E+01
42	C8H18	OCTANE	114.230	398.77	16.865	-10269.143	1.53010E+06	4.3100E-01	2.1560E+00
43	C8H18	2-METHYLHEPTANE	114.230	390.80	15.344	-9220.273	1.37382E+06	7.4740E-01	3.1710E+00
44	C8H18	3-METHYLHEPTANE	114.230	392.07	14.588	-8755.025	1.30450E+06	7.9200E-01	3.1200E+00
45	C8H18	4-METHYLHEPTANE	114.230	390.92	15.358	-9230.278	1.37531E+06	7.4220E-01	3.1560E+00
46	C8H18	3-ETHYLHEXANE	114.230	391.70	15.448	-9296.104	1.38512E+06	7.0910E-01	3.0420E+00
47	C8H18	2,2-DIMETHYLHEXANE	114.230	380.01	14.116	-8327.433	1.24079E+06	1.3940E+00	5.1350E+00
48	C8H18	2,3-DIMETHYLHEXANE	114.230	388.78	15.111	-9051.116	1.34862E+06	8.4070E-01	3.4710E+00
49	C8H18	2,4-DIMETHYLHEXANE	114.230	382.82	14.432	-8557.258	1.27503E+06	1.1870E+00	4.5340E+00
50	C8H18	2,5-DIMETHYLHEXANE	114.230	381.76	14.313	-8470.421	1.26209E+06	1.2610E+00	4.7570E+00
51	C8H18	3,3-DIMETHYLHEXANE	114.230	385.12	14.693	-8746.792	1.30327E+06	1.0390E+00	4.0970E+00
52	C8H18	3,4-DIMETHYLHEXANE	114.230	391.65	15.465	-9308.189	1.38692E+06	7.0300E-01	3.0250E+00
53	C8H18	3-ETHYL-2-METHYLPENTANE	114.230	388.81	15.115	-9053.816	1.34902E+06	8.3920E-01	3.4690E+00
54	C8H18	3-ETHYL-3-METHYLPENTANE	114.230	391.42	15.416	-9272.560	1.38161E+06	7.2080E-01	3.0830E+00
55	C8H18	2,2,3-TRIMETHYLPENTANE	114.230	383.25	14.480	-8592.351	1.28026E+06	1.1580E+00	4.4470E+00
56	C8H18	2,2,4-TRIMETHYLPENTANE	114.230	372.38	13.041	-7565.611	1.12728E+06	2.2210E+00	7.2760E+00
57	C8H18	2,3,3-TRIMETHYLPENTANE	114.230	387.89	15.009	-8976.902	1.33756E+06	8.8530E-01	3.6140E+00
58	C8H18	2,3,4-TRIMETHYLPENTANE	114.230	386.58	12.446	-7338.101	1.09338E+06	1.3600E+00	4.2970E+00
59	C8H18	2,2,3,3-TETRAMETHYLBUTANE	114.230	379.60	14.071	-8294.366	1.23586E+06	1.4270E+00	5.2330E+00
60	C9H12	PROPYLBENZENE	120.194	430.17	11.722	-5962.906	8.88473E+05	5.2160E+01	1.3270E+02
61	C9H12	CUMENE	120.194	426.31	13.052	-6766.120	1.00815E+06	5.0000E+01	1.4450E+02
62	C9H12	M-ETHYLTOLUENE	120.194	434.50	11.458	-5685.778	8.47181E+05	8.2820E+01	2.0190E+02
63	C9H12	O-ETHYLTOLUENE	120.194	438.30	9.107	-4254.669	6.33946E+05	9.3050E+01	1.8110E+02
64	C9H12	P-ETHYLTOLUENE	120.194	435.20	10.245	-4927.887	7.34255E+05	9.4850E+01	2.0520E+02
65	C9H12	1,2,3-TRIMETHYLBENZENE	120.194	449.27	7.011	-3095.900	4.61289E+05	6.5510E+01	1.0650E+02
66	C9H12	1,2,4-TRIMETHYLBENZENE	120.194	442.26	10.827	-5405.927	8.05483E+05	5.7040E+01	1.3320E+02
67	C9H12	MESITYLENE	120.194	437.89	13.768	-7202.428	1.07316E+06	4.8200E+01	1.4910E+02
68	C9H18	BUTYLCYCLOPENTANE	126.241	429.76	20.741	-12579.687	1.87437E+06	3.0920E-01	3.0920E+00
69	C9H18	PROPYLCYCLOHEXANE	126.241	429.90	18.885	-11356.483	1.69212E+06	6.7730E-01	4.0120E+00
70	C9H18	C-C-135TRIMETHYLCYCLOHEXANE	126.241	411.66	16.852	-9869.544	1.47056E+06	1.9620E+00	9.2050E+00
71	C9H18	C-T-135TRIMETHYLCYCLOHEXANE	126.241	413.70	17.089	-10041.877	1.49624E+06	1.7400E+00	8.3870E+00
72	C9H20	NONANE	128.257	423.81	18.222	-11404.793	1.69931E+06	1.2200E-01	7.2870E-01
73	C9H20	2-METHYLOCTANE	128.257	416.19	18.267	-11349.456	1.69107E+06	1.6780E-01	9.9230E-01
74	C9H20	3-METHYLOCTANE	128.257	417.37	18.398	-11445.016	1.70531E+06	1.5660E-01	9.4150E-01
75	C9H20	4-METHYLOCTANE	128.257	415.58	21.024	-13090.267	1.95045E+06	1.1500E-01	8.9390E-01
76	C9H20	3-ETHYLHEPTANE	128.257	416.16	18.264	-11346.784	1.69067E+06	1.6810E-01	9.9530E-01
77	C9H20	4-ETHYLHEPTANE	128.257	414.36	18.060	-11198.056	1.66851E+06	1.8690E-01	1.0800E+00
78	C9H20	2,2-DIMETHYLHEPTANE	128.257	405.85	17.089	-10489.506	1.56294E+06	3.0860E-01	1.5960E+00

400

Table 17-1 SOLUBILITY IN WATER - ORGANIC COMPOUNDS (continued)

								S - solubility, ppm (wt), T - K	
NO	FORMULA	NAME	MW, g/mol	T_B, K	A	B	C	S @ 25 C	S @ 100 C
79	C9H20	2,3-DIMETHYLHEPTANE	128.257	413.66	17.980	-11139.361	1.65976E+06	1.9480E-01	1.1170E+00
80	C9H20	2,4-DIMETHYLHEPTANE	128.257	406.05	17.112	-10505.966	1.56539E+06	3.0500E-01	1.5830E+00
81	C9H20	2,5-DIMETHYLHEPTANE	128.257	409.16	17.469	-10766.279	1.60418E+06	2.5390E-01	1.3720E+00
82	C9H20	2,6-DIMETHYLHEPTANE	128.257	408.37	17.379	-10700.421	1.59436E+06	2.6600E-01	1.4240E+00
83	C9H20	3,3-DIMETHYLHEPTANE	128.257	410.17	17.585	-10850.868	1.61678E+06	2.3920E-01	1.3100E+00
84	C9H20	3,4-DIMETHYLHEPTANE	128.257	413.76	17.993	-11148.951	1.66119E+06	1.9360E-01	1.1100E+00
85	C9H20	3,5-DIMETHYLHEPTANE	128.257	409.16	17.469	-10766.279	1.60418E+06	2.5390E-01	1.3720E+00
86	C9H20	4,4-DIMETHYLHEPTANE	128.257	408.36	17.379	-10700.442	1.59437E+06	2.6610E-01	1.4240E+00
87	C9H20	3-ETHYL-2-METHYLHEXANE	128.257	411.16	17.699	-10934.184	1.62919E+06	2.2560E-01	1.2510E+00
88	C9H20	4-ETHYL-2-METHYLHEXANE	128.257	406.96	17.216	-10581.772	1.57668E+06	2.8910E-01	1.5190E+00
89	C9H20	3-ETHYL-3-METHYLHEXANE	128.257	413.76	17.993	-11148.951	1.66119E+06	1.9360E-01	1.1100E+00
90	C9H20	3-ETHYL-4-METHYLHEXANE	128.257	413.56	17.970	-11132.245	1.65870E+06	1.9590E-01	1.1200E+00
91	C9H20	2,2,3-TRIMETHYLHEXANE	128.257	406.75	17.192	-10564.177	1.57406E+06	2.9270E-01	1.5340E+00
92	C9H20	2,2,4-TRIMETHYLHEXANE	128.257	399.69	16.374	-9969.326	1.48543E+06	4.4380E-01	2.1150E+00
93	C9H20	2,2,5-TRIMETHYLHEXANE	128.257	397.24	15.733	-9536.120	1.42088E+06	5.4010E-01	2.4090E+00
94	C9H20	2,3,3-TRIMETHYLHEXANE	128.257	410.84	17.662	-10907.365	1.62520E+06	2.2990E-01	1.2690E+00
95	C9H20	2,3,4-TRIMETHYLHEXANE	128.257	412.20	17.817	-11020.301	1.64202E+06	2.1220E-01	1.1930E+00
96	C9H20	2,3,5-TRIMETHYLHEXANE	128.257	404.50	16.933	-10375.648	1.54597E+06	3.3420E-01	1.6990E+00
97	C9H20	2,4,4-TRIMETHYLHEXANE	128.257	403.81	16.853	-10317.287	1.53728E+06	3.4810E-01	1.7550E+00
98	C9H20	3,3,4-TRIMETHYLHEXANE	128.257	413.62	17.977	-11137.427	1.65948E+06	1.9520E-01	1.1170E+00
99	C9H20	3,3-DIETHYLPENTANE	128.257	419.30	18.609	-11600.131	1.72842E+06	1.3990E-01	8.6120E-01
100	C9H20	3-ETHYL-2,2-DIMETHYLPENTANE	128.257	406.99	17.221	-10585.565	1.57725E+06	2.8850E-01	1.5150E+00
101	C9H20	3-ETHYL-2,3-DIMETHYLPENTANE	128.257	417.86	18.451	-11483.842	1.71109E+06	1.5220E-01	9.2110E-01
102	C9H20	3-ETHYL-2,4-DIMETHYLPENTANE	128.257	409.85	17.547	-10823.408	1.61269E+06	2.4380E-01	1.3290E+00
103	C9H20	2,2,3,3-TETRAMETHYLPENTANE	128.257	413.44	17.956	-11121.973	1.65717E+06	1.9730E-01	1.1270E+00
104	C9H20	2,2,3,4-TETRAMETHYLPENTANE	128.257	406.18	17.126	-10516.489	1.56696E+06	3.0270E-01	1.5720E+00
105	C9H20	2,2,4,4-TETRAMETHYLPENTANE	128.257	395.44	15.881	-9610.821	1.43201E+06	5.6970E-01	2.5670E+00
106	C9H20	2,3,3,4-TETRAMETHYLPENTANE	128.257	414.72	18.100	-11227.357	1.67288E+06	1.8300E-01	1.0620E+00
107	C10H14	BUTYLBENZENE	134.221	456.43	9.533	-5001.869	7.45278E+05	1.3820E+01	3.0270E+01
108	C10H14	M-DIETHYLBENZENE	134.221	454.25	10.975	-5686.134	8.47234E+05	2.7230E+01	6.6290E+01
109	C10H14	O-DIETHYLBENZENE	134.221	457.02	10.908	-5686.174	8.47240E+05	2.3320E+01	5.6810E+01
110	C10H14	P-DIETHYLBENZENE	134.221	456.94	10.910	-5686.175	8.47240E+05	2.3420E+01	5.7070E+01
111	C10H14	1,2,3,4-TETRAMETHYLBENZENE	134.221	478.25	10.407	-5686.600	8.47303E+05	7.3390E+00	1.7890E+01
112	C10H14	1,2,3,5-TETRAMETHYLBENZENE	134.221	471.25	10.569	-5686.435	8.47279E+05	1.0660E+01	2.6000E+01
113	C10H20	1-CYCLOPENTYLPENTANE	140.268	453.76	22.985	-14259.115	2.12461E+06	1.1500E-01	1.0730E+00
114	C10H20	BUTYLCYCLOHEXANE	140.268	454.10	21.087	-13014.701	1.93919E+06	1.7810E-01	1.3680E+00
115	C10H22	DECANE	142.284	446.86	17.396	-11133.010	1.65882E+06	5.2000E-02	2.9790E-01
116	C10H22	2-METHYLNONANE	142.284	440.16	20.737	-13178.037	1.96353E+06	4.2300E-02	3.3340E-01
117	C10H22	3-METHYLNONANE	142.284	440.96	20.803	-13228.564	1.97106E+06	4.0500E-02	3.2190E-01
118	C10H22	4-METHYLNONANE	142.284	438.86	20.615	-13086.581	1.94906E+06	4.5500E-02	3.5340E-01
119	C10H22	5-METHYLNONANE	142.284	438.26	20.552	-13039.905	1.94295E+06	4.7100E-02	3.6340E-01
120	C10H22	3-ETHYLOCTANE	142.284	439.66	20.691	-13143.350	1.95836E+06	4.3500E-02	3.4110E-01
121	C10H22	4-ETHYLOCTANE	142.284	436.80	20.418	-12939.069	1.92792E+06	5.1100E-02	3.8790E-01
122	C10H22	2,2-DIMETHYLOCTANE	142.284	430.06	19.758	-12446.785	1.85457E+06	7.4900E-02	5.2620E-01
123	C10H22	2,3-DIMETHYLOCTANE	142.284	437.47	20.483	-12987.251	1.93510E+06	4.9200E-02	3.7680E-01
124	C10H22	2,4-DIMETHYLOCTANE	142.284	429.06	19.649	-12366.790	1.84265E+06	7.9400E-02	5.5070E-01
125	C10H22	2,5-DIMETHYLOCTANE	142.284	431.66	19.915	-12563.459	1.87196E+06	6.8400E-02	4.9020E-01
126	C10H22	2,6-DIMETHYLOCTANE	142.284	433.54	20.107	-12706.174	1.89322E+06	6.1400E-02	4.4940E-01
127	C10H22	2,7-DIMETHYLOCTANE	142.284	433.03	20.058	-12669.340	1.88773E+06	6.3200E-02	4.6020E-01
128	C10H22	3,3-DIMETHYLOCTANE	142.284	434.36	20.189	-12766.974	1.90228E+06	5.8600E-02	4.3320E-01
129	C10H22	3,4-DIMETHYLOCTANE	142.284	436.56	20.394	-12921.348	1.92528E+06	5.1800E-02	3.9190E-01
130	C10H22	3,5-DIMETHYLOCTANE	142.284	432.56	20.002	-12628.855	1.88170E+06	6.5000E-02	4.7000E-01
131	C10H22	3,6-DIMETHYLOCTANE	142.284	433.96	20.143	-12733.496	1.89729E+06	6.0000E-02	4.4120E-01
132	C10H22	4,4-DIMETHYLOCTANE	142.284	430.66	19.816	-12490.136	1.86103E+06	7.2400E-02	5.1210E-01
133	C10H22	4,5-DIMETHYLOCTANE	142.284	435.29	20.279	-12834.012	1.91227E+06	5.5600E-02	4.1570E-01
134	C10H22	4-PROPYLHEPTANE	142.284	430.66	19.816	-12490.136	1.86103E+06	7.2400E-02	5.1210E-01
135	C10H22	4-ISOPROPYLHEPTANE	142.284	432.06	19.961	-12596.914	1.87694E+06	6.6800E-02	4.8140E-01
136	C10H22	3-ETHYL-2-METHYLHEPTANE	142.284	434.36	20.189	-12766.974	1.90228E+06	5.8600E-02	4.3320E-01
137	C10H22	4-ETHYL-2-METHYLHEPTANE	142.284	429.36	19.684	-12392.112	1.84642E+06	7.8000E-02	5.4350E-01
138	C10H22	5-ETHYL-2-METHYLHEPTANE	142.284	432.86	20.032	-12650.920	1.88499E+06	6.3900E-02	4.6400E-01
139	C10H22	3-ETHYL-3-METHYLHEPTANE	142.284	436.96	20.439	-12953.822	1.93012E+06	5.0600E-02	3.8540E-01
140	C10H22	4-ETHYL-3-METHYLHEPTANE	142.284	435.36	20.283	-12837.546	1.91279E+06	5.5400E-02	4.1410E-01
141	C10H22	3-ETHYL-5-METHYLHEPTANE	142.284	431.36	19.883	-12539.989	1.86846E+06	6.9600E-02	4.9670E-01
142	C10H22	3-ETHYL-4-METHYLHEPTANE	142.284	436.16	20.354	-12891.362	1.92081E+06	5.3000E-02	3.9950E-01
143	C10H22	4-ETHYL-4-METHYLHEPTANE	142.284	433.96	20.143	-12733.496	1.89729E+06	6.0000E-02	4.4120E-01
144	C10H22	2,2,3-TRIMETHYLHEPTANE	142.284	430.76	19.825	-12496.720	1.86201E+06	7.2000E-02	5.1030E-01
145	C10H22	2,2,4-TRIMETHYLHEPTANE	142.284	421.46	18.846	-11774.072	1.75434E+06	1.2330E-01	7.8000E-01
146	C10H22	2,2,5-TRIMETHYLHEPTANE	142.284	423.96	19.116	-11972.385	1.78389E+06	1.0660E-01	6.9640E-01
147	C10H22	2,2,6-TRIMETHYLHEPTANE	142.284	422.09	18.912	-11822.675	1.76158E+06	1.1890E-01	7.5830E-01
148	C10H22	2,3,3-TRIMETHYLHEPTANE	142.284	433.36	20.082	-12688.453	1.89058E+06	6.2100E-02	4.5300E-01
149	C10H22	2,3,4-TRIMETHYLHEPTANE	142.284	433.06	20.060	-12671.053	1.88799E+06	6.3100E-02	4.5930E-01
150	C10H22	2,3,5-TRIMETHYLHEPTANE	142.284	433.86	20.138	-12729.393	1.89668E+06	6.0300E-02	4.4280E-01
151	C10H22	2,3,6-TRIMETHYLHEPTANE	142.284	429.16	19.664	-12376.871	1.84415E+06	7.8900E-02	5.4920E-01
152	C10H22	2,4,4-TRIMETHYLHEPTANE	142.284	424.16	19.135	-11987.004	1.78606E+06	1.0540E-01	6.8920E-01
153	C10H22	2,4,5-TRIMETHYLHEPTANE	142.284	429.66	19.712	-12412.877	1.84952E+06	7.6700E-02	5.3670E-01
154	C10H22	2,4,6-TRIMETHYLHEPTANE	142.284	420.76	18.772	-11719.532	1.74621E+06	1.2840E-01	8.0520E-01
155	C10H22	2,5,5-TRIMETHYLHEPTANE	142.284	425.96	19.330	-12130.477	1.80744E+06	9.4900E-02	6.3440E-01
156	C10H22	3,3,4-TRIMETHYLHEPTANE	142.284	435.06	20.247	-12811.749	1.90895E+06	5.6400E-02	4.1940E-01

Table 17-1 SOLUBILITY IN WATER - ORGANIC COMPOUNDS (continued)

					$\log_{10} S = A + B/T + C/T^2$			S - solubility, ppm (wt), T - K	
NO	FORMULA	NAME	MW, g/mol	T_B, K	A	B	C	S @ 25 C	S @ 100 C
157	C10H22	3,3,5-TRIMETHYLHEPTANE	142.284	428.85	19.633	-12354.188	1.84077E+06	8.0300E-02	5.5620E-01
158	C10H22	3,4,4-TRIMETHYLHEPTANE	142.284	434.26	20.171	-12754.520	1.90042E+06	5.9000E-02	4.3530E-01
159	C10H22	3,4,5-TRIMETHYLHEPTANE	142.284	435.66	20.308	-12856.587	1.91563E+06	5.4500E-02	4.0880E-01
160	C10H22	3-ISOPROPYL-2-METHYLHEXANE	142.284	439.86	20.712	-13158.846	1.96067E+06	4.3000E-02	3.3800E-01
161	C10H22	3,3-DIETHYLHEXANE	142.284	439.46	20.671	-13128.152	1.95609E+06	4.4000E-02	3.4460E-01
162	C10H22	3,4-DIETHYLHEXANE	142.284	437.06	20.450	-12962.359	1.93139E+06	5.0300E-02	3.8300E-01
163	C10H22	3-ETHYL-2,2-DIMETHYLHEXANE	142.284	429.26	19.669	-12381.538	1.84485E+06	7.8500E-02	5.4600E-01
164	C10H22	4-ETHYL-2,2-DIMETHYLHEXANE	142.284	420.16	18.706	-11670.749	1.73894E+06	1.3300E-01	8.2880E-01
165	C10H22	3-ETHYL-2,3-DIMETHYLHEXANE	142.284	436.86	20.427	-12945.369	1.92886E+06	5.0900E-02	3.8690E-01
166	C10H22	4-ETHYL-2,3-DIMETHYLHEXANE	142.284	434.06	20.160	-12745.277	1.89905E+06	5.9600E-02	4.3920E-01
167	C10H22	3-ETHYL-2,4-DIMETHYLHEXANE	142.284	433.26	20.078	-12684.664	1.89001E+06	6.2400E-02	4.5520E-01
168	C10H22	4-ETHYL-2,4-DIMETHYLHEXANE	142.284	434.26	20.171	-12754.520	1.90042E+06	5.9000E-02	4.3530E-01
169	C10H22	3-ETHYL-2,5-DIMETHYLHEXANE	142.284	427.26	19.461	-12227.633	1.82192E+06	8.8100E-02	5.9840E-01
170	C10H22	4-ETHYL-3,3-DIMETHYLHEXANE	142.284	436.06	20.344	-12883.942	1.91971E+06	5.3300E-02	4.0130E-01
171	C10H22	3-ETHYL-3,4-DIMETHYLHEXANE	142.284	435.26	20.275	-12831.451	1.91189E+06	5.5700E-02	4.1580E-01
172	C10H22	2,2,3,3-TETRAMETHYLHEXANE	142.284	433.48	20.103	-12702.573	1.89268E+06	6.1600E-02	4.5120E-01
173	C10H22	2,2,3,4-TETRAMETHYLHEXANE	142.284	431.96	19.949	-12588.338	1.87566E+06	6.7200E-02	4.8340E-01
174	C10H22	2,2,3,5-TETRAMETHYLHEXANE	142.284	421.56	18.856	-11781.565	1.75545E+06	1.2260E-01	7.7630E-01
175	C10H22	2,2,4,4-TETRAMETHYLHEXANE	142.284	426.96	19.433	-12206.429	1.81876E+06	8.9600E-02	6.0690E-01
176	C10H22	2,2,4,5-TETRAMETHYLHEXANE	142.284	421.04	18.798	-11739.076	1.74912E+06	1.2640E-01	7.9510E-01
177	C10H22	2,2,5,5-TETRAMETHYLHEXANE	142.284	410.63	17.637	-10889.177	1.62249E+06	2.3280E-01	1.2810E+00
178	C10H22	2,3,3,4-TETRAMETHYLHEXANE	142.284	437.75	20.514	-13010.201	1.93852E+06	4.8400E-02	3.7170E-01
179	C10H22	2,3,3,5-TETRAMETHYLHEXANE	142.284	426.26	19.359	-12152.005	1.81065E+06	9.3300E-02	6.2620E-01
180	C10H22	2,3,4,4-TETRAMETHYLHEXANE	142.284	434.76	20.226	-12795.040	1.90646E+06	5.7300E-02	4.2520E-01
181	C10H22	2,3,4,5-TETRAMETHYLHEXANE	142.284	429.36	19.684	-12392.112	1.84642E+06	7.8000E-02	5.4350E-01
182	C10H22	3,3,4,4-TETRAMETHYLHEXANE	142.284	443.16	21.011	-13384.618	1.99431E+06	3.5800E-02	2.9140E-01
183	C10H22	24DIMETHYL3ISOPROPYLPENTANE	142.284	430.20	19.773	-12457.408	1.85615E+06	7.4300E-02	5.2370E-01
184	C10H22	33-DIETHYL-2-METHYLPENTANE	142.284	442.86	20.984	-13363.994	1.99124E+06	3.6400E-02	2.9560E-01
185	C10H22	3ETHYL-223TRIMETHYLPENTANE	142.284	442.66	20.967	-13351.283	1.98934E+06	3.6800E-02	2.9790E-01
186	C10H22	3ETHYL-224TRIMETHYLPENTANE	142.284	428.46	19.587	-12320.464	1.83575E+06	8.2200E-02	5.6690E-01
187	C10H22	3ETHYL-234TRIMETHYLPENTANE	142.284	442.60	20.947	-13338.038	1.98737E+06	3.7000E-02	2.9880E-01
188	C10H22	22334-PENTAMETHYLPENTANE	142.284	439.21	20.650	-13112.457	1.95376E+06	4.4600E-02	3.4800E-01
189	C10H22	22344-PENTAMETHYLPENTANE	142.284	432.45	19.992	-12621.347	1.88058E+06	6.5400E-02	4.7220E-01
190	C11H16	PENTYLBENZENE	148.247	478.62	9.504	-5316.406	7.92144E+05	3.8400E+00	8.8240E+00
191	C11H22	1-CYCLOPENTYLHEXANE	154.295	476.26	24.302	-15330.352	2.28422E+06	3.8000E-02	4.2000E-01
192	C11H22	PENTYLCYCLOHEXANE	154.295	476.87	22.316	-14034.923	2.09120E+06	5.8500E-02	5.2790E-01
193	C11H24	UNDECANE	156.311	468.70	28.051	-18123.096	2.70034E+06	4.4000E-03	7.5240E-02
194	C12H18	HEXYLBENZENE	162.274	499.26	11.076	-6596.944	9.82945E+05	1.0180E+00	2.8590E+00
195	C12H18	1,2,3-TRIETHYLBENZENE	162.274	490.66	10.131	-5686.827	8.47337E+05	3.8860E+00	9.4700E+00
196	C12H18	1,2,4-TRIETHYLBENZENE	162.274	490.66	10.131	-5686.827	8.47337E+05	3.8860E+00	9.4700E+00
197	C12H18	1,3,5-TRIETHYLBENZENE	162.274	489.16	10.164	-5686.808	8.47334E+05	4.1870E+00	1.0220E+01
198	C12H24	1-CYCLOPENTYLHEPTANE	168.322	497.30	24.467	-15649.309	2.33175E+06	1.6200E-02	1.8830E-01
199	C12H24	1-CYCLOHEXYLHEXANE	168.322	497.86	22.412	-14310.589	2.13228E+06	2.5200E-02	2.3700E-01
200	C12H26	DODECANE	170.337	488.61	23.755	-15607.165	2.32547E+06	3.7000E-03	4.2720E-02
201	C13H20	1-PHENYLHEPTANE	176.301	519.16	10.022	-6165.903	9.18720E+05	4.7440E-01	1.2480E+00
202	C13H26	1-CYCLOPENTYLOCTANE	182.348	516.86	23.457	-15199.391	2.26471E+06	9.0000E-03	9.7500E-02
203	C13H26	1-CYCLOHEXYLHEPTANE	182.348	518.06	21.326	-13817.112	2.05875E+06	1.3900E-02	1.2110E-01
204	C13H28	TRIDECANE	184.364	507.77	22.937	-15279.183	2.27660E+06	2.0000E-03	2.1910E-02
205	C14H22	1-PHENYLOCTANE	190.328	537.56	9.743	-6166.179	9.18761E+05	2.4970E-01	6.5570E-01
206	C14H22	1,2,3,4-TETRAETHYLBENZENE	190.328	524.16	9.494	-5687.287	8.47406E+05	8.9390E-01	2.1810E+00
207	C14H22	1,2,3,5-TETRAETHYLBENZENE	190.328	523.66	9.502	-5687.515	8.47440E+05	9.1060E-01	2.2190E+00
208	C14H22	1,2,4,5-TETRAETHYLBENZENE	190.328	523.16	9.510	-5687.282	8.47405E+05	9.2780E-01	2.2630E+00
209	C14H28	1-CYCLOPENTYLNONANE	196.375	535.26	21.217	-13948.638	2.07835E+06	6.5000E-03	5.7870E-02
210	C14H28	1-CYCLOHEXYLOCTANE	196.375	536.76	19.067	-12553.516	1.87047E+06	1.0100E-02	7.2170E-02
211	C14H30	TETRADECANE	198.391	526.14	17.765	-12172.057	1.81364E+06	2.0000E-03	1.4810E-02
212	C15H24	1-PHENYLNONANE	204.355	555.16	9.552	-6165.937	9.18725E+05	1.6090E-01	4.2270E-01
213	C15H30	1-CYCLOPENTYLDECANE	210.402	552.54	17.980	-12044.875	1.79469E+06	5.9000E-03	3.8920E-02
214	C15H30	1-CYCLOHEXYLNONANE	210.402	554.66	15.592	-10498.132	1.56422E+06	9.5000E-03	4.9220E-02
215	C15H32	PENTADECANE	212.418	543.59	18.533	-12808.889	1.90852E+06	1.1000E-03	8.1890E-03
216	C16H34	HEXADECANE	226.445	560.50	16.462	-11626.514	1.73235E+06	9.0000E-04	5.5670E-03
217	C17H36	HEPTADECANE	240.471	576.00	10.581	-8007.275	1.19308E+06	1.4000E-03	4.9080E-03

S - solubility in water, parts per million by weight, ppm (wt)

T - temperature, K

A, B, and C - correlation constants for compound

MW - molecular weight of compound, g/mol

T_B - boiling point temperature of compound, K

Chapter 18

HENRY'S LAW CONSTANT FOR COMPOUND IN WATER

Carl L. Yaws, Sachin Nijhawan, and Li Bu
Lamar University, Beaumont, Texas

ABSTRACT

Results for Henry's law constant are presented for a wide variety of organic chemicals in water. The organic chemicals include hydrocarbon, oxygen, nitrogen, fluorine, chlorine, bromine, iodine, and sulfur compounds. The results are provided in an easy-to-use table that is especially applicable for rapid engineering usage with the personal computer or hand calculator. Representative values for Henry's law constant (atm/mol fraction) are 11,515 for ethylene (C2H4), 1,630 for carbon tetrachloride (CCl4), and 308 for benzene (C6H6) at ambient conditions.

INTRODUCTION

Physical and thermodynamic property data for major organic chemicals are of special value to engineers in the chemical processing and petroleum refining industries. In this article results are presented for Henry's law constant for organic chemicals in water. The compilation is intended for use in initial engineering and environmental impact studies. As an example of such usage, Henry's law constant is helpful in determining the environmental movement and fate of chemicals in air and water.

The results are presented in an easy-to-use tabular format that is especially applicable for rapid engineering usage with the personal computer or hand calculator.

HENRY'S LAW CONSTANT

The results for Henry's law constant for organic chemicals in water are given in Table 18-1. The tabulation is applicable to a wide variety of organic chemicals in contact with water at ambient conditions. The wide variety of substances includes hydrocarbons, fluorocarbons, chlorocarbons, bromocarbons, iodocarbons, alcohols, acids, ketones, aldehydes, ethers, esters, amines, nitriles, sulfides, and thiols.

The tabulation is arranged by carbon number (C1, C2, C3, ... ,C28). This provides ease of use in quickly locating data using the chemical formula.

Henry's law constant may be determined from data for solubility, vapor pressure, and activity coefficient at infinite dilution (see Appendix for equations). Thus, in preparing the tabulation, a literature search was conducted to identify data source publications (1-210) for solubility, vapor pressure, activity coefficient at infinite dilution, and Henry's law constant. Both experimental values for the property under consideration and parameter values for estimation of the property are included in the source publications. The publications were screened and copies of appropriate data were made. These data were then keyed-in to the computer to provide a database.

A comparison of calculated and data (experimental) values for Henry's law constant is shown in Figs. 18-1 and 18-2. The graph in Fig. 18-1 includes many different organic chemicals which contain a variety of functional groups: carbon monoxide, carbon dioxide, trichlorofluoromethane, carbon tetrachloride, bromoform, chloroform, dichloromethane, trichlorotrifluoroethane, tetrachloroethylene, hexachloroethane, trichloroethylene, 1,1-dichloroethylene, cis 1,2-dichloroethylene, trans 1,2-dichloroethylene, 1,1,2,2-tetrachloroethane, vinyl chloride, 1,1,1-trichloroethane, 1,1,2-trichloroethane, 1,2-dibromoethane, 1,1-dichloroethane, ethyl chloride, 1,2,3-trichloropropane, 1,2-dichloropropane, 1,3-dichloropropane, 1-chlorobutane, 2-chlorobutane, 1-chloropentane, 1,2,4-trichlorobenzene, 1,3-dichlorobenzene, 1,2-dichlorobenzene, chlorobenzene, and nitrobenzene. The graph discloses general agreement of calculated values and experimental data.

The graph in Fig. 18-2 includes many different types of hydrocarbons: methane, ethane, hexane, 2-methylpentane, 2-methylhexane, heptane, octane, ethylene, acetylene, cyclopentane, cyclohexane, methyl cyclohexane, benzene, toluene, o-xylene, m-xylene, p-xylene, ethyl benzene, propyl benzene, cumene, 1,3,5-trimethylbenzene, 1,2,4-trimethylbenzene, tetralin, and 1-methylnaphthalene. The graph discloses general agreement of calculations and data.

The compilation for Henry's law constant is usable for engineering and environmental impact studies involving organic compounds in water.

EXAMPLES

The tabulation maybe used for determining Henry's law constant for the compound in water. The use of Henry's law constant in environmental applications is illustrated below.

Example 1 A chemical spill of ethylbenzene (C8H10) occurs into a body of water. The concentration of ethylbenzene in the liquid at the surface of the water is 25 parts per million on a mol basis [x_i = 25 ppm(mol)]. Estimate the concentration of ethylbenzene in the air at the surface of the water.

From thermodynamics at low pressure, the partition coefficient is given by $K_i = H_i/P_t$. Substitution of Henry's law constant (H_i = 452.3 atm/mol fraction for ethylbenzene) from the table and the total pressure (P_t = 1 atm) into this equation yields K_i = 452.3/1 = 452.3.

Since the vapor concentration is given by $y_i = K_i x_l$, substitution of the K value (K_i = 452.3) and liquid concentration (x_i = 25 ppm) yields

$$y_i = K_i x_i = (452.3)(25)$$

$$y_i = 11,308 \text{ ppm (mol)}$$

Example 2 A chemical spill of benzene (C6H6) occurs into a body of water. The concentration of benzene in the air at the surface of the water is measured at 1000 parts per million on a mol basis [y_i = 1000 ppm(mol)]. Estimate the concentration of benzene in the liquid at the surface of the water.

From thermodynamics at low pressure, the partition coefficient is given by $K_i = H_i/P_t$. Substitution of Henry's law constant (H_i = 308.26 atm/mol fraction for benzene) from the table and the total pressure (P_t = 1 atm) into the above equation yields K_i = 308.26/1 = 308.26.

Since the liquid concentration is given by $x_i = y_i / K_i$, substitution of the K value (K_i = 308.26) and vapor concentration (y_i = 1000 ppm) yields

$$x_i = y_i / K_i = 1000/308.26$$

$$x_i = 3.24 \text{ ppm (mol)}$$

Portions of this material appeared in Chem. Eng., 98, 179 (Nov., 1991). These portions are reprinted by special permission.

REFERENCES – ORGANIC COMPOUNDS

1-190. See **REFERENCES – WATER SOLUBILITY - ORGANIC COMPOUNDS** in Chapter 15, **SOLUBILITY IN WATER AND OCTANOL-WATER PARTITION COEFFICIENT**

191. Baker, R. J., B. J. Donelan, L. J. Peterson, W. E. Acree, Jr., and C. C. Tsai, Phys. Chem. Liq., 16, 279 (1987).

192. Acree, W. E., Jr., THERMODYNAMIC PROPERTIES OF NONELECTROLYTE SOLUTIONS, Academic Press, New York, NY (1984).

193. Byers, W. D. and C. M. Morton, "Removing VOC from Groundwater", 1984 Summer National Meeting, A.I.Ch.E., Phil., PA (August, 1984).

194. Roberts, P. V., G. D. Hopkins, C. Munz, and A. H. Riojas, Environ. Sci. Technol., 19, 164 (1985).

195. Roberts, P. V. and J. A. Levy, Journal AWWA, 138 (April, 1985).

196. Groves, F. R., Jr. and R. Doshi, "Prediction of Solubility of Organics in Water via Activity Coefficients", LSU, Chem. Eng. Dept., Baton Rouge, LA (1986).

197. Burkhard, L. P., D. E. Armstrong, and A. W. Andren, Environ. Sci. Technol., 19, 590 (1985).

198. Murphy, T. J., M. D. Mullin, and J. A. Meyer, Environ. Sci. Technol., 21, 155 (1987).

198. Preston, G. T. and J. M. Prausnitz, Ind. Eng. Chem. Fundam., 10, 389 (1971).

199. Dilling, W. L., Environ. Sci. Technol., 11, 405 (1977).

200. TREATABILITY MANUAL, Vol. I, Treatability Data, EPA-600/2-82-001a, Office of Research and Development, U.S. Environmental Protection Agency, Washington, D.C. (Sept., 1981).

201. Ashworth, R. A., G. B. Howe, M. E. Mullins, and T. N. Rogers, "Air-Water Partitioning Coefficients of Organics in Dilute Aqueous Solutions", 1986 Summer National Meeting, AIChE, Boston, Mass. (August, 1986).

202. Ashworth, R. A., G. B. Howe, M. E. Mullins, and T. N. Rogers, J. Hazardous Materials, 18, 25 (1988).

203. Warner, H. P., J. M. Cohen and J. C. Ireland, DETERMINATION OF HENRY'S LAW CONSTANTS OF SELECTED PRIORITY POLLUTANTS, EPA/600/D-87/229 (July, 1987).

204. Leighton, D. T., Jr. and J. M. Calo, J. Chem. Eng. Data, 26, 382 (1981).

205. Gossett, J. M., ANAEROBIC DEGRADATION OF C1 AND C2 CHLORINATED HYDROCARBONS, final report, U.S. Air Force, ESL-TR-85-38, Tyndall AFB, FL (Dec., 1985).

206. Gossett, J. M., Environ. Sci. Technol., 21, 202 (1987).

207. Hansen, K. C., Z. Zhou, C. L. Yaws, and T. M. Aminabhavi, J. Chem. Eng. Data, 38(4), 546 (1993).

208. Hwang, Y. L., G. E. Keller, II, and J. D. Olson, Ind. Eng. Chem. Res., 31(7), 1753 (1992).

209. Hwang, Y. L., G. E. Keller, II, and J. D. Olson, Ind. Eng. Chem. Res., 31(7), 1759 (1992).

210. Yaws, C. L. and others, Chem. Eng., 98, 179 (Nov., 1991).

Figure 18-1 Henry's Law Constant for Organic Chemicals in Water

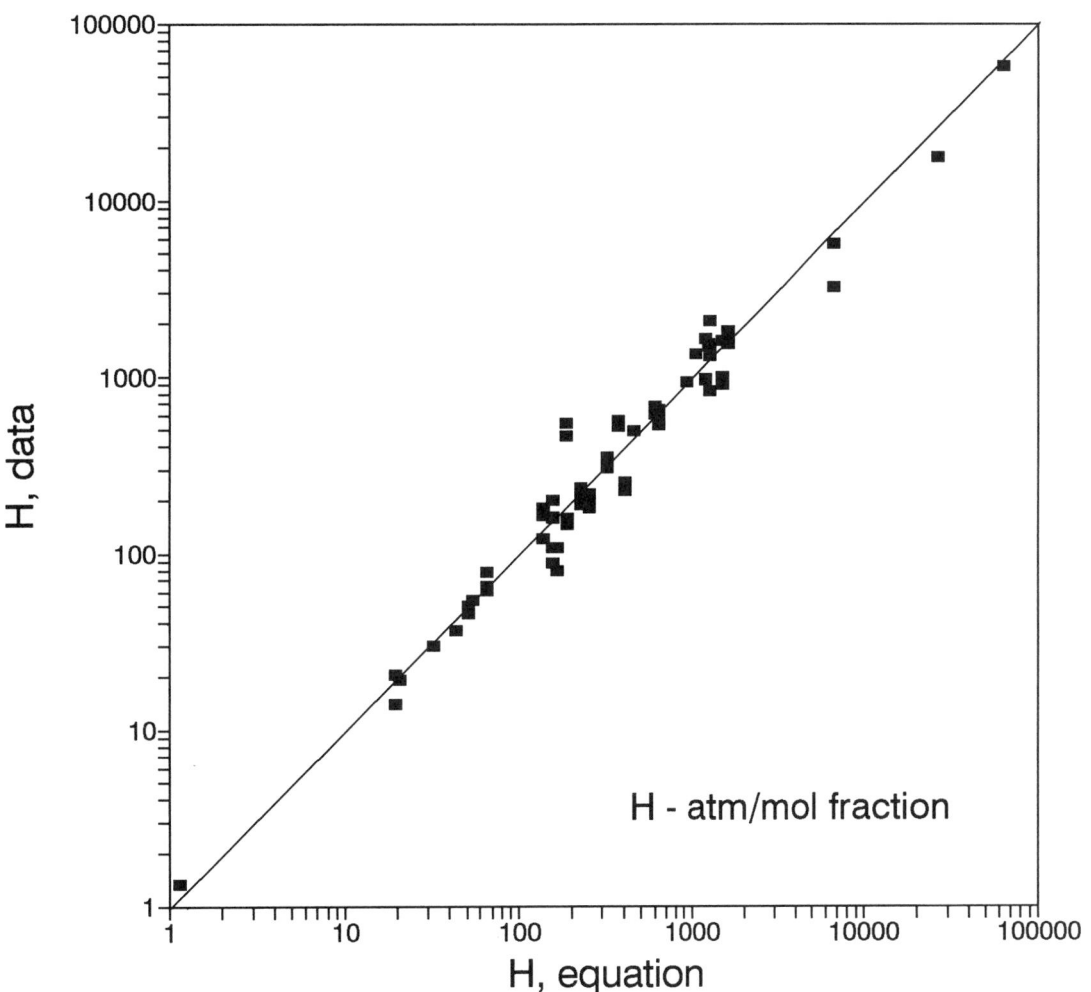

H - atm/mol fraction

Figure 18-2 Henry's Law Constant for Hydrocarbons in Water

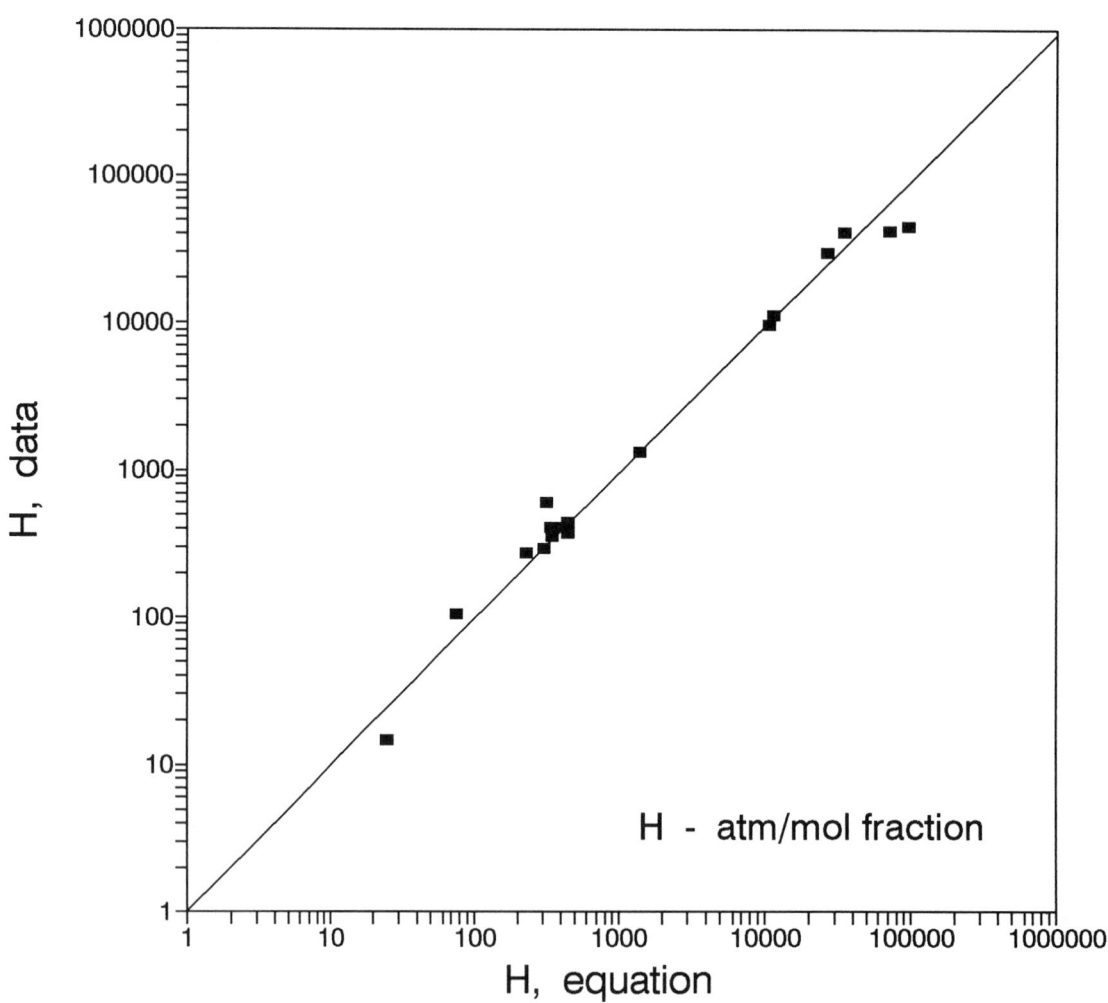

Table 18-1 HENRY'S LAW CONSTANT FOR COMPOUND IN WATER - ORGANIC COMPOUNDS

			Henry's Law Constant, H					
NO	FORMULA	NAME	MW, g/mol	T_F, K	T_B, K	T, C	H @ T, atm/mol frac	H @ T, atm/mol/m³
1	CBrClF2	BROMOCHLORODIFLUOROMETHANE	165.365	113.65	269.14	2.0	9.1899E+03	1.6542E-01
2	CBrCl3	BROMOTRICHLOROMETHANE	198.273	252.15	378.05	------	----------	----------
3	CBrF3	BROMOTRIFLUOROMETHANE	148.910	105.15	215.26	25.0	2.6684E+04	4.8031E-01
4	CBr2F2	DIBROMODIFLUOROMETHANE	209.816	163.05	295.94	------	----------	----------
5	CClF3	CHLOROTRIFLUOROMETHANE	104.459	92.15	191.74	25.0	6.2407E+04	1.1233E+00
6	CClN	CYANOGEN CHLORIDE	61.470	266.65	286.00	25.0	1.0784E+02	1.9412E-03
7	CCl2F2	DICHLORODIFLUOROMETHANE	120.913	115.15	243.36	25.0	2.1667E+04	3.9001E-01
8	CCl2O	PHOSGENE	98.916	145.37	280.71	25.0	7.7493E+02	1.3949E-02
9	CCl3F	TRICHLOROFLUOROMETHANE	137.368	162.04	296.97	30.0	6.7582E+03	1.2165E-01
10	CCl4	CARBON TETRACHLORIDE	153.822	250.33	349.79	25.0	1.6299E+03	2.9338E-02
11	CF2O	CARBONYL FLUORIDE	66.007	161.89	188.58	------	----------	----------
12	CF4	CARBON TETRAFLUORIDE	88.005	89.56	145.09	25.0	2.9576E+05	5.3236E+00
13	CHBr3	TRIBROMOMETHANE	252.731	281.20	422.35	25.0	3.2557E+01	5.8601E-04
14	CHClF2	CHLORODIFLUOROMETHANE	86.468	115.73	232.32	25.0	1.6749E+03	3.0148E-02
15	CHCl2F	DICHLOROFLUOROMETHANE	102.923	138.15	282.05	25.0	2.8982E+02	5.2167E-03
16	CHCl3	CHLOROFORM	119.377	209.63	334.33	25.0	2.2784E+02	4.1011E-03
17	CHF3	TRIFLUOROMETHANE	70.014	117.97	190.99	25.0	4.1804E+03	7.5246E-02
18	CHI3	TRIIODOMETHANE	393.732	396.16	491.16	------	----------	----------
19	CHN	HYDROGEN CYANIDE	27.026	259.91	298.85	25.0	4.9430E+00	8.8973E-05
20	CHNS	ISOTHIOCYANIC-ACID	59.086	------	------	------	----------	----------
21	CH2BrCl	BROMOCHLOROMETHANE	129.384	185.20	341.20	25.0	8.8708E+01	1.5967E-03
22	CH2Br2	DIBROMOMETHANE	173.835	220.60	370.10	25.0	5.1665E+01	9.2995E-04
23	CH2ClF	CHLOROFLUOROMETHANE	68.478	140.16	264.06	25.0	3.4731E+02	6.2516E-03
24	CH2Cl2	DICHLOROMETHANE	84.932	178.01	312.90	25.0	1.3648E+02	2.4567E-03
25	CH2F2	DIFLUOROMETHANE	52.024	137.00	221.50	25.0	6.3570E+02	1.1443E-02
26	CH2I2	DIIODOMETHANE	267.836	279.25	455.15	25.0	1.8963E+01	3.4133E-04
27	CH2O	FORMALDEHYDE	30.026	181.15	254.05	------	----------	----------
28	CH2O2	FORMIC ACID	46.026	281.55	373.71	25.0	4.2568E-02	7.6622E-07
29	CH3Br	METHYL BROMIDE	94.939	179.55	276.71	25.0	3.7656E+02	6.7780E-03
30	CH3Cl	METHYL CHLORIDE	50.488	175.45	248.93	25.0	4.5840E+02	8.2512E-03
31	CH3Cl3Si	METHYL TRICHLOROSILANE	149.478	195.35	339.55	------	----------	----------
32	CH3F	METHYL FLUORIDE	34.033	131.35	194.82	15.0	7.7625E+02	1.3972E-02
33	CH3I	METHYL IODIDE	141.939	206.70	315.58	25.0	1.5740E+02	2.8332E-03
34	CH3NO	FORMAMIDE	45.041	275.70	493.00	------	----------	----------
35	CH3NO2	NITROMETHANE	61.040	244.60	374.35	25.0	1.5907E+01	2.8632E-04
36	CH3NO2	METHYL-NITRITE	61.040	256.16	261.16	------	----------	----------
37	CH3NO3	METHYL-NITRATE	77.040	190.86	339.16	------	----------	----------
38	CH4	METHANE	16.043	90.67	111.66	25.0	3.5356E+04	6.3640E-01
39	CH4Cl2Si	METHYL DICHLOROSILANE	115.034	182.55	314.70	------	----------	----------
40	CH4O	METHANOL	32.042	175.47	337.85	25.0	2.8851E-01	5.1932E-06
41	CH4O3S	METHANESULFONIC ACID	96.107	292.81	561.00	------	----------	----------
42	CH4S	METHYL MERCAPTAN	48.109	150.18	279.11	15.0	1.0775E+02	1.9396E-03
43	CH5ClSi	METHYL CHLOROSILANE	80.589	139.05	281.85	------	----------	----------
44	CH5N	METHYLAMINE	31.057	179.69	266.82	------	----------	----------
45	CH6Si	METHYL SILANE	46.144	116.34	216.25	------	----------	----------
46	CN4O8	TETRANITROMETHANE	196.033	287.05	398.85	------	----------	----------
47	CO	CARBON MONOXIDE	28.010	68.15	81.70	25.0	6.3364E+04	1.1405E+00
48	COS	CARBONYL SULFIDE	60.076	134.35	223.00	25.0	2.8068E+03	5.0523E-02
49	CO2	CARBON DIOXIDE	44.010	216.58	194.67	25.0	1.2122E+03	2.1820E-02
50	CS2	CARBON DISULFIDE	76.143	161.58	319.37	25.0	5.8174E+03	1.0471E-01
51	C2BrF3	BROMOTRIFLUOROETHYLENE	160.921	------	270.65	------	----------	----------
52	C2Br2F4	1,2-DIBROMOTETRAFLUOROETHANE	259.824	162.65	320.41	------	----------	----------
53	C2ClF3	CHLOROTRIFLUOROETHYLENE	116.470	115.00	245.30	------	----------	----------
54	C2ClF5	CHLOROPENTAFLUOROETHANE	154.467	173.71	234.04	25.0	1.4320E+05	2.5776E+00
55	C2Cl2F4	1,2-DICHLOROTETRAFLUOROETHANE	170.921	179.15	276.92	25.0	6.7079E+04	1.2074E+00
56	C2Cl3F3	1,1,2-TRICHLOROTRIFLUOROETHANE	187.375	238.15	320.75	25.0	2.6684E+04	4.8031E-01
57	C2Cl4	TETRACHLOROETHYLENE	165.833	250.80	394.40	25.0	1.4968E+03	2.6942E-02
58	C2Cl4F2	1,1,2,2-TETRACHLORODIFLUOROETHANE	203.830	299.15	366.00	25.0	6.9239E+03	1.2463E-01
59	C2Cl4O	TRICHLOROACETYL CHLORIDE	181.832	------	391.15	------	----------	----------
60	C2Cl6	HEXACHLOROETHANE	236.738	459.95	460.00	25.0	1.4041E+03	2.5274E-02
61	C2F4	TETRAFLUOROETHYLENE	100.016	142.00	197.51	25.0	3.3949E+04	6.1107E-01
62	C2F6	HEXAFLUOROETHANE	138.012	172.45	194.95	25.0	9.3976E+05	1.6915E+01
63	C2HBrClF3	HALOTHANE	197.382	------	323.35	24.0	1.1676E+03	2.1017E-02
64	C2HClF2	2-CHLORO-1,1-DIFLUOROETHYLENE	98.479	134.65	254.55	------	----------	----------
65	C2HCl3	TRICHLOROETHYLENE	131.388	188.40	360.10	25.0	6.4250E+02	1.1565E-02
66	C2HCl3O	DICHLOROACETYL CHLORIDE	147.387	------	382.15	------	----------	----------
67	C2HCl3O	TRICHLOROACETALDEHYDE	147.387	216.00	370.85	------	----------	----------
68	C2HCl5	PENTACHLOROETHANE	202.293	244.15	433.03	25.0	1.0840E+02	1.9511E-03
69	C2HF3	TRIFLUOROETHENE	82.025	94.53	221.01	------	----------	----------
70	C2HF3O2	TRIFLUOROACETIC ACID	114.024	257.90	344.95	------	----------	----------
71	C2HF5	PENTAFLUOROETHANE	120.022	170.15	225.15	------	----------	----------
72	C2H2	ACETYLENE	26.038	192.40	189.00	25.0	1.3997E+03	2.5195E-02
73	C2H2Br4	1,1,2,2-TETRABROMOETHANE	345.654	273.15	516.65	25.0	7.5203E-01	1.3536E-05
74	C2H2Cl2	1,1-DICHLOROETHYLENE	96.943	150.65	304.71	25.0	1.2639E+03	2.2750E-02
75	C2H2Cl2	cis-1,2-DICHLOROETHYLENE	96.943	193.15	333.65	25.0	4.0913E+02	7.3644E-03
76	C2H2Cl2	trans-1,2-DICHLOROETHYLENE	96.943	223.35	320.85	25.0	3.7250E+02	6.7049E-03
77	C2H2Cl2O	CHLOROACETYL CHLORIDE	112.943	251.15	379.15	------	----------	----------
78	C2H2Cl2O	DICHLOROACETALDEHYDE	112.943	223.00	362.00	------	----------	----------

NO	FORMULA	NAME	MW, g/mol	T_F, K	T_B, K	T, C	Henry's Law Constant, H	
							H @ T, atm/mol frac	H @ T, atm/mol/m³
79	C2H2Cl2O2	DICHLOROACETIC ACID	128.942	286.55	467.15	------	----------	----------
80	C2H2Cl3F	1,1,1-TRICHLOROFLUOROETHANE	151.394	-------	366.00	------	----------	----------
81	C2H2Cl4	1,1,1,2-TETRACHLOROETHANE	167.849	202.94	403.65	25.0	1.3422E+02	2.4159E-03
82	C2H2Cl4	1,1,2,2-TETRACHLOROETHANE	167.849	229.35	418.25	25.0	1.9461E+01	3.5030E-04
83	C2H2F2	1,1-DIFLUOROETHYLENE	64.035	129.15	187.50	25.0	2.0879E+04	3.7582E-01
84	C2H2F2	cis-1,2-DIFLUOROETHENE	64.035	107.90	247.86	------	----------	----------
85	C2H2F2	trans-1,2-DIFLUOROETHENE	64.035	107.90	247.86	------	----------	----------
86	C2H2F4	1,1,1,2-TETRAFLUOROETHANE	102.031	172.15	247.15	------	----------	----------
87	C2H2O	KETENE	42.037	122.00	223.34	------	----------	----------
88	C2H2O4	OXALIC ACID	90.036	462.65	569.00	------	----------	----------
89	C2H3Br	VINYL BROMIDE	106.950	135.35	288.95	------	----------	----------
90	C2H3Cl	VINYL CHLORIDE	62.499	119.36	259.78	25.0	1.2437E+03	2.2387E-02
91	C2H3ClF2	1-CHLORO-1,1-DIFLUOROETHANE	100.495	142.35	263.14	------	----------	----------
92	C2H3ClO	ACETYL CHLORIDE	78.498	160.30	323.90	------	----------	----------
93	C2H3ClO	CHLOROACETALDEHYDE	78.498	-------	358.00	------	----------	----------
94	C2H3ClO2	CHLOROACETIC ACID	94.497	333.15	462.50	------	----------	----------
95	C2H3ClO2	METHYL CHLOROFORMATE	94.497	-------	344.00	------	----------	----------
96	C2H3Cl3	1,1,1-TRICHLOROETHANE	133.404	242.75	347.23	25.0	1.2041E+03	2.1674E-02
97	C2H3Cl3	1,1,2-TRICHLOROETHANE	133.404	236.50	387.00	25.0	5.1350E+01	9.2429E-04
98	C2H3F	VINYL FLUORIDE	46.044	112.65	200.95	------	----------	----------
99	C2H3F3	1,1,1-TRIFLUOROETHANE	84.041	161.85	225.75	------	----------	----------
100	C2H3N	ACETONITRILE	41.053	229.32	354.75	25.0	1.1077E+00	1.9938E-05
101	C2H3NO	METHYL ISOCYANATE	57.052	256.15	312.00	------	----------	----------
102	C2H4	ETHYLENE	28.054	104.01	169.47	25.0	1.1515E+04	2.0727E-01
103	C2H4Br2	1,1-DIBROMOETHANE	187.862	210.15	381.15	------	----------	----------
104	C2H4Br2	1,2-DIBROMOETHANE	187.862	282.94	404.51	25.0	4.3697E+01	7.8654E-04
105	C2H4Cl2	1,1-DICHLOROETHANE	98.959	176.19	330.45	25.0	3.2504E+02	5.8508E-03
106	C2H4Cl2	1,2-DICHLOROETHANE	98.959	237.49	356.59	25.0	6.5415E+01	1.1775E-03
107	C2H4Cl2O	BIS(CHLOROMETHYL)ETHER	114.959	231.65	378.00	------	----------	----------
108	C2H4F2	1,1-DIFLUOROETHANE	66.051	156.15	247.35	27.5	1.4108E+03	2.5395E-02
109	C2H4F2	1,2-DIFLUOROETHANE	66.051	-------	303.65	------	----------	----------
110	C2H4I2	1,2-DIIODOETHANE	281.863	356.16	473.16	------	----------	----------
111	C2H4O	ACETALDEHYDE	44.053	150.15	293.55	25.0	5.3431E+00	9.6174E-05
112	C2H4O	ETHYLENE OXIDE	44.053	161.45	283.85	25.0	1.3228E+01	2.3811E-04
113	C2H4OS	THIOACETIC-ACID	76.113	150.16	360.16	------	----------	----------
114	C2H4O2	ACETIC ACID	60.053	289.81	391.05	25.0	6.5146E-02	1.1726E-06
115	C2H4O2	METHYL FORMATE	60.053	174.15	304.90	20.0	1.0611E+02	1.9099E-03
116	C2H4S	THIACYCLOPROPANE	60.114	165.37	328.07	------	----------	----------
117	C2H5Br	BROMOETHANE	108.966	154.55	311.50	25.0	4.1031E+02	7.3855E-03
118	C2H5Cl	ETHYL CHLORIDE	64.514	136.75	285.42	20.0	3.8400E+02	6.9120E-03
119	C2H5ClO	2-CHLOROETHANOL	80.514	205.65	401.75	------	----------	----------
120	C2H5F	ETHYL FLUORIDE	48.060	129.95	235.45	25.0	1.1959E+03	2.1527E-02
121	C2H5I	ETHYL IODIDE	155.966	162.05	345.45	20.0	3.9596E+02	7.1273E-03
122	C2H5N	ETHYLENEIMINE	43.068	195.20	329.00	------	----------	----------
123	C2H5NO	ACETAMIDE	59.068	354.15	494.30	------	----------	----------
124	C2H5NO	N-METHYLFORMAMIDE	59.068	269.35	472.66	------	----------	----------
125	C2H5NO2	NITROETHANE	75.067	183.63	387.22	25.0	2.1892E+00	3.9405E-05
126	C2H5NO3	ETHYL-NITRATE	91.066	178.56	360.36	------	----------	----------
127	C2H6	ETHANE	30.070	90.35	184.55	25.0	2.6770E+04	4.8186E-01
128	C2H6AlCl	DIMETHYLALUMINUM CHLORIDE	92.054	252.15	399.15	------	----------	----------
129	C2H6O	DIMETHYL ETHER	46.069	131.66	248.31	18.0	3.6909E+01	6.6435E-04
130	C2H6O	ETHANOL	46.069	159.05	351.44	25.0	4.5047E-01	8.1083E-06
131	C2H6OS	DIMETHYL SULFOXIDE	78.135	291.67	462.15	------	----------	----------
132	C2H6O2	ETHYLENE GLYCOL	62.068	260.15	470.45	25.0	1.0506E-04	1.8911E-09
133	C2H6O4S	DIMETHYL SULFATE	126.133	241.35	461.95	18.0	2.1715E-01	3.9088E-06
134	C2H6S	DIMETHYL SULFIDE	62.136	174.88	310.48	25.0	1.1194E+02	2.0148E-03
135	C2H6S	ETHYL MERCAPTAN	62.136	125.26	308.15	25.0	1.6077E+02	2.8938E-03
136	C2H6S2	DIMETHYL DISULFIDE	94.202	188.44	382.90	------	----------	----------
137	C2H7N	DIMETHYLAMINE	45.084	180.96	280.03	------	----------	----------
138	C2H7N	ETHYLAMINE	45.084	192.15	289.73	20.0	1.4442E+00	2.5995E-05
139	C2H7NO	MONOETHANOLAMINE	61.084	283.65	444.15	------	----------	----------
140	C2H8N2	ETHYLENEDIAMINE	60.099	284.29	390.41	20.0	2.4318E-03	4.3772E-08
141	C2H8Si	DIMETHYL SILANE	60.171	122.93	253.55	------	----------	----------
142	C2N2	CYANOGEN	52.036	245.25	252.00	20.0	2.9559E+02	5.3206E-03
143	C3F6	HEXAFLUOROPROPYLENE	150.023	116.65	243.55	------	----------	----------
144	C3F6O	HEXAFLUOROACETONE	166.023	151.15	245.88	------	----------	----------
145	C3F8	OCTAFLUOROPROPANE	188.020	125.46	236.40	15.0	1.7555E+06	3.1599E+01
146	C3H2N2	MALONONITRILE	66.062	304.90	491.50	20.0	8.7621E-03	1.5772E-07
147	C3H3Cl	PROPARGYL CHLORIDE	74.510	-------	331.00	------	----------	----------
148	C3H3N	ACRYLONITRILE	53.064	189.63	350.50	20.0	5.4434E+00	9.7980E-05
149	C3H3NO	OXAZOLE	69.063	-------	342.65	------	----------	----------
150	C3H4	METHYLACETYLENE	40.065	170.45	249.94	25.0	5.9069E+02	1.0632E-02
151	C3H4	PROPADIENE	40.065	136.87	238.65	------	----------	----------
152	C3H4Cl2	2,3-DICHLOROPROPENE	110.970	191.50	365.75	25.0	2.3060E+02	4.1507E-03
153	C3H4O	ACROLEIN	56.064	185.45	325.84	20.0	4.5570E+00	8.2025E-05
154	C3H4O	PROPARGYL ALCOHOL	56.064	221.35	386.75	------	----------	----------
155	C3H4O2	ACRYLIC ACID	72.064	286.65	414.15	25.0	2.2397E-02	4.0314E-07
156	C3H4O2	beta-PROPIOLACTONE	72.064	239.75	435.15	------	----------	----------

							Henry's Law Constant, H	
NO	FORMULA	NAME	MW, g/mol	T_F, K	T_B, K	T, C	H @ T, atm/mol frac	H @ T, atm/mol/m³
157	C3H4O2	VINYL FORMATE	72.064	-------	320.00	------	----------	---------
158	C3H4O3	ETHYLENE CARBONATE	88.063	309.55	511.15	------	----------	---------
159	C3H4O3	PYRUVIC ACID	88.063	286.75	438.15	------	----------	---------
160	C3H5Br	3-BROMO-1-PROPENE	120.977	153.76	343.16	25.0	3.2603E+02	5.8685E-03
161	C3H5Cl	2-CHLOROPROPENE	76.525	135.75	295.80	------	----------	---------
162	C3H5Cl	3-CHLOROPROPENE	76.525	138.65	318.11	25.0	5.1336E+02	9.2404E-03
163	C3H5ClO	alpha-EPICHLOROHYDRIN	92.525	215.95	389.26	------	----------	---------
164	C3H5ClO2	METHYL CHLOROACETATE	108.524	241.03	402.97	------	----------	---------
165	C3H5ClO2	ETHYL CHLOROFORMATE	108.524	192.00	366.00	------	----------	---------
166	C3H5Cl3	1,2,3-TRICHLOROPROPANE	147.431	258.45	430.00	25.0	2.0823E+01	3.7481E-04
167	C3H5I	3-IODO-1-PROPENE	167.977	173.86	375.16	------	----------	---------
168	C3H5N	PROPIONITRILE	55.079	180.26	370.50	25.0	1.9098E+00	3.4377E-05
169	C3H5NO	ACRYLAMIDE	71.079	357.65	465.75	------	----------	---------
170	C3H5NO	HYDRACRYLONITRILE	71.079	227.15	494.15	------	----------	---------
171	C3H5NO	LACTONITRILE	71.079	233.00	457.00	------	----------	---------
172	C3H5N3O9	NITROGLYCERINE	227.088	286.15	523.00	20.0	5.4644E-03	9.8359E-08
173	C3H6	CYCLOPROPANE	42.081	145.73	240.37	21.1	4.2356E+03	7.6240E-02
174	C3H6	PROPYLENE	42.081	87.90	225.43	25.0	1.1313E+04	2.0363E-01
175	C3H6Br2	1,2-DIBROMOPROPANE	201.888	217.96	413.16	------	----------	---------
176	C3H6Cl2	1,1-DICHLOROPROPANE	112.986	-------	361.25	------	----------	---------
177	C3H6Cl2	1,2-DICHLOROPROPANE	112.986	172.71	369.52	25.0	1.5594E+02	2.8069E-03
178	C3H6Cl2	1,3-DICHLOROPROPANE	112.987	173.65	393.55	25.0	5.4745E+01	9.8540E-04
179	C3H6Cl2	2,2-DICHLOROPROPANE	112.986	239.36	342.46	------	----------	---------
180	C3H6I2	1,2-DIIODOPROPANE	295.889	253.16	500.16	------	----------	---------
181	C3H6O	ACETONE	58.080	178.45	329.44	25.0	3.6505E+00	6.5709E-05
182	C3H6O	ALLYL ALCOHOL	58.080	144.15	370.23	25.0	3.4201E-01	6.1562E-06
183	C3H6O	METHYL VINYL ETHER	58.080	151.15	278.65	------	----------	---------
184	C3H6O	n-PROPIONALDEHYDE	58.080	193.15	321.15	25.0	2.3948E+00	4.3106E-05
185	C3H6O	1,2-PROPYLENE OXIDE	58.080	161.22	307.05	20.0	4.0240E+00	7.2432E-05
186	C3H6O	1,3-PROPYLENE OXIDE	58.080	-------	321.00	------	----------	---------
187	C3H6O2	ETHYL FORMATE	74.079	193.55	327.46	18.0	1.2227E+01	2.2009E-04
188	C3H6O2	METHYL ACETATE	74.079	175.15	330.09	20.0	3.8824E+00	6.9883E-05
189	C3H6O2	PROPIONIC ACID	74.079	252.45	414.32	25.0	5.1401E-02	9.2521E-07
190	C3H6O2S	3-MERCAPTOPROPIONIC ACID	106.145	290.65	501.00	------	----------	---------
191	C3H6O3	LACTIC ACID	90.079	291.15	447.00	------	----------	---------
192	C3H6O3	METHOXYACETIC ACID	90.079	281.00	478.26	------	----------	---------
193	C3H6O3	TRIOXANE	90.079	334.65	387.65	------	----------	---------
194	C3H6S	THIACYCLOBUTANE	74.140	199.96	368.13	------	----------	---------
195	C3H7Br	1-BROMOPROPANE	122.993	163.15	344.15	20.0	5.0854E+02	9.1536E-03
196	C3H7Br	2-BROMOPROPANE	122.993	184.15	332.56	20.0	6.7498E+02	1.2150E-02
197	C3H7Cl	ISOPROPYL CHLORIDE	78.541	155.97	308.85	20.0	9.7233E+02	1.7502E-02
198	C3H7Cl	n-PROPYL CHLORIDE	78.541	150.35	319.67	20.0	7.2989E+02	1.3138E-02
199	C3H7F	1-FLUOROPROPANE	62.087	114.16	269.95	14.0	8.7726E+02	1.5791E-02
200	C3H7F	2-FLUOROPROPANE	62.087	139.80	263.81	15.0	9.2262E+02	1.6607E-02
201	C3H7I	ISOPROPYL IODIDE	169.993	183.15	362.65	20.0	6.3069E+02	1.1352E-02
202	C3H7I	n-PROPYL IODIDE	169.993	171.85	375.60	23.5	5.0022E+02	9.0038E-03
203	C3H7N	ALLYAMINE	57.095	184.95	326.45	------	----------	---------
204	C3H7N	PROPYLENEIMINE	57.095	229.00	334.00	------	----------	---------
205	C3H7NO	N,N-DIMETHYLFORMAMIDE	73.095	212.72	426.15	------	----------	---------
206	C3H7NO	N-METHYLACETAMIDE	73.095	301.15	478.15	------	----------	---------
207	C3H7NO2	1-NITROPROPANE	89.094	169.16	404.33	20.0	4.7202E+00	8.4962E-05
208	C3H7NO2	2-NITROPROPANE	89.094	181.83	393.40	20.0	6.6466E+00	1.1964E-04
209	C3H7NO3	PROPYL-NITRATE	105.093	173.16	383.16	------	----------	---------
210	C3H7NO3	ISOPROPYL-NITRATE	105.093	173.16	373.66	------	----------	---------
211	C3H8	PROPANE	44.096	85.46	231.11	25.0	3.7998E+04	6.8396E-01
212	C3H8O	ISOPROPANOL	60.096	185.28	355.41	25.0	6.8936E-01	1.2408E-05
213	C3H8O	METHYL ETHYL ETHER	60.096	160.00	280.50	25.0	3.0964E+01	5.5734E-04
214	C3H8O	n-PROPANOL	60.096	146.95	370.35	25.0	4.3467E-01	7.8239E-06
215	C3H8O2	2-METHOXYETHANOL	76.095	188.05	397.55	------	----------	---------
216	C3H8O2	METHYLAL	76.095	168.35	315.00	20.0	7.9384E+00	1.4289E-04
217	C3H8O2	1,2-PROPYLENE GLYCOL	76.095	213.15	460.75	------	----------	---------
218	C3H8O2	1,3-PROPYLENE GLYCOL	76.095	246.45	487.55	------	----------	---------
219	C3H8O3	GLYCEROL	92.095	291.33	563.15	------	----------	---------
220	C3H8S	n-PROPYLMERCAPTAN	76.163	159.95	340.87	25.0	2.5127E+02	4.5228E-03
221	C3H8S	ISOPROPYL MERCAPTAN	76.163	142.61	325.71	25.0	2.3174E+02	4.1714E-03
222	C3H8S	ETHYL-METHYL-SULFIDE	76.156	167.20	340.15	25.0	1.0986E+02	1.9775E-03
223	C3H9N	n-PROPYLAMINE	59.111	190.15	321.65	25.0	1.4208E+00	2.5575E-05
224	C3H9N	ISOPROPYLAMINE	59.111	177.95	305.55	20.0	4.2433E+00	7.6379E-05
225	C3H9N	TRIMETHYLAMINE	59.111	156.08	276.02	------	----------	---------
226	C3H9NO	1-AMINO-2-PROPANOL	75.111	274.89	432.61	------	----------	---------
227	C3H9NO	3-AMINO-1-PROPANOL	75.111	284.15	460.65	------	----------	---------
228	C3H9NO	METHYLETHANOLAMINE	75.111	268.65	431.15	------	----------	---------
229	C3H9O4P	TRIMETHYL PHOSPHATE	140.076	227.00	465.85	25.0	1.0495E-02	1.8890E-07
230	C3H10N2	1,2-PROPANEDIAMINE	74.126	236.53	392.45	------	----------	---------
231	C3H10Si	TRIMETHYL SILANE	74.198	137.26	279.85	------	----------	---------
232	C4Cl4S	TETRACHLOROTHIOPHENE	221.921	301.97	506.54	------	----------	---------
233	C4Cl6	HEXACHLORO-1,3-BUTADIENE	260.760	252.15	488.15	25.0	1.3043E+03	2.3477E-02
234	C4F8	OCTAFLUORO-2-BUTENE	200.031	138.15	270.36	------	----------	---------

							Henry's Law Constant, H	
NO	FORMULA	NAME	MW, g/mol	T_F, K	T_B, K	T, C	H @ T, atm/mol frac	H @ T, atm/mol/m³
235	C4F8	OCTAFLUOROCYCLOBUTANE	200.031	232.96	267.17	26.0	2.1469E+05	3.8644E+00
236	C4F10	DECAFLUOROBUTANE	238.028	144.95	271.15	------	----------	----------
237	C4H2	BUTADIYNE(BIACETYLENE)	50.060	237.16	283.46	25.0	2.8310E+02	5.0958E-03
238	C4H2O3	MALEIC ANHYDRIDE	98.058	326.00	475.15	------	----------	----------
239	C4H4	VINYLACETYLENE	52.076	-------	278.25	30.0	1.5489E+03	2.7880E-02
240	C4H4N2	SUCCINONITRILE	80.089	331.30	540.15	------	----------	----------
241	C4H4O	FURAN	68.075	187.55	304.50	25.0	2.9901E+02	5.3821E-03
242	C4H4O2	DIKETENE	84.075	266.65	399.20	------	----------	----------
243	C4H8O3	SUCCINIC ANHYDRIDE	100.074	393.00	536.58	------	----------	----------
244	C4H4O4	FUMARIC ACID	116.073	560.15	563.15	------	----------	----------
245	C4H4O4	MALEIC ACID	116.073	403.45	565.00	------	----------	----------
246	C4H4S	THIOPHENE	84.142	234.94	357.31	25.0	1.5990E+02	2.8782E-03
247	C4H5Cl	CHLOROPRENE	88.536	143.15	332.55	------	----------	----------
248	C4H5N	trans-CROTONITRILE	67.090	222.00	394.38	------	----------	----------
249	C4H5N	cis-CROTONITRILE	67.090	200.55	380.60	------	----------	----------
250	C4H5N	METHACRYLONITRILE	67.090	237.35	363.45	------	----------	----------
251	C4H5N	PYRROLE	67.090	249.74	403.00	------	----------	----------
252	C4H5N	VINYLACETONITRILE	67.090	186.15	391.67	------	----------	----------
253	C4H5NO2	METHYL CYANOACETATE	99.089	260.08	478.24	------	----------	----------
254	C4H6	CYCLOBUTENE	54.091	153.76	275.75	------	----------	----------
255	C4H6	1,2-BUTADIENE	54.092	136.95	284.00	25.0	4.4566E+03	8.0219E-02
256	C4H6	1,3-BUTADIENE	54.092	164.25	268.74	25.0	3.9555E+03	7.1198E-02
257	C4H6	DIMETHYLACETYLENE	54.092	240.91	300.13	25.0	1.1789E+03	2.1220E-02
258	C4H6	ETHYLACETYLENE	54.092	147.43	281.22	25.0	1.0115E+03	1.8208E-02
259	C4H6Cl2	1,3-DICHLORO-trans-2-BUTENE	124.997	-------	402.00	------	----------	----------
260	C4H6Cl2	1,4-DICHLORO-cis-2-BUTENE	124.997	225.15	425.65	------	----------	----------
261	C4H6Cl2	1,4-DICHLORO-trans-2-BUTENE	124.997	274.15	429.26	------	----------	----------
262	C4H6Cl2	3,4-DICHLORO-1-BUTENE	124.997	212.00	388.00	------	----------	----------
263	C4H6O	trans-CROTONALDEHYDE	70.091	196.65	377.25	20.0	9.3355E-01	1.6804E-05
264	C4H6O	2,5-DIHYDROFURAN	70.091	-------	339.00	------	----------	----------
265	C4H6O	DIVINYL ETHER	70.091	172.05	301.45	------	----------	----------
266	C4H6O	METHACROLEIN	70.091	192.15	341.15	25.0	1.1533E+01	2.0759E-04
267	C4H6O2	2-BUTYNE-1,4-DIOL	86.090	331.00	511.15	------	----------	----------
268	C4H6O2	gamma-BUTYROLACTONE	86.090	229.78	477.15	------	----------	----------
269	C4H6O2	cis-CROTONIC ACID	86.090	288.65	445.05	------	----------	----------
270	C4H6O2	trans-CROTONIC ACID	86.090	344.55	458.15	------	----------	----------
271	C4H6O2	METHACRYLIC ACID	86.090	288.15	434.15	------	----------	----------
272	C4H6O2	METHYL ACRYLATE	86.090	196.32	353.35	23.0	1.0452E+01	1.8813E-04
273	C4H6O2	VINYL ACETATE	86.090	180.35	345.65	25.0	2.7631E+01	4.9735E-04
274	C4H6O3	ACETIC ANHYDRIDE	102.090	200.15	411.78	25.0	2.3906E-01	4.3030E-06
275	C4H6O4	SUCCINIC ACID	118.089	461.15	591.00	------	----------	----------
276	C4H6O5	DIGLYCOLIC ACID	134.089	421.15	610.00	------	----------	----------
277	C4H6O5	MALIC ACID	134.089	403.15	602.00	------	----------	----------
278	C4H6O6	TARTARIC ACID	150.088	479.15	660.00	------	----------	----------
279	C4H7N	n-BUTYRONITRILE	69.106	161.25	390.75	25.0	3.0192E+00	5.4346E-05
280	C4H7N	ISOBUTYRONITRILE	69.106	201.70	376.76	------	----------	----------
281	C4H7NO	ACETONE CYANOHYDRIN	85.106	253.15	463.00	------	----------	----------
282	C4H7NO	2-METHACRYLAMIDE	85.106	383.65	488.00	------	----------	----------
283	C4H7NO	3-METHOXYPROPIONITRILE	85.106	210.12	439.00	------	----------	----------
284	C4H7NO	2-PYRROLIDONE	85.106	298.15	518.15	------	----------	----------
285	C4H8	1-BUTENE	56.107	87.80	266.90	25.0	1.3588E+04	2.4459E-01
286	C4H8	cis-2-BUTENE	56.107	134.26	276.87	25.0	1.3222E+04	2.3800E-01
287	C4H8	trans-2-BUTENE	56.107	167.62	274.03	25.0	1.2834E+04	2.3101E-01
288	C4H8	CYCLOBUTANE	56.107	182.48	285.66	------	----------	----------
289	C4H8	ISOBUTENE	56.107	132.81	266.25	25.0	1.1470E+04	2.0645E-01
290	C4H8Br2	1,2-DIBROMOBUTANE	215.915	207.76	439.46	------	----------	----------
291	C4H8Br2	2,3-DIBROMOBUTANE	215.915	238.66	434.16	------	----------	----------
292	C4H8Cl2	1,4-DICHLOROBUTANE	127.013	235.85	427.05	------	----------	----------
293	C4H8I2	1,2-DIIODOBUTANE	309.916	279.06	476.76	------	----------	----------
294	C4H8O	n-BUTYRALDEHYDE	72.107	176.75	347.95	25.0	6.5655E+00	1.1818E-04
295	C4H8O	ISOBUTYRALDEHYDE	72.107	208.15	337.25	25.0	9.5687E+00	1.7224E-04
296	C4H8O	1,2-EPOXYBUTANE	72.107	123.15	336.57	------	----------	----------
297	C4H8O	METHYL ETHYL KETONE	72.107	186.48	352.79	25.0	1.6483E+00	2.9669E-05
298	C4H8O	ETHYL VINYL ETHER	72.107	157.35	308.70	------	----------	----------
299	C4H8O	TETRAHYDROFURAN	72.107	164.65	338.00	20.0	2.9242E+00	5.2635E-05
300	C4H8O2	cis-2-BUTENE-1,4-DIOL	88.106	284.15	508.15	------	----------	----------
301	C4H8O2	trans-2-BUTENE-1,4-DIOL	88.106	300.45	510.00	------	----------	----------
302	C4H8O2	ISOBUTYRIC ACID	88.106	227.15	427.85	20.0	4.9107E-02	8.8391E-07
303	C4H8O2	n-BUTYRIC ACID	88.106	267.95	436.42	------	----------	----------
304	C4H8O2	1,4-DIOXANE	88.106	284.95	374.47	25.0	3.0799E-01	5.5438E-06
305	C4H8O2	ETHYL ACETATE	88.106	189.60	350.21	25.0	7.6708E+00	1.3807E-04
306	C4H8O2	METHYL PROPIONATE	88.106	185.65	352.60	25.0	1.0540E+01	1.8971E-04
307	C4H8O2	n-PROPYL FORMATE	88.106	180.25	353.97	22.0	2.3715E+01	4.2686E-04
308	C4H8O2S	SULFOLANE	120.172	300.75	558.15	------	----------	----------
309	C4H8S	TETRAHYDROTHIOPHENE	88.173	176.99	394.27	------	----------	----------
310	C4H9Br	1-BROMOBUTANE	137.019	160.75	374.75	25.0	6.8085E+02	1.2255E-02
311	C4H9Br	2-BROMOBUTANE	137.019	161.25	364.37	------	----------	----------
312	C4H9Cl	n-BUTYL CHLORIDE	92.568	150.05	351.58	25.0	9.2757E+02	1.6696E-02

NO	FORMULA	NAME	MW, g/mol	T_F, K	T_B, K	T, C	H @ T, atm/mol frac	H @ T, atm/mol/m³
						Henry's Law Constant, H		
313	C4H9Cl	sec-BUTYL CHLORIDE	92.568	141.85	341.25	25.0	1.0608E+03	1.9094E-02
314	C4H9Cl	tert-BUTYL CHLORIDE	92.568	247.75	323.75	14.9	7.0756E+02	1.2736E-02
315	C4H9I	2-IODO-2-METHYLPROPANE	184.020	234.96	373.16	------	----------	----------
316	C4H9N	PYRROLIDINE	71.122	215.31	359.72	------	----------	----------
317	C4H9NO	N,N-DIMETHYLACETAMIDE	87.122	253.15	439.25	------	----------	----------
318	C4H9NO	MORPHOLINE	87.122	270.05	401.15	20.0	1.3067E-02	2.3520E-07
319	C4H9NO2	1-NITROBUTANE	103.121	191.83	426.05	------	----------	----------
320	C4H9NO2	2-NITROBUTANE	103.121	141.16	412.85	------	----------	----------
321	C4H10	n-BUTANE	58.123	134.86	272.65	25.0	5.0901E+04	9.1621E-01
322	C4H10	ISOBUTANE	58.123	113.54	261.43	25.0	6.3913E+04	1.1504E+00
323	C4H10N2	PIPERAZINE	86.137	379.15	419.15	------	----------	----------
324	C4H10O	n-BUTANOL	74.123	183.85	390.81	25.0	4.8271E-01	8.6887E-06
325	C4H10O	sec-BUTANOL	74.123	158.45	372.70	25.0	4.6543E-01	8.3776E-06
326	C4H10O	tert-BUTANOL	74.123	298.97	355.57	25.0	9.5996E-01	1.7279E-05
327	C4H10O	DIETHYL ETHER	74.123	156.85	307.58	25.0	4.5412E+01	8.1741E-04
328	C4H10O	METHYL-PROPYL-ETHER	74.122	156.87	311.72	25.0	7.9097E+01	1.4237E-03
329	C4H10O	METHYL ISOPROPYL ETHER	74.123	127.93	303.92	25.0	4.7819E+01	8.6074E-04
330	C4H10O	ISOBUTANOL	74.123	165.15	380.81	25.0	6.5536E-01	1.1796E-05
331	C4H10O2	1,3-BUTANEDIOL	90.122	196.15	480.15	------	----------	----------
332	C4H10O2	1,4-BUTANEDIOL	90.122	293.05	501.15	20.0	2.8544E-05	5.1378E-10
333	C4H10O2	2,3-BUTANEDIOL	90.122	280.75	453.85	------	----------	----------
334	C4H10O2	t-BUTYL HYDROPEROXIDE	90.122	277.45	405.50	------	----------	----------
335	C4H10O2	1,2-DIMETHOXYETHANE	90.122	215.15	357.20	------	----------	----------
336	C4H10O2	2-ETHOXYETHANOL	90.122	------	408.15	20.0	2.9101E-02	5.2382E-07
337	C4H10O3	DIETHYLENE GLYCOL	106.122	262.70	518.15	25.0	1.6681E-05	3.0026E-10
338	C4H10O4S	DIETHYL SULFATE	154.187	248.00	483.00	------	----------	----------
339	C4H10S	n-BUTYL MERCAPTAN	90.189	157.46	371.61	25.0	4.9932E+02	8.9877E-03
340	C4H10S	ISOBUTYL MERCAPTAN	90.189	128.31	361.64	25.0	3.8351E+02	6.9030E-03
341	C4H10S	sec-BUTYL MERCAPTAN	90.189	133.02	358.13	25.0	3.6829E+02	6.6291E-03
342	C4H10S	tert-BUTYL MERCAPTAN	90.189	274.26	337.37	25.0	2.9765E+02	5.3576E-03
343	C4H10S	DIETHYL SULFIDE	90.189	169.20	365.25	20.0	1.2336E+02	2.2204E-03
344	C4H10S	ISOPROPYL-METHYL-SULFIDE	90.183	171.65	357.90	25.0	1.5876E+02	2.8577E-03
345	C4H10S	METHYL-PROPYL-SULFIDE	90.183	160.19	368.71	25.0	1.7893E+02	3.2207E-03
346	C4H10S2	DIETHYL DISULFIDE	122.255	171.63	427.13	------	----------	----------
347	C4H11N	n-BUTYLAMINE	73.138	224.05	350.55	25.0	2.4822E+00	4.4679E-05
348	C4H11N	ISOBUTYLAMINE	73.138	188.55	340.88	20.0	1.2336E+00	2.2205E-05
349	C4H11N	sec-BUTYLAMINE	73.138	168.65	336.15	------	----------	----------
350	C4H11N	tert-BUTYLAMINE	73.138	206.19	317.55	------	----------	----------
351	C4H11N	DIETHYLAMINE	73.138	223.35	328.60	25.0	3.9376E+00	7.0875E-05
352	C4H11NO	DIMETHYLETHANOLAMINE	89.137	214.15	407.15	------	----------	----------
353	C4H11NO2	DIETHANOLAMINE	105.137	301.15	542.04	------	----------	----------
354	C4H12N2O	N-AMINOETHYL ETHANOLAMINE	105.137	------	514.00	------	----------	----------
355	C4H11NO2	2-AMINOETHOXYETHANOL	104.152	------	517.00	------	----------	----------
356	C4H12Si	TETRAMETHYLSILANE	88.225	174.07	299.80	------	----------	----------
357	C4H13N3	DIETHYLENE TRIAMINE	103.167	234.15	480.25	------	----------	----------
358	C5Cl6	HEXACHLOROCYCLOPENTADIENE	272.771	284.49	512.15	------	----------	----------
359	C5H4O2	FURFURAL	96.086	236.65	434.85	25.0	1.8365E-01	3.3057E-06
360	C5H5N	PYRIDINE	79.101	231.53	388.41	25.0	5.1700E-01	9.3058E-06
361	C5H6	CYCLOPENTADIENE	66.103	188.15	314.65	------	----------	----------
362	C5H6	2-METHYL-1-BUTENE-3-YNE	66.103	160.15	305.40	------	----------	----------
363	C5H6	1-PENTENE-3-YNE	66.103	------	332.40	------	----------	----------
364	C5H6	1-PENTENE-4-YNE	66.103	------	315.65	------	----------	----------
365	C5H6N2	GLUTARONITRILE	94.116	244.21	559.15	------	----------	----------
366	C5H6O2	FURFURYL ALCOHOL	98.101	258.52	443.15	20.0	5.1698E-03	9.3056E-08
367	C5H6O3	GLUTARIC ANHYDRIDE	114.101	328.00	562.69	------	----------	----------
368	C5H6O4	CITRACONIC ACID	130.100	356.15	607.00	------	----------	----------
369	C5H6O4	ITACONIC ACID	130.100	438.15	601.00	------	----------	----------
370	C5H6S	2-METHYLTHIOPHENE	98.162	209.77	385.71	------	----------	----------
371	C5H6S	3-METHYLTHIOPHENE	98.162	204.18	388.60	------	----------	----------
372	C5H7N	N-METHYLPYRROLE	81.117	216.91	385.89	------	----------	----------
373	C5H7NO2	ETHYL CYANOACETATE	113.116	250.65	479.15	25.0	1.5766E-02	2.8378E-07
374	C5H8	CYCLOPENTENE	68.118	138.13	317.38	25.0	3.5368E+03	6.3661E-02
375	C5H8	ISOPRENE	68.118	127.27	307.21	25.0	5.1555E+03	9.2798E-02
376	C5H8	3-METHYL-1,2-BUTADIENE	68.118	159.53	314.00	25.0	4.9978E+03	8.9961E-02
377	C5H8	2-METHYL-1,3-BUTADIENE	68.118	127.20	307.22	25.0	4.2694E+03	7.6848E-02
378	C5H8	1,2-PENTADIENE	68.118	135.89	318.01	25.0	4.9819E+03	8.9674E-02
379	C5H8	cis-1,3-PENTADIENE	68.118	132.35	317.22	25.0	4.9917E+03	8.9849E-02
380	C5H8	trans-1,3-PENTADIENE	68.118	185.71	315.17	25.0	5.0229E+03	9.0411E-02
381	C5H8	1,4-PENTADIENE	68.118	124.86	299.11	25.0	6.5524E+03	1.1794E-01
382	C5H8	2,3-PENTADIENE	68.118	147.50	321.40	25.0	4.9300E+03	8.8738E-02
383	C5H8	1-PENTYNE	68.118	167.45	313.33	25.0	1.3781E+03	2.4806E-02
384	C5H8	2-PENTYNE	68.118	163.86	329.22	25.0	1.3824E+03	2.4883E-02
385	C5H8	3-METHYL-1-BUTYNE	68.118	183.45	302.15	25.0	1.4647E+03	2.6365E-02
386	C5H8	SPIROPENTANE	68.118	166.11	312.19	------	----------	----------
387	C5H8N4O12	PENTAERYTHRITOL TETRANITRATE	316.138	413.65	543.00	------	----------	----------
388	C5H8O	CYCLOPENTANONE	84.118	221.85	403.80	------	----------	----------
389	C5H8O	METHYL ISOPROPENYL KETONE	84.118	219.55	371.15	25.0	5.5695E+00	1.0025E-04
390	C5H8O2	ACETYLACETONE	100.117	249.65	413.55	20.0	2.5342E-01	4.5615E-06

NO	FORMULA	NAME	MW, g/mol	T_F, K	T_B, K	T, C	H @ T, atm/mol frac	H @ T, atm/mol/m³
							Henry's Law Constant, H	
391	C5H8O2	ALLYL ACETATE	100.117	138.00	377.15	------	----------	----------
392	C5H8O2	ETHYL ACRYLATE	100.117	201.95	372.65	25.0	1.3874E+01	2.4973E-04
393	C5H8O2	METHYL METHACRYLATE	100.117	224.95	373.45	------	----------	----------
394	C5H8O2	VINYL PROPIONATE	100.117	-------	364.35	------	----------	----------
395	C5H8O3	2-HYDROXYETHYL ACRYLATE	116.117	213.00	484.00	------	----------	----------
396	C5H8O3	LEVULINIC ACID	116.117	308.15	518.95	------	----------	----------
397	C5H8O3	METHYL ACETOACETATE	116.117	193.15	444.85	------	----------	----------
398	C5H8O4	GLUTARIC ACID	132.116	370.65	595.54	------	----------	----------
399	C5H9N	VALERONITRILE	83.133	176.95	414.45	------	----------	----------
400	C5H9NO	n-BUTYL ISOCYANATE	99.133	-------	388.15	------	----------	----------
401	C5H9NO	N-METHYL-2-PYRROLIDONE	99.133	249.15	475.15	------	----------	----------
402	C5H9NO4	L-GLUTAMIC ACID	147.131	497.15	670.00	------	----------	----------
403	C5H10	CYCLOPENTANE	70.134	179.31	322.40	25.0	1.0430E+04	1.8774E-01
404	C5H10	2-METHYL-1-BUTENE	70.134	135.58	304.30	25.0	1.6150E+04	2.9069E-01
405	C5H10	2-METHYL-2-BUTENE	70.134	139.39	311.71	25.0	1.5668E+04	2.8202E-01
406	C5H10	3-METHYL-1-BUTENE	70.134	104.66	293.21	25.0	2.9007E+04	5.2213E-01
407	C5H10	1-PENTENE	70.134	107.93	303.11	25.0	2.2190E+04	3.9942E-01
408	C5H10	cis-2-PENTENE	70.134	121.75	310.08	25.0	1.2492E+04	2.2486E-01
409	C5H10	trans-2-PENTENE	70.134	132.89	309.49	25.0	1.2762E+04	2.2972E-01
410	C5H10Br2	2,3-DIBROMO-2-METHYLBUTANE	229.942	288.00	444.01	------	----------	----------
411	C5H10Cl2	1,5-DICHLOROPENTANE	141.040	200.35	453.15	------	----------	----------
412	C5H10O	METHYL ISOPROPYL KETONE	86.134	181.15	367.55	25.0	5.1422E+00	9.2560E-05
413	C5H10O	2-PENTANONE	86.134	196.29	375.46	25.0	4.0515E+00	7.2927E-05
414	C5H10O	DIETHYL KETONE	86.134	234.18	375.14	25.0	6.6596E+00	1.1987E-04
415	C5H10O	VALERALDEHYDE	86.134	182.00	376.15	25.0	1.8292E+01	3.2926E-04
416	C5H10O2	n-BUTYL FORMATE	102.133	181.25	379.25	27.0	2.8327E+01	5.0989E-04
417	C5H10O2	ETHYL PROPIONATE	102.133	199.25	372.25	25.0	1.2218E+01	2.1993E-04
418	C5H10O2	ISOBUTYL FORMATE	102.133	177.35	371.22	25.0	1.6511E+01	2.9720E-04
419	C5H10O2	ISOPROPYL ACETATE	102.133	199.75	361.65	25.0	1.5154E+01	2.7278E-04
420	C5H10O2	n-PROPYL ACETATE	102.133	178.15	374.65	25.0	1.6272E+01	2.9290E-04
421	C5H10O2	METHYL n-BUTYRATE	102.133	187.35	375.90	25.0	1.5849E+01	2.8529E-04
422	C5H10O2	2-METHYLBUTYRIC ACID	102.133	-------	450.15	------	----------	----------
423	C5H10O2	ISOVALERIC ACID	102.133	243.85	448.25	20.0	7.7818E-02	1.4007E-06
424	C5H10O2	VALERIC ACID	102.133	239.15	458.65	25.0	7.4407E-02	1.3393E-06
425	C5H10O2	TETRAHYDROFURFURYL ALCOHOL	102.133	-------	451.15	------	----------	----------
426	C5H10O2S	3-METHYL SULFOLANE	134.199	273.65	549.15	------	----------	----------
427	C5H10O3	DIETHYL CARBONATE	118.133	230.15	399.95	20.0	4.8901E+00	8.8021E-05
428	C5H10O3	ETHYL LACTATE	118.133	247.15	427.65	------	----------	----------
429	C5H10S	THIACYCLOHEXANE	102.194	292.14	414.90	------	----------	----------
430	C5H10S	CYCLOPENTANETHIOL	102.194	155.39	405.33	------	----------	----------
431	C5H11Br	1-BROMOPENTANE	151.046	185.26	402.74	25.0	9.7670E+02	1.7580E-02
432	C5H11Cl	1-CHLOROPENTANE	106.595	174.15	381.54	25.0	2.8627E+03	5.1527E-02
433	C5H11Cl	1-CHLORO-3-METHYLBUTANE	106.595	168.76	371.66	------	----------	----------
434	C5H11Cl	2-CHLORO-2-METHYLBUTANE	106.595	199.66	358.76	25.0	1.7869E+02	3.2164E-03
435	C5H11N	N-METHYLPYRROLIDINE	85.149	183.15	352.30	------	----------	----------
436	C5H11N	PIPERIDINE	85.149	262.65	379.55	------	----------	----------
437	C5H11NO	tert-BUTYLFORMAMIDE	101.148	289.15	475.15	------	----------	----------
438	C5H12	ISOPENTANE	72.150	113.25	300.99	25.0	7.5912E+04	1.3664E+00
439	C5H12	NEOPENTANE	72.150	256.58	282.65	25.0	1.1686E+05	2.1034E+00
440	C5H12	n-PENTANE	72.150	143.42	309.22	25.0	7.0302E+04	1.2654E+00
441	C5H12O	2,2-DIMETHYL-1-PROPANOL	88.150	327.15	386.25	------	----------	----------
442	C5H12O	tert-PENTYL-ALCOHOL	88.149	327.00	386.30	------	----------	----------
443	C5H12O	2-METHYL-1-BUTANOL	88.150	-------	401.85	25.0	6.5537E-01	1.1797E-05
444	C5H12O	2-METHYL-2-BUTANOL	88.150	264.35	375.15	25.0	9.7444E+00	1.7540E-04
445	C5H12O	3-METHYL-1-BUTANOL	88.150	155.95	404.35	25.0	7.4253E-01	1.3365E-05
446	C5H12O	3-METHYL-2-BUTANOL	88.150	-------	384.65	25.0	1.0041E+00	1.8075E-05
447	C5H12O	1-PENTANOL	88.150	195.56	410.95	25.0	7.1022E-01	1.2784E-05
448	C5H12O	2-PENTANOL	88.150	200.00	392.15	25.0	8.4200E-01	1.5156E-05
449	C5H12O	3-PENTANOL	88.150	204.15	388.45	25.0	1.0416E+00	1.8749E-05
450	C5H12O	METHYL sec-BUTYL ETHER	88.150	-------	332.15	25.0	6.4043E+01	1.1528E-03
451	C5H12O	METHYL tert-BUTYL ETHER	88.150	164.55	328.35	25.0	3.0059E+01	5.4106E-04
452	C5H120	METHYL ISOBUTYL ETHER	88.150	-------	331.70	25.0	6.3634E+01	1.1454E-03
453	C5H12O	ETHYL PROPYL ETHER	88.150	145.65	337.01	25.0	6.9216E+01	1.2459E-03
454	C5H12O2	ETHYLENE GLYCOL MONOPROPYL ETHER	104.149	183.15	424.50	------	----------	----------
455	C5H12O2	NEOPENTYL GLYCOL	104.149	400.00	483.00	------	----------	----------
456	C5H12O2	1,5-PENTANEDIOL	104.149	257.15	512.15	------	----------	----------
457	C5H12O3	2-(2-METHOXYETHOXY)ETHANOL	120.148	197.15	466.75	------	----------	----------
458	C5H12O4	PENTAERYTHRITOL	136.148	534.15	631.00	------	----------	----------
459	C5H12S	n-PENTYL MERCAPTAN	104.216	197.45	399.79	25.0	7.6771E+02	1.3819E-02
460	C5H12S	BUTYL-METHYL-SULFIDE	104.210	175.33	396.58	25.0	3.1311E+02	5.6360E-03
461	C5H12S	ETHYL-PROPYL-SULFIDE	104.210	156.15	391.65	25.0	2.9001E+02	5.2202E-03
462	C5H12S	2-METHYL-2-BUTANETHIOL	104.210	169.38	372.28	25.0	5.4414E+02	9.7945E-03
463	C5H13N	n-PENTYLAMINE	87.165	218.15	377.65	------	----------	----------
464	C5H13NO2	METHYL DIETHANOLAMINE	119.164	252.15	520.15	------	----------	----------
465	C6Cl6	HEXACHLOROBENZENE	284.782	501.70	582.55	------	----------	----------
466	C6F6	HEXAFLUOROBENZENE	186.056	278.25	353.41	------	----------	----------
467	C6H3ClN2O4	1-CHLORO-2,4-DINITROBENZENE	202.554	326.55	588.00	------	----------	----------
468	C3H3Cl2NO2	1,2-DICHLORO-4-NITROBENZENE	192.001	315.65	529.00	------	----------	----------

NO	FORMULA	NAME	MW, g/mol	T_F, K	T_B, K	T, C	Henry's Law Constant, H H @ T, atm/mol frac	H @ T, atm/mol/m^3
469	C6H3Cl3	1,2,4-TRICHLOROBENZENE	181.448	290.15	486.15	25.0	1.6494E+02	2.9689E-03
470	C6H3N3O6	1,3,5-TRINITROBENZENE	213.106	398.40	748.00	------	----------	----------
471	C6H4Br2	m-DIBROMOBENZENE	235.906	266.25	491.15	25.0	4.7029E+01	8.4651E-04
472	C6H4ClNO2	m-CHLORONITROBENZENE	157.556	317.65	508.75	20.0	1.6044E+00	2.8880E-05
473	C6H4ClNO2	o-CHLORONITROBENZENE	157.556	306.15	519.00	20.0	5.6816E-01	1.0227E-05
474	C6H4ClNO2	p-CHLORONITROBENZENE	157.556	356.65	515.15	------	----------	----------
475	C6H4Cl2	m-DICHLOROBENZENE	147.003	248.39	446.23	25.0	1.8716E+02	3.3688E-03
476	C6H4Cl2	o-DICHLOROBENZENE	147.003	256.15	453.57	25.0	1.5757E+02	2.8363E-03
477	C6H4Cl2	p-DICHLOROBENZENE	147.003	326.14	447.21	------	----------	----------
478	C6H4F2	m-DIFLUOROBENZENE	114.094	249.16	363.66	25.0	4.3109E+03	7.7596E-02
479	C6H4F2	o-DIFLUOROBENZENE	114.094	239.16	364.66	25.0	3.9053E+02	7.0295E-03
480	C6H4F2	p-DIFLUOROBENZENE	114.094	260.16	362.00	25.0	4.2270E+02	7.6086E-03
481	C6H4N2O4	m-DINITROBENZENE	168.109	364.00	573.00	------	----------	----------
482	C6H4N2O4	o-DINITROBENZENE	168.109	390.08	592.00	------	----------	----------
483	C6H4N2O4	p-DINITROBENZENE	168.109	446.60	572.00	------	----------	----------
484	C6H5Br	BROMOBENZENE	157.010	242.43	429.24	25.0	1.1886E+02	2.1394E-03
485	C6H5Cl	MONOCHLOROBENZENE	112.558	227.95	404.87	25.0	2.5163E+02	4.5293E-03
486	C6H5ClO	m-CHLOROPHENOL	128.558	306.00	487.00	20.0	1.1308E-01	2.0354E-06
487	C6H5ClO	o-CHLOROPHENOL	128.558	282.00	447.53	25.0	2.0731E+00	3.7315E-05
488	C6H5ClO	p-CHLOROPHENOL	128.558	316.00	493.11	20.0	7.0135E-02	1.2624E-06
489	C6H5Cl2N	3,4-DICHLOROANILINE	162.018	344.65	545.00	------	----------	----------
490	C6H5F	FLUOROBENZENE	96.104	230.94	357.88	25.0	3.4893E+02	6.2806E-03
491	C6H5I	IODOBENZENE	204.010	241.83	461.60	25.0	7.8862E+01	1.4195E-03
492	C6H5NO2	NITROBENZENE	123.111	278.91	483.95	25.0	1.1593E+00	2.0867E-05
493	C6H6	BENZENE	78.114	278.68	353.24	25.0	3.0826E+02	5.5486E-03
494	C6H6ClN	m-CHLOROANILINE	127.573	262.75	501.65	20.0	1.1230E-01	2.0214E-06
495	C6H6ClN	o-CHLOROANILINE	127.573	481.99	481.99	20.0	5.0380E-01	9.0683E-06
496	C6H6ClN	p-CHLOROANILINE	127.573	343.05	503.65	------	----------	----------
497	C6H6N2	cis-DICYANO-1-BUTENE	106.127	249.00	501.00	------	----------	----------
498	C6H6N2	trans-DICYANO-1-BUTENE	106.127	260.00	499.00	------	----------	----------
499	C6H6N2	1,4-DICYANO-2-BUTENE	106.127	349.00	547.00	------	----------	----------
500	C6H6N2O2	m-NITROANILINE	138.126	387.15	579.00	------	----------	----------
501	C6H6N2O2	o-NITROANILINE	138.126	344.65	558.00	------	----------	----------
502	C6H6N2O2	p-NITROANILINE	138.126	420.65	609.15	------	----------	----------
503	C6H6O	PHENOL	94.113	314.06	454.99	25.0	4.2199E-02	7.5958E-07
504	C6H6O2	1,2-BENZENEDIOL	110.112	377.60	518.65	------	----------	----------
505	C6H6O2	1,3-BENZENEDIOL	110.112	382.00	549.65	------	----------	----------
506	C6H6O2	p-HYDROQUINONE	110.112	444.65	558.15	------	----------	----------
507	C6H6O3	1,2,3-BENZENETRIOL	126.112	407.00	581.85	------	----------	----------
508	C6H6S	PHENYL MERCAPTAN	110.180	258.26	442.29	------	----------	----------
509	C6H7N	ANILINE	93.128	267.13	457.60	25.0	9.4986E-02	1.7097E-06
510	C6H7N	2-METHYLPYRIDINE	93.128	206.44	402.55	25.0	5.5139E-01	9.9250E-06
511	C6H7N	3-METHYLPYRIDINE	93.128	255.01	417.29	25.0	4.0109E-01	7.2195E-06
512	C6H7N	4-METHYLPYRIDINE	93.128	276.73	418.50	20.0	3.1743E-01	5.7137E-06
513	C6H8	1,3-CYCLOHEXADIENE	80.130	161.00	353.49	25.0	8.1272E+02	1.4629E-02
514	C6H8	METHYLCYCLOPENTADIENE	80.130	-------	345.93	------	----------	----------
515	C6H8N2	ADIPONITRILE	108.143	275.64	568.15	------	----------	----------
516	C6H8N2	METHYLGLUTARONITRILE	108.143	228.15	536.15	------	----------	----------
517	C6H8N2	m-PHENYLENEDIAMINE	108.143	334.00	560.00	------	----------	----------
518	C6H8N2	o-PHENYLENEDIAMINE	108.143	376.95	525.00	------	----------	----------
519	C6H8N2	p-PHENYLENEDIAMINE	108.143	413.00	540.00	------	----------	----------
520	C6H8N2	PHENYLHYDRAZINE	108.143	292.35	516.65	------	----------	----------
521	C6H8N2O	BIS(CYANOETHYL)ETHER	124.142	246.85	579.00	------	----------	----------
522	C6H8O4	DIMETHYL MALEATE	144.127	254.15	478.15	------	----------	----------
523	C6H8O6	ASCORBIC ACID	176.126	465.15	637.00	------	----------	----------
524	C6H8O7	CITRIC ACID	192.125	426.15	659.00	------	----------	----------
525	C6H10	1-METHYLCYCLOPENTENE	82.145	145.96	348.95	------	----------	----------
526	C6H10	3-METHYLCYCLOPENTENE	82.145	130.16	343.16	------	----------	----------
527	C6H10	4-METHYLCYCLOPENTENE	82.145	112.31	348.31	------	----------	----------
528	C6H10	CYCLOHEXENE	82.145	169.67	356.12	25.0	2.4964E+03	4.4935E-02
529	C6H10	2,3-DIMETHYL-1,3-BUTADIENE	82.145	197.15	341.93	25.0	6.8189E+03	1.2274E-01
530	C6H10	1,5-HEXADIENE	82.145	132.47	332.61	25.0	6.5004E+03	1.1701E-01
531	C6H10	cis,trans-2,4-HEXADIENE	82.145	177.05	356.65	25.0	7.7888E+03	1.4020E-01
532	C6H10	trans,trans-2,4-HEXADIENE	82.145	228.25	355.05	25.0	7.5738E+03	1.3633E-01
533	C6H10	1-HEXYNE	82.145	141.25	344.48	25.0	2.2139E+03	3.9850E-02
534	C6H10	2-HEXYNE	82.145	183.65	357.67	25.0	2.1856E+03	3.9340E-02
535	C6H10	3-HEXYNE	82.145	170.05	354.35	25.0	2.0906E+03	3.7631E-02
536	C6H10O	CYCLOHEXANONE	98.145	242.00	428.90	25.0	3.0733E-01	5.5320E-06
537	C6H10O	MESITYL OXIDE	98.145	220.15	402.95	20.0	2.7566E+00	4.9619E-05
538	C6H10O2	epsilon-CAPROLACTONE	114.144	271.85	514.00	------	----------	----------
539	C6H10O2	ETHYL METHACRYLATE	114.144	-------	390.15	------	----------	----------
540	C6H10O2	n-PROPYL ACRYLATE	114.144	-------	392.15	------	----------	----------
541	C6H10O3	ETHYLACETOACETATE	130.144	234.15	453.95	16.5	6.0381E-02	1.0869E-06
542	C6H10O3	PROPIONIC ANHYDRIDE	130.144	228.15	442.15	------	----------	----------
543	C6H10O4	ADIPIC ACID	146.143	425.50	611.00	------	----------	----------
544	C6H10O4	DIETHYL OXALATE	146.143	232.55	458.85	------	----------	----------
545	C6H10O4	ETHYLENE GLYCOL DIACETATE	146.143	242.15	463.65	22.0	5.8979E-03	1.0616E-07
546	C6H10O4	ETHYLIDENE DIACETATE	146.143	292.00	442.15	------	----------	----------

413

NO	FORMULA	NAME	MW, g/mol	T_F, K	T_B, K	T, C	H @ T, atm/mol frac	H @ T, atm/mol/m³
547	C6H11N	HEXANENITRILE	97.160	192.85	436.75	------	----------	----------
548	C6H11NO	epsilon-CAPROLACTAM	113.159	342.36	543.15	------	----------	----------
549	C6H11NO	CYCLOHEXANONE OXIME	113.159	363.15	481.15	------	----------	----------
550	C6H12	CYCLOHEXANE	84.161	279.69	353.87	25.0	1.0785E+04	1.9412E-01
551	C6H12	2,3-DIMETHYL-1-BUTENE	84.161	115.89	328.76	25.0	1.9363E+04	3.4854E-01
552	C6H12	2,3-DIMETHYL-2-BUTENE	84.161	198.82	346.35	25.0	2.1467E+04	3.8640E-01
553	C6H12	3,3-DIMETHYL-1-BUTENE	84.161	157.95	314.40	25.0	1.8960E+04	3.4127E-01
554	C6H12	2-ETHYL-1-BUTENE	84.161	141.61	337.82	25.0	2.0025E+04	3.6044E-01
555	C6H12	trans-3-METHYL-2-PENTENE	84.161	134.70	343.60	25.0	2.0946E+04	3.7702E-01
556	C6H12	1-HEXENE	84.161	133.39	336.63	25.0	1.6404E+04	2.9528E-01
557	C6H12	cis-2-HEXENE	84.161	132.00	342.03	25.0	2.0808E+04	3.7453E-01
558	C6H12	trans-2-HEXENE	84.161	140.17	341.02	25.0	2.0587E+04	3.7057E-01
559	C6H12	cis-3-HEXENE	84.161	135.33	339.60	25.0	2.0444E+04	3.6799E-01
560	C6H12	trans-3-HEXENE	84.161	159.73	340.24	25.0	2.0365E+04	3.6657E-01
561	C6H12	METHYLCYCLOPENTANE	84.161	130.73	344.96	25.0	1.9813E+04	3.5664E-01
562	C6H12	2-METHYL-1-PENTENE	84.161	137.42	335.25	25.0	1.5394E+04	2.7709E-01
563	C6H12	2-METHYL-2-PENTENE	84.161	138.07	340.45	25.0	2.0352E+04	3.6633E-01
564	C6H12	3-METHYL-1-PENTENE	84.161	120.20	327.33	25.0	1.9470E+04	3.5046E-01
565	C6H12	3-METHYL-cis-2-PENTENE	84.161	138.31	340.85	25.0	2.0654E+04	3.7177E-01
566	C6H12	4-METHYL-1-PENTENE	84.161	119.51	327.01	25.0	3.4740E+04	6.2532E-01
567	C6H12	4-METHYL-cis-2-PENTENE	84.161	138.30	329.53	25.0	1.9343E+04	3.4818E-01
568	C6H12	4-METHYL-trans-2-PENTENE	84.161	132.35	331.75	25.0	1.9418E+04	3.4952E-01
569	C6H12N2	TRIETHYLENEDIAMINE	112.175	434.25	447.15	------	----------	----------
570	C6H12O	BUTYL VINYL ETHER	100.161	181.25	366.97	------	----------	----------
571	C6H12O	CYCLOHEXANOL	100.161	296.60	434.00	25.0	1.4029E-01	2.5252E-06
572	C6H12O	1-HEXANAL	100.161	217.15	401.45	30.0	1.4533E+01	2.6159E-04
573	C6H12O	ETHYL ISOPROPYL KETONE	100.161	------	386.55			
574	C6H12O	2-HEXANONE	100.161	217.35	400.85	25.0	4.6863E+00	8.4352E-05
575	C6H12O	3-HEXANONE	100.161	217.50	396.65	25.0	6.8489E+00	1.2328E-04
576	C6H12O	METHYL ISOBUTYL KETONE	100.161	189.15	389.65	25.0	7.5223E+00	1.3540E-04
577	C6H12O2	n-PENTYL FORMATE	116.160	199.65	406.60	25.0	2.2111E+01	3.9799E-04
578	C6H12O2	n-BUTYL ACETATE	116.160	199.65	399.15	25.0	1.4296E+01	2.5732E-04
579	C6H12O2	sec-BUTYL ACETATE	116.160	174.15	385.15	20.0	3.0306E+01	5.4550E-04
580	C6H12O2	tert-BUTYL ACETATE	116.160	------	369.15	25.0	1.8999E+01	3.4198E-04
581	C6H12O2	ETHYL n-BUTYRATE	116.160	175.15	394.65	22.0	2.3220E+01	4.1795E-04
582	C6H12O2	ETHYL ISOBUTYRATE	116.160	185.00	383.00	25.0	2.2797E+01	4.1033E-04
583	C6H12O2	ISOBUTYL ACETATE	116.160	174.30	389.80	25.0	2.3871E+01	4.2967E-04
584	C6H12O2	n-PROPYL PROPIONATE	116.160	197.25	395.65	25.0	2.6314E+01	4.7366E-04
585	C6H12O2	CYCLOHEXYL PEROXIDE	116.160	253.15	490.00	------	----------	----------
586	C6H12O2	DIACETONE ALCOHOL	116.160	229.15	441.00	------	----------	----------
587	C6H12O2	2-ETHYL BUTYRIC ACID	116.160	258.15	466.95	------	----------	----------
588	C6H12O2	n-HEXANOIC ACID	116.160	270.15	478.85	20.0	3.7776E-02	6.7997E-07
589	C6H12O3	2-ETHOXYETHYL ACETATE	132.159	211.45	429.45	20.0	7.8748E-02	1.4174E-06
590	C6H12O3	HYDROXYCAPROIC ACID	132.159	334.00	576.00	------	----------	----------
591	C6H12O3	PARALDEHYDE	132.159	285.75	397.25	30.0	8.5799E-01	1.5444E-05
592	C6H12O3	sec-BUTYL GLYCOLATE	132.160	------	450.65	------	----------	----------
593	C6H12S	THIACYCLOHEPTANE	116.221	292.14	414.90	------	----------	----------
594	C6H13N	CYCLOHEXYLAMINE	99.176	255.45	407.65	------	----------	----------
595	C6H13N	HEXAMETHYLENEIMINE	99.176	236.15	404.85	------	----------	----------
596	C6H14	2,2-DIMETHYLBUTANE	86.177	174.28	322.88	25.0	8.4422E+04	1.5196E+00
597	C6H14	2,3-DIMETHYLBUTANE	86.177	145.19	331.13	25.0	7.2108E+04	1.2979E+00
598	C6H14	n-HEXANE	86.177	177.84	341.88	25.0	7.2173E+04	1.2991E+00
599	C6H14	2-METHYLPENTANE	86.177	119.55	333.41	25.0	9.6367E+04	1.7346E+00
600	C6H14	3-METHYLPENTANE	86.177	110.25	336.42	25.0	6.6721E+04	1.2010E+00
601	C6H14N2O2	LYSINE	146.189	483.00	615.00	------	----------	----------
602	C6H14O	2-ETHYL-1-BUTANOL	102.177	158.75	419.65	25.0	1.1323E+00	2.0382E-05
603	C6H14O	1-HEXANOL	102.177	228.55	430.15	25.0	1.1721E+00	2.1097E-05
604	C6H14O	2-HEXANOL	102.177	223.00	413.04	25.0	1.3101E+00	2.3582E-05
605	C6H14O	2-METHYL-1-PENTANOL	102.177	------	421.15	25.0	1.7548E+00	3.1586E-05
606	C6H14O	4-METHYL-2-PENTANOL	102.177	------	404.85	27.0	2.5778E+00	4.6401E-05
607	C6H14O	n-BUTYL ETHYL ETHER	102.177	170.15	365.35	------	----------	----------
608	C6H14O	DIISOPROPYL ETHER	102.177	187.65	341.45	25.0	9.7738E+01	1.7593E-03
609	C6H14O	DI-n-PROPYL ETHER	102.177	149.95	362.79	25.0	1.2177E+02	2.1919E-03
610	C6H14O	METHYL tert-PENTYL ETHER	102.177	------	359.45	25.0	1.0107E+02	1.8192E-03
611	C6H14O2	ACETAL	118.176	173.15	376.75	25.0	5.2028E+00	9.3649E-05
612	C6H14O2	2-BUTOXYETHANOL	118.176	203.15	444.47	20.0	1.9851E-02	3.5731E-07
613	C6H14O2	1,6-HEXANEDIOL	118.176	315.15	516.15	------	----------	----------
614	C6H14O2	HEXYLENE GLYCOL	118.176	223.15	470.65	------	----------	----------
615	C6H14O2S	DI-n-PROPYL SULFONE	150.242	303.00	543.00	------	----------	----------
616	C6H14O3	DIETHYLENE GLYCOL DIMETHYL ETHER	134.175	203.15	432.91	------	----------	----------
617	C6H14O3	DIPROPYLENE GLYCOL	134.175	233.00	504.95	------	----------	----------
618	C6H14O3	2-(2-ETHOXYETHOXY)ETHANOL	134.175	195.15	475.15	------	----------	----------
619	C6H14O3	TRIMETHYLOLPROPANE	134.175	331.15	562.04	------	----------	----------
620	C6H14O4	TRIETHYLENE GLYCOL	150.175	265.79	551.00	------	----------	----------
621	C6H14O6	SORBITOL	182.174	370.85	777.00	------	----------	----------
622	C6H14S	n-HEXYLMERCAPTAN	118.243	192.62	425.81	25.0	1.2312E+03	2.2161E-02
623	C6H14S	BUTYL-ETHYL-SULFIDE	118.237	178.03	417.41	25.0	4.7176E+02	8.4917E-03
624	C6H14S	ISOPROPYL-SULFIDE	118.237	170.45	393.19	25.0	3.5944E+02	6.4698E-03

							Henry's Law Constant, H	
NO	FORMULA	NAME	MW, g/mol	T_F, K	T_B, K	T, C	H @ T, atm/mol frac	H @ T, atm/mol/m³
625	C6H14S	METHYL-PENTYL-SULFIDE	118.237	179.16	401.16	------	----------	----------
626	C6H14S	PROPYL-SULFIDE	118.237	170.45	416.00	25.0	4.6106E+02	8.2990E-03
627	C6H14S2	PROPYL-DISULFIDE	150.297	187.68	464.65	------	----------	----------
628	C6H15Al	TRIETHYL ALUMINUN	114.167	220.65	458.15	------	----------	----------
629	C6H15Al2Cl3	ETHYL ALUMINUM SESQUICHLORIDE	247.506	253.15	482.15	------	----------	----------
630	C6H15N	DIISOPROPYLAMINE	101.192	176.85	357.05	------	----------	----------
631	C6H15N	DI-n-PROPYLAMINE	101.192	210.15	382.00	36.1	4.9757E+00	8.9562E-05
632	C6H15N	n-HEXYLAMINE	101.192	251.85	404.65	25.0	5.4808E+00	9.8654E-05
633	C6H15N	TRIETHYLAMINE	101.192	158.45	361.92	25.0	6.4591E+00	1.1626E-04
634	C6H15NO	6-AMINOHEXANOL	117.191	331.00	508.00	------	----------	----------
635	C6H15NO2	DIISOPROPANOLAMINE	133.191	318.15	521.90	------	----------	----------
636	C6H15NO3	TRIETHANOLAMINE	149.190	294.35	613.00	------	----------	----------
637	C6H15N3	N-AMINOETHYL PIPERAZINE	129.205	254.15	493.55	------	----------	----------
638	C6H15O4P	TRIETHYL PHOSPHATE	182.156	216.00	484.15	25.0	5.7382E-03	1.0329E-07
639	C6H16N2	HEXAMETHYLENEDIAMINE	116.207	313.95	475.04	------	----------	----------
640	C6H18N3OP	HEXAMETHYL PHOSPHORAMIDE	179.202	280.15	506.15	------	----------	----------
641	C6H18N4	TRIETHYLENE TETRAMINE	146.236	285.15	539.65	------	----------	----------
642	C6H18OSi2	HEXAMETHYLDISILOXANE	162.379	204.93	373.67	------	----------	----------
643	C6H18O3Si3	HEXAMETHYLCYCLOTRISILOXANE	222.464	337.15	408.26	------	----------	----------
644	C6H19NSi2	HEXAMETHYLDISILAZANE	161.395	------	399.15	------	----------	----------
645	C7H3ClF3NO2	4-CHLORO-3-NITROBENZOTRIFLUORIDE	225.554	------	495.15	------	----------	----------
646	C7H3Cl2F3	2,4-DICHLOROBENZOTRIFLUORIDE	215.001	247.55	450.65	------	----------	----------
647	C7H3Cl2NO	3,4-DICHLOROPHENYL ISOCYANATE	188.012	316.15	501.00	------	----------	----------
648	C7H4ClF3	p-CHLOROBENZOTRIFLUORIDE	180.557	237.15	412.15	------	----------	----------
649	C7H4Cl2O	m-CHLOROBENZOYL CHLORIDE	175.014	280.00	498.00	------	----------	----------
650	C7H4F3NO2	3-NITROBENZOTRIFLUORIDE	191.110	272.00	475.93	------	----------	----------
651	C7H5ClO	BENZOYL CHLORIDE	140.569	272.65	470.15	------	----------	----------
652	C7H5ClO2	o-CHLOROBENZOIC ACID	156.568	415.15	560.15	------	----------	----------
653	C7H5Cl3	BENZOTRICHLORIDE	195.475	268.40	486.65	------	----------	----------
654	C7H5F3	BENZOTRIFLUORIDE	146.112	244.14	375.20	25.0	9.2031E+02	1.6565E-02
655	C7H5N	BENZONITRILE	103.123	260.40	464.15	100.0	5.7291E-01	1.0312E-05
656	C7H5NO	PHENYL ISOCYANATE	119.123	243.15	438.75	20.0	3.0447E+04	5.4804E-01
657	C7H5N3O6	2,4,6-TRINITROTOLUENE	227.133	354.00	573.00	------	----------	----------
658	C7H6Cl2	BENZYL DICHLORIDE	161.030	257.00	487.00	------	----------	----------
659	C7H6Cl2	2,4-DICHLOROTOLUENE	161.030	259.65	474.25	------	----------	----------
660	C7H6N2O4	2,4-DINITROTOLUENE	182.136	343.00	590.00	------	----------	----------
661	C7H6N2O4	2,5-DINITROTOLUENE	182.136	325.65	590.00	------	----------	----------
662	C7H6N2O4	2,6-DINITROTOLUENE	182.136	339.00	558.00	------	----------	----------
663	C7H6N2O4	3,4-DINITROTOLUENE	182.136	332.00	610.00	------	----------	----------
664	C7H6N2O4	3,5-DINITROTOLUENE	182.136	365.65	588.00	------	----------	----------
665	C7H6O	BENZALDEHYDE	106.124	247.15	451.90	25.0	1.3849E+00	2.4928E-05
666	C7H6O2	BENZOIC ACID	122.123	395.52	522.40	------	----------	----------
667	C7H6O2	p-HYDROXYBENZALDEHYDE	122.123	390.15	583.15	------	----------	----------
668	C7H6O2	SALICYLALDEHYDE	122.123	266.15	469.65	86.0	3.0656E-01	5.5181E-06
669	C7H6O3	SALICYLIC ACID	138.123	431.75	529.00	------	----------	----------
670	C7H7Br	p-BROMOTOLUENE	171.037	299.65	457.50	------	----------	----------
671	C7H7Cl	BENZYL CHLORIDE	126.585	234.15	452.55	------	----------	----------
672	C7H7Cl	o-CHLOROTOLUENE	126.585	236.65	432.30	------	----------	----------
673	C7H7Cl	p-CHLOROTOLUENE	126.585	280.65	435.65	20.0	2.4264E+02	4.3674E-03
674	C7H7F	p-FLUOROTOLUENE	110.131	216.36	389.76	------	----------	----------
675	C7H7NO	FORMANILIDE	121.139	323.15	544.15	------	----------	----------
676	C7H7NO2	m-NITROTOLUENE	137.138	289.20	505.00	30.0	4.1465E+00	7.4637E-05
677	C7H7NO2	o-NITROTOLUENE	137.138	269.98	495.64	------	----------	----------
678	C7H7NO2	p-NITROTOLUENE	137.138	324.75	511.65	------	----------	----------
679	C7H7NO3	o-NITROANISOLE	153.138	283.60	546.15	------	----------	----------
680	C7H8	TOLUENE	92.141	178.18	383.78	25.0	3.5290E+02	6.3522E-03
681	C7H8	1,3,5-CYCLOHEPTATRIENE	92.140	193.66	388.65	25.0	2.5530E+02	4.5953E-03
682	C7H8O	ANISOLE	108.140	235.65	426.73	25.0	2.0364E+02	3.6655E-03
683	C7H8O	BENZYL ALCOHOL	108.140	257.85	477.85	25.0	1.4405E-02	2.5929E-07
684	C7H8O	m-CRESOL	108.140	285.39	475.43	20.0	4.0198E-02	7.2357E-07
685	C7H8O	o-CRESOL	108.140	304.19	464.15	20.0	9.4952E-02	1.7091E-06
686	C7H8O	p-CRESOL	108.140	307.93	475.13	20.0	4.0616E-02	7.3109E-07
687	C7H8O2	GUAIACOL	124.139	304.65	478.15	37.0	1.6603E+00	2.9885E-05
688	C7H8O2	p-METHOXYPHENOL	124.139	329.00	516.00	37.0	2.7203E-01	4.8965E-06
689	C7H9N	BENZYLAMINE	107.155	227.15	457.65	25.0	1.6201E+00	2.9162E-05
690	C7H9N	2,6-DIMETHYLPYRIDINE	107.155	267.00	417.20	------	----------	----------
691	C7H9N	N-METHYLANILINE	107.155	216.15	469.02	25.0	6.2753E-01	1.1296E-05
692	C7H9N	m-TOLUIDINE	107.155	242.75	476.55	20.0	1.5554E-01	2.7998E-06
693	C7H9N	o-TOLUIDINE	107.155	249.47	473.55	25.0	1.3268E-01	2.3882E-06
694	C7H9N	p-TOLUIDINE	107.155	316.90	473.40	------	----------	----------
695	C7H10	2-NORBORNENE	94.156	319.40	368.65	------	----------	----------
696	C7H10N2	TOLUENEDIAMINE	122.170	371.25	557.15	------	----------	----------
697	C7H11NO	CYCLOHEXYL ISOCYANATE	125.170	------	442.15	------	----------	----------
698	C7H12	1-HEPTYNE	96.172	192.26	372.86	25.0	3.9330E+03	7.0793E-02
699	C7H12O2	n-BUTYL ACRYLATE	128.171	208.55	421.00	20.0	3.1845E+01	5.7321E-04
700	C7H12O2	ISOBUTYL ACRYLATE	128.171	212.00	405.15	23.0	3.7731E+01	6.7915E-04
701	C7H12O2	n-PROPYL METHACRYLATE	128.171	------	414.00	------	----------	----------
702	C7H12O4	DIETHYL MALONATE	160.170	224.25	472.05	37.0	1.3279E-01	2.3901E-06

NO	FORMULA	NAME	MW, g/mol	T_F, K	T_B, K	T, C	H @ T, atm/mol frac	H @ T, atm/mol/m³
							Henry's Law Constant, H	
703	C7H14	CYCLOHEPTANE	98.188	265.15	391.94	25.0	5.1654E+03	9.2976E-02
704	C7H14	1,1-DIMETYLCYCLOPENTANE	98.188	203.36	361.00	25.0	2.4223E+04	4.3601E-01
705	C7H14	cis-1,2-DIMETHYLCYCLOPENTANE	98.188	219.26	372.68	25.0	2.8432E+04	5.1177E-01
706	C7H14	trans-1,2-DIMETHYLCYCLOPENTANE	98.188	155.58	365.02	25.0	2.5282E+04	4.5506E-01
707	C7H14	cis-1,3-DIMETHYLCYCLOPENTANE	98.188	139.45	363.92	25.0	2.4615E+04	4.4307E-01
708	C7H14	trans-1,3-DIMETHYLCYCLOPENTANE	98.188	139.18	364.88	25.0	2.5282E+04	4.5508E-01
709	C7H14	ETHYLCYCLOPENTANE	98.188	134.71	376.62	25.0	2.9902E+04	5.3823E-01
710	C7H14	2-ETHYL-1-PENTENE	98.188	168.00	367.15	25.0	3.3563E+04	6.0413E-01
711	C7H14	3-ETHYL-1-PENTENE	98.188	145.67	357.26	25.0	2.8214E+04	5.0784E-01
712	C7H14	1-HEPTENE	98.188	154.27	366.79	25.0	2.2220E+04	3.9996E-01
713	C7H14	cis-2-HEPTENE	98.188	164.00	371.56	25.0	3.6364E+04	6.5455E-01
714	C7H14	trans-2-HEPTENE	98.188	163.67	371.10	25.0	3.5703E+04	6.4265E-01
715	C7H14	cis-3-HEPTENE	98.188	136.51	368.90	25.0	3.4401E+04	6.1921E-01
716	C7H14	trans-3-HEPTENE	98.188	136.52	368.82	25.0	3.3767E+04	6.0781E-01
717	C7H14	METHYLCYCLOHEXANE	98.188	146.58	374.08	25.0	2.3520E+04	4.2335E-01
718	C7H14	2-METHYL-1-HEXENE	98.188	170.28	364.99	25.0	3.1846E+04	5.7322E-01
719	C7H14	3-METHYL-1-HEXENE	98.188	145.00	357.05	25.0	2.8256E+04	5.0860E-01
720	C7H14	4-METHYL-1-HEXENE	98.188	131.70	359.88	25.0	2.9231E+04	5.2615E-01
721	C7H14	2,3,3-TRIMETHYL-1-BUTENE	98.188	163.30	351.04	25.0	2.8081E+04	5.0545E-01
722	C7H14O	DIISOPROPYL KETONE	114.188	204.81	397.55	25.0	1.9526E+01	3.5147E-04
723	C7H14O	2-HEPTANONE	114.188	238.15	424.05	25.0	7.3369E+00	1.3206E-04
724	C7H14O	1-HEPTANAL	114.188	230.15	425.95	30.0	1.9412E+01	3.4942E-04
725	C7H14O	1-METHYLCYCLOHEXANOL	114.188	299.15	430.15	------	----------	----------
726	C7H14O	cis-2-METHYLCYCLOHEXANOL	114.188	280.15	438.15	------	----------	----------
727	C7H14O	trans-2-METHYLCYCLOHEXANOL	114.188	269.15	439.65	------	----------	----------
728	C7H14O	cis-3-METHYLCYCLOHEXANOL	114.188	267.65	441.15	------	----------	----------
729	C7H14O	trans-3-METHYLCYCLOHEXANOL	114.188	272.65	441.15	------	----------	----------
730	C7H14O	cis-4-METHYLCYCLOHEXANOL	114.188	-------	444.15	------	----------	----------
731	C7H14O	trans-4-METHYLCYCLOHEXANOL	114.188	-------	444.15	------	----------	----------
732	C7H14O	5-METHYL-2-HEXANONE	114.188	199.25	417.95	------	----------	----------
733	C7H14O2	n-BUTYL PROPIONATE	130.187	183.63	419.75	25.0	3.8556E+01	6.9400E-04
734	C7H14O2	ETHYL ISOVALERATE	130.187	173.85	407.45	25.0	3.5199E+01	6.3357E-04
735	C7H14O2	ISOPENTYL ACETATE	130.187	194.65	415.25	25.0	2.6555E+01	4.7799E-04
736	C7H14O2	n-PENTYL ACETATE	130.187	202.35	422.15	25.0	1.9229E+01	3.4613E-04
737	C7H14O2	n-PROPYL n-BUTYRATE	130.187	177.95	416.45	17.0	3.4373E+01	6.1872E-04
738	C7H14O2	n-HEPTANOIC ACID	130.187	265.83	496.15	20.0	2.1562E-02	3.8811E-07
739	C7H14O3	ETHYL-3-ETHOXYPROPIONATE	146.186	-------	438.15	25.0	3.6106E+01	6.4991E-04
740	C7H15Br	1-BROMOHEPTANE	179.100	217.05	452.05	25.0	2.5039E+03	4.5070E-02
741	C7H15N	N-METHYLCYCLOHEXYLAMINE	113.203	264.65	422.00	------	----------	----------
742	C7H16	2,2-DIMETHYLPENTANE	100.204	149.34	352.34	25.0	1.7492E+05	3.1486E+00
743	C7H16	2,3-DIMETHYLPENTANE	100.204	-------	362.93	25.0	9.6001E+04	1.7280E+00
744	C7H16	2,4-DIMETHYLPENTANE	100.204	153.91	353.64	25.0	1.6321E+05	2.9378E+00
745	C7H16	3,3-DIMETHYLPENTANE	100.204	138.70	359.21	25.0	1.0191E+05	1.8343E+00
746	C7H16	3-ETHYLPENTANE	100.204	154.55	366.62	25.0	1.4389E+05	2.5900E+00
747	C7H16	n-HEPTANE	100.204	182.57	371.58	25.0	1.4884E+05	2.6791E+00
748	C7H16	2-METHYLHEXANE	100.204	154.90	363.20	25.0	1.9020E+05	3.4236E+00
749	C7H16	3-METHYLHEXANE	100.204	153.75	365.00	25.0	1.7046E+05	3.0683E+00
750	C7H16	2,2,3-TRIMETHYLBUTANE	100.204	248.57	354.03	25.0	1.3040E+05	2.3472E+00
751	C7H16O	1-HEPTANOL	116.203	239.15	449.45	25.0	1.0500E+00	1.8900E-05
752	C7H16O	2-HEPTANOL	116.203	243.00	432.35	30.0	3.1590E+00	5.6861E-05
753	C7H16O	5-METHYL-1-HEXANOL	116.203	-------	445.15	------	----------	----------
754	C7H16O	ISOPROPYL-TERT-BUTYL-ETHER	116.203	177.80	378.66	25.0	8.2497E+02	1.4849E-02
755	C7H16S	n-HEPTYL MERCAPTAN	132.270	229.92	450.09	25.0	1.6520E+03	2.9735E-02
756	C7H16S	BUTYL-PROPYL-SULFIDE	132.263	206.66	444.16	------	----------	----------
757	C7H16S	ETHYL-PENTYL-SULFIDE	132.263	206.66	444.16	------	----------	----------
758	C7H16S	HEXYL-METHYL-SULFIDE	132.263	206.66	444.16	------	----------	----------
759	C7H17N	1-AMINOHEPTANE	115.219	254.15	430.05	------	----------	----------
760	C8H4Cl2O2	ISOPHTHALOYL CHLORIDE	203.024	317.00	549.00	------	----------	----------
761	C8H4O3	PHTHALIC ANHYDRIDE	148.118	404.26	557.65	------	----------	----------
762	C8H6	ETHYNYLBENZENE	102.135	242.53	418.36	------	----------	----------
763	C8H6O4	ISOPHTHALIC ACID	166.133	619.15	753.00	------	----------	----------
764	C8H6O4	PHTHALIC ACID	166.133	464.15	598.00	------	----------	----------
765	C8H6O4	TEREPHTHALIC ACID	166.133	700.15	832.00	------	----------	----------
766	C8H6S	BENZOTHIOPHENE	134.202	304.50	493.05	20.0	1.7993E+01	3.2386E-04
767	C8H7N	INDOLE	117.150	273.68	526.15	20.0	5.5496E-02	9.9893E-07
768	C8H8	STYRENE	104.152	242.54	418.31	25.0	1.4456E+02	2.6021E-03
769	C8H8	1,3,5,7-CYCLOOCTATETRAENE	104.151	266.16	413.16	------	----------	----------
770	C8H8O	ACETOPHENONE	120.151	293.65	475.15	25.0	5.0557E-01	9.1002E-06
771	C8H8O	p-TOLUALDEHYDE	120.151	-------	477.15	25.0	9.7805E-01	1.7605E-05
772	C8H8O2	METHYL BENZOATE	136.150	260.75	472.65	25.0	1.9326E+00	3.4786E-05
773	C8H8O2	o-TOLUIC ACID	136.150	376.85	532.00	------	----------	----------
774	C8H8O2	p-TOLUIC ACID	136.150	452.75	548.15	------	----------	----------
775	C8H8O3	METHYL SALICYLATE	152.150	265.15	493.65	30.0	5.4338E-01	9.7807E-06
776	C8H8O3	VANILLIN	152.150	355.00	558.00	------	----------	----------
777	C8H9NO	ACETANILIDE	135.166	386.65	576.95	------	----------	----------
778	C8H10	ETHYLBENZENE	106.167	178.20	409.35	25.0	4.5230E+02	8.1413E-03
779	C8H10	m-XYLENE	106.167	225.30	412.27	25.0	3.7653E+02	6.7775E-03
780	C8H10	o-XYLENE	106.167	247.98	417.58	25.0	2.3275E+02	4.1895E-03

						Henry's Law Constant, H		
NO	FORMULA	NAME	MW, g/mol	T_F, K	T_B, K	T, C	H @ T, atm/mol frac	H @ T, atm/mol/m³
781	C8H10	p-XYLENE	106.167	286.41	411.51	25.0	3.4193E+02	6.1547E-03
782	C8H10O	m-ETHYLPHENOL	122.167	-------	477.66	------	----------	----------
783	C8H10O	p-ETHYLPHENOL	122.167	318.23	491.14	------	----------	----------
784	C8H10O	PHENETOLE	122.167	243.63	443.15	------	----------	----------
785	C8H10O	2-PHENYLETHANOL	122.167	247.00	492.05	25.0	1.5686E-01	2.8235E-06
786	C8H10O	2,3-XYLENOL	122.167	345.71	490.07	------	----------	----------
787	C8H10O	2,4-XYLENOL	122.167	297.68	484.13	------	----------	----------
788	C8H10O	2,5-XYLENOL	122.167	347.99	484.33	------	----------	----------
789	C8H10O	2,6-XYLENOL	122.167	318.76	474.22	25.0	3.9039E-01	7.0270E-06
790	C8H10O	3,4-XYLENOL	122.167	338.25	500.15	------	----------	----------
791	C8H10O	3,5-XYLENOL	122.167	336.59	494.89	------	----------	----------
792	C8H11N	N,N-DIMETHYLANILINE	121.182	275.60	466.69	25.0	5.6627E+00	1.0193E-04
793	C8H11N	o-ETHYLANILINE	121.182	226.55	482.65	------	----------	----------
794	C8H11N	2,4,6-TRIMETHYLPYRIDINE	121.182	229.00	444.00	------	----------	----------
795	C8H11NO	p-PHENETIDINE	137.181	277.00	528.00	------	----------	----------
796	C8H12	1,5-CYCLOOCTADIENE	108.183	203.98	423.27	------	----------	----------
797	C8H12	VINYLCYCLOHEXENE	108.183	164.00	401.00	25.0	2.4807E+03	4.4653E-02
798	C8H12O4	1,4-CYCLOHEXANEDICARBOXYLIC ACID	172.181	585.65	669.00	------	----------	----------
799	C8H12O4	DIETHYL MALEATE	172.181	264.35	498.15	------	----------	----------
800	C8H14O2	n-BUTYL METHACRYLATE	142.198	-------	434.00	------	----------	----------
801	C8H14O3	BUTYRIC ANHYDRIDE	158.197	199.85	468.15	------	----------	----------
802	C8H14O4	DIETHYL SUCCINATE	174.197	252.35	489.65	------	----------	----------
803	C8H16	CYCLOOCTANE	112.214	287.60	424.30	------	----------	----------
804	C8H16	1,1-DIMETHYLCYCLOHEXANE	112.215	239.66	392.70	25.0	3.1009E+04	5.5816E-01
805	C8H16	cis-1,2-DIMETHYLCYCLOHEXANE	112.215	223.16	402.94	25.0	1.9779E+04	3.5602E-01
806	C8H16	trans-1,2-DIMETHYLCYCLOHEXANE	112.215	184.99	396.58	25.0	3.3260E+04	5.9868E-01
807	C8H16	cis-1,3-DIMETHYLCYCLOHEXANE	112.215	197.58	393.24	25.0	3.0334E+04	5.4600E-01
808	C8H16	trans-1,3-DIMETHYLCYCLOHEXANE	112.215	183.07	397.61	25.0	3.2157E+04	5.7882E-01
809	C8H16	cis-1,4-DIMETHYLCYCLOHEXANE	112.215	185.72	397.47	25.0	3.2459E+04	5.8425E-01
810	C8H16	trans-1,4-DIMETHYLCYCLOHEXANE	112.215	236.21	392.51	25.0	4.8393E+04	8.7107E-01
811	C8H16	ETHYLCYCLOHEXANE	112.215	161.84	404.95	25.0	3.6064E+04	6.4915E-01
812	C8H16	2-ETHYL-1-HEXENE	112.215	-------	393.15	25.0	5.8442E+04	1.0519E+00
813	C8H16	1-METHYL-1-ETHYLCYCLOPENTANE	112.215	129.35	394.67	25.0	4.8121E+04	8.6618E-01
814	C8H16	1-OCTENE	112.215	171.45	394.44	25.0	3.4837E+04	6.2705E-01
815	C8H16	trans-2-OCTENE	112.215	185.45	398.15	25.0	6.4907E+04	1.1683E+00
816	C8H16	trans-3-OCTENE	112.215	163.15	396.45	25.0	6.2124E+04	1.1182E+00
817	C8H16	trans-4-OCTENE	112.215	179.37	395.41	25.0	6.0189E+04	1.0834E+00
818	C8H16	n-PROPYLCYCLOPENTANE	112.215	155.82	404.11	25.0	4.9654E+04	8.9376E-01
819	C8H16	2,4,4-TRIMETHYL-1-PENTENE	112.215	179.70	374.59	25.0	4.5440E+04	8.1792E-01
820	C8H16	2,4,4-TRIMETHYL-2-PENTENE	112.215	166.84	378.06	25.0	4.4377E+04	7.9879E-01
821	C8H16O	2-ETHYLHEXANAL	128.214	-------	433.80	------	----------	----------
822	C8H16O	1-OCTANAL	128.214	246.00	447.15	30.0	2.9770E+01	5.3585E-04
823	C8H16O	2-OCTANONE	128.214	252.85	445.75	20.0	5.7140E+00	1.0285E-04
824	C8H16O2	n-BUTYL n-BUTYRATE	144.214	181.15	438.15	25.0	5.0473E+01	9.0851E-04
825	C8H16O2	n-HEXYL ACETATE	144.214	192.25	444.65	25.0	2.7298E+01	4.9136E-04
826	C8H16O2	ISOBUTYL ISOBUTYRATE	144.214	192.45	420.65	25.0	8.0024E+01	1.4404E-03
827	C8H16O2	n-OCTANOIC ACID	144.214	289.65	513.05	20.0	5.3203E-02	9.5766E-07
828	C8H16O4	DIETHYLENE GLYCOL ETHYL ETHER ACETATE	176.213	248.15	490.55	------	----------	----------
829	C8H18	2,2-DIMETHYLHEXANE	114.231	151.97	379.99	25.0	2.0352E+05	3.6633E+00
830	C8H18	2,3-DIMETHYLHEXANE	114.231	-------	388.76	25.0	2.3219E+05	4.1795E+00
831	C8H18	2,4-DIMETHYLHEXANE	114.231	-------	382.58	25.0	2.1042E+05	3.7875E+00
832	C8H18	2,5-DIMETHYLHEXANE	114.231	182.00	382.26	25.0	2.0631E+05	3.7135E+00
833	C8H18	3,3-DIMETHYLHEXANE	114.231	147.05	385.12	25.0	2.2963E+05	4.1332E+00
834	C8H18	3,4-DIMETHYLHEXANE	114.231	-------	390.88	25.0	2.4292E+05	4.3726E+00
835	C8H18	3-ETHYLHEXANE	114.231	-------	391.69	25.0	2.3537E+05	4.2366E+00
836	C8H18	3-ETHYL-2-METHYLPENTANE	114.230	158.20	388.81	25.0	2.3757E+05	4.2762E+00
837	C8H18	3-METHYL-3-ETHYLPENTANE	114.231	182.28	391.42	25.0	2.6605E+05	4.7889E+00
838	C8H18	2-METHYLHEPTANE	114.231	164.16	390.80	25.0	2.3052E+05	4.1493E+00
839	C8H18	3-METHYLHEPTANE	114.231	152.60	392.08	25.0	2.0644E+05	3.7159E+00
840	C8H18	4-METHYLHEPTANE	114.231	152.20	390.86	25.0	2.2950E+05	4.1309E+00
841	C8H18	n-OCTANE	114.231	216.38	398.83	25.0	2.7196E+05	4.8952E+00
842	C8H18	2,2,3-TRIMETHYLPENTANE	114.231	160.89	383.00	25.0	2.2782E+05	4.1007E+00
843	C8H18	2,2,4-TRIMETHYLPENTANE	114.231	165.78	372.39	25.0	1.9209E+05	3.4576E+00
844	C8H18	2,3,3-TRIMETHYLPENTANE	114.231	172.22	387.92	25.0	2.5476E+05	4.5856E+00
845	C8H18	2,3,4-TRIMETHYLPENTANE	114.231	163.95	386.62	25.0	9.8283E+04	1.7691E+00
846	C8H18	2,2,3,3-TETRAMETHYLBUTANE	114.230	172.47	379.60	------	----------	----------
847	C8H18O	DI-n-BUTYL ETHER	130.230	177.95	413.44	24.8	3.0899E+01	5.5618E-04
848	C8H18O	DI-sec-BUTYL ETHER	130.230	173.15	394.20	25.0	1.9791E+02	3.5623E-03
849	C8H18O	DI-tert-BUTYL ETHER	130.230	195.00	380.40	25.0	1.8279E+02	3.2902E-03
850	C8H18O	2-ETHYL-1-HEXANOL	130.230	203.15	457.75	25.0	1.2928E+01	2.3271E-04
851	C8H18O	1-OCTANOL	130.230	257.65	468.35	25.0	1.3907E+00	2.5032E-05
852	C8H18O	2-OCTANOL	130.230	241.55	452.95	25.0	5.7244E-01	1.0304E-05
853	C8H18O2	DI-t-BUTYL PEROXIDE	146.230	233.15	384.15	------	----------	----------
854	C8H18O2S	DI-n-BUTYL SULFONE	178.296	318.00	564.00	------	----------	----------
855	C8H18O3	DIETHYLENE GLYCOL DIETHYL ETHER	162.229	228.85	462.15	------	----------	----------
856	C8H18O3	DIETHYLENE GLYCOL MONOBUTYL EHTER	162.229	205.15	504.15	------	----------	----------
857	C8H18O4	TRIETHYLENE GLYCOL DIMETHYL ETHER	178.229	229.35	489.15	------	----------	----------
858	C8H18O5	TETRAETHYLENE GLYCOL	194.228	268.15	581.00	------	----------	----------

417

NO	FORMULA	NAME	MW, g/mol	T_F, K	T_B, K	T, C	Henry's Law Constant, H	
							H @ T, atm/mol frac	H @ T, atm/mol/m^3
859	C8H18S	n-OCTYL MERCAPTAN	146.297	223.95	472.19	25.0	1.8038E+03	3.2468E-02
860	C8H18S	tert-OCTYL MERCAPTAN	146.297	199.00	429.00	25.0	2.1323E+03	3.8381E-02
861	C8H18S	BUTYL-SULFIDE	146.290	209.86	455.15	------	----------	----------
862	C8H18S	ETHYL-HEXYL-SULFIDE	146.290	209.86	468.16	------	----------	----------
863	C8H18S	HEPTYL-METHYL-SULFIDE	146.290	209.86	468.16	------	----------	----------
864	C8H18S	PENTYL-PROPYL-SULFIDE	146.290	209.86	468.16	------	----------	----------
865	C8H18S2	BUTYL-DISULFIDE	178.350	202.16	504.36	------	----------	----------
866	C8H19N	DI-n-BUTYLAMINE	129.246	211.15	432.00	25.0	7.8515E+00	1.4133E-04
867	C8H19N	DIISOBUTYLAMINE	129.246	203.15	412.25	------	----------	----------
868	C8H19N	n-OCTYLAMINE	129.246	272.75	452.75	25.0	4.5641E+01	8.2154E-04
869	C8H23N5	TETRAETHYLENEPENTAMINE	189.304	243.00	606.15	------	----------	----------
870	C8H24O4Si4	OCTAMETHYLCYCLOTETRASILOXANE	296.618	290.80	448.15	------	----------	----------
871	C9H4O5	TRIMELLITIC ANHYDRIDE	192.128	438.15	663.00	------	----------	----------
872	C9H6N2O2	TOLUENE DIISOCYANATE	174.159	287.04	523.15	------	----------	----------
873	C9H7N	ISOQUINOLINE	129.161	299.45	516.40	20.0	1.4532E-01	2.6158E-06
874	C9H7N	QUINOLINE	129.161	258.25	510.75	20.0	8.4257E-02	1.5166E-06
875	C9H7NO	8-HYDROXYQUINOLINE	145.161	346.00	540.00	------	----------	----------
876	C9H8	INDENE	116.163	271.70	455.77	------	----------	----------
877	C9H8O	2-METHYLBENZOFURAN	132.162	------	470.65	------	----------	----------
878	C9H10	INDANE	118.178	221.74	451.12	25.0	1.2170E+02	2.1905E-03
879	C9H10	cis-PROPENYLBENZENE	118.178	211.47	443.16	------	----------	----------
880	C9H10	trans-PROPENYLBENZENE	118.178	243.82	443.16	------	----------	----------
881	C9H1O	alpha-METHYLSTYRENE	118.178	249.95	438.65	------	----------	----------
882	C9H10	m-METHYLSTYRENE	118.178	186.81	444.75	25.0	1.8444E+02	3.3199E-03
883	C9H10	o-METHYLSTYRENE	118.178	204.58	442.96	------	----------	----------
884	C9H10	p-METHYLSTYRENE	118.178	239.02	445.93	25.0	1.7722E+02	3.1899E-03
885	C9H10O2	BENZYL ACETATE	150.177	221.65	486.65	------	----------	----------
886	C9H10O2	ETHYL BENZOATE	150.177	238.45	486.55	25.0	4.0614E+00	7.3105E-05
887	C9H10O3	ETHYL VANILLIN	166.177	350.65	567.00	------	----------	----------
888	C9H11NO	p-DIMETHYLAMINOBENZALDEHYDE	149.192	348.00	588.00	------	----------	----------
889	C9H12	CUMENE	120.194	177.14	425.56	25.0	7.9842E+02	1.4371E-02
890	C9H12	m-ETHYLTOLUENE	120.194	177.61	434.48	25.0	3.2147E+02	5.7865E-03
891	C9H12	o-ETHYLTOLUENE	120.194	192.35	438.33	25.0	2.4608E+02	4.4294E-03
892	C9H12	p-ETHYLTOLUENE	120.194	210.83	435.16	25.0	2.7707E+02	4.9872E-03
893	C9H12	MESITYLENE	120.194	228.46	437.89	25.0	4.5128E+02	8.1230E-03
894	C9H12	n-PROPYLBENZENE	120.194	173.67	432.39	25.0	3.2214E+02	5.7985E-03
895	C9H12	1,2,3-TRIMETHYLBENZENE	120.194	247.79	449.27	25.0	4.1166E+02	7.4098E-03
896	C9H12	1,2,4-TRIMETHYLBENZENE	120.194	229.38	442.53	25.0	3.7707E+02	6.7872E-03
897	C9H12O	BENZYL ETHYL ETHER	136.194	275.65	458.15	------	----------	----------
898	C9H12O	2-PHENYL-2-PROPANOL	136.194	309.15	475.15	------	----------	----------
899	C9H12O2	CUMENE HYDROPEROXIDE	152.193	264.26	442.70	------	----------	----------
900	C9H14O	ISOPHORONE	138.210	265.05	488.35	25.0	3.6407E-01	6.5533E-06
901	C9H14O6	GLYCERYL TRIACETATE	218.207	277.25	532.15	------	----------	----------
902	C9H16	1-NONYNE	124.225	223.16	423.96	------	----------	----------
903	C9H16O4	AZELAIC ACID	188.224	379.65	633.36	------	----------	----------
904	C9H18	BUTYLCYCLOPENTANE	126.241	165.18	429.76	25.0	8.3843E+04	1.5092E+00
905	C9H18	cis,cis-1,3,5-TRIMETHYLCYCLOHEXANE	126.241	223.46	411.66	------	----------	----------
906	C9H18	cis,trans-1,3,5-TRIMETHYLCYCLOHEXANE	126.241	188.76	413.70	------	----------	----------
907	C9H18	ISOPROPYLCYCLOHEXANE	126.242	183.76	427.91	25.0	5.8201E+04	1.0476E+00
908	C9H18	1-NONENE	126.242	191.78	420.02	25.0	4.4450E+04	8.0009E-01
909	C9H18	n-PROPYLCYCLOHEXANE	126.242	178.28	429.90	25.0	5.7042E+04	1.0267E+00
910	C9H18O	DIISOBUTYL KETONE	142.241	227.17	441.41	23.5	6.6086E+00	1.1895E-04
911	C9H18O	1-NONANAL	142.241	255.15	468.15	30.0	3.6621E+01	6.5916E-04
912	C9H18O2	n-BUTYL VALERATE	158.241	180.35	459.65	25.0	3.5613E+01	6.4104E-04
913	C9H18O2	n-NONANOIC ACID	158.241	285.55	528.75	20.0	5.2084E-02	9.3751E-07
914	C9H18O2	n-OCTYL FORMATE	158.241	234.05	471.95	25.0	1.0771E+02	1.9388E-03
915	C9H20	3,3-DIETHYLPENTANE	128.258	240.12	419.34	25.0	4.8928E+05	8.8070E+00
916	C9H20	2,2-DIMETHYL-3-ETHYLPENTANE	128.258	173.68	406.99	25.0	3.6719E+05	6.6093E+00
917	C9H20	3-ETHYL-2,3-DIMETHYLPENTANE	128.257	173.67	417.86	25.0	4.7268E+05	8.5082E+00
918	C9H20	2,4-DIMETHYL-3-ETHYLPENTANE	128.258	150.79	409.87	25.0	3.8656E+05	6.9581E+00
919	C9H20	2,2-DIMETHYLHEPTANE	128.258	160.15	405.84	25.0	3.2665E+05	5.8797E+00
920	C9H20	2,6-DIMETHYLHEPTANE	128.258	170.25	408.36	25.0	3.2777E+05	5.8998E+00
921	C9H20	3-ETHYLHEPTANE	128.258	158.25	416.35	25.0	3.8270E+05	6.8886E+00
922	C9H20	4-ETHYLHEPTANE	128.257	159.96	414.36	25.0	3.7170E+05	6.6906E+00
923	C9H20	2,3-DIMETHYLHEPTANE	128.257	160.16	413.66	25.0	3.7768E+05	6.7981E+00
924	C9H20	2,4-DIMETHYLHEPTANE	128.257	160.16	406.05	25.0	3.1742E+05	5.7135E+00
925	C9H20	2,5-DIMETHYLHEPTANE	128.257	160.16	409.16	25.0	3.3648E+05	6.0566E+00
926	C9H20	3,4-DIMETHYLHEPTANE	128.257	170.26	413.76	25.0	3.8312E+05	6.8961E+00
927	C9H20	3,5-DIMETHYLHEPTANE	128.257	170.26	409.16	25.0	3.4014E+05	6.1224E+00
928	C9H20	4,4-DIMETHYLHEPTANE	128.257	170.26	408.36	25.0	3.6253E+05	6.5255E+00
929	C9H20	3-ETHYL-2-METHYLHEXANE	128.257	160.16	411.16	25.0	3.6655E+05	6.5979E+00
930	C9H20	4-ETHYL-2-METHYLHEXANE	128.257	160.16	406.96	25.0	3.3135E+05	5.9643E+00
931	C9H20	3-ETHYL-3-METHYLHEXANE	128.257	160.16	413.76	25.0	4.0899E+05	7.3617E+00
932	C9H20	3-ETHYL-4-METHYLHEXANE	128.257	160.16	413.56	25.0	3.8508E+05	6.9313E+00
933	C9H20	2,2,3-TRIMETHYLHEXANE	128.257	153.16	406.75	25.0	3.6088E+05	6.4957E+00
934	C9H20	2,2,4-TRIMETHYLHEXANE	128.257	153.00	399.69	25.0	3.1636E+05	5.6944E+00
935	C9H20	2,3,3-TRIMETHYLHEXANE	128.257	156.36	410.84	25.0	4.0056E+05	7.2101E+00
936	C9H20	2,3,4-TRIMETHYLHEXANE	128.257	156.36	412.20	25.0	3.9803E+05	7.1645E+00

NO	FORMULA	NAME	MW, g/mol	T_F, K	T_B, K	T, C	H @ T, atm/mol frac	H @ T, atm/mol/m³
							Henry's Law Constant, H	
937	C9H20	2,3,5-TRIMETHYLHEXANE	128.257	145.36	404.50	25.0	3.2928E+05	5.9270E+00
938	C9H20	2,4,4-TRIMETHYLHEXANE	128.257	159.78	403.81	25.0	3.5950E+05	6.4710E+00
939	C9H20	3,3,4-TRIMETHYLHEXANE	128.257	171.96	413.62	25.0	4.2585E+05	7.6653E+00
940	C9H20	2-METHYLOCTANE	128.258	192.78	416.43	25.0	3.5239E+05	6.3429E+00
941	C9H20	3-METHYLOCTANE	128.258	165.55	417.38	25.0	3.7554E+05	6.7597E+00
942	C9H20	4-METHYLOCTANE	128.258	159.95	415.59	25.0	5.5815E+05	1.0047E+01
943	C9H20	n-NONANE	128.258	219.63	423.97	25.0	3.4166E+05	6.1499E+00
944	C9H20	2,2,3,3-TETRAMETHYLPENTANE	128.258	263.26	413.44	25.0	4.4865E+05	8.0756E+00
945	C9H20	2,2,3,4-TETRAMETHYLPENTANE	128.258	152.06	406.18	25.0	3.9067E+05	7.0320E+00
946	C9H20	2,2,4,4-TETRAMETHYLPENTANE	128.258	206.95	395.44	25.0	3.2903E+05	5.9225E+00
947	C9H20	2,3,3,4-TETRAMETHYLPENTANE	128.257	171.10	414.72	25.0	4.5220E+05	8.1396E+00
948	C9H20	2,2,5-TRIMETHYLHEXANE	128.258	167.39	397.24	25.0	3.0336E+05	5.4605E+00
949	C9H20O	2,6-DIMETHYL-4-HEPTANOL	144.257	208.00	451.00	25.0	3.1883E+00	5.7388E-05
950	C9H20O	1-NONANOL	144.257	268.15	486.25	25.0	1.8541E+00	3.3374E-05
951	C9H20O	2-NONANOL	144.257	238.15	471.65	15.0	2.7387E+00	4.9295E-05
952	C9H20S	n-NONYL MERCAPTAN	160.324	253.05	492.95	25.0	1.4699E+03	2.6459E-02
953	C9H20S	BUTYL-PENTYL-SULFIDE	160.317	231.16	491.16	------	----------	----------
954	C9H20S	ETHYL-HEPTYL-SULFIDE	160.317	231.16	491.16	------	----------	----------
955	C9H20S	HEXYL-PROPYL-SULFIDE	160.317	231.16	491.16	------	----------	----------
956	C9H20S	METHYL-OCTYL-SULFIDE	160.317	231.16	491.16	------	----------	----------
957	C9H21N	n-NONYLAMINE	143.272	273.15	475.35	------	----------	----------
958	C9H21N	TRIPROPYLAMINE	143.272	179.65	429.65	25.0	2.1136E+01	3.8044E-04
959	C10H6O8	PYROMELLITIC ACID	254.153	554.00	722.00	------	----------	----------
960	C10H7Br	1-BROMONAPHTHALENE	207.070	279.35	554.25	21.0	1.5772E+01	2.8389E-04
961	C10H7Cl	1-CHLORONAPHTHALENE	162.618	269.15	532.45	25.0	1.0847E+01	1.9524E-04
962	C10H8	NAPHTHALENE	128.174	353.43	491.14	------	----------	----------
963	C10H8	AZULENE	128.173	173.66	515.16	------	----------	----------
964	C10H9N	QUINALDINE	143.188	272.15	519.75	------	----------	----------
965	C10H10	m-DIVINYLBENZENE	130.189	206.25	472.65	-----	----------	----------
966	C10H10	1-METHYLINDENE	130.189	-------	471.65	-----	----------	----------
967	C10H10	2-METHYLINDENE	130.189	353.15	458.00	-----	----------	----------
968	C10H10O4	DIMETHYL PHTHALATE	194.187	272.15	556.85	20.0	1.0110E-02	1.8198E-07
969	C10H10O4	DIMETHYL TEREPHTHALATE	194.187	413.80	561.15	------	----------	----------
970	C10H12	DICYCLOPENTADIENE	132.205	307.00	443.00	------	----------	----------
971	C10H12	1,2,3,4-TETRAHYDRONAPHTHALENE	132.205	237.40	480.77	28.0	7.6039E+01	1.3687E-03
972	C10H12O	ANETHOLE	148.205	294.50	508.45	25.0	5.4529E+00	9.8152E-05
973	C10H12O4	DIALLYL MALEATE	196.203	226.15	520.00	------	----------	----------
974	C10H14	n-BUTYLBENZENE	134.221	185.30	456.46	25.0	7.2577E+02	1.3064E-02
975	C10H14	sec-BUTYLBENZENE	134.221	197.72	446.48	25.0	4.0737E+02	7.3327E-03
976	C10H14	tert-BUTYLBENZENE	134.221	215.27	442.30	25.0	3.9239E+02	7.0629E-03
977	C10H14	1,2,3,4-TETRAMETHYLBENZENE	134.221	266.91	478.25	------	----------	----------
978	C10H14	m-CYMENE	134.221	209.44	448.23	25.0	4.4262E+02	7.9671E-03
979	C10H14	o-CYMENE	134.221	201.64	451.33	25.0	4.5878E+02	8.2579E-03
980	C10H14	p-CYMENE	134.221	205.25	450.28	25.0	4.2200E+02	7.5959E-03
981	C10H14	m-DIETHYLBENZENE	134.221	189.26	454.29	25.0	4.0843E+02	7.3517E-03
982	C10H14	o-DIETHYLBENZENE	134.221	241.93	456.61	25.0	4.3246E+02	7.7842E-03
983	C10H14	p-DIETHYLBENZENE	134.221	230.32	456.94	25.0	4.4138E+02	7.9448E-03
984	C1OH14	2-ETHYL-m-XYLENE	134.221	256.89	463.19	25.0	4.3441E+02	7.8194E-03
985	C10H14	2-ETHYL-p-XYLENE	134.221	219.52	459.98	25.0	4.6941E+02	8.4493E-03
986	C10H14	3-ETHYL-o-XYLENE	134.221	223.64	467.11	25.0	4.5615E+02	8.2106E-03
987	C10H14	4-ETHYL-m-XYLENE	134.221	210.27	461.59	25.0	4.8194E+02	8.6748E-03
988	C10H14	4-ETHYL-o-XYLENE	134.221	206.22	462.93	25.0	4.3646E+02	7.8563E-03
989	C1OH14	5-ETHYL-m-XYLENE	134.221	188.82	456.93	25.0	4.3083E+02	7.7549E-03
990	C10H14	ISOBUTYLBENZENE	134.221	221.70	445.94	25.0	4.3450E+02	7.8209E-03
991	C10H14	1,2,3,5-TETRAMETHYLBENZENE	134.221	249.46	471.15	25.0	4.5451E+02	8.1811E-03
992	C10H14	1,2,4,5-TETRAMETHYLBENZENE	134.221	352.38	469.99	------	----------	----------
993	C10H14O	p-tert-BUTYLPHENOL	150.221	371.56	512.88	------	----------	----------
994	C10H14O2	p-tert-BUTYLCATECHOL	166.220	325.00	558.00	------	----------	----------
995	C10H15N	N,N-DIETYHLANILINE	149.236	235.15	489.42	12.0	1.0547E-01	1.8984E-06
996	C10H15N	2,6-DIETHYLANILINE	149.236	276.65	508.65	26.7	6.2029E-02	1.1165E-06
997	C10H16	CAMPHENE	136.237	320.15	433.65	------	----------	----------
998	C10H16	D-LIMONENE	136.237	199.00	449.65	25.0	1.4272E+03	2.5689E-02
999	C10H16	alpha-PHELLANDRENE	136.237	-------	448.15	-----	----------	----------
1000	C1OH16	beta-PHELLANDRENE	136.237	-------	447.15	-----	----------	----------
1001	C10H16	alpha-PINENE	136.237	209.15	429.29	-----	----------	----------
1002	C10H16	beta-PINENE	136.237	211.61	439.19	------	----------	----------
1003	C10H16	alpha-TERPINENE	136.237	-------	450.35	-----	----------	----------
1004	C1OH16	gamma-TERPINENE	136.237	-------	456.15	-----	----------	----------
1005	C10H16	TERPINOLENE	136.237	-------	458.15	-----	----------	----------
1006	C10H16O	CAMPHOR	152.236	453.25	480.57	------	----------	----------
1007	C10H18	1-DECYNE	138.252	229.16	447.16	-----	----------	----------
1008	C10H18	cis-DECAHYDRONANPHTALENE	138.253	230.20	468.97	25.0	8.9272E+03	1.6069E-01
1009	C10H18	trans-DECAHYDRONAPHTALENE	138.253	242.79	460.46	25.0	1.3840E+04	2.4912E-01
1010	C10H18O4	SEBACIC ACID	202.251	407.65	642.09	------	----------	----------
1011	C10H20	n-BUTYLCYCLOHEXANE	140.269	198.42	454.13	25.0	7.5701E+04	1.3626E+00
1012	C10H20	1-CYCLOPENTYLPENTANE	140.268	190.16	453.76	------	----------	----------
1013	C10H20	1-DECENE	140.269	206.89	443.75	25.0	1.1590E+05	2.0861E+00
1014	C10H20O	1-DECANAL	156.268	267.15	488.15	------	----------	----------

NO	FORMULA	NAME	MW, g/mol	T_F, K	T_B, K	T, C	Henry's Law Constant, H	
							H @ T, atm/mol frac	H @ T, atm/mol/m³
1015	C10H20O2	n-DECANOIC ACID	172.268	304.75	543.15	20.0	2.9029E-03	5.2252E-08
1016	C10H20O2	2-ETHYLHEXYL ACETATE	172.268	180.15	471.75	25.0	4.3685E+01	7.8632E-04
1017	C10H20O2	ISOPENTYL ISOVALERATE	172.268	215.00	467.15	25.0	1.3570E+02	2.4427E-03
1018	C10H22	n-DECANE	142.285	243.49	447.30	25.0	2.8508E+05	5.1314E+00
1019	C10H22	2-METHYLNONANE	142.285	198.50	440.15	25.0	4.6256E+05	8.3260E+00
1020	C10H22	3-METHYLNONANE	142.285	188.35	440.95	25.0	5.0756E+05	9.1360E+00
1021	C10H22	4-METHYLNONANE	142.285	174.45	438.85	25.0	5.3002E+05	9.5402E+00
1022	C10H22	5-METHYLNONANE	142.285	185.45	438.30	25.0	4.9028E+05	8.8250E+00
1023	C10H22	3-ETHYLOCTANE	142.284	185.46	439.66	------	----------	----------
1024	C10H22	4-ETHYLOCTANE	142.284	185.46	436.80	25.0	5.1647E+05	9.2964E+00
1025	C10H22	2,2-DIMETHYLOCTANE	142.285	------	430.05	25.0	5.0456E+05	9.0821E+00
1026	C10H22	2,3-DIMETHYLOCTANE	142.284	219.16	437.47	25.0	5.2973E+05	9.5351E+00
1027	C10H22	2,4-DIMETHYLOCTANE	142.284	219.16	429.06	25.0	4.5232E+05	8.1417E+00
1028	C10H22	2,5-DIMETHYLOCTANE	142.284	219.16	431.66	25.0	4.8104E+05	8.6587E+00
1029	C10H22	2,6-DIMETHYLOCTANE	142.284	219.16	433.54	25.0	4.9909E+05	8.9835E+00
1030	C10H22	2,7-DIMETHYLOCTANE	142.284	219.16	433.03	25.0	4.6396E+05	8.3513E+00
1031	C10H22	3,3-DIMETHYLOCTANE	142.284	219.16	434.36	25.0	5.4190E+05	9.7542E+00
1032	C10H22	3,4-DIMETHYLOCTANE	142.284	219.16	436.56	25.0	5.3687E+05	9.6635E+00
1033	C10H22	3,5-DIMETHYLOCTANE	142.284	219.16	432.56	25.0	5.0352E+05	9.0633E+00
1034	C10H22	3,6-DIMETHYLOCTANE	142.284	219.16	433.96	25.0	4.9576E+05	8.9236E+00
1035	C10H22	4,4-DIMETHYLOCTANE	142.284	219.16	430.66	25.0	5.2060E+05	9.3707E+00
1036	C10H22	4,5-DIMETHYLOCTANE	142.284	219.16	435.29	25.0	5.3617E+05	9.6510E+00
1037	C10H22	4-PROPYLHEPTANE	142.284	219.16	430.66	25.0	4.5599E+05	8.2077E+00
1038	C10H22	4-ISOPROPYLHEPTANE	142.284	219.16	432.06	25.0	5.2329E+05	9.4191E+00
1039	C10H22	3-ETHYL-2-METHYLHEPTANE	142.284	219.16	434.36	25.0	5.3307E+05	9.5952E+00
1040	C10H22	4-ETHYL-2-METHYLHEPTANE	142.284	219.16	429.36	25.0	4.9032E+05	8.8257E+00
1041	C10H22	5-ETHYL-2-METHYLHEPTANE	142.284	219.16	432.86	25.0	5.0795E+05	9.1430E+00
1042	C10H22	3-ETHYL-3-METHYLHEPTANE	142.284	219.16	436.96	25.0	5.9879E+05	1.0778E+01
1043	C10H22	4-ETHYL-3-METHYLHEPTANE	142.284	219.16	435.36	25.0	5.5893E+05	1.0061E+01
1044	C10H22	3-ETHYL-5-METHYLHEPTANE	142.284	219.16	431.36	25.0	4.9486E+05	8.9074E+00
1045	C10H22	3-ETHYL-4-METHYLHEPTANE	142.284	219.16	436.16	25.0	5.6089E+05	1.0096E+01
1046	C10H22	4-ETHYL-4-METHYLHEPTANE	142.284	219.16	433.96	25.0	5.7509E+05	1.0352E+01
1047	C10H22	2,2,3-TRIMETHYLHEPTANE	142.284	219.16	430.76	25.0	5.4217E+05	9.7589E+00
1048	C10H22	2,2,4-TRIMETHYLHEPTANE	142.284	219.16	421.46	25.0	4.5940E+05	8.2691E+00
1049	C10H22	2,2,5-TRIMETHYLHEPTANE	142.284	219.16	423.96	25.0	4.7399E+05	8.5318E+00
1050	C10H22	2,2,6-TRIMETHYLHEPTANE	142.284	219.16	422.09	25.0	4.3787E+05	7.8817E+00
1051	C10H22	2,3,3-TRIMETHYLHEPTANE	142.284	219.16	433.36	25.0	5.7646E+05	1.0376E+01
1052	C10H22	2,3,4-TRIMETHYLHEPTANE	142.284	219.16	433.06	25.0	5.5257E+05	9.9462E+00
1053	C10H22	2,3,5-TRIMETHYLHEPTANE	142.284	219.16	433.86	25.0	5.7583E+05	1.0365E+01
1054	C10H22	2,3,6-TRIMETHYLHEPTANE	142.284	219.16	429.16	25.0	4.9135E+05	8.8443E+00
1055	C10H22	2,4,4-TRIMETHYLHEPTANE	142.284	219.16	424.16	25.0	5.0124E+05	9.0222E+00
1056	C10H22	2,4,5-TRIMETHYLHEPTANE	142.284	219.16	429.66	25.0	5.1819E+05	9.3273E+00
1057	C10H22	2,4,6-TRIMETHYLHEPTANE	142.284	219.16	420.76	25.0	4.1352E+05	7.4432E+00
1058	C10H22	2,5,5-TRIMETHYLHEPTANE	142.284	219.16	425.96	25.0	4.9440E+05	8.8991E+00
1059	C10H22	3,3,4-TRIMETHYLHEPTANE	142.284	219.16	435.06	25.0	6.0885E+05	1.0959E+01
1060	C10H22	3,3,5-TRIMETHYLHEPTANE	142.284	219.16	428.85	25.0	5.4920E+05	9.8856E+00
1061	C10H22	3,4,4-TRIMETHYLHEPTANE	142.284	219.16	434.26	25.0	6.0671E+05	1.0921E+01
1062	C10H22	3,4,5-TRIMETHYLHEPTANE	142.284	219.16	435.66	25.0	5.8235E+05	1.0482E+01
1063	C10H22	3-ISOPROPYL-2-METHYLHEXANE	142.284	219.16	439.86	25.0	7.2974E+05	1.3135E+01
1064	C10H22	3,3-DIETHYLHEXANE	142.284	219.16	439.46	25.0	6.4521E+05	1.1614E+01
1065	C10H22	3,4-DIETHYLHEXANE	142.284	219.16	437.06	25.0	5.8992E+05	1.0619E+01
1066	C10H22	3-ETHYL-2,2-DIMETHYLHEXANE	142.284	219.16	429.26	25.0	5.5993E+05	1.0079E+01
1067	C10H22	4-ETHYL-2,2-DIMETHYLHEXANE	142.284	219.16	420.16	25.0	4.5789E+05	8.2420E+00
1068	C10H22	3-ETHYL-2,3-DIMETHYLHEXANE	142.284	219.16	436.86	25.0	6.3957E+05	1.1512E+01
1069	C10H22	4-ETHYL-2,3-DIMETHYLHEXANE	142.284	219.16	434.06	25.0	5.7962E+05	1.0433E+01
1070	C10H22	3-ETHYL-2,4-DIMETHYLHEXANE	142.284	219.16	433.26	25.0	5.7773E+05	1.0399E+01
1071	C10H22	4-ETHYL-2,4-DIMETHYLHEXANE	142.284	219.16	434.26	25.0	6.0192E+05	1.0835E+01
1072	C10H22	3-ETHYL-2,5-DIMETHYLHEXANE	142.284	219.16	427.26	25.0	5.0852E+05	9.1533E+00
1073	C10H22	4-ETHYL-3,3-DIMETHYLHEXANE	142.284	219.16	436.06	25.0	6.3880E+05	1.1498E+01
1074	C10H22	3-ETHYL-3,4-DIMETHYLHEXANE	142.284	219.16	435.26	25.0	6.3664E+05	1.1459E+01
1075	C10H22	2,2,3,3-TETRAMETHYLHEXANE	142.284	219.16	433.48	25.0	6.5159E+05	1.1728E+01
1076	C10H22	2,2,3,4-TETRAMETHYLHEXANE	142.284	219.16	431.96	25.0	6.3126E+05	1.1363E+01
1077	C10H22	2,2,3,5-TETRAMETHYLHEXANE	142.284	219.16	421.56	25.0	4.7808E+05	8.6054E+00
1078	C10H22	2,2,4,4-TETRAMETHYLHEXANE	142.284	219.16	426.96	25.0	6.6132E+05	1.1904E+01
1079	C10H22	2,2,4,5-TETRAMETHYLHEXANE	142.284	219.16	421.04	25.0	4.9673E+05	8.9411E+00
1080	C10H22	2,2,5,5-TETRAMETHYLHEXANE	142.284	260.56	410.63	25.0	3.9129E+05	7.0431E+00
1081	C10H22	2,3,3,4-TETRAMETHYLHEXANE	142.284	260.56	437.75	25.0	6.8980E+05	1.2416E+01
1082	C10H22	2,3,3,5-TETRAMETHYLHEXANE	142.284	260.56	426.26	25.0	5.4532E+05	9.8156E+00
1083	C10H22	2,3,4,4-TETRAMETHYLHEXANE	142.284	260.56	434.76	25.0	6.6300E+05	1.1934E+01
1084	C10H22	2,3,4,5-TETRAMETHYLHEXANE	142.284	260.56	429.36	25.0	5.4237E+05	9.7625E+00
1085	C10H22	3,3,4,4-TETRAMETHYLHEXANE	142.284	260.56	443.16	25.0	8.2193E+05	1.4795E+01
1086	C10H22	2,4-DIMETHYL-3-ISOPROPYLPENTANE	142.284	191.46	430.20	25.0	5.9871E+05	1.0777E+01
1087	C10H22	3,3-DIETHYL-2-METHYLPENTANE	142.284	191.46	442.86	25.0	7.2852E+05	1.3113E+01
1088	C10H22	3-ETHYL-2,2,3-TRIMETHYLPENTANE	142.284	191.46	442.66	25.0	8.2603E+05	1.4868E+01
1089	C10H22	3-ETHYL-2,2,4-TRIMETHYLPENTANE	142.284	191.46	428.46	25.0	5.9696E+05	1.0745E+01
1090	C10H22	3-ETHYL-2,3,4-TRIMETHYLPENTANE	142.284	191.46	442.60	25.0	7.8221E+05	1.4080E+01
1091	C10H22	2,2,3,3,4-PENTAMETHYLPENTANE	142.284	236.71	439.21	25.0	8.0376E+05	1.4468E+01
1092	C10H22	2,2,3,4,4-PENTAMETHYLPENTANE	142.284	234.41	432.45	25.0	7.6369E+05	1.3746E+01

NO	FORMULA	NAME	MW, g/mol	T_F, K	T_B, K	T, C	H @ T, atm/mol frac	H @ T, atm/mol/m³
							Henry's Law Constant, H	
1093	C10H22O	1-DECANOL	158.284	280.05	503.35	25.0	2.6559E+00	4.7805E-05
1094	C10H22O	DI-n-PENTYL ETHER	158.284	203.72	459.90	25.0	5.3121E+02	9.5616E-03
1095	C10H22O	ISODECANOL	158.284	213.15	493.00	------	----------	----------
1096	C10H22O5	TETRAETHYLENE GLYCOL DIMETHYL ETHER	222.282	243.45	548.95	------	----------	----------
1097	C10H22S	n-DECYL MERCAPTAN	174.351	247.56	512.35	25.0	8.3931E+02	1.5108E-02
1098	C10H22S	BUTYL-HEXYL-SULFIDE	174.344	238.16	513.16	------	----------	----------
1099	C10H22S	ETHYL-OCTYL-SULFIDE	174.344	238.16	513.16	------	----------	----------
1100	C10H22S	HEPTYL-PROPYL-SULFIDE	174.344	238.16	513.16	------	----------	----------
1101	C10H22S	METHYL-NONYL-SULFIDE	174.344	238.16	513.16	------	----------	----------
1102	C10H22S	PENTYL-SULFIDE	174.344	238.16	513.16	------	----------	----------
1103	C10H22S2	PENTYL-DISULFIDE	206.404	214.16	537.06	------	----------	----------
1104	C10H23N	n-DECYLAMINE	157.299	288.85	493.65	------	----------	----------
1105	C11H10	1-METHYLNAPHTHALENE	142.200	242.67	517.83	25.0	2.5077E+01	4.5139E-04
1106	C11H10	2-METHYLNAPHTHALENE	142.200	307.73	514.20	25.0	2.4135E+01	4.3443E-04
1107	C11H14O2	n-BUTYL BENZOATE	178.231	251.65	523.15	------	----------	----------
1108	C11H16	n-PENTYLBENZENE	148.248	198.15	478.61	25.0	1.2351E+03	2.2231E-02
1109	C11H16O	p-tert-AMYLPHENOL	164.247	366.00	535.15	------	----------	----------
1110	C11H20	1-UNDECYNE	152.279	248.16	468.16	------	----------	----------
1111	C11H20O2	2-ETHYLHEXYL ACRYLATE	184.274	183.15	489.15	23.0	2.3924E+01	4.3063E-04
1112	C11H22	1-UNDECENE	154.296	223.99	465.82	25.0	1.1881E+05	2.1386E+00
1113	C11H22	1-CYCLOPENTYLHEXANE	154.295	200.16	476.26	------	----------	----------
1114	C11H22	PENTYLCYCLOHEXANE	154.295	215.66	476.87	------	----------	----------
1115	C11H22O	1-UNDECANAL	170.295	273.15	506.15	------	----------	----------
1116	C11H24	n-UNDECANE	156.312	247.57	469.08	25.0	1.0681E+05	1.9226E+00
1117	C11H24O	1-UNDECANOL	172.311	289.05	518.15	25.0	1.8188E+00	3.2738E-05
1118	C11H24S	UNDECYL MERCAPTAN	188.378	270.15	530.55	25.0	1.5972E+02	2.8749E-03
1119	C11H24S	BUTYL-HEPTYL-SULFIDE	188.371	254.66	533.16	------	----------	----------
1120	C11H24S	DECYL-METHYL-SULFIDE	188.371	254.66	533.16	------	----------	----------
1121	C11H24S	ETHYL-NONYL-SULFIDE	188.371	254.66	533.16	------	----------	----------
1122	C11H24S	OCTYL-PROPYL-SULFIDE	188.371	254.66	533.16	------	----------	----------
1123	C12H8O	DIBENZOFURAN	168.195	355.65	557.86	------	----------	----------
1124	C12H9N	DIBENZOPYRROLE	167.210	517.95	627.86	------	----------	----------
1125	C12H10	ACENAPHTHENE	154.211	366.56	550.54	------	----------	----------
1126	C12H10	BIPHENYL	154.211	342.37	528.15	------	----------	----------
1127	C12H10O	DIPHENYL ETHER	170.211	300.02	531.46	25.0	2.9535E+01	5.3163E-04
1128	C12H11N	p-AMINODIPHENYL	169.226	326.00	575.00	------	----------	----------
1129	C12H11N	DIPHENYLAMINE	169.226	326.15	575.15	------	----------	----------
1130	C12H11N3	p-AMINOAZOBENZENE	197.240	401.00	633.00	------	----------	----------
1131	C12H11N3	1,3-DIPHENYLTRIAZENE	197.240	372.00	610.00	------	----------	----------
1132	C12H12	1,2-DIMETHYLNAPHTHALENE	156.227	272.16	539.46	------	----------	----------
1133	C12H12	1,3-DIMETHYLNAPHTHALENE	156.227	269.16	538.36	------	----------	----------
1134	C12H12	1,4-DIMETHYLNAPHTHALENE	156.227	280.82	540.46	------	----------	----------
1135	C12H12	1,5-DIMETHYLNAPHTHALENE	156.227	355.16	538.16	------	----------	----------
1136	C12H12	1,6-DIMETHYLNAPHTHALENE	156.227	259.16	536.16	------	----------	----------
1137	C12H12	1,7-DIMETHYLNAPHTHALENE	156.227	260.16	536.16	------	----------	----------
1138	C12H12	2,3-DIMETHYLNAPHTHALENE	156.227	378.16	541.16	------	----------	----------
1139	C12H12	2,6-DIMETHYLNAPHTHALENE	156.227	384.55	535.15	------	----------	----------
1140	C12H12	2,7-DIMETHYLNAPHTHALENE	156.227	370.15	536.15	------	----------	----------
1141	C12H12	1-ETHYLNAPHTHALENE	156.227	259.34	531.48	25.0	2.6828E+01	4.8290E-04
1142	C12H12	2-ETHYLNAPHTHALENE	156.227	265.76	531.49	------	----------	----------
1143	C12H12N2	p-AMINODIPHENYLAMINE	184.241	341.15	627.15	------	----------	----------
1144	C12H12N2	HYDRAZOBENZENE	184.241	404.15	573.00	------	----------	----------
1145	C12H14	1,2,3-TRIMETHYLINDENE	158.243	344.65	509.00	------	----------	----------
1146	C12H14O4	DIETHYL PHTHALATE	222.241	269.15	567.15	25.0	7.9204E-03	1.4257E-07
1147	C12H16	CYCLOHEXYLBENZENE	160.259	280.14	513.27	------	----------	----------
1148	C12H18	m-DIISOPROPYLBENZENE	162.275	210.02	476.33	25.0	5.7371E+02	1.0327E-02
1149	C12H18	p-DIISOPROPYLBENZENE	162.275	256.08	483.65	25.0	5.2581E+02	9.4645E-03
1150	C12H18	n-HEXYLBENZENE	162.275	212.00	499.26	25.0	1.4031E+03	2.5256E-02
1151	C12H18	1,2,3-TRIETHYLBENZENE	162.274	206.66	490.66	------	----------	----------
1152	C12H18	1,2,4-TRIETHYLBENZENE	162.274	206.66	490.66	25.0	7.5649E+02	1.3617E-02
1153	C12H18	1,3,5-TRIETHYLBENZENE	162.274	206.66	489.15	------	----------	----------
1154	C12H18	HEXAMETHYLBENZENE	162.274	438.66	536.60	------	----------	----------
1155	C12H20O4	DIBUTYL MALEATE	228.288	188.15	553.15	------	----------	----------
1156	C12H22	BICYCLOHEXYL	166.307	276.78	512.19	------	----------	----------
1157	C12H22	1-DODECYNE	166.306	254.16	488.16	------	----------	----------
1158	C12H23N	DICYCLOHEXYLAMINE	181.321	273.05	529.00	------	----------	----------
1159	C12H24	1-DODECENE	168.323	237.93	486.50	25.0	1.0598E+05	1.9076E+00
1160	C12H24	1-CYCLOPENTYLHEPTANE	168.322	220.00	497.30	------	----------	----------
1161	C12H24	1-CYCLOHEXYLHEXANE	168.322	263.60	497.86	------	----------	----------
1162	C12H24O	1-DODECANAL	184.322	285.15	523.15	------	----------	----------
1163	C12H24O2	n-DODECANOIC ACID	200.321	317.15	571.85	20.0	6.5641E-03	1.1815E-07
1164	C12H26	n-DODECANE	170.338	263.57	489.47	25.0	4.5619E+05	8.2113E+00
1165	C12H26O	DI-n-HEXYL ETHER	186.338	230.15	498.85	25.0	1.8229E+02	3.2811E-03
1166	C12H26O	1-DODECANOL	186.338	296.95	535.00	25.0	2.8809E+00	5.1856E-05
1167	C12H26O3	DIETHYLENE GLYCOL DI-n-BUTYL ETHER	218.337	212.95	529.15	20.0	1.3809E-01	2.4856E-06
1168	C12H26S	n-DODECYL MERCAPTAN	202.404	265.15	547.75	25.0	3.8593E+02	6.9468E-03
1169	C12H26S	BUTYL-OCTYL-SULFIDE	202.397	259.16	552.16	------	----------	----------
1170	C12H26S	DECYL-ETHYL-SULFIDE	202.397	259.16	552.16	------	----------	----------

NO	FORMULA	NAME	MW, g/mol	T_F, K	T_B, K	T, C	H @ T, atm/mol frac	H @ T, atm/mol/m³
							Henry's Law Constant, H	
1171	C12H26S	HEXYL-SULFIDE	202.397	259.16	552.16	------	----------	----------
1172	C12H26S	METHYL-UNDECYL-SULFIDE	202.397	259.16	552.16	------	----------	----------
1173	C12H26S	NONYL-PROPYL-SULFIDE	202.397	259.16	552.16	------	----------	----------
1174	C12H26S2	HEXYL-DISULFIDE	234.457	225.16	566.66	------	----------	----------
1175	C12H27BO3	TRI-n-BUTYL BORATE	230.156	203.15	506.65	------	----------	----------
1176	C12H27N	DODECYLAMINE	185.353	301.47	532.35	------	----------	----------
1177	C12H27N	TRI-n-BUTYLAMINE	185.353	203.00	487.15	25.0	8.9251E+00	1.6065E-04
1178	C13H10	FLUORENE	166.222	387.94	570.44	------	----------	----------
1179	C13H10O	BENZOPHENONE	182.222	321.35	579.24	25.0	3.2004E-01	5.7607E-06
1180	C13H12	DIPHENYLMETHANE	168.238	298.39	537.42	25.0	1.2626E+01	2.2727E-04
1181	C13H14	1-PROPYLNAPHTHALENE	170.254	264.69	545.96	------	----------	----------
1182	C13H14	2-PROPYLNAPHTHALENE	170.254	270.16	546.66	------	----------	----------
1183	C13H14	2ETHYL-3-METHYLNAPHTHALENE	170.254	344.16	550.16	------	----------	----------
1184	C13H14	2ETHYL-6-METHYLNAPHTHALENE	170.254	318.16	543.16	------	----------	----------
1185	C13H14	2ETHYL-7-METHYLNAPHTHALENE	170.254	318.16	543.16	------	----------	----------
1186	C13H20	n-HEPTYLBENZENE	176.302	225.15	519.25	25.0	3.7009E+02	6.6616E-03
1187	C13H24	1-TRIDECYNE	180.333	268.16	507.16	------	----------	----------
1188	C13H26	1-TRIDECENE	182.349	250.08	505.93	25.0	9.2190E+04	1.6594E+00
1189	C13H26	1-CYCLOPENTYLOCTANE	182.348	229.16	516.86	------	----------	----------
1190	C13H26	1-CYCLOHEXYLHEPTANE	182.348	242.66	518.06	------	----------	----------
1191	C13H26O	1-TRIDECANAL	198.349	288.15	540.15	------	----------	----------
1192	C13H26O2	n-BUTYL NONANOATE	214.348	235.15	503.00	------	----------	----------
1193	C13H26O2	METHYL DODECANOATE	214.348	278.15	540.00	------	----------	----------
1194	C13H28	n-TRIDECANE	184.365	267.76	508.62	25.0	3.7752E+05	6.7954E+00
1195	C13H28O	1-TRIDECANOL	200.365	303.75	547.15	25.0	2.9639E+00	5.3350E-05
1196	C13H28S	BUTYL-NONYL-SULFIDE	216.424	271.16	570.16	------	----------	----------
1197	C13H28S	DECYL-PROPYL-SULFIDE	216.424	271.16	570.16	------	----------	----------
1198	C13H28S	DODECYL-METHYL-SULFIDE	216.424	271.16	570.16	------	----------	----------
1199	C13H28S	ETHYL-UNDECYL-SULFIDE	216.424	271.16	570.16	------	----------	----------
1200	C13H28S	1-TRIDECANETHIOL	216.424	282.04	563.96	------	----------	----------
1201	C14H8O2	ANTHRAQUINONE	208.216	559.15	653.05	------	----------	----------
1202	C14H10	ANTHRACENE	178.233	489.25	615.18	------	----------	----------
1203	C14H10	DIPHENYLACETYLENE	178.233	335.65	573.00	------	----------	----------
1204	C14H10	PHENANTHRENE	178.233	372.38	613.45	------	----------	----------
1205	C14H12	cis-STILBENE	180.249	268.15	535.00	------	----------	----------
1206	C14H12	trans-STILBENE	180.249	397.35	579.65	------	----------	----------
1207	C14H12O2	BENZYL BENZOATE	212.248	292.55	596.65	------	----------	----------
1208	C14H14	1,1-DIPHENYLETHANE	182.265	255.20	545.78	------	----------	----------
1209	C14H14	1,2-DIPHENYLETHANE	182.265	324.34	553.65	------	----------	----------
1210	C14H14O	DIBENZYL ETHER	198.265	276.75	561.45	------	----------	----------
1211	C14H16	1-n-BUTYLNAPHTHALENE	184.281	253.43	562.54	------	----------	----------
1212	C14H16	2-BUTYLNAPHTHALENE	184.280	268.16	562.16	------	----------	----------
1213	C14H22	n-OCTYLBENZENE	190.329	237.15	537.55	25.0	2.7616E+02	4.9708E-03
1214	C14H22	1,2,3,4-TETRAETHYLBENZENE	190.328	284.96	524.16	------	----------	----------
1215	C14H22	1,2,3,5-TETRAETHYLBENZENE	190.328	284.16	523.66	------	----------	----------
1216	C14H22	1,2,4,5-TETRAETHYLBENZENE	190.328	283.16	523.16	------	----------	----------
1217	C14H22O	p-tert-OCTYLPHENOL	206.328	358.55	563.60	------	----------	----------
1218	C14H28	1-TETRADECENE	196.376	260.30	524.25	25.0	3.7034E+04	6.6661E-01
1219	C14H28	1-CYCLOPENTYLNONANE	196.375	244.16	535.26	------	----------	----------
1220	C14H28	1-CYCLOHEXYLOCTANE	196.375	253.46	536.76	------	----------	----------
1221	C14H28O2	n-TETRADECANOIC ACID	228.375	327.55	599.35	------	----------	----------
1222	C14H30	n-TETRADECANE	198.392	279.01	526.73	25.0	7.6139E+04	1.3705E+00
1223	C14H30O	1-TETRADECANOL	214.392	310.65	560.15	25.0	5.7773E+00	1.0399E-04
1224	C14H30S	BUTYL-DECYL-SULFIDE	230.451	276.16	587.16	------	----------	----------
1225	C14H30S	DODECYL-ETHYL-SULFIDE	230.451	276.16	587.16	------	----------	----------
1226	C14H30S	HEPTYL-SULFIDE	230.451	276.16	587.16	------	----------	----------
1227	C14H30S	METHYL-TRIDECYL-SULFIDE	230.451	276.16	587.16	------	----------	----------
1228	C14H30S	PROPYL-UNDECYL-SULFIDE	230.451	276.16	587.16	------	----------	----------
1229	C14H30S	1-TETRADECANETHIOL	230.451	279.26	579.36	------	----------	----------
1230	C14H30S2	HEPTYL-DISULFIDE	262.511	235.16	593.86	------	----------	----------
1231	C14H31N	TETRADECYLAMINE	213.407	311.34	564.45	------	----------	----------
1232	C15H10N2O2	DIPHENYLMETHANE-4,4'-DIISOCYANATE	250.257	311.20	609.00	------	----------	----------
1233	C15H16O	p-CUMYLPHENOL	212.291	346.00	608.15	------	----------	----------
1234	C15H16O2	BISPHENOL A	228.291	426.15	633.65	------	----------	----------
1235	C15H18	1-PENTYLNAPHTHALENE	198.307	251.16	580.16	------	----------	----------
1236	C15H18	2-PENTYLNAPHTHALENE	198.307	269.16	583.16	------	----------	----------
1237	C15H24	n-NONYLBENZENE	204.356	249.00	555.20	25.0	1.6460E+02	2.9627E-03
1238	C15H24O	2,6-DI-tert-BUTYL-p-CRESOL	220.355	344.00	538.00	------	----------	----------
1239	C15H24O	NONYLPHENOL	220.355	------	581.00	25.0	2.1599E-01	3.8877E-06
1240	C15H28	1-PENTADECYNE	208.386	283.16	541.16	------	----------	----------
1241	C15H30	1-PENTADECENE	210.403	269.42	541.61	25.0	1.5154E+04	2.7277E-01
1242	C15H30	1-CYCLOPENTYLDECANE	210.402	251.03	552.54	------	----------	----------
1243	C15H30	1-CYCLOHEXYLNONANE	210.402	262.96	554.66	------	----------	----------
1244	C15H30O2	PENTADECANOIC ACID	242.402	325.68	612.05	------	----------	----------
1245	C15H32	n-PENTADECANE	212.419	283.11	543.83	25.0	4.9916E+04	8.9848E-01
1246	C15H32O	1-PENTADECANOL	228.417	317.04	578.01	------	----------	----------
1247	C15H32S	BUTYL-UNDECYL-SULFIDE	244.478	284.16	603.16	------	----------	----------
1248	C15H32S	DODECYL-PROPYL-SULFIDE	244.478	284.16	603.16	------	----------	----------

Table 18-1 HENRY'S LAW CONSTANT FOR COMPOUND IN WATER - ORGANIC COMPOUNDS (continued)

NO	FORMULA	NAME	MW, g/mol	T_F, K	T_B, K	T, C	H @ T, atm/mol frac	H @ T, atm/mol/m³
1249	C15H32S	ETHYL-TRIDECYL-SULFIDE	244.478	284.16	603.16	------	----------	----------
1250	C15H32S	METHYL-TETRADECYL-SULFIDE	244.478	284.16	603.16	------	----------	----------
1251	C15H32S	1-PENTADECANETHIOL	244.478	290.93	593.86	------	----------	----------
1252	C16H10	FLUORANTHENE	202.255	383.33	655.95	------	----------	----------
1253	C16H10	PYRENE	202.255	423.81	667.95	------	----------	----------
1254	C16H12	1-PHENYLNAPHTHALENE	204.271	318.15	607.15	------	----------	----------
1255	C16H20	1-n-HEXYLNAPHTHALENE	212.335	255.15	595.15	------	----------	----------
1256	C16H22O4	DIBUTYL PHTHALATE	278.348	238.15	613.15	20.0	2.0532E-02	3.6957E-07
1257	C16H26	n-DECYLBENZENE	218.382	258.77	571.04	25.0	6.9163E+01	1.2449E-03
1258	C16H26	PENTAETHYLBENZENE	218.381	327.66	550.16	------	----------	----------
1259	C16H30	1-HEXADECYNE	222.413	288.16	557.16	------	----------	----------
1260	C16H32	n-DECYLCYCLOHEXANE	224.430	271.42	570.75	25.0	1.4945E+03	2.6900E-02
1261	C16H32	1-CYCLOPENTYLUNDECANE	224.429	263.16	568.76	------	----------	----------
1262	C16H32	1-HEXADECENE	224.430	277.51	558.02	25.0	9.4427E+03	1.6997E-01
1263	C16H32O2	n-HEXADECANOIC ACID	256.429	335.95	624.15	------	----------	----------
1264	C16H34	n-HEXADECANE	226.446	291.34	560.01	25.0	2.6327E+04	4.7388E-01
1265	C16H34O	DI-n-OCTYL ETHER	242.445	265.55	559.65	25.0	6.6569E+00	1.1982E-04
1266	C16H34O	1-HEXADECANOL	242.445	322.35	585.15	25.0	5.4271E+00	9.7687E-05
1267	C16H34S	BUTYL-DODECYL-SULFIDE	258.505	288.16	618.16	------	----------	----------
1268	C16H34S	ETHYL-TETRADECYL-SULFIDE	258.505	288.16	618.16	------	----------	----------
1269	C16H34S	METHYL-PENTADECYL-SULFIDE	258.505	288.16	618.16	------	----------	----------
1270	C16H34S	OCTYL-SULFIDE	258.505	288.16	618.16	------	----------	----------
1271	C16H34S	PROPYL-TRIDECYL-SULFIDE	258.505	288.16	618.16	------	----------	----------
1272	C16H34S	1-HEXADECANETHIOL	258.505	290.93	607.16	------	----------	----------
1273	C16H34S2	OCTYL-DISULFIDE	290.565	244.16	619.16	------	----------	----------
1274	C17H28	n-UNDECYLBENZENE	232.409	268.00	586.40	------	----------	----------
1275	C17H32	1-HEPTADECYNE	236.440	295.16	572.16	------	----------	----------
1276	C17H34	1-CYCLOPENTYLDODECANE	238.456	268.16	584.06	------	----------	----------
1277	C17H34	1-CYCLOHEXYLUNDECANE	238.456	278.96	586.26	------	----------	----------
1278	C17H34	1-HEPTADECENE	238.457	284.40	573.48	------	----------	----------
1279	C17H36	n-HEPTADECANE	240.473	295.13	575.30	25.0	2.6619E+03	4.7915E-02
1280	C17H36O	1-HEPTADECANOL	256.472	327.05	597.15	------	----------	----------
1281	C17H36S	BUTYL-TRIDECYL-SULFIDE	272.531	294.16	632.16	------	----------	----------
1282	C17H36S	ETHYL-PENTADECYL-SULFIDE	272.531	294.16	632.16	------	----------	----------
1283	C17H36S	HEXADECYL-METHYL-SULFIDE	272.531	294.16	632.16	------	----------	----------
1284	C17H36S	PROPYL-TETRADECYL-SULFIDE	272.531	294.16	632.16	------	----------	----------
1285	C17H36S	1-HEPTADECANETHIOL	272.531	300.37	621.16	------	----------	----------
1286	C18H12	CHRYSENE	228.293	531.15	714.15	------	----------	----------
1287	C18H14	m-TERPHENYL	230.309	360.00	650.00	------	----------	----------
1288	C18H14	o-TERPHENYL	230.309	329.35	609.00	------	----------	----------
1289	C18H14	p-TERPHENYL	230.309	485.00	649.15	------	----------	----------
1290	C18H15P	TRIPHENYLPHOSPHINE	262.291	354.40	650.15	------	----------	----------
1291	C18H15O4P	TRIPHENYL PHOSPHATE	326.288	323.15	686.65	------	----------	----------
1292	C18H16N2	N,N'-DIPHENYL-p-PHENYLENEDIAMINE	260.339	409.00	688.00	------	----------	----------
1293	C18H22	2,3-DIMETHYL-2,3-DIPHENYLBUTANE	238.373	392.15	589.00	------	----------	----------
1294	C18H22O2	DICUMYL PEROXIDE	270.371	311.15	669.00	------	----------	----------
1295	C18H30	n-DODECYLBENZENE	246.436	275.93	600.76	------	----------	----------
1296	C18H30	HEXAETHYLBENZENE	246.435	401.16	571.16	------	----------	----------
1297	C18H32O2	LINOLEIC ACID	280.451	268.15	628.00	------	----------	----------
1298	C18H34	1-OCTADECYNE	250.467	300.16	586.16	------	----------	----------
1299	C18H34O2	OLEIC ACID	282.467	286.53	633.00	------	----------	----------
1300	C18H34O4	DIBUTYL SEBACATE	314.466	263.95	622.15	------	----------	----------
1301	C18H34O4	DIHEXYL ADIPATE	314.466	259.35	621.15	------	----------	----------
1302	C18H36	1-CYCOPENTYLTRIDECANE	252.482	278.16	598.56	------	----------	----------
1303	C18H36	1-CYCLOHEXYLDODECANE	252.482	285.66	600.86	------	----------	----------
1304	C18H36	1-OCTADECENE	252.484	290.76	587.97	------	----------	----------
1305	C18H36O2	STEARIC ACID	284.483	342.75	648.35	------	----------	----------
1306	C18H38	n-OCTADECANE	254.500	301.33	589.86	25.0	1.5385E+03	2.7693E-02
1307	C18H38O	DINONYL ETHER	270.499	-------	591.00	------	----------	----------
1308	C18H38O	1-OCTADECANOL	270.499	331.05	608.15	34.0	4.6676E+01	8.4016E-04
1309	C18H38S	BUTYL-TETRADECYL-SULFIDE	286.558	298.16	646.16	------	----------	----------
1310	C18H38S	ETHYL-HEXADECYL-SULFIDE	286.558	298.16	646.16	------	----------	----------
1311	C18H38S	HEPTADECYL-METHYL-SULFIDE	286.558	298.16	646.16	------	----------	----------
1312	C18H38S	NONYL-SULFIDE	286.558	298.16	646.16	------	----------	----------
1313	C18H38S	PENTADECYL-PROPYL-SULFIDE	286.558	298.16	646.16	------	----------	----------
1314	C18H38S	1-OCTADECANETHIOL	286.558	300.93	633.16	------	----------	----------
1315	C18H38S2	NONYL-DISULFIDE	318.618	252.16	642.16	------	----------	----------
1316	C19H26	1-n-NONYLNAPHTHALENE	254.415	284.15	639.00	------	----------	----------
1317	C19H32	n-TRIDECYLBENZENE	260.463	283.15	614.43	------	----------	----------
1318	C19H36	1-NONADECYNE	264.493	306.16	600.16	------	----------	----------
1319	C19H36O2	METHYL OLEATE	296.494	293.05	617.00	------	----------	----------
1320	C19H38	1-CYCLOPENTYLTETRADECANE	266.509	282.00	612.16	------	----------	----------
1321	C19H38	1-CYCLOHEXYLTRIDECANE	266.509	291.66	614.66	------	----------	----------
1322	C19H38	1-NONADECENE	266.511	296.55	602.17	------	----------	----------
1323	C19H38O2	NONADECANOIC ACID	298.510	341.23	659.15	------	----------	----------
1324	C19H40	n-NONADECANE	268.527	305.33	603.05	25.0	5.8958E+02	1.0612E-02
1325	C19H40O	1-NONADECANOL	284.524	334.87	631.00	------	----------	----------
1326	C19H40S	BUTYL-PENTADECYL-SULFIDE	300.585	303.16	659.16	------	----------	----------

423

Table 18-1 HENRY'S LAW CONSTANT FOR COMPOUND IN WATER - ORGANIC COMPOUNDS (continued)

NO	FORMULA	NAME	MW, g/mol	T_F, K	T_B, K	T, C	H @ T, atm/mol frac	H @ T, atm/mol/m³
							Henry's Law Constant, H	
1327	C19H40S	ETHYL-HEPTADECYL-SULFIDE	300.585	303.16	659.16	------	----------	----------
1328	C19H40S	HEXADECYL-PROPYL-SULFIDE	300.585	303.16	659.16	------	----------	----------
1329	C19H40S	METHYL-OCTADECYL-SULFIDE	300.585	303.16	659.16	------	----------	----------
1330	C19H40S	1-NONADECANETHIOL	300.585	307.04	645.16	------	----------	----------
1331	C20H16	TRIPHENYLETHYLENE	256.347	342.15	669.00	------	----------	----------
1332	C20H28	1-n-DECYLNAPHTHALENE	268.442	288.15	652.00	------	----------	----------
1333	C20H30O2	ABIETIC ACID	302.457	446.65	649.70	------	----------	----------
1334	C20H31N	DEHYDROABIETYLAMINE	285.473	317.65	660.00	------	----------	----------
1335	C20H34	1-PHENYLTETRADECANE	274.489	289.16	627.16	------	----------	----------
1336	C20H38	1-EICOSYNE	278.520	309.16	613.16	------	----------	----------
1337	C20H40	1-CYCLOPENTYLPENTADECANE	280.536	290.00	625.00	------	----------	----------
1338	C20H40	1-CYCLOHEXYLTETRADECANE	280.536	297.16	627.16	------	----------	----------
1339	C20H40	1-EICOSENE	280.538	301.76	615.54	------	----------	----------
1340	C20H42	n-EICOSANE	282.553	309.59	616.93	25.0	2.3858E+02	4.2944E-03
1341	C20H42O	1-EICOSANOL	298.553	338.55	629.15	------	----------	----------
1342	C20H42S	BUTYL-HEXADECYL-SULFIDE	314.612	308.16	671.16	------	----------	----------
1343	C20H42S	DECYL-SULFIDE	314.612	308.16	671.16	------	----------	----------
1344	C20H42S	ETHYL-OCTADECYL-SULFIDE	314.612	308.16	671.16	------	----------	----------
1345	C20H42S	HEPTADECYL-PROPYL-SULFIDE	314.612	308.16	671.16	------	----------	----------
1346	C20H42S	METHYL-NONADECYL-SULFIDE	314.612	308.16	671.16	------	----------	----------
1347	C20H42S	1-EICOSANETHIOL	314.612	310.37	656.16	------	----------	----------
1348	C20H42S2	DECYL-DISULFIDE	346.672	259.16	663.16	------	----------	----------
1349	C21H21O4P	TRI-o-CRESYL PHOSPHATE	368.369	240.15	-------	------	----------	----------
1350	C21H36	1-PHENYLPENTADECANE	288.515	295.16	639.16	------	----------	----------
1351	C21H42	1-CYCLOPENTYLHEXADECANE	294.563	294.16	637.16	------	----------	----------
1352	C21H42	1-CYCLOHEXYLPENTADECANE	294.563	302.16	640.16	------	----------	----------
1353	C22H38	1-PHENYLHEXADECANE	302.542	300.16	651.16	------	----------	----------
1354	C22H44	1-CYCLOHEXYLHEXADECANE	308.590	306.76	652.16	------	----------	----------
1355	C22H44O2	n-BUTYL STEARATE	340.590	299.45	623.15	------	----------	----------
1356	C24H38O4	DIISOOCTYL PHTHALATE	390.563	-------	694.00	------	----------	----------
1357	C24H38O4	DIOCTYL PHTHALATE	390.563	223.15	657.15	24.0	7.2502E-03	1.3050E-07
1358	C24H42O	DINONYLPHENOL	346.597	-------	722.00	------	----------	----------
1359	C26H20	TETRAPHENYLETHYLENE	332.445	496.15	760.00	------	----------	----------
1360	C28H46O4	DIISODECYL PHTHALATE	446.671	227.59	723.00	24.0	6.1372E-02	1.1047E-06

H - Henry's law constant for compound in water
T - temperature, C
MW - molecular weight of compound, g/mol
T_F - freezing point temperature of compound, K
T_B - boiling point temperature of compound, K

Chapter 19

ADSORPTION ON ACTIVATED CARBON

Carl L. Yaws, Li Bu, and Sachin Nijhawan
Lamar University, Beaumont, Texas

ABSTRACT

Adsorption on activated carbon is an effective method for removing volatile organic compounds (VOC) from gases. In this article, results are presented for adsorption capacity as a function of the VOC concentration in the gas. The correlation constants are displayed in an easy-to-use tabular format that is especially applicable for rapid engineering usage with the personal computer or hand calculator.

The results for adsorption capacity are applicable for conditions (concentrations in parts per million range in gas at 25 C and 1 atm) which are encountered in air pollution control. Correlation and experimental results are in favorable agreement.

INTRODUCTION

Physical and thermodynamic property data for organic compounds are especially helpful to engineers and scientists in industry. In particular, capacity data for adsorption of volatile organic compounds (VOC) on activated carbon is becoming increasingly important in engineering and environmental studies because of more and more stringent regulations regarding air emissions.

In this article, results are presented for adsorption capacity as a function of the VOC concentration in the gas. The results are usable in engineering and environmental studies. As an example of such usage, capacity data issuing from the correlation are useful in the engineering design of carbon adsorption systems to remove trace pollutants from gases.

ADSORPTION CAPACITY CORRELATION

The correlation for adsorption on activated carbon is based on a logarithmic series expansion of concentration in the gas:

$$\log_{10} Q = A + B [\log_{10} y] + C [\log_{10} y]^2 \tag{19-1}$$

where
Q = adsorption capacity at equilibrium, g of compound/100 g of carbon
y = concentration in gas at 25 C and 1 atm, parts per million by volume, ppmv
A, B, and C = correlation constants

The correlation constants (A, B, and C) are given in Table 19-1. The correlation constants in the table were determined from regression of the available data for adsorption on activated carbon. The tabulation is arranged by carbon number (C1, C2, C3, ... ,C14). This provides ease of use in quickly locating data using the chemical formula. The tabulation also gives the adsorption capacity at concentrations of 10, 100, and 1000 parts per million by volume (ppmv) in gas.

A comparison of correlation and experimental data is shown in Figure 19-1 for a representative compound. In the figure, adsorption capacity is for conditions (concentrations in parts per million range in gas at 25 C and 1 atm) which are encountered in air pollution control. The graph discloses favorable agreement of correlation and experimental data.

ESTIMATION EQUATION

In preparing the correlation, a literature search was conducted to identify source publications (1-39) relative to experimental data and property values for estimates. The publications were screened and copies of appropriate data were made. These data were then keyed into the computer to provide a database of adsorption capacity values at different concentrations (partial pressures) for which experimental data are available. The database also served as a basis to check the accuracy of the correlation.

Upon completion of data collection, estimation of adsorption capacity for the remaining compounds was performed. The following equation (developed by Calgon, 7) was used for estimation of the equilibrium adsorption capacity of activated carbon as a fifth order polynomial function of its adsorption potential:

$$\log_{10} Q = A + B \mu + C \mu^2 + D \mu^3 + E \mu^4 + F \mu^5 \tag{19-2}$$

where
Q = adsorption capacity at equilibrium, cm³ of liquid compound/100 g of carbon
μ = adsorption potential of compound, $T/(V_i \, \Gamma_i) \log(P_i^{sat}/p_i)$
T = temperature, K
V_i = liquid molar volume of compound, cm³/g-mol

Γ_i = relative polarizability, $[(n^2-1)/(n^2+1)]_i$ / $[(n^2-1)/(n^2+1)]_{n\text{-heptane}}$

n = refractive index

P_i^{sat} = vapor pressure of compound, atm

p_i = partial pressure of compound, atm

A = 1.71

B = - 1.46E-02

C = - 1.65E-03

D = - 4.11E-04

E = 3.14E-05

F = - 6.75E-07

For the above equation, data for refractive index are from compilations by Yaws (present work), Texas A & M (26,27), and DIPPR (28). Data for vapor pressure and liquid molar volume are from compilations by Yaws (32-36).

The correlation constants (A, B, and C) are given in Table 19-1. The correlation constants in the table were determined from regression of the available data for adsorption on activated carbon. The tabulation is arranged by carbon number (C1, C2, C3, ... ,C14). This provides ease of use in quickly locating data using the chemical formula. The tabulation also gives the adsorption capacity at concentrations of 10, 100, and 1000 parts per million by volume (ppmv) in gas.

A comparison of correlation and experimental data is shown in Figure 19-1 for a representative compound. In the figure, adsorption capacity is for conditions (concentrations in parts per million range in gas at 25 C and 1 atm) which are encountered in air pollution control. The graph discloses favorable agreement of correlation and experimental data.

EXAMPLES

The correlation may be used for determining adsorption capacity of activated carbon for removing compounds from gases. Examples are given below.

Example 1 The air from a paint spraying operation contains 10 ppmv of n-butanol (C4H10O). Estimate the adsorption capacity of activated carbon for removing the compound at 25 C and 1 atm.

Substitution of the coefficients from the tabulation and concentration into the correlation equation yields

$\log_{10} Q = 0.8988 + 0.32534 [\log_{10}(10)] - 0.03648 [\log_{10}(10)]^2 = 1.18767$

$Q = 10^{1.18767}$

Q = 15.41 g of n-butanol/100 g of carbon

Example 1 The air from an industrial operation contains 10 ppmv of cyclohexane (C6H12). Estimate the adsorption capacity of activated carbon for removing the compound at 25 C and 1 atm.

Substitution of the coefficients from the tabulation and concentration into the correlation equation yields

$\log_{10} Q = 0.720 + 0.25698 [\log_{10}(10)] - 0.01550 [\log_{10}(10)]^2 = 0.96142$

$Q = 10^{0.96142}$

Q = 9.15 g of cyclohexane/100 g of carbon

OPERATION AND DESIGN

In actual operation under plant conditions, the capacity of an adsorption bed will seldom achieve equilibrium. Copper and Alley (6) indicate bed capacity at 30 to 40% of equilibrium for plant operating conditions. Damie and Rogers (7) in the EPA design manual suggest a working factor of 3 for design of adsorption beds. The total carbon requirements for an adsorption system is obtained by determining carbon required from equilibrium capacity and then multiplying by the working factor.

Factors affecting adsorption bed capacity are discussed by Copper and Alley (6), Damie and Rogers (7), and Gram and Ramaratnam (25). These include loss due to adsorption zone, loss due to heat wave (adsorption is an exothermic process) and loss due to moisture in entering gas and loss due to residual moisture on the carbon.

Representative adsorption systems for removing organic compounds from gases are shown in Figs. 19-2, 19-3, and 19-4. Figure 19-2 shows an adsorption system with recovery of the organic (such as a solvent) using steam for regeneration. Figure 19-3 applies to an adsorption system with thermal or catalytic oxidation of the organic removed from the gas by carbon adsorption. In Figure 19-4, the organic is initially removed from wastewater by air stripping. The air leaving the stripper contains the organic and is then sent to the adsorption system for recovery of the organic. This last system can be used to recover organics (such as benzenes) from process wastewater encountered in the chemical and petroleum refining industries.

Portions of this material appeared in Pollution Engineering, 27, 34 (1995), Environmental Engineering World, 1, 16 (May-June, 1995), and Oil & Gas Journal, 93, 64 (Feb. 13, 1995). These portions are reprinted by special permission.

REFERENCES – ORGANIC COMPOUNDS

1. Cheremisnoff, P. N. and F. Ellerbusch, CARBON ADSORPTION HANDBOOK, Ann Arbor Science Publishers, Ann Arbor, MI (1980).
2. Yang, R. T., GAS SEPARATION BY ADSORPTION PROCESSES, Butterworth Publishers, Boston, MA (1987).
3. Valenzuela, D. P. and A. L. Myers, ADSORPTION EQUILIBRIUM DATA HANDBOOK, Prentice Hall, Engle Cliffs, NJ (1989).
4. CALGON CARBON ADSORPTION HANDBOOK, Calgon Carbon Corporation, Pittsburg, PA (1994).
5. Adsorption Isotherms, personal communication to Carl L. Yaws, Calgon Carbon Corporation, Pittsburg, PA (1994).
6. Copper, C. D. and F. C. Alley, AIR POLLUTION CONTROL, 2nd ed., Waveland Press, Prospects Heights, IL (1994).
7. Damie, A. S. and T. N. Rogers, AIR STRIPPER DESIGN MANUAL, EPA-450/1-90-003, U. S. Environmental Protection Agency (May,1990).
8. Grant, R. J., M. Manes, and S. B. Smith, AIChE J., 8, 403 (1962).
9. Grant, R. J. and M. Manes, Ind. Eng. Chem. Fundam., 3, 221 (1964).
10. Grant, R. J. and M. Manes, Ind. Eng. Chem. Fundam., 5, 490 (1966).
11. Schenz, T. W. and M. Manes, J. Phy. Chem., 79, 604 (1975).
12. Szepesy, L. and V. Illes, Acta Chim. Hung., 35, 37 (1963).
13. Laukhuf, W. L. S. and C. A. Plank, J. Chem. Eng. Data, 14, 48 (1969).
14. Ray, G. C. and E. O. Box, Ind. Eng. Chem., 42, 1315 (1950).
15. Payne, H. K., G. A. Studervant, and T. W. Leland, Ind. Eng. Chem. Fundam., 7, 363 (1968).
16. Kuro-Oka, M., T. Suzuki, T. Nitta, and T. Katayama, J. Chem. Eng. Japan, 17, 588 (1984).
17. Ritter, J. A. and R. T. Yang, Ind. Eng. Chem. Res., 26, 1679 (1987).
18. Kaul, B. K., Ind. Eng. Chem. Res., 26, 928 (1987).
19. Reich, R., W. T. Ziegler, and K. A. Rogers, Ind. Eng. Chem. Process Des. Dev., 19, 336 (1980).
20. Lewis, W. K., E. R. Gilliland, B. Chertow, and W. Milliken, J. Am. Chem. Soc., 72, 1157 (1950).
21. Lewis, W. K., E. R. Gilliland, B. Chertow, and W. P. Cadogan, Ind. Eng. Chem., 42, 1326 (1950).
22. Maslan, F. D., M. Altman, and E. R. Aberth, J. Phys. Chem., 57, 106 (1953).
23. Cook, W. H. and D. Basmadjian, Can. J. Chem. Eng., 42, 146 (1964).
24. Reich, R., W. T. Zeigler, and K. A. Rogers, Ind. Eng. Chem. Proc. Des. Dev., 19, 336 (1980).
25. Graham, J. R. and M. Ramaratnam, Chem. Eng., 100, 6 (1993).
26. SELECTED VALUES OF PROPERTIES OF HYDROCARBONS AND RELATED COMPOUNDS, Thermodynamics Research Center, TAMU, College Station, TX (1977, 1984).
27. SELECTED VALUES OF PROPERTIES OF CHEMICAL COMPOUNDS, Thermodynamics Research Center, TAMU, College Station, TX (1977, 1987).
28. Daubert, T. E. and R. P. Danner, DATA COMPILATION OF PROPERTIES OF PURE COMPOUNDS, Parts 1, 2, 3, and 4, Supplements 1 and 2, DIPPR Project, AIChE, New York, NY (1985-1992).
29. Lide, D. R. and H. V. Kehianian, CRC HANDBOOK OF THERMOPHYSICAL AND THERMOCHEMICAL DATA, CRC Press, Boca Raton, FL (1994).
30. Howard, P. H. and W. M. Meylan, eds., HANDBOOK OF PHYSICAL PROPERTIES OF ORGANIC CHEMICALS, CRC Press, Boca Raton, FL (1997).
31. Verschueren, K., HANDBOOK OF ENVIRONMENTAL DATA ON ORGANIC CHEMICALS, Van Nostrand Reinhold, New York, NY (1996).
32. Yaws, C. L., PHYSICAL PROPERTIES, McGraw-Hill, New York, NY (1977).
33. Yaws, C. L., THERMODYNAMIC AND PHYSICAL PROPERTY DATA, Gulf Publishing Co., Houston, TX (1992).
34. Yaws, C. L. and R. W. Gallant, PHYSICAL PROPERTIES OF HYDROCARBONS, Vols. 1 (2nd ed.), 2 (3rd ed.), 3 and 4, Gulf Publishing Co., Houston, TX (1992,1993,1993,1995).
35. Yaws, C. L., HANDBOOK OF VAPOR PRESSURE, Vols. 1, 2, 3 and 4, Gulf Publishing Co., Houston, TX (1994,1994,1994,1995).
36. Yaws, C. L., HANDBOOK OF CHEMICAL COMPOUND DATA FOR PROCESS SAFETY, Gulf Publishing Co., Houston, TX (1997).
37. Yaws, C. L. and others, Pollution Engineering, 27 (2), 34 (1995).
38. Yaws, C. L. and others, Environmental Engineering World, 1 (3), 16 (May-June, 1995).
39. Yaws, C. L. and others, Oil & Gas Journal, 93 (7), 64 (Feb. 13, 1995).

Figure 19-1 Adsorption on Activated Carbon

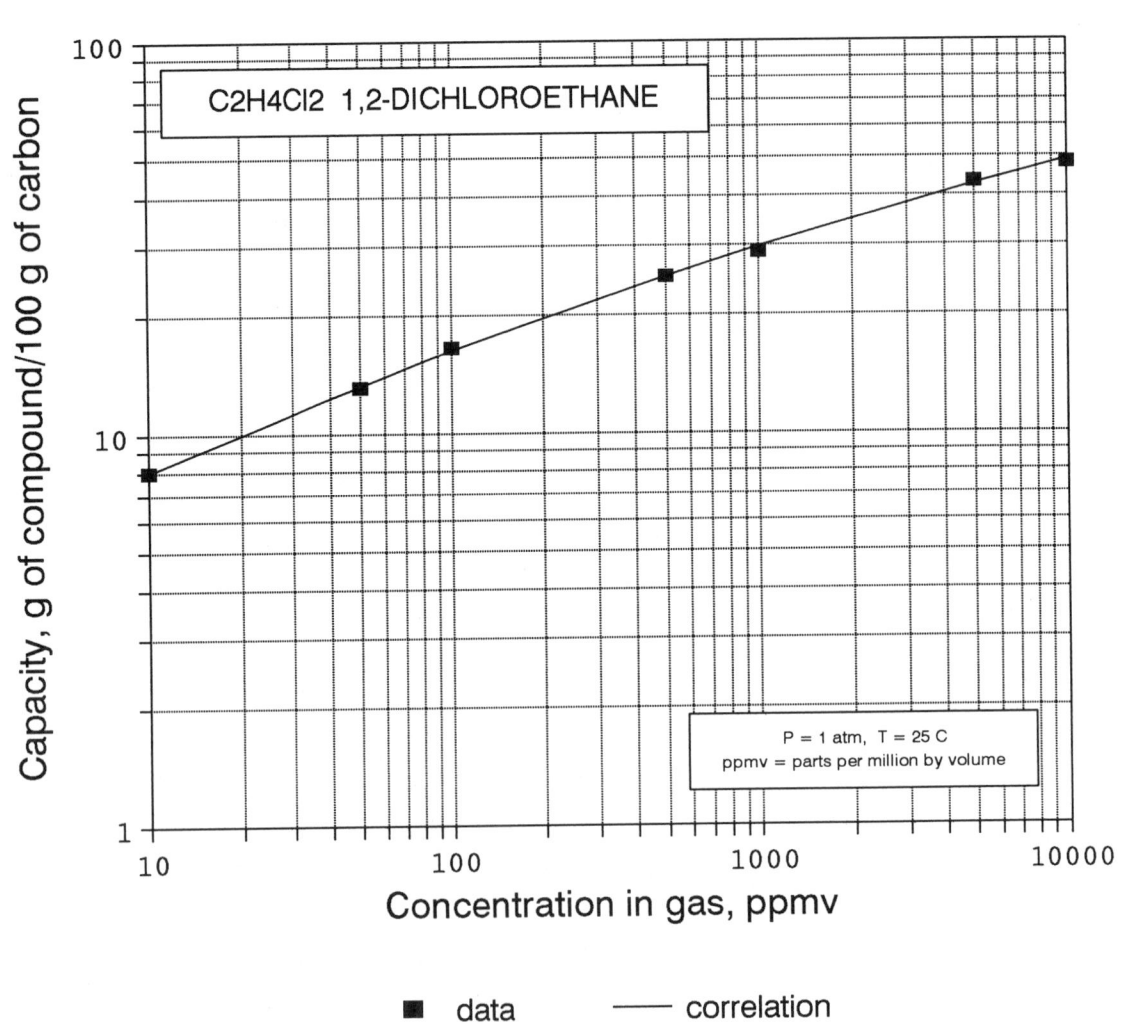

Figure 19-2 Adsorption System with Solvent Recovery

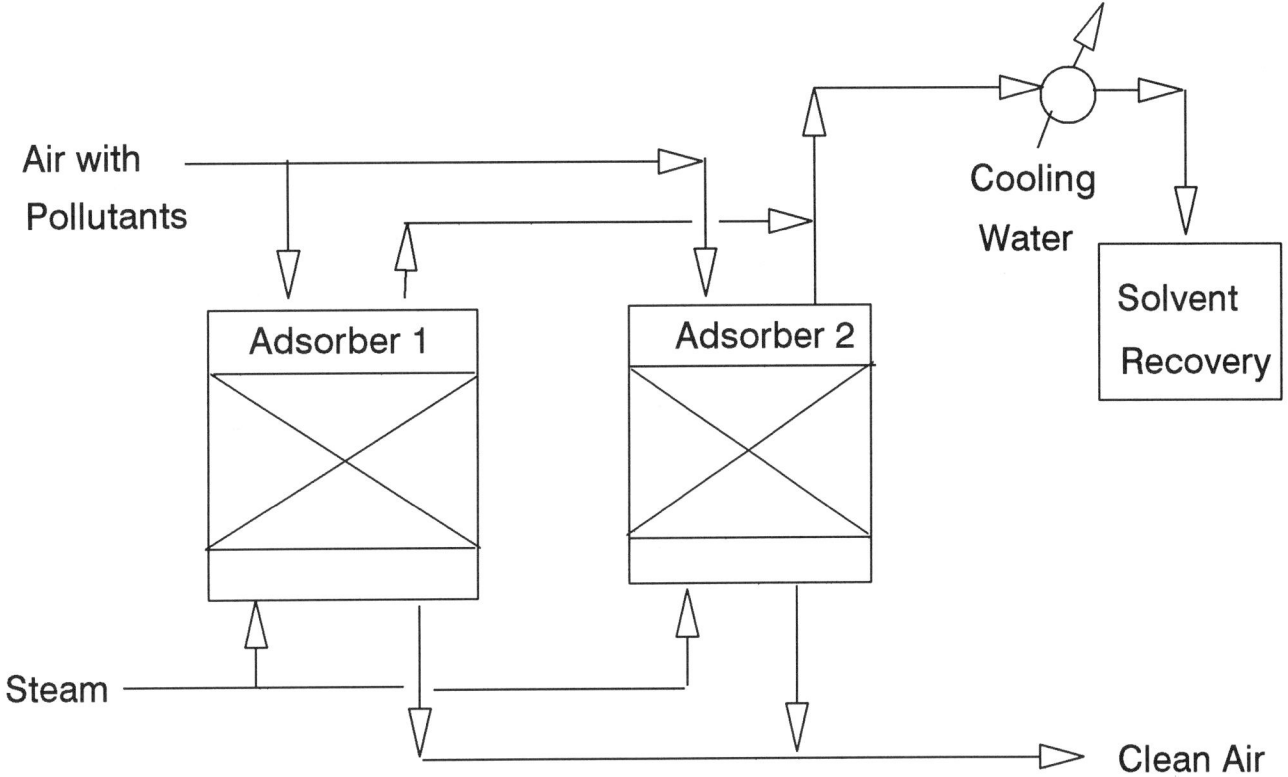

Figure 19-3   Adsorption System with Thermal or Catalytic Oxidation

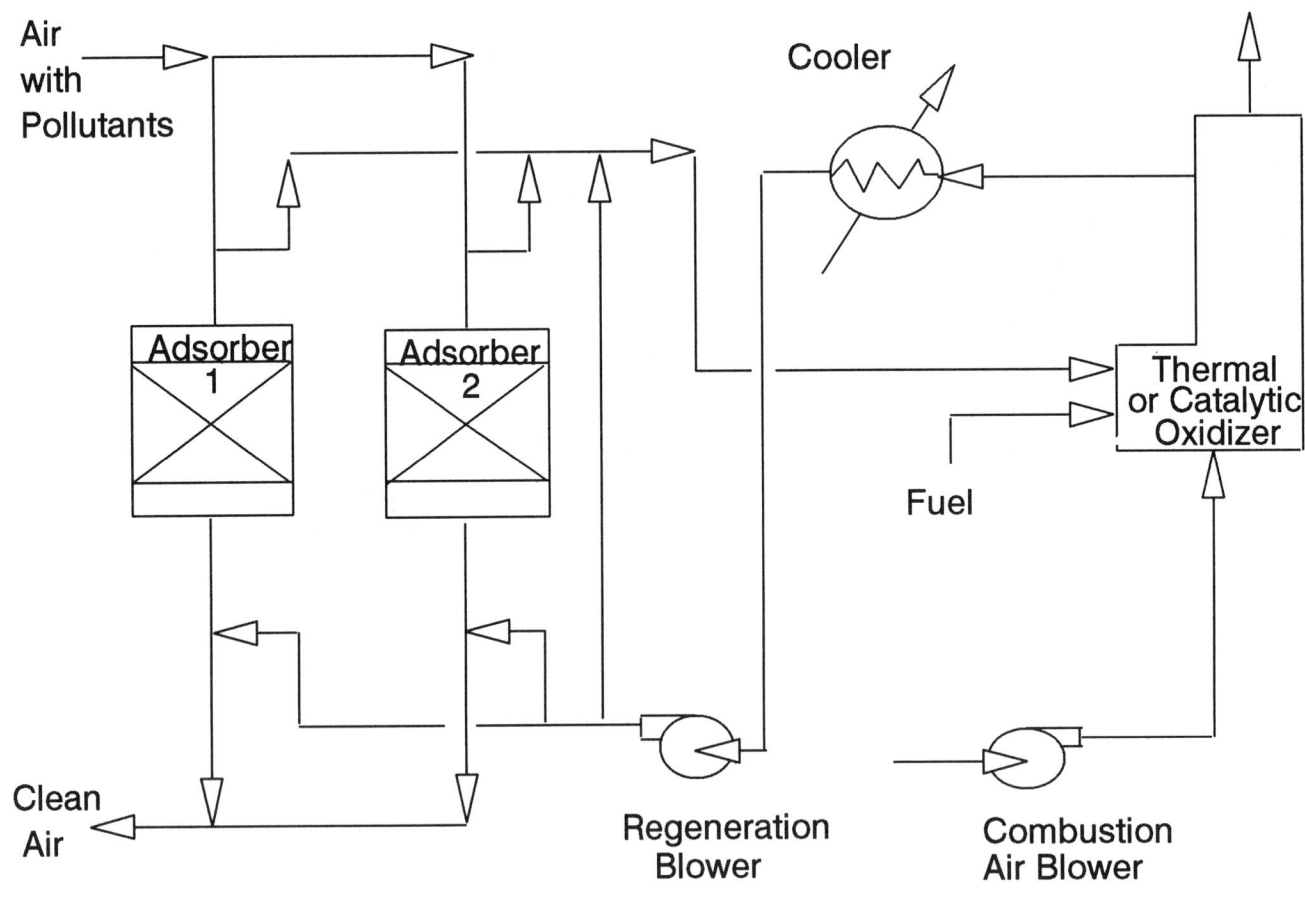

Figure 19-4 Adsorption System with Air Stripper

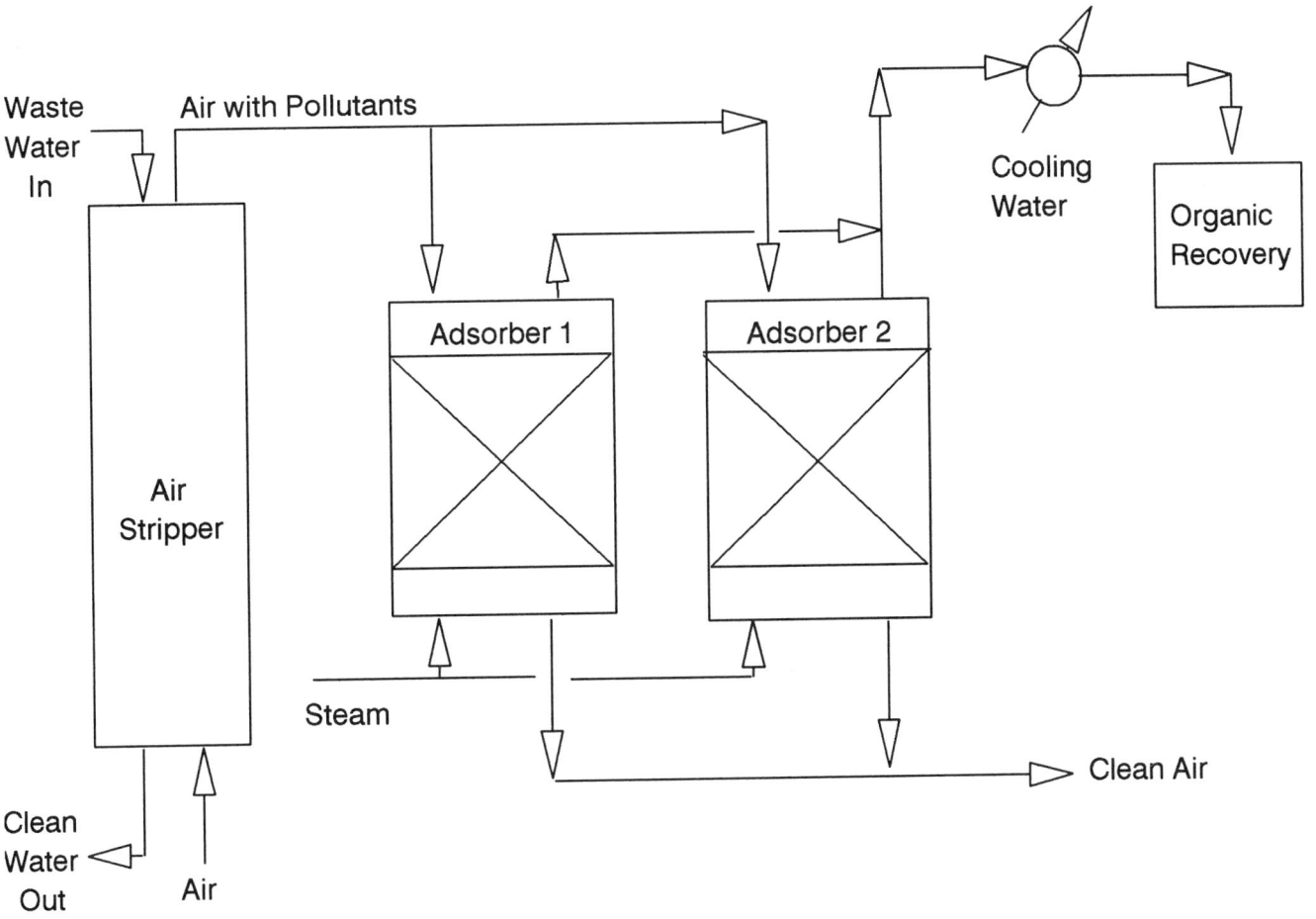

Table 19-1 ADSORPTION CAPACITY OF ACTIVATED CARBON

$$\log_{10} Q = A + B\,[\log_{10} y] + C\,[\log_{10} y]^2 \qquad Q\text{ - g of compound/100 g of carbon, } y\text{ - ppmv}$$

NO	FORMULA	NAME	A	B	C	ymin	ymax	Q @10 ppmv	Q @100 ppmv	Q @1000 ppmv
1	CBrCl3	BROMOTRICHLOROMETHANE	1.3984	2.3228E-01	-2.1840E-02	10	10000	40.63	59.65	79.20
2	CBrF3	BROMOTRIFLUOROMETHANE	-1.4625	5.8361E-01	-1.0440E-02	10	10000	0.13	0.46	1.56
3	CBr2F2	DIBROMODIFLUOROMETHANE	0.8208	3.0701E-01	-1.3840E-02	10	10000	13.00	23.96	41.42
4	CBr3F	TRIBROMOFLUOROMETHANE	-1.4375	5.5503E-01	-4.5000E-03	10	10000	0.13	0.45	1.54
5	CCl2F2	DICHLORODIFLUOROMETHANE	-0.0735	4.0145E-01	-1.4040E-02	10	10000	2.06	4.71	10.10
6	CCl2O	PHOSGENE	-0.6447	6.0428E-01	-2.9860E-02	10	10000	0.85	2.78	7.93
7	CCl3F	TRICHLOROFLUOROMETHANE	0.1731	4.0715E-01	-1.9150E-02	10	10000	3.64	8.14	16.68
8	CCl3NO2	CHLOROPICRIN	1.2675	2.0841E-01	-1.2880E-02	10	10000	29.04	42.93	59.81
9	CCl4	CARBON TETRACHLORIDE	1.0748	2.8186E-01	-2.2730E-02	10	10000	21.57	35.29	51.98
10	CHBr3	TRIBROMOMETHANE	1.7318	1.9948E-01	-2.2460E-02	10	7238	81.07	109.89	134.33
11	CHCl3	CHLOROFORM	0.6710	3.6148E-01	-2.2880E-02	10	10000	10.22	20.07	35.45
12	CHN	HYDROGEN CYANIDE	-4.3925	1.0895E+00	-7.4000E-03	10	10000	4.9E-04	5.7E-03	6.4E-02
13	CH2BrCl	BROMOCHLOROMETHANE	0.6140	4.1353E-01	-2.5310E-02	10	10000	10.05	21.87	42.35
14	CH2BrF	BROMOFLUOROMETHANE	0.4548	3.6332E-01	-1.6060E-02	10	10000	6.34	13.10	25.13
15	CH2Br2	DIBROMOMETHANE	1.0838	3.7211E-01	-3.2380E-02	10	10000	26.52	49.94	81.04
16	CH2Cl2	DICHLOROMETHANE	-0.0704	4.9210E-01	-2.2760E-02	10	10000	2.51	6.65	15.89
17	CH2I2	DIIODOMETHANE	1.9476	1.4984E-01	-1.9470E-02	10	1583	119.66	147.70	166.67
18	CH2O	FORMALDEHYDE	-2.4852	6.9123E-01	-3.7500E-03	10	10000	1.6E-02	7.6E-02	0.36
19	CH2O2	FORMIC ACID	-1.7773	1.0950E+00	-6.3540E-02	10	10000	0.18	1.44	8.63
20	CH3Br	METHYL BROMIDE	-1.2384	7.8564E-01	-5.5210E-02	10	10000	0.31	1.29	4.19
21	CH3Cl	METHYL CHLORIDE	-1.9187	6.2053E-01	-5.4900E-03	10	10000	0.05	0.20	0.78
22	CH3Cl3Si	METHYL TRICHLOROSILANE	1.0720	2.4275E-01	-1.9110E-02	10	10000	19.75	30.27	42.48
23	CH3I	METHYL IODIDE	0.7400	3.2985E-01	-1.3300E-02	10	10000	11.39	22.21	40.72
24	CH3NO	FORMAMIDE	1.3098	2.5274E-01	----------	10	80	36.52	-------	-------
25	CH3NO2	NITROMETHANE	-0.3285	7.0602E-01	-5.1110E-02	10	10000	2.12	7.57	21.36
26	CH4	METHANE	-4.3101	7.7883E-01	-6.2800E-03	10	10000	2.9E-04	1.7E-03	9.3E-03
27	CH4Cl2Si	METHYL DICHLOROSILANE	0.7327	2.9305E-01	-1.8220E-02	10	10000	10.17	17.62	28.04
28	CH4O	METHANOL	-1.9674	8.2107E-01	-1.3930E-02	10	10000	0.07	0.42	2.35
29	CH4S	METHYL MERCAPTAN	-1.1229	6.0573E-01	-2.0940E-02	10	10000	0.29	1.01	3.21
30	CH5N	METHYLAMINE	-1.9355	6.4710E-01	-1.0570E-02	10	10000	0.05	0.21	0.81
31	CN4O8	TETRANITROMETHANE	1.4905	1.8181E-01	-1.8940E-02	10	10000	45.01	60.02	73.35
32	CO	CARBON MONOXIDE	-5.1878	9.0121E-01	-1.3580E-02	10	10000	5.0E-05	3.6E-04	2.5E-03
33	COS	CARBONYL SULFIDE	-1.4288	5.1061E-01	2.8000E-04	10	10000	0.12	0.39	1.27
34	CO2	CARBON DIOXIDE	-3.6522	8.0180E-01	-3.2800E-03	10	10000	1.4E-03	8.7E-03	5.3E-02
35	CS2	CARBON DISULFIDE	-0.1890	4.7093E-01	-1.4810E-02	10	10000	1.85	4.94	12.32
36	C2Br2F4	1,2-DIBROMOTETRAFLUOROETHANE	0.9039	2.5693E-01	-9.7400E-03	10	10000	14.16	23.92	38.64
37	C2ClF5	CHLOROPENTAFLUOROETHANE	0.0826	3.4756E-01	-1.3430E-02	10	10000	2.61	5.30	10.10
38	C2Cl3F3	1,1,2-TRICHLOROTRIFLUOROETHANE	1.2737	1.8656E-01	-1.2310E-02	10	10000	28.05	39.59	52.79
39	C2Cl4	TETRACHLOROETHYLENE	1.4060	2.0802E-01	-2.0970E-02	10	10000	39.17	54.72	69.39
40	C2Cl4F2	1,1,2,2-TETRACHLORODIFLUOROETHANE	1.3731	1.7625E-01	-1.4650E-02	10	10000	34.25	46.45	58.88
41	C2HBrClF3	HALOTHANE	0.9241	3.1204E-01	-2.0040E-02	10	10000	16.45	29.38	47.84
42	C2HCl3	TRICHLOROETHYLENE	1.0241	2.9929E-01	-2.5390E-02	10	10000	19.86	33.20	49.38
43	C2HCl3O	DICHLOROACETYL CHLORIDE	1.2365	2.6219E-01	-2.5960E-02	10	10000	29.70	45.39	61.57
44	C2HCl3O	TRICHLOROACETALDEHYDE	1.1736	2.6971E-01	-2.5130E-02	10	10000	26.19	40.98	57.09
45	C2HCl5	PENTACHLOROETHANE	1.6457	1.3515E-01	-1.5720E-02	10	4829	58.22	71.30	81.22
46	C2HF3O2	TRIFLUOROACETIC ACID	-0.1258	5.9373E-01	-3.4450E-02	10	10000	2.71	8.39	22.15
47	C2H2	ACETYLENE	-2.2418	8.2454E-01	-3.3900E-02	10	10000	3.5E-02	0.19	0.84
48	C2H2Br4	1,1,2,2-TETRABROMOETHANE	------	----------	----------	-----	-------	146.05	-------	-------
49	C2H2Cl2	1,1-DICHLOROETHYLENE	0.4874	3.3282E-01	-1.6220E-02	10	10000	6.37	12.25	21.87
50	C2H2Cl2	cis-1,2-DICHLOROETHYLENE	0.4757	3.9061E-01	-2.5540E-02	10	10000	6.93	14.28	26.16
51	C2H2Cl2	trans-1,2-DICHLOROETHYLENE	0.4757	3.9061E-01	-2.5540E-02	10	10000	6.93	14.28	26.16
52	C2H2Cl2O2	DICHLOROACETIC ACID	1.6924	9.6300E-02	----------	10	235	61.47	76.73	-------
53	C2H2Cl4	1,1,1,2-TETRACHLOROETHANE	1.4410	1.9166E-01	-1.9950E-02	10	10000	40.99	55.52	68.61
54	C2H2Cl4	1,1,2,2-TETRACHLOROETHANE	1.5232	1.7848E-01	-2.0190E-02	10	6073	48.03	63.01	75.33
55	C2H3Cl	VINYL CHLORIDE	-0.9889	6.6564E-01	-4.3200E-02	10	10000	0.43	1.48	4.16
56	C2H3ClO	ACETYL CHLORIDE	0.0363	4.5526E-01	-2.0930E-02	10	10000	2.96	7.30	16.35
57	C2H3ClO2	METHYL CHLOROFORMATE	0.4119	4.2776E-01	-2.7760E-02	10	10000	6.48	14.33	27.88
58	C2H3Cl3	1,1,1-TRICHLOROETHANE	0.9733	2.8737E-01	-2.2770E-02	10	10000	17.29	28.64	42.70
59	C2H3Cl3	1,1,2-TRICHLOROETHANE	1.1716	2.7791E-01	-2.7460E-02	10	10000	26.43	41.46	57.31
60	C2H3N	ACETONITRILE	-0.7967	6.3512E-01	-2.5980E-02	10	10000	0.65	2.34	7.50
61	C2H3NO	METHYL ISOCYANATE	-1.0758	8.5881E-01	-6.8760E-02	10	10000	0.52	2.33	7.62
62	C2H4	ETHYLENE	-2.2710	6.1731E-01	-1.4670E-02	10	10000	2.1E-02	8.0E-02	0.28
63	C2H4Br2	1,1-DIBROMOETHANE	1.3726	2.5671E-01	-2.5160E-02	10	10000	40.19	61.01	82.48
64	C2H4Br2	1,2-DIBROMOETHANE	1.4423	2.4741E-01	-2.6660E-02	10	10000	46.84	70.09	92.77
65	C2H4Cl2	1,1-DICHLOROETHANE	0.5449	3.6091E-01	-2.1920E-02	10	10000	7.65	15.10	26.93
66	C2H4Cl2	1,2-DICHLOROETHANE	0.5534	3.7072E-01	-2.1610E-02	10	10000	7.99	16.16	29.59
67	C2H4Cl2O	BIS(CHLOROMETHYL)ETHER	0.9560	3.3784E-01	-3.2000E-02	10	10000	18.27	31.89	48.04
68	C2H4F2	1,2-DIFLUOROETHANE	-3.9790	2.5186E+00	-3.1617E-01	10	10000	0.02	0.62	5.39
69	C2H4O	ACETALDEHYDE	-1.1705	6.2766E-01	-2.4750E-02	10	10000	0.27	0.97	3.09
70	C2H4O	ETHYLENE OXIDE	-2.4238	9.4878E-01	-4.0620E-02	10	10000	0.03	0.20	1.14
71	C2H4O2	ACETIC ACID	-0.0555	6.8410E-01	-6.0710E-02	10	10000	3.70	11.74	28.21
72	C2H4O2	METHYL FORMATE	-0.9959	6.1693E-01	-1.8470E-02	10	10000	0.40	1.46	4.88
73	C2H4S	THIACYCLOPROPANE	0.0226	4.5520E-01	-2.1540E-02	10	10000	2.86	7.03	15.64
74	C2H5Br	BROMOETHANE	0.3178	4.3594E-01	-3.0720E-02	10	10000	5.28	11.64	22.27
75	C2H5Cl	ETHYL CHLORIDE	-0.5083	5.0364E-01	-2.1790E-02	10	10000	0.94	2.58	6.40
76	C2H5ClO	2-CHLOROETHANOL	0.7416	4.6933E-01	-5.1580E-02	10	9446	14.43	29.78	48.46
77	C2H5I	ETHYL IODIDE	1.0036	3.2123E-01	-2.4050E-02	10	10000	19.99	35.47	56.33
78	C2H5N	ETHYLENEIMINE	-1.1691	9.1238E-01	-7.4000E-02	10	10000	0.47	2.29	7.98

Table 19-1 ADSORPTION CAPACITY OF ACTIVATED CARBON (continued)

NO	FORMULA	NAME	$\log_{10} Q = A + B[\log_{10} y] + C[\log_{10} y]^2$					Q - g of compound/100 g of carbon, y - ppmv		
			A	B	C	ymin	ymax	Q @10 ppmv	Q @100 ppmv	Q @1000 ppmv
79	C2H5NO	N-METHYLFORMAMIDE	1.2333	2.1723E-01	----------	10	333	28.22	46.54	-------
80	C2H5NO2	NITROETHANE	0.4497	4.9708E-01	-4.6120E-02	10	10000	7.96	18.17	33.56
81	C2H6	ETHANE	-2.4039	6.8107E-01	-1.9250E-02	10	10000	1.8E-02	7.6E-02	0.29
82	C2H6O	ETHANOL	-0.5115	6.7525E-01	-4.4730E-02	10	10000	1.32	4.57	12.93
83	C2H6OS	DIMETHYL SULFOXIDE	1.2404	3.1302E-01	-4.7680E-02	10	802	32.05	47.40	-------
84	C2H6O2	ETHYLENE GLYCOL	1.4047	1.8738E-01	-2.6630E-02	10	121	36.77	47.09	-------
85	C2H6O4S	DIMETHYL SULFATE	1.3462	2.1539E-01	-2.3360E-02	10	890	34.53	48.25	-------
86	C2H6S	DIMETHYL SULFIDE	0.4847	3.7358E-01	-2.7700E-02	10	10000	6.77	13.22	22.71
87	C2H6S	ETHYL MERCAPTAN	0.0055	4.0506E-01	-1.8020E-02	10	10000	2.47	5.54	11.44
88	C2H6S2	DIMETHYL DISULFIDE	0.7588	3.5928E-01	-2.9530E-02	10	10000	12.26	22.87	37.23
89	C2H7N	DIMETHYLAMINE	-1.2249	6.3962E-01	-3.2660E-02	10	10000	0.24	0.84	2.51
90	C2H7NO	MONOETHANOLAMINE	1.2157	2.1994E-01	----------	10	485	27.27	45.24	-------
91	C2H8N2	ETHYLENEDIAMINE	0.5650	4.6307E-01	-4.7890E-02	10	10000	9.55	19.94	33.36
92	C3H3Cl	PROPARGYL CHLORIDE	0.2714	4.0480E-01	-2.1350E-02	10	10000	4.52	9.90	19.66
93	C3H3N	ACRYLONITRILE	0.0767	4.9986E-01	-3.5000E-02	10	10000	3.48	8.64	18.25
94	C3H3NO	OXAZOLE	0.6335	3.0620E-01	-2.3500E-02	10	10000	8.25	14.19	21.91
95	C3H4	METHYLACETYLENE	-2.5287	1.7472E+00	-2.1635E-01	10	10000	0.10	1.26	5.83
96	C3H4Cl2	2,3-DICHLOROPROPENE	0.9542	3.0034E-01	-2.6140E-02	10	10000	16.92	28.20	41.68
97	C3H4O	ACROLEIN	-0.2963	4.9437E-01	-2.4710E-02	10	10000	1.49	3.92	9.21
98	C3H4O	PROPARGYL ALCOHOL	0.2297	5.7711E-01	-5.4410E-02	10	10000	5.65	14.67	29.61
99	C3H4O2	ACRYLIC ACID	0.7555	4.7108E-01	-5.6150E-02	10	5221	14.81	29.72	46.06
100	C3H4O3	PYRUVIC ACID	1.0741	4.1414E-01	-5.7680E-02	10	1679	26.95	46.95	62.72
101	C3H5Br	3-BROMO-1-PROPENE	0.8482	3.2392E-01	-2.3980E-02	10	10000	14.06	25.12	40.19
102	C3H5Cl	3-CHLOROPROPENE	0.3279	3.6553E-01	-1.8530E-02	10	10000	4.73	9.66	18.10
103	C3H5ClO	alpha-EPICHLOROHYDRIN	0.8320	3.8983E-01	-3.9320E-02	10	10000	15.22	28.47	44.42
104	C3H5ClO2	METHYL CHLOROACETATE	1.0766	3.2514E-01	-3.6170E-02	10	9959	23.20	38.21	53.27
105	C3H5ClO2	ETHYL CHLOROFORMATE	0.9490	3.2529E-01	-3.2010E-02	10	10000	17.47	29.62	43.33
106	C3H5Cl3	1,2,3-TRICHLOROPROPANE	1.4724	1.8136E-01	-2.1650E-02	10	4843	42.87	56.04	66.32
107	C3H5I	3-IODO-1-PROPENE	1.3363	2.4222E-01	-2.2710E-02	10	10000	35.96	53.70	72.21
108	C3H5N	PROPIONITRILE	0.0593	5.1747E-01	-3.7810E-02	10	10000	3.46	8.77	18.68
109	C3H5NO	HYDRACRYLONITRILE	1.5099	1.1037E-01	----------	10	105	41.72	53.79	-------
110	C3H5NO	LACTONITRILE	1.4416	1.2689E-01	----------	10	157	37.02	49.58	-------
111	C3H6	PROPYLENE	-0.9367	5.7775E-01	-3.8530E-02	10	10000	0.40	1.16	2.82
112	C3H6Cl2	1,1-DICHLOROPROPANE	0.9538	2.8791E-01	-2.4870E-02	10	10000	16.48	26.92	39.24
113	C3H6Cl2	1,2-DICHLOROPROPANE	0.9887	2.8700E-01	-2.5710E-02	10	10000	17.78	28.83	41.53
114	C3H6Cl2	1,3-DICHLOROPROPANE	1.1034	2.7837E-01	-2.8240E-02	10	10000	22.57	35.25	48.35
115	C3H6Cl2	2,2-DICHLOROPROPANE	0.8531	2.9432E-01	-2.2550E-02	10	10000	13.33	22.47	34.13
116	C3H6O	ACETONE	-0.1455	4.7497E-01	-2.2860E-02	10	10000	2.03	5.16	11.85
117	C3H6O	ALLYL ALCOHOL	0.3239	4.9368E-01	-4.3700E-02	10	10000	5.94	13.69	25.80
118	C3H6O	n-PROPIONALDEHYDE	0.0552	4.9738E-01	-4.3310E-02	10	10000	3.23	7.53	14.37
119	C3H6O	1,2-PROPYLENE OXIDE	-0.4283	5.3858E-01	-2.7570E-02	10	10000	1.21	3.46	8.70
120	C3H6O	1,3-PROPYLENE OXIDE	-0.5042	5.1872E-01	-2.2960E-02	10	10000	0.98	2.76	7.00
121	C3H6O2	ETHYL FORMATE	0.1262	4.2260E-01	-2.0900E-02	10	10000	3.37	7.72	16.06
122	C3H6O2	METHYL ACETATE	0.1331	4.2849E-01	-2.1880E-02	10	10000	3.47	7.99	16.66
123	C3H6O2	PROPIONIC ACID	0.7785	4.4570E-01	-5.2090E-02	10	4872	14.86	28.94	44.34
124	C3H6O2S	3-MERCAPTOPROPIONIC ACID	1.6882	5.9160E-02	----------	10	66	55.90	-------	-------
125	C3H6O3	LACTIC ACID	1.6072	9.2250E-02	----------	10	107	50.06	61.90	-------
126	C3H6O3	METHOXYACETIC ACID	1.6189	8.8730E-02	----------	10	71	51.00	-------	-------
127	C3H6S	THIACYCLOBUTANE	0.6742	3.7225E-01	-3.1510E-02	10	10000	10.35	19.62	32.16
128	C3H7Br	1-BROMOPROPANE	0.8360	3.2406E-01	-2.4070E-02	10	10000	13.68	24.43	39.04
129	C3H7Br	2-BROMOPROPANE	0.8114	3.1043E-01	-2.1550E-02	10	10000	12.60	22.18	35.37
130	C3H7Cl	ISOPROPYL CHLORIDE	0.3143	3.4779E-01	-1.6610E-02	10	10000	4.42	8.78	16.15
131	C3H7Cl	n-PROPYL CHLORIDE	0.4013	3.4678E-01	-1.9310E-02	10	10000	5.36	10.41	18.53
132	C3H7I	ISOPROPYL IODIDE	1.2646	2.4157E-01	-2.1220E-02	10	10000	30.54	46.01	62.85
133	C3H7I	n-PROPYL IODIDE	1.3062	2.4227E-01	-2.2500E-02	10	10000	33.57	50.21	67.69
134	C3H7N	ALLYLAMINE	0.1625	3.9815E-01	-2.1050E-02	10	10000	3.46	7.49	14.71
135	C3H7N	PROPYLENEIMINE	0.0692	4.3529E-01	-2.2930E-02	10	10000	3.03	7.05	14.75
136	C3H7NO	N,N-DIMETHYLFORMAMIDE	0.9025	3.7875E-01	-4.5230E-02	10	5220	17.22	30.14	42.83
137	C3H7NO2	1-NITROPROPANE	0.9133	3.4648E-01	-3.7300E-02	10	10000	16.69	28.64	41.40
138	C3H7NO2	2-NITROPROPANE	0.8325	3.5732E-01	-3.6080E-02	10	10000	14.25	25.28	37.99
139	C3H8	PROPANE	-0.7946	4.9029E-01	-2.3980E-02	10	10000	0.47	1.23	2.89
140	C3H8O	ISOPROPANOL	0.2718	4.6419E-01	-3.6820E-02	10	10000	5.00	11.30	21.53
141	C3H8O	n-PROPANOL	0.3864	4.8033E-01	-4.5050E-02	10	10000	6.63	14.69	26.42
142	C3H8O2	2-METHOXYETHANOL	0.7434	4.1792E-01	-4.5360E-02	10	10000	13.06	24.99	38.81
143	C3H8O2	METHYLAL	0.1908	3.8167E-01	-1.7750E-02	10	10000	3.59	7.64	15.00
144	C3H8O2	1,2-PROPYLENE GLYCOL	1.4828	1.1594E-01	----------	10	170	39.69	51.84	-------
145	C3H8O2	1,3-PROPYLENE GLYCOL	1.5856	8.3950E-02	----------	10	58	46.73	-------	-------
146	C3H8S	n-PROPYLMERCAPTAN	0.5903	3.1407E-01	-2.1900E-02	10	10000	7.63	13.52	21.65
147	C3H8S	ISOPROPYL MERCAPTAN	0.5578	3.1539E-01	-2.0510E-02	10	10000	7.12	12.78	20.86
148	C3H8S	ETHYL-METHYL-SULFIDE	0.6283	3.1889E-01	-2.3200E-02	10	10000	8.39	14.90	23.78
149	C3H9N	n-PROPYLAMINE	0.0577	3.4918E-01	-1.2410E-02	10	10000	2.48	5.09	9.85
150	C3H9N	ISOPROPYLAMINE	0.0746	3.7106E-01	-1.5680E-02	10	10000	2.69	5.68	11.14
151	C3H9N	TRIMETHYLAMINE	-0.0942	3.2583E-01	-3.3700E-03	10	10000	1.69	3.50	7.13
152	C3H9NO	1-AMINO-2-PROPANOL	1.2550	2.7456E-01	-4.2540E-02	10	617	30.69	43.04	-------
153	C3H9NO	3-AMINO-1-PROPANOL	1.5373	8.1560E-02	----------	10	101	41.58	50.17	-------
154	C3H9NO	METHYLETHANOLAMINE	1.1475	3.0208E-01	-4.2430E-02	10	1422	25.53	38.18	46.97
155	C3H9O3P	TRIMETHYL-PHOSPHITE	1.0057	2.1001E-01	-1.4020E-02	10	10000	15.91	23.42	32.32
156	C3H9O4P	TRIMETHYL PHOSPHATE	1.4846	1.6933E-01	-2.2900E-02	10	1196	42.76	53.91	61.17

Table 19-1 ADSORPTION CAPACITY OF ACTIVATED CARBON (continued)

| NO | FORMULA | NAME | $\log_{10} Q = A + B[\log_{10} y] + C[\log_{10} y]^2$ | | | | | Q - g of compound/100 g of carbon, y - ppmv | | |
			A	B	C	ymin	ymax	Q @10 ppmv	Q @100 ppmv	Q @1000 ppmv
157	C3H10N2	1,2-PROPANEDIAMINE	0.9024	3.1904E-01	-3.4000E-02	10	10000	15.40	25.38	35.77
158	C4H4O	FURAN	0.0408	4.0613E-01	-1.6200E-02	10	10000	2.70	6.14	12.99
159	C4H4O2	DIKETENE	0.8743	3.7094E-01	-3.9620E-02	10	10000	16.06	28.69	42.71
160	C4H4S	THIOPHENE	0.8075	3.2166E-01	-2.6540E-02	10	10000	12.67	22.12	34.17
161	C4H5Cl	CHLOROPRENE	0.7296	2.9786E-01	-2.1110E-02	10	10000	10.15	17.41	27.11
162	C4H5N	trans-CROTONITRILE	0.7079	3.7284E-01	-3.7560E-02	10	10000	11.05	20.11	30.79
163	C4H5N	cis-CROTONITRILE	0.5813	3.9427E-01	-3.5900E-02	10	10000	8.70	16.84	27.61
164	C4H5N	METHACRYLONITRILE	0.4666	3.8890E-01	-3.0420E-02	10	10000	6.68	13.26	22.88
165	C4H5N	PYRROLE	0.8313	3.8413E-01	-4.2170E-02	10	10000	14.90	26.97	40.19
166	C4H5N	VINYLACETONITRILE	0.6184	4.0844E-01	-4.0320E-02	10	10000	9.70	18.80	30.26
167	C4H5NO2	METHYL CYANOACETATE	1.5659	9.1430E-02	----------	10	180	45.43	56.07	------
168	C4H6	1,3-BUTADIENE	-0.0336	3.4764E-01	-1.2970E-02	10	10000	2.00	4.07	7.81
169	C4H6	DIMETHYLACETYLENE	-0.0667	3.9387E-01	-1.5240E-02	10	10000	2.05	4.57	9.50
170	C4H6	ETHYLACETYLENE	-0.0292	3.3636E-01	-1.0560E-02	10	10000	1.98	3.99	7.67
171	C4H6Cl2	1,3-DICHLORO-trans-2-BUTENE	1.3021	1.9939E-01	-2.0910E-02	10	10000	30.24	41.42	51.53
172	C4H6Cl2	1,4-DICHLORO-cis-2-BUTENE	1.4012	1.7876E-01	-2.0410E-02	10	5383	36.27	47.54	56.73
173	C4H6Cl2	1,4-DICHLORO-trans-2-BUTENE	1.4090	1.8120E-01	-2.1790E-02	10	4503	37.02	48.34	57.09
174	C4H6Cl2	3,4-DICHLORO-1-BUTENE	1.2339	2.1476E-01	-2.1350E-02	10	10000	26.75	37.85	48.54
175	C4H6O	trans-CROTONALDEHYDE	0.6835	3.6560E-01	-3.4000E-02	10	10000	10.35	19.00	29.81
176	C4H6O	2,5-DIHYDROFURAN	0.3399	4.0041E-01	-2.4510E-02	10	10000	5.20	11.03	20.92
177	C4H6O	METHACROLEIN	0.4346	3.7019E-01	-2.4740E-02	10	10000	6.03	11.91	21.01
178	C4H6O2	gamma-BUTYROLACTONE	1.2943	2.9719E-01	-4.6580E-02	10	592	35.07	50.39	------
179	C4H6O2	cis-CROTONIC ACID	1.3087	2.5008E-01	-3.7520E-02	10	773	33.21	45.58	------
180	C4H6O2	METHACRYLIC ACID	1.2310	2.7648E-01	-3.9030E-02	10	1282	29.41	42.44	51.19
181	C4H6O2	METHYL ACRYLATE	0.4587	3.2104E-01	-2.0010E-02	10	10000	5.75	10.49	17.45
182	C4H6O2	VINYL ACETATE	0.6107	3.4797E-01	-2.5950E-02	10	10000	8.56	15.95	26.36
183	C4H6O3	ACETIC ANHYDRIDE	1.0739	3.1083E-01	-3.5750E-02	10	7274	22.33	35.69	48.37
184	C4H7N	n-BUTYRONITRILE	0.6431	3.8787E-01	-3.8220E-02	10	10000	9.83	18.45	29.02
185	C4H7N	ISOBUTYRONITRILE	0.5670	3.8807E-01	-3.5310E-02	10	10000	8.31	15.92	25.90
186	C4H7NO	3-METHOXYPROPIONITRILE	1.1328	2.8534E-01	-3.7320E-02	10	2598	24.04	35.83	44.98
187	C4H8	1-BUTENE	0.0731	3.2701E-01	-1.4520E-02	10	10000	2.43	4.67	8.38
188	C4H8Br2	1,2-DIBROMOBUTANE	1.6923	1.2766E-01	-1.4970E-02	10	4094	63.83	77.23	87.22
189	C4H8Br2	2,3-DIBROMOBUTANE	1.6818	1.2916E-01	-1.4920E-02	10	4934	62.52	75.93	86.10
190	C4H8Cl2	1,4-DICHLOROBUTANE	1.3828	1.7796E-01	-2.0300E-02	10	5427	34.71	45.45	54.20
191	C4H8O	n-BUTYRALDEHYDE	0.4506	3.7372E-01	-2.6890E-02	10	10000	6.27	12.31	21.36
192	C4H8O	ISOBUTYRALDEHYDE	0.3932	3.6715E-01	-2.3790E-02	10	10000	5.45	10.77	19.08
193	C4H8O	1,2-EPOXYBUTANE	0.3672	3.7654E-01	-2.3600E-02	10	10000	5.25	10.61	19.25
194	C4H8O	METHYL ETHYL KETONE	0.4653	3.7688E-01	-2.8010E-02	10	10000	6.52	12.79	22.07
195	C4H8O	ETHYL VINYL ETHER	0.3331	3.3471E-01	-1.7110E-02	10	10000	4.47	8.59	15.25
196	C4H8O	TETRAHYDROFURAN	0.2986	3.5648E-01	-1.5500E-02	10	10000	4.36	8.90	16.93
197	C4H8O2	ISOBUTYRIC ACID	1.1402	2.9004E-01	-3.8330E-02	10	2388	24.66	36.89	46.27
198	C4H8O2	n-BUTYRIC ACID	1.2259	2.6481E-01	-3.7370E-02	10	1244	28.40	40.37	48.30
199	C4H8O2	1,4-DIOXANE	0.6678	3.6208E-01	-3.0340E-02	10	10000	9.99	18.65	30.27
200	C4H8O2	ETHYL ACETATE	0.6361	3.4441E-01	-2.6910E-02	10	10000	8.99	16.49	26.74
201	C4H8O2	METHYL PROPIONATE	0.6427	3.4862E-01	-2.7670E-02	10	10000	9.20	16.96	27.52
202	C4H8O2	n-PROPYL FORMATE	0.6586	3.4340E-01	-2.7500E-02	10	10000	9.43	17.19	27.62
203	C4H8O2S	SULFOLANE	1.7776	1.1800E-03	5.0000E-05	10	10000	60.10	60.28	60.48
204	C4H8S	TETRAHYDROTHIOPHENE	0.9378	3.3197E-01	-3.8770E-02	10	10000	17.02	27.97	38.44
205	C4H9Br	1-BROMOBUTANE	1.1670	2.4380E-01	-2.2700E-02	10	10000	24.44	36.62	49.44
206	C4H9Br	2-BROMOBUTANE	1.1387	2.4481E-01	-2.2030E-02	10	10000	22.99	34.69	47.30
207	C4H9Cl	n-BUTYL CHLORIDE	0.8002	2.9114E-01	-2.3700E-02	10	10000	11.69	19.40	28.87
208	C4H9Cl	sec-BUTYL CHLORIDE	0.7505	2.9132E-01	-2.2120E-02	10	10000	10.46	17.56	26.63
209	C4H9Cl	tert-BUTYL CHLORIDE	0.6867	2.8529E-01	-1.9440E-02	10	10000	8.97	15.12	23.32
210	C4H9N	PYRROLIDINE	0.6069	3.6363E-01	-3.0040E-02	10	10000	8.72	16.37	26.76
211	C4H9NO	N,N-DIMETHYLACETAMIDE	1.2003	2.5124E-01	-3.2630E-02	10	2631	26.23	37.34	45.74
212	C4H9NO	MORPHOLINE	1.0067	3.0572E-01	-3.2940E-02	10	10000	19.03	30.65	42.41
213	C4H10	n-BUTANE	0.0307	3.4304E-01	-1.5960E-02	10	10000	2.28	4.50	8.25
214	C4H10	ISOBUTANE	-0.0168	3.3495E-01	-1.2740E-02	10	10000	2.02	4.00	7.47
215	C4H10O	n-BUTANOL	0.8988	3.2534E-01	-3.6480E-02	10	9276	15.41	25.33	35.20
216	C4H10O	sec-BUTANOL	0.7681	3.4611E-01	-3.4780E-02	10	10000	12.01	20.95	31.15
217	C4H10O	DIETHYL ETHER	0.2348	3.6044E-01	-2.2360E-02	10	10000	3.74	7.35	13.03
218	C4H10O	METHYL-PROPYL-ETHER	0.3676	3.2893E-01	-1.7870E-02	10	10000	4.77	9.00	15.62
219	C4H10O	METHYL ISOPROPYL ETHER	0.3637	3.1940E-01	-1.6470E-02	10	10000	4.64	8.64	14.92
220	C4H10O	ISOBUTANOL	0.8482	3.3155E-01	-3.5590E-02	10	10000	13.94	23.38	33.30
221	C4H10O2	1,3-BUTANEDIOL	------	----------	----------	------	-------	48.89	-------	-------
222	C4H10O2	1,4-BUTANEDIOL	------	----------	----------	------	-------	51.19	-------	-------
223	C4H10O2	2,3-BUTANEDIOL	1.5064	9.2390E-02	----------	10	239	39.70	49.11	-------
224	C4H10O2	t-BUTYL HYDROPEROXIDE	1.0856	2.6496E-01	-3.0350E-02	10	7205	20.90	31.20	40.49
225	C4H10O2	1,2-DIMETHOXYETHANE	0.7498	3.1330E-01	-2.6160E-02	10	10000	10.89	18.70	28.46
226	C4H10O2	2-ETHOXYETHANOL	1.0791	2.7792E-01	-3.1990E-02	10	6983	21.14	32.14	42.16
227	C4H10O4S	DIETHYL SULFATE	1.6480	5.8050E-02	----------	10	278	50.82	58.08	-------
228	C4H10S	n-BUTYL MERCAPTAN	0.9809	2.4388E-01	-2.2510E-02	10	10000	15.93	23.91	32.35
229	C4H10S	ISOBUTYL MERCAPTAN	0.9371	2.4802E-01	-2.1790E-02	10	10000	14.57	22.18	30.55
230	C4H10S	sec-BUTYL MERCAPTAN	0.9229	2.4856E-01	-2.1460E-02	10	10000	14.12	21.59	29.88
231	C4H10S	tert-BUTYL MERCAPTAN	0.8438	2.4937E-01	-1.9390E-02	10	10000	11.85	18.41	26.15
232	C4H10S	DIETHYL SULFIDE	0.9599	2.4465E-01	-2.1950E-02	10	10000	15.23	22.98	31.35
233	C4H10S	ISOPROPYL-METHYL-SULFIDE	0.9277	2.4689E-01	-2.1320E-02	10	10000	14.23	21.69	29.95
234	C4H10S	METHYL-PROPYL-SULFIDE	0.9722	2.4453E-01	-2.2290E-02	10	10000	15.65	23.55	32.00

434

Table 19-1 ADSORPTION CAPACITY OF ACTIVATED CARBON (continued)

			$\log_{10} Q = A + B [\log_{10} y] + C [\log_{10} y]^2$			Q - g of compound/100 g of carbon, y - ppmv				
NO	FORMULA	NAME	A	B	C	ymin	ymax	Q @10 ppmv	Q @100 ppmv	Q @1000 ppmv
235	C4H10S2	DIETHYL DISULFIDE	1.4259	1.2564E-01	-1.3730E-02	10	5627	34.50	41.91	47.79
236	C4H11N	n-BUTYLAMINE	0.6457	3.1857E-01	-2.5700E-02	10	10000	8.68	15.14	23.45
237	C4H11N	ISOBUTYLAMINE	0.6014	3.1538E-01	-2.3690E-02	10	10000	7.82	13.72	21.59
238	C4H11N	sec-BUTYLAMINE	0.5771	3.1254E-01	-2.2500E-02	10	10000	7.36	12.95	20.52
239	C4H11N	tert-BUTYLAMINE	0.5004	3.0334E-01	-1.8970E-02	10	10000	6.09	10.74	17.37
240	C4H11N	DIETHYLAMINE	0.5477	3.0799E-01	-2.1070E-02	10	10000	6.83	12.01	19.14
241	C4H11NO	DIMETHYLETHANOLAMINE	1.1838	2.3249E-01	-2.8570E-02	10	4183	24.42	34.24	42.09
242	C4H12Si	TETRAMETHYLSILANE	0.7087	2.3089E-01	-1.5050E-02	10	10000	8.40	12.89	18.45
243	C4H13N3	DIETHYLENE TRIAMINE	1.5427	6.4790E-02	----------	10	308	40.50	47.02	-------
244	C5Cl6	HEXACHLOROCYCLOPENTADIENE	1.8947	2.4970E-02	----------	10	78	83.12	-------	-------
245	C5H4O2	FURFURAL	1.3022	2.4236E-01	-3.1070E-02	10	2907	32.62	45.99	56.19
246	C5H5N	PYRIDINE	0.9185	3.1579E-01	-3.1430E-02	10	10000	15.96	26.57	38.29
247	C5H6	CYCLOPENTADIENE	0.2800	3.4240E-01	-1.6120E-02	10	10000	4.04	7.95	14.52
248	C5H6	2-METHYL-1-BUTENE-3-YNE	0.3965	3.1192E-01	-1.6680E-02	10	10000	4.92	8.99	15.21
249	C5H6	1-PENTENE-3-YNE	0.5415	3.1586E-01	-2.1500E-02	10	10000	6.85	12.23	19.75
250	C5H6	1-PENTENE-4-YNE	0.3859	3.2858E-01	-1.8390E-02	10	10000	4.97	9.32	16.08
251	C5H6N2	GLUTARONITRILE	------	----------	----------	-----	-------	50.12	-------	-------
252	C5H6O2	FURFURYL ALCOHOL	1.4117	2.0912E-01	-3.0690E-02	10	800	38.92	50.96	-------
253	C5H6S	2-METHYLTHIOPHENE	1.1369	2.2962E-01	-2.2580E-02	10	10000	22.07	32.05	41.93
254	C5H6S	3-METHYLTHIOPHENE	1.1467	2.2873E-01	-2.2740E-02	10	10000	22.52	32.59	42.48
255	C5H7N	N-METHYLPYRROLE	0.9706	2.7959E-01	-2.7790E-02	10	10000	16.69	26.22	36.24
256	C5H7NO2	ETHYL CYANOACETATE	1.6500	4.9490E-02	----------	10	51	50.06	-------	-------
257	C5H8	CYCLOPENTENE	0.3983	3.3418E-01	-1.8850E-02	10	10000	5.17	9.80	17.03
258	C5H8	ISOPRENE	0.3607	3.4086E-01	-2.2060E-02	10	10000	4.78	9.00	15.30
259	C5H8	3-METHYL-1,2-BUTADIENE	0.5129	2.9543E-01	-1.8270E-02	10	10000	6.17	10.73	17.17
260	C5H8	2-METHYL-1,3-BUTADIENE	0.4990	2.9044E-01	-1.7140E-02	10	10000	5.92	10.26	16.44
261	C5H8	1,2-PENTADIENE	0.5198	2.9918E-01	-1.8960E-02	10	10000	6.31	11.02	17.65
262	C5H8	cis-1,3-PENTADIENE	0.5594	2.8894E-01	-1.8650E-02	10	10000	6.76	11.55	18.13
263	C5H8	trans-1,3-PENTADIENE	0.5576	2.8487E-01	-1.8250E-02	10	10000	6.67	11.33	17.70
264	C5H8	1,4-PENTADIENE	0.4063	2.9813E-01	-1.5840E-02	10	10000	4.88	8.69	14.39
265	C5H8	2,3-PENTADIENE	0.5508	2.9638E-01	-1.9490E-02	10	10000	6.73	11.63	18.39
266	C5H8	1-PENTYNE	0.3928	3.2123E-01	-1.8210E-02	10	10000	4.96	9.17	15.58
267	C5H8	2-PENTYNE	0.4914	3.2217E-01	-2.1400E-02	10	10000	6.20	11.22	18.42
268	C5H8	3-METHYL-1-BUTYNE	0.3530	3.1235E-01	-1.6160E-02	10	10000	4.46	8.19	13.95
269	C5H8O	CYCLOPENTANONE	0.9557	3.1398E-01	-3.3410E-02	10	10000	17.23	28.18	39.53
270	C5H8O	METHYL ISOPROPENYL KETONE	0.8522	2.9307E-01	-2.6740E-02	10	10000	13.14	21.45	30.95
271	C5H8O2	ACETYLACETONE	1.2470	2.1191E-01	-2.4530E-02	10	5458	27.19	37.39	45.92
272	C5H8O2	ALLYL ACETATE	0.9691	2.7055E-01	-2.5550E-02	10	10000	16.37	25.59	35.55
273	C5H8O2	ETHYL ACRYLATE	0.5609	2.7017E-01	-8.8900E-03	10	10000	6.64	11.63	19.56
274	C5H8O2	METHYL METHACRYLATE	1.2144	1.0453E-01	-5.7300E-03	10	10000	20.57	25.15	29.96
275	C5H8O3	2-HYDROXYETHYL ACRYLATE	1.6305	4.5910E-02	----------	10	69	47.47	-------	-------
276	C5H8O3	METHYL ACETOACETATE	1.4224	1.7734E-01	-2.4150E-02	10	1175	37.63	47.91	54.58
277	C5H9N	VALERONITRILE	1.0281	2.6173E-01	-2.9150E-02	10	9597	18.23	27.22	35.56
278	C5H9NO	n-BUTYL ISOCYANATE	1.0780	2.3588E-01	-2.3970E-02	10	10000	19.49	28.43	37.14
279	C5H9NO	N-METHYL-2-PYRROLIDONE	1.5139	8.8720E-02	----------	10	448	40.05	49.13	-------
280	C5H10	CYCLOPENTANE	0.4432	3.2910E-01	-1.9910E-02	10	10000	5.65	10.51	17.84
281	C5H10	2-METHYL-1-BUTENE	0.4494	2.9381E-01	-1.6710E-02	10	10000	5.33	9.34	15.15
282	C5H10	2-METHYL-2-BUTENE	0.4876	2.9496E-01	-1.7870E-02	10	10000	5.82	10.14	16.28
283	C5H10	1-PENTENE	0.4416	2.9253E-01	-1.6500E-02	10	10000	5.22	9.13	14.82
284	C5H10	cis-2-PENTENE	0.4803	2.9384E-01	-1.7620E-02	10	10000	5.71	9.94	15.97
285	C5H10	trans-2-PENTENE	0.4776	2.9283E-01	-1.7520E-02	10	10000	5.66	9.84	15.79
286	C5H10Cl2	1,5-DICHLOROPENTANE	1.5262	1.1672E-01	-1.4690E-02	10	1484	42.48	50.21	55.47
287	C5H10O	METHYL ISOPROPYL KETONE	0.8176	2.9264E-01	-2.6140E-02	10	10000	12.14	19.88	28.86
288	C5H10O	2-PENTANONE	0.8638	2.8549E-01	-2.6680E-02	10	10000	13.26	21.29	30.21
289	C5H10O	DIETHYL KETONE	0.8623	2.8778E-01	-2.6900E-02	10	10000	13.28	21.39	30.45
290	C5H10O	VALERALDEHYDE	0.8790	2.8289E-01	-2.6710E-02	10	10000	13.65	21.77	30.70
291	C5H10O2	n-BUTYL FORMATE	1.0142	2.5183E-01	-2.4370E-02	10	10000	17.45	26.33	35.51
292	C5H10O2	ETHYL PROPIONATE	0.9778	2.5928E-01	-2.4430E-02	10	10000	16.32	25.04	34.34
293	C5H10O2	ISOBUTYL FORMATE	0.9792	2.5416E-01	-2.3680E-02	10	10000	16.21	24.71	33.78
294	C5H10O2	ISOPROPYL ACETATE	0.8483	2.2591E-01	-1.1250E-02	10	10000	11.56	17.99	26.60
295	C5H10O2	n-PROPYL ACETATE	0.9925	2.5587E-01	-2.4380E-02	10	10000	16.75	25.51	34.73
296	C5H10O2	METHYL n-BUTYRATE	0.9922	2.5872E-01	-2.4720E-02	10	10000	16.83	25.75	35.14
297	C5H10O2	ISOVALERIC ACID	0.6287	3.1278E-01	-2.6590E-02	10	578	8.22	14.06	-------
298	C5H10O2	VALERIC ACID	1.5005	7.9620E-02	----------	10	321	38.02	45.68	-------
299	C5H10O2	TETRAHYDROFURFURYL ALCOHOL	1.3926	1.9211E-01	-2.6660E-02	10	1059	36.14	46.79	53.57
300	C5H10O3	DIETHYL CARBONATE	1.1953	2.1465E-01	-2.2700E-02	10	10000	24.39	34.18	43.15
301	C5H10O3	ETHYL LACTATE	1.3126	1.9639E-01	-2.3510E-02	10	4934	30.59	40.87	49.00
302	C5H10S	THIACYCLOHEXANE	1.2743	1.8674E-01	-2.0060E-02	10	10000	27.60	36.94	45.08
303	C5H11Cl	1-CHLOROPENTANE	1.0984	2.1414E-01	-2.0490E-02	10	10000	19.59	27.84	36.00
304	C5H11Cl	1-CHLORO-3-METHYLBUTANE	1.0576	2.1782E-01	-1.9910E-02	10	10000	18.01	25.92	34.03
305	C5H11Cl	2-CHLORO-2-METHYLBUTANE	1.0194	2.2273E-01	-1.9540E-02	10	10000	16.69	24.36	32.49
306	C5H11N	N-METHYLPYRROLIDINE	0.8400	2.6710E-01	-2.2190E-02	10	10000	12.16	19.29	27.64
307	C5H11N	PIPERIDINE	0.9580	2.6527E-01	-2.5340E-02	10	10000	15.77	24.39	33.55
308	C5H11NO	tert-BUTYLFORMAMIDE	1.4958	7.3120E-02	----------	10	359	37.06	43.85	-------
309	C5H12	ISOPENTANE	0.4498	2.8626E-01	-1.6210E-02	10	10000	5.25	9.07	14.54
310	C5H12	NEOPENTANE	0.3779	3.1821E-01	-2.3130E-02	10	10000	4.71	8.35	13.32
311	C5H12	n-PENTANE	0.4838	2.8785E-01	-1.7420E-02	10	10000	5.68	9.77	15.51
312	C5H12O	2-METHYL-1-BUTANOL	1.1773	2.1844E-01	-2.6810E-02	10	4116	23.38	32.13	39.02

435

Table 19-1 ADSORPTION CAPACITY OF ACTIVATED CARBON (continued)

NO	FORMULA	NAME	$\log_{10} Q = A + B[\log_{10} y] + C[\log_{10} y]^2$					Q - g of compound/100 g of carbon, y - ppmv		
			A	B	C	ymin	ymax	Q @10 ppmv	Q @100 ppmv	Q @1000 ppmv
313	C5H12O	2-METHYL-2-BUTANOL	1.0064	2.5140E-01	-2.5700E-02	10	10000	17.07	25.49	33.83
314	C5H12O	3-METHYL-1-BUTANOL	1.1713	2.2051E-01	-2.7030E-02	10	4187	23.16	31.93	38.87
315	C5H12O	3-METHYL-2-BUTANOL	1.0736	2.3892E-01	-2.5780E-02	10	10000	19.35	28.07	36.17
316	C5H12O	1-PENTANOL	1.0714	2.7214E-01	-3.4750E-02	10	3250	20.36	29.97	37.59
317	C5H12O	2-PENTANOL	1.1130	2.2907E-01	-2.5800E-02	10	8031	20.71	29.37	36.99
318	C5H12O	3-PENTANOL	1.0758	2.3994E-01	-2.5980E-02	10	10000	19.49	28.30	36.45
319	C5H12O	METHYL sec-BUTYL ETHER	0.7373	2.6744E-01	-1.9980E-02	10	10000	9.65	15.57	22.90
320	C5H12O	METHYL tert-BUTYL ETHER	0.7107	2.6973E-01	-1.9520E-02	10	10000	9.14	14.86	22.09
321	C5H120	METHYL ISOBUTYL ETHER	0.7911	2.4959E-01	-1.9000E-02	10	10000	10.51	16.38	23.38
322	C5H12O	ETHYL PROPYL ETHER	0.7551	2.6369E-01	-2.0220E-02	10	10000	9.97	15.91	23.13
323	C5H12O2	ETHYLENE GLYCOL MONOPROPYL ETHER	1.2746	1.9202E-01	-2.3400E-02	10	4105	27.75	36.73	43.66
324	C5H12O3	2-(2-METHOXYETHOXY)ETHANOL	1.5824	6.1080E-02	----------	10	237	44.01	50.65	-------
325	C5H12S	n-PENTYL MERCAPTAN	1.2216	1.7004E-01	-1.7360E-02	10	10000	23.67	31.06	37.62
326	C5H12S	BUTYL-METHYL-SULFIDE	1.2134	1.7168E-01	-1.7380E-02	10	10000	23.32	30.71	37.33
327	C5H12S	ETHYL-PROPYL-SULFIDE	1.1971	1.7402E-01	-1.7330E-02	10	10000	22.58	29.91	36.57
328	C5H12S	2-METHYL-2-BUTANETHIOL	1.1202	1.8741E-01	-1.7300E-02	10	10000	19.51	26.65	33.63
329	C5H13N	n-PENTYLAMINE	0.9921	2.3123E-01	-2.2320E-02	10	10000	15.89	23.19	30.55
330	C6F6	HEXAFLUOROBENZENE	1.1420	2.7082E-01	-2.2970E-02	10	10000	24.54	39.06	55.94
331	C6H3Cl3	1,2,4-TRICHLOROBENZENE	1.6830	9.4560E-02	-9.9800E-03	10	566	58.56	67.96	-------
332	C6H4Br2	m-DIBROMOBENZENE	1.8986	4.2930E-02	----------	10	353	87.39	96.48	-------
333	C6H4Cl2	m-DICHLOROBENZENE	1.5364	1.2406E-01	-1.3170E-02	10	2821	44.39	53.93	61.67
334	C6H4Cl2	o-DICHLOROBENZENE	1.5364	1.2406E-01	-1.3170E-02	10	1783	44.39	53.93	61.67
335	C6H5Br	BROMOBENZENE	1.5581	1.4840E-01	-1.6560E-02	10	5593	48.97	61.47	71.48
336	C6H5Cl	MONOCHLOROBENZENE	1.0271	3.0619E-01	-3.3530E-02	10	10000	19.94	32.01	44.04
337	C6H5ClO	o-CHLOROPHENOL	1.4985	1.5264E-01	-1.8610E-02	10	3330	42.90	53.62	61.51
338	C6H5F	FLUOROBENZENE	0.8977	2.8858E-01	-2.4490E-02	10	10000	14.51	23.82	34.91
339	C6H5I	IODOBENZENE	1.7808	9.7760E-02	-1.1990E-02	10	1398	73.55	84.79	92.51
340	C6H5NO2	NITROBENZENE	1.6486	6.1090E-02	----------	10	329	51.25	58.99	-------
341	C6H6	BENZENE	0.8112	2.8864E-01	-2.3780E-02	10	10000	11.91	19.65	29.05
342	C6H6ClN	m-CHLOROANILINE	1.7151	4.1520E-02	----------	10	87	57.09	-------	-------
343	C6H6N2	cis-DICYANO-1-BUTENE	------	----------	----------	-----	--------	51.27	-------	-------
344	C6H6N2	trans-DICYANO-1-BUTENE	------	----------	----------	-----	--------	50.52	-------	-------
345	C6H6O	PHENOL	1.4560	1.0349E-01	-1.0860E-02	10	10000	35.37	41.64	46.64
346	C6H6S	PHENYL MERCAPTAN	1.4839	1.3156E-01	-1.6470E-02	10	1988	39.72	47.99	53.74
347	C6H7N	ANILINE	-0.1340	5.0000E-04	----------	10	645	0.74	0.74	-------
348	C6H7N	2-METHYLPYRIDINE	1.1743	2.1689E-01	-2.2860E-02	10	10000	23.35	32.86	41.61
349	C6H7N	4-METHYLPYRIDINE	1.2387	2.0376E-01	-2.2840E-02	10	7761	26.28	35.88	44.10
350	C6H8	1,3-CYCLOHEXADIENE	0.8333	2.7528E-01	-2.2830E-02	10	10000	12.18	19.61	28.43
351	C6H8	METHYLCYCLOPENTADIENE	0.8040	2.7260E-01	-2.1900E-02	10	10000	11.34	18.26	26.59
352	C6H8N2	PHENYLHYDRAZINE	------	----------	----------	-----	--------	53.68	-------	-------
353	C6H8O4	DIMETHYL MALEATE	1.6538	5.2700E-02	----------	10	227	50.88	57.44	-------
354	C6H10	1-METHYLCYCLOPENTENE	0.8107	2.6695E-01	-2.1700E-02	10	10000	11.37	18.10	26.07
355	C6H10	3-METHYLCYCLOPENTENE	0.7632	2.6774E-01	-2.0510E-02	10	10000	10.24	16.47	24.09
356	C6H10	4-METHYLCYCLOPENTENE	0.7965	2.6861E-01	-2.1740E-02	10	10000	11.05	17.65	25.50
357	C6H10	CYCLOHEXENE	0.8394	2.7057E-01	-2.2800E-02	10	10000	12.22	19.47	27.92
358	C6H10	2,3-DIMETHYL-1,3-BUTADIENE	0.8559	2.3724E-01	-1.9080E-02	10	10000	11.86	17.95	24.88
359	C6H10	1,5-HEXADIENE	0.7803	2.4517E-01	-1.8630E-02	10	10000	10.16	15.71	22.29
360	C6H10	cis,trans-2,4-HEXADIENE	0.9506	2.1782E-01	-1.8960E-02	10	10000	14.11	20.44	27.13
361	C6H10	trans,trans-2,4-HEXADIENE	0.9442	2.1718E-01	-1.8800E-02	10	10000	13.88	20.11	26.70
362	C6H10	1-HEXYNE	0.7890	2.5895E-01	-2.0830E-02	10	10000	10.64	16.73	23.90
363	C6H10	2-HEXYNE	0.8579	2.5214E-01	-2.1750E-02	10	10000	12.25	18.84	26.22
364	C6H10	3-HEXYNE	0.8496	2.5011E-01	-2.1260E-02	10	10000	11.98	18.40	25.62
365	C6H10O	CYCLOHEXANONE	1.1217	2.4930E-01	-2.9160E-02	10	5690	21.97	31.89	40.47
366	C6H10O	MESITYL OXIDE	1.1844	1.9382E-01	-2.0340E-02	10	10000	22.80	30.95	38.26
367	C6H10O2	epsilon-CAPROLACTONE	1.5883	6.7840E-02	----------	10	248	45.30	52.96	-------
368	C6H10O2	ETHYL METHACRYLATE	1.1863	1.9341E-01	-1.9260E-02	10	10000	22.93	31.34	39.19
369	C6H10O2	n-PROPYL ACRYLATE	1.2212	1.8302E-01	-1.8720E-02	10	10000	24.30	32.54	39.98
370	C6H10O3	PROPIONIC ANHYDRIDE	1.4411	1.4193E-01	-1.8110E-02	10	1794	36.72	44.93	50.57
371	C6H10O4	DIETHYL OXALATE	1.5623	1.0360E-01	-1.4070E-02	10	544	44.86	51.67	-------
372	C6H10O4	ETHYLENE GLYCOL DIACETATE	1.6681	4.2330E-02	----------	10	102	51.33	56.59	-------
373	C6H10O4	ETHYLIDENE DIACETATE	1.5361	1.1480E-01	-1.5510E-02	10	765	43.19	50.56	-------
374	C6H11N	HEXANENITRILE	1.2533	1.7713E-01	-2.1650E-02	10	3746	25.63	33.19	38.89
375	C6H12	CYCLOHEXANE	0.7200	2.5698E-01	-1.5500E-02	10	10000	9.15	14.86	22.46
376	C6H12	2,3-DIMETHYL-1-BUTENE	0.7814	2.3912E-01	-1.7950E-02	10	10000	10.06	15.41	21.73
377	C6H12	2,3-DIMETHYL-2-BUTENE	0.9294	2.1371E-01	-1.7820E-02	10	10000	13.35	19.30	25.72
378	C6H12	3,3-DIMETHYL-1-BUTENE	0.7268	2.3806E-01	-1.6580E-02	10	10000	8.88	13.70	19.58
379	C6H12	2-ETHYL-1-BUTENE	0.8172	2.3627E-01	-1.8820E-02	10	10000	10.89	16.57	23.12
380	C6H12	1-HEXENE	0.8098	2.3627E-01	-1.8530E-02	10	10000	10.66	16.15	22.49
381	C6H12	cis-2-HEXENE	0.8362	2.3588E-01	-1.9020E-02	10	10000	11.30	17.06	23.59
382	C6H12	trans-2-HEXENE	0.8335	2.3401E-01	-1.8800E-02	10	10000	11.19	16.84	23.24
383	C6H12	cis-3-HEXENE	0.8287	2.3467E-01	-1.8710E-02	10	10000	11.08	16.72	23.14
384	C6H12	trans-3-HEXENE	0.8334	2.3337E-01	-1.8700E-02	10	10000	11.17	16.80	23.19
385	C6H12	METHYLCYCLOPENTANE	0.8069	2.5867E-01	-2.0700E-02	10	10000	11.09	17.43	24.91
386	C6H12	2-METHYL-1-PENTENE	0.8071	2.3746E-01	-1.8480E-02	10	10000	10.62	16.15	22.55
387	C6H12	2-METHYL-2-PENTENE	0.8389	2.3366E-01	-1.8750E-02	10	10000	11.32	17.03	23.51
388	C6H12	3-METHYL-1-PENTENE	0.7724	2.3834E-01	-1.7750E-02	10	10000	9.84	15.07	21.27
389	C6H12	3-METHYL-cis-2-PENTENE	0.8349	2.3643E-01	-1.8940E-02	10	10000	11.28	17.06	23.64
390	C6H12	4-METHYL-1-PENTENE	0.7720	2.3762E-01	-1.7690E-02	10	10000	9.81	15.01	21.16

436

Table 19-1 ADSORPTION CAPACITY OF ACTIVATED CARBON (continued)

NO	FORMULA	NAME	$\log_{10} Q = A + B [\log_{10} y] + C [\log_{10} y]^2$					Q - g of compound/100 g of carbon, y - ppmv			
			A	B	C	ymin	ymax	Q @10 ppmv	Q @100 ppmv	Q @1000 ppmv
391	C6H12	4-METHYL-cis-2-PENTENE	0.7885	2.3617E-01	-1.7870E-02	10	10000	10.16	15.47	21.69
392	C6H12	4-METHYL-trans-2-PENTENE	0.7992	2.3490E-01	-1.8010E-02	10	10000	10.38	15.74	21.97
393	C6H12O	BUTYL VINYL ETHER	1.0495	2.0397E-01	-1.8780E-02	10	10000	17.17	24.12	31.07
394	C6H12O	CYCLOHEXANOL	1.4082	1.6169E-01	-2.2520E-02	10	994	35.27	43.81	-------
395	C6H12O	1-HEXANAL	1.1588	1.9497E-01	-2.0430E-02	10	10000	21.54	29.31	36.29
396	C6H12O	ETHYL ISOPROPYL KETONE	1.1094	2.0644E-01	-2.0850E-02	10	10000	19.72	27.47	34.76
397	C6H12O	2-HEXANONE	1.1602	1.9471E-01	-2.0500E-02	10	10000	21.60	29.35	36.29
398	C6H12O	3-HEXANONE	1.1340	2.0198E-01	-2.0840E-02	10	10000	20.66	28.48	35.67
399	C6H12O	METHYL ISOBUTYL KETONE	1.1020	2.0559E-01	-2.0590E-02	10	10000	19.36	26.97	34.16
400	C6H12O2	n-PENTYL FORMATE	1.2839	1.6558E-01	-1.8180E-02	10	8096	27.00	34.86	41.40
401	C6H12O2	n-BUTYL ACETATE	1.2297	1.7812E-01	-1.8510E-02	10	10000	24.51	32.50	39.58
402	C6H12O2	sec-BUTYL ACETATE	1.1730	1.8935E-01	-1.8710E-02	10	10000	22.06	29.98	37.38
403	C6H12O2	tert-BUTYL ACETATE	1.1092	1.9996E-01	-1.8500E-02	10	10000	19.53	27.24	34.88
404	C6H12O2	ETHYL n-BUTYRATE	1.1966	1.8552E-01	-1.8740E-02	10	10000	23.08	31.09	38.41
405	C6H12O2	ETHYL ISOBUTYRATE	1.1643	1.8958E-01	-1.8530E-02	10	10000	21.65	29.47	36.84
406	C6H12O2	ISOBUTYL ACETATE	1.1913	1.8595E-01	-1.8710E-02	10	10000	22.83	30.78	38.08
407	C6H12O2	n-PROPYL PROPIONATE	1.2143	1.8176E-01	-1.8630E-02	10	10000	23.85	31.86	39.08
408	C6H12O2	CYCLOHEXYL PEROXIDE	-------	----------	----------	-----	-------	49.92	-------	-------
409	C6H12O2	DIACETONE ALCOHOL	1.3904	1.4843E-01	-1.8680E-02	10	2247	33.13	40.98	46.51
410	C6H12O2	2-ETHYL BUTYRIC ACID	1.5452	5.7790E-02	----------	10	246	40.08	45.79	-------
411	C6H12O2	n-HEXANOIC ACID	1.6015	4.1720E-02	----------	10	57	43.98	-------	-------
412	C6H12O3	2-ETHOXYETHYL ACETATE	1.4132	1.3968E-01	-1.6970E-02	10	3081	34.35	42.14	47.80
413	C6H12O3	PARALDEHYDE	1.3114	1.6495E-01	-1.7090E-02	10	10000	28.79	37.41	44.92
414	C6H13N	CYCLOHEXYLAMINE	1.2161	1.8350E-01	-1.9320E-02	10	10000	24.01	32.05	39.15
415	C6H13N	HEXAMETHYLENEIMINE	1.2391	1.7954E-01	-1.9180E-02	10	10000	25.09	33.22	40.28
416	C6H14	2,2-DIMETHYLBUTANE	0.7673	2.3271E-01	-1.7040E-02	10	10000	9.62	14.61	20.52
417	C6H14	2,3-DIMETHYLBUTANE	0.7972	2.3308E-01	-1.7780E-02	10	10000	10.29	15.57	21.70
418	C6H14	n-HEXANE	0.8360	2.3017E-01	-1.8590E-02	10	10000	11.16	16.67	22.87
419	C6H14	2-METHYLPENTANE	0.8069	2.3064E-01	-1.7880E-02	10	10000	10.46	15.73	21.77
420	C6H14	3-METHYLPENTANE	0.8165	2.3234E-01	-1.8230E-02	10	10000	10.73	16.15	22.36
421	C6H14O	2-ETHYL-1-BUTANOL	1.3404	1.4928E-01	-1.8990E-02	10	2013	29.56	36.56	41.43
422	C6H14O	1-HEXANOL	1.2985	1.6289E-01	-2.1060E-02	10	1220	27.56	34.67	39.59
423	C6H14O	2-HEXANOL	1.3008	1.5709E-01	-1.9240E-02	10	3272	27.46	34.52	39.72
424	C6H14O	2-METHYL-1-PENTANOL	1.3215	1.5542E-01	-1.9470E-02	10	2523	28.67	35.84	40.97
425	C6H14O	4-METHYL-2-PENTANOL	1.2466	1.6720E-01	-1.8610E-02	10	6903	24.84	32.11	38.08
426	C6H14O	n-BUTYL ETHYL ETHER	1.0441	1.9895E-01	-1.8230E-02	10	10000	16.78	23.39	29.98
427	C6H14O	DIISOPROPYL ETHER	0.9460	2.0848E-01	-1.7100E-02	10	10000	13.72	19.71	26.16
428	C6H14O	DI-n-PROPYL ETHER	1.0285	2.0034E-01	-1.8020E-02	10	10000	16.25	22.76	29.33
429	C6H14O	METHYL tert-PENTYL ETHER	1.0126	2.0786E-01	-1.8350E-02	10	10000	15.93	22.64	29.58
430	C6H14O2	ACETAL	1.1580	1.8195E-01	-1.7630E-02	10	10000	21.01	28.28	35.09
431	C6H14O2	2-BUTOXYETHANOL	1.4381	1.2089E-01	-1.5610E-02	10	1147	34.95	41.44	45.74
432	C6H14O2	HEXYLENE GLYCOL	-------	----------	----------	-----	-------	46.30	-------	-------
433	C6H14O3	DIETHYLENE GLYCOL DIMETHYL ETHER	1.4094	1.3122E-01	-1.5500E-02	10	3910	33.51	40.73	46.09
434	C6H14O3	DIPROPYLENE GLYCOL	-------	----------	----------	-----	-------	49.63	-------	-------
435	C6H14O3	2-(2-ETHOXYETHOXY)ETHANOL	1.6075	4.4930E-02	----------	10	166	44.92	49.82	-------
436	C6H14S	n-HEXYLMERCAPTAN	1.3741	1.1580E-01	-1.2520E-02	10	5576	30.02	35.95	40.63
437	C6H14S	BUTYL-ETHYL-SULFIDE	1.3533	1.2075E-01	-1.2810E-02	10	7989	28.92	34.95	39.83
438	C6H14S	PROPYL-SULFIDE	1.3498	1.2165E-01	-1.2870E-02	10	8481	28.75	34.80	39.71
439	C6H15Al	TRIETHYL ALUMINUM	-------	----------	----------	-----	-------	41.16	-------	-------
440	C6H15N	DIISOPROPYLAMINE	1.0469	1.8337E-01	-1.6150E-02	10	10000	16.37	22.34	28.29
441	C6H15N	DI-n-PROPYLAMINE	1.1528	1.6979E-01	-1.6830E-02	10	10000	20.22	26.61	32.41
442	C6H15N	n-HEXYLAMINE	1.2235	1.5698E-01	-1.6440E-02	10	10000	23.12	29.63	35.19
443	C6H15N	TRIETHYLAMINE	1.0607	1.8327E-01	-1.6360E-02	10	10000	16.89	23.00	29.06
444	C6H15NO3	TRIETHANOLAMINE	1.0924	1.4626E-01	-1.0970E-02	10	10000	16.89	21.93	27.07
445	C6H15N3	N-AMINOETHYL PIPERAZINE	1.6367	3.5630E-02	----------	10	77	47.03	-------	-------
446	C6H15O4P	TRIETHYL PHOSPHATE	1.6142	6.8620E-02	-8.5000E-03	10	516	47.24	52.17	-------
447	C6H18N3OP	HEXAMETHYL PHOSPHORAMIDE	1.6732	2.5690E-02	----------	10	60	49.99	-------	-------
448	C6H18OSi2	HEXAMETHYLDISILOXANE	1.3481	8.6670E-02	-7.6200E-03	10	10000	26.74	30.97	34.64
449	C6H19NSi2	HEXAMETHYLDISILAZANE	1.4107	7.1460E-02	-6.5300E-03	10	10000	29.90	33.69	36.84
450	C7H3ClF3NO2	4-CHLORO-3-NITROBENZOTRIFLUORIDE	1.8307	3.0790E-02	----------	10	78	72.69	-------	-------
451	C7H4ClF3	p-CHLOROBENZOTRIFLUORIDE	1.5257	1.1492E-01	-1.1730E-02	10	10000	42.55	51.13	58.20
452	C7H4F3NO2	3-NITROBENZOTRIFLUORIDE	1.7519	4.8320E-02	----------	10	338	63.13	70.56	-------
453	C7H5ClO	BENZOYL CHLORIDE	1.6132	9.8160E-02	-1.2800E-02	10	822	49.95	57.32	-------
454	C7H5F3	BENZOTRIFLUORIDE	1.2335	2.0792E-01	-1.9580E-02	10	10000	26.41	37.24	47.97
455	C7H5N	BENZONITRILE	1.4657	1.3529E-01	-1.7970E-02	10	1009	38.28	46.17	51.27
456	C7H5NO	PHENYL ISOCYANATE	1.4563	1.4278E-01	-1.7250E-02	10	3384	38.18	47.08	53.63
457	C7H6Cl2	BENZYL DICHLORIDE	1.6464	8.8700E-02	-1.1490E-02	10	672	52.92	59.96	-------
458	C7H6Cl2	2,4-DICHLOROTOLUENE	1.6660	7.7420E-02	-9.7900E-03	10	603	54.15	60.48	-------
459	C7H6O	BENZALDEHYDE	1.4578	1.4292E-01	-1.8480E-02	10	1553	38.21	46.74	52.51
460	C7H6O2	SALICYLALDEHYDE	1.5738	1.1367E-01	-1.5300E-02	10	780	47.00	54.94	-------
461	C7H7Cl	BENZYL CHLORIDE	1.4357	1.3714E-01	-1.3990E-02	10	1716	36.21	45.08	52.62
462	C7H7Cl	o-CHLOROTOLUENE	1.4590	1.3317E-01	-1.5520E-02	10	4638	37.73	46.06	52.34
463	C7H7Cl	p-CHLOROTOLUENE	1.4704	1.2760E-01	-1.5100E-02	10	3671	38.27	46.25	52.15
464	C7H7NO2	m-NITROTOLUENE	1.6656	4.6740E-02	----------	10	272	51.57	57.43	-------
465	C7H7NO2	o-NITROTOLUENE	1.6713	4.5820E-02	----------	10	243	52.13	57.93	-------
466	C7H8	TOLUENE	1.1147	2.0795E-01	-2.0160E-02	10	10000	20.07	28.18	36.06
467	C7H8O	ANISOLE	1.3747	1.5735E-01	-1.8610E-02	10	4750	32.61	41.20	47.78
468	C7H8O	BENZYL ALCOHOL	1.5675	1.0268E-01	-1.4540E-02	10	107	45.25	51.84	-------

Table 19-1 ADSORPTION CAPACITY OF ACTIVATED CARBON (continued)

NO	FORMULA	NAME	$\log_{10} Q = A + B [\log_{10} y] + C [\log_{10} y]^2$					Q - g of compound/100 g of carbon, y - ppmv		
			A	B	C	ymin	ymax	Q @10 ppmv	Q @100 ppmv	Q @1000 ppmv
469	C7H8O	m-CRESOL	1.6198	4.9260E-02	----------	10	149	46.68	52.28	-------
470	C7H9N	BENZYLAMINE	1.4953	1.1541E-01	-1.5460E-02	10	857	39.38	46.16	-------
471	C7H9N	2,6-DIMETHYLPYRIDINE	1.3417	1.4804E-01	-1.6230E-02	10	7152	29.75	37.40	43.63
472	C7H9N	N-METHYLANILINE	1.5262	1.0091E-01	-1.3540E-02	10	596	41.07	47.19	-------
473	C7H9N	m-TOLUIDINE	1.5709	5.6470E-02	----------	10	398	42.40	48.28	-------
474	C7H9N	o-TOLUIDINE	1.5810	5.4750E-02	----------	10	339	43.23	49.04	-------
475	C7H11NO	CYCLOHEXYL ISOCYANATE	1.5311	1.1077E-01	-1.3930E-02	10	1341	42.45	49.76	54.70
476	C7H12	1-HEPTYNE	1.0477	1.9339E-01	-1.7690E-02	10	10000	16.73	23.11	29.42
477	C7H12O2	n-BUTYL ACRYLATE	1.3677	1.2987E-01	-1.4010E-02	10	7171	30.45	37.27	42.78
478	C7H12O2	n-PROPYL METHACRYLATE	1.3621	1.3127E-01	-1.4030E-02	10	8387	30.16	37.03	42.63
479	C7H12O4	DIETHYL MALONATE	1.6220	4.6640E-02	----------	10	354	46.63	51.91	-------
480	C7H14	CYCLOHEPTANE	1.1510	1.8568E-01	-1.8380E-02	10	10000	20.81	28.11	34.88
481	C7H14	1,1-DIMETYLCYCLOPENTANE	1.0314	1.9751E-01	-1.7430E-02	10	10000	16.27	22.74	29.31
482	C7H14	cis-1,2-DIMETHYLCYCLOPENTANE	1.0733	1.9446E-01	-1.7970E-02	10	10000	17.78	24.57	31.26
483	C7H14	trans-1,2-DIMETHYLCYCLOPENTANE	1.0448	1.9512E-01	-1.7520E-02	10	10000	16.69	23.17	29.68
484	C7H14	cis-1,3-DIMETHYLCYCLOPENTANE	1.0410	1.9451E-01	-1.7410E-02	10	10000	16.52	22.93	29.36
485	C7H14	trans-1,3-DIMETHYLCYCLOPENTANE	1.0435	1.9490E-01	-1.7490E-02	10	10000	16.63	23.09	29.57
486	C7H14	ETHYLCYCLOPENTANE	1.0866	1.9120E-01	-1.7950E-02	10	10000	18.19	24.96	31.52
487	C7H14	2-ETHYL-1-PENTENE	1.0677	1.7905E-01	-1.6260E-02	10	10000	17.00	22.95	28.74
488	C7H14	3-ETHYL-1-PENTENE	1.0374	1.7816E-01	-1.5960E-02	10	10000	15.96	21.72	27.45
489	C7H14	1-HEPTENE	1.0670	1.7691E-01	-1.6080E-02	10	10000	16.90	22.73	28.38
490	C7H14	cis-2-HEPTENE	1.0841	1.7515E-01	-1.6130E-02	10	10000	17.50	23.43	29.13
491	C7H14	trans-2-HEPTENE	1.0832	1.7397E-01	-1.6010E-02	10	10000	17.43	23.29	28.91
492	C7H14	cis-3-HEPTENE	1.0782	1.7492E-01	-1.5980E-02	10	10000	17.27	23.13	28.79
493	C7H14	trans-3-HEPTENE	1.0793	1.7362E-01	-1.5880E-02	10	10000	17.26	23.07	28.66
494	C7H14	METHYLCYCLOHEXANE	1.0785	1.9250E-01	-1.7840E-02	10	10000	17.91	24.67	31.29
495	C7H14	2-METHYL-1-HEXENE	1.0640	1.7808E-01	-1.6070E-02	10	10000	16.83	22.69	28.42
496	C7H14	3-METHYL-1-HEXENE	1.0365	1.8069E-01	-1.5860E-02	10	10000	15.90	21.60	27.28
497	C7H14	4-METHYL-1-HEXENE	1.0467	1.8048E-01	-1.6010E-02	10	10000	16.26	22.06	27.80
498	C7H14	2,3,3-TRIMETHYL-1-BUTENE	1.0159	1.8598E-01	-1.5850E-02	10	10000	15.35	21.11	26.99
499	C7H14O	DIISOPROPYL KETONE	1.1987	1.9686E-01	-2.0350E-02	10	10000	23.73	32.44	40.38
500	C7H14O	2-HEPTANONE	1.3261	1.3700E-01	-1.5910E-02	10	4995	28.00	34.39	39.25
501	C7H14O	1-HEPTANAL	1.3330	1.3507E-01	-1.5760E-02	10	4649	28.34	34.68	39.48
502	C7H14O	cis-2-METHYLCYCLOHEXANOL	1.4273	1.2961E-01	-1.6250E-02	10	1902	34.72	41.83	46.75
503	C7H14O	trans-2-METHYLCYCLOHEXANOL	1.4360	1.2443E-01	-1.5750E-02	10	1575	35.05	41.87	46.51
504	C7H14O	5-METHYL-2-HEXANONE	1.3081	1.3817E-01	-1.5070E-02	10	6844	26.99	33.43	38.63
505	C7H14O2	n-BUTYL PROPIONATE	1.3749	1.2365E-01	-1.3450E-02	10	5806	30.55	37.01	42.14
506	C7H14O2	ETHYL ISOVALERATE	1.3363	1.3158E-01	-1.3590E-02	10	10000	28.46	35.08	40.62
507	C7H14O2	ISOPENTYL ACETATE	1.3593	1.2706E-01	-1.3650E-02	10	7362	29.69	36.21	41.46
508	C7H14O2	n-PENTYL ACETATE	1.3871	1.2300E-01	-1.4210E-02	10	4610	31.33	37.70	42.49
509	C7H14O2	n-PROPYL n-BUTYRATE	1.3558	1.2849E-01	-1.3780E-02	10	7716	29.55	36.12	41.43
510	C7H14O3	ETHYL-3-ETHOXYPROPIONATE	1.4889	9.9430E-02	-1.2060E-02	10	1690	37.69	43.60	47.71
511	C7H15Br	1-BROMOHEPTANE	1.5994	8.2030E-02	-9.6000E-03	10	1674	46.97	53.09	57.42
512	C7H15N	N-METHYLCYCLOHEXYLAMINE	1.3558	1.3191E-01	-1.4510E-02	10	5619	29.73	36.44	41.78
513	C7H16	2,2-DIMETHYLPENTANE	1.0196	1.7959E-01	-1.5410E-02	10	10000	15.27	20.76	26.29
514	C7H16	2,3-DIMETHYLPENTANE	1.0572	1.7698E-01	-1.5800E-02	10	10000	16.54	22.29	27.93
515	C7H16	2,4-DIMETHYLPENTANE	1.0266	1.7702E-01	-1.5290E-02	10	10000	15.43	20.87	26.31
516	C7H16	3,3-DIMETHYLPENTANE	1.0445	1.7773E-01	-1.5600E-02	10	10000	16.09	21.75	27.37
517	C7H16	3-ETHYLPENTANE	1.0699	1.7595E-01	-1.5950E-02	10	10000	16.98	22.80	28.46
518	C7H16	n-HEPTANE	1.0855	1.7044E-01	-1.5770E-02	10	10000	17.39	23.09	28.51
519	C7H16	2-METHYLHEXANE	1.0584	1.7308E-01	-1.5500E-02	10	10000	16.44	22.00	27.42
520	C7H16	3-METHYLHEXANE	1.0640	1.7455E-01	-1.5740E-02	10	10000	16.70	22.40	27.93
521	C7H16	2,2,3-TRIMETHYLBUTANE	1.0261	1.8087E-01	-1.5550E-02	10	10000	15.54	21.16	26.84
522	C7H16O	1-HEPTANOL	1.5133	4.8470E-02	----------	10	284	36.45	40.76	-------
523	C7H16O	2-HEPTANOL	1.4080	1.0927E-01	-1.3510E-02	10	1621	31.89	37.36	41.13
524	C7H16O	5-METHYL-1-HEXANOL	1.4645	8.9210E-02	-1.1790E-02	10	508	34.83	39.42	-------
525	C7H16S	n-HEPTYL MERCAPTAN	1.4677	8.2080E-02	-9.5900E-03	10	1718	34.69	39.22	42.43
526	C7H17N	1-AMINOHEPTANE	1.3636	1.1080E-01	-1.2890E-02	10	3604	28.94	34.17	38.02
527	C8H7N	INDOLE	------	----------	----------	------	-------	55.74	-------	-------
528	C8H8	STYRENE	1.3570	1.3495E-01	-1.4510E-02	10	8044	30.02	37.06	42.78
529	C8H8O	ACETOPHENONE	1.5583	9.3380E-02	-1.2440E-02	10	522	43.57	49.57	-------
530	C8H8O2	METHYL BENZOATE	1.5951	8.6010E-02	-1.1280E-02	10	505	46.76	52.73	-------
531	C8H8O3	METHYL SALICYLATE	------	----------	----------	------	------	57.49	-------	-------
532	C8H10	ETHYLBENZENE	1.3044	1.4449E-01	-1.4930E-02	10	10000	27.16	34.17	40.14
533	C8H10	m-XYLENE	1.3152	1.4019E-01	-1.4570E-02	10	10000	27.60	34.46	40.24
534	C8H10	o-XYLENE	1.3340	1.3931E-01	-1.4940E-02	10	8722	28.73	35.72	41.44
535	C8H10	p-XYLENE	1.3112	1.4069E-01	-1.4580E-02	10	10000	27.37	34.22	40.00
536	C8H10O	PHENETOLE	1.4772	1.0788E-01	-1.3070E-02	10	2055	37.32	43.71	48.21
537	C8H10O	2-PHENYLETHANOL	1.6388	3.8720E-02	----------	10	114	47.59	52.03	-------
538	C8H11N	N,N-DIMETHYLANILINE	1.5309	8.3390E-02	-1.0410E-02	10	931	40.17	45.29	-------
539	C8H11N	o-ETHYLANILINE	1.6083	4.1170E-02	----------	10	225	44.62	49.05	-------
540	C8H11N	2,4,6-TRIMETHYLPYRIDINE	1.4542	1.0528E-01	-1.2470E-02	10	2601	35.24	41.19	45.47
541	C8H11NO	p-PHENETIDINE	------	----------	----------	------	------	53.74	-------	-------
542	C8H12	1,5-CYCLOOCTADIENE	1.3527	1.3455E-01	-1.4680E-02	10	6515	29.69	36.57	42.09
543	C8H12	VINYLCYCLOHEXENE	1.2640	1.4697E-01	-1.4680E-02	10	10000	24.90	31.56	37.39
544	C8H12O4	DIETHYL MALEATE	1.6350	2.7620E-02	----------	10	136	45.98	49.00	-------
545	C8H14O2	n-BUTYL METHACRYLATE	1.4619	9.4570E-02	-1.0950E-02	10	2792	35.11	40.48	44.37
546	C8H14O3	BUTYRIC ANHYDRIDE	1.5923	4.2250E-02	----------	10	372	43.11	47.51	-------

Table 19-1 ADSORPTION CAPACITY OF ACTIVATED CARBON (continued)

NO	FORMULA	NAME	A	B	C	ymin	ymax	Q @10 ppmv	Q @100 ppmv	Q @1000 ppmv
			$\log_{10} Q = A + B [\log_{10} y] + C [\log_{10} y]^2$				Q - g of compound/100 g of carbon, y - ppmv			
547	C8H14O4	DIETHYL SUCCINATE	1.6745	2.8950E-02	----------	10	58	50.52	-------	-------
548	C8H16	1,1-DIMETHYLCYCLOHEXANE	1.2350	1.4297E-01	-1.3860E-02	10	10000	23.13	29.21	34.61
549	C8H16	cis-1,2-DIMETHYLCYCLOHEXANE	1.2675	1.3849E-01	-1.3840E-02	10	10000	24.67	30.84	36.18
550	C8H16	trans-1,2-DIMETHYLCYCLOHEXANE	1.2440	1.4010E-01	-1.3720E-02	10	10000	23.46	29.47	34.74
551	C8H16	cis-1,3-DIMETHYLCYCLOHEXANE	1.2353	1.4003E-01	-1.3610E-02	10	10000	23.00	28.90	34.11
552	C8H16	trans-1,3-DIMETHYLCYCLOHEXANE	1.2521	1.3988E-01	-1.3790E-02	10	10000	23.89	29.97	35.28
553	C8H16	cis-1,4-DIMETHYLCYCLOHEXANE	1.2502	1.4004E-01	-1.3790E-02	10	10000	23.79	29.86	35.17
554	C8H16	trans-1,4-DIMETHYLCYCLOHEXANE	1.2303	1.4069E-01	-1.3620E-02	10	10000	22.77	28.65	33.87
555	C8H16	ETHYLCYCLOHEXANE	1.2737	1.3532E-01	-1.3610E-02	10	10000	24.85	30.89	36.07
556	C8H16	2-ETHYL-1-HEXENE	1.2408	1.2902E-01	-1.2520E-02	10	10000	22.76	28.10	32.75
557	C8H16	1-METHYL-1-ETHYLCYCLOPENTANE	1.2417	1.4190E-01	-1.3890E-02	10	10000	23.43	29.51	34.86
558	C8H16	1-OCTENE	1.2428	1.2652E-01	-1.2360E-02	10	10000	22.75	27.95	32.44
559	C8H16	trans-2-OCTENE	1.2503	1.2524E-01	-1.2280E-02	10	10000	23.08	28.29	32.77
560	C8H16	trans-3-OCTENE	1.2473	1.2470E-01	-1.2170E-02	10	10000	22.90	28.05	32.50
561	C8H16	trans-4-OCTENE	1.2449	1.2509E-01	-1.2190E-02	10	10000	22.79	27.94	32.39
562	C8H16	n-PROPYLCYCLOPENTANE	1.2709	1.3422E-01	-1.3520E-02	10	10000	24.64	30.57	35.64
563	C8H16	2,4,4-TRIMETHYL-1-PENTENE	1.1870	1.3821E-01	-1.2680E-02	10	10000	20.54	25.86	30.73
564	C8H16	2,4,4-TRIMETHYL-2-PENTENE	1.2074	1.3440E-01	-1.2520E-02	10	10000	21.34	26.67	31.47
565	C8H16O	2-ETHYLHEXANAL	1.4181	9.9430E-02	-1.1680E-02	10	2586	32.05	37.18	40.86
566	C8H16O	1-OCTANAL	1.4412	9.1400E-02	-1.0980E-02	10	1549	33.24	38.02	41.36
567	C8H16O	2-OCTANONE	1.4707	8.2160E-02	-1.0410E-02	10	723	34.87	39.21	-------
568	C8H16O2	n-BUTYL n-BUTYRATE	1.4588	9.1250E-02	-1.0620E-02	10	2383	34.63	39.70	43.34
569	C8H16O2	n-HEXYL ACETATE	1.4750	8.6180E-02	-1.0150E-02	10	1742	35.56	40.43	43.86
570	C8H16O2	ISOBUTYL ISOBUTYRATE	1.4189	9.7130E-02	-1.0210E-02	10	5701	32.05	37.35	41.54
571	C8H16O4	DIETHYLENE GLYCOL ETHYL ETHER ACET	1.6402	3.3050E-02	----------	10	195	47.13	50.85	-------
572	C8H18	2,2-DIMETHYLHEXANE	1.2010	1.3127E-01	-1.2250E-02	10	10000	20.89	25.97	30.51
573	C8H18	2,3-DIMETHYLHEXANE	1.2281	1.2866E-01	-1.2330E-02	10	10000	22.10	27.30	31.85
574	C8H18	2,4-DIMETHYLHEXANE	1.2096	1.2945E-01	-1.2170E-02	10	10000	21.23	26.29	30.79
575	C8H18	2,5-DIMETHYLHEXANE	1.2069	1.2974E-01	-1.2210E-02	10	10000	21.11	26.16	30.64
576	C8H18	3,3-DIMETHYLHEXANE	1.2157	1.3114E-01	-1.2400E-02	10	10000	21.60	26.81	31.44
577	C8H18	3,4-DIMETHYLHEXANE	1.2344	1.2900E-01	-1.2430E-02	10	10000	22.44	27.71	32.32
578	C8H18	3-ETHYLHEXANE	1.2373	1.2693E-01	-1.2280E-02	10	10000	22.49	27.67	32.18
579	C8H18	3-METHYL-3-ETHYLPENTANE	1.2336	1.3050E-01	-1.2540E-02	10	10000	22.47	27.83	32.53
580	C8H18	2-METHYLHEPTANE	1.2308	1.2568E-01	-1.2120E-02	10	10000	22.10	27.14	31.53
581	C8H18	3-METHYLHEPTANE	1.2365	1.2554E-01	-1.2150E-02	10	10000	22.38	27.48	31.89
582	C8H18	4-METHYLHEPTANE	1.2312	1.3003E-01	-1.2590E-02	10	10000	22.32	27.60	32.21
583	C8H18	n-OCTANE	1.0673	1.8700E-01	-1.2490E-02	10	10000	17.45	24.62	32.80
584	C8H18	2,2,3-TRIMETHYLPENTANE	1.2111	1.3305E-01	-1.2490E-02	10	10000	21.46	26.74	31.46
585	C8H18	2,2,4-TRIMETHYLPENTANE	1.1771	1.3603E-01	-1.2370E-02	10	10000	19.99	25.10	29.77
586	C8H18	2,3,3-TRIMETHYLPENTANE	1.2242	1.3218E-01	-1.2560E-02	10	10000	22.07	27.43	32.18
587	C8H18	2,3,4-TRIMETHYLPENTANE	1.2216	1.3166E-01	-1.2510E-02	10	10000	21.91	27.22	31.92
588	C8H18O	DI-n-BUTYL ETHER	1.3571	1.0114E-01	-1.0470E-02	10	7920	28.04	32.92	36.83
589	C8H18O	DI-sec-BUTYL ETHER	1.3057	1.1246E-01	-1.0900E-02	10	10000	25.54	30.69	35.07
590	C8H18O	DI-tert-BUTYL ETHER	1.2693	1.2049E-01	-1.1190E-02	10	10000	23.91	29.21	33.89
591	C8H18O	2-ETHYL-1-HEXANOL	1.5488	3.7070E-02	----------	10	179	38.53	41.97	-------
592	C8H18O	1-OCTANOL	1.5600	3.2840E-02	----------	10	104	39.16	42.24	-------
593	C8H18O	2-OCTANOL	1.5261	4.1410E-02	----------	10	318	36.94	40.64	-------
594	C8H18O2	DI-t-BUTYL PEROXIDE	1.3320	1.0436E-01	-9.7300E-03	10	10000	26.71	31.75	36.10
595	C8H18O3	DIETHYLENE GLYCOL DIETHYL ETHER	1.5458	6.4110E-02	-7.6700E-03	10	684	40.01	43.98	-------
596	C8H18O3	DIETHYLENE GLYCOL MONOBUTYL EHTER	------	----------	----------	-----	-------	47.51	-------	-------
597	C8H18O4	TRIETHYLENE GLYCOL DIMETHYL ETHER	1.6564	2.5930E-02	----------	10	53	48.12	-------	-------
598	C8H18S	n-OCTYL MERCAPTAN	1.5275	5.7290E-02	-6.7300E-03	10	558	37.85	41.23	-------
599	C8H19N	DI-n-BUTYLAMINE	1.4010	8.7370E-02	-9.8300E-03	10	3402	30.10	34.39	37.56
600	C8H19N	DIISOBUTYLAMINE	1.3539	9.4890E-02	-9.6100E-03	10	9560	27.49	32.00	35.65
601	C8H19N	n-OCTYLAMINE	1.4501	7.6420E-02	-8.9700E-03	10	1273	32.92	36.90	39.68
602	C8H24O4Si4	OCTAMETHYLCYCLOTETRASILOXANE	1.6239	2.7890E-02	-2.4100E-03	10	1285	44.61	46.78	48.52
603	C9H6N2O2	TOLUENE DIISOCYANATE	1.5119	1.3042E-01	-1.6930E-02	10	18	42.21	-------	-------
604	C9H7N	QUINOLINE	1.6853	3.3280E-02	----------	10	79	52.31	-------	-------
605	C9H8	INDENE	1.5128	1.0037E-01	-1.2350E-02	10	1429	39.88	46.14	50.44
606	C9H8O	2-METHYLBENZOFURAN	1.5824	8.2760E-02	-1.0590E-02	10	640	45.14	50.77	-------
607	C9H10	INDANE	1.4815	1.0534E-01	-1.2730E-02	10	2027	37.50	43.78	48.19
608	C9H10O	alpha-METHYLSTYRENE	1.4516	1.0128E-01	-1.1660E-02	10	3460	34.77	40.51	44.72
609	C9H10	m-METHYLSTYRENE	1.4681	9.6840E-02	-1.1350E-02	10	2503	35.77	41.34	45.33
610	C9H10	o-METHYLSTYRENE	1.4706	9.5790E-02	-1.1230E-02	10	2435	35.90	41.42	45.38
611	C9H10	p-METHYLSTYRENE	1.4712	9.7690E-02	-1.1500E-02	10	2405	36.09	41.74	45.78
612	C9H10O2	BENZYL ACETATE	1.6485	3.5930E-02	----------	10	233	48.36	52.53	-------
613	C9H10O2	ETHYL BENZOATE	1.6328	4.0460E-02	----------	10	351	47.13	51.73	-------
614	C9H12	CUMENE	1.3396	1.3829E-01	-1.7200E-02	10	5984	28.89	35.27	39.79
615	C9H12	m-ETHYLTOLUENE	1.4260	1.0166E-01	-1.1580E-02	10	3995	32.82	38.28	42.34
616	C9H12	o-ETHYLTOLUENE	1.4387	1.0099E-01	-1.1630E-02	10	3432	33.73	39.28	43.35
617	C9H12	p-ETHYLTOLUENE	1.4255	1.0122E-01	-1.1530E-02	10	3939	32.75	38.17	42.20
618	C9H12	MESITYLENE	1.4369	9.8470E-02	-1.1340E-02	10	3260	33.42	38.77	42.69
619	C9H12	n-PROPYLBENZENE	1.4177	1.0439E-01	-1.1820E-02	10	4503	32.38	37.95	42.12
620	C9H12	1,2,3-TRIMETHYLBENZENE	1.4648	9.5550E-02	-1.1270E-02	10	2222	35.40	40.81	44.66
621	C9H12	1,2,4-TRIMETHYLBENZENE	1.4450	9.8070E-02	-1.1370E-02	10	2976	34.02	39.42	43.35
622	C9H12O	BENZYL ETHYL ETHER	1.5275	8.0740E-02	-9.6200E-03	10	1216	39.69	44.72	48.21
623	C9H14O	ISOPHORONE	1.5466	6.9800E-02	-8.6300E-03	10	575	40.53	44.84	-------
624	C9H18	ISOPROPYLCYCLOHEXANE	1.3881	9.7440E-02	-1.0180E-02	10	6307	29.88	34.86	38.80

439

Table 19-1 ADSORPTION CAPACITY OF ACTIVATED CARBON (continued)

			$\log_{10} Q = A + B [\log_{10} y] + C [\log_{10} y]^2$					Q - g of compound/100 g of carbon, y - ppmv		
NO	FORMULA	NAME	A	B	C	ymin	ymax	Q @10 ppmv	Q @100 ppmv	Q @1000 ppmv
625	C9H18	1-NONENE	1.3599	8.9970E-02	-9.2100E-03	10	7085	27.58	31.84	35.23
626	C9H18	n-PROPYLCYCLOHEXANE	1.3913	9.5030E-02	-9.9700E-03	10	5513	29.94	34.79	38.60
627	C9H18O	DIISOBUTYL KETONE	1.4521	7.7370E-02	-8.7700E-03	10	2215	33.17	37.31	40.30
628	C9H18O	1-NONANAL	1.5257	3.9900E-02	----------	10	488	36.78	40.32	-------
629	C9H18O2	n-BUTYL VALERATE	1.5264	6.2690E-02	-7.4000E-03	10	776	38.16	41.89	-------
630	C9H18O2	n-OCTYL FORMATE	1.5278	6.3680E-02	-7.5400E-03	10	801	38.36	42.17	-------
631	C9H20	3,3-DIETHYLPENTANE	1.3581	9.4920E-02	-9.6100E-03	10	9591	27.76	32.32	36.00
632	C9H20	2,2-DIMETHYL-3-ETHYLPENTANE	1.3247	1.0037E-01	-9.8100E-03	10	10000	26.02	30.63	34.48
633	C9H20	2,4-DIMETHYL-3-ETHYLPENTANE	1.3388	9.5950E-02	-9.3800E-03	10	10000	26.63	31.13	34.85
634	C9H20	2,2-DIMETHYLHEPTANE	1.3250	9.4050E-02	-9.1300E-03	10	10000	25.70	29.96	33.49
635	C9H20	2,6-DIMETHYLHEPTANE	1.3306	9.2550E-02	-9.0400E-03	10	10000	25.95	30.17	33.64
636	C9H20	3-ETHYLHEPTANE	1.3510	9.1290E-02	-9.2300E-03	10	8938	27.11	31.38	34.82
637	C9H20	4-ETHYLHEPTANE	1.3476	9.2560E-02	-9.3300E-03	10	9758	26.96	31.29	34.78
638	C9H20	2,3-DIMETHYLHEPTANE	1.3452	9.1840E-02	-9.0500E-03	10	10000	26.79	31.09	34.61
639	C9H20	2,4-DIMETHYLHEPTANE	1.3285	9.4170E-02	-9.1700E-03	10	10000	25.91	30.21	33.77
640	C9H20	2,5-DIMETHYLHEPTANE	1.3344	9.2900E-02	-9.0900E-03	10	10000	26.19	30.47	33.98
641	C9H20	3,4-DIMETHYLHEPTANE	1.3472	9.2370E-02	-9.1100E-03	10	10000	26.94	31.29	34.86
642	C9H20	3,5-DIMETHYLHEPTANE	1.3365	9.3490E-02	-9.1600E-03	10	10000	26.35	30.68	34.24
643	C9H20	4,4-DIMETHYLHEPTANE	1.3324	9.5200E-02	-9.2900E-03	10	10000	26.20	30.59	34.23
644	C9H20	3-ETHYL-2-METHYLHEXANE	1.3430	9.3840E-02	-9.2200E-03	10	10000	26.77	31.18	34.80
645	C9H20	4-ETHYL-2-METHYLHEXANE	1.3321	9.4830E-02	-9.2500E-03	10	10000	26.16	30.53	34.14
646	C9H20	3-ETHYL-3-METHYLHEXANE	1.3471	9.4420E-02	-9.3100E-03	10	10000	27.05	31.53	35.20
647	C9H20	3-ETHYL-4-METHYLHEXANE	1.3499	9.3410E-02	-9.2200E-03	10	10000	27.17	31.61	35.24
648	C9H20	2,2,3-TRIMETHYLHEXANE	1.3307	9.6080E-02	-9.3300E-03	10	10000	26.15	30.59	34.27
649	C9H20	2,2,4-TRIMETHYLHEXANE	1.3116	9.8680E-02	-9.4500E-03	10	10000	25.17	29.59	33.31
650	C9H20	2,3,3-TRIMETHYLHEXANE	1.3401	9.5610E-02	-9.3600E-03	10	10000	26.69	31.18	34.89
651	C9H20	2,3,4-TRIMETHYLHEXANE	1.3441	9.4780E-02	-9.3100E-03	10	10000	26.89	31.36	35.05
652	C9H20	2,3,5-TRIMETHYLHEXANE	1.3250	9.6360E-02	-9.3400E-03	10	10000	25.83	30.23	33.89
653	C9H20	2,4,4-TRIMETHYLHEXANE	1.3204	9.7820E-02	-9.4300E-03	10	10000	25.63	30.09	33.81
654	C9H20	3,3,4-TRIMETHYLHEXANE	1.3477	9.4900E-02	-9.3400E-03	10	10000	27.12	31.63	35.35
655	C9H20	2-METHYLOCTANE	1.3491	8.9290E-02	-9.0500E-03	10	8192	26.88	31.01	34.32
656	C9H20	3-METHYLOCTANE	1.3518	9.0010E-02	-9.1300E-03	10	8257	27.08	31.28	34.64
657	C9H20	4-METHYLOCTANE	1.3479	9.0920E-02	-9.1800E-03	10	9016	26.89	31.12	34.51
658	C9H20	n-NONANE	1.2650	1.0884E-01	-1.1580E-02	10	5855	23.03	27.31	30.71
659	C9H20	2,2,3,3-TETRAMETHYLPENTANE	1.3498	9.6470E-02	-9.4800E-03	10	10000	27.34	31.98	35.80
660	C9H20	2,2,3,4-TETRAMETHYLPENTANE	1.3292	9.8470E-02	-9.5300E-03	10	10000	26.19	30.76	34.58
661	C9H20	2,2,4,4-TETRAMETHYLPENTANE	1.3014	1.0127E-01	-9.5500E-03	10	10000	24.72	29.22	33.06
662	C9H20	2,3,3,4-TETRAMETHYLPENTANE	1.3510	9.1700E-02	-8.9800E-03	10	10000	27.14	31.51	35.10
663	C9H20	2,2,5-TRIMETHYLHEXANE	1.3038	9.9700E-02	-9.5000E-03	10	10000	24.77	29.19	32.91
664	C9H20O	2,6-DIMETHYL-4-HEPTANOL	1.5287	3.6970E-02	----------	10	399	36.77	40.02	-------
665	C9H20O	1-NONANOL	-------	----------	----------	-------	--------	41.13	-------	-------
666	C9H20O	2-NONANOL	1.5681	2.9160E-02	----------	10	89	39.56	-------	-------
667	C9H20S	n-NONYL MERCAPTAN	1.5769	2.6730E-02	----------	10	173	40.14	42.69	-------
668	C9H21N	n-NONYLAMINE	1.5237	3.4130E-02	----------	10	366	36.13	39.08	-------
669	C9H21N	TRIPROPYLAMINE	1.4476	6.6290E-02	-7.3200E-03	10	1989	32.10	35.55	38.07
670	C10H7Br	1-BROMONAPHTHALENE	-------	----------	----------	-------	--------	75.42	-------	-------
671	C10H7Cl	1-CHLORONAPHTHALENE	-------	----------	----------	-------	--------	58.68	-------	-------
672	C10H9N	QUINALDINE	-------	----------	----------	-------	--------	53.84	-------	-------
673	C10H10	m-DIVINYLBENZENE	1.5533	6.4830E-02	-7.7300E-03	10	761	40.78	44.88	
674	C10H12	1,2,3,4-TETRAHYDRONAPHTHALENE	1.5901	4.3280E-02	----------	10	484	42.99	47.50	
675	C10H12O	ANETHOLE	1.6538	2.6650E-02	----------	10	74	47.91	-------	-------
676	C10H14	n-BUTYLBENZENE	1.4998	7.1460E-02	-8.2300E-03	10	1346	36.56	40.72	43.66
677	C10H14	sec-BUTYLBENZENE	1.4809	7.7430E-02	-8.7500E-03	10	2304	35.45	39.88	43.09
678	C10H14	tert-BUTYLBENZENE	1.4754	7.9830E-02	-8.9500E-03	10	2810	35.18	39.74	43.08
679	C10H14	m-CYMENE	1.4824	7.6480E-02	-8.6400E-03	10	2268	35.50	39.88	43.06
680	C10H14	o-CYMENE	1.4936	7.5890E-02	-8.6400E-03	10	1975	36.37	40.81	44.00
681	C10H14	p-CYMENE	1.4865	7.4370E-02	-8.4400E-03	10	1927	35.68	39.94	43.02
682	C10H14	m-DIETHYLBENZENE	1.4989	7.1820E-02	-8.2300E-03	10	1489	36.52	40.71	43.69
683	C10H14	o-DIETHYLBENZENE	1.5078	7.1800E-02	-8.2700E-03	10	1385	37.27	41.53	44.55
684	C10H14	p-DIETHYLBENZENE	1.5006	7.0840E-02	-8.1300E-03	10	1388	36.59	40.72	43.65
685	C1OH14	2-ETHYL-m-XYLENE	1.5242	6.9070E-02	-8.2400E-03	10	965	38.46	42.60	-------
686	C10H14	2-ETHYL-p-XYLENE	1.5112	7.0070E-02	-8.0700E-03	10	1246	37.42	41.59	44.53
687	C10H14	3-ETHYL-o-XYLENE	1.5305	6.7280E-02	-8.0600E-03	10	817	38.87	42.93	-------
688	C10H14	4-ETHYL-m-XYLENE	1.5129	6.9330E-02	-8.0000E-03	10	1170	37.52	41.65	44.56
689	C10H14	4-ETHYL-o-XYLENE	1.5173	6.8390E-02	-8.1300E-03	10	984	37.80	41.83	-------
690	C1OH14	5-ETHYL-m-XYLENE	1.5033	7.0330E-02	-8.0700E-03	10	1355	36.77	40.89	43.82
691	C10H14	ISOBUTYLBENZENE	1.4741	7.7750E-02	-8.7400E-03	10	2534	34.92	39.32	42.53
692	C10H14	1,2,3,5-TETRAMETHYLBENZENE	1.5376	6.4370E-02	-7.7300E-03	10	654	39.28	43.19	-------
693	C10H15N	N,N-DIETYHLANILINE	1.6162	2.8590E-02	----------	10	186	44.13	47.14	-------
694	C10H16	D-LIMONENE	1.4052	6.8190E-02	-7.1400E-03	10	2604	29.26	32.59	35.12
695	C10H16	alpha-PHELLANDRENE	1.4794	7.5400E-02	-8.5600E-03	10	1995	35.17	39.44	42.51
696	C10H16	beta-PHELLANDRENE	1.4827	7.1280E-02	-7.9700E-03	10	2086	35.16	39.21	42.15
697	C10H16	alpha-PINENE	1.4366	8.8960E-02	-9.1600E-03	10	6251	32.84	37.84	41.79
698	C10H16	beta-PINENE	1.4623	8.5050E-02	-9.4300E-03	10	3849	34.51	39.33	42.91
699	C10H16	alpha-TERPINENE	1.4795	7.1690E-02	-8.0600E-03	10	1987	34.92	38.96	41.88
700	C1OH16	gamma-TERPINENE	1.4923	7.1560E-02	-8.2100E-03	10	1430	35.95	40.05	42.96
701	C10H16	TERPINOLENE	1.5201	6.4980E-02	-7.7400E-03	10	778	37.79	41.60	-------
702	C10H18	cis-DECAHYDRONANPHTALENE	1.5213	7.1790E-02	-8.4200E-03	10	1034	38.43	42.77	45.80

Table 19-1 ADSORPTION CAPACITY OF ACTIVATED CARBON (continued)

$$\log_{10} Q = A + B\,[\log_{10} y] + C\,[\log_{10} y]^2 \qquad Q \text{ - g of compound/100 g of carbon,} \quad y \text{ - ppmv}$$

NO	FORMULA	NAME	A	B	C	ymin	ymax	Q @10 ppmv	Q @100 ppmv	Q @1000 ppmv
703	C10H18	trans-DECAHYDRONAPHTHALENE	1.4953	7.5380E-02	-8.6900E-03	10	1603	36.47	40.86	43.98
704	C10H20	n-BUTYLCYCLOHEXANE	1.4679	6.9000E-02	-7.7600E-03	10	1729	33.82	37.57	40.28
705	C10H20	1-DECENE	1.4365	6.6060E-02	-7.2400E-03	10	2200	31.29	34.65	37.12
706	C10H20O	1-DECANAL	1.5673	2.7420E-02	----------	10	135	39.33	41.89	-------
707	C10H20O2	2-ETHYLHEXYL ACETATE	1.5779	3.0400E-02	----------	10	301	40.58	43.52	-------
708	C10H20O2	ISOPENTYL ISOVALERATE	1.5275	5.5780E-02	-6.0900E-03	10	1165	37.77	41.18	43.65
709	C10H22	n-DECANE	1.4382	6.3590E-02	-6.9800E-03	10	1877	31.25	34.47	36.83
710	C10H22	2-METHYLNONANE	1.4272	6.5740E-02	-7.1400E-03	10	2480	30.60	33.89	36.32
711	C10H22	3-METHYLNONANE	1.4286	6.6590E-02	-7.2300E-03	10	2603	30.76	34.11	36.59
712	C10H22	4-METHYLNONANE	1.4229	6.8000E-02	-7.3400E-03	10	3056	30.45	33.85	36.38
713	C10H22	5-METHYLNONANE	1.4247	6.7540E-02	-7.3000E-03	10	2915	30.55	33.93	36.44
714	C10H22	3-ETHYLOCTANE	1.4289	6.7780E-02	-7.3400E-03	10	2860	30.86	34.28	36.83
715	C10H22	4-ETHYLOCTANE	1.4232	6.8980E-02	-7.4200E-03	10	3341	30.53	34.00	36.59
716	C10H22	2,2-DIMETHYLOCTANE	1.4048	7.1540E-02	-7.5600E-03	10	4791	29.43	32.93	35.59
717	C10H22	2,3-DIMETHYLOCTANE	1.4234	6.8950E-02	-7.4200E-03	10	3300	30.54	34.01	36.60
718	C10H22	2,4-DIMETHYLOCTANE	1.4074	7.1180E-02	-7.5400E-03	10	4546	29.58	33.08	35.73
719	C10H22	2,5-DIMETHYLOCTANE	1.4121	7.0590E-02	-7.5100E-03	10	4165	29.87	33.36	36.00
720	C10H22	2,6-DIMETHYLOCTANE	1.4135	6.9580E-02	-7.4200E-03	10	3882	29.90	33.34	35.93
721	C10H22	2,7-DIMETHYLOCTANE	1.4130	6.9100E-02	-7.3800E-03	10	3715	29.83	33.24	35.80
722	C10H22	3,3-DIMETHYLOCTANE	1.4179	7.0500E-02	-7.5200E-03	10	4024	30.26	33.79	36.45
723	C10H22	3,4-DIMETHYLOCTANE	1.4246	6.9980E-02	-7.5200E-03	10	3520	30.69	34.23	36.88
724	C10H22	3,5-DIMETHYLOCTANE	1.4151	7.1060E-02	-7.5800E-03	10	4142	30.10	33.65	36.32
725	C10H22	3,6-DIMETHYLOCTANE	1.4182	7.0150E-02	-7.5100E-03	10	3765	30.26	33.76	36.39
726	C10H22	4,4-DIMETHYLOCTANE	1.4104	7.1800E-02	-7.6000E-03	10	4773	29.83	33.39	36.10
727	C10H22	4,5-DIMETHYLOCTANE	1.4232	7.0800E-02	-7.5900E-03	10	3777	30.65	34.23	36.92
728	C10H22	4-PROPYLHEPTANE	1.4146	7.1060E-02	-7.5700E-03	10	4181	30.06	33.60	36.27
729	C10H22	4-ISOPROPYLHEPTANE	1.4144	7.1680E-02	-7.6200E-03	10	4429	30.09	33.67	36.38
730	C10H22	3-ETHYL-2-METHYLHEPTANE	1.4200	7.0990E-02	-7.5900E-03	10	3958	30.44	34.01	36.70
731	C10H22	4-ETHYL-2-METHYLHEPTANE	1.4097	7.2340E-02	-7.6600E-03	10	4844	29.81	33.40	36.13
732	C10H22	5-ETHYL-2-METHYLHEPTANE	1.4150	7.0880E-02	-7.5600E-03	10	4107	30.08	33.61	36.27
733	C10H22	3-ETHYL-3-METHYLHEPTANE	1.4242	7.1020E-02	-7.6100E-03	10	3838	30.73	34.34	37.04
734	C10H22	4-ETHYL-3-METHYLHEPTANE	1.4235	7.1340E-02	-7.6400E-03	10	3921	30.70	34.32	37.04
735	C10H22	3-ETHYL-5-METHYLHEPTANE	1.4157	7.1560E-02	-7.6100E-03	10	4359	30.17	33.75	36.46
736	C10H22	3-ETHYL-4-METHYLHEPTANE	1.4249	7.0960E-02	-7.6200E-03	10	3761	30.78	34.39	37.09
737	C10H22	4-ETHYL-4-METHYLHEPTANE	1.4202	7.2310E-02	-7.7100E-03	10	4368	30.53	34.20	36.96
738	C10H22	2,2,3-TRIMETHYLHEPTANE	1.4120	7.2880E-02	-7.7100E-03	10	4943	30.01	33.65	36.41
739	C10H22	2,2,4-TRIMETHYLHEPTANE	1.3942	7.3010E-02	-7.1900E-03	10	7172	28.84	32.47	35.36
740	C10H22	2,2,5-TRIMETHYLHEPTANE	1.3988	7.1830E-02	-7.1000E-03	10	6397	29.08	32.67	35.52
741	C10H22	2,2,6-TRIMETHYLHEPTANE	1.3957	7.1890E-02	-7.1000E-03	10	6589	28.87	32.44	35.28
742	C10H22	2,3,3-TRIMETHYLHEPTANE	1.4181	7.2460E-02	-7.7100E-03	10	4530	30.40	34.06	36.83
743	C10H22	2,3,4-TRIMETHYLHEPTANE	1.4186	7.2370E-02	-7.7100E-03	10	4417	30.42	34.08	36.84
744	C10H22	2,3,5-TRIMETHYLHEPTANE	1.4166	7.2330E-02	-7.7100E-03	10	4399	30.28	33.92	36.66
745	C10H22	2,3,6-TRIMETHYLHEPTANE	1.4087	7.2400E-02	-7.6500E-03	10	4910	29.75	33.33	36.06
746	C10H22	2,4,4-TRIMETHYLHEPTANE	1.4009	7.2350E-02	-7.1400E-03	10	6687	29.25	32.89	35.78
747	C10H22	2,4,5-TRIMETHYLHEPTANE	1.4130	7.0390E-02	-7.0200E-03	10	5032	29.95	33.55	36.39
748	C10H22	2,4,6-TRIMETHYLHEPTANE	1.3943	7.2040E-02	-7.1000E-03	10	6725	28.79	32.36	35.20
749	C10H22	2,5,5-TRIMETHYLHEPTANE	1.4064	7.2070E-02	-7.1600E-03	10	5942	29.60	33.26	36.16
750	C10H22	3,3,4-TRIMETHYLHEPTANE	1.4229	7.2600E-02	-7.7500E-03	10	4345	30.74	34.44	37.23
751	C10H22	3,3,5-TRIMETHYLHEPTANE	1.4103	7.1470E-02	-7.1100E-03	10	5587	29.83	33.48	36.37
752	C10H22	3,4,4-TRIMETHYLHEPTANE	1.4216	7.3160E-02	-7.8000E-03	10	4530	30.69	34.41	37.23
753	C10H22	3,4,5-TRIMETHYLHEPTANE	1.4250	7.1950E-02	-7.7100E-03	10	4017	30.85	34.52	37.28
754	C10H22	3-ISOPROPYL-2-METHYLHEXANE	1.4218	7.1220E-02	-7.6200E-03	10	3976	30.58	34.18	36.89
755	C10H22	3,3-DIETHYLHEXANE	1.4313	7.1190E-02	-7.6700E-03	10	3595	31.25	34.92	37.66
756	C10H22	3,4-DIETHYLHEXANE	1.4242	7.1490E-02	-7.6800E-03	10	3760	30.76	34.39	37.11
757	C10H22	3-ETHYL-2,2-DIMETHYLHEXANE	1.4130	7.1750E-02	-7.1500E-03	10	5563	30.03	33.72	36.64
758	C10H22	4-ETHYL-2,2-DIMETHYLHEXANE	1.3951	7.3920E-02	-7.2700E-03	10	7712	28.95	32.64	35.59
759	C10H22	3-ETHYL-2,3-DIMETHYLHEXANE	1.4279	7.2600E-02	-7.7800E-03	10	4123	31.10	34.83	37.64
760	C10H22	4-ETHYL-2,3-DIMETHYLHEXANE	1.4219	7.2760E-02	-7.7600E-03	10	4377	30.68	34.39	37.18
761	C10H22	3-ETHYL-2,4-DIMETHYLHEXANE	1.4204	7.3090E-02	-7.7800E-03	10	4566	30.60	34.31	37.12
762	C10H22	4-ETHYL-2,4-DIMETHYLHEXANE	1.4217	7.2910E-02	-7.7700E-03	10	4495	30.68	34.39	37.19
763	C10H22	3-ETHYL-2,5-DIMETHYLHEXANE	1.4086	7.1540E-02	-7.1100E-03	10	5669	29.72	33.36	36.24
764	C10H22	4-ETHYL-3,3-DIMETHYLHEXANE	1.4264	7.3010E-02	-7.8000E-03	10	4308	31.01	34.77	37.60
765	C10H22	3-ETHYL-3,4-DIMETHYLHEXANE	1.4248	7.3420E-02	-7.8300E-03	10	4492	30.93	34.70	37.55
766	C10H22	2,2,3,3-TETRAMETHYLHEXANE	1.4238	7.1630E-02	-7.1700E-03	10	5086	30.78	34.55	37.52
767	C10H22	2,2,3,4-TETRAMETHYLHEXANE	1.4165	7.2150E-02	-7.2100E-03	10	5373	30.30	34.04	36.99
768	C10H22	2,2,3,5-TETRAMETHYLHEXANE	1.3977	7.3860E-02	-7.2800E-03	10	7420	29.13	32.83	35.79
769	C10H22	2,2,4,4-TETRAMETHYLHEXANE	1.4026	7.3910E-02	-7.2800E-03	10	7502	29.46	33.22	36.21
770	C10H22	2,2,4,5-TETRAMETHYLHEXANE	1.3944	7.4390E-02	-7.3100E-03	10	7947	28.94	32.65	35.62
771	C10H22	2,2,5,5-TETRAMETHYLHEXANE	1.3744	7.5670E-02	-7.1700E-03	10	10000	27.73	31.41	34.42
772	C10H22	2,3,3,4-TETRAMETHYLHEXANE	1.4296	7.3130E-02	-7.8300E-03	10	4230	31.25	35.04	37.89
773	C10H22	2,3,3,5-TETRAMETHYLHEXANE	1.4078	7.3230E-02	-7.2600E-03	10	6441	29.77	33.51	36.49
774	C10H22	2,3,4,4-TETRAMETHYLHEXANE	1.4261	7.7610E-02	-8.3300E-03	10	4812	31.29	35.32	38.37
775	C10H22	2,3,4,5-TETRAMETHYLHEXANE	1.4149	7.1400E-02	-7.1200E-03	10	5358	30.14	33.82	36.73
776	C10H22	3,3,4,4-TETRAMETHYLHEXANE	1.4401	7.2610E-02	-7.8300E-03	10	3730	31.98	35.81	38.68
777	C10H22	2,4-DIMETHYL-3-ISOPROPYLPENTANE	1.4172	7.2390E-02	-7.2200E-03	10	5636	30.36	34.13	37.10
778	C10H22	3,3-DIETHYL-2-METHYLPENTANE	1.4417	7.1710E-02	-7.7700E-03	10	3361	32.04	35.81	38.62
779	C10H22	3-ETHYL-2,2,3-TRIMETHYLPENTANE	1.4413	7.1530E-02	-7.6800E-03	10	3853	32.00	35.78	38.62
780	C10H22	3-ETHYL-2,2,4-TRIMETHYLPENTANE	1.4121	7.3770E-02	-7.3400E-03	10	6210	30.10	33.91	36.93

Table 19-1 ADSORPTION CAPACITY OF ACTIVATED CARBON (continued)

| NO | FORMULA | NAME | $log_{10} Q = A + B [log_{10} y] + C [log_{10} y]^2$ | | | | | Q - g of compound/100 g of carbon, y - ppmv | | |
			A	B	C	ymin	ymax	Q @10 ppmv	Q @100 ppmv	Q @1000 ppmv
781	C10H22	3-ETHYL-2,3,4-TRIMETHYLPENTANE	1.4379	7.2360E-02	-7.8100E-03	10	3661	31.80	35.59	38.43
782	C10H22	2,2,3,3,4-PENTAMETHYLPENTANE	1.4328	7.4150E-02	-7.9200E-03	10	4541	31.55	35.43	38.36
783	C10H22	2,2,3,4,4-PENTAMETHYLPENTANE	1.4182	7.3630E-02	-7.3200E-03	10	6321	30.51	34.37	37.43
784	C10H22O	1-DECANOL	------	----------	----------	-----	-------	42.23	-------	-------
785	C10H22O	DI-n-PENTYL ETHER	1.4906	5.4510E-02	-5.9300E-03	10	1127	34.61	37.67	39.89
786	C10H22O	ISODECANOL	------	----------	----------	-----	-------	41.94	-------	-------
787	C10H22S	n-DECYL MERCAPTAN	------	----------	----------	-----	-------	41.75	-------	-------
788	C10H23N	n-DECYLAMINE	1.5549	2.5390E-02	----------	10	138	38.05	40.34	-------
789	C11H10	1-METHYLNAPHTHALENE	1.6641	2.7980E-02	----------	10	89	49.21	-------	-------
790	C11H14O2	n-BUTYL BENZOATE	------	----------	----------	-----	-------	49.61	-------	-------
791	C11H16	n-PENTYLBENZENE	1.5422	5.2660E-02	-6.0200E-03	10	576	38.80	42.02	-------
792	C11H20O2	2-ETHYLHEXYL ACRYLATE	1.5959	2.5790E-02	----------	10	234	41.85	44.41	-------
793	C11H22	1-UNDECENE	1.4911	4.7280E-02	-5.2000E-03	10	648	34.13	36.71	-------
794	C11H22O	1-UNDECANAL	1.5853	2.2540E-02	----------	10	62	40.54	-------	-------
795	C11H24	n-UNDECANE	1.4859	6.5480E-02	-8.9300E-03	10	542	34.87	38.12	-------
796	C12H12	1-ETHYLNAPHTHALENE	------	----------	----------	-----	-------	50.17	-------	-------
797	C12H16	CYCLOHEXYLBENZENE	1.6430	2.3020E-02	----------	10	52	46.35	-------	-------
798	C12H18	m-DIISOPROPYLBENZENE	1.5569	4.3700E-02	-4.7300E-03	10	517	39.43	42.20	-------
799	C12H18	p-DIISOPROPYLBENZENE	1.5756	2.7610E-02	----------	10	323	40.11	42.74	-------
800	C12H18	n-HEXYLBENZENE	1.5884	2.4860E-02	----------	10	159	41.05	43.47	-------
801	C12H22	BICYCLOHEXYL	1.6033	2.4980E-02	----------	10	100	42.49	45.01	-------
802	C12H23N	DICYCLOHEXYLAMINE	------	----------	----------	-----	-------	45.24	-------	-------
803	C12H24	1-DODECENE	1.5349	2.3850E-02	----------	10	209	36.20	38.24	-------
804	C12H24O	1-DODECANAL	------	----------	----------	-----	-------	41.83	-------	-------
805	C12H26	n-DODECANE	1.5223	2.0710E-02	4.1000E-04	10	179	34.95	36.76	-------
806	C12H26O	DI-n-HEXYL ETHER	1.5725	1.9700E-02	----------	10	61	39.10	-------	-------
807	C12H26O3	DIETHYLENE GLYCOL DI-n-BUTYL ETHER	------	----------	----------	-----	-------	44.23	-------	-------
808	C12H26S	n-DODECYL MERCAPTAN	------	----------	----------	-----	-------	43.09	-------	-------
809	C12H27BO3	TRI-n-BUTYL BORATE	1.5752	3.1920E-02	-3.0500E-03	10	692	40.19	42.35	-------
810	C12H27N	TRI-n-BUTYLAMINE	1.5576	2.0170E-02	----------	10	123	37.83	39.63	-------
811	C13H20	n-HEPTYLBENZENE	------	----------	----------	-----	-------	42.57	-------	-------
812	C13H26	1-TRIDECENE	1.5548	1.9740E-02	----------	10	84	37.54	-------	-------
813	C13H26O2	n-BUTYL NONANOATE	------	----------	----------	-----	-------	42.90	-------	-------
814	C13H28	n-TRIDECANE	1.5520	1.9220E-02	----------	10	73	37.26	-------	-------
815	C14H14	1,1-DIPHENYLETHANE	------	----------	----------	-----	-------	50.69	-------	-------
816	C14H22	n-OCTYLBENZENE	------	----------	----------	-----	-------	43.44	-------	-------
817	C14H28	1-TETRADECENE	------	----------	----------	-----	-------	38.94	-------	-------
818	C14H30	n-TETRADECANE	1.5086	2.2070E-02	2.4000E-04	10	15	33.96	-------	-------

Q - adsorption capacity at equilibrium, g of compound/100 g of carbon

y - concentration of compound in gas at 25 C and 1 atm, parts per million by volume, ppmv

A, B, and C - regression coefficients for chemical compound

ymin - minimum concentration, ppmv

ymax - minimum concentration, ppmv

Chapter 20

SOIL SORPTION COEFFICIENT

Daniel H. Chen and Carl L. Yaws
Lamar University, Beaumont, Texas

ABSTRACT

Results for soil sorption coefficient K_{OC} are presented for 336 hydrocarbon and organic chemicals. The chemicals include hydrocarbon, oxygen, nitrogen, halogen, sulfur, and phosphorus compounds. Representative results for soil adsorption coefficient [(mg sorbed / kg organic carbon in soil) / (mg/L aqueous concentration) or abbreviated as L/kg] are 4300 for trifluralin ($C10H9F3N3O4$), 3910 for hexachlorobenzene ($C6CL6$), 1300 for naphthalene ($C10H8$), and 27 for phenol ($C6H6O$) in a normal soil environment (20 C, pH 4-8, carbon exchange capacity greater than 7 MEQ/100 g, sand composition less than 70%, etc.). The melting point (MP) and molecular weight (MW) data are also provided as they are needed in many environmental property data correlations.

INTRODUCTION

The soil sorption coefficient K_{OC} , which determines the partitioning of an organic chemical between the soil/sediment and the aqueous solution, is an important environmental parameter. K_{OC} affects the physical movement of pollutants, chemical degradation (photolysis and hydrolysis), biodegradation, acidity, and buffered solution-phase concentration. As a result, the soil sorption coefficient is widely used in river, runoff, and soil/ground water models for the assessments of the fate and transport of chemicals (5,24,28,36). K_{OC} is also known as "soil organic carbon partition coefficient" or "soil sorption coefficient standardized with respect to organic carbon." With the value of K_{OC} known, the partition uptake of water contaminants for a particular soil/sediment or the degree of leaching of the pollutants into the ground water can be estimated.

A compilation of the soil sorption coefficient data for 336 compounds is provided in an easy-to-use tabular format that is especially applicable for rapid engineering use with the personal computer or hand calculator.

SOIL SORPTION COEFFICIENT

The amount of chemicals sorbed onto a soil or sediment depends on the concentration of chemicals and their equilibrium distribution coefficient (i.e., K_{OC}). For dilute aqueous solutions, the distribution coefficient can be adequately expressed with the Freundlich equation with 1/n equal to one (8,11,24,25,38):

$$x / m = K_{OC} C \tag{20-1}$$

where x = weight of solute sorbed, mg
C = equilibrium concentration of solute in aqueous phase, mg/L
m = weight of sorbent (organic carbon in soil), kg
K_{OC} = soil sorption coefficient

From the above equation, K_{OC} can be interpreted as the ratio of the solid phase concentration (normalized for the organic carbon content) to the solution phase concentration of the chemical at equilibrium. Therefore, in commonly used units, K_{OC} is

$$K_{OC} = \frac{\text{mg sorbed/ kg organic carbon}}{\text{mg/L aqueous concentration}} \tag{20-2}$$

The unit of K_{OC} can be abbreviated as L/kg. The average organic carbon content of a typical soil is from 0.5% to 3.5%. By basing the sorption coefficient on soil (or sediment) organic carbon, one can eliminate much of the variation between soils due to organic carbon content. Note that the cited experimental or predicted K_{OC} values are intended for a normal environment as stated above. Attempts to extrapolate far beyond these conditions may incur considerable errors (10-12,15,24). In case that the coefficient is expressed in terms of soil organic matter, K_{om}, the following equivalence can be used to obtain K_{OC}:

$$K_{OC} = 1.72 K_{om} \tag{20-3}$$

This assumes that the organic matter contains about 58% C (8).

The results for the soil sorption coefficient for organic chemicals in water are given in Table 20-1. The melting point (MP) and molecular weight (MW) are also provided to facilitate predictions for other environmental properties. The tabulation is applicable to a wide variety of organic chemicals in contact with water at normal ambient conditions. The wide variety of substances includes hydrocarbons, acids, alcohols, esters, ethers, ketones, fluorides, chlorides, bromides, amines, sulfones, nitros, amides, sulfides, and phosphates. The tabulation is arranged by carbon number (C, C2, C3, ..., C21). This provides ease of use in quickly locating data using the chemical formula.

In preparing the tabulation, a literature search was conducted to identify data source publications (1-40). The publications were screened and copies of appropriate data were made. These data were then keyed into the computer to provide a database.

The nonlinear group contribution method (3,13,21) has been used for the estimation of the soil sorption coefficient when experimental K_{OC} values are not available. The method is based on comprehensive (225 compounds) and updated data. Comparison with literature methods yields favorable results (2,3,13,19,32,33). In general, the prediction errors are within ± 0.82 order of magnitude (95% confidence limit). A comparison of calculated and data (experimental) values for the soil sorption coefficient is shown in Fig. 20-1. The graph discloses general agreement of calculated and data values for different organic chemicals.

The compilation for the soil sorption coefficient maybe used in engineering and environmental impact studies involving organic compounds in water.

EXAMPLES

The tabulation may be used for determining the soil sorption coefficient for the compound in water. The use of the soil sorption coefficient in environmental applications involving organic chemicals in water is illustrated below.

Example 1 For an aqueous concentration of benzene (C_6H_6) in contaminated river water of 10 ppm by weight, what will be the maximum uptake of benzene by the bottom sediment? The average organic carbon content of the bottom sediment is 3%.

The equation $x/m = K_{OC} C$ is used in determining the solution. First, calculate the amount of organic carbon per ton (metric) of bottom sediment: m = 1000 x 3% = 30 kg organic carbon. Then substitute the soil sorption coefficient of benzene from the tabulation

K_{OC} = 83 (mg sorbed/kg org carbon) / (mg/L aqueous conc) = 83 (mg sorbed/kg org carbon) / (ppm aqueous conc)

and the aqueous concentration C = 10 ppm into the equation for x/m to obtain

$$x /30 = 83 \times 10 \text{ mg}$$

$$x = 30 \times 83 \times 10 \text{ mg} = 2.5E04 \text{ mg} = 0.025 \text{ kg}$$

Example 2 Atrazine ($C_8H_{14}CLN_5$) is uniformly applied to a field and incorporated into the soil. The soil has a bulk density of 1.25 kg/L, 2% organic carbon and 25% each air and water by volume. Estimate the equilibrium distribution of the pesticide resulting from a 1 kg/hectare (1 hectare = 10,000 m^2) application incorporated to 10-cm depth. Volatilization into the air is assumed to be negligible.

The amount of organic carbon in soil per hectare is m = 10,000 x .1 x 1000 x 1.25 x 2% = 25,000 kg of org carbon. From the tabulation, the soil sorption coefficient of atrazine is

K_{OC} = 149 (mg sorbed/kg organic carbon) / (mg/L aqueous conc)

Determination of the amount of atrazine in the soil phase per hectare requires a trial and error procedure. Trying x = .94 kg = 940000 mg in the equation $x/m = K_{OC} C$ and solving for C gives

C = (940000 /25,000)/149 = .252 mg/L

The amount of atrazine in the aqueous phase per hectare is

.252 x 10,000 x .1 x 1,000 x 25% = 63000 mg = .063 kg

The total amount of atrazine is 1.003 kg/hectare. This is close enough to the application (1.003 vs 1 application).

The equilibrium distribution of atrazine is estimated to be

.063/(.94 + .063) = 6.3% in water

.94/ (.94 + .063) = 93.7% in soil

Portions of this material appeared in Pollution Engineering, 24, 54 (June 15, 1992) and are reprinted by special permission.

REFERENCES – ORGANIC COMPOUNDS

1. Boyd, S. A., M.D. Mikesell, and J.F. Lee, Chlorophenols in soil. Sawhney, B. L. and K. Brown (eds.), REACTION AND MOVEMENT OF ORGANIC CHEMICALS IN SOILS , Soil Science Society of American and American Society Agronomy, SSSA Special Publication no. 22, 209 (1989).

2. Briggs, G. G. J. Agric. Food Chem. 29(5), 1050 (1981).

3. Chen,T. L., D. H. Chen, and C. L. Yaws, Predicting soil adsorption with molecular structure, paper 54b, AIChE National Meeting, August 19-22, 1990, San Diego, California.

4. Chiou, C. T., L. J. Peters, and V. H. Freed, Science, 206, 831 (1979).

5. Chiou, C. T., Soil Science Society of American and American Society Agronomy, SSSA Special Publication no. 22, 1 (1989).

6. Dickson, K. I. (ed.), MODELING THE FATE OF CHEMICALS IN THE AQUATIC ENVIRONMENT, Ann Arbor Sci. Pub., Ann Arbor, MI (1982).

7. Green, R. E. and S. R. Obien, Weed Sci., 17, 514 (1969).

8. Grover, R. and R. J. Soil Sci., 109, 136 (1970).

9. Gschwend, P. M. and S.C. Wu, Environ. Sci. Technol., 19 90 (1985).

10. Hamaker, J. W. and J. M. Thompson, Adsorption. Goring C.A.I. and Hamaker J.W.(eds.), ORGANIC CHEMICALS IN THE SOIL ENVIRONMENT, Vol. 1, Marcel-Dekker , New York, NY (1972).

11. Hance, R. J., J. Agric. Food Chem., 17, 667 (1969).

12. Hodson, J. and N. A. Williams, Chemosphere, 17, No. 1, 67 (1988).

13. Jeng, C. Y., Estimation of soil sorption coefficient and acentric factor with nonlinear group contribution methods. M.S.E. Thesis, Lamar University, Beaumont, TX (August, 1989).

14. Khan, S. V. J. of Environmental Quality, 3 ,202 (1974).

15. Karickhoff, S. W. and Brown, D. S. J. Environ. Qual., 7, 246 (1978).

16. Karickhoff, S. W., D.S. Brown, and T.A. Scott, Water Res., 13, 241 (1979).

17. Karickhoff, S. W., Chemosphere, 10(8), 833 (1981).

18. Karickhoff, S. W., J. of Hydraulic Engineering, ASCE, 110, 707 (1984).

19. Kenaga, E. E. and C. A. I. Goring, AQUATIC TOXICOLOGY, Eaton J. C., Parrish .P. R., Hendricks, A. C. (eds), the American Society for Testing and Materials, Philadelphia, PA, 78 (1980).

20. Ladlie, J. S., W.F. Meggitt, and D. Penner, Weed Science, 24, 477 (1976).

21. Lai, W. Y., D. H. Chen, and R. N. Maddox, Ind. Eng. Chem. Res., 26, 1072 (1987).

22. Lambert, S. M. J. Agric. Food Chem., 15, 572 (1967).

23. Liu, L. C., H. Cibes-Viadw, and F. K. S. Koo, Weed Sci., 18, 470 (1970).

24. Lyman, W. J., W. F. Reehl, and D. H. Rosenblatt, HANDBOOK OF CHEMICAL PROPERTY ESTIMATION METHODS, McGraw-Hill, New York, NY (1982).

25. Means, J. C., S. G. Wood, J. J. Hassett, and W. L. Banwart, Environ. Sci. Technol., 14, 1524 (1980).

26. Means, J. C., S. G. Wood, J. J. Hassett, and W. L. Banwart, Environ. Sci. Technol., 16, 93 (1982).

27. Nearpass, D. C., Soil Sci., 103, 177 (1967).

28. Pickens, J. and W. C. Lennox, Water Resources Research, 12, No. 2, 171 (1976).

29. Pierce, R. H., Jr., C.E. Olney, and G.T. Felbeck, Jr. Geochem. Cosmochim. Acta, 38, 1061 (1974).

30. Poinke, H. B. and G. Chesters, J. Environ. Qual., 2 29 (1973).

31. Reid, R. C., J. M. Prausnitz, and B. E. Poling, THE PROPERTIES OF GASES AND LIQUIDS, 4th ed., McGraw-Hill, New York, NY (1987).

32. Sabljic, A. Environ. Sci. Technol. 21(4), 358 (1987).

33. Sabljic A. J. Argic. Food Chem., 32, 243 (1984).

34. Sax, N. I. and R. J. Lewis, Jr., HAWLEY'S CONDENSED CHEMICAL DICTIONARY, 11th ed., Van Nostrand Reinhold Co., New York, NY (1987).

35. Swann, R. L., D. A. Laskowski, P. J. McCall, K. Vander Kuy, and H. J. Dishburger, Residue Reviews, 85, 17 (1983).

36. Thibodeaux, L. J., CHEMODYNAMICS, John Wiley & Sons, New York, NY (1979).

37. TREATABILITY MANUAL, Vol. I, Treatability Data, EPA-600/2-82-001a, Office of Research and Development, U.S. Environmental Protection Agency, Washington, D.C. Sept., 1981.

38. U. S. Environmental Protection Agency, Toxic substances control act for premanufacture testing of new chemical substances. Fed. Regist., 44, 16257 (1979).

39. Weast, R. C., ed., CRC HANDBOOK OF CHEMISTRY AND PHYSICS, 68th ed., 1987-88, CRC Press, Inc., Boca Raton, FL (1987).

40. Yaws, C. L., THERMODYNAMIC AND PHYSICAL PROPERTY DATA, Gulf Publishing Co., Houston, TX (1992).

41. Chen, D. H. and others, Pollution Engineering, 24, 54 (June 15, 1992).

Figure 20-1 Soil Sorption Coefficient For Compounds in Water

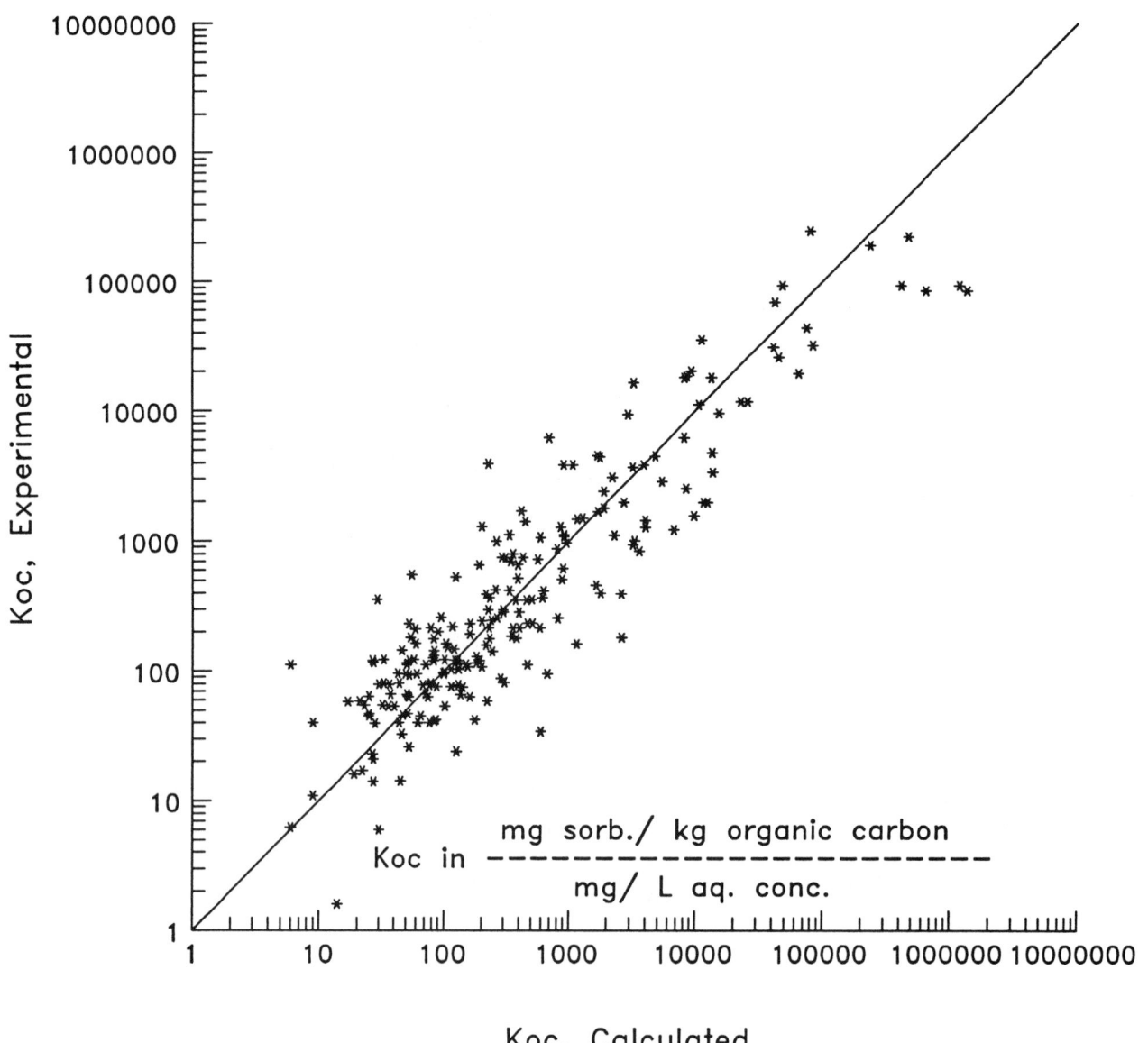

Table 20-1 SOIL SORPTION COEFFICIENT - ORGANIC COMPOUNDS

NO	FORMULA	NAME	Molecular Weight g/mol	Melting Point C	K_{OC} @ 20 C L/kg	CODE
1	CCL4	TETRACHLOROMETHANE	153.838	-23.16	1.22E+02	data
2	CHCL3	TRICHLORO-METHANE	119.389	-111.16	7.68E+01	data
3	CH2CL2	DICHLORO-METHANE	84.940	-95.16	4.74E+01	data
4	CH3NO2	NITROMETHANE	61.040	-28.16	1.76E+02	estimated
5	CH4N2O	UREA	60.058	132.70	1.40E+01	data
6	CH12O3NPS2	DIMETHOATE	181.226	52.00	9.00E+00	data
7	C2CL3F3	1,1,2-TRICHLOROTRIFLUOROETHANE	187.376	-35.16	5.32E+01	estimated
8	C2CL4	TETRACHLOROETHENE	165.848	-22.16	3.59E+02	data
9	C2CL6	HEXACHLOROETHANE	236.740	184.84	4.13E+02	estimated
10	C2HCL3	TRICHLOROETHENE	131.388	-86.16	1.37E+02	estimated
11	C2HCL5	PENTACHLOROETHANE	202.294	-29.16	3.07E+02	estimated
12	C2H2BR2	1,1-DIBROMOETHENE	185.868	------	4.40E+01	data
13	C2H2BR2	CIS-1,2-DIBROMOETHENE	185.868	-53.00	7.68E+01	data
14	C2H2BR2	TRANS-1,2-DIBROMOETHENE	185.868	-6.50	7.68E+01	data
15	C2H2CL2	1,1-DICHLOROETHENE	96.944	-117.16	8.02E+01	estimated
16	C2H2CL2	CIS-1,2-DICHLOROETHENE	96.944	-80.16	8.02E+01	estimated
17	C2H2CL2	TRANS-1,2-DICHLOROETHENE	96.944	-50.16	8.02E+01	estimated
18	C2H2CL4	1,1,2,2-TETRACHLOROETHANE	167.864	-36.16	7.86E+01	data
19	C2H3BR2CL	1,2-DIBROMO-3-CHLOROETHANE	222.333	------	2.22E+02	data
20	C2H3CL3	TRICHLOROETHANE	133.415	-37.16	2.41E+02	estimated
21	C2H3NS	METHYL ISOTHIOCYANATE	73.118	------	6.00E+00	data
22	C2H4BR2	DIBROMOETHANE	187.884	9.84	6.24E+01	data
23	C2H4CL2	1,1-DICHLOROETHANE	98.960	-97.16	8.02E+01	estimated
24	C2H4CL2	1,2-DICHLOROETHANE	98.966	-35.16	3.28E+02	data
25	C3H4CL2	1,1-DICHLOROPROPENE	110.976	------	4.52E+01	data
26	C3H4CL2	1,3-DICHLOROPROPENE	110.976	------	4.63E+01	data
27	C3H4CLN5	TRIETAZINE	145.559	103.00	6.00E+02	data
28	C3H5BR2CL	1,2-DIBROMO-3-CHLOROPROPANE	236.359	------	1.29E+02	data
29	C3H5CL3	1,2,3-TRICHLOROPROPANE	147.431	-14.16	2.42E+02	estimated
30	C3H6BR2	1,2-DIBROMOPROPANE	201.888	-55.16	6.97E+01	estimated
31	C3H6CL2	1,2-DICHLOROPROPANE	112.992	-100.16	4.63E+01	data
32	C3H6CL2	1,3-DICHLOROPROPANE	112.986	-99.16	1.41E+02	estimated
33	C3H6CL2	2,2-DICHLOROPROPANE	112.986	-34.16	1.41E+02	estimated
34	C3H8NO5P	GLYPHOSATE	169.082	------	2.64E+03	data
35	C4H8BR2	1,2-DIBROMOBUTANE	215.915	-65.16	1.22E+02	estimated
36	C4H8BR2	2,3-DIBROMOBUTANE	215.915	-34.16	1.22E+02	estimated
37	C5H2CL3NO	3,5,6-TRICHLORO-2-PYRIDINOL	253.359	------	1.30E+02	data
38	C5H10BR2	2,3-DIBROMO-2-METHYLBUTANE	229.942	14.84	2.13E+02	estimated
39	C5H10N2O2S	METHOMYL	162.212	------	1.60E+02	data
40	C6CL6	HEXACHLOROBENZENE	284.802	229.84	3.91E+03	data
41	C6F6	HEXAFLUOROBENZENE	186.056	4.84	1.83E+02	estimated
42	C6HCL5	PEATACHLOROBENZENE	250.353	-45.60	5.44E+03	data
43	C6HCL6O	PENTACHLOROPHENOL	301.810	174.00	9.00E+02	data
44	C6H2CL4	1,2,3,4-TETRACHLOROBENZENE	215.904	47.50	1.16E+04	data
45	C6H2CL4	1,2,3,5-TETRACHLOROBENZENE	215.904	54.50	2.73E+03	data
46	C6H2CL4	1,2,4,5-TETRACHLOROBENZENE	215.904	140.00	1.25E+04	data
47	C6H3BRCLNO2	3-CHLORO-4-BROMOBENZENE	236.465	61.00	3.94E+02	data
48	C6H3CL2NO2	3,4-DICHLORONITROBENZENE	192.006	43.00	3.35E+02	data
49	C6H3CL2NO2	DICHLOROPICOLINIC ACID	192.006	------	2.00E+00	data
50	C6H3CL3	1,2,3-TRICHLOROBENZENE	181.455	54.00	4.03E+03	data
51	C6H3CL3	1,2,4-TRICHLOROBENZENE	181.455	17.00	8.62E+02	data
52	C6H3CL3N2O2	PICLORAM	241.463	------	1.70E+01	data
53	C6H3CL3O	2,4,5-TRICHLOROPHENOL	198.051	69.00	5.13E+02	data
54	C6H3CL3O	2,4,6-TRICHLOROPHENOL	198.051	69.00	4.68E+02	data
55	C6H3CL4N	NITRAPYRIN	230.930	62.00	4.20E+02	data
56	C6H4BRNO2	2-BROMONITROBENZENE	202.016	43.00	2.60E+02	data
57	C6H4BRNO2	3-BROMONITROBENZENE	202.016	17.00	2.60E+02	data
58	C6H4BRNO2	4-BROMONITROBENZENE	202.016	127.00	2.60E+02	data
59	C6H4CL2	M-DICHLOROBENZENE	147.006	-25.16	3.13E+02	data
60	C6H4CL2	O-DICHLOROBENZENE	147.006	52.84	2.92E+02	data
61	C6H4CL2	P-DICHLOROBENZENE	147.006	-24.16	4.32E+02	data
62	C6H4CL3NO	2-METHOXY-3,5,6-TRICHLOROPYRIDINE	212.471	------	9.20E+02	data
63	C6H4CLNO2	6-CHLOROPICOLINIC ACID	157.557	------	9.00E+00	data
64	C6H4F2	M-DIFLUOROBENZENE	114.094	-24.16	1.83E+02	estimated
65	C6H4F2	O-DIFLUOROBENZENE	114.094	-34.16	1.83E+02	estimated
66	C6H4F2	P-DIFLUOROBENZENE	114.094	-13.16	1.83E+02	estimated
67	C6H5BR	BROMOBENZENE	157.009	-31.16	2.97E+02	estimated
68	C6H5BRCLN	3-METHYL-4-BROMOANILINE	206.481	-251.16	1.80E+02	data
69	C6H5CL	CHLOROBENZENE	112.557	-45.16	2.17E+02	data
70	C6H5CL2N	3,4-DICHLOROANILINE	162.022	72.00	1.93E+02	data
71	C6H5CLN	3,5-DICHLOROANILINE	162.022	52.00	1.28E+02	data
72	C6H5CLO	2-CHLOROPHENOL	129.161	9.00	2.30E+02	data
73	C6H5F	FLUOROBENZENE	96.104	-39.16	1.82E+02	estimated
74	C6H5NO2	NITROBENZENE	123.076	5.84	8.64E+01	data
75	C6H6	BENZENE	78.108	5.84	8.30E+01	data
76	C6H6CL3	METHOXYCHLOR	184.479	89.00	8.00E+04	data
77	C6H6CL6	GAMMA-BHC	290.850	------	9.11E+02	data
78	C6H6CL6	LINDANE	290.850	113.00	1.08E+03	data

447

Table 20-1 SOIL SORPTION COEFFICIENT - ORGANIC COMPOUNDS (continued)

NO	FORMULA	NAME	Molecular Weight g/mol	Melting Point C	K_{OC} @ 20 C L/kg	CODE
79	C6H6N2O2	3-AMINONITROBENZENE	138.124	114.00	5.30E+01	data
80	C6H6N2O2	4-AMINONITROBENZENE	138.124	149.00	7.50E+01	data
81	C6H6O	PHENOL	94.113	40.84	2.70E+01	data
82	C6H7N	2-PICOLINE	93.128	-66.16	4.42E+01	estimated
83	C6H7N	3-PICOLINE	93.128	-18.16	4.42E+01	estimated
84	C6H7N	ANILINE	93.128	-6.16	2.50E+01	data
85	C6H15N	TRIETHYLAMINE	101.191	-115.16	1.43E+02	estimated
86	C7H4CL3NO3	TRICLOPYR	256.481	------	2.70E+01	data
87	C7H4N2O4S	ASULAM	212.184	------	3.00E+02	data
88	C7H5CL2NO2	CHLORAMBEN	206.032	------	2.10E+01	data
89	C7H5CL2NS	CHLORTHIAMID	206.098	------	1.07E+02	data
90	C7H5F3	A,A,A-TRIFLUOROTOLUENE	146.111	-29.16	3.16E+02	estimated
91	C7H6CL2N2O	(3,4-DICHLOROPHENYL)UREA	205.048	------	3.06E+02	data
92	C7H6CL3	3-(TRICHLOROMETHYL)ANILINE	196.489	------	2.27E+02	data
93	C7H6O2	BENZOIC ACID	122.123	122.84	7.91E+00	estimated
94	C7H7BRN2O	(3-BROMOPHENYL)UREA	215.058	------	1.14E+02	data
95	C7H7BRN2O	(4-BROMOPHENYL)UREA	215.058	------	1.31E+02	data
96	C7H7CLN2O	(2-CHLOROPHENYL)UREA	170.599	------	4.03E+01	data
97	C7H7CLN2O	(3-CHLOROPHENYL)UREA	170.599	------	1.02E+02	data
98	C7H7F	P-FLUOROTOLUENE	110.130	-57.16	3.13E+02	estimated
99	C7H7FN2O	(2-FLOROPHENYL)UREA	154.142	------	1.01E+02	data
100	C7H7FN2O	(3-FLOROPHENYL)UREA	154.142	------	5.80E+01	data
101	C7H7FN2O	(4-FLOROPHENYL)UREA	154.142	------	3.28E+01	data
102	C7H8	TOLUENE	92.140	-95.16	3.03E+02	estimated
103	C7H8CLNO	3-CHLORO-4-METHOXYANILINE	157.599	52.00	8.40E+01	data
104	C7H8NO2	PHENYLUREA	138.142	147.00	2.21E+01	data
105	C7H8O	ANISOLE	108.134	-37.80	3.40E+01	data
106	C7H8O	M-CRESOL	108.139	11.84	3.60E+02	estimated
107	C7H8O	O-CRESOL	108.139	30.84	3.60E+02	estimated
108	C7H8O	P-CRESOL	108.139	34.84	3.60E+02	estimated
109	C7H9N	2-METHYLANILINE	107.150	-43.00	4.42E+01	data
110	C7H9N	3-METHYLANILINE	107.150	-64.00	4.42E+01	data
111	C7H9N	4-METHYLANILINE	107.150	-4.87	7.86E+01	data
112	C7H11CLN2O	3-(3-CHLORO-4-METHYLPHENYL)-1-METHYLUREA	174.631	------	1.25E+02	data
113	C7H12CLN5	SIMAZINE	201.663	226.00	1.35E+02	data
114	C7H14N2O2S	ALDICARB	190.967	100.00	3.00E+01	data
115	C7H17O2PS3	PHORATE	260.384	------	3.20E+03	data
116	C8H5CL3O3	2,4,5-TRICHLOROPHENOXYACETIC ACID	255.491	0.00	5.30E+01	data
117	C8H6	ETHYNYLBENZENE	102.135	-30.16	5.19E+02	estimated
118	C8H6CL2O3	2,4-DICHLOROPHENOXYACETIC ACID	221.042	------	6.00E+01	data
119	C8H6CL2O3	DICAMBA	221.844	115.00	4.70E+02	data
120	C8H6CL3NO	3-(TRIFLOROMETHYL)ACETANILIDE	238.507	------	5.56E+01	data
121	C8H7CL2NO	3,4-DICHLOROACETANILIDE	204.058	------	2.17E+02	data
122	C8H7CL2NO2	CHLORAMBEN	220.058	------	5.07E+02	data
123	C8H7CL3N2O	(3-TRIFLOROMETHYL)PHENYLUREA	253.523	------	9.02E+01	data
124	C8H8	1,3,5,7-CYCLOOCTATETRAENE	104.151	------	5.19E+02	estimated
125	C8H8	STYRENE	104.151	-30.16	5.19E+02	estimated
126	C8H8BRNO	2-BROMOACETANILIDE	214.068	99.00	1.01E+02	data
127	C8H8BRNO	3-BROMOACETANILIDE	214.068	87.50	1.01E+02	data
128	C8H8BRNO	4-BROMOACETANILIDE	214.068	168.00	1.01E+02	data
129	C8H8CL2O2	CHLOREONEB	207.058	------	1.16E+03	data
130	C8H8CL2N2O	3-(3,4-DICHLOROPHENYL)-1-METHYLUREA	219.074	------	2.85E+02	data
131	C8H8CLNO	2-CHLOROACETANILIDE	169.609	------	3.76E+01	data
132	C8H8CLNO	3-CHLOROACETANILIDE	169.609	------	7.17E+01	data
133	C8H8CLNO2	METHYL-N-(3-CHLOROPHENYL)CARBAMATE	185.609	------	1.40E+02	data
134	C8H8FNO	2-FLOROACETANILIDE	153.152	------	7.84E+01	estimated
135	C8H8FNO	3-FLOROACETANILIDE	153.152	------	3.67E+01	data
136	C8H8FNO	4-FLOROACETANILIDE	153.152	------	2.99E+01	data
137	C8H8N2O3	2-NITROACETANILIDE	180.160	94.00	8.62E+01	data
138	C8H8N2O3	3-NITROACETANILIDE	180.160	155.00	8.62E+01	data
139	C8H8N2O3	4-NITROACETANILIDE	180.160	216.00	8.62E+01	data
140	C8H8O	ACETOPHENONE	120.144	20.50	4.30E+01	data
141	C8H9BRN2O	(3-METHYL-4-BROMOPHENYL)UREA	229.084	------	2.32E+02	data
142	C8H9CLN2O	3-(3-CHLOROPHENYL)-1-METHYLUREA	184.625	------	8.40E+01	data
143	C8H9CLN2O2	(3-CHLORO-4-METHOXYLPHENYL)UREA	200.625	------	9.90E+01	data
144	C8H9FN2OF	(3-METHYL-4-FLOROPHENYL)UREA	168.168	------	5.96E+01	data
145	C8H9NO	ACETANILIDE	135.160	114.30	2.66E+01	data
146	C8H9NO2	METHYL-N-PHENYLCARBAMATE	151.160	------	5.30E+01	data
147	C8H10	ETHYLBENZENE	106.167	-95.16	5.19E+02	estimated
148	C8H10	M-XYLENE	106.167	-48.16	5.19E+02	estimated
149	C8H10	O-XYLENE	106.167	-25.16	5.19E+02	estimated
150	C8H10	P-XYLENE	106.167	12.84	5.19E+02	estimated
151	C8H10CL3O3PS	METHYL-CHLOROPYRIFOS	323.577	------	3.30E+03	data
152	C8H10N2O	(3-METHYLPHENYL)UREA	150.176	------	3.59E+01	data
153	C8H10NO2	3-PHENYL-1-METHYLUREA	152.168	151.00	1.90E+01	data
154	C8H10O5PS	METHYL PARATHION	249.206	36.00	9.80E+03	data
155	C8H11BRN2O2	ISOCIL	247.100	158.00	1.30E+02	data
156	C8H14CLN5	ATRAZINE	216.697	172.00	1.49E+02	data

Table 20-1 SOIL SORPTION COEFFICIENT - ORGANIC COMPOUNDS (continued)

NO	FORMULA	NAME	Molecular Weight g/mol	Melting Point C	K_{OC} @ 20 C L/kg	CODE
157	C8H14N4OS	METRIBUZIN	214.290	125.00	9.50E+01	data
158	C8H19O2PS3	DISULFOTON	274.410	------	1.78E+03	data
159	C9H6CL2N2O3	METHAZOLE	261.068	------	2.62E+03	data
160	C9H7CL3O3	SILVEX	269.517	181.00	2.60E+03	data
161	C9H7N	QUINOLINE	129.154	15.60	5.70E+02	data
162	C9H10	M-METHYLSTYRENE	118.178	-86.16	8.84E+02	estimated
163	C9H10	O-METHYLSTYRENE	118.178	-68.16	8.84E+02	estimated
164	C9H10	P-METHYLSTYRENE	118.178	-34.16	8.84E+02	estimated
165	C9H10	CIS-PROPENYLBENZENE	118.178	-62.16	8.84E+02	estimated
166	C9H10	TRANS-PROPENYLBENZENE	118.178	-29.16	8.84E+02	estimated
167	C9H10BRCLN2O2	3-(3-CHLORO-4-BROMOPHENYL)-1-METHYL-1-METHOXYU	293.559	------	3.76E+02	data
168	C9H10BRCLN2O2	CHLORBROMURON	293.559	------	4.60E+02	data
169	C9H10CL2N2	DIURON	217.100	158.00	4.00E+02	data
170	C9H10CL2N2O	3-(3,4-DICHLOROPHENYL)-1,1-DIMETHYLUREA	233.100	------	1.61E+02	data
171	C9H10CL2N2O2	3-(3,4-DICHLOROPHENYL)-1-METHYL-1-METHOXYUREA	249.100	------	2.66E+02	data
172	C9H10CL2N2O2	LINURON	249.100	93.00	8.20E+02	data
173	C9H10CLNO2	3-CHLORO-4-METHOXYACETANILIDE	199.635	------	8.23E+01	data
174	C9H11BRN2O2	3-(4-BROMOPHENYL)-1-METHYL-1-METHOXYUREA	259.110	------	1.04E+02	data
175	C9H11BRN2O2	METOBROMURON	259.110	95.00	6.00E+01	data
176	C9H11CL3NO3PS	CHLORPYRIFOS	350.603	43.00	6.10E+03	data
177	C9H11CLN2O	3-(3-CHLOROPHENYL)-1,1-DIMETHYLUREA	198.651	------	6.10E+01	data
178	C9H11CLN2O	3-(4-CHLOROPHENYL)-1,1-DIMETHYLUREA	198.651	170.00	4.96E+01	data
179	C9H11CLN2O	MONURON	198.651	175.00	1.00E+02	data
180	C9H11CLN2O2	3-(3-CHLORO-4-METHOXYPHENYL)-1-METHYLUREA	214.651	------	6.85E+01	data
181	C9H11CLN2O2	3-(4-CHLOROPHENYL)-1-METHYL-1-METHOXYUREA	214.651	------	1.53E+02	data
182	C9H11CLN2O2	MONOINURON	214.651	------	2.00E+02	data
183	C9H11FN2O	3-(3-FLOROPHENYL)-1,1-DIMETHYLUREA	182.194	------	5.31E+01	data
184	C9H11FN2O	3-(4-FLOROPHENYL)-1,1-DIMETHYLUREA	182.194	------	2.66E+01	data
185	C9H11NO	2-METHYLACETANILIDE	149.186	110.00	2.79E+01	data
186	C9H11NO	3-METHYLACETANILIDE	149.186	65.50	2.79E+01	data
187	C9H11NO	4-METHYLACETANILIDE	149.186	148.50	2.79E+01	data
188	C9H11NO2	4-METHOXYACETANILIDE	165.186	131.00	2.49E+01	data
189	C9H11NO2	ETHYL N-PHENYLCARBAMATE	165.186	------	6.54E+01	data
190	C9H12	1,2,3-TRIMETHYLBENZENE	120.193	-25.16	8.84E+02	estimated
191	C9H12	1,2,4-TRIMETHYLBENZENE	120.193	-46.16	8.84E+02	estimated
192	C9H12	CUMENE	120.193	-96.16	8.84E+02	estimated
193	C9H12	M-ETHYLTOLUENE	120.193	-95.16	8.84E+02	estimated
194	C9H12	O-ETHYLTOLUENE	120.193	-81.16	8.84E+02	estimated
195	C9H12	P-ETHYLTOLUENE	120.193	-62.16	8.84E+02	estimated
196	C9H12	PROPYLBENZENE	120.193	-99.16	8.84E+02	estimated
197	C9H12N2O	1,1-DIMETHYL-3-PHENYLUREA	164.202	------	4.63E+01	data
198	C9H12N2O	FENURON	164.202	128.00	2.70E+01	data
199	C9H13BRN2O2	BROMACIL	261.126	158.00	7.20E+01	data
200	C9H13CLN2O2	TERBACIL	216.667	176.00	5.10E+01	data
201	C9H13CLN6	CYANAZINE	241.707	167.00	2.00E+02	data
202	C9H16CLN5	PROPAZINE	229.715	213.00	1.60E+02	data
203	C9H16N4OS	TEBUTHIURON	228.316	------	6.20E+02	data
204	C9H17N5S	AMETRYN	227.332	------	3.92E+02	data
205	C9H19NOS	S-ETHYL DI-N,N-PROPYLTHIOCARBAMATE	189.316	------	2.40E+02	data
206	C9H22O4P2S4	ETHION	384.490	-13.00	1.54E+04	data
207	C10HO5PS	PARATHION	264.154	6.00	4.80E+03	data
208	C10H7N3S	THIABENDAZOLE	201.246	------	1.72E+03	data
209	C10H8	AZULENE	128.173	174.00	1.50E+03	estimated
210	C10H8	NAPHTHALENE	128.164	80.00	1.30E+03	data
211	C10H8CLNO3	PYRAZON	225.629	------	1.20E+02	data
212	C10H8F3N2O	3-(3-(TRIFLOROMETHYL)PHENYL)-1,1-DIMETHYLUREA	229.180	------	2.99E+02	data
213	C10H8O	NAPHTHOL	144.164	96.00	3.81E+02	data
214	C10H9F3N3O4	TRIFLURALIN	292.196	49.00	4.30E+03	data
215	C10H11F3N2O	FLUOMETURON	232.204	164.00	1.75E+02	data
216	C10H12CLNO2	CHLORPROPHAM	213.661	41.00	5.90E+02	data
217	C10H12N2O	3-PHENYL-1-CYCLEPROPYLUREA	176.212	------	5.19E+01	data
218	C10H12N2O5	DINOSEB	240.212	------	1.24E+02	data
219	C10H13CLN2O	3-(3-CHLORO-4-METHYLPHENYL)-1,1-DIMETHYLUREA	212.677	------	1.04E+02	data
220	C10H13CLN2O2	3-(3-CHLORO-4-METHOXYLPHENYL)-1,1-DIMETHYLUREA	228.677	------	5.44E+01	data
221	C10H13NO2	PROPHAM	179.212	------	5.10E+01	data
222	C10H13NO2	iso-PROPYL-N-PHENYLCARBAMATE	179.212	------	8.80E+01	data
223	C10H13NO2	n-PROPYL-N-PHENYLCARBAMATE	179.212	------	1.14E+02	data
224	C10H14	1,2,3,4-TETRAMETHYLBENZENE	134.220	-6.16	1.50E+03	estimated
225	C10H14	1,2,3,5-TETRAMETHYLBENZENE	134.220	-24.16	1.50E+03	estimated
226	C10H14	1,2,4,5-TETRAMETHYLBENZENE	134.220	78.84	1.50E+03	estimated
227	C10H14	BUTYLBENZENE	134.220	-88.16	1.50E+03	estimated
228	C10H14	M-DIETHYLBENZENE	134.220	-84.16	1.50E+03	estimated
229	C10H14	O-DIETHYLBENZENE	134.220	-31.16	1.50E+03	estimated
230	C10H14	P-DIETHYLBENZENE	134.220	-42.16	1.50E+03	estimated
231	C10H14N2O	3-(4-METHYLPHENYL)-1,1-DIMETHYLUREA	178.228	------	3.20E+01	data
232	C10H14N2O2	3-(3-METHOXYPHENYL)-1,1-DIMETHYLUREA	194.228	------	5.19E+01	data
233	C10H14N2O2	3-(4-METHOXYPHENYL)-1,1-DIMETHYLUREA	194.228	------	2.49E+01	data
234	C10H16CL3NOS	TRIALLATE	304.673	------	2.22E+03	data

449

Table 20-1 SOIL SORPTION COEFFICIENT - ORGANIC COMPOUNDS (continued)

NO	FORMULA	NAME	Molecular Weight g/mol	Melting Point C	K$_{OC}$ @ 20 C L/kg	CODE
235	C10H17CL2NOS	DIALLATE	270.224	------	1.90E+03	data
236	C10H18	DECAHYDRONAPHTHALENE,CIS	138.252	-43.16	2.92E+03	estimated
237	C10H18	DECAHYDRONAPHTHALENE,TRANS	138.252	-30.16	2.92E+03	estimated
238	C10H18CLN5	IPAZINE	243.741	87.00	1.66E+03	data
239	C10H19N5O	PROMETON	225.292	------	3.50E+02	data
240	C10H19N5S	PROMETRYNE	241.358	------	8.10E+02	data
241	C10H19N5S	TERBUTYRNE	241.358	------	7.00E+02	data
242	C10H19O6PS2	MALATHION	330.364	3.00	1.80E+03	data
243	C10H21NOS	PEBULATE	203.342	------	6.30E+02	data
244	C11H10	2-METHYLNAPHTHALENE	142.190	34.60	8.50E+03	data
245	C11H13F3N4O4	DINITRAMINE	354.312	------	4.00E+03	data
246	C11H14CLNO	PROPACHLOR	211.687	71.00	2.65E+02	data
247	C11H15BRN2O	3-(3,5-DIMETHYL-4-BROMOPHENYL)-1,1-DIMETHYLUREA	271.162	------	3.35E+02	data
248	C11H15NO	BUTYRANILIDE	177.238	------	5.07E+01	data
249	C11H15NO2	n-BUTYLPHENYLCARBAMATE	193.238	------	1.80E+02	data
250	C11H16	PENTAMETHYLBENZENE	148.247	54.84	2.52E+03	estimated
251	C11H16	PENTYLBENZENE	148.247	-75.16	2.52E+03	estimated
252	C11H16CLO2PS3	CARBOPHENOTHION	341.416	------	4.54E+04	data
253	C11H16N2O	3-(3,5-METHYLPHENYL)-1,1-DIMETHYLUREA	192.254	------	5.32E+01	data
254	C11H21N5S	DIPROPETRYNE	255.384	------	1.17E+03	data
255	C11H21NOS	CYCLOATE	215.352	------	3.45E+02	data
256	C12H4CL6	2,2',4,4',5,5'-HEXACHLOROBIPHENYL	361.902	103.00	1.20E+06	data
257	C12H4CL6	2,2',4,4',6,6'-HEXACHLOROBIPHENYL	361.902	103.00	4.17E+05	data
258	C12H5CL5	2,2',4,5,5'-PENTACHLOROBIPHENYL	326.445	75.00	5.00E+05	data
259	C12H7CL3	2,4,4'-TRICHLOROBIPHENYL	258.753	------	4.14E+04	data
260	C12H8CL2	2,2'-DICHLOROBIPHENYL	224.308	------	8.25E+03	data
261	C12H8CL2	2,4-DICHLOROBIPHENYL	224.308	------	1.34E+04	data
262	C12H8S	DIBENZOTHIOPHENE	184.250	98.00	1.12E+04	data
263	C12H9CL	2-CHLOROBIPHENYL	189.863	34.00	2.93E+03	data
264	C12H9CL	3-CHLOROBIPHENYL	189.863	16.00	2.93E+03	data
265	C12H9CLF3N3O	NORFLUORAZON	303.673	------	1.91E+03	data
266	C12H10	BIPHENYL	154.211	71.00	4.23E+03	estimated
267	C12H11CL2NO	PRONAMIDE	256.130	------	2.00E+02	data
268	C12H11N	DIPHENYLAMINE	169.216	55.00	5.96E+02	data
269	C12H11N02	CARBARYL	201.216	142.00	3.90E+02	data
270	C12H12	1,2-DIMETHYLNAPHTHALENE	156.226	-1.16	4.23E+03	estimated
271	C12H12	1,3-DIMETHYLNAPHTHALENE	156.226	-4.16	4.23E+03	estimated
272	C12H12	1,4-DIMETHYLNAPHTHALENE	156.226	7.84	4.23E+03	estimated
273	C12H12	1,5-DIMETHYLNAPHTHALENE	156.226	81.84	4.23E+03	estimated
274	C12H12	1,6-DIMETHYLNAPHTHALENE	156.226	-14.16	4.23E+03	estimated
275	C12H12	1,7-DIMETHYLNAPHTHALENE	156.226	-13.16	4.23E+03	estimated
276	C12H12	1-ETHYLNAPHTHALENE	156.226	-14.16	4.23E+03	estimated
277	C12H12	2,3-DIMETHYLNAPHTHALENE	156.226	104.84	4.23E+03	estimated
278	C12H12	2,6-DIMETHYLNAPHTHALENE	156.226	111.84	4.23E+03	estimated
279	C12H12	2,7-DIMETHYLNAPHTHALENE	156.226	97.84	4.23E+03	estimated
280	C12H12	2-ETHYLNAPHTHALENE	156.226	-7.16	4.23E+03	estimated
281	C12H12F3N3O4	FLUCHORALIN	319.240	------	3.60E+03	data
282	C12H15NO3	CARBOFURAN	211.168	151.00	1.30E+02	data
283	C12H16CL2N2	NEBURON	275.178	------	2.30E+03	data
284	C12H16N2O	3-PHENYLCYCLOPENTYLUREA	204.264	------	8.40E+01	data
285	C12H17NO2	n-PENTYL PHENYLCARBAMATE	207.264	------	4.03E+02	data
286	C12H18	1,2,3-TRIETHYLBENZENE	162.274	-66.16	5.78E+03	estimated
287	C12H18	1,2,4-TRIETHYLBENZENE	162.274	-66.16	5.78E+03	estimated
288	C12H18	1,3,5-TRIETHYLBENZENE	162.274	-66.16	5.78E+03	estimated
289	C12H18	HEXAMETHYLBENZENE	162.274	165.84	5.78E+03	estimated
290	C12H18	HEXYLBENZENE	162.274	-61.16	5.78E+03	estimated
291	C12H21N2O3PS	DIAZINON	304.350	------	2.27E+02	data
292	C13H10BRCL2O2P5	LEPTOPHOS	499.040	------	9.30E+03	data
293	C13H12N2O2	(4-PHENNOXYPHENYL)UREA	228.242	------	3.59E+02	data
294	C13H14	1-PROPYLNAPHTHALENE	170.253	-8.16	7.06E+03	estimated
295	C13H14	2-PROPYLNAPHTHALENE	170.253	-3.16	7.06E+03	estimated
296	C13H14	2-ETHYL-3-METHYLNAPHTHALENE	170.253	70.84	7.06E+03	estimated
297	C13H14	2-ETHYL-6-METHYLNAPHTHALENE	170.253	44.84	7.06E+03	estimated
298	C13H14	2-ETHYL-7-METHYLNAPHTHALENE	170.253	44.84	7.06E+03	estimated
299	C13H16F3N3O4	BENEFIN	335.282	66.00	1.07E+04	data
300	C13H18N2O	2-PHENYLCYCLOHEXYLUREA	218.290	------	1.16E+03	data
301	C13H19N3O6S	NITRALIN	345.372	------	9.60E+02	data
302	C13H20	1-PHENYLHEPTANE	176.301	-48.16	7.06E+03	estimated
303	C14H9CL5	4,4'-DICHLORODIPHENYLTRICHLOROETHANE	354.497	109.00	1.50E+05	data
304	C14H9CLF2N2O2	DIFLUBENZURON	310.685	------	6.79E+03	data
305	C14H10	ANTHRACENE	178.220	216.00	2.60E+04	data
306	C14H10	PHENANTHRENE	178.220	101.00	2.30E+04	data
307	C14H16	1-BUTYLNAPHTHALENE	184.280	-20.16	1.13E+04	estimated
308	C14H16	2-BUTYLNAPHTHALENE	184.280	-5.16	1.13E+04	estimated
309	C14H16F3N3O4	PROFLURALIN	347.292	------	8.60E+03	data
310	C14H17CLNO2	ALACHLOR	266.741	------	1.90E+02	data
311	C14H20N2O	3-PHENYLCYCLOPENTYLUREA	374.144	------	2.32E+02	data
312	C14H20N2O8S2	PARAQUAT	408.410	------	1.54E+04	data

450

Table 20-1 SOIL SORPTION COEFFICIENT - ORGANIC COMPOUNDS (continued)

NO	FORMULA	NAME	Molecular Weight g/mol	Melting Point C	K_{OC} @ 20 C L/kg	CODE
313	C14H21N3O4	BUTRALIN	295.332	------	8.20E+03	data
314	C14H22	1,2,3,4-TETRAETHYLBENZENE	190.327	11.84	1.13E+04	estimated
315	C14H22	1,2,3,5-TETRAETHYLBENZENE	190.327	10.84	1.13E+04	estimated
316	C14H22	1,2,4,5-TETRAETHYLBENZENE	190.327	9.84	1.13E+04	estimated
317	C14H22	1-PHENYLOCTANE	190.327	-36.16	1.13E+04	estimated
318	C15H10O2	ANTHRACENCE-9-CARBOXYLIC ACID	364.058	------	8.82E+02	data
319	C15H12	9-METHYLANTHRANCENE	192.246	82.00	6.50E+04	data
320	C15H15CLN2O2	CHLOROXURON	290.743	------	3.20E+03	data
321	C15H18	1-PENTYLNAPHTHALENE	198.307	-22.16	1.93E+04	estimated
322	C15H18	2-PENTYLNAPHTHALENE	198.307	-4.16	1.93E+04	estimated
323	C15H18CL2N2O3	OXADIAZON	345.224	------	3.24E+03	data
324	C15H23N3O4	ISOPROPALIN	309.358	------	7.53E+04	data
325	C15H24	1-PHENYLNONANE	204.354	-24.16	1.93E+04	estimated
326	C16H10	PYRENE	202.240	156.00	8.40E+04	data
327	C16H22O4	DIBUTYLPHTHALATE	278.336	------	4.50E+02	data
328	C16H26	1-PHENYLDECANE	218.381	-14.16	3.18E+04	estimated
329	C16H26	PENTAETHYLBENZENE	218.381	54.84	3.18E+04	estimated
330	C17H21NO2	NAPROPAMIDE	271.346	------	6.80E+02	data
331	C17H28	1-PHENYLUNDECANE	232.408	-5.16	5.20E+04	estimated
332	C18H12	1,2-BENZANTHRACENE	228.276	162.00	6.50E+05	data
333	C18H12	TETRACENE	228.276	332.00	6.50E+05	data
334	C18H30	HEXAETHYLBENZENE	246.435	127.84	8.45E+04	estimated
335	C20H16	9,10-DIMETHYLBENZANTHRACENE	256.328	123.00	4.76E+05	data
336	C21H36	1-PHENYLPENTADECANE	288.515	21.84	3.53E+05	estimated

K_{OC} - soil sorption coefficient, (mg sorbed/kg organic carbon) / (mg/L aqueous concentration)

451

Chapter 21

VISCOSITY OF GAS

Carl L. Yaws, Xiaoyan Lin, and Li Bu
Lamar University, Beaumont, Texas

ABSTRACT

Results for gas viscosity as function of temperature are presented for a wide range of organic and inorganic chemicals. The major chemicals include many compound types. The results are provided in easy-to-use tables that are especially applicable for rapid engineering usage with the personal computer or hand calculator. The agreement of correlation and data is quite good.

INTRODUCTION

Gas viscosity data are important in many engineering applications in the chemical processing and petroleum refining industries. The objective of this article is to provide the engineer with such viscosity data. The compilation of data is presented for a wide temperature range to enable the engineer to determine values at desired temperatures of interest.

GAS VISCOSITY CORRELATION

The correlation for gas viscosity as a function of temperature is given by the equation shown below:

$$n_{gas} = A + B\,T + C\,T^2 \tag{21-1}$$

where
n_{gas} = viscosity of gas, micropoise
A, B, and C = regression coefficients for chemical compound
T = temperature, K

The results for gas viscosity at low pressure are given in Tables 21-1 and 21-2. The tabulations are arranged by chemical formula carbon number to provide ease of use in quickly locating data.

In preparing the compilation, a literature search was conducted to identify data source publications for organics (1-37) and inorganics (1-62). Both experimental values for the property under consideration and parameter values for estimation of the property are included in the source publications. The publications were screened for appropriate data. The compilation resulting from the screening is based on both experimental data and estimated values. In the absence of experimental data, estimates were primarily based on modified Chapman-Enskog method (29, Chung, intermolecular forces, collision diameter) and Reichenberg equation (29, corresponding state, group contribution). Experimental data and estimates were then regressed to provide the same equation for all compounds.

Very limited experimental data are available for highly polar and high molecular weight compounds. Thus, the values for these compounds should be considered rough approximations.

A comparison of correlation and experimental data is shown in Fig. 21-1 for a representative chemical. The graph discloses good agreement of correlation and data.

EXAMPLES

The correlation results may be used for prediction and calculation of gas viscosity. Examples are given below.

Example 1 Calculate the gas viscosity of n-hexane (C6H14) at a temperature of 300 K.

Substitution of the coefficients from the table and temperature into the correlation equation yields

$n_{gas} = -\,8.2223 + (2.6229\text{E-}01)\,(300) + (-5.7366\text{E-}05)\,(300^2)$

$n_{gas} = 65.3$ micropoise

The calculated and data values compare favorably (65.3 vs 66.6, deviation = 1.95%).

Example 2 Calculate the gas viscosity of carbon tetrachloride (CCl4) at a temperature of 520 K.

Substitution of the coefficients from the table and temperature into the correlation equation yields

$n_{gas} = -\,7.7453 + (3.9481\text{E-}01)\,(520) + (-1.1150\text{E-}02)\,(520^2)$

$n_{gas} = 169.3$ micropoise

The calculated and data values compare favorably (169.3 vs 167.0, deviation = 1.36%).

REFERENCES – ORGANIC COMPOUNDS

1-34. See **REFERENCES - ORGANIC COMPOUNDS** in **Chapter 1, CRITICAL PROPERTIES AND ACENTRIC FACTOR**

35. Golubev, I. F., <u>VISCOSITY OF GASES AND GAS MIXTURES</u>, translated from Russian, U. S. Dept. of Commerce, Springfield, VA (1970).

36. Stephan, K. and K. Lucas, <u>VISCOSITY OF DENSE FLUIDS</u>, Plenum Press, New York, NY (1979).

37. Yaws, C. L., <u>HANDBOOK OF VISCOSITY</u>, Vols. 1, 2 , 3 and 4, Gulf Publishing Company, Houston, TX (1995, 1995, 1995, 1997).

REFERENCES – INORGANIC COMPOUNDS

1-56. See **REFERENCES - INORGANIC COMPOUNDS** in **Chapter 1, CRITICAL PROPERTIES AND ACENTRIC FACTOR**

57. Golubev, I. F., <u>VISCOSITY OF GASES AND GAS MIXTURES</u>, translated from Russian, U. S. Dept. of Commerce, Springfield, VA (1970).

58. Lyon, R. N., ed., <u>LIQUID-METALS HANDBOOK</u>, Atomic Energy Commission and Dept. of Navy, Washington, DC (1954).

59. Emsley, J., <u>THE ELEMENTS</u>, 2nd ed., Clarendon Press, Oxford University Press, New York, NY (1991).

60. Perry, D. L. and S. L. Phillips, <u>HANDBOOK OF INORGANIC COMPOUNDS</u>, CRC Press, New York, NY (1995).

61. Stephan, K. and K. Lucas, <u>VISCOSITY OF DENSE FLUIDS</u>, Plenum Press, New York, NY (1979).

62. Yaws, C. L., <u>HANDBOOK OF VISCOSITY</u>, Vol. 4, Gulf Publishing Co., Houston, TX (1997).

Figure 21-1 Viscosity of Gas

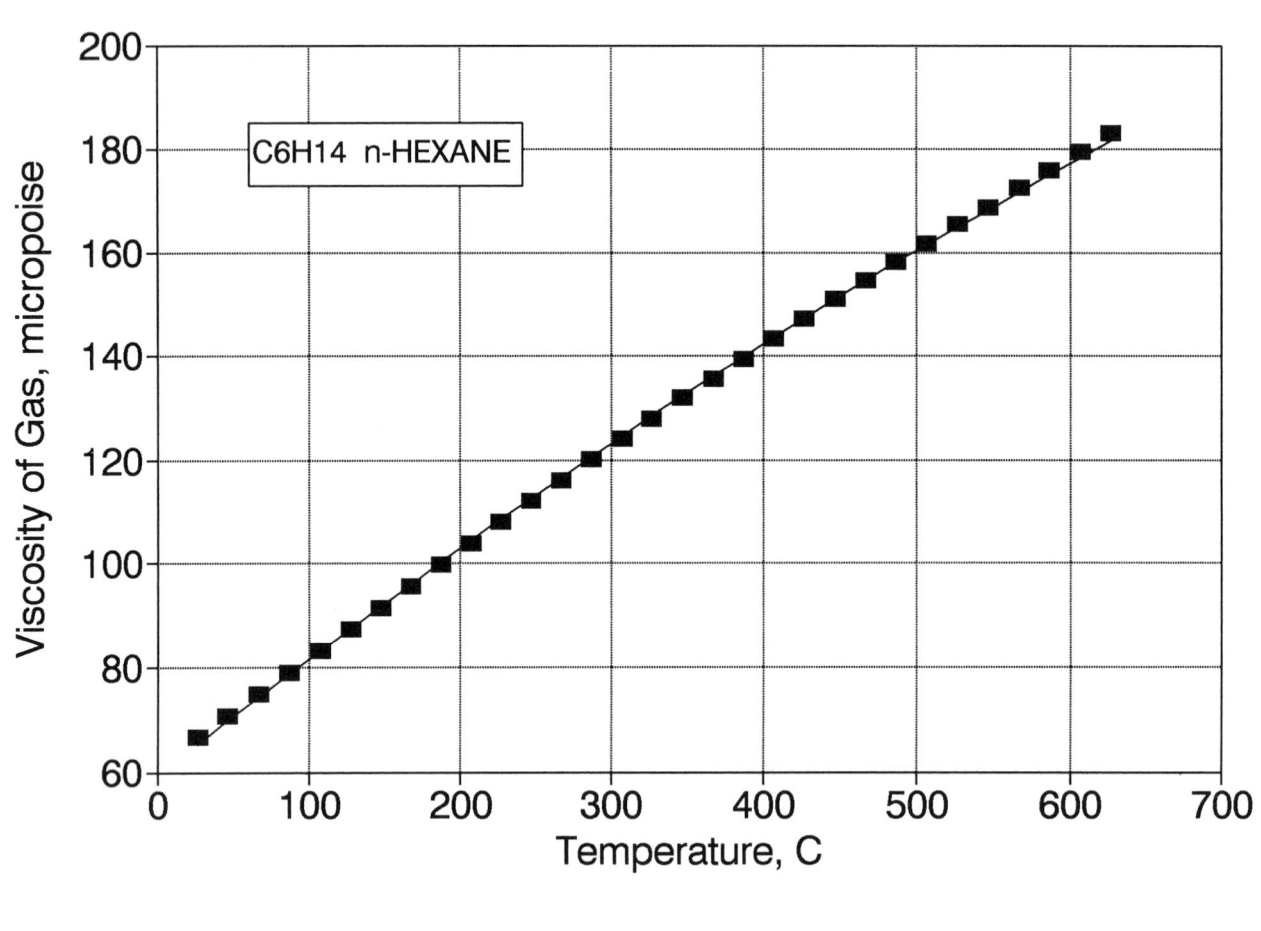

Table 21-1 VISCOSITY OF GAS - ORGANIC COMPOUNDS

$$n_{gas} = A + B\,T + C\,T^2 \qquad (n_{gas} - \text{micropoise}, T - K)$$

NO	FORMULA	NAME	A	B	C	TMIN	TMAX	n_{gas} @25 C	n_{gas} @TMIN	n_{gas} @TMAX
1	CBrClF2	BROMOCHLORODIFLUOROMETHANE	-4.369	4.8220E-01	-1.0333E-04	233	1000	130.21	102.37	374.50
2	CBrCl3	BROMOTRICHLOROMETHANE	-10.274	3.9477E-01	-6.3126E-05	252	1000	101.81	85.20	321.37
3	CBrF3	BROMOTRIFLUOROMETHANE	-4.914	6.0829E-01	-2.2428E-04	230	500	156.51	123.13	243.16
4	CBr2F2	DIBROMODIFLUOROMETHANE	-9.401	5.0711E-01	-1.0180E-04	296	1000	132.74	131.78	395.91
5	CClF3	CHLOROTRIFLUOROMETHANE	21.229	3.9296E-01	6.5918E-05	230	500	144.25	115.10	234.19
6	CClN	CYANOGEN CHLORIDE	-5.236	3.1918E-01	-6.6139E-05	286	1000	84.05	80.64	247.81
7	CCl2F2	DICHLORODIFLUOROMETHANE	13.986	3.7629E-01	-1.3267E-05	250	575	125.00	107.23	225.97
8	CCl2O	PHOSGENE	-37.235	5.2032E-01	-1.9868E-05	200	900	116.13	66.03	414.96
9	CCl3F	TRICHLOROFLUOROMETHANE	-3.391	4.0268E-01	-2.9161E-05	248	997	114.08	94.68	369.09
10	CCl4	CARBON TETRACHLORIDE	-7.745	3.9481E-01	-1.1150E-04	280	800	100.06	94.06	236.74
11	CF2O	CARBONYL FLUORIDE	0.003	4.2190E-01	-1.1407E-04	189	1000	115.65	75.67	307.83
12	CF4	CARBON TETRAFLUORIDE	10.896	5.9490E-01	-1.7559E-04	145	900	172.66	93.46	404.08
13	CHBr3	TRIBROMOMETHANE	-17.026	3.4203E-01	-3.9741E-05	273	573	81.42	73.39	165.91
14	CHClF2	CHLORODIFLUOROMETHANE	-9.517	5.0379E-01	-1.3232E-04	232	500	128.93	100.24	209.30
15	CHCl2F	DICHLOROFLUOROMETHANE	-3.016	4.2016E-01	-7.7903E-05	280	500	115.33	108.52	187.59
16	CHCl3	CHLOROFORM	-4.392	3.7309E-01	-5.1696E-05	250	700	102.25	85.65	231.44
17	CHF3	TRIFLUOROMETHANE	-10.495	5.6948E-01	-1.7587E-04	191	700	143.66	91.86	301.97
18	CHI3	TRIIODOMETHANE	-8.619	4.1553E-01	-2.5315E-05	275	1000	113.02	103.74	381.60
19	CHN	HYDROGEN CYANIDE	-8.486	9.0368E-02	7.9146E-05	300	425	------	25.75	44.22
20	CHNS	ISOTHIOCYANIC-ACID	------	--------	--------	----	----	------	------	------
21	CH2BrCl	BROMOCHLOROMETHANE	-10.798	4.5078E-01	-7.8817E-05	341	1000	------	133.75	361.17
22	CH2Br2	DIBROMOMETHANE	-9.547	4.6829E-01	-6.9599E-05	370	1000	------	154.19	389.14
23	CH2ClF	CHLOROFLUOROMETHANE	-12.138	4.4954E-01	-1.0225E-04	275	1000	112.80	103.75	335.15
24	CH2Cl2	DICHLOROMETHANE	-20.372	4.3745E-01	-7.7549E-05	273	993	103.16	93.27	337.55
25	CH2F2	DIFLUOROMETHANE	-3.807	5.2411E-01	-1.2877E-04	222	1000	141.01	106.20	391.53
26	CH2I2	DIIODOMETHANE	-28.175	5.0959E-01	-6.8132E-05	288	573	117.70	112.94	241.45
27	CH2O	FORMALDEHYDE	-6.439	4.4802E-01	-1.0130E-04	254	1000	118.13	100.82	340.28
28	CH2O2	FORMIC ACID	-13.139	2.7486E-01	1.9189E-04	363	386	------	111.92	121.55
29	CH3Br	METHYL BROMIDE	-27.740	5.5901E-01	-7.0942E-05	260	440	132.62	112.81	204.49
30	CH3Cl	METHYL CHLORIDE	-1.374	3.8627E-01	-4.8650E-05	230	700	109.47	84.90	245.18
31	CH3Cl3Si	METHYL TRICHLOROSILANE	-9.343	3.6088E-01	-5.1146E-05	195	1000	93.71	59.08	300.39
32	CH3F	METHYL FLUORIDE	-0.901	4.7265E-01	-1.2197E-04	195	1000	129.18	86.63	349.78
33	CH3I	METHYL IODIDE	-21.698	5.4959E-01	-9.1954E-05	300	1000	------	134.90	435.94
34	CH3NO	FORMAMIDE	------	--------	--------	----	----	------	------	------
35	CH3NO2	NITROMETHANE	-20.206	3.2616E-01	-4.3498E-05	374	1000	------	95.69	262.46
36	CH3NO2	METHYL-NITRITE	------	--------	--------	----	----	------	------	------
37	CH3NO3	METHYL-NITRATE	------	--------	--------	----	----	------	------	------
38	CH4	METHANE	3.844	4.0112E-01	-1.4303E-04	91	850	110.72	39.16	241.46
39	CH4Cl2Si	METHYL DICHLOROSILANE	-8.885	3.6879E-01	-5.9741E-05	183	1000	95.76	56.60	300.16
40	CH4O	METHANOL	-14.236	3.8935E-01	-6.2762E-05	240	1000	96.27	75.59	312.35
41	CH4O3S	METHANESULFONIC ACID	------	--------	--------	----	----	------	------	------
42	CH4S	METHYL MERCAPTAN	-39.380	4.6695E-01	-6.2465E-05	273	473	94.29	83.44	167.51
43	CH5ClSi	METHYL CHLOROSILANE	-6.691	3.5509E-01	-6.0616E-05	139	1000	93.79	41.50	287.78
44	CH5N	METHYLAMINE	-5.334	3.4181E-01	-7.4297E-05	267	1000	89.97	80.63	262.18
45	CH6Si	METHYL SILANE	-5.429	3.4287E-01	-7.7044E-05	116	1000	89.95	33.31	260.40
46	CN4O8	TETRANITROMETHANE	-15.752	3.4934E-01	-5.9980E-05	287	1000	83.07	79.57	273.61
47	CO	CARBON MONOXIDE	23.811	5.3944E-01	-1.5411E-04	68	1250	170.95	59.78	457.31
48	COS	CARBONYL SULFIDE	-11.490	4.9840E-01	-9.9991E-05	134	1000	128.22	53.50	386.92
49	CO2	CARBON DIOXIDE	11.811	4.9838E-01	-1.0851E-04	195	1500	150.76	104.87	515.23
50	CS2	CARBON DISULFIDE	-7.700	3.6594E-01	-2.5416E-05	273	583	99.15	90.31	197.00
51	C2BrF3	BROMOTRIFLUOROETHYLENE	-7.541	4.8895E-01	-1.0519E-04	271	1000	128.89	117.24	376.22
52	C2Br2F4	1,2-DIBROMOTETRAFLUOROETHANE	-7.132	4.5248E-01	-8.5233E-05	320	1000	------	128.93	360.11
53	C2ClF3	CHLOROTRIFLUOROETHYLENE	-3.948	4.5063E-01	-1.0575E-04	245	1000	121.01	100.11	340.93
54	C2ClF5	CHLOROPENTAFLUOROETHANE	3.736	4.4326E-01	-1.0051E-04	250	500	126.96	108.27	200.24
55	C2Cl2F4	1,2-DICHLOROTETRAFLUOROETHANE	26.642	2.7226E-01	8.7050E-05	230	500	115.55	93.87	184.53
56	C2Cl3F3	1,1,1-TRICHLOROTRIFLUOROETHANE	32.995	2.5082E-01	-5.4159E-05	237	1000	102.96	89.40	229.66
57	C2Cl4	TETRACHLOROETHYLENE	-10.841	3.1582E-01	2.2833E-05	251	570	85.35	69.87	176.59
58	C2Cl4F2	1,1,2,2-TETRACHLORODIFLUOROETHANE	-8.861	3.6067E-01	-4.3826E-05	366	1000	------	114.59	287.98
59	C2Cl4O	TRICHLOROACETYL CHLORIDE	-19.554	4.0312E-01	-6.2634E-05	391	991	------	128.49	318.43
60	C2Cl6	HEXACHLOROETHANE	-19.961	3.6258E-01	-4.5006E-05	460	1000	------	137.30	297.61
61	C2F4	TETRAFLUOROETHYLENE	-0.662	5.1057E-01	-1.3502E-04	198	1000	139.56	95.14	374.89
62	C2F6	HEXAFLUOROETHANE	0.972	5.3473E-01	-1.4324E-04	195	1000	147.67	99.80	392.46
63	C2HBrClF3	HALOTHANE	-7.358	4.2906E-01	-7.5755E-05	223	1000	113.83	84.56	345.95
64	C2HClF2	2-CHLORO-1,1-DIFLUOROETHYLENE	-4.775	4.2835E-01	-9.6753E-05	255	1000	114.34	98.16	326.82
65	C2HCl3	TRICHLOROETHYLENE	-23.621	4.2164E-01	-6.8948E-05	333	990	------	109.14	326.23
66	C2HCl3O	DICHLOROACETYL CHLORIDE	-19.499	4.0357E-01	-6.4804E-05	382	992	------	125.21	317.07
67	C2HCl3O	TRICHLOROACETALDEHYDE	-9.163	3.7547E-01	-4.3356E-05	216	1000	98.93	69.92	322.95
68	C2HCl5	PENTACHLOROETHANE	-4.876	3.0910E-01	-3.8437E-05	244	1000	83.87	68.26	265.79
69	C2HF3	TRIFLUOROETHYLENE	-5.548	4.6672E-01	-1.1472E-04	275	1000	123.41	114.12	346.45
70	C2HF3O2	TRIFLUOROACETIC ACID	-6.823	4.0112E-01	-7.5762E-05	258	1000	106.04	91.62	318.54
71	C2HF5	PENTAFLUOROETHANE	-4.504	5.2704E-01	-1.3533E-04	170	1000	140.60	81.18	387.21
72	C2H2	ACETYLENE	-11.557	4.2363E-01	-1.4174E-04	193	500	102.15	64.92	164.82
73	C2H2Br4	1,1,2,2-TETRABROMOETHANE	-8.309	3.6748E-01	-3.5912E-05	517	1000	------	172.08	323.26
74	C2H2Cl2	1,1-DICHLOROETHYLENE	-8.078	3.8859E-01	-7.9107E-05	305	1000	103.08	103.08	301.41
75	C2H2Cl2	cis-1,2-DICHLOROETHYLENE	-7.574	3.6605E-01	-6.7144E-05	334	1000	------	107.20	291.33
76	C2H2Cl2	trans-1,2-DICHLOROETHYLENE	-7.125	3.6505E-01	-6.8692E-05	321	1000	------	102.98	289.23
77	C2H2Cl2O	CHLOROACETYL CHLORIDE	-18.205	3.8542E-01	-5.7860E-05	350	1000	------	109.60	309.36
78	C2H2Cl2O	DICHLOROACETALDEHYDE	-5.363	3.2751E-01	-5.2553E-05	223	1000	87.61	65.06	269.59

455

Table 21-1 VISCOSITY OF GAS - ORGANIC COMPOUNDS (continued)

| NO | FORMULA | NAME | $n_{gas} = A + B T + C T^2$ | | | | | $(n_{gas}$ - micropoise, T - K) | | |
			A	B	C	TMIN	TMAX	n_{gas} @25 C	n_{gas} @TMIN	n_{gas} @TMAX
79	C2H2Cl2O2	DICHLOROACETIC ACID	-5.508	3.0753E-01	-3.7358E-05	287	1000	82.86	79.68	264.66
80	C2H2Cl3F	1,1,1-TRICHLOROFLUOROETHANE	-4.853	3.4986E-01	-5.4139E-05	173	1000	94.65	54.05	290.87
81	C2H2Cl4	1,1,1,2-TETRACHLOROETHANE	-9.963	3.3385E-01	-5.3254E-05	404	1000	-----	116.22	270.63
82	C2H2Cl4	1,1,2,2-TETRACHLOROETHANE	-20.605	3.7770E-01	-5.4670E-05	418	998	-----	127.72	301.89
83	C2H2F2	1,1-DIFLUOROETHYLENE	2.537	4.6454E-01	-1.1805E-04	188	1000	130.55	85.70	349.03
84	C2H2F2	cis-1,2-DIFLUOROETHENE	-9.608	4.3427E-01	-1.0232E-04	275	1000	110.77	102.08	322.34
85	C2H2F2	trans-1,2-DIFLUOROETHENE	-9.608	4.3427E-01	-1.0232E-04	275	1000	110.77	102.08	322.34
86	C2H2F4	1,1,1,2-TETRAFLUOROETHANE	-5.083	5.0162E-01	-1.1800E-04	172	1000	133.99	77.70	378.54
87	C2H2O	KETENE	-10.924	4.1236E-01	-9.3098E-05	200	1000	103.75	67.82	308.34
88	C2H2O4	OXALIC ACID	-6.842	2.8812E-01	-3.1536E-05	463	1000	-----	119.80	249.74
89	C2H3Br	VINYL BROMIDE	-9.309	4.6189E-01	-9.4500E-05	289	1000	120.00	116.28	358.08
90	C2H3Cl	VINYL CHLORIDE	-6.067	3.9013E-01	-8.3970E-05	260	1000	102.79	89.69	300.09
91	C2H3ClF2	1-CHLORO-1,1-DIFLUOROETHANE	-6.450	4.3456E-01	-9.8180E-05	263	1000	114.39	101.05	329.93
92	C2H3ClO	ACETYL CHLORIDE	1.012	3.1562E-01	-6.2753E-06	277	761	94.56	87.96	237.56
93	C2H3ClO	CHLOROACETALDEHYDE	-6.383	3.3296E-01	-5.6014E-05	293	1000	87.91	86.37	270.56
94	C2H3ClO2	CHLOROACETIC ACID	-6.489	3.1233E-01	-3.8880E-05	333	1000	-----	93.21	266.96
95	C2H3ClO2	METHYL CHLOROFORMATE	-4.742	3.3821E-01	-5.7423E-05	192	1000	90.99	58.08	276.05
96	C2H3Cl3	1,1,1-TRICHLOROETHANE	-19.216	4.0308E-01	-6.8618E-05	347	997	-----	112.39	314.45
97	C2H3Cl3	1,1,2-TRICHLOROETHANE	-8.293	3.3989E-01	-5.3678E-05	387	987	-----	115.21	274.89
98	C2H3F	VINYL FLUORIDE	-1.954	4.2566E-01	-1.0977E-04	201	1000	115.20	79.17	313.94
99	C2H3F3	1,1,1-TRIFLUOROETHANE	-3.539	4.9228E-01	-1.2372E-04	226	1000	132.24	101.40	365.02
100	C2H3N	ACETONITRILE	-1.384	2.5204E-01	-2.4595E-05	308	750	-----	73.91	173.81
101	C2H3NO	METHYL ISOCYANATE	-15.434	3.2530E-01	-5.4292E-05	312	1000	-----	80.77	255.57
102	C2H4	ETHYLENE	-3.985	3.8726E-01	-1.1227E-04	150	1000	101.50	51.58	271.00
103	C2H4Br2	1,1-DIBROMOETHANE	-6.137	4.1000E-01	-5.4252E-05	210	1000	111.28	77.57	349.61
104	C2H4Br2	1,2-DIBROMOETHANE	-21.796	4.5011E-01	-5.9439E-05	405	995	-----	150.75	367.22
105	C2H4Cl2	1,1-DICHLOROETHANE	-12.991	4.0085E-01	-1.1779E-04	320	472	-----	103.22	149.97
106	C2H4Cl2	1,2-DICHLOROETHANE	1.025	3.1792E-01	-4.1853E-05	322	561	-----	99.06	166.21
107	C2H4Cl2O	BIS(CHLOROMETHYL)ETHER	-5.207	3.3051E-01	-5.0373E-05	232	1000	88.86	68.76	274.93
108	C2H4F2	1,1-DIFLUOROETHANE	-2.949	4.3374E-01	-9.2979E-05	247	1000	118.11	98.51	337.81
109	C2H4F2	1,2-DIFLUOROETHANE	-5.951	4.2281E-01	-8.0663E-05	215	1000	112.94	81.22	336.20
110	C2H4I2	1,2-DIIODOETHANE	-9.916	3.8940E-01	-3.2473E-05	275	1000	103.30	94.71	347.01
111	C2H4O	ACETALDEHYDE	0.069	3.0246E-01	-4.2372E-05	294	1000	86.48	85.33	260.16
112	C2H4O	ETHYLENE OXIDE	-12.180	3.7672E-01	-7.7599E-05	250	1000	93.24	77.15	286.94
113	C2H4OS	THIOACETIC-ACID	-11.202	3.1219E-01	-5.2683E-05	275	1000	77.19	70.67	248.30
114	C2H4O2	ACETIC ACID	-28.660	2.3510E-01	2.2087E-04	366	523	-----	86.97	154.71
115	C2H4O2	METHYL FORMATE	-16.031	3.9071E-01	-7.1883E-05	300	1000	-----	94.71	302.80
116	C2H4S	THIACYCLOPROPANE	-13.859	3.8573E-01	-6.8981E-05	275	1000	95.01	87.00	302.89
117	C2H5Br	BROMOETHANE	-10.148	4.3036E-01	-8.1147E-05	312	1000	-----	116.23	339.07
118	C2H5Cl	ETHYL CHLORIDE	0.458	3.2827E-01	-1.2467E-05	213	523	97.22	69.81	168.73
119	C2H5ClO	2-CHLOROETHANOL	-2.529	1.0612E-01	-1.7298E-05	402	1000	-----	37.34	86.29
120	C2H5F	ETHYL FLUORIDE	-4.569	3.9509E-01	-9.5177E-05	235	1000	104.77	83.02	295.34
121	C2H5I	ETHYL IODIDE	-17.071	4.6084E-01	-6.7487E-05	300	1000	-----	115.11	376.28
122	C2H5N	ETHYLENEIMINE	-12.220	3.0761E-01	-4.7922E-05	329	1000	-----	83.80	247.47
123	C2H5NO	ACETAMIDE	-17.090	2.5005E-01	-2.3629E-05	494	994	-----	100.67	208.11
124	C2H5NO	N-METHYLFORMAMIDE	-8.853	3.2988E-01	-4.1861E-05	473	983	-----	137.81	274.97
125	C2H5NO2	NITROETHANE	-16.701	2.9762E-01	-3.7773E-05	387	1000	-----	92.82	243.15
126	C2H5NO3	ETHYL-NITRATE	-----	--------	--------	----	----	-----	-----	-----
127	C2H6	ETHANE	0.514	3.3449E-01	-7.1071E-05	150	1000	93.92	49.09	263.93
128	C2H6AlCl	DIMETHYLALUMINUM CHLORIDE	-7.230	2.6575E-01	-2.7116E-05	252	1000	69.59	58.02	231.40
129	C2H6O	DIMETHYL ETHER	-4.276	3.0262E-01	6.3528E-05	216	373	91.60	64.05	117.44
130	C2H6O	ETHANOL	1.499	3.0741E-01	-4.4479E-05	200	1000	89.20	61.20	264.43
131	C2H6OS	DIMETHYL SULFOXIDE	-14.902	2.7069E-01	-2.0800E-05	462	992	-----	105.72	233.15
132	C2H6O2	ETHYLENE GLYCOL	-7.178	3.1246E-01	-4.4028E-05	260	1000	82.07	71.09	261.25
133	C2H6O4S	DIMETHYL SULFATE	-----	--------	--------	----	----	-----	-----	-----
134	C2H6S	DIMETHYL SULFIDE	-14.076	3.5306E-01	-6.2780E-05	310	990	-----	89.34	273.92
135	C2H6S	ETHYL MERCAPTAN	-15.432	3.5300E-01	-6.6454E-05	308	998	-----	86.99	270.67
136	C2H6S2	DIMETHYL DISULFIDE	-15.432	3.5300E-01	-6.6454E-05	308	998	-----	86.99	270.67
137	C2H7N	DIMETHYLAMINE	-9.275	2.8958E-01	-1.3875E-06	250	450	76.94	63.03	120.75
138	C2H7N	ETHYLAMINE	-5.538	3.0778E-01	-6.4363E-05	290	1000	80.51	78.31	237.88
139	C2H7NO	MONOETHANOLAMINE	-12.592	2.8971E-01	-3.9470E-05	400	1000	-----	96.98	237.65
140	C2H8N2	ETHYLENEDIAMINE	-4.843	2.7125E-01	-4.0319E-05	284	1000	72.45	68.94	226.09
141	C2H8Si	DIMETHYL SILANE	-5.555	3.1359E-01	-6.1179E-05	123	1000	82.50	32.09	246.86
142	C2N2	CYANOGEN	-0.293	3.4461E-01	-1.1345E-05	252	600	101.44	85.83	202.39
143	C3F6	HEXAFLUOROPROPYLENE	-4.971	4.8190E-01	-1.1694E-04	244	1000	128.31	105.65	359.99
144	C3F6O	HEXAFLUOROACETONE	-3.319	4.0569E-01	-9.9426E-05	246	1000	108.80	90.46	302.94
145	C3F8	OCTAFLUOROPROPANE	-1.656	4.8878E-01	-1.2050E-04	236	1000	133.36	106.98	366.62
146	C3H2N2	MALONONITRILE	-4.096	2.1382E-01	-2.5089E-05	305	1000	-----	58.79	184.63
147	C3H3Cl	PROPARGYL CHLORIDE	-7.092	3.4137E-01	-5.9573E-05	293	1000	89.39	87.81	274.70
148	C3H3N	ACRYLONITRILE	-4.783	2.4047E-01	-1.4529E-05	303	1000	-----	66.75	221.16
149	C3H3NO	OXAZOLE	-6.523	2.9548E-01	-3.4972E-05	189	1000	78.47	48.07	253.99
150	C3H4	METHYLACETYLENE	-9.626	3.5605E-01	-1.1504E-04	173	573	86.30	48.53	156.62
151	C3H4	PROPADIENE	-12.431	3.4616E-01	-8.6996E-05	175	600	83.04	45.48	163.95
152	C3H4Cl2	2,3-DICHLOROPROPENE	-4.599	3.2783E-01	-4.8207E-05	192	1000	88.86	56.57	275.02
153	C3H4O	ACROLEIN	-16.910	3.2167E-01	-5.2581E-05	326	996	-----	82.37	251.31
154	C3H4O	PROPARGYL ALCOHOL	-9.381	3.2585E-01	-5.6888E-05	387	1000	-----	108.20	259.58
155	C3H4O2	ACRYLIC ACID	-6.532	3.0600E-01	-4.6620E-05	287	1000	80.56	77.45	252.85
156	C3H4O2	beta-PROPIOLACTONE	-5.165	3.4624E-01	-3.9574E-05	240	1000	94.55	75.65	301.50

456

Table 21-1 VISCOSITY OF GAS - ORGANIC COMPOUNDS (continued)

NO	FORMULA	NAME	$n_{gas} = A + B T + C T^2$ A	B	C	TMIN	TMAX	n_{gas} @25 C	n_{gas} @TMIN	n_{gas} @TMAX
157	C3H4O2	VINYL FORMATE	-9.585	3.4817E-01	-5.4813E-05	200	1000	89.35	57.86	283.77
158	C3H4O3	ETHYLENE CARBONATE	-21.019	2.8757E-01	-1.5204E-05	511	1000	-----	121.96	251.35
159	C3H4O3	PYRUVIC ACID	-4.496	2.6567E-01	-3.5453E-05	287	1000	71.56	68.83	225.72
160	C3H5Br	3-BROMO-1-PROPENE	-13.907	3.8907E-01	-7.2088E-05	275	1000	95.69	87.64	303.08
161	C3H5Cl	2-CHLOROPROPENE	-7.512	3.5502E-01	-7.1618E-05	296	1000	91.97	91.30	275.89
162	C3H5Cl	3-CHLOROPROPENE	-7.035	3.3654E-01	-6.2098E-05	318	1000	-----	93.71	267.41
163	C3H5ClO	alpha-EPICHLOROHYDRIN	-17.221	3.5374E-01	-5.1027E-05	389	999	-----	112.66	285.24
164	C3H5ClO2	METHYL CHLOROACETATE	-5.318	3.0230E-01	-4.3113E-05	241	1000	80.98	65.03	253.87
165	C3H5ClO2	ETHYL CHLOROFORMATE	-4.146	3.0221E-01	-5.3347E-05	192	1000	81.22	51.91	244.72
166	C3H5Cl3	1,2,3-TRICHLOROPROPANE	-5.612	2.9957E-01	-3.9788E-05	430	1000	-----	115.85	254.17
167	C3H5I	3-IODO-1-PROPENE	-14.552	4.0915E-01	-6.5528E-05	275	1000	101.61	93.01	329.07
168	C3H5N	PROPIONITRILE	-15.409	2.5623E-01	-3.0933E-05	371	991	-----	75.39	208.14
169	C3H5NO	ACRYLAMIDE	-6.359	2.7465E-01	-3.4477E-05	358	1000	-----	87.55	233.81
170	C3H5NO	HYDRACRYLONITRILE	-----	--------	--------	----	----	-----	-----	-----
171	C3H5NO	LACTONITRILE	-3.619	2.4863E-01	-3.1666E-05	233	1000	67.70	52.59	213.35
172	C3H5N3O9	NITROGLYCERINE	-7.544	3.1769E-01	-2.0439E-05	286	1000	85.36	81.64	289.71
173	C3H6	CYCLOPROPANE	-9.521	3.7037E-01	-1.3138E-04	240	446	89.23	71.80	129.53
174	C3H6	PROPYLENE	-7.230	3.4180E-01	-9.4516E-05	193	1000	86.28	55.22	240.05
175	C3H6Br2	1,2-DIBROMOPROPANE	-12.319	3.6024E-01	-5.1048E-05	275	1000	90.55	82.89	296.87
176	C3H6Cl2	1,1-DICHLOROPROPANE	-4.966	3.1422E-01	-4.9988E-05	200	1000	84.28	55.88	259.27
177	C3H6Cl2	1,2-DICHLOROPROPANE	-16.802	3.5263E-01	-5.6713E-05	370	1000	-----	105.91	279.12
178	C3H6Cl2	1,3-DICHLOROPROPANE	-3.947	3.0000E-01	-4.1205E-05	174	1000	81.84	47.01	254.85
179	C3H6Cl2	2,2-DICHLOROPROPANE	-12.308	3.4447E-01	-6.3934E-05	275	1000	84.71	77.59	268.23
180	C3H6I2	1,2-DIIODOPROPANE	-7.746	3.4742E-01	-2.3613E-05	275	1000	93.74	86.01	316.06
181	C3H6O	ACETONE	-4.055	2.6655E-01	-5.6936E-06	300	650	-----	75.40	166.80
182	C3H6O	ALLYL ALCOHOL	-9.271	3.2529E-01	-5.9894E-05	370	1000	-----	102.89	256.13
183	C3H6O	METHYL VINYL ETHER	-6.632	3.4394E-01	-7.6282E-05	279	1000	89.13	83.39	261.03
184	C3H6O	n-PROPIONALDEHYDE	-14.885	3.2999E-01	-5.9490E-05	321	991	-----	84.91	253.71
185	C3H6O	1,2-PROPYLENE OXIDE	-15.286	3.6549E-01	-6.7583E-05	300	1000	-----	88.28	282.62
186	C3H6O	1,3-PROPYLENE OXIDE	-16.375	3.5336E-01	-6.1967E-05	321	991	-----	90.67	272.95
187	C3H6O2	ETHYL FORMATE	-15.295	3.5359E-01	-6.3604E-05	300	1000	-----	85.06	274.69
188	C3H6O2	METHYL ACETATE	-14.780	3.3569E-01	-6.4353E-05	200	1000	79.59	49.78	256.56
189	C3H6O2	PROPIONIC ACID	-9.446	2.9112E-01	-4.2258E-05	252	1000	73.59	61.23	239.42
190	C3H6O2S	3-MERCAPTOPROPIONIC ACID	-5.241	2.6958E-01	-1.2110E-05	291	1000	74.06	72.18	252.23
191	C3H6O3	LACTIC ACID	-5.514	2.8171E-01	-4.1031E-05	291	1000	74.83	72.99	235.16
192	C3H6O3	METHOXYACETIC ACID	-5.195	2.8270E-01	-3.4304E-05	281	1000	76.04	71.54	243.20
193	C3H6O3	TRIOXANE	-17.988	3.7799E-01	-5.4524E-05	388	998	-----	120.46	304.94
194	C3H6S	THIACYCLOBUTANE	-11.724	3.3137E-01	-5.1942E-05	275	1000	82.46	75.47	267.70
195	C3H7Br	1-BROMOPROPANE	-7.958	3.6805E-01	-6.3823E-05	344	1000	-----	111.10	296.27
196	C3H7Br	2-BROMOPROPANE	-11.793	5.1644E-01	-9.3057E-05	333	1000	-----	149.86	411.59
197	C3H7Cl	ISOPROPYL CHLORIDE	-5.856	3.3081E-01	-6.3622E-05	309	1000	-----	90.29	261.33
198	C3H7Cl	n-PROPYL CHLORIDE	-6.268	3.1908E-01	-5.9816E-05	320	1000	-----	89.71	253.00
199	C3H7F	1-FLUOROPROPANE	-9.240	3.4756E-01	-7.9346E-05	275	1000	87.33	80.34	258.97
200	C3H7F	2-FLUOROPROPANE	-9.212	3.5924E-01	-8.2656E-05	275	1000	90.55	83.33	267.37
201	C3H7I	ISOPROPYL IODIDE	-16.914	4.1524E-01	-5.9631E-05	363	1000	-----	125.96	338.69
202	C3H7I	n-PROPYL IODIDE	-18.302	4.1521E-01	-5.7130E-05	376	1000	-----	129.74	339.78
203	C3H7N	ALLYAMINE	-5.422	2.9789E-01	-5.5989E-05	326	1000	-----	85.74	236.48
204	C3H7N	PROPYLENEIMINE	-6.684	4.1385E-01	-7.1281E-05	229	1000	110.37	84.35	335.88
205	C3H7NO	N,N-DIMETHYLFORMAMIDE	-17.828	2.7374E-01	-3.5679E-05	426	996	-----	92.31	219.42
206	C3H7NO	N-METHYLACETAMIDE	-4.202	2.7991E-01	-2.9560E-05	301	1000	-----	77.37	246.15
207	C3H7NO2	1-NITROPROPANE	-13.355	2.7955E-01	-3.4102E-05	404	1000	-----	94.02	232.09
208	C3H7NO2	2-NITROPROPANE	-17.523	2.9622E-01	-4.2088E-05	393	1000	-----	92.39	236.61
209	C3H7NO3	PROPYL-NITRATE	-----	--------	--------	----	----	-----	-----	-----
210	C3H7NO3	ISOPROPYL-NITRATE	-----	--------	--------	----	----	-----	-----	-----
211	C3H8	PROPANE	-5.462	3.2722E-01	-1.0672E-04	193	750	82.61	53.72	179.92
212	C3H8O	ISOPROPANOL	-10.859	3.0873E-01	-4.8098E-05	200	1000	76.91	48.96	249.77
213	C3H8O	METHYL ETHYL ETHER	-12.218	3.5144E-01	-7.2869E-05	281	991	86.09	80.78	264.50
214	C3H8O	n-PROPANOL	-14.894	3.2171E-01	-5.8021E-05	200	1000	75.87	47.13	248.80
215	C3H8O2	2-METHOXYETHANOL	-4.620	2.9356E-01	-4.5367E-05	398	1000	-----	105.03	243.57
216	C3H8O2	METHYLAL	-6.319	3.1872E-01	-6.3471E-05	315	975	-----	87.78	244.10
217	C3H8O2	1,2-PROPYLENE GLYCOL	-15.620	2.8898E-01	-3.2128E-05	461	991	-----	110.77	239.21
218	C3H8O2	1,3-PROPYLENE GLYCOL	-18.014	3.2093E-01	-4.1746E-05	450	1000	-----	117.95	261.17
219	C3H8O3	GLYCEROL	-23.119	2.8879E-01	-3.4277E-05	563	993	-----	128.61	229.85
220	C3H8S	n-PROPYLMERCAPTAN	-5.705	2.9923E-01	-3.5958E-05	160	1000	80.31	41.25	257.57
221	C3H8S	ISOPROPYL MERCAPTAN	-5.045	3.0176E-01	-3.5890E-05	143	1000	81.73	37.37	260.83
222	C3H8S	ETHYL-METHYL-SULFIDE	-10.851	3.0497E-01	-5.7467E-05	275	1000	74.97	68.67	236.65
223	C3H9N	n-PROPYLAMINE	-12.339	2.9593E-01	-5.5438E-05	300	1000	-----	71.45	228.15
224	C3H9N	ISOPROPYLAMINE	-5.471	2.9085E-01	-5.8656E-05	306	1000	-----	78.04	226.72
225	C3H9N	TRIMETHYLAMINE	-11.545	3.2107E-01	-6.8081E-05	250	1000	78.13	64.47	241.44
226	C3H9NO	1-AMINO-2-PROPANOL	-9.375	2.8556E-01	-4.6813E-05	275	1000	71.60	65.61	229.37
227	C3H9NO	3-AMINO-1-PROPANOL	-7.770	2.7106E-01	-3.9489E-05	284	1000	69.54	66.03	223.80
228	C3H9NO	METHYLETHANOLAMINE	-5.232	2.9002E-01	-4.0148E-05	269	1000	77.67	69.88	244.64
229	C3H9O4P	TRIMETHYL PHOSPHATE	-----	--------	--------	----	----	-----	-----	-----
230	C3H10N2	1,2-PROPANEDIAMINE	-4.513	2.5972E-01	-3.9437E-05	237	1000	69.42	54.83	215.77
231	C3H10Si	TRIMETHYL SILANE	-5.691	2.9660E-01	-5.3047E-05	137	1000	78.02	33.95	237.86
232	C4Cl4S	TETRACHLOROTHIOPHENE	-5.773	2.9116E-01	-1.2711E-05	302	1000	-----	81.00	272.68
233	C4Cl6	HEXACHLORO-1,3-BUTADIENE	-19.865	3.3082E-01	-4.0579E-05	488	998	-----	131.91	269.88
234	C4F8	OCTAFLUORO-2-BUTENE	-4.269	4.4109E-01	-1.0018E-04	270	1000	118.34	107.52	336.64

457

Table 21-1 VISCOSITY OF GAS - ORGANIC COMPOUNDS (continued)

NO	FORMULA	NAME	$n_{gas} = A + B T + C T^2$					$(n_{gas}$ - micropoise, T - K$)$		
			A	B	C	TMIN	TMAX	n_{gas} @25 C	n_{gas} @TMIN	n_{gas} @TMAX
235	C4F8	OCTAFLUOROCYCLOBUTANE	-1.797	4.2747E-01	-9.2773E-05	267	1000	117.41	105.72	332.90
236	C4F10	DECAFLUOROBUTANE	-5.210	4.5542E-01	-1.0643E-04	271	1000	121.11	110.39	343.78
237	C4H2	BUTADIYNE(BIACETYLENE)	-11.043	3.3678E-01	-7.0731E-05	275	1000	83.08	76.22	255.01
238	C4H2O3	MALEIC ANHYDRIDE	-11.219	2.9181E-01	-1.0579E-05	326	1000	-----	82.79	270.01
239	C4H4	VINYLACETYLENE	-----	--------	--------	----	----	-----	-----	-----
240	C4H4N2	SUCCINONITRILE	-11.239	2.2724E-01	-1.6566E-05	540	990	-----	106.64	197.49
241	C4H4O	FURAN	-13.696	3.5655E-01	-6.6378E-05	305	995	-----	88.88	275.36
242	C4H4O2	DIKETENE	-10.684	3.3267E-01	-5.5054E-05	267	1000	83.61	74.21	266.93
243	C4H8O3	SUCCINIC ANHYDRIDE	-10.125	2.7767E-01	-7.1078E-06	393	1000	-----	97.90	260.44
244	C4H4O4	FUMARIC ACID	-11.333	2.6574E-01	-3.6422E-05	560	1000	-----	126.06	217.99
245	C4H4O4	MALEIC ACID	-5.808	2.5158E-01	-2.7666E-05	403	1000	-----	91.09	218.11
246	C4H4S	THIOPHENE	-23.815	3.6576E-01	-4.9330E-05	293	997	80.85	79.12	291.81
247	C4H5Cl	CHLOROPRENE	-6.106	3.2925E-01	-5.7652E-05	333	1000	-----	97.14	265.49
248	C4H5N	trans-CROTONITRILE	-4.425	2.4447E-01	-3.8374E-05	222	1000	65.05	47.96	201.67
249	C4H5N	cis-CROTONITRILE	-3.997	2.4281E-01	-3.8310E-05	201	1000	64.99	43.26	200.50
250	C4H5N	METHACRYLONITRILE	-14.870	2.7470E-01	-3.9631E-05	237	1000	63.51	48.01	220.20
251	C4H5N	PYRROLE	-14.353	2.9421E-01	-3.9026E-05	403	993	-----	97.88	239.32
252	C4H5N	VINYLACETONITRILE	-3.406	2.4325E-01	-3.5875E-05	186	1000	65.93	40.60	203.97
253	C4H5NO2	METHYL CYANOACETATE	-3.952	2.3617E-01	-2.7783E-05	478	1000	-----	102.59	204.44
254	C4H6	CYCLOBUTENE	-10.008	3.3653E-01	-7.4332E-05	275	1000	83.72	76.92	252.19
255	C4H6	1,2-BUTADIENE	0.879	2.7938E-01	-6.6959E-05	284	994	78.22	74.82	212.42
256	C4H6	1,3-BUTADIENE	10.256	2.6833E-01	-4.1148E-05	250	650	86.60	74.77	167.28
257	C4H6	DIMETHYLACETYLENE	-6.293	2.9596E-01	-7.0767E-05	300	990	-----	76.13	217.35
258	C4H6	ETHYLACETYLENE	-11.051	3.0621E-01	-6.9582E-05	213	573	74.06	51.01	141.56
259	C4H6Cl2	1,3-DICHLORO-trans-2-BUTENE	-5.217	2.9435E-01	-4.1350E-05	276	1000	78.87	72.87	247.78
260	C4H6Cl2	1,4-DICHLORO-cis-2-BUTENE	-3.875	2.8158E-01	-3.5458E-05	225	1000	76.93	57.69	242.25
261	C4H6Cl2	1,4-DICHLORO-trans-2-BUTENE	-9.420	3.3823E-01	-5.0131E-05	429	1000	-----	126.45	278.68
262	C4H6Cl2	3,4-DICHLORO-1-BUTENE	-4.456	2.9664E-01	-4.4188E-05	212	1000	80.06	56.45	248.00
263	C4H6O	trans-CROTONALDEHYDE	-6.563	2.8401E-01	-4.7535E-05	377	1000	-----	93.75	229.91
264	C4H6O	2,5-DIHYDROFURAN	-5.341	3.2501E-01	-5.3301E-05	273	1000	86.82	79.41	266.36
265	C4H6O	DIVINYL ETHER	-5.585	3.2050E-01	-6.5196E-05	301	1000	-----	84.98	249.72
266	C4H6O	METHACROLEIN	-14.103	3.1016E-01	-5.0202E-05	341	991	-----	85.82	243.96
267	C4H6O2	2-BUTYNE-1,4-DIOL	-9.931	2.7828E-01	-4.0329E-05	511	1000	-----	121.74	228.02
268	C4H6O2	gamma-BUTYROLACTONE	-15.630	2.6618E-01	-2.4638E-05	477	1000	-----	105.73	225.91
269	C4H6O2	cis-CROTONIC ACID	-4.245	2.6784E-01	-3.4297E-05	289	1000	72.56	70.30	229.30
270	C4H6O2	trans-CROTONIC ACID	-5.445	2.7036E-01	-3.4454E-05	345	1000	-----	83.73	230.46
271	C4H6O2	METHACRYLIC ACID	-6.079	2.8607E-01	-4.1281E-05	288	1000	75.54	72.89	238.71
272	C4H6O2	METHYL ACRYLATE	-11.173	3.1448E-01	-6.8774E-05	200	1000	76.48	48.97	234.53
273	C4H6O2	VINYL ACETATE	-7.462	3.0466E-01	-5.7544E-05	346	1000	-----	91.06	239.65
274	C4H6O3	ACETIC ANHYDRIDE	-1.485	2.8869E-01	-2.3391E-05	295	993	82.51	81.64	262.12
275	C4H6O4	SUCCINIC ACID	-8.458	2.5493E-01	-3.7515E-05	461	1000	-----	101.09	208.96
276	C4H6O5	DIGLYCOLIC ACID	-7.036	2.4872E-01	-3.2693E-05	421	1000	-----	91.88	208.99
277	C4H6O5	MALIC ACID	-3.713	2.3442E-01	-2.2353E-05	403	1000	-----	87.13	208.35
278	C4H6O6	TARTARIC ACID	-3.727	2.2814E-01	-1.5516E-05	479	1000	-----	101.99	208.90
279	C4H7N	n-BUTYRONITRILE	-2.785	2.3017E-01	-3.3051E-05	161	1000	62.90	33.42	194.33
280	C4H7N	ISOBUTYRONITRILE	-5.884	2.4813E-01	-4.2192E-05	377	977	-----	81.66	196.27
281	C4H7NO	ACETONE CYANOHYDRIN	-4.498	2.4466E-01	-3.2844E-05	252	1000	65.53	55.07	207.32
282	C4H7NO	2-METHACRYLAMIDE	-5.578	2.3845E-01	-2.8600E-05	384	1000	-----	81.77	204.27
283	C4H7NO	3-METHOXYPROPIONITRILE	-3.407	2.3563E-01	-3.0376E-05	210	1000	64.15	44.74	201.85
284	C4H7NO	2-PYRROLIDONE	-11.939	2.5385E-01	-1.8021E-05	518	998	-----	114.72	223.45
285	C4H8	1-BUTENE	-9.143	3.1562E-01	-8.4164E-05	175	800	77.48	43.51	189.49
286	C4H8	cis-2-BUTENE	-9.923	3.2622E-01	-1.0258E-04	277	450	78.22	72.57	116.10
287	C4H8	trans-2-BUTENE	-9.923	3.2622E-01	-1.0258E-04	277	450	78.22	72.57	116.10
288	C4H8	CYCLOBUTANE	0.401	3.0851E-01	-7.9599E-05	286	1000	85.31	82.12	229.31
289	C4H8	ISOBUTENE	-8.630	3.2415E-01	-7.1963E-05	175	1000	81.62	45.89	243.56
290	C4H8Br2	1,2-DIBROMOBUTANE	-10.590	3.2259E-01	-4.1691E-05	275	1000	81.88	74.97	270.31
291	C4H8Br2	2,3-DIBROMOBUTANE	-10.890	3.3031E-01	-4.3071E-05	275	1000	83.76	76.69	276.35
292	C4H8Cl2	1,4-DICHLOROBUTANE	-7.575	2.8464E-01	-4.3227E-05	427	1000	-----	106.08	233.84
293	C4H8I2	1,2-DIIODOBUTANE	-9.446	3.4113E-01	-3.2515E-05	275	1000	89.37	81.91	299.17
294	C4H8O	n-BUTYRALDEHYDE	7.694	2.0543E-01	1.0683E-05	298	523	69.89	69.86	118.06
295	C4H8O	ISOBUTYRALDEHYDE	-14.239	3.1427E-01	-5.5302E-05	300	1000	-----	75.06	244.73
296	C4H8O	1,2-EPOXYBUTANE	-6.710	3.5563E-01	-6.1245E-05	337	1000	-----	106.18	287.68
297	C4H8O	METHYL ETHYL KETONE	3.010	2.2899E-01	7.8953E-06	273	573	71.99	66.11	136.81
298	C4H8O	ETHYL VINYL ETHER	-3.364	2.9504E-01	-5.5005E-05	309	1000	-----	82.55	236.67
299	C4H8O	TETRAHYDROFURAN	-14.222	3.3961E-01	-5.4608E-05	338	998	-----	94.33	270.32
300	C4H8O2	cis-2-BUTENE-1,4-DIOL	-8.512	2.7407E-01	-3.9067E-05	508	1000	-----	120.63	226.49
301	C4H8O2	trans-2-BUTENE-1,4-DIOL	-7.730	2.7278E-01	-3.8322E-05	510	1000	-----	121.42	226.73
302	C4H8O2	ISOBUTYRIC ACID	-4.438	2.6891E-01	-3.8739E-05	227	1000	72.29	54.61	225.73
303	C4H8O2	n-BUTYRIC ACID	-5.781	2.6159E-01	-3.4903E-05	268	1000	69.11	61.82	220.91
304	C4H8O2	1,4-DIOXANE	-16.701	3.4988E-01	-5.3736E-05	374	994	-----	106.64	277.99
305	C4H8O2	ETHYL ACETATE	-9.259	3.0725E-01	-7.1069E-05	190	1000	76.03	46.55	226.92
306	C4H8O2	METHYL PROPIONATE	-6.740	2.8777E-01	-5.2551E-05	353	1000	-----	88.29	228.48
307	C4H8O2	n-PROPYL FORMATE	-15.004	3.2541E-01	-5.5609E-05	350	1000	-----	92.08	254.80
308	C4H8O2S	SULFOLANE	-3.749	2.2530E-01	6.1191E-06	301	1000	-----	64.62	227.67
309	C4H8S	TETRAHYDROTHIOPHENE	-5.612	2.9145E-01	-4.0805E-05	394	1000	-----	102.88	245.03
310	C4H9Br	1-BROMOBUTANE	-6.940	3.2469E-01	-5.4949E-05	375	1000	-----	107.09	262.80
311	C4H9Br	2-BROMOBUTANE	-3.856	3.2515E-01	-4.6707E-05	161	1000	88.94	47.28	274.59
312	C4H9Cl	n-BUTYL CHLORIDE	-5.847	2.8831E-01	-5.0974E-05	352	1000	-----	89.32	231.49

458

Table 21-1 VISCOSITY OF GAS - ORGANIC COMPOUNDS (continued)

| NO | FORMULA | NAME | $n_{gas} = A + B T + C T^2$ | | | | | $(n_{gas}$ - micropoise, T - K$)$ | | |
			A	B	C	TMIN	TMAX	n_{gas} @25 C	n_{gas} @TMIN	n_{gas} @TMAX
313	C4H9Cl	sec-BUTYL CHLORIDE	-7.451	3.0318E-01	-5.8353E-05	341	1000	-----	89.15	237.38
314	C4H9Cl	tert-BUTYL CHLORIDE	-5.254	3.0857E-01	-5.5912E-05	324	1000	-----	88.85	247.40
315	C4H9I	2-IODO-2-METHYLPROPANE	-13.505	3.7797E-01	-6.1936E-05	275	1000	93.68	85.75	302.53
316	C4H9N	PYRROLIDINE	-14.454	3.1056E-01	-4.9770E-05	360	1000	-----	90.90	246.34
317	C4H9NO	N,N-DIMETHYLACETAMIDE	-14.937	2.5775E-01	-3.1476E-05	439	1000	-----	92.15	211.34
318	C4H9NO	MORPHOLINE	-15.402	3.0711E-01	-4.4586E-05	401	1000	-----	100.58	247.12
319	C4H9NO2	1-NITROBUTANE	-9.157	2.6430E-01	-3.8758E-05	275	1000	66.20	60.59	216.39
320	C4H9NO2	2-NITROBUTANE	-9.767	2.7911E-01	-4.2147E-05	275	1000	69.70	63.80	227.20
321	C4H10	n-BUTANE	-4.946	2.9001E-01	-6.9665E-05	150	1200	75.33	36.99	242.75
322	C4H10	ISOBUTANE	-4.731	2.9131E-01	-8.0995E-05	150	1000	74.92	37.14	205.58
323	C4H10N2	PIPERAZINE	-6.782	3.0412E-01	-4.4349E-05	379	1000	-----	102.11	252.99
324	C4H10O	n-BUTANOL	-11.144	2.8790E-01	-5.6275E-05	391	1000	-----	92.82	220.48
325	C4H10O	sec-BUTANOL	-14.992	3.1418E-01	-5.5185E-05	373	993	-----	94.52	242.57
326	C4H10O	tert-BUTANOL	-10.039	2.8178E-01	-7.7623E-05	303	473	-----	68.21	105.88
327	C4H10O	DIETHYL ETHER	-7.932	3.0235E-01	-7.3858E-05	200	1000	75.65	49.58	220.56
328	C4H10O	METHYL-PROPYL-ETHER	-9.772	2.9933E-01	-6.3063E-05	275	1000	73.87	67.77	226.50
329	C4H10O	METHYL ISOPROPYL ETHER	-3.016	2.9496E-01	-6.0300E-05	128	1000	79.57	33.75	231.64
330	C4H10O	ISOBUTANOL	-11.412	2.7821E-01	-2.9510E-06	175	1100	71.27	37.18	291.05
331	C4H10O2	1,3-BUTANEDIOL	-7.553	2.7198E-01	-4.0704E-05	480	990	-----	113.62	221.81
332	C4H10O2	1,4-BUTANEDIOL	-16.116	2.6953E-01	-3.0297E-05	500	1000	-----	111.07	223.12
333	C4H10O2	2,3-BUTANEDIOL	-5.470	2.7493E-01	-4.0547E-05	281	1000	72.90	68.58	228.91
334	C4H10O2	t-BUTYL HYDROPEROXIDE	-5.789	2.9373E-01	-4.8826E-05	277	1000	77.45	71.83	239.11
335	C4H10O2	1,2-DIMETHOXYETHANE	8.641	2.4096E-01	-1.7203E-05	357	1000	-----	92.47	232.40
336	C4H10O2	2-ETHOXYETHANOL	-6.483	2.7545E-01	-4.6641E-05	408	1000	-----	98.14	222.33
337	C4H10O3	DIETHYLENE GLYCOL	-3.863	2.5115E-01	-2.6000E-05	263	1000	68.71	60.39	221.29
338	C4H10O4S	DIETHYL SULFATE	-----	--------	--------	----	----	-----	-----	-----
339	C4H10S	n-BUTYL MERCAPTAN	-13.994	3.0281E-01	-4.7931E-05	372	1000	-----	92.02	240.89
340	C4H10S	ISOBUTYL MERCAPTAN	-3.448	2.7288E-01	-2.5320E-05	128	1000	75.66	31.07	244.11
341	C4H10S	sec-BUTYL MERCAPTAN	-3.678	2.7493E-01	-2.6200E-05	133	1000	75.95	32.42	245.01
342	C4H10S	tert-BUTYL MERCAPTAN	-10.888	3.0565E-01	-4.7129E-05	274	1000	76.05	69.32	247.63
343	C4H10S	DIETHYL SULFIDE	-13.578	3.0021E-01	-4.9223E-05	365	1000	-----	89.44	237.41
344	C4H10S	ISOPROPYL-METHYL-SULFIDE	-10.210	2.8441E-01	-5.1363E-05	275	1000	70.02	64.12	222.84
345	C4H10S	METHYL-PROPYL-SULFIDE	-9.882	2.7483E-01	-4.8169E-05	275	1000	67.78	62.05	216.78
346	C4H10S2	DIETHYL DISULFIDE	-3.418	2.6320E-01	-1.6484E-05	172	1000	73.59	41.36	243.30
347	C4H11N	n-BUTYLAMINE	-12.439	2.8092E-01	-4.8899E-05	351	991	-----	80.14	217.93
348	C4H11N	ISOBUTYLAMINE	-3.639	2.6244E-01	-4.5505E-05	189	1000	70.56	44.34	213.30
349	C4H11N	sec-BUTYLAMINE	-4.850	2.6464E-01	-4.7781E-05	336	1000	-----	78.67	212.01
350	C4H11N	tert-BUTYLAMINE	-5.751	2.8198E-01	-5.6921E-05	318	1000	-----	78.16	219.31
351	C4H11N	DIETHYLAMINE	-6.033	2.8574E-01	-5.6208E-05	329	1000	-----	81.89	223.50
352	C4H11NO	DIMETHYLETHANOLAMINE	-6.392	2.7958E-01	-3.0439E-05	214	1000	74.26	52.04	242.75
353	C4H11NO2	DIETHANOLAMINE	-----	--------	--------	----	----	-----	-----	-----
354	C4H12N2O	N-AMINOETHYL ETHANOLAMINE	-4.750	2.4628E-01	-3.2317E-05	293	1000	65.81	64.64	209.21
355	C4H11NO2	2-AMINOETHOXYETHANOL	-3.969	2.4086E-01	-2.7574E-05	273	1000	65.39	59.73	209.32
356	C4H12Si	TETRAMETHYLSILANE	-6.722	2.9090E-01	-5.0708E-05	174	1000	75.50	42.36	233.47
357	C4H13N3	DIETHYLENE TRIAMINE	-3.995	2.3224E-01	-2.8336E-05	234	1000	62.73	48.80	199.91
358	C5Cl6	HEXACHLOROCYCLOPENTADIENE	-14.581	3.0945E-01	-3.0866E-05	512	992	-----	135.77	262.02
359	C5H4O2	FURFURAL	-17.273	3.1360E-01	-3.4489E-05	435	995	-----	112.62	260.61
360	C5H5N	PYRIDINE	-5.739	2.7135E-01	-1.7202E-05	369	998	-----	92.05	247.93
361	C5H6	CYCLOPENTADIENE	-7.155	3.1587E-01	-7.4264E-05	300	1000	-----	80.92	234.45
362	C5H6	2-METHYL-1-BUTENE-3-YNE	-4.004	2.8654E-01	-7.2314E-05	160	TMAX	75.00	39.99	210.22
363	C5H6	1-PENTENE-3-YNE	-3.901	2.7034E-01	-6.9140E-05	150	1000	70.55	35.09	197.30
364	C5H6	1-PENTENE-4-YNE	-4.849	2.8226E-01	-6.8930E-05	150	1000	73.18	35.94	208.48
365	C5H6N2	GLUTARONITRILE	-2.613	1.9111E-01	-1.7676E-05	244	1000	52.80	42.97	170.82
366	C5H6O2	FURFURYL ALCOHOL	-11.203	3.0249E-01	-5.1315E-05	259	1000	74.42	63.70	239.97
367	C5H6O3	GLUTARIC ANHYDRIDE	-3.518	2.4149E-01	-1.9830E-05	328	1000	-----	73.56	218.14
368	C5H6O4	CITRACONIC ACID	-3.432	2.3255E-01	-1.9261E-05	356	1000	-----	76.91	209.86
369	C5H6O4	ITACONIC ACID	-4.935	2.3961E-01	-2.3471E-05	439	1000	-----	95.73	211.20
370	C5H6S	2-METHYLTHIOPHENE	-10.527	2.9935E-01	-4.5922E-05	275	1000	74.64	68.32	242.90
371	C5H6S	3-METHYLTHIOPHENE	-10.402	2.9724E-01	-4.4884E-05	275	1000	74.23	67.94	241.95
372	C5H7N	N-METHYLPYRROLE	-5.230	2.6377E-01	-2.2975E-05	217	1000	71.37	50.93	235.56
373	C5H7NO2	ETHYL CYANOACETATE	-3.806	2.2299E-01	-2.7623E-05	251	1000	60.22	50.42	191.56
374	C5H8	CYCLOPENTENE	8.952	2.5246E-01	-6.3618E-05	317	1000	-----	82.59	197.79
375	C5H8	ISOPRENE	-5.283	2.9865E-01	-5.7039E-05	307	1000	-----	81.03	236.33
376	C5H8	3-METHYL-1,2-BUTADIENE	-4.745	2.6940E-01	-6.3368E-05	160	1000	69.94	36.74	201.29
377	C5H8	2-METHYL-1,3-BUTADIENE	-9.938	2.9943E-01	-6.2307E-05	275	1000	73.80	67.69	227.18
378	C5H8	1,2-PENTADIENE	-1.743	2.5280E-01	-6.9924E-05	136	1000	67.41	31.34	181.13
379	C5H8	cis-1,3-PENTADIENE	5.050	2.2864E-01	-5.4984E-05	317	1000	-----	72.00	178.71
380	C5H8	trans-1,3-PENTADIENE	4.797	2.2925E-01	-5.6092E-05	315	1000	-----	71.44	177.95
381	C5H8	1,4-PENTADIENE	-7.594	2.7944E-01	-6.2839E-05	299	1000	-----	70.34	209.01
382	C5H8	2,3-PENTADIENE	-4.490	2.6290E-01	-6.2596E-05	148	1000	68.33	33.05	195.81
383	C5H8	1-PENTYNE	-6.582	2.9264E-01	-5.9751E-05	167	1000	75.36	40.62	226.31
384	C5H8	2-PENTYNE	-9.806	2.7813E-01	-5.3664E-05	275	1000	68.35	62.62	214.66
385	C5H8	3-METHYL-1-BUTYNE	-7.410	3.0446E-01	-5.4297E-05	183	1000	78.54	46.49	242.75
386	C5H8	SPIROPENTANE	-10.736	3.1346E-01	-6.3274E-05	275	1000	77.10	70.68	239.45
387	C5H8N4O12	PENTAERYTHRITOL TETRANITRATE	-12.735	2.8265E-01	-3.2378E-05	414	1000	-----	98.73	237.54
388	C5H8O	CYCLOPENTANONE	-14.060	2.9812E-01	-3.5355E-05	404	1000	-----	100.61	248.71
389	C5H8O	METHYL ISOPROPENYL KETONE	-4.354	2.6337E-01	-4.1839E-05	220	1000	70.45	51.56	217.18
390	C5H8O2	ACETYLACETONE	-4.447	2.3407E-01	-3.4740E-05	414	1000	-----	86.50	194.88

459

Table 21-1 VISCOSITY OF GAS - ORGANIC COMPOUNDS (continued)

			$n_{gas} = A + B\,T + C\,T^2$						(n_{gas} - micropoise, T - K)		
NO	FORMULA	NAME	A	B	C	TMIN	TMAX	n_{gas} @25 C	n_{gas} @TMIN	n_{gas} @TMAX
391	C5H8O2	ALLYL ACETATE	-3.659	2.7061E-01	-4.7381E-05	377	1000	------	91.63	219.57
392	C5H8O2	ETHYL ACRYLATE	-12.000	2.9702E-01	-6.1136E-05	293	950	71.12	69.78	214.99
393	C5H8O2	METHYL METHACRYLATE	-14.458	3.1073E-01	-5.0853E-05	373	1000	------	94.37	245.42
394	C5H8O2	VINYL PROPIONATE	-6.404	2.7590E-01	-4.8704E-05	368	998	------	88.53	220.44
395	C5H8O3	2-HYDROXYETHYL ACRYLATE	-3.065	2.5008E-01	-2.9097E-05	213	1000	68.91	48.88	217.92
396	C5H8O3	LEVULINIC ACID	-3.824	2.2768E-01	-2.4762E-05	308	1000	------	63.95	199.09
397	C5H8O3	METHYL ACETOACETATE	-2.907	2.3823E-01	-2.8686E-05	193	1000	65.57	42.00	206.64
398	C5H8O4	GLUTARIC ACID	-4.719	2.2743E-01	-2.3918E-05	371	1000	------	76.37	198.79
399	C5H9N	VALERONITRILE	-4.681	2.2436E-01	-3.4736E-05	414	1000	------	82.25	184.94
400	C5H9NO	n-BUTYL ISOCYANATE	-3.875	2.4837E-01	-3.9247E-05	193	1000	66.69	42.60	205.25
401	C5H9NO	N-METHYL-2-PYRROLIDONE	-14.571	2.5695E-01	-2.4206E-05	475	995	------	101.92	217.03
402	C5H9NO4	L-GLUTAMIC ACID	-2.384	2.0827E-01	-1.3258E-05	497	1000	------	97.85	192.63
403	C5H10	CYCLOPENTANE	-2.321	3.0131E-01	-1.2230E-04	250	550	76.64	65.36	126.40
404	C5H10	2-METHYL-1-BUTENE	2.683	2.4490E-01	-6.1313E-05	304	994	------	71.47	185.53
405	C5H10	2-METHYL-2-BUTENE	2.271	2.4524E-01	-6.1866E-05	312	992	------	72.76	184.67
406	C5H10	3-METHYL-1-BUTENE	-5.811	2.8881E-01	-6.8810E-05	293	993	74.18	72.90	213.13
407	C5H10	1-PENTENE	-2.903	2.7060E-01	-6.8331E-05	303	993	------	72.82	198.43
408	C5H10	cis-2-PENTENE	-4.967	2.7418E-01	-6.2827E-05	300	1000	------	71.63	206.39
409	C5H10	trans-2-PENTENE	-4.873	2.7393E-01	-6.2670E-05	300	1000	------	71.67	206.39
410	C5H10Br2	2,3-DIBROMO-2-METHYLBUTANE	-10.011	3.1037E-01	-3.8699E-05	275	1000	79.09	72.41	261.66
411	C5H10Cl2	1,5-DICHLOROPENTANE	-7.383	2.6287E-01	-3.8178E-05	453	1000	------	103.86	217.31
412	C5H10O	METHYL ISOPROPYL KETONE	-5.547	2.5571E-01	-4.3205E-05	368	1000	------	82.70	206.96
413	C5H10O	2-PENTANONE	-5.868	2.4975E-01	-4.2230E-05	375	1000	------	81.85	201.65
414	C5H10O	DIETHYL KETONE	-3.593	2.4308E-01	-3.7613E-05	234	1000	65.54	51.23	201.87
415	C5H10O	VALERALDEHYDE	-5.084	2.5099E-01	-4.1966E-05	376	1000	------	83.35	203.94
416	C5H10O2	n-BUTYL FORMATE	-12.396	3.0859E-01	-5.0283E-05	300	1250	------	75.66	294.77
417	C5H10O2	ETHYL PROPIONATE	-13.338	3.6573E-01	-5.6002E-05	293	1000	90.73	89.01	296.39
418	C5H10O2	ISOBUTYL FORMATE	-8.430	2.9441E-01	-5.4618E-05	371	1000	------	93.31	231.39
419	C5H10O2	ISOPROPYL ACETATE	-6.725	2.7363E-01	-5.0297E-05	362	1000	------	85.74	216.61
420	C5H10O2	n-PROPYL ACETATE	-4.867	2.5928E-01	-4.2456E-05	375	1000	------	86.39	211.96
421	C5H10O2	METHYL n-BUTYRATE	-6.837	2.6711E-01	-4.8664E-05	376	1000	------	86.72	211.61
422	C5H10O2	2-METHYLBUTYRIC ACID	-6.842	2.5411E-01	-3.6454E-05	450	1000	------	100.13	210.81
423	C5H10O2	ISOVALERIC ACID	-4.687	2.5089E-01	-3.6366E-05	244	1000	66.88	54.36	209.84
424	C5H10O2	VALERIC ACID	-4.273	2.4277E-01	-3.3016E-05	239	1000	65.17	51.86	205.48
425	C5H10O2	TETRAHYDROFURFURYL ALCOHOL	-15.578	3.1183E-01	-4.1258E-05	451	1000	------	116.67	254.99
426	C5H10O2S	3-METHYL SULFOLANE	-12.958	2.4827E-01	-1.2033E-05	549	1000	------	119.72	223.28
427	C5H10O3	DIETHYL CARBONATE	-6.754	2.5352E-01	-4.3576E-05	400	1000	------	87.68	203.19
428	C5H10O3	ETHYL LACTATE	-4.612	2.5496E-01	-3.9865E-05	247	1000	67.86	55.93	210.48
429	C5H10S	THIACYCLOHEXANE	-9.150	2.7762E-01	-3.6178E-05	275	1000	70.41	64.46	232.29
430	C5H10S	CYCLOPENTANETHIOL	-9.447	2.7436E-01	-3.9565E-05	275	1000	68.84	63.01	225.35
431	C5H11Br	1-BROMOPENTANE	-11.008	3.0615E-01	-5.3415E-05	275	1000	75.52	69.14	241.73
432	C5H11Cl	1-CHLOROPENTANE	-7.208	2.6970E-01	-4.7760E-05	382	1000	------	88.85	214.73
433	C5H11Cl	1-CHLORO-3-METHYLBUTANE	-9.912	2.7573E-01	-4.8836E-05	275	1000	67.96	62.22	216.98
434	C5H11Cl	2-CHLORO-2-METHYLBUTANE	-10.282	2.8660E-01	-5.2014E-05	275	1000	70.54	64.60	224.30
435	C5H11N	N-METHYLPYRROLIDINE	------	--------	---------	----	----	----	------	------
436	C5H11N	PIPERIDINE	-14.147	2.8977E-01	-4.5274E-05	380	1000	------	89.43	230.35
437	C5H11NO	tert-BUTYLFORMAMIDE	-6.066	2.1951E-01	-1.2657E-05	289	1000	58.26	56.32	200.79
438	C5H12	ISOPENTANE	-0.842	2.6759E-01	-6.8487E-05	301	1000	------	73.50	198.26
439	C5H12	NEOPENTANE	8.041	2.3255E-01	-4.7189E-05	300	600	------	73.56	130.58
440	C5H12	n-PENTANE	-3.202	2.6746E-01	-6.6178E-05	303	900	------	71.76	183.91
441	C5H12O	2,2-DIMETHYL-1-PROPANOL	-6.340	2.7243E-01	-4.7586E-05	386	1000	------	91.73	218.50
442	C5H12O	tert-PENTYL-ALCOHOL	-8.741	2.4363E-01	-4.4212E-05	275	1000	59.97	54.91	190.68
443	C5H12O	2-METHYL-1-BUTANOL	-5.084	2.5865E-01	-4.2658E-05	402	1000	------	92.00	210.91
444	C5H12O	2-METHYL-2-BUTANOL	-12.171	2.9012E-01	-4.5740E-05	375	1000	------	90.19	232.21
445	C5H12O	3-METHYL-1-BUTANOL	-13.735	2.8560E-01	-4.4938E-05	404	1000	------	94.31	226.93
446	C5H12O	3-METHYL-2-BUTANOL	-6.351	2.6909E-01	-4.4695E-05	385	1000	------	90.62	218.04
447	C5H12O	1-PENTANOL	-4.955	2.4999E-01	-3.8436E-05	411	1000	------	91.30	206.60
448	C5H12O	2-PENTANOL	-6.211	2.6201E-01	-4.5780E-05	392	1000	------	89.46	210.02
449	C5H12O	3-PENTANOL	-6.116	2.6194E-01	-4.6092E-05	388	1000	------	88.58	209.73
450	C5H12O	METHYL sec-BUTYL ETHER	-4.222	2.6931E-01	-5.0079E-05	150	1000	71.62	35.05	215.01
451	C5H12O	METHYL tert-BUTYL ETHER	-12.469	3.0744E-01	-5.7113E-05	300	1000	------	74.62	237.86
452	C5H12O	METHYL ISOBUTYL ETHER	-3.675	2.6780E-01	-4.9221E-05	150	1000	71.79	35.39	214.90
453	C5H12O	ETHYL PROPYL ETHER	-6.463	2.6908E-01	-5.4110E-05	337	1000	------	78.07	208.51
454	C5H12O2	ETHYLENE GLYCOL MONOPROPYL ETHER	-3.152	2.4536E-01	-3.5554E-05	183	1000	66.84	40.56	206.65
455	C5H12O2	NEOPENTYL GLYCOL	-6.703	2.5965E-01	-3.7941E-05	483	1000	------	109.86	215.01
456	C5H12O2	1,5-PENTANEDIOL	-17.120	2.8007E-01	-3.8354E-05	512	992	------	116.22	222.97
457	C5H12O3	2-(2-METHOXYETHOXY)ETHANOL	-7.450	2.5546E-01	-3.9256E-05	467	1000	------	103.29	208.75
458	C5H12O4	PENTAERYTHRITOL	-9.957	2.4582E-01	-3.2316E-05	631	991	------	132.29	201.91
459	C5H12S	n-PENTYL MERCAPTAN	-5.032	2.5938E-01	-2.3821E-05	197	1000	70.19	45.14	230.53
460	C5H12S	BUTYL-METHYL-SULFIDE	-9.007	2.5249E-01	-4.1009E-05	275	1000	62.63	57.33	202.47
461	C5H12S	ETHYL-PROPYL-SULFIDE	-9.120	2.5478E-01	-4.2212E-05	275	1000	63.09	57.75	203.45
462	C5H12S	2-METHYL-2-BUTANETHIOL	-9.791	2.7233E-01	-4.7362E-05	275	1000	67.19	61.52	215.18
463	C5H13N	n-PENTYLAMINE	-12.196	2.6955E-01	-4.4090E-05	378	998	------	83.39	212.90
464	C5H13NO2	METHYL DIETHANOLAMINE	-5.098	2.3734E-01	-1.5569E-05	252	1000	64.28	53.72	216.67
465	C6Cl6	HEXACHLOROBENZENE	-15.496	2.9854E-01	-2.5388E-05	502	1000	------	127.97	257.66
466	C6F6	HEXAFLUOROBENZENE	-6.824	3.2624E-01	-5.9783E-05	353	1000	------	100.89	259.63
467	C6H3ClN2O4	1-CHLORO-2,4-DINITROBENZENE	-4.793	2.4534E-01	-8.3861E-06	327	1000	------	74.54	232.16
468	C3H3Cl2NO2	1,2-DICHLORO-4-NITROBENZENE	-4.861	2.6412E-01	-9.6561E-06	316	1000	------	77.64	249.60

Table 21-1 VISCOSITY OF GAS - ORGANIC COMPOUNDS (continued)

NO	FORMULA	NAME	A	B	C	TMIN	TMAX	n_gas @25 C	n_gas @TMIN	n_gas @TMAX
				$n_{gas} = A + B\,T + C\,T^2$			(n_gas - micropoise, T - K)			
469	C6H3Cl3	1,2,4-TRICHLOROBENZENE	-2.543	2.4820E-01	-2.6755E-05	486	1000	-----	111.76	218.90
470	C6H3N3O6	1,3,5-TRINITROBENZENE	1.596	1.9419E-01	9.3174E-06	398	748	-----	80.36	152.06
471	C6H4Br2	m-DIBROMOBENZENE	-4.001	2.9804E-01	-2.7722E-05	266	1000	82.40	73.32	266.32
472	C6H4ClNO2	m-CHLORONITROBENZENE	-12.534	2.6434E-01	-2.4236E-05	509	1000	-----	115.74	227.57
473	C6H4ClNO2	o-CHLORONITROBENZENE	-15.643	2.6244E-01	-2.5439E-05	519	1000	-----	113.71	221.36
474	C6H4ClNO2	p-CHLORONITROBENZENE	-10.992	2.6119E-01	-2.1184E-05	515	1000	-----	117.90	229.01
475	C6H4Cl2	m-DICHLOROBENZENE	-17.003	3.1334E-01	-3.9564E-05	248	1000	72.90	58.27	256.77
476	C6H4Cl2	o-DICHLOROBENZENE	-15.796	3.0983E-01	-3.5367E-05	256	1000	73.44	61.20	258.67
477	C6H4Cl2	p-DICHLOROBENZENE	-14.728	3.1231E-01	-3.6655E-05	326	1000	-----	83.19	260.93
478	C6H4F2	m-DIFLUOROBENZENE	-11.484	3.1977E-01	-5.7476E-05	275	1000	78.75	72.11	250.81
479	C6H4F2	o-DIFLUOROBENZENE	-11.709	3.2593E-01	-5.8364E-05	275	1000	80.28	73.51	255.86
480	C6H4F2	p-DIFLUOROBENZENE	-11.455	3.1875E-01	-5.6861E-05	275	1000	78.53	71.90	250.43
481	C6H4N2O4	m-DINITROBENZENE	-6.413	2.4020E-01	-8.7331E-06	364	1000	-----	79.86	225.05
482	C6H4N2O4	o-DINITROBENZENE	-9.807	2.1855E-01	-3.3659E-06	390	1000	-----	74.92	205.38
483	C6H4N2O4	p-DINITROBENZENE	-----	--------	--------	----	----	-----	-----	-----
484	C6H5Br	BROMOBENZENE	-8.397	3.1604E-01	-4.5491E-05	429	1000	-----	118.81	262.15
485	C6H5Cl	MONOCHLOROBENZENE	-14.868	3.1615E-01	-4.2899E-05	400	1000	-----	104.73	258.38
486	C6H5ClO	m-CHLOROPHENOL	-4.494	2.5958E-01	-2.9326E-05	306	1000	-----	72.19	225.76
487	C6H5ClO	o-CHLOROPHENOL	-4.519	2.6038E-01	-3.2159E-05	282	1000	70.25	66.35	223.70
488	C6H5ClO	p-CHLOROPHENOL	-4.170	2.5757E-01	-2.6997E-05	316	1000	-----	74.53	226.40
489	C6H5Cl2N	3,4-DICHLOROANILINE	-4.909	2.4359E-01	-2.4153E-05	345	1000	-----	76.25	214.53
490	C6H5F	FLUOROBENZENE	-4.846	2.9660E-01	-4.7452E-05	358	1000	-----	95.26	244.30
491	C6H5I	IODOBENZENE	-18.952	3.6336E-01	-4.1614E-05	462	992	-----	140.04	300.55
492	C6H5NO2	NITROBENZENE	-16.569	2.9184E-01	-2.5523E-05	450	1000	-----	109.59	249.75
493	C6H6	BENZENE	-0.151	2.5706E-01	-8.9797E-06	287	628	75.69	72.89	157.74
494	C6H6ClN	m-CHLOROANILINE	-3.820	2.4716E-01	-2.4566E-05	263	1000	67.69	59.48	218.77
495	C6H6ClN	o-CHLOROANILINE	-7.777	2.5913E-01	-3.3853E-05	482	1000	-----	109.26	217.50
496	C6H6ClN	p-CHLOROANILINE	-4.532	2.4874E-01	-2.5747E-05	343	1000	-----	77.76	218.46
497	C6H6N2	cis-DICYANO-1-BUTENE	-2.402	1.8815E-01	-2.0400E-05	249	1000	51.88	43.18	165.35
498	C6H6N2	trans-DICYANO-1-BUTENE	-3.009	1.9051E-01	-2.2520E-05	260	1000	51.79	45.00	164.98
499	C6H6N2	1,4-DICYANO-2-BUTENE	-3.075	1.8905E-01	-1.8832E-05	349	1000	-----	60.61	167.14
500	C6H6N2O2	m-NITROANILINE	-----	--------	--------	----	----	-----	-----	-----
501	C6H6N2O2	o-NITROANILINE	-----	--------	--------	----	----	-----	-----	-----
502	C6H6N2O2	p-NITROANILINE	-----	--------	--------	----	----	-----	-----	-----
503	C6H6O	PHENOL	-7.185	2.7179E-01	-3.6205E-05	455	1000	-----	108.98	228.40
504	C6H6O2	1,2-BENZENEDIOL	-4.299	2.5050E-01	-2.5609E-05	378	1000	-----	86.73	220.59
505	C6H6O2	1,3-BENZENEDIOL	-5.336	2.5160E-01	-2.4473E-05	382	1000	-----	87.20	221.79
506	C6H6O2	p-HYDROQUINONE	-7.052	2.5544E-01	-2.6997E-05	558	1000	-----	127.08	221.39
507	C6H6O3	1,2,3-BENZENETRIOL	-12.427	2.5742E-01	-3.3513E-05	407	1000	-----	86.79	211.48
508	C6H6S	PHENYL MERCAPTAN	-10.964	2.8411E-01	-2.9115E-05	442	1000	-----	108.92	244.03
509	C6H7N	ANILINE	-6.918	2.5935E-01	-3.4348E-05	458	1000	-----	104.66	218.08
510	C6H7N	2-METHYLPYRIDINE	-8.492	2.7564E-01	-4.4458E-05	403	1000	-----	95.37	222.69
511	C6H7N	3-METHYLPYRIDINE	-8.101	2.7476E-01	-4.3048E-05	417	1000	-----	98.99	223.61
512	C6H7N	4-METHYLPYRIDINE	-4.851	2.6338E-01	-3.4062E-05	419	1000	-----	99.53	224.47
513	C6H8	1,3-CYCLOHEXADIENE	-6.446	2.8814E-01	-5.3550E-05	161	1000	74.70	38.56	228.14
514	C6H8	METHYLCYCLOPENTADIENE	-5.790	2.8810E-01	-6.1646E-05	150	1000	74.63	36.04	220.66
515	C6H8N2	ADIPONITRILE	6.550	1.8368E-01	7.6364E-07	523	998	-----	102.82	190.62
516	C6H8N2	METHYLGLUTARONITRILE	-2.677	1.8831E-01	-1.9003E-05	228	1000	51.78	39.27	166.63
517	C6H8N2	m-PHENYLENEDIAMINE	-3.685	2.2941E-01	-2.1502E-05	560	1000	-----	118.04	204.22
518	C6H8N2	o-PHENYLENEDIAMINE	-9.454	2.4489E-01	-3.2600E-05	525	1000	-----	110.13	202.84
519	C6H8N2	p-PHENYLENEDIAMINE	-13.435	2.5506E-01	-3.8643E-05	540	1000	-----	113.03	202.98
520	C6H8N2	PHENYLHYDRAZINE	-----	--------	--------	----	----	-----	-----	-----
521	C6H8N2O	BIS(CYANOETHYL)ETHER	-2.275	1.8450E-01	-1.5774E-05	247	1000	51.33	42.33	166.45
522	C6H8O4	DIMETHYL MALEATE	-3.447	2.3780E-01	-2.7868E-05	254	1000	64.98	55.16	206.49
523	C6H8O6	ASCORBIC ACID	-3.911	2.2810E-01	-1.1250E-05	465	1000	-----	99.72	212.94
524	C6H8O7	CITRIC ACID	-4.092	2.0164E-01	-1.2766E-05	426	1000	-----	79.49	184.78
525	C6H10	1-METHYLCYCLOPENTENE	-9.810	2.7420E-01	-5.0593E-05	275	1000	67.45	61.77	213.80
526	C6H10	3-METHYLCYCLOPENTENE	-10.114	2.8369E-01	-5.3107E-05	275	1000	69.75	63.89	220.47
527	C6H10	4-METHYLCYCLOPENTENE	-10.069	2.8122E-01	-5.1674E-05	275	1000	69.18	63.36	219.48
528	C6H10	CYCLOHEXENE	-6.359	2.7997E-01	-6.2110E-05	356	1000	-----	85.44	211.50
529	C6H10	2,3-DIMETHYL-1,3-BUTADIENE	-2.031	2.6811E-01	-6.8135E-05	197	1000	71.85	48.14	197.94
530	C6H10	1,5-HEXADIENE	-4.084	2.5831E-01	-6.1663E-05	132	1000	67.45	28.94	192.56
531	C6H10	cis,trans-2,4-HEXADIENE	-0.779	2.3135E-01	-6.1159E-05	177	1000	62.76	38.25	169.41
532	C6H10	trans,trans-2,4-HEXADIENE	-0.120	2.3122E-01	-6.0482E-05	228	1000	63.44	49.45	170.62
533	C6H10	1-HEXYNE	-4.688	2.6760E-01	-6.0693E-05	141	1000	69.70	31.84	202.22
534	C6H10	2-HEXYNE	-2.939	2.4188E-01	-6.1529E-05	184	1000	63.71	39.48	177.41
535	C6H10	3-HEXYNE	-3.402	2.4569E-01	-6.2042E-05	170	1000	64.34	36.57	180.25
536	C6H10O	CYCLOHEXANONE	-17.560	3.1862E-01	-4.3723E-05	400	1000	-----	102.89	257.34
537	C6H10O	MESITYL OXIDE	-14.841	2.7693E-01	-4.1290E-05	403	993	-----	90.06	219.44
538	C6H10O2	epsilon-CAPROLACTONE	-7.494	2.5759E-01	-3.3447E-05	488	1000	-----	110.24	216.65
539	C6H10O2	ETHYL METHACRYLATE	-13.468	2.9461E-01	-4.5224E-05	390	990	-----	94.55	233.87
540	C6H10O2	n-PROPYL ACRYLATE	-3.778	2.4715E-01	-3.7146E-05	273	1000	66.61	60.93	206.23
541	C6H10O3	ETHYLACETOACETATE	-7.124	2.3699E-01	-3.6844E-05	454	1000	-----	92.87	193.02
542	C6H10O3	PROPIONIC ANHYDRIDE	-5.800	2.2542E-01	-3.4762E-05	442	1000	-----	87.04	184.86
543	C6H10O4	ADIPIC ACID	-5.803	2.1821E-01	-2.4224E-05	426	1000	-----	82.76	188.18
544	C6H10O4	DIETHYL OXALATE	-4.232	2.2030E-01	-3.0144E-05	459	1000	-----	90.54	185.92
545	C6H10O4	ETHYLENE GLYCOL DIACETATE	-6.292	2.2434E-01	-3.1732E-05	464	1000	-----	90.97	186.32
546	C6H10O4	ETHYLIDENE DIACETATE	-4.890	2.4323E-01	-3.4780E-05	292	1000	64.54	63.17	203.56

Table 21-1 VISCOSITY OF GAS - ORGANIC COMPOUNDS (continued)

| NO | FORMULA | NAME | $n_{gas} = A + B T + C T^2$ | | | | | n_{gas} @25 C | n_{gas} @TMIN | n_{gas} @TMAX |
			A	B	C	TMIN	TMAX	(n_{gas} - micropoise, T - K)		
547	C6H11N	HEXANENITRILE	-5.831	2.1602E-01	-3.3487E-05	467	977	------	87.75	173.26
548	C6H11NO	epsilon-CAPROLACTAM	-13.339	2.2699E-01	-2.0402E-05	500	1000	------	95.06	193.25
549	C6H11NO	CYCLOHEXANONE OXIME	-8.820	2.4991E-01	-2.1506E-05	363	1000	------	79.06	219.58
550	C6H12	CYCLOHEXANE	1.190	2.4542E-01	-3.8334E-05	315	600	------	74.69	134.64
551	C6H12	2,3-DIMETHYL-1-BUTENE	-3.359	2.5778E-01	-6.1806E-05	329	1000	------	74.76	192.62
552	C6H12	2,3-DIMETHYL-2-BUTENE	-2.243	2.3736E-01	-5.1998E-05	346	1000	------	73.66	183.12
553	C6H12	3,3-DIMETHYL-1-BUTENE	-5.082	2.7805E-01	-6.6157E-05	158	1000	71.94	37.20	206.81
554	C6H12	2-ETHYL-1-BUTENE	-2.457	2.4415E-01	-5.4956E-05	338	1000	------	73.79	186.74
555	C6H12	trans-3-METHYL-2-PENTENE	-9.277	2.6339E-01	-5.0928E-05	275	1000	64.73	59.30	203.18
556	C6H12	1-HEXENE	-2.118	2.4787E-01	-6.0615E-05	337	997	------	74.53	184.76
557	C6H12	cis-2-HEXENE	-2.222	2.4415E-01	-5.8163E-05	342	1000	------	74.47	183.76
558	C6H12	trans-2-HEXENE	-1.539	2.4127E-01	-5.4824E-05	341	1000	------	74.36	184.91
559	C6H12	cis-3-HEXENE	-3.267	2.4898E-01	-6.1922E-05	135	1000	65.46	29.22	183.79
560	C6H12	trans-3-HEXENE	-3.763	2.5109E-01	-6.3728E-05	160	1000	65.43	34.78	183.60
561	C6H12	METHYLCYCLOPENTANE	-4.640	2.6892E-01	-6.0039E-05	345	1000	------	80.99	204.24
562	C6H12	2-METHYL-1-PENTENE	-3.045	2.5006E-01	-5.9757E-05	335	1000	------	74.02	187.26
563	C6H12	2-METHYL-2-PENTENE	-1.934	2.4210E-01	-5.5304E-05	340	1000	------	73.99	184.86
564	C6H12	3-METHYL-1-PENTENE	-4.167	2.6539E-01	-6.1247E-05	120	1000	69.51	26.80	199.98
565	C6H12	3-METHYL-cis-2-PENTENE	-3.616	2.5295E-01	-6.2531E-05	138	1000	66.24	30.10	186.80
566	C6H12	4-METHYL-1-PENTENE	-1.873	2.5473E-01	-5.8960E-05	327	1000	------	75.12	193.90
567	C6H12	4-METHYL-cis-2-PENTENE	-1.231	2.5082E-01	-5.7548E-05	330	1000	------	75.27	192.04
568	C6H12	4-METHYL-trans-2-PENTENE	-1.399	2.5087E-01	-5.8880E-05	332	1000	------	75.40	190.59
569	C6H12N2	TRIETHYLENEDIAMINE	------	--------	--------	----	----	------	------	------
570	C6H12O	BUTYL VINYL ETHER	3.743	2.2933E-01	-3.0055E-05	367	1000	------	83.86	203.02
571	C6H12O	CYCLOHEXANOL	-13.542	2.9086E-01	-3.9472E-05	400	1000	------	96.49	237.85
572	C6H12O	1-HEXANAL	-4.826	2.3443E-01	-3.7349E-05	401	1000	------	83.17	192.26
573	C6H12O	ETHYL ISOPROPYL KETONE	-3.558	2.3413E-01	-3.6351E-05	200	1000	63.02	41.81	194.22
574	C6H12O	2-HEXANONE	-5.732	2.3496E-01	-3.8931E-05	401	1000	------	82.23	190.30
575	C6H12O	3-HEXANONE	-3.443	2.2788E-01	-3.4144E-05	218	1000	61.46	44.61	190.29
576	C6H12O	METHYL ISOBUTYL KETONE	-3.237	2.3310E-01	-3.5612E-05	189	1000	63.10	39.55	194.25
577	C6H12O2	n-PENTYL FORMATE	-3.821	2.5213E-01	-3.8996E-05	200	1000	67.89	45.05	209.31
578	C6H12O2	n-BUTYL ACETATE	-5.140	2.4347E-01	-3.9118E-05	399	1000	------	85.78	199.21
579	C6H12O2	sec-BUTYL ACETATE	-4.319	2.4664E-01	-3.9369E-05	385	1000	------	84.80	202.95
580	C6H12O2	tert-BUTYL ACETATE	-5.817	2.6124E-01	-4.7341E-05	283	1000	67.86	64.32	208.08
581	C6H12O2	ETHYL n-BUTYRATE	-4.606	2.4198E-01	-3.8677E-05	395	1000	------	84.94	198.70
582	C6H12O2	ETHYL ISOBUTYRATE	-3.319	2.4379E-01	-3.8102E-05	185	1000	65.98	40.48	202.37
583	C6H12O2	ISOBUTYL ACETATE	-27.832	3.1735E-01	1.5127E-04	273	500	80.23	70.08	168.66
584	C6H12O2	n-PROPYL PROPIONATE	-5.416	2.4462E-01	-4.0452E-05	396	1000	------	85.11	198.75
585	C6H12O2	CYCLOHEXYL PEROXIDE	-3.726	2.5302E-01	-2.9000E-05	253	1000	69.13	58.43	220.29
586	C6H12O2	DIACETONE ALCOHOL	-14.358	2.8019E-01	-3.8947E-05	441	991	------	101.63	225.06
587	C6H12O2	2-ETHYL BUTYRIC ACID	-4.233	2.3280E-01	-3.1326E-05	258	1000	62.39	53.74	197.24
588	C6H12O2	n-HEXANOIC ACID	-3.291	2.2319E-01	-2.5978E-05	270	1000	60.94	55.08	193.92
589	C6H12O3	2-ETHOXYETHYL ACETATE	-6.371	2.4712E-01	-4.2895E-05	211	1000	63.49	43.86	197.85
590	C6H12O3	HYDROXYCAPROIC ACID	-4.577	2.2348E-01	-2.4462E-05	334	1000	------	67.34	194.44
591	C6H12O3	PARALDEHYDE	-15.800	3.2591E-01	-5.1607E-05	397	997	------	105.45	257.83
592	C6H12O3	sec-BUTYL GLYCOLATE	------	--------	--------	----	----	------	------	------
593	C6H12S	THIACYCLOHEPTANE	-8.308	2.4504E-01	-3.4005E-05	275	1000	61.73	56.51	202.73
594	C6H13N	CYCLOHEXYLAMINE	-4.631	2.4967E-01	-3.7020E-05	255	1000	66.52	56.63	208.02
595	C6H13N	HEXAMETHYLENEIMINE	2.283	2.3223E-01	-1.4755E-05	298	513	70.21	70.18	117.53
596	C6H14	2,2-DIMETHYLBUTANE	3.201	2.4239E-01	-5.0887E-05	323	859	------	76.18	173.87
597	C6H14	2,3-DIMETHYLBUTANE	-2.819	2.5392E-01	-6.0124E-05	331	991	------	74.64	189.77
598	C6H14	n-HEXANE	-8.222	2.6229E-01	-5.7366E-05	300	1000	------	65.30	196.70
599	C6H14	2-METHYLPENTANE	-2.338	2.4813E-01	-5.9665E-05	333	1000	------	73.67	186.13
600	C6H14	3-METHYLPENTANE	-7.066	2.6489E-01	-6.6209E-05	110	1000	66.03	21.27	191.62
601	C6H14N2O2	LYSINE	------	--------	--------	----	----	------	------	------
602	C6H14O	2-ETHYL-1-BUTANOL	-7.532	2.4905E-01	-4.4285E-05	420	1000	------	89.26	197.23
603	C6H14O	1-HEXANOL	-4.207	2.3268E-01	-3.4103E-05	430	1000	------	89.54	194.37
604	C6H14O	2-HEXANOL	-3.687	2.3713E-01	-3.5075E-05	223	1000	63.89	47.45	198.37
605	C6H14O	2-METHYL-1-PENTANOL	-6.124	2.4421E-01	-4.0468E-05	421	1000	------	89.52	197.62
606	C6H14O	4-METHYL-2-PENTANOL	-6.300	2.6484E-01	-4.3962E-05	405	1000	------	93.75	214.58
607	C6H14O	n-BUTYL ETHYL ETHER	7.689	2.0548E-01	-1.4958E-05	365	1000	------	80.70	198.21
608	C6H14O	DIISOPROPYL ETHER	-14.251	2.9988E-01	-5.4923E-05	300	1000	------	70.77	230.71
609	C6H14O	DI-n-PROPYL ETHER	7.525	2.0534E-01	-1.4766E-05	363	1000	------	80.12	198.10
610	C6H14O	METHYL tert-PENTYL ETHER	-4.074	2.5734E-01	-4.3814E-05	160	1000	68.76	35.98	209.45
611	C6H14O2	ACETAL	-5.631	2.5270E-01	-4.4889E-05	377	1000	------	83.26	202.18
612	C6H14O2	2-BUTOXYETHANOL	-7.234	2.4131E-01	-4.0046E-05	444	1000	------	92.01	194.03
613	C6H14O2	1,6-HEXANEDIOL	-17.742	2.7624E-01	-4.0936E-05	516	996	------	113.90	216.78
614	C6H14O2	HEXYLENE GLYCOL	-16.757	2.8373E-01	-4.3375E-05	471	1000	------	107.26	223.60
615	C6H14O2S	DI-n-PROPYL SULFONE	-5.044	2.1872E-01	-6.8294E-06	303	1000	------	60.60	206.85
616	C6H14O3	DIETHYLENE GLYCOL DIMETHYL ETHER	------	--------	--------	----	----	------	------	------
617	C6H14O3	DIPROPYLENE GLYCOL	-4.673	2.3591E-01	-3.2645E-05	495	1000	------	104.10	198.59
618	C6H14O3	2-(2-ETHOXYETHOXY)ETHANOL	-5.502	2.3453E-01	-3.3199E-05	475	1000	------	98.41	195.83
619	C6H14O3	TRIMETHYLOLPROPANE	-7.121	2.3476E-01	-3.1603E-05	562	982	------	114.83	192.94
620	C6H14O4	TRIETHYLENE GLYCOL	-6.792	2.2915E-01	-3.0399E-05	561	1000	------	112.19	191.96
621	C6H14O6	SORBITOL	-6.200	1.7820E-01	-1.6463E-05	777	987	------	122.32	153.65
622	C6H14S	n-HEXYLMERCAPTAN	------	--------	--------	----	----	------	------	------
623	C6H14S	BUTYL-ETHYL-SULFIDE	-8.326	2.3653E-01	-3.6398E-05	275	1000	58.96	53.97	191.81
624	C6H14S	ISOPROPYL-SULFIDE	-9.020	2.5216E-01	-4.1578E-05	275	1000	62.47	57.18	201.56

462

Table 21-1 VISCOSITY OF GAS - ORGANIC COMPOUNDS (continued)

NO	FORMULA	NAME	$n_{gas} = A + B\,T + C\,T^2$					(n_{gas} - micropoise, T - K)		
			A	B	C	TMIN	TMAX	n_{gas} @25 C	n_{gas} @TMIN	n_{gas} @TMAX
625	C6H14S	METHYL-PENTYL-SULFIDE	-8.656	2.4227E-01	-3.9690E-05	275	1000	60.05	54.97	193.92
626	C6H14S	PROPYL-SULFIDE	-8.308	2.3618E-01	-3.6262E-05	275	1000	58.89	53.90	191.61
627	C6H14S2	PROPYL-DISULFIDE	-7.293	2.2831E-01	-2.7937E-05	275	1000	58.29	53.38	193.08
628	C6H15Al	TRIETHYL ALUMINUN	------	--------	--------	----	----	-----	-----	-----
629	C6H15Al2Cl3	ETHYL ALUMINUM SESQUICHLORIDE	------	--------	--------	----	----	-----	-----	-----
630	C6H15N	DIISOPROPYLAMINE	-6.631	2.6157E-01	-4.9666E-05	357	1000	-----	80.42	205.27
631	C6H15N	DI-n-PROPYLAMINE	-4.362	2.4094E-01	-3.9967E-05	382	1000	-----	81.84	196.61
632	C6H15N	n-HEXYLAMINE	-3.799	2.2315E-01	-3.4251E-05	252	1000	59.69	50.26	185.10
633	C6H15N	TRIETHYLAMINE	-9.642	2.7518E-01	-4.2748E-05	363	1000	-----	84.62	222.79
634	C6H15NO	6-AMINOHEXANOL	-3.990	2.1542E-01	-2.6031E-05	331	1000	64.46	185.40	
635	C6H15NO2	DIISOPROPANOLAMINE	-5.544	2.3707E-01	-3.3267E-05	318	1000	-----	66.48	198.26
636	C6H15NO3	TRIETHANOLAMINE	-15.056	2.4627E-01	-2.6625E-05	600	523	-----	123.12	106.46
637	C6H15N3	N-AMINOETHYL PIPERAZINE	-2.185	2.2451E-01	-1.8214E-05	254	1000	63.13	53.67	204.11
638	C6H15O4P	TRIETHYL PHOSPHATE	------	--------	--------	----	----	-----	-----	-----
639	C6H16N2	HEXAMETHYLENEDIAMINE	-5.902	2.1437E-01	-3.0763E-05	475	985	-----	88.98	175.41
640	C6H18N3OP	HEXAMETHYL PHOSPHORAMIDE	------	--------	--------	----	----	-----	-----	-----
641	C6H18N4	TRIETHYLENE TETRAMINE	-3.396	2.0540E-01	-2.3077E-05	285	1000	55.79	53.27	178.93
642	C6H18OSi2	HEXAMETHYLDISILOXANE	-6.880	2.5931E-01	-3.8932E-05	205	1000	66.97	44.64	213.50
643	C6H18O3Si3	HEXAMETHYLCYCLOTRISILOXANE	-12.178	2.7177E-01	-4.4673E-05	337	1000	-----	74.33	214.92
644	C6H19NSi2	HEXAMETHYLDISILAZANE	-10.245	2.5854E-01	-4.1826E-05	293	1000	63.12	61.92	206.47
645	C7H3ClF3NO2	4-CHLORO-3-NITROBENZOTRIFLUORIDE	------	--------	--------	----	----	-----	-----	-----
646	C7H3Cl2F3	2,4-DICHLOROBENZOTRIFLUORIDE	-7.528	2.8105E-01	-4.1446E-05	451	991	-----	110.80	230.29
647	C7H3Cl2NO	3,4-DICHLOROPHENYL ISOCYANATE	------	--------	--------	----	----	-----	-----	-----
648	C7H4ClF3	p-CHLOROBENZOTRIFLUORIDE	-15.751	3.4622E-01	-4.9106E-05	412	1000	-----	118.56	281.36
649	C7H4Cl2O	m-CHLOROBENZOYL CHLORIDE	-7.344	2.4344E-01	-3.1881E-05	489	998	-----	104.07	203.86
650	C7H4F3NO2	3-NITROBENZOTRIFLUORIDE	------	--------	--------	----	----	-----	-----	-----
651	C7H5ClO	BENZOYL CHLORIDE	-10.832	2.8031E-01	-2.3901E-05	470	990	-----	115.63	243.25
652	C7H5ClO2	o-CHLOROBENZOIC ACID	------	--------	--------	----	----	-----	-----	-----
653	C7H5Cl3	BENZOTRICHLORIDE	-15.523	2.9859E-01	-3.3118E-05	494	1000	-----	123.90	249.95
654	C7H5F3	BENZOTRIFLUORIDE	-16.442	3.4665E-01	-5.4376E-05	377	997	-----	106.52	275.12
655	C7H5N	BENZONITRILE	-13.343	2.5305E-01	-2.4503E-05	464	994	-----	98.80	213.98
656	C7H5NO	PHENYL ISOCYANATE	-3.456	2.4204E-01	-2.9816E-05	243	1000	66.06	53.60	208.77
657	C7H5N3O6	2,4,6-TRINITROTOLUENE	2.310	2.3305E-01	2.0170E-05	354	513	-----	87.34	127.17
658	C7H6Cl2	BENZYL DICHLORIDE	-4.192	2.5224E-01	-2.6342E-05	257	1000	68.67	58.89	221.71
659	C7H6Cl2	2,4-DICHLOROTOLUENE	-15.138	2.9593E-01	-3.3312E-05	474	994	-----	117.65	246.10
660	C7H6N2O4	2,4-DINITROTOLUENE	0.445	2.0438E-01	1.3809E-05	343	743	-----	72.17	159.92
661	C7H6N2O4	2,5-DINITROTOLUENE	-3.001	2.3074E-01	-4.1815E-05	326	1000	-----	71.78	223.56
662	C7H6N2O4	2,6-DINITROTOLUENE	-5.409	2.4877E-01	-9.2886E-06	339	1000	-----	77.86	234.07
663	C7H6N2O4	3,4-DINITROTOLUENE	-5.728	2.0275E-01	3.2242E-06	332	1000	-----	61.94	200.25
664	C7H6N2O4	3,5-DINITROTOLUENE	-6.368	2.3345E-01	-7.9257E-06	366	1000	-----	78.01	219.16
665	C7H6O	BENZALDEHYDE	-4.551	2.4474E-01	-2.8486E-05	452	1000	-----	100.25	211.70
666	C7H6O2	BENZOIC ACID	-3.930	2.3677E-01	-2.4221E-05	396	1000	-----	86.03	208.62
667	C7H6O2	p-HYDROXYBENZALDEHYDE	-5.502	2.3573E-01	-2.4232E-05	390	1000	-----	82.75	206.00
668	C7H6O2	SALICYLALDEHYDE	-4.990	2.3816E-01	-3.1302E-05	470	1000	-----	100.03	201.87
669	C7H6O3	SALICYLIC ACID	-5.473	2.3230E-01	-2.7265E-05	432	1000	-----	89.79	199.56
670	C7H7Br	p-BROMOTOLUENE	-8.694	2.8932E-01	-4.0248E-05	458	1000	-----	115.37	240.38
671	C7H7Cl	BENZYL CHLORIDE	-9.249	2.6773E-01	-3.9915E-05	453	1000	-----	103.84	218.57
672	C7H7Cl	o-CHLOROTOLUENE	-14.068	2.9696E-01	-3.6040E-05	237	1000	71.27	54.29	246.85
673	C7H7Cl	p-CHLOROTOLUENE	-14.442	2.9298E-01	-3.5493E-05	281	1000	69.75	65.08	243.05
674	C7H7F	p-FLUOROTOLUENE	-10.001	2.8027E-01	-4.5589E-05	275	1000	69.51	63.63	224.68
675	C7H7NO	FORMANILIDE	-12.403	2.6197E-01	-3.7292E-05	544	1000	-----	119.07	212.27
676	C7H7NO2	m-NITROTOLUENE	-4.131	2.1907E-01	-5.2718E-06	289	1000	60.72	58.74	209.67
677	C7H7NO2	o-NITROTOLUENE	-4.418	2.2526E-01	-9.6088E-06	270	1000	61.89	55.70	211.23
678	C7H7NO2	p-NITROTOLUENE	-7.336	2.2670E-01	-1.2930E-05	325	1000	-----	64.98	206.43
679	C7H7NO3	o-NITROANISOLE	-3.193	2.1792E-01	2.4631E-06	284	1000	62.00	58.89	217.19
680	C7H8	TOLUENE	1.787	2.3566E-01	-9.3508E-06	275	600	71.22	65.89	139.82
681	C7H8	1,3,5-CYCLOHEPTATRIENE	-9.446	2.6526E-01	-4.2720E-05	275	1000	65.84	60.27	213.09
682	C7H8O	ANISOLE	-6.401	2.5885E-01	-3.7145E-05	427	997	-----	97.35	214.75
683	C7H8O	BENZYL ALCOHOL	-9.062	2.5842E-01	-3.8883E-05	479	1000	-----	105.80	210.47
684	C7H8O	m-CRESOL	-6.367	2.4928E-01	-3.2222E-05	475	1000	-----	104.77	210.69
685	C7H8O	o-CRESOL	-15.575	3.0914E-01	-3.4443E-05	450	1000	-----	116.56	259.12
686	C7H8O	p-CRESOL	-7.806	2.5588E-01	-3.6266E-05	308	1000	67.56	211.81	
687	C7H8O2	GUAIACOL	-3.906	2.4101E-01	-2.6965E-05	305	1000	67.09	210.14	
688	C7H8O2	p-METHOXYPHENOL	-8.848	2.5339E-01	-3.2747E-05	516	1000	-----	113.58	211.80
689	C7H9N	BENZYLAMINE	-9.498	2.5128E-01	-3.9342E-05	227	1000	61.92	45.52	202.44
690	C7H9N	2,6-DIMETHYLPYRIDINE	-3.730	2.4165E-01	-3.2236E-05	267	1000	65.45	58.49	205.68
691	C7H9N	N-METHYLANILINE	-4.107	2.5167E-01	-2.8701E-05	216	1000	68.38	48.91	218.86
692	C7H9N	m-TOLUIDINE	-5.778	2.3808E-01	-2.9576E-05	477	1000	-----	101.06	202.73
693	C7H9N	o-TOLUIDINE	-9.201	2.4867E-01	-3.7188E-05	474	1000	-----	100.31	202.28
694	C7H9N	p-TOLUIDINE	-5.071	2.3751E-01	-3.0731E-05	317	1000	67.13	201.71	
695	C7H10	2-NORBORNENE	-7.488	2.7092E-01	-5.9603E-05	319	1000	-----	72.87	203.83
696	C7H10N2	TOLUENEDIAMINE	-6.350	2.2185E-01	-2.4575E-05	557	977	-----	109.60	186.94
697	C7H11NO	CYCLOHEXYL ISOCYANATE	-3.134	2.3618E-01	-3.0393E-06	193	1000	64.58	41.32	202.65
698	C7H12	1-HEPTYNE	-8.810	2.4506E-01	-4.3313E-05	275	1000	60.40	55.31	192.94
699	C7H12O2	n-BUTYL ACRYLATE	-6.121	2.3892E-01	-3.8503E-05	421	991	-----	87.64	192.84
700	C7H12O2	ISOBUTYL ACRYLATE	-3.795	2.3794E-01	-3.6529E-05	212	1000	63.90	45.01	197.62
701	C7H12O2	n-PROPYL METHACRYLATE	-13.626	2.8190E-01	-4.2333E-05	414	994	-----	95.82	224.76
702	C7H12O4	DIETHYL MALONATE	-6.683	2.1674E-01	-3.3894E-05	472	1000	-----	88.07	176.16

Table 21-1 VISCOSITY OF GAS - ORGANIC COMPOUNDS (continued)

| NO | FORMULA | NAME | $n_{gas} = A + B T + C T^2$ | | | | | $(n_{gas}$ - micropoise, T - K$)$ | | |
			A	B	C	TMIN	TMAX	n_{gas} @25 C	n_{gas} @TMIN	n_{gas} @TMAX
703	C7H14	CYCLOHEPTANE	-7.438	2.6108E-01	-5.4889E-05	392	1000	------	86.47	198.75
704	C7H14	1,1-DIMETYLCYCLOPENTANE	-7.026	2.7282E-01	-6.2906E-05	203	1000	68.72	45.76	202.89
705	C7H14	cis-1,2-DIMETHYLCYCLOPENTANE	-4.725	2.5583E-01	-5.5687E-05	373	1000	------	82.95	195.42
706	C7H14	trans-1,2-DIMETHYLCYCLOPENTANE	-5.409	2.6456E-01	-5.9782E-05	365	1000	------	83.19	199.37
707	C7H14	cis-1,3-DIMETHYLCYCLOPENTANE	-4.777	2.6247E-01	-5.7086E-05	139	1000	68.40	30.60	200.61
708	C7H14	trans-1,3-DIMETHYLCYCLOPENTANE	-5.014	2.6336E-01	-5.9223E-05	139	1000	68.24	30.45	199.12
709	C7H14	ETHYLCYCLOPENTANE	-5.163	2.5421E-01	-5.7884E-05	377	1000	------	82.45	191.16
710	C7H14	2-ETHYL-1-PENTENE	-3.250	2.3857E-01	-5.8878E-05	168	1000	62.65	35.17	176.44
711	C7H14	3-ETHYL-1-PENTENE	-4.820	2.5287E-01	-5.5637E-05	146	1000	65.63	30.91	192.41
712	C7H14	1-HEPTENE	-0.160	2.2630E-01	-5.1029E-05	367	997	------	76.02	174.74
713	C7H14	cis-2-HEPTENE	-3.122	2.3118E-01	-5.5157E-05	372	1000	------	75.24	172.90
714	C7H14	trans-2-HEPTENE	-3.319	2.3451E-01	-6.1537E-05	164	1000	61.13	33.49	169.65
715	C7H14	cis-3-HEPTENE	-2.338	2.2994E-01	-5.3715E-05	369	1000	------	75.20	173.89
716	C7H14	trans-3-HEPTENE	-3.347	2.3615E-01	-6.0997E-05	137	1000	61.64	27.86	171.81
717	C7H14	METHYLCYCLOHEXANE	-3.574	2.4991E-01	-5.3652E-05	374	994	------	82.39	191.83
718	C7H14	2-METHYL-1-HEXENE	-2.433	2.3440E-01	-6.0899E-05	170	1000	62.04	35.65	171.07
719	C7H14	3-METHYL-1-HEXENE	-4.257	2.4902E-01	-5.9706E-05	145	1000	64.68	30.60	185.06
720	C7H14	4-METHYL-1-HEXENE	-4.452	2.5068E-01	-5.5183E-05	132	1000	65.38	27.68	191.04
721	C7H14	2,3,3-TRIMETHYL-1-BUTENE	-5.246	2.6000E-01	-5.9728E-05	163	1000	66.96	35.55	195.03
722	C7H14O	DIISOPROPYL KETONE	-3.146	2.2431E-01	-3.3432E-05	205	1000	60.76	41.43	187.73
723	C7H14O	2-HEPTANONE	-4.179	2.1570E-01	-3.0591E-05	424	1000	------	81.78	180.93
724	C7H14O	1-HEPTANAL	-3.797	2.1786E-01	-3.1445E-05	426	1000	------	83.30	182.62
725	C7H14O	1-METHYLCYCLOHEXANOL	-6.692	2.3801E-01	-3.5259E-05	430	1000	------	89.13	196.06
726	C7H14O	cis-2-METHYLCYCLOHEXANOL	-6.682	2.6537E-01	-4.1341E-05	438	1000	------	101.62	217.35
727	C7H14O	trans-2-METHYLCYCLOHEXANOL	-6.168	2.6314E-01	-3.9040E-05	440	1000	------	102.06	217.93
728	C7H14O	cis-3-METHYLCYCLOHEXANOL	-4.209	2.5283E-01	-3.3080E-05	441	1000	------	100.86	215.54
729	C7H14O	trans-3-METHYLCYCLOHEXANOL	-9.451	2.7138E-01	-4.6195E-05	440	1000	------	101.01	215.73
730	C7H14O	cis-4-METHYLCYCLOHEXANOL	8.219	2.2310E-01	-1.5838E-05	444	1000	------	104.15	215.48
731	C7H14O	trans-4-METHYLCYCLOHEXANOL	8.161	2.2152E-01	-1.5725E-05	444	1000	------	103.42	213.96
732	C7H14O	5-METHYL-2-HEXANONE	-7.312	2.3209E-01	-4.0528E-05	418	1000	------	82.62	184.25
733	C7H14O2	n-BUTYL PROPIONATE	-3.381	2.2311E-01	-3.1675E-05	420	1000	------	84.74	188.05
734	C7H14O2	ETHYL ISOVALERATE	-5.613	2.3574E-01	-3.8287E-05	407	1000	------	83.99	191.84
735	C7H14O2	ISOPENTYL ACETATE	-6.783	2.4016E-01	-4.0559E-05	415	1000	------	85.90	192.82
736	C7H14O2	n-PENTYL ACETATE	-5.773	2.3055E-01	-3.6543E-05	422	1000	------	85.01	188.23
737	C7H14O2	n-PROPYL n-BUTYRATE	-5.773	2.3055E-01	-3.6543E-05	422	1000	------	85.01	188.23
738	C7H14O2	n-HEPTANOIC ACID	-3.235	2.1129E-01	-2.4882E-05	266	1000	57.55	51.21	183.17
739	C7H14O3	ETHYL-3-ETHOXYPROPIONATE	-3.458	2.2290E-01	-3.1016E-05	223	1000	60.24	44.71	188.43
740	C7H15Br	1-BROMOHEPTANE	-3.554	2.4585E-01	-3.0768E-05	217	1000	67.01	48.35	211.53
741	C7H15N	N-METHYLCYCLOHEXYLAMINE	-3.864	2.4602E-01	-3.3136E-05	265	1000	66.54	59.00	209.02
742	C7H16	2,2-DIMETHYLPENTANE	-0.362	2.3381E-01	-5.0753E-05	352	1000	------	75.65	182.70
743	C7H16	2,3-DIMETHYLPENTANE	1.938	2.2586E-01	-5.1563E-05	363	993	------	77.13	175.37
744	C7H16	2,4-DIMETHYLPENTANE	5.790	2.1603E-01	-4.1541E-05	343	994	------	75.00	179.48
745	C7H16	3,3-DIMETHYLPENTANE	-4.688	2.4742E-01	-5.4788E-05	359	1000	------	77.07	187.94
746	C7H16	3-ETHYLPENTANE	-2.472	2.3698E-01	-5.2486E-05	367	1000	------	77.43	182.02
747	C7H16	n-HEPTANE	-10.378	2.4401E-01	-5.4003E-05	338	700	------	65.93	133.97
748	C7H16	2-METHYLHEXANE	-0.928	2.2971E-01	-5.5133E-05	363	993	------	75.19	172.81
749	C7H16	3-METHYLHEXANE	2.752	2.1590E-01	-4.9642E-05	365	995	------	76.10	171.57
750	C7H16	2,2,3-TRIMETHYLBUTANE	34.054	7.8678E-02	9.6635E-05	343	535	------	72.41	103.81
751	C7H16O	1-HEPTANOL	-6.640	2.2701E-01	-3.6633E-05	449	1000	------	87.90	183.74
752	C7H16O	2-HEPTANOL	-6.021	2.3027E-01	-3.8208E-05	432	1000	------	86.33	186.04
753	C7H16O	5-METHYL-1-HEXANOL	-3.640	2.2222E-01	-3.1463E-05	293	1000	59.82	58.77	187.12
754	C7H16O	ISOPROPYL-TERT-BUTYL-ETHER	-9.069	2.5230E-01	-4.4759E-05	275	1000	62.18	56.93	198.47
755	C7H16S	n-HEPTYL MERCAPTAN	-4.487	2.3471E-01	-1.8025E-05	230	1000	63.89	48.54	212.20
756	C7H16S	BUTYL-PROPYL-SULFIDE	-7.370	2.2215E-01	-2.9351E-05	275	1000	56.26	51.50	185.43
757	C7H16S	ETHYL-PENTYL-SULFIDE	-7.453	2.1926E-01	-3.0611E-05	275	1000	55.20	50.53	181.20
758	C7H16S	HEXYL-METHYL-SULFIDE	-7.453	2.1926E-01	-3.0611E-05	275	1000	55.20	50.53	181.20
759	C7H17N	1-AMINOHEPTANE	-3.730	2.1059E-01	-2.9832E-05	254	1000	56.41	47.83	177.03
760	C8H4Cl2O2	ISOPHTHALOYL CHLORIDE	-3.407	2.1022E-01	-2.1042E-05	317	1000	------	61.12	185.77
761	C8H4O3	PHTHALIC ANHYDRIDE	-10.366	2.2977E-01	-1.0424E-05	550	1000	------	112.85	208.98
762	C8H6	ETHYNYLBENZENE	-8.347	2.5247E-01	-3.3114E-05	275	1000	63.98	58.58	211.01
763	C8H6O4	ISOPHTHALIC ACID	-15.789	2.3729E-01	-3.2320E-05	619	1000	------	118.71	189.18
764	C8H6O4	PHTHALIC ACID	-1.927	2.0550E-01	-1.7011E-05	464	1000	------	89.76	186.56
765	C8H6O4	TEREPHTHALIC ACID	8.169	1.7644E-01	6.1619E-06	700	1000	------	134.70	190.77
766	C8H6S	BENZOTHIOPHENE	-3.784	2.4817E-01	-2.4206E-05	305	1000	------	69.66	220.18
767	C8H7N	INDOLE	-1.519	2.2460E-01	-1.3076E-06	274	1000	65.33	59.92	221.77
768	C8H8	STYRENE	-10.035	2.5191E-01	-3.7932E-05	243	1000	61.70	48.94	203.94
769	C8H8	1,3,5,7-CYCLOOCTATETRAENE	-8.597	2.5453E-01	-3.5010E-05	275	1000	64.18	58.75	210.92
770	C8H8O	ACETOPHENONE	-15.968	2.7182E-01	-3.2956E-05	475	1000	------	105.71	222.90
771	C8H8O	p-TOLUALDEHYDE	-15.375	2.5587E-01	-3.1881E-05	477	997	------	99.42	208.04
772	C8H8O2	METHYL BENZOATE	-3.763	2.3371E-01	-2.6845E-05	473	1000	------	100.78	203.10
773	C8H8O2	o-TOLUIC ACID	-5.470	2.2720E-01	-2.7335E-05	377	1000	------	76.30	194.39
774	C8H8O2	p-TOLUIC ACID	-4.764	2.2386E-01	-2.3427E-05	453	1000	------	91.84	195.67
775	C8H8O3	METHYL SALICYLATE	-3.840	2.2518E-01	-2.5988E-05	265	1000	60.99	54.01	195.35
776	C8H8O3	VANILLIN	-4.085	2.1743E-01	-2.1385E-05	355	1000	------	70.41	191.96
777	C8H9NO	ACETANILIDE	-4.945	2.2579E-01	-2.3026E-05	387	1000	------	78.99	197.82
778	C8H10	ETHYLBENZENE	-4.267	2.4735E-01	-5.4264E-05	409	1000	------	87.82	188.82
779	C8H10	m-XYLENE	-21.620	2.7820E-01	-6.0531E-05	250	1000	55.94	44.15	196.05
780	C8H10	o-XYLENE	-19.763	2.8022E-01	-5.9293E-05	250	1000	58.51	46.59	201.16

Table 21-1 VISCOSITY OF GAS - ORGANIC COMPOUNDS (continued)

NO	FORMULA	NAME	$n_{gas} = A + B\,T + C\,T^2$			$(n_{gas}$ - micropoise, T - K)				
			A	B	C	TMIN	TMAX	n_{gas} @25 C	n_{gas} @TMIN	n_{gas} @TMAX
781	C8H10	p-XYLENE	-17.226	2.5098E-01	-2.8232E-05	286	1000	55.09	52.25	205.52
782	C8H10O	m-ETHYLPHENOL	-----	--------	--------	----	----	-----	-----	-----
783	C8H10O	p-ETHYLPHENOL	-7.023	2.3683E-01	-3.2455E-05	491	1000	-----	101.44	197.35
784	C8H10O	PHENETOLE	-5.185	2.4696E-01	-3.9071E-05	443	993	-----	96.55	201.52
785	C8H10O	2-PHENYLETHANOL	-3.941	2.2946E-01	-2.7667E-05	247	1000	62.01	51.05	197.85
786	C8H10O	2,3-XYLENOL	-6.830	2.5325E-01	-3.2357E-05	490	1000	-----	109.49	214.06
787	C8H10O	2,4-XYLENOL	-8.590	2.5771E-01	-3.5700E-05	484	1000	-----	107.78	213.42
788	C8H10O	2,5-XYLENOL	-11.075	2.6488E-01	-4.0537E-05	484	1000	-----	107.63	213.27
789	C8H10O	2,6-XYLENOL	-7.918	2.5778E-01	-3.6971E-05	474	1000	-----	105.96	212.89
790	C8H10O	3,4-XYLENOL	-6.212	2.5085E-01	-2.9900E-05	500	1000	-----	111.74	214.74
791	C8H10O	3,5-XYLENOL	-6.082	2.5092E-01	-3.0962E-05	495	1000	-----	110.54	213.88
792	C8H11N	N,N-DIMETHYLANILINE	-11.427	2.3666E-01	-2.7800E-05	467	1000	-----	93.03	197.43
793	C8H11N	o-ETHYLANILINE	-3.086	2.1729E-01	-2.3428E-05	227	1000	59.62	45.03	190.78
794	C8H11N	2,4,6-TRIMETHYLPYRIDINE	-3.488	2.2503E-01	-2.8166E-05	229	1000	61.10	46.57	193.38
795	C8H11NO	p-PHENETIDINE	-3.442	2.1740E-01	-2.2122E-05	277	1000	59.41	55.08	191.84
796	C8H12	1,5-CYCLOOCTADIENE	-7.018	2.4696E-01	-5.5632E-05	204	1000	64.63	43.07	194.24
797	C8H12	VINYLCYCLOHEXENE	-3.056	2.4666E-01	-5.5458E-05	401	1000	-----	86.94	188.15
798	C8H12O4	1,4-CYCLOHEXANEDICARBOXYLIC ACID	-1.884	2.1013E-01	-1.4686E-05	586	1000	-----	116.21	193.56
799	C8H12O4	DIETHYL MALEATE	-3.791	2.1444E-01	-2.7014E-05	264	1000	57.74	50.94	183.64
800	C8H14O2	n-BUTYL METHACRYLATE	-13.415	2.7028E-01	-3.8766E-05	434	994	-----	96.58	216.94
801	C8H14O3	BUTYRIC ANHYDRIDE	-5.611	2.0629E-01	-3.0923E-05	468	978	-----	84.16	166.56
802	C8H14O4	DIETHYL SUCCINATE	-2.551	2.0547E-01	-2.3707E-05	252	1000	56.60	47.72	179.21
803	C8H16	CYCLOOCTANE	-7.878	2.3495E-01	-3.1777E-05	275	1000	59.35	54.33	195.30
804	C8H16	1,1-DIMETHYLCYCLOHEXANE	-4.456	2.3353E-01	-4.9690E-05	393	1000	-----	79.65	179.38
805	C8H16	cis-1,2-DIMETHYLCYCLOHEXANE	-4.676	2.2762E-01	-4.8251E-05	403	1000	-----	79.22	174.69
806	C8H16	trans-1,2-DIMETHYLCYCLOHEXANE	-5.784	2.3542E-01	-4.9707E-05	397	1000	-----	79.84	179.93
807	C8H16	cis-1,3-DIMETHYLCYCLOHEXANE	-5.235	2.3630E-01	-5.1512E-05	393	1000	-----	79.67	179.55
808	C8H16	trans-1,3-DIMETHYLCYCLOHEXANE	-6.527	2.3630E-01	-5.0689E-05	398	1000	-----	79.49	179.08
809	C8H16	cis-1,4-DIMETHYLCYCLOHEXANE	-6.541	2.3635E-01	-5.0685E-05	397	1000	-----	79.30	179.12
810	C8H16	trans-1,4-DIMETHYLCYCLOHEXANE	-5.458	2.3662E-01	-5.1728E-05	393	1000	-----	79.54	179.43
811	C8H16	ETHYLCYCLOHEXANE	-6.237	2.3574E-01	-5.0889E-05	405	995	-----	80.89	177.94
812	C8H16	2-ETHYL-1-HEXENE	-0.255	2.3722E-01	-5.4208E-05	393	1000	-----	84.60	182.76
813	C8H16	1-METHYL-1-ETHYLCYCLOPENTANE	-4.149	2.4039E-01	-5.1894E-05	129	1000	62.91	26.00	184.35
814	C8H16	1-OCTENE	2.722	2.0327E-01	-4.3879E-05	394	994	-----	76.00	161.42
815	C8H16	trans-2-OCTENE	-0.235	2.0966E-01	-4.5808E-05	398	1000	-----	75.95	163.62
816	C8H16	trans-3-OCTENE	-1.983	2.1735E-01	-5.1136E-05	396	1000	-----	76.07	164.23
817	C8H16	trans-4-OCTENE	-4.363	2.2443E-01	-5.5480E-05	179	1000	57.62	34.03	164.59
818	C8H16	n-PROPYLCYCLOPENTANE	1.611	2.2775E-01	-5.8607E-05	404	1000	-----	84.06	170.75
819	C8H16	2,4,4-TRIMETHYL-1-PENTENE	-2.985	2.3159E-01	-5.3140E-05	375	1000	-----	76.39	175.47
820	C8H16	2,4,4-TRIMETHYL-2-PENTENE	-2.125	2.2611E-01	-5.0084E-05	378	1000	-----	76.19	173.90
821	C8H16O	2-ETHYLHEXANAL	-6.088	2.1785E-01	-3.4958E-05	434	1000	-----	81.87	176.80
822	C8H16O	1-OCTANAL	-4.191	2.0954E-01	-3.1582E-05	246	1000	55.48	45.44	173.77
823	C8H16O	2-OCTANONE	-3.232	2.0349E-01	-2.7728E-05	253	1000	54.97	46.48	172.53
824	C8H16O2	n-BUTYL n-BUTYRATE	-6.645	2.2136E-01	-3.6247E-05	438	1000	-----	83.36	178.47
825	C8H16O2	n-HEXYL ACETATE	-2.776	2.0956E-01	-2.7615E-05	192	1000	57.25	36.44	179.17
826	C8H16O2	ISOBUTYL ISOBUTYRATE	-2.059	2.1602E-01	-2.7891E-05	421	1000	-----	83.94	186.07
827	C8H16O2	n-OCTANOIC ACID	-4.091	2.0435E-01	-2.6127E-05	290	1000	54.51	52.97	174.13
828	C8H16O4	DIETHYLENE GLYCOL ETHYL ETHER ACETATE	-5.954	2.1682E-01	-3.1524E-05	491	1000	-----	92.90	179.34
829	C8H18	2,2-DIMETHYLHEXANE	-3.105	2.2995E-01	-5.4122E-05	380	1000	-----	76.46	172.72
830	C8H18	2,3-DIMETHYLHEXANE	-0.354	2.1977E-01	-4.6587E-05	389	1000	-----	78.09	172.83
831	C8H18	2,4-DIMETHYLHEXANE	-2.240	2.2738E-01	-5.3458E-05	383	1000	-----	77.01	171.68
832	C8H18	2,5-DIMETHYLHEXANE	-1.204	2.2182E-01	-5.1017E-05	382	1000	-----	76.09	169.60
833	C8H18	3,3-DIMETHYLHEXANE	4.922	2.0565E-01	-4.3444E-05	385	1000	-----	77.66	167.13
834	C8H18	3,4-DIMETHYLHEXANE	-1.990	2.2616E-01	-4.9251E-05	391	1000	-----	78.91	174.92
835	C8H18	3-ETHYLHEXANE	3.212	2.0721E-01	-4.4757E-05	392	1000	-----	77.56	165.67
836	C8H18	3-ETHYL-2-METHYLPENTANE	-8.532	2.3733E-01	-4.1168E-05	275	1000	58.57	53.62	187.63
837	C8H18	3-METHYL-3-ETHYLPENTANE	-5.654	2.4024E-01	-5.5917E-05	182	1000	61.00	36.22	178.67
838	C8H18	2-METHYLHEPTANE	-1.925	2.1984E-01	-5.1869E-05	391	1000	-----	76.10	166.05
839	C8H18	3-METHYLHEPTANE	2.097	2.0867E-01	-4.8059E-05	392	1000	-----	76.51	162.71
840	C8H18	4-METHYLHEPTANE	-0.947	2.1869E-01	-5.0959E-05	391	1000	-----	76.77	166.78
841	C8H18	n-OCTANE	3.940	1.6640E-01	1.4470E-05	374	670	-----	68.20	121.92
842	C8H18	2,2,3-TRIMETHYLPENTANE	2.213	2.1794E-01	-4.8665E-05	383	993	-----	78.55	170.64
843	C8H18	2,2,4-TRIMETHYLPENTANE	4.636	2.0622E-01	-3.2545E-05	355	992	-----	73.74	177.18
844	C8H18	2,3,3-TRIMETHYLPENTANE	-4.079	2.3612E-01	-5.1078E-05	388	1000	-----	79.85	180.96
845	C8H18	2,3,4-TRIMETHYLPENTANE	-2.823	2.3136E-01	-5.0133E-05	387	1000	-----	79.21	178.40
846	C8H18	2,2,3,3-TETRAMETHYLBUTANE	-8.600	2.3922E-01	-4.1411E-05	275	1000	59.04	54.05	189.21
847	C8H18O	DI-n-BUTYL ETHER	-6.436	2.2225E-01	-3.8560E-05	413	1000	-----	78.78	177.25
848	C8H18O	DI-sec-BUTYL ETHER	-2.403	2.1823E-01	-3.1737E-05	173	1000	59.84	34.40	184.09
849	C8H18O	DI-tert-BUTYL ETHER	-3.400	2.3528E-01	-3.7970E-05	195	1000	63.37	41.04	193.91
850	C8H18O	2-ETHYL-1-HEXANOL	-15.161	2.5809E-01	-3.8227E-05	450	1000	-----	93.24	204.70
851	C8H18O	1-OCTANOL	-5.734	2.1231E-01	-3.1198E-05	468	1000	-----	86.79	175.38
852	C8H18O	2-OCTANOL	-7.609	2.2266E-01	-3.6824E-05	453	1000	-----	85.70	178.23
853	C8H18O2	DI-t-BUTYL PEROXIDE	-3.445	2.3372E-01	-3.7286E-05	233	1000	62.92	48.99	192.99
854	C8H18O2S	DI-n-BUTYL SULFONE	-3.870	2.0602E-01	-4.3349E-06	318	1000	-----	61.21	197.82
855	C8H18O3	DIETHYLENE GLYCOL DIETHYL ETHER	-3.450	2.1209E-01	-2.9382E-05	229	1000	57.17	43.58	179.26
856	C8H18O3	DIETHYLENE GLYCOL MONOBUTYL EHTER	-8.537	2.1984E-01	-3.5753E-05	205	1000	53.83	35.03	175.55
857	C8H18O4	TRIETHYLENE GLYCOL DIMETHYL ETHER	-3.318	2.1021E-01	-2.7608E-05	229	1000	56.90	43.37	179.28
858	C8H18O5	TETRAETHYLENE GLYCOL	-3.068	1.9637E-01	-1.9716E-05	600	1000	-----	107.66	173.59

465

Table 21-1 VISCOSITY OF GAS - ORGANIC COMPOUNDS (continued)

NO	FORMULA	NAME	$n_{gas} = A + B\,T + C\,T^2$					$(n_{gas}$ - micropoise, T - K)		
			A	B	C	TMIN	TMAX	n_{gas} @25 C	n_{gas} @TMIN	n_{gas} @TMAX
859	C8H18S	n-OCTYL MERCAPTAN	-2.619	2.1932E-01	-1.1098E-05	224	1000	61.78	45.95	205.60
860	C8H18S	tert-OCTYL MERCAPTAN	-----	--------	--------	----	----	-----	-----	-----
861	C8H18S	BUTYL-SULFIDE	-7.108	2.1298E-01	-2.8510E-05	275	1000	53.86	49.31	177.36
862	C8H18S	ETHYL-HEXYL-SULFIDE	-6.733	2.0564E-01	-2.6429E-05	275	1000	52.23	47.82	172.48
863	C8H18S	HEPTYL-METHYL-SULFIDE	-6.733	2.0564E-01	-2.6429E-05	275	1000	52.23	47.82	172.48
864	C8H18S	PENTYL-PROPYL-SULFIDE	-6.733	2.0564E-01	-2.6429E-05	275	1000	52.23	47.82	172.48
865	C8H18S2	BUTYL-DISULFIDE	-5.863	1.9802E-01	-2.1111E-05	275	1000	51.30	47.00	171.05
866	C8H19N	DI-n-BUTYLAMINE	-4.793	2.1273E-01	-2.8463E-05	432	1000	-----	81.79	179.47
867	C8H19N	DIISOBUTYLAMINE	-3.262	2.2006E-01	-3.2461E-05	203	1000	59.46	40.07	184.34
868	C8H19N	n-OCTYLAMINE	-4.389	2.0340E-01	-2.9847E-05	273	1000	53.60	48.91	169.16
869	C8H23N5	TETRAETHYLENEPENTAMINE	-3.025	1.8476E-01	-1.7424E-05	243	1000	50.51	40.84	164.31
870	C8H24O4Si4	OCTAMETHYLCYCLOTETRASILOXANE	-8.561	2.4239E-01	-3.2718E-05	291	1000	60.80	59.20	201.11
871	C9H4O5	TRIMELLITIC ANHYDRIDE	-8.497	2.7296E-01	-2.6923E-05	663	993	-----	160.64	236.01
872	C9H6N2O2	TOLUENE DIISOCYANATE	-6.020	2.1233E-01	-2.6552E-05	523	973	-----	97.77	175.44
873	C9H7N	ISOQUINOLINE	-3.491	2.3319E-01	-2.0356E-05	299	1000	-----	64.41	209.34
874	C9H7N	QUINOLINE	-6.172	2.3955E-01	-2.4770E-05	511	1000	-----	109.77	208.61
875	C9H7NO	8-HYDROXYQUINOLINE	-2.965	1.9989E-01	-1.7754E-05	348	1000	-----	64.45	179.17
876	C9H8	INDENE	-6.357	2.4242E-01	-5.0608E-05	456	996	-----	93.66	184.89
877	C9H8O	2-METHYLBENZOFURAN	-5.700	2.4788E-01	-3.2055E-05	290	1000	65.36	63.49	210.12
878	C9H10	INDANE	-7.579	2.5439E-01	-5.4799E-05	222	1000	63.40	46.20	192.01
879	C9H10	cis-PROPENYLBENZENE	-7.474	2.3000E-01	-2.9113E-05	275	1000	58.51	53.57	193.41
880	C9H10	trans-PROPENYLBENZENE	-7.474	2.3000E-01	-2.9113E-05	275	1000	58.51	53.57	193.41
881	C9H10	alpha-METHYLSTYRENE	-3.020	2.2996E-01	-4.5027E-05	439	1000	-----	89.25	181.91
882	C9H10	m-METHYLSTYRENE	-3.712	2.2924E-01	-5.9757E-05	187	1000	59.32	37.07	165.77
883	C9H10	o-METHYLSTYRENE	-5.576	2.4030E-01	-5.4707E-05	205	1000	61.21	41.39	180.02
884	C9H10	p-METHYLSTYRENE	-6.238	2.3464E-01	-5.1790E-05	239	1000	59.12	46.88	176.61
885	C9H10O2	BENZYL ACETATE	-3.100	2.4378E-01	-2.7324E-05	487	1000	-----	109.14	213.36
886	C9H10O2	ETHYL BENZOATE	-7.390	2.3088E-01	-3.2553E-05	486	1000	-----	97.13	190.94
887	C9H10O3	ETHYL VANILLIN	-4.216	2.0860E-01	-2.3777E-05	351	1000	-----	66.07	180.61
888	C9H11NO	p-DIMETHYLAMINOBENZALDEHYDE	-5.021	1.9649E-01	-3.7363E-07	348	1000	-----	63.31	191.10
889	C9H12	CUMENE	-12.027	2.5591E-01	-4.3606E-05	150	1000	60.40	25.38	200.28
890	C9H12	m-ETHYLTOLUENE	-4.327	2.1775E-01	-4.7848E-05	434	1000	-----	81.16	165.57
891	C9H12	o-ETHYLTOLUENE	-1.335	2.1358E-01	-4.4230E-05	438	1000	-----	83.73	168.02
892	C9H12	p-ETHYLTOLUENE	-1.133	2.1196E-01	-4.3377E-05	435	1000	-----	82.86	167.45
893	C9H12	MESITYLENE	-15.901	2.5143E-01	-4.7254E-05	374	750	-----	71.52	146.09
894	C9H12	n-PROPYLBENZENE	-5.389	2.3737E-01	-5.1146E-05	432	992	-----	87.61	179.75
895	C9H12	1,2,3-TRIMETHYLBENZENE	-1.909	2.2938E-01	-4.7290E-05	449	1000	-----	91.55	180.18
896	C9H12	1,2,4-TRIMETHYLBENZENE	-3.350	2.3022E-01	-5.1265E-05	443	1000	-----	88.58	175.61
897	C9H12O	BENZYL ETHYL ETHER	-5.439	2.2580E-01	-3.1859E-05	458	1000	-----	91.30	188.50
898	C9H12O	2-PHENYL-2-PROPANOL	-5.211	2.3193E-01	-3.2288E-05	309	1000	-----	63.37	194.43
899	C9H12O2	CUMENE HYDROPEROXIDE	-----	--------	--------	----	----	-----	-----	-----
900	C9H14O	ISOPHORONE	-13.415	2.4722E-01	-2.6941E-05	488	1000	-----	100.81	206.86
901	C9H14O6	GLYCERYL TRIACETATE	-3.353	1.9886E-01	-2.2907E-05	277	1000	53.90	49.97	172.60
902	C9H16	1-NONYNE	-7.589	2.1597E-01	-3.3048E-05	275	1000	53.86	49.30	175.33
903	C9H16O4	AZELAIC ACID	-3.008	1.8282E-01	-1.6326E-05	380	1000	-----	64.11	163.49
904	C9H18	BUTYLCYCLOPENTANE	-7.688	2.2218E-01	-3.2468E-05	275	1000	55.67	50.96	182.02
905	C9H18	cis,cis-1,3,5-TRIMETHYLCYCLOHEXANE	-8.300	2.3555E-01	-3.6376E-05	275	1000	58.70	53.73	190.87
906	C9H18	cis,trans-1,3,5-TRIMETHYLCYCLOHEXANE	-8.247	2.3295E-01	-3.6603E-05	275	1000	57.95	53.05	188.10
907	C9H18	ISOPROPYLCYCLOHEXANE	-4.373	2.2904E-01	-5.6324E-05	184	1000	58.91	35.86	168.34
908	C9H18	1-NONENE	2.045	1.9481E-01	-4.4875E-05	400	1000	-----	72.79	151.98
909	C9H18	n-PROPYLCYCLOHEXANE	-0.749	2.1149E-01	-4.8009E-05	430	1000	-----	81.32	162.73
910	C9H18O	DIISOBUTYL KETONE	-3.235	2.0203E-01	-2.8206E-05	227	1000	54.49	41.17	170.59
911	C9H18O	1-NONANAL	-3.028	1.9523E-01	-2.5569E-05	255	1000	52.91	45.09	166.63
912	C9H18O2	n-BUTYL VALERATE	-2.561	1.9888E-01	-2.5595E-05	180	1000	54.46	32.41	170.72
913	C9H18O2	n-NONANOIC ACID	-3.887	1.9474E-01	-2.3857E-05	286	1000	52.05	49.86	167.00
914	C9H18O2	n-OCTYL FORMATE	-2.972	2.0679E-01	-2.5911E-05	234	1000	56.38	44.00	177.91
915	C9H20	3,3-DIETHYLPENTANE	4.011	2.0336E-01	-4.3531E-05	419	1000	-----	81.58	163.84
916	C9H20	2,2-DIMETHYL-3-ETHYLPENTANE	-4.929	2.3066E-01	-5.0451E-05	174	1000	59.36	33.68	175.28
917	C9H20	3-ETHYL-2,3-DIMETHYLPENTANE	-8.122	2.3029E-01	-3.5680E-05	275	1000	57.37	52.51	186.49
918	C9H20	2,4-DIMETHYL-3-ETHYLPENTANE	-4.321	2.2703E-01	-5.1733E-05	151	1000	58.77	28.78	170.98
919	C9H20	2,2-DIMETHYLHEPTANE	-3.730	2.2218E-01	-5.7903E-05	160	1000	57.37	30.34	160.55
920	C9H20	2,6-DIMETHYLHEPTANE	-2.022	2.1185E-01	-5.5730E-05	170	1000	56.19	32.38	154.10
921	C9H20	3-ETHYLHEPTANE	-3.932	2.1932E-01	-5.4820E-05	158	1000	56.58	29.35	160.57
922	C9H20	4-ETHYLHEPTANE	-7.684	2.1474E-01	-3.5484E-05	275	1000	53.19	48.69	171.57
923	C9H20	2,3-DIMETHYLHEPTANE	-7.882	2.2079E-01	-3.6004E-05	275	1000	54.75	50.11	176.90
924	C9H20	2,4-DIMETHYLHEPTANE	-8.008	2.2314E-01	-3.7711E-05	275	1000	55.17	50.50	177.42
925	C9H20	2,5-DIMETHYLHEPTANE	-7.899	2.2042E-01	-3.6815E-05	275	1000	54.55	49.93	175.71
926	C9H20	3,4-DIMETHYLHEPTANE	-7.990	2.2411E-01	-3.6304E-05	275	1000	55.60	50.89	179.82
927	C9H20	3,5-DIMETHYLHEPTANE	-8.017	2.2388E-01	-3.7175E-05	275	1000	55.43	50.74	178.69
928	C9H20	4,4-DIMETHYLHEPTANE	-8.137	2.2744E-01	-3.7534E-05	275	1000	56.34	51.57	181.77
929	C9H20	3-ETHYL-2-METHYLHEXANE	-8.109	2.2697E-01	-3.7171E-05	275	1000	56.26	51.50	181.69
930	C9H20	4-ETHYL-2-METHYLHEXANE	-8.117	2.2641E-01	-3.7930E-05	275	1000	56.02	51.28	180.36
931	C9H20	3-ETHYL-3-METHYLHEXANE	-8.150	2.2939E-01	-3.6556E-05	275	1000	56.99	52.17	184.68
932	C9H20	3-ETHYL-4-METHYLHEXANE	-8.123	2.2807E-01	-3.6753E-05	275	1000	56.61	51.82	183.19
933	C9H20	2,2,3-TRIMETHYLHEXANE	-8.204	2.2962E-01	-3.7617E-05	275	1000	56.91	52.10	183.80
934	C9H20	2,2,4-TRIMETHYLHEXANE	-8.328	2.3187E-01	-3.9537E-05	275	1000	57.29	52.45	184.00
935	C9H20	2,3,3-TRIMETHYLHEXANE	-8.175	2.2987E-01	-3.6794E-05	275	1000	57.09	52.26	184.90
936	C9H20	2,3,4-TRIMETHYLHEXANE	-8.114	2.2795E-01	-3.6647E-05	275	1000	56.59	51.80	183.19

Table 21-1 VISCOSITY OF GAS - ORGANIC COMPOUNDS (continued)

NO	FORMULA	NAME	$n_{gas} = A + BT + CT^2$					n_{gas} - micropoise, T - K		
			A	B	C	TMIN	TMAX	n_{gas} @25 C	n_{gas} @TMIN	n_{gas} @TMAX
937	C9H20	2,3,5-TRIMETHYLHEXANE	-8.128	2.2666E-01	-3.8055E-05	275	1000	56.07	51.33	180.48
938	C9H20	2,4,4-TRIMETHYLHEXANE	-8.248	2.3017E-01	-3.8400E-05	275	1000	56.96	52.15	183.52
939	C9H20	3,3,4-TRIMETHYLHEXANE	-8.168	2.3073E-01	-3.6244E-05	275	1000	57.40	52.54	186.32
940	C9H20	2-METHYLOCTANE	1.581	1.9878E-01	-4.4337E-05	416	1000	-----	76.60	156.02
941	C9H20	3-METHYLOCTANE	2.472	1.9729E-01	-4.3816E-05	417	1000	-----	77.12	155.95
942	C9H20	4-METHYLOCTANE	3.957	1.9343E-01	-4.0662E-05	416	1000	-----	77.39	156.73
943	C9H20	n-NONANE	-6.802	1.8688E-01	3.4929E-07	273	773	48.95	44.24	137.86
944	C9H20	2,2,3,3-TETRAMETHYLPENTANE	-1.922	2.2330E-01	-5.0470E-05	413	1000	-----	81.69	170.91
945	C9H20	2,2,3,4-TETRAMETHYLPENTANE	-0.622	2.1865E-01	-5.0811E-05	406	1000	-----	79.77	167.22
946	C9H20	2,2,4,4-TETRAMETHYLPENTANE	3.584	2.0199E-01	-4.5209E-05	395	1000	-----	76.32	160.37
947	C9H20	2,3,3,4-TETRAMETHYLPENTANE	-7.776	2.2061E-01	-3.4107E-05	275	1000	54.97	50.31	178.73
948	C9H20	2,2,5-TRIMETHYLHEXANE	-1.028	2.1683E-01	-5.2617E-05	397	1000	-----	76.76	163.19
949	C9H20O	2,6-DIMETHYL-4-HEPTANOL	-5.674	2.1542E-01	-3.5338E-05	451	1000	-----	84.29	174.41
950	C9H20O	1-NONANOL	-6.044	2.0031E-01	-2.5555E-05	486	1000	-----	85.27	168.71
951	C9H20O	2-NONANOL	-3.625	2.0151E-01	-2.8409E-05	238	1000	53.93	42.73	169.48
952	C9H20S	n-NONYL MERCAPTAN	-----	--------	--------	----	----	-----	-----	-----
953	C9H20S	BUTYL-PENTYL-SULFIDE	-6.075	1.9381E-01	-2.2879E-05	275	1000	49.68	45.49	164.86
954	C9H20S	ETHYL-HEPTYL-SULFIDE	-6.075	1.9381E-01	-2.2879E-05	275	1000	49.68	45.49	164.86
955	C9H20S	HEXYL-PROPYL-SULFIDE	-6.075	1.9381E-01	-2.2879E-05	275	1000	49.68	45.49	164.86
956	C9H20S	METHYL-OCTYL-SULFIDE	-6.075	1.9381E-01	-2.2879E-05	275	1000	49.68	45.49	164.86
957	C9H21N	n-NONYLAMINE	-3.637	1.9185E-01	-2.5462E-05	273	1000	51.30	46.84	162.75
958	C9H21N	TRIPROPYLAMINE	-3.986	2.2556E-01	-2.1556E-05	180	1000	61.35	35.92	200.02
959	C10H6O8	PYROMELLITIC ACID	-5.610	1.8093E-01	-1.8344E-05	554	1000	-----	89.00	156.98
960	C10H7Br	1-BROMONAPHTHALENE	-7.252	2.5256E-01	-2.6628E-05	554	1000	-----	124.49	218.68
961	C10H7Cl	1-CHLORONAPHTHALENE	-6.855	2.3520E-01	-2.6843E-05	532	1000	-----	110.67	201.50
962	C10H8	NAPHTHALENE	-16.789	2.5406E-01	-3.5495E-05	353	1000	-----	68.47	201.78
963	C10H8	AZULENE	-4.802	2.0834E-01	-1.4894E-05	275	1000	55.99	51.37	188.64
964	C10H9N	QUINALDINE	-3.783	2.2094E-01	-2.2246E-05	272	1000	60.11	54.67	194.91
965	C10H10	m-DIVINYLBENZENE	-6.025	2.3547E-01	-5.2396E-05	206	1000	59.52	40.26	177.05
966	C10H10	1-METHYLINDENE	-6.755	2.3609E-01	-5.1219E-05	350	1000	-----	69.60	178.12
967	C10H10	2-METHYLINDENE	-4.375	2.3584E-01	-4.7501E-05	353	1000	-----	72.96	183.96
968	C10H10O4	DIMETHYL PHTHALATE	-2.094	2.0161E-01	-1.9452E-05	557	1000	-----	104.17	180.06
969	C10H10O4	DIMETHYL TEREPHTHALATE	-0.783	2.1295E-01	-1.6980E-05	414	1000	-----	84.47	195.19
970	C10H12	DICYCLOPENTADIENE	-----	--------	--------	----	----	-----	-----	-----
971	C10H12	1,2,3,4-TETRAHYDRONAPHTHALENE	-15.394	2.5754E-01	-3.2295E-05	481	991	-----	101.01	208.11
972	C10H12O	ANETHOLE	-3.775	2.1393E-01	-2.3706E-05	295	1000	57.90	57.27	186.45
973	C10H12O4	DIALLYL MALEATE	-3.163	2.0247E-01	-2.4597E-05	226	1000	55.02	41.34	174.71
974	C10H14	n-BUTYLBENZENE	-3.667	2.2078E-01	-4.6696E-05	456	996	-----	87.30	169.91
975	C10H14	sec-BUTYLBENZENE	-5.272	2.2665E-01	-4.7376E-05	446	1000	-----	86.39	174.00
976	C10H14	tert-BUTYLBENZENE	-5.023	2.2769E-01	-4.5804E-05	442	1000	-----	86.67	176.86
977	C10H14	1,2,3,4-TETRAMETHYLBENZENE	-6.300	2.0771E-01	-2.3098E-05	275	1000	53.58	49.07	178.31
978	C10H14	m-CYMENE	-3.958	2.2572E-01	-4.8946E-05	448	1000	-----	87.34	172.82
979	C10H14	o-CYMENE	-4.700	2.2501E-01	-4.8782E-05	273	1000	58.05	53.09	171.53
980	C10H14	p-CYMENE	-2.692	2.2025E-01	-4.9525E-05	450	1000	-----	86.39	168.03
981	C10H14	m-DIETHYLBENZENE	-2.462	2.1821E-01	-4.9306E-05	454	994	-----	86.44	165.72
982	C10H14	o-DIETHYLBENZENE	11.390	1.7425E-01	-1.6743E-05	457	1000	-----	87.53	168.90
983	C10H14	p-DIETHYLBENZENE	12.885	1.7073E-01	-1.8158E-05	457	1000	-----	87.12	165.46
984	C1OH14	2-ETHYL-m-XYLENE	-6.430	2.3108E-01	-5.1367E-05	257	1000	57.90	49.56	173.28
985	C10H14	2-ETHYL-p-XYLENE	-4.831	2.2485E-01	-5.5172E-05	220	1000	57.30	41.97	164.85
986	C10H14	3-ETHYL-o-XYLENE	-1.450	2.0767E-01	-4.2096E-05	467	1000	-----	86.35	164.12
987	C10H14	4-ETHYL-m-XYLENE	-4.813	2.2400E-01	-5.5594E-05	210	1000	57.03	39.78	163.59
988	C10H14	4-ETHYL-o-XYLENE	-4.526	2.2104E-01	-5.1668E-05	206	1000	56.78	38.82	164.85
989	C1OH14	5-ETHYL-m-XYLENE	-2.232	2.1240E-01	-5.3828E-05	189	1000	56.31	35.99	156.34
990	C10H14	ISOBUTYLBENZENE	13.342	1.8277E-01	-1.9304E-05	446	1000	-----	91.02	176.81
991	C10H14	1,2,3,5-TETRAMETHYLBENZENE	-4.899	2.2250E-01	-5.1723E-05	249	1000	56.84	47.30	165.88
992	C10H14	1,2,4,5-TETRAMETHYLBENZENE	2.071	2.0350E-01	-4.1091E-05	470	1000	-----	88.64	164.48
993	C10H14O	p-tert-BUTYLPHENOL	-8.575	2.2637E-01	-3.2804E-05	513	1000	-----	98.92	184.99
994	C10H14O2	p-tert-BUTYLCATECHOL	-7.536	2.1498E-01	-2.7042E-05	558	1000	-----	104.00	180.40
995	C10H15N	N,N-DIETYHLANILINE	-----	--------	--------	----	----	-----	-----	-----
996	C10H15N	2,6-DIETHYLANILINE	-3.450	1.9811E-01	-2.4274E-05	277	1000	53.46	49.56	170.39
997	C10H16	CAMPHENE	-4.616	2.2730E-01	-5.1828E-05	320	1000	-----	62.81	170.86
998	C10H16	D-LIMONENE	-6.323	2.2421E-01	-5.0292E-05	199	1000	56.05	36.30	167.60
999	C10H16	alpha-PHELLANDRENE	-6.307	2.3170E-01	-5.3718E-05	220	1000	58.00	42.07	171.68
1000	C1OH16	beta-PHELLANDRENE	-4.197	2.2534E-01	-5.1869E-05	220	1000	58.38	42.87	169.27
1001	C10H16	alpha-PINENE	-1.923	2.2208E-01	-4.4463E-05	429	1000	-----	85.17	175.69
1002	C10H16	beta-PINENE	-3.139	2.2078E-01	-4.6522E-05	439	1000	-----	84.82	171.12
1003	C10H16	alpha-TERPINENE	-4.982	2.2231E-01	-4.8022E-05	220	1000	57.03	41.60	169.31
1004	C1OH16	gamma-TERPINENE	-4.646	2.1949E-01	-4.9340E-05	220	1000	56.41	41.25	165.50
1005	C10H16	TERPINOLENE	-3.203	2.0925E-01	-4.5597E-05	458	1000	-----	83.07	160.45
1006	C10H16O	CAMPHOR	-10.210	2.5187E-01	-2.2342E-05	481	1000	-----	105.77	219.32
1007	C10H18	1-DECYNE	-6.996	2.0413E-01	-2.9089E-05	275	1000	51.28	46.94	168.04
1008	C10H18	cis-DECAHYDRONANPHTALENE	-6.213	2.3052E-01	-4.4417E-05	469	1000	-----	92.13	179.89
1009	C10H18	trans-DECAHYDRONAPHTHALENE	6.503	1.8310E-01	-3.6662E-05	460	1000	-----	82.97	152.94
1010	C10H18O4	SEBACIC ACID	-5.372	1.8420E-01	-2.2453E-05	408	1000	-----	66.04	156.38
1011	C10H20	n-BUTYLCYCLOHEXANE	-3.361	2.0736E-01	-5.1499E-05	198	1000	53.89	35.68	152.50
1012	C10H20	1-CYCLOPENTYLPENTANE	-7.005	2.0901E-01	-2.8239E-05	275	1000	52.80	48.34	173.77
1013	C10H20	1-DECENE	4.015	1.8083E-01	-3.8216E-05	444	1000	-----	76.77	146.63
1014	C10H20O	1-DECANAL	-4.467	1.9082E-01	-2.5974E-05	488	1000	-----	82.47	160.38

467

Table 21-1 VISCOSITY OF GAS - ORGANIC COMPOUNDS (continued)

			$n_{gas} = A + B T + C T^2$					(n_{gas} - micropoise, T - K)		
NO	FORMULA	NAME	A	B	C	TMIN	TMAX	n_{gas} @25 C	n_{gas} @TMIN	n_{gas} @TMAX
1015	C10H20O2	n-DECANOIC ACID	-4.079	1.8754E-01	-2.3182E-05	305	1000	-----	50.96	160.28
1016	C10H20O2	2-ETHYLHEXYL ACETATE	-6.044	2.0364E-01	-3.1611E-05	472	1000	-----	83.03	165.99
1017	C10H20O2	ISOPENTYL ISOVALERATE	-5.173	2.0522E-01	-3.2170E-05	467	1000	-----	83.65	167.88
1018	C10H22	n-DECANE	-7.297	1.8506E-01	-4.8008E-06	278	783	47.45	43.78	134.66
1019	C10H22	2-METHYLNONANE	-2.147	2.0076E-01	-5.3880E-05	199	1000	52.92	35.67	144.73
1020	C10H22	3-METHYLNONANE	-2.383	2.0236E-01	-5.2486E-05	188	1000	53.29	33.81	147.49
1021	C10H22	4-METHYLNONANE	-2.468	2.0337E-01	-5.2018E-05	174	1000	53.54	31.34	148.88
1022	C10H22	5-METHYLNONANE	-2.521	2.0343E-01	-5.1960E-05	185	1000	53.51	33.34	148.95
1023	C10H22	3-ETHYLOCTANE	-7.344	2.0957E-01	-3.1788E-05	275	1000	52.31	47.88	170.44
1024	C10H22	4-ETHYLOCTANE	-7.495	2.1304E-01	-3.2723E-05	275	1000	53.11	48.62	172.82
1025	C10H22	2,2-DIMETHYLOCTANE	-2.070	2.0582E-01	-5.3005E-05	225	1000	54.58	41.56	150.75
1026	C10H22	2,3-DIMETHYLOCTANE	-7.349	2.0963E-01	-3.1836E-05	275	1000	52.32	47.89	170.44
1027	C10H22	2,4-DIMETHYLOCTANE	-7.553	2.1290E-01	-3.3735E-05	275	1000	52.92	48.44	171.61
1028	C10H22	2,5-DIMETHYLOCTANE	-7.470	2.1113E-01	-3.3094E-05	275	1000	52.54	48.09	170.57
1029	C10H22	2,6-DIMETHYLOCTANE	-7.359	2.0801E-01	-3.2596E-05	275	1000	51.76	47.38	168.06
1030	C10H22	2,7-DIMETHYLOCTANE	-7.279	2.0572E-01	-3.2257E-05	275	1000	51.19	46.85	166.18
1031	C10H22	3,3-DIMETHYLOCTANE	-7.501	2.1373E-01	-3.2572E-05	275	1000	53.33	48.81	173.66
1032	C10H22	3,4-DIMETHYLOCTANE	-7.494	2.1393E-01	-3.2408E-05	275	1000	53.41	48.89	174.03
1033	C10H22	3,5-DIMETHYLOCTANE	-7.564	2.1437E-01	-3.3265E-05	275	1000	53.39	48.87	173.54
1034	C10H22	3,6-DIMETHYLOCTANE	-7.470	2.1209E-01	-3.2709E-05	275	1000	52.86	48.38	171.91
1035	C10H22	4,4-DIMETHYLOCTANE	-7.671	2.1752E-01	-3.3692E-05	275	1000	54.19	49.60	176.16
1036	C10H22	4,5-DIMETHYLOCTANE	-7.567	2.1563E-01	-3.2851E-05	275	1000	53.80	49.25	175.21
1037	C10H22	4-PROPYLHEPTANE	-7.689	2.1697E-01	-3.4216E-05	275	1000	53.96	49.39	175.07
1038	C10H22	4-ISOPROPYLHEPTANE	-7.731	2.1935E-01	-3.3901E-05	275	1000	54.66	50.03	177.72
1039	C10H22	3-ETHYL-2-METHYLHEPTANE	-7.614	2.1669E-01	-3.3148E-05	275	1000	54.05	49.47	175.93
1040	C10H22	4-ETHYL-2-METHYLHEPTANE	-7.725	2.1813E-01	-3.4316E-05	275	1000	54.26	49.67	176.09
1041	C10H22	5-ETHYL-2-METHYLHEPTANE	-7.558	2.1427E-01	-3.3209E-05	275	1000	53.37	48.86	173.50
1042	C10H22	3-ETHYL-3-METHYLHEPTANE	-7.635	2.1935E-01	-3.2592E-05	275	1000	54.87	50.22	179.12
1043	C10H22	4-ETHYL-3-METHYLHEPTANE	-7.703	2.1995E-01	-3.3287E-05	275	1000	54.92	50.27	178.96
1044	C10H22	3-ETHYL-5-METHYLHEPTANE	-7.708	2.1852E-01	-3.3878E-05	275	1000	54.43	49.82	176.93
1045	C10H22	3-ETHYL-4-METHYLHEPTANE	-7.656	2.1888E-01	-3.2999E-05	275	1000	54.67	50.04	178.23
1046	C10H22	4-ETHYL-4-METHYLHEPTANE	-7.776	2.2235E-01	-3.3501E-05	275	1000	55.54	50.84	181.07
1047	C10H22	2,2,3-TRIMETHYLHEPTANE	-7.670	2.1844E-01	-3.3332E-05	275	1000	54.50	49.88	177.44
1048	C10H22	2,2,4-TRIMETHYLHEPTANE	-7.850	2.2051E-01	-3.5452E-05	275	1000	54.74	50.11	177.21
1049	C10H22	2,2,5-TRIMETHYLHEPTANE	-7.734	2.1778E-01	-3.4653E-05	275	1000	54.12	49.53	175.39
1050	C10H22	2,2,6-TRIMETHYLHEPTANE	-7.655	2.1490E-01	-3.4661E-05	275	1000	53.34	48.82	172.58
1051	C10H22	2,3,3-TRIMETHYLHEPTANE	-7.664	2.1957E-01	-3.2891E-05	275	1000	54.88	50.23	179.02
1052	C10H22	2,3,4-TRIMETHYLHEPTANE	-7.681	2.1920E-01	-3.3238E-05	275	1000	54.72	50.09	178.28
1053	C10H22	2,3,5-TRIMETHYLHEPTANE	-7.597	2.1662E-01	-3.2940E-05	275	1000	54.06	49.48	176.08
1054	C10H22	2,3,6-TRIMETHYLHEPTANE	-7.602	2.1504E-01	-3.3595E-05	275	1000	53.53	48.99	173.84
1055	C10H22	2,4,4-TRIMETHYLHEPTANE	-7.885	2.2240E-01	-3.5146E-05	275	1000	55.30	50.62	179.37
1056	C10H22	2,4,5-TRIMETHYLHEPTANE	-7.721	2.1893E-01	-3.3910E-05	275	1000	54.54	49.92	177.30
1057	C10H22	2,4,6-TRIMETHYLHEPTANE	-7.779	2.1798E-01	-3.5475E-05	275	1000	54.06	49.48	174.73
1058	C10H22	2,5,5-TRIMETHYLHEPTANE	-7.760	2.1930E-01	-3.4385E-05	275	1000	54.57	49.95	177.16
1059	C10H22	3,3,4-TRIMETHYLHEPTANE	-7.712	2.2210E-01	-3.2775E-05	275	1000	55.59	50.89	181.61
1060	C10H22	3,3,5-TRIMETHYLHEPTANE	-7.405	2.1047E-01	-3.2337E-05	275	1000	52.47	48.03	170.73
1061	C10H22	3,4,4-TRIMETHYLHEPTANE	-7.753	2.2298E-01	-3.3034E-05	275	1000	55.79	51.07	182.19
1062	C10H22	3,4,5-TRIMETHYLHEPTANE	-7.550	2.1528E-01	-3.2736E-05	275	1000	53.73	49.18	174.99
1063	C10H22	3-ISOPROPYL-2-METHYLHEXANE	-7.598	2.1916E-01	-3.2202E-05	275	1000	54.88	50.24	179.36
1064	C10H22	3,3-DIETHYLHEXANE	-7.737	2.2436E-01	-3.2486E-05	275	1000	56.27	51.51	184.14
1065	C10H22	3,4-DIETHYLHEXANE	-7.750	2.2235E-01	-3.3167E-05	275	1000	55.60	50.89	181.43
1066	C10H22	3-ETHYL-2,2-DIMETHYLHEXANE	-7.885	2.2456E-01	-3.4267E-05	275	1000	56.02	51.28	182.41
1067	C10H22	4-ETHYL-2,2-DIMETHYLHEXANE	-7.986	2.2436E-01	-3.6061E-05	275	1000	55.70	50.99	180.31
1068	C10H22	3-ETHYL-2,3-DIMETHYLHEXANE	-7.738	2.2412E-01	-3.2560E-05	275	1000	56.19	51.43	183.82
1069	C10H22	4-ETHYL-2,3-DIMETHYLHEXANE	-7.772	2.2263E-01	-3.3370E-05	275	1000	55.64	50.93	181.49
1070	C10H22	3-ETHYL-2,4-DIMETHYLHEXANE	-7.810	2.2344E-01	-3.3622E-05	275	1000	55.82	51.09	182.01
1071	C10H22	4-ETHYL-2,4-DIMETHYLHEXANE	-7.711	2.2175E-01	-3.2852E-05	275	1000	55.48	50.79	181.19
1072	C10H22	3-ETHYL-2,5-DIMETHYLHEXANE	-7.841	2.2171E-01	-3.4700E-05	275	1000	55.18	50.51	179.17
1073	C10H22	4-ETHYL-3,3-DIMETHYLHEXANE	-7.792	2.2536E-01	-3.2861E-05	275	1000	56.48	51.70	184.71
1074	C10H22	3-ETHYL-3,4-DIMETHYLHEXANE	-7.832	2.2620E-01	-3.3115E-05	275	1000	56.67	51.87	185.25
1075	C10H22	2,2,3,3-TETRAMETHYLHEXANE	-7.267	2.0950E-01	-3.0824E-05	275	1000	52.46	48.01	171.41
1076	C10H22	2,2,3,4-TETRAMETHYLHEXANE	-7.800	2.2419E-01	-3.3269E-05	275	1000	56.08	51.34	183.12
1077	C10H22	2,2,3,5-TETRAMETHYLHEXANE	-7.943	2.2419E-01	-3.5324E-05	275	1000	55.76	51.04	180.92
1078	C10H22	2,2,4,4-TETRAMETHYLHEXANE	-7.881	2.2414E-01	-3.4363E-05	275	1000	55.89	51.16	181.90
1079	C10H22	2,2,4,5-TETRAMETHYLHEXANE	-7.931	2.2341E-01	-3.5497E-05	275	1000	55.52	50.82	179.98
1080	C10H22	2,2,5,5-TETRAMETHYLHEXANE	-7.854	2.1918E-01	-3.6578E-05	275	1000	54.24	49.65	174.75
1081	C10H22	2,3,3,4-TETRAMETHYLHEXANE	-7.696	2.2473E-01	-3.1957E-05	275	1000	56.47	51.69	185.08
1082	C10H22	2,3,3,5-TETRAMETHYLHEXANE	-7.905	2.2481E-01	-3.4476E-05	275	1000	56.06	51.31	182.43
1083	C10H22	2,3,4,4-TETRAMETHYLHEXANE	-7.769	2.2496E-01	-3.2705E-05	275	1000	56.40	51.62	184.49
1084	C10H22	2,3,4,5-TETRAMETHYLHEXANE	-7.829	2.2331E-01	-3.3915E-05	275	1000	55.74	51.02	181.57
1085	C10H22	3,3,4,4-TETRAMETHYLHEXANE	-7.550	2.2499E-01	-3.0486E-05	275	1000	56.82	52.02	186.95
1086	C10H22	2,4-DIMETHYL-3-ISOPROPYLPENTANE	-7.898	2.2556E-01	-3.4125E-05	275	1000	56.32	51.55	183.54
1087	C10H22	3,3-DIETHYL-2-METHYLPENTANE	-7.662	2.2593E-01	-3.1372E-05	275	1000	56.91	52.10	186.90
1088	C10H22	3-ETHYL-2,2,3-TRIMETHYLPENTANE	-7.595	2.2608E-01	-3.0711E-05	275	1000	57.08	52.25	187.77
1089	C10H22	3-ETHYL-2,2,4-TRIMETHYLPENTANE	-7.959	2.2751E-01	-3.4321E-05	275	1000	56.82	52.01	185.23
1090	C10H22	3-ETHYL-2,3,4-TRIMETHYLPENTANE	-7.601	2.2495E-01	-3.0970E-05	275	1000	56.71	51.92	186.38
1091	C10H22	2,2,3,3,4-PENTAMETHYLPENTANE	-7.630	2.2634E-01	-3.0993E-05	275	1000	57.10	52.27	187.72
1092	C10H22	2,2,3,4,4-PENTAMETHYLPENTANE	-7.799	2.2602E-01	-3.2782E-05	275	1000	56.67	51.88	185.44

Table 21-1 VISCOSITY OF GAS - ORGANIC COMPOUNDS (continued)

			$n_{gas} = A + B T + C T^2$				(n_{gas} - micropoise, T - K)		
NO	FORMULA	NAME	A	B	C	TMIN	TMAX	n_{gas} @25 C	n_{gas} @TMIN	n_{gas} @TMAX
1093	C10H22O	1-DECANOL	-1.075	1.8019E-01	-1.7007E-05	503	1000	-----	85.26	162.11
1094	C10H22O	DI-n-PENTYL ETHER	-2.796	1.9169E-01	-2.5704E-05	204	1000	52.07	35.24	163.19
1095	C10H22O	ISODECANOL	-2.995	1.8974E-01	-2.3764E-05	493	1000	-----	84.77	162.98
1096	C10H22O5	TETRAETHYLENE GLYCOL DIMETHYL ETHER	-3.435	1.9092E-01	-2.3102E-05	243	1000	51.43	41.59	164.38
1097	C10H22S	n-DECYL MERCAPTAN	-3.177	2.0833E-01	-1.0827E-05	248	1000	57.97	47.82	194.33
1098	C10H22S	BUTYL-HEXYL-SULFIDE	-5.474	1.8331E-01	-1.9835E-05	275	1000	47.42	43.44	158.00
1099	C10H22S	ETHYL-OCTYL-SULFIDE	-5.474	1.8331E-01	-1.9835E-05	275	1000	47.42	43.44	158.00
1100	C10H22S	HEPTYL-PROPYL-SULFIDE	-5.474	1.8331E-01	-1.9835E-05	275	1000	47.42	43.44	158.00
1101	C10H22S	METHYL-NONYL-SULFIDE	-5.474	1.8331E-01	-1.9835E-05	275	1000	47.42	43.44	158.00
1102	C10H22S	PENTYL-SULFIDE	-5.474	1.8331E-01	-1.9835E-05	275	1000	47.42	43.44	158.00
1103	C10H22S2	PENTYL-DISULFIDE	-4.966	1.7967E-01	-1.7077E-05	275	1000	47.08	43.15	157.63
1104	C10H23N	n-DECYLAMINE	-3.838	1.8489E-01	-2.4366E-05	289	1000	49.12	47.56	156.69
1105	C11H10	1-METHYLNAPHTHALENE	-7.548	2.1310E-01	-4.0458E-05	518	998	-----	91.98	164.83
1106	C11H10	2-METHYLNAPHTHALENE	-6.045	2.1481E-01	-4.2980E-05	514	994	-----	93.01	165.01
1107	C11H14O2	n-BUTYL BENZOATE	-3.279	1.9850E-01	-2.1886E-05	252	1000	53.96	45.35	173.34
1108	C11H16	n-PENTYLBENZENE	-4.385	2.1258E-01	-4.9519E-05	198	1000	54.59	35.76	158.68
1109	C11H16O	p-tert-AMYLPHENOL	-4.529	2.0306E-01	-2.1569E-05	535	1000	-----	97.93	176.96
1110	C11H20	1-UNDECYNE	-6.460	1.9390E-01	-2.5860E-05	275	1000	49.05	44.91	161.58
1111	C11H20O2	2-ETHYLHEXYL ACRYLATE	-0.522	1.8662E-01	-2.2695E-05	489	1000	-----	85.31	163.40
1112	C11H22	1-UNDECENE	-1.935	1.9180E-01	-4.7790E-05	224	1000	51.00	38.63	142.08
1113	C11H22	1-CYCLOPENTYLHEXANE	-6.380	1.9753E-01	-2.4699E-05	275	1000	50.32	46.07	166.45
1114	C11H22	PENTYLCYCLOHEXANE	-6.363	1.9961E-01	-2.4324E-05	275	1000	50.99	46.69	168.92
1115	C11H22O	1-UNDECANAL	-2.568	1.7785E-01	-2.0333E-05	273	1000	48.65	44.47	154.95
1116	C11H24	n-UNDECANE	-10.044	1.8311E-01	-6.9882E-06	316	784	-----	47.12	129.22
1117	C11H24O	1-UNDECANOL	-4.268	1.8323E-01	-2.3458E-05	518	1000	-----	84.35	155.50
1118	C11H24S	UNDECYL MERCAPTAN	-2.270	1.7520E-01	-1.5256E-05	270	1000	48.61	43.92	157.67
1119	C11H24S	BUTYL-HEPTYL-SULFIDE	-4.959	1.7431E-01	-1.7368E-05	275	1000	45.47	41.66	151.98
1120	C11H24S	DECYL-METHYL-SULFIDE	-4.959	1.7431E-01	-1.7368E-05	275	1000	45.47	41.66	151.98
1121	C11H24S	ETHYL-NONYL-SULFIDE	-4.959	1.7431E-01	-1.7368E-05	275	1000	45.47	41.66	151.98
1122	C11H24S	OCTYL-PROPYL-SULFIDE	-4.959	1.7431E-01	-1.7368E-05	275	1000	45.47	41.66	151.98
1123	C12H8O	DIBENZOFURAN	-3.545	2.0368E-01	-4.7881E-06	356	1000	-----	68.36	195.35
1124	C12H9N	DIBENZOPYRROLE	-7.636	2.1807E-01	-9.1469E-06	628	1000	-----	125.70	201.29
1125	C12H10	ACENAPHTHENE	-5.508	2.0308E-01	-3.9619E-05	551	1000	-----	94.36	157.95
1126	C12H10	BIPHENYL	-13.498	2.4098E-01	-2.9320E-05	500	1000	-----	99.66	198.16
1127	C12H10O	DIPHENYL ETHER	-2.645	2.0518E-01	-1.9891E-05	531	1000	-----	100.70	182.64
1128	C12H11N	p-AMINODIPHENYL	-3.206	1.9469E-01	-1.7328E-05	326	1000	-----	58.42	174.16
1129	C12H11N	DIPHENYLAMINE	-3.913	2.0782E-01	-2.0583E-05	575	1000	-----	108.78	183.32
1130	C12H11N3	p-AMINOAZOBENZENE	-4.302	1.9234E-01	-5.3037E-05	401	1000	-----	71.97	182.73
1131	C12H11N3	1,3-DIPHENYLTRIAZENE	-2.246	1.8949E-01	-1.2739E-05	372	1000	-----	66.48	174.51
1132	C12H12	1,2-DIMETHYLNAPHTHALENE	-4.343	1.9009E-01	-1.3411E-05	275	1000	51.14	46.92	172.34
1133	C12H12	1,3-DIMETHYLNAPHTHALENE	-4.384	1.9044E-01	-1.3588E-05	275	1000	51.19	46.96	172.47
1134	C12H12	1,4-DIMETHYLNAPHTHALENE	-4.307	1.8976E-01	-1.3251E-05	275	1000	51.09	46.88	172.20
1135	C12H12	1,5-DIMETHYLNAPHTHALENE	-4.391	1.9051E-01	-1.3620E-05	275	1000	51.20	46.97	172.50
1136	C12H12	1,6-DIMETHYLNAPHTHALENE	-4.464	1.9116E-01	-1.3943E-05	275	1000	51.29	47.05	172.75
1137	C12H12	1,7-DIMETHYLNAPHTHALENE	-4.464	1.9116E-01	-1.3943E-05	275	1000	51.29	47.05	172.75
1138	C12H12	2,3-DIMETHYLNAPHTHALENE	-4.281	1.8954E-01	-1.3140E-05	275	1000	51.06	46.85	172.12
1139	C12H12	2,6-DIMETHYLNAPHTHALENE	-6.244	2.1670E-01	-4.4807E-05	383	1000	-----	70.18	165.65
1140	C12H12	2,7-DIMETHYLNAPHTHALENE	-6.613	2.1830E-01	-4.6669E-05	369	1000	-----	67.59	165.02
1141	C12H12	1-ETHYLNAPHTHALENE	-3.445	2.0356E-01	-4.6596E-05	259	1000	53.10	46.15	153.52
1142	C12H12	2-ETHYLNAPHTHALENE	-4.424	1.9323E-01	-1.3676E-05	275	1000	51.97	47.68	175.13
1143	C12H12N2	p-AMINODIPHENYLAMINE	-0.565	1.8695E-01	-1.2372E-05	627	1000	-----	111.79	174.01
1144	C12H12N2	HYDRAZOBENZENE	-5.376	2.1015E-01	-2.3784E-05	404	1000	-----	75.64	180.99
1145	C12H14	1,2,3-TRIMETHYLINDENE	-0.567	2.0121E-01	-4.4388E-05	345	1000	-----	63.57	156.25
1146	C12H14O4	DIETHYL PHTHALATE	-6.267	1.9349E-01	-2.2181E-05	567	1000	-----	96.31	165.04
1147	C12H16	CYCLOHEXYLBENZENE	-4.913	2.1010E-01	-4.6325E-05	280	1000	53.61	50.28	158.86
1148	C12H18	m-DIISOPROPYLBENZENE	-3.377	2.0894E-01	-4.4368E-05	476	1000	-----	86.03	161.19
1149	C12H18	p-DIISOPROPYLBENZENE	-2.144	2.0446E-01	-4.2702E-05	484	1000	-----	86.81	159.61
1150	C12H18	n-HEXYLBENZENE	-1.470	1.9567E-01	-4.1289E-05	499	1000	-----	85.89	152.91
1151	C12H18	1,2,3-TRIETHYLBENZENE	-6.064	1.9470E-01	-2.2709E-05	275	1000	49.97	45.76	165.93
1152	C12H18	1,2,4-TRIETHYLBENZENE	-6.064	1.9470E-01	-2.2709E-05	275	1000	49.97	45.76	165.93
1153	C12H18	1,3,5-TRIETHYLBENZENE	-6.109	1.9519E-01	-2.2972E-05	275	1000	50.05	45.83	166.11
1154	C12H18	HEXAMETHYLBENZENE	-4.509	1.8288E-01	-1.4502E-05	275	1000	48.73	44.69	163.87
1155	C12H20O4	DIBUTYL MALEATE	-2.357	1.7626E-01	-1.9256E-05	188	1000	48.48	30.10	154.65
1156	C12H22	BICYCLOHEXYL	1.871	1.8841E-01	-3.3600E-05	512	1000	-----	89.53	156.68
1157	C12H22	1-DODECYNE	-5.962	1.8477E-01	-2.3058E-05	275	1000	47.08	43.11	155.75
1158	C12H23N	DICYCLOHEXYLAMINE	-3.862	2.0157E-01	-2.3298E-05	273	1000	54.16	49.43	174.41
1159	C12H24	1-DODECENE	-1.293	1.8285E-01	-4.6080E-05	238	1000	49.13	39.62	135.48
1160	C12H24	1-CYCLOPENTYLHEPTANE	-5.903	1.8724E-01	-2.2345E-05	275	1000	47.94	43.90	158.99
1161	C12H24	1-CYCLOHEXYLHEXANE	-5.794	1.8942E-01	-2.1380E-05	275	1000	48.78	44.68	162.25
1162	C12H24O	1-DODECANAL	-6.860	1.8436E-01	-2.8455E-05	523	1000	-----	81.78	149.05
1163	C12H24O2	n-DODECANOIC ACID	-2.410	1.6834E-01	-1.6470E-05	317	1000	49.30	49.30	149.46
1164	C12H26	n-DODECANE	-12.217	1.8099E-01	-8.9949E-06	311	751	-----	43.20	118.63
1165	C12H26O	DI-n-HEXYL ETHER	-4.727	1.8283E-01	-2.6794E-05	499	1000	-----	79.83	151.31
1166	C12H26O	1-DODECANOL	-4.397	1.7627E-01	-2.1280E-05	535	1000	-----	83.82	150.59
1167	C12H26O3	DIETHYLENE GLYCOL DI-n-BUTYL ETHER	-2.688	1.7631E-01	-2.1329E-05	213	1000	47.98	33.90	152.29
1168	C12H26S	n-DODECYL MERCAPTAN	-3.470	1.9845E-01	-9.1469E-06	265	1000	54.88	48.48	185.83
1169	C12H26S	BUTYL-OCTYL-SULFIDE	-4.494	1.6631E-01	-1.5238E-05	275	1000	43.74	40.09	146.58
1170	C12H26S	DECYL-ETHYL-SULFIDE	-4.494	1.6631E-01	-1.5238E-05	275	1000	43.74	40.09	146.58

Table 21-1 VISCOSITY OF GAS - ORGANIC COMPOUNDS (continued)

NO	FORMULA	NAME	$n_{gas} = A + BT + CT^2$					$(n_{gas}$ - micropoise, T - K)		
			A	B	C	TMIN	TMAX	n_{gas} @25 C	n_{gas} @TMIN	n_{gas} @TMAX
1171	C12H26S	HEXYL-SULFIDE	-4.494	1.6631E-01	-1.5238E-05	275	1000	43.74	40.09	146.58
1172	C12H26S	METHYL-UNDECYL-SULFIDE	-4.494	1.6631E-01	-1.5238E-05	275	1000	43.74	40.09	146.58
1173	C12H26S	NONYL-PROPYL-SULFIDE	-4.494	1.6631E-01	-1.5238E-05	275	1000	43.74	40.09	146.58
1174	C12H26S2	HEXYL-DISULFIDE	-4.245	1.6494E-01	-1.3989E-05	275	1000	43.69	40.06	146.71
1175	C12H27BO3	TRI-n-BUTYL BORATE	0.198	1.7520E-01	-3.7453E-07	203	1000	52.40	35.75	175.02
1176	C12H27N	DODECYLAMINE	-3.686	1.7200E-01	-2.2140E-05	301	1000	-----	46.08	146.17
1177	C12H27N	TRI-n-BUTYLAMINE	-11.149	2.2597E-01	-3.0110E-05	487	1000	-----	91.76	184.71
1178	C13H10	FLUORENE	-7.732	2.5212E-01	-4.2730E-05	570	1000	-----	122.09	201.66
1179	C13H10O	BENZOPHENONE	-11.279	2.1167E-01	-3.0610E-05	579	1000	-----	101.02	169.78
1180	C13H12	DIPHENYLMETHANE	12.726	1.6606E-01	-1.5965E-05	537	1000	-----	97.30	162.82
1181	C13H14	1-PROPYLNAPHTHALENE	-4.271	1.8362E-01	-1.3314E-05	275	1000	49.29	45.22	166.03
1182	C13H14	2-PROPYLNAPHTHALENE	-4.247	1.8340E-01	-1.3207E-05	275	1000	49.26	45.19	165.95
1183	C13H14	2ETHYL-3-METHYLNAPHTHALENE	-4.149	1.8254E-01	-1.2778E-05	275	1000	49.14	45.08	165.61
1184	C13H14	2ETHYL-6-METHYLNAPHTHALENE	-4.391	1.8469E-01	-1.3847E-05	275	1000	49.44	45.35	166.45
1185	C13H14	2ETHYL-7-METHYLNAPHTHALENE	-4.391	1.8469E-01	-1.3847E-05	275	1000	49.44	45.35	166.45
1186	C13H20	n-HEPTYLBENZENE	-1.764	1.8810E-01	-4.6445E-05	225	1000	50.19	38.21	139.89
1187	C13H24	1-TRIDECYNE	-5.502	1.7655E-01	-2.0614E-05	275	1000	45.30	41.49	150.43
1188	C13H26	1-TRIDECENE	-1.112	1.7632E-01	-4.5756E-05	250	1000	47.39	40.11	129.45
1189	C13H26	1-CYCLOPENTYLOCTANE	-5.322	1.7871E-01	-1.9244E-05	275	1000	46.25	42.37	154.14
1190	C13H26	1-CYCLOHEXYLHEPTANE	-5.271	1.8027E-01	-1.8810E-05	275	1000	46.80	42.88	156.19
1191	C13H26O	1-TRIDECANAL	-3.544	1.6984E-01	-2.1948E-05	288	1000	45.14	43.55	144.35
1192	C13H26O2	n-BUTYL NONANOATE	-2.714	1.7040E-01	-2.1592E-05	235	1000	46.17	36.14	146.09
1193	C13H26O2	METHYL DODECANOATE	-1.922	1.6635E-01	-1.6536E-05	278	1000	46.21	43.05	147.89
1194	C13H28	n-TRIDECANE	-10.691	1.6482E-01	-1.8752E-06	273	773	38.28	34.17	115.59
1195	C13H28O	1-TRIDECANOL	-3.624	1.6920E-01	-2.0340E-05	547	1000	-----	82.84	145.24
1196	C13H28S	BUTYL-NONYL-SULFIDE	-4.077	1.5921E-01	-1.3397E-05	275	1000	42.20	38.69	141.74
1197	C13H28S	DECYL-PROPYL-SULFIDE	-4.077	1.5921E-01	-1.3397E-05	275	1000	42.20	38.69	141.74
1198	C13H28S	DODECYL-METHYL-SULFIDE	-4.077	1.5921E-01	-1.3397E-05	275	1000	42.20	38.69	141.74
1199	C13H28S	ETHYL-UNDECYL-SULFIDE	-4.077	1.5921E-01	-1.3397E-05	275	1000	42.20	38.69	141.74
1200	C13H28S	1-TRIDECANETHIOL	-4.198	1.6009E-01	-1.3982E-05	275	1000	42.29	38.77	141.91
1201	C14H8O2	ANTHRAQUINONE	-11.195	2.3260E-01	-1.3976E-05	653	1000	-----	134.73	207.43
1202	C14H10	ANTHRACENE	16.949	1.3230E-01	-1.6366E-05	615	1000	-----	92.12	132.88
1203	C14H10	DIPHENYLACETYLENE	-5.522	2.0061E-01	-3.8600E-05	573	1000	-----	96.75	156.49
1204	C14H10	PHENANTHRENE	2.085	1.7139E-01	-3.9740E-05	372	1000	-----	60.34	133.74
1205	C14H12	cis-STILBENE	-5.762	2.1705E-01	-4.6739E-05	268	1000	54.80	49.05	164.55
1206	C14H12	trans-STILBENE	6.605	1.7245E-01	-3.0224E-05	580	1000	-----	96.46	148.83
1207	C14H12O2	BENZYL BENZOATE	-8.059	2.0011E-01	-2.5588E-05	597	1000	-----	102.29	166.46
1208	C14H14	1,1-DIPHENYLETHANE	-3.628	2.0407E-01	-4.1976E-05	546	1000	-----	95.28	158.47
1209	C14H14	1,2-DIPHENYLETHANE	-0.578	1.9519E-01	-3.9807E-05	554	1000	-----	95.34	154.80
1210	C14H14O	DIBENZYL ETHER	-6.552	1.9743E-01	-2.2715E-05	561	1000	-----	97.06	168.16
1211	C14H16	1-n-BUTYLNAPHTHALENE	-6.676	2.0784E-01	-4.2598E-05	253	1000	51.51	43.18	158.57
1212	C14H16	2-BUTYLNAPHTHALENE	-3.894	1.7504E-01	-1.1856E-05	275	1000	47.24	43.35	159.29
1213	C14H22	n-OCTYLBENZENE	-1.269	1.8033E-01	-4.6111E-05	237	1000	48.40	38.88	132.95
1214	C14H22	1,2,3,4-TETRAETHYLBENZENE	-5.210	1.7795E-01	-1.8608E-05	275	1000	46.19	42.32	154.13
1215	C14H22	1,2,3,5-TETRAETHYLBENZENE	-5.224	1.7810E-01	-1.8684E-05	275	1000	46.22	42.34	154.19
1216	C14H22	1,2,4,5-TETRAETHYLBENZENE	-5.238	1.7824E-01	-1.8760E-05	275	1000	46.24	42.36	154.24
1217	C14H22O	p-tert-OCTYLPHENOL	-4.771	1.8702E-01	-2.1326E-05	359	1000	-----	59.62	160.92
1218	C14H28	1-TETRADECENE	-0.689	1.6935E-01	-4.3424E-05	260	1000	45.94	40.41	125.24
1219	C14H28	1-CYCLOPENTYLNONANE	-4.875	1.7087E-01	-1.7108E-05	275	1000	44.55	40.82	148.89
1220	C14H28	1-CYCLOHEXYLOCTANE	-4.813	1.7226E-01	-1.6664E-05	275	1000	45.06	41.30	150.78
1221	C14H28O2	n-TETRADECANOIC ACID	-3.143	1.6062E-01	-1.7213E-05	328	1000	-----	47.69	140.26
1222	C14H30	n-TETRADECANE	-10.397	1.5700E-01	1.0229E-06	279	773	36.50	33.49	111.58
1223	C14H30O	1-TETRADECANOL	-4.530	1.6609E-01	-2.0507E-05	560	1000	-----	82.05	141.05
1224	C14H30S	BUTYL-DECYL-SULFIDE	-3.703	1.5289E-01	-1.1797E-05	275	1000	40.83	37.45	137.39
1225	C14H30S	DODECYL-ETHYL-SULFIDE	-3.703	1.5289E-01	-1.1797E-05	275	1000	40.83	37.45	137.39
1226	C14H30S	HEPTYL-SULFIDE	-3.703	1.5289E-01	-1.1797E-05	275	1000	40.83	37.45	137.39
1227	C14H30S	METHYL-TRIDECYL-SULFIDE	-3.703	1.5289E-01	-1.1797E-05	275	1000	40.83	37.45	137.39
1228	C14H30S	PROPYL-UNDECYL-SULFIDE	-3.703	1.5289E-01	-1.1797E-05	275	1000	40.83	37.45	137.39
1229	C14H30S	1-TETRADECANETHIOL	-3.863	1.5407E-01	-1.2544E-05	275	1000	40.96	37.56	137.66
1230	C14H30S2	HEPTYL-DISULFIDE	-3.648	1.5306E-01	-1.1522E-05	275	1000	40.96	37.57	137.89
1231	C14H31N	TETRADECYLAMINE	-6.676	1.6894E-01	-2.4467E-05	311	1000	-----	43.50	137.80
1232	C15H10N2O2	DIPHENYLMETHANE-4,4'-DIISOCYANATE	-2.735	1.7081E-01	-1.5594E-05	311	1000	-----	48.88	152.48
1233	C15H16O	p-CUMYLPHENOL	-12.935	2.0610E-01	-3.0535E-05	608	1000	-----	101.09	162.63
1234	C15H16O2	BISPHENOL A	-11.811	1.9992E-01	-2.9746E-05	634	1000	-----	102.98	158.36
1235	C15H18	1-PENTYLNAPHTHALENE	-3.485	1.6684E-01	-1.0269E-05	275	1000	45.35	41.62	153.09
1236	C15H18	2-PENTYLNAPHTHALENE	-3.390	1.6602E-01	-9.8748E-06	275	1000	45.23	41.52	152.75
1237	C15H24	n-NONYLBENZENE	-2.878	1.8068E-01	-4.2965E-05	249	1000	47.17	39.45	134.84
1238	C15H24O	2,6-DI-tert-BUTYL-p-CRESOL	-6.529	1.7816E-01	-2.6546E-05	538	1000	-----	81.64	145.09
1239	C15H24O	NONYLPHENOL	-8.368	1.8029E-01	-2.6549E-05	581	1000	-----	87.42	145.37
1240	C15H28	1-PENTADECYNE	-4.722	1.6280E-01	-1.6758E-05	275	1000	42.33	38.78	141.32
1241	C15H30	1-PENTADECENE	-0.080	1.6236E-01	-4.1530E-05	269	1000	44.64	40.59	120.75
1242	C15H30	1-CYCLOPENTYLDECANE	-4.474	1.6387E-01	-1.5267E-05	275	1000	43.03	39.44	144.13
1243	C15H30	1-CYCLOHEXYLNONANE	-4.395	1.6499E-01	-1.4773E-05	275	1000	43.48	39.86	145.82
1244	C15H30O2	PENTADECANOIC ACID	-2.533	1.5421E-01	-1.5453E-05	326	1000	-----	46.10	136.22
1245	C15H32	n-PENTADECANE	-11.516	1.5643E-01	-6.6776E-07	298	773	35.06	35.04	109.01
1246	C15H32O	1-PENTADECANOL	-4.311	1.5375E-01	-1.4958E-05	275	1000	40.20	36.84	134.48
1247	C15H32S	BUTYL-UNDECYL-SULFIDE	-3.369	1.4730E-01	-1.0407E-05	275	1000	39.62	36.35	133.52
1248	C15H32S	DODECYL-PROPYL-SULFIDE	-3.369	1.4730E-01	-1.0407E-05	275	1000	39.62	36.35	133.52

Table 21-1 VISCOSITY OF GAS - ORGANIC COMPOUNDS (continued)

NO	FORMULA	NAME	$n_{gas} = A + BT + CT^2$				$(n_{gas}$ - micropoise, T - K)			
			A	B	C	TMIN	TMAX	n_{gas} @25 C	n_{gas} @TMIN	n_{gas} @TMAX
1249	C15H32S	ETHYL-TRIDECYL-SULFIDE	-3.369	1.4730E-01	-1.0407E-05	275	1000	39.62	36.35	133.52
1250	C15H32S	METHYL-TETRADECYL-SULFIDE	-3.369	1.4730E-01	-1.0407E-05	275	1000	39.62	36.35	133.52
1251	C15H32S	1-PENTADECANETHIOL	-3.563	1.4872E-01	-1.1284E-05	275	1000	39.77	36.48	133.87
1252	C16H10	FLUORANTHENE	-1.199	1.7289E-01	-4.1287E-05	383	1000	-----	58.96	130.40
1253	C16H10	PYRENE	-0.966	1.6484E-01	-2.7841E-05	668	998	-----	96.72	135.82
1254	C16H12	1-PHENYLNAPHTHALENE	-1.806	1.8963E-01	-3.9275E-05	607	998	-----	98.83	148.33
1255	C16H20	1-n-HEXYLNAPHTHALENE	-5.499	1.9295E-01	-4.4139E-05	255	1000	48.10	40.83	143.31
1256	C16H22O4	DIBUTYL PHTHALATE	-5.916	1.7495E-01	-2.0380E-05	613	1000	-----	93.67	148.65
1257	C16H26	n-DECYLBENZENE	-4.129	1.7898E-01	-4.1240E-05	259	1000	45.57	39.46	133.61
1258	C16H26	PENTAETHYLBENZENE	-4.626	1.6557E-01	-1.6015E-05	275	1000	43.32	39.69	144.93
1259	C16H30	1-HEXADECYNE	-4.374	1.5686E-01	-1.5123E-05	275	1000	41.05	37.62	137.36
1260	C16H32	n-DECYLCYCLOHEXANE	1.008	1.5802E-01	-3.8927E-05	271	1000	44.66	40.97	120.10
1261	C16H32	1-CYCLOPENTYLUNDECANE	-4.117	1.5768E-01	-1.3678E-05	275	1000	41.68	38.21	139.89
1262	C16H32	1-HEXADECENE	0.960	1.5408E-01	-3.8317E-05	278	1000	43.49	40.83	116.72
1263	C16H32O2	n-HEXADECANOIC ACID	-3.002	1.5158E-01	-1.6018E-05	336	1000	-----	46.12	132.56
1264	C16H34	n-HEXADECANE	-13.585	1.6007E-01	-5.5846E-06	291	773	33.64	32.52	106.81
1265	C16H34O	DI-n-OCTYL ETHER	-2.291	1.5370E-01	-1.6928E-05	266	1000	42.03	37.40	134.48
1266	C16H34O	1-HEXADECANOL	-4.826	1.5734E-01	-1.8946E-05	585	1000	-----	80.73	133.57
1267	C16H34S	BUTYL-DODECYL-SULFIDE	-3.069	1.4236E-01	-9.1923E-06	275	1000	38.56	35.39	130.10
1268	C16H34S	ETHYL-TETRADECYL-SULFIDE	-3.172	1.5052E-01	-9.3878E-06	275	1000	40.87	37.51	137.96
1269	C16H34S	METHYL-PENTADECYL-SULFIDE	-3.069	1.4236E-01	-9.1923E-06	275	1000	38.56	35.39	130.10
1270	C16H34S	OCTYL-SULFIDE	-3.069	1.4236E-01	-9.1923E-06	275	1000	38.56	35.39	130.10
1271	C16H34S	PROPYL-TRIDECYL-SULFIDE	-3.069	1.4236E-01	-9.1923E-06	275	1000	38.56	35.39	130.10
1272	C16H34S	1-HEXADECANETHIOL	-3.301	1.4404E-01	-1.0211E-05	275	1000	38.74	35.54	130.53
1273	C16H34S2	OCTYL-DISULFIDE	-3.139	1.4354E-01	-9.4705E-06	275	1000	38.82	35.62	130.93
1274	C17H28	n-UNDECYLBENZENE	-2.631	1.7109E-01	-4.0704E-05	268	1000	44.76	40.30	127.76
1275	C17H32	1-HEPTADECYNE	-4.061	1.5160E-01	-1.3695E-05	275	1000	39.92	36.59	133.84
1276	C17H34	1-CYCLOPENTYLDODECANE	-3.794	1.5214E-01	-1.2282E-05	275	1000	40.47	37.12	136.06
1277	C17H34	1-CYCLOHEXYLUNDECANE	-3.714	1.5300E-01	-1.1843E-05	275	1000	40.85	37.47	137.44
1278	C17H34	1-HEPTADECENE	3.237	1.4164E-01	-3.5052E-05	284	1000	42.35	40.64	109.82
1279	C17H36	n-HEPTADECANE	-6.166	1.2965E-01	1.9105E-05	575	775	-----	74.70	105.79
1280	C17H36O	1-HEPTADECANOL	-5.072	1.5457E-01	-1.9519E-05	597	1000	-----	80.25	129.98
1281	C17H36S	BUTYL-TRIDECYL-SULFIDE	-2.801	1.3804E-01	-8.1315E-06	275	1000	37.63	34.55	127.11
1282	C17H36S	ETHYL-PENTADECYL-SULFIDE	-2.801	1.3804E-01	-8.1315E-06	275	1000	37.63	34.55	127.11
1283	C17H36S	HEXADECYL-METHYL-SULFIDE	-2.801	1.3804E-01	-8.1315E-06	275	1000	37.63	34.55	127.11
1284	C17H36S	PROPYL-TETRADECYL-SULFIDE	-2.801	1.3804E-01	-8.1315E-06	275	1000	37.63	34.55	127.11
1285	C17H36S	1-HEPTADECANETHIOL	-3.030	1.3966E-01	-9.1064E-06	275	1000	37.80	34.69	127.52
1286	C18H12	CHRYSENE	5.851	1.3767E-01	-2.7934E-05	531	1000	-----	71.08	115.59
1287	C18H14	m-TERPHENYL	-5.315	1.7735E-01	-1.6308E-05	638	1000	-----	101.20	155.73
1288	C18H14	o-TERPHENYL	-5.046	1.7719E-01	-1.6852E-05	605	1000	-----	95.99	155.29
1289	C18H14	p-TERPHENYL	-5.080	1.7661E-01	-1.5866E-05	649	1000	-----	102.86	155.66
1290	C18H15P	TRIPHENYLPHOSPHINE	2.069	2.0532E-01	1.0449E-05	354	1000	-----	76.06	217.84
1291	C18H15O4P	TRIPHENYL PHOSPHATE	-----	--------	--------	----	----	-----	-----	-----
1292	C18H16N2	N,N'-DIPHENYL-p-PHENYLENEDIAMINE	-3.958	1.7305E-01	-1.5762E-05	409	1000	-----	64.18	153.33
1293	C18H22	2,3-DIMETHYL-2,3-DIPHENYLBUTANE	1.531	1.7034E-01	-3.7516E-05	392	1000	-----	62.54	134.35
1294	C18H22O2	DICUMYL PEROXIDE	-2.252	1.6758E-01	-1.2252E-05	311	1000	-----	48.68	153.08
1295	C18H30	n-DODECYLBENZENE	3.178	1.5222E-01	-2.8495E-05	600	1000	-----	84.25	126.90
1296	C18H30	HEXAETHYLBENZENE	-4.203	1.5615E-01	-1.4220E-05	275	1000	41.09	37.66	137.73
1297	C18H32O2	LINOLEIC ACID	-2.954	1.6129E-01	-1.6972E-05	268	1000	43.63	39.05	141.36
1298	C18H34	1-OCTADECYNE	-3.779	1.4693E-01	-1.2445E-05	275	1000	38.92	35.69	130.71
1299	C18H34O2	OLEIC ACID	-2.483	1.5050E-01	-1.4741E-05	287	1000	41.08	39.50	133.28
1300	C18H34O4	DIBUTYL SEBACATE	-5.738	1.5358E-01	-1.9799E-05	622	1000	-----	82.13	128.04
1301	C18H34O4	DIHEXYL ADIPATE	-2.421	1.4529E-01	-1.4709E-05	259	1000	39.59	34.22	128.16
1302	C18H36	1-CYCOPENTYLTRIDECANE	-3.499	1.4713E-01	-1.1038E-05	275	1000	39.39	36.13	132.59
1303	C18H36	1-CYCLOHEXYLDODECANE	-3.419	1.4792E-01	-1.0623E-05	275	1000	39.74	36.46	133.88
1304	C18H36	1-OCTADECENE	1.068	1.4653E-01	-3.7442E-05	291	1000	41.43	40.54	110.16
1305	C18H36O2	STEARIC ACID	-3.566	1.4606E-01	-1.6454E-05	343	1000	-----	44.60	126.04
1306	C18H38	n-OCTADECANE	-6.947	1.2597E-01	2.2320E-05	590	775	-----	75.15	104.09
1307	C18H38O	DINONYL ETHER	-1.968	1.4495E-01	-1.4970E-05	591	1000	-----	78.47	128.01
1308	C18H38O	1-OCTADECANOL	-2.025	1.4238E-01	-1.3308E-05	608	1000	-----	79.62	127.05
1309	C18H38S	BUTYL-TETRADECYL-SULFIDE	-2.540	1.3419E-01	-7.1132E-06	275	1000	36.84	33.82	124.54
1310	C18H38S	ETHYL-HEXADECYL-SULFIDE	-2.540	1.3419E-01	-7.1132E-06	275	1000	36.84	33.82	124.54
1311	C18H38S	HEPTADECYL-METHYL-SULFIDE	-2.540	1.3419E-01	-7.1132E-06	275	1000	36.84	33.82	124.54
1312	C18H38S	NONYL-SULFIDE	-2.540	1.3419E-01	-7.1132E-06	275	1000	36.84	33.82	124.54
1313	C18H38S	PENTADECYL-PROPYL-SULFIDE	-2.540	1.3419E-01	-7.1132E-06	275	1000	36.84	33.82	124.54
1314	C18H38S	1-OCTADECANETHIOL	-2.811	1.3606E-01	-8.2363E-06	275	1000	37.02	33.98	125.01
1315	C18H38S2	NONYL-DISULFIDE	-2.707	1.3620E-01	-7.7749E-06	275	1000	37.21	34.16	125.72
1316	C19H26	1-n-NONYLNAPHTHALENE	-3.668	1.6282E-01	-3.8584E-05	284	1000	41.45	39.46	120.57
1317	C19H32	n-TRIDECYLBENZENE	3.774	1.4787E-01	-2.7163E-05	614	1000	-----	84.33	124.48
1318	C19H36	1-NONADECYNE	-3.503	1.4264E-01	-1.1243E-05	275	1000	38.03	34.87	127.89
1319	C19H36O2	METHYL OLEATE	-6.186	1.5358E-01	-2.1018E-05	617	1000	-----	80.57	126.38
1320	C19H38	1-CYCLOPENTYLTETRADECANE	-3.253	1.4019E-01	-1.0127E-05	275	1000	37.64	34.53	126.81
1321	C19H38	1-CYCLOHEXYLTRIDECANE	-3.152	1.4339E-01	-9.5379E-06	275	1000	38.75	35.56	130.70
1322	C19H38	1-NONADECENE	1.734	1.4112E-01	-3.5773E-05	297	1000	40.63	40.49	107.08
1323	C19H38O2	NONADECANOIC ACID	-3.238	1.4167E-01	-1.5160E-05	341	1000	-----	43.31	123.27
1324	C19H40	n-NONADECANE	-7.682	1.2181E-01	2.5813E-05	603	775	-----	75.16	102.23
1325	C19H40O	1-NONADECANOL	-3.214	1.4062E-01	-9.9244E-06	275	1000	37.83	34.71	127.48
1326	C19H40S	BUTYL-PENTADECYL-SULFIDE	-2.306	1.3090E-01	-6.2165E-06	275	1000	36.17	33.22	122.38

Table 21-1 VISCOSITY OF GAS - ORGANIC COMPOUNDS (continued)

| NO | FORMULA | NAME | $n_{gas} = A + BT + CT^2$ | | | | | $(n_{gas}$ - micropoise, T - K) | | |
			A	B	C	TMIN	TMAX	n_{gas} @25 C	n_{gas} @TMIN	n_{gas} @TMAX
1327	C19H40S	ETHYL-HEPTADECYL-SULFIDE	-2.306	1.3090E-01	-6.2165E-06	275	1000	36.17	33.22	122.38
1328	C19H40S	HEXADECYL-PROPYL-SULFIDE	-2.306	1.3090E-01	-6.2165E-06	275	1000	36.17	33.22	122.38
1329	C19H40S	METHYL-OCTADECYL-SULFIDE	-2.306	1.3090E-01	-6.2165E-06	275	1000	36.17	33.22	122.38
1330	C19H40S	1-NONADECANETHIOL	-2.595	1.3282E-01	-7.3849E-06	275	1000	36.35	33.37	122.84
1331	C20H16	TRIPHENYLETHYLENE	-3.987	1.7431E-01	-3.6757E-05	342	1000	-----	51.33	133.57
1332	C20H28	1-n-DECYLNAPHTHALENE	4.605	1.3680E-01	-2.3578E-05	652	1000	-----	83.78	117.83
1333	C20H30O2	ABIETIC ACID	-5.968	1.6617E-01	-1.7915E-05	447	1000	-----	64.73	142.29
1334	C20H31N	DEHYDROABIETYLAMINE	-2.809	1.5592E-01	-1.3263E-05	318	1000	-----	45.43	139.85
1335	C20H34	1-PHENYLTETRADECANE	-2.721	1.2934E-01	-8.0461E-06	275	1000	35.13	32.24	118.57
1336	C20H38	1-EICOSYNE	-3.255	1.3890E-01	-1.0188E-05	275	1000	37.25	34.17	125.46
1337	C20H40	1-CYCLOPENTYLPENTADECANE	-3.027	1.3545E-01	-9.2396E-06	275	1000	36.54	33.52	123.18
1338	C20H40	1-CYCLOHEXYLTETRADECANE	-2.922	1.3954E-01	-8.6235E-06	275	1000	37.91	34.80	127.99
1339	C20H40	1-EICOSENE	2.083	1.3646E-01	-3.4264E-05	302	1000	-----	40.17	104.28
1340	C20H42	n-EICOSANE	-7.990	1.1866E-01	2.7167E-05	617	775	-----	75.57	100.29
1341	C20H42O	1-EICOSANOL	-6.015	1.4613E-01	-1.8942E-05	629	1000	-----	78.41	121.17
1342	C20H42S	BUTYL-HEXADECYL-SULFIDE	-2.095	1.2813E-01	-5.4231E-06	275	1000	35.62	32.73	120.61
1343	C20H42S	DECYL-SULFIDE	-2.095	1.2813E-01	-5.4231E-06	275	1000	35.62	32.73	120.61
1344	C20H42S	ETHYL-OCTADECYL-SULFIDE	-2.095	1.2813E-01	-5.4231E-06	275	1000	35.62	32.73	120.61
1345	C20H42S	HEPTADECYL-PROPYL-SULFIDE	-2.095	1.2813E-01	-5.4231E-06	275	1000	35.62	32.73	120.61
1346	C20H42S	METHYL-NONADECYL-SULFIDE	-2.095	1.2813E-01	-5.4231E-06	275	1000	35.62	32.73	120.61
1347	C20H42S	1-EICOSANETHIOL	-2.403	1.3015E-01	-6.6397E-06	275	1000	35.81	32.89	121.11
1348	C20H42S2	DECYL-DISULFIDE	-2.332	1.3089E-01	-6.3233E-06	275	1000	36.13	33.18	122.23
1349	C21H21O4P	TRI-o-CRESYL PHOSPHATE	-----	--------	--------	----	----	-----	-----	-----
1350	C21H36	1-PHENYLPENTADECANE	-2.556	1.2694E-01	-7.3914E-06	275	1000	34.63	31.79	116.99
1351	C21H42	1-CYCLOPENTYLHEXADECANE	-2.772	1.3556E-01	-8.0785E-06	275	1000	36.93	33.90	124.71
1352	C21H42	1-CYCLOHEXYLPENTADECANE	-2.683	1.3591E-01	-7.6800E-06	275	1000	37.16	34.11	125.55
1353	C22H38	1-PHENYLHEXADECANE	-2.348	1.2216E-01	-6.6284E-06	275	1000	33.48	30.74	113.18
1354	C22H44	1-CYCLOHEXYLHEXADECANE	-2.470	1.3282E-01	-6.8516E-06	275	1000	36.52	33.54	123.50
1355	C22H44O2	n-BUTYL STEARATE	-2.150	1.3327E-01	-1.4512E-05	623	1000	-----	75.24	116.61
1356	C24H38O4	DIISOOCTYL PHTHALATE	-1.743	1.3161E-01	-1.0812E-05	223	1000	36.54	27.07	119.06
1357	C24H38O4	DIOCTYL PHTHALATE	-14.000	1.6564E-01	-3.3834E-05	223	1000	32.38	21.26	117.81
1358	C24H42O	DINONYLPHENOL	3.061	1.1889E-01	-5.1703E-06	722	1000	-----	86.20	116.78
1359	C26H20	TETRAPHENYLETHYLENE	5.223	1.3078E-01	-2.8031E-05	496	1000	-----	63.19	107.97
1360	C28H46O4	DIISODECYL PHTHALATE	-4.094	1.2702E-01	-1.5011E-05	643	973	-----	71.37	105.29

n_{gas} - viscosity of gas, micropoise

A, B, and C - regression coefficients for chemical compound

T - temperature, K

TMIN - minimum temperature, K

TMAX - maximum temperature, K

Table 21-2 VISCOSITY OF GAS - INORGANIC COMPOUNDS

			$n_{gas} = A + BT + CT^2$					$(n_{gas}$ - micropoise, T - K$)$		
NO	FORMULA	NAME	A	B	C	TMIN	TMAX	n_{gas} @25 C	n_{gas} @TMIN	n_{gas} @TMAX
1	Ag	SILVER	73.744	1.8017E-01	7.3069E-06	2485	2885	-----	566.59	654.35
2	AgCl	SILVER CHLORIDE	-6.149	8.0109E-02	-3.8336E-07	1837	2237	-----	139.72	171.14
3	AgI	SILVER IODIDE	-7.923	1.0437E-01	-5.6252E-07	1779	2179	-----	175.97	216.83
4	Al	ALUMINUM	47.517	1.2002E-01	5.1703E-06	2329	2729	-----	355.09	413.56
5	AlB3H12	ALUMINUM BOROHYDRIDE	-18.132	5.4504E-01	-1.0539E-04	250	1000	135.00	111.54	421.52
6	AlBr3	ALUMINUN BROMIDE	-25.267	4.4842E-01	-5.8813E-05	529	929	-----	195.49	340.56
7	AlCl3	ALUMINUM CHLORIDE	-20.021	3.9745E-01	-6.7400E-05	453	853	-----	146.19	269.96
8	AlF3	ALUMINUM FLUORIDE	-4.721	6.1826E-02	-3.1278E-07	1810	2210	-----	106.16	130.39
9	AlI3	ALUMINUM IODIDE	-24.944	3.9595E-01	-3.5175E-05	659	1059	-----	220.71	354.92
10	Al2O3	ALUMINUM OXIDE	-21.809	2.4206E-01	3.4367E-07	3253	3653	-----	769.25	867.02
11	Al2S3O12	ALUMINUM SULFATE	---------	----------	----------	----	----	-----	------	------
12	Ar	ARGON	44.997	6.3892E-01	-1.2455E-04	150	1500	224.42	138.03	723.14
13	As	ARSENIC	-14.744	5.7697E-01	2.9027E-06	885	1285	-----	498.15	731.46
14	AsBr3	ARSENIC TRIBROMIDE	-24.988	5.0869E-01	-5.5767E-05	493	893	-----	212.24	384.80
15	AsCl3	ARSENIC TRICHLORIDE	-19.566	4.5049E-01	-6.4140E-05	404	804	-----	151.96	301.17
16	AsF3	ARSENIC TRIFLUORIDE	-21.586	6.4134E-01	-1.1942E-04	250	1000	159.01	131.29	500.33
17	AsF5	ARSENIC PENTAFLUORIDE	-10.274	6.6832E-01	-1.6394E-04	250	1000	174.41	146.56	494.11
18	AsH3	ARSINE	-11.170	6.2070E-01	-1.6254E-04	156	1000	159.44	81.70	446.99
19	AsI3	ARSENIC TRIIODIDE	-26.889	4.9106E-01	-2.9281E-05	676	1076	-----	291.69	467.59
20	As2O3	ARSENIC TRIOXIDE	-22.960	4.0747E-01	-2.0972E-05	730	1130	-----	263.32	410.70
21	At	ASTATINE	-67.704	1.7296E+00	-6.4452E-05	607	1007	-----	958.42	1608.65
22	Au	GOLD	-97.513	4.9518E-01	-5.9892E-06	3120	3520	-----	1389.15	1571.31
23	B	BORON	11.398	1.1909E-01	1.4266E-06	4133	4533	-----	527.97	580.55
24	BBr3	BORON TRIBROMIDE	-17.739	5.2932E-01	-8.6114E-05	250	1000	132.42	109.21	425.47
25	BCl3	BORON TRICHLORIDE	47.391	1.8060E-01	1.9496E-04	286	600	118.57	114.99	225.94
26	BF3	BORON TRIFLUORIDE	14.681	5.6153E-01	-1.3794E-04	172	700	169.84	107.18	340.16
27	BH2CO	BORINE CARBONYL	-4.732	4.1633E-01	-1.0371E-04	250	1000	110.18	92.87	307.89
28	BH3O3	BORIC ACID	---------	----------	----------	----	----	-----	------	------
29	B2D6	DEUTERODIBORANE	0.395	2.9927E-01	-7.6983E-05	250	1000	82.78	70.40	222.68
30	B2H5Br	DIBORANE HYDROBROMIDE	-12.949	4.2049E-01	-8.9232E-05	250	1000	104.49	86.60	318.31
31	B2H6	DIBORANE	1.439	2.6553E-01	-1.3774E-05	108	1000	79.38	29.96	253.19
32	B3N3H6	BORINE TRIAMINE	-9.514	2.8423E-01	-5.4051E-05	250	1000	70.42	58.17	220.67
33	B4H10	TETRABORANE	-8.480	2.7557E-01	-5.8511E-05	250	1000	68.48	56.76	208.58
34	B5H9	PENTABORANE	-8.812	2.6130E-01	-4.4076E-05	250	1000	65.18	53.76	208.41
35	B5H11	TETRAHYDROPENTABORANE	-8.858	2.6188E-01	-4.6770E-05	250	1000	65.06	53.69	206.25
36	B10H14	DECABORANE	-13.009	2.7320E-01	-2.8891E-05	486	886	-----	112.94	206.37
37	Ba	BARIUM	-0.110	1.4287E-01	2.1467E-06	1907	2307	-----	280.15	340.92
38	Be	BERYLLIUM	0.978	2.9418E-02	4.2605E-07	2744	3144	-----	84.91	97.68
39	BeB2H8	BERYLLIUM BOROHYDRIDE	-6.366	1.9143E-01	-3.0176E-05	250	1000	48.03	39.61	154.89
40	BeBr2	BERYLLIUM BROMIDE	-15.101	2.6581E-01	-1.3087E-05	747	1147	-----	176.16	272.57
41	BeCl2	BERYLLIUM CHLORIDE	-10.348	1.8102E-01	-8.6146E-06	760	1160	-----	122.25	188.04
42	BeF2	BERYLLIUM FLUORIDE	---------	----------	----------	----	----	-----	------	------
43	BeI2	BERYLLIUM IODIDE	-18.766	3.2828E-01	-1.5622E-05	760	1160	-----	221.70	341.02
44	Bi	BISMUTH	57.468	2.6858E-01	1.4445E-05	1698	2098	-----	555.16	684.53
45	BiBr3	BISMUTH TRIBROMIDE	-22.946	4.2955E-01	-1.9314E-05	734	1134	-----	281.94	439.33
46	BiCl3	BISMUTH TRICHLORIDE	-21.943	4.0581E-01	-2.0254E-05	714	1114	-----	257.48	404.99
47	BrF5	BROMINE PENTAFLUORIDE	-19.868	6.4047E-01	-1.3518E-04	250	1000	159.07	131.80	485.42
48	Br2	BROMINE	19.086	4.3931E-01	4.8205E-05	266	600	154.35	139.35	300.03
49	C	CARBON	-19.805	1.7342E-01	2.0401E-07	4203	4603	-----	712.68	782.77
50	CCl2O	PHOSGENE	-37.045	5.1974E-01	-1.9479E-05	200	900	116.18	66.12	414.94
51	CF2O	CARBONYL FLUORIDE	-0.145	4.2223E-01	-1.1419E-04	189	1000	115.59	75.58	307.90
52	CH4N2O	UREA	-13.895	2.7802E-01	-3.8420E-05	465	865	-----	107.08	197.85
53	CH4N2S	THIOUREA	-9.887	2.0456E-01	-5.2055E-07	454	1000	-----	82.88	194.15
54	CNBr	CYANOGEN BROMIDE	-15.199	4.4944E-01	-8.0699E-05	250	1000	111.63	92.12	353.54
55	CNCl	CYANOGEN CHLORIDE	-5.287	3.1928E-01	-6.6168E-05	286	1000	84.02	80.61	247.82
56	CNF	CYANOGEN FLUORIDE	-9.099	5.1725E-01	-1.2558E-04	250	1000	133.96	112.37	382.57
57	CO	CARBON MONOXIDE	35.086	5.0651E-01	-1.3314E-04	150	1250	174.27	108.07	460.19
58	COS	CARBONYL SULFIDE	-11.409	4.9811E-01	-9.9754E-05	134	1000	128.23	53.55	386.95
59	COSe	CARBON OXYSELENIDE	-18.568	7.6929E-01	-1.7890E-04	250	1000	194.89	162.57	571.82
60	CO2	CARBON DIOXIDE	11.336	4.9918E-01	-1.0876E-04	195	1500	150.50	104.54	515.40
61	CS2	CARBON DISULFIDE	-7.702	3.6594E-01	-2.5416E-05	273	583	99.14	90.31	197.00
62	CSeS	CARBON SELENOSULFIDE	-16.674	4.9624E-01	-8.1796E-05	250	1000	124.01	102.27	397.77
63	C2N2	CYANOGEN	-10.232	4.4240E-01	-1.0374E-04	250	1000	112.45	93.88	328.43
64	C3S2	CARBON SUBSULFIDE	---------	----------	----------	----	----	-----	------	------
65	Ca	CALCIUM	-0.506	8.1000E-02	1.1906E-06	1762	2162	-----	145.91	180.18
66	CaF2	CALCIUM FLUORIDE	-7.713	8.5508E-02	6.1125E-09	2807	3207	-----	232.36	266.57
67	CbF5	COLUMBIUM FLUORIDE	-17.202	3.5711E-01	-3.6309E-05	498	898	-----	151.63	274.20
68	Cd	CADMIUM	16.576	5.1705E-01	2.2389E-05	1043	1443	-----	580.21	809.30
69	CdCl2	CADMIUM CHLORIDE	-13.867	2.0603E-01	-3.2740E-06	1240	1640	-----	236.58	315.22
70	CdF2	CADMIUM FLUORIDE	-11.266	1.4177E-01	-4.5419E-07	2024	2424	-----	273.82	329.72
71	CdI2	CADMIUM IODIDE	-20.346	3.1720E-01	-7.2498E-06	1069	1469	-----	310.46	429.98
72	CdO	CADMIUM OXIDE	-10.614	1.3840E-01	-6.6921E-07	1832	2232	-----	240.69	294.96
73	ClF	CHLORINE MONOFLUORIDE	3.539	7.1487E-01	-1.8508E-04	250	1000	200.22	170.69	533.33
74	ClFO3	PERCHLORYL FLUORIDE	-18.769	5.9256E-01	-1.2994E-04	125	1000	146.35	53.27	443.85
75	ClF3	CHLORINE TRIFLUORIDE	-17.711	5.8672E-01	-1.2616E-04	250	1000	146.00	121.08	442.85
76	ClF5	CHLORINE PENTAFLUORIDE	-12.890	5.0651E-01	-1.1636E-04	250	1000	127.78	106.47	377.26
77	ClHO3S	CHLOROSULFONIC ACID	-18.175	4.1053E-01	-5.2079E-05	427	827	-----	147.63	285.72
78	ClHO4	PERCHLORIC ACID	-13.788	4.3539E-01	-6.0076E-05	250	1000	110.68	91.30	361.53

473

Table 21-2 VISCOSITY OF GAS - INORGANIC COMPOUNDS (continued)

			$n_{gas} = A + B T + C T^2$					(n_{gas} - micropoise, T - K)		
NO	FORMULA	NAME	A	B	C	TMIN	TMAX	n_{gas} @25 C	n_{gas} @TMIN	n_{gas} @TMAX
79	ClO2	CHLORINE DIOXIDE	-18.886	6.1600E-01	-1.3100E-04	250	1000	153.21	127.00	466.20
80	Cl2	CHLORINE	-3.571	4.8700E-01	-8.5300E-05	200	1000	134.03	90.41	398.09
81	Cl2O	CHLORINE MONOXIDE	-16.577	5.7500E-01	-1.2700E-04	250	1000	143.60	119.26	431.77
82	Cl2O7	CHLORINE HEPTOXIDE	-16.299	4.8300E-01	-8.2100E-05	250	1000	120.37	99.29	384.50
83	Co	COBALT	73.507	1.8600E-01	7.3900E-06	2528	2928	-----	590.99	681.53
84	CoCl2	COBALT CHLORIDE	-11.486	1.6700E-01	-2.2400E-05	1323	1723	-----	205.67	269.78
85	CoNC3O4	COBALT NITROSYL TRICARBONYL	-14.162	4.2000E-01	-7.1000E-05	250	1000	104.70	86.36	334.68
86	Cr	CHROMIUM	74.686	1.5800E-01	5.6700E-06	2840	3240	-----	569.25	646.25
87	CrC6O6	CHROMIUM CARBONYL	-18.887	4.2500E-01	-5.5500E-05	424	824	-----	151.35	293.66
88	CrO2Cl2	CHROMIUM OXYCHLORIDE	-13.758	4.3100E-01	-6.0500E-05	250	1000	109.44	90.27	356.98
89	Cs	CESIUM	-176.607	5.4100E-01	-1.3300E-04	950	1300	-----	217.42	301.92
90	CsBr	CESIUM BROMIDE	-14.119	1.9400E-01	-1.5800E-06	1573	1973	-----	287.12	362.48
91	CsCl	CESIUM CHLORIDE	-12.558	1.7300E-01	-1.4000E-06	1573	1973	-----	255.37	322.40
92	CsF	CESIUM FLUORIDE	-12.015	1.6700E-01	-1.5000E-06	1524	1924	-----	238.76	303.43
93	CsI	CESIUM IODIDE	-15.646	2.1600E-01	-1.8300E-06	1553	1953	-----	315.22	399.01
94	Cu	COPPER	-20.261	2.1100E-01	1.2000E-07	3150	3550	-----	646.21	731.01
95	CuBr	CUPROUS BROMIDE	-11.504	1.5600E-01	-1.1400E-06	1628	2028	-----	239.82	300.64
96	CuCl	CUPROUS CHLORIDE	-22.237	1.6400E-01	-4.8700E-06	1763	2163	-----	252.57	310.71
97	CuCl2	CUPRIC CHLORIDE	-13.722	1.8200E-01	-3.6700E-06	1266	1666	-----	211.11	279.70
98	CuI	COPPER IODIDE	-13.289	1.8100E-01	-1.3700E-06	1609	2009	-----	274.71	345.20
99	DCN	DEUTERIUM CYANIDE	-12.525	3.9300E-01	-8.1100E-05	250	1000	97.48	80.69	299.55
100	D2	DEUTERIUM	18.778	4.2100E-01	-2.0700E-04	60	480	126.01	43.32	173.32
101	D2O	DEUTERIUM OXIDE	99.754	-1.5800E-01	5.9800E-04	290	800	105.94	104.36	356.53
102	Eu	EUROPIUM	96.085	3.5700E-01	1.9800E-05	1742	2142	-----	778.90	952.66
103	F2	FLUORINE	-0.811	8.9800E-01	-3.9600E-04	90	500	231.59	76.76	348.97
104	F2O	FLUORINE OXIDE	19.793	6.4900E-01	-1.7300E-04	250	1000	197.84	171.17	495.49
105	Fe	IRON	99.062	1.8700E-01	6.3700E-06	3000	3400	-----	718.67	809.95
106	FeC5O5	IRON PENTACARBONYL	-11.631	3.5500E-01	-5.3300E-05	250	1000	89.57	73.87	290.43
107	FeCl2	FERROUS CHLORIDE	-11.400	1.6700E-01	-2.3500E-06	1299	1699	-----	201.35	265.27
108	FeCl3	FERRIC CHLORIDE	-15.552	2.9900E-01	-2.2700E-05	592	992	-----	153.66	258.98
109	Fr	FRANCIUM	-48.139	1.3900E+00	-6.1500E-06	879	1279	-----	1172.35	1724.60
110	Ga	GALLIUM	50.609	1.2100E-01	4.8400E-06	2517	2917	-----	385.10	443.91
111	GaCl3	GALLIUM TRICHLORIDE	-20.386	3.4200E-01	-1.7600E-05	351	1000	-----	97.31	303.52
112	Gd	GADOLINIUM	-4.471	7.5200E-01	1.1100E-05	1770	2170	-----	1360.71	1678.83
113	Ge	GERMANIUM	86.056	2.0900E-01	6.6000E-06	3125	3525	-----	802.99	904.07
114	GeBr4	GERMANIUM BROMIDE	-22.236	4.6900E-01	-5.6600E-05	462	862	-----	182.34	339.95
115	GeCl4	GERMANIUM CHLORIDE	-14.589	4.3400E-01	-7.2000E-05	250	1000	108.30	89.32	347.05
116	GeHCl3	TRICHLORO GERMANE	-15.397	4.5500E-01	-7.8700E-05	250	1000	113.40	93.55	361.37
117	GeH4	GERMANE	-1.711	5.8300E-01	-1.4900E-04	250	1000	158.93	134.78	432.72
118	Ge2H6	DIGERMANE	-15.993	4.9500E-01	-1.0000E-04	250	1000	122.58	101.41	378.37
119	Ge3H8	TRIGERMANE	-14.661	4.5300E-01	-6.5800E-05	250	1000	114.56	94.48	372.53
120	HBr	HYDROGEN BROMIDE	10.125	5.8500E-01	-4.3100E-05	273	673	180.71	166.62	384.31
121	HCN	HYDROGEN CYANIDE	-8.482	9.0400E-02	7.9200E-05	300	425	25.75	44.22
122	HCl	HYDROGEN CHLORIDE	-9.118	5.5500E-01	-1.1100E-04	200	1000	146.44	97.42	434.55
123	HF	HYDROGEN FLUORIDE	-186.394	1.3500E+00	-1.1300E-03	286	473	114.61	106.27	197.73
124	HI	HYDROGEN IODIDE	-0.961	6.6800E-01	-9.2600E-05	250	650	189.90	160.19	393.96
125	HNO3	NITRIC ACID	-14.473	4.3300E-01	-8.2500E-05	250	1000	107.22	88.56	335.78
126	H2	HYDROGEN	27.758	2.1200E-01	-3.2800E-05	150	1500	88.03	58.81	271.76
127	H2O	WATER	-36.826	4.2900E-01	-1.6200E-05	280	1073	89.68	82.07	404.97
128	H2O2	HYDROGEN PEROXIDE	8.039	2.7000E-01	8.2900E-05	373	600	-----	120.24	199.81
129	H2S	HYDROGEN SULFIDE	-14.839	5.1000E-01	-1.2600E-04	230	570	125.96	95.75	234.84
130	H2SO4	SULFURIC ACID	-20.619	3.4900E-01	-3.3000E-05	610	1010	-----	180.12	298.43
131	H2S2	HYDROGEN DISULFIDE	-14.978	4.4300E-01	-8.0100E-05	250	1000	110.02	90.80	348.05
132	H2Se	HYDROGEN SELENIDE	-16.042	6.4700E-01	-1.5000E-04	250	1000	163.62	136.41	481.53
133	H2Te	HYDROGEN TELLURIDE	-19.538	6.9400E-01	-1.5500E-04	250	1000	173.75	144.40	520.30
134	H3NO3S	SULFAMIC ACID	----------	----------	----------	----	----	-----	-----	-----
135	He	HELIUM-3	51.443	4.2900E-01	-1.0700E-04	150	1000	169.90	113.42	373.41
136	He	HELIUM-4	71.094	4.4300E-01	-5.1800E-05	150	2000	198.63	136.41	750.41
137	Hf	HAFNIUM	205.918	1.5900E-01	2.5300E-06	5960	6360	-----	1243.35	1319.40
138	Hg	MERCURY	-15.995	1.0200E+00	-1.6100E-05	473	1073	-----	462.16	1058.35
139	HgBr2	MERCURIC BROMIDE	-23.182	4.4600E-01	-3.3900E-05	592	992	-----	229.05	386.04
140	HgCl2	MERCURIC CHLORIDE	-20.215	3.9300E-01	-3.1200E-05	577	977	-----	196.26	334.13
141	HgI2	MERCURIC IODIDE	-23.047	4.9600E-01	-2.5100E-05	627	1027	-----	278.19	460.04
142	IF7	IODINE HEPTAFLUORIDE	-19.310	6.6300E-01	-1.4500E-04	250	1000	165.53	137.43	498.95
143	I2	IODINE	-8.163	5.2000E-01	-8.9200E-05	387	750	-----	179.74	331.68
144	In	INDIUM	55.358	1.5700E-01	6.7000E-06	2323	2723	-----	455.34	531.51
145	Ir	IRIDIUM	741.090	8.3400E-01	1.8800E-05	4450	4850	-----	4825.70	5229.29
146	K	POTASSIUM	-128.200	3.3900E-01	-5.0000E-05	1100	1500	-----	184.20	267.80
147	KBr	POTASSIUM BROMIDE	-10.438	1.4100E-01	-9.7300E-07	1656	2056	-----	220.29	275.22
148	KCl	POTASSIUM CHLORIDE	2.376	5.3000E-02	1.4800E-06	1689	2089	-----	96.06	119.48
149	KF	POTASSIUM FLUORIDE	-7.186	9.4700E-02	-5.1500E-07	1775	2175	-----	159.36	196.44
150	KI	POTASSIUM IODIDE	-12.428	1.7000E-01	-1.3200E-06	1597	1997	-----	255.51	321.56
151	KOH	POTASSIUM HYDROXIDE	-7.222	9.8700E-02	-7.6100E-07	1600	2000	-----	148.69	187.06
152	Kr	KRYPTON	31.096	7.9800E-01	-1.7900E-04	150	1500	253.15	146.78	826.31
153	La	LANTHANUM	115.282	2.5500E-01	6.8800E-06	3643	4043	-----	1135.05	1258.14
154	Li	LITHIUM	216.200	-1.0400E-01	4.2900E-05	1273	1673	-----	152.89	161.68
155	LiBr	LITHIUM BROMIDE	-9.007	1.2300E-01	-9.8500E-07	1583	1983	-----	183.99	231.98
156	LiCl	LITHIUM CHLORIDE	-6.231	8.4200E-02	-5.8200E-07	1655	2055	-----	131.45	164.24

Table 21-2 VISCOSITY OF GAS - INORGANIC COMPOUNDS (continued)

NO	FORMULA	NAME	A	B	C	TMIN	TMAX	n_{gas} @25 C	n_{gas} @TMIN	n_{gas} @TMAX
					$n_{gas} = A + B T + C T^2$			(n_{gas} - micropoise, T - K)		
157	LiF	LITHIUM FLUORIDE	-4.710	6.0000E-02	-2.2400E-07	1954	2354	-----	111.73	135.35
158	LiI	LITHIUM IODIDE	-11.420	1.6100E-01	-1.7000E-06	1444	1844	-----	218.16	280.50
159	Lu	LUTECIUM	-71.708	8.2900E-01	-5.5800E-07	2535	2935	-----	2025.03	2355.22
160	Mg	MAGNESIUM	-30.160	4.3300E-01	-5.2200E-06	1376	1776	-----	556.03	722.72
161	MgCl2	MAGNESIUM CHLORIDE	-9.295	1.2500E-01	-8.0200E-07	1691	2091	-----	199.13	247.76
162	MgO	MAGNESIUM OXIDE	-11.384	7.0400E-02	-2.0200E-07	3873	4273	-----	258.40	285.92
163	Mn	MANGANESE	68.202	1.8800E-01	7.8400E-06	2392	2792	-----	563.87	655.52
164	MnCl2	MANGANESE CHLORIDE	-11.040	1.5500E-01	-1.5700E-06	1463	1863	-----	212.93	272.99
165	Mo	MOLYBDENUM	25.163	2.3300E-01	2.2400E-06	5081	5481	-----	1268.87	1371.68
166	MoF6	MOLYBDENUM FLUORIDE	-19.680	6.0200E-01	-1.2000E-04	250	1000	149.20	123.37	462.22
167	MoO3	MOLYBDENUM OXIDE	-11.882	1.6900E-01	-1.8500E-06	1424	1824	-----	224.66	289.76
168	NCl3	NITROGEN TRICHLORIDE	-15.168	4.4900E-01	-7.6700E-05	250	1000	111.92	92.32	357.25
169	ND3	HEAVY AMMONIA	-9.516	4.4800E-01	-1.0600E-04	250	1000	114.52	95.77	331.80
170	NF3	NITROGEN TRIFLUORIDE	-1.996	7.1000E-01	-2.0100E-04	66	1000	192.00	44.02	507.95
171	NH3	AMMONIA	-7.874	3.6700E-01	-4.4700E-05	195	1000	101.28	63.61	355.10
172	NH3O	HYDROXYLAMINE	-7.975	3.7500E-01	-6.1300E-05	306	1000	-----	100.91	305.31
173	NH4Br	AMMONIUM BROMIDE	-4.028	7.3800E-02	-4.4900E-06	669	1069	-----	43.36	69.78
174	NH4Cl	AMMONIUM CHLORIDE	-4.344	6.9700E-02	-7.5800E-06	612	1012	-----	35.48	58.44
175	NH4I	AMMONIUM IODIDE	-4.886	8.9100E-02	-5.2900E-06	678	1078	-----	53.11	85.05
176	NH5O	AMMONIUM HYDROXIDE	---------	----------	----------	----	----	-----	-----	-----
177	NH5S	AMMONIUM HYDROGENSULFIDE	-2.728	8.3400E-02	-1.6600E-05	250	1000	20.66	17.08	64.03
178	NO	NITRIC OXIDE	39.921	5.3700E-01	-1.2400E-04	150	1500	189.04	117.70	566.08
179	NOCl	NITROSYL CHLORIDE	-15.159	4.7300E-01	-8.5900E-05	214	1000	118.33	82.20	372.25
180	NOF	NITROSYL FLUORIDE	-10.167	7.1200E-01	-1.7600E-04	250	1000	186.59	156.93	526.57
181	NO2	NITROGEN DIOXIDE	-372.375	2.3300E+00	-2.1500E-03	295	460	130.84	127.52	243.97
182	N2	NITROGEN	42.606	4.7500E-01	-9.8800E-05	150	1500	175.52	111.67	533.12
183	N2F4	TETRAFLUOROHYDRAZINE	-6.882	5.3100E-01	-1.3000E-04	112	1000	139.75	50.92	393.43
184	N2H4	HYDRAZINE	-9.979	2.4900E-01	-2.6800E-05	387	997	-----	82.18	211.15
185	N2H4C	AMMONIUM CYANIDE	-15.234	4.7100E-01	-9.5500E-05	250	1000	116.70	96.55	360.28
186	N2H6CO2	AMMONIUM CARBAMATE	-9.750	2.8900E-01	-5.2500E-05	250	1000	71.64	59.13	226.37
187	N2O	NITROUS OXIDE	-5.680	5.5600E-01	-1.5200E-04	182	1000	146.55	90.46	398.12
188	N2O3	NITROGEN TRIOXIDE	-11.415	4.2900E-01	-9.7300E-05	250	1000	107.77	89.69	320.06
189	N2O4	NITROGEN TETRAOXIDE	-22.781	8.3300E-01	-1.8700E-04	250	1000	208.80	173.65	622.54
190	N2O5	NITROGEN PENTOXIDE	-16.271	4.8800E-01	-9.4100E-05	250	1000	120.97	99.93	378.00
191	Na	SODIUM	181.000	1.5000E-02	5.0000E-05	1200	1500	-----	271.00	316.00
192	NaBr	SODIUM BROMIDE	12.073	6.8100E-02	3.5100E-06	1664	2064	-----	135.12	167.60
193	NaCN	SODIUM CYANIDE	-6.280	8.6200E-02	-3.9100E-07	1769	2169	-----	144.99	178.86
194	NaCl	SODIUM CHLORIDE	1.624	8.5900E-02	1.8200E-06	1738	2138	-----	156.41	193.58
195	NaF	SODIUM FLUORIDE	16.862	6.1900E-02	2.9800E-06	1983	2383	-----	151.26	181.21
196	NaI	SODIUM IODIDE	-15.662	2.1500E-01	-1.7400E-06	1577	1977	-----	319.08	402.61
197	NaOH	SODIUM HYDROXIDE	-5.820	1.0300E-01	-4.5500E-08	1663	2063	-----	165.06	206.12
198	Na2SO4	SODIUM SULFATE	---------	----------	----------	----	----	-----	-----	-----
199	Nb	NIOBIUM	138.221	1.3000E-01	2.4800E-06	5115	5515	-----	866.53	928.96
200	Nd	NEODYMIUM	138.495	2.2700E-01	6.8600E-06	3384	3784	-----	985.00	1095.45
201	Ne	NEON	102.964	7.4600E-01	-1.3600E-04	150	1500	313.31	211.82	915.45
202	Ni	NICKEL	70.991	1.9300E-01	7.9800E-06	2415	2815	-----	584.02	677.97
203	NiC4O4	NICKEL CARBONYL	-13.078	3.9500E-01	-7.7300E-05	250	1000	97.89	80.90	304.84
204	NiF2	NICKEL FLUORIDE	-9.042	1.1400E-01	-3.7400E-07	2013	2413	-----	218.94	263.88
205	Np	NEPTUNIUM	---------	----------	----------	----	----	-----	-----	-----
206	O2	OXYGEN	44.224	5.6200E-01	-1.1300E-04	150	1500	201.85	126.04	633.08
207	O3	OZONE	48.384	3.3200E-01	-1.6000E-04	162	350	133.09	97.93	144.92
208	Os	OSMIUM	193.295	1.9300E-01	3.9000E-06	4880	5280	-----	1226.86	1319.82
209	OsOF5	OSMIUM OXIDE PENTAFLUORIDE	-17.104	5.2300E-01	-7.8100E-05	250	1000	131.99	108.85	428.13
210	OsO4	OSMIUM TETROXIDE - YELLOW	-20.337	4.7100E-01	-6.6200E-05	403	803	-----	158.58	314.89
211	OsO4	OSMIUM TETROXIDE - WHITE	-20.337	4.7100E-01	-6.6200E-05	403	803	-----	158.58	314.89
212	P	PHOSPHORUS - WHITE	-9.199	2.4500E-01	-2.2700E-05	553	1000	-----	119.49	213.37
213	PBr3	PHOSPHORUS TRIBROMIDE	-22.117	4.6900E-01	-6.0900E-05	448	848	-----	175.98	332.20
214	PCl2F3	PHOSPHORUS DICHLORIDE TRIFLUORIDE	-15.006	5.0000E-01	-1.0800E-04	250	1000	124.60	103.36	377.41
215	PCl3	PHOSPHORUS TRICHLORIDE	-5.014	3.8600E-01	-5.6100E-05	200	1000	105.16	69.99	325.10
216	PCl5	PHOSPHORUS PENTACHLORIDE	-22.302	4.6500E-01	-7.2900E-05	433	833	-----	165.32	314.35
217	PH3	PHOSPHINE	-2.849	4.4500E-01	-1.0400E-04	139	1000	120.53	56.96	338.19
218	PH4Br	PHOSPHONIUM BROMIDE	-16.721	5.0900E-01	-1.0100E-04	250	1000	126.14	104.29	391.51
219	PH4Cl	PHOSPHONIUM CHLORIDE	-3.659	5.3100E-01	-1.3400E-04	250	1000	142.77	120.74	393.34
220	PH4I	PHOSPHONIUM IODIDE	-21.615	6.4000E-01	-1.1600E-04	250	1000	158.82	131.08	501.83
221	POCl3	PHOSPHORUS OXYCHLORIDE	-13.509	4.1000E-01	-6.2500E-05	250	1000	103.29	85.18	334.34
222	PSBr3	PHOSPHORUS THIOBROMIDE	-22.075	4.8300E-01	-5.8000E-05	448	848	-----	182.51	345.51
223	PSCl3	PHOSPHORUS THIOCHLORIDE	-11.901	3.8100E-01	-5.1000E-05	250	1000	97.11	80.12	317.91
224	P4O6	PHOSPHORUS TRIOXIDE	-19.024	4.0900E-01	-5.1900E-05	446	846	-----	153.13	289.96
225	P4O10	PHOSPHORUS PENTOXIDE	---------	----------	----------	----	----	-----	-----	-----
226	P4S10	PHOSPHORUS PENTASULFIDE	14.570	6.0400E-01	2.3200E-05	561	1000	-----	360.50	641.38
227	Pb	LEAD	55.375	2.2200E-01	1.0200E-05	2024	2424	-----	547.42	654.55
228	PbBr2	LEAD BROMIDE	-19.836	2.9900E-01	-5.3000E-06	1187	1587	-----	327.51	441.19
229	PbCl2	LEAD CHLORIDE	-17.125	2.5500E-01	-4.1700E-06	1227	1627	-----	289.87	387.23
230	PbF2	LEAD FLUORIDE	-15.171	2.0900E-01	-1.7200E-06	1566	1966	-----	307.51	388.57
231	PbI2	LEAD IODIDE	-22.434	3.4200E-01	-6.6300E-06	1145	1545	-----	360.50	490.18
232	PbO	LEAD OXIDE	-14.136	1.8700E-01	-1.0800E-06	1745	2145	-----	309.70	383.01
233	PbS	LEAD SULFIDE	-15.012	2.0700E-01	-1.7500E-06	1554	1954	-----	302.59	382.98
234	Pd	PALLADIUM	118.981	1.9500E-01	5.8900E-06	3385	3785	-----	846.14	940.98

Table 21-2 VISCOSITY OF GAS - INORGANIC COMPOUNDS (continued)

NO	FORMULA	NAME	$n_{gas} = A + BT + CT^2$					$(n_{gas}$ - micropoise, T - K)$		
			A	B	C	TMIN	TMAX	n_{gas} @25 C	n_{gas} @TMIN	n_{gas} @TMAX
235	Po	POLONIUM	64.353	6.7500E-01	3.8500E-05	1235	1635	-----	957.22	1271.60
236	Pt	PLATINUM	-1.208	5.6000E-02	3.6700E-07	3980	4380	-----	227.63	251.27
237	Ra	RADIUM	109.368	4.9000E-01	2.4800E-05	1809	2209	-----	1077.77	1313.82
238	Rb	RUBIDIUM	-295.000	6.3700E-01	-1.2100E-04	1100	1500	-----	259.09	387.72
239	RbBr	RUBIDIUM BROMIDE	-12.356	1.6800E-01	-1.2300E-06	1625	2025	-----	257.26	322.63
240	RbCl	RUBIDIUM CHLORIDE	-10.525	1.4200E-01	-9.8500E-07	1654	2054	-----	221.93	277.34
241	RbF	RUBIDIUM FLUORIDE	-9.748	1.3100E-01	-8.6000E-07	1681	2081	-----	207.97	259.06
242	RbI	RUBIDIUM IODIDE	-14.097	1.9400E-01	-1.5600E-06	1577	1977	-----	287.18	362.37
243	Re	RHENIUM	260.509	2.0300E-01	3.2600E-06	5915	6315	-----	1575.74	1672.91
244	Re2O7	RHENIUM HEPTOXIDE	-26.508	4.9600E-01	-3.3100E-05	636	1036	-----	275.41	451.56
245	Rh	RHODIUM	127.361	1.6900E-01	4.3500E-06	3940	4340	-----	860.32	942.29
246	Rn	RADON	-18.323	9.5000E-01	-2.2800E-04	250	1000	244.51	204.81	702.87
247	Ru	RUTHENIUM	135.297	1.5000E-01	3.3400E-06	4500	4900	-----	878.21	950.78
248	RuF5	RUTHENIUM PENTAFLUORIDE	-17.056	3.2600E-01	-2.4200E-05	600	1000	-----	170.08	285.16
249	S	SULFUR	-5.897	1.6200E-01	-2.8000E-06	718	1118	-----	109.14	171.98
250	SF4	SULFUR TETRAFLUORIDE	-10.261	6.1500E-01	-1.5000E-04	250	1000	159.80	134.14	454.86
251	SF6	SULFUR HEXAFLUORIDE	-12.176	5.9800E-01	-1.7800E-04	205	1000	150.35	102.97	408.09
252	SOBr2	THIONYL BROMIDE	-22.523	5.0700E-01	-7.1900E-05	413	813	-----	174.66	342.27
253	SOCl2	THIONYL CHLORIDE	-7.701	4.1800E-01	-4.4000E-05	172	1000	113.02	62.90	366.35
254	SOF2	SULFUROUS OXYFLUORIDE	-10.632	5.8700E-01	-1.4200E-04	250	1000	151.62	127.13	433.91
255	SO2	SULFUR DIOXIDE	-11.103	5.0200E-01	-1.0800E-04	200	1000	129.10	85.07	383.30
256	SO2Cl2	SULFURYL CHLORIDE	-20.529	4.6500E-01	-7.5800E-05	343	1000	-----	129.89	368.19
257	SO3	SULFUR TRIOXIDE	-12.039	5.4300E-01	-1.6000E-04	298	694	135.74	135.67	288.01
258	S2Cl2	SULFUR MONOCHLORIDE	-17.792	4.0200E-01	-5.7200E-05	411	811	-----	137.56	270.21
259	Sb	ANTIMONY	82.105	3.5400E-01	1.7100E-05	1898	2298	-----	816.14	986.58
260	SbBr3	ANTIMONY TRIBROMIDE	-23.531	4.6800E-01	-4.0600E-05	548	948	-----	220.67	383.53
261	SbCl3	ANTIMONY TRICHLORIDE	-21.042	4.3200E-01	-4.6400E-05	493	893	-----	180.42	327.31
262	SbCl5	ANTIMONY PENTACHLORIDE	-19.363	4.3600E-01	-6.1600E-05	413	813	-----	150.05	294.08
263	SbH3	STIBINE	-17.754	6.2600E-01	-1.3900E-04	250	1000	156.41	129.95	469.04
264	SbI3	ANTIMONY TRIIODIDE	-26.665	4.8800E-01	-2.9200E-05	674	1074	-----	288.64	463.21
265	Sb2O3	ANTIMONY TRIOXIDE	-16.249	2.1800E-01	-1.3800E-06	1698	2098	-----	349.14	434.05
266	Sc	SCANDIUM	67.172	1.5300E-01	5.7600E-06	2700	3100	-----	523.54	598.29
267	Se	SELENIUM	-8.981	3.8900E-01	3.1800E-06	930	1330	-----	355.44	513.87
268	SeCl4	SELENIUM TETRACHLORIDE	-20.624	4.3400E-01	-5.1900E-05	465	865	-----	169.80	315.64
269	SeF6	SELENIUM HEXAFLUORIDE	-12.928	7.3300E-01	-1.7800E-04	250	1000	189.69	159.11	541.87
270	SeOCl2	SELENIUM OXYCHLORIDE	-21.694	4.7000E-01	-6.0600E-05	441	841	-----	173.65	330.44
271	SeO2	SELENIUM DIOXIDE	-12.871	2.4800E-01	-1.8900E-05	590	990	-----	126.87	214.12
272	Si	SILICON	-11.338	6.0500E-02	-4.3100E-07	3514	3914	-----	196.08	219.01
273	SiBrCl2F	BROMODICHLOROFLUOROSILANE	-15.940	4.8800E-01	-9.7900E-05	250	1000	121.00	100.07	374.69
274	SiBrF3	TRIFLUOROBROMOSILANE	-11.203	5.9300E-01	-1.4300E-04	250	1000	152.84	128.07	438.68
275	SiBr2ClF	DIBROMOCHLOROFLUOROSILANE	-16.957	5.0300E-01	-9.2500E-05	250	1000	124.70	102.94	393.27
276	SiClF3	TRIFLUOROCHLOROSILANE	-4.919	5.4500E-01	-1.3700E-04	250	1000	145.34	122.72	403.09
277	SiCl2F2	DICHLORODIFLUOROSILANE	-10.944	5.0500E-01	-1.2000E-04	250	1000	128.92	107.77	374.21
278	SiCl3F	TRICHLOROFLUOROSILANE	-13.967	4.6100E-01	-9.9000E-05	250	1000	114.77	95.17	348.32
279	SiCl4	SILICON TETRACHLORIDE	0.103	3.8700E-01	-1.5300E-04	273	573	102.03	94.49	171.91
280	SiF4	SILICON TETRAFLUORIDE	24.103	4.5600E-01	-7.8100E-05	296	407	153.21	152.33	196.89
281	SiHBr3	TRIBROMOSILANE	-14.609	4.4800E-01	-6.6500E-05	250	1000	112.99	93.18	366.68
282	SiHCl3	TRICHLOROSILANE	6.109	3.7500E-01	-3.3400E-05	305	500	-----	117.36	185.24
283	SiHF3	TRIFLUOROSILANE	1.206	5.5700E-01	-1.4300E-04	250	1000	154.44	131.42	414.48
284	SiH2Br2	DIBROMOSILANE	-17.281	5.1100E-01	-9.0600E-05	250	1000	126.96	104.75	402.95
285	SiH2Cl2	DICHLOROSILANE	-18.511	4.7300E-01	-1.1600E-04	273	600	112.32	102.08	223.81
286	SiH2F2	DIFLUOROSILANE	-2.992	5.1700E-01	-1.3100E-04	250	1000	139.43	118.00	382.86
287	SiH2I2	DIIODOSILANE	-27.518	5.9800E-01	-8.7600E-05	423	823	-----	209.75	405.28
288	SiH3Br	MONOBROMOSILANE	-15.994	5.3800E-01	-1.1700E-04	250	1000	134.09	111.26	405.44
289	SiH3Cl	MONOCHLOROSILANE	-10.025	4.4300E-01	-1.0400E-04	250	1000	112.85	94.27	328.88
290	SiH3F	MONOFLUOROSILANE	1.757	4.7900E-01	-1.2400E-04	250	1000	133.51	113.72	356.86
291	SiH3I	IODOSILANE	-21.491	6.4500E-01	-1.2400E-04	250	1000	159.84	132.06	499.37
292	SiH4	SILANE	5.831	3.8300E-01	-4.5400E-05	288	373	115.97	112.35	142.35
293	SiO2	SILICON DIOXIDE	---------	----------	----------	----	----	-----	-----	-----
294	Si2Cl6	HEXACHLORODISILANE	-15.023	3.3900E-01	-4.8100E-05	412	812	-----	116.28	228.15
295	Si2F6	HEXAFLUORODISILANE	-11.810	4.7600E-01	-1.1000E-04	250	1000	120.29	100.28	354.05
296	Si2H5Cl	DISILANYL CHLORIDE	-11.645	3.5300E-01	-6.9200E-05	250	1000	87.30	72.16	271.67
297	Si2H6	DISILANE	-10.479	3.8200E-01	-8.5700E-05	250	1000	95.68	79.56	285.42
298	Si2OCl3F3	TRICHLOROTRIFLUORODISILOXANE	-13.519	4.0800E-01	-7.9800E-05	250	1000	101.15	83.60	315.06
299	Si2OCl6	HEXACHLORODISILOXANE	-15.111	3.4200E-01	-4.9100E-05	409	809	-----	116.63	229.57
300	Si2OH6	DISILOXANE	-9.191	3.5900E-01	-8.2400E-05	250	1000	90.62	75.49	267.70
301	Si3Cl8	OCTACHLOROTRISILANE	-14.151	2.9100E-01	-3.2700E-05	485	885	-----	119.19	217.58
302	Si3H8	TRISILANE	-9.625	2.8700E-01	-5.4000E-05	250	1000	71.07	58.69	223.11
303	Si3H9N	TRISILAZANE	-10.069	3.0200E-01	-5.7700E-05	250	1000	74.69	61.70	233.70
304	Si4H10	TETRASILANE	-8.422	2.5500E-01	-3.9200E-05	250	1000	64.15	52.90	207.44
305	Sm	SAMARIUM	92.251	3.8700E-01	1.9200E-05	1874	2274	-----	885.67	1072.47
306	Sn	TIN	39.390	1.2700E-01	3.8500E-06	2995	3395	-----	454.40	515.05
307	SnBr4	STANNIC BROMIDE	-22.720	4.7000E-01	-5.4100E-05	478	878	-----	189.80	348.65
308	SnCl2	STANNOUS CHLORIDE	-15.310	2.5300E-01	-8.5800E-06	896	1296	-----	204.50	298.17
309	SnCl4	STANNIC CHLORIDE	-14.239	4.4200E-01	-6.3500E-05	250	1000	111.93	92.32	364.38
310	SnH4	STANNIC HYDRIDE	-10.487	6.7500E-01	-1.6500E-04	250	1000	176.01	147.89	498.94
311	SnI4	STANNIC IODIDE	-30.281	5.7200E-01	-3.9800E-05	621	1021	-----	309.28	511.74
312	Sr	STRONTIUM	61.141	3.3600E-01	1.7900E-05	1630	2030	-----	656.56	817.23

Table 21-2 VISCOSITY OF GAS - INORGANIC COMPOUNDS (continued)

NO	FORMULA	NAME	$n_{gas} = A + BT + CT^2$					$(n_{gas}$ - micropoise, T - K$)$		
			A	B	C	TMIN	TMAX	n_{gas} @25 C	n_{gas} @TMIN	n_{gas} @TMAX
313	SrO	STRONTIUM OXIDE	---------	----------	----------	----	----	-----	-----	-----
314	Ta	TANTALUM	200.818	1.6900E-01	2.9200E-06	5565	5965	------	1232.53	1313.66
315	Tc	TECNNETIUM	140.400	1.3600E-01	2.6700E-06	5000	5400	------	885.39	950.76
316	Te	TELLURIUM	79.742	2.7900E-01	1.9500E-05	1285	1685	------	470.45	605.20
317	TeCl4	TELLURIUM TETRACHLORIDE	-19.581	3.6000E-01	-2.2100E-05	665	1065	------	209.93	338.55
318	TeF6	TELLURIUM HEXAFLUORIDE	-13.259	6.6900E-01	-1.6100E-04	250	1000	172.07	144.07	495.64
319	Ti	TITANIUM	10.245	2.9200E-01	3.3000E-06	3442	3842	------	1053.35	1179.65
320	TiCl4	TITANIUM TETRACHLORIDE	-17.416	3.8600E-01	-5.9300E-05	409	809	------	130.67	256.32
321	Tl	THALLIUM	100.360	4.8200E-01	2.4900E-05	1745	2145	------	1017.19	1248.68
322	TlBr	THALLOUS BROMIDE	-17.829	2.7600E-01	-6.0000E-06	1092	1492	------	276.46	380.68
323	TlI	THALLOUS IODIDE	-19.229	2.9700E-01	-6.4100E-06	1096	1496	------	298.99	411.30
324	Tm	THULIUM	112.968	3.5300E-01	1.5600E-05	2219	2619	------	973.01	1144.36
325	U	URANIUM	198.746	2.4700E-01	6.0300E-06	4135	4535	------	1322.56	1442.21
326	UF6	URANIUM FLUORIDE	-23.834	7.2200E-01	-1.4200E-04	250	1000	178.92	147.89	556.45
327	V	VANADIUM	86.128	1.2600E-01	3.5100E-06	3665	4065	------	595.74	657.07
328	VCl4	VANADIUM TETRACHLORIDE	-3.652	3.7700E-01	-3.1700E-05	247	1000	105.97	87.56	341.80
329	VOCl3	VANADIUM OXYTRICHLORIDE	-12.405	3.9500E-01	-5.3500E-05	250	1000	100.61	83.00	329.09
330	W	TUNGSTEN	177.786	2.4600E-01	4.4200E-06	5645	6045	------	1706.58	1825.60
331	WF6	TUNGSTEN FLUORIDE	-22.675	7.3300E-01	-1.5500E-04	250	1000	182.22	151.00	555.59
332	Xe	XENON	7.386	7.8700E-01	-1.5100E-04	100	1600	228.53	84.54	880.48
333	Yb	YTTERBIUM	87.650	4.6400E-01	2.4600E-05	1660	2060	------	926.20	1148.50
334	Yt	YTTRIUM	102.271	1.9400E-01	6.5000E-06	3055	3455	------	756.25	850.86
335	Zn	ZINC	43.470	3.3200E-01	2.3000E-05	1181	1581	------	467.99	626.32
336	ZnCl2	ZINC CHLORIDE	-12.609	2.0100E-01	-5.2900E-06	1005	1405	------	183.64	258.77
337	ZnF2	ZINC FLUORIDE	-9.592	1.2700E-01	-6.9500E-07	1770	2170	------	212.30	261.84
338	ZnO	ZINC OXIDE	---------	----------	----------	----	----	-----	-----	-----
339	ZnSO4	ZINC SULFATE	---------	----------	----------	----	----	-----	-----	-----
340	Zr	ZIRCONIUM	35.675	3.2900E-01	3.5900E-06	4598	4998	------	1623.16	1768.43
341	ZrBr4	ZIRCONIUM BROMIDE	-24.454	4.5900E-01	-3.1200E-05	630	1030	------	252.31	415.19
342	ZrCl4	ZIRCONIUM CHLORIDE	-18.571	3.5400E-01	-2.6000E-05	604	1004	------	186.05	311.13
343	ZrI4	ZIRCONIUM IODIDE	-28.831	5.1900E-01	-2.8600E-05	704	1104	------	322.06	508.78

n_{gas} - viscosity of gas, micropoise

A, B, and C - regression coefficients for chemical compound

T - temperature, K

TMIN - minimum temperature, K

TMAX - maximum temperature, K

477

Chapter 22

VISCOSITY OF LIQUID

Carl L. Yaws, Xiaoyan Lin, and Li Bu
Lamar University, Beaumont, Texas

ABSTRACT

Results for liquid viscosity as function of temperature are presented for a wide range of organic and inorganic chemicals. The major chemicals include many compound types. The results are provided in easy-to-use tables that are especially applicable for rapid engineering usage with the personal computer or hand calculator. The agreement of correlation and data is quite good.

INTRODUCTION

Liquid viscosity data are important in many engineering applications in the chemical processing and petroleum refining industries. The objective of this article is to provide the engineer with such viscosity data. The compilation of data is presented for a wide temperature range to enable the engineer to determine values at temperatures of interest.

LIQUID VISCOSITY CORRELATION

The correlation for liquid viscosity as a function of temperature is given by the equation shown below:

$$\log_{10} n_{liq} = A + B/T + C\,T + D\,T^2 \tag{22-1}$$

where
n_{liq} = viscosity of liquid, centipoise
A, B, C, and D = regression coefficients for chemical compound
T = temperature, K

The results for liquid viscosity are given in Tables 22-1 and 22-2. The tabulations are arranged by chemical formula to provide ease of use in quickly locating data. Many of the values for the liquid cover the full range from melting to critical point.

In preparing the compilation, a literature search was conducted to identify data source publications for organics (1-40) and inorganics (1-126). Both experimental values for the property under consideration and parameter values for estimation of the property are included in the source publications. The publications were screened for appropriate data. The compilation resulting from the screening is based on both experimental data and estimated values.

For organic compounds, liquid viscosities at low temperatures were primarily estimated using the Van Velzen method (29, group and structural contributions). The Przezdziecki and Sridhar equation (29, corresponding states) and boiling point method (empirical) were also used for selected compounds. For liquid viscosities at high temperatures, both experimental data and estimates were extended using a modified Letsou and Stiel equation (29, corresponding states) for saturated liquids. Experimental data and estimates were then regressed to provide the same equation for all compounds.

For inorganic compounds, liquid viscosities for metals were primarily estimated using the Grosse method (64, melting point, liquid volume). For inorganics that are solids at ambient conditions, a modified Letsou and Stiel method (29, corresponding states, melting point, boiling point) was used. For inorganics that are gases and liquids at ambient conditions, a modified Letsou and Stiel method was also used. Experimental data and estimates were then regressed to provide the same equation for all compounds.

For gas and liquid viscosities, the experimental data for inorganics is very limited or scarce when compared to that available for organics. The estimation methods for inorganics are also very limited or scarce in comparison to organics. Thus, in the absence of experimental data and the scarcity of estimation methods, the estimates for inorganics should be considered as very rough approximations.

Very limited experimental data for liquid viscosities are available at temperatures in the region of the melting and critical point temperatures. Thus, the values in the regions of melting and critical point temperatures should be considered rough approximations. The values in the intermediate region (above melting and below critical point) are more accurate.

A comparison of correlation and experimental data for liquid viscosity is shown in Fig. 22-1 for a representative chemical. The graph discloses good agreement of correlation and data.

EXAMPLES

The correlation results may be used for prediction and calculation of liquid viscosity. Examples are given below.

Example 1 Calculate the liquid viscosity of cyclohexane (C6H12) at a temperature of 353.85 K (80.7 C).

Substitution of the coefficients from the table and temperature into the correlation equation yields

$\log_{10} n_{liq}$ = + 4.7423 - 2.5322E+02/353.85 + (-1.6927E-02) (353.85) + (1.2472E-05) (353.85^2) = -.4012

n_{liq} = $10^{-.4012}$

n_{liq} = 0.397 centipoise

The calculated and data values compare favorably (0.397 vs 0.413, deviation = 3.87%).

Example 2 Calculate the liquid viscosity of benzene (C6H6) at a temperature of 343.35 K (70.2 C).

Substitution of the coefficients from the table and temperature into the correlation equation yields

$\log_{10} n_{liq}$ = - 7.4005 + 1.1815E+03/343.85 + (1.4888E-02) (343.85) + (-1.3713E-05) (343.85^2) = -.4647

n_{liq} = $10^{-.4647}$

n_{liq} = 0.343 centipoise

The calculated and data values compare favorably (0.343 vs 0.3507, deviation = 2.2%).

Portions of this material appeared in Chem. Eng., 101 (4), 119 (April, 1994) and is reprinted by special permission.

REFERENCES - ORGANIC COMPOUNDS

1-34. See **REFERENCES - ORGANIC COMPOUNDS** in **Chapter 1, CRITICAL PROPERTIES AND ACENTRIC FACTOR**
35. Golubev, I. F., VISCOSITY OF GASES AND GAS MIXTURES, translated from Russian, U. S. Dept. of Commerce, Springfield, VA (1970).
36. Stephan, K. and K. Lucas, VISCOSITY OF DENSE FLUIDS, Plenum Press, New York, NY (1979).
37. Viswanath, D. S. and G. Natarajan, DATA BOOK ON THE VISCOSITY OF LIQUIDS, Hemisphere Publishing Corporation, New York, NY (1989).
38. Yaws, C. L., Xiaoyan Lin, and Li Bu, Chem. Eng., 101 (4), 119 (April, 1994).
39. Yaws, C. L., HANDBOOK OF TRANSPORT PROPERTY DATA, Gulf Publishing Co., Houston, TX (1995).
40. Yaws, C. L., HANDBOOK OF VISCOSITY, Vols. 1, 2, 3 and 4, Gulf Publishing Company, Houston, TX (1995, 1995, 1995, 1997).

REFERENCES – INORGANIC COMPOUNDS

1-56. See **REFERENCES - INORGANIC COMPOUNDS** in **Chapter 1, CRITICAL PROPERTIES AND ACENTRIC FACTOR**
57. Golubev, I. F., VISCOSITY OF GASES AND GAS MIXTURES, translated from Russian, U. S. Dept. of Commerce, Springfield, VA (1970).
58. Viswanath, D. S. and G. Natarajan, DATA BOOK ON THE VISCOSITY OF LIQUIDS, Hemisphere Publishing Corporation, New York, NY (1989).
59. Lyon, R. N., ed., LIQUID-METALS HANDBOOK, Atomic Energy Commission and Dept. of Navy, Washington, DC (1954).
60. Emsley, J., THE ELEMENTS, 2nd ed., Clarendon Press, Oxford University Press, New York, NY (1991).
61. Perry, D. L. and S. L. Phillips, HANDBOOK OF INORGANIC COMPOUNDS, CRC Press, New York, NY (1995).
62. Stephan, K. and K. Lucas, VISCOSITY OF DENSE FLUIDS, Plenum Press, New York, NY (1979).
63. Van Horn, K. R., ed., ALUMINUM, Vol. 1, American Society for Metals, Metals Park, Ohio (1967).
64. Grosse, A. V., J. Inorg. Nucl. Chem., 23, 333 (1961).
65. Chapman, T. W., AIChE J., 12, No. 2, 395 (1966).
66. Bacon, R. F. and R. Fanelli, J. Amer. Chem. Soc., 15, 639 (1943).
67. Bacon, J. F. and A. A. Hasapis, J. Appl. Phys., 30 (9), 1470 (1959).
68. Niselson, L. A. and T. D. Sokolova, Russ. J. Inorg. Chem., 10, 827 (1965).
69. Saji, Y. and S. Kobayashi, Cryogenics, 136 (1964).
70. Maitland, G. C. and E. B. Smith, J. Chem. Eng. Data, 17 (2), 150 (1972).
71. Kestin, J. and E. A., Knierrim, J. Phys. Chem. Ref. Data, 13 (1), 229 (1984).
72. Runovskaya, I. V., A. D. Zorin, and G. G. Devyatykh, Russ. J. Inorg. Chem., 15, 1338 (1970).
73. Reichenburg, D., AICHE J., 21, 181 (1975).
74. Stiel, L.T. and G. Thodos, AICHE J., 10, 266 (1964).
75. Boon, J. P. and J. C. Thomas, J. Physica, 33, 547 (1967).
76. Rao, R. V. G. and K. N. Swamy, Z. Phys. Chem. (Leipzig), 2, 250 (1974).
77. Rudenko, N. S. and L. W. Schubrukow, Phys Zeit. der Sowjetunion, 6, 470 (1934).
78. Hetteman, W., W. Grevendork, and A. DeBock, J. Chem. Phys., 53(1), 185 (1970).
79. Kulifeev, V. K., V. I. Panchishnyi, and G. P. Standevich, Isv. Vyssh. Ucheb. Zaved. Tsvet. Met., 11(2), 116 (1968).
80. Simkin, J. and R. L. Jarry, J. Phys. Chem., 61, 503 (1957).
81. Usanovich, M., T. Sumarokova, and V. Udovenko, Acta Physicochim. USSR, 11, 505 (1939).
82. Kestin, J., J. V. Sengers, B. Kamgar-Parsi, and J.M. H. Levelt Sengers, J. Phys. Chem. Ref. Data, 13 (2), 601 (1984).
83. Matsunaga, N. and A. Nagashima, J. Phys. Chem. Ref. Data, 12 (4), 933 (1983).
84. Hanley, H.J.M. and R. Prydz, J. Phys. Chem. Ref. Data, 1 (4), 1101 (1972).
85. Moore, G. A. and T. R. Shives, IRON, Metals Handbook, 8th ed., 1206 (1961).

86. Greenwood, N.N. and K. Wade, J. Inorg. Nucl. Chem., 3, 349 (1957).

87. Janz, G. J., A. T. Ward, and R. D. Reeves, MOLTEN SALT DATA, Technical Bulletin Series, Rensselaer Polytechnic Institue, Troy, NY (1964).

88. Baker, C. E., J. Chem. Phys., 46, 2846 (1967).

89. Stern, S. A., J. L. Mullhauupt and W. B. Kay, Chem. Rev., 60, 185 (1960).

90. Mason, D. M., I. Petker, and S. P. Vango, J. Phys. Chem., 59, 511 (1955).

91. Naumova, A. S., Zh, Obshch. Khim., 19, 1228 (1949).

92. Taylor, E. G., L. M. Lynne, and A. G. Follous, Can. J. Chem., 29, 439 (1951).

93. Haar, L., J. S. Gallagher, and G. S. Kell, NBS/NRC STEAM TABLES. THERMODYNAMIC AND TRANSPORT PROPERTIES AND COMPUTER PROGRAMS FOR VAPOR AND LIQUID STATES OF WATER IN SI UNITS, Hemisphere Publish Corporation, Washington, DC (1984).

94. Misra, S. C. and K. N. Parida, Ind. J. Pure Appl. Phys., 7, 772 (1969).

95. Janz, G. L., C. B. Bansal, N. P. Bansal, R. M. Murphy, and R. P. T. Tompkins, PHYSICAL PROPERTIES DATA COMPILATIONS RELEVANT TOR ENERGY STORAGE. II. MOLTEN SALTS: DATA ON SINGLE AND MULTICOMPONENT SALT SYSTEMS, Nat. Bur. Stand., Molten Salts Data Center, Troy, NY (April, 1979).

96. Morozov, I. R., J. Appl. Chem. (USSR), 24, 975 (1951).

97. Davison, H. W., NASA Tech. Note D-4650 (1968).

98. Andrade, E. N., C. Da, and E. R. Dobbs, Proc. Roy. Soc. London, 211A, 12 (1952).

99. Leu, A. L., S. M. Ma, and H. Eyring, Proc. Nat. Acad. Sci. USA, 72 (3), 1026 (1975).

100. Krynicki, K. and J. W. Hennel, Acta Physica Polonica, 24 (8), 269 (1963).

101. Hanley, H. J. M. and J. F. Ely, J. Phys. Chem. Ref. Data, 2 (4), 735 (1973).

102. Mason, D. M., O. W. Wilcox, and B. H. Sage, J. Phys. Chem., 56, 1008 (1952).

103. Janz, G. J., J. Phys. Chem. Ref. Data, 9 (4), 791 (1980).

104. Janz, G. J., G. L. Gardner, U. Krebs, and R. P. T. Tomkins, J. Phys. Chem. Ref. Data, 3 (1), 1 (1974).

105. Morozov, I. R., J. Appl. Chem. (USSR), 24, 975 (1951).

106. Gossink, R. G. and J. M. Stevels, Inorg. Chem., 11 (9), 2180 (1982).

107. Forster, S. (translation), Cryogenics, 3, 176 (1963).

108. McCarty, R. D. and L. A. Weber, National Bureau of Standards Technical Note 384, Washington, DC (1971).

109. Hersh, C. K., A. W. Berger, and J. R. C. Brown, Adv. Chem. Ser. No. 21, Am. Chem. Soc., Washington, DC (1959).

110. Streng A. G., J. Chem. Eng. Data, 6 (3), 43 (1961).

111. Jenkins, A. C. and F. S. Dipaolo, J. Chem. Phys., 29 (4), 905 (1958).

112. Mole, M. F., W. S. Holmes, and J. C. McCoubrey, J. Chem. Soc., 81, 5082 (1959).

113. Gutmann, V., Monatshofte Fur Chemie, 83, 164 (1952).

114. Yoon, P. and G. Thodos, AICHE J., 16, 300 (1970).

115. Murgulescu, I.G. and M. Serban, Rev. Roum. Chim. 19, 1417 (1974).

116. Veda, K. and K. Kigoshi, J. Inorg. Nucl. Chem., 36, 989 (1974).

117. Hyne, R. A. and P. F. Tiley, J. Chem. Soc., 2348 (1961).

118. Niselson, L. A., P. P. Pugachevich, T. D. Sokolova, and R. A. Bederdinov, Russ. J. Inorg. Chem., 10 (6), 705 (1965).

119. Ellis, C. P. and J. G. Raw, J. Chem. Soc., 3765 (1956).

120. Runovskaya, I. V., A. D. Zorin, and G. G. Devyatykh, Russ. J. Inorg. Chem., 15 (9), (1970).

121. Waseda, Y. and K. Suzuki, Phys. Status Solidi B: Basic Research, 57, 351 (1973).

122. Rudenko, N. S and V. G. Konareva, Zh. Fiz. Khim., 38, 270 (1964).

123. Culpin, M. F., Proc. Phys. Soc., 70, 1079 (1957).

124. Spells, K. E, Proc. Phys. Soc., 48, 299 (1936).

125. Yaws, C. L., HANDBOOK OF TRANSPORT PROPERTY DATA, Gulf Publishing Co., Houston, TX (1995).

126. Yaws, C. L., HANDBOOK OF VISCOSITY, Vol. 4, Gulf Publishing Co., Houston, TX (1997).

Figure 22-1 Viscosity of Liquid

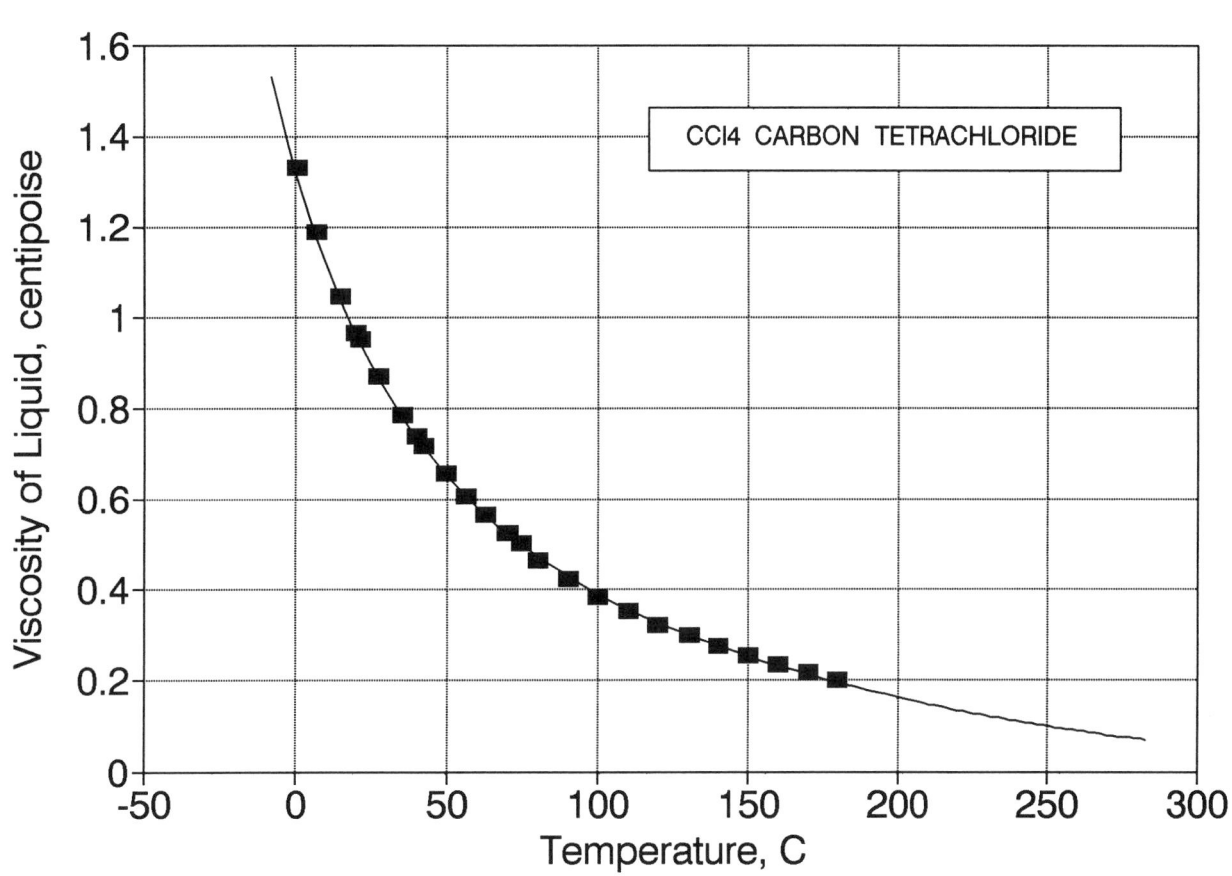

CCl4 CARBON TETRACHLORIDE

■ data —— equation

Table 22-1 VISCOSITY OF LIQUID - ORGANIC COMPOUNDS

			$\log_{10} n_{liq} = A + B/T + C\,T + C\,T^2$						(n_{liq} - centipoise, T - K)	
NO	FORMULA	NAME	A	B	C	D	TMIN	TMAX	n_{liq} @25 C	n_{liq} @TMAX
1	CBrClF2	BROMOCHLORODIFLUOROMETHANE	-5.3203	5.2567E+02	1.8649E-02	-2.7670E-05	160	426	0.349	0.069
2	CBrCl3	BROMOTRICHLOROMETHANE	-2.4807	6.5384E+02	2.4106E-03	-2.9900E-06	252	606	1.463	0.091
3	CBrF3	BROMOTRIFLUOROMETHANE	-17.3630	1.4685E+03	6.7833E-02	-9.6658E-05	170	340	0.157	0.070
4	CBr2F2	DIBROMODIFLUOROMETHANE	-2.3825	3.4361E+02	6.1053E-03	-9.9243E-06	220	478	0.511	0.097
5	CClF3	CHLOROTRIFLUOROMETHANE	-15.2276	1.1396E+03	6.6727E-02	-1.0913E-04	170	302	0.061	0.056
6	CClN	CYANOGEN CHLORIDE	-34.0692	4.3759E+03	8.8017E-02	-8.1600E-05	267	449	0.395	0.056
7	CCl2F2	DICHLORODIFLUOROMETHANE	-14.1271	1.2812E+03	5.1192E-02	-6.8214E-05	170	385	0.234	0.063
8	CCl2O	PHOSGENE	-5.9900	8.9328E+02	1.2942E-02	-1.4515E-05	253	455	0.375	0.072
9	CCl3F	TRICHLOROFLUOROMETHANE	-8.7050	9.7314E+02	2.6505E-02	-3.1615E-05	170	471	0.448	0.068
10	CCl4	CARBON TETRACHLORIDE	-6.4564	1.0379E+03	1.4021E-02	-1.4107E-05	265	556	0.893	0.070
11	CF2O	CARBONYL FLUORIDE	15.3424	-1.0000E+03	-6.0892E-02	5.3802E-05	162	297	-----	0.043
12	CF4	CARBON TETRAFLUORIDE	-8.1062	4.7871E+02	5.0987E-02	-1.3379E-04	90	228	-----	0.046
13	CHBr3	TRIBROMOMETHANE	-3.3401	7.2801E+02	5.4337E-03	-5.0412E-06	281	696	1.878	0.111
14	CHClF2	CHLORODIFLUOROMETHANE	-10.8934	9.7972E+02	3.8730E-02	-5.2025E-05	170	369	0.207	0.093
15	CHCl2F	DICHLOROFLUOROMETHANE	-9.3552	9.7434E+02	2.9611E-02	-3.6067E-05	170	452	0.343	0.065
16	CHCl3	CHLOROFORM	-4.7831	6.9902E+02	1.0929E-02	-1.2244E-05	210	536	0.539	0.073
17	CHF3	TRIFLUOROMETHANE	-21.3082	1.5503E+03	9.5945E-02	-1.5489E-04	170	299	0.054	0.052
18	CHI3	TRIIODOMETHANE	-----	--------	--------	--------	----	----	-----	-----
19	CHN	HYDROGEN CYANIDE	-12.0812	1.3183E+03	3.5234E-02	-4.0185E-05	260	457	0.188	0.033
20	CHNS	ISOTHIOCYANIC-ACID	-----	--------	--------	--------	----	----	-----	-----
21	CH2BrCl	BROMOCHLOROMETHANE	-5.2060	7.7028E+02	1.1905E-02	-1.2455E-05	185	557	0.660	0.088
22	CH2Br2	DIBROMOMETHANE	-5.4865	8.6380E+02	1.2128E-02	-1.1614E-05	230	611	0.987	0.100
23	CH2ClF	CHLOROFLUOROMETHANE	-9.6115	9.7434E+02	2.9611E-02	-3.6067E-05	170	424	0.190	0.057
24	CH2Cl2	DICHLOROMETHANE	-5.1043	6.8653E+02	1.2459E-02	-1.4540E-05	208	510	0.417	0.065
25	CH2F2	DIFLUOROMETHANE	-0.9739	1.3685E+02	2.7097E-03	-1.3376E-05	200	352	0.127	0.051
26	CH2I2	DIIODOMETHANE	-1.4610	5.0743E+02	1.1256E-03	-1.8470E-06	279	747	2.584	0.107
27	CH2O	FORMALDEHYDE	-6.3673	6.5848E+02	1.9414E-02	-2.7279E-05	193	408	0.160	0.042
28	CH2O2	FORMIC ACID	-4.2125	9.7953E+02	5.5520E-03	-5.7723E-06	281	580	1.641	0.057
29	CH3Br	METHYL BROMIDE	-9.5533	1.0306E+03	2.8322E-02	-3.1920E-05	193	467	0.324	0.083
30	CH3Cl	METHYL CHLORIDE	-7.3473	8.5395E+02	1.9485E-02	-2.3484E-05	249	416	0.173	0.056
31	CH3Cl3Si	METHYL TRICHLOROSILANE	-5.0787	7.1071E+02	1.2214E-02	-1.4351E-05	225	517	0.469	0.060
32	CH3F	METHYL FLUORIDE	-7.1229	5.0487E+02	3.0365E-02	-5.4345E-05	131	318	0.062	0.042
33	CH3I	METHYL IODIDE	-9.3737	1.1262E+03	2.5513E-02	-2.6102E-05	250	528	0.490	0.090
34	CH3NO	FORMAMIDE	-10.3646	1.9650E+03	1.8169E-02	-1.2609E-05	276	771	3.329	0.050
35	CH3NO2	NITROMETHANE	-7.1521	1.0567E+03	1.5983E-02	-1.5345E-05	245	588	0.621	0.055
36	CH3NO2	METHYL-NITRITE	-----	--------	--------	--------	----	----	-----	-----
37	CH3NO3	METHYL-NITRATE	-----	--------	--------	--------	----	----	-----	-----
38	CH4	METHANE	-7.3801	3.1925E+02	4.7934E-02	-1.4120E-04	91	191	-----	0.020
39	CH4Cl2Si	METHYL DICHLOROSILANE	-7.2040	8.6650E+02	1.9542E-02	-2.2621E-05	275	483	0.329	0.056
40	CH4O	METHANOL	-9.0562	1.2542E+03	2.2383E-02	-2.3538E-05	230	513	0.539	0.047
41	CH4O3S	METHANESULFONIC ACID	-----	--------	--------	--------	----	----	-----	-----
42	CH4S	METHYL MERCAPTAN	-3.8298	4.4874E+02	9.7914E-03	-1.3437E-05	150	470	0.251	0.057
43	CH5ClSi	METHYL CHLOROSILANE	-12.7087	1.3825E+03	3.7365E-02	-4.2205E-05	220	442	0.207	0.049
44	CH5N	METHYLAMINE	-9.4670	9.8286E+02	2.8918E-02	-3.5672E-05	180	430	0.191	0.045
45	CH6Si	METHYL SILANE	-2.3551	1.5725E+02	8.7028E-03	-2.0236E-05	180	353	0.093	0.044
46	CN4O8	TETRANITROMETHANE	1.5505	4.2561E+02	-1.2276E-02	1.0222E-05	287	540	1.685	0.049
47	CO	CARBON MONOXIDE	-1.1224	5.7858E+01	-4.9174E-03	8.2233E-06	69	133	-----	0.064
48	COS	CARBONYL SULFIDE	-2.7939	3.0912E+02	7.0546E-03	-1.3233E-05	134	379	0.148	0.062
49	CO2	CARBON DIOXIDE	-19.4921	1.5948E+03	7.9274E-02	-1.2025E-04	219	304	0.064	0.055
50	CS2	CARBON DISULFIDE	-9.1108	1.1216E+03	2.3216E-02	-2.2648E-05	235	552	0.363	0.068
51	C2BrF3	BROMOTRIFLUOROETHYLENE	12.0582	-5.7499E+02	-5.2673E-02	5.9107E-05	220	432	0.478	0.101
52	C2Br2F4	1,2-DIBROMOTETRAFLUOROETHANE	1.4494	1.4794E+01	-5.6542E-03	6.9522E-07	227	488	0.750	0.077
53	C2ClF3	CHLOROTRIFLUOROETHYLENE	-14.1754	2.2800E+03	2.2548E-02	-1.0524E-05	220	379	0.182	0.075
54	C2ClF5	CHLOROPENTAFLUOROETHANE	-21.2668	1.8746E+03	8.1455E-02	-1.1282E-04	190	353	0.189	0.055
55	C2Cl2F4	1,2-DICHLOROTETRAFLUOROETHANE	-13.2921	1.3844E+03	4.4079E-02	-5.5267E-05	179	419	0.381	0.060
56	C2Cl3F3	1,1,2-TRICHLOROTRIFLUOROETHANE	-1.8516	4.1245E+02	3.2446E-03	-7.4593E-06	237	487	0.686	0.064
57	C2Cl4	TETRACHLOROETHYLENE	-7.4654	1.1063E+03	1.6888E-02	-1.5458E-05	251	620	0.806	0.070
58	C2Cl4F2	1,1,2,2-TETRACHLORODIFLUOROETHANE	-4.4583	1.0258E+03	5.4537E-03	-5.1203E-06	299	551	-----	0.071
59	C2Cl4O	TRICHLOROACETYL CHLORIDE	-3.2061	7.3624E+02	4.6638E-03	-5.5161E-06	273	590	1.457	0.075
60	C2Cl6	HEXACHLOROETHANE	-5.0439	8.4518E+02	9.8272E-03	-8.6017E-06	460	698	-----	0.068
61	C2F4	TETRAFLUOROETHYLENE	18.2028	-1.3315E+03	-6.6186E-02	5.3425E-05	174	306	0.057	0.040
62	C2F6	HEXAFLUOROETHANE	-14.9996	1.0011E+03	7.0996E-02	-1.2231E-04	173	293	-----	0.052
63	C2HBrClF3	HALOTHANE	2.8293	5.5081E+01	-1.4157E-02	1.2453E-05	223	521	0.795	0.087
64	C2HClF2	2-CHLORO-1,1-DIFLUOROETHYLENE	13.0230	-7.5837E+02	-5.6827E-02	6.5532E-05	200	401	0.230	0.076
65	C2HCl3	TRICHLOROETHYLENE	-5.5389	7.8313E+02	1.2849E-02	-1.3292E-05	250	571	0.546	0.068
66	C2HCl3O	DICHLOROACETYL CHLORIDE	-18.9816	2.5504E+03	4.5842E-02	-3.9281E-05	298	579	0.560	0.063
67	C2HCl3O	TRICHLOROACETALDEHYDE	-3.9777	7.8167E+02	6.8624E-03	-7.6177E-06	225	565	1.030	0.071
68	C2HCl5	PENTACHLOROETHANE	-2.2339	6.8941E+02	1.5280E-03	-2.1471E-06	244	665	2.203	0.074
69	C2HF3	TRIFLUOROETHENE	18.5468	-1.3315E+03	-6.6186E-02	5.3425E-05	174	306	0.125	0.088
70	C2HF3O2	TRIFLUOROACETIC ACID	7.9900	-5.8864E+02	-2.6509E-02	2.0573E-05	258	491	0.873	0.054
71	C2HF5	PENTAFLUOROETHANE	5.4452	-1.8170E+02	-3.0751E-02	3.7720E-05	170	342	0.105	0.064
72	C2H2	ACETYLENE	-0.0709	2.8381E+01	-4.6617E-03	3.1151E-06	193	308	0.082	0.076
73	C2H2Br4	1,1,2,2-TETRABROMOETHANE	-12.1956	2.4476E+03	2.0163E-02	-1.2385E-05	300	824	-----	0.095
74	C2H2Cl2	1,1-DICHLOROETHYLENE	-2.8187	4.7865E+02	5.0534E-03	-7.6546E-06	151	482	0.410	0.068
75	C2H2Cl2	cis-1,2-DICHLOROETHYLENE	-5.4151	7.2994E+02	1.3225E-02	-1.4921E-05	208	527	0.446	0.062
76	C2H2Cl2	trans-1,2-DICHLOROETHYLENE	-7.5792	9.4638E+02	1.9835E-02	-2.1586E-05	223	508	0.389	0.062
77	C2H2Cl2O	CHLOROACETYL CHLORIDE	-6.4016	1.1523E+03	1.2139E-02	-1.1268E-05	251	581	1.205	0.068
78	C2H2Cl2O	DICHLOROACETALDEHYDE	-10.5217	1.8824E+03	1.9229E-02	-1.5230E-05	223	555	1.483	0.071

Table 22-1 VISCOSITY OF LIQUID - ORGANIC COMPOUNDS (continued)

			$\log_{10} n_{liq} = A + B/T + C\,T + C\,T^2$					(n_{liq} - centipoise, T - K)		
NO	FORMULA	NAME	A	B	C	D	TMIN	TMAX	n_{liq} @25 C	n_{liq} @TMAX
79	C2H2Cl2O2	DICHLOROACETIC ACID	-22.2651	3.8370E+03	4.3150E-02	-3.0014E-05	287	686	6.330	0.064
80	C2H2Cl3F	1,1,1-TRICHLOROFLUOROETHANE	-1.1714	4.5197E+02	-9.0334E-04	-8.6437E-07	220	565	0.996	0.070
81	C2H2Cl4	1,1,1,2-TETRACHLOROETHANE	-2.5068	6.5745E+02	2.4956E-03	-3.2075E-06	203	624	1.436	0.072
82	C2H2Cl4	1,1,2,2-TETRACHLOROETHANE	-3.5146	7.7224E+02	5.5571E-03	-5.8524E-06	273	645	1.630	0.068
83	C2H2F2	1,1-DIFLUOROETHYLENE	12.1301	-5.3605E+02	-7.1633E-02	1.1068E-04	129	303	0.065	0.066
84	C2H2F2	cis-1,2-DIFLUOROETHENE	12.4757	-5.3605E+02	-7.1633E-02	1.1068E-04	129	300	0.144	0.145
85	C2H2F2	trans-1,2-DIFLUOROETHENE	12.4757	-5.3605E+02	-7.1633E-02	1.1068E-04	129	300	0.144	0.145
86	C2H2F4	1,1,1,2-TETRAFLUOROETHANE	-14.4406	1.2563E+02	5.2393E-02	-6.9771E-05	172	380	0.156	0.050
87	C2H2O	KETENE	4.3585	-1.4009E+02	-2.5364E-02	3.0329E-05	130	370	0.105	0.056
88	C2H2O4	OXALIC ACID	-----	-------	-------	-------	----	----	-----	-----
89	C2H3Br	VINYL BROMIDE	-7.3663	8.1930E+02	2.2026E-02	-2.6270E-05	180	473	0.411	0.081
90	C2H3Cl	VINYL CHLORIDE	-1.1063	2.1454E+02	-8.5045E-04	-1.3519E-06	130	432	0.174	0.059
91	C2H3ClF2	1-CHLORO-1,1-DIFLUOROETHANE	-16.8159	1.5115E+02	6.1839E-02	-8.0549E-05	200	410	0.339	0.048
92	C2H3ClO	ACETYL CHLORIDE	-10.9887	1.3155E+02	2.0872E-02	-3.1074E-05	275	508	0.369	0.057
93	C2H3ClO	CHLOROACETALDEHYDE	-43.4494	5.7337E+03	1.0624E-01	-8.8273E-05	293	555	0.407	0.045
94	C2H3ClO2	CHLOROACETIC ACID	-8.5505	1.8169E+03	1.3987E-02	-1.0355E-05	333	686	-----	0.066
95	C2H3ClO2	METHYL CHLOROFORMATE	-8.3513	1.1333E+03	2.0644E-02	-2.1205E-05	192	525	0.525	0.063
96	C2H3Cl3	1,1,1-TRICHLOROETHANE	-3.9096	7.0709E+02	7.5847E-03	-9.1662E-06	243	545	0.810	0.063
97	C2H3Cl3	1,1,2-TRICHLOROETHANE	-3.2716	6.8810E+02	4.8932E-03	-5.4671E-06	237	602	1.021	0.069
98	C2H3F	VINYL FLUORIDE	-1.1547	1.6886E+02	3.4703E-04	-7.2507E-06	113	328	0.074	0.049
99	C2H3F3	1,1,1-TRIFLUOROETHANE	-2.2123	2.4044E+02	7.7406E-04	1.6026E-07	162	346	0.069	0.059
100	C2H3N	ACETONITRILE	-2.9528	4.1475E+02	6.4299E-03	-9.1660E-06	288	546	0.347	0.038
101	C2H3NO	METHYL ISOCYANATE	-2.3960	2.3790E+02	6.4147E-03	-1.0316E-05	273	505	0.250	0.048
102	C2H4	ETHYLENE	-4.5611	3.0811E+02	1.8030E-02	-3.8145E-05	105	282	---	0.038
103	C2H4Br2	1,1-DIBROMOETHANE	-4.1302	7.5049E+02	7.8279E-03	-7.5923E-06	210	628	1.112	0.097
104	C2H4Br2	1,2-DIBROMOETHANE	-5.4223	1.0377E+03	9.6953E-03	-8.3417E-06	283	650	1.612	0.089
105	C2H4Cl2	1,1-DICHLOROETHANE	-3.8388	5.9046E+02	8.0953E-03	-9.9210E-06	176	523	0.471	0.065
106	C2H4Cl2	1,2-DICHLOROETHANE	-0.1656	2.7576E+02	-3.3493E-03	1.4093E-06	245	561	0.769	0.078
107	C2H4Cl2O	BIS(CHLOROMETHYL)ETHER	-2.4635	5.7215E+02	3.2060E-03	-4.6903E-06	232	579	0.987	0.064
108	C2H4F2	1,1-DIFLUOROETHANE	-37.3585	3.4509E+03	1.3343E-01	-1.6427E-04	243	387	0.249	0.039
109	C2H4F2	1,2-DIFLUOROETHANE	-10.3352	1.1272E+03	2.9491E-02	-3.2763E-05	215	476	0.212	0.044
110	C2H4I2	1,2-DIIODOETHANE	-----	-------	-------	-------	----	----	-----	-----
111	C2H4O	ACETALDEHYDE	-6.6171	6.8123E+02	1.9979E-02	-2.5563E-05	260	461	0.225	0.043
112	C2H4O	ETHYLENE OXIDE	-5.7794	6.7020E+02	1.5686E-02	-1.9462E-05	190	469	0.260	0.053
113	C2H4OS	THIOACETIC-ACID	-4.4348	7.8482E+02	6.6650E-03	-7.5606E-06	290	577	0.326	0.018
114	C2H4O2	ACETIC ACID	-3.8937	7.8482E+02	6.6650E-03	-7.5606E-06	290	593	1.132	0.053
115	C2H4O2	METHYL FORMATE	-8.0637	1.0137E+03	2.0884E-02	-2.2997E-05	250	487	0.330	0.054
116	C2H4S	THIACYCLOPROPANE	-4.6797	7.8482E+02	6.6650E-03	-7.5606E-06	290	555	0.185	0.013
117	C2H5Br	BROMOETHANE	-5.3844	6.7418E+02	1.4140E-02	-1.6501E-05	155	504	0.422	0.077
118	C2H5Cl	ETHYL CHLORIDE	-4.4279	5.1891E+02	1.2035E-02	-1.6620E-05	150	460	0.265	0.052
119	C2H5ClO	2-CHLOROETHANOL	-10.3253	1.8994E+03	1.9820E-02	-1.6723E-05	250	585	2.938	0.062
120	C2H5F	ETHYL FLUORIDE	-5.4713	4.8451E+02	1.8733E-02	-2.9742E-05	130	375	0.124	0.046
121	C2H5I	ETHYL IODIDE	-10.4954	1.3679E+03	2.6346E-02	-2.4827E-05	273	561	0.550	0.081
122	C2H5N	ETHYLENEIMINE	-4.7646	7.3673E+02	9.5205E-03	-1.0431E-05	250	537	0.415	0.052
123	C2H5NO	ACETAMIDE	-15.0576	3.0478E+03	2.4646E-02	-1.5506E-05	354	761	-----	0.053
124	C2H5NO	N-METHYLFORMAMIDE	-6.9983	1.3266E+03	1.2046E-02	-9.2805E-06	269	721	1.651	0.050
125	C2H5NO2	NITROETHANE	-3.7814	6.3484E+02	7.5441E-03	-8.5933E-06	200	593	0.681	0.055
126	C2H5NO3	ETHYL-NITRATE	-----	-------	-------	-------	----	----	-----	-----
127	C2H6	ETHANE	-4.2694	2.8954E+02	1.7111E-02	-3.6092E-05	98	305	0.039	0.035
128	C2H6AlCl	DIMETHYLALUMINUM CHLORIDE	-----	-------	-------	-------	----	----	-----	-----
129	C2H6O	DIMETHYL ETHER	-7.4844	5.8392E+02	2.7815E-02	-4.0433E-05	132	400	0.149	0.043
130	C2H6O	ETHANOL	-6.4406	1.1176E+03	1.3721E-02	-1.5465E-05	240	516	1.057	0.049
131	C2H6OS	DIMETHYL SULFOXIDE	-3.6341	8.5487E+02	4.8721E-03	-4.4070E-06	292	726	1.968	0.057
132	C2H6O2	ETHYLENE GLYCOL	-16.9728	3.1886E+03	3.2537E-02	-2.4480E-05	261	645	17.645	0.059
133	C2H6O4S	DIMETHYL SULFATE	-----	-------	-------	-------	----	----	-----	-----
134	C2H6S	DIMETHYL SULFIDE	-7.8503	9.4330E+02	2.0507E-02	-2.2249E-05	273	503	0.282	0.051
135	C2H6S	ETHYL MERCAPTAN	-3.0781	4.2896E+02	6.4239E-03	-9.0060E-06	125	499	0.299	0.056
136	C2H6S2	DIMETHYL DISULFIDE	-6.8447	1.0217E+03	1.4599E-02	-1.3298E-05	188	606	0.566	0.064
137	C2H7N	DIMETHYLAMINE	-11.5558	1.2126E+03	3.4999E-02	-4.1253E-05	240	438	0.190	0.042
138	C2H7N	ETHYLAMINE	-7.0668	9.0544E+02	1.7675E-02	-2.0701E-05	192	456	0.251	0.047
139	C2H7NO	MONOETHANOLAMINE	-13.1818	2.8596E+03	2.0826E-02	-1.4230E-05	288	638	22.577	0.062
140	C2H8N2	ETHYLENEDIAMINE	-18.3052	2.9617E+03	3.7865E-02	-2.9650E-05	303	593	-----	0.052
141	C2H8Si	DIMETHYL SILANE	-----	-------	-------	-------	----	----	-----	-----
142	C2N2	CYANOGEN	-79.0878	1.0824E+04	1.8829E-01	-1.5317E-04	269	400	0.548	0.060
143	C3F6	HEXAFLUOROPROPYLENE	17.0240	-9.7487E+02	-7.8331E-02	9.8244E-05	160	368	0.136	0.071
144	C3F6O	HEXAFLUOROACETONE	7.0539	-2.5115E+02	-3.6300E-02	4.2474E-05	148	357	0.146	0.064
145	C3F8	OCTAFLUOROPROPANE	4.8406	-2.8807E+01	-2.9852E-02	3.6536E-05	125	345	0.123	0.064
146	C3H2N2	MALONONITRILE	-15.5463	2.8953E+03	2.7321E-02	-1.8425E-05	305	715	-----	0.042
147	C3H3Cl	PROPARGYL CHLORIDE	-2.0693	3.0412E+02	4.3357E-03	-7.1639E-06	293	541	0.404	0.055
148	C3H3N	ACRYLONITRILE	-6.3470	8.1502E+02	1.5664E-02	-1.7275E-05	240	535	0.332	0.041
149	C3H3NO	OXAZOLE	-----	-------	-------	-------	----	----	-----	-----
150	C3H4	METHYLACETYLENE	-8.4493	7.7571E+02	2.8408E-02	-3.8708E-05	170	402	0.152	0.044
151	C3H4	PROPADIENE	-4.0226	3.5646E+02	1.3512E-02	-2.3072E-05	173	393	0.141	0.043
152	C3H4Cl2	2,3-DICHLOROPROPENE	-1.9545	5.4794E+02	7.6738E-04	-1.9049E-06	192	577	0.877	0.064
153	C3H4O	ACROLEIN	-5.5517	7.0871E+02	1.4056E-02	-1.6788E-05	223	506	0.334	0.046
154	C3H4O	PROPARGYL ALCOHOL	-12.5948	2.0520E+03	2.5944E-02	-2.1497E-05	221	580	1.294	0.057
155	C3H4O2	ACRYLIC ACID	-15.9215	2.4408E+03	3.4383E-02	-2.7677E-05	293	615	1.138	0.053
156	C3H4O2	beta-PROPIOLACTONE	-6.6127	1.0109E+03	1.2263E-02	-9.3813E-06	240	686	0.398	0.072

Table 22-1 VISCOSITY OF LIQUID - ORGANIC COMPOUNDS (continued)

			$\log_{10} n_{liq} = A + B/T + C\,T + C\,T^2$						$(n_{liq}$ - centipoise, T - K$)$	
NO	FORMULA	NAME	A	B	C	D	TMIN	TMAX	n_{liq} @25 C	n_{liq} @TMAX
157	C3H4O2	VINYL FORMATE	-4.4042	5.9523E+02	1.0606E-02	-1.3477E-05	200	498	0.360	0.054
158	C3H4O3	ETHYLENE CARBONATE	-8.5203	1.3364E+03	1.7273E-02	-1.2911E-05	311	790	-----	0.057
159	C3H4O3	PYRUVIC ACID	-17.9536	2.9060E+03	3.7391E-02	-2.8775E-05	287	635	2.418	0.058
160	C3H5Br	3-BROMO-1-PROPENE	-11.5038	1.4831E+03	2.9203E-02	-2.8268E-05	250	532	0.462	0.066
161	C3H5Cl	2-CHLOROPROPENE	-2.1425	3.7179E+02	2.8199E-03	-5.4004E-06	136	478	0.292	0.056
162	C3H5Cl	3-CHLOROPROPENE	-6.0433	8.0040E+02	1.4186E-02	-1.5455E-05	250	514	0.314	0.053
163	C3H5ClO	alpha-EPICHLOROHYDRIN	-2.3159	5.6462E+02	2.7982E-03	-4.1693E-06	223	610	1.100	0.058
164	C3H5ClO2	METHYL CHLOROACETATE	-6.3947	1.0712E+03	1.3156E-02	-1.2533E-05	241	600	1.015	0.059
165	C3H5ClO2	ETHYL CHLOROFORMATE	-12.4292	1.5619E+03	3.3551E-02	-3.4501E-05	192	508	0.557	0.061
166	C3H5Cl3	1,2,3-TRICHLOROPROPANE	-1.7913	6.4440E+02	3.8924E-04	-1.4969E-06	258	652	2.254	0.065
167	C3H5I	3-IODO-1-PROPENE	-6.5801	9.4055E+02	1.5487E-02	-1.5127E-05	225	578	0.704	0.088
168	C3H5N	PROPIONITRILE	-5.6142	8.0233E+02	1.2446E-02	-1.3286E-05	250	564	0.404	0.040
169	C3H5NO	ACRYLAMIDE	2.7157	2.5375E+02	-1.1286E-02	7.4172E-06	358	710	-----	0.063
170	C3H5NO	HYDRACRYLONITRILE	-13.0827	1.9718E+03	2.6820E-02	-2.0038E-05	227	690	0.557	0.055
171	C3H5NO	LACTONITRILE	-13.0054	2.6255E+03	2.1406E-02	-1.4674E-05	233	643	7.557	0.060
172	C3H5N3O9	NITROGLYCERINE	-30.0495	5.6062E+03	5.2649E-02	-3.2609E-05	286	680	35.675	0.083
173	C3H6	CYCLOPROPANE	-3.2541	3.2192E+02	9.9766E-03	-1.8191E-05	146	398	0.152	0.044
174	C3H6	PROPYLENE	-5.1758	4.2982E+02	1.8611E-02	-3.1662E-05	90	365	0.100	0.038
175	C3H6Br2	1,2-DIBROMOPROPANE	-2.6053	5.9927E+02	3.7360E-03	-4.8006E-06	218	580	1.235	0.095
176	C3H6Cl2	1,1-DICHLOROPROPANE	-3.5820	6.1920E+02	6.6598E-03	-7.8836E-06	200	560	0.602	0.060
177	C3H6Cl2	1,2-DICHLOROPROPANE	-2.8218	5.9927E+02	3.7360E-03	-4.8006E-06	173	572	0.750	0.062
178	C3H6Cl2	1,3-DICHLOROPROPANE	-2.8361	6.0541E+02	4.0515E-03	-5.0439E-06	174	603	0.900	0.060
179	C3H6Cl2	2,2-DICHLOROPROPANE	-3.6167	6.1920E+02	6.6598E-03	-7.8836E-06	240	539	0.556	0.068
180	C3H6I2	1,2-DIIODOPROPANE	-2.2369	5.9927E+02	3.7360E-03	-4.8006E-06	254	700	2.885	0.076
181	C3H6O	ACETONE	-7.2126	9.0305E+02	1.8385E-02	-2.0353E-05	223	508	0.308	0.045
182	C3H6O	ALLYL ALCOHOL	-11.8248	1.9173E+03	2.5034E-02	-2.2322E-05	281	545	1.217	0.051
183	C3H6O	METHYL VINYL ETHER	-5.8282	5.4577E+02	1.8540E-02	-2.5458E-05	151	437	0.185	0.046
184	C3H6O	n-PROPIONALDEHYDE	-9.8172	1.2714E+03	2.4587E-02	-2.5572E-05	280	496	0.320	0.045
185	C3H6O	1,2-PROPYLENE OXIDE	-7.2842	9.7539E+02	1.7425E-02	-1.9160E-05	200	482	0.302	0.049
186	C3H6O	1,3-PROPYLENE OXIDE	-10.8675	1.2901E+03	2.7838E-02	-2.7265E-05	255	520	0.217	0.052
187	C3H6O2	ETHYL FORMATE	-6.3477	8.5383E+02	1.5404E-02	-1.7222E-05	245	508	0.378	0.052
188	C3H6O2	METHYL ACETATE	-7.0933	9.3074E+02	1.7481E-02	-1.9038E-05	250	507	0.353	0.051
189	C3H6O2	PROPIONIC ACID	-5.0177	8.7365E+02	1.0302E-02	-1.0883E-05	252	604	1.039	0.048
190	C3H6O2S	3-MERCAPTOPROPIONIC ACID	-18.6383	3.2116E+03	3.5095E-02	-2.3670E-05	291	729	3.111	0.059
191	C3H6O3	LACTIC ACID	-----	--------	-------	--------	----	----	-----	-----
192	C3H6O3	METHOXYACETIC ACID	-16.1627	2.5857E+03	3.2785E-02	-2.4094E-05	281	691	1.389	0.054
193	C3H6O3	TRIOXANE	-----	--------	-------	--------	----	----	-----	-----
194	C3H6S	THIACYCLOBUTANE	-4.3146	7.8482E+02	6.6650E-03	-7.5606E-06	290	555	0.429	0.029
195	C3H7Br	1-BROMOPROPANE	-6.3524	8.3429E+02	1.5963E-02	-1.6934E-05	200	544	0.501	0.071
196	C3H7Br	2-BROMOPROPANE	-11.4833	1.4831E+03	2.9203E-02	-2.8268E-05	250	532	0.484	0.069
197	C3H7Cl	ISOPROPYL CHLORIDE	-8.3740	1.0583E+03	2.1296E-02	-2.2963E-05	250	489	0.305	0.052
198	C3H7Cl	n-PROPYL CHLORIDE	-6.4801	8.5514E+02	1.5738E-02	-1.7463E-05	250	503	0.337	0.052
199	C3H7F	1-FLUOROPROPANE	-6.7921	8.5514E+02	1.5738E-02	-1.7463E-05	250	422	0.164	0.058
200	C3H7F	2-FLUOROPROPANE	-6.7749	8.5514E+02	1.5738E-02	-1.7463E-05	250	415	0.171	0.064
201	C3H7I	ISOPROPYL IODIDE	-6.6205	9.4055E+02	1.5487E-02	-1.5127E-05	225	578	0.641	0.080
202	C3H7I	n-PROPYL IODIDE	-7.8304	1.1203E+03	1.8135E-02	-1.6845E-05	240	593	0.687	0.078
203	C3H7N	ALLYLAMINE	-6.2650	9.0217E+02	1.4036E-02	-1.5335E-05	185	505	0.382	0.050
204	C3H7N	PROPYLENEIMINE	-22.2367	2.5333E+03	6.0497E-02	-5.6947E-05	229	529	0.172	0.042
205	C3H7NO	N,N-DIMETHYLFORMAMIDE	-5.3292	8.9547E+02	1.0559E-02	-1.0088E-05	240	647	0.843	0.046
206	C3H7NO	N-METHYLACETAMIDE	-12.0806	2.3614E+03	2.0484E-02	-1.3984E-05	301	718	-----	0.051
207	C3H7NO2	1-NITROPROPANE	-6.5870	1.0437E+03	1.4009E-02	-1.3397E-05	240	605	0.793	0.051
208	C3H7NO2	2-NITROPROPANE	-3.5405	6.1915E+02	6.7366E-03	-7.8310E-06	182	594	0.706	0.055
209	C3H7NO3	PROPYL-NITRATE	-----	--------	-------	--------	----	----	-----	-----
210	C3H7NO3	ISOPROPYL-NITRATE	-----	--------	-------	--------	----	----	-----	-----
211	C3H8	PROPANE	-3.1759	2.9712E+02	9.5453E-03	-1.8781E-05	85	370	0.099	0.039
212	C3H8O	ISOPROPANOL	-0.7009	8.4150E+02	-8.6068E-03	8.2964E-06	187	508	1.963	0.053
213	C3H8O	METHYL ETHYL ETHER	-23.8621	2.2331E+03	8.0478E-02	-9.3416E-05	233	438	0.208	0.037
214	C3H8O	n-PROPANOL	-3.7702	9.9151E+02	4.0836E-03	-5.4586E-06	220	537	1.939	0.050
215	C3H8O2	2-METHOXYETHANOL	-15.1858	2.4914E+03	3.0625E-02	-2.4291E-05	250	564	1.387	0.060
216	C3H8O2	METHYLAL	-3.9255	5.1395E+02	1.0108E-02	-1.4478E-05	168	481	0.335	0.045
217	C3H8O2	1,2-PROPYLENE GLYCOL	-29.4920	5.2456E+03	5.8169E-02	-4.2343E-05	233	626	47.962	0.051
218	C3H8O2	1,3-PROPYLENE GLYCOL	-7.9787	1.9800E+03	1.1850E-02	-9.3205E-06	246	658	23.270	0.062
219	C3H8O3	GLYCEROL	-18.2152	4.2305E+03	2.8705E-02	-1.8648E-05	293	723	749.338	0.044
220	C3H8S	n-PROPYLMERCAPTAN	-4.1495	5.8084E+02	9.1470E-03	-1.0824E-05	160	536	0.365	0.053
221	C3H8S	ISOPROPYL MERCAPTAN	-2.7739	4.3005E+02	5.2653E-03	-7.6377E-06	143	517	0.363	0.055
222	C3H8S	ETHYL-METHYL-SULFIDE	-2.7867	4.3005E+02	5.2653E-03	-7.6377E-06	168	517	0.352	0.053
223	C3H9N	n-PROPYLAMINE	-4.9620	7.9761E+02	9.5072E-03	-1.0844E-05	190	497	0.384	0.049
224	C3H9N	ISOPROPYLAMINE	-3.6769	6.8343E+02	3.4740E-03	-1.5653E-06	273	472	0.325	0.115
225	C3H9N	TRIMETHYLAMINE	-3.9726	4.4221E+02	1.0657E-02	-1.6134E-05	200	433	0.179	0.043
226	C3H9NO	1-AMINO-2-PROPANOL	-31.3874	5.5522E+03	5.9514E-02	-4.0875E-05	275	614	22.149	0.061
227	C3H9NO	3-AMINO-1-PROPANOL	-26.7477	4.7518E+03	5.0500E-02	-3.4638E-05	284	649	14.703	0.057
228	C3H9NO	METHYLETHANOLAMINE	-10.5527	2.3490E+03	1.5490E-02	-1.0492E-05	269	630	10.270	0.059
229	C3H9O4P	TRIMETHYL PHOSPHATE	-6.0196	1.2096E+03	9.3929E-03	-6.3749E-06	240	764	1.867	0.104
230	C3H10N2	1,2-PROPANEDIAMINE	-13.3632	2.2430E+03	2.6624E-02	-2.1292E-05	237	587	1.604	0.056
231	C3H10Si	TRIMETHYL SILANE	-----	--------	-------	--------	----	----	-----	-----
232	C4Cl4S	TETRACHLOROTHIOPHENE	-----	--------	-------	--------	----	----	-----	-----
233	C4Cl6	HEXACHLORO-1,3-BUTADIENE	-0.1976	5.7506E+02	-3.8913E-03	2.1134E-06	252	741	5.739	0.072
234	C4F8	OCTAFLUORO-2-BUTENE	-1.9732	4.6632E+02	3.5001E-05	-2.1423E-08	174	392	0.398	0.169

Table 22-1 VISCOSITY OF LIQUID - ORGANIC COMPOUNDS (continued)

| NO | FORMULA | NAME | $\log_{10} n_{liq} = A + B/T + C\,T + C\,T^2$ | | | | | | $(n_{liq}$ - centipoise, T - K$)$ | |
			A	B	C	D	TMIN	TMAX	n_{liq} @25 C	n_{liq} @TMAX
235	C4F8	OCTAFLUOROCYCLOBUTANE	-2.0637	4.9325E+02	3.8652E-05	-2.0508E-08	245	388	0.398	0.166
236	C4F10	DECAFLUOROBUTANE	-2.0446	6.0604E+02	-2.3464E-05	1.3468E-08	210	386	0.960	0.330
237	C4H2	BUTADIYNE(BIACETYLENE)	0.6495	7.9658E+01	-5.8889E-03	2.9221E-06	250	425	0.263	0.073
238	C4H2O3	MALEIC ANHYDRIDE	-1.0811	5.5616E+02	-1.2536E-03	4.1553E-07	326	721	-----	0.101
239	C4H4	VINYLACETYLENE	-----	-------	-------	-------	----	----	-----	-----
240	C4H4N2	SUCCINONITRILE	-1.1564	5.9038E+02	-6.9203E-04	2.1818E-07	331	770	-----	0.161
241	C4H4O	FURAN	-3.6715	5.5958E+02	7.4026E-03	-9.7493E-06	220	490	0.351	0.057
242	C4H4O2	DIKETENE	-----	-------	-------	-------	----	----	-----	-----
243	C4H4O3	SUCCINIC ANHYDRIDE	-1.7466	5.0881E+02	-9.6474E-05	2.7683E-08	393	811	-----	0.066
244	C4H4O4	FUMARIC ACID	-3.0627	1.1080E+03	6.7047E-04	-1.6982E-07	560	771	-----	0.062
245	C4H4O4	MALEIC ACID	-52.9808	9.9837E+03	9.2110E-02	-5.4257E-05	403	773	-----	0.052
246	C4H4S	THIOPHENE	-6.5928	1.0206E+03	1.3826E-02	-1.3089E-05	273	579	0.615	0.061
247	C4H5Cl	CHLOROPRENE	-1.8266	4.2271E+02	2.5963E-05	-1.2880E-08	174	525	0.396	0.097
248	C4H5N	trans-CROTONITRILE	-5.6497	8.2828E+02	1.2430E-02	-1.2891E-05	222	586	0.488	0.042
249	C4H5N	cis-CROTONITRILE	-4.1893	6.5278E+02	8.7208E-03	-1.0178E-05	201	568	0.496	0.043
250	C4H5N	METHACRYLONITRILE	-9.9472	1.3376E+03	2.3416E-02	-2.2227E-05	237	554	0.351	0.041
251	C4H5N	PYRROLE	-7.8869	1.3872E+03	1.4837E-02	-1.2263E-05	250	640	1.257	0.057
252	C4H5N	VINYLACETONITRILE	-4.3272	6.6740E+02	8.9861E-03	-1.0067E-05	186	584	0.496	0.043
253	C4H5NO2	METHYL CYANOACETATE	-8.0928	1.5829E+03	1.5829E-03	-1.0951E-05	260	687	2.727	0.051
254	C4H6	CYCLOBUTENE	0.4832	7.9658E+01	-5.8889E-03	2.9221E-06	250	425	0.180	0.050
255	C4H6	1,2-BUTADIENE	-3.5214	4.0958E+02	7.8148E-03	-1.0143E-05	181	444	0.191	0.074
256	C4H6	1,3-BUTADIENE	0.3772	7.9658E+01	-5.8889E-03	2.9221E-06	250	425	0.141	0.039
257	C4H6	DIMETHYLACETYLENE	-----	-------	-------	-------	----	----	-----	-----
258	C4H6	ETHYLACETYLENE	-6.9326	7.0136E+02	2.1415E-02	-2.7943E-05	193	443	0.209	0.045
259	C4H6Cl2	1,3-DICHLORO-trans-2-BUTENE	-6.0849	1.0588E+03	1.1364E-02	-1.0201E-05	276	618	0.887	0.057
260	C4H6Cl2	1,4-DICHLORO-cis-2-BUTENE	-6.4731	1.0734E+03	1.2744E-02	-1.1246E-05	225	640	0.845	0.057
261	C4H6Cl2	1,4-DICHLORO-trans-2-BUTENE	-9.9634	1.5750E+03	2.0497E-02	-1.6731E-05	274	646	0.877	0.054
262	C4H6Cl2	3,4-DICHLORO-1-BUTENE	-3.9073	7.5340E+02	6.5569E-03	-7.0911E-06	212	589	0.879	0.059
263	C4H6O	trans-CROTONALDEHYDE	-5.2285	7.3316E+02	1.1862E-02	-1.2777E-05	197	571	0.428	0.046
264	C4H6O	2,5-DIHYDROFURAN	-25.4792	3.3299E+03	6.1867E-02	-5.1933E-05	273	542	0.152	0.057
265	C4H6O	DIVINYL ETHER	-7.2676	7.4786E+02	2.1778E-02	-2.6886E-05	172	463	0.221	0.046
266	C4H6O	METHACROLEIN	-2.0243	4.1152E+02	2.2500E-03	-4.4668E-06	192	530	0.426	0.049
267	C4H6O2	2-BUTYNE-1,4-DIOL	-----	-------	-------	-------	----	----	-----	-----
268	C4H6O2	gamma-BUTYROLACTONE	-----	-------	-------	-------	----	----	-----	-----
269	C4H6O2	cis-CROTONIC ACID	-13.7877	2.2066E+03	2.8706E-02	-2.2671E-05	289	647	1.434	0.051
270	C4H6O2	trans-CROTONIC ACID	-23.6229	3.8043E+03	4.8126E-02	-3.4867E-05	345	666	-----	0.047
271	C4H6O2	METHACRYLIC ACID	-4.2607	8.4654E+02	7.6050E-03	-7.7782E-06	295	643	1.428	0.054
272	C4H6O2	METHYL ACRYLATE	-12.1755	1.6859E+03	2.8551E-02	-2.6324E-05	273	536	0.448	0.051
273	C4H6O2	VINYL ACETATE	-9.0671	1.1863E+03	2.2663E-02	-2.3208E-05	250	524	0.403	0.050
274	C4H6O3	ACETIC ANHYDRIDE	-17.3580	2.3611E+03	4.2734E-02	-3.8202E-05	265	569	0.806	0.055
275	C4H6O4	SUCCINIC ACID	-62.8791	1.2466E+04	1.0404E-01	-5.8132E-05	461	806	-----	0.048
276	C4H6O5	DIGLYCOLIC ACID	-55.7461	1.1066E+04	9.1535E-02	-5.0702E-05	421	820	-----	0.052
277	C4H6O5	MALIC ACID	-3.4665	1.2435E+03	-1.4020E-03	4.0266E-07	403	781	-----	0.002
278	C4H6O6	TARTARIC ACID	-----	-------	-------	-------	----	----	-----	-----
279	C4H7N	n-BUTYRONITRILE	-3.4336	5.8884E+02	6.3219E-03	-7.7312E-06	161	582	0.548	0.044
280	C4H7N	ISOBUTYRONITRILE	-4.9492	7.8421E+02	9.9363E-03	-1.0712E-05	202	565	0.491	0.043
281	C4H7NO	ACETONE CYANOHYDRIN	-----	-------	-------	-------	----	----	-----	-----
282	C4H7NO	2-METHACRYLAMIDE	-2.3289	1.0046E+03	2.8308E-04	-8.1490E-07	384	741	-----	0.062
283	C4H7NO	3-METHOXYPROPIONITRILE	-7.4121	1.1519E+03	1.5220E-02	-1.3361E-05	210	638	0.633	0.046
284	C4H7NO	2-PYRROLIDONE	-14.3974	2.9106E+03	2.3755E-02	-1.4858E-05	298	792	13.384	0.059
285	C4H8	1-BUTENE	-4.9218	4.9503E+02	1.4390E-02	-2.0853E-05	160	420	0.150	0.042
286	C4H8	cis-2-BUTENE	-6.3837	6.8524E+02	1.8356E-02	-2.3949E-05	200	439	0.181	0.042
287	C4H8	trans-2-BUTENE	-7.9461	8.1334E+02	2.4583E-02	-3.2036E-05	200	429	0.183	0.040
288	C4H8	CYCLOBUTANE	-3.3295	3.8110E+02	9.4773E-03	-1.5146E-05	182	460	0.268	0.045
289	C4H8	ISOBUTENE	-5.1190	3.4126E+02	1.9893E-02	-3.1057E-05	240	418	0.157	0.039
290	C4H8Br2	1,2-DIBROMOBUTANE	-2.2194	5.7867E+02	2.4157E-03	-3.6109E-06	236	641	1.320	0.056
291	C4H8Br2	2,3-DIBROMOBUTANE	-2.1848	5.7867E+02	2.4157E-03	-3.6109E-06	239	641	1.430	0.061
292	C4H8Cl2	1,4-DICHLOROBUTANE	-2.2175	5.7867E+02	2.4157E-03	-3.6109E-06	236	641	1.326	0.056
293	C4H8I2	1,2-DIIODOBUTANE	-1.9185	5.7867E+02	2.4157E-03	-3.6109E-06	280	660	2.640	0.095
294	C4H8O	n-BUTYRALDEHYDE	-4.6882	6.8181E+02	1.0648E-02	-1.2871E-05	275	525	0.426	0.045
295	C4H8O	ISOBUTYRALDEHYDE	-4.9534	7.1084E+02	1.1385E-02	-1.3812E-05	208	507	0.396	0.047
296	C4H8O	1,2-EPOXYBUTANE	0.1130	3.6921E+01	-1.3654E-03	-2.8143E-06	180	526	0.380	0.049
297	C4H8O	METHYL ETHYL KETONE	-0.8761	2.9257E+02	-1.9833E-03	9.4354E-07	189	536	0.396	0.075
298	C4H8O	ETHYL VINYL ETHER	-5.7800	6.6206E+02	1.5513E-02	-1.9072E-05	157	475	0.235	0.048
299	C4H8O	TETRAHYDROFURAN	-2.7860	4.7681E+02	4.9173E-03	-6.8815E-06	165	540	0.465	0.056
300	C4H8O2	cis-2-BUTENE-1,4-DIOL	-13.7194	2.7695E+03	2.4403E-02	-1.7709E-05	284	678	18.668	0.059
301	C4H8O2	trans-2-BUTENE-1,4-DIOL	-----	-------	-------	-------	----	----	-----	-----
302	C4H8O2	ISOBUTYRIC ACID	-6.5789	1.1041E+03	1.3998E-02	-1.3690E-05	250	609	1.205	0.048
303	C4H8O2	n-BUTYRIC ACID	-7.9846	1.3636E+03	1.6315E-02	-1.4511E-05	268	628	1.457	0.051
304	C4H8O2	1,4-DIOXANE	-7.5724	1.3813E+03	1.3556E-02	-1.1464E-05	288	587	1.211	0.061
305	C4H8O2	ETHYL ACETATE	-3.6861	5.5228E+02	8.0018E-03	-1.0439E-05	220	523	0.421	0.050
306	C4H8O2	METHYL PROPIONATE	-9.4254	1.2618E+03	2.3072E-02	-2.3116E-05	274	531	0.427	0.048
307	C4H8O2	n-PROPYL FORMATE	-5.9614	8.7189E+02	1.3431E-02	-1.4452E-05	250	538	0.482	0.050
308	C4H8O2S	SULFOLANE	-2.6246	9.0825E+02	2.7366E-03	-2.7441E-06	301	849	-----	0.062
309	C4H8S	TETRAHYDROTHIOPHENE	-20.0809	3.0608E+03	4.2486E-02	-3.2296E-05	293	632	0.958	0.052
310	C4H9Br	1-BROMOBUTANE	-8.1638	1.1342E+03	1.9376E-02	-1.8526E-05	240	577	0.589	0.065
311	C4H9Br	2-BROMOBUTANE	-6.8873	9.5838E+02	1.6202E-02	-1.6077E-05	220	567	0.547	0.068
312	C4H9Cl	n-BUTYL CHLORIDE	-5.5792	7.8795E+02	1.2714E-02	-1.3909E-05	250	537	0.415	0.051

485

Table 22-1 VISCOSITY OF LIQUID - ORGANIC COMPOUNDS (continued)

									n_{liq} - centipoise, T - K)	
			log_{10} n_{liq} = A + B/T + C T + C T^2							
NO	FORMULA	NAME	A	B	C	D	TMIN	TMAX	n_{liq} @25 C	n_{liq} @TMAX
313	C4H9Cl	sec-BUTYL CHLORIDE	-9.9788	1.3237E+03	2.4470E-02	-2.4390E-05	250	521	0.388	0.049
314	C4H9Cl	tert-BUTYL CHLORIDE	-10.2240	1.5480E+03	2.1671E-02	-1.9828E-05	280	507	0.464	0.052
315	C4H9I	2-IODO-2-METHYLPROPANE	-6.6956	9.5838E+02	1.6202E-02	-1.6077E-05	235	570	0.832	0.099
316	C4H9N	PYRROLIDINE	-7.7495	1.2386E+03	1.6051E-02	-1.4879E-05	215	569	0.737	0.055
317	C4H9NO	N,N-DIMETHYLACETAMIDE	-4.6530	8.3643E+02	8.3151E-03	-7.8250E-06	253	658	0.863	0.050
318	C4H9NO	MORPHOLINE	-5.0308	1.1150E+03	7.3378E-03	-6.5955E-06	280	618	2.044	0.062
319	C4H9NO2	1-NITROBUTANE	-6.6823	1.0437E+03	1.4009E-02	-1.3397E-05	240	605	0.637	0.041
320	C4H9NO2	2-NITROBUTANE	-3.6817	6.1915E+02	6.7366E-03	-7.8310E-06	182	594	0.510	0.040
321	C4H10	n-BUTANE	-6.8590	6.7393E+02	2.1973E-02	-3.0686E-05	180	425	0.168	0.033
322	C4H10	ISOBUTANE	-13.4207	1.3131E+03	4.4329E-02	-5.5793E-05	190	408	0.174	0.039
323	C4H10N2	PIPERAZINE	-19.4532	2.7745E+03	4.3501E-02	-3.4220E-05	379	638	------	0.052
324	C4H10O	n-BUTANOL	-5.3970	1.3256E+03	6.2223E-03	-5.5062E-06	250	563	2.599	0.052
325	C4H10O	sec-BUTANOL	-20.6736	3.5493E+03	4.0352E-02	-3.0937E-05	288	536	3.248	0.049
326	C4H10O	tert-BUTANOL	-35.2655	5.4737E+03	7.7742E-02	-6.3499E-05	298	506	4.241	0.043
327	C4H10O	DIETHYL ETHER	-8.5060	1.0020E+03	2.2753E-02	-2.5780E-05	233	467	0.222	0.044
328	C4H10O	METHYL-PROPYL-ETHER	-8.4519	1.0020E+03	2.2753E-02	-2.5780E-05	233	467	0.252	0.050
329	C4H10O	METHYL ISOPROPYL ETHER	-5.1341	5.5878E+02	1.4090E-02	-1.8212E-05	128	465	0.210	0.048
330	C4H10O	ISOBUTANOL	-11.9687	2.1770E+03	2.3767E-02	-2.1427E-05	211	548	3.269	0.039
331	C4H10O2	1,3-BUTANEDIOL	-24.0994	4.6841E+03	4.4009E-02	-3.0862E-05	293	643	97.500	0.053
332	C4H10O2	1,4-BUTANEDIOL	-14.3559	3.1423E+03	2.4253E-02	-1.7544E-05	293	667	71.600	0.053
333	C4H10O2	2,3-BUTANEDIOL	-6.0949	2.1560E+03	2.8597E-03	-1.0083E-06	281	611	79.314	0.064
334	C4H10O2	t-BUTYL HYDROPEROXIDE	------	--------	--------	--------	----	----	------	-----
335	C4H10O2	1,2-DIMETHOXYETHANE	-5.7926	8.1316E+02	1.3579E-02	-1.5006E-05	215	536	0.446	0.049
336	C4H10O2	2-ETHOXYETHANOL	-10.1080	1.8207E+03	1.9143E-02	-1.6143E-05	210	569	1.867	0.057
337	C4H10O3	DIETHYLENE GLYCOL	-14.7942	3.1502E+03	2.3543E-02	-1.4786E-05	270	745	29.963	0.059
338	C4H10O4S	DIETHYL SULFATE	------	--------	--------	--------	----	----	------	-----
339	C4H10S	n-BUTYL MERCAPTAN	-6.3227	8.9344E+02	1.4252E-02	-1.4396E-05	225	569	0.440	0.050
340	C4H10S	ISOBUTYL MERCAPTAN	-4.3170	6.6031E+02	8.9698E-03	-1.0136E-05	225	559	0.469	0.051
341	C4H10S	sec-BUTYL MERCAPTAN	-4.1544	6.4242E+02	8.5141E-03	-9.8036E-06	225	554	0.465	0.052
342	C4H10S	tert-BUTYL MERCAPTAN	-1.7534	5.2688E+02	-6.2337E-04	-5.9282E-07	274	530	0.596	0.055
343	C4H10S	DIETHYL SULFIDE	-6.1635	8.4801E+02	1.4306E-02	-1.4925E-05	225	557	0.416	0.050
344	C4H10S	ISOPROPYL-METHYL-SULFIDE	-6.1506	8.4801E+02	1.4306E-02	-1.4925E-05	225	551	0.429	0.055
345	C4H10S	METHYL-PROPYL-SULFIDE	-6.0984	8.4801E+02	1.4306E-02	-1.4925E-05	225	557	0.484	0.058
346	C4H10S2	DIETHYL DISULFIDE	-8.6196	1.3497E+03	1.7682E-02	-1.4750E-05	225	642	0.738	0.057
347	C4H11N	n-BUTYLAMINE	-7.6575	1.2106E+03	1.5854E-02	-1.5448E-05	278	532	0.571	0.048
348	C4H11N	ISOBUTYLAMINE	-6.3000	1.0270E+03	1.2630E-02	-1.3256E-05	250	514	0.539	0.049
349	C4H11N	sec-BUTYLAMINE	-4.7287	8.5139E+02	7.7286E-03	-8.3765E-06	225	514	0.486	0.049
350	C4H11N	tert-BUTYLAMINE	-8.3011	1.3490E+03	1.5924E-02	-1.5016E-05	260	484	0.433	0.047
351	C4H11N	DIETHYLAMINE	-9.2189	1.2133E+03	2.2638E-02	-2.3633E-05	240	497	0.316	0.043
352	C4H11NO	DIMETHYLETHANOLAMINE	------	--------	--------	--------	----	----	------	-----
353	C4H11NO2	DIETHANOLAMINE	-27.9385	5.9547E+03	4.4120E-02	-2.5871E-05	301	715	------	0.051
354	C4H12N2O	N-AMINOETHYL ETHANOLAMINE	-30.0369	5.6570E+03	5.2553E-02	-3.2716E-05	293	699	49.798	0.064
355	C4H11NO2	2-AMINOETHOXYETHANOL	-32.0709	6.2908E+03	5.4172E-02	-3.2693E-05	298	698	187.817	0.067
356	C4H12Si	TETRAMETHYLSILANE	------	--------	--------	--------	----	----	------	-----
357	C4H13N3	DIETHYLENE TRIAMINE	-6.9597	1.4821E+03	1.1547E-02	-9.3746E-06	234	676	4.175	0.057
358	C5Cl6	HEXACHLOROCYCLOPENTADIENE	-10.4172	2.1941E+03	1.6514E-02	-1.0745E-05	298	746	8.135	0.073
359	C5H4O2	FURFURAL	-0.6087	2.8604E+02	4.5345E-04	-3.0939E-06	273	657	1.625	0.062
360	C5H5N	PYRIDINE	-6.8100	1.1496E+03	1.3229E-02	-1.1661E-05	232	620	0.898	0.058
361	C5H6	CYCLOPENTADIENE	-5.6949	6.5093E+02	1.5245E-02	-1.7940E-05	188	507	0.275	0.051
362	C5H6	2-METHYL-1-BUTENE-3-YNE	-2.9024	3.8325E+02	6.6626E-03	-1.0295E-05	160	492	0.285	0.046
363	C5H6	1-PENTENE-3-YNE	-3.6138	4.8914E+02	7.2390E-03	-8.8128E-06	150	520	0.252	0.051
364	C5H6	1-PENTENE-4-YNE	-3.0207	3.9788E+02	7.0631E-03	-1.0509E-05	150	503	0.306	0.046
365	C5H6N2	GLUTARONITRILE	-8.1374	1.8110E+03	1.1719E-02	-7.6393E-06	244	782	5.645	0.047
366	C5H6O2	FURFURYL ALCOHOL	-12.3697	2.2704E+03	2.3717E-02	-1.8564E-05	259	632	4.637	0.063
367	C5H6O3	GLUTARIC ANHYDRIDE	-8.9237	1.7641E+03	1.2843E-02	-7.1214E-06	328	838	------	0.088
368	C5H6O4	CITRACONIC ACID	-26.2612	5.2278E+03	4.2998E-02	-2.4625E-05	356	829	------	0.058
369	C5H6O4	ITACONIC ACID	-10.8903	2.8916E+03	1.1869E-02	-5.2006E-06	439	821	------	0.074
370	C5H6S	2-METHYLTHIOPHENE	-6.6049	1.0206E+03	1.3826E-02	-1.3089E-05	273	579	0.598	0.060
371	C5H6S	3-METHYLTHIOPHENE	-6.6134	1.0206E+03	1.3826E-02	-1.3089E-05	273	579	0.587	0.058
372	C5H7N	N-METHYLPYRROLE	-0.0619	3.6968E+02	-1.1876E-03	-3.1716E-06	217	610	3.483	0.044
373	C5H7NO2	ETHYL CYANOACETATE	-14.7933	2.5859E+03	2.7355E-02	-1.9234E-05	251	679	2.118	0.053
374	C5H8	CYCLOPENTENE	-3.1576	3.8500E+02	6.8491E-03	-9.1013E-06	138	507	0.233	0.054
375	C5H8	ISOPRENE	-9.9164	1.0172E+03	2.9353E-02	-3.2692E-05	273	484	0.219	0.054
376	C5H8	3-METHYL-1,2-BUTADIENE	-5.0944	5.6342E+02	1.3413E-02	-1.6605E-05	160	490	0.208	0.044
377	C5H8	2-METHYL-1,3-BUTADIENE	-5.0725	5.6342E+02	1.3413E-02	-1.6605E-05	160	483	0.219	0.050
378	C5H8	1,2-PENTADIENE	-3.7920	4.2763E+02	9.2560E-03	-1.2181E-05	136	500	0.209	0.044
379	C5H8	cis-1,3-PENTADIENE	-5.3378	5.8984E+02	1.3932E-02	-1.6748E-05	186	499	0.202	0.042
380	C5H8	trans-1,3-PENTADIENE	-5.8162	6.5547E+02	1.4971E-02	-1.7425E-05	222	500	0.198	0.042
381	C5H8	1,4-PENTADIENE	-2.6680	3.3442E+02	5.2518E-03	-8.2626E-06	125	479	0.193	0.045
382	C5H8	2,3-PENTADIENE	-4.7505	5.1167E+02	1.2667E-02	-1.5930E-05	148	497	0.212	0.044
383	C5H8	1-PENTYNE	-2.8855	3.9387E+02	7.4684E-03	-1.2463E-05	167	481	0.358	0.044
384	C5H8	2-PENTYNE	-2.9691	3.9387E+02	7.4684E-03	-1.2463E-05	167	481	0.296	0.036
385	C5H8	3-METHYL-1-BUTYNE	-2.6844	3.6528E+02	7.3057E-03	-1.3270E-05	183	463	0.346	0.044
386	C5H8	SPIROPENTANE	-2.7924	3.6528E+02	7.3057E-03	-1.3270E-05	183	463	0.270	0.034
387	C5H8N4O12	PENTAERYTHRITOL TETRANITRATE	------	--------	--------	--------	----	----	------	-----
388	C5H8O	CYCLOPENTANONE	-53.1880	8.1650E+03	1.1041E-01	-7.9251E-05	288	626	1.179	0.008
389	C5H8O	METHYL ISOPROPENYL KETONE	-7.3581	1.0035E+03	1.7021E-02	-1.6781E-05	220	566	0.390	0.047
390	C5H8O2	ACETYLACETONE	-8.8988	1.3244E+03	2.0022E-02	-1.8365E-05	273	602	0.759	0.050

Table 22-1 VISCOSITY OF LIQUID - ORGANIC COMPOUNDS (continued)

$$\log_{10} n_{liq} = A + B/T + C\,T + C\,T^2 \qquad (n_{liq} \text{ - centipoise, T - K})$$

NO	FORMULA	NAME	A	B	C	D	TMIN	TMAX	n_{liq} @25 C	n_{liq} @TMAX
391	C5H8O2	ALLYL ACETATE	-8.8594	1.2199E+03	2.1012E-02	-2.0417E-05	240	559	0.481	0.049
392	C5H8O2	ETHYL ACRYLATE	-5.5960	8.1055E+02	1.2995E-02	-1.4271E-05	250	553	0.535	0.049
393	C5H8O2	METHYL METHACRYLATE	-4.7825	7.3478E+02	1.0258E-02	-1.1343E-05	260	564	0.540	0.050
394	C5H8O2	VINYL PROPIONATE	-6.3902	9.2664E+02	1.4476E-02	-1.5129E-05	230	546	0.489	0.050
395	C5H8O3	2-HYDROXYETHYL ACRYLATE	-20.5844	3.8501E+03	3.6430E-02	-2.4036E-05	260	662	11.320	0.065
396	C5H8O3	LEVULINIC ACID	-21.8785	3.8805E+03	4.0538E-02	-2.6906E-05	308	723	-----	0.054
397	C5H8O3	METHYL ACETOACETATE	-4.8619	9.5371E+02	8.8042E-03	-8.6041E-06	210	642	1.574	0.054
398	C5H8O4	GLUTARIC ACID	-23.2118	4.5522E+03	3.9304E-02	-2.3667E-05	371	807	-----	0.054
399	C5H9N	VALERONITRILE	-4.9993	8.1288E+02	1.0342E-02	-1.0939E-05	275	603	0.689	0.041
400	C5H9NO	n-BUTYL ISOCYANATE	-----	-------	-------	-------	----	----	-----	-----
401	C5H9NO	N-METHYL-2-PYRROLIDONE	-11.7394	2.0342E+03	2.1569E-02	-1.5109E-05	249	724	1.483	0.058
402	C5H9NO4	L-GLUTAMIC ACID	-60.7700	1.4128E+04	8.4987E-02	-4.0318E-05	497	886	-----	0.067
403	C5H10	CYCLOPENTANE	-0.5385	2.5255E+02	-2.3007E-03	-2.3062E-07	225	512	0.400	0.052
404	C5H10	2-METHYL-1-BUTENE	-3.3641	4.2134E+02	7.7836E-03	-1.1733E-05	136	465	0.212	0.042
405	C5H10	2-METHYL-2-BUTENE	-3.9960	4.7505E+02	9.9954E-03	-1.3953E-05	139	471	0.217	0.042
406	C5H10	3-METHYL-1-BUTENE	-8.7648	1.0250E+03	2.3732E-02	-2.7510E-05	273	450	0.201	0.042
407	C5H10	1-PENTENE	-5.5640	6.1906E+02	1.5123E-02	-1.9316E-05	175	465	0.201	0.042
408	C5H10	cis-2-PENTENE	-3.4667	4.1022E+02	8.4927E-03	-1.2319E-05	122	476	0.222	0.044
409	C5H10	trans-2-PENTENE	-3.7334	4.3487E+02	9.3927E-03	-1.3291E-05	133	475	0.221	0.044
410	C5H10Br2	2,3-DIBROMO-2-METHYLBUTANE	-2.7815	6.8497E+02	3.8366E-03	-4.5639E-06	289	663	1.795	0.062
411	C5H10Cl2	1,5-DICHLOROPENTANE	-2.8380	6.8497E+02	3.8366E-03	-4.5639E-06	200	663	1.576	0.054
412	C5H10O	METHYL ISOPROPYL KETONE	-5.5433	7.9650E+02	1.2368E-02	-1.3239E-05	181	553	0.435	0.049
413	C5H10O	2-PENTANONE	-7.4145	1.0492E+03	1.6999E-02	-1.6941E-05	250	561	0.464	0.046
414	C5H10O	DIETHYL KETONE	-9.2905	1.2716E+03	2.1925E-02	-2.1036E-05	274	561	0.438	0.045
415	C5H10O	VALERALDEHYDE	-2.4006	4.6830E+02	4.0666E-03	-6.6593E-06	182	554	0.617	0.045
416	C5H10O2	n-BUTYL FORMATE	-6.2873	9.9161E+02	1.3347E-02	-1.3623E-05	250	559	0.641	0.049
417	C5H10O2	ETHYL PROPIONATE	-7.3410	1.0413E+03	1.7121E-02	-1.7582E-05	250	546	0.493	0.047
418	C5H10O2	ISOBUTYL FORMATE	-9.8983	1.4374E+03	2.2751E-02	-2.1524E-05	289	551	0.620	0.051
419	C5H10O2	ISOPROPYL ACETATE	-5.3287	8.2209E+02	1.1484E-02	-1.2697E-05	216	538	0.530	0.050
420	C5H10O2	n-PROPYL ACETATE	-8.7395	1.3076E+03	1.9248E-02	-1.8348E-05	293	549	0.568	0.048
421	C5H10O2	METHYL n-BUTYRATE	-5.2565	8.0396E+02	1.1391E-02	-1.2415E-05	200	555	0.540	0.049
422	C5H10O2	2-METHYLBUTYRIC ACID	-9.1382	1.6081E+03	1.7951E-02	-1.5001E-05	298	643	1.879	0.050
423	C5H10O2	ISOVALERIC ACID	-6.7507	1.2740E+03	1.2636E-02	-1.1318E-05	244	634	1.922	0.053
424	C5H10O2	VALERIC ACID	-7.9425	1.4583E+03	1.5088E-02	-1.2781E-05	288	651	2.046	0.050
425	C5H10O2	TETRAHYDROFURFURYL ALCOHOL	-19.7007	3.4266E+03	3.7490E-02	-2.6468E-05	250	639	4.140	0.065
426	C5H10O2S	3-METHYL SULFOLANE	-5.1223	1.4211E+03	5.9806E-03	-4.0278E-06	290	817	11.726	0.065
427	C5H10O3	DIETHYL CARBONATE	-6.9038	1.0458E+03	1.5802E-02	-1.6047E-05	250	576	0.774	0.049
428	C5H10O3	ETHYL LACTATE	-20.0105	3.2123E+03	4.1891E-02	-3.2733E-05	247	588	2.206	0.059
429	C5H10S	THIACYCLOHEXANE	-5.7132	8.4801E+02	1.4306E-02	-1.4925E-05	293	600	1.174	0.081
430	C5H10S	CYCLOPENTANETHIOL	-6.0385	8.4801E+02	1.4306E-02	-1.4925E-05	225	557	0.555	0.066
431	C5H11Br	1-BROMOPENTANE	-3.9431	6.5614E+02	7.6907E-03	-8.8986E-06	186	564	0.575	0.053
432	C5H11Cl	1-CHLOROPENTANE	-3.9553	6.5614E+02	7.6907E-03	-8.8986E-06	174	568	0.559	0.050
433	C5H11Cl	1-CHLORO-3-METHYLBUTANE	-4.1014	6.5614E+02	7.6907E-03	-8.8986E-06	174	558	0.399	0.039
434	C5H11Cl	2-CHLORO-2-METHYLBUTANE	-4.0734	6.5614E+02	7.6907E-03	-8.8986E-06	200	548	0.426	0.046
435	C5H11N	N-METHYLPYRROLIDINE	-5.2640	8.7099E+02	9.7493E-03	-9.7536E-06	183	550	0.498	0.054
436	C5H11N	PIPERIDINE	-0.5054	4.9531E+02	-4.1258E-03	2.5391E-06	285	594	1.417	0.059
437	C5H11NO	tert-BUTYLFORMAMIDE	11.3399	-1.0643E+03	-2.8136E-02	1.7528E-05	289	692	8.702	0.053
438	C5H12	ISOPENTANE	-5.8108	7.0656E+02	1.4826E-02	-1.8547E-05	210	460	0.214	0.042
439	C5H12	NEOPENTANE	-4.6261	7.8569E+02	6.8775E-03	-8.1947E-06	258	434	0.214	0.042
440	C5H12	n-PENTANE	-7.1711	7.4736E+02	2.1697E-02	-2.7176E-05	143	470	0.245	0.041
441	C5H12O	2,2-DIMETHYL-1-PROPANOL	-21.3537	4.0439E+03	3.7271E-02	-2.5735E-05	327	550	-----	0.052
442	C5H12O	tert-PENTYL-ALCOHOL	-----	-------	-------	-------	----	----	-----	-----
443	C5H12O	2-METHYL-1-BUTANOL	-13.3823	2.5243E+03	2.4256E-02	-1.9091E-05	273	565	4.160	0.050
444	C5H12O	2-METHYL-2-BUTANOL	-25.3557	4.2312E+03	5.0802E-02	-3.8373E-05	264	545	3.727	0.050
445	C5H12O	3-METHYL-1-BUTANOL	-7.9943	1.7691E+03	1.1475E-02	-8.9300E-06	273	579	3.688	0.051
446	C5H12O	3-METHYL-2-BUTANOL	-4.7191	1.3883E+03	1.6744E-03	2.9612E-07	250	574	2.903	0.057
447	C5H12O	1-PENTANOL	-6.9286	1.5609E+03	9.9633E-03	-8.3534E-06	253	586	3.425	0.051
448	C5H12O	2-PENTANOL	-21.4761	3.5740E+03	4.3957E-02	-3.4712E-05	250	552	3.398	0.049
449	C5H12O	3-PENTANOL	-2.8804	1.1077E+03	-1.8779E-03	2.1199E-06	222	547	2.907	0.056
450	C5H12O	METHYL sec-BUTYL ETHER	-5.0497	6.4203E+02	1.2303E-02	-1.4886E-05	150	498	0.281	0.047
451	C5H12O	METHYL tert-BUTYL ETHER	-9.7896	1.2002E+03	2.6168E-02	-2.8297E-05	219	497	0.333	0.044
452	C5H12O	METHYL ISOBUTYL ETHER	-4.3736	5.8318E+02	1.0287E-02	-1.3123E-05	150	497	0.304	0.047
453	C5H12O	ETHYL PROPYL ETHER	-5.7475	7.2196E+02	1.4621E-02	-1.7401E-05	200	500	0.306	0.045
454	C5H12O2	ETHYLENE GLYCOL MONOPROPYL ETHER	-7.7167	1.6490E+03	1.1781E-02	-9.5068E-06	224	582	3.030	0.057
455	C5H12O2	NEOPENTYL GLYCOL	-70.1674	1.2205E+04	1.3211E-01	-8.4759E-05	400	643	-----	0.052
456	C5H12O2	1,5-PENTANEDIOL	-7.0464	2.3070E+03	5.2107E-03	-2.3581E-06	263	673	108.459	0.066
457	C5H12O3	2-(2-METHOXYETHOXY)ETHANOL	-9.0628	1.7101E+03	1.7174E-02	-1.4454E-05	240	630	3.225	0.054
458	C5H12O4	PENTAERYTHRITOL	50.8517	-1.1053E+04	-7.2356E-02	3.0441E-05	534	780	-----	0.058
459	C5H12S	n-PENTYL MERCAPTAN	-4.6076	7.4223E+02	9.1728E-03	-9.5706E-06	197	598	0.583	0.050
460	C5H12S	BUTYL-METHYL-SULFIDE	-4.5547	7.4223E+02	9.1728E-03	-9.5706E-06	197	591	0.659	0.060
461	C5H12S	ETHYL-PROPYL-SULFIDE	-4.5784	7.4223E+02	9.1728E-03	-9.5706E-06	197	584	0.624	0.061
462	C5H12S	2-METHYL-2-BUTANETHIOL	-4.7261	7.4223E+02	9.1728E-03	-9.5706E-06	197	566	0.444	0.051
463	C5H13N	n-PENTYLAMINE	-6.7617	1.1650E+03	1.3126E-02	-1.2846E-05	218	555	0.827	0.046
464	C5H13NO2	METHYL DIETHANOLAMINE	-12.0930	2.9675E+03	1.6516E-02	-1.0129E-05	252	678	76.540	0.067
465	C6Cl6	HEXACHLOROBENZENE	-33.4785	7.1609E+03	5.2053E-02	-2.8454E-05	502	825	-----	0.060
466	C6F6	HEXAFLUOROBENZENE	26.0147	-2.4756E+03	-7.7839E-02	6.7076E-05	278	517	2.927	0.082
467	C6H3ClN2O4	1-CHLORO-2,4-DINITROBENZENE	-18.8604	3.8454E+03	3.0491E-02	-1.7891E-05	327	814	-----	0.067
468	C3H3Cl2NO2	1,2-DICHLORO-4-NITROBENZENE	-16.9196	3.2771E+03	2.9025E-02	-1.8435E-05	316	758	-----	0.065

Table 22-1 VISCOSITY OF LIQUID - ORGANIC COMPOUNDS (continued)

NO	FORMULA	NAME	$\log_{10} n_{liq} = A + B/T + C\,T + C\,T^2$						$(n_{liq}$ - centipoise, T - K)	
			A	B	C	D	TMIN	TMAX	n_{liq} @25 C	n_{liq} @TMAX
469	C6H3Cl3	1,2,4-TRICHLOROBENZENE	-7.5691	1.4901E+03	1.3010E-02	-9.7071E-06	290	725	2.785	0.065
470	C6H3N3O6	1,3,5-TRINITROBENZENE	-25.8759	6.0592E+03	3.4686E-02	-1.5994E-05	398	1005	-----	0.072
471	C6H4Br2	m-DIBROMOBENZENE	-11.7673	2.1521E+03	2.0981E-02	-1.3972E-05	266	761	2.913	0.086
472	C6H4ClNO2	m-CHLORONITROBENZENE	-7.4607	1.4381E+03	1.3542E-02	-1.0431E-05	318	742	-----	0.061
473	C6H4ClNO2	o-CHLORONITROBENZENE	-15.4421	2.8372E+03	2.7503E-02	-1.8034E-05	306	757	-----	0.062
474	C6H4ClNO2	p-CHLORONITROBENZENE	-20.9286	3.7707E+03	3.8037E-02	-2.4642E-05	357	751	-----	0.058
475	C6H4Cl2	m-DICHLOROBENZENE	-4.6414	8.2076E+02	8.8567E-03	-8.1810E-06	248	684	1.059	0.062
476	C6H4Cl2	o-DICHLOROBENZENE	-3.8623	7.8030E+02	6.3271E-03	-5.8508E-06	256	705	1.322	0.063
477	C6H4Cl2	p-DICHLOROBENZENE	-6.6362	1.1967E+03	1.2224E-02	-9.9914E-06	326	685	-----	0.063
478	C6H4F2	m-DIFLUOROBENZENE	-4.8543	8.2076E+02	8.8567E-03	-8.1810E-06	250	552	0.649	0.107
479	C6H4F2	o-DIFLUOROBENZENE	-4.1929	7.8030E+02	6.3271E-03	-5.8508E-06	256	554	0.617	0.084
480	C6H4F2	p-DIFLUOROBENZENE	-4.1583	7.8030E+02	6.3271E-03	-5.8508E-06	261	556	0.669	0.090
481	C6H4N2O4	m-DINITROBENZENE	-8.5112	1.8823E+03	1.3972E-02	-9.7139E-06	364	805	-----	0.060
482	C6H4N2O4	o-DINITROBENZENE	-22.2948	4.6876E+03	3.5439E-02	-2.0308E-05	390	831	-----	0.059
483	C6H4N2O4	p-DINITROBENZENE	-25.9514	5.3986E+03	4.2022E-02	-2.4436E-05	447	803	-----	0.057
484	C6H5Br	BROMOBENZENE	-6.5135	1.0984E+03	1.2749E-02	-1.0749E-05	242	670	1.038	0.070
485	C6H5Cl	MONOCHLOROBENZENE	-4.8717	8.2340E+02	9.1981E-03	-8.6530E-06	250	632	0.730	0.061
486	C6H5ClO	m-CHLOROPHENOL	-39.6221	6.8879E+03	7.3543E-02	-4.6375E-05	306	729	-----	0.062
487	C6H5ClO	o-CHLOROPHENOL	-15.0741	2.6314E+03	2.9170E-02	-2.1317E-05	282	675	3.579	0.063
488	C6H5ClO	p-CHLOROPHENOL	-16.6697	3.2044E+03	2.9330E-02	-1.9280E-05	316	738	-----	0.066
489	C6H5Cl2N	3,4-DICHLOROANILINE	-15.9808	3.1009E+03	2.4746E-02	-1.3793E-05	345	800	-----	0.073
490	C6H5F	FLUOROBENZENE	-6.5062	9.7585E+02	1.4152E-02	-1.4056E-05	232	560	0.545	0.057
491	C6H5I	IODOBENZENE	-6.7338	1.2140E+03	1.2409E-02	-9.6198E-06	242	721	1.523	0.079
492	C6H5NO2	NITROBENZENE	-7.7710	1.4019E+03	1.4653E-02	-1.1512E-05	273	719	1.890	0.058
493	C6H6	BENZENE	-7.4005	1.1815E+03	1.4888E-02	-1.3713E-05	285	562	0.606	0.055
494	C6H6ClN	m-CHLOROANILINE	-14.2377	2.6141E+03	2.4952E-02	-1.6249E-05	263	751	3.350	0.066
495	C6H6ClN	o-CHLOROANILINE	-12.7917	2.3163E+03	2.3202E-02	-1.6047E-05	271	722	2.940	0.064
496	C6H6ClN	p-CHLOROANILINE	-10.8969	2.0918E+03	1.9036E-02	-1.3102E-05	328	754	-----	0.061
497	C6H6N2	cis-DICYANO-1-BUTENE	-9.4027	2.0397E+03	1.4323E-02	-9.9881E-06	249	691	6.622	0.048
498	C6H6N2	trans-DICYANO-1-BUTENE	-10.3176	2.1809E+03	1.6236E-02	-1.1299E-05	260	689	6.816	0.047
499	C6H6N2	1,4-DICYANO-2-BUTENE	-30.6811	5.7439E+03	5.3148E-02	-3.2357E-05	349	755	-----	0.041
500	C6H6N2O2	m-NITROANILINE	-21.4002	4.5473E+03	3.0755E-02	-1.5638E-05	387	815	-----	0.072
501	C6H6N2O2	o-NITROANILINE	-15.2760	3.2868E+03	2.1290E-02	-1.0953E-05	345	784	-----	0.075
502	C6H6N2O2	p-NITROANILINE	-16.5136	3.7808E+03	2.1340E-02	-1.0131E-05	421	851	-----	0.057
503	C6H6O	PHENOL	1.5349	4.2620E+02	-9.1577E-03	6.2322E-06	318	694	-----	0.062
504	C6H6O2	1,2-BENZENEDIOL	-24.7533	4.8589E+03	4.2834E-02	-2.6506E-05	385	764	-----	0.072
505	C6H6O2	1,3-BENZENEDIOL	-26.2943	5.6833E+03	4.1514E-02	-2.3604E-05	400	810	-----	0.073
506	C6H6O2	p-HYDROQUINONE	-67.2454	1.5146E+04	9.9970E-02	-5.1078E-05	445	822	-----	0.070
507	C6H6O3	1,2,3-BENZENETRIOL	2.2220	1.2238E+03	-1.0975E-02	6.4587E-06	407	830	-----	0.109
508	C6H6S	PHENYL MERCAPTAN	-3.7030	7.2932E+02	6.1906E-03	-5.9690E-06	258	689	1.144	0.061
509	C6H7N	ANILINE	-13.8625	2.5109E+03	2.5681E-02	-1.8281E-05	268	699	3.898	0.056
510	C6H7N	2-METHYLPYRIDINE	-5.4649	9.1344E+02	1.0641E-02	-1.0072E-05	220	621	0.752	0.054
511	C6H7N	3-METHYLPYRIDINE	-10.9953	1.7608E+03	2.2185E-02	-1.7629E-05	293	645	0.907	0.051
512	C6H7N	4-METHYLPYRIDINE	-8.2356	1.3239E+03	1.6712E-02	-1.4085E-05	277	646	0.862	0.054
513	C6H8	1,3-CYCLOHEXADIENE	-3.1367	5.9143E+02	4.8065E-03	-5.9644E-06	180	558	0.562	0.056
514	C6H8	METHYLCYCLOPENTADIENE	-3.3654	5.2704E+02	5.7171E-03	-6.6472E-06	150	541	0.328	0.057
515	C6H8N2	ADIPONITRILE	0.4949	4.1398E+02	-4.2058E-03	1.5475E-06	276	781	5.848	0.048
516	C6H8N2	METHYLGLUTARONITRILE	-4.3436	1.0468E+03	6.5095E-03	-5.9407E-06	228	742	3.803	0.042
517	C6H8N2	m-PHENYLENEDIAMINE	-44.9990	8.9345E+03	7.3255E-02	-4.0464E-05	363	824	-----	0.054
518	C6H8N2	o-PHENYLENEDIAMINE	-12.6956	2.8298E+03	1.9504E-02	-1.2084E-05	377	781	-----	0.062
519	C6H8N2	p-PHENYLENEDIAMINE	-18.6583	3.9059E+03	3.0083E-02	-1.8039E-05	413	796	-----	0.058
520	C6H8N2	PHENYLHYDRAZINE	-29.5908	5.2240E+03	5.4755E-02	-3.5011E-05	292	761	13.917	0.046
521	C6H8N2O	BIS(CYANOETHYL)ETHER	-4.7689	1.2262E+03	7.0649E-03	-6.0124E-06	247	783	8.236	0.044
522	C6H8O4	DIMETHYL MALEATE	-2.1332	6.7918E+02	2.3965E-03	-3.8790E-06	254	675	3.269	0.053
523	C6H8O6	ASCORBIC ACID	-----	-------	-------	-------	----	----	-----	-----
524	C6H8O7	CITRIC ACID	-----	-------	-------	-------	----	----	-----	-----
525	C6H10	1-METHYLCYCLOPENTENE	-4.7432	8.1472E+02	7.3925E-03	-7.6551E-06	273	541	0.326	0.033
526	C6H10	3-METHYLCYCLOPENTENE	-4.7806	8.1472E+02	7.3925E-03	-7.6551E-06	273	535	0.299	0.032
527	C6H10	4-METHYLCYCLOPENTENE	-4.7918	8.1472E+02	7.3925E-03	-7.6551E-06	273	543	0.291	0.029
528	C6H10	CYCLOHEXENE	-4.4683	8.1472E+02	7.3925E-03	-7.6551E-06	273	560	0.614	0.053
529	C6H10	2,3-DIMETHYL-1,3-BUTADIENE	-5.6382	6.9686E+02	1.3722E-02	-1.5374E-05	197	526	0.265	0.045
530	C6H10	1,5-HEXADIENE	-3.6271	4.6186E+02	8.2645E-03	-1.0973E-05	132	507	0.257	0.045
531	C6H10	cis,trans-2,4-HEXADIENE	-6.1222	7.2326E+02	1.5438E-02	-1.6909E-05	177	538	0.253	0.043
532	C6H10	trans,trans-2,4-HEXADIENE	-9.7857	1.1547E+03	2.5456E-02	-2.5786E-05	228	535	0.242	0.041
533	C6H10	1-HEXYNE	-3.0390	4.3910E+02	7.8829E-03	-1.2220E-05	141	516	0.499	0.042
534	C6H10	2-HEXYNE	-3.4983	5.3680E+02	7.0308E-03	-8.9311E-06	184	549	0.402	0.044
535	C6H10	3-HEXYNE	-3.2573	5.0738E+02	6.4200E-03	-8.5118E-06	170	544	0.400	0.045
536	C6H10O	CYCLOHEXANONE	-3.5761	9.3118E+02	3.6856E-03	-3.7510E-06	275	629	2.054	0.055
537	C6H10O	MESITYL OXIDE	-5.4305	8.4932E+02	1.1426E-02	-1.1624E-05	220	600	0.619	0.045
538	C6H10O2	epsilon-CAPROLACTONE	-7.5834	1.3332E+03	1.2604E-02	-8.4577E-06	272	771	0.784	0.069
539	C6H10O2	ETHYL METHACRYLATE	-8.0496	1.1661E+03	1.7964E-02	-1.6967E-05	223	577	0.512	0.049
540	C6H10O2	n-PROPYL ACRYLATE	-11.6479	1.6509E+03	2.7040E-02	-2.4607E-05	273	569	0.581	0.047
541	C6H10O3	ETHYLACETOACETATE	-7.6190	1.3438E+03	1.4739E-02	-1.2640E-05	234	643	1.442	0.053
542	C6H10O3	PROPIONIC ANHYDRIDE	-11.7549	1.7933E+03	2.5883E-02	-2.2100E-05	273	618	1.029	0.050
543	C6H10O4	ADIPIC ACID	-54.8285	1.1789E+04	8.5563E-02	-4.6275E-05	433	809	-----	0.048
544	C6H10O4	DIETHYL OXALATE	-14.8918	2.4136E+03	3.0683E-02	-2.3887E-05	260	646	1.691	0.050
545	C6H10O4	ETHYLENE GLYCOL DIACETATE	-3.1566	8.0484E+02	4.5706E-03	-5.4906E-06	242	653	2.615	0.052
546	C6H10O4	ETHYLIDENE DIACETATE	-14.9263	2.3871E+03	3.0890E-02	-2.4126E-05	292	635	1.397	0.052

Table 22-1 VISCOSITY OF LIQUID - ORGANIC COMPOUNDS (continued)

			$\log_{10} n_{liq} = A + B/T + C\,T + C\,T^2$						$(n_{liq}$ - centipoise, T - K)	
NO	FORMULA	NAME	A	B	C	D	TMIN	TMAX	n_{liq} @25 C	n_{liq} @TMAX
547	C6H11N	HEXANENITRILE	-4.8341	8.6235E+02	9.0581E-03	-9.2062E-06	193	622	0.872	0.042
548	C6H11NO	epsilon-CAPROLACTAM	-21.8927	4.6844E+03	3.4249E-02	-1.9662E-05	342	806	-----	0.056
549	C6H11NO	CYCLOHEXANONE OXIME	-----	-------	-------	-------	----	----	-----	-----
550	C6H12	CYCLOHEXANE	4.7423	-2.5322E+02	-1.6927E-02	1.2472E-05	285	554	0.901	0.054
551	C6H12	2,3-DIMETHYL-1-BUTENE	-3.0632	4.3429E+02	6.0597E-03	-8.7246E-06	116	500	0.266	0.045
552	C6H12	2,3-DIMETHYL-2-BUTENE	-6.1147	7.5638E+02	1.4993E-02	-1.6622E-05	199	524	0.260	0.042
553	C6H12	3,3-DIMETHYL-1-BUTENE	-3.5056	4.7789E+02	7.6202E-03	-1.0831E-05	158	480	0.255	0.045
554	C6H12	2-ETHYL-1-BUTENE	-3.6762	4.9545E+02	7.9274E-03	-1.0337E-05	142	512	0.269	0.044
555	C6H12	trans-3-METHYL-2-PENTENE	-3.6700	4.9545E+02	7.9274E-03	-1.0337E-05	142	512	0.273	0.044
556	C6H12	1-HEXENE	-6.3469	7.8976E+02	1.5694E-02	-1.7747E-05	218	504	0.253	0.042
557	C6H12	cis-2-HEXENE	-3.5211	4.7446E+02	7.7214E-03	-1.0351E-05	132	513	0.283	0.044
558	C6H12	trans-2-HEXENE	-3.6519	4.8838E+02	8.0942E-03	-1.0694E-05	140	513	0.281	0.043
559	C6H12	cis-3-HEXENE	-3.6154	4.8261E+02	8.0850E-03	-1.0837E-05	135	509	0.282	0.044
560	C6H12	trans-3-HEXENE	-4.3388	5.5486E+02	1.0354E-02	-1.3080E-05	160	509	0.280	0.043
561	C6H12	METHYLCYCLOPENTANE	-1.0639	3.1769E+02	-4.7988E-04	-2.0504E-06	248	533	0.475	0.049
562	C6H12	2-METHYL-1-PENTENE	-8.1875	1.0565E+03	1.9948E-02	-2.0963E-05	230	507	0.275	0.042
563	C6H12	2-METHYL-2-PENTENE	-3.7409	4.9959E+02	8.1854E-03	-1.0587E-05	138	514	0.272	0.044
564	C6H12	3-METHYL-1-PENTENE	-3.3186	4.5588E+02	7.0761E-03	-9.9805E-06	120	495	0.271	0.046
565	C6H12	3-METHYL-cis-2-PENTENE	-3.9620	5.1828E+02	8.9503E-03	-1.1310E-05	138	515	0.275	0.045
566	C6H12	4-METHYL-1-PENTENE	-3.1323	4.4203E+02	6.3382E-03	-9.1455E-06	120	496	0.267	0.045
567	C6H12	4-METHYL-cis-2-PENTENE	-3.6002	4.8538E+02	7.8742E-03	-1.0661E-05	138	499	0.268	0.044
568	C6H12	4-METHYL-trans-2-PENTENE	-3.3633	4.6465E+02	7.0194E-03	-9.6896E-06	132	501	0.267	0.045
569	C6H12N2	TRIETHYLENEDIAMINE	11.5928	-2.4836E+03	-1.7095E-02	4.9208E-06	434	655	-----	0.052
570	C6H12O	BUTYL VINYL ETHER	-4.7429	7.1709E+02	1.0233E-02	-1.1890E-05	181	536	0.453	0.046
571	C6H12O	CYCLOHEXANOL	-5.3792	1.8793E+03	1.7011E-03	1.0187E-07	303	625	-----	0.054
572	C6H12O	1-HEXANAL	-1.5906	3.8705E+02	2.1713E-03	-5.0434E-06	217	579	0.807	0.044
573	C6H12O	ETHYL ISOPROPYL KETONE	-6.0045	9.1123E+02	1.2923E-02	-1.3230E-05	200	567	0.535	0.047
574	C6H12O	2-HEXANONE	-6.9812	1.0500E+03	1.5094E-02	-1.4504E-05	217	587	0.564	0.047
575	C6H12O	3-HEXANONE	-6.7371	1.0210E+03	1.4470E-02	-1.4069E-05	218	583	0.564	0.047
576	C6H12O	METHYL ISOBUTYL KETONE	-3.0570	5.0050E+02	6.5038E-03	-8.8243E-06	246	571	0.598	0.045
577	C6H12O2	n-PENTYL FORMATE	-5.7440	9.1130E+02	1.2435E-02	-1.3004E-05	200	576	0.731	0.049
578	C6H12O2	n-BUTYL ACETATE	-8.3884	1.3075E+03	1.7671E-02	-1.6145E-05	275	579	0.677	0.049
579	C6H12O2	sec-BUTYL ACETATE	-4.2484	7.3709E+02	8.1159E-03	-9.2626E-06	174	561	0.661	0.051
580	C6H12O2	tert-BUTYL ACETATE	-8.1645	1.2261E+03	1.8000E-02	-1.7534E-05	283	545	0.570	0.049
581	C6H12O2	ETHYL n-BUTYRATE	-6.5990	1.0010E+03	1.4580E-02	-1.4750E-05	250	571	0.623	0.047
582	C6H12O2	ETHYL ISOBUTYRATE	-5.4028	8.4827E+02	1.1442E-02	-1.2326E-05	185	553	0.573	0.049
583	C6H12O2	ISOBUTYL ACETATE	-7.4697	1.1287E+03	1.6664E-02	-1.6546E-05	275	561	0.651	0.048
584	C6H12O2	n-PROPYL PROPIONATE	-5.4369	8.7969E+02	1.1064E-02	-1.1348E-05	197	578	0.636	0.049
585	C6H12O2	CYCLOHEXYL PEROXIDE	-----	-------	-------	-------	----	----	-----	-----
586	C6H12O2	DIACETONE ALCOHOL	-16.5815	2.7668E+03	3.4806E-02	-2.8348E-05	229	606	3.596	0.046
587	C6H12O2	2-ETHYL BUTYRIC ACID	-7.1034	1.3882E+03	1.2840E-02	-1.1012E-05	258	655	2.523	0.050
588	C6H12O2	n-HEXANOIC ACID	-6.2881	1.3092E+03	1.0624E-02	-9.1112E-06	275	667	2.888	0.051
589	C6H12O3	2-ETHOXYETHYL ACETATE	-7.3105	1.2304E+03	1.4721E-02	-1.3742E-05	225	597	0.963	0.044
590	C6H12O3	HYDROXYCAPROIC ACID	-3.7600	1.5809E+03	4.4909E-05	-1.4217E-08	334	758	-----	0.022
591	C6H12O3	PARALDEHYDE	-14.2350	2.1910E+03	3.1012E-02	-2.6181E-05	286	579	1.078	0.053
592	C6H12O3	sec-BUTYL GLYCOLATE	-----	-------	-------	-------	----	----	-----	-----
593	C6H12S	THIACYCLOHEPTANE	-3.7574	7.2932E+02	6.1906E-03	-5.9690E-06	293	640	1.009	0.079
594	C6H13N	CYCLOHEXYLAMINE	-4.4924	1.0272E+03	6.1025E-03	-5.7699E-06	288	615	1.817	0.056
595	C6H13N	HEXAMETHYLENEIMINE	-3.5971	9.3397E+02	3.6596E-03	-3.7383E-06	303	615	-----	0.057
596	C6H14	2,2-DIMETHYLBUTANE	-3.6841	6.5672E+02	5.5289E-03	-7.1890E-06	269	489	0.337	0.044
597	C6H14	2,3-DIMETHYLBUTANE	-12.6145	1.7240E+03	2.9725E-02	-2.8274E-05	273	500	0.329	0.042
598	C6H14	n-HEXANE	-5.0715	6.5536E+02	1.2349E-02	-1.5042E-05	178	507	0.296	0.041
599	C6H14	2-METHYLPENTANE	-5.1822	6.7047E+02	1.2528E-02	-1.5278E-05	258	498	0.278	0.041
600	C6H14	3-METHYLPENTANE	-5.2757	6.8590E+02	1.2664E-02	-1.5109E-05	220	504	0.287	0.043
601	C6H14N2O2	LYSINE	-108.5082	2.2844E+04	1.6947E-01	-8.8781E-05	498	821	-----	0.041
602	C6H14O	2-ETHYL-1-BUTANOL	-6.9398	1.7143E+03	8.2026E-03	-6.1032E-06	237	580	5.165	0.053
603	C6H14O	1-HEXANOL	-9.7166	2.0060E+03	1.5908E-02	-1.2286E-05	278	611	4.596	0.050
604	C6H14O	2-HEXANOL	-19.7738	3.4234E+03	3.8599E-02	-2.9152E-05	230	586	4.219	0.047
605	C6H14O	2-METHYL-1-PENTANOL	-18.4079	3.2619E+03	3.5729E-02	-2.7509E-05	270	582	5.493	0.047
606	C6H14O	4-METHYL-2-PENTANOL	-25.3841	4.1810E+03	5.2001E-02	-3.9695E-05	280	574	4.116	0.047
607	C6H14O	n-BUTYL ETHYL ETHER	-4.7316	7.1301E+02	1.0253E-02	-1.2054E-05	170	531	0.442	0.045
608	C6H14O	DIISOPROPYL ETHER	-5.5350	7.4894E+02	1.3145E-02	-1.5557E-05	188	500	0.326	0.044
609	C6H14O	DI-n-PROPYL ETHER	-3.7969	5.8653E+02	7.7642E-03	-9.7923E-06	150	531	0.412	0.047
610	C6H14O	METHYL tert-PENTYL ETHER	-5.1638	7.1242E+02	1.1515E-02	-1.2790E-05	160	534	0.333	0.047
611	C6H14O2	ACETAL	-6.7319	9.0418E+02	1.6397E-02	-1.7543E-05	200	541	0.427	0.047
612	C6H14O2	2-BUTOXYETHANOL	-7.3676	1.5830E+03	1.1138E-02	-8.8708E-06	203	600	2.979	0.058
613	C6H14O2	1,6-HEXANEDIOL	-72.9906	1.3107E+04	1.3170E-01	-8.0122E-05	315	670	-----	0.070
614	C6H14O2	HEXYLENE GLYCOL	-69.7580	1.0040E+04	1.6420E-01	-1.3020E-04	230	621	19.890	0.015
615	C6H14O2S	DI-n-PROPYL SULFONE	-12.8643	2.5541E+03	2.1501E-02	-1.4018E-05	303	763	-----	0.053
616	C6H14O3	DIETHYLENE GLYCOL DIMETHYL ETHER	-5.9960	1.0158E+03	1.2239E-02	-1.2021E-05	203	604	0.981	0.049
617	C6H14O3	DIPROPYLENE GLYCOL	-13.6865	3.3862E+03	1.6049E-02	-7.1661E-06	233	654	65.896	0.084
618	C6H14O3	2-(2-ETHOXYETHOXY)ETHANOL	-9.7859	1.8448E+03	1.8441E-02	-1.5177E-05	240	632	3.553	0.053
619	C6H14O3	TRIMETHYLOLPROPANE	-----	-------	-------	-------	----	----	-----	-----
620	C6H14O4	TRIETHYLENE GLYCOL	-13.8883	3.0642E+03	2.0531E-02	-1.0500E-05	285	700	37.758	0.520
621	C6H14O6	SORBITOL	-26.2502	5.7927E+03	3.4769E-02	-1.6250E-05	371	959	-----	0.015
622	C6H14S	n-HEXYLMERCAPTAN	-5.2857	8.5597E+02	1.0430E-02	-1.0034E-05	193	623	0.635	0.049
623	C6H14S	BUTYL-ETHYL-SULFIDE	-5.1693	8.5597E+02	1.0430E-02	-1.0034E-05	179	609	0.831	0.074
624	C6H14S	ISOPROPYL-SULFIDE	-5.2862	8.5597E+02	1.0430E-02	-1.0034E-05	171	585	0.635	0.070

489

Table 22-1 VISCOSITY OF LIQUID - ORGANIC COMPOUNDS (continued)

			$\log_{10} n_{liq} = A + B/T + C\,T + C\,T^2$						(n_{liq} - centipoise, T - K)	
NO	FORMULA	NAME	A	B	C	D	TMIN	TMAX	n_{liq} @25 C	n_{liq} @TMAX
625	C6H14S	METHYL-PENTYL-SULFIDE	-5.2477	8.5597E+02	1.0430E-02	-1.0034E-05	180	587	0.693	0.075
626	C6H14S	PROPYL-SULFIDE	-5.1761	8.5597E+02	1.0430E-02	-1.0034E-05	171	609	0.818	0.072
627	C6H14S2	PROPYL-DISULFIDE	-5.1929	8.5597E+02	1.0430E-02	-1.0034E-05	188	623	0.787	0.061
628	C6H15Al	TRIETHYL ALUMINUN	-4.3627	9.9686E+02	5.9002E-03	-4.1150E-06	221	720	2.367	0.137
629	C6H15Al2Cl3	ETHYL ALUMINUM SESQUICHLORIDE	-----	-------	-------	-------	----	----	-----	-----
630	C6H15N	DIISOPROPYLAMINE	-9.4766	1.2451E+03	2.3557E-02	-2.4018E-05	250	523	0.387	0.045
631	C6H15N	DI-n-PROPYLAMINE	-15.3095	2.0644E+03	3.7073E-02	-3.3467E-05	260	556	0.493	0.047
632	C6H15N	n-HEXYLAMINE	-4.3636	8.3506E+02	7.3533E-03	-7.8672E-06	252	583	0.852	0.048
633	C6H15N	TRIETHYLAMINE	-8.2302	1.0707E+03	2.0144E-02	-2.0640E-05	250	535	0.341	0.044
634	C6H15NO	6-AMINOHEXANOL	-42.5962	7.8271E+03	7.7049E-02	-4.8866E-05	331	681	-----	0.051
635	C6H15NO2	DIISOPROPANOLAMINE	-110.6700	1.8305E+04	2.2070E-01	-1.4706E-04	303	672	-----	0.030
636	C6H15NO3	TRIETHANOLAMINE	-21.0078	4.7935E+03	3.1279E-02	-1.7784E-05	294	787	652.576	0.048
637	C6H15N3	N-AMINOETHYL PIPERAZINE	-0.4594	7.1166E+02	-3.0976E-03	8.2669E-07	254	708	11.952	0.058
638	C6H15O4P	TRIETHYL PHOSPHATE	-10.0212	1.7337E+03	1.8103E-02	-1.1778E-05	250	794	1.393	0.129
639	C6H16N2	HEXAMETHYLENEDIAMINE	-24.6252	4.1044E+03	4.8655E-02	-3.4491E-05	314	663	-----	0.046
640	C6H18N3OP	HEXAMETHYL PHOSPHORAMIDE	-5.4288	1.2492E+03	6.6845E-03	-2.8892E-06	280	506	3.142	0.481
641	C6H18N4	TRIETHYLENE TETRAMINE	-40.9603	7.3562E+03	7.6974E-02	-5.0374E-05	285	718	152.896	0.038
642	C6H18OSi2	HEXAMETHYLDISILOXANE	-3.3863	5.7418E+02	6.7648E-03	-9.7088E-06	240	519	0.494	0.041
643	C6H18O3Si3	HEXAMETHYLCYCLOTRISILOXANE	-4.3150	8.3634E+02	7.6570E-03	-9.1028E-06	337	554	-----	0.044
644	C6H19NSi2	HEXAMETHYLDISILAZANE	-3.3927	6.0328E+02	6.8022E-03	-9.4801E-06	290	544	0.655	0.041
645	C7H3ClF3NO2	4-CHLORO-3-NITROBENZOTRIFLUORIDE	-13.2679	2.5485E+03	2.3563E-02	-1.6631E-05	293	686	6.710	0.061
646	C7H3Cl2F3	2,4-DICHLOROBENZOTRIFLUORIDE	-6.6149	1.3248E+03	1.1407E-02	-9.6282E-06	267	646	2.364	0.061
647	C7H3Cl2NO	3,4-DICHLOROPHENYL ISOCYANATE	-23.6469	4.1728E+03	4.3080E-02	-2.7652E-05	316	733	-----	0.058
648	C7H4ClF3	p-CHLOROBENZOTRIFLUORIDE	-8.3109	1.3727E+03	1.7034E-02	-1.5091E-05	254	601	1.072	0.058
649	C7H4Cl2O	m-CHLOROBENZOYL CHLORIDE	-7.4045	1.5604E+03	1.2133E-02	-9.0082E-06	280	724	4.424	0.065
650	C7H4F3NO2	3-NITROBENZOTRIFLUORIDE	-11.9324	2.1272E+03	2.2447E-02	-1.6797E-05	272	667	2.522	0.057
651	C7H5ClO	BENZOYL CHLORIDE	-11.9830	1.9357E+03	2.3900E-02	-1.7861E-05	273	697	1.115	0.060
652	C7H5ClO2	o-CHLOROBENZOIC ACID	-35.6472	7.0126E+03	5.9808E-02	-3.4812E-05	415	792	-----	0.055
653	C7H5Cl3	BENZOTRICHLORIDE	-5.7549	1.2014E+03	8.9628E-03	-6.7458E-06	273	737	2.224	0.066
654	C7H5F3	BENZOTRIFLUORIDE	-8.3782	1.1826E+03	1.9281E-02	-1.8350E-05	244	565	0.508	0.056
655	C7H5N	BENZONITRILE	-6.5416	1.1609E+03	1.2160E-02	-1.0015E-05	270	699	1.223	0.053
656	C7H5NO	PHENYL ISOCYANATE	-----	-------	-------	-------	----	----	-----	-----
657	C7H5N3O6	2,4,6-TRINITROTOLUENE	-14.9246	2.8645E+03	2.8227E-02	-1.9563E-05	354	795	-----	0.057
658	C7H6Cl2	BENZYL DICHLORIDE	-5.2934	1.1077E+03	7.6332E-03	-5.5580E-06	257	731	1.598	0.068
659	C7H6Cl2	2,4-DICHLOROTOLUENE	-7.5511	1.4411E+03	1.3164E-02	-1.0027E-05	260	705	2.070	0.062
660	C7H6N2O4	2,4-DINITROTOLUENE	-7.2273	1.5200E+03	1.2525E-02	-9.1730E-06	343	814	-----	0.057
661	C7H6N2O4	2,5-DINITROTOLUENE	-21.6018	4.2975E+03	3.5182E-02	-2.0389E-05	326	814	-----	0.064
662	C7H6N2O4	2,6-DINITROTOLUENE	-2.9240	8.1054E+02	4.7543E-03	-5.0883E-06	339	770	-----	0.059
663	C7H6N2O4	3,4-DINITROTOLUENE	-7.1438	1.5239E+03	1.1952E-02	-8.4090E-06	332	842	-----	0.059
664	C7H6N2O4	3,5-DINITROTOLUENE	-31.5919	6.1804E+03	5.2284E-02	-2.9883E-05	366	814	-----	0.057
665	C7H6O	BENZALDEHYDE	1.2039	2.2414E+01	-3.9285E-03	5.5651E-07	247	695	1.436	0.060
666	C7H6O2	BENZOIC ACID	-67.6079	1.3482E+04	1.1017E-01	-6.2162E-05	396	751	-----	0.011
667	C7H6O2	p-HYDROXYBENZALDEHYDE	-34.5840	7.6927E+03	4.9247E-02	-2.4193E-05	390	844	-----	0.073
668	C7H6O2	SALICYLALDEHYDE	-9.9310	1.7245E+03	1.9801E-02	-1.5708E-05	266	680	2.293	0.064
669	C7H6O3	SALICYLIC ACID	------	-------	-------	-------	----	----	-----	-----
670	C7H7Br	p-BROMOTOLUENE	-9.1633	1.5029E+03	1.8358E-02	-1.4308E-05	300	699	-----	0.067
671	C7H7Cl	BENZYL CHLORIDE	-6.4178	1.1281E+03	1.1724E-02	-9.5598E-06	234	686	1.027	0.059
672	C7H7Cl	o-CHLOROTOLUENE	-5.0510	8.8285E+02	9.5294E-03	-8.7952E-06	237	656	0.932	0.058
673	C7H7Cl	p-CHLOROTOLUENE	-8.5917	1.3598E+03	1.7594E-02	-1.4550E-05	281	660	0.834	0.055
674	C7H7F	p-FLUOROTOLUENE	-8.7463	1.3598E+03	1.7594E-02	-1.4550E-05	281	590	0.584	0.075
675	C7H7NO	FORMANILIDE	-21.7761	4.4373E+03	3.5166E-02	-2.0639E-05	328	787	-----	0.057
676	C7H7NO2	m-NITROTOLUENE	-20.0665	3.4718E+03	3.6803E-02	-2.3950E-05	289	734	2.641	0.059
677	C7H7NO2	o-NITROTOLUENE	-9.9050	1.7941E+03	1.8107E-02	-1.3234E-05	270	720	2.161	0.058
678	C7H7NO2	p-NITROTOLUENE	-11.3921	2.0234E+03	2.1371E-02	-1.5415E-05	325	736	-----	0.054
679	C7H7NO3	o-NITROANISOLE	-6.0024	1.2301E+03	1.0354E-02	-8.0089E-06	284	782	3.151	0.059
680	C7H8	TOLUENE	-5.1649	8.1068E+02	1.0454E-02	-1.0488E-05	200	592	0.548	0.052
681	C7H8	1,3,5-CYCLOHEPTATRIENE	-5.2048	8.1068E+02	1.0454E-02	-1.0488E-05	200	592	0.500	0.048
682	C7H8O	ANISOLE	-8.8307	1.4337E+03	1.7992E-02	-1.5096E-05	287	642	1.001	0.054
683	C7H8O	BENZYL ALCOHOL	-25.0704	4.2124E+03	4.9668E-02	-3.5036E-05	288	677	5.651	0.052
684	C7H8O	m-CRESOL	-18.3829	3.4649E+03	3.2787E-02	-2.1943E-05	298	706	11.569	0.054
685	C7H8O	o-CRESOL	-18.6396	3.3896E+03	3.4286E-02	-2.3373E-05	304	698	-----	0.058
686	C7H8O	p-CRESOL	-19.0715	3.5560E+03	3.4524E-02	-2.3222E-05	308	705	-----	0.059
687	C7H8O2	GUAIACOL	-16.0239	2.8581E+03	3.0360E-02	-2.1530E-05	305	697	-----	0.060
688	C7H8O2	p-METHOXYPHENOL	-16.2546	3.3208E+03	2.7118E-02	-1.7208E-05	345	758	-----	0.062
689	C7H9N	BENZYLAMINE	-6.0952	1.1500E+03	1.0800E-02	-9.0063E-06	227	684	1.518	0.057
690	C7H9N	2,6-DIMETHYLPYRIDINE	-7.1740	1.1515E+03	1.4740E-02	-1.3258E-05	267	624	0.802	0.051
691	C7H9N	N-METHYLANILINE	-13.7299	2.2913E+03	2.7208E-02	-2.0030E-05	250	702	1.935	0.058
692	C7H9N	m-TOLUIDINE	-13.5859	2.4293E+03	2.5395E-02	-1.8204E-05	273	709	3.276	0.050
693	C7H9N	o-TOLUIDINE	-11.1346	2.1223E+03	1.9780E-02	-1.4453E-05	273	694	3.947	0.049
694	C7H9N	p-TOLUIDINE	-12.3169	2.2349E+03	2.2874E-02	-1.6774E-05	317	693	-----	0.051
695	C7H10	2-NORBORNENE	-9.3004	1.5318E+03	1.8049E-02	-1.5132E-05	319	583	-----	0.051
696	C7H10N2	TOLUENEDIAMINE	-33.0277	6.1042E+03	5.7240E-02	-3.3853E-05	371	804	-----	0.050
697	C7H11NO	CYCLOHEXYL ISOCYANATE	-----	-------	-------	-------	----	----	-----	-----
698	C7H12	1-HEPTYNE	-5.2222	8.1068E+02	1.0454E-02	-1.0488E-05	200	559	0.480	0.062
699	C7H12O2	n-BUTYL ACRYLATE	-6.9308	1.1689E+03	1.3471E-02	-1.2339E-05	250	598	0.811	0.046
700	C7H12O2	ISOBUTYL ACRYLATE	-7.4193	1.1066E+03	1.6403E-02	-1.5787E-05	212	580	0.602	0.049
701	C7H12O2	n-PROPYL METHACRYLATE	-7.6284	1.1672E+03	1.6216E-02	-1.4907E-05	223	599	0.625	0.048
702	C7H12O4	DIETHYL MALONATE	-9.1440	1.6103E+03	1.7730E-02	-1.4503E-05	240	653	1.795	0.052

490

Table 22-1 VISCOSITY OF LIQUID - ORGANIC COMPOUNDS (continued)

| NO | FORMULA | NAME | $\log_{10} n_{liq} = A + B/T + C\,T + C\,T^2$ | | | | | | $(n_{liq}$ - centipoise, T - K) | |
			A	B	C	D	TMIN	TMAX	n_{liq} @25 C	n_{liq} @TMAX
703	C7H14	CYCLOHEPTANE	2.2571	-8.2344E+01	-6.5186E-03	1.3401E-06	265	604	1.434	0.047
704	C7H14	1,1-DIMETYLCYCLOPENTANE	-2.9826	5.2928E+02	5.0996E-03	-6.9685E-06	203	547	0.494	0.049
705	C7H14	cis-1,2-DIMETHYLCYCLOPENTANE	-3.9448	6.3812E+02	7.6699E-03	-8.8791E-06	219	565	0.493	0.048
706	C7H14	trans-1,2-DIMETHYLCYCLOPENTANE	-2.8753	5.1637E+02	4.7706E-03	-6.5528E-06	156	553	0.497	0.049
707	C7H14	cis-1,3-DIMETHYLCYCLOPENTANE	-2.7490	5.0276E+02	4.4211E-03	-6.2776E-06	139	551	0.498	0.049
708	C7H14	trans-1,3-DIMETHYLCYCLOPENTANE	-2.7501	5.0315E+02	4.4009E-03	-6.2117E-06	139	553	0.498	0.049
709	C7H14	ETHYLCYCLOPENTANE	-3.1046	5.4840E+02	5.4107E-03	-6.9666E-06	253	570	0.535	0.048
710	C7H14	2-ETHYL-1-PENTENE	-5.0433	6.7571E+02	1.1550E-02	-1.2981E-05	168	543	0.326	0.044
711	C7H14	3-ETHYL-1-PENTENE	-4.1071	5.7416E+02	8.9597E-03	-1.0906E-05	146	530	0.332	0.046
712	C7H14	1-HEPTENE	-8.3002	1.0763E+03	2.0357E-02	-2.0898E-05	250	537	0.332	0.041
713	C7H14	cis-2-HEPTENE	-4.1620	5.9928E+02	8.8279E-03	-1.0428E-05	164	549	0.357	0.043
714	C7H14	trans-2-HEPTENE	-4.4221	6.2108E+02	9.8018E-03	-1.1566E-05	164	543	0.359	0.043
715	C7H14	cis-3-HEPTENE	-3.4630	5.2468E+02	6.8238E-03	-8.6790E-06	137	545	0.363	0.044
716	C7H14	trans-3-HEPTENE	-3.6419	5.3868E+02	7.5309E-03	-9.5346E-06	137	540	0.365	0.044
717	C7H14	METHYLCYCLOHEXANE	-1.9879	5.0806E+02	1.2152E-03	-2.7318E-06	248	572	0.685	0.050
718	C7H14	2-METHYL-1-HEXENE	-4.9412	6.6561E+02	1.1296E-02	-1.2911E-05	170	538	0.325	0.043
719	C7H14	3-METHYL-1-HEXENE	-4.0259	5.6592E+02	8.7407E-03	-1.0790E-05	145	528	0.330	0.045
720	C7H14	4-METHYL-1-HEXENE	-3.4368	5.1940E+02	6.8729E-03	-8.9153E-06	132	534	0.365	0.046
721	C7H14	2,3,3-TRIMETHYL-1-BUTENE	-3.3759	4.5743E+02	7.5119E-03	-9.9892E-06	163	531	0.324	0.045
722	C7H14O	DIISOPROPYL KETONE	-6.0698	9.6613E+02	1.2535E-02	-1.2504E-05	205	576	0.626	0.048
723	C7H14O	2-HEPTANONE	-6.3072	1.0186E+03	1.2947E-02	-1.2369E-05	250	612	0.741	0.044
724	C7H14O	1-HEPTANAL	-1.5042	3.8309E+02	2.2617E-03	-5.1258E-06	230	603	0.999	0.043
725	C7H14O	1-METHYLCYCLOHEXANOL	2.8090	8.1555E+02	-1.8028E-02	1.5145E-05	299	603	-----	0.063
726	C7H14O	cis-2-METHYLCYCLOHEXANOL	-61.1284	9.7604E+03	1.2667E-01	-8.9897E-05	298	614	24.185	0.045
727	C7H14O	trans-2-METHYLCYCLOHEXANOL	-69.3044	1.1008E+04	1.4418E-01	-1.0213E-04	280	616	33.512	0.042
728	C7H14O	cis-3-METHYLCYCLOHEXANOL	-47.3336	7.8174E+03	9.5221E-02	-6.6657E-05	289	618	22.432	0.051
729	C7H14O	trans-3-METHYLCYCLOHEXANOL	-56.2704	9.1353E+03	1.1490E-01	-8.0775E-05	289	617	27.965	0.048
730	C7H14O	cis-4-METHYLCYCLOHEXANOL	-1.1605	1.2470E+03	-7.1483E-03	6.1622E-06	298	622	27.446	0.061
731	C7H14O	trans-4-METHYLCYCLOHEXANOL	2.2649	8.3276E+02	-1.5347E-02	1.2225E-05	298	622	37.068	0.061
732	C7H14O	5-METHYL-2-HEXANONE	-6.0147	9.7149E+02	1.2143E-02	-1.1693E-05	199	601	0.668	0.047
733	C7H14O2	n-BUTYL PROPIONATE	-5.3815	9.0843E+02	1.0671E-02	-1.0766E-05	184	594	0.776	0.049
734	C7H14O2	ETHYL ISOVALERATE	-4.5111	8.0507E+02	8.2414E-03	-8.6925E-06	174	588	0.747	0.050
735	C7H14O2	ISOPENTYL ACETATE	-4.6434	8.2935E+02	8.6638E-03	-9.0386E-06	195	599	0.828	0.049
736	C7H14O2	n-PENTYL ACETATE	-6.3879	1.0379E+03	1.3281E-02	-1.2904E-05	223	598	0.805	0.047
737	C7H14O2	n-PROPYL n-BUTYRATE	-5.3652	9.0990E+02	1.0578E-02	-1.0705E-05	273	594	0.774	0.047
738	C7H14O2	n-HEPTANOIC ACID	-7.2247	1.4530E+03	1.3212E-02	-1.1316E-05	280	680	3.819	0.046
739	C7H14O3	ETHYL-3-ETHOXYPROPIONATE	-8.0517	1.3544E+03	1.6245E-02	-1.4468E-05	223	609	1.118	0.050
740	C7H15Br	1-BROMOHEPTANE	-5.9893	1.0290E+03	1.1969E-02	-1.0905E-05	240	651	1.151	0.058
741	C7H15N	N-METHYLCYCLOHEXYLAMINE	-7.3265	1.4620E+03	1.2053E-02	-9.8197E-06	265	622	1.985	0.053
742	C7H16	2,2-DIMETHYLPENTANE	-2.8909	5.7125E+02	3.5976E-03	-5.2719E-06	149	521	0.426	0.045
743	C7H16	2,3-DIMETHYLPENTANE	-3.4160	5.9455E+02	5.8631E-03	-7.5893E-06	160	537	0.448	0.045
744	C7H16	2,4-DIMETHYLPENTANE	-4.2276	6.8593E+02	8.0689E-03	-9.8333E-06	170	520	0.402	0.043
745	C7H16	3,3-DIMETHYLPENTANE	-2.9875	5.5438E+02	4.4350E-03	-6.1370E-06	139	536	0.445	0.046
746	C7H16	3-ETHYLPENTANE	-4.3369	6.3373E+02	9.0780E-03	-1.0576E-05	155	541	0.359	0.045
747	C7H16	n-HEPTANE	-5.7782	8.0587E+02	1.3355E-02	-1.4794E-05	183	540	0.390	0.041
748	C7H16	2-METHYLHEXANE	-7.2033	9.7969E+02	1.6986E-02	-1.7924E-05	273	530	0.358	0.041
749	C7H16	3-METHYLHEXANE	-4.2439	6.2293E+02	8.8962E-03	-1.0606E-05	154	535	0.359	0.044
750	C7H16	2,2,3-TRIMETHYLBUTANE	1.0062	1.0680E+02	-6.3684E-03	3.1024E-06	249	531	0.551	0.050
751	C7H16O	1-HEPTANOL	-13.9237	2.6296E+03	2.5331E-02	-1.8972E-05	288	632	5.781	0.047
752	C7H16O	2-HEPTANOL	-29.5002	4.7658E+03	6.1363E-02	-4.6360E-05	275	588	4.557	0.045
753	C7H16O	5-METHYL-1-HEXANOL	-16.5523	3.0409E+03	3.0976E-02	-2.3331E-05	293	605	6.434	0.047
754	C7H16O	ISOPROPYL-TERT-BUTYL-ETHER	-6.7539	1.1616E+03	1.1379E-02	-1.0664E-05	254	558	0.386	0.023
755	C7H16S	n-HEPTYL MERCAPTAN	-7.0230	1.1355E+03	1.4051E-02	-1.2325E-05	230	645	0.757	0.047
756	C7H16S	BUTYL-PROPYL-SULFIDE	-6.8537	1.1355E+03	1.4051E-02	-1.2325E-05	230	645	1.118	0.070
757	C7H16S	ETHYL-PENTYL-SULFIDE	-6.8537	1.1355E+03	1.4051E-02	-1.2325E-05	230	638	1.118	0.075
758	C7H16S	HEXYL-METHYL-SULFIDE	-6.8537	1.1355E+03	1.4051E-02	-1.2325E-05	230	638	1.118	0.075
759	C7H17N	1-AMINOHEPTANE	-6.2343	1.1616E+03	1.1379E-02	-1.0664E-05	254	607	1.278	0.045
760	C8H4Cl2O2	ISOPHTHALOYL CHLORIDE	-----	--------	--------	--------	----	----	-----	-----
761	C8H4O3	PHTHALIC ANHYDRIDE	-25.1512	5.0387E+03	4.1950E-02	-2.4927E-05	405	791	-----	0.064
762	C8H6	ETHYNYLBENZENE	-8.0291	1.2666E+03	1.6127E-02	-1.3475E-05	243	648	0.675	0.052
763	C8H6O4	ISOPHTHALIC ACID	-----	--------	--------	--------	----	----	-----	-----
764	C8H6O4	PHTHALIC ACID	-47.1881	1.0175E+04	7.4008E-02	-4.0603E-05	464	800	-----	0.056
765	C8H6O4	TEREPHTHALIC ACID	-----	--------	--------	--------	----	----	-----	-----
766	C8H6S	BENZOTHIOPHENE	-11.4042	1.7823E+03	2.6024E-02	-2.0696E-05	305	754	-----	0.065
767	C8H7N	INDOLE	-14.8844	3.0324E+03	2.3499E-02	-1.3990E-05	300	790	-----	0.061
768	C8H8	STYRENE	-8.0291	1.2666E+03	1.6127E-02	-1.3475E-05	243	648	0.675	0.052
769	C8H8	1,3,5,7-CYCLOOCTATETRAENE	-7.9184	1.2666E+03	1.6127E-02	-1.3475E-05	267	642	0.871	0.071
770	C8H8O	ACETOPHENONE	-8.5218	1.5262E+03	1.5887E-02	-1.2340E-05	293	701	1.725	0.053
771	C8H8O	p-TOLUALDEHYDE	2.3677	-1.7942E+02	-5.7057E-03	1.2297E-06	293	698	1.493	0.053
772	C8H8O2	METHYL BENZOATE	-13.6724	2.3439E+03	2.6095E-02	-1.8870E-05	288	693	1.958	0.054
773	C8H8O2	o-TOLUIC ACID	7.1162	-8.2578E+02	-1.4902E-02	7.0339E-06	385	751	-----	0.062
774	C8H8O2	p-TOLUIC ACID	-37.6832	7.4837E+03	6.3063E-02	-3.6895E-05	453	773	-----	0.050
775	C8H8O3	METHYL SALICYLATE	-9.5757	1.6932E+03	1.8609E-02	-1.4419E-05	265	701	2.343	0.063
776	C8H8O3	VANILLIN	-28.4395	5.3459E+03	5.0055E-02	-3.0772E-05	355	777	-----	0.057
777	C8H9NO	ACETANILIDE	-11.5657	2.6635E+03	1.7255E-02	-1.0524E-05	387	825	-----	0.054
778	C8H10	ETHYLBENZENE	-5.2585	8.3065E+02	1.0784E-02	-1.0618E-05	210	617	0.629	0.050
779	C8H10	m-XYLENE	-6.0517	9.2460E+02	1.2583E-02	-1.1850E-05	225	617	0.559	0.050
780	C8H10	o-XYLENE	-7.8805	1.2500E+03	1.6116E-02	-1.3993E-05	268	630	0.747	0.050

Table 22-1 VISCOSITY OF LIQUID - ORGANIC COMPOUNDS (continued)

NO	FORMULA	NAME	$\log_{10} n_{liq} = A + B/T + C\,T + C\,T^2$				(n_liq - centipoise, T - K)			
			A	B	C	D	TMIN	TMAX	n_{liq} @25 C	n_{liq} @TMAX
781	C8H10	p-XYLENE	-9.4655	1.4400E+03	1.9910E-02	-1.6994E-05	288	616	0.616	0.049
782	C8H10O	m-ETHYLPHENOL	-----	-------	-------	-------	----	----	-----	-----
783	C8H10O	p-ETHYLPHENOL	-21.4968	4.8257E+03	3.1773E-02	-1.7902E-05	318	716	-----	0.065
784	C8H10O	PHENETOLE	-8.7231	1.4354E+03	1.7844E-02	-1.5176E-05	244	647	1.155	0.049
785	C8H10O	2-PHENYLETHANOL	-18.5077	3.4385E+03	3.2236E-02	-2.0797E-05	247	684	6.131	0.069
786	C8H10O	2,3-XYLENOL	-16.7336	3.8194E+03	2.4742E-02	-1.4554E-05	346	723	-----	0.068
787	C8H10O	2,4-XYLENOL	-11.1854	2.8959E+03	1.3956E-02	-7.9259E-06	346	708	-----	0.065
788	C8H10O	2,5-XYLENOL	-21.4155	3.8112E+03	4.0173E-02	-2.7215E-05	348	707	-----	0.059
789	C8H10O	2,6-XYLENOL	-3.2859	1.7038E+03	-2.6415E-03	3.1503E-06	319	701	-----	0.069
790	C8H10O	3,4-XYLENOL	-15.4495	3.0275E+03	2.7037E-02	-1.8095E-05	338	730	-----	0.062
791	C8H10O	3,5-XYLENOL	-14.1287	2.8523E+03	2.3707E-02	-1.5830E-05	337	716	-----	0.052
792	C8H11N	N,N-DIMETHYLANILINE	-7.4875	1.3067E+03	1.4255E-02	-1.1636E-05	276	687	1.291	0.052
793	C8H11N	o-ETHYLANILINE	-7.6282	1.6589E+03	1.1381E-02	-7.9949E-06	227	704	4.153	0.060
794	C8H11N	2,4,6-TRIMETHYLPYRIDINE	-----	-------	-------	-------	----	----	-----	-----
795	C8H11NO	p-PHENETIDINE	-13.2792	2.6147E+03	2.2098E-02	-1.4215E-05	277	754	6.538	0.059
796	C8H12	1,5-CYCLOOCTADIENE	-4.5175	8.2100E+02	6.6319E-03	-5.3783E-06	204	645	0.544	0.062
797	C8H12	VINYLCYCLOHEXENE	-3.9510	6.0496E+02	6.6571E-03	-6.3538E-06	164	599	0.315	0.058
798	C8H12O4	1,4-CYCLOHEXANEDICARBOXYLIC ACID	-81.3130	2.0380E+04	1.0529E-01	-4.6131E-05	586	889	-----	0.057
799	C8H12O4	DIETHYL MALEATE	-6.6406	1.3199E+03	1.2048E-02	-1.0365E-05	264	680	2.865	0.050
800	C8H14O2	n-BUTYL METHACRYLATE	-5.1723	9.5710E+02	8.6671E-03	-7.8972E-06	223	616	0.832	0.053
801	C8H14O3	BUTYRIC ANHYDRIDE	-10.9571	1.7847E+03	2.3032E-02	-1.9292E-05	298	639	1.517	0.047
802	C8H14O4	DIETHYL SUCCINATE	-20.6926	3.4669E+03	4.0016E-02	-2.8085E-05	270	660	2.342	0.055
803	C8H16	CYCLOOCTANE	-4.2651	8.2100E+02	6.6319E-03	-5.3783E-06	288	647	0.972	0.110
804	C8H16	1,1-DIMETHYLCYCLOHEXANE	0.2327	2.2589E+02	-4.0384E-03	1.2952E-06	240	591	0.797	0.048
805	C8H16	cis-1,2-DIMETHYLCYCLOHEXANE	0.6194	2.2882E+02	-5.0881E-03	2.0513E-06	250	606	1.128	0.046
806	C8H16	trans-1,2-DIMETHYLCYCLOHEXANE	-1.7206	4.7198E+02	8.7743E-04	-2.6044E-06	185	596	0.781	0.047
807	C8H16	cis-1,3-DIMETHYLCYCLOHEXANE	-1.0155	3.9208E+02	-1.0214E-03	-1.0588E-06	198	591	0.796	0.047
808	C8H16	trans-1,3-DIMETHYLCYCLOHEXANE	-4.9313	8.1772E+02	9.5354E-03	-9.7698E-06	200	598	0.611	0.044
809	C8H16	cis-1,4-DIMETHYLCYCLOHEXANE	-1.2997	4.2648E+02	-2.8726E-04	-1.5898E-06	186	598	0.801	0.047
810	C8H16	trans-1,4-DIMETHYLCYCLOHEXANE	-2.1405	4.9124E+02	2.2554E-03	-3.8951E-06	250	590	0.681	0.046
811	C8H16	ETHYLCYCLOHEXANE	-1.6818	4.5374E+02	1.0537E-03	-2.7699E-06	248	609	0.809	0.048
812	C8H16	2-ETHYL-1-HEXENE	-5.9271	8.3237E+02	1.3442E-02	-1.3847E-05	200	574	0.438	0.047
813	C8H16	1-METHYL-1-ETHYLCYCLOPENTANE	-2.7779	5.5107E+02	4.2266E-03	-5.7514E-06	129	582	0.660	0.048
814	C8H16	1-OCTENE	-5.6209	8.1305E+02	1.2523E-02	-1.3384E-05	250	567	0.447	0.041
815	C8H16	trans-2-OCTENE	-4.6713	7.1254E+02	9.6242E-03	-1.0481E-05	185	577	0.453	0.042
816	C8H16	trans-3-OCTENE	-4.0401	6.3834E+02	7.9623E-03	-9.1313E-06	163	574	0.460	0.043
817	C8H16	trans-4-OCTENE	-4.4295	6.8365E+02	9.0187E-03	-1.0058E-05	179	573	0.455	0.043
818	C8H16	n-PROPYLCYCLOPENTANE	-3.9428	6.9107E+02	7.1909E-03	-7.9215E-06	200	603	0.653	0.046
819	C8H16	2,4,4-TRIMETHYL-1-PENTENE	-5.6110	7.2824E+02	1.3001E-02	-1.3928E-05	180	553	0.295	0.043
820	C8H16	2,4,4-TRIMETHYL-2-PENTENE	-5.0453	6.6543E+02	1.1348E-02	-1.2325E-05	167	558	0.298	0.044
821	C8H16O	2-ETHYLHEXANAL	-2.5673	5.3666E+02	4.8215E-03	-7.1064E-06	200	607	1.093	0.042
822	C8H16O	1-OCTANAL	-1.0519	3.2257E+02	1.5830E-03	-4.7462E-06	246	621	1.202	0.042
823	C8H16O	2-OCTANONE	-9.4604	1.5010E+03	1.9707E-02	-1.6941E-05	253	624	0.878	0.044
824	C8H16O2	n-BUTYL n-BUTYRATE	-6.5225	1.1072E+03	1.2711E-02	-1.1661E-05	240	616	0.880	0.048
825	C8H16O2	n-HEXYL ACETATE	-5.2445	9.4658E+02	9.9025E-03	-9.7602E-06	192	618	1.036	0.048
826	C8H16O2	ISOBUTYL ISOBUTYRATE	-4.9934	9.0376E+02	8.9765E-03	-8.8800E-06	192	602	0.841	0.049
827	C8H16O2	n-OCTANOIC ACID	-10.4823	2.0670E+03	1.8423E-02	-1.3722E-05	290	692	5.290	0.048
828	C8H16O4	DIETHYLENE GLYCOL ETHYL ETHER ACETATE	-9.2498	1.7071E+03	1.7219E-02	-1.3768E-05	248	660	2.431	0.051
829	C8H18	2,2-DIMETHYLHEXANE	-3.6086	6.2134E+02	6.5706E-03	-8.2845E-06	175	550	0.499	0.043
830	C8H18	2,3-DIMETHYLHEXANE	-6.5588	9.5743E+02	1.4545E-02	-1.4843E-05	272	563	0.467	0.042
831	C8H18	2,4-DIMETHYLHEXANE	-3.1002	5.8604E+02	5.0527E-03	-6.9337E-06	272	554	0.569	0.043
832	C8H18	2,5-DIMETHYLHEXANE	-4.4617	6.8825E+02	9.3610E-03	-1.0979E-05	200	550	0.459	0.041
833	C8H18	3,3-DIMETHYLHEXANE	-3.9015	6.2872E+02	7.6331E-03	-9.0752E-06	180	562	0.475	0.044
834	C8H18	3,4-DIMETHYLHEXANE	-6.8074	9.8501E+02	1.5166E-02	-1.5214E-05	272	569	0.463	0.042
835	C8H18	3-ETHYLHEXANE	-10.4972	1.4539E+03	2.4344E-02	-2.2604E-05	272	565	0.425	0.041
836	C8H18	3-ETHYL-2-METHYLPENTANE	-6.8711	9.8501E+02	1.5166E-02	-1.5214E-05	272	567	0.400	0.038
837	C8H18	3-METHYL-3-ETHYLPENTANE	-3.8409	6.4987E+02	7.0411E-03	-8.0724E-06	182	577	0.525	0.046
838	C8H18	2-METHYLHEPTANE	-4.1606	6.5646E+02	8.2941E-03	-9.6539E-06	164	560	0.453	0.043
839	C8H18	3-METHYLHEPTANE	-4.0095	6.3962E+02	7.8131E-03	-9.0866E-06	153	564	0.454	0.044
840	C8H18	4-METHYLHEPTANE	-4.1486	6.5297E+02	8.2245E-03	-9.4852E-06	152	562	0.447	0.044
841	C8H18	n-OCTANE	-5.9245	8.8809E+02	1.2955E-02	-1.3596E-05	216	569	0.511	0.040
842	C8H18	2,2,3-TRIMETHYLPENTANE	-3.2959	6.1877E+02	5.2212E-03	-6.5755E-06	161	564	0.564	0.045
843	C8H18	2,2,4-TRIMETHYLPENTANE	-15.0420	2.0021E+03	3.7102E-02	-3.4486E-05	273	544	0.467	0.041
844	C8H18	2,3,3-TRIMETHYLPENTANE	-3.3709	6.6781E+02	4.6835E-03	-5.4639E-06	172	574	0.602	0.048
845	C8H18	2,3,4-TRIMETHYLPENTANE	-2.9927	5.6484E+02	4.7581E-03	-6.3758E-06	164	566	0.567	0.045
846	C8H18	2,2,3,3-TETRAMETHYLBUTANE	-3.1680	5.6484E+02	4.7581E-03	-6.3758E-06	173	566	0.379	0.030
847	C8H18O	DI-n-BUTYL ETHER	-7.5007	1.1344E+03	1.6448E-02	-1.5927E-05	247	581	0.620	0.043
848	C8H18O	DI-sec-BUTYL ETHER	-4.1967	7.4040E+02	7.8708E-03	-9.1766E-06	173	559	0.657	0.046
849	C8H18O	DI-tert-BUTYL ETHER	0.3030	3.2030E+02	-5.5073E-03	2.4004E-06	220	550	1.061	0.046
850	C8H18O	2-ETHYL-1-HEXANOL	-5.5092	1.5321E+03	4.6767E-03	-2.7905E-06	203	640	5.967	0.054
851	C8H18O	1-OCTANOL	-13.7751	2.7042E+03	2.3738E-02	-1.6852E-05	288	653	7.487	0.048
852	C8H18O	2-OCTANOL	-43.7680	7.2244E+03	8.6417E-02	-5.9079E-05	300	637	-----	0.045
853	C8H18O2	DI-t-BUTYL PEROXIDE	-----	-------	-------	-------	----	----	-----	-----
854	C8H18O2S	DI-n-BUTYL SULFONE	-14.6560	2.9668E+03	2.4207E-02	-1.5428E-05	323	767	-----	0.050
855	C8H18O3	DIETHYLENE GLYCOL DIETHYL ETHER	-5.6470	9.4671E+02	1.2405E-02	-1.2745E-05	229	624	1.241	0.044
856	C8H18O3	DIETHYLENE GLYCOL MONOBUTYL EHTER	-8.8165	1.8848E+03	1.2060E-02	-7.2713E-06	205	654	2.848	0.070
857	C8H18O4	TRIETHYLENE GLYCOL DIMETHYL ETHER	-5.3563	1.1244E+03	9.7557E-03	-9.5530E-06	229	651	2.981	0.047
858	C8H18O5	TETRAETHYLENE GLYCOL	-11.5630	2.8807E+03	1.3907E-02	-6.8703E-06	268	722	43.108	0.077

Table 22-1 VISCOSITY OF LIQUID - ORGANIC COMPOUNDS (continued)

$$\log_{10} n_{liq} = A + B/T + C\,T + C\,T^2 \qquad (n_{liq}\text{ - centipoise, T - K})$$

NO	FORMULA	NAME	A	B	C	D	TMIN	TMAX	n_{liq} @25 C	n_{liq} @TMAX
859	C8H18S	n-OCTYL MERCAPTAN	-6.6459	1.1271E+03	1.2766E-02	-1.1016E-05	224	664	0.915	0.047
860	C8H18S	tert-OCTYL MERCAPTAN	1.2458	1.6401E+02	-5.8299E-03	2.0729E-06	199	627	1.746	0.046
861	C8H18S	BUTYL-SULFIDE	1.1054	1.6401E+02	-5.8299E-03	2.0729E-06	210	627	1.263	0.034
862	C8H18S	ETHYL-HEXYL-SULFIDE	1.1682	1.6401E+02	-5.8299E-03	2.0729E-06	210	627	1.460	0.039
863	C8H18S	HEPTYL-METHYL-SULFIDE	1.1682	1.6401E+02	-5.8299E-03	2.0729E-06	210	627	1.460	0.039
864	C8H18S	PENTYL-PROPYL-SULFIDE	1.1682	1.6401E+02	-5.8299E-03	2.0729E-06	210	627	1.460	0.039
865	C8H18S2	BUTYL-DISULFIDE	0.9683	1.6401E+02	-5.8299E-03	2.0729E-06	203	627	0.921	0.025
866	C8H19N	DI-n-BUTYLAMINE	-12.4705	1.8784E+03	2.7406E-02	-2.3254E-05	250	608	0.858	0.048
867	C8H19N	DIISOBUTYLAMINE	-7.8148	1.2106E+03	1.6928E-02	-1.6193E-05	220	580	0.713	0.044
868	C8H19N	n-OCTYLAMINE	-9.4368	1.6234E+03	1.8687E-02	-1.5856E-05	273	627	1.480	0.043
869	C8H23N5	TETRAETHYLENEPENTAMINE	-5.1543	1.6519E+03	7.3118E-03	-6.6283E-06	243	774	94.842	0.047
870	C8H24O4Si4	OCTAMETHYLCYCLOTETRASILOXANE	4.9229	-3.4936E+02	-1.3404E-02	6.2710E-06	291	587	2.052	0.042
871	C9H4O5	TRIMELLITIC ANHYDRIDE	-5.9519	1.4849E+03	6.5746E-03	-3.4602E-06	438	890	-----	0.067
872	C9H6N2O2	TOLUENE DIISOCYANATE	-----	--------	--------	--------	----	----	-----	-----
873	C9H7N	ISOQUINOLINE	-7.6994	1.5750E+03	1.2537E-02	-8.5640E-06	299	803	-----	0.064
874	C9H7N	QUINOLINE	-10.9841	2.0787E+03	1.8981E-02	-1.2647E-05	274	782	3.333	0.061
875	C9H7NO	8-HYDROXYQUINOLINE	-----	--------	--------	--------	----	----	-----	-----
876	C9H8	INDENE	-7.0231	1.3236E+03	1.2332E-02	-9.8399E-06	272	687	1.653	0.054
877	C9H8O	2-METHYLBENZOFURAN	-----	--------	--------	--------	----	----	-----	-----
878	C9H10	INDANE	-7.3304	1.3306E+03	1.2617E-02	-8.6008E-06	250	865	1.348	0.049
879	C9H10	cis-PROPENYLBENZENE	-7.5064	1.3306E+03	1.2617E-02	-8.6008E-06	250	664	0.899	0.121
880	C9H10	trans-PROPENYLBENZENE	-7.5064	1.3306E+03	1.2617E-02	-8.6008E-06	250	664	0.899	0.121
881	C9H1O	alpha-METHYLSTYRENE	-8.0605	1.2867E+03	1.6119E-02	-1.3444E-05	250	654	0.734	0.050
882	C9H10	m-METHYLSTYRENE	-5.2814	9.1365E+02	9.5793E-03	-8.5309E-06	187	657	0.760	0.053
883	C9H10	o-METHYLSTYRENE	-6.5858	1.1184E+03	1.2275E-02	-1.0288E-05	205	659	0.814	0.054
884	C9H10	p-METHYLSTYRENE	-6.6270	1.0834E+03	1.2912E-02	-1.1056E-05	239	665	0.748	0.050
885	C9H10O2	BENZYL ACETATE	-10.3224	1.8519E+03	1.8807E-02	-1.3764E-05	248	699	1.874	0.056
886	C9H10O2	ETHYL BENZOATE	-8.5065	1.5478E+03	1.5816E-02	-1.2374E-05	284	698	1.997	0.053
887	C9H10O3	ETHYL VANILLIN	-52.2191	1.0531E+04	8.5019E-02	-4.7611E-05	351	748	-----	0.065
888	C9H11NO	p-DIMETHYLAMINOBENZALDEHYDE	-16.6480	3.3619E+03	2.6709E-02	-1.5770E-05	348	832	-----	0.050
889	C9H12	CUMENE	-5.9339	9.6384E+02	1.1916E-02	-1.1108E-05	200	631	0.731	0.049
890	C9H12	m-ETHYLTOLUENE	-4.1591	7.8018E+02	6.8385E-03	-6.7586E-06	178	637	0.787	0.048
891	C9H12	o-ETHYLTOLUENE	-5.0194	9.2620E+02	8.3900E-03	-7.4638E-06	192	651	0.842	0.050
892	C9H12	p-ETHYLTOLUENE	-6.4219	1.0175E+03	1.3010E-02	-1.1808E-05	283	640	0.661	0.045
893	C9H12	MESITYLENE	-15.0790	2.3442E+03	3.1356E-02	-2.4418E-05	288	637	0.916	0.046
894	C9H12	n-PROPYLBENZENE	-6.9452	1.1276E+03	1.3933E-02	-1.2344E-05	248	638	0.783	0.049
895	C9H12	1,2,3-TRIMETHYLBENZENE	-13.6106	2.1392E+03	2.7632E-02	-2.0999E-05	268	665	0.863	0.050
896	C9H12	1,2,4-TRIMETHYLBENZENE	-8.7592	1.4166E+03	1.7480E-02	-1.4420E-05	250	649	0.835	0.049
897	C9H12O	BENZYL ETHYL ETHER	-10.5203	1.7552E+03	2.0716E-02	-1.6305E-05	276	662	1.241	0.050
898	C9H12O	2-PHENYL-2-PROPANOL	-26.4916	5.0837E+03	4.2602E-02	-2.4306E-05	309	660	-----	0.055
899	C9H12O2	CUMENE HYDROPEROXIDE	-----	--------	--------	--------	----	----	-----	-----
900	C9H14O	ISOPHORONE	-9.8841	1.8200E+03	1.7704E-02	-1.2906E-05	265	715	2.246	0.053
901	C9H14O6	GLYCERYL TRIACETATE	-12.8264	2.3323E+03	2.3698E-02	-1.7016E-05	277	704	3.541	0.055
902	C9H16	1-NONYNE	-6.3942	9.9350E+02	1.4232E-02	-1.4097E-05	273	593	0.848	0.058
903	C9H16O4	AZELAIC ACID	-19.9279	4.3483E+03	3.1484E-02	-1.8655E-05	380	811	-----	0.050
904	C9H18	BUTYLCYCLOPENTANE	-3.1274	6.6560E+02	4.2080E-03	-5.0128E-06	184	625	0.820	0.041
905	C9H18	cis,cis-1,3,5-TRIMETHYLCYCLOHEXANE	-3.1598	6.6560E+02	4.2080E-03	-5.0128E-06	224	607	0.761	0.044
906	C9H18	cis,trans-1,3,5-TRIMETHYLCYCLOHEXANE	-3.1499	6.6560E+02	4.2080E-03	-5.0128E-06	189	602	0.779	0.047
907	C9H18	ISOPROPYLCYCLOHEXANE	-3.0456	6.6560E+02	4.2080E-03	-5.0128E-06	184	627	0.990	0.048
908	C9H18	1-NONENE	-6.5557	9.9350E+02	1.4232E-02	-1.4097E-05	273	593	0.584	0.040
909	C9H18	n-PROPYLCYCLOHEXANE	-3.0155	6.4844E+02	4.2748E-03	-5.0585E-06	248	639	0.964	0.046
910	C9H18O	DIISOBUTYL KETONE	-10.3632	1.5878E+03	2.2574E-02	-1.9757E-05	240	615	0.864	0.043
911	C9H18O	1-NONANAL	-1.1460	3.3324E+02	2.1306E-03	-5.2064E-06	255	640	1.394	0.040
912	C9H18O2	n-BUTYL VALERATE	-5.4173	9.9333E+02	1.0045E-02	-9.6026E-06	180	629	1.137	0.048
913	C9H18O2	n-NONANOIC ACID	-14.3522	2.7639E+03	2.5363E-02	-1.7668E-05	293	703	8.117	0.048
914	C9H18O2	n-OCTYL FORMATE	-6.5800	1.1842E+03	1.2581E-02	-1.1323E-05	234	645	1.369	0.046
915	C9H20	3,3-DIETHYLPENTANE	-10.3759	1.4524E+03	2.3128E-02	-2.0065E-05	240	610	0.405	0.044
916	C9H20	2,2-DIMETHYL-3-ETHYLPENTANE	-3.6172	6.8704E+02	5.9251E-03	-6.8455E-06	174	590	0.700	0.046
917	C9H20	3-ETHYL-2,3-DIMETHYLPENTANE	-3.6642	6.8704E+02	5.9251E-03	-6.8455E-06	174	590	0.628	0.041
918	C9H20	2,4-DIMETHYL-3-ETHYLPENTANE	-3.3642	6.5573E+02	5.2943E-03	-6.3412E-06	151	591	0.708	0.046
919	C9H20	2,2-DIMETHYLHEPTANE	-3.3832	6.5551E+02	5.5498E-03	-6.9853E-06	160	577	0.707	0.043
920	C9H20	2,6-DIMETHYLHEPTANE	-3.5056	6.6931E+02	5.8862E-03	-7.2680E-06	170	579	0.705	0.042
921	C9H20	3-ETHYLHEPTANE	-4.2966	7.0894E+02	8.2216E-03	-8.9310E-06	158	590	0.548	0.044
922	C9H20	4-ETHYLHEPTANE	-3.5725	6.6931E+02	5.8862E-03	-7.2680E-06	170	579	0.604	0.036
923	C9H20	2,3-DIMETHYLHEPTANE	-3.5758	6.6931E+02	5.8862E-03	-7.2680E-06	170	579	0.600	0.036
924	C9H20	2,4-DIMETHYLHEPTANE	-3.6126	6.6931E+02	5.8862E-03	-7.2680E-06	170	576	0.551	0.034
925	C9H20	2,5-DIMETHYLHEPTANE	-3.5976	6.6931E+02	5.8862E-03	-7.2680E-06	170	579	0.570	0.034
926	C9H20	3,4-DIMETHYLHEPTANE	-3.5754	6.6931E+02	5.8862E-03	-7.2680E-06	171	579	0.600	0.036
927	C9H20	3,5-DIMETHYLHEPTANE	-3.5976	6.6931E+02	5.8862E-03	-7.2680E-06	171	579	0.570	0.034
928	C9H20	4,4-DIMETHYLHEPTANE	-3.6014	6.6931E+02	5.8862E-03	-7.2680E-06	171	579	0.565	0.034
929	C9H20	3-ETHYL-2-METHYLHEXANE	-3.5879	6.6931E+02	5.8862E-03	-7.2680E-06	170	579	0.583	0.035
930	C9H20	4-ETHYL-2-METHYLHEXANE	-3.6082	6.6931E+02	5.8862E-03	-7.2680E-06	170	579	0.557	0.033
931	C9H20	3-ETHYL-3-METHYLHEXANE	-3.5754	6.6931E+02	5.8862E-03	-7.2680E-06	170	579	0.600	0.036
932	C9H20	3-ETHYL-4-METHYLHEXANE	-3.5763	6.6931E+02	5.8862E-03	-7.2680E-06	170	579	0.599	0.036
933	C9H20	2,2,3-TRIMETHYLHEXANE	-3.6092	6.6931E+02	5.8862E-03	-7.2680E-06	170	579	0.555	0.033
934	C9H20	2,2,4-TRIMETHYLHEXANE	-3.6433	6.6931E+02	5.8862E-03	-7.2680E-06	170	573	0.513	0.032
935	C9H20	2,3,3-TRIMETHYLHEXANE	-3.5895	6.6931E+02	5.8862E-03	-7.2680E-06	170	579	0.581	0.035
936	C9H20	2,3,4-TRIMETHYLHEXANE	-3.5829	6.6931E+02	5.8862E-03	-7.2680E-06	170	579	0.590	0.035

Table 22-1 VISCOSITY OF LIQUID - ORGANIC COMPOUNDS (continued)

$$\log_{10} n_{liq} = A + B/T + C\,T + C\,T^2 \qquad (n_{liq} - \text{centipoise}, T - K)$$

NO	FORMULA	NAME	A	B	C	D	TMIN	TMAX	n_{liq} @25 C	n_{liq} @TMAX
937	C9H20	2,3,5-TRIMETHYLHEXANE	-3.6201	6.6931E+02	5.8862E-03	-7.2680E-06	170	579	0.542	0.032
938	C9H20	2,4,4-TRIMETHYLHEXANE	-3.6234	6.6931E+02	5.8862E-03	-7.2680E-06	170	579	0.537	0.032
939	C9H20	3,3,4-TRIMETHYLHEXANE	-3.5760	6.6931E+02	5.8862E-03	-7.2680E-06	172	579	0.599	0.036
940	C9H20	2-METHYLOCTANE	-5.3995	8.4005E+02	1.1185E-02	-1.1542E-05	193	587	0.533	0.042
941	C9H20	3-METHYLOCTANE	-4.4887	7.3148E+02	8.7404E-03	-9.3922E-06	166	590	0.544	0.044
942	C9H20	4-METHYLOCTANE	-4.2975	7.0922E+02	8.2517E-03	-9.0323E-06	160	588	0.548	0.043
943	C9H20	n-NONANE	-6.0742	9.6861E+02	1.2677E-02	-1.2675E-05	220	596	0.672	0.040
944	C9H20	2,2,3,3-TETRAMETHYLPENTANE	-10.3834	1.7503E+03	1.8921E-02	-1.4297E-05	263	611	0.720	0.051
945	C9H20	2,2,3,4-TETRAMETHYLPENTANE	-3.0746	6.7833E+02	3.5028E-03	-4.1661E-06	152	592	0.749	0.048
946	C9H20	2,2,4,4-TETRAMETHYLPENTANE	-5.5019	1.0202E+03	8.9431E-03	-8.3908E-06	207	571	0.692	0.045
947	C9H20	2,3,3,4-TETRAMETHYLPENTANE	-5.5594	1.0202E+03	8.9431E-03	-8.3908E-06	207	571	0.607	0.040
948	C9H20	2,2,5-TRIMETHYLHEXANE	-5.3624	7.4791E+02	1.1724E-02	-1.2311E-05	167	568	0.353	0.044
949	C9H20O	2,6-DIMETHYL-4-HEPTANOL	-37.0693	6.0617E+03	7.6138E-02	-5.5731E-05	273	603	10.190	0.043
950	C9H20O	1-NONANOL	-16.3943	3.1795E+03	2.8251E-02	-1.9118E-05	288	673	9.848	0.048
951	C9H20O	2-NONANOL	-6.3355	1.5855E+03	8.8205E-03	-7.8222E-06	238	623	8.256	0.047
952	C9H20S	n-NONYL MERCAPTAN	-8.2997	1.4136E+03	1.5962E-02	-1.2914E-05	253	681	1.129	0.045
953	C9H20S	BUTYL-PENTYL-SULFIDE	-8.0771	1.4136E+03	1.5962E-02	-1.2914E-05	253	681	1.885	0.076
954	C9H20S	ETHYL-HEPTYL-SULFIDE	-8.0771	1.4136E+03	1.5962E-02	-1.2914E-05	253	681	1.885	0.076
955	C9H20S	HEXYL-PROPYL-SULFIDE	-8.0771	1.4136E+03	1.5962E-02	-1.2914E-05	253	681	1.885	0.076
956	C9H20S	METHYL-OCTYL-SULFIDE	-8.0771	1.4136E+03	1.5962E-02	-1.2914E-05	253	681	1.885	0.076
957	C9H21N	n-NONYLAMINE	-5.7257	1.2815E+03	8.9626E-03	-8.1432E-06	273	648	3.317	0.044
958	C9H21N	TRIPROPYLAMINE	-7.9459	1.1326E+03	1.8727E-02	-1.8597E-05	200	578	0.607	0.042
959	C10H6O8	PYROMELLITIC ACID	13.7921	-1.7081E+03	-2.2239E-02	8.4878E-06	554	893	-----	0.061
960	C10H7Br	1-BROMONAPHTHALENE	-----	-------	-------	-------	----	----	-----	-----
961	C10H7Cl	1-CHLORONAPHTHALENE	-14.5862	2.6360E+03	2.5825E-02	-1.6742E-05	274	785	2.927	0.053
962	C10H8	NAPHTHALENE	-10.3716	1.8572E+03	1.9320E-02	-1.4012E-05	353	748	-----	0.053
963	C10H8	AZULENE	-8.8695	1.7618E+03	1.2453E-02	-8.1383E-06	272	773	1.069	0.015
964	C10H9N	QUINALDINE	-8.3254	1.7618E+03	1.2453E-02	-8.1383E-06	272	773	3.742	0.052
965	C10H10	m-DIVINYLBENZENE	-6.8961	1.1314E+03	1.2966E-02	-1.0417E-05	206	692	0.689	0.053
966	C10H10	1-METHYLINDENE	-----	-------	-------	-------	----	----	-----	-----
967	C10H10	2-METHYLINDENE	-----	-------	-------	-------	----	----	-----	-----
968	C10H10O4	DIMETHYL PHTHALATE	-15.0712	3.0137E+03	2.5581E-02	-1.6595E-05	272	766	15.438	0.053
969	C10H10O4	DIMETHYL TEREPHTHALATE	-8.8688	1.9149E+03	1.4612E-02	-1.0350E-05	414	772	-----	0.053
970	C10H12	DICYCLOPENTADIENE	-7.9390	1.3455E+03	1.5146E-02	-1.2413E-05	307	660	-----	0.049
971	C10H12	1,2,3,4-TETRAHYDRONAPHTHALENE	-6.3710	1.2740E+03	1.0494E-02	-8.1163E-06	273	720	2.039	0.056
972	C10H12O	ANETHOLE	-17.9835	3.2145E+03	3.2021E-02	-2.0822E-05	295	723	3.120	0.054
973	C10H12O4	DIALLYL MALEATE	-5.6029	1.2056E+03	9.7937E-03	-8.7961E-06	226	693	3.791	0.050
974	C10H14	n-BUTYLBENZENE	-7.2536	1.2071E+03	1.4327E-02	-1.2296E-05	248	661	0.941	0.047
975	C10H14	sec-BUTYLBENZENE	-8.0972	1.3543E+03	1.5755E-02	-1.3018E-05	198	665	0.967	0.046
976	C10H14	tert-BUTYLBENZENE	-4.4917	8.6627E+02	7.2702E-03	-6.6859E-06	215	660	0.971	0.051
977	C10H14	1,2,3,4-TETRAMETHYLBENZENE	-4.3557	8.6627E+02	7.2702E-03	-6.6859E-06	267	660	1.328	0.070
978	C10H14	m-CYMENE	-3.7837	6.2624E+02	7.7037E-03	-8.2406E-06	209	657	0.760	0.047
979	C10H14	o-CYMENE	-4.5397	7.5849E+02	9.5321E-03	-9.7293E-06	250	662	0.958	0.045
980	C10H14	p-CYMENE	-3.8139	6.2626E+02	7.8756E-03	-8.4952E-06	205	653	0.758	0.046
981	C10H14	m-DIETHYLBENZENE	-4.2172	8.0549E+02	7.0455E-03	-6.7549E-06	189	663	0.965	0.050
982	C10H14	o-DIETHYLBENZENE	-7.4901	1.3219E+03	1.3631E-02	-1.0985E-05	242	668	1.074	0.049
983	C10H14	p-DIETHYLBENZENE	-5.8058	1.0140E+03	1.0909E-02	-9.7803E-06	230	658	0.951	0.048
984	C1OH14	2-ETHYL-m-XYLENE	-10.8133	1.8029E+03	2.0906E-02	-1.5999E-05	257	671	1.108	0.050
985	C10H14	2-ETHYL-p-XYLENE	-6.5040	1.1567E+03	1.1867E-02	-1.0032E-05	220	663	1.052	0.050
986	C10H14	3-ETHYL-o-XYLENE	-7.3375	1.3118E+03	1.3018E-02	-1.0250E-05	224	680	1.078	0.051
987	C10H14	4-ETHYL-m-XYLENE	-5.8408	1.0314E+03	1.0713E-02	-9.3524E-06	210	665	0.958	0.050
988	C10H14	4-ETHYL-o-XYLENE	-5.8776	1.0709E+03	1.0393E-02	-8.8903E-06	206	667	1.053	0.051
989	C1OH14	5-ETHYL-m-XYLENE	-5.2785	9.8458E+02	9.0236E-03	-8.0075E-06	189	655	1.005	0.050
990	C10H14	ISOBUTYLBENZENE	-5.9116	1.0337E+03	1.1068E-02	-9.8554E-06	222	650	0.953	0.051
991	C10H14	1,2,3,5-TETRAMETHYLBENZENE	-10.9140	1.8343E+03	2.0790E-02	-1.5622E-05	249	679	1.117	0.050
992	C10H14	1,2,4,5-TETRAMETHYLBENZENE	-25.5606	4.1829E+03	5.0616E-02	-3.5474E-05	352	675	-----	0.044
993	C10H14O	p-tert-BUTYLPHENOL	-47.3907	9.3735E+03	7.9337E-02	-4.6220E-05	372	734	-----	0.051
994	C10H14O2	p-tert-BUTYLCATECHOL	-28.2625	5.6522E+03	4.6362E-02	-2.6862E-05	325	776	-----	0.066
995	C10H15N	N,N-DIETYHLANILINE	-8.2367	1.5091E+03	1.5104E-02	-1.1837E-05	250	702	1.887	0.048
996	C10H15N	2,6-DIETHYLANILINE	-16.1868	2.9213E+03	3.0305E-02	-2.1579E-05	277	678	5.352	0.056
997	C10H16	CAMPHENE	4.2772	-2.0051E+02	-1.3642E-02	8.4866E-06	320	638	-----	0.052
998	C10H16	D-LIMONENE	-3.3667	6.6580E+02	5.4395E-03	-5.8645E-06	199	660	0.927	0.048
999	C10H16	alpha-PHELLANDRENE	-6.0555	9.8032E+02	1.2018E-02	-1.0848E-05	220	649	0.710	0.048
1000	C1OH16	beta-PHELLANDRENE	-5.9733	9.7015E+02	1.1817E-02	-1.0707E-05	220	648	0.711	0.048
1001	C10H16	alpha-PINENE	-3.0249	6.6321E+02	4.9209E-03	-6.2076E-06	209	632	1.303	0.045
1002	C10H16	beta-PINENE	-3.7170	8.3963E+02	5.3086E-03	-5.6180E-06	212	643	1.522	0.048
1003	C10H16	alpha-TERPINENE	-6.0176	9.7751E+02	1.1868E-02	-1.0671E-05	220	652	0.709	0.048
1004	C1OH16	gamma-TERPINENE	-6.1321	9.9555E+02	1.2023E-02	-1.0614E-05	220	661	0.705	0.048
1005	C10H16	TERPINOLENE	-5.4542	9.1919E+02	1.0115E-02	-8.9234E-06	220	672	0.710	0.048
1006	C10H16O	CAMPHOR	-14.2809	3.5410E+03	1.8549E-02	-1.0210E-05	453	709	-----	0.054
1007	C10H18	1-DECYNE	-6.7242	1.1003E+03	1.4341E-02	-1.3520E-05	273	617	1.097	0.058
1008	C10H18	cis-DECAHYDRONANPHTALENE	-3.9921	1.0210E+03	4.9544E-03	-4.5071E-06	243	702	3.227	0.052
1009	C10H18	trans-DECAHYDRONAPHTHALENE	-2.4871	7.2432E+02	2.0298E-03	-2.6970E-06	243	687	2.031	0.049
1010	C10H18O4	SEBACIC ACID	-22.7000	5.0113E+03	3.5521E-02	-2.0661E-05	408	815	-----	0.047
1011	C10H20	n-BUTYLCYCLOHEXANE	-2.9752	6.7445E+02	4.1686E-03	-4.8706E-06	253	667	1.250	0.045
1012	C10H20	1-CYCLOPENTYLPENTANE	-3.0422	6.7445E+02	4.1686E-03	-4.8706E-06	253	647	1.071	0.046
1013	C10H20	1-DECENE	-6.8845	1.1003E+03	1.4341E-02	-1.3520E-05	273	617	0.758	0.040
1014	C10H20O	1-DECANAL	-1.0851	3.1235E+02	2.4212E-03	-5.5349E-06	267	657	1.557	0.039

494

Table 22-1 VISCOSITY OF LIQUID - ORGANIC COMPOUNDS (continued)

			$\log_{10} n_{liq} = A + B/T + C\,T + C\,T^2$						$(n_{liq}$ - centipoise, T - K)	
NO	FORMULA	NAME	A	B	C	D	TMIN	TMAX	n_{liq} @25 C	n_{liq} @TMAX
1015	C10H20O2	n-DECANOIC ACID	-9.9842	2.0737E+03	1.7187E-02	-1.2840E-05	305	713	-----	0.045
1016	C10H20O2	2-ETHYLHEXYL ACETATE	-6.5103	1.1823E+03	1.2249E-02	-1.1009E-05	209	639	1.345	0.047
1017	C10H20O2	ISOPENTYL ISOVALERATE	-6.9331	1.2322E+03	1.3086E-02	-1.1479E-05	215	637	1.205	0.048
1018	C10H22	n-DECANE	-6.0716	1.0177E+03	1.2247E-02	-1.1892E-05	243	618	0.863	0.040
1019	C10H22	2-METHYLNONANE	-6.6483	1.0765E+03	1.3719E-02	-1.2961E-05	199	610	0.795	0.046
1020	C10H22	3-METHYLNONANE	-4.3139	7.6340E+02	8.1597E-03	-8.8183E-06	188	613	0.786	0.042
1021	C10H22	4-METHYLNONANE	-4.1949	7.4214E+02	7.8148E-03	-8.4958E-06	174	610	0.740	0.042
1022	C10H22	5-METHYLNONANE	-4.3301	7.6483E+02	8.2897E-03	-9.0456E-06	185	610	0.799	0.041
1023	C10H22	3-ETHYLOCTANE	-4.3293	7.6483E+02	8.2897E-03	-9.0456E-06	186	610	0.801	0.041
1024	C10H22	4-ETHYLOCTANE	-4.3431	7.6483E+02	8.2897E-03	-9.0456E-06	186	609	0.776	0.040
1025	C10H22	2,2-DIMETHYLOCTANE	-4.5297	8.2587E+02	8.3200E-03	-8.9247E-06	225	602	0.846	0.041
1026	C10H22	2,3-DIMETHYLOCTANE	-4.5644	8.2587E+02	8.3200E-03	-8.9247E-06	225	602	0.781	0.038
1027	C10H22	2,4-DIMETHYLOCTANE	-4.6050	8.2587E+02	8.3200E-03	-8.9247E-06	225	599	0.712	0.036
1028	C10H22	2,5-DIMETHYLOCTANE	-4.5925	8.2587E+02	8.3200E-03	-8.9247E-06	225	602	0.732	0.036
1029	C10H22	2,6-DIMETHYLOCTANE	-4.5834	8.2587E+02	8.3200E-03	-8.9247E-06	225	602	0.748	0.037
1030	C10H22	2,7-DIMETHYLOCTANE	-4.5858	8.2587E+02	8.3200E-03	-8.9247E-06	225	602	0.744	0.036
1031	C10H22	3,3-DIMETHYLOCTANE	-4.5794	8.2587E+02	8.3200E-03	-8.9247E-06	225	602	0.755	0.037
1032	C10H22	3,4-DIMETHYLOCTANE	-4.5688	8.2587E+02	8.3200E-03	-8.9247E-06	225	602	0.773	0.038
1033	C10H22	3,5-DIMETHYLOCTANE	-4.5881	8.2587E+02	8.3200E-03	-8.9247E-06	225	602	0.740	0.036
1034	C10H22	3,6-DIMETHYLOCTANE	-4.5814	8.2587E+02	8.3200E-03	-8.9247E-06	225	602	0.751	0.037
1035	C10H22	4,4-DIMETHYLOCTANE	-4.5973	8.2587E+02	8.3200E-03	-8.9247E-06	225	602	0.724	0.035
1036	C10H22	4,5-DIMETHYLOCTANE	-4.5749	8.2587E+02	8.3200E-03	-8.9247E-06	225	602	0.763	0.037
1037	C10H22	4-PROPYLHEPTANE	-4.5973	8.2587E+02	8.3200E-03	-8.9247E-06	225	601	0.724	0.036
1038	C10H22	4-ISOPROPYLHEPTANE	-4.5905	8.2587E+02	8.3200E-03	-8.9247E-06	225	602	0.736	0.036
1039	C10H22	3-ETHYL-2-METHYLHEPTANE	-4.5794	8.2587E+02	8.3200E-03	-8.9247E-06	225	602	0.755	0.037
1040	C10H22	4-ETHYL-2-METHYLHEPTANE	-4.6036	8.2587E+02	8.3200E-03	-8.9247E-06	225	601	0.714	0.035
1041	C10H22	5-ETHYL-2-METHYLHEPTANE	-4.5867	8.2587E+02	8.3200E-03	-8.9247E-06	225	602	0.742	0.036
1042	C10H22	3-ETHYL-3-METHYLHEPTANE	-4.5669	8.2587E+02	8.3200E-03	-8.9247E-06	225	602	0.777	0.038
1043	C10H22	4-ETHYL-3-METHYLHEPTANE	-4.5746	8.2587E+02	8.3200E-03	-8.9247E-06	225	602	0.763	0.037
1044	C10H22	3-ETHYL-5-METHYLHEPTANE	-4.5939	8.2587E+02	8.3200E-03	-8.9247E-06	225	602	0.730	0.036
1045	C10H22	3-ETHYL-4-METHYLHEPTANE	-4.5707	8.2587E+02	8.3200E-03	-8.9247E-06	225	602	0.770	0.038
1046	C10H22	4-ETHYL-4-METHYLHEPTANE	-4.5814	8.2587E+02	8.3200E-03	-8.9247E-06	225	602	0.751	0.037
1047	C10H22	2,2,3-TRIMETHYLHEPTANE	-4.5968	8.2587E+02	8.3200E-03	-8.9247E-06	225	602	0.725	0.035
1048	C10H22	2,2,4-TRIMETHYLHEPTANE	-4.6417	8.2587E+02	8.3200E-03	-8.9247E-06	225	594	0.654	0.035
1049	C10H22	2,2,5-TRIMETHYLHEPTANE	-4.6296	8.2587E+02	8.3200E-03	-8.9247E-06	225	593	0.672	0.036
1050	C10H22	2,2,6-TRIMETHYLHEPTANE	-4.6386	8.2587E+02	8.3200E-03	-8.9247E-06	225	593	0.659	0.035
1051	C10H22	2,3,3-TRIMETHYLHEPTANE	-4.5843	8.2587E+02	8.3200E-03	-8.9247E-06	225	593	0.746	0.040
1052	C10H22	2,3,4-TRIMETHYLHEPTANE	-4.5857	8.2587E+02	8.3200E-03	-8.9247E-06	225	602	0.744	0.036
1053	C10H22	2,3,5-TRIMETHYLHEPTANE	-4.5818	8.2587E+02	8.3200E-03	-8.9247E-06	225	602	0.751	0.037
1054	C10H22	2,3,6-TRIMETHYLHEPTANE	-4.6045	8.2587E+02	8.3200E-03	-8.9247E-06	225	602	0.712	0.035
1055	C10H22	2,4,4-TRIMETHYLHEPTANE	-4.6287	8.2587E+02	8.3200E-03	-8.9247E-06	225	600	0.674	0.034
1056	C10H22	2,4,5-TRIMETHYLHEPTANE	-4.6021	8.2587E+02	8.3200E-03	-8.9247E-06	225	602	0.716	0.035
1057	C10H22	2,4,6-TRIMETHYLHEPTANE	-4.6451	8.2587E+02	8.3200E-03	-8.9247E-06	225	590	0.649	0.036
1058	C10H22	2,5,5-TRIMETHYLHEPTANE	-4.6200	8.2587E+02	8.3200E-03	-8.9247E-06	225	602	0.687	0.034
1059	C10H22	3,3,4-TRIMETHYLHEPTANE	-4.5760	8.2587E+02	8.3200E-03	-8.9247E-06	225	602	0.761	0.037
1060	C10H22	3,3,5-TRIMETHYLHEPTANE	-4.6060	8.2587E+02	8.3200E-03	-8.9247E-06	225	602	0.710	0.035
1061	C10H22	3,4,4-TRIMETHYLHEPTANE	-4.5799	8.2587E+02	8.3200E-03	-8.9247E-06	225	602	0.754	0.037
1062	C10H22	3,4,5-TRIMETHYLHEPTANE	-4.5732	8.2587E+02	8.3200E-03	-8.9247E-06	225	602	0.766	0.037
1063	C10H22	3-ISOPROPYL-2-METHYLHEXANE	-4.5529	8.2587E+02	8.3200E-03	-8.9247E-06	225	602	0.802	0.039
1064	C10H22	3,3-DIETHYLHEXANE	-4.5548	8.2587E+02	8.3200E-03	-8.9247E-06	225	602	0.799	0.039
1065	C10H22	3,4-DIETHYLHEXANE	-4.5664	8.2587E+02	8.3200E-03	-8.9247E-06	225	602	0.778	0.038
1066	C10H22	3-ETHYL-2,2-DIMETHYLHEXANE	-4.6040	8.2587E+02	8.3200E-03	-8.9247E-06	225	602	0.713	0.035
1067	C10H22	4-ETHYL-2,2-DIMETHYLHEXANE	-4.6480	8.2587E+02	8.3200E-03	-8.9247E-06	225	594	0.645	0.034
1068	C10H22	3-ETHYL-2,3-DIMETHYLHEXANE	-4.5674	8.2587E+02	8.3200E-03	-8.9247E-06	225	602	0.776	0.038
1069	C10H22	4-ETHYL-2,3-DIMETHYLHEXANE	-4.5809	8.2587E+02	8.3200E-03	-8.9247E-06	225	602	0.752	0.037
1070	C10H22	3-ETHYL-2,4-DIMETHYLHEXANE	-4.5847	8.2587E+02	8.3200E-03	-8.9247E-06	225	602	0.746	0.036
1071	C10H22	4-ETHYL-2,4-DIMETHYLHEXANE	-4.5799	8.2587E+02	8.3200E-03	-8.9247E-06	225	602	0.754	0.037
1072	C10H22	3-ETHYL-2,5-DIMETHYLHEXANE	-4.6137	8.2587E+02	8.3200E-03	-8.9247E-06	225	602	0.697	0.034
1073	C10H22	4-ETHYL-3,3-DIMETHYLHEXANE	-4.5712	8.2587E+02	8.3200E-03	-8.9247E-06	225	602	0.769	0.038
1074	C10H22	3-ETHYL-3,4-DIMETHYLHEXANE	-4.5751	8.2587E+02	8.3200E-03	-8.9247E-06	225	602	0.762	0.037
1075	C10H22	2,2,3,3-TETRAMETHYLHEXANE	-4.5837	8.2587E+02	8.3200E-03	-8.9247E-06	225	602	0.747	0.037
1076	C10H22	2,2,3,4-TETRAMETHYLHEXANE	-4.5910	8.2587E+02	8.3200E-03	-8.9247E-06	225	602	0.735	0.036
1077	C10H22	2,2,3,5-TETRAMETHYLHEXANE	-4.6412	8.2587E+02	8.3200E-03	-8.9247E-06	225	601	0.655	0.032
1078	C10H22	2,2,4,4-TETRAMETHYLHEXANE	-4.6151	8.2587E+02	8.3200E-03	-8.9247E-06	225	602	0.695	0.034
1079	C10H22	2,2,4,5-TETRAMETHYLHEXANE	-4.6437	8.2587E+02	8.3200E-03	-8.9247E-06	225	598	0.651	0.033
1080	C10H22	2,2,5,5-TETRAMETHYLHEXANE	-4.6939	8.2587E+02	8.3200E-03	-8.9247E-06	261	581	0.580	0.035
1081	C10H22	2,3,3,4-TETRAMETHYLHEXANE	-4.5631	8.2587E+02	8.3200E-03	-8.9247E-06	261	602	0.784	0.038
1082	C10H22	2,3,3,5-TETRAMETHYLHEXANE	-4.6185	8.2587E+02	8.3200E-03	-8.9247E-06	261	602	0.690	0.034
1083	C10H22	2,3,4,4-TETRAMETHYLHEXANE	-4.5775	8.2587E+02	8.3200E-03	-8.9247E-06	261	602	0.758	0.037
1084	C10H22	2,3,4,5-TETRAMETHYLHEXANE	-4.6036	8.2587E+02	8.3200E-03	-8.9247E-06	261	602	0.714	0.035
1085	C10H22	3,3,4,4-TETRAMETHYLHEXANE	-4.5370	8.2587E+02	8.3200E-03	-8.9247E-06	261	602	0.832	0.041
1086	C10H22	2,4-DIMETHYL-3-ISOPROPYLPENTANE	-4.5995	8.2587E+02	8.3200E-03	-8.9247E-06	225	602	0.721	0.035
1087	C10H22	3,3-DIETHYL-2-METHYLPENTANE	-4.5384	8.2587E+02	8.3200E-03	-8.9247E-06	225	602	0.830	0.041
1088	C10H22	3-ETHYL-2,2,3-TRIMETHYLPENTANE	-4.5394	8.2587E+02	8.3200E-03	-8.9247E-06	225	602	0.828	0.040
1089	C10H22	3-ETHYL-2,2,4-TRIMETHYLPENTANE	-4.6079	8.2587E+02	8.3200E-03	-8.9247E-06	225	602	0.707	0.035
1090	C10H22	3-ETHYL-2,3,4-TRIMETHYLPENTANE	-4.5397	8.2587E+02	8.3200E-03	-8.9247E-06	225	602	0.827	0.040
1091	C10H22	2,2,3,3,4-PENTAMETHYLPENTANE	-4.5560	8.2587E+02	8.3200E-03	-8.9247E-06	237	602	0.797	0.039
1092	C10H22	2,2,3,4,4-PENTAMETHYLPENTANE	-4.5886	8.2587E+02	8.3200E-03	-8.9247E-06	237	602	0.739	0.036

Table 22-1 VISCOSITY OF LIQUID - ORGANIC COMPOUNDS (continued)

NO	FORMULA	NAME	$\log_{10} n_{liq} = A + B/T + C\,T + C\,T^2$						$(n_{liq}$ - centipoise, T - K)	
			A	B	C	D	TMIN	TMAX	n_{liq} @25 C	n_{liq} @TMAX
1093	C10H22O	1-DECANOL	-12.9966	2.6249E+03	2.2319E-02	-1.5904E-05	288	690	11.169	0.043
1094	C10H22O	DI-n-PENTYL ETHER	-6.7681	1.1594E+03	1.3073E-02	-1.1825E-05	204	622	0.927	0.045
1095	C10H22O	ISODECANOL	-10.4212	2.4141E+03	1.4388E-02	-9.3282E-06	240	644	13.687	0.053
1096	C10H22O5	TETRAETHYLENE GLYCOL DIMETHYL ETHER	-8.8707	1.7050E+03	1.5949E-02	-1.2257E-05	243	705	3.262	0.050
1097	C10H22S	n-DECYL MERCAPTAN	-7.8372	1.3858E+03	1.4624E-02	-1.1718E-05	248	696	1.347	0.045
1098	C10H22S	BUTYL-HEXYL-SULFIDE	-7.5851	1.3858E+03	1.4624E-02	-1.1718E-05	248	696	2.407	0.081
1099	C10H22S	ETHYL-OCTYL-SULFIDE	-7.5851	1.3858E+03	1.4624E-02	-1.1718E-05	248	696	2.407	0.081
1100	C10H22S	HEPTYL-PROPYL-SULFIDE	-7.5851	1.3858E+03	1.4624E-02	-1.1718E-05	248	696	2.407	0.081
1101	C10H22S	METHYL-NONYL-SULFIDE	-7.5851	1.3858E+03	1.4624E-02	-1.1718E-05	248	696	2.407	0.081
1102	C10H22S	PENTYL-SULFIDE	-7.5851	1.3858E+03	1.4624E-02	-1.1718E-05	248	696	2.407	0.081
1103	C10H22S2	PENTYL-DISULFIDE	-7.9471	1.3858E+03	1.4624E-02	-1.1718E-05	248	696	1.046	0.035
1104	C10H23N	n-DECYLAMINE	-6.2418	1.3980E+03	9.9630E-03	-8.7388E-06	289	663	4.373	0.043
1105	C11H10	1-METHYLNAPHTHALENE	-7.8389	1.5609E+03	1.3221E-02	-9.4950E-06	250	772	3.120	0.054
1106	C11H10	2-METHYLNAPHTHALENE	-5.3837	1.0715E+03	9.2405E-03	-7.5290E-06	308	761	------	0.050
1107	C11H14O2	n-BUTYL BENZOATE	-9.5453	1.8976E+03	1.5779E-02	-1.0980E-05	252	724	3.530	0.055
1108	C11H16	n-PENTYLBENZENE	-7.8348	1.3476E+03	1.5192E-02	-1.2594E-05	220	680	1.245	0.045
1109	C11H16O	p-tert-AMYLPHENOL	-48.7375	9.6443E+03	8.1016E-02	-4.6531E-05	366	751	------	0.051
1110	C11H20	1-UNDECYNE	-6.9901	1.1713E+03	1.4798E-02	-1.3604E-05	349	638	------	0.056
1111	C11H20O2	2-ETHYLHEXYL ACRYLATE	-5.4142	1.0721E+03	9.2332E-03	-8.3421E-06	183	655	1.559	0.049
1112	C11H22	1-UNDECENE	-7.1496	1.1713E+03	1.4798E-02	-1.3604E-05	224	638	0.959	0.039
1113	C11H22	1-CYCLOPENTYLHEXANE	-6.9930	1.1713E+03	1.4798E-02	-1.3604E-05	224	638	1.375	0.056
1114	C11H22	PENTYLCYCLOHEXANE	-6.9350	1.1713E+03	1.4798E-02	-1.3604E-05	224	638	1.571	0.064
1115	C11H22O	1-UNDECANAL	-1.6299	3.7917E+02	3.9856E-03	-6.7204E-06	273	672	1.709	0.038
1116	C11H24	n-UNDECANE	-6.7868	1.1682E+03	1.3438E-02	-1.2334E-05	248	639	1.100	0.039
1117	C11H24O	1-UNDECANOL	-12.0055	2.6247E+03	1.8181E-02	-1.1814E-05	293	704	14.732	0.046
1118	C11H24S	UNDECYL MERCAPTAN	-9.2294	1.6449E+03	1.7120E-02	-1.3088E-05	270	710	1.692	0.044
1119	C11H24S	BUTYL-HEPTYL-SULFIDE	-8.9799	1.6449E+03	1.7120E-02	-1.3088E-05	270	710	3.006	0.078
1120	C11H24S	DECYL-METHYL-SULFIDE	-8.9799	1.6449E+03	1.7120E-02	-1.3088E-05	270	710	3.006	0.078
1121	C11H24S	ETHYL-NONYL-SULFIDE	-8.9799	1.6449E+03	1.7120E-02	-1.3088E-05	270	710	3.006	0.078
1122	C11H24S	OCTYL-PROPYL-SULFIDE	-8.9799	1.6449E+03	1.7120E-02	-1.3088E-05	270	710	3.006	0.078
1123	C12H8O	DIBENZOFURAN	------	--------	--------	--------	----	----	------	------
1124	C12H9N	DIBENZOPYRROLE	------	--------	--------	--------	----	----	------	------
1125	C12H10	ACENAPHTHENE	-11.4875	2.4585E+03	1.7704E-02	-1.0967E-05	367	803	------	0.052
1126	C12H10	BIPHENYL	-9.9122	2.0514E+03	1.5545E-02	-9.9043E-06	373	789	------	0.061
1127	C12H10O	DIPHENYL ETHER	-7.6018	1.5551E+03	1.2621E-02	-9.1364E-06	300	763	------	0.056
1128	C12H11N	p-AMINODIPHENYL	-15.0833	3.0302E+03	2.4527E-02	-1.4842E-05	326	817	------	0.057
1129	C12H11N	DIPHENYLAMINE	-16.0796	3.2265E+03	2.6580E-02	-1.6289E-05	328	817	------	0.052
1130	C12H11N3	p-AMINOAZOBENZENE	------	--------	--------	--------	----	----	------	------
1131	C12H11N3	1,3-DIPHENYLTRIAZENE	------	--------	--------	--------	----	----	------	------
1132	C12H12	1,2-DIMETHYLNAPHTHALENE	-7.7787	1.5212E+03	1.3895E-02	-1.0523E-05	273	770	3.395	0.045
1133	C12H12	1,3-DIMETHYLNAPHTHALENE	-7.7840	1.5212E+03	1.3895E-02	-1.0523E-05	270	770	3.354	0.045
1134	C12H12	1,4-DIMETHYLNAPHTHALENE	-7.7739	1.5212E+03	1.3895E-02	-1.0523E-05	281	770	3.432	0.046
1135	C12H12	1,5-DIMETHYLNAPHTHALENE	-7.7850	1.5212E+03	1.3895E-02	-1.0523E-05	356	770	------	0.045
1136	C12H12	1,6-DIMETHYLNAPHTHALENE	-7.7946	1.5212E+03	1.3895E-02	-1.0523E-05	270	770	3.273	0.044
1137	C12H12	1,7-DIMETHYLNAPHTHALENE	-7.7946	1.5212E+03	1.3895E-02	-1.0523E-05	270	770	3.273	0.044
1138	C12H12	2,3-DIMETHYLNAPHTHALENE	-7.7705	1.5212E+03	1.3895E-02	-1.0523E-05	379	770	------	0.046
1139	C12H12	2,6-DIMETHYLNAPHTHALENE	-19.5550	3.7700E+03	3.3206E-02	-2.0537E-05	383	777	------	0.050
1140	C12H12	2,7-DIMETHYLNAPHTHALENE	-17.5672	3.3933E+03	2.9777E-02	-1.8597E-05	369	778	------	0.051
1141	C12H12	1-ETHYLNAPHTHALENE	-7.7627	1.5212E+03	1.3895E-02	-1.0523E-05	270	776	3.522	0.044
1142	C12H12	2-ETHYLNAPHTHALENE	-7.8172	1.5212E+03	1.3895E-02	-1.0523E-05	270	774	3.107	0.040
1143	C12H12N2	p-AMINODIPHENYLAMINE	-14.8162	3.3378E+03	2.1060E-02	-1.1304E-05	341	867	------	0.062
1144	C12H12N2	HYDRAZOBENZENE	-25.3140	5.4179E+03	3.9904E-02	-2.2948E-05	404	792	------	0.054
1145	C12H14	1,2,3-TRIMETHYLINDENE	-17.0891	3.0665E+03	3.0112E-02	-1.9502E-05	345	726	------	0.052
1146	C12H14O4	DIETHYL PHTHALATE	-12.1587	2.5728E+03	1.8732E-02	-1.1566E-05	269	757	10.649	0.062
1147	C12H16	CYCLOHEXYLBENZENE	-8.5159	1.6746E+03	1.4146E-02	-1.0006E-05	280	744	2.685	0.053
1148	C12H18	m-DIISOPROPYLBENZENE	-3.5276	8.2582E+02	4.8244E-03	-4.9109E-06	225	684	1.754	0.048
1149	C12H18	p-DIISOPROPYLBENZENE	-8.8585	1.4249E+03	1.7340E-02	-1.3658E-05	256	689	0.752	0.047
1150	C12H18	n-HEXYLBENZENE	-5.4643	1.1674E+03	7.5583E-03	-5.6485E-06	253	698	1.594	0.054
1151	C12H18	1,2,3-TRIETHYLBENZENE	-3.5887	8.2582E+02	4.8244E-03	-4.9109E-06	225	684	1.524	0.042
1152	C12H18	1,2,4-TRIETHYLBENZENE	-3.5887	8.2582E+02	4.8244E-03	-4.9109E-06	225	684	1.524	0.042
1153	C12H18	1,3,5-TRIETHYLBENZENE	-3.5960	8.2582E+02	4.8244E-03	-4.9109E-06	225	682	1.499	0.042
1154	C12H18	HEXAMETHYLBENZENE	-3.3670	8.2582E+02	4.8244E-03	-4.9109E-06	225	684	2.539	0.070
1155	C12H20O4	DIBUTYL MALEATE	-4.9604	1.1559E+03	7.9908E-03	-7.1881E-06	188	716	4.571	0.049
1156	C12H22	BICYCLOHEXYL	-14.0181	2.5425E+03	2.5246E-02	-1.7328E-05	277	727	3.135	0.047
1157	C12H22	1-DODECYNE	-7.3005	1.2603E+03	1.5116E-02	-1.3452E-05	255	657	1.728	0.055
1158	C12H23N	DICYCLOHEXYLAMINE	-7.1261	1.6284E+03	1.0548E-02	-7.6204E-06	273	737	6.354	0.052
1159	C12H24	1-DODECENE	-7.4600	1.2603E+03	1.5116E-02	-1.3452E-05	238	657	1.197	0.038
1160	C12H24	1-CYCLOPENTYLHEPTANE	-6.8914	1.1713E+03	1.4798E-02	-1.3604E-05	224	638	1.737	0.071
1161	C12H24	1-CYCLOHEXYLHEXANE	-6.8337	1.1713E+03	1.4798E-02	-1.3604E-05	264	638	1.984	0.081
1162	C12H24O	1-DODECANAL	-1.8002	3.8496E+02	4.7835E-03	-7.4064E-06	285	685	1.815	0.037
1163	C12H24O2	n-DODECANOIC ACID	-8.6022	1.8713E+03	1.5184E-02	-1.2043E-05	317	734	------	0.040
1164	C12H26	n-DODECANE	-7.0687	1.2530E+03	1.3735E-02	-1.2215E-05	262	658	1.390	0.038
1165	C12H26O	DI-n-HEXYL ETHER	-5.2939	1.0885E+03	9.6168E-03	-9.4315E-06	230	658	2.431	0.040
1166	C12H26O	1-DODECANOL	-14.5095	2.9618E+03	2.4288E-02	-1.6342E-05	297	721	16.338	0.041
1167	C12H26O3	DIETHYLENE GLYCOL DI-n-BUTYL ETHER	-5.2831	1.0464E+03	9.5588E-03	-8.8557E-06	213	680	1.947	0.046
1168	C12H26S	n-DODECYL MERCAPTAN	-6.9334	1.3524E+03	1.2394E-02	-1.0073E-05	265	724	2.526	0.042
1169	C12H26S	BUTYL-OCTYL-SULFIDE	-6.7658	1.3524E+03	1.2394E-02	-1.0073E-05	265	724	3.716	0.062
1170	C12H26S	DECYL-ETHYL-SULFIDE	-6.7658	1.3524E+03	1.2394E-02	-1.0073E-05	265	724	3.716	0.062

Table 22-1 VISCOSITY OF LIQUID - ORGANIC COMPOUNDS (continued)

			$\log_{10} n_{liq} = A + B/T + C\,T + C\,T^2$						$(n_{liq}$ - centipoise, T - K)	
NO	FORMULA	NAME	A	B	C	D	TMIN	TMAX	n_{liq} @25 C	n_{liq} @TMAX
1171	C12H26S	HEXYL-SULFIDE	-6.7658	1.3524E+03	1.2394E-02	-1.0073E-05	265	724	3.716	0.062
1172	C12H26S	METHYL-UNDECYL-SULFIDE	-6.7658	1.3524E+03	1.2394E-02	-1.0073E-05	265	724	3.716	0.062
1173	C12H26S	NONYL-PROPYL-SULFIDE	-6.7658	1.3524E+03	1.2394E-02	-1.0073E-05	265	724	3.716	0.062
1174	C12H26S2	HEXYL-DISULFIDE	-7.2601	1.3524E+03	1.2394E-02	-1.0073E-05	265	724	1.190	0.020
1175	C12H27BO3	TRI-n-BUTYL BORATE	-4.1309	9.2087E+02	5.5619E-03	-4.6086E-06	250	743	1.608	0.050
1176	C12H27N	DODECYLAMINE	-7.3328	1.6322E+03	1.2027E-02	-9.8498E-06	301	696	-----	0.041
1177	C12H27N	TRI-n-BUTYLAMINE	-9.5150	1.5646E+03	1.9693E-02	-1.6870E-05	240	644	1.272	0.040
1178	C13H10	FLUORENE	-13.2138	2.7388E+03	2.0893E-02	-1.2270E-05	388	870	-----	0.067
1179	C13H10O	BENZOPHENONE	-15.2815	3.0748E+03	2.5300E-02	-1.5645E-05	325	816	-----	0.052
1180	C13H12	DIPHENYLMETHANE	-8.6295	1.6730E+03	1.4894E-02	-1.0640E-05	298	768	2.996	0.051
1181	C13H14	1-PROPYLNAPHTHALENE	-7.7473	1.5212E+03	1.3895E-02	-1.0523E-05	270	770	3.649	0.049
1182	C13H14	2-PROPYLNAPHTHALENE	-7.7439	1.5212E+03	1.3895E-02	-1.0523E-05	271	770	3.678	0.049
1183	C13H14	2ETHYL-3-METHYLNAPHTHALENE	-7.7271	1.5212E+03	1.3895E-02	-1.0523E-05	345	770	-----	0.051
1184	C13H14	2ETHYL-6-METHYLNAPHTHALENE	-7.7608	1.5212E+03	1.3895E-02	-1.0523E-05	320	766	-----	0.049
1185	C13H14	2ETHYL-7-METHYLNAPHTHALENE	-7.7608	1.5212E+03	1.3895E-02	-1.0523E-05	320	766	-----	0.049
1186	C13H20	n-HEPTYLBENZENE	-9.0835	1.5999E+03	1.7555E-02	-1.3886E-05	230	714	1.915	0.041
1187	C13H24	1-TRIDECYNE	-3.6815	9.2942E+02	3.3679E-03	-1.2409E-06	250	675	2.136	0.253
1188	C13H26	1-TRIDECENE	-3.8333	9.2942E+02	3.3679E-03	-1.2409E-06	250	675	1.506	0.178
1189	C13H26	1-CYCLOPENTYLOCTANE	-6.7970	1.1713E+03	1.4798E-02	-1.3604E-05	264	638	2.159	0.088
1190	C13H26	1-CYCLOHEXYLHEPTANE	-6.7362	1.1713E+03	1.4798E-02	-1.3604E-05	264	638	2.484	0.101
1191	C13H26O	1-TRIDECANAL	-2.3833	4.6257E+02	6.2560E-03	-8.3851E-06	288	700	1.941	0.035
1192	C13H26O2	n-BUTYL NONANOATE	-7.7968	1.4649E+03	1.4685E-02	-1.2657E-05	235	652	2.343	0.044
1193	C13H26O2	METHYL DODECANOATE	-8.9487	1.7269E+03	1.5804E-02	-1.2024E-05	278	712	3.065	0.043
1194	C13H28	n-TRIDECANE	-7.2994	1.3248E+03	1.3974E-02	-1.2097E-05	268	676	1.718	0.038
1195	C13H28O	1-TRIDECANOL	-18.9853	3.9869E+03	2.9275E-02	-1.7181E-05	304	731	-----	0.049
1196	C13H28S	BUTYL-NONYL-SULFIDE	-6.6790	1.3524E+03	1.2394E-02	-1.0073E-05	272	724	4.538	0.076
1197	C13H28S	DECYL-PROPYL-SULFIDE	-6.6790	1.3524E+03	1.2394E-02	-1.0073E-05	272	724	4.538	0.076
1198	C13H28S	DODECYL-METHYL-SULFIDE	-6.6790	1.3524E+03	1.2394E-02	-1.0073E-05	272	724	4.538	0.076
1199	C13H28S	ETHYL-UNDECYL-SULFIDE	-6.6790	1.3524E+03	1.2394E-02	-1.0073E-05	272	724	4.538	0.076
1200	C13H28S	1-TRIDECANETHIOL	-7.0753	1.3524E+03	1.2394E-02	-1.0073E-05	283	724	1.822	0.031
1201	C14H8O2	ANTHRAQUINONE	-----	-------	-------	-------	----	----	-----	-----
1202	C14H10	ANTHRACENE	-2.5527	7.8859E+02	3.1219E-03	-3.0878E-06	489	873	-----	0.053
1203	C14H10	DIPHENYLACETYLENE	-7.2174	1.6312E+03	1.0668E-02	-7.0635E-06	336	832	-----	0.054
1204	C14H10	PHENANTHRENE	-8.0216	1.8233E+03	1.2068E-02	-7.7480E-06	372	869	-----	0.052
1205	C14H12	cis-STILBENE	-----	-------	-------	-------	----	----	-----	-----
1206	C14H12	trans-STILBENE	-----	-------	-------	-------	----	----	-----	-----
1207	C14H12O2	BENZYL BENZOATE	-13.0757	2.6140E+03	2.1898E-02	-1.3920E-05	293	820	9.620	0.051
1208	C14H14	1,1-DIPHENYLETHANE	-8.6891	1.7683E+03	1.4252E-02	-9.8599E-06	265	775	4.117	0.052
1209	C14H14	1,2-DIPHENYLETHANE	-9.8294	1.9762E+03	1.6301E-02	-1.1032E-05	324	780	-----	0.051
1210	C14H14O	DIBENZYL ETHER	-14.3261	2.6523E+03	2.5534E-02	-1.6954E-05	277	777	4.738	0.049
1211	C14H16	1-n-BUTYLNAPHTHALENE	-9.3474	1.9174E+03	1.5182E-02	-1.0166E-05	275	792	5.086	0.053
1212	C14H16	2-BUTYLNAPHTHALENE	-9.4131	1.9174E+03	1.5182E-02	-1.0166E-05	275	780	4.372	0.050
1213	C14H22	n-OCTYLBENZENE	-9.0954	1.6429E+03	1.7317E-02	-1.3531E-05	230	729	2.372	0.039
1214	C14H22	1,2,3,4-TETRAETHYLBENZENE	-9.1259	1.6429E+03	1.7317E-02	-1.3531E-05	285	708	2.211	0.047
1215	C14H22	1,2,3,5-TETRAETHYLBENZENE	-9.1283	1.6429E+03	1.7317E-02	-1.3531E-05	285	707	2.199	0.047
1216	C14H22	1,2,4,5-TETRAETHYLBENZENE	-9.1307	1.6429E+03	1.7317E-02	-1.3531E-05	284	706	2.187	0.048
1217	C14H22O	p-tert-OCTYLPHENOL	-33.4808	6.4935E+03	5.4349E-02	-3.0605E-05	359	765	-----	0.047
1218	C14H28	1-TETRADECENE	-6.7310	1.2666E+03	1.2388E-02	-1.0611E-05	260	692	1.851	0.039
1219	C14H28	1-CYCLOPENTYLNONANE	-6.7082	1.1713E+03	1.4798E-02	-1.3604E-05	264	650	2.649	0.092
1220	C14H28	1-CYCLOHEXYLOCTANE	-6.6460	1.1713E+03	1.4798E-02	-1.3604E-05	264	660	3.057	0.093
1221	C14H28O2	n-TETRADECANOIC ACID	-6.7310	1.2666E+03	1.2388E-02	-1.0611E-05	328	756	-----	0.018
1222	C14H30	n-TETRADECANE	-7.8717	1.4467E+03	1.4940E-02	-1.2495E-05	279	692	2.110	0.037
1223	C14H30O	1-TETRADECANOL	-20.2470	4.2845E+03	3.0943E-02	-1.7801E-05	311	741	-----	0.049
1224	C14H30S	BUTYL-DECYL-SULFIDE	-6.5969	1.3524E+03	1.2394E-02	-1.0073E-05	277	724	5.482	0.092
1225	C14H30S	DODECYL-ETHYL-SULFIDE	-6.5969	1.3524E+03	1.2394E-02	-1.0073E-05	277	724	5.482	0.092
1226	C14H30S	HEPTYL-SULFIDE	-6.5969	1.3524E+03	1.2394E-02	-1.0073E-05	277	724	5.482	0.092
1227	C14H30S	METHYL-TRIDECYL-SULFIDE	-6.5969	1.3524E+03	1.2394E-02	-1.0073E-05	277	724	5.482	0.092
1228	C14H30S	PROPYL-UNDECYL-SULFIDE	-6.5969	1.3524E+03	1.2394E-02	-1.0073E-05	277	724	5.482	0.092
1229	C14H30S	1-TETRADECANETHIOL	-7.0534	1.3524E+03	1.2394E-02	-1.0073E-05	280	724	1.916	0.032
1230	C14H30S2	HEPTYL-DISULFIDE	-7.2007	1.3524E+03	1.2394E-02	-1.0073E-05	272	724	1.365	0.023
1231	C14H31N	TETRADECYLAMINE	-7.7009	1.7421E+03	1.2714E-02	-1.0158E-05	311	722	-----	0.039
1232	C15H10N2O2	DIPHENYLMETHANE-4,4'-DIISOCYANATE	-----	-------	-------	-------	----	----	-----	-----
1233	C15H16O	p-CUMYLPHENOL	-26.9629	6.0753E+03	3.7190E-02	-1.8052E-05	346	834	-----	0.061
1234	C15H16O2	BISPHENOL A	-1.2831	1.1047E+03	-8.9645E-04	2.3595E-07	433	849	-----	0.267
1235	C15H18	1-PENTYLNAPHTHALENE	-9.3262	1.9174E+03	1.5182E-02	-1.0166E-05	275	792	5.341	0.055
1236	C15H18	2-PENTYLNAPHTHALENE	-9.3117	1.9174E+03	1.5182E-02	-1.0166E-05	275	792	5.522	0.057
1237	C15H24	n-NONYLBENZENE	-10.1069	1.8159E+03	1.9433E-02	-1.4862E-05	249	741	2.861	0.038
1238	C15H24O	2,6-Di-tert-BUTYL-p-CRESOL	-1.3289	7.1070E+02	2.2805E-04	-2.1941E-06	344	720	-----	0.048
1239	C15H24O	NONYLPHENOL	-44.4391	8.2245E+03	8.1683E-02	-5.2033E-05	279	757	748.864	0.028
1240	C15H28	1-PENTADECYNE	-8.8647	1.5853E+03	1.7951E-02	-1.4746E-05	284	708	3.117	0.049
1241	C15H30	1-PENTADECENE	-9.0123	1.5853E+03	1.7951E-02	-1.4746E-05	269	708	2.219	0.035
1242	C15H30	1-CYCLOPENTYLDECANE	-6.6248	1.1713E+03	1.4798E-02	-1.3604E-05	264	660	3.210	0.098
1243	C15H30	1-CYCLOHEXYLNONANE	-6.5596	1.1713E+03	1.4798E-02	-1.3604E-05	264	670	3.730	0.099
1244	C15H30O2	PENTADECANOIC ACID	-10.2431	2.2863E+03	1.6842E-02	-1.1988E-05	326	766	-----	0.041
1245	C15H32	n-PENTADECANE	-7.8643	1.4798E+03	1.4720E-02	-1.2148E-05	283	707	2.558	0.037
1246	C15H32O	1-PENTADECANOL	-----	-------	-------	-------	----	----	-----	-----
1247	C15H32S	BUTYL-UNDECYL-SULFIDE	-6.5197	1.3524E+03	1.2394E-02	-1.0073E-05	285	740	6.548	0.092
1248	C15H32S	DODECYL-PROPYL-SULFIDE	-6.5197	1.3524E+03	1.2394E-02	-1.0073E-05	285	740	6.548	0.092

Table 22-1 VISCOSITY OF LIQUID - ORGANIC COMPOUNDS (continued)

| NO | FORMULA | NAME | $\log_{10} n_{liq} = A + B/T + C\,T + C\,T^2$ | | | | | | $(n_{liq}$ - centipoise, T - K) | |
			A	B	C	D	TMIN	TMAX	n_{liq} @25 C	n_{liq} @TMAX
1249	C15H32S	ETHYL-TRIDECYL-SULFIDE	-6.5197	1.3524E+03	1.2394E-02	-1.0073E-05	285	740	6.548	0.092
1250	C15H32S	METHYL-TETRADECYL-SULFIDE	-6.5197	1.3524E+03	1.2394E-02	-1.0073E-05	285	740	6.548	0.092
1251	C15H32S	1-PENTADECANETHIOL	-6.9217	1.3524E+03	1.2394E-02	-1.0073E-05	291	740	2.595	0.036
1252	C16H10	FLUORANTHENE	-4.3644	1.2936E+03	5.6806E-03	-4.2783E-06	383	905	-----	0.050
1253	C16H10	PYRENE	-9.2925	2.4028E+03	1.2287E-02	-6.9222E-06	424	936	-----	0.051
1254	C16H12	1-PHENYLNAPHTHALENE	-2.6291	7.5395E+02	3.6472E-03	-3.6558E-06	318	849	-----	0.053
1255	C16H20	1-n-HEXYLNAPHTHALENE	-11.0297	2.2299E+03	1.8534E-02	-1.2297E-05	270	813	7.624	0.045
1256	C16H22O4	DIBUTYL PHTHALATE	-16.7220	3.2816E+03	2.9364E-02	-1.9331E-05	238	781	20.942	0.042
1257	C16H26	n-DECYLBENZENE	-9.6815	1.7942E+03	1.8161E-02	-1.3740E-05	259	753	3.385	0.039
1258	C16H26	PENTAETHYLBENZENE	-9.7411	1.7942E+03	1.8161E-02	-1.3740E-05	328	723	-----	0.049
1259	C16H30	1-HEXADECYNE	-6.5008	1.3190E+03	1.1821E-02	-9.8620E-06	289	722	3.723	0.052
1260	C16H32	n-DECYLCYCLOHEXANE	-8.9041	1.8268E+03	1.4648E-02	-1.0453E-05	271	751	4.583	0.043
1261	C16H32	1-CYCLOPENTYLUNDECANE	-8.9804	1.8268E+03	1.4648E-02	-1.0453E-05	271	743	3.844	0.039
1262	C16H32	1-HEXADECENE	-6.6407	1.3190E+03	1.1821E-02	-9.8620E-06	278	722	2.698	0.038
1263	C16H32O2	n-HEXADECANOIC ACID	-9.4484	2.0983E+03	1.6578E-02	-1.2551E-05	336	776	-----	0.036
1264	C16H34	n-HEXADECANE	-8.1894	1.5571E+03	1.5270E-02	-1.2371E-05	291	721	3.063	0.035
1265	C16H34O	DI-n-OCTYL ETHER	-7.0351	1.4420E+03	1.3198E-02	-1.1536E-05	266	707	5.139	0.037
1266	C16H34O	1-HEXADECANOL	-27.3347	5.3133E+03	4.6664E-02	-2.8638E-05	322	761	-----	0.037
1267	C16H34S	BUTYL-DODECYL-SULFIDE	-6.4473	1.3524E+03	1.2394E-02	-1.0073E-05	289	750	7.736	0.097
1268	C16H34S	ETHYL-TETRADECYL-SULFIDE	-6.4473	1.3524E+03	1.2394E-02	-1.0073E-05	289	750	7.736	0.097
1269	C16H34S	METHYL-PENTADECYL-SULFIDE	-6.4473	1.3524E+03	1.2394E-02	-1.0073E-05	289	750	7.736	0.097
1270	C16H34S	OCTYL-SULFIDE	-6.4473	1.3524E+03	1.2394E-02	-1.0073E-05	289	750	7.736	0.097
1271	C16H34S	PROPYL-TRIDECYL-SULFIDE	-6.4473	1.3524E+03	1.2394E-02	-1.0073E-05	289	750	7.736	0.097
1272	C16H34S	1-HEXADECANETHIOL	-6.8739	1.3524E+03	1.2394E-02	-1.0073E-05	291	750	2.897	0.036
1273	C16H34S2	OCTYL-DISULFIDE	-7.1340	1.3524E+03	1.2394E-02	-1.0073E-05	285	724	1.592	0.027
1274	C17H28	n-UNDECYLBENZENE	-10.2163	1.9150E+03	1.8950E-02	-1.4024E-05	268	764	4.073	0.038
1275	C17H32	1-HEPTADECYNE	-9.6382	1.7675E+03	1.9082E-02	-1.5029E-05	296	736	4.399	0.046
1276	C17H34	1-CYCLOPENTYLDODECANE	-8.9066	1.8268E+03	1.4648E-02	-1.0453E-05	271	743	4.556	0.046
1277	C17H34	1-CYCLOHEXYLUNDECANE	-8.8410	1.8268E+03	1.4648E-02	-1.0453E-05	279	743	5.299	0.054
1278	C17H34	1-HEPTADECENE	-9.7772	1.7675E+03	1.9082E-02	-1.5029E-05	284	736	3.194	0.034
1279	C17H36	n-HEPTADECANE	-8.1307	1.5791E+04	1.4949E-02	-1.1987E-05	295	733	3.607	0.035
1280	C17H36O	1-HEPTADECANOL	-21.9653	4.8551E+03	3.1884E-02	-1.7191E-05	327	770	-----	0.050
1281	C17H36S	BUTYL-TRIDECYL-SULFIDE	-6.3797	1.3524E+03	1.2394E-02	-1.0073E-05	295	770	9.039	0.089
1282	C17H36S	ETHYL-PENTADECYL-SULFIDE	-6.3797	1.3524E+03	1.2394E-02	-1.0073E-05	295	770	9.039	0.089
1283	C17H36S	HEXADECYL-METHYL-SULFIDE	-6.3797	1.3524E+03	1.2394E-02	-1.0073E-05	295	770	9.039	0.089
1284	C17H36S	PROPYL-TETRADECYL-SULFIDE	-6.3797	1.3524E+03	1.2394E-02	-1.0073E-05	295	770	9.039	0.089
1285	C17H36S	1-HEPTADECANETHIOL	-----	-------	-------	-------	----	----	-----	-----
1286	C18H12	CHRYSENE	-43.1838	1.1296E+04	5.3288E-02	-2.2743E-05	589	979	-----	0.053
1287	C18H14	m-TERPHENYL	-11.2713	2.7227E+03	1.5526E-02	-8.3974E-06	373	925	-----	0.071
1288	C18H14	o-TERPHENYL	-13.2893	3.2056E+03	1.7962E-02	-9.3402E-06	373	891	-----	0.079
1289	C18H14	p-TERPHENYL	-22.3505	5.2148E+03	3.1682E-02	-1.6127E-05	498	926	-----	0.062
1290	C18H15P	TRIPHENYLPHOSPHINE	-----	-------	-------	-------	----	----	-----	-----
1291	C18H15O4P	TRIPHENYL PHOSPHATE	-27.1995	5.1831E+03	4.6451E-02	-2.4354E-05	323	400	-----	2.767
1292	C18H16N2	N,N'-DIPHENYL-p-PHENYLENEDIAMINE	-15.8283	4.2642E+03	2.0363E-02	-1.0448E-05	409	906	-----	0.056
1293	C18H22	2,3-DIMETHYL-2,3-DIPHENYLBUTANE	2.4742	2.9167E+01	-6.4163E-03	2.0758E-06	392	805	-----	0.049
1294	C18H22O2	DICUMYL PEROXIDE	-25.4883	4.6893E+03	4.6096E-02	-2.8240E-05	311	884	-----	0.031
1295	C18H30	n-DODECYLBENZENE	-8.6157	1.7664E+03	1.4453E-02	-1.0385E-05	283	774	4.953	0.043
1296	C18H30	HEXAETHYLBENZENE	-8.7390	1.7664E+03	1.4453E-02	-1.0385E-05	402	734	-----	0.048
1297	C18H32O2	LINOLEIC ACID	-2.5389	1.2657E+03	-2.6625E-04	-4.7215E-07	268	775	38.453	0.040
1298	C18H34	1-OCTADECYNE	-6.8855	1.4654E+03	1.1707E-02	-9.0997E-06	301	747	-----	0.055
1299	C18H34O2	OLEIC ACID	-6.1303	1.6893E+03	8.3740E-03	-6.4505E-06	293	781	28.770	0.043
1300	C18H34O4	DIBUTYL SEBACATE	-9.2401	1.9231E+03	1.5798E-02	-1.1453E-05	264	768	7.982	0.044
1301	C18H34O4	DIHEXYL ADIPATE	-8.7851	1.8614E+03	1.3867E-02	-9.5074E-06	259	767	5.589	0.048
1302	C18H36	1-CYCOPENTYLTRIDECANE	-8.8366	1.8268E+03	1.4648E-02	-1.0453E-05	279	743	5.353	0.054
1303	C18H36	1-CYCLOHEXYLDODECANE	-8.7705	1.8268E+03	1.4648E-02	-1.0453E-05	286	743	6.233	0.063
1304	C18H36	1-OCTADECENE	-7.0295	1.4654E+03	1.1707E-02	-9.0997E-06	291	748	3.690	0.039
1305	C18H36O2	STEARIC ACID	-3.5929	1.3465E+03	2.9104E-03	-2.7617E-06	343	799	-----	0.045
1306	C18H38	n-OCTADECANE	-8.5505	1.6708E+03	1.5675E-02	-1.2341E-05	301	745	-----	0.033
1307	C18H38O	DINONYL ETHER	-6.7960	1.6057E+03	1.0044E-02	-7.6457E-06	297	736	8.026	0.043
1308	C18H38O	1-OCTADECANOL	-22.0941	4.9629E+03	3.1506E-02	-1.6680E-05	331	777	-----	0.050
1309	C18H38S	BUTYL-TETRADECYL-SULFIDE	-6.3122	1.3524E+03	1.2394E-02	-1.0073E-05	300	780	-----	0.091
1310	C18H38S	ETHYL-HEXADECYL-SULFIDE	-6.3122	1.3524E+03	1.2394E-02	-1.0073E-05	300	780	-----	0.091
1311	C18H38S	HEPTADECYL-METHYL-SULFIDE	-6.3122	1.3524E+03	1.2394E-02	-1.0073E-05	300	780	-----	0.091
1312	C18H38S	NONYL-SULFIDE	-6.3122	1.3524E+03	1.2394E-02	-1.0073E-05	300	780	-----	0.091
1313	C18H38S	PENTADECYL-PROPYL-SULFIDE	-6.3122	1.3524E+03	1.2394E-02	-1.0073E-05	300	780	-----	0.091
1314	C18H38S	1-OCTADECANETHIOL	-----	-------	-------	-------	----	----	-----	-----
1315	C18H38S2	NONYL-DISULFIDE	-7.0601	1.3524E+03	1.2394E-02	-1.0073E-05	266	724	1.887	0.032
1316	C19H26	1-n-NONYLNAPHTHALENE	-5.5160	1.3030E+03	7.9138E-03	-5.6525E-06	284	849	5.144	0.046
1317	C19H32	n-TRIDECYLBENZENE	-9.0086	1.8273E+03	1.5565E-02	-1.1269E-05	283	783	5.743	0.040
1318	C19H36	1-NONADECYNE	-10.6099	1.9747E+03	2.0654E-02	-1.5668E-05	307	758	-----	0.045
1319	C19H36O2	METHYL OLEATE	-11.3485	2.3310E+03	1.8224E-02	-1.1939E-05	293	764	6.949	0.045
1320	C19H38	1-CYCLOPENTYLTETRADECANE	-8.7710	1.8268E+03	1.4648E-02	-1.0453E-05	286	743	6.226	0.063
1321	C19H38	1-CYCLOHEXYLTRIDECANE	-8.7039	1.8268E+03	1.4648E-02	-1.0453E-05	292	743	7.266	0.074
1322	C19H38	1-NONADECENE	-10.7334	1.9747E+03	2.0654E-02	-1.5668E-05	297	760	4.518	0.033
1323	C19H38O2	NONADECANOIC ACID	-10.6370	2.6099E+03	1.5791E-02	-1.0329E-05	341	810	-----	0.040
1324	C19H40	n-NONADECANE	-8.4648	1.6797E+03	1.5428E-02	-1.2108E-05	305	756	-----	0.032
1325	C19H40O	1-NONADECANOL	-----	-------	-------	-------	----	----	-----	-----
1326	C19H40S	BUTYL-PENTADECYL-SULFIDE	-6.2494	1.3524E+03	1.2394E-02	-1.0073E-05	304	790	-----	0.093

Table 22-1 VISCOSITY OF LIQUID - ORGANIC COMPOUNDS (continued)

NO	FORMULA	NAME	$\log_{10} n_{liq} = A + B/T + C\,T + C\,T^2$						$(n_{liq}$ - centipoise, T - K)$	
			A	B	C	D	TMIN	TMAX	n_{liq} @25 C	n_{liq} @TMAX
1327	C19H40S	ETHYL-HEPTADECYL-SULFIDE	-6.2494	1.3524E+03	1.2394E-02	-1.0073E-05	304	790	-----	0.093
1328	C19H40S	HEXADECYL-PROPYL-SULFIDE	-6.2494	1.3524E+03	1.2394E-02	-1.0073E-05	304	790	-----	0.093
1329	C19H40S	METHYL-OCTADECYL-SULFIDE	-6.2494	1.3524E+03	1.2394E-02	-1.0073E-05	304	790	-----	0.093
1330	C19H40S	1-NONADECANETHIOL	-----	-------	-------	-------	----	----	-----	-----
1331	C20H16	TRIPHENYLETHYLENE	-5.9039	1.6667E+03	7.5221E-03	-4.9183E-06	342	908	-----	0.051
1332	C20H28	1-n-DECYLNAPHTHALENE	-8.7454	2.0201E+03	1.3015E-02	-8.3498E-06	288	859	14.731	0.042
1333	C20H30O2	ABIETIC ACID	-37.4215	8.3940E+03	5.3218E-02	-2.6322E-05	447	832	-----	0.053
1334	C20H31N	DEHYDROABIETYLAMINE	-6.1760	1.6565E+03	8.5784E-03	-6.0199E-06	318	863	-----	0.046
1335	C20H34	1-PHENYLTETRADECANE	-8.9260	1.8273E+03	1.5565E-02	-1.1269E-05	290	730	6.946	0.086
1336	C20H38	1-EICOSYNE	-9.4932	1.8835E+03	1.7307E-02	-1.2860E-05	310	769	-----	0.046
1337	C20H40	1-CYCLOPENTYLPENTADECANE	-8.7090	1.8268E+03	1.4648E-02	-1.0453E-05	292	743	7.182	0.073
1338	C20H40	1-CYCLOHEXYLTETRADECANE	-8.6436	1.8268E+03	1.4648E-02	-1.0453E-05	298	743	8.349	0.085
1339	C20H40	1-EICOSENE	-9.5998	1.8835E+03	1.7307E-02	-1.2860E-05	302	771	-----	0.035
1340	C20H42	n-EICOSANE	-9.2095	1.8221E+03	1.6798E-02	-1.2861E-05	310	767	-----	0.030
1341	C20H42O	1-EICOSANOL	-21.2803	4.9924E+03	2.8849E-02	-1.4606E-05	339	792	-----	0.051
1342	C20H42S	BUTYL-HEXADECYL-SULFIDE	-6.1915	1.3524E+03	1.2394E-02	-1.0073E-05	310	800	-----	0.093
1343	C20H42S	DECYL-SULFIDE	-6.1915	1.3524E+03	1.2394E-02	-1.0073E-05	310	800	-----	0.093
1344	C20H42S	ETHYL-OCTADECYL-SULFIDE	-6.1915	1.3524E+03	1.2394E-02	-1.0073E-05	310	800	-----	0.093
1345	C20H42S	HEPTADECYL-PROPYL-SULFIDE	-6.1915	1.3524E+03	1.2394E-02	-1.0073E-05	310	800	-----	0.093
1346	C20H42S	METHYL-NONADECYL-SULFIDE	-6.1915	1.3524E+03	1.2394E-02	-1.0073E-05	310	800	-----	0.093
1347	C20H42S	1-EICOSANETHIOL	-----	-------	-------	-------	----	----	-----	-----
1348	C20H42S2	DECYL-DISULFIDE	-6.9834	1.3524E+03	1.2394E-02	-1.0073E-05	295	724	2.251	0.038
1349	C21H21O4P	TRI-o-CRESYL PHOSPHATE	-27.1995	5.1831E+03	4.6451E-02	-2.4354E-05	260	375	73.987	4.135
1350	C21H36	1-PHENYLPENTADECANE	-8.8681	1.8273E+03	1.5565E-02	-1.1269E-05	296	783	7.937	0.055
1351	C21H42	1-CYCLOPENTYLHEXADECANE	-8.6503	1.8268E+03	1.4648E-02	-1.0453E-05	296	743	8.221	0.083
1352	C21H42	1-CYCLOHEXYLPENTADECANE	-8.5809	1.8268E+03	1.4648E-02	-1.0453E-05	303	743	-----	0.098
1353	C22H38	1-PHENYLHEXADECANE	-8.8102	1.8273E+03	1.5565E-02	-1.1269E-05	302	783	-----	0.063
1354	C22H44	1-CYCLOHEXYLHEXADECANE	-8.5230	1.8268E+03	1.4648E-02	-1.0453E-05	307	760	-----	0.095
1355	C22H44O2	n-BUTYL STEARATE	-15.2751	2.9345E+03	2.6926E-02	-1.8079E-05	298	764	9.731	0.038
1356	C24H38O4	DIISOOCTYL PHTHALATE	-28.0228	5.3736E+03	4.7818E-02	-2.8343E-05	273	851	54.673	0.029
1357	C24H38O4	DIOCTYL PHTHALATE	-24.8674	4.8847E+03	4.2201E-02	-2.5219E-05	230	851	71.844	0.033
1358	C24H42O	DINONYLPHENOL	-16.9590	3.9003E+03	2.2397E-02	-1.0860E-05	350	886	-----	0.058
1359	C26H20	TETRAPHENYLETHYLENE	-6.0905	1.9939E+03	7.0976E-03	-4.3352E-06	496	996	-----	0.048
1360	C28H46O4	DIISODECYL PHTHALATE	-16.9629	3.7058E+03	2.6036E-02	-1.4866E-05	275	887	80.827	0.041

n_{liq} - viscosity of liquid, centipoise

A, B, and C - regression coefficients for chemical compound

T - temperature, K

TMIN - minimum temperature, K

TMAX - maximum temperature, K

Table 22-2 VISCOSITY OF LIQUID - INORGANIC COMPOUNDS

NO	FORMULA	NAME	$\log_{10} n_{liq} = A + B/T + C\,T + C\,T^2$						n_{liq} @25 C	n_{liq} @TMAX
			A	B	C	D	TMIN	TMAX		
1	Ag	SILVER	1.5885	-3.7613E+01	-8.6402E-04	7.9672E-08	1234	2485	-----	0.828
2	AgCl	SILVER CHLORIDE	-0.5912	6.6508E+02	2.3019E-16	-6.4297E-20	728	1837	-----	0.590
3	AgI	SILVER IODIDE	-0.5214	8.4847E+02	-3.5867E-16	9.1142E-20	825	1779	-----	0.903
4	Al	ALUMINUM	-0.5859	6.8230E+02	-4.9031E-06	1.0432E-09	933	2306	-----	0.506
5	AlB3H12	ALUMINUM BOROHYDRIDE	2.7856	-1.2911E+02	-1.0870E-02	6.6001E-06	209	488	0.499	0.061
6	AlBr3	ALUMINUN BROMIDE	5.4170	-3.1616E+02	-1.5599E-02	1.1700E-05	373	523	-----	0.715
7	AlCl3	ALUMINUM CHLORIDE	-4.2733	1.2552E+03	2.8802E-03	-9.5172E-07	461	549	-----	0.203
8	AlF3	ALUMINUM FLUORIDE	-1.6859	2.6370E+03	-1.0012E-12	2.1435E-16	1313	1810	-----	0.590
9	AlI3	ALUMINUM IODIDE	-4.0815	1.3653E+03	4.5037E-03	-2.2992E-06	473	673	-----	0.865
10	Al2O3	ALUMINUM OXIDE	---------	----------	----------	----------	----	-----	-----	-----
11	Al2S3O12	ALUMINUM SULFATE	---------	----------	----------	----------	----	----	-----	-----
12	Ar	ARGON	-13.4194	4.9102E+02	1.1657E-01	-3.8874E-04	84	150	-----	0.039
13	As	ARSENIC	-0.4125	8.0101E+02	3.0015E-11	-8.4101E-15	1090	1290	-----	1.616
14	AsBr3	ARSENIC TRIBROMIDE	-1.3540	3.3244E+02	1.8906E-03	-2.8457E-06	306	743	-----	0.085
15	AsCl3	ARSENIC TRICHLORIDE	2.3246	-1.2293E+02	-7.2888E-03	3.0636E-06	255	621	1.027	0.060
16	AsF3	ARSENIC TRIFLUORIDE	2.0024	1.3535E+02	-1.1178E-02	8.7640E-06	267	504	0.799	0.073
17	AsF5	ARSENIC PENTAFLUORIDE	-29.8696	3.0285E+03	9.6855E-02	-1.1466E-04	193	340	0.094	0.052
18	AsH3	ARSINE	-1.1627	2.6699E+02	-4.4640E-03	5.1834E-06	165	354	0.073	0.046
19	AsI3	ARSENIC TRIIODIDE	-1.1284	6.0802E+02	3.6047E-14	-2.1883E-17	419	676	-----	0.590
20	As2O3	ARSENIC TRIOXIDE	-1.2657	9.3043E+02	-2.9644E-12	1.4442E-15	586	786	-----	0.828
21	At	ASTATINE	---------	----------	----------	----------	----	----	-----	-----
22	Au	GOLD	-135.7401	6.6888E+04	9.3033E-02	-2.1218E-05	1373	1637	-----	3.595
23	B	BORON	-0.0531	1.3341E+03	-1.0422E-14	1.0838E-18	2348	4133	-----	1.861
24	BBr3	BORON TRIBROMIDE	2.1355	-1.1170E+02	-7.0872E-03	2.5614E-06	228	552	0.751	0.063
25	BCl3	BORON TRICHLORIDE	0.7902	1.2104E+01	-4.4149E-03	-1.8390E-06	166	429	0.224	0.039
26	BF3	BORON TRIFLUORIDE	-4.1942	5.7368E+02	3.3621E-03	-5.2236E-06	146	248	-----	0.043
27	BH2CO	BORINE CARBONYL	3.4997	-1.7416E+02	-1.9810E-02	1.8470E-05	136	323	0.045	0.031
28	BH3O3	BORIC ACID	---------	----------	----------	----------	----	----	-----	-----
29	B2D6	DEUTERODIBORANE	5.2180	-3.0926E+02	-3.3123E-02	4.4076E-05	130	279	-----	0.020
30	B2H5Br	DIBORANE HYDROBROMIDE	1.5099	-5.2076E+01	-6.6170E-03	7.0141E-07	169	444	0.266	0.039
31	B2H6	DIBORANE	0.3145	3.5890E+01	-8.5233E-03	2.8667E-06	108	275	-----	0.021
32	B3N3H6	BORINE TRIAMINE	3.8496	-3.0260E+02	-1.3885E-02	8.6668E-06	215	495	0.292	0.031
33	B4H10	TETRABORANE	0.0398	2.0065E+00	-1.6278E-03	-4.7305E-06	153	443	0.138	0.025
34	B5H9	PENTABORANE	3.8805	-2.8383E+02	-1.4311E-02	9.8515E-06	226	540	0.345	0.032
35	B5H11	TETRAHYDROPENTABORANE	16.0089	-1.8521E+03	-4.4671E-02	3.4334E-05	290	520	0.339	0.032
36	B10H14	DECABORANE	-1.9377	8.4239E+02	4.5194E-13	-3.1881E-16	373	573	-----	0.341
37	Ba	BARIUM	-0.5000	7.4760E+02	5.4990E-15	-1.2786E-18	1000	1907	-----	0.780
38	Be	BERYLLIUM	-0.2959	1.0406E+03	-2.1490E-14	3.3682E-18	1560	2744	-----	1.212
39	BeB2H8	BERYLLIUM BOROHYDRIDE	-1.8042	8.4239E+02	-2.3406E-12	1.5801E-15	396	596	-----	0.407
40	BeBr2	BERYLLIUM BROMIDE	-0.8970	9.3043E+02	9.4994E-12	-3.6780E-15	763	963	-----	1.173
41	BeCl2	BERYLLIUM CHLORIDE	-1.0498	9.3043E+02	-2.6120E-12	1.1231E-15	678	878	-----	1.023
42	BeF2	BERYLLIUM FLUORIDE	---------	----------	----------	----------	----	----	-----	-----
43	BeI2	BERYLLIUM IODIDE	-0.9002	9.3043E+02	4.4768E-12	-1.7369E-15	761	961	-----	1.169
44	Bi	BISMUTH	-2.2282	8.6699E+02	2.0642E-03	-7.1676E-07	535	981	-----	0.979
45	BiBr3	BISMUTH TRIBROMIDE	-1.3436	8.1815E+02	3.9014E-15	-1.5928E-18	491	734	-----	0.590
46	BiCl3	BISMUTH TRICHLORIDE	-1.5439	9.3896E+02	-2.3639E-12	1.3000E-15	503	714	-----	0.590
47	BrF5	BROMINE PENTAFLUORIDE	2.6826	-1.2405E+02	-9.9355E-03	4.1913E-06	212	447	0.475	0.063
48	Br2	BROMINE	-1.4000	3.8750E+02	4.1570E-04	-5.2140E-07	266	588	0.949	0.210
49	C	CARBON	---------	----------	----------	----------	----	----	-----	-----
50	CCl2O	PHOSGENE	28.4916	-2.8631E+03	-8.9606E-02	8.3415E-05	253	432	0.387	0.053
51	CF2O	CARBONYL FLUORIDE	-15.0660	1.4437E+03	5.0410E-02	-7.0750E-05	162	282	-----	0.044
52	CH4N2O	UREA	-1.7534	8.4239E+02	-8.2438E-13	5.4719E-16	406	606	-----	0.433
53	CH4N2S	THIOUREA	-1.5327	8.4239E+02	1.9332E-12	-1.1672E-15	454	654	-----	0.569
54	CNBr	CYANOGEN BROMIDE	-2.2216	8.4239E+02	1.6322E-13	-1.2632E-16	331	531	-----	0.232
55	CNCl	CYANOGEN CHLORIDE	-83.1777	1.0139E+04	2.2490E-01	-2.0923E-04	267	427	0.192	0.028
56	CNF	CYANOGEN FLUORIDE	5.9598	-2.8206E+02	-3.0561E-02	3.4093E-05	177	350	0.086	0.043
57	CO	CARBON MONOXIDE	-2.3460	1.0520E+02	4.6130E-03	-1.9640E-05	68	133	-----	0.051
58	COS	CARBONYL SULFIDE	2.4566	-9.1718E+01	-1.3880E-02	1.0731E-05	134	360	0.092	0.039
59	COSe	CARBON OXYSELENIDE	5.8909	-3.2369E+02	-2.6595E-02	2.7158E-05	201	386	0.195	0.068
60	CO2	CARBON DIOXIDE	-17.9151	1.4605E+03	7.3127E-02	-1.1230E-04	219	304	0.064	0.055
61	CS2	CARBON DISULFIDE	4.3002	-3.2854E+02	-1.5733E-02	1.1895E-05	235	524	0.367	0.050
62	CSeS	CARBON SELENOSULFIDE	0.4976	4.2639E+01	-2.2843E-03	-1.8981E-06	198	548	0.618	0.057
63	C2N2	CYANOGEN	-97.5433	1.0496E+04	2.9925E-01	-3.1328E-04	239	380	0.108	0.036
64	C3S2	CARBON SUBSULFIDE	---------	----------	----------	----------	----	----	-----	-----
65	Ca	CALCIUM	-170.0004	6.4164E+04	1.5131E-01	-4.5206E-05	1085	1173	-----	0.970
66	CaF2	CALCIUM FLUORIDE	-18.3286	1.7915E+04	5.4022E-03	-4.8473E-07	1720	2000	-----	0.312
67	CbF5	COLUMBIUM FLUORIDE	-2.0939	8.4239E+02	6.0639E-13	-4.5370E-16	349	549	-----	0.276
68	Cd	CADMIUM	-3.2376	9.8440E+02	4.8330E-03	-2.5217E-06	623	873	-----	1.539
69	CdCl2	CADMIUM CHLORIDE	41.0999	-1.1109E+04	-4.8207E-02	1.8470E-05	842	1005	-----	1.792
70	CdF2	CADMIUM FLUORIDE	-0.5844	7.1909E+02	8.1123E-16	-1.9891E-19	793	2024	-----	0.590
71	CdI2	CADMIUM IODIDE	-1.1121	9.4398E+02	-1.7991E-14	6.7885E-18	658	1069	-----	0.590
72	CdO	CADMIUM OXIDE	-1.0627	2.2605E+03	5.9983E-10	-1.1549E-13	1632	1832	-----	1.483
73	ClF	CHLORINE MONOFLUORIDE	3.1587	-2.5059E+01	-2.6369E-02	3.7470E-05	128	268	-----	0.049
74	ClFO3	PERCHLORYL FLUORIDE	-1.5292	2.7504E+02	-6.3016E-04	4.8326E-07	125	333	0.177	0.138
75	ClF3	CHLORINE TRIFLUORIDE	-18.4280	2.0326E+03	5.7151E-02	-6.5406E-05	290	321	0.412	0.324
76	ClF5	CHLORINE PENTAFLUORIDE	9.2179	-6.6358E+02	-3.6683E-02	3.6480E-05	210	395	0.199	0.055
77	ClHO3S	CHLOROSULFONIC ACID	-0.5472	1.2693E+02	1.0574E-03	-3.6166E-06	193	665	0.745	0.056
78	ClHO4	PERCHLORIC ACID	-1.9874	5.6067E+02	5.1727E-14	-5.9370E-17	200	385	0.782	0.294

Table 22-2 VISCOSITY OF LIQUID - INORGANIC COMPOUNDS (continued)

| NO | FORMULA | NAME | $\log_{10} n_{llq} = A + B/T + C\,T + C\,T^2$ | | | | $(n_{llq}$ - centipoise, T - K$)$ | | | |
			A	B	C	D	TMIN	TMAX	n_{llq} @25 C	n_{llq} @TMAX
79	ClO2	CHLORINE DIOXIDE	1.9595	8.7671E+01	-1.2417E-02	1.0822E-05	214	442	0.326	0.061
80	Cl2	CHLORINE	-0.7681	1.5140E+02	-8.0650E-04	4.0750E-07	172	417	0.343	0.214
81	Cl2O	CHLORINE MONOXIDE	0.9863	6.3169E+00	-5.1457E-03	-6.7430E-07	157	422	0.259	0.051
82	Cl2O7	CHLORINE HEPTOXIDE	-0.0996	8.9745E+01	-3.4213E-04	-4.0018E-06	182	537	0.554	0.054
83	Co	COBALT	-1.7980	2.8803E+03	6.0798E-04	-9.1452E-08	1723	2023	-----	3.030
84	CoCl2	COBALT CHLORIDE	-1.9938	2.3349E+03	-2.7926E-12	7.9857E-16	1008	1323	-----	0.590
85	CoNC3O4	COBALT NITROSYL TRICARBONYL	2.9661	-2.2022E+01	8.6248E-06	0.590	262	539	0.910	0.055
86	Cr	CHROMIUM	0.0920	1.2803E+03	-2.8151E-12	3.7478E-16	2180	2840	-----	3.490
87	CrC6O6	CHROMIUM CARBONYL	-1.6662	8.4239E+02	1.1481E-12	-7.3381E-16	424	624	-----	0.483
88	CrO2Cl2	CHROMIUM OXYCHLORIDE	-0.9174	1.5519E+02	2.3739E-03	-5.7504E-06	177	595	0.631	0.052
89	Cs	CESIUM	-0.8647	2.3202E+02	-2.2032E-04	3.8720E-08	302	1400	-----	0.117
90	CsBr	CESIUM BROMIDE	-0.9841	1.1876E+03	-1.0722E-14	2.8340E-18	909	1573	-----	0.590
91	CsCl	CESIUM CHLORIDE	-1.2528	1.2190E+03	1.0471E-13	-2.8321E-17	919	1573	-----	0.333
92	CsF	CESIUM FLUORIDE	-1.1573	1.4146E+03	1.2378E-13	-3.3464E-17	956	1524	-----	0.590
93	CsI	CESIUM IODIDE	-0.9773	1.1619E+03	4.8950E-14	-1.3453E-17	894	1553	-----	0.590
94	Cu	COPPER	-41.4454	2.3433E+04	2.5263E-02	-5.1354E-06	1438	1823	-----	2.492
95	CuBr	CUPROUS BROMIDE	-0.7327	8.1981E+02	1.0347E-14	-2.9195E-18	777	1628	-----	0.590
96	CuCl	CUPROUS CHLORIDE	-0.4443	6.4464E+02	-7.8736E-16	2.1800E-19	703	1763	-----	0.834
97	CuCl2	CUPRIC CHLORIDE	-1.6170	1.7572E+03	2.1141E-12	-6.5097E-16	906	1266	-----	0.590
98	CuI	COPPER IODIDE	-0.8915	1.0658E+03	7.1140E-14	-1.9322E-17	878	1609	-----	0.590
99	DCN	DEUTERIUM CYANIDE	-21.3016	3.0568E+03	4.8211E-02	-4.2161E-05	261	458	0.378	0.041
100	D2	DEUTERIUM	-23.2435	1.8361E+02	8.7699E-01	-1.2136E-02	19	26	-----	0.026
101	D2O	DEUTERIUM OXIDE	-3.4931	9.3628E+02	1.7670E-03	-1.1438E-06	277	612	1.181	0.049
102	Eu	EUROPIUM	-0.3423	8.0390E+02	5.2201E-14	-1.2352E-17	1095	1742	-----	1.316
103	F2	FLUORINE	-1.5760	8.5630E+01	-4.0730E-04	-2.7250E-06	54	145	-----	0.079
104	F2O	FLUORINE OXIDE	-0.9093	3.5507E+01	6.5834E-03	-5.0636E-05	49	204	-----	0.032
105	Fe	IRON	-0.5131	2.3416E+03	-7.7036E-15	1.0844E-18	1808	3000	-----	1.851
106	FeC5O5	IRON PENTACARBONYL	3.4808	-2.2186E+02	-1.1133E-02	5.9877E-06	252	577	0.891	0.046
107	FeCl2	FERROUS CHLORIDE	-1.7013	1.9125E+03	9.6262E-13	-2.8714E-16	945	1299	-----	0.590
108	FeCl3	FERRIC CHLORIDE	-1.2899	9.3043E+02	6.2336E-12	-3.0767E-15	577	777	-----	0.808
109	Fr	FRANCIUM	---------	----------	----------	----------	----	----	-----	-----
110	Ga	GALLIUM	-0.2008	1.7463E+02	-1.7838E-04	4.1968E-08	326	1373	-----	0.576
111	GaCl3	GALLIUM TRICHLORIDE	-1.3055	8.8743E+02	-3.5967E-03	2.3234E-06	340	659	-----	0.048
112	Gd	GADOLINIUM	-0.0747	1.0526E+03	-1.1233E-10	2.2313E-14	1587	1770	-----	3.311
113	Ge	GERMANIUM	2.2947	2.1761E+02	-3.5627E-03	1.1586E-06	1223	1473	-----	0.511
114	GeBr4	GERMANIUM BROMIDE	3.8266	-4.4201E+02	-8.4955E-03	3.2984E-06	299	703	-----	0.072
115	GeCl4	GERMANIUM CHLORIDE	2.2138	-9.0657E+01	-7.9188E-03	3.3082E-06	224	545	0.696	0.052
116	GeHCl3	TRICHLORO GERMANE	1.1233	-1.8915E+00	-4.5060E-03	-4.4242E-08	202	532	0.588	0.051
117	GeH4	GERMANE	1.7757	-4.0884E+01	-1.2479E-02	7.4301E-06	107	293	-----	0.041
118	Ge2H6	DIGERMANE	0.3472	4.8040E+01	-2.1103E-03	-3.6266E-06	164	466	0.360	0.048
119	Ge3H8	TRIGERMANE	-1.0331	1.5572E+02	2.8517E-03	-6.3370E-06	168	586	0.597	0.053
120	HBr	HYDROGEN BROMIDE	-9.2380	8.6670E+02	3.4320E-02	-5.1720E-05	186	363	0.201	0.062
121	HCN	HYDROGEN CYANIDE	23.1072	-2.3977E+03	-7.2644E-02	6.5607E-05	260	434	0.173	0.026
122	HCl	HYDROGEN CHLORIDE	-1.5150	1.9460E+02	3.0670E-03	-1.3760E-05	159	325	0.067	0.042
123	HF	HYDROGEN FLUORIDE	-6.0990	8.1600E+02	1.2920E-02	-1.3290E-05	190	461	0.204	0.064
124	HI	HYDROGEN IODIDE	-9.3730	1.0150E+03	3.1860E-02	-4.2200E-05	222	424	0.601	0.088
125	HNO3	NITRIC ACID	-3.5221	7.2948E+02	3.9634E-03	-2.2372E-06	240	356	0.808	0.451
126	H2	HYDROGEN	-7.0154	4.0791E+01	2.3714E-01	-4.0830E-03	14	33	-----	0.004
127	H2O	WATER	-10.2158	1.7925E+03	1.7730E-02	-1.2631E-05	273	643	0.911	0.056
128	H2O2	HYDROGEN PEROXIDE	-1.6150	5.0380E+02	5.3010E-04	-1.1680E-06	273	728	1.189	0.041
129	H2S	HYDROGEN SULFIDE	16.3303	-1.2250E+03	-6.8722E-02	8.0406E-05	190	355	0.076	0.041
130	H2SO4	SULFURIC ACID	-18.7045	3.4962E+03	3.3080E-02	-1.7018E-05	283	367	23.541	4.679
131	H2S2	HYDROGEN DISULFIDE	0.4177	4.8162E+01	-2.1390E-03	-2.7604E-06	183	515	0.497	0.048
132	H2Se	HYDROGEN SELENIDE	-36.7366	3.9462E+03	1.1122E-01	-1.1861E-04	209	391	0.131	0.051
133	H2Te	HYDROGEN TELLURIDE	-6.2197	1.0412E+03	1.0098E-02	-9.6551E-06	224	416	0.266	0.065
134	H3NO3S	SULFAMIC ACID	---------	----------	----------	----------	----	----	-----	-----
135	He	HELIUM-3	-3.4013	4.9866E-01	3.5450E-01	-6.7520E-02	1	3	-----	0.0017
136	He	HELIUM-4	-2.5419	-8.2540E-02	1.1831E-01	-2.4514E-02	2	5	-----	0.0026
137	Hf	HAFNIUM	0.2485	1.3811E+03	4.4735E-16	-3.6010E-20	2506	5960	-----	3.022
138	Hg	MERCURY	-0.2748	1.3697E+02	4.1785E-06	-1.1995E-09	234	1073	1.534	0.718
139	HgBr2	MERCURIC BROMIDE	-1.5016	9.3043E+02	4.2416E-12	-2.3266E-15	510	710	-----	0.644
140	HgCl2	MERCURIC CHLORIDE	-1.4201	9.3043E+02	-3.3310E-13	1.7316E-16	550	750	-----	0.661
141	HgI2	MERCURIC IODIDE	-1.4262	9.3043E+02	-1.5854E-12	8.3978E-16	532	732	-----	0.700
142	IF7	IODINE HEPTAFLUORIDE	-17.3376	2.4977E+03	4.3864E-02	-4.6953E-05	279	424	0.879	0.051
143	I2	IODINE	-0.9048	5.1900E+02	-1.9830E-04	4.6760E-08	387	819	-----	0.396
144	In	INDIUM	-1.0541	3.9062E+02	1.3601E-03	-9.1852E-07	432	723	-----	0.976
145	Ir	IRIDIUM	0.4304	1.4396E+03	2.0882E-14	-1.9538E-18	2719	4450	-----	5.674
146	K	POTASSIUM	-0.9780	2.5961E+02	-2.0679E-04	3.3831E-08	340	2000	-----	0.075
147	KBr	POTASSIUM BROMIDE	-1.2966	1.4028E+03	4.0885E-14	-1.0358E-17	1003	1656	-----	0.355
148	KCl	POTASSIUM CHLORIDE	-1.2073	1.2595E+03	4.1889E-05	-5.0883E-09	1070	1710	-----	0.385
149	KF	POTASSIUM FLUORIDE	-1.2513	1.8146E+03	-4.1237E-13	9.4412E-17	1153	1775	-----	0.590
150	KI	POTASSIUM IODIDE	-1.1430	1.4596E+03	-7.8064E-14	2.0159E-17	996	1597	-----	0.590
151	KOH	POTASSIUM HYDROXIDE	-0.0007	7.8935E+02	-1.3943E-03	2.8348E-07	679	973	-----	0.528
152	Kr	KRYPTON	4.5330	5.2335E+01	-6.5203E-02	1.6402E-04	116	199	-----	0.021
153	La	LANTHANUM	-0.2571	8.5889E+02	-3.6388E-16	5.2455E-20	1193	3643	-----	0.952
154	Li	LITHIUM	-0.6442	2.3326E+02	-2.0400E-04	2.1555E-08	460	3400	-----	0.095
155	LiBr	LITHIUM BROMIDE	-0.9195	9.3828E+02	1.2409E-14	-3.5057E-18	820	1583	-----	0.471
156	LiCl	LITHIUM CHLORIDE	-1.2463	1.0542E+03	-6.3253E-16	1.6586E-19	887	1655	-----	0.246

Table 22-2 VISCOSITY OF LIQUID - INORGANIC COMPOUNDS (continued)

NO	FORMULA	NAME	$\log_{10} n_{liq} = A + B/T + C\,T + C\,T^2$						$(n_{liq}$ - centipoise, T - K$)$	
			A	B	C	D	TMIN	TMAX	n_{liq} @25 C	n_{liq} @TMAX
157	LiF	LITHIUM FLUORIDE	-1.0063	1.5187E+03	-2.7959E-14	6.0617E-18	1143	1954	-----	0.590
158	LiI	LITHIUM IODIDE	-0.7761	7.8983E+02	-2.3974E-15	7.3984E-19	719	1444	-----	0.590
159	Lu	LUTECIUM	0.0706	1.1944E+03	-1.2657E-12	1.8912E-16	1936	2535	-----	3.481
160	Mg	MAGNESIUM	9.0202	-2.5686E+03	-9.1580E-03	2.7173E-06	973	1173	-----	0.671
161	MgCl2	MAGNESIUM CHLORIDE	-0.9985	1.3011E+03	6.9688E-14	-1.7500E-17	985	1691	-----	0.590
162	MgO	MAGNESIUM OXIDE	---------	----------	----------	----------	----	----	-----	-----
163	Mn	MANGANESE	-0.0772	1.0220E+03	-2.9541E-14	5.0678E-18	1519	2392	-----	2.239
164	MnCl2	MANGANESE CHLORIDE	-1.1717	1.3791E+03	3.8518E-13	-1.0844E-16	923	1463	-----	0.590
165	Mo	MOLYBDENUM	0.2903	1.4841E+03	1.1277E-14	-9.5257E-19	2895	5081	-----	3.823
166	MoF6	MOLYBDENUM FLUORIDE	-84.6129	1.1248E+04	2.1014E-01	-1.7788E-04	290	473	0.899	0.058
167	MoO3	MOLYBDENUM OXIDE	-1.8835	2.3560E+03	1.1319E-12	-3.0228E-16	1068	1424	-----	0.590
168	NCl3	NITROGEN TRICHLORIDE	3.1175	-8.5075E+01	-1.2499E-02	8.6126E-06	246	536	0.743	0.054
169	ND3	HEAVY AMMONIA	-4.2790	7.4777E+02	3.7842E-03	-4.0909E-06	199	369	0.099	0.039
170	NF3	NITROGEN TRIFLUORIDE	-0.6966	4.2367E+01	4.4355E-03	-3.9292E-05	66	222	-----	0.035
171	NH3	AMMONIA	-8.5910	8.7640E+02	2.6810E-02	-3.6120E-05	195	406	0.135	0.032
172	NH3O	HYDROXYLAMINE	1.5212	3.3060E+02	-9.8009E-03	6.7070E-06	306	545	-----	0.060
173	NH4Br	AMMONIUM BROMIDE	-1.4880	8.4239E+02	-1.3278E-13	7.5834E-17	469	669	-----	0.590
174	NH4Cl	AMMONIUM CHLORIDE	-0.8508	9.3043E+02	-4.7565E-12	1.7769E-15	793	993	-----	1.220
175	NH4I	AMMONIUM IODIDE	-1.4399	8.4239E+02	7.9273E-14	-4.3778E-17	478	678	-----	0.635
176	NH5O	AMMONIUM HYDROXIDE	---------	----------	----------	----------	----	----	-----	-----
177	NH5S	AMMONIUM HYDROGENSULFIDE	-1.8314	8.4239E+02	9.8094E-13	-6.6686E-16	391	591	-----	0.393
178	NO	NITRIC OXIDE	-6.8816	3.2461E+02	5.0147E-02	-1.6563E-04	110	180	-----	0.038
179	NOCl	NITROSYL CHLORIDE	-2.5251	6.2166E+02	-8.9116E-04	8.4659E-07	214	419	0.234	0.054
180	NOF	NITROSYL FLUORIDE	2.5313	-6.8091E+01	-1.4952E-02	1.2600E-05	139	335	0.092	0.054
181	NO2	NITROGEN DIOXIDE	-8.4310	9.3260E+02	2.7590E-02	-3.7540E-05	262	431	0.385	0.045
182	N2	NITROGEN	-15.6104	4.6505E+02	1.6259E-01	-6.3353E-04	63	125	-----	0.034
183	N2F4	TETRAFLUOROHYDRAZINE	0.0995	3.0613E+01	-9.7653E-04	-1.5585E-05	112	294	-----	0.037
184	N2H4	HYDRAZINE	-8.0240	1.2990E+03	1.6110E-02	-1.3300E-05	275	653	0.899	0.065
185	N2H4C	AMMONIUM CYANIDE	12.3268	-1.1944E+03	-3.1178E-02	1.4489E-05	309	465	-----	0.025
186	N2H6CO2	AMMONIUM CARBAMATE	15.4658	-1.7205E+03	-4.4131E-02	3.4552E-05	281	513	0.406	0.037
187	N2O	NITROUS OXIDE	0.4733	2.1800E+01	-4.9230E-03	-4.2740E-06	171	310	0.050	0.040
188	N2O3	NITROGEN TRIOXIDE	1.4767	-2.6041E+01	-6.6284E-03	-3.5983E-07	170	404	0.240	0.047
189	N2O4	NITROGEN TETRAOXIDE	-24.1632	2.4294E+03	8.2309E-02	-1.0044E-04	253	293	-----	0.419
190	N2O5	NITROGEN PENTOXIDE	-102.8008	1.4106E+04	2.4736E-01	-2.0203E-04	303	490	-----	0.049
191	Na	SODIUM	-0.8963	2.9461E+02	-1.6338E-04	2.3166E-08	380	2300	-----	0.095
192	NaBr	SODIUM BROMIDE	-0.9551	1.1218E+03	8.8113E-11	-2.5942E-14	1054	1213	-----	0.933
193	NaCN	SODIUM CYANIDE	-0.7241	8.7558E+02	-4.4033E-15	1.1455E-18	837	1769	-----	0.590
194	NaCl	SODIUM CHLORIDE	-0.9169	1.0789E+03	-7.6231E-05	1.1105E-08	1080	1210	-----	0.792
195	NaF	SODIUM FLUORIDE	-1.5525	2.2477E+03	6.6921E-05	-6.7942E-09	1269	2060	-----	0.444
196	NaI	SODIUM IODIDE	-1.0095	1.2307E+03	2.5450E-14	-6.8176E-18	924	1577	-----	0.590
197	NaOH	SODIUM HYDROXIDE	-4.1939	2.0515E+03	2.7917E-03	-6.1590E-07	623	900	-----	1.257
198	Na2SO4	SODIUM SULFATE	11.2905	-4.5769E+03	-6.7848E-03	9.2443E-07	1180	1267	-----	3.679
199	Nb	NIOBIUM	0.2170	1.4477E+03	-2.5440E-16	2.0408E-20	2750	5115	-----	3.162
200	Nd	NEODYMIUM	-0.1910	9.0993E+02	-1.6741E-16	2.4602E-20	1289	3384	-----	1.196
201	Ne	NEON	-12.8696	1.4174E+02	3.5221E-01	-3.8226E-03	25	44	-----	0.028
202	Ni	NICKEL	60.6143	-3.4747E+04	-3.3857E-02	6.2505E-06	1727	2023	-----	3.357
203	NiC4O4	NICKEL CARBONYL	1.9163	1.6605E+02	-1.2282E-02	9.9219E-06	248	483	0.494	0.044
204	NiF2	NICKEL FLUORIDE	-3.5053	6.5954E+03	7.7331E-11	-1.3803E-14	1723	2013	-----	0.590
205	Np	NEPTUNIUM	-0.0479	6.9326E+02	3.8094E-17	-5.2755E-21	913	4175	-----	1.313
206	O2	OXYGEN	-5.0957	1.7983E+02	3.9779E-02	-1.4664E-04	54	150	-----	0.059
207	O3	OZONE	-1.1716	1.3645E+02	9.6397E-04	-1.7734E-05	80	248	-----	0.034
208	Os	OSMIUM	0.5276	1.5764E+03	1.0422E-14	-8.3841E-19	3306	4880	-----	7.090
209	OsOF5	OSMIUM OXIDE PENTAFLUORIDE	-2.2079	8.4239E+02	1.6199E-13	-1.2324E-16	333	533	-----	0.236
210	OsO4	OSMIUM TETROXIDE - YELLOW	-2.2371	8.4239E+02	-2.6198E-15	2.9102E-18	329	529	-----	0.227
211	OsO4	OSMIUM TETROXIDE - WHITE	2.7477	4.6966E+02	-1.0303E-02	6.4757E-06	315	624	-----	0.070
212	P	PHOSPHORUS - WHITE	2.9047	-1.2032E+02	-8.5106E-03	4.1988E-06	317	944	-----	0.031
213	PBr3	PHOSPHORUS TRIBROMIDE	-0.1249	7.6136E+01	4.3426E-04	-3.2230E-06	233	675	0.941	0.065
214	PCl2F3	PHOSPHORUS DICHLORIDE TRIFLUORIDE	-87.0377	1.0623E+04	2.3538E-01	-2.1737E-04	265	434	0.280	0.045
215	PCl3	PHOSPHORUS TRICHLORIDE	0.5598	8.0528E+01	-3.5561E-03	-3.4200E-07	200	535	0.549	0.051
216	PCl5	PHOSPHORUS PENTACHLORIDE	-1.6226	8.4239E+02	-2.8701E-12	1.8031E-15	433	633	-----	0.511
217	PH3	PHOSPHINE	4.3805	-2.0015E+02	-2.8762E-02	3.8322E-05	148	309	0.035	0.032
218	PH4Br	PHOSPHONIUM BROMIDE	13.6469	-1.3645E+03	-4.1988E-02	3.5118E-05	261	477	0.471	0.056
219	PH4Cl	PHOSPHONIUM CHLORIDE	-741.3525	6.9484E+04	2.6365E+00	-3.1381E-03	245	306	0.065	0.045
220	PH4I	PHOSPHONIUM IODIDE	-10.1513	1.8445E+03	1.7634E-02	-1.3774E-05	292	513	1.170	0.073
221	POCl3	PHOSPHORUS OXYCHLORIDE	9.4372	-1.0421E+03	-2.4294E-02	1.5352E-05	275	572	1.157	0.055
222	PSBr3	PHOSPHORUS THIOBROMIDE	8.0397	-1.1154E+03	-1.6755E-02	8.5008E-06	311	693	-----	0.080
223	PSCl3	PHOSPHORUS THIOCHLORIDE	1.1948	8.8775E+01	-5.1347E-03	1.3311E-06	242	607	1.202	0.052
224	P4O6	PHOSPHORUS TRIOXIDE	2.9013	-1.8887E+02	-8.0812E-03	3.5814E-06	296	679	1.502	0.061
225	P4O10	PHOSPHORUS PENTOXIDE	---------	----------	----------	----------	----	----	-----	-----
226	P4S10	PHOSPHORUS PENTASULFIDE	-1.5982	1.0776E+03	6.1092E-12	-3.0308E-15	561	787	-----	0.590
227	Pb	LEAD	1.4174	-3.5466E+01	-2.0867E-03	8.3570E-07	600	1173	-----	1.228
228	PbBr2	LEAD BROMIDE	-0.4266	7.8178E+02	8.4465E-14	-3.1055E-17	646	1187	-----	1.706
229	PbCl2	LEAD CHLORIDE	-0.8819	1.1563E+03	2.0992E-13	-7.0255E-17	774	1227	-----	1.149
230	PbF2	LEAD FLUORIDE	-1.6493	2.2242E+03	6.9535E-13	-1.7259E-16	1128	1566	-----	0.590
231	PbI2	LEAD IODIDE	-1.0212	9.0700E+02	-1.7090E-13	6.3172E-17	675	1145	-----	0.590
232	PbO	LEAD OXIDE	-1.3311	1.9230E+03	-2.6958E-13	6.1979E-17	1163	1745	-----	0.590
233	PbS	LEAD SULFIDE	-4.8090	7.1177E+03	-2.7695E-09	6.2797E-13	1387	1554	-----	0.591
234	Pd	PALLADIUM	0.0957	1.1530E+03	5.8668E-16	-7.8347E-20	1828	3385	-----	2.731

Table 22-2 VISCOSITY OF LIQUID - INORGANIC COMPOUNDS (continued)

			$\log_{10} n_{liq} = A + B/T + C\,T + C\,T^2$						(n_{liq} - centipoise, T - K)	
NO	FORMULA	NAME	A	B	C	D	TMIN	TMAX	n_{liq} @25 C	n_{liq} @TMAX
235	Po	POLONIUM	-0.4120	4.1562E+02	-4.3258E-15	1.6715E-18	527	1235	-----	0.840
236	Pt	PLATINUM	0.3130	1.2327E+03	4.3822E-15	-4.9221E-19	2042	3980	-----	4.195
237	Ra	RADIUM	-0.4276	7.3103E+02	-1.5347E-14	3.7306E-18	973	1809	-----	0.947
238	Rb	RUBIDIUM	-0.8363	2.0900E+02	-2.4619E-04	4.0545E-08	320	1900	-----	0.090
239	RbBr	RUBIDIUM BROMIDE	-1.1491	1.2774E+03	2.7266E-14	-7.1331E-18	955	1625	-----	0.434
240	RbCl	RUBIDIUM CHLORIDE	-1.1368	1.3532E+03	1.2391E-13	-3.1589E-17	988	1654	-----	0.480
241	RbF	RUBIDIUM FLUORIDE	-1.1082	1.4779E+03	-1.2003E-13	2.9778E-17	1033	1681	-----	0.590
242	RbI	RUBIDIUM IODIDE	-0.9914	1.2021E+03	-4.8992E-14	1.3203E-17	915	1577	-----	0.590
243	Re	RHENIUM	0.5334	1.6071E+03	6.0193E-15	-4.3283E-19	3459	5915	-----	6.384
244	Re2O7	RHENIUM HEPTOXIDE	-1.3126	9.3043E+02	3.0772E-12	-1.5393E-15	569	769	-----	0.789
245	Rh	RHODIUM	0.1996	1.2990E+03	3.1577E-14	-3.4477E-18	2237	3940	-----	3.383
246	Rn	RADON	-27.4776	2.7997E+03	8.8259E-02	-1.0196E-04	202	359	0.146	0.073
247	Ru	RUTHENIUM	0.2808	1.4095E+03	1.4204E-14	-1.3468E-18	2607	4500	-----	3.927
248	RuF5	RUTHENIUM PENTAFLUORIDE	-1.0537	4.9484E+02	-1.8102E-13	1.2698E-16	360	600	-----	0.590
249	S	SULFUR	326.3558	-5.5393E+04	-6.0794E-01	3.7191E-04	460	579	-----	2332.918
250	SF4	SULFUR TETRAFLUORIDE	2.3876	-7.3676E+01	-1.2442E-02	6.7929E-06	149	346	0.108	0.048
251	SF6	SULFUR HEXAFLUORIDE	-0.0511	1.0013E+02	-3.4646E-03	2.1475E-06	223	319	0.277	0.238
252	SOBr2	THIONYL BROMIDE	0.4017	1.6327E+01	-1.0601E-02	-2.3978E-06	221	629	0.846	0.065
253	SOCl2	THIONYL CHLORIDE	3.8942	-3.2103E+02	-1.1986E-02	6.4907E-06	225	539	0.662	0.053
254	SOF2	SULFUROUS OXYFLUORIDE	3.8925	-1.2787E+02	-2.1522E-02	2.2044E-05	163	353	0.101	0.048
255	SO2	SULFUR DIOXIDE	-2.6700	4.0670E+02	6.1410E-03	-1.2540E-05	200	431	0.257	0.039
256	SO2Cl2	SULFURYL CHLORIDE	8.8303	-8.6708E+02	-2.5704E-02	1.8056E-05	250	518	0.730	0.049
257	SO3	SULFUR TRIOXIDE	12.5700	-9.8890E+02	-4.0790E-02	3.5020E-05	290	491	1.602	0.093
258	S2Cl2	SULFUR MONOCHLORIDE	-0.7765	1.5768E+02	1.7889E-03	-4.8243E-06	193	626	0.719	0.051
259	Sb	ANTIMONY	-31.7771	1.2258E+04	2.7784E-02	-8.1861E-06	923	1275	-----	0.900
260	SbBr3	ANTIMONY TRIBROMIDE	-1.3719	6.2640E+02	-2.4652E-13	1.8020E-16	370	548	-----	0.590
261	SbCl3	ANTIMONY TRICHLORIDE	8.4581	-1.1108E+03	-1.6733E-02	7.7803E-06	347	754	-----	0.062
262	SbCl5	ANTIMONY PENTACHLORIDE	2.9722	-1.7332E+02	-9.0240E-03	4.4500E-06	276	629	1.247	0.060
263	SbH3	STIBINE	2.0337	2.7207E+01	-1.2998E-02	1.2211E-05	185	418	0.216	0.063
264	SbI3	ANTIMONY TRIIODIDE	-1.2663	6.9917E+02	-3.9065E-13	2.3452E-16	440	674	-----	0.590
265	Sb2O3	ANTIMONY TRIOXIDE	-0.8953	1.1313E+03	-1.1744E-14	3.0022E-18	929	1698	-----	0.590
266	Sc	SCANDIUM	-0.2058	1.1476E+03	-9.1051E-14	1.3519E-17	1814	2700	-----	1.657
267	Se	SELENIUM	-0.5999	3.8887E+02	4.4937E-16	-1.6133E-19	494	930	-----	0.658
268	SeCl4	SELENIUM TETRACHLORIDE	21.2641	-3.4907E+03	-4.1213E-02	2.3344E-05	415	707	-----	0.072
269	SeF6	SELENIUM HEXAFLUORIDE	-38.6146	4.0136E+03	1.2932E-01	-1.5693E-04	238	351	0.284	0.075
270	SeOCl2	SELENIUM OXYCHLORIDE	2.9746	-2.7349E+02	-7.4501E-03	2.8136E-06	282	671	1.219	0.068
271	SeO2	SELENIUM DIOXIDE	-1.1952	9.3043E+02	1.4284E-12	-6.7000E-16	613	813	-----	0.890
272	Si	SILICON	-0.5968	2.4339E+03	-7.3444E-04	1.1871E-07	1685	3151	-----	0.110
273	SiBrCl2F	BROMODICHLOROFLUOROSILANE	0.0289	6.9421E+01	-8.4711E-04	-4.9434E-06	161	472	0.371	0.047
274	SiBrF3	TRIFLUOROBROMOSILANE	-27.5315	2.9820E+03	8.3985E-02	-9.4711E-05	203	357	0.106	0.048
275	SiBr2ClF	DIBROMOCHLOROFLUOROSILANE	-0.0013	7.8784E+01	-7.3177E-04	-4.1095E-06	174	509	0.478	0.052
276	SiClF3	TRIFLUOROCHLOROSILANE	3.2982	-1.2878E+02	-1.9331E-02	1.8065E-05	131	314	0.051	0.040
277	SiCl2F2	DICHLORODIFLUOROSILANE	0.8253	-1.1455E+00	-4.7875E-03	-3.2269E-06	133	371	0.128	0.040
278	SiCl3F	TRICHLOROFLUOROSILANE	0.3391	3.6068E+01	-2.0809E-03	-4.6663E-06	152	437	0.266	0.042
279	SiCl4	SILICON TETRACHLORIDE	14.9167	-1.6353E+03	-4.2676E-02	3.3228E-05	258	482	0.459	0.047
280	SiF4	SILICON TETRAFLUORIDE	-516.9862	3.4964E+04	2.5511E+00	-4.2016E-03	186	246	-----	0.028
281	SiHBr3	TRIBROMOSILANE	-0.2847	9.9899E+01	5.8749E-04	-4.3067E-06	200	580	0.696	0.060
282	SiHCl3	TRICHLOROSILANE	18.7411	-1.9605E+03	-5.7360E-02	4.9953E-05	266	455	0.319	0.047
283	SiHF3	TRIFLUOROSILANE	1.2537	1.3402E+02	-1.8829E-02	2.6948E-05	142	276	-----	0.039
284	SiH2Br2	DIBROMOSILANE	1.2409	-1.5985E+01	-4.8862E-03	3.9120E-07	203	523	0.582	0.058
285	SiH2Cl2	DICHLOROSILANE	-0.3898	1.0536E+02	-9.9271E-04	-4.6930E-06	151	427	0.178	0.038
286	SiH2F2	DIFLUOROSILANE	1.8709	8.1044E+01	-1.8984E-02	2.3893E-05	145	302	0.040	0.038
287	SiH2I2	DIIODOSILANE	-0.8524	2.1847E+02	1.3070E-03	-3.5798E-06	272	623	0.895	0.084
288	SiH3Br	MONOBROMOSILANE	2.9311	-1.1425E+02	-1.3458E-02	9.8506E-06	179	431	0.258	0.050
289	SiH3Cl	MONOCHLOROSILANE	3.2255	-1.6592E+02	-1.5788E-02	1.2042E-05	155	377	0.108	0.035
290	SiH3F	MONOFLUOROSILANE	-1.0218	2.7482E+02	-8.5591E-03	1.1568E-05	125	272	-----	0.033
291	SiH3I	IODOSILANE	3.5969	-2.1526E+02	-1.3256E-02	8.9865E-06	216	489	0.527	0.067
292	SiH4	SILANE	0.1193	3.5928E+00	-4.3007E-03	-9.8142E-06	88	256	-----	0.025
293	SiO2	SILICON DIOXIDE	-7.6771	4.1923E+04	-6.2876E-03	2.1394E-06	1883	2673	-----	3061132.284
294	Si2Cl6	HEXACHLORODISILANE	3.2696	-2.1949E+02	-9.6366E-03	4.5737E-06	272	628	1.166	0.047
295	Si2F6	HEXAFLUORODISILANE	-12.1723	1.6244E+03	3.5843E-02	-4.9291E-05	255	390	0.381	0.030
296	Si2H5Cl	DISILANYL CHLORIDE	15.6168	-1.6271E+03	-4.7109E-02	3.9014E-05	265	482	0.382	0.040
297	Si2H6	DISILANE	1.0739	-3.5984E+01	-5.8945E-03	-2.9257E-07	141	410	0.148	0.033
298	Si2OCl3F3	TRICHLOROTRIFLUORODISILOXANE	15.8119	-1.6495E+03	-4.7267E-02	3.9004E-05	266	483	0.451	0.046
299	Si2OCl6	HEXACHLORODISILOXANE	1.3798	-3.8058E+01	-4.4084E-03	2.1402E-07	240	623	0.905	0.045
300	Si2OH6	DISILOXANE	-0.2942	3.7983E+01	2.1375E-04	-9.0929E-06	129	396	0.123	0.029
301	Si3Cl8	OCTACHLOROTRISILANE	25.4608	-4.4198E+03	-4.6962E-02	2.5519E-05	435	737	-----	0.052
302	Si3H8	TRISILANE	-0.7161	7.8248E+01	1.7409E-03	-7.4486E-06	156	499	0.253	0.028
303	Si3H9N	TRISILAZANE	-0.0139	3.1184E+01	-7.3036E-04	-4.9945E-06	167	492	0.269	0.030
304	Si4H10	TETRASILANE	-0.7004	9.3239E+01	1.5569E-03	-5.7948E-06	180	569	0.364	0.030
305	Sm	SAMARIUM	-0.1561	9.3849E+02	7.5703E-13	-1.5731E-16	1345	1874	-----	2.212
306	Sn	TIN	-0.1453	2.4520E+02	-2.0340E-04	7.2413E-08	505	1573	-----	0.741
307	SnBr4	STANNIC BROMIDE	4.0293	-5.2959E+02	-8.0939E-03	2.7468E-06	304	727	-----	0.074
308	SnCl2	STANNOUS CHLORIDE	-0.9912	6.8291E+02	-4.5992E-14	2.1875E-17	520	896	-----	0.590
309	SnCl4	STANNIC CHLORIDE	2.1553	-9.3791E+01	-7.1670E-03	2.8398E-06	243	589	0.904	0.058
310	SnH4	STANNIC HYDRIDE	0.8203	1.2270E+01	-5.5589E-03	-2.2955E-06	123	341	0.100	0.049
311	SnI4	STANNIC IODIDE	-1.3607	7.0289E+02	2.4892E-13	-1.6068E-16	418	621	-----	0.590
312	Sr	STRONTIUM	-0.5473	7.7761E+02	1.9923E-13	-4.9894E-17	1050	1630	-----	0.851

Table 22-2 VISCOSITY OF LIQUID - INORGANIC COMPOUNDS (continued)

NO	FORMULA	NAME	$\log_{10} n_{liq} = A + B/T + C\,T + C\,T^2$						(n_{liq} - centipoise, T - K)	
			A	B	C	D	TMIN	TMAX	n_{liq} @25 C	n_{liq} @TMAX
313	SrO	STRONTIUM OXIDE	---------	---------	---------	---------	----	----	-----	-----
314	Ta	TANTALUM	0.4442	1.5731E+03	-1.2515E-14	9.5163E-19	3290	5565	-----	5.332
315	Tc	TECNNETIUM	0.2648	1.3589E+03	6.8694E-16	-6.2439E-20	2430	5000	-----	3.440
316	Te	TELLURIUM	-0.4375	5.6414E+02	-4.1436E-14	1.3903E-17	723	1373	-----	0.941
317	TeCl4	TELLURIUM TETRACHLORIDE	-1.8608	1.0853E+03	1.1141E-11	-6.4100E-15	497	665	-----	0.591
318	TeF6	TELLURIUM HEXAFLUORIDE	-40.6498	4.1427E+03	1.3925E-01	-1.7224E-04	235	363	0.283	0.041
319	Ti	TITANIUM	-0.0872	1.1963E+03	-1.9444E-14	2.4331E-18	1941	3442	-----	1.821
320	TiCl4	TITANIUM TETRACHLORIDE	1.1919	-2.3230E+01	-3.9580E-03	-9.1688E-08	249	606	0.843	0.053
321	Tl	THALLIUM	-1.1773	6.3806E+02	1.1252E-03	-4.7356E-07	577	730	-----	1.844
322	TlBr	THALLOUS BROMIDE	-1.3552	1.2298E+03	-7.2136E-13	2.6460E-16	733	1092	-----	0.590
323	TlI	THALLOUS IODIDE	-1.2558	1.1254E+03	4.1745E-13	-1.5488E-16	713	1096	-----	0.590
324	Tm	THULIUM	0.0287	1.1492E+03	7.9093E-12	-1.3071E-15	1818	2219	-----	3.520
325	U	URANIUM	0.0809	9.6962E+02	-2.4806E-16	3.1141E-20	1408	4135	-----	2.067
326	UF6	URANIUM FLUORIDE	-2.1384	8.4239E+02	6.1642E-14	-4.7680E-17	342	542	-----	0.261
327	V	VANADIUM	0.0502	1.2813E+03	2.5757E-14	-2.9588E-18	2183	3665	-----	2.511
328	VCl4	VANADIUM TETRACHLORIDE	1.2623	-5.4791E+01	-4.4547E-03	8.5998E-07	247	662	0.671	0.040
329	VOCl3	VANADIUM OXYTRICHLORIDE	-0.7835	1.5921E+02	1.8844E-03	-5.2628E-06	194	604	0.699	0.050
330	W	TUNGSTEN	0.5475	1.6510E+03	7.1972E-14	-5.1729E-18	3695	5645	-----	6.918
331	WF6	TUNGSTEN FLUORIDE	-61.0247	7.7331E+03	1.5928E-01	-1.4340E-04	273	445	0.451	0.069
332	Xe	XENON	4.2337	7.5934E+01	-4.4511E-02	8.1137E-05	161	275	-----	0.025
333	Yb	YTTERBIUM	-0.2688	8.0505E+02	-6.7993E-14	1.6482E-17	1097	1660	-----	1.645
334	Yt	YTTRIUM	-0.1425	1.1416E+03	9.1199E-15	-1.2539E-18	1799	3055	-----	1.703
335	Zn	ZINC	-0.2870	6.0271E+02	-7.2405E-05	1.3098E-08	693	1181	-----	1.433
336	ZnCl2	ZINC CHLORIDE	-1.1879	9.6367E+02	3.8216E-13	-1.5591E-16	638	1005	-----	0.590
337	ZnF2	ZINC FLUORIDE	-1.2394	1.7883E+03	-1.7990E-13	4.1345E-17	1145	1770	-----	0.590
338	ZnO	ZINC OXIDE	---------	---------	---------	---------	----	----	-----	-----
339	ZnSO4	ZINC SULFATE	---------	---------	---------	---------	----	----	-----	-----
340	Zr	ZIRCONIUM	0.0088	1.2628E+03	-1.7517E-15	1.7682E-19	2128	4598	-----	1.921
341	ZrBr4	ZIRCONIUM BROMIDE	-0.9644	9.3043E+02	1.2122E-11	-4.9169E-15	723	923	-----	1.106
342	ZrCl4	ZIRCONIUM CHLORIDE	-0.9880	9.3043E+02	1.3089E-11	-5.3963E-15	710	910	-----	1.083
343	ZrI4	ZIRCONIUM IODIDE	-0.8828	9.3043E+02	9.7652E-12	-3.7407E-15	772	972	-----	1.187

n_{liq} - viscosity of liquid, centipoise

A, B, and C - regression coefficients for chemical compound

T - temperature, K

TMIN - minimum temperature, K

TMAX - maximum temperature, K

Chapter 23

THERMAL CONDUCTIVITY OF GAS

Carl L. Yaws, Xiaoyan Lin, and Li Bu
Lamar University, Beaumont, Texas

ABSTRACT

Results for gas thermal conductivity as function of temperature are presented for a wide range of organic and inorganic chemicals. The major chemicals include many compound types. The results are provided in easy-to-use tables that are especially applicable for rapid engineering usage with the personal computer or hand calculator. The agreement of correlation and data is quite good.

INTRODUCTION

Gas thermal conductivity data are important in many engineering applications in the chemical processing and petroleum refining industries. The objective of this article is to provide the engineer with such data. The compilation of data is presented for a wide temperature range to enable the engineer to determine values at the temperatures of interest.

THERMAL CONDUCTIVITY CORRELATION

The correlation for thermal conductivity of gas as a function of temperature is given by the equation shown below:

$$k_{gas} = A + B\,T + C\,T^2 \tag{23-1}$$

where
k_{gas} = thermal conductivity of gas, W/(m K)
A, B, and C = regression coefficients for chemical compound
T = temperature, K

The results for gas thermal conductivity at low pressure are given in Tables 23-1 and 23-2. The tabulation is arranged by chemical formula to provides ease of use in quickly locating data.

In preparing the compilation, a literature search was conducted to identify data source publications for organics (1-38) and inorganics (1-99). Both experimental values for the property under consideration and parameter values for estimation of the property are included in the source publications. The publications were screened for appropriate data. The compilation resulting from the screening is based on both experimental data and estimated values.

In the absence of experimental data for organic compounds, estimates were primarily based on correlations (29) of Roy and Thodos, Misic and Thodos, Stiel and Thodos, and modified Eucken models. For inorganic compounds, estimates were primarily based on modified Eucken models. Experimental data and estimates were then regressed to provide the same equation for all compounds.

Very limited experimental data are available for highly polar and high molecular weight compounds. Also, very few experimental data are available at high temperatures above 600 K. Thus, the values for these compounds and high temperatures should be considered rough approximations.

A comparison of correlation and experimental data is shown in Fig. 23-1 for a representative chemical. The graph discloses good agreement of correlation and data.

EXAMPLES

The correlation results may be used for prediction and calculation of gas thermal conductivity. Examples are given below.

Example 1 Calculate the gas thermal conductivity of n-hexane (C6H14) at a temperature of 300 K.

Substitution of the coefficients from the table and temperature into the correlation equation yields

k_{gas} = - 0.00200 + (7.7788E-06) (300) + (1.3824E-07) (300^2)

k_{gas} = 0.01261 W/(m K)

The calculated and data values compare favorably (0.01261 vs 0.01280, deviation = 1.5%).

Example 2 Calculate the gas thermal conductivity of carbon dioxide (CO2) at a temperature of 550 K.

Substitution of the coefficients from the table and temperature into the correlation equation yields

$k_{gas} = -0.01200 + (1.0208E\text{-}04)(550) - (2.2403E\text{-}08)(550^2)$

$k_{gas} = 0.03344 \ W/(m \ K)$

The calculated and data values compare favorably (0.03344 vs 0.03228, deviation = 3.59%).

Portions of this material appeared in Oil & Gas Journal, <u>92</u>, 43 (April, 1994) and are reprinted by special permission.

REFERENCES – ORGANIC COMPOUNDS

1-34. See **REFERENCES - ORGANIC COMPOUNDS** in **Chapter 1, CRITICAL PROPERTIES AND ACENTRIC FACTOR**

35. Tsederberg, N. V., <u>THERMAL CONDUCTIVITY OF GASES AND LIQUIDS</u>, R. D. Cess, editor, MIT Press, Cambridge, MA (1965).

36. Vargaftik, N. B., L. P. Filippov, A. A. Tarzimanov, and E. E. Totskiy, <u>THERMAL CONDUCTIVITY OF GASES AND LIQUIDS</u>, Standards Press, Moscow, USSR (1978).

37. Yaws, C. L., X. Lin, Li Bu, and Sachin Nijhawan, Oil & Gas Journal, <u>92</u> (16), 43 (April 18, 1994).

38. Yaws, C. L., <u>HANDBOOK OF THERMAL CONDUCTIVITY</u>, Vols. 1, 2, 3, and 4, Gulf Publishing Co., Houston, TX (1995, 1995, 1995, 1997).

REFERENCES – INORGANIC COMPOUNDS

1-64. See **REFERENCES - INORGANIC COMPOUNDS** in **Chapter 2, HEAT CAPACITY OF GAS**

65. Tsederberg, N. V., <u>THERMAL CONDUCTIVITY OF GASES AND LIQUIDS</u>, MIT Press, Cambridge, MA (1965).

66. Vargaftik, N. B. and others, <u>HANDBOOK OF THERMAL CONDUCTIVITY OF LIQUIDS AND GASES</u>, CRC Press, Inc., Boca Raton, FL (1994).

67. Janz, G. J., C. B. Allan, N. P. Bansal, and R. M. Murphy, <u>PHYSICAL PROPERTIES DATA COMPILATIONS RELEVANT TO ENERGY STORAGE</u>, Nat. Bur. Stand., Molten Salts Data Center, Troy, NY (1979).

68. Lyon, R. N., ed., <u>LIQUID-METALS HANDBOOK</u>, 2nd ed., U. S. Government Printing Office, Washington, DC (1954).

69. Ho, C. Y., R. W. Powell, and P. E. Liley, J., Phys. Chem. Ref. Data, <u>1</u> (2), 279 (1972).

70. Saxena, S. C., High Temp. Sci., <u>3</u>, 168 (1971).

71. Stiel, L. I. and G. Thodos, AICHE J., <u>10</u>, 266 (1964).

72. Choy, P. and C. J. G. Raw, J. Chem. Phys., <u>45</u> (5), 1413 (1966).

73. Chapman, S.and T. G. Cowling, AICHE J., <u>10</u>, 266 (1964).

74. Eucken A., Z. Phys., <u>14</u>, 324 (1913).

75. Matsunaga, N. and A. Nagashima, J. Phys. Chem. Ref. Data, <u>12</u> (4), 933 (1983).

76. Kestin, J. and J. V. Sengers, J. Phys. Chem. Ref. Data, <u>13</u> (2), 601 (1984).

77. Baker, C. E., J. Chem. Phys., <u>46</u>, 2846 (1967).

78. Jones, L. W., Int. J. Heat Mass Transfer, <u>10</u>, 745 (1967).

79. Kestin, J. and E. A. Knierim, J. Phys. Chem. Ref. Data, <u>13</u> (1), 229 (1984).

80. Hanley, H. J. M. and J. F. Ely, J. Phys. Chem. Ref. Data, <u>2</u> (4), 735 (1973).

81. Streng, A. G., J. Chem. Eng. Data, <u>6</u> (3), 43 (1961).

82. Bakalin, S. S. and S. A. Olbybin, Teplofiz Vys. Temp., <u>14</u>, 391 (1976).

83. Palmer, G., Ind. Eng., <u>40</u>, 89 (1948).

84. Fedorov, V. I. and V. I. Machev, High Temp., <u>8</u>, 858 (1970).

85. Sengers, J. V., J. T. R. Watson, and R. S. Basu, J. Phys. Chem. Ref. Data, <u>13(3)</u>, 893 (1984).

86. Ewing, C. T., J. A. Grand, and R. R. Miller, J. Amer. Chem. Soc., <u>74</u>, 11 (1952).

87. Davison, H. W., U. S. NASA Tech. Note D-4650, (1968).

88. Andrade, E. N., C. Da, and E. R. Dobbs, Proc. Roy. Soc. London, 211A <u>12</u>, (1952).

89. Schaefer, C. A. and G. Thodos, AIChE J., <u>5(3)</u>, 367 (1959).

90. Richter, G. N. and B. H. Sage, J. Chem. Eng. Data, <u>8(2)</u>, 221 (1963).

91. McDonald, J. and H. T. Davis, Phys. Chem. Liquids, <u>2(3)</u>, 119 (1971).

92. Federov, V. I. and V. I. Machuev, Teplofiz. Vys. Temp., <u>8(4)</u>, 912 (1970).

93. Waterman, T. E., D. P. Kirsh, and R. I. Brabets, J. Chem. Phys., <u>29(4)</u>, 905 (1958).

94. Sladkov, I. B. and T. G. Kotina, Russ. J. Phys. Chem., <u>48(7)</u>, 1115 (1974).

95. Sugawara, A., J. Appl. Phys., <u>36</u>, 2375 (1965).

96. Ivannikov, P. S., I. V. Litvinenko, and I. V. Radchenko, J. Eng. Phys. USSR, <u>23(5)</u>, 1397 (1972).

97. Dushin, Yu. A., J. Eng. Phys., <u>10(4)</u>, 538 (1966).

98. Yaws, C. L., <u>HANDBOOK OF TRANSPORT PROPERTY DATA</u>, Gulf Publishing Co., Houston, TX (1995).

99. Yaws, C. L., <u>HANDBOOK OF THERMAL CONDUCTIVITY</u>, Vol. 4, Gulf Publishing Co., Houston, TX (1997).

Figure 23-1 Thermal Conductivity of Gas

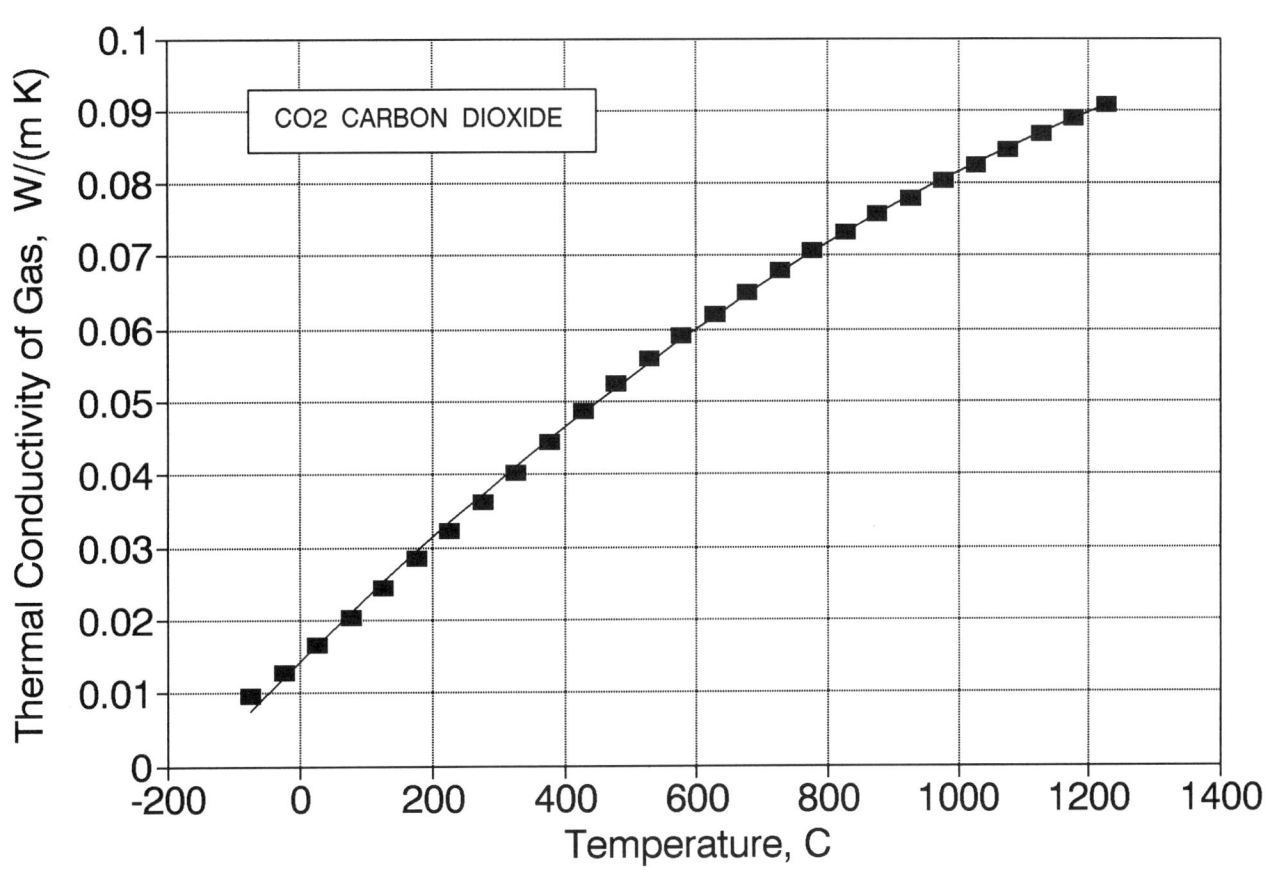

Table 23-1 THERMAL CONDUCTIVITY OF GAS - ORGANIC COMPOUNDS

			$k_{gas} = A + B T + C T^2$					$(k_{gas} - W/(m\ K),\ T - K)$		
NO	FORMULA	NAME	A	B	C	TMIN	TMAX	k_{gas} @25 C	k_{gas} @TMIN	k_{gas} @TMAX
1	CBrClF2	BROMOCHLORODIFLUOROMETHANE	-0.00408	3.4053E-05	2.1046E-08	173	1000	0.00794	0.00244	0.05102
2	CBrCl3	BROMOTRICHLOROMETHANE	-0.00407	2.3399E-05	2.0915E-08	378	1000	------	0.00776	0.04024
3	CBrF3	BROMOTRIFLUOROMETHANE	-0.00528	4.7159E-05	1.1929E-08	215	1000	0.00984	0.00541	0.05381
4	CBr2F2	DIBROMODIFLUOROMETHANE	-0.00507	3.3401E-05	2.0218E-08	296	996	0.00669	0.00659	0.04825
5	CCIF3	CHLOROTRIFLUOROMETHANE	-0.00381	4.8679E-05	1.7208E-08	192	600	0.01223	0.00617	0.03159
6	CCIN	CYANOGEN CHLORIDE	-0.00239	3.3400E-05	-3.7346E-09	286	1000	0.00724	0.00686	0.02728
7	CCl2F2	DICHLORODIFLUOROMETHANE	-0.00079	2.1035E-05	4.6311E-08	115	385	0.00960	0.00224	0.01417
8	CCl2O	PHOSGENE	-0.00442	3.8294E-05	1.9223E-08	250	800	0.00871	0.00635	0.03852
9	CCl3F	TRICHLOROFLUOROMETHANE	-0.00252	3.1158E-05	1.2304E-08	250	500	0.00786	0.00604	0.01614
10	CCl4	CARBON TETRACHLORIDE	-0.00070	2.2065E-05	6.7913E-09	255	556	0.00648	0.00537	0.01367
11	CF2O	CARBONYL FLUORIDE	-0.00560	6.4614E-05	-1.0689E-08	189	1000	0.01271	0.00623	0.04833
12	CF4	CARBON TETRAFLUORIDE	-0.00406	6.3276E-05	2.5435E-09	145	750	0.01503	0.00517	0.04483
13	CHBr3	TRIBROMOMETHANE	-0.00169	1.2862E-05	1.7430E-08	422	1000	------	0.00684	0.02860
14	CHClF2	CHLORODIFLUOROMETHANE	-0.00011	1.6709E-05	6.3391E-08	232	500	0.01051	0.00718	0.02409
15	CHCl2F	DICHLOROFLUOROMETHANE	-0.00123	1.8714E-05	4.7235E-08	250	450	0.00855	0.00640	0.01676
16	CHCl3	CHLOROFORM	-0.00019	2.2269E-05	1.2257E-09	273	573	0.00656	0.00598	0.01297
17	CHF3	TRIFLUOROMETHANE	-0.00431	5.5962E-05	-1.0820E-09	191	1000	0.01228	0.00634	0.05057
18	CHI3	TRIIODOMETHANE	-0.00099	1.1538E-05	-6.7781E-11	275	1000	0.00244	0.00218	0.01048
19	CHN	HYDROGEN CYANIDE	-0.00136	3.7886E-05	2.6807E-08	260	360	0.01232	0.01030	0.01575
20	CHNS	ISOTHIOCYANIC-ACID	------	--------	--------	----	----	------	------	------
21	CH2BrCl	BROMOCHLOROMETHANE	-0.00391	2.8128E-05	1.8518E-08	341	1000	------	0.00783	0.04274
22	CH2Br2	DIBROMOMETHANE	-0.00373	2.3061E-05	1.4326E-08	370	1000	------	0.00676	0.03366
23	CH2ClF	CHLOROFLUOROMETHANE	-0.00716	5.5521E-05	-1.2176E-09	275	1000	0.00929	0.00802	0.04714
24	CH2Cl2	DICHLOROMETHANE	-0.00122	1.8440E-05	3.5385E-08	250	500	0.00742	0.00560	0.01685
25	CH2F2	DIFLUOROMETHANE	-0.00413	5.0088E-05	2.0672E-08	222	1000	0.01099	0.00709	0.04803
26	CH2I2	DIIODOMETHANE	-0.00311	1.9559E-05	-5.0194E-10	455	1000	------	0.00569	0.01595
27	CH2O	FORMALDEHYDE	0.00171	1.9431E-05	9.5287E-08	254	994	0.01597	0.01279	0.11517
28	CH2O2	FORMIC ACID	------	--------	--------	----	----	------	------	------
29	CH3Br	METHYL BROMIDE	-0.00421	2.7182E-05	3.9758E-08	273	825	0.00743	0.00617	0.04528
30	CH3Cl	METHYL CHLORIDE	-0.00185	2.0296E-05	7.3234E-08	213	750	0.01071	0.00580	0.05457
31	CH3Cl3Si	METHYL TRICHLOROSILANE	-0.00439	4.2665E-05	-1.3646E-10	340	1000	------	0.01010	0.03814
32	CH3F	METHYL FLUORIDE	-0.00436	6.3911E-05	-3.4987E-09	195	1000	0.01438	0.00797	0.05605
33	CH3I	METHYL IODIDE	-0.00774	4.3432E-05	4.6420E-10	220	1500	0.00525	0.00184	0.05845
34	CH3NO	FORMAMIDE	------	--------	--------	----	----	------	------	------
35	CH3NO2	NITROMETHANE	-0.00377	3.4946E-05	2.8700E-08	374	994	------	0.01331	0.05932
36	CH3NO2	METHYL-NITRITE	------	--------	--------	----	----	------	------	------
37	CH3NO3	METHYL-NITRATE	------	--------	--------	----	----	------	------	------
38	CH4	METHANE	-0.00935	1.4028E-04	3.3180E-08	97	1400	0.03542	0.00457	0.25207
39	CH4Cl2Si	METHYL DICHLOROSILANE	-0.00554	5.1703E-05	2.0656E-09	315	1000	------	0.01095	0.04823
40	CH4O	METHANOL	0.00234	5.4340E-06	1.3154E-07	273	684	0.01565	0.01363	0.06760
41	CH4O3S	METHANESULFONIC ACID	------	--------	--------	----	----	------	------	------
42	CH4S	METHYL MERCAPTAN	-0.00561	5.3048E-05	2.7049E-08	279	639	0.01261	0.01130	0.03933
43	CH5ClSi	METHYL CHLOROSILANE	-0.00716	6.3113E-05	7.8172E-09	282	1000	0.01235	0.01126	0.06377
44	CH5N	METHYLAMINE	-0.01136	8.0400E-05	5.8607E-08	267	800	0.01782	0.01428	0.09047
45	CH6Si	METHYL SILANE	-0.01097	9.1895E-05	1.6306E-08	216	1000	0.01788	0.00964	0.09723
46	CN4O8	TETRANITROMETHANE	-0.01219	6.7062E-05	-1.1177E-08	399	1000	------	0.01279	0.04370
47	CO	CARBON MONOXIDE	0.00158	8.2511E-05	-1.9081E-08	70	1250	0.02448	0.00726	0.07490
48	COS	CARBONYL SULFIDE	-0.00452	6.3569E-05	-5.6376E-09	223	1000	0.01393	0.00938	0.05341
49	CO2	CARBON DIOXIDE	-0.01200	1.0208E-04	-2.2403E-08	195	1500	0.01644	0.00705	0.09071
50	CS2	CARBON DISULFIDE	-0.00239	3.2574E-05	2.1193E-10	273	999	0.00734	0.00652	0.03036
51	C2BrF3	BROMOTRIFLUOROETHYLENE	-0.00515	4.5808E-05	-7.4219E-09	271	1000	0.00785	0.00672	0.03324
52	C2Br2F4	1,2-DIBROMOTETRAFLUOROETHANE	-0.00462	2.7959E-05	1.9955E-08	320	1000	------	0.00637	0.04329
53	C2ClF3	CHLOROTRIFLUOROETHYLENE	-0.00987	6.5826E-05	1.8776E-08	245	1000	0.01143	0.00738	0.07473
54	C2ClF5	CHLOROPENTAFLUOROETHANE	-0.00591	5.9903E-05	6.0256E-09	234	1000	0.01249	0.00844	0.06002
55	C2Cl2F4	1,2-DICHLOROTETRAFLUOROETHANE	0.01638	-7.7593E-05	2.0028E-07	277	500	0.01105	0.01025	0.02765
56	C2Cl3F3	1,1,2-TRICHLOROTRIFLUOROETHANE	------	--------	--------	----	----	------	------	------
57	C2Cl4	TETRACHLOROETHYLENE	-0.00129	1.2069E-05	3.4803E-08	277	825	0.00540	0.00472	0.03235
58	C2Cl4F2	1,1,2,2-TETRACHLORODIFLUOROETHANE	-0.00566	3.0538E-05	2.7170E-08	366	1000	------	0.00916	0.05205
59	C2Cl4O	TRICHLOROACETYL CHLORIDE	-0.00313	3.2817E-05	-1.8896E-09	391	991	------	0.00941	0.02754
60	C2Cl6	HEXACHLOROETHANE	-0.00379	3.1150E-05	-3.7944E-09	460	1000	------	0.00974	0.02357
61	C2F4	TETRAFLUOROETHYLENE	-0.00736	8.0682E-05	-1.4513E-08	198	1000	0.01541	0.00805	0.05881
62	C2F6	HEXAFLUOROETHANE	-0.01821	1.4585E-04	1.1582E-07	195	700	0.03557	0.01463	0.14064
63	C2HBrClF3	HALOTHANE	-0.00605	4.7638E-05	-5.6613E-09	323	1000	------	0.00875	0.03593
64	C2HClF2	2-CHLORO-1,1-DIFLUOROETHYLENE	-0.00820	5.7024E-05	1.9550E-08	255	1000	0.01054	0.00761	0.06837
65	C2HCl3	TRICHLOROETHYLENE	-0.00142	1.5596E-05	3.3673E-08	250	800	0.00622	0.00458	0.03261
66	C2HCl3O	DICHLOROACETYL CHLORIDE	-0.00377	3.6287E-05	-7.8238E-10	382	1000	------	0.00998	0.03173
67	C2HCl3O	TRICHLOROACETALDEHYDE	-0.00463	4.3971E-05	-4.6342E-10	371	1000	------	0.01162	0.03888
68	C2HCl5	PENTACHLOROETHANE	-0.00244	1.7034E-05	3.4894E-08	433	1000	------	0.01148	0.04949
69	C2HF3	TRIFLUOROETHENE	-0.00864	7.1787E-05	-1.0373E-08	275	1000	0.01184	0.01032	0.05277
70	C2HF3O2	TRIFLUOROACETIC ACID	-0.01011	7.3715E-05	-8.4874E-09	345	1000	------	0.01431	0.05512
71	C2HF5	PENTAFLUOROETHANE	-0.00926	6.4844E-05	8.6106E-09	225	1000	0.01084	0.00577	0.06419
72	C2H2	ACETYLENE	-0.00358	6.2542E-05	7.0646E-08	200	600	0.02135	0.01175	0.05938
73	C2H2Br4	1,1,2,2-TETRABROMOETHANE	-0.00255	2.1836E-05	-5.8239E-10	517	1000	------	0.00858	0.01870
74	C2H2Cl2	1,1-DICHLOROETHYLENE	-0.00331	2.9439E-05	2.8676E-08	273	673	0.00802	0.00686	0.02949
75	C2H2Cl2	cis-1,2-DICHLOROETHYLENE	-0.00438	3.2220E-05	3.6668E-08	334	700	------	0.01047	0.03614
76	C2H2Cl2	trans-1,2-DICHLOROETHYLENE	-0.00473	3.4990E-05	3.8396E-08	321	700	------	0.01046	0.03858
77	C2H2Cl2O	CHLOROACETYL CHLORIDE	-0.00582	4.7691E-05	3.3316E-09	379	1000	------	0.01273	0.04520
78	C2H2Cl2O	DICHLOROACETALDEHYDE	-0.00485	4.3837E-05	8.9786E-10	362	1000	------	0.01114	0.03988

Table 23-1 THERMAL CONDUCTIVITY OF GAS - ORGANIC COMPOUNDS (continued)

			$k_{gas} = A + B T + C T^2$					$(k_{gas} - W/(m\ K),\ T - K)$		
NO	FORMULA	NAME	A	B	C	TMIN	TMAX	k_{gas} @25 C	k_{gas} @TMIN	k_{gas} @TMAX
79	C2H2Cl2O2	DICHLOROACETIC ACID	-0.00519	4.2625E-05	4.2170E-10	467	1000	------	0.01481	0.03786
80	C2H2Cl3F	1,1,1-TRICHLOROFLUOROETHANE	-0.00628	3.4037E-05	3.0005E-08	366	1000	------	0.01020	0.05776
81	C2H2Cl4	1,1,1,2-TETRACHLOROETHANE	-0.00279	2.0578E-05	3.6861E-08	404	1000	------	0.01154	0.05465
82	C2H2Cl4	1,1,2,2-TETRACHLOROETHANE	-0.00268	1.9386E-05	3.5334E-08	418	998	------	0.01160	0.05186
83	C2H2F2	1,1-DIFLUOROETHYLENE	-0.00591	7.1299E-05	1.1278E-09	188	1000	0.01545	0.00753	0.06652
84	C2H2F2	cis-1,2-DIFLUOROETHENE	-0.01023	7.5179E-05	-5.4967E-09	275	1000	0.01170	0.01003	0.05945
85	C2H2F2	trans-1,2-DIFLUOROETHENE	-0.00975	7.4602E-05	-5.2134E-09	275	1000	0.01203	0.01037	0.05964
86	C2H2F4	1,1,1,2-TETRAFLUOROETHANE	-0.00668	4.8544E-05	1.2108E-08	247	1000	0.00887	0.00605	0.05397
87	C2H2O	KETENE	-0.00652	7.2924E-05	3.6631E-09	200	1000	0.01555	0.00821	0.07007
88	C2H2O4	OXALIC ACID	-0.00204	2.3825E-05	3.5957E-09	569	1000	------	0.01268	0.02538
89	C2H3Br	VINYL BROMIDE	-0.00682	4.6167E-05	6.3199E-10	289	1000	0.00700	0.00658	0.03998
90	C2H3Cl	VINYL CHLORIDE	-0.00764	5.8427E-05	2.4051E-08	260	1000	0.01192	0.00918	0.07484
91	C2H3ClF2	1-CHLORO-1,1-DIFLUOROETHANE	-0.00972	6.2984E-05	2.5156E-08	263	1000	0.01129	0.00858	0.07842
92	C2H3ClO	ACETYL CHLORIDE	-0.00302	3.6914E-05	1.8003E-08	324	994	------	0.01083	0.05146
93	C2H3ClO	CHLOROACETALDEHYDE	-0.00688	5.4997E-05	6.1014E-09	358	1000	------	0.01359	0.05422
94	C2H3ClO2	CHLOROACETIC ACID	-0.00684	4.7752E-05	5.6251E-09	463	1000	------	0.01648	0.04654
95	C2H3ClO2	METHYL CHLOROFORMATE	-0.00878	6.3484E-05	-2.5154E-09	344	1000	------	0.01276	0.05219
96	C2H3Cl3	1,1,1-TRICHLOROETHANE	-0.00070	3.9347E-06	3.3468E-09	347	997	------	0.00107	0.00655
97	C2H3Cl3	1,1,2-TRICHLOROETHANE	-0.00609	3.2555E-05	3.0529E-08	387	997	------	0.01108	0.05671
98	C2H3F	VINYL FLUORIDE	-0.00758	7.3243E-05	2.3394E-09	201	1000	0.01447	0.00724	0.06800
99	C2H3F3	1,1,1-TRIFLUOROETHANE	-0.00666	5.8603E-05	1.4415E-08	226	1000	0.01209	0.00732	0.06636
100	C2H3N	ACETONITRILE	0.01203	-4.4320E-05	1.3710E-07	355	995	------	0.01357	0.10366
101	C2H3NO	METHYL ISOCYANATE	-0.00947	5.7305E-05	1.9060E-08	312	1000	------	0.01026	0.06690
102	C2H4	ETHYLENE	-0.00123	3.6219E-05	1.2459E-07	150	750	0.02064	0.00701	0.09602
103	C2H4Br2	1,1-DIBROMOETHANE	-0.00438	3.4738E-05	3.2113E-09	381	1000	------	0.00932	0.03357
104	C2H4Br2	1,2-DIBROMOETHANE	-0.00433	3.0693E-05	5.2748E-09	405	995	------	0.00897	0.03143
105	C2H4Cl2	1,1-DICHLOROETHANE	-0.00739	4.4752E-05	3.3174E-08	330	990	------	0.01099	0.06943
106	C2H4Cl2	1,2-DICHLOROETHANE	-0.00682	4.0081E-05	3.1925E-08	357	997	------	0.01156	0.06487
107	C2H4Cl2O	BIS(CHLOROMETHYL)ETHER	-0.00626	4.9757E-05	4.6350E-09	378	1000	------	0.01321	0.04813
108	C2H4F2	1,1-DIFLUOROETHANE	-0.00315	4.2328E-05	2.2875E-08	247	1000	0.01150	0.00870	0.06205
109	C2H4F2	1,2-DIFLUOROETHANE	-0.00461	3.5650E-05	1.6149E-08	304	1000	------	0.00772	0.04719
110	C2H4I2	1,2-DIIODOETHANE	-0.00219	1.7577E-05	3.4121E-09	275	1000	0.00335	0.00290	0.01880
111	C2H4O	ACETALDEHYDE	-0.00181	2.1187E-05	8.0192E-08	200	700	0.01164	0.00564	0.05231
112	C2H4O	ETHYLENE OXIDE	0.01612	-7.3460E-05	2.1215E-07	273	673	0.01308	0.01188	0.06277
113	C2H4OS	THIOACETIC-ACID	-0.00614	4.9734E-05	4.6291E-09	275	1000	0.00910	0.00789	0.04822
114	C2H4O2	ACETIC ACID	0.00234	-6.5956E-06	1.1569E-07	295	687	0.01066	0.01046	0.05241
115	C2H4O2	METHYL FORMATE	0.00085	6.0228E-06	1.2439E-07	300	1000	------	0.01385	0.13126
116	C2H4S	THIACYCLOPROPANE	-0.01161	7.0163E-05	6.5663E-09	275	1000	0.00989	0.00818	0.06512
117	C2H5Br	BROMOETHANE	-0.00565	4.1720E-05	1.2067E-08	312	992	------	0.00854	0.04761
118	C2H5Cl	ETHYL CHLORIDE	-0.00291	3.1284E-05	5.5316E-08	273	773	0.01133	0.00975	0.05433
119	C2H5ClO	2-CHLOROETHANOL	-0.00305	2.0681E-05	7.5149E-08	402	1000	------	0.01741	0.09278
120	C2H5F	ETHYL FLUORIDE	-0.00772	6.7262E-05	1.6128E-08	235	1000	0.01377	0.00898	0.07567
121	C2H5I	ETHYL IODIDE	-0.00468	3.0149E-05	2.7090E-08	273	1000	0.00672	0.00557	0.05256
122	C2H5N	ETHYLENEIMINE	-0.02282	1.0564E-04	1.7972E-08	329	999	------	0.01388	0.10065
123	C2H5NO	ACETAMIDE	-0.00804	4.5522E-05	1.5108E-08	494	994	------	0.01813	0.05214
124	C2H5NO	N-METHYLFORMAMIDE	-0.00827	4.8526E-05	3.0067E-08	473	993	------	0.02141	0.06956
125	C2H5NO2	NITROETHANE	-0.00830	5.0042E-05	2.2752E-08	387	997	------	0.01447	0.06421
126	C2H5NO3	ETHYL-NITRATE	------	---------	---------	----	----	------	------	------
127	C2H6	ETHANE	-0.01936	1.2547E-04	3.8298E-08	225	825	0.02145	0.01081	0.11022
128	C2H6AlCl	DIMETHYLALUMINUM CHLORIDE	-0.00734	5.0132E-05	1.0924E-08	399	1000	------	0.01440	0.05372
129	C2H6O	DIMETHYL ETHER	-0.03150	1.5032E-04	1.3879E-09	273	1500	0.01344	0.00964	0.19710
130	C2H6O	ETHANOL	-0.00556	4.3620E-05	8.5033E-08	351	991	------	0.02023	0.12118
131	C2H6OS	DIMETHYL SULFOXIDE	-0.01136	6.3395E-05	1.1846E-08	462	1000	------	0.02046	0.06388
132	C2H6O2	ETHYLENE GLYCOL	-0.01076	7.9631E-05	5.7243E-09	470	990	------	0.02793	0.07369
133	C2H6O4S	DIMETHYL SULFATE	------	---------	---------	----	----	------	------	------
134	C2H6S	DIMETHYL SULFIDE	-0.00827	6.0624E-05	1.8376E-08	310	990	------	0.01229	0.06976
135	C2H6S	ETHYL MERCAPTAN	-0.00817	6.0155E-05	1.7688E-08	308	998	------	0.01204	0.06948
136	C2H6S2	DIMETHYL DISULFIDE	-0.00779	5.1893E-05	1.4514E-08	383	1000	------	0.01421	0.05862
137	C2H7N	DIMETHYLAMINE	-0.01559	8.3374E-05	6.3822E-08	273	990	0.01494	0.01193	0.12950
138	C2H7N	ETHYLAMINE	-0.01204	7.8124E-05	5.6706E-08	274	800	0.01629	0.01362	0.08675
139	C2H7NO	MONOETHANOLAMINE	0.00878	-2.9523E-05	8.6097E-08	400	1000	------	0.01075	0.06535
140	C2H8N2	ETHYLENEDIAMINE	-0.01120	7.2328E-05	2.2662E-08	390	1000	------	0.02045	0.08379
141	C2H8Si	DIMETHYL SILANE	-0.00965	8.0952E-05	2.2082E-08	254	1000	0.01645	0.01234	0.09338
142	C2N2	CYANOGEN	-0.00399	5.4443E-05	6.4633E-09	252	1000	0.01282	0.01014	0.05692
143	C3F6	HEXAFLUOROPROPYLENE	-0.00978	6.2257E-05	1.5800E-08	244	1000	0.01019	0.00635	0.06828
144	C3F6O	HEXAFLUOROACETONE	-0.00916	6.9596E-05	3.0262E-08	246	1000	0.01428	0.00979	0.09070
145	C3F8	OCTAFLUOROPROPANE	-0.01201	7.4793E-05	1.1958E-08	236	1000	0.01135	0.00631	0.07474
146	C3H2N2	MALONONITRILE	0.01003	-3.4680E-05	9.0857E-08	492	1000	------	0.01496	0.06621
147	C3H3Cl	PROPARGYL CHLORIDE	-0.00708	6.2447E-05	3.9000E-09	331	1000	------	0.01402	0.05927
148	C3H3N	ACRYLONITRILE	0.01204	-4.9047E-05	1.4193E-07	298	1000	0.01003	0.01003	0.10492
149	C3H3NO	OXAZOLE	-0.00671	2.8038E-05	9.5600E-08	343	1000	------	0.01415	0.11693
150	C3H4	METHYLACETYLENE	-0.01650	1.0715E-04	6.5108E-09	250	1000	0.01603	0.01069	0.09716
151	C3H4	PROPADIENE	-0.01973	1.1713E-04	9.5083E-10	315	999	------	0.01726	0.09823
152	C3H4Cl2	2,3-DICHLOROPROPENE	-0.00682	3.7519E-05	3.3414E-08	366	1000	------	0.01139	0.06411
153	C3H4O	ACROLEIN	-0.00827	4.9529E-05	3.5739E-08	326	996	------	0.01167	0.07651
154	C3H4O	PROPARGYL ALCOHOL	-0.01123	8.0885E-05	1.4120E-08	387	1000	------	0.02219	0.08378
155	C3H4O2	ACRYLIC ACID	-0.00889	6.0453E-05	1.2049E-08	414	1000	------	0.01820	0.06361
156	C3H4O2	beta-PROPIOLACTONE	-0.01833	8.8775E-05	6.6444E-09	435	1000	------	0.02154	0.07709

Table 23-1 THERMAL CONDUCTIVITY OF GAS - ORGANIC COMPOUNDS (continued)

NO	FORMULA	NAME	$k_{gas} = A + B\,T + C\,T^2$					$(k_{gas}$ - W/(m K), T - K)		
			A	B	C	TMIN	TMAX	k_{gas} @25 C	k_{gas} @TMIN	k_{gas} @TMAX
157	C3H4O2	VINYL FORMATE	0.00182	-4.4946E-07	1.2522E-07	320	1000	------	0.01450	0.12659
158	C3H4O3	ETHYLENE CARBONATE	------	--------	--------	----	----	-----	-----	-----
159	C3H4O3	PYRUVIC ACID	-0.00799	5.7126E-05	6.8910E-09	438	1000	------	0.01835	0.05603
160	C3H5Br	3-BROMO-1-PROPENE	-0.00651	4.3412E-05	4.3390E-09	275	1000	0.00682	0.00576	0.04124
161	C3H5Cl	2-CHLOROPROPENE	-0.00925	5.9051E-05	4.0074E-08	296	996	0.01192	0.01174	0.08932
162	C3H5Cl	3-CHLOROPROPENE	-0.00787	4.8625E-05	3.4407E-08	318	1000	------	0.01107	0.07516
163	C3H5ClO	alpha-EPICHLOROHYDRIN	-0.00785	5.1666E-05	1.4771E-08	389	999	------	0.01448	0.05851
164	C3H5ClO2	METHYL CHLOROACETATE	-0.00761	5.6206E-05	7.3097E-09	403	1000	------	0.01623	0.05591
165	C3H5ClO2	ETHYL CHLOROFORMATE	-0.00720	5.3967E-05	8.7189E-09	366	1000	------	0.01372	0.05549
166	C3H5Cl3	1,2,3-TRICHLOROPROPANE	-0.00278	1.9819E-05	3.6275E-08	430	1000	------	0.01245	0.05331
167	C3H5I	3-IODO-1-PROPENE	-0.00494	3.4049E-05	3.5917E-09	275	1000	0.00553	0.00470	0.03270
168	C3H5N	PROPIONITRILE	0.00997	-4.5594E-05	1.5030E-07	273	1000	0.00974	0.00872	0.11468
169	C3H5NO	ACRYLAMIDE	-0.01602	7.7591E-05	5.0142E-09	466	1000	------	0.02123	0.06659
170	C3H5NO	HYDRACRYLONITRILE	------	--------	--------	----	----	------	-----	-----
171	C3H5NO	LACTONITRILE	-0.00989	6.4376E-05	1.2371E-08	457	1000	------	0.02211	0.06686
172	C3H5N3O9	NITROGLYCERINE	------	--------	--------	----	----	-----	-----	-----
173	C3H6	CYCLOPROPANE	-0.00430	3.7668E-05	1.0953E-07	240	1000	0.01667	0.01105	0.14290
174	C3H6	PROPYLENE	-0.01116	7.5155E-05	6.5558E-08	250	1000	0.01708	0.01173	0.12955
175	C3H6Br2	1,2-DIBROMOPROPANE	-0.00455	3.0748E-05	4.2464E-09	275	1000	0.00499	0.00423	0.03044
176	C3H6Cl2	1,1-DICHLOROPROPANE	-0.00746	4.0892E-05	3.5593E-08	361	1000	------	0.01194	0.06903
177	C3H6Cl2	1,2-DICHLOROPROPANE	-0.00567	3.2384E-05	2.6457E-08	370	1000	------	0.00993	0.05317
178	C3H6Cl2	1,3-DICHLOROPROPANE	-0.00649	3.4301E-05	3.2037E-08	394	1000	------	0.01200	0.05985
179	C3H6Cl2	2,2-DICHLOROPROPANE	-0.00809	5.5402E-05	1.6310E-09	275	1000	0.00857	0.00727	0.04894
180	C3H6I2	1,2-DIIODOPROPANE	-0.00273	1.9471E-05	5.3579E-09	275	1000	0.00355	0.00303	0.02210
181	C3H6O	ACETONE	-0.00084	8.7475E-06	1.0678E-07	273	572	0.01126	0.00951	0.03910
182	C3H6O	ALLYL ALCOHOL	0.00014	1.4038E-05	9.2237E-08	370	1000	------	0.01796	0.10642
183	C3H6O	METHYL VINYL ETHER	-0.00648	5.8563E-05	3.7240E-08	279	650	0.01429	0.01276	0.04732
184	C3H6O	n-PROPIONALDEHYDE	0.00155	2.5422E-06	1.0588E-07	321	991	------	0.01328	0.10805
185	C3H6O	1,2-PROPYLENE OXIDE	-0.01000	6.9207E-05	2.8343E-08	307	997	------	0.01392	0.08717
186	C3H6O	1,3-PROPYLENE OXIDE	-0.00675	3.1581E-05	9.5229E-08	321	991	------	0.01320	0.11807
187	C3H6O2	ETHYL FORMATE	-0.00184	1.1677E-05	1.1060E-07	273	768	0.01147	0.00959	0.07236
188	C3H6O2	METHYL ACETATE	-0.00524	4.0236E-05	5.7371E-08	277	771	0.01186	0.01031	0.05989
189	C3H6O2	PROPIONIC ACID	0.08924	-2.5339E-04	2.9793E-07	559	722	------	0.04069	0.06160
190	C3H6O2S	3-MERCAPTOPROPIONIC ACID	-0.00982	5.9243E-05	1.3336E-08	501	1000	------	0.02321	0.06276
191	C3H6O3	LACTIC ACID	-0.00909	6.0750E-05	1.0657E-08	447	1000	------	0.02019	0.06232
192	C3H6O3	METHOXYACETIC ACID	-0.01052	6.6126E-05	1.3535E-08	478	1000	------	0.02418	0.06914
193	C3H6O3	TRIOXANE	-0.00645	1.9795E-05	1.1199E-07	388	998	------	0.01809	0.12485
194	C3H6S	THIACYCLOBUTANE	-0.01125	6.1963E-05	1.5015E-08	275	1000	0.00856	0.00693	0.06573
195	C3H7Br	1-BROMOPROPANE	-0.00984	5.4683E-05	6.3166E-09	344	1000	------	0.00972	0.05116
196	C3H7Br	2-BROMOPROPANE	-0.01411	7.9402E-05	6.6943E-09	333	1000	------	0.01307	0.07199
197	C3H7Cl	ISOPROPYL CHLORIDE	-0.00888	5.5302E-05	3.7804E-08	309	1000	------	0.01182	0.08423
198	C3H7Cl	n-PROPYL CHLORIDE	-0.00865	5.2286E-05	3.7301E-08	320	1000	------	0.01190	0.08094
199	C3H7F	1-FLUOROPROPANE	-0.01305	8.3174E-05	1.0626E-08	275	1000	0.01269	0.01063	0.08075
200	C3H7F	2-FLUOROPROPANE	-0.01444	8.9879E-05	8.2057E-09	275	1000	0.01309	0.01090	0.08364
201	C3H7I	ISOPROPYL IODIDE	-0.00546	3.5717E-05	1.2961E-08	363	993	------	0.00921	0.04279
202	C3H7I	n-PROPYL IODIDE	-0.00552	3.5041E-05	1.3485E-08	376	996	------	0.00956	0.04276
203	C3H7N	ALLYLAMINE	0.02130	-9.5355E-05	2.5635E-07	326	1000	------	0.01746	0.18230
204	C3H7N	PROPYLENEIMINE	-0.02813	1.5217E-04	1.1791E-08	334	1000	------	0.02401	0.13583
205	C3H7NO	N,N-DIMETHYLFORMAMIDE	-0.00906	5.4289E-05	1.9961E-08	426	996	------	0.01769	0.06481
206	C3H7NO	N-METHYLACETAMIDE	-0.01740	8.0569E-05	1.5950E-08	478	1000	------	0.02476	0.07912
207	C3H7NO2	1-NITROPROPANE	-0.00895	5.3414E-05	2.3131E-08	404	994	------	0.01640	0.06700
208	C3H7NO2	2-NITROPROPANE	-0.01667	7.7246E-05	7.9691E-09	393	1000	------	0.01492	0.06855
209	C3H7NO3	PROPYL-NITRATE	------	--------	--------	----	----	------	------	-----
210	C3H7NO3	ISOPROPYL-NITRATE	------	--------	--------	----	----	-----	-----	-----
211	C3H8	PROPANE	-0.00869	6.6409E-05	7.8760E-08	233	773	0.01811	0.01106	0.08971
212	C3H8O	ISOPROPANOL	0.07775	-3.6017E-04	5.7593E-07	355	450	------	0.02247	0.03230
213	C3H8O	METHYL ETHYL ETHER	-0.00108	7.9704E-06	1.5933E-07	273	1000	0.01546	0.01297	0.16622
214	C3H8O	n-PROPANOL	-0.00333	2.8691E-05	1.0222E-07	372	720	------	0.02149	0.07032
215	C3H8O2	2-METHOXYETHANOL	-0.01071	6.8951E-05	2.0904E-08	398	1000	------	0.02004	0.07915
216	C3H8O2	METHYLAL	-0.01773	9.8899E-05	2.9645E-09	315	1000	------	0.01372	0.08413
217	C3H8O2	1,2-PROPYLENE GLYCOL	-0.01227	7.2936E-05	2.2476E-08	461	1000	------	0.02613	0.08314
218	C3H8O2	1,3-PROPYLENE GLYCOL	0.00040	1.0202E-05	8.9545E-08	450	1000	------	0.02312	0.10015
219	C3H8O3	GLYCEROL	-0.00842	4.4977E-05	2.4839E-08	563	993	------	0.02478	0.06073
220	C3H8S	n-PROPYLMERCAPTAN	-0.00019	6.8622E-06	-9.5609E-10	341	1000	------	0.00204	0.00572
221	C3H8S	ISOPROPYL MERCAPTAN	-0.01039	7.2423E-05	2.4480E-08	326	1000	------	0.01582	0.08651
222	C3H8S	ETHYL-METHYL-SULFIDE	-0.00928	6.0706E-05	1.4628E-08	275	1000	0.01012	0.00852	0.06605
223	C3H9N	n-PROPYLAMINE	-0.01367	8.0487E-05	4.6285E-08	322	750	------	0.01705	0.07273
224	C3H9N	ISOPROPYLAMINE	-0.01430	8.9570E-05	3.6544E-08	306	750	------	0.01653	0.07343
225	C3H9N	TRIMETHYLAMINE	-0.02411	1.2442E-04	1.6155E-08	273	996	0.01442	0.01106	0.11584
226	C3H9NO	1-AMINO-2-PROPANOL	-0.01161	7.3773E-05	1.8971E-08	433	1000	------	0.02389	0.08113
227	C3H9NO	3-AMINO-1-PROPANOL	-0.01163	7.0354E-05	1.9736E-08	461	1000	------	0.02500	0.07846
228	C3H9NO	METHYLETHANOLAMINE	-0.01313	8.3075E-05	2.2122E-08	431	1000	------	0.02678	0.09207
229	C3H9O4P	TRIMETHYL PHOSPHATE	------	--------	--------	----	----	------	------	-----
230	C3H10N2	1,2-PROPANEDIAMINE	0.02083	-8.7351E-05	2.2027E-07	392	1000	------	0.02044	0.15375
231	C3H10Si	TRIMETHYL SILANE	-0.01166	8.1852E-05	2.0546E-08	280	1000	0.01457	0.01287	0.09074
232	C4Cl4S	TETRACHLOROTHIOPHENE	------	--------	--------	----	----	------	-----	-----
233	C4Cl6	HEXACHLORO-1,3-BUTADIENE	-0.00197	1.2558E-05	3.3506E-08	488	998	------	0.01214	0.04393
234	C4F8	OCTAFLUORO-2-BUTENE	-0.01057	6.1000E-05	2.1327E-08	270	1000	0.00951	0.00745	0.07176

Table 23-1 THERMAL CONDUCTIVITY OF GAS - ORGANIC COMPOUNDS (continued)

NO	FORMULA	NAME	$k_{gas} = A + B T + C T^2$					$(k_{gas} - W/(m K), T - K)$		
			A	B	C	TMIN	TMAX	k_{gas} @25 C	k_{gas} @TMIN	k_{gas} @TMAX
235	C4F8	OCTAFLUOROCYCLOBUTANE	-0.00180	3.8498E-05	3.2808E-09	267	600	0.00997	0.00871	0.02248
236	C4F10	DECAFLUOROBUTANE	-0.01123	6.3948E-05	2.0767E-08	271	1000	0.00968	0.00763	0.07349
237	C4H2	BUTADIYNE(BIACETYLENE)	-0.00870	7.7983E-05	-7.2142E-09	275	1000	0.01391	0.01220	0.06207
238	C4H2O3	MALEIC ANHYDRIDE	-0.01006	6.7349E-05	9.6585E-09	475	1000	-----	0.02411	0.06695
239	C4H4	VINYLACETYLENE	-------	--------	-------	----	----	-----	-----	-----
240	C4H4N2	SUCCINONITRILE	-0.00768	4.4693E-05	1.0167E-08	540	990	-----	0.01942	0.04653
241	C4H4O	FURAN	-0.00805	3.6512E-05	1.0838E-07	305	995	-----	0.01317	0.13558
242	C4H4O2	DIKETENE	-0.01127	6.9980E-05	2.6027E-08	399	1000	-----	0.02080	0.08474
243	C4H8O3	SUCCINIC ANHYDRIDE	-0.01000	5.8323E-05	1.2919E-08	537	1000	-----	0.02504	0.06124
244	C4H4O4	FUMARIC ACID	-0.00846	6.0688E-05	2.0178E-09	563	1000	-----	0.02635	0.05425
245	C4H4O4	MALEIC ACID	-0.00889	5.7067E-05	-7.6083E-10	565	1000	-----	0.02311	0.04742
246	C4H4S	THIOPHENE	-0.00640	2.2517E-05	1.0225E-07	357	997	-----	0.01467	0.11769
247	C4H5Cl	CHLOROPRENE	-0.00442	4.0609E-05	4.1938E-08	333	1000	-----	0.01375	0.07813
248	C4H5N	trans-CROTONITRILE	0.01323	-5.2954E-05	1.3968E-07	394	1000	-----	0.01405	0.09996
249	C4H5N	cis-CROTONITRILE	0.01472	-6.0739E-05	1.5873E-07	381	1000	-----	0.01462	0.11271
250	C4H5N	METHACRYLONITRILE	0.01165	-4.7234E-05	1.3161E-07	363	1000	-----	0.01185	0.09603
251	C4H5N	PYRROLE	-0.00905	5.3595E-05	2.4407E-08	403	993	-----	0.01651	0.06824
252	C4H5N	VINYLACETONITRILE	0.00816	-3.7331E-05	1.3829E-07	392	800	-----	0.01478	0.06680
253	C4H5NO2	METHYL CYANOACETATE	-0.00774	4.8250E-05	1.0354E-08	478	1000	-----	0.01769	0.05086
254	C4H6	CYCLOBUTENE	-0.01682	9.2597E-05	8.8624E-09	275	1000	0.01158	0.00931	0.08464
255	C4H6	1,2-BUTADIENE	-0.01059	7.7710E-05	2.9026E-08	284	994	0.01516	0.01382	0.09533
256	C4H6	1,3-BUTADIENE	-0.00085	7.1537E-06	1.6202E-07	250	850	0.01569	0.01106	0.12229
257	C4H6	DIMETHYLACETYLENE	-0.01032	7.0697E-05	3.3110E-08	300	990	-----	0.01387	0.09212
258	C4H6	ETHYLACETYLENE	-0.01990	1.1278E-04	7.8857E-09	281	1000	0.01443	0.01241	0.10077
259	C4H6Cl2	1,3-DICHLORO-trans-2-BUTENE	-0.00689	5.0851E-05	6.6399E-09	402	1000	-----	0.01463	0.05060
260	C4H6Cl2	1,4-DICHLORO-cis-2-BUTENE	-0.00314	2.2503E-05	4.1277E-08	426	1000	-----	0.01394	0.06064
261	C4H6Cl2	1,4-DICHLORO-trans-2-BUTENE	-0.00823	5.7352E-05	8.8581E-09	429	1000	-----	0.01800	0.05798
262	C4H6Cl2	3,4-DICHLORO-1-BUTENE	-0.00705	5.0428E-05	9.1966E-09	388	1000	-----	0.01390	0.05257
263	C4H6O	trans-CROTONALDEHYDE	0.00146	-4.7110E-07	9.5147E-08	377	1000	-----	0.01481	0.09614
264	C4H6O	2,5-DIHYDROFURAN	-0.01873	9.8507E-05	1.0017E-08	339	1000	-----	0.01582	0.08979
265	C4H6O	DIVINYL ETHER	-0.00093	6.6325E-06	1.3838E-07	301	1000	-----	0.01360	0.14408
266	C4H6O	METHACROLEIN	-0.01809	8.1438E-05	1.3203E-08	341	1000	-----	0.01122	0.07655
267	C4H6O2	2-BUTYNE-1,4-DIOL	-0.01163	6.5134E-05	7.8504E-09	511	1000	-----	0.02370	0.06135
268	C4H6O2	gamma-BUTYROLACTONE	-0.01761	7.1605E-05	1.0108E-08	477	1000	-----	0.01885	0.06410
269	C4H6O2	cis-CROTONIC ACID	-0.01022	6.9486E-05	1.1123E-08	445	1000	-----	0.02290	0.07039
270	C4H6O2	trans-CROTONIC ACID	-0.00945	5.8963E-05	1.4303E-08	458	1000	-----	0.02056	0.06382
271	C4H6O2	METHACRYLIC ACID	-0.00951	5.8900E-05	1.7241E-08	434	1000	-----	0.01930	0.06663
272	C4H6O2	METHYL ACRYLATE	-0.00757	5.7245E-05	1.0008E-08	353	993	-----	0.01388	0.05914
273	C4H6O2	VINYL ACETATE	-0.00846	5.8704E-05	1.7678E-08	346	996	-----	0.01397	0.06755
274	C4H6O3	ACETIC ANHYDRIDE	-0.00846	5.2818E-05	1.7355E-08	413	993	-----	0.01631	0.06110
275	C4H6O4	SUCCINIC ACID	-0.00891	5.7738E-05	4.6079E-09	591	1000	-----	0.02682	0.05344
276	C4H6O5	DIGLYCOLIC ACID	-0.00866	5.4393E-05	4.8789E-09	610	1000	-----	0.02640	0.05061
277	C4H6O5	MALIC ACID	-0.00845	5.4754E-05	4.0264E-09	602	1000	-----	0.02597	0.05033
278	C4H6O6	TARTARIC ACID	-0.00852	5.3276E-05	3.4807E-09	660	1000	-----	0.02816	0.04824
279	C4H7N	n-BUTYRONITRILE	-0.01638	8.3032E-05	9.6206E-09	273	1000	0.00923	0.00700	0.07627
280	C4H7N	ISOBUTYRONITRILE	-0.01418	8.0980E-05	5.8675E-09	377	1000	-----	0.01718	0.07267
281	C4H7NO	ACETONE CYANOHYDRIN	-0.00990	6.6596E-05	1.0078E-08	463	1000	-----	0.02309	0.06677
282	C4H7NO	2-METHACRYLAMIDE	-0.00533	4.8497E-05	2.6917E-08	488	1000	-----	0.02475	0.07008
283	C4H7NO	3-METHOXYPROPIONITRILE	0.01433	-5.6999E-05	1.4063E-07	439	1000	-----	0.01641	0.09796
284	C4H7NO	2-PYRROLIDONE	-0.00961	4.9586E-05	2.2637E-08	518	998	-----	0.02215	0.06242
285	C4H8	1-BUTENE	-0.00293	3.0205E-05	1.0192E-07	225	800	0.01514	0.00903	0.08646
286	C4H8	cis-2-BUTENE	-0.02545	1.2682E-04	2.2968E-09	273	1273	0.01257	0.00934	0.13971
287	C4H8	trans-2-BUTENE	-0.02331	1.2197E-04	4.7243E-09	285	1257	0.01348	0.01184	0.13747
288	C4H8	CYCLOBUTANE	-0.00569	2.7532E-05	1.3399E-07	286	1000	0.01443	0.01314	0.15583
289	C4H8	ISOBUTENE	-0.00327	3.0146E-05	1.2529E-07	250	850	0.01686	0.01210	0.11288
290	C4H8Br2	1,2-DIBROMOBUTANE	-0.00464	3.0955E-05	5.8210E-09	275	1000	0.00511	0.00431	0.03214
291	C4H8Br2	2,3-DIBROMOBUTANE	-0.00516	3.2898E-05	5.2383E-09	275	1000	0.00511	0.00428	0.03298
292	C4H8Cl2	1,4-DICHLOROBUTANE	-0.00306	2.1837E-05	4.0406E-08	427	1000	-----	0.01363	0.05918
293	C4H8I2	1,2-DIIODOBUTANE	-0.00320	2.2100E-05	5.8717E-09	275	1000	0.00391	0.00332	0.02477
294	C4H8O	n-BUTYRALDEHYDE	-0.00088	5.9120E-06	1.0642E-07	348	998	-----	0.01407	0.11101
295	C4H8O	ISOBUTYRALDEHYDE	0.00200	-2.0700E-06	1.2185E-07	300	1000	-----	0.01235	0.12178
296	C4H8O	1,2-EPOXYBUTANE	-0.01132	7.7341E-05	2.6647E-08	337	1000	-----	0.01777	0.09267
297	C4H8O	METHYL ETHYL KETONE	0.00169	-2.1055E-06	1.1764E-07	353	993	-----	0.01561	0.11560
298	C4H8O	ETHYL VINYL ETHER	0.00250	-6.1279E-06	1.6422E-07	309	1000	-----	0.01629	0.16059
299	C4H8O	TETRAHYDROFURAN	-0.00752	2.8645E-05	1.1344E-07	338	998	-----	0.01512	0.13405
300	C4H8O2	cis-2-BUTENE-1,4-DIOL	-0.01089	6.2735E-05	1.7348E-08	508	1000	-----	0.02546	0.06919
301	C4H8O2	trans-2-BUTENE-1,4-DIOL	-0.01103	6.4484E-05	1.6323E-08	510	1000	-----	0.02610	0.06978
302	C4H8O2	ISOBUTYRIC ACID	-0.01589	7.9419E-05	6.6228E-09	428	1000	-----	0.01931	0.07015
303	C4H8O2	n-BUTYRIC ACID	0.12421	-3.6238E-04	3.7750E-07	523	707	-----	0.03794	0.05670
304	C4H8O2	1,4-DIOXANE	-0.01642	8.0095E-05	1.7582E-08	285	1000	0.00902	0.00784	0.08126
305	C4H8O2	ETHYL ACETATE	0.00207	-4.8558E-06	1.1222E-07	273	1000	0.01060	0.00911	0.10943
306	C4H8O2	METHYL PROPIONATE	0.00133	-1.5412E-06	1.0895E-07	350	1000	-----	0.01414	0.10874
307	C4H8O2	n-PROPYL FORMATE	0.00137	-1.7055E-06	1.0998E-07	350	1000	-----	0.01425	0.10964
308	C4H8O2S	SULFOLANE	-0.00950	4.8342E-05	1.9246E-08	558	1000	-----	0.02347	0.05809
309	C4H8S	TETRAHYDROTHIOPHENE	-0.00872	5.4337E-05	1.9864E-08	394	1000	-----	0.01577	0.06548
310	C4H9Br	1-BROMOBUTANE	-0.00683	4.5241E-05	1.4105E-08	375	995	-----	0.01212	0.05215
311	C4H9Br	2-BROMOBUTANE	-0.00759	5.2507E-05	1.3669E-08	364	1000	-----	0.01333	0.05859
312	C4H9Cl	n-BUTYL CHLORIDE	-0.00720	4.3604E-05	4.4541E-08	273	1000	0.00976	0.00802	0.08095

Table 23-1 THERMAL CONDUCTIVITY OF GAS - ORGANIC COMPOUNDS (continued)

			$k_{gas} = A + B T + C T^2$					$(k_{gas} - W/(m\ K),\ T - K)$		
NO	FORMULA	NAME	A	B	C	TMIN	TMAX	k_{gas} @25 C	k_{gas} @TMIN	k_{gas} @TMAX
313	C4H9Cl	sec-BUTYL CHLORIDE	-0.00920	5.2142E-05	4.3379E-08	341	1000	-----	0.01362	0.08632
314	C4H9Cl	tert-BUTYL CHLORIDE	-0.00934	5.4477E-05	4.4848E-08	324	1000	-----	0.01302	0.08999
315	C4H9I	2-IODO-2-METHYLPROPANE	-0.00704	4.3502E-05	5.3335E-09	275	1000	0.00640	0.00533	0.04180
316	C4H9N	PYRROLIDINE	-0.00761	2.6565E-05	1.2103E-07	360	1000	-----	0.01764	0.13999
317	C4H9NO	N,N-DIMETHYLACETAMIDE	-0.01632	7.1845E-05	1.7748E-08	439	1000	-----	0.01864	0.07327
318	C4H9NO	MORPHOLINE	-0.00627	1.7533E-05	1.1623E-07	401	1000	-----	0.01945	0.12749
319	C4H9NO2	1-NITROBUTANE	-0.00947	5.6026E-05	1.3489E-08	275	1000	0.00843	0.00696	0.06005
320	C4H9NO2	2-NITROBUTANE	-0.01032	5.9918E-05	1.3962E-08	275	1000	0.00879	0.00721	0.06356
321	C4H10	n-BUTANE	-0.00182	1.9396E-05	1.3818E-07	225	675	0.01625	0.00954	0.07423
322	C4H10	ISOBUTANE	-0.00115	1.4943E-05	1.4921E-07	261	673	0.01657	0.01291	0.07649
323	C4H10N2	PIPERAZINE	-0.03129	1.4026E-04	-1.4536E-08	419	1000	-----	0.02493	0.09443
324	C4H10O	n-BUTANOL	0.01783	-4.8291E-05	1.6334E-07	371	713	-----	0.02240	0.06644
325	C4H10O	sec-BUTANOL	-0.01103	7.0807E-05	2.6401E-08	373	993	-----	0.01905	0.08531
326	C4H10O	tert-BUTANOL	0.01023	-1.3515E-05	1.2679E-07	356	1000	-----	0.02149	0.12351
327	C4H10O	DIETHYL ETHER	-0.00032	1.6530E-05	1.1709E-07	200	600	0.01502	0.00767	0.05175
328	C4H10O	METHYL-PROPYL-ETHER	-0.01159	7.3721E-05	1.6595E-08	275	1000	0.01187	0.00994	0.07873
329	C4H10O	METHYL ISOPROPYL ETHER	0.00225	-4.9306E-06	1.5452E-07	304	1000	-----	0.01503	0.15184
330	C4H10O	ISOBUTANOL	-0.00278	1.8899E-05	9.5375E-08	381	991	-----	0.01827	0.10961
331	C4H10O2	1,3-BUTANEDIOL	-0.01033	6.4637E-05	1.3921E-08	480	990	-----	0.02390	0.06730
332	C4H10O2	1,4-BUTANEDIOL	-0.00319	1.6128E-05	8.4206E-08	500	1000	-----	0.02593	0.09714
333	C4H10O2	2,3-BUTANEDIOL	-0.01215	7.9859E-05	1.4777E-08	454	1000	-----	0.02715	0.08249
334	C4H10O2	t-BUTYL HYDROPEROXIDE	-0.01208	8.4197E-05	1.5494E-08	406	1000	-----	0.02466	0.08761
335	C4H10O2	1,2-DIMETHOXYETHANE	-0.00452	4.7532E-05	2.8883E-08	357	1000	-----	0.01613	0.07190
336	C4H10O2	2-ETHOXYETHANOL	-0.01043	6.6755E-05	1.9427E-08	408	1000	-----	0.02004	0.07575
337	C4H10O3	DIETHYLENE GLYCOL	0.00003	6.4926E-06	7.7523E-08	518	1000	-----	0.02419	0.08405
338	C4H10O4S	DIETHYL SULFATE	-----	--------	--------	---	---	-----	-----	-----
339	C4H10S	n-BUTYL MERCAPTAN	-0.01008	5.9889E-05	3.2200E-08	372	1000	-----	0.01665	0.08201
340	C4H10S	ISOBUTYL MERCAPTAN	-0.01070	7.0253E-05	2.4641E-08	362	1000	-----	0.01796	0.08419
341	C4H10S	sec-BUTYL MERCAPTAN	-0.01462	8.1642E-05	1.8126E-08	358	1000	-----	0.01693	0.08515
342	C4H10S	tert-BUTYL MERCAPTAN	-0.01599	8.5348E-05	1.2343E-08	337	1000	-----	0.01417	0.08170
343	C4H10S	DIETHYL SULFIDE	-0.01485	7.5068E-05	1.8733E-08	365	1000	-----	0.01505	0.07895
344	C4H10S	ISOPROPYL-METHYL-SULFIDE	-0.00858	5.5002E-05	2.0988E-08	275	1000	0.00968	0.00813	0.06741
345	C4H10S	METHYL-PROPYL-SULFIDE	-0.00881	5.5653E-05	1.7640E-08	275	1000	0.00935	0.00783	0.06448
346	C4H10S2	DIETHYL DISULFIDE	-0.00948	6.0975E-05	1.5211E-08	427	1000	-----	0.01933	0.06671
347	C4H11N	n-BUTYLAMINE	-0.00898	6.0958E-05	6.0534E-08	273	700	0.01458	0.01217	0.06335
348	C4H11N	ISOBUTYLAMINE	-0.01865	1.0231E-04	1.7910E-08	341	991	-----	0.01832	0.10033
349	C4H11N	sec-BUTYLAMINE	-0.01375	8.1713E-05	3.8756E-08	336	700	-----	0.01808	0.06244
350	C4H11N	tert-BUTYLAMINE	-0.02218	1.2023E-04	8.5934E-09	318	998	-----	0.01692	0.10637
351	C4H11N	DIETHYLAMINE	-0.02191	1.1401E-04	1.5079E-08	273	1000	0.01342	0.01034	0.10718
352	C4H11NO	DIMETHYLETHANOLAMINE	-0.01852	9.6357E-05	1.5390E-08	407	1000	-----	0.02325	0.09323
353	C4H11NO2	DIETHANOLAMINE	-----	--------	--------	----	----	-----	-----	-----
354	C4H11NO2	2-AMINOETHOXYETHANOL	-0.01141	6.9314E-05	1.3928E-08	514	1000	-----	0.02790	0.07183
355	C4H12N2O	N-AMINOETHYL ETHANOLAMINE	-0.01666	8.1012E-05	8.5233E-09	517	1000	-----	0.02750	0.07288
356	C4H12Si	TETRAMETHYLSILANE	-0.01305	8.9038E-05	1.5212E-08	300	1000	-----	0.01503	0.09120
357	C4H13N3	DIETHYLENE TRIAMINE	-0.01657	8.1317E-05	9.0637E-09	480	1000	-----	0.02455	0.07381
358	C5Cl6	HEXACHLOROCYCLOPENTADIENE	-0.00249	2.8374E-05	1.1725E-09	512	1000	-----	0.01234	0.02706
359	C5H4O2	FURFURAL	-0.00850	5.3869E-05	1.4154E-08	435	995	-----	0.01761	0.05911
360	C5H5N	PYRIDINE	-0.00469	1.6151E-05	9.4006E-08	388	998	-----	0.01573	0.10506
361	C5H6	CYCLOPENTADIENE	-0.02157	1.1092E-04	4.9200E-09	315	995	-----	0.01386	0.09367
362	C5H6	2-METHYL-1-BUTENE-3-YNE	-0.02061	1.0795E-04	-3.8523E-10	305	1000	-----	0.01228	0.08695
363	C5H6	1-PENTENE-3-YNE	-0.01948	9.7606E-05	2.9124E-09	332	1000	-----	0.01325	0.08104
364	C5H6	1-PENTENE-4-YNE	-0.01895	9.9571E-05	3.1222E-09	316	1000	-----	0.01283	0.08374
365	C5H6N2	GLUTARONITRILE	0.00651	-2.5737E-05	8.0925E-08	559	800	-----	0.01741	0.03771
366	C5H6O2	FURFURYL ALCOHOL	-0.00980	6.6130E-05	1.1295E-08	443	1000	-----	0.02171	0.06763
367	C5H6O3	GLUTARIC ANHYDRIDE	-0.01623	6.8241E-05	1.3318E-08	537	1000	-----	0.02426	0.06533
368	C5H6O4	CITRACONIC ACID	-0.00950	5.9892E-05	5.3306E-09	607	1000	-----	0.02882	0.05572
369	C5H6O4	ITACONIC ACID	-0.00930	5.8805E-05	5.3028E-09	601	1000	-----	0.02796	0.05481
370	C5H6S	2-METHYLTHIOPHENE	-0.00972	5.5850E-05	9.9456E-09	275	1000	0.00782	0.00639	0.05608
371	C5H6S	3-METHYLTHIOPHENE	-0.00982	5.6366E-05	8.5607E-09	275	1000	0.00775	0.00633	0.05511
372	C5H7N	N-METHYLPYRROLE	-0.01610	8.0996E-05	2.0850E-08	386	1000	-----	0.01827	0.08575
373	C5H7NO2	ETHYL CYANOACETATE	-0.00781	5.0032E-05	9.1160E-09	479	1000	-----	0.01825	0.05134
374	C5H8	CYCLOPENTENE	-0.02478	1.0653E-04	1.9323E-08	317	1000	-----	0.01093	0.10107
375	C5H8	ISOPRENE	-0.00963	7.4045E-05	1.7069E-08	307	1000	-----	0.01471	0.08148
376	C5H8	3-METHYL-1,2-BUTADIENE	-0.01830	9.3896E-05	1.3853E-08	314	1000	-----	0.01255	0.08945
377	C5H8	2-METHYL-1,3-BUTADIENE	-0.01402	8.6464E-05	5.2556E-09	275	1000	0.01223	0.01016	0.07770
378	C5H8	1,2-PENTADIENE	-0.01814	9.4388E-05	1.2776E-08	318	1000	-----	0.01317	0.08902
379	C5H8	cis-1,3-PENTADIENE	-0.02272	1.0735E-04	2.3430E-09	317	1000	-----	0.01155	0.08697
380	C5H8	trans-1,3-PENTADIENE	-0.02171	1.0827E-04	9.7348E-10	315	1000	-----	0.01249	0.08753
381	C5H8	1,4-PENTADIENE	-0.02050	1.0732E-04	4.4840E-09	299	1000	-----	0.01199	0.09130
382	C5H8	2,3-PENTADIENE	-0.01808	9.5064E-05	1.1518E-08	321	1000	-----	0.01362	0.08850
383	C5H8	1-PENTYNE	-0.02135	1.1353E-04	4.1053E-09	313	1000	-----	0.01459	0.09629
384	C5H8	2-PENTYNE	-0.01047	6.5991E-05	1.6124E-08	275	1000	0.01064	0.00890	0.07165
385	C5H8	3-METHYL-1-BUTYNE	-0.02238	1.1883E-04	5.0355E-09	302	1000	-----	0.01397	0.10149
386	C5H8	SPIROPENTANE	-0.01655	8.8186E-05	1.1391E-08	275	1000	0.01076	0.00856	0.08303
387	C5H8N4O12	PENTAERYTHRITOL TETRANITRATE	-----	--------	--------	---	---	-----	-----	-----
388	C5H8O	CYCLOPENTANONE	-0.01474	6.5495E-05	2.2503E-08	404	1000	-----	0.01539	0.07326
389	C5H8O	METHYL ISOPROPENYL KETONE	0.00191	-5.0402E-06	1.2275E-07	371	1000	-----	0.01694	0.11962
390	C5H8O2	ACETYLACETONE	0.00202	-6.4635E-06	1.1971E-07	414	1000	-----	0.01986	0.11527

512

Table 23-1 THERMAL CONDUCTIVITY OF GAS - ORGANIC COMPOUNDS (continued)

			$k_{gas} = A + B T + C T^2$						(k_{gas} - W/(m K), T - K)		
NO	FORMULA	NAME	A	B	C	TMIN	TMAX	k_{gas} @25 C	k_{gas} @TMIN	k_{gas} @TMAX
391	C5H8O2	ALLYL ACETATE	0.00167	-4.3717E-06	1.1046E-07	377	1000	------	0.01572	0.10776
392	C5H8O2	ETHYL ACRYLATE	-0.01596	8.5844E-05	6.3310E-10	373	1000	------	0.01615	0.07052
393	C5H8O2	METHYL METHACRYLATE	-0.00858	5.6352E-05	1.8538E-08	373	993	------	0.01502	0.06566
394	C5H8O2	VINYL PROPIONATE	-0.00769	5.3910E-05	1.3380E-08	364	994	------	0.01371	0.05912
395	C5H8O3	2-HYDROXYETHYL ACRYLATE	-0.00958	6.6063E-05	8.3262E-09	483	1000	------	0.02427	0.06481
396	C5H8O3	LEVULINIC ACID	-0.00876	5.3118E-05	1.0459E-08	519	1000	------	0.02163	0.05482
397	C5H8O3	METHYL ACETOACETATE	-0.01100	5.9144E-05	8.6685E-09	445	1000	------	0.01704	0.05681
398	C5H8O4	GLUTARIC ACID	-0.00932	5.6714E-05	7.0118E-09	596	1000	------	0.02697	0.05441
399	C5H9N	VALERONITRILE	-0.00829	5.3602E-05	1.4187E-08	414	1000	------	0.01633	0.05950
400	C5H9NO	n-BUTYL ISOCYANATE	-0.00907	5.8175E-05	1.9087E-08	388	1000	------	0.01638	0.06819
401	C5H9NO	N-METHYL-2-PYRROLIDONE	-0.01854	7.7242E-05	1.6547E-08	475	1000	------	0.02188	0.07525
402	C5H9NO4	L-GLUTAMIC ACID	-0.00933	5.2703E-05	6.9953E-09	670	1000	------	0.02912	0.05037
403	C5H10	CYCLOPENTANE	-0.02062	9.4739E-05	3.5560E-08	273	1000	0.01079	0.00789	0.10968
404	C5H10	2-METHYL-1-BUTENE	-0.02204	1.1084E-04	9.4377E-09	304	994	------	0.01253	0.09746
405	C5H10	2-METHYL-2-BUTENE	-0.01074	7.0325E-05	3.7385E-08	312	992	------	0.01484	0.09581
406	C5H10	3-METHYL-1-BUTENE	-0.01347	9.2089E-05	2.6520E-08	363	993	------	0.02345	0.10412
407	C5H10	1-PENTENE	-0.01898	1.0348E-04	1.9775E-08	230	900	0.01363	0.00587	0.09017
408	C5H10	cis-2-PENTENE	-0.02130	1.0595E-04	1.5701E-08	300	1000	------	0.01190	0.10035
409	C5H10	trans-2-PENTENE	-0.02255	1.1262E-04	1.0063E-08	300	1000	------	0.01214	0.10013
410	C5H10Br2	2,3-DIBROMO-2-METHYLBUTANE	-0.00586	3.5837E-05	5.6388E-09	275	1000	0.00533	0.00442	0.03562
411	C5H10Cl2	1,5-DICHLOROPENTANE	-0.00306	2.0818E-05	4.3956E-08	453	1000	------	0.01539	0.06171
412	C5H10O	METHYL ISOPROPYL KETONE	0.00210	-6.1498E-06	1.3239E-07	368	1000	------	0.01777	0.12834
413	C5H10O	2-PENTANONE	-0.00010	2.9234E-07	1.1956E-07	273	995	0.01062	0.00889	0.11856
414	C5H10O	DIETHYL KETONE	0.00004	-2.7408E-07	1.2105E-07	273	1000	0.01072	0.00899	0.12082
415	C5H10O	VALERALDEHYDE	0.00195	-5.2241E-06	1.1664E-07	376	1000	------	0.01648	0.11337
416	C5H10O2	n-BUTYL FORMATE	-0.00003	1.9886E-07	1.0678E-07	379	999	------	0.01538	0.10674
417	C5H10O2	ETHYL PROPIONATE	0.00169	-3.9463E-06	9.7439E-08	350	1000	------	0.01225	0.09518
418	C5H10O2	ISOBUTYL FORMATE	-0.01149	7.6532E-05	2.3296E-08	371	1000	------	0.02011	0.08834
419	C5H10O2	ISOPROPYL ACETATE	0.00186	-5.5341E-06	1.2208E-07	362	1000	------	0.01585	0.11841
420	C5H10O2	n-PROPYL ACETATE	------	1.2015E-11	1.0832E-07	375	1000	------	0.01523	0.10832
421	C5H10O2	METHYL n-BUTYRATE	-0.00008	5.3061E-07	1.0498E-07	376	1000	------	0.01496	0.10543
422	C5H10O2	2-METHYLBUTYRIC ACID	-0.01063	6.5791E-05	1.7472E-08	450	1000	------	0.02251	0.07263
423	C5H10O2	ISOVALERIC ACID	-0.01009	6.3363E-05	1.5718E-08	448	1000	------	0.02145	0.06899
424	C5H10O2	VALERIC ACID	-0.00993	6.2466E-05	1.4445E-08	459	1000	------	0.02179	0.06698
425	C5H10O2	TETRAHYDROFURFURYL ALCOHOL	-0.01256	7.4885E-05	2.4009E-08	451	1000	------	0.02610	0.08633
426	C5H10O2S	3-METHYL SULFOLANE	-0.00995	5.2079E-05	1.8925E-08	549	1000	------	0.02435	0.06105
427	C5H10O3	DIETHYL CARBONATE	------	------	--------	----	----	------	------	------
428	C5H10O3	ETHYL LACTATE	-0.01357	7.3304E-05	3.4967E-09	428	1000	------	0.01844	0.06323
429	C5H10S	THIACYCLOHEXANE	-0.01234	6.0384E-05	2.3795E-08	275	1000	0.00778	0.00607	0.07184
430	C5H10S	CYCLOPENTANETHIOL	-0.01072	5.6324E-05	1.7970E-08	275	1000	0.00767	0.00613	0.06357
431	C5H11Br	1-BROMOPENTANE	-0.00755	4.5704E-05	9.6435E-09	275	1000	0.00693	0.00575	0.04780
432	C5H11Cl	1-CHLOROPENTANE	-0.00691	3.9991E-05	5.0394E-08	273	1000	0.00949	0.00776	0.08348
433	C5H11Cl	1-CHLORO-3-METHYLBUTANE	-0.00983	5.9717E-05	1.1167E-08	275	1000	0.00897	0.00744	0.06105
434	C5H11Cl	2-CHLORO-2-METHYLBUTANE	-0.01090	6.4071E-05	1.0375E-08	275	1000	0.00913	0.00750	0.06355
435	C5H11N	N-METHYLPYRROLIDINE	-0.00622	2.2417E-05	9.6605E-08	352	1000	------	0.01364	0.11280
436	C5H11N	PIPERIDINE	-0.03828	1.6072E-04	-1.7430E-08	380	1000	------	0.02028	0.10501
437	C5H11NO	tert-BUTYLFORMAMIDE	-0.01047	5.9203E-05	2.1578E-08	475	1000	------	0.02252	0.07031
438	C5H12	ISOPENTANE	-0.00389	3.1816E-05	1.0396E-07	250	475	0.01484	0.01056	0.03468
439	C5H12	NEOPENTANE	0.00071	-8.1028E-06	1.8759E-07	270	425	0.01497	0.01220	0.03115
440	C5H12	n-PENTANE	-0.00137	1.8081E-05	1.2136E-07	225	480	0.01481	0.00884	0.03527
441	C5H12O	2,2-DIMETHYL-1-PROPANOL	-0.00247	1.6408E-05	9.9920E-08	386	1000	------	0.01875	0.11386
442	C5H12O	tert-PENTYL-ALCOHOL	-0.01141	6.5724E-05	1.3064E-08	275	1000	0.00935	0.00765	0.06738
443	C5H12O	2-METHYL-1-BUTANOL	-0.02047	1.0324E-04	1.1287E-08	402	1000	------	0.02286	0.09406
444	C5H12O	2-METHYL-2-BUTANOL	-0.01059	6.7601E-05	2.5324E-08	375	995	------	0.01832	0.08174
445	C5H12O	3-METHYL-1-BUTANOL	-0.00240	1.5505E-05	1.0343E-07	404	1000	------	0.02075	0.11654
446	C5H12O	3-METHYL-2-BUTANOL	-0.01024	6.7212E-05	2.0039E-08	385	1000	------	0.01861	0.07701
447	C5H12O	1-PENTANOL	-0.00237	1.5340E-05	9.9882E-08	411	991	------	0.02081	0.11092
448	C5H12O	2-PENTANOL	-0.01815	9.7025E-05	6.6841E-09	392	1000	------	0.02091	0.08556
449	C5H12O	3-PENTANOL	-0.00498	4.6859E-05	3.4060E-08	388	1000	------	0.01833	0.07594
450	C5H12O	METHYL sec-BUTYL ETHER	-0.01627	9.4309E-05	1.0734E-08	332	1000	------	0.01622	0.08877
451	C5H12O	METHYL tert-BUTYL ETHER	0.00443	-1.1115E-05	1.1739E-07	273	1000	0.01155	0.01014	0.11071
452	C5H120	METHYL ISOBUTYL ETHER	0.00229	-7.2080E-06	1.4524E-07	332	1000	------	0.01591	0.14032
453	C5H12O	ETHYL PROPYL ETHER	-0.01461	9.8182E-05	-8.8708E-09	273	550	0.01387	0.01153	0.03671
454	C5H12O2	ETHYLENE GLYCOL MONOPROPYL ETHER	-0.01074	6.8825E-05	1.7860E-08	425	1000	------	0.02174	0.07595
455	C5H12O2	NEOPENTYL GLYCOL	-0.02009	9.4129E-05	2.0098E-09	483	1000	------	0.02584	0.07605
456	C5H12O2	1,5-PENTANEDIOL	-0.00222	1.2905E-05	8.7672E-08	512	992	------	0.02737	0.09686
457	C5H12O3	2-(2-METHOXYETHOXY)ETHANOL	-0.01010	6.1490E-05	1.6098E-08	467	1000	------	0.02213	0.06749
458	C5H12O4	PENTAERYTHRITOL	0.00004	5.5260E-06	7.1811E-08	631	1000	------	0.03212	0.07738
459	C5H12S	n-PENTYL MERCAPTAN	-0.01410	7.0973E-05	2.3670E-08	400	1000	------	0.01808	0.08054
460	C5H12S	BUTYL-METHYL-SULFIDE	-0.00826	5.1219E-05	2.0589E-08	275	1000	0.00884	0.00738	0.06355
461	C5H12S	ETHYL-PROPYL-SULFIDE	-0.00841	5.1418E-05	2.1121E-08	275	1000	0.00880	0.00733	0.06413
462	C5H12S	2-METHYL-2-BUTANETHIOL	-0.01071	6.4668E-05	1.2025E-08	275	1000	0.00964	0.00798	0.06598
463	C5H13N	n-PENTYLAMINE	-0.01089	6.4183E-05	5.5586E-08	280	700	0.01319	0.01144	0.06128
464	C5H13NO2	METHYL DIETHANOLAMINE	-0.01745	8.0243E-05	1.0917E-08	520	1000	------	0.02723	0.07371
465	C6Cl6	HEXACHLOROBENZENE	-0.00182	1.0469E-05	3.1217E-08	583	1000	------	0.01489	0.03987
466	C6F6	HEXAFLUOROBENZENE	-0.00405	2.7373E-05	4.8096E-08	353	600	------	0.01161	0.02969
467	C6H3ClN2O4	1-CHLORO-2,4-DINITROBENZENE	------	------	--------	----	----	------	------	------
468	C3H3Cl2NO2	1,2-DICHLORO-4-NITROBENZENE	------	------	--------	----	----	------	------	------

Table 23-1 THERMAL CONDUCTIVITY OF GAS - ORGANIC COMPOUNDS (continued)

			$k_{gas} = A + BT + CT^2$						$(k_{gas}$ - W/(m K), T - K)		
NO	FORMULA	NAME	A	B	C	TMIN	TMAX	k_{gas} @25 C	k_{gas} @TMIN	k_{gas} @TMAX
469	C6H3Cl3	1,2,4-TRICHLOROBENZENE	-0.00267	1.7256E-05	4.1140E-08	486	1000	-----	0.01543	0.05573
470	C6H3N3O6	1,3,5-TRINITROBENZENE	------	--------	--------	----	----	-----	-----	-----
471	C6H4Br2	m-DIBROMOBENZENE	-0.00485	3.2423E-05	4.2040E-09	491	1000	-----	0.01208	0.03178
472	C6H4ClNO2	m-CHLORONITROBENZENE	-0.01272	5.8834E-05	-1.0453E-09	509	1000	-----	0.01696	0.04507
473	C6H4ClNO2	o-CHLORONITROBENZENE	-0.01285	5.7825E-05	-1.1171E-09	519	1000	-----	0.01686	0.04386
474	C6H4ClNO2	p-CHLORONITROBENZENE	-0.01253	5.8543E-05	-5.8007E-10	515	1000	-----	0.01747	0.04543
475	C6H4Cl2	m-DICHLOROBENZENE	-0.00312	2.1250E-05	4.6201E-08	446	1000	-----	0.01555	0.06433
476	C6H4Cl2	o-DICHLOROBENZENE	-0.00291	1.9569E-05	4.4032E-08	454	1000	-----	0.01505	0.06069
477	C6H4Cl2	p-DICHLOROBENZENE	-0.00311	2.1221E-05	4.6100E-08	447	1000	-----	0.01559	0.06421
478	C6H4F2	m-DIFLUOROBENZENE	-0.01051	6.1043E-05	1.7135E-09	275	1000	0.00784	0.00641	0.05225
479	C6H4F2	o-DIFLUOROBENZENE	-0.01049	6.1008E-05	3.5257E-09	275	1000	0.00801	0.00655	0.05404
480	C6H4F2	p-DIFLUOROBENZENE	-0.01040	6.0704E-05	1.8982E-09	275	1000	0.00787	0.00644	0.05220
481	C6H4N2O4	m-DINITROBENZENE	------	--------	--------	----	----	-----	-----	-----
482	C6H4N2O4	o-DINITROBENZENE	------	--------	--------	----	----	-----	-----	-----
483	C6H4N2O4	p-DINITROBENZENE	------	--------	--------	----	----	-----	-----	-----
484	C6H5Br	BROMOBENZENE	-0.01148	5.3994E-05	-1.7338E-10	429	1000	-----	0.01165	0.04234
485	C6H5Cl	MONOCHLOROBENZENE	-0.00974	4.5774E-05	4.3529E-08	400	1000	-----	0.01553	0.07956
486	C6H5ClO	m-CHLOROPHENOL	-0.00789	5.3233E-05	6.6551E-09	487	1000	-----	0.01961	0.05200
487	C6H5ClO	o-CHLOROPHENOL	-0.00748	5.1974E-05	7.1317E-09	448	1000	-----	0.01724	0.05163
488	C6H5ClO	p-CHLOROPHENOL	-0.00798	5.3247E-05	6.8750E-09	493	1000	-----	0.01994	0.05214
489	C6H5Cl2N	3,4-DICHLOROANILINE	-0.00678	4.5736E-05	3.6748E-09	545	1000	-----	0.01924	0.04263
490	C6H5F	FLUOROBENZENE	-0.00518	3.4460E-05	6.4643E-08	358	600	-----	0.01544	0.03877
491	C6H5I	IODOBENZENE	-0.00509	3.0753E-05	8.9177E-09	462	992	-----	0.01102	0.03419
492	C6H5NO2	NITROBENZENE	------	--------	--------	----	----	-----	-----	-----
493	C6H6	BENZENE	-0.00565	3.4493E-05	6.9298E-08	325	700	-----	0.01288	0.05245
494	C6H6ClN	m-CHLOROANILINE	-0.00835	5.4258E-05	7.8223E-09	502	1000	-----	0.02086	0.05373
495	C6H6ClN	o-CHLOROANILINE	-0.00775	4.9775E-05	8.8297E-09	482	1000	-----	0.01829	0.05085
496	C6H6ClN	p-CHLOROANILINE	-0.00835	5.4268E-05	7.7382E-09	504	1000	-----	0.02097	0.05366
497	C6H6N2	cis-DICYANO-1-BUTENE	0.01389	-5.3103E-05	1.2187E-07	501	1000	-----	0.01787	0.08266
498	C6H6N2	trans-DICYANO-1-BUTENE	0.01386	-5.3189E-05	1.2253E-07	499	1000	-----	0.01783	0.08320
499	C6H6N2	1,4-DICYANO-2-BUTENE	0.01259	-4.6044E-05	1.0455E-07	547	1000	-----	0.01869	0.07110
500	C6H6N2O2	m-NITROANILINE	------	--------	--------	----	----	-----	-----	-----
501	C6H6N2O2	o-NITROANILINE	------	--------	--------	----	----	-----	-----	-----
502	C6H6N2O2	p-NITROANILINE	------	--------	--------	----	----	-----	-----	-----
503	C6H6O	PHENOL	-0.00552	4.4952E-05	3.9900E-08	455	995	-----	0.02319	0.07871
504	C6H6O2	1,2-BENZENEDIOL	-0.00992	6.4254E-05	8.4716E-09	519	1000	-----	0.02571	0.06281
505	C6H6O2	1,3-BENZENEDIOL	-0.01036	6.4433E-05	9.0642E-09	550	1000	-----	0.02782	0.06314
506	C6H6O2	p-HYDROQUINONE	-0.01029	6.3928E-05	8.5918E-09	558	1000	-----	0.02806	0.06223
507	C6H6O3	1,2,3-BENZENETRIOL	-0.00892	5.7198E-05	5.2779E-09	582	1000	-----	0.02616	0.05356
508	C6H6S	PHENYL MERCAPTAN	-0.01559	7.4463E-05	5.5203E-09	442	1000	-----	0.01840	0.06439
509	C6H7N	ANILINE	-0.01796	8.3464E-05	1.5022E-09	458	1000	-----	0.02058	0.06701
510	C6H7N	2-METHYLPYRIDINE	-0.01825	8.1741E-05	5.2735E-09	403	1000	-----	0.01555	0.06876
511	C6H7N	3-METHYLPYRIDINE	-0.01774	8.4519E-05	5.7390E-09	417	1000	-----	0.01850	0.07252
512	C6H7N	4-METHYLPYRIDINE	-0.01891	8.4421E-05	7.4076E-09	419	1000	-----	0.01776	0.07292
513	C6H8	1,3-CYCLOHEXADIENE	-0.02301	1.0251E-04	1.5448E-08	353	1000	-----	0.01510	0.09495
514	C6H8	METHYLCYCLOPENTADIENE	-0.02496	1.1665E-04	1.8289E-09	346	1000	-----	0.01562	0.09352
515	C6H8N2	ADIPONITRILE	0.01165	-4.1640E-05	9.3712E-08	568	1000	-----	0.01823	0.06372
516	C6H8N2	METHYLGLUTARONITRILE	0.00799	-3.3574E-05	1.0027E-07	536	800	-----	0.01880	0.04530
517	C6H8N2	m-PHENYLENEDIAMINE	-0.01683	7.9123E-05	-6.3338E-10	560	1000	-----	0.02728	0.06166
518	C6H8N2	o-PHENYLENEDIAMINE	-0.01783	8.2039E-05	-2.9441E-09	525	1000	-----	0.02443	0.06126
519	C6H8N2	p-PHENYLENEDIAMINE	-0.01912	8.5298E-05	-4.8802E-09	540	1000	-----	0.02552	0.06130
520	C6H8N2	PHENYLHYDRAZINE	------	--------	--------	----	----	-----	-----	-----
521	C6H8N2O	BIS(CYANOETHYL)ETHER	-0.00825	4.9618E-05	7.2838E-09	579	1000	-----	0.02292	0.04865
522	C6H8O4	DIMETHYL MALEATE	0.00138	-4.7128E-06	7.8611E-08	478	1000	-----	0.01709	0.07528
523	C6H8O6	ASCORBIC ACID	-0.00773	5.3920E-05	7.2142E-09	637	1000	-----	0.02954	0.05340
524	C6H8O7	CITRIC ACID	-0.00783	4.7615E-05	3.9996E-09	659	1000	-----	0.02529	0.04378
525	C6H10	1-METHYLCYCLOPENTENE	-0.01386	7.0524E-05	1.6833E-08	275	1000	0.00866	0.00681	0.07350
526	C6H10	3-METHYLCYCLOPENTENE	-0.01423	7.2551E-05	1.7620E-08	275	1000	0.00897	0.00705	0.07594
527	C6H10	4-METHYLCYCLOPENTENE	-0.01397	7.1249E-05	1.8253E-08	275	1000	0.00890	0.00700	0.07553
528	C6H10	CYCLOHEXENE	-0.01193	7.1179E-05	4.2228E-08	356	996	-----	0.01876	0.10086
529	C6H10	2,3-DIMETHYL-1,3-BUTADIENE	-0.02297	1.1246E-04	7.3288E-10	342	1000	-----	0.01558	0.09022
530	C6H10	1,5-HEXADIENE	-0.02265	1.0750E-04	4.4254E-09	333	1000	-----	0.01364	0.08928
531	C6H10	cis,trans-2,4-HEXADIENE	-0.02297	1.0411E-04	2.7109E-09	357	1000	-----	0.01454	0.08385
532	C6H10	trans,trans-2,4-HEXADIENE	-0.02246	1.0474E-04	2.5653E-09	355	1000	-----	0.01505	0.08485
533	C6H10	1-HEXYNE	-0.02300	1.1387E-04	8.6721E-10	344	1000	-----	0.01627	0.09174
534	C6H10	2-HEXYNE	-0.02195	1.0011E-04	6.6097E-09	358	1000	-----	0.01474	0.08477
535	C6H10	3-HEXYNE	-0.02073	9.5896E-05	1.0364E-08	354	1000	-----	0.01452	0.08553
536	C6H10O	CYCLOHEXANONE	0.00192	-7.2593E-06	1.3921E-07	400	1000	-----	0.02129	0.13387
537	C6H10O	MESITYL OXIDE	-0.01110	7.0738E-05	2.1760E-08	403	1000	-----	0.02094	0.08140
538	C6H10O2	epsilon-CAPROLACTONE	-0.02399	9.8072E-05	4.6009E-10	488	1000	-----	0.02398	0.07454
539	C6H10O2	ETHYL METHACRYLATE	-0.00879	5.6001E-05	1.9424E-08	390	990	-----	0.01600	0.06569
540	C6H10O2	n-PROPYL ACRYLATE	-0.00907	6.2865E-05	1.3193E-08	392	1000	-----	0.01760	0.06699
541	C6H10O3	ETHYLACETOACETATE	-0.01382	6.7076E-05	2.7945E-09	454	1000	-----	0.01721	0.05605
542	C6H10O3	PROPIONIC ANHYDRIDE	-0.00879	5.4626E-05	1.5361E-08	442	992	-----	0.01836	0.06052
543	C6H10O4	ADIPIC ACID	-0.00902	5.3113E-05	7.4185E-09	611	1000	-----	0.02620	0.05151
544	C6H10O4	DIETHYL OXALATE	-0.00762	4.9563E-05	9.4162E-09	459	1000	-----	0.01711	0.05136
545	C6H10O4	ETHYLENE GLYCOL DIACETATE	-0.00769	4.9373E-05	9.8343E-09	464	1000	-----	0.01734	0.05152
546	C6H10O4	ETHYLIDENE DIACETATE	-0.01361	7.2743E-05	-1.7770E-09	442	1000	-----	0.01820	0.05736

Table 23-1 THERMAL CONDUCTIVITY OF GAS - ORGANIC COMPOUNDS (continued)

NO	FORMULA	NAME	$k_{gas} = A + B\,T + C\,T^2$					$(k_{gas}$ - W/(m K), T - K)		
			A	B	C	TMIN	TMAX	k_{gas} @25 C	k_{gas} @TMIN	k_{gas} @TMAX
547	C6H11N	HEXANENITRILE	-0.01361	7.2743E-05	-1.7770E-09	442	1000	-----	0.01820	0.05736
548	C6H11NO	epsilon-CAPROLACTAM	-0.00212	-1.6616E-06	8.6366E-08	500	1000	-----	0.01864	0.08258
549	C6H11NO	CYCLOHEXANONE OXIME	-0.02483	9.8579E-05	8.4898E-09	481	1000	-----	0.02455	0.08224
550	C6H12	CYCLOHEXANE	-0.00159	-1.7494E-07	1.4588E-07	325	650	-----	0.01376	0.05993
551	C6H12	2,3-DIMETHYL-1-BUTENE	-0.02233	1.1326E-04	6.7224E-09	329	1000	-----	0.01566	0.09765
552	C6H12	2,3-DIMETHYL-2-BUTENE	-0.02251	9.9283E-05	1.4181E-08	346	1000	-----	0.01354	0.09095
553	C6H12	3,3-DIMETHYL-1-BUTENE	-0.02536	1.2004E-04	6.7777E-09	314	1000	-----	0.01300	0.10146
554	C6H12	2-ETHYL-1-BUTENE	-0.02345	1.1104E-04	6.2519E-09	338	1000	-----	0.01480	0.09384
555	C6H12	trans-3-METHYL-2-PENTENE	-0.01252	7.1459E-05	1.5980E-08	275	1000	0.01021	0.00834	0.07492
556	C6H12	1-HEXENE	-0.00705	4.4426E-05	7.7958E-08	273	800	0.01313	0.01089	0.07838
557	C6H12	cis-2-HEXENE	-0.02390	1.0827E-04	8.4714E-09	342	1000	-----	0.01412	0.09284
558	C6H12	trans-2-HEXENE	-0.02284	1.0704E-04	8.6641E-09	341	1000	-----	0.01467	0.09286
559	C6H12	cis-3-HEXENE	-0.02485	1.1197E-04	6.6069E-09	340	1000	-----	0.01398	0.09373
560	C6H12	trans-3-HEXENE	-0.02318	1.0980E-04	7.3222E-09	340	1000	-----	0.01500	0.09394
561	C6H12	METHYLCYCLOPENTANE	-0.00746	2.5087E-05	1.3442E-07	345	1000	-----	0.01719	0.15205
562	C6H12	2-METHYL-1-PENTENE	-0.02322	1.1108E-04	6.8641E-09	335	1000	-----	0.01476	0.09472
563	C6H12	2-METHYL-2-PENTENE	-0.02355	1.0791E-04	8.1723E-09	340	1000	-----	0.01408	0.09253
564	C6H12	3-METHYL-1-PENTENE	-0.02358	1.1861E-04	5.2451E-09	327	1000	-----	0.01577	0.10028
565	C6H12	3-METHYL-cis-2-PENTENE	-0.02422	1.1082E-04	8.3013E-09	341	1000	-----	0.01453	0.09490
566	C6H12	4-METHYL-1-PENTENE	-0.02365	1.1116E-04	9.0111E-09	327	997	-----	0.01366	0.09613
567	C6H12	4-METHYL-cis-2-PENTENE	-0.02325	1.1237E-04	7.8684E-09	330	1000	-----	0.01469	0.09699
568	C6H12	4-METHYL-trans-2-PENTENE	-0.02227	1.1154E-04	7.7384E-09	332	1000	-----	0.01561	0.09701
569	C6H12N2	TRIETHYLENEDIAMINE	------	--------	--------	----	----	-----	-----	-----
570	C6H12O	BUTYL VINYL ETHER	0.00040	-2.5246E-06	1.2832E-07	367	1000	-----	0.01676	0.12620
571	C6H12O	CYCLOHEXANOL	-0.02696	1.1162E-04	1.2497E-08	434	1000	-----	0.02384	0.09716
572	C6H12O	1-HEXANAL	0.00201	-6.9706E-06	1.1641E-07	401	1000	-----	0.01793	0.11145
573	C6H12O	ETHYL ISOPROPYL KETONE	0.00223	-8.3013E-06	1.3364E-07	387	1000	-----	0.01903	0.12757
574	C6H12O	2-HEXANONE	-0.00018	-1.1096E-06	1.1858E-07	273	1000	0.01003	0.00835	0.11729
575	C6H12O	3-HEXANONE	0.00046	-2.8354E-06	1.2146E-07	397	1000	-----	0.01848	0.11908
576	C6H12O	METHYL ISOBUTYL KETONE	0.00209	-7.7952E-06	1.2993E-07	390	1000	-----	0.01881	0.12422
577	C6H12O2	n-PENTYL FORMATE	0.00193	-7.0093E-06	1.1429E-07	407	1000	-----	0.01801	0.10921
578	C6H12O2	n-BUTYL ACETATE	-0.01347	6.3061E-05	5.5881E-08	273	600	0.01030	0.00791	0.04448
579	C6H12O2	sec-BUTYL ACETATE	-0.01064	7.0457E-05	1.9982E-08	385	1000	-----	0.01945	0.07980
580	C6H12O2	tert-BUTYL ACETATE	-0.01558	8.6536E-05	2.7575E-09	369	1000	-----	0.01673	0.07371
581	C6H12O2	ETHYL n-BUTYRATE	0.00044	-2.7311E-06	1.1199E-07	395	995	-----	0.01683	0.10860
582	C6H12O2	ETHYL ISOBUTYRATE	-0.01336	7.1037E-05	1.1391E-08	383	810	-----	0.01552	0.05165
583	C6H12O2	ISOBUTYL ACETATE	-0.00255	2.2339E-05	9.5719E-08	390	650	-----	0.02072	0.05241
584	C6H12O2	n-PROPYL PROPIONATE	-0.01456	7.9880E-05	4.0996E-09	396	1000	-----	0.01772	0.06942
585	C6H12O2	CYCLOHEXYL PEROXIDE	-0.01456	7.9880E-05	4.0996E-09	396	1000	-----	0.01772	0.06942
586	C6H12O2	DIACETONE ALCOHOL	-0.01456	7.9880E-05	4.0996E-09	396	1000	-----	0.01772	0.06942
587	C6H12O2	2-ETHYL BUTYRIC ACID	-0.01001	5.9638E-05	1.7463E-08	467	1000	-----	0.02165	0.06709
588	C6H12O2	n-HEXANOIC ACID	-0.00999	6.0544E-05	1.5093E-08	479	1000	-----	0.02247	0.06565
589	C6H12O3	2-ETHOXYETHYL ACETATE	-0.01431	7.3519E-05	2.8216E-09	429	1000	-----	0.01775	0.06203
590	C6H12O3	HYDROXYCAPROIC ACID	-0.01082	6.3540E-05	1.0893E-08	576	1000	-----	0.02939	0.06361
591	C6H12O3	PARALDEHYDE	-0.00986	6.3529E-05	1.9397E-08	397	997	-----	0.01842	0.07276
592	C6H12O3	sec-BUTYL GLYCOLATE	------	--------	--------	----	----	-----	-----	-----
593	C6H12S	THIACYCLOHEPTANE	-0.01420	6.4940E-05	1.6240E-08	275	1000	0.00661	0.00489	0.06698
594	C6H13N	CYCLOHEXYLAMINE	-0.02518	1.1000E-04	1.3973E-08	408	1000	-----	0.02203	0.09879
595	C6H13N	HEXAMETHYLENEIMINE	-0.01545	7.0179E-05	3.6497E-08	405	1000	-----	0.01896	0.09123
596	C6H14	2,2-DIMETHYLBUTANE	-0.02495	1.1807E-04	1.5286E-08	323	993	-----	0.01478	0.10737
597	C6H14	2,3-DIMETHYLBUTANE	-0.02527	1.1748E-04	1.3227E-08	331	991	-----	0.01507	0.10414
598	C6H14	n-HEXANE	-0.00200	7.7788E-06	1.3824E-07	290	480	0.01261	0.01188	0.03358
599	C6H14	2-METHYLPENTANE	-0.02517	1.1820E-04	9.7793E-09	333	993	-----	0.01528	0.10185
600	C6H14	3-METHYLPENTANE	-0.02525	1.1734E-04	1.1464E-08	336	1000	-----	0.01547	0.10355
601	C6H14N2O2	LYSINE	------	--------	--------	----	----	-----	-----	-----
602	C6H14O	2-ETHYL-1-BUTANOL	-0.02106	1.0226E-04	7.8607E-09	420	1000	-----	0.02328	0.08906
603	C6H14O	1-HEXANOL	-0.00003	7.1095E-06	1.0391E-07	430	990	-----	0.02224	0.10885
604	C6H14O	2-HEXANOL	0.00009	6.8974E-06	1.1173E-07	413	1000	-----	0.02200	0.11872
605	C6H14O	2-METHYL-1-PENTANOL	0.00003	7.1723E-06	1.1403E-07	421	1000	-----	0.02326	0.12123
606	C6H14O	4-METHYL-2-PENTANOL	-0.01046	6.7260E-05	1.9467E-08	405	1000	-----	0.01997	0.07627
607	C6H14O	n-BUTYL ETHYL ETHER	0.00213	-7.9371E-06	1.3400E-07	365	1000	-----	0.01709	0.12819
608	C6H14O	DIISOPROPYL ETHER	-0.00466	5.0338E-05	3.6643E-08	300	1000	-----	0.01374	0.08232
609	C6H14O	DI-n-PROPYL ETHER	-0.00914	5.7163E-05	2.5520E-08	363	1000	-----	0.01497	0.07354
610	C6H14O	METHYL tert-PENTYL ETHER	-0.01703	9.4155E-05	9.9660E-09	359	1000	-----	0.01806	0.08709
611	C6H14O2	ACETAL	-0.01761	9.0102E-05	2.2962E-09	377	1000	-----	0.01668	0.07479
612	C6H14O2	2-BUTOXYETHANOL	-0.01015	6.2014E-05	1.8278E-08	444	1000	-----	0.02099	0.07014
613	C6H14O2	1,6-HEXANEDIOL	-0.00195	1.1148E-05	9.2023E-08	516	996	-----	0.02830	0.10044
614	C6H14O2	HEXYLENE GLYCOL	-0.01094	6.5187E-05	1.8708E-08	471	1000	-----	0.02391	0.07296
615	C6H14O2S	DI-n-PROPYL SULFONE	-0.00546	4.2561E-05	3.0821E-08	543	1000	-----	0.02674	0.06792
616	C6H14O3	DIETHYLENE GLYCOL DIMETHYL ETHER	------	--------	--------	----	----	-----	-----	-----
617	C6H14O3	DIPROPYLENE GLYCOL	-0.01036	6.3705E-05	1.3322E-08	495	1000	-----	0.02444	0.06667
618	C6H14O3	2-(2-ETHOXYETHOXY)ETHANOL	-0.00989	5.9411E-05	1.5932E-08	475	1000	-----	0.02192	0.06545
619	C6H14O3	TRIMETHYLOLPROPANE	0.00003	5.4634E-06	8.6657E-08	562	1000	-----	0.03047	0.09215
620	C6H14O4	TRIETHYLENE GLYCOL	-0.00382	2.7242E-05	4.4887E-08	266	700	0.00829	0.00660	0.03724
621	C6H14O6	SORBITOL	-0.00791	4.1821E-05	5.0914E-09	777	997	-----	0.02766	0.03885
622	C6H14S	n-HEXYLMERCAPTAN	------	--------	--------	----	----	-----	-----	-----
623	C6H14S	BUTYL-ETHYL-SULFIDE	-0.00815	4.9201E-05	2.0664E-08	275	1000	0.00836	0.00694	0.06172
624	C6H14S	ISOPROPYL-SULFIDE	-0.01013	6.1638E-05	1.1453E-08	275	1000	0.00927	0.00769	0.06296

Table 23-1 THERMAL CONDUCTIVITY OF GAS - ORGANIC COMPOUNDS (continued)

NO	FORMULA	NAME	$k_{gas} = A + B\,T + C\,T^2$					$(k_{gas}$ - W/(m K), T - K)		
			A	B	C	TMIN	TMAX	k_{gas} @25 C	k_{gas} @TMIN	k_{gas} @TMAX
625	C6H14S	METHYL-PENTYL-SULFIDE	-0.00839	5.1161E-05	1.9389E-08	275	1000	0.00859	0.00715	0.06216
626	C6H14S	PROPYL-SULFIDE	-0.00787	4.7365E-05	2.3049E-08	275	1000	0.00830	0.00690	0.06254
627	C6H14S2	PROPYL-DISULFIDE	-0.00678	4.2940E-05	1.5620E-08	275	1000	0.00741	0.00621	0.05178
628	C6H15Al	TRIETHYL ALUMINUN	------	--------	---------	----	----	-----	-----	-----
629	C6H15Al2Cl3	ETHYL ALUMINUM SESQUICHLORIDE	------	--------	----/---	----	----	-----	-----	-----
630	C6H15N	DIISOPROPYLAMINE	-0.02243	1.1290E-04	9.0082E-09	357	1000	-----	0.01902	0.09948
631	C6H15N	DI-n-PROPYLAMINE	-0.01864	9.7340E-05	1.5899E-08	280	992	0.01180	0.00986	0.09357
632	C6H15N	n-HEXYLAMINE	-0.01101	6.9097E-05	2.2740E-08	405	1000	-----	0.02070	0.08083
633	C6H15N	TRIETHYLAMINE	0.06339	-2.8473E-04	4.4125E-07	273	1000	0.01772	0.01854	0.21991
634	C6H15NO	6-AMINOHEXANOL	-0.01143	6.7029E-05	1.6847E-08	508	1000	-----	0.02697	0.07245
635	C6H15NO2	DIISOPROPANOLAMINE	-0.01157	7.0707E-05	1.3092E-08	522	1000	-----	0.02891	0.07223
636	C6H15NO3	TRIETHANOLAMINE	-0.01074	5.2692E-05	1.8196E-08	633	993	-----	0.02990	0.05953
637	C6H15N3	N-AMINOETHYL PIPERAZINE	-0.01191	6.7419E-05	2.2113E-08	494	1000	-----	0.02679	0.07762
638	C6H15O4P	TRIETHYL PHOSPHATE	------	--------	---------	----	----	-----	-----	-----
639	C6H16N2	HEXAMETHYLENEDIAMINE	-0.01217	7.3189E-05	1.9829E-08	475	995	-----	0.02707	0.08028
640	C6H18N3OP	HEXAMETHYL PHOSPHORAMIDE	------	--------	---------	----	----	-----	-----	-----
641	C6H18N4	TRIETHYLENE TETRAMINE	-0.01123	6.4536E-05	1.5212E-08	540	1000	-----	0.02806	0.06852
642	C6H18OSi2	HEXAMETHYLDISILOXANE	-0.00955	6.6223E-05	1.5880E-08	374	1000	-----	0.01744	0.07255
643	C6H18O3Si3	HEXAMETHYLCYCLOTRISILOXANE	-0.01486	7.2287E-05	3.3299E-09	408	1000	-----	0.01519	0.06076
644	C6H19NSi2	HEXAMETHYLDISILAZANE	-0.01670	7.9936E-05	9.6457E-09	399	1000	-----	0.01673	0.07288
645	C7H3ClF3NO2	4-CHLORO-3-NITROBENZOTRIFLUORIDE	------	--------	---------	----	----	-----	-----	-----
646	C7H3Cl2F3	2,4-DICHLOROBENZOTRIFLUORIDE	-0.00204	1.4209E-05	2.5948E-08	451	1000	-----	0.00965	0.03812
647	C7H3Cl2NO	3,4-DICHLOROPHENYL ISOCYANATE	------	--------	---------	----	----	-----	-----	-----
648	C7H4ClF3	p-CHLOROBENZOTRIFLUORIDE	-0.00470	2.3913E-05	2.1583E-08	412	1000	-----	0.00882	0.04080
649	C7H4Cl2O	m-CHLOROBENZOYL CHLORIDE	-0.00581	3.8952E-05	4.6627E-09	498	1000	-----	0.01474	0.03780
650	C7H4F3NO2	3-NITROBENZOTRIFLUORIDE	------	--------	---------	----	----	-----	-----	-----
651	C7H5ClO	BENZOYL CHLORIDE	-0.00821	4.9147E-05	1.3800E-08	470	1000	-----	0.01794	0.05474
652	C7H5ClO2	o-CHLOROBENZOIC ACID	-0.00724	4.0174E-05	1.0286E-08	560	1000	-----	0.01848	0.04322
653	C7H5Cl3	BENZOTRICHLORIDE	-0.00196	1.2735E-05	2.6872E-08	494	1000	-----	0.01089	0.03765
654	C7H5F3	BENZOTRIFLUORIDE	-0.00480	2.6729E-05	2.1257E-08	377	1000	-----	0.00830	0.04319
655	C7H5N	BENZONITRILE	-0.00777	4.5588E-05	1.4657E-08	464	1000	-----	0.01654	0.05248
656	C7H5NO	PHENYL ISOCYANATE	-0.01296	6.1228E-05	3.3193E-09	439	1000	-----	0.01456	0.05159
657	C7H5N3O6	2,4,6-TRINITROTOLUENE	------	--------	---------	----	----	-----	-----	-----
658	C7H6Cl2	BENZYL DICHLORIDE	-0.00716	4.4714E-05	8.9822E-09	487	1000	-----	0.01675	0.04654
659	C7H6Cl2	2,4-DICHLOROTOLUENE	-0.00216	1.4670E-05	2.7213E-08	474	994	-----	0.01091	0.03931
660	C7H6N2O4	2,4-DINITROTOLUENE	------	--------	---------	----	----	-----	-----	-----
661	C7H6N2O4	2,5-DINITROTOLUENE	------	--------	---------	----	----	-----	-----	-----
662	C7H6N2O4	2,6-DINITROTOLUENE	------	--------	---------	----	----	-----	-----	-----
663	C7H6N2O4	3,4-DINITROTOLUENE	------	--------	---------	----	----	-----	-----	-----
664	C7H6N2O4	3,5-DINITROTOLUENE	------	--------	---------	----	----	-----	-----	-----
665	C7H6O	BENZALDEHYDE	-0.01407	6.5719E-05	1.2910E-09	452	1000	-----	0.01590	0.05294
666	C7H6O2	BENZOIC ACID	-0.00824	4.5025E-05	1.5092E-08	522	1000	-----	0.01938	0.05188
667	C7H6O2	p-HYDROXYBENZALDEHYDE	-0.01035	6.1233E-05	9.6772E-09	583	1000	-----	0.02864	0.06056
668	C7H6O2	SALICYLALDEHYDE	-0.01585	7.3309E-05	-1.2729E-09	470	1000	-----	0.01832	0.05619
669	C7H6O3	SALICYLIC ACID	------	--------	---------	----	----	-----	-----	-----
670	C7H7Br	p-BROMOTOLUENE	-0.00651	3.9659E-05	1.0960E-08	458	1000	-----	0.01395	0.04411
671	C7H7Cl	BENZYL CHLORIDE	-0.00676	4.8275E-05	2.1168E-08	453	700	-----	0.01945	0.03740
672	C7H7Cl	o-CHLOROTOLUENE	-0.00246	1.7966E-05	2.8211E-08	432	1000	-----	0.01057	0.04372
673	C7H7Cl	p-CHLOROTOLUENE	-0.00244	1.7780E-05	2.7905E-08	436	1000	-----	0.01062	0.04325
674	C7H7F	p-FLUOROTOLUENE	-0.01066	5.8768E-05	9.8030E-09	275	1000	0.00773	0.00624	0.05791
675	C7H7NO	FORMANILIDE	-0.00918	5.2720E-05	1.2195E-08	544	1000	-----	0.02311	0.05574
676	C7H7NO2	m-NITROTOLUENE	------	--------	---------	----	----	-----	-----	-----
677	C7H7NO2	o-NITROTOLUENE	------	--------	---------	----	----	-----	-----	-----
678	C7H7NO2	p-NITROTOLUENE	------	--------	---------	----	----	-----	-----	-----
679	C7H7NO3	o-NITROANISOLE	------	--------	---------	----	----	-----	-----	-----
680	C7H8	TOLUENE	-0.00776	4.4905E-05	6.4514E-08	350	800	-----	0.01586	0.06945
681	C7H8	1,3,5-CYCLOHEPTATRIENE	-0.01236	6.8696E-05	8.9758E-09	275	1000	0.00892	0.00721	0.06531
682	C7H8O	ANISOLE	-0.00841	5.4312E-05	1.3274E-08	427	1000	-----	0.01720	0.05918
683	C7H8O	BENZYL ALCOHOL	-0.02239	9.3867E-05	1.1774E-09	478	1000	-----	0.02275	0.07265
684	C7H8O	m-CRESOL	-0.01019	6.3156E-05	1.4276E-08	475	1000	-----	0.02303	0.06724
685	C7H8O	o-CRESOL	-0.01453	7.3771E-05	1.1445E-08	460	960	-----	0.02183	0.06684
686	C7H8O	p-CRESOL	-0.01026	6.3162E-05	1.4789E-08	475	1000	-----	0.02308	0.06769
687	C7H8O2	GUAIACOL	-0.00948	5.8177E-05	1.3631E-08	478	1000	-----	0.02144	0.06233
688	C7H8O2	p-METHOXYPHENOL	-0.01778	7.7713E-05	-2.6925E-10	516	1000	-----	0.02225	0.05966
689	C7H9N	BENZYLAMINE	-0.01746	7.8683E-05	3.1880E-09	458	1000	-----	0.01925	0.06441
690	C7H9N	2,6-DIMETHYLPYRIDINE	-0.01496	7.4216E-05	9.6792E-09	417	1000	-----	0.01767	0.06894
691	C7H9N	N-METHYLANILINE	-0.01939	8.7524E-05	2.1511E-08	469	1000	-----	0.02213	0.07029
692	C7H9N	m-TOLUIDINE	-0.01130	6.6264E-05	1.9972E-08	477	1000	-----	0.02485	0.07494
693	C7H9N	o-TOLUIDINE	-0.01126	6.6401E-05	1.9757E-08	474	1000	-----	0.02465	0.07490
694	C7H9N	p-TOLUIDINE	-0.01122	6.6479E-05	1.9337E-08	473	1000	-----	0.02455	0.07460
695	C7H10	2-NORBORNENE	-0.02632	1.0831E-04	1.1393E-08	369	1000	-----	0.01520	0.09338
696	C7H10N2	TOLUENEDIAMINE	-0.00459	3.7823E-05	2.2289E-08	557	997	-----	0.02339	0.05527
697	C7H11NO	CYCLOHEXYL ISOCYANATE	-0.02086	8.5930E-05	4.3256E-09	442	1000	-----	0.01797	0.06940
698	C7H12	1-HEPTYNE	-0.01029	6.3474E-05	1.3957E-08	275	1000	0.00988	0.00822	0.06714
699	C7H12O2	n-BUTYL ACRYLATE	-0.00810	5.5918E-05	1.0107E-08	421	991	-----	0.01723	0.05724
700	C7H12O2	ISOBUTYL ACRYLATE	-0.01292	7.3271E-05	5.7666E-09	405	1000	-----	0.01770	0.06612
701	C7H12O2	n-PROPYL METHACRYLATE	-0.00903	5.5800E-05	1.9106E-08	414	994	-----	0.01735	0.06531
702	C7H12O4	DIETHYL MALONATE	-0.00770	4.8737E-05	9.9882E-09	472	1000	-----	0.01753	0.05103

Table 23-1 THERMAL CONDUCTIVITY OF GAS - ORGANIC COMPOUNDS (continued)

			$k_{gas} = A + B\,T + C\,T^2$						$(k_{gas}$ - W/(m K), T - K$)$		
NO	FORMULA	NAME	A	B	C	TMIN	TMAX	k_{gas} @25 C	k_{gas} @TMIN	k_{gas} @TMAX
703	C7H14	CYCLOHEPTANE	-0.01815	8.2733E-05	3.8757E-08	273	1000	0.00996	0.00732	0.10334
704	C7H14	1,1-DIMETYLCYCLOPENTANE	-0.02729	1.1701E-04	1.3570E-08	361	1000	-----	0.01672	0.10329
705	C7H14	cis-1,2-DIMETHYLCYCLOPENTANE	-0.02637	1.1233E-04	1.3745E-08	373	1000	-----	0.01744	0.09971
706	C7H14	trans-1,2-DIMETHYLCYCLOPENTANE	-0.02645	1.1404E-04	1.4169E-08	365	1000	-----	0.01706	0.10176
707	C7H14	cis-1,3-DIMETHYLCYCLOPENTANE	-0.02673	1.1533E-04	1.3413E-08	364	1000	-----	0.01703	0.10201
708	C7H14	trans-1,3-DIMETHYLCYCLOPENTANE	-0.02657	1.1456E-04	1.3656E-08	365	1000	-----	0.01706	0.10165
709	C7H14	ETHYLCYCLOPENTANE	-0.02464	1.0590E-04	1.5918E-08	377	1000	-----	0.01755	0.09718
710	C7H14	2-ETHYL-1-PENTENE	-0.02622	1.1564E-04	2.3883E-09	367	1000	-----	0.01654	0.09181
711	C7H14	3-ETHYL-1-PENTENE	-0.02688	1.1958E-04	3.0658E-09	357	1000	-----	0.01620	0.09577
712	C7H14	1-HEPTENE	-0.00771	4.6067E-05	6.8648E-08	325	800	-----	0.01451	0.07308
713	C7H14	cis-2-HEPTENE	-0.02483	1.0727E-04	5.1453E-09	372	1000	-----	0.01579	0.08759
714	C7H14	trans-2-HEPTENE	-0.02478	1.0907E-04	4.8661E-09	371	1000	-----	0.01635	0.08916
715	C7H14	cis-3-HEPTENE	-0.02529	1.0925E-04	4.3971E-09	369	1000	-----	0.01562	0.08836
716	C7H14	trans-3-HEPTENE	-0.02553	1.1168E-04	3.6230E-09	369	1000	-----	0.01617	0.08977
717	C7H14	METHYLCYCLOHEXANE	-0.01221	6.8869E-05	4.7632E-08	374	994	-----	0.02021	0.10331
718	C7H14	2-METHYL-1-HEXENE	-0.02504	1.1217E-04	3.7454E-08	365	1000	-----	0.01640	0.09088
719	C7H14	3-METHYL-1-HEXENE	-0.02647	1.1790E-04	2.9719E-09	357	1000	-----	0.01600	0.09440
720	C7H14	4-METHYL-1-HEXENE	-0.02691	1.1911E-04	3.0242E-09	360	1000	-----	0.01636	0.09522
721	C7H14	2,3,3-TRIMETHYL-1-BUTENE	-0.02859	1.2805E-04	-1.2852E-10	351	1000	-----	0.01634	0.09933
722	C7H14O	DIISOPROPYL KETONE	0.00236	-1.0037E-05	1.3841E-07	398	1000	-----	0.02029	0.13073
723	C7H14O	2-HEPTANONE	0.00080	-4.7604E-06	1.1455E-07	424	1000	-----	0.01937	0.11059
724	C7H14O	1-HEPTANAL	0.00193	-7.7435E-06	1.1335E-07	426	1000	-----	0.01920	0.10754
725	C7H14O	1-METHYLCYCLOHEXANOL	-0.01073	6.0731E-05	2.8618E-08	430	1000	-----	0.02068	0.07862
726	C7H14O	cis-2-METHYLCYCLOHEXANOL	-0.01211	6.8635E-05	3.0581E-08	438	1000	-----	0.02382	0.08711
727	C7H14O	trans-2-METHYLCYCLOHEXANOL	-0.01214	6.8477E-05	3.0950E-08	440	1000	-----	0.02398	0.08729
728	C7H14O	cis-3-METHYLCYCLOHEXANOL	-0.02360	1.0023E-04	8.7735E-09	441	1000	-----	0.02231	0.08540
729	C7H14O	trans-3-METHYLCYCLOHEXANOL	-0.01208	6.8142E-05	3.0508E-08	441	1000	-----	0.02390	0.08657
730	C7H14O	cis-4-METHYLCYCLOHEXANOL	-0.01212	6.8970E-05	2.9148E-08	444	1000	-----	0.02425	0.08600
731	C7H14O	trans-4-METHYLCYCLOHEXANOL	-0.01204	6.8489E-05	2.8939E-08	444	1000	-----	0.02407	0.08539
732	C7H14O	5-METHYL-2-HEXANONE	0.00089	-5.2671E-06	1.2134E-07	418	1000	-----	0.01989	0.11696
733	C7H14O2	n-BUTYL PROPIONATE	-0.00967	6.0969E-05	1.7773E-08	420	1000	-----	0.01907	0.06907
734	C7H14O2	ETHYL ISOVALERATE	0.00163	-6.9483E-06	1.1341E-07	407	1000	-----	0.01759	0.10809
735	C7H14O2	ISOPENTYL ACETATE	-0.00536	7.3541E-05	3.0480E-08	415	800	-----	0.03041	0.07298
736	C7H14O2	n-PENTYL ACETATE	-0.00027	2.6117E-06	1.1495E-07	273	600	0.01073	0.00901	0.04268
737	C7H14O2	n-PROPYL n-BUTYRATE	-0.00954	6.0418E-05	1.7843E-08	416	996	-----	0.01868	0.06834
738	C7H14O2	n-HEPTANOIC ACID	-0.00992	5.8552E-05	1.5206E-08	496	1000	-----	0.02286	0.06384
739	C7H14O3	ETHYL-3-ETHOXYPROPIONATE	0.00188	-7.8909E-06	1.0588E-07	438	1000	-----	0.01874	0.09987
740	C7H15Br	1-BROMOHEPTANE	-0.00812	4.9825E-05	1.3677E-08	452	1000	-----	0.01720	0.05538
741	C7H15N	N-METHYLCYCLOHEXYLAMINE	-0.02345	1.0309E-04	1.0640E-08	422	1000	-----	0.02195	0.09028
742	C7H16	2,2-DIMETHYLPENTANE	-0.02542	1.1603E-04	7.9145E-09	352	1000	-----	0.01640	0.09852
743	C7H16	2,3-DIMETHYLPENTANE	-0.01223	7.4018E-05	3.9510E-08	363	993	-----	0.01984	0.10023
744	C7H16	2,4-DIMETHYLPENTANE	-0.01212	7.5158E-05	3.7926E-08	354	994	-----	0.01924	0.10006
745	C7H16	3,3-DIMETHYLPENTANE	-0.02638	1.1824E-04	7.6535E-09	359	1000	-----	0.01705	0.09951
746	C7H16	3-ETHYLPENTANE	-0.02624	1.1672E-04	7.0036E-09	367	1000	-----	0.01754	0.09748
747	C7H16	n-HEPTANE	-0.00172	1.6565E-05	1.0525E-07	250	750	0.01257	0.00900	0.06991
748	C7H16	2-METHYLHEXANE	-0.01193	7.3057E-05	3.6736E-08	363	993	-----	0.01943	0.09684
749	C7H16	3-METHYLHEXANE	-0.01210	7.3744E-05	3.7221E-08	365	995	-----	0.01978	0.09813
750	C7H16	2,2,3-TRIMETHYLBUTANE	-0.02639	1.1615E-04	1.6588E-08	354	1000	-----	0.01681	0.10635
751	C7H16O	1-HEPTANOL	-0.00089	1.0487E-05	9.7505E-08	449	1000	-----	0.02348	0.10710
752	C7H16O	2-HEPTANOL	-0.00003	6.3709E-06	1.1568E-07	432	1000	-----	0.02431	0.12202
753	C7H16O	5-METHYL-1-HEXANOL	-0.00184	1.1187E-05	1.0576E-07	445	1000	-----	0.02408	0.11511
754	C7H16O	ISOPROPYL-TERT-BUTYL-ETHER	-0.01046	6.3059E-05	1.8080E-08	275	1000	0.00995	0.00825	0.07068
755	C7H16S	n-HEPTYL MERCAPTAN	-0.01462	7.4335E-05	1.9050E-08	450	1000	-----	0.02269	0.07877
756	C7H16S	BUTYL-PROPYL-SULFIDE	-0.00759	4.5374E-05	2.3593E-08	275	1000	0.00804	0.00667	0.06138
757	C7H16S	ETHYL-PENTYL-SULFIDE	-0.00778	4.6593E-05	2.0463E-08	275	1000	0.00792	0.00657	0.05923
758	C7H16S	HEXYL-METHYL-SULFIDE	-0.00775	4.6845E-05	1.9938E-08	275	1000	0.00799	0.00664	0.05903
759	C7H17N	1-AMINOHEPTANE	-0.00880	6.1045E-05	3.4374E-08	430	700	-----	0.02381	0.05077
760	C8H4Cl2O2	ISOPHTHALOYL CHLORIDE	-0.00542	3.4898E-05	3.9077E-09	549	1000	-----	0.01492	0.03339
761	C8H4O3	PHTHALIC ANHYDRIDE	-0.00753	3.8226E-05	1.5493E-08	558	1000	-----	0.01862	0.04619
762	C8H6	ETHYNYLBENZENE	-0.01012	5.7144E-05	8.3282E-09	275	1000	0.00766	0.00622	0.05535
763	C8H6O4	ISOPHTHALIC ACID	-0.00858	4.4907E-05	6.3743E-09	753	1000	-----	0.02885	0.04270
764	C8H6O4	PHTHALIC ACID	-0.00739	4.0704E-05	8.9383E-09	598	1000	-----	0.02015	0.04225
765	C8H6O4	TEREPHTHALIC ACID	-0.00924	4.5635E-05	6.5815E-09	832	1000	-----	0.03328	0.04298
766	C8H6S	BENZOTHIOPHENE	-0.00886	5.6047E-05	1.0132E-08	493	1000	-----	0.02123	0.05732
767	C8H7N	INDOLE	-0.01544	6.7583E-05	1.1291E-08	526	1000	-----	0.02323	0.06343
768	C8H8	STYRENE	-0.00712	4.5538E-05	3.9529E-08	273	973	0.00997	0.00826	0.07461
769	C8H8	1,3,5,7-CYCLOOCTATETRAENE	-0.01093	5.9475E-05	1.2339E-08	275	1000	0.00790	0.00636	0.06088
770	C8H8O	ACETOPHENONE	-0.00815	4.3566E-05	2.0894E-08	475	1000	-----	0.01726	0.05631
771	C8H8O	p-TOLUALDEHYDE	-0.00947	5.3038E-05	2.0083E-08	477	1000	-----	0.02040	0.06365
772	C8H8O2	METHYL BENZOATE	-0.00804	4.5774E-05	1.6365E-08	473	1000	-----	0.01727	0.05410
773	C8H8O2	o-TOLUIC ACID	-0.00835	4.5361E-05	1.4871E-08	532	1000	-----	0.01999	0.05188
774	C8H8O2	p-TOLUIC ACID	-0.00855	4.5799E-05	1.4900E-08	548	1000	-----	0.02102	0.05215
775	C8H8O3	METHYL SALICYLATE	-0.01382	6.2711E-05	3.3082E-10	494	1000	-----	0.01724	0.04922
776	C8H8O3	VANILLIN	-0.00926	5.4095E-05	1.0583E-08	558	1000	-----	0.02422	0.05542
777	C8H9NO	ACETANILIDE	-0.00954	5.0738E-05	1.4864E-08	577	1000	-----	0.02468	0.05606
778	C8H10	ETHYLBENZENE	-0.00797	4.0572E-05	6.7289E-08	400	825	-----	0.01903	0.07130
779	C8H10	m-XYLENE	-0.00375	2.9995E-05	7.4603E-08	400	825	-----	0.02018	0.07177
780	C8H10	o-XYLENE	-0.00979	7.4087E-05	1.8418E-08	400	825	-----	0.02279	0.06387

517

Table 23-1 THERMAL CONDUCTIVITY OF GAS - ORGANIC COMPOUNDS (continued)

NO	FORMULA	NAME	$k_{gas} = A + B\,T + C\,T^2$					$(k_{gas} - W/(m\,K),\ T - K)$		
			A	B	C	TMIN	TMAX	k_{gas} @25 C	k_{gas} @TMIN	k_{gas} @TMAX
781	C8H10	p-XYLENE	-0.00870	4.7349E-05	5.8829E-08	400	825	------	0.01965	0.07040
782	C8H10O	m-ETHYLPHENOL	------	--------	--------	----	----	------	------	------
783	C8H10O	p-ETHYLPHENOL	-0.01729	7.9086E-05	-3.9212E-11	491	1000	------	0.02153	0.06176
784	C8H10O	PHENETOLE	-0.00831	5.0314E-05	1.5670E-08	443	1000	------	0.01705	0.05767
785	C8H10O	2-PHENYLETHANOL	-0.01601	7.3080E-05	4.7158E-09	492	1000	------	0.02109	0.06179
786	C8H10O	2,3-XYLENOL	-0.01621	8.0847E-05	4.6664E-09	490	1000	------	0.02453	0.06930
787	C8H10O	2,4-XYLENOL	-0.01061	6.3622E-05	1.6317E-08	484	1000	------	0.02401	0.06933
788	C8H10O	2,5-XYLENOL	-0.01062	6.3888E-05	1.6071E-08	484	1000	------	0.02407	0.06934
789	C8H10O	2,6-XYLENOL	-0.01043	6.3338E-05	1.6044E-08	474	1000	------	0.02320	0.06895
790	C8H10O	3,4-XYLENOL	-0.01142	6.9766E-05	1.4701E-08	500	1000	------	0.02714	0.07305
791	C8H10O	3,5-XYLENOL	-0.01074	6.3474E-05	1.6542E-08	495	1000	------	0.02473	0.06928
792	C8H11N	N,N-DIMETHYLANILINE	-0.02003	8.2062E-05	3.8364E-09	467	1000	------	0.01913	0.06587
793	C8H11N	o-ETHYLANILINE	-0.01014	6.1355E-05	1.5104E-08	483	1000	------	0.02302	0.06632
794	C8H11N	2,4,6-TRIMETHYLPYRIDINE	-0.01734	7.9239E-05	5.3151E-09	444	1000	------	0.01889	0.06721
795	C8H11NO	p-PHENETIDINE	-0.01005	5.8755E-05	1.3434E-08	528	1000	------	0.02472	0.06214
796	C8H12	1,5-CYCLOOCTADIENE	-0.02554	1.0401E-04	1.2248E-08	423	1000	------	0.02065	0.09072
797	C8H12	VINYLCYCLOHEXENE	-0.02497	1.0672E-04	6.0506E-09	401	1000	------	0.01880	0.08780
798	C8H12O4	1,4-CYCLOHEXANEDICARBOXYLIC ACID	-0.01112	5.8457E-05	1.1846E-08	669	1000	------	0.03329	0.05918
799	C8H12O4	DIETHYL MALEATE	0.00165	-7.1000E-06	8.6929E-08	498	1000	------	0.01967	0.08148
800	C8H14O2	n-BUTYL METHACRYLATE	-0.01848	8.7139E-05	6.5835E-09	434	1000	------	0.02058	0.07524
801	C8H14O3	BUTYRIC ANHYDRIDE	-0.00710	4.5385E-05	9.0705E-09	468	998	------	0.01613	0.04723
802	C8H14O4	DIETHYL SUCCINATE	-0.00826	5.1970E-05	9.8994E-09	490	1000	------	0.01958	0.05361
803	C8H16	CYCLOOCTANE	-0.01347	6.3151E-05	2.4928E-08	275	1000	0.00757	0.00578	0.07461
804	C8H16	1,1-DIMETHYLCYCLOHEXANE	-0.02802	1.1255E-04	1.1180E-08	393	1000	------	0.01794	0.09571
805	C8H16	cis-1,2-DIMETHYLCYCLOHEXANE	-0.02680	1.0843E-04	1.0955E-08	403	1000	------	0.01868	0.09259
806	C8H16	trans-1,2-DIMETHYLCYCLOHEXANE	-0.02699	1.1109E-04	1.0310E-08	397	1000	------	0.01874	0.09441
807	C8H16	cis-1,3-DIMETHYLCYCLOHEXANE	-0.02736	1.1131E-04	1.1533E-08	393	1000	------	0.01817	0.09548
808	C8H16	trans-1,3-DIMETHYLCYCLOHEXANE	-0.02652	1.0839E-04	1.1691E-08	398	1000	------	0.01847	0.09356
809	C8H16	cis-1,4-DIMETHYLCYCLOHEXANE	-0.02652	1.0841E-04	1.1697E-08	397	1000	------	0.01836	0.09359
810	C8H16	trans-1,4-DIMETHYLCYCLOHEXANE	-0.02750	1.1269E-04	1.0360E-08	393	1000	------	0.01839	0.09555
811	C8H16	ETHYLCYCLOHEXANE	-0.01194	6.6023E-05	4.1195E-08	405	995	------	0.02156	0.09454
812	C8H16	2-ETHYL-1-HEXENE	-0.02838	1.2204E-04	1.8230E-09	393	1000	------	0.01986	0.09548
813	C8H16	1-METHYL-1-ETHYLCYCLOPENTANE	-0.02598	1.1067E-04	1.0210E-08	395	1000	------	0.01933	0.09490
814	C8H16	1-OCTENE	-0.00463	2.8829E-05	8.4175E-08	394	500	------	0.01980	0.03083
815	C8H16	trans-2-OCTENE	-0.02481	1.0586E-04	2.9507E-09	398	1000	------	0.01779	0.08400
816	C8H16	trans-3-OCTENE	-0.02522	1.0720E-04	2.5380E-09	396	1000	------	0.01763	0.08452
817	C8H16	trans-4-OCTENE	-0.02519	1.0724E-04	2.6386E-09	395	1000	------	0.01758	0.08469
818	C8H16	n-PROPYLCYCLOPENTANE	-0.01185	6.7767E-05	3.5487E-08	404	1000	------	0.02132	0.09140
819	C8H16	2,4,4-TRIMETHYL-1-PENTENE	-0.02665	1.1719E-04	-5.7671E-12	375	1000	------	0.01730	0.09053
820	C8H16	2,4,4-TRIMETHYL-2-PENTENE	-0.02613	1.1450E-04	9.0816E-10	378	1000	------	0.01728	0.08928
821	C8H16O	2-ETHYLHEXANAL	0.00209	-9.5110E-06	1.2283E-07	434	1000	------	0.02110	0.11541
822	C8H16O	1-OCTANAL	0.00204	-8.8077E-06	1.1152E-07	447	1000	------	0.02039	0.10475
823	C8H16O	2-OCTANONE	0.00205	-9.0803E-06	1.1603E-07	446	1000	------	0.02108	0.10900
824	C8H16O2	n-BUTYL n-BUTYRATE	-0.00805	5.3875E-05	1.0085E-08	438	998	------	0.01748	0.05576
825	C8H16O2	n-HEXYL ACETATE	0.00190	-8.2258E-06	1.0590E-07	445	1000	------	0.01921	0.09957
826	C8H16O2	ISOBUTYL ISOBUTYRATE	-0.01063	6.5754E-05	2.1014E-08	421	1000	------	0.02078	0.07614
827	C8H16O2	n-OCTANOIC ACID	-0.00981	5.8314E-05	1.3250E-08	513	1000	------	0.02359	0.06175
828	C8H16O4	DIETHYLENE GLYCOL ETHYL ETHER ACETATE	-0.01399	6.7839E-05	2.5453E-09	491	1000	------	0.01993	0.05639
829	C8H18	2,2-DIMETHYLHEXANE	-0.02694	1.1625E-04	7.4267E-09	380	1000	------	0.01831	0.09674
830	C8H18	2,3-DIMETHYLHEXANE	-0.02737	1.1702E-04	5.5484E-09	389	1000	------	0.01899	0.09520
831	C8H18	2,4-DIMETHYLHEXANE	-0.02625	1.1629E-04	5.4923E-09	383	1000	------	0.01909	0.09553
832	C8H18	2,5-DIMETHYLHEXANE	-0.02571	1.1255E-04	7.2788E-09	382	1000	------	0.01835	0.09412
833	C8H18	3,3-DIMETHYLHEXANE	-0.02788	1.2026E-04	5.8458E-09	385	1000	------	0.01929	0.09823
834	C8H18	3,4-DIMETHYLHEXANE	-0.02737	1.1807E-04	5.3388E-09	391	1000	------	0.01915	0.09558
835	C8H18	3-ETHYLHEXANE	-0.02738	1.1743E-04	3.1424E-09	392	1000	------	0.01914	0.09319
836	C8H18	3-ETHYL-2-METHYLPENTANE	-0.01159	6.6507E-05	1.8595E-08	275	1000	0.00989	0.00811	0.07351
837	C8H18	3-METHYL-3-ETHYLPENTANE	-0.02632	1.1337E-04	1.2309E-08	391	1000	------	0.01989	0.09936
838	C8H18	2-METHYLHEPTANE	-0.02706	1.1601E-04	2.3365E-09	391	1000	------	0.01866	0.09129
839	C8H18	3-METHYLHEPTANE	-0.02729	1.1640E-04	2.9533E-09	392	1000	------	0.01879	0.09206
840	C8H18	4-METHYLHEPTANE	-0.02760	1.1834E-04	1.8233E-09	391	1000	------	0.01895	0.09256
841	C8H18	n-OCTANE	-0.00213	1.8456E-05	9.4775E-08	300	800	------	0.01194	0.07329
842	C8H18	2,2,3-TRIMETHYLPENTANE	-0.01237	7.2537E-05	3.9419E-08	383	993	------	0.02119	0.09853
843	C8H18	2,2,4-TRIMETHYLPENTANE	-0.00622	4.1978E-05	7.7251E-08	355	580	------	0.01842	0.04411
844	C8H18	2,3,3-TRIMETHYLPENTANE	-0.02800	1.1853E-04	1.0470E-08	388	1000	------	0.01957	0.10100
845	C8H18	2,3,4-TRIMETHYLPENTANE	-0.02631	1.1542E-04	9.2518E-09	387	1000	------	0.01974	0.09836
846	C8H18	2,2,3,3-TETRAMETHYLBUTANE	-0.01398	7.7000E-05	1.3185E-08	275	1000	0.01015	0.00819	0.07621
847	C8H18O	DI-n-BUTYL ETHER	-0.00439	4.2683E-05	3.8415E-08	323	995	------	0.01340	0.07611
848	C8H18O	DI-sec-BUTYL ETHER	0.00129	-7.8611E-06	1.3732E-07	394	1000	------	0.01951	0.13075
849	C8H18O	DI-tert-BUTYL ETHER	-0.01061	6.8838E-05	2.2520E-08	380	1000	------	0.01880	0.08075
850	C8H18O	2-ETHYL-1-HEXANOL	0.00206	-8.9440E-07	1.0488E-07	450	1000	------	0.02290	0.10605
851	C8H18O	1-OCTANOL	-0.00005	5.1932E-06	9.7714E-08	468	1000	------	0.02378	0.10286
852	C8H18O	2-OCTANOL	-0.01093	6.6858E-05	1.8531E-08	453	1000	------	0.02316	0.07446
853	C8H18O2	DI-t-BUTYL PEROXIDE	-0.01004	6.7967E-05	1.7277E-08	384	1000	------	0.01861	0.07520
854	C8H18O2S	DI-n-BUTYL SULFONE	-0.01331	6.2631E-05	1.8296E-08	564	1000	------	0.02783	0.06762
855	C8H18O3	DIETHYLENE GLYCOL DIETHYL ETHER	-0.00965	5.9025E-05	1.5504E-08	462	1000	------	0.02093	0.06488
856	C8H18O3	DIETHYLENE GLYCOL MONOBUTYL EHTER	-0.00965	5.6473E-05	1.4669E-08	504	1000	------	0.02254	0.06149
857	C8H18O4	TRIETHYLENE GLYCOL DIMETHYL ETHER	0.00189	-8.9510E-06	1.0149E-07	489	1000	------	0.02178	0.09443
858	C8H18O5	TETRAETHYLENE GLYCOL	-0.01018	5.5741E-05	1.2880E-08	600	990	------	0.02790	0.05763

518

Table 23-1 THERMAL CONDUCTIVITY OF GAS - ORGANIC COMPOUNDS (continued)

| | | | $k_{gas} = A + B\,T + C\,T^2$ | | | | | $(k_{gas}$ - W/(m K), T - K) | | |
NO	FORMULA	NAME	A	B	C	TMIN	TMAX	k_{gas} @25 C	k_{gas} @TMIN	k_{gas} @TMAX
859	C8H18S	n-OCTYL MERCAPTAN	-0.00576	4.7991E-05	3.6269E-08	472	1000	------	0.02497	0.07850
860	C8H18S	tert-OCTYL MERCAPTAN	------	--------	--------	----	----	------	------	------
861	C8H18S	BUTYL-SULFIDE	-0.00756	4.4836E-05	2.2113E-08	275	1000	0.00777	0.00644	0.05939
862	C8H18S	ETHYL-HEXYL-SULFIDE	-0.00746	4.4403E-05	2.0160E-08	275	1000	0.00757	0.00628	0.05710
863	C8H18S	HEPTYL-METHYL-SULFIDE	-0.00744	4.4659E-05	1.9706E-08	275	1000	0.00763	0.00633	0.05693
864	C8H18S	PENTYL-PROPYL-SULFIDE	-0.00726	4.3120E-05	2.1891E-08	275	1000	0.00754	0.00625	0.05775
865	C8H18S2	BUTYL-DISULFIDE	-0.00634	3.9338E-05	1.5622E-08	275	1000	0.00678	0.00566	0.04862
866	C8H19N	DI-n-BUTYLAMINE	-0.01218	7.1511E-05	2.7791E-08	432	1000	------	0.02390	0.08712
867	C8H19N	DIISOBUTYLAMINE	-0.01109	6.8773E-05	2.2919E-08	412	1000	------	0.02113	0.08060
868	C8H19N	n-OCTYLAMINE	-0.00877	5.9869E-05	3.1692E-08	453	700	------	0.02485	0.04867
869	C8H23N5	TETRAETHYLENEPENTAMINE	-0.01101	6.0553E-05	1.2889E-08	606	1000	------	0.03042	0.06243
870	C8H24O4Si4	OCTAMETHYLCYCLOTETRASILOXANE	-0.01415	6.7033E-05	3.7115E-09	448	1000	------	0.01663	0.05659
871	C9H4O5	TRIMELLITIC ANHYDRIDE	-0.00903	4.8872E-05	8.6629E-09	663	1000	------	0.02718	0.04850
872	C9H6N2O2	TOLUENE DIISOCYANATE	-0.00663	4.0463E-05	7.6806E-09	523	993	------	0.01663	0.04112
873	C9H7N	ISOQUINOLINE	-0.01670	7.2615E-05	2.8204E-08	516	1000	------	0.02152	0.05874
874	C9H7N	QUINOLINE	-0.01068	6.0360E-05	1.8156E-08	511	1000	------	0.02490	0.06784
875	C9H7NO	8-HYDROXYQUINOLINE	-0.00821	5.0777E-05	7.8718E-09	540	1000	------	0.02150	0.05044
876	C9H8	INDENE	-0.00980	5.4597E-05	2.4290E-08	456	996	------	0.02015	0.06867
877	C9H8O	2-METHYLBENZOFURAN	-0.00952	5.8741E-05	1.4022E-08	471	1000	------	0.02126	0.06324
878	C9H10	INDANE	-0.02053	8.6309E-05	1.0067E-08	451	1000	------	0.02044	0.07585
879	C9H10	cis-PROPENYLBENZENE	-0.00910	5.1208E-05	1.4653E-08	275	1000	0.00747	0.00609	0.05676
880	C9H10	trans-PROPENYLBENZENE	-0.00940	5.2667E-05	1.3633E-08	275	1000	0.00751	0.00611	0.05690
881	C9H1O	alpha-METHYLSTYRENE	-0.00992	5.5077E-05	2.7139E-08	439	999	------	0.01949	0.07219
882	C9H10	m-METHYLSTYRENE	-0.01828	8.0813E-05	7.8454E-09	445	1000	------	0.01924	0.07038
883	C9H10	o-METHYLSTYRENE	-0.01876	8.3088E-05	8.3762E-09	443	1000	------	0.01969	0.07270
884	C9H10	p-METHYLSTYRENE	-0.01006	5.7370E-05	2.3743E-08	446	1000	------	0.02025	0.07105
885	C9H10O2	BENZYL ACETATE	-0.00855	4.9702E-05	1.4755E-08	487	1000	------	0.01915	0.05591
886	C9H10O2	ETHYL BENZOATE	-0.00817	4.6374E-05	1.5547E-08	487	1000	------	0.01810	0.05375
887	C9H10O3	ETHYL VANILLIN	-0.00914	5.3952E-05	9.4382E-09	567	1000	------	0.02449	0.05425
888	C9H11NO	p-DIMETHYLAMINOBENZALDEHYDE	-0.01078	5.4566E-05	1.9441E-08	588	1000	------	0.02803	0.06323
889	C9H12	CUMENE	-0.00803	4.2071E-05	1.1791E-07	400	650	------	0.02766	0.06913
890	C9H12	m-ETHYLTOLUENE	-0.00984	5.5910E-05	2.5606E-08	434	994	------	0.01925	0.07103
891	C9H12	o-ETHYLTOLUENE	-0.01023	5.8520E-05	2.5095E-08	438	998	------	0.02022	0.07317
892	C9H12	p-ETHYLTOLUENE	-0.00998	5.6322E-05	2.6487E-08	435	995	------	0.01953	0.07228
893	C9H12	MESITYLENE	-0.01943	8.3184E-05	1.1735E-08	438	1000	------	0.01926	0.07549
894	C9H12	n-PROPYLBENZENE	-0.02709	1.0607E-04	-2.0873E-09	432	1000	------	0.01834	0.07689
895	C9H12	1,2,3-TRIMETHYLBENZENE	-0.02000	8.5193E-05	1.1559E-08	449	1000	------	0.02058	0.07675
896	C9H12	1,2,4-TRIMETHYLBENZENE	-0.01052	5.8378E-05	2.8098E-08	443	1000	------	0.02086	0.07596
897	C9H12O	BENZYL ETHYL ETHER	0.00114	-2.9858E-06	6.6643E-08	458	998	------	0.01375	0.06454
898	C9H12O	2-PHENYL-2-PROPANOL	-0.01001	6.0550E-05	1.5573E-08	475	1000	------	0.02226	0.06611
899	C9H12O2	CUMENE HYDROPEROXIDE	------	--------	--------	----	----	------	------	------
900	C9H14O	ISOPHORONE	-0.01114	6.0706E-05	2.4761E-08	488	1000	------	0.02438	0.07433
901	C9H14O6	GLYCERYL TRIACETATE	-0.01174	5.8935E-05	-4.2145E-10	532	1000	------	0.01949	0.04677
902	C9H16	1-NONYNE	-0.00924	5.5717E-05	1.5442E-08	275	1000	0.00874	0.00725	0.06192
903	C9H16O4	AZELAIC ACID	-0.00935	5.3205E-05	8.1020E-09	633	1000	------	0.02758	0.05196
904	C9H18	BUTYLCYCLOPENTANE	-0.00854	4.6965E-05	3.1737E-08	275	1000	0.00828	0.00678	0.07016
905	C9H18	cis,cis-1,3,5-TRIMETHYLCYCLOHEXANE	-0.01322	6.5795E-05	2.2665E-08	275	1000	0.00841	0.00659	0.07524
906	C9H18	cis,trans-1,3,5-TRIMETHYLCYCLOHEXANE	-0.01297	6.5247E-05	2.0700E-08	275	1000	0.00832	0.00654	0.07298
907	C9H18	ISOPROPYLCYCLOHEXANE	-0.02592	1.0534E-04	1.3165E-08	428	1000	------	0.02158	0.09259
908	C9H18	1-NONENE	-0.01089	6.3813E-05	3.0482E-08	400	1000	------	0.01951	0.08341
909	C9H18	n-PROPYLCYCLOHEXANE	-0.00575	4.4891E-05	5.2081E-08	430	1000	------	0.02318	0.09122
910	C9H18O	DIISOBUTYL KETONE	0.00213	-1.0359E-05	1.2639E-07	441	1000	------	0.02214	0.11816
911	C9H18O	1-NONANAL	0.00198	-9.0458E-06	1.0749E-07	468	1000	------	0.02129	0.10042
912	C9H18O2	n-BUTYL VALERATE	0.00186	-8.5085E-06	1.0316E-07	460	1000	------	0.01977	0.09651
913	C9H18O2	n-NONANOIC ACID	-0.00971	5.8249E-05	1.1540E-08	529	1000	------	0.02433	0.06008
914	C9H18O2	n-OCTYL FORMATE	0.00189	-8.7169E-06	1.0248E-07	472	1000	------	0.02061	0.09565
915	C9H20	3,3-DIETHYLPENTANE	-0.01277	7.2251E-05	3.6371E-08	419	999	------	0.02389	0.09571
916	C9H20	2,2-DIMETHYL-3-ETHYLPENTANE	-0.02651	1.1035E-04	1.2089E-08	407	1000	------	0.02400	0.09593
917	C9H20	3-ETHYL-2,3-DIMETHYLPENTANE	-0.01293	6.9939E-05	1.6430E-08	275	1000	0.00938	0.00755	0.07344
918	C9H20	2,4-DIMETHYL-3-ETHYLPENTANE	-0.02585	1.0958E-04	1.0014E-08	410	1000	------	0.02076	0.09374
919	C9H20	2,2-DIMETHYLHEPTANE	-0.02574	1.1018E-04	7.7586E-09	406	1000	------	0.02027	0.09220
920	C9H20	2,6-DIMETHYLHEPTANE	-0.02543	1.0889E-04	5.8666E-09	408	1000	------	0.01997	0.08933
921	C9H20	3-ETHYLHEPTANE	-0.02608	1.1067E-04	3.8539E-09	416	1000	------	0.02063	0.08844
922	C9H20	4-ETHYLHEPTANE	-0.01107	6.1643E-05	1.6284E-08	275	1000	0.00876	0.00711	0.06686
923	C9H20	2,3-DIMETHYLHEPTANE	-0.01124	6.3210E-05	1.7005E-08	275	1000	0.00912	0.00743	0.06898
924	C9H20	2,4-DIMETHYLHEPTANE	-0.01181	6.5026E-05	1.5989E-08	275	1000	0.00900	0.00728	0.06921
925	C9H20	2,5-DIMETHYLHEPTANE	-0.01165	6.4145E-05	1.6042E-08	275	1000	0.00890	0.00720	0.06854
926	C9H20	3,4-DIMETHYLHEPTANE	-0.01180	6.4921E-05	1.6981E-08	275	1000	0.00907	0.00734	0.07010
927	C9H20	3,5-DIMETHYLHEPTANE	-0.01223	6.5913E-05	1.6005E-08	275	1000	0.00884	0.00711	0.06969
928	C9H20	4,4-DIMETHYLHEPTANE	-0.01215	6.7921E-05	1.5794E-08	275	1000	0.00950	0.00772	0.07157
929	C9H20	3-ETHYL-2-METHYLHEXANE	-0.01196	6.5828E-05	1.6971E-08	275	1000	0.00918	0.00743	0.07084
930	C9H20	4-ETHYL-2-METHYLHEXANE	-0.01238	6.6725E-05	1.5996E-08	275	1000	0.00894	0.00718	0.07034
931	C9H20	3-ETHYL-3-METHYLHEXANE	-0.01263	6.9091E-05	1.6242E-08	275	1000	0.00941	0.00760	0.07270
932	C9H20	3-ETHYL-4-METHYLHEXANE	-0.01241	6.6855E-05	1.6969E-08	275	1000	0.00903	0.00726	0.07141
933	C9H20	2,2,3-TRIMETHYLHEXANE	-0.01254	6.9270E-05	1.5666E-08	275	1000	0.00951	0.00769	0.07240
934	C9H20	2,2,4-TRIMETHYLHEXANE	-0.01313	7.1160E-05	1.4484E-08	275	1000	0.00937	0.00753	0.07251
935	C9H20	2,3,3-TRIMETHYLHEXANE	-0.01252	6.9186E-05	1.6168E-08	275	1000	0.00955	0.00773	0.07283
936	C9H20	2,3,4-TRIMETHYLHEXANE	-0.01227	6.6731E-05	1.6986E-08	275	1000	0.00914	0.00737	0.07145

Table 23-1 THERMAL CONDUCTIVITY OF GAS - ORGANIC COMPOUNDS (continued)

			$k_{gas} = A + B\,T + C\,T^2$					$(k_{gas}$ - W/(m K), T - K)		
NO	FORMULA	NAME	A	B	C	TMIN	TMAX	k_{gas} @25 C	k_{gas} @TMIN	k_{gas} @TMAX
937	C9H20	2,3,5-TRIMETHYLHEXANE	-0.01226	6.6737E-05	1.5958E-08	275	1000	0.00906	0.00730	0.07044
938	C9H20	2,4,4-TRIMETHYLHEXANE	-0.01301	7.0471E-05	1.4859E-08	275	1000	0.00932	0.00749	0.07232
939	C9H20	3,3,4-TRIMETHYLHEXANE	-0.01297	7.0159E-05	1.6186E-08	275	1000	0.00939	0.00755	0.07338
940	C9H20	2-METHYLOCTANE	-0.02750	1.1432E-04	-1.2209E-10	416	1000	-----	0.02004	0.08670
941	C9H20	3-METHYLOCTANE	-0.02782	1.1521E-04	-2.2227E-11	417	1000	-----	0.02022	0.08737
942	C9H20	4-METHYLOCTANE	-0.02793	1.1621E-04	-4.2077E-10	416	1000	-----	0.02034	0.08786
943	C9H20	n-NONANE	-0.00655	3.2637E-05	7.7150E-08	449	678	-----	0.02366	0.05104
944	C9H20	2,2,3,3-TETRAMETHYLPENTANE	-0.02921	1.1850E-04	1.0144E-08	413	1000	-----	0.02146	0.09943
945	C9H20	2,2,3,4-TETRAMETHYLPENTANE	-0.02850	1.1698E-04	8.0693E-09	406	1000	-----	0.02032	0.09655
946	C9H20	2,2,4,4-TETRAMETHYLPENTANE	-0.02637	1.1114E-04	1.1595E-08	395	1000	-----	0.01934	0.09637
947	C9H20	2,3,3,4-TETRAMETHYLPENTANE	-0.01226	6.6914E-05	1.5764E-08	275	1000	0.00909	0.00733	0.07042
948	C9H20	2,2,5-TRIMETHYLHEXANE	-0.01211	6.9649E-05	3.7423E-08	397	997	-----	0.02144	0.09453
949	C9H20O	2,6-DIMETHYL-4-HEPTANOL	-0.01820	8.5102E-05	3.7637E-09	451	1000	-----	0.02095	0.07067
950	C9H20O	1-NONANOL	-0.01187	6.7488E-05	2.2469E-08	486	1000	-----	0.02624	0.07809
951	C9H20O	2-NONANOL	-0.00147	8.6692E-06	1.0418E-07	472	1000	-----	0.02583	0.11138
952	C9H20S	n-NONYL MERCAPTAN	-----	--------	--------	----	----	-----	-----	-----
953	C9H20S	BUTYL-PENTYL-SULFIDE	-0.00698	4.1287E-05	2.1392E-08	275	1000	0.00723	0.00599	0.05570
954	C9H20S	ETHYL-HEPTYL-SULFIDE	-0.00715	4.2404E-05	1.9891E-08	275	1000	0.00726	0.00602	0.05515
955	C9H20S	HEXYL-PROPYL-SULFIDE	-0.00698	4.1287E-05	2.1392E-08	275	1000	0.00723	0.00599	0.05570
956	C9H20S	METHYL-OCTYL-SULFIDE	-0.00731	4.2632E-05	1.9493E-08	275	1000	0.00731	0.00607	0.05500
957	C9H21N	n-NONYLAMINE	-0.00924	5.9820E-05	2.8719E-08	475	750	-----	0.02565	0.05178
958	C9H21N	TRIPROPYLAMINE	-0.02328	1.0622E-04	1.3649E-08	430	1000	-----	0.02492	0.09659
959	C10H6O8	PYROMELLITIC ACID	-0.00588	3.1496E-05	4.4893E-09	722	1000	-----	0.01920	0.03011
960	C10H7Br	1-BROMONAPHTHALENE	-0.00686	3.9597E-05	8.4459E-09	554	1000	-----	0.01767	0.04118
961	C10H7Cl	1-CHLORONAPHTHALENE	-0.00786	4.6122E-05	1.0111E-08	532	1000	-----	0.01954	0.04837
962	C10H8	NAPHTHALENE	-0.02306	9.2610E-05	4.4577E-10	353	1259	-----	0.00969	0.09424
963	C10H8	AZULENE	-0.00880	4.5043E-05	1.3831E-08	275	1000	0.00586	0.00463	0.05007
964	C10H9N	QUINALDINE	-0.00929	5.3175E-05	1.4263E-08	520	1000	-----	0.02222	0.05815
965	C10H10	m-DIVINYLBENZENE	-0.00988	5.7353E-05	1.8576E-08	473	1000	-----	0.02140	0.06605
966	C10H10	1-METHYLINDENE	-0.01990	8.8360E-05	2.4675E-09	472	1000	-----	0.02236	0.07093
967	C10H10	2-METHYLINDENE	-0.02021	9.0063E-05	3.1453E-09	458	1000	-----	0.02170	0.07300
968	C10H10O4	DIMETHYL PHTHALATE	-0.00767	4.3884E-05	9.6831E-09	557	1000	-----	0.01978	0.04590
969	C10H10O4	DIMETHYL TEREPHTHALATE	-0.00361	3.5675E-05	1.3239E-08	561	1000	-----	0.02057	0.04530
970	C10H12	DICYCLOPENTADIENE	-----	--------	--------	----	----	-----	-----	-----
971	C10H12	1,2,3,4-TETRAHYDRONAPHTHALENE	-0.00942	5.0729E-05	2.3456E-08	481	991	-----	0.02041	0.06389
972	C10H12O	ANETHOLE	-0.00963	5.5759E-05	1.5008E-08	508	1000	-----	0.02257	0.06114
973	C10H12O4	DIALLYL MALEATE	-0.00806	5.3705E-05	5.7459E-09	520	1000	-----	0.02142	0.05139
974	C10H14	n-BUTYLBENZENE	-0.02091	8.8748E-05	7.1867E-09	456	996	-----	0.02105	0.07461
975	C10H14	sec-BUTYLBENZENE	-0.01071	6.0505E-05	2.6424E-08	446	996	-----	0.02153	0.07577
976	C10H14	tert-BUTYLBENZENE	-0.01094	6.2733E-05	2.6379E-08	442	992	-----	0.02194	0.07725
977	C10H14	1,2,3,4-TETRAMETHYLBENZENE	-0.00798	4.7539E-05	1.7532E-08	275	1000	0.00775	0.00642	0.05709
978	C10H14	m-CYMENE	-0.01087	6.0806E-05	2.7371E-08	448	998	-----	0.02186	0.07708
979	C10H14	o-CYMENE	-0.01075	5.9879E-05	2.7733E-08	451	991	-----	0.02190	0.07583
980	C10H14	p-CYMENE	-0.01029	6.7187E-05	1.3232E-08	450	1000	-----	0.02262	0.07013
981	C10H14	m-DIETHYLBENZENE	-0.02017	8.6688E-05	8.0428E-09	454	994	-----	0.02084	0.07394
982	C10H14	o-DIETHYLBENZENE	-0.01075	6.0773E-05	2.4600E-08	457	1000	-----	0.02216	0.07462
983	C10H14	p-DIETHYLBENZENE	-0.01063	5.9356E-05	2.5488E-08	457	1000	-----	0.02182	0.07421
984	C10H14	2-ETHYL-m-XYLENE	-0.01119	6.3211E-05	2.4836E-08	463	1000	-----	0.02340	0.07686
985	C10H14	2-ETHYL-p-XYLENE	-0.01078	6.0393E-05	2.5169E-08	460	1000	-----	0.02233	0.07478
986	C10H14	3-ETHYL-o-XYLENE	-0.01877	8.2876E-05	8.7574E-09	467	1000	-----	0.02184	0.07286
987	C10H14	4-ETHYL-m-XYLENE	-0.01077	6.0066E-05	2.5338E-08	462	1000	-----	0.02239	0.07463
988	C10H14	4-ETHYL-o-XYLENE	-0.01076	5.9922E-05	2.5219E-08	463	1000	-----	0.02239	0.07438
989	C10H14	5-ETHYL-m-XYLENE	-0.01925	8.1961E-05	9.8622E-09	457	1000	-----	0.02027	0.07257
990	C10H14	ISOBUTYLBENZENE	-0.01124	6.3606E-05	2.7317E-08	446	1000	-----	0.02256	0.07968
991	C10H14	1,2,3,5-TETRAMETHYLBENZENE	-0.01078	5.8559E-05	2.6562E-08	471	1000	-----	0.02269	0.07434
992	C10H14	1,2,4,5-TETRAMETHYLBENZENE	-0.00373	4.7379E-06	9.8524E-08	470	1000	-----	0.02026	0.09953
993	C10H14O	p-tert-BUTYLPHENOL	-0.01007	5.9304E-05	1.4237E-08	513	1000	-----	0.02410	0.06347
994	C10H14O2	p-tert-BUTYLCATECHOL	-0.00974	5.7770E-05	1.0368E-08	558	1000	-----	0.02572	0.05840
995	C10H15N	N,N-DIETYHLANILINE	-----	--------	--------	----	----	-----	-----	-----
996	C10H15N	2,6-DIETHYLANILINE	-0.00972	5.8191E-05	1.3059E-08	509	1000	-----	0.02328	0.06153
997	C10H16	CAMPHENE	-0.02672	1.0380E-04	8.2702E-09	434	1000	-----	0.01989	0.08535
998	C10H16	D-LIMONENE	-0.02197	9.3562E-05	7.4279E-09	450	1000	-----	0.02164	0.07902
999	C10H16	alpha-PHELLANDRENE	-0.02222	9.6145E-05	7.8464E-09	448	1000	-----	0.02243	0.08177
1000	C10H16	beta-PHELLANDRENE	-0.02301	9.9718E-05	5.9927E-09	447	1000	-----	0.02276	0.08270
1001	C10H16	alpha-PINENE	-0.02514	1.0151E-04	-7.1878E-09	429	1000	-----	0.01708	0.06918
1002	C10H16	beta-PINENE	-0.02797	1.0908E-04	3.9401E-09	439	1000	-----	0.02068	0.08505
1003	C10H16	alpha-TERPINENE	-0.01190	7.0296E-05	2.3739E-08	450	1000	-----	0.02454	0.08214
1004	C1OH16	gamma-TERPINENE	-0.01176	6.9037E-05	2.3256E-08	456	1000	-----	0.02456	0.08053
1005	C10H16	TERPINOLENE	-0.02133	9.0600E-05	8.4117E-08	458	1000	-----	0.02193	0.07768
1006	C10H16O	CAMPHOR	-0.02563	9.7273E-05	1.0104E-08	481	1000	-----	0.02350	0.08175
1007	C10H18	1-DECYNE	-0.00880	5.2643E-05	1.5781E-08	275	1000	0.00830	0.00687	0.05962
1008	C10H18	cis-DECAHYDRONANPHTALENE	-0.00457	-1.5454E-05	2.9327E-07	469	1000	-----	0.05269	0.27325
1009	C10H18	trans-DECAHYDRONAPHTHALENE	-0.00480	-1.2283E-05	2.7948E-07	460	1000	-----	0.04869	0.26240
1010	C10H18O4	SEBACIC ACID	-0.00925	5.2244E-05	7.9754E-09	642	1000	-----	0.02758	0.05097
1011	C10H20	n-BUTYLCYCLOHEXANE	-0.01194	6.4176E-05	3.3729E-08	454	1000	-----	0.02415	0.08597
1012	C10H20	1-CYCLOPENTYLPENTANE	-0.01044	5.4533E-05	2.1040E-08	275	1000	0.00769	0.00615	0.06513
1013	C10H20	1-DECENE	-0.01118	6.2751E-05	2.8360E-08	444	1000	-----	0.02227	0.07993
1014	C10H20O	1-DECANAL	0.00195	-9.2739E-06	1.0395E-07	488	1000	-----	0.02218	0.09663

520

Table 23-1 THERMAL CONDUCTIVITY OF GAS - ORGANIC COMPOUNDS (continued)

NO	FORMULA	NAME	$k_{gas} = A + BT + CT^2$					$(k_{gas}$ - W/(m K), T - K)		
			A	B	C	TMIN	TMAX	k_{gas} @25 C	k_{gas} @TMIN	k_{gas} @TMAX
1015	C10H20O2	n-DECANOIC ACID	-0.00965	5.5721E-05	1.2593E-08	543	1000	-----	0.02432	0.05866
1016	C10H20O2	2-ETHYLHEXYL ACETATE	-0.00816	4.9835E-05	1.2459E-08	472	1000	-----	0.01814	0.05413
1017	C10H20O2	ISOPENTYL ISOVALERATE	-0.01063	6.4970E-05	1.6592E-08	467	1000	-----	0.02333	0.07093
1018	C10H22	n-DECANE	-0.00113	8.1090E-06	9.6092E-08	470	700	-----	0.02391	0.05163
1019	C10H22	2-METHYLNONANE	-0.02612	1.0783E-04	1.3143E-09	440	1000	-----	0.02158	0.08302
1020	C10H22	3-METHYLNONANE	-0.02629	1.0816E-04	1.7111E-09	441	1000	-----	0.02174	0.08358
1021	C10H22	4-METHYLNONANE	-0.02649	1.0941E-04	1.2206E-09	439	1000	-----	0.02178	0.08414
1022	C10H22	5-METHYLNONANE	-0.02646	1.0917E-04	1.3130E-09	438	1000	-----	0.02161	0.08402
1023	C10H22	3-ETHYLOCTANE	-0.01052	5.9140E-05	1.7492E-08	275	1000	0.00867	0.00707	0.06611
1024	C10H22	4-ETHYLOCTANE	-0.01071	6.0191E-05	1.7554E-08	275	1000	0.00880	0.00717	0.06704
1025	C10H22	2,2-DIMETHYLOCTANE	-0.02572	1.0727E-04	5.6114E-09	430	1000	-----	0.02144	0.08716
1026	C10H22	2,3-DIMETHYLOCTANE	-0.01077	5.9849E-05	1.7080E-08	275	1000	0.00859	0.00698	0.06616
1027	C10H22	2,4-DIMETHYLOCTANE	-0.01062	6.0273E-05	1.6950E-08	275	1000	0.00886	0.00724	0.06660
1028	C10H22	2,5-DIMETHYLOCTANE	-0.01088	6.0458E-05	1.6629E-08	275	1000	0.00862	0.00700	0.06621
1029	C10H22	2,6-DIMETHYLOCTANE	-0.01072	5.9564E-05	1.6389E-08	275	1000	0.00850	0.00690	0.06523
1030	C10H22	2,7-DIMETHYLOCTANE	-0.01027	5.8245E-05	1.6537E-08	275	1000	0.00857	0.00700	0.06451
1031	C10H22	3,3-DIMETHYLOCTANE	-0.01110	6.2595E-05	1.6493E-08	275	1000	0.00903	0.00736	0.06799
1032	C10H22	3,4-DIMETHYLOCTANE	-0.01098	6.0999E-05	1.7502E-08	275	1000	0.00876	0.00712	0.06752
1033	C10H22	3,5-DIMETHYLOCTANE	-0.01138	6.2025E-05	1.6719E-08	275	1000	0.00860	0.00694	0.06736
1034	C10H22	3,6-DIMETHYLOCTANE	-0.01126	6.1329E-05	1.6654E-08	275	1000	0.00851	0.00686	0.06672
1035	C10H22	4,4-DIMETHYLOCTANE	-0.01132	6.3799E-05	1.6485E-08	275	1000	0.00917	0.00747	0.06896
1036	C10H22	4,5-DIMETHYLOCTANE	-0.01107	6.1514E-05	1.7537E-08	275	1000	0.00883	0.00717	0.06798
1037	C10H22	4-PROPYLHEPTANE	-0.01093	6.1460E-05	1.7381E-08	275	1000	0.00894	0.00729	0.06791
1038	C10H22	4-ISOPROPYLHEPTANE	-0.01127	6.2660E-05	1.7571E-08	275	1000	0.00897	0.00729	0.06896
1039	C10H22	3-ETHYL-2-METHYLHEPTANE	-0.01113	6.1840E-05	1.7549E-08	275	1000	0.00887	0.00720	0.06826
1040	C10H22	4-ETHYL-2-METHYLHEPTANE	-0.01160	6.3196E-05	1.6753E-08	275	1000	0.00873	0.00705	0.06835
1041	C10H22	5-ETHYL-2-METHYLHEPTANE	-0.01138	6.1989E-05	1.6734E-08	275	1000	0.00859	0.00693	0.06734
1042	C10H22	3-ETHYL-3-METHYLHEPTANE	-0.01173	6.4813E-05	1.7030E-08	275	1000	0.00911	0.00738	0.07011
1043	C10H22	4-ETHYL-3-METHYLHEPTANE	-0.01164	6.3423E-05	1.7655E-08	275	1000	0.00884	0.00714	0.06944
1044	C10H22	3-ETHYL-5-METHYLHEPTANE	-0.01196	6.3930E-05	1.6700E-08	275	1000	0.00859	0.00688	0.06867
1045	C10H22	3-ETHYL-4-METHYLHEPTANE	-0.01158	6.3094E-05	1.7639E-08	275	1000	0.00880	0.00710	0.06915
1046	C10H22	4-ETHYL-4-METHYLHEPTANE	-0.01190	6.5776E-05	1.7007E-08	275	1000	0.00922	0.00747	0.07088
1047	C10H22	2,2,3-TRIMETHYLHEPTANE	-0.01159	6.4628E-05	1.6458E-08	275	1000	0.00914	0.00743	0.06950
1048	C10H22	2,2,4-TRIMETHYLHEPTANE	-0.01212	6.6339E-05	1.5214E-08	275	1000	0.00901	0.00727	0.06943
1049	C10H22	2,2,5-TRIMETHYLHEPTANE	-0.01196	6.5453E-05	1.5226E-08	275	1000	0.00891	0.00719	0.06872
1050	C10H22	2,2,6-TRIMETHYLHEPTANE	-0.01147	6.3972E-05	1.5120E-08	275	1000	0.00895	0.00727	0.06762
1051	C10H22	2,3,3-TRIMETHYLHEPTANE	-0.01163	6.4861E-05	1.6882E-08	275	1000	0.00921	0.00748	0.07011
1052	C10H22	2,3,4-TRIMETHYLHEPTANE	-0.01149	6.3155E-05	1.7542E-08	275	1000	0.00890	0.00720	0.06921
1053	C10H22	2,3,5-TRIMETHYLHEPTANE	-0.01172	6.3198E-05	1.6898E-08	275	1000	0.00862	0.00694	0.06838
1054	C10H22	2,3,6-TRIMETHYLHEPTANE	-0.01132	6.2194E-05	1.6630E-08	275	1000	0.00870	0.00704	0.06750
1055	C10H22	2,4,4-TRIMETHYLHEPTANE	-0.01221	6.6799E-05	1.5684E-08	275	1000	0.00910	0.00735	0.07027
1056	C10H22	2,4,5-TRIMETHYLHEPTANE	-0.01186	6.3977E-05	1.6735E-08	275	1000	0.00870	0.00700	0.06885
1057	C10H22	2,4,6-TRIMETHYLHEPTANE	-0.01188	6.4079E-05	1.5678E-08	275	1000	0.00862	0.00693	0.06788
1058	C10H22	2,5,5-TRIMETHYLHEPTANE	-0.01203	6.5821E-05	1.5616E-08	275	1000	0.00898	0.00725	0.06941
1059	C10H22	3,3,4-TRIMETHYLHEPTANE	-0.01211	6.6255E-05	1.6974E-08	275	1000	0.00915	0.00739	0.07112
1060	C10H22	3,3,5-TRIMETHYLHEPTANE	-0.01188	6.3797E-05	1.4958E-08	275	1000	0.00847	0.00680	0.06688
1061	C10H22	3,4,4-TRIMETHYLHEPTANE	-0.01216	6.6539E-05	1.6970E-08	275	1000	0.00919	0.00742	0.07135
1062	C10H22	3,4,5-TRIMETHYLHEPTANE	-0.01163	6.2746E-05	1.6811E-08	275	1000	0.00857	0.00690	0.06793
1063	C10H22	3-ISOPROPYL-2-METHYLHEXANE	-0.01146	6.2971E-05	1.8105E-08	275	1000	0.00892	0.00723	0.06962
1064	C10H22	3,3-DIETHYLHEXANE	-0.01234	6.6902E-05	1.7506E-08	275	1000	0.00916	0.00738	0.07207
1065	C10H22	3,4-DIETHYLHEXANE	-0.01211	6.4763E-05	1.7737E-08	275	1000	0.00878	0.00704	0.07039
1066	C10H22	3-ETHYL-2,2-DIMETHYLHEXANE	-0.01228	6.7173E-05	1.6540E-08	275	1000	0.00922	0.00744	0.07143
1067	C10H22	4-ETHYL-2,2-DIMETHYLHEXANE	-0.01270	6.8225E-05	1.5114E-08	275	1000	0.00898	0.00720	0.07064
1068	C10H22	3-ETHYL-2,3-DIMETHYLHEXANE	-0.01221	6.6779E-05	1.7410E-08	275	1000	0.00925	0.00747	0.07198
1069	C10H22	4-ETHYL-2,3-DIMETHYLHEXANE	-0.01202	6.4806E-05	1.7652E-08	275	1000	0.00887	0.00714	0.07044
1070	C10H22	3-ETHYL-2,4-DIMETHYLHEXANE	-0.01207	6.5065E-05	1.7645E-08	275	1000	0.00890	0.00716	0.07064
1071	C10H22	4-ETHYL-2,4-DIMETHYLHEXANE	-0.01247	6.6964E-05	1.6483E-08	275	1000	0.00896	0.00719	0.07098
1072	C10H22	3-ETHYL-2,5-DIMETHYLHEXANE	-0.01203	6.4855E-05	1.6745E-08	275	1000	0.00880	0.00707	0.06957
1073	C10H22	4-ETHYL-3,3-DIMETHYLHEXANE	-0.01264	6.7901E-05	1.7071E-08	275	1000	0.00912	0.00732	0.07233
1074	C10H22	3-ETHYL-3,4-DIMETHYLHEXANE	-0.01270	6.8174E-05	1.7062E-08	275	1000	0.00914	0.00734	0.07254
1075	C10H22	2,2,3,3-TETRAMETHYLHEXANE	-0.01166	6.4304E-05	1.4993E-08	275	1000	0.00885	0.00716	0.06764
1076	C10H22	2,2,3,4-TETRAMETHYLHEXANE	-0.01248	6.7573E-05	1.6649E-08	275	1000	0.00915	0.00736	0.07174
1077	C10H22	2,2,3,5-TETRAMETHYLHEXANE	-0.01255	6.7983E-05	1.5480E-08	275	1000	0.00910	0.00732	0.07091
1078	C10H22	2,2,4,4-TETRAMETHYLHEXANE	-0.01302	7.0232E-05	1.4710E-08	275	1000	0.00923	0.00741	0.07192
1079	C10H22	2,2,4,5-TETRAMETHYLHEXANE	-0.01252	6.7798E-05	1.5262E-08	275	1000	0.00905	0.00728	0.07054
1080	C10H22	2,2,5,5-TETRAMETHYLHEXANE	-0.01246	6.8481E-05	1.3084E-08	275	1000	0.00912	0.00736	0.06911
1081	C10H22	2,3,3,4-TETRAMETHYLHEXANE	-0.01247	6.7523E-05	1.7451E-08	275	1000	0.00921	0.00742	0.07250
1082	C10H22	2,3,3,5-TETRAMETHYLHEXANE	-0.01256	6.8010E-05	1.6046E-08	275	1000	0.00914	0.00736	0.07150
1083	C10H22	2,3,4,4-TETRAMETHYLHEXANE	-0.01250	6.7701E-05	1.7078E-08	275	1000	0.00920	0.00741	0.07228
1084	C10H22	2,3,4,5-TETRAMETHYLHEXANE	-0.01195	6.5010E-05	1.7454E-08	275	1000	0.00898	0.00725	0.07051
1085	C10H22	3,3,4,4-TETRAMETHYLHEXANE	-0.01282	6.9432E-05	1.7146E-08	275	1000	0.00941	0.00757	0.07376
1086	C10H22	2,4-DIMETHYL-3-ISOPROPYLPENTANE	-0.01257	6.8089E-05	1.6390E-08	275	1000	0.00919	0.00739	0.07191
1087	C10H22	3,3-DIETHYL-2-METHYLPENTANE	-0.01264	6.7841E-05	1.7973E-08	275	1000	0.00918	0.00738	0.07317
1088	C10H22	3-ETHYL-2,2,3-TRIMETHYLPENTANE	-0.01288	6.9771E-05	1.7190E-08	275	1000	0.00945	0.00761	0.07408
1089	C10H22	3-ETHYL-2,2,4-TRIMETHYLPENTANE	-0.01268	6.8661E-05	1.6586E-08	275	1000	0.00927	0.00746	0.07257
1090	C10H22	3-ETHYL-2,3,4-TRIMETHYLPENTANE	-0.01246	6.7444E-05	1.8022E-08	275	1000	0.00925	0.00745	0.07301
1091	C10H22	2,2,3,3,4-PENTAMETHYLPENTANE	-0.01251	6.8963E-05	1.7404E-08	275	1000	0.00960	0.00777	0.07386
1092	C10H22	2,2,3,4,4-PENTAMETHYLPENTANE	-0.01295	7.0402E-05	1.5872E-08	275	1000	0.00945	0.00761	0.07332

521

Table 23-1 THERMAL CONDUCTIVITY OF GAS - ORGANIC COMPOUNDS (continued)

NO	FORMULA	NAME	$k_{gas} = A + B\,T + C\,T^2$					$(k_{gas} - W/(m\,K),\ T - K)$		
			A	B	C	TMIN	TMAX	k_{gas} @25 C	k_{gas} @TMIN	k_{gas} @TMAX
1093	C10H22O	1-DECANOL	0.03262	-1.1884E-04	2.0994E-07	503	702	-----	0.02596	0.05265
1094	C10H22O	DI-n-PENTYL ETHER	-0.01420	7.1760E-05	1.0355E-08	460	1000	-----	0.02100	0.06792
1095	C10H22O	ISODECANOL	-0.01047	6.1440E-05	1.6754E-08	493	1000	-----	0.02389	0.06772
1096	C10H22O5	TETRAETHYLENE GLYCOL DIMETHYL ETHER	-0.00913	5.3148E-05	1.1123E-08	549	1000	-----	0.02340	0.05514
1097	C10H22S	n-DECYL MERCAPTAN	-0.01171	6.3915E-05	2.2689E-08	512	1000	-----	0.02696	0.07489
1098	C10H22S	BUTYL-HEXYL-SULFIDE	-0.00670	3.9552E-05	2.0950E-08	275	1000	0.00695	0.00576	0.05380
1099	C10H22S	ETHYL-OCTYL-SULFIDE	-0.00686	4.0563E-05	1.9598E-08	275	1000	0.00698	0.00578	0.05330
1100	C10H22S	HEPTYL-PROPYL-SULFIDE	-0.00670	3.9552E-05	2.0950E-08	275	1000	0.00695	0.00576	0.05380
1101	C10H22S	METHYL-NONYL-SULFIDE	-0.00684	4.0763E-05	1.9247E-08	275	1000	0.00702	0.00583	0.05317
1102	C10H22S	PENTYL-SULFIDE	-0.00670	3.9552E-05	2.0950E-08	275	1000	0.00695	0.00576	0.05380
1103	C10H22S2	PENTYL-DISULFIDE	-0.00606	3.7069E-05	1.5630E-08	275	1000	0.00638	0.00532	0.04664
1104	C10H23N	n-DECYLAMINE	-0.00917	5.8694E-05	2.6803E-08	494	750	-----	0.02637	0.04993
1105	C11H10	1-METHYLNAPHTHALENE	-0.01703	7.1429E-05	6.7407E-09	518	998	-----	0.02178	0.06097
1106	C11H10	2-METHYLNAPHTHALENE	-0.01706	7.1614E-05	7.3389E-09	514	994	-----	0.02169	0.06138
1107	C11H14O2	n-BUTYL BENZOATE	0.00115	-4.3515E-06	6.2142E-08	523	1000	-----	0.01587	0.05894
1108	C11H16	n-PENTYLBENZENE	-0.01096	6.0321E-05	2.4626E-08	479	1000	-----	0.02358	0.07399
1109	C11H16O	p-tert-AMYLPHENOL	-0.01008	5.7404E-05	1.4538E-08	535	1000	-----	0.02479	0.06186
1110	C11H20	1-UNDECYNE	-0.00840	4.9997E-05	1.5963E-08	275	1000	0.00793	0.00656	0.05756
1111	C11H20O2	2-ETHYLHEXYL ACRYLATE	0.00187	-9.1216E-06	9.9909E-08	489	999	-----	0.02130	0.09247
1112	C11H22	1-UNDECENE	-0.02584	1.0261E-04	-2.1510E-10	466	1000	-----	0.02193	0.07655
1113	C11H22	1-CYCLOPENTYLHEXANE	-0.00978	5.1386E-05	2.0695E-08	275	1000	0.00738	0.00592	0.06230
1114	C11H22	PENTYLCYCLOHEXANE	-0.01071	5.5518E-05	2.0159E-08	275	1000	0.00763	0.00608	0.06497
1115	C11H22O	1-UNDECANAL	0.00191	-9.3691E-06	1.0081E-07	506	1000	-----	0.02298	0.09335
1116	C11H24	n-UNDECANE	0.01364	-4.8303E-05	1.4396E-07	470	800	-----	0.02274	0.06713
1117	C11H24O	1-UNDECANOL	-0.01150	6.4324E-05	1.9854E-08	516	1000	-----	0.02698	0.07268
1118	C11H24S	UNDECYL MERCAPTAN	-0.00988	5.5448E-05	1.5534E-08	531	1000	-----	0.02394	0.06110
1119	C11H24S	BUTYL-HEPTYL-SULFIDE	-0.00646	3.8048E-05	2.0491E-08	275	1000	0.00671	0.00555	0.05208
1120	C11H24S	DECYL-METHYL-SULFIDE	-0.00658	3.9114E-05	1.8993E-08	275	1000	0.00677	0.00561	0.05153
1121	C11H24S	ETHYL-NONYL-SULFIDE	-0.00660	3.8935E-05	1.9302E-08	275	1000	0.00672	0.00557	0.05164
1122	C11H24S	OCTYL-PROPYL-SULFIDE	-0.00646	3.8048E-05	2.0491E-08	275	1000	0.00671	0.00555	0.05208
1123	C12H8O	DIBENZOFURAN	-0.00911	4.9183E-05	1.4632E-08	558	1000	-----	0.02289	0.05471
1124	C12H9N	DIBENZOPYRROLE	-0.01782	7.3507E-05	2.4511E-09	628	1000	-----	0.02931	0.05814
1125	C12H10	ACENAPHTHENE	-0.00954	5.0025E-05	1.7866E-08	551	1000	-----	0.02345	0.05835
1126	C12H10	BIPHENYL	-0.00788	4.2910E-05	3.4569E-08	373	1000	-----	0.01293	0.06960
1127	C12H10O	DIPHENYL ETHER	-0.00885	5.1306E-05	1.2076E-08	531	1000	-----	0.02180	0.05453
1128	C12H11N	p-AMINODIPHENYL	-0.00918	5.2745E-05	1.0297E-08	575	1000	-----	0.02455	0.05386
1129	C12H11N	DIPHENYLAMINE	-0.00972	5.4414E-05	1.2286E-08	575	1000	-----	0.02563	0.05698
1130	C12H11N3	p-AMINOAZOBENZENE	-0.00975	5.4448E-05	9.3225E-09	633	1000	-----	0.02845	0.05402
1131	C12H11N3	1,3-DIPHENYLTRIAZENE	-0.00903	5.2512E-05	7.9461E-09	610	1000	-----	0.02596	0.05143
1132	C12H12	1,2-DIMETHYLNAPHTHALENE	-0.00800	4.3484E-05	1.3915E-08	275	1000	0.00620	0.00501	0.04940
1133	C12H12	1,3-DIMETHYLNAPHTHALENE	-0.00777	4.2610E-05	1.4444E-08	275	1000	0.00622	0.00504	0.04928
1134	C12H12	1,4-DIMETHYLNAPHTHALENE	-0.00799	4.3405E-05	1.3941E-08	275	1000	0.00619	0.00500	0.04936
1135	C12H12	1,5-DIMETHYLNAPHTHALENE	-0.00802	4.3587E-05	1.3880E-08	275	1000	0.00621	0.00502	0.04945
1136	C12H12	1,6-DIMETHYLNAPHTHALENE	-0.00781	4.2782E-05	1.4387E-08	275	1000	0.00622	0.00504	0.04936
1137	C12H12	1,7-DIMETHYLNAPHTHALENE	-0.00781	4.2782E-05	1.4387E-08	275	1000	0.00622	0.00504	0.04936
1138	C12H12	2,3-DIMETHYLNAPHTHALENE	-0.00718	4.0210E-05	1.5838E-08	275	1000	0.00622	0.00508	0.04887
1139	C12H12	2,6-DIMETHYLNAPHTHALENE	-0.01677	7.1801E-05	9.5201E-09	535	1000	-----	0.02437	0.06455
1140	C12H12	2,7-DIMETHYLNAPHTHALENE	-0.01682	7.1879E-05	9.4156E-09	536	1000	-----	0.02441	0.06447
1141	C12H12	1-ETHYLNAPHTHALENE	-0.01007	5.3746E-05	1.9325E-08	531	1000	-----	0.02392	0.06300
1142	C12H12	2-ETHYLNAPHTHALENE	-0.00794	4.3407E-05	1.4556E-08	275	1000	0.00630	0.00510	0.05002
1143	C12H12N2	p-AMINODIPHENYLAMINE	-0.01522	6.6481E-05	5.2547E-10	627	1000	-----	0.02667	0.05179
1144	C12H12N2	HYDRAZOBENZENE	-0.00927	5.3347E-05	1.0456E-08	573	1000	-----	0.02473	0.05453
1145	C12H14	1,2,3-TRIMETHYLINDENE	-0.01086	6.1319E-05	1.8696E-08	509	1000	-----	0.02520	0.06916
1146	C12H14O4	DIETHYL PHTHALATE	-0.00863	4.7444E-05	1.2248E-08	567	1000	-----	0.02221	0.05106
1147	C12H16	CYCLOHEXYLBENZENE	-0.01111	5.6012E-05	2.9171E-08	513	1000	-----	0.02530	0.07407
1148	C12H18	m-DIISOPROPYLBENZENE	-0.02095	8.8293E-05	6.1091E-09	476	1000	-----	0.02246	0.07345
1149	C12H18	p-DIISOPROPYLBENZENE	-0.02100	8.8014E-05	5.8660E-09	484	1000	-----	0.02297	0.07288
1150	C12H18	n-HEXYLBENZENE	-0.02036	8.4338E-05	6.6954E-09	499	1000	-----	0.02339	0.07067
1151	C12H18	1,2,3-TRIETHYLBENZENE	-0.00805	4.6007E-05	1.6409E-08	275	1000	0.00713	0.00584	0.05437
1152	C12H18	1,2,4-TRIETHYLBENZENE	-0.00817	4.6432E-05	1.6195E-08	275	1000	0.00711	0.00582	0.05446
1153	C12H18	1,3,5-TRIETHYLBENZENE	-0.00840	4.6889E-05	1.6008E-08	275	1000	0.00700	0.00571	0.05450
1154	C12H18	HEXAMETHYLBENZENE	-0.00730	4.4977E-05	1.6745E-08	275	1000	0.00760	0.00634	0.05442
1155	C12H20O4	DIBUTYL MALEATE	-0.00826	4.7750E-05	1.0165E-08	553	1000	-----	0.02125	0.04966
1156	C12H22	BICYCLOHEXYL	-0.01229	6.0636E-05	3.6019E-08	512	992	-----	0.02820	0.08331
1157	C12H22	1-DODECYNE	-0.00800	4.7468E-05	1.6213E-08	275	1000	0.00759	0.00628	0.05568
1158	C12H23N	DICYCLOHEXYLAMINE	-0.01184	6.1839E-05	2.5082E-08	529	1000	-----	0.02789	0.07508
1159	C12H24	1-DODECENE	-0.02600	1.0107E-04	-1.0231E-09	487	1000	-----	0.02298	0.07405
1160	C12H24	1-CYCLOPENTYLHEPTANE	-0.00921	4.8689E-05	1.9969E-08	275	1000	0.00708	0.00569	0.05945
1161	C12H24	1-CYCLOHEXYLHEXANE	-0.01002	5.2257E-05	1.9946E-08	275	1000	0.00733	0.00586	0.06218
1162	C12H24O	1-DODECANAL	0.00187	-9.4540E-06	9.7853E-08	523	1000	-----	0.02369	0.09027
1163	C12H24O2	n-DODECANOIC ACID	-0.00951	5.2199E-05	1.3328E-08	572	1000	-----	0.02471	0.05602
1164	C12H26	n-DODECANE	-0.00812	2.9150E-05	7.2085E-08	489	1000	-----	0.02337	0.09312
1165	C12H26O	DI-n-HEXYL ETHER	-0.00942	5.7682E-05	1.2128E-08	499	1000	-----	0.02238	0.06039
1166	C12H26O	1-DODECANOL	-0.01135	6.3614E-05	1.7549E-08	535	1000	-----	0.02771	0.06981
1167	C12H26O3	DIETHYLENE GLYCOL DI-n-BUTYL ETHER	-0.00930	5.3413E-05	1.3380E-08	529	1000	-----	0.02270	0.05749
1168	C12H26S	n-DODECYL MERCAPTAN	-0.00575	4.4400E-05	3.2638E-08	548	1000	-----	0.02838	0.07129
1169	C12H26S	BUTYL-OCTYL-SULFIDE	-0.00622	3.6645E-05	2.0087E-08	275	1000	0.00649	0.00538	0.05051
1170	C12H26S	DECYL-ETHYL-SULFIDE	-0.00635	3.7435E-05	1.9027E-08	275	1000	0.00650	0.00538	0.05011

522

Table 23-1 THERMAL CONDUCTIVITY OF GAS - ORGANIC COMPOUNDS (continued)

| NO | FORMULA | NAME | $k_{gas} = A + BT + CT^2$ | | | | | $(k_{gas}$ - W/(m K), T - K) | | |
			A	B	C	TMIN	TMAX	k_{gas} @25 C	k_{gas} @TMIN	k_{gas} @TMAX
1171	C12H26S	HEXYL-SULFIDE	-0.00622	3.6645E-05	2.0087E-08	275	1000	0.00649	0.00538	0.05051
1172	C12H26S	METHYL-UNDECYL-SULFIDE	-0.00633	3.7597E-05	1.8746E-08	275	1000	0.00655	0.00543	0.05001
1173	C12H26S	NONYL-PROPYL-SULFIDE	-0.00622	3.6645E-05	2.0087E-08	275	1000	0.00649	0.00538	0.05051
1174	C12H26S2	HEXYL-DISULFIDE	-0.00576	3.4987E-05	1.5493E-08	275	1000	0.00605	0.00503	0.04472
1175	C12H27BO3	TRI-n-BUTYL BORATE	------	--------	--------	----	----	-----	-----	-----
1176	C12H27N	DODECYLAMINE	-0.00905	5.6320E-05	2.3942E-08	532	750	-----	0.02769	0.04666
1177	C12H27N	TRI-n-BUTYLAMINE	-0.01351	7.4453E-05	2.9099E-08	487	997	-----	0.02965	0.08964
1178	C13H10	FLUORENE	-0.00904	5.6671E-05	6.6790E-09	570	1000	-----	0.02543	0.05431
1179	C13H10O	BENZOPHENONE	-0.00813	4.3772E-05	1.1875E-08	579	1000	-----	0.02119	0.04752
1180	C13H12	DIPHENYLMETHANE	-0.01004	5.3011E-05	1.9575E-08	537	1000	-----	0.02407	0.06255
1181	C13H14	1-PROPYLNAPHTHALENE	-0.00780	4.2550E-05	1.4006E-08	275	1000	0.00613	0.00496	0.04876
1182	C13H14	2-PROPYLNAPHTHALENE	-0.00760	4.1771E-05	1.4401E-08	275	1000	0.00613	0.00498	0.04857
1183	C13H14	2ETHYL-3-METHYLNAPHTHALENE	-0.00715	4.0167E-05	1.5258E-08	275	1000	0.00618	0.00505	0.04828
1184	C13H14	2ETHYL-6-METHYLNAPHTHALENE	-0.00686	3.9673E-05	1.5647E-08	275	1000	0.00636	0.00523	0.04846
1185	C13H14	2ETHYL-7-METHYLNAPHTHALENE	-0.00686	3.9673E-05	1.5647E-08	275	1000	0.00636	0.00523	0.04846
1186	C13H20	n-HEPTYLBENZENE	-0.01079	5.7422E-05	2.2320E-08	519	1000	-----	0.02502	0.06895
1187	C13H24	1-TRIDECYNE	-0.00766	4.5330E-05	1.6254E-08	275	1000	0.00730	0.00603	0.05392
1188	C13H26	1-TRIDECENE	-0.02607	9.9453E-05	-1.5725E-09	506	1000	-----	0.02385	0.07181
1189	C13H26	1-CYCLOPENTYLOCTANE	-0.00868	4.6179E-05	2.0113E-08	275	1000	0.00688	0.00554	0.05761
1190	C13H26	1-CYCLOHEXYLHEPTANE	-0.00942	4.9431E-05	1.9685E-08	275	1000	0.00707	0.00566	0.05970
1191	C13H26O	1-TRIDECANAL	0.00184	-9.4112E-06	9.4178E-08	540	1000	-----	0.02422	0.08661
1192	C13H26O2	n-BUTYL NONANOATE	0.00202	-1.0410E-05	1.0444E-07	503	1000	-----	0.02321	0.09605
1193	C13H26O2	METHYL DODECANOATE	-0.00914	5.1369E-05	1.3643E-08	540	1000	-----	0.02258	0.05587
1194	C13H28	n-TRIDECANE	-0.00784	2.7116E-05	7.0226E-08	509	1000	-----	0.02416	0.08950
1195	C13H28O	1-TRIDECANOL	-0.00098	5.3568E-06	8.6625E-08	547	1000	-----	0.02787	0.09100
1196	C13H28S	BUTYL-NONYL-SULFIDE	-0.00601	3.5360E-05	1.9721E-08	275	1000	0.00629	0.00521	0.04907
1197	C13H28S	DECYL-PROPYL-SULFIDE	-0.00601	3.5360E-05	1.9721E-08	275	1000	0.00629	0.00521	0.04907
1198	C13H28S	DODECYL-METHYL-SULFIDE	-0.00611	3.6228E-05	1.8501E-08	275	1000	0.00634	0.00525	0.04862
1199	C13H28S	ETHYL-UNDECYL-SULFIDE	-0.00612	3.6083E-05	1.8751E-08	275	1000	0.00630	0.00522	0.04871
1200	C13H28S	1-TRIDECANETHIOL	-0.00619	3.6651E-05	1.8291E-08	275	1000	0.00636	0.00527	0.04875
1201	C14H8O2	ANTHRAQUINONE	-0.01018	5.3038E-05	1.2081E-08	653	1000	-----	0.02961	0.05494
1202	C14H10	ANTHRACENE	-0.00508	3.4417E-05	2.3646E-08	489	615	-----	0.01740	0.02503
1203	C14H10	DIPHENYLACETYLENE	-0.00913	4.8508E-05	1.4880E-08	573	993	-----	0.02355	0.05371
1204	C14H10	PHENANTHRENE	-0.00923	4.7388E-05	1.3940E-08	613	1000	-----	0.02506	0.05210
1205	C14H12	cis-STILBENE	-0.00981	5.1510E-05	1.9775E-08	535	1000	-----	0.02341	0.06148
1206	C14H12	trans-STILBENE	-0.00980	5.1139E-05	1.6241E-08	580	1000	-----	0.02532	0.05758
1207	C14H12O2	BENZYL BENZOATE	-0.00798	4.2392E-05	1.1341E-08	597	1000	-----	0.02137	0.04575
1208	C14H14	1,1-DIPHENYLETHANE	-0.02834	9.7961E-05	-7.5061E-09	546	1000	-----	0.02291	0.06211
1209	C14H14	1,2-DIPHENYLETHANE	-0.01012	5.2760E-05	1.9062E-08	554	1000	-----	0.02496	0.06170
1210	C14H14O	DIBENZYL ETHER	-0.00858	4.5472E-05	1.5047E-08	561	991	-----	0.02167	0.05126
1211	C14H16	1-n-BUTYLNAPHTHALENE	-0.01718	7.2875E-05	9.1628E-09	563	1000	-----	0.02675	0.06486
1212	C14H16	2-BUTYLNAPHTHALENE	-0.00730	4.0286E-05	1.4581E-08	275	1000	0.00601	0.00488	0.04757
1213	C14H22	n-OCTYLBENZENE	-0.01078	5.6575E-05	2.1438E-08	538	1000	-----	0.02586	0.06723
1214	C14H22	1,2,3,4-TETRAETHYLBENZENE	-0.00757	4.4608E-05	1.4899E-08	275	1000	0.00705	0.00582	0.05194
1215	C14H22	1,2,3,5-TETRAETHYLBENZENE	-0.00751	4.3935E-05	1.5383E-08	275	1000	0.00696	0.00574	0.05181
1216	C14H22	1,2,4,5-TETRAETHYLBENZENE	-0.00717	4.2765E-05	1.6160E-08	275	1000	0.00702	0.00581	0.05176
1217	C14H22O	p-tert-OCTYLPHENOL	-0.01018	5.8348E-05	1.2288E-08	564	1000	-----	0.02664	0.06046
1218	C14H28	1-TETRADECENE	-0.02610	9.7799E-05	-2.0865E-09	524	1000	-----	0.02457	0.06961
1219	C14H28	1-CYCLOPENTYLNONANE	-0.00825	4.4090E-05	1.9755E-08	275	1000	0.00665	0.00537	0.05560
1220	C14H28	1-CYCLOHEXYLOCTANE	-0.00890	4.6963E-05	1.9449E-08	275	1000	0.00683	0.00549	0.05751
1221	C14H28O2	n-TETRADECANOIC ACID	-0.00936	5.0289E-05	1.2546E-08	599	1000	-----	0.02526	0.05348
1222	C14H30	n-TETRADECANE	-0.00180	1.0242E-05	7.7727E-08	527	1000	-----	0.02518	0.08617
1223	C14H30O	1-TETRADECANOL	-0.00003	2.4670E-06	7.6134E-08	560	1000	-----	0.02523	0.07857
1224	C14H30S	BUTYL-DECYL-SULFIDE	-0.00581	3.4197E-05	1.9378E-08	275	1000	0.00611	0.00506	0.04777
1225	C14H30S	DODECYL-ETHYL-SULFIDE	-0.00591	3.4860E-05	1.8494E-08	275	1000	0.00613	0.00508	0.04744
1226	C14H30S	HEPTYL-SULFIDE	-0.00581	3.4197E-05	1.9378E-08	275	1000	0.00611	0.00506	0.04777
1227	C14H30S	METHYL-TRIDECYL-SULFIDE	-0.00590	3.4994E-05	1.8263E-08	275	1000	0.00616	0.00510	0.04736
1228	C14H30S	PROPYL-UNDECYL-SULFIDE	-0.00581	3.4197E-05	1.9378E-08	275	1000	0.00611	0.00506	0.04777
1229	C14H30S	1-TETRADECANETHIOL	-0.00600	3.5483E-05	1.8026E-08	275	1000	0.00618	0.00512	0.04751
1230	C14H30S2	HEPTYL-DISULFIDE	-0.00549	3.3137E-05	1.5340E-08	275	1000	0.00575	0.00478	0.04299
1231	C14H31N	TETRADECYLAMINE	-0.01117	6.0842E-05	1.7026E-08	564	994	-----	0.02856	0.06613
1232	C15H10N2O2	DIPHENYLMETHANE-4,4'-DIISOCYANATE	-0.00684	3.8456E-05	7.1593E-09	609	1000	-----	0.01923	0.03878
1233	C15H16O	p-CUMYLPHENOL	-0.01833	7.2980E-05	-3.4325E-09	608	1000	-----	0.02477	0.05122
1234	C15H16O2	BISPHENOL A	-0.00886	5.0729E-05	7.4757E-09	634	1000	-----	0.02631	0.04934
1235	C15H18	1-PENTYLNAPHTHALENE	-0.00713	3.9249E-05	1.4497E-08	275	1000	0.00586	0.00476	0.04662
1236	C15H18	2-PENTYLNAPHTHALENE	-0.00693	3.8427E-05	1.4899E-08	275	1000	0.00585	0.00476	0.04640
1237	C15H24	n-NONYLBENZENE	-0.01084	5.6173E-05	2.0794E-08	555	1000	-----	0.02674	0.06613
1238	C15H24O	2,6-DI-tert-BUTYL-p-CRESOL	-0.01633	7.1176E-05	-1.4771E-09	538	1000	-----	0.02154	0.05337
1239	C15H24O	NONYLPHENOL	-0.00932	5.1812E-05	1.1798E-08	581	1000	-----	0.02477	0.05429
1240	C15H28	1-PENTADECYNE	-0.00707	4.1684E-05	1.6268E-08	275	1000	0.00680	0.00562	0.05088
1241	C15H30	1-PENTADECENE	-0.02698	1.0116E-04	-7.5469E-09	542	1000	-----	0.02563	0.06663
1242	C15H30	1-CYCLOPENTYLDECANE	-0.00787	4.2243E-05	1.9406E-08	275	1000	0.00645	0.00521	0.05378
1243	C15H30	1-CYCLOHEXYLNONANE	-0.00843	4.4698E-05	1.9246E-08	275	1000	0.00661	0.00532	0.05551
1244	C15H30O2	PENTADECANOIC ACID	-0.00927	4.9669E-05	1.1808E-08	612	1000	-----	0.02555	0.05221
1245	C15H32	n-PENTADECANE	-0.00723	2.3158E-05	6.7125E-08	544	1000	-----	0.02523	0.08305
1246	C15H32O	1-PENTADECANOL	-0.00673	3.8972E-05	1.6494E-08	275	1000	0.00636	0.00523	0.04874
1247	C15H32S	BUTYL-UNDECYL-SULFIDE	-0.00563	3.3158E-05	1.9067E-08	275	1000	0.00595	0.00493	0.04660
1248	C15H32S	DODECYL-PROPYL-SULFIDE	-0.00563	3.3158E-05	1.9067E-08	275	1000	0.00595	0.00493	0.04660

Table 23-1 THERMAL CONDUCTIVITY OF GAS - ORGANIC COMPOUNDS (continued)

| NO | FORMULA | NAME | $k_{gas} = A + BT + CT^2$ | | | | | $(k_{gas}$ - W/(m K), T - K) | | |
			A	B	C	TMIN	TMAX	k_{gas} @25 C	k_{gas} @TMIN	k_{gas} @TMAX
1249	C15H32S	ETHYL-TRIDECYL-SULFIDE	-0.00573	3.3765E-05	1.8258E-08	275	1000	0.00596	0.00494	0.04629
1250	C15H32S	METHYL-TETRADECYL-SULFIDE	-0.00572	3.3886E-05	1.8048E-08	275	1000	0.00599	0.00496	0.04621
1251	C15H32S	1-PENTADECANETHIOL	-0.00583	3.3429E-05	1.7791E-08	275	1000	0.00602	0.00498	0.04639
1252	C16H10	FLUORANTHENE	-0.00899	4.6986E-05	1.0387E-08	656	1000	-----	0.02630	0.04838
1253	C16H10	PYRENE	-0.00860	4.3319E-05	1.1141E-08	668	998	-----	0.02531	0.04573
1254	C16H12	1-PHENYLNAPHTHALENE	-0.00955	4.9187E-05	1.4702E-08	607	1000	-----	0.02572	0.05434
1255	C16H20	1-n-HEXYLNAPHTHALENE	-0.01676	7.0285E-05	8.6552E-09	595	1000	-----	0.02812	0.06218
1256	C16H22O4	DIBUTYL PHTHALATE	-0.00911	4.8916E-05	1.1470E-08	613	1000	-----	0.02519	0.05128
1257	C16H26	n-DECYLBENZENE	-0.01081	5.5409E-05	2.0073E-08	571	1000	-----	0.02737	0.06467
1258	C16H26	PENTAETHYLBENZENE	-0.00727	4.2920E-05	1.3998E-08	275	1000	0.00677	0.00559	0.04965
1259	C16H30	1-HEXADECYNE	-0.00682	4.0159E-05	1.6205E-08	275	1000	0.00659	0.00545	0.04954
1260	C16H32	n-DECYLCYCLOHEXANE	-0.01164	5.8255E-05	2.3755E-08	571	1000	-----	0.02937	0.07037
1261	C16H32	1-CYCLOPENTYLUNDECANE	-0.00751	4.0505E-05	1.9183E-08	275	1000	0.00627	0.00508	0.05218
1262	C16H32	1-HEXADECENE	-0.01088	5.5158E-05	2.2391E-08	558	1000	-----	0.02687	0.06667
1263	C16H32O2	n-HEXADECANOIC ACID	-0.00934	4.9840E-05	1.1461E-08	624	1000	-----	0.02622	0.05196
1264	C16H34	n-HEXADECANE	-0.00671	2.0080E-05	6.7235E-08	560	1000	-----	0.02562	0.08061
1265	C16H34O	DI-n-OCTYL ETHER	-0.00945	5.1927E-05	1.3988E-08	560	1000	-----	0.02402	0.05647
1266	C16H34O	1-HEXADECANOL	0.00001	2.1149E-06	8.0125E-08	585	1000	-----	0.02867	0.08225
1267	C16H34S	BUTYL-DODECYL-SULFIDE	-0.00547	3.2226E-05	1.8790E-08	275	1000	0.00581	0.00481	0.04555
1268	C16H34S	ETHYL-TETRADECYL-SULFIDE	-0.00587	3.4634E-05	1.9238E-08	275	1000	0.00617	0.00511	0.04800
1269	C16H34S	METHYL-PENTADECYL-SULFIDE	-0.00555	3.2895E-05	1.7851E-08	275	1000	0.00584	0.00485	0.04520
1270	C16H34S	OCTYL-SULFIDE	-0.00547	3.2226E-05	1.8790E-08	275	1000	0.00581	0.00481	0.04555
1271	C16H34S	PROPYL-TRIDECYL-SULFIDE	-0.00547	3.2226E-05	1.8790E-08	275	1000	0.00581	0.00481	0.04555
1272	C16H34S	1-HEXADECANETHIOL	-0.00567	3.3497E-05	1.7577E-08	275	1000	0.00588	0.00487	0.04540
1273	C16H34S2	OCTYL-DISULFIDE	-0.00524	3.1577E-05	1.5227E-08	275	1000	0.00553	0.00460	0.04156
1274	C17H28	n-UNDECYLBENZENE	-0.01083	5.5002E-05	1.9427E-08	586	1000	-----	0.02807	0.06360
1275	C17H32	1-HEPTADECYNE	-0.00659	3.8793E-05	1.6142E-08	275	1000	0.00641	0.00530	0.04835
1276	C17H34	1-CYCLOPENTYLDODECANE	-0.00721	3.9056E-05	1.8880E-08	275	1000	0.00611	0.00496	0.05073
1277	C17H34	1-CYCLOHEXYLUNDECANE	-0.00768	4.1133E-05	1.8689E-08	275	1000	0.00625	0.00504	0.05214
1278	C17H34	1-HEPTADECENE	-0.01081	5.6622E-05	1.8218E-08	573	1000	-----	0.02762	0.06403
1279	C17H36	n-HEPTADECANE	0.00124	-6.3091E-06	8.1047E-08	575	1000	-----	0.02441	0.07598
1280	C17H36O	1-HEPTADECANOL	-0.00006	2.1078E-06	7.7468E-08	597	1000	-----	0.02881	0.07952
1281	C17H36S	BUTYL-TRIDECYL-SULFIDE	-0.00533	3.1406E-05	1.8549E-08	275	1000	0.00568	0.00471	0.04463
1282	C17H36S	ETHYL-PENTADECYL-SULFIDE	-0.00541	3.1927E-05	1.7856E-08	275	1000	0.00570	0.00472	0.04437
1283	C17H36S	HEXADECYL-METHYL-SULFIDE	-0.00540	3.2031E-05	1.7679E-08	275	1000	0.00572	0.00475	0.04431
1284	C17H36S	PROPYL-TETRADECYL-SULFIDE	-0.00533	3.1406E-05	1.8549E-08	275	1000	0.00568	0.00471	0.04463
1285	C17H36S	1-HEPTADECANETHIOL	-0.00552	3.2605E-05	1.7415E-08	275	1000	0.00575	0.00476	0.04450
1286	C18H12	CHRYSENE	-0.00870	4.3518E-05	9.6092E-09	714	994	-----	0.02727	0.04405
1287	C18H14	m-TERPHENYL	-0.00741	4.0043E-05	8.0074E-09	638	1000	-----	0.02140	0.04064
1288	C18H14	o-TERPHENYL	-0.00715	3.9304E-05	8.4426E-09	605	1000	-----	0.01972	0.04060
1289	C18H14	p-TERPHENYL	-0.00748	4.0276E-05	7.8174E-09	649	1000	-----	0.02195	0.04061
1290	C18H15P	TRIPHENYLPHOSPHINE	-0.00570	4.3077E-05	2.4635E-08	650	1000	-----	. 0.03271	0.06201
1291	C18H15O4P	TRIPHENYL PHOSPHATE	-----	--------	--------	----	----	-----	-----	-----
1292	C18H16N2	N,N'-DIPHENYL-p-PHENYLENEDIAMINE	-0.00882	4.7619E-05	7.5718E-09	688	1000	-----	0.02753	0.04637
1293	C18H22	2,3-DIMETHYL-2,3-DIPHENYLBUTANE	-0.01060	5.4719E-05	1.7566E-08	589	1000	-----	0.02772	0.06169
1294	C18H22O2	DICUMYL PEROXIDE	-----	--------	--------	----	----	-----	-----	-----
1295	C18H30	n-DODECYLBENZENE	-0.01081	5.2783E-05	2.1190E-08	600	1000	-----	0.02849	0.06316
1296	C18H30	HEXAETHYLBENZENE	-0.00694	4.1560E-05	1.3234E-08	275	1000	0.00663	0.00549	0.04785
1297	C18H32O2	LINOLEIC ACID	-0.00938	5.1481E-05	9.9426E-09	628	1000	-----	0.02687	0.05204
1298	C18H34	1-OCTADECYNE	-0.00638	3.7513E-05	1.6147E-08	275	1000	0.00624	0.00516	0.04728
1299	C18H34O2	OLEIC ACID	-0.00917	4.8698E-05	1.1022E-08	633	1000	-----	0.02607	0.05055
1300	C18H34O4	DIBUTYL SEBACATE	-0.00826	4.5607E-05	8.7811E-09	622	1000	-----	0.02350	0.04613
1301	C18H34O4	DIHEXYL ADIPATE	-0.00727	3.9979E-05	7.9202E-09	621	1000	-----	0.02061	0.04063
1302	C18H36	1-CYCOPENTYLTRIDECANE	-0.00694	3.7752E-05	1.8603E-08	275	1000	0.00597	0.00485	0.04942
1303	C18H36	1-CYCLOHEXYLDODECANE	-0.00735	3.9570E-05	1.8513E-08	275	1000	0.00609	0.00493	0.05073
1304	C18H36	1-OCTADECENE	-0.01079	5.3470E-05	2.1104E-08	588	1000	-----	0.02795	0.06378
1305	C18H36O2	STEARIC ACID	-0.00916	4.9075E-05	9.8658E-09	648	1000	-----	0.02678	0.04978
1306	C18H38	n-OCTADECANE	-0.00172	6.6775E-07	7.2881E-08	590	1000	-----	0.02404	0.07183
1307	C18H38O	DINONYL ETHER	-0.00933	5.0232E-05	1.2875E-08	591	1000	-----	0.02485	0.05378
1308	C18H38O	1-OCTADECANOL	-0.01063	5.5997E-05	1.4819E-08	608	1000	-----	0.02889	0.06019
1309	C18H38S	BUTYL-TETRADECYL-SULFIDE	-0.00519	3.0664E-05	1.8372E-08	275	1000	0.00559	0.00463	0.04385
1310	C18H38S	ETHYL-HEXADECYL-SULFIDE	-0.00527	3.1147E-05	1.7726E-08	275	1000	0.00559	0.00464	0.04360
1311	C18H38S	HEPTADECYL-METHYL-SULFIDE	-0.00526	3.1244E-05	1.7560E-08	275	1000	0.00562	0.00466	0.04354
1312	C18H38S	NONYL-SULFIDE	-0.00519	3.0664E-05	1.8372E-08	275	1000	0.00559	0.00463	0.04385
1313	C18H38S	PENTADECYL-PROPYL-SULFIDE	-0.00519	3.0664E-05	1.8372E-08	275	1000	0.00559	0.00463	0.04385
1314	C18H38S	1-OCTADECANETHIOL	-0.00539	3.1877E-05	1.7271E-08	275	1000	0.00565	0.00468	0.04376
1315	C18H38S2	NONYL-DISULFIDE	-0.00505	3.0347E-05	1.5188E-08	275	1000	0.00535	0.00444	0.04049
1316	C19H26	1-n-NONYLNAPHTHALENE	-0.00998	4.9182E-05	1.5845E-08	639	1000	-----	0.02792	0.05505
1317	C19H32	n-TRIDECYLBENZENE	-0.01882	7.4211E-05	5.9657E-09	614	1000	-----	0.02899	0.06136
1318	C19H36	1-NONADECYNE	-0.00619	3.6402E-05	1.6103E-08	275	1000	0.00609	0.00504	0.04632
1319	C19H36O2	METHYL OLEATE	-0.00860	4.6444E-05	1.0352E-08	617	1000	-----	0.02400	0.04820
1320	C19H38	1-CYCLOPENTYLTETRADECANE	-0.00658	3.5903E-05	1.7923E-08	275	1000	0.00572	0.00465	0.04725
1321	C19H38	1-CYCLOHEXYLTRIDECANE	-0.00709	3.8280E-05	1.8262E-08	275	1000	0.00595	0.00482	0.04945
1322	C19H38	1-NONADECENE	-0.01069	5.3678E-05	1.8616E-08	602	1000	-----	0.02837	0.06160
1323	C19H38O2	NONADECANOIC ACID	-0.00909	4.8001E-05	9.8948E-09	659	1000	-----	0.02684	0.04881
1324	C19H40	n-NONADECANE	0.00153	-7.5609E-06	7.4184E-08	603	1000	-----	0.02394	0.06815
1325	C19H40O	1-NONADECANOL	-0.00610	3.5474E-05	1.7023E-08	275	1000	0.00599	0.00494	0.04640
1326	C19H40S	BUTYL-PENTADECYL-SULFIDE	-0.00508	3.0032E-05	1.8232E-08	275	1000	0.00549	0.00456	0.04318

524

Table 23-1 THERMAL CONDUCTIVITY OF GAS - ORGANIC COMPOUNDS (continued)

NO	FORMULA	NAME	$k_{gas} = A + B T + C T^2$ A	B	C	TMIN	TMAX	k_{gas} @25 C	k_{gas} @TMIN	k_{gas} @TMAX
								(k_{gas} - W/(m K), T - K)		
1327	C19H40S	ETHYL-HEPTADECYL-SULFIDE	-0.00515	3.0482E-05	1.7631E-08	275	1000	0.00551	0.00457	0.04296
1328	C19H40S	HEXADECYL-PROPYL-SULFIDE	-0.00508	3.0032E-05	1.8232E-08	275	1000	0.00549	0.00456	0.04318
1329	C19H40S	METHYL-OCTADECYL-SULFIDE	-0.00515	3.0573E-05	1.7476E-08	275	1000	0.00552	0.00458	0.04290
1330	C19H40S	1-NONADECANETHIOL	-0.00527	3.1189E-05	1.7192E-08	275	1000	0.00556	0.00461	0.04311
1331	C20H16	TRIPHENYLETHYLENE	-0.00930	4.6613E-05	1.2116E-08	669	1000	-----	0.02731	0.04943
1332	C20H28	1-n-DECYLNAPHTHALENE	-0.00994	4.8645E-05	1.5353E-08	652	1000	-----	0.02830	0.05406
1333	C20H30O2	ABIETIC ACID	-0.00981	5.1249E-05	1.1691E-08	650	1000	-----	0.02844	0.05313
1334	C20H31N	DEHYDROABIETYLAMINE	-0.00001	8.9713E-07	-6.1569E-11	660	1000	-----	0.00056	0.00083
1335	C20H34	1-PHENYLTETRADECANE	-0.00573	3.2099E-05	1.4631E-08	275	1000	0.00514	0.00420	0.04100
1336	C20H38	1-EICOSYNE	-0.00603	3.5443E-05	1.6063E-08	275	1000	0.00597	0.00493	0.04548
1337	C20H40	1-CYCLOPENTYLPENTADECANE	-0.00634	3.4690E-05	1.7525E-08	275	1000	0.00556	0.00453	0.04588
1338	C20H40	1-CYCLOHEXYLTETRADECANE	-0.00686	3.7189E-05	1.8046E-08	275	1000	0.00583	0.00473	0.04838
1339	C20H40	1-EICOSENE	-0.01067	5.1723E-05	2.0085E-08	616	1000	-----	0.02881	0.06114
1340	C20H42	n-EICOSANE	0.00154	-7.5268E-06	7.0837E-08	617	1000	-----	0.02386	0.06485
1341	C20H42O	1-EICOSANOL	0.00006	1.4370E-06	7.0530E-08	629	1000	-----	0.02887	0.07203
1342	C20H42S	BUTYL-HEXADECYL-SULFIDE	-0.00499	2.9501E-05	1.8139E-08	275	1000	0.00542	0.00449	0.04265
1343	C20H42S	DECYL-SULFIDE	-0.00499	2.9501E-05	1.8139E-08	275	1000	0.00542	0.00449	0.04265
1344	C20H42S	ETHYL-OCTADECYL-SULFIDE	-0.00506	2.9935E-05	1.7562E-08	275	1000	0.00543	0.00450	0.04244
1345	C20H42S	HEPTADECYL-PROPYL-SULFIDE	-0.00499	2.9501E-05	1.8139E-08	275	1000	0.00542	0.00449	0.04265
1346	C20H42S	METHYL-NONADECYL-SULFIDE	-0.00505	3.0020E-05	1.7416E-08	275	1000	0.00545	0.00452	0.04239
1347	C20H42S	1-EICOSANETHIOL	-0.00518	3.0674E-05	1.7113E-08	275	1000	0.00549	0.00455	0.04261
1348	C20H42S2	DECYL-DISULFIDE	-0.00490	2.9459E-05	1.5285E-08	275	1000	0.00524	0.00436	0.03984
1349	C21H21O4P	TRI-o-CRESYL PHOSPHATE	-----	--------	--------	----	----	-----	-----	-----
1350	C21H36	1-PHENYLPENTADECANE	-0.00560	3.1494E-05	1.4713E-08	275	1000	0.00510	0.00417	0.04061
1351	C21H42	1-CYCLOPENTYLHEXADECANE	-0.00631	3.4659E-05	1.8072E-08	275	1000	0.00563	0.00459	0.04642
1352	C21H42	1-CYCLOHEXYLPENTADECANE	-0.00663	3.6076E-05	1.7962E-08	275	1000	0.00572	0.00465	0.04741
1353	C22H38	1-PHENYLHEXADECANE	-0.00539	3.0355E-05	1.4433E-08	275	1000	0.00494	0.00405	0.03940
1354	C22H44	1-CYCLOHEXYLHEXADECANE	-0.00643	3.5135E-05	1.7893E-08	275	1000	0.00564	0.00459	0.04660
1355	C22H44O2	n-BUTYL STEARATE	-0.00719	4.0258E-05	7.1090E-09	623	1000	-----	0.02065	0.04018
1356	C24H38O4	DIISOOCTYL PHTHALATE	-0.00777	4.0239E-05	7.8945E-09	694	1000	-----	0.02396	0.04036
1357	C24H38O4	DIOCTYL PHTHALATE	-0.00575	6.5471E-05	-7.3091E-09	657	1000	-----	0.03411	0.05241
1358	C24H42O	DINONYLPHENOL	-0.01218	5.3914E-05	3.2714E-09	722	1000	-----	0.02845	0.04501
1359	C26H20	TETRAPHENYLETHYLENE	-0.00895	4.3161E-05	9.4152E-09	760	1000	-----	0.02929	0.04363
1360	C28H46O4	DIISODECYL PHTHALATE	-0.00746	3.8410E-05	6.8581E-09	723	1000	-----	0.02390	0.03781

k_{gas} - thermal conductivity of gas, W/(m K)

A, B, and C - regression coefficients for chemical compound

T - temperature, K

TMIN - minimum temperature, K

TMAX - maximum temperature, K

Table 23-2 THERMAL CONDUCTIVITY OF GAS - INORGANIC COMPOUNDS

NO	FORMULA	NAME	$k_{gas} = A + B\,T + C\,T^2$					$(k_{gas} - W/(m\,K),\ T - K)$		
			A	B	C	TMIN	TMAX	k_{gas} @25 C	k_{gas} @TMIN	k_{gas} @TMAX
1	Ag	SILVER	0.00213	5.2096E-06	2.1128E-10	2485	2885	-----	0.01638	0.01892
2	AgCl	SILVER CHLORIDE	-0.00149	4.3307E-06	-3.8455E-10	1837	2000	-----	0.00517	0.00563
3	AgI	SILVER IODIDE	-0.00050	2.7137E-06	-1.0906E-10	1779	2000	-----	0.00398	0.00449
4	Al	ALUMINUM	0.00571	1.3607E-05	6.6450E-10	2329	2729	-----	0.04101	0.04779
5	AlB3H12	ALUMINUM BOROHYDRIDE	--------	----------	----------	----	----	-----	-----	-----
6	AlBr3	ALUMINUN BROMIDE	-0.00198	2.1725E-05	-3.7297E-09	529	929	-----	0.00847	0.01498
7	AlCl3	ALUMINUM CHLORIDE	-0.00399	3.8899E-05	-9.3405E-09	453	853	-----	0.01171	0.02239
8	AlF3	ALUMINUM FLUORIDE	-0.00312	1.1025E-05	-7.0783E-10	1810	2210	-----	0.01452	0.01779
9	AlI3	ALUMINUM IODIDE	-0.00106	1.1045E-05	-7.3749E-10	659	1059	-----	0.00590	0.00981
10	Al2O3	ALUMINUM OXIDE	--------	----------	----------	----	----	-----	-----	-----
11	Al2S3O12	ALUMINUM SULFATE	--------	----------	----------	----	----	-----	-----	-----
12	Ar	ARGON	0.00548	4.3869E-05	-6.8141E-09	150	1500	0.01795	0.01191	0.05595
13	As	ARSENIC	-0.00055	2.3762E-05	3.7646E-10	885	1285	-----	0.02077	0.03061
14	AsBr3	ARSENIC TRIBROMIDE	-0.00176	2.0825E-05	-2.8637E-09	493	893	-----	0.00781	0.01455
15	AsCl3	ARSENIC TRICHLORIDE	-0.00228	2.9795E-05	-4.7013E-09	404	804	-----	0.00899	0.01864
16	AsF3	ARSENIC TRIFLUORIDE	-0.00599	6.0184E-05	-1.2305E-08	298	1000	0.01086	0.01085	0.04189
17	AsF5	ARSENIC PENTAFLUORIDE	-0.00459	6.7397E-05	-1.7395E-08	298	1000	0.01396	0.01395	0.04541
18	AsH3	ARSINE	-0.00434	5.9571E-05	-4.1392E-09	211	1000	0.01305	0.00805	0.05109
19	AsI3	ARSENIC TRIIODIDE	-0.00086	1.2585E-05	-9.1863E-10	676	1076	-----	0.00723	0.01162
20	As2O3	ARSENIC TRIOXIDE	--------	----------	----------	----	----	-----	-----	-----
21	At	ASTATINE	--------	----------	----------	----	----	-----	-----	-----
22	Au	GOLD	-0.00296	7.9743E-06	4.0978E-10	3120	3520	-----	0.02591	0.03019
23	B	BORON	0.01655	2.7140E-05	1.4093E-09	4133	4533	-----	0.15279	0.16853
24	BBr3	BORON TRIBROMIDE	-0.00228	2.5377E-05	-4.3096E-09	298	1000	0.00490	0.00490	0.01879
25	BCl3	BORON TRICHLORIDE	-0.00476	6.1881E-05	-2.5706E-08	298	632	0.01140	0.01140	0.02408
26	BF3	BORON TRIFLUORIDE	-0.00486	7.2166E-05	-7.5235E-10	261	700	0.01659	0.01392	0.04529
27	BH2CO	BORINE CARBONYL	-0.01854	1.0477E-04	2.7415E-08	298	1000	0.01513	0.01512	0.11365
28	BH3O3	BORIC ACID	--------	----------	----------	----	----	-----	-----	-----
29	B2D6	DEUTERODIBORANE	-0.01391	1.1534E-04	1.9026E-08	250	1000	0.02217	0.01611	0.12046
30	B2H5Br	DIBORANE HYDROBROMIDE	-0.00650	5.2310E-05	1.2892E-08	250	1000	0.01024	0.00738	0.05870
31	B2H6	DIBORANE	-0.01217	1.0306E-04	6.9756E-08	181	1000	0.02476	0.00877	0.16065
32	B3N3H6	BORINE TRIAMINE	-0.01734	9.3406E-05	3.7802E-09	298	1000	0.01085	0.01083	0.07985
33	B4H10	TETRABORANE	-0.03114	1.3806E-04	1.7477E-08	298	1000	0.01158	0.01155	0.12440
34	B5H9	PENTABORANE	-0.02474	1.1641E-04	2.1444E-08	298	1000	0.01187	0.01185	0.11311
35	B5H11	TETRAHYDROPENTABORANE	-0.02415	1.1436E-04	1.8690E-08	298	1000	0.01161	0.01159	0.10890
36	B10H14	DECABORANE	--------	----------	----------	----	----	-----	-----	-----
37	Ba	BARIUM	0.00025	2.8816E-06	4.6727E-10	1907	2307	-----	0.00744	0.00938
38	Be	BERYLLIUM	0.00804	3.9301E-06	1.4051E-09	2744	3144	-----	0.02940	0.03429
39	BeB2H8	BERYLLIUM BOROHYDRIDE	--------	----------	----------	----	----	-----	-----	-----
40	BeBr2	BERYLLIUM BROMIDE	-0.00147	1.3648E-05	-3.7029E-10	747	1147	-----	0.00852	0.01370
41	BeCl2	BERYLLIUM CHLORIDE	-0.00246	1.9594E-05	-3.4768E-10	760	1160	-----	0.01223	0.01980
42	BeF2	BERYLLIUM FLUORIDE	--------	----------	----------	----	----	-----	-----	-----
43	BeI2	BERYLLIUM IODIDE	-0.00111	1.0896E-05	-3.4475E-10	760	1160	-----	0.00697	0.01107
44	Bi	BISMUTH	0.00086	4.0084E-06	2.1558E-10	1698	2098	-----	0.00829	0.01022
45	BiBr3	BISMUTH TRIBROMIDE	--------	----------	----------	----	----	-----	-----	-----
46	BiCl3	BISMUTH TRICHLORIDE	-0.00109	1.5093E-05	-9.7617E-10	714	1000	-----	0.00919	0.01303
47	BrF5	BROMINE PENTAFLUORIDE	-0.00792	7.3671E-05	-1.7395E-08	298	1000	0.01250	0.01249	0.04836
48	Br2	BROMINE	0.00514	-1.7001E-05	5.2592E-08	275	500	0.00475	0.00444	0.00979
49	C	CARBON	-0.01914	4.9735E-05	4.5834E-10	4203	4603	-----	0.19799	0.21950
50	CCl2O	PHOSGENE	-0.00440	3.8240E-05	1.9254E-08	250	800	0.00871	0.00636	0.03851
51	CF2O	CARBONYL FLUORIDE	-0.00556	6.4485E-05	-1.0611E-08	189	1000	0.01272	0.00625	0.04831
52	CH4N2O	UREA	-0.01444	7.2675E-05	1.2011E-08	465	865	-----	0.02195	0.05741
53	CH4N2S	THIOUREA	-0.00348	3.1111E-05	1.5603E-08	536	1000	-----	0.01768	0.04323
54	CNBr	CYANOGEN BROMIDE	-0.00249	3.2496E-05	-3.1172E-09	298	1000	0.00692	0.00692	0.02689
55	CNCl	CYANOGEN CHLORIDE	-0.00238	3.3381E-05	-3.7228E-09	286	1000	0.00724	0.00686	0.02728
56	CNF	CYANOGEN FLUORIDE	-0.00711	8.6799E-05	-1.3904E-08	298	1000	0.01753	0.01752	0.06579
57	CO	CARBON MONOXIDE	0.00150	8.2713E-05	-1.9171E-08	70	1250	0.02446	0.00720	0.07494
58	COS	CARBONYL SULFIDE	-0.00450	6.3507E-05	-5.5987E-09	223	1000	0.01394	0.00938	0.05341
59	COSe	CARBON OXYSELENIDE	-0.00550	5.7272E-05	-8.6637E-09	250	1000	0.01081	0.00828	0.04311
60	CO2	CARBON DIOXIDE	-0.01183	1.0174E-04	-2.2242E-08	195	1500	0.01653	0.00716	0.09074
61	CS2	CARBON DISULFIDE	-0.00239	3.2557E-05	2.2176E-10	273	999	0.00734	0.00651	0.03036
62	CSeS	CARBON SELENOSULFIDE	-0.00342	3.4679E-05	-4.3416E-09	250	1000	0.00653	0.00498	0.02692
63	C2N2	CYANOGEN	-0.00732	8.4273E-05	-1.0877E-08	298	1000	0.01684	0.01683	0.06608
64	C3S2	CARBON SUBSULFIDE	--------	----------	----------	----	----	-----	-----	-----
65	Ca	CALCIUM	0.00479	8.0623E-07	1.6046E-09	1762	2162	-----	0.01119	0.01403
66	CaF2	CALCIUM FLUORIDE	-0.00079	8.7281E-06	6.2392E-13	2807	3207	-----	0.02371	0.02721
67	CbF5	COLUMBIUM FLUORIDE	--------	----------	----------	----	----	-----	-----	-----
68	Cd	CADMIUM	0.00046	1.4346E-05	6.2121E-10	1043	1443	-----	0.01610	0.02245
69	CdCl2	CADMIUM CHLORIDE	--------	----------	----------	----	----	-----	-----	-----
70	CdF2	CADMIUM FLUORIDE	--------	----------	----------	----	----	-----	-----	-----
71	CdI2	CADMIUM IODIDE	--------	----------	----------	----	----	-----	-----	-----
72	CdO	CADMIUM OXIDE	--------	----------	----------	----	----	-----	-----	-----
73	ClF	CHLORINE MONOFLUORIDE	-0.00352	7.4127E-05	-1.9053E-08	298	1000	0.01689	0.01688	0.05155
74	ClFO3	PERCHLORYL FLUORIDE	-0.00740	7.1874E-05	-4.8101E-09	125	1000	0.01360	0.00151	0.05966
75	ClF3	CHLORINE TRIFLUORIDE	-0.00825	7.9705E-05	-1.8660E-08	298	1000	0.01386	0.01385	0.05280
76	ClF5	CHLORINE PENTAFLUORIDE	-0.00893	7.9234E-05	-2.0167E-08	298	1000	0.01290	0.01289	0.05014
77	ClHO3S	CHLOROSULFONIC ACID	-0.00815	5.7719E-05	-3.3985E-09	427	827	-----	0.01588	0.03726
78	ClHO4	PERCHLORIC ACID	-0.00855	6.6086E-05	-3.4013E-09	298	1000	0.01085	0.01084	0.05413

526

Table 23-2 THERMAL CONDUCTIVITY OF GAS - INORGANIC COMPOUNDS (continued)

			$k_{gas} = A + B\,T + C\,T^2$					$(k_{gas}$ - W/(m K), T - K)$		
NO	FORMULA	NAME	A	B	C	TMIN	TMAX	k_{gas} @25 C	k_{gas} @TMIN	k_{gas} @TMAX
79	ClO2	CHLORINE DIOXIDE	-0.00667	7.0078E-05	-9.8436E-09	250	1000	0.01335	0.01023	0.05356
80	Cl2	CHLORINE	-0.00194	3.8300E-05	-6.3523E-09	200	1000	0.00891	0.00547	0.03001
81	Cl2O	CHLORINE MONOXIDE	-0.00494	5.7501E-05	-1.3425E-08	298	1000	0.01101	0.01100	0.03914
82	Cl2O7	CHLORINE HEPTOXIDE	-0.00289	4.7938E-05	-6.5547E-09	250	1000	0.01082	0.00868	0.03849
83	Co	COBALT	0.00981	9.3221E-06	8.5338E-10	2528	2928	-----	0.03883	0.04442
84	CoCl2	COBALT CHLORIDE	-0.00081	1.1854E-05	-1.5893E-10	1323	1723	-----	0.01459	0.01914
85	CoNC3O4	COBALT NITROSYL TRICARBONYL	-0.00346	4.0031E-05	-3.1017E-09	250	1000	0.00820	0.00635	0.03347
86	Cr	CHROMIUM	0.00367	8.1796E-06	2.1710E-09	2840	3240	-----	0.04441	0.05296
87	CrC6O6	CHROMIUM CARBONYL	--------	----------	----------	----	----	-----	-----	-----
88	CrO2Cl2	CHROMIUM OXYCHLORIDE	-0.00372	4.0256E-05	-1.2360E-09	250	1000	0.00817	0.00627	0.03530
89	Cs	CESIUM	0.00231	3.5257E-06	8.1273E-10	700	1500	-----	0.00518	0.00943
90	CsBr	CESIUM BROMIDE	--------	----------	----------	----	----	-----	-----	-----
91	CsCl	CESIUM CHLORIDE	-0.00057	5.6482E-06	2.1864E-12	1573	1973	-----	0.00832	0.01058
92	CsF	CESIUM FLUORIDE	-0.00083	6.3242E-06	-1.0181E-10	1524	1924	-----	0.00857	0.01096
93	CsI	CESIUM IODIDE	-0.00042	4.5295E-06	-6.9594E-11	1553	1953	-----	0.00645	0.00816
94	Cu	COPPER	0.00270	4.4160E-06	2.2054E-09	3150	3550	-----	0.03849	0.04617
95	CuBr	CUPROUS BROMIDE	-0.00147	6.9368E-06	-3.6469E-10	1628	2000	-----	0.00886	0.01094
96	CuCl	CUPROUS CHLORIDE	-0.00121	8.9161E-06	-2.6395E-10	1763	2163	-----	0.01369	0.01684
97	CuCl2	CUPRIC CHLORIDE	--------	----------	----------	----	----	-----	-----	-----
98	CuI	COPPER IODIDE	-0.00104	5.7983E-06	-2.4613E-10	1609	2000	-----	0.00765	0.00957
99	DCN	DEUTERIUM CYANIDE	-0.00692	8.3019E-05	-4.2818E-10	250	1000	0.01779	0.01381	0.07567
100	D2	DEUTERIUM	0.05930	2.6600E-04	-6.1759E-09	233	1500	0.13806	0.12094	0.44440
101	D2O	DEUTERIUM OXIDE	0.20802	-7.6855E-04	8.5518E-07	374	1000	-----	0.04020	0.29465
102	Eu	EUROPIUM	0.00198	7.3498E-06	4.0725E-10	1742	2142	-----	0.01602	0.01959
103	F2	FLUORINE	-0.00076	9.5961E-05	-2.1800E-08	70	700	0.02591	0.00585	0.05573
104	F2O	FLUORINE OXIDE	-0.00763	1.0940E-04	-3.1124E-08	298	1000	0.02222	0.02221	0.07065
105	Fe	IRON	0.02350	-4.3192E-06	4.1782E-09	3000	3400	-----	0.04815	0.05711
106	FeC5O5	IRON PENTACARBONYL	-0.00263	3.2159E-05	-1.9895E-09	250	1000	0.00678	0.00529	0.02754
107	FeCl2	FERROUS CHLORIDE	-0.00229	1.4039E-05	-8.1659E-10	1299	1699	-----	0.01457	0.01921
108	FeCl3	FERRIC CHLORIDE	-0.00160	2.0728E-05	-1.0126E-09	592	992	-----	0.01032	0.01797
109	Fr	FRANCIUM	-0.00067	1.9531E-05	-8.6129E-11	879	1279	-----	0.01643	0.02417
110	Ga	GALLIUM	0.00240	8.0397E-06	-1.2371E-09	2517	2917	-----	0.01480	0.01533
111	GaCl3	GALLIUM TRICHLORIDE	-0.00210	2.3204E-05	-1.7078E-09	474	1000	-----	0.00851	0.01940
112	Gd	GADOLINIUM	-0.00221	2.0309E-05	-3.4197E-10	1770	2170	-----	0.03267	0.04025
113	Ge	GERMANIUM	0.00339	8.2281E-06	2.6028E-10	3125	3525	-----	0.03164	0.03563
114	GeBr4	GERMANIUM BROMIDE	-0.00130	1.8376E-05	-2.6173E-09	462	862	-----	0.00663	0.01260
115	GeCl4	GERMANIUM CHLORIDE	-0.00234	3.1764E-05	-6.0290E-09	298	1000	0.00659	0.00659	0.02340
116	GeHCl3	TRICHLORO GERMANE	-0.00404	3.6875E-05	-5.6104E-09	250	1000	0.00646	0.00483	0.02722
117	GeH4	GERMANE	-0.01240	9.1289E-05	-1.5424E-08	250	1000	0.01345	0.00946	0.06347
118	Ge2H6	DIGERMANE	-0.00597	5.2916E-05	-6.5667E-09	250	1000	0.00922	0.00685	0.04038
119	Ge3H8	TRIGERMANE	-0.00399	4.3100E-05	-3.4375E-09	250	1000	0.00855	0.00657	0.03567
120	HBr	HYDROGEN BROMIDE	-0.00081	3.2275E-05	-1.2108E-09	273	473	0.00871	0.00791	0.01419
121	HCN	HYDROGEN CYANIDE	-0.01140	7.7207E-05	2.8371E-09	270	680	0.01187	0.00965	0.04241
122	HCl	HYDROGEN CHLORIDE	0.00119	4.4775E-05	2.0997E-10	159	1000	0.01456	0.00831	0.04617
123	HF	HYDROGEN FLUORIDE	0.02860	-8.3485E-05	2.1679E-07	275	1000	0.02298	0.02204	0.16191
124	HI	HYDROGEN IODIDE	-0.00026	2.1969E-05	-1.6856E-09	230	1000	0.00614	0.00470	0.02002
125	HNO3	NITRIC ACID	-0.01060	7.8123E-05	-5.0028E-10	250	1000	0.01265	0.00890	0.06702
126	H2	HYDROGEN	0.03951	4.5918E-04	-6.4933E-08	150	1500	0.17064	0.10693	0.58218
127	H2O	WATER	0.00053	4.7093E-05	4.9551E-08	275	1073	0.01898	0.01723	0.10811
128	H2O2	HYDROGEN PEROXIDE	-0.00858	8.6933E-05	-6.2970E-09	275	1200	0.01678	0.01485	0.08667
129	H2S	HYDROGEN SULFIDE	-0.00931	8.3043E-05	-1.9514E-08	180	600	0.01371	0.00501	0.03349
130	H2SO4	SULFURIC ACID	-0.01341	7.1828E-05	-4.4286E-09	610	1010	-----	0.02876	0.05462
131	H2S2	HYDROGEN DISULFIDE	-0.00194	5.1423E-05	1.6764E-09	250	1000	0.01354	0.01102	0.05116
132	H2Se	HYDROGEN SELENIDE	-0.00132	3.7431E-05	3.0942E-09	250	1000	0.01012	0.00823	0.03921
133	H2Te	HYDROGEN TELLURIDE	-0.00095	2.4944E-05	2.4644E-09	250	1000	0.00671	0.00544	0.02646
134	H3NO3S	SULFAMIC ACID	--------	----------	----------	----	----	-----	-----	-----
135	He	HELIUM-3	0.04848	4.6034E-04	-1.2371E-07	100	1000	0.17473	0.09328	0.38511
136	He	HELIUM-4	0.05516	3.2540E-04	-2.2723E-08	100	2000	0.15016	0.08747	0.61507
137	Hf	HAFNIUM	0.21341	-7.3828E-05	7.5266E-09	5960	6000	-----	0.04075	0.04140
138	Hg	MERCURY	0.00602	3.8279E-06	-1.2957E-09	473	880	-----	0.00754	0.00839
139	HgBr2	MERCURIC BROMIDE	-0.00069	1.0600E-05	-6.5878E-10	592	992	-----	0.00535	0.00918
140	HgCl2	MERCURIC CHLORIDE	-0.00092	1.2276E-05	-6.2036E-10	577	977	-----	0.00596	0.01048
141	HgI2	MERCURIC IODIDE	-0.00052	9.3934E-06	-4.0777E-10	627	1027	-----	0.00521	0.00870
142	IF7	IODINE HEPTAFLUORIDE	-0.00793	7.0635E-05	-1.7066E-08	298	1000	0.01161	0.01160	0.04564
143	I2	IODINE	-0.00001	1.0959E-05	3.2555E-11	298	819	0.00326	0.00326	0.00899
144	In	INDIUM	0.00157	7.1635E-06	-6.1699E-10	2323	2723	-----	0.01488	0.01650
145	Ir	IRIDIUM	0.01789	2.0141E-05	4.5342E-10	4450	4850	-----	0.11650	0.12624
146	K	POTASSIUM	0.00816	2.3387E-06	5.1685E-09	725	1500	-----	0.01257	0.02330
147	KBr	POTASSIUM BROMIDE	-0.00067	6.5359E-06	1.0407E-11	1656	2056	-----	0.01018	0.01281
148	KCl	POTASSIUM CHLORIDE	0.00017	3.8444E-06	1.0755E-10	1689	2089	-----	0.00697	0.00867
149	KF	POTASSIUM FLUORIDE	-0.00143	9.5818E-06	-1.4474E-10	1775	2175	-----	0.01512	0.01873
150	KI	POTASSIUM IODIDE	-0.00052	5.6095E-06	1.0388E-11	1597	1997	-----	0.00846	0.01072
151	KOH	POTASSIUM HYDROXIDE	-0.00100	1.3659E-05	-1.0540E-10	1600	2000	-----	0.02058	0.02590
152	Kr	KRYPTON	0.00168	2.7493E-05	-4.7254E-09	120	2000	0.00946	0.00491	0.03776
153	La	LANTHANUM	0.01410	2.5120E-06	1.1481E-09	3643	4043	-----	0.03849	0.04302
154	Li	LITHIUM	0.20076	-1.7178E-04	6.1898E-08	1150	2000	-----	0.08507	0.10479
155	LiBr	LITHIUM BROMIDE	-0.00153	8.6678E-06	-2.3008E-10	1583	1983	-----	0.01161	0.01475
156	LiCl	LITHIUM CHLORIDE	-0.00232	1.2269E-05	-3.6710E-10	1655	2055	-----	0.01698	0.02134

Table 23-2 THERMAL CONDUCTIVITY OF GAS - INORGANIC COMPOUNDS (continued)

| NO | FORMULA | NAME | $k_{gas} = A + BT + CT^2$ | | | | | | | $(k_{gas}$ - W/(m K), T - K$)$ | | |
			A	B	C	TMIN	TMAX	k_{gas} @25 C	k_{gas} @TMIN	k_{gas} @TMAX
157	LiF	LITHIUM FLUORIDE	-0.00302	1.4370E-05	-3.7856E-10	1954	2354	-----	0.02361	0.02871
158	LiI	LITHIUM IODIDE	-0.00115	7.2163E-06	-1.6895E-10	1444	1844	-----	0.00892	0.01158
159	Lu	LUTECIUM	-0.03994	4.6789E-05	-5.4978E-09	2535	2935	-----	0.04334	0.05003
160	Mg	MAGNESIUM	-0.00343	5.5750E-05	-9.8113E-10	1376	1776	-----	0.07142	0.09249
161	MgCl2	MAGNESIUM CHLORIDE	-0.00218	1.2820E-05	-4.7620E-10	1691	2091	-----	0.01814	0.02254
162	MgO	MAGNESIUM OXIDE	0.00429	5.9951E-06	5.1782E-10	3873	4273	-----	0.03528	0.03936
163	Mn	MANGANESE	0.00387	1.0700E-05	4.4526E-10	2392	2792	-----	0.03201	0.03722
164	MnCl2	MANGANESE CHLORIDE	-0.00083	6.8824E-06	-1.2806E-10	1463	1863	-----	0.00896	0.01155
165	Mo	MOLYBDENUM	-0.20267	7.6284E-05	-3.3409E-09	5081	5481	-----	0.09868	0.11508
166	MoF6	MOLYBDENUM FLUORIDE	-0.00716	6.6967E-05	-1.4512E-08	298	1000	0.01152	0.01151	0.04530
167	MoO3	MOLYBDENUM OXIDE	-0.00502	1.7598E-05	-1.1926E-09	1424	1824	-----	0.01762	0.02311
168	NCl3	NITROGEN TRICHLORIDE	-0.00519	4.8811E-05	-1.0609E-08	250	1000	0.00842	0.00635	0.03301
169	ND3	HEAVY AMMONIA	-0.00641	1.1345E-04	2.6124E-08	250	1000	0.02974	0.02359	0.13316
170	NF3	NITROGEN TRIFLUORIDE	-0.00889	1.1086E-04	-2.4596E-08	144	1000	0.02198	0.00656	0.07737
171	NH3	AMMONIA	0.00457	2.3239E-05	1.4810E-07	200	700	0.02466	0.01514	0.09341
172	NH3O	HYDROXYLAMINE	-0.01228	1.1043E-04	1.2568E-09	383	1000	-----	0.03020	0.09941
173	NH4Br	AMMONIUM BROMIDE	---------	----------	----------	----	----	-----	------	-----
174	NH4Cl	AMMONIUM CHLORIDE	---------	----------	----------	----	----	-----	------	-----
175	NH4I	AMMONIUM IODIDE	---------	----------	----------	----	----	-----	------	-----
176	NH5O	AMMONIUM HYDROXIDE	---------	----------	----------	----	----	-----	------	-----
177	NH5S	AMMONIUM HYDROGENSULFIDE	---------	----------	----------	----	----	-----	------	-----
178	NO	NITRIC OXIDE	0.00176	8.2369E-05	-1.2527E-08	110	1000	0.02520	0.01067	0.07160
179	NOCl	NITROSYL CHLORIDE	-0.00353	4.4714E-05	-4.4300E-09	268	1000	0.00941	0.00814	0.03675
180	NOF	NITROSYL FLUORIDE	-0.00736	1.0820E-04	-1.9433E-08	250	1000	0.02317	0.01848	0.08141
181	NO2	NITROGEN DIOXIDE	-0.01289	1.0390E-04	-2.1445E-08	298	1300	0.01618	0.01617	0.08594
182	N2	NITROGEN	0.00309	7.5930E-05	-1.1014E-08	78	1500	0.02475	0.00895	0.09220
183	N2F4	TETRAFLUOROHYDRAZINE	-0.01031	8.9801E-05	-1.8640E-08	199	1000	0.01481	0.00682	0.06085
184	N2H4	HYDRAZINE	-0.00926	6.3765E-05	1.4671E-08	387	1000	-----	0.01761	0.06918
185	N2H4C	AMMONIUM CYANIDE	-0.01575	1.4752E-04	8.8181E-09	250	1000	0.02902	0.02168	0.14059
186	N2H6CO2	AMMONIUM CARBAMATE	---------	----------	----------	----	----	-----	------	-----
187	N2O	NITROUS OXIDE	-0.00603	7.9837E-05	-2.2582E-09	182	1000	0.01757	0.00843	0.07155
188	N2O3	NITROGEN TRIOXIDE	-0.00749	7.0445E-05	-9.1558E-09	250	1000	0.01270	0.00955	0.05380
189	N2O4	NITROGEN TETRAOXIDE	-0.01736	1.4761E-04	-2.2481E-08	250	1000	0.02465	0.01814	0.10777
190	N2O5	NITROGEN PENTOXIDE	-0.00641	6.9189E-05	2.6245E-09	250	1000	0.01445	0.01105	0.06540
191	Na	SODIUM	0.07327	-8.2947E-05	4.0996E-08	825	1500	-----	0.03274	0.04109
192	NaBr	SODIUM BROMIDE	0.00041	3.7847E-06	1.9125E-10	1664	2064	-----	0.00724	0.00904
193	NaCN	SODIUM CYANIDE	-0.00096	1.3135E-05	-5.9560E-11	1769	2169	-----	0.02209	0.02725
194	NaCl	SODIUM CHLORIDE	0.00015	7.9407E-06	1.6779E-10	1738	2138	-----	0.01446	0.01789
195	NaF	SODIUM FLUORIDE	0.00217	7.9637E-06	3.8411E-10	1983	2383	-----	0.01947	0.02333
196	NaI	SODIUM IODIDE	---------	----------	----------	----	----	-----	------	-----
197	NaOH	SODIUM HYDROXIDE	-0.00113	1.9961E-05	-8.8245E-12	1663	2063	-----	0.03204	0.04001
198	Na2SO4	SODIUM SULFATE	---------	----------	----------	----	----	-----	------	-----
199	Nb	NIOBIUM	-0.00851	9.0443E-06	4.4032E-10	5115	5515	-----	0.04927	0.05476
200	Nd	NEODYMIUM	0.00359	5.8764E-06	1.7775E-10	3384	3784	-----	0.02551	0.02837
201	Ne	NEON	0.01379	1.2156E-04	-2.3590E-08	100	1500	0.04794	0.02571	0.14305
202	Ni	NICKEL	0.00912	8.7850E-06	6.2342E-10	2415	2815	-----	0.03397	0.03879
203	NiC4O4	NICKEL CARBONYL	-0.00323	3.6380E-05	-3.7936E-09	250	1000	0.00728	0.00563	0.02936
204	NiF2	NICKEL FLUORIDE	---------	----------	----------	----	----	-----	------	-----
205	Np	NEPTUNIUM	---------	----------	----------	----	----	-----	------	-----
206	O2	OXYGEN	0.00121	8.6157E-05	-1.3346E-08	80	1500	0.02571	0.00802	0.10042
207	O3	OZONE	0.00242	4.1109E-05	-7.7769E-09	162	1000	0.01399	0.00888	0.03575
208	Os	OSMIUM	0.00466	4.6428E-06	9.3953E-11	4880	5280	-----	0.02955	0.03179
209	OsOF5	OSMIUM OXIDE PENTAFLUORIDE	---------	----------	----------	----	----	-----	------	-----
210	OsO4	OSMIUM TETROXIDE - YELLOW	-0.00339	2.6992E-05	-2.2724E-09	403	803	-----	0.00712	0.01682
211	OsO4	OSMIUM TETROXIDE - WHITE	-0.00339	2.6992E-05	-2.2724E-09	403	803	-----	0.00712	0.01682
212	P	PHOSPHORUS - WHITE	-0.00047	2.3973E-05	-2.1465E-09	200	1600	0.00649	0.00424	0.03239
213	PBr3	PHOSPHORUS TRIBROMIDE	-0.00165	2.0988E-05	-3.3365E-09	448	848	-----	0.00708	0.01375
214	PCl2F3	PHOSPHORUS DICHLORIDE TRIFLUORIDE	-0.00641	6.4986E-05	-1.8973E-08	250	1000	0.01128	0.00865	0.03960
215	PCl3	PHOSPHORUS TRICHLORIDE	-0.00190	3.3080E-05	-4.5926E-09	200	1000	0.00755	0.00453	0.02659
216	PCl5	PHOSPHORUS PENTACHLORIDE	-0.00574	4.9726E-05	-1.3684E-08	433	833	-----	0.01323	0.02619
217	PH3	PHOSPHINE	-0.00455	6.3782E-05	1.6295E-08	185	1000	0.01592	0.00781	0.07553
218	PH4Br	PHOSPHONIUM BROMIDE	-0.00305	4.2385E-05	4.4068E-09	250	1000	0.00998	0.00782	0.04374
219	PH4Cl	PHOSPHONIUM CHLORIDE	-0.00407	7.5254E-05	4.6148E-10	250	1000	0.01841	0.01477	0.07165
220	PH4I	PHOSPHONIUM IODIDE	-0.00270	3.7465E-05	5.0195E-09	250	1000	0.00892	0.00698	0.03978
221	POCl3	PHOSPHORUS OXYCHLORIDE	-0.00452	4.3932E-05	-8.9116E-09	250	1000	0.00779	0.00591	0.03050
222	PSBr3	PHOSPHORUS THIOBROMIDE	-0.00225	2.5241E-05	-3.8730E-09	448	848	-----	0.00828	0.01637
223	PSCl3	PHOSPHORUS THIOCHLORIDE	-0.00371	3.8145E-05	-7.8275E-09	250	1000	0.00697	0.00534	0.02661
224	P4O6	PHOSPHORUS TRIOXIDE	-0.01107	6.8049E-05	-1.1990E-08	446	846	-----	0.01689	0.03792
225	P4O10	PHOSPHORUS PENTOXIDE	---------	----------	----------	----	----	-----	------	-----
226	P4S10	PHOSPHORUS PENTASULFIDE	-0.00568	6.0729E-05	-4.5433E-10	787	1000	-----	0.04183	0.05459
227	Pb	LEAD	0.00626	-4.0665E-06	2.8092E-09	2024	2424	-----	0.00954	0.01291
228	PbBr2	LEAD BROMIDE	---------	----------	----------	----	----	-----	------	-----
229	PbCl2	LEAD CHLORIDE	-0.00087	7.8433E-06	-2.3343E-10	1227	1627	-----	0.00840	0.01127
230	PbF2	LEAD FLUORIDE	-0.00144	7.9785E-06	-3.4913E-10	1566	1966	-----	0.01020	0.01290
231	PbI2	LEAD IODIDE	-0.00047	6.0732E-06	-1.3930E-10	1145	1545	-----	0.00630	0.00858
232	PbO	LEAD OXIDE	-0.00098	5.0338E-06	-1.1718E-10	1745	2145	-----	0.00745	0.00928
233	PbS	LEAD SULFIDE	-0.00123	5.5830E-06	-3.1548E-10	1554	1954	-----	0.00668	0.00847
234	Pd	PALLADIUM	0.00756	1.6614E-05	-1.5062E-09	3385	3785	-----	0.04654	0.04887

Table 23-2 THERMAL CONDUCTIVITY OF GAS - INORGANIC COMPOUNDS (continued)

NO	FORMULA	NAME	$k_{gas} = A + B\,T + C\,T^2$					k_{gas} - W/(m K), T - K		
			A	B	C	TMIN	TMAX	k_{gas} @25 C	k_{gas} @TMIN	k_{gas} @TMAX
235	Po	POLONIUM	0.00096	1.0079E-05	5.7480E-10	1235	1635	-----	0.01428	0.01898
236	Pt	PLATINUM	0.00000	8.6140E-07	4.6936E-11	3980	4380	-----	0.00417	0.00467
237	Ra	RADIUM	0.00151	6.7688E-06	3.4229E-10	1809	2000	-----	0.01487	0.01642
238	Rb	RUBIDIUM	0.00344	4.0964E-06	1.9453E-09	700	1500	-----	0.00726	0.01396
239	RbBr	RUBIDIUM BROMIDE	--------	----------	----------	----	----	-----	-----	-----
240	RbCl	RUBIDIUM CHLORIDE	-0.00143	7.3356E-06	-2.2852E-10	1654	2000	-----	0.01008	0.01233
241	RbF	RUBIDIUM FLUORIDE	--------	----------	----------	----	----	-----	-----	-----
242	RbI	RUBIDIUM IODIDE	--------	----------	----------	----	----	-----	-----	-----
243	Re	RHENIUM	0.00532	4.1469E-06	6.6505E-11	5915	6315	-----	0.03218	0.03416
244	Re2O7	RHENIUM HEPTOXIDE	--------	----------	----------	----	----	-----	-----	-----
245	Rh	RHODIUM	0.00497	6.2712E-06	3.0683E-10	3940	4340	-----	0.03444	0.03797
246	Rn	RADON	-0.00026	1.3342E-05	-3.2099E-09	298	1000	0.00343	0.00343	0.00987
247	Ru	RUTHENIUM	0.00556	6.1651E-06	1.3703E-10	4500	4900	-----	0.03608	0.03906
248	RuF5	RUTHENIUM PENTAFLUORIDE	--------	----------	----------	----	----	-----	-----	-----
249	S	SULFUR	0.00004	1.5923E-05	-5.0751E-10	718	1118	-----	0.01121	0.01721
250	SF4	SULFUR TETRAFLUORIDE	-0.01038	9.3508E-05	-2.3977E-08	298	1000	0.01537	0.01536	0.05915
251	SF6	SULFUR HEXAFLUORIDE	-0.00833	8.2437E-05	-2.8655E-08	273	1000	0.01370	0.01204	0.04545
252	SOBr2	THIONYL BROMIDE	-0.00254	2.8353E-05	-3.8276E-09	413	813	-----	0.00852	0.01798
253	SOCl2	THIONYL CHLORIDE	-0.00277	3.5059E-05	-3.5463E-09	349	1000	-----	0.00903	0.02874
254	SOF2	SULFUROUS OXYFLUORIDE	-0.00893	8.2467E-05	-1.9178E-08	298	1000	0.01395	0.01394	0.05436
255	SO2	SULFUR DIOXIDE	-0.00394	4.4847E-05	2.1066E-09	198	1000	0.00962	0.00502	0.04301
256	SO2Cl2	SULFURYL CHLORIDE	-0.00677	6.7459E-05	1.9374E-09	270	680	0.01352	0.01159	0.04000
257	SO3	SULFUR TRIOXIDE	-0.00354	5.0217E-05	-5.5547E-09	275	998	0.01094	0.00985	0.04104
258	S2Cl2	SULFUR MONOCHLORIDE	-0.00332	3.6918E-05	-6.0280E-09	411	811	-----	0.01084	0.02266
259	Sb	ANTIMONY	0.00210	9.0743E-06	4.3880E-10	1898	2000	-----	0.02090	0.02200
260	SbBr3	ANTIMONY TRIBROMIDE	-0.00093	1.4756E-05	-1.2353E-09	548	948	-----	0.00679	0.01195
261	SbCl3	ANTIMONY TRICHLORIDE	-0.00176	2.2840E-05	-3.1749E-09	493	893	-----	0.00873	0.01610
262	SbCl5	ANTIMONY PENTACHLORIDE	-0.00168	2.6004E-05	-4.2431E-09	413	813	-----	0.00834	0.01666
263	SbH3	STIBINE	-0.00539	4.3243E-05	-3.4260E-09	298	1000	0.00720	0.00719	0.03443
264	SbI3	ANTIMONY TRIIODIDE	-0.00088	1.1406E-05	-8.4752E-10	674	1000	-----	0.00642	0.00968
265	Sb2O3	ANTIMONY TRIOXIDE	--------	----------	----------	----	----	-----	-----	-----
266	Sc	SCANDIUM	0.01739	-4.3366E-06	5.1462E-09	2700	3100	-----	0.04320	0.05340
267	Se	SELENIUM	-0.00150	1.7811E-05	2.4971E-10	930	1330	-----	0.01528	0.02263
268	SeCl4	SELENIUM TETRACHLORIDE	-0.00376	3.4855E-05	-7.1702E-09	465	865	-----	0.01090	0.02102
269	SeF6	SELENIUM HEXAFLUORIDE	-0.01258	9.2755E-05	-2.3304E-08	250	1000	0.01300	0.00915	0.05687
270	SeOCl2	SELENIUM OXYCHLORIDE	-0.00461	3.7375E-05	-6.8113E-09	441	841	-----	0.01055	0.02200
271	SeO2	SELENIUM DIOXIDE	--------	----------	----------	----	----	-----	-----	-----
272	Si	SILICON	-0.00864	1.1161E-05	-5.3147E-10	3514	3914	-----	0.02402	0.02690
273	SiBrCl2F	BROMODICHLOROFLUOROSILANE	-0.00327	3.7741E-05	-8.0510E-09	298	1000	0.00727	0.00726	0.02642
274	SiBrF3	TRIFLUOROBROMOSILANE	-0.00596	5.7644E-05	-1.4282E-08	298	1000	0.00996	0.00995	0.03740
275	SiBr2ClF	DIBROMOCHLOROFLUOROSILANE	-0.00288	3.4677E-05	-6.8123E-09	298	1000	0.00685	0.00685	0.02498
276	SiClF3	TRIFLUOROCHLOROSILANE	-0.00716	7.3263E-05	-1.9005E-08	298	1000	0.01299	0.01298	0.04710
277	SiCl2F2	DICHLORODIFLUOROSILANE	-0.00441	5.3871E-05	-1.3504E-08	298	1000	0.01045	0.01044	0.03596
278	SiCl3F	TRICHLOROFLUOROSILANE	-0.00395	4.7170E-05	-1.0707E-08	298	1000	0.00916	0.00916	0.03251
279	SiCl4	SILICON TETRACHLORIDE	0.00905	-1.7720E-05	5.4336E-08	275	650	0.00860	0.00829	0.02049
280	SiF4	SILICON TETRAFLUORIDE	-0.00189	5.9473E-05	-3.0812E-09	275	702	0.01557	0.01423	0.03834
281	SiHBr3	TRIBROMOSILANE	-0.00268	2.5207E-05	-3.2482E-09	250	1000	0.00455	0.00342	0.01928
282	SiHCl3	TRICHLOROSILANE	-0.00635	4.9468E-05	-7.7224E-09	275	1229	0.00771	0.00667	0.04278
283	SiHF3	TRIFLUOROSILANE	-0.01093	9.3880E-05	-1.9359E-08	298	1000	0.01534	0.01533	0.06359
284	SiH2Br2	DIBROMOSILANE	-0.00422	3.4935E-05	-1.9892E-09	250	1000	0.00602	0.00439	0.02873
285	SiH2Cl2	DICHLOROSILANE	0.00340	8.5800E-06	4.6426E-08	275	624	0.01009	0.00927	0.02683
286	SiH2F2	DIFLUOROSILANE	-0.01379	9.6915E-05	-1.2666E-08	298	1000	0.01398	0.01397	0.07046
287	SiH2I2	DIIODOSILANE	-0.00416	2.8849E-05	-1.2490E-09	423	823	-----	0.00782	0.01874
288	SiH3Br	MONOBROMOSILANE	-0.00795	5.6923E-05	-2.8594E-09	298	1000	0.00877	0.00876	0.04611
289	SiH3Cl	MONOCHLOROSILANE	-0.01076	7.6935E-05	-5.6791E-09	298	1000	0.01167	0.01166	0.06050
290	SiH3F	MONOFLUOROSILANE	-0.01602	1.1442E-04	-1.1983E-08	298	1000	0.01703	0.01701	0.08642
291	SiH3I	IODOSILANE	-0.00666	4.6587E-05	-5.4486E-10	298	1000	0.00718	0.00717	0.03938
292	SiH4	SILANE	-0.00346	7.5853E-05	4.0964E-08	275	543	0.02280	0.02050	0.04981
293	SiO2	SILICON DIOXIDE	--------	----------	----------	----	----	-----	-----	-----
294	Si2Cl6	HEXACHLORODISILANE	-0.00306	3.4287E-05	-7.1457E-09	412	812	-----	0.00985	0.02007
295	Si2F6	HEXAFLUORODISILANE	-0.00551	6.5085E-05	-1.4546E-08	250	1000	0.01260	0.00985	0.04503
296	Si2H5Cl	DISILANYL CHLORIDE	-0.00658	6.0637E-05	-4.4215E-09	298	1000	0.01111	0.01110	0.04964
297	Si2H6	DISILANE	-0.01362	9.9007E-05	4.9324E-09	250	1000	0.01634	0.01144	0.09032
298	Si2OCl3F3	TRICHLOROTRIFLUORODISILOXANE	-0.00272	3.9020E-05	-7.9771E-09	298	1000	0.00820	0.00820	0.02832
299	Si2OCl6	HEXACHLORODISILOXANE	-0.00287	3.1599E-05	-6.7224E-09	409	809	-----	0.00893	0.01829
300	Si2OH6	DISILOXANE	-0.01035	8.0601E-05	1.6236E-09	250	1000	0.01383	0.00990	0.07187
301	Si3Cl8	OCTACHLOROTRISILANE	-0.00199	2.4790E-05	-3.8632E-09	485	885	-----	0.00912	0.01692
302	Si3H8	TRISILANE	-0.00713	5.6886E-05	4.1692E-09	250	1000	0.01020	0.00735	0.05393
303	Si3H9N	TRISILAZANE	-0.00687	6.4052E-05	1.1155E-09	250	1000	0.01233	0.00921	0.05830
304	Si4H10	TETRASILANE	-0.00511	5.0150E-05	2.9820E-09	250	1000	0.01011	0.00761	0.04802
305	Sm	SAMARIUM	0.00410	1.0986E-05	-3.6715E-10	1874	2274	-----	0.02340	0.02718
306	Sn	TIN	0.00873	5.4559E-07	4.8172E-10	2995	3395	-----	0.01469	0.01613
307	SnBr4	STANNIC BROMIDE	-0.00208	2.0360E-05	-3.8607E-09	478	878	-----	0.00677	0.01282
308	SnCl2	STANNOUS CHLORIDE	-0.00101	1.6712E-05	-5.6693E-10	925	1296	-----	0.01396	0.01970
309	SnCl4	STANNIC CHLORIDE	-0.00276	3.0087E-05	-6.1585E-09	250	1000	0.00566	0.00438	0.02117
310	SnH4	STANNIC HYDRIDE	-0.00597	5.2409E-05	5.8176E-09	250	1000	0.01017	0.00750	0.05226
311	SnI4	STANNIC IODIDE	-0.00080	1.3470E-05	-9.7534E-10	621	1000	-----	0.00719	0.01169
312	Sr	STRONTIUM	0.00218	1.1963E-05	6.3803E-10	1630	2030	-----	0.02337	0.02909

Table 23-2 THERMAL CONDUCTIVITY OF GAS - INORGANIC COMPOUNDS (continued)

NO	FORMULA	NAME	$k_{gas} = A + B T + C T^2$					$(k_{gas}$ - W/(m K), T - K)		
			A	B	C	TMIN	TMAX	k_{gas} @25 C	k_{gas} @TMIN	k_{gas} @TMAX
313	SrO	STRONTIUM OXIDE	--------	----------	----------	----	----	-----	-----	-----
314	Ta	TANTALUM	0.00582	4.9055E-06	8.4815E-11	5565	5965	-----	0.03575	0.03810
315	Tc	TECNNETIUM	0.00666	6.4359E-06	1.2656E-10	5000	5400	-----	0.04200	0.04510
316	Te	TELLURIUM	0.00196	6.2899E-06	1.1778E-09	1285	1685	-----	0.01199	0.01590
317	TeCl4	TELLURIUM TETRACHLORIDE	-0.00107	1.9679E-05	-1.2099E-09	665	1065	-----	0.01148	0.01852
318	TeF6	TELLURIUM HEXAFLUORIDE	-0.00922	6.7842E-05	-1.6803E-08	250	1000	0.00951	0.00669	0.04182
319	Ti	TITANIUM	-0.03118	2.9012E-05	2.9894E-09	3442	3842	-----	0.10410	0.12441
320	TiCl4	TITANIUM TETRACHLORIDE	-0.00231	2.6334E-05	-2.6851E-10	409	809	-----	0.00842	0.01882
321	Tl	THALLIUM	0.00092	7.4529E-06	8.2689E-10	1745	2145	-----	0.01644	0.02071
322	TlBr	THALLOUS BROMIDE	-0.00029	5.0866E-06	-8.4499E-11	1092	1492	-----	0.00516	0.00711
323	TlI	THALLOUS IODIDE	-0.00028	4.7253E-06	-8.6258E-11	1096	1496	-----	0.00480	0.00660
324	Tm	THULIUM	--------	----------	----------	----	----	-----	-----	-----
325	U	URANIUM	0.00358	4.4496E-06	1.0877E-10	4135	4535	-----	0.02384	0.02600
326	UF6	URANIUM FLUORIDE	-0.00700	4.9626E-05	-1.0019E-08	250	1000	0.00691	0.00478	0.03261
327	V	VANADIUM	0.00610	8.9815E-06	3.4649E-10	3665	4065	-----	0.04367	0.04834
328	VCl4	VANADIUM TETRACHLORIDE	-0.00126	2.8736E-05	-2.6809E-09	298	1000	0.00707	0.00707	0.02480
329	VOCl3	VANADIUM OXYTRICHLORIDE	-0.00379	3.9109E-05	-8.6371E-09	250	1000	0.00710	0.00545	0.02668
330	W	TUNGSTEN	-0.50306	1.8809E-04	-1.5842E-08	5645	6000	-----	0.05389	0.05517
331	WF6	TUNGSTEN FLUORIDE	-0.00630	5.7971E-05	-1.3328E-08	298	1000	0.00980	0.00979	0.03834
332	Xe	XENON	0.00034	1.8809E-05	-3.0072E-09	165	1500	0.00568	0.00336	0.02179
333	Yb	YTTERBIUM	0.00158	8.3703E-06	4.4268E-10	1660	2060	-----	0.01669	0.02070
334	Yt	YTTRIUM	-0.00654	9.1709E-06	1.7625E-09	3055	3455	-----	0.03793	0.04618
335	Zn	ZINC	0.00242	1.5100E-05	1.1654E-09	1181	1581	-----	0.02188	0.02921
336	ZnCl2	ZINC CHLORIDE	-0.00078	1.2263E-05	-1.7583E-10	1005	1405	-----	0.01137	0.01610
337	ZnF2	ZINC FLUORIDE	--------	----------	----------	----	----	-----	-----	-----
338	ZnO	ZINC OXIDE	--------	----------	----------	----	----	-----	-----	-----
339	ZnSO4	ZINC SULFATE	--------	----------	----------	----	----	-----	-----	-----
340	Zr	ZIRCONIUM	-0.02960	2.6914E-05	7.1497E-13	4598	4998	-----	0.09417	0.10493
341	ZrBr4	ZIRCONIUM BROMIDE	-0.00123	1.6328E-05	-8.1814E-10	630	1030	-----	0.00873	0.01472
342	ZrCl4	ZIRCONIUM CHLORIDE	-0.00199	2.1687E-05	-6.9774E-10	604	1004	-----	0.01085	0.01908
343	ZrI4	ZIRCONIUM IODIDE	-0.00091	1.2772E-05	-6.2272E-10	704	1104	-----	0.00777	0.01243

k_{gas} - thermal conductivity of gas, W/(m K)

A, B, and C - regression coefficients for chemical compound

T - temperature, K

TMIN - minimum temperature, K

TMAX - maximum temperature, K

Chapter 24

THERMAL CONDUCTIVITY OF LIQUID AND SOLID

Carl L. Yaws, Xiaoyan Lin, Li Bu, and Sachin Nijhawan
Lamar University, Beaumont, Texas

ABSTRACT

Results for thermal conductivity of liquid as a function of temperature are presented for a wide range of organic chemicals. Results are also given for thermal conductivity of liquid and solid as a function of temperature of inorganic chemicals. The chemicals include many compound types. The results are provided in easy-to-use tabulations that are especially applicable for rapid engineering usage with the personal computer or hand calculator. The agreement of correlation and data is quite good.

INTRODUCTION

Thermal conductivity of liquid and solid is important in many engineering applications in the chemical processing and petroleum refining industries. The objective of this article is to provide the engineer with such data. The compilation of data is presented for a wide temperature range to enable the engineer to determine values at the desired temperatures of interest.

THERMAL CONDUCTIVITY CORRELATION

For organic compounds, the correlation for thermal conductivity of liquid as a function of temperature is given by the equation shown below:

$$\log_{10} k_{liq} = A + B [1-T/C]^{2/7} \qquad (24\text{-}1)$$

where
k_{liq} = thermal conductivity of liquid, W/(m K)
A, B, and C = regression coefficients for chemical compound
T = temperature, K

For inorganic compounds, the correlation for thermal conductivity of liquid and solid as a function of temperature is given by the equation shown below:

$$k = A + B T + C T^2 \qquad (24\text{-}2)$$

where
k = thermal conductivity of liquid or solid, W/(m K)
A, B, and C = regression coefficients for chemical compound
T = temperature, K

The results for thermal conductivity of liquid and solid are given in Tables 24-1 and 24-2. The tabulation is arranged by chemical formula to provide ease of use in quickly locating data.

In preparing the compilation, a literature search was conducted to identify data source publications for organics (1-37) and inorganics (1-99). Both experimental values for the property under consideration and parameter values for estimation of the property are included in the source publications. The publications were screened for appropriate data. The compilation resulting from the screening is based on both experimental data and estimated values. In the absence of experimental data for organic compounds, estimates of liquids were primarily based on modified Missenard and Pachaiyappan methods (29) and the Sato equation (29). For inorganic compounds, estimates of liquids were primarily based on modified methods of Sato, Reidel, and Pachaiyappan (29). For inorganic compounds, estimates of solids were primarily based on the work of Ho, Powell, and Liley (23, 24, and 69). Experimental data and estimates were then regressed to provide the same equation for all compounds.

Very limited experimental data for liquid thermal conductivities are available at temperatures in the region of the melting point. Also, there are very few reliable data at temperatures above a reduced temperature of $T_r = 0.65$. Thus, the values in the regions of melting point and reduced temperatures above 0.65 should be considered rough approximations. The values in the intermediate region (above melting point and below reduced temperature of 0.65) are more accurate.

A comparison of correlation and experimental data is shown in Fig. 24-1 for a representative chemical. The graph discloses good agreement of correlation and data.

EXAMPLES

The correlation results may be used for prediction and calculation of thermal conductivity. Examples are given below.

Example 1 Calculate the liquid thermal conductivity of methanol (CH4O) at a temperature of 370 K.

Substitution of the coefficients from the table and temperature into the correlation equation yields

$\log_{10} k_{liq} = -1.1793 + 0.6191 [1-370/512.58]^{2/7} = -.7498$

$k_{liq} = 10^{-.7498}$

$k_{liq} = 0.1779$ W/(m K)

The calculated and data values compare favorably (0.1779 vs 0.1822, deviation = 2.4%).

Example 2 Calculate the liquid thermal conductivity of toluene (C7H8) at a temperature of 360 K.

Substitution of the coefficients from the table and temperature into the correlation equation yields

$\log_{10} \lambda_{liq} = -1.6735 + 0.9773 [1-360/591.79]^{2/7} = -.9258$

$k_{liq} = 10^{-.9258}$

$k_{liq} = 0.11863$ W/m K

The calculated and data values compare favorably (0.11863 vs 0.11860, deviation = 0.1%).

REFERENCES – ORGANIC COMPOUNDS

1-34. See **REFERENCES - ORGANIC COMPOUNDS** in Chapter 1, **CRITICAL PROPERTIES AND ACENTRIC FACTOR**

35. Tsederberg, N. V., THERMAL CONDUCTIVITY OF GASES AND LIQUIDS, R. D. Cess, editor, MIT Press, Cambridge, MA (1965).

36. Vargaftik, N. B., L. P. Filippov, A. A. Tarzimanov, and E. E. Totskiy, THERMAL CONDUCTIVITY OF GASES AND LIQUIDS, Standards Press, Moscow, USSR (1978).

37. Yaws, C. L., HANDBOOK OF THERMAL CONDUCTIVITY, Vols. 1, 2, 3, and 4, Gulf Publishing Co., Houston, TX (1995, 1995, 1995, 1997).

REFERENCES – INORGANIC COMPOUNDS

1-66. See **REFERENCES - INORGANIC COMPOUNDS** in Chapter 3, **HEAT CAPACITY OF LIQUID**

67. Tsederberg, N. V., THERMAL CONDUCTIVITY OF GASES AND LIQUIDS, MIT Press, Cambridge, MA (1965).

68. Vargaftik, N. B. and others, HANDBOOK OF THERMAL CONDUCTIVITY OF LIQUIDS AND GASES, CRC Press, Inc., Boca Raton, FL (1994).

69. Ho. C. Y., R. W. Powell, and P. E. Liley, J., Phys. Chem. Ref. Data, 1 (2), 279 (1972).

70. Saxena, S. C., High Temp. Sci., 3, 168 (1971).

71. Stiel, L. I. and G. Thodos, AIChE J., 10, 266 (1964).

72. Choy, P. and C. J. G. Raw, J. Chem. Phys., 45 (5), 1413 (1966).

73. Chapman, S.and T. G. Cowling, AIChE J., 10, 266 (1964).

74. Eucken A., Z. Phys., 14, 324 (1913).

75. Matsunaga, N. and A. Nagashima, J. Phys. Chem. Ref. Data, 12 (4), 933 (1983).

76. Kestin, J. and J. V. Sengers, J. Phys. Chem. Ref. Data, 13 (2), 601 (1984).

77. Baker, C. E., J. Chem. Phys., 46, 2846 (1967).

78. Jones, L. W., Int. J. Heat Mass Transfer, 10, 745 (1967).

79. Kestin, J. and E. A. Knierim, J. Phys. Chem. Ref. Data, 13 (1), 229 (1984).

80. Hanley, H. J. M. and J. F. Ely, J. Phys. Chem. Ref. Data, 2 (4), 735 (1973).

81. Streng, A. G., J. Chem. Eng. Data, 6 (3), 43 (1961).

82. Bakalin, S. S. and S. A. Olbybin, Teplofiz Vys. Temp., 14, 391 (1976).

83. Palmer, G., Ind. Eng., 40, 89 (1948).

84. Fedorov, V. I. and V. I. Machev, High Temp., 8, 858 (1970).

85. Sengers, J. V., J. T. R. Watson, and R. S. Basu, J. Phys. Chem. Ref. Data, 13(3), 893 (1984).

86. Ewing, C. T., J. A. Grand, and R. R. Miller, J. Amer. Chem. Soc., 74, 11 (1952).

87. Davison, H. W., U. S. NASA Tech. Note D-4650, (1968).

88. Andrade, E. N., C. Da, and E. R. Dobbs, Proc. Roy. Soc. London, 211A 12, (1952).

89. Schaefer, C. A. and G. Thodos, AIChE J., 5(3), 367 (1959).

90. Richter, G. N. and B. H. Sage, J. Chem. Eng. Data, 8(2), 221 (1963).

91. McDonald, J. and H. T. Davis, Phys. Chem. Liquids, 2(3), 119 (1971).

92. Federov, V. I. and V. I. Machuev, Teplofiz. Vys. Temp., 8(4), 912 (1970).

93. Waterman, T. E., D. P. Kirsh, and R. I. Brabets, J. Chem. Phys., 29(4), 905 (1958).

94. Sladkov, I. B. and T. G. Kotina, Russ. J. Phys. Chem., 48(7), 1115 (1974).

95. Sugawara, A., J. Appl. Phys., 36, 2375 (1965).

96. Ivannikov, P. S., I. V. Litvinenko, and I. V. Radchenko, J. Eng. Phys. USSR, 23(5), 1397 (1972).

97. Dushin, Yu. A., J. Eng. Phys., 10(4), 538 (1966).

98. Yaws, C. L., HANDBOOK OF TRANSPORT PROPERTY DATA, Gulf Publishing Co., Houston, TX (1995).

99. Yaws, C. L., HANDBOOK OF THERMAL CONDUCTIVITY, Vols. 1 and 4, Gulf Publishing Co., Houston, TX (1995, 1997).

Figure 24-1 Thermal Conductivity of Liquid

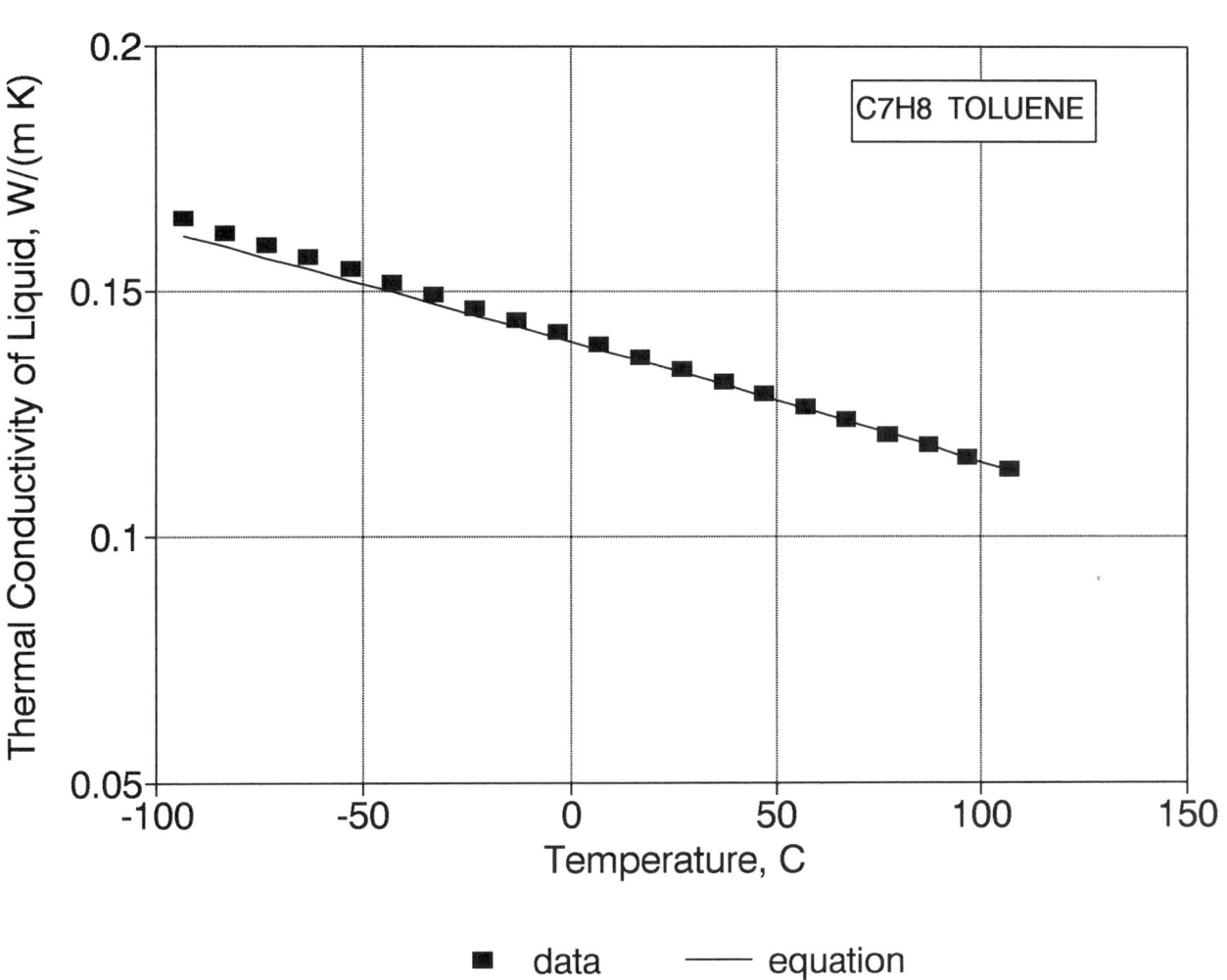

Table 24-1 THERMAL CONDUCTIVITY OF LIQUID - ORGANIC COMPOUNDS

			log₁₀ k_liq = A + B [1 - T/C]^(2/7)					(k_liq - W/(m K), T - K)		
NO	FORMULA	NAME	A	B	C	TMIN	TMAX	k_liq @25 C	k_liq @TMIN	k_liq @TMAX
1	CBrClF2	BROMOCHLORODIFLUOROMETHANE	-1.1059	0.2320	426.15	114	405	0.1145	0.1274	0.0784
2	CBrCl3	BROMOTRICHLOROMETHANE	-1.8873	1.0168	606.00	252	576	0.0893	0.0945	0.0130
3	CBrF3	BROMOTRIFLUOROMETHANE	-2.0074	1.1951	340.15	105	323	0.0447	0.1150	0.0098
4	CBr2F2	DIBROMODIFLUOROMETHANE	-1.6881	0.9967	478.00	163	454	0.1163	0.1548	0.0205
5	CClF3	CHLOROTRIFLUOROMETHANE	-1.6922	0.9350	301.96	92	287	-----	0.1397	0.0203
6	CClN	CYANOGEN CHLORIDE	-1.5416	1.0380	449.00	267	427	0.1654	0.1748	0.0287
7	CCl2F2	DICHLORODIFLUOROMETHANE	-1.8990	1.1855	384.95	115	366	0.0751	0.1463	0.0126
8	CCl2O	PHOSGENE	-1.4013	0.7045	455.00	145	432	0.1313	0.1681	0.0397
9	CCl3F	TRICHLOROFLUOROMETHANE	-1.9663	1.2354	471.20	162	448	0.0915	0.1320	0.0108
10	CCl4	CARBON TETRACHLORIDE	-1.8791	1.0875	556.35	250	529	0.0987	0.1063	0.0132
11	CF2O	CARBONYL FLUORIDE	-1.4494	1.0257	297.00	162	282	-----	0.2260	0.0355
12	CF4	CARBON TETRAFLUORIDE	-1.7559	0.9890	227.50	90	216	-----	0.1236	0.0175
13	CHBr3	TRIBROMOMETHANE	-1.9473	1.1088	696.00	281	661	0.0995	0.0998	0.0113
14	CHClF2	CHLORODIFLUOROMETHANE	-1.6896	1.0163	369.30	116	351	0.0882	0.1647	0.0204
15	CHCl2F	DICHLOROFLUOROMETHANE	-1.8416	1.1592	451.58	138	429	0.1023	0.1570	0.0144
16	CHCl3	CHLOROFORM	-1.5271	0.7577	536.40	210	510	0.1185	0.1330	0.0297
17	CHF3	TRIFLUOROMETHANE	-1.6073	0.9937	298.89	118	284	-----	0.1758	0.0247
18	CHI3	TRIIODOMETHANE	-1.7076	0.5740	794.55	396	755	-----	0.0571	0.0196
19	CHN	HYDROGEN CYANIDE	-1.4117	1.0351	456.65	260	434	0.2256	0.2429	0.0388
20	CHNS	ISOTHIOCYANIC-ACID	-------	-------	-------	-----	-----	-----	-----	-----
21	CH2BrCl	BROMOCHLOROMETHANE	-1.7805	0.9975	557.00	185	529	0.1049	0.1264	0.0166
22	CH2Br2	DIBROMOMETHANE	-1.9470	1.1902	611.00	221	580	0.1086	0.1232	0.0113
23	CH2ClF	CHLOROFLUOROMETHANE	-1.1872	0.3845	424.91	140	404	0.1216	0.1423	0.0650
24	CH2Cl2	DICHLOROMETHANE	-1.8069	1.2216	510.00	178	485	0.1392	0.1841	0.0156
25	CH2F2	DIFLUOROMETHANE	-1.5190	1.0688	351.60	137	334	0.1273	0.2513	0.0303
26	CH2I2	DIIODOMETHANE	-2.3257	1.5226	747.00	279	710	0.0979	0.0987	0.0047
27	CH2O	FORMALDEHYDE	-------	-------	-------	-----	-----	-----	-----	-----
28	CH2O2	FORMIC ACID	-0.8626	0.3692	580.00	282	551	0.2740	0.2743	0.1372
29	CH3Br	METHYL BROMIDE	-1.7379	1.0082	467.00	180	444	0.1038	0.1352	0.0183
30	CH3Cl	METHYL CHLORIDE	-1.7528	1.3686	416.25	175	395	0.1593	0.2541	0.0177
31	CH3Cl3Si	METHYL TRICHLOROSILANE	-1.6182	0.9960	517.00	195	491	0.1448	0.1753	0.0241
32	CH3F	METHYL FLUORIDE	-1.2953	0.9173	317.70	131	302	0.1313	0.3051	0.0507
33	CH3I	METHYL IODIDE	-1.8201	0.9896	528.00	207	502	0.0912	0.1072	0.0151
34	CH3NO	FORMAMIDE	-0.6492	0.2262	771.00	276	732	0.3528	0.3535	0.2243
35	CH3NO2	NITROMETHANE	-1.5364	1.0441	588.15	245	559	0.2074	0.2234	0.0291
36	CH3NO2	METHYL-NITRITE	-------	-------	-------	-----	-----	-----	-----	-----
37	CH3NO3	METHYL-NITRATE	-------	-------	-------	-----	-----	-----	-----	-----
38	CH4	METHANE	-1.0976	0.5387	190.58	91	181	-----	0.2206	0.0799
39	CH4Cl2Si	METHYL DICHLOROSILANE	-1.6664	0.9997	483.00	183	459	0.1240	0.1578	0.0216
40	CH4O	METHANOL	-1.1793	0.6191	512.58	175	487	0.2011	0.2322	0.0662
41	CH4O3S	METHANESULFONIC ACID	-------	-------	-------	-----	-----	-----	-----	-----
42	CH4S	METHYL MERCAPTAN	-------	-------	-------	-----	-----	-----	-----	-----
43	CH5ClSi	METHYL CHLOROSILANE	-1.6000	0.9860	442.00	139	420	0.1305	0.1901	0.0251
44	CH5N	METHYLAMINE	-1.0947	0.5539	430.05	180	409	0.1997	0.2369	0.0804
45	CH6Si	METHYL SILANE	-1.5494	0.9862	352.50	116	335	0.1068	0.2109	0.0282
46	CN4O8	TETRANITROMETHANE	-1.7461	1.0399	540.00	287	513	0.1204	0.1193	0.0179
47	CO	CARBON MONOXIDE	-1.7115	1.1359	132.92	68	126	-----	0.1580	0.0194
48	COS	CARBONYL SULFIDE	-1.5784	0.9882	378.80	134	360	0.1140	0.1935	0.0264
49	CO2	CARBON DIOXIDE	-1.3679	0.8092	304.19	217	289	-----	0.1500	0.0429
50	CS2	CARBON DISULFIDE	-1.2917	0.5809	552.00	162	524	0.1491	0.1702	0.0511
51	C2BrF3	BROMOTRIFLUOROETHYLENE	-1.8388	0.9861	432.00	162	410	0.0736	0.1036	0.0145
52	C2Br2F4	1,2-DIBROMOTETRAFLUOROETHANE	-1.9334	0.9638	487.80	163	463	0.0634	0.0828	0.0117
53	C2ClF3	CHLOROTRIFLUOROETHYLENE	-1.3362	0.5001	379.15	115	360	0.0967	0.1294	0.0461
54	C2ClF5	CHLOROPENTAFLUOROETHANE	-1.8980	1.0488	353.15	174	335	0.0523	0.0897	0.0126
55	C2Cl2F4	1,2-DICHLOROTETRAFLUOROETHANE	-1.7806	0.8820	418.85	179	398	0.0688	0.0918	0.0166
56	C2Cl3F3	1,1,2-TRICHLOROTRIFLUOROETHANE	-1.9265	1.0580	487.25	238	463	0.0760	0.0860	0.0118
57	C2Cl4	TETRACHLOROETHYLENE	-2.0968	1.3739	620.00	251	589	0.1103	0.1190	0.0080
58	C2Cl4F2	1,1,2,2-TETRACHLORODIFLUOROETHANE	-1.6650	0.7165	551.00	299	523	-----	0.0789	0.0216
59	C2Cl4O	TRICHLOROACETYL CHLORIDE	-1.7653	1.0299	590.00	273	561	0.1194	0.1219	0.0172
60	C2Cl6	HEXACHLOROETHANE	-------	-------	-------	-----	-----	-----	-----	-----
61	C2F4	TETRAFLUOROETHYLENE	-------	-------	-------	-----	-----	-----	-----	-----
62	C2F6	HEXAFLUOROETHANE	-1.7571	0.9052	292.80	172	278	-----	0.0851	0.0175
63	C2HBrClF3	HALOTHANE	-1.8115	1.0047	521.00	223	495	0.0948	0.1084	0.0154
64	C2HClF2	2-CHLORO-1,1-DIFLUOROETHYLENE	-1.7408	0.9926	400.55	135	381	0.0854	0.1365	0.0182
65	C2HCl3	TRICHLOROETHYLENE	-2.1042	1.4428	571.00	188	542	0.1159	0.1490	0.0079
66	C2HCl3O	DICHLOROACETYL CHLORIDE	-1.7322	1.0211	579.00	298	550	0.1254	0.1216	0.0185
67	C2HCl3O	TRICHLOROACETALDEHYDE	-1.7155	0.9989	565.00	216	537	0.1232	0.1402	0.0193
68	C2HCl5	PENTACHLOROETHANE	-1.8582	0.9887	665.00	244	632	0.0946	0.1004	0.0139
69	C2HF3	TRIFLUOROETHENE	-1.2321	0.3890	347.22	95	330	0.0978	0.1321	0.0586
70	C2HF3O2	TRIFLUOROACETIC ACID	-1.5757	1.0271	491.25	258	467	0.1625	0.1740	0.0266
71	C2HF5	PENTAFLUOROETHANE	-1.7216	0.9650	342.00	170	325	0.0653	0.1147	0.0190
72	C2H2	ACETYLENE	-------	-------	-------	-----	-----	-----	-----	-----
73	C2H2Br4	1,1,2,2-TETRABROMOETHANE	-1.5070	0.4990	824.00	273	783	0.0855	0.0860	0.0311
74	C2H2Cl2	1,1-DICHLOROETHYLENE	-1.7266	0.9964	482.00	151	458	0.1071	0.1453	0.0188
75	C2H2Cl2	cis-1,2-DICHLOROETHYLENE	-1.7165	0.9901	527.00	193	501	0.1158	0.1397	0.0192
76	C2H2Cl2	trans-1,2-DICHLOROETHYLENE	-1.7117	0.9782	508.00	223	483	0.1117	0.1282	0.0194
77	C2H2Cl2O	CHLOROACETYL CHLORIDE	-1.6852	1.0377	581.00	251	552	0.1444	0.1540	0.0206
78	C2H2Cl2O	DICHLOROACETALDEHYDE	-1.6851	0.9984	555.00	223	527	0.1306	0.1473	0.0206

Table 24-1 THERMAL CONDUCTIVITY OF LIQUID - ORGANIC COMPOUNDS (continued)

NO	FORMULA	NAME	$\log_{10} k_{liq} = A + B [1 - T/C]^{2/7}$					$(k_{liq} - W/(m\ K),\ T - K)$		
			A	B	C	TMIN	TMAX	k_{liq} @25 C	k_{liq} @TMIN	k_{liq} @TMAX
79	C2H2Cl2O2	DICHLOROACETIC ACID	-1.5872	1.0111	686.00	287	652	0.1870	0.1860	0.0259
80	C2H2Cl3F	1,1,1-TRICHLOROFLUOROETHANE	-1.8131	0.9980	565.00	220	537	0.0983	0.1110	0.0154
81	C2H2Cl4	1,1,1,2-TETRACHLOROETHANE	-1.9394	1.1331	624.00	203	593	0.1004	0.1164	0.0115
82	C2H2Cl4	1,1,2,2-TETRACHLOROETHANE	-1.7916	1.0073	645.00	229	613	0.1127	0.1229	0.0162
83	C2H2F2	1,1-DIFLUOROETHYLENE	-1.6721	0.9990	302.80	129	288	-----	0.1482	0.0213
84	C2H2F2	cis-1,2-DIFLUOROETHENE	-1.1268	0.3281	394.67	108	375	0.1238	0.1482	0.0747
85	C2H2F2	trans-1,2-DIFLUOROETHENE	-1.1268	0.3281	394.67	108	375	0.1238	0.1482	0.0747
86	C2H2F4	1,1,1,2-TETRAFLUOROETHANE	-1.7425	0.9769	380.00	172	361	0.0772	0.1174	0.0181
87	C2H2O	KETENE	-1.5301	0.9860	370.00	122	352	0.1222	0.2203	0.0295
88	C2H2O4	OXALIC ACID	-------	-------	-------	-----	-----	-----	-----	-----
89	C2H3Br	VINYL BROMIDE	-1.7532	1.0010	473.00	135	449	0.1000	0.1414	0.0177
90	C2H3Cl	VINYL CHLORIDE	-1.7110	0.9777	432.00	119	410	0.0974	0.1498	0.0195
91	C2H3ClF2	1-CHLORO-1,1-DIFLUOROETHANE	-1.7037	0.9591	410.20	142	390	0.0908	0.1377	0.0198
92	C2H3ClO	ACETYL CHLORIDE	-1.6413	0.9858	508.00	160	483	0.1332	0.1728	0.0228
93	C2H3ClO	CHLOROACETALDEHYDE	-1.6434	1.0212	555.00	293	527	0.1500	0.1467	0.0227
94	C2H3ClO2	CHLOROACETIC ACID	-1.3157	0.5349	686.00	333	652	-----	0.1319	0.0483
95	C2H3ClO2	METHYL CHLOROFORMATE	-1.6454	0.9973	525.00	192	499	0.1378	0.1670	0.0226
96	C2H3Cl3	1,1,1-TRICHLOROETHANE	-1.7352	0.9286	545.00	243	518	0.1012	0.1096	0.0184
97	C2H3Cl3	1,1,2-TRICHLOROETHANE	-2.6567	2.1686	602.00	237	572	0.1340	0.1601	0.0022
98	C2H3F	VINYL FLUORIDE	-1.1468	0.4921	327.80	113	311	0.1261	0.1931	0.0713
99	C2H3F3	1,1,1-TRIFLUOROETHANE	-1.7049	0.9726	346.25	162	329	0.0705	0.1249	0.0197
100	C2H3N	ACETONITRILE	-1.5248	1.0023	545.50	229	518	0.1883	0.2107	0.0299
101	C2H3NO	METHYL ISOCYANATE	-1.6639	1.0134	505.00	256	480	0.1322	0.1416	0.0217
102	C2H4	ETHYLENE	-1.3314	0.8527	282.36	104	268	-----	0.2568	0.0466
103	C2H4Br2	1,1-DIBROMOETHANE	-1.8696	1.0151	628.00	210	597	0.0944	0.1065	0.0135
104	C2H4Br2	1,2-DIBROMOETHANE	-1.5732	0.6869	650.15	283	618	0.1008	0.1008	0.0267
105	C2H4Cl2	1,1-DICHLOROETHANE	-1.7265	0.9930	523.00	176	497	0.1132	0.1412	0.0188
106	C2H4Cl2	1,2-DICHLOROETHANE	-1.6509	0.9701	561.00	237	533	0.1350	0.1476	0.0223
107	C2H4Cl2O	BIS(CHLOROMETHYL)ETHER	-1.7290	0.9999	579.00	232	550	0.1214	0.1337	0.0187
108	C2H4F2	1,1-DIFLUOROETHANE	-1.6971	1.0889	386.60	156	367	0.1041	0.1708	0.0201
109	C2H4F2	1,2-DIFLUOROETHANE	-1.6318	0.9762	476.00	215	452	0.1274	0.1513	0.0233
110	C2H4I2	1,2-DIIODOETHANE	-1.4948	0.3995	749.91	356	712	-----	0.0681	0.0320
111	C2H4O	ACETALDEHYDE	-1.4826	0.9821	461.00	150	438	0.1766	0.2447	0.0329
112	C2H4O	ETHYLENE OXIDE	-1.4656	0.8777	469.15	161	446	0.1557	0.2026	0.0342
113	C2H4OS	THIOACETIC-ACID	-1.0457	0.1872	577.34	150	548	0.1278	0.1334	0.0900
114	C2H4O2	ACETIC ACID	-1.2836	0.5893	592.71	290	563	0.1581	0.1569	0.0520
115	C2H4O2	METHYL FORMATE	-1.4729	0.9778	487.20	174	463	0.1876	0.2408	0.0337
116	C2H4S	THIACYCLOPROPANE	-1.0471	0.2475	555.00	165	527	0.1417	0.1497	0.0897
117	C2H5Br	BROMOETHANE	-1.7584	1.0024	503.80	155	479	0.1041	0.1374	0.0174
118	C2H5Cl	ETHYL CHLORIDE	-2.0001	1.4496	460.35	137	437	0.1191	0.2003	0.0100
119	C2H5ClO	2-CHLOROETHANOL	-1.6860	0.9969	585.00	206	556	0.1340	0.1540	0.0206
120	C2H5F	ETHYL FLUORIDE	-1.5942	0.9970	375.31	130	357	0.1097	0.1913	0.0255
121	C2H5I	ETHYL IODIDE	-1.8611	1.0025	561.00	162	533	0.0883	0.1104	0.0138
122	C2H5N	ETHYLENEIMINE	-1.5062	0.9900	537.00	195	510	0.1902	0.2272	0.0312
123	C2H5NO	ACETAMIDE	-2.1785	1.9059	761.00	354	723	-----	0.2478	0.0066
124	C2H5NO	N-METHYLFORMAMIDE	-0.9309	0.2775	721.00	269	685	0.2029	0.2041	0.1172
125	C2H5NO2	NITROETHANE	-1.5543	0.9394	593.00	184	563	0.1641	0.1926	0.0279
126	C2H5NO3	ETHYL-NITRATE	-------	-------	-------	-----	-----	-----	-----	-----
127	C2H6	ETHANE	-1.3474	0.7003	305.42	90	290	-----	0.1916	0.0449
128	C2H6AlCl	DIMETHYLALUMINUM CHLORIDE	-1.6842	0.9995	619.00	252	588	0.1394	0.1471	0.0207
129	C2H6O	DIMETHYL ETHER	-1.5099	0.9936	400.10	132	380	0.1453	0.2342	0.0309
130	C2H6O	ETHANOL	-1.3172	0.6987	516.25	159	490	0.1694	0.2030	0.0482
131	C2H6OS	DIMETHYL SULFOXIDE	-2.3306	1.7746	726.00	292	690	0.1567	0.1534	0.0047
132	C2H6O2	ETHYLENE GLYCOL	-0.5918	-------	645.00	260	613	0.2560	0.2560	0.2560
133	C2H6O4S	DIMETHYL SULFATE	-1.6460	0.9860	758.00	241	720	0.1617	0.1705	0.0226
134	C2H6S	DIMETHYL SULFIDE	-1.6145	0.9868	503.04	175	478	0.1409	0.1786	0.0243
135	C2H6S	ETHYL MERCAPTAN	-------	-------	-------	-----	-----	-----	-----	-----
136	C2H6S2	DIMETHYL DISULFIDE	-1.6601	0.9837	606.00	188	576	0.1414	0.1654	0.0219
137	C2H7N	DIMETHYLAMINE	-1.2557	0.6069	437.65	181	416	0.1521	0.1819	0.0555
138	C2H7N	ETHYLAMINE	-0.7418	0.0838	456.15	192	433	0.2090	0.2133	0.1812
139	C2H7NO	MONOETHANOLAMINE	-1.3743	1.0185	638.00	284	606	0.2995	0.2991	0.0422
140	C2H8N2	ETHYLENEDIAMINE	-1.4651	1.0149	593.00	284	563	0.2323	0.2319	0.0343
141	C2H8Si	DIMETHYL SILANE	-1.5839	0.9852	402.00	123	382	0.1217	0.1985	0.0261
142	C2N2	CYANOGEN	-1.3833	1.0380	400.15	245	380	0.2085	0.2449	0.0414
143	C3F6	HEXAFLUOROPROPYLENE	-1.8426	0.9978	368.00	117	350	0.0600	0.1111	0.0144
144	C3F6O	HEXAFLUOROACETONE	-1.8604	0.9867	357.14	151	339	0.0536	0.0940	0.0138
145	C3F8	OCTAFLUOROPROPANE	-1.8913	0.9905	345.05	125	328	0.0466	0.0938	0.0128
146	C3H2N2	MALONONITRILE	-0.9680	0.4645	715.00	305	679	-----	0.2653	0.1076
147	C3H3Cl	PROPARGYL CHLORIDE	-1.6852	1.0209	541.00	293	514	0.1339	0.1309	0.0206
148	C3H3N	ACRYLONITRILE	-2.1221	1.7052	535.00	190	508	0.1694	0.2341	0.0075
149	C3H3NO	OXAZOLE	-1.7525	0.9870	554.00	200	526	0.1094	0.1284	0.0177
150	C3H4	METHYLACETYLENE	-1.6189	1.0028	402.39	170	382	0.1156	0.1693	0.0240
151	C3H4	PROPADIENE	-1.6287	0.9897	393.15	137	373	0.1074	0.1736	0.0235
152	C3H4Cl2	2,3-DICHLOROPROPENE	-1.7388	0.9897	577.00	192	548	0.1162	0.1368	0.0182
153	C3H4O	ACROLEIN	-1.5489	0.9977	506.00	185	481	0.1678	0.2087	0.0283
154	C3H4O	PROPARGYL ALCOHOL	-1.5717	0.9908	580.00	221	551	0.1716	0.1924	0.0268
155	C3H4O2	ACRYLIC ACID	-1.6101	0.9742	615.00	287	584	0.1570	0.1559	0.0245
156	C3H4O2	beta-PROPIOLACTONE	-1.5242	0.9924	686.00	240	652	0.2084	0.2219	0.0299

Table 24-1 THERMAL CONDUCTIVITY OF LIQUID - ORGANIC COMPOUNDS (continued)

			$\log_{10} k_{liq} = A + B[1 - T/C]^{2/7}$					$(k_{liq} - W/(m\ K),\ T - K)$		
NO	FORMULA	NAME	A	B	C	TMIN	TMAX	k_{liq} @25 C	k_{liq} @TMIN	k_{liq} @TMAX
157	C3H4O2	VINYL FORMATE	-1.5429	0.9761	498.00	200	473	0.1618	0.1956	0.0286
158	C3H4O3	ETHYLENE CARBONATE	-------	-------	-------	-----	-----	-----	-----	-----
159	C3H4O3	PYRUVIC ACID	-1.6684	1.0168	634.52	287	603	0.1513	0.1503	0.0215
160	C3H5Br	3-BROMO-1-PROPENE	-1.1684	0.2156	540.20	154	513	0.1007	0.1062	0.0679
161	C3H5Cl	2-CHLOROPROPENE	-1.6818	1.0038	478.00	136	454	0.1195	0.1678	0.0208
162	C3H5Cl	3-CHLOROPROPENE	-1.6775	1.0055	514.15	139	488	0.1280	0.1722	0.0210
163	C3H5ClO	alpha-EPICHLOROHYDRIN	-1.7020	0.9936	610.00	216	580	0.1313	0.1472	0.0199
164	C3H5ClO2	METHYL CHLOROACETATE	-1.6783	1.0052	600.00	241	570	0.1405	0.1516	0.0210
165	C3H5ClO2	ETHYL CHLOROFORMATE	-1.5969	1.0052	508.15	192	483	0.1528	0.1874	0.0253
166	C3H5Cl3	1,2,3-TRICHLOROPROPANE	-1.7903	0.9867	652.00	258	619	0.1092	0.1136	0.0162
167	C3H5I	3-IODO-1-PROPENE	-1.2243	0.1967	595.81	174	566	0.0865	0.0897	0.0597
168	C3H5N	PROPIONITRILE	-1.5733	0.9918	564.40	180	536	0.1686	0.2037	0.0267
169	C3H5NO	ACRYLAMIDE	-1.7139	1.0196	710.00	358	675	-----	0.1281	0.0193
170	C3H5NO	HYDRACRYLONITRILE	-1.3772	0.9962	690.00	227	656	0.2953	0.3200	0.0420
171	C3H5NO	LACTONITRILE	-1.4913	1.0030	643.00	233	611	0.2229	0.2416	0.0323
172	C3H5N3O9	NITROGLYCERINE	-------	-------	-------	-----	-----	-----	-----	-----
173	C3H6	CYCLOPROPANE	-1.5958	0.9876	397.91	146	378	0.1173	0.1833	0.0254
174	C3H6	PROPYLENE	-1.4376	0.7718	364.76	88	347	0.1089	0.1872	0.0365
175	C3H6Br2	1,2-DIBROMOPROPANE	-1.2694	0.2067	634.11	218	602	0.0800	0.0817	0.0538
176	C3H6Cl2	1,1-DICHLOROPROPANE	-1.7409	0.9837	560.00	200	532	0.1124	0.1315	0.0182
177	C3H6Cl2	1,2-DICHLOROPROPANE	-2.3301	1.7507	572.00	173	543	0.1226	0.1733	0.0047
178	C3H6Cl2	1,3-DICHLOROPROPANE	-1.7461	1.0000	603.00	174	573	0.1194	0.1431	0.0179
179	C3H6Cl2	2,2-DICHLOROPROPANE	-1.3921	0.5137	539.46	239	512	0.1038	0.1089	0.0405
180	C3H6I2	1,2-DIIODOPROPANE	-1.3090	0.1522	780.49	253	741	0.0666	0.0670	0.0491
181	C3H6O	ACETONE	-1.3857	0.7643	508.20	178	483	0.1615	0.1925	0.0411
182	C3H6O	ALLYL ALCOHOL	-1.5961	0.9943	545.05	144	518	0.1574	0.2041	0.0253
183	C3H6O	METHYL VINYL ETHER	-1.5601	0.9963	437.00	151	415	0.1438	0.2067	0.0275
184	C3H6O	n-PROPIONALDEHYDE	-1.5067	0.9267	496.00	193	471	0.1607	0.1952	0.0311
185	C3H6O	1,2-PROPYLENE OXIDE	-1.5287	1.0014	482.25	161	458	0.1705	0.2271	0.0296
186	C3H6O	1,3-PROPYLENE OXIDE	-1.6834	1.0446	520.00	255	494	0.1366	0.1464	0.0207
187	C3H6O2	ETHYL FORMATE	-1.4474	0.8436	508.40	194	483	0.1615	0.1910	0.0357
188	C3H6O2	METHYL ACETATE	-1.6616	1.0979	506.80	175	481	0.1550	0.2010	0.0218
189	C3H6O2	PROPIONIC ACID	-1.2207	0.4764	604.00	252	574	0.1484	0.1525	0.0602
190	C3H6O2S	3-MERCAPTOPROPIONIC ACID	-1.5614	1.0081	729.00	291	693	0.2023	0.2002	0.0275
191	C3H6O3	LACTIC ACID	-0.3904	0.0057	616.00	291	585	0.4115	0.4114	0.4070
192	C3H6O3	METHOXYACETIC ACID	-1.5621	1.0070	691.00	281	656	0.1972	0.1978	0.0274
193	C3H6O3	TRIOXANE	-1.6457	1.0284	604.00	335	574	-----	0.1429	0.0226
194	C3H6S	THIACYCLOBUTANE	-1.0826	0.2390	603.00	200	573	0.1300	0.1345	0.0827
195	C3H7Br	1-BROMOPROPANE	-1.6989	0.8779	544.00	163	517	0.1002	0.1227	0.0200
196	C3H7Br	2-BROMOPROPANE	-1.8087	0.9932	532.00	184	505	0.0948	0.1158	0.0155
197	C3H7Cl	ISOPROPYL CHLORIDE	-1.6811	0.9975	489.00	156	465	0.1206	0.1609	0.0208
198	C3H7Cl	n-PROPYL CHLORIDE	-1.4818	0.7346	503.15	150	478	0.1221	0.1506	0.0330
199	C3H7F	1-FLUOROPROPANE	-1.0825	0.2855	422.00	114	401	0.1314	0.1503	0.0827
200	C3H7F	2-FLUOROPROPANE	-1.1685	0.3901	415.68	140	395	0.1269	0.1499	0.0678
201	C3H7I	ISOPROPYL IODIDE	-1.8432	0.9338	578.00	183	549	0.0824	0.0974	0.0143
202	C3H7I	n-PROPYL IODIDE	-1.8930	1.0239	593.00	172	563	0.0882	0.1071	0.0128
203	C3H7N	ALLYLAMINE	-1.1389	0.5366	505.00	185	480	0.1892	0.2128	0.0726
204	C3H7N	PROPYLENEIMINE	-1.6207	1.0022	529.00	229	503	0.1479	0.1667	0.0239
205	C3H7NO	N,N-DIMETHYLFORMAMIDE	-1.4326	0.8321	647.00	213	615	0.1840	0.2015	0.0369
206	C3H7NO	N-METHYLACETAMIDE	-1.6399	1.0055	718.00	301	682	-----	0.1627	0.0229
207	C3H7NO2	1-NITROPROPANE	-1.5785	0.9348	605.00	169	575	0.1554	0.1853	0.0264
208	C3H7NO2	2-NITROPROPANE	-1.5223	0.8227	594.00	182	564	0.1419	0.1636	0.0300
209	C3H7NO3	PROPYL-NITRATE	-------	-------	-------	-----	-----	-----	-----	-----
210	C3H7NO3	ISOPROPYL-NITRATE	-------	-------	-------	-----	-----	-----	-----	-----
211	C3H8	PROPANE	-1.2127	0.6611	369.82	85	351	0.1588	0.2501	0.0613
212	C3H8O	ISOPROPANOL	-1.3721	0.6580	508.31	185	483	0.1378	0.1589	0.0425
213	C3H8O	METHYL ETHYL ETHER	-1.5603	0.9786	437.80	160	416	0.1399	0.1957	0.0275
214	C3H8O	n-PROPANOL	-1.2131	0.5097	536.71	147	510	0.1553	0.1776	0.0612
215	C3H8O2	2-METHOXYETHANOL	-1.0528	0.4122	564.00	188	536	0.1904	0.2049	0.0886
216	C3H8O2	METHYLAL	-1.5966	0.9909	480.60	168	457	0.1428	0.1874	0.0253
217	C3H8O2	1,2-PROPYLENE GLYCOL	-0.8118	0.1372	626.00	213	595	0.2006	0.2037	0.1542
218	C3H8O2	1,3-PROPYLENE GLYCOL	-0.4653	-0.2405	658.00	250	390	0.2149	0.2266	0.3425
219	C3H8O3	GLYCEROL	-0.3550	-0.2097	723.00	293	550	0.2916	0.2994	0.4416
220	C3H8S	n-PROPYLMERCAPTAN	-1.6489	0.9858	536.00	160	509	0.1357	0.1722	0.0224
221	C3H8S	ISOPROPYL MERCAPTAN	-1.6600	0.9791	517.00	143	491	0.1276	0.1688	0.0219
222	C3H8S	ETHYL-METHYL-SULFIDE	-1.0885	0.2416	532.80	167	506	0.1267	0.1340	0.0816
223	C3H9N	n-PROPYLAMINE	-1.1416	0.4932	496.95	190	472	0.1730	0.1924	0.0722
224	C3H9N	ISOPROPYLAMINE	-1.5363	0.9005	471.85	178	448	0.1382	0.1749	0.0291
225	C3H9N	TRIMETHYLAMINE	-1.6174	0.9865	433.25	156	412	0.1230	0.1753	0.0241
226	C3H9NO	1-AMINO-2-PROPANOL	-1.5902	1.0148	614.00	275	583	0.1774	0.1801	0.0257
227	C3H9NO	3-AMINO-1-PROPANOL	-1.5366	1.0157	649.00	284	617	0.2067	0.2065	0.0291
228	C3H9NO	METHYLETHANOLAMINE	-1.5791	1.0121	630.00	269	599	0.1835	0.1881	0.0264
229	C3H9O4P	TRIMETHYL PHOSPHATE	-------	-------	-------	-----	-----	-----	-----	-----
230	C3H10N2	1,2-PROPANEDIAMINE	-1.5008	1.0033	587.00	237	558	0.2082	0.2269	0.0316
231	C3H10Si	TRIMETHYL SILANE	-1.6641	0.9872	432.00	137	410	0.1102	0.1640	0.0217
232	C4Cl4S	TETRACHLOROTHIOPHENE	-------	-------	-------	-----	-----	-----	-----	-----
233	C4Cl6	HEXACHLORO-1,3-BUTADIENE	-1.9958	0.9923	741.00	252	704	0.0726	0.0756	0.0101
234	C4F8	OCTAFLUORO-2-BUTENE	-1.8993	0.9952	392.00	138	372	0.0578	0.0939	0.0126

Table 24-1 THERMAL CONDUCTIVITY OF LIQUID - ORGANIC COMPOUNDS (continued)

			$\log_{10} k_{liq} = A + B [1 - T/C]^{2/7}$					$(k_{liq} - W/(m K), T - K)$		
NO	FORMULA	NAME	A	B	C	TMIN	TMAX	k_{liq} @25 C	k_{liq} @TMIN	k_{liq} @TMAX
235	C4F8	OCTAFLUOROCYCLOBUTANE	-1.9271	1.1242	388.37	233	369	0.0651	0.0828	0.0118
236	C4F10	DECAFLUOROBUTANE	-1.9336	0.9892	386.35	145	367	0.0519	0.0838	0.0117
237	C4H2	BUTADIYNE(BIACETYLENE)	-1.8537	1.2964	478.02	237	454	0.1339	0.1571	0.0140
238	C4H2O3	MALEIC ANHYDRIDE	-1.6158	1.0110	721.00	326	685	-----	0.1678	0.0242
239	C4H4	VINYLACETYLENE	-------	------	------	-----	-----	-----	-----	-----
240	C4H4N2	SUCCINONITRILE	-1.4849	1.0114	770.00	331	732	-----	0.2327	0.0327
241	C4H4O	FURAN	-1.6602	0.9948	490.15	188	466	0.1262	0.1578	0.0219
242	C4H4O2	DIKETENE	-1.6106	1.0085	616.00	267	585	0.1676	0.1725	0.0245
243	C4H4O3	SUCCINIC ANHYDRIDE	-1.5742	1.0180	811.00	393	770	-----	0.1803	0.0267
244	C4H4O4	FUMARIC ACID	-1.6471	0.8844	771.00	560	732	-----	0.0866	0.0225
245	C4H4O4	MALEIC ACID	-1.5672	0.9720	773.00	403	734	-----	0.1611	0.0271
246	C4H4S	THIOPHENE	-1.4069	0.7030	579.35	235	550	0.1462	0.1558	0.0392
247	C4H5Cl	CHLOROPRENE	-2.5064	2.0553	525.00	143	499	0.1291	0.2290	0.0031
248	C4H5N	trans-CROTONITRILE	-1.6587	1.0017	586.00	222	557	0.1442	0.1613	0.0219
249	C4H5N	cis-CROTONITRILE	-1.6238	1.0033	568.00	201	540	0.1539	0.1797	0.0238
250	C4H5N	METHACRYLONITRILE	-1.5773	1.0053	554.00	237	526	0.1694	0.1862	0.0265
251	C4H5N	PYRROLE	-1.6178	0.9977	639.75	250	608	0.1645	0.1737	0.0241
252	C4H5N	VINYLACETONITRILE	-1.6149	0.9895	584.00	186	555	0.1556	0.1844	0.0243
253	C4H5NO2	METHYL CYANOACETATE	-1.5042	0.9878	687.00	260	653	0.2164	0.2240	0.0313
254	C4H6	CYCLOBUTENE	-1.1401	0.3900	446.33	154	424	0.1395	0.1595	0.0724
255	C4H6	1,2-BUTADIENE	-1.6517	0.9876	444.00	137	422	0.1166	0.1703	0.0223
256	C4H6	1,3-BUTADIENE	-1.6512	0.9899	425.37	164	404	0.1122	0.1592	0.0223
257	C4H6	DIMETHYLACETYLENE	-1.6900	1.0114	488.15	241	464	0.1209	0.1350	0.0204
258	C4H6	ETHYLACETYLENE	-1.6463	0.9884	443.20	147	421	0.1180	0.1690	0.0226
259	C4H6Cl2	1,3-DICHLORO-trans-2-BUTENE	-1.7525	0.9775	618.00	276	587	0.1141	0.1155	0.0177
260	C4H6Cl2	1,4-DICHLORO-cis-2-BUTENE	-1.7539	0.9849	640.00	225	608	0.1173	0.1286	0.0176
261	C4H6Cl2	1,4-DICHLORO-trans-2-BUTENE	-1.7509	0.9805	646.00	274	614	0.1177	0.1195	0.0177
262	C4H6Cl2	3,4-DICHLORO-1-BUTENE	-1.7639	0.9906	589.00	212	560	0.1111	0.1261	0.0172
263	C4H6O	trans-CROTONALDEHYDE	-1.9879	1.6110	571.00	197	542	0.2073	0.2679	0.0103
264	C4H6O	2,5-DIHYDROFURAN	-1.6927	1.0187	542.00	273	515	0.1313	0.1344	0.0203
265	C4H6O	DIVINYL ETHER	-1.5983	0.9879	463.00	172	440	0.1371	0.1816	0.0252
266	C4H6O	METHACROLEIN	-1.6470	0.9896	530.00	192	504	0.1363	0.1644	0.0225
267	C4H6O2	2-BUTYNE-1,4-DIOL	-1.5373	1.0296	695.00	331	660	-----	0.2026	0.0290
268	C4H6O2	gamma-BUTYROLACTONE	-1.6401	0.9843	739.00	230	702	0.1619	0.1732	0.0229
269	C4H6O2	cis-CROTONIC ACID	-1.6538	1.0211	647.00	289	615	0.1593	0.1577	0.0222
270	C4H6O2	trans-CROTONIC ACID	-1.5922	0.9597	666.00	345	633	-----	0.1493	0.0256
271	C4H6O2	METHACRYLIC ACID	-1.6272	0.9747	643.00	288	611	0.1544	0.1532	0.0236
272	C4H6O2	METHYL ACRYLATE	-1.5867	0.9925	536.00	196	509	0.1586	0.1893	0.0259
273	C4H6O2	VINYL ACETATE	-1.7519	1.1895	524.00	180	498	0.1525	0.1970	0.0177
274	C4H6O3	ACETIC ANHYDRIDE	-1.3593	0.7106	569.15	200	541	0.1643	0.1834	0.0437
275	C4H6O4	SUCCINIC ACID	-1.5555	0.9783	806.00	461	766	-----	0.1572	0.0278
276	C4H6O5	DIGLYCOLIC ACID	-1.5871	1.0334	820.00	421	779	-----	0.1740	0.0259
277	C4H6O5	MALIC ACID	-1.5921	1.0366	781.00	403	742	-----	0.1724	0.0256
278	C4H6O6	TARTARIC ACID	-1.5788	1.0606	828.00	479	787	-----	0.1708	0.0264
279	C4H7N	n-BUTYRONITRILE	-1.5713	0.9796	582.25	161	553	0.1685	0.2073	0.0268
280	C4H7N	ISOBUTYRONITRILE	-1.5949	0.9952	565.00	202	537	0.1616	0.1883	0.0254
281	C4H7NO	ACETONE CYANOHYDRIN	-1.4284	0.9965	647.00	253	615	0.2552	0.2680	0.0373
282	C4H7NO	2-METHACRYLAMIDE	-1.6405	1.0253	741.00	384	704	-----	0.1507	0.0229
283	C4H7NO	3-METHOXYPROPIONITRILE	-1.6418	0.9917	638.00	210	606	0.1537	0.1723	0.0228
284	C4H7NO	2-PYRROLIDONE	-1.5852	1.0000	792.00	298	752	0.1943	0.1908	0.0260
285	C4H8	1-BUTENE	-1.6539	0.9786	419.59	88	399	0.1078	0.1809	0.0222
286	C4H8	cis-2-BUTENE	-1.6584	0.9867	435.58	134	414	0.1125	0.1675	0.0220
287	C4H8	trans-2-BUTENE	-1.6736	0.9990	428.63	168	407	0.1090	0.1529	0.0212
288	C4H8	CYCLOBUTANE	-1.6584	0.9979	459.93	182	437	0.1208	0.1575	0.0220
289	C4H8	ISOBUTENE	-1.4902	0.8491	417.90	133	397	0.1270	0.1843	0.0323
290	C4H8Br2	1,2-DIBROMOBUTANE	-1.2556	0.1727	659.28	208	626	0.0776	0.0791	0.0555
291	C4H8Br2	2,3-DIBROMOBUTANE	-1.2832	0.2074	656.96	239	624	0.0779	0.0790	0.0521
292	C4H8Cl2	1,4-DICHLOROBUTANE	-1.7598	0.9887	641.00	236	609	0.1167	0.1258	0.0174
293	C4H8I2	1,2-DIIODOBUTANE	-1.3561	0.2007	726.41	279	690	0.0655	0.0656	0.0440
294	C4H8O	n-BUTYRALDEHYDE	-1.4504	0.7790	525.00	177	499	0.1454	0.1726	0.0354
295	C4H8O	ISOBUTYRALDEHYDE	-2.4296	2.0609	507.00	208	482	0.1479	0.2109	0.0037
296	C4H8O	1,2-EPOXYBUTANE	-1.5694	0.9987	526.00	123	500	0.1648	0.2249	0.0270
297	C4H8O	METHYL ETHYL KETONE	-1.4647	0.7938	535.50	186	509	0.1460	0.1708	0.0343
298	C4H8O	ETHYL VINYL ETHER	-1.6073	0.9976	475.15	157	451	0.1397	0.1886	0.0247
299	C4H8O	TETRAHYDROFURAN	-1.7003	0.9827	540.15	165	513	0.1205	0.1511	0.0199
300	C4H8O2	cis-2-BUTENE-1,4-DIOL	-1.5334	0.9829	677.88	284	644	0.1993	0.1991	0.0293
301	C4H8O2	trans-2-BUTENE-1,4-DIOL	-1.5265	0.9756	681.00	300	647	-----	0.1950	0.0298
302	C4H8O2	ISOBUTYRIC ACID	-1.6358	0.9707	609.15	227	579	0.1463	0.1608	0.0231
303	C4H8O2	n-BUTYRIC ACID	-1.3420	0.6161	628.00	268	597	0.1481	0.1505	0.0455
304	C4H8O2	1,4-DIOXANE	-2.1607	1.6668	587.00	285	558	0.1586	0.1578	0.0069
305	C4H8O2	ETHYL ACETATE	-1.6938	1.0862	523.30	190	497	0.1445	0.1789	0.0202
306	C4H8O2	METHYL PROPIONATE	-1.4504	0.7831	530.60	186	504	0.1473	0.1722	0.0354
307	C4H8O2	n-PROPYL FORMATE	-1.4604	0.7949	538.00	180	511	0.1481	0.1745	0.0346
308	C4H8O2S	SULFOLANE	-1.4179	0.8122	849.00	301	807	-----	0.1963	0.0382
309	C4H8S	TETRAHYDROTHIOPHENE	-1.6093	0.9128	631.95	177	600	0.1417	0.1647	0.0246
310	C4H9Br	1-BROMOBUTANE	-1.7651	0.9645	577.00	161	548	0.1043	0.1283	0.0172
311	C4H9Br	2-BROMOBUTANE	-1.7934	1.0012	567.00	161	539	0.1036	0.1292	0.0161
312	C4H9Cl	n-BUTYL CHLORIDE	-1.7147	1.0040	537.00	150	510	0.1207	0.1564	0.0193

Table 24-1 THERMAL CONDUCTIVITY OF LIQUID - ORGANIC COMPOUNDS (continued)

NO	FORMULA	NAME	$\log_{10} k_{liq} = A + B [1 - T/C]^{2/7}$					$(k_{liq} - W/(m\ K),\ T - K)$		
			A	B	C	TMIN	TMAX	k_{liq} @25 C	k_{liq} @TMIN	k_{liq} @TMAX
313	C4H9Cl	sec-BUTYL CHLORIDE	-1.7136	1.0008	520.60	142	495	0.1179	0.1567	0.0193
314	C4H9Cl	tert-BUTYL CHLORIDE	-1.6946	0.9670	507.00	248	482	0.1137	0.1236	0.0202
315	C4H9I	2-IODO-2-METHYLPROPANE	-1.3411	0.3191	587.90	235	559	0.0831	0.0855	0.0456
316	C4H9N	PYRROLIDINE	-1.2920	0.6432	568.55	215	540	0.1691	0.1838	0.0511
317	C4H9NO	N,N-DIMETHYLACETAMIDE	-2.1127	1.5857	658.00	253	625	0.1667	0.1797	0.0077
318	C4H9NO	MORPHOLINE	-1.6196	1.0089	618.00	270	587	0.1645	0.1684	0.0240
319	C4H9NO2	1-NITROBUTANE	-1.0965	0.1760	624.00	192	593	0.1121	0.1150	0.0801
320	C4H9NO2	2-NITROBUTANE	-1.0771	0.1493	615.00	141	584	0.1113	0.1150	0.0837
321	C4H10	n-BUTANE	-1.8929	1.2885	425.18	135	404	0.1046	0.1796	0.0128
322	C4H10	ISOBUTANE	-1.6862	0.9802	408.14	114	388	0.0972	0.1589	0.0206
323	C4H10N2	PIPERAZINE	-1.7185	1.0393	638.00	379	606	-----	0.1166	0.0191
324	C4H10O	n-BUTANOL	-1.3120	0.6190	562.93	184	535	0.1538	0.1725	0.0488
325	C4H10O	sec-BUTANOL	-1.4633	0.7473	536.01	158	509	0.1346	0.1617	0.0344
326	C4H10O	tert-BUTANOL	-1.2018	0.3521	506.20	299	481	-----	0.1161	0.0628
327	C4H10O	DIETHYL ETHER	-1.5629	0.9357	466.70	157	443	0.1369	0.1832	0.0274
328	C4H10O	METHYL-PROPYL-ETHER	-1.1184	0.2872	476.20	157	452	0.1254	0.1367	0.0761
329	C4H10O	METHYL ISOPROPYL ETHER	-1.6271	1.0013	464.50	128	441	0.1317	0.1909	0.0236
330	C4H10O	ISOBUTANOL	-1.3936	0.6487	547.73	165	520	0.1332	0.1542	0.0404
331	C4H10O2	1,3-BUTANEDIOL	-1.5828	1.0140	643.00	196	611	0.1844	0.2114	0.0261
332	C4H10O2	1,4-BUTANEDIOL	-0.7958	0.1404	667.00	293	634	0.2103	0.2098	0.1600
333	C4H10O2	2,3-BUTANEDIOL	-1.5826	0.9730	611.00	281	580	0.1664	0.1670	0.0261
334	C4H10O2	t-BUTYL HYDROPEROXIDE	-1.7927	1.0205	576.00	277	547	0.1086	0.1100	0.0161
335	C4H10O2	1,2-DIMETHOXYETHANE	-1.6301	0.9871	536.15	215	509	0.1421	0.1636	0.0234
336	C4H10O2	2-ETHOXYETHANOL	-1.2079	0.5565	569.00	210	541	0.1747	0.1887	0.0620
337	C4H10O3	DIETHYLENE GLYCOL	-0.4818	-0.2372	744.60	263	400	0.2057	0.2206	0.3298
338	C4H10O4S	DIETHYL SULFATE	-------	-------	-------	-----	-----	-----	-----	-----
339	C4H10S	n-BUTYL MERCAPTAN	-1.6638	0.9832	569.00	157	541	0.1354	0.1689	0.0217
340	C4H10S	ISOBUTYL MERCAPTAN	-1.6629	0.9755	559.00	128	531	0.1323	0.1733	0.0217
341	C4H10S	sec-BUTYL MERCAPTAN	-1.6637	0.9747	554.00	133	526	0.1312	0.1710	0.0217
342	C4H10S	tert-BUTYL MERCAPTAN	-1.7045	1.0228	530.00	274	504	0.1268	0.1297	0.0197
343	C4H10S	DIETHYL SULFIDE	-1.6613	0.9782	557.15	169	529	0.1332	0.1641	0.0218
344	C4H10S	ISOPROPYL-METHYL-SULFIDE	-1.1085	0.2226	551.00	172	523	0.1174	0.1231	0.0779
345	C4H10S	METHYL-PROPYL-SULFIDE	-1.0868	0.1962	563.00	160	535	0.1179	0.1232	0.0819
346	C4H10S2	DIETHYL DISULFIDE	-1.6821	0.9805	642.00	172	610	0.1375	0.1621	0.0208
347	C4H11N	n-BUTYLAMINE	-1.5519	0.9377	531.90	224	505	0.1547	0.1743	0.0281
348	C4H11N	ISOBUTYLAMINE	-1.5874	0.9991	513.73	189	488	0.1557	0.1911	0.0259
349	C4H11N	sec-BUTYLAMINE	-1.6162	0.9862	514.30	169	489	0.1424	0.1809	0.0242
350	C4H11N	tert-BUTYLAMINE	-1.5830	1.0085	483.90	206	460	0.1528	0.1854	0.0261
351	C4H11N	DIETHYLAMINE	-1.5768	0.9190	496.60	223	472	0.1350	0.1544	0.0265
352	C4H11NO	DIMETHYLETHANOLAMINE	-1.5964	1.0056	571.82	214	543	0.1653	0.1884	0.0253
353	C4H11NO2	DIETHANOLAMINE	-1.6192	1.0212	715.00	301	679	-----	0.1756	0.0240
354	C4H11NO2	2-AMINOETHOXYETHANOL	-1.5931	1.0189	699.00	293	664	0.1889	0.1861	0.0255
355	C4H12N2O	N-AMINOETHYL ETHANOLAMINE	-1.5861	1.0126	698.00	273	663	0.1894	0.1923	0.0259
356	C4H12Si	TETRAMETHYLSILANE	-1.6714	0.9974	450.40	174	428	0.1149	0.1541	0.0213
357	C4H13N3	DIETHYLENE TRIAMINE	-1.6950	0.9997	676.00	234	642	0.1418	0.1525	0.0202
358	C5Cl6	HEXACHLOROCYCLOPENTADIENE	-2.0979	1.0017	746.00	284	709	0.0586	0.0586	0.0080
359	C5H4O2	FURFURAL	-1.3650	0.7132	657.00	237	624	0.1718	0.1808	0.0432
360	C5H5N	PYRIDINE	-1.2083	0.5146	619.95	232	589	0.1653	0.1729	0.0619
361	C5H6	CYCLOPENTADIENE	-1.6475	0.9973	507.00	188	482	0.1338	0.1654	0.0225
362	C5H6	2-METHYL-1-BUTENE-3-YNE	-1.6815	0.9844	492.00	160	467	0.1183	0.1555	0.0208
363	C5H6	1-PENTENE-3-YNE	-1.6450	0.9830	520.00	150	494	0.1336	0.1744	0.0226
364	C5H6	1-PENTENE-4-YNE	-1.6388	0.9803	503.00	150	478	0.1317	0.1744	0.0230
365	C5H6N2	GLUTARONITRILE	-1.5769	0.9950	782.00	244	743	0.1952	0.2047	0.0265
366	C5H6O2	FURFURYL ALCOHOL	-1.5466	0.9860	632.00	259	600	0.1884	0.1961	0.0284
367	C5H6O3	GLUTARIC ANHYDRIDE	-1.6387	1.0022	838.00	328	796	-----	0.1669	0.0230
368	C5H6O4	CITRACONIC ACID	-1.5583	0.9722	829.00	356	788	-----	0.1822	0.0277
369	C5H6O4	ITACONIC ACID	-1.5617	0.9687	821.00	439	780	-----	0.1596	0.0274
370	C5H6S	2-METHYLTHIOPHENE	-1.1341	0.2305	610.00	210	580	0.1138	0.1171	0.0734
371	C5H6S	3-METHYLTHIOPHENE	-1.1236	0.2174	615.00	204	584	0.1138	0.1171	0.0752
372	C5H7N	N-METHYLPYRROLE	-1.6575	0.9879	610.00	217	580	0.1439	0.1609	0.0220
373	C5H7NO2	ETHYL CYANOACETATE	-1.6092	1.0012	679.00	251	645	0.1736	0.1822	0.0246
374	C5H8	CYCLOPENTENE	-1.6455	0.9742	507.00	138	482	0.1290	0.1735	0.0226
375	C5H8	ISOPRENE	-1.7942	1.1542	484.00	127	460	0.1213	0.1812	0.0161
376	C5H8	3-METHYL-1,2-BUTADIENE	-1.6967	0.9872	490.00	160	466	0.1144	0.1509	0.0201
377	C5H8	2-METHYL-1,3-BUTADIENE	-1.0588	0.2330	483.30	127	459	0.1313	0.1424	0.0873
378	C5H8	1,2-PENTADIENE	-1.6591	0.9799	500.00	136	475	0.1250	0.1701	0.0219
379	C5H8	cis-1,3-PENTADIENE	-1.6566	0.9769	499.00	132	474	0.1249	0.1711	0.0220
380	C5H8	trans-1,3-PENTADIENE	-1.6774	0.9895	500.00	186	475	0.1220	0.1518	0.0210
381	C5H8	1,4-PENTADIENE	-1.6611	0.9748	479.00	125	455	0.1194	0.1691	0.0218
382	C5H8	2,3-PENTADIENE	-1.6589	0.9840	497.00	148	472	0.1255	0.1678	0.0219
383	C5H8	1-PENTYNE	-1.6530	0.9898	481.20	167	457	0.1253	0.1645	0.0222
384	C5H8	2-PENTYNE	-1.0755	0.2541	521.99	164	496	0.1330	0.1416	0.0840
385	C5H8	3-METHYL-1-BUTYNE	-1.6968	0.9963	463.20	183	440	0.1109	0.1438	0.0201
386	C5H8	SPIROPENTANE	-1.1127	0.2986	499.74	166	475	0.1311	0.1417	0.0771
387	C5H8N4O12	PENTAERYTHRITOL TETRANITRATE	-----	-----	-----	-----	-----	-----	-----	-----
388	C5H8O	CYCLOPENTANONE	-1.5890	0.9910	626.00	222	595	0.1717	0.1898	0.0258
389	C5H8O	METHYL ISOPROPENYL KETONE	-1.6154	0.9767	566.00	220	538	0.1490	0.1679	0.0242
390	C5H8O2	ACETYLACETONE	-1.5987	0.9872	602.00	250	572	0.1634	0.1734	0.0252

Table 24-1 THERMAL CONDUCTIVITY OF LIQUID - ORGANIC COMPOUNDS (continued)

NO	FORMULA	NAME	$\log_{10} k_{liq} = A + B [1 - T/C]^{2/7}$					$(k_{liq} - W/(m\ K),\ T - K)$		
			A	B	C	TMIN	TMAX	k_{liq} @25 C	k_{liq} @TMIN	k_{liq} @TMAX
391	C5H8O2	ALLYL ACETATE	-1.6316	1.0083	559.00	138	531	0.1511	0.1966	0.0234
392	C5H8O2	ETHYL ACRYLATE	-1.6195	0.9830	553.00	202	525	0.1473	0.1722	0.0240
393	C5H8O2	METHYL METHACRYLATE	-1.8863	1.3001	564.00	225	536	0.1453	0.1685	0.0130
394	C5H8O2	VINYL PROPIONATE	-1.6247	0.9733	546.00	250	519	0.1419	0.1520	0.0237
395	C5H8O3	2-HYDROXYETHYL ACRYLATE	-1.6645	0.9995	662.00	213	629	0.1506	0.1674	0.0217
396	C5H8O3	LEVULINIC ACID	-1.6097	0.9783	723.00	308	687	-----	0.1643	0.0246
397	C5H8O3	METHYL ACETOACETATE	-2.0207	1.4479	642.00	193	610	0.1551	0.1897	0.0095
398	C5H8O4	GLUTARIC ACID	-1.5710	0.9737	807.00	371	767	-----	0.1718	0.0269
399	C5H9N	VALERONITRILE	-1.5914	0.9869	603.00	177	573	0.1662	0.1980	0.0256
400	C5H9NO	n-BUTYL ISOCYANATE	-1.6928	0.9947	568.00	193	540	0.1292	0.1527	0.0203
401	C5H9NO	N-METHYL-2-PYRROLIDONE	-1.1284	0.4817	724.00	249	688	0.1930	0.1974	0.0744
402	C5H9NO4	L-GLUTAMIC ACID	-------	-------	-------	-----	-----	-----	-----	-----
403	C5H10	CYCLOPENTANE	-1.6175	0.9240	511.76	179	486	0.1266	0.1559	0.0241
404	C5H10	2-METHYL-1-BUTENE	-1.7016	0.9847	465.00	136	442	0.1079	0.1531	0.0199
405	C5H10	2-METHYL-2-BUTENE	-1.7011	0.9879	471.00	139	447	0.1098	0.1538	0.0199
406	C5H10	3-METHYL-1-BUTENE	-1.7016	0.9808	450.37	105	428	0.1042	0.1597	0.0199
407	C5H10	1-PENTENE	-1.6663	0.9787	464.78	108	442	0.1158	0.1726	0.0216
408	C5H10	cis-2-PENTENE	-1.2021	0.4769	475.93	122	452	0.1438	0.1713	0.0628
409	C5H10	trans-2-PENTENE	-1.6694	0.9777	475.37	133	452	0.1170	0.1643	0.0214
410	C5H10Br2	2,3-DIBROMO-2-METHYLBUTANE	-1.3483	0.2744	668.37	288	635	0.0765	0.0763	0.0448
411	C5H10Cl2	1,5-DICHLOROPENTANE	-1.7054	0.8846	663.00	200	630	0.1098	0.1224	0.0197
412	C5H10O	METHYL ISOPROPYL KETONE	-1.6608	1.0197	553.00	181	525	0.1434	0.1749	0.0218
413	C5H10O	2-PENTANONE	-1.5427	0.8664	561.08	196	533	0.1429	0.1649	0.0287
414	C5H10O	DIETHYL KETONE	-1.5868	0.9252	560.95	234	533	0.1439	0.1576	0.0259
415	C5H10O	VALERALDEHYDE	-1.3874	0.6935	554.00	182	526	0.1475	0.1686	0.0410
416	C5H10O2	n-BUTYL FORMATE	-1.4694	0.7764	559.00	181	531	0.1429	0.1659	0.0339
417	C5H10O2	ETHYL PROPIONATE	-1.4559	0.7611	546.00	199	519	0.1417	0.1611	0.0350
418	C5H10O2	ISOBUTYL FORMATE	-1.6401	0.9975	551.35	177	524	0.1441	0.1764	0.0229
419	C5H10O2	ISOPROPYL ACETATE	-1.6674	1.0117	538.00	200	511	0.1367	0.1624	0.0215
420	C5H10O2	n-PROPYL ACETATE	-1.6554	1.0143	549.40	178	522	0.1431	0.1758	0.0221
421	C5H10O2	METHYL n-BUTYRATE	-1.6416	0.9912	554.50	187	527	0.1424	0.1710	0.0228
422	C5H10O2	2-METHYLBUTYRIC ACID	-1.6505	0.9761	643.00	357	611	-----	0.1286	0.0224
423	C5H10O2	ISOVALERIC ACID	-1.8111	1.1217	634.00	244	602	0.1332	0.1432	0.0154
424	C5H10O2	VALERIC ACID	-1.3306	0.5788	651.00	239	618	0.1430	0.1488	0.0467
425	C5H10O2	TETRAHYDROFURFURYL ALCOHOL	-1.6621	0.9895	639.00	193	607	0.1461	0.1678	0.0218
426	C5H10O2S	3-METHYL SULFOLANE	-1.6930	0.9946	817.00	274	776	0.1516	0.1532	0.0203
427	C5H10O3	DIETHYL CARBONATE	-------	-------	-------	-----	-----	-----	-----	-----
428	C5H10O3	ETHYL LACTATE	-1.5767	1.0176	588.00	247	559	0.1798	0.1926	0.0265
429	C5H10S	THIACYCLOHEXANE	-1.2567	0.3748	657.12	292	624	0.1145	0.1138	0.0554
430	C5H10S	CYCLOPENTANETHIOL	-1.0828	0.1565	629.00	155	598	0.1115	0.1150	0.0826
431	C5H11Br	1-BROMOPENTANE	-1.1971	0.2023	564.76	185	537	0.0925	0.0960	0.0635
432	C5H11Cl	1-CHLOROPENTANE	-1.7334	0.9954	568.00	174	540	0.1178	0.1437	0.0185
433	C5H11Cl	1-CHLORO-3-METHYLBUTANE	-1.1290	0.2050	558.87	169	531	0.1086	0.1134	0.0743
434	C5H11Cl	2-CHLORO-2-METHYLBUTANE	-1.1829	0.2715	548.97	200	522	0.1082	0.1131	0.0656
435	C5H11N	N-METHYLPYRROLIDINE	-1.7236	0.9865	550.00	183	523	0.1163	0.1408	0.0189
436	C5H11N	PIPERIDINE	-1.5720	1.0072	594.05	263	564	0.1792	0.1861	0.0268
437	C5H11NO	tert-BUTYLFORMAMIDE	-1.6969	1.0088	692.00	289	657	0.1452	0.1439	0.0201
438	C5H12	ISOPENTANE	-1.6824	0.9955	460.43	113	437	0.1139	0.1704	0.0208
439	C5H12	NEOPENTANE	-1.7534	1.0306	433.78	257	412	0.0968	0.1062	0.0176
440	C5H12	n-PENTANE	-1.2287	0.5322	469.65	143	446	0.1480	0.1769	0.0591
441	C5H12O	2,2-DIMETHYL-1-PROPANOL	-1.6332	0.9395	550.00	327	523	-----	0.1193	0.0233
442	C5H12O	tert-PENTYL-ALCOHOL	-3.1715	2.9426	549.00	327	522	-----	0.1121	0.0007
443	C5H12O	2-METHYL-1-BUTANOL	-1.6474	0.9911	565.00	203	537	0.1421	0.1652	0.0225
444	C5H12O	2-METHYL-2-BUTANOL	-1.6654	0.9673	545.15	264	518	0.1277	0.1330	0.0216
445	C5H12O	3-METHYL-1-BUTANOL	-0.8268	-------	579.45	156	550	0.1490	0.1490	0.1490
446	C5H12O	3-METHYL-2-BUTANOL	-1.6753	0.9967	574.00	188	545	0.1359	0.1614	0.0211
447	C5H12O	1-PENTANOL	-1.2628	0.5481	586.15	196	557	0.1530	0.1665	0.0546
448	C5H12O	2-PENTANOL	-1.6591	0.9915	552.00	200	524	0.1365	0.1604	0.0219
449	C5H12O	3-PENTANOL	-1.6616	0.9893	547.00	204	520	0.1344	0.1572	0.0218
450	C5H12O	METHYL sec-BUTYL ETHER	-1.6565	0.9938	498.00	150	473	0.1286	0.1717	0.0221
451	C5H12O	METHYL tert-BUTYL ETHER	-1.3554	0.5475	497.10	165	472	0.1164	0.1345	0.0441
452	C5H12O	METHYL ISOBUTYL ETHER	-1.6565	0.9932	497.00	150	472	0.1282	0.1714	0.0221
453	C5H12O	ETHYL PROPYL ETHER	-1.6520	0.9987	500.23	146	475	0.1315	0.1767	0.0223
454	C5H12O2	ETHYLENE GLYCOL MONOPROPYL ETHER	-1.7081	0.9992	582.00	183	553	0.1276	0.1523	0.0196
455	C5H12O2	NEOPENTYL GLYCOL	-1.6595	1.0611	643.00	400	611	-----	0.1329	0.0219
456	C5H12O2	1,5-PENTANEDIOL	-1.5818	1.0119	673.00	257	639	0.1881	0.1958	0.0262
457	C5H12O3	2-(2-METHOXYETHOXY)ETHANOL	-1.0049	0.3242	630.00	197	599	0.1841	0.1925	0.0989
458	C5H12O4	PENTAERYTHRITOL	-1.4744	0.9034	780.00	534	741	-----	0.1423	0.0335
459	C5H12S	n-PENTYL MERCAPTAN	-1.6804	0.9923	598.00	197	568	0.1362	0.1578	0.0209
460	C5H12S	BUTYL-METHYL-SULFIDE	-1.1081	0.1860	591.00	175	561	0.1107	0.1146	0.0780
461	C5H12S	ETHYL-PROPYL-SULFIDE	-1.0992	0.1743	584.00	156	555	0.1104	0.1147	0.0796
462	C5H12S	2-METHYL-2-BUTANETHIOL	-1.1231	0.2028	566.00	169	538	0.1098	0.1145	0.0753
463	C5H13N	n-PENTYLAMINE	-1.2926	0.6093	555.00	218	527	0.1571	0.1700	0.0510
464	C5H13NO2	METHYL DIETHANOLAMINE	-1.5918	1.0074	678.00	252	644	0.1828	0.1916	0.0256
465	C6Cl6	HEXACHLOROBENZENE	-1.9240	1.1177	825.00	502	784	-----	0.0814	0.0119
466	C6F6	HEXAFLUOROBENZENE	-1.8030	0.9591	516.73	278	491	0.0885	0.0896	0.0157
467	C6H3ClN2O4	1-CHLORO-2,4-DINITROBENZENE	-------	-------	-------	-----	-----	-----	-----	-----
468	C6H3Cl2NO2	1,2-DICHLORO-4-NITROBENZENE	-------	-------	-------	-----	-----	-----	-----	-----

Table 24-1 THERMAL CONDUCTIVITY OF LIQUID - ORGANIC COMPOUNDS (continued)

NO	FORMULA	NAME	$\log_{10} k_{liq} = A + B[1 - T/C]^{2/7}$					$(k_{liq} - W/(m\ K),\ T - K)$		
			A	B	C	TMIN	TMAX	k_{liq} @25 C	k_{liq} @TMIN	k_{liq} @TMAX
469	C6H3Cl3	1,2,4-TRICHLOROBENZENE	-1.4962	0.6264	725.00	290	689	0.1102	0.1096	0.0319
470	C6H3N3O6	1,3,5-TRINITROBENZENE	-------	-------	-------	-----	-----	-----	-----	-----
471	C6H4Br2	m-DIBROMOBENZENE	-1.6606	0.7907	761.00	266	723	0.1060	0.1079	0.0218
472	C6H4ClNO2	m-CHLORONITROBENZENE	-1.7455	1.0121	742.00	318	705	-----	0.1280	0.0180
473	C6H4ClNO2	o-CHLORONITROBENZENE	-1.7359	1.0095	757.00	306	719	-----	0.1336	0.0184
474	C6H4ClNO2	p-CHLORONITROBENZENE	-1.7481	1.0167	751.00	357	713	-----	0.1218	0.0179
475	C6H4Cl2	m-DICHLOROBENZENE	-1.7344	0.9478	683.95	248	650	0.1176	0.1235	0.0184
476	C6H4Cl2	o-DICHLOROBENZENE	-1.7480	0.9729	705.00	256	670	0.1212	0.1259	0.0179
477	C6H4Cl2	p-DICHLOROBENZENE	-1.7572	0.9659	684.75	326	651	-----	0.1083	0.0175
478	C6H4F2	m-DIFLUOROBENZENE	-1.3276	0.4374	552.94	249	525	0.1054	0.1087	0.0470
479	C6H4F2	o-DIFLUOROBENZENE	-1.2813	0.3786	554.46	239	527	0.1053	0.1090	0.0523
480	C6H4F2	p-DIFLUOROBENZENE	-1.4037	0.5319	556.00	260	528	0.1055	0.1083	0.0395
481	C6H4N2O4	m-DINITROBENZENE	-1.5895	1.0176	805.00	364	765	-----	0.1806	0.0257
482	C6H4N2O4	o-DINITROBENZENE	-1.6661	1.0237	831.00	390	789	-----	0.1501	0.0216
483	C6H4N2O4	p-DINITROBENZENE	-1.7864	1.0389	803.00	447	763	-----	0.1050	0.0164
484	C6H5Br	BROMOBENZENE	-1.9541	1.1821	670.15	242	637	0.1109	0.1194	0.0111
485	C6H5Cl	MONOCHLOROBENZENE	-1.6502	0.9051	632.35	228	601	0.1271	0.1379	0.0224
486	C6H5ClO	m-CHLOROPHENOL	-1.7797	1.0094	729.00	306	693	-----	0.1188	0.0166
487	C6H5ClO	o-CHLOROPHENOL	-1.6755	0.9728	675.00	282	641	0.1406	0.1409	0.0211
488	C6H5ClO	p-CHLOROPHENOL	-1.7607	1.0072	738.00	316	701	-----	0.1225	0.0174
489	C6H5Cl2N	3,4-DICHLOROANILINE	-1.7220	1.0139	800.00	345	760	-----	0.1352	0.0190
490	C6H5F	FLUOROBENZENE	-1.7815	1.0969	560.09	231	532	0.1263	0.1415	0.0165
491	C6H5I	IODOBENZENE	-2.8339	2.2319	721.15	242	685	0.1209	0.1370	0.0015
492	C6H5NO2	NITROBENZENE	-1.3942	0.6571	719.00	279	683	0.1478	0.1484	0.0403
493	C6H6	BENZENE	-1.6846	1.0520	562.16	279	534	0.1456	0.1469	0.0207
494	C6H6ClN	m-CHLOROANILINE	-1.6657	0.9933	751.00	263	713	0.1563	0.1604	0.0216
495	C6H6ClN	o-CHLOROANILINE	-1.7082	1.0368	722.00	482	686	-----	0.1059	0.0196
496	C6H6ClN	p-CHLOROANILINE	-1.7198	1.0152	754.00	343	716	-----	0.1327	0.0191
497	C6H6N2	cis-DICYANO-1-BUTENE	-1.5501	1.0055	691.00	249	656	0.2021	0.2124	0.0282
498	C6H6N2	trans-DICYANO-1-BUTENE	-1.5408	1.0035	689.00	260	655	0.2054	0.2127	0.0288
499	C6H6N2	1,4-DICYANO-2-BUTENE	-1.5254	1.0202	755.00	349	717	-----	0.2079	0.0298
500	C6H6N2O2	m-NITROANILINE	-1.6637	1.0250	815.00	387	774	-----	0.1504	0.0217
501	C6H6N2O2	o-NITROANILINE	-1.6626	1.0190	784.00	345	745	-----	0.1551	0.0217
502	C6H6N2O2	p-NITROANILINE	-1.6108	1.0267	851.00	421	808	-----	0.1664	0.0245
503	C6H6O	PHENOL	-1.1489	0.4091	694.25	314	660	-----	0.1553	0.0710
504	C6H6O2	1,2-BENZENEDIOL	-1.5704	0.9675	764.00	378	726	-----	0.1636	0.0269
505	C6H6O2	1,3-BENZENEDIOL	-1.5486	0.9736	810.00	382	770	-----	0.1786	0.0283
506	C6H6O2	p-HYDROQUINONE	-1.6580	1.0313	822.00	445	781	-----	0.1421	0.0220
507	C6H6O3	1,2,3-BENZENETRIOL	-1.5445	0.9665	830.00	407	789	-----	0.1742	0.0285
508	C6H6S	PHENYL MERCAPTAN	-1.6817	0.9958	689.00	258	655	0.1463	0.1518	0.0208
509	C6H7N	ANILINE	-1.3485	0.6888	699.00	267	664	0.1734	0.1763	0.0448
510	C6H7N	2-METHYLPYRIDINE	-1.6901	0.9887	621.00	206	590	0.1349	0.1529	0.0204
511	C6H7N	3-METHYLPYRIDINE	-1.6957	1.0009	645.00	255	613	0.1389	0.1454	0.0202
512	C6H7N	4-METHYLPYRIDINE	-1.7337	1.0058	646.15	277	614	0.1286	0.1299	0.0185
513	C6H8	1,3-CYCLOHEXADIENE	-1.6374	0.9577	558.00	161	530	0.1357	0.1683	0.0230
514	C6H8	METHYLCYCLOPENTADIENE	-1.6423	0.9558	541.00	150	514	0.1312	0.1674	0.0228
515	C6H8N2	ADIPONITRILE	-1.3732	0.7055	781.00	276	742	0.1745	0.1756	0.0423
516	C6H8N2	METHYLGLUTARONITRILE	-1.6111	0.9928	742.00	228	705	0.1763	0.1892	0.0245
517	C6H8N2	m-PHENYLENEDIAMINE	-1.6545	1.0061	824.00	334	783	-----	0.1599	0.0222
518	C6H8N2	o-PHENYLENEDIAMINE	-1.6232	1.0159	781.00	377	742	-----	0.1608	0.0238
519	C6H8N2	p-PHENYLENEDIAMINE	-1.6192	1.0242	796.00	413	756	-----	0.1578	0.0240
520	C6H8N2	PHENYLHYDRAZINE	-1.7253	1.0032	761.00	292	723	0.1396	0.1381	0.0188
521	C6H8N2O	BIS(CYANOETHYL)ETHER	-1.6256	0.9993	783.00	247	744	0.1761	0.1840	0.0237
522	C6H8O4	DIMETHYL MALEATE	-1.6586	0.9783	675.00	254	641	0.1478	0.1543	0.0219
523	C6H8O6	ASCORBIC ACID	-1.6122	1.0655	783.00	465	744	-----	0.1559	0.0244
524	C6H8O7	CITRIC ACID	-1.6225	1.0458	822.00	426	781	-----	0.1631	0.0239
525	C6H10	1-METHYLCYCLOPENTENE	-1.0717	0.2009	541.99	146	515	0.1225	0.1291	0.0848
526	C6H10	3-METHYLCYCLOPENTENE	-1.0636	0.1901	535.71	130	509	0.1222	0.1292	0.0864
527	C6H10	4-METHYLCYCLOPENTENE	-1.0473	0.1698	543.75	112	517	0.1225	0.1291	0.0897
528	C6H10	CYCLOHEXENE	-1.6375	0.9374	560.40	170	532	0.1309	0.1593	0.0230
529	C6H10	2,3-DIMETHYL-1,3-BUTADIENE	-1.7201	0.9969	526.00	197	500	0.1161	0.1393	0.0191
530	C6H10	1,5-HEXADIENE	-1.6655	0.9826	507.00	132	482	0.1251	0.1703	0.0216
531	C6H10	cis,trans-2,4-HEXADIENE	-1.6672	0.9862	538.00	177	511	0.1305	0.1608	0.0215
532	C6H10	trans,trans-2,4-HEXADIENE	-1.6870	1.0056	535.00	228	508	0.1287	0.1450	0.0206
533	C6H10	1-HEXYNE	-1.6575	0.9838	516.20	141	490	0.1293	0.1719	0.0220
534	C6H10	2-HEXYNE	-1.6719	0.9908	549.00	184	522	0.1319	0.1597	0.0213
535	C6H10	3-HEXYNE	-1.6709	0.9864	544.00	170	517	0.1304	0.1619	0.0213
536	C6H10O	CYCLOHEXANONE	-1.7647	1.0954	629.15	242	598	0.1403	0.1513	0.0172
537	C6H10O	MESITYL OXIDE	-1.6270	0.9997	600.00	220	570	0.1565	0.1749	0.0236
538	C6H10O2	epsilon-CAPROLACTONE	-1.7751	0.9963	771.00	272	732	0.1234	0.1251	0.0168
539	C6H10O2	ETHYL METHACRYLATE	-1.6607	0.9890	577.00	223	548	0.1389	0.1553	0.0218
540	C6H10O2	n-PROPYL ACRYLATE	-1.6397	0.9643	569.00	273	541	0.1381	0.1410	0.0229
541	C6H10O3	ETHYLACETOACETATE	-1.6494	0.9958	643.00	234	611	0.1528	0.1652	0.0224
542	C6H10O3	PROPIONIC ANHYDRIDE	-1.6727	1.0046	618.00	228	587	0.1444	0.1586	0.0212
543	C6H10O4	ADIPIC ACID	-1.5701	0.9580	809.00	426	769	-----	0.1551	0.0269
544	C6H10O4	DIETHYL OXALATE	-1.6294	0.9880	646.00	233	614	0.1579	0.1709	0.0235
545	C6H10O4	ETHYLENE GLYCOL DIACETATE	-1.6875	0.9903	653.00	242	620	0.1395	0.1487	0.0205
546	C6H10O4	ETHYLIDENE DIACETATE	-1.7337	1.0146	635.00	292	603	0.1297	0.1277	0.0185

Table 24-1 THERMAL CONDUCTIVITY OF LIQUID - ORGANIC COMPOUNDS (continued)

| NO | FORMULA | NAME | $\log_{10} k_{liq} = A + B [1 - T/C]^{2/7}$ | | | | | (k_{liq} - W/(m K), T - K) | | |
			A	B	C	TMIN	TMAX	k_{liq} @25 C	k_{liq} @TMIN	k_{liq} @TMAX
547	C6H11N	HEXANENITRILE	-1.6095	0.9930	622.05	193	591	0.1639	0.1894	0.0246
548	C6H11NO	epsilon-CAPROLACTAM	-1.6509	1.0120	806.00	342	766	-----	0.1599	0.0223
549	C6H11NO	CYCLOHEXANONE OXIME	-1.7615	1.0232	715.00	363	679	-----	0.1150	0.0173
550	C6H12	CYCLOHEXANE	-1.6817	0.9649	553.54	280	526	0.1236	0.1244	0.0208
551	C6H12	2,3-DIMETHYL-1-BUTENE	-1.7230	0.9800	500.00	116	475	0.1080	0.1519	0.0189
552	C6H12	2,3-DIMETHYL-2-BUTENE	-1.7317	0.9951	524.00	199	498	0.1124	0.1344	0.0185
553	C6H12	3,3-DIMETHYL-1-BUTENE	-1.7315	0.9866	480.00	158	456	0.1038	0.1387	0.0186
554	C6H12	2-ETHYL-1-BUTENE	-1.7244	0.9812	512.00	142	486	0.1097	0.1461	0.0189
555	C6H12	trans-3-METHYL-2-PENTENE	-1.0747	0.1996	521.00	135	495	0.1208	0.1281	0.0842
556	C6H12	1-HEXENE	-1.7827	1.1185	504.03	133	479	0.1212	0.1724	0.0165
557	C6H12	cis-2-HEXENE	-1.6730	0.9828	513.00	132	487	0.1240	0.1678	0.0212
558	C6H12	trans-2-HEXENE	-1.6761	0.9799	513.00	140	487	0.1225	0.1634	0.0211
559	C6H12	cis-3-HEXENE	-1.6726	0.9803	509.00	135	484	0.1229	0.1661	0.0213
560	C6H12	trans-3-HEXENE	-1.6842	0.9919	509.00	160	484	0.1222	0.1586	0.0207
561	C6H12	METHYLCYCLOPENTANE	-1.6949	0.9753	532.79	131	506	0.1193	0.1586	0.0202
562	C6H12	2-METHYL-1-PENTENE	-1.7259	0.9792	507.00	137	482	0.1082	0.1459	0.0188
563	C6H12	2-METHYL-2-PENTENE	-1.7292	0.9827	514.00	138	488	0.1091	0.1460	0.0187
564	C6H12	3-METHYL-1-PENTENE	-1.7256	0.9789	495.00	120	470	0.1063	0.1494	0.0188
565	C6H12	3-METHYL-cis-2-PENTENE	-1.7247	0.9822	515.00	138	489	0.1103	0.1475	0.0188
566	C6H12	4-METHYL-1-PENTENE	-1.7293	0.9796	496.00	120	471	0.1057	0.1484	0.0187
567	C6H12	4-METHYL-cis-2-PENTENE	-1.7321	0.9846	499.00	138	474	0.1064	0.1446	0.0185
568	C6H12	4-METHYL-trans-2-PENTENE	-1.7268	0.9765	501.00	132	476	0.1065	0.1456	0.0188
569	C6H12N2	TRIETHYLENEDIAMINE	-1.8102	1.0587	655.00	434	622	-----	0.0875	0.0155
570	C6H12O	BUTYL VINYL ETHER	-1.6689	0.9976	536.00	181	509	0.1324	0.1626	0.0214
571	C6H12O	CYCLOHEXANOL	-1.3475	0.5719	625.15	297	594	0.1342	0.1323	0.0449
572	C6H12O	1-HEXANAL	-1.4172	0.7037	579.00	217	550	0.1429	0.1558	0.0383
573	C6H12O	ETHYL ISOPROPYL KETONE	-1.6367	0.9833	567.00	200	539	0.1438	0.1677	0.0231
574	C6H12O	2-HEXANONE	-1.5936	0.9060	587.05	217	558	0.1400	0.1561	0.0255
575	C6H12O	3-HEXANONE	-1.6210	0.9777	582.82	218	554	0.1499	0.1685	0.0239
576	C6H12O	METHYL ISOBUTYL KETONE	-1.6520	0.9961	571.40	189	543	0.1428	0.1696	0.0223
577	C6H12O2	n-PENTYL FORMATE	-1.3693	0.6420	576.00	200	547	0.1419	0.1565	0.0427
578	C6H12O2	n-BUTYL ACETATE	-1.6072	0.9107	579.65	200	551	0.1360	0.1561	0.0247
579	C6H12O2	sec-BUTYL ACETATE	-1.6762	0.9960	561.00	174	533	0.1336	0.1635	0.0211
580	C6H12O2	tert-BUTYL ACETATE	-1.6708	0.9562	545.00	283	518	0.1235	0.1236	0.0213
581	C6H12O2	ETHYL n-BUTYRATE	-1.6755	1.0136	571.00	175	542	0.1397	0.1703	0.0211
582	C6H12O2	ETHYL ISOBUTYRATE	-1.6789	0.9958	553.15	185	525	0.1316	0.1587	0.0209
583	C6H12O2	ISOBUTYL ACETATE	-1.6695	0.9962	561.00	174	533	0.1357	0.1661	0.0214
584	C6H12O2	n-PROPYL PROPIONATE	-1.6581	0.9859	578.00	197	549	0.1391	0.1623	0.0220
585	C6H12O2	CYCLOHEXYL PEROXIDE	-1.6212	1.0027	685.00	253	651	0.1700	0.1778	0.0239
586	C6H12O2	DIACETONE ALCOHOL	-1.6610	0.9873	606.00	229	576	0.1421	0.1560	0.0218
587	C6H12O2	2-ETHYL BUTYRIC ACID	-1.6634	0.9864	655.00	258	622	0.1465	0.1524	0.0217
588	C6H12O2	n-HEXANOIC ACID	-1.3784	0.6393	667.00	270	634	0.1450	0.1470	0.0418
589	C6H12O3	2-ETHOXYETHYL ACETATE	-1.7101	0.9985	597.00	211	567	0.1286	0.1459	0.0195
590	C6H12O3	HYDROXYCAPROIC ACID	-1.6086	1.0244	758.00	334	720	-----	0.1773	0.0246
591	C6H12O3	PARALDEHYDE	-1.2305	0.4847	579.00	286	550	0.1458	0.1454	0.0588
592	C6H12O3	sec-BUTYL GLYCOLATE	-------	-------	------	----	----	------	------	------
593	C6H12S	THIACYCLOHEPTANE	-1.2873	0.3818	640.07	292	608	0.1076	0.1070	0.0516
594	C6H13N	CYCLOHEXYLAMINE	-1.7144	0.9971	615.00	255	584	0.1290	0.1355	0.0193
595	C6H13N	HEXAMETHYLENEIMINE	-1.7388	1.1970	615.00	236	584	0.1785	0.1966	0.0182
596	C6H14	2,2-DIMETHYLBUTANE	-1.7773	1.0273	488.78	174	464	0.1018	0.1321	0.0167
597	C6H14	2,3-DIMETHYLBUTANE	-1.7202	0.9624	499.98	145	475	0.1053	0.1403	0.0190
598	C6H14	n-HEXANE	-1.8389	1.1860	507.43	178	482	0.1208	0.1588	0.0145
599	C6H14	2-METHYLPENTANE	-1.7102	0.9861	497.50	120	473	0.1120	0.1573	0.0195
600	C6H14	3-METHYLPENTANE	-1.7207	1.0050	504.43	110	479	0.1142	0.1630	0.0190
601	C6H14N2O2	LYSINE	-1.7028	1.0544	821.00	483	780	-----	0.1251	0.0198
602	C6H14O	2-ETHYL-1-BUTANOL	-1.6669	1.0023	580.00	159	551	0.1408	0.1748	0.0215
603	C6H14O	1-HEXANOL	-1.3521	0.6421	611.35	229	581	0.1508	0.1601	0.0445
604	C6H14O	2-HEXANOL	-1.6584	0.9779	586.20	223	557	0.1380	0.1537	0.0220
605	C6H14O	2-METHYL-1-PENTANOL	-1.6532	0.9855	582.00	223	553	0.1411	0.1575	0.0222
606	C6H14O	4-METHYL-2-PENTANOL	-1.6816	0.9980	574.40	183	546	0.1343	0.1609	0.0208
607	C6H14O	n-BUTYL ETHYL ETHER	-1.6657	0.9849	531.00	170	504	0.1296	0.1622	0.0216
608	C6H14O	DIISOPROPYL ETHER	-2.2301	1.6397	500.05	188	475	0.1085	0.1548	0.0059
609	C6H14O	DI-n-PROPYL ETHER	-1.4463	0.6954	530.60	150	504	0.1268	0.1522	0.0358
610	C6H14O	METHYL tert-PENTYL ETHER	-1.6858	0.9965	534.00	160	507	0.1268	0.1616	0.0206
611	C6H14O2	ACETAL	-1.7534	1.0740	541.00	173	514	0.1262	0.1591	0.0176
612	C6H14O2	2-BUTOXYETHANOL	-1.2084	0.5059	600.00	203	570	0.1612	0.1729	0.0619
613	C6H14O2	1,6-HEXANEDIOL	-1.5811	0.9722	670.00	315	637	-----	0.1655	0.0262
614	C6H14O2	HEXYLENE GLYCOL	-1.6339	0.9892	621.00	223	590	0.1537	0.1698	0.0232
615	C6H14O2S	DI-n-PROPYL SULFONE	-1.7132	1.0123	763.00	303	725	-----	0.1426	0.0194
616	C6H14O3	DIETHYLENE GLYCOL DIMETHYL ETHER	-1.7449	1.0276	604.00	203	574	0.1262	0.1453	0.0180
617	C6H14O3	DIPROPYLENE GLYCOL	-1.6412	0.9914	654.00	233	621	0.1556	0.1681	0.0228
618	C6H14O3	2-(2-ETHOXYETHOXY)ETHANOL	-1.2811	0.6107	632.00	195	600	0.1690	0.1840	0.0523
619	C6H14O3	TRIMETHYLOLPROPANE	-1.5678	0.9715	709.00	331	674	-----	0.1711	0.0271
620	C6H14O4	TRIETHYLENE GLYCOL	-0.7913	0.0995	700.00	266	665	0.1966	0.1971	0.1617
621	C6H14O6	SORBITOL	-1.4973	0.9872	959.00	371	911	-----	0.2254	0.0318
622	C6H14S	n-HEXYLMERCAPTAN	-1.6845	0.9908	623.00	193	592	0.1374	0.1587	0.0207
623	C6H14S	BUTYL-ETHYL-SULFIDE	-1.1242	0.1734	609.00	178	579	0.1044	0.1076	0.0751
624	C6H14S	ISOPROPYL-SULFIDE	-1.1349	0.1850	585.71	170	556	0.1038	0.1076	0.0733

Table 24-1 THERMAL CONDUCTIVITY OF LIQUID - ORGANIC COMPOUNDS (continued)

			$\log_{10} k_{liq} = A + B [1 - T/C]^{2/7}$					$(k_{liq} - W/(m K), T - K)$		
NO	FORMULA	NAME	A	B	C	TMIN	TMAX	k_{liq} @25 C	k_{liq} @TMIN	k_{liq} @TMAX
625	C6H14S	METHYL-PENTYL-SULFIDE	-1.1362	0.1882	587.98	179	559	0.1041	0.1077	0.0731
626	C6H14S	PROPYL-SULFIDE	-1.1204	0.1681	609.73	170	579	0.1043	0.1076	0.0758
627	C6H14S2	PROPYL-DISULFIDE	-1.1555	0.1477	673.00	188	639	0.0932	0.0951	0.0699
628	C6H15Al	TRIETHYL ALUMINUM	-------	-------	-------	-----	-----	-----	-----	-----
629	C6H15Al2Cl3	ETHYL ALUMINUM SESQUICHLORIDE	-------	-------	-------	-----	-----	-----	-----	-----
630	C6H15N	DIISOPROPYLAMINE	-1.6751	0.9314	523.10	177	497	0.1140	0.1400	0.0211
631	C6H15N	DI-n-PROPYLAMINE	-1.8360	1.1801	555.80	210	528	0.1292	0.1531	0.0146
632	C6H15N	n-HEXYLAMINE	-1.1828	0.4509	583.00	252	554	0.1530	0.1572	0.0656
633	C6H15N	TRIETHYLAMINE	-1.8052	1.1067	535.15	158	508	0.1180	0.1548	0.0157
634	C6H15NO	6-AMINOHEXANOL	-1.7323	1.0289	681.00	331	647	-----	0.1277	0.0185
635	C6H15NO2	DIISOPROPANOLAMINE	-1.6411	1.0325	672.00	318	638	-----	0.1609	0.0229
636	C6H15NO3	TRIETHANOLAMINE	-1.0922	0.4389	787.00	294	748	0.1954	0.1942	0.0809
637	C6H15N3	N-AMINOETHYL PIPERAZINE	-1.6857	0.9969	708.00	254	673	0.1469	0.1531	0.0206
638	C6H15O4P	TRIETHYL PHOSPHATE	-------	-------	-------	-----	-----	-----	-----	-----
639	C6H16N2	HEXAMETHYLENEDIAMINE	-1.5005	0.7359	663.00	314	630	-----	0.1270	0.0316
640	C6H18N3OP	HEXAMETHYL PHOSPHORAMIDE	-------	-------	-------	-----	-----	-----	-----	-----
641	C6H18N4	TRIETHYLENE TETRAMINE	-1.6641	1.0139	718.00	285	682	0.1606	0.1602	0.0217
642	C6H18OSi2	HEXAMETHYLDISILOXANE	-1.7621	0.9977	518.70	205	493	0.1045	0.1240	0.0173
643	C6H18O3Si3	HEXAMETHYLCYCLOTRISILOXANE	-1.8092	1.0460	554.20	337	526	-----	0.0937	0.0155
644	C6H19NSi2	HEXAMETHYLDISILAZANE	-1.8517	1.0367	544.00	293	517	0.0943	0.0922	0.0141
645	C7H3ClF3NO2	4-CHLORO-3-NITROBENZOTRIFLUORIDE	-------	-------	-------	-----	-----	-----	-----	-----
646	C7H3Cl2F3	2,4-DICHLOROBENZOTRIFLUORIDE	-1.8722	0.9864	646.00	248	614	0.0900	0.0952	0.0134
647	C7H3Cl2NO	3,4-DICHLOROPHENYL ISOCYANATE	-1.7899	1.0120	733.00	316	696	-----	0.1152	0.0162
648	C7H4ClF3	p-CHLOROBENZOTRIFLUORIDE	-1.8375	0.9837	601.00	237	571	0.0936	0.1015	0.0145
649	C7H4Cl2O	m-CHLOROBENZOYL CHLORIDE	-1.8183	0.9852	724.00	280	688	0.1067	0.1072	0.0152
650	C7H4F3NO2	3-NITROBENZOTRIFLUORIDE	-------	-------	-------	-----	-----	-----	-----	-----
651	C7H5ClO	BENZOYL CHLORIDE	-1.8355	1.0012	697.00	273	662	0.1043	0.1058	0.0146
652	C7H5ClO2	o-CHLOROBENZOIC ACID	-1.6299	0.9595	792.00	415	752	-----	0.1358	0.0234
653	C7H5Cl3	BENZOTRICHLORIDE	-1.8450	0.9906	737.00	268	700	0.1021	0.1042	0.0143
654	C7H5F3	BENZOTRIFLUORIDE	-1.7754	0.9962	565.00	244	537	0.1068	0.1155	0.0168
655	C7H5N	BENZONITRILE	-------	-------	-------	-----	-----	-----	-----	-----
656	C7H5NO	PHENYL ISOCYANATE	-1.8002	0.9996	648.00	243	616	0.1091	0.1164	0.0158
657	C7H5N3O6	2,4,6-TRINITROTOLUENE	3.6003	-5.8533	795.00	357	393	-----	4.4017	3983.8227
658	C7H6Cl2	BENZYL DICHLORIDE	-1.7975	0.9832	731.00	257	694	0.1119	0.1159	0.0159
659	C7H6Cl2	2,4-DICHLOROTOLUENE	-1.8047	0.9890	705.00	260	670	0.1098	0.1135	0.0157
660	C7H6N2O4	2,4-DINITROTOLUENE	-------	-------	-------	-----	-----	-----	-----	-----
661	C7H6N2O4	2,5-DINITROTOLUENE	-------	-------	-------	-----	-----	-----	-----	-----
662	C7H6N2O4	2,6-DINITROTOLUENE	-------	-------	-------	-----	-----	-----	-----	-----
663	C7H6N2O4	3,4-DINITROTOLUENE	-------	-------	-------	-----	-----	-----	-----	-----
664	C7H6N2O4	3,5-DINITROTOLUENE	-------	-------	-------	-----	-----	-----	-----	-----
665	C7H6O	BENZALDEHYDE	-2.4889	1.9557	695.00	247	660	0.1505	0.1666	0.0032
666	C7H6O2	BENZOIC ACID	-1.6074	0.9579	751.00	396	713	-----	0.1421	0.0247
667	C7H6O2	p-HYDROXYBENZALDEHYDE	-1.6961	1.0193	844.00	390	802	-----	0.1401	0.0201
668	C7H6O2	SALICYLALDEHYDE	-1.6495	1.0064	680.00	266	646	0.1599	0.1642	0.0224
669	C7H6O3	SALICYLIC ACID	-------	-------	-------	-----	-----	-----	-----	-----
670	C7H7Br	p-BROMOTOLUENE	-1.8197	1.0055	699.00	300	664	-----	0.1064	0.0151
671	C7H7Cl	BENZYL CHLORIDE	-1.5601	0.8204	686.00	234	652	0.1371	0.1454	0.0275
672	C7H7Cl	o-CHLOROTOLUENE	-1.7397	1.0002	656.00	237	623	0.1263	0.1357	0.0182
673	C7H7Cl	p-CHLOROTOLUENE	-1.7240	0.9684	660.00	281	627	0.1235	0.1239	0.0189
674	C7H7F	p-FLUOROTOLUENE	-1.1658	0.2429	590.48	216	561	0.1079	0.1111	0.0683
675	C7H7NO	FORMANILIDE	-1.6665	1.0115	787.00	323	748	-----	0.1564	0.0216
676	C7H7NO2	m-NITROTOLUENE	-1.7136	0.9973	734.00	289	697	0.1399	0.1387	0.0193
677	C7H7NO2	o-NITROTOLUENE	-1.6935	0.9850	720.00	270	684	0.1419	0.1445	0.0203
678	C7H7NO2	p-NITROTOLUENE	-1.6940	0.9779	736.00	325	699	-----	0.1330	0.0202
679	C7H7NO3	o-NITROANISOLE	-1.6759	1.0014	782.00	284	743	0.1575	0.1573	0.0211
680	C7H8	TOLUENE	-1.6735	0.9773	591.79	178	562	0.1338	0.1596	0.0212
681	C7H8	1,3,5-CYCLOHEPTATRIENE	-1.1017	0.2098	593.90	194	564	0.1175	0.1214	0.0791
682	C7H8O	ANISOLE	-1.6295	0.9887	641.65	236	610	0.1576	0.1699	0.0235
683	C7H8O	BENZYL ALCOHOL	-0.9950	0.2347	677.00	258	643	0.1599	0.1613	0.1012
684	C7H8O	m-CRESOL	-1.2227	0.4627	705.85	285	671	0.1489	0.1487	0.0599
685	C7H8O	o-CRESOL	-1.2791	0.5437	697.55	304	663	-----	0.1504	0.0526
686	C7H8O	p-CRESOL	-1.3021	0.5377	704.65	308	669	-----	0.1408	0.0499
687	C7H8O2	GUAIACOL	-1.7827	1.0133	697.00	305	662	-----	0.1166	0.0165
688	C7H8O2	p-METHOXYPHENOL	-1.7008	1.0103	758.00	329	720	-----	0.1405	0.0199
689	C7H9N	BENZYLAMINE	-1.6799	0.9933	683.50	227	649	0.1457	0.1579	0.0209
690	C7H9N	2,6-DIMETHYLPYRIDINE	-1.7385	1.0120	623.75	267	593	0.1265	0.1302	0.0183
691	C7H9N	N-METHYLANILINE	-1.3638	0.6685	701.55	216	666	0.1610	0.1713	0.0433
692	C7H9N	m-TOLUIDINE	-1.6508	0.9978	709.15	243	674	0.1596	0.1688	0.0223
693	C7H9N	o-TOLUIDINE	-1.6342	0.9995	694.15	249	659	0.1649	0.1732	0.0232
694	C7H9N	p-TOLUIDINE	-1.6736	1.0521	693.15	317	658	-----	0.1579	0.0212
695	C7H10	2-NORBORNENE	-1.7378	1.0260	583.00	319	554	-----	0.1162	0.0183
696	C7H10N2	TOLUENEDIAMINE	-1.5750	1.0187	804.00	371	764	-----	0.1851	0.0266
697	C7H11NO	CYCLOHEXYL ISOCYANATE	-1.7389	0.9882	633.00	193	601	0.1216	0.1399	0.0182
698	C7H12	1-HEPTYNE	-1.1299	0.2348	559.69	192	532	0.1146	0.1193	0.0741
699	C7H12O2	n-BUTYL ACRYLATE	-1.6774	0.9946	598.00	209	568	0.1378	0.1567	0.0210
700	C7H12O2	ISOBUTYL ACRYLATE	-1.6941	0.9926	580.00	212	551	0.1299	0.1479	0.0202
701	C7H12O2	n-PROPYL METHACRYLATE	-1.6853	0.9925	599.00	223	569	0.1349	0.1499	0.0206
702	C7H12O4	DIETHYL MALONATE	-1.7202	0.9940	653.00	224	620	0.1303	0.1427	0.0190

Table 24-1 THERMAL CONDUCTIVITY OF LIQUID - ORGANIC COMPOUNDS (continued)

			$\log_{10} k_{liq} = A + B [1 - T/C]^{2/7}$					$(k_{liq} - W/(m\ K),\ T - K)$		
NO	FORMULA	NAME	A	B	C	TMIN	TMAX	k_{liq} @25 C	k_{liq} @TMIN	k_{liq} @TMAX
703	C7H14	CYCLOHEPTANE	-1.7334	1.0054	604.30	265	574	0.1243	0.1285	0.0185
704	C7H14	1,1-DIMETYLCYCLOPENTANE	-1.7319	0.9958	547.00	203	520	0.1157	0.1357	0.0185
705	C7H14	cis-1,2-DIMETHYLCYCLOPENTANE	-1.7337	1.0000	565.15	219	537	0.1184	0.1341	0.0185
706	C7H14	trans-1,2-DIMETHYLCYCLOPENTANE	-1.7228	0.9796	553.15	156	525	0.1154	0.1455	0.0189
707	C7H14	cis-1,3-DIMETHYLCYCLOPENTANE	-1.7250	0.9787	551.00	139	523	0.1144	0.1483	0.0188
708	C7H14	trans-1,3-DIMETHYLCYCLOPENTANE	-1.7217	0.9766	553.00	139	525	0.1151	0.1488	0.0190
709	C7H14	ETHYLCYCLOPENTANE	-1.7230	0.9811	569.52	135	541	0.1177	0.1517	0.0189
710	C7H14	2-ETHYL-1-PENTENE	-1.7490	0.9860	543.00	168	516	0.1087	0.1355	0.0178
711	C7H14	3-ETHYL-1-PENTENE	-1.7495	0.9851	530.00	146	504	0.1067	0.1393	0.0178
712	C7H14	1-HEPTENE	-1.7662	1.0883	537.29	154	510	0.1251	0.1644	0.0171
713	C7H14	cis-2-HEPTENE	-1.6750	0.9768	549.00	164	522	0.1276	0.1592	0.0211
714	C7H14	trans-2-HEPTENE	-1.6834	0.9869	543.00	164	516	0.1267	0.1590	0.0207
715	C7H14	cis-3-HEPTENE	-1.6782	0.9805	545.00	137	518	0.1270	0.1659	0.0210
716	C7H14	trans-3-HEPTENE	-1.6808	0.9832	540.00	137	513	0.1261	0.1655	0.0209
717	C7H14	METHYLCYCLOHEXANE	-1.7992	1.0419	572.19	147	544	0.1109	0.1422	0.0159
718	C7H14	2-METHYL-1-HEXENE	-1.7526	0.9893	538.00	170	511	0.1078	0.1345	0.0177
719	C7H14	3-METHYL-1-HEXENE	-1.7510	0.9853	528.00	145	502	0.1061	0.1390	0.0177
720	C7H14	4-METHYL-1-HEXENE	-1.7438	0.9784	534.00	132	507	0.1074	0.1425	0.0180
721	C7H14	2,3,3-TRIMETHYL-1-BUTENE	-1.7435	0.9842	531.00	163	504	0.1082	0.1370	0.0181
722	C7H14O	DIISOPROPYL KETONE	-1.6464	0.9770	576.00	205	547	0.1402	0.1614	0.0226
723	C7H14O	2-HEPTANONE	-1.6861	1.0000	611.55	238	581	0.1380	0.1493	0.0206
724	C7H14O	1-HEPTANAL	-1.4274	0.7062	603.00	230	573	0.1425	0.1522	0.0374
725	C7H14O	1-METHYLCYCLOHEXANOL	-1.6767	0.9979	603.00	299	573	-----	0.1354	0.0211
726	C7H14O	cis-2-METHYLCYCLOHEXANOL	-1.6551	0.9757	614.00	280	583	0.1418	0.1426	0.0221
727	C7H14O	trans-2-METHYLCYCLOHEXANOL	-1.6697	1.0008	616.00	269	585	0.1441	0.1478	0.0214
728	C7H14O	cis-3-METHYLCYCLOHEXANOL	-1.6686	1.0013	618.00	268	587	0.1449	0.1488	0.0214
729	C7H14O	trans-3-METHYLCYCLOHEXANOL	-1.6702	1.0019	617.00	273	586	0.1444	0.1470	0.0214
730	C7H14O	cis-4-METHYLCYCLOHEXANOL	-1.6624	0.9980	622.00	187	591	0.1465	0.1709	0.0218
731	C7H14O	trans-4-METHYLCYCLOHEXANOL	-1.6624	0.9980	622.00	187	591	0.1465	0.1709	0.0218
732	C7H14O	5-METHYL-2-HEXANONE	-1.6685	1.0299	601.00	199	571	0.1507	0.1749	0.0215
733	C7H14O2	n-BUTYL PROPIONATE	-1.6733	0.9959	594.00	184	564	0.1389	0.1646	0.0212
734	C7H14O2	ETHYL ISOVALERATE	-1.7025	0.9995	587.95	174	559	0.1300	0.1570	0.0198
735	C7H14O2	ISOPENTYL ACETATE	-1.6874	0.9819	599.00	195	569	0.1316	0.1526	0.0205
736	C7H14O2	n-PENTYL ACETATE	-1.6758	0.9850	598.00	202	568	0.1358	0.1559	0.0211
737	C7H14O2	n-PROPYL n-BUTYRATE	-1.6748	0.9809	594.00	178	564	0.1346	0.1605	0.0211
738	C7H14O2	n-HEPTANOIC ACID	-1.4366	0.6964	680.00	266	646	0.1425	0.1452	0.0366
739	C7H14O3	ETHYL-3-ETHOXYPROPIONATE	-1.6832	1.0032	609.00	223	579	0.1395	0.1548	0.0207
740	C7H15Br	1-BROMOHEPTANE	-1.6210	0.7940	651.00	217	618	0.1111	0.1204	0.0239
741	C7H15N	N-METHYLCYCLOHEXYLAMINE	-1.7499	1.0072	622.00	265	591	0.1219	0.1259	0.0178
742	C7H16	2,2-DIMETHYLPENTANE	-1.7590	0.9870	520.50	149	494	0.1035	0.1355	0.0174
743	C7H16	2,3-DIMETHYLPENTANE	-1.8184	1.0680	537.35	160	510	0.1069	0.1383	0.0152
744	C7H16	2,4-DIMETHYLPENTANE	-1.7527	0.9762	519.79	154	494	0.1029	0.1333	0.0177
745	C7H16	3,3-DIMETHYLPENTANE	-1.7559	0.9833	536.40	139	510	0.1057	0.1387	0.0175
746	C7H16	3-ETHYLPENTANE	-1.7563	0.9874	540.64	155	514	0.1069	0.1364	0.0175
747	C7H16	n-HEPTANE	-1.8482	1.1843	540.26	183	513	0.1240	0.1570	0.0142
748	C7H16	2-METHYLHEXANE	-1.7110	0.9951	530.37	155	504	0.1188	0.1531	0.0195
749	C7H16	3-METHYLHEXANE	-1.7622	0.9895	535.25	154	508	0.1052	0.1350	0.0173
750	C7H16	2,2,3-TRIMETHYLBUTANE	-1.7761	1.0083	531.17	249	505	0.1049	0.1133	0.0167
751	C7H16O	1-HEPTANOL	-1.5652	0.9223	631.90	239	600	0.1597	0.1708	0.0272
752	C7H16O	2-HEPTANOL	-1.6662	0.9817	588.00	243	559	0.1367	0.1472	0.0216
753	C7H16O	5-METHYL-1-HEXANOL	-1.6450	0.9695	605.00	293	575	0.1424	0.1399	0.0226
754	C7H16O	ISOPROPYL-TERT-BUTYL-ETHER	-1.1509	0.2109	558.21	178	530	0.1044	0.1088	0.0706
755	C7H16S	n-HEPTYL MERCAPTAN	-1.7017	1.0018	645.00	230	613	0.1372	0.1493	0.0199
756	C7H16S	BUTYL-PROPYL-SULFIDE	-1.1469	0.1713	653.50	207	621	0.0993	0.1013	0.0713
757	C7H16S	ETHYL-PENTYL-SULFIDE	-1.1490	0.1758	638.37	207	606	0.0995	0.1016	0.0710
758	C7H16S	HEXYL-METHYL-SULFIDE	-1.1490	0.1758	638.37	207	606	0.0995	0.1016	0.0710
759	C7H17N	1-AMINOHEPTANE	-1.1421	0.3945	607.00	254	577	0.1525	0.1556	0.0721
760	C8H4Cl2O2	ISOPHTHALOYL CHLORIDE	-1.7803	1.0149	768.00	317	730	-----	0.1208	0.0166
761	C8H4O3	PHTHALIC ANHYDRIDE	-1.6831	1.0293	791.00	404	751	-----	0.1388	0.0207
762	C8H6	ETHYNYLBENZENE	-1.1408	0.2298	655.43	243	623	0.1128	0.1145	0.0723
763	C8H6O4	ISOPHTHALIC ACID	-------	-------	-------	-----	-----	-----	-----	-----
764	C8H6O4	PHTHALIC ACID	-------	-------	-------	-----	-----	-----	-----	-----
765	C8H6O4	TEREPHTHALIC ACID	-------	-------	-------	-----	-----	-----	-----	-----
766	C8H6S	BENZOTHIOPHENE	-1.7276	1.0008	754.00	305	716	-----	0.1338	0.0187
767	C8H7N	INDOLE	-1.6837	0.9961	790.00	274	751	0.1536	0.1553	0.0207
768	C8H8	STYRENE	-1.7023	1.0002	648.00	243	616	0.1369	0.1460	0.0198
769	C8H8	1,3,5,7-CYCLOOCTATETRAENE	-1.1937	0.2921	642.55	266	610	0.1124	0.1133	0.0640
770	C8H8O	ACETOPHENONE	-2.0849	1.5494	701.00	294	666	0.1729	0.1687	0.0082
771	C8H8O	p-TOLUALDEHYDE	-1.7596	1.0074	698.00	290	663	0.1258	0.1245	0.0174
772	C8H8O2	METHYL BENZOATE	-1.6414	0.9769	693.00	261	658	0.1550	0.1600	0.0228
773	C8H8O2	o-TOLUIC ACID	-1.6254	0.9641	751.00	377	713	-----	0.1420	0.0237
774	C8H8O2	p-TOLUIC ACID	-1.5980	0.9423	773.00	453	734	-----	0.1313	0.0252
775	C8H8O3	METHYL SALICYLATE	-1.6717	0.9822	701.00	265	666	0.1468	0.1506	0.0213
776	C8H8O3	VANILLIN	-1.7080	1.0230	777.00	355	738	-----	0.1381	0.0196
777	C8H9NO	ACETANILIDE	-1.7347	1.0237	825.00	387	784	-----	0.1283	0.0184
778	C8H10	ETHYLBENZENE	-1.7498	1.0437	617.17	178	586	0.1302	0.1554	0.0178
779	C8H10	m-XYLENE	-1.7286	1.0193	617.05	225	586	0.1305	0.1442	0.0187
780	C8H10	o-XYLENE	-1.7372	1.0282	630.37	248	599	0.1315	0.1398	0.0183

Table 24-1 THERMAL CONDUCTIVITY OF LIQUID - ORGANIC COMPOUNDS (continued)

| NO | FORMULA | NAME | $\log_{10} k_{liq} = A + B [1 - T/C]^{2/7}$ | | | | | $(k_{liq} - W/(m\ K),\ T - K)$ | | |
			A	B	C	TMIN	TMAX	k_{liq} @25 C	k_{liq} @TMIN	k_{liq} @TMAX
781	C8H10	p-XYLENE	-1.7354	1.0254	616.26	286	585	0.1299	0.1291	0.0184
782	C8H10O	m-ETHYLPHENOL	-------	-------	-------	-----	-----	-----	-----	-----
783	C8H10O	p-ETHYLPHENOL	-1.6225	0.9776	716.45	318	681	-----	0.1564	0.0239
784	C8H10O	PHENETOLE	-1.4237	0.6814	647.15	244	615	0.1404	0.1466	0.0377
785	C8H10O	2-PHENYLETHANOL	-1.6255	0.9898	684.00	247	650	0.1640	0.1730	0.0237
786	C8H10O	2,3-XYLENOL	-1.1910	0.4541	722.95	346	687	-----	0.1516	0.0644
787	C8H10O	2,4-XYLENOL	-1.1954	0.4501	707.65	298	672	0.1547	0.1532	0.0638
788	C8H10O	2,5-XYLENOL	-1.1950	0.4487	707.05	348	672	-----	0.1477	0.0638
789	C8H10O	2,6-XYLENOL	-1.2065	0.4624	701.05	319	666	-----	0.1504	0.0622
790	C8H10O	3,4-XYLENOL	-1.6115	0.9738	729.95	338	693	-----	0.1559	0.0245
791	C8H10O	3,5-XYLENOL	-1.1839	0.4551	715.65	337	680	-----	0.1550	0.0655
792	C8H11N	N,N-DIMETHYLANILINE	-1.5076	0.7769	687.15	276	653	0.1421	0.1434	0.0311
793	C8H11N	o-ETHYLANILINE	-1.6969	0.9915	704.00	227	669	0.1413	0.1527	0.0201
794	C8H11N	2,4,6-TRIMETHYLPYRIDINE	-1.7344	0.9957	653.00	229	620	0.1265	0.1375	0.0184
795	C8H11NO	p-PHENETIDINE	-1.7085	1.0051	754.00	277	716	0.1452	0.1464	0.0196
796	C8H12	1,5-CYCLOOCTADIENE	-1.6838	0.9633	645.00	204	613	0.1328	0.1494	0.0207
797	C8H12	VINYLCYCLOHEXENE	-1.7094	0.9809	599.00	164	569	0.1248	0.1516	0.0195
798	C8H12O4	1,4-CYCLOHEXANEDICARBOXYLIC ACID	-1.7081	1.0707	889.00	586	845	-----	0.1137	0.0196
799	C8H12O4	DIETHYL MALEATE	-1.7114	0.9736	680.00	264	646	0.1301	0.1338	0.0194
800	C8H14O2	n-BUTYL METHACRYLATE	-1.7073	0.9923	616.00	223	585	0.1300	0.1438	0.0196
801	C8H14O3	BUTYRIC ANHYDRIDE	-1.6905	0.9930	639.00	200	607	0.1378	0.1568	0.0204
802	C8H14O4	DIETHYL SUCCINATE	-1.7317	0.9849	660.00	252	627	0.1253	0.1314	0.0185
803	C8H16	CYCLOOCTANE	-1.2354	0.3272	647.20	288	615	0.1094	0.1091	0.0582
804	C8H16	1,1-DIMETHYLCYCLOHEXANE	-1.7558	1.0033	591.15	240	562	0.1162	0.1259	0.0175
805	C8H16	cis-1,2-DIMETHYLCYCLOHEXANE	-1.7459	0.9949	606.15	223	576	0.1186	0.1316	0.0180
806	C8H16	trans-1,2-DIMETHYLCYCLOHEXANE	-1.2763	0.4639	596.15	185	566	0.1271	0.1374	0.0529
807	C8H16	cis-1,3-DIMETHYLCYCLOHEXANE	-1.7537	0.9909	591.15	198	562	0.1141	0.1323	0.0176
808	C8H16	trans-1,3-DIMETHYLCYCLOHEXANE	-1.7435	0.9882	598.00	183	568	0.1169	0.1383	0.0181
809	C8H16	cis-1,4-DIMETHYLCYCLOHEXANE	-1.7459	0.9896	598.15	186	568	0.1166	0.1373	0.0180
810	C8H16	trans-1,4-DIMETHYLCYCLOHEXANE	-1.7647	1.0023	590.15	236	561	0.1135	0.1238	0.0172
811	C8H16	ETHYLCYCLOHEXANE	-1.7405	0.9818	609.15	162	579	0.1174	0.1424	0.0182
812	C8H16	2-ETHYL-1-HEXENE	-1.7709	0.9929	574.00	173	545	0.1083	0.1316	0.0169
813	C8H16	1-METHYL-1-ETHYLCYCLOPENTANE	-1.7315	0.9795	582.00	129	553	0.1165	0.1501	0.0186
814	C8H16	1-OCTENE	-1.6274	0.9136	566.60	171	538	0.1290	0.1555	0.0236
815	C8H16	trans-2-OCTENE	-1.6935	0.9937	577.00	185	548	0.1300	0.1548	0.0203
816	C8H16	trans-3-OCTENE	-1.6905	0.9884	574.00	163	545	0.1292	0.1594	0.0204
817	C8H16	trans-4-OCTENE	-1.6921	0.9901	573.00	179	544	0.1290	0.1554	0.0203
818	C8H16	n-PROPYLCYCLOPENTANE	-1.7409	0.9779	603.00	156	573	0.1158	0.1419	0.0182
819	C8H16	2,4,4-TRIMETHYL-1-PENTENE	-1.7673	0.9910	553.00	180	525	0.1064	0.1293	0.0171
820	C8H16	2,4,4-TRIMETHYL-2-PENTENE	-1.7646	0.9898	558.00	167	530	0.1074	0.1330	0.0172
821	C8H16O	2-ETHYLHEXANAL	-1.7220	0.9981	607.00	200	577	0.1261	0.1452	0.0190
822	C8H16O	1-OCTANAL	-1.7252	1.0125	621.00	246	590	0.1302	0.1389	0.0188
823	C8H16O	2-OCTANONE	-1.6779	0.9911	624.00	253	593	0.1397	0.1471	0.0210
824	C8H16O2	n-BUTYL n-BUTYRATE	-1.7168	0.9953	616.00	181	585	0.1280	0.1509	0.0192
825	C8H16O2	n-HEXYL ACETATE	-1.3859	0.6458	618.00	192	587	0.1410	0.1552	0.0411
826	C8H16O2	ISOBUTYL ISOBUTYRATE	-1.7293	0.9957	602.00	192	572	0.1229	0.1434	0.0187
827	C8H16O2	n-OCTANOIC ACID	-1.6624	0.9409	692.00	290	657	0.1491	0.1476	0.0218
828	C8H16O4	DIETHYLENE GLYCOL ETHYL ETHER ACETATE	-1.7930	1.0113	660.00	248	627	0.1145	0.1210	0.0161
829	C8H18	2,2-DIMETHYLHEXANE	-1.7714	0.9821	549.80	152	522	0.1033	0.1314	0.0169
830	C8H18	2,3-DIMETHYLHEXANE	-1.7995	1.0203	563.40	272	535	0.1055	0.1080	0.0159
831	C8H18	2,4-DIMETHYLHEXANE	-1.8086	1.0259	553.50	272	526	0.1032	0.1058	0.0155
832	C8H18	2,5-DIMETHYLHEXANE	-1.7869	0.9989	550.00	182	523	0.1028	0.1251	0.0163
833	C8H18	3,3-DIMETHYLHEXANE	-1.7657	0.9782	562.00	147	534	0.1053	0.1338	0.0172
834	C8H18	3,4-DIMETHYLHEXANE	-1.7937	1.0161	568.80	272	540	0.1067	0.1092	0.0161
835	C8H18	3-ETHYLHEXANE	-1.7971	1.0170	565.40	272	537	0.1057	0.1082	0.0160
836	C8H18	3-ETHYL-2-METHYLPENTANE	-1.1245	0.1828	567.00	158	539	0.1055	0.1099	0.0751
837	C8H18	3-METHYL-3-ETHYLPENTANE	-1.7687	0.9873	576.50	182	548	0.1079	0.1291	0.0170
838	C8H18	2-METHYLHEPTANE	-1.7035	0.9932	559.64	164	532	0.1246	0.1550	0.0198
839	C8H18	3-METHYLHEPTANE	-1.7146	0.9986	563.67	153	535	0.1232	0.1557	0.0193
840	C8H18	4-METHYLHEPTANE	-1.7083	0.9985	561.74	152	534	0.1248	0.1581	0.0196
841	C8H18	n-OCTANE	-1.8388	1.1699	568.83	216	540	0.1281	0.1487	0.0145
842	C8H18	2,2,3-TRIMETHYLPENTANE	-1.7705	0.9879	563.50	161	535	0.1062	0.1322	0.0170
843	C8H18	2,2,4-TRIMETHYLPENTANE	-1.6144	0.7698	543.96	166	517	0.0998	0.1188	0.0243
844	C8H18	2,3,3-TRIMETHYLPENTANE	-1.7681	0.9870	573.50	172	545	0.1077	0.1311	0.0171
845	C8H18	2,3,4-TRIMETHYLPENTANE	-1.7695	0.9879	566.30	164	538	0.1068	0.1321	0.0170
846	C8H18	2,2,3,3-TETRAMETHYLBUTANE	-1.1401	0.2007	567.80	172	539	0.1052	0.1096	0.0724
847	C8H18O	DI-n-BUTYL ETHER	-1.4074	0.6132	581.00	178	552	0.1235	0.1384	0.0391
848	C8H18O	DI-sec-BUTYL ETHER	-1.7430	0.9913	559.00	173	531	0.1133	0.1389	0.0181
849	C8H18O	DI-tert-BUTYL ETHER	-1.7549	0.9830	550.00	195	523	0.1075	0.1275	0.0176
850	C8H18O	2-ETHYL-1-HEXANOL	-1.6754	0.9952	640.25	203	608	0.1434	0.1625	0.0211
851	C8H18O	1-OCTANOL	-1.6972	1.0775	652.50	258	620	0.1614	0.1686	0.0201
852	C8H18O	2-OCTANOL	-1.2061	0.4359	637.15	242	605	0.1438	0.1481	0.0622
853	C8H18O2	DI-t-BUTYL PEROXIDE	-1.7630	1.0153	547.00	233	520	0.1116	0.1241	0.0173
854	C8H18O2S	DI-n-BUTYL SULFONE	-1.7352	1.0149	767.00	318	729	-----	0.1338	0.0184
855	C8H18O3	DIETHYLENE GLYCOL DIETHYL ETHER	-1.4158	0.7047	624.00	229	593	0.1477	0.1575	0.0384
856	C8H18O3	DIETHYLENE GLYCOL MONOBUTYL EHTER	-1.3443	0.6670	654.00	205	621	0.1645	0.1781	0.0453
857	C8H18O4	TRIETHYLENE GLYCOL DIMETHYL ETHER	-2.2323	1.7319	651.00	229	618	0.1666	0.1928	0.0059
858	C8H18O5	TETRAETHYLENE GLYCOL	-1.6415	0.9871	722.00	268	686	0.1608	0.1642	0.0228

Table 24-1 THERMAL CONDUCTIVITY OF LIQUID - ORGANIC COMPOUNDS (continued)

NO	FORMULA	NAME	$\log_{10} k_{liq} = A + B [1 - T/C]^{2/7}$					$(k_{liq} - W/(m\ K),\ T - K)$		
			A	B	C	TMIN	TMAX	k_{liq} @25 C	k_{liq} @TMIN	k_{liq} @TMAX
859	C8H18S	n-OCTYL MERCAPTAN	-1.7402	0.9950	664.00	224	631	0.1256	0.1373	0.0182
860	C8H18S	tert-OCTYL MERCAPTAN	-1.7613	0.9948	627.00	199	596	0.1164	0.1332	0.0173
861	C8H18S	BUTYL-SULFIDE	-1.1662	0.1705	650.00	210	618	0.0948	0.0967	0.0682
862	C8H18S	ETHYL-HEXYL-SULFIDE	-1.1598	0.1631	660.72	210	628	0.0950	0.0967	0.0692
863	C8H18S	HEPTYL-METHYL-SULFIDE	-1.1598	0.1631	660.72	210	628	0.0950	0.0967	0.0692
864	C8H18S	PENTYL-PROPYL-SULFIDE	-1.1598	0.1631	660.72	210	628	0.0950	0.0967	0.0692
865	C8H18S2	BUTYL-DISULFIDE	-1.1842	0.1398	704.16	202	669	0.0861	0.0875	0.0654
866	C8H19N	DI-n-BUTYLAMINE	-1.6042	0.8870	607.50	211	577	0.1340	0.1495	0.0249
867	C8H19N	DIISOBUTYLAMINE	-1.7621	0.9963	580.00	203	551	0.1118	0.1293	0.0173
868	C8H19N	n-OCTYLAMINE	-1.0950	0.3289	627.00	273	596	0.1508	0.1517	0.0804
869	C8H23N5	TETRAETHYLENEPENTAMINE	-1.6786	1.0022	774.00	243	735	0.1562	0.1641	0.0210
870	C8H24O4Si4	OCTAMETHYLCYCLOTETRASILOXANE	-1.8019	1.0388	586.50	291	557	0.1112	0.1094	0.0158
871	C9H4O5	TRIMELLITIC ANHYDRIDE	-1.5385	1.0319	890.00	438	846	-----	0.1992	0.0289
872	C9H6N2O2	TOLUENE DIISOCYANATE	-1.7078	0.9876	737.00	287	700	0.1393	0.1385	0.0196
873	C9H7N	ISOQUINOLINE	-1.7402	0.9953	803.15	299	763	-----	0.1328	0.0182
874	C9H7N	QUINOLINE	-1.8878	1.2178	782.15	258	743	0.1493	0.1550	0.0129
875	C9H7NO	8-HYDROXYQUINOLINE	-1.7414	1.0149	788.00	346	749	-----	0.1285	0.0181
876	C9H8	INDENE	-1.6863	1.0061	687.00	272	653	0.1475	0.1501	0.0206
877	C9H8O	2-METHYLBENZOFURAN	-1.7593	1.0104	698.00	290	663	0.1266	0.1253	0.0174
878	C9H10	INDANE	-1.7070	0.9855	684.90	222	651	0.1349	0.1472	0.0196
879	C9H10	cis-PROPENYLBENZENE	-1.1245	0.1725	664.60	211	631	0.1050	0.1069	0.0751
880	C9H10	trans-PROPENYLBENZENE	-1.1488	0.2039	664.60	244	631	0.1055	0.1068	0.0710
881	C9H1O	alpha-METHYLSTYRENE	-1.7068	0.9915	654.00	250	621	0.1338	0.1409	0.0196
882	C9H10	m-METHYLSTYRENE	-1.6988	0.9833	657.00	187	624	0.1344	0.1546	0.0200
883	C9H10	o-METHYLSTYRENE	-1.7026	0.9871	659.00	205	626	0.1344	0.1509	0.0198
884	C9H10	p-METHYLSTYRENE	-1.7070	0.9953	665.00	239	632	0.1357	0.1452	0.0196
885	C9H10O2	BENZYL ACETATE	-1.6849	0.9991	699.00	222	664	0.1470	0.1601	0.0207
886	C9H10O2	ETHYL BENZOATE	-1.6750	0.9835	698.00	238	663	0.1458	0.1553	0.0211
887	C9H10O3	ETHYL VANILLIN	-1.6942	1.0286	748.00	351	711	-----	0.1421	0.0202
888	C9H11NO	p-DIMETHYLAMINOBENZALDEHYDE	-1.7582	1.0148	832.00	348	790	-----	0.1263	0.0175
889	C9H12	CUMENE	-1.1835	0.3543	631.15	177	600	0.1293	0.1371	0.0655
890	C9H12	m-ETHYLTOLUENE	-1.7189	0.9845	637.15	178	605	0.1268	0.1487	0.0191
891	C9H12	o-ETHYLTOLUENE	-1.7176	0.9874	651.15	192	619	0.1292	0.1481	0.0192
892	C9H12	p-ETHYLTOLUENE	-1.7291	0.9935	640.15	211	608	0.1263	0.1414	0.0187
893	C9H12	MESITYLENE	-1.2339	0.4698	637.36	228	605	0.1440	0.1501	0.0584
894	C9H12	n-PROPYLBENZENE	-1.7052	0.9688	638.38	174	606	0.1271	0.1494	0.0197
895	C9H12	1,2,3-TRIMETHYLBENZENE	-1.6931	0.9586	664.53	248	631	0.1305	0.1374	0.0203
896	C9H12	1,2,4-TRIMETHYLBENZENE	-1.7273	1.0028	649.13	229	617	0.1300	0.1416	0.0187
897	C9H12O	BENZYL ETHYL ETHER	-1.6898	0.9805	662.00	276	629	0.1370	0.1385	0.0204
898	C9H12O	2-PHENYL-2-PROPANOL	-1.6545	0.9738	660.00	309	627	-----	0.1405	0.0222
899	C9H12O2	CUMENE HYDROPEROXIDE	-1.8736	1.0150	605.00	264	575	0.0917	0.0950	0.0134
900	C9H14O	ISOPHORONE	-1.6075	0.9905	715.00	265	679	0.1744	0.1788	0.0247
901	C9H14O6	GLYCERYL TRIACETATE	-1.7907	0.9843	704.00	277	669	0.1123	0.1133	0.0162
902	C9H16	1-NONYNE	-1.1635	0.2120	610.81	223	580	0.1027	0.1050	0.0686
903	C9H16O4	AZELAIC ACID	-1.6152	0.9721	811.00	380	770	-----	0.1532	0.0243
904	C9H18	BUTYLCYCLOPENTANE	-1.1249	0.1562	625.05	165	594	0.1011	0.1041	0.0750
905	C9H18	cis,cis-1,3,5-TRIMETHYLCYCLOHEXANE	-1.1781	0.2237	607.86	223	577	0.1015	0.1039	0.0664
906	C9H18	cis,trans-1,3,5-TRIMETHYLCYCLOHEXANE	-1.1479	0.1859	602.20	189	572	0.1012	0.1042	0.0711
907	C9H18	ISOPROPYLCYCLOHEXANE	-1.7546	0.9867	627.00	184	596	0.1164	0.1359	0.0176
908	C9H18	1-NONENE	-1.8036	1.0743	593.25	192	564	0.1192	0.1413	0.0157
909	C9H18	n-PROPYLCYCLOHEXANE	-1.7622	0.9859	639.15	178	607	0.1153	0.1351	0.0173
910	C9H18O	DIISOBUTYL KETONE	-1.6430	0.9749	615.00	227	584	0.1458	0.1600	0.0228
911	C9H18O	1-NONANAL	-1.3939	0.6593	640.00	255	608	0.1436	0.1481	0.0404
912	C9H18O2	n-BUTYL VALERATE	-1.7291	0.9996	629.00	180	598	0.1267	0.1490	0.0187
913	C9H18O2	n-NONANOIC ACID	-1.4696	0.7282	703.00	286	668	0.1420	0.1416	0.0339
914	C9H18O2	n-OCTYL FORMATE	-1.3800	0.6256	645.00	234	613	0.1393	0.1463	0.0417
915	C9H20	3,3-DIETHYLPENTANE	-1.7892	1.0016	610.05	240	580	0.1091	0.1177	0.0162
916	C9H20	2,2-DIMETHYL-3-ETHYLPENTANE	-1.7833	0.9898	590.00	174	561	0.1062	0.1279	0.0165
917	C9H20	3-ETHYL-2,3-DIMETHYLPENTANE	-2.6081	2.0588	606.80	174	576	0.1228	0.1777	0.0025
918	C9H20	2,4-DIMETHYL-3-ETHYLPENTANE	-1.7726	0.9791	591.00	151	561	0.1068	0.1326	0.0169
919	C9H20	2,2-DIMETHYLHEPTANE	-1.7894	0.9861	576.80	160	548	0.1027	0.1271	0.0162
920	C9H20	2,6-DIMETHYLHEPTANE	-1.7923	0.9880	579.00	170	550	0.1026	0.1249	0.0161
921	C9H20	3-ETHYLHEPTANE	-1.7835	0.9841	590.00	158	561	0.1050	0.1294	0.0165
922	C9H20	4-ETHYLHEPTANE	-2.6081	2.0588	584.95	160	556	0.1179	0.1822	0.0025
923	C9H20	2,3-DIMETHYLHEPTANE	-2.6081	2.0588	589.60	160	560	0.1189	0.1828	0.0025
924	C9H20	2,4-DIMETHYLHEPTANE	-2.6081	2.0588	576.80	160	548	0.1160	0.1808	0.0025
925	C9H20	2,5-DIMETHYLHEPTANE	-2.6081	2.0588	581.10	160	552	0.1170	0.1815	0.0025
926	C9H20	3,4-DIMETHYLHEPTANE	-2.6081	2.0588	591.90	170	562	0.1195	0.1776	0.0025
927	C9H20	3,5-DIMETHYLHEPTANE	-2.6081	2.0588	583.20	170	554	0.1175	0.1762	0.0025
928	C9H20	4,4-DIMETHYLHEPTANE	-2.6081	2.0588	585.40	170	556	0.1180	0.1765	0.0025
929	C9H20	3-ETHYL-2-METHYLHEXANE	-2.6081	2.0588	588.10	160	559	0.1186	0.1826	0.0025
930	C9H20	4-ETHYL-2-METHYLHEXANE	-2.6081	2.0588	580.00	160	551	0.1167	0.1813	0.0025
931	C9H20	3-ETHYL-3-METHYLHEXANE	-2.6081	2.0588	597.50	160	568	0.1207	0.1841	0.0025
932	C9H20	3-ETHYL-4-METHYLHEXANE	-2.6081	2.0588	593.70	160	564	0.1199	0.1834	0.0025
933	C9H20	2,2,3-TRIMETHYLHEXANE	-2.6081	2.0588	588.00	153	559	0.1186	0.1866	0.0025
934	C9H20	2,2,4-TRIMETHYLHEXANE	-2.6081	2.0588	573.50	153	545	0.1152	0.1844	0.0025
935	C9H20	2,3,3-TRIMETHYLHEXANE	-2.6081	2.0588	596.00	156	566	0.1204	0.1860	0.0025
936	C9H20	2,3,4-TRIMETHYLHEXANE	-2.6081	2.0588	594.50	156	565	0.1200	0.1858	0.0025

545

Table 24-1 THERMAL CONDUCTIVITY OF LIQUID - ORGANIC COMPOUNDS (continued)

NO	FORMULA	NAME	$\log_{10} k_{liq} = A + B [1 - T/C]^{2/7}$					$(k_{liq} - W/(m\ K),\ T - K)$		
			A	B	C	TMIN	TMAX	k_{liq} @25 C	k_{liq} @TMIN	k_{liq} @TMAX
937	C9H20	2,3,5-TRIMETHYLHEXANE	-2.6081	2.0588	579.20	145	550	0.1165	0.1898	0.0025
938	C9H20	2,4,4-TRIMETHYLHEXANE	-2.6081	2.0588	581.50	160	552	0.1171	0.1815	0.0025
939	C9H20	3,3,4-TRIMETHYLHEXANE	-2.6081	2.0588	602.30	172	572	0.1218	0.1781	0.0025
940	C9H20	2-METHYLOCTANE	-1.6942	0.9857	586.75	193	557	0.1290	0.1509	0.0202
941	C9H20	3-METHYLOCTANE	-1.7918	0.9901	590.15	166	561	0.1042	0.1270	0.0162
942	C9H20	4-METHYLOCTANE	-1.7871	0.9856	587.65	160	558	0.1042	0.1282	0.0163
943	C9H20	n-NONANE	-1.7865	1.1033	595.65	220	566	0.1313	0.1486	0.0163
944	C9H20	2,2,3,3-TETRAMETHYLPENTANE	-1.7988	1.0142	610.85	263	580	0.1093	0.1134	0.0159
945	C9H20	2,2,3,4-TETRAMETHYLPENTANE	-1.7758	0.9829	592.15	152	563	0.1069	0.1326	0.0168
946	C9H20	2,2,4,4-TETRAMETHYLPENTANE	-1.7906	0.9966	571.35	207	543	0.1039	0.1198	0.0162
947	C9H20	2,3,3,4-TETRAMETHYLPENTANE	-2.6081	2.0588	607.50	171	577	0.1229	0.1795	0.0025
948	C9H20	2,2,5-TRIMETHYLHEXANE	-2.6081	2.0588	568.05	167	540	0.1139	0.1755	0.0025
949	C9H20O	2,6-DIMETHYL-4-HEPTANOL	-1.6993	0.9921	603.00	208	573	0.1310	0.1489	0.0200
950	C9H20O	1-NONANOL	-1.4848	0.8197	673.00	268	639	0.1617	0.1648	0.0327
951	C9H20O	2-NONANOL	-1.6670	0.9763	623.00	238	592	0.1392	0.1499	0.0215
952	C9H20S	n-NONYL MERCAPTAN	-1.7464	1.0012	681.00	253	647	0.1267	0.1326	0.0179
953	C9H20S	BUTYL-PENTYL-SULFIDE	-1.1790	0.1641	681.56	231	647	0.0912	0.0924	0.0662
954	C9H20S	ETHYL-HEPTYL-SULFIDE	-1.1790	0.1641	681.56	231	647	0.0912	0.0924	0.0662
955	C9H20S	HEXYL-PROPYL-SULFIDE	-1.1790	0.1641	681.56	231	647	0.0912	0.0924	0.0662
956	C9H20S	METHYL-OCTYL-SULFIDE	-1.1790	0.1641	681.56	231	647	0.0912	0.0924	0.0662
957	C9H21N	n-NONYLAMINE	-1.0924	0.3278	648.00	273	616	0.1522	0.1531	0.0808
958	C9H21N	TRIPROPYLAMINE	-1.5650	0.7943	577.50	180	549	0.1204	0.1393	0.0272
959	C10H6O8	PYROMELLITIC ACID	-1.8637	1.0653	893.00	554	848	-----	0.0838	0.0137
960	C10H7Br	1-BROMONAPHTHALENE	-1.8040	0.9932	824.00	279	783	0.1174	0.1180	0.0157
961	C10H7Cl	1-CHLORONAPHTHALENE	-1.7189	0.9531	785.00	269	746	0.1296	0.1318	0.0191
962	C10H8	NAPHTHALENE	-1.0304	0.1860	748.35	353	711	-----	0.1326	0.0932
963	C10H8	AZULENE	-1.0959	0.1136	773.48	174	735	0.1007	0.1022	0.0802
964	C10H9N	QUINALDINE	-1.7357	0.9944	773.00	272	734	0.1347	0.1366	0.0184
965	C10H10	m-DIVINYLBENZENE	-1.7132	0.9876	692.00	206	657	0.1341	0.1492	0.0194
966	C10H10	1-METHYLINDENE	-1.7259	1.0236	703.00	350	668	-----	0.1265	0.0188
967	C10H10	2-METHYLINDENE	-1.7191	1.0236	684.00	353	650	-----	0.1257	0.0191
968	C10H10O4	DIMETHYL PHTHALATE	-1.6451	0.9331	766.00	272	728	0.1464	0.1483	0.0226
969	C10H10O4	DIMETHYL TEREPHTHALATE	-1.7191	1.0367	772.00	414	733	-----	0.1254	0.0191
970	C10H12	DICYCLOPENTADIENE	-1.6755	0.9683	660.00	307	627	-----	0.1329	0.0211
971	C10H12	1,2,3,4-TETRAHYDRONAPHTHALENE	-1.1461	0.3017	720.15	237	684	0.1297	0.1322	0.0714
972	C10H12O	ANETHOLE	-------	-------	-------	----	----	-----	-----	-----
973	C10H12O4	DIALLYL MALEATE	-1.7974	1.0330	693.00	226	658	0.1208	0.1314	0.0159
974	C10H14	n-BUTYLBENZENE	-1.8105	1.0843	660.55	185	628	0.1267	0.1482	0.0155
975	C10H14	sec-BUTYLBENZENE	-1.7383	0.9837	664.54	198	631	0.1235	0.1397	0.0183
976	C10H14	tert-BUTYLBENZENE	-1.5954	0.7911	660.00	215	627	0.1177	0.1277	0.0254
977	C10H14	1,2,3,4-TETRAMETHYLBENZENE	-1.1665	0.1945	695.10	267	660	0.0998	0.1003	0.0682
978	C10H14	m-CYMENE	-1.7415	0.9895	657.00	209	624	0.1233	0.1378	0.0181
979	C10H14	o-CYMENE	-1.7317	0.9853	662.00	202	629	0.1255	0.1414	0.0185
980	C10H14	p-CYMENE	-1.7448	0.9921	653.15	205	620	0.1227	0.1380	0.0180
981	C10H14	m-DIETHYLBENZENE	-1.7413	0.9888	663.00	189	630	0.1237	0.1418	0.0181
982	C10H14	o-DIETHYLBENZENE	-1.7429	0.9984	668.00	242	635	0.1260	0.1342	0.0181
983	C10H14	p-DIETHYLBENZENE	-1.7474	0.9969	657.96	230	625	0.1235	0.1340	0.0179
984	C1OH14	2-ETHYL-m-XYLENE	-1.7417	1.0025	671.00	257	637	0.1276	0.1328	0.0181
985	C10H14	2-ETHYL-p-XYLENE	-1.7425	0.9982	663.00	220	630	0.1256	0.1382	0.0181
986	C10H14	3-ETHYL-o-XYLENE	-1.7371	0.9951	680.00	224	646	0.1279	0.1393	0.0183
987	C10H14	4-ETHYL-m-XYLENE	-1.7409	0.9952	665.00	210	632	0.1255	0.1399	0.0182
988	C10H14	4-ETHYL-o-XYLENE	-1.7381	0.9903	667.00	206	634	0.1253	0.1403	0.0183
989	C1OH14	5-ETHYL-m-XYLENE	-1.7354	0.9854	655.00	189	622	0.1239	0.1423	0.0184
990	C10H14	ISOBUTYLBENZENE	-1.7490	0.9961	650.15	222	618	0.1222	0.1343	0.0178
991	C10H14	1,2,3,5-TETRAMETHYLBENZENE	-1.7441	1.0028	679.00	249	645	0.1276	0.1343	0.0180
992	C10H14	1,2,4,5-TETRAMETHYLBENZENE	-1.3032	0.4911	675.15	352	641	-----	0.1224	0.0498
993	C10H14O	p-tert-BUTYLPHENOL	-1.6377	0.9639	734.00	372	697	-----	0.1372	0.0230
994	C10H14O2	p-tert-BUTYLCATECHOL	-1.7032	1.0160	776.00	325	737	-----	0.1436	0.0198
995	C10H15N	N,N-DIETHYLANILINE	-1.6971	0.9932	702.00	235	667	0.1416	0.1514	0.0201
996	C10H15N	2,6-DIETHYLANILINE	-1.7114	1.0163	678.00	277	644	0.1412	0.1426	0.0194
997	C10H16	CAMPHENE	-1.7932	1.0634	638.00	320	606	-----	0.1161	0.0161
998	C10H16	D-LIMONENE	-1.7644	1.0003	660.00	199	627	0.1197	0.1357	0.0172
999	C10H16	alpha-PHELLANDRENE	-1.7685	1.0155	649.00	220	617	0.1212	0.1339	0.0170
1000	C1OH16	beta-PHELLANDRENE	-1.7421	0.9796	648.00	220	616	0.1200	0.1322	0.0181
1001	C10H16	alpha-PINENE	-1.7430	0.9964	632.00	209	600	0.1223	0.1376	0.0181
1002	C10H16	beta-PINENE	-1.7360	0.9914	643.00	212	611	0.1241	0.1386	0.0184
1003	C10H16	alpha-TERPINENE	-1.7476	0.9807	652.00	220	619	0.1191	0.1311	0.0179
1004	C1OH16	gamma-TERPINENE	-1.7441	0.9824	661.00	220	628	0.1212	0.1332	0.0180
1005	C10H16	TERPINOLENE	-1.7885	1.0404	672.00	220	638	0.1234	0.1360	0.0163
1006	C10H16O	CAMPHOR	-1.5928	0.9230	709.00	453	674	-----	0.1198	0.0255
1007	C10H18	1-DECYNE	-1.1727	0.1959	632.49	229	601	0.0979	0.0996	0.0672
1008	C10H18	cis-DECAHYDRONANPHTALENE	-1.3879	0.5151	702.25	230	667	0.1127	0.1171	0.0409
1009	C10H18	trans-DECAHYDRONAPHTHALENE	-1.3710	0.4977	687.05	243	653	0.1127	0.1161	0.0426
1010	C10H18O4	SEBACIC ACID	-1.6287	0.9779	815.00	408	774	-----	0.1448	0.0235
1011	C10H20	n-BUTYLCYCLOHEXANE	-1.7790	0.9871	667.00	198	634	0.1133	0.1282	0.0166
1012	C10H20	1-CYCLOPENTYLPENTANE	-1.1482	0.1587	647.49	190	615	0.0966	0.0988	0.0711
1013	C10H20	1-DECENE	-1.7491	1.0443	617.05	207	586	0.1305	0.1489	0.0178
1014	C10H20O	1-DECANAL	-1.5334	0.8250	657.00	267	624	0.1448	0.1479	0.0293

Table 24-1 THERMAL CONDUCTIVITY OF LIQUID - ORGANIC COMPOUNDS (continued)

			$\log_{10} k_{liq} = A + B [1 - T/C]^{2/7}$						$(k_{liq} - W/(m\ K),\ T - K)$		
NO	FORMULA	NAME	A	B	C	TMIN	TMAX	k_{liq} @25 C	k_{liq} @TMIN	k_{liq} @TMAX
1015	C10H20O2	n-DECANOIC ACID	-1.4112	0.6591	713.00	305	677	-----	0.1394	0.0388
1016	C10H20O2	2-ETHYLHEXYL ACETATE	-1.7576	1.0042	639.00	180	607	0.1207	0.1414	0.0175
1017	C10H20O2	ISOPENTYL ISOVALERATE	-1.7570	0.9960	637.00	215	605	0.1188	0.1323	0.0175
1018	C10H22	n-DECANE	-1.7768	1.0839	618.45	243	588	0.1322	0.1425	0.0167
1019	C10H22	2-METHYLNONANE	-1.6900	0.9850	610.00	199	580	0.1328	0.1526	0.0204
1020	C10H22	3-METHYLNONANE	-1.6908	0.9873	613.00	188	582	0.1335	0.1557	0.0204
1021	C10H22	4-METHYLNONANE	-1.6992	0.9962	610.00	174	580	0.1328	0.1587	0.0200
1022	C10H22	5-METHYLNONANE	-1.6942	0.9883	610.00	185	580	0.1323	0.1554	0.0202
1023	C10H22	3-ETHYLOCTANE	-1.7768	1.0839	613.60	185	583	0.1317	0.1567	0.0167
1024	C10H22	4-ETHYLOCTANE	-1.7768	1.0839	609.60	185	579	0.1312	0.1564	0.0167
1025	C10H22	2,2-DIMETHYLOCTANE	-1.8164	1.0045	602.00	225	572	0.1023	0.1134	0.0153
1026	C10H22	2,3-DIMETHYLOCTANE	-1.7768	1.0839	613.20	219	583	0.1316	0.1481	0.0167
1027	C10H22	2,4-DIMETHYLOCTANE	-1.7768	1.0839	599.40	219	569	0.1299	0.1468	0.0167
1028	C10H22	2,5-DIMETHYLOCTANE	-1.7768	1.0839	603.00	219	573	0.1304	0.1472	0.0167
1029	C10H22	2,6-DIMETHYLOCTANE	-1.7768	1.0839	603.10	219	573	0.1304	0.1472	0.0167
1030	C10H22	2,7-DIMETHYLOCTANE	-1.7768	1.0839	602.90	219	573	0.1304	0.1472	0.0167
1031	C10H22	3,3-DIMETHYLOCTANE	-1.7768	1.0839	612.10	219	581	0.1315	0.1479	0.0167
1032	C10H22	3,4-DIMETHYLOCTANE	-1.7768	1.0839	614.00	219	583	0.1317	0.1481	0.0167
1033	C10H22	3,5-DIMETHYLOCTANE	-1.7768	1.0839	606.30	219	576	0.1308	0.1474	0.0167
1034	C10H22	3,6-DIMETHYLOCTANE	-1.7768	1.0839	608.30	219	578	0.1310	0.1476	0.0167
1035	C10H22	4,4-DIMETHYLOCTANE	-1.7768	1.0839	606.90	219	577	0.1308	0.1475	0.0167
1036	C10H22	4,5-DIMETHYLOCTANE	-1.7768	1.0839	612.20	219	582	0.1315	0.1480	0.0167
1037	C10H22	4-PROPYLHEPTANE	-1.7768	1.0839	601.00	219	571	0.1301	0.1470	0.0167
1038	C10H22	4-ISOPROPYLHEPTANE	-1.7768	1.0839	607.60	219	577	0.1309	0.1475	0.0167
1039	C10H22	3-ETHYL-2-METHYLHEPTANE	-1.7768	1.0839	610.90	219	580	0.1313	0.1478	0.0167
1040	C10H22	4-ETHYL-2-METHYLHEPTANE	-1.7768	1.0839	601.80	219	572	0.1302	0.1471	0.0167
1041	C10H22	5-ETHYL-2-METHYLHEPTANE	-1.7768	1.0839	606.70	219	576	0.1308	0.1474	0.0167
1042	C10H22	3-ETHYL-3-METHYLHEPTANE	-1.7768	1.0839	620.00	219	589	0.1324	0.1487	0.0167
1043	C10H22	4-ETHYL-3-METHYLHEPTANE	-1.7768	1.0839	614.30	219	584	0.1317	0.1482	0.0167
1044	C10H22	3-ETHYL-5-METHYLHEPTANE	-1.7768	1.0839	606.60	219	576	0.1308	0.1474	0.0167
1045	C10H22	3-ETHYL-4-METHYLHEPTANE	-1.7768	1.0839	615.50	219	585	0.1319	0.1483	0.0167
1046	C10H22	4-ETHYL-4-METHYLHEPTANE	-1.7768	1.0839	615.70	219	585	0.1319	0.1483	0.0167
1047	C10H22	2,2,3-TRIMETHYLHEPTANE	-1.7768	1.0839	611.70	219	581	0.1314	0.1479	0.0167
1048	C10H22	2,2,4-TRIMETHYLHEPTANE	-1.7768	1.0839	594.50	219	565	0.1293	0.1464	0.0167
1049	C10H22	2,2,5-TRIMETHYLHEPTANE	-1.7768	1.0839	598.00	219	568	0.1297	0.1467	0.0167
1050	C10H22	2,2,6-TRIMETHYLHEPTANE	-1.7768	1.0839	593.40	219	564	0.1292	0.1462	0.0167
1051	C10H22	2,3,3-TRIMETHYLHEPTANE	-1.7768	1.0839	617.50	219	587	0.1321	0.1485	0.0167
1052	C10H22	2,3,4-TRIMETHYLHEPTANE	-1.7768	1.0839	613.70	219	583	0.1317	0.1481	0.0167
1053	C10H22	2,3,5-TRIMETHYLHEPTANE	-1.7768	1.0839	612.80	219	582	0.1316	0.1480	0.0167
1054	C10H22	2,3,6-TRIMETHYLHEPTANE	-1.7768	1.0839	604.10	219	574	0.1305	0.1473	0.0167
1055	C10H22	2,4,4-TRIMETHYLHEPTANE	-1.7768	1.0839	600.30	219	570	0.1300	0.1469	0.0167
1056	C10H22	2,4,5-TRIMETHYLHEPTANE	-1.7768	1.0839	606.90	219	577	0.1308	0.1475	0.0167
1057	C10H22	2,4,6-TRIMETHYLHEPTANE	-1.7768	1.0839	590.30	219	561	0.1288	0.1459	0.0167
1058	C10H22	2,5,5-TRIMETHYLHEPTANE	-1.7768	1.0839	602.90	219	573	0.1304	0.1472	0.0167
1059	C10H22	3,3,4-TRIMETHYLHEPTANE	-1.7768	1.0839	622.10	219	591	0.1327	0.1489	0.0167
1060	C10H22	3,3,5-TRIMETHYLHEPTANE	-1.7768	1.0839	609.50	219	579	0.1312	0.1477	0.0167
1061	C10H22	3,4,4-TRIMETHYLHEPTANE	-1.7768	1.0839	620.90	219	590	0.1325	0.1488	0.0167
1062	C10H22	3,4,5-TRIMETHYLHEPTANE	-1.7768	1.0839	612.80	219	582	0.1316	0.1480	0.0167
1063	C10H22	3-ISOPROPYL-2-METHYLHEXANE	-1.7768	1.0839	623.40	219	592	0.1328	0.1490	0.0167
1064	C10H22	3,3-DIETHYLHEXANE	-1.7768	1.0839	627.80	219	596	0.1333	0.1493	0.0167
1065	C10H22	3,4-DIETHYLHEXANE	-1.7768	1.0839	618.80	219	588	0.1323	0.1486	0.0167
1066	C10H22	3-ETHYL-2,2-DIMETHYLHEXANE	-1.7768	1.0839	611.70	219	581	0.1314	0.1479	0.0167
1067	C10H22	4-ETHYL-2,2-DIMETHYLHEXANE	-1.7768	1.0839	594.60	219	565	0.1293	0.1464	0.0167
1068	C10H22	3-ETHYL-2,3-DIMETHYLHEXANE	-1.7768	1.0839	626.80	219	595	0.1332	0.1492	0.0167
1069	C10H22	4-ETHYL-2,3-DIMETHYLHEXANE	-1.7768	1.0839	617.30	219	586	0.1321	0.1484	0.0167
1070	C10H22	3-ETHYL-2,4-DIMETHYLHEXANE	-1.7768	1.0839	616.10	219	585	0.1320	0.1483	0.0167
1071	C10H22	4-ETHYL-2,4-DIMETHYLHEXANE	-1.7768	1.0839	620.90	219	590	0.1325	0.1488	0.0167
1072	C10H22	3-ETHYL-2,5-DIMETHYLHEXANE	-1.7768	1.0839	603.50	219	573	0.1304	0.1472	0.0167
1073	C10H22	4-ETHYL-3,3-DIMETHYLHEXANE	-1.7768	1.0839	625.70	219	594	0.1331	0.1491	0.0167
1074	C10H22	3-ETHYL-3,4-DIMETHYLHEXANE	-1.7768	1.0839	624.50	219	593	0.1329	0.1491	0.0167
1075	C10H22	2,2,3,3-TETRAMETHYLHEXANE	-1.7768	1.0839	623.00	219	592	0.1328	0.1490	0.0167
1076	C10H22	2,2,3,4-TETRAMETHYLHEXANE	-1.7768	1.0839	620.40	219	589	0.1325	0.1487	0.0167
1077	C10H22	2,2,3,5-TETRAMETHYLHEXANE	-1.7768	1.0839	601.30	219	571	0.1302	0.1470	0.0167
1078	C10H22	2,2,4,4-TETRAMETHYLHEXANE	-1.7768	1.0839	610.20	219	580	0.1313	0.1478	0.0167
1079	C10H22	2,2,4,5-TETRAMETHYLHEXANE	-1.7768	1.0839	598.50	219	569	0.1298	0.1468	0.0167
1080	C10H22	2,2,5,5-TETRAMETHYLHEXANE	-1.7768	1.0839	581.40	261	552	0.1276	0.1336	0.0167
1081	C10H22	2,3,3,4-TETRAMETHYLHEXANE	-1.7768	1.0839	633.10	261	601	0.1339	0.1394	0.0167
1082	C10H22	2,3,3,5-TETRAMETHYLHEXANE	-1.7768	1.0839	610.10	261	580	0.1312	0.1371	0.0167
1083	C10H22	2,3,4,4-TETRAMETHYLHEXANE	-1.7768	1.0839	626.60	261	595	0.1332	0.1388	0.0167
1084	C10H22	2,3,4,5-TETRAMETHYLHEXANE	-1.7768	1.0839	613.20	261	583	0.1316	0.1374	0.0167
1085	C10H22	3,3,4,4-TETRAMETHYLHEXANE	-1.7768	1.0839	646.70	261	614	0.1354	0.1408	0.0167
1086	C10H22	2,4-DIMETHYL-3-ISOPROPYLPENTANE	-1.7768	1.0839	614.40	191	584	0.1318	0.1553	0.0167
1087	C10H22	3,3-DIETHYL-2-METHYLPENTANE	-1.7768	1.0839	639.90	191	608	0.1347	0.1572	0.0167
1088	C10H22	3-ETHYL-2,2,3-TRIMETHYLPENTANE	-1.7768	1.0839	646.00	191	614	0.1353	0.1576	0.0167
1089	C10H22	3-ETHYL-2,2,4-TRIMETHYLPENTANE	-1.7768	1.0839	615.30	191	585	0.1319	0.1554	0.0167
1090	C10H22	3-ETHYL-2,3,4-TRIMETHYLPENTANE	-1.7768	1.0839	642.30	191	610	0.1349	0.1573	0.0167
1091	C10H22	2,2,3,3,4-PENTAMETHYLPENTANE	-1.7768	1.0839	643.80	237	612	0.1351	0.1464	0.0167
1092	C10H22	2,2,3,4,4-PENTAMETHYLPENTANE	-1.7768	1.0839	627.30	234	596	0.1333	0.1456	0.0167

Table 24-1 THERMAL CONDUCTIVITY OF LIQUID - ORGANIC COMPOUNDS (continued)

			$\log_{10} k_{liq} = A + B [1 - T/C]^{2/7}$					$(k_{liq}$ - W/(m K), T - K)		
NO	FORMULA	NAME	A	B	C	TMIN	TMAX	k_{liq} @25 C	k_{liq} @TMIN	k_{liq} @TMAX
1093	C10H22O	1-DECANOL	-1.5529	0.8957	690.00	280	656	0.1618	0.1626	0.0280
1094	C10H22O	DI-n-PENTYL ETHER	-1.7493	0.9842	622.00	204	591	0.1168	0.1327	0.0178
1095	C10H22O	ISODECANOL	-1.6789	0.9937	644.00	213	612	0.1423	0.1587	0.0209
1096	C10H22O5	TETRAETHYLENE GLYCOL DIMETHYL ETHER	-1.7110	1.0058	705.00	243	670	0.1408	0.1490	0.0195
1097	C10H22S	n-DECYL MERCAPTAN	-1.7376	1.0045	696.00	248	661	0.1314	0.1382	0.0183
1098	C10H22S	BUTYL-HEXYL-SULFIDE	-1.1908	0.1571	701.03	238	666	0.0878	0.0886	0.0644
1099	C10H22S	ETHYL-OCTYL-SULFIDE	-1.1908	0.1571	701.03	238	666	0.0878	0.0886	0.0644
1100	C10H22S	HEPTYL-PROPYL-SULFIDE	-1.1908	0.1571	701.03	238	666	0.0878	0.0886	0.0644
1101	C10H22S	METHYL-NONYL-SULFIDE	-1.1908	0.1571	701.03	238	666	0.0878	0.0886	0.0644
1102	C10H22S	PENTYL-SULFIDE	-1.1908	0.1571	701.03	238	666	0.0878	0.0886	0.0644
1103	C10H22S2	PENTYL-DISULFIDE	-1.2107	0.1354	726.94	214	691	0.0805	0.0815	0.0616
1104	C10H23N	n-DECYLAMINE	-1.1406	0.3816	663.00	289	630	0.1518	0.1512	0.0723
1105	C11H10	1-METHYLNAPHTHALENE	-1.7031	0.9920	772.04	243	733	0.1445	0.1517	0.0198
1106	C11H10	2-METHYLNAPHTHALENE	-1.7240	1.0073	761.00	308	723	-----	0.1366	0.0189
1107	C11H14O2	n-BUTYL BENZOATE	-1.7114	0.9826	724.00	252	688	0.1358	0.1416	0.0194
1108	C11H16	n-PENTYLBENZENE	-1.7786	1.0126	679.90	198	646	0.1202	0.1360	0.0166
1109	C11H16O	p-tert-AMYLPHENOL	-1.6402	0.9691	751.00	366	713	-----	0.1408	0.0229
1110	C11H20	1-UNDECYNE	-1.1924	0.1964	650.99	248	618	0.0939	0.0949	0.0642
1111	C11H20O2	2-ETHYLHEXYL ACRYLATE	-1.7687	1.0030	655.00	183	622	0.1187	0.1378	0.0170
1112	C11H22	1-UNDECENE	-1.7475	1.0119	638.00	224	606	0.1252	0.1378	0.0179
1113	C11H22	1-CYCLOPENTYLHEXANE	-1.1636	0.1535	667.67	200	634	0.0925	0.0942	0.0686
1114	C11H22	PENTYLCYCLOHEXANE	-1.1693	0.1604	674.01	216	640	0.0926	0.0940	0.0677
1115	C11H22O	1-UNDECANAL	-1.7583	1.0167	672.00	273	638	0.1263	0.1284	0.0174
1116	C11H24	n-UNDECANE	-1.6318	0.9325	638.76	248	607	0.1404	0.1482	0.0233
1117	C11H24O	1-UNDECANOL	-1.6798	0.9957	704.00	289	669	0.1482	0.1470	0.0209
1118	C11H24S	UNDECYL MERCAPTAN	-1.7714	1.0137	710.00	270	675	0.1248	0.1272	0.0169
1119	C11H24S	BUTYL-HEPTYL-SULFIDE	-1.2064	0.1572	717.91	255	682	0.0848	0.0853	0.0622
1120	C11H24S	DECYL-METHYL-SULFIDE	-1.2064	0.1572	717.91	255	682	0.0848	0.0853	0.0622
1121	C11H24S	ETHYL-NONYL-SULFIDE	-1.2064	0.1572	717.91	255	682	0.0848	0.0853	0.0622
1122	C11H24S	OCTYL-PROPYL-SULFIDE	-1.2064	0.1572	717.91	255	682	0.0848	0.0853	0.0622
1123	C12H8O	DIBENZOFURAN	-1.7205	0.9746	837.80	356	796	-----	0.1265	0.0190
1124	C12H9N	DIBENZOPYRROLE	-1.6626	1.0379	899.00	518	854	-----	0.1357	0.0217
1125	C12H10	ACENAPHTHENE	-1.2648	0.4524	803.15	367	763	-----	0.1289	0.0544
1126	C12H10	BIPHENYL	-1.4292	0.6647	789.26	342	750	-----	0.1347	0.0372
1127	C12H10O	DIPHENYL ETHER	-1.5285	0.7853	763.00	300	725	-----	0.1398	0.0296
1128	C12H11N	p-AMINODIPHENYL	-1.7466	1.0067	817.00	326	776	-----	0.1303	0.0179
1129	C12H11N	DIPHENYLAMINE	-1.2631	0.4669	817.00	326	776	-----	0.1369	0.0546
1130	C12H11N3	p-AMINOAZOBENZENE	-1.7397	1.0227	877.00	401	833	-----	0.1283	0.0182
1131	C12H11N3	1,3-DIPHENYLTRIAZENE	-1.7757	1.0175	845.00	372	803	-----	0.1191	0.0168
1132	C12H12	1,2-DIMETHYLNAPHTHALENE	-1.1645	0.1502	775.34	272	737	0.0925	0.0927	0.0685
1133	C12H12	1,3-DIMETHYLNAPHTHALENE	-1.1637	0.1491	773.76	269	735	0.0925	0.0927	0.0686
1134	C12H12	1,4-DIMETHYLNAPHTHALENE	-1.1678	0.1546	776.78	281	738	0.0926	0.0927	0.0680
1135	C12H12	1,5-DIMETHYLNAPHTHALENE	-1.2233	0.2283	773.47	355	735	-----	0.0924	0.0598
1136	C12H12	1,6-DIMETHYLNAPHTHALENE	-1.1606	0.1450	770.60	259	732	0.0924	0.0928	0.0691
1137	C12H12	1,7-DIMETHYLNAPHTHALENE	-1.1610	0.1455	770.60	260	732	0.0924	0.0928	0.0690
1138	C12H12	2,3-DIMETHYLNAPHTHALENE	-1.2497	0.2636	777.78	378	739	-----	0.0923	0.0563
1139	C12H12	2,6-DIMETHYLNAPHTHALENE	-1.7473	1.0248	777.00	385	738	-----	0.1210	0.0179
1140	C12H12	2,7-DIMETHYLNAPHTHALENE	-1.7439	1.0203	778.00	370	739	-----	0.1238	0.0180
1141	C12H12	1-ETHYLNAPHTHALENE	-1.7225	0.9971	776.00	259	737	0.1398	0.1441	0.0189
1142	C12H12	2-ETHYLNAPHTHALENE	-1.1645	0.1491	774.90	266	736	0.0923 C	0.0926	0.0685
1143	C12H12N2	p-AMINODIPHENYLAMINE	-1.7000	1.0101	867.00	341	824	-----	0.1469	0.0200
1144	C12H12N2	HYDRAZOBENZENE	-------	------	------	----	----	-----	-----	-----
1145	C12H14	1,2,3-TRIMETHYLINDENE	-1.7299	1.0200	726.00	345	690	-----	0.1279	0.0186
1146	C12H14O4	DIETHYL PHTHALATE	-1.3265	0.5567	757.00	269	719	0.1432	0.1447	0.0472
1147	C12H16	CYCLOHEXYLBENZENE	-1.7275	0.9671	744.00	280	707	0.1282	0.1288	0.0187
1148	C12H18	m-DIISOPROPYLBENZENE	-1.7699	0.9902	684.00	210	650	0.1177	0.1306	0.0170
1149	C12H18	p-DIISOPROPYLBENZENE	-1.7814	1.0018	689.00	256	655	0.1177	0.1225	0.0165
1150	C12H18	n-HEXYLBENZENE	-1.7737	0.9910	698.00	212	663	0.1179	0.1300	0.0168
1151	C12H18	1,2,3-TRIETHYLBENZENE	-1.1710	0.1494	684.37	207	650	0.0903	0.0918	0.0675
1152	C12H18	1,2,4-TRIETHYLBENZENE	-1.1710	0.1494	684.37	207	650	0.0903	0.0918	0.0675
1153	C12H18	1,3,5-TRIETHYLBENZENE	-1.1717	0.1502	682.28	207	648	0.0903	0.0918	0.0673
1154	C12H18	HEXAMETHYLBENZENE	-1.4998	0.5901	758.00	439	720	-----	0.0894	0.0316
1155	C12H20O4	DIBUTYL MALEATE	-1.8319	1.0337	716.00	188	680	0.1133	0.1290	0.0147
1156	C12H22	BICYCLOHEXYL	-1.7808	1.0044	727.00	277	691	0.1211	0.1221	0.0166
1157	C12H22	1-DODECYNE	-1.2025	0.1863	668.16	254	635	0.0901	0.0909	0.0627
1158	C12H23N	DICYCLOHEXYLAMINE	-1.8152	1.0039	737.00	273	700	0.1123	0.1139	0.0153
1159	C12H24	1-DODECENE	-1.6991	0.9948	657.00	238	624	0.1374	0.1473	0.0200
1160	C12H24	1-CYCLOPENTYLHEPTANE	-1.1832	0.1570	675.00	220	645	0.0891	0.0904	0.0656
1161	C12H24	1-CYCLOHEXYLHEXANE	-1.2028	0.1818	691.81	264	657	0.0895	0.0900	0.0627
1162	C12H24O	1-DODECANAL	-1.7316	1.0818	685.00	285	651	0.1539	0.1535	0.0186
1163	C12H24O2	n-DODECANOIC ACID	-1.4322	0.7077	734.00	317	697	-----	0.1455	0.0370
1164	C12H26	n-DODECANE	-1.7989	1.1109	658.20	264	625	0.1368	0.1415	0.0159
1165	C12H26O	DI-n-HEXYL ETHER	-1.5209	0.7658	658.00	230	625	0.1329	0.1415	0.0301
1166	C12H26O	1-DODECANOL	-1.3911	0.6474	721.00	297	685	0.1461	0.1443	0.0406
1167	C12H26O3	DIETHYLENE GLYCOL DI-n-BUTYL ETHER	-1.4951	0.7795	680.00	213	646	0.1465	0.1586	0.0320
1168	C12H26S	n-DODECYL MERCAPTAN	-1.7831	1.0100	724.00	265	688	0.1216	0.1247	0.0165
1169	C12H26S	BUTYL-OCTYL-SULFIDE	-1.2169	0.1517	733.68	259	697	0.0820	0.0824	0.0607
1170	C12H26S	DECYL-ETHYL-SULFIDE	-1.2169	0.1517	733.68	259	697	0.0820	0.0824	0.0607

Table 24-1 THERMAL CONDUCTIVITY OF LIQUID - ORGANIC COMPOUNDS (continued)

			$\log_{10} k_{liq} = A + B [1 - T/C]^{2/7}$					$(k_{liq} - W/(m\ K),\ T - K)$		
NO	FORMULA	NAME	A	B	C	TMIN	TMAX	k_{liq} @25 C	k_{liq} @TMIN	k_{liq} @TMAX
1171	C12H26S	HEXYL-SULFIDE	-1.2169	0.1517	733.68	259	697	0.0820	0.0824	0.0607
1172	C12H26S	METHYL-UNDECYL-SULFIDE	-1.2169	0.1517	733.68	259	697	0.0820	0.0824	0.0607
1173	C12H26S	NONYL-PROPYL-SULFIDE	-1.2169	0.1517	733.68	259	697	0.0820	0.0824	0.0607
1174	C12H26S2	HEXYL-DISULFIDE	-1.2345	0.1326	747.10	225	710	0.0759	0.0766	0.0583
1175	C12H27BO3	TRI-n-BUTYL BORATE	-------	-------	-------	-----	-----	-----	-----	-----
1176	C12H27N	DODECYLAMINE	-1.1301	0.3850	696.00	301	661	-----	0.1561	0.0741
1177	C12H27N	TRI-n-BUTYLAMINE	-1.4764	0.6715	644.00	203	612	0.1218	0.1325	0.0334
1178	C13H10	FLUORENE	-1.6208	1.0116	870.00	388	827	-----	0.1672	0.0239
1179	C13H10O	BENZOPHENONE	-1.5854	0.9864	816.00	321	775	-----	0.1825	0.0260
1180	C13H12	DIPHENYLMETHANE	-1.4751	0.7029	768.00	298	730	0.1367	0.1349	0.0335
1181	C13H14	1-PROPYLNAPHTHALENE	-1.1784	0.1452	771.45	265	733	0.0887	0.0890	0.0663
1182	C13H14	2-PROPYLNAPHTHALENE	-1.1803	0.1477	772.44	270	734	0.0888	0.0890	0.0660
1183	C13H14	2ETHYL-3-METHYLNAPHTHALENE	-1.2206	0.2020	776.44	344	738	-----	0.0888	0.0602
1184	C13H14	2ETHYL-6-METHYLNAPHTHALENE	-1.2069	0.1835	766.56	318	728	-----	0.0889	0.0621
1185	C13H14	2ETHYL-7-METHYLNAPHTHALENE	-1.2069	0.1835	766.56	318	728	-----	0.0889	0.0621
1186	C13H20	n-HEPTYLBENZENE	-1.7909	0.9965	714.00	225	678	0.1156	0.1251	0.0162
1187	C13H24	1-TRIDECYNE	-1.2174	0.1846	684.11	268	650	0.0870	0.0873	0.0606
1188	C13H26	1-TRIDECENE	-1.7092	1.0068	675.00	250	641	0.1390	0.1462	0.0195
1189	C13H26	1-CYCLOPENTYLOCTANE	-1.1959	0.1514	702.06	229	667	0.0858	0.0868	0.0637
1190	C13H26	1-CYCLOHEXYLHEPTANE	-1.1999	0.1563	708.63	243	673	0.0859	0.0866	0.0631
1191	C13H26O	1-TRIDECANAL	-1.7559	1.0201	700.00	288	665	0.1302	0.1293	0.0175
1192	C13H26O2	n-BUTYL NONANOATE	-1.8021	0.9790	652.00	235	619	0.1047	0.1127	0.0158
1193	C13H26O2	METHYL DODECANOATE	-1.7783	0.9870	712.00	278	676	0.1167	0.1175	0.0167
1194	C13H28	n-TRIDECANE	-1.6564	0.9372	675.80	268	642	0.1372	0.1402	0.0221
1195	C13H28O	1-TRIDECANOL	-1.8051	1.1252	731.00	304	694	-----	0.1410	0.0157
1196	C13H28S	BUTYL-NONYL-SULFIDE	-1.2297	0.1507	748.42	271	711	0.0795	0.0797	0.0589
1197	C13H28S	DECYL-PROPYL-SULFIDE	-1.2297	0.1507	748.42	271	711	0.0795	0.0797	0.0589
1198	C13H28S	DODECYL-METHYL-SULFIDE	-1.2297	0.1507	748.42	271	711	0.0795	0.0797	0.0589
1199	C13H28S	ETHYL-UNDECYL-SULFIDE	-1.2297	0.1507	748.42	271	711	0.0795	0.0797	0.0589
1200	C13H28S	1-TRIDECANETHIOL	-1.2359	0.1591	742.13	282	705	0.0797	0.0797	0.0581
1201	C14H8O2	ANTHRAQUINONE	-1.5523	0.9688	900.00	559	855	-----	0.1456	0.0280
1202	C14H10	ANTHRACENE	-1.2472	0.4397	873.00	489	829	-----	0.1241	0.0566
1203	C14H10	DIPHENYLACETYLENE	-1.7656	1.0030	832.00	336	790	-----	0.1232	0.0172
1204	C14H10	PHENANTHRENE	-1.0147	0.1435	869.25	372	826	-----	0.1277	0.0967
1205	C14H12	cis-STILBENE	-1.7315	0.9996	757.00	268	719	0.1364	0.1391	0.0186
1206	C14H12	trans-STILBENE	-1.7801	1.0841	820.00	397	779	-----	0.1271	0.0166
1207	C14H12O2	BENZYL BENZOATE	-1.7335	0.9898	820.00	293	779	0.1369	0.1354	0.0185
1208	C14H14	1,1-DIPHENYLETHANE	-1.7429	0.9996	775.00	255	736	0.1340	0.1388	0.0181
1209	C14H14	1,2-DIPHENYLETHANE	-1.7400	1.0102	780.00	324	741	-----	0.1310	0.0182
1210	C14H14O	DIBENZYL ETHER	-1.7666	0.9923	777.00	277	738	0.1252	0.1261	0.0171
1211	C14H16	1-n-BUTYLNAPHTHALENE	-1.7347	0.9739	792.00	253	752	0.1307	0.1354	0.0184
1212	C14H16	2-BUTYLNAPHTHALENE	-1.1916	0.1413	780.96	268	742	0.0854	0.0856	0.0643
1213	C14H22	n-OCTYLBENZENE	-1.8063	1.0017	729.00	237	693	0.1137	0.1209	0.0156
1214	C14H22	1,2,3,4-TETRAETHYLBENZENE	-1.2277	0.1829	708.20	285	673	0.0849	0.0848	0.0592
1215	C14H22	1,2,3,5-TETRAETHYLBENZENE	-1.2276	0.1827	707.52	284	672	0.0849	0.0848	0.0592
1216	C14H22	1,2,4,5-TETRAETHYLBENZENE	-1.2273	0.1823	706.85	283	672	0.0848	0.0848	0.0593
1217	C14H22O	p-tert-OCTYLPHENOL	-1.6666	0.9739	765.00	359	727	-----	0.1365	0.0215
1218	C14H28	1-TETRADECENE	-1.6917	0.9840	692.00	260	657	0.1399	0.1447	0.0203
1219	C14H28	1-CYCLOPENTYLNONANE	-1.2111	0.1517	716.95	244	681	0.0830	0.0837	0.0615
1220	C14H28	1-CYCLOHEXYLOCTANE	-1.2133	0.1542	723.61	253	687	0.0830	0.0835	0.0612
1221	C14H28O2	n-TETRADECANOIC ACID	-1.3018	0.6091	756.00	328	718	-----	0.1621	0.0499
1222	C14H30	n-TETRADECANE	-1.6813	0.9584	692.40	279	658	0.1363	0.1372	0.0208
1223	C14H30O	1-TETRADECANOL	-1.4537	0.7984	741.00	311	704	-----	0.1668	0.0352
1224	C14H30S	BUTYL-DECYL-SULFIDE	-1.2398	0.1471	762.23	276	724	0.0772	0.0773	0.0576
1225	C14H30S	DODECYL-ETHYL-SULFIDE	-1.2398	0.1471	762.23	276	724	0.0772	0.0773	0.0576
1226	C14H30S	HEPTYL-SULFIDE	-1.2398	0.1471	762.23	276	724	0.0772	0.0773	0.0576
1227	C14H30S	METHYL-TRIDECYL-SULFIDE	-1.2398	0.1471	762.23	276	724	0.0772	0.0773	0.0576
1228	C14H30S	PROPYL-UNDECYL-SULFIDE	-1.2398	0.1471	762.23	276	724	0.0772	0.0773	0.0576
1229	C14H30S	1-TETRADECANETHIOL	-1.2434	0.1519	753.80	279	716	0.0773	0.0774	0.0571
1230	C14H30S2	HEPTYL-DISULFIDE	-1.2559	0.1305	765.96	235	728	0.0720	0.0726	0.0555
1231	C14H31N	TETRADECYLAMINE	-1.7823	1.0245	722.30	311	686	-----	0.1202	0.0165
1232	C15H10N2O2	DIPHENYLMETHANE-4,4'-DIISOCYANATE	-1.8483	1.0108	802.00	311	762	-----	0.1052	0.0142
1233	C15H16O	p-CUMYLPHENOL	-1.7658	1.0169	834.00	346	792	-----	0.1251	0.0171
1234	C15H16O2	BISPHENOL A	-1.6286	0.9637	849.00	426	807	-----	0.1410	0.0235
1235	C15H18	1-PENTYLNAPHTHALENE	-1.1974	0.1289	793.32	251	754	0.0823	0.0827	0.0635
1236	C15H18	2-PENTYLNAPHTHALENE	-1.2015	0.1345	797.48	269	758	0.0824	0.0826	0.0629
1237	C15H24	n-NONYLBENZENE	-1.8155	1.0011	741.00	249	704	0.1119	0.1170	0.0153
1238	C15H24O	2,6-DI-tert-BUTYL-p-CRESOL	-1.6963	0.9674	720.00	344	684	-----	0.1247	0.0201
1239	C15H24O	NONYLPHENOL	-1.6850	1.0169	757.00	279	719	0.1524	0.1532	0.0207
1240	C15H28	1-PENTADECYNE	-1.2395	0.1753	711.41	283	676	0.0814	0.0814	0.0576
1241	C15H30	1-PENTADECENE	-1.7166	1.0132	708.00	269	673	0.1413	0.1443	0.0192
1242	C15H30	1-CYCLOPENTYLDECANE	-1.2229	0.1488	730.64	251	694	0.0804	0.0809	0.0599
1243	C15H30	1-CYCLOHEXYLNONANE	-1.2255	0.1519	737.79	263	701	0.0804	0.0808	0.0595
1244	C15H30O2	PENTADECANOIC ACID	-1.6873	0.9778	766.00	326	728	-----	0.1374	0.0205
1245	C15H32	n-PENTADECANE	-1.7650	1.0668	706.80	283	671	0.1404	0.1404	0.0172
1246	C15H32O	1-PENTADECANOL	-1.2626	0.1871	722.53	317	686	-----	0.0784	0.0546
1247	C15H32S	BUTYL-UNDECYL-SULFIDE	-1.2506	0.1455	775.15	284	736	0.0752	0.0752	0.0562
1248	C15H32S	DODECYL-PROPYL-SULFIDE	-1.2506	0.1455	775.15	284	736	0.0752	0.0752	0.0562

549

Table 24-1 THERMAL CONDUCTIVITY OF LIQUID - ORGANIC COMPOUNDS (continued)

NO	FORMULA	NAME	A	B	C	TMIN	TMAX	k_{liq} @25 C	k_{liq} @TMIN	k_{liq} @TMAX
								$\log_{10} k_{liq} = A + B [1 - T/C]^{2/7}$	(k_{liq} - W/(m K), T - K)	
1249	C15H32S	ETHYL-TRIDECYL-SULFIDE	-1.2506	0.1455	775.15	284	736	0.0752	0.0752	0.0562
1250	C15H32S	METHYL-TETRADECYL-SULFIDE	-1.2506	0.1455	775.15	284	736	0.0752	0.0752	0.0562
1251	C15H32S	1-PENTADECANETHIOL	-1.2558	0.1525	764.77	291	727	0.0753	0.0752	0.0555
1252	C16H10	FLUORANTHENE	-1.7217	1.0163	905.00	383	860	------	0.1371	0.0190
1253	C16H10	PYRENE	-1.1478	0.2832	936.00	424	889	------	0.1223	0.0712
1254	C16H12	1-PHENYLNAPHTHALENE	-1.7308	1.0036	849.00	318	807	------	0.1377	0.0186
1255	C16H20	1-n-HEXYLNAPHTHALENE	-1.7818	0.9943	813.00	255	772	0.1233	0.1273	0.0165
1256	C16H22O4	DIBUTYL PHTHALATE	-1.6470	0.8980	781.00	238	742	0.1367	0.1436	0.0225
1257	C16H26	n-DECYLBENZENE	-1.8290	1.0063	753.00	259	715	0.1102	0.1137	0.0148
1258	C16H26	PENTAETHYLBENZENE	-1.2696	0.2033	723.64	328	687	------	0.0793	0.0538
1259	C16H30	1-HEXADECYNE	-1.2490	0.1704	724.26	288	688	0.0790	0.0789	0.0564
1260	C16H32	n-DECYLCYCLOHEXANE	-1.8492	1.0049	751.25	271	714	0.1048	0.1066	0.0142
1261	C16H32	1-CYCLOPENTYLUNDECANE	-1.2359	0.1488	743.30	263	706	0.0781	0.0784	0.0581
1262	C16H32	1-HEXADECENE	-1.7204	1.0172	722.00	278	686	0.1423	0.1434	0.0190
1263	C16H32O2	n-HEXADECANOIC ACID	-1.6139	0.9634	776.00	336	737	------	0.1569	0.0243
1264	C16H34	n-HEXADECANE	-1.8486	1.1616	720.60	291	685	0.1408	0.1391	0.0142
1265	C16H34O	DI-n-OCTYL ETHER	-1.8375	0.9775	707.00	266	672	0.0996	0.1021	0.0145
1266	C16H34O	1-HEXADECANOL	-1.7942	1.1033	761.00	322	723	------	0.1374	0.0161
1267	C16H34S	BUTYL-DODECYL-SULFIDE	-1.2599	0.1427	787.27	288	748	0.0732	0.0732	0.0550
1268	C16H34S	ETHYL-TETRADECYL-SULFIDE	-1.2593	0.1412	791.68	288	752	0.0731	0.0731	0.0550
1269	C16H34S	METHYL-PENTADECYL-SULFIDE	-1.2599	0.1427	787.27	288	748	0.0732	0.0732	0.0550
1270	C16H34S	OCTYL-SULFIDE	-1.2599	0.1427	787.27	288	748	0.0732	0.0732	0.0550
1271	C16H34S	PROPYL-TRIDECYL-SULFIDE	-1.2599	0.1427	787.27	288	748	0.0732	0.0732	0.0550
1272	C16H34S	1-HEXADECANETHIOL	-1.2641	0.1484	774.68	291	736	0.0733	0.0732	0.0544
1273	C16H34S2	OCTYL-DISULFIDE	-1.2753	0.1288	784.46	244	745	0.0687	0.0691	0.0531
1274	C17H28	n-UNDECYLBENZENE	-1.8347	1.0032	764.00	268	726	0.1087	0.1109	0.0146
1275	C17H32	1-HEPTADECYNE	-1.2593	0.1676	736.21	295	699	0.0768	0.0766	0.0550
1276	C17H34	1-CYCLOPENTYLDODECANE	-1.2463	0.1462	755.17	268	717	0.0759	0.0761	0.0567
1277	C17H34	1-CYCLOHEXYLUNDECANE	-1.2483	0.1486	761.74	279	724	0.0760	0.0760	0.0565
1278	C17H34	1-HEPTADECENE	-1.7147	1.0094	736.00	284	699	0.1431	0.1429	0.0193
1279	C17H36	n-HEPTADECANE	-1.5446	0.8191	733.37	295	697	0.1449	0.1430	0.0285
1280	C17H36O	1-HEPTADECANOL	-1.7864	1.0920	770.00	327	732	------	0.1367	0.0164
1281	C17H36S	BUTYL-TRIDECYL-SULFIDE	-1.2696	0.1411	798.63	294	759	0.0714	0.0713	0.0538
1282	C17H36S	ETHYL-PENTADECYL-SULFIDE	-1.2696	0.1411	798.63	294	759	0.0714	0.0713	0.0538
1283	C17H36S	HEXADECYL-METHYL-SULFIDE	-1.2696	0.1411	798.63	294	759	0.0714	0.0713	0.0538
1284	C17H36S	PROPYL-TETRADECYL-SULFIDE	-1.2696	0.1411	798.63	294	759	0.0714	0.0713	0.0538
1285	C17H36S	1-HEPTADECANETHIOL	-1.2746	0.1481	786.01	300	747	------	0.0713	0.0531
1286	C18H12	CHRYSENE	-1.7230	1.0425	979.00	531	930	------	0.1246	0.0189
1287	C18H14	m-TERPHENYL	-1.2834	0.4800	924.85	360	879	------	0.1347	0.0521
1288	C18H14	o-TERPHENYL	-1.4433	0.6429	890.95	329	846	------	0.1304	0.0360
1289	C18H14	p-TERPHENYL	-1.5067	0.7626	925.95	485	880	------	0.1259	0.0311
1290	C18H15P	TRIPHENYLPHOSPHINE	-1.7693	0.9930	1008.00	354	958	------	0.1262	0.0170
1291	C18H15O4P	TRIPHENYL PHOSPHATE	------	------	------	------	------	------	------	------
1292	C18H16N2	N,N'-DIPHENYL-p-PHENYLENEDIAMINE	-1.7812	1.0231	906.00	409	861	------	0.1175	0.0166
1293	C18H22	2,3-DIMETHYL-2,3-DIPHENYLBUTANE	-1.7541	1.0268	805.00	392	765	------	0.1208	0.0176
1294	C18H22O2	DICUMYL PEROXIDE	-1.7138	1.0042	884.00	311	840	------	0.1466	0.0193
1295	C18H30	n-DODECYLBENZENE	-1.8479	1.0078	774.26	276	736	0.1069	0.1080	0.0142
1296	C18H30	HEXAETHYLBENZENE	-1.3499	0.2837	734.78	401	698	------	0.0745	0.0447
1297	C18H32O2	LINOLEIC ACID	-1.7073	0.9823	775.00	268	736	0.1405	0.1432	0.0196
1298	C18H34	1-OCTADECYNE	-1.2687	0.1644	747.33	300	710	------	0.0744	0.0539
1299	C18H34O2	OLEIC ACID	-2.9905	2.6266	781.00	287	742	0.1990	0.1966	0.0010
1300	C18H34O4	DIBUTYL SEBACATE	-1.8794	0.9921	768.00	264	730	0.0961	0.0985	0.0132
1301	C18H34O4	DIHEXYL ADIPATE	-1.8860	0.9986	767.00	259	729	0.0959	0.0988	0.0130
1302	C18H36	1-CYCOPENTYLTRIDECANE	-1.2578	0.1461	766.47	278	728	0.0740	0.0741	0.0552
1303	C18H36	1-CYCLOHEXYLDODECANE	-1.2587	0.1470	772.83	286	734	0.0740	0.0740	0.0551
1304	C18H36	1-OCTADECENE	-1.7179	1.0127	748.00	291	711	0.1438	0.1424	0.0191
1305	C18H36O2	STEARIC ACID	-1.8215	1.1522	799.00	343	759	------	0.1409	0.0151
1306	C18H38	n-OCTADECANE	-1.5198	0.8067	745.26	301	708	------	0.1475	0.0302
1307	C18H38O	DINONYL ETHER	-1.8634	0.9832	736.00	273	699	0.0964	0.0977	0.0137
1308	C18H38O	1-OCTADECANOL	-1.6609	1.0778	777.00	331	738	------	0.1772	0.0218
1309	C18H38S	BUTYL-TETRADECYL-SULFIDE	-1.2782	0.1388	810.53	298	770	0.0698	0.0696	0.0527
1310	C18H38S	ETHYL-HEXADECYL-SULFIDE	-1.2782	0.1388	810.53	298	770	0.0698	0.0696	0.0527
1311	C18H38S	HEPTADECYL-METHYL-SULFIDE	-1.2782	0.1388	810.53	298	770	0.0698	0.0696	0.0527
1312	C18H38S	NONYL-SULFIDE	-1.2782	0.1388	810.53	298	770	0.0698	0.0696	0.0527
1313	C18H38S	PENTADECYL-PROPYL-SULFIDE	-1.2782	0.1388	810.53	298	770	0.0698	0.0696	0.0527
1314	C18H38S	1-OCTADECANETHIOL	-1.2826	0.1449	795.36	301	756	------	0.0696	0.0522
1315	C18H38S2	NONYL-DISULFIDE	-1.2930	0.1271	802.30	252	762	0.0658	0.0661	0.0509
1316	C19H26	1-n-NONYLNAPHTHALENE	-1.8182	1.0024	849.00	284	807	0.1169	0.1168	0.0152
1317	C19H32	n-TRIDECYLBENZENE	-1.8571	1.0126	783.00	283	744	0.1061	0.1062	0.0139
1318	C19H36	1-NONADECYNE	-1.2779	0.1619	758.94	306	721	------	0.0725	0.0527
1319	C19H36O2	METHYL OLEATE	-1.8537	0.9858	764.00	293	726	0.1005	0.0993	0.0140
1320	C19H38	1-CYCLOPENTYLTETRADECANE	-1.2680	0.1457	772.00	282	733	0.0722	0.0722	0.0540
1321	C19H38	1-CYCLOHEXYLTRIDECANE	-1.2685	0.1454	783.38	292	744	0.0722	0.0720	0.0539
1322	C19H38	1-NONADECENE	-1.7206	1.0155	760.00	297	722	0.1446	0.1420	0.0190
1323	C19H38O2	NONADECANOIC ACID	-1.6897	0.9683	810.00	341	770	------	0.1348	0.0204
1324	C19H40	n-NONADECANE	-1.5053	0.7902	755.93	305	718	------	0.1477	0.0312
1325	C19H40O	1-NONADECANOL	-1.2955	0.1689	775.30	335	737	------	0.0702	0.0506
1326	C19H40S	BUTYL-PENTADECYL-SULFIDE	-1.2869	0.1372	821.75	303	781	------	0.0680	0.0517

Table 24-1 THERMAL CONDUCTIVITY OF LIQUID - ORGANIC COMPOUNDS (continued)

NO	FORMULA	NAME	$\log_{10} k_{liq} = A + B [1 - T/C]^{2/7}$					$(k_{liq} - W/(m\ K),\ T - K)$		
			A	B	C	TMIN	TMAX	k_{liq} @25 C	k_{liq} @TMIN	k_{liq} @TMAX
1327	C19H40S	ETHYL-HEPTADECYL-SULFIDE	-1.2869	0.1372	821.75	303	781	-----	0.0681	0.0591
1328	C19H40S	HEXADECYL-PROPYL-SULFIDE	-1.2869	0.1372	821.75	303	781	-----	0.0681	0.0591
1329	C19H40S	METHYL-OCTADECYL-SULFIDE	-1.2869	0.1372	821.75	303	781	-----	0.0681	0.0591
1330	C19H40S	1-NONADECANETHIOL	-1.2917	0.1440	805.29	307	765	-----	0.0682	0.0588
1331	C20H16	TRIPHENYLETHYLENE	-1.7755	1.0075	908.00	342	863	-----	0.1273	0.0448
1332	C20H28	1-n-DECYLNAPHTHALENE	-1.8282	1.0039	859.00	288	816	0.1150	0.1162	0.0397
1333	C20H30O2	ABIETIC ACID	-1.6935	0.9517	832.00	447	790	-----	0.1175	0.0515
1334	C20H31N	DEHYDROABIETYLAMINE	-2.6513	1.0122	863.00	318	820	-----	0.0172	0.0060
1335	C20H34	1-PHENYLTETRADECANE	-1.2714	0.1415	792.00	289	752	0.0712	0.0713	0.0615
1336	C20H38	1-EICOSYNE	-1.2859	0.1585	769.79	309	731	-----	0.0710	0.0605
1337	C20H40	1-CYCLOPENTYLPENTADECANE	-1.2785	0.1463	780.00	290	741	0.0706	0.0707	0.0608
1338	C20H40	1-CYCLOHEXYLTETRADECANE	-1.2781	0.1443	792.82	297	753	0.0705	0.0705	0.0607
1339	C20H40	1-EICOSENE	-1.7233	1.0181	771.00	302	732	-----	0.1445	0.0514
1340	C20H42	n-EICOSANE	-1.8447	1.1788	767.04	310	729	-----	0.1486	0.0452
1341	C20H42O	1-EICOSANOL	-1.7565	1.0513	792.00	339	752	-----	0.1379	0.0491
1342	C20H42S	BUTYL-HEXADECYL-SULFIDE	-1.2954	0.1359	832.33	308	791	-----	0.0666	0.0578
1343	C20H42S	DECYL-SULFIDE	-1.2954	0.1359	832.33	308	791	-----	0.0666	0.0578
1344	C20H42S	ETHYL-OCTADECYL-SULFIDE	-1.2954	0.1359	832.33	308	791	-----	0.0666	0.0578
1345	C20H42S	HEPTADECYL-PROPYL-SULFIDE	-1.2954	0.1359	832.33	308	791	-----	0.0666	0.0578
1346	C20H42S	METHYL-NONADECYL-SULFIDE	-1.2954	0.1359	832.33	308	791	-----	0.0666	0.0578
1347	C20H42S	1-EICOSANETHIOL	-1.2999	0.1422	814.57	310	774	-----	0.0667	0.0576
1348	C20H42S2	DECYL-DISULFIDE	-1.3092	0.1253	820.08	259	779	0.0632	0.0636	0.0555
1349	C21H21O4P	TRI-o-CRESYL PHOSPHATE	-1.3800	0.5756	806.00	240	680	0.1332	0.1381	0.0909
1350	C21H36	1-PHENYLPENTADECANE	-1.2813	0.1413	800.00	295	760	0.0696	0.0696	0.0601
1351	C21H42	1-CYCLOPENTYLHEXADECANE	-1.2862	0.1418	797.25	294	757	0.0688	0.0689	0.0595
1352	C21H42	1-CYCLOHEXYLPENTADECANE	-1.2869	0.1426	803.46	302	763	-----	0.0688	0.0594
1353	C22H38	1-PHENYLHEXADECANE	-1.2905	0.1408	808.00	300	768	-----	0.0680	0.0588
1354	C22H44	1-CYCLOHEXYLHEXADECANE	-1.2954	0.1411	813.42	307	773	-----	0.0673	0.0581
1355	C22H44O2	n-BUTYL STEARATE	-1.4902	0.7884	764.00	299	726	-----	0.1563	0.0699
1356	C24H38O4	DIISOOCTYL PHTHALATE	-1.7906	0.9976	851.00	223	808	0.1234	0.1331	0.0431
1357	C24H38O4	DIOCTYL PHTHALATE	-1.7804	1.0238	806.00	223	766	0.1309	0.1422	0.0450
1358	C24H42O	DINONYLPHENOL	-1.6584	0.9739	886.00	350	842	-----	0.1532	0.0568
1359	C26H20	TETRAPHENYLETHYLENE	-1.7345	1.0370	996.00	496	946	-----	0.1310	0.0509
1360	C28H46O4	DIISODECYL PHTHALATE	-1.9715	1.0041	887.00	228	843	0.0835	0.0893	0.0285

k_{liq} - thermal conductivity of liquid, W/(m K)

A, B, and C - regression coefficients for chemical compound

T - temperature, K

TMIN - minimum temperature, K

TMAX - maximum temperature, K

Table 24-2 THERMAL CONDUCTIVITY OF LIQUID AND SOLID - INORGANIC COMPOUNDS

NO	FORMULA	NAME	\multicolumn{3}{c}{$k = A + B T + C T^2$}					\multicolumn{3}{c}{(k - W/(m K), T - K)}		
			A	B	C	TMIN	TMAX	Phase	k @25 C	k @TMIN	k @TMAX
1	Ag	SILVER	438.2178	-2.2947E-02	-3.5429E-05	200	1200	solid	428.227	432.211	359.664
2	AgCl	SILVER CHLORIDE	2.1134	-5.0343E-03	5.8694E-06	221	373	solid	1.134	1.287	1.052
3	AgI	SILVER IODIDE	0.1110	1.7467E-03	-2.5859E-06	290	310	solid	0.402	0.400	0.404
4	Al	ALUMINUM	228.2103	5.7999E-02	-8.6806E-05	200	934	solid	237.786	236.338	206.656
5	AlB3H12	ALUMINUM BOROHYDRIDE	0.2804	-1.4584E-04	-5.3593E-07	209	462	liquid	0.189	0.227	0.099
6	AlBr3	ALUMINUN BROMIDE	0.1529	-2.3596E-05	-1.6993E-07	390	687	liquid	------	0.118	0.056
7	AlCl3	ALUMINUM CHLORIDE	--------	----------	----------	----	----	------	-----	-----	-----
8	AlF3	ALUMINUM FLUORIDE	--------	----------	----------	----	----	------	-----	-----	-----
9	AlI3	ALUMINUM IODIDE	--------	----------	----------	----	----	------	-----	-----	-----
10	Al2O3	ALUMINUM OXIDE	79.6671	-1.6625E-01	9.6265E-05	200	1000	solid	38.657	50.268	9.682
11	Al2S3O12	ALUMINUM SULFATE	--------	----------	----------	----	----	------	-----	-----	-----
12	Ar	ARGON	0.1819	-3.1760E-04	-4.1100E-06	84	150	liquid	------	0.126	0.042
13	As	ARSENIC	122.7520	-3.4038E-01	3.3153E-04	200	500	solid	50.739	67.937	35.445
14	AsBr3	ARSENIC TRIBROMIDE	0.1354	-4.9347E-05	-1.0551E-07	306	710	liquid	------	0.110	0.047
15	AsCl3	ARSENIC TRICHLORIDE	0.1764	-7.6933E-05	-2.0083E-07	255	589	liquid	0.136	0.144	0.061
16	AsF3	ARSENIC TRIFLUORIDE	0.1960	-4.8443E-05	-4.4249E-07	267	477	liquid	0.142	0.152	0.072
17	AsF5	ARSENIC PENTAFLUORIDE	0.1668	-2.7463E-05	-9.1562E-07	193	322	liquid	0.077	0.127	0.063
18	AsH3	ARSINE	0.2485	-1.6905E-04	-9.2454E-07	156	336	liquid	0.116	0.200	0.087
19	AsI3	ARSENIC TRIIODIDE	0.1112	-2.9772E-05	-4.3803E-08	419	991	liquid	------	0.091	0.039
20	As2O3	ARSENIC TRIOXIDE	--------	----------	----------	----	----	------	-----	-----	-----
21	At	ASTATINE	-8.6579	6.5794E-02	-1.0425E-04	298	310	solid	1.691	1.691	1.720
22	Au	GOLD	335.4544	-5.8253E-02	-7.4773E-06	200	1300	solid	317.422	323.505	247.089
23	B	BORON	--------	----------	----------	----	----	------	-----	-----	-----
24	BBr3	BORON TRIBROMIDE	0.1521	-7.4015E-05	-2.2066E-07	228	523	liquid	0.110	0.124	0.053
25	BCl3	BORON TRICHLORIDE	0.2258	-1.5356E-04	-5.1475E-07	166	407	liquid	0.134	0.186	0.078
26	BF3	BORON TRIFLUORIDE	0.2781	-1.4752E-05	-3.0436E-06	146	235	liquid	------	0.211	0.107
27	BH2CO	BORINE CARBONYL	0.3691	-2.9796E-04	-1.5882E-06	136	306	liquid	0.139	0.299	0.129
28	BH3O3	BORIC ACID	--------	----------	----------	----	----	------	-----	-----	-----
29	B2D6	DEUTERODIBORANE	0.3928	-2.6308E-04	-2.6052E-06	136	264	liquid	------	0.309	0.142
30	B2H5Br	DIBORANE HYDROBROMIDE	0.2333	-1.5623E-04	-4.9250E-07	169	420	liquid	0.143	0.193	0.081
31	B2H6	DIBORANE	0.4585	-4.8053E-04	-2.5611E-06	108	261	liquid	------	0.377	0.159
32	B3N3H6	BORINE TRIAMINE	0.2637	-1.3220E-04	-4.9540E-07	215	469	liquid	0.180	0.212	0.093
33	B4H10	TETRABORANE	0.3332	-2.4373E-04	-6.6177E-07	153	420	liquid	0.202	0.280	0.114
34	B5H9	PENTABORANE	0.2850	-1.3884E-04	-4.3683E-07	226	512	liquid	0.205	0.231	0.099
35	B5H11	TETRAHYDROPENTABORANE	0.2933	-1.3966E-04	-5.0069E-07	226	492	liquid	0.207	0.236	0.103
36	B10H14	DECABORANE	0.2059	-4.8187E-05	-1.9149E-07	373	713	liquid	------	0.161	0.074
37	Ba	BARIUM	26.5122	-5.3067E-02	8.6615E-05	150	300	solid	18.390	20.501	18.387
38	Be	BERYLLIUM	318.8202	-4.7499E-01	2.5262E-04	273	1000	solid	199.658	207.975	96.450
39	BeB2H8	BERYLLIUM BOROHYDRIDE	0.2938	2.0204E-04	-9.5365E-07	396	532	liquid	------	0.224	0.131
40	BeBr2	BERYLLIUM BROMIDE	--------	----------	----------	----	----	------	-----	-----	-----
41	BeCl2	BERYLLIUM CHLORIDE	--------	----------	----------	----	----	------	-----	-----	-----
42	BeF2	BERYLLIUM FLUORIDE	--------	----------	----------	----	----	------	-----	-----	-----
43	BeI2	BERYLLIUM IODIDE	--------	----------	----------	----	----	------	-----	-----	-----
44	Bi	BISMUTH	14.7277	-3.2047E-02	3.1392E-05	200	545	solid	7.963	9.574	6.586
45	BiBr3	BISMUTH TRIBROMIDE	--------	----------	----------	----	----	------	-----	-----	-----
46	BiCl3	BISMUTH TRICHLORIDE	--------	----------	----------	----	----	------	-----	-----	-----
47	BrF5	BROMINE PENTAFLUORIDE	0.1878	-8.5540E-05	-4.7230E-07	212	423	liquid	0.120	0.148	0.067
48	Br2	BROMINE	0.1325	1.0331E-04	-4.5514E-07	266	584	liquid	0.123	0.128	0.038
49	C	CARBON	168.9106	-1.3343E-01	3.7650E-05	200	2000	solid	132.475	143.731	52.651
50	CCl2O	PHOSGENE	0.1682	1.1873E-04	-8.4939E-07	145	410	liquid	0.128	0.168	0.074
51	CF2O	CARBONYL FLUORIDE	0.3069	-5.3980E-05	-2.4657E-06	162	267	liquid	------	0.233	0.117
52	CH4N2O	UREA	0.2888	1.1326E-05	-4.5537E-07	406	635	liquid	------	0.218	0.112
53	CH4N2S	THIOUREA	0.1934	-1.9223E-05	-1.7982E-07	454	769	liquid	------	0.148	0.072
54	CNBr	CYANOGEN BROMIDE	0.1964	4.5432E-05	-5.7894E-07	331	491	liquid	------	0.148	0.079
55	CNCl	CYANOGEN CHLORIDE	0.2423	3.7906E-05	-9.9081E-07	267	404	liquid	0.166	0.182	0.096
56	CNF	CYANOGEN FLUORIDE	0.3512	-2.5473E-04	-1.3047E-06	150	332	liquid	0.159	0.284	0.123
57	CO	CARBON MONOXIDE	0.2855	-1.7840E-03	-1.6136E-17	70	125	liquid	------	0.161	0.062
58	COS	CARBONYL SULFIDE	0.2354	-1.9881E-04	-7.4048E-07	134	341	liquid	0.110	0.195	0.082
59	COSe	CARBON OXYSELENIDE	0.2346	-1.9639E-04	-6.1505E-07	134	366	liquid	0.121	0.197	0.080
60	CO2	CARBON DIOXIDE	0.4320	-1.1929E-03	-6.5352E-17	250	300	liquid	0.076	0.134	0.074
61	CS2	CARBON DISULFIDE	0.2181	-1.4539E-04	-2.9062E-07	162	497	liquid	0.149	0.187	0.074
62	CSeS	CARBON SELENOSULFIDE	0.2191	-1.2504E-04	-2.9305E-07	198	519	liquid	0.156	0.183	0.075
63	C2N2	CYANOGEN	0.2905	6.6083E-05	-1.5309E-06	239	360	liquid	0.174	0.219	0.116
64	C3S2	CARBON SUBSULFIDE	--------	----------	----------	----	----	------	-----	-----	-----
65	Ca	CALCIUM	244.4135	-1.5732E-01	6.5607E-05	200	800	solid	203.341	215.574	160.546
66	CaF2	CALCIUM FLUORIDE	51.9961	-2.2344E-01	2.7388E-04	229	421	solid	9.724	15.191	6.471
67	CbF5	COLUMBIUM FLUORIDE	0.1696	-5.0359E-05	-1.3656E-07	349	730	liquid	------	0.135	0.060
68	Cd	CADMIUM	101.2521	-5.5398E-03	-2.7578E-05	200	594	solid	97.149	99.041	88.231
69	CdCl2	CADMIUM CHLORIDE	--------	----------	----------	----	----	------	-----	-----	-----
70	CdF2	CADMIUM FLUORIDE	--------	----------	----------	----	----	------	-----	-----	-----
71	CdI2	CADMIUM IODIDE	--------	----------	----------	----	----	------	-----	-----	-----
72	CdO	CADMIUM OXIDE	--------	----------	----------	----	----	------	-----	-----	-----
73	ClF	CHLORINE MONOFLUORIDE	0.3103	-2.3025E-04	-2.1802E-06	128	254	liquid	------	0.245	0.111
74	ClFO3	PERCHLORYL FLUORIDE	0.2379	-2.1407E-04	-7.7506E-07	125	332	liquid	0.105	0.199	0.081
75	ClF3	CHLORINE TRIFLUORIDE	0.2454	-1.3866E-04	-5.9536E-07	190	413	liquid	0.151	0.198	0.087
76	ClF5	CHLORINE PENTAFLUORIDE	0.2040	-1.0074E-04	-6.6485E-07	190	374	liquid	0.115	0.161	0.073
77	ClHO3S	CHLOROSULFONIC ACID	0.2080	-1.1086E-04	-1.6832E-07	193	630	liquid	0.160	0.180	0.071
78	ClHO4	PERCHLORIC ACID	0.2178	-1.3010E-04	-2.1479E-07	172	568	liquid	0.160	0.189	0.075

Table 24-2 THERMAL CONDUCTIVITY OF LIQUID AND SOLID - INORGANIC COMPOUNDS (continued)

NO	FORMULA	NAME	A	B	C	TMIN	TMAX	Phase	k @25 C	k @TMIN	k @TMAX
				$k = A + B T + C T^2$						(k - W/(m K), T - K)	
79	ClO2	CHLORINE DIOXIDE	0.2778	-1.2070E-04	-7.2849E-07	214	419	liquid	0.177	0.219	0.099
80	Cl2	CHLORINE	0.2246	-6.4000E-05	-7.8800E-07	172	410	liquid	0.135	0.190	0.066
81	Cl2O	CHLORINE MONOXIDE	0.2588	-1.8641E-04	-5.9294E-07	157	400	liquid	0.151	0.215	0.089
82	Cl2O7	CHLORINE HEPTOXIDE	0.1808	-1.1071E-04	-2.4145E-07	182	509	liquid	0.126	0.153	0.062
83	Co	COBALT	150.4892	-1.9030E-01	8.9428E-05	200	1200	solid	101.701	116.006	50.906
84	CoCl2	COBALT CHLORIDE	--------	----------	----------	----	----	------	------	------	-----
85	CoNC3O4	COBALT NITROSYL TRICARBONYL	0.1759	-6.1478E-05	-3.1136E-07	262	511	liquid	0.130	0.138	0.063
86	Cr	CHROMIUM	116.0771	-7.4968E-02	2.4602E-05	250	1200	solid	95.912	98.873	61.542
87	CrC6O6	CHROMIUM CARBONYL	0.1353	2.8624E-05	-2.5348E-07	424	622	liquid		0.102	0.055
88	CrO2Cl2	CHROMIUM OXYCHLORIDE	0.1988	-1.1915E-04	-2.0211E-07	177	564	liquid	0.145	0.171	0.067
89	Cs	CESIUM	20.3124	3.1566E-03	-6.1694E-06	350	1800	liquid	------	20.661	6.005
90	CsBr	CESIUM BROMIDE	3.8030	-1.7239E-02	2.4591E-05	228	368	solid	0.849	1.151	0.789
91	CsCl	CESIUM CHLORIDE	--------	----------	----------	----	----	------	------	------	-----
92	CsF	CESIUM FLUORIDE	--------	----------	----------	----	----	------	------	------	-----
93	CsI	CESIUM IODIDE	3.3449	-1.1618E-02	1.3715E-05	227	361	solid	1.100	1.414	0.938
94	Cu	COPPER	426.2970	-8.3932E-02	9.3782E-06	200	1300	solid	402.106	409.886	333.035
95	CuBr	CUPROUS BROMIDE	1.1573	-2.2601E-03	-8.5227E-09	220	400	solid	0.483	0.660	0.252
96	CuCl	CUPROUS CHLORIDE	0.0647	3.3955E-04	-1.4748E-07	703	2192	liquid	------	0.231	0.100
97	CuCl2	CUPRIC CHLORIDE	--------	----------	----------	----	----	------	------	------	-----
98	CuI	COPPER IODIDE	-0.0465	1.1344E-02	-1.8350E-05	290	310	solid	1.705	1.700	1.707
99	DCN	DEUTERIUM CYANIDE	0.4128	-4.9224E-05	-1.2473E-06	261	434	liquid	0.287	0.315	0.157
100	D2	DEUTERIUM	-0.0953	1.7754E-02	-3.3575E-04	19	38	liquid	------	0.121	0.095
101	D2O	DEUTERIUM OXIDE	-0.0138	3.6838E-03	-5.4096E-06	277	580	liquid	0.604	0.592	0.303
102	Eu	EUROPIUM	27.6243	-1.0058E-01	1.8378E-04	123	300	solid	13.973	18.033	13.991
103	F2	FLUORINE	0.2758	-1.6297E-03	-3.7475E-18	53	130	liquid	------	0.189	0.064
104	F2O	FLUORINE OXIDE	0.3292	-6.2445E-04	-2.6048E-06	49	194	liquid	------	0.292	0.110
105	Fe	IRON	117.3180	-1.3759E-01	5.4170E-05	200	1000	solid	81.111	91.967	33.898
106	FeC5O5	IRON PENTACARBONYL	0.1692	-7.1908E-05	-2.3565E-07	252	546	liquid	0.127	0.136	0.060
107	FeCl2	FERROUS CHLORIDE	--------	----------	----------	----	----	------	------	------	-----
108	FeCl3	FERRIC CHLORIDE	--------	----------	----------	----	----	------	------	------	-----
109	Fr	FRANCIUM	24.6434	-6.1069E-02	9.6549E-05	298	310	solid	15.018	15.019	14.990
110	Ga	GALLIUM	6.6484	7.5930E-02	-1.6369E-05	303	700	liquid	------	28.152	51.779
111	GaCl3	GALLIUM TRICHLORIDE	0.1857	-3.4299E-05	-2.4565E-07	351	625	liquid	------	0.143	0.068
112	Gd	GADOLINIUM	18.9802	-5.4808E-02	8.9115E-05	100	310	solid	10.561	14.391	10.554
113	Ge	GERMANIUM	122.6462	-2.4844E-01	1.4443E-04	250	1000	solid	61.413	69.563	18.636
114	GeBr4	GERMANIUM BROMIDE	0.1301	-1.4000E-04	5.3235E-18	300	360	liquid	------	0.088	0.080
115	GeCl4	GERMANIUM CHLORIDE	0.1183	1.1086E-04	-5.7143E-07	260	360	liquid	0.101	0.108	0.084
116	GeHCl3	TRICHLORO GERMANE	0.1801	-1.0085E-04	-2.6414E-07	202	504	liquid	0.127	0.149	0.062
117	GeH4	GERMANE	0.2693	-2.8399E-04	-1.2736E-06	107	277	liquid	------	0.224	0.093
118	Ge2H6	DIGERMANE	0.1977	-1.3548E-04	-3.5845E-07	164	442	liquid	0.125	0.166	0.068
119	Ge3H8	TRIGERMANE	0.1650	-1.0229E-04	-1.7027E-07	168	555	liquid	0.119	0.143	0.056
120	HBr	HYDROGEN BROMIDE	0.1649	1.1789E-04	-1.1999E-06	194	327	liquid	0.093	0.143	0.075
121	HCN	HYDROGEN CYANIDE	0.3333	1.5737E-06	-1.2143E-06	260	411	liquid	0.226	0.252	0.129
122	HCl	HYDROGEN CHLORIDE	0.8045	-2.1020E-03	-2.3238E-16	273	323	liquid	0.178	0.231	0.126
123	HF	HYDROGEN FLUORIDE	0.6678	-4.7979E-04	-1.0548E-06	204	415	liquid	0.431	0.526	0.287
124	HI	HYDROGEN IODIDE	0.0483	1.1854E-04	-4.5588E-07	225	420	liquid	0.043	0.052	0.018
125	HNO3	NITRIC ACID	-0.2535	2.9368E-03	-3.6854E-06	233	468	liquid	0.294	0.231	0.314
126	H2	HYDROGEN	-0.1433	2.3627E-02	-5.1480E-04	14	33	liquid	------	0.0866	0.0737
127	H2O	WATER	-0.2758	4.6120E-03	-5.5391E-06	273	633	liquid	0.607	0.570	0.424
128	H2O2	HYDROGEN PEROXIDE	0.4425	-1.8406E-04	-3.8824E-07	273	657	liquid	0.353	0.363	0.154
129	H2S	HYDROGEN SULFIDE	0.5719	-1.9273E-03	1.5450E-06	193	336	liquid	0.135	0.257	0.099
130	H2SO4	SULFURIC ACID	0.1553	1.0699E-03	-1.2858E-06	283	833	liquid	0.360	0.355	0.154
131	H2S2	HYDROGEN DISULFIDE	0.2990	-1.8376E-04	-4.4762E-07	183	488	liquid	0.204	0.250	0.103
132	H2Se	HYDROGEN SELENIDE	0.2318	-6.9072E-05	-8.8127E-07	209	370	liquid	0.133	0.179	0.086
133	H2Te	HYDROGEN TELLURIDE	0.1961	-5.2407E-05	-6.6190E-07	224	394	liquid	0.122	0.151	0.073
134	H3NO3S	SULFAMIC ACID	--------	----------	----------	----	----	------	------	------	-----
135	He	HELIUM-3	0.0050	5.1690E-03	-3.3214E-16	1	3	liquid	------	0.0102	0.0179
136	He	HELIUM-4	0.0078	4.1000E-03	-2.9000E-04	2	5	liquid	------	0.0154	0.0213
137	Hf	HAFNIUM	26.0954	-1.1689E-02	6.1749E-06	200	1200	solid	23.159	24.005	20.960
138	Hg	MERCURY	0.9230	2.8887E-02	-1.5499E-05	234	1562	liquid	8.158	6.834	8.229
139	HgBr2	MERCURIC BROMIDE	--------	----------	----------	----	----	------	------	------	-----
140	HgCl2	MERCURIC CHLORIDE	--------	----------	----------	----	----	------	------	------	-----
141	HgI2	MERCURIC IODIDE	--------	----------	----------	----	----	------	------	------	-----
142	IF7	IODINE HEPTAFLUORIDE	0.1238	5.0118E-05	-5.7268E-07	279	403	liquid	0.088	0.093	0.051
143	I2	IODINE	0.0965	1.7970E-04	-3.4020E-07	387	785	liquid	------	0.115	0.028
144	In	INDIUM	108.8836	-1.0543E-01	4.8374E-05	200	400	solid	81.750	89.733	74.451
145	Ir	IRIDIUM	158.8514	-3.7903E-02	4.8752E-06	200	1200	solid	147.984	151.466	120.388
146	K	POTASSIUM	69.3316	-4.7493E-02	9.4295E-06	350	1600	liquid	------	53.864	17.482
147	KBr	POTASSIUM BROMIDE	13.8899	-5.4091E-02	8.0015E-05	241	372	solid	4.875	5.501	4.841
148	KCl	POTASSIUM CHLORIDE	39.3670	-1.9735E-01	2.8892E-04	120	360	solid	6.210	19.845	5.765
149	KF	POTASSIUM FLUORIDE	--------	----------	----------	----	----	------	------	------	-----
150	KI	POTASSIUM IODIDE	--------	----------	----------	----	----	------	------	------	-----
151	KOH	POTASSIUM HYDROXIDE	--------	----------	----------	----	----	------	------	------	-----
152	Kr	KRYPTON	0.1756	-7.0050E-04	1.3789E-18	116	210	liquid	------	0.0943	0.0285
153	La	LANTHANUM	7.6462	2.1036E-02	-5.9037E-06	200	1100	solid	13.393	11.617	23.642
154	Li	LITHIUM	28.9331	3.5692E-02	-1.0087E-05	454	3677	liquid	------	43.058	23.793
155	LiBr	LITHIUM BROMIDE	--------	----------	----------	----	----	------	------	------	-----
156	LiCl	LITHIUM CHLORIDE	--------	----------	----------	----	----	------	------	------	-----

Table 24-2 THERMAL CONDUCTIVITY OF LIQUID AND SOLID - INORGANIC COMPOUNDS (continued)

| NO | FORMULA | NAME | k = A + B T + C T² | | | | | | (k - W/(m K), T - K) | | |
			A	B	C	TMIN	TMAX	Phase	k @25 C	k @TMIN	k @TMAX
157	LiF	LITHIUM FLUORIDE	-2.3678	1.5310E-02	-6.1414E-06	378	772	solid	------	2.542	5.791
158	LiI	LITHIUM IODIDE	--------	----------	----------	----	----	-------	------	------	------
159	Lu	LUTECIUM	18.4078	-1.5138E-04	-2.1934E-05	200	300	solid	16.413	17.500	16.388
160	Mg	MAGNESIUM	165.8119	-3.7721E-02	1.6112E-05	200	900	solid	155.998	158.912	144.914
161	MgCl2	MAGNESIUM CHLORIDE	--------	----------	----------	----	----	-------	------	------	------
162	MgO	MAGNESIUM OXIDE	90.1725	-1.6014E-01	7.7161E-05	250	1200	solid	49.286	54.960	9.116
163	Mn	MANGANESE	2.6275	3.5723E-02	-6.2094E-05	50	300	solid	7.759	4.258	7.756
164	MnCl2	MANGANESE CHLORIDE	--------	----------	----------	----	----	-------	------	------	------
165	Mo	MOLYBDENUM	152.6278	-5.0955E-02	9.8237E-06	200	2000	solid	138.309	142.830	90.013
166	MoF6	MOLYBDENUM FLUORIDE	0.1450	1.3643E-05	-4.6861E-07	290	448	liquid	0.107	0.110	0.057
167	MoO3	MOLYBDENUM OXIDE	--------	----------	----------	----	----	-------	------	------	------
168	NCl3	NITROGEN TRICHLORIDE	0.2102	-8.6655E-05	-3.5545E-07	246	508	liquid	0.153	0.167	0.074
169	ND3	HEAVY AMMONIA	0.4969	-1.4733E-04	-2.1397E-06	199	350	liquid	0.263	0.383	0.183
170	NF3	NITROGEN TRIFLUORIDE	0.2908	-4.6529E-04	-2.1267E-06	66	210	liquid	------	0.251	0.099
171	NH3	AMMONIA	1.1606	-2.2840E-03	3.1245E-18	220	400	liquid	0.480	0.658	0.247
172	NH3O	HYDROXYLAMINE	0.4095	-5.2700E-05	-8.5576E-07	306	517	liquid	------	0.313	0.154
173	NH4Br	AMMONIUM BROMIDE	--------	----------	----------	----	----	-------	------	------	------
174	NH4Cl	AMMONIUM CHLORIDE	--------	----------	----------	----	----	-------	------	------	------
175	NH4I	AMMONIUM IODIDE	--------	----------	----------	----	----	-------	------	------	------
176	NH5O	AMMONIUM HYDROXIDE	--------	----------	----------	----	----	-------	------	------	------
177	NH5S	AMMONIUM HYDROGENSULFIDE	0.1661	6.3855E-04	-1.6809E-06	391	449	liquid	------	0.159	0.114
178	NO	NITRIC OXIDE	0.1878	1.0293E-03	-9.4300E-06	110	176	liquid	------	0.187	0.077
179	NOCl	NITROSYL CHLORIDE	0.2767	-1.0288E-04	-8.5984E-07	214	397	liquid	0.170	0.215	0.100
180	NOF	NITROSYL FLUORIDE	0.3380	-2.6904E-04	-1.3357E-06	139	317	liquid	0.139	0.275	0.118
181	NO2	NITROGEN DIOXIDE	-0.0120	1.5191E-03	-3.5073E-06	262	388	liquid	0.129	0.145	0.049
182	N2	NITROGEN	0.2130	-4.2050E-04	-7.2951E-06	70	126	liquid	------	0.148	0.044
183	N2F4	TETRAFLUOROHYDRAZINE	0.2547	-2.3893E-04	-1.2702E-06	112	278	liquid	------	0.212	0.090
184	N2H4	HYDRAZINE	0.4008	-1.5493E-04	-4.8625E-07	275	588	liquid	0.311	0.321	0.142
185	N2H4C	AMMONIUM CYANIDE	0.2982	1.2592E-04	-1.1749E-06	309	442	liquid	------	0.225	0.124
186	N2H6CO2	AMMONIUM CARBAMATE	0.2382	8.2829E-06	-6.3390E-07	309	486	liquid	------	0.180	0.093
187	N2O	NITROUS OXIDE	0.2539	5.4127E-05	-2.1713E-06	182	279	liquid	------	0.192	0.100
188	N2O3	NITROGEN TRIOXIDE	0.2831	-1.8315E-04	-7.7909E-07	170	383	liquid	0.159	0.229	0.099
189	N2O4	NITROGEN TETRAOXIDE	0.2401	6.9938E-05	-1.1302E-06	262	388	liquid	0.160	0.181	0.097
190	N2O5	NITROGEN PENTOXIDE	0.2011	2.4661E-05	-6.1811E-07	303	464	liquid	------	0.152	0.079
191	Na	SODIUM	98.8908	-4.3158E-02	5.2876E-06	371	2316	liquid	------	83.607	27.299
192	NaBr	SODIUM BROMIDE	0.3616	-3.3985E-05	-7.3045E-09	1020	3858	liquid	------	0.319	0.122
193	NaCN	SODIUM CYANIDE	--------	----------	----------	----	----	-------	------	------	------
194	NaCl	SODIUM CHLORIDE	51.6119	-2.9610E-01	4.7053E-04	80	380	solid	5.157	30.935	7.038
195	NaF	SODIUM FLUORIDE	7082.4870	-2.6107E+02	2.4011E+00	12	68	solid	------	4295.405	432.413
196	NaI	SODIUM IODIDE	--------	----------	----------	----	----	-------	------	------	------
197	NaOH	SODIUM HYDROXIDE	-3.2252	4.0045E-03	5.0633E-06	592	592	solid	------	0.920	0.920
198	Na2SO4	SODIUM SULFATE	--------	----------	----------	----	----	-------	------	------	------
199	Nb	NIOBIUM	49.1361	1.5425E-02	-1.5966E-07	200	2000	solid	53.721	52.215	79.347
200	Nd	NEODYMIUM	15.6243	2.1652E-03	3.1777E-06	200	1200	solid	16.552	16.184	22.798
201	Ne	NEON	0.0080	8.6895E-03	-1.7697E-04	25	44	liquid	------	0.115	0.048
202	Ni	NICKEL	144.4370	-2.2793E-01	1.6434E-04	250	800	solid	91.088	97.726	67.271
203	NiC4O4	NICKEL CARBONYL	0.1740	-5.4156E-05	-4.0987E-07	248	458	liquid	0.121	0.135	0.063
204	NiF2	NICKEL FLUORIDE	--------	----------	----------	----	----	-------	------	------	------
205	Np	NEPTUNIUM	-4.0579	6.5794E-02	-1.0425E-04	298	310	solid	6.291	6.291	6.320
206	O2	OXYGEN	0.2320	-5.6357E-04	-3.8093E-06	60	155	liquid	------	0.184	0.053
207	O3	OZONE	0.1076	2.0786E-03	-8.7941E-06	80	235	liquid	------	0.218	0.110
208	Os	OSMIUM	88.4000	-3.1565E-03	1.4658E-06	250	1600	solid	87.589	87.702	87.102
209	OsOF5	OSMIUM OXIDE PENTAFLUORIDE	0.1242	-8.9873E-06	-2.4035E-07	333	548	liquid	------	0.0946	0.0471
210	OsO4	OSMIUM TETROXIDE - YELLOW	0.1401	-2.8986E-05	-2.0468E-07	329	591	liquid	------	0.108	0.051
211	OsO4	OSMIUM TETROXIDE - WHITE	0.1420	-3.7029E-05	-1.9643E-07	315	591	liquid	------	0.111	0.052
212	P	PHOSPHORUS - WHITE	0.3917	2.2121E-03	-2.5062E-06	317	894	liquid	------	0.841	0.366
213	PBr3	PHOSPHORUS TRIBROMIDE	0.1502	-7.2169E-05	-1.2836E-07	233	640	liquid	0.117	0.126	0.051
214	PCl2F3	PHOSPHORUS DICHLORIDE TRIFLUORIDE	0.1669	1.4694E-05	-6.3564E-07	265	411	liquid	0.115	0.126	0.066
215	PCl3	PHOSPHORUS TRICHLORIDE	0.1839	-1.1322E-04	-2.4781E-07	181	507	liquid	0.128	0.155	0.063
216	PCl5	PHOSPHORUS PENTACHLORIDE	0.1371	8.7024E-05	-3.7396E-07	433	582	liquid	------	0.105	0.061
217	PH3	PHOSPHINE	0.2693	-2.0269E-04	-1.3427E-06	139	292	liquid	------	0.215	0.096
218	PH4Br	PHOSPHONIUM BROMIDE	0.1966	1.5528E-04	-6.2075E-07	291	452	liquid	0.146	0.149	0.077
219	PH4Cl	PHOSPHONIUM CHLORIDE	0.2092	8.6905E-04	-4.0096E-06	245	290	liquid	------	0.181	0.124
220	PH4I	PHOSPHONIUM IODIDE	0.1722	-1.8856E-05	-4.1517E-07	292	486	liquid	0.130	0.131	0.065
221	POCl3	PHOSPHORUS OXYCHLORIDE	0.1892	-6.5087E-05	-2.9332E-07	274	542	liquid	0.144	0.149	0.068
222	PSBr3	PHOSPHORUS THIOBROMIDE	0.1338	-4.4935E-05	-1.3213E-07	311	657	liquid	------	0.107	0.047
223	PSCl3	PHOSPHORUS THIOCHLORIDE	0.1853	-8.8258E-05	-2.1280E-07	237	575	liquid	0.140	0.152	0.064
224	P4O6	PHOSPHORUS TRIOXIDE	0.1602	-5.8241E-05	-1.6033E-07	296	643	liquid	0.129	0.129	0.056
225	P4O10	PHOSPHORUS PENTOXIDE	--------	----------	----------	----	----	-------	------	------	------
226	P4S10	PHOSPHORUS PENTASULFIDE	0.2202	-4.0677E-05	-7.0155E-08	561	1162	liquid	------	0.175	0.078
227	Pb	LEAD	39.3335	-1.3469E-02	5.1500E-07	200	600	solid	35.363	36.660	31.438
228	PbBr2	LEAD BROMIDE	--------	----------	----------	----	----	-------	------	------	------
229	PbCl2	LEAD CHLORIDE	--------	----------	----------	----	----	-------	------	------	------
230	PbF2	LEAD FLUORIDE	--------	----------	----------	----	----	-------	------	------	------
231	PbI2	LEAD IODIDE	--------	----------	----------	----	----	-------	------	------	------
232	PbO	LEAD OXIDE	--------	----------	----------	----	----	-------	------	------	------
233	PbS	LEAD SULFIDE	--------	----------	----------	----	----	-------	------	------	------
234	Pd	PALLADIUM	62.3067	3.2174E-02	-9.5280E-07	200	1800	solid	71.815	68.703	117.133

NO	FORMULA	NAME	$k = A + B T + C T^2$						(k - W/(m K), T - K)		
			A	B	C	TMIN	TMAX	Phase	k @25 C	k @TMIN	k @TMAX
235	Po	POLONIUM	9.6421	6.5794E-02	-1.0425E-04	298	310	solid	19.991	19.991	20.020
236	Pt	PLATINUM	70.5776	1.3686E-03	6.9316E-06	200	2000	solid	71.602	71.129	101.041
237	Ra	RADIUM	18.6000	-3.4976E-11	5.7938E-14	293	310	solid	18.600	18.600	18.600
238	Rb	RUBIDIUM	42.1670	-2.5394E-02	3.8426E-06	350	1500	liquid	------	33.750	12.722
239	RbBr	RUBIDIUM BROMIDE	--------	----------	----------	----	----	------	------	------	------
240	RbCl	RUBIDIUM CHLORIDE	--------	----------	----------	----	----	------	------	------	------
241	RbF	RUBIDIUM FLUORIDE	--------	----------	----------	----	----	------	------	------	------
242	RbI	RUBIDIUM IODIDE	--------	----------	----------	----	----	------	------	------	------
243	Re	RHENIUM	50.3091	-1.2751E-02	6.9863E-06	298	2000	solid	47.128	47.130	52.752
244	Re2O7	RHENIUM HEPTOXIDE	--------	----------	----------	----	----	------	------	------	------
245	Rh	RHODIUM	167.9688	-6.5530E-02	1.8568E-05	200	2000	solid	150.082	155.606	111.181
246	Rn	RADON	0.0943	-1.4286E-04	-1.3437E-07	202	360	liquid	0.040	0.060	0.025
247	Ru	RUTHENIUM	126.1698	-3.5146E-02	7.0257E-06	200	2000	solid	116.316	119.422	83.981
248	RuF5	RUTHENIUM PENTAFLUORIDE	0.1703	-5.3488E-05	-8.3035E-08	360	880	liquid	------	0.140	0.059
249	S	SULFUR	0.0262	4.1738E-04	-3.2471E-07	388	1182	liquid	------	0.139	0.066
250	SF4	SULFUR TETRAFLUORIDE	0.2341	-1.6998E-04	-8.9665E-07	149	328	liquid	0.104	0.189	0.082
251	SF6	SULFUR HEXAFLUORIDE	0.2544	-6.5950E-04	7.0217E-18	223	319	liquid	0.058	0.107	0.044
252	SOBr2	THIONYL BROMIDE	0.1694	-8.6221E-05	-1.6895E-07	221	596	liquid	0.129	0.142	0.058
253	SOCl2	THIONYL CHLORIDE	0.1974	-1.2157E-04	-2.5795E-07	172	510	liquid	0.138	0.169	0.068
254	SOF2	SULFUROUS OXYFLUORIDE	0.2505	-1.5546E-04	-9.8143E-07	163	334	liquid	0.117	0.199	0.089
255	SO2	SULFUR DIOXIDE	0.3822	-6.2540E-04	-5.6891E-19	200	400	liquid	0.196	0.257	0.132
256	SO2Cl2	SULFURYL CHLORIDE	0.2062	-1.0100E-04	-3.5044E-07	222	491	liquid	0.145	0.167	0.072
257	SO3	SULFUR TRIOXIDE	0.9288	-3.0803E-03	2.6600E-06	290	481	liquid	0.247	0.259	0.063
258	S2Cl2	SULFUR MONOCHLORIDE	0.2125	-1.1856E-04	-1.9856E-07	193	593	liquid	0.160	0.182	0.072
259	Sb	ANTIMONY	39.6013	-6.0319E-02	3.9401E-05	200	900	solid	25.120	29.114	17.229
260	SbBr3	ANTIMONY TRIBROMIDE	0.1232	-3.5768E-05	-7.9151E-08	370	803	liquid	------	0.099	0.043
261	SbCl3	ANTIMONY TRICHLORIDE	0.1550	-4.5388E-05	-1.3227E-07	347	715	liquid	------	0.123	0.055
262	SbCl5	ANTIMONY PENTACHLORIDE	0.1370	-5.3077E-05	-1.6073E-07	276	596	liquid	0.107	0.110	0.048
263	SbH3	STIBINE	0.1999	-1.1437E-04	-5.3519E-07	185	396	liquid	0.118	0.160	0.071
264	SbI3	ANTIMONY TRIIODIDE	0.1050	-2.6214E-05	-4.3393E-08	440	988	liquid	------	0.085	0.037
265	Sb2O3	ANTIMONY TRIOXIDE	--------	----------	----------	----	----	------	------	------	------
266	Sc	SCANDIUM	13.1216	1.4801E-02	-1.9547E-05	200	300	solid	15.797	15.300	15.803
267	Se	SELENIUM	4.3030	-1.8659E-02	2.9082E-05	200	490	solid	1.325	1.734	2.143
268	SeCl4	SELENIUM TETRACHLORIDE	0.1615	-6.1998E-05	-1.4201E-07	290	670	liquid	0.130	0.132	0.056
269	SeF6	SELENIUM HEXAFLUORIDE	0.1378	1.0612E-04	-1.0358E-06	238	332	liquid	0.077	0.104	0.059
270	SeOCl2	SELENIUM OXYCHLORIDE	0.1855	-7.2606E-05	-1.8405E-07	282	636	liquid	0.147	0.150	0.065
271	SeO2	SELENIUM DIOXIDE	--------	----------	----------	----	----	------	------	------	------
272	Si	SILICON	436.7820	-1.2856E+00	1.1110E-03	250	600	solid	152.241	184.820	65.382
273	SiBrCl2F	BROMODICHLOROFLUOROSILANE	0.1734	-1.2039E-04	-3.0074E-07	161	447	liquid	0.111	0.146	0.059
274	SiBrF3	TRIFLUOROBROMOSILANE	0.1695	-2.7093E-05	-8.4434E-07	203	338	liquid	0.086	0.129	0.064
275	SiBr2ClF	DIBROMOCHLOROFLUOROSILANE	0.1568	-1.0084E-04	-2.3517E-07	174	482	liquid	0.106	0.132	0.054
276	SiClF3	TRIFLUOROCHLOROSILANE	0.2150	-1.8107E-04	-9.7165E-07	131	297	liquid	------	0.175	0.076
277	SiCl2F2	DICHLORODIFLUOROSILANE	0.2064	-1.7459E-04	-5.9819E-07	133	352	liquid	0.101	0.173	0.071
278	SiCl3F	TRICHLOROFLUOROSILANE	0.1963	-1.4463E-04	-4.0204E-07	152	414	liquid	0.117	0.165	0.068
279	SiCl4	SILICON TETRACHLORIDE	0.1510	-8.0951E-05	-2.9387E-07	204	456	liquid	0.101	0.122	0.053
280	SiF4	SILICON TETRAFLUORIDE	0.1727	-2.9350E-04	1.2500E-07	260	340	liquid	0.096	0.105	0.087
281	SiHBr3	TRIBROMOSILANE	0.1508	-8.4590E-05	-1.7490E-07	200	549	liquid	0.110	0.127	0.052
282	SiHCl3	TRICHLOROSILANE	0.2132	-2.3826E-04	-1.7829E-07	145	431	liquid	0.126	0.175	0.077
283	SiHF3	TRIFLUOROSILANE	0.2423	-1.3314E-04	-1.7375E-06	142	262	liquid	------	0.188	0.088
284	SiH2Br2	DIBROMOSILANE	0.1755	-9.7668E-05	-2.7088E-07	203	495	liquid	0.122	0.145	0.061
285	SiH2Cl2	DICHLOROSILANE	0.3105	-5.1350E-04	-1.3364E-19	151	413	liquid	0.157	0.233	0.098
286	SiH2F2	DIFLUOROSILANE	0.2752	-1.5536E-04	-1.5992E-06	151	286	liquid	------	0.215	0.100
287	SiH2I2	DIIODOSILANE	0.1443	-5.7123E-05	-1.6897E-07	272	594	liquid	0.112	0.116	0.051
288	SiH3Br	MONOBROMOSILANE	0.2219	-1.3708E-04	-5.2949E-07	179	409	liquid	0.134	0.180	0.077
289	SiH3Cl	MONOCHLOROSILANE	0.2890	-2.0728E-04	-8.9603E-07	155	357	liquid	0.148	0.235	0.101
290	SiH3F	MONOFLUOROSILANE	0.3051	-6.0560E-05	-2.6235E-06	155	258	liquid	------	0.233	0.115
291	SiH3I	IODOSILANE	0.1870	-9.1803E-05	-3.6543E-07	216	464	liquid	0.127	0.150	0.066
292	SiH4	SILANE	0.2977	-1.3786E-03	1.7321E-06	88	243	liquid	------	0.190	0.065
293	SiO2	SILICON DIOXIDE	2.1744	-3.7847E-03	4.6353E-06	273	1400	solid	1.458	1.487	5.961
294	Si2Cl6	HEXACHLORODISILANE	0.1448	-5.7491E-05	-1.6883E-07	272	595	liquid	0.113	0.117	0.051
295	Si2F6	HEXAFLUORODISILANE	0.1535	6.2233E-05	-8.2874E-07	255	370	liquid	0.098	0.115	0.063
296	Si2H5Cl	DISILANYL CHLORIDE	0.2293	-6.1631E-05	-5.6190E-07	254	456	liquid	0.161	0.177	0.084
297	Si2H6	DISILANE	0.3005	-2.3899E-04	-6.9282E-07	141	389	liquid	0.168	0.253	0.103
298	Si2OCl3F3	TRICHLOROTRIFLUORODISILOXANE	0.1471	-4.0323E-05	-3.5605E-07	254	458	liquid	0.103	0.114	0.054
299	Si2OCl6	HEXACHLORODISILOXANE	0.1432	-6.7458E-05	-1.5464E-07	240	590	liquid	0.109	0.118	0.050
300	Si2OH6	DISILOXANE	0.2760	-2.3576E-04	-6.6397E-07	129	375	liquid	0.147	0.235	0.094
301	Si3Cl8	OCTACHLOROTRISILANE	0.1278	-5.6357E-05	-9.1772E-08	254	698	liquid	0.103	0.108	0.044
302	Si3H8	TRISILANE	0.2559	-1.7790E-04	-3.7962E-07	156	473	liquid	0.169	0.219	0.087
303	Si3H9N	TRISILAZANE	0.2356	-1.5710E-04	-3.7591E-07	167	466	liquid	0.155	0.199	0.081
304	Si4H10	TETRASILANE	0.2225	-1.3484E-04	-2.5461E-07	180	539	liquid	0.160	0.190	0.076
305	Sm	SAMARIUM	14.2476	-6.3556E-03	1.0067E-05	250	600	solid	13.248	13.288	14.058
306	Sn	TIN	92.3073	-1.1537E-01	1.0000E-04	200	500	solid	66.799	73.233	59.622
307	SnBr4	STANNIC BROMIDE	0.1980	-2.0000E-04	3.3437E-18	260	340	liquid	0.138	0.146	0.130
308	SnCl2	STANNOUS CHLORIDE	0.1616	-1.1786E-04	-1.7857E-07	260	340	liquid	0.111	0.119	0.101
309	SnCl4	STANNIC CHLORIDE	0.1482	-6.7765E-05	-1.8857E-07	243	558	liquid	0.111	0.121	0.052
310	SnH4	STANNIC HYDRIDE	0.2175	-1.9936E-04	-7.5278E-07	123	323	liquid	0.091	0.182	0.075
311	SnI4	STANNIC IODIDE	0.0936	-2.4132E-05	-4.6721E-08	418	910	liquid	------	0.0753	0.0330
312	Sr	STRONTIUM	54.6626	-8.3896E-02	6.4179E-05	200	800	solid	35.354	40.451	28.620

Table 24-2 THERMAL CONDUCTIVITY OF LIQUID AND SOLID - INORGANIC COMPOUNDS (continued)

NO	FORMULA	NAME	$k = A + B T + C T^2$						$(k - W/(m\ K),\ T - K)$		
			A	B	C	TMIN	TMAX	Phase	k @25 C	k @TMIN	k @TMAX
313	SrO	STRONTIUM OXIDE	205.8964	-6.6390E-01	7.0558E-04	493	1003	solid	-----	50.084	249.825
314	Ta	TANTALUM	56.4764	3.5218E-03	1.7006E-07	200	2000	solid	57.542	57.188	64.200
315	Tc	TECNNETIUM	55.2837	-2.2322E-02	2.2516E-05	273	900	solid	50.630	50.868	53.432
316	Te	TELLURIUM	3.8891	-8.2606E-03	6.8317E-06	200	700	solid	2.033	2.510	1.454
317	TeCl4	TELLURIUM TETRACHLORIDE	--------	----------	----------	----	----	-------	-----	-----	-----
318	TeF6	TELLURIUM HEXAFLUORIDE	0.1286	5.6649E-05	-8.1322E-07	235	342	liquid	0.073	0.0970	0.0529
319	Ti	TITANIUM	23.1703	-9.1462E-03	6.4557E-06	298	1800	solid	21.017	21.018	27.624
320	TiCl4	TITANIUM TETRACHLORIDE	0.0942	3.9437E-04	-7.4801E-07	249	574	liquid	0.145	0.146	0.074
321	Tl	THALLIUM	58.2977	-5.3024E-02	4.1437E-05	200	500	solid	46.172	49.350	42.145
322	TlBr	THALLOUS BROMIDE	1.2859	-4.4118E-03	6.9124E-06	298	343	solid	0.585	0.585	0.586
323	TlI	THALLOUS IODIDE	--------	----------	----------	----	----	-------	-----	-----	-----
324	Tm	THULIUM	11.8032	3.1724E-02	-4.9077E-05	200	300	solid	16.899	16.185	16.903
325	U	URANIUM	21.3170	1.9475E-02	3.0155E-06	200	1200	solid	27.392	25.333	49.029
326	UF6	URANIUM FLUORIDE	--------	----------	----------	----	----	-------	-----	-----	-----
327	V	VANADIUM	27.7557	8.9910E-03	1.3750E-06	250	2000	solid	30.559	30.089	51.238
328	VCl4	VANADIUM TETRACHLORIDE	0.1639	-7.4795E-05	-1.5272E-07	247	627	liquid	0.128	0.136	0.057
329	VOCl3	VANADIUM OXYTRICHLORIDE	0.1884	-1.0654E-04	-1.9305E-07	194	572	liquid	0.139	0.160	0.064
330	W	TUNGSTEN	192.9791	-7.8532E-02	1.4383E-05	250	3400	solid	170.843	174.245	92.238
331	WF6	TUNGSTEN FLUORIDE	0.1217	1.1726E-05	-4.4358E-07	273	422	liquid	0.086	0.092	0.048
332	Xe	XENON	0.1376	-3.9765E-04	-3.2519E-19	161	270	liquid	-----	0.074	0.030
333	Yb	YTTERBIUM	48.6283	-6.8259E-02	7.7496E-05	200	500	solid	35.166	38.076	33.873
334	Yt	YTTRIUM	12.8950	1.4996E-02	-3.2810E-06	250	1173	solid	17.074	16.439	25.971
335	Zn	ZINC	125.5025	-3.0320E-02	-1.1471E-05	200	693	solid	115.443	118.980	98.982
336	ZnCl2	ZINC CHLORIDE	2.2616	-6.2681E-03	4.9822E-06	563	641	solid	-----	0.312	0.291
337	ZnF2	ZINC FLUORIDE	--------	----------	----------	----	----	-------	-----	-----	-----
338	ZnO	ZINC OXIDE	71.4550	-1.8504E-01	1.6306E-04	319	418	solid	-----	29.020	22.599
339	ZnSO4	ZINC SULFATE	--------	----------	----------	----	----	-------	-----	-----	-----
340	Zr	ZIRCONIUM	27.4961	-2.0812E-02	1.6586E-05	250	1200	solid	22.765	23.330	26.406
341	ZrBr4	ZIRCONIUM BROMIDE	--------	----------	----------	----	----	-------	-----	-----	-----
342	ZrCl4	ZIRCONIUM CHLORIDE	--------	----------	----------	----	----	-------	-----	-----	-----
343	ZrI4	ZIRCONIUM IODIDE	--------	----------	----------	----	----	-------	-----	-----	-----

k - thermal conductivity, W/(m K)

A, B, and C - regression coefficients for chemical compound

T - temperature, K

TMIN - minimum temperature, K

TMAX - maximum temperature, K

Chapter 25

EXPLOSIVE LIMITS IN AIR, FLASH POINT, AND AUTOIGNITION TEMPERATURE

Carl L. Yaws, Sachin D. Sheth, and Mei Han
Lamar University, Beaumont, Texas

ABSTRACT

Results for explosive (lower and upper flammable) limits in air, flash point, and autoignition temperature are presented for organic compounds. The results are displayed in an easy-to-use table that is especially applicable for rapid engineering usage. The organic compounds encompass hydrocarbon, oxygen, nitrogen, halogen, silicon, sulfur, and other chemical types.

EXPLOSIVE LIMITS IN AIR

The results for lower (LEL) and upper (UEL) explosive limits in air are presented in Tables 25-1 and 25-2. The LEL and UEL values are the lower and upper concentrations (expressed as volume %) for flammability in air. The tabulation is based on both experimental data and estimated values.

In the data collection, a literature search was conducted to identify data source publications (1-56) for explosive limits in air. Both experimental values for the property under consideration and parameter values for estimation of the property are included in the source publications. The publications were screened and copies of appropriate data were made. These data were then keyed into the computer to provide a database for which experimental values are available. The database also served as a basis to check the accuracy of the estimation methods.

Upon completion of data collection, estimation of values for explosive limits in air was performed. The estimates are primarily based on the methods of Shebeko (17) and Jones (2). The Jones method (regression of the stoichiometric concentrations for volume % fuel in fuel plus air) is shown below:

$$C_mH_xO_y + z\,O_2 \longrightarrow m\,CO_2 + x/2\,H_2O$$

$$\text{LEL, \%} = 0.55\,(100)/(4.76m + 1.19x + 1 - 2.38y) \tag{25-1}$$

$$\text{UEL, \%} = 3.50\,(100)/(4.76m + 1.19x + 1 - 2.38y) \tag{25-2}$$

Evaluation of these equations with normal alkanes disclosed favorable agreement of estimates and data. For lower explosive limit, very favorable agreement was obtained for small, intermediate, and large size alkanes. For upper explosive limit, rough agreement was experienced for small and large size alkanes. More favorable agreement was exhibited for intermediate size alkanes.

A comparison of experimental data and estimates for lower explosion limit in air and data is shown in Fig. 25-1 for normal alkanes. The graph discloses favorable agreement of data and estimates.

EXPLOSIVE LIMITS FOR MIXTURES

The lower and upper explosive limits in air are often needed for gas mixtures. The Le Chatelier equation (2) for gas mixtures is

$$\text{LEL}_{mixture}, \% = 1 / \Sigma\,(y_i/\text{LEL}_i) \tag{25-3}$$

$$\text{UEL}_{mixture}, \% = 1 / \Sigma\,(y_i/\text{UEL}_i) \tag{25-4}$$

where y_i = mole fraction of component i on a combustible basis

FLASH POINT AND AUTOIGNITION TEMPERATURE

The results for flash point and autoignition temperatures are also given in Tables 25-1 and 25-2. The flash point represents the temperature at which the liquid gives off enough vapor to flash (combust) when exposed to an external ignition source. The autoignition temperature is the temperature at which the substance will automatically ignite (combust) without an external ignition source. The tabulation is based on both experimental data and estimated values.

In the data collection, a literature search was conducted to identify data source publications (1-83) for flash point and autoignition temperature. The publications were screened and copies of appropriate data were made. These data were then keyed into the computer to provide a database for which experimental data are available. The database also served as a basis to check the accuracy of the estimation methods.

Upon completion of data collection, estimation of values for the remaining compounds was performed. The estimates are primarily based on the methods of Shebeko (22), Gmehling and Rasmussen (23), and vapor pressure methods. The vapor pressure method is based on determining the temperature at which the vapor pressure will provide an equilibrium concentration that is equal to the lower explosive limit (LEL) concentration in air. The equations are briefly given below:

$$y_i = P_i/P = LEL_i/100 \qquad\qquad (25\text{-}5)$$

where
y_i = vapor concentration of component i, mole fraction
P_i = vapor pressure of component i, atm
P = total pressure, atm
P_i/P = equilibrium concentration of component i, mole fraction
$LEL_i/100$ = lower explosive limit concentration of component i, mole fraction

Evaluation of the vapor pressure method with normal alkanes disclosed favorable agreement of estimates and data for small, intermediate, and large size molecules. Evaluation with other compound types was not performed. If the lower explosive limit (LEL) used in the calculations is estimated, the estimates for flashpoint should be considered as rough values.

A comparison of experimental data and estimates for flash point and data is shown in Fig. 25-2 for normal alkanes. The graph discloses favorable agreement of data and estimates.

EXAMPLES

The tabulated values may be used in engineering applications involving pure components and mixtures in air. Examples are given below.

Example 1 A process vessel contains n-pentane (C_5H_{12}) at a concentration of 2 vol % in air. Are the contents of the vessel flammable?

Inspection of the table discloses that LEL = 1.4 vol % for n-pentane. Since the vessel contents exceed the LEL for n-pentane, the contents are flammable. This is shown below:

Vessel contents of 2 vol % > LEL of 1.4 vol %

Vessel contents are flammable.

Example 2 Estimate the lower (LEL) and upper (UEL) explosive limits in air for the gas mixture below:

	vol %	y_i (combustible basis)	LEL_i	UEL_i
Methane	1	0.2	5.0	15.0
Ethane	2	0.4	3.0	12.5
Propane	2	0.4	2.1	9.5
Air	95	---	---	---

Substitution of y_i, LEL_i, and UEL_i into the equations for gas mixtures yields

$$LEL_{mixture} = 1 / \Sigma(y_i/LEL_i) = 1/(0.2/5 + 0.4/3 + 0.4/2.1) = 2.75 \text{ vol \%}$$

$$UEL_{mixture} = 1 / \Sigma(y_i/UEL_i) = 1/(0.2/15 + 0.4/12.5 + 0.4/9.5) = 11.4 \text{ vol \%}$$

Example 3 A process vessel at a temperature of 80 F contains liquid toluene (C_7H_8) in contact with air. Is the vapor in the process vessel flammable?

Inspection of the table discloses that the flash point is 40 F for toluene. Since the temperature of the process vessel contents exceeds the flash point for toluene, the vapor is flammable. This is shown below:

Vessel temperature of 80 F > Flash point of 40 F

Vapor in vessel is flammable.

Example 4 A small quantity of residual n-tetradecane ($C_{14}H_{30}$) is in the piston bore of a piston-type compressor. If air at ambient conditions is compressed to 570 psia and 420 F, will the n-tetradecane undergo autoignition?

Inspection of the table discloses that the autoignition temperature of n-tetradecane is 392 F. Since the temperature of 420 F at the end of the compression exceeds the autoignition temperature, autoignition of the n-tetradecane will occur. This is shown below:

Compressor temp. of 420 F > Autoignition temp. of 392 F

Autoignition of n-tetradecane will occur.

REFERENCES – EXPLOSIVE LIMITS IN AIR – ORGANIC AND INORGANIC COMPOUNDS

1. Daubert, T. E. and R. P. Danner, <u>DATA COMPILATION OF PROPERTIES OF PURE COMPOUNDS</u>, Parts 1, 2, 3, and 4, Supplements 1 and 2, DIPPR Project, AIChE, New York, NY (1985-1992).
2. Crowl, D. A. and J. F. Louvar, <u>CHEMICAL PROCESS SAFETY</u>, Prentice Hall, Inc., Englewood Cliffs, NJ (1990).
3. <u>NIOSH POCKET GUIDE TO CHEMICAL HAZARDS</u>, U. S. Dept. of Health and Human Services, Superintendent of Documents, Washington, DC (June, 1994).
4. Lees, F. P., <u>LOSS PREVENTION IN THE PROCESS INDUSTRIES</u>, Vols. 1 and 2, Butterworth-Heinemann, London, England (1992).
5. <u>SELECTED VALUES OF PROPERTIES OF HYDROCARBONS AND RELATED COMPOUNDS</u>, Thermodynamics Research Center, TAMU, College Station, TX (1977, 1984).
6. <u>SELECTED VALUES OF PROPERTIES OF CHEMICAL COMPOUNDS</u>, Thermodynamics Research Center, TAMU, College Station, TX (1977, 1987).
7. <u>TECHNICAL DATA BOOK - PETROLEUM REFINING</u>, Vols. I and II, American Petroleum Institute, Washington, DC (1972, 1977, 1982).
8. <u>CONDENSED CHEMICAL DICTIONARY</u>, 10th and 11th eds., G. G. Hawley (10th) and Sax, N. I. and R. J. Lewis, Jr. (11th), Van Nostrand Reinhold Co., New York, NY (1981, 1987).
9. Braker, W. and A. L. Mossman, <u>MATHESON GAS DATA BOOK</u>, 6th ed., Matheson Gas Products, Secaucus, NJ (1980).
10. <u>CRC HANDBOOK OF CHEMISTRY AND PHYSICS</u>, 75th - 78th eds., CRC Press, Inc., Boca Raton, FL (1994-1997).
11. <u>ENCYCLOPEDIA OF CHEMICAL TECHNOLOGY</u>, 3rd and 4th eds., John Wiley and Sons, Inc., New York, NY (1978-1997).
12. Riddick, J. A., W. B. Bunger, and T. K. Sakano, <u>ORGANIC SOLVENTS: PHYSICAL PROPERTIES AND METHODS OF PURIFICATION</u>, 3rd and 4th eds., Wiley Interscience, New York. NY (1970, 1986).
13. Tryon, G. H., ed., <u>FIRE PROTECTION HANDBOOK</u>, 12th ed., National Fire Protection Association, Boston, MA (1962).
14. Weiss, G., <u>HAZARDOUS CHEMICALS DATA BOOK</u>, Noyes Data Corp., Park Ridge, NJ (1986).
15. <u>FIRE POTECTION GUIDE ON HAZARDOUS MATERIALS</u>, 7th ed., National Fire Protection Association, Quincy, MA (1978).
16. <u>SFPE HANDBOOK OF FIRE POTECTION ENGINEERING</u>, 1st ed., Society of Fire Protection Engineers, National Fire Protection Association, Quincy, MA (1988).
17. Shebeko, Y. N., A. V. Ivanov, and T. M. Dmitrieva, Sov. Chem. Ind. 15(3), 311 (1983).
18. Steere, N. V., ed., <u>HANDBOOK OF LABORATORY SAFETY</u>, 2nd ed., CRC Press Inc., Boca Raton, FL (1982).
19. Zabetakis, M. G., U. S. Bureau of Mines Bulletin No. 627 (1965).
20. Gas Processors Association, Publication No. 2145-84, Tulsa, OK (1984).
21. Halls, E. W., H. H. Liebhafsky, and D. H. Getz, Ind. Eng. Chem., 41, 1959 (1949).
22. Gmehling, J. and P. Rasmussen, Ind. Eng. Chem., Fundam, 21, 186 (1982).
23. Hercules, Inc., Bulettin HE-109A, HE-110, HE-120A, Cumberland, MD.
24. DeMicheli, S. and V. Tartari, J. Chem. Eng. Data, 27(3), 273 (1982).
25. Coward, H. F. and G. W. Jones, Bureau of Mines Bulletin 503 (1952).
26. Eastman Chemicals, Publication No. C103-A, Kingsport, TN (1977).
27. Pfanstiehl Laboratories, "MALEIC ACID", Material Safety Data Sheet, Waukegan, IL (1972).
28. Huffman, H. M., S. S. Todd, and G. D. Oliver, J. Am. Chem. Soc., 71, 584 (1971).
29. Eastman Chemicals, Publication No. A1013, Kingsport, TN (1976).
30. DuPont, "METHACRYLATE MONOMERS", Wilmington, DE (1979).
31. McGlashan, M. L. and I. R. McKinnon, J. Chem. Thermo., 9(10), 1205 (1977).
32. GAF Corporation, "M-PYROL HANDBOOK", New York, NY (1972).
33. Eastman Chemicals, Publication No. M-144C, Kingsport, TN (1978).
34. Petro-Tex Chemical Corp., "METHYL TERT-BUTYL ETHER", Material Safety Data Sheet, Houston, TX (1979).
35. Eastman Chemicals, Publication Number M-198D, Kingsport, TN (1985).
36. Eastman Chemicals, "NEOPENTYL GLYCOL", Material Safety Data Sheet, Kingsport, TN (1973).
37. Thompson, W. H., Ph.D. Thesis, Pennsylavania State University, University Park, PA (1966).
38. DuPont, "3,4-DICHLOROANILINE", Material Safety Data Sheet, Wilmington, DE (Oct. 1985).
39. Thiokol Ventra Division, "4-CHLOROANILINE", Material Safety Data Sheet, Danvers, MA (1986).
40. Eastman Chemicals, "m-NITROANILINE", Material Safety Data Sheet, Kingsport, TN (1978).
41. Exxon Chemical Company, "MESITYL OXIDE", Technical Brochure, Houston, TX (1976).
42. Eastman Chemical Products, Technical Data Publication No. N-135, USA (August, 1979).
43. Union Carbide, "2-METHYLPENTANOL", Material Safety Data Sheet, Danbury, CT (1981).
44. Eastman Chemicals, "TRIETHYLPHOSPHATE", Material Safety Data Sheet, Kingsport, TN (1975).
45. DuPont, "o-NITROTOLUENE", Material Safety Data Sheet, DuPont Company, Wilmington, DE (1985).
46. DuPont, "p-NITROTOLUENE", Material Safety Data Sheet, DuPont Company, Wilmington, DE (1985).
47. Rohm and Haas, "BUTYL ACRYLATE", Material Safety Data Sheet, Philadelphia (Oct., 1978).
48. Eastman Chemicals, "ETHYL 3-ETHOXYPROPIONATE", Kingsport, TN (1988).
49. Eastman Chemicals, Publication No. A-111-2B, Kingsport, TN (1977).
50. Eastman Chemicals, Technical Data M-143D, Kingsport, TN (1978).
51. EM Science, "2-(2-ETHOXYETHOXY)ETHYL ACETATE", Material Safety Data Sheet, Gibbstown, NJ (1984).
52. Eastman Chemicals, Publication No. B-115-B, Kingsport, TN (1977).
53. Fisher Scientific Company, "2-OCTANOL", Material Safety Data Sheet, Fairlawn, NJ (1980).
54. DuPont, "N,N-DIETHYLANILINE", Material Safety Data Sheet, Wilmington, DE (Oct. 1985).

55. Eastman Chemical Products, Publication No. L-142C, Kingsport, TN (Jan. 1980).
56. Yaws, C. L., HANDBOOK OF CHEMICAL COMPOUND DATA FOR PROCESS SAFETY, Gulf Publishing Co., Houston, TX (1997).

REFERENCES - FLASH POINT AND AUTOIGNITION TEMPERATURE – ORGANIC AND INORGANIC COMPOUNDS

1-21. See above REFERENCES – EXPLOSIVE LIMITS IN AIR - ORGANIC COMPOUNDS
22. Shebeko, N. Y., A. Y. Korolchenko, A. V. Ivanov, and E. N. Alekhina, Sov. Chem. Ind., 16, 1371 (1984).
23. Gmehling, J. and P. Rasmussen, Ind. Eng. Chem., Fundam, 21, 186 (1982).
24. HANDBOOK OF FINE CHEMICALS, Aldrich Chemical Co., Milwaukee, WI (1984-96).
25. Wilson, A. L., Ind. Eng. Chem., 27, 867 (1935).
26. Bailey, A. S., G. B. Pickering, and J. C. Smith, Inst. Petrol., 35, 103 (1949).
27. Fishbein, L., J. A. Gallaghan, and J. Arner, Chem. Soc., 78, 1218 (1956).
28. Amundsen, L. H., R. H. Mayer, L. S. Pitts, and L. A. Malentacchi, J. Amer. Chem. Soc., 73, 2118 (1951).
29. Hatch, G., J. Org. Chem., 24, 1881 (1959).
30. Riddle, E. H., MONOMERIC ACRYLIC ESTERS, Reinhold Publishing Corp., New York, NY (1954).
31. Wohl, A. and B. Mylo, Chem. Ber., 45, 322 (1912).
32. Medcalf, E. C., A. G. Hill, and G. N. Vriens, Petrol. Refiner, 31(7), 97 (1950).
33. Farkas, A., G. A. Gerner, W. E. Erner, and J. B. Maerker, J. Chem. Eng. Data, 4(4), 334 (1959).
34. Stephenson, R. M., FLASH POINTS OF ORGANIC AND ORGANOMETALLIC COMPOUNDS, Elsevier, New York, NY (1987).
35. Roth, C. A., Ind. Eng. Chem. Res. Dev., 11, 134 (1972).
36. Smutney, E. J. and A. Bondi, J. Phys. Chem., 69, 1214 (1965).
37. Dickey, F. H., J. H. Raley, F. F. Rust, R. S. Treseder, and W. E. Vaughan, Ind. Eng. Chem. 41, 1673 (1949).
38. Lipowitz, J. and M. Ziemelis, Fire Flamm., 7, 504 (1976).
39. Carswell, T. S. and H. L. Morrill, Ind. Eng. Chem. 29, 1247 (1937).
40. Curme, G. O. J., ed., GLYCOLS, Reinhold Publishing Corp., New York, NY (1952).
41. Rieth, V. H. and H. Eckardt, Frieberger. Forsch., 164A, 146 (1960).
42. Kratzke, H., S. Muller, M. Bohn, and R. Kohler, J. Chem. Thermo., 17, 283 (1985).
43. Griffith, S. T. and R. R. Wilson, Combustion Flame, 2, 244 (1958).
44. Eastman Chemical Products, Publication No. L-142C, Kingsport, TN (Jan. 1980).
45. Burner, W. M. and L. T. Sherwood, Ind. Eng. Chem., 41, 1654 (1949).
46. Reuther, H., Chem. Tech., 5, 330 (1953).
47. Reuther, H., Chem. Tech., 17, 752 (1965).
48. Lipowitz, J. and M. Ziemelis, Fire Flamm., 7, 504 (1976).
49. Hercules, Inc., Bulettin HE-109A, HE-110, HE-120A, Cumberland, MD.
50. DeMicheli, S. and V. Tartari, J. Chem. Eng. Data, 27(3), 273 (1982).
51. Coward, H. F. and G. W. Jones, Bureau of Mines Bulletin 503 (1952).
52. Eastman Chemicals, Publication No. C103-A, Kingsport, TN (1977).
53. Pfanstiehl Laboratories, "MALEIC ACID", Material Safety Data Sheet, Waukegan, IL (1972).
54. Huffman, H. M., S. S. Todd, and G. D. Oliver, J. Am. Chem. Soc., 71, 584 (1971).
55. Eastman Chemicals, Publication No. A1013, Kingsport, TN (1976).
56. DuPont, "METHACRYLATE MONOMERS", Wilmington, DE (1979).
57. McGlashan, M. L. and I. R. McKinnon, J. Chem. Thermo., 9(10), 1205 (1977).
58. GAF Corporation, "M-PYROL HANDBOOK", New York, NY (1972).
59. Eastman Chemicals, Publication No. M-144C, Kingsport, TN (1978).
60. Petro-Tex Chemical Corp., "METHYL TERT-BUTYL ETHER", Material Safety Data Sheet, Houston, TX (1979).
61. Eastman Chemicals, Publication Number M-198D, Kingsport, TN (1985).
62. Eastman Chemicals, "NEOPENTYL GLYCOL", Material Safety Data Sheet, Kingsport, TN (1973).
63. Thompson, W. H., Ph.D. Thesis, Pennsylvania State University, University Park, PA (1966).
64. DuPont, "3,4-DICHLOROANILINE", Material Safety Data Sheet, Wilmington, DE (Oct. 1985).
65. Thiokol Ventra Division, "4-CHLOROANILINE", Material Safety Data Sheet, Danvers, MA (1986).
66. Eastman Chemicals, "m-NITROANILINE", Material Safety Data Sheet, Kingsport, TN (1978).
67. Exxon Chemical Company, "MESITYL OXIDE", Technical Brochure, Houston, TX (1976).
68. Eastman Chemical Products, Technical Data Publication No. N-135, USA (August, 1979).
69. Union Carbide, "2-METHYLPENTANOL", Material Safety Data Sheet, Danbury, CT (1981).
70. Eastman Chemicals, "TRIETHYLPHOSPHATE", Material Safety Data Sheet, Kingsport, TN (1975).
71. DuPont, "o-NITROTOLUENE", Material Safety Data Sheet, DuPont Company, Wilmington, DE (1985).
72. DuPont, "p-NITROTOLUENE", Material Safety Data Sheet, DuPont Company, Wilmington, DE (1985).
73. Rohm and Haas, "BUTYL ACRYLATE", Material Safety Data Sheet, Philadelphia (Oct., 1978).
74. Eastman Chemicals, "ETHYL 3-ETHOXYPROPIONATE", Kingsport, TN (1988).
75. Eastman Chemicals, Publication No. A-111-2B, Kingsport, TN (1977).
76. Eastman Chemicals, Technical Data M-143D, Kingsport, TN (1978).
77. EM Science, "2-(2-ETHOXYETHOXY)ETHYL ACETATE", Material Safety Data Sheet, Gibbstown, NJ (1984).
78. Eastman Chemicals, Publication No. B-115-B, Kingsport, TN (1977).
79. Fisher Scientific Company, "2-OCTANOL", Material Safety Data Sheet, Fairlawn, NJ (1980).
80. DuPont, "N,N-DIETHYLANILINE", Material Safety Data Sheet, Wilmington, DE (Oct. 1985).
81. Hilado, C. J. and C. W. Clark, Chem. Eng., 75 (1972).
82. Stull, D. R., FUNDAMENTALS OF FIRE AND EXPLOSION, AIChE, New York, NY (1977).
83. Yaws, C. L., HANDBOOK OF CHEMICAL COMPOUND DATA FOR PROCESS SAFETY, Gulf Publishing Co., Houston, TX (1997).

Figure 25-1 Lower Explosive Limit for Normal Alkanes

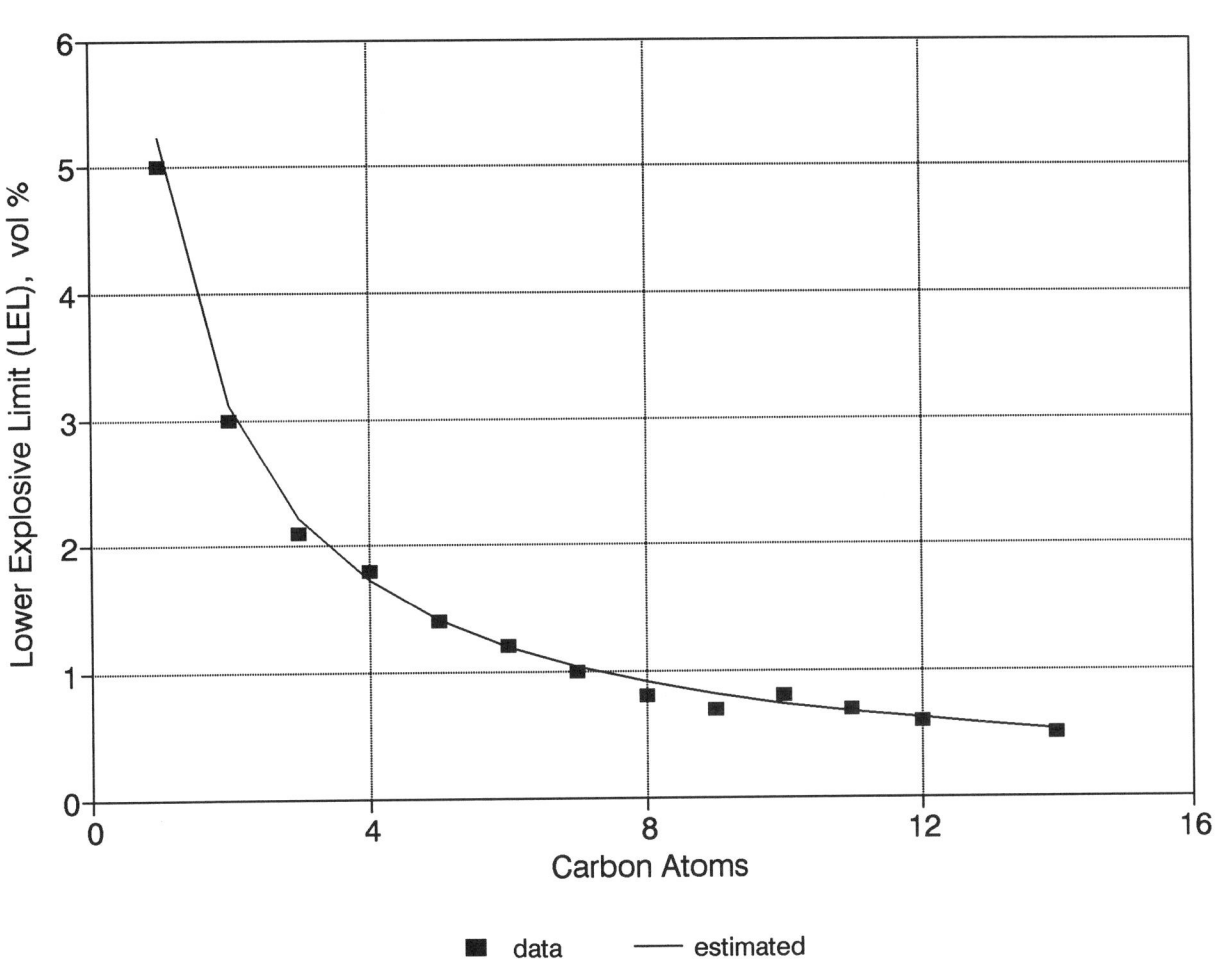

Figure 25-2 Flash Point for Normal Alkanes

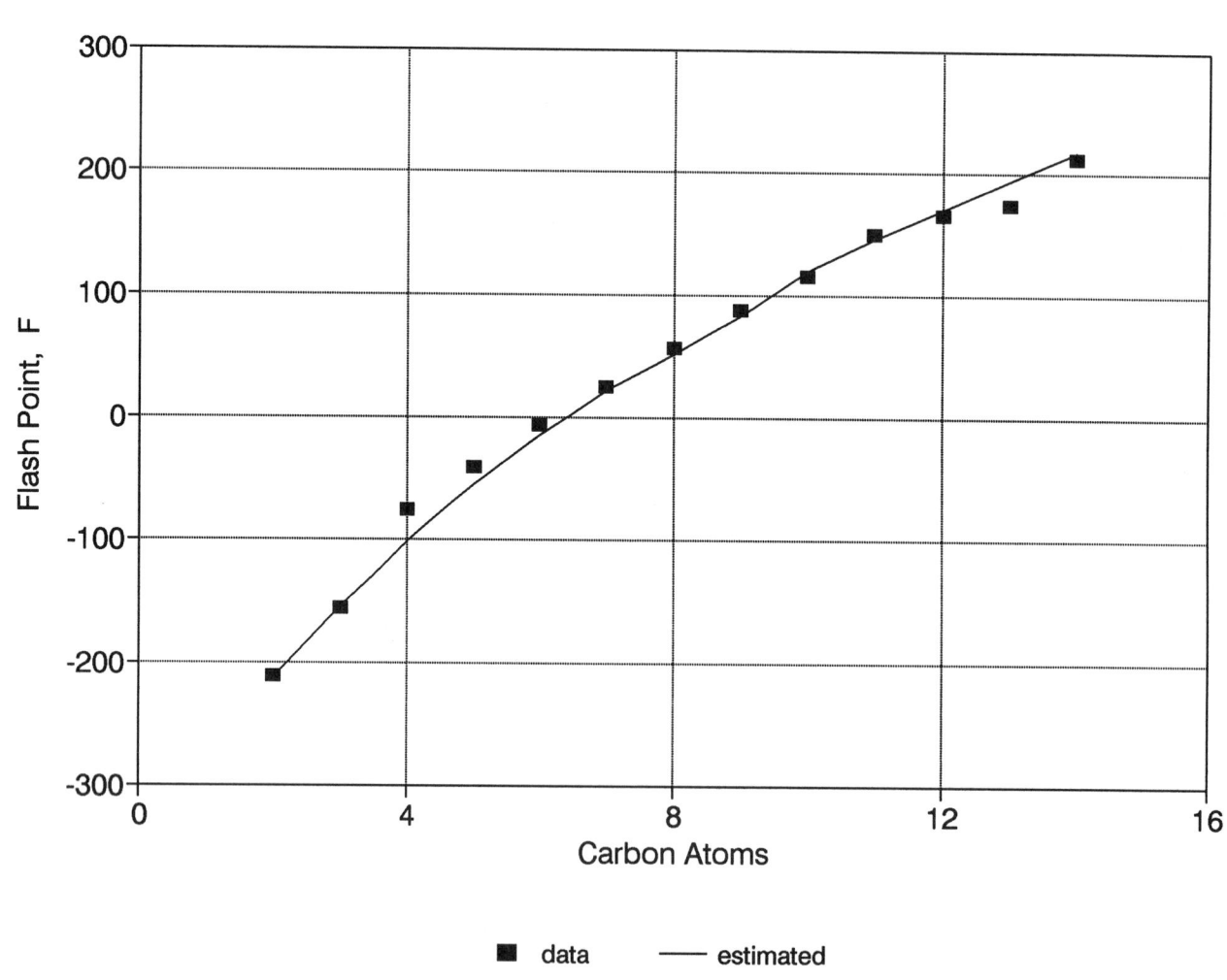

Table 25-1 EXPLOSION LIMITS IN AIR, FLASH POINT, AND AUTOIGNITION TEMPERATURE - ORGANIC COMPOUNDS

NO	FORMULA	NAME	Code	Lower Explosive Limit LEL, vol %	Upper Explosive Limit UEL, vol %	Flash Point Code	Flash Point F	Autoignition Temperature Code	Autoignition Temperature F
1	CBrClF2	BROMOCHLORODIFLUOROMETHANE	---	----	----	---	-----	---	-----
2	CBrCl3	BROMOTRICHLOROMETHANE	---	----	----	---	-----	---	-----
3	CBrF3	BROMOTRIFLUOROMETHANE	---	----	----	---	-----	---	-----
4	CBr2F2	DIBROMODIFLUOROMETHANE	---	----	----	---	-----	---	-----
5	CClF3	CHLOROTRIFLUOROMETHANE	---	----	----	---	-----	---	-----
6	CClN	CYANOGEN CHLORIDE	2	----	23.5	---	-----	---	-----
7	CCl2F2	DICHLORODIFLUOROMETHANE	---	----	----	---	-----	---	-----
8	CCl2O	PHOSGENE	---	----	----	---	-----	---	-----
9	CCl3F	TRICHLOROFLUOROMETHANE	---	----	----	---	-----	---	-----
10	CCl4	CARBON TETRACHLORIDE	---	----	----	---	-----	---	-----
11	CF2O	CARBONYL FLUORIDE	---	----	----	---	-----	---	-----
12	CF4	CARBON TETRAFLUORIDE	---	----	----	---	-----	---	-----
13	CHBr3	TRIBROMOMETHANE	2	----	35.3	2	181	---	-----
14	CHClF2	CHLORODIFLUOROMETHANE	2	----	26.9	2	-109	1	1170
15	CHCl2F	DICHLOROFLUOROMETHANE	2	----	54.7	2	-33	1	1026
16	CHCl3	CHLOROFORM	2	----	12.9	---	-----	---	-----
17	CHF3	TRIFLUOROMETHANE	2	----	35.3	2	-170	---	-----
18	CHI3	TRIIODOMETHANE	---	----	----	---	-----	---	-----
19	CHN	HYDROGEN CYANIDE	1	6.0	41.0	---	-----	1	1000
20	CHNS	ISOTHIOCYANIC-ACID	---	----	----	---	-----	---	-----
21	CH2BrCl	BROMOCHLOROMETHANE	2	----	22.6	2	52	---	-----
22	CH2Br2	DIBROMOMETHANE	2	----	27.2	2	93	---	-----
23	CH2ClF	CHLOROFLUOROMETHANE	---	----	----	---	-----	---	-----
24	CH2Cl2	DICHLOROMETHANE	1	15.5	66.0	2	25	1	1224
25	CH2F2	DIFLUOROMETHANE	2	----	27.2	2	-128	---	-----
26	CH2I2	DIIODOMETHANE	---	----	----	2	219	---	-----
27	CH2O	FORMALDEHYDE	1	7.0	73.0	2	-64	1	806
28	CH2O2	FORMIC ACID	1	18.0	57.0	1	156	1	1114
29	CH3Br	METHYL BROMIDE	1	10.0	16.0	2	-47	1	999
30	CH3Cl	METHYL CHLORIDE	1	10.7	17.4	2	-87	1	1170
31	CH3Cl3Si	METHYL TRICHLOROSILANE	1	5.1	----	1	5	---	-----
32	CH3F	METHYL FLUORIDE	2	----	22.2	---	-----	---	-----
33	CH3I	METHYL IODIDE	---	----	----	2	1	---	-----
34	CH3NO	FORMAMIDE	2	7.0	29.3	1	347	---	-----
35	CH3NO2	NITROMETHANE	1	7.3	22.2	1	95	1	714
36	CH3NO2	METHYL-NITRITE	---	----	----	---	-----	---	-----
37	CH3NO3	METHYL-NITRATE	---	----	----	---	-----	---	-----
38	CH4	METHANE	1	5.0	15.0	1	-306	1	1112
39	CH4Cl2Si	METHYL DICHLOROSILANE	1	3.4	----	1	-26	1	446
40	CH4O	METHANOL	1	7.3	36.0	1	52	1	867
41	CH4O3S	METHANESULFONIC ACID	---	----	----	---	-----	---	-----
42	CH4S	METHYL MERCAPTAN	1	3.9	21.8	2	-64	---	-----
43	CH5ClSi	METHYL CHLOROSILANE	2	5.9	----	2	-58	---	-----
44	CH5N	METHYLAMINE	1	4.9	20.7	2	-72	1	806
45	CH6Si	METHYL SILANE	2	4.3	----	2	-162	1	266
46	CN4O8	TETRANITROMETHANE	---	----	----	---	-----	---	-----
47	CO	CARBON MONOXIDE	1	12.5	74.0	---	-----	1	1128
48	COS	CARBONYL SULFIDE	1	12.0	29.0	2	-125	---	-----
49	CO2	CARBON DIOXIDE	---	----	----	---	-----	---	-----
50	CS2	CARBON DISULFIDE	1	1.3	50.0	1	-22	1	194
51	C2BrF3	BROMOTRIFLUOROETHYLENE	2	----	37.0	2	-71	---	-----
52	C2Br2F4	1,2-DIBROMOTETRAFLUOROETHANE	---	----	----	---	-----	---	-----
53	C2ClF3	CHLOROTRIFLUOROETHYLENE	1	8.4	38.7	2	-100	---	-----
54	C2ClF5	CHLOROPENTAFLUOROETHANE	---	----	----	---	-----	---	-----
55	C2Cl2F4	1,2-DICHLOROTETRAFLUOROETHANE	---	----	----	---	-----	---	-----
56	C2Cl3F3	1,1,2-TRICHLOROTRIFLUOROETHANE	---	----	----	---	-----	---	-----
57	C2Cl4	TETRACHLOROETHYLENE	---	----	----	2	113	---	-----
58	C2Cl4F2	1,1,2,2-TETRACHLORODIFLUOROETHANE	---	----	----	---	-----	---	-----
59	C2Cl4O	TRICHLOROACETYL CHLORIDE	---	----	----	---	-----	---	-----
60	C2Cl6	HEXACHLOROETHANE	---	----	----	---	-----	---	-----
61	C2F4	TETRAFLUOROETHYLENE	1	11.0	60.0	2	-163	1	392
62	C2F6	HEXAFLUOROETHANE	---	----	----	---	-----	---	-----
63	C2HBrClF3	HALOTHANE	---	----	----	---	-----	---	-----
64	C2HClF2	2-CHLORO-1,1-DIFLUOROETHYLENE	2	----	26.2	---	-----	---	-----
65	C2HCl3	TRICHLOROETHYLENE	1	8.0	10.5	1	90	1	770
66	C2HCl3O	DICHLOROACETYL CHLORIDE	---	----	31.7	1	151	---	-----
67	C2HCl3O	TRICHLOROACETALDEHYDE	2	7.8	14.7	2	91	---	-----
68	C2HCl5	PENTACHLOROETHANE	---	----	----	---	-----	---	-----
69	C2HF3	TRIFLUOROETHENE	---	----	----	---	-----	---	-----
70	C2HF3O2	TRIFLUOROACETIC ACID	2	----	28.0	2	70	---	-----
71	C2HF5	PENTAFLUOROETHANE	---	----	----	---	-----	---	-----
72	C2H2	ACETYLENE	1	2.5	80.0	---	-----	1	581
73	C2H2Br4	1,1,2,2-TETRABROMOETHANE	2	----	24.5	2	287	1	635
74	C2H2Cl2	1,1-DICHLOROETHYLENE	1	7.3	16.0	1	-1	1	1058
75	C2H2Cl2	cis-1,2-DICHLOROETHYLENE	1	5.6	12.8	1	39	2	860
76	C2H2Cl2	trans-1,2-DICHLOROETHYLENE	1	5.6	12.8	1	35	1	860
77	C2H2Cl2O	CHLOROACETYL CHLORIDE	2	----	16.3	2	98	---	-----
78	C2H2Cl2O	DICHLOROACETALDEHYDE	2	9.0	29.0	1	140	---	-----

NO	FORMULA	NAME	Code	Lower Explosive Limit LEL, vol %	Upper Explosive Limit UEL, vol %	Flash Point Code	F	Autoignition Temperature Code	F
79	C2H2Cl2O2	DICHLOROACETIC ACID	2	11.9	43.3	1	230	---	-----
80	C2H2Cl3F	1,1,1-TRICHLOROFLUOROETHANE	---	----	----	2	61	---	-----
81	C2H2Cl4	1,1,1,2-TETRACHLOROETHANE	2	4.9	12.1	2	116	---	-----
82	C2H2Cl4	1,1,2,2-TETRACHLOROETHANE	---	----	----	2	143	---	-----
83	C2H2F2	1,1-DIFLUOROETHYLENE	1	5.5	21.3	2	-195	1	1184
84	C2H2F2	cis-1,2-DIFLUOROETHENE	---	----	----	---	-----	---	-----
85	C2H2F2	trans-1,2-DIFLUOROETHENE	---	----	----	---	-----	---	-----
86	C2H2F4	1,1,1,2-TETRAFLUOROETHANE	---	----	----	2	-110	---	-----
87	C2H2O	KETENE	---	----	----	---	-----	---	-----
88	C2H2O4	OXALIC ACID	2	8.5	28.0	2	431	---	-----
89	C2H3Br	VINYL BROMIDE	1	9.0	15.0	2	-60	---	-----
90	C2H3Cl	VINYL CHLORIDE	1	3.6	33.0	1	-108	1	882
91	C2H3ClF2	1-CHLORO-1,1-DIFLUOROETHANE	1	6.2	17.9	2	-85	1	1169
92	C2H3ClO	ACETYL CHLORIDE	2	5.7	15.6	1	40	1	734
93	C2H3ClO	CHLOROACETALDEHYDE	2	5.7	18.4	---	-----	---	-----
94	C2H3ClO2	CHLOROACETIC ACID	---	----	----	---	-----	---	-----
95	C2H3ClO2	METHYL CHLOROFORMATE	1	6.7	15.6	1	54	1	939
96	C2H3Cl3	1,1,1-TRICHLOROETHANE	1	8.0	10.5	2	30	1	999
97	C2H3Cl3	1,1,2-TRICHLOROETHANE	2	8.6	27.8	2	89	1	860
98	C2H3F	VINYL FLUORIDE	1	2.6	21.7	2	-194	1	725
99	C2H3F3	1,1,1-TRIFLUOROETHANE	2	----	20.5	---	-----	---	-----
100	C2H3N	ACETONITRILE	1	4.4	16.0	1	42	1	975
101	C2H3NO	METHYL ISOCYANATE	1	5.3	26.0	1	19	1	993
102	C2H4	ETHYLENE	1	2.7	36.0	1	-213	1	842
103	C2H4Br2	1,1-DIBROMOETHANE	2	3.9	20.5	2	82	---	-----
104	C2H4Br2	1,2-DIBROMOETHANE	---	----	----	---	-----	---	-----
105	C2H4Cl2	1,1-DICHLOROETHANE	1	5.4	11.4	1	10	1	856
106	C2H4Cl2	1,2-DICHLOROETHANE	1	6.2	16.0	1	56	1	775
107	C2H4Cl2O	BIS(CHLOROMETHYL)ETHER	2	6.5	21.9	2	95	---	-----
108	C2H4F2	1,1-DIFLUOROETHANE	1	3.7	18.0	2	-114	---	-----
109	C2H4F2	1,2-DIFLUOROETHANE	2	4.6	21.5	2	-35	---	-----
110	C2H4I2	1,2-DIIODOETHANE	---	----	----	---	-----	---	-----
111	C2H4O	ACETALDEHYDE	1	1.6	10.4	1	-36	1	365
112	C2H4O	ETHYLENE OXIDE	1	3.0	----	2	-67	1	804
113	C2H4OS	THIOACETIC-ACID	---	----	----	---	-----	---	-----
114	C2H4O2	ACETIC ACID	1	5.4	16.0	1	109	1	800
115	C2H4O2	METHYL FORMATE	1	5.9	20.0	1	-2	1	853
116	C2H4S	THIACYCLOPROPANE	---	----	----	---	-----	---	-----
117	C2H5Br	BROMOETHANE	1	6.7	11.3	2	-28	1	952
118	C2H5Cl	ETHYL CHLORIDE	1	3.8	15.4	1	-58	1	966
119	C2H5ClO	2-CHLOROETHANOL	1	4.9	15.9	1	105	1	797
120	C2H5F	ETHYL FLUORIDE	2	----	17.3	---	-----	---	-----
121	C2H5I	ETHYL IODIDE	---	----	----	---	-----	---	-----
122	C2H5N	ETHYLENEIMINE	1	3.3	5.4	1	12	1	608
123	C2H5NO	ACETAMIDE	2	3.6	20.5	2	259	---	-----
124	C2H5NO	N-METHYLFORMAMIDE	2	3.6	18.6	2	217	---	-----
125	C2H5NO2	NITROETHANE	1	3.4	17.3	1	82	1	680
126	C2H5NO3	ETHYL-NITRATE	---	----	----	---	-----	---	-----
127	C2H6	ETHANE	1	3.0	12.5	1	-211	1	959
128	C2H6AlCl	DIMETHYLALUMINUM CHLORIDE	---	----	----	---	-----	---	-----
129	C2H6O	DIMETHYL ETHER	1	3.4	18.0	1	-42	1	662
130	C2H6O	ETHANOL	1	4.3	19.0	1	55	1	793
131	C2H6OS	DIMETHYL SULFOXIDE	1	2.6	28.5	1	190	1	419
132	C2H6O2	ETHYLENE GLYCOL	1	3.2	21.6	1	232	1	752
133	C2H6O4S	DIMETHYL SULFATE	---	----	----	1	181	1	370
134	C2H6S	DIMETHYL SULFIDE	1	2.2	19.7	1	-29	1	403
135	C2H6S	ETHYL MERCAPTAN	1	2.8	18.0	2	-55	1	570
136	C2H6S2	DIMETHYL DISULFIDE	2	1.9	----	1	44	---	-----
137	C2H7N	DIMETHYLAMINE	1	2.8	14.4	1	-58	1	752
138	C2H7N	ETHYLAMINE	1	3.5	14.0	2	-51	1	723
139	C2H7NO	MONOETHANOLAMINE	2	3.1	21.6	1	185	3	-----
140	C2H8N2	ETHYLENEDIAMINE	1	4.2	14.4	1	93	1	725
141	C2H8Si	DIMETHYL SILANE	2	2.7	83.0	2	-119	1	446
142	C2N2	CYANOGEN	1	6.0	32.0	---	-----	---	-----
143	C3F6	HEXAFLUOROPROPYLENE	2	----	28.3	---	-----	---	-----
144	C3F6O	HEXAFLUOROACETONE	---	----	----	---	-----	---	-----
145	C3F8	OCTAFLUOROPROPANE	---	----	----	---	-----	---	-----
146	C3H2N2	MALONONITRILE	2	2.9	19.0	1	234	---	-----
147	C3H3Cl	PROPARGYL CHLORIDE	2	3.4	41.0	1	64	---	-----
148	C3H3N	ACRYLONITRILE	1	2.4	17.3	1	32	1	898
149	C3H3NO	OXAZOLE	---	----	----	1	66	---	-----
150	C3H4	METHYLACETYLENE	1	1.7	39.9	2	-125	---	-----
151	C3H4	PROPADIENE	1	2.1	----	2	-140	---	-----
152	C3H4Cl2	2,3-DICHLOROPROPENE	1	2.6	7.8	1	50	---	-----
153	C3H4O	ACROLEIN	1	2.8	31.0	1	-15	1	453
154	C3H4O	PROPARGYL ALCOHOL	1	2.4	50.3	1	97	1	239
155	C3H4O2	ACRYLIC ACID	1	2.4	20.2	1	123	1	820
156	C3H4O2	beta-PROPIOLACTONE	1	2.9	25.1	1	165	---	-----

NO	FORMULA	NAME	Code	Lower Explosive Limit LEL, vol %	Upper Explosive Limit UEL, vol %	Flash Point Code	Flash Point F	Autoignition Temperature Code	Autoignition Temperature F
157	C3H4O2	VINYL FORMATE	2	3.3	18.8	2	-2	---	-----
158	C3H4O3	ETHYLENE CARBONATE	2	3.6	25.1	1	305	---	-----
159	C3H4O3	PYRUVIC ACID	2	3.6	16.6	1	183	---	-----
160	C3H5Br	3-BROMO-1-PROPENE	---	----	----	---	-----	---	-----
161	C3H5Cl	2-CHLOROPROPENE	1	4.5	16.0	1	25	---	-----
162	C3H5Cl	3-CHLOROPROPENE	1	2.9	11.1	1	-26	1	905
163	C3H5ClO	alpha-EPICHLOROHYDRIN	1	3.8	21.0	1	88	1	781
164	C3H5ClO2	METHYL CHLOROACETATE	2	3.5	12.8	1	125	---	-----
165	C3H5ClO2	ETHYL CHLOROFORMATE	2	3.5	12.8	1	36	1	932
166	C3H5Cl3	1,2,3-TRICHLOROPROPANE	1	3.2	12.6	1	165	1	579
167	C3H5I	3-IODO-1-PROPENE	---	----	----	---	-----	---	-----
168	C3H5N	PROPIONITRILE	1	3.1	14.0	1	36	2	953
169	C3H5NO	ACRYLAMIDE	2	2.7	20.6	2	252	---	-----
170	C3H5NO	HYDRACRYLONITRILE	1	2.3	12.1	1	265	1	922
171	C3H5NO	LACTONITRILE	2	2.7	17.9	1	171	---	-----
172	C3H5N3O9	NITROGLYCERINE	---	----	----	---	-----	1	518
173	C3H6	CYCLOPROPANE	1	2.4	10.4	2	-138	1	928
174	C3H6	PROPYLENE	1	2.0	11.0	1	-162	1	851
175	C3H6Br2	1,2-DIBROMOPROPANE	---	----	----	---	-----	---	-----
176	C3H6Cl2	1,1-DICHLOROPROPANE	1	3.1	14.5	1	70	2	1034
177	C3H6Cl2	1,2-DICHLOROPROPANE	1	3.4	14.5	1	60	1	1035
178	C3H6Cl2	1,3-DICHLOROPROPANE	2	3.4	14.5	1	70	---	-----
179	C3H6Cl2	2,2-DICHLOROPROPANE	---	----	----	---	-----	---	-----
180	C3H6I2	1,2-DIIODOPROPANE	---	----	----	---	-----	---	-----
181	C3H6O	ACETONE	1	2.6	12.8	1	0	1	1000
182	C3H6O	ALLYL ALCOHOL	1	2.5	18.0	1	70	1	713
183	C3H6O	METHYL VINYL ETHER	1	2.6	39.0	1	-69	1	549
184	C3H6O	n-PROPIONALDEHYDE	1	2.6	16.1	1	-22	1	405
185	C3H6O	1,2-PROPYLENE OXIDE	1	2.1	21.5	1	-35	1	869
186	C3H6O	1,3-PROPYLENE OXIDE	1	2.8	37.0	2	-19	---	-----
187	C3H6O2	ETHYL FORMATE	1	2.7	13.5	1	25	1	851
188	C3H6O2	METHYL ACETATE	1	3.1	16.0	1	14	1	935
189	C3H6O2	PROPIONIC ACID	2	2.9	14.8	1	131	1	887
190	C3H6O2S	3-MERCAPTOPROPIONIC ACID	2	2.2	----	1	199	---	-----
191	C3H6O3	LACTIC ACID	2	3.1	18.0	2	209	---	-----
192	C3H6O3	METHOXYACETIC ACID	2	3.1	20.2	2	241	---	-----
193	C3H6O3	TRIOXANE	1	3.6	29.0	1	113	1	777
194	C3H6S	THIACYCLOBUTANE	---	----	----	---	-----	---	-----
195	C3H7Br	1-BROMOPROPANE	2	----	13.8	2	7	1	914
196	C3H7Br	2-BROMOPROPANE	2	----	13.8	2	-11	---	-----
197	C3H7Cl	ISOPROPYL CHLORIDE	1	2.8	10.7	1	-26	1	1100
198	C3H7Cl	n-PROPYL CHLORIDE	1	2.6	11.1	1	-24	1	968
199	C3H7F	1-FLUOROPROPANE	---	----	----	---	-----	---	-----
200	C3H7F	2-FLUOROPROPANE	---	----	----	---	-----	---	-----
201	C3H7I	ISOPROPYL IODIDE	---	----	----	2	28	---	-----
202	C3H7I	n-PROPYL IODIDE	---	----	----	2	46	---	-----
203	C3H7N	ALLYAMINE	1	2.2	22.0	1	-20	1	705
204	C3H7N	PROPYLENEIMINE	2	2.3	15.1	1	14	---	-----
205	C3H7NO	N,N-DIMETHYLFORMAMIDE	1	2.2	15.2	1	136	1	833
206	C3H7NO	N-METHYLACETAMIDE	2	2.4	13.9	1	226	---	-----
207	C3H7NO2	1-NITROPROPANE	1	2.2	13.8	1	97	1	790
208	C3H7NO2	2-NITROPROPANE	1	2.6	11.0	1	75	1	802
209	C3H7NO3	PROPYL-NITRATE	1	2.0	----	2	63	---	-----
210	C3H7NO3	ISOPROPYL-NITRATE	---	----	----	---	-----	---	-----
211	C3H8	PROPANE	1	2.1	9.5	1	-156	1	842
212	C3H8O	ISOPROPANOL	1	2.0	12.0	1	53	1	750
213	C3H8O	METHYL ETHYL ETHER	1	2.0	10.1	1	-35	1	374
214	C3H8O	n-PROPANOL	1	2.0	12.0	1	59	1	700
215	C3H8O2	2-METHOXYETHANOL	1	2.3	24.5	1	102	1	545
216	C3H8O2	METHYLAL	1	1.6	17.6	1	0	1	459
217	C3H8O2	1,2-PROPYLENE GLYCOL	1	2.6	12.5	1	210	1	790
218	C3H8O2	1,3-PROPYLENE GLYCOL	1	2.6	16.6	2	251	2	712
219	C3H8O3	GLYCEROL	2	2.7	19.0	1	320	1	739
220	C3H8S	n-PROPYLMERCAPTAN	2	1.8	----	1	-4	---	-----
221	C3H8S	ISOPROPYL MERCAPTAN	2	1.8	----	1	-31	---	-----
222	C3H8S	ETHYL-METHYL-SULFIDE	---	----	----	---	-----	---	-----
223	C3H9N	n-PROPYLAMINE	1	2.0	10.4	1	10	1	604
224	C3H9N	ISOPROPYLAMINE	1	2.0	10.4	1	-35	1	756
225	C3H9N	TRIMETHYLAMINE	1	2.0	11.6	1	20	1	374
226	C3H9NO	1-AMINO-2-PROPANOL	2	2.2	16.9	1	165	1	705
227	C3H9NO	3-AMINO-1-PROPANOL	2	2.2	16.5	1	175	---	-----
228	C3H9NO	METHYLETHANOLAMINE	2	2.2	15.2	1	162	1	662
229	C3H9O4P	TRIMETHYL PHOSPHATE	---	----	----	---	-----	---	-----
230	C3H10N2	1,2-PROPANEDIAMINE		1.9	16.9	1	91	---	-----
231	C3H10Si	TRIMETHYL SILANE	2	2.0	----	2	-92	1	608
232	C4Cl4S	TETRACHLOROTHIOPHENE	---	----	----	2	241	---	-----
233	C4Cl6	HEXACHLORO-1,3-BUTADIENE	2	----	15.7	2	217	1	1130
234	C4F8	OCTAFLUORO-2-BUTENE	2	----	18.7	---	-----	---	-----

565

NO	FORMULA	NAME	Code	Lower Explosive Limit LEL, vol %	Upper Explosive Limit UEL, vol %	Flash Point Code	Flash Point F	Autoignition Temperature Code	Autoignition Temperature F
235	C4F8	OCTAFLUOROCYCLOBUTANE	---	----	----	---	-----	---	-----
236	C4F10	DECAFLUOROBUTANE	---	----	----	---	-----	---	-----
237	C4H2	BUTADIYNE(BIACETYLENE)	2	2.5	15.6	2	-68	---	-----
238	C4H2O3	MALEIC ANHYDRIDE	1	1.4	7.1	1	216	1	890
239	C4H4	VINYLACETYLENE	2	2.2	31.7	2	-82	---	-----
240	C4H4N2	SUCCINONITRILE	2	2.1	14.4	1	270	---	-----
241	C4H4O	FURAN	1	2.3	14.3	1	-32	---	-----
242	C4H4O2	DIKETENE	2	2.5	16.8	1	93	1	590
243	C4H4O3	SUCCINIC ANHYDRIDE	2	2.7	20.3	2	304	---	-----
244	C4H4O4	FUMARIC ACID	2	2.9	18.7	---	-----	---	1364
245	C4H4O4	MALEIC ACID	---	----	----	---	-----	---	-----
246	C4H4S	THIOPHENE	---	----	----	1	30	---	-----
247	C4H5Cl	CHLOROPRENE	1	4.0	20.0	1	-4	---	-----
248	C4H5N	trans-CROTONITRILE	2	2.1	15.5	2	73	---	-----
249	C4H5N	cis-CROTONITRILE	2	2.1	15.5	2	53	---	-----
250	C4H5N	METHACRYLONITRILE	2	2.1	15.5	1	55	---	-----
251	C4H5N	PYRROLE	1	2.0	12.0	1	102	---	-----
252	C4H5N	VINYLACETONITRILE	2	2.0	15.6	1	73	---	-----
253	C4H5NO2	METHYL CYANOACETATE	2	----	13.0	1	230	---	-----
254	C4H6	CYCLOBUTENE	2	2.0	12.9	2	-95	---	-----
255	C4H6	1,2-BUTADIENE	2	2.0	12.0	2	-105	---	-----
256	C4H6	1,3-BUTADIENE	1	2.0	11.5	1	-105	1	804
257	C4H6	DIMETHYLACETYLENE	2	2.0	41.8	2	-56	---	-----
258	C4H6	ETHYLACETYLENE	2	2.0	32.9	2	-81	---	-----
259	C4H6Cl2	1,3-DICHLORO-trans-2-BUTENE	2	2.4	12.2	1	81	---	-----
260	C4H6Cl2	1,4-DICHLORO-cis-2-BUTENE	2	2.5	12.7	1	131	---	-----
261	C4H6Cl2	1,4-DICHLORO-trans-2-BUTENE	1	1.5	4.0	2	124	---	-----
262	C4H6Cl2	3,4-DICHLORO-1-BUTENE	2	2.4	13.3	1	82	---	-----
263	C4H6O	trans-CROTONALDEHYDE	1	2.1	15.5	1	55	1	450
264	C4H6O	2,5-DIHYDROFURAN	2	2.1	20.0	1	3	---	-----
265	C4H6O	DIVINYL ETHER	1	1.7	27.0	1	-53	1	680
266	C4H6O	METHACROLEIN	2	2.1	14.6	1	36	1	453
267	C4H6O2	2-BUTYNE-1,4-DIOL	2	2.3	35.7	2	305	---	-----
268	C4H6O2	gamma-BUTYROLACTONE	1	2.0	12.6	1	209	---	-----
269	C4H6O2	cis-CROTONIC ACID	2	2.3	14.6	2	178	2	745
270	C4H6O2	trans-CROTONIC ACID	1	2.2	15.1	1	202	1	745
271	C4H6O2	METHACRYLIC ACID	1	1.6	8.7	1	171	---	-----
272	C4H6O2	METHYL ACRYLATE	1	2.8	25.0	1	26	1	874
273	C4H6O2	VINYL ACETATE	1	2.6	13.4	1	17	1	800
274	C4H6O3	ACETIC ANHYDRIDE	1	2.9	10.3	1	129	1	734
275	C4H6O4	SUCCINIC ACID	2	2.6	14.4	2	411	---	-----
276	C4H6O5	DIGLYCOLIC ACID	2	2.8	20.3	1	440	1	446
277	C4H6O5	MALIC ACID	2	2.8	16.8	2	473	---	-----
278	C4H6O6	TARTARIC ACID	2	3.0	18.7	1	410	1	802
279	C4H7N	n-BUTYRONITRILE	1	1.6	11.4	1	79	1	935
280	C4H7N	ISOBUTYRONITRILE	2	1.9	11.5	1	47	---	-----
281	C4H7NO	ACETONE CYANOHYDRIN	1	2.2	12.0	1	165	1	1270
282	C4H7NO	2-METHACRYLAMIDE	2	2.0	15.1	2	217	---	-----
283	C4H7NO	3-METHOXYPROPIONITRILE	2	2.0	16.8	1	149	---	-----
284	C4H7NO	2-PYRROLIDONE	2	----	13.1	1	265	---	-----
285	C4H8	1-BUTENE	1	1.6	9.3	1	-110	1	723
286	C4H8	cis-2-BUTENE	1	1.6	9.7	1	-100	1	617
287	C4H8	trans-2-BUTENE	1	1.8	9.7	1	-100	1	615
288	C4H8	CYCLOBUTANE	2	1.8	11.1	2	-83	2	800
289	C4H8	ISOBUTENE	1	1.8	8.8	1	-105	1	869
290	C4H8Br2	1,2-DIBROMOBUTANE	---	----	----	---	-----	---	-----
291	C4H8Br2	2,3-DIBROMOBUTANE	---	----	----	---	-----	---	-----
292	C4H8Cl2	1,4-DICHLOROBUTANE	2	2.2	10.1	1	126	---	-----
293	C4H8I2	1,2-DIIODOBUTANE	---	----	----	---	-----	---	-----
294	C4H8O	n-BUTYRALDEHYDE	1	2.5	12.5	1	20	1	446
295	C4H8O	ISOBUTYRALDEHYDE	1	1.6	10.6	1	19	1	490
296	C4H8O	1,2-EPOXYBUTANE	1	1.5	18.3	1	5	1	822
297	C4H8O	METHYL ETHYL KETONE	1	1.8	10.0	1	21	1	960
298	C4H8O	ETHYL VINYL ETHER	1	1.7	28.0	1	-50	1	395
299	C4H8O	TETRAHYDROFURAN	1	2.0	11.8	1	6	1	610
300	C4H8O2	cis-2-BUTENE-1,4-DIOL	2	2.1	13.7	1	262	---	-----
301	C4H8O2	trans-2-BUTENE-1,4-DIOL	2	2.1	16.1	2	278	---	-----
302	C4H8O2	ISOBUTYRIC ACID	1	2.0	9.2	1	170	1	935
303	C4H8O2	n-BUTYRIC ACID	1	2.2	13.4	1	161	1	846
304	C4H8O2	1,4-DIOXANE	1	2.0	22.0	1	54	1	356
305	C4H8O2	ETHYL ACETATE	1	2.2	11.4	1	25	1	800
306	C4H8O2	METHYL PROPIONATE	1	2.5	13.0	1	28	1	876
307	C4H8O2	n-PROPYL FORMATE	2	2.1	11.3	1	27	1	851
308	C4H8O2S	SULFOLANE	---	----	----	1	350	---	-----
309	C4H8S	TETRAHYDROTHIOPHENE	1	1.5	9.0	1	65	---	-----
310	C4H9Br	1-BROMOBUTANE	1	2.6	6.6	1	64	1	509
311	C4H9Br	2-BROMOBUTANE	2	----	11.6	1	70	---	-----
312	C4H9Cl	n-BUTYL CHLORIDE	1	1.8	10.1	1	-18	1	860

NO	FORMULA	NAME	Code	Lower Explosive Limit LEL, vol %	Upper Explosive Limit UEL, vol %	Flash Point Code	Flash Point F	Autoignition Temperature Code	Autoignition Temperature F
313	C4H9Cl	sec-BUTYL CHLORIDE	2	1.9	9.9	1	-20	---	-----
314	C4H9Cl	tert-BUTYL CHLORIDE	2	1.9	9.1	1	23	---	-----
315	C4H9I	2-IODO-2-METHYLPROPANE	---	----	----	---	-----	---	-----
316	C4H9N	PYRROLIDINE	2	----	12.0	1	37	---	-----
317	C4H9NO	N,N-DIMETHYLACETAMIDE	1	1.8	13.8	1	145	1	669
318	C4H9NO	MORPHOLINE	1	1.8	10.8	1	100	1	590
319	C4H9NO2	1-NITROBUTANE	---	----	----	---	-----	---	-----
320	C4H9NO2	2-NITROBUTANE	---	----	----	---	-----	---	-----
321	C4H10	n-BUTANE	1	1.8	8.5	1	-76	1	761
322	C4H10	ISOBUTANE	1	1.8	8.4	1	-117	1	860
323	C4H10N2	PIPERAZINE	2	1.6	12.5	1	178	1	851
324	C4H10O	n-BUTANOL	1	1.4	11.2	1	84	1	650
325	C4H10O	sec-BUTANOL	1	1.7	9.8	1	75	1	763
326	C4H10O	tert-BUTANOL	1	2.4	8.0	1	52	1	892
327	C4H10O	DIETHYL ETHER	1	1.9	48.0	1	-49	1	356
328	C4H10O	METHYL-PROPYL-ETHER	2	1.9	11.8	---	-----	---	-----
329	C4H10O	METHYL ISOPROPYL ETHER	2	1.8	14.3	2	-53	---	-----
330	C4H10O	ISOBUTANOL	1	1.7	10.9	1	82	1	800
331	C4H10O2	1,3-BUTANEDIOL	2	1.9	12.6	1	228	1	741
332	C4H10O2	1,4-BUTANEDIOL	2	1.9	13.2	1	273	2	674
333	C4H10O2	2,3-BUTANEDIOL	2	1.9	13.7	1	185	1	756
334	C4H10O2	t-BUTYL HYDROPEROXIDE	---	----	----	1	80	---	-----
335	C4H10O2	1,2-DIMETHOXYETHANE	2	1.9	18.7	1	29	1	395
336	C4H10O2	2-ETHOXYETHANOL	1	1.7	15.6	1	110	1	455
337	C4H10O3	DIETHYLENE GLYCOL	2	2.0	17.1	1	255	1	444
338	C4H10O4S	DIETHYL SULFATE	---	----	----	1	219	1	817
339	C4H10S	n-BUTYL MERCAPTAN	---	----	----	1	35	---	-----
340	C4H10S	ISOBUTYL MERCAPTAN	2	1.4	----	1	16	---	-----
341	C4H10S	sec-BUTYL MERCAPTAN	2	1.4	----	1	-9	---	-----
342	C4H10S	tert-BUTYL MERCAPTAN	---	----	----	1	-15	---	-----
343	C4H10S	DIETHYL SULFIDE	---	----	----	1	14	---	-----
344	C4H10S	ISOPROPYL-METHYL-SULFIDE	---	----	----	---	-----	---	-----
345	C4H10S	METHYL-PROPYL-SULFIDE	---	----	----	---	-----	---	-----
346	C4H10S2	DIETHYL DISULFIDE	2	1.2	----	2	102	---	-----
347	C4H11N	n-BUTYLAMINE	1	1.7	9.8	1	10	1	594
348	C4H11N	ISOBUTYLAMINE	2	1.6	11.6	1	0	1	712
349	C4H11N	sec-BUTYLAMINE	2	1.6	11.6	1	-20	2	712
350	C4H11N	tert-BUTYLAMINE	1	1.7	8.9	1	16	---	-----
351	C4H11N	DIETHYLAMINE	1	1.8	10.1	1	-15	1	594
352	C4H11NO	DIMETHYLETHANOLAMINE	2	1.7	11.5	1	106	1	563
353	C4H11NO2	DIETHANOLAMINE	2	1.8	13.4	1	305	1	1224
354	C4H11NO2	2-AMINOETHOXYETHANOL	2	1.8	17.1	2	282	---	-----
355	C4H12N2O	N-AMINOETHYL ETHANOLAMINE	2	1.0	8.0	1	215	1	694
356	C4H12Si	TETRAMETHYLSILANE	1	1.5	----	1	-17	1	350
357	C4H13N3	DIETHYLENE TRIAMINE	1	2.0	6.7	1	210	1	676
358	C5Cl6	HEXACHLOROCYCLOPENTADIENE	---	----	----	---	-----	---	-----
359	C5H4O2	FURFURAL	1	2.1	19.3	1	140	1	600
360	C5H5N	PYRIDINE	1	1.8	12.4	1	68	1	900
361	C5H6	CYCLOPENTADIENE	2	1.7	14.6	2	-51	1	1184
362	C5H6	2-METHYL-1-BUTENE-3-YNE	2	1.7	25.1	2	-62	---	-----
363	C5H6	1-PENTENE-3-YNE	2	1.7	25.1	2	-22	---	-----
364	C5H6	1-PENTENE-4-YNE	2	1.7	22.9	2	-47	---	-----
365	C5H6N2	GLUTARONITRILE	2	1.6	11.5	2	278	---	-----
366	C5H6O2	FURFURYL ALCOHOL	1	1.8	16.3	1	149	1	736
367	C5H6O3	GLUTARIC ANHYDRIDE	2	2.0	13.1	1	235	---	-----
368	C5H6O4	CITRACONIC ACID	2	2.1	14.6	2	428	---	-----
369	C5H6O4	ITACONIC ACID	2	2.1	14.4	2	419	---	-----
370	C5H6S	2-METHYLTHIOPHENE	---	----	----	---	-----	---	-----
371	C5H6S	3-METHYLTHIOPHENE	---	----	----	---	-----	---	-----
372	C5H7N	N-METHYLPYRROLE	2	1.6	14.0	1	59	1	1094
373	C5H7NO2	ETHYL CYANOACETATE	2	1.7	10.6	1	230	---	-----
374	C5H8	CYCLOPENTENE	2	1.5	12.1	1	-20	1	743
375	C5H8	ISOPRENE	1	2.0	9.0	1	-65	1	428
376	C5H8	3-METHYL-1,2-BUTADIENE	2	1.6	15.2	2	-46	---	-----
377	C5H8	2-METHYL-1,3-BUTADIENE	2	1.6	10.2	---	-----	---	-----
378	C5H8	1,2-PENTADIENE	2	1.5	12.3	2	-40	---	-----
379	C5H8	cis-1,3-PENTADIENE	2	1.6	13.1	2	-42	---	-----
380	C5H8	trans-1,3-PENTADIENE	2	1.6	13.1	2	-46	---	-----
381	C5H8	1,4-PENTADIENE	2	1.6	13.1	2	-67	---	-----
382	C5H8	2,3-PENTADIENE	2	1.6	12.1	2	-35	---	-----
383	C5H8	1-PENTYNE	2	1.6	22.3	1	-29	---	-----
384	C5H8	2-PENTYNE	2	1.6	10.2	2	-22	---	-----
385	C5H8	3-METHYL-1-BUTYNE	2	1.6	22.8	2	-62	---	-----
386	C5H8	SPIROPENTANE	2	1.6	10.2	---	-----	---	-----
387	C5H8N4O12	PENTAERYTHRITOL TETRANITRATE	---	----	----	---	-----	---	-----
388	C5H8O	CYCLOPENTANONE	2	1.7	10.4	1	79	---	-----
389	C5H8O	METHYL ISOPROPENYL KETONE	1	1.8	9.0	1	70	---	-----
390	C5H8O2	ACETYLACETONE	1	2.4	11.6	1	93	1	644

NO	FORMULA	NAME	Code	Lower Explosive Limit LEL, vol %	Upper Explosive Limit UEL, vol %	Flash Point Code	F	Autoignition Temperature Code	F
391	C5H8O2	ALLYL ACETATE	2	1.7	11.8	1	72	1	705
392	C5H8O2	ETHYL ACRYLATE	2	1.8	9.5	1	60	1	721
393	C5H8O2	METHYL METHACRYLATE	1	2.1	12.5	1	52	---	-----
394	C5H8O2	VINYL PROPIONATE	2	1.7	11.8	1	34	---	-----
395	C5H8O3	2-HYDROXYETHYL ACRYLATE	2	1.8	12.9	1	208	---	-----
396	C5H8O3	LEVULINIC ACID	2	1.8	10.6	2	278	---	-----
397	C5H8O3	METHYL ACETOACETATE	2	1.8	8.8	1	170	1	536
398	C5H8O4	GLUTARIC ACID	2	1.9	11.5	2	408	---	-----
399	C5H9N	VALERONITRILE	2	1.5	9.6	2	89	---	-----
400	C5H9NO	n-BUTYL ISOCYANATE	2	1.5	8.9	2	64	---	-----
401	C5H9NO	N-METHYL-2-PYRROLIDONE	1	2.2	12.2	1	204	1	655
402	C5H9NO4	L-GLUTAMIC ACID	2	1.7	13.2	2	520	---	-----
403	C5H10	CYCLOPENTANE	2	1.4	9.4	2	-38	1	682
404	C5H10	2-METHYL-1-BUTENE	1	1.4	9.6	1	20	2	689
405	C5H10	2-METHYL-2-BUTENE	1	1.4	9.6	1	20	---	-----
406	C5H10	3-METHYL-1-BUTENE	2	1.5	9.1	2	-80	2	689
407	C5H10	1-PENTENE	1	1.5	8.7	1	0	1	523
408	C5H10	cis-2-PENTENE	2	1.4	10.6	2	-50	2	550
409	C5H10	trans-2-PENTENE	2	1.4	10.6	2	-55	2	545
410	C5H10Br2	2,3-DIBROMO-2-METHYLBUTANE	---	----	----	---	-----	---	-----
411	C5H10Cl2	1,5-DICHLOROPENTANE	2	1.7	8.6	2	145	---	-----
412	C5H10O	METHYL ISOPROPYL KETONE	2	1.5	9.0	2	32	---	-----
413	C5H10O	2-PENTANONE	1	1.5	8.2	1	45	1	846
414	C5H10O	DIETHYL KETONE	1	1.5	8.0	1	55	1	846
415	C5H10O	VALERALDEHYDE	2	1.5	9.5	1	54	1	432
416	C5H10O2	n-BUTYL FORMATE	1	1.7	8.0	1	64	1	612
417	C5H10O2	ETHYL PROPIONATE	1	1.9	11.0	1	54	1	890
418	C5H10O2	ISOBUTYL FORMATE	1	1.7	8.0	2	39	1	608
419	C5H10O2	ISOPROPYL ACETATE	2	1.8	7.2	1	36	2	894
420	C5H10O2	n-PROPYL ACETATE	1	2.0	8.0	1	59	1	842
421	C5H10O2	METHYL n-BUTYRATE	2	1.6	8.8	1	57	---	-----
422	C5H10O2	2-METHYLBUTYRIC ACID	2	1.6	9.8	2	167	---	-----
423	C5H10O2	ISOVALERIC ACID	2	1.6	9.8	2	167	1	781
424	C5H10O2	VALERIC ACID	2	1.6	9.6	1	205	1	752
425	C5H10O2	TETRAHYDROFURFURYL ALCOHOL	1	1.5	9.7	1	183	1	540
426	C5H10O2S	3-METHYL SULFOLANE	---	----	----	2	284	---	-----
427	C5H10O3	DIETHYL CARBONATE	2	1.7	12.4	1	77	---	-----
428	C5H10O3	ETHYL LACTATE	1	1.5	10.6	1	115	1	752
429	C5H10S	THIACYCLOHEXANE	---	----	----	---	-----	---	-----
430	C5H10S	CYCLOPENTANETHIOL	---	----	----	---	-----	---	-----
431	C5H11Br	1-BROMOPENTANE	---	----	----	---	-----	---	-----
432	C5H11Cl	1-CHLOROPENTANE	1	1.6	8.6	1	55	1	500
433	C5H11Cl	1-CHLORO-3-METHYLBUTANE	---	----	----	---	-----	---	-----
434	C5H11Cl	2-CHLORO-2-METHYLBUTANE	---	----	----	---	-----	---	-----
435	C5H11N	N-METHYLPYRROLIDINE	2	----	9.5	1	7	---	-----
436	C5H11N	PIPERIDINE	2	1.4	10.0	1	37	---	-----
437	C5H11NO	tert-BUTYLFORMAMIDE	2	1.5	9.7	1	203	---	-----
438	C5H12	ISOPENTANE	1	1.4	7.6	2	-71	1	788
439	C5H12	NEOPENTANE	1	1.4	7.5	1	-85	1	842
440	C5H12	n-PENTANE	1	1.4	7.8	1	-40	1	500
441	C5H12O	2,2-DIMETHYL-1-PROPANOL	2	1.5	9.1	2	99	---	-----
442	C5H12O	tert-PENTYL-ALCOHOL	2	1.5	9.5	---	-----	---	-----
443	C5H12O	2-METHYL-1-BUTANOL	1	1.4	9.0	1	122	1	725
444	C5H12O	2-METHYL-2-BUTANOL	1	1.2	9.0	1	105	1	819
445	C5H12O	3-METHYL-1-BUTANOL	1	1.2	9.0	1	109	1	662
446	C5H12O	3-METHYL-2-BUTANOL	2	1.5	9.9	1	103	2	818
447	C5H12O	1-PENTANOL	1	1.2	10.0	1	91	1	572
448	C5H12O	2-PENTANOL	2	1.5	9.7	1	94	1	650
449	C5H12O	3-PENTANOL	1	1.2	9.0	1	105	1	815
450	C5H12O	METHYL sec-BUTYL ETHER	2	1.4	11.5	2	-20	---	-----
451	C5H12O	METHYL tert-BUTYL ETHER	1	2.0	15.1	1	-18	---	-----
452	C5H12O	METHYL ISOBUTYL ETHER	2	1.5	15.3	2	-20	---	-----
453	C5H12O	ETHYL PROPYL ETHER	1	1.7	9.0	2	-13	---	-----
454	C5H12O2	ETHYLENE GLYCOL MONOPROPYL ETHER	1	1.3	15.8	1	120	1	455
455	C5H12O2	NEOPENTYL GLYCOL	2	1.4	22.0	1	264	1	750
456	C5H12O2	1,5-PENTANEDIOL	2	1.5	10.9	1	265	1	635
457	C5H12O3	2-(2-METHOXYETHOXY)ETHANOL	1	1.4	22.7	1	205	1	379
458	C5H12O4	PENTAERYTHRITOL	2	1.6	12.5	2	500	---	-----
459	C5H12S	n-PENTYL MERCAPTAN	---	----	----	1	64	---	-----
460	C5H12S	BUTYL-METHYL-SULFIDE	---	----	----	---	-----	---	-----
461	C5H12S	ETHYL-PROPYL-SULFIDE	---	----	----	---	-----	---	-----
462	C5H12S	2-METHYL-2-BUTANETHIOL	---	----	----	---	-----	---	-----
463	C5H13N	n-PENTYLAMINE	2	1.3	9.5	1	45	---	-----
464	C5H13NO2	METHYL DIETHANOLAMINE	2	1.4	10.8	1	260	---	-----
465	C6Cl6	HEXACHLOROBENZENE	2	3.5	6.7	1	467	---	-----
466	C6F6	HEXAFLUOROBENZENE	2	----	13.6	1	50	---	-----
467	C6H3ClN2O4	1-CHLORO-2,4-DINITROBENZENE	1	2.0	22.0	1	381	1	810
468	C6H3Cl2NO2	1,2-DICHLORO-4-NITROBENZENE	2	2.1	8.4	2	278	---	-----

568

NO	FORMULA	NAME	Code	Lower Explosive Limit LEL, vol %	Upper Explosive Limit UEL, vol %	Flash Point Code	Flash Point F	Autoignition Temperature Code	Autoignition Temperature F
469	C6H3Cl3	1,2,4-TRICHLOROBENZENE	1	2.5	6.6	1	222	1	1060
470	C6H3N3O6	1,3,5-TRINITROBENZENE	---	----	----	---	-----	---	-----
471	C6H4Br2	m-DIBROMOBENZENE	2	1.9	9.8	1	199	---	-----
472	C6H4ClNO2	m-CHLORONITROBENZENE	2	----	8.8	1	261	---	-----
473	C6H4ClNO2	o-CHLORONITROBENZENE	2	----	8.8	1	261	---	-----
474	C6H4ClNO2	p-CHLORONITROBENZENE	2	----	8.8	1	261	---	-----
475	C6H4Cl2	m-DICHLOROBENZENE	2	1.8	7.8	1	161	2	1196
476	C6H4Cl2	o-DICHLOROBENZENE	1	2.2	9.2	1	151	1	1198
477	C6H4Cl2	p-DICHLOROBENZENE	2	1.8	7.8	1	151	2	1196
478	C6H4F2	m-DIFLUOROBENZENE	---	----	----	---	-----	---	-----
479	C6H4F2	o-DIFLUOROBENZENE	---	----	----	---	-----	---	-----
480	C6H4F2	p-DIFLUOROBENZENE	---	----	----	---	-----	---	-----
481	C6H4N2O4	m-DINITROBENZENE	---	----	----	1	302	---	-----
482	C6H4N2O4	o-DINITROBENZENE	2	1.8	9.8	1	302	---	-----
483	C6H4N2O4	p-DINITROBENZENE	2	1.8	9.8	2	347	---	-----
484	C6H5Br	BROMOBENZENE	2	1.5	9.1	1	124	1	1049
485	C6H5Cl	MONOCHLOROBENZENE	1	1.3	7.1	1	90	1	1180
486	C6H5ClO	m-CHLOROPHENOL	2	1.7	8.8	2	194	---	-----
487	C6H5ClO	o-CHLOROPHENOL	2	1.7	8.8	1	147	---	-----
488	C6H5ClO	p-CHLOROPHENOL	2	1.7	8.8	1	241	---	-----
489	C6H5Cl2N	3,4-DICHLOROANILINE	1	2.8	7.2	1	331	1	509
490	C6H5F	FLUOROBENZENE	2	1.6	9.1	1	5	---	-----
491	C6H5I	IODOBENZENE	---	----	----	2	151	---	-----
492	C6H5NO2	NITROBENZENE	1	1.8	9.1	1	190	1	899
493	C6H6	BENZENE	1	1.4	7.1	1	12	1	1044
494	C6H6ClN	m-CHLOROANILINE	2	1.5	8.8	2	221	---	-----
495	C6H6ClN	o-CHLOROANILINE	2	1.5	8.8	1	195	---	-----
496	C6H6ClN	p-CHLOROANILINE	1	2.2	8.8	1	235	---	-----
497	C6H6N2	cis-DICYANO-1-BUTENE	2	1.4	11.6	2	232	---	-----
498	C6H6N2	trans-DICYANO-1-BUTENE	2	1.4	11.6	2	228	---	-----
499	C6H6N2	1,4-DICYANO-2-BUTENE	2	1.4	11.6	2	295	---	-----
500	C6H6N2O2	m-NITROANILINE	2	1.7	9.8	1	390	2	1070
501	C6H6N2O2	o-NITROANILINE	2	1.5	9.8	1	334	1	970
502	C6H6N2O2	p-NITROANILINE	2	1.5	9.8	1	390	2	1070
503	C6H6O	PHENOL	1	1.8	8.6	1	175	1	1319
504	C6H6O2	1,2-BENZENEDIOL	2	1.6	9.8	1	260	2	1052
505	C6H6O2	1,3-BENZENEDIOL	1	1.6	9.8	1	260	1	1052
506	C6H6O2	p-HYDROQUINONE	2	1.6	15.3	1	329	1	960
507	C6H6O3	1,2,3-BENZENETRIOL	2	1.7	10.8	2	365	---	-----
508	C6H6S	PHENYL MERCAPTAN	2	1.2	----	1	163	---	-----
509	C6H7N	ANILINE	1	1.3	11.0	1	158	1	1143
510	C6H7N	2-METHYLPYRIDINE	1	----	11.9	1	102	1	1000
511	C6H7N	3-METHYLPYRIDINE	2	----	11.9	2	102	2	998
512	C6H7N	4-METHYLPYRIDINE	2	----	11.9	1	135	2	998
513	C6H8	1,3-CYCLOHEXADIENE	2	1.4	11.8	2	3	---	-----
514	C6H8	METHYLCYCLOPENTADIENE	1	1.3	7.6	1	120	1	835
515	C6H8N2	ADIPONITRILE	1	1.7	5.0	1	199	1	1022
516	C6H8N2	METHYLGLUTARONITRILE	1	0.3	3.3	1	208	---	-----
517	C6H8N2	m-PHENYLENEDIAMINE	2	1.3	9.8	1	280	---	-----
518	C6H8N2	o-PHENYLENEDIAMINE	1	1.5	9.8	1	313	---	-----
519	C6H8N2	p-PHENYLENEDIAMINE	2	1.3	9.8	1	312	---	-----
520	C6H8N2	PHENYLHYDRAZINE	2	1.3	9.5	1	192	1	345
521	C6H8N2O	BIS(CYANOETHYL)ETHER	2	1.4	12.4	2	343	---	-----
522	C6H8O4	DIMETHYL MALEATE	2	1.6	10.4	1	235	---	-----
523	C6H8O6	ASCORBIC ACID	2	1.7	14.5	2	529	---	-----
524	C6H8O7	CITRIC ACID	2	1.8	4.8	2	547	---	1850
525	C6H10	1-METHYLCYCLOPENTENE	2	1.3	8.4	---	-----	---	-----
526	C6H10	3-METHYLCYCLOPENTENE	2	1.3	8.4	---	-----	---	-----
527	C6H10	4-METHYLCYCLOPENTENE	2	1.3	8.4	---	-----	---	-----
528	C6H10	CYCLOHEXENE	2	1.2	10.1	1	20	1	509
529	C6H10	2,3-DIMETHYL-1,3-BUTADIENE	---	----	----	1	-8	---	-----
530	C6H10	1,5-HEXADIENE	2	1.3	10.9	1	-51	---	-----
531	C6H10	cis,trans-2,4-HEXADIENE	2	1.3	17.1	1	19	---	-----
532	C6H10	trans,trans-2,4-HEXADIENE	2	1.3	17.1	1	19	---	-----
533	C6H10	1-HEXYNE	2	1.3	16.6	1	-6	---	-----
534	C6H10	2-HEXYNE	2	1.3	17.8	1	12	---	-----
535	C6H10	3-HEXYNE	2	1.3	17.8	2	1	---	-----
536	C6H10O	CYCLOHEXANONE	1	1.0	8.8	1	111	1	788
537	C6H10O	MESITYL OXIDE	1	1.3	8.8	1	83	1	652
538	C6H10O2	epsilon-CAPROLACTONE	1	----	9.6	1	228	1	640
539	C6H10O2	ETHYL METHACRYLATE	1	1.8	9.6	1	70	---	-----
540	C6H10O2	n-PROPYL ACRYLATE	2	1.4	9.9	2	68	2	647
541	C6H10O3	ETHYLACETOACETATE	1	1.4	9.5	1	135	1	563
542	C6H10O3	PROPIONIC ANHYDRIDE	1	1.5	11.9	1	145	1	545
543	C6H10O4	ADIPIC ACID	1	1.6	9.6	1	325	1	792
544	C6H10O4	DIETHYL OXALATE	2	1.5	8.4	2	252	---	-----
545	C6H10O4	ETHYLENE GLYCOL DIACETATE	1	1.6	8.4	1	190	1	900
546	C6H10O4	ETHYLIDENE DIACETATE	2	1.5	8.5	1	151	---	-----

NO	FORMULA	NAME	Code	Lower Explosive Limit LEL, vol %	Upper Explosive Limit UEL, vol %	Flash Point Code	F	Autoignition Temperature Code	F
547	C6H11N	HEXANENITRILE	2	1.3	8.2	2	116	---	-----
548	C6H11NO	epsilon-CAPROLACTAM	1	1.4	8.0	2	257	---	-----
549	C6H11NO	CYCLOHEXANONE OXIME	2	1.3	9.8	2	192	---	-----
550	C6H12	CYCLOHEXANE	1	1.3	8.0	1	-4	1	500
551	C6H12	2,3-DIMETHYL-1-BUTENE	2	1.2	9.1	2	-35	1	680
552	C6H12	2,3-DIMETHYL-2-BUTENE	2	1.2	8.1	2	-10	1	753
553	C6H12	3,3-DIMETHYL-1-BUTENE	2	1.2	9.0	1	-19	---	-----
554	C6H12	2-ETHYL-1-BUTENE	2	1.2	9.0	2	-22	1	599
555	C6H12	trans-3-METHYL-2-PENTENE	2	1.3	8.0	---	-----	---	-----
556	C6H12	1-HEXENE	2	1.2	9.2	2	-24	1	487
557	C6H12	cis-2-HEXENE	2	1.2	9.0	2	-17	1	487
558	C6H12	trans-2-HEXENE	2	1.2	9.0	2	-17	1	473
559	C6H12	cis-3-HEXENE	2	1.2	9.0	2	-19	2	539
560	C6H12	trans-3-HEXENE	2	1.2	9.0	1	10	2	529
561	C6H12	METHYLCYCLOPENTANE	2	1.2	8.4	1	-17	1	624
562	C6H12	2-METHYL-1-PENTENE	2	1.2	9.0	2	-26	1	572
563	C6H12	2-METHYL-2-PENTENE	2	1.2	9.4	2	-17	---	-----
564	C6H12	3-METHYL-1-PENTENE	2	1.2	9.4	1	-18	---	-----
565	C6H12	3-METHYL-cis-2-PENTENE	2	1.2	8.6	2	-19	---	-----
566	C6H12	4-METHYL-1-PENTENE	2	1.2	9.4	2	-38	1	572
567	C6H12	4-METHYL-cis-2-PENTENE	2	1.2	9.1	2	-33	---	-----
568	C6H12	4-METHYL-trans-2-PENTENE	2	1.2	9.1	2	-29	---	-----
569	C6H12N2	TRIETHYLENEDIAMINE	2	1.2	9.2	2	143	---	-----
570	C6H12O	BUTYL VINYL ETHER	2	1.3	11.3	1	30	1	437
571	C6H12O	CYCLOHEXANOL	1	1.2	9.3	1	154	1	572
572	C6H12O	1-HEXANAL	2	1.3	8.1	1	80	---	-----
573	C6H12O	ETHYL ISOPROPYL KETONE	2	1.3	7.8	2	61	---	-----
574	C6H12O	2-HEXANONE	1	1.2	8.0	1	77	1	795
575	C6H12O	3-HEXANONE	1	1.0	8.0	1	57	---	-----
576	C6H12O	METHYL ISOBUTYL KETONE	1	1.4	7.5	1	60	1	858
577	C6H12O2	n-PENTYL FORMATE	2	1.3	8.1	2	91	---	-----
578	C6H12O2	n-BUTYL ACETATE	1	1.7	7.6	1	71	1	790
579	C6H12O2	sec-BUTYL ACETATE	1	1.7	7.6	1	60	2	791
580	C6H12O2	tert-BUTYL ACETATE	2	1.3	7.3	1	61	---	-----
581	C6H12O2	ETHYL n-BUTYRATE	2	1.3	7.7	1	78	1	865
582	C6H12O2	ETHYL ISOBUTYRATE	2	1.3	7.8	1	57	---	-----
583	C6H12O2	ISOBUTYL ACETATE	1	1.3	10.5	1	64	1	793
584	C6H12O2	n-PROPYL PROPIONATE	2	1.3	7.7	1	175	2	831
585	C6H12O2	CYCLOHEXYL PEROXIDE	2	1.3	11.7	2	228	---	-----
586	C6H12O2	DIACETONE ALCOHOL	1	1.8	6.9	1	116	1	1118
587	C6H12O2	2-ETHYL BUTYRIC ACID	2	1.3	8.4	1	188	1	752
588	C6H12O2	n-HEXANOIC ACID	2	1.3	8.2	1	215	1	716
589	C6H12O3	2-ETHOXYETHYL ACETATE	1	1.7	12.7	1	130	1	715
590	C6H12O3	HYDROXYCAPROIC ACID	2	1.4	8.7	2	372	---	-----
591	C6H12O3	PARALDEHYDE	1	1.3	16.2	1	96	1	460
592	C6H12O3	sec-BUTYL GLYCOLATE	---	----	----	---	-----	---	-----
593	C6H12S	THIACYCLOHEPTANE	---	----	----	---	-----	---	-----
594	C6H13N	CYCLOHEXYLAMINE	2	1.2	9.3	1	88	1	559
595	C6H13N	HEXAMETHYLENEIMINE	1	1.6	2.3	1	99	1	626
596	C6H14	2,2-DIMETHYLBUTANE	1	1.2	7.0	1	-54	1	797
597	C6H14	2,3-DIMETHYLBUTANE	1	1.2	7.0	1	-20	1	788
598	C6H14	n-HEXANE	1	1.2	7.5	1	-7	1	453
599	C6H14	2-METHYLPENTANE	1	1.2	7.0	1	-31	1	583
600	C6H14	3-METHYLPENTANE	2	1.2	7.7	2	-26	1	532
601	C6H14N2O2	LYSINE	2	1.2	10.1	2	412	---	-----
602	C6H14O	2-ETHYL-1-BUTANOL	2	1.2	8.3	1	135	---	-----
603	C6H14O	1-HEXANOL	2	1.2	8.2	1	145	2	545
604	C6H14O	2-HEXANOL	2	1.2	8.3	1	106	---	-----
605	C6H14O	2-METHYL-1-PENTANOL	2	1.0	7.7	1	124	---	-----
606	C6H14O	4-METHYL-2-PENTANOL	1	1.0	5.5	1	106	---	-----
607	C6H14O	n-BUTYL ETHYL ETHER	2	1.2	9.5	1	39	---	-----
608	C6H14O	DIISOPROPYL ETHER	1	1.4	21.0	1	-18	1	830
609	C6H14O	DI-n-PROPYL ETHER	2	1.2	9.5	1	70	1	419
610	C6H14O	METHYL tert-PENTYL ETHER	2	1.2	9.1	2	8	---	-----
611	C6H14O2	ACETAL	1	1.6	10.4	1	-5	1	446
612	C6H14O2	2-BUTOXYETHANOL	1	1.1	12.7	1	157	1	460
613	C6H14O2	1,6-HEXANEDIOL	2	1.3	9.2	1	210	---	-----
614	C6H14O2	HEXYLENE GLYCOL	2	1.3	9.0	1	199	2	583
615	C6H14O2S	DI-n-PROPYL SULFONE	2	1.1	----	1	259	---	-----
616	C6H14O3	DIETHYLENE GLYCOL DIMETHYL ETHER	2	1.3	14.2	1	145	1	392
617	C6H14O3	DIPROPYLENE GLYCOL	1	2.2	11.5	1	244	---	-----
618	C6H14O3	2-(2-ETHOXYETHOXY)ETHANOL	1	1.2	10.4	1	182	1	399
619	C6H14O3	TRIMETHYLOLPROPANE	2	1.3	9.7	1	355	---	-----
620	C6H14O4	TRIETHYLENE GLYCOL	1	0.9	9.2	1	350	1	700
621	C6H14O6	SORBITOL	2	1.5	12.4	2	725	---	-----
622	C6H14S	n-HEXYLMERCAPTAN	2	0.4	----	1	68	---	-----
623	C6H14S	BUTYL-ETHYL-SULFIDE	---	----	----	---	-----	---	-----
624	C6H14S	ISOPROPYL-SULFIDE	---	----	----	---	-----	---	-----

NO	FORMULA	NAME	Code	Lower Explosive Limit LEL, vol %	Upper Explosive Limit UEL, vol %	Flash Point Code	Flash Point F	Autoignition Temperature Code	Autoignition Temperature F
625	C6H14S	METHYL-PENTYL-SULFIDE	---	----	----	---	----	---	----
626	C6H14S	PROPYL-SULFIDE	---	----	----	---	----	---	----
627	C6H14S2	PROPYL-DISULFIDE	---	----	----	---	----	---	----
628	C6H15Al	TRIETHYL ALUMINUN	---	----	----	---	----	---	----
629	C6H15Al2Cl3	ETHYL ALUMINUM SESQUICHLORIDE	---	----	----	1	-4	---	----
630	C6H15N	DIISOPROPYLAMINE	1	0.8	7.1	1	19	1	600
631	C6H15N	DI-n-PROPYLAMINE	2	1.1	7.7	1	63	1	570
632	C6H15N	n-HEXYLAMINE	2	1.1	8.2	1	85	---	----
633	C6H15N	TRIETHYLAMINE	1	1.2	8.0	1	10	---	----
634	C6H15NO	6-AMINOHEXANOL	2	1.2	9.2	2	251	---	----
635	C6H15NO2	DIISOPROPANOLAMINE	2	1.2	9.8	1	255	1	705
636	C6H15NO3	TRIETHANOLAMINE	2	1.2	9.9	1	355	---	----
637	C6H15N3	N-AMINOETHYL PIPERAZINE	2	1.1	9.4	1	199	---	----
638	C6H15O4P	TRIETHYL PHOSPHATE	1	1.7	10.0	1	210	1	851
639	C6H16N2	HEXAMETHYLENEDIAMINE	1	0.7	6.3	1	200	---	----
640	C6H18N3OP	HEXAMETHYL PHOSPHORAMIDE	---	----	----	1	222	---	----
641	C6H18N4	TRIETHYLENE TETRAMINE	2	1.0	9.5	1	275	1	640
642	C6H18OSi2	HEXAMETHYLDISILOXANE	1	1.3	18.6	1	28	1	646
643	C6H18O3Si3	HEXAMETHYLCYCLOTRISILOXANE	2	1.1	----	1	95	---	----
644	C6H19NSi2	HEXAMETHYLDISILAZANE	1	0.8	16.3	1	46	---	----
645	C7H3ClF3NO2	4-CHLORO-3-NITROBENZOTRIFLUORIDE	2	1.8	10.1	1	214	---	----
646	C7H3Cl2F3	2,4-DICHLOROBENZOTRIFLUORIDE	2	----	7.9	---	----	---	----
647	C7H3Cl2NO	3,4-DICHLOROPHENYL ISOCYANATE	2	1.6	8.6	1	288	---	----
648	C7H4ClF3	p-CHLOROBENZOTRIFLUORIDE	---	----	8.1	1	109	---	----
649	C7H4Cl2O	m-CHLOROBENZOYL CHLORIDE	2	----	7.3	2	208	---	----
650	C7H4F3NO2	3-NITROBENZOTRIFLUORIDE	2	1.6	9.0	1	190	---	----
651	C7H5ClO	BENZOYL CHLORIDE	1	1.2	4.9	1	162	1	185
652	C7H5ClO2	o-CHLOROBENZOIC ACID	2	1.5	7.7	2	322	---	----
653	C7H5Cl3	BENZOTRICHLORIDE	2	1.6	6.5	2	174	---	----
654	C7H5F3	BENZOTRIFLUORIDE	2	----	8.4	1	54	---	----
655	C7H5N	BENZONITRILE	2	1.3	8.0	1	167	---	----
656	C7H5NO	PHENYL ISOCYANATE	2	1.3	8.4	1	132	---	----
657	C7H5N3O6	2,4,6-TRINITROTOLUENE	---	----	----	---	----	---	----
658	C7H6Cl2	BENZYL DICHLORIDE	2	1.5	9.9	2	172	---	----
659	C7H6Cl2	2,4-DICHLOROTOLUENE	2	1.5	6.5	1	205	---	----
660	C7H6N2O4	2,4-DINITROTOLUENE	2	1.4	8.2	1	405	---	----
661	C7H6N2O4	2,5-DINITROTOLUENE	2	1.5	8.2	2	365	---	----
662	C7H6N2O4	2,6-DINITROTOLUENE	2	1.4	8.2	2	318	---	----
663	C7H6N2O4	3,4-DINITROTOLUENE	2	1.4	8.2	1	404	---	----
664	C7H6N2O4	3,5-DINITROTOLUENE	2	1.5	8.2	2	340	---	----
665	C7H6O	BENZALDEHYDE	2	1.4	7.8	1	148	1	377
666	C7H6O2	BENZOIC ACID	2	1.4	8.0	1	250	1	1065
667	C7H6O2	p-HYDROXYBENZALDEHYDE	2	1.4	8.4	2	341	---	----
668	C7H6O2	SALICYLALDEHYDE	2	1.4	8.4	1	172	---	----
669	C7H6O3	SALICYLIC ACID	2	1.5	8.6	1	315	1	1013
670	C7H7Br	p-BROMOTOLUENE	2	----	7.6	1	185	---	----
671	C7H7Cl	BENZYL CHLORIDE	1	1.1	7.1	1	140	1	1085
672	C7H7Cl	o-CHLOROTOLUENE	2	1.3	6.7	1	124	---	----
673	C7H7Cl	p-CHLOROTOLUENE	2	1.3	6.7	1	125	---	----
674	C7H7F	p-FLUOROTOLUENE	---	----	----	---	----	---	----
675	C7H7NO	FORMANILIDE	2	----	8.0	---	----	---	----
676	C7H7NO2	m-NITROTOLUENE	1	1.6	7.6	1	215	2	734
677	C7H7NO2	o-NITROTOLUENE	1	2.2	7.6	1	223	1	581
678	C7H7NO2	p-NITROTOLUENE	1	1.6	7.6	1	223	1	734
679	C7H7NO3	o-NITROANISOLE	---	----	----	2	223	---	----
680	C7H8	TOLUENE	1	1.2	7.1	1	40	1	997
681	C7H8	1,3,5-CYCLOHEPTATRIENE	2	1.3	8.0	2	49	---	----
682	C7H8O	ANISOLE	2	1.3	9.0	1	126	---	----
683	C7H8O	BENZYL ALCOHOL	2	1.3	7.9	1	199	1	817
684	C7H8O	m-CRESOL	1	1.1	7.6	1	202	1	1038
685	C7H8O	o-CRESOL	1	1.4	7.6	1	178	1	1110
686	C7H8O	p-CRESOL	1	1.1	7.6	1	202	1	1038
687	C7H8O2	GUAIACOL	2	1.3	9.6	1	179	---	----
688	C7H8O2	p-METHOXYPHENOL	2	1.3	9.6	1	257	---	----
689	C7H9N	BENZYLAMINE	2	1.2	7.8	1	140	---	----
690	C7H9N	2,6-DIMETHYLPYRIDINE	---	----	----	1	91	---	----
691	C7H9N	N-METHYLANILINE	2	1.2	7.4	1	172	---	----
692	C7H9N	m-TOLUIDINE	2	1.2	7.6	1	187	1	900
693	C7H9N	o-TOLUIDINE	2	1.2	7.6	1	185	1	900
694	C7H9N	p-TOLUIDINE	2	1.2	7.6	1	188	1	900
695	C7H10	2-NORBORNENE	2	1.2	9.8	1	5	1	941
696	C7H10N2	TOLUENEDIAMINE	---	----	----	1	300	---	----
697	C7H11NO	CYCLOHEXYL ISOCYANATE	2	1.1	8.6	1	120	---	----
698	C7H12	1-HEPTYNE	2	1.1	7.2	---	----	---	----
699	C7H12O2	n-BUTYL ACRYLATE	1	1.5	9.9	1	102	1	559
700	C7H12O2	ISOBUTYL ACRYLATE	2	2.0	8.0	1	88	1	644
701	C7H12O2	n-PROPYL METHACRYLATE	2	1.2	8.3	2	88	---	----
702	C7H12O4	DIETHYL MALONATE	2	1.3	7.3	1	199	---	----

NO	FORMULA	NAME	Code	Lower Explosive Limit LEL, vol %	Upper Explosive Limit UEL, vol %	Flash Point Code	Flash Point F	Autoignition Temperature Code	Autoignition Temperature F
703	C7H14	CYCLOHEPTANE	2	1.1	7.1	2	44	---	-----
704	C7H14	1,1-DIMETYLCYCLOPENTANE	2	1.1	6.8	2	3	---	-----
705	C7H14	cis-1,2-DIMETHYLCYCLOPENTANE	2	1.1	7.3	2	25	---	-----
706	C7H14	trans-1,2-DIMETHYLCYCLOPENTANE	2	1.1	7.3	2	14	---	-----
707	C7H14	cis-1,3-DIMETHYLCYCLOPENTANE	2	1.1	7.3	2	7	---	-----
708	C7H14	trans-1,3-DIMETHYLCYCLOPENTANE	2	1.1	7.3	2	8	---	-----
709	C7H14	ETHYLCYCLOPENTANE	1	1.1	6.7	2	25	1	500
710	C7H14	2-ETHYL-1-PENTENE	2	1.0	7.8	2	14	---	-----
711	C7H14	3-ETHYL-1-PENTENE	2	1.0	8.1	2	1	---	-----
712	C7H14	1-HEPTENE	2	1.0	8.0	1	32	1	505
713	C7H14	cis-2-HEPTENE	2	1.1	7.8	2	17	---	-----
714	C7H14	trans-2-HEPTENE	2	1.1	7.8	1	30	---	-----
715	C7H14	cis-3-HEPTENE	2	1.1	7.8	2	16	---	-----
716	C7H14	trans-3-HEPTENE	2	1.1	7.8	1	21	---	-----
717	C7H14	METHYLCYCLOHEXANE	1	1.2	7.2	1	25	1	545
718	C7H14	2-METHYL-1-HEXENE	2	1.0	7.8	1	21	---	-----
719	C7H14	3-METHYL-1-HEXENE	2	1.0	8.2	1	21	---	-----
720	C7H14	4-METHYL-1-HEXENE	2	1.1	8.1	2	5	---	-----
721	C7H14	2,3,3-TRIMETHYL-1-BUTENE	2	1.0	7.4	1	1	1	219
722	C7H14O	DIISOPROPYL KETONE	2	1.1	7.0	1	59	---	-----
723	C7H14O	2-HEPTANONE	1	1.1	7.9	1	102	1	739
724	C7H14O	1-HEPTANAL	2	1.1	7.1	1	100	---	-----
725	C7H14O	1-METHYLCYCLOHEXANOL	2	1.1	7.7	2	129	---	-----
726	C7H14O	cis-2-METHYLCYCLOHEXANOL	2	1.1	8.2	2	127	1	565
727	C7H14O	trans-2-METHYLCYCLOHEXANOL	2	1.1	8.2	2	131	1	565
728	C7H14O	cis-3-METHYLCYCLOHEXANOL	2	1.1	8.2	2	147	---	-----
729	C7H14O	trans-3-METHYLCYCLOHEXANOL	2	1.1	8.2	2	145	---	-----
730	C7H14O	cis-4-METHYLCYCLOHEXANOL	2	1.1	8.2	2	154	1	567
731	C7H14O	trans-4-METHYLCYCLOHEXANOL	2	1.1	8.2	2	158	1	567
732	C7H14O	5-METHYL-2-HEXANONE	1	1.0	8.2	1	97	1	376
733	C7H14O2	n-BUTYL PROPIONATE	2	1.1	6.8	1	90	1	800
734	C7H14O2	ETHYL ISOVALERATE	2	1.1	6.9	1	95	---	-----
735	C7H14O2	ISOPENTYL ACETATE	1	1.0	7.5	1	77	1	680
736	C7H14O2	n-PENTYL ACETATE	2	1.1	6.8	1	77	2	674
737	C7H14O2	n-PROPYL n-BUTYRATE	2	1.1	6.8	2	88	---	-----
738	C7H14O2	n-HEPTANOIC ACID	2	1.1	7.2	2	230	2	710
739	C7H14O3	ETHYL-3-ETHOXYPROPIONATE	1	1.1	8.7	1	136	1	711
740	C7H15Br	1-BROMOHEPTANE	2	1.0	7.2	2	129	---	-----
741	C7H15N	N-METHYLCYCLOHEXYLAMINE	2	1.0	7.6	1	84	---	-----
742	C7H16	2,2-DIMETHYLPENTANE	2	1.0	6.0	2	-10	2	638
743	C7H16	2,3-DIMETHYLPENTANE	1	1.1	6.7	2	5	1	639
744	C7H16	2,4-DIMETHYLPENTANE	2	1.0	6.5	1	10	2	638
745	C7H16	3,3-DIMETHYLPENTANE	2	1.0	7.0	2	-2	2	638
746	C7H16	3-ETHYLPENTANE	2	1.0	7.0	2	10	---	-----
747	C7H16	n-HEPTANE	1	1.0	7.0	1	25	1	433
748	C7H16	2-METHYLHEXANE	2	1.0	6.0	1	25	1	536
749	C7H16	3-METHYLHEXANE	2	1.0	7.0	1	25	1	536
750	C7H16	2,2,3-TRIMETHYLBUTANE	2	1.0	6.1	2	-11	1	842
751	C7H16O	1-HEPTANOL	2	1.0	7.2	1	170	2	539
752	C7H16O	2-HEPTANOL	2	1.0	7.3	1	160	---	-----
753	C7H16O	5-METHYL-1-HEXANOL	2	1.0	7.3	2	151	---	-----
754	C7H16O	ISOPROPYL-TERT-BUTYL-ETHER	2	1.1	6.9	---	-----	---	-----
755	C7H16S	n-HEPTYL MERCAPTAN	2	0.9	----	1	115	---	-----
756	C7H16S	BUTYL-PROPYL-SULFIDE	---	----	----	---	-----	---	-----
757	C7H16S	ETHYL-PENTYL-SULFIDE	---	----	----	---	-----	---	-----
758	C7H16S	HEXYL-METHYL-SULFIDE	---	----	----	---	-----	---	-----
759	C7H17N	1-AMINOHEPTANE	2	1.0	7.2	1	131	---	-----
760	C8H4Cl2O2	ISOPHTHALOYL CHLORIDE	2	1.5	6.9	1	356	---	-----
761	C8H4O3	PHTHALIC ANHYDRIDE	1	1.7	----	1	305	1	-441
762	C8H6	ETHYNYLBENZENE	2	1.2	7.6	---	-----	---	-----
763	C8H6O4	ISOPHTHALIC ACID	2	1.3	7.7	2	624	2	925
764	C8H6O4	PHTHALIC ACID	2	1.3	7.7	2	401	2	925
765	C8H6O4	TEREPHTHALIC ACID	2	1.3	7.7	---	-----	1	925
766	C8H6S	BENZOTHIOPHENE	---	----	----	---	-----	---	-----
767	C8H7N	INDOLE	2	----	8.1	---	-----	---	-----
768	C8H8	STYRENE	1	1.1	6.1	1	90	1	914
769	C8H8	1,3,5,7-CYCLOOCTATETRAENE	2	1.1	7.2	2	79	---	-----
770	C8H8O	ACETOPHENONE	2	1.1	6.7	1	180	1	1060
771	C8H8O	p-TOLUALDEHYDE	2	1.1	6.7	2	178	---	-----
772	C8H8O2	METHYL BENZOATE	2	1.2	6.7	1	181	---	-----
773	C8H8O2	o-TOLUIC ACID	2	1.2	6.8	2	273	---	-----
774	C8H8O2	p-TOLUIC ACID	2	1.2	6.8	2	273	---	-----
775	C8H8O3	METHYL SALICYLATE	2	1.2	7.2	1	205	1	851
776	C8H8O3	VANILLIN	2	1.2	8.8	2	318	---	-----
777	C8H9NO	ACETANILIDE	2	1.1	6.9	1	343	1	986
778	C8H10	ETHYLBENZENE	1	1.0	6.7	1	59	1	810
779	C8H10	m-XYLENE	1	1.1	7.0	1	77	1	982
780	C8H10	o-XYLENE	1	1.0	6.0	1	63	1	867

NO	FORMULA	NAME	Code	Lower Explosive Limit LEL, vol %	Upper Explosive Limit UEL, vol %	Flash Point Code	Flash Point F	Autoignition Temperature Code	Autoignition Temperature F
781	C8H10	p-XYLENE	1	1.1	7.0	1	77	1	984
782	C8H10O	m-ETHYLPHENOL	2	1.1	6.7	2	77	---	-----
783	C8H10O	p-ETHYLPHENOL	2	1.1	6.7	1	219	---	-----
784	C8H10O	PHENETOLE	2	1.1	7.8	2	125	---	-----
785	C8H10O	2-PHENYLETHANOL	2	1.1	7.0	1	205	---	-----
786	C8H10O	2,3-XYLENOL	2	1.1	6.4	2	199	2	1110
787	C8H10O	2,4-XYLENOL	2	1.1	6.4	2	192	2	1110
788	C8H10O	2,5-XYLENOL	2	1.1	6.4	2	192	2	1110
789	C8H10O	2,6-XYLENOL	1	1.4	6.4	1	186	1	1110
790	C8H10O	3,4-XYLENOL	2	1.1	6.4	2	219	2	1110
791	C8H10O	3,5-XYLENOL	2	1.1	6.4	2	212	2	1110
792	C8H11N	N,N-DIMETHYLANILINE	2	1.0	6.4	1	145	1	700
793	C8H11N	o-ETHYLANILINE	2	1.0	6.7	1	207	---	-----
794	C8H11N	2,4,6-TRIMETHYLPYRIDINE	2	1.0	8.9	1	135	---	-----
795	C8H11NO	p-PHENETIDINE	2	1.0	8.3	1	241	---	-----
796	C8H12	1,5-CYCLOOCTADIENE	2	1.0	8.6	1	95	---	-----
797	C8H12	VINYLCYCLOHEXENE	2	1.0	8.8	1	61	1	518
798	C8H12O4	1,4-CYCLOHEXANEDICARBOXYLIC ACID	2	1.0	8.2	2	489	---	-----
799	C8H12O4	DIETHYL MALEATE	2	1.1	7.7	1	199	1	662
800	C8H14O2	n-BUTYL METHACRYLATE	1	2.0	8.0	1	126	---	-----
801	C8H14O3	BUTYRIC ANHYDRIDE	1	1.1	7.6	1	180	1	535
802	C8H14O4	DIETHYL SUCCINATE	2	1.1	6.5	1	194	---	-----
803	C8H16	CYCLOOCTANE	2	0.9	6.0	---	-----	---	-----
804	C8H16	1,1-DIMETHYLCYCLOHEXANE	2	0.9	6.1	2	37	2	579
805	C8H16	cis-1,2-DIMETHYLCYCLOHEXANE	2	0.9	6.5	2	71	1	579
806	C8H16	trans-1,2-DIMETHYLCYCLOHEXANE	2	0.9	6.5	2	62	1	579
807	C8H16	cis-1,3-DIMETHYLCYCLOHEXANE	2	0.9	6.5	2	59	1	583
808	C8H16	trans-1,3-DIMETHYLCYCLOHEXANE	2	0.9	6.5	2	46	1	583
809	C8H16	cis-1,4-DIMETHYLCYCLOHEXANE	2	0.9	6.5	2	61	1	579
810	C8H16	trans-1,4-DIMETHYLCYCLOHEXANE	2	0.9	6.5	2	53	1	579
811	C8H16	ETHYLCYCLOHEXANE	1	0.9	6.6	1	95	1	504
812	C8H16	2-ETHYL-1-HEXENE	2	0.9	6.9	2	43	---	-----
813	C8H16	1-METHYL-1-ETHYLCYCLOPENTANE	2	0.9	6.1	2	43	---	-----
814	C8H16	1-OCTENE	2	0.9	7.1	1	70	1	446
815	C8H16	trans-2-OCTENE	2	0.9	6.9	1	70	---	-----
816	C8H16	trans-3-OCTENE	2	0.9	6.9	2	48	---	-----
817	C8H16	trans-4-OCTENE	2	0.9	6.9	2	46	---	-----
818	C8H16	n-PROPYLCYCLOPENTANE	1	0.9	6.4	2	61	1	516
819	C8H16	2,4,4-TRIMETHYL-1-PENTENE	2	0.9	6.7	2	1	1	788
820	C8H16	2,4,4-TRIMETHYL-2-PENTENE	2	0.9	6.4	1	1	1	586
821	C8H16O	2-ETHYLHEXANAL	1	0.9	7.2	1	111	1	374
822	C8H16O	1-OCTANAL	2	1.0	6.4	1	125	---	-----
823	C8H16O	2-OCTANONE	2	1.0	6.1	1	160	---	-----
824	C8H16O2	n-BUTYL n-BUTYRATE	2	1.0	6.1	1	128	---	-----
825	C8H16O2	n-HEXYL ACETATE	2	1.0	6.1	1	138	---	-----
826	C8H16O2	ISOBUTYL ISOBUTYRATE	1	1.0	7.6	1	100	1	810
827	C8H16O2	n-OCTANOIC ACID	2	1.0	6.4	2	248	2	710
828	C8H16O4	DIETHYLENE GLYCOL ETHYL ETHER ACETATE	1	1.0	19.4	1	224	1	680
829	C8H18	2,2-DIMETHYLHEXANE	2	0.9	5.5	2	25	2	818
830	C8H18	2,3-DIMETHYLHEXANE	2	0.9	5.9	1	42	1	820
831	C8H18	2,4-DIMETHYLHEXANE	2	0.9	5.9	1	50	2	818
832	C8H18	2,5-DIMETHYLHEXANE	2	0.9	5.9	2	28	2	818
833	C8H18	3,3-DIMETHYLHEXANE	2	0.9	5.5	2	30	2	818
834	C8H18	3,4-DIMETHYLHEXANE	2	0.9	5.9	2	39	2	818
835	C8H18	3-ETHYLHEXANE	2	0.9	5.8	2	43	---	-----
836	C8H18	3-ETHYL-2-METHYLPENTANE	2	0.9	5.8	2	37	---	-----
837	C8H18	3-METHYL-3-ETHYLPENTANE	2	0.9	5.5	2	37	---	-----
838	C8H18	2-METHYLHEPTANE	2	1.0	5.8	2	41	2	476
839	C8H18	3-METHYLHEPTANE	2	1.0	5.8	2	43	2	462
840	C8H18	4-METHYLHEPTANE	2	1.0	5.8	2	41	---	-----
841	C8H18	n-OCTANE	1	0.8	6.5	1	56	1	428
842	C8H18	2,2,3-TRIMETHYLPENTANE	2	1.0	5.6	2	26	1	806
843	C8H18	2,2,4-TRIMETHYLPENTANE	1	1.1	6.0	1	10	1	784
844	C8H18	2,3,3-TRIMETHYLPENTANE	2	1.0	5.6	2	32	1	806
845	C8H18	2,3,4-TRIMETHYLPENTANE	2	1.0	6.0	2	32	2	800
846	C8H18	2,2,3,3-TETRAMETHYLBUTANE	2	0.9	5.8	2	32	---	-----
847	C8H18O	DI-n-BUTYL ETHER	1	1.5	7.6	1	77	1	382
848	C8H18O	DI-sec-BUTYL ETHER	2	0.9	7.4	1	59	---	-----
849	C8H18O	DI-tert-BUTYL ETHER	2	0.9	6.8	2	26	---	-----
850	C8H18O	2-ETHYL-1-HEXANOL	1	0.9	9.7	1	164	1	550
851	C8H18O	1-OCTANOL	2	0.9	6.4	1	178	2	539
852	C8H18O	2-OCTANOL	1	0.8	6.5	1	140	---	-----
853	C8H18O2	DI-t-BUTYL PEROXIDE	2	0.9	8.2	1	64	---	-----
854	C8H18O2S	DI-n-BUTYL SULFONE	2	0.8	----	1	289	---	-----
855	C8H18O3	DIETHYLENE GLYCOL DIETHYL ETHER	2	0.9	10.1	1	180	1	421
856	C8H18O3	DIETHYLENE GLYCOL MONOBUTYL EHTER	1	0.9	24.6	1	172	1	400
857	C8H18O4	TRIETHYLENE GLYCOL DIMETHYL ETHER	2	1.0	11.4	1	230	1	376
858	C8H18O5	TETRAETHYLENE GLYCOL	2	1.0	11.0	1	385	---	-----

NO	FORMULA	NAME	Code	Lower Explosive Limit LEL, vol %	Upper Explosive Limit UEL, vol %	Flash Point Code	Flash Point F	Autoignition Temperature Code	Autoignition Temperature F
859	C8H18S	n-OCTYL MERCAPTAN	2	0.8	----	1	154	---	-----
860	C8H18S	tert-OCTYL MERCAPTAN	---	----	----	1	88	---	-----
861	C8H18S	BUTYL-SULFIDE	---	----	----	---	-----	---	-----
862	C8H18S	ETHYL-HEXYL-SULFIDE	---	----	----	---	-----	---	-----
863	C8H18S	HEPTYL-METHYL-SULFIDE	---	----	----	---	-----	---	-----
864	C8H18S	PENTYL-PROPYL-SULFIDE	---	----	----	---	-----	---	-----
865	C8H18S2	BUTYL-DISULFIDE	---	----	----	---	-----	---	-----
866	C8H19N	DI-n-BUTYLAMINE	1	1.1	6.1	1	117	---	-----
867	C8H19N	DIISOBUTYLAMINE	2	0.9	6.3	1	70	---	-----
868	C8H19N	n-OCTYLAMINE	2	0.9	6.4	1	140	---	-----
869	C8H23N5	TETRAETHYLENEPENTAMINE	2	0.8	4.6	1	325	1	572
870	C8H24O4Si4	OCTAMETHYLCYCLOTETRASILOXANE	1	0.8	7.4	1	135	1	752
871	C9H4O5	TRIMELLITIC ANHYDRIDE	2	1.3	9.5	2	493	---	-----
872	C9H6N2O2	TOLUENE DIISOCYANATE	1	0.9	9.5	1	261	---	-----
873	C9H7N	ISOQUINOLINE	2	1.0	7.8	1	225	---	-----
874	C9H7N	QUINOLINE	2	1.0	7.8	1	214	1	896
875	C9H7NO	8-HYDROXYQUINOLINE	2	1.0	7.0	2	268	---	-----
876	C9H8	INDENE	2	1.0	7.2	2	131	---	-----
877	C9H8O	2-METHYLBENZOFURAN	2	1.0	8.6	2	154	---	-----
878	C9H10	INDANE	2	1.0	6.1	1	122	---	-----
879	C9H10	cis-PROPENYLBENZENE	2	1.0	6.3	2	125	---	-----
880	C9H10	trans-PROPENYLBENZENE	2	1.0	6.3	---	-----	---	-----
881	C9H1O	alpha-METHYLSTYRENE	1	1.9	6.1	1	129	1	1066
882	C9H10	m-METHYLSTYRENE	1	0.7	11.0	1	124	2	912
883	C9H10	o-METHYLSTYRENE	2	1.0	6.7	1	117	---	-----
884	C9H10	p-METHYLSTYRENE	1	1.9	6.1	1	115	1	1067
885	C9H10O2	BENZYL ACETATE	2	1.0	6.1	1	215	1	862
886	C9H10O2	ETHYL BENZOATE	2	1.0	6.1	1	190	1	914
887	C9H10O3	ETHYL VANILLIN	2	1.0	7.8	2	341	---	-----
888	C9H11NO	p-DIMETHYLAMINOBENZALDEHYDE	2	1.0	6.1	1	297	---	-----
889	C9H12	CUMENE	1	0.9	6.5	1	111	1	795
890	C9H12	m-ETHYLTOLUENE	2	0.9	5.5	2	100	1	896
891	C9H12	o-ETHYLTOLUENE	2	0.9	5.5	2	106	1	824
892	C9H12	p-ETHYLTOLUENE	2	0.9	5.5	2	108	1	887
893	C9H12	MESITYLENE	2	0.9	5.2	1	112	1	1022
894	C9H12	n-PROPYLBENZENE	2	0.9	5.7	1	86	1	853
895	C9H12	1,2,3-TRIMETHYLBENZENE	2	0.9	5.2	1	124	1	878
896	C9H12	1,2,4-TRIMETHYLBENZENE	2	0.9	5.2	1	114	1	959
897	C9H12O	BENZYL ETHYL ETHER	2	0.9	6.7	2	138	---	-----
898	C9H12O	2-PHENYL-2-PROPANOL	2	1.0	6.1	2	174	---	-----
899	C9H12O2	CUMENE HYDROPEROXIDE	1	0.9	6.5	1	120	1	300
900	C9H14O	ISOPHORONE	1	0.8	3.8	1	183	1	860
901	C9H14O6	GLYCERYL TRIACETATE	1	1.0	6.4	1	280	1	811
902	C9H16	1-NONYNE	2	0.9	5.6	---	-----	---	-----
903	C9H16O4	AZELAIC ACID	2	0.9	6.5	2	426	---	-----
904	C9H18	BUTYLCYCLOPENTANE	2	0.8	5.4	---	-----	---	-----
905	C9H18	cis,cis-1,3,5-TRIMETHYLCYCLOHEXANE	2	0.8	5.4	---	-----	---	-----
906	C9H18	cis,trans-1,3,5-TRIMETHYLCYCLOHEXANE	2	0.8	5.4	---	-----	---	-----
907	C9H18	ISOPROPYLCYCLOHEXANE	2	0.8	5.9	1	95	1	541
908	C9H18	1-NONENE	2	0.8	6.4	2	80	2	458
909	C9H18	n-PROPYLCYCLOHEXANE	2	0.9	5.9	2	88	1	478
910	C9H18O	DIISOBUTYL KETONE	1	0.8	6.2	1	120	1	745
911	C9H18O	1-NONANAL	2	0.8	5.8	2	152	---	-----
912	C9H18O2	n-BUTYL VALERATE	2	0.9	5.6	2	145	---	-----
913	C9H18O2	n-NONANOIC ACID	2	0.9	5.9	2	262	2	710
914	C9H18O2	n-OCTYL FORMATE	2	0.9	5.8	2	147	---	-----
915	C9H20	3,3-DIETHYLPENTANE	1	0.7	5.7	2	70	1	554
916	C9H20	2,2-DIMETHYL-3-ETHYLPENTANE	2	0.8	5.2	2	55	---	-----
917	C9H20	3-ETHYL-2,3-DIMETHYLPENTANE	2	0.8	5.2	2	59	---	-----
918	C9H20	2,4-DIMETHYL-3-ETHYLPENTANE	2	0.8	5.5	2	59	1	734
919	C9H20	2,2-DIMETHYLHEPTANE	2	0.8	5.1	1	75	---	-----
920	C9H20	2,6-DIMETHYLHEPTANE	2	0.8	5.4	1	79	---	-----
921	C9H20	3-ETHYLHEPTANE	2	0.8	5.4	2	71	---	-----
922	C9H20	4-ETHYLHEPTANE	2	0.8	5.2	2	59	---	-----
923	C9H20	2,3-DIMETHYLHEPTANE	2	0.8	5.2	2	59	---	-----
924	C9H20	2,4-DIMETHYLHEPTANE	2	0.8	5.2	2	59	---	-----
925	C9H20	2,5-DIMETHYLHEPTANE	2	0.8	5.2	2	59	---	-----
926	C9H20	3,4-DIMETHYLHEPTANE	2	0.8	5.2	2	59	---	-----
927	C9H20	3,5-DIMETHYLHEPTANE	2	0.8	5.2	2	59	---	-----
928	C9H20	4,4-DIMETHYLHEPTANE	2	0.8	5.2	2	59	---	-----
929	C9H20	3-ETHYL-2-METHYLHEXANE	2	0.8	5.2	2	59	---	-----
930	C9H20	4-ETHYL-2-METHYLHEXANE	2	0.8	5.2	2	59	---	-----
931	C9H20	3-ETHYL-3-METHYLHEXANE	2	0.8	5.2	2	59	---	-----
932	C9H20	3-ETHYL-4-METHYLHEXANE	2	0.8	5.2	2	59	---	-----
933	C9H20	2,2,3-TRIMETHYLHEXANE	2	0.8	5.2	2	59	---	-----
934	C9H20	2,2,4-TRIMETHYLHEXANE	2	0.8	5.2	2	59	---	-----
935	C9H20	2,3,3-TRIMETHYLHEXANE	2	0.8	5.2	2	59	---	-----
936	C9H20	2,3,4-TRIMETHYLHEXANE	2	0.8	5.2	2	59	---	-----

NO	FORMULA	NAME	Code	Lower Explosive Limit LEL, vol %	Upper Explosive Limit UEL, vol %	Flash Point Code	Flash Point F	Autoignition Temperature Code	Autoignition Temperature F
937	C9H20	2,3,5-TRIMETHYLHEXANE	2	0.8	5.2	2	59	---	-----
938	C9H20	2,4,4-TRIMETHYLHEXANE	2	0.8	5.2	2	59	---	-----
939	C9H20	3,3,4-TRIMETHYLHEXANE	2	0.8	5.2	2	59	---	-----
940	C9H20	2-METHYLOCTANE	2	0.9	5.4	2	75	1	428
941	C9H20	3-METHYLOCTANE	2	0.9	5.4	2	75	1	428
942	C9H20	4-METHYLOCTANE	2	0.9	5.4	2	71	1	428
943	C9H20	n-NONANE	1	0.7	5.6	1	88	1	403
944	C9H20	2,2,3,3-TETRAMETHYLPENTANE	1	0.8	4.9	2	61	1	806
945	C9H20	2,2,3,4-TETRAMETHYLPENTANE	2	0.9	5.3	2	52	2	806
946	C9H20	2,2,4,4-TETRAMETHYLPENTANE	2	0.9	5.0	2	37	2	806
947	C9H20	2,3,3,4-TETRAMETHYLPENTANE	2	0.8	5.2	2	52	---	-----
948	C9H20	2,2,5-TRIMETHYLHEXANE	2	0.8	5.2	1	55	---	-----
949	C9H20O	2,6-DIMETHYL-4-HEPTANOL	2	0.8	5.7	2	151	---	-----
950	C9H20O	1-NONANOL	1	0.8	6.1	1	165	2	530
951	C9H20O	2-NONANOL	2	0.8	5.9	1	179	---	-----
952	C9H20S	n-NONYL MERCAPTAN	2	0.7	----	1	172	---	-----
953	C9H20S	BUTYL-PENTYL-SULFIDE	---	----	----	---	-----	---	-----
954	C9H20S	ETHYL-HEPTYL-SULFIDE	---	----	----	---	-----	---	-----
955	C9H20S	HEXYL-PROPYL-SULFIDE	---	----	----	---	-----	---	-----
956	C9H20S	METHYL-OCTYL-SULFIDE	---	----	----	---	-----	---	-----
957	C9H21N	n-NONYLAMINE	2	0.8	5.9	2	160	---	-----
958	C9H21N	TRIPROPYLAMINE	2	0.8	5.4	1	84	---	-----
959	C10H6O8	PYROMELLITIC ACID	2	1.2	7.8	2	617	---	-----
960	C10H7Br	1-BROMONAPHTHALENE	2	----	7.8	2	264	---	-----
961	C10H7Cl	1-CHLORONAPHTHALENE	1	1.0	5.7	1	250	---	-----
962	C10H8	NAPHTHALENE	1	0.9	5.9	1	176	1	1089
963	C10H8	AZULENE	2	0.9	6.0	---	-----	---	-----
964	C10H9N	QUINALDINE	2	0.9	6.8	1	175	---	-----
965	C10H10	m-DIVINYLBENZENE	1	0.3	6.9	2	149	---	-----
966	C10H10	1-METHYLINDENE	2	0.9	6.6	2	149	---	-----
967	C10H10	2-METHYLINDENE	2	0.9	6.4	2	149	---	-----
968	C10H10O4	DIMETHYL PHTHALATE	1	0.9	5.8	1	295	1	1033
969	C10H10O4	DIMETHYL TEREPHTHALATE	1	----	5.5	1	313	1	1058
970	C10H12	DICYCLOPENTADIENE	1	1.0	8.3	1	90	1	950
971	C10H12	1,2,3,4-TETRAHYDRONAPHTHALENE	1	0.8	5.0	1	160	1	723
972	C10H12O	ANETHOLE	2	0.9	7.2	1	192	---	-----
973	C10H12O4	DIALLYL MALEATE	2	0.9	7.9	2	248	---	-----
974	C10H14	n-BUTYLBENZENE	1	0.8	5.8	1	160	1	774
975	C10H14	sec-BUTYLBENZENE	1	0.8	6.9	1	126	1	784
976	C10H14	tert-BUTYLBENZENE	1	0.7	5.7	1	140	1	842
977	C10H14	1,2,3,4-TETRAMETHYLBENZENE	2	0.8	5.4	---	-----	---	-----
978	C10H14	m-CYMENE	2	0.8	5.2	2	122	2	817
979	C10H14	o-CYMENE	2	0.8	5.2	2	127	2	710
980	C10H14	p-CYMENE	1	0.7	5.6	1	117	1	817
981	C10H14	m-DIETHYLBENZENE	2	0.8	5.1	1	133	1	842
982	C10H14	o-DIETHYLBENZENE	2	0.8	5.1	1	135	1	743
983	C10H14	p-DIETHYLBENZENE	1	0.8	6.1	1	134	1	806
984	C1OH14	2-ETHYL-m-XYLENE	2	0.8	4.9	2	136	---	-----
985	C10H14	2-ETHYL-p-XYLENE	2	0.8	4.9	2	133	---	-----
986	C10H14	3-ETHYL-o-XYLENE	2	0.8	4.9	2	149	---	-----
987	C10H14	4-ETHYL-m-XYLENE	2	0.8	4.9	2	134	---	-----
988	C10H14	4-ETHYL-o-XYLENE	2	0.8	4.9	2	136	---	-----
989	C1OH14	5-ETHYL-m-XYLENE	2	0.8	4.9	2	127	---	-----
990	C10H14	ISOBUTYLBENZENE	1	0.8	6.0	1	131	1	802
991	C10H14	1,2,3,5-TETRAMETHYLBENZENE	2	0.8	4.6	1	146	2	801
992	C10H14	1,2,4,5-TETRAMETHYLBENZENE	2	0.8	4.6	1	130	---	-----
993	C10H14O	p-tert-BUTYLPHENOL	2	0.8	5.5	1	235	---	-----
994	C10H14O2	p-tert-BUTYLCATECHOL	2	0.9	5.8	1	266	---	-----
995	C10H15N	N,N-DIETYHLANILINE	1	1.6	9.5	1	185	1	630
996	C10H15N	2,6-DIETHYLANILINE	2	0.8	5.4	2	242	---	-----
997	C10H16	CAMPHENE	2	0.8	6.7	1	97	---	-----
998	C10H16	D-LIMONENE	1	0.7	6.1	1	113	1	458
999	C10H16	alpha-PHELLANDRENE	2	0.8	7.1	2	116	---	-----
1000	C1OH16	beta-PHELLANDRENE	2	0.8	7.1	1	120	---	-----
1001	C10H16	alpha-PINENE	2	0.8	6.6	1	91	1	491
1002	C10H16	beta-PINENE	2	0.8	6.7	2	103	2	491
1003	C10H16	alpha-TERPINENE	2	0.8	6.8	1	115	---	-----
1004	C1OH16	gamma-TERPINENE	2	0.8	6.8	1	124	---	-----
1005	C10H16	TERPINOLENE	2	0.8	6.8	1	100	---	-----
1006	C10H16O	CAMPHOR	1	0.6	3.5	1	150	1	871
1007	C10H18	1-DECYNE	2	0.8	5.0	---	-----	---	-----
1008	C10H18	cis-DECAHYDRONANPHTALENE	1	0.7	4.9	1	136	1	482
1009	C10H18	trans-DECAHYDRONAPHTHALENE	1	0.7	4.9	1	136	1	482
1010	C10H18O4	SEBACIC ACID	2	0.8	6.0	2	433	---	-----
1011	C10H20	n-BUTYLCYCLOHEXANE	2	0.9	5.5	2	118	1	475
1012	C10H20	1-CYCLOPENTYLPENTANE	2	0.8	4.8	---	-----	---	-----
1013	C10H20	1-DECENE	2	0.7	5.9	1	128	1	455
1014	C10H20O	1-DECANAL	2	0.8	5.4	1	185	---	-----

NO	FORMULA	NAME	Code	Lower Explosive Limit LEL, vol %	Upper Explosive Limit UEL, vol %	Flash Point Code	Flash Point F	Autoignition Temperature Code	Autoignition Temperature F
1015	C10H20O2	n-DECANOIC ACID	2	0.8	5.5	2	287	2	710
1016	C10H20O2	2-ETHYLHEXYL ACETATE	1	0.8	8.1	1	190	---	-----
1017	C10H20O2	ISOPENTYL ISOVALERATE	2	0.8	5.4	2	136	---	-----
1018	C10H22	n-DECANE	1	0.8	5.4	1	115	1	406
1019	C10H22	2-METHYLNONANE	2	0.7	5.0	2	106	---	-----
1020	C10H22	3-METHYLNONANE	2	0.7	5.0	1	106	1	417
1021	C10H22	4-METHYLNONANE	2	0.7	5.0	2	101	---	-----
1022	C10H22	5-METHYLNONANE	2	0.7	5.0	2	102	---	-----
1023	C10H22	3-ETHYLOCTANE	2	0.7	4.7	2	106	---	-----
1024	C10H22	4-ETHYLOCTANE	2	0.7	4.7	2	106	---	-----
1025	C10H22	2,2-DIMETHYLOCTANE	2	0.7	4.8	2	88	---	-----
1026	C10H22	2,3-DIMETHYLOCTANE	2	0.7	4.7	2	106	---	-----
1027	C10H22	2,4-DIMETHYLOCTANE	2	0.7	4.7	2	106	---	-----
1028	C10H22	2,5-DIMETHYLOCTANE	2	0.7	4.7	2	106	---	-----
1029	C10H22	2,6-DIMETHYLOCTANE	2	0.7	4.7	2	106	---	-----
1030	C10H22	2,7-DIMETHYLOCTANE	2	0.7	4.7	2	106	---	-----
1031	C10H22	3,3-DIMETHYLOCTANE	2	0.7	4.7	2	106	---	-----
1032	C10H22	3,4-DIMETHYLOCTANE	2	0.7	4.7	2	106	---	-----
1033	C10H22	3,5-DIMETHYLOCTANE	2	0.7	4.7	2	106	---	-----
1034	C10H22	3,6-DIMETHYLOCTANE	2	0.7	4.7	2	106	---	-----
1035	C10H22	4,4-DIMETHYLOCTANE	2	0.7	4.7	2	106	---	-----
1036	C10H22	4,5-DIMETHYLOCTANE	2	0.7	4.7	2	106	---	-----
1037	C10H22	4-PROPYLHEPTANE	2	0.7	4.7	2	106	---	-----
1038	C10H22	4-ISOPROPYLHEPTANE	2	0.7	4.7	2	106	---	-----
1039	C10H22	3-ETHYL-2-METHYLHEPTANE	2	0.7	4.7	2	106	---	-----
1040	C10H22	4-ETHYL-2-METHYLHEPTANE	2	0.7	4.7	2	106	---	-----
1041	C10H22	5-ETHYL-2-METHYLHEPTANE	2	0.7	4.7	2	106	---	-----
1042	C10H22	3-ETHYL-3-METHYLHEPTANE	2	0.7	4.7	2	106	---	-----
1043	C10H22	4-ETHYL-3-METHYLHEPTANE	2	0.7	4.7	2	106	---	-----
1044	C10H22	3-ETHYL-5-METHYLHEPTANE	2	0.7	4.7	2	88	---	-----
1045	C10H22	3-ETHYL-4-METHYLHEPTANE	2	0.7	4.7	2	88	---	-----
1046	C10H22	4-ETHYL-4-METHYLHEPTANE	2	0.7	4.7	1	100	2	413
1047	C10H22	2,2,3-TRIMETHYLHEPTANE	2	0.7	4.7	2	88	---	-----
1048	C10H22	2,2,4-TRIMETHYLHEPTANE	2	0.7	4.7	2	88	---	-----
1049	C10H22	2,2,5-TRIMETHYLHEPTANE	2	0.7	4.7	2	88	---	-----
1050	C10H22	2,2,6-TRIMETHYLHEPTANE	2	0.7	4.7	2	88	---	-----
1051	C10H22	2,3,3-TRIMETHYLHEPTANE	2	0.7	4.7	2	88	---	-----
1052	C10H22	2,3,4-TRIMETHYLHEPTANE	2	0.7	4.7	2	88	---	-----
1053	C10H22	2,3,5-TRIMETHYLHEPTANE	2	0.7	4.7	2	88	---	-----
1054	C10H22	2,3,6-TRIMETHYLHEPTANE	2	0.7	4.7	2	88	---	-----
1055	C10H22	2,4,4-TRIMETHYLHEPTANE	2	0.7	4.7	2	88	---	-----
1056	C10H22	2,4,5-TRIMETHYLHEPTANE	2	0.7	4.7	2	88	---	-----
1057	C10H22	2,4,6-TRIMETHYLHEPTANE	2	0.7	4.7	2	88	---	-----
1058	C10H22	2,5,5-TRIMETHYLHEPTANE	2	0.7	4.7	2	88	---	-----
1059	C10H22	3,3,4-TRIMETHYLHEPTANE	2	0.7	4.7	2	88	---	-----
1060	C10H22	3,3,5-TRIMETHYLHEPTANE	2	0.7	4.7	2	88	---	-----
1061	C10H22	3,4,4-TRIMETHYLHEPTANE	2	0.7	4.7	2	88	---	-----
1062	C10H22	3,4,5-TRIMETHYLHEPTANE	2	0.7	4.7	2	88	---	-----
1063	C10H22	3-ISOPROPYL-2-METHYLHEXANE	2	0.7	4.7	2	88	---	-----
1064	C10H22	3,3-DIETHYLHEXANE	2	0.7	4.7	2	100	---	-----
1065	C10H22	3,4-DIETHYLHEXANE	2	0.7	4.7	2	100	---	-----
1066	C10H22	3-ETHYL-2,2-DIMETHYLHEXANE	2	0.7	4.7	2	100	---	-----
1067	C10H22	4-ETHYL-2,2-DIMETHYLHEXANE	2	0.7	4.7	2	100	---	-----
1068	C10H22	3-ETHYL-2,3-DIMETHYLHEXANE	2	0.7	4.7	2	100	---	-----
1069	C10H22	4-ETHYL-2,3-DIMETHYLHEXANE	2	0.7	4.7	2	100	---	-----
1070	C10H22	3-ETHYL-2,4-DIMETHYLHEXANE	2	0.7	4.7	2	100	---	-----
1071	C10H22	4-ETHYL-2,4-DIMETHYLHEXANE	2	0.7	4.7	2	100	---	-----
1072	C10H22	3-ETHYL-2,5-DIMETHYLHEXANE	2	0.7	4.7	2	100	---	-----
1073	C10H22	4-ETHYL-3,3-DIMETHYLHEXANE	2	0.7	4.7	2	100	2	413
1074	C10H22	3-ETHYL-3,4-DIMETHYLHEXANE	2	0.7	4.7	2	88	---	-----
1075	C10H22	2,2,3,3-TETRAMETHYLHEXANE	2	0.7	4.7	2	88	---	-----
1076	C10H22	2,2,3,4-TETRAMETHYLHEXANE	2	0.7	4.7	2	88	---	-----
1077	C10H22	2,2,3,5-TETRAMETHYLHEXANE	2	0.7	4.7	2	88	---	-----
1078	C10H22	2,2,4,4-TETRAMETHYLHEXANE	2	0.7	4.7	2	88	---	-----
1079	C10H22	2,2,4,5-TETRAMETHYLHEXANE	2	0.7	4.7	2	88	---	-----
1080	C10H22	2,2,5,5-TETRAMETHYLHEXANE	2	0.7	4.7	2	88	---	-----
1081	C10H22	2,3,3,4-TETRAMETHYLHEXANE	2	0.7	4.7	2	88	---	-----
1082	C10H22	2,3,3,5-TETRAMETHYLHEXANE	2	0.7	4.7	2	88	---	-----
1083	C10H22	2,3,4,4-TETRAMETHYLHEXANE	2	0.7	4.7	2	88	---	-----
1084	C10H22	2,3,4,5-TETRAMETHYLHEXANE	2	0.7	4.7	2	88	---	-----
1085	C10H22	3,3,4,4-TETRAMETHYLHEXANE	2	0.7	4.7	2	88	---	-----
1086	C10H22	2,4-DIMETHYL-3-ISOPROPYLPENTANE	2	0.7	4.7	2	88	---	-----
1087	C10H22	3,3-DIETHYL-2-METHYLPENTANE	2	0.7	4.7	2	88	---	-----
1088	C10H22	3-ETHYL-2,2,3-TRIMETHYLPENTANE	2	0.7	4.7	2	88	---	-----
1089	C10H22	3-ETHYL-2,2,4-TRIMETHYLPENTANE	2	0.7	4.7	2	88	---	-----
1090	C10H22	3-ETHYL-2,3,4-TRIMETHYLPENTANE	2	0.7	4.7	2	88	---	-----
1091	C10H22	2,2,3,3,4-PENTAMETHYLPENTANE	2	0.7	4.7	2	88	---	-----
1092	C10H22	2,2,3,4,4-PENTAMETHYLPENTANE	2	0.7	4.7	2	100	2	413

NO	FORMULA	NAME	Code	Lower Explosive Limit LEL, vol %	Upper Explosive Limit UEL, vol %	Flash Point Code	Flash Point F	Autoignition Temperature Code	Autoignition Temperature F
1093	C10H22O	1-DECANOL	2	0.7	5.5	1	180	1	550
1094	C10H22O	DI-n-PENTYL ETHER	2	0.7	6.0	1	135	1	340
1095	C10H22O	ISODECANOL	2	0.7	5.5	1	220	---	-----
1096	C10H22O5	TETRAETHYLENE GLYCOL DIMETHYL ETHER	2	0.8	9.9	1	286	---	-----
1097	C10H22S	n-DECYL MERCAPTAN	2	0.6	----	2	203	---	-----
1098	C10H22S	BUTYL-HEXYL-SULFIDE	---	-----	-----	---	-----	---	-----
1099	C10H22S	ETHYL-OCTYL-SULFIDE	---	----	----	---	----	---	-----
1100	C10H22S	HEPTYL-PROPYL-SULFIDE	---	----	----	---	----	---	-----
1101	C10H22S	METHYL-NONYL-SULFIDE	---	----	----	---	-----	---	-----
1102	C10H22S	PENTYL-SULFIDE	---	----	----	---	-----	---	-----
1103	C10H22S2	PENTYL-DISULFIDE	---	----	----	---	-----	---	-----
1104	C10H23N	n-DECYLAMINE	2	0.7	5.5	1	210	---	-----
1105	C11H10	1-METHYLNAPHTHALENE	2	0.8	5.3	2	206	1	984
1106	C11H10	2-METHYLNAPHTHALENE	2	0.8	5.3	2	203	2	984
1107	C11H14O2	n-BUTYL BENZOATE	2	0.8	5.4	1	223	---	-----
1108	C11H16	n-PENTYLBENZENE	2	0.7	5.1	1	151	---	-----
1109	C11H16O	p-tert-AMYLPHENOL	2	0.8	5.2	1	232	---	-----
1110	C11H20	1-UNDECYNE	2	0.7	4.5	---	-----	---	-----
1111	C11H20O2	2-ETHYLHEXYL ACRYLATE	1	0.7	8.2	1	160	1	496
1112	C11H22	1-UNDECENE	2	0.7	5.6	1	160	2	458
1113	C11H22	1-CYCLOPENTYLHEXANE	2	0.7	4.4	---	-----	---	-----
1114	C11H22	PENTYLCYCLOHEXANE	2	0.7	4.4	---	-----	---	-----
1115	C11H22O	1-UNDECANAL	2	0.7	5.2	2	199	---	-----
1116	C11H24	n-UNDECANE	2	0.7	4.8	1	149	2	395
1117	C11H24O	1-UNDECANOL	2	0.7	5.2	1	200	2	530
1118	C11H24S	UNDECYL MERCAPTAN	2	0.6	----	2	228	---	-----
1119	C11H24S	BUTYL-HEPTYL-SULFIDE	---	-----	----	---	-----	---	-----
1120	C11H24S	DECYL-METHYL-SULFIDE	---	----	----	---	-----	---	-----
1121	C11H24S	ETHYL-NONYL-SULFIDE	---	----	----	---	-----	---	-----
1122	C11H24S	OCTYL-PROPYL-SULFIDE	---	----	----	---	-----	---	-----
1123	C12H8O	DIBENZOFURAN	2	0.8	6.4	2	233	---	-----
1124	C12H9N	DIBENZOPYRROLE	2	----	5.5	---	-----	---	-----
1125	C12H10	ACENAPHTHENE	2	0.8	6.7	2	248	---	-----
1126	C12H10	BIPHENYL	1	0.6	5.8	1	235	1	1004
1127	C12H10O	DIPHENYL ETHER	1	0.8	1.5	1	239	1	1144
1128	C12H11N	p-AMINODIPHENYL	2	0.7	5.5	2	296	1	842
1129	C12H11N	DIPHENYLAMINE	2	0.7	5.4	1	307	1	1173
1130	C12H11N3	p-AMINOAZOBENZENE	2	0.7	----	2	379	---	-----
1131	C12H11N3	1,3-DIPHENYLTRIAZENE	---	----	----	---	-----	---	-----
1132	C12H12	1,2-DIMETHYLNAPHTHALENE	2	0.8	4.8	---	-----	---	-----
1133	C12H12	1,3-DIMETHYLNAPHTHALENE	2	0.8	4.8	---	-----	---	-----
1134	C12H12	1,4-DIMETHYLNAPHTHALENE	2	0.8	4.8	---	-----	---	-----
1135	C12H12	1,5-DIMETHYLNAPHTHALENE	2	0.8	4.8	---	-----	---	-----
1136	C12H12	1,6-DIMETHYLNAPHTHALENE	2	0.8	4.8	---	-----	---	-----
1137	C12H12	1,7-DIMETHYLNAPHTHALENE	2	0.8	4.8	---	-----	---	-----
1138	C12H12	2,3-DIMETHYLNAPHTHALENE	2	0.8	4.8	---	-----	---	-----
1139	C12H12	2,6-DIMETHYLNAPHTHALENE	2	0.7	5.0	2	228	---	-----
1140	C12H12	2,7-DIMETHYLNAPHTHALENE	2	0.7	5.0	2	228	---	-----
1141	C12H12	1-ETHYLNAPHTHALENE	2	0.7	5.2	2	224	1	896
1142	C12H12	2-ETHYLNAPHTHALENE	2	0.8	4.8	---	-----	---	-----
1143	C12H12N2	p-AMINODIPHENYLAMINE	2	0.7	5.5	1	380	---	-----
1144	C12H12N2	HYDRAZOBENZENE	2	0.7	5.5	2	305	---	-----
1145	C12H14	1,2,3-TRIMETHYLINDENE	2	0.7	5.7	1	187	---	-----
1146	C12H14O4	DIETHYL PHTHALATE	1	0.7	5.3	1	322	1	855
1147	C12H16	CYCLOHEXYLBENZENE	2	0.7	5.4	1	210	---	-----
1148	C12H18	m-DIISOPROPYLBENZENE	2	0.7	4.9	2	170	1	170
1149	C12H18	p-DIISOPROPYLBENZENE	2	0.7	4.9	2	178	1	170
1150	C12H18	n-HEXYLBENZENE	2	0.7	5.0	2	181	---	-----
1151	C12H18	1,2,3-TRIETHYLBENZENE	2	0.7	4.4	---	-----	---	-----
1152	C12H18	1,2,4-TRIETHYLBENZENE	2	0.7	4.4	2	168	---	-----
1153	C12H18	1,3,5-TRIETHYLBENZENE	2	0.7	4.4	---	-----	---	-----
1154	C12H18	HEXAMETHYLBENZENE	2	0.7	4.4	---	-----	---	-----
1155	C12H20O4	DIBUTYL MALEATE	2	0.7	5.7	1	284	---	-----
1156	C12H22	BICYCLOHEXYL	1	0.7	5.1	1	165	1	473
1157	C12H22	1-DODECYNE	2	0.7	4.2	---	-----	---	-----
1158	C12H23N	DICYCLOHEXYLAMINE	2	0.6	5.6	1	219	---	-----
1159	C12H24	1-DODECENE	2	0.6	5.4	1	120	1	491
1160	C12H24	1-CYCLOPENTYLHEPTANE	2	0.6	4.0	---	-----	---	-----
1161	C12H24	1-CYCLOHEXYLHEXANE	2	0.6	4.0	---	-----	---	-----
1162	C12H24O	1-DODECANAL	2	0.6	5.0	2	221	---	-----
1163	C12H24O2	n-DODECANOIC ACID	2	0.6	8.4	2	314	2	710
1164	C12H26	n-DODECANE	1	0.6	4.7	1	165	1	399
1165	C12H26O	DI-n-HEXYL ETHER	2	0.6	5.5	1	171	1	365
1166	C12H26O	1-DODECANOL	2	0.6	5.1	1	260	1	527
1167	C12H26O3	DIETHYLENE GLYCOL DI-n-BUTYL ETHER	2	0.6	7.0	1	244	1	397
1168	C12H26S	n-DODECYL MERCAPTAN	---	----	----	1	262	---	-----
1169	C12H26S	BUTYL-OCTYL-SULFIDE	---	----	----	---	-----	---	-----
1170	C12H26S	DECYL-ETHYL-SULFIDE	---	----	----	---	-----	---	-----

NO	FORMULA	NAME	Code	Lower Explosive Limit LEL, vol %	Upper Explosive Limit UEL, vol %	Flash Point Code	Flash Point F	Autoignition Temperature Code	Autoignition Temperature F
1171	C12H26S	HEXYL-SULFIDE	---	-----	-----	---	-----	---	-----
1172	C12H26S	METHYL-UNDECYL-SULFIDE	---	-----	-----	---	-----	---	-----
1173	C12H26S	NONYL-PROPYL-SULFIDE	---	-----	-----	---	-----	---	-----
1174	C12H26S2	HEXYL-DISULFIDE	---	-----	-----	---	-----	---	-----
1175	C12H27BO3	TRI-n-BUTYL BORATE	---	-----	-----	1	199	---	-----
1176	C12H27N	DODECYLAMINE	2	0.6	5.1	2	232	---	-----
1177	C12H27N	TRI-n-BUTYLAMINE	2	0.6	4.9	1	187	---	-----
1178	C13H10	FLUORENE	2	0.7	5.5	2	274	---	-----
1179	C13H10O	BENZOPHENONE	2	0.7	5.4	2	289	---	-----
1180	C13H12	DIPHENYLMETHANE	2	0.7	5.2	1	266	1	905
1181	C13H14	1-PROPYLNAPHTHALENE	2	0.7	4.4	---	-----	---	-----
1182	C13H14	2-PROPYLNAPHTHALENE	2	0.7	4.4	---	-----	---	-----
1183	C13H14	2ETHYL-3-METHYLNAPHTHALENE	2	0.7	4.4	---	-----	---	-----
1184	C13H14	2ETHYL-6-METHYLNAPHTHALENE	2	0.7	4.4	---	-----	---	-----
1185	C13H14	2ETHYL-7-METHYLNAPHTHALENE	2	0.7	4.4	---	-----	---	-----
1186	C13H20	n-HEPTYLBENZENE	2	0.6	4.9	1	203	---	-----
1187	C13H24	1-TRIDECYNE	2	0.6	3.8	---	-----	---	-----
1188	C13H26	1-TRIDECENE	2	0.6	5.4	2	175	2	458
1189	C13H26	1-CYCLOPENTYLOCTANE	2	0.6	3.7	---	-----	---	-----
1190	C13H26	1-CYCLOHEXYLHEPTANE	2	0.6	3.7	---	-----	---	-----
1191	C13H26O	1-TRIDECANAL	2	0.6	5.0	2	242	---	-----
1192	C13H26O2	n-BUTYL NONANOATE	2	0.6	4.9	2	205	---	-----
1193	C13H26O2	METHYL DODECANOATE	2	0.6	4.9	2	244	---	-----
1194	C13H28	n-TRIDECANE	2	0.6	4.7	1	174	2	395
1195	C13H28O	1-TRIDECANOL	2	0.6	5.0	1	250	---	-----
1196	C13H28S	BUTYL-NONYL-SULFIDE	---	-----	-----	---	-----	---	-----
1197	C13H28S	DECYL-PROPYL-SULFIDE	---	-----	-----	---	-----	---	-----
1198	C13H28S	DODECYL-METHYL-SULFIDE	---	-----	-----	---	-----	---	-----
1199	C13H28S	ETHYL-UNDECYL-SULFIDE	---	-----	-----	---	-----	---	-----
1200	C13H28S	1-TRIDECANETHIOL	---	-----	-----	---	-----	---	-----
1201	C14H8O2	ANTHRAQUINONE	2	0.8	4.8	1	365	---	-----
1202	C14H10	ANTHRACENE	1	0.6	5.3	1	250	1	1004
1203	C14H10	DIPHENYLACETYLENE	2	0.7	5.3	2	269	---	-----
1204	C14H10	PHENANTHRENE	2	0.7	5.2	1	340	---	-----
1205	C14H12	cis-STILBENE	2	0.7	4.3	2	251	---	-----
1206	C14H12	trans-STILBENE	2	0.7	4.3	2	293	---	-----
1207	C14H12O2	BENZYL BENZOATE	2	0.7	4.5	1	311	1	896
1208	C14H14	1,1-DIPHENYLETHANE	2	0.6	4.2	1	264	1	824
1209	C14H14	1,2-DIPHENYLETHANE	2	0.6	4.2	1	264	1	896
1210	C14H14O	DIBENZYL ETHER	2	0.6	6.0	1	275	---	-----
1211	C14H16	1-n-BUTYLNAPHTHALENE	2	0.6	5.2	2	262	2	680
1212	C14H16	2-BUTYLNAPHTHALENE	2	0.6	4.0	---	-----	---	-----
1213	C14H22	n-OCTYLBENZENE	2	0.6	4.9	1	225	---	-----
1214	C14H22	1,2,3,4-TETRAETHYLBENZENE	2	0.6	3.7	2	202	---	-----
1215	C14H22	1,2,3,5-TETRAETHYLBENZENE	2	0.6	3.7	---	-----	---	-----
1216	C14H22	1,2,4,5-TETRAETHYLBENZENE	2	0.6	3.7	---	-----	---	-----
1217	C14H22O	p-tert-OCTYLPHENOL	2	0.6	5.0	2	271	---	-----
1218	C14H28	1-TETRADECENE	2	0.5	5.4	1	230	1	455
1219	C14H28	1-CYCLOPENTYLNONANE	2	0.5	3.5	---	-----	---	-----
1220	C14H28	1-CYCLOHEXYLOCTANE	2	0.5	3.5	---	-----	---	-----
1221	C14H28O2	n-TETRADECANOIC ACID	2	0.5	5.8	2	341	2	710
1222	C14H30	n-TETRADECANE	1	0.5	4.7	1	212	1	392
1223	C14H30O	1-TETRADECANOL	2	0.5	3.5	1	286	---	-----
1224	C14H30S	BUTYL-DECYL-SULFIDE	---	-----	-----	---	-----	---	-----
1225	C14H30S	DODECYL-ETHYL-SULFIDE	---	-----	-----	---	-----	---	-----
1226	C14H30S	HEPTYL-SULFIDE	---	-----	-----	---	-----	---	-----
1227	C14H30S	METHYL-TRIDECYL-SULFIDE	---	-----	-----	---	-----	---	-----
1228	C14H30S	PROPYL-UNDECYL-SULFIDE	---	-----	-----	---	-----	---	-----
1229	C14H30S	1-TETRADECANETHIOL	---	-----	-----	---	-----	---	-----
1230	C14H30S2	HEPTYL-DISULFIDE	---	-----	-----	---	-----	---	-----
1231	C14H31N	TETRADECYLAMINE	2	0.5	-----	2	271	---	-----
1232	C15H10N2O2	DIPHENYLMETHANE-4,4'-DIISOCYANATE	2	0.6	-----	1	394	---	-----
1233	C15H16O	p-CUMYLPHENOL	2	0.6	3.9	2	320	---	-----
1234	C15H16O2	BISPHENOL A	2	0.6	4.0	1	415	---	-----
1235	C15H18	1-PENTYLNAPHTHALENE	2	0.6	3.7	---	-----	---	-----
1236	C15H18	2-PENTYLNAPHTHALENE	2	0.6	3.7	---	-----	---	-----
1237	C15H24	n-NONYLBENZENE	2	0.5	4.9	2	250	---	-----
1238	C15H24O	2,6-DI-tert-BUTYL-p-CRESOL	2	0.5	4.9	2	241	---	-----
1239	C15H24O	NONYLPHENOL	1	1.0	5.0	1	280	---	-----
1240	C15H28	1-PENTADECYNE	2	0.5	3.3	---	-----	---	-----
1241	C15H30	1-PENTADECENE	2	0.4	5.4	2	235	2	458
1242	C15H30	1-CYCLOPENTYLDECANE	2	0.5	3.2	---	-----	---	-----
1243	C15H30	1-CYCLOHEXYLNONANE	2	0.5	3.2	---	-----	---	-----
1244	C15H30O2	PENTADECANOIC ACID	2	0.5	3.3	2	359	2	710
1245	C15H32	n-PENTADECANE	2	0.5	4.7	2	239	2	395
1246	C15H32O	1-PENTADECANOL	2	0.5	3.2	---	-----	---	-----
1247	C15H32S	BUTYL-UNDECYL-SULFIDE	---	-----	-----	---	-----	---	-----
1248	C15H32S	DODECYL-PROPYL-SULFIDE	---	-----	-----	---	-----	---	-----

NO	FORMULA	NAME	Code	Lower Explosive Limit LEL, vol %	Upper Explosive Limit UEL, vol %	Flash Point Code	Flash Point F	Autoignition Temperature Code	Autoignition Temperature F
1249	C15H32S	ETHYL-TRIDECYL-SULFIDE	---	----	----	---	-----	---	-----
1250	C15H32S	METHYL-TETRADECYL-SULFIDE	---	----	----	---	-----	---	-----
1251	C15H32S	1-PENTADECANETHIOL	---	----	----	---	-----	---	-----
1252	C16H10	FLUORANTHENE	2	0.6	3.9	2	363	---	-----
1253	C16H10	PYRENE	2	0.6	3.9	1	390	---	-----
1254	C16H12	1-PHENYLNAPHTHALENE	2	0.6	3.8	2	323	---	-----
1255	C16H20	1-n-HEXYLNAPHTHALENE	2	0.5	5.2	2	305	---	-----
1256	C16H22O4	DIBUTYL PHTHALATE	1	0.5	----	1	315	1	756
1257	C16H26	n-DECYLBENZENE	2	0.5	3.2	1	225	---	-----
1258	C16H26	PENTAETHYLBENZENE	2	0.5	3.2	2	236	---	-----
1259	C16H30	1-HEXADECYNE	2	0.5	3.1	---	-----	---	-----
1260	C16H32	n-DECYLCYCLOHEXANE	2	0.5	4.7	2	268	---	-----
1261	C16H32	1-CYCLOPENTYLUNDECANE	2	0.5	3.0	---	-----	---	-----
1262	C16H32	1-HEXADECENE	2	0.5	3.0	2	255	1	464
1263	C16H32O2	n-HEXADECANOIC ACID	2	0.5	4.9	2	356	2	710
1264	C16H34	n-HEXADECANE	2	0.5	4.7	2	259	1	396
1265	C16H34O	DI-n-OCTYL ETHER	2	0.5	5.5	1	234	1	401
1266	C16H34O	1-HEXADECANOL	2	0.5	3.0	2	311	---	-----
1267	C16H34S	BUTYL-DODECYL-SULFIDE	---	----	----	---	-----	---	-----
1268	C16H34S	ETHYL-TETRADECYL-SULFIDE	---	----	----	---	-----	---	-----
1269	C16H34S	METHYL-PENTADECYL-SULFIDE	---	----	----	---	-----	---	-----
1270	C16H34S	OCTYL-SULFIDE	---	----	----	---	-----	---	-----
1271	C16H34S	PROPYL-TRIDECYL-SULFIDE	---	----	----	---	-----	---	-----
1272	C16H34S	1-HEXADECANETHIOL	---	----	----	---	-----	---	-----
1273	C16H34S2	OCTYL-DISULFIDE	---	----	----	---	-----	---	-----
1274	C17H28	n-UNDECYLBENZENE	2	0.5	4.9	2	293	---	-----
1275	C17H32	1-HEPTADECYNE	2	0.5	2.9	---	-----	---	-----
1276	C17H34	1-CYCLOPENTYLDODECANE	2	0.4	2.9	---	-----	---	-----
1277	C17H34	1-CYCLOHEXYLUNDECANE	2	0.4	2.9	---	-----	---	-----
1278	C17H34	1-HEPTADECENE	2	0.4	5.4	2	275	2	458
1279	C17H36	n-HEPTADECANE	2	0.4	2.8	2	277	2	395
1280	C17H36O	1-HEPTADECANOL	2	0.4	2.9	1	310	---	-----
1281	C17H36S	BUTYL-TRIDECYL-SULFIDE	---	----	----	---	-----	---	-----
1282	C17H36S	ETHYL-PENTADECYL-SULFIDE	---	----	----	---	-----	---	-----
1283	C17H36S	HEXADECYL-METHYL-SULFIDE	---	----	----	---	-----	---	-----
1284	C17H36S	PROPYL-TETRADECYL-SULFIDE	---	----	----	---	-----	---	-----
1285	C17H36S	1-HEPTADECANETHIOL	---	----	----	---	-----	---	-----
1286	C18H12	CHRYSENE	2	0.5	3.5	---	-----	---	-----
1287	C18H14	m-TERPHENYL	2	0.5	3.4	1	375	---	-----
1288	C18H14	o-TERPHENYL	2	0.5	3.4	1	325	---	-----
1289	C18H14	p-TERPHENYL	2	0.5	3.4	1	405	---	-----
1290	C18H15P	TRIPHENYLPHOSPHINE	2	0.5	----	1	356	---	-----
1291	C18H15O4P	TRIPHENYL PHOSPHATE	2	0.5	----	1	428	---	-----
1292	C18H16N2	N,N'-DIPHENYL-p-PHENYLENEDIAMINE	2	0.5	----	1	450	---	-----
1293	C18H22	2,3-DIMETHYL-2,3-DIPHENYLBUTANE	2	0.5	5.2	2	282	---	-----
1294	C18H22O2	DICUMYL PEROXIDE	2	0.5	3.2	1	261	---	-----
1295	C18H30	n-DODECYLBENZENE	2	0.4	2.9	1	285	---	-----
1296	C18H30	HEXAETHYLBENZENE	2	0.4	2.9	---	-----	---	-----
1297	C18H32O2	LINOLEIC ACID	2	0.4	2.9	2	381	---	-----
1298	C18H34	1-OCTADECYNE	2	0.4	2.8	---	-----	---	-----
1299	C18H34O2	OLEIC ACID	2	0.4	2.9	1	372	1	685
1300	C18H34O4	DIBUTYL SEBACATE	1	0.4	5.1	1	352	1	689
1301	C18H34O4	DIHEXYL ADIPATE	2	0.4	3.0	1	376	---	-----
1302	C18H36	1-CYCOPENTYLTRIDECANE	2	0.4	2.7	---	-----	---	-----
1303	C18H36	1-CYCLOHEXYLDODECANE	2	0.4	2.7	---	-----	---	-----
1304	C18H36	1-OCTADECENE	2	0.4	2.7	2	295	1	482
1305	C18H36O2	STEARIC ACID	2	0.4	4.9	1	385	1	743
1306	C18H38	n-OCTADECANE	2	0.4	2.7	2	296	1	455
1307	C18H38O	DINONYL ETHER	2	0.4	2.7	2	314	---	-----
1308	C18H38O	1-OCTADECANOL	2	0.4	2.7	2	350	---	-----
1309	C18H38S	BUTYL-TETRADECYL-SULFIDE	---	----	----	---	-----	---	-----
1310	C18H38S	ETHYL-HEXADECYL-SULFIDE	---	----	----	---	-----	---	-----
1311	C18H38S	HEPTADECYL-METHYL-SULFIDE	---	----	----	---	-----	---	-----
1312	C18H38S	NONYL-SULFIDE	---	----	----	---	-----	---	-----
1313	C18H38S	PENTADECYL-PROPYL-SULFIDE	---	----	----	---	-----	---	-----
1314	C18H38S	1-OCTADECANETHIOL	---	----	----	---	-----	---	-----
1315	C18H38S2	NONYL-DISULFIDE	---	----	----	---	-----	---	-----
1316	C19H26	1-n-NONYLNAPHTHALENE	2	0.4	2.9	2	349	---	-----
1317	C19H32	n-TRIDECYLBENZENE	2	0.4	2.7	2	331	---	-----
1318	C19H36	1-NONADECYNE	2	0.4	2.6	---	-----	---	-----
1319	C19H36O2	METHYL OLEATE	2	0.4	7.8	2	350	---	-----
1320	C19H38	1-CYCLOPENTYLTETRADECANE	2	0.4	2.6	---	-----	---	-----
1321	C19H38	1-CYCLOHEXYLTRIDECANE	2	0.4	2.6	---	-----	---	-----
1322	C19H38	1-NONADECENE	2	0.4	5.4	2	314	2	458
1323	C19H38O2	NONADECANOIC ACID	2	0.4	4.9	2	415	2	710
1324	C19H40	n-NONADECANE	2	0.4	2.5	2	313	1	446
1325	C19H40O	1-NONADECANOL	2	0.4	2.6	---	-----	---	-----
1326	C19H40S	BUTYL-PENTADECYL-SULFIDE	---	----	----	---	-----	---	-----

Table 25-1 EXPLOSION LIMITS IN AIR, FLASH POINT, AND AUTOIGNITION TEMPERATURE - ORGANIC COMPOUNDS (continued)

				Lower Explosive Limit	Upper Explosive Limit	Flash Point		Autoignition Temperature	
NO	FORMULA	NAME	Code	LEL, vol %	UEL, vol %	Code	F	Code	F
1327	C19H40S	ETHYL-HEPTADECYL-SULFIDE	---	----	----	---	-----	---	-----
1328	C19H40S	HEXADECYL-PROPYL-SULFIDE	---	----	----	---	-----	---	-----
1329	C19H40S	METHYL-OCTADECYL-SULFIDE	---	----	----	---	-----	---	-----
1330	C19H40S	1-NONADECANETHIOL	---	----	----	---	-----	---	-----
1331	C20H16	TRIPHENYLETHYLENE	2	0.4	5.3	2	394	---	-----
1332	C20H28	1-n-DECYLNAPHTHALENE	2	0.4	2.7	2	365	---	-----
1333	C20H30O2	ABIETIC ACID	2	0.4	2.8	2	435	---	-----
1334	C20H31N	DEHYDROABIETYLAMINE	2	0.4	----	1	376	1	430
1335	C20H34	1-PHENYLTETRADECANE	2	0.4	2.6	---	-----	---	-----
1336	C20H38	1-EICOSYNE	2	0.4	2.5	---	-----	---	-----
1337	C20H40	1-CYCLOPENTYLPENTADECANE	2	0.4	2.4	---	-----	---	-----
1338	C20H40	1-CYCLOHEXYLTETRADECANE	2	0.4	2.4	---	-----	---	-----
1339	C20H40	1-EICOSENE	2	0.4	2.4	2	331	2	458
1340	C20H42	n-EICOSANE	2	0.4	2.4	2	336	2	395
1341	C20H42O	1-EICOSANOL	2	0.4	2.4	2	377	---	-----
1342	C20H42S	BUTYL-HEXADECYL-SULFIDE	---	----	----	---	-----	---	-----
1343	C20H42S	DECYL-SULFIDE	---	----	----	---	-----	---	-----
1344	C20H42S	ETHYL-OCTADECYL-SULFIDE	---	----	----	---	-----	---	-----
1345	C20H42S	HEPTADECYL-PROPYL-SULFIDE	---	----	----	---	-----	---	-----
1346	C20H42S	METHYL-NONADECYL-SULFIDE	---	----	----	---	-----	---	-----
1347	C20H42S	1-EICOSANETHIOL	---	----	----	---	-----	---	-----
1348	C20H42S2	DECYL-DISULFIDE	---	----	----	---	-----	---	-----
1349	C21H21O4P	TRI-o-CRESYL PHOSPHATE	2	0.4	----	1	437	1	725
1350	C21H36	1-PHENYLPENTADECANE	2	0.4	2.4	---	-----	---	-----
1351	C21H42	1-CYCLOPENTYLHEXADECANE	2	0.4	2.3	---	-----	---	-----
1352	C21H42	1-CYCLOHEXYLPENTADECANE	2	0.4	2.3	---	-----	---	-----
1353	C22H38	1-PHENYLHEXADECANE	2	0.4	2.3	---	-----	---	-----
1354	C22H44	1-CYCLOHEXYLHEXADECANE	2	0.3	2.2	---	-----	---	-----
1355	C22H44O2	n-BUTYL STEARATE	2	0.3	2.3	1	320	1	671
1356	C24H38O4	DIISOOCTYL PHTHALATE	2	0.3	5.3	2	448	---	-----
1357	C24H38O4	DIOCTYL PHTHALATE	1	0.3	5.3	1	421	1	736
1358	C24H42O	DINONYLPHENOL	2	0.3	2.1	2	487	---	-----
1359	C26H20	TETRAPHENYLETHYLENE	2	0.3	5.3	2	509	---	-----
1360	C28H46O4	DIISODECYL PHTHALATE	1	0.3	----	1	450	1	755

Code: 1 - experimental, 2 - estimate
LEL - Lower Explosive Limit
UEL - Upper Explosive Limit
Flash Point - Flash Point Temperature
F - Fahrenheit

Table 25-2 EXPLOSION LIMITS IN AIR, FLASH POINT, AND AUTOIGNITION TEMPERATURE - INORGANIC COMPOUNDS

NO	FORMULA	NAME	Code	Lower Explosive Limit LEL, vol %	Upper Explosive Limit UEL, vol %	Flash Point Code	Flash Point F	Autoignition Temperature Code	Autoignition Temperature F
1	B2H6	DIBORANE	1	0.9	98.0	1	-130	1	125
2	CH4N2O	UREA	2	5.6	35.3	---	-----	---	-----
3	CNCl	CYANOGEN CHLORIDE	2	----	23.5	---	-----	---	-----
4	CO	CARBON MONOXIDE	1	12.5	74.0	---	-----	1	1128
5	COS	CARBONYL SULFIDE	1	12.0	29.0	---	-----	---	-----
6	CS2	CARBON DISULFIDE	1	1.3	50.0	1	-22	1	194
7	D2	DEUTERIUM	1	5.0	75.0	---	-----	2	752
8	HCN	HYDROGEN CYANIDE	3	6.0	41.0	3	0	3	1000
9	H2	HYDROGEN	1	4.0	75.0	---	-----	1	752
10	H2S	HYDROGEN SULFIDE	1	4.3	45.0	---	-----	1	500
11	H3NO3S	SULFAMIC ACID	2	9.3	----	---	-----	---	-----
12	NH3	AMMONIA	1	16.0	25.0	---	-----	1	1204
13	NH3O	HYDROXYLAMINE	---	----	----	---	-----	---	264
14	N2H4	HYDRAZINE	1	4.7	99.9	1	100	1	518
15	P	PHOSPHORUS(WHITE)	---	----	----	---	-----	1	86
16	PH3	PHOSPHINE	---	----	----	---	-----	1	212
17	P4S10	PHOSPHORUS PENTASULFIDE	---	----	----	---	-----	1	287
18	S	SULFUR	1	2.0	----	1	405	1	450
19	SiHCl3	TRICHLOROSILANE	1	1.2	90.5	1	-18	1	220
20	SiH2Cl2	DICHLOROSILANE	1	4.0	96.0	---	-----	1	136

Code: 1 - experimental, 2 - estimate, 3 - 96 % hydrocyanic acid

LEL - Lower Explosive Limit

UEL - Upper Explosive Limit

Flash Point - Flash Point Temperature

F - Fahrenheit

Chapter 26

ENTHALPY OF COMBUSTION

Carl L. Yaws, Sachin D. Sheth, and Mei Han
Lamar University, Beaumont, Texas

ABSTRACT

Results for enthalpy of combustion are presented for major organic chemicals. The results are displayed in an easy-to-use table that is especially applicable for rapid engineering usage. The organic chemicals encompass hydrocarbon, oxygen, nitrogen, halogen, silicon, sulfur, and other type compounds.

ENTHALPY OF COMBUSTION

The results for enthalpy of combustion are presented in Table 26-1 for organic chemicals. The enthalpy of combustion is the net increase in heat content when a substance in its standard state at ambient conditions (77 F, 1 atm) undergoes complete oxidation.

The tabulated values are the negative of the enthalpy of combustion. A positive value as shown means that heat is released in the combustion. A negative value means that heat is required for the combustion. For substances in the table, the products of combustion are CO_2 (gas), H_2O (gas), F_2 (gas), Cl_2 (gas), Br_2 (gas), I_2 (gas), N_2 (gas), SO_2 (gas), H_3PO_4 (solid), and SiO_2 (crystobalite).

In the data collection, a literature search was conducted to identify data source publications (1-94) for the table. The publications were screened and copies of appropriate data were made. These data were then keyed into the computer to provide a database for use in preparing the table.

EXAMPLES

The tabulated values may be used in engineering applications involving combustion. Examples are given below.

Example 1 Combustion of propane (C_3H_8, 50 kg/h) occurs at ambient conditions (77 F, 1 atm). Estimate the quantity of heat released in the combustion.

Substitution of the tabulated value for propane into the equation below provides the quantity of heat released:

$$\Delta H = (-\Delta H_{combustion}) \, (mass) = (46,333 \text{ kjoule/kg}) \, (50 \text{ kg/h})$$

$$\Delta H = 2.32 \text{ million kjoule/h}$$

Example 2 Combustion of n-hexane (C_6H_{14}, 150 lb/h) occurs at ambient conditions (77 F, 1 atm). Estimate the quantity of heat released in the combustion.

Substitution of the tabulated value for n-hexane into the equation below provides the quantity of heat released:

$$\Delta H = (-\Delta H_{combustion}) \, (mass) = (19,236.4 \text{ BTU/lb}) \, (150 \text{ lb/h})$$

$$\Delta H = 2.89 \text{ million BTU/h}$$

REFERENCES – ORGANIC COMPOUNDS

1. API Research Project No. 44, SELECTED VALUES OF PHYSICAL AND THERMODYNAMIC PROPERTIES OF HYDROCARBONS AND RELATED COMPOUNDS, Carnegie Press, Carnegie Institute of Technology, Pittsburgh, PA (1953).
2. SELECTED VALUES OF PROPERTIES OF HYDROCARBONS AND RELATED COMPOUNDS, Thermodynamics Research Center, TAMU, College Station, TX (1977, 1984).
3. SELECTED VALUES OF PROPERTIES OF CHEMICAL COMPOUNDS, Thermodynamics Research Center, TAMU, College Station, TX (1977, 1987).
4. TECHNICAL DATA BOOK - PETROLEUM REFINING, Vols. I and II, American Petroleum Institute, Washington, DC (1972, 1977, 1982).
5. Daubert, T. E. and R. P. Danner, DATA COMPILATION OF PROPERTIES OF PURE COMPOUNDS, Parts 1, 2, 3 and 4, Supplements 1 and 2, DIPPR Project, AIChE, New York, NY (1985-1992).
6. Braker, W. and A. L. Mossman, MATHESON GAS DATA BOOK, 6th ed., Matheson Gas Products, Secaucus, NJ (1980).
7. CRC HANDBOOK OF CHEMISTRY AND PHYSICS, 75th - 78th eds., CRC Press, Inc., Boca Raton, FL (1994-1997).
8. PERRY'S CHEMICAL ENGINEERING HANDBOOK, 6th ed., McGraw-Hill, New York, NY (1984).
9. Lees, F. P., LOSS PREVENTION IN THE PROCESS INDUSTRIES, Vols. 1 and 2, Butterworth-Heinemann, London, England (1992).
10. CONDENSED CHEMICAL DICTIONARY, 10th and 11th eds., G. G. Hawley (10th) and Sax, N. I. and R. J. Lewis, Jr. (11th), Van Nostrand Reinhold Co., New York, NY (1981, 1987).
11. Sax, N. I., DANGEROUS PROPERTIES OF INDUSTRIAL MATERIALS, 6th ed., Van Nostrand Reinhold Co., New York, NY

(1984).

12. Driesbach, R. R., PHYSICAL PROPERTIES OF CHEMICAL COMPOUNDS, Vols. I (No. 15), II (No. 22), and III (No. 29), Advances in Chemistry Series, American Chemical Society, Washington, DC (1955,1959,1961).

13. Vargaftik, N. B., TABLES ON THE THERMOPHYSICAL PROPERTIES OF LIQUIDS AND GASES, 2nd ed., English translation, Hemisphere Publishing Corporation, New York, NY (1975, 1983).

14. Timmermans, J., PHYSICO-CHEMICAL CONSTANTS OF PURE ORGANIC COMPOUNDS, Vols. 1 and 2, Elsevier, New York, NY (1950,1965).

15. ENCYCLOPEDIA OF CHEMICAL TECHNOLOGY, 3rd and 4th eds., John Wiley and Sons, Inc., New York, NY (1978-1997).

16. Riddick, J. A. and W. B. Bunger, ORGANIC SOLVENTS: PHYSICAL PROPERTIES AND METHODS OF PURIFICATION, 3rd ed., Wiley Interscience, New York, NY (1970).

17. Cox, J. D. and G. Pilcher, THERMOCHEMISTRY OF ORGANIC AND ORGANOMETALLIC COMPOUNDS, Academic Press, New York, NY (1970).

18. American Institute of Chemical Engineers, FIRE AND EXPLOSION INDEX - HAZARD CLASSIFICATION GUIDE, 5th ed., New York, NY (1981).

19. Weiss, G., ed., HAZARDOUS CHEMICALS DATA BOOK, Noyes Data Corporation, Park Ridge, NJ (1980).

20. Pedley, J. B., R. D. Naylor, and S. P. Kirby, THERMOCHEMICAL DATA OF ORGANIC COMPUNDS, Chapman and Hall, London, England (1986).

21. Riddick, J. A., W. B. Bunger, and T. K. Sakano, ORGANIC SOLVENTS: PHYSICAL PROPERTIES AND METHODS OF PURIFICATION, 4th ed., Wiley Interscience, New York, NY (1986).

22. Stull, D. R., E. F. Westrum, and G. C. Sinke, THE CHEMICAL THERMODYNAMICS OF ORGANIC COMPOUNDS, John Wiley and Sons, New York, NY (1969).

23. Kudchadker, S. A., A. P. Kudchadker, R. C. Wilhoit, and B. J. Zwolinski, "KEY CHEMICALS DATA BOOKS - PHENOL, Thermodynamics Research Center, Texas Eng. Expt. Station, Texas A&M University, College Station, TX (1977).

24. Kudchadker, A. P., S. A. Kudchadker, and R. C. Wilhoit, KEY CHEMICALS DATA BOOKS - FURAN, DIHYDROFURAN AND TETRAHYDROFURAN, Thermodynamics Research Center, Texas Eng. Expt. Station, Texas A&M University, College Station, TX (1984).

25. Kudchadker, A. P., S. A. Kudchadker, and R. C. Wilhoit, KEY CHEMICALS DATA BOOKS - CRESOLS, Thermodynamics Research Center, Texas Eng. Expt. Station, Texas A&M University, College Station, TX (1978).

26. Kudchadker, A. P. and S. A. Kudchadker, KEY CHEMICALS DATA BOOKS - XYLENOLS, Thermodynamics Research Center, Texas Eng. Expt. Station, Texas A&M University, College Station, TX (1978).

27. Wilhoit, R. C. and B. J. Zwolinski, J. Phys. Chem. Ref. Data, 2 (1), (1973).

28. Domalski, E. S., J. Phys. Chem. Ref. Data, 1(2), 221 (1972).

29. U.S. Industrial Chemicals Co., VINYL ACETATE MONOMERS HANDBOOK, National Distillers and Chemical Corporation, New York, NY (1978).

30. Morrison, G. O. and T. P. G. Shaw, Trans. Electrochem. Soc., 63, 425 (1933).

31. Mansson, M., J. Chem. Thermo., 4, 865 (1972).

32. Kobe, K. A. and R. E. Pennington, Petroleum Refiner, 29, 135 (1950).

33. Delafontaine, J., R. Sabbah, and M. Lafitte, Zert fur Phys. Chem. Neise Folge, 84, 157 (1973).

34. American Society of Testing and Materials, ASTM Data Series Publication DS 51, Philadelphia, PA (1974).

35. Minadakis, C. and R. Sabbah, Thermochim. Acta, 55(2), 147 (1982).

36. Fenwick, J. O., D. Harrop, and A. J. Head, J. Chem. Thermo., 7, 1173 (1975).

37. Gas Processors Association, Publication No. 2145-82, Tulsa, OK (1982).

38. An, Xu-Wu and M. Mansson, J. Chem. Thermo., 15, 287 (1983).

39. Steele, W. V., J. Chem. Thermo., 11, 1185 (1979).

40. Vilcu, R. and S. Perisanu, Rev. Roum. Chim., 24 (1), 237 (1979).

41. Good, W. D., J. Chem. Eng. Data, 17, 28 (1972).

42. McCormick, D. G. and W. S. Hamilton, J. Chem. Thermo., 10, 275 (1978).

43. Hutchens, J. O., A. G. Cole, R. A. Robie, and J. W. Stout, J. Biol. Chem., 57, 359 (1953).

44. Bell, E. R., F. H. Dickey, J. H. Raley, F. F. Rust, and W. E. Vaughan, Ind. Eng. Chem., 41, 2597 (1949).

45. Tannerbaum, S., S. Kaye, and G. F. Lowenz, J. Amer. Chem. Soc., 75, 3753 (1953).

46. Mansson, M., N. Yoshiaki, and S. Sunner, Acta Chem. Scand., 22, 171 (1968).

47. Wilhoit, R. C. and I. Lei, J. Chem. Eng. Data, 10, 166 (1965).

48. Lebedeva, N. D., Russ. J. Phys. Chem., 38, 11 (1964).

49. Bruylarts, P. and A. Christiaen, Chem. Zentrablatt II, 538 (1925).

50. Roth, W. A. and K. Isecke, Chem. Ber., 77, 537 (1944).

51. Bartolo, H. F. and F. D. Rossini, J. Phys. Chem., 64, 1685 (1960).

52. Lebedeva, V. P. and others, Anal. Chem., 22, 871 (1950).

53. Steele, W. V., A. Chirico, T. A. Nguyen, and I. A. Hossenlopp, Topical Report NIPER-319, National Institute for Petroleum and Energy Research, Bartlesville, OK (Jan. 1988).

54. Wilhoit, R. C. and D. Shiao, J. Chem. Eng. Data, 9, 595 (1964).

55. Mukaiyama, T., Bull. Chem. Soc. Japan, 28, 253 (1955).

56. Sinke, G. C. and D. R. Stull, J. Phys. Chem., 62, 397 (1958).

57. Parks, G. S., J. R. Mosley, and P. V. Peterson, J. Chem. Phys., 18, 152 (1950).

58. Lebedeva, N. D., Y. A. Katin, and G. Y. Akhmedova, Russ. J. Phys. Chem., 45, 771 (1971).

59. Fenwick, J. O., D. Harrop, and A. J. Head, J. Chem. Thermo., 7, 943 (1975).

60. Furukawa, J., M. Sakujama, S. Seki, Y. Saito, and Kusano, Bull. Chem. Soc. Japan, 3329 (1982).

61. Petit, M., Am. Chem. Phys., 18(6), 145 (1889).

62. Ethyl Corporation, TRIETHYLALUMINUM, Baton Rouge, LA (1982).

63. Garner, W. E. and C. L. Abernethy, Proc. Roy. Soc. London, A99, 213 (1921).

64. Bedford, A. F. and C. T. Mortimer, J. Chem. Soc., 163, 1622 (1960).

65. Tanaka, T., J. Chem. Phys., 22, 957 (1954).

66. Osthoff, R. L., W. J. Grubb, and C. A. Burkhard, J. Amer. Chem. Soc., 75, 227 (1958).

67. Baker, G., J. H. Littlefair, R. Shaw, and J. C. J. Thynne, J. Chem. Soc., 6970 (1965).

68. Lenchnitz, C., R. W. Velicky, G. Silvestro, and L. P. Schlosberg, J. Chem. Thermo., 3, 689 (1971).

69. Good, W. D., J. Chem. Thermo., 3, 97 (1971).

70. American Petroleum Institute, API Publication No. 705, Washington, DC (October, 1978).

71. American Petroleum Institute, API Publication No. 706, Washington, DC (October, 1978).

72. American Petroleum Institute, API Publication No. 707, Washington, DC (October, 1978).

73. American Petroleum Institute, API Publication No. 708, Washington, DC (January, 1979).

74. American Petroleum Institute, API Publication No. 709, Washington, DC (March, 1979).

75. American Petroleum Institute, API Publication No. 710, Washington, DC (1979).

76. American Petroleum Institute, API Publication No. 714, Washington, DC (April, 1980).

77. American Petroleum Institute, API Publication No. 715, Washington, DC (January, 1981).

78. American Petroleum Institute, API Publication No. 717, Washington, DC (November, 1981).

79. American Petroleum Institute, API Publication No. 718, Washington, DC (Jan. 1982).

80. American Petroleum Institute, API Publication No. 719, Washington, DC (April, 1982).

81. American Petroleum Institute, API Publication No. 720, Washington, DC (Jan. 1983).

82. American Petroleum Institute, API Publication No. 722, Washington, DC (Sept. 1984).

83. American Petroleum Institute, API Publication No. 723, Washington, DC (1984).

84. American Petroleum Institute, API Publication No. 724, Washington, DC (February, 1985).

85. Serijan, K. T. and P. H. Wise, J. Amer. Chem. Soc., 73, 4766 (1951).

86. Hickman, K. and W. Weyerts, J. Amer. Chem. Soc., 52, 4714 (1930).

87. Hipsher, H. F. and P. H. Wise, J. Amer. Chem. Soc., 76, 1747 (1954).

88. Good, W. D., J. Chem. Eng. Data, 14 (2), 231 (1969).

89. Labbauf, A., J. B. Greenshields, and F. D. Rossini, J. Chem. Eng. Data, 6, 261 (1961).

90. Good, W. D., J. Chem. Thermo., 5, 715 (1973).

91. Kirklin, D. R. and E. S. Domalski, J. Chem. Thermo., 20, 743 (1988).

92. Mortimer, C. T., Pure Appl. Chem., 2, 71 (1961).

93. Tavernier, P. and M. Lamouroux, Mem. Pondres, 37, 197 (1955).

94. Yaws, C. L., HANDBOOK OF CHEMICAL COMPOUND DATA FOR PROCESS SAFETY, Gulf Publishing Co., Houston, TX (1997).

Table 26-1 ENTHALPY OF COMBUSTION - ORGANIC COMPOUNDS

NO	FORMULA	NAME	State	- Enthalpy of Combustion @ 77 F		
				kjoule/mol	kjoule/kg	BTU/lb
1	CBrClF2	BROMOCHLORODIFLUOROMETHANE	gas	-53.3	-322.3	-138.6
2	CBrCl3	BROMOTRICHLOROMETHANE	liquid	303.0	1528.2	657.1
3	CBrF3	BROMOTRIFLUOROMETHANE	gas	-270.9	-1819.0	-782.2
4	CBr2F2	DIBROMODIFLUOROMETHANE	gas	-34.4	-164.0	-70.5
5	CClF3	CHLOROTRIFLUOROMETHANE	gas	-314.4	-3010.0	-1294.3
6	CClN	CYANOGEN CHLORIDE	gas	531.0	8638.4	3714.5
7	CCl2F2	DICHLORODIFLUOROMETHANE	gas	-98.1	-811.4	-348.9
8	CCl2O	PHOSGENE	gas	174.9	1768.1	760.3
9	CCl3F	TRICHLOROFLUOROMETHANE	gas	104.8	763.1	328.1
10	CCl4	CARBON TETRACHLORIDE	liquid	258.1	1677.7	721.4
11	CF2O	CARBONYL FLUORIDE	gas	-245.0	-3711.7	-1596.0
12	CF4	CARBON TETRAFLUORIDE	gas	-539.7	-6132.5	-2637.0
13	CHBr3	TRIBROMOMETHANE	liquid	439.7	1740.0	748.2
14	CHClF2	CHLORODIFLUOROMETHANE	gas	65.7	759.9	326.7
15	CHCl2F	DICHLOROFLUOROMETHANE	gas	231.2	2246.0	965.8
16	CHCl3	CHLOROFORM	liquid	380.0	3182.8	1368.6
17	CHF3	TRIFLUOROMETHANE	gas	-178.9	-2554.5	-1098.4
18	CHI3	TRIIODOMETHANE	solid	-------	-------	-------
19	CHN	HYDROGEN CYANIDE	liquid	623.3	23062.6	9916.9
20	CHNS	ISOTHIOCYANIC-ACID	gas	940.1	15911.1	6841.8
21	CH2BrCl	BROMOCHLOROMETHANE	liquid	544.0	4204.5	1808.0
22	CH2Br2	DIBROMOMETHANE	liquid	-------	-------	-------
23	CH2ClF	CHLOROFLUOROMETHANE	gas	371.5	5424.6	2332.6
24	CH2Cl2	DICHLOROMETHANE	liquid	513.9	6050.5	2601.7
25	CH2F2	DIFLUOROMETHANE	gas	182.6	3510.5	1509.5
26	CH2I2	DIIODOMETHANE	liquid	639.7	2388.5	1027.1
27	CH2O	FORMALDEHYDE	gas	519.4	17299.7	7438.9
28	CH2O2	FORMIC ACID	liquid	211.5	4594.4	1975.6
29	CH3Br	METHYL BROMIDE	gas	705.4	7430.2	3195.0
30	CH3Cl	METHYL CHLORIDE	gas	675.4	13377.0	5752.1
31	CH3Cl3Si	METHYL TRICHLOROSILANE	liquid	1064.0	7118.1	3060.8
32	CH3F	METHYL FLUORIDE	gas	521.9	15336.3	6594.6
33	CH3I	METHYL IODIDE	liquid	709.6	4999.4	2149.7
34	CH3NO	FORMAMIDE	liquid	506.0	11235.1	4831.1
35	CH3NO2	NITROMETHANE	liquid	643.2	10536.7	4530.8
36	CH3NO2	METHYL-NITRITE	gas	693.1	11354.4	4882.4
37	CH3NO3	METHYL-NITRATE	liquid	636.6	8263.1	3553.1
38	CH4	METHANE	gas	802.3	50009.3	21504.0
39	CH4Cl2Si	METHYL DICHLOROSILANE	liquid	1357.0	11796.5	5072.5
40	CH4O	METHANOL	liquid	638.1	19914.5	8563.2
41	CH4O3S	METHANESULFONIC ACID	liquid	591.0	6149.4	2644.2
42	CH4S	METHYL MERCAPTAN	gas	1151.7	23939.4	10293.9
43	CH5ClSi	METHYL CHLOROSILANE	gas	1693.0	21007.8	9033.4
44	CH5N	METHYLAMINE	gas	975.1	31396.5	13500.5
45	CH6Si	METHYL SILANE	gas	1999.0	43320.9	18628.0
46	CN4O8	TETRANITROMETHANE	liquid	431.8	2202.7	947.2
47	CO	CARBON MONOXIDE	gas	283.0	10103.9	4344.7
48	COS	CARBONYL SULFIDE	gas	548.3	9126.3	3924.3
49	CO2	CARBON DIOXIDE	gas	0.0	0.0	0.0
50	CS2	CARBON DISULFIDE	liquid	1104.2	14501.7	6235.7
51	C2BrF3	BROMOTRIFLUOROETHYLENE	gas	316.6	1967.2	845.9
52	C2Br2F4	1,2-DIBROMOTETRAFLUOROETHANE	liquid	194.9	750.1	322.5
53	C2ClF3	CHLOROTRIFLUOROETHYLENE	gas	214.0	1837.4	790.1
54	C2ClF5	CHLOROPENTAFLUOROETHANE	gas	-322.0	-2084.6	-896.4
55	C2Cl2F4	1,2-DICHLOROTETRAFLUOROETHANE	gas	-100.4	-587.4	-252.6
56	C2Cl3F3	1,1,2-TRICHLOROTRIFLUOROETHANE	liquid	64.8	346.0	148.8
57	C2Cl4	TETRACHLOROETHYLENE	liquid	735.5	4435.0	1907.1
58	C2Cl4F2	1,1,2,2-TETRACHLORODIFLUOROETHANE	solid	51.6	253.2	108.9
59	C2Cl4O	TRICHLOROACETYL CHLORIDE	liquid	445.1	2448.0	1052.7
60	C2Cl6	HEXACHLOROETHANE	solid	580.3	2451.3	1054.1
61	C2F4	TETRAFLUOROETHYLENE	gas	128.0	1279.8	550.3
62	C2F6	HEXAFLUOROETHANE	gas	-557.0	-4035.9	-1735.4
63	C2HBrClF3	HALOTHANE	liquid	158.5	803.0	345.3
64	C2HClF2	2-CHLORO-1,1-DIFLUOROETHYLENE	gas	579.0	5879.4	2528.2
65	C2HCl3	TRICHLOROETHYLENE	liquid	864.1	6576.8	2828.0
66	C2HCl3O	DICHLOROACETYL CHLORIDE	liquid	627.1	4254.8	1829.6
67	C2HCl3O	TRICHLOROACETALDEHYDE	liquid	672.0	4559.4	1960.6
68	C2HCl5	PENTACHLOROETHANE	liquid	720.0	3559.4	1530.5
69	C2HF3	TRIFLUOROETHENE	gas	417.4	5089.0	2188.3
70	C2HF3O2	TRIFLUOROACETIC ACID	liquid	-152.0	-1333.1	-573.2
71	C2HF5	PENTAFLUOROETHANE	gas	-196.6	-1638.0	-704.4
72	C2H2	ACETYLENE	gas	1255.6	48221.8	20735.4
73	C2H2Br4	1,1,2,2-TETRABROMOETHANE	liquid	911.0	2635.6	1133.3
74	C2H2Cl2	1,1-DICHLOROETHYLENE	liquid	1003.7	10353.5	4452.0
75	C2H2Cl2	cis-1,2-DICHLOROETHYLENE	liquid	994.5	10258.5	4411.2
76	C2H2Cl2	trans-1,2-DICHLOROETHYLENE	liquid	999.1	10306.5	4431.8
77	C2H2Cl2O	CHLOROACETYL CHLORIDE	liquid	744.3	6590.4	2833.9
78	C2H2Cl2O	DICHLOROACETALDEHYDE	liquid	790.0	6994.7	3007.7

585

Table 26-1 ENTHALPY OF COMBUSTION - ORGANIC COMPOUNDS (continued)

NO	FORMULA	NAME	State	Enthalpy of Combustion @ 77 F		
				kjoule/mol	kjoule/kg	BTU/lb
79	C2H2Cl2O2	DICHLOROACETIC ACID	liquid	530.9	4117.4	1770.5
80	C2H2Cl3F	1,1,1-TRICHLOROFLUOROETHANE	liquid	692.0	4570.9	1965.5
81	C2H2Cl4	1,1,1,2-TETRACHLOROETHANE	liquid	837.8	4991.6	2146.4
82	C2H2Cl4	1,1,2,2-TETRACHLOROETHANE	liquid	834.6	4972.6	2138.2
83	C2H2F2	1,1-DIFLUOROETHYLENE	gas	692.0	10806.6	4646.8
84	C2H2F2	cis-1,2-DIFLUOROETHENE	gas	707.2	11044.5	4749.1
85	C2H2F2	trans-1,2-DIFLUOROETHENE	gas	707.2	11044.5	4749.1
86	C2H2F4	1,1,1,2-TETRAFLUOROETHANE	gas	133.1	1304.0	560.7
87	C2H2O	KETENE	gas	967.8	23021.6	9899.3
88	C2H2O4	OXALIC ACID	solid	199.0	2210.2	950.4
89	C2H3Br	VINYL BROMIDE	gas	1212.7	11338.9	4875.7
90	C2H3Cl	VINYL CHLORIDE	gas	1158.0	18528.3	7967.2
91	C2H3ClF2	1-CHLORO-1,1-DIFLUOROETHANE	gas	663.0	6597.3	2836.9
92	C2H3ClO	ACETYL CHLORIDE	liquid	876.9	11170.3	4803.2
93	C2H3ClO	CHLOROACETALDEHYDE	liquid	921.0	11732.8	5045.1
94	C2H3ClO2	CHLOROACETIC ACID	solid	663.0	7016.1	3016.9
95	C2H3ClO2	METHYL CHLOROFORMATE	liquid	689.0	7291.2	3135.2
96	C2H3Cl3	1,1,1-TRICHLOROETHANE	liquid	975.0	7308.3	3142.5
97	C2H3Cl3	1,1,2-TRICHLOROETHANE	liquid	963.9	7225.0	3106.8
98	C2H3F	VINYL FLUORIDE	gas	1011.7	21972.2	9448.0
99	C2H3F3	1,1,1-TRIFLUOROETHANE	gas	404.0	4807.2	2067.1
100	C2H3N	ACETONITRILE	liquid	1190.4	28996.7	12468.6
101	C2H3NO	METHYL ISOCYANATE	liquid	1064.5	18658.4	8023.1
102	C2H4	ETHYLENE	gas	1322.6	47144.8	20272.3
103	C2H4Br2	1,1-DIBROMOETHANE	liquid	1160.0	6174.7	2655.1
104	C2H4Br2	1,2-DIBROMOETHANE	liquid	1176.9	6264.7	2693.8
105	C2H4Cl2	1,1-DICHLOROETHANE	liquid	1109.7	11213.7	4821.9
106	C2H4Cl2	1,2-DICHLOROETHANE	liquid	1105.0	11166.2	4801.5
107	C2H4Cl2O	BIS(CHLOROMETHYL)ETHER	liquid	990.0	8611.8	3703.1
108	C2H4F2	1,1-DIFLUOROETHANE	gas	758.5	11483.6	4937.9
109	C2H4F2	1,2-DIFLUOROETHANE	liquid	813.0	12308.7	5292.7
110	C2H4I2	1,2-DIIODOETHANE	solid	-------	-------	-------
111	C2H4O	ACETALDEHYDE	gas	1104.6	25074.3	10782.0
112	C2H4O	ETHYLENE OXIDE	gas	1218.0	27648.5	11888.9
113	C2H4OS	THIOACETIC-ACID	liquid	1353.2	17778.5	7644.8
114	C2H4O2	ACETIC ACID	liquid	786.4	13095.6	5631.1
115	C2H4O2	METHYL FORMATE	liquid	920.9	15334.8	6594.0
116	C2H4S	THIACYCLOPROPANE	liquid	1622.6	26992.2	11606.7
117	C2H5Br	BROMOETHANE	liquid	1284.4	11787.2	5068.5
118	C2H5Cl	ETHYL CHLORIDE	gas	1284.9	19916.6	8564.1
119	C2H5ClO	2-CHLOROETHANOL	liquid	1080.0	13413.8	5767.9
120	C2H5F	ETHYL FLUORIDE	gas	1110.0	23096.1	9931.3
121	C2H5I	ETHYL IODIDE	liquid	1356.1	8694.8	3738.8
122	C2H5N	ETHYLENEIMINE	liquid	1481.0	34387.5	14786.6
123	C2H5NO	ACETAMIDE	solid	1074.1	18184.1	7819.2
124	C2H5NO	N-METHYLFORMAMIDE	liquid	1149.6	19462.3	8368.8
125	C2H5NO2	NITROETHANE	liquid	1249.8	16649.1	7159.1
126	C2H5NO3	ETHYL-NITRATE	liquid	1239.0	13605.7	5850.4
127	C2H6	ETHANE	gas	1428.6	47509.1	20428.9
128	C2H6AlCl	DIMETHYLALUMINUM CHLORIDE	liquid	-------	-------	-------
129	C2H6O	DIMETHYL ETHER	gas	1328.4	28835.0	12399.1
130	C2H6O	ETHANOL	liquid	1235.5	26818.5	11531.9
131	C2H6OS	DIMETHYL SULFOXIDE	liquid	1547.3	19802.9	8515.2
132	C2H6O2	ETHYLENE GLYCOL	liquid	1057.6	17039.4	7326.9
133	C2H6O4S	DIMETHYL SULFATE	liquid	1070.0	8483.1	3647.7
134	C2H6S	DIMETHYL SULFIDE	liquid	1744.0	28067.5	12069.0
135	C2H6S	ETHYL MERCAPTAN	liquid	1735.7	27933.9	12011.6
136	C2H6S2	DIMETHYL DISULFIDE	liquid	2043.8	21695.9	9329.2
137	C2H7N	DIMETHYLAMINE	gas	1614.6	35813.1	15399.7
138	C2H7N	ETHYLAMINE	gas	1587.4	35209.8	15140.2
139	C2H7NO	MONOETHANOLAMINE	liquid	1363.1	22315.2	9595.5
140	C2H8N2	ETHYLENEDIAMINE	liquid	1691.0	28136.9	12098.9
141	C2H8Si	DIMETHYL SILANE	gas	2569.0	42695.0	18358.8
142	C2N2	CYANOGEN	gas	1096.1	21064.3	9057.6
143	C3F6	HEXAFLUOROPROPYLENE	gas	100.0	666.6	286.6
144	C3F6O	HEXAFLUOROACETONE	gas	-279.0	-1680.5	-722.6
145	C3F8	OCTAFLUOROPROPANE	gas	-523.0	-2781.6	-1196.1
146	C3H2N2	MALONONITRILE	solid	1609.0	24355.9	10473.0
147	C3H3Cl	PROPARGYL CHLORIDE	liquid	1670.0	22413.1	9637.6
148	C3H3N	ACRYLONITRILE	liquid	1690.0	31848.3	13694.8
149	C3H3NO	OXAZOLE	liquid	1495.2	21649.8	9309.4
150	C3H4	METHYLACETYLENE	gas	1849.6	46165.0	19850.9
151	C3H4	PROPADIENE	gas	1856.3	46332.2	19922.9
152	C3H4Cl2	2,3-DICHLOROPROPENE	liquid	1580.0	14238.1	6122.4
153	C3H4O	ACROLEIN	liquid	1553.6	27711.2	11915.8
154	C3H4O	PROPARGYL ALCOHOL	liquid	1656.9	29553.7	12708.1
155	C3H4O2	ACRYLIC ACID	liquid	1280.2	17764.8	7638.8
156	C3H4O2	beta-PROPIOLACTONE	liquid	1329.0	18441.9	7930.0

Table 26-1 ENTHALPY OF COMBUSTION - ORGANIC COMPOUNDS (continued)

NO	FORMULA	NAME	State	kjoule/mol	kjoule/kg	BTU/lb
				- Enthalpy of Combustion @ 77 F		
157	C3H4O2	VINYL FORMATE	liquid	1370.0	19010.9	8174.7
158	C3H4O3	ETHYLENE CARBONATE	solid	1083.0	12298.0	5288.1
159	C3H4O3	PYRUVIC ACID	liquid	987.0	11207.9	4819.4
160	C3H5Br	3-BROMO-1-PROPENE	liquid	-------	-------	-------
161	C3H5Cl	2-CHLOROPROPENE	gas	1760.0	22999.0	9889.6
162	C3H5Cl	3-CHLOROPROPENE	liquid	1760.0	22999.0	9889.6
163	C3H5ClO	alpha-EPICHLOROHYDRIN	liquid	1660.5	17946.5	7717.0
164	C3H5ClO2	METHYL CHLOROACETATE	liquid	1290.0	11886.8	5111.3
165	C3H5ClO2	ETHYL CHLOROFORMATE	liquid	1280.0	11794.6	5071.7
166	C3H5Cl3	1,2,3-TRICHLOROPROPANE	liquid	1553.6	10537.8	4531.3
167	C3H5I	3-IODO-1-PROPENE	liquid	-------	-------	-------
168	C3H5N	PROPIONITRILE	liquid	1800.7	32693.0	14058.0
169	C3H5NO	ACRYLAMIDE	solid	1620.0	22791.5	9800.4
170	C3H5NO	HYDRACRYLONITRILE	liquid	1628.8	22915.3	9853.6
171	C3H5NO	LACTONITRILE	liquid	1650.0	23213.6	9981.9
172	C3H5N3O9	NITROGLYCERINE	liquid	1422.0	6261.9	2692.6
173	C3H6	CYCLOPROPANE	gas	1959.3	46560.2	20020.9
174	C3H6	PROPYLENE	gas	1925.7	45761.7	19677.6
175	C3H6Br2	1,2-DIBROMOPROPANE	liquid	-------	-------	-------
176	C3H6Cl2	1,1-DICHLOROPROPANE	liquid	1720.0	15223.1	6545.9
177	C3H6Cl2	1,2-DICHLOROPROPANE	liquid	1704.6	15086.8	6487.3
178	C3H6Cl2	1,3-DICHLOROPROPANE	liquid	1744.0	15435.4	6637.2
179	C3H6Cl2	2,2-DICHLOROPROPANE	liquid	1702.7	15069.7	6480.0
180	C3H6I2	1,2-DIIODOPROPANE	liquid	-------	-------	-------
181	C3H6O	ACETONE	liquid	1659.2	28567.5	12284.0
182	C3H6O	ALLYL ALCOHOL	liquid	1731.9	29819.2	12822.3
183	C3H6O	METHYL VINYL ETHER	gas	1774.3	30549.2	13136.2
184	C3H6O	n-PROPIONALDEHYDE	liquid	1685.7	29023.8	12480.2
185	C3H6O	1,2-PROPYLENE OXIDE	liquid	1785.3	30738.6	13217.6
186	C3H6O	1,3-PROPYLENE OXIDE	liquid	1801.1	31010.7	13334.6
187	C3H6O2	ETHYL FORMATE	liquid	1507.0	20343.1	8747.6
188	C3H6O2	METHYL ACETATE	liquid	1461.0	19722.2	8480.5
189	C3H6O2	PROPIONIC ACID	liquid	1395.0	18831.2	8097.4
190	C3H6O2S	3-MERCAPTOPROPIONIC ACID	liquid	1734.5	16340.9	7026.6
191	C3H6O3	LACTIC ACID	liquid	1235.0	13710.2	5895.4
192	C3H6O3	METHOXYACETIC ACID	liquid	1270.0	14098.7	6062.5
193	C3H6O3	TRIOXANE	solid	1383.8	15362.1	6605.7
194	C3H6S	THIACYCLOBUTANE	liquid	2234.4	30137.7	12959.2
195	C3H7Br	1-BROMOPROPANE	liquid	1891.2	15376.5	6611.9
196	C3H7Br	2-BROMOPROPANE	liquid	1886.1	15335.0	6594.1
197	C3H7Cl	ISOPROPYL CHLORIDE	liquid	1854.8	23615.7	10154.7
198	C3H7Cl	n-PROPYL CHLORIDE	liquid	1864.6	23740.5	10208.4
199	C3H7F	1-FLUOROPROPANE	gas	1747.7	28149.5	12104.3
200	C3H7F	2-FLUOROPROPANE	gas	1740.2	28028.2	12052.1
201	C3H7I	ISOPROPYL IODIDE	liquid	1919.7	11292.8	4855.9
202	C3H7I	n-PROPYL IODIDE	liquid	1929.8	11352.2	4881.5
203	C3H7N	ALLYAMINE	liquid	2050.0	35905.1	15439.2
204	C3H7N	PROPYLENEIMINE	liquid	2080.0	36430.5	15665.1
205	C3H7NO	N,N-DIMETHYLFORMAMIDE	liquid	1788.7	24470.9	10522.5
206	C3H7NO	N-METHYLACETAMIDE	solid	1710.0	23394.2	10059.5
207	C3H7NO2	1-NITROPROPANE	liquid	1858.9	20864.5	8971.7
208	C3H7NO2	2-NITROPROPANE	liquid	1845.6	20715.2	8907.5
209	C3H7NO3	PROPYL-NITRATE	liquid	1854.8	17649.4	7589.2
210	C3H7NO3	ISOPROPYL-NITRATE	liquid	1837.9	17488.1	7519.9
211	C3H8	PROPANE	gas	2043.1	46333.0	19923.2
212	C3H8O	ISOPROPANOL	liquid	1830.0	30451.3	13094.0
213	C3H8O	METHYL ETHYL ETHER	gas	1931.4	32138.6	13819.6
214	C3H8O	n-PROPANOL	liquid	1843.8	30680.9	13192.8
215	C3H8O2	2-METHOXYETHANOL	liquid	1670.0	21946.3	9436.9
216	C3H8O2	METHYLAL	liquid	1799.8	23652.0	10170.4
217	C3H8O2	1,2-PROPYLENE GLYCOL	liquid	1647.6	21651.9	9310.3
218	C3H8O2	1,3-PROPYLENE GLYCOL	liquid	1683.1	22118.4	9510.9
219	C3H8O3	GLYCEROL	liquid	1477.0	16037.8	6896.2
220	C3H8S	n-PROPYLMERCAPTAN	liquid	2345.0	30789.2	13239.4
221	C3H8S	ISOPROPYL MERCAPTAN	liquid	2339.8	30721.0	13210.0
222	C3H8S	ETHYL-METHYL-SULFIDE	liquid	2358.8	30972.8	13318.3
223	C3H9N	n-PROPYLAMINE	liquid	2164.8	36622.6	15747.7
224	C3H9N	ISOPROPYLAMINE	liquid	2156.6	36483.9	15688.1
225	C3H9N	TRIMETHYLAMINE	gas	2244.9	37977.7	16330.4
226	C3H9NO	1-AMINO-2-PROPANOL	liquid	1970.0	26227.8	11278.0
227	C3H9NO	3-AMINO-1-PROPANOL	liquid	1980.0	26361.0	11335.2
228	C3H9NO	METHYLETHANOLAMINE	liquid	2010.0	26760.4	11507.0
229	C3H9O4P	TRIMETHYL PHOSPHATE	liquid	2200.0	15705.8	6753.5
230	C3H10N2	1,2-PROPANEDIAMINE	liquid	2290.0	30893.3	13284.1
231	C3H10Si	TRIMETHYL SILANE	gas	3142.5	42352.9	18211.7
232	C4Cl4S	TETRACHLOROTHIOPHENE	solid	-------	-------	-------
233	C4Cl6	HEXACHLORO-1,3-BUTADIENE	liquid	1477.4	5665.7	2436.3
234	C4F8	OCTAFLUORO-2-BUTENE	gas	-75.9	-379.4	-163.2

587

Table 26-1 ENTHALPY OF COMBUSTION - ORGANIC COMPOUNDS (continued)

NO	FORMULA	NAME	State	- Enthalpy of Combustion @ 77 F		
				kjoule/mol	kjoule/kg	BTU/lb
235	C4F8	OCTAFLUOROCYCLOBUTANE	gas	46.1	230.5	99.1
236	C4F10	DECAFLUOROBUTANE	gas	-570.0	-2394.7	-1029.7
237	C4H2	BUTADIYNE(BIACETYLENE)	gas	2289.2	45729.2	19663.5
238	C4H2O3	MALEIC ANHYDRIDE	solid	1389.5	14170.2	6093.2
239	C4H4	VINYLACETYLENE	gas	2362.0	45356.8	19503.4
240	C4H4N2	SUCCINONITRILE	solid	2197.4	27437.0	11797.9
241	C4H4O	FURAN	liquid	1995.9	29319.1	12607.2
242	C4H4O2	DIKETENE	liquid	1824.0	21694.9	9328.8
243	C4H4O3	SUCCINIC ANHYDRIDE	solid	1460.0	14589.2	6273.4
244	C4H4O4	FUMARIC ACID	solid	1247.0	10743.2	4619.6
245	C4H4O4	MALEIC ACID	solid	1268.4	10927.6	4698.9
246	C4H4S	THIOPHENE	liquid	2435.2	28941.6	12444.9
247	C4H5Cl	CHLOROPRENE	liquid	2222.1	25098.3	10792.3
248	C4H5N	trans-CROTONITRILE	liquid	2279.0	33969.3	14606.8
249	C4H5N	cis-CROTONITRILE	liquid	2283.0	34028.9	14632.4
250	C4H5N	METHACRYLONITRILE	liquid	2243.0	33432.7	14376.1
251	C4H5N	PYRROLE	liquid	2241.8	33414.8	14368.4
252	C4H5N	VINYLACETONITRILE	liquid	2296.0	34222.7	14715.8
253	C4H5NO2	METHYL CYANOACETATE	liquid	1881.0	18982.9	8162.7
254	C4H6	CYCLOBUTENE	gas	2430.9	44940.8	19324.6
255	C4H6	1,2-BUTADIENE	gas	2461.7	45509.5	19569.1
256	C4H6	1,3-BUTADIENE	gas	2409.7	44548.2	19155.7
257	C4H6	DIMETHYLACETYLENE	liquid	2418.9	44718.3	19228.9
258	C4H6	ETHYLACETYLENE	gas	2464.7	45565.0	19592.9
259	C4H6Cl2	1,3-DICHLORO-trans-2-BUTENE	liquid	2180.0	17440.4	7499.4
260	C4H6Cl2	1,4-DICHLORO-cis-2-BUTENE	liquid	2332.0	18656.4	8022.3
261	C4H6Cl2	1,4-DICHLORO-trans-2-BUTENE	liquid	2187.0	17496.4	7523.5
262	C4H6Cl2	3,4-DICHLORO-1-BUTENE	liquid	2200.0	17600.4	7568.2
263	C4H6O	trans-CROTONALDEHYDE	liquid	2155.5	30752.9	13223.7
264	C4H6O	2,5-DIHYDROFURAN	liquid	2160.0	30817.1	13251.3
265	C4H6O	DIVINYL ETHER	liquid	2260.0	32243.8	13864.8
266	C4H6O	METHACROLEIN	liquid	2150.0	30674.4	13190.0
267	C4H6O2	2-BUTYNE-1,4-DIOL	solid	2050.0	23812.3	10239.3
268	C4H6O2	gamma-BUTYROLACTONE	liquid	1867.9	21697.1	9329.7
269	C4H6O2	cis-CROTONIC ACID	liquid	1952.0	22673.9	9749.8
270	C4H6O2	trans-CROTONIC ACID	solid	1884.3	21887.6	9411.7
271	C4H6O2	METHACRYLIC ACID	liquid	1931.6	22437.0	9647.9
272	C4H6O2	METHYL ACRYLATE	liquid	1937.4	22504.4	9676.9
273	C4H6O2	VINYL ACETATE	liquid	1949.0	22639.1	9734.8
274	C4H6O3	ACETIC ANHYDRIDE	liquid	1675.4	16411.0	7056.7
275	C4H6O4	SUCCINIC ACID	solid	1359.1	11509.1	4948.9
276	C4H6O5	DIGLYCOLIC ACID	solid	1354.8	10103.7	4344.6
277	C4H6O5	MALIC ACID	solid	1190.0	8874.7	3816.1
278	C4H6O6	TARTARIC ACID	solid	1009.0	6722.7	2890.8
279	C4H7N	n-BUTYRONITRILE	liquid	2414.8	34943.4	15025.7
280	C4H7N	ISOBUTYRONITRILE	liquid	2408.4	34850.8	14985.8
281	C4H7NO	ACETONE CYANOHYDRIN	liquid	2239.1	26309.5	11313.1
282	C4H7NO	2-METHACRYLAMIDE	solid	1850.0	21737.6	9347.2
283	C4H7NO	3-METHOXYPROPIONITRILE	liquid	2290.0	26907.6	11570.3
284	C4H7NO	2-PYRROLIDONE	liquid	2156.9	25343.7	10897.8
285	C4H8	1-BUTENE	gas	2541.2	45292.0	19475.6
286	C4H8	cis-2-BUTENE	gas	2534.4	45170.8	19423.5
287	C4H8	trans-2-BUTENE	gas	2528.7	45069.2	19379.8
288	C4H8	CYCLOBUTANE	gas	2567.8	45766.1	19679.4
289	C4H8	ISOBUTENE	gas	2524.0	44985.5	19343.8
290	C4H8Br2	1,2-DIBROMOBUTANE	liquid	-------	-------	-------
291	C4H8Br2	2,3-DIBROMOBUTANE	liquid	-------	-------	-------
292	C4H8Cl2	1,4-DICHLOROBUTANE	liquid	2320.0	18265.8	7854.3
293	C4H8I2	1,2-DIIODOBUTANE	liquid	-------	-------	-------
294	C4H8O	n-BUTYRALDEHYDE	liquid	2303.5	31945.6	13736.6
295	C4H8O	ISOBUTYRALDEHYDE	liquid	2291.3	31776.4	13663.8
296	C4H8O	1,2-EPOXYBUTANE	liquid	2400.1	33285.3	14312.7
297	C4H8O	METHYL ETHYL KETONE	liquid	2261.6	31364.5	13486.7
298	C4H8O	ETHYL VINYL ETHER	liquid	2372.9	32908.0	14150.5
299	C4H8O	TETRAHYDROFURAN	liquid	2325.0	32243.7	13864.8
300	C4H8O2	cis-2-BUTENE-1,4-DIOL	liquid	2170.0	24629.4	10590.7
301	C4H8O2	trans-2-BUTENE-1,4-DIOL	solid	2170.0	24629.4	10590.7
302	C4H8O2	ISOBUTYRIC ACID	liquid	2000.1	22701.1	9761.5
303	C4H8O2	n-BUTYRIC ACID	liquid	2006.2	22770.3	9791.2
304	C4H8O2	1,4-DIOXANE	liquid	2187.8	24831.5	10677.5
305	C4H8O2	ETHYL ACETATE	liquid	2061.0	23392.3	10058.7
306	C4H8O2	METHYL PROPIONATE	liquid	2078.0	23585.2	10141.6
307	C4H8O2	n-PROPYL FORMATE	liquid	2041.0	23165.3	9961.1
308	C4H8O2S	SULFOLANE	solid	2380.0	19804.9	8516.1
309	C4H8S	TETRAHYDROTHIOPHENE	liquid	2764.6	31354.3	13482.3
310	C4H9Br	1-BROMOBUTANE	liquid	2500.3	18247.8	7846.6
311	C4H9Br	2-BROMOBUTANE	liquid	2507.3	18298.9	7868.5
312	C4H9Cl	n-BUTYL CHLORIDE	liquid	2474.2	26728.5	11493.2

Table 26-1 ENTHALPY OF COMBUSTION - ORGANIC COMPOUNDS (continued)

NO	FORMULA	NAME	State	Enthalpy of Combustion @ 77 F kjoule/mol	kjoule/kg	BTU/lb
313	C4H9Cl	sec-BUTYL CHLORIDE	liquid	2465.2	26631.2	11451.4
314	C4H9Cl	tert-BUTYL CHLORIDE	liquid	2449.1	26457.3	11376.6
315	C4H9I	2-IODO-2-METHYLPROPANE	liquid	2560.8	13915.6	5983.7
316	C4H9N	PYRROLIDINE	liquid	2621.4	36857.8	15848.9
317	C4H9NO	N,N-DIMETHYLACETAMIDE	liquid	2380.0	27318.0	11746.7
318	C4H9NO	MORPHOLINE	liquid	2460.0	28236.3	12141.6
319	C4H9NO2	1-NITROBUTANE	liquid	2481.9	24068.2	10349.3
320	C4H9NO2	2-NITROBUTANE	liquid	2464.4	23897.8	10276.0
321	C4H10	n-BUTANE	gas	2657.5	45722.0	19660.5
322	C4H10	ISOBUTANE	gas	2649.0	45575.8	19597.6
323	C4H10N2	PIPERAZINE	solid	2738.0	31786.6	13668.2
324	C4H10O	n-BUTANOL	liquid	2456.0	33134.1	14247.7
325	C4H10O	sec-BUTANOL	liquid	2440.5	32925.0	14157.8
326	C4H10O	tert-BUTANOL	solid	2423.9	32701.1	14061.5
327	C4H10O	DIETHYL ETHER	liquid	2503.5	33774.9	14523.2
328	C4H10O	METHYL-PROPYL-ETHER	liquid	2517.2	33960.7	14603.1
329	C4H10O	METHYL ISOPROPYL ETHER	liquid	2531.1	34147.3	14683.3
330	C4H10O	ISOBUTANOL	liquid	2449.0	33039.7	14207.1
331	C4H10O2	1,3-BUTANEDIOL	liquid	2268.4	25170.3	10823.2
332	C4H10O2	1,4-BUTANEDIOL	liquid	2280.0	25299.0	10878.6
333	C4H10O2	2,3-BUTANEDIOL	liquid	2237.0	24821.9	10673.4
334	C4H10O2	t-BUTYL HYDROPEROXIDE	liquid	2736.3	30362.2	13055.7
335	C4H10O2	1,2-DIMETHOXYETHANE	liquid	2401.9	26651.6	11460.2
336	C4H10O2	2-ETHOXYETHANOL	liquid	2340.0	25964.8	11164.9
337	C4H10O3	DIETHYLENE GLYCOL	liquid	2155.0	20306.8	8731.9
338	C4H10O4S	DIETHYL SULFATE	liquid	2270.0	14722.4	6330.6
339	C4H10S	n-BUTYL MERCAPTAN	liquid	2955.4	32769.0	14090.7
340	C4H10S	ISOBUTYL MERCAPTAN	liquid	2949.0	32698.0	14060.1
341	C4H10S	sec-BUTYL MERCAPTAN	liquid	2950.0	32709.1	14064.9
342	C4H10S	tert-BUTYL MERCAPTAN	liquid	2939.5	32592.7	14014.8
343	C4H10S	DIETHYL SULFIDE	liquid	2960.7	32827.7	14115.9
344	C4H10S	ISOPROPYL-METHYL-SULFIDE	liquid	2962.7	32851.7	14126.2
345	C4H10S	METHYL-PROPYL-SULFIDE	liquid	2969.9	32932.4	14161.0
346	C4H10S2	DIETHYL DISULFIDE	liquid	3257.0	26641.0	11455.6
347	C4H11N	n-BUTYLAMINE	liquid	2776.3	37959.7	16322.7
348	C4H11N	ISOBUTYLAMINE	liquid	2771.7	37896.9	16295.6
349	C4H11N	sec-BUTYLAMINE	liquid	2766.5	37825.8	16265.1
350	C4H11N	tert-BUTYLAMINE	liquid	2753.6	37649.4	16189.2
351	C4H11N	DIETHYLAMINE	liquid	2800.3	38287.9	16463.8
352	C4H11NO	DIMETHYLETHANOLAMINE	liquid	2650.0	29729.5	12783.7
353	C4H11NO2	DIETHANOLAMINE	solid	2410.5	22927.2	9858.7
354	C4H11NO2	2-AMINOETHOXYETHANOL	liquid	2450.0	23302.9	10020.3
355	C4H12N2O	N-AMINOETHYL ETHANOLAMINE	liquid	2740.0	26307.7	11312.3
356	C4H12Si	TETRAMETHYLSILANE	liquid	3680.0	41711.5	17936.0
357	C4H13N3	DIETHYLENE TRIAMINE	liquid	3080.0	29854.5	12837.4
358	C5Cl6	HEXACHLOROCYCLOPENTADIENE	liquid	1800.0	6598.9	2837.5
359	C5H4O2	FURFURAL	liquid	2249.7	23413.4	10067.8
360	C5H5N	PYRIDINE	liquid	2672.1	33780.9	14525.8
361	C5H6	CYCLOPENTADIENE	liquid	2795.4	42288.5	18184.1
362	C5H6	2-METHYL-1-BUTENE-3-YNE	liquid	2930.0	44324.8	19059.6
363	C5H6	1-PENTENE-3-YNE	liquid	2910.0	44022.2	18929.5
364	C5H6	1-PENTENE-4-YNE	liquid	2930.0	44324.8	19059.6
365	C5H6N2	GLUTARONITRILE	liquid	2800.0	29750.5	12792.7
366	C5H6O2	FURFURYL ALCOHOL	liquid	2416.8	24635.8	10593.4
367	C5H6O3	GLUTARIC ANHYDRIDE	solid	2270.0	19894.7	8554.7
368	C5H6O4	CITRACONIC ACID	solid	1870.0	14373.6	6180.6
369	C5H6O4	ITACONIC ACID	solid	1853.2	14244.4	6125.1
370	C5H6S	2-METHYLTHIOPHENE	liquid	3042.3	30992.5	13326.8
371	C5H6S	3-METHYLTHIOPHENE	liquid	3041.1	30980.0	13321.4
372	C5H7N	N-METHYLPYRROLE	liquid	2876.0	35455.0	15245.6
373	C5H7NO2	ETHYL CYANOACETATE	liquid	2415.9	21357.7	9183.8
374	C5H8	CYCLOPENTENE	liquid	2939.1	43147.2	18553.3
375	C5H8	ISOPRENE	liquid	2984.2	43809.3	18838.0
376	C5H8	3-METHYL-1,2-BUTADIENE	liquid	3032.0	44511.0	19139.7
377	C5H8	2-METHYL-1,3-BUTADIENE	liquid	2986.8	43846.7	18854.1
378	C5H8	1,2-PENTADIENE	liquid	3046.7	44726.8	19232.5
379	C5H8	cis-1,3-PENTADIENE	liquid	2989.1	43881.2	18868.9
380	C5H8	trans-1,3-PENTADIENE	liquid	2982.7	43787.3	18828.5
381	C5H8	1,4-PENTADIENE	liquid	3015.7	44271.7	19036.8
382	C5H8	2,3-PENTADIENE	liquid	3038.0	44599.1	19177.6
383	C5H8	1-PENTYNE	liquid	3051.0	44789.9	19259.7
384	C5H8	2-PENTYNE	liquid	3036.7	44580.1	19169.4
385	C5H8	3-METHYL-1-BUTYNE	liquid	3046.0	44716.5	19228.1
386	C5H8	SPIROPENTANE	liquid	3095.6	45444.3	19541.1
387	C5H8N4O12	PENTAERYTHRITOL TETRANITRATE	solid	2348.0	7427.1	3193.7
388	C5H8O	CYCLOPENTANONE	liquid	2698.0	32074.0	13791.8
389	C5H8O	METHYL ISOPROPENYL KETONE	liquid	2720.0	32335.5	13904.3
390	C5H8O2	ACETYLACETONE	liquid	2511.0	25080.7	10784.7

Table 26-1 ENTHALPY OF COMBUSTION - ORGANIC COMPOUNDS (continued)

NO	FORMULA	NAME	State	- Enthalpy of Combustion @ 77 F		
				kjoule/mol	kjoule/kg	BTU/lb
391	C5H8O2	ALLYL ACETATE	liquid	2561.7	25587.1	11002.4
392	C5H8O2	ETHYL ACRYLATE	liquid	2550.6	25476.2	10954.8
393	C5H8O2	METHYL METHACRYLATE	liquid	2546.8	25438.2	10938.4
394	C5H8O2	VINYL PROPIONATE	liquid	2587.5	25844.8	11113.2
395	C5H8O3	2-HYDROXYETHYL ACRYLATE	liquid	2370.0	20410.4	8776.5
396	C5H8O3	LEVULINIC ACID	solid	2238.0	19273.7	8287.7
397	C5H8O3	METHYL ACETOACETATE	liquid	2310.0	19893.7	8554.3
398	C5H8O4	GLUTARIC ACID	solid	1975.3	14951.3	6429.0
399	C5H9N	VALERONITRILE	liquid	3022.9	36362.2	15635.8
400	C5H9NO	n-BUTYL ISOCYANATE	liquid	2890.0	29152.8	12535.7
401	C5H9NO	N-METHYL-2-PYRROLIDONE	liquid	2805.1	28296.3	12167.4
402	C5H9NO4	L-GLUTAMIC ACID	solid	1848.4	12563.0	5402.1
403	C5H10	CYCLOPENTANE	liquid	3070.9	43786.2	18828.1
404	C5H10	2-METHYL-1-BUTENE	liquid	3114.9	44413.6	19097.8
405	C5H10	2-METHYL-2-BUTENE	liquid	3107.6	44309.5	19053.1
406	C5H10	3-METHYL-1-BUTENE	gas	3124.1	44544.7	19154.2
407	C5H10	1-PENTENE	liquid	3129.7	44624.6	19188.6
408	C5H10	cis-2-PENTENE	liquid	3122.4	44520.5	19143.8
409	C5H10	trans-2-PENTENE	liquid	3118.0	44457.8	19116.8
410	C5H10Br2	2,3-DIBROMO-2-METHYLBUTANE	liquid	-------	-------	-------
411	C5H10Cl2	1,5-DICHLOROPENTANE	liquid	2926.1	20746.6	8921.0
412	C5H10O	METHYL ISOPROPYL KETONE	liquid	2877.0	33401.4	14362.6
413	C5H10O	2-PENTANONE	liquid	2880.0	33436.3	14377.6
414	C5H10O	DIETHYL KETONE	liquid	2880.4	33440.9	14379.6
415	C5H10O	VALERALDEHYDE	liquid	2910.5	33790.4	14529.9
416	C5H10O2	n-BUTYL FORMATE	liquid	2803.9	27453.4	11805.0
417	C5H10O2	ETHYL PROPIONATE	liquid	2674.0	26181.5	11258.1
418	C5H10O2	ISOBUTYL FORMATE	liquid	2700.8	26444.0	11370.9
419	C5H10O2	ISOPROPYL ACETATE	liquid	2658.1	26025.9	11191.1
420	C5H10O2	n-PROPYL ACETATE	liquid	2672.0	26162.0	11249.6
421	C5H10O2	METHYL n-BUTYRATE	liquid	2690.0	26338.2	11325.4
422	C5H10O2	2-METHYLBUTYRIC ACID	liquid	2680.0	26240.3	11283.3
423	C5H10O2	ISOVALERIC ACID	liquid	2615.3	25606.8	11010.9
424	C5H10O2	VALERIC ACID	liquid	2616.5	25618.6	11016.0
425	C5H10O2	TETRAHYDROFURFURYL ALCOHOL	liquid	2741.2	26839.5	11541.0
426	C5H10O2S	3-METHYL SULFOLANE	liquid	2990.0	22280.3	9580.5
427	C5H10O3	DIETHYL CARBONATE	liquid	2495.4	21123.6	9083.2
428	C5H10O3	ETHYL LACTATE	liquid	2481.6	21006.8	9032.9
429	C5H10S	THIACYCLOHEXANE	liquid	3378.0	33055.2	14213.7
430	C5H10S	CYCLOPENTANETHIOL	liquid	3394.1	33211.9	14281.1
431	C5H11Br	1-BROMOPENTANE	liquid	-------	-------	-------
432	C5H11Cl	1-CHLOROPENTANE	liquid	3085.2	28943.2	12445.6
433	C5H11Cl	1-CHLORO-3-METHYLBUTANE	liquid	3088.5	28974.0	12458.8
434	C5H11Cl	2-CHLORO-2-METHYLBUTANE	liquid	3068.5	28787.0	12378.4
435	C5H11N	N-METHYLPYRROLIDINE	liquid	3256.5	38244.7	16445.2
436	C5H11N	PIPERIDINE	liquid	3211.0	37710.4	16215.5
437	C5H11NO	tert-BUTYLFORMAMIDE	liquid	2960.0	29264.0	12583.5
438	C5H12	ISOPENTANE	liquid	3240.3	44910.6	19311.6
439	C5H12	NEOPENTANE	gas	3250.4	45050.6	19371.8
440	C5H12	n-PENTANE	liquid	3245.0	44975.7	19339.6
441	C5H12O	2,2-DIMETHYL-1-PROPANOL	solid	3099.5	35161.7	15119.5
442	C5H12O	tert-PENTYL-ALCOHOL	solid	3049.7	34597.1	14876.8
443	C5H12O	2-METHYL-1-BUTANOL	liquid	3062.0	34736.2	14936.6
444	C5H12O	2-METHYL-2-BUTANOL	liquid	3039.1	34476.5	14824.9
445	C5H12O	3-METHYL-1-BUTANOL	liquid	3062.3	34739.6	14938.0
446	C5H12O	3-METHYL-2-BUTANOL	liquid	3052.0	34622.8	14887.8
447	C5H12O	1-PENTANOL	liquid	3060.5	34719.2	14929.3
448	C5H12O	2-PENTANOL	liquid	3051.5	34617.1	14885.4
449	C5H12O	3-PENTANOL	liquid	3048.3	34580.8	14869.8
450	C5H12O	METHYL sec-BUTYL ETHER	liquid	3110.0	35280.8	15170.7
451	C5H12O	METHYL tert-BUTYL ETHER	liquid	3099.9	35166.2	15121.5
452	C5H12O	METHYL ISOBUTYL ETHER	liquid	3122.0	35416.9	15229.3
453	C5H12O	ETHYL PROPYL ETHER	liquid	3120.0	35394.2	15219.5
454	C5H12O2	ETHYLENE GLYCOL MONOPROPYL ETHER	liquid	2949.0	28315.2	12175.5
455	C5H12O2	NEOPENTYL GLYCOL	solid	2868.0	27537.5	11841.1
456	C5H12O2	1,5-PENTANEDIOL	liquid	2887.3	27722.8	11920.8
457	C5H12O3	2-(2-METHOXYETHOXY)ETHANOL	liquid	2790.0	23221.4	9985.2
458	C5H12O4	PENTAERYTHRITOL	solid	2498.1	18348.4	7889.8
459	C5H12S	n-PENTYL MERCAPTAN	liquid	3565.7	34214.5	14712.2
460	C5H12S	BUTYL-METHYL-SULFIDE	liquid	3583.2	34384.4	14785.3
461	C5H12S	ETHYL-PROPYL-SULFIDE	liquid	3580.3	34357.1	14773.5
462	C5H12S	2-METHYL-2-BUTANETHIOL	liquid	3561.3	34174.1	14694.9
463	C5H13N	n-PENTYLAMINE	liquid	3387.0	38857.3	16708.7
464	C5H13NO2	METHYL DIETHANOLAMINE	liquid	3060.0	25678.9	11041.9
465	C6Cl6	HEXACHLOROBENZENE	solid	2200.0	7725.2	3321.8
466	C6F6	HEXAFLUOROBENZENE	liquid	1369.4	7360.1	3164.9
467	C6H3ClN2O4	1-CHLORO-2,4-DINITROBENZENE	solid	2760.0	13626.0	5859.2
468	C6H3Cl2NO2	1,2-DICHLORO-4-NITROBENZENE	solid	2740.0	14270.8	6136.4

Table 26-1 ENTHALPY OF COMBUSTION - ORGANIC COMPOUNDS (continued)

NO	FORMULA	NAME	State	- Enthalpy of Combustion @ 77 F		
				kjoule/mol	kjoule/kg	BTU/lb
469	C6H3Cl3	1,2,4-TRICHLOROBENZENE	liquid	2656.3	14639.5	6295.0
470	C6H3N3O6	1,3,5-TRINITROBENZENE	solid	2679.9	12575.4	5407.4
471	C6H4Br2	m-DIBROMOBENZENE	liquid	2886.3	12235.0	5261.0
472	C6H4ClNO2	m-CHLORONITROBENZENE	solid	2820.0	17898.4	7696.3
473	C6H4ClNO2	o-CHLORONITROBENZENE	solid	2820.0	17898.4	7696.3
474	C6H4ClNO2	p-CHLORONITROBENZENE	solid	2820.0	17898.4	7696.3
475	C6H4Cl2	m-DICHLOROBENZENE	liquid	2825.0	19217.3	8263.4
476	C6H4Cl2	o-DICHLOROBENZENE	liquid	2826.0	19224.1	8266.4
477	C6H4Cl2	p-DICHLOROBENZENE	solid	2802.0	19060.8	8196.2
478	C6H4F2	m-DIFLUOROBENZENE	liquid	2503.1	21939.3	9433.9
479	C6H4F2	o-DIFLUOROBENZENE	liquid	2519.1	22079.0	9494.0
480	C6H4F2	p-DIFLUOROBENZENE	liquid	2505.9	21963.5	9444.3
481	C6H4N2O4	m-DINITROBENZENE	solid	2813.5	16736.2	7196.6
482	C6H4N2O4	o-DINITROBENZENE	solid	2843.0	16911.6	7272.0
483	C6H4N2O4	p-DINITROBENZENE	solid	2806.2	16692.7	7177.9
484	C6H5Br	BROMOBENZENE	liquid	3019.2	19229.3	8268.6
485	C6H5Cl	MONOCHLOROBENZENE	liquid	2976.1	26440.6	11369.5
486	C6H5ClO	m-CHLOROPHENOL	solid	2760.0	21468.9	9231.6
487	C6H5ClO	o-CHLOROPHENOL	liquid	2790.0	21702.3	9332.0
488	C6H5ClO	p-CHLOROPHENOL	solid	2780.0	21624.5	9298.5
489	C6H5Cl2N	3,4-DICHLOROANILINE	solid	3000.0	18516.5	7962.1
490	C6H5F	FLUOROBENZENE	liquid	2814.5	29286.0	12593.0
491	C6H5I	IODOBENZENE	liquid	3047.7	14939.0	6423.8
492	C6H5NO2	NITROBENZENE	liquid	2978.2	24191.2	10402.2
493	C6H6	BENZENE	liquid	3135.6	40141.3	17260.8
494	C6H6ClN	m-CHLOROANILINE	liquid	3079.1	24136.0	10378.5
495	C6H6ClN	o-CHLOROANILINE	solid	3086.5	24194.0	10403.4
496	C6H6ClN	p-CHLOROANILINE	solid	3060.0	23986.3	10314.1
497	C6H6N2	cis-DICYANO-1-BUTENE	liquid	3290.0	31000.6	13330.3
498	C6H6N2	trans-DICYANO-1-BUTENE	liquid	3290.0	31000.6	13330.3
499	C6H6N2	1,4-DICYANO-2-BUTENE	solid	3350.0	31566.0	13573.4
500	C6H6N2O2	m-NITROANILINE	solid	3060.0	22153.7	9526.1
501	C6H6N2O2	o-NITROANILINE	solid	3060.0	22153.7	9526.1
502	C6H6N2O2	p-NITROANILINE	solid	3050.0	22081.3	9495.0
503	C6H6O	PHENOL	solid	2921.4	31041.4	13347.8
504	C6H6O2	1,2-BENZENEDIOL	solid	2733.0	24820.2	10672.7
505	C6H6O2	1,3-BENZENEDIOL	solid	2719.0	24693.0	10618.0
506	C6H6O2	p-HYDROQUINONE	solid	2740.0	24883.8	10700.0
507	C6H6O3	1,2,3-BENZENETRIOL	solid	2540.0	20140.8	8660.6
508	C6H6S	PHENYL MERCAPTAN	liquid	3447.4	31288.8	13454.2
509	C6H7N	ANILINE	liquid	3238.5	34774.7	14953.1
510	C6H7N	2-METHYLPYRIDINE	liquid	3263.9	35047.5	15070.4
511	C6H7N	3-METHYLPYRIDINE	liquid	3269.1	35103.3	15094.4
512	C6H7N	4-METHYLPYRIDINE	liquid	3265.3	35062.5	15076.9
513	C6H8	1,3-CYCLOHEXADIENE	liquid	3399.9	42429.8	18244.8
514	C6H8	METHYLCYCLOPENTADIENE	liquid	3400.0	42431.0	18245.4
515	C6H8N2	ADIPONITRILE	liquid	3413.6	31565.6	13573.2
516	C6H8N2	METHYLGLUTARONITRILE	liquid	3400.0	31439.9	13519.1
517	C6H8N2	m-PHENYLENEDIAMINE	solid	3320.0	30700.1	13201.0
518	C6H8N2	o-PHENYLENEDIAMINE	solid	3330.0	30792.6	13240.8
519	C6H8N2	p-PHENYLENEDIAMINE	solid	3330.0	30792.6	13240.8
520	C6H8N2	PHENYLHYDRAZINE	liquid	3470.9	32095.5	13801.1
521	C6H8N2O	BIS(CYANOETHYL)ETHER	liquid	3260.0	26260.3	11291.9
522	C6H8O4	DIMETHYL MALEATE	liquid	2647.5	18369.2	7898.8
523	C6H8O6	ASCORBIC ACID	solid	2163.9	12286.1	5283.0
524	C6H8O7	CITRIC ACID	solid	1784.0	9285.6	3992.8
525	C6H10	1-METHYLCYCLOPENTENE	liquid	3538.5	43076.1	18522.7
526	C6H10	3-METHYLCYCLOPENTENE	liquid	3552.8	43250.0	18597.5
527	C6H10	4-METHYLCYCLOPENTENE	liquid	3558.4	43319.1	18627.2
528	C6H10	CYCLOHEXENE	liquid	3531.4	42989.8	18485.6
529	C6H10	2,3-DIMETHYL-1,3-BUTADIENE	liquid	3584.0	43630.2	18761.0
530	C6H10	1,5-HEXADIENE	liquid	3620.0	44068.4	18949.4
531	C6H10	cis,trans-2,4-HEXADIENE	liquid	3488.0	42461.5	18258.4
532	C6H10	trans,trans-2,4-HEXADIENE	liquid	3580.0	43581.5	18740.0
533	C6H10	1-HEXYNE	liquid	3661.0	44567.5	19164.0
534	C6H10	2-HEXYNE	liquid	3640.0	44311.9	19054.1
535	C6H10	3-HEXYNE	liquid	3640.0	44311.9	19054.1
536	C6H10O	CYCLOHEXANONE	liquid	3298.8	33611.5	14452.9
537	C6H10O	MESITYL OXIDE	liquid	3331.0	33939.6	14594.0
538	C6H10O2	epsilon-CAPROLACTONE	liquid	3083.8	27016.8	11617.2
539	C6H10O2	ETHYL METHACRYLATE	liquid	3150.0	27596.7	11866.6
540	C6H10O2	n-PROPYL ACRYLATE	liquid	3160.0	27684.3	11904.3
541	C6H10O3	ETHYLACETOACETATE	liquid	2960.0	22744.0	9779.9
542	C6H10O3	PROPIONIC ANHYDRIDE	liquid	2891.1	22214.6	9552.3
543	C6H10O4	ADIPIC ACID	solid	2580.0	17653.9	7591.2
544	C6H10O4	DIETHYL OXALATE	liquid	2723.0	18632.4	8011.9
545	C6H10O4	ETHYLENE GLYCOL DIACETATE	liquid	2704.7	18507.2	7958.1
546	C6H10O4	ETHYLIDENE DIACETATE	liquid	2700.0	18475.1	7944.3

Table 26-1 ENTHALPY OF COMBUSTION - ORGANIC COMPOUNDS (continued)

NO	FORMULA	NAME	State	kjoule/mol	kjoule/kg	BTU/lb
				- Enthalpy of Combustion @ 77 F		
547	C6H11N	HEXANENITRILE	liquid	3637.0	37433.1	16096.2
548	C6H11NO	epsilon-CAPROLACTAM	solid	3362.1	29711.3	12775.9
549	C6H11NO	CYCLOHEXANONE OXIME	solid	3610.0	31902.0	13717.9
550	C6H12	CYCLOHEXANE	liquid	3655.8	43438.2	18678.4
551	C6H12	2,3-DIMETHYL-1-BUTENE	liquid	3717.9	44176.0	18995.7
552	C6H12	2,3-DIMETHYL-2-BUTENE	liquid	3710.6	44089.3	18958.4
553	C6H12	3,3-DIMETHYL-1-BUTENE	liquid	3725.0	44260.4	19032.0
554	C6H12	2-ETHYL-1-BUTENE	liquid	3724.9	44259.2	19031.5
555	C6H12	trans-3-METHYL-2-PENTENE	liquid	3727.4	44289.2	19044.4
556	C6H12	1-HEXENE	liquid	3739.4	44431.5	19105.5
557	C6H12	cis-2-HEXENE	liquid	3728.1	44297.2	19047.8
558	C6H12	trans-2-HEXENE	liquid	3726.5	44278.2	19039.6
559	C6H12	cis-3-HEXENE	liquid	3733.0	44355.5	19072.8
560	C6H12	trans-3-HEXENE	liquid	3726.0	44272.3	19037.1
561	C6H12	METHYLCYCLOPENTANE	liquid	3673.7	43650.9	18769.9
562	C6H12	2-METHYL-1-PENTENE	liquid	3722.1	44225.9	19017.2
563	C6H12	2-METHYL-2-PENTENE	liquid	3713.5	44123.8	18973.2
564	C6H12	3-METHYL-1-PENTENE	liquid	3734.0	44367.3	19078.0
565	C6H12	3-METHYL-cis-2-PENTENE	liquid	3718.0	44177.2	18996.2
566	C6H12	4-METHYL-1-PENTENE	liquid	3732.0	44343.6	19067.7
567	C6H12	4-METHYL-cis-2-PENTENE	liquid	3725.0	44260.4	19032.0
568	C6H12	4-METHYL-trans-2-PENTENE	liquid	3720.5	44206.9	19009.0
569	C6H12N2	TRIETHYLENEDIAMINE	solid	3760.0	33519.1	14413.2
570	C6H12O	BUTYL VINYL ETHER	liquid	3591.4	35856.3	15418.2
571	C6H12O	CYCLOHEXANOL	liquid	3463.8	34582.3	14870.4
572	C6H12O	1-HEXANAL	liquid	3521.8	35161.4	15119.4
573	C6H12O	ETHYL ISOPROPYL KETONE	liquid	3486.0	34804.0	14965.7
574	C6H12O	2-HEXANONE	liquid	3490.0	34843.9	14982.9
575	C6H12O	3-HEXANONE	liquid	3492.0	34863.9	14991.5
576	C6H12O	METHYL ISOBUTYL KETONE	liquid	3487.2	34815.9	14970.9
577	C6H12O2	n-PENTYL FORMATE	liquid	3314.0	28529.6	12267.7
578	C6H12O2	n-BUTYL ACETATE	liquid	3283.0	28262.7	12153.0
579	C6H12O2	sec-BUTYL ACETATE	liquid	3267.1	28125.9	12094.1
580	C6H12O2	tert-BUTYL ACETATE	liquid	3250.0	27978.7	12030.8
581	C6H12O2	ETHYL n-BUTYRATE	liquid	3284.5	28275.1	12158.5
582	C6H12O2	ETHYL ISOBUTYRATE	liquid	3270.0	28150.8	12104.9
583	C6H12O2	ISOBUTYL ACETATE	liquid	3276.0	28202.5	12127.1
584	C6H12O2	n-PROPYL PROPIONATE	liquid	3280.0	28236.9	12141.9
585	C6H12O2	CYCLOHEXYL PEROXIDE	liquid	3540.0	30475.2	13104.3
586	C6H12O2	DIACETONE ALCOHOL	liquid	3219.0	27711.8	11916.1
587	C6H12O2	2-ETHYL BUTYRIC ACID	liquid	3295.9	28373.8	12200.7
588	C6H12O2	n-HEXANOIC ACID	liquid	3228.7	27795.3	11952.0
589	C6H12O3	2-ETHOXYETHYL ACETATE	liquid	3150.0	23834.9	10249.0
590	C6H12O3	HYDROXYCAPROIC ACID	solid	3130.0	23683.6	10183.9
591	C6H12O3	PARALDEHYDE	liquid	3125.2	23647.3	10168.3
592	C6H12O3	sec-BUTYL GLYCOLATE	liquid	-------	-------	-------
593	C6H12S	THIACYCLOHEPTANE	liquid	3717.0	31982.5	13752.5
594	C6H13N	CYCLOHEXYLAMINE	liquid	3785.5	38169.5	16412.9
595	C6H13N	HEXAMETHYLENEIMINE	liquid	3853.9	38859.2	16709.5
596	C6H14	2,2-DIMETHYLBUTANE	liquid	3842.0	44582.7	19170.5
597	C6H14	2,3-DIMETHYLBUTANE	liquid	3848.7	44660.4	19204.0
598	C6H14	n-HEXANE	liquid	3855.2	44735.8	19236.4
599	C6H14	2-METHYLPENTANE	liquid	3849.1	44665.0	19206.0
600	C6H14	3-METHYLPENTANE	liquid	3851.4	44691.7	19217.4
601	C6H14N2O2	LYSINE	solid	3590.0	24557.3	10559.6
602	C6H14O	2-ETHYL-1-BUTANOL	liquid	3671.4	35931.8	15450.7
603	C6H14O	1-HEXANOL	liquid	3674.4	35961.1	15463.3
604	C6H14O	2-HEXANOL	liquid	3666.0	35878.9	15427.9
605	C6H14O	2-METHYL-1-PENTANOL	liquid	3673.4	35951.3	15459.1
606	C6H14O	4-METHYL-2-PENTANOL	liquid	3661.3	35832.9	15408.2
607	C6H14O	n-BUTYL ETHYL ETHER	liquid	3725.1	36457.3	15676.6
608	C6H14O	DIISOPROPYL ETHER	liquid	3702.3	36234.2	15580.7
609	C6H14O	DI-n-PROPYL ETHER	liquid	3725.0	36456.3	15676.2
610	C6H14O	METHYL tert-PENTYL ETHER	liquid	3710.0	36309.5	15613.1
611	C6H14O2	ACETAL	liquid	3562.7	30147.4	12963.4
612	C6H14O2	2-BUTOXYETHANOL	liquid	3550.0	30039.9	12917.2
613	C6H14O2	1,6-HEXANEDIOL	solid	3487.5	29511.1	12689.8
614	C6H14O2	HEXYLENE GLYCOL	liquid	3435.6	29071.9	12500.9
615	C6H14O2S	DI-n-PROPYL SULFONE	solid	3800.0	25292.5	10875.8
616	C6H14O3	DIETHYLENE GLYCOL DIMETHYL ETHER	liquid	3489.0	26003.4	11181.4
617	C6H14O3	DIPROPYLENE GLYCOL	liquid	3340.0	24892.9	10703.9
618	C6H14O3	2-(2-ETHOXYETHOXY)ETHANOL	liquid	3420.0	25489.1	10960.3
619	C6H14O3	TRIMETHYLOLPROPANE	solid	3413.7	25442.1	10940.1
620	C6H14O4	TRIETHYLENE GLYCOL	liquid	3249.8	21640.1	9305.2
621	C6H14O6	SORBITOL	solid	2914.7	15999.5	6879.8
622	C6H14S	n-HEXYLMERCAPTAN	liquid	4176.0	35317.1	15186.4
623	C6H14S	BUTYL-ETHYL-SULFIDE	liquid	4193.6	35468.2	15251.3
624	C6H14S	ISOPROPYL-SULFIDE	liquid	4179.1	35345.0	15198.4

592

Table 26-1 ENTHALPY OF COMBUSTION - ORGANIC COMPOUNDS (continued)

NO	FORMULA	NAME	State	- Enthalpy of Combustion @ 77 F		
				kjoule/mol	kjoule/kg	BTU/lb
625	C6H14S	METHYL-PENTYL-SULFIDE	liquid	4196.9	35495.5	15263.0
626	C6H14S	PROPYL-SULFIDE	liquid	4193.6	35467.5	15251.0
627	C6H14S2	PROPYL-DISULFIDE	liquid	4494.3	29902.5	12858.1
628	C6H15Al	TRIETHYL ALUMINUN	liquid	4820.2	42220.6	18154.9
629	C6H15Al2Cl3	ETHYL ALUMINUM SESQUICHLORIDE	liquid	-------	-------	-------
630	C6H15N	DIISOPROPYLAMINE	liquid	3990.0	39430.0	16954.9
631	C6H15N	DI-n-PROPYLAMINE	liquid	4018.9	39715.6	17077.7
632	C6H15N	n-HEXYLAMINE	liquid	4000.0	39528.8	16997.4
633	C6H15N	TRIETHYLAMINE	liquid	4040.5	39929.0	17169.5
634	C6H15NO	6-AMINOHEXANOL	solid	3850.0	32852.4	14126.5
635	C6H15NO2	DIISOPROPANOLAMINE	solid	3720.0	27929.8	12009.8
636	C6H15NO3	TRIETHANOLAMINE	liquid	3510.8	23532.4	10118.9
637	C6H15N3	N-AMINOETHYL PIPERAZINE	liquid	4135.0	32003.4	13761.5
638	C6H15O4P	TRIETHYL PHOSPHATE	liquid	3848.6	21128.0	9085.1
639	C6H16N2	HEXAMETHYLENEDIAMINE	solid	4196.9	36115.7	15529.8
640	C6H18N3OP	HEXAMETHYL PHOSPHORAMIDE	liquid	4920.0	27455.1	11805.7
641	C6H18N4	TRIETHYLENE TETRAMINE	liquid	4450.0	30430.3	13085.0
642	C6H18OSi2	HEXAMETHYLDISILOXANE	liquid	5542.4	34132.5	14677.0
643	C6H18O3Si3	HEXAMETHYLCYCLOTRISILOXANE	solid	5490.0	24678.2	10611.6
644	C6H19NSi2	HEXAMETHYLDISILAZANE	liquid	5960.0	36928.0	15879.1
645	C7H3ClF3NO2	4-CHLORO-3-NITROBENZOTRIFLUORIDE	liquid	2418.7	10723.4	4611.1
646	C7H3Cl2F3	2,4-DICHLOROBENZOTRIFLUORIDE	liquid	2414.7	11231.1	4829.4
647	C7H3Cl2NO	3,4-DICHLOROPHENYL ISOCYANATE	solid	3110.0	16541.5	7112.8
648	C7H4ClF3	p-CHLOROBENZOTRIFLUORIDE	liquid	2574.9	14260.9	6132.2
649	C7H4Cl2O	m-CHLOROBENZOYL CHLORIDE	liquid	3413.7	19505.3	8387.3
650	C7H4F3NO2	3-NITROBENZOTRIFLUORIDE	liquid	2580.0	13500.1	5805.0
651	C7H5ClO	BENZOYL CHLORIDE	liquid	3203.8	22791.7	9800.4
652	C7H5ClO2	o-CHLOROBENZOIC ACID	solid	3037.1	19398.0	8341.1
653	C7H5Cl3	BENZOTRICHLORIDE	liquid	3291.8	16840.0	7241.2
654	C7H5F3	BENZOTRIFLUORIDE	liquid	2741.2	18761.0	8067.2
655	C7H5N	BENZONITRILE	liquid	3522.4	34157.3	14687.6
656	C7H5NO	PHENYL ISOCYANATE	liquid	3360.0	28206.1	12128.6
657	C7H5N3O6	2,4,6-TRINITROTOLUENE	solid	3291.9	14493.3	6232.1
658	C7H6Cl2	BENZYL DICHLORIDE	liquid	3440.0	21362.5	9185.9
659	C7H6Cl2	2,4-DICHLOROTOLUENE	liquid	3413.4	21197.3	9114.8
660	C7H6N2O4	2,4-DINITROTOLUENE	solid	3416.0	18755.2	8064.7
661	C7H6N2O4	2,5-DINITROTOLUENE	solid	3446.0	18919.9	8135.6
662	C7H6N2O4	2,6-DINITROTOLUENE	solid	3429.0	18826.6	8095.4
663	C7H6N2O4	3,4-DINITROTOLUENE	solid	3466.0	19029.7	8182.8
664	C7H6N2O4	3,5-DINITROTOLUENE	solid	3437.0	18870.5	8114.3
665	C7H6O	BENZALDEHYDE	liquid	3393.1	31973.0	13748.4
666	C7H6O2	BENZOIC ACID	solid	3095.1	25344.1	10898.0
667	C7H6O2	p-HYDROXYBENZALDEHYDE	solid	3317.0	27161.1	11679.3
668	C7H6O2	SALICYLALDEHYDE	liquid	3200.2	26204.7	11268.0
669	C7H6O3	SALICYLIC ACID	solid	2894.7	20957.4	9011.7
670	C7H7Br	p-BROMOTOLUENE	solid	3620.0	21165.0	9101.0
671	C7H7Cl	BENZYL CHLORIDE	liquid	3570.4	28205.6	12128.4
672	C7H7Cl	o-CHLOROTOLUENE	liquid	3570.0	28202.4	12127.0
673	C7H7Cl	p-CHLOROTOLUENE	liquid	3570.0	28202.4	12127.0
674	C7H7F	p-FLUOROTOLUENE	liquid	3416.2	31019.1	13338.2
675	C7H7NO	FORMANILIDE	solid	3450.0	28479.7	12246.3
676	C7H7NO2	m-NITROTOLUENE	liquid	3570.0	26032.2	11193.8
677	C7H7NO2	o-NITROTOLUENE	liquid	3590.0	26178.0	11256.5
678	C7H7NO2	p-NITROTOLUENE	solid	3550.0	25886.3	11131.1
679	C7H7NO3	o-NITROANISOLE	liquid	3440.0	22463.4	9659.3
680	C7H8	TOLUENE	liquid	3733.9	40523.8	17425.2
681	C7H8	1,3,5-CYCLOHEPTATRIENE	liquid	3870.7	42008.4	18063.6
682	C7H8O	ANISOLE	liquid	3601.8	33306.8	14321.9
683	C7H8O	BENZYL ALCOHOL	liquid	3561.3	32932.3	14160.9
684	C7H8O	m-CRESOL	liquid	3527.8	32622.5	14027.7
685	C7H8O	o-CRESOL	solid	3517.4	32526.4	13986.3
686	C7H8O	p-CRESOL	solid	3522.6	32574.4	14007.0
687	C7H8O2	GUAIACOL	solid	3470.0	27952.5	12019.6
688	C7H8O2	p-METHOXYPHENOL	solid	3400.0	27388.7	11777.1
689	C7H9N	BENZYLAMINE	liquid	3850.0	35929.3	15449.6
690	C7H9N	2,6-DIMETHYLPYRIDINE	liquid	3855.6	35981.5	15472.1
691	C7H9N	N-METHYLANILINE	liquid	3900.0	36395.9	15650.2
692	C7H9N	m-TOLUIDINE	liquid	3840.0	35835.9	15409.5
693	C7H9N	o-TOLUIDINE	liquid	3840.0	35835.9	15409.5
694	C7H9N	p-TOLUIDINE	solid	3830.0	35742.6	15369.3
695	C7H10	2-NORBORNENE	solid	4020.0	42695.1	18358.9
696	C7H10N2	TOLUENEDIAMINE	solid	4022.1	32922.2	14156.5
697	C7H11NO	CYCLOHEXYL ISOCYANATE	liquid	3850.0	30758.2	13226.0
698	C7H12	1-HEPTYNE	liquid	4280.0	44503.7	19136.6
699	C7H12O2	n-BUTYL ACRYLATE	liquid	3766.3	29385.0	12635.5
700	C7H12O2	ISOBUTYL ACRYLATE	liquid	3770.0	29413.8	12647.9
701	C7H12O2	n-PROPYL METHACRYLATE	liquid	3760.0	29335.8	12614.4
702	C7H12O4	DIETHYL MALONATE	liquid	3362.0	20990.2	9025.8

Table 26-1 ENTHALPY OF COMBUSTION - ORGANIC COMPOUNDS (continued)

NO	FORMULA	NAME	State	- Enthalpy of Combustion @ 77 F		
				kjoule/mol	kjoule/kg	BTU/lb
703	C7H14	CYCLOHEPTANE	liquid	4289.7	43688.6	18786.1
704	C7H14	1,1-DIMETYLCYCLOPENTANE	liquid	4275.2	43541.0	18722.6
705	C7H14	cis-1,2-DIMETHYLCYCLOPENTANE	liquid	4282.0	43610.2	18752.4
706	C7H14	trans-1,2-DIMETHYLCYCLOPENTANE	liquid	4276.1	43550.1	18726.6
707	C7H14	cis-1,3-DIMETHYLCYCLOPENTANE	liquid	4277.1	43560.3	18730.9
708	C7H14	trans-1,3-DIMETHYLCYCLOPENTANE	liquid	4279.0	43579.7	18739.3
709	C7H14	ETHYLCYCLOPENTANE	liquid	4283.9	43629.6	18760.7
710	C7H14	2-ETHYL-1-PENTENE	liquid	4337.0	44170.4	18993.3
711	C7H14	3-ETHYL-1-PENTENE	liquid	4349.0	44292.6	19045.8
712	C7H14	1-HEPTENE	liquid	4349.5	44297.7	19048.0
713	C7H14	cis-2-HEPTENE	liquid	4342.4	44225.4	19016.9
714	C7H14	trans-2-HEPTENE	liquid	4340.0	44200.9	19006.4
715	C7H14	cis-3-HEPTENE	liquid	4343.3	44234.5	19020.8
716	C7H14	trans-3-HEPTENE	liquid	4340.0	44200.9	19006.4
717	C7H14	METHYLCYCLOHEXANE	liquid	4257.1	43356.6	18643.3
718	C7H14	2-METHYL-1-HEXENE	liquid	4335.0	44150.0	18984.5
719	C7H14	3-METHYL-1-HEXENE	liquid	4346.0	44262.0	19032.7
720	C7H14	4-METHYL-1-HEXENE	liquid	4350.0	44302.8	19050.2
721	C7H14	2,3,3-TRIMETHYL-1-BUTENE	liquid	4329.9	44098.1	18962.2
722	C7H14O	DIISOPROPYL KETONE	liquid	4095.0	35861.9	15420.6
723	C7H14O	2-HEPTANONE	liquid	4099.5	35901.3	15437.6
724	C7H14O	1-HEPTANAL	liquid	4136.0	36221.0	15575.0
725	C7H14O	1-METHYLCYCLOHEXANOL	solid	4057.8	35536.1	15280.5
726	C7H14O	cis-2-METHYLCYCLOHEXANOL	liquid	4057.4	35532.6	15279.0
727	C7H14O	trans-2-METHYLCYCLOHEXANOL	liquid	4031.8	35308.4	15182.6
728	C7H14O	cis-3-METHYLCYCLOHEXANOL	liquid	4031.4	35304.9	15181.1
729	C7H14O	trans-3-METHYLCYCLOHEXANOL	liquid	4053.2	35495.8	15263.2
730	C7H14O	cis-4-METHYLCYCLOHEXANOL	liquid	4034.4	35331.2	15192.4
731	C7H14O	trans-4-METHYLCYCLOHEXANOL	liquid	4014.3	35155.2	15116.7
732	C7H14O	5-METHYL-2-HEXANONE	liquid	4100.0	35905.7	15439.5
733	C7H14O2	n-BUTYL PROPIONATE	liquid	3900.0	29956.9	12881.5
734	C7H14O2	ETHYL ISOVALERATE	liquid	3876.5	29776.4	12803.9
735	C7H14O2	ISOPENTYL ACETATE	liquid	3889.9	29879.3	12848.1
736	C7H14O2	n-PENTYL ACETATE	liquid	3893.1	29903.9	12858.7
737	C7H14O2	n-PROPYL n-BUTYRATE	liquid	3904.8	29993.8	12897.3
738	C7H14O2	n-HEPTANOIC ACID	liquid	3839.3	29490.7	12681.0
739	C7H14O3	ETHYL-3-ETHOXYPROPIONATE	liquid	3760.0	25720.7	11059.9
740	C7H15Br	1-BROMOHEPTANE	liquid	4330.0	24176.4	10395.9
741	C7H15N	N-METHYLCYCLOHEXYLAMINE	liquid	4420.0	39044.9	16789.3
742	C7H16	2,2-DIMETHYLPENTANE	liquid	4450.8	44417.4	19099.5
743	C7H16	2,3-DIMETHYLPENTANE	liquid	4460.7	44516.2	19142.0
744	C7H16	2,4-DIMETHYLPENTANE	liquid	4454.5	44454.3	19115.4
745	C7H16	3,3-DIMETHYLPENTANE	liquid	4456.3	44472.3	19123.1
746	C7H16	3-ETHYLPENTANE	liquid	4464.6	44555.1	19158.7
747	C7H16	n-HEPTANE	liquid	4464.9	44558.1	19160.0
748	C7H16	2-METHYLHEXANE	liquid	4459.6	44505.2	19137.2
749	C7H16	3-METHYLHEXANE	liquid	4462.7	44536.1	19150.5
750	C7H16	2,2,3-TRIMETHYLBUTANE	liquid	4452.6	44435.4	19107.2
751	C7H16O	1-HEPTANOL	liquid	4288.7	36907.0	15870.0
752	C7H16O	2-HEPTANOL	liquid	4330.0	37262.4	16022.8
753	C7H16O	5-METHYL-1-HEXANOL	liquid	4290.0	36918.2	15874.8
754	C7H16O	ISOPROPYL-TERT-BUTYL-ETHER	liquid	4303.7	37035.6	15925.3
755	C7H16S	n-HEPTYL MERCAPTAN	liquid	4786.0	36183.6	15558.9
756	C7H16S	BUTYL-PROPYL-SULFIDE	liquid	4807.8	36350.7	15630.8
757	C7H16S	ETHYL-PENTYL-SULFIDE	liquid	4805.4	36332.0	15622.8
758	C7H16S	HEXYL-METHYL-SULFIDE	liquid	4807.8	36350.3	15630.6
759	C7H17N	1-AMINOHEPTANE	liquid	4610.0	40010.8	17204.6
760	C8H4Cl2O2	ISOPHTHALOYL CHLORIDE	solid	3330.0	16402.0	7052.9
761	C8H4O3	PHTHALIC ANHYDRIDE	solid	3171.5	21412.0	9207.2
762	C8H6	ETHYNYLBENZENE	liquid	4166.4	40792.7	17540.8
763	C8H6O4	ISOPHTHALIC ACID	solid	3070.7	18483.4	7947.9
764	C8H6O4	PHTHALIC ACID	solid	3091.5	18608.6	8001.7
765	C8H6O4	TEREPHTHALIC ACID	solid	3057.6	18404.5	7913.9
766	C8H6S	BENZOTHIOPHENE	solid	4336.0	32309.5	13893.1
767	C8H7N	INDOLE	liquid	4081.1	34836.5	14979.7
768	C8H8	STYRENE	liquid	4219.3	40511.0	17419.7
769	C8H8	1,3,5,7-CYCLOOCTATETRAENE	liquid	4379.2	42046.9	18080.2
770	C8H8O	ACETOPHENONE	liquid	3973.0	33066.7	14218.7
771	C8H8O	p-TOLUALDEHYDE	liquid	3990.0	33208.2	14279.5
772	C8H8O2	METHYL BENZOATE	liquid	3772.0	27704.7	11913.0
773	C8H8O2	o-TOLUIC ACID	solid	3699.0	27168.6	11682.5
774	C8H8O2	p-TOLUIC ACID	solid	3692.2	27118.6	11661.0
775	C8H8O3	METHYL SALICYLATE	liquid	3583.8	23554.4	10128.4
776	C8H8O3	VANILLIN	solid	3660.0	24055.2	10343.7
777	C8H9NO	ACETANILIDE	solid	4026.0	29785.6	12807.8
778	C8H10	ETHYLBENZENE	liquid	4344.8	40924.2	17597.4
779	C8H10	m-XYLENE	liquid	4331.8	40801.8	17544.8
780	C8H10	o-XYLENE	liquid	4332.8	40811.2	17548.8

Table 26-1 ENTHALPY OF COMBUSTION - ORGANIC COMPOUNDS (continued)

NO	FORMULA	NAME	State	- Enthalpy of Combustion @ 77 F		
				kjoule/mol	kjoule/kg	BTU/lb
781	C8H10	p-XYLENE	liquid	4332.8	40811.2	17548.8
782	C8H10O	m-ETHYLPHENOL	liquid	-------	-------	-------
783	C8H10O	p-ETHYLPHENOL	solid	4133.0	33830.7	14547.2
784	C8H10O	PHENETOLE	liquid	4204.8	34418.5	14799.9
785	C8H10O	2-PHENYLETHANOL	liquid	4180.0	34215.5	14712.6
786	C8H10O	2,3-XYLENOL	solid	4116.0	33691.6	14487.4
787	C8H10O	2,4-XYLENOL	liquid	4128.4	33793.1	14531.0
788	C8H10O	2,5-XYLENOL	solid	4110.6	33647.4	14468.4
789	C8H10O	2,6-XYLENOL	solid	4119.8	33722.7	14500.8
790	C8H10O	3,4-XYLENOL	solid	4114.9	33682.6	14483.5
791	C8H10O	3,5-XYLENOL	solid	4112.8	33665.4	14476.1
792	C8H11N	N,N-DIMETHYLANILINE	liquid	4525.0	37340.5	16056.4
793	C8H11N	o-ETHYLANILINE	liquid	4450.0	36721.6	15790.3
794	C8H11N	2,4,6-TRIMETHYLPYRIDINE	liquid	4450.0	36721.6	15790.3
795	C8H11NO	p-PHENETIDINE	liquid	4310.0	31418.3	13509.9
796	C8H12	1,5-CYCLOOCTADIENE	liquid	4660.0	43075.2	18522.3
797	C8H12	VINYLCYCLOHEXENE	liquid	4626.0	42760.9	18387.2
798	C8H12O4	1,4-CYCLOHEXANEDICARBOXYLIC ACID	solid	3623.0	21041.8	9048.0
799	C8H12O4	DIETHYL MALEATE	liquid	3800.0	22069.8	9490.0
800	C8H14O2	n-BUTYL METHACRYLATE	liquid	4420.0	31083.4	13365.9
801	C8H14O3	BUTYRIC ANHYDRIDE	liquid	4118.7	26035.3	11195.2
802	C8H14O4	DIETHYL SUCCINATE	liquid	3920.0	22503.3	9676.4
803	C8H16	CYCLOOCTANE	liquid	4925.5	43893.5	18874.2
804	C8H16	1,1-DIMETHYLCYCLOHEXANE	liquid	4863.9	43344.5	18638.1
805	C8H16	cis-1,2-DIMETHYLCYCLOHEXANE	liquid	4870.8	43406.0	18664.6
806	C8H16	trans-1,2-DIMETHYLCYCLOHEXANE	liquid	4864.4	43348.9	18640.0
807	C8H16	cis-1,3-DIMETHYLCYCLOHEXANE	liquid	4859.7	43307.0	18622.0
808	C8H16	trans-1,3-DIMETHYLCYCLOHEXANE	liquid	4866.9	43371.2	18649.6
809	C8H16	cis-1,4-DIMETHYLCYCLOHEXANE	liquid	4867.0	43372.1	18650.0
810	C8H16	trans-1,4-DIMETHYLCYCLOHEXANE	liquid	4860.3	43312.4	18624.3
811	C8H16	ETHYLCYCLOHEXANE	liquid	4870.5	43403.3	18663.4
812	C8H16	2-ETHYL-1-HEXENE	liquid	4950.0	44111.7	18968.1
813	C8H16	1-METHYL-1-ETHYLCYCLOPENTANE	liquid	4889.2	43569.9	18735.1
814	C8H16	1-OCTENE	liquid	4960.6	44206.2	19008.7
815	C8H16	trans-2-OCTENE	liquid	4950.0	44111.7	18968.1
816	C8H16	trans-3-OCTENE	liquid	4950.0	44111.7	18968.1
817	C8H16	trans-4-OCTENE	liquid	4950.0	44111.7	18968.1
818	C8H16	n-PROPYLCYCLOPENTANE	liquid	4893.6	43609.1	18751.9
819	C8H16	2,4,4-TRIMETHYL-1-PENTENE	liquid	4937.3	43998.6	18919.4
820	C8H16	2,4,4-TRIMETHYL-2-PENTENE	liquid	4940.5	44027.1	18931.6
821	C8H16O	2-ETHYLHEXANAL	liquid	4734.4	36925.8	15878.1
822	C8H16O	1-OCTANAL	liquid	4745.4	37011.6	15915.0
823	C8H16O	2-OCTANONE	liquid	4701.0	36665.3	15766.1
824	C8H16O2	n-BUTYL n-BUTYRATE	liquid	4502.6	31221.7	13425.3
825	C8H16O2	n-HEXYL ACETATE	liquid	4505.0	31238.3	13432.5
826	C8H16O2	ISOBUTYL ISOBUTYRATE	liquid	4488.6	31124.6	13383.6
827	C8H16O2	n-OCTANOIC ACID	liquid	4448.3	30845.1	13263.4
828	C8H16O4	DIETHYLENE GLYCOL ETHYL ETHER ACETATE	liquid	4240.0	24061.8	10346.6
829	C8H18	2,2-DIMETHYLHEXANE	liquid	5062.6	44319.0	19057.2
830	C8H18	2,3-DIMETHYLHEXANE	liquid	5071.9	44400.4	19092.2
831	C8H18	2,4-DIMETHYLHEXANE	liquid	5067.3	44360.1	19074.8
832	C8H18	2,5-DIMETHYLHEXANE	liquid	5064.1	44332.1	19062.8
833	C8H18	3,3-DIMETHYLHEXANE	liquid	5066.9	44356.6	19073.3
834	C8H18	3,4-DIMETHYLHEXANE	liquid	5072.8	44408.3	19095.6
835	C8H18	3-ETHYLHEXANE	liquid	5074.1	44419.6	19100.4
836	C8H18	3-ETHYL-2-METHYLPENTANE	liquid	5085.4	44518.8	19143.1
837	C8H18	3-METHYL-3-ETHYLPENTANE	liquid	5071.6	44397.8	19091.0
838	C8H18	2-METHYLHEPTANE	liquid	5069.4	44378.5	19082.8
839	C8H18	3-METHYLHEPTANE	liquid	5072.1	44402.1	19092.9
840	C8H18	4-METHYLHEPTANE	liquid	5072.8	44408.3	19095.6
841	C8H18	n-OCTANE	liquid	5074.1	44419.6	19100.4
842	C8H18	2,2,3-TRIMETHYLPENTANE	liquid	5067.5	44361.9	19075.6
843	C8H18	2,2,4-TRIMETHYLPENTANE	liquid	5065.3	44342.6	19067.3
844	C8H18	2,3,3-TRIMETHYLPENTANE	liquid	5068.8	44373.2	19080.5
845	C8H18	2,3,4-TRIMETHYLPENTANE	liquid	5069.4	44378.5	19082.8
846	C8H18	2,2,3,3-TETRAMETHYLBUTANE	liquid	5072.2	44403.7	19093.6
847	C8H18O	DI-n-BUTYL ETHER	liquid	4947.2	37988.2	16334.9
848	C8H18O	DI-sec-BUTYL ETHER	liquid	4923.3	37804.7	16256.0
849	C8H18O	DI-tert-BUTYL ETHER	liquid	4925.0	37817.7	16261.6
850	C8H18O	2-ETHYL-1-HEXANOL	liquid	4892.1	37565.1	16153.0
851	C8H18O	1-OCTANOL	liquid	4899.5	37621.9	16177.4
852	C8H18O	2-OCTANOL	liquid	4881.4	37482.9	16117.7
853	C8H18O2	DI-t-BUTYL PEROXIDE	liquid	4940.0	33782.4	14526.4
854	C8H18O2S	DI-n-BUTYL SULFONE	solid	5010.0	28099.3	12082.7
855	C8H18O3	DIETHYLENE GLYCOL DIETHYL ETHER	liquid	4690.0	28909.8	12431.2
856	C8H18O3	DIETHYLENE GLYCOL MONOBUTYL EHTER	liquid	4640.0	28601.5	12298.7
857	C8H18O4	TRIETHYLENE GLYCOL DIMETHYL ETHER	liquid	4580.0	25697.3	11049.8
858	C8H18O5	TETRAETHYLENE GLYCOL	liquid	4380.0	22550.8	9696.9

Table 26-1 ENTHALPY OF COMBUSTION - ORGANIC COMPOUNDS (continued)

NO	FORMULA	NAME	State	- Enthalpy of Combustion @ 77 F		
				kjoule/mol	kjoule/kg	BTU/lb
859	C8H18S	n-OCTYL MERCAPTAN	liquid	5397.0	36890.7	15863.0
860	C8H18S	tert-OCTYL MERCAPTAN	liquid	5370.0	36706.2	15783.6
861	C8H18S	BUTYL-SULFIDE	liquid	5420.3	37052.0	15932.4
862	C8H18S	ETHYL-HEXYL-SULFIDE	liquid	5418.0	37036.0	15925.5
863	C8H18S	HEPTYL-METHYL-SULFIDE	liquid	5420.4	37052.5	15932.6
864	C8H18S	PENTYL-PROPYL-SULFIDE	liquid	5417.8	37034.8	15925.0
865	C8H18S2	BUTYL-DISULFIDE	liquid	5719.9	32071.2	13790.6
866	C8H19N	DI-n-BUTYLAMINE	liquid	5239.7	40540.5	17432.4
867	C8H19N	DIISOBUTYLAMINE	liquid	5227.0	40442.3	17390.2
868	C8H19N	n-OCTYLAMINE	liquid	5220.0	40388.1	17366.9
869	C8H23N5	TETRAETHYLENEPENTAMINE	liquid	5830.0	30797.0	13242.7
870	C8H24O4Si4	OCTAMETHYLCYCLOTETRASILOXANE	liquid	7368.0	24840.0	10681.2
871	C9H4O5	TRIMELLITIC ANHYDRIDE	solid	3120.0	16239.2	6982.8
872	C9H6N2O2	TOLUENE DIISOCYANATE	liquid	3970.3	22797.0	9802.7
873	C9H7N	ISOQUINOLINE	solid	4533.3	35098.1	15092.2
874	C9H7N	QUINOLINE	liquid	4544.2	35182.4	15128.5
875	C9H7NO	8-HYDROXYQUINOLINE	solid	4300.0	29622.3	12737.6
876	C9H8	INDENE	liquid	4619.5	39767.4	17100.0
877	C9H8O	2-METHYLBENZOFURAN	liquid	4450.0	33670.8	14478.4
878	C9H10	INDANE	liquid	4762.0	40295.1	17326.9
879	C9H10	cis-PROPENYLBENZENE	liquid	4836.1	40922.1	17596.5
880	C9H10	trans-PROPENYLBENZENE	liquid	4831.9	40886.7	17581.3
881	C9H10	alpha-METHYLSTYRENE	liquid	4817.9	40768.2	17530.3
882	C9H10	m-METHYLSTYRENE	liquid	4818.0	40769.0	17530.7
883	C9H10	o-METHYLSTYRENE	liquid	4820.0	40785.9	17538.0
884	C9H10	p-METHYLSTYRENE	liquid	4822.9	40810.5	17548.5
885	C9H10O2	BENZYL ACETATE	liquid	4381.9	29178.2	12546.6
886	C9H10O2	ETHYL BENZOATE	liquid	4410.0	29365.3	12627.1
887	C9H10O3	ETHYL VANILLIN	solid	4250.0	25575.1	10997.3
888	C9H11NO	p-DIMETHYLAMINOBENZALDEHYDE	solid	4730.0	31704.1	13632.8
889	C9H12	CUMENE	liquid	4951.3	41194.2	17713.5
890	C9H12	m-ETHYLTOLUENE	liquid	4943.8	41131.8	17686.7
891	C9H12	o-ETHYLTOLUENE	liquid	4946.1	41151.0	17694.9
892	C9H12	p-ETHYLTOLUENE	liquid	4942.7	41122.7	17682.8
893	C9H12	MESITYLENE	liquid	4929.1	41009.5	17634.1
894	C9H12	n-PROPYLBENZENE	liquid	4954.1	41217.5	17723.5
895	C9H12	1,2,3-TRIMETHYLBENZENE	liquid	4934.0	41050.3	17651.6
896	C9H12	1,2,4-TRIMETHYLBENZENE	liquid	4930.7	41022.8	17639.8
897	C9H12O	BENZYL ETHYL ETHER	liquid	4826.8	35440.6	15239.5
898	C9H12O	2-PHENYL-2-PROPANOL	solid	4690.0	34436.2	14807.6
899	C9H12O2	CUMENE HYDROPEROXIDE	liquid	4844.4	31830.6	13687.2
900	C9H14O	ISOPHORONE	liquid	4932.9	35691.3	15347.3
901	C9H14O6	GLYCERYL TRIACETATE	liquid	3903.0	17886.7	7691.3
902	C9H16	1-NONYNE	liquid	5505.3	44317.5	19056.5
903	C9H16O4	AZELAIC ACID	solid	4422.4	23495.4	10103.0
904	C9H18	BUTYLCYCLOPENTANE	liquid	5518.4	43713.4	18796.7
905	C9H18	cis,cis-1,3,5-TRIMETHYLCYCLOHEXANE	liquid	5474.2	43362.8	18646.0
906	C9H18	cis,trans-1,3,5-TRIMETHYLCYCLOHEXANE	liquid	5481.5	43420.7	18670.9
907	C9H18	ISOPROPYLCYCLOHEXANE	liquid	5480.0	43408.7	18665.7
908	C9H18	1-NONENE	liquid	5569.1	44114.5	18969.2
909	C9H18	n-PROPYLCYCLOHEXANE	liquid	5479.7	43406.3	18664.7
910	C9H18O	DIISOBUTYL KETONE	liquid	5310.0	37331.0	16052.3
911	C9H18O	1-NONANAL	liquid	5355.8	37653.0	16190.8
912	C9H18O2	n-BUTYL VALERATE	liquid	5105.0	32260.9	13872.2
913	C9H18O2	n-NONANOIC ACID	liquid	5060.6	31980.3	13751.5
914	C9H18O2	n-OCTYL FORMATE	liquid	5157.0	32589.5	14013.5
915	C9H20	3,3-DIETHYLPENTANE	liquid	5684.4	44320.0	19057.6
916	C9H20	2,2-DIMETHYL-3-ETHYLPENTANE	liquid	5687.1	44341.1	19066.7
917	C9H20	3-ETHYL-2,3-DIMETHYLPENTANE	liquid	5696.6	44415.4	19098.6
918	C9H20	2,4-DIMETHYL-3-ETHYLPENTANE	liquid	5690.0	44363.7	19076.4
919	C9H20	2,2-DIMETHYLHEPTANE	liquid	5672.0	44223.4	19016.0
920	C9H20	2,6-DIMETHYLHEPTANE	liquid	5673.8	44237.4	19022.1
921	C9H20	3-ETHYLHEPTANE	liquid	5683.8	44315.4	19055.6
922	C9H20	4-ETHYLHEPTANE	liquid	5698.3	44429.1	19104.5
923	C9H20	2,3-DIMETHYLHEPTANE	liquid	5693.7	44392.9	19088.9
924	C9H20	2,4-DIMETHYLHEPTANE	liquid	5689.6	44360.9	19075.2
925	C9H20	2,5-DIMETHYLHEPTANE	liquid	5689.3	44358.9	19074.3
926	C9H20	3,4-DIMETHYLHEPTANE	liquid	5696.3	44413.4	19097.8
927	C9H20	3,5-DIMETHYLHEPTANE	liquid	5692.2	44381.1	19083.9
928	C9H20	4,4-DIMETHYLHEPTANE	liquid	5688.5	44352.4	19071.5
929	C9H20	3-ETHYL-2-METHYLHEXANE	liquid	5696.7	44416.4	19099.0
930	C9H20	4-ETHYL-2-METHYLHEXANE	liquid	5692.2	44381.1	19083.9
931	C9H20	3-ETHYL-3-METHYLHEXANE	liquid	5693.4	44390.6	19088.0
932	C9H20	3-ETHYL-4-METHYLHEXANE	liquid	5699.2	44435.6	19107.3
933	C9H20	2,2,3-TRIMETHYLHEXANE	liquid	5689.5	44359.9	19074.8
934	C9H20	2,2,4-TRIMETHYLHEXANE	liquid	5688.2	44350.1	19070.6
935	C9H20	2,3,3-TRIMETHYLHEXANE	liquid	5691.6	44376.9	19082.1
936	C9H20	2,3,4-TRIMETHYLHEXANE	liquid	5694.7	44400.7	19092.3

Table 26-1 ENTHALPY OF COMBUSTION - ORGANIC COMPOUNDS (continued)

NO	FORMULA	NAME	State	- Enthalpy of Combustion @ 77 F		
				kjoule/mol	kjoule/kg	BTU/lb
937	C9H20	2,3,5-TRIMETHYLHEXANE	liquid	5687.8	44347.2	19069.3
938	C9H20	2,4,4-TRIMETHYLHEXANE	liquid	5690.3	44366.4	19077.6
939	C9H20	3,3,4-TRIMETHYLHEXANE	liquid	5694.4	44398.1	19091.2
940	C9H20	2-METHYLOCTANE	liquid	5679.1	44278.7	19039.8
941	C9H20	3-METHYLOCTANE	liquid	5681.2	44295.1	19046.9
942	C9H20	4-METHYLOCTANE	liquid	5680.2	44287.3	19043.5
943	C9H20	n-NONANE	liquid	5685.1	44325.5	19060.0
944	C9H20	2,2,3,3-TETRAMETHYLPENTANE	liquid	5681.5	44297.4	19047.9
945	C9H20	2,2,3,4-TETRAMETHYLPENTANE	liquid	5684.0	44316.9	19056.3
946	C9H20	2,2,4,4-TETRAMETHYLPENTANE	liquid	5679.8	44284.2	19042.2
947	C9H20	2,3,3,4-TETRAMETHYLPENTANE	liquid	5694.3	44397.4	19090.9
948	C9H20	2,2,5-TRIMETHYLHEXANE	liquid	5666.4	44179.7	18997.3
949	C9H20O	2,6-DIMETHYL-4-HEPTANOL	liquid	5490.0	38057.1	16364.5
950	C9H20O	1-NONANOL	liquid	5500.7	38131.3	16396.4
951	C9H20O	2-NONANOL	liquid	5490.0	38057.1	16364.5
952	C9H20S	n-NONYL MERCAPTAN	liquid	6006.1	37462.3	16108.8
953	C9H20S	BUTYL-PENTYL-SULFIDE	liquid	6029.7	37610.8	16172.6
954	C9H20S	ETHYL-HEPTYL-SULFIDE	liquid	6030.6	37616.5	16175.1
955	C9H20S	HEXYL-PROPYL-SULFIDE	liquid	6030.4	37615.8	16174.8
956	C9H20S	METHYL-OCTYL-SULFIDE	liquid	6033.0	37631.7	16181.6
957	C9H21N	n-NONYLAMINE	liquid	5830.0	40691.8	17497.5
958	C9H21N	TRIPROPYLAMINE	liquid	5874.0	40998.9	17629.5
959	C10H6O8	PYROMELLITIC ACID	solid	3190.0	12551.5	5397.1
960	C10H7Br	1-BROMONAPHTHALENE	liquid	4882.4	23578.5	10138.8
961	C10H7Cl	1-CHLORONAPHTHALENE	liquid	4836.4	29740.9	12788.6
962	C10H8	NAPHTHALENE	solid	4980.9	38860.5	16710.0
963	C10H8	AZULENE	liquid	5129.0	40016.6	17207.1
964	C10H9N	QUINALDINE	liquid	5232.0	36539.4	15711.9
965	C10H10	m-DIVINYLBENZENE	liquid	5300.0	40710.0	17505.3
966	C10H10	1-METHYLINDENE	liquid	5210.0	40018.7	17208.1
967	C10H10	2-METHYLINDENE	solid	5260.0	40402.8	17373.2
968	C10H10O4	DIMETHYL PHTHALATE	liquid	4408.9	22704.4	9762.9
969	C10H10O4	DIMETHYL TEREPHTHALATE	solid	4411.5	22717.8	9768.7
970	C10H12	DICYCLOPENTADIENE	solid	5540.0	41904.6	18019.0
971	C10H12	1,2,3,4-TETRAHYDRONAPHTHALENE	liquid	5357.5	40524.2	17425.4
972	C10H12O	ANETHOLE	liquid	5280.0	35626.3	15319.3
973	C10H12O4	DIALLYL MALEATE	liquid	4790.0	24413.5	10497.8
974	C10H14	n-BUTYLBENZENE	liquid	5564.4	41457.0	17826.5
975	C10H14	sec-BUTYLBENZENE	liquid	5561.4	41434.6	17816.9
976	C10H14	tert-BUTYLBENZENE	liquid	5557.1	41402.6	17803.1
977	C10H14	1,2,3,4-TETRAMETHYLBENZENE	liquid	5544.8	41311.3	17763.8
978	C10H14	m-CYMENE	liquid	5549.2	41343.8	17777.8
979	C10H14	o-CYMENE	liquid	5554.5	41383.2	17794.8
980	C10H14	p-CYMENE	liquid	5549.8	41348.2	17779.7
981	C10H14	m-DIETHYLBENZENE	liquid	5554.3	41381.8	17794.2
982	C10H14	o-DIETHYLBENZENE	liquid	5559.3	41419.0	17810.2
983	C10H14	p-DIETHYLBENZENE	liquid	5550.0	41349.7	17780.4
984	C1OH14	2-ETHYL-m-XYLENE	liquid	5548.0	41334.8	17774.0
985	C10H14	2-ETHYL-p-XYLENE	liquid	5543.0	41297.6	17758.0
986	C10H14	3-ETHYL-o-XYLENE	liquid	5547.3	41329.6	17771.7
987	C10H14	4-ETHYL-m-XYLENE	liquid	5544.0	41305.0	17761.2
988	C10H14	4-ETHYL-o-XYLENE	liquid	5542.0	41290.1	17754.7
989	C1OH14	5-ETHYL-m-XYLENE	liquid	5540.0	41275.2	17748.3
990	C10H14	ISOBUTYLBENZENE	liquid	5558.0	41409.3	17806.0
991	C10H14	1,2,3,5-TETRAMETHYLBENZENE	liquid	5532.0	41215.6	17722.7
992	C10H14	1,2,4,5-TETRAMETHYLBENZENE	solid	5529.8	41199.2	17715.7
993	C10H14O	p-tert-BUTYLPHENOL	solid	5360.0	35680.8	15342.7
994	C10H14O2	p-tert-BUTYLCATECHOL	solid	5170.0	31103.4	13374.4
995	C10H15N	N,N-DIETYHLANILINE	liquid	5730.0	38395.6	16510.1
996	C10H15N	2,6-DIETHYLANILINE	liquid	5650.0	37859.5	16279.6
997	C10H16	CAMPHENE	solid	5790.0	42499.5	18274.8
998	C10H16	D-LIMONENE	liquid	5815.4	42685.9	18354.9
999	C10H16	alpha-PHELLANDRENE	liquid	5810.0	42646.3	18337.9
1000	C1OH16	beta-PHELLANDRENE	liquid	5810.0	42646.3	18337.9
1001	C10H16	alpha-PINENE	liquid	5850.0	42939.9	18464.1
1002	C10H16	beta-PINENE	liquid	5860.0	43013.3	18495.7
1003	C10H16	alpha-TERPINENE	liquid	5800.0	42572.9	18306.3
1004	C1OH16	gamma-TERPINENE	liquid	5810.0	42646.3	18337.9
1005	C10H16	TERPINOLENE	liquid	-------	-------	-------
1006	C10H16O	CAMPHOR	solid	5560.0	36522.2	15704.6
1007	C10H18	1-DECYNE	liquid	6118.1	44253.4	19029.0
1008	C10H18	cis-DECAHYDRONANPHTALENE	liquid	5892.1	42618.2	18325.8
1009	C10H18	trans-DECAHYDRONAPHTHALENE	liquid	5880.9	42537.2	18291.0
1010	C10H18O4	SEBACIC ACID	solid	5029.0	24865.1	10692.0
1011	C10H20	n-BUTYLCYCLOHEXANE	liquid	6090.2	43418.0	18669.7
1012	C10H20	1-CYCLOPENTYLPENTANE	liquid	6131.1	43709.8	18795.2
1013	C10H20	1-DECENE	liquid	6178.2	44045.4	18939.5
1014	C10H20O	1-DECANAL	liquid	5964.4	38167.8	16412.1

Table 26-1 ENTHALPY OF COMBUSTION - ORGANIC COMPOUNDS (continued)

NO	FORMULA	NAME	State	Enthalpy of Combustion @ 77 F		
				kjoule/mol	kjoule/kg	BTU/lb
1015	C10H20O2	n-DECANOIC ACID	solid	5752.0	33389.8	14357.6
1016	C10H20O2	2-ETHYLHEXYL ACETATE	liquid	5720.0	33204.1	14277.8
1017	C10H20O2	ISOPENTYL ISOVALERATE	liquid	5713.1	33164.0	14260.5
1018	C10H22	n-DECANE	liquid	6294.2	44236.6	19021.7
1019	C10H22	2,2-DIMETHYLOCTANE	liquid	6281.0	44143.8	18981.8
1020	C10H22	2-METHYLNONANE	liquid	6288.8	44198.6	19005.4
1021	C10H22	3-ETHYLOCTANE	liquid	6308.9	44339.7	19066.1
1022	C10H22	4-ETHYLOCTANE	liquid	6309.4	44343.3	19067.6
1023	C10H22	2,3-DIMETHYLOCTANE	liquid	6311.0	44355.2	19072.7
1024	C10H22	2,4-DIMETHYLOCTANE	liquid	6311.3	44357.0	19073.5
1025	C10H22	2,5-DIMETHYLOCTANE	liquid	6296.7	44254.2	19029.3
1026	C10H22	2,6-DIMETHYLOCTANE	liquid	6302.0	44291.8	19045.5
1027	C10H22	2,7-DIMETHYLOCTANE	liquid	6307.2	44328.6	19061.3
1028	C10H22	3,3-DIMETHYLOCTANE	liquid	6302.4	44294.5	19046.6
1029	C10H22	3,4-DIMETHYLOCTANE	liquid	6302.0	44291.8	19045.5
1030	C10H22	3,5-DIMETHYLOCTANE	liquid	6299.1	44271.2	19036.6
1031	C10H22	3,6-DIMETHYLOCTANE	liquid	6301.5	44288.3	19044.0
1032	C10H22	4,4-DIMETHYLOCTANE	liquid	6309.9	44347.4	19069.4
1033	C10H22	4,5-DIMETHYLOCTANE	liquid	6305.4	44315.4	19055.6
1034	C10H22	4-PROPYLHEPTANE	liquid	6305.0	44313.0	19054.6
1035	C10H22	4-ISOPROPYLHEPTANE	liquid	6301.9	44290.9	19045.1
1036	C10H22	3-ETHYL-2-METHYLHEPTANE	liquid	6310.1	44348.6	19069.9
1037	C10H22	4-ETHYL-2-METHYLHEPTANE	liquid	6312.1	44363.0	19076.1
1038	C10H22	5-ETHYL-2-METHYLHEPTANE	liquid	6312.8	44367.4	19078.0
1039	C10H22	3-ETHYL-3-METHYLHEPTANE	liquid	6310.2	44349.5	19070.3
1040	C10H22	4-ETHYL-3-METHYLHEPTANE	liquid	6305.7	44318.0	19056.7
1041	C10H22	3-ETHYL-5-METHYLHEPTANE	liquid	6305.4	44315.4	19055.6
1042	C10H22	3-ETHYL-4-METHYLHEPTANE	liquid	6306.9	44325.9	19060.2
1043	C10H22	4-ETHYL-4-METHYLHEPTANE	liquid	6313.2	44370.3	19079.2
1044	C10H22	2,2,3-TRIMETHYLHEPTANE	liquid	6308.3	44335.6	19064.3
1045	C10H22	2,2,4-TRIMETHYLHEPTANE	liquid	6313.1	44369.8	19079.0
1046	C10H22	3-METHYLNONANE	liquid	6291.0	44214.4	19012.2
1047	C10H22	2,2,5-TRIMETHYLHEPTANE	liquid	6302.4	44294.5	19046.6
1048	C10H22	2,2,6-TRIMETHYLHEPTANE	liquid	6301.1	44285.4	19042.7
1049	C10H22	2,3,3-TRIMETHYLHEPTANE	liquid	6292.9	44228.0	19018.0
1050	C10H22	2,3,4-TRIMETHYLHEPTANE	liquid	6290.0	44207.1	19009.1
1051	C10H22	2,3,5-TRIMETHYLHEPTANE	liquid	6304.7	44310.6	19053.6
1052	C10H22	2,3,6-TRIMETHYLHEPTANE	liquid	6308.3	44336.2	19064.6
1053	C10H22	2,4,4-TRIMETHYLHEPTANE	liquid	6303.7	44303.3	19050.4
1054	C10H22	2,4,5-TRIMETHYLHEPTANE	liquid	6300.6	44281.8	19041.2
1055	C10H22	2,4,6-TRIMETHYLHEPTANE	liquid	6303.5	44302.4	19050.0
1056	C10H22	2,5,5-TRIMETHYLHEPTANE	liquid	6303.7	44303.6	19050.5
1057	C10H22	3,3,4-TRIMETHYLHEPTANE	liquid	6311.2	44356.5	19073.3
1058	C10H22	3,3,5-TRIMETHYLHEPTANE	liquid	6295.2	44243.9	19024.9
1059	C10H22	3,4,4-TRIMETHYLHEPTANE	liquid	6307.6	44330.9	19062.3
1060	C10H22	3,4,5-TRIMETHYLHEPTANE	liquid	6306.1	44320.6	19057.9
1061	C10H22	3-ISOPROPYL-2-METHYLHEXANE	liquid	6307.7	44331.8	19062.7
1062	C10H22	3,3-DIETHYLHEXANE	liquid	6311.1	44355.3	19072.8
1063	C10H22	3,4-DIETHYLHEXANE	liquid	6309.0	44341.2	19066.7
1064	C10H22	3-ETHYL-2,2-DIMETHYLHEXANE	liquid	6312.1	44362.4	19075.8
1065	C10H22	4-ETHYL-2,2-DIMETHYLHEXANE	liquid	6316.1	44390.9	19088.1
1066	C10H22	3-ETHYL-2,3-DIMETHYLHEXANE	liquid	6305.7	44317.7	19056.6
1067	C10H22	4-ETHYL-2,3-DIMETHYLHEXANE	liquid	6304.2	44306.8	19051.9
1068	C10H22	3-ETHYL-2,4-DIMETHYLHEXANE	liquid	6310.0	44347.7	19069.5
1069	C10H22	4-ETHYL-2,4-DIMETHYLHEXANE	liquid	6311.3	44356.8	19073.4
1070	C10H22	3-ETHYL-2,5-DIMETHYLHEXANE	liquid	6311.3	44357.4	19073.7
1071	C10H22	4-ETHYL-3,3-DIMETHYLHEXANE	liquid	6308.1	44334.5	19063.8
1072	C10H22	3-ETHYL-3,4-DIMETHYLHEXANE	liquid	6304.0	44305.7	19051.4
1073	C10H22	4-METHYLNONANE	liquid	6291.0	44214.4	19012.2
1074	C10H22	2,2,3,3-TETRAMETHYLHEXANE	liquid	6313.1	44369.5	19078.9
1075	C10H22	2,2,3,4-TETRAMETHYLHEXANE	liquid	6307.1	44327.7	19060.9
1076	C10H22	2,2,3,5-TETRAMETHYLHEXANE	liquid	6311.6	44359.2	19074.4
1077	C10H22	2,2,4,4-TETRAMETHYLHEXANE	liquid	6296.1	44250.4	19027.7
1078	C10H22	2,2,4,5-TETRAMETHYLHEXANE	liquid	6309.0	44340.6	19066.5
1079	C10H22	2,2,5,5-TETRAMETHYLHEXANE	liquid	6299.4	44273.3	19037.5
1080	C10H22	2,3,3,4-TETRAMETHYLHEXANE	liquid	6280.8	44142.7	18981.4
1081	C10H22	2,3,3,5-TETRAMETHYLHEXANE	liquid	6310.6	44351.8	19071.3
1082	C10H22	2,3,4,4-TETRAMETHYLHEXANE	liquid	6306.4	44322.4	19058.6
1083	C10H22	2,3,4,5-TETRAMETHYLHEXANE	liquid	6313.7	44374.2	19080.9
1084	C10H22	3,3,4,4-TETRAMETHYLHEXANE	liquid	6306.5	44323.0	19058.9
1085	C10H22	2,4-DIMETHYL-3-ISOPROPYLPENTANE	liquid	6313.7	44374.2	19080.9
1086	C10H22	3,3-DIETHYL-2-METHYLPENTANE	liquid	6307.0	44326.5	19060.4
1087	C10H22	3-ETHYL-2,2,3-TRIMETHYLPENTANE	liquid	6314.9	44382.4	19084.4
1088	C10H22	3-ETHYL-2,2,4-TRIMETHYLPENTANE	liquid	6312.3	44363.9	19076.5
1089	C10H22	3-ETHYL-2,3,4-TRIMETHYLPENTANE	liquid	6311.9	44360.9	19075.2
1090	C10H22	2,2,3,3,4-PENTAMETHYLPENTANE	liquid	6312.9	44368.3	19078.4
1091	C10H22	2,2,3,4,4-PENTAMETHYLPENTANE	liquid	6317.8	44402.7	19093.2
1092	C10H22	5-METHYLNONANE	liquid	6291.0	44214.4	19012.2

598

Table 26-1 ENTHALPY OF COMBUSTION - ORGANIC COMPOUNDS (continued)

NO	FORMULA	NAME	State	- Enthalpy of Combustion @ 77 F		
				kjoule/mol	kjoule/kg	BTU/lb
1093	C10H22O	1-DECANOL	liquid	6117.0	38645.7	16617.7
1094	C10H22O	DI-n-PENTYL ETHER	liquid	6170.0	38980.6	16761.6
1095	C10H22O	ISODECANOL	liquid	6110.0	38601.5	16598.6
1096	C10H22O5	TETRAETHYLENE GLYCOL DIMETHYL ETHER	liquid	5640.0	25373.2	10910.5
1097	C10H22S	n-DECYL MERCAPTAN	liquid	6616.0	37946.4	16317.0
1098	C10H22S	BUTYL-HEXYL-SULFIDE	liquid	6642.3	38098.9	16382.5
1099	C10H22S	ETHYL-OCTYL-SULFIDE	liquid	6643.2	38104.0	16384.7
1100	C10H22S	HEPTYL-PROPYL-SULFIDE	liquid	6643.1	38103.3	16384.4
1101	C10H22S	METHYL-NONYL-SULFIDE	liquid	6645.6	38117.9	16390.7
1102	C10H22S	PENTYL-SULFIDE	liquid	6642.3	38098.9	16382.5
1103	C10H22S2	PENTYL-DISULFIDE	liquid	6946.3	33653.9	14471.2
1104	C10H23N	n-DECYLAMINE	liquid	6440.0	40941.1	17604.7
1105	C11H10	1-METHYLNAPHTHALENE	liquid	5594.0	39339.0	16915.8
1106	C11H10	2-METHYLNAPHTHALENE	solid	5582.7	39259.5	16881.6
1107	C11H14O2	n-BUTYL BENZOATE	liquid	5590.0	31363.8	13486.4
1108	C11H16	n-PENTYLBENZENE	liquid	6177.0	41666.7	17916.7
1109	C11H16O	p-tert-AMYLPHENOL	solid	5980.0	36408.6	15655.7
1110	C11H20	1-UNDECYNE	liquid	6731.0	44201.8	19006.8
1111	C11H20O2	2-ETHYLHEXYL ACRYLATE	liquid	6220.3	33754.8	14514.6
1112	C11H22	1-UNDECENE	liquid	6787.9	43992.7	18916.9
1113	C11H22	1-CYCLOPENTYLHEXANE	liquid	6744.2	43709.7	18795.2
1114	C11H22	PENTYLCYCLOHEXANE	liquid	6720.2	43554.3	18728.3
1115	C11H22O	1-UNDECANAL	liquid	6570.0	38580.1	16589.4
1116	C11H24	n-UNDECANE	liquid	6903.8	44166.8	18991.7
1117	C11H24O	1-UNDECANOL	liquid	6739.1	39110.1	16817.3
1118	C11H24S	UNDECYL MERCAPTAN	liquid	7225.4	38355.9	16493.0
1119	C11H24S	BUTYL-HEPTYL-SULFIDE	liquid	7255.1	38514.9	16561.4
1120	C11H24S	DECYL-METHYL-SULFIDE	liquid	7258.4	38532.7	16569.1
1121	C11H24S	ETHYL-NONYL-SULFIDE	liquid	7256.0	38519.8	16563.5
1122	C11H24S	OCTYL-PROPYL-SULFIDE	liquid	7256.0	38519.6	16563.4
1123	C12H8O	DIBENZOFURAN	solid	5680.0	33770.3	14521.2
1124	C12H9N	DIBENZOPYRROLE	solid	5935.6	35497.9	15264.1
1125	C12H10	ACENAPHTHENE	solid	6001.4	38916.8	16734.2
1126	C12H10	BIPHENYL	solid	6031.7	39113.3	16818.7
1127	C12H10O	DIPHENYL ETHER	solid	5920.0	34780.4	14955.6
1128	C12H11N	p-AMINODIPHENYL	solid	6160.0	36401.0	15652.4
1129	C12H11N	DIPHENYLAMINE	solid	6180.0	36519.2	15703.3
1130	C12H11N3	p-AMINOAZOBENZENE	solid	6380.0	32346.4	13908.9
1131	C12H11N3	1,3-DIPHENYLTRIAZENE	solid	6384.0	32366.7	13917.7
1132	C12H12	1,2-DIMETHYLNAPHTHALENE	liquid	6209.7	39748.2	17091.7
1133	C12H12	1,3-DIMETHYLNAPHTHALENE	liquid	6208.3	39738.8	17087.7
1134	C12H12	1,4-DIMETHYLNAPHTHALENE	liquid	6208.8	39742.1	17089.1
1135	C12H12	1,5-DIMETHYLNAPHTHALENE	solid	6208.3	39738.9	17087.7
1136	C12H12	1,6-DIMETHYLNAPHTHALENE	liquid	6209.2	39744.7	17090.2
1137	C12H12	1,7-DIMETHYLNAPHTHALENE	liquid	6208.5	39740.2	17088.3
1138	C12H12	2,3-DIMETHYLNAPHTHALENE	solid	6209.8	39748.4	17091.8
1139	C12H12	2,6-DIMETHYLNAPHTHALENE	solid	6167.6	39478.5	16975.7
1140	C12H12	2,7-DIMETHYLNAPHTHALENE	solid	6167.9	39480.4	16976.6
1141	C12H12	1-ETHYLNAPHTHALENE	liquid	6171.0	39500.2	16985.1
1142	C12H12	2-ETHYLNAPHTHALENE	liquid	6224.2	39840.6	17131.5
1143	C12H12N2	p-AMINODIPHENYLAMINE	solid	6286.4	34120.5	14671.8
1144	C12H12N2	HYDRAZOBENZENE	solid	6394.6	34707.8	14924.4
1145	C12H14	1,2,3-TRIMETHYLINDENE	solid	6470.0	40886.5	17581.2
1146	C12H14O4	DIETHYL PHTHALATE	liquid	5636.0	25359.9	10904.7
1147	C12H16	CYCLOHEXYLBENZENE	liquid	6574.8	41026.1	17641.2
1148	C12H18	m-DIISOPROPYLBENZENE	liquid	6770.0	41719.3	17939.3
1149	C12H18	p-DIISOPROPYLBENZENE	liquid	6770.0	41719.3	17939.3
1150	C12H18	n-HEXYLBENZENE	liquid	6783.7	41803.7	17975.6
1151	C12H18	1,2,3-TRIETHYLBENZENE	liquid	6790.8	41848.0	17994.6
1152	C12H18	1,2,4-TRIETHYLBENZENE	liquid	6787.7	41828.9	17986.4
1153	C12H18	1,3,5-TRIETHYLBENZENE	liquid	6784.2	41807.3	17977.1
1154	C12H18	HEXAMETHYLBENZENE	solid	6751.9	41607.9	17891.4
1155	C12H20O4	DIBUTYL MALEATE	liquid	6360.0	27859.5	11979.6
1156	C12H22	BICYCLOHEXYL	liquid	7053.0	42409.5	18236.1
1157	C12H22	1-DODECYNE	liquid	7343.9	44159.1	18988.4
1158	C12H23N	DICYCLOHEXYLAMINE	liquid	7260.0	40039.5	17217.0
1159	C12H24	1-DODECENE	liquid	7397.7	43949.4	18898.3
1160	C12H24	1-CYCLOPENTYLHEPTANE	liquid	7357.3	43709.5	18795.1
1161	C12H24	1-CYCLOHEXYLHEXANE	liquid	7333.5	43568.3	18734.4
1162	C12H24O	1-DODECANAL	liquid	7181.5	38961.7	16753.5
1163	C12H24O2	n-DODECANOIC ACID	solid	6849.9	34194.6	14703.7
1164	C12H26	n-DODECANE	liquid	7513.6	44109.9	18967.3
1165	C12H26O	DI-n-HEXYL ETHER	liquid	7383.8	39625.8	17039.1
1166	C12H26O	1-DODECANOL	liquid	7338.0	39380.1	16933.4
1167	C12H26O3	DIETHYLENE GLYCOL DI-n-BUTYL ETHER	liquid	7140.0	32701.7	14061.7
1168	C12H26S	n-DODECYL MERCAPTAN	liquid	7840.0	38734.4	16655.8
1169	C12H26S	BUTYL-OCTYL-SULFIDE	liquid	7868.0	38874.2	16715.9
1170	C12H26S	DECYL-ETHYL-SULFIDE	liquid	7867.4	38871.1	16714.6

Table 26-1 ENTHALPY OF COMBUSTION - ORGANIC COMPOUNDS (continued)

NO	FORMULA	NAME	State	- Enthalpy of Combustion @ 77 F		
				kjoule/mol	kjoule/kg	BTU/lb
1171	C12H26S	HEXYL-SULFIDE	liquid	7868.0	38874.2	16715.9
1172	C12H26S	METHYL-UNDECYL-SULFIDE	liquid	7871.3	38890.5	16722.9
1173	C12H26S	NONYL-PROPYL-SULFIDE	liquid	7868.8	38877.9	16717.5
1174	C12H26S2	HEXYL-DISULFIDE	liquid	8173.1	34859.7	14989.7
1175	C12H27BO3	TRI-n-BUTYL BORATE	liquid	-------	-------	-------
1176	C12H27N	DODECYLAMINE	solid	7660.0	41326.5	17770.4
1177	C12H27N	TRI-n-BUTYLAMINE	liquid	7690.0	41488.4	17840.0
1178	C13H10	FLUORENE	solid	6425.1	38653.7	16621.1
1179	C13H10O	BENZOPHENONE	solid	6292.0	34529.3	14847.6
1180	C13H12	DIPHENYLMETHANE	solid	6658.9	39580.2	17019.5
1181	C13H14	1-PROPYLNAPHTHALENE	liquid	6835.6	40149.3	17264.2
1182	C13H14	2-PROPYLNAPHTHALENE	liquid	6834.3	40141.8	17261.0
1183	C13H14	2ETHYL-3-METHYLNAPHTHALENE	solid	6826.4	40095.5	17241.1
1184	C13H14	2ETHYL-6-METHYLNAPHTHALENE	solid	6822.6	40073.1	17231.4
1185	C13H14	2ETHYL-7-METHYLNAPHTHALENE	solid	6822.2	40070.9	17230.5
1186	C13H20	n-HEPTYLBENZENE	liquid	7393.4	41936.0	18032.5
1187	C13H24	1-TRIDECYNE	liquid	7956.9	44123.6	18973.2
1188	C13H26	1-TRIDECENE	liquid	8007.3	43911.9	18882.1
1189	C13H26	1-CYCLOPENTYLOCTANE	liquid	7970.5	43710.6	18795.5
1190	C13H26	1-CYCLOHEXYLHEPTANE	liquid	7946.7	43580.0	18739.4
1191	C13H26O	1-TRIDECANAL	liquid	7790.0	39274.2	16887.9
1192	C13H26O2	n-BUTYL NONANOATE	liquid	7561.0	35274.4	15168.0
1193	C13H26O2	METHYL DODECANOATE	liquid	7566.8	35301.5	15179.6
1194	C13H28	n-TRIDECANE	liquid	8123.5	44062.1	18946.7
1195	C13H28O	1-TRIDECANOL	solid	7953.0	39692.6	17067.8
1196	C13H28S	BUTYL-NONYL-SULFIDE	liquid	8481.0	39187.0	16850.4
1197	C13H28S	DECYL-PROPYL-SULFIDE	liquid	8481.8	39190.4	16851.9
1198	C13H28S	DODECYL-METHYL-SULFIDE	liquid	8484.3	39202.4	16857.0
1199	C13H28S	ETHYL-UNDECYL-SULFIDE	liquid	8481.9	39191.2	16852.2
1200	C13H28S	1-TRIDECANETHIOL	liquid	8480.0	39182.5	16848.5
1201	C14H8O2	ANTHRAQUINONE	solid	6292.0	30218.6	12994.0
1202	C14H10	ANTHRACENE	solid	6847.4	38418.3	16519.8
1203	C14H10	DIPHENYLACETYLENE	solid	7148.4	40107.1	17246.0
1204	C14H10	PHENANTHRENE	solid	6834.4	38345.3	16488.5
1205	C14H12	cis-STILBENE	liquid	7196.9	39927.5	17168.8
1206	C14H12	trans-STILBENE	solid	7200.0	39944.7	17176.2
1207	C14H12O2	BENZYL BENZOATE	liquid	6690.0	31519.7	13553.5
1208	C14H14	1,1-DIPHENYLETHANE	liquid	7250.0	39777.2	17104.2
1209	C14H14	1,2-DIPHENYLETHANE	solid	7280.2	39942.9	17175.5
1210	C14H14O	DIBENZYL ETHER	liquid	7145.1	36038.1	15496.4
1211	C14H16	1-n-BUTYLNAPHTHALENE	liquid	7425.0	40291.7	17325.4
1212	C14H16	2-BUTYLNAPHTHALENE	liquid	7446.9	40410.8	17376.6
1213	C14H22	n-OCTYLBENZENE	liquid	8003.0	42048.2	18080.7
1214	C14H22	1,2,3,4-TETRAETHYLBENZENE	liquid	8003.3	42049.8	18081.4
1215	C14H22	1,2,3,5-TETRAETHYLBENZENE	liquid	8003.7	42052.3	18082.5
1216	C14H22	1,2,4,5-TETRAETHYLBENZENE	liquid	8003.4	42050.3	18081.6
1217	C14H22O	p-tert-OCTYLPHENOL	solid	8460.0	41002.7	17631.2
1218	C14H28	1-TETRADECENE	liquid	8616.9	43879.6	18868.2
1219	C14H28	1-CYCLOPENTYLNONANE	liquid	8584.0	43712.1	18796.2
1220	C14H28	1-CYCLOHEXYLOCTANE	liquid	8560.1	43590.7	18744.0
1221	C14H28O2	n-TETRADECANOIC ACID	solid	8060.0	35292.8	15175.9
1222	C14H30	n-TETRADECANE	liquid	8733.3	44020.4	18928.8
1223	C14H30O	1-TETRADECANOL	solid	8562.7	39939.5	17174.0
1224	C14H30S	BUTYL-DECYL-SULFIDE	liquid	9094.1	39462.3	16968.8
1225	C14H30S	DODECYL-ETHYL-SULFIDE	liquid	9095.1	39466.3	16970.5
1226	C14H30S	HEPTYL-SULFIDE	liquid	9094.2	39462.5	16968.9
1227	C14H30S	METHYL-TRIDECYL-SULFIDE	liquid	9097.5	39476.9	16975.0
1228	C14H30S	PROPYL-UNDECYL-SULFIDE	liquid	9094.9	39465.8	16970.3
1229	C14H30S	1-TETRADECANETHIOL	liquid	9093.6	39460.1	16967.8
1230	C14H30S2	HEPTYL-DISULFIDE	liquid	9399.9	35807.7	15397.3
1231	C14H31N	TETRADECYLAMINE	solid	8830.0	41376.3	17791.8
1232	C15H10N2O2	DIPHENYLMETHANE-4,4'-DIISOCYANATE	solid	7056.0	28195.0	12123.9
1233	C15H16O	p-CUMYLPHENOL	solid	7690.0	36223.9	15576.3
1234	C15H16O2	BISPHENOL A	solid	7590.4	33248.8	14297.0
1235	C15H18	1-PENTYLNAPHTHALENE	liquid	8060.5	40646.6	17478.0
1236	C15H18	2-PENTYLNAPHTHALENE	liquid	8059.5	40641.4	17475.8
1237	C15H24	n-NONYLBENZENE	liquid	8611.3	42138.7	18119.6
1238	C15H24O	2,6-DI-tert-BUTYL-p-CRESOL	solid	8390.0	38074.9	16372.2
1239	C15H24O	NONYLPHENOL	liquid	8420.0	38211.1	16430.8
1240	C15H28	1-PENTADECYNE	liquid	9183.3	44068.8	18949.6
1241	C15H30	1-PENTADECENE	liquid	9230.0	43868.2	18863.3
1242	C15H30	1-CYCLOPENTYLDECANE	liquid	9197.5	43713.8	18796.9
1243	C15H30	1-CYCLOHEXYLNONANE	liquid	9173.6	43600.3	18748.1
1244	C15H30O2	PENTADECANOIC ACID	solid	8668.9	35762.5	15377.9
1245	C15H32	n-PENTADECANE	liquid	9343.0	43983.8	18913.0
1246	C15H32O	1-PENTADECANOL	solid	9207.2	40308.6	17332.7
1247	C15H32S	BUTYL-UNDECYL-SULFIDE	liquid	9707.5	39707.1	17074.0
1248	C15H32S	DODECYL-PROPYL-SULFIDE	liquid	9708.3	39710.1	17075.4

Table 26-1 ENTHALPY OF COMBUSTION - ORGANIC COMPOUNDS (continued)

NO	FORMULA	NAME	State	Enthalpy of Combustion @ 77 F		
				kjoule/mol	kjoule/kg	BTU/lb
1249	C15H32S	ETHYL-TRIDECYL-SULFIDE	liquid	9708.4	39710.7	17075.6
1250	C15H32S	METHYL-TETRADECYL-SULFIDE	liquid	9710.8	39720.6	17079.8
1251	C15H32S	1-PENTADECANETHIOL	liquid	9704.9	39696.2	17069.4
1252	C16H10	FLUORANTHENE	solid	7695.0	38046.0	16359.8
1253	C16H10	PYRENE	solid	7620.1	37675.7	16200.6
1254	C16H12	1-PHENYLNAPHTHALENE	solid	7920.0	38772.0	16672.0
1255	C16H20	1-n-HEXYLNAPHTHALENE	liquid	8648.0	40728.1	17513.1
1256	C16H22O4	DIBUTYL PHTHALATE	liquid	8106.6	29124.0	12523.3
1257	C16H26	n-DECYLBENZENE	liquid	9222.3	42230.1	18159.0
1258	C16H26	PENTAETHYLBENZENE	solid	9219.7	42218.6	18154.0
1259	C16H30	1-HEXADECYNE	liquid	9796.7	44047.2	18940.3
1260	C16H32	n-DECYLCYCLOHEXANE	liquid	9749.0	43438.9	18678.7
1261	C16H32	1-CYCLOPENTYLUNDECANE	liquid	9810.9	43714.9	18797.4
1262	C16H32	1-HEXADECENE	liquid	9836.3	43827.9	18846.0
1263	C16H32O2	n-HEXADECANOIC ACID	solid	9274.7	36168.7	15552.5
1264	C16H34	n-HEXADECANE	liquid	9952.8	43952.2	18899.4
1265	C16H34O	DI-n-OCTYL ETHER	liquid	9820.0	40504.0	17416.7
1266	C16H34O	1-HEXADECANOL	solid	9797.0	40409.2	17375.9
1267	C16H34S	BUTYL-DODECYL-SULFIDE	liquid	10321.1	39926.0	17168.2
1268	C16H34S	ETHYL-TETRADECYL-SULFIDE	liquid	10325.1	39941.8	17175.0
1269	C16H34S	METHYL-PENTADECYL-SULFIDE	liquid	10324.4	39939.0	17173.8
1270	C16H34S	OCTYL-SULFIDE	liquid	10321.1	39926.2	17168.3
1271	C16H34S	PROPYL-TRIDECYL-SULFIDE	liquid	10321.8	39929.0	17169.4
1272	C16H34S	1-HEXADECANETHIOL	liquid	10318.6	39916.3	17164.0
1273	C16H34S2	OCTYL-DISULFIDE	liquid	10625.4	36568.1	15724.3
1274	C17H28	n-UNDECYLBENZENE	liquid	9832.0	42304.7	18191.0
1275	C17H32	1-HEPTADECYNE	liquid	10410.2	44028.7	18932.4
1276	C17H34	1-CYCLOPENTYLDODECANE	liquid	10421.7	43705.1	18793.2
1277	C17H34	1-CYCLOHEXYLUNDECANE	liquid	10400.7	43617.0	18755.3
1278	C17H34	1-HEPTADECENE	liquid	10446.0	43806.6	18836.9
1279	C17H36	n-HEPTADECANE	liquid	10568.0	43946.7	18897.1
1280	C17H36O	1-HEPTADECANOL	solid	10396.0	40534.6	17429.9
1281	C17H36S	BUTYL-TRIDECYL-SULFIDE	liquid	10935.0	40123.8	17253.3
1282	C17H36S	ETHYL-PENTADECYL-SULFIDE	liquid	10935.9	40127.1	17254.6
1283	C17H36S	HEXADECYL-METHYL-SULFIDE	liquid	10938.3	40136.0	17258.5
1284	C17H36S	PROPYL-TETRADECYL-SULFIDE	liquid	10935.7	40126.6	17254.4
1285	C17H36S	1-HEPTADECANETHIOL	solid	10932.4	40114.3	17249.1
1286	C18H12	CHRYSENE	solid	8679.4	38018.7	16348.0
1287	C18H14	m-TERPHENYL	solid	9050.0	39295.0	16896.9
1288	C18H14	o-TERPHENYL	solid	9050.0	39295.0	16896.9
1289	C18H14	p-TERPHENYL	solid	9052.5	39305.9	16901.5
1290	C18H15P	TRIPHENYLPHOSPHINE	solid	10032.0	38247.6	16446.5
1291	C18H15O4P	TRIPHENYL PHOSPHATE	solid	9059.0	27763.8	11938.4
1292	C18H16N2	N,N'-DIPHENYL-p-PHENYLENEDIAMINE	solid	9340.0	35876.3	15426.8
1293	C18H22	2,3-DIMETHYL-2,3-DIPHENYLBUTANE	solid	9800.0	41112.0	17678.2
1294	C18H22O2	DICUMYL PEROXIDE	solid	9630.0	35617.7	15315.6
1295	C18H30	n-DODECYLBENZENE	liquid	10442.0	42372.1	18220.0
1296	C18H30	HEXAETHYLBENZENE	solid	10439.8	42363.5	18216.3
1297	C18H32O2	LINOLEIC ACID	liquid	10400.0	37083.1	15945.7
1298	C18H34	1-OCTADECYNE	solid	11023.9	44013.6	18925.8
1299	C18H34O2	OLEIC ACID	liquid	10523.0	37253.9	16019.2
1300	C18H34O4	DIBUTYL SEBACATE	liquid	10033.0	31904.9	13719.1
1301	C18H34O4	DIHEXYL ADIPATE	liquid	10100.0	32117.9	13810.7
1302	C18H36	1-CYCOPENTYLTRIDECANE	liquid	11038.0	43718.2	18798.8
1303	C18H36	1-CYCLOHEXYLDODECANE	liquid	11014.5	43624.9	18758.7
1304	C18H36	1-OCTADECENE	liquid	11056.0	43788.9	18829.2
1305	C18H36O2	STEARIC ACID	solid	10489.0	36870.4	15854.3
1306	C18H38	n-OCTADECANE	solid	11173.0	43901.8	18877.8
1307	C18H38O	DINONYL ETHER	liquid	11100.0	41035.3	17645.2
1308	C18H38O	1-OCTADECANOL	solid	10998.0	40658.2	17483.0
1309	C18H38S	BUTYL-TETRADECYL-SULFIDE	solid	11549.1	40302.8	17330.2
1310	C18H38S	ETHYL-HEXADECYL-SULFIDE	solid	11550.0	40306.0	17331.6
1311	C18H38S	HEPTADECYL-METHYL-SULFIDE	solid	11552.4	40314.5	17335.2
1312	C18H38S	NONYL-SULFIDE	solid	11549.1	40302.8	17330.2
1313	C18H38S	PENTADECYL-PROPYL-SULFIDE	solid	11549.8	40305.4	17331.3
1314	C18H38S	1-OCTADECANETHIOL	solid	11546.6	40294.2	17326.5
1315	C18H38S2	NONYL-DISULFIDE	liquid	11854.1	37204.9	15998.1
1316	C19H26	1-n-NONYLNAPHTHALENE	liquid	10500.0	41271.2	17746.6
1317	C19H32	n-TRIDECYLBENZENE	liquid	11051.0	42428.3	18244.2
1318	C19H36	1-NONADECYNE	solid	11637.8	44000.6	18920.3
1319	C19H36O2	METHYL OLEATE	liquid	11100.0	37437.5	16098.1
1320	C19H38	1-CYCLOPENTYLTETRADECANE	liquid	11651.6	43719.5	18799.4
1321	C19H38	1-CYCLOHEXYLTRIDECANE	liquid	11628.3	43631.8	18761.7
1322	C19H38	1-NONADECENE	liquid	11700.0	43900.6	18877.3
1323	C19H38O2	NONADECANOIC ACID	solid	11088.0	37144.5	15972.1
1324	C19H40	n-NONADECANE	solid	11782.0	43876.4	18866.9
1325	C19H40O	1-NONADECANOL	solid	11666.8	41004.5	17631.9
1326	C19H40S	BUTYL-PENTADECYL-SULFIDE	solid	12163.6	40466.4	17400.5

601

Table 26-1 ENTHALPY OF COMBUSTION - ORGANIC COMPOUNDS (continued)

NO	FORMULA	NAME	State	- Enthalpy of Combustion @ 77 F		
				kjoule/mol	kjoule/kg	BTU/lb
1311	C18H38S	HEPTADECYL-METHYL-SULFIDE	solid	11552.4	40314.5	17335.2
1312	C18H38S	NONYL-SULFIDE	solid	11549.1	40302.8	17330.2
1313	C18H38S	PENTADECYL-PROPYL-SULFIDE	solid	11549.8	40305.4	17331.3
1314	C18H38S	1-OCTADECANETHIOL	solid	11546.6	40294.2	17326.5
1315	C18H38S2	NONYL-DISULFIDE	liquid	11854.1	37204.9	15998.1
1316	C19H26	1-n-NONYLNAPHTHALENE	liquid	10500.0	41271.2	17746.6
1317	C19H32	n-TRIDECYLBENZENE	liquid	11051.0	42428.3	18244.2
1318	C19H36	1-NONADECYNE	solid	11637.8	44000.6	18920.3
1319	C19H36O2	METHYL OLEATE	liquid	11100.0	37437.5	16098.1
1320	C19H38	1-CYCLOPENTYLTETRADECANE	liquid	11651.6	43719.5	18799.4
1321	C19H38	1-CYCLOHEXYLTRIDECANE	liquid	11628.3	43631.8	18761.7
1322	C19H38	1-NONADECENE	liquid	11700.0	43900.6	18877.3
1323	C19H38O2	NONADECANOIC ACID	solid	11088.0	37144.5	15972.1
1324	C19H40	n-NONADECANE	solid	11782.0	43876.4	18866.9
1325	C19H40O	1-NONADECANOL	solid	11666.8	41004.5	17631.9
1326	C19H40S	BUTYL-PENTADECYL-SULFIDE	solid	12163.6	40466.4	17400.5
1327	C19H40S	ETHYL-HEPTADECYL-SULFIDE	solid	12164.5	40469.4	17401.9
1328	C19H40S	HEXADECYL-PROPYL-SULFIDE	solid	12164.4	40469.0	17401.7
1329	C19H40S	METHYL-OCTADECYL-SULFIDE	solid	12166.9	40477.5	17405.3
1330	C19H40S	1-NONADECANETHIOL	solid	12161.1	40458.2	17397.0
1331	C20H16	TRIPHENYLETHYLENE	solid	10000.0	39009.6	16774.1
1332	C20H28	1-n-DECYLNAPHTHALENE	liquid	11100.0	41349.7	17780.4
1333	C20H30O2	ABIETIC ACID	solid	10900.0	36038.2	15496.4
1334	C20H31N	DEHYDROABIETYLAMINE	solid	11440.2	40074.4	17232.0
1335	C20H34	1-PHENYLTETRADECANE	liquid	11713.1	42672.3	18349.1
1336	C20H38	1-EICOSYNE	solid	12252.1	43990.0	18915.7
1337	C20H40	1-CYCLOPENTYLPENTADECANE	liquid	12265.2	43720.7	18799.9
1338	C20H40	1-CYCLOHEXYLTETRADECANE	liquid	12242.0	43637.9	18764.3
1339	C20H40	1-EICOSENE	solid	12241.0	43634.0	18762.6
1340	C20H42	n-EICOSANE	solid	12392.0	43857.3	18858.6
1341	C20H42O	1-EICOSANOL	solid	12216.0	40917.4	17594.5
1342	C20H42S	BUTYL-HEXADECYL-SULFIDE	solid	12778.5	40616.8	17465.2
1343	C20H42S	DECYL-SULFIDE	solid	12778.5	40616.8	17465.2
1344	C20H42S	ETHYL-OCTADECYL-SULFIDE	solid	12779.4	40619.6	17466.4
1345	C20H42S	HEPTADECYL-PROPYL-SULFIDE	solid	12779.3	40619.2	17466.3
1346	C20H42S	METHYL-NONADECYL-SULFIDE	solid	12781.8	40627.3	17469.8
1347	C20H42S	1-EICOSANETHIOL	solid	12776.1	40609.0	17461.9
1348	C20H42S2	DECYL-DISULFIDE	liquid	13084.3	37742.6	16229.3
1349	C21H21O4P	TRI-o-CRESYL PHOSPHATE	liquid	11000.0	29861.4	12840.4
1350	C21H36	1-PHENYLPENTADECANE	liquid	12326.7	42724.6	18371.6
1351	C21H42	1-CYCLOPENTYLHEXADECANE	liquid	12878.9	43722.0	18800.4
1352	C21H42	1-CYCLOHEXYLPENTADECANE	solid	12855.9	43644.0	18766.9
1353	C22H38	1-PHENYLHEXADECANE	solid	12940.3	42772.0	18391.9
1354	C22H44	1-CYCLOHEXYLHEXADECANE	solid	13469.9	43649.7	18769.4
1355	C22H44O2	n-BUTYL STEARATE	solid	13158.0	38633.0	16612.2
1356	C24H38O4	DIISOOCTYL PHTHALATE	liquid	12950.0	33157.3	14257.6
1357	C24H38O4	DIOCTYL PHTHALATE	liquid	12956.0	33172.6	14264.2
1358	C24H42O	DINONYLPHENOL	liquid	14000.0	40392.7	17368.9
1359	C26H20	TETRAPHENYLETHYLENE	solid	13000.0	39104.2	16814.8
1360	C28H46O4	DIISODECYL PHTHALATE	liquid	15538.0	34786.2	14958.1

Enthalpy of combustion applies at 77 F and 1 atm.

The numerical values as shown are the negative of the ethalpy of combustion.

A positive value means that heat is released in the combustion.

A negative value means that heat is required in the combustion.

The products of combustion are CO2(gas), H2O(gas), F2(gas), Cl2(gas), Br2(gas), I2(gas), N2(gas), H3PO4(solid), and SiO2(crystobalite).

Chapter 27

EXPOSURE LIMITS FOR SAFEGUARDING HEALTH

Carl L. Yaws and Eric L. Jaycox
Lamar University, Beaumont, Texas

ABSTRACT

Results for exposure limits in air for safeguarding health are presented for organic and inorganic chemicals. The results include threshold limit value (TLV of ACGIH), permissible exposure limit (PEL of OSHA), recommended exposure limit (REL of NIOSH), and maximum concentration value in the workplace (MAK of DFG). The results are displayed in easy-to-use tabulations that are especially applicable for rapid engineering usage. The organic chemicals encompass hydrocarbon, oxygen, nitrogen, halogen, silicon, sulfur, and other compound types. The results are useful in engineering applications involving exposure of chemicals and mixtures in the workplace.

EXPOSURE LIMITS FOR SAFEGUARDING HEALTH

The results for exposure limits in air for safeguarding health in the workplace are presented in Tables 27-1 and 27-2 for organic and inorganic chemicals. The tabulated values that apply to exposure in the workplace in a 40-hour week are summarized below:

- TLV (ACGIH) – Threshold limit value in air in workplace of the American Conference of Governmental Industrial Hygienists.

- PEL (OSHA) – Permissible exposure limit in air in workplace of the Occupational Safety and Health Administration.

- REL (NIOSH) – Recommended exposure limit in air in workplace of the National Institute for Occupational Safety and Health.

- MAK (DFG) – Maximum concentration value in air in workplace of the Federal Republic of Germany.

In the data collection, a literature search was conducted to identify data source publications for organics (1-13) and inorganics (1-12). The publications were screened and copies of appropriate data were made. These data were then keyed into the computer to provide a database for use in preparing the tabulations.

MIXTURES

If more than one substance is present in the workplace, then exposure limits are needed for gas mixtures. The following equation (1) maybe used for exposure limits of gas mixtures:

$$PEL_{mixture} = \Sigma \, y_i \, / \, \Sigma \, (y_i/PEL_i) \tag{27-1}$$

where

$PEL_{mixture}$ = permissible exposure limit of mixture, ppm (vol)
y_i = mole fraction of component i, ppm

EXAMPLES

The tabulated values maybe used in engineering applications involving exposure of pure components and mixtures in the workplace. Examples are given below.

Example 1 Due to a small leak, the workplace contains methylamine (CH5N) at a concentration of 13.7 parts per million by volume in air.

Are the workers overexposed?

Inspection of the table discloses that the permissible exposure level (PEL) = 10 ppm (vol) for methylamine. Since the workplace concentration exceed the PEL for methylamine, the workers are overexposed. This is shown below:

Workplace concentration of 13.7 ppm (vol) > PEL of 10 ppm (vol)

Workers are overexposed.

Example 2 Estimate the permissible exposure level (PEL) for the gas mixture below:

	y_i	PEL_i
	ppm	ppm (vol)
Acetonitrile (C2H3N)	10	40
Ethylamine (C2H7N)	2	10
Monoethanolamine (C2H7NO)	2	3

Substitution of y_i and PEL_i into the equations for gas mixtures yields

$$PEL_{mixture} = \Sigma\, y_i\, /\, \Sigma\, (y_i/PEL_i) = (10 + 2 + 2\,)/(10/40 + 2/10 + 2/3)$$

$$PEL_{mixture} = 12.5 \text{ ppm (vol)}$$

REFERENCES – ORGANIC COMPOUNDS

1. Crowl, D. A. and J. F. Louvar, CHEMICAL PROCESS SAFETY, Prentice Hall, Inc., Englewood Cliffs, NJ (1990).
2. NIOSH POCKET GUIDE TO CHEMICAL HAZARDS, U. S. Dept. of Health and Human Services, Superintendent of Documents, Washington, DC (June, 1994).
3. GUIDE TO OCCUPATIONAL EXPOSURE VALUES – 1997, American Conference of Governmental Industrial Hygienists, ACGIH, Inc., Cincinnati, OH (1997).
4. 1997 TLVs and BEIs, American Conference of Governmental Industrial Hygienists, ACGIH, Inc., Cincinnati, OH (1997).
5. June 1993 Air Contaminants Final Rule, specified in Tables Z-1, Z-2, and Z-3 (Federal Register, 58:35388-35351, June 30, 1993; corrected in Federal Register, 58:40191, July 27, 1993; and subsequent amendments).
6. Lees, F. P., LOSS PREVENTION IN THE PROCESS INDUSTRIES, Vols. 1 and 2, Butterworth-Heinemann, London, England (1992).
7. CONDENSED CHEMICAL DICTIONARY, 10th and 11th eds., G. G. Hawley (10th) and Sax, N. I. and R. J. Lewis, Jr. (11th), Van Nostrand Reinhold Co., New York, NY (1981, 1987).
8. Sax, N. I., DANGEROUS PROPERTIES OF INDUSTRIAL MATERIALS, 9th ed., Vols. 1, 2, and 3, Van Nostrand Reinhold Company, New York, NY (1996).
9. CRC HANDBOOK OF CHEMISTRY AND PHYSICS, 75th - 77th ed., CRC Press, Inc., Boca Raton, FL (1994-1996).
10. Springer, C. and J. R. Welker, INDUSTRIAL HYGIENE: AN INTRODUCTION FOR CHEMICAL ENGINEERS, American Institute of Chemical Engineers, New York, NY (1995).
11. Fawcett, H. H. and W. C. Wood, eds., SAFETY AND ACCIDENT PREVENTION IN CHEMICAL OPERATIONS, 2nd ed., John Wiley and Sons, New York, NY (1982).
12. Williams, P. L. and J. L. Burson, eds., INDUSTRIAL TOXICOLOGY, SAFETY AND HEALTH APPLICATIONS IN THE WORKPLACE, Van Nostrand Reinhold Company, New York, NY (1985).
13. de la Cruz, P. L. and D. G. Sarvadi, Am. Ind. Hyg. Assoc. J., 55(10), 894 (1994).

REFERENCES – INORGANIC COMPOUNDS

1. Crowl, D. A. and J. F. Louvar, CHEMICAL PROCESS SAFETY, Prentice Hall, Inc., Englewood Cliffs, NJ (1990).
2. NIOSH POCKET GUIDE TO CHEMICAL HAZARDS, U. S. Dept. of Health and Human Services, Superintendent of Documents, Washington, DC (June, 1994).
3. GUIDE TO OCCUPATIONAL EXPOSURE VALUES – 1997, American Conference of Governmental Industrial Hygienists, ACGIH, Inc., Cincinnati, OH (1997).
4. 1997 TLVs and BEIs, American Conference of Governmental Industrial Hygienists, ACGIH, Inc., Cincinnati, OH (1997).
5. June 1993 Air Contaminants Final Rule, specified in Tables Z-1, Z-2 and Z-3 (Federal Register, 58:35388-35351, June 30, 1993; corrected in Federal Register, 58:40191, July 27, 1993; and subsequent amendments).
6. Lees, F. P., LOSS PREVENTION IN THE PROCESS INDUSTRIES, Vols. 1 and 2, Butterworth-Heinemann, London, England (1992).
7. CONDENSED CHEMICAL DICTIONARY, 10th and 11th eds., G. G. Hawley (10th) and Sax, N. I. and R. J. Lewis, Jr. (11th), Van Nostrand Reinhold Co., New York, NY (1981, 1987).
8. Sax, N. I., DANGEROUS PROPERTIES OF INDUSTRIAL MATERIALS, 9th ed., Vols. 1, 2 and 3, Van Nostrand Reinhold Company, New York, NY (1996).
9. CRC HANDBOOK OF CHEMISTRY AND PHYSICS, 75th - 77th ed., CRC Press, Inc., Boca Raton, FL (1994-1996).
10. Springer, C. and J. R. Welker, INDUSTRIAL HYGIENE: AN INTRODUCTION FOR CHEMICAL ENGINEERS, American Institute of Chemical Engineers, New York, NY (1995).
11. Fawcett, H. H. and W. C. Wood, eds., SAFETY AND ACCIDENT PREVENTION IN CHEMICAL OPERATIONS, 2nd ed., John Wiley and Sons, New York, NY (1982).
12. Williams, P. L. and J. L. Burson, eds., INDUSTRIAL TOXICOLOGY, SAFETY AND HEALTH APPLICATIONS IN THE WORKPLACE, Van Nostrand Reinhold Company, New York, NY (1985).

Table 27-1 EXPOSURE LIMITS - ORGANIC COMPOUNDS

NO	FORMULA	NAME	CAS No	TLV (ACGIH) Threshold Limit Value		PEL (OSHA) Permissible Exposure Limit		REL (NIOSH) Recommended Exposure Limit		MAK (DFG) Maximum Concentration Value	
				ppm (vol)	mg/m³	ppm (vol)	mg/m³	ppm (vol)	mg/m³	ppm (vol)	mg/m³
1	C	GRAPHITE	7782-42-5	-----	2	-----	-----	-----	2.5	-----	6
2	CBrF2	DIFLUORODIBROMOMETHANE	75-61-6	100	858	100	860	100	860	100	860
3	CBrF3	BROMOTRIFLUOROMETHANE	75-63-8	1000	6090	1000	6100	1000	6100	1000	6100
4	CBrF3	TRIFLUOROBROMOMETHANE	75-63-8	1000	6090	1000	6100	1000	6100	1000	6100
5	CBr4	CARBON TETRABROMIDE	558-13-4	0.1	1.4	-----	-----	0.1	1.4	-----	-----
6	CCIF3	CHLOROTRIFLUOROMETHANE	75-72-9	-----	-----	-----	-----	-----	-----	1000	4330
7	CCl2F2	DICHLORODIFLUOROMETHANE	75-71-8	1000	4950	1000	4950	1000	4950	1000	5000
8	CCl2O	CARBONYL CHLORIDE	75-44-5	0.1	0.4	0.1	0.4	0.1	0.4	0.02	0.08
9	CCl2O	PHOSGENE	75-44-5	0.1	0.4	0.1	0.4	0.1	0.4	0.02	0.08
10	CCl3F	FLUOROTRICHLOROMETHANE	75-69-4	-----	-----	1000	5600	-----	-----	1000	5600
11	CCl3F	TRICHLOROFLUOROMETHANE	75-69-4	-----	-----	1000	5600	-----	-----	1000	5600
12	CCl3NO2	CHLOROPICRIN	76-06-2	0.1	0.67	0.1	0.7	0.1	0.7	0.1	0.7
13	CCl3NO2	NITROTRICHLOROMETHANE	76-06-2	0.1	0.67	0.1	0.7	0.1	0.7	0.1	0.7
14	CCl3NO2	TRICHLORONITROMETHANE	76-06-2	0.1	0.67	0.1	0.7	0.1	0.7	0.1	0.7
15	CCl4	CARBON TETRACHLORIDE	56-23-5	5	31	10	-----	-----	-----	10	65
16	CCl4	TETRACHLOROMETHANE	56-23-5	5	31	10	-----	-----	-----	10	65
17	CCl4S	PERCHLOROMETHYL MERCAPTAN	594-42-3	0.1	0.76	0.1	0.8	0.1	0.8	-----	-----
18	CF2O	CARBONYL FLUORIDE	353-50-4	2	5.4	-----	-----	2	5	-----	-----
19	CHBr3	BROMOFORM	75-25-2	0.5	5.2	0.5	5	0.5	5	-----	-----
20	CHBr3	TRIBROMOMETHANE	75-25-2	0.5	5.2	0.5	5	0.5	5	-----	-----
21	CHCIF2	CHLORODIFLUOROMETHANE	75-45-6	1000	3540	-----	-----	1000	3500	500	1800
22	CHCl2F	DICHLOROFLUOROMETHANE	75-43-4	10	42	1000	4200	10	40	10	45
23	CHCl3	CHLOROFORM	67-66-3	10	49	-----	-----	-----	-----	10	50
24	CHCl3	TRICHLOROMETHANE	67-66-3	10	49	-----	-----	-----	-----	10	50
25	CHI3	IODOFORM	75-47-8	0.6	10	-----	-----	0.6	10	-----	-----
26	CHN	HYDROGEN CYANIDE	74-90-8	-----	-----	10	11	-----	-----	10	11
27	CH2BrCl	BROMOCHLOROMETHANE	74-97-5	200	1060	200	1050	200	1050	200	1050
28	CH2BrCl	CHLOROBROMOMETHANE	74-97-5	200	1060	200	1050	200	1050	200	1500
29	CH2Cl2	DICHLOROMETHANE	75-09-2	50	174	25	-----	-----	-----	100	350
30	CH2Cl2	METHYLENE CHLORIDE	75-09-2	50	174	25	-----	-----	-----	100	350
31	CH2N2	CYNAMIDE	420-04-2	-----	2	-----	-----	-----	2	-----	-----
32	CH2N2	DIAZOMETHANE	334-88-3	0.2	0.34	0.2	0.4	0.2	0.4	-----	-----
33	CH2O	FORMALDEHYDE	50-00-0	-----	-----	0.75	-----	0.016	-----	0.5	0.6
34	CH2O2	FORMIC ACID	64-18-6	5	9.4	5	9	5	9	5	9
35	CH3Br	METHYL BROMIDE	74-83-9	1	3.9	-----	-----	-----	-----	-----	-----
36	CH3Cl	METHYL CHLORIDE	74-87-3	50	103	100	205	-----	-----	50	105
37	CH3Hg	METHYL MERCURY (as Hg)	22967-92-6	-----	0.01	-----	0.01	-----	0.01	-----	0.01
38	CH3I	METHYL IODIDE	74-88-4	2	12	5	28	2	10	-----	-----
39	CH3NO	FORMAMIDE	75-12-7	10	18	-----	-----	10	15	-----	-----
40	CH3NO2	NITROMETHANE	75-52-5	20	50	100	250	-----	-----	100	250
41	CH4O	METHANOL	67-56-1	200	262	200	260	200	260	200	260
42	CH4O	METHYL ALCOHOL	67-56-1	200	262	200	260	200	260	200	260
43	CH4S	METHANETHIOL	74-93-1	0.5	0.98	-----	-----	-----	-----	0.5	1
44	CH4S	METHYL MERCAPTAN	74-93-1	0.5	0.98	-----	-----	-----	-----	0.5	1
45	CH5N	METHYLAMINE	74-89-5	5	6.4	10	12	10	12	10	12
46	CH6N2	METHYL HYDRAZINE	60-34-4	0.01	0.019	-----	-----	-----	-----	-----	-----
47	CNK	POTASSIUM CYANIDE (as CN)	151-50-8	-----	-----	-----	5	-----	-----	-----	5
48	CNNa	SODIUM CYANIDE (as CN)	143-33-9	-----	-----	-----	5	-----	-----	-----	5
49	CN4O8	TETRANITROMETHANE	509-14-8	0.005	0.04	1	8	1	8	-----	-----
50	CO	CARBON MONOXIDE	630-08-0	25	29	50	55	35	40	30	33
51	CO2	CARBON DIOXIDE	124-38-9	5000	9000	5000	9000	5000	9000	5000	9000
52	CS2	CARBON DISULFIDE	75-15-0	10	31	20	-----	1	3	10	30
53	C2ClF4	DICHLOROTETRAFLUOROETHANE	76-14-2	1000	6990	1000	7000	1000	7000	1000	7000
54	C2ClF5	CHLOROPENTAFLUOROETHANE	76-15-3	1000	6320	-----	-----	1000	6320	-----	-----
55	C2Cl3F3	1,1,2-TRICHLORO-1,2,2-TRIFLUOROETHANE	76-13-1	1000	7670	1000	7600	1000	7600	500	3800
56	C2Cl4	PERCHLOROETHYLENE	127-18-4	25	170	100	-----	-----	-----	50	345
57	C2Cl4	TETRACHLOROETHYLENE	127-18-4	25	170	100	-----	-----	-----	50	345
58	C2Cl4F2	1,1,1,2-TETRACHLORO-2,2-DIFLUOROETHANE	76-11-9	500	4170	500	4170	500	4170	1000	8340
59	C2Cl4F2	1,1,2,2-TETRACHLORO-1,2-DIFLUOROETHANE	76-12-0	500	4170	500	4170	500	4170	200	1690
60	C2Cl6	HEXACHLOROETHANE	67-72-1	1	9.7	1	10	1	10	1	10
61	C2HBrClF3	HALOTHANE	151-67-7	50	404	-----	-----	-----	-----	5	40
62	C2HCl3	TRICHLOROETHYLENE	79-01-6	50	269	100	-----	-----	-----	-----	-----
63	C2HCl3O2	TRICHLOROACETIC ACID	76-03-9	1	6.7	-----	-----	1	7	-----	-----
64	C2HCl5	PENTACHLOROETHANE	76-01-7	-----	-----	-----	-----	-----	-----	5	40
65	C2H2Br4	ACETYLENE TETRABROMIDE	79-27-6	1	14	1	14	-----	-----	1	14
66	C2H2Br4	1,1,2,2-TETRABROMOETHANE	79-27-6	-----	-----	-----	-----	-----	-----	1	14
67	C2H2Cl2	sym-1,2,-DICHLOROETHYLENE	540-59-0	200	793	200	790	200	790	200	790
68	C2H2Cl2	1,1-DICHLOROETHYLENE	75-35-4	5	20	-----	-----	-----	-----	2	8
69	C2H2Cl2	cis-1,2,-DICHLOROETHYLENE	156-59-2	200	793	200	790	200	790	200	790
70	C2H2Cl2	trans-1,2,-DICHLOROETHYLENE	156-60-5	200	793	200	790	200	790	200	790
71	C2H2Cl2	VINYLIDENE CHLORIDE	75-35-4	5	20	-----	-----	-----	-----	2	8
72	C2H2Cl4	1,1,2,2-TETRACHLOROETHANE	79-34-5	1	6.9	5	35	1	7	1	7
73	C2H2ClO	CHLOROACETYL CHLORIDE	79-04-9	0.05	0.23	-----	-----	0.05	0.2	-----	-----
74	C2H2FO2Na	SODIUM FLUOROACETATE	62-74-8	-----	0.05	-----	0.05	-----	0.05	-----	0.05
75	C2H2F2	1,1-DIFLUOROETHYLENE	75-38-7	-----	-----	-----	-----	-----	1	-----	-----
76	C2H2F2	VINYLIDENE FLUORIDE	75-38-7	-----	-----	-----	-----	-----	1	-----	-----
77	C2H2F4	1,1,2,2-TETRAFLUOROETHANE	811-97-2	-----	-----	-----	-----	-----	-----	1000	4200

605

Table 27-1 EXPOSURE LIMITS - ORGANIC COMPOUNDS (continued)

NO	FORMULA	NAME	CAS No	TLV (ACGIH) Threshold Limit Value		PEL (OSHA) Permissible Exposure Limit		REL (NIOSH) Recommended Exposure Limit		MAK (DFG) Maximum Concentration Value	
				ppm (vol)	mg/m^3	ppm (vol)	mg/m^3	ppm (vol)	mg/m^3	ppm (vol)	mg/m^3
78	C2H2O	KETENE	463-51-4	0.5	0.86	0.5	0.9	0.5	0.9	0.5	0.9
79	C2H2O4	OXALIC ACID	144-62-7	-----	1	-----	1	-----	1	-----	-----
80	C2H3Br	VINYL BROMIDE	593-60-2	5	22	-----	-----	-----	-----	-----	-----
81	C2H3Cl	VINYL CHLORIDE	75-01-4	5	13	1	-----	-----	-----	-----	-----
82	C2H3ClF2	1-CHLORO-1,1-DIFLUOROETHANE	75-68-3	-----	-----	-----	-----	-----	-----	1000	4170
83	C2H3ClO	CHLOROACETALDEHYDE	107-20-0	-----	-----	-----	-----	-----	-----	1	3
84	C2H3Cl2NO2	1,1-DICHLORO-1-NITROETHANE	594-72-9	2	12	-----	-----	2	10	10	60
85	C2H3Cl3	1,1,1-TRICHLOROETHANE	71-55-6	350	1910	350	1900	-----	-----	200	1080
86	C2H3F	VINYL FLUORIDE	72-02-5	-----	-----	-----	-----	1	-----	-----	-----
87	C2H3N	ACETONITRILE	75-05-08	40	67	40	70	20	34	40	70
88	C2H3NO	METHYL ISOCYANATE	624-83-9	0.02	0.047	0.02	0.05	0.02	0.05	0.01	0.024
89	C2H4O	ACETALDEHYDE	75-07-0	-----	-----	200	360	-----	-----	50	90
90	C2H4Br2	1,2-DIBROMOETHANE	106-93-4	-----	-----	20	-----	0.045	-----	-----	-----
91	C2H4Br2	ETHYLENE DIBROMIDE	106-93-4	-----	-----	20	-----	0.045	-----	-----	-----
92	C2H4Cl2	1,1-DICHLOROETHANE	75-34-3	100	405	100	400	100	400	100	400
93	C2H4Cl2	1,2-DICHLOROETHANE	107-06-2	10	40	50	-----	1	4	-----	-----
94	C2H4Cl2	ETHYLIDENE CHLORIDE	75-34-3	100	405	100	400	100	400	100	400
95	C2H4Cl2	ETHYLENE DICHLORIDE	107-06-2	10	40	50	-----	1	4	-----	-----
96	C2H4Cl2O	BIS(CHLOROMETHYL)ETHER	542-88-1	0.001	0.0047	-----	-----	-----	-----	-----	-----
97	C2H4N2O6	ETHYLENE GLYCOL DINITRATE	628-96-6	0.05	0.31	-----	-----	-----	-----	0.05	0.3
98	C2H4N4	AMITROLE	61-82-5	-----	0.2	-----	-----	-----	0.2	-----	0.2
99	C2H4O	ETHYLENE OXIDE	75-21-8	1	1.8	1	-----	<0.1	<0.18	-----	-----
100	C2H4O2	ACETIC ACID	64-19-7	10	25	10	25	10	25	10	25
101	C2H4O2	METHYL FORMATE	107-31-3	100	246	100	250	100	250	50	125
102	C2H4O2S	THIOGLYCOLIC ACID	68-11-1	1	3.8	-----	-----	1	4	-----	-----
103	C2H5Br	ETHYL BROMIDE	74-96-4	5	22	200	890	-----	-----	-----	-----
104	C2H5Cl	ETHYL CHLORIDE	75-00-3	100	264	1000	2600	-----	-----	-----	-----
105	C2H5ClO	2-CHLOROETHANOL	107-07-3	-----	-----	5	16	-----	-----	1	3
106	C2H5ClO	ETHYLENE CHLOROHYDRIN	107-07-3	-----	-----	5	16	-----	-----	1	3
107	C2H5N	ETHYLENIMINE	151-56-4	0.5	0.88	-----	-----	-----	-----	-----	-----
108	C2H5NO2	NITROEHTANE	79-24-3	100	307	100	310	100	310	100	310
109	C2H6O	DIMETHYL ETHER	115-10-6	-----	-----	-----	-----	-----	-----	1000	1910
110	C2H6O	ETHANOL	64-17-5	1000	1880	1000	1900	1000	1900	1000	1900
111	C2H6O	ETHYL ALCOHOL	64-17-5	1000	1880	1000	1900	1000	1900	1000	1900
112	C2H6O2	ETHYLENE GLYCOL	107-21-1	-----	-----	-----	-----	-----	-----	10	26
113	C2H6O4S	DIMETHYL SULFATE	77-78-1	0.1	0.52	1	5	0.1	0.5	-----	-----
114	C2H6S	ETHANETHIOL	75-08-1	0.5	1.3	-----	-----	-----	-----	0.5	1
115	C2H6S	ETHYL MERCAPTAN	75-08-1	0.5	1.3	-----	-----	-----	-----	0.5	1
116	C2H7N	DIMETHYLAMINE	124-40-3	5	9.2	10	18	10	18	2	4
117	C2H7N	ETHYLAMINE	75-04-7	5	9.2	10	18	10	18	5	9
118	C2H7NO	ETHANOLAMINE	141-43-5	3	7.5	3	6	3	8	2	5
119	C2H8N2	ETHYLENE DIAMINE	107-15-3	10	25	10	25	10	25	10	25
120	C2H8N2	1,1-DIMETHYLHYDRAZINE	57-14-7	0.01	0.025	0.5	1	-----	-----	-----	-----
121	C2N2	CYANOGEN	460-19-5	10	21	-----	-----	10	20	10	22
122	C2N2Ca	CALCIUM CYANIDE (as CN)	592-01-8	-----	-----	-----	5	-----	-----	-----	5
123	C3F6O	HEXAFLUOROACETONE	684-16-2	0.1	0.68	-----	-----	0.1	0.7	-----	-----
124	C3H2ClF5O	ENFLURANE	13838-16-9	75	566	-----	-----	-----	-----	20	150
125	C3H2N2	MALONONITRILE	109-77-3	-----	-----	-----	-----	3	8	-----	-----
126	C3H3Cl2O2Na	2,2-DICHLOROPROPIONIC ACID SODIUM SALT	127-20-8	-----	-----	-----	-----	-----	-----	1	6
127	C3H3Cl3	1,1,2-TRICHLOROETHANE	79-00-5	10	55	10	45	10	45	10	55
128	C3H3N	ACRYLONITRILE	107-13-1	2	4.3	2	-----	1	-----	-----	-----
129	C3H3N	VINYL CYANIDE	107-13-1	2	4.3	2	-----	1	-----	-----	-----
130	C3H4	METHYL ACETYLENE	74-99-7	1000	1640	1000	1650	1000	1650	1000	1650
131	C3H4	PROPYNE	74-99-7	1000	1640	1000	1650	1000	1650	1000	1650
132	C3H4Cl2	1,3-DICHLOROPROPENE	542-75-6	1	4.5	-----	-----	1	5	-----	-----
133	C3H4Cl2O2	2,2-DICHLOROPROPIONIC ACID	75-99-0	1	5.8	-----	-----	1	6	1	6
134	C3H4O	ACROLEIN	107-02-8	0.1	0.23	0.1	0.25	0.1	0.25	0.1	0.23
135	C3H4O	PROPARGYL ALCOHOL	107-19-7	1	2.3	-----	-----	1	2	2	5
136	C3H4O2	ACRYLIC ACID	79-10-7	2	5.9	-----	-----	2	6	-----	-----
137	C3H4O2	beta-PROPIOLACTONE	57-57-8	0.5	1.5	-----	-----	-----	-----	-----	-----
138	C3H5Br2Cl	1,2-DIBROMO-3-CHLOROPROPANE	96-12-8	-----	-----	0.001	-----	-----	-----	-----	-----
139	C3H5Cl	ALLYL CHLORIDE	107-05-1	1	3	1	3	1	3	1	3
140	C3H5ClO	1-CHLORO,2,3-EPOXYPROPANE	106-89-8	0.5	1.9	5	19	-----	-----	-----	-----
141	C3H5ClO	EPICHLOROHYDRIN	106-89-8	0.5	1.9	5	19	-----	-----	-----	-----
142	C3H5ClO2	METHYL CHLOROACETATE	96-34-4	-----	-----	-----	-----	-----	-----	1	5
143	C3H5ClO2	2-CHLOROPROPIONIC ACID	598-78-7	0.1	0.44	-----	-----	-----	-----	-----	-----
144	C3H5ClO2	METHYL CHLOROACETATE	96-34-4	-----	-----	-----	-----	-----	-----	1	5
145	C3H5Cl3	1,2,3-TRICHLOROPROPANE	96-18-4	10	60	50	300	10	60	-----	-----
146	C3H5N	PROPIONITRILE	107-12-0	-----	-----	-----	-----	6	14	-----	-----
147	C3H5NO	ACRYLAMIDE	79-06-1	-----	0.03	-----	0.3	-----	0.03	-----	-----
148	C3H5N3O9	NITROGLYCERIN	55-63-0	0.05	0.46	-----	-----	-----	-----	0.05	0.5
149	C3H6ClNO2	1-CHLORO-1-NITROPROPANE	600-25-9	2	10	20	100	2	10	-----	-----
150	C3H6Cl2	1,2-DICHLOROPROPANE	78-87-5	75	347	75	350	-----	-----	-----	-----
151	C3H6Cl2	PROPYLENE DICHLORIDE	78-87-5	75	347	75	350	-----	-----	-----	-----
152	C3H6N2O6	PROPYLENE GLYCOL DINITRATE	6423-43-4	0.05	0.34	-----	-----	0.05	0.3	0.05	0.3
153	C3H6N6O6	CYCLONITE	121-82-4	-----	0.5	-----	-----	-----	1.5	-----	-----
154	C3H6N6O6	RDX	121-82-4	-----	0.5	-----	-----	-----	1.5	-----	-----

Table 27-1 EXPOSURE LIMITS - ORGANIC COMPOUNDS (continued)

NO	FORMULA	NAME	CAS No	TLV (ACGIH) Threshold Limit Value		PEL (OSHA) Permissible Exposure Limit		REL (NIOSH) Recommended Exposure Limit		MAK (DFG) Maximum Concentration Value	
				ppm (vol)	mg/m³	ppm (vol)	mg/m³	ppm (vol)	mg/m³	ppm (vol)	mg/m³
155	C3H6O	ACETONE	67-64-1	500	1188	1000	2400	250	590	500	1200
156	C3H6O	ALLYL ALCOHOL	107-18-6	2	4.8	2	5	2	5	2	5
157	C3H6O	1,2-EPOXYPROPANE	75-56-9	20	48	100	240	-----	-----	-----	-----
158	C3H6O	PROPYLENE OXIDE	75-56-9	20	48	100	240	-----	-----	-----	-----
159	C3H6O2	2,3-EPOXY-1-PROPANOL	556-52-5	2	6.1	50	150	25	75	50	150
160	C3H6O2	ETHYL FORMATE	109-94-4	100	303	100	300	100	300	100	300
161	C3H6O2	GLYCIDOL	556-52-5	2	6.1	50	150	25	75	50	150
162	C3H6O2	METHYL ACETATE	79-20-9	200	606	200	610	200	610	200	610
163	C3H6O2	PROPIONIC ACID	79-09-4	10	30	-----	-----	10	30	10	30
164	C3H7N	PROPYLENE IMINE	75-55-8	2	4.7	2	5	2	5	-----	-----
165	C3H7NO	DIMETHYLFORMAMIDE	68-12-2	10	30	10	30	10	30	10	30
166	C3H7NO	2-METHYLAZIRIDINE	75-55-8	2	4.7	2	5	2	5	-----	-----
167	C3H7NO2	1-NITROPROPANE	108-03-2	25	91	25	90	25	90	25	90
168	C3H7NO2	2-NITROPROPANE	79-46-9	10	36	25	90	-----	-----	25	90
169	C3H7NO3	n-PROPYL NITRATE	627-13-4	25	107	25	110	25	105	25	110
170	C3H8	PROPANE	74-98-6	2500	4508	1000	1800	1000	1800	1000	1800
171	C3H8O	ISOPROPYL ALCOHOL	67-63-0	400	983	400	980	400	980	200	490
172	C3H8O	n-PROPYL ALCOHOL	71-23-8	200	492	200	500	200	500	-----	-----
173	C3H8O2	DIMETHOXY METHANE	109-87-5	1000	3110	1000	3100	1000	3100	1000	3100
174	C3H8O2	ETHYLENE GLYCOL MONOMETHYL ETHER	109-86-4	5	16	25	80	0.1	0.3	5	15
175	C3H8O2	2-METHOXYETHANOL	109-86-4	5	16	25	80	0.1	0.3	5	15
176	C3H8O2	METHYLAL	109-87-5	1000	3110	1000	3100	1000	3100	1000	3100
177	C3H8O2	2-METHOXYETHANOL	109-86-4	5	16	25	80	0.1	0.3	5	15
178	C3H8O3	GLYCERIN MIST	56-81-5	10	-----	-----	15*/5**	-----	-----	-----	-----
179	C3H9N	ISOPROPYLAMINE	75-31-0	5	12	5	12	-----	-----	5	12
180	C3H9N	TRIMETHYLAMINE	75-50-3	5	12	-----	-----	10	24	-----	-----
181	C3H9O3P	TRIMETHYL PHOSPHITE	121-45-9	2	10	-----	-----	2	10	-----	-----
182	C4Cl6	HEXACHLOROBUTADIENE	87-68-3	0.02	0.21	-----	-----	0.02	0.24	-----	-----
183	C4HO4Co	COBALT HYDROCARBONYL (as Co)	16842-03-8	-----	0.1	-----	-----	-----	0.1	-----	-----
184	C4H2O3	MALEIC ANHYDRIDE	108-31-6	0.25	1	0.25	1	0.25	1	0.1	0.4
185	C4H4N2	SUCCINONITRILE	110-61-2	-----	-----	-----	-----	6	20	-----	-----
186	C4H5Cl	2-CHLORO-1,3-BUTADIENE	126-99-8	10	36	25	90	-----	-----	5	8
187	C4H5Cl	beta-CHLOROPRENE	126-99-8	10	36	25	90	-----	-----	5	18
188	C4H5N	METHYLACRYLONITRILE	126-98-7	1	2.7	-----	-----	1	3	-----	-----
189	C4H6	1,3-BUTADIENE	106-99-0	2	4.4	1	5	-----	-----	-----	-----
190	C4H6Cl2	1,4-DICHLORO-2-BUTENE	764-41-0	0.005	0.025	-----	-----	-----	-----	-----	-----
191	C4H6O	CROTONALDEHYDE	4170-30-3	2	5.7	2	6	2	6	-----	-----
192	C4H6O2	METHYL ACRYLATE	96-33-3	2	7	10	35	10	35	5	18
193	C4H6O2	METHACRYLIC ACID	79-41-4	20	70	-----	-----	20	70	-----	-----
194	C4H6O2	VINYL ACETATE	108-05-4	10	35	-----	-----	-----	-----	10	35
195	C4H6O3	ACETIC ANHYDRIDE	108-24-7	5	21	5	20	-----	-----	5	20
196	C4H7Br2Cl2O4P	DIMETHYL-1,2-DIBROMO-2,2-DICHLOROETHYL PHOSPHATE	300-76-5	-----	3	-----	3	-----	3	-----	3
197	C4H7Br2Cl2O4P	NALED (DIBROM)	300-76-5	-----	3	-----	3	-----	3	-----	3
198	C4H7Cl2O4P	DICHLORVOS (DDVP)	62-73-7	0.1	0.9	-----	1	-----	1	0.1	1
199	C4H7N	n-BUTYRONITRILE	109-74-0	-----	-----	-----	-----	8	22	-----	-----
200	C4H7N	ISOBUTYRONITRILE	78-82-0	-----	-----	-----	-----	8	22	-----	-----
201	C4H8Cl2O	DICHLOROETHYL ETHER	111-44-4	5	29	-----	-----	5	30	10	60
202	C4H8O	2-BUTANONE	78-93-3	200	590	200	590	200	590	200	590
203	C4H8O	METHYL ETHYL KETONE	78-93-3	200	590	200	590	200	590	200	590
204	C4H8O	TETRAHYDROFURAN	109-99-9	200	590	200	590	200	590	50	150
205	C4H8O2	DIETHYLENE DIOXIDE	123-91-1	25	90	100	360	-----	-----	20	72
206	C4H8O2	DIOXANE	123-91-1	25	90	100	360	-----	-----	20	72
207	C4H8O2	ETHYL ACETATE	141-78-6	400	1440	400	1400	400	1400	400	1400
208	C4H9NO	N,N-DIMETHYLACETAMIDE	127-19-5	10	36	10	35	10	35	10	35
209	C4H9NO	MORPHOLINE	110-91-8	20	71	20	70	20	70	10	36
210	C4H10	BUTANE	106-97-8	800	1900	-----	-----	800	1900	1000	2350
211	C4H10	ISOBUTANE	75-28-5	-----	-----	-----	-----	800	1900	1000	2350
212	C4H10O	n-BUTANOL	71-36-3	-----	-----	100	300	-----	-----	100	300
213	C4H10O	sec-BUTANOL	78-92-2	100	300	150	450	100	305	100	300
214	C4H10O	tert-BUTANOL	75-65-0	100	303	100	300	100	300	100	300
215	C4H10O	DIETHYL ETHER	60-29-7	400	1210	400	1200	400	1200	400	1200
216	C4H10O	ETHYL ETHER	60-29-7	400	1210	400	1200	-----	-----	400	1200
217	C4H10O	ISOBUTYL ALCOHOL	78-83-1	50	152	100	300	50	150	100	300
218	C4H10O2	2-ETHOXYETHANOL	110-80-5	5	18	200	740	0.5	1.8	5	19
219	C4H10O2	ETHYLENE GLYCOL MONOETHYL ETHER	110-80-5	5	18	200	740	0.5	1.8	5	19
220	C4H10O2	GLYCOL MONOETHYL ETHER	110-80-5	5	18	200	740	0.5	1.8	5	19
221	C4H10O2	1-METHOXY-2-PROPANOL	107-98-2	-----	-----	-----	-----	-----	-----	100	375
222	C4H10O2	2-METHOXY-1-PROPANOL	1589-47-5	-----	-----	-----	-----	-----	-----	20	75
223	C4H10O2	PROPYLENE GLYCOL-2-METHYL ETHER	1589-47-5	-----	-----	-----	-----	-----	-----	20	75
224	C4H10O2	PROPYLENE GLYCOL MONOMETHYL ETHER	107-98-2	100	369	-----	-----	100	360	100	375
225	C4H10O3	DIETHYLENE GLYCOL	111-46-6	-----	-----	-----	-----	-----	-----	10	44
226	C4H10S	BUTYL MERCAPTAN	109-79-5	0.5	1.8	10	35	0.5	1.5	0.5	1.5
227	C4H11N	n-BUTYLAMINE	109-73-9	-----	-----	-----	-----	-----	-----	5	15
228	C4H11N	sec-BUTYLAMINE	13952-84-6	-----	-----	-----	-----	-----	-----	5	15
229	C4H11N	tert-BUTYLAMINE	75-64-9	-----	-----	-----	-----	-----	-----	5	15
230	C4H11N	DIETHYLAMINE	109-89-7	5	15	25	75	10	30	5	15
231	C4H11N	N,N-DIMETHYL ETHYLAMINE	598-56-1	-----	-----	-----	-----	-----	-----	25	75

Table 27-1 EXPOSURE LIMITS - ORGANIC COMPOUNDS (continued)

NO	FORMULA	NAME	CAS No	TLV (ACGIH) Threshold Limit Value		PEL (OSHA) Permissible Exposure Limit		REL (NIOSH) Recommended Exposure Limit		MAK (DFG) Maximum Concentration Value	
				ppm (vol)	mg/m³	ppm (vol)	mg/m³	ppm (vol)	mg/m³	ppm (vol)	mg/m³
232	C4H11N	ISOBUTYLAMINE	78-81-9	-----	-----	-----	-----	-----	-----	5	15
233	C4H11NO2	DIETHANOLAMINE	111-42-2	0.46	2	-----	-----	3	15	-----	-----
234	C4H12Cl2N2	PIPERAZINE DIHYDROCHLORIDE	142-64-3	-----	5	-----	-----	-----	5	-----	-----
235	C4H12OSi	DIMETHYLETHOXYSILANE	14857-34-2	0.5	2.1	-----	-----	-----	-----	-----	-----
236	C4H12O4Si	METHYL SILICATE	681-84-5	1	6	-----	-----	1	6	-----	-----
237	C4H13N3	DIETHYLENE TRIAMINE	111-40-0	1	4.2	-----	-----	1	4	-----	-----
238	C4H12Pb	TETRAMETHYL LEAD (as Pb)	75-74-1	-----	0.15	-----	0.075	-----	0.075	-----	0.05
239	C4O4Ni	NICKEL CARBONYL (as Ni)	13463-39-3	0.05	0.12	0.001	0.007	0.001	0.007	-----	-----
240	C5Cl6	HEXACHLOROCYCLOPENTADIENE	77-47-4	0.01	0.11	-----	-----	0.01	0.1	-----	-----
241	C5H4O2	FURFURAL	98-01-1	2	7.9	5	20	-----	-----	-----	-----
242	C5H5N	PYRIDINE	110-86-1	5	16	5	15	5	15	5	15
243	C5H5NO2	METHYL-2-CYANO-ACRYLATE	137-05-3	2	9.1	-----	-----	2	8	2	8
244	C5H5NOSNa	SODIUM PYRITHIONE	3811-73-2	-----	-----	-----	-----	-----	-----	-----	1
245	C5H6Cl2N2O2	1,3-DICHLORO-5,5-DIMETHYL HYDANTOIN	118-52-5	-----	0.2	-----	0.2	-----	0.2	-----	-----
246	C5H6N2	2-AMINOPYRIDINE	504-29-0	0.5	2	0.5	2	0.5	2	0.5	2
247	C5H6N2	CYCLOPENTADIENE	542-92-7	75	203	75	200	75	200	75	200
248	C5H6O2	FURFURYL ALCOHOL	98-00-0	10	40	50	200	10	40	10	40
249	C5H8O2	ETHYL ACRYLATE	140-88-5	5	20	25	100	-----	-----	5	20
250	C5H8O2	GLUTARALDEHYDE	111-30-8	-----	-----	-----	-----	-----	-----	0.1	0.4
251	C5H8O2	METHACRYLIC ACID, METHYL ESTER	80-62-6	100	410	100	410	100	410	50	210
252	C5H8O2	METHYL METHACRYLATE	80-62-6	100	410	100	410	100	410	50	210
253	C5H9NO	N-METHYL-2-PYRROLIDONE	872-50-4	-----	-----	-----	-----	-----	-----	20	80
254	C5H10	CYCLOPENTANE	287-92-3	600	1720	-----	-----	600	1720	-----	-----
255	C5H10N2O2S	METHOMYL	16752-77-5	-----	2.5	-----	-----	-----	2.5	-----	-----
256	C5H10O	DIETHYL KETONE	96-22-0	200	705	-----	-----	200	705	-----	-----
257	C5H10O	METHYL ISOPROPYL KETONE	563-80-4	200	705	-----	-----	200	705	-----	-----
258	C5H10O	2-PENTANONE	107-87-9	200	705	200	700	150	530	200	700
259	C5H10O	n-VALERALDEHYDE	110-62-3	50	176	-----	-----	50	175	-----	-----
260	C5H10O2	ISOPROPYL ACETATE	108-21-4	250	1040	250	950	-----	-----	200	840
261	C5H10O2	n-PROPYL ACETATE	109-60-4	200	835	200	840	200	840	200	840
262	C5H10O3	ETHYLENE GLYCOL METHYL ETHER ACETATE	110-49-6	5	24	25	120	0.1	0.5	5	25
263	C5H10O3	ETHYLENE GLYCOL MONOMETHYL ETHER ACETATE	110-49-6	5	24	25	120	0.1	0.5	5	25
264	C5H10O3	2-METHOXYETHYL ACETATE	110-49-6	5	24	25	120	0.1	0.5	5	25
265	C5H12	ISOPENTANE	78-78-4	600	1770	1000	2950	120	350	1000	2950
266	C5H12	n-PENTANE	109-66-0	600	1770	1000	2950	120	350	1000	2950
267	C5H12	NEOPENTANE	463-82-1	600	1770	1000	2950	120	350	1000	2950
268	C5H12O	ISOAMYL ALCOHOL	123-51-3	100	361	100	360	100	360	100	360
269	C5H12O	METHYL-TERT-BUTYL ETHER	1634-04-4	40	144	-----	-----	-----	-----	-----	-----
270	C5H12O2	ETHYLENE GLYCOL MONO N-PROPYL ETHER	2807-30-9	-----	-----	-----	-----	-----	-----	20	85
271	C5H12O2	ISOPROPOXYETHANOL	109-59-1	25	106	-----	-----	-----	-----	5	22
272	C5H12O2	2-PROPOXYETHANOL	2807-30-9	-----	-----	-----	-----	-----	-----	20	85
273	C5H12O4	PENTAERYTHRITOL	115-77-5	-----	10	-----	15*/5**	-----	10*/5**	-----	-----
274	C5O5Fe	IRON PENTACARBONYL (as Fe)	13463-40-6	0.1	0.8	-----	-----	0.1	0.8	0.1	0.8
275	C6Cl5NO2	PENTACHLORONITROBENZENE	82-68-8	-----	0.5	-----	-----	-----	-----	-----	-----
276	C6Cl6	HEXACHLOROBENZENE	118-74-1	-----	0.002	-----	-----	-----	-----	-----	-----
277	C6HCl5O	PENTACHLOROPHENOL	87-86-5	-----	0.5	-----	0.5	-----	0.5	-----	-----
278	C6H3Cl3	TRICHLOROBENZENE	12002-48-1	-----	-----	-----	-----	-----	-----	5	38
279	C6H3Cl3N2O2	PICLORAM	1918-02-1	-----	10	-----	15*/5**	-----	10*/5**	-----	-----
280	C6H3Cl4N	2-CHLORO-6-TRICHLOROMETHYL PYRIDINE	1929-82-4	-----	10	-----	15*/5**	-----	10*/5**	-----	-----
281	C6H3N3O7	PICRIC ACID	88-89-1	-----	0.1	-----	0.1	-----	0.1	-----	0.1
282	C6H4Cl2	o-DICHLOROBENZENE	95-50-1	25	150	-----	-----	-----	-----	50	300
283	C6H4Cl2	p-DICHLOROBENZENE	106-46-7	10	60	75	450	-----	-----	50	300
284	C6H4ClNO2	o-CHLORONITROBENZENE	88-73-3	-----	-----	-----	-----	-----	-----	-----	-----
285	C6H4ClNO2	m-CHLORONITROBENZENE	121-73-3	-----	-----	-----	-----	-----	-----	-----	-----
286	C6H4ClNO2	p-CHLORONITROBENZENE	100-00-5	0.1	0.64	-----	1	-----	-----	-----	-----
287	C6H4ClNO2	p-NITROCHLOROBENZENE	100-00-5	0.1	0.64	-----	1	-----	-----	-----	-----
288	C6H4N2O4	o-DINITROBENZENE	528-29-0	0.15	1	-----	1	-----	1	-----	-----
289	C6H4N2O4	m-DINITROBENZENE	99-65-0	0.15	1	-----	1	-----	1	-----	-----
290	C6H4N2O4	p-DINITROBENZENE	100-25-4	0.15	1	-----	1	-----	1	-----	-----
291	C6H4O2	p-BENZOQUINONE	106-51-4	0.1	0.44	0.1	0.4	0.1	0.4	0.1	0.4
292	C6H4O2	QUINONE	106-51-4	0.1	0.44	0.1	0.4	0.1	0.4	0.1	0.4
293	C6H5Cl	CHLOROBENZENE	108-90-7	10	46	75	350	-----	-----	10	46
294	C6H5Cl	MONOCHLOROBENZENE	108-90-7	10	46	75	350	-----	-----	10	46
295	C6H5NO2	NITROBENZENE	98-95-3	1	5	1	5	1	5	1	5
296	C6H6	BENZENE	71-43-2	0.5	1.6	1	3	0.1	0.32	-----	-----
297	C6H6Cl6	gamma-HEXACHLOROCYCLOHEXANE (LINDANE)	58-89-9	-----	0.5	-----	0.5	-----	0.5	-----	0.5
298	C6H6Cl6	LINDANE	58-89-9	-----	0.5	-----	0.5	-----	0.5	-----	0.5
299	C6H6N2O2	p-NITROANILINE	100-01-6	-----	3	1	6	-----	3	1	6
300	C6H6O	PHENOL	108-95-2	5	19	5	19	5	19	5	19
301	C6H6O2	CATECHOL	120-80-9	5	23	-----	-----	5	20	-----	-----
302	C6H6O2	PYROCATECHOL	120-80-9	5	23	-----	-----	5	20	-----	-----
303	C6H6O2	DIHYDROXYBENZENE	123-31-9	-----	2	-----	2	-----	-----	-----	-----
304	C6H6O2	HYDROQUINONE	123-31-9	-----	2	-----	2	-----	-----	-----	-----
305	C6H6O2	RESORCINOL	108-46-3	10	45	-----	-----	10	45	-----	-----
306	C6H6S	PHENYL MERCAPTAN	108-98-5	0.5	2.3	-----	-----	-----	-----	-----	-----
307	C6H7N	ANILINE	62-53-3	2	7.6	5	19	-----	-----	2	8
308	C6H7NO2	ETHYL CYANOACRYLATE	7085-85-0	0.2	1	-----	-----	-----	-----	-----	-----

Table 27-1 EXPOSURE LIMITS - ORGANIC COMPOUNDS (continued)

NO	FORMULA	NAME	CAS No	TLV (ACGIH) Threshold Limit Value		PEL (OSHA) Permissible Exposure Limit		REL (NIOSH) Recommended Exposure Limit		MAK (DFG) Maximum Concentration Value	
				ppm (vol)	mg/m³	ppm (vol)	mg/m³	ppm (vol)	mg/m³	ppm (vol)	mg/m³
309	C6H8N2	ADIPONITRILE	111-69-3	2	8.8	-----	-----	4	18	-----	-----
310	C6H8N2	o-PHENYLENEDIAMINE	95-54-5	-----	0.1	-----	-----	-----	-----	-----	-----
311	C6H8N2	m-PHENYLENEDIAMINE	108-45-2	-----	0.1	-----	-----	-----	-----	-----	-----
312	C6H8N2	p-PHENYLENEDIAMINE	106-50-3	-----	0.1	-----	0.1	-----	0.1	-----	0.1
313	C6H8N2	PHENYLHYDRAZINE	100-63-0	0.1	0.44	5	22	-----	-----	-----	-----
314	C6H10	CYCLOHEXENE	110-83-8	300	1010	300	1015	300	1015	300	1015
315	C6H10O	CYCLOHEXANONE	108-94-1	25	100	50	200	25	100	-----	-----
316	C6H10O2	ALLYL GLYCIDYL ETHER	106-92-3	5	23	-----	-----	5	22	-----	-----
317	C6H10O3	DIGLYCIDYL ETHER	2238-07-5	0.1	0.53	-----	-----	0.1	0.5	0.1	0.5
318	C6H10O3	2-HYDROXYPROPYL ACRYLATE	999-61-1	0.5	2.8	-----	-----	0.5	3	-----	-----
319	C6H10O4	ADIPIC ACID	124-04-9	-----	5	-----	-----	-----	-----	-----	-----
320	C6H11NO	CAPROLACTAM (dust)	105-60-2	-----	1	-----	-----	-----	1	-----	5
321	C6H11NO	CAPROLACTAM (vapor)	105-60-2	5	23	-----	-----	0.22	1	-----	5
322	C6H12	CYCLOHEXANE	110-82-7	300	1030	300	1050	300	1050	200	700
323	C6H12N2S4	THIRAM	137-26-8	-----	1	-----	5	-----	5	-----	5
324	C6H12O	CYCLOHEXANOL	108-93-0	50	206	50	200	50	200	50	200
325	C6H12O	HEXONE	108-10-1	50	205	100	410	50	205	20	82
326	C6H12O	METHYL N-BUTYL KETONE	591-78-6	5	20	100	410	1	4	5	21
327	C6H12O	METHYL ISOBUTYL KETONE	108-10-1	50	205	100	410	50	205	20	82
328	C6H12O2	n-BUTYL ACETATE	123-86-4	150	713	150	710	150	710	200	950
329	C6H12O2	sec-BUTYL ACETATE	105-46-4	200	950	200	950	200	950	200	950
330	C6H12O2	tert-BUTYL ACETATE	540-88-5	200	950	200	950	200	950	200	950
331	C6H12O2	DIACETONE ALCOHOL	123-42-2	50	238	50	240	50	240	50	240
332	C6H12O2	2-HEXANONE	591-78-6	5	20	100	410	1	4	5	21
333	C6H12O2	1-HEXENE	592-41-6	30	130	-----	-----	-----	-----	-----	-----
334	C6H12O2	4-HYDROXY-4-METHYL-2-PENTANONE	123-42-2	50	238	50	240	50	240	50	240
335	C6H12O2	ISOBUTYL ACETATE	110-19-0	150	713	150	700	150	700	200	950
336	C6H12O2	ISOPROPYL GLYCIDYL ETHER	4016-14-2	50	238	50	240	-----	-----	-----	-----
337	C6H12O3	2-ETHOXYETHYL ACETATE	111-15-9	5	27	100	540	0.5	2.7	5	27
338	C6H12O3	ETHYLENE GLYCOL MONOETHYL ETHER ACETATE	111-15-9	5	27	100	540	0.5	2.7	5	27
339	C6H12O3	1-METHOXYPROPYL-2-ACETATE	108-65-6	-----	-----	-----	-----	-----	-----	50	275
340	C6H12O3	2-METHOXYPROPYL-1-ACETATE	70657-70-4	-----	-----	-----	-----	-----	-----	20	110
341	C6H12O3	PROPYLENE GLYCOL-1-METHYL ETHER-2-ACETATE	108-65-6	-----	-----	-----	-----	-----	-----	50	275
342	C6H12O3	PROPYLENE GLYCOL-2-METHYL ETHER-1-ACETATE	70657-70-4	-----	-----	-----	-----	-----	-----	20	110
343	C6H12S2	ALLYL PROPYL DISULFIDE	2179-59-1	2	12	2	12	2	12	2	12
344	C6H13N	CYCLOHEXYLAMINE	108-91-8	10	41	-----	-----	10	40	10	40
345	C6H13NO	N-ETHYLMORPHOLINE	100-74-3	5	24	20	94	5	23	-----	-----
346	C6H14	2,2-DIMETHYL BUTANE	75-83-2	500	1760	-----	-----	100	350	200	700
347	C6H14	2,3-DIMETHYL BUTANE	79-29-8	500	1760	-----	-----	100	350	200	700
348	C6H14	HEXANE	110-54-3	50	176	500	1800	50	180	50	180
349	C6H14	2-METHYLPENTANE	107-83-5	500	1760	-----	-----	100	350	200	700
350	C6H14	3-METHYLPENTANE	96-14-0	500	1760	-----	-----	-----	-----	200	700
351	C6H14O	ISOPROPYL ETHER	108-20-3	250	1040	500	2100	500	2100	500	2100
352	C6H14O	METHYL AMYL ALCOHOL	108-11-2	25	104	25	100	25	100	25	100
353	C6H14O	METHYL ISOBUTYL CARBINOL	108-11-2	25	104	25	100	25	100	25	100
354	C6H14O	4-METHYL-2-PENTANOL	108-11-2	25	104	25	100	25	100	25	100
355	C6H14O2	2-BUTOXYETHANOL	111-76-2	25	121	50	240	5	24	20	100
356	C6H14O2	ETHYLENE GLYCOL MONOBUTYL ETHER	111-76-2	25	121	50	240	5	24	20	100
357	C6H14O3	DIETHYLENE GLYCOL DIMETHYL ETHER	111-96-6	-----	-----	-----	-----	-----	-----	5	27
358	C6H15N	DIISOPROPYLAMINE	108-18-9	5	21	5	20	5	20	-----	-----
359	C6H15N	TRIETHYLAMINE	121-44-8	1	4.1	25	100	-----	-----	1	4.2
360	C6H15NO	2-DIETHYLAMINOETHANOL	100-37-8	2	9.6	10	50	10	50	10	50
361	C6H15NO3	TRIETHANOLAMINE	102-71-6	-----	5	-----	-----	-----	-----	-----	-----
362	C6H15O3PS2	METHYL DEMETON	8022-00-2	-----	0.5	-----	-----	-----	0.5	0.5	5
363	C6H16N2	1,6-HEXANEDIAMINE	124-09-4	0.5	2.3	-----	-----	-----	-----	-----	-----
364	C7H5N3O6	2,4,6-TRINITROPHENOL	88-89-1	-----	0.1	-----	0.1	-----	0.1	-----	0.1
365	C7H5N3O6	2,4,6-TRINITROTOLUENE (TNT)	118-96-7	-----	0.1	-----	1.5	-----	0.5	0.01	0.09
366	C7H5N5O8	N-METHYL-N,2,4,6-TETRANITROANILINE	-----	-----	1.5	-----	1.5	-----	1.5	-----	1.5
367	C7H5N5O8	TETRYL	-----	-----	1.5	-----	1.5	-----	1.5	-----	1.5
368	C7H5N5O8	2,4,6-TRINITROPHENYLMETHYLNITRAMINE	-----	-----	1.5	-----	1.5	-----	1.5	-----	1.5
369	C7H6N2O4	DINITROTOLUENE	25321-14-6	-----	0.2	-----	1.5	-----	1.5	-----	-----
370	C7H6N2O5	4,6-DINITRO-o-CRESOL	534-52-1	-----	0.2	-----	0.2	-----	0.2	-----	0.2
371	C7H7Cl	BENZYL CHLORIDE	100-44-7	1	5.2	1	5	-----	-----	-----	-----
372	C7H7Cl	o-CHLOROTOLUENE	95-49-8	50	259	-----	-----	50	250	-----	-----
373	C7H7Cl2NO	CLOPIDOL	2971-90-6	-----	10	-----	15*/5**	-----	10*/5**	-----	-----
374	C7H7NO2	2-NITROTOLUENE	88-72-2	2	11	5	30	2	11	-----	-----
375	C7H7NO2	3-NITROTOLUENE	99-08-1	2	11	5	30	2	11	5	28
376	C7H7NO2	4-NITROTOLUENE	99-99-0	2	11	5	30	2	11	5	28
377	C7H8	TOLUENE	108-88-3	50	188	200	-----	100	375	50	190
378	C7H8	TOLUOL	108-88-3	50	188	200	-----	100	375	50	190
379	C7H8O	CRESOL	1319-77-3	5	22	5	22	2.3	10	5	22
380	C7H8O2	4-METHOXYPHENOL	150-76-5	-----	5	-----	-----	-----	5	-----	-----
381	C7H9N	N-METHYL ANILINE	100-61-8	0.5	2.2	2	9	0.5	2	0.5	2
382	C7H9N	o-TOLUIDINE	95-53-4	2	8.8	5	22	-----	-----	-----	-----
383	C7H9N	m-TOLUIDINE	108-44-1	2	8.8	-----	-----	-----	-----	-----	-----
384	C7H9N	p-TOLUIDINE	106-49-0	2	8.8	-----	-----	-----	-----	-----	-----
385	C7H9NO	o-ANISIDINE	90-04-0	0.1	0.5	-----	0.5	-----	0.5	-----	-----

Table 27-1 EXPOSURE LIMITS - ORGANIC COMPOUNDS (continued)

NO	FORMULA	NAME	CAS No	TLV (ACGIH) Threshold Limit Value		PEL (OSHA) Permissible Exposure Limit		REL (NIOSH) Recommended Exposure Limit		MAK (DFG) Maximum Concentration Value	
				ppm (vol)	mg/m³	ppm (vol)	mg/m³	ppm (vol)	mg/m³	ppm (vol)	mg/m³
386	C7H9NO	p-ANISIDINE	104-94-9	0.1	0.5	-----	0.5	-----	0.5	0.1	0.5
387	C7H12O2	n-BUTYL ACRYLATE	141-32-2	10	52	-----	-----	10	55	2	11
388	C7H12O2	o-METHYLCYCLOHEXANONE	583-60-8	50	229	100	460	50	230	50	230
389	C7H13O6P	MEVINPHOS	7786-34-7	0.01	0.092	-----	0.1	0.01	0.1	0.01	0.09
390	C7H13O6P	PHOSDRIN	7786-34-7	0.01	0.092	-----	0.1	0.01	0.1	0.01	0.09
391	C7H14	METHYLCYCLOHEXANE	108-87-2	400	1610	500	2000	400	1600	500	2000
392	C7H14O2	n-BUTYL GLYCIDYL ETHER	2426-08-6	25	133	50	270	-----	-----	-----	-----
393	C7H14NO5P	MONOCHROTOPHOS	6923-22-4	-----	0.25	-----	-----	-----	0.25	-----	-----
394	C7H14O	DIPROPYL KETONE	123-19-3	50	233	-----	-----	50	235	-----	-----
395	C7H14O	ETHYL BUTYL KETONE	106-35-4	50	-----	50	230	50	230	-----	-----
396	C7H14O	2-HEPTANONE	110-43-0	50	233	100	465	100	465	-----	-----
397	C7H14O	3-HEPTANONE	106-35-4	50	234	50	230	50	230	-----	-----
398	C7H14O	METHYL N-AMYL KETONE	110-43-0	50	233	100	465	100	465	-----	-----
399	C7H14O	METHYLCYCLOHEXANOL	25639-42-3	50	234	100	470	50	235	50	235
400	C7H14O	METHYL IOSOAMYL KETONE	110-12-3	50	234	100	475	50	240	-----	-----
401	C7H14O2	n-AMYL ACETATE	628-63-7	100	532	100	525	100	525	50	260
402	C7H14O2	sec-AMYL ACETATE	626-38-0	125	665	125	650	125	650	50	260
403	C7H14O2	ISOAMYL ACETATE	123-92-2	100	532	100	525	100	525	-----	-----
404	C7H14O3	n-BUTYL LACTATE	138-22-7	5	30	-----	-----	5	25	-----	-----
405	C7H14O3	2-PROPOXYETHYL ACETATE	20706-25-6	-----	-----	-----	-----	-----	-----	20	120
406	C7H16	HEPTANE	142-82-5	400	1640	500	2000	85	350	500	2000
407	C7H16O3	DIPROPYLENE GLYCOL METHYL ETHER	34590-94-8	100	606	100	600	100	600	50	300
408	C7H17O2PS3	PHORATE	298-02-2	-----	0.05	-----	-----	-----	0.05	-----	-----
409	C8H4F15NO2	AMMONIUM PERFLUOROOCTANOATE	3825-26-1	-----	0.01	-----	-----	-----	-----	-----	-----
410	C8H4N2	m-PHTHALODINITRILE	626-17-5	-----	5	-----	-----	-----	5	-----	-----
411	C8H4O3	PHTHALIC ANHYDRIDE	85-44-9	1	6.1	2	12	1	6	-----	1
412	C8H5Cl3O3	2,4,5-TRICHLOROPHENOXYACETIC ACID	93-76-5	-----	10	-----	10	-----	10	-----	10
413	C8H6Cl2O3	2,4-DICHLOROPHENOXYACETIC ACID	94-75-7	-----	10	-----	10	-----	10	-----	1
414	C8H6O4	TEREPHTHALIC ACID	100-21-0	-----	10	-----	-----	-----	-----	-----	-----
415	C8H7Cl	o-CHLOROSTYRENE	2039-87-4	50	283	-----	-----	50	285	-----	-----
416	C8H7Cl2O5SNa	SESONE	136-78-7	-----	10	-----	15*/5**	-----	10*/5**	-----	-----
417	C8H7ClO	alpha-CHLOROACETOPHENONE	532-27-4	0.05	0.32	0.05	0.3	0.05	0.3	-----	-----
418	C8H7ClO	PHENACYL CHLORIDE	532-27-4	0.05	0.32	0.05	0.3	0.05	0.3	-----	-----
419	C8H7N3O5	DINITOLMIDE	148-01-6	-----	5	-----	-----	-----	5	-----	-----
420	C8H7N3O5	3,5-DINITRO-o-TOLUAMIDE	148-01-6	-----	5	-----	-----	-----	5	-----	-----
421	C8H8	PHENYLETHYLENE	100-42-5	20	85	100	-----	50	215	20	85
422	C8H8	STYRENE	100-42-5	20	85	100	-----	50	215	20	85
423	C8H8	VINYL BENZENE	100-42-5	20	85	100	-----	50	215	20	85
424	C8H8O	ACETOPHENONE	98-86-2	10	49	-----	-----	-----	-----	-----	-----
425	C8H10	DIMETHYLBENZENE	1330-20-7	100	434	100	435	100	435	100	440
426	C8H10	ETHYL BENZENE	100-41-4	100	434	100	435	100	435	100	440
427	C8H10	XYLENE	1330-20-7	100	434	100	435	100	435	100	440
428	C8H10	o-XYLENE	95-47-6	100	434	100	435	100	435	100	440
429	C8H10	m-XYLENE	108-38-3	100	434	100	435	100	435	100	440
430	C8H10	p-XYLENE	106-42-3	100	434	100	435	100	435	100	440
431	C8H10NO5PS	METHYL PARATHION	298-00-0	-----	0.2	-----	-----	-----	0.2	-----	-----
432	C8H11N	DIMETHYLAMINOBENZENE	1300-73-8	0.5	2.5	5	25	2	10	5	25
433	C8H11N	DIMETHYLANILINE	121-69-7	5	25	5	25	5	25	5	25
434	C8H11N	XYLIDINE	1300-73-8	0.5	2.5	5	25	2	10	5	25
435	C8H11N	2,4-XYLIDINE	95-68-1	-----	-----	-----	-----	-----	-----	5	25
436	C8H12	4-VINYL CYCLOHEXENE	100-40-3	0.1	0.44	-----	-----	-----	-----	-----	-----
437	C8H12N2	TETRAMETHYL SUCCINONITRILE	3333-52-6	0.5	2.8	0.5	3	0.5	3	0.5	3
438	C8H12N2O2	1,6-HEXAMETHYLENE DIISOCYANATE	822-06-0	0.005	0.034	-----	-----	0.005	0.035	0.005	0.035
439	C8H12O2	VINYL CYCLOHEXENE DIOXIDE	106-87-6	0.1	0.57	-----	-----	10	60	-----	-----
440	C8H14ClN5	ATRAZINE	1912-24-9	-----	5	-----	-----	-----	5	-----	2
441	C8H14N4OS	METRIBUZIN	21087-64-9	-----	5	-----	-----	-----	5	-----	-----
442	C8H16NO5P	DICROTOPHOS	141-66-2	-----	0.25	-----	-----	-----	0.25	-----	-----
443	C8H16O	ETHYL AMYL KETONE	541-85-5	25	131	25	130	25	130	-----	-----
444	C8H16O	5-METHYL-3-HEPTANONE	541-85-5	25	131	25	130	25	130	-----	-----
445	C8H16O2	sec-HEXYL ACETATE	108-84-9	50	295	50	300	50	300	50	300
446	C8H16O3	2-BUTOXYETHYL ACETATE	112-07-2	-----	-----	-----	-----	5	31	20	135
447	C8H16O3	ETHYLENE GLYCOL MONOBUTYL ETHER ACETATE	112-07-2	-----	-----	-----	-----	5	33	20	135
448	C8H18	OCTANE	111-65-9	300	1400	500	2350	75	350	500	2350
449	C8H18CrO4	tert-BUTYL CHROMATE	1189-85-1	-----	-----	-----	-----	-----	0.001	-----	-----
450	C8H18O	ISOOCTYL ALCOHOL	26952-21-6	50	266	-----	-----	50	270	-----	-----
451	C8H18O3	DIETHYLENE GLYCOL MONOBUTYL ETHER	112-34-5	-----	-----	-----	-----	-----	-----	-----	100
452	C8H19O2PS3	DISULFOTON	298-04-4	-----	0.1	-----	-----	-----	0.1	-----	-----
453	C16H38O6P2S4	DEMETON	8065-48-3	0.01	0.11	-----	0.1	-----	0.1	0.01	0.1
454	C8H19O4P	DIBUTYL PHOSPHATE	107-66-4	1	8.6	1	5	1	5	-----	-----
455	C8H20O4Si	ETHYL SILICATE	78-10-4	10	85	100	850	10	85	20	170
456	C8H20O5P2S2	SULFOTEP	3689-24-5	-----	0.2	-----	0.2	-----	0.2	0.015	0.2
457	C8H20O7P2	TETRAETHYL PHOSPHATE	107-49-3	0.004	0.047	-----	0.05	-----	0.05	0.005	0.05
458	C8H20O7P2	TETRAETHYL PYROPHOSPHATE	107-49-3	0.004	0.047	-----	0.05	-----	0.05	0.005	0.05
459	C8H20Pb	TETRAETHYL LEAD (as Pb)	78-00-2	-----	0.1	-----	0.075	-----	0.075	-----	0.05
460	C8O8Co2	COBALT CARBONYL (as Co)	10210-68-1	-----	0.1	-----	-----	-----	0.1	-----	-----
461	C9H4O5	TRIMELLITIC ANHYDRIDE	552-30-7	-----	-----	-----	-----	0.005	0.04	-----	0.04
462	C9H6Cl6O3S	ENDOSULFAN	115-29-7	-----	0.1	-----	-----	-----	0.1	-----	-----

Table 27-1 EXPOSURE LIMITS - ORGANIC COMPOUNDS (continued)

NO	FORMULA	NAME	CAS No	TLV (ACGIH) Threshold Limit Value ppm (vol)	mg/m^3	PEL (OSHA) Permissible Exposure Limit ppm (vol)	mg/m^3	REL (NIOSH) Recommended Exposure Limit ppm (vol)	mg/m^3	MAK (DFG) Maximum Concentration Value ppm (vol)	mg/m^3
463	C9H6N2O2	TOLUENE-2,4-DIISOCYANATE	584-84-9	0.005	0.036	-----	-----	-----	-----	0.01	0.07
464	C9H6N2O2	TOLUENE-2,6-DIISOCYANATE	91-08-7	-----	-----	-----	-----	-----	-----	0.01	0.07
465	C9H8Cl3NO2S	CAPTAN	133-06-2	-----	5	-----	-----	-----	5	-----	-----
466	C9H8O4	ACETYLSALICYLIC ACID (ASPIRIN)	50-78-2	-----	5	-----	-----	-----	5	-----	-----
467	C9H10	alpha-METHYL STYRENE	98-83-9	50	242	-----	-----	50	240	-----	-----
468	C9H10	METHYL STYRENE	25013-15-4	50	242	100	480	100	480	100	480
469	C9H10	VINYL TOLUENE	25013-15-4	50	242	100	480	100	480	100	480
470	C9H10Cl2N2O	DIURON	330-54-1	-----	10	-----	-----	-----	10	-----	-----
471	C9H10O2	BENZYL ACETATE	140-11-4	10	61	-----	-----	-----	-----	-----	-----
472	C9H10O2	PHENYL GLYCIDYL ETHER	122-60-1	0.1	0.6	10	60	-----	-----	-----	-----
473	C9H11Cl3NO3PS	CHLORPYRIFOS	2921-88-2	-----	0.2	-----	-----	-----	0.2	-----	-----
474	C9H12	CUMENE	98-82-8	50	-----	50	245	50	245	50	245
475	C9H12	TRIMETHYL BENZENE	25551-13-7	25	123	-----	-----	25	125	-----	-----
476	C9H13BrN2O2	BROMACIL	314-40-9	-----	10	-----	-----	1	10	-----	-----
477	C9H13N	n-ISOPROPYLANILINE	768-52-5	2	-----	-----	-----	2	10	-----	-----
478	C9H14O	ISOPHORONE	78-59-1	-----	-----	25	140	4	23	2	11
479	C9H18N3S6Fe	FERBAM	14484-64-1	-----	10	-----	15	-----	10	-----	15
480	C9H18O	DIISOBUTYL KETONE	108-83-8	25	145	50	290	25	150	50	290
481	C9H20	NONANE	111-84-2	200	1050	-----	-----	200	1050	-----	-----
482	C9H22O4P2S4	ETHION	563-12-2	-----	0.4	-----	-----	-----	0.4	-----	-----
483	C10Cl8	OCTACHLORONAPHTHALENE	2234-13-1	0.1	0.3	-----	0.1	-----	0.1	-----	-----
484	C10Cl10O	CHLORDECONE	143-50-0	-----	-----	-----	-----	-----	0.001	-----	-----
485	C10Cl10O	KEPONE	143-50-0	-----	-----	-----	-----	-----	0.001	-----	-----
486	C10H2Cl6	HEXACHLORONAPHTHALENE	1335-87-1	-----	0.2	-----	0.2	-----	0.2	-----	-----
487	C10H3Cl5	PENTACHLORONAPHTHALENE	1321-64-8	-----	0.5	-----	0.5	-----	0.5	-----	-----
488	C10H4Cl4	TETRACHLORONAPHTHALENE	1335-88-2	-----	2	-----	2	-----	2	-----	0.5
489	C10H5Cl3	TRICHLORONAPHTHALENE	1321-65-9	-----	5	-----	5	-----	5	-----	-----
490	C10H5Cl7	HEPTACHLOR	76-44-8	-----	0.5	-----	0.5	-----	0.5	-----	0.5
491	C10H5Cl7O	HEPTACHLOR EPOXIDE	1024-57-3	-----	0.5	-----	0.5	-----	0.5	-----	0.5
492	C10H5ClN2	o-CHLOROBENZYLIDENE MALONONITRILE	2698-41-1	-----	-----	0.05	0.4	-----	-----	-----	-----
493	C10H6Cl8	CHLORDANE	57-74-9	-----	0.5	-----	0.5	-----	0.5	-----	0.5
494	C10H8	NAPHTHALENE	91-20-3	10	52	10	50	10	50	-----	-----
495	C10H9Cl4NO2S	CAPTAFOL	2425-06-1	-----	0.1	-----	-----	-----	0.1	-----	-----
496	C10H10	DIVINYL BENZENE	1321-74-0	10	53	-----	-----	10	50	-----	-----
497	C10H10Cl8	CHLORINATED CAMPHENE	8001-35-2	-----	0.5	-----	0.5	-----	-----	-----	0.5
498	C10H10Cl8	TOXAPHENE	8001-35-2	-----	0.5	-----	0.5	-----	-----	-----	0.5
499	C10H10Fe	DICYCLOPENTADIENYL IRON	102-54-5	-----	10	-----	15	-----	10	-----	-----
500	C10H10Fe	FERROCENE	102-54-5	-----	10	-----	15*/5**	-----	10*/5**	-----	-----
501	C10H10O4	DIMETHYLPHTHALATE	131-11-3	-----	5	-----	5	-----	5	-----	-----
502	C10H12	DICYCLOPENTADIENE	77-73-6	5	27	-----	-----	5	30	0.5	3
503	C10H12N3O3PS2	GUTHION	86-50-0	-----	0.2	-----	0.2	-----	0.2	-----	0.2
504	C10H12N3O3PS2	METHYL AZINPHOS	86-50-0	-----	0.2	-----	0.2	-----	0.2	-----	0.2
505	C10H14N2	NICOTINE	54-11-5	-----	0.5	-----	0.5	-----	0.5	0.07	0.5
506	C10H14NO5P	PARATHION	56-38-2	-----	0.1	-----	0.1	-----	0.05	-----	0.1
507	C10H14O	o-sec-BUTYLPHENOL	89-72-5	5	31	-----	-----	5	30	-----	-----
508	C10H14O	p-tert-BUTYLPHENOL	98-54-4	-----	-----	-----	-----	-----	-----	0.08	0.5
509	C10H15OPS2	FONOFOS	944-22-9	-----	0.1	-----	-----	-----	0.1	-----	-----
510	C10H15O3PS2	FENTHION	55-38-9	-----	0.2	-----	-----	-----	-----	-----	0.2
511	C10H16O	SYNTHETIC CAMPHOR	76-22-2	2	12	-----	2	-----	2	2	13
512	C10H19O6PS2	MALATHION	121-75-5	-----	10	-----	15	-----	10	-----	15
513	C10H20N2S4	DISULFIRAM	97-77-8	-----	2	-----	-----	-----	2	-----	2
514	C10H23NO	2-N-DIBUTYLAMINOETHANOL	102-81-8	0.5	3.5	-----	-----	2	14	-----	-----
515	C11H10N2S	alpha-NAPHTHYLTHIOUREA	86-88-4	-----	0.3	-----	0.3	-----	0.3	-----	0.3
516	C11H15NO3	PROPOXUR	114-26-1	-----	0.5	-----	-----	-----	0.5	-----	2
517	C11H16	p-tert-BUTYLTOLUENE	98-51-1	1	6.1	10	60	10	60	10	60
518	C11H17O4PS2	FENSULFOTHION	115-90-2	-----	0.1	-----	-----	-----	0.1	-----	-----
519	C12H4Cl6O	CHLORINATED DIPHENYL OXIDE	31242-93-0	-----	0.5	-----	0.5	-----	0.5	-----	0.5
520	C12H6N2O2	NAPHTHALENE DIISOCYANATE	3173-72-6	-----	-----	-----	-----	0.005	0.04	-----	-----
521	C12H6N2O2	1,5-NAPHTHYLENE DIISOCYANATE	3173-72-6	-----	-----	-----	-----	-----	-----	0.01	0.09
522	C12H8Cl6O	DIELDRIN	60-57-1	-----	0.25	-----	0.25	-----	0.25	-----	0.25
523	C12H8Cl6O	ENDRIN	72-20-8	-----	0.1	-----	0.1	-----	0.1	-----	0.1
524	C12H9NS	PHENOTHIAZINE	92-84-2	-----	5	-----	-----	-----	5	-----	-----
525	C12H10	BIPHENYL	92-52-4	0.2	1.3	0.2	1	0.2	1	0.2	1
526	C12H10	DIPHENYL	92-52-4	0.2	1.3	0.2	1	0.2	1	0.2	1
527	C12H10O	DIPHENYL ETHER	101-84-8	1	7	1	7	1	7	1	7
528	C12H10O	PHENYL ETHER	101-84-8	1	7	1	7	1	7	1	7
529	C12H11N	DIPHENYLAMINE	122-39-4	-----	10	-----	-----	-----	10	-----	-----
530	C12H11NO2	CARBARYL	63-25-2	-----	5	-----	5	-----	5	-----	5
531	C12H12N2	DIQUAT	2764-72-9	-----	0.5	-----	-----	-----	0.5	-----	-----
532	C12H14N2	PARAQUAT	4685-14-7	-----	0.5	-----	0.5	-----	0.1	-----	0.1
533	C12H14O4	DIETHYL PHTHALATE	84-66-2	-----	5	-----	-----	-----	5	-----	-----
534	C12H15N3O6	1,3,5-TRIGLYCIDYL-S-TRIAZINETRIONE	2451-62-9	-----	0.05	-----	-----	-----	-----	-----	-----
535	C12H15NO3	CARBOFURAN	1563-66-2	-----	0.1	-----	-----	-----	0.1	-----	-----
536	C12H18Cl6	ALDRIN	309-00-2	-----	0.25	-----	0.25	-----	0.25	-----	0.25
537	C12H18N2O2	ISOPHORONE DIISOCYANATE	4098-71-9	0.005	0.045	-----	-----	0.005	0.045	0.01	0.09
538	C12H19ClNO3P	CRUFOMATE	299-86-5	-----	5	-----	-----	-----	5	-----	-----
539	C12H19O2PS3	SULPROFOS	35400-43-2	-----	1	-----	-----	-----	1	-----	-----

Table 27-1 EXPOSURE LIMITS - ORGANIC COMPOUNDS (continued)

NO	FORMULA	NAME	CAS No	TLV (ACGIH) Threshold Limit Value		PEL (OSHA) Permissible Exposure Limit		REL (NIOSH) Recommended Exposure Limit		MAK (DFG) Maximum Concentration Value	
				ppm (vol)	mg/m³	ppm (vol)	mg/m³	ppm (vol)	mg/m³	ppm (vol)	mg/m³
540	C12H21N2O3PS	DIAZINON	333-41-5	-----	0.1	-----	-----	-----	0.1	-----	0.1
541	C12H22O11	SUCROSE	57-50-1	-----	10	-----	15*/5**	-----	10*/5**	-----	-----
542	C12H26O6P2S4	DIOXATHION	78-34-2	-----	0.2	-----	-----	-----	0.2	-----	-----
543	C12H27O4P	TRIBUTYL PHOSPHATE	126-73-8	0.2	2.2	-----	5	0.2	2.5	-----	-----
544	C13H12Cl2N2	4,4'-METHYLENE BIS(2-CHLOROANILINE)	101-14-4	0.01	0.11	-----	-----	-----	0.003	-----	-----
545	C13H14N2	4,4'-METHYLENE DIANILINE	101-77-9	0.1	0.81	0.01	-----	-----	-----	-----	-----
546	C13H14N2	4,4'-DIAMINOPHENYLMETHANE	101-77-9	0.1	0.81	0.01	-----	-----	-----	-----	-----
547	C13H22NO3PS	FENAMIPHOS	22224-92-6	-----	0.1	-----	-----	-----	0.1	-----	-----
548	C14H9Cl5	DICHLORODIPHENYLTRICHLOROETHANE (DDT)	50-29-3	-----	1	-----	1	-----	0.5	-----	1
549	C14H10O4	BENZOYL PEROXIDE	94-36-0	-----	5	-----	5	-----	5	-----	5
550	C14H10O4	DIBENZOYL PEROXIDE	94-36-0	-----	5	-----	5	-----	5	-----	0.5
551	C14H14NO4PS	EPN	2104-64-5	-----	0.1	-----	0.5	-----	0.5	-----	-----
552	C14H14O3	PINDONE	83-26-1	-----	0.1	-----	0.1	-----	0.1	-----	-----
553	C14H18N4O3	BENOMYL	17804-35-2	0.84	10	-----	15*/5**	-----	-----	-----	-----
554	C14H23O4P	DIBUTYL PHENYL PHOSPHATE	2528-36-1	0.3	3.5	-----	-----	-----	-----	-----	-----
555	C15H10N2O2	METHYLENE BISPHENYL ISOCYNATE (MDI)	101-68-8	0.005	0.051	-----	-----	0.005	0.05	0.005	0.05
556	C15H16N4O5S	METHYL SULFOMETURON	74222-97-2	-----	5	-----	-----	-----	-----	-----	-----
557	C15H16O2	BISPHENOL A	80-05-7	-----	-----	-----	-----	-----	-----	-----	5
558	C15H16O2	4,4'-ISOPROPYLIDENE DIPHENOL	80-05-7	-----	-----	-----	-----	-----	-----	-----	5
559	C15H22N2O2	METHYLENE BIS(4-CYCLO-HEXYLISOCYANATE)	5124-30-1	0.005	0.054	-----	-----	-----	-----	-----	-----
560	C15H24O	2,6-DI-tert-BUTYL-p-CRESOL	128-37-0	-----	10	-----	-----	-----	10	-----	-----
561	C16H15Cl3O2	METHOXYCHLOR	72-43-5	-----	10	-----	15	-----	-----	-----	15
562	C16H20O6P2S	TEMEPHOS	3383-96-8	-----	10	-----	15*/5**	-----	10*/5**	-----	-----
563	C16H22O4	DIBUTYL PHTHALATE	84-74-2	-----	5	-----	5	-----	5	-----	-----
564	C18H12	CHRYSENE	218-01-9	-----	-----	-----	0.2	-----	0.1	-----	-----
565	C18H15N	TRIPHENYL AMINE	603-34-9	-----	5	-----	-----	-----	5	-----	-----
566	C18H15O4P	TRIPHENYL PHOSPHATE	115-86-6	-----	3	-----	3	-----	3	-----	-----
567	C18H34OSn	CYHEXATIN	13121-70-5	-----	5	-----	-----	-----	5	-----	-----
568	C18H34OSn	TRICYCLOHEXYLTIN HYDROXIDE	13121-70-5	-----	5	-----	-----	-----	5	-----	-----
569	C19H16O4	WARFARIN	81-81-2	-----	0.1	-----	0.1	-----	0.1	-----	0.5
570	C20H12	BENZO(a)PYRENE	50-32-8	-----	-----	-----	0.2	-----	0.1	-----	-----
571	C21H21O4P	TRIORTHOCRESYL PHOSPHATE	78-30-8	-----	0.1	-----	0.1	-----	0.1	-----	-----
572	C21H22N2O2	STRYCHNINE	57-24-9	-----	0.15	-----	0.15	-----	0.15	-----	0.15
573	C22H30O2S	4,4'-THIOBIS	96-69-5	-----	10	-----	15*/5**	-----	10*/5**	-----	-----
574	C23H22O6	ROTENONE	83-79-4	-----	5	-----	5	-----	5	-----	5
575	C24H38O4	DI(2-ETHYLHEXYL)PHTHALATE	117-81-7	-----	5	-----	5	-----	5	-----	10
576	C24H38O4	DI-sec-OCTYL-PHTHALATE	117-81-7	-----	5	-----	5	-----	5	-----	10
577	C36H70O4Zn	ZINC STEARATE	557-05-1	-----	10	-----	15*/5**	-----	10*/5**	-----	-----

* total dust

* respirable dust

TLV (ACGIH) - Threshold Limit Value of ACGIH

ACGIH - American Conference of Governmental Hygienists

PEL (OSHA) - Permissible Exposure Limit of OSHA

OSHA - Occupational Safety and Health Aministration

REL (NIOSH) - Recommended Exposure Limit of NIOSH

NIOSH - National Institute for Occupational Safety and Health

MAK (DFG) - Maximum Concentration Value in Workplace of DFG

DFG - Federal Republic of Germany

ppm (vol) - parts per million by volume of air

mg/m³ - milligrams per cubic meter of air

Table 27-2 EXPOSURE LIMITS - INORGANIC COMPOUNDS

NO	FORMULA	NAME	CAS No	TLV (ACGIH) Threshold Limit Value		PEL (OSHA) Permissible Exposure Limit		REL (NIOSH) Recommended Exposure Limit		MAK (DFG) Maximum Concentration Value	
				ppm (vol)	mg/m³	ppm (vol)	mg/m³	ppm (vol)	mg/m³	ppm (vol)	mg/m³
1	Ag	SILVER	7440-22-4	-----	0.1	-----	0.01	-----	0.01	-----	0.1
2	Al	ALUMINIUM	7429-90-5	-----	10	-----	15*/5**	-----	10*/5**	-----	6
3	AlO3H3	ALUMINIUM HYDROXIDE	21645-51-2	-----	-----	-----	-----	-----	-----	-----	6
4	Al2O3	alpha-ALUMINA	1344-28-1	-----	10	-----	15*/5**	-----	-----	-----	6
5	Al2O3	ALUMINIUM OXIDE	1344-28-1	-----	10	-----	15*/5**	-----	-----	-----	6
6	Al2O3	EMERY	1302-74-5	-----	10	-----	15	-----	-----	-----	-----
7	As	ARSENIC	7440-38-2	-----	0.01	-----	0.01	-----	-----	-----	-----
8	AsH3	ARSINE	7784-42-1	0.05	0.16	0.05	0.2	-----	-----	0.05	0.2
9	AsH3O4	ARSENIC ACID	7778-39-4	-----	0.01	-----	-----	-----	-----	-----	-----
10	BF3	BORON TRIFLUORIDE	7637-07-2	-----	-----	-----	-----	-----	-----	1	3
11	B2H6	DIBORANE	19287-45-7	0.1	0.11	0.1	0.1	0.1	0.1	0.1	0.1
12	B2O3	BORON OXIDE	1303-86-2	-----	10	-----	15*/5**	-----	10	-----	15
13	B5H9	PENTABORANE	19624-22-7	0.005	0.013	0.005	0.01	0.005	0.01	0.005	0.01
14	B10H14	DECABORANE	17702-41-9	0.05	0.25	0.05	0.3	0.05	0.3	0.05	0.3
15	Ba	BARIUM	7440-39-3	-----	0.5	-----	0.5	-----	0.5	-----	0.5
16	BaSO4	BARIUM SULFATE	7727-43-7	-----	10	-----	15*/5**	-----	10*/5**	-----	-----
17	Be	BERYLLIUM	7440-41-7	-----	0.002	-----	0.002	-----	-----	-----	-----
18	Bi2Te3	BISMUTH TELLURIDE, UNDOPED	1304-82-1	-----	10	-----	15*/5**	-----	10*/5**	-----	-----
19	Bi2Te3	BISMUTH TELLURIDE, Se-DOPED	1304-82-1	-----	5	-----	-----	-----	5	-----	-----
20	Br2	BROMINE	7726-95-6	0.1	0.66	0.1	0.7	0.1	0.7	0.1	0.7
21	BrF5	BROMINE PENTAFLUORIDE	7789-30-2	0.1	0.72	-----	-----	0.1	0.7	-----	-----
22	C	GRAPHITE (NATURAL)	7782-42-5	-----	2	-----	-----	-----	2.5	-----	6
23	CCl2O	PHOSGENE	75-44-5	0.1	0.4	0.1	0.4	0.1	0/4	0.02	0.08
24	CO	CARBON MONOXIDE	630-08-0	25	29	50	55	35	40	30	33
25	CO2	CARBON DIOXIDE	124-38-9	5000	9000	5000	9000	5000	9000	5000	9000
26	CS2	CARBON DISULFIDE	75-15-0	10	31	20	-----	1	3	10	30
27	CaCN2	CALCIUM CYANAMIDE	156-62-7	-----	0.5	-----	-----	-----	0.5	-----	1
28	CaCO3	CALCIUM CARBONATE	471-34-1	-----	10	-----	15*/5**	-----	10*/5**	-----	-----
29	CaCO3	LIMESTONE	1317-65-3	-----	10	-----	15*/5**	-----	10*/5**	-----	-----
30	CaCO3	MARBLE	1317-65-3	-----	10	-----	15*/5**	-----	10*/5**	-----	-----
31	CaCrO4	CALCIUM CHROMATE	13756-19-0	-----	0.001	-----	-----	-----	0.001	-----	-----
32	CaH4SO6	GYPSUM	13397-24-5	-----	10	-----	15*/5**	-----	10*/5**	-----	6
33	CaO	CALCIUM OXIDE	1305-78-8	-----	2	-----	5	-----	2	-----	5
34	CaO2H2	CALCIUM HYDROXIDE	1305-62-0	-----	5	-----	15*/5**	-----	5	-----	-----
35	CaSiO3	CALCIUM SILICATE	1344-95-2	-----	10	-----	15*/5**	-----	10*/5**	-----	-----
36	CaSO4	CALCIUM SULFATE	7778-18-9	-----	10	-----	15*/5**	-----	10*/5**	-----	6
37	Cd	CADMIUM	7440-43-9	-----	0.01	-----	0.005	-----	-----	-----	-----
38	CdO	CADMIUM OXIDE	1306-19-0	-----	-----	-----	0.005	-----	-----	-----	-----
39	ClFO3	PERCHLORYL FLUORIDE	7616-94-6	3	13	3	13.5	3	14	-----	-----
40	ClF3	CHLORINE TRIFLUORIDE	7790-91-2	-----	-----	-----	-----	-----	-----	0.1	0.4
41	ClO2	CHLORINE DIOXIDE	10049-04-4	0.1	0.28	0.1	0.3	0.1	0.3	0.1	0.3
42	Cl2	CHLORINE	7782-50-5	0.5	1.5	-----	-----	-----	-----	0.5	1.5
43	Co	COBALT	7440-48-4	-----	0.02	-----	0.1	-----	0.05	-----	-----
44	Cr	CHROMIUM (METAL)	7440-47-3	-----	0.5	-----	1	-----	0.5	-----	-----
45	CrC6H9O6	CHROMIC ACID	1066-30-4	-----	0.05	-----	-----	-----	0.001	-----	-----
46	CrCi2O2	CHROMYL CHLORIDE	14977-61-8	0.025	0.16	-----	-----	-----	0.001	-----	-----
47	CrO3	CHROMIUM TRIOXIDE	1333-82-0	-----	0.05	-----	-----	-----	0.001	-----	-----
48	CsOH	CESIUM HYDROXIDE	21351-79-1	-----	2	-----	-----	-----	2	-----	-----
49	Cu	COPPER AND INOGRANIC COMPOUNDS	7440-50-8	-----	-----	-----	-----	-----	-----	-----	1
50	Cu	COPPER, FUME	7440-50-8	-----	0.2	-----	0.1	-----	0.1	-----	0.1
51	Cu	COPPER, DUSTS & MISTS	7440-50-8	-----	-----	-----	1	-----	1	-----	1
52	F2	FLUORINE	7782-41-4	1	1.6	0.1	0.2	0.1	0.2	0.1	0.2
53	F2O	FLUORINE OXIDE	7783-41-7	-----	-----	0.05	0.1	-----	-----	-----	-----
54	FeC5O5	IRON PENTACARBONYL	13463-40-6	0.1	0.8	-----	-----	0.1	0.8	0.1	0.8
55	FeCl2	FERROUS CHLORIDE	7758-94-3	-----	-----	-----	-----	-----	1	-----	-----
56	FeCl3	FERRIC CHLORIDE	7705-08-0	-----	-----	-----	-----	-----	1	-----	-----
57	Fe2O3	IRON OXIDE, DUST & FUME	1309-37-1	-----	5	-----	10	-----	5	-----	6
58	GeH4	GERMANIUM TETRAHYDRIDE	7782-65-2	0.2	0.63	-----	-----	0.2	0.6	-----	-----
59	HBr	HYDROGEN BROMIDE	10035-10-6	-----	-----	3	10	-----	-----	2	6.7
60	HCN	HYDROGEN CYANIDE	74-90-8	-----	-----	10	11	-----	-----	10	11
61	HCl	HYDROGEN CHLORIDE	7647-01-0	-----	-----	-----	-----	-----	-----	5	7
62	HF	HYDROGEN FLUORIDE	7664-39-3	-----	-----	3	-----	3	2.5	3	2
63	HNO3	NITRIC ACID	7697-37-2	2	5.2	2	5	2	5	2	5
64	H2O2	HYDROGEN PEROXIDE	7722-84-1	1	1.4	1	1.4	1	1.4	1	1.4
65	H2S	HYDROGEN SULFIDE	7783-06-4	10	14	-----	-----	-----	-----	10	15
66	H2SO4	SULFURIC ACID	7664-93-9	-----	1	-----	1	-----	1	-----	1
67	H2Se	HYDROGEN SELENIDE	7783-07-5	0.05	0.16	0.05	0.2	0.05	0.2	0.05	0.2
68	H3PO4	PHOSPHORIC ACID	7664-38-2	-----	1	-----	1	-----	1	-----	-----
69	Hf	HAFNIUM	7440-58-6	-----	0.5	-----	0.5	-----	0.5	-----	0.5
70	I 2	IODINE	7553-56-2	-----	-----	-----	-----	-----	-----	0.1	1
71	In	INDIUM	7440-74-6	-----	0.1	-----	-----	-----	0.1	-----	-----
72	KCN	POTASSIUM CYANIDE	151-50-8	-----	-----	-----	5	-----	-----	-----	5
73	KOH	POTASSIUM HYDROXIDE	1310-58-3	-----	-----	-----	-----	-----	2	-----	-----
74	K2H2S2O8	POTASSIUM PERSULFATE	7727-21-1	-----	0.1	-----	-----	-----	-----	-----	-----
75	LiH	LITHIUM HYDRIDE	7580-67-8	-----	0.025	-----	0.025	-----	0.025	-----	-----
76	Mg	MANGANESE, FUME	7439-96-5	-----	0.2	-----	-----	-----	1	-----	0.5
77	MgCO3	MAGNESITE	546-93-0	-----	10	-----	15*/5**	-----	10*/5**	-----	-----

Table 27-2 EXPOSURE LIMITS - INORGANIC COMPOUNDS (continued)

NO	FORMULA	NAME	CAS No	TLV (ACGIH) Threshold Limit Value		PEL (OSHA) Permissible Exposure Limit		REL (NIOSH) Recommended Exposure Limit		MAK (DFG) Maximum Concentration Value	
				ppm (vol)	mg/m³	ppm (vol)	mg/m³	ppm (vol)	mg/m³	ppm (vol)	mg/m³
78	MgO	MAGNESIUM OXIDE, FUME	1309-48-4	-----	10	-----	15	-----	-----	-----	6
79	Mn	MANGANESE	7439-96-5	-----	0.2	-----	-----	-----	1	-----	0.5
80	Mn3O4	MANGANESE TETROXIDE	1317-35-7	-----	-----	-----	-----	-----	-----	-----	0.5
81	Mo	MOLYBDENUM (soluble compounds, as Mo)	7439-98-7	-----	5	-----	5	-----	-----	-----	5
82	Mo	MOLYBDENUM (insoluble compounds, as Mo)	7439-98-7	-----	10	-----	15	-----	-----	-----	15
83	NF3	NITROGEN TRIFLUORIDE	7783-54-2	10	29	10	29	10	29	-----	-----
84	NH3	AMMONIA	7664-41-7	25	17	50	35	25	18	20	14
85	NH3	HYDRAZOIC ACID	7782-79-8	-----	-----	-----	-----	-----	-----	0.1	0.18
86	NH4Cl	AMMONIUM CHLORIDE, FUME	12125-02-9	-----	10	-----	-----	-----	10	-----	-----
87	NO	NITRIC OXIDE	10102-43-9	25	-----	25	30	25	30	-----	-----
88	NO2	NITROGEN DIOXIDE	10102-44-0	3	5.6	-----	-----	-----	-----	5	9
89	N2H4	HYDRAZINE	302-01-2	0.01	0.013	1	1.3	-----	-----	-----	-----
90	N2H6NSO3	AMMONIUM SULFAMATE	7773-06-0	-----	10	-----	15*/5*	-----	10*/5*	-----	15
91	N2H8S2O8	AMMONIUM PERSULFATE	7727-54-0	-----	0.1	-----	-----	-----	-----	-----	-----
92	N2O	NITROUS OXIDE	10024-97-2	50	90	-----	-----	25	46	100	200
93	NaCN	SODIUM CYANIDE	143-33-9	-----	-----	-----	5	-----	-----	-----	5
94	NaF	SODIUM FLUORIDE	7681-49-4	-----	-----	-----	2.5	-----	2.5	-----	-----
95	NaHO3S	SODIUM BISULFITE	7631-90-5	-----	5	-----	-----	-----	5	-----	-----
96	NaN3	SODIUM AZIDE	26628-22-8	-----	-----	-----	-----	-----	-----	-----	0.2
97	NaOH	SODIUM HYDROXIDE	1310-73-2	-----	-----	-----	2	-----	-----	-----	2
98	Na2B4H20O17	SODIUM TETRABORATE, DECAHYDRATE	1303-96-4	-----	5	-----	-----	-----	5	-----	-----
99	Na2S2O5	SODIUM METABISULFITE	7681-57-4	-----	5	-----	-----	-----	5	-----	-----
100	Na2S2O8	SODIUM PERSULFATE	7775-27-1	-----	0.1	-----	-----	-----	-----	-----	-----
101	Na4P2O7	TETRASODIUM PYROPHOSPHATE	7722-88-5	-----	5	-----	-----	-----	5	-----	-----
102	Ni	NICKEL	7440-02-0	-----	1	-----	1	-----	0.015	-----	-----
103	NiC4O4	NICKEL CARBONYL (as Ni)	13463-39-3	-----	-----	0.001	0.007	0.001	0.007	-----	-----
104	OF2	OXYGEN DIFLUORIDE	7783-41-7	-----	-----	0.05	0.1	-----	-----	-----	-----
105	O3	OZONE	10028-15-6	-----	-----	0.1	0.2	-----	-----	-----	-----
106	OsO4	OSMIUM TETROXIDE (as Os)	20816-12-0	0.0002	0.0016	-----	0.002	0.0002	0.002	0.0002	0.002
107	P	PHOSPHORUS - YELLOW	7723-14-0	0.02	0.1	-----	0.1	-----	0.1	-----	0.1
108	PCl3	PHOSPHORUS TRICHLORIDE	7719-12-2	0.2	1.1	0.5	3	0.2	1.5	0.5	3
109	PCl5	PHOSPHORUS PENTACHLORIDE	10026-13-8	0.1	0.85	-----	1	-----	1	-----	1
110	PH3	PHOSPHINE	7803-51-2	0.3	0.42	0.3	0.4	0.3	0.4	0.1	0.14
111	POCl3	PHOSPHORUS OXYCHLORIDE	10025-87-3	0.1	0.63	-----	-----	0.1	0.6	0.2	1
112	P2O5	PHOSPHORUS PENTOXIDE	1314-56-3	-----	-----	-----	-----	-----	-----	-----	1
113	P2S5	PHOSPHORUS PENTASULFIDE	1314-80-3	-----	1	-----	1	-----	1	-----	1
114	Pb	LEAD	7439-92-1	-----	0.05	-----	0.05	-----	<0.1	-----	0.1
115	PbCrO4	LEAD CHROMATE	7758-97-6	-----	0.012	-----	-----	-----	0.001	-----	-----
116	Pb3As2O8	LEAD ARSENATE	3687-31-8	-----	0.15	-----	0.01	-----	-----	-----	-----
117	Pb3P2O8	LEAD PHOSPHATE	7446-27-7	-----	0.05	-----	0.05	-----	<0.1	-----	-----
118	Pt	PLATINUM	7440-06-4	-----	1	-----	-----	-----	1	-----	-----
119	Rh	RHODIUM	7440-16-6	-----	1	-----	0.1	-----	0.1	-----	-----
120	SF6	SULFUR HEXAFLUORIDE	2551-62-4	1000	-----	1000	6000	1000	6000	1000	6000
121	SO2	SULFUR DIOXIDE	7446-09-5	2	5.2	5	13	2	5	2	5
122	SO2F2	SULFURYL FLUORIDE	2699-79-8	5	21	5	20	5	20	-----	-----
123	S2Cl2	SULFUR MONOCHLORIDE	10025-67-9	-----	-----	1	6	-----	-----	1	6
124	S2F10	SULFUR PENTAFLUORIDE	5714-22-7	-----	-----	0.025	0.25	-----	-----	0.025	0.25
125	Sb	ANTIMONY	7440-36-0	-----	0.5	-----	0.5	-----	0.5	-----	0.5
126	SbH3	STIBINE	7803-52-3	0.1	0.51	0.1	0.5	0.1	0.5	0.1	0.5
127	Sb2O3	ANTIMONY TRIOXIDE	1309-64-4	-----	0.5	-----	0.5	-----	0.5	-----	-----
128	Se	SELENIUM	7782-49-2	-----	0.2	-----	0.2	-----	0.2	-----	0.1
129	SeF6	SELENIUM HEXAFLUORIDE	7783-79-1	0.05	0.16	0.05	0.16	0.05	0.16	-----	-----
130	Si	SILICON	7440-21-3	-----	10	-----	15*/5**	-----	10*/5**	-----	-----
131	SiC	SILICON CARBIDE	409-21-2	-----	10	-----	15*/5**	-----	10*/5**	-----	4
132	SiC8H20O4	ETHYL SILICATE	78-10-4	10	85	100	850	10	85	20	170
133	SiH4	SILANE	7803-62-5	5	-----	-----	-----	5	7	-----	-----
134	SiH4	SILICON TETRAHYDRIDE	7803-62-5	5	-----	-----	-----	5	7	-----	-----
135	SiO2	SILICA-AMORPHOUS SILICA, FUSED	60676-86-0	-----	0.1	-----	-----	-----	0.05	-----	0.3
136	SiO2	SILICA-AMORPHOUS SILICA GEL	112926-00-8	-----	10	-----	-----	-----	6	-----	4
137	SiO2	SILICA-CRYSTALLINE CRISTOBALITE	14464-46-1	-----	0.05	-----	-----	-----	0.05	-----	0.15
138	SiO2	SILICA-CRYSTALLINE QUARTZ	14808-60-7	-----	0.1	-----	-----	-----	0.05	-----	0.15
139	SiO2	SILICA-CRYSTALLINE TRIDYMITE	15468-32-3	-----	0.05	-----	-----	-----	0.05	-----	0.15
140	SiO2	SILICA-CRYSTALLINE TRIPOLI	1317-95-9	-----	0.1	-----	-----	-----	0.05	-----	0.15
141	SiO2	SILICA, FUME	69012-64-2	-----	2	-----	-----	-----	-----	-----	0.3
142	SiO2	SILICA, FUSED	60676-86-0	-----	0.1	-----	-----	-----	0.05	-----	0.3
143	SiO2	SILICA GEL	112926-00-8	-----	10	-----	-----	-----	6	-----	4
144	SiO2	SILICA, PRECIPITATED	112926-00-8	-----	10	-----	-----	-----	6	-----	4
145	SiO2	CRISTOBALITE	14464-46-1	-----	0.05	-----	-----	-----	0.05	-----	0.15
146	SiO2	QUARTZ	14808-60-7	-----	0.1	-----	-----	-----	0.05	-----	0.15
147	SiO2	TRIDYMITE	15468-32-3	-----	0.05	-----	-----	-----	0.05	-----	0.15
148	Sn	TIN	7440-31-5	-----	2	-----	2	-----	2	-----	-----
149	SnO2	TIN OXIDE	21651-19-4	-----	2	-----	-----	-----	2	-----	2
150	SrCrO4	STRONTIUM CHROMATE	7789-06-2	-----	0.0005	-----	-----	-----	0.001	-----	-----
151	Ta	TANTALUM	7440-25-7	-----	5	-----	5	-----	5	-----	5
152	Ta2O5	TANTALUM OXIDE, DUST	1314-61-0	-----	5	-----	5	-----	5	-----	-----
153	Te	TELLURIUM	13494-80-9	-----	0.1	-----	0.1	-----	0.1	-----	0.1
154	TeF6	TELLURIUM HEXAFLUORIDE (as Te)	7783-80-4	0.02	0.2	0.02	0.2	0.02	0.2	-----	-----

Table 27-2 EXPOSURE LIMITS - INORGANIC COMPOUNDS (continued)

NO	FORMULA	NAME	CAS No	TLV (ACGIH) Threshold Limit Value		PEL (OSHA) Permissible Exposure Limit		REL (NIOSH) Recommended Exposure Limit		MAK (DFG) Maximum Concentration Value	
				ppm (vol)	mg/m³	ppm (vol)	mg/m³	ppm (vol)	mg/m³	ppm (vol)	mg/m³
155	TiO2	TITANIUM DIOXIDE	13463-67-7	-----	10	-----	15	-----	-----	-----	6
156	Tl	THALLIUM	7440-28-0	-----	0.1	-----	0.1	-----	0.1	-----	0.1
157	U	URANIUM	7440-61-1	-----	0.2	-----	0.05	-----	0.05	-----	0.25
158	V2O5	VANADIUM PENTOXIDE	1314-62-1	-----	0.05	-----	-----	-----	-----	-----	0.05
159	W	TUNGSTEN	7440-33-7	-----	5	-----	-----	-----	5	-----	-----
160	Yt	YTTRIUM	7440-65-5	-----	1	-----	1	-----	1	-----	5
161	ZnCl2	ZINC CHLORIDE, FUME	7646-85-7	-----	1	-----	1	-----	1	-----	-----
162	ZnO	ZINC OXIDE, DUST	1314-13-2	-----	10	-----	15*/5**	-----	5	-----	-----
163	ZnO	ZINC OXIDE, FUME	1314-13-2	-----	5	-----	5	-----	5	-----'	5
164	Zr	ZIRCONIUM	7440-67-7	-----	-----	-----	5	-----	5	-----	-----

* total dust
* respirable dust
TLV (ACGIH) - Threshold Limit Value of ACGIH
ACGIH - American Conference of Governmental Hygienists
PEL (OSHA) - Permissible Exposure Limit of OSHA
OSHA - Occupational Safety and Health Aministration
REL (NIOSH) - Recommended Exposure Limit of NIOSH
NIOSH - National Institute for Occupational Safety and Health
MAK (DFG) - Maximum Concentration Value in Workplace of DFG
DFG - Federal Republic of Germany
ppm (vol) - parts per million by volume
mg/m³ - milligrams per cubic meter

Chapter 28

COEFFICIENT OF THERMAL EXPANSION OF LIQUID

Carl L. Yaws
Lamar University, Beaumont, Texas

ABSTRACT

Results for thermal expansion coefficient of liquids are presented for organic and inorganic chemicals. The results are especially helpful in the design of relief systems for process equipment containing liquids that are subject to thermal expansion. The regression coefficients are displayed in easy-to-use tabulations. Correlation and experimental results are in favorable agreement.

INTRODUCTION

Physical and thermodynamic property data, such as thermal expansion coefficient, are important in process engineering. The following brief discussion illustrates such importance. Liquids contained in process equipment will expand with an increase in temperature. To accommodate such expansion, it is necessary to design a relief system which will relieve (or vent) the thermally expanding liquid and prevent pressure build-up from the expansion. If provisions are not made for a relief system, the pressure will increase from the thermally expanding liquid. If the pressure increase is excessive, damage to the process equipment will occur.

THERMAL EXPANSION COEFFICIENT

The following equation was selected for correlation of thermal expansion coefficient of liquid as a function of temperature:

$$B_{liq} = a \, (1 - T/T_C)^m \qquad (28\text{-}1)$$

where

B_{liq} = thermal expansion coefficient of liquid, 1/C
a and m = regression coefficients for chemical compound
T = temperature, K
T_C = critical temperature, K

The results for thermal expansion coefficient are given in Tables 28-1 and 28-2. The values are applicable to a wide variety of substances. The tabulations also disclose the temperature range for which the equation is useable. The respective minimum and maximum temperatures are denoted by TMIN and TMAX. Spot values at ambient temperature (25 C) are provided for both thermal expansion coefficient and liquid density.

For the tabulations, a literature search was conducted to identify data source publications for organics (1-42) and inorganics (1-120). Both experimental values for the property under consideration and parameter values for estimation of the property are included in the source publications. The publications were screened and copies of appropriate data were made. These data were next keyed into the computer to provide a database of liquid volume values for which experimental data are available. These data were then regressed for volume and change of volume with temperature as a function of temperature.

The coefficient of thermal expansion involves both volume and change of volume with temperature. The variation of volume with temperature is shown in Fig. 28-1 for a representative compound. Inspection of the figure discloses that the curve at constant pressure (P=29.6 atm) is very similar in shape to the curve at saturation (P=saturation). In fact, the curves are roughly parallel for the range shown. Also, the closeness of the curves indicates that the volume is about the same for both saturation and constant pressure as shown. These observations of similar shape and closeness suggest that the coefficient of thermal expansion at constant pressure is approximately equal to that at saturation:

$$B_{liq} = (1/v) \, (\partial v/\partial T)_P \approx (1/v) \, (\partial v/\partial T)_{saturation} \qquad (28\text{-}2)$$

This equation was used in preparing the tabulated results. The equation is applicable to the liquid at conditions below the critical point (temperatures and pressures below critical).

A comparison of calculated and actual data values for thermal expansion coefficient of liquid in Fig. 28-2 for a representative compound. The graph indicates good agreement of calculated and data values.

VOLUMETRIC EXPANSION RATE

Crowl and Louvar (41) have shown that the volumetric expansion (flow) rate for a liquid contained in process equipment that undergoes thermal expansion from heat input is given by

$$Q_v = \frac{B_{liq}}{\rho_{liq} \, C_P} UA \, (T_{ext} - T) \tag{28-3}$$

where
Q_v = volumetric expansion rate
ρ_{liq} = density of liquid
C_P = heat capacity of liquid
U = overall heat transfer coefficient
A = area for heat transfer
T_{ext} = external temperature
T = temperature of liquid

This equation describes the volumetric expansion rate at the beginning of the heat transfer and is applicable for the design of relief systems. The relief system should be sized to accommodate this volumetric flow (Crowl and Louvar). Property data for use in the equation are available from Yaws (32-34).

EXAMPLES

The correlation results may be used for calculation of the thermal expansion coefficient of liquid and volumetric flow from thermal expansion. Examples are given below.

Example 1 Estimate the thermal expansion coefficient of liquid for n-pentane (C5H12) at 40 C.

Substitution of the correlation constants from the table and temperature into the correlation equation yields

$$B_{liq} = (7.883E\text{-}04) \, (1-(40+273.15)/469.65)^{-.7179}$$

$$B_{liq} = 0.00174 \; C^{-1}$$

Example 2 Estimate the thermal expansion coefficient of liquid for n-butane (C4H10) at 40 C.

Substitution of the correlation constants from the table and temperature into the correlation equation yields

$$B_{liq} = (8.757E\text{-}04) \, (1-(40+273.15)/425.18)^{-.7137}$$

$$B_{liq} = 0.00227 \; C^{-1}$$

Example 3 The tubing in a reactor contains benzene (C6H6) at 25 C (76.7 F). Other data are:

heat capacity of liquid (C_P)	0.413 BTU/lb-F
overall heat transfer coefficient (U)	40 BTU/hr-ft²-F
surface area of tubing (A)	500 ft²

Estimate the volumetric expansion rate if the tubing is exposed to 500 F superheated steam.

Substitution of the spot values at 25 C (B_{liq}=1.137E-03 C^{-1} and ρ_{liq}=0.873 g/cm³) from the table into the volumetric expansion equation yields

$$Q_v = \frac{1.137E\text{-}03 \; C^{-1}/(1.8 \; F/C)}{(0.873 \; g/cm^3 \; 62.4 \; lb/ft^3/g/cm^3)(0.413 \; BTU/lb\text{-}F)} \; (40 \; BTU/hr\text{-}ft^2\text{-}F) \, (500 \; ft^2) \, (500\text{-}76.7 \; f)$$

$$Q_v = 237.69 \; ft^3/h = 29.63 \; gal/min$$

The relief system should be designed to accommodate this volumetric flow.

Portions of this material appeared in Chem. Eng., <u>102</u>, 98 (Aug., 1995) and are reprinted by special permission.

REFERENCES – ORGANIC COMPOUNDS

1-40. See **REFERENCES – ORGANIC COMPOUNDS** in **Chapter 8, DENSITY OF LIQUID**
41. Crowl, D. A. and J. F. Louvar, <u>CHEMICAL PROCESS SAFETY</u>, Prentice Hall, Inc., Englewood Cliffs, NJ (1990).
42. Yaws, C. L. and others, Chem. Eng., <u>102</u> (8), 98 (Aug., 1995).

REFERENCES – INORGANIC COMPOUNDS

1-120. See **REFERENCES – INORGANIC COMPOUNDS** in **Chapter 8, DENSITY OF LIQUID**

Figure 28-1 Volume of Liquid

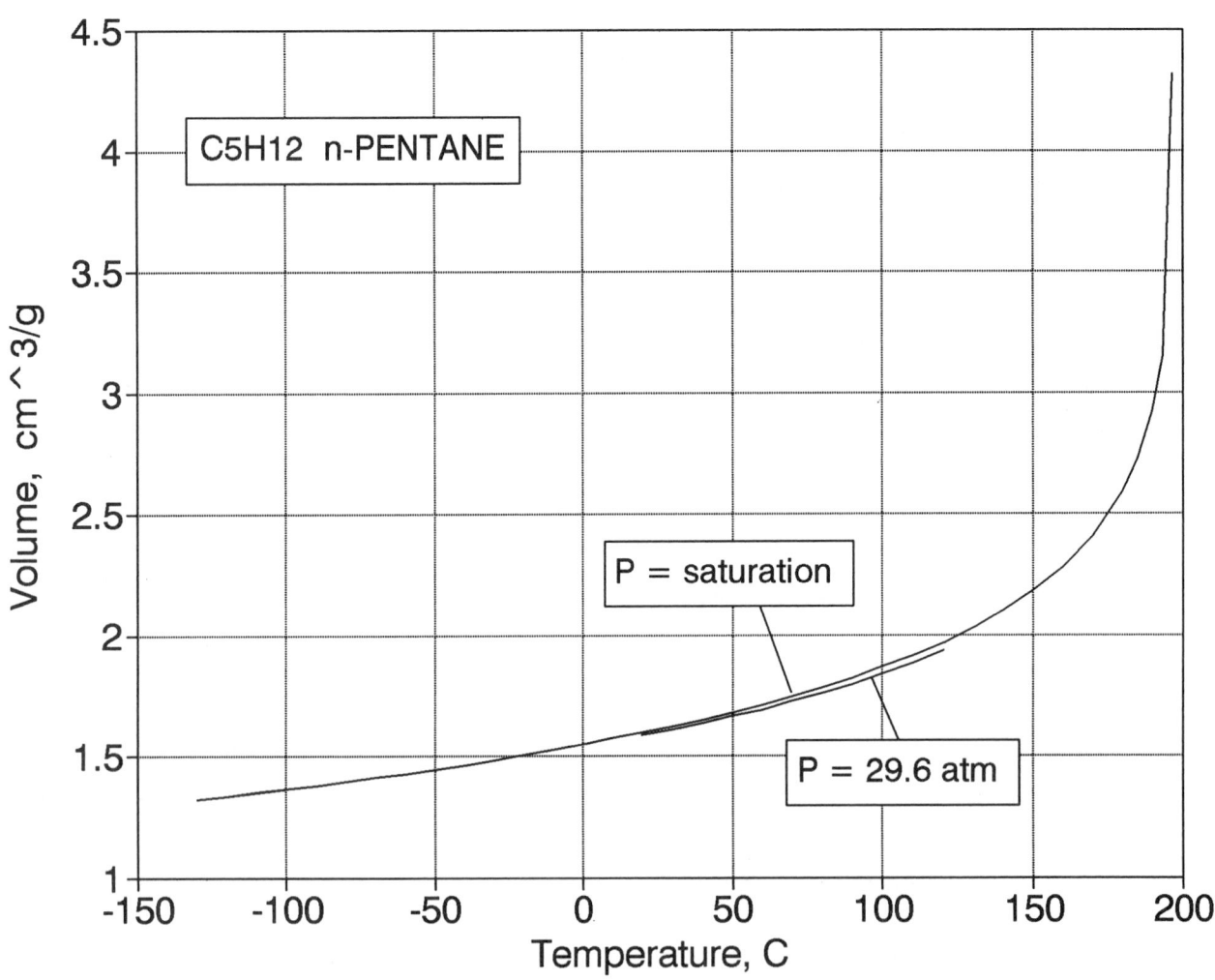

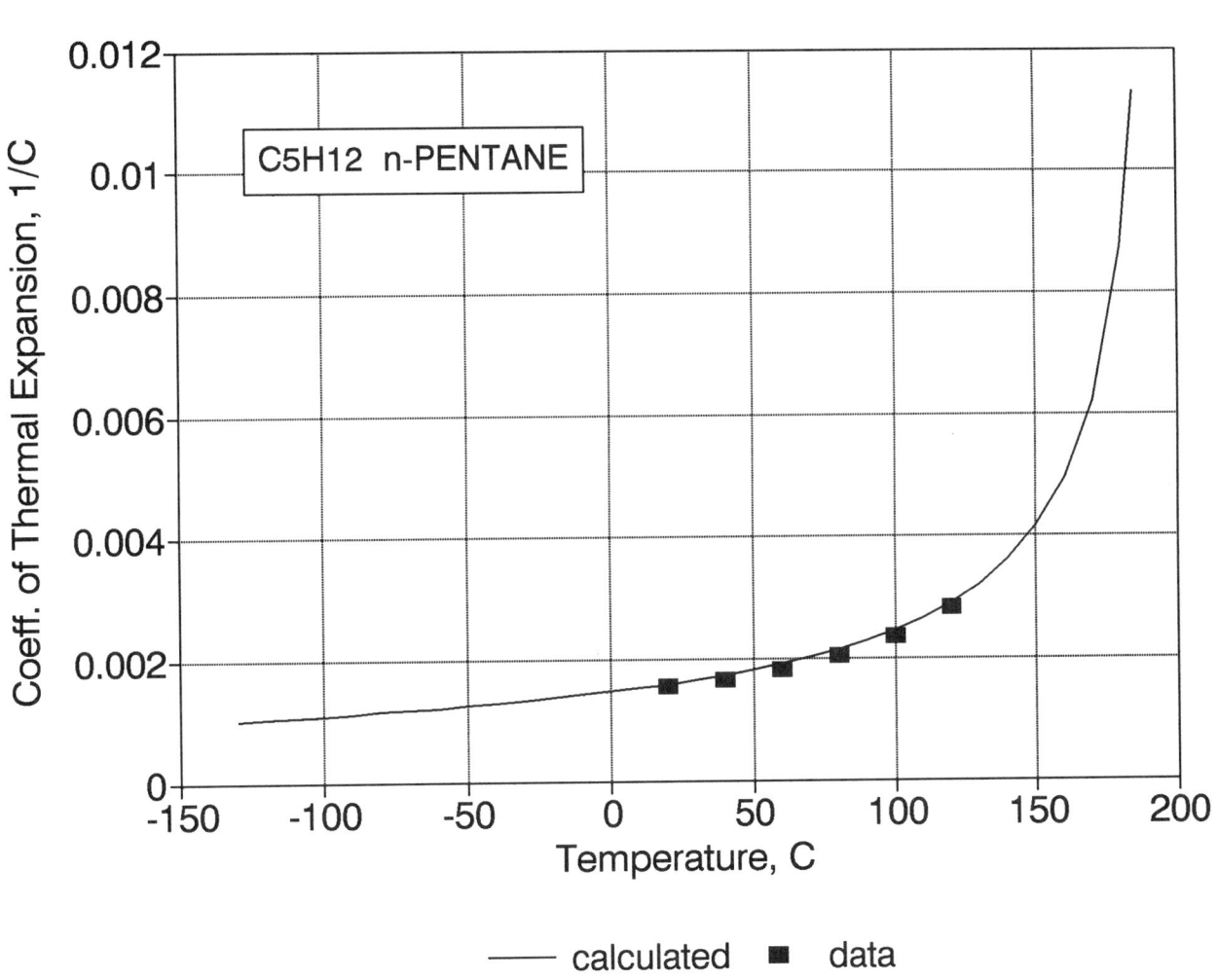

Figure 28-2 Coefficient of Thermal Expansion

Table 28-1 COEFFICIENT OF THERMAL EXPANSION - ORGANIC COMPOUNDS

			$B_{liq} = a \ (1 - T/T_C)^m$				$(B_{liq} - 1/C, \ T - K, \ density_{liq} - g/cm^3)$		
NO	FORMULA	NAME	a	T_C	m	TMIN	TMAX	B_{liq} @ 25 C	density$_{liq}$ @ 25 C
1	CBrClF2	BROMOCHLORODIFLUOROMETHANE	8.4480E-04	426.15	-0.7356	113.65	404.84	2.0460E-03	1.810
2	CBrCl3	BROMOTRICHLOROMETHANE	6.0010E-04	606.00	-0.7143	252.15	575.70	9.7350E-04	1.994
3	CBrF3	BROMOTRIFLUOROMETHANE	1.0780E-03	340.15	-0.7199	105.15	323.14	4.8610E-03	1.536
4	CBr2F2	DIBROMODIFLUOROMETHANE	7.0470E-04	478.00	-0.7374	163.05	454.10	1.4490E-03	2.274
5	CClF3	CHLOROTRIFLUOROMETHANE	1.2290E-03	301.96	-0.7093	92.15	286.86	--------	0.841
6	CClN	CYANOGEN CHLORIDE	7.8160E-04	449.00	-0.7614	266.65	426.55	1.7930E-03	1.172
7	CCl2F2	DICHLORODIFLUOROMETHANE	9.8380E-04	384.95	-0.7035	115.15	365.70	2.8050E-03	1.307
8	CCl2O	PHOSGENE	7.8010E-04	455.00	-0.7280	145.37	432.25	1.6940E-03	1.363
9	CCl3F	TRICHLOROFLUOROMETHANE	7.8150E-04	471.20	-0.7143	162.04	447.64	1.5980E-03	1.477
10	CCl4	CARBON TETRACHLORIDE	6.6990E-04	556.35	-0.7100	250.33	528.53	1.1550E-03	1.583
11	CF2O	CARBONYL FLUORIDE	1.0690E-03	297.00	-0.7143	161.89	282.15	--------	------
12	CF4	CARBON TETRAFLUORIDE	1.6100E-03	227.50	-0.7091	89.56	216.13	--------	------
13	CHBr3	TRIBROMOMETHANE	6.0300E-04	696.00	-0.6994	281.20	661.20	8.9160E-04	2.876
14	CHClF2	CHLORODIFLUOROMETHANE	1.0080E-03	369.30	-0.7188	115.73	350.84	3.2940E-03	1.193
15	CHCl2F	DICHLOROFLUOROMETHANE	8.2580E-04	451.58	-0.7143	138.15	429.00	1.7860E-03	1.367
16	CHCl3	CHLOROFORM	7.3760E-04	536.40	-0.7123	209.63	509.58	1.3150E-03	1.480
17	CHF3	TRIFLUOROMETHANE	1.3220E-03	298.89	-0.7109	117.97	283.95	--------	0.667
18	CHI3	TRIIODOMETHANE	4.5650E-04	794.55	-0.7143	396.16	754.82	--------	------
19	CHN	HYDROGEN CYANIDE	1.0390E-03	456.65	-0.7179	259.91	433.82	2.2220E-03	0.680
20	CHNS	ISOTHIOCYANIC-ACID	--------	------	------	------	------	--------	------
21	CH2BrCl	BROMOCHLOROMETHANE	6.5850E-04	557.00	-0.7143	185.20	529.15	1.1380E-03	1.926
22	CH2Br2	DIBROMOMETHANE	6.2880E-04	611.00	-0.7242	220.60	580.45	1.0210E-03	2.482
23	CH2ClF	CHLOROFLUOROMETHANE	9.8820E-04	424.91	-0.7143	140.16	403.66	2.3450E-03	1.256
24	CH2Cl2	DICHLOROMETHANE	7.7360E-04	510.00	-0.7098	178.01	484.50	1.4430E-03	1.318
25	CH2F2	DIFLUOROMETHANE	1.1010E-03	351.60	-0.7190	137.00	334.02	4.2660E-03	0.957
26	CH2I2	DIIODOMETHANE	4.9260E-04	747.00	-0.7346	279.25	709.65	7.1610E-04	3.306
27	CH2O	FORMALDEHYDE	1.0530E-03	408.00	-0.7143	181.15	387.60	2.6870E-03	0.736
28	CH2O2	FORMIC ACID	5.7720E-04	580.00	-0.7634	281.55	551.00	1.0010E-03	1.214
29	CH3Br	METHYL BROMIDE	8.0180E-04	467.00	-0.7197	179.55	443.65	1.6670E-03	1.662
30	CH3Cl	METHYL CHLORIDE	9.2560E-04	416.25	-0.7131	175.45	395.44	2.2730E-03	0.913
31	CH3Cl3Si	METHYL TRICHLOROSILANE	7.2170E-04	517.00	-0.7221	195.35	491.15	1.3430E-03	1.266
32	CH3F	METHYL FLUORIDE	1.2760E-03	317.70	-0.7146	131.35	301.82	9.3590E-03	0.566
33	CH3I	METHYL IODIDE	6.8630E-04	528.00	-0.7321	206.70	501.60	1.2620E-03	2.265
34	CH3NO	FORMAMIDE	5.1990E-04	771.00	-0.7482	275.70	732.45	7.4950E-04	1.129
35	CH3NO2	NITROMETHANE	7.0610E-04	588.15	-0.7097	244.60	558.74	1.1660E-03	1.129
36	CH3NO2	METHYL-NITRITE	--------	------	------	------	------	--------	------
37	CH3NO3	METHYL-NITRATE	--------	------	------	------	------	--------	------
38	CH4	METHANE	1.8090E-03	190.58	-0.7230	90.67	181.05	--------	------
39	CH4Cl2Si	METHYL DICHLOROSILANE	5.9030E-04	483.00	-0.7747	182.55	458.85	1.2420E-03	1.103
40	CH4O	METHANOL	5.9220E-04	512.58	-0.7669	175.47	486.95	1.1550E-03	0.787
41	CH4O3S	METHANESULFONIC ACID	--------	------	------	------	------	--------	1.477
42	CH4S	METHYL MERCAPTAN	7.7220E-04	469.95	-0.7148	150.18	446.45	1.5850E-03	0.862
43	CH5ClSi	METHYL CHLOROSILANE	8.0380E-04	442.00	-0.7343	139.05	419.90	1.8330E-03	0.884
44	CH5N	METHYLAMINE	8.1550E-04	430.05	-0.7725	179.69	408.55	2.0320E-03	0.655
45	CH6Si	METHYL SILANE	1.0770E-03	352.50	-0.7120	116.34	334.88	4.0770E-03	0.486
46	CN4O8	TETRANITROMETHANE	9.0300E-04	540.00	-0.7143	287.05	513.00	1.6030E-03	1.626
47	CO	CARBON MONOXIDE	2.8090E-03	132.92	-0.7095	68.15	126.27	--------	------
48	COS	CARBONYL SULFIDE	8.8830E-04	378.80	-0.7286	134.35	359.86	2.7420E-03	1.005
49	CO2	CARBON DIOXIDE	1.2800E-03	304.19	-0.7097	216.58	288.98	--------	0.713
50	CS2	CARBON DISULFIDE	7.2850E-04	552.00	-0.6774	161.58	524.40	1.2330E-03	1.256
51	C2BrF3	BROMOTRIFLUOROETHYLENE	9.9110E-04	432.00	-0.7000	173.00	410.40	2.2510E-03	1.830
52	C2Br2F4	1,2-DIBROMOTETRAFLUOROETHANE	7.2950E-04	487.80	-0.7345	162.65	463.41	1.4600E-03	2.162
53	C2ClF3	CHLOROTRIFLUOROETHYLENE	1.0940E-03	379.15	-0.6926	115.00	360.19	3.1870E-03	1.275
54	C2ClF5	CHLOROPENTAFLUOROETHANE	1.0150E-03	353.15	-0.7147	173.71	335.49	3.8360E-03	1.287
55	C2Cl2F4	1,2-DICHLOROTETRAFLUOROETHANE	8.6000E-04	418.85	-0.7255	179.15	397.91	2.1210E-03	1.455
56	C2Cl3F3	1,1,2-TRICHLOROTRIFLUOROETHANE	7.4970E-04	487.25	-0.7196	238.15	462.89	1.4810E-03	1.564
57	C2Cl4	TETRACHLOROETHYLENE	6.4140E-04	620.00	-0.6437	250.80	589.00	9.7810E-04	1.613
58	C2Cl4F2	1,1,2,2-TETRACHLORODIFLUOROETHANE	6.8390E-04	551.00	-0.7143	299.15	523.45	--------	------
59	C2Cl4O	TRICHLOROACETYL CHLORIDE	6.2920E-04	590.00	-0.7143	273.15	560.50	1.0400E-03	1.613
60	C2Cl6	HEXACHLOROETHANE	5.8930E-04	698.00	-0.7143	459.95	663.00	--------	------
61	C2F4	TETRAFLUOROETHYLENE	1.2250E-03	306.45	-0.7143	142.00	291.13	--------	0.920
62	C2F6	HEXAFLUOROETHANE	1.3440E-03	292.80	-0.7021	172.45	278.16	--------	------
63	C2HBrClF3	HALOTHANE	7.2150E-04	521.00	-0.7143	223.15	494.95	1.3230E-03	1.869
64	C2HClF2	2-CHLORO-1,1-DIFLUOROETHYLENE	9.4840E-04	400.55	-0.7118	134.65	380.52	2.5040E-03	1.217
65	C2HCl3	TRICHLOROETHYLENE	6.5600E-04	571.00	-0.7143	188.40	542.45	1.1120E-03	1.458
66	C2HCl3O	DICHLOROACETYL CHLORIDE	6.4660E-04	579.00	-0.7143	298.15	550.05	--------	1.519
67	C2HCl3O	TRICHLOROACETALDEHYDE	6.5260E-04	565.00	-0.7216	216.00	536.75	1.1210E-03	1.499
68	C2HCl5	PENTACHLOROETHANE	5.6690E-04	665.00	-0.7151	244.15	631.75	8.6750E-04	1.675
69	C2HF3	TRIFLUOROETHENE	1.0300E-03	347.22	-0.7143	94.53	329.86	4.1670E-03	0.919
70	C2HF3O2	TRIFLUOROACETIC ACID	7.9660E-04	491.25	-0.6971	257.90	466.69	1.5270E-03	1.480
71	C2HF5	PENTAFLUOROETHANE	1.1220E-03	342.00	-0.7143	170.15	324.90	4.8670E-03	1.174
72	C2H2	ACETYLENE	1.2100E-03	308.32	-0.7143	192.40	292.90	--------	0.377
73	C2H2Br4	1,1,2,2-TETRABROMOETHANE	3.4950E-04	824.00	-0.7860	273.15	782.80	4.9750E-04	2.927
74	C2H2Cl2	1,1-DICHLOROETHYLENE	7.3380E-04	482.00	-0.7143	150.65	457.90	1.4610E-03	1.117
75	C2H2Cl2	cis-1,2-DICHLOROETHYLENE	7.2760E-04	527.00	-0.7143	193.15	500.65	1.3200E-03	1.265
76	C2H2Cl2	trans-1,2-DICHLOROETHYLENE	7.4270E-04	508.00	-0.7143	223.35	482.60	1.3970E-03	1.244
77	C2H2Cl2O	CHLOROACETYL CHLORIDE	6.7350E-04	581.00	-0.7143	251.15	551.95	1.1260E-03	1.434
78	C2H2Cl2O	DICHLOROACETALDEHYDE	7.0560E-04	555.00	-0.7143	223.00	527.25	1.2230E-03	1.433

Table 28-1 COEFFICIENT OF THERMAL EXPANSION - ORGANIC COMPOUNDS (continued)

			$B_{liq} = a \ (1 - T/T_C)^m$				$(B_{liq} - 1/C, \ T - K, \ density_{liq} - g/cm^3)$	
NO	FORMULA	NAME	a	T_C	m	TMIN	TMAX	B_{liq} @ 25 C	$density_{liq}$ @ 25 C
79	C2H2Cl2O2	DICHLOROACETIC ACID	6.6310E-04	686.00	-0.6745	286.55	651.70	9.7420E-04	1.553
80	C2H2Cl3F	1,1,1-TRICHLOROFLUOROETHANE	7.0100E-04	565.00	-0.7143	173.00	536.75	1.1980E-03	1.575
81	C2H2Cl4	1,1,1,2-TETRACHLOROETHANE	6.1730E-04	624.00	-0.7143	202.94	592.80	9.8180E-04	1.535
82	C2H2Cl4	1,1,2,2-TETRACHLOROETHANE	6.1890E-04	645.00	-0.7041	229.35	612.75	9.5780E-04	1.587
83	C2H2F2	1,1-DIFLUOROETHYLENE	1.4660E-03	302.80	-0.6906	129.15	287.66	--------	0.594
84	C2H2F2	cis-1,2-DIFLUOROETHENE	1.0390E-03	394.67	-0.7143	107.90	374.94	2.8420E-03	1.023
85	C2H2F2	trans-1,2-DIFLUOROETHENE	1.0390E-03	394.67	-0.7143	107.90	374.94	2.8420E-03	1.023
86	C2H2F4	1,1,1,2-TETRAFLUOROETHANE	1.1730E-03	380.00	-0.6863	172.15	361.00	3.3650E-03	1.199
87	C2H2O	KETENE	1.0050E-03	370.00	-0.7143	122.00	351.50	3.2410E-03	0.660
88	C2H2O4	OXALIC ACID	5.4620E-04	804.00	-0.7143	462.65	763.80	--------	------
89	C2H3Br	VINYL BROMIDE	8.3920E-04	473.00	-0.7110	135.35	449.35	1.7030E-03	1.499
90	C2H3Cl	VINYL CHLORIDE	8.2160E-04	432.00	-0.7284	119.36	410.40	1.9290E-03	0.903
91	C2H3ClF2	1-CHLORO-1,1-DIFLUOROETHANE	1.0210E-03	410.20	-0.7020	142.35	389.69	2.5380E-03	1.107
92	C2H3ClO	ACETYL CHLORIDE	7.3910E-04	508.00	-0.7143	160.30	482.60	1.3900E-03	1.102
93	C2H3ClO	CHLOROACETALDEHYDE	7.3550E-04	555.00	-0.7143	293.00	527.25	1.2750E-03	1.200
94	C2H3ClO2	CHLOROACETIC ACID	6.0760E-04	686.00	-0.7143	333.15	651.70	--------	------
95	C2H3ClO2	METHYL CHLOROFORMATE	6.9780E-04	525.00	-0.7219	192.00	498.75	1.2790E-03	1.213
96	C2H3Cl3	1,1,1-TRICHLOROETHANE	6.9960E-04	545.00	-0.7067	242.75	517.75	1.2240E-03	1.330
97	C2H3Cl3	1,1,2-TRICHLOROETHANE	7.0420E-04	602.00	-0.6900	236.50	571.90	1.1290E-03	1.435
98	C2H3F	VINYL FLUORIDE	1.1290E-03	327.80	-0.7143	112.65	311.41	6.2810E-03	0.620
99	C2H3F3	1,1,1-TRIFLUOROETHANE	1.0150E-03	346.25	-0.7375	161.85	328.94	4.3530E-03	0.953
100	C2H3N	ACETONITRILE	7.6590E-04	545.50	-0.7187	229.32	518.23	1.3520E-03	0.779
101	C2H3NO	METHYL ISOCYANATE	8.2070E-04	505.00	-0.7143	256.15	479.75	1.5530E-03	0.926
102	C2H4	ETHYLENE	1.2860E-03	282.36	-0.7143	104.01	268.24	--------	------
103	C2H4Br2	1,1-DIBROMOETHANE	6.3310E-04	628.00	-0.7018	210.15	596.60	9.9480E-04	2.045
104	C2H4Br2	1,2-DIBROMOETHANE	5.8140E-04	650.15	-0.7143	282.94	617.64	9.0120E-04	2.169
105	C2H4Cl2	1,1-DICHLOROETHANE	7.2810E-04	523.00	-0.7130	176.19	496.85	1.3290E-03	1.168
106	C2H4Cl2	1,2-DICHLOROETHANE	6.8990E-04	561.00	-0.6896	237.49	532.95	1.1640E-03	1.246
107	C2H4Cl2O	BIS(CHLOROMETHYL)ETHER	6.7600E-04	579.00	-0.7143	231.65	550.05	1.1330E-03	1.312
108	C2H4F2	1,1-DIFLUOROETHANE	9.8470E-04	386.60	-0.7203	156.15	367.27	2.8490E-03	0.898
109	C2H4F2	1,2-DIFLUOROETHANE	9.0310E-04	476.00	-0.7143	215.00	452.20	1.8240E-03	1.016
110	C2H4I2	1,2-DIIODOETHANE	5.3590E-04	749.91	-0.7143	356.16	712.41	--------	------
111	C2H4O	ACETALDEHYDE	8.1110E-04	461.00	-0.7224	150.15	437.95	1.7200E-03	0.774
112	C2H4O	ETHYLENE OXIDE	8.0920E-04	469.15	-0.7175	161.45	445.69	1.6690E-03	0.862
113	C2H4OS	THIOACETIC-ACID	5.6850E-04	577.34	-0.7143	150.16	548.47	9.5530E-04	1.059
114	C2H4O2	ACETIC ACID	5.9380E-04	592.71	-0.7316	289.81	563.07	9.9030E-04	1.043
115	C2H4O2	METHYL FORMATE	7.6890E-04	487.20	-0.7232	174.15	462.84	1.5250E-03	0.967
116	C2H4S	THIACYCLOPROPANE	7.9370E-04	555.00	-0.7143	165.37	527.25	1.3760E-03	1.007
117	C2H5Br	BROMOETHANE	7.4990E-04	503.80	-0.7202	154.55	478.61	1.4300E-03	1.450
118	C2H5Cl	ETHYL CHLORIDE	6.4960E-04	460.35	-0.7686	136.75	437.33	1.4480E-03	0.890
119	C2H5ClO	2-CHLOROETHANOL	5.0180E-04	585.00	-0.7811	205.65	555.75	8.7560E-04	1.196
120	C2H5F	ETHYL FLUORIDE	8.4960E-04	375.31	-0.7558	129.95	356.54	2.8080E-03	0.712
121	C2H5I	ETHYL IODIDE	7.1860E-04	561.00	-0.7015	162.05	532.95	1.2230E-03	1.920
122	C2H5N	ETHYLENEIMINE	6.3380E-04	537.00	-0.7664	195.20	510.15	1.1790E-03	0.831
123	C2H5NO	ACETAMIDE	5.7010E-04	761.00	-0.7143	354.15	722.95	--------	------
124	C2H5NO	N-METHYLFORMAMIDE	5.6960E-04	721.00	-0.7253	269.35	684.95	8.3870E-04	0.999
125	C2H5NO2	NITROETHANE	6.7580E-04	593.00	-0.7220	183.63	563.35	1.1190E-03	1.043
126	C2H5NO3	ETHYL-NITRATE	--------	------	------	------	------	--------	------
127	C2H6	ETHANE	1.2030E-03	305.42	-0.7167	90.35	290.15	--------	0.315
128	C2H6AlCl	DIMETHYLALUMINUM CHLORIDE	6.8780E-04	619.00	-0.7143	252.15	588.05	1.1000E-03	0.988
129	C2H6O	DIMETHYL ETHER	9.3600E-04	400.10	-0.7194	131.66	380.10	2.5030E-03	0.655
130	C2H6O	ETHANOL	6.1070E-04	516.25	-0.7633	159.05	490.44	1.1790E-03	0.787
131	C2H6OS	DIMETHYL SULFOXIDE	6.0870E-04	726.00	-0.6780	291.67	689.70	8.7120E-04	1.095
132	C2H6O2	ETHYLENE GLYCOL	3.6440E-04	645.00	-0.8280	260.15	612.75	6.0910E-04	1.110
133	C2H6O4S	DIMETHYL SULFATE	6.5930E-04	758.00	-0.6298	241.35	720.10	9.0320E-04	1.322
134	C2H6S	DIMETHYL SULFIDE	7.4830E-04	503.04	-0.7143	174.88	477.89	1.4210E-03	0.850
135	C2H6S	ETHYL MERCAPTAN	7.3220E-04	499.15	-0.7213	125.26	474.19	1.4110E-03	0.833
136	C2H6S2	DIMETHYL DISULFIDE	6.5960E-04	606.00	-0.6886	188.44	575.70	1.0520E-03	1.057
137	C2H7N	DIMETHYLAMINE	7.4650E-04	437.65	-0.7520	180.96	415.77	1.7640E-03	0.650
138	C2H7N	ETHYLAMINE	8.5270E-04	456.15	-0.7141	192.15	433.34	1.8180E-03	0.677
139	C2H7NO	MONOETHANOLAMINE	4.7240E-04	638.00	-0.7985	283.65	606.10	7.8110E-04	1.014
140	C2H8N2	ETHYLENEDIAMINE	5.3540E-04	593.00	-0.7983	284.29	563.35	9.3520E-04	0.893
141	C2H8Si	DIMETHYL SILANE	9.4290E-04	402.00	-0.7158	122.93	381.90	2.4840E-03	0.578
142	C2N2	CYANOGEN	8.0520E-04	400.15	-0.7937	245.25	380.14	2.3820E-03	0.866
143	C3F6	HEXAFLUOROPROPYLENE	1.2000E-03	368.00	-0.6887	116.65	349.60	3.7700E-03	1.304
144	C3F6O	HEXAFLUOROACETONE	9.3150E-04	357.14	-0.7710	151.15	339.28	3.7340E-03	1.321
145	C3F8	OCTAFLUOROPROPANE	1.0620E-03	345.05	-0.7183	125.46	327.80	4.4520E-03	1.317
146	C3H2N2	MALONONITRILE	5.7900E-04	715.00	-0.7391	304.90	679.25	--------	------
147	C3H3Cl	PROPARGYL CHLORIDE	7.9200E-04	541.00	-0.6868	293.00	513.95	1.3730E-03	1.024
148	C3H3N	ACRYLONITRILE	7.9660E-04	535.00	-0.7106	189.63	508.25	1.4210E-03	0.801
149	C3H3NO	OXAZOLE	5.7960E-04	554.00	-0.7143	189.15	526.30	1.0070E-03	0.718
150	C3H4	METHYLACETYLENE	9.2220E-04	402.39	-0.7210	170.45	382.27	2.4420E-03	0.607
151	C3H4	PROPADIENE	1.0090E-03	393.15	-0.6970	136.87	373.49	2.7140E-03	0.579
152	C3H4Cl2	2,3-DICHLOROPROPENE	7.2820E-04	577.00	-0.6937	191.50	548.15	1.2060E-03	1.201
153	C3H4O	ACROLEIN	6.6030E-04	506.00	-0.7511	185.45	480.70	1.2880E-03	0.834
154	C3H4O	PROPARGYL ALCOHOL	6.8480E-04	580.00	-0.7143	253.15	551.00	1.1470E-03	0.945
155	C3H4O2	ACRYLIC ACID	6.7590E-04	615.00	-0.6930	286.65	584.25	1.0700E-03	1.046
156	C3H4O2	beta-PROPIOLACTONE	6.0140E-04	686.00	-0.7143	239.75	651.70	9.0370E-04	1.262

621

NO	FORMULA	NAME	$B_{liq} = a \ (1 - T/T_C)^m$				$(B_{liq} - 1/C, \ T - K, \ density_{liq} - g/cm^3)$	
			a	T_C	m	TMIN	TMAX	B_{liq} @ 25 C	$density_{liq}$ @ 25 C
157	C3H4O2	VINYL FORMATE	8.1640E-04	498.00	-0.7055	200.00	473.10	1.5550E-03	0.954
158	C3H4O3	ETHYLENE CARBONATE	5.2120E-04	790.00	-0.7140	309.55	750.50	-------	------
159	C3H4O3	PYRUVIC ACID	5.2770E-04	634.52	-0.7660	286.75	602.79	8.5800E-04	1.265
160	C3H5Br	3-BROMO-1-PROPENE	6.6950E-04	540.20	-0.7143	153.76	513.19	1.1880E-03	1.389
161	C3H5Cl	2-CHLOROPROPENE	7.7900E-04	478.00	-0.7143	135.75	454.10	1.5660E-03	0.895
162	C3H5Cl	3-CHLOROPROPENE	7.4950E-04	514.15	-0.7143	138.65	488.44	1.3920E-03	0.931
163	C3H5ClO	alpha-EPICHLOROHYDRIN	6.6030E-04	610.00	-0.6969	215.95	579.50	1.0540E-03	1.174
164	C3H5ClO2	METHYL CHLOROACETATE	7.1290E-04	600.00	-0.7143	241.03	570.00	1.1460E-03	1.229
165	C3H5ClO2	ETHYL CHLOROFORMATE	7.4830E-04	508.15	-0.7170	192.00	482.74	1.4100E-03	1.127
166	C3H5Cl3	1,2,3-TRICHLOROPROPANE	6.1360E-04	652.00	-0.7143	258.45	619.40	9.4950E-04	1.384
167	C3H5I	3-IODO-1-PROPENE	6.6670E-04	595.81	-0.7143	173.86	566.02	1.0940E-03	1.839
168	C3H5N	PROPIONITRILE	7.2050E-04	564.40	-0.7196	180.26	536.18	1.2370E-03	0.777
169	C3H5NO	ACRYLAMIDE	5.5470E-04	710.00	-0.7143	357.65	674.50	--------	------
170	C3H5NO	HYDRACRYLONITRILE	4.6880E-04	690.00	-0.7755	227.15	655.50	7.2700E-04	1.040
171	C3H5NO	LACTONITRILE	5.0900E-04	643.00	-0.7666	233.00	610.85	8.2060E-04	0.983
172	C3H5N3O9	NITROGLYCERINE	5.6330E-04	680.00	-0.7001	286.15	646.00	8.4370E-04	1.586
173	C3H6	CYCLOPROPANE	9.2960E-04	397.91	-0.7143	145.73	378.01	2.4970E-03	0.619
174	C3H6	PROPYLENE	1.0700E-03	364.76	-0.6975	87.90	346.52	3.5030E-03	0.504
175	C3H6Br2	1,2-DIBROMOPROPANE	4.9950E-04	634.11	-0.7143	217.96	602.40	7.8640E-04	1.925
176	C3H6Cl2	1,1-DICHLOROPROPANE	6.7640E-04	560.00	-0.7143	200.00	532.00	1.1640E-03	1.126
177	C3H6Cl2	1,2-DICHLOROPROPANE	6.7180E-04	572.00	-0.7143	172.71	543.40	1.1370E-03	1.150
178	C3H6Cl2	1,3-DICHLOROPROPANE	6.0150E-04	603.00	-0.7290	173.65	572.85	9.8910E-04	1.181
179	C3H6Cl2	2,2-DICHLOROPROPANE	7.0140E-04	539.46	-0.7143	239.36	512.49	1.2460E-03	1.106
180	C3H6I2	1,2-DIIODOPROPANE	5.1940E-04	780.49	-0.7143	253.16	741.47	7.3250E-04	2.566
181	C3H6O	ACETONE	7.9810E-04	508.20	-0.7010	178.45	482.79	1.4830E-03	0.786
182	C3H6O	ALLYL ALCOHOL	7.1820E-04	545.05	-0.7143	144.15	517.80	1.2640E-03	0.845
183	C3H6O	METHYL VINYL ETHER	7.8560E-04	437.00	-0.7420	151.15	415.15	1.8390E-03	0.744
184	C3H6O	n-PROPIONALDEHYDE	8.1360E-04	496.00	-0.7140	193.15	471.20	1.5680E-03	0.796
185	C3H6O	1,2-PROPYLENE OXIDE	7.8280E-04	482.25	-0.7065	161.22	458.14	1.5460E-03	0.823
186	C3H6O	1,3-PROPYLENE OXIDE	7.5410E-04	520.00	-0.7143	255.00	494.00	1.3860E-03	0.894
187	C3H6O2	ETHYL FORMATE	7.5730E-04	508.40	-0.7065	193.55	482.98	1.4130E-03	0.917
188	C3H6O2	METHYL ACETATE	7.3260E-04	506.80	-0.7255	175.15	481.46	1.3950E-03	0.927
189	C3H6O2	PROPIONIC ACID	6.1800E-04	604.00	-0.7236	252.45	573.80	1.0110E-03	0.988
190	C3H6O2S	3-MERCAPTOPROPIONIC ACID	8.0100E-04	729.00	-0.5953	290.65	692.55	1.0960E-03	1.213
191	C3H6O3	LACTIC ACID	6.1860E-04	616.00	-0.7143	291.15	585.20	9.9230E-04	1.201
192	C3H6O3	METHOXYACETIC ACID	5.4830E-04	691.00	-0.7255	281.00	656.45	8.2590E-04	1.170
193	C3H6O3	TRIOXANE	6.3970E-04	604.00	-0.7143	334.65	573.80	--------	------
194	C3H6S	THIACYCLOBUTANE	7.0050E-04	603.00	-0.7143	199.96	572.85	1.1400E-03	1.014
195	C3H7Br	1-BROMOPROPANE	7.2180E-04	544.00	-0.7084	163.15	516.80	1.2670E-03	1.345
196	C3H7Br	2-BROMOPROPANE	9.9590E-04	532.00	-0.6200	184.15	505.40	1.6580E-03	1.282
197	C3H7Cl	ISOPROPYL CHLORIDE	7.5420E-04	489.00	-0.7143	155.97	464.55	1.4770E-03	0.855
198	C3H7Cl	n-PROPYL CHLORIDE	7.3580E-04	503.15	-0.7143	150.35	477.99	1.3970E-03	0.856
199	C3H7F	1-FLUOROPROPANE	9.0430E-04	422.00	-0.7143	114.16	400.90	2.1710E-03	0.787
200	C3H7F	2-FLUOROPROPANE	9.2060E-04	415.68	-0.7143	139.80	394.90	2.2700E-03	0.733
201	C3H7I	ISOPROPYL IODIDE	6.9140E-04	578.00	-0.6977	183.15	549.10	1.1470E-03	1.695
202	C3H7I	n-PROPYL IODIDE	6.8170E-04	593.00	-0.6988	171.85	563.35	1.1110E-03	1.739
203	C3H7N	ALLYLAMINE	5.8730E-04	505.00	-0.7924	184.95	479.75	1.1910E-03	0.757
204	C3H7N	PROPYLENEIMINE	7.3540E-04	529.00	-0.7140	229.00	502.55	1.3290E-03	0.802
205	C3H7NO	N,N-DIMETHYLFORMAMIDE	6.2740E-04	647.00	-0.7237	212.72	614.65	9.8100E-04	0.945
206	C3H7NO	N-METHYLACETAMIDE	5.5110E-04	718.00	-0.7262	301.15	682.10	--------	------
207	C3H7NO2	1-NITROPROPANE	6.3670E-04	605.00	-0.7264	169.16	574.75	1.0430E-03	0.996
208	C3H7NO2	2-NITROPROPANE	6.4490E-04	594.00	-0.7263	181.83	564.30	1.0700E-03	0.983
209	C3H7NO3	PROPYL-NITRATE	--------	------	------	------	------	--------	------
210	C3H7NO3	ISOPROPYL-NITRATE	--------	------	------	------	------	--------	------
211	C3H8	PROPANE	9.9500E-04	369.82	-0.7130	85.46	351.33	3.2060E-03	0.493
212	C3H8O	ISOPROPANOL	6.3530E-04	508.31	-0.7570	185.28	482.89	1.2400E-03	0.783
213	C3H8O	METHYL ETHYL ETHER	8.2040E-04	437.80	-0.7105	160.00	415.91	1.8480E-03	0.692
214	C3H8O	n-PROPANOL	6.0500E-04	536.71	-0.7506	146.95	509.87	1.1120E-03	0.802
215	C3H8O2	2-METHOXYETHANOL	6.9210E-04	564.00	-0.7143	188.05	535.80	1.1840E-03	0.960
216	C3H8O2	METHYLAL	7.8290E-04	480.60	-0.6825	168.35	456.57	1.5160E-03	0.854
217	C3H8O2	1,2-PROPYLENE GLYCOL	4.3890E-04	626.00	-0.7954	213.15	594.70	7.3420E-04	1.033
218	C3H8O2	1,3-PROPYLENE GLYCOL	6.0050E-04	658.00	-0.7143	246.45	625.10	9.2420E-04	1.052
219	C3H8O3	GLYCEROL	2.9630E-04	723.00	-0.8459	291.33	686.85	4.6460E-04	1.257
220	C3H8S	n-PROPYLMERCAPTAN	6.4110E-04	536.00	-0.7308	159.95	509.20	1.1610E-03	0.836
221	C3H8S	ISOPROPYL MERCAPTAN	7.0160E-04	517.00	-0.7143	142.61	491.15	1.2970E-03	0.809
222	C3H8S	ETHYL-METHYL-SULFIDE	7.1830E-04	532.80	-0.7143	167.20	506.16	1.2900E-03	0.832
223	C3H9N	n-PROPYLAMINE	7.0930E-04	496.95	-0.7539	190.15	472.10	1.4150E-03	0.714
224	C3H9N	ISOPROPYLAMINE	7.9550E-04	471.85	-0.7028	177.95	448.26	1.6060E-03	0.684
225	C3H9N	TRIMETHYLAMINE	8.4260E-04	433.25	-0.7313	156.08	411.59	1.9760E-03	0.629
226	C3H9NO	1-AMINO-2-PROPANOL	5.2810E-04	614.00	-0.7787	274.89	583.30	8.8620E-04	0.957
227	C3H9NO	3-AMINO-1-PROPANOL	6.0830E-04	649.00	-0.7143	284.15	616.55	9.4400E-04	0.972
228	C3H9NO	METHYLETHANOLAMINE	6.2450E-04	630.00	-0.7143	268.65	598.50	9.8710E-04	0.934
229	C3H9O4P	TRIMETHYL PHOSPHATE	--------	------	------	------	------	--------	1.202
230	C3H10N2	1,2-PROPANEDIAMINE	6.7730E-04	587.00	-0.7438	236.53	557.65	1.1480E-03	0.856
231	C3H10Si	TRIMETHYL SILANE	9.0040E-04	432.00	-0.7077	137.26	410.40	2.0630E-03	0.614
232	C4Cl4S	TETRACHLOROTHIOPHENE	5.2450E-04	753.00	-0.7143	301.97	715.35	--------	------
233	C4Cl6	HEXACHLORO-1,3-BUTADIENE	5.2990E-04	741.00	-0.7143	252.15	703.95	7.6540E-04	1.556
234	C4F8	OCTAFLUORO-2-BUTENE	9.8300E-04	392.00	-0.7187	138.15	372.40	2.7470E-03	1.442

Table 28-1 COEFFICIENT OF THERMAL EXPANSION - ORGANIC COMPOUNDS (continued)

NO	FORMULA	NAME	$B_{liq} = a\,(1 - T/T_C)^m$			$(B_{liq}$ - 1/C, T - K, density$_{liq}$ - g/cm$^3)$		
			a	T_C	m	TMIN	TMAX	B_{liq} @ 25 C	density$_{liq}$ @ 25 C
235	C4F8	OCTAFLUOROCYCLOBUTANE	9.6280E-04	388.37	-0.7223	232.96	368.95	2.7630E-03	1.495
236	C4F10	DECAFLUOROBUTANE	9.4170E-04	386.35	-0.7330	144.95	367.03	2.7800E-03	1.497
237	C4H2	BUTADIYNE(BIACETYLENE)	7.8040E-04	478.02	-0.7143	237.16	454.12	1.5690E-03	0.709
238	C4H2O3	MALEIC ANHYDRIDE	6.6220E-04	721.00	-0.6442	326.00	684.95	--------	------
239	C4H4	VINYLACETYLENE	8.3940E-04	454.00	-0.7143	179.95	431.30	1.8010E-03	0.680
240	C4H4N2	SUCCINONITRILE	5.3470E-04	770.00	-0.7298	331.30	731.50	--------	------
241	C4H4O	FURAN	7.4270E-04	490.15	-0.7395	187.55	465.64	1.4850E-03	0.935
242	C4H4O2	DIKETENE	6.0220E-04	616.00	-0.7143	266.65	585.20	9.6600E-04	1.050
243	C4H8O3	SUCCINIC ANHYDRIDE	5.2860E-04	811.00	-0.7143	393.00	770.45	--------	------
244	C4H4O4	FUMARIC ACID	5.4300E-04	771.00	-0.7143	560.15	732.45	--------	------
245	C4H4O4	MALEIC ACID	5.4970E-04	773.00	-0.7100	403.45	734.35	--------	------
246	C4H4S	THIOPHENE	6.7240E-04	579.35	-0.6923	234.94	550.38	1.1090E-03	1.059
247	C4H5Cl	CHLOROPRENE	7.0620E-04	525.00	-0.7214	143.15	498.75	1.2940E-03	0.950
248	C4H5N	trans-CROTONITRILE	7.2880E-04	586.00	-0.7143	222.00	556.70	1.2110E-03	0.807
249	C4H5N	cis-CROTONITRILE	7.1970E-04	568.00	-0.7178	200.55	539.60	1.2280E-03	0.819
250	C4H5N	METHACRYLONITRILE	7.6940E-04	554.00	-0.7037	237.35	526.30	1.3250E-03	0.795
251	C4H5N	PYRROLE	5.4190E-04	639.75	-0.7521	249.74	607.76	8.6860E-04	0.965
252	C4H5N	VINYLACETONITRILE	7.1020E-04	584.00	-0.7101	186.15	554.80	1.1800E-03	0.829
253	C4H5NO2	METHYL CYANOACETATE	6.8960E-04	687.00	-0.6805	260.08	652.65	1.0160E-03	1.119
254	C4H6	CYCLOBUTENE	8.1950E-04	446.33	-0.7143	153.76	424.01	1.8010E-03	0.704
255	C4H6	1,2-BUTADIENE	8.4980E-04	444.00	-0.7143	136.95	421.80	1.8820E-03	0.646
256	C4H6	1,3-BUTADIENE	8.8920E-04	425.37	-0.7093	164.25	404.10	2.0930E-03	0.615
257	C4H6	DIMETHYLACETYLENE	7.6800E-04	488.15	-0.7143	240.91	463.74	1.5070E-03	0.686
258	C4H6	ETHYLACETYLENE	8.5690E-04	443.20	-0.7169	147.43	421.04	1.9080E-03	0.648
259	C4H6Cl2	1,3-DICHLORO-trans-2-BUTENE	5.9520E-04	618.00	-0.7212	276.00	587.10	9.5710E-04	1.153
260	C4H6Cl2	1,4-DICHLORO-cis-2-BUTENE	6.3070E-04	640.00	-0.7143	225.15	608.00	9.8700E-04	1.188
261	C4H6Cl2	1,4-DICHLORO-trans-2-BUTENE	6.2620E-04	646.00	-0.7143	274.15	613.70	9.7440E-04	1.187
262	C4H6Cl2	3,4-DICHLORO-1-BUTENE	6.5600E-04	589.00	-0.7143	212.00	559.55	1.0860E-03	1.148
263	C4H6O	trans-CROTONALDEHYDE	7.9510E-04	571.00	-0.6765	196.65	542.45	1.3100E-03	0.847
264	C4H6O	2,5-DIHYDROFURAN	7.0310E-04	542.00	-0.7143	273.00	514.90	1.2440E-03	0.939
265	C4H6O	DIVINYL ETHER	7.9440E-04	463.00	-0.7143	172.05	439.85	1.6610E-03	0.731
266	C4H6O	METHACROLEIN	7.5910E-04	530.00	-0.7143	192.15	503.50	1.3700E-03	0.840
267	C4H6O2	2-BUTYNE-1,4-DIOL	5.5380E-04	695.00	-0.7143	331.00	660.25	--------	------
268	C4H6O2	gamma-BUTYROLACTONE	5.1070E-04	739.00	-0.7350	229.78	702.05	7.4660E-04	1.125
269	C4H6O2	cis-CROTONIC ACID	6.2700E-04	647.00	-0.7143	288.65	614.65	9.7470E-04	1.023
270	C4H6O2	trans-CROTONIC ACID	6.1240E-04	666.00	-0.7143	344.55	632.70	--------	------
271	C4H6O2	METHACRYLIC ACID	6.2710E-04	643.00	-0.7143	288.15	610.85	9.7860E-04	1.012
272	C4H6O2	METHYL ACRYLATE	7.2770E-04	536.00	-0.7143	196.32	509.20	1.3000E-03	0.949
273	C4H6O2	VINYL ACETATE	7.3090E-04	524.00	-0.7173	180.35	497.80	1.3370E-03	0.926
274	C4H6O3	ACETIC ANHYDRIDE	6.7520E-04	569.15	-0.7301	200.15	540.69	1.1610E-03	1.077
275	C4H6O4	SUCCINIC ACID	5.5170E-04	806.00	-0.7143	461.15	765.70	--------	------
276	C4H6O5	DIGLYCOLIC ACID	5.3560E-04	820.00	-0.7143	421.15	779.00	--------	------
277	C4H6O5	MALIC ACID	4.9560E-04	781.00	-0.7143	403.15	741.95	--------	------
278	C4H6O6	TARTARIC ACID	5.0710E-04	828.00	-0.7143	479.15	786.60	--------	------
279	C4H7N	n-BUTYRONITRILE	6.9390E-04	582.25	-0.7141	161.25	553.14	1.1580E-03	0.786
280	C4H7N	ISOBUTYRONITRILE	7.4750E-04	565.00	-0.7002	201.70	536.75	1.2640E-03	0.766
281	C4H7NO	ACETONE CYANOHYDRIN	6.3050E-04	647.00	-0.7143	253.15	614.65	9.8010E-04	0.928
282	C4H7NO	2-METHACRYLAMIDE	5.2120E-04	741.00	-0.7100	383.65	703.95	--------	------
283	C4H7NO	3-METHOXYPROPIONITRILE	6.7440E-04	638.00	-0.7143	210.12	606.10	1.0580E-03	0.924
284	C4H7NO	2-PYRROLIDONE	5.3140E-04	792.00	-0.7036	298.15	752.40	--------	1.108
285	C4H8	1-BUTENE	8.9970E-04	419.59	-0.7147	87.80	398.61	2.1820E-03	0.588
286	C4H8	cis-2-BUTENE	8.5750E-04	435.58	-0.7143	134.26	413.80	1.9550E-03	0.617
287	C4H8	trans-2-BUTENE	8.6740E-04	428.63	-0.7143	167.62	407.20	2.0290E-03	0.599
288	C4H8	CYCLOBUTANE	6.2970E-04	459.93	-0.7619	182.48	436.93	1.3960E-03	0.689
289	C4H8	ISOBUTENE	8.8460E-04	417.90	-0.7204	132.81	397.01	2.1770E-03	0.589
290	C4H8Br2	1,2-DIBROMOBUTANE	4.8710E-04	659.28	-0.7143	207.76	626.32	7.4870E-04	1.785
291	C4H8Br2	2,3-DIBROMOBUTANE	4.9010E-04	656.96	-0.7143	238.66	624.11	7.5500E-04	1.774
292	C4H8Cl2	1,4-DICHLOROBUTANE	6.2580E-04	641.00	-0.7027	235.85	608.95	9.7150E-04	1.135
293	C4H8I2	1,2-DIIODOBUTANE	5.2230E-04	726.41	-0.7143	279.06	690.09	7.6180E-04	2.280
294	C4H8O	n-BUTYRALDEHYDE	7.5830E-04	525.00	-0.7143	176.75	498.75	1.3810E-03	0.797
295	C4H8O	ISOBUTYRALDEHYDE	7.6580E-04	507.00	-0.7143	208.15	481.65	1.4430E-03	0.784
296	C4H8O	1,2-EPOXYBUTANE	7.3900E-04	526.00	-0.7143	123.15	499.70	1.3430E-03	0.824
297	C4H8O	METHYL ETHYL KETONE	7.3660E-04	535.50	-0.7143	186.48	508.73	1.3170E-03	0.799
298	C4H8O	ETHYL VINYL ETHER	7.9190E-04	475.15	-0.7143	157.35	451.39	1.6030E-03	0.749
299	C4H8O	TETRAHYDROFURAN	6.8470E-04	540.15	-0.7088	164.65	513.14	1.2100E-03	0.880
300	C4H8O2	cis-2-BUTENE-1,4-DIOL	5.8800E-04	677.88	-0.7143	284.15	643.99	8.8950E-04	1.070
301	C4H8O2	trans-2-BUTENE-1,4-DIOL	5.8560E-04	681.00	-0.7143	300.45	646.95	--------	------
302	C4H8O2	ISOBUTYRIC ACID	6.0270E-04	609.15	-0.7314	227.15	578.69	9.8550E-04	0.946
303	C4H8O2	n-BUTYRIC ACID	5.9730E-04	628.00	-0.7200	267.95	596.60	9.4950E-04	0.953
304	C4H8O2	1,4-DIOXANE	6.5840E-04	587.00	-0.6953	284.95	557.65	1.0780E-03	1.029
305	C4H8O2	ETHYL ACETATE	7.1860E-04	523.30	-0.7220	189.60	497.14	1.3210E-03	0.894
306	C4H8O2	METHYL PROPIONATE	7.0600E-04	530.60	-0.7230	185.65	504.07	1.2820E-03	0.909
307	C4H8O2	n-PROPYL FORMATE	6.9840E-04	538.00	-0.7200	180.25	511.10	1.2490E-03	0.900
308	C4H8O2S	SULFOLANE	4.6910E-04	849.00	-0.6960	300.75	806.55	--------	------
309	C4H8S	TETRAHYDROTHIOPHENE	4.7770E-04	631.95	-0.7512	176.99	600.35	7.7150E-04	0.997
310	C4H9Br	1-BROMOBUTANE	6.6960E-04	577.00	-0.7109	160.75	548.15	1.1230E-03	1.269
311	C4H9Br	2-BROMOBUTANE	5.4350E-04	567.00	-0.7600	161.25	538.65	9.5820E-04	1.253
312	C4H9Cl	n-BUTYL CHLORIDE	7.3200E-04	537.00	-0.7046	150.05	510.15	1.2950E-03	0.880

623

Table 28-1 COEFFICIENT OF THERMAL EXPANSION - ORGANIC COMPOUNDS (continued)

			$B_{liq} = a \ (1 - T/T_C)^m$			$(B_{liq} - 1/C, \ T - K, \ density_{liq} - g/cm^3)$			
NO	FORMULA	NAME	a	T_C	m	TMIN	TMAX	B_{liq} @ 25 C	density$_{liq}$ @ 25 C
313	C4H9Cl	sec-BUTYL CHLORIDE	7.0310E-04	520.60	-0.7209	141.85	494.57	1.2980E-03	0.868
314	C4H9Cl	tert-BUTYL CHLORIDE	7.2290E-04	507.00	-0.7143	247.75	481.65	1.3620E-03	0.836
315	C4H9I	2-IODO-2-METHYLPROPANE	6.4180E-04	587.90	-0.7143	234.96	558.51	1.0640E-03	1.536
316	C4H9N	PYRROLIDINE	6.1970E-04	568.55	-0.7367	215.31	540.12	1.0710E-03	0.860
317	C4H9NO	N,N-DIMETHYLACETAMIDE	5.9870E-04	658.00	-0.7298	253.15	625.10	9.3000E-04	0.937
318	C4H9NO	MORPHOLINE	5.6440E-04	618.00	-0.7436	270.05	587.10	9.2100E-04	0.996
319	C4H9NO2	1-NITROBUTANE	6.1680E-04	624.00	-0.7143	191.83	592.80	9.8100E-04	0.968
320	C4H9NO2	2-NITROBUTANE	6.1870E-04	615.00	-0.7143	141.16	584.25	9.9360E-04	0.978
321	C4H10	n-BUTANE	8.7570E-04	425.18	-0.7137	134.86	403.92	2.0740E-03	0.573
322	C4H10	ISOBUTANE	8.6860E-04	408.14	-0.7270	113.54	387.73	2.2530E-03	0.552
323	C4H10N2	PIPERAZINE	5.0610E-04	638.00	-0.7143	379.15	606.10	--------	------
324	C4H10O	n-BUTANOL	5.7680E-04	562.93	-0.7143	183.85	534.78	1.0190E-03	0.806
325	C4H10O	sec-BUTANOL	6.4790E-04	536.01	-0.7396	158.45	509.21	1.1820E-03	0.805
326	C4H10O	tert-BUTANOL	7.3570E-04	506.20	-0.7263	298.97	480.89	--------	------
327	C4H10O	DIETHYL ETHER	8.0960E-04	466.70	-0.7064	156.85	443.37	1.6620E-03	0.708
328	C4H10O	METHYL-PROPYL-ETHER	8.0130E-04	476.20	-0.7143	156.87	452.39	1.6180E-03	0.723
329	C4H10O	METHYL ISOPROPYL ETHER	6.6090E-04	464.50	-0.7556	127.93	441.28	1.4360E-03	0.714
330	C4H10O	ISOBUTANOL	5.5700E-04	547.73	-0.7657	165.15	520.34	1.0170E-03	0.797
331	C4H10O2	1,3-BUTANEDIOL	5.9890E-04	643.00	-0.7143	196.15	610.85	9.3460E-04	1.002
332	C4H10O2	1,4-BUTANEDIOL	5.9160E-04	667.00	-0.7143	293.05	633.65	9.0320E-04	1.013
333	C4H10O2	2,3-BUTANEDIOL	6.1310E-04	611.00	-0.7140	280.75	580.45	9.8870E-04	0.994
334	C4H10O2	t-BUTYL HYDROPEROXIDE	6.5270E-04	576.00	-0.7143	277.45	547.20	1.0990E-03	0.886
335	C4H10O2	1,2-DIMETHOXYETHANE	7.1410E-04	536.15	-0.7143	215.15	509.34	1.2760E-03	0.865
336	C4H10O2	2-ETHOXYETHANOL	6.7670E-04	569.00	-0.7143	183.00	540.55	1.1500E-03	0.925
337	C4H10O3	DIETHYLENE GLYCOL	4.3680E-04	744.60	-0.7578	262.70	707.37	6.4360E-04	1.114
338	C4H10O4S	DIETHYL SULFATE	6.9200E-04	792.00	-0.5916	248.00	752.40	9.1510E-04	1.172
339	C4H10S	n-BUTYL MERCAPTAN	6.8800E-04	569.00	-0.7006	157.46	540.55	1.1570E-03	0.837
340	C4H10S	ISOBUTYL MERCAPTAN	5.8730E-04	559.00	-0.7405	128.31	531.05	1.0330E-03	0.830
341	C4H10S	sec-BUTYL MERCAPTAN	7.2470E-04	554.00	-0.6932	133.02	526.30	1.2380E-03	0.825
342	C4H10S	tert-BUTYL MERCAPTAN	7.8280E-04	530.00	-0.6800	274.26	503.50	1.3730E-03	0.795
343	C4H10S	DIETHYL SULFIDE	6.5730E-04	557.15	-0.7256	169.20	529.29	1.1460E-03	0.832
344	C4H10S	ISOPROPYL-METHYL-SULFIDE	6.8090E-04	551.00	-0.7143	171.65	523.45	1.1880E-03	0.825
345	C4H10S	METHYL-PROPYL-SULFIDE	6.7200E-04	563.00	-0.7143	160.19	534.85	1.1520E-03	0.837
346	C4H10S2	DIETHYL DISULFIDE	6.0000E-04	642.00	-0.6994	171.63	609.90	9.2850E-04	0.988
347	C4H11N	n-BUTYLAMINE	6.4250E-04	531.90	-0.7572	224.05	505.31	1.1980E-03	0.741
348	C4H11N	ISOBUTYLAMINE	6.4210E-04	513.73	-0.7635	188.55	488.04	1.2460E-03	0.730
349	C4H11N	sec-BUTYLAMINE	6.6420E-04	514.30	-0.7529	168.65	488.59	1.2760E-03	0.720
350	C4H11N	tert-BUTYLAMINE	6.7880E-04	483.90	-0.7460	206.19	459.71	1.3870E-03	0.688
351	C4H11N	DIETHYLAMINE	7.5360E-04	496.60	-0.7272	223.35	471.77	1.4680E-03	0.702
352	C4H11NO	DIMETHYLETHANOLAMINE	6.7080E-04	571.82	-0.7143	214.15	543.23	1.1350E-03	0.882
353	C4H11NO2	DIETHANOLAMINE	3.7800E-04	715.00	-0.8108	301.15	679.25	--------	------
354	C4H11NO2	2-AMINOETHOXYETHANOL	5.7120E-04	699.00	-0.7143	293.15	664.05	8.4970E-04	1.051
355	C4H12N2O	N-AMINOETHYL ETHANOLAMINE	5.6250E-04	698.00	-0.7143	273.15	663.10	8.3750E-04	1.022
356	C4H12Si	TETRAMETHYLSILANE	8.5460E-04	450.40	-0.7062	174.07	427.88	1.8380E-03	0.641
357	C4H13N3	DIETHYLENE TRIAMINE	5.7470E-04	676.00	-0.7143	234.15	642.20	8.7080E-04	0.954
358	C5Cl6	HEXACHLOROCYCLOPENTADIENE	5.2480E-04	746.00	-0.7143	284.49	708.70	7.5560E-04	1.703
359	C5H4O2	FURFURAL	5.8530E-04	657.00	-0.7143	236.65	624.15	9.0150E-04	1.155
360	C5H5N	PYRIDINE	6.9420E-04	619.95	-0.6955	231.53	588.95	1.0950E-03	0.979
361	C5H6	CYCLOPENTADIENE	7.2640E-04	507.00	-0.7143	188.15	481.65	1.3690E-03	0.797
362	C5H6	2-METHYL-1-BUTENE-3-YNE	8.0480E-04	492.00	-0.6918	160.15	467.40	1.5330E-03	0.699
363	C5H6	1-PENTENE-3-YNE	9.9880E-04	520.00	-0.6355	150.00	494.00	1.7160E-03	0.734
364	C5H6	1-PENTENE-4-YNE	7.5050E-04	503.00	-0.7143	150.00	477.85	1.4260E-03	0.724
365	C5H6N2	GLUTARONITRILE	6.7610E-04	782.00	-0.6552	244.21	742.90	9.2610E-04	0.981
366	C5H6O2	FURFURYL ALCOHOL	4.7010E-04	632.00	-0.7682	258.52	600.40	7.6760E-04	1.127
367	C5H6O3	GLUTARIC ANHYDRIDE	5.0260E-04	838.00	-0.7143	328.00	796.10	--------	------
368	C5H6O4	CITRACONIC ACID	5.3000E-04	829.00	-0.7143	356.15	787.55	--------	------
369	C5H6O4	ITACONIC ACID	5.3220E-04	821.00	-0.7143	438.75	779.95	--------	------
370	C5H6S	2-METHYLTHIOPHENE	6.2200E-04	610.00	-0.7143	209.77	579.50	1.0040E-03	1.014
371	C5H6S	3-METHYLTHIOPHENE	6.2050E-04	615.00	-0.7143	204.18	584.25	9.9650E-04	1.016
372	C5H7N	N-METHYLPYRROLE	8.1460E-04	610.00	-0.6552	216.91	579.50	1.2650E-03	0.903
373	C5H7NO2	ETHYL CYANOACETATE	6.1770E-04	679.00	-0.7070	250.65	645.05	9.2970E-04	1.058
374	C5H8	CYCLOPENTENE	7.2760E-04	507.00	-0.7143	138.13	481.65	1.3710E-03	0.767
375	C5H8	ISOPRENE	7.8420E-04	484.00	-0.7143	127.27	459.80	1.5540E-03	0.676
376	C5H8	3-METHYL-1,2-BUTADIENE	8.3060E-04	490.00	-0.7096	159.53	465.50	1.6160E-03	0.681
377	C5H8	2-METHYL-1,3-BUTADIENE	8.2430E-04	483.30	-0.7143	127.20	459.14	1.6360E-03	0.675
378	C5H8	1,2-PENTADIENE	7.6590E-04	500.00	-0.7121	135.89	475.00	1.4610E-03	0.688
379	C5H8	cis-1,3-PENTADIENE	7.8030E-04	499.00	-0.7143	132.35	474.05	1.4950E-03	0.686
380	C5H8	trans-1,3-PENTADIENE	8.0120E-04	500.00	-0.6960	185.71	475.00	1.5060E-03	0.671
381	C5H8	1,4-PENTADIENE	1.0440E-03	479.00	-0.6627	124.86	455.05	1.9900E-03	0.653
382	C5H8	2,3-PENTADIENE	7.6190E-04	497.00	-0.7299	147.50	472.15	1.4870E-03	0.690
383	C5H8	1-PENTYNE	1.0620E-03	481.20	-0.6470	167.45	457.14	1.9840E-03	0.688
384	C5H8	2-PENTYNE	7.1670E-04	521.99	-0.7143	163.86	495.89	1.3120E-03	0.705
385	C5H8	3-METHYL-1-BUTYNE	8.9570E-04	463.20	-0.6919	183.45	440.04	1.8290E-03	0.660
386	C5H8	SPIROPENTANE	6.9410E-04	499.74	-0.7143	166.11	474.75	1.3280E-03	0.735
387	C5H8N4O12	PENTAERYTHRITOL TETRANITRATE	5.2170E-04	676.00	-0.7143	413.65	642.20	--------	------
388	C5H8O	CYCLOPENTANONE	6.5200E-04	626.00	-0.6867	221.85	594.70	1.0170E-03	0.945
389	C5H8O	METHYL ISOPROPENYL KETONE	6.9790E-04	566.00	-0.7143	219.55	537.70	1.1910E-03	0.846
390	C5H8O2	ACETYLACETONE	5.6400E-04	602.00	-0.7495	249.65	571.90	9.4140E-04	0.971

Table 28-1 COEFFICIENT OF THERMAL EXPANSION - ORGANIC COMPOUNDS (continued)

$$B_{liq} = a\,(1 - T/T_C)^m \qquad (B_{liq} - 1/C,\ T - K,\ density_{liq} - g/cm^3)$$

NO	FORMULA	NAME	a	T_C	m	TMIN	TMAX	B_{liq} @ 25 C	$density_{liq}$ @ 25 C
391	C5H8O2	ALLYL ACETATE	6.9520E-04	559.00	-0.7143	138.00	531.05	1.1980E-03	0.922
392	C5H8O2	ETHYL ACRYLATE	6.9930E-04	553.00	-0.7143	201.95	525.35	1.2160E-03	0.918
393	C5H8O2	METHYL METHACRYLATE	6.9510E-04	564.00	-0.7143	224.95	535.80	1.1890E-03	0.937
394	C5H8O2	VINYL PROPIONATE	7.0090E-04	546.00	-0.7143	364.35	518.70	--------	------
395	C5H8O3	2-HYDROXYETHYL ACRYLATE	5.8200E-04	662.00	-0.7143	213.00	628.90	8.9250E-04	1.008
396	C5H8O3	LEVULINIC ACID	4.9350E-04	723.00	-0.7428	308.15	686.85	--------	------
397	C5H8O3	METHYL ACETOACETATE	4.3680E-04	642.00	-0.7870	193.15	609.90	7.1400E-04	1.072
398	C5H8O4	GLUTARIC ACID	4.4430E-04	807.00	-0.7448	370.65	766.65	--------	------
399	C5H9N	VALERONITRILE	6.4630E-04	603.00	-0.7203	176.95	572.85	1.0560E-03	0.794
400	C5H9NO	n-BUTYL ISOCYANATE	1.1420E-03	568.00	-0.5880	193.00	539.60	1.7690E-03	0.877
401	C5H9NO	N-METHYL-2-PYRROLIDONE	5.1740E-04	724.00	-0.7264	249.15	687.80	7.6080E-04	1.025
402	C5H9NO4	L-GLUTAMIC ACID	4.9570E-04	886.00	-0.7143	497.15	841.70	--------	------
403	C5H10	CYCLOPENTANE	7.2590E-04	511.76	-0.7143	179.31	486.17	1.3550E-03	0.750
404	C5H10	2-METHYL-1-BUTENE	8.2610E-04	465.00	-0.7143	135.58	441.75	1.7180E-03	0.645
405	C5H10	2-METHYL-2-BUTENE	8.2420E-04	471.00	-0.7143	139.39	447.45	1.6870E-03	0.657
406	C5H10	3-METHYL-1-BUTENE	7.6880E-04	450.37	-0.7361	104.66	427.85	1.7080E-03	0.622
407	C5H10	1-PENTENE	8.2660E-04	464.78	-0.7095	107.93	441.54	1.7110E-03	0.635
408	C5H10	cis-2-PENTENE	8.6640E-04	475.93	-0.7014	121.75	452.13	1.7280E-03	0.650
409	C5H10	trans-2-PENTENE	7.8220E-04	475.37	-0.7224	132.89	451.60	1.5950E-03	0.643
410	C5H10Br2	2,3-DIBROMO-2-METHYLBUTANE	4.8180E-04	668.37	-0.7143	288.00	634.95	7.3470E-04	1.410
411	C5H10Cl2	1,5-DICHLOROPENTANE	6.0720E-04	663.00	-0.7143	200.35	629.85	9.3030E-04	1.096
412	C5H10O	METHYL ISOPROPYL KETONE	6.9190E-04	553.00	-0.7143	181.15	525.35	1.2030E-03	0.805
413	C5H10O	2-PENTANONE	7.7400E-04	561.08	-0.6715	196.29	533.03	1.2880E-03	0.802
414	C5H10O	DIETHYL KETONE	6.9030E-04	560.95	-0.7264	234.18	532.90	1.1970E-03	0.810
415	C5H10O	VALERALDEHYDE	5.7760E-04	554.00	-0.7554	182.00	526.30	1.0350E-03	0.805
416	C5H10O2	n-BUTYL FORMATE	6.5960E-04	559.00	-0.7215	181.25	531.05	1.1430E-03	0.887
417	C5H10O2	ETHYL PROPIONATE	7.2010E-04	546.00	-0.7058	199.25	518.70	1.2570E-03	0.884
418	C5H10O2	ISOBUTYL FORMATE	6.6170E-04	551.35	-0.7317	177.35	523.78	1.1690E-03	0.875
419	C5H10O2	ISOPROPYL ACETATE	7.0130E-04	538.00	-0.7140	199.75	511.10	1.2490E-03	0.871
420	C5H10O2	n-PROPYL ACETATE	6.9020E-04	549.40	-0.7217	178.15	521.93	1.2140E-03	0.883
421	C5H10O2	METHYL n-BUTYRATE	6.7920E-04	554.50	-0.7228	187.35	526.78	1.1860E-03	0.893
422	C5H10O2	2-METHYLBUTYRIC ACID	6.1170E-04	643.00	-0.7143	193.00	610.85	9.5450E-04	0.932
423	C5H10O2	ISOVALERIC ACID	5.7280E-04	634.00	-0.7257	243.85	602.30	9.0830E-04	0.926
424	C5H10O2	VALERIC ACID	5.7650E-04	651.00	-0.7185	239.15	618.45	8.9520E-04	0.934
425	C5H10O2	TETRAHYDROFURFURYL ALCOHOL	4.6110E-04	639.00	-0.7667	193.00	607.05	7.4660E-04	1.048
426	C5H10O2S	3-METHYL SULFOLANE	5.0560E-04	817.00	-0.6856	273.65	776.15	6.9020E-04	1.188
427	C5H10O3	DIETHYL CARBONATE	6.5550E-04	576.00	-0.7143	230.15	547.20	1.1030E-03	0.970
428	C5H10O3	ETHYL LACTATE	1.2010E-03	588.00	-0.5447	247.15	558.60	1.7660E-03	1.027
429	C5H10S	THIACYCLOHEXANE	5.9930E-04	657.12	-0.7143	292.14	624.26	9.2300E-04	0.981
430	C5H10S	CYCLOPENTANETHIOL	5.8480E-04	629.00	-0.7143	155.39	597.55	9.2530E-04	0.961
431	C5H11Br	1-BROMOPENTANE	6.0410E-04	564.76	-0.7143	185.26	536.52	1.0330E-03	1.212
432	C5H11Cl	1-CHLOROPENTANE	6.4730E-04	568.00	-0.7196	174.15	539.60	1.1060E-03	0.878
433	C5H11Cl	1-CHLORO-3-METHYLBUTANE	6.9060E-04	558.87	-0.7143	168.76	530.93	1.1910E-03	0.865
434	C5H11Cl	2-CHLORO-2-METHYLBUTANE	6.9510E-04	548.97	-0.7143	199.66	521.52	1.2160E-03	0.860
435	C5H11N	N-METHYLPYRROLIDINE	6.7400E-04	550.00	-0.7140	183.15	522.50	1.1770E-03	0.806
436	C5H11N	PIPERIDINE	6.2500E-04	594.05	-0.7286	262.65	564.35	1.0390E-03	0.858
437	C5H11NO	tert-BUTYLFORMAMIDE	5.9430E-04	692.00	-0.7143	289.15	657.40	8.8890E-04	0.899
438	C5H12	ISOPENTANE	8.0150E-04	460.43	-0.7133	113.25	437.41	1.6860E-03	0.616
439	C5H12	NEOPENTANE	8.7180E-04	433.78	-0.7081	256.58	412.09	1.9860E-03	0.586
440	C5H12	n-PENTANE	7.8830E-04	469.65	-0.7179	143.42	446.17	1.6250E-03	0.621
441	C5H12O	2,2-DIMETHYL-1-PROPANOL	6.6690E-04	550.00	-0.7143	327.15	522.50	--------	------
442	C5H12O	tert-PENTYL-ALCOHOL	6.6250E-04	549.00	-0.7143	327.00	521.55	--------	------
443	C5H12O	2-METHYL-1-BUTANOL	5.4070E-04	565.00	-0.7678	203.00	536.75	9.6180E-04	0.814
444	C5H12O	2-METHYL-2-BUTANOL	6.5480E-04	545.15	-0.7353	264.35	517.89	1.1720E-03	0.805
445	C5H12O	3-METHYL-1-BUTANOL	6.6270E-04	579.45	-0.7143	155.95	550.48	1.1100E-03	0.812
446	C5H12O	3-METHYL-2-BUTANOL	6.6210E-04	574.00	-0.7143	188.00	545.30	1.1180E-03	0.814
447	C5H12O	1-PENTANOL	5.6410E-04	586.15	-0.7494	195.56	556.84	9.6080E-04	0.812
448	C5H12O	2-PENTANOL	6.8400E-04	552.00	-0.7143	200.00	524.40	1.1910E-03	0.805
449	C5H12O	3-PENTANOL	6.9320E-04	547.00	-0.7143	204.15	519.65	1.2170E-03	0.818
450	C5H12O	METHYL sec-BUTYL ETHER	7.5120E-04	498.00	-0.7143	150.00	473.10	1.4420E-03	0.737
451	C5H12O	METHYL tert-BUTYL ETHER	7.4450E-04	497.10	-0.7171	164.55	472.25	1.4360E-03	0.735
452	C5H12O	METHYL ISOBUTYL ETHER	7.4670E-04	497.00	-0.7143	150.00	472.15	1.4370E-03	0.725
453	C5H12O	ETHYL PROPYL ETHER	7.7300E-04	500.23	-0.7080	145.65	475.22	1.4690E-03	0.724
454	C5H12O2	ETHYLENE GLYCOL MONOPROPYL ETHER	6.6000E-04	582.00	-0.7143	183.15	552.90	1.1020E-03	0.906
455	C5H12O2	NEOPENTYL GLYCOL	5.7530E-04	643.00	-0.7143	400.00	610.85	--------	------
456	C5H12O2	1,5-PENTANEDIOL	5.8740E-04	673.00	-0.7143	257.15	639.35	8.9230E-04	0.994
457	C5H12O3	2-(2-METHOXYETHOXY)ETHANOL	6.2540E-04	630.00	-0.7143	197.15	598.50	9.8860E-04	1.017
458	C5H12O4	PENTAERYTHRITOL	4.7200E-04	780.00	-0.7100	534.15	741.00	--------	------
459	C5H12S	n-PENTYL MERCAPTAN	6.5740E-04	598.00	-0.6987	197.45	568.10	1.0650E-03	0.838
460	C5H12S	BUTYL-METHYL-SULFIDE	6.4200E-04	591.00	-0.7143	175.33	561.45	1.0600E-03	0.838
461	C5H12S	ETHYL-PROPYL-SULFIDE	6.4420E-04	584.00	-0.7143	156.15	554.80	1.0730E-03	0.832
462	C5H12S	2-METHYL-2-BUTANETHIOL	6.2840E-04	566.00	-0.7143	169.38	537.70	1.0720E-03	0.821
463	C5H13N	n-PENTYLAMINE	6.3150E-04	555.00	-0.7481	218.15	527.25	1.1240E-03	0.751
464	C5H13NO2	METHYL DIETHANOLAMINE	5.7700E-04	678.00	-0.7143	252.15	644.10	8.7280E-04	1.029
465	C6Cl6	HEXACHLOROBENZENE	5.2590E-04	825.00	-0.7143	501.70	783.75	--------	------
466	C6F6	HEXAFLUOROBENZENE	7.2720E-04	516.73	-0.7215	278.25	490.89	1.3530E-03	1.606
467	C6H3ClN2O4	1-CHLORO-2,4-DINITROBENZENE	4.3000E-04	813.77	-0.7578	326.55	773.08	--------	------
468	C6H3Cl2NO2	1,2-DICHLORO-4-NITROBENZENE	5.1190E-04	758.00	-0.7257	315.65	720.10	--------	------

Table 28-1 COEFFICIENT OF THERMAL EXPANSION - ORGANIC COMPOUNDS (continued)

			$B_{liq} = a \ (1 - T/T_C)^m$					$(B_{liq}$ - 1/C, T - K, density$_{liq}$ - g/cm^3)	
NO	FORMULA	NAME	a	T_C	m	TMIN	TMAX	B_{liq} @ 25 C	density$_{liq}$ @ 25 C
469	C6H3Cl3	1,2,4-TRICHLOROBENZENE	5.4250E-04	725.00	-0.7143	290.15	688.75	7.9210E-04	1.449
470	C6H3N3O6	1,3,5-TRINITROBENZENE	4.4230E-04	1005.00	-0.7143	398.40	954.75	--------	------
471	C6H4Br2	m-DIBROMOBENZENE	4.8580E-04	761.00	-0.7143	266.25	722.95	6.9290E-04	1.947
472	C6H4ClNO2	m-CHLORONITROBENZENE	4.9490E-04	742.00	-0.7550	317.65	704.90	--------	------
473	C6H4ClNO2	o-CHLORONITROBENZENE	5.2850E-04	757.00	-0.7143	306.15	719.15	--------	------
474	C6H4ClNO2	p-CHLORONITROBENZENE	4.4350E-04	751.00	-0.7732	356.65	713.45	--------	------
475	C6H4Cl2	m-DICHLOROBENZENE	6.1830E-04	683.95	-0.6847	248.39	649.75	9.1510E-04	1.283
476	C6H4Cl2	o-DICHLOROBENZENE	5.8690E-04	705.00	-0.6919	256.15	669.75	8.5850E-04	1.301
477	C6H4Cl2	p-DICHLOROBENZENE	6.0090E-04	684.75	-0.6921	326.14	650.51	--------	------
478	C6H4F2	m-DIFLUOROBENZENE	6.8620E-04	552.94	-0.7143	249.16	525.29	1.1930E-03	1.162
479	C6H4F2	o-DIFLUOROBENZENE	6.8630E-04	554.46	-0.7143	239.16	526.74	1.1910E-03	1.150
480	C6H4F2	p-DIFLUOROBENZENE	6.8050E-04	556.00	-0.7143	260.16	528.20	1.1780E-03	1.162
481	C6H4N2O4	m-DINITROBENZENE	4.0600E-04	805.00	-0.7728	364.00	764.75	--------	------
482	C6H4N2O4	o-DINITROBENZENE	4.8780E-04	831.00	-0.7143	390.08	789.45	--------	------
483	C6H4N2O4	p-DINITROBENZENE	4.9320E-04	803.00	-0.7143	446.60	762.85	--------	------
484	C6H5Br	BROMOBENZENE	5.5690E-04	670.15	-0.7179	242.43	636.64	8.4980E-04	1.487
485	C6H5Cl	MONOCHLOROBENZENE	5.9030E-04	632.35	-0.7096	227.95	600.73	9.2820E-04	1.101
486	C6H5ClO	m-CHLOROPHENOL	5.0880E-04	729.00	-0.7143	306.00	692.55	--------	------
487	C6H5ClO	o-CHLOROPHENOL	5.4370E-04	675.00	-0.7286	282.00	641.25	8.3140E-04	1.255
488	C6H5ClO	p-CHLOROPHENOL	5.5200E-04	738.00	-0.7046	316.00	701.10	--------	------
489	C6H5Cl2N	3,4-DICHLOROANILINE	5.0070E-04	800.00	-0.7143	344.65	760.00	--------	------
490	C6H5F	FLUOROBENZENE	6.5620E-04	560.09	-0.7171	230.94	532.09	1.1320E-03	1.019
491	C6H5I	IODOBENZENE	5.3490E-04	721.15	-0.7105	241.83	685.09	7.8150E-04	1.822
492	C6H5NO2	NITROBENZENE	5.5520E-04	719.00	-0.7143	278.91	683.05	8.1390E-04	1.199
493	C6H6	BENZENE	6.6060E-04	562.16	-0.7182	278.68	534.05	1.1370E-03	0.873
494	C6H6ClN	m-CHLOROANILINE	6.3050E-04	751.00	-0.6759	262.75	713.45	8.8750E-04	1.211
495	C6H6ClN	o-CHLOROANILINE	5.1150E-04	722.00	-0.7396	481.99	685.90	--------	------
496	C6H6ClN	p-CHLOROANILINE	5.4360E-04	754.00	-0.7150	343.05	716.30	--------	------
497	C6H6N2	cis-DICYANO-1-BUTENE	6.6340E-04	691.00	-0.7143	249.00	656.45	9.9300E-04	1.062
498	C6H6N2	trans-DICYANO-1-BUTENE	6.6330E-04	689.00	-0.7143	260.00	654.55	9.9440E-04	1.054
499	C6H6N2	1,4-DICYANO-2-BUTENE	6.0830E-04	755.00	-0.7143	349.00	717.25	--------	------
500	C6H6N2O2	m-NITROANILINE	4.6690E-04	815.00	-0.7143	387.15	774.25	--------	------
501	C6H6N2O2	o-NITROANILINE	4.7050E-04	784.00	-0.7143	344.65	744.80	--------	------
502	C6H6N2O2	p-NITROANILINE	4.6280E-04	851.00	-0.7143	420.65	808.45	--------	------
503	C6H6O	PHENOL	5.2480E-04	694.25	-0.6788	314.06	659.54	--------	------
504	C6H6O2	1,2-BENZENEDIOL	4.8070E-04	764.00	-0.7331	377.60	725.80	--------	------
505	C6H6O2	1,3-BENZENEDIOL	4.0860E-04	810.00	-0.7567	382.00	769.50	--------	------
506	C6H6O2	p-HYDROQUINONE	3.8850E-04	822.00	-0.7143	444.65	780.90	--------	------
507	C6H6O3	1,2,3-BENZENETRIOL	3.1110E-04	830.00	-0.7143	407.00	788.50	--------	------
508	C6H6S	PHENYL MERCAPTAN	5.9660E-04	689.00	-0.6920	258.26	654.55	8.8320E-04	1.073
509	C6H7N	ANILINE	5.6660E-04	699.00	-0.7143	267.13	664.05	8.4290E-04	1.018
510	C6H7N	2-METHYLPYRIDINE	6.1300E-04	621.00	-0.7282	206.44	589.95	9.8700E-04	0.940
511	C6H7N	3-METHYLPYRIDINE	6.2090E-04	645.00	-0.7166	255.01	612.75	9.6840E-04	0.952
512	C6H7N	4-METHYLPYRIDINE	6.0610E-04	646.15	-0.7248	276.73	613.84	9.4910E-04	0.950
513	C6H8	1,3-CYCLOHEXADIENE	7.6290E-04	558.00	-0.6851	161.00	530.10	1.2880E-03	0.837
514	C6H8	METHYLCYCLOPENTADIENE	6.8310E-04	541.00	-0.7143	150.00	513.95	1.2110E-03	0.805
515	C6H8N2	ADIPONITRILE	5.3390E-04	781.00	-0.7162	275.64	741.95	7.5340E-04	0.960
516	C6H8N2	METHYLGLUTARONITRILE	7.2630E-04	742.00	-0.6454	228.15	704.90	1.0120E-03	0.950
517	C6H8N2	m-PHENYLENEDIAMINE	4.7420E-04	824.00	-0.7464	334.00	782.80	--------	------
518	C6H8N2	o-PHENYLENEDIAMINE	5.0580E-04	781.00	-0.7143	376.95	741.95	--------	------
519	C6H8N2	p-PHENYLENEDIAMINE	5.0050E-04	796.00	-0.7143	413.00	756.20	--------	------
520	C6H8N2	PHENYLHYDRAZINE	5.1360E-04	761.00	-0.7595	292.35	722.95	7.4930E-04	1.094
521	C6H8N2O	BIS(CYANOETHYL)ETHER	5.1410E-04	783.00	-0.6981	246.85	743.85	7.1840E-04	1.044
522	C6H8O4	DIMETHYL MALEATE	5.9640E-04	675.00	-0.7088	254.15	641.25	9.0140E-04	1.148
523	C6H8O6	ASCORBIC ACID	4.7090E-04	783.00	-0.7143	465.15	743.85	--------	------
524	C6H8O7	CITRIC ACID	4.9730E-04	822.00	-0.7143	426.15	780.90	--------	------
525	C6H10	1-METHYLCYCLOPENTENE	6.8440E-04	541.99	-0.7143	145.96	514.89	1.2110E-03	0.776
526	C6H10	3-METHYLCYCLOPENTENE	7.0030E-04	535.71	-0.7143	130.16	508.92	1.2520E-03	0.759
527	C6H10	4-METHYLCYCLOPENTENE	6.9780E-04	543.75	-0.7143	112.31	516.56	1.2310E-03	0.763
528	C6H10	CYCLOHEXENE	6.6450E-04	560.40	-0.7143	169.67	532.38	1.1430E-03	0.806
529	C6H10	2,3-DIMETHYL-1,3-BUTADIENE	7.1550E-04	526.00	-0.7110	197.15	499.70	1.2970E-03	0.723
530	C6H10	1,5-HEXADIENE	6.9880E-04	507.00	-0.7323	132.47	481.65	1.3380E-03	0.688
531	C6H10	cis,trans-2,4-HEXADIENE	7.2350E-04	538.00	-0.7109	177.05	511.10	1.2850E-03	0.719
532	C6H10	trans,trans-2,4-HEXADIENE	6.9770E-04	535.00	-0.7179	228.25	508.25	1.2520E-03	0.710
533	C6H10	1-HEXYNE	6.9920E-04	516.20	-0.7229	141.25	490.39	1.3040E-03	0.712
534	C6H10	2-HEXYNE	7.9250E-04	549.00	-0.6839	183.65	521.55	1.3540E-03	0.727
535	C6H10	3-HEXYNE	7.0620E-04	544.00	-0.7143	170.05	516.80	1.2450E-03	0.718
536	C6H10O	CYCLOHEXANONE	5.6250E-04	629.15	-0.7283	242.00	597.69	8.9790E-04	0.942
537	C6H10O	MESITYL OXIDE	6.4990E-04	600.00	-0.7152	220.15	570.00	1.0620E-03	0.852
538	C6H10O2	epsilon-CAPROLACTONE	5.0750E-04	771.00	-0.7143	271.85	732.45	7.1970E-04	1.067
539	C6H10O2	ETHYL METHACRYLATE	6.7300E-04	577.00	-0.7143	223.15	548.15	1.1310E-03	0.908
540	C6H10O2	n-PROPYL ACRYLATE	6.7730E-04	569.00	-0.7143	273.15	540.55	1.1510E-03	0.900
541	C6H10O3	ETHYLACETOACETATE	6.1760E-04	643.00	-0.7143	234.15	610.85	9.6370E-04	1.023
542	C6H10O3	PROPIONIC ANHYDRIDE	6.2620E-04	618.00	-0.7143	228.15	587.10	1.0020E-03	1.007
543	C6H10O4	ADIPIC ACID	4.7070E-04	809.00	-0.7180	425.50	768.55	--------	------
544	C6H10O4	DIETHYL OXALATE	7.1030E-04	646.00	-0.6658	232.55	613.70	1.0730E-03	1.073
545	C6H10O4	ETHYLENE GLYCOL DIACETATE	6.1420E-04	653.00	-0.7143	242.15	620.35	9.4950E-04	1.101
546	C6H10O4	ETHYLIDENE DIACETATE	6.1100E-04	635.00	-0.7046	292.00	603.25	9.5500E-04	1.069

Table 28-1 COEFFICIENT OF THERMAL EXPANSION - ORGANIC COMPOUNDS (continued)

NO	FORMULA	NAME	$B_{liq} = a \ (1 - T/T_C)^m$					$(B_{liq} - 1/C, \ T - K, \ density_{liq} - g/cm^3)$	
			a	T_C	m	TMIN	TMAX	B_{liq} @ 25 C	$density_{liq}$ @ 25 C
547	C6H11N	HEXANENITRILE	6.3970E-04	622.05	-0.7136	192.85	590.95	1.0190E-03	0.801
548	C6H11NO	epsilon-CAPROLACTAM	4.8640E-04	806.00	-0.7143	342.36	765.70	--------	------
549	C6H11NO	CYCLOHEXANONE OXIME	4.9330E-04	715.00	-0.7143	363.15	679.25	--------	------
550	C6H12	CYCLOHEXANE	6.6670E-04	553.54	-0.7149	279.69	525.86	1.1590E-03	0.773
551	C6H12	2,3-DIMETHYL-1-BUTENE	7.5300E-04	500.00	-0.7143	115.89	475.00	1.4390E-03	0.673
552	C6H12	2,3-DIMETHYL-2-BUTENE	7.1380E-04	524.00	-0.7360	198.82	497.80	1.3260E-03	0.703
553	C6H12	3,3-DIMETHYL-1-BUTENE	7.8420E-04	480.00	-0.7009	157.95	456.00	1.5480E-03	0.648
554	C6H12	2-ETHYL-1-BUTENE	7.4220E-04	512.00	-0.7248	141.61	486.40	1.3970E-03	0.685
555	C6H12	trans-3-METHYL-2-PENTENE	7.1800E-04	521.00	-0.7143	134.70	494.95	1.3170E-03	0.693
556	C6H12	1-HEXENE	7.5060E-04	504.03	-0.7143	133.39	478.83	1.4230E-03	0.667
557	C6H12	cis-2-HEXENE	7.3890E-04	513.00	-0.7216	132.00	487.35	1.3850E-03	0.683
558	C6H12	trans-2-HEXENE	7.4440E-04	513.00	-0.7143	140.17	487.35	1.3860E-03	0.673
559	C6H12	cis-3-HEXENE	7.4890E-04	509.00	-0.7143	135.33	483.55	1.4050E-03	0.675
560	C6H12	trans-3-HEXENE	7.7690E-04	509.00	-0.7045	159.73	483.55	1.4450E-03	0.673
561	C6H12	METHYLCYCLOPENTANE	6.9410E-04	532.79	-0.7172	130.73	506.15	1.2500E-03	0.745
562	C6H12	2-METHYL-1-PENTENE	7.5190E-04	507.00	-0.7143	137.42	481.65	1.4170E-03	0.675
563	C6H12	2-METHYL-2-PENTENE	7.4710E-04	514.00	-0.7143	138.07	488.30	1.3890E-03	0.681
564	C6H12	3-METHYL-1-PENTENE	7.4280E-04	495.00	-0.7156	120.20	470.25	1.4370E-03	0.663
565	C6H12	3-METHYL-cis-2-PENTENE	7.2720E-04	515.00	-0.7161	138.31	489.25	1.3510E-03	0.689
566	C6H12	4-METHYL-1-PENTENE	7.5070E-04	496.00	-0.7143	119.51	471.20	1.4470E-03	0.659
567	C6H12	4-METHYL-cis-2-PENTENE	7.5000E-04	499.00	-0.7143	138.30	474.05	1.4370E-03	0.665
568	C6H12	4-METHYL-trans-2-PENTENE	7.4750E-04	501.00	-0.7143	132.35	475.95	1.4260E-03	0.664
569	C6H12N2	TRIETHYLENEDIAMINE	5.6470E-04	655.00	-0.7143	434.25	622.25	--------	------
570	C6H12O	BUTYL VINYL ETHER	7.1390E-04	536.00	-0.7143	181.25	509.20	1.2760E-03	0.774
571	C6H12O	CYCLOHEXANOL	6.4580E-04	625.15	-0.7143	296.60	593.89	1.0260E-03	0.960
572	C6H12O	1-HEXANAL	5.5080E-04	579.00	-0.7555	217.15	550.05	9.5140E-04	0.810
573	C6H12O	ETHYL ISOPROPYL KETONE	8.5030E-04	567.00	-0.6570	200.00	538.65	1.3880E-03	0.806
574	C6H12O	2-HEXANONE	6.7850E-04	587.05	-0.7037	217.35	557.70	1.1170E-03	0.807
575	C6H12O	3-HEXANONE	6.5200E-04	582.82	-0.7134	217.50	553.68	1.0870E-03	0.810
576	C6H12O	METHYL ISOBUTYL KETONE	6.7570E-04	571.40	-0.7143	189.15	542.83	1.1450E-03	0.796
577	C6H12O2	n-PENTYL FORMATE	7.2640E-04	576.00	-0.6914	199.65	547.20	1.2020E-03	0.881
578	C6H12O2	n-BUTYL ACETATE	7.1750E-04	579.65	-0.6910	199.65	550.67	1.1820E-03	0.876
579	C6H12O2	sec-BUTYL ACETATE	7.2860E-04	561.00	-0.6958	174.15	532.95	1.2350E-03	0.868
580	C6H12O2	tert-BUTYL ACETATE	6.4990E-04	545.00	-0.7300	283.15	517.75	1.1590E-03	0.861
581	C6H12O2	ETHYL n-BUTYRATE	6.4760E-04	571.00	-0.7361	175.15	542.45	1.1150E-03	0.874
582	C6H12O2	ETHYL ISOBUTYRATE	6.9370E-04	553.15	-0.7143	185.00	525.49	1.2060E-03	0.863
583	C6H12O2	ISOBUTYL ACETATE	6.7010E-04	561.00	-0.7163	174.30	532.95	1.1530E-03	0.869
584	C6H12O2	n-PROPYL PROPIONATE	7.0160E-04	578.00	-0.6977	197.25	549.10	1.1640E-03	0.877
585	C6H12O2	CYCLOHEXYL PEROXIDE	5.5690E-04	685.00	-0.7143	253.15	650.75	8.3760E-04	1.015
586	C6H12O2	DIACETONE ALCOHOL	5.5770E-04	606.00	-0.7489	229.15	575.70	9.2610E-04	0.934
587	C6H12O2	2-ETHYL BUTYRIC ACID	6.4500E-04	655.00	-0.6890	258.15	622.25	9.8020E-04	0.919
588	C6H12O2	n-HEXANOIC ACID	5.4930E-04	667.00	-0.7235	270.15	633.65	8.4320E-04	0.921
589	C6H12O3	2-ETHOXYETHYL ACETATE	7.0830E-04	597.00	-0.7143	211.45	567.15	1.1610E-03	0.970
590	C6H12O3	HYDROXYCAPROIC ACID	5.5070E-04	758.00	-0.7143	334.00	720.10	--------	------
591	C6H12O3	PARALDEHYDE	6.7970E-04	579.00	-0.6866	285.75	550.05	1.1170E-03	0.985
592	C6H12O3	sec-BUTYL GLYCOLATE	--------	------	------	------	------	--------	------
593	C6H12S	THIACYCLOHEPTANE	5.0580E-04	640.07	-0.7143	292.14	608.07	7.9160E-04	0.766
594	C6H13N	CYCLOHEXYLAMINE	6.0120E-04	615.00	-0.7293	255.45	584.25	9.7510E-04	0.863
595	C6H13N	HEXAMETHYLENEIMINE	5.7190E-04	615.00	-0.7438	236.15	584.25	9.3660E-04	0.875
596	C6H14	2,2-DIMETHYLBUTANE	7.3580E-04	488.78	-0.7201	174.28	464.34	1.4490E-03	0.644
597	C6H14	2,3-DIMETHYLBUTANE	7.4730E-04	499.98	-0.7143	145.19	474.98	1.4290E-03	0.658
598	C6H14	n-HEXANE	7.2780E-04	507.43	-0.7219	177.84	482.06	1.3790E-03	0.656
599	C6H14	2-METHYLPENTANE	7.5780E-04	497.50	-0.7143	119.55	472.63	1.4560E-03	0.648
600	C6H14	3-METHYLPENTANE	7.3290E-04	504.43	-0.7208	110.25	479.21	1.3960E-03	0.660
601	C6H14N2O2	LYSINE	4.6880E-04	821.00	-0.7143	483.00	779.95	--------	------
602	C6H14O	2-ETHYL-1-BUTANOL	5.7380E-04	580.00	-0.7527	158.75	551.00	9.8780E-04	0.829
603	C6H14O	1-HEXANOL	5.4820E-04	611.35	-0.7460	228.55	580.78	9.0300E-04	0.816
604	C6H14O	2-HEXANOL	6.1440E-04	586.20	-0.7306	223.00	556.89	1.0320E-03	0.810
605	C6H14O	2-METHYL-1-PENTANOL	6.6120E-04	582.00	-0.7143	223.00	552.90	1.1040E-03	0.827
606	C6H14O	4-METHYL-2-PENTANOL	6.5450E-04	574.40	-0.7143	183.00	545.68	1.1040E-03	0.805
607	C6H14O	n-BUTYL ETHYL ETHER	7.1440E-04	531.00	-0.7143	170.15	504.45	1.2870E-03	0.745
608	C6H14O	DIISOPROPYL ETHER	7.4870E-04	500.05	-0.7143	187.65	475.05	1.4310E-03	0.721
609	C6H14O	DI-n-PROPYL ETHER	7.0940E-04	530.60	-0.7143	149.95	504.07	1.2790E-03	0.741
610	C6H14O	METHYL tert-PENTYL ETHER	7.1460E-04	534.00	-0.7143	160.00	507.30	1.2810E-03	0.766
611	C6H14O2	ACETAL	6.9140E-04	541.00	-0.7143	173.15	513.95	1.2250E-03	0.821
612	C6H14O2	2-BUTOXYETHANOL	5.6080E-04	600.00	-0.7455	203.15	570.00	9.3600E-04	0.896
613	C6H14O2	1,6-HEXANEDIOL	5.8910E-04	670.00	-0.7143	315.15	636.50	--------	------
614	C6H14O2	HEXYLENE GLYCOL	5.7910E-04	621.00	-0.7143	223.15	589.95	9.2400E-04	0.918
615	C6H14O2S	DI-n-PROPYL SULFONE	5.4520E-04	763.00	-0.6943	303.00	724.85	--------	------
616	C6H14O3	DIETHYLENE GLYCOL DIMETHYL ETHER	6.5170E-04	604.00	-0.7143	203.15	573.80	1.0600E-03	0.942
617	C6H14O3	DIPROPYLENE GLYCOL	5.8010E-04	654.00	-0.7143	233.00	621.30	8.9610E-04	1.018
618	C6H14O3	2-(2-ETHOXYETHOXY)ETHANOL	6.1840E-04	632.00	-0.7143	195.15	600.40	9.7550E-04	0.984
619	C6H14O3	TRIMETHYLOLPROPANE	3.0030E-04	709.00	-0.8424	331.15	673.55	--------	------
620	C6H14O4	TRIETHYLENE GLYCOL	4.0250E-04	700.00	-0.7904	265.79	665.00	6.2420E-04	1.122
621	C6H14O6	SORBITOL	3.3050E-04	959.00	-0.7827	370.85	911.05	--------	------
622	C6H14S	n-HEXYLMERCAPTAN	8.1520E-04	623.00	-0.6278	192.62	591.85	1.2270E-03	0.837
623	C6H14S	BUTYL-ETHYL-SULFIDE	6.2300E-04	609.00	-0.7143	178.03	578.55	1.0070E-03	0.833
624	C6H14S	ISOPROPYL-SULFIDE	6.3150E-04	585.71	-0.7143	170.45	556.42	1.0500E-03	0.822

Table 28-1 COEFFICIENT OF THERMAL EXPANSION - ORGANIC COMPOUNDS (continued)

$B_{liq} = a \ (1 - T/T_C)^m$ $(B_{liq} - 1/C, \ T - K, \ density_{liq} - g/cm^3)$

NO	FORMULA	NAME	a	T_C	m	TMIN	TMAX	B_{liq} @ 25 C	$density_{liq}$ @ 25 C
625	C6H14S	METHYL-PENTYL-SULFIDE	6.2560E-04	587.98	-0.7143	179.16	558.58	1.0370E-03	0.839
626	C6H14S	PROPYL-SULFIDE	6.2050E-04	609.73	-0.7143	170.45	579.24	1.0020E-03	0.833
627	C6H14S2	PROPYL-DISULFIDE	5.4500E-04	673.00	-0.7143	187.68	639.35	8.2780E-04	0.955
628	C6H15Al	TRIETHYL ALUMINUM	6.6830E-04	720.15	-0.3445	220.65	684.14	8.0340E-04	0.833
629	C6H15Al2Cl3	ETHYL ALUMINUM SESQUICHLORIDE	--------	------	-------	------	------	--------	1.092
630	C6H15N	DIISOPROPYLAMINE	7.0220E-04	523.10	-0.7290	176.85	496.95	1.2990E-03	0.713
631	C6H15N	DI-n-PROPYLAMINE	6.7720E-04	555.80	-0.7257	210.15	528.01	1.1830E-03	0.737
632	C6H15N	n-HEXYLAMINE	5.9310E-04	583.00	-0.7477	251.85	553.85	1.0130E-03	0.761
633	C6H15N	TRIETHYLAMINE	6.9510E-04	535.15	-0.7128	158.45	508.39	1.2420E-03	0.724
634	C6H15NO	6-AMINOHEXANOL	5.5720E-04	681.00	-0.7143	331.00	646.95	--------	------
635	C6H15NO2	DIISOPROPANOLAMINE	4.6480E-04	672.00	-0.7771	318.15	638.40	--------	------
636	C6H15NO3	TRIETHANOLAMINE	3.6050E-04	787.00	-0.7965	294.35	747.65	5.2680E-04	1.120
637	C6H15N3	N-AMINOETHYL PIPERAZINE	5.3370E-04	708.00	-0.7143	254.15	672.60	7.8870E-04	0.983
638	C6H15O4P	TRIETHYL PHOSPHATE	6.4740E-04	794.00	-0.7434	216.00	754.30	9.1880E-04	1.066
639	C6H16N2	HEXAMETHYLENEDIAMINE	5.4400E-04	663.00	-0.7143	313.95	629.85	--------	------
640	C6H18N3OP	HEXAMETHYL PHOSPHORAMIDE	--------	------	-------	------	------	--------	1.020
641	C6H18N4	TRIETHYLENE TETRAMINE	5.4260E-04	718.00	-0.7143	285.15	682.10	7.9600E-04	0.978
642	C6H18OSi2	HEXAMETHYLDISILOXANE	7.2770E-04	518.70	-0.7143	204.93	492.77	1.3410E-03	0.760
643	C6H18O3Si3	HEXAMETHYLCYCLOTRISILOXANE	6.9430E-04	554.20	-0.6907	337.15	526.49	--------	------
644	C6H19NSi2	HEXAMETHYLDISILAZANE	7.0790E-04	544.00	-0.7143	293.15	516.80	1.2480E-03	0.772
645	C7H3ClF3NO2	4-CHLORO-3-NITROBENZOTRIFLUORIDE	5.9260E-04	686.00	-0.7143	293.15	651.70	8.9060E-04	1.506
646	C7H3Cl2F3	2,4-DICHLOROBENZOTRIFLUORIDE	6.2200E-04	646.00	-0.7143	247.55	613.70	9.6790E-04	1.492
647	C7H3Cl2NO	3,4-DICHLOROPHENYL ISOCYANATE	5.3570E-04	733.00	-0.7255	316.15	696.35	--------	------
648	C7H4ClF3	p-CHLOROBENZOTRIFLUORIDE	6.3180E-04	601.00	-0.7143	237.15	570.95	1.0310E-03	1.226
649	C7H4Cl2O	m-CHLOROBENZOYL CHLORIDE	5.5020E-04	724.00	-0.7143	280.00	687.80	8.0390E-04	1.430
650	C7H4F3NO2	3-NITROBENZOTRIFLUORIDE	6.2550E-04	667.00	-0.7143	272.00	633.65	9.5500E-04	1.426
651	C7H5ClO	BENZOYL CHLORIDE	5.5490E-04	697.00	-0.7140	272.65	662.15	8.2670E-04	1.206
652	C7H5ClO2	o-CHLOROBENZOIC ACID	5.3420E-04	792.00	-0.7143	415.15	752.40	--------	------
653	C7H5Cl3	BENZOTRICHLORIDE	5.3140E-04	737.00	-0.7143	268.40	700.15	7.6960E-04	1.369
654	C7H5F3	BENZOTRIFLUORIDE	6.7520E-04	565.00	-0.7143	244.14	536.75	1.1540E-03	1.178
655	C7H5N	BENZONITRILE	5.6650E-04	699.35	-0.7159	260.40	664.38	8.4330E-04	1.001
656	C7H5NO	PHENYL ISOCYANATE	5.9940E-04	648.00	-0.7143	243.15	615.60	9.3090E-04	1.093
657	C7H5N3O6	2,4,6-TRINITROTOLUENE	4.7260E-04	795.00	-0.7183	354.00	755.25	--------	------
658	C7H6Cl2	BENZYL DICHLORIDE	5.8910E-04	731.00	-0.6806	257.00	694.45	8.4150E-04	1.247
659	C7H6Cl2	2,4-DICHLOROTOLUENE	5.5430E-04	705.00	-0.7143	259.65	669.75	8.2090E-04	1.247
660	C7H6N2O4	2,4-DINITROTOLUENE	5.3200E-04	814.00	-0.7085	343.00	773.30	--------	------
661	C7H6N2O4	2,5-DINITROTOLUENE	5.0540E-04	814.00	-0.7143	325.65	773.30	--------	------
662	C7H6N2O4	2,6-DINITROTOLUENE	5.2670E-04	770.00	-0.7143	339.00	731.50	--------	------
663	C7H6N2O4	3,4-DINITROTOLUENE	4.8970E-04	842.00	-0.7143	332.00	799.90	--------	------
664	C7H6N2O4	3,5-DINITROTOLUENE	5.0470E-04	814.00	-0.7143	365.65	773.30	--------	------
665	C7H6O	BENZALDEHYDE	5.5590E-04	695.00	-0.7150	247.15	660.25	8.2980E-04	1.040
666	C7H6O2	BENZOIC ACID	5.3030E-04	751.00	-0.7143	395.52	713.45	--------	------
667	C7H6O2	p-HYDROXYBENZALDEHYDE	4.0530E-04	844.00	-0.7577	390.15	801.80	--------	------
668	C7H6O2	SALICYLALDEHYDE	5.2990E-04	680.00	-0.7375	266.15	646.00	8.1090E-04	1.162
669	C7H6O3	SALICYLIC ACID	--------	------	-------	------	------	--------	------
670	C7H7Br	p-BROMOTOLUENE	5.7570E-04	699.00	-0.6992	299.95	664.05	--------	------
671	C7H7Cl	BENZYL CHLORIDE	5.7120E-04	686.00	-0.7143	234.15	651.70	8.5840E-04	1.097
672	C7H7Cl	o-CHLOROTOLUENE	5.6700E-04	656.00	-0.7160	236.65	623.20	8.7510E-04	1.077
673	C7H7Cl	p-CHLOROTOLUENE	5.7270E-04	660.00	-0.7125	280.65	627.00	8.7880E-04	1.063
674	C7H7F	p-FLUOROTOLUENE	6.4810E-04	590.48	-0.7143	216.36	560.96	1.0710E-03	0.991
675	C7H7NO	FORMANILIDE	5.2590E-04	787.00	-0.7100	323.15	747.65	--------	------
676	C7H7NO2	m-NITROTOLUENE	5.5950E-04	734.00	-0.7279	289.20	697.30	8.1760E-04	1.152
677	C7H7NO2	o-NITROTOLUENE	5.7170E-04	720.00	-0.7291	269.98	684.00	8.4430E-04	1.158
678	C7H7NO2	p-NITROTOLUENE	5.1980E-04	736.00	-0.7437	324.75	699.20	--------	------
679	C7H7NO3	o-NITROANISOLE	5.2180E-04	782.00	-0.7117	283.60	742.90	7.3430E-04	1.244
680	C7H8	TOLUENE	6.5930E-04	591.79	-0.7011	178.18	562.20	1.0780E-03	0.865
681	C7H8	1,3,5-CYCLOHEPTATRIENE	6.3710E-04	593.90	-0.7143	193.66	564.21	1.0480E-03	0.882
682	C7H8O	ANISOLE	5.8600E-04	641.65	-0.7199	235.65	609.57	9.1890E-04	0.990
683	C7H8O	BENZYL ALCOHOL	4.4060E-04	677.00	-0.7760	257.85	643.15	6.9130E-04	1.041
684	C7H8O	m-CRESOL	4.8450E-04	705.85	-0.7293	285.39	670.56	7.2310E-04	1.030
685	C7H8O	o-CRESOL	5.2730E-04	697.55	-0.6901	304.19	662.67	--------	------
686	C7H8O	p-CRESOL	5.7210E-04	704.65	-0.6659	307.93	669.42	--------	------
687	C7H8O2	GUAIACOL	5.2690E-04	697.00	-0.7288	304.65	662.15	--------	------
688	C7H8O2	p-METHOXYPHENOL	4.9810E-04	758.00	-0.7155	329.00	720.10	--------	------
689	C7H9N	BENZYLAMINE	2.6110E-04	683.50	-0.8655	227.15	649.33	4.2870E-04	0.981
690	C7H9N	2,6-DIMETHYLPYRIDINE	5.1740E-04	623.75	-0.7283	267.00	592.56	8.3070E-04	0.918
691	C7H9N	N-METHYLANILINE	5.1130E-04	701.55	-0.7463	216.15	666.47	7.7270E-04	0.982
692	C7H9N	m-TOLUIDINE	5.4160E-04	709.15	-0.7310	242.75	673.69	8.0690E-04	0.985
693	C7H9N	o-TOLUIDINE	5.3180E-04	694.15	-0.7427	249.47	659.44	8.0680E-04	0.994
694	C7H9N	p-TOLUIDINE	5.3730E-04	693.15	-0.7384	316.90	658.49	--------	------
695	C7H10	2-NORBORNENE	6.3620E-04	583.00	-0.7143	319.40	553.85	--------	------
696	C7H10N2	TOLUENEDIAMINE	4.9740E-04	804.00	-0.7143	371.25	763.80	--------	------
697	C7H11NO	CYCLOHEXYL ISOCYANATE	6.3220E-04	633.00	-0.7143	193.00	601.35	9.9640E-04	1.077
698	C7H12	1-HEPTYNE	6.5720E-04	559.69	-0.7143	192.26	531.71	1.1320E-03	0.728
699	C7H12O2	n-BUTYL ACRYLATE	6.9800E-04	598.00	-0.6916	208.55	568.10	1.1250E-03	0.894
700	C7H12O2	ISOBUTYL ACRYLATE	6.5830E-04	580.00	-0.7143	212.00	551.00	1.1020E-03	0.885
701	C7H12O2	n-PROPYL METHACRYLATE	6.5030E-04	599.00	-0.7143	223.00	569.05	1.0630E-03	0.897
702	C7H12O4	DIETHYL MALONATE	6.5930E-04	653.00	-0.6839	224.25	620.35	1.0000E-03	1.050

NO	FORMULA	NAME	$B_{liq} = a$	$(1 - T/T_C)^m$		$(B_{liq}$ - 1/C,	T - K,	$density_{liq}$ - g/cm^3)	
			a	T_C	m	TMIN	TMAX	B_{liq} @ 25 C	$density_{liq}$ @ 25 C
703	C7H14	CYCLOHEPTANE	6.1560E-04	604.30	-0.7143	265.15	574.09	1.0010E-03	0.806
704	C7H14	1,1-DIMETYLCYCLOPENTANE	7.0730E-04	547.00	-0.6984	203.36	519.65	1.2260E-03	0.750
705	C7H14	cis-1,2-DIMETHYLCYCLOPENTANE	6.6230E-04	565.15	-0.7143	219.26	536.89	1.1320E-03	0.768
706	C7H14	trans-1,2-DIMETHYLCYCLOPENTANE	7.1700E-04	553.15	-0.6902	155.58	525.49	1.2240E-03	0.747
707	C7H14	cis-1,3-DIMETHYLCYCLOPENTANE	7.2670E-04	551.00	-0.6860	139.45	523.45	1.2400E-03	0.740
708	C7H14	trans-1,3-DIMETHYLCYCLOPENTANE	7.0030E-04	553.00	-0.6953	139.18	525.35	1.2000E-03	0.744
709	C7H14	ETHYLCYCLOPENTANE	6.3960E-04	569.52	-0.7223	134.71	541.04	1.0930E-03	0.763
710	C7H14	2-ETHYL-1-PENTENE	6.3580E-04	543.00	-0.7337	168.00	515.85	1.1410E-03	0.704
711	C7H14	3-ETHYL-1-PENTENE	5.7500E-04	530.00	-0.7580	145.67	503.50	1.0760E-03	0.692
712	C7H14	1-HEPTENE	6.8150E-04	537.29	-0.7258	154.27	510.43	1.2260E-03	0.693
713	C7H14	cis-2-HEPTENE	7.3150E-04	549.00	-0.7114	164.00	521.55	1.2770E-03	0.703
714	C7H14	trans-2-HEPTENE	7.0220E-04	543.00	-0.7133	163.67	515.85	1.2390E-03	0.697
715	C7H14	cis-3-HEPTENE	7.0170E-04	545.00	-0.7206	136.51	517.75	1.2420E-03	0.698
716	C7H14	trans-3-HEPTENE	7.0150E-04	540.00	-0.7143	136.52	513.00	1.2450E-03	0.694
717	C7H14	METHYLCYCLOHEXANE	6.6900E-04	572.19	-0.7073	146.58	543.58	1.1260E-03	0.766
718	C7H14	2-METHYL-1-HEXENE	7.1550E-04	538.00	-0.7079	170.28	511.10	1.2680E-03	0.698
719	C7H14	3-METHYL-1-HEXENE	6.9230E-04	528.00	-0.7178	145.00	501.60	1.2580E-03	0.687
720	C7H14	4-METHYL-1-HEXENE	6.9310E-04	534.00	-0.7163	131.70	507.30	1.2450E-03	0.694
721	C7H14	2,3,3-TRIMETHYL-1-BUTENE	7.1150E-04	531.00	-0.7040	163.30	504.45	1.2710E-03	0.701
722	C7H14O	DIISOPROPYL KETONE	6.6040E-04	576.00	-0.7382	204.81	547.20	1.1310E-03	0.912
723	C7H14O	2-HEPTANONE	6.6850E-04	611.55	-0.6955	238.15	580.97	1.0640E-03	0.811
724	C7H14O	1-HEPTANAL	6.4220E-04	603.00	-0.7105	230.15	572.85	1.0430E-03	0.813
725	C7H14O	1-METHYLCYCLOHEXANOL	7.7820E-04	603.00	-0.6867	299.15	572.85	--------	------
726	C7H14O	cis-2-METHYLCYCLOHEXANOL	5.4970E-04	614.00	-0.7631	280.15	583.30	9.1300E-04	0.932
727	C7H14O	trans-2-METHYLCYCLOHEXANOL	5.1430E-04	616.00	-0.7738	269.15	585.20	8.5810E-04	0.921
728	C7H14O	cis-3-METHYLCYCLOHEXANOL	6.5620E-04	618.00	-0.7182	267.65	587.10	1.0530E-03	0.911
729	C7H14O	trans-3-METHYLCYCLOHEXANOL	4.9430E-04	617.00	-0.7805	272.65	586.15	8.2750E-04	0.918
730	C7H14O	cis-4-METHYLCYCLOHEXANOL	5.9970E-04	622.00	-0.7143	294.85	590.90	9.5590E-04	0.913
731	C7H14O	trans-4-METHYLCYCLOHEXANOL	5.6520E-04	622.00	-0.7495	293.00	590.90	9.2180E-04	0.908
732	C7H14O	5-METHYL-2-HEXANONE	6.4630E-04	601.00	-0.7143	199.25	570.95	1.0540E-03	0.808
733	C7H14O2	n-BUTYL PROPIONATE	6.5280E-04	594.00	-0.7143	183.63	564.30	1.0740E-03	0.872
734	C7H14O2	ETHYL ISOVALERATE	6.5160E-04	587.95	-0.7143	173.85	558.55	1.0800E-03	0.865
735	C7H14O2	ISOPENTYL ACETATE	6.6700E-04	599.00	-0.6991	194.65	569.05	1.0800E-03	0.867
736	C7H14O2	n-PENTYL ACETATE	6.5150E-04	598.00	-0.7068	202.35	568.10	1.0610E-03	0.872
737	C7H14O2	n-PROPYL n-BUTYRATE	6.6710E-04	594.00	-0.7021	177.95	564.30	1.0880E-03	0.868
738	C7H14O2	n-HEPTANOIC ACID	6.0740E-04	680.00	-0.6941	265.83	646.00	9.0660E-04	0.913
739	C7H14O3	ETHYL-3-ETHOXYPROPIONATE	7.1690E-04	609.00	-0.6785	223.00	578.55	1.1310E-03	0.945
740	C7H15Br	1-BROMOHEPTANE	5.7510E-04	651.00	-0.7143	217.05	618.45	8.9070E-04	1.135
741	C7H15N	N-METHYLCYCLOHEXYLAMINE	6.0930E-04	622.00	-0.7143	264.65	590.90	9.7110E-04	0.865
742	C7H16	2,2-DIMETHYLPENTANE	6.9860E-04	520.50	-0.7170	149.34	494.48	1.2850E-03	0.673
743	C7H16	2,3-DIMETHYLPENTANE	6.3170E-04	537.35	-0.7287	160.00	510.48	1.1390E-03	0.691
744	C7H16	2,4-DIMETHYLPENTANE	6.6880E-04	519.79	-0.7305	153.91	493.80	1.2470E-03	0.668
745	C7H16	3,3-DIMETHYLPENTANE	7.5260E-04	536.40	-0.6916	138.70	509.58	1.3190E-03	0.687
746	C7H16	3-ETHYLPENTANE	7.1690E-04	540.64	-0.7086	154.55	513.61	1.2650E-03	0.695
747	C7H16	n-HEPTANE	6.9550E-04	540.26	-0.7209	182.57	513.25	1.2410E-03	0.682
748	C7H16	2-METHYLHEXANE	6.8940E-04	530.37	-0.7210	154.90	503.85	1.2500E-03	0.674
749	C7H16	3-METHYLHEXANE	6.2320E-04	535.25	-0.7375	153.75	508.49	1.1360E-03	0.684
750	C7H16	2,2,3-TRIMETHYLBUTANE	6.7250E-04	531.17	-0.7200	248.57	504.61	1.2170E-03	0.687
751	C7H16O	1-HEPTANOL	5.7670E-04	631.90	-0.7270	239.15	600.31	9.1730E-04	0.820
752	C7H16O	2-HEPTANOL	5.7100E-04	588.00	-0.7466	243.00	558.60	9.6820E-04	0.814
753	C7H16O	5-METHYL-1-HEXANOL	7.5730E-04	605.00	-0.6688	293.15	574.75	1.1930E-03	0.812
754	C7H16O	ISOPROPYL-TERT-BUTYL-ETHER	6.8750E-04	558.21	-0.7143	177.80	530.30	1.1860E-03	0.750
755	C7H16S	n-HEPTYL MERCAPTAN	6.1400E-04	645.00	-0.6967	229.92	612.75	9.4600E-04	0.839
756	C7H16S	BUTYL-PROPYL-SULFIDE	5.9060E-04	653.50	-0.7143	206.66	620.83	9.1270E-04	0.839
757	C7H16S	ETHYL-PENTYL-SULFIDE	6.0460E-04	638.37	-0.7143	206.66	606.45	9.4780E-04	0.839
758	C7H16S	HEXYL-METHYL-SULFIDE	6.0460E-04	638.37	-0.7143	206.66	606.45	9.4780E-04	0.839
759	C7H17N	1-AMINOHEPTANE	6.0220E-04	607.00	-0.7341	254.15	576.65	9.8890E-04	0.772
760	C8H4Cl2O2	ISOPHTHALOYL CHLORIDE	5.1550E-04	768.00	-0.7143	317.00	729.60	--------	------
761	C8H4O3	PHTHALIC ANHYDRIDE	4.3310E-04	791.00	-0.7654	404.26	751.45	--------	------
762	C8H6	ETHYNYLBENZENE	5.6590E-04	655.43	-0.7143	242.53	622.66	8.7300E-04	0.901
763	C8H6O4	ISOPHTHALIC ACID	4.5660E-04	1007.00	-0.7143	619.15	956.65	--------	------
764	C8H6O4	PHTHALIC ACID	4.9230E-04	800.00	-0.7143	464.15	760.00	--------	------
765	C8H6O4	TEREPHTHALIC ACID	4.3880E-04	1113.00	-0.7143	700.15	1057.35	--------	------
766	C8H6S	BENZOTHIOPHENE	5.9830E-04	754.00	-0.6670	304.50	716.30	--------	------
767	C8H7N	INDOLE	4.5560E-04	790.00	-0.7676	273.68	750.50	6.5550E-04	1.102
768	C8H8	STYRENE	5.8860E-04	648.00	-0.7143	242.54	615.60	9.1420E-04	0.900
769	C8H8	1,3,5,7-CYCLOOCTATETRAENE	5.8550E-04	642.55	-0.7143	266.16	610.42	9.1410E-04	0.907
770	C8H8O	ACETOPHENONE	5.6700E-04	701.00	-0.7094	293.65	665.95	8.3990E-04	1.024
771	C8H8O	p-TOLUALDEHYDE	5.7140E-04	698.00	-0.7143	298.85	663.10	--------	------
772	C8H8O2	METHYL BENZOATE	5.5890E-04	693.00	-0.7324	260.75	658.35	8.4380E-04	1.085
773	C8H8O2	o-TOLUIC ACID	5.3220E-04	751.00	-0.7143	376.85	713.45	--------	------
774	C8H8O2	p-TOLUIC ACID	5.3060E-04	773.00	-0.7143	452.75	734.35	--------	------
775	C8H8O3	METHYL SALICYLATE	4.9600E-04	701.00	-0.7388	265.15	665.95	7.4690E-04	1.175
776	C8H8O3	VANILLIN	4.9820E-04	777.00	-0.7143	355.00	738.15	--------	------
777	C8H9NO	ACETANILIDE	4.9860E-04	825.00	-0.7143	386.65	783.75	--------	------
778	C8H10	ETHYLBENZENE	6.2960E-04	617.17	-0.7079	178.20	586.31	1.0050E-03	0.865
779	C8H10	m-XYLENE	5.9600E-04	617.05	-0.7276	225.30	586.20	9.6340E-04	0.861
780	C8H10	o-XYLENE	5.8440E-04	630.37	-0.7259	247.98	598.85	9.3020E-04	0.876

$$B_{liq} = a \ (1 - T/T_C)^m \qquad (B_{liq} - 1/C, \ T - K, \ density_{liq} - g/cm^3)$$

NO	FORMULA	NAME	a	T_C	m	TMIN	TMAX	B_{liq} @ 25 C	density$_{liq}$ @ 25 C
781	C8H10	p-XYLENE	6.0980E-04	616.26	-0.7210	286.41	585.45	9.8230E-04	0.858
782	C8H10O	m-ETHYLPHENOL	--------	------	------	------	------	--------	------
783	C8H10O	p-ETHYLPHENOL	5.2320E-04	716.45	-0.7143	318.23	680.63	--------	------
784	C8H10O	PHENETOLE	6.0390E-04	647.15	-0.7143	243.63	614.79	9.3860E-04	0.961
785	C8H10O	2-PHENYLETHANOL	6.1540E-04	684.00	-0.6972	247.00	649.80	9.1740E-04	1.016
786	C8H10O	2,3-XYLENOL	4.8510E-04	722.95	-0.7143	345.71	686.80	--------	------
787	C8H10O	2,4-XYLENOL	4.7350E-04	707.65	-0.7513	297.68	672.27	7.1420E-04	1.015
788	C8H10O	2,5-XYLENOL	5.2740E-04	707.05	-0.7030	347.99	671.70	--------	------
789	C8H10O	2,6-XYLENOL	5.0730E-04	701.05	-0.7143	318.76	666.00	--------	------
790	C8H10O	3,4-XYLENOL	4.8080E-04	729.95	-0.7193	338.25	693.45	--------	------
791	C8H10O	3,5-XYLENOL	4.9370E-04	715.65	-0.7737	336.59	679.87	--------	------
792	C8H11N	N,N-DIMETHYLANILINE	5.0140E-04	687.15	-0.7665	275.60	652.79	7.7550E-04	0.949
793	C8H11N	o-ETHYLANILINE	5.5480E-04	704.00	-0.7143	226.55	668.80	8.2220E-04	0.977
794	C8H11N	2,4,6-TRIMETHYLPYRIDINE	5.9630E-04	653.00	-0.7143	229.00	620.35	9.2190E-04	0.913
795	C8H11NO	p-PHENETIDINE	5.2890E-04	754.00	-0.7143	277.00	716.30	7.5760E-04	1.057
796	C8H12	1,5-CYCLOOCTADIENE	7.2760E-04	645.00	-0.6525	203.98	612.75	1.0910E-03	0.878
797	C8H12	VINYLCYCLOHEXENE	6.3010E-04	599.00	-0.7143	164.00	569.05	1.0300E-03	0.826
798	C8H12O4	1,4-CYCLOHEXANEDICARBOXYLIC ACID	4.9400E-04	889.00	-0.7143	585.65	844.55	--------	------
799	C8H12O4	DIETHYL MALEATE	6.1960E-04	680.00	-0.6664	264.35	646.00	9.1020E-04	0.961
800	C8H14O2	n-BUTYL METHACRYLATE	6.3470E-04	616.00	-0.7143	223.00	585.20	1.0180E-03	0.891
801	C8H14O3	BUTYRIC ANHYDRIDE	6.0490E-04	639.00	-0.7143	199.85	607.05	9.4770E-04	0.960
802	C8H14O4	DIETHYL SUCCINATE	6.0490E-04	660.00	-0.7048	252.35	627.00	9.2390E-04	1.036
803	C8H16	CYCLOOCTANE	5.7640E-04	647.20	-0.7143	287.60	614.84	8.9590E-04	0.830
804	C8H16	1,1-DIMETHYLCYCLOHEXANE	6.4830E-04	591.15	-0.7224	239.66	561.59	1.0760E-03	0.777
805	C8H16	cis-1,2-DIMETHYLCYCLOHEXANE	6.2460E-04	606.15	-0.7319	223.16	575.84	1.0250E-03	0.792
806	C8H16	trans-1,2-DIMETHYLCYCLOHEXANE	6.1760E-04	596.15	-0.7342	184.99	566.34	1.0280E-03	0.772
807	C8H16	cis-1,3-DIMETHYLCYCLOHEXANE	6.5480E-04	591.15	-0.7161	197.58	561.59	1.0820E-03	0.762
808	C8H16	trans-1,3-DIMETHYLCYCLOHEXANE	6.5490E-04	598.00	-0.7220	183.07	568.10	1.0780E-03	0.781
809	C8H16	cis-1,4-DIMETHYLCYCLOHEXANE	6.5100E-04	598.15	-0.7229	185.72	568.24	1.0720E-03	0.779
810	C8H16	trans-1,4-DIMETHYLCYCLOHEXANE	6.3600E-04	590.15	-0.7225	236.21	560.64	1.0570E-03	0.759
811	C8H16	ETHYLCYCLOHEXANE	6.0210E-04	609.15	-0.7325	161.84	578.69	9.8520E-04	0.784
812	C8H16	2-ETHYL-1-HEXENE	7.2220E-04	574.00	-0.6581	173.00	545.30	1.1700E-03	0.723
813	C8H16	1-METHYL-1-ETHYLCYCLOPENTANE	6.5090E-04	582.00	-0.7143	129.35	552.90	1.0870E-03	0.777
814	C8H16	1-OCTENE	6.8610E-04	566.60	-0.7143	171.45	538.27	1.1700E-03	0.711
815	C8H16	trans-2-OCTENE	6.7560E-04	577.00	-0.7143	185.45	548.15	1.1360E-03	0.716
816	C8H16	trans-3-OCTENE	6.7650E-04	574.00	-0.7143	163.15	545.30	1.1420E-03	0.711
817	C8H16	trans-4-OCTENE	6.7690E-04	573.00	-0.7143	179.37	544.35	1.1440E-03	0.710
818	C8H16	n-PROPYLCYCLOPENTANE	6.3030E-04	603.00	-0.7097	155.82	572.85	1.0230E-03	0.773
819	C8H16	2,4,4-TRIMETHYL-1-PENTENE	6.8950E-04	553.00	-0.7143	179.70	525.35	1.1990E-03	0.711
820	C8H16	2,4,4-TRIMETHYL-2-PENTENE	6.8770E-04	558.00	-0.7143	166.84	530.10	1.1870E-03	0.717
821	C8H16O	2-ETHYLHEXANAL	6.5160E-04	607.00	-0.7143	200.00	576.65	1.0560E-03	0.819
822	C8H16O	1-OCTANAL	5.2290E-04	621.00	-0.7504	246.00	589.95	8.5420E-04	0.816
823	C8H16O	2-OCTANONE	6.2850E-04	624.00	-0.7037	252.85	592.80	9.9280E-04	0.815
824	C8H16O2	n-BUTYL n-BUTYRATE	5.8590E-04	616.00	-0.7235	181.15	585.20	9.4570E-04	0.866
825	C8H16O2	n-HEXYL ACETATE	6.3980E-04	618.00	-0.7019	192.25	587.10	1.0160E-03	0.868
826	C8H16O2	ISOBUTYL ISOBUTYRATE	6.3200E-04	602.00	-0.7143	192.45	571.90	1.0300E-03	0.852
827	C8H16O2	n-OCTANOIC ACID	5.3510E-04	692.00	-0.7198	289.65	657.40	8.0280E-04	0.903
828	C8H16O4	DIETHYLENE GLYCOL ETHYL ETHER ACETATE	6.0180E-04	660.00	-0.7143	248.15	627.00	9.2450E-04	1.006
829	C8H18	2,2-DIMETHYLHEXANE	6.7310E-04	549.80	-0.7223	151.97	522.31	1.1840E-03	0.692
830	C8H18	2,3-DIMETHYLHEXANE	6.2960E-04	563.40	-0.7305	272.04	535.23	1.0920E-03	0.708
831	C8H18	2,4-DIMETHYLHEXANE	7.0060E-04	553.50	-0.7015	272.04	525.83	1.2050E-03	0.693
832	C8H18	2,5-DIMETHYLHEXANE	6.7500E-04	550.00	-0.7218	182.00	522.50	1.1860E-03	0.690
833	C8H18	3,3-DIMETHYLHEXANE	5.6540E-04	562.00	-0.7432	147.05	533.90	9.9180E-04	0.707
834	C8H18	3,4-DIMETHYLHEXANE	6.2340E-04	568.80	-0.7324	272.04	540.36	1.0740E-03	0.716
835	C8H18	3-ETHYLHEXANE	6.0280E-04	565.40	-0.7343	272.04	537.13	1.0450E-03	0.710
836	C8H18	3-ETHYL-2-METHYLPENTANE	6.9060E-04	567.00	-0.7143	158.20	538.65	1.1770E-03	0.711
837	C8H18	3-METHYL-3-ETHYLPENTANE	6.3520E-04	576.50	-0.7183	182.28	547.68	1.0720E-03	0.724
838	C8H18	2-METHYLHEPTANE	6.5490E-04	559.64	-0.7287	164.16	531.66	1.1400E-03	0.696
839	C8H18	3-METHYLHEPTANE	6.2240E-04	563.67	-0.7287	152.60	535.49	1.0770E-03	0.702
840	C8H18	4-METHYLHEPTANE	6.8150E-04	561.74	-0.7168	152.20	533.65	1.1720E-03	0.713
841	C8H18	n-OCTANE	6.4760E-04	568.83	-0.7306	216.38	540.39	1.1140E-03	0.699
842	C8H18	2,2,3-TRIMETHYLPENTANE	5.9940E-04	563.50	-0.7266	160.89	535.33	1.0360E-03	0.712
843	C8H18	2,2,4-TRIMETHYLPENTANE	6.7790E-04	543.96	-0.7154	165.78	516.76	1.1970E-03	0.690
844	C8H18	2,3,3-TRIMETHYLPENTANE	6.1800E-04	573.50	-0.7259	172.22	544.83	1.0530E-03	0.722
845	C8H18	2,3,4-TRIMETHYLPENTANE	6.1570E-04	566.30	-0.7353	163.95	537.99	1.0670E-03	0.716
846	C8H18	2,2,3,3-TETRAMETHYLBUTANE	6.4060E-04	567.80	-0.7143	372.47	539.41	1.0900E-03	0.692
847	C8H18O	DI-n-BUTYL ETHER	6.6270E-04	581.00	-0.7143	177.95	551.95	1.1080E-03	0.764
848	C8H18O	DI-sec-BUTYL ETHER	6.7160E-04	559.00	-0.7143	173.15	531.05	1.1580E-03	0.759
849	C8H18O	DI-tert-BUTYL ETHER	6.8000E-04	550.00	-0.7143	195.00	522.50	1.1880E-03	0.760
850	C8H18O	2-ETHYL-1-HEXANOL	5.8130E-04	640.25	-0.7227	203.15	608.24	9.1440E-04	0.830
851	C8H18O	1-OCTANOL	5.7780E-04	652.50	-0.7191	257.65	619.88	8.9630E-04	0.823
852	C8H18O	2-OCTANOL	6.0030E-04	637.15	-0.7057	241.55	605.29	9.3700E-04	0.817
853	C8H18O2	DI-t-BUTYL PEROXIDE	6.6600E-04	547.00	-0.7143	233.15	519.65	1.1690E-03	0.790
854	C8H18O2S	DI-n-BUTYL SULFONE	5.3900E-04	767.00	-0.6948	318.00	728.65	--------	------
855	C8H18O3	DIETHYLENE GLYCOL DIETHYL ETHER	6.2560E-04	624.00	-0.7143	228.85	592.80	9.9500E-04	0.904
856	C8H18O3	DIETHYLENE GLYCOL MONOBUTYL EHTER	5.9880E-04	654.00	-0.7143	205.15	621.30	9.2490E-04	0.952
857	C8H18O4	TRIETHYLENE GLYCOL DIMETHYL ETHER	8.4350E-04	651.00	-0.6082	229.35	618.45	1.2240E-03	0.980
858	C8H18O5	TETRAETHYLENE GLYCOL	5.5230E-04	722.00	-0.7143	268.15	685.90	8.0790E-04	1.122

			$B_{liq} = a \ (1 - T/T_C)^m$			$(B_{liq} - 1/C, \ T - K, \ density_{liq} - g/cm^3)$			
NO	FORMULA	NAME	a	T_C	m	TMIN	TMAX	B_{liq} @ 25 C	$density_{liq}$ @ 25 C
859	C8H18S	n-OCTYL MERCAPTAN	5.8830E-04	664.00	-0.7002	223.95	630.80	8.9300E-04	0.840
860	C8H18S	tert-OCTYL MERCAPTAN	6.0920E-04	627.00	-0.7143	199.00	595.65	9.6600E-04	0.841
861	C8H18S	BUTYL-SULFIDE	5.9210E-04	650.00	-0.7143	209.86	617.50	9.1790E-04	0.840
862	C8H18S	ETHYL-HEXYL-SULFIDE	5.9600E-04	660.72	-0.7143	209.86	627.68	9.1500E-04	0.840
863	C8H18S	HEPTYL-METHYL-SULFIDE	5.9600E-04	660.72	-0.7143	209.86	627.68	9.1500E-04	0.840
864	C8H18S	PENTYL-PROPYL-SULFIDE	5.9600E-04	660.72	-0.7143	209.86	627.68	9.1500E-04	0.840
865	C8H18S2	BUTYL-DISULFIDE	5.3880E-04	704.16	-0.7143	202.16	668.95	7.9850E-04	0.934
866	C8H19N	DI-n-BUTYLAMINE	6.3000E-04	607.50	-0.7173	211.15	577.13	1.0220E-03	0.757
867	C8H19N	DIISOBUTYLAMINE	6.2730E-04	580.00	-0.7280	203.15	551.00	1.0610E-03	0.743
868	C8H19N	n-OCTYLAMINE	6.2180E-04	627.00	-0.7143	272.75	595.65	9.8590E-04	0.779
869	C8H23N5	TETRAETHYLENEPENTAMINE	5.1170E-04	774.00	-0.7143	243.00	735.30	7.2430E-04	0.994
870	C8H24O4Si4	OCTAMETHYLCYCLOTETRASILOXANE	6.5020E-04	586.50	-0.7231	290.80	557.18	1.0860E-03	0.949
871	C9H4O5	TRIMELLITIC ANHYDRIDE	4.3870E-04	890.00	-0.7143	438.15	845.50	--------	------
872	C9H6N2O2	TOLUENE DIISOCYANATE	2.7400E-04	737.00	-0.8561	287.04	700.15	4.2700E-04	1.220
873	C9H7N	ISOQUINOLINE	4.8490E-04	803.15	-0.7211	299.45	762.99	--------	------
874	C9H7N	QUINOLINE	4.6500E-04	782.15	-0.7638	258.25	743.04	6.7090E-04	1.090
875	C9H7NO	8-HYDROXYQUINOLINE	4.7380E-04	788.00	-0.7100	346.00	748.60	--------	------
876	C9H8	INDENE	6.1840E-04	687.00	-0.6900	271.70	652.65	9.1580E-04	0.994
877	C9H8O	2-METHYLBENZOFURAN	5.6110E-04	698.00	-0.7143	290.00	663.10	8.3530E-04	1.051
878	C9H10	INDANE	5.9250E-04	684.90	-0.6978	221.74	650.66	8.8270E-04	0.960
879	C9H10	cis-PROPENYLBENZENE	5.8240E-04	664.60	-0.7143	211.47	631.37	8.9110E-04	0.904
880	C9H10	trans-PROPENYLBENZENE	5.8240E-04	664.60	-0.7143	243.82	631.37	8.9110E-04	0.902
881	C9H1O	alpha-METHYLSTYRENE	6.3450E-04	654.00	-0.7049	249.95	621.30	9.7440E-04	0.905
882	C9H10	m-METHYLSTYRENE	6.0130E-04	657.00	-0.7093	186.81	624.15	9.2340E-04	0.908
883	C9H10	o-METHYLSTYRENE	6.0470E-04	659.00	-0.7069	204.58	626.05	9.2560E-04	0.908
884	C9H10	p-METHYLSTYRENE	7.7230E-04	665.00	-0.6534	239.02	631.75	1.1390E-03	0.916
885	C9H10O2	BENZYL ACETATE	6.5460E-04	699.00	-0.6664	221.65	664.05	9.4820E-04	1.045
886	C9H10O2	ETHYL BENZOATE	5.8450E-04	698.00	-0.7151	238.45	663.10	8.7060E-04	1.042
887	C9H10O3	ETHYL VANILLIN	5.3570E-04	748.00	-0.7143	350.65	710.60	--------	------
888	C9H11NO	p-DIMETHYLAMINOBENZALDEHYDE	5.1100E-04	832.00	-0.7143	348.00	790.40	--------	------
889	C9H12	CUMENE	6.1580E-04	631.15	-0.7100	177.14	599.59	9.6960E-04	0.860
890	C9H12	m-ETHYLTOLUENE	6.0220E-04	637.15	-0.7404	177.61	605.29	9.6090E-04	0.860
891	C9H12	o-ETHYLTOLUENE	5.9400E-04	651.15	-0.7293	192.35	618.59	9.2840E-04	0.877
892	C9H12	p-ETHYLTOLUENE	6.0930E-04	640.15	-0.7280	210.83	608.14	9.6170E-04	0.857
893	C9H12	MESITYLENE	5.9290E-04	637.36	-0.7202	228.46	605.49	9.3390E-04	0.861
894	C9H12	n-PROPYLBENZENE	6.2770E-04	638.38	-0.7087	173.67	606.46	9.8050E-04	0.860
895	C9H12	1,2,3-TRIMETHYLBENZENE	5.1350E-04	664.53	-0.7392	247.79	631.30	7.9740E-04	0.891
896	C9H12	1,2,4-TRIMETHYLBENZENE	5.7620E-04	649.13	-0.7228	229.38	616.67	8.9860E-04	0.872
897	C9H12O	BENZYL ETHYL ETHER	5.2170E-04	662.00	-0.7368	275.65	628.90	8.1080E-04	0.945
898	C9H12O	2-PHENYL-2-PROPANOL	5.3630E-04	660.00	-0.7398	309.15	627.00		------
899	C9H12O2	CUMENE HYDROPEROXIDE	6.0520E-04	605.00	-0.7140	264.26	574.75	9.8260E-04	1.043
900	C9H14O	ISOPHORONE	5.3250E-04	715.00	-0.7143	265.05	679.25	7.8280E-04	0.920
901	C9H14O6	GLYCERYL TRIACETATE	5.6860E-04	704.00	-0.7143	277.25	668.80	8.4280E-04	1.158
902	C9H16	1-NONYNE	6.2300E-04	610.81	-0.7143	223.16	580.27	1.0050E-03	0.752
903	C9H16O4	AZELAIC ACID	5.1530E-04	811.00	-0.7143	379.65	770.45	--------	------
904	C9H18	BUTYLCYCLOPENTANE	6.2280E-04	625.05	-0.7143	165.18	593.80	9.8960E-04	0.781
905	C9H18	cis,cis-1,3,5-TRIMETHYLCYCLOHEXANE	6.5730E-04	607.86	-0.7143	223.46	577.47	1.0640E-03	0.767
906	C9H18	cis,trans-1,3,5-TRIMETHYLCYCLOHEXANE	6.6730E-04	602.20	-0.7143	188.76	572.09	1.0870E-03	0.718
907	C9H18	ISOPROPYLCYCLOHEXANE	6.2020E-04	627.00	-0.7020	183.76	595.65	9.7560E-04	0.798
908	C9H18	1-NONENE	6.6150E-04	593.25	-0.7143	191.78	563.59	1.0890E-03	0.725
909	C9H18	n-PROPYLCYCLOHEXANE	6.2030E-04	639.15	-0.6998	178.28	607.19	9.6290E-04	0.790
910	C9H18O	DIISOBUTYL KETONE	5.9480E-04	615.00	-0.7187	227.17	584.25	9.5810E-04	0.802
911	C9H18O	1-NONANAL	5.0750E-04	640.00	-0.7499	255.15	608.00	8.1220E-04	0.819
912	C9H18O2	n-BUTYL VALERATE	6.2340E-04	629.00	-0.7035	180.35	597.55	9.7960E-04	0.863
913	C9H18O2	n-NONANOIC ACID	6.0880E-04	703.00	-0.6839	285.55	667.85	8.8800E-04	0.902
914	C9H18O2	n-OCTYL FORMATE	6.3870E-04	645.00	-0.6911	234.05	612.75	9.8060E-04	0.870
915	C9H20	3,3-DIETHYLPENTANE	5.9570E-04	610.05	-0.7066	240.12	579.55	9.5700E-04	0.750
916	C9H20	2,2-DIMETHYL-3-ETHYLPENTANE	6.3590E-04	590.00	-0.7143	173.68	560.50	1.0510E-03	0.731
917	C9H20	3-ETHYL-2,3-DIMETHYLPENTANE	6.4530E-04	606.80	-0.7143	173.67	576.46	1.0460E-03	0.751
918	C9H20	2,4-DIMETHYL-3-ETHYLPENTANE	6.4040E-04	591.00	-0.7143	150.79	561.45	1.0570E-03	0.734
919	C9H20	2,2-DIMETHYLHEPTANE	6.6250E-04	576.80	-0.7062	160.15	547.96	1.1080E-03	0.707
920	C9H20	2,6-DIMETHYLHEPTANE	6.5570E-04	579.00	-0.7077	170.25	550.05	1.0940E-03	0.706
921	C9H20	3-ETHYLHEPTANE	6.2740E-04	590.00	-0.7210	158.25	560.50	1.0420E-03	0.723
922	C9H20	4-ETHYLHEPTANE	6.6750E-04	584.95	-0.7143	159.96	555.70	1.1110E-03	0.724
923	C9H20	2,3-DIMETHYLHEPTANE	6.6790E-04	589.60	-0.7143	160.16	560.12	1.1050E-03	0.722
924	C9H20	2,4-DIMETHYLHEPTANE	6.8270E-04	576.80	-0.7143	160.16	547.96	1.1480E-03	0.711
925	C9H20	2,5-DIMETHYLHEPTANE	6.7380E-04	581.10	-0.7143	160.16	552.05	1.1270E-03	0.713
926	C9H20	3,4-DIMETHYLHEPTANE	6.6530E-04	591.90	-0.7143	170.26	562.31	1.0970E-03	0.727
927	C9H20	3,5-DIMETHYLHEPTANE	6.7330E-04	583.20	-0.7143	170.26	554.04	1.1230E-03	0.719
928	C9H20	4,4-DIMETHYLHEPTANE	6.7660E-04	585.40	-0.7143	170.26	556.13	1.1250E-03	0.721
929	C9H20	3-ETHYL-2-METHYLHEXANE	6.7540E-04	588.10	-0.7143	160.16	558.70	1.1190E-03	0.729
930	C9H20	4-ETHYL-2-METHYLHEXANE	6.8090E-04	580.00	-0.7143	160.16	551.00	1.1400E-03	0.719
931	C9H20	3-ETHYL-3-METHYLHEXANE	6.2690E-04	597.50	-0.7143	160.16	567.63	1.0860E-03	0.737
932	C9H20	3-ETHYL-4-METHYLHEXANE	6.6910E-04	593.70	-0.7143	160.16	564.02	1.1010E-03	0.736
933	C9H20	2,2,3-TRIMETHYLHEXANE	6.6590E-04	588.00	-0.7143	153.16	558.60	1.1040E-03	0.725
934	C9H20	2,2,4-TRIMETHYLHEXANE	6.8470E-04	573.50	-0.7143	153.00	544.83	1.1560E-03	0.713
935	C9H20	2,3,3-TRIMETHYLHEXANE	6.5890E-04	596.00	-0.7143	156.36	566.20	1.0810E-03	0.734
936	C9H20	2,3,4-TRIMETHYLHEXANE	6.6240E-04	594.50	-0.7143	156.36	564.78	1.0890E-03	0.735

631

NO	FORMULA	NAME	$B_{liq} = a \ (1 - T/T_C)^m$					$(B_{liq} - 1/C, \ T - K, \ density_{liq} - g/cm^3)$	
			a	T_C	m	TMIN	TMAX	B_{liq} @ 25 C	$density_{liq}$ @ 25 C
937	C9H20	2,3,5-TRIMETHYLHEXANE	6.7600E-04	579.20	-0.7143	145.36	550.24	1.1330E-03	0.718
938	C9H20	2,4,4-TRIMETHYLHEXANE	6.7920E-04	581.50	-0.7143	159.78	552.43	1.1350E-03	0.720
939	C9H20	3,3,4-TRIMETHYLHEXANE	6.5010E-04	602.30	-0.7143	171.96	572.19	1.0590E-03	0.741
940	C9H20	2-METHYLOCTANE	6.6130E-04	586.75	-0.7143	192.78	557.41	1.0980E-03	0.710
941	C9H20	3-METHYLOCTANE	6.2650E-04	590.15	-0.7199	165.55	560.64	1.0400E-03	0.717
942	C9H20	4-METHYLOCTANE	6.7620E-04	587.65	-0.7002	159.95	558.27	1.1100E-03	0.716
943	C9H20	n-NONANE	6.5440E-04	595.65	-0.7143	219.63	565.87	1.0740E-03	0.715
944	C9H20	2,2,3,3-TETRAMETHYLPENTANE	6.1240E-04	610.85	-0.7029	263.26	580.31	9.8040E-04	0.753
945	C9H20	2,2,3,4-TETRAMETHYLPENTANE	6.2480E-04	592.15	-0.7080	152.06	562.54	1.0260E-03	0.735
946	C9H20	2,2,4,4-TETRAMETHYLPENTANE	6.4000E-04	571.35	-0.7137	206.95	542.78	1.0840E-03	0.716
947	C9H20	2,3,3,4-TETRAMETHYLPENTANE	6.6650E-04	592.60	-0.7143	152.00	562.97	1.0980E-03	0.735
948	C9H20	2,2,5-TRIMETHYLHEXANE	6.8570E-04	568.05	-0.7046	167.39	539.65	1.1580E-03	0.707
949	C9H20O	2,6-DIMETHYL-4-HEPTANOL	6.2320E-04	603.00	-0.7143	208.00	572.85	1.0150E-03	0.807
950	C9H20O	1-NONANOL	5.1930E-04	673.00	-0.7369	268.15	639.35	7.9930E-04	0.824
951	C9H20O	2-NONANOL	5.8290E-04	623.00	-0.7278	238.15	591.85	9.3630E-04	0.820
952	C9H20S	n-NONYL MERCAPTAN	5.8730E-04	681.00	-0.6943	253.05	646.95	8.7600E-04	0.841
953	C9H20S	BUTYL-PENTYL-SULFIDE	5.8790E-04	681.56	-0.7143	231.16	647.48	8.8670E-04	0.840
954	C9H20S	ETHYL-HEPTYL-SULFIDE	5.8790E-04	681.56	-0.7143	231.16	647.48	8.8670E-04	0.840
955	C9H20S	HEXYL-PROPYL-SULFIDE	5.8790E-04	681.56	-0.7143	231.16	647.48	8.8670E-04	0.840
956	C9H20S	METHYL-OCTYL-SULFIDE	5.8790E-04	681.56	-0.7143	231.16	647.48	8.8670E-04	0.840
957	C9H21N	n-NONYLAMINE	6.0600E-04	648.00	-0.7143	273.15	615.60	9.4120E-04	0.785
958	C9H21N	TRIPROPYLAMINE	5.6070E-04	577.50	-0.7550	179.65	548.63	9.7020E-04	0.754
959	C10H6O8	PYROMELLITIC ACID	4.4740E-04	893.00	-0.7143	554.00	848.35	--------	------
960	C10H7Br	1-BROMONAPHTHALENE	4.9450E-04	824.00	-0.6970	279.35	782.80	6.7630E-04	1.478
961	C10H7Cl	1-CHLORONAPHTHALENE	5.3070E-04	785.00	-0.6854	269.15	745.75	7.3640E-04	1.171
962	C10H8	NAPHTHALENE	5.0520E-04	748.35	-0.7270	353.43	710.93	--------	------
963	C10H8	AZULENE	5.1500E-04	773.48	-0.7143	173.66	734.81	7.2930E-04	1.053
964	C10H9N	QUINALDINE	5.4800E-04	773.00	-0.7143	272.15	734.35	7.7610E-04	1.055
965	C10H10	m-DIVINYLBENZENE	5.9210E-04	692.00	-0.6971	206.25	657.40	8.7710E-04	0.925
966	C10H10	1-METHYLINDENE	4.0320E-04	703.00	-0.7877	350.00	667.85	--------	------
967	C10H10	2-METHYLINDENE	4.6830E-04	684.00	-0.7605	353.15	649.80	--------	------
968	C10H10O4	DIMETHYL PHTHALATE	5.4920E-04	766.00	-0.6928	272.15	727.70	7.7280E-04	1.189
969	C10H10O4	DIMETHYL TEREPHTHALATE	4.4440E-04	772.00	-0.7388	413.80	733.40	--------	------
970	C10H12	DICYCLOPENTADIENE	6.0360E-04	660.00	-0.7143	307.00	627.00	--------	------
971	C10H12	1,2,3,4-TETRAHYDRONAPHTHALENE	5.0430E-04	720.15	-0.7323	237.40	684.14	7.4590E-04	0.967
972	C10H12O	ANETHOLE	5.5550E-04	723.00	-0.7048	294.50	686.85	8.0810E-04	0.984
973	C10H12O4	DIALLYL MALEATE	5.7990E-04	693.00	-0.7143	226.15	658.35	8.6660E-04	1.073
974	C10H14	n-BUTYLBENZENE	6.0210E-04	660.55	-0.7104	185.30	627.52	9.2240E-04	0.858
975	C10H14	sec-BUTYLBENZENE	5.7890E-04	664.54	-0.7143	197.72	631.31	8.8580E-04	0.858
976	C10H14	tert-BUTYLBENZENE	5.8140E-04	660.00	-0.7143	215.27	627.00	8.9310E-04	0.863
977	C10H14	1,2,3,4-TETRAMETHYLBENZENE	5.9180E-04	695.10	-0.7143	266.91	660.35	8.8300E-04	0.901
978	C10H14	m-CYMENE	6.0480E-04	657.00	-0.7058	209.44	624.15	9.2690E-04	0.857
979	C10H14	o-CYMENE	5.9790E-04	662.00	-0.7120	201.64	628.90	9.1570E-04	0.873
980	C10H14	p-CYMENE	5.9730E-04	653.15	-0.7125	205.25	620.49	9.2230E-04	0.852
981	C10H14	m-DIETHYLBENZENE	5.8610E-04	663.00	-0.7143	189.26	629.85	8.9800E-04	0.860
982	C10H14	o-DIETHYLBENZENE	5.8710E-04	668.00	-0.7143	241.93	634.60	8.9560E-04	0.876
983	C10H14	p-DIETHYLBENZENE	5.9510E-04	657.96	-0.7143	230.32	625.06	9.1590E-04	0.858
984	C1OH14	2-ETHYL-m-XYLENE	5.9660E-04	671.00	-0.7087	256.89	637.45	9.0470E-04	0.886
985	C10H14	2-ETHYL-p-XYLENE	6.0210E-04	663.00	-0.7068	219.52	629.85	9.1840E-04	0.873
986	C10H14	3-ETHYL-o-XYLENE	5.9760E-04	680.00	-0.7152	223.64	646.00	9.0290E-04	0.888
987	C10H14	4-ETHYL-m-XYLENE	6.0460E-04	665.00	-0.7046	210.27	631.75	9.1940E-04	0.872
988	C10H14	4-ETHYL-o-XYLENE	5.8640E-04	667.00	-0.7143	206.22	633.65	8.9530E-04	0.871
989	C1OH14	5-ETHYL-m-XYLENE	6.0710E-04	655.00	-0.7054	188.82	622.25	9.3170E-04	0.861
990	C10H14	ISOBUTYLBENZENE	6.4040E-04	650.15	-0.6776	221.70	617.64	9.7040E-04	0.849
991	C10H14	1,2,3,5-TETRAMETHYLBENZENE	5.7620E-04	679.00	-0.7143	249.46	645.05	8.7080E-04	0.887
992	C10H14	1,2,4,5-TETRAMETHYLBENZENE	5.8020E-04	675.15	-0.7143	352.38	641.39	--------	------
993	C10H14O	p-tert-BUTYLPHENOL	5.1710E-04	734.00	-0.7143	371.56	697.30	--------	------
994	C10H14O2	p-tert-BUTYLCATECHOL	4.7220E-04	776.00	-0.7143	325.00	737.20	--------	------
995	C10H15N	N,N-DIETHYLANILINE	5.3570E-04	702.00	-0.7380	235.15	666.90	8.0550E-04	0.931
996	C10H15N	2,6-DIETHYLANILINE	5.4540E-04	678.00	-0.7143	276.65	644.10	8.2490E-04	0.902
997	C10H16	CAMPHENE	5.6520E-04	638.00	-0.7353	320.15	606.10	--------	------
998	C10H16	D-LIMONENE	5.6900E-04	660.00	-0.7279	199.00	627.00	8.8120E-04	0.839
999	C10H16	alpha-PHELLANDRENE	5.9180E-04	649.00	-0.7143	220.00	616.55	9.1820E-04	0.843
1000	C1OH16	beta-PHELLANDRENE	6.4670E-04	648.00	-0.6850	220.00	615.60	9.8640E-04	0.837
1001	C10H16	alpha-PINENE	5.6140E-04	632.00	-0.7396	209.15	600.40	9.0000E-04	0.857
1002	C10H16	beta-PINENE	4.2810E-04	643.00	-0.7932	211.61	610.85	7.0180E-04	0.867
1003	C10H16	alpha-TERPINENE	5.8790E-04	652.00	-0.7143	220.00	619.40	9.0970E-04	0.830
1004	C1OH16	gamma-TERPINENE	5.8630E-04	661.00	-0.7143	220.00	627.95	8.9990E-04	0.845
1005	C10H16	TERPINOLENE	5.8560E-04	672.00	-0.7143	200.00	638.40	8.9020E-04	0.858
1006	C10H16O	CAMPHOR	5.8700E-04	709.00	-0.7143	453.25	673.55	--------	------
1007	C10H18	1-DECYNE	6.1380E-04	632.49	-0.7143	229.16	600.87	9.6770E-04	0.762
1008	C10H18	cis-DECAHYDRONANPHTALENE	5.4140E-04	702.25	-0.7128	230.20	667.14	8.0270E-04	0.894
1009	C10H18	trans-DECAHYDRONAPHTHALENE	5.6270E-04	687.05	-0.7048	242.79	652.70	8.4040E-04	0.868
1010	C10H18O4	SEBACIC ACID	5.1770E-04	815.00	-0.7143	407.65	774.25	--------	------
1011	C10H20	n-BUTYLCYCLOHEXANE	6.0110E-04	667.00	-0.6976	198.42	633.65	9.0870E-04	0.796
1012	C10H20	1-CYCLOPENTYLPENTANE	6.1170E-04	647.49	-0.7143	190.16	615.12	9.5050E-04	0.787
1013	C10H20	1-DECENE	6.2750E-04	617.05	-0.7144	206.89	586.20	1.0060E-03	0.737
1014	C10H20O	1-DECANAL	5.3560E-04	657.00	-0.7316	267.15	624.15	8.3360E-04	0.821

			$B_{liq} = a \ (1 - T/T_C)^m$					$(B_{liq} - 1/C, \ T - K, \ density_{liq} - g/cm^3)$	
NO	FORMULA	NAME	a	T_C	m	TMIN	TMAX	B_{liq} @ 25 C	$density_{liq}$ @ 25 C
1015	C10H20O2	n-DECANOIC ACID	4.9220E-04	713.00	-0.7314	304.75	677.35	------	------
1016	C10H20O2	2-ETHYLHEXYL ACETATE	6.1220E-04	639.00	-0.7143	180.15	607.05	9.5910E-04	0.869
1017	C10H20O2	ISOPENTYL ISOVALERATE	4.8690E-04	637.00	-0.7561	215.00	605.15	7.8470E-04	0.854
1018	C10H22	n-DECANE	6.3600E-04	618.45	-0.7143	243.49	587.53	1.0180E-03	0.728
1019	C10H22	2-METHYLNONANE	6.5540E-04	610.00	-0.6992	198.50	579.50	1.0480E-03	0.723
1020	C10H22	3-METHYLNONANE	6.2040E-04	613.00	-0.7131	188.35	582.35	9.9770E-04	0.729
1021	C10H22	4-METHYLNONANE	6.6960E-04	610.00	-0.6954	174.45	579.50	1.0680E-03	0.728
1022	C10H22	5-METHYLNONANE	6.4340E-04	610.00	-0.7055	185.45	579.50	1.0330E-03	0.729
1023	C10H22	3-ETHYLOCTANE	6.6260E-04	613.60	-0.7143	185.46	582.92	1.0660E-03	0.736
1024	C10H22	4-ETHYLOCTANE	6.7480E-04	609.60	-0.7143	185.46	579.12	1.0900E-03	0.734
1025	C10H22	2,2-DIMETHYLOCTANE	6.3510E-04	602.00	-0.7069	225.00	571.90	1.0300E-03	0.721
1026	C10H22	2,3-DIMETHYLOCTANE	6.5920E-04	613.20	-0.7143	219.16	582.54	1.0610E-03	0.734
1027	C10H22	2,4-DIMETHYLOCTANE	6.7430E-04	599.40	-0.7143	219.16	569.43	1.1020E-03	0.723
1028	C10H22	2,5-DIMETHYLOCTANE	6.6840E-04	603.00	-0.7143	219.16	572.85	1.0880E-03	0.726
1029	C10H22	2,6-DIMETHYLOCTANE	6.6250E-04	603.10	-0.7143	219.16	572.95	1.0780E-03	0.723
1030	C10H22	2,7-DIMETHYLOCTANE	6.6270E-04	602.90	-0.7143	219.16	572.76	1.0790E-03	0.720
1031	C10H22	3,3-DIMETHYLOCTANE	6.6030E-04	612.10	-0.7143	219.16	581.50	1.0640E-03	0.735
1032	C10H22	3,4-DIMETHYLOCTANE	6.6020E-04	614.00	-0.7143	219.16	583.30	1.0610E-03	0.741
1033	C10H22	3,5-DIMETHYLOCTANE	6.7060E-04	606.30	-0.7143	219.16	575.99	1.0870E-03	0.733
1034	C10H22	3,6-DIMETHYLOCTANE	6.6450E-04	608.30	-0.7143	219.16	577.89	1.0750E-03	0.732
1035	C10H22	4,4-DIMETHYLOCTANE	6.7190E-04	606.90	-0.7143	219.16	576.56	1.0890E-03	0.731
1036	C10H22	4,5-DIMETHYLOCTANE	6.6600E-04	612.20	-0.7143	219.16	581.59	1.0730E-03	0.743
1037	C10H22	4-PROPYLHEPTANE	6.8240E-04	601.00	-0.7143	219.16	570.95	1.1130E-03	0.732
1038	C10H22	4-ISOPROPYLHEPTANE	6.7700E-04	607.60	-0.7143	219.16	577.22	1.0960E-03	0.735
1039	C10H22	3-ETHYL-2-METHYLHEPTANE	6.6940E-04	610.90	-0.7143	219.16	580.36	1.0800E-03	0.740
1040	C10H22	4-ETHYL-2-METHYLHEPTANE	6.8150E-04	601.80	-0.7143	219.16	571.71	1.1110E-03	0.732
1041	C10H22	5-ETHYL-2-METHYLHEPTANE	6.7010E-04	606.70	-0.7143	219.16	576.37	1.0860E-03	0.732
1042	C10H22	3-ETHYL-3-METHYLHEPTANE	6.6150E-04	620.00	-0.7143	219.16	589.00	1.0570E-03	0.746
1043	C10H22	4-ETHYL-3-METHYLHEPTANE	6.6960E-04	614.30	-0.7143	219.16	583.59	1.0760E-03	0.746
1044	C10H22	3-ETHYL-5-METHYLHEPTANE	6.7420E-04	606.60	-0.7143	219.16	576.27	1.0930E-03	0.737
1045	C10H22	3-ETHYL-4-METHYLHEPTANE	6.6830E-04	615.50	-0.7143	219.16	584.73	1.0730E-03	0.747
1046	C10H22	4-ETHYL-4-METHYLHEPTANE	6.6810E-04	615.70	-0.7143	219.16	584.92	1.0720E-03	0.747
1047	C10H22	2,2,3-TRIMETHYLHEPTANE	6.5890E-04	611.70	-0.7143	219.16	581.12	1.0620E-03	0.738
1048	C10H22	2,2,4-TRIMETHYLHEPTANE	6.8190E-04	594.50	-0.7143	219.16	564.78	1.1210E-03	0.724
1049	C10H22	2,2,5-TRIMETHYLHEPTANE	6.7400E-04	598.00	-0.7143	219.16	568.10	1.1030E-03	0.724
1050	C10H22	2,2,6-TRIMETHYLHEPTANE	6.7330E-04	593.40	-0.7143	219.16	563.73	1.1090E-03	0.720
1051	C10H22	2,3,3-TRIMETHYLHEPTANE	6.5460E-04	617.50	-0.7143	219.16	586.63	1.0480E-03	0.745
1052	C10H22	2,3,4-TRIMETHYLHEPTANE	6.6250E-04	613.70	-0.7143	219.16	583.02	1.0650E-03	0.745
1053	C10H22	2,3,5-TRIMETHYLHEPTANE	6.6540E-04	612.80	-0.7143	219.16	582.16	1.0710E-03	0.741
1054	C10H22	2,3,6-TRIMETHYLHEPTANE	6.6710E-04	604.10	-0.7143	219.16	573.90	1.0850E-03	0.731
1055	C10H22	2,4,4-TRIMETHYLHEPTANE	6.7730E-04	600.30	-0.7143	219.16	570.29	1.1060E-03	0.731
1056	C10H22	2,4,5-TRIMETHYLHEPTANE	6.6990E-04	606.90	-0.7143	219.16	576.56	1.0860E-03	0.737
1057	C10H22	2,4,6-TRIMETHYLHEPTANE	6.8080E-04	590.30	-0.7143	219.16	560.79	1.1250E-03	0.719
1058	C10H22	2,5,5-TRIMETHYLHEPTANE	6.7040E-04	602.90	-0.7143	219.16	572.76	1.0910E-03	0.736
1059	C10H22	3,3,4-TRIMETHYLHEPTANE	6.5350E-04	622.10	-0.7143	219.16	591.00	1.0420E-03	0.752
1060	C10H22	3,3,5-TRIMETHYLHEPTANE	6.5360E-04	609.50	-0.7143	219.16	579.03	1.0560E-03	0.739
1061	C10H22	3,4,4-TRIMETHYLHEPTANE	6.5480E-04	620.90	-0.7143	219.16	589.86	1.0450E-03	0.753
1062	C10H22	3,4,5-TRIMETHYLHEPTANE	6.6540E-04	612.80	-0.7143	219.16	582.16	1.0710E-03	0.752
1063	C10H22	3-ISOPROPYL-2-METHYLHEXANE	6.6570E-04	623.40	-0.7143	219.16	592.23	1.0590E-03	0.744
1064	C10H22	3,3-DIETHYLHEXANE	6.5710E-04	627.80	-0.7143	219.16	596.41	1.0410E-03	0.757
1065	C10H22	3,4-DIETHYLHEXANE	6.6870E-04	618.80	-0.7143	219.16	587.86	1.0690E-03	0.747
1066	C10H22	3-ETHYL-2,2-DIMETHYLHEXANE	6.6850E-04	611.70	-0.7143	219.16	581.12	1.0780E-03	0.745
1067	C10H22	4-ETHYL-2,2-DIMETHYLHEXANE	6.8180E-04	594.60	-0.7143	219.16	564.87	1.1210E-03	0.730
1068	C10H22	3-ETHYL-2,3-DIMETHYLHEXANE	6.5050E-04	626.80	-0.7143	219.16	595.46	1.0320E-03	0.760
1069	C10H22	4-ETHYL-2,3-DIMETHYLHEXANE	6.6250E-04	617.30	-0.7143	219.16	586.44	1.0610E-03	0.752
1070	C10H22	3-ETHYL-2,4-DIMETHYLHEXANE	6.6380E-04	616.10	-0.7143	219.16	585.30	1.0650E-03	0.751
1071	C10H22	4-ETHYL-2,4-DIMETHYLHEXANE	6.5480E-04	620.90	-0.7143	219.16	589.86	1.0450E-03	0.752
1072	C10H22	3-ETHYL-2,5-DIMETHYLHEXANE	6.7560E-04	603.50	-0.7143	219.16	573.33	1.0990E-03	0.737
1073	C10H22	4-ETHYL-3,3-DIMETHYLHEXANE	6.5360E-04	625.70	-0.7143	219.16	594.42	1.0380E-03	0.760
1074	C10H22	3-ETHYL-3,4-DIMETHYLHEXANE	6.5670E-04	624.50	-0.7143	219.16	593.28	1.0440E-03	0.760
1075	C10H22	2,2,3,3-TETRAMETHYLHEXANE	6.5070E-04	623.00	-0.7143	219.16	591.85	1.0360E-03	0.761
1076	C10H22	2,2,3,4-TETRAMETHYLHEXANE	6.5530E-04	620.40	-0.7143	219.16	589.38	1.0460E-03	0.751
1077	C10H22	2,2,3,5-TETRAMETHYLHEXANE	6.6830E-04	601.30	-0.7143	219.16	571.24	1.0900E-03	0.733
1078	C10H22	2,2,4,4-TETRAMETHYLHEXANE	6.7410E-04	610.20	-0.7143	219.16	579.69	1.0880E-03	0.742
1079	C10H22	2,2,4,5-TETRAMETHYLHEXANE	6.7540E-04	598.50	-0.7143	219.16	568.58	1.1050E-03	0.731
1080	C10H22	2,2,5,5-TETRAMETHYLHEXANE	6.6960E-04	581.40	-0.7143	260.56	552.33	1.1190E-03	0.715
1081	C10H22	2,3,3,4-TETRAMETHYLHEXANE	6.4030E-04	633.10	-0.7143	260.56	601.45	1.0090E-03	0.765
1082	C10H22	2,3,3,5-TETRAMETHYLHEXANE	6.6250E-04	610.00	-0.7143	260.56	579.60	1.0700E-03	0.745
1083	C10H22	2,3,4,4-TETRAMETHYLHEXANE	6.4880E-04	626.60	-0.7143	260.56	595.27	1.0290E-03	0.779
1084	C10H22	2,3,4,5-TETRAMETHYLHEXANE	6.5920E-04	613.20	-0.7143	260.56	582.54	1.0610E-03	0.746
1085	C10H22	3,3,4,4-TETRAMETHYLHEXANE	6.2680E-04	646.70	-0.7143	260.56	614.37	9.7480E-04	0.779
1086	C10H22	2,4-DIMETHYL-3-ISOPROPYLPENTANE	6.6560E-04	614.40	-0.7143	191.46	583.68	1.0700E-03	0.754
1087	C10H22	3,3-DIETHYL-2-METHYLPENTANE	6.3910E-04	639.90	-0.7143	191.46	607.91	1.0000E-03	0.775
1088	C10H22	3-ETHYL-2,2,3-TRIMETHYLPENTANE	6.2930E-04	646.00	-0.7143	191.46	613.70	9.7930E-04	0.778
1089	C10H22	3-ETHYL-2,2,4-TRIMETHYLPENTANE	6.6270E-04	615.30	-0.7143	191.46	584.54	1.0640E-03	0.753
1090	C10H22	3-ETHYL-2,3,4-TRIMETHYLPENTANE	6.3300E-04	642.30	-0.7143	191.46	610.19	9.8840E-04	0.773
1091	C10H22	2,2,3,3,4-PENTAMETHYLPENTANE	6.2420E-04	643.80	-0.7143	236.71	611.61	9.7330E-04	0.777
1092	C10H22	2,2,3,4,4-PENTAMETHYLPENTANE	6.5000E-04	627.30	-0.7143	234.41	595.94	1.0300E-03	0.764

Table 28-1 COEFFICIENT OF THERMAL EXPANSION - ORGANIC COMPOUNDS (continued)

NO	FORMULA	NAME	$B_{liq} = a$ $(1 - T/T_C)^m$				(B_{liq} - 1/C, T - K, density$_{liq}$ - g/cm^3)	
			a	T_C	m	TMIN	TMAX	B_{liq} @ 25 C	density$_{liq}$ @ 25 C
1093	C10H22O	1-DECANOL	5.6510E-04	690.00	-0.7096	280.05	655.50	8.4430E-04	0.825
1094	C10H22O	DI-n-PENTYL ETHER	6.2750E-04	622.00	-0.7006	203.72	590.90	9.9130E-04	0.780
1095	C10H22O	ISODECANOL	6.0820E-04	644.00	-0.7143	213.15	611.80	9.4820E-04	0.837
1096	C10H22O5	TETRAETHYLENE GLYCOL DIMETHYL ETHER	6.7710E-04	705.00	-0.6473	243.45	669.75	9.6640E-04	1.006
1097	C10H22S	n-DECYL MERCAPTAN	4.7860E-04	696.00	-0.7388	247.56	661.20	7.2350E-04	0.841
1098	C10H22S	BUTYL-HEXYL-SULFIDE	5.8330E-04	701.03	-0.7143	238.16	665.98	8.6650E-04	0.841
1099	C10H22S	ETHYL-OCTYL-SULFIDE	5.8330E-04	701.03	-0.7143	238.16	665.98	8.6650E-04	0.841
1100	C10H22S	HEPTYL-PROPYL-SULFIDE	5.8330E-04	701.03	-0.7143	238.16	665.98	8.6650E-04	0.841
1101	C10H22S	METHYL-NONYL-SULFIDE	5.8330E-04	701.03	-0.7143	238.16	665.98	8.6650E-04	0.841
1102	C10H22S	PENTYL-SULFIDE	5.8330E-04	701.03	-0.7143	238.16	665.98	8.6650E-04	0.841
1103	C10H22S2	PENTYL-DISULFIDE	5.4020E-04	726.94	-0.7143	214.16	690.59	7.8760E-04	0.918
1104	C10H23N	n-DECYLAMINE	5.9450E-04	663.00	-0.7143	288.85	629.85	9.1080E-04	0.791
1105	C11H10	1-METHYLNAPHTHALENE	4.9810E-04	772.04	-0.7429	242.67	733.44	7.1580E-04	1.017
1106	C11H10	2-METHYLNAPHTHALENE	4.8210E-04	761.00	-0.7441	307.73	722.95	---------	------
1107	C11H14O2	n-BUTYL BENZOATE	5.4600E-04	724.00	-0.7029	251.65	687.80	7.9290E-04	1.001
1108	C11H16	n-PENTYLBENZENE	5.6140E-04	679.90	-0.7187	198.15	645.91	8.5000E-04	0.855
1109	C11H16O	p-tert-AMYLPHENOL	5.1100E-04	751.00	-0.7143	366.00	713.45	---------	------
1110	C11H20	1-UNDECYNE	6.0670E-04	650.99	-0.7143	248.16	618.44	9.3960E-04	0.769
1111	C11H20O2	2-ETHYLHEXYL ACRYLATE	8.5570E-04	655.00	-0.6048	183.15	622.25	1.2360E-03	0.880
1112	C11H22	1-UNDECENE	6.1930E-04	638.00	-0.7097	223.99	606.10	9.6830E-04	0.747
1113	C11H22	1-CYCLOPENTYLHEXANE	6.0540E-04	667.67	-0.7143	200.16	634.29	9.2370E-04	0.793
1114	C11H22	PENTYLCYCLOHEXANE	5.9800E-04	674.01	-0.7143	215.66	640.31	9.0750E-04	0.800
1115	C11H22O	1-UNDECANAL	6.1700E-04	672.00	-0.6904	273.15	638.40	9.2490E-04	0.823
1116	C11H24	n-UNDECANE	6.2010E-04	638.76	-0.7143	247.57	606.82	9.7170E-04	0.737
1117	C11H24O	1-UNDECANOL	5.7310E-04	704.00	-0.6980	289.05	668.80	8.4170E-04	0.831
1118	C11H24S	UNDECYL MERCAPTAN	4.6010E-04	710.00	-0.7430	270.15	674.50	6.8950E-04	0.841
1119	C11H24S	BUTYL-HEPTYL-SULFIDE	5.8150E-04	717.91	-0.7143	254.66	682.01	8.5310E-04	0.841
1120	C11H24S	DECYL-METHYL-SULFIDE	5.8150E-04	717.91	-0.7143	254.66	682.01	8.5310E-04	0.841
1121	C11H24S	ETHYL-NONYL-SULFIDE	5.8150E-04	717.91	-0.7143	254.66	682.01	8.5310E-04	0.841
1122	C11H24S	OCTYL-PROPYL-SULFIDE	5.8150E-04	717.91	-0.7143	254.66	682.01	8.5310E-04	0.841
1123	C12H8O	DIBENZOFURAN	5.1780E-04	837.80	-0.7055	355.65	795.91	---------	------
1124	C12H9N	DIBENZOPYRROLE	4.9600E-04	899.00	-0.7143	517.95	854.05	---------	------
1125	C12H10	ACENAPHTHENE	4.5220E-04	803.15	-0.7598	366.56	762.99	---------	------
1126	C12H10	BIPHENYL	4.8460E-04	789.26	-0.7211	342.37	749.80	---------	------
1127	C12H10O	DIPHENYL ETHER	4.4980E-04	763.00	-0.7334	300.02	724.85	---------	------
1128	C12H11N	p-AMINODIPHENYL	4.6970E-04	817.00	-0.7143	326.00	776.15	---------	------
1129	C12H11N	DIPHENYLAMINE	4.8710E-04	817.00	-0.7167	326.15	776.15	---------	------
1130	C12H11N3	p-AMINOAZOBENZENE	4.4520E-04	877.00	-0.7143	401.00	833.15	---------	------
1131	C12H11N3	1,3-DIPHENYLTRIAZENE	4.5810E-04	845.00	-0.7143	372.00	802.75	---------	------
1132	C12H12	1,2-DIMETHYLNAPHTHALENE	5.2130E-04	775.34	-0.7143	272.16	736.57	7.3740E-04	1.014
1133	C12H12	1,3-DIMETHYLNAPHTHALENE	5.2090E-04	773.76	-0.7143	269.16	735.07	7.3740E-04	1.003
1134	C12H12	1,4-DIMETHYLNAPHTHALENE	5.2040E-04	776.78	-0.7143	280.82	737.94	7.3540E-04	1.013
1135	C12H12	1,5-DIMETHYLNAPHTHALENE	5.2110E-04	773.47	-0.7143	355.16	734.80	---------	------
1136	C12H12	1,6-DIMETHYLNAPHTHALENE	5.2150E-04	770.60	-0.7143	259.16	732.07	7.3960E-04	0.999
1137	C12H12	1,7-DIMETHYLNAPHTHALENE	5.2150E-04	770.60	-0.7143	260.16	732.07	7.3960E-04	0.999
1138	C12H12	2,3-DIMETHYLNAPHTHALENE	5.2120E-04	777.78	-0.7143	378.16	738.89	---------	------
1139	C12H12	2,6-DIMETHYLNAPHTHALENE	5.7590E-04	777.00	-0.6809	384.55	738.15	---------	------
1140	C12H12	2,7-DIMETHYLNAPHTHALENE	5.7660E-04	778.00	-0.6801	370.15	739.10	---------	------
1141	C12H12	1-ETHYLNAPHTHALENE	4.6660E-04	776.00	-0.7360	259.34	737.20	6.6660E-04	1.004
1142	C12H12	2-ETHYLNAPHTHALENE	5.0530E-04	774.90	-0.7143	265.76	736.16	7.1490E-04	0.988
1143	C12H12N2	p-AMINODIPHENYLAMINE	4.3890E-04	867.00	-0.7143	341.15	823.65	---------	------
1144	C12H12N2	HYDRAZOBENZENE	4.8460E-04	792.00	-0.7143	404.15	752.40	---------	------
1145	C12H14	1,2,3-TRIMETHYLINDENE	5.5710E-04	726.00	-0.7143	344.65	689.70	---------	------
1146	C12H14O4	DIETHYL PHTHALATE	6.4270E-04	757.00	-0.6475	269.15	719.15	8.8870E-04	1.113
1147	C12H16	CYCLOHEXYLBENZENE	5.9960E-04	744.00	-0.6773	280.14	706.80	8.4820E-04	0.939
1148	C12H18	m-DIISOPROPYLBENZENE	8.5450E-04	684.00	-0.5960	210.02	649.80	1.2020E-03	0.852
1149	C12H18	p-DIISOPROPYLBENZENE	5.6220E-04	689.00	-0.7143	256.08	654.55	8.4290E-04	0.853
1150	C12H18	n-HEXYLBENZENE	5.6390E-04	698.00	-0.7143	212.00	663.10	8.3950E-04	0.855
1151	C12H18	1,2,3-TRIETHYLBENZENE	5.8550E-04	684.37	-0.7143	206.66	650.15	8.8100E-04	0.870
1152	C12H18	1,2,4-TRIETHYLBENZENE	5.8550E-04	684.37	-0.7143	206.66	650.15	8.8100E-04	0.870
1153	C12H18	1,3,5-TRIETHYLBENZENE	5.8560E-04	682.28	-0.7143	206.66	648.17	8.8270E-04	0.887
1154	C12H18	HEXAMETHYLBENZENE	5.7760E-04	758.00	-0.7143	438.66	720.10	---------	------
1155	C12H20O4	DIBUTYL MALEATE	5.7640E-04	716.00	-0.6926	188.15	680.20	8.3700E-04	0.991
1156	C12H22	BICYCLOHEXYL	5.1850E-04	727.00	-0.7187	276.78	690.65	7.5770E-04	0.883
1157	C12H22	1-DODECYNE	6.0320E-04	668.16	-0.7143	254.16	634.75	9.2000E-04	0.775
1158	C12H23N	DICYCLOHEXYLAMINE	5.3450E-04	737.00	-0.7020	273.05	700.15	7.6920E-04	0.909
1159	C12H24	1-DODECENE	6.1840E-04	657.00	-0.7034	237.93	624.15	9.4630E-04	0.756
1160	C12H24	1-CYCLOPENTYLHEPTANE	6.0580E-04	679.00	-0.7143	220.00	645.05	9.1560E-04	0.806
1161	C12H24	1-CYCLOHEXYLHEXANE	5.9280E-04	691.81	-0.7143	263.60	657.22	8.8680E-04	0.892
1162	C12H24O	1-DODECANAL	5.9070E-04	685.00	-0.7143	285.15	650.75	8.8850E-04	0.826
1163	C12H24O2	n-DODECANOIC ACID	5.2990E-04	734.00	-0.7067	317.15	697.30	---------	------
1164	C12H26	n-DODECANE	6.0670E-04	658.20	-0.7104	263.57	625.29	9.3140E-04	0.745
1165	C12H26O	DI-n-HEXYL ETHER	5.9840E-04	658.00	-0.6975	230.15	625.10	9.1160E-04	0.790
1166	C12H26O	1-DODECANOL	5.7750E-04	721.00	-0.6890	296.95	684.95	8.3400E-04	0.830
1167	C12H26O3	DIETHYLENE GLYCOL DI-n-BUTYL ETHER	5.8260E-04	680.00	-0.7143	212.95	646.00	8.7980E-04	0.881
1168	C12H26S	n-DODECYL MERCAPTAN	5.6370E-04	724.00	-0.6881	265.15	687.80	8.1220E-04	0.842
1169	C12H26S	BUTYL-OCTYL-SULFIDE	5.8090E-04	733.68	-0.7143	259.16	697.00	8.4310E-04	0.842
1170	C12H26S	DECYL-ETHYL-SULFIDE	5.8090E-04	733.68	-0.7143	259.16	697.00	8.4310E-04	0.842

Table 28-1 COEFFICIENT OF THERMAL EXPANSION - ORGANIC COMPOUNDS (continued)

NO	FORMULA	NAME	$B_{liq} = a$ $(1 - T/T_C)^m$				$(B_{liq} - 1/C,$ T - K, $density_{liq}$ - $g/cm^3)$	
			a	T_C	m	TMIN	TMAX	B_{liq} @ 25 C	$density_{liq}$ @ 25 C
1171	C12H26S	HEXYL-SULFIDE	5.8090E-04	733.68	-0.7143	259.16	697.00	8.4310E-04	0.842
1172	C12H26S	METHYL-UNDECYL-SULFIDE	5.8090E-04	733.68	-0.7143	259.16	697.00	8.4310E-04	0.842
1173	C12H26S	NONYL-PROPYL-SULFIDE	5.8090E-04	733.68	-0.7143	259.16	697.00	8.4310E-04	0.842
1174	C12H26S2	HEXYL-DISULFIDE	5.4420E-04	747.10	-0.7143	225.16	709.75	7.8290E-04	0.908
1175	C12H27BO3	TRI-n-BUTYL BORATE	7.4690E-04	743.15	-0.6097	203.15	705.99	1.0210E-03	0.854
1176	C12H27N	DODECYLAMINE	5.3110E-04	696.00	-0.7226	301.47	661.20	--------	------
1177	C12H27N	TRI-n-BUTYLAMINE	5.6530E-04	644.00	-0.7257	203.00	611.80	8.8760E-04	0.775
1178	C13H10	FLUORENE	4.4240E-04	870.00	-0.7143	387.94	826.50	--------	------
1179	C13H10O	BENZOPHENONE	4.8280E-04	816.00	-0.7269	321.35	775.20	--------	------
1180	C13H12	DIPHENYLMETHANE	5.1100E-04	768.00	-0.7143	298.39	729.60	--------	------
1181	C13H14	1-PROPYLNAPHTHALENE	5.1640E-04	771.45	-0.7143	264.69	732.88	7.3200E-04	0.987
1182	C13H14	2-PROPYLNAPHTHALENE	5.1570E-04	772.44	-0.7143	270.16	733.82	7.3070E-04	0.973
1183	C13H14	2ETHYL-3-METHYLNAPHTHALENE	5.2060E-04	776.44	-0.7143	344.16	737.62	--------	------
1184	C13H14	2ETHYL-6-METHYLNAPHTHALENE	5.2270E-04	766.56	-0.7143	318.16	728.23	--------	------
1185	C13H14	2ETHYL-7-METHYLNAPHTHALENE	5.2270E-04	766.56	-0.7143	318.16	728.23	--------	------
1186	C13H20	n-HEPTYLBENZENE	6.1100E-04	714.00	-0.6791	225.16	678.30	8.8200E-04	0.853
1187	C13H24	1-TRIDECYNE	6.0130E-04	684.11	-0.7143	268.16	649.90	9.0500E-04	0.781
1188	C13H26	1-TRIDECENE	6.1880E-04	675.00	-0.6960	250.08	641.25	9.2840E-04	0.762
1189	C13H26	1-CYCLOPENTYLOCTANE	5.9630E-04	702.06	-0.7143	229.16	666.96	8.8510E-04	0.801
1190	C13H26	1-CYCLOHEXYLHEPTANE	5.9080E-04	708.63	-0.7143	242.66	673.20	8.7260E-04	0.807
1191	C13H26O	1-TRIDECANAL	6.1720E-04	700.00	-0.6783	288.15	665.00	8.9930E-04	0.827
1192	C13H26O2	n-BUTYL NONANOATE	5.9910E-04	652.00	-0.7143	235.15	619.40	9.2700E-04	0.851
1193	C13H26O2	METHYL DODECANOATE	6.0580E-04	712.00	-0.7143	278.15	676.40	8.9250E-04	1.039
1194	C13H28	n-TRIDECANE	6.3930E-04	675.80	-0.6880	267.76	642.01	9.5400E-04	0.754
1195	C13H28O	1-TRIDECANOL	5.5450E-04	731.00	-0.7143	303.75	694.45	--------	------
1196	C13H28S	BUTYL-NONYL-SULFIDE	5.7980E-04	748.42	-0.7143	271.16	711.00	8.3340E-04	0.842
1197	C13H28S	DECYL-PROPYL-SULFIDE	5.7980E-04	748.42	-0.7143	271.16	711.00	8.3340E-04	0.842
1198	C13H28S	DODECYL-METHYL-SULFIDE	5.7980E-04	748.42	-0.7143	271.16	711.00	8.3340E-04	0.842
1199	C13H28S	ETHYL-UNDECYL-SULFIDE	5.7980E-04	748.42	-0.7143	271.16	711.00	8.3340E-04	0.842
1200	C13H28S	1-TRIDECANETHIOL	5.6580E-04	742.13	-0.7143	282.04	705.02	8.1670E-04	0.842
1201	C14H8O2	ANTHRAQUINONE	4.5280E-04	900.00	-0.7143	559.15	855.00	--------	------
1202	C14H10	ANTHRACENE	3.6240E-04	873.00	-0.7647	489.25	829.35	--------	------
1203	C14H10	DIPHENYLACETYLENE	4.6790E-04	832.00	-0.7143	335.65	790.40	--------	------
1204	C14H10	PHENANTHRENE	3.9390E-04	869.25	-0.7514	372.38	825.79	--------	------
1205	C14H12	cis-STILBENE	5.1670E-04	757.00	-0.7143	268.15	719.15	7.3880E-04	1.011
1206	C14H12	trans-STILBENE	5.0910E-04	820.00	-0.7143	397.35	779.00	--------	------
1207	C14H12O2	BENZYL BENZOATE	5.9800E-04	820.00	-0.6730	292.55	779.00	8.1060E-04	1.115
1208	C14H14	1,1-DIPHENYLETHANE	5.4190E-04	775.00	-0.6965	255.20	736.25	7.6000E-04	0.996
1209	C14H14	1,2-DIPHENYLETHANE	5.1840E-04	780.00	-0.7100	324.34	741.00	--------	------
1210	C14H14O	DIBENZYL ETHER	5.0800E-04	777.00	-0.7143	276.75	738.15	7.1780E-04	1.042
1211	C14H16	1-n-BUTYLNAPHTHALENE	4.9240E-04	792.00	-0.7166	253.43	752.40	6.9070E-04	0.973
1212	C14H16	2-BUTYLNAPHTHALENE	5.1610E-04	780.96	-0.7143	268.16	741.91	7.2760E-04	0.962
1213	C14H22	n-OCTYLBENZENE	6.0960E-04	729.00	-0.6737	237.15	692.55	8.6880E-04	0.853
1214	C14H22	1,2,3,4-TETRAETHYLBENZENE	5.8770E-04	708.20	-0.7143	284.96	672.79	8.6830E-04	0.883
1215	C14H22	1,2,3,5-TETRAETHYLBENZENE	5.8650E-04	707.52	-0.7143	284.16	672.14	8.6700E-04	0.878
1216	C14H22	1,2,4,5-TETRAETHYLBENZENE	5.8710E-04	706.85	-0.7143	283.16	671.51	8.6830E-04	0.875
1217	C14H22O	p-tert-OCTYLPHENOL	5.2250E-04	765.00	-0.7100	358.55	726.75	--------	------
1218	C14H28	1-TETRADECENE	6.0660E-04	692.00	-0.6955	260.30	657.40	8.9770E-04	0.768
1219	C14H28	1-CYCLOPENTYLNONANE	5.9440E-04	716.95	-0.7143	244.16	681.10	8.7270E-04	0.804
1220	C14H28	1-CYCLOHEXYLOCTANE	5.8900E-04	723.61	-0.7143	253.46	687.43	8.6070E-04	0.810
1221	C14H28O2	n-TETRADECANOIC ACID	4.9170E-04	756.00	-0.7164	327.55	718.20	--------	------
1222	C14H30	n-TETRADECANE	5.3840E-04	692.40	-0.7265	279.01	657.78	8.1050E-04	0.758
1223	C14H30O	1-TETRADECANOL	5.7370E-04	741.00	-0.6818	310.65	703.95	--------	------
1224	C14H30S	BUTYL-DECYL-SULFIDE	5.7970E-04	762.23	-0.7143	276.16	724.12	8.2620E-04	0.843
1225	C14H30S	DODECYL-ETHYL-SULFIDE	5.7970E-04	762.23	-0.7143	276.16	724.12	8.2620E-04	0.843
1226	C14H30S	HEPTYL-SULFIDE	5.7970E-04	762.23	-0.7143	276.16	724.12	8.2620E-04	0.843
1227	C14H30S	METHYL-TRIDECYL-SULFIDE	5.7970E-04	762.23	-0.7143	276.16	724.12	8.2620E-04	0.843
1228	C14H30S	PROPYL-UNDECYL-SULFIDE	5.7970E-04	762.23	-0.7143	276.16	724.12	8.2620E-04	0.843
1229	C14H30S	1-TETRADECANETHIOL	5.6880E-04	753.80	-0.7143	279.26	716.11	8.1490E-04	0.842
1230	C14H30S2	HEPTYL-DISULFIDE	5.5150E-04	765.96	-0.7143	235.16	727.66	7.8430E-04	0.900
1231	C14H31N	TETRADECYLAMINE	5.5620E-04	722.30	-0.7143	311.34	686.19	--------	------
1232	C15H10N2O2	DIPHENYLMETHANE-4,4'-DIISOCYANATE	5.0340E-04	802.00	-0.7143	311.20	761.90	--------	------
1233	C15H16O	p-CUMYLPHENOL	4.6810E-04	834.00	-0.7143	346.00	792.30	--------	------
1234	C15H16O2	BISPHENOL A	4.2720E-04	849.00	-0.7143	426.15	806.55	--------	------
1235	C15H18	1-PENTYLNAPHTHALENE	5.1700E-04	793.32	-0.7143	251.16	753.65	7.2390E-04	0.962
1236	C15H18	2-PENTYLNAPHTHALENE	5.1730E-04	797.48	-0.7143	269.16	757.61	7.2280E-04	0.953
1237	C15H24	n-NONYLBENZENE	5.6620E-04	741.00	-0.6986	249.00	703.95	8.1130E-04	0.852
1238	C15H24O	2,6-DI-tert-BUTYL-p-CRESOL	5.3000E-04	720.00	-0.7143	344.00	684.00	--------	------
1239	C15H24O	NONYLPHENOL	5.2850E-04	757.00	-0.7076	279.15	719.15	7.5320E-04	0.949
1240	C15H28	1-PENTADECYNE	5.9910E-04	711.41	-0.7143	283.16	675.84	8.8300E-04	0.789
1241	C15H30	1-PENTADECENE	6.0820E-04	708.00	-0.6889	269.42	672.60	8.8630E-04	0.773
1242	C15H30	1-CYCLOPENTYLDECANE	5.9390E-04	730.64	-0.7143	251.03	694.11	8.6370E-04	0.807
1243	C15H30	1-CYCLOHEXYLNONANE	5.8810E-04	737.79	-0.7143	262.96	700.90	8.5130E-04	0.813
1244	C15H30O2	PENTADECANOIC ACID	5.9010E-04	766.00	-0.6657	325.68	727.70	--------	------
1245	C15H32	n-PENTADECANE	6.1270E-04	706.80	-0.6842	283.11	671.46	8.9140E-04	0.765
1246	C15H32O	1-PENTADECANOL	5.8810E-04	722.53	-0.7143	317.04	686.40	--------	------
1247	C15H32S	BUTYL-UNDECYL-SULFIDE	5.8050E-04	775.15	-0.7143	284.16	736.39	8.2120E-04	0.843
1248	C15H32S	DODECYL-PROPYL-SULFIDE	5.8050E-04	775.15	-0.7143	284.16	736.39	8.2120E-04	0.843

Table 28-1 COEFFICIENT OF THERMAL EXPANSION - ORGANIC COMPOUNDS (continued)

NO	FORMULA	NAME	$B_{liq} = a\,(1 - T/T_C)^m$			$(B_{liq} - 1/C,\ T - K,\ density_{liq} - g/cm^3)$			
			a	T_C	m	TMIN	TMAX	B_{liq} @ 25 C	$density_{liq}$ @ 25 C
1249	C15H32S	ETHYL-TRIDECYL-SULFIDE	5.8050E-04	775.15	-0.7143	284.16	736.39	8.2120E-04	0.843
1250	C15H32S	METHYL-TETRADECYL-SULFIDE	5.8050E-04	775.15	-0.7143	284.16	736.39	8.2120E-04	0.843
1251	C15H32S	1-PENTADECANETHIOL	5.7080E-04	764.77	-0.7143	290.93	726.53	8.1240E-04	0.843
1252	C16H10	FLUORANTHENE	4.6780E-04	905.00	-0.7143	383.33	859.75	--------	------
1253	C16H10	PYRENE	4.7120E-04	936.00	-0.7143	423.81	889.20	--------	------
1254	C16H12	1-PHENYLNAPHTHALENE	4.8610E-04	849.00	-0.7095	318.15	806.55	--------	------
1255	C16H20	1-n-HEXYLNAPHTHALENE	5.0770E-04	813.00	-0.6991	255.15	772.35	6.9880E-04	0.947
1256	C16H22O4	DIBUTYL PHTHALATE	6.6040E-04	781.00	-0.6263	238.15	741.95	8.9260E-04	1.043
1257	C16H26	n-DECYLBENZENE	5.6370E-04	753.00	-0.7038	258.77	715.35	8.0380E-04	0.852
1258	C16H26	PENTAETHYLBENZENE	5.9420E-04	723.64	-0.7143	327.66	687.46	--------	------
1259	C16H30	1-HEXADECYNE	5.9910E-04	724.26	-0.7143	288.16	688.05	8.7510E-04	0.793
1260	C16H32	n-DECYLCYCLOHEXANE	5.7360E-04	751.25	-0.6779	271.42	713.69	8.0810E-04	0.815
1261	C16H32	1-CYCLOPENTYLUNDECANE	5.9440E-04	743.30	-0.7143	263.16	706.14	8.5730E-04	0.810
1262	C16H32	1-HEXADECENE	6.0690E-04	722.00	-0.6843	277.51	685.90	8.7390E-04	0.777
1263	C16H32O2	n-HEXADECANOIC ACID	5.0000E-04	776.00	-0.7053	335.95	737.20	--------	------
1264	C16H34	n-HEXADECANE	6.1510E-04	720.60	-0.6762	291.34	684.57	8.8250E-04	0.770
1265	C16H34O	DI-n-OCTYL ETHER	5.6010E-04	707.00	-0.6961	265.55	671.65	8.2000E-04	0.803
1266	C16H34O	1-HEXADECANOL	5.6790E-04	761.00	-0.6764	322.35	722.95	--------	------
1267	C16H34S	BUTYL-DODECYL-SULFIDE	5.8230E-04	787.27	-0.7143	288.16	747.91	8.1810E-04	0.843
1268	C16H34S	ETHYL-TETRADECYL-SULFIDE	6.1490E-04	791.68	-0.7143	288.16	752.10	8.6180E-04	0.844
1269	C16H34S	METHYL-PENTADECYL-SULFIDE	5.8230E-04	787.27	-0.7143	288.16	747.91	8.1810E-04	0.843
1270	C16H34S	OCTYL-SULFIDE	5.8230E-04	787.27	-0.7143	288.16	747.91	8.1810E-04	0.843
1271	C16H34S	PROPYL-TRIDECYL-SULFIDE	5.8230E-04	787.27	-0.7143	288.16	747.91	8.1810E-04	0.843
1272	C16H34S	1-HEXADECANETHIOL	5.7380E-04	774.68	-0.7143	290.93	735.95	8.1190E-04	0.843
1273	C16H34S2	OCTYL-DISULFIDE	5.5980E-04	784.46	-0.7143	244.16	745.24	7.8780E-04	0.894
1274	C17H28	n-UNDECYLBENZENE	5.6610E-04	764.00	-0.6916	268.00	725.80	7.9700E-04	0.851
1275	C17H32	1-HEPTADECYNE	6.0200E-04	736.21	-0.7143	295.16	699.40	8.7220E-04	0.796
1276	C17H34	1-CYCLOPENTYLDODECANE	5.9590E-04	755.17	-0.7143	268.16	717.41	8.5300E-04	0.812
1277	C17H34	1-CYCLOHEXYLUNDECANE	5.8900E-04	761.74	-0.7143	278.96	723.65	8.3970E-04	0.817
1278	C17H34	1-HEPTADECENE	6.1660E-04	736.00	-0.6657	284.40	699.20	8.7120E-04	0.782
1279	C17H36	n-HEPTADECANE	5.7170E-04	733.37	-0.6948	295.13	696.70	8.2150E-04	0.773
1280	C17H36O	1-HEPTADECANOL	5.9500E-04	770.00	-0.6610	327.05	731.50	--------	------
1281	C17H36S	BUTYL-TRIDECYL-SULFIDE	5.8480E-04	798.63	-0.7143	294.16	758.70	8.1660E-04	0.844
1282	C17H36S	ETHYL-PENTADECYL-SULFIDE	5.8480E-04	798.63	-0.7143	294.16	758.70	8.1660E-04	0.844
1283	C17H36S	HEXADECYL-METHYL-SULFIDE	5.8480E-04	798.63	-0.7143	294.16	758.70	8.1660E-04	0.844
1284	C17H36S	PROPYL-TETRADECYL-SULFIDE	5.8480E-04	798.63	-0.7143	294.16	758.70	8.1660E-04	0.844
1285	C17H36S	1-HEPTADECANETHIOL	5.7610E-04	786.01	-0.7143	300.37	746.71	--------	------
1286	C18H12	CHRYSENE	4.5150E-04	979.00	-0.7143	531.15	930.05	--------	------
1287	C18H14	m-TERPHENYL	4.6220E-04	924.85	-0.7032	360.00	878.61	--------	------
1288	C18H14	o-TERPHENYL	5.0760E-04	890.95	-0.6828	329.35	846.40	--------	------
1289	C18H14	p-TERPHENYL	4.5820E-04	925.95	-0.7057	485.00	879.65	--------	------
1290	C18H15P	TRIPHENYLPHOSPHINE	5.7270E-04	1008.00	-0.4453	354.40	957.60	--------	------
1291	C18H15O4P	TRIPHENYL PHOSPHATE	--------	------	------	------	------	--------	------
1292	C18H16N2	N,N'-DIPHENYL-p-PHENYLENEDIAMINE	4.3590E-04	906.00	-0.7143	409.00	860.70	--------	------
1293	C18H22	2,3-DIMETHYL-2,3-DIPHENYLBUTANE	5.1850E-04	805.00	-0.7143	392.15	764.75	--------	------
1294	C18H22O2	DICUMYL PEROXIDE	4.3660E-04	884.00	-0.7143	311.15	839.80	--------	------
1295	C18H30	n-DODECYLBENZENE	5.2120E-04	774.26	-0.7143	275.93	735.55	7.3770E-04	0.849
1296	C18H30	HEXAETHYLBENZENE	6.0320E-04	734.78	-0.7143	401.16	698.04	--------	------
1297	C18H32O2	LINOLEIC ACID	5.3040E-04	775.00	-0.7140	268.15	736.25	7.5030E-04	0.902
1298	C18H34	1-OCTADECYNE	6.0220E-04	747.33	-0.7143	300.16	709.96	--------	------
1299	C18H34O2	OLEIC ACID	4.8830E-04	781.00	-0.7103	286.53	741.95	6.8710E-04	0.888
1300	C18H34O4	DIBUTYL SEBACATE	6.7420E-04	768.00	-0.6215	263.95	729.60	9.1490E-04	0.932
1301	C18H34O4	DIHEXYL ADIPATE	6.8460E-04	767.00	-0.6115	259.35	728.65	9.2510E-04	0.932
1302	C18H36	1-CYCOPENTYLTRIDECANE	5.9620E-04	766.47	-0.7143	278.16	728.15	8.4770E-04	0.814
1303	C18H36	1-CYCLOHEXYLDODECANE	5.9130E-04	772.83	-0.7143	285.66	734.19	8.3760E-04	0.819
1304	C18H36	1-OCTADECENE	6.1140E-04	748.00	-0.6728	290.76	710.60	8.6080E-04	0.785
1305	C18H36O2	STEARIC ACID	4.9770E-04	799.00	-0.6975	342.75	759.05	--------	------
1306	C18H38	n-OCTADECANE	4.9860E-04	745.26	-0.7260	301.33	708.00	--------	------
1307	C18H38O	DINONYL ETHER	5.9360E-04	736.00	-0.6697	273.00	699.20	8.4050E-04	0.808
1308	C18H38O	1-OCTADECANOL	5.5930E-04	777.00	-0.6728	331.05	738.15	--------	------
1309	C18H38S	BUTYL-TETRADECYL-SULFIDE	5.8540E-04	810.53	-0.7143	298.16	770.00	--------	------
1310	C18H38S	ETHYL-HEXADECYL-SULFIDE	5.8540E-04	810.53	-0.7143	298.16	770.00	--------	------
1311	C18H38S	HEPTADECYL-METHYL-SULFIDE	5.8540E-04	810.53	-0.7143	298.16	770.00	--------	------
1312	C18H38S	NONYL-SULFIDE	5.8540E-04	810.53	-0.7143	298.16	770.00	--------	------
1313	C18H38S	PENTADECYL-PROPYL-SULFIDE	5.8540E-04	810.53	-0.7143	298.16	770.00	--------	------
1314	C18H38S	1-OCTADECANETHIOL	5.7820E-04	795.36	-0.7143	300.93	755.59	--------	------
1315	C18H38S2	NONYL-DISULFIDE	5.6610E-04	802.30	-0.7143	252.16	762.19	7.8890E-04	0.889
1316	C19H26	1-n-NONYLNAPHTHALENE	5.2630E-04	849.00	-0.6984	284.15	806.55	7.1190E-04	0.934
1317	C19H32	n-TRIDECYLBENZENE	5.9170E-04	783.00	-0.6802	283.15	743.85	8.1980E-04	0.851
1318	C19H36	1-NONADECYNE	6.0400E-04	758.94	-0.7143	306.16	720.99	--------	------
1319	C19H36O2	METHYL OLEATE	5.8220E-04	764.00	-0.6675	293.05	725.80	8.1000E-04	0.870
1320	C19H38	1-CYCLOPENTYLTETRADECANE	5.9380E-04	772.00	-0.7143	282.00	733.40	8.4150E-04	0.816
1321	C19H38	1-CYCLOHEXYLTRIDECANE	5.9250E-04	783.38	-0.7143	291.66	744.21	8.3420E-04	0.821
1322	C19H38	1-NONADECENE	6.1100E-04	760.00	-0.6666	296.55	722.00	8.5160E-04	0.788
1323	C19H38O2	NONADECANOIC ACID	5.3080E-04	810.00	-0.6784	341.23	769.50	--------	------
1324	C19H40	n-NONADECANE	5.6190E-04	755.93	-0.6935	305.33	718.13	--------	------
1325	C19H40O	1-NONADECANOL	5.9500E-04	775.30	-0.7143	334.87	736.54	--------	------
1326	C19H40S	BUTYL-PENTADECYL-SULFIDE	5.8670E-04	821.75	-0.7143	303.16	780.66	--------	------

Table 28-1 COEFFICIENT OF THERMAL EXPANSION - ORGANIC COMPOUNDS (continued)

NO	FORMULA	NAME	$B_{liq} = a \, (1 - T/T_C)^m$					$(B_{liq} - 1/C, \; T - K, \; density_{liq} - g/cm^3)$	
			a	T_C	m	TMIN	TMAX	B_{liq} @ 25 C	$density_{liq}$ @ 25 C
1327	C19H40S	ETHYL-HEPTADECYL-SULFIDE	5.8670E-04	821.75	-0.7143	303.16	780.66	--------	-------
1328	C19H40S	HEXADECYL-PROPYL-SULFIDE	5.8670E-04	821.75	-0.7143	303.16	780.66	--------	-------
1329	C19H40S	METHYL-OCTADECYL-SULFIDE	5.8670E-04	821.75	-0.7143	303.16	780.66	--------	-------
1330	C19H40S	1-NONADECANETHIOL	5.8180E-04	805.29	-0.7143	307.04	765.03	--------	-------
1331	C20H16	TRIPHENYLETHYLENE	4.5040E-04	908.00	-0.7143	342.15	862.60	--------	-------
1332	C20H28	1-n-DECYLNAPHTHALENE	5.3390E-04	859.00	-0.6925	288.15	816.05	7.1720E-04	0.928
1333	C20H30O2	ABIETIC ACID	5.1070E-04	832.00	-0.7143	446.65	790.40	--------	-------
1334	C20H31N	DEHYDROABIETYLAMINE	4.6930E-04	863.00	-0.7143	317.65	819.85	--------	-------
1335	C20H34	1-PHENYLTETRADECANE	5.1480E-04	792.00	-0.7143	289.16	752.40	7.2140E-04	0.851
1336	C20H38	1-EICOSYNE	6.0680E-04	769.79	-0.7143	309.16	731.30	--------	-------
1337	C20H40	1-CYCLOPENTYLPENTADECANE	6.0640E-04	780.00	-0.7143	290.00	741.00	8.5540E-04	0.818
1338	C20H40	1-CYCLOHEXYLTETRADECANE	5.9470E-04	792.82	-0.7143	297.16	753.18	8.3300E-04	0.822
1339	C20H40	1-EICOSENE	6.0840E-04	771.00	-0.6641	301.76	732.45	--------	-------
1340	C20H42	n-EICOSANE	5.5920E-04	767.04	-0.6912	309.59	728.69	--------	-------
1341	C20H42O	1-EICOSANOL	4.7900E-04	792.00	-0.7113	338.55	752.40	--------	-------
1342	C20H42S	BUTYL-HEXADECYL-SULFIDE	5.9060E-04	832.33	-0.7143	308.16	790.71	--------	-------
1343	C20H42S	DECYL-SULFIDE	5.9060E-04	832.33	-0.7143	308.16	790.71	--------	-------
1344	C20H42S	ETHYL-OCTADECYL-SULFIDE	5.9060E-04	832.33	-0.7143	308.16	790.71	--------	-------
1345	C20H42S	HEPTADECYL-PROPYL-SULFIDE	5.9060E-04	832.33	-0.7143	308.16	790.71	--------	-------
1346	C20H42S	METHYL-NONADECYL-SULFIDE	5.9060E-04	832.33	-0.7143	308.16	790.71	--------	-------
1347	C20H42S	1-EICOSANETHIOL	5.8440E-04	814.57	-0.7143	310.37	773.84	--------	-------
1348	C20H42S2	DECYL-DISULFIDE	5.7490E-04	820.08	-0.7143	259.16	779.08	7.9400E-04	0.885
1349	C21H21O4P	TRI-o-CRESYL PHOSPHATE	--------	------	------	------	------		1.165
1350	C21H36	1-PHENYLPENTADECANE	5.2490E-04	800.00	-0.7143	295.16	760.00	7.3230E-04	0.851
1351	C21H42	1-CYCLOPENTYLHEXADECANE	6.0280E-04	797.25	-0.7143	294.16	757.39	8.4230E-04	0.819
1352	C21H42	1-CYCLOHEXYLPENTADECANE	5.9620E-04	803.46	-0.7143	302.16	763.29	--------	-------
1353	C22H38	1-PHENYLHEXADECANE	5.1970E-04	808.00	-0.7143	300.16	767.60	--------	-------
1354	C22H44	1-CYCLOHEXYLHEXADECANE	5.9840E-04	813.42	-0.7143	306.76	772.75	--------	-------
1355	C22H44O2	n-BUTYL STEARATE	5.3370E-04	764.00	-0.7143	299.45	725.80	--------	-------
1356	C24H38O4	DIISOOCTYL PHTHALATE	4.8370E-04	851.00	-0.7143	254.00	808.45	6.5820E-04	0.983
1357	C24H38O4	DIOCTYL PHTHALATE	6.6040E-04	806.00	-0.6155	223.15	765.70	8.7750E-04	0.980
1358	C24H42O	DINONYLPHENOL	5.1100E-04	886.00	-0.7143	350.00	841.70	--------	-------
1359	C26H20	TETRAPHENYLETHYLENE	4.4770E-04	996.00	-0.7143	496.15	946.20	--------	-------
1360	C28H46O4	DIISODECYL PHTHALATE	4.7390E-04	887.00	-0.7140	227.59	842.65	6.3490E-04	0.973

B_{liq} - coefficient of thermal expansion of liquid, 1/C

T - temperature, K

$density_{liq}$ - density of liquid, g/cm^3

TMIN - minimum temperature, K

TMAX - maximum temperature, K

Table 28-2 COEFFICIENT OF THERMAL EXPANSION - INORGANIC COMPOUNDS

			$B_{liq} = a \ (1 - T/T_c)^m$					$(B_{liq} - 1/C, \ T - K, \ density_{liq} - g/cm^3)$	
NO	FORMULA	NAME	a	T_c	m	TMIN	TMAX	B_{liq} @ 25 C	$density_{liq}$ @ 25 C
1	Ag	SILVER	7.3980E-05	6410.00	-0.8635	1234.00	6089.50	-------	-------
2	AgCl	SILVER CHLORIDE	1.1500E-04	2992.10	-0.7143	728.15	881.74	-------	-------
3	AgI	SILVER IODIDE	1.1870E-04	2897.64	-0.7143	825.15	973.89	-------	-------
4	Al	ALUMINUM	1.0270E-04	7151.00	-0.7143	933.00	1304.49	-------	-------
5	AlB3H12	ALUMINUM BOROHYDRIDE	---------	-------	-------	-------	-------	-------	-------
6	AlBr3	ALUMINUN BROMIDE	7.3290E-04	763.00	-0.7143	390.15	436.81	-------	-------
7	AlCl3	ALUMINUM CHLORIDE	7.0170E-04	629.00	-0.6876	465.75	597.55	-------	-------
8	AlF3	ALUMINUM FLUORIDE	1.1670E-04	2948.13	-0.7143	1313.15	1437.49	-------	-------
9	AlI3	ALUMINUM IODIDE	3.4990E-04	983.00	-0.7143	464.15	533.33	-------	-------
10	Al2O3	ALUMINUM OXIDE	6.4480E-05	5335.00	-0.7143	2325.00	2398.75	-------	-------
11	Al2S3O12	ALUMINUM SULFATE	----------	-------	-------	-------	-------	-------	-------
12	Ar	ARGON	2.4760E-03	150.86	-0.7016	83.78	143.32	-------	-------
13	As	ARSENIC	4.9240E-04	1673.15	-0.7143	1090.15	1225.64	-------	-------
14	AsBr3	ARSENIC TRIBROMIDE	4.6880E-04	789.01	-0.7143	306.15	379.67	-------	-------
15	AsCl3	ARSENIC TRICHLORIDE	5.6560E-04	654.00	-0.7143	255.15	621.30	8.7360E-04	2.150
16	AsF3	ARSENIC TRIFLUORIDE	6.9760E-04	530.21	-0.7143	267.25	503.70	1.2590E-03	2.649
17	AsF5	ARSENIC PENTAFLUORIDE	1.0340E-03	357.73	-0.7143	193.35	339.85	3.7200E-03	1.890
18	AsH3	ARSINE	1.2040E-03	373.00	-0.5612	156.23	354.35	2.9640E-03	1.321
19	AsI3	ARSENIC TRIIODIDE	3.1240E-04	1101.22	-0.7143	419.15	588.19	-------	-------
20	As2O3	ARSENIC TRIOXIDE	2.8920E-04	1189.50	-0.7143	585.95	625.24	-------	-------
21	At	ASTATINE	----------	-------	-------	-------	-------	-------	-------
22	Au	GOLD	5.2360E-05	4398.00	-0.7143	1337.33	1460.46	-------	-------
23	B	BORON	4.3350E-05	7934.59	-0.7143	2348.15	2420.74	-------	-------
24	BBr3	BORON TRIBROMIDE	6.3660E-04	581.00	-0.7143	228.15	551.95	1.0650E-03	2.625
25	BCl3	BORON TRICHLORIDE	1.0040E-03	451.95	-0.4889	166.15	429.35	1.7010E-03	1.318
26	BF3	BORON TRIFLUORIDE	1.1350E-03	260.90	-0.7790	146.05	247.86	-------	-------
27	BH2CO	BORINE CARBONYL	----------	-------	-------	-------	-------	-------	-------
28	BH3O3	BORIC ACID	----------	-------	-------	-------	-------	-------	-------
29	B2D6	DEUTERODIBORANE	----------	-------	-------	-------	-------	-------	-------
30	B2H5Br	DIBORANE HYDROBROMIDE	----------	-------	-------	-------	-------	-------	-------
31	B2H6	DIBORANE	1.2990E-03	289.80	-0.7121	107.65	275.31	-------	-------
32	B3N3H6	BORINE TRIAMINE	7.0970E-04	521.20	-0.7143	214.95	495.14	1.3010E-03	0.902
33	B4H10	TETRABORANE	7.9260E-04	466.66	-0.7143	153.25	443.33	1.6410E-03	0.513
34	B5H9	PENTABORANE	6.3990E-04	568.45	-0.7143	226.35	540.03	1.0880E-03	0.643
35	B5H11	TETRAHYDROPENTABORANE	6.7610E-04	547.13	-0.7143	230.00	519.77	1.1860E-03	0.741
36	B10H14	DECABORANE	4.3450E-04	791.78	-0.7143	372.75	407.98	-------	-------
37	Ba	BARIUM	9.6300E-05	3572.13	-0.7143	1000.15	1140.14	-------	-------
38	Be	BERYLLIUM	6.6150E-05	5199.80	-0.7143	1560.15	1672.14	-------	-------
39	BeB2H8	BERYLLIUM BOROHYDRIDE	----------	-------	-------	-------	-------	-------	-------
40	BeBr2	BERYLLIUM BROMIDE	2.8270E-04	1216.86	-0.7143	763.15	873.73	-------	-------
41	BeCl2	BERYLLIUM CHLORIDE	2.7790E-04	1238.03	-0.7143	678.15	760.00	-------	-------
42	BeF2	BERYLLIUM FLUORIDE	----------	-------	-------	-------	-------	-------	-------
43	BeI2	BERYLLIUM IODIDE	2.7790E-04	1238.03	-0.7143	761.15	874.42	-------	-------
44	Bi	BISMUTH	1.1190E-04	4620.00	-0.7143	544.15	1173.39	-------	-------
45	BiBr3	BISMUTH TRIBROMIDE	2.8200E-04	1220.00	-0.7143	491.15	1045.00	-------	-------
46	BiCl3	BISMUTH TRICHLORIDE	2.9200E-04	1178.00	-0.7143	503.15	1045.00	-------	-------
47	BrF5	BROMINE PENTAFLUORIDE	7.8700E-04	470.00	-0.7143	211.75	446.50	1.6150E-03	2.466
48	Br2	BROMINE	6.8810E-04	584.15	-0.6705	265.85	554.94	1.1110E-03	3.104
49	C	CARBON	9.6610E-05	6810.00	-0.7143	4765.00	4716.75	-------	-------
50	CCl2O	PHOSGENE	7.8010E-04	455.00	-0.7280	145.37	432.25	1.6940E-03	1.363
51	CF2O	CARBONYL FLUORIDE	1.0690E-03	297.00	-0.7143	161.89	282.15	-------	-------
52	CH4N2O	UREA	4.4080E-04	705.00	-0.7143	405.85	480.56	-------	-------
53	CH4N2S	THIOUREA	4.1760E-04	854.00	-0.7143	454.15	811.30	-------	-------
54	CNBr	CYANOGEN BROMIDE	6.3110E-04	545.03	-0.7143	331.15	316.26	-------	-------
55	CNCl	CYANOGEN CHLORIDE	7.8160E-04	449.00	-0.7614	266.65	426.55	1.7930E-03	1.172
56	CNF	CYANOGEN FLUORIDE	1.0040E-03	368.51	-0.7143	200.00	350.09	3.2750E-03	1.144
57	CO	CARBON MONOXIDE	2.8090E-03	132.92	-0.7095	68.15	126.27	-------	-------
58	COS	CARBONYL SULFIDE	8.8830E-04	378.80	-0.7286	134.30	359.86	2.7420E-03	1.005
59	COSe	CARBON OXYSELENIDE	9.0980E-04	406.58	-0.7143	200.00	386.25	2.3380E-03	1.731
60	CO2	CARBON DIOXIDE	1.2800E-03	304.19	-0.7097	216.58	288.98	-------	0.713
61	CS2	CARBON DISULFIDE	7.2850E-04	552.00	-0.6774	161.11	524.40	1.2330E-03	1.256
62	CSeS	CARBON SELENOSULFIDE	6.4160E-04	576.53	-0.7143	197.95	547.71	1.0790E-03	1.977
63	C2N2	CYANOGEN	9.2500E-04	399.90	-0.7143	238.75	379.91	2.4590E-03	0.868
64	C3S2	CARBON SUBSULFIDE	----------	-------	-------	-------	-------	-------	-------
65	Ca	CALCIUM	1.0450E-04	3292.23	-0.7143	1115.15	1249.39	-------	-------
66	CaF2	CALCIUM FLUORIDE	1.4390E-04	4570.85	-0.7143	1691.00	2850.00	-------	-------
67	CbF5	COLUMBIUM FLUORIDE	4.2400E-04	811.32	-0.7143	348.65	402.23	-------	-------
68	Cd	CADMIUM	1.0410E-04	2291.00	-0.7143	594.05	829.49	-------	-------
69	CdCl2	CADMIUM CHLORIDE	1.7030E-04	2019.79	-0.7143	841.15	1140.00	-------	-------
70	CdF2	CADMIUM FLUORIDE	1.0430E-04	3296.66	-0.7143	793.15	943.49	-------	-------
71	CdI2	CADMIUM IODIDE	1.9760E-04	1741.29	-0.7143	658.15	1045.00	-------	-------
72	CdO	CADMIUM OXIDE	1.1530E-04	2983.96	-0.7143	1000.00	1140.00	-------	-------
73	ClF	CHLORINE MONOFLUORIDE	1.3100E-03	282.32	-0.7143	128.15	268.21	-------	-------
74	ClFO3	PERCHLORYL FLUORIDE	1.0560E-03	368.40	-0.7013	125.41	349.98	3.3750E-03	1.408
75	ClF3	CHLORINE TRIFLUORIDE	8.0520E-04	459.39	-0.7143	190.15	436.42	1.7010E-03	1.785
76	ClF5	CHLORINE PENTAFLUORIDE	7.2020E-04	415.90	-0.7143	190.00	395.11	1.7740E-03	1.774
77	ClHO3S	CHLOROSULFONIC ACID	5.1210E-04	700.00	-0.7143	193.15	665.00	7.6120E-04	1.741
78	ClHO4	PERCHLORIC ACID	7.3500E-04	631.00	-0.6548	171.95	599.45	1.1170E-03	1.760

			$B_{liq} = a \ (1 - T/T_C)^m$				$(B_{liq} - 1/C, \ T - K, \ density_{liq} - g/cm^3)$		
NO	FORMULA	NAME	a	T_C	m	TMIN	TMAX	B_{liq} @ 25 C	$density_{liq}$ @ 25 C
79	ClO2	CHLORINE DIOXIDE	7.9550E-04	465.00	-0.7143	213.55	441.75	1.6540E-03	1.579
80	Cl2	CHLORINE	8.9690E-04	417.15	-0.7117	172.12	396.29	2.1900E-03	1.398
81	Cl2O	CHLORINE MONOXIDE	8.3180E-04	444.68	-0.7143	157.15	422.45	1.8380E-03	1.549
82	Cl2O7	CHLORINE HEPTOXIDE	6.5380E-04	565.78	-0.7143	182.15	537.49	1.1160E-03	1.850
83	Co	COBALT	4.6490E-05	7398.48	-0.7143	1768.15	1869.74	-------	-------
84	CoCl2	COBALT CHLORIDE	1.5960E-04	2154.97	-0.7143	1008.15	1147.74	-------	-------
85	CoNC3O4	COBALT NITROSYL TRICARBONYL	----------	-------	-------	-------	-------	-------	-------
86	Cr	CHROMIUM	4.0180E-05	8560.93	-0.7143	2180.15	2261.14	-------	-------
87	CrC6O6	CHROMIUM CARBONYL	4.9800E-04	690.80	-0.7143	423.65	656.26	-------	-------
88	CrO2Cl2	CHROMIUM OXYCHLORIDE	5.9060E-04	626.33	-0.7143	176.65	595.02	9.3710E-04	1.902
89	Cs	CESIUM	2.7400E-04	2048.10	-0.7143	301.65	1928.50	-------	-------
90	CsBr	CESIUM BROMIDE	1.3430E-04	2562.13	-0.7143	909.15	1425.00	-------	-------
91	CsCl	CESIUM CHLORIDE	1.3430E-04	2562.13	-0.7143	919.15	1425.00	-------	-------
92	CsF	CESIUM FLUORIDE	1.3860E-04	2482.33	-0.7143	976.00	1425.00	-------	-------
93	CsI	CESIUM IODIDE	1.3600E-04	2529.56	-0.7143	894.15	1425.00	-------	-------
94	Cu	COPPER	6.7150E-05	5123.00	-0.7143	1357.77	1479.88	-------	-------
95	CuBr	CUPROUS BROMIDE	1.2970E-04	2651.71	-0.7143	777.15	928.29	-------	-------
96	CuCl	CUPROUS CHLORIDE	1.8560E-04	2435.00	-0.4319	703.00	1674.95	-------	-------
97	CuCl2	CUPRIC CHLORIDE	1.7110E-04	2010.00	-0.7143	906.15	1050.84	-------	-------
98	CuI	COPPER IODIDE	1.3130E-04	2620.77	-0.7143	878.15	1024.24	-------	-------
99	DCN	DEUTERIUM CYANIDE	----------	-------	-------	-------	-------	-------	-------
100	D2	DEUTERIUM	8.6060E-03	38.35	-0.7143	18.73	36.43	-------	-------
101	D2O	DEUTERIUM OXIDE	3.9540E-04	643.89	-0.8000	276.96	611.70	6.5030E-04	1.090
102	Eu	EUROPIUM	6.6790E-05	5150.00	-0.7143	1095.15	1230.39	-------	-------
103	F2	FLUORINE	2.5210E-03	144.31	-0.7100	53.48	137.09	-------	-------
104	F2O	FLUORINE OXIDE	1.7380E-03	215.10	-0.7143	49.25	204.35	-------	-------
105	Fe	IRON	8.1350E-05	9340.00	-0.7143	1808.10	2375.00	-------	-------
106	FeC5O5	IRON PENTACARBONYL	6.0920E-04	607.20	-0.7143	252.15	576.84	9.8680E-04	1.451
107	FeCl2	FERROUS CHLORIDE	1.6260E-04	2115.88	-0.7143	945.15	1087.89	-------	-------
108	FeCl3	FERRIC CHLORIDE	3.5670E-04	964.41	-0.7143	577.15	555.42	-------	-------
109	Fr	FRANCIUM	----------	-------	-------	-------	-------	-------	-------
110	Ga	GALLIUM	9.6500E-05	7620.00	-0.7143	302.91	1304.49	-------	-------
111	GaCl3	GALLIUM TRICHLORIDE	5.9780E-04	694.00	-0.7002	350.90	659.30	-------	-------
112	Gd	GADOLINIUM	1.0400E-04	3307.66	-0.7143	1587.15	1594.65	-------	-------
113	Ge	GERMANIUM	4.0950E-05	8400.00	-0.7143	1211.40	1340.83	-------	-------
114	GeBr4	GERMANIUM BROMIDE	4.9990E-04	740.00	-0.7143	299.25	361.67	-------	-------
115	GeCl4	GERMANIUM CHLORIDE	6.4440E-04	574.00	-0.7143	223.65	545.30	1.0880E-03	1.854
116	GeHCl3	TRICHLORO GERMANE	6.6080E-04	559.78	-0.7143	202.05	531.79	1.1380E-03	1.919
117	GeH4	GERMANE	1.1090E-03	308.00	-0.7143	107.26	292.60	-------	0.859
118	Ge2H6	DIGERMANE	7.5330E-04	491.01	-0.7143	164.15	466.46	1.4690E-03	1.701
119	Ge3H8	TRIGERMANE	6.0010E-04	616.37	-0.7143	167.55	585.55	9.6230E-04	2.211
120	HBr	HYDROGEN BROMIDE	9.8630E-04	363.15	-0.7143	185.15	344.99	3.3710E-03	1.728
121	HCN	HYDROGEN CYANIDE	1.0390E-03	456.65	-0.7179	259.83	433.82	2.2220E-03	0.680
122	HCl	HYDROGEN CHLORIDE	1.2870E-03	324.65	-0.6813	158.97	308.42	7.0940E-03	0.796
123	HF	HYDROGEN FLUORIDE	1.4040E-03	461.15	-0.6267	189.79	438.09	2.6930E-03	0.941
124	HI	HYDROGEN IODIDE	8.3850E-04	423.85	-0.7143	222.38	402.66	1.9980E-03	2.520
125	HNO3	NITRIC ACID	5.4000E-04	520.00	-0.8083	231.55	354.49	1.0750E-03	1.509
126	H2	HYDROGEN	8.7840E-03	33.18	-0.7244	13.95	31.52	-------	-------
127	H2O	WATER	5.7160E-04	647.13	-0.7143	273.16	614.77	8.8850E-04	1.027
128	H2O2	HYDROGEN PEROXIDE	5.4650E-04	730.15	-0.7123	272.74	693.64	7.9430E-04	1.443
129	H2S	HYDROGEN SULFIDE	9.6930E-04	373.53	-0.7143	187.68	354.85	3.0400E-03	0.777
130	H2SO4	SULFURIC ACID	5.0720E-04	925.00	-0.7143	283.46	345.32	6.6970E-04	1.833
131	H2S2	HYDROGEN DISULFIDE	6.8200E-04	542.39	-0.7143	183.45	515.27	1.2060E-03	1.326
132	H2Se	HYDROGEN SELENIDE	8.9980E-04	411.10	-0.7143	209.15	390.55	2.2640E-03	1.766
133	H2Te	HYDROGEN TELLURIDE	8.4440E-04	438.04	-0.7143	224.15	416.14	1.9080E-03	2.378
134	H3NO3S	SULFAMIC ACID	----------	-------	-------	-------	-------	-------	-------
135	He	HELIUM-3	5.1360E-02	3.31	-0.7770	1.01	3.14	-------	-------
136	He	HELIUM-4	4.0360E-02	5.20	-0.7590	1.76	4.94	-------	-------
137	Hf	HAFNIUM	1.5860E-05	21688.00	-0.7143	2506.15	2570.84	-------	-------
138	Hg	MERCURY	1.2770E-04	1735.00	-0.8385	234.31	1648.25	1.4960E-04	13.487
139	HgBr2	MERCURIC BROMIDE	3.5670E-04	964.41	-0.7143	510.15	760.00	-------	-------
140	HgCl2	MERCURIC CHLORIDE	3.6600E-04	939.98	-0.7143	550.15	760.00	-------	-------
141	HgI2	MERCURIC IODIDE	3.1910E-04	1078.10	-0.7143	532.15	712.50	-------	-------
142	IF7	IODINE HEPTAFLUORIDE	8.2650E-04	447.53	-0.7143	278.65	425.15	1.8100E-03	2.709
143	I2	IODINE	4.5020E-04	819.15	-0.6645	386.75	778.19	-------	-------
144	In	INDIUM	1.0960E-04	6730.00	-0.7143	429.75	544.49	-------	-------
145	Ir	IRIDIUM	2.2880E-05	15035.00	-0.7143	2719.15	2773.19	-------	-------
146	K	POTASSIUM	1.9680E-04	2223.00	-0.7143	336.35	2109.00	-------	-------
147	KBr	POTASSIUM BROMIDE	1.2750E-04	2697.31	-0.7143	1003.15	1425.00	-------	-------
148	KCl	POTASSIUM CHLORIDE	3.2260E-04	3470.00	-0.6448	1044.00	2850.00	-------	-------
149	KF	POTASSIUM FLUORIDE	1.1900E-04	2891.12	-0.7143	1153.15	1425.00	-------	-------
150	KI	POTASSIUM IODIDE	1.3220E-04	2601.22	-0.7143	996.15	1425.00	-------	-------
151	KOH	POTASSIUM HYDROXIDE	2.0510E-04	2605.86	-0.7143	679.00	1425.00	-------	-------
152	Kr	KRYPTON	1.7460E-03	209.35	-0.7061	115.78	198.88	-------	-------
153	La	LANTHANUM	4.1400E-05	9511.00	-0.7143	1193.15	1323.49	-------	-------
154	Li	LITHIUM	2.0140E-04	4085.00	-0.7143	453.69	3486.50	-------	-------
155	LiBr	LITHIUM BROMIDE	1.3340E-04	2578.42	-0.7143	820.15	1425.00	-------	-------
156	LiCl	LITHIUM CHLORIDE	1.2760E-04	2695.68	-0.7143	887.15	1425.00	-------	-------

| NO | FORMULA | NAME | $B_{liq} = a \ (1 - T/T_C)^m$ | | | | | $(B_{liq} - 1/C, \ T - K, \ density_{liq} - g/cm^3)$ | |
			a	T_C	m	TMIN	TMAX	B_{liq} @ 25 C	$density_{liq}$ @ 25 C
157	LiF	LITHIUM FLUORIDE	1.0810E-04	3182.65	-0.7143	1143.15	1615.00	-------	-------
158	LiI	LITHIUM IODIDE	1.4630E-04	2352.04	-0.7143	719.15	1520.00	-------	-------
159	Lu	LUTECIUM	8.3320E-05	4128.66	-0.7143	1936.15	2029.34	-------	-------
160	Mg	MAGNESIUM	4.4660E-04	2241.04	-0.7143	923.15	971.99	-------	-------
161	MgCl2	MAGNESIUM CHLORIDE	1.2490E-04	2754.32	-0.7143	985.15	1520.00	-------	-------
162	MgO	MAGNESIUM OXIDE	2.0360E-04	5950.00	-0.7143	3105.00	5652.50	-------	-------
163	Mn	MANGANESE	4.9830E-05	6902.82	-0.7143	1519.15	1633.19	-------	-------
164	MnCl2	MANGANESE CHLORIDE	1.4440E-04	2382.98	-0.7143	923.15	1066.99	-------	-------
165	Mo	MOLYBDENUM	3.5760E-05	9620.00	-0.7143	2895.15	2940.39	-------	-------
166	MoF6	MOLYBDENUM FLUORIDE	7.4260E-04	498.12	-0.7143	290.15	473.21	1.4250E-03	2.523
167	MoO3	MOLYBDENUM OXIDE	1.4830E-04	2319.46	-0.7143	1068.15	1204.74	-------	-------
168	NCl3	NITROGEN TRICHLORIDE	6.5580E-04	564.00	-0.7143	293.00	535.80	1.1220E-03	1.641
169	ND3	HEAVY AMMONIA	----------	-------	-------	-------	-------	-------	-------
170	NF3	NITROGEN TRIFLUORIDE	1.5760E-03	233.85	-0.7086	66.36	222.16	-------	-------
171	NH3	AMMONIA	9.7330E-04	405.65	-0.7113	195.41	385.37	2.5030E-03	0.602
172	NH3O	HYDROXYLAMINE	6.4440E-04	574.00	-0.7143	306.15	545.30	-------	-------
173	NH4Br	AMMONIUM BROMIDE	3.1560E-04	1089.82	-0.7143	800.00	950.00	-------	-------
174	NH4Cl	AMMONIUM CHLORIDE	3.9000E-04	882.00	-0.7143	793.20	839.04	-------	-------
175	NH4I	AMMONIUM IODIDE	3.1150E-04	1104.32	-0.7143	800.00	950.00	-------	-------
176	NH5O	AMMONIUM HYDROXIDE	----------	-------	-------	-------	-------	-------	-------
177	NH5S	AMMONIUM HYDROGENSULFIDE	6.8920E-04	499.10	-0.7143	391.15	432.60	-------	-------
178	NO	NITRIC OXIDE	1.5980E-03	180.15	-0.7580	109.50	171.14	-------	-------
179	NOCl	NITROSYL CHLORIDE	7.3880E-04	440.65	-0.7512	213.55	418.62	1.7250E-03	1.262
180	NOF	NITROSYL FLUORIDE	1.0490E-03	352.67	-0.7143	139.15	335.04	3.9800E-03	1.058
181	NO2	NITROGEN DIOXIDE	7.3380E-04	431.35	-0.7568	293.15	409.78	1.7860E-03	1.442
182	N2	NITROGEN	2.9130E-03	126.10	-0.7075	63.15	119.80	-------	-------
183	N2F4	TETRAFLUOROHYDRAZINE	1.1840E-03	309.35	-0.7143	111.65	300.00	1.2670E-02	0.888
184	N2H4	HYDRAZINE	5.2160E-04	653.15	-0.8102	274.68	620.49	8.5480E-04	1.004
185	N2H4C	AMMONIUM CYANIDE	----------	-------	-------	-------	-------	-------	-------
186	N2H6CO2	AMMONIUM CARBAMATE	----------	-------	-------	-------	-------	-------	-------
187	N2O	NITROUS OXIDE	1.2110E-03	309.57	-0.7118	182.30	300.00	1.2680E-02	0.742
188	N2O3	NITROGEN TRIOXIDE	1.1530E-03	425.00	-0.7143	263.15	403.75	2.7340E-03	1.331
189	N2O4	NITROGEN TETRAOXIDE	9.6480E-04	431.15	-0.7143	261.90	409.59	2.2350E-03	1.434
190	N2O5	NITROGEN PENTOXIDE	7.1750E-04	515.51	-0.7143	303.15	296.07	-------	-------
191	Na	SODIUM	2.5330E-04	2573.00	-0.7143	370.98	2375.00	-------	-------
192	NaBr	SODIUM BROMIDE	3.1410E-04	4287.00	-0.4709	1020.00	2375.00	-------	-------
193	NaCN	SODIUM CYANIDE	2.1750E-04	2900.00	-0.7143	973.15	1114.49	-------	-------
194	NaCl	SODIUM CHLORIDE	2.4780E-04	3400.00	-0.6247	1073.90	3230.00	-------	-------
195	NaF	SODIUM FLUORIDE	3.0310E-04	5530.00	-0.3430	1269.00	2375.00	-------	-------
196	NaI	SODIUM IODIDE	1.3390E-04	2568.65	-0.7143	924.15	1520.00	-------	-------
197	NaOH	SODIUM HYDROXIDE	2.0910E-04	2820.00	-0.7462	596.00	1900.00	-------	-------
198	Na2SO4	SODIUM SULFATE	1.7780E-04	3700.00	-0.7143	1157.00	1900.00	-------	-------
199	Nb	NIOBIUM	1.9210E-05	17904.10	-0.7143	2750.15	2802.64	-------	-------
200	Nd	NEODYMIUM	3.2250E-05	10665.10	-0.7143	1289.15	1414.69	-------	-------
201	Ne	NEON	7.4160E-03	44.40	-0.7214	24.56	42.18	-------	-------
202	Ni	NICKEL	4.9240E-05	6986.15	-0.7143	1728.15	1831.74	-------	-------
203	NiC4O4	NICKEL CARBONYL	7.2760E-04	508.40	-0.7143	248.15	482.98	1.3670E-03	1.306
204	NiF2	NICKEL FLUORIDE	1.0490E-04	3278.75	-0.7143	1723.15	1826.99	-------	-------
205	Np	NEPTUNIUM	----------	-------	-------	-------	-------	-------	-------
206	O2	OXYGEN	2.3560E-03	154.58	-0.7076	54.35	146.85	-------	-------
207	O3	OZONE	1.3350E-03	261.00	-0.7120	80.15	247.95	-------	-------
208	Os	OSMIUM	2.0380E-05	16878.70	-0.7143	3306.15	3330.84	-------	-------
209	OsOF5	OSMIUM OXIDE PENTAFLUORIDE	----------	-------	-------	-------	-------	-------	-------
210	OsO4	OSMIUM TETROXIDE - YELLOW	5.2390E-04	656.60	-0.7143	329.15	347.84	-------	-------
211	OsO4	OSMIUM TETROXIDE - WHITE	5.2390E-04	656.60	-0.7143	315.15	341.19	-------	-------
212	P	PHOSPHORUS - WHITE	4.6210E-04	993.75	-0.3364	317.55	525.49	-------	-------
213	PBr3	PHOSPHORUS TRIBROMIDE	5.2020E-04	711.00	-0.7143	233.15	675.45	7.6710E-04	2.830
214	PCl2F3	PHOSPHORUS DICHLORIDE TRIFLUORIDE	----------	-------	-------	-------	-------	-------	-------
215	PCl3	PHOSPHORUS TRICHLORIDE	7.1670E-04	563.15	-0.7031	200.00	331.79	1.2180E-03	1.566
216	PCl5	PHOSPHORUS PENTACHLORIDE	5.3240E-04	646.15	-0.7143	433.15	478.99	-------	-------
217	PH3	PHOSPHINE	1.4110E-03	324.75	-0.6308	139.37	308.51	6.8370E-03	0.491
218	PH4Br	PHOSPHONIUM BROMIDE	----------	-------	-------	-------	-------	-------	-------
219	PH4Cl	PHOSPHONIUM CHLORIDE	----------	-------	-------	-------	-------	-------	-------
220	PH4I	PHOSPHONIUM IODIDE	6.8540E-04	539.70	-0.7143	291.65	512.71	1.2170E-03	2.843
221	POCl3	PHOSPHORUS OXYCHLORIDE	7.4090E-04	602.15	-0.5741	274.33	359.72	1.0970E-03	1.667
222	PSBr3	PHOSPHORUS THIOBROMIDE	4.7130E-04	729.89	-0.7143	311.15	360.67	-------	-------
223	PSCl3	PHOSPHORUS THIOCHLORIDE	5.7900E-04	638.82	-0.7143	242.75	353.07	9.0730E-04	1.643
224	P4O6	PHOSPHORUS TRIOXIDE	5.1740E-04	714.86	-0.7143	295.65	679.12	7.6080E-04	2.128
225	P4O10	PHOSPHORUS PENTOXIDE	----------	-------	-------	-------	-------	-------	-------
226	P4S10	PHOSPHORUS PENTASULFIDE	2.6650E-04	1291.00	-0.7143	596.15	1226.45	-------	-------
227	Pb	LEAD	9.9390E-05	5400.00	-0.7143	600.64	1209.49	-------	-------
228	PbBr2	LEAD BROMIDE	1.7790E-04	1933.47	-0.7143	646.15	1425.00	-------	-------
229	PbCl2	LEAD CHLORIDE	1.7210E-04	1998.62	-0.7143	774.15	1425.00	-------	-------
230	PbF2	LEAD FLUORIDE	1.3490E-04	2550.73	-0.7143	1128.15	1261.74	-------	-------
231	PbI2	LEAD IODIDE	1.8440E-04	1865.07	-0.7143	675.15	831.39	-------	-------
232	PbO	LEAD OXIDE	1.2100E-04	2842.26	-0.7143	1163.15	1294.99	-------	-------
233	PbS	LEAD SULFIDE	1.3590E-04	2531.19	-0.7143	1387.15	1397.12	-------	-------
234	Pd	PALLADIUM	3.2240E-05	10669.10	-0.7143	1828.05	1926.65	-------	-------

| NO | FORMULA | NAME | $B_{liq} = a \ (1 - T/T_C)^m$ | | | | | $(B_{liq} - 1/C, \ T - K, \ density_{liq} - g/cm^3)$ | |
			a	T_C	m	TMIN	TMAX	B_{liq} @ 25 C	$density_{liq}$ @ 25 C
235	Po	POLONIUM	1.1420E-04	3013.08	-0.7143	527.15	690.79	-------	-------
236	Pt	PLATINUM	4.9260E-05	6983.00	-0.7143	2041.55	2129.47	-------	-------
237	Ra	RADIUM	7.0740E-05	4862.82	-0.7143	973.15	1114.49	-------	-------
238	Rb	RUBIDIUM	3.1370E-04	2111.10	-0.7143	312.46	1928.50	-------	-------
239	RbBr	RUBIDIUM BROMIDE	1.3000E-04	2646.82	-0.7143	955.15	1710.00	-------	-------
240	RbCl	RUBIDIUM CHLORIDE	1.2770E-04	2694.06	-0.7143	988.15	1710.00	-------	-------
241	RbF	RUBIDIUM FLUORIDE	1.2560E-04	2738.03	-0.7143	1033.15	1171.49	-------	-------
242	RbI	RUBIDIUM IODIDE	1.3390E-04	2568.65	-0.7143	915.15	1710.00	-------	-------
243	Re	RHENIUM	1.6010E-05	21482.80	-0.7143	3459.15	3476.19	-------	-------
244	Re2O7	RHENIUM HEPTOXIDE	3.3230E-04	1035.10	-0.7143	569.15	572.23	-------	-------
245	Rh	RHODIUM	2.6650E-05	12906.60	-0.7143	2237.15	2315.29	-------	-------
246	Rn	RADON	9.6070E-04	377.40	-0.7143	202.15	358.53	2.9290E-03	3.578
247	Ru	RUTHENIUM	2.2560E-05	15247.10	-0.7143	2607.15	2666.79	-------	-------
248	RuF5	RUTHENIUM PENTAFLUORIDE	3.5190E-04	977.44	-0.7143	359.65	531.67	-------	-------
249	S	SULFUR	1.9620E-04	1313.00	-0.8860	388.36	1247.35	-------	-------
250	SF4	SULFUR TETRAFLUORIDE	1.0160E-03	364.00	-0.7143	149.15	345.80	3.4460E-03	1.526
251	SF6	SULFUR HEXAFLUORIDE	1.2000E-03	318.69	-0.7079	223.15	302.76	8.3570E-03	1.322
252	SOBr2	THIONYL BROMIDE	5.5900E-04	661.75	-0.7143	220.55	628.66	8.5730E-04	2.659
253	SOCl2	THIONYL CHLORIDE	6.9430E-04	567.00	-0.6937	172.00	580.11	1.1650E-03	1.630
254	SOF2	SULFUROUS OXYFLUORIDE	9.9630E-04	371.25	-0.7143	162.65	352.69	3.1810E-03	1.361
255	SO2	SULFUR DIOXIDE	9.1740E-04	430.75	-0.7107	197.67	409.21	2.1190E-03	1.366
256	SO2Cl2	SULFURYL CHLORIDE	7.3090E-04	545.00	-0.6917	222.00	517.75	1.2640E-03	1.658
257	SO3	SULFUR TRIOXIDE	1.3870E-03	490.85	-0.5821	289.95	466.31	2.3910E-03	1.897
258	S2Cl2	SULFUR MONOCHLORIDE	5.6100E-04	659.37	-0.7143	193.15	626.41	8.6220E-04	1.671
259	Sb	ANTIMONY	7.9880E-05	5070.00	-0.7143	903.78	1180.99	-------	-------
260	SbBr3	ANTIMONY TRIBROMIDE	3.8530E-04	892.75	-0.7143	369.75	436.00	-------	-------
261	SbCl3	ANTIMONY TRICHLORIDE	5.5110E-04	794.00	-0.7143	346.55	754.30	-------	-------
262	SbCl5	ANTIMONY PENTACHLORIDE	5.5830E-04	662.54	-0.7143	275.95	629.41	8.5570E-04	2.326
263	SbH3	STIBINE	7.5220E-04	440.35	-0.7143	185.15	418.33	1.6860E-03	2.096
264	SbI3	ANTIMONY TRIIODIDE	3.1330E-04	1097.96	-0.7143	440.15	608.14	-------	-------
265	Sb2O3	ANTIMONY TRIOXIDE	1.2440E-04	2765.72	-0.7143	929.15	1072.69	-------	-------
266	Sc	SCANDIUM	4.2810E-05	8035.08	-0.7143	1814.15	1913.44	-------	-------
267	Se	SELENIUM	2.9520E-04	1766.00	-0.7143	494.15	659.44	-------	-------
268	SeCl4	SELENIUM TETRACHLORIDE	4.9720E-04	743.95	-0.7143	400.00	570.00	-------	-------
269	SeF6	SELENIUM HEXAFLUORIDE	1.0030E-03	368.80	-0.7143	238.45	350.36	3.2650E-03	1.911
270	SeOCl2	SELENIUM OXYCHLORIDE	5.2330E-04	706.80	-0.7143	281.65	671.46	7.7400E-04	2.414
271	SeO2	SELENIUM DIOXIDE	3.5790E-04	961.16	-0.7143	613.15	699.98	-------	-------
272	Si	SILICON	6.8230E-05	5159.00	-0.7143	1685.00	1790.75	-------	-------
273	SiBrCl2F	BROMODICHLOROFLUOROSILANE	7.4400E-04	497.17	-0.7143	160.85	472.31	1.4310E-03	1.812
274	SiBrF3	TRIFLUOROBROMOSILANE	9.8560E-04	375.28	-0.7143	202.65	356.52	3.0520E-03	1.651
275	SiBr2ClF	DIBROMOCHLOROFLUOROSILANE	6.9100E-04	535.27	-0.7143	173.85	508.51	1.2360E-03	2.286
276	SiClF3	TRIFLUOROCHLOROSILANE	1.1190E-03	330.54	-0.7143	131.15	314.01	5.8810E-03	1.099
277	SiCl2F2	DICHLORODIFLUOROSILANE	9.4620E-04	390.93	-0.7143	133.45	371.39	2.6430E-03	1.305
278	SiCl3F	TRICHLOROFLUOROSILANE	8.0330E-04	460.49	-0.7143	152.35	437.47	1.6920E-03	1.424
279	SiCl4	SILICON TETRACHLORIDE	7.3910E-04	507.00	-0.7185	204.30	481.65	1.3980E-03	1.470
280	SiF4	SILICON TETRAFLUORIDE	1.3850E-03	259.00	-0.7143	186.35	246.05	-------	-------
281	SiHBr3	TRIBROMOSILANE	5.2720E-04	610.00	-0.7143	199.65	579.50	8.5130E-04	2.682
282	SiHCl3	TRICHLOROSILANE	9.8880E-04	479.00	-0.6510	144.95	455.05	1.8640E-03	1.328
283	SiHF3	TRIFLUOROSILANE	1.2710E-03	291.02	-0.7143	141.75	276.47	-------	-------
284	SiH2Br2	DIBROMOSILANE	6.5190E-04	550.00	-0.7143	202.95	522.50	1.1390E-03	2.111
285	SiH2Cl2	DICHLOROSILANE	8.3580E-04	449.00	-0.7143	151.15	426.55	1.8220E-03	1.167
286	SiH2F2	DIFLUOROSILANE	1.1620E-03	318.21	-0.7143	150.00	302.30	8.3710E-03	0.805
287	SiH2I2	DIIODOSILANE	5.4680E-04	660.00	-0.7143	272.15	627.00	8.4000E-04	2.825
288	SiH3Br	MONOBROMOSILANE	8.3650E-04	454.00	-0.7143	179.25	431.30	1.7950E-03	1.473
289	SiH3Cl	MONOCHLOROSILANE	9.8280E-04	396.65	-0.7143	155.05	376.82	2.6580E-03	0.882
290	SiH3F	MONOFLUOROSILANE	1.2920E-03	286.28	-0.7143	150.00	271.96	-------	-------
291	SiH3I	IODOSILANE	7.4860E-04	515.00	-0.7143	216.15	489.25	1.3890E-03	2.031
292	SiH4	SILANE	1.8470E-03	269.70	-0.6042	88.48	256.22	-------	-------
293	SiO2	SILICON DIOXIDE	1.6140E-04	4076.87	-0.7143	2000.00	2090.00	-------	-------
294	Si2Cl6	HEXACHLORODISILANE	5.5960E-04	660.96	-0.7143	271.95	627.91	8.5900E-04	1.547
295	Si2F6	HEXAFLUORODISILANE	8.9930E-04	411.33	-0.7143	254.55	390.76	2.2600E-03	1.341
296	Si2H5Cl	DISILANYL CHLORIDE	7.2970E-04	506.89	-0.7143	250.00	481.55	1.3750E-03	0.951
297	Si2H6	DISILANE	8.8060E-04	432.00	-0.7049	143.85	410.40	2.0110E-03	0.782
298	Si2OCl3F3	TRICHLOROTRIFLUORODISILOXANE	7.2700E-04	508.78	-0.7143	150.00	483.34	1.3650E-03	1.488
299	Si2OCl6	HEXACHLORODISILOXANE	5.6420E-04	655.58	-0.7143	239.95	622.80	8.7020E-04	1.581
300	Si2OH6	DISILOXANE	8.8730E-04	416.86	-0.7143	128.95	396.02	2.1760E-03	0.772
301	Si3Cl8	OCTACHLOROTRISILANE	4.7700E-04	775.41	-0.7143	250.00	736.64	6.7470E-04	1.588
302	Si3H8	TRISILANE	7.0440E-04	525.15	-0.7143	155.95	498.90	1.2820E-03	0.720
303	Si3H9N	TRISILAZANE	7.1380E-04	518.20	-0.7143	167.45	492.29	1.3160E-03	0.858
304	Si4H10	TETRASILANE	6.1720E-04	599.30	-0.7143	179.55	569.33	1.0090E-03	0.771
305	Sm	SAMARIUM	6.7680E-05	5082.92	-0.7143	1345.15	1467.89	-------	-------
306	Sn	TIN	9.1070E-05	7400.00	-0.7143	505.08	928.29	-------	-------
307	SnBr4	STANNIC BROMIDE	4.8360E-04	764.82	-0.7143	304.15	371.45	-------	-------
308	SnCl2	STANNOUS CHLORIDE	2.3570E-04	1459.53	-0.7143	519.95	1140.00	-------	-------
309	SnCl4	STANNIC CHLORIDE	5.9670E-04	619.85	-0.7143	242.95	588.86	9.5330E-04	2.215
310	SnH4	STANNIC HYDRIDE	---------	-------	-------	-------	-------	-------	-------
311	SnI4	STANNIC IODIDE	3.4000E-04	1011.64	-0.7143	417.65	586.77	-------	-------
312	Sr	STRONTIUM	8.0610E-05	4267.20	-0.7143	1050.15	1187.64	-------	-------

NO	FORMULA	NAME	$B_{liq} = a \ (1 - T/T_C)^m$			$(B_{liq} - 1/C, \ T - K, \ density_{liq} - g/cm^3)$			
			a	T_C	m	TMIN	TMAX	B_{liq} @ 25 C	$density_{liq}$ @ 25 C
313	SrO	STRONTIUM OXIDE	----------	-------	-------	-------	-------	-------	-------
314	Ta	TANTALUM	1.7290E-05	19900.90	-0.7143	3290.15	3315.64	-------	-------
315	Tc	TECNNETIUM	1.9770E-05	17400.80	-0.7143	2430.15	2498.64	-------	-------
316	Te	TELLURIUM	7.1070E-05	4840.00	-0.7143	722.66	876.53	-------	-------
317	TeCl4	TELLURIUM TETRACHLORIDE	3.1750E-04	1083.31	-0.7143	497.15	552.09	-------	-------
318	TeF6	TELLURIUM HEXAFLUORIDE	9.7290E-04	380.18	-0.7143	235.35	361.17	2.9090E-03	2.684
319	Ti	TITANIUM	5.3750E-05	6400.00	-0.7143	1941.15	2034.09	-------	-------
320	TiCl4	TITANIUM TETRACHLORIDE	5.7030E-04	638.00	-0.7274	249.05	606.10	9.0170E-04	1.714
321	Tl	THALLIUM	1.1750E-04	4648.06	-0.7143	577.15	738.15	-------	-------
322	TlBr	THALLOUS BROMIDE	1.9340E-04	1778.75	-0.7143	733.15	886.49	-------	-------
323	TlI	THALLOUS IODIDE	1.9270E-04	1785.26	-0.7143	713.15	867.49	-------	-------
324	Tm	THULIUM	5.4750E-05	6283.16	-0.7143	1818.15	1917.24	-------	-------
325	U	URANIUM	2.5090E-05	13712.60	-0.7143	1408.15	1527.74	-------	-------
326	UF6	URANIUM FLUORIDE	7.2510E-04	505.80	-0.7143	342.35	381.27	-------	-------
327	V	VANADIUM	2.9180E-05	11787.10	-0.7143	2183.15	2263.99	-------	-------
328	VCl4	VANADIUM TETRACHLORIDE	4.9490E-04	697.00	-0.6884	247.45	662.15	7.2680E-04	1.821
329	VOCl3	VANADIUM OXYTRICHLORIDE	6.1400E-04	636.00	-0.6830	193.65	380.00	9.4580E-04	1.637
330	W	TUNGSTEN	2.3310E-05	14756.00	-0.7143	3695.15	3700.39	-------	-------
331	WF6	TUNGSTEN FLUORIDE	7.8940E-04	468.56	-0.7143	272.65	445.13	1.6260E-03	3.387
332	Xe	XENON	1.2530E-03	289.74	-0.7103	161.36	275.25	-------	-------
333	Yb	YTTERBIUM	7.8790E-05	4365.92	-0.7143	1097.15	1232.29	-------	-------
334	Yt	YTTRIUM	3.6670E-05	9381.32	-0.7143	1799.15	1899.19	-------	-------
335	Zn	ZINC	1.0880E-04	3170.00	-0.7143	692.70	1019.49	-------	-------
336	ZnCl2	ZINC CHLORIDE	2.1010E-04	1637.05	-0.7143	638.15	1140.00	-------	-------
337	ZnF2	ZINC FLUORIDE	1.1930E-04	2882.98	-0.7143	1145.15	1277.89	-------	-------
338	ZnO	ZINC OXIDE	----------	-------	-------	-------	-------	-------	-------
339	ZnSO4	ZINC SULFATE	----------	-------	-------	-------	-------	-------	-------
340	Zr	ZIRCONIUM	3.9080E-05	8802.00	-0.7143	2128.15	2211.74	-------	-------
341	ZrBr4	ZIRCONIUM BROMIDE	----------	-------	-------	-------	-------	-------	-------
342	ZrCl4	ZIRCONIUM CHLORIDE	3.4960E-04	983.96	-0.7143	710.15	794.91	-------	-------
343	ZrI4	ZIRCONIUM IODIDE	----------	-------	-------	-------	-------	-------	-------

B_{liq} - coefficient of thermal expansion of liquid, 1/C

T - temperature, K

$density_{liq}$ - density of liquid, g/cm³

TMIN - minimum temperature, K

TMAX - maximum temperature, K

Appendix A

CONVERSION TABLE

1. Temperature
To convert from Centigrade to:
Kelvin, add 273.15
Rankine, multiply Kelvin by 1.8
Fahrenheit, multiply Centigrade by 1.8 and add 32

2. Pressure
To convert from psia to:
kPa, multiply by 6.895
psig, subtract 14.7
mm Hg, multiply by 51.71
atmospheres, divide by 14.7
bars, divide by 14.508

3. Heat of Vaporization
To convert from kJ/kg to:
BTU/lb, multiply by 0.43
cal/gram, multiply by 0.239

4. Density
To convert from g/ml to:
lb/ft^3, multiply by 62.43
lb/gallon, multiply by 8.345

5. Surface Tension
To convert from dynes/cm to:
N/m, multiply by 0.001

6. Heat Capacity
To convert from J/g K to:
BTU/lb R, multiply by 0.239
cal/gram K, multiply by 0.239

7. Viscosity
To convert from micropoise to:
lb/ft s, multiply by 0.0672E-06
centipoise, multiply by 1.0E-04
poise, multiply by 1.0E-06
Pa s (Pascal seconds), multiply by 1.0E-07

To convert from centipoise to:
lb/ft s, multiply by 0.000672
micropoise, multiply by 10,000
poise, multiply by 0.01
Pa s (Pascal seconds), multiply by 0.001

8. Thermal Conductivity
To convert from W/m K to:
BTU/hr ft R, multiply by 0.5770
calorie/cm s K, multiply by .002388

9. Enthalpy of Formation
To convert from kJ/mol to:
kcal/mol, multiply by 0.239

10. Gibbs Energy of Formation
To convert from kJ/mol to:
kcal/mol, multiply by 0.239

11. Henry's Law Constant for Compound in Water
To convert from atm/mol fraction to:
atm/(mol/m^3), divide by 55,556
kPa/(mol/m^3), divide by 548.295

Appendix B

HENRY'S LAW CONSTANT - EQUATIONS

Carl L. Yaws

Lamar University, Beaumont, Texas

The calculation of Henry's law constant for a component in water may be achieved using data for solubility, vapor pressure, and activity coefficient at infinite dilution. The derivation of the appropriate equations is briefly given in the following discussion.

LIQUIDS (PARTIAL SOLUBILITY)

For organic chemicals that are liquids at ambient conditions and have partial solubility in water, there are three phases when the organic chemical is in contact with water. These are vapor, organic, and water phases. Such a three-phase system consisting of vapor, liquid I and liquid II is shown in Fig. B-1a. At equilibrium, the fugacity of the component in each liquid phase is

$$f_i^{liq\ I} = f_i^{liq\ II} \tag{B-1}$$

For the organic phase (liquid I), the fugacity of the component is γ_i * mol fraction$_i$ * vapor pressure$_i$ (where γ_i is the activity coefficient). Since the organic phase has only very small concentration of water (ppm level or less), the mol fraction of the organic chemical is approximately equal to 1 (mol fraction$_i \approx 1$). This is also true for the activity coefficient of the organic chemical ($\gamma_i \approx 1$). Thus

$$f_i^{liq\ I} = P_i^{SAT} \tag{B-2}$$

For the water phase (liquid II), the fugacity of the component is given by Henry's law which is applicable at very small concentration. The equation is

$$f_i^{liq\ II} = H_i\, x_i^{liq\ II} \qquad (x_i << 1) \tag{B-3}$$

Substitution of Equations (B-2) and (B-3) into Equation (B-1) yields

$$P_i^{SAT} = H_i\, x_i^{liq\ II} \tag{B-4}$$

Solving for Henry's law constant yields the following equation which is applicable to organic chemicals which are liquids at ambient conditions (25 C, 1 atm) and have only small partial solubility in water:

$$H_i = (\ 1\ /\ x_i^{liq\ II}\ )\, P_i^{SAT} \tag{B-5}$$

where H_i = Henry's law constant, atm/mol fraction
   $x_i^{liq\ II}$ = solubility of organic chemical in water, mol fraction
   P_i^{SAT} = vapor pressure of organic chemical, atm

LIQUIDS (TOTAL SOLUBILITY)

For organic chemicals that are liquids at ambient conditions and have total solubility in water, there are two phases when the organic chemical is in contact with water. These are vapor and liquid phases. Fig. B-1b shows such a two-phase system.

For the liquid phase, the fugacity of the organic chemical is γ_i * mol fraction$_i$ * vapor pressure$_i$ (where γ_i is the activity coefficient). Since the liquid phase has only very small concentration of organic chemical (ppm level or less) in the region where Henry's law is applicable, the activity coefficient is the activity coefficient at infinite dilution ($\gamma_i = \gamma_i^\infty$). Thus

$$f_i^{liq} = \gamma_i^\infty\, x_i\, P_i^{SAT} \tag{B-6}$$

For the liquid phase, the fugacity of the component is given by Henry's law that is applicable at very small concentration. The equation is

$$f_i^{liq} = H_i\, x_i \qquad (x_i << 1) \tag{B-7}$$

Substitution of Equation (B-6) into Equation (B-7) yields

$$\gamma_i^\infty\, x_i\, P_i^{SAT} = H_i\, x_i \tag{B-8}$$

Solving for Henry's law constant yields the following equation which is applicable to organic chemicals which are liquids at ambient conditions (25 C, 1 atm) and have total solubility in water:

$$H_i = \gamma_i^\infty\, P_i^{SAT} \tag{B-9}$$

where H_i = Henry's law constant, atm/mol fraction
γ_i^∞ = activity coefficient at infinite dilution
P_i^{SAT} = vapor pressure of organic chemical, atm

GASES

For organic chemicals that are gases at ambient conditions, there are two phases when the organic chemical is in contact with water. These are vapor and liquid phases. Such a two-phase system consisting of vapor and liquid is shown in Fig. B-1b. At equilibrium, the fugacity of the component in each phase is given by

$$f_i^{vap} = f_i^{liq} \tag{B-10}$$

For the vapor phase, the fugacity of the organic chemical is

$$f_i^{vap} = y_i\, P_t \tag{B-11}$$

Substitution of $y_i = 1 - y_{H2O}$ and $P_t = 1$ atm into the equation yields

$$f_i^{vap} = 1 - y_{H2O} \tag{B-12}$$

For the liquid phase, the fugacity of the component is given by Henry's law that is applicable at very small concentration. The equation is

$$f_i^{liq} = H_i\, x_i \qquad (x_i \ll 1) \tag{B-13}$$

Substitution of Equations (B-12) and (B-13) into Equation (B-10) yields

$$1 - y_{H2O} = H_i\, x_i \tag{B-14}$$

Solving for Henry's law constant yields the following equation which is applicable to organic chemicals which are gases at ambient conditions (25 C, 1 atm):

$$H_i = (1 - y_{H2O}) / x_i \tag{B-15}$$

where H_i = Henry's law constant, atm/mol fraction
x_i = solubility of organic chemical in water, mol fraction
y_{H2O} = mol fraction of water in vapor phase at ambient conditions (at 25 C, $y_{H2O} \approx 0.03117$)

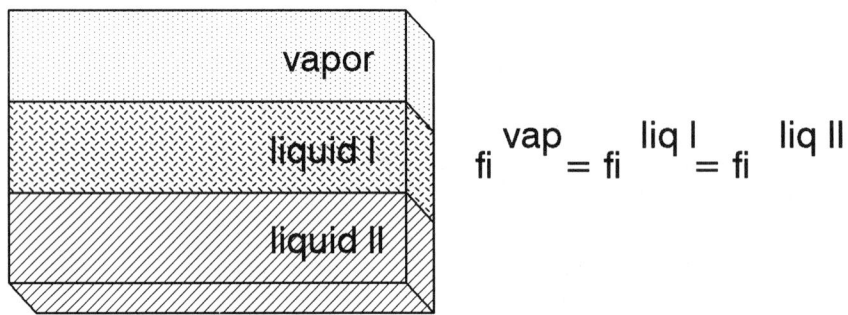

$$fi^{vap} = fi^{liq\,I} = fi^{liq\,II}$$

Figure B-1a Three Phase System

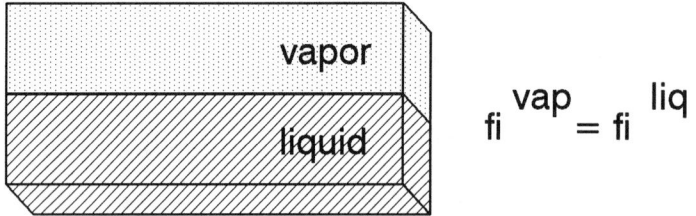

$$fi^{vap} = fi^{liq}$$

Figure B-1b Two Phase System

Appendix C

COMPOUND LIST BY CHEMICAL FORMULA

ORGANICS The following compilation for organics provides the compound list by chemical formula, name, and CAS registry number.

Formula	Name	CAS
CBrClF2	BROMOCHLORODIFLUOROMETHANE	353-59-3
CBrCl3	BROMOTRICHLOROMETHANE	75-62-7
CBrF3	BROMOTRIFLUOROMETHANE	75-63-8
CBr2F2	DIBROMODIFLUOROMETHANE	75-61-6
CClF3	CHLOROTRIFLUOROMETHANE	75-72-9
CClN	CYANOGEN CHLORIDE	506-77-4
CCl2F2	DICHLORODIFLUOROMETHANE	75-71-8
CCl2O	PHOSGENE	75-44-5
CCl3F	TRICHLOROFLUOROMETHANE	75-69-4
CCl4	CARBON TETRACHLORIDE	56-23-5
CF2O	CARBONYL FLUORIDE	353-50-4
CF4	CARBON TETRAFLUORIDE	75-73-0
CHBr3	TRIBROMOMETHANE	75-25-2
CHClF2	CHLORODIFLUOROMETHANE	75-45-6
CHCl2F	DICHLOROFLUOROMETHANE	75-43-4
CHCl3	CHLOROFORM	67-66-3
CHF3	TRIFLUOROMETHANE	75-46-7
CHI3	TRIIODOMETHANE	75-47-8
CHN	HYDROGEN CYANIDE	74-90-8
CHNS	ISOTHIOCYANIC-ACID	3129-90-6
CH2BrCl	BROMOCHLOROMETHANE	74-97-5
CH2Br2	DIBROMOMETHANE	74-95-3
CH2ClF	CHLOROFLUOROMETHANE	593-70-4
CH2Cl2	DICHLOROMETHANE	75-09-2
CH2F2	DIFLUOROMETHANE	75-10-5
CH2I2	DIIODOMETHANE	75-11-6
CH2O	FORMALDEHYDE	50-00-0
CH2O2	FORMIC ACID	64-18-6
CH3Br	METHYL BROMIDE	74-83-9
CH3Cl	METHYL CHLORIDE	74-87-3
CH3Cl3Si	METHYL TRICHLOROSILANE	75-79-6
CH3F	METHYL FLUORIDE	593-53-3
CH3I	METHYL IODIDE	74-88-4
CH3NO	FORMAMIDE	75-12-7
CH3NO2	NITROMETHANE	75-52-5
CH3NO2	METHYL-NITRITE	624-91-9
CH3NO3	METHYL-NITRATE	598-58-3
CH4	METHANE	74-82-8
CH4Cl2Si	METHYL DICHLOROSILANE	75-54-7
CH4O	METHANOL	67-56-1
CH4O3S	METHANESULFONIC ACID	75-75-2
CH4S	METHYL MERCAPTAN	74-93-1
CH5ClSi	METHYL CHLOROSILANE	993-00-0
CH5N	METHYLAMINE	74-89-5
CH6Si	METHYL SILANE	992-94-9
CN4O8	TETRANITROMETHANE	509-14-8
CO	CARBON MONOXIDE	630-08-0
COS	CARBONYL SULFIDE	463-58-1
CO2	CARBON DIOXIDE	124-38-9
CS2	CARBON DISULFIDE	75-15-0
C2BrF3	BROMOTRIFLUOROETHYLENE	598-73-2
C2Br2F4	1,2-DIBROMOTETRAFLUOROETHANE	124-73-2
C2ClF3	CHLOROTRIFLUOROETHYLENE	79-38-9
C2ClF5	CHLOROPENTAFLUOROETHANE	76-15-3
C2Cl2F4	1,2-DICHLOROTETRAFLUOROETHANE	76-14-2
C2Cl3F3	1,1,2-TRICHLOROTRIFLUOROETHANE	76-13-1
C2Cl4	TETRACHLOROETHYLENE	127-18-4
C2Cl4F2	1,1,2-TETRACHLORODIFLUOROETHANE	76-12-0
C2Cl4O	TRICHLOROACETYL CHLORIDE	76-02-8
C2Cl6	HEXACHLOROETHANE	67-72-1
C2F4	TETRAFLUOROETHYLENE	116-14-3
C2F6	HEXAFLUOROETHANE	76-16-4
C2HBrClF3	HALOTHANE	151-67-7
C2HClF2	2-CHLORO-1,1-DIFLUOROETHYLENE	359-10-4
C2HCl3	TRICHLOROETHYLENE	79-01-6
C2HCl3O	DICHLOROACETYL CHLORIDE	79-36-7
C2HCl3O	TRICHLOROACETALDEHYDE	75-87-6
C2HCl5	PENTACHLOROETHANE	76-01-7
C2HF3	TRIFLUOROETHENE	359-11-5
C2HF3O2	TRIFLUOROACETIC ACID	76-05-1
C2HF5	PENTAFLUOROETHANE	354-33-6
C2H2	ACETYLENE	74-86-2
C2H2Br4	1,1,2,2-TETRABROMOETHANE	79-27-6
C2H2Cl2	1,1-DICHLOROETHYLENE	75-35-4
C2H2Cl2	cis-1,2-DICHLOROETHYLENE	156-59-2
C2H2Cl2	trans-1,2-DICHLOROETHYLENE	156-60-5
C2H2Cl2O	CHLOROACETYL CHLORIDE	79-04-9
C2H2Cl2O	DICHLOROACETALDEHYDE	79-02-7
C2H2Cl2O2	DICHLOROACETIC ACID	79-43-6
C2H2Cl3F	1,1,1-TRICHLOROFLUOROETHANE	27154-33-2
C2H2Cl4	1,1,1,2-TETRACHLOROETHANE	630-20-6
C2H2Cl4	1,1,2,2-TETRACHLOROETHANE	79-34-5
C2H2F2	1,1-DIFLUOROETHYLENE	75-38-7
C2H2F2	cis-1,2-DIFLUOROETHENE	1630-78-0
C2H2F2	trans-1,2-DIFLUOROETHENE	1630-77-9
C2H2F4	1,1,1,2-TETRAFLUOROETHANE	811-97-2
C2H2O	KETENE	463-51-4
C2H2O4	OXALIC ACID	144-62-7
C2H3Br	VINYL BROMIDE	593-60-2
C2H3Cl	VINYL CHLORIDE	75-01-4
C2H3ClF2	1-CHLORO-1,1-DIFLUOROETHANE	75-68-3
C2H3ClO	ACETYL CHLORIDE	75-36-5
C2H3ClO	CHLOROACETALDEHYDE	107-20-0
C2H3ClO2	CHLOROACETIC ACID	79-11-8
C2H3ClO2	METHYL CHLOROFORMATE	79-22-1
C2H3Cl3	1,1,1-TRICHLOROETHANE	71-55-6
C2H3Cl3	1,1,2-TRICHLOROETHANE	79-00-5
C2H3F	VINYL FLUORIDE	75-02-5
C2H3F3	1,1,1-TRIFLUOROETHANE	420-46-2
C2H3N	ACETONITRILE	75-05-8
C2H3NO	METHYL ISOCYANATE	624-83-9
C2H4	ETHYLENE	74-85-1
C2H4Br2	1,1-DIBROMOETHANE	557-91-5
C2H4Br2	1,2-DIBROMOETHANE	106-93-4
C2H4Cl2	1,1-DICHLOROETHANE	75-34-3
C2H4Cl2	1,2-DICHLOROETHANE	107-06-2
C2H4Cl2O	BIS(CHLOROMETHYL)ETHER	542-88-1
C2H4F2	1,1-DIFLUOROETHANE	75-37-6
C2H4F2	1,2-DIFLUOROETHANE	624-72-6
C2H4I2	1,2-DIIODOETHANE	624-73-7
C2H4O	ACETALDEHYDE	75-07-0
C2H4O	ETHYLENE OXIDE	75-21-8
C2H4OS	THIOACETIC-ACID	507-09-5
C2H4O2	ACETIC ACID	64-19-7
C2H4O2	METHYL FORMATE	107-31-3
C2H4S	THIACYCLOPROPANE	420-12-2
C2H5Br	BROMOETHANE	74-96-4
C2H5Cl	ETHYL CHLORIDE	75-00-3
C2H5ClO	2-CHLOROETHANOL	107-07-3
C2H5F	ETHYL FLUORIDE	353-36-6
C2H5I	ETHYL IODIDE	75-03-6
C2H5N	ETHYLENEIMINE	151-56-4
C2H5NO	ACETAMIDE	60-35-5
C2H5NO	N-METHYLFORMAMIDE	123-39-7
C2H5NO2	NITROETHANE	79-24-3
C2H5NO3	ETHYL-NITRATE	625-58-1
C2H6	ETHANE	74-84-0
C2H6AlCl	DIMETHYLALUMINUM CHLORIDE	1184-58-3
C2H6O	DIMETHYL ETHER	115-10-6
C2H6O	ETHANOL	64-17-5
C2H6OS	DIMETHYL SULFOXIDE	67-68-5
C2H6O2	ETHYLENE GLYCOL	107-21-1
C2H6O4S	DIMETHYL SULFATE	77-78-1
C2H6S	DIMETHYL SULFIDE	75-18-3
C2H6S	ETHYL MERCAPTAN	75-08-1
C2H6S2	DIMETHYL DISULFIDE	624-92-0
C2H7N	DIMETHYLAMINE	124-40-3
C2H7N	ETHYLAMINE	75-04-7
C2H7NO	MONOETHANOLAMINE	141-43-5
C2H8N2	ETHYLENEDIAMINE	107-15-3
C2H8Si	DIMETHYL SILANE	1111-74-6
C2N2	CYANOGEN	460-19-5
C3F6	HEXAFLUOROPROPYLENE	116-15-4
C3F6O	HEXAFLUOROACETONE	684-16-2
C3F8	OCTAFLUOROPROPANE	76-19-7
C3H2N2	MALONONITRILE	109-77-3
C3H3Cl	PROPARGYL CHLORIDE	624-65-7
C3H3N	ACRYLONITRILE	107-13-1
C3H3NO	OXAZOLE	288-42-6
C3H4	METHYLACETYLENE	74-99-7

Formula	Name	CAS Number
C3H4	PROPADIENE	463-49-0
C3H4Cl2	2,3-DICHLOROPROPENE	78-88-6
C3H4O	ACROLEIN	107-02-8
C3H4O	PROPARGYL ALCOHOL	107-19-7
C3H4O2	ACRYLIC ACID	79-10-7
C3H4O2	beta-PROPIOLACTONE	57-57-8
C3H4O2	VINYL FORMATE	692-45-5
C3H4O3	ETHYLENE CARBONATE	96-49-1
C3H4O3	PYRUVIC ACID	127-17-3
C3H5Br	3-BROMO-1-PROPENE	106-95-6
C3H5Cl	2-CHLOROPROPENE	557-98-2
C3H5Cl	3-CHLOROPROPENE	107-05-1
C3H5ClO	alpha-EPICHLOROHYDRIN	106-89-8
C3H5ClO2	METHYL CHLOROACETATE	96-34-4
C3H5ClO2	ETHYL CHLOROFORMATE	541-41-3
C3H5Cl3	1,2,3-TRICHLOROPROPANE	96-18-4
C3H5I	3-IODO-1-PROPENE	556-56-9
C3H5N	PROPIONITRILE	107-12-0
C3H5NO	ACRYLAMIDE	79-06-1
C3H5NO	HYDRACRYLONITRILE	109-78-4
C3H5NO	LACTONITRILE	78-97-7
C3H5N3O9	NITROGLYCERINE	55-63-0
C3H6	CYCLOPROPANE	75-19-4
C3H6	PROPYLENE	115-07-1
C3H6Br2	1,2-DIBROMOPROPANE	78-75-1
C3H6Cl2	1,1-DICHLOROPROPANE	78-99-9
C3H6Cl2	1,2-DICHLOROPROPANE	78-87-5
C3H6Cl2	1,3-DICHLOROPROPANE	142-28-9
C3H6Cl2	2,2-DICHLOROPROPANE	594-20-7
C3H6I2	1,2-DIIODOPROPANE	598-29-8
C3H6O	ACETONE	67-64-1
C3H6O	ALLYL ALCOHOL	107-18-6
C3H6O	METHYL VINYL ETHER	107-25-5
C3H6O	n-PROPIONALDEHYDE	123-38-6
C3H6O	1,2-PROPYLENE OXIDE	75-56-9
C3H6O	1,3-PROPYLENE OXIDE	503-30-0
C3H6O2	ETHYL FORMATE	109-94-4
C3H6O2	METHYL ACETATE	79-20-9
C3H6O2	PROPIONIC ACID	79-09-4
C3H6O2S	3-MERCAPTOPROPIONIC ACID	107-96-0
C3H6O3	LACTIC ACID	598-82-3
C3H6O3	METHOXYACETIC ACID	625-45-6
C3H6O3	TRIOXANE	110-88-3
C3H6S	THIACYCLOBUTANE	287-27-4
C3H7Br	1-BROMOPROPANE	106-94-5
C3H7Br	2-BROMOPROPANE	75-26-3
C3H7Cl	ISOPROPYL CHLORIDE	75-29-6
C3H7Cl	n-PROPYL CHLORIDE	540-54-5
C3H7F	1-FLUOROPROPANE	460-13-9
C3H7F	2-FLUOROPROPANE	420-26-8
C3H7I	ISOPROPYL IODIDE	75-30-9
C3H7I	n-PROPYL IODIDE	107-08-4
C3H7N	ALLYLAMINE	107-11-9
C3H7N	PROPYLENEIMINE	75-55-8
C3H7NO	N,N-DIMETHYLFORMAMIDE	68-12-2
C3H7NO	N-METHYLACETAMIDE	79-16-3
C3H7NO2	1-NITROPROPANE	103-03-2
C3H7NO2	2-NITROPROPANE	79-46-9
C3H7NO3	PROPYL-NITRATE	627-13-4
C3H7NO3	ISOPROPYL-NITRATE	2902-96-7
C3H8	PROPANE	74-98-6
C3H8O	ISOPROPANOL	67-63-0
C3H8O	METHYL ETHYL ETHER	540-67-0
C3H8O	n-PROPANOL	71-23-8
C3H8O2	2-METHOXYETHANOL	109-86-4
C3H8O2	METHYLAL	109-87-5
C3H8O2	1,2-PROPYLENE GLYCOL	57-55-6
C3H8O2	1,3-PROPYLENE GLYCOL	504-63-2
C3H8O3	GLYCEROL	56-81-5
C3H8S	n-PROPYLMERCAPTAN	107-03-9
C3H8S	ISOPROPYL MERCAPTAN	75-33-2
C3H8S	ETHYL-METHYL-SULFIDE	624-89-5
C3H9N	n-PROPYLAMINE	107-10-8
C3H9N	ISOPROPYLAMINE	75-31-0
C3H9N	TRIMETHYLAMINE	75-50-3
C3H9NO	1-AMINO-2-PROPANOL	78-96-6
C3H9NO	3-AMINO-1-PROPANOL	156-87-6
C3H9NO	METHYLETHANOLAMINE	109-83-1
C3H9O4P	TRIMETHYL PHOSPHATE	512-56-1
C3H10N2	1,2-PROPANEDIAMINE	78-90-0
C3H10Si	TRIMETHYL SILANE	993-07-7
C4Cl4S	TETRACHLOROTHIOPHENE	6012-97-1
C4Cl6	HEXACHLORO-1,3-BUTADIENE	87-68-3
C4F8	OCTAFLUORO-2-BUTENE	360-89-4
C4F8	OCTAFLUOROCYCLOBUTANE	115-25-3
C4F10	DECAFLUOROBUTANE	355-25-9
C4H2	BUTADIYNE(BIACETYLENE)	460-12-8
C4H2O3	MALEIC ANHYDRIDE	108-31-6
C4H4	VINYLACETYLENE	689-97-4
C4H4N2	SUCCINONITRILE	110-61-2
C4H4O	FURAN	110-00-9
C4H4O2	DIKETENE	674-82-8
C4H4O3	SUCCINIC ANHYDRIDE	108-30-5
C4H4O4	FUMARIC ACID	110-17-8
C4H4O4	MALEIC ACID	110-16-7
C4H4S	THIOPHENE	110-02-1
C4H5Cl	CHLOROPRENE	126-99-8
C4H5N	trans-CROTONITRILE	627-26-9
C4H5N	cis-CROTONITRILE	1190-76-7
C4H5N	METHACRYLONITRILE	126-98-7
C4H5N	PYRROLE	109-97-7
C4H5N	VINYLACETONITRILE	109-75-1
C4H5NO2	METHYL CYANOACETATE	105-34-0
C4H6	CYCLOBUTENE	822-35-5
C4H6	1,2-BUTADIENE	590-19-2
C4H6	1,3-BUTADIENE	106-99-0
C4H6	DIMETHYLACETYLENE	503-17-3
C4H6	ETHYLACETYLENE	107-00-6
C4H6Cl2	1,3-DICHLORO-trans-2-BUTENE	7415-31-8
C4H6Cl2	1,4-DICHLORO-cis-2-BUTENE	1476-11-5
C4H6Cl2	1,4-DICHLORO-trans-2-BUTENE	110-57-6
C4H6Cl2	3,4-DICHLORO-1-BUTENE	760-23-6
C4H6O	trans-CROTONALDEHYDE	123-73-9
C4H6O	2,5-DIHYDROFURAN	1708-29-8
C4H6O	DIVINYL ETHER	109-93-3
C4H6O	METHACROLEIN	78-85-3
C4H6O2	2-BUTYNE-1,4-DIOL	110-65-6
C4H6O2	gamma-BUTYROLACTONE	96-48-0
C4H6O2	cis-CROTONIC ACID	503-64-0
C4H6O2	trans-CROTONIC ACID	107-93-7
C4H6O2	METHACRYLIC ACID	79-41-4
C4H6O2	METHYL ACRYLATE	96-33-3
C4H6O2	VINYL ACETATE	108-05-4
C4H6O3	ACETIC ANHYDRIDE	108-24-7
C4H6O4	SUCCINIC ACID	110-15-6
C4H6O5	DIGLYCOLIC ACID	110-99-6
C4H6O5	MALIC ACID	617-48-1
C4H6O6	TARTARIC ACID	133-37-9
C4H7N	n-BUTYRONITRILE	109-74-0
C4H7N	ISOBUTYRONITRILE	78-82-0
C4H7NO	ACETONE CYANOHYDRIN	75-86-5
C4H7NO	2-METHACRYLAMIDE	79-39-0
C4H7NO	3-METHOXYPROPIONITRILE	110-67-8
C4H7NO	2-PYRROLIDONE	616-45-5
C4H8	1-BUTENE	106-98-9
C4H8	cis-2-BUTENE	590-18-1
C4H8	trans-2-BUTENE	624-64-6
C4H8	CYCLOBUTANE	287-23-0
C4H8	ISOBUTENE	115-11-7
C4H8Br2	1,2-DIBROMOBUTANE	533-98-2
C4H8Br2	2,3-DIBROMOBUTANE	5408-86-6
C4H8Cl2	1,4-DICHLOROBUTANE	110-56-5
C4H8I2	1,2-DIIODOBUTANE	628-21-7
C4H8O	n-BUTYRALDEHYDE	123-72-8
C4H8O	ISOBUTYRALDEHYDE	78-84-2
C4H8O	1,2-EPOXYBUTANE	106-88-7
C4H8O	METHYL ETHYL KETONE	78-93-3
C4H8O	ETHYL VINYL ETHER	109-92-2
C4H8O	TETRAHYDROFURAN	109-99-9
C4H8O2	cis-2-BUTENE-1,4-DIOL	6117-80-2
C4H8O2	trans-2-BUTENE-1,4-DIOL	821-11-4
C4H8O2	ISOBUTYRIC ACID	79-31-2
C4H8O2	n-BUTYRIC ACID	107-92-6
C4H8O2	1,4-DIOXANE	123-91-1
C4H8O2	ETHYL ACETATE	141-78-6
C4H8O2	METHYL PROPIONATE	554-12-1
C4H8O2	n-PROPYL FORMATE	110-74-7
C4H8O2S	SULFOLANE	126-33-0
C4H8S	TETRAHYDROTHIOPHENE	110-01-0
C4H9Br	1-BROMOBUTANE	109-65-9
C4H9Br	2-BROMOBUTANE	78-76-2
C4H9Cl	n-BUTYL CHLORIDE	109-69-3
C4H9Cl	sec-BUTYL CHLORIDE	78-86-4
C4H9Cl	tert-BUTYL CHLORIDE	507-20-0
C4H9I	2-IODO-2-METHYLPROPANE	558-17-8
C4H9N	PYRROLIDINE	123-75-1

Formula	Name	CAS		Formula	Name	CAS
C4H9NO	N,N-DIMETHYLACETAMIDE	127-19-5		C5H9NO	n-BUTYL ISOCYANATE	111-36-4
C4H9NO	MORPHOLINE	110-91-8		C5H9NO	N-METHYL-2-PYRROLIDONE	872-50-4
C4H9NO2	1-NITROBUTANE	627-05-4		C5H9NO4	L-GLUTAMIC ACID	56-86-0
C4H9NO2	2-NITROBUTANE	600-24-8		C5H10	CYCLOPENTANE	287-92-3
C4H10	n-BUTANE	106-97-8		C5H10	2-METHYL-1-BUTENE	563-46-2
C4H10	ISOBUTANE	75-28-5		C5H10	2-METHYL-2-BUTENE	513-35-9
C4H10N2	PIPERAZINE	110-85-0		C5H10	3-METHYL-1-BUTENE	563-45-1
C4H10O	n-BUTANOL	71-36-3		C5H10	1-PENTENE	109-67-1
C4H10O	sec-BUTANOL	78-92-2		C5H10	cis-2-PENTENE	627-20-3
C4H10O	tert-BUTANOL	75-65-0		C5H10	trans-2-PENTENE	646-04-8
C4H10O	DIETHYL ETHER	60-29-7		C5H10Br2	2,3-DIBROMO-2-METHYLBUTANE	594-51-4
C4H10O	METHYL-PROPYL-ETHER	557-17-5		C5H10Cl2	1,5-DICHLOROPENTANE	628-76-2
C4H10O	METHYL ISOPROPYL ETHER	598-53-8		C5H10O	METHYL ISOPROPYL KETONE	563-80-4
C4H10O	ISOBUTANOL	78-83-1		C5H10O	2-PENTANONE	107-87-9
C4H10O2	1,3-BUTANEDIOL	107-88-0		C5H10O	DIETHYL KETONE	96-22-0
C4H10O2	1,4-BUTANEDIOL	110-63-4		C5H10O	VALERALDEHYDE	110-62-3
C4H10O2	2,3-BUTANEDIOL	6982-25-8		C5H10O2	n-BUTYL FORMATE	592-84-7
C4H10O2	t-BUTYL HYDROPEROXIDE	75-91-2		C5H10O2	ETHYL PROPIONATE	105-37-3
C4H10O2	1,2-DIMETHOXYETHANE	110-71-4		C5H10O2	ISOBUTYL FORMATE	542-55-2
C4H10O2	2-ETHOXYETHANOL	110-80-5		C5H10O2	ISOPROPYL ACETATE	108-21-4
C4H10O3	DIETHYLENE GLYCOL	111-46-6		C5H10O2	n-PROPYL ACETATE	109-60-4
C4H10O4S	DIETHYL SULFATE	64-67-5		C5H10O2	METHYL n-BUTYRATE	623-42-7
C4H10S	n-BUTYL MERCAPTAN	109-79-5		C5H10O2	2-METHYLBUTYRIC ACID	600-07-7
C4H10S	ISOBUTYL MERCAPTAN	513-44-0		C5H10O2	ISOVALERIC ACID	503-74-2
C4H10S	sec-BUTYL MERCAPTAN	513-53-1		C5H10O2	VALERIC ACID	109-52-4
C4H10S	tert-BUTYL MERCAPTAN	75-66-1		C5H10O2	TETRAHYDROFURFURYL ALCOHOL	97-99-4
C4H10S	DIETHYL SULFIDE	352-93-2		C5H10O2S	3-METHYL SULFOLANE	872-93-5
C4H10S	ISOPROPYL-METHYL-SULFIDE	1551-21-9		C5H10O3	DIETHYL CARBONATE	105-58-8
C4H10S	METHYL-PROPYL-SULFIDE	3877-15-4		C5H10O3	ETHYL LACTATE	97-64-3
C4H10S2	DIETHYL DISULFIDE	110-81-6		C5H10S	THIACYCLOHEXANE	1613-51-0
C4H11N	n-BUTYLAMINE	109-73-9		C5H10S	CYCLOPENTANETHIOL	1679-07-8
C4H11N	ISOBUTYLAMINE	78-81-9		C5H11Br	1-BROMOPENTANE	110-53-2
C4H11N	sec-BUTYLAMINE	13952-84-6		C5H11Cl	1-CHLOROPENTANE	543-59-9
C4H11N	tert-BUTYLAMINE	75-64-9		C5H11Cl	1-CHLORO-3-METHYLBUTANE	107-84-6
C4H11N	DIETHYLAMINE	109-89-7		C5H11Cl	2-CHLORO-2-METHYLBUTANE	594-36-5
C4H11NO	DIMETHYLETHANOLAMINE	108-01-0		C5H11N	N-METHYLPYRROLIDINE	120-94-5
C4H11NO2	DIETHANOLAMINE	111-42-2		C5H11N	PIPERIDINE	110-89-4
C4H11NO2	2-AMINOETHOXYETHANOL	929-06-6		C5H11NO	tert-BUTYLFORMAMIDE	2425-74-3
C4H12N2O	N-AMINOETHYL ETHANOLAMINE	111-41-1		C5H12	ISOPENTANE	78-78-4
C4H12Si	TETRAMETHYLSILANE	75-76-3		C5H12	NEOPENTANE	463-82-1
C4H13N3	DIETHYLENE TRIAMINE	111-40-0		C5H12	n-PENTANE	109-66-0
C5Cl6	HEXACHLOROCYCLOPENTADIENE	77-47-4		C5H12O	2,2-DIMETHYL-1-PROPANOL	75-84-3
C5H4O2	FURFURAL	98-01-1		C5H12O	tert-PENTYL-ALCOHOL	75-85-4
C5H5N	PYRIDINE	110-86-1		C5H12O	2-METHYL-1-BUTANOL	137-32-6
C5H6	CYCLOPENTADIENE	542-92-7		C5H12O	2-METHYL-2-BUTANOL	75-85-4
C5H6	2-METHYL-1-BUTENE-3-YNE	78-80-8		C5H12O	3-METHYL-1-BUTANOL	123-51-3
C5H6	1-PENTENE-3-YNE	646-05-9		C5H12O	3-METHYL-2-BUTANOL	598-75-4
C5H6	1-PENTENE-4-YNE	871-28-3		C5H12O	1-PENTANOL	71-41-0
C5H6N2	GLUTARONITRILE	544-13-8		C5H12O	2-PENTANOL	6032-29-7
C5H6O2	FURFURYL ALCOHOL	98-00-0		C5H12O	3-PENTANOL	584-02-1
C5H6O3	GLUTARIC ANHYDRIDE	108-55-4		C5H12O	METHYL sec-BUTYL ETHER	6795-87-5
C5H6O4	CITRACONIC ACID	498-23-7		C5H12O	METHYL tert-BUTYL ETHER	1634-04-4
C5H6O4	ITACONIC ACID	97-65-4		C5H12O	METHYL ISOBUTYL ETHER	625-44-5
C5H6S	2-METHYLTHIOPHENE	554-14-3		C5H12O	ETHYL PROPYL ETHER	628-32-0
C5H6S	3-METHYLTHIOPHENE	616-44-4		C5H12O2	ETHYLENE GLYCOL MONOPROPYL ETHER	2807-30-9
C5H7N	N-METHYLPYRROLE	96-54-8		C5H12O2	NEOPENTYL GLYCOL	126-30-7
C5H7NO2	ETHYL CYANOACETATE	105-56-6		C5H12O2	1,5-PENTANEDIOL	111-29-5
C5H8	CYCLOPENTENE	142-29-0		C5H12O3	2-(2-METHOXYETHOXY)ETHANOL	111-77-3
C5H8	ISOPRENE	78-79-5		C5H12O4	PENTAERYTHRITOL	115-77-5
C5H8	3-METHYL-1,2-BUTADIENE	598-25-4		C5H12S	n-PENTYL MERCAPTAN	110-66-7
C5H8	2-METHYL-1,3-BUTADIENE	78-79-5		C5H12S	BUTYL-METHYL-SULFIDE	628-29-5
C5H8	1,2-PENTADIENE	591-95-7		C5H12S	ETHYL-PROPYL-SULFIDE	4110-50-3
C5H8	cis-1,3-PENTADIENE	1574-41-0		C5H12S	2-METHYL-2-BUTANETHIOL	1679-09-0
C5H8	trans-1,3-PENTADIENE	2004-70-8		C5H13N	n-PENTYLAMINE	110-58-7
C5H8	1,4-PENTADIENE	591-93-5		C5H13NO2	METHYL DIETHANOLAMINE	105-59-9
C5H8	2,3-PENTADIENE	591-96-8		C6Cl6	HEXACHLOROBENZENE	118-74-1
C5H8	1-PENTYNE	627-19-0		C6F6	HEXAFLUOROBENZENE	392-56-3
C5H8	2-PENTYNE	627-21-4		C6H3ClN2O4	1-CHLORO-2,4-DINITROBENZENE	97-00-7
C5H8	3-METHYL-1-BUTYNE	598-23-2		C6H3Cl2NO2	1,2-DICHLORO-4-NITROBENZENE	99-54-7
C5H8	SPIROPENTANE	157-40-4		C6H3Cl3	1,2,4-TRICHLOROBENZENE	120-82-1
C5H8N4O12	PENTAERYTHRITOL TETRANITRATE	78-11-5		C6H3N3O6	1,3,5-TRINITROBENZENE	99-35-4
C5H8O	CYCLOPENTANONE	120-92-3		C6H4Br2	m-DIBROMOBENZENE	108-36-1
C5H8O	METHYL ISOPROPENYL KETONE	814-78-8		C6H4ClNO2	m-CHLORONITROBENZENE	121-73-3
C5H8O2	ACETYLACETONE	123-54-6		C6H4ClNO2	o-CHLORONITROBENZENE	88-73-3
C5H8O2	ALLYL ACETATE	591-87-7		C6H4ClNO2	p-CHLORONITROBENZENE	100-00-5
C5H8O2	ETHYL ACRYLATE	140-88-5		C6H4Cl2	m-DICHLOROBENZENE	541-73-1
C5H8O2	METHYL METHACRYLATE	80-62-6		C6H4Cl2	o-DICHLOROBENZENE	95-50-1
C5H8O2	VINYL PROPIONATE	105-38-4		C6H4Cl2	p-DICHLOROBENZENE	106-46-7
C5H8O3	2-HYDROXYETHYL ACRYLATE	818-61-1		C6H4F2	m-DIFLUOROBENZENE	372-18-9
C5H8O3	LEVULINIC ACID	123-76-2		C6H4F2	o-DIFLUOROBENZENE	367-11-3
C5H8O3	METHYL ACETOACETATE	105-45-3		C6H4F2	p-DIFLUOROBENZENE	540-36-3
C5H8O4	GLUTARIC ACID	110-94-1		C6H4N2O4	m-DINITROBENZENE	99-65-0
C5H9N	VALERONITRILE	110-59-8		C6H4N2O4	o-DINITROBENZENE	528-29-0

C6H4N2O4	p-DINITROBENZENE	100-25-4	
C6H5Br	BROMOBENZENE	108-86-1	
C6H5Cl	MONOCHLOROBENZENE	108-90-7	
C6H5ClO	m-CHLOROPHENOL	108-43-0	
C6H5ClO	o-CHLOROPHENOL	95-57-8	
C6H5ClO	p-CHLOROPHENOL	106-48-9	
C6H5Cl2N	3,4-DICHLOROANILINE	95-76-1	
C6H5F	FLUOROBENZENE	462-06-6	
C6H5I	IODOBENZENE	591-50-4	
C6H5NO2	NITROBENZENE	98-95-3	
C6H6	BENZENE	71-43-2	
C6H6ClN	m-CHLOROANILINE	108-42-9	
C6H6ClN	o-CHLOROANILINE	95-51-2	
C6H6ClN	p-CHLOROANILINE	106-47-8	
C6H6N2	cis-DICYANO-1-BUTENE	2141-58-4	
C6H6N2	trans-DICYANO-1-BUTENE	2141-59-5	
C6H6N2	1,4-DICYANO-2-BUTENE	1119-85-3	
C6H6N2O2	m-NITROANILINE	99-09-2	
C6H6N2O2	o-NITROANILINE	88-74-4	
C6H6N2O2	p-NITROANILINE	100-01-6	
C6H6O	PHENOL	108-95-2	
C6H6O2	1,2-BENZENEDIOL	120-80-9	
C6H6O2	1,3-BENZENEDIOL	108-46-3	
C6H6O2	p-HYDROQUINONE	123-31-9	
C6H6O3	1,2,3-BENZENETRIOL	87-66-1	
C6H6S	PHENYL MERCAPTAN	108-98-5	
C6H7N	ANILINE	62-53-3	
C6H7N	2-METHYLPYRIDINE	109-06-8	
C6H7N	3-METHYLPYRIDINE	108-99-6	
C6H7N	4-METHYLPYRIDINE	108-89-4	
C6H8	1,3-CYCLOHEXADIENE	592-57-4	
C6H8	METHYLCYCLOPENTADIENE	26519-91-5	
C6H8N2	ADIPONITRILE	111-69-3	
C6H8N2	METHYLGLUTARONITRILE	4553-62-2	
C6H8N2	m-PHENYLENEDIAMINE	108-45-2	
C6H8N2	o-PHENYLENEDIAMINE	95-54-5	
C6H8N2	p-PHENYLENEDIAMINE	106-50-3	
C6H8N2	PHENYLHYDRAZINE	100-63-0	
C6H8N2O	BIS(CYANOETHYL)ETHER	1656-48-0	
C6H8O4	DIMETHYL MALEATE	624-58-6	
C6H8O6	ASCORBIC ACID	50-81-7	
C6H8O7	CITRIC ACID	77-92-9	
C6H10	1-METHYLCYCLOPENTENE	693-89-0	
C6H10	3-METHYLCYCLOPENTENE	1120-62-3	
C6H10	4-METHYLCYCLOPENTENE	1759-81-5	
C6H10	CYCLOHEXENE	110-83-8	
C6H10	2,3-DIMETHYL-1,3-BUTADIENE	513-81-5	
C6H10	1,5-HEXADIENE	592-42-7	
C6H10	cis,trans-2,4-HEXADIENE	5194-50-3	
C6H10	trans,trans-2,4-HEXADIENE	5194-51-4	
C6H10	1-HEXYNE	693-02-7	
C6H10	2-HEXYNE	764-35-2	
C6H10	3-HEXYNE	928-49-4	
C6H10O	CYCLOHEXANONE	108-94-1	
C6H10O	MESITYL OXIDE	141-79-7	
C6H10O2	epsilon-CAPROLACTONE	502-44-3	
C6H10O2	ETHYL METHACRYLATE	97-63-2	
C6H10O2	n-PROPYL ACRYLATE	925-60-0	
C6H10O3	ETHYLACETOACETATE	141-97-9	
C6H10O3	PROPIONIC ANHYDRIDE	123-62-6	
C6H10O4	ADIPIC ACID	124-04-9	
C6H10O4	DIETHYL OXALATE	95-92-1	
C6H10O4	ETHYLENE GLYCOL DIACETATE	111-55-7	
C6H10O4	ETHYLIDENE DIACETATE	542-10-9	
C6H11N	HEXANENITRILE	628-73-9	
C6H11NO	epsilon-CAPROLACTAM	105-60-2	
C6H11NO	CYCLOHEXANONE OXIME	100-64-1	
C6H12	CYCLOHEXANE	110-82-7	
C6H12	2,3-DIMETHYL-1-BUTENE	563-78-0	
C6H12	2,3-DIMETHYL-2-BUTENE	563-79-1	
C6H12	3,3-DIMETHYL-1-BUTENE	558-37-2	
C6H12	2-ETHYL-1-BUTENE	760-21-4	
C6H12	trans-3-METHYL-2-PENTENE	616-12-6	
C6H12	1-HEXENE	592-41-6	
C6H12	cis-2-HEXENE	7688-21-3	
C6H12	trans-2-HEXENE	4050-45-7	
C6H12	cis-3-HEXENE	7642-09-3	
C6H12	trans-3-HEXENE	13269-52-8	
C6H12	METHYLCYCLOPENTANE	96-37-7	
C6H12	2-METHYL-1-PENTENE	763-29-1	
C6H12	2-METHYL-2-PENTENE	625-27-4	
C6H12	3-METHYL-1-PENTENE	760-20-3	
C6H12	3-METHYL-cis-2-PENTENE	922-62-3	
C6H12	4-METHYL-1-PENTENE	691-37-2	
C6H12	4-METHYL-cis-2-PENTENE	691-38-3	
C6H12	4-METHYL-trans-2-PENTENE	674-76-0	
C6H12N2	TRIETHYLENEDIAMINE	280-57-9	
C6H12O	BUTYL VINYL ETHER	111-34-2	
C6H12O	CYCLOHEXANOL	108-93-0	
C6H12O	1-HEXANAL	66-25-1	
C6H12O	ETHYL ISOPROPYL KETONE	565-69-5	
C6H12O	2-HEXANONE	591-78-6	
C6H12O	3-HEXANONE	589-38-8	
C6H12O	METHYL ISOBUTYL KETONE	108-10-1	
C6H12O2	n-PENTYL FORMATE	638-49-3	
C6H12O2	n-BUTYL ACETATE	123-86-4	
C6H12O2	sec-BUTYL ACETATE	105-46-4	
C6H12O2	tert-BUTYL ACETATE	540-88-5	
C6H12O2	ETHYL n-BUTYRATE	105-54-4	
C6H12O2	ETHYL ISOBUTYRATE	97-62-1	
C6H12O2	ISOBUTYL ACETATE	110-19-0	
C6H12O2	n-PROPYL PROPIONATE	106-36-5	
C6H12O2	CYCLOHEXYL PEROXIDE	766-07-4	
C6H12O2	DIACETONE ALCOHOL	123-42-2	
C6H12O2	2-ETHYL BUTYRIC ACID	88-09-5	
C6H12O2	n-HEXANOIC ACID	142-62-1	
C6H12O3	2-ETHOXYETHYL ACETATE	111-15-9	
C6H12O3	HYDROXYCAPROIC ACID	6064-63-7	
C6H12O3	PARALDEHYDE	123-63-7	
C6H12O3	sec-BUTYL GLYCOLATE	----------	
C6H12S	THIACYCLOHEPTANE	----------	
C6H13N	CYCLOHEXYLAMINE	108-91-8	
C6H13N	HEXAMETHYLENEIMINE	111-49-9	
C6H14	2,2-DIMETHYLBUTANE	75-83-2	
C6H14	2,3-DIMETHYLBUTANE	79-29-8	
C6H14	n-HEXANE	110-54-3	
C6H14	2-METHYLPENTANE	107-83-5	
C6H14	3-METHYLPENTANE	96-14-0	
C6H14N2O2	LYSINE	56-87-1	
C6H14O	2-ETHYL-1-BUTANOL	97-95-0	
C6H14O	1-HEXANOL	111-27-3	
C6H14O	2-HEXANOL	626-93-7	
C6H14O	2-METHYL-1-PENTANOL	105-30-6	
C6H14O	4-METHYL-2-PENTANOL	108-11-2	
C6H14O	n-BUTYL ETHYL ETHER	628-81-9	
C6H14O	DIISOPROPYL ETHER	108-20-3	
C6H14O	DI-n-PROPYL ETHER	111-43-3	
C6H14O	METHYL tert-PENTYL ETHER	994-05-8	
C6H14O2	ACETAL	105-57-7	
C6H14O2	2-BUTOXYETHANOL	111-76-2	
C6H14O2	1,6-HEXANEDIOL	629-11-8	
C6H14O2	HEXYLENE GLYCOL	107-41-5	
C6H14O2S	DI-n-PROPYL SULFONE	598-03-8	
C6H14O3	DIETHYLENE GLYCOL DIMETHYL ETHER	111-96-6	
C6H14O3	DIPROPYLENE GLYCOL	25265-71-8	
C6H14O3	2-(2-ETHOXYETHOXY)ETHANOL	111-90-0	
C6H14O3	TRIMETHYLOLPROPANE	77-99-6	
C6H14O4	TRIETHYLENE GLYCOL	112-27-6	
C6H14O6	SORBITOL	50-70-4	
C6H14S	n-HEXYLMERCAPTAN	111-31-9	
C6H14S	BUTYL-ETHYL-SULFIDE	638-46-0	
C6H14S	DIISOPROPYL-SULFIDE	625-80-9	
C6H14S	METHYL-PENTYL-SULFIDE	1741-83-9	
C6H14S	DIPROPYL-SULFIDE	111-47-7	
C6H14S2	DIPROPYL-DISULFIDE	629-19-6	
C6H15Al	TRIETHYL ALUMINUN	97-93-8	
C6H15Al2Cl3	ETHYL ALUMINUM SESQUICHLORIDE	12075-68-2	
C6H15N	DIISOPROPYLAMINE	108-18-9	
C6H15N	DI-n-PROPYLAMINE	142-84-7	
C6H15N	n-HEXYLAMINE	111-26-2	
C6H15N	TRIETHYLAMINE	121-44-8	
C6H15NO	6-AMINOHEXANOL	4048-33-3	
C6H15NO2	DIISOPROPANOLAMINE	110-97-4	
C6H15NO3	TRIETHANOLAMINE	102-71-6	
C6H15N3	N-AMINOETHYL PIPERAZINE	140-31-8	
C6H15O4P	TRIETHYL PHOSPHATE	78-40-0	
C6H16N2	HEXAMETHYLENEDIAMINE	124-09-4	
C6H18N3OP	HEXAMETHYL PHOSPHORAMIDE	680-31-9	
C6H18N4	TRIETHYLENE TETRAMINE	112-24-3	
C6H18OSi2	HEXAMETHYLDISILOXANE	107-46-0	
C6H18O3Si3	HEXAMETHYLCYCLOTRISILOXANE	541-05-9	
C6H19NSi2	HEXAMETHYLDISILAZANE	999-97-3	
C7H3ClF3NO2	4-CHLORO-3-NITROBENZOTRIFLUORIDE	121-71-5	
C7H3Cl2F3	2,4-DICHLOROBENZOTRIFLUORIDE	320-60-5	
C7H3Cl2NO	3,4-DICHLOROPHENYL ISOCYANATE	102-36-3	
C7H4ClF3	p-CHLOROBENZOTRIFLUORIDE	98-56-6	

C7H4Cl2O	m-CHLOROBENZOYL CHLORIDE	618-46-2
C7H4F3NO2	3-NITROBENZOTRIFLUORIDE	98-46-4
C7H5ClO	BENZOYL CHLORIDE	98-88-4
C7H5ClO2	o-CHLOROBENZOIC ACID	118-91-2
C7H5Cl3	BENZOTRICHLORIDE	98-07-7
C7H5F3	BENZOTRIFLUORIDE	98-08-8
C7H5N	BENZONITRILE	100-47-0
C7H5NO	PHENYL ISOCYANATE	103-71-9
C7H5N3O6	2,4,6-TRINITROTOLUENE	118-96-7
C7H6Cl2	BENZYL DICHLORIDE	98-87-3
C7H6Cl2	2,4-DICHLOROTOLUENE	95-73-8
C7H6N2O4	2,4-DINITROTOLUENE	121-14-2
C7H6N2O4	2,5-DINITROTOLUENE	619-15-8
C7H6N2O4	2,6-DINITROTOLUENE	606-20-2
C7H6N2O4	3,4-DINITROTOLUENE	610-39-9
C7H6N2O4	3,5-DINITROTOLUENE	618-85-9
C7H6O	BENZALDEHYDE	100-52-7
C7H6O2	BENZOIC ACID	65-85-0
C7H6O2	p-HYDROXYBENZALDEHYDE	123-08-0
C7H6O2	SALICYLALDEHYDE	90-02-8
C7H6O3	SALICYLIC ACID	69-72-7
C7H7Br	p-BROMOTOLUENE	106-38-7
C7H7Cl	BENZYL CHLORIDE	100-44-7
C7H7Cl	o-CHLOROTOLUENE	95-49-8
C7H7Cl	p-CHLOROTOLUENE	106-43-4
C7H7F	p-FLUOROTOLUENE	352-32-9
C7H7NO	FORMANILIDE	103-70-8
C7H7NO2	m-NITROTOLUENE	99-08-1
C7H7NO2	o-NITROTOLUENE	88-72-2
C7H7NO2	p-NITROTOLUENE	99-99-0
C7H7NO3	o-NITROANISOLE	91-23-6
C7H8	TOLUENE	108-88-3
C7H8	1,3,5-CYCLOHEPTATRIENE	544-55-2
C7H8O	ANISOLE	100-66-3
C7H8O	BENZYL ALCOHOL	100-51-6
C7H8O	m-CRESOL	108-39-4
C7H8O	o-CRESOL	95-48-7
C7H8O	p-CRESOL	106-44-5
C7H8O2	GUAIACOL	90-05-1
C7H8O2	p-METHOXYPHENOL	150-76-5
C7H9N	BENZYLAMINE	100-46-9
C7H9N	2,6-DIMETHYLPYRIDINE	108-48-5
C7H9N	N-METHYLANILINE	100-61-8
C7H9N	m-TOLUIDINE	108-44-1
C7H9N	o-TOLUIDINE	95-53-4
C7H9N	p-TOLUIDINE	106-49-0
C7H10	2-NORBORNENE	498-66-8
C7H10N2	TOLUENEDIAMINE	95-80-7
C7H11NO	CYCLOHEXYL ISOCYANATE	3173-53-3
C7H12	1-HEPTYNE	628-71-7
C7H12O2	n-BUTYL ACRYLATE	141-32-2
C7H12O2	ISOBUTYL ACRYLATE	106-63-8
C7H12O2	n-PROPYL METHACRYLATE	2210-28-8
C7H12O4	DIETHYL MALONATE	105-53-3
C7H14	CYCLOHEPTANE	291-64-5
C7H14	1,1-DIMETYLCYCLOPENTANE	1638-26-2
C7H14	cis-1,2-DIMETHYLCYCLOPENTANE	1192-18-3
C7H14	trans-1,2-DIMETHYLCYCLOPENTANE	822-50-4
C7H14	cis-1,3-DIMETHYLCYCLOPENTANE	2532-58-3
C7H14	trans-1,3-DIMETHYLCYCLOPENTANE	1759-58-6
C7H14	ETHYLCYCLOPENTANE	1640-89-7
C7H14	2-ETHYL-1-PENTENE	3404-71-5
C7H14	3-ETHYL-1-PENTENE	4038-04-4
C7H14	1-HEPTENE	592-76-7
C7H14	cis-2-HEPTENE	6443-92-1
C7H14	trans-2-HEPTENE	14686-13-6
C7H14	cis-3-HEPTENE	7642-10-6
C7H14	trans-3-HEPTENE	14686-14-7
C7H14	METHYLCYCLOHEXANE	108-87-2
C7H14	2-METHYL-1-HEXENE	6094-02-6
C7H14	3-METHYL-1-HEXENE	3404-61-3
C7H14	4-METHYL-1-HEXENE	3769-23-1
C7H14	2,3,3-TRIMETHYL-1-BUTENE	594-56-9
C7H14O	DIISOPROPYL KETONE	565-80-0
C7H14O	2-HEPTANONE	110-43-0
C7H14O	1-HEPTANAL	111-71-7
C7H14O	1-METHYLCYCLOHEXANOL	590-67-0
C7H14O	cis-2-METHYLCYCLOHEXANOL	7443-70-1
C7H14O	trans-2-METHYLCYCLOHEXANOL	7443-52-9
C7H14O	cis-3-METHYLCYCLOHEXANOL	5454-79-5
C7H14O	trans-3-METHYLCYCLOHEXANOL	7443-55-2
C7H14O	cis-4-METHYLCYCLOHEXANOL	7731-28-4
C7H14O	trans-4-METHYLCYCLOHEXANOL	7731-29-5
C7H14O	5-METHYL-2-HEXANONE	110-12-3
C7H14O2	n-BUTYL PROPIONATE	590-01-2
C7H14O2	ETHYL ISOVALERATE	108-64-5
C7H14O2	ISOPENTYL ACETATE	123-92-2
C7H14O2	n-PENTYL ACETATE	628-63-7
C7H14O2	n-PROPYL n-BUTYRATE	105-66-8
C7H14O2	n-HEPTANOIC ACID	111-14-8
C7H14O3	ETHYL-3-ETHOXYPROPIONATE	763-69-9
C7H15Br	1-BROMOHEPTANE	629-04-9
C7H15N	N-METHYLCYCLOHEXYLAMINE	100-60-7
C7H16	2,2-DIMETHYLPENTANE	590-35-2
C7H16	2,3-DIMETHYLPENTANE	565-59-3
C7H16	2,4-DIMETHYLPENTANE	108-08-7
C7H16	3,3-DIMETHYLPENTANE	562-49-2
C7H16	3-ETHYLPENTANE	617-78-7
C7H16	n-HEPTANE	142-82-5
C7H16	2-METHYLHEXANE	591-76-4
C7H16	3-METHYLHEXANE	589-34-4
C7H16	2,2,3-TRIMETHYLBUTANE	464-06-2
C7H16O	1-HEPTANOL	111-70-6
C7H16O	2-HEPTANOL	543-49-7
C7H16O	5-METHYL-1-HEXANOL	627-98-5
C7H16O	ISOPROPYL-TERT-BUTYL-ETHER	1860-27-1
C7H16S	n-HEPTYL MERCAPTAN	1639-09-4
C7H16S	BUTYL-PROPYL-SULFIDE	----------
C7H16S	ETHYL-PENTYL-SULFIDE	----------
C7H16S	HEXYL-METHYL-SULFIDE	20291-60-5
C7H17N	1-AMINOHEPTANE	111-68-2
C8H4Cl2O2	ISOPHTHALOYL CHLORIDE	99-63-8
C8H4O3	PHTHALIC ANHYDRIDE	85-44-9
C8H6	ETHYNYLBENZENE	536-74-3
C8H6O4	ISOPHTHALIC ACID	121-91-5
C8H6O4	PHTHALIC ACID	88-99-3
C8H6O4	TEREPHTHALIC ACID	100-21-0
C8H6S	BENZOTHIOPHENE	95-15-8
C8H7N	INDOLE	120-72-9
C8H8	STYRENE	100-42-5
C8H8	1,3,5,7-CYCLOOCTATETRAENE	629-20-9
C8H8O	ACETOPHENONE	98-86-2
C8H8O	p-TOLUALDEHYDE	104-87-0
C8H8O2	METHYL BENZOATE	93-58-3
C8H8O2	o-TOLUIC ACID	118-90-1
C8H8O2	p-TOLUIC ACID	99-94-5
C8H8O3	METHYL SALICYLATE	119-36-8
C8H8O3	VANILLIN	121-33-5
C8H9NO	ACETANILIDE	103-84-4
C8H10	ETHYLBENZENE	100-41-4
C8H10	m-XYLENE	108-38-3
C8H10	o-XYLENE	95-47-6
C8H10	p-XYLENE	106-42-3
C8H10O	m-ETHYLPHENOL	620-17-7
C8H10O	p-ETHYLPHENOL	123-07-9
C8H10O	PHENETOLE	103-73-1
C8H10O	2-PHENYLETHANOL	60-12-8
C8H10O	2,3-XYLENOL	526-75-0
C8H10O	2,4-XYLENOL	105-67-9
C8H10O	2,5-XYLENOL	95-87-4
C8H10O	2,6-XYLENOL	576-26-1
C8H10O	3,4-XYLENOL	95-65-8
C8H10O	3,5-XYLENOL	108-68-9
C8H11N	N,N-DIMETHYLANILINE	121-69-7
C8H11N	o-ETHYLANILINE	578-54-1
C8H11N	2,4,6-TRIMETHYLPYRIDINE	108-75-8
C8H11NO	p-PHENETIDINE	156-43-3
C8H12	1,5-CYCLOOCTADIENE	111-78-4
C8H12	VINYLCYCLOHEXENE	100-40-3
C8H12O4	1,4-CYCLOHEXANEDICARBOXYLIC ACID	619-82-9
C8H12O4	DIETHYL MALEATE	141-05-9
C8H14O2	n-BUTYL METHACRYLATE	97-88-1
C8H14O3	BUTYRIC ANHYDRIDE	106-31-0
C8H14O4	DIETHYL SUCCINATE	123-25-1
C8H16	CYCLOOCTANE	292-64-8
C8H16	1,1-DIMETHYLCYCLOHEXANE	590-66-9
C8H16	cis-1,2-DIMETHYLCYCLOHEXANE	2207-01-4
C8H16	trans-1,2-DIMETHYLCYCLOHEXANE	6876-23-9
C8H16	cis-1,3-DIMETHYLCYCLOHEXANE	638-04-0
C8H16	trans-1,3-DIMETHYLCYCLOHEXANE	2207-03-6
C8H16	cis-1,4-DIMETHYLCYCLOHEXANE	624-29-3
C8H16	trans-1,4-DIMETHYLCYCLOHEXANE	2207-04-7
C8H16	ETHYLCYCLOHEXANE	1678-91-7
C8H16	2-ETHYL-1-HEXENE	1632-16-2
C8H16	1-METHYL-1-ETHYLCYCLOPENTANE	16747-50-5
C8H16	1-OCTENE	111-66-0

651

Formula	Name	CAS
C8H16	trans-2-OCTENE	13389-42-9
C8H16	trans-3-OCTENE	14919-01-8
C8H16	trans-4-OCTENE	14850-23-8
C8H16	n-PROPYLCYCLOPENTANE	2040-96-2
C8H16	2,4,4-TRIMETHYL-1-PENTENE	107-39-1
C8H16	2,4,4-TRIMETHYL-2-PENTENE	107-40-4
C8H16O	2-ETHYLHEXANAL	123-05-7
C8H16O	1-OCTANAL	124-13-0
C8H16O	2-OCTANONE	111-13-7
C8H16O2	n-BUTYL n-BUTYRATE	109-21-7
C8H16O2	n-HEXYL ACETATE	142-92-7
C8H16O2	ISOBUTYL ISOBUTYRATE	97-85-8
C8H16O2	n-OCTANOIC ACID	124-07-2
C8H16O4	DIETHYLENE GLYCOL ETHYL ETHER ACETATE	112-15-2
C8H18	2,2-DIMETHYLHEXANE	590-73-8
C8H18	2,3-DIMETHYLHEXANE	584-94-1
C8H18	2,4-DIMETHYLHEXANE	589-43-5
C8H18	2,5-DIMETHYLHEXANE	592-13-2
C8H18	3,3-DIMETHYLHEXANE	563-16-6
C8H18	3,4-DIMETHYLHEXANE	583-48-2
C8H18	3-ETHYLHEXANE	619-99-8
C8H18	3-ETHYL-2-METHYLPENTANE	609-26-7
C8H18	3-METHYL-3-ETHYLPENTANE	1067-08-9
C8H18	2-METHYLHEPTANE	592-27-8
C8H18	3-METHYLHEPTANE	589-81-1
C8H18	4-METHYLHEPTANE	589-53-7
C8H18	n-OCTANE	111-65-9
C8H18	2,2,3-TRIMETHYLPENTANE	564-02-3
C8H18	2,2,4-TRIMETHYLPENTANE	540-84-1
C8H18	2,3,3-TRIMETHYLPENTANE	560-21-4
C8H18	2,3,4-TRIMETHYLPENTANE	565-75-3
C8H18	2,2,3,3-TETRAMETHYLBUTANE	594-82-1
C8H18O	DI-n-BUTYL ETHER	142-96-1
C8H18O	DI-sec-BUTYL ETHER	6863-58-7
C8H18O	DI-tert-BUTYL ETHER	6163-66-2
C8H18O	2-ETHYL-1-HEXANOL	104-76-7
C8H18O	1-OCTANOL	111-87-5
C8H18O	2-OCTANOL	123-96-6
C8H18O2	DI-t-BUTYL PEROXIDE	110-05-4
C8H18O2S	DI-n-BUTYL SULFONE	598-04-9
C8H18O3	DIETHYLENE GLYCOL DIETHYL ETHER	112-36-7
C8H18O3	DIETHYLENE GLYCOL MONOBUTYL ETHER	112-34-5
C8H18O4	TRIETHYLENE GLYCOL DIMETHYL ETHER	112-49-2
C8H18O5	TETRAETHYLENE GLYCOL	112-60-7
C8H18S	n-OCTYL MERCAPTAN	111-88-6
C8H18S	tert-OCTYL MERCAPTAN	141-59-3
C8H18S	DIBUTYL-SULFIDE	544-40-1
C8H18S	ETHYL-HEXYL-SULFIDE	---------
C8H18S	HEPTYL-METHYL-SULFIDE	---------
C8H18S	PENTYL-PROPYL-SULFIDE	---------
C8H18S2	DIBUTYL-DISULFIDE	629-45-8
C8H19N	DI-n-BUTYLAMINE	111-92-2
C8H19N	DIISOBUTYLAMINE	110-96-3
C8H19N	n-OCTYLAMINE	111-86-4
C8H23N5	TETRAETHYLENEPENTAMINE	112-57-2
C8H24O4Si4	OCTAMETHYLCYCLOTETRASILOXANE	556-67-2
C9H4O5	TRIMELLITIC ANHYDRIDE	552-30-7
C9H6N2O2	TOLUENE DIISOCYANATE	26471-62-5
C9H7N	ISOQUINOLINE	119-65-3
C9H7N	QUINOLINE	91-22-5
C9H7NO	8-HYDROXYQUINOLINE	148-24-3
C9H8	INDENE	95-13-6
C9H8O	2-METHYLBENZOFURAN	4265-25-2
C9H10	INDANE	496-11-7
C9H10	cis-PROPENYLBENZENE	766-90-5
C9H10	trans-PROPENYLBENZENE	873-66-5
C9H10	alpha-METHYLSTYRENE	98-83-9
C9H10	m-METHYLSTYRENE	100-80-1
C9H10	o-METHYLSTYRENE	611-15-4
C9H10	p-METHYLSTYRENE	622-97-9
C9H10O2	BENZYL ACETATE	140-11-4
C9H10O2	ETHYL BENZOATE	93-89-0
C9H10O3	ETHYL VANILLIN	121-32-4
C9H11NO	p-DIMETHYLAMINOBENZALDEHYDE	100-10-7
C9H12	CUMENE	98-82-8
C9H12	m-ETHYLTOLUENE	620-14-4
C9H12	o-ETHYLTOLUENE	611-14-3
C9H12	p-ETHYLTOLUENE	622-96-8
C9H12	MESITYLENE	108-67-8
C9H12	n-PROPYLBENZENE	103-65-1
C9H12	1,2,3-TRIMETHYLBENZENE	526-73-8
C9H12	1,2,4-TRIMETHYLBENZENE	95-63-6
C9H12O	BENZYL ETHYL ETHER	539-30-0
C9H12O	2-PHENYL-2-PROPANOL	617-94-7
C9H12O2	CUMENE HYDROPEROXIDE	80-15-9
C9H14O	ISOPHORONE	78-59-1
C9H14O6	GLYCERYL TRIACETATE	102-76-1
C9H16	1-NONYNE	3452-09-3
C9H16O4	AZELAIC ACID	123-99-9
C9H18	BUTYLCYCLOPENTANE	2040-95-1
C9H18	cis,cis-1,3,5-TRIMETHYLCYCLOHEXANE	1795-27-3
C9H18	cis,trans-1,3,5-TRIMETHYLCYCLOHEXANE	1795-26-2
C9H18	ISOPROPYLCYCLOHEXANE	696-29-7
C9H18	1-NONENE	124-11-8
C9H18	n-PROPYLCYCLOHEXANE	1678-92-8
C9H18O	DIISOBUTYL KETONE	108-83-8
C9H18O	1-NONANAL	124-19-6
C9H18O2	n-BUTYL VALERATE	591-68-4
C9H18O2	n-NONANOIC ACID	112-05-0
C9H18O2	n-OCTYL FORMATE	112-32-3
C9H20	3,3-DIETHYLPENTANE	1067-20-5
C9H20	2,2-DIMETHYL-3-ETHYLPENTANE	16747-32-3
C9H20	3-ETHYL-2,3-DIMETHYLPENTANE	16747-33-4
C9H20	2,4-DIMETHYL-3-ETHYLPENTANE	1068-87-7
C9H20	2,2-DIMETHYLHEPTANE	1071-26-7
C9H20	2,6-DIMETHYLHEPTANE	1072-05-5
C9H20	3-ETHYLHEPTANE	15869-80-4
C9H20	4-ETHYLHEPTANE	2216-32-2
C9H20	2,3-DIMETHYLHEPTANE	3074-71-3
C9H20	2,4-DIMETHYLHEPTANE	2213-23-2
C9H20	2,5-DIMETHYLHEPTANE	2216-30-0
C9H20	3,4-DIMETHYLHEPTANE	922-28-1
C9H20	3,5-DIMETHYLHEPTANE	926-82-9
C9H20	4,4-DIMETHYLHEPTANE	1068-19-5
C9H20	3-ETHYL-2-METHYLHEXANE	16789-46-1
C9H20	4-ETHYL-2-METHYLHEXANE	3074-75-7
C9H20	3-ETHYL-3-METHYLHEXANE	3074-76-8
C9H20	3-ETHYL-4-METHYLHEXANE	3074-77-9
C9H20	2,2,3-TRIMETHYLHEXANE	16747-25-4
C9H20	2,2,4-TRIMETHYLHEXANE	16747-26-5
C9H20	2,3,3-TRIMETHYLHEXANE	16747-28-7
C9H20	2,3,4-TRIMETHYLHEXANE	921-47-1
C9H20	2,3,5-TRIMETHYLHEXANE	1069-53-0
C9H20	2,4,4-TRIMETHYLHEXANE	16747-30-1
C9H20	3,3,4-TRIMETHYLHEXANE	16747-31-2
C9H20	2-METHYLOCTANE	3221-61-2
C9H20	3-METHYLOCTANE	2216-33-3
C9H20	4-METHYLOCTANE	2216-34-4
C9H20	n-NONANE	111-84-2
C9H20	2,2,3,3-TETRAMETHYLPENTANE	7154-79-2
C9H20	2,2,3,4-TETRAMETHYLPENTANE	1186-53-4
C9H20	2,2,4,4-TETRAMETHYLPENTANE	1070-87-7
C9H20	2,3,3,4-TETRAMETHYLPENTANE	16747-38-9
C9H20	2,2,5-TRIMETHYLHEXANE	3522-94-9
C9H20O	2,6-DIMETHYL-4-HEPTANOL	108-82-7
C9H20O	1-NONANOL	143-08-8
C9H20O	2-NONANOL	70419-06-6
C9H20S	n-NONYL MERCAPTAN	1455-21-6
C9H20S	BUTYL-PENTYL-SULFIDE	---------
C9H20S	ETHYL-HEPTYL-SULFIDE	---------
C9H20S	HEXYL-PROPYL-SULFIDE	---------
C9H20S	METHYL-OCTYL-SULFIDE	---------
C9H21N	n-NONYLAMINE	112-20-9
C9H21N	TRIPROPYLAMINE	102-69-2
C10H6O8	PYROMELLITIC ACID	89-05-4
C10H7Br	1-BROMONAPHTHALENE	90-11-9
C10H7Cl	1-CHLORONAPHTHALENE	90-13-1
C10H8	NAPHTHALENE	91-20-3
C10H8	AZULENE	275-51-4
C10H9N	QUINALDINE	91-63-4
C10H10	m-DIVINYLBENZENE	108-57-6
C10H10	1-METHYLINDENE	767-59-9
C10H10	2-METHYLINDENE	2177-47-1
C10H10O4	DIMETHYL PHTHALATE	131-11-3
C10H10O4	DIMETHYL TEREPHTHALATE	120-61-6
C10H12	DICYCLOPENTADIENE	77-73-6
C10H12	1,2,3,4-TETRAHYDRONAPHTHALENE	119-64-2
C10H12O	ANETHOLE	104-46-1
C10H12O4	DIALLYL MALEATE	999-21-3
C10H14	n-BUTYLBENZENE	104-51-8
C10H14	sec-BUTYLBENZENE	135-98-8
C10H14	tert-BUTYLBENZENE	98-06-6
C10H14	1,2,3,4-TETRAMETHYLBENZENE	488-23-3
C10H14	m-CYMENE	535-77-3
C10H14	o-CYMENE	527-84-4
C10H14	p-CYMENE	99-87-6

Formula	Name	CAS
C10H14	m-DIETHYLBENZENE	141-93-5
C10H14	o-DIETHYLBENZENE	135-01-3
C10H14	p-DIETHYLBENZENE	105-05-5
C10H14	2-ETHYL-m-XYLENE	2870-04-4
C10H14	2-ETHYL-p-XYLENE	1758-88-9
C10H14	3-ETHYL-o-XYLENE	933-98-2
C10H14	4-ETHYL-m-XYLENE	874-41-9
C10H14	4-ETHYL-o-XYLENE	934-80-5
C10H14	5-ETHYL-m-XYLENE	934-74-7
C10H14	ISOBUTYLBENZENE	538-93-2
C10H14	1,2,3,5-TETRAMETHYLBENZENE	527-53-7
C10H14	1,2,4,5-TETRAMETHYLBENZENE	95-93-2
C10H14O	p-tert-BUTYLPHENOL	98-54-4
C10H14O2	p-tert-BUTYLCATECHOL	98-29-3
C10H15N	N,N-DIETHYLANILINE	91-66-7
C10H15N	2,6-DIETHYLANILINE	579-66-8
C10H16	CAMPHENE	79-92-5
C10H16	D-LIMONENE	5989-27-5
C10H16	alpha-PHELLANDRENE	99-83-2
C10H16	beta-PHELLANDRENE	555-10-2
C10H16	alpha-PINENE	80-56-8
C10H16	beta-PINENE	127-91-3
C10H16	alpha-TERPINENE	99-86-5
C10H16	gamma-TERPINENE	99-85-4
C10H16	TERPINOLENE	586-62-9
C10H16O	CAMPHOR	76-22-2
C10H18	1-DECYNE	7064-93-2
C10H18	cis-DECAHYDRONANPHTALENE	493-01-6
C10H18	trans-DECAHYDRONAPHTHALENE	493-02-7
C10H18O4	SEBACIC ACID	111-20-6
C10H20	n-BUTYLCYCLOHEXANE	1678-93-9
C10H20	1-CYCLOPENTYLPENTANE	3741-00-2
C10H20	1-DECENE	872-05-9
C10H20O	1-DECANAL	112-31-2
C10H20O2	n-DECANOIC ACID	334-48-5
C10H20O2	2-ETHYLHEXYL ACETATE	103-09-3
C10H20O2	ISOPENTYL ISOVALERATE	659-70-1
C10H22	n-DECANE	124-18-5
C10H22	2-METHYLNONANE	871-83-0
C10H22	3-METHYLNONANE	5911-04-6
C10H22	4-METHYLNONANE	17301-94-9
C10H22	5-METHYLNONANE	15869-85-9
C10H22	3-ETHYLOCTANE	5881-17-4
C10H22	4-ETHYLOCTANE	15869-86-0
C10H22	2,2-DIMETHYLOCTANE	15869-87-1
C10H22	2,3-DIMETHYLOCTANE	7146-60-3
C10H22	2,4-DIMETHYLOCTANE	4032-94-4
C10H22	2,5-DIMETHYLOCTANE	15869-89-3
C10H22	2,6-DIMETHYLOCTANE	2051-30-1
C10H22	2,7-DIMETHYLOCTANE	1072-16-8
C10H22	3,3-DIMETHYLOCTANE	4110-44-5
C10H22	3,4-DIMETHYLOCTANE	15869-92-8
C10H22	3,5-DIMETHYLOCTANE	15869-93-9
C10H22	3,6-DIMETHYLOCTANE	15869-94-0
C10H22	4,4-DIMETHYLOCTANE	15869-95-1
C10H22	4,5-DIMETHYLOCTANE	15869-96-2
C10H22	4-PROPYLHEPTANE	3178-29-8
C10H22	4-ISOPROPYLHEPTANE	---------
C10H22	3-ETHYL-2-METHYLHEPTANE	---------
C10H22	4-ETHYL-2-METHYLHEPTANE	52896-88-5
C10H22	5-ETHYL-2-METHYLHEPTANE	13475-78-0
C10H22	3-ETHYL-3-METHYLHEPTANE	17302-01-1
C10H22	4-ETHYL-3-METHYLHEPTANE	52896-89-6
C10H22	3-ETHYL-5-METHYLHEPTANE	52896-90-9
C10H22	3-ETHYL-4-METHYLHEPTANE	15896-91-0
C10H22	4-ETHYL-4-METHYLHEPTANE	17302-4-4
C10H22	2,2,3-TRIMETHYLHEPTANE	52896-92-1
C10H22	2,2,4-TRIMETHYLHEPTANE	14720-74-2
C10H22	2,2,5-TRIMETHYLHEPTANE	20291-95-6
C10H22	2,2,6-TRIMETHYLHEPTANE	1190-83-6
C10H22	2,3,3-TRIMETHYLHEPTANE	52896-93-2
C10H22	2,3,4-TRIMETHYLHEPTANE	52896-95-4
C10H22	2,3,5-TRIMETHYLHEPTANE	20278-85-7
C10H22	2,3,6-TRIMETHYLHEPTANE	4032-93-3
C10H22	2,4,4-TRIMETHYLHEPTANE	4032-92-2
C10H22	2,4,5-TRIMETHYLHEPTANE	20278-84-6
C10H22	2,4,6-TRIMETHYLHEPTANE	2613-61-8
C10H22	2,5,5-TRIMETHYLHEPTANE	1189-99-7
C10H22	3,3,4-TRIMETHYLHEPTANE	2278-87-9
C10H22	3,3,5-TRIMETHYLHEPTANE	7154-80-5
C10H22	3,4,4-TRIMETHYLHEPTANE	2278-88-0
C10H22	3,4,5-TRIMETHYLHEPTANE	2278-89-1
C10H22	3-ISOPROPYL-2-METHYLHEXANE	---------
C10H22	3,3-DIETHYLHEXANE	17302-02-2
C10H22	3,4-DIETHYLHEXANE	19398-77-7
C10H22	3-ETHYL-2,2-DIMETHYLHEXANE	20291-91-2
C10H22	4-ETHYL-2,2-DIMETHYLHEXANE	---------
C10H22	3-ETHYL-2,3-DIMETHYLHEXANE	52897-00-4
C10H22	4-ETHYL-2,3-DIMETHYLHEXANE	52897-01-5
C10H22	3-ETHYL-2,4-DIMETHYLHEXANE	7220-26-0
C10H22	4-ETHYL-2,4-DIMETHYLHEXANE	52897-03-7
C10H22	3-ETHYL-2,5-DIMETHYLHEXANE	52897-04-8
C10H22	4-ETHYL-3,3-DIMETHYLHEXANE	52897-05-9
C10H22	3-ETHYL-3,4-DIMETHYLHEXANE	52897-06-0
C10H22	2,2,3,3-TETRAMETHYLHEXANE	13475-81-5
C10H22	2,2,3,4-TETRAMETHYLHEXANE	52897-08-2
C10H22	2,2,3,5-TETRAMETHYLHEXANE	52897-09-3
C10H22	2,2,4,4-TETRAMETHYLHEXANE	51750-65-3
C10H22	2,2,4,5-TETRAMETHYLHEXANE	16747-42-5
C10H22	2,2,5,5-TETRAMETHYLHEXANE	1071-81-4
C10H22	2,3,3,4-TETRAMETHYLHEXANE	52897-10-6
C10H22	2,3,3,5-TETRAMETHYLHEXANE	52897-11-7
C10H22	2,3,4,4-TETRAMETHYLHEXANE	52897-12-8
C10H22	2,3,4,5-TETRAMETHYLHEXANE	52897-15-1
C10H22	3,3,4,4-TETRAMETHYLHEXANE	5171-84-6
C10H22	2,4-DIMETHYL-3-ISOPROPYLPENTANE	---------
C10H22	3,3-DIETHYL-2-METHYLPENTANE	52897-16-2
C10H22	3-ETHYL-2,2,3-TRIMETHYLPENTANE	52897-17-3
C10H22	3-ETHYL-2,2,4-TRIMETHYLPENTANE	---------
C10H22	3-ETHYL-2,3,4-TRIMETHYLPENTANE	52897-19-5
C10H22	2,2,3,3,4-PENTAMETHYLPENTANE	16747-44-7
C10H22	2,2,3,4,4-PENTAMETHYLPENTANE	16747-45-8
C10H22O	1-DECANOL	112-30-1
C10H22O	DI-n-PENTYL ETHER	693-65-2
C10H22O	ISODECANOL	55505-26-5
C10H22O5	TETRAETHYLENE GLYCOL DIMETHYL ETHER	143-24-8
C10H22S	n-DECYL MERCAPTAN	143-10-2
C10H22S	BUTYL-HEXYL-SULFIDE	---------
C10H22S	ETHYL-OCTYL-SULFIDE	---------
C10H22S	HEPTYL-PROPYL-SULFIDE	---------
C10H22S	METHYL-NONYL-SULFIDE	---------
C10H22S	PENTYL-SULFIDE	---------
C10H22S2	DIPENTYL-DISULFIDE	112-51-6
C10H23N	n-DECYLAMINE	2016-57-1
C11H10	1-METHYLNAPHTHALENE	90-12-0
C11H10	2-METHYLNAPHTHALENE	91-57-6
C11H14O2	n-BUTYL BENZOATE	136-60-7
C11H16	n-PENTYLBENZENE	538-68-1
C11H16O	p-tert-AMYLPHENOL	80-46-6
C11H20	1-UNDECYNE	2243-98-3
C11H20O2	2-ETHYLHEXYL ACRYLATE	103-11-7
C11H22	1-UNDECENE	821-95-4
C11H22	1-CYCLOPENTYLHEXANE	---------
C11H22	PENTYLCYCLOHEXANE	4292-92-6
C11H22O	1-UNDECANAL	112-44-7
C11H24	n-UNDECANE	1120-21-4
C11H24O	1-UNDECANOL	112-42-5
C11H24S	UNDECYL MERCAPTAN	5332-52-5
C11H24S	BUTYL-HEPTYL-SULFIDE	---------
C11H24S	DECYL-METHYL-SULFIDE	---------
C11H24S	ETHYL-NONYL-SULFIDE	---------
C11H24S	OCTYL-PROPYL-SULFIDE	---------
C12H8O	DIBENZOFURAN	132-64-9
C12H9N	DIBENZOPYRROLE	86-74-8
C12H10	ACENAPHTHENE	83-32-9
C12H10	BIPHENYL	92-54-4
C12H10O	DIPHENYL ETHER	101-84-8
C12H11N	p-AMINODIPHENYL	92-67-1
C12H11N	DIPHENYLAMINE	122-39-4
C12H11N3	p-AMINOAZOBENZENE	60-09-3
C12H11N3	1,3-DIPHENYLTRIAZENE	136-35-6
C12H12	1,2-DIMETHYLNAPHTHALENE	573-98-8
C12H12	1,3-DIMETHYLNAPHTHALENE	575-41-7
C12H12	1,4-DIMETHYLNAPHTHALENE	571-58-4
C12H12	1,5-DIMETHYLNAPHTHALENE	571-61-9
C12H12	1,6-DIMETHYLNAPHTHALENE	575-43-9
C12H12	1,7-DIMETHYLNAPHTHALENE	575-37-1
C12H12	2,3-DIMETHYLNAPHTHALENE	581-40-8
C12H12	2,6-DIMETHYLNAPHTHALENE	581-42-0
C12H12	2,7-DIMETHYLNAPHTHALENE	582-16-1
C12H12	1-ETHYLNAPHTHALENE	1127-76-0
C12H12	2-ETHYLNAPHTHALENE	939-27-5
C12H12N2	p-AMINODIPHENYLAMINE	101-54-2
C12H12N2	HYDRAZOBENZENE	122-66-7
C12H14	1,2,3-TRIMETHYLINDENE	4773-83-5
C12H14O4	DIETHYL PHTHALATE	84-66-2

Formula	Name	CAS
C12H16	CYCLOHEXYLBENZENE	827-52-1
C12H18	m-DIISOPROPYLBENZENE	99-62-7
C12H18	p-DIISOPROPYLBENZENE	100-18-5
C12H18	n-HEXYLBENZENE	1077-16-3
C12H18	1,2,3-TRIETHYLBENZENE	---------
C12H18	1,2,4-TRIETHYLBENZENE	877-44-1
C12H18	1,3,5-TRIETHYLBENZENE	102-25-0
C12H18	HEXAMETHYLBENZENE	87-85-4
C12H20O4	DIBUTYL MALEATE	105-76-0
C12H22	BICYCLOHEXYL	92-51-3
C12H22	1-DODECYNE	765-03-7
C12H23N	DICYCLOHEXYLAMINE	101-83-7
C12H24	1-DODECENE	112-41-4
C12H24	1-CYCLOPENTYLHEPTANE	---------
C12H24	1-CYCLOHEXYLHEXANE	4292-75-5
C12H24O	1-DODECANAL	112-54-9
C12H24O2	n-DODECANOIC ACID	143-07-7
C12H26	n-DODECANE	112-40-3
C12H26O	DI-n-HEXYL ETHER	112-58-3
C12H26O	1-DODECANOL	112-53-8
C12H26O3	DIETHYLENE GLYCOL DI-n-BUTYL ETHER	112-73-2
C12H26S	n-DODECYL MERCAPTAN	112-55-0
C12H26S	BUTYL-OCTYL-SULFIDE	---------
C12H26S	DECYL-ETHYL-SULFIDE	---------
C12H26S	HEXYL-SULFIDE	6294-31-1
C12H26S	METHYL-UNDECYL-SULFIDE	10059-13-9
C12H26S	NONYL-PROPYL-SULFIDE	---------
C12H26S2	HEXYL-DISULFIDE	---------
C12H27BO3	TRI-n-BUTYL BORATE	688-74-4
C12H27N	DODECYLAMINE	124-22-1
C12H27N	TRI-n-BUTYLAMINE	102-82-9
C13H10	FLUORENE	86-73-7
C13H10O	BENZOPHENONE	119-61-9
C13H12	DIPHENYLMETHANE	101-81-5
C13H14	1-PROPYLNAPHTHALENE	2765-18-6
C13H14	2-PROPYLNAPHTHALENE	2027-19-2
C13H14	2ETHYL-3-METHYLNAPHTHALENE	---------
C13H14	2ETHYL-6-METHYLNAPHTHALENE	7372-86-3
C13H14	2ETHYL-7-METHYLNAPHTHALENE	---------
C13H20	n-HEPTYLBENZENE	1078-71-3
C13H24	1-TRIDECYNE	26186-02-7
C13H26	1-TRIDECENE	2437-56-1
C13H26	1-CYCLOPENTYLOCTANE	1795-20-6
C13H26	1-CYCLOHEXYLHEPTANE	5617-41-4
C13H26O	1-TRIDECANAL	10486-19-8
C13H26O2	n-BUTYL NONANOATE	50623-57-9
C13H26O2	METHYL DODECANOATE	111-82-0
C13H28	n-TRIDECANE	629-50-5
C13H28O	1-TRIDECANOL	112-70-9
C13H28S	BUTYL-NONYL-SULFIDE	---------
C13H28S	DECYL-PROPYL-SULFIDE	---------
C13H28S	DODECYL-METHYL-SULFIDE	---------
C13H28S	ETHYL-UNDECYL-SULFIDE	---------
C13H28S	1-TRIDECANETHIOL	---------
C14H8O2	ANTHRAQUINONE	84-65-1
C14H10	ANTHRACENE	120-12-7
C14H10	DIPHENYLACETYLENE	501-65-5
C14H10	PHENANTHRENE	85-01-8
C14H12	cis-STILBENE	645-49-8
C14H12	trans-STILBENE	103-30-0
C14H12O2	BENZYL BENZOATE	120-51-4
C14H14	1,1-DIPHENYLETHANE	612-00-0
C14H14	1,2-DIPHENYLETHANE	103-29-7
C14H14O	DIBENZYL ETHER	103-50-4
C14H16	1-n-BUTYLNAPHTHALENE	1634-09-9
C14H16	2-BUTYLNAPHTHALENE	---------
C14H22	n-OCTYLBENZENE	2189-60-8
C14H22	1,2,3,4-TETRAETHYLBENZENE	642-32-0
C14H22	1,2,3,5-TETRAETHYLBENZENE	---------
C14H22	1,2,4,5-TETRAETHYLBENZENE	635-81-4
C14H22O	p-tert-OCTYLPHENOL	140-66-9
C14H28	1-TETRADECENE	1120-36-1
C14H28	1-CYCLOPENTYLNONANE	2882-98-6
C14H28	1-CYCLOHEXYLOCTANE	1795-15-9
C14H28O2	n-TETRADECANOIC ACID	544-63-8
C14H30	n-TETRADECANE	629-59-4
C14H30O	1-TETRADECANOL	112-72-1
C14H30S	BUTYL-DECYL-SULFIDE	---------
C14H30S	DODECYL-ETHYL-SULFIDE	---------
C14H30S	HEPTYL-SULFIDE	---------
C14H30S	METHYL-TRIDECYL-SULFIDE	---------
C14H30S	PROPYL-UNDECYL-SULFIDE	---------
C14H30S	1-TETRADECANETHIOL	2079-95-0
C14H30S2	HEPTYL-DISULFIDE	---------
C14H31N	TETRADECYLAMINE	2016-42-4
C15H10N2O2	DIPHENYLMETHANE-4,4'-DIISOCYANATE	101-68-8
C15H16O	p-CUMYLPHENOL	599-64-4
C15H16O2	BISPHENOL A	80-05-7
C15H18	1-PENTYLNAPHTHALENE	86-89-5
C15H18	2-PENTYLNAPHTHALENE	93-22-1
C15H24	n-NONYLBENZENE	1081-77-2
C15H24O	2,6-DI-tert-BUTYL-p-CRESOL	128-37-0
C15H24O	NONYLPHENOL	25154-52-3
C15H28	1-PENTADECYNE	765-13-9
C15H30	1-PENTADECENE	13360-61-7
C15H30	1-CYCLOPENTYLDECANE	295-48-7
C15H30	1-CYCLOHEXYLNONANE	2883-02-5
C15H30O2	PENTADECANOIC ACID	1002-84-2
C15H32	n-PENTADECANE	629-62-9
C15H32O	1-PENTADECANOL	629-76-5
C15H32S	BUTYL-UNDECYL-SULFIDE	---------
C15H32S	DODECYL-PROPYL-SULFIDE	---------
C15H32S	ETHYL-TRIDECYL-SULFIDE	---------
C15H32S	METHYL-TETRADECYL-SULFIDE	---------
C15H32S	1-PENTADECANETHIOL	---------
C16H10	FLUORANTHENE	206-44-0
C16H10	PYRENE	129-00-0
C16H12	1-PHENYLNAPHTHALENE	605-02-7
C16H20	1-n-HEXYLNAPHTHALENE	2876-53-1
C16H22O4	DIBUTYL PHTHALATE	84-74-2
C16H26	n-DECYLBENZENE	104-72-3
C16H26	PENTAETHYLBENZENE	605-01-6
C16H30	1-HEXADECYNE	629-74-3
C16H32	n-DECYLCYCLOHEXANE	1795-16-0
C16H32	1-CYCLOPENTYLUNDECANE	6758-23-5
C16H32	1-HEXADECENE	629-73-2
C16H32O2	n-HEXADECANOIC ACID	57-10-3
C16H34	n-HEXADECANE	544-76-3
C16H34O	DI-n-OCTYL ETHER	629-82-3
C16H34O	1-HEXADECANOL	36653-82-4
C16H34S	BUTYL-DODECYL-SULFIDE	---------
C16H34S	ETHYL-TETRADECYL-SULFIDE	---------
C16H34S	METHYL-PENTADECYL-SULFIDE	---------
C16H34S	OCTYL-SULFIDE	---------
C16H34S	PROPYL-TRIDECYL-SULFIDE	---------
C16H34S	1-HEXADECANETHIOL	2917-26-2
C16H34S2	OCTYL-DISULFIDE	---------
C17H28	n-UNDECYLBENZENE	6742-54-7
C17H32	1-HEPTADECYNE	6765-39-5
C17H34	1-CYCLOPENTYLDODECANE	5634-30-0
C17H34	1-CYCLOHEXYLUNDECANE	---------
C17H34	1-HEPTADECENE	6765-39-5
C17H36	n-HEPTADECANE	629-78-7
C17H36O	1-HEPTADECANOL	1454-85-9
C17H36S	BUTYL-TRIDECYL-SULFIDE	---------
C17H36S	ETHYL-PENTADECYL-SULFIDE	---------
C17H36S	HEXADECYL-METHYL-SULFIDE	---------
C17H36S	PROPYL-TETRADECYL-SULFIDE	---------
C17H36S	1-HEPTADECANETHIOL	---------
C18H12	CHRYSENE	218-01-9
C18H14	m-TERPHENYL	92-06-8
C18H14	o-TERPHENYL	84-15-1
C18H14	p-TERPHENYL	92-94-4
C18H15P	TRIPHENYLPHOSPHINE	603-35-0
C18H15O4P	TRIPHENYL PHOSPHATE	115-86-6
C18H16N2	N,N'-DIPHENYL-p-PHENYLENEDIAMINE	74-31-7
C18H22	2,3-DIMETHYL-2,3-DIPHENYLBUTANE	1889-67-4
C18H22O2	DICUMYL PEROXIDE	80-43-3
C18H30	n-DODECYLBENZENE	123-01-3
C18H30	HEXAETHYLBENZENE	604-88-6
C18H32O2	LINOLEIC ACID	60-33-3
C18H34	1-OCTADECYNE	629-89-0
C18H34O2	OLEIC ACID	112-80-1
C18H34O4	DIBUTYL SEBACATE	109-43-3
C18H34O4	DIHEXYL ADIPATE	110-33-8
C18H36	1-CYCOPENTYLTRIDECANE	6006-34-4
C18H36	1-CYCLOHEXYLDODECANE	1795-17-1
C18H36	1-OCTADECENE	112-88-9
C18H36O2	STEARIC ACID	57-11-4
C18H38	n-OCTADECANE	593-45-3
C18H38O	DINONYL ETHER	2456-27-1
C18H38O	1-OCTADECANOL	112-92-5
C18H38S	BUTYL-TETRADECYL-SULFIDE	---------
C18H38S	ETHYL-HEXADECYL-SULFIDE	---------
C18H38S	HEPTADECYL-METHYL-SULFIDE	---------
C18H38S	NONYL-SULFIDE	---------

Formula	Name	CAS
C18H38S	PENTADECYL-PROPYL-SULFIDE	----------
C18H38S	1-OCTADECANETHIOL	2885-00-9
C18H38S2	NONYL-DISULFIDE	----------
C19H26	1-n-NONYLNAPHTHALENE	26438-26-6
C19H32	n-TRIDECYLBENZENE	123-02-4
C19H36	1-NONADECYNE	26186-01-6
C19H3602	METHYL OLEATE	112-62-9
C19H38	1-CYCLOPENTYLTETRADECANE	1795-22-8
C19H38	1-CYCLOHEXYLTRIDECANE	6006-33-3
C19H38	1-NONADECENE	18435-45-5
C19H3802	NONADECANOIC ACID	646-30-0
C19H40	n-NONADECANE	629-92-5
C19H400	1-NONADECANOL	1454-84-8
C19H40S	BUTYL-PENTADECYL-SULFIDE	----------
C19H40S	ETHYL-HEPTADECYL-SULFIDE	----------
C19H40S	HEXADECYL-PROPYL-SULFIDE	----------
C19H40S	METHYL-OCTADECYL-SULFIDE	----------
C19H40S	1-NONADECANETHIOL	----------
C20H16	TRIPHENYLETHYLENE	58-72-0
C20H28	1-n-DECYLNAPHTHALENE	26438-27-7
C20H3002	ABIETIC ACID	514-10-3
C20H31N	DEHYDROABIETYLAMINE	1446-61-3
C20H34	1-PHENYLTETRADECANE	----------
C20H38	1-EICOSYNE	765-27-5
C20H40	1-CYCLOPENTYLPENTADECANE	4669-01-6
C20H40	1-CYCLOHEXYLTETRADECANE	1795-18-2
C20H40	1-EICOSENE	3452-07-1
C20H42	n-EICOSANE	112-95-8
C20H420	1-EICOSANOL	629-96-9
C20H42S	BUTYL-HEXADECYL-SULFIDE	----------
C20H42S	DECYL-SULFIDE	----------
C20H42S	ETHYL-OCTADECYL-SULFIDE	----------
C20H42S	HEPTADECYL-PROPYL-SULFIDE	----------
C20H42S	METHYL-NONADECYL-SULFIDE	----------
C20H42S	1-EICOSANETHIOL	----------
C20H42S2	DECYL-DISULFIDE	----------
C21H2104P	TRI-o-CRESYL PHOSPHATE	78-30-8
C21H36	1-PHENYLPENTADECANE	2131-18-2
C21H42	1-CYCLOPENTYLHEXADECANE	606-95-7
C21H42	1-CYCLOHEXYLPENTADECANE	6812-39-1
C22H38	1-PHENYLHEXADECANE	1459-09-2
C22H44	1-CYCLOHEXYLHEXADECANE	6892-38-0
C22H4402	n-BUTYL STEARATE	123-95-5
C24H3804	DIISOOCTYL PHTHALATE	27554-26-3
C24H3804	DIOCTYL PHTHALATE	117-81-7
C24H420	DINONYLPHENOL	1323-65-5
C26H20	TETRAPHENYLETHYLENE	632-51-9
C28H4604	DIISODECYL PHTHALATE	26761-40-0

INORGANICS

The following compilation for inorganics provides the compound list by chemical formula, name, and CAS registry number.

Formula	Name	CAS
Ag	SILVER	7440-22-4
AgCl	SILVER CHLORIDE	7783-90-6
AgI	SILVER IODIDE	7783-96-2
Al	ALUMINUM	7429-90-5
AlB3H12	ALUMINUM BOROHYDRIDE	16962-07-5
AlBr3	ALUMINUN BROMIDE	7727-15-3
AlCl3	ALUMINUM CHLORIDE	7446-70-0
AlF3	ALUMINUM FLUORIDE	7784-18-1
AlI3	ALUMINUM IODIDE	7784-23-8
Al2O3	ALUMINUM OXIDE	1344-28-1
Al2S3O12	ALUMINUM SULFATE	10043-01-3
Ar	ARGON	7440-37-1
As	ARSENIC	7440-37-1
AsBr3	ARSENIC TRIBROMIDE	7784-33-0
AsCl3	ARSENIC TRICHLORIDE	7784-34-1
AsF3	ARSENIC TRIFLUORIDE	7784-35-2
AsF5	ARSENIC PENTAFLUORIDE	7784-36-3
AsH3	ARSINE	7784-42-1
AsI3	ARSENIC TRIIODIDE	7784-45-4
As2O3	ARSENIC TRIOXIDE	1327-53-3
At	ASTATINE	7440-68-8
Au	GOLD	7440-57-5
B	BORON	7440-42-8
BBr3	BORON TRIBROMIDE	10294-33-4
BCl3	BORON TRICHLORIDE	10294-34-5
BF3	BORON TRIFLUORIDE	7637-07-2
BH2CO	BORINE CARBONYL	----------
BH3O3	BORIC ACID	10043-35-3
B2D6	DEUTERODIBORANE	----------
B2H5Br	DIBORANE HYDROBROMIDE	----------
B2H6	DIBORANE	19287-45-7
B3N3H6	BORINE TRIAMINE	----------
B4H10	TETRABORANE	18283-93-7
B5H9	PENTABORANE	18433-84-6
B5H11	TETRAHYDROPENTABORANE	17702-41-9
B10H14	DECABORANE	----------
Ba	BARIUM	7440-39-3
Be	BERYLLIUM	7440-41-7
BeB2H8	BERYLLIUM BOROHYDRIDE	17440-85-6
BeBr2	BERYLLIUM BROMIDE	7787-46-4
BeCl2	BERYLLIUM CHLORIDE	7787-47-5
BeF2	BERYLLIUM FLUORIDE	7787-49-7
BeI2	BERYLLIUM IODIDE	7787-53-3
Bi	BISMUTH	7440-69-9
BiBr3	BISMUTH TRIBROMIDE	7787-58-8
BiCl3	BISMUTH TRICHLORIDE	7787-60-2
BrF5	BROMINE PENTAFLUORIDE	7789-30-2
Br2	BROMINE	7726-95-6
C	CARBON	7440-44-0
CCl2O	PHOSGENE	75-44-5
CF2O	CARBONYL FLUORIDE	353-50-4
CH4N2O	UREA	57-13-6
CH4N2S	THIOUREA	62-56-6
CNBr	CYANOGEN BROMIDE	506-68-3
CNCl	CYANOGEN CHLORIDE	506-77-4
CNF	CYANOGEN FLUORIDE	1495-50-7
CO	CARBON MONOXIDE	630-08-0
COS	CARBONYL SULFIDE	463-58-1
COSe	CARBON OXYSELENIDE	1603-84-5
CO2	CARBON DIOXIDE	124-38-9
CS2	CARBON DISULFIDE	75-15-0
CSeS	CARBON SELENOSULFIDE	5951-19-9
C2N2	CYANOGEN	460-19-5
C3S2	CARBON SUBSULFIDE	----------
Ca	CALCIUM	7440-70-2
CaF2	CALCIUM FLUORIDE	7789-75-5
CbF5	COLUMBIUM FLUORIDE	----------
Cd	CADMIUM	7440-43-9
CdCl2	CADMIUM CHLORIDE	10108-64-2
CdF2	CADMIUM FLUORIDE	7790-79-6
CdI2	CADMIUM IODIDE	7790-80-9
CdO	CADMIUM OXIDE	1306-19-0
ClF	CHLORINE MONOFLUORIDE	7790-89-8
ClFO3	PERCHLORYL FLUORIDE	7616-94-6
ClF3	CHLORINE TRIFLUORIDE	7790-91-2
ClF5	CHLORINE PENTAFLUORIDE	13637-63-3
ClHO3S	CHLOROSULFONIC ACID	7790-94-5
ClHO4	PERCHLORIC ACID	7601-90-3
ClO2	CHLORINE DIOXIDE	10049-04-4
Cl2	CHLORINE	7782-50-5
Cl2O	CHLORINE MONOXIDE	7791-21-1
Cl2O7	CHLORINE HEPTOXIDE	10294-48-1
Co	COBALT	7440-48-4
CoCl2	COBALT CHLORIDE	7646-79-9
CoNC3O4	COBALT NITROSYL TRICARBONYL	14096-82-3
Cr	CHROMIUM	7440-47-3
CrC6O6	CHROMIUM CARBONYL	13007-92-6
CrO2Cl2	CHROMIUM OXYCHLORIDE	14977-61-8
Cs	CESIUM	7440-46-2
CsBr	CESIUM BROMIDE	7787-69-1
CsCl	CESIUM CHLORIDE	7647-17-8
CsF	CESIUM FLUORIDE	13400-13-0
CsI	CESIUM IODIDE	7789-17-5
Cu	COPPER	7440-50-8
CuBr	CUPROUS BROMIDE	7787-70-4
CuCl	CUPROUS CHLORIDE	7758-89-6
CuCl2	CUPRIC CHLORIDE	7447-39-4
CuI	COPPER IODIDE	7681-65-4
DCN	DEUTERIUM CYANIDE	----------
D2	DEUTERIUM	7782-39-0
D2O	DEUTERIUM OXIDE	7789-20-0
Eu	EUROPIUM	7440-53-1
F2	FLUORINE	7782-41-4
F2O	FLUORINE OXIDE	7783-41-7
Fe	IRON	7439-89-6
FeC5O5	IRON PENTACARBONYL	13463-40-6
FeCl2	FERROUS CHLORIDE	7758-94-3
FeCl3	FERRIC CHLORIDE	7705-08-0
Fr	FRANCIUM	7440-73-5
Ga	GALLIUM	7440-55-3
GaCl3	GALLIUM TRICHLORIDE	13450-90-3
Gd	GADOLINIUM	7440-54-2

Formula	Name	CAS	Formula	Name	CAS
Ge	GERMANIUM	7440-56-4	NaI	SODIUM IODIDE	7681-82-5
GeBr4	GERMANIUM BROMIDE	13450-92-5	NaOH	SODIUM HYDROXIDE	1310-73-2
GeCl4	GERMANIUM CHLORIDE	10038-98-9	Na2SO4	SODIUM SULFATE	7757-82-6
GeHCl3	TRICHLORO GERMANE	----------	Nb	NIOBIUM	7440-03-1
GeH4	GERMANIUM TETRAHYDRIDE	7782-65-2	Nd	NEODYMIUM	7440-00-8
Ge2H6	DIGERMANE	13818-89-8	Ne	NEON	7440-01-9
Ge3H8	TRIGERMANE	14691-44-2	Ni	NICKEL	7440-02-0
HBr	HYDROGEN BROMIDE	10035-10-6	NiC4O4	NICKEL CARBONYL	13436-39-3
HCN	HYDROGEN CYANIDE	74-90-8	NiF2	NICKEL FLUORIDE	10028-18-9
HCl	HYDROGEN CHLORIDE	7647-01-0	Np	NEPTUNIUM	7439-99-8
HF	HYDROGEN FLUORIDE	7664-39-3	O2	OXYGEN	7782-44-7
HI	HYDROGEN IODIDE	10034-85-2	O3	OZONE	10028-15-6
HNO3	NITRIC ACID	7697-37-2	Os	OSMIUM	7440-04-2
H2	HYDROGEN	1333-74-0	OsOF5	OSMIUM OXIDE PENTAFLUORIDE	----------
H2O	WATER	7732-18-5	OsO4	OSMIUM TETROXIDE - YELLOW	20816-12-0
H2O2	HYDROGEN PEROXIDE	7722-84-1	OsO4	OSMIUM TETROXIDE - WHITE	----------
H2S	HYDROGEN SULFIDE	7783-06-4	P	PHOSPHORUS - WHITE	7723-14-0
H2SO4	SULFURIC ACID	7664-93-9	PBr3	PHOSPHORUS TRIBROMIDE	7789-60-8
H2S2	HYDROGEN DISULFIDE	----------	PCl2F3	PHOSPHORUS DICHLORIDE TRIFLUORIDE	----------
H2Se	HYDROGEN SELENIDE	7783-07-5	PCl3	PHOSPHORUS TRICHLORIDE	7719-12-2
H2Te	HYDROGEN TELLURIDE	7783-09-7	PCl5	PHOSPHORUS PENTACHLORIDE	10026-13-8
H3NO3S	SULFAMIC ACID	5329-14-6	PH3	PHOSPHINE	7803-51-2
He	HELIUM-3	14762-55-1	PH4Br	PHOSPHONIUM BROMIDE	----------
He	HELIUM-4	7440-59-7	PH4Cl	PHOSPHONIUM CHLORIDE	24567-53-1
Hf	HAFNIUM	7440-58-6	PH4I	PHOSPHONIUM IODIDE	12125-09-6
Hg	MERCURY	7439-97-6	POCl3	PHOSPHORUS OXYCHLORIDE	10025-87-3
HgBr2	MERCURIC BROMIDE	7789-47-1	PSBr3	PHOSPHORUS THIOBROMIDE	7789-60-8
HgCl2	MERCURIC CHLORIDE	7487-94-7	PSCl3	PHOSPHORUS THIOCHLORIDE	3982-91-0
HgI2	MERCURIC IODIDE	7774-29-0	P4O6	PHOSPHORUS TRIOXIDE	1314-24-5
IF7	IODINE HEPTAFLUORIDE	16921-96-3	P4O10	PHOSPHORUS PENTOXIDE	16752-60-6
I2	IODINE	7553-56-2	P4S10	PHOSPHORUS PENTASULFIDE	1314-80-3
In	INDIUM	7440-74-6	Pb	LEAD	7439-92-1
Ir	IRIDIUM	7439-88-5	PbBr2	LEAD BROMIDE	10031-22-8
K	POTASSIUM	7440-09-7	PbCl2	LEAD CHLORIDE	7758-95-4
KBr	POTASSIUM BROMIDE	7758-02-3	PbF2	LEAD FLUORIDE	7783-46-2
KCl	POTASSIUM CHLORIDE	7447-40-7	PbI2	LEAD IODIDE	10101-63-0
KF	POTASSIUM FLUORIDE	7789-23-3	PbO	LEAD OXIDE	1317-36-8
KI	POTASSIUM IODIDE	7681-11-0	PbS	LEAD SULFIDE	1314-87-0
KOH	POTASSIUM HYDROXIDE	1310-58-3	Pd	PALLADIUM	7440-05-3
Kr	KRYPTON	7439-90-9	Po	POLONIUM	7440-08-6
La	LANTHANUM	7439-91-0	Pt	PLATINUM	7440-06-4
Li	LITHIUM	7439-93-2	Ra	RADIUM	7440-14-4
LiBr	LITHIUM BROMIDE	7550-35-8	Rb	RUBIDIUM	7440-17-7
LiCl	LITHIUM CHLORIDE	7447-41-8	RbBr	RUBIDIUM BROMIDE	7789-39-1
LiF	LITHIUM FLUORIDE	7789-24-4	RbCl	RUBIDIUM CHLORIDE	7791-11-9
LiI	LITHIUM IODIDE	10377-51-2	RbF	RUBIDIUM FLUORIDE	13446-74-7
Lu	LUTECIUM	----------	RbI	RUBIDIUM IODIDE	7790-29-6
Mg	MAGNESIUM	7439-95-4	Re	RHENIUM	7440-15-5
MgCl2	MAGNESIUM CHLORIDE	7786-30-3	Re2O7	RHENIUM HEPTOXIDE	1314-68-7
MgO	MAGNESIUM OXIDE	1309-48-4	Rh	RHODIUM	7440-16-6
Mn	MANGANESE	7439-96-5	Rn	RADON	10043-92-2
MnCl2	MANGANESE CHLORIDE	7773-01-5	Ru	RUTHENIUM	7440-18-8
Mo	MOLYBDENUM	7439-98-7	RuF5	RUTHENIUM PENTAFLUORIDE	----------
MoF6	MOLYBDENUM FLUORIDE	7783-77-9	S	SULFUR	7704-34-9
MoO3	MOLYBDENUM OXIDE	1313-27-5	SF4	SULFUR TETRAFLUORIDE	7783-60-0
NCl3	NITROGEN TRICHLORIDE	10025-85-1	SF6	SULFUR HEXAFLUORIDE	2551-62-4
ND3	HEAVY AMMONIA	----------	SOBr2	THIONYL BROMIDE	507-16-4
NF3	NITROGEN TRIFLUORIDE	7783-54-2	SOCl2	THIONYL CHLORIDE	7719-09-7
NH3	AMMONIA	7664-41-7	SOF2	SULFUROUS OXYFLUORIDE	----------
NH3O	HYDROXYLAMINE	7803-49-8	SO2	SULFUR DIOXIDE	7446-09-5
NH4Br	AMMONIUM BROMIDE	12124-97-9	SO2Cl2	SULFURYL CHLORIDE	7791-25-5
NH4Cl	AMMONIUM CHLORIDE	12125-02-9	SO3	SULFUR TRIOXIDE	7446-11-9
NH4I	AMMONIUM IODIDE	12027-06-4	S2Cl2	SULFUR MONOCHLORIDE	10025-67-9
NH5O	AMMONIUM HYDROXIDE	1336-21-6	Sb	ANTIMONY	7440-36-0
NH5S	AMMONIUM HYDROGENSULFIDE	12124-99-1	SbBr3	ANTIMONY TRIBROMIDE	7789-61-9
NO	NITRIC OXIDE	10102-43-9	SbCl3	ANTIMONY TRICHLORIDE	10025-91-9
NOCl	NITROSYL CHLORIDE	2696-92-6	SbCl5	ANTIMONY PENTACHLORIDE	7647-18-9
NOF	NITROSYL FLUORIDE	7789-25-5	SbH3	STIBINE	7803-52-3
NO2	NITROGEN DIOXIDE	10102-44-0	SbI3	ANTIMONY TRIIODIDE	7790-44-5
N2	NITROGEN	7727-37-9	Sb2O3	ANTIMONY TRIOXIDE	1317-98-2
N2F4	TETRAFLUOROHYDRAZINE	10036-47-2	Sc	SCANDIUM	7440-20-2
N2H4	HYDRAZINE	302-01-2	Se	SELENIUM	7782-49-2
N2H4C	AMMONIUM CYANIDE	12211-52-8	SeCl4	SELENIUM TETRACHLORIDE	10026-03-6
N2H6CO2	AMMONIUM CARBAMATE	1111-78-0	SeF6	SELENIUM HEXAFLUORIDE	7783-79-1
N2O	NITROUS OXIDE	10024-97-2	SeOCl2	SELENIUM OXYCHLORIDE	7791-23-3
N2O3	NITROGEN TRIOXIDE	10544-73-7	SeO2	SELENIUM DIOXIDE	7446-08-4
N2O4	NITROGEN TETRAOXIDE	10544-72-6	Si	SILICON	7440-21-3
N2O5	NITROGEN PENTOXIDE	10102-03-1	SiBrCl2F	BROMODICHLOROFLUOROSILANE	----------
Na	SODIUM	7440-23-5	SiBrF3	TRIFLUOROBROMOSILANE	----------
NaBr	SODIUM BROMIDE	7647-15-6	SiBr2ClF	DIBROMOCHLOROFLUOROSILANE	----------
NaCN	SODIUM CYANIDE	143-33-9	SiClF3	TRIFLUOROCHLOROSILANE	----------
NaCl	SODIUM CHLORIDE	7647-14-5	SiCl2F2	DICHLORODIFLUOROSILANE	18356-71-3
NaF	SODIUM FLUORIDE	7681-49-4	SiCl3F	TRICHLOROFLUOROSILANE	14965-52-7

SiCl4	SILICON TETRACHLORIDE	10026-04-7
SiF4	SILICON TETRAFLUORIDE	7783-61-1
SiHBr3	TRIBROMOSILANE	7789-57-3
SiHCl3	TRICHLOROSILANE	10025-78-2
SiHF3	TRIFLUOROSILANE	13465-71-9
SiH2Br2	DIBROMOSILANE	13768-94-0
SiH2Cl2	DICHLOROSILANE	4109-96-0
SiH2F2	DIFLUOROSILANE	13824-36-7
SiH2I2	DIIODOSILANE	13760-02-6
SiH3Br	MONOBROMOSILANE	13465-73-1
SiH3Cl	MONOCHLOROSILANE	13465-78-6
SiH3F	MONOFLUOROSILANE	13537-33-2
SiH3I	IODOSILANE	13598-42-0
SiH4	SILANE	7803-62-5
SiO2	SILICON DIOXIDE	14808-60-7
Si2Cl6	HEXACHLORODISILANE	13465-77-5
Si2F6	HEXAFLUORODISILANE	----------
Si2H5Cl	DISILANYL CHLORIDE	----------
Si2H6	DISILANE	1590-87-0
Si2OCl3F3	TRICHLOROTRIFLUORODISILOXANE	----------
Si2OCl6	HEXACHLORODISILOXANE	----------
Si2OH6	DISILOXANE	13597-73-4
Si3Cl8	OCTACHLOROTRISILANE	----------
Si3H8	TRISILANE	7783-26-8
Si3H9N	TRISILAZANE	----------
Si4H10	TETRASILANE	7783-29-1
Sm	SAMARIUM	7440-19-9
Sn	TIN	7440-31-5
SnBr4	STANNIC BROMIDE	7789-67-5
SnCl2	STANNOUS CHLORIDE	7772-99-8
SnCl4	STANNIC CHLORIDE	7646-78-8
SnH4	STANNIC HYDRIDE	----------
SnI4	STANNIC IODIDE	7790-47-8
Sr	STRONTIUM	7440-24-6
SrO	STRONTIUM OXIDE	1314-11-0
Ta	TANTALUM	7440-25-7
Tc	TECNNETIUM	7440-26-8
Te	TELLURIUM	13494-80-9
TeCl4	TELLURIUM TETRACHLORIDE	10026-07-0
TeF6	TELLURIUM HEXAFLUORIDE	7783-80-4
Ti	TITANIUM	7440-32-6
TiCl4	TITANIUM TETRACHLORIDE	7550-45-0
Tl	THALLIUM	7440-28-0
TlBr	THALLOUS BROMIDE	7789-40-4
TlI	THALLOUS IODIDE	7790-30-9
Tm	THULIUM	7440-30-4
U	URANIUM	7440-61-1
UF6	URANIUM FLUORIDE	7783-81-5
V	VANADIUM	7440-62-2
VCl4	VANADIUM TETRACHLORIDE	7632-51-1
VOCl3	VANADIUM OXYTRICHLORIDE	7727-18-6
W	TUNGSTEN	7440-33-7
WF6	TUNGSTEN FLUORIDE	7783-82-6
Xe	XENON	7440-63-3
Yb	YTTERBIUM	7440-64-4
Yt	YTTRIUM	7440-65-5
Zn	ZINC	7440-66-6
ZnCl2	ZINC CHLORIDE	7646-85-7
ZnF2	ZINC FLUORIDE	7783-49-5
ZnO	ZINC OXIDE	1314-13-2
ZnSO4	ZINC SULFATE	7733-02-0
Zr	ZIRCONIUM	7440-67-7
ZrBr4	ZIRCONIUM BROMIDE	13777-25-8
ZrCl4	ZIRCONIUM CHLORIDE	10026-11-6
ZrI4	ZIRCONIUM IODIDE	13986-26-0

Appendix D

COMPOUND LIST BY CAS REGISTRY NUMBER

ORGANICS The following compilation for organics provides the compound list by CAS registry number, name, and chemical formula.

CAS	Name	Formula
50-00-0	FORMALDEHYDE	CH2O
50-70-4	SORBITOL	C6H14O6
50-81-7	ASCORBIC ACID	C6H8O6
55-63-0	NITROGLYCERINE	C3H5N3O9
56-23-5	CARBON TETRACHLORIDE	CCl4
56-81-5	GLYCEROL	C3H8O3
56-86-0	L-GLUTAMIC ACID	C5H9NO4
56-87-1	LYSINE	C6H14N2O2
57-10-3	n-HEXADECANOIC ACID	C16H32O2
57-11-4	STEARIC ACID	C18H36O2
57-55-6	1,2-PROPYLENE GLYCOL	C3H8O2
57-57-8	beta-PROPIOLACTONE	C3H4O2
58-72-0	TRIPHENYLETHYLENE	C20H16
60-09-3	p-AMINOAZOBENZENE	C12H11N3
60-12-8	2-PHENYLETHANOL	C8H10O
60-29-7	DIETHYL ETHER	C4H10O
60-33-3	LINOLEIC ACID	C18H32O2
60-35-5	ACETAMIDE	C2H5NO
62-53-3	ANILINE	C6H7N
64-17-5	ETHANOL	C2H6O
64-18-6	FORMIC ACID	CH2O2
64-19-7	ACETIC ACID	C2H4O2
64-67-5	DIETHYL SULFATE	C4H10O4S
65-85-0	BENZOIC ACID	C7H6O2
66-25-1	1-HEXANAL	C6H12O
67-56-1	METHANOL	CH4O
67-63-0	ISOPROPANOL	C3H8O
67-64-1	ACETONE	C3H6O
67-66-3	CHLOROFORM	CHCl3
67-68-5	DIMETHYL SULFOXIDE	C2H6OS
67-72-1	HEXACHLOROETHANE	C2Cl6
68-12-2	N,N-DIMETHYLFORMAMIDE	C3H7NO
69-72-7	SALICYLIC ACID	C7H6O3
71-23-8	n-PROPANOL	C3H8O
71-36-3	n-BUTANOL	C4H10O
71-41-0	1-PENTANOL	C5H12O
71-43-2	BENZENE	C6H6
71-55-6	1,1,1-TRICHLOROETHANE	C2H3Cl3
74-31-7	N,N'-DIPHENYL-p-PHENYLENEDIAMINE	C18H16N2
74-82-8	METHANE	CH4
74-83-9	METHYL BROMIDE	CH3Br
74-84-0	ETHANE	C2H6
74-85-1	ETHYLENE	C2H4
74-86-2	ACETYLENE	C2H2
74-87-3	METHYL CHLORIDE	CH3Cl
74-88-4	METHYL IODIDE	CH3I
74-89-5	METHYLAMINE	CH5N
74-90-8	HYDROGEN CYANIDE	CHN
74-93-1	METHYL MERCAPTAN	CH4S
74-95-3	DIBROMOMETHANE	CH2Br2
74-96-4	BROMOETHANE	C2H5Br
74-97-5	BROMOCHLOROMETHANE	CH2BrCl
74-98-6	PROPANE	C3H8
74-99-7	METHYLACETYLENE	C3H4
75-00-3	ETHYL CHLORIDE	C2H5Cl
75-01-4	VINYL CHLORIDE	C2H3Cl
75-02-5	VINYL FLUORIDE	C2H3F
75-03-6	ETHYL IODIDE	C2H5I
75-04-7	ETHYLAMINE	C2H7N
75-05-8	ACETONITRILE	C2H3N
75-07-0	ACETALDEHYDE	C2H4O
75-08-1	ETHYL MERCAPTAN	C2H6S
75-09-2	DICHLOROMETHANE	CH2Cl2
75-10-5	DIFLUOROMETHANE	CH2F2
75-11-6	DIIODOMETHANE	CH2I2
75-12-7	FORMAMIDE	CH3NO
75-15-0	CARBON DISULFIDE	CS2
75-18-3	DIMETHYL SULFIDE	C2H6S
75-19-4	CYCLOPROPANE	C3H6
75-21-8	ETHYLENE OXIDE	C2H4O
75-25-2	TRIBROMOMETHANE	CHBr3
75-26-3	2-BROMOPROPANE	C3H7Br
75-28-5	ISOBUTANE	C4H10
75-29-6	ISOPROPYL CHLORIDE	C3H7Cl
75-30-9	ISOPROPYL IODIDE	C3H7I
75-31-0	ISOPROPYLAMINE	C3H9N
75-33-2	ISOPROPYL MERCAPTAN	C3H8S
75-34-3	1,1-DICHLOROETHANE	C2H4Cl2
75-35-4	1,1-DICHLOROETHYLENE	C2H2Cl2
75-36-5	ACETYL CHLORIDE	C2H3ClO
75-37-6	1,1-DIFLUOROETHANE	C2H4F2
75-38-7	1,1-DIFLUOROETHYLENE	C2H2F2
75-43-4	DICHLOROFLUOROMETHANE	CHCl2F
75-44-5	PHOSGENE	CCl2O
75-45-6	CHLORODIFLUOROMETHANE	CHClF2
75-46-7	TRIFLUOROMETHANE	CHF3
75-47-8	TRIIODOMETHANE	CHI3
75-50-3	TRIMETHYLAMINE	C3H9N
75-52-5	NITROMETHANE	CH3NO2
75-54-7	METHYL DICHLOROSILANE	CH4Cl2Si
75-55-8	PROPYLENEIMINE	C3H7N
75-56-9	1,2-PROPYLENE OXIDE	C3H6O
75-61-6	DIBROMODIFLUOROMETHANE	CBr2F2
75-62-7	BROMOTRICHLOROMETHANE	CBrCl3
75-63-8	BROMOTRIFLUOROMETHANE	CBrF3
75-64-9	tert-BUTYLAMINE	C4H11N
75-65-0	tert-BUTANOL	C4H10O
75-66-1	tert-BUTYL MERCAPTAN	C4H10S
75-68-3	1-CHLORO-1,1-DIFLUOROETHANE	C2H3ClF2
75-69-4	TRICHLOROFLUOROMETHANE	CCl3F
75-71-8	DICHLORODIFLUOROMETHANE	CCl2F2
75-72-9	CHLOROTRIFLUOROMETHANE	CClF3
75-73-0	CARBON TETRAFLUORIDE	CF4
75-75-2	METHANESULFONIC ACID	CH4O3S
75-76-3	TETRAMETHYLSILANE	C4H12Si
75-79-6	METHYL TRICHLOROSILANE	CH3Cl3Si
75-83-2	2,2-DIMETHYLBUTANE	C6H14
75-84-3	2,2-DIMETHYL-1-PROPANOL	C5H12O
75-85-4	2-METHYL-2-BUTANOL	C5H12O
75-85-4	tert-PENTYL-ALCOHOL	C5H12O
75-86-5	ACETONE CYANOHYDRIN	C4H7NO
75-87-6	TRICHLOROACETALDEHYDE	C2HCl3O
75-91-2	t-BUTYL HYDROPEROXIDE	C4H10O2
76-01-7	PENTACHLOROETHANE	C2HCl5
76-02-8	TRICHLOROACETYL CHLORIDE	C2Cl4O
76-05-1	TRIFLUOROACETIC ACID	C2HF3O2
76-12-0	1,1,2,2-TETRACHLORODIFLUOROETHANE	C2Cl4F2
76-13-1	1,1,2-TRICHLOROTRIFLUOROETHANE	C2Cl3F3
76-14-2	1,2-DICHLOROTETRAFLUOROETHANE	C2Cl2F4
76-15-3	CHLOROPENTAFLUOROETHANE	C2ClF5
76-16-4	HEXAFLUOROETHANE	C2F6
76-19-7	OCTAFLUOROPROPANE	C3F8
76-22-2	CAMPHOR	C10H16O
77-47-4	HEXACHLOROCYCLOPENTADIENE	C5Cl6
77-73-6	DICYCLOPENTADIENE	C10H12
77-78-1	DIMETHYL SULFATE	C2H6O4S
77-92-9	CITRIC ACID	C6H8O7
77-99-9	TRIMETHYLOLPROPANE	C6H14O3
78-11-5	PENTAERYTHRITOL TETRANITRATE	C5H8N4O12
78-30-8	TRI-o-CRESYL PHOSPHATE	C21H21O4P
78-40-0	TRIETHYL PHOSPHATE	C6H15O4P
78-59-1	ISOPHORONE	C9H14O
78-75-1	1,2-DIBROMOPROPANE	C3H6Br2
78-76-2	2-BROMOBUTANE	C4H9Br
78-78-4	ISOPENTANE	C5H12
78-79-5	2-METHYL-1,3-BUTADIENE	C5H8
78-79-5	ISOPRENE	C5H8
78-80-8	2-METHYL-1-BUTENE-3-YNE	C5H6
78-81-9	ISOBUTYLAMINE	C4H11N
78-82-0	ISOBUTYRONITRILE	C4H7N
78-83-1	ISOBUTANOL	C4H10O
78-84-2	ISOBUTYRALDEHYDE	C4H8O
78-85-3	METHACROLEIN	C4H6O
78-86-4	sec-BUTYL CHLORIDE	C4H9Cl
78-87-5	1,2-DICHLOROPROPANE	C3H6Cl2
78-88-6	2,3-DICHLOROPROPENE	C3H4Cl2
78-90-0	1,2-PROPANEDIAMINE	C3H10N2
78-92-2	sec-BUTANOL	C4H10O
78-93-3	METHYL ETHYL KETONE	C4H8O
78-96-6	1-AMINO-2-PROPANOL	C3H9NO

78-97-7	LACTONITRILE	C3H5NO
78-99-9	1,1-DICHLOROPROPANE	C3H6Cl2
79-00-5	1,1,2-TRICHLOROETHANE	C2H3Cl3
79-01-6	TRICHLOROETHYLENE	C2HCl3
79-02-7	DICHLOROACETALDEHYDE	C2H2Cl2O
79-04-9	CHLOROACETYL CHLORIDE	C2H2Cl2O
79-06-1	ACRYLAMIDE	C3H5NO
79-09-4	PROPIONIC ACID	C3H6O2
79-10-7	ACRYLIC ACID	C3H4O2
79-11-8	CHLOROACETIC ACID	C2H3ClO2
79-16-3	N-METHYLACETAMIDE	C3H7NO
79-20-9	METHYL ACETATE	C3H6O2
79-22-1	METHYL CHLOROFORMATE	C2H3ClO2
79-24-3	NITROETHANE	C2H5NO2
79-27-6	1,1,2,2-TETRABROMOETHANE	C2H2Br4
79-29-8	2,3-DIMETHYLBUTANE	C6H14
79-31-2	ISOBUTYRIC ACID	C4H8O2
79-34-5	1,1,2,2-TETRACHLOROETHANE	C2H2Cl4
79-36-7	DICHLOROACETYL CHLORIDE	C2HCl3O
79-38-9	CHLOROTRIFLUOROETHYLENE	C2ClF3
79-39-0	2-METHACRYLAMIDE	C4H7NO
79-41-4	METHACRYLIC ACID	C4H6O2
79-43-6	DICHLOROACETIC ACID	C2H2Cl2O2
79-46-9	2-NITROPROPANE	C3H7NO2
79-92-5	CAMPHENE	C10H16
80-05-7	BISPHENOL A	C15H16O2
80-15-9	CUMENE HYDROPEROXIDE	C9H12O2
80-43-3	DICUMYL PEROXIDE	C18H22O2
80-46-6	p-tert-AMYLPHENOL	C11H16O
80-56-8	alpha-PINENE	C10H16
80-62-6	METHYL METHACRYLATE	C5H8O2
83-32-9	ACENAPHTHENE	C12H10
84-15-1	o-TERPHENYL	C18H14
84-65-1	ANTHRAQUINONE	C14H8O2
84-66-2	DIETHYL PHTHALATE	C12H14O4
84-74-2	DIBUTYL PHTHALATE	C16H22O4
85-01-8	PHENANTHRENE	C14H10
85-44-9	PHTHALIC ANHYDRIDE	C8H4O3
86-73-7	FLUORENE	C13H10
86-74-8	DIBENZOPYRROLE	C12H9N
86-89-5	1-PENTYLNAPHTHALENE	C15H18
87-66-1	1,2,3-BENZENETRIOL	C6H6O3
87-68-3	HEXACHLORO-1,3-BUTADIENE	C4Cl6
87-85-4	HEXAMETHYLBENZENE	C12H18
88-09-5	2-ETHYL BUTYRIC ACID	C6H12O2
88-72-2	o-NITROTOLUENE	C7H7NO2
88-73-3	o-CHLORONITROBENZENE	C6H4ClNO2
88-74-4	o-NITROANILINE	C6H6N2O2
88-99-3	PHTHALIC ACID	C8H6O4
89-05-4	PYROMELLITIC ACID	C10H6O8
90-02-8	SALICYLALDEHYDE	C7H6O2
90-05-1	GUAIACOL	C7H8O2
90-11-9	1-BROMONAPHTHALENE	C10H7Br
90-12-0	1-METHYLNAPHTHALENE	C11H10
90-13-1	1-CHLORONAPHTHALENE	C10H7Cl
91-20-3	NAPHTHALENE	C10H8
91-22-5	QUINOLINE	C9H7N
91-23-6	o-NITROANISOLE	C7H7NO3
91-57-6	2-METHYLNAPHTHALENE	C11H10
91-63-4	QUINALDINE	C10H9N
91-66-7	N,N-DIETHYLANILINE	C10H15N
92-06-8	m-TERPHENYL	C18H14
92-51-3	BICYCLOHEXYL	C12H22
92-54-4	BIPHENYL	C12H10
92-67-1	p-AMINODIPHENYL	C12H11N
92-94-4	p-TERPHENYL	C18H14
93-22-1	2-PENTYLNAPHTHALENE	C15H18
93-58-3	METHYL BENZOATE	C8H8O2
93-89-0	ETHYL BENZOATE	C9H10O2
95-13-6	INDENE	C9H8
95-15-8	BENZOTHIOPHENE	C8H6S
95-47-6	o-XYLENE	C8H10
95-48-7	o-CRESOL	C7H8O
95-49-8	o-CHLOROTOLUENE	C7H7Cl
95-50-1	o-DICHLOROBENZENE	C6H4Cl2
95-51-2	o-CHLOROANILINE	C6H6ClN
95-53-4	o-TOLUIDINE	C7H9N
95-54-5	o-PHENYLENEDIAMINE	C6H8N2
95-57-8	o-CHLOROPHENOL	C6H5ClO
95-63-6	1,2,4-TRIMETHYLBENZENE	C9H12
95-65-8	3,4-XYLENOL	C8H10O
95-73-8	2,4-DICHLOROTOLUENE	C7H6Cl2
95-76-1	3,4-DICHLOROANILINE	C6H5Cl2N

95-80-7	TOLUENEDIAMINE	C7H10N2
95-87-4	2,5-XYLENOL	C8H10O
95-92-1	DIETHYL OXALATE	C6H10O4
95-93-2	1,2,4,5-TETRAMETHYLBENZENE	C10H14
96-14-0	3-METHYLPENTANE	C6H14
96-18-4	1,2,3-TRICHLOROPROPANE	C3H5Cl3
96-22-0	DIETHYL KETONE	C5H10O
96-33-3	METHYL ACRYLATE	C4H6O2
96-34-4	METHYL CHLOROACETATE	C3H5ClO2
96-37-7	METHYLCYCLOPENTANE	C6H12
96-48-0	gamma-BUTYROLACTONE	C4H6O2
96-49-1	ETHYLENE CARBONATE	C3H4O3
96-54-8	N-METHYLPYRROLE	C5H7N
97-00-7	1-CHLORO-2,4-DINITROBENZENE	C6H3ClN2O4
97-62-1	ETHYL ISOBUTYRATE	C6H12O2
97-63-2	ETHYL METHACRYLATE	C6H10O2
97-64-3	ETHYL LACTATE	C5H10O3
97-65-4	ITACONIC ACID	C5H6O4
97-85-8	ISOBUTYL ISOBUTYRATE	C8H16O2
97-88-1	n-BUTYL METHACRYLATE	C8H14O2
97-93-8	TRIETHYL ALUMINUN	C6H15Al
97-95-0	2-ETHYL-1-BUTANOL	C6H14O
97-99-4	TETRAHYDROFURFURYL ALCOHOL	C5H10O2
98-00-0	FURFURYL ALCOHOL	C5H6O2
98-01-1	FURFURAL	C5H4O2
98-06-6	tert-BUTYLBENZENE	C10H14
98-07-7	BENZOTRICHLORIDE	C7H5Cl3
98-08-8	BENZOTRIFLUORIDE	C7H5F3
98-29-3	p-tert-BUTYLCATECHOL	C10H14O2
98-46-4	3-NITROBENZOTRIFLUORIDE	C7H4F3NO2
98-54-4	p-tert-BUTYLPHENOL	C10H14O
98-56-6	p-CHLOROBENZOTRIFLUORIDE	C7H4ClF3
98-82-8	CUMENE	C9H12
98-83-9	alpha-METHYLSTYRENE	C9H10
98-86-2	ACETOPHENONE	C8H8O
98-87-3	BENZYL DICHLORIDE	C7H6Cl2
98-88-4	BENZOYL CHLORIDE	C7H5ClO
98-95-3	NITROBENZENE	C6H5NO2
99-08-1	m-NITROTOLUENE	C7H7NO2
99-09-2	m-NITROANILINE	C6H6N2O2
99-35-4	1,3,5-TRINITROBENZENE	C6H3N3O6
99-54-7	1,2-DICHLORO-4-NITROBENZENE	C6H3Cl2NO2
99-62-9	m-DIISOPROPYLBENZENE	C12H18
99-63-8	ISOPHTHALOYL CHLORIDE	C8H4Cl2O2
99-65-0	m-DINITROBENZENE	C6H4N2O4
99-83-2	alpha-PHELLANDRENE	C10H16
99-85-4	gamma-TERPINENE	C10H16
99-86-5	alpha-TERPINENE	C10H16
99-87-6	p-CYMENE	C10H14
99-94-5	p-TOLUIC ACID	C8H8O2
99-99-0	p-NITROTOLUENE	C7H7NO2
100-00-5	p-CHLORONITROBENZENE	C6H4ClNO2
100-01-6	p-NITROANILINE	C6H6N2O2
100-10-7	p-DIMETHYLAMINOBENZALDEHYDE	C9H11NO
100-18-5	p-DIISOPROPYLBENZENE	C12H18
100-21-0	TEREPHTHALIC ACID	C8H6O4
100-25-4	p-DINITROBENZENE	C6H4N2O4
100-40-3	VINYLCYCLOHEXENE	C8H12
100-41-4	ETHYLBENZENE	C8H10
100-42-5	STYRENE	C8H8
100-44-7	BENZYL CHLORIDE	C7H7Cl
100-46-9	BENZYLAMINE	C7H9N
100-47-0	BENZONITRILE	C7H5N
100-51-6	BENZYL ALCOHOL	C7H8O
100-52-7	BENZALDEHYDE	C7H6O
100-60-7	N-METHYLCYCLOHEXYLAMINE	C7H15N
100-61-8	N-METHYLANILINE	C7H9N
100-63-0	PHENYLHYDRAZINE	C6H8N2
100-64-1	CYCLOHEXANONE OXIME	C6H11NO
100-66-3	ANISOLE	C7H8O
100-80-1	m-METHYLSTYRENE	C9H10
101-14-2	p-AMINODIPHENYLAMINE	C12H12N2
101-68-8	DIPHENYLMETHANE-4,4'-DIISOCYANATE	C15H10N2O2
101-81-5	DIPHENYLMETHANE	C13H12
101-83-7	DICYCLOHEXYLAMINE	C12H23N
101-84-8	DIPHENYL ETHER	C12H10O
102-25-0	1,3,5-TRIETHYLBENZENE	C12H18
102-36-3	3,4-DICHLOROPHENYL ISOCYANATE	C7H3Cl2NO
102-69-2	TRIPROPYLAMINE	C9H21N
102-71-6	TRIETHANOLAMINE	C6H15NO3
102-76-1	GLYCERYL TRIACETATE	C9H14O6
102-82-9	TRI-n-BUTYLAMINE	C12H27N
103-03-2	1-NITROPROPANE	C3H7NO2

103-09-3	2-ETHYLHEXYL ACETATE	C10H20O2	108-08-7	2,4-DIMETHYLPENTANE	C7H16
103-11-7	2-ETHYLHEXYL ACRYLATE	C11H20O2	108-10-1	METHYL ISOBUTYL KETONE	C6H12O
103-29-7	1,2-DIPHENYLETHANE	C14H14	108-11-2	4-METHYL-2-PENTANOL	C6H14O
103-30-0	trans-STILBENE	C14H12	108-18-9	DIISOPROPYLAMINE	C6H15N
103-50-4	DIBENZYL ETHER	C14H14O	108-20-3	DIISOPROPYL ETHER	C6H14O
103-65-1	n-PROPYLBENZENE	C9H12	108-21-4	ISOPROPYL ACETATE	C5H10O2
103-70-8	FORMANILIDE	C7H7NO	108-24-7	ACETIC ANHYDRIDE	C4H6O3
103-71-9	PHENYL ISOCYANATE	C7H5NO	108-30-5	SUCCINIC ANHYDRIDE	C4H4O3
103-73-1	PHENETOLE	C8H10O	108-31-6	MALEIC ANHYDRIDE	C4H2O3
103-84-4	ACETANILIDE	C8H9NO	108-36-1	m-DIBROMOBENZENE	C6H4Br2
104-46-1	ANETHOLE	C10H12O	108-38-3	m-XYLENE	C8H10
104-51-8	n-BUTYLBENZENE	C10H14	108-39-4	m-CRESOL	C7H8O
104-72-3	n-DECYLBENZENE	C16H26	108-42-9	m-CHLOROANILINE	C6H6ClN
104-76-7	2-ETHYL-1-HEXANOL	C8H18O	108-43-0	m-CHLOROPHENOL	C6H5ClO
104-87-0	p-TOLUALDEHYDE	C8H8O	108-44-1	m-TOLUIDINE	C7H9N
105-05-5	p-DIETHYLBENZENE	C10H14	108-45-2	m-PHENYLENEDIAMINE	C6H8N2
105-30-6	2-METHYL-1-PENTANOL	C6H14O	108-46-3	1,3-BENZENEDIOL	C6H6O2
105-34-0	METHYL CYANOACETATE	C4H5NO2	108-48-5	2,6-DIMETHYLPYRIDINE	C7H9N
105-37-3	ETHYL PROPIONATE	C5H10O2	108-55-4	GLUTARIC ANHYDRIDE	C5H6O3
105-38-4	VINYL PROPIONATE	C5H8O2	108-57-6	m-DIVINYLBENZENE	C10H10
105-45-3	METHYL ACETOACETATE	C5H8O3	108-64-5	ETHYL ISOVALERATE	C7H14O2
105-46-4	sec-BUTYL ACETATE	C6H12O2	108-67-8	MESITYLENE	C9H12
105-53-3	DIETHYL MALONATE	C7H12O4	108-68-9	3,5-XYLENOL	C8H10O
105-54-4	ETHYL n-BUTYRATE	C6H12O2	108-75-8	2,4,6-TRIMETHYLPYRIDINE	C8H11N
105-56-6	ETHYL CYANOACETATE	C5H7NO2	108-82-7	2,6-DIMETHYL-4-HEPTANOL	C9H20O
105-57-7	ACETAL	C6H14O2	108-83-8	DIISOBUTYL KETONE	C9H18O
105-58-8	DIETHYL CARBONATE	C5H10O3	108-86-1	BROMOBENZENE	C6H5Br
105-59-9	METHYL DIETHANOLAMINE	C5H13NO2	108-87-2	METHYLCYCLOHEXANE	C7H14
105-60-2	epsilon-CAPROLACTAM	C6H11NO	108-88-3	TOLUENE	C7H8
105-66-8	n-PROPYL n-BUTYRATE	C7H14O2	108-89-4	4-METHYLPYRIDINE	C6H7N
105-67-9	2,4-XYLENOL	C8H10O	108-90-7	MONOCHLOROBENZENE	C6H5Cl
105-76-0	DIBUTYL MALEATE	C12H20O4	108-91-8	CYCLOHEXYLAMINE	C6H13N
106-31-0	BUTYRIC ANHYDRIDE	C8H14O3	108-93-0	CYCLOHEXANOL	C6H12O
106-36-5	n-PROPYL PROPIONATE	C6H12O2	108-94-1	CYCLOHEXANONE	C6H10O
106-38-7	p-BROMOTOLUENE	C7H7Br	108-95-2	PHENOL	C6H6O
106-42-3	p-XYLENE	C8H10	108-98-5	PHENYL MERCAPTAN	C6H6S
106-43-4	p-CHLOROTOLUENE	C7H7Cl	108-99-6	3-METHYLPYRIDINE	C6H7N
106-44-5	p-CRESOL	C7H8O	109-06-8	2-METHYLPYRIDINE	C6H7N
106-46-7	p-DICHLOROBENZENE	C6H4Cl2	109-21-7	n-BUTYL n-BUTYRATE	C8H16O2
106-47-8	p-CHLOROANILINE	C6H6ClN	109-43-3	DIBUTYL SEBACATE	C18H34O4
106-48-9	p-CHLOROPHENOL	C6H5ClO	109-52-4	VALERIC ACID	C5H10O2
106-49-0	p-TOLUIDINE	C7H9N	109-60-4	n-PROPYL ACETATE	C5H10O2
106-50-3	p-PHENYLENEDIAMINE	C6H8N2	109-65-9	1-BROMOBUTANE	C4H9Br
106-63-8	ISOBUTYL ACRYLATE	C7H12O2	109-66-0	n-PENTANE	C5H12
106-88-7	1,2-EPOXYBUTANE	C4H8O	109-67-1	1-PENTENE	C5H10
106-89-8	alpha-EPICHLOROHYDRIN	C3H5ClO	109-69-3	n-BUTYL CHLORIDE	C4H9Cl
106-93-4	1,2-DIBROMOETHANE	C2H4Br2	109-73-9	n-BUTYLAMINE	C4H11N
106-94-5	1-BROMOPROPANE	C3H7Br	109-74-0	n-BUTYRONITRILE	C4H7N
106-95-6	3-BROMO-1-PROPENE	C3H5Br	109-75-1	VINYLACETONITRILE	C4H5N
106-97-8	n-BUTANE	C4H10	109-77-3	MALONONITRILE	C3H2N2
106-98-9	1-BUTENE	C4H8	109-78-4	HYDRACRYLONITRILE	C3H5NO
106-99-0	1,3-BUTADIENE	C4H6	109-79-5	n-BUTYL MERCAPTAN	C4H10S
107-00-6	ETHYLACETYLENE	C4H6	109-83-1	METHYLETHANOLAMINE	C3H9NO
107-02-8	ACROLEIN	C3H4O	109-86-4	2-METHOXYETHANOL	C3H8O2
107-03-9	n-PROPYLMERCAPTAN	C3H8S	109-87-5	METHYLAL	C3H8O2
107-05-1	3-CHLOROPROPENE	C3H5Cl	109-89-7	DIETHYLAMINE	C4H11N
107-06-2	1,2-DICHLOROETHANE	C2H4Cl2	109-92-2	ETHYL VINYL ETHER	C4H8O
107-07-3	2-CHLOROETHANOL	C2H5ClO	109-93-3	DIVINYL ETHER	C4H6O
107-08-4	n-PROPYL IODIDE	C3H7I	109-94-4	ETHYL FORMATE	C3H6O2
107-10-8	n-PROPYLAMINE	C3H9N	109-97-7	PYRROLE	C4H5N
107-11-9	ALLYLAMINE	C3H7N	109-99-9	TETRAHYDROFURAN	C4H8O
107-12-0	PROPIONITRILE	C3H5N	110-00-9	FURAN	C4H4O
107-13-1	ACRYLONITRILE	C3H3N	110-01-0	TETRAHYDROTHIOPHENE	C4H8S
107-15-3	ETHYLENEDIAMINE	C2H8N2	110-02-1	THIOPHENE	C4H4S
107-18-6	ALLYL ALCOHOL	C3H6O	110-05-4	DI-t-BUTYL PEROXIDE	C8H18O2
107-19-7	PROPARGYL ALCOHOL	C3H4O	110-12-3	5-METHYL-2-HEXANONE	C7H14O
107-20-0	CHLOROACETALDEHYDE	C2H3ClO	110-15-6	SUCCINIC ACID	C4H6O4
107-21-1	ETHYLENE GLYCOL	C2H6O2	110-16-7	MALEIC ACID	C4H4O4
107-25-5	METHYL VINYL ETHER	C3H6O	110-17-8	FUMARIC ACID	C4H4O4
107-31-3	METHYL FORMATE	C2H4O2	110-19-0	ISOBUTYL ACETATE	C6H12O2
107-39-1	2,4,4-TRIMETHYL-1-PENTENE	C8H16	110-33-8	DIHEXYL ADIPATE	C18H34O4
107-40-4	2,4,4-TRIMETHYL-2-PENTENE	C8H16	110-43-0	2-HEPTANONE	C7H14O
107-41-5	HEXYLENE GLYCOL	C6H14O2	110-53-2	1-BROMOPENTANE	C5H11Br
107-46-0	HEXAMETHYLDISILOXANE	C6H18OSi2	110-54-3	n-HEXANE	C6H14
107-83-5	2-METHYLPENTANE	C6H14	110-56-5	1,4-DICHLOROBUTANE	C4H8Cl2
107-84-6	1-CHLORO-3-METHYLBUTANE	C5H11Cl	110-57-6	1,4-DICHLORO-trans-2-BUTENE	C4H6Cl2
107-87-9	2-PENTANONE	C5H10O	110-58-7	n-PENTYLAMINE	C5H13N
107-88-0	1,3-BUTANEDIOL	C4H10O2	110-59-8	VALERONITRILE	C5H9N
107-92-6	n-BUTYRIC ACID	C4H8O2	110-61-2	SUCCINONITRILE	C4H4N2
107-93-7	trans-CROTONIC ACID	C4H6O2	110-62-3	VALERALDEHYDE	C5H10O
107-96-0	3-MERCAPTOPROPIONIC ACID	C3H6O2S	110-63-4	1,4-BUTANEDIOL	C4H10O2
108-01-0	DIMETHYLETHANOLAMINE	C4H11NO	110-65-6	2-BUTYNE-1,4-DIOL	C4H6O2
108-05-4	VINYL ACETATE	C4H6O2	110-66-7	n-PENTYL MERCAPTAN	C5H12S

110-67-8	3-METHOXYPROPIONITRILE	C4H7NO		115-11-7	ISOBUTENE	C4H8
110-71-4	1,2-DIMETHOXYETHANE	C4H10O2		115-25-3	OCTAFLUOROCYCLOBUTANE	C4F8
110-74-7	n-PROPYL FORMATE	C4H8O2		115-77-5	PENTAERYTHRITOL	C5H12O4
110-80-5	2-ETHOXYETHANOL	C4H10O2		115-86-6	TRIPHENYL PHOSPHATE	C18H15O4P
110-81-6	DIETHYL DISULFIDE	C4H10S2		116-14-3	TETRAFLUOROETHYLENE	C2F4
110-82-7	CYCLOHEXANE	C6H12		116-15-4	HEXAFLUOROPROPYLENE	C3F6
110-83-8	CYCLOHEXENE	C6H10		117-81-7	DIOCTYL PHTHALATE	C24H38O4
110-85-0	PIPERAZINE	C4H10N2		118-74-1	HEXACHLOROBENZENE	C6Cl6
110-86-1	PYRIDINE	C5H5N		118-90-1	o-TOLUIC ACID	C8H8O2
110-88-3	TRIOXANE	C3H6O3		118-91-2	o-CHLOROBENZOIC ACID	C7H5ClO2
110-89-4	PIPERIDINE	C5H11N		118-96-7	2,4,6-TRINITROTOLUENE	C7H5N3O6
110-91-8	MORPHOLINE	C4H9NO		119-36-8	METHYL SALICYLATE	C8H8O3
110-94-1	GLUTARIC ACID	C5H8O4		119-61-9	BENZOPHENONE	C13H10O
110-96-3	DIISOBUTYLAMINE	C8H19N		119-64-2	1,2,3,4-TETRAHYDRONAPHTHALENE	C10H12
110-97-4	DIISOPROPANOLAMINE	C6H15NO2		119-65-3	ISOQUINOLINE	C9H7N
110-99-6	DIGLYCOLIC ACID	C4H6O5		120-12-7	ANTHRACENE	C14H10
111-13-7	2-OCTANONE	C8H16O		120-51-4	BENZYL BENZOATE	C14H12O2
111-14-8	n-HEPTANOIC ACID	C7H14O2		120-61-6	DIMETHYL TEREPHTHALATE	C10H10O4
111-15-9	2-ETHOXYETHYL ACETATE	C6H12O3		120-72-9	INDOLE	C8H7N
111-20-6	SEBACIC ACID	C10H18O4		120-80-9	1,2-BENZENEDIOL	C6H6O2
111-26-2	n-HEXYLAMINE	C6H15N		120-82-1	1,2,4-TRICHLOROBENZENE	C6H3Cl3
111-27-3	1-HEXANOL	C6H14O		120-92-3	CYCLOPENTANONE	C5H8O
111-29-5	1,5-PENTANEDIOL	C5H12O2		120-94-5	N-METHYLPYRROLIDINE	C5H11N
111-31-9	n-HEXYLMERCAPTAN	C6H14S		121-14-2	2,4-DINITROTOLUENE	C7H6N2O4
111-34-2	BUTYL VINYL ETHER	C6H12O		121-32-4	ETHYL VANILLIN	C9H10O3
111-36-4	n-BUTYL ISOCYANATE	C5H9NO		121-33-5	VANILLIN	C8H8O3
111-40-0	DIETHYLENE TRIAMINE	C4H13N3		121-44-8	TRIETHYLAMINE	C6H15N
111-41-1	N-AMINOETHYL ETHANOLAMINE	C4H12N2O		121-69-7	N,N-DIMETHYLANILINE	C8H11N
111-42-2	DIETHANOLAMINE	C4H11NO2		121-71-5	4-CHLORO-3-NITROBENZOTRIFLUORIDE	C7H3ClF3NO2
111-43-3	DI-n-PROPYL ETHER	C6H14O		121-73-3	m-CHLORONITROBENZENE	C6H4ClNO2
111-46-6	DIETHYLENE GLYCOL	C4H10O3		121-91-5	ISOPHTHALIC ACID	C8H6O4
111-47-7	DIPROPYL-SULFIDE	C6H14S		122-39-4	DIPHENYLAMINE	C12H11N
111-49-9	HEXAMETHYLENEIMINE	C6H13N		122-66-7	HYDRAZOBENZENE	C12H12N2
111-55-7	ETHYLENE GLYCOL DIACETATE	C6H10O4		123-01-3	n-DODECYLBENZENE	C18H30
111-65-9	n-OCTANE	C8H18		123-02-4	n-TRIDECYLBENZENE	C19H32
111-66-0	1-OCTENE	C8H16		123-05-7	2-ETHYLHEXANAL	C8H16O
111-68-2	1-AMINOHEPTANE	C7H17N		123-07-9	p-ETHYLPHENOL	C8H10O
111-69-3	ADIPONITRILE	C6H8N2		123-08-0	p-HYDROXYBENZALDEHYDE	C7H6O2
111-70-6	1-HEPTANOL	C7H16O		123-25-1	DIETHYL SUCCINATE	C8H14O4
111-71-7	1-HEPTANAL	C7H14O		123-31-9	p-HYDROQUINONE	C6H6O2
111-76-2	2-BUTOXYETHANOL	C6H14O2		123-38-6	n-PROPIONALDEHYDE	C3H6O
111-77-3	2-(2-METHOXYETHOXY)ETHANOL	C5H12O3		123-39-7	N-METHYLFORMAMIDE	C2H5NO
111-78-4	1,5-CYCLOOCTADIENE	C8H12		123-42-2	DIACETONE ALCOHOL	C6H12O2
111-82-0	METHYL DODECANOATE	C13H26O2		123-51-3	3-METHYL-1-BUTANOL	C5H12O
111-84-2	n-NONANE	C9H20		123-54-6	ACETYLACETONE	C5H8O2
111-86-4	n-OCTYLAMINE	C8H19N		123-62-6	PROPIONIC ANHYDRIDE	C6H10O3
111-87-5	1-OCTANOL	C8H18O		123-63-7	PARALDEHYDE	C6H12O3
111-88-6	n-OCTYL MERCAPTAN	C8H18S		123-72-8	n-BUTYRALDEHYDE	C4H8O
111-90-0	2-(2-ETHOXYETHOXY)ETHANOL	C6H14O3		123-73-9	trans-CROTONALDEHYDE	C4H6O
111-92-2	DI-n-BUTYLAMINE	C8H19N		123-75-1	PYRROLIDINE	C4H9N
111-96-6	DIETHYLENE GLYCOL DIMETHYL ETHER	C6H14O3		123-76-2	LEVULINIC ACID	C5H8O3
112-05-0	n-NONANOIC ACID	C9H18O2		123-86-4	n-BUTYL ACETATE	C6H12O2
112-15-2	DIETHYLENE GLYCOL ETHYL ETHER ACETATE	C8H16O4		123-91-1	1,4-DIOXANE	C4H8O2
112-20-9	n-NONYLAMINE	C9H21N		123-92-2	ISOPENTYL ACETATE	C7H14O2
112-24-3	TRIETHYLENE TETRAMINE	C6H18N4		123-95-5	n-BUTYL STEARATE	C22H44O2
112-27-6	TRIETHYLENE GLYCOL	C6H14O4		123-96-6	2-OCTANOL	C8H18O
112-30-1	1-DECANOL	C10H22O		123-99-9	AZELAIC ACID	C9H16O4
112-31-2	1-DECANAL	C10H20O		124-04-9	ADIPIC ACID	C6H10O4
112-32-3	n-OCTYL FORMATE	C9H18O2		124-07-2	n-OCTANOIC ACID	C8H16O2
112-34-5	DIETHYLENE GLYCOL MONOBUTYL ETHER	C8H18O3		124-09-4	HEXAMETHYLENEDIAMINE	C6H16N2
112-36-7	DIETHYLENE GLYCOL DIETHYL ETHER	C8H18O3		124-11-8	1-NONENE	C9H18
112-40-3	n-DODECANE	C12H26		124-13-0	1-OCTANAL	C8H16O
112-41-4	1-DODECENE	C12H24		124-18-5	n-DECANE	C10H22
112-42-5	1-UNDECANOL	C11H24O		124-19-6	1-NONANAL	C9H18O
112-44-7	1-UNDECANAL	C11H22O		124-22-1	DODECYLAMINE	C12H27N
112-49-2	TRIETHYLENE GLYCOL DIMETHYL ETHER	C8H18O4		124-38-9	CARBON DIOXIDE	CO2
112-51-6	DIPENTYL-DISULFIDE	C10H22S2		124-40-3	DIMETHYLAMINE	C2H7N
112-53-8	1-DODECANOL	C12H26O		124-73-2	1,2-DIBROMOTETRAFLUOROETHANE	C2Br2F4
112-54-9	1-DODECANAL	C12H24O		126-30-7	NEOPENTYL GLYCOL	C5H12O2
112-55-0	n-DODECYL MERCAPTAN	C12H26S		126-33-0	SULFOLANE	C4H8O2S
112-57-2	TETRAETHYLENEPENTAMINE	C8H23N5		126-98-7	METHACRYLONITRILE	C4H5N
112-58-3	DI-n-HEXYL ETHER	C12H26O		126-99-8	CHLOROPRENE	C4H5Cl
112-60-7	TETRAETHYLENE GLYCOL	C8H18O5		127-17-3	PYRUVIC ACID	C3H4O3
112-62-9	METHYL OLEATE	C19H36O2		127-18-4	TETRACHLOROETHYLENE	C2Cl4
112-70-9	1-TRIDECANOL	C13H28O		127-19-5	N,N-DIMETHYLACETAMIDE	C4H9NO
112-72-1	1-TETRADECANOL	C14H30O		127-91-3	beta-PINENE	C10H16
112-73-2	DIETHYLENE GLYCOL DI-n-BUTYL ETHER	C12H26O3		128-37-0	2,6-DI-tert-BUTYL-p-CRESOL	C15H24O
112-80-1	OLEIC ACID	C18H34O2		129-00-0	PYRENE	C16H10
112-88-9	1-OCTADECENE	C18H36		131-11-3	DIMETHYL PHTHALATE	C10H10O4
112-92-5	1-OCTADECANOL	C18H38O		132-64-9	DIBENZOFURAN	C12H8O
112-95-8	n-EICOSANE	C20H42		133-37-9	TARTARIC ACID	C4H6O6
115-07-1	PROPYLENE	C3H6		135-01-3	o-DIETHYLBENZENE	C10H14
115-10-6	DIMETHYL ETHER	C2H6O		135-98-8	sec-BUTYLBENZENE	C10H14

136-35-6	1,3-DIPHENYLTRIAZENE	C12H11N3
136-60-7	n-BUTYL BENZOATE	C11H14O2
137-32-6	2-METHYL-1-BUTANOL	C5H12O
140-11-4	BENZYL ACETATE	C9H10O2
140-31-8	N-AMINOETHYL PIPERAZINE	C6H15N3
140-66-9	p-tert-OCTYLPHENOL	C14H22O
140-88-5	ETHYL ACRYLATE	C5H8O2
141-05-9	DIETHYL MALEATE	C8H12O4
141-32-2	n-BUTYL ACRYLATE	C7H12O2
141-43-5	MONOETHANOLAMINE	C2H7NO
141-59-3	tert-OCTYL MERCAPTAN	C8H18S
141-78-6	ETHYL ACETATE	C4H8O2
141-79-7	MESITYL OXIDE	C6H10O
141-93-5	m-DIETHYLBENZENE	C10H14
141-97-9	ETHYLACETOACETATE	C6H10O3
142-28-9	1,3-DICHLOROPROPANE	C3H6Cl2
142-29-0	CYCLOPENTENE	C5H8
142-62-1	n-HEXANOIC ACID	C6H12O2
142-82-5	n-HEPTANE	C7H16
142-84-7	DI-n-PROPYLAMINE	C6H15N
142-92-7	n-HEXYL ACETATE	C8H16O2
142-96-1	DI-n-BUTYL ETHER	C8H18O
143-07-7	n-DODECANOIC ACID	C12H24O2
143-08-8	1-NONANOL	C9H20O
143-10-2	n-DECYL MERCAPTAN	C10H22S
143-24-8	TETRAETHYLENE GLYCOL DIMETHYL ETHER	C10H22O5
144-62-7	OXALIC ACID	C2H2O4
148-24-3	8-HYDROXYQUINOLINE	C9H7NO
150-76-5	p-METHOXYPHENOL	C7H8O2
151-56-4	ETHYLENEIMINE	C2H5N
151-67-7	HALOTHANE	C2HBrClF3
156-43-4	p-PHENETIDINE	C8H11NO
156-59-2	cis-1,2-DICHLOROETHYLENE	C2H2Cl2
156-60-5	trans-1,2-DICHLOROETHYLENE	C2H2Cl2
156-87-6	3-AMINO-1-PROPANOL	C3H9NO
157-40-4	SPIROPENTANE	C5H8
206-44-0	FLUORANTHENE	C16H10
218-01-9	CHRYSENE	C18H12
275-51-4	AZULENE	C10H8
280-57-9	TRIETHYLENEDIAMINE	C6H12N2
287-23-0	CYCLOBUTANE	C4H8
287-27-4	THIACYCLOBUTANE	C3H6S
287-92-3	CYCLOPENTANE	C5H10
288-42-6	OXAZOLE	C3H3NO
291-64-5	CYCLOHEPTANE	C7H14
292-64-8	CYCLOOCTANE	C8H16
295-48-7	1-CYCLOPENTYLDECANE	C15H30
320-60-5	2,4-DICHLOROBENZOTRIFLUORIDE	C7H3Cl2F3
334-48-5	n-DECANOIC ACID	C10H20O2
352-32-9	p-FLUOROTOLUENE	C7H7F
352-93-2	DIETHYL SULFIDE	C4H10S
353-36-6	ETHYL FLUORIDE	C2H5F
353-50-4	CARBONYL FLUORIDE	CF2O
353-59-3	BROMOCHLORODIFLUOROMETHANE	CBrClF2
354-33-6	PENTAFLUOROETHANE	C2HF5
355-25-9	DECAFLUOROBUTANE	C4F10
359-10-4	2-CHLORO-1,1-DIFLUOROETHYLENE	C2HClF2
359-11-5	TRIFLUOROETHENE	C2HF3
360-89-4	OCTAFLUORO-2-BUTENE	C4F8
367-11-3	o-DIFLUOROBENZENE	C6H4F2
372-18-9	m-DIFLUOROBENZENE	C6H4F2
392-56-3	HEXAFLUOROBENZENE	C6F6
420-12-2	THIACYCLOPROPANE	C2H4S
420-26-8	2-FLUOROPROPANE	C3H7F
420-46-2	1,1,1-TRIFLUOROETHANE	C2H3F3
460-12-8	BUTADIYNE(BIACETYLENE)	C4H2
460-13-9	1-FLUOROPROPANE	C3H7F
460-19-5	CYANOGEN	C2N2
462-06-6	FLUOROBENZENE	C6H5F
463-49-0	PROPADIENE	C3H4
463-51-4	KETENE	C2H2O
463-58-1	CARBONYL SULFIDE	COS
463-82-1	NEOPENTANE	C5H12
464-06-2	2,2,3-TRIMETHYLBUTANE	C7H16
488-23-3	1,2,3,4-TETRAMETHYLBENZENE	C10H14
493-01-6	cis-DECAHYDRONANPHTALENE	C10H18
493-02-7	trans-DECAHYDRONAPHTHALENE	C10H18
496-11-7	INDANE	C9H10
498-23-7	CITRACONIC ACID	C5H6O4
498-66-8	2-NORBORNENE	C7H10
501-65-5	DIPHENYLACETYLENE	C14H10
502-44-3	epsilon-CAPROLACTONE	C6H10O2
503-17-3	DIMETHYLACETYLENE	C4H6
503-30-0	1,3-PROPYLENE OXIDE	C3H6O
503-64-0	cis-CROTONIC ACID	C4H6O2
503-74-2	ISOVALERIC ACID	C5H10O2
504-63-2	1,3-PROPYLENE GLYCOL	C3H8O2
506-77-4	CYANOGEN CHLORIDE	CClN
507-09-5	THIOACETIC-ACID	C2H4OS
507-20-0	tert-BUTYL CHLORIDE	C4H9Cl
509-14-8	TETRANITROMETHANE	CN4O8
512-56-1	TRIMETHYL PHOSPHATE	C3H9O4P
513-35-9	2-METHYL-2-BUTENE	C5H10
513-44-0	ISOBUTYL MERCAPTAN	C4H10S
513-53-3	sec-BUTYL MERCAPTAN	C4H10S
513-81-5	2,3-DIMETHYL-1,3-BUTADIENE	C6H10
514-10-3	ABIETIC ACID	C20H30O2
526-73-8	1,2,3-TRIMETHYLBENZENE	C9H12
526-75-0	2,3-XYLENOL	C8H10O
527-53-7	1,2,3,5-TETRAMETHYLBENZENE	C10H14
527-84-4	o-CYMENE	C10H14
528-29-0	o-DINITROBENZENE	C6H4N2O4
533-98-2	1,2-DIBROMOBUTANE	C4H8Br2
535-77-3	m-CYMENE	C10H14
536-74-3	ETHYNYLBENZENE	C8H6
538-68-1	n-PENTYLBENZENE	C11H16
538-93-2	ISOBUTYLBENZENE	C10H14
539-30-0	BENZYL ETHYL ETHER	C9H12O
540-36-3	p-DIFLUOROBENZENE	C6H4F2
540-54-5	n-PROPYL CHLORIDE	C3H7Cl
540-67-0	METHYL ETHYL ETHER	C3H8O
540-84-1	2,2,4-TRIMETHYLPENTANE	C8H18
540-88-5	tert-BUTYL ACETATE	C6H12O2
541-05-9	HEXAMETHYLCYCLOTRISILOXANE	C6H18O3Si3
541-41-3	ETHYL CHLOROFORMATE	C3H5ClO2
541-73-1	m-DICHLOROBENZENE	C6H4Cl2
542-10-9	ETHYLIDENE DIACETATE	C6H10O4
542-55-2	ISOBUTYL FORMATE	C5H10O2
542-88-1	BIS(CHLOROMETHYL)ETHER	C2H4Cl2O
542-92-7	CYCLOPENTADIENE	C5H6
543-49-7	2-HEPTANOL	C7H16O
543-59-9	1-CHLOROPENTANE	C5H11Cl
544-13-8	GLUTARONITRILE	C5H6N2
544-40-1	DIBUTYL-SULFIDE	C8H18S
544-55-2	1,3,5-CYCLOHEPTATRIENE	C7H8
544-63-8	n-TETRADECANOIC ACID	C14H28O2
544-76-3	n-HEXADECANE	C16H34
552-30-7	TRIMELLITIC ANHYDRIDE	C9H4O5
554-12-1	METHYL PROPIONATE	C4H8O2
554-14-3	2-METHYLTHIOPHENE	C5H6S
555-10-2	beta-PHELLANDRENE	C10H16
556-56-9	3-IODO-1-PROPENE	C3H5I
556-67-2	OCTAMETHYLCYCLOTETRASILOXANE	C8H24O4Si4
557-17-5	METHYL-PROPYL-ETHER	C4H10O
557-91-5	1,1-DIBROMOETHANE	C2H4Br2
557-98-2	2-CHLOROPROPENE	C3H5Cl
558-17-8	2-IODO-2-METHYLPROPANE	C4H9I
558-37-2	3,3-DIMETHYL-1-BUTENE	C6H12
560-21-4	2,3,3-TRIMETHYLPENTANE	C8H18
562-49-2	3,3-DIMETHYLPENTANE	C7H16
563-16-6	3,3-DIMETHYLHEXANE	C8H18
563-45-1	3-METHYL-1-BUTENE	C5H10
563-46-2	2-METHYL-1-BUTENE	C5H10
563-78-0	2,3-DIMETHYL-1-BUTENE	C6H12
563-79-1	2,3-DIMETHYL-2-BUTENE	C6H12
563-80-4	METHYL ISOPROPYL KETONE	C5H10O
564-02-3	2,2,3-TRIMETHYLPENTANE	C8H18
565-59-3	2,3-DIMETHYLPENTANE	C7H16
565-69-5	ETHYL ISOPROPYL KETONE	C6H12O
565-75-3	2,3,4-TRIMETHYLPENTANE	C8H18
565-80-0	DIISOPROPYL KETONE	C7H14O
571-58-4	1,4-DIMETHYLNAPHTHALENE	C12H12
571-61-9	1,5-DIMETHYLNAPHTHALENE	C12H12
573-98-8	1,2-DIMETHYLNAPHTHALENE	C12H12
575-37-1	1,7-DIMETHYLNAPHTHALENE	C12H12
575-41-7	1,3-DIMETHYLNAPHTHALENE	C12H12
575-43-9	1,6-DIMETHYLNAPHTHALENE	C12H12
576-26-1	2,6-XYLENOL	C8H10O
578-54-1	o-ETHYLANILINE	C8H11N
579-66-8	2,6-DIETHYLANILINE	C10H15N
581-40-8	2,3-DIMETHYLNAPHTHALENE	C12H12
581-42-0	2,6-DIMETHYLNAPHTHALENE	C12H12
582-16-1	2,7-DIMETHYLNAPHTHALENE	C12H12
583-48-2	3,4-DIMETHYLHEXANE	C8H18
584-02-1	3-PENTANOL	C5H12O
584-94-1	2,3-DIMETHYLHEXANE	C8H18

6-62-9	TERPINOLENE	C10H16		624-83-9	METHYL ISOCYANATE	C2H3NO
9-34-4	3-METHYLHEXANE	C7H16		624-89-5	ETHYL-METHYL-SULFIDE	C3H8S
9-38-8	3-HEXANONE	C6H12O		624-91-9	METHYL-NITRITE	CH3NO2
9-43-5	2,4-DIMETHYLHEXANE	C8H18		624-92-0	DIMETHYL DISULFIDE	C2H6S2
9-53-7	4-METHYLHEPTANE	C8H18		625-27-4	2-METHYL-2-PENTENE	C6H12
9-81-1	3-METHYLHEPTANE	C8H18		625-44-5	METHYL ISOBUTYL ETHER	C5H12O
0-01-2	n-BUTYL PROPIONATE	C7H14O2		625-45-6	METHOXYACETIC ACID	C3H6O3
0-18-1	cis-2-BUTENE	C4H8		625-58-1	ETHYL-NITRATE	C2H5NO3
0-19-2	1,2-BUTADIENE	C4H6		625-80-9	DIISOPROPYL-SULFIDE	C6H14S
0-35-2	2,2-DIMETHYLPENTANE	C7H16		626-93-7	2-HEXANOL	C6H14O
0-66-9	1,1-DIMETHYLCYCLOHEXANE	C8H16		627-05-4	1-NITROBUTANE	C4H9NO2
0-67-0	1-METHYLCYCLOHEXANOL	C7H14O		627-13-4	PROPYL-NITRATE	C3H7NO3
0-73-8	2,2-DIMETHYLHEXANE	C8H18		627-19-0	1-PENTYNE	C5H8
1-50-4	IODOBENZENE	C6H5I		627-20-3	cis-2-PENTENE	C5H10
1-68-4	n-BUTYL VALERATE	C9H18O2		627-21-4	2-PENTYNE	C5H8
1-76-4	2-METHYLHEXANE	C7H16		627-26-9	trans-CROTONITRILE	C4H5N
1-78-6	2-HEXANONE	C6H12O		627-98-5	5-METHYL-1-HEXANOL	C7H16O
1-87-7	ALLYL ACETATE	C5H8O2		628-21-7	1,2-DIIODOBUTANE	C4H8I2
1-93-5	1,4-PENTADIENE	C5H8		628-29-5	BUTYL-METHYL-SULFIDE	C5H12S
1-95-7	1,2-PENTADIENE	C5H8		628-32-0	ETHYL PROPYL ETHER	C5H12O
1-96-8	2,3-PENTADIENE	C5H8		628-63-7	n-PENTYL ACETATE	C7H14O2
2-13-2	2,5-DIMETHYLHEXANE	C8H18		628-71-7	1-HEPTYNE	C7H12
2-27-8	2-METHYLHEPTANE	C8H18		628-73-9	HEXANENITRILE	C6H11N
2-41-6	1-HEXENE	C6H12		628-76-2	1,5-DICHLOROPENTANE	C5H10Cl2
2-42-7	1,5-HEXADIENE	C6H10		628-81-9	n-BUTYL ETHYL ETHER	C6H14O
2-57-4	1,3-CYCLOHEXADIENE	C6H8		629-04-9	1-BROMOHEPTANE	C7H15Br
2-76-7	1-HEPTENE	C7H14		629-11-8	1,6-HEXANEDIOL	C6H14O2
2-84-7	n-BUTYL FORMATE	C5H10O2		629-19-6	DIPROPYL-DISULFIDE	C6H14S2
3-45-3	n-OCTADECANE	C18H38		629-20-9	1,3,5,7-CYCLOOCTATETRAENE	C8H8
3-53-3	METHYL FLUORIDE	CH3F		629-45-8	DIBUTYL-DISULFIDE	C8H18S2
3-60-2	VINYL BROMIDE	C2H3Br		629-50-5	n-TRIDECANE	C13H28
3-70-0	CHLOROFLUOROMETHANE	CH2ClF		629-59-4	n-TETRADECANE	C14H30
4-20-7	2,2-DICHLOROPROPANE	C3H6Cl2		629-62-9	n-PENTADECANE	C15H32
4-36-5	2-CHLORO-2-METHYLBUTANE	C5H11Cl		629-73-2	1-HEXADECENE	C16H32
4-51-4	2,3-DIBROMO-2-METHYLBUTANE	C5H10Br2		629-74-3	1-HEXADECYNE	C16H30
4-56-9	2,3,3-TRIMETHYL-1-BUTENE	C7H14		629-76-5	1-PENTADECANOL	C15H32O
4-82-1	2,2,3,3-TETRAMETHYLBUTANE	C8H18		629-78-7	n-HEPTADECANE	C17H36
8-03-8	DI-n-PROPYL SULFONE	C6H14O2S		629-82-3	DI-n-OCTYL ETHER	C16H34O
8-04-9	DI-n-BUTYL SULFONE	C8H18O2S		629-89-0	1-OCTADECYNE	C18H34
8-23-2	3-METHYL-1-BUTYNE	C5H8		629-92-5	n-NONADECANE	C19H40
8-25-4	3-METHYL-1,2-BUTADIENE	C5H8		629-96-9	1-EICOSANOL	C20H42O
8-29-8	1,2-DIIODOPROPANE	C3H6I2		630-08-0	CARBON MONOXIDE	CO
8-53-8	METHYL ISOPROPYL ETHER	C4H10O		630-20-6	1,1,1,2-TETRACHLOROETHANE	C2H2Cl4
8-58-3	METHYL-NITRATE	CH3NO3		632-51-9	TETRAPHENYLETHYLENE	C26H20
8-73-2	BROMOTRIFLUOROETHYLENE	C2BrF3		635-81-4	1,2,4,5-TETRAETHYLBENZENE	C14H22
8-75-4	3-METHYL-2-BUTANOL	C5H12O		638-04-0	cis-1,3-DIMETHYLCYCLOHEXANE	C8H16
8-82-3	LACTIC ACID	C3H6O3		638-46-0	BUTYL-ETHYL-SULFIDE	C6H14S
9-64-4	p-CUMYLPHENOL	C15H16O		638-49-3	n-PENTYL FORMATE	C6H12O2
0-07-7	2-METHYLBUTYRIC ACID	C5H10O2		642-32-0	1,2,3,4-TETRAETHYLBENZENE	C14H22
0-24-8	2-NITROBUTANE	C4H9NO2		645-49-8	cis-STILBENE	C14H12
3-35-0	TRIPHENYLPHOSPHINE	C18H15P		646-04-8	trans-2-PENTENE	C5H10
4-88-6	HEXAETHYLBENZENE	C18H30		646-05-9	1-PENTENE-3-YNE	C5H6
5-01-6	PENTAETHYLBENZENE	C16H26		646-30-0	NONADECANOIC ACID	C19H38O2
5-02-7	1-PHENYLNAPHTHALENE	C16H12		659-70-1	ISOPENTYL ISOVALERATE	C10H20O2
5-20-2	2,6-DINITROTOLUENE	C7H6N2O4		674-76-0	4-METHYL-trans-2-PENTENE	C6H12
6-95-7	1-CYCLOPENTYLHEXADECANE	C21H42		674-82-8	DIKETENE	C4H4O2
9-26-7	3-ETHYL-2-METHYLPENTANE	C8H18		680-31-9	HEXAMETHYL PHOSPHORAMIDE	C6H18N3OP
0-39-9	3,4-DINITROTOLUENE	C7H6N2O4		684-16-2	HEXAFLUOROACETONE	C3F6O
1-14-3	o-ETHYLTOLUENE	C9H12		688-74-4	TRI-n-BUTYL BORATE	C12H27BO3
1-15-4	o-METHYLSTYRENE	C9H10		689-97-4	VINYLACETYLENE	C4H4
2-00-0	1,1-DIPHENYLETHANE	C14H14		691-37-2	4-METHYL-1-PENTENE	C6H12
6-12-6	trans-3-METHYL-2-PENTENE	C6H12		691-38-3	4-METHYL-cis-2-PENTENE	C6H12
6-44-4	3-METHYLTHIOPHENE	C5H6S		692-45-5	VINYL FORMATE	C3H4O2
6-45-5	2-PYRROLIDONE	C4H7NO		693-02-7	1-HEXYNE	C6H10
7-48-1	MALIC ACID	C4H6O5		693-65-2	DI-n-PENTYL ETHER	C10H22O
7-78-1	3-ETHYLPENTANE	C7H16		693-89-0	1-METHYLCYCLOPENTENE	C6H10
7-94-7	2-PHENYL-2-PROPANOL	C9H12O		696-29-7	ISOPROPYLCYCLOHEXANE	C9H18
8-46-2	m-CHLOROBENZOYL CHLORIDE	C7H4Cl2O		760-20-3	3-METHYL-1-PENTENE	C6H12
8-85-9	3,5-DINITROTOLUENE	C7H6N2O4		760-21-4	2-ETHYL-1-BUTENE	C6H12
9-15-8	2,5-DINITROTOLUENE	C7H6N2O4		760-23-6	3,4-DICHLORO-1-BUTENE	C4H6Cl2
9-82-0	1,4-CYCLOHEXANEDICARBOXYLIC ACID	C8H12O4		763-29-1	2-METHYL-1-PENTENE	C6H12
9-99-8	3-ETHYLHEXANE	C8H18		763-69-9	ETHYL-3-ETHOXYPROPIONATE	C7H14O3
0-14-4	m-ETHYLTOLUENE	C9H12		764-35-2	2-HEXYNE	C6H10
0-17-7	m-ETHYLPHENOL	C8H10O		765-03-7	1-DODECYNE	C12H22
2-96-8	p-ETHYLTOLUENE	C9H12		765-13-9	1-PENTADECYNE	C15H28
2-97-7	p-METHYLSTYRENE	C9H10		765-27-5	1-EICOSENE	C20H38
3-42-7	METHYL n-BUTYRATE	C5H10O2		766-07-4	CYCLOHEXYL PEROXIDE	C6H12O2
3-29-3	cis-1,4-DIMETHYLCYCLOHEXANE	C8H16		766-90-5	cis-PROPENYLBENZENE	C9H10
4-58-6	DIMETHYL MALEATE	C6H8O4		767-59-9	1-METHYLINDENE	C10H10
4-64-6	trans-2-BUTENE	C4H8		811-97-2	1,1,1,2-TETRAFLUOROETHANE	C2H2F4
4-65-1	PROPARGYL CHLORIDE	C3H3Cl		814-78-8	METHYL ISOPROPENYL KETONE	C5H8O
4-72-6	1,2-DIFLUOROETHANE	C2H4F2		818-61-1	2-HYDROXYETHYL ACRYLATE	C5H8O3
4-73-7	1,2-DIIODOETHANE	C2H4I2		821-11-4	trans-2-BUTENE-1,4-DIOL	C4H8O2

821-95-4	1-UNDECENE	C11H22
822-35-5	CYCLOBUTENE	C4H6
822-50-4	trans-1,2-DIMETHYLCYCLOPENTANE	C7H14
827-52-1	CYCLOHEXYLBENZENE	C12H16
871-28-3	1-PENTENE-4-YNE	C5H6
871-83-0	2-METHYLNONANE	C10H22
872-05-9	1-DECENE	C10H20
872-50-4	N-METHYL-2-PYRROLIDONE	C5H9NO
872-93-5	3-METHYL SULFOLANE	C5H10O2S
873-66-5	trans-PROPENYLBENZENE	C9H10
874-41-9	4-ETHYL-m-XYLENE	C10H14
877-44-1	1,2,4-TRIETHYLBENZENE	C12H18
921-47-1	2,3,4-TRIMETHYLHEXANE	C9H20
922-28-1	3,4-DIMETHYLHEPTANE	C9H20
922-62-3	3-METHYL-cis-2-PENTENE	C6H12
925-60-0	n-PROPYL ACRYLATE	C6H10O2
926-82-9	3,5-DIMETHYLHEPTANE	C9H20
928-49-4	3-HEXYNE	C6H10
929-06-6	2-AMINOETHOXYETHANOL	C4H11NO2
933-98-2	3-ETHYL-o-XYLENE	C10H14
934-74-7	5-ETHYL-m-XYLENE	C10H14
934-80-5	4-ETHYL-o-XYLENE	C10H14
939-27-5	2-ETHYLNAPHTHALENE	C12H12
993-00-0	METHYL CHLOROSILANE	CH5ClSi
993-07-7	TRIMETHYL SILANE	C3H10Si
994-05-8	METHYL tert-PENTYL ETHER	C6H14O
992-94-9	METHYL SILANE	CH6Si
999-21-3	DIALLYL MALEATE	C10H12O4
999-97-3	HEXAMETHYLDISILAZANE	C6H19NSi2
1002-84-2	PENTADECANOIC ACID	C15H30O2
1067-08-9	3-METHYL-3-ETHYLPENTANE	C8H18
1067-20-5	3,3-DIETHYLPENTANE	C9H20
1068-19-5	4,4-DIMETHYLHEPTANE	C9H20
1068-87-7	2,4-DIMETHYL-3-ETHYLPENTANE	C9H20
1069-53-0	2,3,5-TRIMETHYLHEXANE	C9H20
1070-87-7	2,2,4,4-TETRAMETHYLPENTANE	C9H20
1071-26-7	2,2-DIMETHYLHEPTANE	C9H20
1071-81-4	2,2,5,5-TETRAMETHYLHEXANE	C10H22
1072-05-5	2,6-DIMETHYLHEPTANE	C9H20
1072-16-8	2,7-DIMETHYLOCTANE	C10H22
1077-16-3	n-HEXYLBENZENE	C12H18
1078-71-3	n-HEPTYLBENZENE	C13H20
1081-77-2	n-NONYLBENZENE	C15H24
1111-74-6	DIMETHYL SILANE	C2H8Si
1119-85-3	1,4-DICYANO-2-BUTENE	C6H6N2
1120-21-4	n-UNDECANE	C11H24
1120-36-1	1-TETRADECENE	C14H28
1120-62-3	3-METHYLCYCLOPENTENE	C6H10
1127-76-0	1-ETHYLNAPHTHALENE	C12H12
1184-58-3	DIMETHYLALUMINUM CHLORIDE	C2H6AlCl
1186-53-4	2,2,3,4-TETRAMETHYLPENTANE	C9H20
1189-99-7	2,5,5-TRIMETHYLHEPTANE	C10H22
1190-76-7	cis-CROTONITRILE	C4H5N
1190-83-6	2,2,6-TRIMETHYLHEPTANE	C10H22
1192-18-3	cis-1,2-DIMETHYLCYCLOPENTANE	C7H14
1323-65-5	DINONYLPHENOL	C24H42O
1446-61-3	DEHYDROABIETYLAMINE	C20H31N
1454-84-8	1-NONADECANOL	C19H40O
1454-85-9	1-HEPTADECANOL	C17H36O
1455-21-6	n-NONYL MERCAPTAN	C9H20S
1459-09-2	1-PHENYLHEXADECANE	C22H38
1476-11-5	1,4-DICHLORO-cis-2-BUTENE	C4H6Cl2
1551-21-9	ISOPROPYL-METHYL-SULFIDE	C4H10S
1574-41-0	cis-1,3-PENTADIENE	C5H8
1613-51-0	THIACYCLOHEXANE	C5H10S
1630-77-9	trans-1,2-DIFLUOROETHENE	C2H2F2
1630-78-0	cis-1,2-DIFLUOROETHENE	C2H2F2
1632-16-2	2-ETHYL-1-HEXENE	C8H16
1634-04-4	METHYL tert-BUTYL ETHER	C5H12O
1634-09-9	1-n-BUTYLNAPHTHALENE	C14H16
1638-26-2	1,1-DIMETYLCYCLOPENTANE	C7H14
1639-09-4	n-HEPTYL MERCAPTAN	C7H16S
1640-89-7	ETHYLCYCLOPENTANE	C7H14
1656-48-0	BIS(CYANOETHYL)ETHER	C6H8N2O
1678-91-7	ETHYLCYCLOHEXANE	C8H16
1678-92-8	n-PROPYLCYCLOHEXANE	C9H18
1678-93-9	n-BUTYLCYCLOHEXANE	C10H20
1679-07-8	CYCLOPENTANETHIOL	C5H10S
1679-09-0	2-METHYL-2-BUTANETHIOL	C5H12S
1708-29-8	2,5-DIHYDROFURAN	C4H6O
1741-83-9	METHYL-PENTYL-SULFIDE	C6H14S
1758-88-9	2-ETHYL-p-XYLENE	C10H14
1759-58-6	trans-1,3-DIMETHYLCYCLOPENTANE	C7H14

1759-81-5	4-METHYLCYCLOPENTENE	C6H10
1795-15-9	1-CYCLOHEXYLOCTANE	C14H28
1795-16-0	n-DECYLCYCLOHEXANE	C16H32
1795-17-1	1-CYCLOHEXYLDODECANE	C18H36
1795-18-2	1-CYCLOHEXYLTETRADECANE	C20H40
1795-20-6	1-CYCLOPENTYLOCTANE	C13H26
1795-22-8	1-CYCLOPENTYLTETRADECANE	C19H38
1795-26-2	cis,trans-1,3,5-TRIMETHYLCYCLOHEXANE	C9H18
1795-27-3	cis,cis-1,3,5-TRIMETHYLCYCLOHEXANE	C9H18
1860-27-1	ISOPROPYL-TERT-BUTYL-ETHER	C7H16O
1889-67-4	2,3-DIMETHYL-2,3-DIPHENYLBUTANE	C18H22
2004-70-8	trans-1,3-PENTADIENE	C5H8
2016-42-4	TETRADECYLAMINE	C14H31N
2016-57-1	n-DECYLAMINE	C10H23N
2027-19-2	2-PROPYLNAPHTHALENE	C13H14
2040-95-1	BUTYLCYCLOPENTANE	C9H18
2040-96-2	n-PROPYLCYCLOPENTANE	C8H16
2051-30-1	2,6-DIMETHYLOCTANE	C10H22
2079-95-0	1-TETRADECANETHIOL	C14H30S
2131-18-2	1-PHENYLPENTADECANE	C21H36
2141-58-4	cis-DICYANO-1-BUTENE	C6H6N2
2141-59-5	trans-DICYANO-1-BUTENE	C6H6N2
2177-47-1	2-METHYLINDENE	C10H10
2189-60-8	n-OCTYLBENZENE	C14H22
2207-01-4	cis-1,2-DIMETHYLCYCLOHEXANE	C8H16
2207-03-6	trans-1,3-DIMETHYLCYCLOHEXANE	C8H16
2207-04-7	trans-1,4-DIMETHYLCYCLOHEXANE	C8H16
2210-28-8	n-PROPYL METHACRYLATE	C7H12O2
2213-23-2	2,4-DIMETHYLHEPTANE	C9H20
2216-30-0	2,5-DIMETHYLHEPTANE	C9H20
2216-32-2	4-ETHYLHEPTANE	C9H20
2216-33-3	3-METHYLOCTANE	C9H20
2216-34-4	4-METHYLOCTANE	C9H20
2243-98-3	1-UNDECYNE	C11H20
2278-87-9	3,3,4-TRIMETHYLHEPTANE	C10H22
2278-88-0	3,4,4-TRIMETHYLHEPTANE	C10H22
2278-89-1	3,4,5-TRIMETHYLHEPTANE	C10H22
2425-74-3	tert-BUTYLFORMAMIDE	C5H11NO
2437-56-1	1-TRIDECENE	C13H26
2456-27-1	DINONYL ETHER	C18H38O
2532-58-3	cis-1,3-DIMETHYLCYCLOPENTANE	C7H14
2613-61-8	2,4,6-TRIMETHYLHEPTANE	C10H22
2765-18-6	n-PROPYLNAPHTHALENE	C13H14
2807-30-9	ETHYLENE GLYCOL MONOPROPYL ETHER	C5H12O2
2870-04-4	2-ETHYL-m-XYLENE	C10H14
2876-53-1	1-n-HEXYLNAPHTHALENE	C16H20
2882-98-6	1-CYCLOPENTYLNONANE	C14H28
2883-02-5	1-CYCLOHEXYLNONANE	C15H30
2885-00-9	1-OCTADECANETHIOL	C18H38S
2902-96-7	ISOPROPYL-NITRATE	C3H7NO3
2917-26-2	1-HEXADECANETHIOL	C16H34S
3074-71-3	2,3-DIMETHYLHEPTANE	C9H20
3074-75-7	4-ETHYL-2-METHYLHEXANE	C9H20
3074-76-8	3-ETHYL-3-METHYLHEXANE	C9H20
3074-77-9	3-ETHYL-4-METHYLHEXANE	C9H20
3129-90-6	ISOTHIOCYANIC-ACID	CHNS
3173-53-3	CYCLOHEXYL ISOCYANATE	C7H11NO
3178-29-8	4-PROPYLHEPTANE	C10H22
3221-61-2	2-METHYLOCTANE	C9H20
3404-61-3	3-METHYL-1-HEXENE	C7H14
3404-71-5	2-ETHYL-1-PENTENE	C7H14
3452-07-1	1-EICOSENE	C20H40
3452-09-3	1-NONYNE	C9H16
3522-94-9	2,2,5-TRIMETHYLHEXANE	C9H20
3741-00-2	1-CYCLOPENTYLPENTANE	C10H20
3769-23-1	4-METHYL-1-HEXENE	C7H14
3877-15-4	METHYL-PROPYL-SULFIDE	C4H10S
4032-92-2	2,4,4-TRIMETHYLHEPTANE	C10H22
4032-93-3	2,3,6-TRIMETHYLHEPTANE	C10H22
4032-94-4	2,4-DIMETHYLOCTANE	C10H22
4038-04-4	3-ETHYL-1-PENTENE	C7H14
4048-33-3	6-AMINOHEXANOL	C6H15NO
4050-45-7	trans-2-HEXENE	C6H12
4110-44-5	3,3-DIMETHYLOCTANE	C10H22
4110-50-3	ETHYL-PROPYL-SULFIDE	C5H12S
4265-25-2	2-METHYLBENZOFURAN	C9H8O
4292-75-5	1-CYCLOHEXYLHEXANE	C12H24
4292-92-6	PENTYLCYCLOHEXANE	C11H22
4553-62-2	METHYLGLUTARONITRILE	C6H8N2
4669-01-6	1-CYCLOPENTYLPENTADECANE	C20H40
4773-83-5	1,2,3-TRIMETHYLINDENE	C12H14
5171-84-6	3,3,4,4-TETRAMETHYLHEXANE	C10H22
5194-50-3	cis,trans-2,4-HEXADIENE	C6H10

5194-51-4	trans,trans-2,4-HEXADIENE	C6H10
5332-52-5	UNDECYL MERCAPTAN	C11H24S
5408-86-6	2,3-DIBROMOBUTANE	C4H8Br2
5454-79-5	cis-3-METHYLCYCLOHEXANOL	C7H14O
5617-41-4	1-CYCLOHEXYLHEPTANE	C13H26
5634-30-0	1-CYCLOPENTYLDODECANE	C17H34
5881-17-4	3-ETHYLOCTANE	C10H22
5911-04-6	3-METHYLNONANE	C10H22
5989-27-5	D-LIMONENE	C10H16
6006-33-3	1-CYCLOHEXYLTRIDECANE	C19H38
6006-34-4	1-CYCOPENTYLTRIDECANE	C18H36
6012-97-1	TETRACHLOROTHIOPHENE	C4Cl4S
6032-29-7	2-PENTANOL	C5H12O
6064-63-7	HYDROXYCAPROIC ACID	C6H12O3
6094-02-6	2-METHYL-1-HEXENE	C7H14
6117-80-2	cis-2-BUTENE-1,4-DIOL	C4H8O2
6163-66-2	DI-tert-BUTYL ETHER	C8H18O
6294-31-1	HEXYL-SULFIDE	C12H26S
6443-92-1	cis-2-HEPTENE	C7H14
6742-54-7	n-UNDECYLBENZENE	C17H28
6758-23-5	1-CYCLOPENTYLUNDECANE	C16H32
6765-39-5	1-HEPTADECENE	C17H34
6765-39-5	1-HEPTADECYNE	C17H32
6795-87-5	METHYL sec-BUTYL ETHER	C5H12O
6812-39-1	1-CYCLOHEXYLPENTADECANE	C21H42
6863-58-7	DI-sec-BUTYL ETHER	C8H18O
6876-23-9	trans-1,2-DIMETHYLCYCLOHEXANE	C8H16
6892-38-0	1-CYCLOHEXYLHEXADECANE	C22H44
6982-25-8	2,3-BUTANEDIOL	C4H10O2
7064-93-2	1-DECYNE	C10H18
7146-60-3	2,3-DIMETHYLOCTANE	C10H22
7154-79-2	2,2,3,3-TETRAMETHYLPENTANE	C9H20
7154-80-5	3,3,5-TRIMETHYLHEPTANE	C10H22
7220-26-0	3-ETHYL-2,4-DIMETHYLHEXANE	C10H22
7372-86-3	2ETHYL-6-METHYLNAPHTHALENE	C13H14
7415-31-8	1,3-DICHLORO-trans-2-BUTENE	C4H6Cl2
7443-52-9	trans-2-METHYLCYCLOHEXANOL	C7H14O
7443-55-2	trans-3-METHYLCYCLOHEXANOL	C7H14O
7443-70-1	cis-2-METHYLCYCLOHEXANOL	C7H14O
7642-09-3	cis-3-HEXENE	C6H12
7642-10-6	cis-3-HEPTENE	C7H14
7688-21-3	cis-2-HEXENE	C6H12
7731-28-4	cis-4-METHYLCYCLOHEXANOL	C7H14O
7731-29-5	trans-4-METHYLCYCLOHEXANOL	C7H14O
10059-13-9	METHYL-UNDECYL-SULFIDE	C12H26S
10486-19-8	1-TRIDECANAL	C13H26O
12075-68-2	ETHYL ALUMINUM SESQUICHLORIDE	C6H15Al2Cl3
13269-52-8	trans-3-HEXENE	C6H12
13360-61-7	1-PENTADECENE	C15H30
13389-42-9	trans-2-OCTENE	C8H16
13475-78-0	5-ETHYL-2-METHYLHEPTANE	C10H22
13475-81-5	2,2,3,3-TETRAMETHYLHEXANE	C10H22
13952-84-6	sec-BUTYLAMINE	C4H11N
14686-13-6	trans-2-HEPTENE	C7H14
14686-14-7	trans-3-HEPTENE	C7H14
14720-74-2	2,2,4-TRIMETHYLHEPTANE	C10H22
14850-23-8	trans-4-OCTENE	C8H16
14919-01-8	trans-3-OCTENE	C8H16
15869-80-4	3-ETHYLHEPTANE	C9H20
15869-85-9	5-METHYLNONANE	C10H22
15869-86-0	4-ETHYLOCTANE	C10H22
15869-87-1	2,2-DIMETHYLOCTANE	C10H22
15869-89-3	2,5-DIMETHYLOCTANE	C10H22
15869-92-8	3,4-DIMETHYLOCTANE	C10H22
15869-93-9	3,5-DIMETHYLOCTANE	C10H22
15869-94-0	3,6-DIMETHYLOCTANE	C10H22
15869-95-1	4,4-DIMETHYLOCTANE	C10H22
15869-96-2	4,5-DIMETHYLOCTANE	C10H22
15896-91-0	3-ETHYL-4-METHYLHEPTANE	C10H22
16747-25-4	2,2,3-TRIMETHYLHEXANE	C9H20
16747-26-5	2,2,4-TRIMETHYLHEXANE	C9H20
16747-28-7	2,3,3-TRIMETHYLHEXANE	C9H20
16747-30-1	2,4,4-TRIMETHYLHEXANE	C9H20
16747-31-2	3,3,4-TRIMETHYLHEXANE	C9H20
16747-32-3	2,2-DIMETHYL-3-ETHYLPENTANE	C9H20
16747-33-4	3-ETHYL-2,3-DIMETHYLPENTANE	C9H20
16747-38-9	2,3,3,4-TETRAMETHYLPENTANE	C9H20
16747-42-5	2,2,4,5-TETRAMETHYLHEXANE	C10H22
16747-44-7	2,2,3,3,4-PENTAMETHYLPENTANE	C10H22
16747-45-8	2,2,3,4,4-PENTAMETHYLPENTANE	C10H22
16747-50-5	1-METHYL-1-ETHYLCYCLOPENTANE	C8H16
16789-46-1	3-ETHYL-2-METHYLHEXANE	C9H20
17301-94-9	4-METHYLNONANE	C10H22

17302-01-1	3-ETHYL-3-METHYLHEPTANE	C10H22
17302-02-2	3,3-DIETHYLHEXANE	C10H22
17302-4-4	4-ETHYL-4-METHYLHEPTANE	C10H22
18435-45-5	1-NONADECENE	C19H38
19398-77-7	3,4-DIETHYLHEXANE	C10H22
20278-84-6	2,4,5-TRIMETHYLHEPTANE	C10H22
20278-85-7	2,3,5-TRIMETHYLHEPTANE	C10H22
20291-60-5	HEXYL-METHYL-SULFIDE	C7H16S
20291-91-2	3-ETHYL-2,2-DIMETHYLHEXANE	C10H22
20291-95-6	2,2,5-TRIMETHYLHEXANE	C10H22
25154-52-3	NONYLPHENOL	C15H24O
25265-71-8	DIPROPYLENE GLYCOL	C6H14O3
26186-01-6	1-NONADECYNE	C19H36
26186-02-7	1-TRIDECYNE	C13H24
26438-26-6	1-n-NONYLNAPHTHALENE	C19H26
26438-27-7	1-n-DECYLNAPHTHALENE	C20H28
26471-62-5	TOLUENE DIISOCYANATE	C9H6N2O2
26519-91-5	METHYLCYCLOPENTADIENE	C6H8
26761-40-0	DIISODECYL PHTHALATE	C28H46O4
27154-33-2	1,1,1-TRICHLOROFLUOROETHANE	C2H2Cl3F
27554-26-3	DIISOOCTYL PHTHALATE	C24H38O4
36653-82-4	1-HEXADECANOL	C16H34O
50623-57-9	n-BUTYL NONANOATE	C13H26O2
51750-65-3	2,2,4,4-TETRAMETHYLHEXANE	C10H22
52896-88-5	4-ETHYL-2-METHYLHEPTANE	C10H22
52896-89-6	4-ETHYL-3-METHYLHEPTANE	C10H22
52896-90-9	3-ETHYL-5-METHYLHEPTANE	C10H22
52896-92-1	2,2,3-TRIMETHYLHEPTANE	C10H22
52896-93-2	2,3,3-TRIMETHYLHEPTANE	C10H22
52896-95-4	2,3,4-TRIMETHYLHEPTANE	C10H22
52897-00-4	3-ETHYL-2,3-DIMETHYLHEXANE	C10H22
52897-01-5	4-ETHYL-2,3-DIMETHYLHEXANE	C10H22
52897-03-7	4-ETHYL-2,4-DIMETHYLHEXANE	C10H22
52897-04-8	3-ETHYL-2,5-DIMETHYLHEXANE	C10H22
52897-05-9	4-ETHYL-3,3-DIMETHYLHEXANE	C10H22
52897-06-0	3-ETHYL-3,4-DIMETHYLHEXANE	C10H22
52897-08-2	2,2,3,4-TETRAMETHYLHEXANE	C10H22
52897-09-3	2,2,3,5-TETRAMETHYLHEXANE	C10H22
52897-10-6	2,3,3,4-TETRAMETHYLHEXANE	C10H22
52897-11-7	2,3,3,5-TETRAMETHYLHEXANE	C10H22
52897-12-8	2,3,4,4-TETRAMETHYLHEXANE	C10H22
52897-15-1	2,3,4,5-TETRAMETHYLHEXANE	C10H22
52897-16-2	3,3-DIETHYL-2-METHYLPENTANE	C10H22
52897-17-3	3-ETHYL-2,2,3-TRIMETHYLPENTANE	C10H22
52897-19-5	3-ETHYL-2,3,4-TRIMETHYLPENTANE	C10H22
55505-26-5	ISODECANOL	C10H22O
70419-06-6	2-NONANOL	C9H20O
---------	sec-BUTYL GLYCOLATE	C6H12O3
---------	THIACYCLOHEPTANE	C6H12S
---------	BUTYL-PROPYL-SULFIDE	C7H16S
---------	ETHYL-PENTYL-SULFIDE	C7H16S
---------	ETHYL-HEXYL-SULFIDE	C8H18S
---------	HEPTYL-METHYL-SULFIDE	C8H18S
---------	PENTYL-PROPYL-SULFIDE	C8H18S
---------	BUTYL-PENTYL-SULFIDE	C9H20S
---------	ETHYL-HEPTYL-SULFIDE	C9H20S
---------	HEXYL-PROPYL-SULFIDE	C9H20S
---------	METHYL-OCTYL-SULFIDE	C9H20S
---------	4-ISOPROPYLHEPTANE	C10H22
---------	3-ETHYL-2-METHYLHEPTANE	C10H22
---------	3-ISOPROPYL-2-METHYLHEXANE	C10H22
---------	4-ETHYL-2,2-DIMETHYLHEXANE	C10H22
---------	2,4-DIMETHYL-3-ISOPROPYLPENTANE	C10H22
---------	3-ETHYL-2,2,4-TRIMETHYLPENTANE	C10H22
---------	BUTYL-HEXYL-SULFIDE	C10H22S
---------	ETHYL-OCTYL-SULFIDE	C10H22S
---------	HEPTYL-PROPYL-SULFIDE	C10H22S
---------	METHYL-NONYL-SULFIDE	C10H22S
---------	PENTYL-SULFIDE	C10H22S
---------	1-CYCLOPENTYLHEXANE	C11H22
---------	BUTYL-HEPTYL-SULFIDE	C11H24S
---------	DECYL-METHYL-SULFIDE	C11H24S
---------	ETHYL-NONYL-SULFIDE	C11H24S
---------	OCTYL-PROPYL-SULFIDE	C11H24S
---------	1,2,3-TRIETHYLBENZENE	C12H18
---------	1-CYCLOPENTYLHEPTANE	C12H24
---------	BUTYL-OCTYL-SULFIDE	C12H26S
---------	DECYL-ETHYL-SULFIDE	C12H26S
---------	NONYL-PROPYL-SULFIDE	C12H26S
---------	HEXYL-DISULFIDE	C12H26S2
---------	2ETHYL-3-METHYLNAPHTHALENE	C13H14
---------	2ETHYL-7-METHYLNAPHTHALENE	C13H14
---------	BUTYL-NONYL-SULFIDE	C13H28S

---------	DECYL-PROPYL-SULFIDE	C13H28S
---------	DODECYL-METHYL-SULFIDE	C13H28S
---------	ETHYL-UNDECYL-SULFIDE	C13H28S
---------	1-TRIDECANETHIOL	C13H28S
---------	2-BUTYLNAPHTHALENE	C14H16
---------	1,2,3,5-TETRAETHYLBENZENE	C14H22
---------	BUTYL-DECYL-SULFIDE	C14H30S
---------	DODECYL-ETHYL-SULFIDE	C14H30S
---------	HEPTYL-SULFIDE	C14H30S
---------	METHYL-TRIDECYL-SULFIDE	C14H30S
---------	PROPYL-UNDECYL-SULFIDE	C14H30S
---------	HEPTYL-DISULFIDE	C14H30S2
---------	BUTYL-UNDECYL-SULFIDE	C15H32S
---------	DODECYL-PROPYL-SULFIDE	C15H32S
---------	ETHYL-TRIDECYL-SULFIDE	C15H32S
---------	METHYL-TETRADECYL-SULFIDE	C15H32S
---------	1-PENTADECANETHIOL	C15H32S
---------	BUTYL-DODECYL-SULFIDE	C16H34S
---------	ETHYL-TETRADECYL-SULFIDE	C16H34S
---------	METHYL-PENTADECYL-SULFIDE	C16H34S
---------	OCTYL-SULFIDE	C16H34S
---------	PROPYL-TRIDECYL-SULFIDE	C16H34S
---------	OCTYL-DISULFIDE	C16H34S2
---------	1-CYCLOHEXYLUNDECANE	C17H34
---------	BUTYL-TRIDECYL-SULFIDE	C17H36S
---------	ETHYL-PENTADECYL-SULFIDE	C17H36S
---------	HEXADECYL-METHYL-SULFIDE	C17H36S
---------	PROPYL-TETRADECYL-SULFIDE	C17H36S
---------	1-HEPTADECANETHIOL	C17H36S
---------	BUTYL-TETRADECYL-SULFIDE	C18H38S
---------	ETHYL-HEXADECYL-SULFIDE	C18H38S
---------	HEPTADECYL-METHYL-SULFIDE	C18H38S
---------	NONYL-SULFIDE	C18H38S
---------	PENTADECYL-PROPYL-SULFIDE	C18H38S
---------	NONYL-DISULFIDE	C18H38S2
---------	BUTYL-PENTADECYL-SULFIDE	C19H40S
---------	ETHYL-HEPTADECYL-SULFIDE	C19H40S
---------	HEXADECYL-PROPYL-SULFIDE	C19H40S
---------	METHYL-OCTADECYL-SULFIDE	C19H40S
---------	1-NONADECANETHIOL	C19H40S
---------	1-PHENYLTETRADECANE	C20H34
---------	BUTYL-HEXADECYL-SULFIDE	C20H42S
---------	DECYL-SULFIDE	C20H42S
---------	ETHYL-OCTADECYL-SULFIDE	C20H42S
---------	HEPTADECYL-PROPYL-SULFIDE	C20H42S
---------	METHYL-NONADECYL-SULFIDE	C20H42S
---------	1-EICOSANETHIOL	C20H42S
---------	DECYL-DISULFIDE	C20H42S2

INORGANICS

The following compilation for inorganics provides the compound list by CAS registry number, name, and chemical formula.

CAS	Name	Formula
57-13-6	UREA	CH4N2O
62-56-6	THIOUREA	CH4N2S
74-90-8	HYDROGEN CYANIDE	HCN
75-44-5	PHOSGENE	CCl2O
75-15-0	CARBON DISULFIDE	CS2
124-38-9	CARBON DIOXIDE	CO2
143-33-9	SODIUM CYANIDE	NaCN
302-01-2	HYDRAZINE	N2H4
353-50-4	CARBONYL FLUORIDE	CF2O
460-19-5	CYANOGEN	C2N2
463-58-1	CARBONYL SULFIDE	COS
506-77-4	CYANOGEN CHLORIDE	CNCl
506-68-3	CYANOGEN BROMIDE	CNBr
507-16-4	THIONYL BROMIDE	SOBr2
630-08-0	CARBON MONOXIDE	CO
1111-78-0	AMMONIUM CARBAMATE	N2H6CO2
1306-19-0	CADMIUM OXIDE	CdO
1309-48-4	MAGNESIUM OXIDE	MgO
1310-58-3	POTASSIUM HYDROXIDE	KOH
1310-73-2	SODIUM HYDROXIDE	NaOH
1313-27-5	MOLYBDENUM OXIDE	MoO3
1314-13-2	ZINC OXIDE	ZnO
1314-80-3	PHOSPHORUS PENTASULFIDE	P4S10
1314-87-0	LEAD SULFIDE	PbS
1314-11-0	STRONTIUM OXIDE	SrO
1314-24-5	PHOSPHORUS TRIOXIDE	P4O6
1314-68-7	RHENIUM HEPTOXIDE	Re2O7
1317-36-8	LEAD OXIDE	PbO

CAS	Name	Formula
1317-98-2	ANTIMONY TRIOXIDE	Sb2O3
1327-53-3	ARSENIC TRIOXIDE	As2O3
1333-74-0	HYDROGEN	H2
1336-21-6	AMMONIUM HYDROXIDE	NH5O
1344-28-1	ALUMINUM OXIDE	Al2O3
1495-50-7	CYANOGEN FLUORIDE	CNF
1590-87-0	DISILANE	Si2H6
1603-84-5	CARBON OXYSELENIDE	COSe
2551-62-4	SULFUR HEXAFLUORIDE	SF6
2696-92-6	NITROSYL CHLORIDE	NOCl
3982-91-0	PHOSPHORUS THIOCHLORIDE	PSCl3
4109-96-0	DICHLOROSILANE	SiH2Cl2
5329-14-6	SULFAMIC ACID	H3NO3S
5951-19-9	CARBON SELENOSULFIDE	CSeS
7429-90-5	ALUMINUM	Al
7439-96-5	MANGANESE	Mn
7439-95-4	MAGNESIUM	Mg
7439-99-8	NEPTUNIUM	Np
7439-97-6	MERCURY	Hg
7439-98-7	MOLYBDENUM	Mo
7439-89-6	IRON	Fe
7439-90-9	KRYPTON	Kr
7439-91-0	LANTHANUM	La
7439-93-2	LITHIUM	Li
7439-92-1	LEAD	Pb
7439-88-5	IRIDIUM	Ir
7440-66-6	ZINC	Zn
7440-67-7	ZIRCONIUM	Zr
7440-65-5	YTTRIUM	Yt
7440-63-3	XENON	Xe
7440-64-4	YTTERBIUM	Yb
7440-73-5	FRANCIUM	Fr
7440-74-6	INDIUM	In
7440-70-2	CALCIUM	Ca
7440-68-8	ASTATINE	At
7440-69-9	BISMUTH	Bi
7440-62-2	VANADIUM	V
7440-24-6	STRONTIUM	Sr
7440-25-7	TANTALUM	Ta
7440-22-4	SILVER	Ag
7440-23-5	SODIUM	Na
7440-26-8	TECNNETIUM	Tc
7440-01-9	NEON	Ne
7440-02-0	NICKEL	Ni
7440-28-0	THALLIUM	Tl
7440-00-8	NEODYMIUM	Nd
7440-38-2	ARSENIC	As
7440-39-3	BARIUM	Ba
7440-36-0	ANTIMONY	Sb
7440-37-1	ARGON	Ar
7440-30-4	THULIUM	Tm
7440-20-2	SCANDIUM	Sc
7440-21-3	SILICON	Si
7440-31-5	TIN	Sn
7440-32-6	TITANIUM	Ti
7440-47-3	CHROMIUM	Cr
7440-19-9	SAMARIUM	Sm
7440-17-7	RUBIDIUM	Rb
7440-18-8	RUTHENIUM	Ru
7440-46-2	CESIUM	Cs
7440-42-8	BORON	B
7440-41-7	BERYLLIUM	Be
7440-44-0	CARBON	C
7440-43-9	CADMIUM	Cd
7440-05-3	PALLADIUM	Pd
7440-06-4	PLATINUM	Pt
7440-03-1	NIOBIUM	Nb
7440-04-2	OSMIUM	Os
7440-08-6	POLONIUM	Po
7440-15-5	RHENIUM	Re
7440-16-6	RHODIUM	Rh
7440-09-7	POTASSIUM	K
7440-14-4	RADIUM	Ra
7440-54-2	GADOLINIUM	Gd
7440-55-3	GALLIUM	Ga
7440-53-1	EUROPIUM	Eu
7440-48-4	COBALT	Co
7440-50-8	COPPER	Cu
7440-56-4	GERMANIUM	Ge
7440-61-1	URANIUM	U
7440-33-7	TUNGSTEN	W
7440-59-7	HELIUM-4	He
7440-57-5	GOLD	Au

CAS Number	Name	Formula	CAS Number	Name	Formula
7440-58-6	HAFNIUM	Hf	7784-18-1	ALUMINUM FLUORIDE	AlF3
7446-11-9	SULFUR TRIOXIDE	SO3	7786-30-3	MAGNESIUM CHLORIDE	MgCl2
7446-70-0	ALUMINUM CHLORIDE	AlCl3	7787-70-4	CUPROUS BROMIDE	CuBr
7446-08-4	SELENIUM DIOXIDE	SeO2	7787-60-2	BISMUTH TRICHLORIDE	BiCl3
7446-09-5	SULFUR DIOXIDE	SO2	7787-69-1	CESIUM BROMIDE	CsBr
7447-41-8	LITHIUM CHLORIDE	LiCl	7787-46-4	BERYLLIUM BROMIDE	BeBr2
7447-40-7	POTASSIUM CHLORIDE	KCl	7787-47-5	BERYLLIUM CHLORIDE	BeCl2
7447-39-4	CUPRIC CHLORIDE	CuCl2	7787-49-7	BERYLLIUM FLUORIDE	BeF2
7487-94-7	MERCURIC CHLORIDE	HgCl2	7787-53-3	BERYLLIUM IODIDE	BeI2
7550-45-0	TITANIUM TETRACHLORIDE	TiCl4	7787-58-8	BISMUTH TRIBROMIDE	BiBr3
7550-35-8	LITHIUM BROMIDE	LiBr	7789-67-5	STANNIC BROMIDE	SnBr4
7553-56-2	IODINE	I2	7789-61-9	ANTIMONY TRIBROMIDE	SbBr3
7601-90-3	PERCHLORIC ACID	ClHO4	7789-60-8	PHOSPHORUS TRIBROMIDE	PBr3
7616-94-6	PERCHLORYL FLUORIDE	ClFO3	7789-75-5	CALCIUM FLUORIDE	CaF2
7632-51-1	VANADIUM TETRACHLORIDE	VCl4	7789-40-4	THALLOUS BROMIDE	TlBr
7637-07-2	BORON TRIFLUORIDE	BF3	7789-60-8	PHOSPHORUS THIOBROMIDE	PSBr3
7646-85-7	ZINC CHLORIDE	ZnCl2	7789-30-2	BROMINE PENTAFLUORIDE	BrF5
7646-79-9	COBALT CHLORIDE	CoCl2	7789-57-3	TRIBROMOSILANE	SiHBr3
7646-78-8	STANNIC CHLORIDE	SnCl4	7789-20-0	DEUTERIUM OXIDE	D2O
7647-15-6	SODIUM BROMIDE	NaBr	7789-39-1	RUBIDIUM BROMIDE	RbBr
7647-17-8	CESIUM CHLORIDE	CsCl	7789-17-5	CESIUM IODIDE	CsI
7647-18-9	ANTIMONY PENTACHLORIDE	SbCl5	7789-23-3	POTASSIUM FLUORIDE	KF
7647-01-0	HYDROGEN CHLORIDE	HCl	7789-47-1	MERCURIC BROMIDE	HgBr2
7647-14-5	SODIUM CHLORIDE	NaCl	7789-25-5	NITROSYL FLUORIDE	NOF
7664-93-9	SULFURIC ACID	H2SO4	7789-24-4	LITHIUM FLUORIDE	LiF
7664-41-7	AMMONIA	NH3	7790-80-9	CADMIUM IODIDE	CdI2
7664-39-3	HYDROGEN FLUORIDE	HF	7790-89-8	CHLORINE MONOFLUORIDE	ClF
7681-49-4	SODIUM FLUORIDE	NaF	7790-94-5	CHLOROSULFONIC ACID	ClHO3S
7681-11-0	POTASSIUM IODIDE	KI	7790-91-2	CHLORINE TRIFLUORIDE	ClF3
7681-65-4	COPPER IODIDE	CuI	7790-30-9	THALLOUS IODIDE	TlI
7681-82-5	SODIUM IODIDE	NaI	7790-29-6	RUBIDIUM IODIDE	RbI
7697-37-2	NITRIC ACID	HNO3	7790-44-5	ANTIMONY TRIIODIDE	SbI3
7704-34-9	SULFUR	S	7790-79-6	CADMIUM FLUORIDE	CdF2
7705-08-0	FERRIC CHLORIDE	FeCl3	7790-47-8	STANNIC IODIDE	SnI4
7719-12-2	PHOSPHORUS TRICHLORIDE	PCl3	7791-23-3	SELENIUM OXYCHLORIDE	SeOCl2
7719-09-7	THIONYL CHLORIDE	SOCl2	7791-21-1	CHLORINE MONOXIDE	Cl2O
7722-84-1	HYDROGEN PEROXIDE	H2O2	7791-11-9	RUBIDIUM CHLORIDE	RbCl
7723-14-0	PHOSPHORUS - WHITE	P	7791-25-5	SULFURYL CHLORIDE	SO2Cl2
7726-95-6	BROMINE	Br2	7803-52-3	STIBINE	SbH3
7727-15-3	ALUMINUN BROMIDE	AlBr3	7803-62-5	SILANE	SiH4
7727-37-9	NITROGEN	N2	7803-51-2	PHOSPHINE	PH3
7727-18-6	VANADIUM OXYTRICHLORIDE	VOCl3	7803-49-8	HYDROXYLAMINE	NH3O
7732-18-5	WATER	H2O	10024-97-2	NITROUS OXIDE	N2O
7733-02-0	ZINC SULFATE	ZnSO4	10025-85-1	NITROGEN TRICHLORIDE	NCl3
7757-82-6	SODIUM SULFATE	Na2SO4	10025-87-3	PHOSPHORUS OXYCHLORIDE	POCl3
7758-94-3	FERROUS CHLORIDE	FeCl2	10025-91-9	ANTIMONY TRICHLORIDE	SbCl3
7758-95-4	LEAD CHLORIDE	PbCl2	10025-67-9	SULFUR MONOCHLORIDE	S2Cl2
7758-02-3	POTASSIUM BROMIDE	KBr	10025-78-2	TRICHLOROSILANE	SiHCl3
7758-89-6	CUPROUS CHLORIDE	CuCl	10026-03-6	SELENIUM TETRACHLORIDE	SeCl4
7772-99-8	STANNOUS CHLORIDE	SnCl2	10026-11-6	ZIRCONIUM CHLORIDE	ZrCl4
7773-01-5	MANGANESE CHLORIDE	MnCl2	10026-13-8	PHOSPHORUS PENTACHLORIDE	PCl5
7774-29-0	MERCURIC IODIDE	HgI2	10026-04-7	SILICON TETRACHLORIDE	SiCl4
7782-49-2	SELENIUM	Se	10026-07-0	TELLURIUM TETRACHLORIDE	TeCl4
7782-50-5	CHLORINE	Cl2	10028-18-9	NICKEL FLUORIDE	NiF2
7782-65-2	GERMANIUM TETRAHYDRIDE	GeH4	10028-15-6	OZONE	O3
7782-39-0	DEUTERIUM	D2	10031-22-8	LEAD BROMIDE	PbBr2
7782-41-4	FLUORINE	F2	10034-85-2	HYDROGEN IODIDE	HI
7782-44-7	OXYGEN	O2	10035-10-6	HYDROGEN BROMIDE	HBr
7783-29-1	TETRASILANE	Si4H10	10036-47-2	TETRAFLUOROHYDRAZINE	N2F4
7783-90-6	SILVER CHLORIDE	AgCl	10038-98-9	GERMANIUM CHLORIDE	GeCl4
7783-82-6	TUNGSTEN FLUORIDE	WF6	10043-92-2	RADON	Rn
7783-96-2	SILVER IODIDE	AgI	10043-35-3	BORIC ACID	BH3O3
7783-07-5	HYDROGEN SELENIDE	H2Se	10043-01-3	ALUMINUM SULFATE	Al2S3O12
7783-06-4	HYDROGEN SULFIDE	H2S	10049-04-4	CHLORINE DIOXIDE	ClO2
7783-79-1	SELENIUM HEXAFLUORIDE	SeF6	10101-63-0	LEAD IODIDE	PbI2
7783-77-9	MOLYBDENUM FLUORIDE	MoF6	10102-43-9	NITRIC OXIDE	NO
7783-61-1	SILICON TETRAFLUORIDE	SiF4	10102-03-1	NITROGEN PENTOXIDE	N2O5
7783-81-5	URANIUM FLUORIDE	UF6	10102-44-0	NITROGEN DIOXIDE	NO2
7783-80-4	TELLURIUM HEXAFLUORIDE	TeF6	10108-64-2	CADMIUM CHLORIDE	CdCl2
7783-49-5	ZINC FLUORIDE	ZnF2	10294-48-1	CHLORINE HEPTOXIDE	Cl2O7
7783-54-2	NITROGEN TRIFLUORIDE	NF3	10294-34-5	BORON TRICHLORIDE	BCl3
7783-41-7	FLUORINE OXIDE	F2O	10294-33-4	BORON TRIBROMIDE	BBr3
7783-46-2	LEAD FLUORIDE	PbF2	10377-51-2	LITHIUM IODIDE	LiI
7783-09-1	HYDROGEN TELLURIDE	H2Te	10544-73-7	NITROGEN TRIOXIDE	N2O3
7783-26-8	TRISILANE	Si3H8	10544-72-6	NITROGEN TETRAOXIDE	N2O4
7783-60-0	SULFUR TETRAFLUORIDE	SF4	12027-06-4	AMMONIUM IODIDE	NH4I
7784-36-3	ARSENIC PENTAFLUORIDE	AsF5	12124-97-9	AMMONIUM BROMIDE	NH4Br
7784-34-1	ARSENIC TRICHLORIDE	AsCl3	12124-99-1	AMMONIUM HYDROGENSULFIDE	NH5S
7784-35-2	ARSENIC TRIFLUORIDE	AsF3	12125-02-9	AMMONIUM CHLORIDE	NH4Cl
7784-42-1	ARSINE	AsH3	12125-09-6	PHOSPHONIUM IODIDE	PH4I
7784-45-4	ARSENIC TRIIODIDE	AsI3	12211-52-8	AMMONIUM CYANIDE	N2H4C
7784-33-0	ARSENIC TRIBROMIDE	AsBr3	13007-92-6	CHROMIUM CARBONYL	CrC6O6
7784-23-8	ALUMINUM IODIDE	AlI3	13400-13-0	CESIUM FLUORIDE	CsF

13446-74-7	RUBIDIUM FLUORIDE	RbF
13450-90-3	GALLIUM TRICHLORIDE	GaCl3
13450-92-5	GERMANIUM BROMIDE	GeBr4
13463-39-3	NICKEL CARBONYL	NiC4O4
13463-40-6	IRON PENTACARBONYL	FeC5O5
13465-71-9	TRIFLUOROSILANE	SiHF3
13465-77-5	HEXACHLORODISILANE	Si2Cl6
13465-78-6	MONOCHLOROSILANE	SiH3Cl
13465-73-1	MONOBROMOSILANE	SiH3Br
13494-80-9	TELLURIUM	Te
13537-33-2	MONOFLUOROSILANE	SiH3F
13597-73-4	DISILOXANE	Si2OH6
13598-42-0	IODOSILANE	SiH3I
13637-63-3	CHLORINE PENTAFLUORIDE	ClF5
13760-02-6	DIIODOSILANE	SiH2I2
13768-94-0	DIBROMOSILANE	SiH2Br2
13777-25-8	ZIRCONIUM BROMIDE	ZrBr4
13818-89-8	DIGERMANE	Ge2H6
13824-36-7	DIFLUOROSILANE	SiH2F2
13986-26-0	ZIRCONIUM IODIDE	ZrI4
14096-82-3	COBALT NITROSYL TRICARBONYL	CoNC3O4
14691-44-2	TRIGERMANE	Ge3H8
14762-55-1	HELIUM-3	He
14808-60-7	SILICON DIOXIDE	SiO2
14965-52-7	TRICHLOROFLUOROSILANE	SiCl3F
14977-61-8	CHROMIUM OXYCHLORIDE	CrO2Cl2
16752-60-6	PHOSPHORUS PENTOXIDE	P4O10
16921-96-3	IODINE HEPTAFLUORIDE	IF7
16962-07-5	ALUMINUM BOROHYDRIDE	AlB3H12
17440-85-6	BERYLLIUM BOROHYDRIDE	BeB2H8
17702-41-9	DECABORANE	B10H14
18283-93-7	TETRABORANE	B4H10
18356-71-3	DICHLORODIFLUOROSILANE	SiCl2F2
18433-84-6	PENTABORANE	B5H9
19287-45-7	DIBORANE	B2H6
20816-12-0	OSMIUM TETROXIDE - YELLOW	OsO4
24567-53-1	PHOSPHONIUM CHLORIDE	PH4Cl
----------	BORINE CARBONYL	BH2CO
----------	DEUTERODIBORANE	B2D6
----------	DIBORANE HYDROBROMIDE	B2H5Br
----------	BORINE TRIAMINE	B3N3H6
----------	TETRAHYDROPENTABORANE	B5H11
----------	CARBON SUBSULFIDE	C3S2
----------	COLUMBIUM FLUORIDE	CbF5
----------	DEUTERIUM CYANIDE	DCN
----------	TRICHLORO GERMANE	GeHCl3
----------	HYDROGEN DISULFIDE	H2S2
----------	LUTECIUM	Lu
----------	HEAVY AMMONIA	ND3
----------	OSMIUM OXIDE PENTAFLUORIDE	OsOF5
----------	OSMIUM TETROXIDE - WHITE	OsO4
----------	PHOSPHORUS DICHLORIDE TRIFLUORIDE	PCl2F3
----------	PHOSPHONIUM BROMIDE	PH4Br
----------	RUTHENIUM PENTAFLUORIDE	RuF5
----------	SULFUROUS OXYFLUORIDE	SOF2
----------	BROMODICHLOROFLUOROSILANE	SiBrCl2F
----------	TRIFLUOROBROMOSILANE	SiBrF3
----------	DIBROMOCHLOROFLUOROSILANE	SiBr2ClF
----------	TRIFLUOROCHLOROSILANE	SiClF3
----------	HEXAFLUORODISILANE	Si2F6
----------	DISILANYL CHLORIDE	Si2H5Cl
----------	TRICHLOROTRIFLUORODISILOXANE	Si2OCl3F3
----------	HEXACHLORODISILOXANE	Si2OCl6
----------	OCTACHLOROTRISILANE	Si3Cl8
----------	TRISILAZANE	Si3H9N
----------	STANNIC HYDRIDE	SnH4

Appendix E

COMPOUND LIST BY NAME AND SYNONYM

ORGANICS The following compilation for organics provides the compound list by name (synonym), formula in table, name in table, and CAS registry number. To locate property data for a specific compound, the user should use the name (synonym) to identify the formula and name. This formula and name are then used in the data tabulations to locate the property data of interest.

Name (Synonym)	Formula	Name in Tables	CAS Reg. No
2-AB	C4H11N	sec-BUTYLAMINE	13952-84-6
ABIETIC ACID	C20H30O2	ABIETIC ACID	514-10-3
ACENAPHTHENE	C12H10	ACENAPHTHENE	83-32-9
ACENAPHTHYLENE, 1,2-DIHYDRO-	C12H10	ACENAPHTHENE	83-32-9
ACETAAL	C6H14O2	ACETAL	105-57-7
ACETAL	C6H14O2	ACETAL	105-57-7
ACETALDEHYD	C2H4O	ACETALDEHYDE	75-07-0
ACETALDEHYDE	C2H4O	ACETALDEHYDE	75-07-0
ACETALDEHYDE, TRICHLORO	C2HCl3O	TRICHLOROACETALDEHYDE	75-87-6
ACETALDEHYDE, TRIMER	C6H12O3	PARALDEHYDE	123-63-7
ACETAL DIETHYLIQUE	C6H14O2	ACETAL	105-57-7
ACETALE	C6H14O2	ACETAL	105-57-7
ACETAMIDE	C2H5NO	ACETAMIDE	60-35-5
ACETAMIDE, N-PHENYL-	C8H9NO	ACETANILIDE	103-84-4
ACETAMIDOBENZENE	C8H9NO	ACETANILIDE	103-84-4
ACETANHYDRIDE	C4H6O3	ACETIC ANHYDRIDE	108-24-7
ACETANIL	C8H9NO	ACETANILIDE	103-84-4
ACETANILID	C8H9NO	ACETANILIDE	103-84-4
ACETANILIDE	C8H9NO	ACETANILIDE	103-84-4
ACETATE dAMYLE	C7H14O2	n-PENTYL ACETATE	628-63-7
ACETATE de BUTYLE	C6H12O2	n-BUTYL ACETATE	123-86-4
ACETATE de BUTYLE SECONDAIRE	C6H12O2	sec-BUTYL ACETATE	105-46-4
ACETATE de CELLOSOLVE	C6H12O3	2-ETHOXYETHYL ACETATE	111-15-9
ACETATE de METHYLE	C3H6O2	METHYL ACETATE	79-20-9
ACETATE de PROPYLE NORMAL	C5H10O2	n-PROPYL ACETATE	109-60-4
ACETATE dETHYLGLYCOL	C6H12O3	2-ETHOXYETHYL ACETATE	111-15-9
ACETATE de VINYLE	C4H6O2	VINYL ACETATE	108-05-4
ACETATE dISOBUTYLE	C6H12O2	ISOBUTYL ACETATE	110-19-0
ACETATE dISOPROPYLE	C5H10O2	ISOPROPYL ACETATE	108-21-4
ACETDIMETHYLAMIDE	C4H9NO	N,N-DIMETHYLACETAMIDE	127-19-5
ACETENE	C2H4	ETHYLENE	74-85-1
ACETIC ACID	C2H4O2	ACETIC ACID	64-19-7
ACETIC ACID ALLYL ESTER	C5H8O2	ALLYL ACETATE	591-87-7
ACETIC ACID AMIDE	C2H5NO	ACETAMIDE	60-35-5
ACETIC ACID, AMYL ESTER	C7H14O2	n-PENTYL ACETATE	628-63-7
ACETIC ACID, ANHYDRIDE	C4H6O3	ACETIC ANHYDRIDE	108-24-7
ACETIC ACID ANILIDE	C8H9NO	ACETANILIDE	103-84-4
ACETIC ACID-2-BUTOXY ESTER	C6H12O2	sec-BUTYL ACETATE	105-46-4
ACETIC ACID n-BUTYL ESTER	C6H12O2	n-BUTYL ACETATE	123-86-4
ACETIC ACID-tert-BUTYL ESTER	C6H12O2	tert-BUTYL ACETATE	540-88-5
ACETIC ACID CHLORIDE	C2H3ClO	ACETYL CHLORIDE	75-36-5
ACETIC ACID, DIMETHYL-	C4H8O2	ISOBUTYRIC ACID	79-31-2
ACETIC ACID DIMETHYLAMIDE	C4H9NO	N,N-DIMETHYLACETAMIDE	127-19-5
ACETIC ACID-1,1-DIMETHYLETHYL ESTER	C6H12O2	tert-BUTYL ACETATE	540-88-5
ACETIC ACID, ETHENYL ESTER	C4H6O2	VINYL ACETATE	108-05-4
ACETIC ACID-2-ETHOXYETHYL ESTER	C6H12O3	2-ETHOXYETHYL ACETATE	111-15-9
ACETIC ACID a-ETHYHEXYL ESTER	C10H20O2	2-ETHYLHEXYL ACETATE	103-09-3
ACETIC ACID, ETHYLENE ETHER	C4H6O2	VINYL ACETATE	108-05-4
ACETIC ACID HEXYL ESTER	C8H16O2	n-HEXYL ACETATE	142-92-7
ACETIC ACID, ISOBUTYL ESTER	C6H12O2	ISOBUTYL ACETATE	110-19-0
ACETIC ACID, ISOPENTYL ESTER	C7H14O2	ISOPENTYL ACETATE	123-92-2
ACETIC ACID ISOPROPYL ESTER	C5H10O2	ISOPROPYL ACETATE	108-21-4
ACETIC ACID METHYL ESTER	C3H6O2	METHYL ACETATE	79-20-9
ACETIC ACID-1-METHYLETHYL ESTER	C5H10O2	ISOPROPYL ACETATE	108-21-4
ACETIC ACID-1-METHYLPROPYL ESTER	C6H12O2	sec-BUTYL ACETATE	105-46-4
ACETIC ACID-2-METHYLPROPYL ESTER	C6H12O2	ISOBUTYL ACETATE	110-19-0
ACETIC ACID, 2,2'-OXYBIS-	C4H6O5	DIGLYCOLIC ACID	110-99-6
ACETIC ACID, OXYDI-	C4H6O5	DIGLYCOLIC ACID	110-99-6
ACETIC ACID-2-PROPENYL ESTER	C5H8O2	ALLYL ACETATE	591-87-7
ACETIC ACID, n-PROPYL ESTER	C5H10O2	n-PROPYL ACETATE	109-60-4
ACETIC ACID VINYL ESTER	C4H6O2	VINYL ACETATE	108-05-4
ACETIC ALDEHYDE	C2H4O	ACETALDEHYDE	75-07-0
ACETIC ANHYDRIDE	C4H6O3	ACETIC ANHYDRIDE	108-24-7
ACETIC CHLORIDE	C2H3ClO	ACETYL CHLORIDE	75-36-5

Name (Synonym)	Formula	Name in Tables	CAS Reg. No
ACETIC ETHER	C4H8O2	ETHYL ACETATE	141-78-6
ACETIC OXIDE	C4H6O3	ACETIC ANHYDRIDE	108-24-7
ACETIDIN	C4H8O2	ETHYL ACETATE	141-78-6
ACETIMIDIC ACID	C2H5NO	ACETAMIDE	60-35-5
ACETOACETIC ACID, ETHYL ESTER	C6H10O2	n-PROPYL ACRYLATE	925-60-0
ACETOACETIC ESTER	C6H10O2	n-PROPYL ACRYLATE	925-60-0
ACETOACETIC METHYL ESTER	C5H8O3	METHYL ACETOACETATE	105-45-3
ACETOACETONE	C5H8O2	ACETYLACETONE	123-54-6
ACETOANILIDE	C8H9NO	ACETANILIDE	103-84-4
ACETON	C3H6O	ACETONE	67-64-1
ACETONE	C3H6O	ACETONE	67-64-1
ACETONE CYANOHYDRIN	C4H7NO	ACETONE CYANOHYDRIN	75-86-5
ACETONE OILS	C3H6O	ACETONE	67-64-1
ACETONITRILE	C2H3N	ACETONITRILE	75-05-8
ACETOPHENONE	C8H8O	ACETOPHENONE	98-86-2
ACETOPROPIONIC ACID	C5H8O3	LEVULINIC ACID	123-76-2
ACETOXYETHANE	C4H8O2	ETHYL ACETATE	141-78-6
1-ACETOXYETHYLENE	C4H6O2	VINYL ACETATE	108-05-4
1-ACETOXYPROPANE	C5H10O2	n-PROPYL ACETATE	109-60-4
2-ACETOXYPROPANE	C5H10O2	ISOPROPYL ACETATE	108-21-4
3-ACETOXYPROPENE	C5H8O2	ALLYL ACETATE	591-87-7
ACETYLACETONE	C5H8O2	ACETYLACETONE	123-54-6
ACETYLAMINOBENZENE	C8H9NO	ACETANILIDE	103-84-4
ACETYL ANHYDRIDE	C4H6O3	ACETIC ANHYDRIDE	108-24-7
ACETYLANILINE	C8H9NO	ACETANILIDE	103-84-4
N-ACETYLANILINE	C8H9NO	ACETANILIDE	103-84-4
ACETYLBENZENE	C8H8O	ACETOPHENONE	98-86-2
ACETYL CHLORIDE	C2H3ClO	ACETYL CHLORIDE	75-36-5
ACETYLEN	C2H2	ACETYLENE	74-86-2
ACETYLENE	C2H2	ACETYLENE	74-86-2
ACETYLENE, METHYL-	C3H4	METHYLACETYLENE	74-99-7
ACETYLENE, PHENYL-	C8H6	ETHYNYLBENZENE	536-74-3
ACETYLENE TETRACHLORIDE	C2H2Cl4	1,1,2,2-TETRACHLOROETHANE	79-34-5
ACETYLENE TRICHLORIDE	C2HCl3	TRICHLOROETHYLENE	79-01-6
ACETYL ETHER	C4H6O3	ACETIC ANHYDRIDE	108-24-7
ACETYLFORMIC ACID	C3H4O3	PYRUVIC ACID	127-17-3
ACETYL MERCAPTAN	C2H4OS	THIOACETIC-ACID	507-09-5
ACETYL OXIDE	C4H6O3	ACETIC ANHYDRIDE	108-24-7
b-ACETYLPROPIONIC ACID	C5H8O3	LEVULINIC ACID	123-76-2
ACIDE ACETIQUE	C2H4O2	ACETIC ACID	64-19-7
ACIDE ANISIQUE	C8H8O3	METHYL SALICYLATE	119-36-8
ACIDE BENZOIQUE	C7H6O2	BENZOIC ACID	65-85-0
ACIDE CARBOLIQUE	C6H6O	PHENOL	108-95-2
ACIDE FORMIQUE	CH2O2	FORMIC ACID	64-18-6
ACIDE ISOPHTALIQUE	C8H6O4	ISOPHTHALIC ACID	121-91-5
ACIDE METHYL-o-BENZOIQUE	C8H8O3	METHYL SALICYLATE	119-36-8
ACIDE PHTHALIQUE	C8H6O4	PHTHALIC ACID	88-99-3
ACIDE PROPIONIQUE	C3H6O2	PROPIONIC ACID	79-09-4
ACIDE TEREPHTHALIQUE	C8H6O4	TEREPHTHALIC ACID	100-21-0
ACIDO ACETICO	C2H4O2	ACETIC ACID	64-19-7
ACIDO FORMICO	CH2O2	FORMIC ACID	64-18-6
ACIDO SALICILICO	C7H6O3	SALICYLIC ACID	69-72-7
ACIFLOCTIN	C6H10O4	ADIPIC ACID	124-04-9
ACILETTEN	C6H8O7	CITRIC ACID	77-92-9
ACINETTEN	C6H10O4	ADIPIC ACID	124-04-9
ACINTENE A	C10H16	alpha-PINENE	80-56-8
ACINTENE O	C10H12O	ANETHOLE	104-46-1
ACNA BLACK DF BASE	C12H12N2	p-AMINODIPHENYLAMINE	101-54-2
ACQUINITE	C3H4O	ACROLEIN	107-02-8
ACRALDEHYDE	C3H4O	ACROLEIN	107-02-8
ACRITET	C3H3N	ACRYLONITRILE	107-13-1
ACROLEIC ACID	C3H4O2	ACRYLIC ACID	79-10-7
ACROLEIN	C3H4O	ACROLEIN	107-02-8
ACROLEIN, 2-METHYL-	C4H6O	METHACROLEIN	78-85-3
ACRYLALDEHYDE	C3H4O	ACROLEIN	107-02-8
ACRYLAMIDE	C3H5NO	ACRYLAMIDE	79-06-1
ACRYLATE dETHYLE	C5H8O2	ETHYL ACRYLATE	140-88-5
ACRYLIC ACID	C3H4O2	ACRYLIC ACID	79-10-7
ACRYLIC ACID BUTYL ESTER	C7H12O2	n-BUTYL ACRYLATE	141-32-2
ACRYLIC ACID n-BUTYL ESTER	C7H12O2	n-BUTYL ACRYLATE	141-32-2
ACRYLIC ACID ETHYL ESTER	C5H8O2	ETHYL ACRYLATE	140-88-5
ACRYLIC ACID, GLACIAL	C3H4O2	ACRYLIC ACID	79-10-7
ACRYLIC ACID ISOBUTYL ESTER	C7H12O2	ISOBUTYL ACRYLATE	106-63-8
ACRYLIC ACID, 2-METHYL-	C4H6O2	METHACRYLIC ACID	79-41-4
ACRYLIC ACID, 2-METHYL-, METHYL ESTER	C5H8O2	METHYL METHACRYLATE	80-62-6
ACRYLIC ALDEHYDE	C3H4O	ACROLEIN	107-02-8

Name (Synonym)	Formula	Name in Tables	CAS Reg. No
ACRYLIC AMIDE	C3H5NO	ACRYLAMIDE	79-06-1
ACRYLNITRIL	C3H3N	ACRYLONITRILE	107-13-1
ACRYLON	C3H3N	ACRYLONITRILE	107-13-1
ACRYLONITRILE	C3H3N	ACRYLONITRILE	107-13-1
ACRYLONITRILE MONOMER	C3H3N	ACRYLONITRILE	107-13-1
2-(ACRYLOYLOXY)ETHANOL	C5H8O3	2-HYDROXYETHYL ACRYLATE	818-61-1
ACRYLSAEUREAETHYLESTER	C5H8O2	ETHYL ACRYLATE	140-88-5
ACTIVE ACETYL ACETATE	C6H10O2	n-PROPYL ACRYLATE	925-60-0
ACTIVE DICUMYL PEROXIDE	C18H22O2	DICUMYL PEROXIDE	80-43-3
ACTYLOL	C5H10O3	ETHYL LACTATE	97-64-3
ACYTOL	C5H10O3	ETHYL LACTATE	97-64-3
ADAKANE 12	C12H26	n-DODECANE	112-40-3
ADILACTETTEN	C6H10O4	ADIPIC ACID	124-04-9
ADIPIC ACID	C6H10O4	ADIPIC ACID	124-04-9
ADIPIC ACID, DIHEXYL ESTER	C18H34O4	DIHEXYL ADIPATE	110-33-8
ADIPIC ACID DINITRILE	C6H8N2	ADIPONITRILE	111-69-3
ADIPIC ACID NITRILE	C6H8N2	ADIPONITRILE	111-69-3
ADIPIC KETONE	C5H8O	CYCLOPENTANONE	120-92-3
ADIPINIC ACID	C6H10O4	ADIPIC ACID	124-04-9
ADIPODINITRILE	C6H8N2	ADIPONITRILE	111-69-3
ADIPONITRILE	C6H8N2	ADIPONITRILE	111-69-3
ADOL	C16H34O	1-HEXADECANOL	36653-82-4
ADOL 68	C18H38O	1-OCTADECANOL	112-92-5
ADRONAL	C6H12O	CYCLOHEXANOL	108-93-0
ADVASTAB 401	C15H24O	2,6-DI-tert-BUTYL-p-CRESOL	128-37-0
AEROTHENE TT	C2H3Cl3	1,1,1-TRICHLOROETHANE	71-55-6
AETHALDIAMIN	C2H8N2	ETHYLENEDIAMINE	107-15-3
AETHANETHIOL	C2H6S	ETHYL MERCAPTAN	75-08-1
AETHANOL	C2H6O	ETHANOL	64-17-5
AETHANOLAMIN	C2H7NO	MONOETHANOLAMINE	141-43-5
AETHER	C4H10O	DIETHYL ETHER	60-29-7
2-AETHOXY-AETHYLACETAT	C6H12O3	2-ETHOXYETHYL ACETATE	111-15-9
AETHYLACETAT	C4H8O2	ETHYL ACETATE	141-78-6
AETHYLACRYLAT	C5H8O2	ETHYL ACRYLATE	140-88-5
AETHYLALKOHOL	C2H6O	ETHANOL	64-17-5
AETHYLAMINE	C2H7N	ETHYLAMINE	75-04-7
AETHYLBENZOL	C8H10	ETHYLBENZENE	100-41-4
AETHYLENECHLORHYDRIN	C2H5ClO	2-CHLOROETHANOL	107-07-3
AETHYLENEDIAMIN	C2H8N2	ETHYLENEDIAMINE	107-15-3
AETHYLENGLYKOLAETHERACETAT	C6H12O3	2-ETHOXYETHYL ACETATE	111-15-9
AETHYLENIMIN	C2H5N	ETHYLENEIMINE	151-56-4
AETHYLENSULFID	C2H4S	THIACYCLOPROPANE	420-12-2
AETHYLFORMIAT	C3H6O2	ETHYL FORMATE	109-94-4
1-AETHYLHEXANOL	C8H18O	2-ETHYL-1-HEXANOL	104-76-7
AETHYLIDENCHLORID	C2H4Cl2	1,1-DICHLOROETHANE	75-34-3
AETHYLIS	C2H5Cl	ETHYL CHLORIDE	75-00-3
AETHYLIS CHLORIDUM	C2H5Cl	ETHYL CHLORIDE	75-00-3
AETHYLMERCAPTAN	C2H6S	ETHYL MERCAPTAN	75-08-1
AETHYLMETHYLKETON	C4H8O	METHYL ETHYL KETONE	78-93-3
AETHYLPROPYLKETON	C6H12O	3-HEXANONE	589-38-8
AGENT 504	C10H22O	1-DECANOL	112-30-1
AGERITE	C18H16N2	N,N'-DIPHENYL-p-PHENYLENEDIAMINE	74-31-7
AGERITEDPPD	C18H16N2	N,N'-DIPHENYL-p-PHENYLENEDIAMINE	74-31-7
AGIDOL	C15H24O	2,6-DI-tert-BUTYL-p-CRESOL	128-37-0
AI3-25449	C4H7NO	3-METHOXYPROPIONITRILE	110-67-8
AKROLEIN	C3H4O	ACROLEIN	107-02-8
AKRYLAMID	C3H5NO	ACRYLAMIDE	79-06-1
ALAMINE 4	C12H27N	DODECYLAMINE	124-22-1
b-ALANINOL	C3H9NO	3-AMINO-1-PROPANOL	156-87-6
ALCOHOL	C2H6O	ETHANOL	64-17-5
ALCOHOL C-10	C10H22O	1-DECANOL	112-30-1
ALCOHOL C-11	C11H24O	1-UNDECANOL	112-42-5
ALCOHOL C-12	C12H26O	1-DODECANOL	112-53-8
ALCOHOL C-16	C16H34O	1-HEXADECANOL	36653-82-4
ALCOHOL C-8	C8H18O	1-OCTANOL	111-87-5
ALCOHOL C-9	C9H20O	1-NONANOL	143-08-8
ALCOHOLS	C2H6O	ETHANOL	64-17-5
ALCOOL	C3H6O	ALLYL ALCOHOL	107-18-6
ALCOOL ALLYLIQUE	C3H6O	ALLYL ALCOHOL	107-18-6
ALCOOL AMILICO	C5H12O	3-METHYL-1-BUTANOL	123-51-3
ALCOOL AMYLIQUE	C5H12O	1-PENTANOL	71-41-0
ALCOOL BUTYLIQUE	C4H10O	n-BUTANOL	71-36-3
ALCOOL BUTYLIQUE SECONDAIRE	C4H10O	sec-BUTANOL	78-92-2
ALCOOL BUTYLIQUE TERTIAIRE	C4H10O	tert-BUTANOL	75-65-0
ALCOOL ETHYLIQUE	C2H6O	ETHANOL	64-17-5
ALCOOL ETILICO	C2H6O	ETHANOL	64-17-5

Name (Synonym)	Formula	Name in Tables	CAS Reg. No
ALCOOL ISOAMYLIQUE	C5H12O	3-METHYL-1-BUTANOL	123-51-3
ALCOOL ISOBUTYLIQUE	C4H10O	ISOBUTANOL	78-83-1
ALCOOL ISOPROPILICO	C3H8O	ISOPROPANOL	67-63-0
ALCOOL ISOPROPYLIQUE	C3H8O	ISOPROPANOL	67-63-0
ALCOOL METHYL AMYLIQUE	C6H14O	4-METHYL-2-PENTANOL	108-11-2
ALCOOL METHYLIQUE	CH4O	METHANOL	67-56-1
ALCOOL METILICO	CH4O	METHANOL	67-56-1
ALCOOL PROPILICO	C3H8O	n-PROPANOL	71-23-8
ALCOOL PROPYLIQUE	C3H8O	n-PROPANOL	71-23-8
ALDEHYDE-14	C11H22O	1-UNDECANAL	112-44-7
ALDEHYDE ACETIQUE	C2H4O	ACETALDEHYDE	75-07-0
ALDEHYDE ACRYLIQUE	C3H4O	ACROLEIN	107-02-8
ALDEHYDE BUTYRIQUE	C4H8O	n-BUTYRALDEHYDE	123-72-8
ALDEHYDE C-10	C10H20O	1-DECANAL	112-31-2
ALDEHYDE C-6	C6H12O	1-HEXANAL	66-25-1
ALDEHYDE C-8	C8H16O	1-OCTANAL	124-13-0
ALDEHYDE C-9	C9H18O	1-NONANAL	124-19-6
ALDEHYDE PROPIONIQUE	C3H6O	n-PROPIONALDEHYDE	123-38-6
ALDEIDE ACETICA	C2H4O	ACETALDEHYDE	75-07-0
ALDEIDE ACRILICA	C3H4O	ACROLEIN	107-02-8
ALDEIDE BUTIRRICA	C4H8O	n-BUTYRALDEHYDE	123-72-8
ALFOL 12	C12H26O	1-DODECANOL	112-53-8
ALFOL 8	C8H18O	1-OCTANOL	111-87-5
ALGEON 22	CHClF2	CHLORODIFLUOROMETHANE	75-45-6
ALGOFRENE TYPE 1	CCl3F	TRICHLOROFLUOROMETHANE	75-69-4
ALGOFRENE TYPE 2	CCl2F2	DICHLORODIFLUOROMETHANE	75-71-8
ALGOFRENE TYPE 5	CHCl2F	DICHLOROFLUOROMETHANE	75-43-4
ALGOFRENE TYPE 6	CHClF2	CHLORODIFLUOROMETHANE	75-45-6
ALGOFRENE TYPE 67	C2H4F2	1,1-DIFLUOROETHANE	75-37-6
ALGRAIN	C2H6O	ETHANOL	64-17-5
ALGYLEN	C2HCl3	TRICHLOROETHYLENE	79-01-6
ALKOHOL	C2H6O	ETHANOL	64-17-5
ALKOHOLU ETYLOWEGO	C2H6O	ETHANOL	64-17-5
ALLILE	C3H5Cl	3-CHLOROPROPENE	107-05-1
ALLILOWY ALKOHOL	C3H6O	ALLYL ALCOHOL	107-18-6
ALLOMALEIC ACID	C4H4O4	FUMARIC ACID	110-17-8
ALLYL ACETATE	C5H8O2	ALLYL ACETATE	591-87-7
ALLYL AL	C3H6O	ALLYL ALCOHOL	107-18-6
ALLYL ALCOHOL	C3H6O	ALLYL ALCOHOL	107-18-6
ALLYL ALDEHYDE	C3H4O	ACROLEIN	107-02-8
ALLYLALKOHOL	C3H6O	ALLYL ALCOHOL	107-18-6
ALLYLAMINE	C3H7N	ALLYLAMINE	107-11-9
ALLYLCHLORID	C3H5Cl	3-CHLOROPROPENE	107-05-1
ALLYL CYANIDE	C4H5N	VINYLACETONITRILE	109-75-1
ALLYLE	C3H5Cl	3-CHLOROPROPENE	107-05-1
ALLYLENE	C3H4	METHYLACETYLENE	74-99-7
ALLYLIC ALCOHOL	C3H6O	ALLYL ALCOHOL	107-18-6
ALLYLNITRILE	C4H5N	VINYLACETONITRILE	109-75-1
ALLYL TRICHLORIDE	C3H5Cl3	1,2,3-TRICHLOROPROPANE	96-18-4
ALMOND ARTIFICIAL ESSENTIAL OIL	C7H6O	BENZALDEHYDE	100-52-7
ALUMINUM, TRIETHYL-	C6H15Al	TRIETHYL ALUMINUN	97-93-8
AMATIN	C6Cl6	HEXACHLOROBENZENE	118-74-1
AMBER ACID	C4H6O4	SUCCINIC ACID	110-15-6
AMEISENSAEURE	CH2O2	FORMIC ACID	64-18-6
AMID KYSELINY AKRYLOVE	C3H5NO	ACRYLAMIDE	79-06-1
AMID KYSELINY OCTOVE	C2H5NO	ACETAMIDE	60-35-5
AMILPHENOL	C11H16O	p-tert-AMYLPHENOL	80-46-6
AMINE BB	C12H27N	DODECYLAMINE	124-22-1
AMINIC ACID	CH2O2	FORMIC ACID	64-18-6
2-AMINOETHANOL	C2H7NO	MONOETHANOLAMINE	141-43-5
m-AMINOANILINE	C6H8N2	m-PHENYLENEDIAMINE	108-45-2
p-AMINOANILINE	C6H8N2	p-PHENYLENEDIAMINE	106-50-3
2-AMINOANILINE	C6H8N2	o-PHENYLENEDIAMINE	95-54-5
3-AMINOANILINE	C6H8N2	m-PHENYLENEDIAMINE	108-45-2
4-AMINOANILINE	C6H8N2	p-PHENYLENEDIAMINE	106-50-3
AMINOAZOBENZENE	C12H11N3	p-AMINOAZOBENZENE	60-09-3
p-AMINOAZOBENZENE	C12H11N3	p-AMINOAZOBENZENE	60-09-3
4-AMINOAZOBENZENE	C12H11N3	p-AMINOAZOBENZENE	60-09-3
4-AMINO-1,1'-AZOBENZENE	C12H11N3	p-AMINOAZOBENZENE	60-09-3
p-AMINOAZOBENZOL	C12H11N3	p-AMINOAZOBENZENE	60-09-3
4-AMINOAZOBENZOL	C12H11N3	p-AMINOAZOBENZENE	60-09-3
AMINOBENZENE	C6H7N	ANILINE	62-53-3
p-AMINOBIPHENYL	C12H11N	p-AMINODIPHENYL	92-67-1
4-AMINOBIPHENYL	C12H11N	p-AMINODIPHENYL	92-67-1
1-AMINO-BUTAAN	C4H11N	n-BUTYLAMINE	109-73-9
1-AMINOBUTAN	C4H11N	n-BUTYLAMINE	109-73-9

Name (Synonym)	Formula	Name in Tables	CAS Reg. No
1-AMINOBUTANE	C4H11N	n-BUTYLAMINE	109-73-9
2-AMINOBUTANE	C4H11N	sec-BUTYLAMINE	13952-84-6
gamma-AMINOBUTYRIC ACID LACTAM	C4H7NO	2-PYRROLIDONE	616-45-5
4-AMINOBUTYRIC ACID LACTAM	C4H7NO	2-PYRROLIDONE	616-45-5
gamma-AMINOBUTYRIC LACTAM	C4H7NO	2-PYRROLIDONE	616-45-5
gamma-AMINOBUTYROLACTAM	C4H7NO	2-PYRROLIDONE	616-45-5
AMINOCAPROIC LACTAM	C6H11NO	epsilon-CAPROLACTAM	105-60-2
m-AMINOCHLOROBENZENE	C6H6ClN	m-CHLOROANILINE	108-42-9
1-AMINO-2-CHLOROBENZENE	C6H6ClN	o-CHLOROANILINE	95-51-2
1-AMINO-3-CHLOROBENZENE	C6H6ClN	m-CHLOROANILINE	108-42-9
1-AMINO-4-CHLOROBENZENE	C6H6ClN	p-CHLOROANILINE	106-47-8
AMINOCYCLOHEXANE	C6H13N	CYCLOHEXYLAMINE	108-91-8
1-AMINODECANE	C10H23N	n-DECYLAMINE	2016-57-1
1-AMINO-3,4-DICHLOROBENZENE	C6H5Cl2N	3,4-DICHLOROANILINE	95-76-1
4-AMINODIFENIL	C12H11N	p-AMINODIPHENYL	92-67-1
p-AMINODIFENYLAMIN	C12H12N2	p-AMINODIPHENYLAMINE	101-54-2
p-AMINODIPHENYL	C12H11N	p-AMINODIPHENYL	92-67-1
p-AMINODIPHENYLAMINE	C12H12N2	p-AMINODIPHENYLAMINE	101-54-2
4-AMINODIPHENYLAMINE	C12H12N2	p-AMINODIPHENYLAMINE	101-54-2
p-AMINODIPHENYLIMIDE	C12H11N3	p-AMINOAZOBENZENE	60-09-3
1-AMINODODECANE	C12H27N	DODECYLAMINE	124-22-1
2-AMINOETANOLO	C2H7NO	MONOETHANOLAMINE	141-43-5
AMINOETHANE	C2H7N	ETHYLAMINE	75-04-7
1-AMINOETHANE	C2H7N	ETHYLAMINE	75-04-7
2-AMINOETHANOL	C2H7NO	MONOETHANOLAMINE	141-43-5
2-AMINOETHOXYETHANOL	C4H11NO2	2-AMINOETHOXYETHANOL	929-06-6
2-(2-AMINOETHOXY)ETHANOL	C4H11NO2	2-AMINOETHOXYETHANOL	929-06-6
b-AMINOETHYL ALCOHOL	C2H7NO	MONOETHANOLAMINE	141-43-5
o-AMINOETHYLBENZENE	C8H11N	o-ETHYLANILINE	578-54-1
AMINOETHYLENE	C2H5N	ETHYLENEIMINE	151-56-4
AMINOETHYLETHANEDIAMINE	C4H13N3	DIETHYLENE TRIAMINE	111-40-0
AMINOETHYL ETHANOLAMINE	C4H12N2O	N-AMINOETHYL ETHANOLAMINE	111-41-1
N-AMINOETHYL ETHANOLAMINE	C4H12N2O	N-AMINOETHYL ETHANOLAMINE	111-41-1
N-(2-AMINOETHYL)ETHYLENEDIAMINE	C4H13N3	DIETHYLENE TRIAMINE	111-40-0
AMINOETHYLPIPERAZINE	C6H15N3	N-AMINOETHYL PIPERAZINE	140-31-8
N-AMINOETHYL PIPERAZINE	C6H15N3	N-AMINOETHYL PIPERAZINE	140-31-8
N-(b-AMINOETHYL)PIPERAZINE	C6H15N3	N-AMINOETHYL PIPERAZINE	140-31-8
N-(2-AMINOETHYL)PIPERAZINE	C6H15N3	N-AMINOETHYL PIPERAZINE	140-31-8
1-(2-AMINOETHYL)PIPERAZINE	C6H15N3	N-AMINOETHYL PIPERAZINE	140-31-8
a-AMINOGLUTARIC ACID	C5H9NO4	L-GLUTAMIC ACID	56-86-0
l-2-AMINOGLUTARIC ACID	C5H9NO4	L-GLUTAMIC ACID	56-86-0
1-AMINOHEPTANE	C7H17N	1-AMINOHEPTANE	111-68-2
AMINOHEXAHYDROBENZENE	C6H13N	CYCLOHEXYLAMINE	108-91-8
1-AMINOHEXANE	C6H15N	n-HEXYLAMINE	111-26-2
6-AMINOHEXANOIC ACID CYCLIC LACTAM	C6H11NO	epsilon-CAPROLACTAM	105-60-2
6-AMINOHEXANOL	C6H15NO	6-AMINOHEXANOL	4048-33-3
2-AMINOISOBUTANE	C4H11N	tert-BUTYLAMINE	75-64-9
a-AMINOISOPROPYL ALCOHOL	C3H9NO	1-AMINO-2-PROPANOL	78-96-6
AMINOMETHANE	CH5N	METHYLAMINE	74-89-5
(AMINOMETHYL)BENZENE	C7H9N	BENZYLAMINE	100-46-9
1-AMINO-2-METHYLBENZENE	C7H9N	o-TOLUIDINE	95-53-4
2-AMINO-1-METHYLBENZENE	C7H9N	o-TOLUIDINE	95-53-4
3-AMINO-1-METHYLBENZENE	C7H9N	m-TOLUIDINE	108-44-1
4-AMINO-1-METHYLBENZENE	C7H9N	p-TOLUIDINE	106-49-0
1-AMINO-2-METHYLPROPANE	C4H11N	ISOBUTYLAMINE	78-81-9
2-AMINO-2-METHYLPROPANE	C4H11N	tert-BUTYLAMINE	75-64-9
m-AMINONITROBENZENE	C6H6N2O2	m-NITROANILINE	99-09-2
p-AMINONITROBENZENE	C6H6N2O2	p-NITROANILINE	100-01-6
1-AMINO-2-NITROBENZENE	C6H6N2O2	o-NITROANILINE	88-74-4
1-AMINO-3-NITROBENZENE	C6H6N2O2	m-NITROANILINE	99-09-2
1-AMINO-4-NITROBENZENE	C6H6N2O2	p-NITROANILINE	100-01-6
1-AMINOPENTANE	C5H13N	n-PENTYLAMINE	110-58-7
2-AMINOPENTANE	C4H11N	DIETHYLAMINE	109-89-7
2-AMINOPENTANEDIOIC ACID	C5H9NO4	L-GLUTAMIC ACID	56-86-0
AMINOPHEN	C6H7N	ANILINE	62-53-3
3-AMINOPHENYLMETHANE	C7H9N	m-TOLUIDINE	108-44-1
2-AMINO-PROPAAN	C3H9N	ISOPROPYLAMINE	75-31-0
2-AMINOPROPAN	C3H9N	ISOPROPYLAMINE	75-31-0
1-AMINOPROPANE	C3H9N	n-PROPYLAMINE	107-10-8
2-AMINOPROPANE	C3H9N	ISOPROPYLAMINE	75-31-0
1-AMINOPROPANE-1,3-DICARBOXYLIC ACID	C5H9NO4	L-GLUTAMIC ACID	56-86-0
2-AMINO-PROPANO	C3H9N	ISOPROPYLAMINE	75-31-0
g-AMINOPROPANOL	C3H9NO	3-AMINO-1-PROPANOL	156-87-6
1-AMINO-2-PROPANOL	C3H9NO	1-AMINO-2-PROPANOL	78-96-6
3-AMINOPROPANOL	C3H9NO	3-AMINO-1-PROPANOL	156-87-6
3-AMINO-1-PROPANOL	C3H9NO	3-AMINO-1-PROPANOL	156-87-6

Name (Synonym)	Formula	Name in Tables	CAS Reg. No
3-AMINOPROPENE	C3H7N	ALLYLAMINE	107-11-9
3-AMINOPROPYL ALCOHOL	C3H9NO	3-AMINO-1-PROPANOL	156-87-6
3-AMINOPROPYLENE	C3H7N	ALLYLAMINE	107-11-9
3-AMINOTOLUEN	C7H9N	m-TOLUIDINE	108-44-1
4-AMINOTOLUEN	C7H9N	p-TOLUIDINE	106-49-0
a-AMINOTOLUENE	C7H9N	BENZYLAMINE	100-46-9
m-AMINOTOLUENE	C7H9N	m-TOLUIDINE	108-44-1
o-AMINOTOLUENE	C7H9N	o-TOLUIDINE	95-53-4
p-AMINOTOLUENE	C7H9N	p-TOLUIDINE	106-49-0
w-AMINOTOLUENE	C7H9N	BENZYLAMINE	100-46-9
2-AMINOTOLUENE	C7H9N	o-TOLUIDINE	95-53-4
3-AMINOTOLUENE	C7H9N	m-TOLUIDINE	108-44-1
4-AMINOTOLUENE	C7H9N	p-TOLUIDINE	106-49-0
3-AMINO-p-TOLUIDINE	C7H10N2	TOLUENEDIAMINE	95-80-7
5-AMINO-o-TOLUIDINE	C7H10N2	TOLUENEDIAMINE	95-80-7
AMINUTRIN	C6H14N2O2	LYSINE	56-87-1
AMPROLENE	C2H4O	ETHYLENE OXIDE	75-21-8
AMYL ACETATE	C7H14O2	n-PENTYL ACETATE	628-63-7
AMYL ACETIC ESTER	C7H14O2	n-PENTYL ACETATE	628-63-7
N-AMYL ALCOHOL	C5H12O	1-PENTANOL	71-41-0
sec-AMYL ALCOHOL	C5H12O	2-PENTANOL	6032-29-7
tert-AMYL ALCOHOL	C5H12O	tert-PENTYL-ALCOHOL	75-85-4
AMYL ALCOHOL, NORMAL	C5H12O	1-PENTANOL	71-41-0
AMYL ALDEHYDE	C5H10O	VALERALDEHYDE	110-62-3
N-AMYLALKOHOL	C5H12O	1-PENTANOL	71-41-0
AMYLAMINE	C5H13N	n-PENTYLAMINE	110-58-7
n-AMYLAMINE	C5H13N	n-PENTYLAMINE	110-58-7
AMYLAZETAT	C7H14O2	n-PENTYL ACETATE	628-63-7
AMYL BROMIDE	C5H11Br	1-BROMOPENTANE	110-53-2
AMYLCARBINOL	C6H14O	1-HEXANOL	111-27-3
AMYL CHLORIDE	C5H11Cl	1-CHLOROPENTANE	543-59-9
n-AMYL CHLORIDE	C5H11Cl	1-CHLOROPENTANE	543-59-9
AMYLENE HYDRATE	C5H12O	tert-PENTYL-ALCOHOL	75-85-4
AMYLESTER KYSELINY OCTOVE	C7H14O2	n-PENTYL ACETATE	628-63-7
AMYL ETHER	C10H22O	DI-n-PENTYL ETHER	693-65-2
n-AMYL ETHER	C10H22O	DI-n-PENTYL ETHER	693-65-2
AMYL FORMATE	C6H12O2	n-PENTYL FORMATE	638-49-3
AMYL HYDRIDE	C5H12	n-PENTANE	109-66-0
AMYL HYDROSULFIDE	C5H12S	n-PENTYL MERCAPTAN	110-66-7
AMYL MERCAPTAN	C5H12S	n-PENTYL MERCAPTAN	110-66-7
n-AMYL MERCAPTAN	C5H12S	n-PENTYL MERCAPTAN	110-66-7
AMYL METHYL CARBINOL	C7H16O	2-HEPTANOL	543-49-7
AMYL-METHYL-CETONE	C7H14O	2-HEPTANONE	110-43-0
AMYL METHYL KETONE	C7H14O	2-HEPTANONE	110-43-0
n-AMYL METHYL KETONE	C7H14O	2-HEPTANONE	110-43-0
AMYLOWY ALKOHOL	C5H12O	3-METHYL-1-BUTANOL	123-51-3
p-tert-AMYLPHENOL	C11H16O	p-tert-AMYLPHENOL	80-46-6
AMYL PHENOL 4T	C11H16O	p-tert-AMYLPHENOL	80-46-6
AMYL SULFHYDRATE	C5H12S	n-PENTYL MERCAPTAN	110-66-7
AMYL THIOALCOHOL	C5H12S	n-PENTYL MERCAPTAN	110-66-7
AN	C8H9NO	ACETANILIDE	103-84-4
ANAESTHETIC ETHER	C4H10O	DIETHYL ETHER	60-29-7
ANAMENTH	C2HCl3	TRICHLOROETHYLENE	79-01-6
ANCHOIC ACID	C9H16O4	AZELAIC ACID	123-99-9
ANESTHENYL	C3H8O2	METHYLAL	109-87-5
ANESTHESIA ETHER	C4H10O	DIETHYL ETHER	60-29-7
ANESTHETIC ETHER	C4H10O	DIETHYL ETHER	60-29-7
ANETHOLE	C10H12O	ANETHOLE	104-46-1
ANGIBID	C3H5N3O9	NITROGLYCERINE	55-63-0
ANGICAP	C5H8N4O12	PENTAERYTHRITOL TETRANITRATE	78-11-5
ANGININE	C3H5N3O9	NITROGLYCERINE	55-63-0
ANGIOLINGUAL	C3H5N3O9	NITROGLYCERINE	55-63-0
ANGITET	C5H8N4O12	PENTAERYTHRITOL TETRANITRATE	78-11-5
ANGORIN	C3H5N3O9	NITROGLYCERINE	55-63-0
ANHYDRIDE CARBONIQUE	CO2	CARBON DIOXIDE	124-38-9
ANHYDRIDE PHTHALIQUE	C8H4O3	PHTHALIC ANHYDRIDE	85-44-9
ANHYDRID KYSELINY MASELNE	C8H14O3	BUTYRIC ANHYDRIDE	106-31-0
ANHYDRID KYSELINY OCTOVE	C4H6O3	ACETIC ANHYDRIDE	108-24-7
ANHYDROL	C2H6O	ETHANOL	64-17-5
ANHYDROTRIMELLIC ACID	C9H4O5	TRIMELLITIC ANHYDRIDE	552-30-7
ANHYDROUS CHLORAL	C2HCl3O	TRICHLOROACETALDEHYDE	75-87-6
ANIDRIDE FTALICA	C8H4O3	PHTHALIC ANHYDRIDE	85-44-9
ANILIN	C6H7N	ANILINE	62-53-3
ANILINA	C6H7N	ANILINE	62-53-3
ANILINE	C6H7N	ANILINE	62-53-3
ANILINE, o-ETHYL-	C8H11N	o-ETHYLANILINE	578-54-1

Name (Synonym)	Formula	Name in Tables	CAS Reg. No
ANILINE, N-ACETYL-	C8H9NO	ACETANILIDE	103-84-4
ANILINE, 4-NITRO-	C6H6N2O2	p-NITROANILINE	100-01-6
ANILINE OIL	C6H7N	ANILINE	62-53-3
ANILINE YELLOW	C12H11N3	p-AMINOAZOBENZENE	60-09-3
p-ANILINOANILINE	C12H12N2	p-AMINODIPHENYLAMINE	101-54-2
ANILINOBENZENE	C12H11N	DIPHENYLAMINE	122-39-4
ANILINOMETHANE	C7H9N	N-METHYLANILINE	100-61-8
ANISE CAMPHOR	C10H12O	ANETHOLE	104-46-1
o-ANISIC ACID	C8H8O3	METHYL SALICYLATE	119-36-8
ANISOLE	C7H8O	ANISOLE	100-66-3
ANKILOSTIN	C2Cl4	TETRACHLOROETHYLENE	127-18-4
(6)ANNULENE	C6H6	BENZENE	71-43-2
ANODYNON	C2H5Cl	ETHYL CHLORIDE	75-00-3
ANOL	C6H12O	CYCLOHEXANOL	108-93-0
ANOZOL	C12H14O4	DIETHYL PHTHALATE	84-66-2
ANPROLENE	C2H4O	ETHYLENE OXIDE	75-21-8
ANPROLINE	C2H4O	ETHYLENE OXIDE	75-21-8
ANSUL ETHER 181AT	C10H22O5	TETRAETHYLENE GLYCOL DIMETHYL ETHER	143-24-8
ANTAK	C10H22O	1-DECANOL	112-30-1
ANTHRACEN	C14H10	ANTHRACENE	120-12-7
ANTHRACENE	C14H10	ANTHRACENE	120-12-7
9,10-ANTHRACENEDIONE	C14H8O2	ANTHRAQUINONE	84-65-1
ANTHRACIN	C14H10	ANTHRACENE	120-12-7
ANTHRADIONE	C14H8O2	ANTHRAQUINONE	84-65-1
ANTHRAPOLE AZ	C11H14O2	n-BUTYL BENZOATE	136-60-7
ANTHRAQUINONE	C14H8O2	ANTHRAQUINONE	84-65-1
9,10-ANTHRAQUINONE	C14H8O2	ANTHRAQUINONE	84-65-1
ANTICARIE	C6Cl6	HEXACHLOROBENZENE	118-74-1
ANTIFEBRIN	C8H9NO	ACETANILIDE	103-84-4
ANTIOXIDANT 29	C15H24O	2,6-DI-tert-BUTYL-p-CRESOL	128-37-0
ANTIOXIDANT D	C6H11NO	CYCLOHEXANONE OXIME	100-64-1
ANTIOXIDANT DBPC	C15H24O	2,6-DI-tert-BUTYL-p-CRESOL	128-37-0
ANTIREN	C4H10N2	PIPERAZINE	110-85-0
ANTISAL 1a	C7H8	TOLUENE	108-88-3
ANTISOL 1	C2Cl4	TETRACHLOROETHYLENE	127-18-4
ANTORA	C5H8N4O12	PENTAERYTHRITOL TETRANITRATE	78-11-5
AO 29	C15H24O	2,6-DI-tert-BUTYL-p-CRESOL	128-37-0
AO 4K	C15H24O	2,6-DI-tert-BUTYL-p-CRESOL	128-37-0
APCO 2330	C6H8N2	m-PHENYLENEDIAMINE	108-45-2
APEX 4	C22H44O2	n-BUTYL STEARATE	123-95-5
APV	C6H14O3	2-(2-ETHOXYETHOXY)ETHANOL	111-90-0
AQUALINE	C3H4O	ACROLEIN	107-02-8
ARALDITE HARDENER HY 951	C6H18N4	TRIETHYLENE TETRAMINE	112-24-3
ARALDITE HY 951	C6H18N4	TRIETHYLENE TETRAMINE	112-24-3
ARCOTRATE	C5H8N4O12	PENTAERYTHRITOL TETRANITRATE	78-11-5
ARCTON	CHF3	TRIFLUOROMETHANE	75-46-7
ARCTON 22	CHClF2	CHLORODIFLUOROMETHANE	75-45-6
ARCTON 3	CClF3	CHLOROTRIFLUOROMETHANE	75-72-9
ARCTON 6	CCl2F2	DICHLORODIFLUOROMETHANE	75-71-8
ARCTON 63	C2Cl3F3	1,1,2-TRICHLOROTRIFLUOROETHANE	76-13-1
ARCTON 7	CHCl2F	DICHLOROFLUOROMETHANE	75-43-4
ARCTON 9	CCl3F	TRICHLOROFLUOROMETHANE	75-69-4
ARCTUVIN	C6H6O2	p-HYDROQUINONE	123-31-9
AREGINAL	C3H6O2	ETHYL FORMATE	109-94-4
ARIZOLE	C10H12O	ANETHOLE	104-46-1
ARKLONE P	C2Cl3F3	1,1,2-TRICHLOROTRIFLUOROETHANE	76-13-1
ARMEEN 12D	C12H27N	DODECYLAMINE	124-22-1
ARTIC	CH3Cl	METHYL CHLORIDE	74-87-3
ARTIFICIAL ALMOND OIL	C7H6O	BENZALDEHYDE	100-52-7
ARTIFICIAL ANT OIL	C5H4O2	FURFURAL	98-01-1
ASCABIN	C14H12O2	BENZYL BENZOATE	120-51-4
ASCABIOL	C14H12O2	BENZYL BENZOATE	120-51-4
ASCORBIC ACID	C6H8O6	ASCORBIC ACID	50-81-7
l(+)-ASCORBIC ACID	C6H8O6	ASCORBIC ACID	50-81-7
ASCORBUTINA	C6H8O6	ASCORBIC ACID	50-81-7
ASUCCIN	C4H6O4	SUCCINIC ACID	110-15-6
ASYMMETRICAL TRIMETHYL BENZENE	C9H12	1,2,4-TRIMETHYLBENZENE	95-63-6
ATALCO C	C16H34O	1-HEXADECANOL	36653-82-4
ATALCO S	C18H38O	1-OCTADECANOL	112-92-5
ATHYLENGLYKOL	C2H6O2	ETHYLENE GLYCOL	107-21-1
ATHYLENGLYKOL-MONOATHYLATHER	C4H10O2	2-ETHOXYETHANOL	110-80-5
AUXINUTRIL	C5H12O4	PENTAERYTHRITOL	115-77-5
AVLOTANE	C2Cl6	HEXACHLOROETHANE	67-72-1
AVOLIN	C10H10O4	DIMETHYL PHTHALATE	131-11-3
AZABENZENE	C5H5N	PYRIDINE	110-86-1
AZACYCLOHEPTANE	C6H13N	HEXAMETHYLENEIMINE	111-49-9

Name (Synonym)	Formula	Name in Tables	CAS Reg. No
1-AZACYCLOHEPTANE	C6H13N	HEXAMETHYLENEIMINE	111-49-9
2-AZACYCLOHEPTANONE	C6H11NO	epsilon-CAPROLACTAM	105-60-2
AZACYCLOHEXANE	C5H11N	PIPERIDINE	110-89-4
1-AZA-2,4-CYCLOPENTADIENE	C4H5N	PYRROLE	109-97-7
AZACYCLOPENTANE	C4H9N	PYRROLIDINE	123-75-1
AZACYCLOPROPANE	C2H5N	ETHYLENEIMINE	151-56-4
9-AZAFLUORENE	C12H9N	DIBENZOPYRROLE	86-74-8
1-AZAINDENE	C8H7N	INDOLE	120-72-9
1-AZANAPHTHALENE	C9H7N	QUINOLINE	91-22-5
2-AZANAPHTHALENE	C9H7N	ISOQUINOLINE	119-65-3
3-AZAPENTANE-1,5-DIAMINE	C4H13N3	DIETHYLENE TRIAMINE	111-40-0
AZELAIC ACID	C9H16O4	AZELAIC ACID	123-99-9
AZIJNZUUR	C2H4O2	ACETIC ACID	64-19-7
AZINE	C5H5N	PYRIDINE	110-86-1
AZIRANE	C2H5N	ETHYLENEIMINE	151-56-4
AZIRIDIN	C2H5N	ETHYLENEIMINE	151-56-4
AZIRIDINE	C2H5N	ETHYLENEIMINE	151-56-4
AZOAMINE RED ZH	C6H6N2O2	p-NITROANILINE	100-01-6
AZOBASE MNA	C6H6N2O2	m-NITROANILINE	99-09-2
AZOENE FAST ORANGE GR BASE	C6H6N2O2	o-NITROANILINE	88-74-4
AZOGEN DEVELOPER H	C7H10N2	TOLUENEDIAMINE	95-80-7
AZOGENE FAST ORANGE GR	C6H6N2O2	o-NITROANILINE	88-74-4
AZOIC DIAZO COMPONENT 37	C6H6N2O2	p-NITROANILINE	100-01-6
AZOIC DIAZO COMPONENT 6	C6H6N2O2	o-NITROANILINE	88-74-4
AZOLE	C4H5N	PYRROLE	109-97-7
AZOSALT R	C12H12N2	p-AMINODIPHENYLAMINE	101-54-2
AZULENE	C10H8	AZULENE	275-51-4
B-500	C9H7N	QUINOLINE	91-22-5
BAKERS P AND S LIQUID and OINTMENT	C6H6O	PHENOL	108-95-2
BANANA OIL	C7H14O2	ISOPENTYL ACETATE	123-92-2
BARITRATE	C5H8N4O12	PENTAERYTHRITOL TETRANITRATE	78-11-5
BASF URSOL D	C6H8N2	p-PHENYLENEDIAMINE	106-50-3
BCME	C2H4Cl2O	BIS(CHLOROMETHYL)ETHER	542-88-1
BEHP	C24H38O4	DIOCTYL PHTHALATE	117-81-7
BENYLATE	C14H12O2	BENZYL BENZOATE	120-51-4
1,2-BENZACENAPHTHENE	C16H10	FLUORANTHENE	206-44-0
BENZAL ALCOHOL	C7H8O	BENZYL ALCOHOL	100-51-6
BENZALDEHYDE	C7H6O	BENZALDEHYDE	100-52-7
1-BENZAZINE	C9H7N	QUINOLINE	91-22-5
2-BENZAZINE	C9H7N	ISOQUINOLINE	119-65-3
1-BENZAZOLE	C8H7N	INDOLE	120-72-9
BENZEEN	C6H6	BENZENE	71-43-2
BENZEN	C6H6	BENZENE	71-43-2
BENZENAMINE	C6H7N	ANILINE	62-53-3
BENZENAMINE, N,N-DIETHYL-	C10H15N	N,N-DIETYHLANILINE	91-66-7
BENZENAMINE, 2,6-DIETHYL-	C10H15N	2,6-DIETHYLANILINE	579-66-8
BENZENAMINE, N,N,-DIMETHYL-	C8H11N	N,N-DIMETHYLANILINE	121-69-7
BENZENAMINE, 2-ETHYL-	C8H11N	o-ETHYLANILINE	578-54-1
BENZENAMINE, N-METHYL-(9CI)	C7H9N	N-METHYLANILINE	100-61-8
BENZENAMINE, 4-NITRO-	C6H6N2O2	p-NITROANILINE	100-01-6
BENZENE	C6H6	BENZENE	71-43-2
4-BENZENEAZOANILINE	C12H11N3	p-AMINOAZOBENZENE	60-09-3
BENZENE, 1,4-BIS(1-METHYLETHYL)-	C12H18	p-DIISOPROPYLBENZENE	100-18-5
BENZENECARBALDEHYDE	C7H6O	BENZALDEHYDE	100-52-7
BENZENECARBINOL	C7H8O	BENZYL ALCOHOL	100-51-6
BENZENECARBONAL	C7H6O	BENZALDEHYDE	100-52-7
BENZENECARBONYL CHLORIDE	C7H5ClO	BENZOYL CHLORIDE	98-88-4
BENZENECARBOXYLIC ACID	C7H6O2	BENZOIC ACID	65-85-0
BENZENE CHLORIDE	C6H5Cl	MONOCHLOROBENZENE	108-90-7
BENZENE, 1-CHLORO-4-METHYL-	C7H7Cl	p-CHLOROTOLUENE	106-43-4
m-BENZENEDIAMINE	C6H8N2	m-PHENYLENEDIAMINE	108-45-2
o-BENZENEDIAMINE	C6H8N2	o-PHENYLENEDIAMINE	95-54-5
p-BENZENEDIAMINE	C6H8N2	p-PHENYLENEDIAMINE	106-50-3
1,2-BENZENEDIAMINE	C6H8N2	o-PHENYLENEDIAMINE	95-54-5
1,3-BENZENEDIAMINE	C6H8N2	m-PHENYLENEDIAMINE	108-45-2
1,4-BENZENEDIAMINE	C6H8N2	p-PHENYLENEDIAMINE	106-50-3
1,4-BENZENEDIAMINE, N-PHENYL-	C12H12N2	p-AMINODIPHENYLAMINE	101-54-2
BENZENE, 1,3-DIBROMO-	C6H4Br2	m-DIBROMOBENZENE	108-36-1
1,3-BENZENEDICARBONYL CHLORIDE	C8H4Cl2O2	ISOPHTHALOYL CHLORIDE	99-63-8
m-BENZENEDICARBOXYLIC ACID	C8H6O4	ISOPHTHALIC ACID	121-91-5
o-BENZENEDICARBOXYLIC ACID	C8H6O4	PHTHALIC ACID	88-99-3
p-BENZENEDICARBOXYLIC ACID	C8H6O4	TEREPHTHALIC ACID	100-21-0
BENZENE-1,2-DICARBOXYLIC ACID	C8H6O4	PHTHALIC ACID	88-99-3
BENZENE-1,3-DICARBOXYLIC ACID	C8H6O4	TEREPHTHALIC ACID	100-21-0
1,2-BENZENEDICARBOXYLIC ACID	C8H6O4	ISOPHTHALIC ACID	121-91-5
1,4-BENZENEDICARBOXYLIC ACID	C8H6O4	PHTHALIC ACID	88-99-3

Name (Synonym)	Formula	Name in Tables	CAS Reg. No
1,2-BENZENEDICARBOXYLIC ACID ANHYDRIDE	C8H4O3	PHTHALIC ANHYDRIDE	85-44-9
BENZENE-o-DICARBOXYLIC ACID DI-n-BUTYL ESTER	C16H22O4	DIBUTYL PHTHALATE	84-74-2
o-BENZENEDICARBOXYLIC ACID, DIBUTYL ESTER	C16H22O4	DIBUTYL PHTHALATE	84-74-2
1,2-BENZENEDICARBOXYLIC ACID, DIETHYL ESTER	C12H14O4	DIETHYL PHTHALATE	84-66-2
1,2-BENZENEDICARBOXYLIC ACID, DIISOOCTYL ESTER	C24H38O4	DIISOOCTYL PHTHALATE	27554-26-3
1,2-BENZENEDICARBOXYLIC ACID DIMETHYL ESTER	C10H10O4	DIMETHYL PHTHALATE	131-11-3
1,4-BENZENE DICARBOXYLIC ACID DIMETHYL ESTER	C10H10O4	DIMETHYL TEREPHTHALATE	120-61-6
BENZENE, 1,2-DICHLORO-	C6H4Cl2	o-DICHLOROBENZENE	95-50-1
BENZENE, 2,4-DICHLORO-1-METHYL-(9Cl)	C7H6Cl2	2,4-DICHLOROTOLUENE	95-73-8
BENZENE, p-DIHYDROXY-	C6H6O2	p-HYDROQUINONE	123-31-9
BENZENE-, 1,3-DIISOCYANATOMETHYL-	C9H6N2O2	TOLUENE DIISOCYANATE	26471-62-5
BENZENE, p-DIISOPROPYL-	C12H18	p-DIISOPROPYLBENZENE	100-18-5
m-BENZENEDIOL	C6H6O2	1,3-BENZENEDIOL	108-46-3
o-BENZENEDIOL	C6H6O2	1,2-BENZENEDIOL	120-80-9
p-BENZENEDIOL	C6H6O2	p-HYDROQUINONE	123-31-9
1,2-BENZENEDIOL	C6H6O2	1,2-BENZENEDIOL	120-80-9
1,3-BENZENEDIOL	C6H6O2	1,3-BENZENEDIOL	108-46-3
1,4-BENZENEDIOL	C6H6O2	p-HYDROQUINONE	123-31-9
BENZENE, 1,1'-(1,2-ETHENEDIYL)BIS-	C14H12	trans-STILBENE	103-30-0
BENZENE, 1-ETHENYL-3-METHYL-	C9H10	m-METHYLSTYRENE	100-80-1
BENZENE, 1-ETHENYL-4-METHYL-	C9H10	p-METHYLSTYRENE	622-97-9
BENZENE, ETHOXY-	C8H10O	PHENETOLE	103-73-1
BENZENEFORMIC ACID	C7H6O2	BENZOIC ACID	65-85-0
BENZENEIODIDE	C6H5I	IODOBENZENE	591-50-4
BENZENE ISOPROPYL	C9H12	CUMENE	98-82-8
BENZENEMETHANAMINE	C7H9N	BENZYLAMINE	100-46-9
BENZENEMETHANOIC ACID	C7H6O2	BENZOIC ACID	65-85-0
BENZENEMETHANOL	C7H8O	BENZYL ALCOHOL	100-51-6
BENZENE, METHOXY	C7H8O	ANISOLE	100-66-3
BENZENE, METHYL-	C7H8	TOLUENE	108-88-3
BENZENE, 1-METHYL-2-(1-METHYLETHYL)-	C10H14	o-CYMENE	527-84-4
BENZENE, 1-METHYL-3-(1-METHYLETHYL)-	C10H14	m-CYMENE	535-77-3
BENZENE, 2-METHYL-1,3,5-TRINITRO-	C7H5N3O6	2,4,6-TRINITROTOLUENE	118-96-7
BENZENENITRILE	C7H5N	BENZONITRILE	100-47-0
1,2,4,5-BENZENETETRACARBOXYLIC ACID	C10H6O8	PYROMELLITIC ACID	89-05-4
BENZENETETRAHYDRIDE	C6H10	CYCLOHEXENE	110-83-8
1,2,4-BENZENETRICARBOXYLIC ANHYDRIDE	C9H4O5	TRIMELLITIC ANHYDRIDE	552-30-7
BENZENE, 1,3,5-TRIMETHYL-	C9H12	MESITYLENE	108-67-8
1,2,3-BENZENETRIOL	C6H6O3	1,2,3-BENZENETRIOL	87-66-1
BENZENOL	C6H6O	PHENOL	108-95-2
BENZENYL CHLORIDE	C7H5Cl3	BENZOTRICHLORIDE	98-07-7
BENZENYL FLUORIDE	C7H5F3	BENZOTRIFLUORIDE	98-08-8
BENZENYL TRICHLORIDE	C7H5Cl3	BENZOTRICHLORIDE	98-07-7
BENZILE (CLORURO di)	C7H7Cl	BENZYL CHLORIDE	100-44-7
BENZIN	C6H6	BENZENE	71-43-2
BENZINE	C6H6	BENZENE	71-43-2
1-BENZINE	C9H7N	QUINOLINE	91-22-5
BENZINOFORM	CCl4	CARBON TETRACHLORIDE	56-23-5
BENZINOL	C2HCl3	TRICHLOROETHYLENE	79-01-6
BENZOATE	C7H6O2	BENZOIC ACID	65-85-0
BENZOESAEURE	C7H6O2	BENZOIC ACID	65-85-0
BENZO(jk)FLUORENE	C16H10	FLUORANTHENE	206-44-0
BENZOFUR MT	C7H10N2	TOLUENEDIAMINE	95-80-7
BENZOHYDROQUINONE	C6H6O2	p-HYDROQUINONE	123-31-9
BENZOIC ACID	C7H6O2	BENZOIC ACID	65-85-0
BENZOIC ACID-n-BUTYL ESTER	C11H14O2	n-BUTYL BENZOATE	136-60-7
BENZOIC ACID, CHLORIDE	C7H5ClO	BENZOYL CHLORIDE	98-88-4
BENZOIC ACID NITRILE	C7H5N	BENZONITRILE	100-47-0
BENZOIC ACID, PHENYLMETHYL ESTER	C14H12O2	BENZYL BENZOATE	120-51-4
BENZOIC ALDEHYDE	C7H6O	BENZALDEHYDE	100-52-7
BENZOIC ETHER	C9H10O2	ETHYL BENZOATE	93-89-0
BENZOIC TRICHLORIDE	C7H5Cl3	BENZOTRICHLORIDE	98-07-7
BENZOL	C6H6	BENZENE	71-43-2
BENZOLE	C6H6	BENZENE	71-43-2
BENZOLENE	C6H6	BENZENE	71-43-2
BENZOLO	C6H6	BENZENE	71-43-2
BENZONITRILE	C7H5N	BENZONITRILE	100-47-0
BENZO(a)PHENANTHRENE	C18H12	CHRYSENE	218-01-9
BENZO(def)PHENANTHRENE	C16H10	PYRENE	129-00-0
1,2-BENZOPHENANTHRENE	C18H12	CHRYSENE	218-01-9
BENZOPHENONE	C13H10O	BENZOPHENONE	119-61-9
BENZO(b)PYRIDINE	C9H7N	QUINOLINE	91-22-5
BENZO(c)PYRIDINE	C9H7N	ISOQUINOLINE	119-65-3
BENZOPYRROLE	C8H7N	INDOLE	120-72-9
2,3-BENZOPYRROLE	C8H7N	INDOLE	120-72-9
BENZOQUINOL	C6H6O2	p-HYDROQUINONE	123-31-9

677

Name (Synonym)	Formula	Name in Tables	CAS Reg. No
BENZOTHIOPHENE	C8H6S	BENZOTHIOPHENE	95-15-8
BENZOTRICHLORIDE	C7H5Cl3	BENZOTRICHLORIDE	98-07-7
BENZOTRIFLUORIDE	C7H5F3	BENZOTRIFLUORIDE	98-08-8
BENZOYL ALCOHOL	C7H8O	BENZYL ALCOHOL	100-51-6
BENZOYLBENZENE	C13H10O	BENZOPHENONE	119-61-9
BENZOYL CHLORIDE	C7H5ClO	BENZOYL CHLORIDE	98-88-4
BENZOYL METHIDE	C8H8O	ACETOPHENONE	98-86-2
BENZ(a)PHENANTHRENE	C18H12	CHRYSENE	218-01-9
1,2-BENZPHENANTHRENE	C18H12	CHRYSENE	218-01-9
BENZYL ACETATE	C9H10O2	BENZYL ACETATE	140-11-4
BENZYL ALCOHOL	C7H8O	BENZYL ALCOHOL	100-51-6
BENZYL ALCOHOL BENZOIC ESTER	C14H12O2	BENZYL BENZOATE	120-51-4
BENZYLAMINE	C7H9N	BENZYLAMINE	100-46-9
BENZYL BENZENECARBOXYLATE	C14H12O2	BENZYL BENZOATE	120-51-4
BENZYL BENZOATE	C14H12O2	BENZYL BENZOATE	120-51-4
BENZYL CARBINOL	C8H10O	2-PHENYLETHANOL	60-12-8
BENZYLCHLORID	C7H7Cl	BENZYL CHLORIDE	100-44-7
BENZYL CHLORIDE	C7H7Cl	BENZYL CHLORIDE	100-44-7
BENZYL DICHLORIDE	C7H6Cl2	BENZYL DICHLORIDE	98-87-3
BENZYLE (CHLORURE de)	C7H7Cl	BENZYL CHLORIDE	100-44-7
BENZYLENE CHLORIDE	C7H6Cl2	BENZYL DICHLORIDE	98-87-3
BENZYL ETHYL ETHER	C9H12O	BENZYL ETHYL ETHER	539-30-0
BENZYLETS	C14H12O2	BENZYL BENZOATE	120-51-4
BENZYLIDENE CHLORIDE	C7H6Cl2	BENZYL DICHLORIDE	98-87-3
BENZYLIDYNE CHLORIDE	C7H5Cl3	BENZOTRICHLORIDE	98-07-7
BENZYLIDYNE FLUORIDE	C7H5F3	BENZOTRIFLUORIDE	98-08-8
BENZYL OXIDE	C14H14O	DIBENZYL ETHER	103-50-4
BENZYL PHENYLFORMATE	C14H12O2	BENZYL BENZOATE	120-51-4
BETAPRONE	C3H4O2	beta-PROPIOLACTONE	57-57-8
BETULA OIL	C8H8O3	METHYL SALICYLATE	119-36-8
BIALLYL	C6H10	1,5-HEXADIENE	592-42-7
N,N'-BIANILINE	C12H12N2	HYDRAZOBENZENE	122-66-7
BIC	C5H9NO	n-BUTYL ISOCYANATE	111-36-4
BICARBURET of HYDROGEN	C6H6	BENZENE	71-43-2
BICARBURETTED HYDROGEN	C2H4	ETHYLENE	74-85-1
BICHLORACETIC ACID	C2H2Cl2O2	DICHLOROACETIC ACID	79-43-6
BICHLORURE de PROPYLENE	C3H6Cl2	1,2-DICHLOROPROPANE	78-87-5
BICYCLO(0.3.5)DECA-1,3,5,7,9-PENTAENE	C10H8	AZULENE	275-51-4
BICYCLO(5.3.0)DECAPENTAENE	C10H8	AZULENE	275-51-4
BICYCLO(5.3.0)-DECA-2,4,6,8,10-PENTAENE	C10H8	AZULENE	275-51-4
BICYCLO(2,2,2)-1,4-DIAZAOCTANE	C6H12N2	TRIETHYLENEDIAMINE	280-57-9
BICYCLO(2.2.1)HEPTANE, 2,2-DIMETHYL-3-METHYLENE-	C10H16	CAMPHENE	79-92-5
BICYCLOHEXYL	C12H22	BICYCLOHEXYL	92-51-3
BICYCLOPENTADIENE	C10H12	DICYCLOPENTADIENE	77-73-6
BIETHYLENE	C4H6	1,3-BUTADIENE	106-99-0
BIG DIPPER	C12H11N	DIPHENYLAMINE	122-39-4
BIHEXYL	C12H26	n-DODECANE	112-40-3
BIMETHYL	C2H6	ETHANE	74-84-0
BINITROBENZENE	C6H4N2O4	m-DINITROBENZENE	99-65-0
BIOCIDE	C3H4O	ACROLEIN	107-02-8
BIOQUIN	C9H7NO	8-HYDROXYQUINOLINE	148-24-3
BIPHENYL	C12H10	BIPHENYL	92-54-4
BIPHENYLAMINE	C12H11N	p-AMINODIPHENYL	92-67-1
p-BIPHENYLAMINE	C12H11N	p-AMINODIPHENYL	92-67-1
(1,1'-BIPHENYL)-4-AMINE	C12H11N	p-AMINODIPHENYL	92-67-1
4-BIPHENYLAMINE	C12H11N	p-AMINODIPHENYL	92-67-1
2,2'-BIPHENYLENE OXIDE	C12H8O	DIBENZOFURAN	132-64-9
o-BIPHENYLENEMETHANE	C13H10	FLUORENE	86-73-7
o-BIPHENYLMETHANE	C13H10	FLUORENE	86-73-7
BIPHENYL OXIDE	C12H10O	DIPHENYL ETHER	101-84-8
BIRNENOEL	C7H14O2	n-PENTYL ACETATE	628-63-7
BIS(b-AMINOETHYL)AMINE	C4H13N3	DIETHYLENE TRIAMINE	111-40-0
BIS(2-AMINOETHYL)AMINE	C4H13N3	DIETHYLENE TRIAMINE	111-40-0
N,N'-BIS(2-AMINOETHYL)-1,2-DIAMINOETHANE	C6H18N4	TRIETHYLENE TETRAMINE	112-24-3
N,N'-BIS(2-AMINOETHYL)ETHYLENEDIAMINE	C6H18N4	TRIETHYLENE TETRAMINE	112-24-3
N,N'-BIS(2-AMINOETHYL)-1,2-ETHYLENEDIAMINE	C6H18N4	TRIETHYLENE TETRAMINE	112-24-3
BIS(BUTOXYETHYL) ETHER	C12H26O3	DIETHYLENE GLYCOL DI-n-BUTYL ETHER	112-73-2
BIS(2-BUTOXYETHYL) ETHER	C12H26O3	DIETHYLENE GLYCOL DI-n-BUTYL ETHER	112-73-2
BIS(2-BUTYL)ETHER	C8H18O	DI-sec-BUTYL ETHER	6863-58-7
BIS(n-BUTYL)SEBACATE	C18H34O4	DIBUTYL SEBACATE	109-43-3
BIS(CARBOXYMETHYL)ETHER	C4H6O5	DIGLYCOLIC ACID	110-99-6
BIS(CHLOROMETHYL)ETHER	C2H4Cl2O	BIS(CHLOROMETHYL)ETHER	542-88-1
BIS-CME	C2H4Cl2O	BIS(CHLOROMETHYL)ETHER	542-88-1
BIS(CYANOETHYL)ETHER	C6H8N2O	BIS(CYANOETHYL)ETHER	1656-48-0
BISCYCLOPENTADIENE	C10H12	DICYCLOPENTADIENE	77-73-6
2,2-BISDIHYDROXYMETHYL-1,3-PROPANEDIOL TETRANITRATE	C5H8N4O12	PENTAERYTHRITOL TETRANITRATE	78-11-5

678

Name (Synonym)	Formula	Name in Tables	CAS Reg. No
BIS(a,a-DIMETHYLBENZYL)PEROXIDE	C18H22O2	DICUMYL PEROXIDE	80-43-3
2,6-BIS(1,1-DIMETHYLETHYL)-4-METHYLPHENOL	C15H24O	2,6-DI-tert-BUTYL-p-CRESOL	128-37-0
BIS(2-ETHOXYETHYL)ETHER	C8H18O3	DIETHYLENE GLYCOL DIETHYL ETHER	112-36-7
BIS(2-ETHYLHEXYL)-1,2-BENZENEDICARBOXYLATE	C24H38O4	DIOCTYL PHTHALATE	117-81-7
BIS(2-ETHYLHEXYL)PHTHALATE	C24H38O4	DIOCTYL PHTHALATE	117-81-7
BISFEROL A	C15H16O2	BISPHENOL A	80-05-7
BIS(2-HYDROXYETHYL)AMINE	C4H11NO2	DIETHANOLAMINE	111-42-2
BIS(2-HYDROXYETHYL) ETHER	C4H10O3	DIETHYLENE GLYCOL	111-46-6
BIS(2-HYDROXYETHYL)METHYLAMINE	C5H13NO2	METHYL DIETHANOLAMINE	105-59-9
2,2-BIS-4'-HYDROXYFENYLPROPAN	C15H16O2	BISPHENOL A	80-05-7
2,2-BIS(HYDROXYMETHYL)-1,3-PROPANEDIOL	C5H12O4	PENTAERYTHRITOL	115-77-5
2,2-BIS(HYDROXYMETHYL)-1,3-PROPANEDIOL TETRANITRATE	C5H8N4O12	PENTAERYTHRITOL TETRANITRATE	78-11-5
BIS(4-HYDROXYPHENYL) DIMETHYLMETHANE	C15H16O2	BISPHENOL A	80-05-7
BIS(4-HYDROXYPHENYL)PROPANE	C15H16O2	BISPHENOL A	80-05-7
2,2-BIS(p-HYDROXYPHENYL)PROPANE	C15H16O2	BISPHENOL A	80-05-7
2,2-BIS(4-HYDROXYPHENYL)PROPANE	C15H16O2	BISPHENOL A	80-05-7
BIS(2-HYDROXYPROPYL)AMINE	C6H15NO2	DIISOPROPANOLAMINE	110-97-4
BIS(p-ISOCYANATOPHENYL)METHANE	C15H10N2O2	DIPHENYLMETHANE-4,4'-DIISOCYANATE	101-68-8
BIS(1,4-ISOCYANATOPHENYL)METHANE	C15H10N2O2	DIPHENYLMETHANE-4,4'-DIISOCYANATE	101-68-8
BIS(4-ISOCYANATOPHENYL)METHANE	C15H10N2O2	DIPHENYLMETHANE-4,4'-DIISOCYANATE	101-68-8
BIS(2-(2-METHOXYETHOXY)ETHYL) ETHER	C10H22O5	TETRAETHYLENE GLYCOL DIMETHYL ETHER	143-24-8
BIS(2-METHOXYETHYL)ETHER	C10H22O5	TETRAETHYLENE GLYCOL DIMETHYL ETHER	143-24-8
1,4-BIS(1-METHYLETHYL)BENZENE	C12H18	p-DIISOPROPYLBENZENE	100-18-5
BIS(6-METHYLHEPTYL)ESTER of PHTHALIC ACID	C24H38O4	DIISOOCTYL PHTHALATE	27554-26-3
BISOFLEX 81	C24H38O4	DIOCTYL PHTHALATE	117-81-7
BISOFLEX DOP	C24H38O4	DIOCTYL PHTHALATE	117-81-7
BISOMER 2HEA	C5H8O3	2-HYDROXYETHYL ACRYLATE	818-61-1
BISPHENOL A	C15H16O2	BISPHENOL A	80-05-7
BIS(2-PROPANOL)AMINE	C6H15NO2	DIISOPROPANOLAMINE	110-97-4
BIS(TRIMETHYLSILYL)AMINE	C6H19NSi2	HEXAMETHYLDISILAZANE	999-97-3
BIVINYL	C4H6	1,3-BUTADIENE	106-99-0
gamma-BL	C4H6O2	gamma-BUTYROLACTONE	96-48-0
BLACOSOLV	C2HCl3	TRICHLOROETHYLENE	79-01-6
BLASTING OIL	C3H5N3O9	NITROGLYCERINE	55-63-0
BLASTING GELATIN	C3H5N3O9	NITROGLYCERINE	55-63-0
BLO	C4H6O2	gamma-BUTYROLACTONE	96-48-0
BLON	C4H6O2	gamma-BUTYROLACTONE	96-48-0
BLUE OIL	C6H7N	ANILINE	62-53-3
BOLETIC ACID	C4H4O4	FUMARIC ACID	110-17-8
BONOFORM	C2H2Cl4	1,1,2,2-TETRACHLOROETHANE	79-34-5
BORER SOL	C2H4Cl2	1,2-DICHLOROETHANE	107-06-2
BORESTER 2	C12H27BO3	TRI-n-BUTYL BORATE	688-74-4
BORIC ACID, TRI-sec-BUTYL ESTER	C12H27BO3	TRI-n-BUTYL BORATE	688-74-4
2-BORNANONE	C10H16O	CAMPHOR	76-22-2
BPL	C3H4O2	beta-PROPIOLACTONE	57-57-8
BRASILAZINA OIL YELLOW G	C12H11N3	p-AMINOAZOBENZENE	60-09-3
BRECOLANE NDG	C4H10O3	DIETHYLENE GLYCOL	111-46-6
BRENTAMINE FAST ORANGE GR BASE	C6H6N2O2	o-NITROANILINE	88-74-4
BROCIDE	C2H4Cl2	1,2-DICHLOROETHANE	107-06-2
BROMALLYLENE	C3H5Br	3-BROMO-1-PROPENE	106-95-6
BROMIC ETHER	C2H5Br	BROMOETHANE	74-96-4
BROMOBENZENE	C6H5Br	BROMOBENZENE	108-86-1
1-BROMOBUTANE	C4H9Br	1-BROMOBUTANE	109-65-9
2-BROMOBUTANE	C4H9Br	2-BROMOBUTANE	78-76-2
BROMOCHLORODIFLUOROMETHANE	CBrClF2	BROMOCHLORODIFLUOROMETHANE	353-59-3
BROMOCHLOROMETHANE	CH2BrCl	BROMOCHLOROMETHANE	74-97-5
BROMOCHLOROTRIFLUOROETHANE	C2HBrClF3	HALOTHANE	151-67-7
2-BROMO-2-CHLORO-1,1,1-TRIFLUOROETHANE	C2HBrClF3	HALOTHANE	151-67-7
BROMOETHANE	C2H5Br	BROMOETHANE	74-96-4
BROMOETHENE	C2H3Br	VINYL BROMIDE	593-60-2
BROMOETHYLENE	C2H3Br	VINYL BROMIDE	593-60-2
BROMOFLUOROFORM	CBrF3	BROMOTRIFLUOROMETHANE	75-63-8
BROMOFORME	CHBr3	TRIBROMOMETHANE	75-25-2
BROMOFORMIO	CHBr3	TRIBROMOMETHANE	75-25-2
BROMOFUME	C2H4Br2	1,2-DIBROMOETHANE	106-93-4
1-BROMOHEPTANE	C7H15Br	1-BROMOHEPTANE	629-04-9
BROMOMETHANE	CH3Br	METHYL BROMIDE	74-83-9
a-BROMONAPHTHALENE	C10H7Br	1-BROMONAPHTHALENE	90-11-9
1-BROMONAPHTHALENE	C10H7Br	1-BROMONAPHTHALENE	90-11-9
BROMO-O-GAS	CH3Br	METHYL BROMIDE	74-83-9
1-BROMOPENTANE	C5H11Br	1-BROMOPENTANE	110-53-2
1-BROMOPROPANE	C3H7Br	1-BROMOPROPANE	106-94-5
2-BROMOPROPANE	C3H7Br	2-BROMOPROPANE	75-26-3
1-BROMO-2-PROPENE	C3H5Br	3-BROMO-1-PROPENE	106-95-6
3-BROMOPROPENE	C3H5Br	3-BROMO-1-PROPENE	106-95-6
3-BROMO-1-PROPENE	C3H5Br	3-BROMO-1-PROPENE	106-95-6

Name (Synonym)	Formula	Name in Tables	CAS Reg. No
3-BROMOPROPYLENE	C3H5Br	3-BROMO-1-PROPENE	106-95-6
p-BROMOTOLUENE	C7H7Br	p-BROMOTOLUENE	106-38-7
BROMOTRICHLOROMETHANE	CBrCl3	BROMOTRICHLOROMETHANE	75-62-7
BROMOTRIFLUOROETHENE	C2BrF3	BROMOTRIFLUOROETHYLENE	598-73-2
BROMOTRIFLUOROETHYLENE	C2BrF3	BROMOTRIFLUOROETHYLENE	598-73-2
BROMOTRIFLUOROMETHANE	CBrF3	BROMOTRIFLUOROMETHANE	75-63-8
BROMURE dETHYLE	C2H5Br	BROMOETHANE	74-96-4
BS	C22H44O2	n-BUTYL STEARATE	123-95-5
BTS	C3H4O3	PYRUVIC ACID	127-17-3
BUCB	C8H18O3	DIETHYLENE GLYCOL MONOBUTYL ETHER	112-34-5
BUCS	C6H14O2	2-BUTOXYETHANOL	111-76-2
BUKS	C15H24O	2,6-DI-tert-BUTYL-p-CRESOL	128-37-0
BUNT-CURE	C6Cl6	HEXACHLOROBENZENE	118-74-1
BUNT-NO-MORE	C6Cl6	HEXACHLOROBENZENE	118-74-1
a-g-BUTADIENE	C4H6	1,3-BUTADIENE	106-99-0
BUTA-1,3-DIENE	C4H6	1,3-BUTADIENE	106-99-0
1,2-BUTADIENE	C4H6	1,2-BUTADIENE	590-19-2
1,3-BUTADIENE	C4H6	1,3-BUTADIENE	106-99-0
BUTADIENE DIMER	C8H12	VINYLCYCLOHEXENE	100-40-3
1,3-BUTADIYNE	C4H2	BUTADIYNE(BIACETYLENE)	460-12-8
BUTADIYNE(BIACETYLENE)	C4H2	BUTADIYNE(BIACETYLENE)	460-12-8
BUTAFUME	C4H11N	sec-BUTYLAMINE	13952-84-6
BUTAL	C4H8O	n-BUTYRALDEHYDE	123-72-8
BUTALDEHYDE	C4H8O	n-BUTYRALDEHYDE	123-72-8
BUTALYDE	C4H8O	n-BUTYRALDEHYDE	123-72-8
BUTANAL	C4H8O	n-BUTYRALDEHYDE	123-72-8
n-BUTANAL	C4H8O	n-BUTYRALDEHYDE	123-72-8
1-BUTANAMINE	C4H11N	n-BUTYLAMINE	109-73-9
2-BUTANAMINE	C4H11N	sec-BUTYLAMINE	13952-84-6
1,3-BUTANDIOL	C4H10O2	1,3-BUTANEDIOL	107-88-0
2,3-BUTANDIOL	C4H10O2	2,3-BUTANEDIOL	6982-25-8
n-BUTANE	C4H10	n-BUTANE	106-97-8
BUTANECARBOXYLIC ACID	C5H10O2	VALERIC ACID	109-52-4
1-BUTANECARBOXYLIC ACID	C5H10O2	VALERIC ACID	109-52-4
1,4-BUTANEDICARBOXYLIC ACID	C6H10O4	ADIPIC ACID	124-04-9
BUTANE, 1,4-DIIODO-	C4H8I2	1,2-DIIODOBUTANE	628-21-7
1,4-BUTANEDINITRILE	C4H4N2	SUCCINONITRILE	110-61-2
BUTANEDIOIC ACID	C4H6O4	SUCCINIC ACID	110-15-6
BUTANEDIOIC ACID, DIETHYL ESTER	C8H14O4	DIETHYL SUCCINATE	123-25-1
BUTANEDIOIC ANHYDRIDE	C4H4O3	SUCCINIC ANHYDRIDE	108-30-5
BUTANE-1,3-DIOL	C4H10O2	1,3-BUTANEDIOL	107-88-0
BUTANE-1,4-DIOL	C4H10O2	1,4-BUTANEDIOL	110-63-4
BUTANE-2,3-DIOL	C4H10O2	2,3-BUTANEDIOL	6982-25-8
1,3-BUTANEDIOL	C4H10O2	1,3-BUTANEDIOL	107-88-0
1,4-BUTANEDIOL	C4H10O2	1,4-BUTANEDIOL	110-63-4
2,3-BUTANEDIOL	C4H10O2	2,3-BUTANEDIOL	6982-25-8
BUTANE MIXTURES	C4H10	n-BUTANE	106-97-8
BUTANEN	C4H10	n-BUTANE	106-97-8
BUTANENITRILE	C4H7N	n-BUTYRONITRILE	109-74-0
n-BUTANENITRILE	C4H7N	n-BUTYRONITRILE	109-74-0
BUTANE, 2,2'-OXYBIS-	C8H18O	DI-sec-BUTYL ETHER	6863-58-7
BUTANETHIOL	C4H10S	n-BUTYL MERCAPTAN	109-79-5
tert-BUTANETHIOL	C4H10S	tert-BUTYL MERCAPTAN	75-66-1
BUTANI	C4H10	n-BUTANE	106-97-8
BUTANOIC ACID	C4H8O2	n-BUTYRIC ACID	107-92-6
BUTANOIC ACID, ANHYDRIDE	C8H14O3	BUTYRIC ANHYDRIDE	106-31-0
BUTANOIC ACID ETHYL ESTER	C6H12O2	ETHYL n-BUTYRATE	105-54-4
BUTANOIC ACID, PROPYL ESTER	C7H14O2	n-PROPYL n-BUTYRATE	105-66-8
BUTANOIC ANHYDRIDE	C8H14O3	BUTYRIC ANHYDRIDE	106-31-0
BUTANOL	C4H10O	n-BUTANOL	71-36-3
n-BUTANOL	C4H10O	n-BUTANOL	71-36-3
sec-BUTANOL	C4H10O	sec-BUTANOL	78-92-2
tert-BUTANOL	C4H10O	tert-BUTANOL	75-65-0
BUTAN-1-OL	C4H10O	n-BUTANOL	71-36-3
BUTAN-2-OL	C4H10O	sec-BUTANOL	78-92-2
1-BUTANOL	C4H10O	n-BUTANOL	71-36-3
2-BUTANOL	C4H10O	sec-BUTANOL	78-92-2
2-BUTANOL ACETATE	C6H12O2	sec-BUTYL ACETATE	105-46-4
BUTANOLEN	C4H10O	n-BUTANOL	71-36-3
BUTANOLO	C4H10O	n-BUTANOL	71-36-3
BUTANOL SECONDAIRE	C4H10O	sec-BUTANOL	78-92-2
BUTANOL TERTIAIRE	C4H10O	tert-BUTANOL	75-65-0
2-BUTANONE	C4H8O	METHYL ETHYL KETONE	78-93-3
BUTANONE 2	C4H8O	METHYL ETHYL KETONE	78-93-3
(E)-2-BUTENAL	C4H6O	trans-CROTONALDEHYDE	123-73-9
trans-2-BUTENAL	C4H6O	trans-CROTONALDEHYDE	123-73-9

680

Name (Synonym)	Formula	Name in Tables	CAS Reg. No
1-BUTENE	C4H8	1-BUTENE	106-98-9
cis-2-BUTENE	C4H8	cis-2-BUTENE	590-18-1
trans-2-BUTENE	C4H8	trans-2-BUTENE	624-64-6
1-BUTENE, 3,4-DICHLORO-	C4H6Cl2	3,4-DICHLORO-1-BUTENE	760-23-6
1-BUTENE-cis-DICYANO	C6H6N2	cis-DICYANO-1-BUTENE	2141-58-4
1-BUTENE-trans-DICYANO	C6H6N2	trans-DICYANO-1-BUTENE	2141-59-5
cis-BUTENEDIOIC ACID	C4H4O4	MALEIC ACID	110-16-7
(E)-BUTENEDIOIC ACID	C4H4O4	FUMARIC ACID	110-17-8
trans-BUTENEDIOIC ACID	C4H4O4	FUMARIC ACID	110-17-8
(Z)-BUTENEDIOIC ACID	C4H4O4	MALEIC ACID	110-16-7
2-BUTENEDIOIC ACID, DIBUTYL ESTER	C12H18	HEXAMETHYLBENZENE	87-85-4
(Z)-2-BUTENEDIOIC ACID DIETHYL ESTER	C8H12O4	DIETHYL MALEATE	141-05-9
cis-BUTENEDIOIC ANHYDRIDE	C4H2O3	MALEIC ANHYDRIDE	108-31-6
cis-2-BUTENE-1,4-DIOL	C4H8O2	cis-2-BUTENE-1,4-DIOL	6117-80-2
trans-2-BUTENE-1,4-DIOL	C4H8O2	trans-2-BUTENE-1,4-DIOL	821-11-4
1-BUTENE-4-NITRILE	C4H5N	VINYLACETONITRILE	109-75-1
2-BUTENENITRILE	C4H5N	trans-CROTONITRILE	627-26-9
3-BUTENO-b-LACTONE	C4H4O2	DIKETENE	674-82-8
b-BUTENONITRILE	C4H5N	VINYLACETONITRILE	109-75-1
BUTEN-3-YNE	C4H4	VINYLACETYLENE	689-97-4
n-BUTILAMINA	C4H11N	n-BUTYLAMINE	109-73-9
BUTILE (ACETATI di)	C6H12O2	n-BUTYL ACETATE	123-86-4
BUTIL METACRILATO	C8H14O2	n-BUTYL METHACRYLATE	97-88-1
BUTOKSYETYLOWY ALKOHOL	C6H14O2	2-BUTOXYETHANOL	111-76-2
2-BUTOSSI-ETANOLO	C6H14O2	2-BUTOXYETHANOL	111-76-2
2-BUTOXY-AETHANOL	C6H14O2	2-BUTOXYETHANOL	111-76-2
1-BUTOXYBUTANE	C8H18O	DI-n-BUTYL ETHER	142-96-1
BUTOXYDIETHYLENE GLYCOL	C8H18O3	DIETHYLENE GLYCOL MONOBUTYL ETHER	112-34-5
BUTOXYDIGLYCOL	C8H18O3	DIETHYLENE GLYCOL MONOBUTYL ETHER	112-34-5
BUTOXYETHANOL	C6H14O2	2-BUTOXYETHANOL	111-76-2
n-BUTOXYETHANOL	C6H14O2	2-BUTOXYETHANOL	111-76-2
2-BUTOXYETHANOL	C6H14O2	2-BUTOXYETHANOL	111-76-2
2-BUTOXY-1-ETHANOL	C6H14O2	2-BUTOXYETHANOL	111-76-2
BUTOXYETHENE	C6H12O	BUTYL VINYL ETHER	111-34-2
2-(2-BUTOXYETHOXY)ETHANOL	C8H18O3	DIETHYLENE GLYCOL MONOBUTYL ETHER	112-34-5
BUTTERSAEURE	C4H8O2	n-BUTYRIC ACID	107-92-6
BUTYLACETAT	C6H12O2	n-BUTYL ACETATE	123-86-4
n-BUTYL ACETATE	C6H12O2	n-BUTYL ACETATE	123-86-4
sec-BUTYL ACETATE	C6H12O2	sec-BUTYL ACETATE	105-46-4
tert-BUTYL ACETATE	C6H12O2	tert-BUTYL ACETATE	540-88-5
BUTYL ACETATE	C6H12O2	n-BUTYL ACETATE	123-86-4
1-BUTYL ACETATE	C6H12O2	n-BUTYL ACETATE	123-86-4
2-BUTYL ACETATE	C6H12O2	sec-BUTYL ACETATE	105-46-4
BUTYLACETATEN	C6H12O2	n-BUTYL ACETATE	123-86-4
BUTYLACETIC ACID	C6H12O2	n-HEXANOIC ACID	142-62-1
BUTYL ACRYLATE	C7H12O2	n-BUTYL ACRYLATE	141-32-2
n-BUTYL ACRYLATE	C7H12O2	n-BUTYL ACRYLATE	141-32-2
BUTYLACRYLATE, INHIBITED	C7H12O2	n-BUTYL ACRYLATE	141-32-2
BUTYL ALCOHOL	C4H10O	n-BUTANOL	71-36-3
2-BUTYL ALCOHOL	C4H10O	sec-BUTANOL	78-92-2
sec-BUTYL ALCOHOL ACETATE	C6H12O2	sec-BUTYL ACETATE	105-46-4
n-BUTYL ALDEHYDE	C4H8O	n-BUTYRALDEHYDE	123-72-8
n-BUTYLAMIN	C4H11N	n-BUTYLAMINE	109-73-9
BUTYLAMINE	C4H11N	n-BUTYLAMINE	109-73-9
BUTYLAMINE, tertiary	C4H11N	tert-BUTYLAMINE	75-64-9
n-BUTYLAMINE	C4H11N	n-BUTYLAMINE	109-73-9
sec-BUTYLAMINE	C4H11N	sec-BUTYLAMINE	13952-84-6
tert-BUTYLAMINE	C4H11N	tert-BUTYLAMINE	75-64-9
BUTYLATED HYDROXYTOLUENE	C15H24O	2,6-DI-tert-BUTYL-p-CRESOL	128-37-0
n-BUTYLBENZENE	C10H14	n-BUTYLBENZENE	104-51-8
sec-BUTYLBENZENE	C10H14	sec-BUTYLBENZENE	135-98-8
tert-BUTYLBENZENE	C10H14	tert-BUTYLBENZENE	98-06-6
n-BUTYL BENZOATE	C11H14O2	n-BUTYL BENZOATE	136-60-7
BUTYL BORATE	C12H27BO3	TRI-n-BUTYL BORATE	688-74-4
n-BUTYL BORATE	C12H27BO3	TRI-n-BUTYL BORATE	688-74-4
BUTYL BROMIDE	C4H9Br	1-BROMOBUTANE	109-65-9
n-BUTYL BROMIDE	C4H9Br	1-BROMOBUTANE	109-65-9
sec-BUTYL BROMIDE	C4H9Br	2-BROMOBUTANE	78-76-2
N-BUTYL-1-BUTANAMINE	C8H19N	DI-n-BUTYLAMINE	111-92-2
BUTYL BUTYRATE	C8H16O2	n-BUTYL n-BUTYRATE	109-21-7
n-BUTYL BUTYRATE	C8H16O2	n-BUTYL n-BUTYRATE	109-21-7
n-BUTYL n-BUTYRATE	C8H16O2	n-BUTYL n-BUTYRATE	109-21-7
dl-sec-BUTYLCARBINOL	C5H12O	2-METHYL-1-BUTANOL	137-32-6
N-BUTYLCARBINOL	C5H12O	1-PENTANOL	71-41-0
BUTYL CARBITOL	C8H18O3	DIETHYLENE GLYCOL MONOBUTYL ETHER	112-34-5
p-tert-BUTYLCATECHOL	C10H14O2	p-tert-BUTYLCATECHOL	98-29-3

Name (Synonym)	Formula	Name in Tables	CAS Reg. No
4-tert-BUTYLCATECHOL	C10H14O2	p-tert-BUTYLCATECHOL	98-29-3
BUTYL CELLOSOLVE	C6H14O2	2-BUTOXYETHANOL	111-76-2
BUTYL CHLORIDE	C4H9Cl	n-BUTYL CHLORIDE	109-69-3
n-BUTYL CHLORIDE	C4H9Cl	n-BUTYL CHLORIDE	109-69-3
sec-BUTYL CHLORIDE	C4H9Cl	sec-BUTYL CHLORIDE	78-86-4
tert-BUTYL CHLORIDE	C4H9Cl	tert-BUTYL CHLORIDE	507-20-0
n-BUTYLCYCLOHEXANE	C10H20	n-BUTYLCYCLOHEXANE	1678-93-9
BUTYLCYCLOPENTANE	C9H18	BUTYLCYCLOPENTANE	2040-95-1
BUTYL-DECYL-SULFIDE	C14H30S	BUTYL-DECYL-SULFIDE	----------
o-BUTYL DIETHYLENE GLYCOL	C8H18O3	DIETHYLENE GLYCOL MONOBUTYL ETHER	112-34-5
BUTYL DIGLYME	C12H26O3	DIETHYLENE GLYCOL DI-n-BUTYL ETHER	112-73-2
BUTYL DIOXITOL	C8H18O3	DIETHYLENE GLYCOL MONOBUTYL ETHER	112-34-5
BUTYL-DODECYL-SULFIDE	C16H34S	BUTYL-DODECYL-SULFIDE	----------
BUTYLE (ACETATE de)	C6H12O2	n-BUTYL ACETATE	123-86-4
BUTYLENE	C4H8	1-BUTENE	106-98-9
a-BUTYLENE	C4H8	1-BUTENE	106-98-9
cis-2-BUTYLENE	C4H8	cis-2-BUTENE	590-18-1
g-BUTYLENE	C4H8	ISOBUTENE	115-11-7
trans-2-BUTYLENE	C4H8	trans-2-BUTENE	624-64-6
b-BUTYLENE GLYCOL	C4H10O2	1,3-BUTANEDIOL	107-88-0
1,3-BUTYLENE GLYCOL	C4H10O2	1,3-BUTANEDIOL	107-88-0
1,4-BUTYLENE GLYCOL	C4H10O2	1,4-BUTANEDIOL	110-63-4
2,3-BUTYLENE GLYCOL	C4H10O2	2,3-BUTANEDIOL	6982-25-8
BUTYLENE HYDRATE	C4H10O	sec-BUTANOL	78-92-2
BUTYLENE OXIDE	C4H8O	1,2-EPOXYBUTANE	106-88-7
1,2-BUTYLENE OXIDE	C4H8O	1,2-EPOXYBUTANE	106-88-7
BUTYLESTER KYSELINY MRAVENCI	C5H10O2	n-BUTYL FORMATE	592-84-7
BUTYL ETHANOATE	C6H12O2	n-BUTYL ACETATE	123-86-4
BUTYL ETHER	C8H18O	DI-n-BUTYL ETHER	142-96-1
BUTYL ETHYLENE	C6H12	1-HEXENE	592-41-6
o-BUTYL ETHYLENE GLYCOL	C6H14O2	2-BUTOXYETHANOL	111-76-2
n-BUTYL ETHYL ETHER	C6H14O	n-BUTYL ETHYL ETHER	628-81-9
BUTYL-ETHYL-SULFIDE	C6H14S	BUTYL-ETHYL-SULFIDE	638-46-0
p-tert-BUTYLFENOL	C10H14O	p-tert-BUTYLPHENOL	98-54-4
BUTYL FORMAL	C5H10O	VALERALDEHYDE	110-62-3
tert-BUTYLFORMAMIDE	C5H11NO	tert-BUTYLFORMAMIDE	2425-74-3
BUTYL FORMATE	C5H10O2	n-BUTYL FORMATE	592-84-7
n-BUTYL FORMATE	C5H10O2	n-BUTYL FORMATE	592-84-7
BUTYL GLYCOL	C6H14O2	2-BUTOXYETHANOL	111-76-2
sec-BUTYL GLYCOLATE	C6H12O3	sec-BUTYL GLYCOLATE	----------
BUTYL-HEPTYL-SULFIDE	C11H24S	BUTYL-HEPTYL-SULFIDE	----------
BUTYL-HEXADECYL-SULFIDE	C20H42S	BUTYL-HEXADECYL-SULFIDE	----------
BUTYL-HEXYL-SULFIDE	C10H22S	BUTYL-HEXYL-SULFIDE	----------
terc. BUTYLHYDROPEROXID	C4H10O2	t-BUTYL HYDROPEROXIDE	75-91-2
t-BUTYL HYDROPEROXIDE	C4H10O2	t-BUTYL HYDROPEROXIDE	75-91-2
BUTYL HYDROXIDE	C4H10O	n-BUTANOL	71-36-3
tert-BUTYL HYDROXIDE	C4H10O	tert-BUTANOL	75-65-0
BUTYLHYDROXYTOLUENE	C15H24O	2,6-DI-tert-BUTYL-p-CRESOL	128-37-0
tert-BUTYL IODIDE	C4H9I	2-IODO-2-METHYLPROPANE	558-17-8
n-BUTYL ISOCYANATE	C5H9NO	n-BUTYL ISOCYANATE	111-36-4
BUTYL MERCAPTAN	C4H10S	n-BUTYL MERCAPTAN	109-79-5
n-BUTYL MERCAPTAN	C4H10S	n-BUTYL MERCAPTAN	109-79-5
sec-BUTYL MERCAPTAN	C4H10S	sec-BUTYL MERCAPTAN	513-53-1
tert-BUTYL MERCAPTAN	C4H10S	tert-BUTYL MERCAPTAN	75-66-1
BUTYLMETHACRYLAAT	C8H14O2	n-BUTYL METHACRYLATE	97-88-1
BUTYL-2-METHACRYLATE	C8H14O2	n-BUTYL METHACRYLATE	97-88-1
n-BUTYL METHACRYLATE	C8H14O2	n-BUTYL METHACRYLATE	97-88-1
BUTYL METHYL KETONE	C6H12O	2-HEXANONE	591-78-6
n-BUTYL METHYL KETONE	C6H12O	2-HEXANONE	591-78-6
BUTYL-2-METHYL-2-PROPENOATE	C8H14O2	n-BUTYL METHACRYLATE	97-88-1
BUTYL-METHYL-SULFIDE	C5H12S	BUTYL-METHYL-SULFIDE	628-29-5
BUTYL MONOSULFIDE	C8H18S	DIBUTYL-SULFIDE	544-40-1
1-n-BUTYLNAPHTHALENE	C14H16	1-n-BUTYLNAPHTHALENE	1634-09-9
2-BUTYLNAPHTHALENE	C14H16	2-BUTYLNAPHTHALENE	----------
n-BUTYL NONANOATE	C13H26O2	n-BUTYL NONANOATE	50623-57-9
BUTYL-NONYL-SULFIDE	C13H28S	BUTYL-NONYL-SULFIDE	----------
BUTYL OCTADECANOATE	C22H44O2	n-BUTYL STEARATE	123-95-5
n-BUTYL OCTADECANOATE	C22H44O2	n-BUTYL STEARATE	123-95-5
BUTYL-OCTYL-SULFIDE	C12H26S	BUTYL-OCTYL-SULFIDE	----------
BUTYLOWY ALKOHOL	C4H10O	n-BUTANOL	71-36-3
BUTYL OXITOL	C6H14O2	2-BUTOXYETHANOL	111-76-2
BUTYL-PENTADECYL-SULFIDE	C19H40S	BUTYL-PENTADECYL-SULFIDE	----------
BUTYL-PENTYL-SULFIDE	C9H20S	BUTYL-PENTYL-SULFIDE	----------
BUTYLPHEN	C10H14O	p-tert-BUTYLPHENOL	98-54-4
p-tert-BUTYLPHENOL	C10H14O	p-tert-BUTYLPHENOL	98-54-4
n-BUTYL PHTHALATE	C16H22O4	DIBUTYL PHTHALATE	84-74-2

Name (Synonym)	Formula	Name in Tables	CAS Reg. No
BUTYL-2-PROPENOATE	C7H12O2	n-BUTYL ACRYLATE	141-32-2
BUTYL PROPIONATE	C7H14O2	n-BUTYL PROPIONATE	590-01-2
n-BUTYL PROPIONATE	C7H14O2	n-BUTYL PROPIONATE	590-01-2
BUTYL-PROPYL-SULFIDE	C7H16S	BUTYL-PROPYL-SULFIDE	----------
p-tert-BUTYLPYROCATECHOL	C10H14O2	p-tert-BUTYLCATECHOL	98-29-3
4-tert-BUTYLPYROKATECHIN	C10H14O2	p-tert-BUTYLCATECHOL	98-29-3
n-BUTYL STEARATE	C22H44O2	n-BUTYL STEARATE	123-95-5
n-BUTYL-SULFIDE	C8H18S	DIBUTYL-SULFIDE	544-40-1
BUTYL-TETRADECYL-SULFIDE	C18H38S	BUTYL-TETRADECYL-SULFIDE	----------
BUTYLTHIOBUTANE	C8H18S	DIBUTYL-SULFIDE	544-40-1
BUTYL-TRIDECYL-SULFIDE	C17H36S	BUTYL-TRIDECYL-SULFIDE	----------
BUTYL-UNDECYL-SULFIDE	C15H32S	BUTYL-UNDECYL-SULFIDE	----------
n-BUTYL VALERATE	C9H18O2	n-BUTYL VALERATE	591-68-4
BUTYL VINYL ETHER	C6H12O	BUTYL VINYL ETHER	111-34-2
1-BUTYNE	C4H6	ETHYLACETYLENE	107-00-6
2-BUTYNE	C4H6	DIMETHYLACETYLENE	503-17-3
1,4-BUTYNEDIOL	C4H6O2	2-BUTYNE-1,4-DIOL	110-65-6
2-BUTYNE-1,4-DIOL	C4H6O2	2-BUTYNE-1,4-DIOL	110-65-6
BUTYRAL	C4H8O	n-BUTYRALDEHYDE	123-72-8
BUTYRALDEHYD	C4H8O	n-BUTYRALDEHYDE	123-72-8
BUTYRALDEHYDE	C4H8O	n-BUTYRALDEHYDE	123-72-8
n-BUTYRALDEHYDE	C4H8O	n-BUTYRALDEHYDE	123-72-8
BUTYRANHYDRID	C8H14O3	BUTYRIC ANHYDRIDE	106-31-0
BUTYRIC or NORMAL PRIMARY BUTYL ALCOHOL	C4H10O	n-BUTANOL	71-36-3
n-BUTYRIC ACID	C4H8O2	n-BUTYRIC ACID	107-92-6
BUTYRIC ACID ANHYDRIDE	C8H14O3	BUTYRIC ANHYDRIDE	106-31-0
n-BUTYRIC ACID ANHYDRIDE	C8H14O3	BUTYRIC ANHYDRIDE	106-31-0
BUTYRIC ACID LACTONE	C4H6O2	gamma-BUTYROLACTONE	96-48-0
BUTYRIC ACID NITRILE	C4H7N	n-BUTYRONITRILE	109-74-0
BUTYRIC ACID, PROPYL ESTER	C7H14O2	n-PROPYL n-BUTYRATE	105-66-8
BUTYRIC ALDEHYDE	C4H8O	n-BUTYRALDEHYDE	123-72-8
BUTYRIC ANHYDRIDE	C8H14O3	BUTYRIC ANHYDRIDE	106-31-0
n-BUTYRIC ANHYDRIDE	C8H14O3	BUTYRIC ANHYDRIDE	106-31-0
BUTYRIC ETHER	C6H12O2	ETHYL n-BUTYRATE	105-54-4
BUTYROLACTAM	C4H7NO	2-PYRROLIDONE	616-45-5
gamma-BUTYROLACTAM	C4H7NO	2-PYRROLIDONE	616-45-5
alpha-BUTYROLACTONE	C4H6O2	gamma-BUTYROLACTONE	96-48-0
gamma-BUTYROLACTONE	C4H6O2	gamma-BUTYROLACTONE	96-48-0
BUTYRONITRILE	C4H7N	n-BUTYRONITRILE	109-74-0
n-BUTYRONITRILE	C4H7N	n-BUTYRONITRILE	109-74-0
BUTYRYL LACTONE	C4H6O2	gamma-BUTYROLACTONE	96-48-0
BUTYRYL OXIDE	C8H14O3	BUTYRIC ANHYDRIDE	106-31-0
C 10 ALCOHOL	C10H22O	1-DECANOL	112-30-1
C 10 ALDEHYDE	C10H20O	1-DECANAL	112-31-2
C 12 ALDEHYDE, LAURIC	C12H24O	1-DODECANAL	112-54-9
C 56	C5Cl6	HEXACHLOROCYCLOPENTADIENE	77-47-4
C 8 ACID	C8H16O2	n-OCTANOIC ACID	124-07-2
C 8 ALDEHYDE	C8H16O	1-OCTANAL	124-13-0
C 9 ALDEHYDE	C9H18O	1-NONANAL	124-19-6
CACHALOT C-50	C16H34O	1-HEXADECANOL	36653-82-4
CACHALOT L-50	C12H26O	1-DODECANOL	112-53-8
CADOX	C8H18O2	DI-t-BUTYL PEROXIDE	110-05-4
CADOX TBH	C4H10O2	t-BUTYL HYDROPEROXIDE	75-91-2
2-CAMPHANONE	C10H16O	CAMPHOR	76-22-2
CAMPHENE	C10H16	CAMPHENE	79-92-5
CAMPHOGEN	C10H14	p-CYMENE	99-87-6
CAMPHOR	C10H16O	CAMPHOR	76-22-2
CAMPHOR TAR	C10H8	NAPHTHALENE	91-20-3
CAO 1	C15H24O	2,6-DI-tert-BUTYL-p-CRESOL	128-37-0
CAO 3	C15H24O	2,6-DI-tert-BUTYL-p-CRESOL	128-37-0
CAPRALDEHYDE	C10H20O	1-DECANAL	112-31-2
CAPRIC ACID	C10H20O2	n-DECANOIC ACID	334-48-5
n-CAPRIC ACID	C10H20O2	n-DECANOIC ACID	334-48-5
CAPRIC ALCOHOL	C10H22O	1-DECANOL	112-30-1
CAPRIC ALDEHYDE	C10H20O	1-DECANAL	112-31-2
CAPRINALDEHYDE	C10H20O	1-DECANAL	112-31-2
CAPRINIC ACID	C10H20O2	n-DECANOIC ACID	334-48-5
CAPRINIC ALCOHOL	C10H22O	1-DECANOL	112-30-1
CAPRINIC ALDEHYDE	C10H20O	1-DECANAL	112-31-2
CAPROALDEHYDE	C6H12O	1-HEXANAL	66-25-1
CAPROIC ACID	C6H12O2	n-HEXANOIC ACID	142-62-1
n-CAPROIC ACID	C6H12O2	n-HEXANOIC ACID	142-62-1
CAPROIC ALDEHYDE	C6H12O	1-HEXANAL	66-25-1
epsilon-CAPROLACTAM	C6H11NO	epsilon-CAPROLACTAM	105-60-2
w-CAPROLACTAM	C6H11NO	epsilon-CAPROLACTAM	105-60-2
6-CAPROLACTAM	C6H11NO	epsilon-CAPROLACTAM	105-60-2

Name (Synonym)	Formula	Name in Tables	CAS Reg. No
CAPROLACTONE	C6H10O2	epsilon-CAPROLACTONE	502-44-3
e-CAPROLACTONE	C6H10O2	epsilon-CAPROLACTONE	502-44-3
epsilon-CAPROLACTONE	C6H10O2	epsilon-CAPROLACTONE	502-44-3
CAPROLATTAME	C6H11NO	epsilon-CAPROLACTAM	105-60-2
CAPRONALDEHYDE	C6H12O	1-HEXANAL	66-25-1
CAPRONIC ACID	C6H12O2	n-HEXANOIC ACID	142-62-1
CAPRONITRILE	C6H11N	HEXANENITRILE	628-73-9
CAPROYL ALCOHOL	C6H14O	1-HEXANOL	111-27-3
n-CAPROYLALDEHYDE	C6H12O	1-HEXANAL	66-25-1
CAPRYL ALCOHOL	C8H18O	1-OCTANOL	111-87-5
CAPRYLAMINE	C8H19N	n-OCTYLAMINE	111-86-4
CAPRYLIC ACID	C8H16O2	n-OCTANOIC ACID	124-07-2
n-CAPRYLIC ACID	C8H16O2	n-OCTANOIC ACID	124-07-2
CAPRYLIC ALCOHOL	C8H18O	1-OCTANOL	111-87-5
CAPRYLIC ETHER	C16H34O	DI-n-OCTYL ETHER	629-82-3
CAPRYLYLAMINE	C8H19N	n-OCTYLAMINE	111-86-4
CAPRYNIC ACID	C10H20O2	n-DECANOIC ACID	334-48-5
CARADATE 30	C15H10N2O2	DIPHENYLMETHANE-4,4'-DIISOCYANATE	101-68-8
CARBACRYL	C3H3N	ACRYLONITRILE	107-13-1
CARBAMALDEHYDE	CH3NO	FORMAMIDE	75-12-7
CARBANIL	C7H5NO	PHENYL ISOCYANATE	103-71-9
CARBANILALDEHYDE	C7H7NO	FORMANILIDE	103-70-8
9H-CARBAZOLE	C12H9N	DIBENZOPYRROLE	86-74-8
CARBETHOXYACETIC ESTER	C7H12O4	DIETHYL MALONATE	105-53-3
CARBINAMINE	CH5N	METHYLAMINE	74-89-5
CARBINOL	CH4O	METHANOL	67-56-1
CARBITOL	C6H14O3	2-(2-ETHOXYETHOXY)ETHANOL	111-90-0
CARBITOL SOLVENT	C6H14O3	2-(2-ETHOXYETHOXY)ETHANOL	111-90-0
CARBOLIC ACID	C6H6O	PHENOL	108-95-2
CARBOLSAEURE	C6H6O	PHENOL	108-95-2
CARBOMETHENE	C2H2O	KETENE	463-51-4
CARBON OXIDE	CO2	CARBON DIOXIDE	124-38-9
CARBONA	CCl4	CARBON TETRACHLORIDE	56-23-5
CARBON BICHLORIDE	C2Cl4	TETRACHLOROETHYLENE	127-18-4
CARBON BISULFIDE	CS2	CARBON DISULFIDE	75-15-0
CARBON BISULPHIDE	CS2	CARBON DISULFIDE	75-15-0
CARBON CHLORIDE	CCl4	CARBON TETRACHLORIDE	56-23-5
CARBON DICHLORIDE	C2Cl4	TETRACHLOROETHYLENE	127-18-4
CARBON DIFLUORIDE OXIDE	CF2O	CARBONYL FLUORIDE	353-50-4
CARBON DIOXIDE	CO2	CARBON DIOXIDE	124-38-9
CARBON DISULFIDE	CS2	CARBON DISULFIDE	75-15-0
CARBON DISULPHIDE	CS2	CARBON DISULFIDE	75-15-0
CARBONE	CCl2O	PHOSGENE	75-44-5
CARBON FLUORIDE	CF4	CARBON TETRAFLUORIDE	75-73-0
CARBON FLUORIDE OXIDE	CF2O	CARBONYL FLUORIDE	353-50-4
CARBON HEXACHLORIDE	C2Cl6	HEXACHLOROETHANE	67-72-1
CARBON HYDRIDE NITRIDE	CHN	HYDROGEN CYANIDE	74-90-8
CARBONIC ACID ANHYDRIDE	CO2	CARBON DIOXIDE	124-38-9
CARBONIC ACID, CYCLIC ETHYLENE ESTER	C3H4O3	ETHYLENE CARBONATE	96-49-1
CARBONIC ACID GAS	CO2	CARBON DIOXIDE	124-38-9
CARBONIC ANHYDRIDE	CO2	CARBON DIOXIDE	124-38-9
CARBONIC DIFLUORIDE	CF2O	CARBONYL FLUORIDE	353-50-4
CARBONIC OXIDE	CO	CARBON MONOXIDE	630-08-0
CARBONIO	CCl2O	PHOSGENE	75-44-5
CARBON MONOXIDE	CO	CARBON MONOXIDE	630-08-0
CARBON NITRIDE	C2N2	CYANOGEN	460-19-5
CARBON OIL	C6H6	BENZENE	71-43-2
CARBON OXIDE	CO	CARBON MONOXIDE	630-08-0
CARBON OXIDE SULFIDE	COS	CARBONYL SULFIDE	463-58-1
CARBON OXYCHLORIDE	CCl2O	PHOSGENE	75-44-5
CARBON OXYFLUORIDE	CF2O	CARBONYL FLUORIDE	353-50-4
CARBON OXYSULFIDE	COS	CARBONYL SULFIDE	463-58-1
CARBON SULFIDE	CS2	CARBON DISULFIDE	75-15-0
CARBON SULPHIDE	CS2	CARBON DISULFIDE	75-15-0
CARBON TET	CCl4	CARBON TETRACHLORIDE	56-23-5
CARBON TETRACHLORIDE	CCl4	CARBON TETRACHLORIDE	56-23-5
CARBON TETRAFLUORIDE	CF4	CARBON TETRAFLUORIDE	75-73-0
CARBON TRIFLUORIDE	CHF3	TRIFLUOROMETHANE	75-46-7
CARBONYLCHLORID	CCl2O	PHOSGENE	75-44-5
CARBONYL CHLORIDE	CCl2O	PHOSGENE	75-44-5
CARBONYL DIFLUORIDE	CF2O	CARBONYL FLUORIDE	353-50-4
CARBONYL FLUORIDE	CF2O	CARBONYL FLUORIDE	353-50-4
CARBONYL SULFIDE	COS	CARBONYL SULFIDE	463-58-1
CARBONYL SULFIDE-32S	COS	CARBONYL SULFIDE	463-58-1
CARBOXYBENZENE	C7H6O2	BENZOIC ACID	65-85-0
CARBOXYETHANE	C3H6O2	PROPIONIC ACID	79-09-4

Name (Synonym)	Formula	Name in Tables	CAS Reg. No
4-CARBOXYPHTHALIC ANHYDRIDE	C9H4O5	TRIMELLITIC ANHYDRIDE	552-30-7
CARDAMIST	C3H5N3O9	NITROGLYCERINE	55-63-0
CATALIN CAO-3	C15H24O	2,6-DI-tert-BUTYL-p-CRESOL	128-37-0
CATECHIN	C6H6O2	1,2-BENZENEDIOL	120-80-9
2-CBA	C7H5ClO2	o-CHLOROBENZOIC ACID	118-91-2
CCS 203	C4H10O	n-BUTANOL	71-36-3
CCS 301	C4H10O	sec-BUTANOL	78-92-2
CECOLENE	C2HCl3	TRICHLOROETHYLENE	79-01-6
CEKU C.B.	C6Cl6	HEXACHLOROBENZENE	118-74-1
CELLOFOR	C12H11N3	1,3-DIPHENYLTRIAZENE	136-35-6
CELLON	C2H2Cl4	1,1,2,2-TETRACHLOROETHANE	79-34-5
CELLOSOLVE	C4H10O2	2-ETHOXYETHANOL	110-80-5
CELLOSOLVE ACETATE	C6H12O3	2-ETHOXYETHYL ACETATE	111-15-9
CELLOSOLVE SOLVENT	C4H10O2	2-ETHOXYETHANOL	110-80-5
CELLUFLEX DPB	C16H22O4	DIBUTYL PHTHALATE	84-74-2
CELLUFLEX TPP	C18H15O4P	TRIPHENYL PHOSPHATE	115-86-6
CELMIDE	C2H4Br2	1,2-DIBROMOETHANE	106-93-4
CENTURY 1240	C18H36O2	STEARIC ACID	57-11-4
CENTURY CD FATTY ACID	C18H34O2	OLEIC ACID	112-80-1
CERES YELLOW R	C12H11N3	p-AMINOAZOBENZENE	60-09-3
CETAFFINE	C16H34O	1-HEXADECANOL	36653-82-4
CETAL	C16H34O	1-HEXADECANOL	36653-82-4
CETALOL CA	C16H34O	1-HEXADECANOL	36653-82-4
CETANE	C16H34	n-HEXADECANE	544-76-3
n-CETANE	C16H34	n-HEXADECANE	544-76-3
CETYL ALCOHOL	C16H34O	1-HEXADECANOL	36653-82-4
CETYLIC ACID	C16H32O2	n-HEXADECANOIC ACID	57-10-3
CETYLIC ALCOHOL	C16H34O	1-HEXADECANOL	36653-82-4
CETYLOL	C16H34O	1-HEXADECANOL	36653-82-4
CEVITAMIC ACID	C6H8O6	ASCORBIC ACID	50-81-7
CEVITAMIN	C6H8O6	ASCORBIC ACID	50-81-7
CFC-112	C2Cl4F2	1,1,2,2-TETRACHLORODIFLUOROETHANE	76-12-0
CFC 142b	C2H3ClF2	1-CHLORO-1,1-DIFLUOROETHANE	75-68-3
CFC 31	CH2ClF	CHLOROFLUOROMETHANE	593-70-4
CHA	C6H13N	CYCLOHEXYLAMINE	108-91-8
CHALOTHANE	C2HBrClF3	HALOTHANE	151-67-7
CHELEN	C2H5Cl	ETHYL CHLORIDE	75-00-3
CHEMANOX 11	C15H24O	2,6-DI-tert-BUTYL-p-CRESOL	128-37-0
CHEVRON ACETONE	C3H6O	ACETONE	67-64-1
CHINALDINE	C10H9N	QUINALDINE	91-63-4
CHINOLEINE	C9H7N	QUINOLINE	91-22-5
CHINOLIN	C9H7N	QUINOLINE	91-22-5
CHINOLINE	C9H7N	QUINOLINE	91-22-5
3-CHLOORANILINEN	C6H6ClN	m-CHLOROANILINE	108-42-9
CHLOORBENZEEN	C6H5Cl	MONOCHLOROBENZENE	108-90-7
1-CHLOOR-2,4-DINITROBENZEEN	C6H3ClN2O4	1-CHLORO-2,4-DINITROBENZENE	97-00-7
1-CHLOOR-2,3-EPOXY-PROPAAN	C3H5ClO	alpha-EPICHLOROHYDRIN	106-89-8
2-CHLOORETHANOL	C2H5ClO	2-CHLOROETHANOL	107-07-3
CHLOOR-METHAAN	CH3Cl	METHYL CHLORIDE	74-87-3
1-CHLOOR-4-NITROBENZEEN	C6H4ClNO2	p-CHLORONITROBENZENE	100-00-5
CHLORACETYL CHLORIDE	C2H2Cl2O	CHLOROACETYL CHLORIDE	79-04-9
2-CHLORAETHANOL	C2H5ClO	2-CHLOROETHANOL	107-07-3
CHLORAL	C2HCl3O	TRICHLOROACETALDEHYDE	75-87-6
CHLORALDEHYDE	C2H2Cl2O	DICHLOROACETALDEHYDE	79-02-7
CHLORALLYLENE	C3H5Cl	3-CHLOROPROPENE	107-05-1
CHLORAMEISENSAEUREAETHYLESTER	C3H5ClO2	ETHYL CHLOROFORMATE	541-41-3
4-CHLORANILIN	C6H6ClN	p-CHLOROANILINE	106-47-8
m-CHLORANILINE	C6H6ClN	m-CHLOROANILINE	108-42-9
o-CHLORANILINE	C6H6ClN	o-CHLOROANILINE	95-51-2
p-CHLORANILINE	C6H6ClN	p-CHLOROANILINE	106-47-8
CHLORBENZENE	C6H5Cl	MONOCHLOROBENZENE	108-90-7
CHLORBENZOL	C6H5Cl	MONOCHLOROBENZENE	108-90-7
CHLORCYAN	CClN	CYANOGEN CHLORIDE	506-77-4
1-CHLOR-2,4-DINITROBENZENE	C6H3ClN2O4	1-CHLORO-2,4-DINITROBENZENE	97-00-7
1-CHLOR-2,3-EPOXY-PROPAN	C3H5ClO	alpha-EPICHLOROHYDRIN	106-89-8
2-CHLORETHANOL	C2H5ClO	2-CHLOROETHANOL	107-07-3
CHLORETHENE	C2H3Cl	VINYL CHLORIDE	75-01-4
CHLORETHYL	C2H5Cl	ETHYL CHLORIDE	75-00-3
CHLORETHYLENE	C2H3Cl	VINYL CHLORIDE	75-01-4
p-CHLORFENOL	C6H5ClO	p-CHLOROPHENOL	106-48-9
CHLORID KYSELINY CHLOROCTOVE	C2H2Cl2O	CHLOROACETYL CHLORIDE	79-04-9
CHLORID KYSELINY DICHLOROCTOVE	C2HCl3O	DICHLOROACETYL CHLORIDE	79-36-7
CHLORIDUM	C2H5Cl	ETHYL CHLORIDE	75-00-3
CHLORINATED HYDROCHLORIC ETHER	C2H4Cl2	1,1-DICHLOROETHANE	75-34-3
CHLORINE CYANIDE	CClN	CYANOGEN CHLORIDE	506-77-4
CHLOR-METHAN	CH3Cl	METHYL CHLORIDE	74-87-3

Name (Synonym)	Formula	Name in Tables	CAS Reg. No
a-CHLORNAPHTHALENE	C10H7Cl	1-CHLORONAPHTHALENE	90-13-1
1-CHLOR-4-NITROBENZOL	C6H4ClNO2	p-CHLORONITROBENZENE	100-00-5
CHLOROACETALDEHYDE	C2H3ClO	CHLOROACETALDEHYDE	107-20-0
2-CHLOROACETALDEHYDE	C2H3ClO	CHLOROACETALDEHYDE	107-20-0
CHLOROACETALDEHYDE MONOMER	C2H3ClO	CHLOROACETALDEHYDE	107-20-0
CHLOROACETIC ACID	C2H3ClO2	CHLOROACETIC ACID	79-11-8
a-CHLOROACETIC ACID	C2H3ClO2	CHLOROACETIC ACID	79-11-8
CHLOROACETIC ACID CHLORIDE	C2H2Cl2O	CHLOROACETYL CHLORIDE	79-04-9
CHLOROACETIC ACID METHYL ESTER	C3H5ClO2	METHYL CHLOROACETATE	96-34-4
CHLOROACETIC CHLORIDE	C2H2Cl2O	CHLOROACETYL CHLORIDE	79-04-9
CHLOROACETYL CHLORIDE	C2H2Cl2O	CHLOROACETYL CHLORIDE	79-04-9
CHLOROALLYLENE	C3H5Cl	3-CHLOROPROPENE	107-05-1
m-CHLOROANILINE	C6H6ClN	m-CHLOROANILINE	108-42-9
o-CHLOROANILINE	C6H6ClN	o-CHLOROANILINE	95-51-2
p-CHLOROANILINE	C6H6ClN	p-CHLOROANILINE	106-47-8
3-CHLOROANILINE	C6H6ClN	m-CHLOROANILINE	108-42-9
CHLOROBEN	C6H4Cl2	o-DICHLOROBENZENE	95-50-1
CHLOROBENZAL	C7H6Cl2	BENZYL DICHLORIDE	98-87-3
a-CHLOROBENZALDEHYDE	C7H5ClO	BENZOYL CHLORIDE	98-88-4
CHLOROBENZEN	C6H5Cl	MONOCHLOROBENZENE	108-90-7
2-CHLORO-BENZENAMINE	C6H6ClN	o-CHLOROANILINE	95-51-2
3-CHLOROBENZENAMINE	C6H6ClN	m-CHLOROANILINE	108-42-9
4-CHLOROBENZENAMINE	C6H6ClN	p-CHLOROANILINE	106-47-8
4-CHLOROBENZENEAMINE	C6H6ClN	p-CHLOROANILINE	106-47-8
o-CHLOROBENZOIC ACID	C7H5ClO2	o-CHLOROBENZOIC ACID	118-91-2
CHLOROBENZOL	C6H5Cl	MONOCHLOROBENZENE	108-90-7
p-CHLOROBENZOTRIFLUORIDE	C7H4ClF3	p-CHLOROBENZOTRIFLUORIDE	98-56-6
m-CHLOROBENZOYL CHLORIDE	C7H4Cl2O	m-CHLOROBENZOYL CHLORIDE	618-46-2
CHLOROBUTADIENE	C4H5Cl	CHLOROPRENE	126-99-8
2-CHLORO-1,3-BUTADIENE	C4H5Cl	CHLOROPRENE	126-99-8
2-CHLOROBUTA-1,3-DIENE	C4H5Cl	CHLOROPRENE	126-99-8
1-CHLOROBUTANE	C4H9Cl	n-BUTYL CHLORIDE	109-69-3
CHLOROCARBONATE DETHYLE	C3H5ClO2	ETHYL CHLOROFORMATE	541-41-3
CHLOROCARBONIC ACID METHYL ESTER	C2H3ClO2	METHYL CHLOROFORMATE	79-22-1
CHLORO(CHLOROMETHOXY)METHANE	C2H4Cl2O	BIS(CHLOROMETHYL)ETHER	542-88-1
CHLOROCYAN	CClN	CYANOGEN CHLORIDE	506-77-4
CHLOROCYANIDE	CClN	CYANOGEN CHLORIDE	506-77-4
CHLOROCYANOGEN	CClN	CYANOGEN CHLORIDE	506-77-4
CHLORODEN	C6H4Cl2	o-DICHLOROBENZENE	95-50-1
1-CHLORO-2,2-DICHLOROETHYLENE	C2HCl3	TRICHLOROETHYLENE	79-01-6
CHLORODIFLUOROBROMOMETHANE	CBrClF2	BROMOCHLORODIFLUOROMETHANE	353-59-3
1-CHLORO-1,1-DIFLUOROETHANE	C2H3ClF2	1-CHLORO-1,1-DIFLUOROETHANE	75-68-3
CHLORODIFLUOROETHANES	C2H3ClF2	1-CHLORO-1,1-DIFLUOROETHANE	75-68-3
2-CHLORO-1,1-DIFLUOROETHYLENE	C2HClF2	2-CHLORO-1,1-DIFLUOROETHYLENE	359-10-4
CHLORODIFLUOROMETHANE	CHClF2	CHLORODIFLUOROMETHANE	75-45-6
CHLORODIFLUOROMONOBROMOMETHANE	CBrClF2	BROMOCHLORODIFLUOROMETHANE	353-59-3
1-CHLORO-2,4-DINITROBENZENE	C6H3ClN2O4	1-CHLORO-2,4-DINITROBENZENE	97-00-7
4-CHLORO-1,3-DINITROBENZENE	C6H3ClN2O4	1-CHLORO-2,4-DINITROBENZENE	97-00-7
6-CHLORO-1,3-DINITROBENZENE	C6H3ClN2O4	1-CHLORO-2,4-DINITROBENZENE	97-00-7
1-CHLORO-2,4-DINITROBENZOL	C6H3ClN2O4	1-CHLORO-2,4-DINITROBENZENE	97-00-7
1-CHLORO-2,3-EPOXYPROPANE	C3H5ClO	alpha-EPICHLOROHYDRIN	106-89-8
3-CHLORO-1,2-EPOXYPROPANE	C3H5ClO	alpha-EPICHLOROHYDRIN	106-89-8
CHLOROETENE	C2H3Cl3	1,1,1-TRICHLOROETHANE	71-55-6
2-CHLOROETHANAL	C2H3ClO	CHLOROACETALDEHYDE	107-20-0
2-CHLORO-1-ETHANAL	C2H3ClO	CHLOROACETALDEHYDE	107-20-0
CHLOROETHANE	C2H5Cl	ETHYL CHLORIDE	75-00-3
CHLOROETHANOIC ACID	C2H3ClO2	CHLOROACETIC ACID	79-11-8
D-CHLOROETHANOL	C2H5ClO	2-CHLOROETHANOL	107-07-3
2-CHLOROETHANOL	C2H5ClO	2-CHLOROETHANOL	107-07-3
CHLOROETHENE	C2H3Cl	VINYL CHLORIDE	75-01-4
b-CHLOROETHYL ALCOHOL	C2H5ClO	2-CHLOROETHANOL	107-07-3
2-CHLOROETHYL ALCOHOL	C2H5ClO	2-CHLOROETHANOL	107-07-3
CHLOROETHYLENE	C2H3Cl	VINYL CHLORIDE	75-01-4
CHLOROETHYLIDENE FLUORIDE	C2H3ClF2	1-CHLORO-1,1-DIFLUOROETHANE	75-68-3
a-CHLOROETHYLIDENE FLUORIDE	C2H3ClF2	1-CHLORO-1,1-DIFLUOROETHANE	75-68-3
CHLOROETHYLOWY ALKOHOL	C2H5ClO	2-CHLOROETHANOL	107-07-3
CHLOROFLUOROMETHANE	CH2ClF	CHLOROFLUOROMETHANE	593-70-4
CHLOROFORM	CHCl3	CHLOROFORM	67-66-3
CHLOROFORME	CHCl3	CHLOROFORM	67-66-3
CHLOROFORMIC ACID ETHYL ESTER	C3H5ClO2	ETHYL CHLOROFORMATE	541-41-3
CHLOROFORMIC ACID METHYL ESTER	C2H3ClO2	METHYL CHLOROFORMATE	79-22-1
CHLOROFORMYL CHLORIDE	CCl2O	PHOSGENE	75-44-5
epi-CHLOROHYDRIN	C3H5ClO	alpha-EPICHLOROHYDRIN	106-89-8
2-CHLOROISOBUTANE	C4H9Cl	tert-BUTYL CHLORIDE	507-20-0
CHLOROMETHANE	CH3Cl	METHYL CHLORIDE	74-87-3
CHLOROMETHYLBENZENE	C7H7Cl	BENZYL CHLORIDE	100-44-7

686

Name (Synonym)	Formula	Name in Tables	CAS Reg. No
2-CHLORO-1-METHYLBENZENE	C7H7Cl	o-CHLOROTOLUENE	95-49-8
4-CHLORO-1-METHYLBENZENE	C7H7Cl	p-CHLOROTOLUENE	106-43-4
1-CHLORO-3-METHYLBUTANE	C5H11Cl	1-CHLORO-3-METHYLBUTANE	107-84-6
2-CHLORO-2-METHYLBUTANE	C5H11Cl	2-CHLORO-2-METHYLBUTANE	594-36-5
(CHLOROMETHYL)ETHYLENE OXIDE	C3H5ClO	alpha-EPICHLOROHYDRIN	106-89-8
CHLOROMETHYLOXIRANE	C3H5ClO	alpha-EPICHLOROHYDRIN	106-89-8
2-(CHLOROMETHYL)OXIRANE	C3H5ClO	alpha-EPICHLOROHYDRIN	106-89-8
2-CHLORO-2-METHYLPROPANE	C4H9Cl	tert-BUTYL CHLORIDE	507-20-0
CHLOROMETHYLSILANE	CH5ClSi	METHYL CHLOROSILANE	993-00-0
a-CHLORONAPHTHALENE	C10H7Cl	1-CHLORONAPHTHALENE	90-13-1
CHLORO-m-NITROBENZENE	C6H4ClNO2	m-CHLORONITROBENZENE	121-73-3
m-CHLORONITROBENZENE	C6H4ClNO2	m-CHLORONITROBENZENE	121-73-3
o-CHLORONITROBENZENE	C6H4ClNO2	o-CHLORONITROBENZENE	88-73-3
p-CHLORONITROBENZENE	C6H4ClNO2	p-CHLORONITROBENZENE	100-00-5
1-CHLORONAPHTHALENE	C10H7Cl	1-CHLORONAPHTHALENE	90-13-1
1-CHLORO-2-NITROBENZENE	C6H4ClNO2	o-CHLORONITROBENZENE	88-73-3
1-CHLORO-4-NITROBENZENE	C6H4ClNO2	p-CHLORONITROBENZENE	100-00-5
2-CHLORONITROBENZENE	C6H4ClNO2	o-CHLORONITROBENZENE	88-73-3
2-CHLORO-1-NITROBENZENE	C6H4ClNO2	o-CHLORONITROBENZENE	88-73-3
4-CHLORONITROBENZENE	C6H4ClNO2	p-CHLORONITROBENZENE	100-00-5
4-CHLORO-1-NITROBENZENE	C6H4ClNO2	p-CHLORONITROBENZENE	100-00-5
4-CHLORO-3-NITROBENZOTRIFLUORIDE	C7H3ClF3NO2	4-CHLORO-3-NITROBENZOTRIFLUORIDE	121-71-5
CHLOROPENTAFLUOROETHANE	C2ClF5	CHLOROPENTAFLUOROETHANE	76-15-3
1-CHLOROPENTANE	C5H11Cl	1-CHLOROPENTANE	543-59-9
m-CHLOROPHENOL	C6H5ClO	m-CHLOROPHENOL	108-43-0
o-CHLOROPHENOL	C6H5ClO	o-CHLOROPHENOL	95-57-8
p-CHLOROPHENOL	C6H5ClO	p-CHLOROPHENOL	106-48-9
m-CHLOROPHENYLAMINE	C6H6ClN	m-CHLOROANILINE	108-42-9
3-CHLOROPHENYLAMINE	C6H6ClN	m-CHLOROANILINE	108-42-9
4-CHLOROPHENYLAMINE	C6H6ClN	p-CHLOROANILINE	106-47-8
p-CHLOROPHENYL CHLORIDE	C6H4Cl2	p-DICHLOROBENZENE	106-46-7
CHLOROPHENYLMETHANE	C7H7Cl	BENZYL CHLORIDE	100-44-7
(p-CHLOROPHENYL)TRIFLUOROMETHANE	C7H4ClF3	p-CHLOROBENZOTRIFLUORIDE	98-56-6
b-CHLOROPRENE	C4H5Cl	CHLOROPRENE	126-99-8
CHLOROPRENE	C4H5Cl	CHLOROPRENE	126-99-8
3-CHLOROPRENE	C3H5Cl	3-CHLOROPROPENE	107-05-1
1-CHLORO-2-PROPENE	C3H5Cl	3-CHLOROPROPENE	107-05-1
2-CHLOROPROPENE	C3H5Cl	2-CHLOROPROPENE	557-98-2
3-CHLOROPROPENE	C3H5Cl	3-CHLOROPROPENE	107-05-1
3-CHLORO-1-PROPENE	C3H5Cl	3-CHLOROPROPENE	107-05-1
1-CHLORO PROPENE-2	C3H5Cl	3-CHLOROPROPENE	107-05-1
a-CHLOROPROPYLENE	C3H5Cl	3-CHLOROPROPENE	107-05-1
3-CHLOROPROPYLENE	C3H5Cl	3-CHLOROPROPENE	107-05-1
3-CHLORO-1-PROPYLENE	C3H5Cl	3-CHLOROPROPENE	107-05-1
CHLOROPROPYLENE OXIDE	C3H5ClO	alpha-EPICHLOROHYDRIN	106-89-8
g-CHLOROPROPYLENE OXIDE	C3H5ClO	alpha-EPICHLOROHYDRIN	106-89-8
3-CHLORO-1,2-PROPYLENE OXIDE	C3H5ClO	alpha-EPICHLOROHYDRIN	106-89-8
6-CHLORO-1H-PURINE	C3H3Cl	PROPARGYL CHLORIDE	624-65-7
6-CHLORO-9H-PURINE	C3H3Cl	PROPARGYL CHLORIDE	624-65-7
CHLOROTHANE NU	C2H3Cl3	1,1,1-TRICHLOROETHANE	71-55-6
CHLOROTHENE	C2H3Cl3	1,1,1-TRICHLOROETHANE	71-55-6
CHLOROTHENE VG	C2H3Cl3	1,1,1-TRICHLOROETHANE	71-55-6
2,3,4,5-CHLOROTHIOPHENE	C4Cl4S	TETRACHLOROTHIOPHENE	6012-97-1
a-CHLOROTOLUENE	C7H7Cl	BENZYL CHLORIDE	100-44-7
o-CHLOROTOLUENE	C7H7Cl	o-CHLOROTOLUENE	95-49-8
p-CHLOROTOLUENE	C7H7Cl	p-CHLOROTOLUENE	106-43-4
w-CHLOROTOLUENE	C7H7Cl	BENZYL CHLORIDE	100-44-7
2-CHLOROTOLUENE	C7H7Cl	o-CHLOROTOLUENE	95-49-8
4-CHLOROTOLUENE	C7H7Cl	p-CHLOROTOLUENE	106-43-4
CHLOROTRIFLUOROETHYLENE	C2ClF3	CHLOROTRIFLUOROETHYLENE	79-38-9
1-CHLORO-1,2,2-TRIFLUOROETHYLENE	C2ClF3	CHLOROTRIFLUOROETHYLENE	79-38-9
2-CHLORO-1,1,2-TRIFLUOROETHYLENE	C2ClF3	CHLOROTRIFLUOROETHYLENE	79-38-9
CHLOROTRIFLUOROMETHANE	CClF3	CHLOROTRIFLUOROMETHANE	75-72-9
p-CHLOROTRIFLUOROMETHYLBENZENE	C7H4ClF3	p-CHLOROBENZOTRIFLUORIDE	98-56-6
4-CHLOROTRIFLUOROMETHYLBENZENE	C7H4ClF3	p-CHLOROBENZOTRIFLUORIDE	98-56-6
1-CHLORO-4-(TRIMETHYL)-BENZENE (9CI)	C7H4ClF3	p-CHLOROBENZOTRIFLUORIDE	98-56-6
3-CHLORPROPEN	C3H5Cl	3-CHLOROPROPENE	107-05-1
CHLORTEN	C2H3Cl3	1,1,1-TRICHLOROETHANE	71-55-6
a-CHLORTOLUOL	C7H7Cl	BENZYL CHLORIDE	100-44-7
CHLORTRIFLUORAETHYLEN	C2ClF3	CHLOROTRIFLUOROETHYLENE	79-38-9
CHLORURE de BENZENYLE	C7H5Cl3	BENZOTRICHLORIDE	98-07-7
CHLORURE de BENZYLE	C7H7Cl	BENZYL CHLORIDE	100-44-7
CHLORURE DE BENZYLIDENE	C7H6Cl2	BENZYL DICHLORIDE	98-87-3
CHLORURE de BUTYLE	C4H9Cl	n-BUTYL CHLORIDE	109-69-3
CHLORURE de CHLORACETYLE	C2H2Cl2O	CHLOROACETYL CHLORIDE	79-04-9
CHLORURE DE CYANOGENE	CClN	CYANOGEN CHLORIDE	506-77-4

687

Name (Synonym)	Formula	Name in Tables	CAS Reg. No
CHLORURE de DICHLORACETYLE	C2HCl3O	DICHLOROACETYL CHLORIDE	79-36-7
CHLORURE de METHYLE	CH3Cl	METHYL CHLORIDE	74-87-3
CHLORURE de METHYLENE	CH2Cl2	DICHLOROMETHANE	75-09-2
CHLORURE d'ETHYLIDENE	C2H4Cl2	1,1-DICHLOROETHANE	75-34-3
CHLORURE de VINYLIDENE	C2H2Cl2	1,1-DICHLOROETHYLENE	75-35-4
CHLORYL	C2H5Cl	ETHYL CHLORIDE	75-00-3
CHLORYL ANESTHETIC	C2H5Cl	ETHYL CHLORIDE	75-00-3
CHLORYLEA	C2HCl3	TRICHLOROETHYLENE	79-01-6
CHOLAXINE	C6H14O6	SORBITOL	50-70-4
CHOT	C5H8N4O12	PENTAERYTHRITOL TETRANITRATE	78-11-5
CHROMAR	C8H10	p-XYLENE	106-42-3
CHRYSENE	C18H12	CHRYSENE	218-01-9
C.I. 10355	C12H11N	DIPHENYLAMINE	122-39-4
C.I. 11000	C12H11N3	p-AMINOAZOBENZENE	60-09-3
C.I. 37025	C6H6N2O2	o-NITROANILINE	88-74-4
C.I. 37030	C6H6N2O2	m-NITROANILINE	99-09-2
C.I. 37035	C6H6N2O2	p-NITROANILINE	100-01-6
C.I. 37077	C7H9N	o-TOLUIDINE	95-53-4
C.I. 37107	C7H9N	p-TOLUIDINE	106-49-0
C.I. 37240	C12H12N2	p-AMINODIPHENYLAMINE	101-54-2
C.I. 76000	C6H7N	ANILINE	62-53-3
C.I. 76010	C6H8N2	o-PHENYLENEDIAMINE	95-54-5
C.I. 76025	C6H8N2	m-PHENYLENEDIAMINE	108-45-2
C.I. 76085	C12H12N2	p-AMINODIPHENYLAMINE	101-54-2
C.I. 76505	C6H6O2	1,3-BENZENEDIOL	108-46-3
C.I. 76515	C6H6O3	1,2,3-BENZENETRIOL	87-66-1
C.I. AZOIC COUPLING COMPONENT 107	C7H9N	p-TOLUIDINE	106-49-0
C.I. AZOIC DIAZO COMPONENT 22	C12H12N2	p-AMINODIPHENYLAMINE	101-54-2
C.I. AZOIC DIAZO COMPONENT 37	C6H6N2O2	p-NITROANILINE	100-01-6
C.I. AZOIC DIAZO COMPONENT 6	C6H6N2O2	o-NITROANILINE	88-74-4
C.I. AZOIC DIAZO COMPONENT 7	C6H6N2O2	m-NITROANILINE	99-09-2
CICLOESANO	C6H12	CYCLOHEXANE	110-82-7
CICLOESANOLO	C6H12O	CYCLOHEXANOL	108-93-0
CICLOESANONE	C6H10O	CYCLOHEXANONE	108-94-1
C.I. DEVELOPER 15	C12H12N2	p-AMINODIPHENYLAMINE	101-54-2
C.I. DEVELOPER 17	C6H6N2O2	p-NITROANILINE	100-01-6
C.I. DEVELOPER 4	C6H6O2	1,3-BENZENEDIOL	108-46-3
CINNAMENE	C8H8	STYRENE	100-42-5
CINNAMENOL	C8H8	STYRENE	100-42-5
C.I. OXIDATION BASE 16	C6H8N2	o-PHENYLENEDIAMINE	95-54-5
C.I. OXIDATION BASE 2	C12H12N2	p-AMINODIPHENYLAMINE	101-54-2
C.I. OXIDATION BASE 26	C6H6O2	1,2-BENZENEDIOL	120-80-9
C.I. OXIDATION BASE 31	C6H6O2	1,3-BENZENEDIOL	108-46-3
C.I. OXIDATION BASE 32	C6H6O3	1,2,3-BENZENETRIOL	87-66-1
CIRRASOL 185A	C9H18O2	n-NONANOIC ACID	112-05-0
C.I. SOLVENT BLUE 7	C12H11N3	p-AMINOAZOBENZENE	60-09-3
C.I. SOLVENT YELLOW 1	C12H11N3	p-AMINOAZOBENZENE	60-09-3
CITRACONIC ACID	C5H6O4	CITRACONIC ACID	498-23-7
CITRETTEN	C6H8O7	CITRIC ACID	77-92-9
CITRIC ACID	C6H8O7	CITRIC ACID	77-92-9
CITRO	C6H8O7	CITRIC ACID	77-92-9
CLORALIO	C2HCl3O	TRICHLOROACETALDEHYDE	75-87-6
CLOROBEN	C6H4Cl2	o-DICHLOROBENZENE	95-50-1
CLOROBENZENE	C6H5Cl	MONOCHLOROBENZENE	108-90-7
1-CLORO-2,4-DINITROBENZENE	C6H3ClN2O4	1-CHLORO-2,4-DINITROBENZENE	97-00-7
CLOROFORMIO	CHCl3	CHLOROFORM	67-66-3
CLOROMETANO	CH3Cl	METHYL CHLORIDE	74-87-3
1-CLORO-4-NITROBENZENE	C6H4ClNO2	p-CHLORONITROBENZENE	100-00-5
CLORURO di ETHENE	C2H4Cl2	1,2-DICHLOROETHANE	107-06-2
CLORURO di ETILIDENE	C2H4Cl2	1,1-DICHLOROETHANE	75-34-3
CLORURO di METILE	CH3Cl	METHYL CHLORIDE	74-87-3
CIP	C3H3Cl	PROPARGYL CHLORIDE	624-65-7
CO 12	C12H26O	1-DODECANOL	112-53-8
CO 1214	C12H26O	1-DODECANOL	112-53-8
CO 1670	C16H34O	1-HEXADECANOL	36653-82-4
CO 1895	C18H38O	1-OCTADECANOL	112-92-5
CO 1897	C18H38O	1-OCTADECANOL	112-92-5
COAL NAPHTHA	C6H6	BENZENE	71-43-2
COAL TAR PITCH VOLATILES: PHENANTHRENE	C14H10	PHENANTHRENE	85-01-8
COLAMINE	C2H7NO	MONOETHANOLAMINE	141-43-5
COLEBENZ	C14H12O2	BENZYL BENZOATE	120-51-4
a-g,a'-COLLIDINE	C8H11N	2,4,6-TRIMETHYLPYRIDINE	108-75-8
g-COLLIDINE	C8H11N	2,4,6-TRIMETHYLPYRIDINE	108-75-8
s-COLLIDINE	C8H11N	2,4,6-TRIMETHYLPYRIDINE	108-75-8
sym-COLLIDINE	C8H11N	2,4,6-TRIMETHYLPYRIDINE	108-75-8
2,4,6-COLLIDINE	C8H11N	2,4,6-TRIMETHYLPYRIDINE	108-75-8

Name (Synonym)	Formula	Name in Tables	CAS Reg. No
COLOGNE SPIRIT	C2H6O	ETHANOL	64-17-5
COLONIAL SPIRIT	CH4O	METHANOL	67-56-1
COLUMBIAN SPIRITS	CH4O	METHANOL	67-56-1
COMPOUND 889	C24H38O4	DIOCTYL PHTHALATE	117-81-7
CO-OP HEXA	C6Cl6	HEXACHLOROBENZENE	118-74-1
CORFLEX 880	C24H38O4	DIISOOCTYL PHTHALATE	27554-26-3
CP 34	C4H4S	THIOPHENE	110-02-1
m-CRESOL	C7H8O	m-CRESOL	108-39-4
o-CRESOL	C7H8O	o-CRESOL	95-48-7
p-CRESOL	C7H8O	p-CRESOL	106-44-5
2-CRESOL	C7H8O	o-CRESOL	95-48-7
3-CRESOL	C7H8O	m-CRESOL	108-39-4
4-CRESOL	C7H8O	p-CRESOL	106-44-5
m-CRESYLIC ACID	C7H8O	m-CRESOL	108-39-4
o-CRESYLIC ACID	C7H8O	o-CRESOL	95-48-7
p-CRESYLIC ACID	C7H8O	p-CRESOL	106-44-5
o-CRESYL PHOSPHATE	C21H21O4P	TRI-o-CRESYL PHOSPHATE	78-30-8
CRODACID	C14H28O2	n-TETRADECANOIC ACID	544-63-8
CRODACOL-CAS	C16H34O	1-HEXADECANOL	36653-82-4
CRODACOL-S	C18H38O	1-OCTADECANOL	112-92-5
CROLEAN	C3H4O	ACROLEIN	107-02-8
CROTONAL	C4H6O	trans-CROTONALDEHYDE	123-73-9
CROTONALDEHYDE	C4H6O	trans-CROTONALDEHYDE	123-73-9
trans-CROTONALDEHYDE	C4H6O	trans-CROTONALDEHYDE	123-73-9
cis-CROTONIC ACID	C4H6O2	cis-CROTONIC ACID	503-64-0
trans-CROTONIC ACID	C4H6O2	trans-CROTONIC ACID	107-93-7
CROTONIC ALDEHYDE	C4H6O	trans-CROTONALDEHYDE	123-73-9
cis-CROTONITRILE	C4H5N	cis-CROTONITRILE	1190-76-7
trans-CROTONITRILE	C4H5N	trans-CROTONITRILE	627-26-9
CRYOFLUORAN	C2Cl2F4	1,2-DICHLOROTETRAFLUOROETHANE	76-14-2
CRYOFLUORANE	C2Cl2F4	1,2-DICHLOROTETRAFLUOROETHANE	76-14-2
CSAC	C6H12O3	2-ETHOXYETHYL ACETATE	111-15-9
CTFE	C2ClF3	CHLOROTRIFLUOROETHYLENE	79-38-9
CUM	C9H12	CUMENE	98-82-8
CUMEEN	C9H12	CUMENE	98-82-8
CUMEENHYDROPEROXYDE	C9H12O2	CUMENE HYDROPEROXIDE	80-15-9
CUMENE	C9H12	CUMENE	98-82-8
psi-CUMENE	C9H12	1,2,4-TRIMETHYLBENZENE	95-63-6
CUMENE HYDROPEROXIDE	C9H12O2	CUMENE HYDROPEROXIDE	80-15-9
CUMENE PEROXIDE	C18H22O2	DICUMYL PEROXIDE	80-43-3
CUMENT HYDROPEROXIDE	C9H12O2	CUMENE HYDROPEROXIDE	80-15-9
CUMENYL HYDROPEROXIDE	C9H12O2	CUMENE HYDROPEROXIDE	80-15-9
CUMOLHYDROPEROXID	C9H12O2	CUMENE HYDROPEROXIDE	80-15-9
a-CUMYL ALCOHOL	C9H12O	2-PHENYL-2-PROPANOL	617-94-7
CUMYL HYDROPEROXIDE	C9H12O2	CUMENE HYDROPEROXIDE	80-15-9
a-CUMYL HYDROPEROXIDE	C9H12O2	CUMENE HYDROPEROXIDE	80-15-9
CUMYL PEROXIDE	C18H22O2	DICUMYL PEROXIDE	80-43-3
p-(a-CUMYL)PHENOL	C15H16O	p-CUMYLPHENOL	599-64-4
p-CUMYLPHENOL	C15H16O	p-CUMYLPHENOL	599-64-4
CURITHANE 103	C4H6O2	METHYL ACRYLATE	96-33-3
CYANACETATE ETHYLE	C5H7NO2	ETHYL CYANOACETATE	105-56-6
ISO-CYANATOMETHANE	C2H3NO	METHYL ISOCYANATE	624-83-9
CYANOACETIC ACID ETHYL ESTER	C5H7NO2	ETHYL CYANOACETATE	105-56-6
CYANOACETIC ACID METHYL ESTER	C4H5NO2	METHYL CYANOACETATE	105-34-0
CYANOACETIC ESTER	C5H7NO2	ETHYL CYANOACETATE	105-56-6
CYANOACETONITRILE	C3H2N2	MALONONITRILE	109-77-3
CYANOBENZENE	C7H5N	BENZONITRILE	100-47-0
1-CYANOBUTANE	C5H9N	VALERONITRILE	110-59-8
CYANOETHANE	C3H5N	PROPIONITRILE	107-12-0
2-CYANOETHANOL	C3H5NO	HYDRACRYLONITRILE	109-78-4
2-CYANOETHYL ALCOHOL	C3H5NO	HYDRACRYLONITRILE	109-78-4
CYANOETHYLENE	C3H3N	ACRYLONITRILE	107-13-1
CYANOGEN	C2N2	CYANOGEN	460-19-5
CYANOGEN CHLORIDE	CClN	CYANOGEN CHLORIDE	506-77-4
CYANOGEN CHLORIDE,inhibited	CClN	CYANOGEN CHLORIDE	506-77-4
CYANOGENE	C2N2	CYANOGEN	460-19-5
CYANOGEN GAS	C2N2	CYANOGEN	460-19-5
CYANOMETHANE	C2H3N	ACETONITRILE	75-05-8
1-CYANO-2-METHOXYETHANE	C4H7NO	3-METHOXYPROPIONITRILE	110-67-8
1-CYANOPROPANE	C4H7N	n-BUTYRONITRILE	109-74-0
2-CYANOPROPANE	C4H7N	ISOBUTYRONITRILE	78-82-0
2-CYANOPROPENE-1	C4H5N	METHACRYLONITRILE	126-98-7
CYANURE de VINYLE	C3H3N	ACRYLONITRILE	107-13-1
CYCLAL CETYL ALCOHOL	C16H34O	1-HEXADECANOL	36653-82-4
CYCLIC DIMETHYLSILOXANE TETRAMER	C8H24O4Si4	OCTAMETHYLCYCLOTETRASILOXANE	556-67-2
CYCLIC ETHYLENE CARBONATE	C3H4O3	ETHYLENE CARBONATE	96-49-1

Name (Synonym)	Formula	Name in Tables	CAS Reg. No
CYCLIC TETRAMETHYLENE SULFONE	C4H8O2S	SULFOLANE	126-33-0
CYCLOBUTANE	C4H8	CYCLOBUTANE	287-23-0
CYCLOBUTENE	C4H6	CYCLOBUTENE	822-35-5
CYCLOHEPTANE	C7H14	CYCLOHEPTANE	291-64-5
1,3,5-CYCLOHEPTATRIENE	C7H8	1,3,5-CYCLOHEPTATRIENE	544-55-2
CYCLOHEXAAN	C6H12	CYCLOHEXANE	110-82-7
1,3-CYCLOHEXADIENE	C6H8	1,3-CYCLOHEXADIENE	592-57-4
CYCLOHEXAMETHYLENIMINE	C6H13N	HEXAMETHYLENEIMINE	111-49-9
CYCLOHEXAN	C6H12	CYCLOHEXANE	110-82-7
CYCLOHEXANAMINE	C6H13N	CYCLOHEXYLAMINE	108-91-8
CYCLOHEXANAMINE, N-METHYL-	C7H15N	N-METHYLCYCLOHEXYLAMINE	100-60-7
CYCLOHEXANE	C6H12	CYCLOHEXANE	110-82-7
CYCLOHEXANE, ISOCYANATO-(9CI)	C7H11NO	CYCLOHEXYL ISOCYANATE	3173-53-3
1,4-CYCLOHEXANEDICARBOXYLIC ACID	C8H12O4	1,4-CYCLOHEXANEDICARBOXYLIC ACID	619-82-9
CYCLOHEXANOL	C6H12O	CYCLOHEXANOL	108-93-0
CYCLOHEXANON	C6H10O	CYCLOHEXANONE	108-94-1
CYCLOHEXANONE	C6H10O	CYCLOHEXANONE	108-94-1
CYCLOHEXANONE ISO-OXIME	C6H11NO	epsilon-CAPROLACTAM	105-60-2
CYCLOHEXANONE OXIME	C6H11NO	CYCLOHEXANONE OXIME	100-64-1
CYCLOHEXATRIENE	C6H6	BENZENE	71-43-2
CYCLOHEXENE	C6H10	CYCLOHEXENE	110-83-8
CYCLOHEXYL ALCOHOL	C6H12O	CYCLOHEXANOL	108-93-0
CYCLOHEXYLAMINE	C6H13N	CYCLOHEXYLAMINE	108-91-8
CYCLOHEXYLBENZENE	C12H16	CYCLOHEXYLBENZENE	827-52-1
N-CYCLOHEXYLCYCLOHEXANAMINE	C12H23N	DICYCLOHEXYLAMINE	101-83-7
1-CYCLOHEXYLDODECANE	C18H36	1-CYCLOHEXYLDODECANE	1795-17-1
1-CYCLOHEXYLHEPTANE	C13H26	1-CYCLOHEXYLHEPTANE	5617-41-4
1-CYCLOHEXYLHEXADECANE	C22H44	1-CYCLOHEXYLHEXADECANE	6892-38-0
1-CYCLOHEXYLHEXANE	C12H24	1-CYCLOHEXYLHEXANE	4292-75-5
CYCLOHEXYL ISOCYANATE	C7H11NO	CYCLOHEXYL ISOCYANATE	3173-53-3
CYCLOHEXYLMETHANE	C7H14	METHYLCYCLOHEXANE	108-87-2
CYCLOHEXYLMETHYLAMINE	C7H15N	N-METHYLCYCLOHEXYLAMINE	100-60-7
1-CYCLOHEXYLNONANE	C15H30	1-CYCLOHEXYLNONANE	2883-02-5
1-CYCLOHEXYLOCTANE	C14H28	1-CYCLOHEXYLOCTANE	1795-15-9
1-CYCLOHEXYLPENTADECANE	C21H42	1-CYCLOHEXYLPENTADECANE	6812-39-1
CYCLOHEXYL PEROXIDE	C6H12O2	CYCLOHEXYL PEROXIDE	766-07-4
1-CYCLOHEXYLTETRADECANE	C20H40	1-CYCLOHEXYLTETRADECANE	1795-18-2
1-CYCLOHEXYLTRIDECANE	C19H38	1-CYCLOHEXYLTRIDECANE	6006-33-3
1-CYCLOHEXYLUNDECANE	C17H34	1-CYCLOHEXYLUNDECANE	----------
CYCLON	CHN	HYDROGEN CYANIDE	74-90-8
CYCLONE B	CHN	HYDROGEN CYANIDE	74-90-8
1,5-CYCLOOCTADIENE	C8H12	1,5-CYCLOOCTADIENE	111-78-4
CYCLOOCTANE	C8H16	CYCLOOCTANE	292-64-8
1,3,5,7-CYCLOOCTATETRAENE	C8H8	1,3,5,7-CYCLOOCTATETRAENE	629-20-9
CYCLOOXABUTANE	C3H6O	1,3-PROPYLENE OXIDE	503-30-0
CYCLOPENTACYCLOHEPTENE	C10H8	AZULENE	275-51-4
CYCLOPENTADIENE	C5H6	CYCLOPENTADIENE	542-92-7
1,3-CYCLOPENTADIENE, DIMER	C10H12	DICYCLOPENTADIENE	77-73-6
CYCLOPENTANE	C5H10	CYCLOPENTANE	287-92-3
CYCLOPENTANETHIOL	C5H10S	CYCLOPENTANETHIOL	1679-07-8
CYCLOPENTANONE	C5H8O	CYCLOPENTANONE	120-92-3
CYCLOPENTENE	C5H8	CYCLOPENTENE	142-29-0
CYCLOPENTIMINE	C5H11N	PIPERIDINE	110-89-4
1-CYCLOPENTYLDECANE	C15H30	1-CYCLOPENTYLDECANE	295-48-7
1-CYCLOPENTYLDODECANE	C17H34	1-CYCLOPENTYLDODECANE	5634-30-0
1-CYCLOPENTYLHEPTANE	C12H24	1-CYCLOPENTYLHEPTANE	----------
1-CYCLOPENTYLHEXADECANE	C21H42	1-CYCLOPENTYLHEXADECANE	606-95-7
1-CYCLOPENTYLHEXANE	C11H22	1-CYCLOPENTYLHEXANE	----------
CYCLOPENTYL MERCAPTAN	C5H10S	CYCLOPENTANETHIOL	1679-07-8
1-CYCLOPENTYLNONANE	C14H28	1-CYCLOPENTYLNONANE	2882-98-6
1-CYCLOPENTYLOCTANE	C13H26	1-CYCLOPENTYLOCTANE	1795-20-6
1-CYCLOPENTYLPENTADECANE	C20H40	1-CYCLOPENTYLPENTADECANE	4669-01-6
1-CYCLOPENTYLPENTANE	C10H20	1-CYCLOPENTYLPENTANE	3741-00-2
1-CYCLOPENTYLTETRADECANE	C19H38	1-CYCLOPENTYLTETRADECANE	1795-22-8
1-CYCLOPENTYLUNDECANE	C16H32	1-CYCLOPENTYLUNDECANE	6758-23-5
CYCLOPROPANE	C3H6	CYCLOPROPANE	75-19-4
CYCLOTETRAMETHYLENE OXIDE	C4H8O	TETRAHYDROFURAN	109-99-9
CYCLOTETRAMETHYLENE SULFONE	C4H8O2S	SULFOLANE	126-33-0
1-CYCOPENTYLTRIDECANE	C18H36	1-CYCOPENTYLTRIDECANE	6006-34-4
CYKLOHEKSAN	C6H12	CYCLOHEXANE	110-82-7
CYKLOHEKSANOL	C6H12O	CYCLOHEXANOL	108-93-0
CYKLOHEKSANON	C6H10O	CYCLOHEXANONE	108-94-1
CYKLOHEKSEN	C6H10	CYCLOHEXENE	110-83-8
b-CYMENE	C10H14	m-CYMENE	535-77-3
CYMENE	C10H14	p-CYMENE	99-87-6
m-CYMENE	C10H14	m-CYMENE	535-77-3

Name (Synonym)	Formula	Name in Tables	CAS Reg. No
o-CYMENE	C10H14	o-CYMENE	527-84-4
p-CYMENE	C10H14	p-CYMENE	99-87-6
CYMOL	C10H14	p-CYMENE	99-87-6
m-CYMOL	C10H14	m-CYMENE	535-77-3
o-CYMOL	C10H14	o-CYMENE	527-84-4
CYPENTIL	C5H11N	PIPERIDINE	110-89-4
1,1,2,2-CZTEROCHLOROETAN	C2H2Cl4	1,1,2,2-TETRACHLOROETHANE	79-34-5
D 33LV	C6H12N2	TRIETHYLENEDIAMINE	280-57-9
DAAB	C12H11N3	1,3-DIPHENYLTRIAZENE	136-35-6
DABCO	C6H12N2	TRIETHYLENEDIAMINE	280-57-9
DABCO 33LV	C6H12N2	TRIETHYLENEDIAMINE	280-57-9
DABCO CRYSTAL	C6H12N2	TRIETHYLENEDIAMINE	280-57-9
DABCO EG	C6H12N2	TRIETHYLENEDIAMINE	280-57-9
DABCO R-8020	C6H12N2	TRIETHYLENEDIAMINE	280-57-9
DABCO S-25	C6H12N2	TRIETHYLENEDIAMINE	280-57-9
DAF 68	C24H38O4	DIOCTYL PHTHALATE	117-81-7
DAI CARI XBN	C11H14O2	n-BUTYL BENZOATE	136-60-7
DAIFLON	C2ClF3	CHLOROTRIFLUOROETHYLENE	79-38-9
DAIFLON S 3	C2Cl3F3	1,1,2-TRICHLOROTRIFLUOROETHANE	76-13-1
DAITO ORANGE BASE R	C6H6N2O2	m-NITROANILINE	99-09-2
DALTOGEN	C6H15NO3	TRIETHANOLAMINE	102-71-6
DAR-CHEM 14	C18H36O2	STEARIC ACID	57-11-4
DBE	C2H4Br2	1,2-DIBROMOETHANE	106-93-4
DBM	C12H18	HEXAMETHYLBENZENE	87-85-4
DBMP	C15H24O	2,6-DI-tert-BUTYL-p-CRESOL	128-37-0
DBP	C16H22O4	DIBUTYL PHTHALATE	84-74-2
DBPC	C15H24O	2,6-DI-tert-BUTYL-p-CRESOL	128-37-0
DCA	C6H5Cl2N	3,4-DICHLOROANILINE	95-76-1
3,4-DCA	C6H5Cl2N	3,4-DICHLOROANILINE	95-76-1
DCB	C6H4Cl2	o-DICHLOROBENZENE	95-50-1
1,2-DCE	C2H4Cl2	1,2-DICHLOROETHANE	107-06-2
DCHA	C12H23N	DICYCLOHEXYLAMINE	101-83-7
DCNB	C6H3Cl2NO2	1,2-DICHLORO-4-NITROBENZENE	99-54-7
D-D	C3H6Cl2	2,2-DICHLOROPROPANE	594-20-7
DD MIXTURE	C3H6Cl2	2,2-DICHLOROPROPANE	594-20-7
DD SOIL FUMIGANT	C3H6Cl2	2,2-DICHLOROPROPANE	594-20-7
DEA	C10H15N	N,N-DIETYHLANILINE	91-66-7
DEACTIVATOR E	C4H10O3	DIETHYLENE GLYCOL	111-46-6
DEACTIVATOR H	C4H10O3	DIETHYLENE GLYCOL	111-46-6
DEANOL	C4H11NO	DIMETHYLETHANOLAMINE	108-01-0
DEC	C5H10O3	DIETHYL CARBONATE	105-58-8
DECAFLUOROBUTANE	C4F10	DECAFLUOROBUTANE	355-25-9
cis-DECAHYDRONANPHTALENE	C10H18	cis-DECAHYDRONANPHTALENE	493-01-6
trans-DECAHYDRONAPHTHALENE	C10H18	trans-DECAHYDRONAPHTHALENE	493-02-7
DECALDEHYDE	C10H20O	1-DECANAL	112-31-2
n-DECALDEHYDE	C10H20O	1-DECANAL	112-31-2
DECANAL	C10H20O	1-DECANAL	112-31-2
n-DECANAL	C10H20O	1-DECANAL	112-31-2
1-DECANAL	C10H20O	1-DECANAL	112-31-2
DECANALDEHYDE	C10H20O	1-DECANAL	112-31-2
DECANAL DIMETHYL ACETAL	C10H22O	1-DECANOL	112-30-1
n-DECANE	C10H22	n-DECANE	124-18-5
DECANEDIOIC ACID	C10H18O4	SEBACIC ACID	111-20-6
DECANEDIOIC ACID, DIBUTYL ESTER	C18H34O4	DIBUTYL SEBACATE	109-43-3
n-DECANOIC ACID	C10H20O2	n-DECANOIC ACID	334-48-5
DECANOL	C10H22O	1-DECANOL	112-30-1
n-DECANOL	C10H22O	1-DECANOL	112-30-1
1-DECANOL	C10H22O	1-DECANOL	112-30-1
n-DECATYL ALCOHOL	C10H22O	1-DECANOL	112-30-1
DECCOTANE	C4H11N	sec-BUTYLAMINE	13952-84-6
1-DECENE	C10H20	1-DECENE	872-05-9
n-DECOIC ACID	C10H20O2	n-DECANOIC ACID	334-48-5
DECYL ALDEHYDE	C10H20O	1-DECANAL	112-31-2
n-DECYL ALDEHYDE	C10H20O	1-DECANAL	112-31-2
1-DECYL ALDEHYDE	C11H22O	1-UNDECANAL	112-44-7
DECYLAMINE	C10H23N	n-DECYLAMINE	2016-57-1
n-DECYLAMINE	C10H23N	n-DECYLAMINE	2016-57-1
n-DECYLBENZENE	C16H26	n-DECYLBENZENE	104-72-3
n-DECYLCYCLOHEXANE	C16H32	n-DECYLCYCLOHEXANE	1795-16-0
DECYL-DISULFIDE	C20H42S2	DECYL-DISULFIDE	----------
DECYL-ETHYL-SULFIDE	C12H26S	DECYL-ETHYL-SULFIDE	----------
DECYLIC ACID	C10H20O2	n-DECANOIC ACID	334-48-5
n-DECYLIC ACID	C10H20O2	n-DECANOIC ACID	334-48-5
DECYLIC ALCOHOL	C10H22O	1-DECANOL	112-30-1
DECYLIC ALDEHYDE	C10H20O	1-DECANAL	112-31-2
n-DECYL MERCAPTAN	C10H22S	n-DECYL MERCAPTAN	143-10-2

Name (Synonym)	Formula	Name in Tables	CAS Reg. No
DECYL-METHYL-SULFIDE	C11H24S	DECYL-METHYL-SULFIDE	----------
1-n-DECYLNAPHTHALENE	C20H28	1-n-DECYLNAPHTHALENE	26438-27-7
DECYL OCTYL ALCOHOL	C18H38O	1-OCTADECANOL	112-92-5
DECYL-PROPYL-SULFIDE	C13H28S	DECYL-PROPYL-SULFIDE	----------
DECYL-SULFIDE	C20H42S	DECYL-SULFIDE	----------
1-DECYNE	C10H18	1-DECYNE	7064-93-2
DEG	C4H10O3	DIETHYLENE GLYCOL	111-46-6
D.E.H. 20	C4H13N3	DIETHYLENE TRIAMINE	111-40-0
DEH 24	C6H18N4	TRIETHYLENE TETRAMINE	112-24-3
D.E.H. 26	C8H23N5	TETRAETHYLENEPENTAMINE	112-57-2
DEHP	C24H38O4	DIOCTYL PHTHALATE	117-81-7
DEHYDROABIETYLAMINE	C20H31N	DEHYDROABIETYLAMINE	1446-61-3
DEK	C5H10O	DIETHYL KETONE	96-22-0
DELPHINIC ACID	C5H10O2	ISOVALERIC ACID	503-74-2
DELTAN	C2H6OS	DIMETHYL SULFOXIDE	67-68-5
DELTRATE-20	C5H8N4O12	PENTAERYTHRITOL TETRANITRATE	78-11-5
DEMASORB	C2H6OS	DIMETHYL SULFOXIDE	67-68-5
DEMAVET	C2H6OS	DIMETHYL SULFOXIDE	67-68-5
DEMESO	C2H6OS	DIMETHYL SULFOXIDE	67-68-5
DEMSODROX	C2H6OS	DIMETHYL SULFOXIDE	67-68-5
2-DEOXYGLYCEROL	C3H8O2	1,3-PROPYLENE GLYCOL	504-63-2
4-DEOXYTETRONIC ACID	C4H6O2	gamma-BUTYROLACTONE	96-48-0
DEPRELIN	C4H4N2	SUCCINONITRILE	110-61-2
DERMASORB	C2H6OS	DIMETHYL SULFOXIDE	67-68-5
DESMODUR 44	C15H10N2O2	DIPHENYLMETHANE-4,4'-DIISOCYANATE	101-68-8
DESMODUR T100	C9H6N2O2	TOLUENE DIISOCYANATE	26471-62-5
DETA	C4H13N3	DIETHYLENE TRIAMINE	111-40-0
DETERGENT ALKYLATE	C18H30	n-DODECYLBENZENE	123-01-3
DEVELOPER 11	C6H8N2	m-PHENYLENEDIAMINE	108-45-2
DEVELOPER P	C6H6N2O2	p-NITROANILINE	100-01-6
DEVELOPER R	C6H6O2	1,3-BENZENEDIOL	108-46-3
DEVOL ORANGE B	C6H6N2O2	o-NITROANILINE	88-74-4
DEVOL ORANGE R	C6H6N2O2	m-NITROANILINE	99-09-2
DEVOL RED GG	C6H6N2O2	p-NITROANILINE	100-01-6
DEVOTON	C3H6O2	METHYL ACETATE	79-20-9
DFA	C12H11N	DIPHENYLAMINE	122-39-4
DIACETIC ETHER	C6H10O2	n-PROPYL ACRYLATE	925-60-0
DIACETONALCOHOL	C6H12O2	DIACETONE ALCOHOL	123-42-2
DIACETONALCOOL	C6H12O2	DIACETONE ALCOHOL	123-42-2
DIACETONALKOHOL	C6H12O2	DIACETONE ALCOHOL	123-42-2
DIACETONE	C6H12O2	DIACETONE ALCOHOL	123-42-2
DIACETONE ALCOHOL	C6H12O2	DIACETONE ALCOHOL	123-42-2
DIACETONE-ALCOOL	C6H12O2	DIACETONE ALCOHOL	123-42-2
DIACETYLMETHANE	C5H8O2	ACETYLACETONE	123-54-6
DIAETHANOLAMIN	C4H11NO2	DIETHANOLAMINE	111-42-2
1,1-DIAETHOXY-AETHAN	C6H14O2	ACETAL	105-57-7
DIAETHYLACETAL	C6H14O2	ACETAL	105-57-7
DIAETHYLAETHER	C4H10O	DIETHYL ETHER	60-29-7
DIAETHYLAMIN	C4H11N	DIETHYLAMINE	109-89-7
DIAETHYLANILIN	C10H15N	N,N-DIETHYLANILINE	91-66-7
DIAETHYLCARBONAT	C5H10O3	DIETHYL CARBONATE	105-58-8
DIAETHYLSULFAT	C4H10O4S	DIETHYL SULFATE	64-67-5
DIAKARMON	C6H14O6	SORBITOL	50-70-4
DIAKON	C5H8O2	METHYL METHACRYLATE	80-62-6
DIALLYL	C6H10	1,5-HEXADIENE	592-42-7
DIALLYL MALEATE	C10H12O4	DIALLYL MALEATE	999-21-3
1,2-DIAMINOAETHAN	C2H8N2	ETHYLENEDIAMINE	107-15-3
m-DIAMINOBENZENE	C6H8N2	m-PHENYLENEDIAMINE	108-45-2
o-DIAMINOBENZENE	C6H8N2	o-PHENYLENEDIAMINE	95-54-5
1,2-DIAMINOBENZENE	C6H8N2	o-PHENYLENEDIAMINE	95-54-5
1,3-DIAMINOBENZENE	C6H8N2	m-PHENYLENEDIAMINE	108-45-2
a,e-DIAMINOCAPROIC ACID	C6H14N2O2	LYSINE	56-87-1
2,2'-DIAMINODIETHYLAMINE	C4H13N3	DIETHYLENE TRIAMINE	111-40-0
1,2-DIAMINO-ETHAAN	C2H8N2	ETHYLENEDIAMINE	107-15-3
1,2-DIAMINOETHANE	C2H8N2	ETHYLENEDIAMINE	107-15-3
1,2-DIAMINO-ETHANO	C2H8N2	ETHYLENEDIAMINE	107-15-3
1,6-DIAMINOHEXANE	C6H16N2	HEXAMETHYLENEDIAMINE	124-09-4
2,6-DIAMINOHEXANOIC ACID	C6H14N2O2	LYSINE	56-87-1
1,3-DIAMINO-4-METHYLBENZENE	C7H10N2	TOLUENEDIAMINE	95-80-7
2,4-DIAMINO-1-METHYLBENZENE	C7H10N2	TOLUENEDIAMINE	95-80-7
1,2-DIAMINOPROPANE	C3H10N2	1,2-PROPANEDIAMINE	78-90-0
2,4-DIAMINOTOLUEN	C7H10N2	TOLUENEDIAMINE	95-80-7
DIAMINOTOLUENE	C7H10N2	TOLUENEDIAMINE	95-80-7
2,4-DIAMINO-1-TOLUENE	C7H10N2	TOLUENEDIAMINE	95-80-7
2,4-DIAMINOTOLUOL	C7H10N2	TOLUENEDIAMINE	95-80-7
DIAMYL ETHER	C10H22O	DI-n-PENTYL ETHER	693-65-2

692

Name (Synonym)	Formula	Name in Tables	CAS Reg. No
DI-n-AMYL ETHER	C10H22O	DI-n-PENTYL ETHER	693-65-2
DIAN	C15H16O2	BISPHENOL A	80-05-7
DIAREX HF 77	C8H8	STYRENE	100-42-5
3,6-DIAZAOCTANE-1,8-DIAMINE	C6H18N4	TRIETHYLENE TETRAMINE	112-24-3
DIAZOAMINOBENZEN	C12H11N3	1,3-DIPHENYLTRIAZENE	136-35-6
DIAZOAMINOBENZENE	C12H11N3	1,3-DIPHENYLTRIAZENE	136-35-6
p-DIAZOAMINOBENZENE	C12H11N3	1,3-DIPHENYLTRIAZENE	136-35-6
DIAZOAMINOBENZOL	C12H11N3	1,3-DIPHENYLTRIAZENE	136-35-6
DIAZO FAST ORANGE GR	C6H6N2O2	o-NITROANILINE	88-74-4
DIAZO FAST ORANGE R	C6H6N2O2	m-NITROANILINE	99-09-2
DIAZO FAST RED GG	C6H6N2O2	p-NITROANILINE	100-01-6
DIBENZOFURAN	C12H8O	DIBENZOFURAN	132-64-9
DIBENZO(b,d)FURAN	C12H8O	DIBENZOFURAN	132-64-9
1,2,5,6-DIBENZONAPHTHALENE	C18H12	CHRYSENE	218-01-9
DIBENZOPYRROLE	C12H9N	DIBENZOPYRROLE	86-74-8
DIBENZO(b,d)PYRROLE	C12H9N	DIBENZOPYRROLE	86-74-8
DIBENZYL	C14H14	1,2-DIPHENYLETHANE	103-29-7
DIBENZYL ETHER	C14H14O	DIBENZYL ETHER	103-50-4
m-DIBROMOBENZENE	C6H4Br2	m-DIBROMOBENZENE	108-36-1
1,2-DIBROMOBUTANE	C4H8Br2	1,2-DIBROMOBUTANE	533-98-2
2,3-DIBROMOBUTANE	C4H8Br2	2,3-DIBROMOBUTANE	5408-86-6
DIBROMODIFLUOROMETHANE	CBr2F2	DIBROMODIFLUOROMETHANE	75-61-6
a,b-DIBROMOETHANE	C2H4Br2	1,2-DIBROMOETHANE	106-93-4
sym-DIBROMOETHANE	C2H4Br2	1,2-DIBROMOETHANE	106-93-4
1,1-DIBROMOETHANE	C2H4Br2	1,1-DIBROMOETHANE	557-91-5
1,2-DIBROMOETHANE	C2H4Br2	1,2-DIBROMOETHANE	106-93-4
DIBROMOMETHANE	CH2Br2	DIBROMOMETHANE	74-95-3
2,3-DIBROMO-2-METHYLBUTANE	C5H10Br2	2,3-DIBROMO-2-METHYLBUTANE	594-51-4
1,2-DIBROMOPERFLUOROETHANE	C2Br2F4	1,2-DIBROMOTETRAFLUOROETHANE	124-73-2
1,2-DIBROMOPROPANE	C3H6Br2	1,2-DIBROMOPROPANE	78-75-1
sym-DIBROMOTETRAFLUOROETHANE	C2Br2F4	1,2-DIBROMOTETRAFLUOROETHANE	124-73-2
1,2-DIBROMOTETRAFLUOROETHANE	C2Br2F4	1,2-DIBROMOTETRAFLUOROETHANE	124-73-2
1,2-DIBROMO-1,1,2,2-TETRAFLUOROETHANE	C2Br2F4	1,2-DIBROMOTETRAFLUOROETHANE	124-73-2
2,2'-DIBUTOXYETHYL ETHER	C12H26O3	DIETHYLENE GLYCOL DI-n-BUTYL ETHER	112-73-2
DI-n-BUTYLAMINE	C8H19N	DI-n-BUTYLAMINE	111-92-2
DIBUTYLATED HYDROXYTOLUENE	C15H24O	2,6-DI-tert-BUTYL-p-CRESOL	128-37-0
DIBUTYL-1,2-BENZENEDICARBOXYLATE	C16H22O4	DIBUTYL PHTHALATE	84-74-2
2,6-DI-tert-BUTYL-p-CRESOL	C15H24O	2,6-DI-tert-BUTYL-p-CRESOL	128-37-0
DIBUTYL-DISULFIDE	C8H18S2	DIBUTYL-DISULFIDE	629-45-8
DI-n-BUTYL ETHER	C8H18O	DI-n-BUTYL ETHER	142-96-1
DI-sec-BUTYL ETHER	C8H18O	DI-sec-BUTYL ETHER	6863-58-7
DI-tert-BUTYL ETHER	C8H18O	DI-tert-BUTYL ETHER	6163-66-2
2,6-DI-tert-BUTYL-1-HYDROXY-4-METHYLBENZENE	C15H24O	2,6-DI-tert-BUTYL-p-CRESOL	128-37-0
DIBUTYL MALEATE	C12H20O4	DIBUTYL MALEATE	105-76-0
DIBUTYL OXIDE	C8H18O	DI-n-BUTYL ETHER	142-96-1
DI-tert-BUTYLPEROXID	C8H18O2	DI-t-BUTYL PEROXIDE	110-05-4
DI-t-BUTYL PEROXIDE	C8H18O2	DI-t-BUTYL PEROXIDE	110-05-4
DIBUTYL PHTHALATE	C16H22O4	DIBUTYL PHTHALATE	84-74-2
DI-n-BUTYL PHTHALATE	C16H22O4	DIBUTYL PHTHALATE	84-74-2
DIBUTYL SEBACATE	C18H34O4	DIBUTYL SEBACATE	109-43-3
DI-n-BUTYL SEBACATE	C18H34O4	DIBUTYL SEBACATE	109-43-3
DIBUTYL-SULFIDE	C8H18S	DIBUTYL-SULFIDE	544-40-1
DI-n-BUTYLSULFIDE	C8H18S	DIBUTYL-SULFIDE	544-40-1
n-DIBUTYL SULFIDE	C8H18S	DIBUTYL-SULFIDE	544-40-1
DI-n-BUTYL SULFONE	C8H18O2S	DI-n-BUTYL SULFONE	598-04-9
DIBUTYL SULPHIDE	C8H18S	DIBUTYL-SULFIDE	544-40-1
DIBUTYL THIOETHER	C8H18S	DIBUTYL-SULFIDE	544-40-1
DICARBETHOXYMETHANE	C7H12O4	DIETHYL MALONATE	105-53-3
o-DICARBOXYBENZENE	C8H6O4	PHTHALIC ACID	88-99-3
DICHA	C12H23N	DICYCLOHEXYLAMINE	101-83-7
p-DICHLOORBENZEEN	C6H4Cl2	p-DICHLOROBENZENE	106-46-7
1,4-DICHLOORBENZEEN	C6H4Cl2	p-DICHLOROBENZENE	106-46-7
1,1-DICHLOORETHAAN	C2H4Cl2	1,1-DICHLOROETHANE	75-34-3
1,2-DICHLOORETHAAN	C2H4Cl2	1,2-DICHLOROETHANE	107-06-2
DICHLORACETYL CHLORIDE	C2HCl3O	DICHLOROACETYL CHLORIDE	79-36-7
1,1-DICHLORAETHAN	C2H4Cl2	1,1-DICHLOROETHANE	75-34-3
3,4-DICHLORANILIN	C6H5Cl2N	3,4-DICHLOROANILINE	95-76-1
3,4-DICHLORANILINE	C6H5Cl2N	3,4-DICHLOROANILINE	95-76-1
1,4-DICHLOR-BENZOL	C6H4Cl2	p-DICHLOROBENZENE	106-46-7
o-DICHLOR BENZOL	C6H4Cl2	o-DICHLOROBENZENE	95-50-1
p-DICHLORBENZOL	C6H4Cl2	p-DICHLOROBENZENE	106-46-7
DICHLORDIMETHYLAETHER	C2H4Cl2O	BIS(CHLOROMETHYL)ETHER	542-88-1
DICHLOREMULSION	C2H4Cl2	1,2-DICHLOROETHANE	107-06-2
DICHLORETHANOIC ACID	C2H2Cl2O2	DICHLOROACETIC ACID	79-43-6
3,4-DICHLORFENYLISOKYANAT	C7H3Cl2NO	3,4-DICHLOROPHENYL ISOCYANATE	102-36-3
DI-CHLORICIDE	C6H4Cl2	p-DICHLOROBENZENE	106-46-7

Name (Synonym)	Formula	Name in Tables	CAS Reg. No
DI-CHLOR-MULSION	C2H4Cl2	1,2-DICHLOROETHANE	107-06-2
3,4-DICHLORNITROBENZEN	C6H3Cl2NO2	1,2-DICHLORO-4-NITROBENZENE	99-54-7
DICHLOROACETALDEHYDE	C2H2Cl2O	DICHLOROACETALDEHYDE	79-02-7
a,a-DICHLOROACETALDEHYDE	C2H2Cl2O	DICHLOROACETALDEHYDE	79-02-7
DICHLOROACETIC ACID	C2H2Cl2O2	DICHLOROACETIC ACID	79-43-6
2,2-DICHLOROACETIC ACID	C2H2Cl2O2	DICHLOROACETIC ACID	79-43-6
DICHLOROACETYL CHLORIDE	C2HCl3O	DICHLOROACETYL CHLORIDE	79-36-7
a,a-DICHLOROACETYL CHLORIDE	C2HCl3O	DICHLOROACETYL CHLORIDE	79-36-7
2,2-DICHLOROACETYL CHLORIDE	C2HCl3O	DICHLOROACETYL CHLORIDE	79-36-7
3,4-DICHLOROANILINE	C6H5Cl2N	3,4-DICHLOROANILINE	95-76-1
4,5-DICHLOROANILINE	C6H5Cl2N	3,4-DICHLOROANILINE	95-76-1
3,4-DICHLOROBENZENAMINE	C6H5Cl2N	3,4-DICHLOROANILINE	95-76-1
m-DICHLOROBENZENE	C6H4Cl2	m-DICHLOROBENZENE	541-73-1
o-DICHLOROBENZENE	C6H4Cl2	o-DICHLOROBENZENE	95-50-1
p-DICHLOROBENZENE	C6H4Cl2	p-DICHLOROBENZENE	106-46-7
1,2-DICHLOROBENZENE	C6H4Cl2	o-DICHLOROBENZENE	95-50-1
1,3-DICHLOROBENZENE	C6H4Cl2	m-DICHLOROBENZENE	541-73-1
1,4-DICHLOROBENZENE	C6H4Cl2	p-DICHLOROBENZENE	106-46-7
DICHLOROBENZENE, PARA	C6H4Cl2	p-DICHLOROBENZENE	106-46-7
p-DICHLOROBENZOL	C6H4Cl2	p-DICHLOROBENZENE	106-46-7
2,4-DICHLOROBENZOTRIFLUORIDE	C7H3Cl2F3	2,4-DICHLOROBENZOTRIFLUORIDE	320-60-5
1,4-DICHLOROBUTANE	C4H8Cl2	1,4-DICHLOROBUTANE	110-56-5
1,3-DICHLORO-trans-2-BUTENE	C4H6Cl2	1,3-DICHLORO-trans-2-BUTENE	7415-31-8
1,4-DICHLORO-cis-2-BUTENE	C4H6Cl2	1,4-DICHLORO-cis-2-BUTENE	1476-11-5
1,4-DICHLORO-trans-2-BUTENE	C4H6Cl2	1,4-DICHLORO-trans-2-BUTENE	110-57-6
1,4-DICHLORO-2-BUTENE	C4H6Cl2	1,4-DICHLORO-trans-2-BUTENE	110-57-6
3,4-DICHLORO-1-BUTENE	C4H6Cl2	3,4-DICHLORO-1-BUTENE	760-23-6
1,4-DICHLOROBUTENE-2	C4H6Cl2	1,4-DICHLORO-trans-2-BUTENE	110-57-6
1,1-DICHLORO-2,2-DICHLOROETHANE	C2H2Cl4	1,1,2,2-TETRACHLOROETHANE	79-34-5
DICHLORODIFLUOROMETHANE	CCl2F2	DICHLORODIFLUOROMETHANE	75-71-8
sym-DICHLORODIMETHYL ETHER	C2H4Cl2O	BIS(CHLOROMETHYL)ETHER	542-88-1
a,b-DICHLOROETHANE	C2H4Cl2	1,2-DICHLOROETHANE	107-06-2
sym-DICHLOROETHANE	C2H4Cl2	1,2-DICHLOROETHANE	107-06-2
1,1-DICHLOROETHANE	C2H4Cl2	1,1-DICHLOROETHANE	75-34-3
DICHLORO-1,2-ETHANE	C2H4Cl2	1,2-DICHLOROETHANE	107-06-2
1,2-DICHLOROETHANE	C2H4Cl2	1,2-DICHLOROETHANE	107-06-2
DICHLOROETHANOIC ACID	C2H2Cl2O2	DICHLOROACETIC ACID	79-43-6
DICHLOROETHANOYL CHLORIDE	C2HCl3O	DICHLOROACETYL CHLORIDE	79-36-7
1,1-DICHLOROETHENE	C2H2Cl2	1,1-DICHLOROETHYLENE	75-35-4
DICHLOROETHYLENE	C2H4Cl2	1,2-DICHLOROETHANE	107-06-2
cis-1,2-DICHLOROETHYLENE	C2H2Cl2	cis-1,2-DICHLOROETHYLENE	156-59-2
trans-DICHLOROETHYLENE	C2H2Cl2	trans-1,2-DICHLOROETHYLENE	156-60-5
trans-1,2-DICHLOROETHYLENE	C2H2Cl2	trans-1,2-DICHLOROETHYLENE	156-60-5
1,1-DICHLOROETHYLENE	C2H2Cl2	1,1-DICHLOROETHYLENE	75-35-4
1,2-DICHLOROETHYLENE	C2H2Cl2	cis-1,2-DICHLOROETHYLENE	156-59-2
DICHLOROFLUOROMETHANE	CHCl2F	DICHLOROFLUOROMETHANE	75-43-4
DICHLOROMETHANE	CH2Cl2	DICHLOROMETHANE	75-09-2
sym-DICHLOROMETHYL ETHER	C2H4Cl2O	BIS(CHLOROMETHYL)ETHER	542-88-1
(DICHLOROMETHYL)BENZENE	C7H6Cl2	BENZYL DICHLORIDE	98-87-3
2,4-DICHLORO-1-METHYLBENZENE	C7H6Cl2	2,4-DICHLOROTOLUENE	95-73-8
DICHLOROMETHYLSILANE	CH4Cl2Si	METHYL DICHLOROSILANE	75-54-7
DICHLOROMONOFLUOROMETHANE	CHCl2F	DICHLOROFLUOROMETHANE	75-43-4
1,2-DICHLORO-4-NITROBENZENE	C6H3Cl2NO2	1,2-DICHLORO-4-NITROBENZENE	99-54-7
3,4-DICHLORONITROBENZENE	C6H3Cl2NO2	1,2-DICHLORO-4-NITROBENZENE	99-54-7
1,5-DICHLOROPENTANE	C5H10Cl2	1,5-DICHLOROPENTANE	628-76-2
3,4-DICHLOROPHENYL ISOCYANATE	C7H3Cl2NO	3,4-DICHLOROPHENYL ISOCYANATE	102-36-3
a,b-DICHLOROPROPANE	C3H6Cl2	1,2-DICHLOROPROPANE	78-87-5
1,1-DICHLOROPROPANE	C3H6Cl2	1,1-DICHLOROPROPANE	78-99-9
1,2-DICHLOROPROPANE	C3H6Cl2	1,2-DICHLOROPROPANE	78-87-5
1,3-DICHLOROPROPANE	C3H6Cl2	1,3-DICHLOROPROPANE	142-28-9
2,2-DICHLOROPROPANE	C3H6Cl2	2,2-DICHLOROPROPANE	594-20-7
2,3-DICHLOROPROPENE	C3H4Cl2	2,3-DICHLOROPROPENE	78-88-6
2,3-DICHLORO-1-PROPENE	C3H4Cl2	2,3-DICHLOROPROPENE	78-88-6
2,3-DICHLOROPROPYLENE	C3H4Cl2	2,3-DICHLOROPROPENE	78-88-6
DICHLOROTETRAFLUOROETHANE	C2Cl2F4	1,2-DICHLOROTETRAFLUOROETHANE	76-14-2
sym-DICHLOROTETRAFLUOROETHANE	C2Cl2F4	1,2-DICHLOROTETRAFLUOROETHANE	76-14-2
1,2-DICHLOROTETRAFLUOROETHANE	C2Cl2F4	1,2-DICHLOROTETRAFLUOROETHANE	76-14-2
1,2-DICHLORO-1,1,2,2-TETRAFLUOROETHANE	C2Cl2F4	1,2-DICHLOROTETRAFLUOROETHANE	76-14-2
a-a-DICHLOROTOLUENE	C7H6Cl2	BENZYL DICHLORIDE	98-87-3
2,4-DICHLOROTOLUENE	C7H6Cl2	2,4-DICHLOROTOLUENE	95-73-8
DICHLORPROPAN-DICHLORPROPENGEMISCH	C3H6Cl2	2,2-DICHLOROPROPANE	594-20-7
p-DICLOROBENZENE	C6H4Cl2	p-DICHLOROBENZENE	106-46-7
1,4-DICLOROBENZENE	C6H4Cl2	p-DICHLOROBENZENE	106-46-7
1,1-DICLOROETANO	C2H4Cl2	1,1-DICHLOROETHANE	75-34-3
DICOL	C4H10O3	DIETHYLENE GLYCOL	111-46-6
DICUMYL PEROXIDE	C18H22O2	DICUMYL PEROXIDE	80-43-3

Name (Synonym)	Formula	Name in Tables	CAS Reg. No
DI-CUP	C18H22O2	DICUMYL PEROXIDE	80-43-3
DI-CUP 40 KF	C18H22O2	DICUMYL PEROXIDE	80-43-3
DI-CUPR	C18H22O2	DICUMYL PEROXIDE	80-43-3
1,4-DICYANOBUTANE	C6H8N2	ADIPONITRILE	111-69-3
cis-DICYANO-1-BUTENE	C6H6N2	cis-DICYANO-1-BUTENE	2141-58-4
trans-DICYANO-1-BUTENE	C6H6N2	trans-DICYANO-1-BUTENE	2141-59-5
1,4-DICYANO-2-BUTENE	C6H6N2	1,4-DICYANO-2-BUTENE	1119-85-3
b,b'-DICYANODIETHYL ETHER	C6H8N2O	BIS(CYANOETHYL)ETHER	1656-48-0
s-DICYANOETHANE	C4H4N2	SUCCINONITRILE	110-61-2
1,2-DICYANOETHANE	C4H4N2	SUCCINONITRILE	110-61-2
DICYANOGEN	C2N2	CYANOGEN	460-19-5
DICYANOMETHANE	C3H2N2	MALONONITRILE	109-77-3
1,3-DICYANOPROPANE	C5H6N2	GLUTARONITRILE	544-13-8
DICYCLOHEXYLAMINE	C12H23N	DICYCLOHEXYLAMINE	101-83-7
DICYCLOPENTADIENE	C10H12	DICYCLOPENTADIENE	77-73-6
DICYKLOHEXYLAMIN	C12H23N	DICYCLOHEXYLAMINE	101-83-7
DICYKLOPENTADIEN	C10H12	DICYCLOPENTADIENE	77-73-6
DIDAKENE	C2Cl4	TETRACHLOROETHYLENE	127-18-4
DIETHANOLAMIN	C4H11NO2	DIETHANOLAMINE	111-42-2
DIETHANOLAMINE	C4H11NO2	DIETHANOLAMINE	111-42-2
DIETHANOLMETHYLAMINE	C5H13NO2	METHYL DIETHANOLAMINE	105-59-9
1,1-DIETHOXY-ETHAAN	C6H14O2	ACETAL	105-57-7
1,1-DIETHOXYETHANE	C6H14O2	ACETAL	105-57-7
DIETHYL	C4H10	n-BUTANE	106-97-8
DIETHYL ACETAL	C6H14O2	ACETAL	105-57-7
DIETHYLAMINE	C4H11N	DIETHYLAMINE	109-89-7
N,N-DIETHYLAMINE	C4H11N	DIETHYLAMINE	109-89-7
N,N-DIETHYLAMINOBENZENE	C10H15N	N,N-DIETYHLANILINE	91-66-7
(DIETHYLAMINO)ETHANE	C6H15N	TRIETHYLAMINE	121-44-8
N,N-DIETHYLANILIN	C10H15N	N,N-DIETYHLANILINE	91-66-7
DIETHYLANILINE	C10H15N	N,N-DIETYHLANILINE	91-66-7
2,6-DIETHYLANILINE	C10H15N	2,6-DIETHYLANILINE	579-66-8
N,N-DIETHYLBENZENAMINE	C10H15N	N,N-DIETYHLANILINE	91-66-7
2,6-DIETHYLBENZENAMINE	C10H15N	2,6-DIETHYLANILINE	579-66-8
m-DIETHYLBENZENE	C10H14	m-DIETHYLBENZENE	141-93-5
o-DIETHYLBENZENE	C10H14	o-DIETHYLBENZENE	135-01-3
p-DIETHYLBENZENE	C10H14	p-DIETHYLBENZENE	105-05-5
DIETHYL CARBINOL	C5H12O	3-PENTANOL	584-02-1
DIETHYL CARBONATE	C5H10O3	DIETHYL CARBONATE	105-58-8
DIETHYLCETONE	C5H10O	DIETHYL KETONE	96-22-0
DIETHYLDISULFID	C4H10S2	DIETHYL DISULFIDE	110-81-6
DIETHYL DISULFIDE	C4H10S2	DIETHYL DISULFIDE	110-81-6
N,N-DIETHYLENE DIAMINE	C4H10N2	PIPERAZINE	110-85-0
1,4-DIETHYLENEDIAMINE	C4H10N2	PIPERAZINE	110-85-0
DIETHYLENE DIOXIDE	C4H8O2	1,4-DIOXANE	123-91-1
1,4-DIETHYLENE DIOXIDE	C4H8O2	1,4-DIOXANE	123-91-1
DIETHYLENE ETHER	C4H8O2	1,4-DIOXANE	123-91-1
DIETHYLENE GLYCOL	C4H10O3	DIETHYLENE GLYCOL	111-46-6
DIETHYLENEGLYCOL DIBUTYL ETHER	C12H26O3	DIETHYLENE GLYCOL DI-n-BUTYL ETHER	112-73-2
DIETHYLENE GLYCOL DIETHYL ETHER	C8H18O3	DIETHYLENE GLYCOL DIETHYL ETHER	112-36-7
DIETHYLENE GLYCOL DIMETHYL ETHER	C6H14O3	DIETHYLENE GLYCOL DIMETHYL ETHER	111-96-6
DIETHYLENE GLYCOL DI-n-BUTYL ETHER	C12H26O3	DIETHYLENE GLYCOL DI-n-BUTYL ETHER	112-73-2
DIETHYLENE GLYCOL ETHYL ETHER	C6H14O3	2-(2-ETHOXYETHOXY)ETHANOL	111-90-0
DIETHYLENE GLYCOL ETHYL ETHER ACETATE	C8H16O4	DIETHYLENE GLYCOL ETHYL ETHER ACETATE	112-15-2
DIETHYLENE GLYCOL METHYL ETHER	C5H12O3	2-(2-METHOXYETHOXY)ETHANOL	111-77-3
DIETHYLENE GLYCOL MONOBUTYL ETHER	C8H18O3	DIETHYLENE GLYCOL MONOBUTYL ETHER	112-34-5
DIETHYLENE GLYCOL MONOETHYL	C6H14O3	2-(2-ETHOXYETHOXY)ETHANOL	111-90-0
DIETHYLENE GLYCOL MONOETHYL ETHER ACETATE	C8H16O4	DIETHYLENE GLYCOL ETHYL ETHER ACETATE	112-15-2
DIETHYLENE GLYCOL-n-BUTYL ETHER	C8H18O3	DIETHYLENE GLYCOL MONOBUTYL ETHER	112-34-5
DIETHYLENE IMIDE OXIDE	C4H9NO	MORPHOLINE	110-91-8
DIETHYLENE IMIDOXIDE	C4H9NO	MORPHOLINE	110-91-8
DI(ETHYLENE OXIDE)	C4H8O2	1,4-DIOXANE	123-91-1
DIETHYLENE OXIDE	C4H8O	TETRAHYDROFURAN	109-99-9
DIETHYLENE OXIMIDE	C4H9NO	MORPHOLINE	110-91-8
DIETHYLENE TRIAMINE	C4H13N3	DIETHYLENE TRIAMINE	111-40-0
DIETHYLENIMIDE OXIDE	C4H9NO	MORPHOLINE	110-91-8
DIETHYL ESTER SULFURIC ACID	C4H10O4S	DIETHYL SULFATE	64-67-5
DIETHYLESTER KYSELINY SIROVE	C4H10O4S	DIETHYL SULFATE	64-67-5
N,N-DIETHYLETHANAMINE	C6H15N	TRIETHYLAMINE	121-44-8
DIETHYL ETHANEDIOATE	C6H10O4	DIETHYL OXALATE	95-92-1
DIETHYL ETHER	C4H10O	DIETHYL ETHER	60-29-7
DIETHYL GLYCOL DIMETHYL ETHER	C6H14O3	DIETHYLENE GLYCOL DIMETHYL ETHER	111-96-6
3,3-DIETHYLHEXANE	C10H22	3,3-DIETHYLHEXANE	17302-02-2
3,4-DIETHYLHEXANE	C10H22	3,4-DIETHYLHEXANE	19398-77-7
DI(2-ETHYLHEXYL)ORTHOPHTHALATE	C24H38O4	DIOCTYL PHTHALATE	117-81-7
DIETHYL KETONE	C5H10O	DIETHYL KETONE	96-22-0

Name (Synonym)	Formula	Name in Tables	CAS Reg. No
DIETHYL MALEATE	C8H12O4	DIETHYL MALEATE	141-05-9
DIETHYL MALONATE	C7H12O4	DIETHYL MALONATE	105-53-3
DIETHYLMETHYL METHANE	C6H14	3-METHYLPENTANE	96-14-0
3,3-DIETHYL-2-METHYLPENTANE	C10H22	3,3-DIETHYL-2-METHYLPENTANE	52897-16-2
DIETHYLOLAMINE	C4H11NO2	DIETHANOLAMINE	111-42-2
DIETHYL-o-PHTHALATE	C12H14O4	DIETHYL PHTHALATE	84-66-2
DIETHYL OXALATE	C6H10O4	DIETHYL OXALATE	95-92-1
DIETHYL OXIDE	C4H10O	DIETHYL ETHER	60-29-7
3,3-DIETHYLPENTANE	C9H20	3,3-DIETHYLPENTANE	1067-20-5
DIETHYLPHENYLAMINE	C10H15N	N,N-DIETYHLANILINE	91-66-7
DIETHYL PHTHALATE	C12H14O4	DIETHYL PHTHALATE	84-66-2
DIETHYL PROPANEDIOATE	C7H12O4	DIETHYL MALONATE	105-53-3
DIETHYL SUCCINATE	C8H14O4	DIETHYL SUCCINATE	123-25-1
DIETHYL SULFATE	C4H10O4S	DIETHYL SULFATE	64-67-5
DIETHYL SULFIDE	C4H10S	DIETHYL SULFIDE	352-93-2
DIETHYL TETRAOXOSULFATE	C4H10O4S	DIETHYL SULFATE	64-67-5
DIETILAMINA	C4H11N	DIETHYLAMINE	109-89-7
1,1-DIETOSSIETANO	C6H14O2	ACETAL	105-57-7
N,N-DIETYHLANILINE	C10H15N	N,N-DIETYHLANILINE	91-66-7
DIFENIL-METAN-DIISOCIANATO	C15H10N2O2	DIPHENYLMETHANE-4,4'-DIISOCYANATE	101-68-8
N,N'-DIFENYL-p-FENYLENDIAMIN	C18H16N2	N,N'-DIPHENYL-p-PHENYLENEDIAMINE	74-31-7
DIFENYLMETHAAN-DIISSOCYANAAT	C15H10N2O2	DIPHENYLMETHANE-4,4'-DIISOCYANATE	101-68-8
m-DIFLUOROBENZENE	C6H4F2	m-DIFLUOROBENZENE	372-18-9
o-DIFLUOROBENZENE	C6H4F2	o-DIFLUOROBENZENE	367-11-3
p-DIFLUOROBENZENE	C6H4F2	p-DIFLUOROBENZENE	540-36-3
1,1-DIFLUORO-1-CHLOROETHANE	C2H3ClF2	1-CHLORO-1,1-DIFLUOROETHANE	75-68-3
DIFLUOROCHLOROETHANES	C2H3ClF2	1-CHLORO-1,1-DIFLUOROETHANE	75-68-3
1,1-DIFLUORO-2-CHLOROETHYLENE	C2HClF2	2-CHLORO-1,1-DIFLUOROETHYLENE	359-10-4
DIFLUOROCHLOROMETHANE	CHClF2	CHLORODIFLUOROMETHANE	75-45-6
DIFLUORODICHLOROMETHANE	CCl2F2	DICHLORODIFLUOROMETHANE	75-71-8
1,1-DIFLUOROETHENE	C2H2F2	1,1-DIFLUOROETHYLENE	75-38-7
cis-1,2-DIFLUOROETHENE	C2H2F2	cis-1,2-DIFLUOROETHENE	1630-78-0
trans-1,2-DIFLUOROETHENE	C2H2F2	trans-1,2-DIFLUOROETHENE	1630-77-9
DIFLUOROETHANE	C2H4F2	1,1-DIFLUOROETHANE	75-37-6
1,1-DIFLUOROETHANE	C2H4F2	1,1-DIFLUOROETHANE	75-37-6
1,2-DIFLUOROETHANE	C2H4F2	1,2-DIFLUOROETHANE	624-72-6
1,1-DIFLUOROETHYLENE	C2H2F2	1,1-DIFLUOROETHYLENE	75-38-7
DIFLUOROFORMALDEHYDE	CF2O	CARBONYL FLUORIDE	353-50-4
DIFLUOROMETHANE	CH2F2	DIFLUOROMETHANE	75-10-5
DIFLUOROMONOCHLOROMETHANE	CHClF2	CHLORODIFLUOROMETHANE	75-45-6
1,2-DIFLUORO-1,1,2,2-TETRACHLOROETHANE	C2Cl4F2	1,1,2,2-TETRACHLORODIFLUOROETHANE	76-12-0
DIGLYCOL	C4H10O3	DIETHYLENE GLYCOL	111-46-6
DIGLYCOLAMINE	C4H11NO2	2-AMINOETHOXYETHANOL	929-06-6
DIGLYCOLIC ACID	C4H6O5	DIGLYCOLIC ACID	110-99-6
DIGLYCOL MONOBUTYL ETHER	C8H18O3	DIETHYLENE GLYCOL MONOBUTYL ETHER	112-34-5
DIGLYCOL MONOETHYL ETHER	C6H14O3	2-(2-ETHOXYETHOXY)ETHANOL	111-90-0
DIGLYCOL MONOETHYL ETHER ACETATE	C8H16O4	DIETHYLENE GLYCOL ETHYL ETHER ACETATE	112-15-2
DIGLYCOL MONOMETHYL ETHER	C5H12O3	2-(2-METHOXYETHOXY)ETHANOL	111-77-3
DIGLYME	C6H14O3	DIETHYLENE GLYCOL DIMETHYL ETHER	111-96-6
DIHEXYL	C12H26	n-DODECANE	112-40-3
DIHEXYL ADIPATE	C18H34O4	DIHEXYL ADIPATE	110-33-8
DI-n-HEXYL ETHER	C12H26O	DI-n-HEXYL ETHER	112-58-3
DIHEXYL HEXANEDIOATE	C18H34O4	DIHEXYL ADIPATE	110-33-8
DIHYDROAZIRENE	C2H5N	ETHYLENEIMINE	151-56-4
DIHYDRO-1H-AZIRINE	C2H5N	ETHYLENEIMINE	151-56-4
DIHYDROBUTADIENE SULPHONE	C4H8O2S	SULFOLANE	126-33-0
1,3-DIHYDRO-1,3-DIOXO-5-ISOBENZOFURANCARBOXYLIC ACID	C9H4O5	TRIMELLITIC ANHYDRIDE	552-30-7
2,5-DIHYDROFURAN	C4H6O	2,5-DIHYDROFURAN	1708-29-8
DIHYDRO-2,5-FURANDIONE	C4H4O3	SUCCINIC ANHYDRIDE	108-30-5
2,5-DIHYDROFURAN-2,5-DIONE	C4H2O3	MALEIC ANHYDRIDE	108-31-6
DIHYDRO-2(3H)-FURANONE	C4H6O2	gamma-BUTYROLACTONE	96-48-0
2,3-DIHYDROINDENE	C9H10	INDANE	496-11-7
DIHYDROOXIRENE	C2H4O	ETHYLENE OXIDE	75-21-8
2,3-DIHYDROTHIIRENE	C2H4S	THIACYCLOPROPANE	420-12-2
m-DIHYDROXYBENZENE	C6H6O2	1,3-BENZENEDIOL	108-46-3
o-DIHYDROXYBENZENE	C6H6O2	1,2-BENZENEDIOL	120-80-9
p-DIHYDROXYBENZENE	C6H6O2	p-HYDROQUINONE	123-31-9
DIHYDROXYBENZENE	C6H6O2	p-HYDROQUINONE	123-31-9
1,2-DIHYDROXYBENZENE	C6H6O2	1,2-BENZENEDIOL	120-80-9
1,3-DIHYDROXYBENZENE	C6H6O2	1,3-BENZENEDIOL	108-46-3
1,4-DIHYDROXYBENZENE	C6H6O2	p-HYDROQUINONE	123-31-9
1,3-DIHYDROXYBUTANE	C4H10O2	1,3-BUTANEDIOL	107-88-0
1,4-DIHYDROXYBUTANE	C4H10O2	1,4-BUTANEDIOL	110-63-4
2,3-DIHYDROXYBUTANE	C4H10O2	2,3-BUTANEDIOL	6982-25-8
DIHYDROXYDIETHYL ETHER	C4H10O3	DIETHYLENE GLYCOL	111-46-6
b,b'-DIHYDROXYDIETHYL ETHER	C4H10O3	DIETHYLENE GLYCOL	111-46-6

Name (Synonym)	Formula	Name in Tables	CAS Reg. No
2,2'-DIHYDROXYDIETHYLAMINE	C4H11NO2	DIETHANOLAMINE	111-42-2
p,p'-DIHYDROXYDIPHENYLDIMETHYLMETHANE	C15H16O2	BISPHENOL A	80-05-7
4,4'-DIHYDROXYDIPHENYLDIMETHYLMETHANE	C15H16O2	BISPHENOL A	80-05-7
p,p'-DIHYDROXYDIPHENYLPROPANE	C15H16O2	BISPHENOL A	80-05-7
2,2-(4,4'-DIHYDROXYDIPHENYL)PROPANE	C15H16O2	BISPHENOL A	80-05-7
4,4'-DIHYDROXYDIPHENYLPROPANE	C15H16O2	BISPHENOL A	80-05-7
4,4'-DIHYDROXYDIPHENYL-2,2-PROPANE	C15H16O2	BISPHENOL A	80-05-7
4,4'-DIHYDROXY-2,2-DIPHENYLPROPANE	C15H16O2	BISPHENOL A	80-05-7
1,2-DIHYDROXYETHANE	C2H6O2	ETHYLENE GLYCOL	107-21-1
DI-b-HYDROXYETHOXYETHANE	C6H14O4	TRIETHYLENE GLYCOL	112-27-6
2,2'-DIHYDROXYETHYL ETHER	C4H10O3	DIETHYLENE GLYCOL	111-46-6
DI(2-HYDROXYETHYL)AMINE	C4H11NO2	DIETHANOLAMINE	111-42-2
1,6-DIHYDROXYHEXANE	C6H14O2	1,6-HEXANEDIOL	629-11-8
2,4-DIHYDROXY-2-METHYLPENTANE	C6H14O2	HEXYLENE GLYCOL	107-41-5
1,5-DIHYDROXYPENTANE	C5H12O2	1,5-PENTANEDIOL	111-29-5
b-DI-p-HYDROXYPHENYLPROPANE	C15H16O2	BISPHENOL A	80-05-7
1,2-DIHYDROXYPROPANE	C3H8O2	1,2-PROPYLENE GLYCOL	57-55-6
1,3-DIHYDROXYPROPANE	C3H8O2	1,3-PROPYLENE GLYCOL	504-63-2
1,2-DIIODOBUTANE	C4H8I2	1,2-DIIODOBUTANE	628-21-7
1,4-DIIODOBUTANE	C4H8I2	1,2-DIIODOBUTANE	628-21-7
1,2-DIIODOETHANE	C2H4I2	1,2-DIIODOETHANE	624-73-7
DIIODOMETHANE	CH2I2	DIIODOMETHANE	75-11-6
1,2-DIIODOPROPANE	C3H6I2	1,2-DIIODOPROPANE	598-29-8
DIISOBUTILCHETONE	C9H18O	DIISOBUTYL KETONE	108-83-8
DIISOBUTYLAMINE	C8H19N	DIISOBUTYLAMINE	110-96-3
DI-ISOBUTYLCETONE	C9H18O	DIISOBUTYL KETONE	108-83-8
DIISOBUTYLKETON	C9H18O	DIISOBUTYL KETONE	108-83-8
DIISOBUTYL KETONE	C9H18O	DIISOBUTYL KETONE	108-83-8
4-4'-DIISOCYANATE de DIPHENYLMETHANE	C15H10N2O2	DIPHENYLMETHANE-4,4'-DIISOCYANATE	101-68-8
4,4'-DIISOCYANATODIPHENYLMETHANE	C15H10N2O2	DIPHENYLMETHANE-4,4'-DIISOCYANATE	101-68-8
DIISOCYANATOMETHYLBENZENE	C9H6N2O2	TOLUENE DIISOCYANATE	26471-62-5
DIISOCYANATOTOLUENE	C9H6N2O2	TOLUENE DIISOCYANATE	26471-62-5
DIISODECYL PHTHALATE	C28H46O4	DIISODECYL PHTHALATE	26761-40-0
DIISOOCTYL PHTHALATE	C24H38O4	DIISOOCTYL PHTHALATE	27554-26-3
DIISOPROPANOLAMINE	C6H15NO2	DIISOPROPANOLAMINE	110-97-4
s-DIISOPROPYLACETONE	C9H18O	DIISOBUTYL KETONE	108-83-8
DIISOPROPYLAMINE	C6H15N	DIISOPROPYLAMINE	108-18-9
m-DIISOPROPYLBENZENE	C12H18	m-DIISOPROPYLBENZENE	99-62-7
p-DIISOPROPYLBENZENE	C12H18	p-DIISOPROPYLBENZENE	100-18-5
DIISOPROPYLBENZENE PEROXIDE	C18H22O2	DICUMYL PEROXIDE	80-43-3
p-DIISOPROPYLBENZOL	C12H18	p-DIISOPROPYLBENZENE	100-18-5
DIISOPROPYL ETHER	C6H14O	DIISOPROPYL ETHER	108-20-3
DIISOPROPYL KETONE	C7H14O	DIISOPROPYL KETONE	565-80-0
DIISOPROPYL OXIDE	C6H14O	DIISOPROPYL ETHER	108-20-3
DIISOPROPYL-SULFIDE	C6H14S	DIISOPROPYL-SULFIDE	625-80-9
DIKETENE	C4H4O2	DIKETENE	674-82-8
DIKETONE ALCOHOL	C6H12O2	DIACETONE ALCOHOL	123-42-2
2,5-DIKETOTETRAHYDROFURAN	C4H4O3	SUCCINIC ANHYDRIDE	108-30-5
DILANTIN DB	C6H4Cl2	o-DICHLOROBENZENE	95-50-1
DILATIN DB	C6H4Cl2	o-DICHLOROBENZENE	95-50-1
DIMER CYKLOPENTADIENU	C10H12	DICYCLOPENTADIENE	77-73-6
a,b-DIMETHOXYETHANE	C4H10O2	1,2-DIMETHOXYETHANE	110-71-4
DIMETHOXYETHANE	C4H10O2	1,2-DIMETHOXYETHANE	110-71-4
1,2-DIMETHOXYETHANE	C4H10O2	1,2-DIMETHOXYETHANE	110-71-4
DIMETHOXYMETHANE	C3H8O2	METHYLAL	109-87-5
DIMETHOXYTETRAETHYLENE GLYCOL	C10H22O5	TETRAETHYLENE GLYCOL DIMETHYL ETHER	143-24-8
DIMETHOXYTETRAGLYCOL	C10H22O5	TETRAETHYLENE GLYCOL DIMETHYL ETHER	143-24-8
DIMETHYL	C2H6	ETHANE	74-84-0
DIMETHYLACETAMIDE	C4H9NO	N,N-DIMETHYLACETAMIDE	127-19-5
N,N-DIMETHYLACETAMIDE	C4H9NO	N,N-DIMETHYLACETAMIDE	127-19-5
DIMETHYLACETIC ACID	C4H8O2	ISOBUTYRIC ACID	79-31-2
DIMETHYLACETONE	C5H10O	DIETHYL KETONE	96-22-0
DIMETHYLACETONE AMIDE	C4H9NO	N,N-DIMETHYLACETAMIDE	127-19-5
DIMETHYLACETONITRILE	C4H7N	ISOBUTYRONITRILE	78-82-0
DIMETHYLACETYLENE	C4H6	DIMETHYLACETYLENE	503-17-3
DIMETHYLAETHANOLAMIN	C4H11NO	DIMETHYLETHANOLAMINE	108-01-0
DIMETHYLALUMINUM CHLORIDE	C2H6AlCl	DIMETHYLALUMINUM CHLORIDE	1184-58-3
DIMETHYLAMIDE ACETATE	C4H9NO	N,N-DIMETHYLACETAMIDE	127-19-5
DIMETHYLAMINE	C2H7N	DIMETHYLAMINE	124-40-3
2-(DIMETHYLAMINO)ETHANOL	C4H11NO	DIMETHYLETHANOLAMINE	108-01-0
DIMETHYLAMINOAETHANOL	C4H11NO	DIMETHYLETHANOLAMINE	108-01-0
p-DIMETHYLAMINOBENZALDEHYDE	C9H11NO	p-DIMETHYLAMINOBENZALDEHYDE	100-10-7
4-(DIMETHYLAMINO) BENZALDEHYDE	C9H11NO	p-DIMETHYLAMINOBENZALDEHYDE	100-10-7
(DIMETHYLAMINO)BENZENE	C8H11N	N,N-DIMETHYLANILINE	121-69-7
4-DIMETHYLAMINOBENZENECARBONAL	C9H11NO	p-DIMETHYLAMINOBENZALDEHYDE	100-10-7
b-DIMETHYLAMINOETHANOL	C4H11NO	DIMETHYLETHANOLAMINE	108-01-0

697

Name (Synonym)	Formula	Name in Tables	CAS Reg. No
DIMETHYLAMINOETHANOL	C4H11NO	DIMETHYLETHANOLAMINE	108-01-0
N,N-DIMETHYLAMINOETHANOL	C4H11NO	DIMETHYLETHANOLAMINE	108-01-0
b-DIMETHYLAMINOETHYL ALCOHOL	C4H11NO	DIMETHYLETHANOLAMINE	108-01-0
N,N-DIMETHYLANILINE	C8H11N	N,N-DIMETHYLANILINE	121-69-7
m-DIMETHYLBENZENE	C8H10	m-XYLENE	108-38-3
o-DIMETHYLBENZENE	C8H10	o-XYLENE	95-47-6
p-DIMETHYLBENZENE	C8H10	p-XYLENE	106-42-3
1,2-DIMETHYLBENZENE	C8H10	o-XYLENE	95-47-6
1,3-DIMETHYLBENZENE	C8H10	m-XYLENE	108-38-3
1,4-DIMETHYLBENZENE	C8H10	p-XYLENE	106-42-3
DIMETHYL-1,2-BENZENEDICARBOXYLATE	C10H10O4	DIMETHYL PHTHALATE	131-11-3
DIMETHYL-1,4-BENZENE DICARBOXYLATE	C10H10O4	DIMETHYL TEREPHTHALATE	120-61-6
N,N-DIMETHYLBENZENEAMINE	C8H11N	N,N-DIMETHYLANILINE	121-69-7
a,a-DIMETHYLBENZENEMETHANOL	C9H12O	2-PHENYL-2-PROPANOL	617-94-7
DIMETHYL BENZENEORTHODICARBOXYLATE	C10H10O4	DIMETHYL PHTHALATE	131-11-3
a,a-DIMETHYLBENZYL ALCOHOL	C9H12O	2-PHENYL-2-PROPANOL	617-94-7
a,a-DIMETHYLBENZYL HYDROPEROXIDE	C9H12O2	CUMENE HYDROPEROXIDE	80-15-9
p-(a-a-DIMETHYLBENZYL)PHENOL	C15H16O	p-CUMYLPHENOL	599-64-4
DIMETHYL BIS(p-HYDROXYPHENYL)METHANE	C15H16O2	BISPHENOL A	80-05-7
2,3-DIMETHYL-1,3-BUTADIENE	C6H10	2,3-DIMETHYL-1,3-BUTADIENE	513-81-5
1,2-DIMETHYLBUTANE	C6H14	2-METHYLPENTANE	107-83-5
2,2-DIMETHYLBUTANE	C6H14	2,2-DIMETHYLBUTANE	75-83-2
2,3-DIMETHYLBUTANE	C6H14	2,3-DIMETHYLBUTANE	79-29-8
1,3-DIMETHYL BUTANOL	C6H14O	2-METHYL-1-PENTANOL	105-30-6
2,3-DIMETHYL-1-BUTENE	C6H12	2,3-DIMETHYL-1-BUTENE	563-78-0
2,3-DIMETHYL-2-BUTENE	C6H12	2,3-DIMETHYL-2-BUTENE	563-79-1
3,3-DIMETHYL-1-BUTENE	C6H12	3,3-DIMETHYL-1-BUTENE	558-37-2
DIMETHYLCARBINOL	C3H8O	ISOPROPANOL	67-63-0
DIMETHYLCELLOSOLVE	C4H10O2	1,2-DIMETHOXYETHANE	110-71-4
cis-1,2-DIMETHYLCYCLOHEXANE	C8H16	cis-1,2-DIMETHYLCYCLOHEXANE	2207-01-4
trans-1,2-DIMETHYLCYCLOHEXANE	C8H16	trans-1,2-DIMETHYLCYCLOHEXANE	6876-23-9
cis-1,3-DIMETHYLCYCLOHEXANE	C8H16	cis-1,3-DIMETHYLCYCLOHEXANE	638-04-0
trans-1,3-DIMETHYLCYCLOHEXANE	C8H16	trans-1,3-DIMETHYLCYCLOHEXANE	2207-03-6
cis-1,4-DIMETHYLCYCLOHEXANE	C8H16	cis-1,4-DIMETHYLCYCLOHEXANE	624-29-3
trans-1,4-DIMETHYLCYCLOHEXANE	C8H16	trans-1,4-DIMETHYLCYCLOHEXANE	2207-03-6
1,1-DIMETHYLCYCLOHEXANE	C8H16	1,1-DIMETHYLCYCLOHEXANE	590-66-9
cis-1,2-DIMETHYLCYCLOPENTANE	C7H14	cis-1,2-DIMETHYLCYCLOPENTANE	1192-18-3
trans-1,2-DIMETHYLCYCLOPENTANE	C7H14	trans-1,2-DIMETHYLCYCLOPENTANE	822-50-4
cis-1,3-DIMETHYLCYCLOPENTANE	C7H14	cis-1,3-DIMETHYLCYCLOPENTANE	2532-58-3
trans-1,3-DIMETHYLCYCLOPENTANE	C7H14	trans-1,3-DIMETHYLCYCLOPENTANE	1759-58-6
DIMETHYL-1,1'-DICHLOROETHER	C2H4Cl2O	BIS(CHLOROMETHYL)ETHER	542-88-1
2,3-DIMETHYL-2,3-DIPHENYLBUTANE	C18H22	2,3-DIMETHYL-2,3-DIPHENYLBUTANE	1889-67-4
DIMETHYL DISULFIDE	C2H6S2	DIMETHYL DISULFIDE	624-92-0
DIMETHYLENEDIAMINE	C2H8N2	ETHYLENEDIAMINE	107-15-3
DIMETHYLENEIMINE	C2H5N	ETHYLENEIMINE	151-56-4
DIMETHYLENE OXIDE	C2H4O	ETHYLENE OXIDE	75-21-8
DIMETHYLENIMINE	C2H5N	ETHYLENEIMINE	151-56-4
DIMETHYLESTER KYSELINY SIROVE	C2H6O4S	DIMETHYL SULFATE	77-78-1
1,1-DIMETHYLETHANOL	C4H10O	tert-BUTANOL	75-65-0
DIMETHYLETHANOLAMINE	C4H11NO	DIMETHYLETHANOLAMINE	108-01-0
N,N-DIMETHYLETHANOLAMINE	C4H11NO	DIMETHYLETHANOLAMINE	108-01-0
DIMETHYL ETHER	C2H6O	DIMETHYL ETHER	115-10-6
1,1-DIMETHYLETHYLAMINE	C4H11N	tert-BUTYLAMINE	75-64-9
4-(1,1-DIMETHYLETHYL)-1,2-BENZENEDIOL	C10H14O2	p-tert-BUTYLCATECHOL	98-29-3
DIMETHYLETHYLCARBINOL	C5H12O	tert-PENTYL-ALCOHOL	75-85-4
1,1-DIMETHYLETHYL HYDROPEROXIDE	C4H10O2	t-BUTYL HYDROPEROXIDE	75-91-2
2,2-DIMETHYL-3-ETHYLPENTANE	C9H20	2,2-DIMETHYL-3-ETHYLPENTANE	16747-32-3
2,4-DIMETHYL-3-ETHYLPENTANE	C9H20	2,4-DIMETHYL-3-ETHYLPENTANE	1068-87-7
4-(1,1-DIMETHYLETHYL)PHENOL	C10H14O	p-tert-BUTYLPHENOL	98-54-4
DIMETHYL FORMAL	C3H8O2	METHYLAL	109-87-5
DIMETHYLFORMALDEHYDE	C3H6O	ACETONE	67-64-1
DIMETHYLFORMAMID	C3H7NO	N,N-DIMETHYLFORMAMIDE	68-12-2
N,N-DIMETHYLFORMAMIDE	C3H7NO	N,N-DIMETHYLFORMAMIDE	68-12-2
2,2-DIMETHYLHEPTANE	C9H20	2,2-DIMETHYLHEPTANE	1071-26-7
2,3-DIMETHYLHEPTANE	C9H20	2,3-DIMETHYLHEPTANE	3074-71-3
2,4-DIMETHYLHEPTANE	C9H20	2,4-DIMETHYLHEPTANE	2213-23-2
2,5-DIMETHYLHEPTANE	C9H20	2,5-DIMETHYLHEPTANE	2216-30-0
2,6-DIMETHYLHEPTANE	C9H20	2,6-DIMETHYLHEPTANE	1072-05-5
3,4-DIMETHYLHEPTANE	C9H20	3,4-DIMETHYLHEPTANE	922-28-1
3,5-DIMETHYLHEPTANE	C9H20	3,5-DIMETHYLHEPTANE	926-82-9
4,4-DIMETHYLHEPTANE	C9H20	4,4-DIMETHYLHEPTANE	1068-19-5
2,6-DIMETHYL-4-HEPTANOL	C9H20O	2,6-DIMETHYL-4-HEPTANOL	108-82-7
2,6-DIMETHYL HEPTANOL-4	C9H20O	2,6-DIMETHYL-4-HEPTANOL	108-82-7
2,6-DIMETHYL-HEPTAN-4-ON	C9H18O	DIISOBUTYL KETONE	108-83-8
2,6-DIMETHYLHEPTAN-4-ONE	C9H18O	DIISOBUTYL KETONE	108-83-8
2,6-DIMETHYL-4-HEPTANONE	C9H18O	DIISOBUTYL KETONE	108-83-8

Name (Synonym)	Formula	Name in Tables	CAS Reg. No
2,6-DIMETHYL-4-HEPTYLPHENOL,	C15H24O	NONYLPHENOL	25154-52-3
2,2-DIMETHYLHEXANE	C8H18	2,2-DIMETHYLHEXANE	590-73-8
2,3-DIMETHYLHEXANE	C8H18	2,3-DIMETHYLHEXANE	584-94-1
2,4-DIMETHYLHEXANE	C8H18	2,4-DIMETHYLHEXANE	589-43-5
2,5-DIMETHYLHEXANE	C8H18	2,5-DIMETHYLHEXANE	592-13-2
3,3-DIMETHYLHEXANE	C8H18	3,3-DIMETHYLHEXANE	563-16-6
3,4-DIMETHYLHEXANE	C8H18	3,4-DIMETHYLHEXANE	583-48-2
N,N-DIMETHYL-N-(2-HYDROXYETHYL)AMINE	C4H11NO	DIMETHYLETHANOLAMINE	108-01-0
N,N-DIMETHYL-2-HYDROXYETHYLAMINE	C4H11NO	DIMETHYLETHANOLAMINE	108-01-0
2,4-DIMETHYL-3-ISOPROPYLPENTANE	C10H22	2,4-DIMETHYL-3-ISOPROPYLPENTANE	----------
DIMETHYLKETAL	C3H6O	ACETONE	67-64-1
DIMETHYL KETONE	C3H6O	ACETONE	67-64-1
DIMETHYL MALEATE	C6H8O4	DIMETHYL MALEATE	624-58-6
DIMETHYLMETHANE	C3H8	PROPANE	74-98-6
6,6-DIMETHYL-2-METHYLENEBICYCLO(3.1.1)HEPTANE	C10H16	beta-PINENE	127-91-3
DIMETHYLMETHYLENE-p,p'-DIPHENOL	C15H16O2	BISPHENOL A	80-05-7
DIMETHYL MONOSULFATE	C2H6O4S	DIMETHYL SULFATE	77-78-1
1,2-DIMETHYLNAPHTHALENE	C12H12	1,2-DIMETHYLNAPHTHALENE	573-98-8
1,3-DIMETHYLNAPHTHALENE	C12H12	1,3-DIMETHYLNAPHTHALENE	575-41-7
1,4-DIMETHYLNAPHTHALENE	C12H12	1,4-DIMETHYLNAPHTHALENE	571-58-4
1,5-DIMETHYLNAPHTHALENE	C12H12	1,5-DIMETHYLNAPHTHALENE	571-61-9
1,6-DIMETHYLNAPHTHALENE	C12H12	1,6-DIMETHYLNAPHTHALENE	575-43-9
1,7-DIMETHYLNAPHTHALENE	C12H12	1,7-DIMETHYLNAPHTHALENE	575-37-1
2,3-DIMETHYLNAPHTHALENE	C12H12	2,3-DIMETHYLNAPHTHALENE	581-40-8
2,6-DIMETHYLNAPHTHALENE	C12H12	2,6-DIMETHYLNAPHTHALENE	581-42-0
2,7-DIMETHYLNAPHTHALENE	C12H12	2,7-DIMETHYLNAPHTHALENE	582-16-1
DIMETHYLNITROMETHANE	C3H7NO2	2-NITROPROPANE	79-46-9
2,2-DIMETHYLOCTANE	C10H22	2,2-DIMETHYLOCTANE	15869-87-1
2,3-DIMETHYLOCTANE	C10H22	2,3-DIMETHYLOCTANE	7146-60-3
2,4-DIMETHYLOCTANE	C10H22	2,4-DIMETHYLOCTANE	4032-94-4
2,5-DIMETHYLOCTANE	C10H22	2,5-DIMETHYLOCTANE	15869-89-3
2,6-DIMETHYLOCTANE	C10H22	2,6-DIMETHYLOCTANE	2051-30-1
2,7-DIMETHYLOCTANE	C10H22	2,7-DIMETHYLOCTANE	1072-16-8
3,3-DIMETHYLOCTANE	C10H22	3,3-DIMETHYLOCTANE	4110-44-5
3,4-DIMETHYLOCTANE	C10H22	3,4-DIMETHYLOCTANE	15869-92-8
3,5-DIMETHYLOCTANE	C10H22	3,5-DIMETHYLOCTANE	15869-93-9
3,6-DIMETHYLOCTANE	C10H22	3,6-DIMETHYLOCTANE	15869-94-0
4,4-DIMETHYLOCTANE	C10H22	4,4-DIMETHYLOCTANE	15869-95-1
4,5-DIMETHYLOCTANE	C10H22	4,5-DIMETHYLOCTANE	15869-96-2
2,2-DIMETHYLPENTANE	C7H16	2,2-DIMETHYLPENTANE	590-35-2
2,3-DIMETHYLPENTANE	C7H16	2,3-DIMETHYLPENTANE	565-59-3
2,4-DIMETHYLPENTANE	C7H16	2,4-DIMETHYLPENTANE	108-08-7
3,3-DIMETHYLPENTANE	C7H16	3,3-DIMETHYLPENTANE	562-49-2
2,3-DIMETHYLPHENOL	C8H10O	2,3-XYLENOL	526-75-0
2,4-DIMETHYLPHENOL	C8H10O	2,4-XYLENOL	105-67-9
2,5-DIMETHYLPHENOL	C8H10O	2,5-XYLENOL	95-87-4
2,6-DIMETHYLPHENOL	C8H10O	2,6-XYLENOL	576-26-1
3,4-DIMETHYLPHENOL	C8H10O	3,4-XYLENOL	95-65-8
3,5-DIMETHYLPHENOL	C8H10O	3,5-XYLENOL	108-68-9
3,6-DIMETHYLPHENOL	C8H10O	2,5-XYLENOL	95-87-4
4,5-DIMETHYLPHENOL	C8H10O	3,4-XYLENOL	95-65-8
4,6-DIMETHYLPHENOL	C8H10O	2,4-XYLENOL	105-67-9
DIMETHYLPHENYLAMINE	C8H11N	N,N-DIMETHYLANILINE	121-69-7
N,N-DIMETHYLPHENYLAMINE	C8H11N	N,N-DIMETHYLANILINE	121-69-7
DIMETHYLPHENYLCARBINOL	C9H12O	2-PHENYL-2-PROPANOL	617-94-7
4-(DIMETHYLPHENYLMETHYL)PHENOL	C15H16O	p-CUMYLPHENOL	599-64-4
DIMETHYL PHTHALATE	C10H10O4	DIMETHYL PHTHALATE	131-11-3
2,2-DIMETHYLPROPANE	C5H12	NEOPENTANE	463-82-1
2,2-DIMETHYL-1,3-PROPANEDIOL	C5H12O2	NEOPENTYL GLYCOL	126-30-7
2,2-DIMETHYL-1-PROPANOL	C5H12O	2,2-DIMETHYL-1-PROPANOL	75-84-3
p-(a,a-DIMETHYLPROPYL)PHENOL	C11H16O	p-tert-AMYLPHENOL	80-46-6
p-(1,1-DIMETHYLPROPYL)PHENOL	C11H16O	p-tert-AMYLPHENOL	80-46-6
a-a'-DIMETHYLPYRIDINE	C7H9N	2,6-DIMETHYLPYRIDINE	108-48-5
2,6-DIMETHYLPYRIDINE	C7H9N	2,6-DIMETHYLPYRIDINE	108-48-5
DIMETHYL SILANE	C2H8Si	DIMETHYL SILANE	1111-74-6
DIMETHYLSULFAAT	C2H6O4S	DIMETHYL SULFATE	77-78-1
DIMETHYLSULFAT	C2H6O4S	DIMETHYL SULFATE	77-78-1
DIMETHYL SULFATE	C2H6O4S	DIMETHYL SULFATE	77-78-1
DIMETHYLSULFID	C2H6S	DIMETHYL SULFIDE	75-18-3
DIMETHYL SULFIDE	C2H6S	DIMETHYL SULFIDE	75-18-3
DIMETHYL SULFOXIDE	C2H6OS	DIMETHYL SULFOXIDE	67-68-5
DIMETHYL SULPHIDE	C2H6S	DIMETHYL SULFIDE	75-18-3
DIMETHYL SULPHOXIDE	C2H6OS	DIMETHYL SULFOXIDE	67-68-5
DIMETHYL TEREPHTHALATE	C10H10O4	DIMETHYL TEREPHTHALATE	120-61-6
DIMETHYLTRIMETHYLENE GLYCOL	C5H12O2	NEOPENTYL GLYCOL	126-30-7
2,6-DIMETIL-EPTAN-4-ONE	C9H18O	DIISOBUTYL KETONE	108-83-8

Name (Synonym)	Formula	Name in Tables	CAS Reg. No
DIMETILFORMAMIDE	C3H7NO	N,N-DIMETHYLFORMAMIDE	68-12-2
DIMETILSOLFATO	C2H6O4S	DIMETHYL SULFATE	77-78-1
1,1-DIMETYLCYCLOPENTANE	C7H14	1,1-DIMETHYLCYCLOPENTANE	1638-26-2
DIMETYLFORMAMIDU	C3H7NO	N,N-DIMETHYLFORMAMIDE	68-12-2
DIMEXIDE	C2H6OS	DIMETHYL SULFOXIDE	67-68-5
DINILE	C4H4N2	SUCCINONITRILE	110-61-2
1,3-DINITRATO-2,2-BIS(NITRATOMETHYL)PROPANE	C5H8N4O12	PENTAERYTHRITOL TETRANITRATE	78-11-5
m-DINITROBENZENE	C6H4N2O4	m-DINITROBENZENE	99-65-0
o-DINITROBENZENE	C6H4N2O4	o-DINITROBENZENE	528-29-0
p-DINITROBENZENE	C6H4N2O4	p-DINITROBENZENE	100-25-4
1,2-DINITROBENZENE	C6H4N2O4	o-DINITROBENZENE	528-29-0
1,3-DINITROBENZENE	C6H4N2O4	m-DINITROBENZENE	99-65-0
2,4-DINITROBENZENE	C6H4N2O4	m-DINITROBENZENE	99-65-0
1,3-DINITROBENZOL	C6H4N2O4	m-DINITROBENZENE	99-65-0
1,3-DINITRO-4-CHLOROBENZENE	C6H3ClN2O4	1-CHLORO-2,4-DINITROBENZENE	97-00-7
2,4-DINITROCHLOROBENZENE	C6H3ClN2O4	1-CHLORO-2,4-DINITROBENZENE	97-00-7
2,4-DINITRO-1-CHLOROBENZENE	C6H3ClN2O4	1-CHLORO-2,4-DINITROBENZENE	97-00-7
DINITROCHLOROBENZOL	C6H3ClN2O4	1-CHLORO-2,4-DINITROBENZENE	97-00-7
2,4-DINITROTOLUENE	C7H6N2O4	2,4-DINITROTOLUENE	121-14-2
2,5-DINITROTOLUENE	C7H6N2O4	2,5-DINITROTOLUENE	619-15-8
2,6-DINITROTOLUENE	C7H6N2O4	2,6-DINITROTOLUENE	606-20-2
3,4-DINITROTOLUENE	C7H6N2O4	3,4-DINITROTOLUENE	610-39-9
3,5-DINITROTOLUENE	C7H6N2O4	3,5-DINITROTOLUENE	618-85-9
2,4-DINITROTOLUOL	C7H6N2O4	2,4-DINITROTOLUENE	121-14-2
DINONYL ETHER	C18H38O	DINONYL ETHER	2456-27-1
DINONYLPHENOL	C24H42O	DINONYLPHENOL	1323-65-5
DIOCTYL ETHER	C16H34O	DI-n-OCTYL ETHER	629-82-3
DI-n-OCTYL ETHER	C16H34O	DI-n-OCTYL ETHER	629-82-3
DIOCTYL PHTHALATE	C24H38O4	DIOCTYL PHTHALATE	117-81-7
DIOKAN	C4H8O2	1,4-DIOXANE	123-91-1
DIOKSAN	C4H8O2	1,4-DIOXANE	123-91-1
DIOLAMINE	C4H11NO2	DIETHANOLAMINE	111-42-2
DIOLANE	C6H14O2	HEXYLENE GLYCOL	107-41-5
DIOSSANO-1,4	C4H8O2	1,4-DIOXANE	123-91-1
DIOXAAN-1,4	C4H8O2	1,4-DIOXANE	123-91-1
2,5-DIOXAHEXANE	C4H10O2	1,2-DIMETHOXYETHANE	110-71-4
p-DIOXAN	C4H8O2	1,4-DIOXANE	123-91-1
1,4-DIOXANE	C4H8O2	1,4-DIOXANE	123-91-1
DIOXANNE	C4H8O2	1,4-DIOXANE	123-91-1
3,6-DIOXAOCTANE-1,8-DIOL	C6H14O4	TRIETHYLENE GLYCOL	112-27-6
1,1-DIOXIDETETRAHYDROTHIOFURAN	C4H8O2S	SULFOLANE	126-33-0
1,1-DIOXIDETETRAHYDROTHIOPHENE	C4H8O2S	SULFOLANE	126-33-0
DIOXITOL	C6H14O3	2-(2-ETHOXYETHOXY)ETHANOL	111-90-0
9,10-DIOXOANTHRACENE	C14H8O2	ANTHRAQUINONE	84-65-1
p-DIOXOBENZENE	C6H6O2	p-HYDROQUINONE	123-31-9
1,3-DIOXOLAN-2-ONE	C3H4O3	ETHYLENE CARBONATE	96-49-1
DIOXOLONE-2	C3H4O3	ETHYLENE CARBONATE	96-49-1
1,3-DIOXOPHTHALAN	C8H4O3	PHTHALIC ANHYDRIDE	85-44-9
1,3-DIOXO-5-PHTHALANCARBOXYLIC ACID	C9H4O5	TRIMELLITIC ANHYDRIDE	552-30-7
DIOXOTHIOLAN	C4H8O2S	SULFOLANE	126-33-0
1,1-DIOXOTHIOLAN	C4H8O2S	SULFOLANE	126-33-0
m-DIOXYBENZENE	C6H6O2	1,3-BENZENEDIOL	108-46-3
o-DIOXYBENZENE	C6H6O2	1,2-BENZENEDIOL	120-80-9
p-DIOXYBENZENE	C6H6O2	p-HYDROQUINONE	123-31-9
DIOXYETHYLENE ETHER	C4H8O2	1,4-DIOXANE	123-91-1
DIPA	C6H15N	DIISOPROPYLAMINE	108-18-9
DIPENTYL-DISULFIDE	C10H22S2	DIPENTYL-DISULFIDE	112-51-6
DIPENTYL ETHER	C10H22O	DI-n-PENTYL ETHER	693-65-2
DI-n-PENTYL ETHER	C10H22O	DI-n-PENTYL ETHER	693-65-2
o-DIPHENOL	C6H6O2	1,2-BENZENEDIOL	120-80-9
DIPHENYLACETYLENE	C14H10	DIPHENYLACETYLENE	501-65-5
DIPHENYLAMINE	C12H11N	DIPHENYLAMINE	122-39-4
N,N-DIPHENYLAMINE	C12H11N	DIPHENYLAMINE	122-39-4
DIPHENYLAMINE, 4-AMINO-	C12H12N2	p-AMINODIPHENYLAMINE	101-54-2
m-DIPHENYLBENZENE	C18H14	m-TERPHENYL	92-06-8
p-DIPHENYLBENZENE	C18H14	p-TERPHENYL	92-94-4
1,2-DIPHENYLBENZENE	C18H14	o-TERPHENYL	84-15-1
1,4-DIPHENYLBENZENE	C18H14	p-TERPHENYL	92-94-4
DIPHENYL BLACK	C12H12N2	p-AMINODIPHENYLAMINE	101-54-2
DIPHENYLENE OXIDE	C12H8O	DIBENZOFURAN	132-64-9
DIPHENYLENEIMINE	C12H9N	DIBENZOPYRROLE	86-74-8
DIPHENYLENEMETHANE	C13H10	FLUORENE	86-73-7
DIPHENYLENIMIDE	C12H9N	DIBENZOPYRROLE	86-74-8
DIPHENYLENIMINE	C12H9N	DIBENZOPYRROLE	86-74-8
1,1-DIPHENYLETHANE	C14H14	1,1-DIPHENYLETHANE	612-00-0
1,2-DIPHENYLETHANE	C14H14	1,2-DIPHENYLETHANE	103-29-7

Name (Synonym)	Formula	Name in Tables	CAS Reg. No
trans-DIPHENYLETHENE	C14H12	trans-STILBENE	103-30-0
trans-1,2-DIPHENYLETHENE	C14H12	trans-STILBENE	103-30-0
DIPHENYL ETHER	C12H10O	DIPHENYL ETHER	101-84-8
(E)-1,2-DIPHENYLETHYLENE	C14H12	trans-STILBENE	103-30-0
trans-a-b-DIPHENYLETHYLENE	C14H12	trans-STILBENE	103-30-0
trans-1,2-DIPHENYLETHYLENE	C14H12	trans-STILBENE	103-30-0
sym-DIPHENYLHYDRAZINE	C12H12N2	HYDRAZOBENZENE	122-66-7
1,2-DIPHENYLHYDRAZINE	C12H12N2	HYDRAZOBENZENE	122-66-7
DIPHENYL KETONE	C13H10O	BENZOPHENONE	119-61-9
DIPHENYLMETHANE	C13H12	DIPHENYLMETHANE	101-81-5
DIPHENYL METHANE DIISOCYANATE	C15H10N2O2	DIPHENYLMETHANE-4,4'-DIISOCYANATE	101-68-8
p,p'-DIPHENYLMETHANE DIISOCYANATE	C15H10N2O2	DIPHENYLMETHANE-4,4'-DIISOCYANATE	101-68-8
DIPHENYLMETHANE 4,4'-DIISOCYANATE	C15H10N2O2	DIPHENYLMETHANE-4,4'-DIISOCYANATE	101-68-8
4,4'-DIPHENYLMETHANE DIISOCYANATE	C15H10N2O2	DIPHENYLMETHANE-4,4'-DIISOCYANATE	101-68-8
DIPHENYLMETHANE-4,4'-DIISOCYANATE-TRIMELLIC ANHYDRID	C9H4O5	TRIMELLITIC ANHYDRIDE	552-30-7
DIPHENYLMETHANONE	C13H10O	BENZOPHENONE	119-61-9
2,2-DI(4-PHENYLOL)PROPANE	C15H16O2	BISPHENOL A	80-05-7
DIPHENYL OXIDE	C12H10O	DIPHENYL ETHER	101-84-8
DIPHENYL-p-PHENYLENEDIAMINE	C18H16N2	N,N'-DIPHENYL-p-PHENYLENEDIAMINE	74-31-7
N,N'-DIPHENYL-p-PHENYLENEDIAMINE	C18H16N2	N,N'-DIPHENYL-p-PHENYLENEDIAMINE	74-31-7
1,3-DIPHENYLTRIAZENE	C12H11N3	1,3-DIPHENYLTRIAZENE	136-35-6
DIPHOSGENE	CCl2O	PHOSGENE	75-44-5
DIPIRARTRIL-TROPICO	C2H6OS	DIMETHYL SULFOXIDE	67-68-5
DI-n-PROPYLAMINE	C6H15N	DI-n-PROPYLAMINE	142-84-7
n-DIPROPYLAMINE	C6H15N	DI-n-PROPYLAMINE	142-84-7
DIPROPYL-2,2'-DIHYDROXYAMINE	C6H15NO2	DIISOPROPANOLAMINE	110-97-4
DIPROPYL-DISULFIDE	C6H14S2	DIPROPYL-DISULFIDE	629-19-6
DIPROPYLENE GLYCOL	C6H14O3	DIPROPYLENE GLYCOL	25265-71-8
DIPROPYL ETHER	C6H14O	DI-n-PROPYL ETHER	111-43-3
DI-n-PROPYL ETHER	C6H14O	DI-n-PROPYL ETHER	111-43-3
DIPROPYL METHANE	C7H16	n-HEPTANE	142-82-5
DIPROPYL OXIDE	C6H14O	DI-n-PROPYL ETHER	111-43-3
N,N-DIPROPYL-1-PROPANAMINE	C9H21N	TRIPROPYLAMINE	102-69-2
DIPROPYL-SULFIDE	C6H14S	DIPROPYL-SULFIDE	111-47-7
DI-n-PROPYL SULFONE	C6H14O2S	DI-n-PROPYL SULFONE	598-03-8
DIRECT BROWN BR	C6H8N2	m-PHENYLENEDIAMINE	108-45-2
DISPERMINE	C4H10N2	PIPERAZINE	110-85-0
DISSOLVANT APV	C4H10O3	DIETHYLENE GLYCOL	111-46-6
DISTOKAL	C2Cl6	HEXACHLOROETHANE	67-72-1
DISTOPAN	C2Cl6	HEXACHLOROETHANE	67-72-1
DISTOPIN	C2Cl6	HEXACHLOROETHANE	67-72-1
DITHANE A-4	C6H4N2O4	p-DINITROBENZENE	100-25-4
DITHIOCARBONIC ANHYDRIDE	CS2	CARBON DISULFIDE	75-15-0
DIVINYL	C4H6	1,3-BUTADIENE	106-99-0
m-DIVINYLBENZEN	C10H10	m-DIVINYLBENZENE	108-57-6
DIVINYLENE OXIDE	C4H4O	FURAN	110-00-9
DIVINYLENE SULFIDE	C4H4S	THIOPHENE	110-02-1
DIVINYLENIMINE	C4H5N	PYRROLE	109-97-7
DIVINYL ETHER	C4H6O	DIVINYL ETHER	109-93-3
DIVYNYL OXIDE	C4H6O	DIVINYL ETHER	109-93-3
DIZENE	C6H4Cl2	o-DICHLOROBENZENE	95-50-1
DMA	C2H7N	DIMETHYLAMINE	124-40-3
DMAC	C4H9NO	N,N-DIMETHYLACETAMIDE	127-19-5
DMAE	C4H11NO	DIMETHYLETHANOLAMINE	108-01-0
DMF	C3H7NO	N,N-DIMETHYLFORMAMIDE	68-12-2
DMFA	C3H7NO	N,N-DIMETHYLFORMAMIDE	68-12-2
DMP	C10H10O4	DIMETHYL PHTHALATE	131-11-3
2,5-DMP	C8H10O	2,5-XYLENOL	95-87-4
2,6-DMP	C8H10O	2,6-XYLENOL	576-26-1
3,4-DMP	C8H10O	3,4-XYLENOL	95-65-8
3,5-DMP	C8H10O	3,5-XYLENOL	108-68-9
DMS	C2H6O4S	DIMETHYL SULFATE	77-78-1
DMS-70	C2H6OS	DIMETHYL SULFOXIDE	67-68-5
DMS-90	C2H6OS	DIMETHYL SULFOXIDE	67-68-5
DMSO	C2H6OS	DIMETHYL SULFOXIDE	67-68-5
DNCB	C6H3ClN2O4	1-CHLORO-2,4-DINITROBENZENE	97-00-7
2,5-DNT	C7H6N2O4	2,5-DINITROTOLUENE	619-15-8
2,6-DNT	C7H6N2O4	2,6-DINITROTOLUENE	606-20-2
3,4-DNT	C7H6N2O4	3,4-DINITROTOLUENE	610-39-9
3,5-DNT	C7H6N2O4	3,5-DINITROTOLUENE	618-85-9
DNT	C7H6N2O4	2,4-DINITROTOLUENE	121-14-2
2,4-DNT	C7H6N2O4	2,4-DINITROTOLUENE	121-14-2
DODECAHYDRODIPHENYLAMINE	C12H23N	DICYCLOHEXYLAMINE	101-83-7
n-DODECAN	C12H26	n-DODECANE	112-40-3
1-DODECANAL	C12H24O	1-DODECANAL	112-54-9
1-DODECANAMINE	C12H27N	DODECYLAMINE	124-22-1

Name (Synonym)	Formula	Name in Tables	CAS Reg. No
n-DODECANE	C12H26	n-DODECANE	112-40-3
1-DODECANETHIOL	C12H26S	n-DODECYL MERCAPTAN	112-55-0
DODECANOIC ACID	C12H24O2	n-DODECANOIC ACID	143-07-7
n-DODECANOIC ACID	C12H24O2	n-DODECANOIC ACID	143-07-7
n-DODECANOL	C12H26O	1-DODECANOL	112-53-8
1-DODECANOL	C12H26O	1-DODECANOL	112-53-8
1-DODECENE	C12H24	1-DODECENE	112-41-4
DODECOIC ACID	C12H24O2	n-DODECANOIC ACID	143-07-7
n-DODECYL ALCOHOL	C12H26O	1-DODECANOL	112-53-8
1-DODECYL ALDEHYDE	C12H24O	1-DODECANAL	112-54-9
DODECYLAMINE	C12H27N	DODECYLAMINE	124-22-1
n-DODECYLAMINE	C12H27N	DODECYLAMINE	124-22-1
1-DODECYLAMINE	C12H27N	DODECYLAMINE	124-22-1
DODECYLBENZENE	C18H30	n-DODECYLBENZENE	123-01-3
n-DODECYLBENZENE	C18H30	n-DODECYLBENZENE	123-01-3
DODECYL-ETHYL-SULFIDE	C14H30S	DODECYL-ETHYL-SULFIDE	----------
DODECYL MERCAPTAN	C12H26S	n-DODECYL MERCAPTAN	112-55-0
m-DODECYL MERCAPTAN	C12H26S	n-DODECYL MERCAPTAN	112-55-0
n-DODECYL MERCAPTAN	C12H26S	n-DODECYL MERCAPTAN	112-55-0
1-DODECYL MERCAPTAN	C12H26S	n-DODECYL MERCAPTAN	112-55-0
DODECYL-METHYL-SULFIDE	C13H28S	DODECYL-METHYL-SULFIDE	----------
DODECYL-PROPYL-SULFIDE	C15H32S	DODECYL-PROPYL-SULFIDE	----------
1-DODECYNE	C12H22	1-DODECYNE	765-03-7
DOLCYMENE	C10H14	p-CYMENE	99-87-6
DOLEN-PUR	C4Cl6	HEXACHLORO-1,3-BUTADIENE	87-68-3
DOLICUR	C2H6OS	DIMETHYL SULFOXIDE	67-68-5
DOLIGUR	C2H6OS	DIMETHYL SULFOXIDE	67-68-5
DOMOSO	C2H6OS	DIMETHYL SULFOXIDE	67-68-5
DOP	C24H38O4	DIOCTYL PHTHALATE	117-81-7
DOW CORNING 200	C6H18OSi2	HEXAMETHYLDISILOXANE	107-46-0
DOWANOL	C6H14O3	2-(2-ETHOXYETHOXY)ETHANOL	111-90-0
DOWANOL DB	C8H18O3	DIETHYLENE GLYCOL MONOBUTYL ETHER	112-34-5
DOWANOL DE	C6H14O3	2-(2-ETHOXYETHOXY)ETHANOL	111-90-0
DOWANOL DM	C5H12O3	2-(2-METHOXYETHOXY)ETHANOL	111-77-3
DOWANOL EB	C6H14O2	2-BUTOXYETHANOL	111-76-2
DOWANOL EE	C4H10O2	2-ETHOXYETHANOL	110-80-5
DOWANOL EM	C3H8O2	2-METHOXYETHANOL	109-86-4
DOWFROST	C3H8O2	1,2-PROPYLENE GLYCOL	57-55-6
DOWFUME 40	C2H4Br2	1,2-DIBROMOETHANE	106-93-4
DOWFUME EDB	C2H4Br2	1,2-DIBROMOETHANE	106-93-4
DOWFUME N	C3H6Cl2	2,2-DICHLOROPROPANE	594-20-7
DOWFUME W-8	C2H4Br2	1,2-DIBROMOETHANE	106-93-4
DOW-PER	C2Cl4	TETRACHLOROETHYLENE	127-18-4
DOWTHERM E	C6H4Cl2	o-DICHLOROBENZENE	95-50-1
DOWTHERM SR 1	C2H6O2	ETHYLENE GLYCOL	107-21-1
DOW-TRI	C2HCl3	TRICHLOROETHYLENE	79-01-6
DPA	C12H11N	DIPHENYLAMINE	122-39-4
DPPD	C18H16N2	N,N'-DIPHENYL-p-PHENYLENEDIAMINE	74-31-7
DRACYLIC ACID	C7H6O2	BENZOIC ACID	65-85-0
DROMISOL	C2H6OS	DIMETHYL SULFOXIDE	67-68-5
DRY ICE	CO2	CARBON DIOXIDE	124-38-9
DS	C4H10O4S	DIETHYL SULFATE	64-67-5
DTBP	C8H18O2	DI-t-BUTYL PEROXIDE	110-05-4
DUKERON	C2HCl3	TRICHLOROETHYLENE	79-01-6
DUMASIN	C5H8O	CYCLOPENTANONE	120-92-3
DUODECANE	C12H26	n-DODECANE	112-40-3
DUODECYL ALCOHOL	C12H26O	1-DODECANOL	112-53-8
DUODECYLIC ACID	C12H24O2	n-DODECANOIC ACID	143-07-7
DUODECYLIC ALDEHYDE	C12H24O	1-DODECANAL	112-54-9
DUOTRATE	C5H8N4O12	PENTAERYTHRITOL TETRANITRATE	78-11-5
DURAFUR DEVELOPER G	C6H6O2	1,3-BENZENEDIOL	108-46-3
DURASORB	C2H6OS	DIMETHYL SULFOXIDE	67-68-5
DWUCHLOROFLUOROMETAN	CHCl2F	DICHLOROFLUOROMETHANE	75-43-4
DWUCHLOROPROPAN	C3H6Cl2	1,2-DICHLOROPROPANE	78-87-5
DWUETYLOAMINA	C4H11N	DIETHYLAMINE	109-89-7
DWUETYLOWY ETER	C4H10O	DIETHYL ETHER	60-29-7
DWUMETHYLOFORMAMID	C3H7NO	N,N-DIMETHYLFORMAMIDE	68-12-2
DWUMETYLOANILINA	C8H11N	N,N-DIMETHYLANILINE	121-69-7
DWUMETYLOSULFOTLENKU	C3H2N2	MALONONITRILE	109-77-3
DWUMETYLOWY SIARCZAN	C2H6O4S	DIMETHYL SULFATE	77-78-1
DWUNITROBENZEN	C6H4N2O4	m-DINITROBENZENE	99-65-0
DYMEX	C8H8O	ACETOPHENONE	98-86-2
DYTOL E-46	C18H38O	1-OCTADECANOL	112-92-5
DYTOL F-11	C16H34O	1-HEXADECANOL	36653-82-4
DYTOL J-68	C12H26O	1-DODECANOL	112-53-8
DYTOL M-83	C8H18O	1-OCTANOL	111-87-5

Name (Synonym)	Formula	Name in Tables	CAS Reg. No
DYTOL S-91	C10H22O	1-DECANOL	112-30-1
EAA	C6H10O2	n-PROPYL ACRYLATE	925-60-0
EASTMAN INHIBITOR HPT	C6H18N3OP	HEXAMETHYL PHOSPHORAMIDE	680-31-9
EB	C8H10	ETHYLBENZENE	100-41-4
ECF	C3H5ClO2	ETHYL CHLOROFORMATE	541-41-3
ECH	C3H5ClO	alpha-EPICHLOROHYDRIN	106-89-8
EDB	C2H4Br2	1,2-DIBROMOETHANE	106-93-4
EDB-85	C2H4Br2	1,2-DIBROMOETHANE	106-93-4
E-D-BEE	C2H4Br2	1,2-DIBROMOETHANE	106-93-4
EDC	C2H4Cl2	1,2-DICHLOROETHANE	107-06-2
EGDME	C4H10O2	1,2-DIMETHOXYETHANE	110-71-4
EGITOL	C2Cl6	HEXACHLOROETHANE	67-72-1
EGM	C3H8O2	2-METHOXYETHANOL	109-86-4
EGME	C3H8O2	2-METHOXYETHANOL	109-86-4
EHRLICHS REAGENT	C9H11NO	p-DIMETHYLAMINOBENZALDEHYDE	100-10-7
EI	C2H5N	ETHYLENEIMINE	151-56-4
n-EICOSANE	C20H42	n-EICOSANE	112-95-8
1-EICOSANETHIOL	C20H42S	1-EICOSANETHIOL	----------
1-EICOSANOL	C20H42O	1-EICOSANOL	629-96-9
1-EICOSENE	C20H40	1-EICOSENE	3452-07-1
1-EICOSYNE	C20H38	1-EICOSYNE	765-27-5
EK 1700	C6H8N2	o-PHENYLENEDIAMINE	95-54-5
EKTASOLVE DB	C8H18O3	DIETHYLENE GLYCOL MONOBUTYL ETHER	112-34-5
EKTASOLVE de ACETATE	C8H16O4	DIETHYLENE GLYCOL ETHYL ETHER ACETATE	112-15-2
EKTASOLVE EB	C6H14O2	2-BUTOXYETHANOL	111-76-2
EKTASOLVE EE	C4H10O2	2-ETHOXYETHANOL	110-80-5
EKTASOLVE EE ACETATE SOLVENT	C6H12O3	2-ETHOXYETHYL ACETATE	111-15-9
EKTASOLVE EP	C5H12O2	ETHYLENE GLYCOL MONOPROPYL ETHER	2807-30-9
ELALDEHYDE	C6H12O3	PARALDEHYDE	123-63-7
ELAOL	C16H22O4	DIBUTYL PHTHALATE	84-74-2
ELAYL	C2H4	ETHYLENE	74-85-1
ELECTRO-CF 11	CCl3F	TRICHLOROFLUOROMETHANE	75-69-4
ELECTRO-CF 12	CCl2F2	DICHLORODIFLUOROMETHANE	75-71-8
ELECTRO-CF 22	CHClF2	CHLORODIFLUOROMETHANE	75-45-6
EL P.E.T.N.	C5H8N4O12	PENTAERYTHRITOL TETRANITRATE	78-11-5
EMEREST 2301	C19H36O2	METHYL OLEATE	112-62-9
EMEREST 2325	C22H44O2	n-BUTYL STEARATE	123-95-5
EMEREST 2801	C19H36O2	METHYL OLEATE	112-62-9
EMERSOL 120	C18H36O2	STEARIC ACID	57-11-4
EMERSOL 140	C16H32O2	n-HEXADECANOIC ACID	57-10-3
EMERSOL 143	C16H32O2	n-HEXADECANOIC ACID	57-10-3
EMERSOL 210	C18H34O2	OLEIC ACID	112-80-1
EMERSOL 213	C18H34O2	OLEIC ACID	112-80-1
EMERSOL 233LL	C18H34O2	OLEIC ACID	112-80-1
EMERSOL 6321	C18H34O2	OLEIC ACID	112-80-1
EMERSOL 221 LOW TITER WHITE OLEIC ACID	C18H34O2	OLEIC ACID	112-80-1
EMERSOL 220 WHITE OLEIC ACID	C18H34O2	OLEIC ACID	112-80-1
EMERY 2219	C19H36O2	METHYL OLEATE	112-62-9
EMERY 2310	C19H36O2	METHYL OLEATE	112-62-9
EMERY 655	C14H28O2	n-TETRADECANOIC ACID	544-63-8
EMERY OLEIC ACID ESTER 2301	C19H36O2	METHYL OLEATE	112-62-9
EMFAC 1202	C9H18O2	n-NONANOIC ACID	112-05-0
ENANTHAL	C7H14O	1-HEPTANAL	111-71-7
ENANTHALDEHYDE	C7H14O	1-HEPTANAL	111-71-7
ENANTHIC ACID	C7H14O2	n-HEPTANOIC ACID	111-14-8
ENANTHIC ALCOHOL	C7H16O	1-HEPTANOL	111-70-6
ENANTHOLE	C7H14O	1-HEPTANAL	111-71-7
ENANTHYLIC ACID	C7H14O2	n-HEPTANOIC ACID	111-14-8
ENT 15,406	C3H6Cl2	1,2-DICHLOROPROPANE	78-87-5
ENT 1,860	C2Cl4	TETRACHLOROETHYLENE	127-18-4
ENT 25,764	C4Cl4S	TETRACHLOROTHIOPHENE	6012-97-1
ENT 262	C10H10O4	DIMETHYL PHTHALATE	131-11-3
ENT 26,263	C2H4O	ETHYLENE OXIDE	75-21-8
ENT 50,324	C2H5N	ETHYLENEIMINE	151-56-4
ENT 50,882	C6H18N3OP	HEXAMETHYL PHOSPHORAMIDE	680-31-9
ENT 54	C3H3N	ACRYLONITRILE	107-13-1
ENT 8,420	C3H6Cl2	2,2-DICHLOROPROPANE	594-20-7
ENTSUFON	C7H5N3O6	2,4,6-TRINITROTOLUENE	118-96-7
ENZACTIN	C9H14O6	GLYCERYL TRIACETATE	102-76-1
EPAL 10	C10H22O	1-DECANOL	112-30-1
EPAL 12	C12H26O	1-DODECANOL	112-53-8
EPAL 16NF	C16H34O	1-HEXADECANOL	36653-82-4
EPAL 6	C6H14O	1-HEXANOL	111-27-3
EPAL 8	C8H18O	1-OCTANOL	111-87-5
alpha-EPICHLOROHYDRIN	C3H5ClO	alpha-EPICHLOROHYDRIN	106-89-8
(dl)-a-EPICHLOROHYDRIN	C3H5ClO	alpha-EPICHLOROHYDRIN	106-89-8

Name (Synonym)	Formula	Name in Tables	CAS Reg. No
EPICHLOROHYDRIN	C3H5ClO	alpha-EPICHLOROHYDRIN	106-89-8
EPOXYBUTANE	C4H8O	1,2-EPOXYBUTANE	106-88-7
1,2-EPOXYBUTANE	C4H8O	1,2-EPOXYBUTANE	106-88-7
1,4-EPOXYBUTANE	C4H8O	TETRAHYDROFURAN	109-99-9
1,2-EPOXY-3-CHLOROPROPANE	C3H5ClO	alpha-EPICHLOROHYDRIN	106-89-8
EPOXYETHANE	C2H4O	ETHYLENE OXIDE	75-21-8
1,2-EPOXYETHANE	C2H4O	ETHYLENE OXIDE	75-21-8
EPOXYPROPANE	C3H6O	1,2-PROPYLENE OXIDE	75-56-9
1,2-EPOXYPROPANE	C3H6O	1,2-PROPYLENE OXIDE	75-56-9
1,3-EPOXYPROPANE	C3H6O	1,3-PROPYLENE OXIDE	503-30-0
2,3-EPOXYPROPANE	C3H6O	1,2-PROPYLENE OXIDE	75-56-9
2,3-EPOXYPROPYL CHLORIDE	C3H5ClO	alpha-EPICHLOROHYDRIN	106-89-8
EPSYLON KAPROLAKTAM	C6H11NO	epsilon-CAPROLACTAM	105-60-2
EPTANI	C7H16	n-HEPTANE	142-82-5
ERGOPLAST FDO	C24H38O4	DIOCTYL PHTHALATE	117-81-7
ERINIT	C5H8N4O12	PENTAERYTHRITOL TETRANITRATE	78-11-5
ERYTHRENE	C4H6	1,3-BUTADIENE	106-99-0
ESACHLOROBENZENE	C6Cl6	HEXACHLOROBENZENE	118-74-1
ESANI	C6H14	n-HEXANE	110-54-3
ESEN	C8H4O3	PHTHALIC ANHYDRIDE	85-44-9
ESKIMON 11	CCl3F	TRICHLOROFLUOROMETHANE	75-69-4
ESKIMON 12	CCl2F2	DICHLORODIFLUOROMETHANE	75-71-8
ESSENCE of MIRBANE	C6H5NO2	NITROBENZENE	98-95-3
ESSENCE of MYRBANE	C6H5NO2	NITROBENZENE	98-95-3
ESSENCE of NIOBE	C9H10O2	ETHYL BENZOATE	93-89-0
ESSENCE OF NIOBE	C8H8O2	METHYL BENZOATE	93-58-3
ESSIGESTER	C4H8O2	ETHYL ACETATE	141-78-6
ESSIGSAEURE	C2H4O2	ACETIC ACID	64-19-7
ESTERE CIANOACETICO	C5H7NO2	ETHYL CYANOACETATE	105-56-6
ESTOL 1550	C12H14O4	DIETHYL PHTHALATE	84-66-2
ETANOLAMINA	C2H7NO	MONOETHANOLAMINE	141-43-5
ETANOLO	C2H6O	ETHANOL	64-17-5
ETANTIOLO	C2H6S	ETHYL MERCAPTAN	75-08-1
ETERE ETILICO	C4H10O	DIETHYL ETHER	60-29-7
ETHAANTHIOL	C2H6S	ETHYL MERCAPTAN	75-08-1
ETHAL	C16H34O	1-HEXADECANOL	36653-82-4
ETHANAL	C2H4O	ACETALDEHYDE	75-07-0
ETHANAMIDE	C2H5NO	ACETAMIDE	60-35-5
ETHANAMINE	C2H7N	ETHYLAMINE	75-04-7
ETHANE	C2H6	ETHANE	74-84-0
ETHANECARBOXYLIC ACID	C3H6O2	PROPIONIC ACID	79-09-4
ETHANE, 1,2-DIBROMOTETRAFLUORO-	C2Br2F4	1,2-DIBROMOTETRAFLUOROETHANE	124-73-2
ETHANE, 1,2-DIBROMO-1,1,2,2-TETRAFLUORO	C2Br2F4	1,2-DIBROMOTETRAFLUOROETHANE	124-73-2
1,2-ETHANEDICARBOXYLIC ACID	C4H6O4	SUCCINIC ACID	110-15-6
ETHANE DICHLORIDE	C2H4Cl2	1,2-DICHLOROETHANE	107-06-2
ETHANE, 1,2-DIFLUORO-	C2H4F2	1,2-DIFLUOROETHANE	624-72-6
ETHANEDINITRILE	C2N2	CYANOGEN	460-19-5
ETHANEDIOIC ACID	C2H2O4	OXALIC ACID	144-62-7
1,2-ETHANEDIOL	C2H6O2	ETHYLENE GLYCOL	107-21-1
1,2-ETHANEDIOL DIACETATE	C6H10O4	ETHYLENE GLYCOL DIACETATE	111-55-7
1,2-ETHANEDIOL DIPROPANOATE	C4H6O	trans-CROTONALDEHYDE	123-73-9
ETHANEDIONIC ACID	C2H2O4	OXALIC ACID	144-62-7
2,2'-(1,2-ETHANEDIYLBIS(OXY))BISETHANOL	C6H14O4	TRIETHYLENE GLYCOL	112-27-6
ETHANE HEXACHLORIDE	C2Cl6	HEXACHLOROETHANE	67-72-1
ETHANENITRILE	C2H3N	ACETONITRILE	75-05-8
ETHANE PENTACHLORIDE	C2HCl5	PENTACHLOROETHANE	76-01-7
ETHANE, 1,1,2,2-TETRACHLORO-1,2-DIFLUORO-	C2Cl4F2	1,1,2,2-TETRACHLORODIFLUOROETHANE	76-12-0
ETHANETHIOIC ACID	C2H4OS	THIOACETIC-ACID	507-09-5
ETHANETHIOL	C2H6S	ETHYL MERCAPTAN	75-08-1
ETHANETHIOLIC ACID	C2H4OS	THIOACETIC-ACID	507-09-5
ETHANE TRICHLORIDE	C2H3Cl3	1,1,2-TRICHLOROETHANE	79-00-5
ETHANE, 1,1,1-TRIFLUORO-	C2H3F3	1,1,1-TRIFLUOROETHANE	420-46-2
ETHANOIC ACID	C2H4O2	ACETIC ACID	64-19-7
ETHANOIC ACID, ETHENYL ESTER	C4H6O2	VINYL ACETATE	108-05-4
ETHANOIC ANHYDRATE	C4H6O3	ACETIC ANHYDRIDE	108-24-7
ETHANOL	C2H6O	ETHANOL	64-17-5
b-ETHANOLAMINE	C2H7NO	MONOETHANOLAMINE	141-43-5
ETHANOLAMINE	C2H7NO	MONOETHANOLAMINE	141-43-5
ETHANOLETHYLENE DIAMINE	C4H12N2O	N-AMINOETHYL ETHANOLAMINE	111-41-1
ETHANOL, 2,2'-(METHYLIMINO)BIS-	C5H13NO2	METHYL DIETHANOLAMINE	105-59-9
ETHANOL, 2-PHENYL-	C8H10O	2-PHENYLETHANOL	60-12-8
ETHANOL 200 PROOF	C2H6O	ETHANOL	64-17-5
ETHANOL SOLUTIONS	C2H6O	ETHANOL	64-17-5
ETHANOYL CHLORIDE	C2H3ClO	ACETYL CHLORIDE	75-36-5
ETHENE	C2H4	ETHYLENE	74-85-1
ETHENE, 1,1-DIFLUORO-	C2H2F2	1,1-DIFLUOROETHYLENE	75-38-7

Name (Synonym)	Formula	Name in Tables	CAS Reg. No
(E)-1,1'-(1,2-ETHENEDIYL)BISBENZENE	C14H12	trans-STILBENE	103-30-0
ETHENE, FLUORO-	C2H3F	VINYL FLUORIDE	75-02-5
ETHENE OXIDE	C2H4O	ETHYLENE OXIDE	75-21-8
ETHENONE	C2H2O	KETENE	463-51-4
ETHENYL ACETATE	C4H6O2	VINYL ACETATE	108-05-4
ETHENYLBENZENE	C8H8	STYRENE	100-42-5
4-ETHENYL-1-CYCLOHEXENE	C8H12	VINYLCYCLOHEXENE	100-40-3
ETHENYL ETHANOATE	C4H6O2	VINYL ACETATE	108-05-4
1-ETHENYL-3-METHYLBENZENE	C9H10	m-METHYLSTYRENE	100-80-1
1-ETHENYL-4-METHYLBENZENE	C9H10	p-METHYLSTYRENE	622-97-9
1-(ETHENYLOXY) BUTANE	C6H12O	BUTYL VINYL ETHER	111-34-2
ETHENYLOXYETHENE	C4H6O	DIVINYL ETHER	109-93-3
ETHER	C4H10O	DIETHYL ETHER	60-29-7
ETHER, BIS(2-CYANOETHYL)	C6H8N2O	BIS(CYANOETHYL)ETHER	1656-48-0
ETHER, BIS(2-(2-METHOXYETHOXY)ETHYL)	C10H22O5	TETRAETHYLENE GLYCOL DIMETHYL ETHER	143-24-8
ETHER BUTYLIQUE	C8H18O	DI-n-BUTYL ETHER	142-96-1
ETHER CHLORATUS	C2H5Cl	ETHYL CHLORIDE	75-00-3
ETHER CYANATUS	C3H5N	PROPIONITRILE	107-12-0
ETHER, DI-n-PENTYL-	C10H22O	DI-n-PENTYL ETHER	693-65-2
ETHER, DI-n-PROPYL-	C6H14O	DI-n-PROPYL ETHER	111-43-3
ETHER ETHYLBUTYLIQUE	C6H14O	n-BUTYL ETHYL ETHER	628-81-9
ETHER ETHYLIQUE	C4H10O	DIETHYL ETHER	60-29-7
ETHER, ETHYL PHENYL	C8H10O	PHENETOLE	103-73-1
ETHER HYDROCHLORIC	C2H5Cl	ETHYL CHLORIDE	75-00-3
ETHER ISOPROPYLIQUE	C6H14O	DIISOPROPYL ETHER	108-20-3
ETHER, METHYL PROPYL	C4H10O	METHYL-PROPYL-ETHER	557-17-5
ETHER MONOETHYLIQUE de L'ETHYLENE-GLYCOL	C4H10O2	2-ETHOXYETHANOL	110-80-5
ETHER MURIATIC	C2H5Cl	ETHYL CHLORIDE	75-00-3
ETHINE	C2H2	ACETYLENE	74-86-2
ETHINYLBENZENE	C8H6	ETHYNYLBENZENE	536-74-3
ETHINYL TRICHLORIDE	C2HCl3	TRICHLOROETHYLENE	79-01-6
ETHOL	C16H34O	1-HEXADECANOL	36653-82-4
ETHOXY ACETATE	C6H12O3	2-ETHOXYETHYL ACETATE	111-15-9
ETHOXYBENZENE	C8H10O	PHENETOLE	103-73-1
ETHOXYCARBONYLETHYLENE	C5H8O2	ETHYL ACRYLATE	140-88-5
ETHOXY DIGLYCOL	C6H14O3	2-(2-ETHOXYETHOXY)ETHANOL	111-90-0
ETHOXYETHANE	C4H10O	DIETHYL ETHER	60-29-7
2-ETHOXYETHANOL	C4H10O2	2-ETHOXYETHANOL	110-80-5
2-ETHOXYETHANOL ACETATE	C6H12O3	2-ETHOXYETHYL ACETATE	111-15-9
2-ETHOXYETHANOL, ESTER with ACETIC ACID	C6H12O3	2-ETHOXYETHYL ACETATE	111-15-9
2-(2-ETHOXYETHOXY)ETHANOL	C6H14O3	2-(2-ETHOXYETHOXY)ETHANOL	111-90-0
2-(2-ETHOXYETHOXY)ETHANOL ACETATE	C8H16O4	DIETHYLENE GLYCOL ETHYL ETHER ACETATE	112-15-2
1-ETHOXY-2-(b-ETHOXYETHOXY)ETHANE	C8H18O3	DIETHYLENE GLYCOL DIETHYL ETHER	112-36-7
2-ETHOXYETHYL ACETATE	C6H12O3	2-ETHOXYETHYL ACETATE	111-15-9
ETHOXYFORMIC ANHYDRIDE	C5H10O3	DIETHYL CARBONATE	105-58-8
ETHOXYMETHANE	C3H8O	METHYL ETHYL ETHER	540-67-0
ETHOXYPROPIONIC ACID, ETHYL ESTER	C7H14O3	ETHYL-3-ETHOXYPROPIONATE	763-69-9
3-ETHOXYPROPIONIC ACID, ETHYL ESTER	C7H14O3	ETHYL-3-ETHOXYPROPIONATE	763-69-9
ETHRIOL	C6H14O3	TRIMETHYLOLPROPANE	77-99-6
ETHYLACETAAT	C4H8O2	ETHYL ACETATE	141-78-6
ETHYL ACETATE	C4H8O2	ETHYL ACETATE	141-78-6
ETHYLACETIC ACID	C4H8O2	n-BUTYRIC ACID	107-92-6
ETHYL ACETIC ESTER	C4H8O2	ETHYL ACETATE	141-78-6
ETHYLACETOACETATE	C6H10O3	ETHYLACETOACETATE	141-97-9
ETHYL ACETOACETATE	C6H10O2	n-PROPYL ACRYLATE	925-60-0
ETHYL ACETONE	C5H10O	2-PENTANONE	107-87-9
ETHYL ACETYLACETONATE	C6H10O2	n-PROPYL ACRYLATE	925-60-0
ETHYLACETYLENE	C4H6	ETHYLACETYLENE	107-00-6
ETHYLACRYLAAT	C5H8O2	ETHYL ACRYLATE	140-88-5
ETHYL ACRYLATE	C5H8O2	ETHYL ACRYLATE	140-88-5
ETHYLAKRYLAT	C5H8O2	ETHYL ACRYLATE	140-88-5
ETHYL ALDEHYDE	C2H4O	ACETALDEHYDE	75-07-0
ETHYLALUMINUM SESQUICHLORIDE	C6H15Al2Cl3	ETHYL ALUMINUM SESQUICHLORIDE	12075-68-2
ETHYLAMINE	C2H7N	ETHYLAMINE	75-04-7
o-ETHYLANILINE	C8H11N	o-ETHYLANILINE	578-54-1
2-ETHYLANILINE	C8H11N	o-ETHYLANILINE	578-54-1
ETHYLBENZEEN	C8H10	ETHYLBENZENE	100-41-4
2-ETHYLBENZENAMINE	C8H11N	o-ETHYLANILINE	578-54-1
ETHYLBENZENE	C8H10	ETHYLBENZENE	100-41-4
ETHYL BENZOATE	C9H10O2	ETHYL BENZOATE	93-89-0
ETHYLBENZOL	C8H10	ETHYLBENZENE	100-41-4
ETHYL BENZYL ACETOACETATE	C6H10O2	n-PROPYL ACRYLATE	925-60-0
ETHYL BUTANOATE	C6H12O2	ETHYL n-BUTYRATE	105-54-4
2-ETHYL BUTANOIC ACID	C6H12O2	2-ETHYL BUTYRIC ACID	88-09-5
2-ETHYL-1-BUTANOL	C6H14O	2-ETHYL-1-BUTANOL	97-95-0
2-ETHYLBUTANOL-1	C6H14O	2-ETHYL-1-BUTANOL	97-95-0

Name (Synonym)	Formula	Name in Tables	CAS Reg. No
2-ETHYL-1-BUTENE	C6H12	2-ETHYL-1-BUTENE	760-21-4
2-ETHYLBUTYL ALCOHOL	C6H14O	2-ETHYL-1-BUTANOL	97-95-0
ETHYLBUTYLACETALDEHYDE	C8H16O	2-ETHYLHEXANAL	123-05-7
ETHYL BUTYRATE	C6H12O2	ETHYL n-BUTYRATE	105-54-4
ETHYL n-BUTYRATE	C6H12O2	ETHYL n-BUTYRATE	105-54-4
a-ETHYLBUTYRIC ACID	C6H12O2	2-ETHYL BUTYRIC ACID	88-09-5
2-ETHYL BUTYRIC ACID	C6H12O2	2-ETHYL BUTYRIC ACID	88-09-5
a-ETHYLCAPROALDEHYDE	C8H16O	2-ETHYLHEXANAL	123-05-7
ETHYL CARBINOL	C3H8O	n-PROPANOL	71-23-8
ETHYL CARBITOL	C6H14O3	2-(2-ETHOXYETHOXY)ETHANOL	111-90-0
ETHYL CARBONATE	C5H10O3	DIETHYL CARBONATE	105-58-8
ETHYL CELLOSOLVE	C4H10O2	2-ETHOXYETHANOL	110-80-5
ETHYLCHLOORFORMIAAT	C3H5ClO2	ETHYL CHLOROFORMATE	541-41-3
ETHYL CHLORIDE	C2H5Cl	ETHYL CHLORIDE	75-00-3
ETHYL CHLOROCARBONATE	C3H5ClO2	ETHYL CHLOROFORMATE	541-41-3
ETHYL CHLOROFORMATE	C3H5ClO2	ETHYL CHLOROFORMATE	541-41-3
ETHYL CYANIDE	C3H5N	PROPIONITRILE	107-12-0
ETHYL CYANOACETATE	C5H7NO2	ETHYL CYANOACETATE	105-56-6
ETHYL CYANOETHANOATE	C5H7NO2	ETHYL CYANOACETATE	105-56-6
ETHYLCYCLOHEXANE	C8H16	ETHYLCYCLOHEXANE	1678-91-7
ETHYLCYCLOPENTANE	C7H14	ETHYLCYCLOPENTANE	1640-89-7
ETHYL DIETHYLENE GLYCOL	C6H14O3	2-(2-ETHOXYETHOXY)ETHANOL	111-90-0
ETHYL DIGLYME	C8H18O3	DIETHYLENE GLYCOL DIETHYL ETHER	112-36-7
3-ETHYL-2,2-DIMETHYLHEXANE	C10H22	3-ETHYL-2,2-DIMETHYLHEXANE	20291-91-2
3-ETHYL-2,3-DIMETHYLHEXANE	C10H22	3-ETHYL-2,3-DIMETHYLHEXANE	52897-00-4
3-ETHYL-2,4-DIMETHYLHEXANE	C10H22	3-ETHYL-2,4-DIMETHYLHEXANE	7220-26-0
3-ETHYL-2,5-DIMETHYLHEXANE	C10H22	3-ETHYL-2,5-DIMETHYLHEXANE	52897-04-8
3-ETHYL-3,4-DIMETHYLHEXANE	C10H22	3-ETHYL-3,4-DIMETHYLHEXANE	52897-06-0
4-ETHYL-2,2-DIMETHYLHEXANE	C10H22	4-ETHYL-2,2-DIMETHYLHEXANE	----------
4-ETHYL-2,3-DIMETHYLHEXANE	C10H22	4-ETHYL-2,3-DIMETHYLHEXANE	52897-01-5
4-ETHYL-2,4-DIMETHYLHEXANE	C10H22	4-ETHYL-2,4-DIMETHYLHEXANE	52897-03-7
4-ETHYL-3,3-DIMETHYLHEXANE	C10H22	4-ETHYL-3,3-DIMETHYLHEXANE	52897-05-9
3-ETHYL-2,3-DIMETHYLPENTANE	C9H20	3-ETHYL-2,3-DIMETHYLPENTANE	16747-33-4
ETHYLE	C3H6O2	ETHYL FORMATE	109-94-4
ETHYLE (ACETATE d)	C4H8O2	ETHYL ACETATE	141-78-6
ETHYLE, CHLOROFORMIAT D	C3H5ClO2	ETHYL CHLOROFORMATE	541-41-3
ETHYLEEN-CHLOORHYDRINE	C2H5ClO	2-CHLOROETHANOL	107-07-3
ETHYLEENDIAMINE	C2H8N2	ETHYLENEDIAMINE	107-15-3
ETHYLEENIMINE	C2H5N	ETHYLENEIMINE	151-56-4
ETHYLENE	C2H4	ETHYLENE	74-85-1
ETHYLENE ACETATE	C6H10O4	ETHYLENE GLYCOL DIACETATE	111-55-7
ETHYLENE ALCOHOL	C2H6O2	ETHYLENE GLYCOL	107-21-1
ETHYLENE ALDEHYDE	C3H4O	ACROLEIN	107-02-8
ETHYLENE BROMIDE	C2H4Br2	1,2-DIBROMOETHANE	106-93-4
ETHYLENE CARBONATE	C3H4O3	ETHYLENE CARBONATE	96-49-1
ETHYLENE CARBONIC ACID	C3H4O3	ETHYLENE CARBONATE	96-49-1
ETHYLENECARBOXAMIDE	C3H5NO	ACRYLAMIDE	79-06-1
ETHYLENECARBOXYLIC ACID	C3H4O2	ACRYLIC ACID	79-10-7
ETHYLENE CHLORIDE	C2H4Cl2	1,2-DICHLOROETHANE	107-06-2
ETHYLENE CYANIDE	C4H4N2	SUCCINONITRILE	110-61-2
ETHYLENE CYANOHYDRIN	C3H5NO	HYDRACRYLONITRILE	109-78-4
ETHYLENEDIAMINE	C2H8N2	ETHYLENEDIAMINE	107-15-3
1,2-ETHYLENEDIAMINE	C2H8N2	ETHYLENEDIAMINE	107-15-3
cis-1,2-ETHYLENEDICARBOXYLIC ACID	C4H4O4	MALEIC ACID	110-16-7
(E)1,2-ETHYLENEDICARBOXYLIC ACID	C4H4O4	FUMARIC ACID	110-17-8
trans-1,2-ETHYLENEDICARBOXYLIC ACID	C4H4O4	FUMARIC ACID	110-17-8
1,2-ETHYLENE DICHLORIDE	C2H4Cl2	1,2-DICHLOROETHANE	107-06-2
ETHYLENE DICYANIDE	C4H4N2	SUCCINONITRILE	110-61-2
ETHYLENE DIGLYCOL	C4H10O3	DIETHYLENE GLYCOL	111-46-6
ETHYLENE DIGLYCOL MONOETHYL ETHER	C6H14O3	2-(2-ETHOXYETHOXY)ETHANOL	111-90-0
ETHYLENE DIGLYCOL MONOMETHYL ETHER E	C5H12O3	2-(2-METHOXYETHOXY)ETHANOL	111-77-3
ETHYLENE DIHYDRATE	C2H6O2	ETHYLENE GLYCOL	107-21-1
ETHYLENE DIMETHYL ETHER	C4H10O2	1,2-DIMETHOXYETHANE	110-71-4
2,2'-ETHYLENEDIOXYDIETHANOL	C6H14O4	TRIETHYLENE GLYCOL	112-27-6
2,2'-ETHYLENEDIOXYETHANOL	C6H14O4	TRIETHYLENE GLYCOL	112-27-6
ETHYLENE EPISULFIDE	C2H4S	THIACYCLOPROPANE	420-12-2
ETHYLENE EPISULPHIDE	C2H4S	THIACYCLOPROPANE	420-12-2
ETHYLENE FLUORIDE	C2H4F2	1,1-DIFLUOROETHANE	75-37-6
ETHYLENE, FLUORO-	C2H3F	VINYL FLUORIDE	75-02-5
ETHYLENE GLYCOL	C2H6O2	ETHYLENE GLYCOL	107-21-1
ETHYLENE GLYCOL ACETATE	C6H10O4	ETHYLENE GLYCOL DIACETATE	111-55-7
ETHYLENE GLYCOL ACRYLATE	C5H8O3	2-HYDROXYETHYL ACRYLATE	818-61-1
ETHYLENE GLYCOL-BIS-(2-HYDROXYETHYL ETHER)	C6H14O4	TRIETHYLENE GLYCOL	112-27-6
ETHYLENE GLYCOL-n-BUTYL ETHER	C6H14O2	2-BUTOXYETHANOL	111-76-2
ETHYLENE GLYCOL CARBONATE	C3H4O3	ETHYLENE CARBONATE	96-49-1
ETHYLENE GLYCOL, CHLOROHYDRIN	C2H5ClO	2-CHLOROETHANOL	107-07-3

Name (Synonym)	Formula	Name in Tables	CAS Reg. No
ETHYLENE GLYCOL, CYCLIC CARBONATE	C3H4O3	ETHYLENE CARBONATE	96-49-1
ETHYLENE GLYCOL DIACETATE	C6H10O4	ETHYLENE GLYCOL DIACETATE	111-55-7
ETHYLENE GLYCOL DIHYDROXYDIETHYL ETHER	C6H14O4	TRIETHYLENE GLYCOL	112-27-6
ETHYLENE GLYCOL DIMETHYL ETHER	C4H10O2	1,2-DIMETHOXYETHANE	110-71-4
ETHYLENE GLYCOL DIPROPIONATE	C4H6O	trans-CROTONALDEHYDE	123-73-9
ETHYLENE GLYCOL ETHYL ETHER	C4H10O2	2-ETHOXYETHANOL	110-80-5
ETHYLENE GLYCOL MONOACRYLATE	C5H8O3	2-HYDROXYETHYL ACRYLATE	818-61-1
ETHYLENE GLYCOL MONOBUTYL ETHER	C6H14O2	2-BUTOXYETHANOL	111-76-2
ETHYLENE GLYCOL MONOETHYL ETHER	C4H10O2	2-ETHOXYETHANOL	110-80-5
ETHYLENE GLYCOL MONOETHYL ETHER ACETATE	C6H12O3	2-ETHOXYETHYL ACETATE	111-15-9
ETHYLENE GLYCOL MONOMETHYL ETHER	C3H8O2	2-METHOXYETHANOL	109-86-4
ETHYLENE GLYCOL-MONO-PROPYL	C5H12O2	ETHYLENE GLYCOL MONOPROPYL ETHER	2807-30-9
ETHYLENE GLYCOL-MONO-n-PROPYL ETHER	C5H12O2	ETHYLENE GLYCOL MONOPROPYL ETHER	2807-30-9
ETHYLENE HEXACHLORIDE	C2Cl6	HEXACHLOROETHANE	67-72-1
ETHYLENEIMINE	C2H5N	ETHYLENEIMINE	151-56-4
ETHYLENE MONOCHLORIDE	C2H3Cl	VINYL CHLORIDE	75-01-4
1,8-ETHYLENENAPHTHALENE	C12H10	ACENAPHTHENE	83-32-9
ETHYLENE OXIDE	C2H4O	ETHYLENE OXIDE	75-21-8
ETHYLENE OXIDE, ETHYL-	C4H8O	1,2-EPOXYBUTANE	106-88-7
1,4-ETHYLENEPIPERAZINE	C6H12N2	TRIETHYLENEDIAMINE	280-57-9
ETHYLENE PROPIONATE	C4H6O	trans-CROTONALDEHYDE	123-73-9
ETHYLENESUCCINIC ACID	C4H6O4	SUCCINIC ACID	110-15-6
ETHYLENE SULPHIDE	C2H4S	THIACYCLOPROPANE	420-12-2
ETHYLENE TETRACHLORIDE	C2Cl4	TETRACHLOROETHYLENE	127-18-4
ETHYLENE TRICHLORIDE	C2HCl3	TRICHLOROETHYLENE	79-01-6
ETHYLENIMINE	C2H5N	ETHYLENEIMINE	151-56-4
ETHYLESTER KYSELINY KYANOCTOVE	C5H7NO2	ETHYL CYANOACETATE	105-56-6
ETHYLESTER KYSELINY MLECNE	C5H10O3	ETHYL LACTATE	97-64-3
ETHYL ESTER of MONOACETIC ACID	C3H6O2	METHYL ACETATE	79-20-9
N-ETHYL-ETHANAMINE	C4H11N	DIETHYLAMINE	109-89-7
ETHYL ETHANOATE	C4H8O2	ETHYL ACETATE	141-78-6
ETHYL-3-ETHOXYPROPIONATE	C7H14O3	ETHYL-3-ETHOXYPROPIONATE	763-69-9
ETHYL ETHYLENE OXIDE	C4H8O	1,2-EPOXYBUTANE	106-88-7
ETHYLETHYNE	C4H6	ETHYLACETYLENE	107-00-6
ETHYL FLUORIDE	C2H5F	ETHYL FLUORIDE	353-36-6
ETHYL FORMATE	C3H6O2	ETHYL FORMATE	109-94-4
ETHYLFORMIAAT	C3H6O2	ETHYL FORMATE	109-94-4
ETHYLFORMIC ACID	C3H6O2	PROPIONIC ACID	79-09-4
ETHYL FORMIC ESTER	C3H6O2	ETHYL FORMATE	109-94-4
ETHYLGLYKOLACETAT	C6H12O3	2-ETHOXYETHYL ACETATE	111-15-9
ETHYL-HEPTADECYL-SULFIDE	C19H40S	ETHYL-HEPTADECYL-SULFIDE	----------
3-ETHYLHEPTANE	C9H20	3-ETHYLHEPTANE	15869-80-4
4-ETHYLHEPTANE	C9H20	4-ETHYLHEPTANE	2216-32-2
ETHYL-HEPTYL-SULFIDE	C9H20S	ETHYL-HEPTYL-SULFIDE	----------
ETHYL-HEXADECYL-SULFIDE	C18H38S	ETHYL-HEXADECYL-SULFIDE	----------
2-ETHYLHEXALDEHYDE	C8H16O	2-ETHYLHEXANAL	123-05-7
2-ETHYLHEXANAL	C8H16O	2-ETHYLHEXANAL	123-05-7
3-ETHYLHEXANE	C8H18	3-ETHYLHEXANE	619-99-8
2-ETHYL-1-HEXANOL	C8H18O	2-ETHYL-1-HEXANOL	104-76-7
2-ETHYLHEXANYL ACETATE	C10H20O2	2-ETHYLHEXYL ACETATE	103-09-3
2-ETHYL-1-HEXENE	C8H16	2-ETHYL-1-HEXENE	1632-16-2
b-ETHYLHEXYL ACETATE	C10H20O2	2-ETHYLHEXYL ACETATE	103-09-3
2-ETHYLHEXYL ACETATE	C10H20O2	2-ETHYLHEXYL ACETATE	103-09-3
2-ETHYLHEXYL ACRYLATE	C11H20O2	2-ETHYLHEXYL ACRYLATE	103-11-7
2-ETHYLHEXYL ALCOHOL	C8H18O	2-ETHYL-1-HEXANOL	104-76-7
2-ETHYLHEXYL ETHANOATE	C10H20O2	2-ETHYLHEXYL ACETATE	103-09-3
ETHYLHEXYL PHTHALATE	C24H38O4	DIOCTYL PHTHALATE	117-81-7
2-ETHYLHEXYL PHTHALATE	C24H38O4	DIOCTYL PHTHALATE	117-81-7
2-ETHYLHEXYL-2-PROPENOATE	C11H20O2	2-ETHYLHEXYL ACRYLATE	103-11-7
ETHYL-HEXYL-SULFIDE	C8H18S	ETHYL-HEXYL-SULFIDE	----------
ETHYL HYDRIDE	C2H6	ETHANE	74-84-0
ETHYL HYDROSULFIDE	C2H6S	ETHYL MERCAPTAN	75-08-1
ETHYL a-HYDROXYPROPIONATE	C5H10O3	ETHYL LACTATE	97-64-3
ETHYL 2-HYDROXYPROPIONATE	C5H10O3	ETHYL LACTATE	97-64-3
ETHYLIC ACID	C2H4O2	ACETIC ACID	64-19-7
ETHYLIDENE BROMIDE	C2H4Br2	1,1-DIBROMOETHANE	557-91-5
ETHYLIDENE CHLORIDE	C2H4Cl2	1,1-DICHLOROETHANE	75-34-3
ETHYLIDENE DIACETATE	C6H10O4	ETHYLIDENE DIACETATE	542-10-9
ETHYLIDENE DIBROMIDE	C2H4Br2	1,1-DIBROMOETHANE	557-91-5
ETHYLIDENE DICHLORIDE	C2H4Cl2	1,1-DICHLOROETHANE	75-34-3
ETHYLIDENE DIETHYL ETHER	C6H14O2	ACETAL	105-57-7
ETHYLIDENE FLUORIDE	C2H4F2	1,1-DIFLUOROETHANE	75-37-6
ETHYLIMINE	C2H5N	ETHYLENEIMINE	151-56-4
ETHYL IODIDE	C2H5I	ETHYL IODIDE	75-03-6
ETHYL ISOBUTANOATE	C6H12O2	ETHYL ISOBUTYRATE	97-62-1
ETHYLISOBUTYLMETHANE, ISOHEPTANE	C7H16	2-METHYLHEXANE	591-76-4

Name (Synonym)	Formula	Name in Tables	CAS Reg. No
ETHYL ISOBUTYRATE	C6H12O2	ETHYL ISOBUTYRATE	97-62-1
ETHYL ISOPROPYL KETONE	C6H12O	ETHYL ISOPROPYL KETONE	565-69-5
ETHYL ISOVALERATE	C7H14O2	ETHYL ISOVALERATE	108-64-5
ETHYLJODID	C2H5I	ETHYL IODIDE	75-03-6
ETHYL LACTATE	C5H10O3	ETHYL LACTATE	97-64-3
ETHYL MALEATE	C8H12O4	DIETHYL MALEATE	141-05-9
ETHYLMERCAPTAAN	C2H6S	ETHYL MERCAPTAN	75-08-1
ETHYLMERKAPTAN	C2H6S	ETHYL MERCAPTAN	75-08-1
ETHYL METHACRYLATE	C6H10O2	ETHYL METHACRYLATE	97-63-2
ETHYL METHACRYLATE, INHIBITED	C6H10O2	ETHYL METHACRYLATE	97-63-2
ETHYL METHANOATE	C3H6O2	ETHYL FORMATE	109-94-4
ETHYLMETHYL CARBINOL	C4H10O	sec-BUTANOL	78-92-2
ETHYL-a-METHYL ACRYLATE	C6H10O2	ETHYL METHACRYLATE	97-63-2
ETHYL-2-METHYLACRYLATE	C6H10O2	ETHYL METHACRYLATE	97-63-2
ETHYL METHYL CETONE	C4H8O	METHYL ETHYL KETONE	78-93-3
ETHYL METHYL ETHER	C3H8O	METHYL ETHYL ETHER	540-67-0
3-ETHYL-2-METHYLHEPTANE	C10H22	3-ETHYL-2-METHYLHEPTANE	----------
3-ETHYL-3-METHYLHEPTANE	C10H22	3-ETHYL-3-METHYLHEPTANE	17302-01-1
3-ETHYL-4-METHYLHEPTANE	C10H22	3-ETHYL-4-METHYLHEPTANE	15896-91-0
3-ETHYL-5-METHYLHEPTANE	C10H22	3-ETHYL-5-METHYLHEPTANE	52896-90-9
4-ETHYL-2-METHYLHEPTANE	C10H22	4-ETHYL-2-METHYLHEPTANE	52896-88-5
4-ETHYL-3-METHYLHEPTANE	C10H22	4-ETHYL-3-METHYLHEPTANE	52896-89-6
4-ETHYL-4-METHYLHEPTANE	C10H22	4-ETHYL-4-METHYLHEPTANE	17302-4-4
5-ETHYL-2-METHYLHEPTANE	C10H22	5-ETHYL-2-METHYLHEPTANE	13475-78-0
3-ETHYL-2-METHYLHEXANE	C9H20	3-ETHYL-2-METHYLHEXANE	16789-46-1
3-ETHYL-3-METHYLHEXANE	C9H20	3-ETHYL-3-METHYLHEXANE	3074-76-8
4-ETHYL-2-METHYLHEXANE	C9H20	4-ETHYL-2-METHYLHEXANE	3074-75-7
3-ETHYL-4-METHYLHEXANE	C9H20	3-ETHYL-4-METHYLHEXANE	3074-77-9
ETHYLMETHYLKETON	C4H8O	METHYL ETHYL KETONE	78-93-3
ETHYL METHYL KETONE	C4H8O	METHYL ETHYL KETONE	78-93-3
2ETHYL-3-METHYLNAPHTHALENE	C13H14	2ETHYL-3-METHYLNAPHTHALENE	----------
2ETHYL-6-METHYLNAPHTHALENE	C13H14	2ETHYL-6-METHYLNAPHTHALENE	7372-86-3
2ETHYL-7-METHYLNAPHTHALENE	C13H14	2ETHYL-7-METHYLNAPHTHALENE	----------
3-ETHYL-2-METHYLPENTANE	C8H18	3-ETHYL-2-METHYLPENTANE	609-26-7
ETHYL-2-METHYLPROPANOATE	C6H12O2	ETHYL ISOBUTYRATE	97-62-1
ETHYL-2-METHYL-2-PROPENOATE	C6H10O2	ETHYL METHACRYLATE	97-63-2
ETHYL-2-METHYLPROPIONATE	C6H12O2	ETHYL ISOBUTYRATE	97-62-1
ETHYL-METHYL-SULFIDE	C3H8S	ETHYL-METHYL-SULFIDE	624-89-5
1-ETHYLNAPHTHALENE	C12H12	1-ETHYLNAPHTHALENE	1127-76-0
2-ETHYLNAPHTHALENE	C12H12	2-ETHYLNAPHTHALENE	939-27-5
ETHYL-NITRATE	C2H5NO3	ETHYL-NITRATE	625-58-1
ETHYL NITRILE	C2H3N	ACETONITRILE	75-05-8
ETHYL-NONYL-SULFIDE	C11H24S	ETHYL-NONYL-SULFIDE	----------
ETHYL-OCTADECYL-SULFIDE	C20H42S	ETHYL-OCTADECYL-SULFIDE	----------
3-ETHYLOCTANE	C10H22	3-ETHYLOCTANE	5881-17-4
4-ETHYLOCTANE	C10H22	4-ETHYLOCTANE	15869-86-0
ETHYL-OCTYL-SULFIDE	C10H22S	ETHYL-OCTYL-SULFIDE	----------
ETHYLOLAMINE	C2H7NO	MONOETHANOLAMINE	141-43-5
ETHYL OXALATE	C6H10O4	DIETHYL OXALATE	95-92-1
ETHYLOXIRANE	C4H8O	1,2-EPOXYBUTANE	106-88-7
ETHYL-3-OXOBUTANOATE	C6H10O3	ETHYLACETOACETATE	141-97-9
ETHYL-3-OXOBUTYRATE	C6H10O3	ETHYLACETOACETATE	141-97-9
ETHYL-PENTADECYL-SULFIDE	C17H36S	ETHYL-PENTADECYL-SULFIDE	----------
2-ETHYL-1-PENTENE	C7H14	2-ETHYL-1-PENTENE	3404-71-5
3-ETHYLPENTANE	C7H16	3-ETHYLPENTANE	617-78-7
3-ETHYL-1-PENTENE	C7H14	3-ETHYL-1-PENTENE	4038-04-4
ETHYL-PENTYL-SULFIDE	C7H16S	ETHYL-PENTYL-SULFIDE	----------
m-ETHYLPHENOL	C8H10O	m-ETHYLPHENOL	620-17-7
p-ETHYLPHENOL	C8H10O	p-ETHYLPHENOL	123-07-9
ETHYL PHENYL ETHER	C8H10O	PHENETOLE	103-73-1
ETHYL PHOSPHATE	C6H15O4P	TRIETHYL PHOSPHATE	78-40-0
ETHYL PHTHALATE	C12H14O4	DIETHYL PHTHALATE	84-66-2
ETHYL PROPENOATE	C5H8O2	ETHYL ACRYLATE	140-88-5
ETHYL-2-PROPENOATE	C5H8O2	ETHYL ACRYLATE	140-88-5
ETHYL PROPIONATE	C5H10O2	ETHYL PROPIONATE	105-37-3
ETHYL PROPYL ETHER	C5H12O	ETHYL PROPYL ETHER	628-32-0
ETHYL PROPYL KETONE	C6H12O	3-HEXANONE	589-38-8
ETHYL-PROPYL-SULFIDE	C5H12S	ETHYL-PROPYL-SULFIDE	4110-50-3
ETHYL SUCCINATE	C8H14O4	DIETHYL SUCCINATE	123-25-1
ETHYL SULFATE	C4H10O4S	DIETHYL SULFATE	64-67-5
ETHYL SULFHYDRATE	C2H6S	ETHYL MERCAPTAN	75-08-1
ETHYL-TETRADECYL-SULFIDE	C16H34S	ETHYL-TETRADECYL-SULFIDE	----------
ETHYL THIOALCOHOL	C2H6S	ETHYL MERCAPTAN	75-08-1
m-ETHYLTOLUENE	C9H12	m-ETHYLTOLUENE	620-14-4
o-ETHYLTOLUENE	C9H12	o-ETHYLTOLUENE	611-14-3
p-ETHYLTOLUENE	C9H12	p-ETHYLTOLUENE	622-96-8

Name (Synonym)	Formula	Name in Tables	CAS Reg. No
ETHYL-TRIDECYL-SULFIDE	C15H32S	ETHYL-TRIDECYL-SULFIDE	----------
ETHYLTRIMETHYLOLMETHANE	C6H14O3	TRIMETHYLOLPROPANE	77-99-6
3-ETHYL-2,2,3-TRIMETHYLPENTANE	C10H22	3-ETHYL-2,2,3-TRIMETHYLPENTANE	52897-17-3
3-ETHYL-2,2,4-TRIMETHYLPENTANE	C10H22	3-ETHYL-2,2,4-TRIMETHYLPENTANE	----------
3-ETHYL-2,3,4-TRIMETHYLPENTANE	C10H22	3-ETHYL-2,3,4-TRIMETHYLPENTANE	52897-19-5
ETHYL-UNDECYL-SULFIDE	C13H28S	ETHYL-UNDECYL-SULFIDE	----------
ETHYL VANILLIN	C9H10O3	ETHYL VANILLIN	121-32-4
ETHYL VINYL ETHER	C4H8O	ETHYL VINYL ETHER	109-92-2
2-ETHYL-m-XYLENE	C1OH14	2-ETHYL-m-XYLENE	874-41-9
2-ETHYL-p-XYLENE	C10H14	2-ETHYL-p-XYLENE	1758-88-9
3-ETHYL-o-XYLENE	C10H14	3-ETHYL-o-XYLENE	933-98-2
4-ETHYL-m-XYLENE	C10H14	4-ETHYL-m-XYLENE	874-41-9
4-ETHYL-o-XYLENE	C10H14	4-ETHYL-o-XYLENE	934-80-5
5-ETHYL-m-XYLENE	C1OH14	5-ETHYL-m-XYLENE	934-74-7
ETHYNE	C2H2	ACETYLENE	74-86-2
ETHYNYLBENZENE	C8H6	ETHYNYLBENZENE	536-74-3
ETIL ACRILATO	C5H8O2	ETHYL ACRYLATE	140-88-5
ETILACRILATULUI	C5H8O2	ETHYL ACRYLATE	140-88-5
ETILAMINA	C2H7N	ETHYLAMINE	75-04-7
ETILBENZENE	C8H10	ETHYLBENZENE	100-41-4
ETIL CLOROCARBONATO	C3H5ClO2	ETHYL CHLOROFORMATE	541-41-3
ETIL CLOROFORMIATO	C3H5ClO2	ETHYL CHLOROFORMATE	541-41-3
ETILE	C3H6O2	ETHYL FORMATE	109-94-4
ETILE (ACETATO di)	C4H8O2	ETHYL ACETATE	141-78-6
ETILENIMINA	C2H5N	ETHYLENEIMINE	151-56-4
ETILMERCAPTANO	C2H6S	ETHYL MERCAPTAN	75-08-1
ETOKSYETYLOWY ALKOHOL	C4H10O2	2-ETHOXYETHANOL	110-80-5
2-ETOSSIETIL-ACETATO	C6H12O3	2-ETHOXYETHYL ACETATE	111-15-9
ETRIOL	C6H14O3	TRIMETHYLOLPROPANE	77-99-6
ETTRIOL	C6H14O3	TRIMETHYLOLPROPANE	77-99-6
ETYLOAMINA	C2H7N	ETHYLAMINE	75-04-7
ETYLOBENZEN	C8H10	ETHYLBENZENE	100-41-4
ETYLU BROMEK	C2H5Br	BROMOETHANE	74-96-4
ETYLU CHLOREK	C2H5Cl	ETHYL CHLORIDE	75-00-3
EUCANINE GB	C7H10N2	TOLUENEDIAMINE	95-80-7
EUFIN	C5H10O3	DIETHYL CARBONATE	105-58-8
EVERCYN	CHN	HYDROGEN CYANIDE	74-90-8
EVIPLAST 80	C24H38O4	DIOCTYL PHTHALATE	117-81-7
EVIPLAST 81	C24H38O4	DIOCTYL PHTHALATE	117-81-7
EVOLA	C6H4Cl2	p-DICHLOROBENZENE	106-46-7
EXACT-S	C2H6S	DIMETHYL SULFIDE	75-18-3
F 112	C2Cl4F2	1,1,2,2-TETRACHLORODIFLUOROETHANE	76-12-0
F 114	C2Cl2F4	1,2-DICHLOROTETRAFLUOROETHANE	76-14-2
F 114B2	C2Br2F4	1,2-DIBROMOTETRAFLUOROETHANE	124-73-2
F 115	C2ClF5	CHLOROPENTAFLUOROETHANE	76-15-3
F 116	C2F6	HEXAFLUOROETHANE	76-16-4
F 12	CCl2F2	DICHLORODIFLUOROMETHANE	75-71-8
F 13	CClF3	CHLOROTRIFLUOROMETHANE	75-72-9
F 13B1	CBrF3	BROMOTRIFLUOROMETHANE	75-63-8
FALKITOL	C2Cl6	HEXACHLOROETHANE	67-72-1
FANNOFORM	CH2O	FORMALDEHYDE	50-00-0
FASCIOLIN	CCl4	CARBON TETRACHLORIDE	56-23-5
FAST BLUE R SALT	C12H12N2	p-AMINODIPHENYLAMINE	101-54-2
FAST ORANGE BASE GR	C6H6N2O2	o-NITROANILINE	88-74-4
FAST ORANGE BASE JR	C6H6N2O2	o-NITROANILINE	88-74-4
FAST ORANGE GC BASE	C6H6ClN	m-CHLOROANILINE	108-42-9
FAST ORANGE GR BASE	C6H6N2O2	o-NITROANILINE	88-74-4
FAST ORANGE O BASE	C6H6N2O2	o-NITROANILINE	88-74-4
FAST ORANGE R SALT	C6H6N2O2	m-NITROANILINE	99-09-2
FAST RED 2G BASE	C6H6N2O2	p-NITROANILINE	100-01-6
FAST RED BASE 2J	C6H6N2O2	p-NITROANILINE	100-01-6
FAST RED BASE GG	C6H6N2O2	p-NITROANILINE	100-01-6
FAST RED GG BASE	C6H6N2O2	p-NITROANILINE	100-01-6
FAST RED MP BASE	C6H6N2O2	p-NITROANILINE	100-01-6
FAST RED P BASE	C6H6N2O2	p-NITROANILINE	100-01-6
FAST SPIRIT YELLOW AAB	C12H11N3	p-AMINOAZOBENZENE	60-09-3
FAST YELLOW GC BASE	C6H6ClN	o-CHLOROANILINE	95-51-2
FC 112	C2Cl4F2	1,1,2,2-TETRACHLORODIFLUOROETHANE	76-12-0
FC 114	C2Cl2F4	1,2-DICHLOROTETRAFLUOROETHANE	76-14-2
FC 114B2	C2Br2F4	1,2-DIBROMOTETRAFLUOROETHANE	124-73-2
FC 12	CCl2F2	DICHLORODIFLUOROMETHANE	75-71-8
FC 1318	C4F8	OCTAFLUORO-2-BUTENE	360-89-4
FC 142b	C2H3ClF2	1-CHLORO-1,1-DIFLUOROETHANE	75-68-3
FC 143	C2H4F2	1,2-DIFLUOROETHANE	624-72-6
FC 143a	C2H3F3	1,1,1-TRIFLUOROETHANE	420-46-2
FC 152a	C2H4F2	1,1-DIFLUOROETHANE	75-37-6

709

Name (Synonym)	Formula	Name in Tables	CAS Reg. No
FC 31	CH2ClF	CHLOROFLUOROMETHANE	593-70-4
FC-C 318	C4F8	OCTAFLUOROCYCLOBUTANE	115-25-3
FEDAL-UN	C2Cl4	TETRACHLOROETHYLENE	127-18-4
a-FELLANDRENE	C10H16	alpha-PHELLANDRENE	99-83-2
FEMA No. 2003	C2H4O	ACETALDEHYDE	75-07-0
FEMA No. 2006	C2H4O2	ACETIC ACID	64-19-7
FEMA No. 2007	C9H14O6	GLYCERYL TRIACETATE	102-76-1
FEMA No. 2009	C8H8O	ACETOPHENONE	98-86-2
FEMA No. 2011	C6H10O4	ADIPIC ACID	124-04-9
FEMA No. 2055	C7H14O2	ISOPENTYL ACETATE	123-92-2
FEMA No. 2085	C10H20O2	ISOPENTYL ISOVALERATE	659-70-1
FEMA No. 2086	C10H12O	ANETHOLE	104-46-1
FEMA No. 2097	C7H8O	ANISOLE	100-66-3
FEMA No. 2109	C6H8O6	ASCORBIC ACID	50-81-7
FEMA No. 2127	C7H6O	BENZALDEHYDE	100-52-7
FEMA No. 2134	C13H10O	BENZOPHENONE	119-61-9
FEMA No. 2137	C7H8O	BENZYL ALCOHOL	100-51-6
FEMA No. 2138	C14H12O2	BENZYL BENZOATE	120-51-4
FEMA No. 2170	C4H8O	METHYL ETHYL KETONE	78-93-3
FEMA No. 2174	C6H12O2	n-BUTYL ACETATE	123-86-4
FEMA No. 2175	C6H12O2	ISOBUTYL ACETATE	110-19-0
FEMA No. 2178	C4H10O	n-BUTANOL	71-36-3
FEMA No. 2179	C4H10O	ISOBUTANOL	78-83-1
FEMA No. 2186	C8H16O2	n-BUTYL n-BUTYRATE	109-21-7
FEMA No. 2219	C4H8O	n-BUTYRALDEHYDE	123-72-8
FEMA No. 2220	C4H8O	ISOBUTYRALDEHYDE	78-84-2
FEMA No. 2221	C4H8O2	n-BUTYRIC ACID	107-92-6
FEMA No. 2222	C4H8O2	ISOBUTYRIC ACID	79-31-2
FEMA No. 2229	C10H16	CAMPHENE	79-92-5
FEMA No. 2306	C6H8O7	CITRIC ACID	77-92-9
FEMA No. 2356	C10H14	p-CYMENE	99-87-6
FEMA No. 2362	C10H20O	1-DECANAL	112-31-2
FEMA No. 2365	C10H22O	1-DECANOL	112-30-1
FEMA No. 2371	C14H14O	DIBENZYL ETHER	103-50-4
FEMA No. 2375	C7H12O4	DIETHYL MALONATE	105-53-3
FEMA No. 2377	C8H14O4	DIETHYL SUCCINATE	123-25-1
FEMA No. 2414	C4H8O2	ETHYL ACETATE	141-78-6
FEMA No. 2415	C6H10O3	ETHYLACETOACETATE	141-97-9
FEMA No. 2418	C5H8O2	ETHYL ACRYLATE	140-88-5
FEMA No. 2422	C9H10O2	ETHYL BENZOATE	93-89-0
FEMA No. 2427	C6H12O2	ETHYL n-BUTYRATE	105-54-4
FEMA No. 2428	C6H12O2	ETHYL ISOBUTYRATE	97-62-1
FEMA No. 2429	C6H12O2	2-ETHYL BUTYRIC ACID	88-09-5
FEMA No. 2434	C3H6O2	ETHYL FORMATE	109-94-4
FEMA No. 2440	C5H10O3	ETHYL LACTATE	97-64-3
FEMA No. 2463	C7H14O2	ETHYL ISOVALERATE	108-64-5
FEMA No. 2489	C5H4O2	FURFURAL	98-01-1
FEMA No. 2540	C7H14O	1-HEPTANAL	111-71-7
FEMA No. 2544	C7H14O	2-HEPTANONE	110-43-0
FEMA No. 2548	C7H16O	1-HEPTANOL	111-70-6
FEMA No. 2557	C6H12O	1-HEXANAL	66-25-1
FEMA No. 2559	C6H12O2	n-HEXANOIC ACID	142-62-1
FEMA No. 2565	C8H16O2	n-HEXYL ACETATE	142-92-7
FEMA No. 2567	C6H14O	1-HEXANOL	111-27-3
FEMA No. 2593	C8H7N	INDOLE	120-72-9
FEMA No. 2615	C12H24O	1-DODECANAL	112-54-9
FEMA No. 2617	C12H26O	1-DODECANOL	112-53-8
FEMA No. 2633	C10H16	D-LIMONENE	5989-27-5
FEMA No. 2683	C8H8O2	METHYL BENZOATE	93-58-3
FEMA No. 2731	C6H12O	METHYL ISOBUTYL KETONE	108-10-1
FEMA No. 2745	C8H8O3	METHYL SALICYLATE	119-36-8
FEMA No. 2782	C9H18O	1-NONANAL	124-19-6
FEMA No. 2789	C9H20O	1-NONANOL	143-08-8
FEMA No. 2797	C8H16O	1-OCTANAL	124-13-0
FEMA No. 2800	C8H18O	1-OCTANOL	111-87-5
FEMA No. 2802	C8H16O	2-OCTANONE	111-13-7
FEMA No. 2809	C9H18O2	n-OCTYL FORMATE	112-32-3
FEMA No. 2841	C5H8O2	ACETYLACETONE	123-54-6
FEMA No. 2842	C5H10O	2-PENTANONE	107-87-9
FEMA No. 2856	C10H16	alpha-PHELLANDRENE	99-83-2
FEMA No. 2858	C8H10O	2-PHENYLETHANOL	60-12-8
FEMA No. 2902	C10H16	alpha-PINENE	80-56-8
FEMA No. 2903	C10H16	beta-PINENE	127-91-3
FEMA No. 2923	C3H6O	n-PROPIONALDEHYDE	123-38-6
FEMA No. 2926	C5H10O2	ISOPROPYL ACETATE	108-21-4
FEMA No. 3092	C11H22O	1-UNDECANAL	112-44-7

Name (Synonym)	Formula	Name in Tables	CAS Reg. No
FEMA No. 3097	C11H24O	1-UNDECANOL	112-42-5
FEMA No. 3101	C5H10O2	VALERIC ACID	109-52-4
FEMA No. 3102	C5H10O2	ISOVALERIC ACID	503-74-2
FEMA No. 3107	C8H8O3	VANILLIN	121-33-5
FEMA No. 3291	C4H6O2	gamma-BUTYROLACTONE	96-48-0
FEMA No. 3326	C3H6O	ACETONE	67-64-1
FEMA No. 3386	C4H5N	PYRROLE	109-97-7
FEMA No. 3558	C10H16	alpha-TERPINENE	99-86-5
FEMA No. 3559	C1OH16	gamma-TERPINENE	99-85-4
b-FENETHYLALKOHOL	C8H10O	2-PHENYLETHANOL	60-12-8
FENILIDRAZINA	C6H8N2	PHENYLHYDRAZINE	100-63-0
2-FENILPROPANO	C9H12	CUMENE	98-82-8
FENNOSAN	C9H7NO	8-HYDROXYQUINOLINE	148-24-3
FENOL	C6H6O	PHENOL	108-95-2
FENOLO	C6H6O	PHENOL	108-95-2
m-FENYLENDIAMIN	C6H8N2	m-PHENYLENEDIAMINE	108-45-2
FENYLENODWUAMINA	C6H8N2	p-PHENYLENEDIAMINE	106-50-3
b-FENYLETHANOL	C8H10O	2-PHENYLETHANOL	60-12-8
N-FENYL-p-FENYLENDIAMIN	C12H12N2	p-AMINODIPHENYLAMINE	101-54-2
FENYLHYDRAZINE	C6H8N2	PHENYLHYDRAZINE	100-63-0
FENYLISOKYANAT	C7H5NO	PHENYL ISOCYANATE	103-71-9
FENYLKYANID	C7H5N	BENZONITRILE	100-47-0
2-FENYL-PROPAAN	C9H12	CUMENE	98-82-8
FENZEN	C6H6	BENZENE	71-43-2
FERMENTATION AMYL ALCOHOL	C5H12O	3-METHYL-1-BUTANOL	123-51-3
FERMENTATION BUTYL ALCOHOL	C4H10O	ISOBUTANOL	78-83-1
FERMINE	C10H10O4	DIMETHYL PHTHALATE	131-11-3
FLECK-FLIP	C2HCl3	TRICHLOROETHYLENE	79-01-6
FLEET-X	C9H12	MESITYLENE	108-67-8
FLEXAMINE G	C18H16N2	N,N'-DIPHENYL-p-PHENYLENEDIAMINE	74-31-7
FLEXIMEL	C24H38O4	DIOCTYL PHTHALATE	117-81-7
FLEXOL DOP	C24H38O4	DIOCTYL PHTHALATE	117-81-7
FLEXOL PLASTICIZER DIP	C24H38O4	DIISOOCTYL PHTHALATE	27554-26-3
FLUATE	C2HCl3	TRICHLOROETHYLENE	79-01-6
FLUGEX 12B1	CBrClF2	BROMOCHLORODIFLUOROMETHANE	353-59-3
FLUKOIDS	CCl4	CARBON TETRACHLORIDE	56-23-5
FLUOBRENE	C2Br2F4	1,2-DIBROMOTETRAFLUOROETHANE	124-73-2
FLUOPHOSGENE	CF2O	CARBONYL FLUORIDE	353-50-4
FLUORANE 114	C2Cl2F4	1,2-DICHLOROTETRAFLUOROETHANE	76-14-2
FLUORANTHENE	C16H10	FLUORANTHENE	206-44-0
FLUORENE	C13H10	FLUORENE	86-73-7
FLUOROBENZENE	C6H5F	FLUOROBENZENE	462-06-6
FLUOROCARBON 113	C2Cl3F3	1,1,2-TRICHLOROTRIFLUOROETHANE	76-13-1
FLUOROCARBON 114	C2Cl2F4	1,2-DICHLOROTETRAFLUOROETHANE	76-14-2
FLUOROCARBON 115	C2ClF5	CHLOROPENTAFLUOROETHANE	76-15-3
FLUOROCARBON 12	CCl2F2	DICHLORODIFLUOROMETHANE	75-71-8
FLUOROCARBON 1211	CBrClF2	BROMOCHLORODIFLUOROMETHANE	353-59-3
FLUOROCARBON 22	CHClF2	CHLORODIFLUOROMETHANE	75-45-6
FLUOROCARBON FC142b	C2H3ClF2	1-CHLORO-1,1-DIFLUOROETHANE	75-68-3
FLUOROCARBON FC143	C2H4F2	1,2-DIFLUOROETHANE	624-72-6
FLUOROCARBON FC143a	C2H3F3	1,1,1-TRIFLUOROETHANE	420-46-2
FLUOROCARBON No. 11	CCl3F	TRICHLOROFLUOROMETHANE	75-69-4
FLUORODICHLOROMETHANE	CHCl2F	DICHLOROFLUOROMETHANE	75-43-4
FLUOROETHENE	C2H3F	VINYL FLUORIDE	75-02-5
FLUOROETHYLENE	C2H3F	VINYL FLUORIDE	75-02-5
FLUOROFORM	CHF3	TRIFLUOROMETHANE	75-46-7
FLUOROFORMYL FLUORIDE	CF2O	CARBONYL FLUORIDE	353-50-4
FLUOROPHOSGENE	CF2O	CARBONYL FLUORIDE	353-50-4
FLUOROPLAST 3	C2ClF3	CHLOROTRIFLUOROETHYLENE	79-38-9
FLUOROPLAST 4	C2F4	TETRAFLUOROETHYLENE	116-14-3
1-FLUOROPROPANE	C3H7F	1-FLUOROPROPANE	460-13-9
2-FLUOROPROPANE	C3H7F	2-FLUOROPROPANE	420-26-8
FLUOROTANE	C2HBrClF3	HALOTHANE	151-67-7
p-FLUOROTOLUENE	C7H7F	p-FLUOROTOLUENE	352-32-9
FLUORO-1,1,1-TRICHLOROETHANE	C2H2Cl3F	1,1,1-TRICHLOROFLUOROETHANE	27154-33-2
FLUOROTRICHLOROMETHANE	CCl3F	TRICHLOROFLUOROMETHANE	75-69-4
FLUORYL	CHF3	TRIFLUOROMETHANE	75-46-7
FLUOTHANE	C2HBrClF3	HALOTHANE	151-67-7
FORMAL	C3H8O2	METHYLAL	109-87-5
FORMALDEHYDE	CH2O	FORMALDEHYDE	50-00-0
FORMALDEHYDE DIMETHYLACETAL	C3H8O2	METHYLAL	109-87-5
FORMALIN	CH2O	FORMALDEHYDE	50-00-0
FORMAMIDE	CH3NO	FORMAMIDE	75-12-7
FORMAMIDOBENZENE	C7H7NO	FORMANILIDE	103-70-8
FORMANILIDE	C7H7NO	FORMANILIDE	103-70-8
FORMIATE de METHYLE	C2H4O2	METHYL FORMATE	107-31-3

Name (Synonym)	Formula	Name in Tables	CAS Reg. No
FORMIATE de PROPYLE	C4H8O2	n-PROPYL FORMATE	110-74-7
FORMIC ACID	CH2O2	FORMIC ACID	64-18-6
FORMIC ACID, ETHYL ESTER	C3H6O2	ETHYL FORMATE	109-94-4
FORMIC ACID, ISOBUTYL ESTER	C5H10O2	ISOBUTYL FORMATE	542-55-2
FORMIC ACID, OCTYL ESTER	C9H18O2	n-OCTYL FORMATE	112-32-3
FORMIC ALDEHYDE	CH2O	FORMALDEHYDE	50-00-0
FORMIC ANAMMONIDE	CHN	HYDROGEN CYANIDE	74-90-8
FORMIC ETHER	C3H6O2	ETHYL FORMATE	109-94-4
FORMOL	CH2O	FORMALDEHYDE	50-00-0
FORMONITRILE	CHN	HYDROGEN CYANIDE	74-90-8
FORMOSA CAMPHOR	C10H16O	CAMPHOR	76-22-2
N-FORMYLANILINE	C7H7NO	FORMANILIDE	103-70-8
N-FORMYLDIMETHYLAMINE	C3H7NO	N,N-DIMETHYLFORMAMIDE	68-12-2
p-FORMYLDIMETHYLANILINE	C9H11NO	p-DIMETHYLAMINOBENZALDEHYDE	100-10-7
FORMYLIC ACID	CH2O2	FORMIC ACID	64-18-6
o-FORMYLPHENOL	C7H6O2	SALICYLALDEHYDE	90-02-8
2-FORMYLPHENOL	C7H6O2	SALICYLALDEHYDE	90-02-8
4-FORMYLPHENOL	C7H6O2	p-HYDROXYBENZALDEHYDE	123-08-0
FORMYL TRICHLORIDE	CHCl3	CHLOROFORM	67-66-3
FOSGEEN	CCl2O	PHOSGENE	75-44-5
FOSGEN	CCl2O	PHOSGENE	75-44-5
FOSGENE	CCl2O	PHOSGENE	75-44-5
FOURAMINE	C7H10N2	TOLUENEDIAMINE	95-80-7
FOURAMINE BROWN AP	C6H6O3	1,2,3-BENZENETRIOL	87-66-1
FOURAMINE D	C6H8N2	p-PHENYLENEDIAMINE	106-50-3
FOURAMINE J	C7H10N2	TOLUENEDIAMINE	95-80-7
FOURAMINE PCH	C6H6O2	1,2-BENZENEDIOL	120-80-9
FOURAMINE RS	C6H6O2	1,3-BENZENEDIOL	108-46-3
FOURRINE 1	C6H8N2	p-PHENYLENEDIAMINE	106-50-3
FOURRINE 68	C6H6O2	1,2-BENZENEDIOL	120-80-9
FOURRINE 79	C6H6O2	1,3-BENZENEDIOL	108-46-3
FOURRINE D	C6H8N2	p-PHENYLENEDIAMINE	106-50-3
FOURRINE PG	C6H6O3	1,2,3-BENZENETRIOL	87-66-1
FREON 11	CCl3F	TRICHLOROFLUOROMETHANE	75-69-4
FREON 112	C2Cl4F2	1,1,2,2-TETRACHLORODIFLUOROETHANE	76-12-0
FREON 113TR-T	C2Cl3F3	1,1,2-TRICHLOROTRIFLUOROETHANE	76-13-1
FREON 115	C2ClF5	CHLOROPENTAFLUOROETHANE	76-15-3
FREON 116	C2F6	HEXAFLUOROETHANE	76-16-4
FREON 12	CCl2F2	DICHLORODIFLUOROMETHANE	75-71-8
FREON 12B1	CBrClF2	BROMOCHLORODIFLUOROMETHANE	353-59-3
FREON 12B2	CBr2F2	DIBROMODIFLUOROMETHANE	75-61-6
FREON 13	CClF3	CHLOROTRIFLUOROMETHANE	75-72-9
FREON 13B1	CBrF3	BROMOTRIFLUOROMETHANE	75-63-8
FREON 14	CF4	CARBON TETRAFLUORIDE	75-73-0
FREON 142	C2H3ClF2	1-CHLORO-1,1-DIFLUOROETHANE	75-68-3
FREON 142b	C2H3ClF2	1-CHLORO-1,1-DIFLUOROETHANE	75-68-3
FREON 152	C2H4F2	1,1-DIFLUOROETHANE	75-37-6
FREON 21	CHCl2F	DICHLOROFLUOROMETHANE	75-43-4
FREON 22	CHClF2	CHLORODIFLUOROMETHANE	75-45-6
FREON 23	CHF3	TRIFLUOROMETHANE	75-46-7
FREON 30	CH2Cl2	DICHLOROMETHANE	75-09-2
FREON 31	CH2ClF	CHLOROFLUOROMETHANE	593-70-4
FREON 41	CH3F	METHYL FLUORIDE	593-53-3
FREON C 318	C4F8	OCTAFLUOROCYCLOBUTANE	115-25-3
FREON F 12	CCl2F2	DICHLORODIFLUOROMETHANE	75-71-8
FREON F 23	CHF3	TRIFLUOROMETHANE	75-46-7
FREON MF	CCl3F	TRICHLOROFLUOROMETHANE	75-69-4
FREON R 112	C2Cl4F2	1,1,2,2-TETRACHLORODIFLUOROETHANE	76-12-0
FRIGEN 11	CCl3F	TRICHLOROFLUOROMETHANE	75-69-4
FRIGEN 113a	C2Cl3F3	1,1,2-TRICHLOROTRIFLUOROETHANE	76-13-1
FRIGEN 114	C2Cl2F4	1,2-DICHLOROTETRAFLUOROETHANE	76-14-2
FRIGEN 12	CCl2F2	DICHLORODIFLUOROMETHANE	75-71-8
FRIGEN 22	CHClF2	CHLORODIFLUOROMETHANE	75-45-6
FRIGIDERM	C2Cl2F4	1,2-DICHLOROTETRAFLUOROETHANE	76-14-2
FRUCOTE	C4H11N	sec-BUTYLAMINE	13952-84-6
FTAALZUURANHYDRIDE	C8H4O3	PHTHALIC ANHYDRIDE	85-44-9
FTALOWY BEZWODNIK	C8H4O3	PHTHALIC ANHYDRIDE	85-44-9
FTOROTAN	C2HBrClF3	HALOTHANE	151-67-7
FUMARIC ACID	C4H4O4	FUMARIC ACID	110-17-8
FUMIGRAIN	C3H3N	ACRYLONITRILE	107-13-1
FUMO-GAS	C2H4Br2	1,2-DIBROMOETHANE	106-93-4
FUNGACETIN	C9H14O6	GLYCERYL TRIACETATE	102-76-1
FURAL	C5H4O2	FURFURAL	98-01-1
2-FURALDEHYDE	C5H4O2	FURFURAL	98-01-1
FURALE	C5H4O2	FURFURAL	98-01-1
FURAN	C4H4O	FURAN	110-00-9

Name (Synonym)	Formula	Name in Tables	CAS Reg. No
2-FURANALDEHYDE	C5H4O2	FURFURAL	98-01-1
2-FURANCARBINOL	C5H6O2	FURFURYL ALCOHOL	98-00-0
2-FURANCARBONAL	C5H4O2	FURFURAL	98-01-1
2-FURANCARBOXALDEHYDE	C5H4O2	FURFURAL	98-01-1
2,5-FURANDIONE	C4H2O3	MALEIC ANHYDRIDE	108-31-6
FURANIDINE	C4H8O	TETRAHYDROFURAN	109-99-9
2-FURANMETHANOL	C5H6O2	FURFURYL ALCOHOL	98-00-0
FURFURAL	C5H4O2	FURFURAL	98-01-1
2-FURFURAL	C5H4O2	FURFURAL	98-01-1
FURFURAL ALCOHOL	C5H6O2	FURFURYL ALCOHOL	98-00-0
FURFURALDEHYDE	C5H4O2	FURFURAL	98-01-1
FURFURAN	C4H4O	FURAN	110-00-9
FURFUROL	C5H4O2	FURFURAL	98-01-1
FURFUROLE	C5H4O2	FURFURAL	98-01-1
FURFURYL ALCOHOL	C5H6O2	FURFURYL ALCOHOL	98-00-0
2-FURFURYLALKOHOL	C5H6O2	FURFURYL ALCOHOL	98-00-0
2-FURIL-METANALE	C5H4O2	FURFURAL	98-01-1
a-FUROLE	C5H4O2	FURFURAL	98-01-1
FUROLE	C5H4O2	FURFURAL	98-01-1
FURYL ALCOHOL	C5H6O2	FURFURYL ALCOHOL	98-00-0
2-FURYLCARBINOL	C5H6O2	FURFURYL ALCOHOL	98-00-0
a-FURYLCARBINOL	C5H6O2	FURFURYL ALCOHOL	98-00-0
2-FURYL-METHANAL	C5H4O2	FURFURAL	98-01-1
(2-FURYL)METHANOL	C5H6O2	FURFURYL ALCOHOL	98-00-0
GAFCOL EB	C6H14O2	2-BUTOXYETHANOL	111-76-2
GAMASOL 90	C2H6OS	DIMETHYL SULFOXIDE	67-68-5
gamma-6480	C4H6O2	gamma-BUTYROLACTONE	96-48-0
GAULTHERIA OIL, ARTIFICIAL	C8H8O3	METHYL SALICYLATE	119-36-8
GENETRON	C2Cl2F4	1,2-DICHLOROTETRAFLUOROETHANE	76-14-2
GENETRON 100	C2H4F2	1,1-DIFLUOROETHANE	75-37-6
GENETRON 101	C2H3ClF2	1-CHLORO-1,1-DIFLUOROETHANE	75-68-3
GENETRON 11	CCl3F	TRICHLOROFLUOROMETHANE	75-69-4
GENETRON 1113	C2ClF3	CHLOROTRIFLUOROETHYLENE	79-38-9
GENETRON 112	C2Cl4F2	1,1,2,2-TETRACHLORODIFLUOROETHANE	76-12-0
GENETRON 113	C2Cl3F3	1,1,2-TRICHLOROTRIFLUOROETHANE	76-13-1
GENETRON 115	C2ClF5	CHLOROPENTAFLUOROETHANE	76-15-3
GENETRON 13	CClF3	CHLOROTRIFLUOROMETHANE	75-72-9
GENETRON 142b	C2H3ClF2	1-CHLORO-1,1-DIFLUOROETHANE	75-68-3
GENETRON 21	CHCl2F	DICHLOROFLUOROMETHANE	75-43-4
GENETRON 22	CHClF2	CHLORODIFLUOROMETHANE	75-45-6
GENETRON 23	CHF3	TRIFLUOROMETHANE	75-46-7
GENTRON 142B	C2H3ClF2	1-CHLORO-1,1-DIFLUOROETHANE	75-68-3
GERANIUM CRYSTALS	C12H10O	DIPHENYL ETHER	101-84-8
GETTYSOLVE-B	C6H14	n-HEXANE	110-54-3
GETTYSOLVE-C	C7H16	n-HEPTANE	142-82-5
GILUCOR NITRO	C3H5N3O9	NITROGLYCERINE	55-63-0
GLACIAL ACETIC ACID	C2H4O2	ACETIC ACID	64-19-7
GLACIAL ACRYLIC ACID	C3H4O2	ACRYLIC ACID	79-10-7
GLICOL MONOCLORIDRINA	C2H5ClO	2-CHLOROETHANOL	107-07-3
GLONOIN	C3H5N3O9	NITROGLYCERINE	55-63-0
GLUCITOL	C6H14O6	SORBITOL	50-70-4
d-GLUCITOL	C6H14O6	SORBITOL	50-70-4
GLUSATE	C5H9NO4	L-GLUTAMIC ACID	56-86-0
GLUTACID	C5H9NO4	L-GLUTAMIC ACID	56-86-0
a-GLUTAMIC ACID	C5H9NO4	L-GLUTAMIC ACID	56-86-0
GLUTAMIC ACID	C5H9NO4	L-GLUTAMIC ACID	56-86-0
L-GLUTAMIC ACID	C5H9NO4	L-GLUTAMIC ACID	56-86-0
d-GLUTAMIENSUUR	C5H9NO4	L-GLUTAMIC ACID	56-86-0
GLUTAMINIC ACID	C5H9NO4	L-GLUTAMIC ACID	56-86-0
l-GLUTAMINIC ACID	C5H9NO4	L-GLUTAMIC ACID	56-86-0
GLUTAMINOL	C5H9NO4	L-GLUTAMIC ACID	56-86-0
GLUTARIC ACID	C5H8O4	GLUTARIC ACID	110-94-1
GLUTARIC ACID DINITRILE	C5H6N2	GLUTARONITRILE	544-13-8
GLUTARIC ANHYDRIDE	C5H6O3	GLUTARIC ANHYDRIDE	108-55-4
GLUTARODINITRILE	C5H6N2	GLUTARONITRILE	544-13-8
GLUTARONITRILE	C5H6N2	GLUTARONITRILE	544-13-8
GLUTATON	C5H9NO4	L-GLUTAMIC ACID	56-86-0
GLYCERIN	C3H8O3	GLYCEROL	56-81-5
GLYCERINE	C3H8O3	GLYCEROL	56-81-5
GLYCERINE TRIACETATE	C9H14O6	GLYCERYL TRIACETATE	102-76-1
GLYCERITOL	C3H8O3	GLYCEROL	56-81-5
GLYCEROL	C3H8O3	GLYCEROL	56-81-5
GLYCEROL EPICHLORHYDRIN	C3H5ClO	alpha-EPICHLOROHYDRIN	106-89-8
GLYCEROL, NITRIC ACID TRIESTER	C3H5N3O9	NITROGLYCERINE	55-63-0
GLYCEROL TRIACETATE	C9H14O6	GLYCERYL TRIACETATE	102-76-1
GLYCEROL TRICHLOROHYDRIN	C3H5Cl3	1,2,3-TRICHLOROPROPANE	96-18-4

Name (Synonym)	Formula	Name in Tables	CAS Reg. No
GLYCERYL NITRATE	C3H5N3O9	NITROGLYCERINE	55-63-0
GLYCERYL TRIACETATE	C9H14O6	GLYCERYL TRIACETATE	102-76-1
GLYCERYL TRICHLOROHYDRIN	C3H5Cl3	1,2,3-TRICHLOROPROPANE	96-18-4
GLYCERYL TRINITRATE	C3H5N3O9	NITROGLYCERINE	55-63-0
GLYCINOL	C2H7NO	MONOETHANOLAMINE	141-43-5
GLYCOL	C2H6O2	ETHYLENE GLYCOL	107-21-1
GLYCOL ALCOHOL	C2H6O2	ETHYLENE GLYCOL	107-21-1
GLYCOL BIS(HYDROXYETHYL) ETHER	C6H14O4	TRIETHYLENE GLYCOL	112-27-6
GLYCOL BROMIDE	C2H4Br2	1,2-DIBROMOETHANE	106-93-4
GLYCOL BUTYL ETHER	C6H14O2	2-BUTOXYETHANOL	111-76-2
GLYCOL CHLOROHYDRIN	C2H5ClO	2-CHLOROETHANOL	107-07-3
GLYCOL CYANOHYDRIN	C3H5NO	HYDRACRYLONITRILE	109-78-4
GLYCOL DIACETATE	C6H10O4	ETHYLENE GLYCOL DIACETATE	111-55-7
GLYCOL DICHLORIDE	C2H4Cl2	1,2-DICHLOROETHANE	107-06-2
GLYCOL DIMETHYL ETHER	C4H10O2	1,2-DIMETHOXYETHANE	110-71-4
GLYCOL ETHER	C4H10O3	DIETHYLENE GLYCOL	111-46-6
GLYCOL ETHER DB	C8H18O3	DIETHYLENE GLYCOL MONOBUTYL ETHER	112-34-5
GLYCOL ETHER de ACETATE	C8H16O4	DIETHYLENE GLYCOL ETHYL ETHER ACETATE	112-15-2
GLYCOL ETHER EB	C6H14O2	2-BUTOXYETHANOL	111-76-2
GLYCOL ETHER EE	C4H10O2	2-ETHOXYETHANOL	110-80-5
GLYCOL ETHER EE ACETATE	C6H12O3	2-ETHOXYETHYL ACETATE	111-15-9
GLYCOL ETHER EM	C3H8O2	2-METHOXYETHANOL	109-86-4
GLYCOL ETHYLENE ETHER	C4H8O2	1,4-DIOXANE	123-91-1
GLYCOL ETHYL ETHER	C4H10O2	2-ETHOXYETHANOL	110-80-5
GLYCOLMETHYL ETHER	C3H8O2	2-METHOXYETHANOL	109-86-4
GLYCOL MONOCHLOROHYDRIN	C2H5ClO	2-CHLOROETHANOL	107-07-3
GLYCOL MONOETHYL ETHER	C4H10O2	2-ETHOXYETHANOL	110-80-5
GLYCOL MONOETHYL ETHER ACETATE	C6H12O3	2-ETHOXYETHYL ACETATE	111-15-9
GLYCOL MONOMETHYL ETHER	C3H8O2	2-METHOXYETHANOL	109-86-4
GLYCOMONOCHLORHYDRIN	C2H5ClO	2-CHLOROETHANOL	107-07-3
GLYCON DP	C18H36O2	STEARIC ACID	57-11-4
GLYCON RO	C18H34O2	OLEIC ACID	112-80-1
GLYCON S-70	C18H36O2	STEARIC ACID	57-11-4
GLYCON TP	C18H36O2	STEARIC ACID	57-11-4
GLYCON WO	C18H34O2	OLEIC ACID	112-80-1
GLYCYL ALCOHOL	C3H8O3	GLYCEROL	56-81-5
GLYME	C4H10O2	1,2-DIMETHOXYETHANE	110-71-4
GLYME-3	C8H18O4	TRIETHYLENE GLYCOL DIMETHYL ETHER	112-49-2
GLYPED	C9H14O6	GLYCERYL TRIACETATE	102-76-1
GP-40-66:120	C4Cl6	HEXACHLORO-1,3-BUTADIENE	87-68-3
GRANOX NM	C6Cl6	HEXACHLOROBENZENE	118-74-1
GRAPHLOX	C5Cl6	HEXACHLOROCYCLOPENTADIENE	77-47-4
GRASEX	C2HCl3O	TRICHLOROACETALDEHYDE	75-87-6
GREEN OIL	C14H10	ANTHRACENE	120-12-7
GROCO 2	C18H34O2	OLEIC ACID	112-80-1
GROCO 4	C18H34O2	OLEIC ACID	112-80-1
GROCO 54	C18H36O2	STEARIC ACID	57-11-4
GROCO 5810	C22H44O2	n-BUTYL STEARATE	123-95-5
GROCO 5L	C18H34O2	OLEIC ACID	112-80-1
GROCOLENE	C3H8O3	GLYCEROL	56-81-5
GUAIACOL	C7H8O2	GUAIACOL	90-05-1
GUAICOL	C7H8O2	GUAIACOL	90-05-1
GULITOL	C6H14O6	SORBITOL	50-70-4
I-GULITOL	C6H14O6	SORBITOL	50-70-4
GUM CAMPHOR	C10H16O	CAMPHOR	76-22-2
HALOCARBON 11	CCl3F	TRICHLOROFLUOROMETHANE	75-69-4
HALOCARBON 112	C2Cl4F2	1,1,2,2-TETRACHLORODIFLUOROETHANE	76-12-0
HALOCARBON 113	C2Cl3F3	1,1,2-TRICHLOROTRIFLUOROETHANE	76-13-1
HALOCARBON 1132A	C2H2F2	1,1-DIFLUOROETHYLENE	75-38-7
HALOCARBON 114	C2Cl2F4	1,2-DICHLOROTETRAFLUOROETHANE	76-14-2
HALOCARBON 115	C2ClF5	CHLOROPENTAFLUOROETHANE	76-15-3
HALOCARBON 13/UCON 13	CClF3	CHLOROTRIFLUOROMETHANE	75-72-9
HALOCARBON 14	CF4	CARBON TETRAFLUORIDE	75-73-0
HALOCARBON 152A	C2H4F2	1,1-DIFLUOROETHANE	75-37-6
HALOCARBON 23	CHF3	TRIFLUOROMETHANE	75-46-7
HALOCARBON C-138	C4F8	OCTAFLUOROCYCLOBUTANE	115-25-3
HALON	CCl2F2	DICHLORODIFLUOROMETHANE	75-71-8
HALON 10001	CH3I	METHYL IODIDE	74-88-4
HALON 1001	CH3Br	METHYL BROMIDE	74-83-9
HALON 1011	CH2BrCl	BROMOCHLOROMETHANE	74-97-5
HALON 1202	CBr2F2	DIBROMODIFLUOROMETHANE	75-61-6
HALON 1211	CBrClF2	BROMOCHLORODIFLUOROMETHANE	353-59-3
HALON 1301	CBrF3	BROMOTRIFLUOROMETHANE	75-63-8
HALON 14	CF4	CARBON TETRAFLUORIDE	75-73-0
HALON 2001	C2H5Br	BROMOETHANE	74-96-4
HALON 2402	C2Br2F4	1,2-DIBROMOTETRAFLUOROETHANE	124-73-2

Name (Synonym)	Formula	Name in Tables	CAS Reg. No
HALOTAN	C2HBrClF3	HALOTHANE	151-67-7
HALOTHANE	C2HBrClF3	HALOTHANE	151-67-7
HALSAN	C2HBrClF3	HALOTHANE	151-67-7
HALSO 99	C7H7Cl	o-CHLOROTOLUENE	95-49-8
HASETHROL	C5H8N4O12	PENTAERYTHRITOL TETRANITRATE	78-11-5
HCB	C6Cl6	HEXACHLOROBENZENE	118-74-1
HCBD	C4Cl6	HEXACHLORO-1,3-BUTADIENE	87-68-3
HCCPD	C5Cl6	HEXACHLOROCYCLOPENTADIENE	77-47-4
HDO	C6H14O2	1,6-HEXANEDIOL	629-11-8
HEKSAN	C6H14	n-HEXANE	110-54-3
HEMIMELLITENE	C9H12	1,2,3-TRIMETHYLBENZENE	526-73-8
HEMPA	C6H18N3OP	HEXAMETHYL PHOSPHORAMIDE	680-31-9
HENDECANAL	C11H22O	1-UNDECANAL	112-44-7
HENDECANALDEHYDE	C11H22O	1-UNDECANAL	112-44-7
HENDECANE	C11H24	n-UNDECANE	1120-21-4
HENDECANOIC ALCOHOL	C11H24O	1-UNDECANOL	112-42-5
1-HENDECANOL	C11H24O	1-UNDECANOL	112-42-5
HENDECYL ALCOHOL	C11H24O	1-UNDECANOL	112-42-5
n-HENDECYLENIC ALCOHOL	C11H24O	1-UNDECANOL	112-42-5
n-HEPTADECANE	C17H36	n-HEPTADECANE	629-78-7
1-HEPTADECANECARBOXYLIC ACID	C18H36O2	STEARIC ACID	57-11-4
1-HEPTADECANETHIOL	C17H36S	1-HEPTADECANETHIOL	----------
1-HEPTADECANOL	C17H36O	1-HEPTADECANOL	1454-85-9
1-HEPTADECENE	C17H34	1-HEPTADECENE	6765-39-5
HEPTADECYL-METHYL-SULFIDE	C18H38S	HEPTADECYL-METHYL-SULFIDE	----------
HEPTADECYL-PROPYL-SULFIDE	C20H42S	HEPTADECYL-PROPYL-SULFIDE	----------
1-HEPTADECYNE	C17H32	1-HEPTADECYNE	6765-39-5
HEPTALDEHYDE	C7H14O	1-HEPTANAL	111-71-7
n-HEPTALDEHYDE	C7H14O	1-HEPTANAL	111-71-7
HEPTAN	C7H16	n-HEPTANE	142-82-5
1-HEPTANAL	C7H14O	1-HEPTANAL	111-71-7
1-HEPTANAMINE	C7H17N	1-AMINOHEPTANE	111-68-2
n-HEPTANE	C7H16	n-HEPTANE	142-82-5
HEPTANE, 1-BROMO-	C7H15Br	1-BROMOHEPTANE	629-04-9
1-HEPTANECARBOXYLIC ACID	C8H16O2	n-OCTANOIC ACID	124-07-2
HEPTANEDICARBOXYLIC ACID	C9H16O4	AZELAIC ACID	123-99-9
1,7-HEPTANEDICARBOXYLIC ACID	C9H16O4	AZELAIC ACID	123-99-9
HEPTANEN	C7H16	n-HEPTANE	142-82-5
n-HEPTANOIC ACID	C7H14O2	n-HEPTANOIC ACID	111-14-8
n-HEPTANOL	C7H16O	1-HEPTANOL	111-70-6
1-HEPTANOL	C7H16O	1-HEPTANOL	111-70-6
2-HEPTANOL	C7H16O	2-HEPTANOL	543-49-7
n-HEPTANOL-1	C7H16O	1-HEPTANOL	111-70-6
HEPTANOL-2	C7H16O	2-HEPTANOL	543-49-7
2-HEPTANONE	C7H14O	2-HEPTANONE	110-43-0
cis-2-HEPTENE	C7H14	cis-2-HEPTENE	6443-92-1
cis-3-HEPTENE	C7H14	cis-3-HEPTENE	7642-10-6
trans-2-HEPTENE	C7H14	trans-2-HEPTENE	14686-13-6
trans-3-HEPTENE	C7H14	trans-3-HEPTENE	14686-14-7
1-HEPTENE	C7H14	1-HEPTENE	592-76-7
1-n-HEPTENE	C7H14	1-HEPTENE	592-76-7
HEPTHLIC ACID	C7H14O2	n-HEPTANOIC ACID	111-14-8
n-HEPTOIC ACID	C7H14O2	n-HEPTANOIC ACID	111-14-8
HEPTYLAMINE	C7H17N	1-AMINOHEPTANE	111-68-2
n-HEPTYLAMINE	C7H17N	1-AMINOHEPTANE	111-68-2
n-HEPTYLBENZENE	C13H20	n-HEPTYLBENZENE	1078-71-3
HEPTYL BROMIDE	C7H15Br	1-BROMOHEPTANE	629-04-9
HEPTYL CARBINOL	C8H18O	1-OCTANOL	111-87-5
HEPTYL-DISULFIDE	C14H30S2	HEPTYL-DISULFIDE	----------
1-HEPTYLENE	C7H14	1-HEPTENE	592-76-7
HEPTYL HYDRIDE	C7H16	n-HEPTANE	142-82-5
n-HEPTYLIC ACID	C7H14O2	n-HEPTANOIC ACID	111-14-8
HEPTYL MERCAPTAN	C7H16S	n-HEPTYL MERCAPTAN	1639-09-4
n-HEPTYL MERCAPTAN	C7H16S	n-HEPTYL MERCAPTAN	1639-09-4
HEPTYL-METHYL-SULFIDE	C8H18S	HEPTYL-METHYL-SULFIDE	----------
HEPTYL-PROPYL-SULFIDE	C10H22S	HEPTYL-PROPYL-SULFIDE	----------
HEPTYL-SULFIDE	C14H30S	HEPTYL-SULFIDE	----------
1-HEPTYNE	C7H12	1-HEPTYNE	628-71-7
HERCULES P6	C5H12O4	PENTAERYTHRITOL	115-77-5
HEXA C.B.	C6Cl6	HEXACHLOROBENZENE	118-74-1
HEXACHLOR-AETHAN	C2Cl6	HEXACHLOROETHANE	67-72-1
HEXACHLORBENZOL	C6Cl6	HEXACHLOROBENZENE	118-74-1
HEXACHLOR-1,3-BUTADIEN	C4Cl6	HEXACHLORO-1,3-BUTADIENE	87-68-3
HEXACHLORCYKLOPENTADIEN	C5Cl6	HEXACHLOROCYCLOPENTADIENE	77-47-4
HEXACHLOROBENZENE	C6Cl6	HEXACHLOROBENZENE	118-74-1
HEXACHLORO-1,3-BUTADIENE	C4Cl6	HEXACHLORO-1,3-BUTADIENE	87-68-3

Name (Synonym)	Formula	Name in Tables	CAS Reg. No
1,1,2,3,4,4-HEXACHLORO-1,3-BUTADIENE	C4Cl6	HEXACHLORO-1,3-BUTADIENE	87-68-3
HEXACHLOROCYCLOPENTADIENE	C5Cl6	HEXACHLOROCYCLOPENTADIENE	77-47-4
HEXACHLORO-1,3-CYCLOPENTADIENE	C5Cl6	HEXACHLOROCYCLOPENTADIENE	77-47-4
1,2,3,4,5,5-HEXACHLORO-1,3-CYCLOPENTADIENE	C5Cl6	HEXACHLOROCYCLOPENTADIENE	77-47-4
HEXACHLOROETHANE	C2Cl6	HEXACHLOROETHANE	67-72-1
1,1,1,2,2,2-HEXACHLOROETHANE	C2Cl6	HEXACHLOROETHANE	67-72-1
HEXACHLOROETHYLENE	C2Cl6	HEXACHLOROETHANE	67-72-1
HEXACID 1095	C10H20O2	n-DECANOIC ACID	334-48-5
HEXACID 698	C6H12O2	n-HEXANOIC ACID	142-62-1
HEXACID 898	C8H16O2	n-OCTANOIC ACID	124-07-2
HEXACID C-7	C7H14O2	n-HEPTANOIC ACID	111-14-8
HEXACID C-9	C9H18O2	n-NONANOIC ACID	112-05-0
n-HEXADECANE	C16H34	n-HEXADECANE	544-76-3
1-HEXADECANETHIOL	C16H34S	1-HEXADECANETHIOL	2917-26-2
HEXADECANOIC ACID	C16H32O2	n-HEXADECANOIC ACID	57-10-3
n-HEXADECANOIC ACID	C16H32O2	n-HEXADECANOIC ACID	57-10-3
HEXADECANOL	C16H34O	1-HEXADECANOL	36653-82-4
n-HEXADECANOL	C16H34O	1-HEXADECANOL	36653-82-4
1-HEXADECANOL	C16H34O	1-HEXADECANOL	36653-82-4
1-HEXADECENE	C16H32	1-HEXADECENE	629-73-2
n-HEXADECOIC ACID	C16H32O2	n-HEXADECANOIC ACID	57-10-3
HEXADECYLIC ACID	C16H32O2	n-HEXADECANOIC ACID	57-10-3
HEXADECYL-METHYL-SULFIDE	C17H36S	HEXADECYL-METHYL-SULFIDE	----------
HEXADECYL-PROPYL-SULFIDE	C19H40S	HEXADECYL-PROPYL-SULFIDE	----------
1-HEXADECYNE	C16H30	1-HEXADECYNE	629-74-3
1,5-HEXADIENE	C6H10	1,5-HEXADIENE	592-42-7
cis,trans-2,4-HEXADIENE	C6H10	cis,trans-2,4-HEXADIENE	5194-50-3
HEXA-1,5-DIENE	C6H10	1,5-HEXADIENE	592-42-7
trans,trans-2,4-HEXADIENE	C6H10	trans,trans-2,4-HEXADIENE	5194-51-4
HEXAETHYLBENZENE	C18H30	HEXAETHYLBENZENE	604-88-6
HEXAFLUOROACETONE	C3F6O	HEXAFLUOROACETONE	684-16-2
HEXAFLUOROBENZENE	C6F6	HEXAFLUOROBENZENE	392-56-3
HEXAFLUOROETHANE	C2F6	HEXAFLUOROETHANE	76-16-4
HEXAFLUOROPROPYLENE	C3F6	HEXAFLUOROPROPYLENE	116-15-4
HEXAHYDROANILINE	C6H13N	CYCLOHEXYLAMINE	108-91-8
HEXAHYDRO-2H-AZEPIN-2-ONE	C6H11NO	epsilon-CAPROLACTAM	105-60-2
HEXAHYDROAZEPINE	C6H13N	HEXAMETHYLENEIMINE	111-49-9
HEXAHYDRO-2-AZEPINONE	C6H11NO	epsilon-CAPROLACTAM	105-60-2
HEXAHYDROBENZENAMINE	C6H13N	CYCLOHEXYLAMINE	108-91-8
HEXAHYDROBENZENE	C6H12	CYCLOHEXANE	110-82-7
HEXAHYDRO-1,4-DIAZINE	C4H10N2	PIPERAZINE	110-85-0
HEXAHYDROPHENOL	C6H12O	CYCLOHEXANOL	108-93-0
HEXAHYDROPYRAZINE	C4H10N2	PIPERAZINE	110-85-0
HEXAHYDROPYRIDINE	C5H11N	PIPERIDINE	110-89-4
HEXAHYDROTOLUENE	C7H14	METHYLCYCLOHEXANE	108-87-2
HEXALDEHYDE	C6H12O	1-HEXANAL	66-25-1
HEXALIN	C6H12O	CYCLOHEXANOL	108-93-0
HEXAMETAPOL	C6H18N3OP	HEXAMETHYL PHOSPHORAMIDE	680-31-9
HEXAMETHYLBENZENE	C12H18	HEXAMETHYLBENZENE	87-85-4
HEXAMETHYLCYCLOTRISILOXANE	C6H18O3Si3	HEXAMETHYLCYCLOTRISILOXANE	541-05-9
HEXAMETHYLDISILAZANE	C6H19NSi2	HEXAMETHYLDISILAZANE	999-97-3
HEXAMETHYLDISILOXANE	C6H18OSi2	HEXAMETHYLDISILOXANE	107-46-0
HEXAMETHYLENE	C6H12	CYCLOHEXANE	110-82-7
HEXAMETHYLENEDIAMINE	C6H16N2	HEXAMETHYLENEDIAMINE	124-09-4
1,6-HEXAMETHYLENEDIAMINE	C6H16N2	HEXAMETHYLENEDIAMINE	124-09-4
HEXAMETHYLENEDIOL	C6H14O2	1,6-HEXANEDIOL	629-11-8
HEXAMETHYLENE GLYCOL	C6H14O2	1,6-HEXANEDIOL	629-11-8
HEXAMETHYLENEIMINE	C6H13N	HEXAMETHYLENEIMINE	111-49-9
HEXAMETHYLENIMINE	C6H13N	HEXAMETHYLENEIMINE	111-49-9
HEXAMETHYL PHOSPHORAMIDE	C6H18N3OP	HEXAMETHYL PHOSPHORAMIDE	680-31-9
HEXAMETHYLPHOSPHORIC ACID TRIAMIDE	C6H18N3OP	HEXAMETHYL PHOSPHORAMIDE	680-31-9
HEXAMETHYLPHOSPHORIC TRIAMIDE	C6H18N3OP	HEXAMETHYL PHOSPHORAMIDE	680-31-9
N,N,N,N,N-HEXAMETHYLPHOSPHORIC TRIAMIDE	C6H18N3OP	HEXAMETHYL PHOSPHORAMIDE	680-31-9
HEXAMETHYLPHOSPHOTRIAMIDE	C6H18N3OP	HEXAMETHYL PHOSPHORAMIDE	680-31-9
HEXAMETHYLSILAZANE	C6H19NSi2	HEXAMETHYLDISILAZANE	999-97-3
HEXANAL	C6H12O	1-HEXANAL	66-25-1
1-HEXANAL	C6H12O	1-HEXANAL	66-25-1
1-HEXANAMINE	C6H15N	n-HEXYLAMINE	111-26-2
HEXANAPHTHENE	C6H12	CYCLOHEXANE	110-82-7
HEXANE	C6H14	n-HEXANE	110-54-3
n-HEXANE	C6H14	n-HEXANE	110-54-3
1-HEXANECARBOXYLIC ACID	C7H14O2	n-HEPTANOIC ACID	111-14-8
HEXANEDINITRILE	C6H8N2	ADIPONITRILE	111-69-3
1,6-HEXANEDIOIC ACID	C6H10O4	ADIPIC ACID	124-04-9
HEXANEDIOIC ACID DINITRILE	C6H8N2	ADIPONITRILE	111-69-3
a-w-HEXANEDIOL	C6H14O2	1,6-HEXANEDIOL	629-11-8

Name (Synonym)	Formula	Name in Tables	CAS Reg. No
w-HEXANEDIOL	C6H14O2	1,6-HEXANEDIOL	629-11-8
1,2-HEXANEDIOL	C6H14O2	HEXYLENE GLYCOL	107-41-5
1,6-HEXANEDIOL	C6H14O2	1,6-HEXANEDIOL	629-11-8
6-HEXANELACTAM	C6H11NO	epsilon-CAPROLACTAM	105-60-2
HEXANEN	C6H14	n-HEXANE	110-54-3
HEXANENITRILE	C6H11N	HEXANENITRILE	628-73-9
n-HEXANOIC ACID	C6H12O2	n-HEXANOIC ACID	142-62-1
HEXANOL	C6H14O	1-HEXANOL	111-27-3
n-HEXANOL	C6H14O	1-HEXANOL	111-27-3
sec-HEXANOL	C6H14O	2-ETHYL-1-BUTANOL	97-95-0
1-HEXANOL	C6H14O	1-HEXANOL	111-27-3
2-HEXANOL	C6H14O	2-HEXANOL	626-93-7
6-HEXANOLACTONE	C6H10O2	epsilon-CAPROLACTONE	502-44-3
1,6-HEXANOLIDE	C6H10O2	epsilon-CAPROLACTONE	502-44-3
HEXANON	C6H10O	CYCLOHEXANONE	108-94-1
2-HEXANONE	C6H12O	2-HEXANONE	591-78-6
3-HEXANONE	C6H12O	3-HEXANONE	589-38-8
HEXANONE ISOXIME	C6H11NO	epsilon-CAPROLACTAM	105-60-2
HEXANONE-2	C6H12O	2-HEXANONE	591-78-6
HEXANONISOXIM	C6H11NO	epsilon-CAPROLACTAM	105-60-2
HEXAPLAS M/B	C16H22O4	DIBUTYL PHTHALATE	84-74-2
HEXAPLAS M/O	C24H38O4	DIISOOCTYL PHTHALATE	27554-26-3
HEXAZANE	C5H11N	PIPERIDINE	110-89-4
HEXENE	C6H12	1-HEXENE	592-41-6
cis-2-HEXENE	C6H12	cis-2-HEXENE	7688-21-3
cis-3-HEXENE	C6H12	cis-3-HEXENE	7642-09-3
trans-2-HEXENE	C6H12	trans-2-HEXENE	4050-45-7
trans-3-HEXENE	C6H12	trans-3-HEXENE	13269-52-8
1-HEXENE	C6H12	1-HEXENE	592-41-6
HEXMETHYLPHOSPHORAMIDE	C6H18N3OP	HEXAMETHYL PHOSPHORAMIDE	680-31-9
n-HEXOIC ACID	C6H12O2	n-HEXANOIC ACID	142-62-1
1,6-HEXOLACTAM	C6H11NO	epsilon-CAPROLACTAM	105-60-2
HEXON	C6H12O	METHYL ISOBUTYL KETONE	108-10-1
n-HEXYL ACETATE	C8H16O2	n-HEXYL ACETATE	142-92-7
1-HEXYL ACETATE	C8H16O2	n-HEXYL ACETATE	142-92-7
HEXYL ALCOHOL	C6H14O	1-HEXANOL	111-27-3
sec-HEXYL ALCOHOL	C6H14O	2-ETHYL-1-BUTANOL	97-95-0
HEXYL ALCOHOL, ACETATE	C8H16O2	n-HEXYL ACETATE	142-92-7
N-HEXYLAMINE	C6H15N	n-HEXYLAMINE	111-26-2
n-HEXYLBENZENE	C12H18	n-HEXYLBENZENE	1077-16-3
HEXYL-DISULFIDE	C12H26S2	HEXYL-DISULFIDE	----------
HEXYLENE	C6H12	1-HEXENE	592-41-6
HEXYLENE GLYCOL	C6H14O2	HEXYLENE GLYCOL	107-41-5
HEXYL ETHANOATE	C8H16O2	n-HEXYL ACETATE	142-92-7
HEXYL ETHER	C12H26O	DI-n-HEXYL ETHER	112-58-3
N-HEXYL ETHER	C12H26O	DI-n-HEXYL ETHER	112-58-3
HEXYL MERCAPTAN	C6H14S	n-HEXYLMERCAPTAN	111-31-9
n-HEXYLMERCAPTAN	C6H14S	n-HEXYLMERCAPTAN	111-31-9
HEXYL-METHYL-SULFIDE	C7H16S	HEXYL-METHYL-SULFIDE	20291-60-5
1-n-HEXYLNAPHTHALENE	C16H20	1-n-HEXYLNAPHTHALENE	2876-53-1
HEXYL-PROPYL-SULFIDE	C9H20S	HEXYL-PROPYL-SULFIDE	----------
HEXYL-SULFIDE	C12H26S	HEXYL-SULFIDE	6294-31-1
1-HEXYNE	C6H10	1-HEXYNE	693-02-7
2-HEXYNE	C6H10	2-HEXYNE	764-35-2
3-HEXYNE	C6H10	3-HEXYNE	928-49-4
HI-DRY	C8H18O5	TETRAETHYLENE GLYCOL	112-60-7
HILTONIL FAST ORANGE GR BASE	C6H6N2O2	o-NITROANILINE	88-74-4
HILTONIL FAST ORANGE R BASE	C6H6N2O2	m-NITROANILINE	99-09-2
HMDA	C6H16N2	HEXAMETHYLENEDIAMINE	124-09-4
HMDS	C6H19NSi2	HEXAMETHYLDISILAZANE	999-97-3
HMPA	C6H18N3OP	HEXAMETHYL PHOSPHORAMIDE	680-31-9
HMPT	C6H18N3OP	HEXAMETHYL PHOSPHORAMIDE	680-31-9
HOCH	CH2O	FORMALDEHYDE	50-00-0
HOMOPIPERIDINE	C6H13N	HEXAMETHYLENEIMINE	111-49-9
HOSTETEX L-PEC	C6H3Cl3	1,2,4-TRICHLOROBENZENE	120-82-1
b-HPN	C3H5NO	HYDRACRYLONITRILE	109-78-4
HPT	C6H18N3OP	HEXAMETHYL PHOSPHORAMIDE	680-31-9
HRS 1655	C5Cl6	HEXACHLOROCYCLOPENTADIENE	77-47-4
HUILE dANILINE	C6H7N	ANILINE	62-53-3
HUILE de CAMPHRE	C10H16O	CAMPHOR	76-22-2
HUILE H50	C4H4S	THIOPHENE	110-02-1
HY 951	C6H18N4	TRIETHYLENE TETRAMINE	112-24-3
HYDRACRYLIC ACID b-LACTONE	C3H4O2	beta-PROPIOLACTONE	57-57-8
HYDRACRYLONITRILE	C3H5NO	HYDRACRYLONITRILE	109-78-4
HYDRALIN	C6H12O	CYCLOHEXANOL	108-93-0
HYDRAZINE-BENZENE	C6H8N2	PHENYLHYDRAZINE	100-63-0

Name (Synonym)	Formula	Name in Tables	CAS Reg. No
HYDRAZINOBENEZENE	C6H8N2	PHENYLHYDRAZINE	100-63-0
HYDRAZOBENZEN	C12H12N2	HYDRAZOBENZENE	122-66-7
HYDRAZOBENZENE	C12H12N2	HYDRAZOBENZENE	122-66-7
1,2-HYDRINDENE	C9H10	INDANE	496-11-7
HYDRINDONAPHTHENE	C9H10	INDANE	496-11-7
HYDRIODIC ETHER	C2H5I	ETHYL IODIDE	75-03-6
HYDROBROMIC ETHER	C2H5Br	BROMOETHANE	74-96-4
HYDROCHLORIC ETHER	C2H5Cl	ETHYL CHLORIDE	75-00-3
HYDROCHLOROFLUOROCARBON 142b	C2H3ClF2	1-CHLORO-1,1-DIFLUOROETHANE	75-68-3
HYDROCYANIC ACID	CHN	HYDROGEN CYANIDE	74-90-8
HYDROCYANIC ETHER	C3H5N	PROPIONITRILE	107-12-0
HYDROFOL	C16H32O2	n-HEXADECANOIC ACID	57-10-3
HYDROFOL ACID 1255	C12H24O2	n-DODECANOIC ACID	143-07-7
HYDROFOL ACID 1495	C14H28O2	n-TETRADECANOIC ACID	544-63-8
HYDROFOL ACID 1655	C18H36O2	STEARIC ACID	57-11-4
HYDROFURAN	C4H8O	TETRAHYDROFURAN	109-99-9
HYDROGEN CARBOXYLIC ACID	CH2O2	FORMIC ACID	64-18-6
HYDROGEN CYANIDE	CHN	HYDROGEN CYANIDE	74-90-8
HYDROGEN ISOTHIOCYANATE	CHNS	ISOTHIOCYANIC-ACID	3129-90-6
HYDROPEROXYDE de BUTYLE TERTIAIRE	C4H10O2	t-BUTYL HYDROPEROXIDE	75-91-2
HYDROPEROXYDE de CUMENE	C9H12O2	CUMENE HYDROPEROXIDE	80-15-9
HYDROPEROXYDE de CUMYLE	C9H12O2	CUMENE HYDROPEROXIDE	80-15-9
2-HYDROPEROXY-2-METHYLPROPANE	C4H10O2	t-BUTYL HYDROPEROXIDE	75-91-2
HYDROPHENOL	C6H12O	CYCLOHEXANOL	108-93-0
a-HYDROQUINONE	C6H6O2	p-HYDROQUINONE	123-31-9
HYDROQUINOLE	C6H6O2	p-HYDROQUINONE	123-31-9
m-HYDROQUINONE	C6H6O2	1,3-BENZENEDIOL	108-46-3
o-HYDROQUINONE	C6H6O2	1,2-BENZENEDIOL	120-80-9
p-HYDROQUINONE	C6H6O2	p-HYDROQUINONE	123-31-9
HYDROQUINONE MONOMETHYL ETHER	C7H8O2	p-METHOXYPHENOL	150-76-5
4-HYDROXY-m-ANISALDEHYDE	C8H8O3	VANILLIN	121-33-5
o-HYDROXYANISOLE	C7H8O2	GUAIACOL	90-05-1
2-HYDROXYANISOLE	C7H8O2	GUAIACOL	90-05-1
o-HYDROXYBENZALDEHYDE	C7H6O2	SALICYLALDEHYDE	90-02-8
p-HYDROXYBENZALDEHYDE	C7H6O2	p-HYDROXYBENZALDEHYDE	123-08-0
2-HYDROXYBENZALDEHYDE	C7H6O2	SALICYLALDEHYDE	90-02-8
4-HYDROXYBENZALDEHYDE	C7H6O2	p-HYDROXYBENZALDEHYDE	123-08-0
HYDROXYBENZENE	C6H6O	PHENOL	108-95-2
o-HYDROXYBENZOIC ACID	C7H6O3	SALICYLIC ACID	69-72-7
2-HYDROXYBENZOIC ACID	C7H6O3	SALICYLIC ACID	69-72-7
o-HYDROXYBENZOIC ACID, METHYL ESTER	C8H8O3	METHYL SALICYLATE	119-36-8
2-HYDROXYBENZOIC ACID METHYL ESTER	C8H8O3	METHYL SALICYLATE	119-36-8
HYDROXYBENZOPYRIDINE	C9H7NO	8-HYDROXYQUINOLINE	148-24-3
1-HYDROXYBUTANE	C4H10O	n-BUTANOL	71-36-3
2-HYDROXYBUTANE	C4H10O	sec-BUTANOL	78-92-2
4-HYDROXYBUTANOIC ACID LACTONE	C4H6O2	gamma-BUTYROLACTONE	96-48-0
1-HYDROXY-4-tert-BUTYLBENZENE	C10H14O	p-tert-BUTYLPHENOL	98-54-4
gamma-HYDROXYBUTYRIC ACID CYCLIC ESTER	C4H6O2	gamma-BUTYROLACTONE	96-48-0
4-HYDROXYBUTYRIC ACID g-LACTONE	C4H6O2	gamma-BUTYROLACTONE	96-48-0
gamma-HYDROXYBUTYROLACTONE	C4H6O2	gamma-BUTYROLACTONE	96-48-0
HYDROXYCAPROIC ACID	C6H12O3	HYDROXYCAPROIC ACID	6064-63-7
8-HYDROXY-CHINOLIN	C9H7NO	8-HYDROXYQUINOLINE	148-24-3
1-HYDROXYCUMENE	C9H12O	2-PHENYL-2-PROPANOL	617-94-7
3-HYDROXYCYCLOHEXADIEN-1-ONE	C6H6O2	1,3-BENZENEDIOL	108-46-3
HYDROXYCYCLOHEXANE	C6H12O	CYCLOHEXANOL	108-93-0
1-HYDROXY-2,4-DIMETHYLBENZENE	C8H10O	2,4-XYLENOL	105-67-9
4-HYDROXYDIPHENYLDIMETHYLMETHANE	C15H16O	p-CUMYLPHENOL	599-64-4
HYDROXY ETHER	C4H10O2	2-ETHOXYETHANOL	110-80-5
b-HYDROXYETHYL ACRYLATE	C5H8O3	2-HYDROXYETHYL ACRYLATE	818-61-1
HYDROXYETHYL ACRYLATE	C5H8O3	2-HYDROXYETHYL ACRYLATE	818-61-1
2-HYDROXYETHYL ACRYLATE	C5H8O3	2-HYDROXYETHYL ACRYLATE	818-61-1
b-HYDROXYETHYLAMINE	C2H7NO	MONOETHANOLAMINE	141-43-5
2-HYDROXYETHYLAMINE	C2H7NO	MONOETHANOLAMINE	141-43-5
b-HYDROXYETHYLBENZENE	C8H10O	2-PHENYLETHANOL	60-12-8
N-HYDROXYETHYL-1,2-ETHANEDIAMINE	C4H12N2O	N-AMINOETHYL ETHANOLAMINE	111-41-1
b-HYDROXYETHYLDIMETHYLAMINE	C4H11NO	DIMETHYLETHANOLAMINE	108-01-0
N-(b-HYDROXYETHYL)ETHYLENEDIAMINE	C4H12N2O	N-AMINOETHYL ETHANOLAMINE	111-41-1
N-(2-HYDROXYETHYL)ETHYLENEDIAMINE	C4H12N2O	N-AMINOETHYL ETHANOLAMINE	111-41-1
2-(N-2-HYDROXYETHYL-N-METHYLAMINO)ETHANOL	C5H13NO2	METHYL DIETHANOLAMINE	105-59-9
1-HYDROXYHEPTANE	C7H16O	1-HEPTANOL	111-70-6
2-HYDROXYHEPTANE	C7H16O	2-HEPTANOL	543-49-7
1-HYDROXYHEXANE	C6H14O	1-HEXANOL	111-27-3
6-HYDROXYHEXANOIC ACID LACTONE	C6H10O2	epsilon-CAPROLACTONE	502-44-3
alpha-HYDROXYISOBUTYRONITRILE	C4H7NO	ACETONE CYANOHYDRIN	75-86-5
4-HYDROXY-2-KETO-4-METHYLPENTANE	C6H12O2	DIACETONE ALCOHOL	123-42-2
4-HYDROXY-3-METHOXYBENZALDEHYDE	C8H8O3	VANILLIN	121-33-5

Name (Synonym)	Formula	Name in Tables	CAS Reg. No
1-HYDROXY-2-METHOXYBENZENE	C7H8O2	GUAIACOL	90-05-1
1-HYDROXY-2-METHYLBENZENE	C7H8O	o-CRESOL	95-48-7
1-HYDROXY-3-METHYLBENZENE	C7H8O	m-CRESOL	108-39-4
1-HYDROXY-4-METHYLBENZENE	C7H8O	p-CRESOL	106-44-5
2-(HYDROXYMETHYL)ETHANOL	C3H8O2	1,3-PROPYLENE GLYCOL	504-63-2
2-HYDROXYMETHYLFURAN	C5H6O2	FURFURYL ALCOHOL	98-00-0
4-HYDROXY-4-METHYL-PENTAN-2-ON	C6H12O2	DIACETONE ALCOHOL	123-42-2
4-HYDROXY-4-METHYL-2-PENTANONE	C6H12O2	DIACETONE ALCOHOL	123-42-2
4-HYDROXY-4-METHYLPENTANONE-2	C6H12O2	DIACETONE ALCOHOL	123-42-2
1-HYDROXYMETHYLPROPANE	C4H10O	ISOBUTANOL	78-83-1
2-HYDROXY-2-METHYLPROPIONITRILE	C4H7NO	ACETONE CYANOHYDRIN	75-86-5
HYDROXY No. 253	C15H24O	NONYLPHENOL	25154-52-3
1-HYDROXYOCTANE	C8H18O	1-OCTANOL	111-87-5
o-HYDROXYPHENOL	C6H6O2	1,2-BENZENEDIOL	120-80-9
2-HYDROXYPHENOL	C6H6O2	1,2-BENZENEDIOL	120-80-9
1-HYDROXYPROPANE	C3H8O	n-PROPANOL	71-23-8
3-HYDROXYPROPANENITRILE	C3H5NO	HYDRACRYLONITRILE	109-78-4
2-HYDROXY-1,2,3-PROPANETRICARBOXYLIC ACID	C6H8O7	CITRIC ACID	77-92-9
2-HYDROXYPROPANNITRIL	C3H5NO	LACTONITRILE	78-97-7
3-HYDROXYPROPENE	C3H6O	ALLYL ALCOHOL	107-18-6
3-HYDROXYPROPIONIC ACID LACTONE	C3H4O2	beta-PROPIOLACTONE	57-57-8
b-HYDROXYPROPIONITRILE	C3H5NO	HYDRACRYLONITRILE	109-78-4
2-HYDROXYPROPIONITRILE	C3H5NO	LACTONITRILE	78-97-7
3-HYDROXYPROPIONITRILE	C3H5NO	HYDRACRYLONITRILE	109-78-4
2-HYDROXYPROPYLAMINE	C3H9NO	1-AMINO-2-PROPANOL	78-96-6
3-HYDROXYPROPYLAMINE	C3H9NO	3-AMINO-1-PROPANOL	156-87-6
8-HYDROXYQUINOLINE	C9H7NO	8-HYDROXYQUINOLINE	148-24-3
a-HYDROXYTOLUENE	C7H8O	BENZYL ALCOHOL	100-51-6
HYDROXYTOLUENE	C7H8O	BENZYL ALCOHOL	100-51-6
m-HYDROXYTOLUENE	C7H8O	m-CRESOL	108-39-4
o-HYDROXYTOLUENE	C7H8O	o-CRESOL	95-48-7
p-HYDROXYTOLUENE	C7H8O	p-CRESOL	106-44-5
4-HYDROXYTOLUENE	C7H8O	p-CRESOL	106-44-5
b-HYDROXYTRICARBALLYLIC ACID	C6H8O7	CITRIC ACID	77-92-9
HYLENE M50	C15H10N2O2	DIPHENYLMETHANE-4,4'-DIISOCYANATE	101-68-8
HYLENE-T	C9H6N2O2	TOLUENE DIISOCYANATE	26471-62-5
HY-PHI 1055	C18H34O2	OLEIC ACID	112-80-1
HY-PHI 1088	C18H34O2	OLEIC ACID	112-80-1
HY-PHI 1199	C18H36O2	STEARIC ACID	57-11-4
HY-PHI 2066	C18H34O2	OLEIC ACID	112-80-1
HY-PHI 2088	C18H34O2	OLEIC ACID	112-80-1
HY-PHI 2102	C18H34O2	OLEIC ACID	112-80-1
HYPNONE	C8H8O	ACETOPHENONE	98-86-2
HYSTRENE 80	C18H36O2	STEARIC ACID	57-11-4
HYSTRENE 8016	C16H32O2	n-HEXADECANOIC ACID	57-10-3
HYSTRENE 9014	C14H28O2	n-TETRADECANOIC ACID	544-63-8
HYSTRENE 9512	C12H24O2	n-DODECANOIC ACID	143-07-7
IDROPEROSSIDO di CUMENE	C9H12O2	CUMENE HYDROPEROXIDE	80-15-9
IDROPEROSSIDO di CUMOLO	C9H12O2	CUMENE HYDROPEROXIDE	80-15-9
4-IDROSSI-4-METIL-PENTAN-2-ONE	C6H12O2	DIACETONE ALCOHOL	123-42-2
IDRYL	C16H10	FLUORANTHENE	206-44-0
IF	C4CI4S	TETRACHLOROTHIOPHENE	6012-97-1
IMIDOLE	C4H5N	PYRROLE	109-97-7
2,2'-IMINOBISETHANOL	C4H11NO2	DIETHANOLAMINE	111-42-2
2,2'-IMINOBISETHYLAMINE	C4H13N3	DIETHYLENE TRIAMINE	111-40-0
2,2'-IMINODIETHANOL	C4H11NO2	DIETHANOLAMINE	111-42-2
1,1'-IMINODI-2-PROPANOL	C6H15NO2	DIISOPROPANOLAMINE	110-97-4
INDANE	C9H10	INDANE	496-11-7
INDENE	C9H8	INDENE	95-13-6
INDOL	C8H7N	INDOLE	120-72-9
INDOLE	C8H7N	INDOLE	120-72-9
INDONAPHTHENE	C9H8	INDENE	95-13-6
INDUSTRENE 105	C18H34O2	OLEIC ACID	112-80-1
INDUSTRENE 205	C18H34O2	OLEIC ACID	112-80-1
INDUSTRENE 4516	C16H32O2	n-HEXADECANOIC ACID	57-10-3
INDUSTRENE 5016	C18H36O2	STEARIC ACID	57-11-4
INHIBISOL	C2H3CI3	1,1,1-TRICHLOROETHANE	71-55-6
INITIATING EXPLOSIVE PENTAERYTHRITE TETRANITRATE	C5H8N4O12	PENTAERYTHRITOL TETRANITRATE	78-11-5
IODINEBENZOL	C6H5I	IODOBENZENE	591-50-4
IODOBENZENE	C6H5I	IODOBENZENE	591-50-4
IODOETHANE	C2H5I	ETHYL IODIDE	75-03-6
IODOMETANO	CH3I	METHYL IODIDE	74-88-4
IODOMETHANE	CH3I	METHYL IODIDE	74-88-4
2-IODO-2-METHYLPROPANE	C4H9I	2-IODO-2-METHYLPROPANE	558-17-8
1-IODOPROPANE	C3H7I	n-PROPYL IODIDE	107-08-4
2-IODOPROPANE	C3H7I	ISOPROPYL IODIDE	75-30-9

Name (Synonym)	Formula	Name in Tables	CAS Reg. No
3-IODOPROPENE	C3H5I	3-IODO-1-PROPENE	556-56-9
3-IODO-1-PROPENE	C3H5I	3-IODO-1-PROPENE	556-56-9
3-IODOPROPYLENE	C3H5I	3-IODO-1-PROPENE	556-56-9
IODURE de METHYLE	CH3I	METHYL IODIDE	74-88-4
IPA	C8H6O4	ISOPHTHALIC ACID	121-91-5
ISCEON 113	C2Cl3F3	1,1,2-TRICHLOROTRIFLUOROETHANE	76-13-1
ISCEON 131	CCl3F	TRICHLOROFLUOROMETHANE	75-69-4
ISCOBROME	CH3Br	METHYL BROMIDE	74-83-9
ISOACETOPHORONE	C9H14O	ISOPHORONE	78-59-1
ISOAMYL ALKOHOL	C5H12O	3-METHYL-1-BUTANOL	123-51-3
ISO-AMYLALKOHOL	C5H12O	3-METHYL-1-BUTANOL	123-51-3
ISOAMYL ETHANOATE	C7H14O2	ISOPENTYL ACETATE	123-92-2
ISOAMYLHYDRIDE	C5H12	ISOPENTANE	78-78-4
ISOAMYL ISOVALERATE	C10H20O2	ISOPENTYL ISOVALERATE	659-70-1
ISOAMYL METHYL KETONE	C7H14O	5-METHYL-2-HEXANONE	110-12-3
ISOAMYLOL ISOAMYLOL	C5H12O	3-METHYL-1-BUTANOL	123-51-3
1,3-ISOBENZOFURANDIONE	C8H4O3	PHTHALIC ANHYDRIDE	85-44-9
ISOBUTANAL	C4H8O	ISOBUTYRALDEHYDE	78-84-2
ISOBUTANE	C4H10	ISOBUTANE	75-28-5
ISOBUTANE MIXTURES	C4H10	ISOBUTANE	75-28-5
ISOBUTANOL	C4H10O	ISOBUTANOL	78-83-1
ISOBUTENAL	C4H6O	METHACROLEIN	78-85-3
ISOBUTENE	C4H8	ISOBUTENE	115-11-7
ISOBUTENYL METHYL KETONE	C6H10O	MESITYL OXIDE	141-79-7
ISOBUTYL ACETATE	C6H12O2	ISOBUTYL ACETATE	110-19-0
ISOBUTYL ACRYLATE	C7H12O2	ISOBUTYL ACRYLATE	106-63-8
ISOBUTYLALDEHYDE	C4H8O	ISOBUTYRALDEHYDE	78-84-2
ISOBUTYLALKOHOL	C4H10O	ISOBUTANOL	78-83-1
ISOBUTYLAMINE	C4H11N	ISOBUTYLAMINE	78-81-9
ISOBUTYLBENZENE	C10H14	ISOBUTYLBENZENE	538-93-2
ISOBUTYLCARBINOL	C5H12O	3-METHYL-1-BUTANOL	123-51-3
ISOBUTYLENE	C4H8	ISOBUTENE	115-11-7
ISOBUTYLESTER KYSELINY MRAVENCI	C5H10O2	ISOBUTYL FORMATE	542-55-2
ISOBUTYLESTER KYSELINY OCTOVE	C6H12O2	ISOBUTYL ACETATE	110-19-0
ISOBUTYL FORMATE	C5H10O2	ISOBUTYL FORMATE	542-55-2
ISOBUTYL ISOBUTYRATE	C8H16O2	ISOBUTYL ISOBUTYRATE	97-85-8
ISOBUTYL KETONE	C9H18O	DIISOBUTYL KETONE	108-83-8
ISOBUTYL MERCAPTAN	C4H10S	ISOBUTYL MERCAPTAN	513-44-0
ISOBUTYL METHYL CARBINOL	C6H14O	4-METHYL-2-PENTANOL	108-11-2
ISOBUTYL METHYL KETONE	C6H12O	METHYL ISOBUTYL KETONE	108-10-1
ISOBUTYL-METHYLKETON	C6H12O	METHYL ISOBUTYL KETONE	108-10-1
ISOBUTYLMETHYLMETHANOL	C6H14O	4-METHYL-2-PENTANOL	108-11-2
ISOBUTYL PROPENOATE	C7H12O2	ISOBUTYL ACRYLATE	106-63-8
ISOBUTYL-2-PROPENOATE	C7H12O2	ISOBUTYL ACRYLATE	106-63-8
ISOBUTYLTRIMETHYLETHANE	C8H18	2,2,4-TRIMETHYLPENTANE	540-84-1
ISOBUTYRALDEHYD	C4H8O	ISOBUTYRALDEHYDE	78-84-2
ISOBUTYRALDEHYDE	C4H8O	ISOBUTYRALDEHYDE	78-84-2
ISOBUTYRIC ACID	C4H8O2	ISOBUTYRIC ACID	79-31-2
ISOBUTYRIC ACID, ETHYL ESTER	C6H12O2	ETHYL ISOBUTYRATE	97-62-1
ISOBUTYRIC ACID, ISOBUTYL ESTER	C8H16O2	ISOBUTYL ISOBUTYRATE	97-85-8
ISOBUTYRIC ALDEHYDE	C4H8O	ISOBUTYRALDEHYDE	78-84-2
ISOBUTYRONE	C7H14O	DIISOPROPYL KETONE	565-80-0
ISOBUTYRONITRILE	C4H7N	ISOBUTYRONITRILE	78-82-0
ISOCYANATOCYCLOHEXANE	C7H11NO	CYCLOHEXYL ISOCYANATE	3173-53-3
ISOCYANIC ACID, BUTYL ESTER	C5H9NO	n-BUTYL ISOCYANATE	111-36-4
ISOCYANIC ACID, CYCLOHEXYL ESTER	C7H11NO	CYCLOHEXYL ISOCYANATE	3173-53-3
ISOCYANIC ACID, METHYL ESTER	C2H3NO	METHYL ISOCYANATE	624-83-9
ISOCYANIC ACID, METHYLPHENYLENE ESTER	C9H6N2O2	TOLUENE DIISOCYANATE	26471-62-5
ISOCYANIC ACID, PHENYL ESTER	C7H5NO	PHENYL ISOCYANATE	103-71-9
ISODECANOL	C10H22O	ISODECANOL	55505-26-5
ISODIPHENYLBENZENE	C18H14	m-TERPHENYL	92-06-8
ISOESTRAGOLE	C10H12O	ANETHOLE	104-46-1
ISOFORON	C9H14O	ISOPHORONE	78-59-1
ISOFORONE	C9H14O	ISOPHORONE	78-59-1
ISOHEXYL ALCOHOL	C6H14O	2-METHYL-1-PENTANOL	105-30-6
ISOHOL	C3H8O	ISOPROPANOL	67-63-0
ISOL	C6H14O2	HEXYLENE GLYCOL	107-41-5
ISONATE	C15H10N2O2	DIPHENYLMETHANE-4,4'-DIISOCYANATE	101-68-8
ISONITROPROPANE	C3H7NO2	2-NITROPROPANE	79-46-9
ISOOCTANE	C8H18	2,2,4-TRIMETHYLPENTANE	540-84-1
ISOPENTANE	C5H12	ISOPENTANE	78-78-4
ISOPENTANOIC ACID	C5H10O2	ISOVALERIC ACID	503-74-2
ISOPENTANOL	C5H12O	3-METHYL-1-BUTANOL	123-51-3
ISOPENTYL ACETATE	C7H14O2	ISOPENTYL ACETATE	123-92-2
ISOPENTYL ALCOHOL	C5H12O	3-METHYL-1-BUTANOL	123-51-3
ISOPENTYL ALCOHOL ACETATE	C7H14O2	ISOPENTYL ACETATE	123-92-2

Name (Synonym)	Formula	Name in Tables	CAS Reg. No
ISOPENTYL ISOVALERATE	C10H20O2	ISOPENTYL ISOVALERATE	659-70-1
ISOPENTYL METHYL KETONE	C7H14O	5-METHYL-2-HEXANONE	110-12-3
ISOPHORONE	C9H14O	ISOPHORONE	78-59-1
ISOPHTHALIC ACID	C8H6O4	ISOPHTHALIC ACID	121-91-5
ISOPHTHALIC ACID CHLORIDE	C8H4Cl2O2	ISOPHTHALOYL CHLORIDE	99-63-8
ISOPHTHALIC ACID DICHLORIDE	C8H4Cl2O2	ISOPHTHALOYL CHLORIDE	99-63-8
ISOPHTHALOYL CHLORIDE	C8H4Cl2O2	ISOPHTHALOYL CHLORIDE	99-63-8
ISOPHTHALOYL DICHLORIDE	C8H4Cl2O2	ISOPHTHALOYL CHLORIDE	99-63-8
ISOPHTHALYL DICHLORIDE	C8H4Cl2O2	ISOPHTHALOYL CHLORIDE	99-63-8
ISOPRENE	C5H8	ISOPRENE	78-79-5
ISOPRENE, INHIBITED	C5H8	2-METHYL-1,3-BUTADIENE	78-79-5
ISOPROPANOL	C3H8O	ISOPROPANOL	67-63-0
ISOPROPANOLAMINE	C3H9NO	1-AMINO-2-PROPANOL	78-96-6
ISOPROPENE CYANIDE	C4H5N	METHACRYLONITRILE	126-98-7
ISOPROPENIL-BENZOLO	C9H1O	alpha-METHYLSTYRENE	98-83-9
ISOPROPENYL-BENZEEN	C9H1O	alpha-METHYLSTYRENE	98-83-9
ISOPROPENYLBENZENE	C9H1O	alpha-METHYLSTYRENE	98-83-9
ISOPROPENYL-BENZOL	C9H1O	alpha-METHYLSTYRENE	98-83-9
(+)-4-ISOPROPENYL-1-METHYLCYCLOHEXENE	C10H16	D-LIMONENE	5989-27-5
ISOPROPENYLNITRILE	C4H5N	METHACRYLONITRILE	126-98-7
ISOPRILAMINA	C3H9N	ISOPROPYLAMINE	75-31-0
ISOPROPILBENZENE	C9H12	CUMENE	98-82-8
ISOPROPILE (ACETATO di)	C5H10O2	ISOPROPYL ACETATE	108-21-4
2-ISOPROPOXYPROPANE	C6H14O	DIISOPROPYL ETHER	108-20-3
ISOPROPYLACETAAT	C5H10O2	ISOPROPYL ACETATE	108-21-4
ISOPROPYLACETAT	C5H10O2	ISOPROPYL ACETATE	108-21-4
ISOPROPYL ACETATE	C5H10O2	ISOPROPYL ACETATE	108-21-4
ISOPROPYL (ACETATE d)	C5H10O2	ISOPROPYL ACETATE	108-21-4
ISOPROPYLACETIC ACID	C5H10O2	ISOVALERIC ACID	503-74-2
ISOPROPYLACETONE	C6H12O	METHYL ISOBUTYL KETONE	108-10-1
ISO-PROPYLALKOHOL	C3H8O	ISOPROPANOL	67-63-0
ISOPROPYLAMINE	C3H9N	ISOPROPYLAMINE	75-31-0
ISOPROPYLBENZEEN	C9H12	CUMENE	98-82-8
ISOPROPYL BENZENE	C9H12	CUMENE	98-82-8
ISOPROPYLBENZENE PEROXIDE	C18H22O2	DICUMYL PEROXIDE	80-43-3
ISOPROPYLBENZOL	C9H12	CUMENE	98-82-8
ISOPROPYL BROMIDE	C3H7Br	2-BROMOPROPANE	75-26-3
ISOPROPYLCARBINOL	C4H10O	ISOBUTANOL	78-83-1
ISOPROPYL CHLORIDE	C3H7Cl	ISOPROPYL CHLORIDE	75-29-6
ISOPROPYL CYANIDE	C4H7N	ISOBUTYRONITRILE	78-82-0
ISOPROPYLCYCLOHEXANE	C9H18	ISOPROPYLCYCLOHEXANE	696-29-7
ISOPROPYL DIMETHYL CARBINOL	C6H14O	2-METHYL-1-PENTANOL	105-30-6
ISOPROPYLESTER KYSELINY OCTOVE	C5H10O2	ISOPROPYL ACETATE	108-21-4
ISOPROPYLFORMIC ACID	C4H8O2	ISOBUTYRIC ACID	79-31-2
4-ISOPROPYLHEPTANE	C10H22	4-ISOPROPYLHEPTANE	----------
ISOPROPYLIDENEACETONE	C6H10O	MESITYL OXIDE	141-79-7
ISOPROPYL IODIDE	C3H7I	ISOPROPYL IODIDE	75-30-9
ISOPROPYL KETONE	C7H14O	DIISOPROPYL KETONE	565-80-0
ISOPROPYLKYANID	C4H7N	ISOBUTYRONITRILE	78-82-0
ISOPROPYL MERCAPTAN	C3H8S	ISOPROPYL MERCAPTAN	75-33-2
4-ISOPROPYL-1-METHYL-1,5-CYCLOHEXADIENE	C10H16	alpha-PHELLANDRENE	99-83-2
5-ISOPROPYL-2-METHYL-1,3-CYCLOHEXADIENE	C10H16	alpha-PHELLANDRENE	99-83-2
ISOPROPYL METHYL ETHER	C4H10O	METHYL ISOPROPYL ETHER	598-53-8
ISOPROPYL METHYL KETONE	C5H10O	METHYL ISOPROPYL KETONE	563-80-4
4-ISOPROPYL-1-METHYLBENZENE	C10H14	p-CYMENE	99-87-6
3-ISOPROPYL-2-METHYLHEXANE	C10H22	3-ISOPROPYL-2-METHYLHEXANE	----------
ISOPROPYL-METHYL-SULFIDE	C4H10S	ISOPROPYL-METHYL-SULFIDE	1551-21-9
ISOPROPYL-NITRATE	C3H7NO3	ISOPROPYL-NITRATE	2902-96-7
ISOPROPYL NITRILE	C4H7N	ISOBUTYRONITRILE	78-82-0
13-ISOPROPYLPODOCARPA-7,13-DIEN-15-OIC ACID	C20H30O2	ABIETIC ACID	514-10-3
ISOPROPYL-TERT-BUTYL-ETHER	C7H16O	ISOPROPYL-TERT-BUTYL-ETHER	1860-27-1
ISOPROPYLTHIOL	C3H8S	ISOPROPYL MERCAPTAN	75-33-2
m-ISOPROPYLTOLUENE	C10H14	m-CYMENE	535-77-3
o-ISOPROPYLTOLUENE	C10H14	o-CYMENE	527-84-4
p-ISOPROPYLTOLUENE	C10H14	p-CYMENE	99-87-6
ISOQUINOLINE	C9H7N	ISOQUINOLINE	119-65-3
ISOTHIOCYANIC-ACID	CHNS	ISOTHIOCYANIC-ACID	3129-90-6
ISOTRON 11	CCl3F	TRICHLOROFLUOROMETHANE	75-69-4
ISOVALERIANIC AICD	C5H10O2	ISOVALERIC ACID	503-74-2
ISOVALERIC ACID	C5H10O2	ISOVALERIC ACID	503-74-2
ISOVALERONE	C9H18O	DIISOBUTYL KETONE	108-83-8
ITACONIC ACID	C5H6O4	ITACONIC ACID	97-65-4
IVALON	CH2O	FORMALDEHYDE	50-00-0
IZOFORON	C9H14O	ISOPHORONE	78-59-1
IZOPROPYLOWY ETER	C6H14O	DIISOPROPYL ETHER	108-20-3
JAPAN CAMPHOR	C10H16O	CAMPHOR	76-22-2

Name (Synonym)	Formula	Name in Tables	CAS Reg. No
JEFFERSOL DB	C8H18O3	DIETHYLENE GLYCOL MONOBUTYL ETHER	112-34-5
JEFFERSOL EE	C4H10O2	2-ETHOXYETHANOL	110-80-5
JEFFERSOL EM	C3H8O2	2-METHOXYETHANOL	109-86-4
JODETHAN	C2H5I	ETHYL IODIDE	75-03-6
JOD-METHAN	CH3I	METHYL IODIDE	74-88-4
JOODMETHAAN	CH3I	METHYL IODIDE	74-88-4
JULINS CARBON CHLORIDE	C6Cl6	HEXACHLOROBENZENE	118-74-1
JZF	C18H16N2	N,N'-DIPHENYL-p-PHENYLENEDIAMINE	74-31-7
KAISER CHEMICALS 11	C2Cl3F3	1,1,2-TRICHLOROTRIFLUOROETHANE	76-13-1
KAISER CHEMICALS 12	CCl2F2	DICHLORODIFLUOROMETHANE	75-71-8
KAM 1000	C18H36O2	STEARIC ACID	57-11-4
KAM 2000	C18H36O2	STEARIC ACID	57-11-4
KAM 3000	C18H36O2	STEARIC ACID	57-11-4
KAMPFER	C10H16O	CAMPHOR	76-22-2
e-KAPROLAKTAM	C6H11NO	epsilon-CAPROLACTAM	105-60-2
KARBANIL	C7H5NO	PHENYL ISOCYANATE	103-71-9
KARION	C6H14O6	SORBITOL	50-70-4
KARSAN	CH2O	FORMALDEHYDE	50-00-0
KAYAFUME	CH3Br	METHYL BROMIDE	74-83-9
KAYTRATE	C5H8N4O12	PENTAERYTHRITOL TETRANITRATE	78-11-5
KELENE	C2H5Cl	ETHYL CHLORIDE	75-00-3
KEMAMINE P690	C12H27N	DODECYLAMINE	124-22-1
KEMESTER 105	C19H36O2	METHYL OLEATE	112-62-9
KEMESTER 115	C19H36O2	METHYL OLEATE	112-62-9
KEMESTER 205	C19H36O2	METHYL OLEATE	112-62-9
KEMESTER 213	C19H36O2	METHYL OLEATE	112-62-9
KERALYT	C7H6O3	SALICYLIC ACID	69-72-7
KESSCO BSC	C22H44O2	n-BUTYL STEARATE	123-95-5
KESSCOFLEX BS	C22H44O2	n-BUTYL STEARATE	123-95-5
KESSCOFLEX TRA	C9H14O6	GLYCERYL TRIACETATE	102-76-1
KETENE	C2H2O	KETENE	463-51-4
KETOCYCLOPENTANE	C5H8O	CYCLOPENTANONE	120-92-3
KETO-ETHYLENE	C2H2O	KETENE	463-51-4
3-KETO-I-GULOFURANOLACTONE	C6H8O6	ASCORBIC ACID	50-81-7
KETOHEXAMETHYLENE	C6H10O	CYCLOHEXANONE	108-94-1
2-KETOHEXAMETHYLENIMINE	C6H11NO	epsilon-CAPROLACTAM	105-60-2
KETOLE	C8H7N	INDOLE	120-72-9
KETONE, DIMETHYL	C3H6O	ACETONE	67-64-1
KETONE, METHYL ISOAMYL	C7H14O	5-METHYL-2-HEXANONE	110-12-3
KETONE METHYL PHENYL	C8H8O	ACETOPHENONE	98-86-2
KETONE PROPANE	C3H6O	ACETONE	67-64-1
KETOPENTAMETHYLENE	C5H8O	CYCLOPENTANONE	120-92-3
b-KETOPROPANE	C3H6O	ACETONE	67-64-1
a-KETOPROPIONIC ACID	C3H4O3	PYRUVIC ACID	127-17-3
I-3-KETOTHREOHEXURONIC ACID LACTONE	C6H8O6	ASCORBIC ACID	50-81-7
2-KETO-1,7,7-TRIMETHYLNORCAMPHANE	C10H16O	CAMPHOR	76-22-2
g-KETOVALERIC ACID	C5H8O3	LEVULINIC ACID	123-76-2
4-KETOVALERIC ACID	C5H8O3	LEVULINIC ACID	123-76-2
KF 994	C8H24O4Si4	OCTAMETHYLCYCLOTETRASILOXANE	556-67-2
KHLADON 113	C2Cl3F3	1,1,2-TRICHLOROTRIFLUOROETHANE	76-13-1
KHLADON 114B2	C2Br2F4	1,2-DIBROMOTETRAFLUOROETHANE	124-73-2
KHLADON 744	CO2	CARBON DIOXIDE	124-38-9
KODAFLEX DBS	C18H34O4	DIBUTYL SEBACATE	109-43-3
KODAFLEX TRIACETIN	C9H14O6	GLYCERYL TRIACETATE	102-76-1
KOHLENDIOXYD	CO2	CARBON DIOXIDE	124-38-9
KOHLENDISULFID	CS2	CARBON DISULFIDE	75-15-0
KOHLENMONOXID	CO	CARBON MONOXIDE	630-08-0
KOHLENOXYD	CO	CARBON MONOXIDE	630-08-0
KOHLENSAEURE	CO2	CARBON DIOXIDE	124-38-9
2,4,6-KOLLIDIN	C8H11N	2,4,6-TRIMETHYLPYRIDINE	108-75-8
KOOLMONOXYDE	CO	CARBON MONOXIDE	630-08-0
KOOLSTOFDISULFIDE	CS2	CARBON DISULFIDE	75-15-0
KOOLSTOFOXYCHLORIDE	CCl2O	PHOSGENE	75-44-5
m-KRESOL	C7H8O	m-CRESOL	108-39-4
o-KRESOL	C7H8O	o-CRESOL	95-48-7
p-KRESOL	C7H8O	p-CRESOL	106-44-5
KWAS METANIOWY	CH2O2	FORMIC ACID	64-18-6
KYSELINA ADIPOVA	C6H10O4	ADIPIC ACID	124-04-9
KYSELINA AKRYLOVA	C3H4O2	ACRYLIC ACID	79-10-7
KYSELINA BENZOOVA	C7H6O2	BENZOIC ACID	65-85-0
KYSELINA o-CHLORBENZOOVA	C7H5ClO2	o-CHLOROBENZOIC ACID	118-91-2
KYSELINA CHLOROCTOVA	C2H3ClO2	CHLOROACETIC ACID	79-11-8
KYSELINA CITRONOVA	C6H8O7	CITRIC ACID	77-92-9
KYSELINA DICHLOROCTOVA	C2H2Cl2O2	DICHLOROACETIC ACID	79-43-6
KYSELINA ISOFTALOVA	C8H6O4	ISOPHTHALIC ACID	121-91-5
KYSELINA ISOMASELNA	C4H8O2	ISOBUTYRIC ACID	79-31-2

Name (Synonym)	Formula	Name in Tables	CAS Reg. No
KYSELINA JANTAROVA	C4H6O4	SUCCINIC ACID	110-15-6
KYSELINA METHAKRYLOVA	C4H6O2	METHACRYLIC ACID	79-41-4
KYSELINA METHANSULFONOVA	CH4O3S	METHANESULFONIC ACID	75-75-2
KYSELINA PROPIONOVA	C3H6O2	PROPIONIC ACID	79-09-4
KYSELINA TERFTALOVA	C8H6O4	TEREPHTHALIC ACID	100-21-0
KYSELINA THIOOCTOVA	C2H4OS	THIOACETIC-ACID	507-09-5
KYSELINA TRIFLUOROCTOVA	C2HF3O2	TRIFLUOROACETIC ACID	76-05-1
LACTATE dETHYLE	C5H10O3	ETHYL LACTATE	97-64-3
LACTIC ACID	C3H6O3	LACTIC ACID	598-82-3
LACTONITRILE	C3H5NO	LACTONITRILE	78-97-7
LAEVULIC ACID	C5H8O3	LEVULINIC ACID	123-76-2
LAEVULINIC ACID	C5H8O3	LEVULINIC ACID	123-76-2
lALCOOL n-HEPTYLIQUE PRIMAIRE	C7H16O	1-HEPTANOL	111-70-6
LAM	C4H7NO	2-PYRROLIDONE	616-45-5
LAUREL CAMPHOR	C10H16O	CAMPHOR	76-22-2
LAURIC ACID, METHYL ESTER	C13H26O2	METHYL DODECANOATE	111-82-0
LAURIC ALCOHOL	C12H26O	1-DODECANOL	112-53-8
LAURINAMINE	C12H27N	DODECYLAMINE	124-22-1
LAURINIC ALCOHOL	C12H26O	1-DODECANOL	112-53-8
LAUROSTEARIC ACID	C12H24O2	n-DODECANOIC ACID	143-07-7
LAURYL ALCOHOL	C12H26O	1-DODECANOL	112-53-8
LAURYL 24	C12H26O	1-DODECANOL	112-53-8
n-LAURYL ALCOHOL, PRIMARY	C12H26O	1-DODECANOL	112-53-8
LAURYL ALDEHYDE	C12H24O	1-DODECANAL	112-54-9
LAURYLAMINE	C12H27N	DODECYLAMINE	124-22-1
n-LAURYLAMINE	C12H27N	DODECYLAMINE	124-22-1
m-LAURYL MERCAPTAN	C12H26S	n-DODECYL MERCAPTAN	112-55-0
LEDON 11	CCl3F	TRICHLOROFLUOROMETHANE	75-69-4
LEDON 114	C2Cl2F4	1,2-DICHLOROTETRAFLUOROETHANE	76-14-2
LEINOLEIC ACID	C18H32O2	LINOLEIC ACID	60-33-3
LEPARGYLIC ACID	C9H16O4	AZELAIC ACID	123-99-9
LEUCOL	C9H7N	QUINOLINE	91-22-5
LEUCOLINE	C9H7N	QUINOLINE	91-22-5
LEUKOL	C9H7N	QUINOLINE	91-22-5
LEVULIC ACID	C5H8O3	LEVULINIC ACID	123-76-2
LEVULINIC ACID	C5H8O3	LEVULINIC ACID	123-76-2
LICHENIC ACID	C4H4O4	FUMARIC ACID	110-17-8
D-LIMONENE	C10H16	D-LIMONENE	5989-27-5
d-(+)-LIMONENE	C10H16	D-LIMONENE	5989-27-5
(+)-R-LIMONENE	C10H16	D-LIMONENE	5989-27-5
LINOLEIC ACID	C18H32O2	LINOLEIC ACID	60-33-3
9,12-LINOLEIC ACID	C18H32O2	LINOLEIC ACID	60-33-3
LIOXIN	C8H8O3	VANILLIN	121-33-5
LIQUEFIED PETROLEUM GAS	C4H8	ISOBUTENE	115-11-7
LIQUID ETHYENE	C2H4	ETHYLENE	74-85-1
LOROL	C12H26O	1-DODECANOL	112-53-8
LOROL 20	C8H18O	1-OCTANOL	111-87-5
LOROL 22	C10H22O	1-DECANOL	112-30-1
LOROL 28	C18H38O	1-OCTADECANOL	112-92-5
LOSUNGSMITTEL APV	C6H14O3	2-(2-ETHOXYETHOXY)ETHANOL	111-90-0
LOWETRATE	C5H8N4O12	PENTAERYTHRITOL TETRANITRATE	78-11-5
LPG ETHYL MERCAPTAN 1010	C2H6S	ETHYL MERCAPTAN	75-08-1
LUMBRICAL	C4H10N2	PIPERAZINE	110-85-0
LUPERCO	C18H22O2	DICUMYL PEROXIDE	80-43-3
LUPEROX	C18H22O2	DICUMYL PEROXIDE	80-43-3
LUPEROX 500R	C18H22O2	DICUMYL PEROXIDE	80-43-3
LUPEROX 500T	C18H22O2	DICUMYL PEROXIDE	80-43-3
LUPROSIL	C3H6O2	PROPIONIC ACID	79-09-4
a-a'-LUTIDINE	C7H9N	2,6-DIMETHYLPYRIDINE	108-48-5
LUTOSOL	C3H8O	ISOPROPANOL	67-63-0
LUTROL-9	C2H6O2	ETHYLENE GLYCOL	107-21-1
LUXAN BLACK R	C12H12N2	p-AMINODIPHENYLAMINE	101-54-2
l-LYSINE	C6H14N2O2	LYSINE	56-87-1
l-(+)-LYSINE	C6H14N2O2	LYSINE	56-87-1
LYSINE	C6H14N2O2	LYSINE	56-87-1
LYSINE ACID	C6H14N2O2	LYSINE	56-87-1
MA-1214	C12H26O	1-DODECANOL	112-53-8
MACROGOL 400 BPC	C2H6O2	ETHYLENE GLYCOL	107-21-1
MAGNACIDE H	C3H4O	ACROLEIN	107-02-8
MALEIC ACID	C4H4O4	MALEIC ACID	110-16-7
MALEIC ACID ANHYDRIDE	C4H2O3	MALEIC ANHYDRIDE	108-31-6
MALEIC ACID, DIALLYL ESTER	C10H12O4	DIALLYL MALEATE	999-21-3
MALEIC ACID, DIBUTYL ESTER	C12H18	HEXAMETHYLBENZENE	87-85-4
MALEIC ACID, DIETHYL ESTER	C8H12O4	DIETHYL MALEATE	141-05-9
MALEIC ANHYDRIDE	C4H2O3	MALEIC ANHYDRIDE	108-31-6
MALEINIC ACID	C4H4O4	MALEIC ACID	110-16-7

Name (Synonym)	Formula	Name in Tables	CAS Reg. No
MALENIC ACID	C4H4O4	MALEIC ACID	110-16-7
MALIC ACID	C4H6O5	MALIC ACID	617-48-1
MALONIC ACID, DIETHYL ESTER	C7H12O4	DIETHYL MALONATE	105-53-3
MALONIC ACID ETHYL ESTER NITRILE	C5H7NO2	ETHYL CYANOACETATE	105-56-6
MALONIC DINITRILE	C3H2N2	MALONONITRILE	109-77-3
MALONIC ESTER	C7H12O4	DIETHYL MALONATE	105-53-3
MALONONITRILE	C3H2N2	MALONONITRILE	109-77-3
MAOH	C6H14O	4-METHYL-2-PENTANOL	108-11-2
MARTRATE-45	C5H8N4O12	PENTAERYTHRITOL TETRANITRATE	78-11-5
MATRICARIA CAMPHOR	C10H16O	CAMPHOR	76-22-2
MBK	C6H12O	2-HEXANONE	591-78-6
MCA	C2H3ClO2	CHLOROACETIC ACID	79-11-8
MCB	C6H5Cl	MONOCHLOROBENZENE	108-90-7
MCF	C2H3ClO2	METHYL CHLOROFORMATE	79-22-1
MDEA	C5H13NO2	METHYL DIETHANOLAMINE	105-59-9
MDI	C15H10N2O2	DIPHENYLMETHANE-4,4'-DIISOCYANATE	101-68-8
MEA	C2H7NO	MONOETHANOLAMINE	141-43-5
MECB	C5H12O3	2-(2-METHOXYETHOXY)ETHANOL	111-77-3
MECS	C3H8O2	2-METHOXYETHANOL	109-86-4
M.E.G.	C2H6O2	ETHYLENE GLYCOL	107-21-1
MEK	C4H8O	METHYL ETHYL KETONE	78-93-3
MEMPA	C6H18N3OP	HEXAMETHYL PHOSPHORAMIDE	680-31-9
d-p-MENTHA-1,8-DIENE	C10H16	D-LIMONENE	5989-27-5
p-MENTHA-1,3-DIENE	C10H16	alpha-TERPINENE	99-86-5
p-MENTHA-1,8-DIENE	C10H16	D-LIMONENE	5989-27-5
MEQUINOL	C7H8O2	p-METHOXYPHENOL	150-76-5
MERCAPTAN AMYLIQUE	C5H12S	n-PENTYL MERCAPTAN	110-66-7
MERCAPTAN METHYLIQUE	CH4S	METHYL MERCAPTAN	74-93-1
MERCAPTOCYCLOPENTANE	C5H10S	CYCLOPENTANETHIOL	1679-07-8
1-MERCAPTODODECANE	C12H26S	n-DODECYL MERCAPTAN	112-55-0
2-MERCAPTOPROPANE	C3H8S	ISOPROPYL MERCAPTAN	75-33-2
b-MERCAPTOPROPANOIC ACID	C3H6O2S	3-MERCAPTOPROPIONIC ACID	107-96-0
3-MERCAPTOPROPIONIC ACID	C3H6O2S	3-MERCAPTOPROPIONIC ACID	107-96-0
MERCURIALIN	CH5N	METHYLAMINE	74-89-5
MESITYLENE	C9H12	MESITYLENE	108-67-8
MESITYLOXID	C6H10O	MESITYL OXIDE	141-79-7
MESITYL OXIDE	C6H10O	MESITYL OXIDE	141-79-7
MESITYLOXYDE	C6H10O	MESITYL OXIDE	141-79-7
META TOLUYLENE DIAMINE	C7H10N2	TOLUENEDIAMINE	95-80-7
METACETONE	C5H10O	DIETHYL KETONE	96-22-0
METACETONIC ACID	C3H6O2	PROPIONIC ACID	79-09-4
METAKRYLAN METYLU	C5H8O2	METHYL METHACRYLATE	80-62-6
METANOLO	CH4O	METHANOL	67-56-1
METANTIOLO	CH4S	METHYL MERCAPTAN	74-93-1
METAPHENYLENEDIAMINE	C6H8N2	m-PHENYLENEDIAMINE	108-45-2
METHAANTHIOL	CH4S	METHYL MERCAPTAN	74-93-1
METHACETONE	C5H10O	DIETHYL KETONE	96-22-0
METHACIDE	C7H8	TOLUENE	108-88-3
METHACRALDEHYDE	C4H6O	METHACROLEIN	78-85-3
METHACROLEIN	C4H6O	METHACROLEIN	78-85-3
2-METHACRYLAMIDE	C4H7NO	2-METHACRYLAMIDE	79-39-0
METHACRYLATE de BUTYLE	C8H14O2	n-BUTYL METHACRYLATE	97-88-1
METHACRYLATE de METHYLE	C5H8O2	METHYL METHACRYLATE	80-62-6
METHACRYLIC ACID	C4H6O2	METHACRYLIC ACID	79-41-4
METHACRYLIC ACID, METHYL ESTER	C5H8O2	METHYL METHACRYLATE	80-62-6
METHACRYLIC ALDEHYDE	C4H6O	METHACROLEIN	78-85-3
METHACRYLIC AMIDE	C4H7NO	2-METHACRYLAMIDE	79-39-0
METHACRYLONITRILE	C4H5N	METHACRYLONITRILE	126-98-7
METHACRYLSAEUREBUTYLESTER	C8H14O2	n-BUTYL METHACRYLATE	97-88-1
METHACRYLSAEUREMETHYL ESTER	C5H8O2	METHYL METHACRYLATE	80-62-6
METHAKRYLALDEHYD	C4H6O	METHACROLEIN	78-85-3
METHANAL	CH2O	FORMALDEHYDE	50-00-0
METHANAMIDE	CH3NO	FORMAMIDE	75-12-7
METHANAMINE	CH5N	METHYLAMINE	74-89-5
METHANE	CH4	METHANE	74-82-8
METHANECARBONITRILE	C2H3N	ACETONITRILE	75-05-8
METHANECARBOTHIOLIC ACID	C2H4OS	THIOACETIC-ACID	507-09-5
METHANECARBOXAMIDE	C2H5NO	ACETAMIDE	60-35-5
METHANECARBOXYLIC ACID	C2H4O2	ACETIC ACID	64-19-7
METHANE, CYANO-	C2H3N	ACETONITRILE	75-05-8
METHANEDICARBOXYLIC ACID, DIETHYL ESTER	C7H12O4	DIETHYL MALONATE	105-53-3
METHANE DICHLORIDE	CH2Cl2	DICHLOROMETHANE	75-09-2
METHANE, PHENYL-	C7H8	TOLUENE	108-88-3
METHANESULFONIC ACID	CH4O3S	METHANESULFONIC ACID	75-75-2
METHANE TETRACHLORIDE	CCl4	CARBON TETRACHLORIDE	56-23-5
METHANE, TETRAFLUORO-	CF4	CARBON TETRAFLUORIDE	75-73-0

724

Name (Synonym)	Formula	Name in Tables	CAS Reg. No
METHANE TETRAMETHYLOL	C5H12O4	PENTAERYTHRITOL	115-77-5
METHANETHIOL	CH4S	METHYL MERCAPTAN	74-93-1
METHANE TRICHLORIDE	CHCl3	CHLOROFORM	67-66-3
METHANOIC ACID	CH2O2	FORMIC ACID	64-18-6
METHANOL	CH4O	METHANOL	67-56-1
METHANOLACETONITRILE	C3H5NO	HYDRACRYLONITRILE	109-78-4
METHANOL, BENZYL-	C8H10O	2-PHENYLETHANOL	60-12-8
METHANTHIOL	CH4S	METHYL MERCAPTAN	74-93-1
METHENYL TRIBROMIDE	CHBr3	TRIBROMOMETHANE	75-25-2
METHENYL TRICHLORIDE	CHCl3	CHLOROFORM	67-66-3
METHOLENE 2296	C13H26O2	METHYL DODECANOATE	111-82-0
METHOXYACETIC ACID	C3H6O3	METHOXYACETIC ACID	625-45-6
2-METHOXYACETIC ACID	C3H6O3	METHOXYACETIC ACID	625-45-6
METHOXYBENZENE	C7H8O	ANISOLE	100-66-3
2-METHOXYBENZOIC ACID	C8H8O3	METHYL SALICYLATE	119-36-8
o-METHOXYBENZOIC ACID	C8H8O3	METHYL SALICYLATE	119-36-8
METHOXYCARBONYL CHLORIDE	C2H3ClO2	METHYL CHLOROFORMATE	79-22-1
METHOXYCARBONYLETHYLENE	C4H6O2	METHYL ACRYLATE	96-33-3
METHOXYDIGLYCOL	C5H12O3	2-(2-METHOXYETHOXY)ETHANOL	111-77-3
METHOXYETHANE	C3H8O	METHYL ETHYL ETHER	540-67-0
2-METHOXYETHANOL	C3H8O2	2-METHOXYETHANOL	109-86-4
METHOXYETHENE	C3H6O	METHYL VINYL ETHER	107-25-5
2-(2-METHOXYETHOXY)ETHANOL	C5H12O3	2-(2-METHOXYETHOXY)ETHANOL	111-77-3
3-METHOXY-4-HYDROXYBENZALDEHYDE	C8H8O3	VANILLIN	121-33-5
b-METHOXY-b'-HYDROXYDIETHYL ETHER	C5H12O3	2-(2-METHOXYETHOXY)ETHANOL	111-77-3
2-METHOXY-2-METHYLPROPANE	C5H12O	METHYL tert-BUTYL ETHER	1634-04-4
p-METHOXY-b-METHYLSTYRENE	C10H12O	ANETHOLE	104-46-1
1-METHOXY-2-NITROBENZENE	C7H7NO3	o-NITROANISOLE	91-23-6
2-METHOXYNITROBENZENE	C7H7NO3	o-NITROANISOLE	91-23-6
o-METHOXYPHENOL	C7H8O2	GUAIACOL	90-05-1
p-METHOXYPHENOL	C7H8O2	p-METHOXYPHENOL	150-76-5
2-METHOXYPHENOL	C7H8O2	GUAIACOL	90-05-1
1-(p-METHOXYPHENYL)PROPENE	C10H12O	ANETHOLE	104-46-1
a-METHOXY PROPANE	C4H10O	METHYL-PROPYL-ETHER	557-17-5
1-METHOXYPROPANE	C4H10O	METHYL-PROPYL-ETHER	557-17-5
3-METHOXYPROPANENITRILE	C4H7NO	3-METHOXYPROPIONITRILE	110-67-8
3-METHOXYPROPANNITRIL	C4H7NO	3-METHOXYPROPIONITRILE	110-67-8
1-METHOXY-4-PROPENYLBENZENE	C10H12O	ANETHOLE	104-46-1
4-METHOXYPROPENYLBENZENE	C10H12O	ANETHOLE	104-46-1
3-METHOXYPROPIONITRILE	C4H7NO	3-METHOXYPROPIONITRILE	110-67-8
METHYLACETAAT	C3H6O2	METHYL ACETATE	79-20-9
METHYLACETALDEHYDE	C3H6O	n-PROPIONALDEHYDE	123-38-6
N-METHYLACETAMIDE	C3H7NO	N-METHYLACETAMIDE	79-16-3
METHYLACETAT	C3H6O2	METHYL ACETATE	79-20-9
METHYL ACETATE	C3H6O2	METHYL ACETATE	79-20-9
METHYL ACETIC ACID	C3H6O2	PROPIONIC ACID	79-09-4
METHYLACETIC ANHYDRIDE	C6H10O3	PROPIONIC ANHYDRIDE	123-62-6
METHYL ACETOACETATE	C5H8O3	METHYL ACETOACETATE	105-45-3
METHYL ACETONE	C4H8O	METHYL ETHYL KETONE	78-93-3
METHYL ACETYLACETONATE	C5H8O3	METHYL ACETOACETATE	105-45-3
METHYLACETYLENE	C3H4	METHYLACETYLENE	74-99-7
a-METHYLACROLEIN	C4H6O	METHACROLEIN	78-85-3
b-METHYL ACROLEIN	C4H6O	trans-CROTONALDEHYDE	123-73-9
2-METHYLACROLEIN	C4H6O	METHACROLEIN	78-85-3
METHYLACRYLALDEHYDE	C4H6O	METHACROLEIN	78-85-3
2-METHYLACRYLAMIDE	C4H7NO	2-METHACRYLAMIDE	79-39-0
METHYL ACRYLATE	C4H6O2	METHYL ACRYLATE	96-33-3
a-METHYL ACRYLIC AMIDE	C4H7NO	2-METHACRYLAMIDE	79-39-0
a-METHYLACRYLIC ACID	C4H6O2	METHACRYLIC ACID	79-41-4
a-METHYLACRYLONITRILE	C4H5N	METHACRYLONITRILE	126-98-7
METHYLAL	C3H8O2	METHYLAL	109-87-5
METHYL ALDEHYDE	CH2O	FORMALDEHYDE	50-00-0
METHYLALKOHOL	CH4O	METHANOL	67-56-1
METHYLAMINE	CH5N	METHYLAMINE	74-89-5
METHYLAMINEN	CH5N	METHYLAMINE	74-89-5
(METHYLAMINO)BENZENE	C7H9N	N-METHYLANILINE	100-61-8
N-METHYLAMINOBENZENE	C7H9N	N-METHYLANILINE	100-61-8
1-METHYL-2-AMINOBENZENE	C7H9N	o-TOLUIDINE	95-53-4
2-METHYL-1-AMINOBENZENE	C7H9N	o-TOLUIDINE	95-53-4
N-METHYLAMINODIGLYCOL	C5H13NO2	METHYL DIETHANOLAMINE	105-59-9
N-METHYLAMINOETHANOL	C3H9NO	METHYLETHANOLAMINE	109-83-1
b-(METHYLAMINO)ETHANOL	C3H9NO	METHYLETHANOLAMINE	109-83-1
METHYLAMYL ALCOHOL	C6H14O	2-METHYL-1-PENTANOL	105-30-6
METHYL AMYL ALCOHOL	C6H14O	4-METHYL-2-PENTANOL	108-11-2
METHYL AMYL CARBINOL	C7H16O	2-HEPTANOL	543-49-7
METHYL-AMYL-CETONE	C7H14O	2-HEPTANONE	110-43-0

Name (Synonym)	Formula	Name in Tables	CAS Reg. No
METHYL AMYL KETONE	C7H14O	2-HEPTANONE	110-43-0
m-METHYLANILINE	C7H9N	m-TOLUIDINE	108-44-1
N-METHYLANILINE	C7H9N	N-METHYLANILINE	100-61-8
o-METHYLANILINE	C7H9N	o-TOLUIDINE	95-53-4
p-METHYLANILINE	C7H9N	p-TOLUIDINE	106-49-0
2-METHYLANILINE	C7H9N	o-TOLUIDINE	95-53-4
3-METHYLANILINE	C7H9N	m-TOLUIDINE	108-44-1
4-METHYLANILINE	C7H9N	p-TOLUIDINE	106-49-0
2-METHYLAZACYCLOPROPANE	C3H7N	PROPYLENEIMINE	75-55-8
2-METHYLAZIRIDINE	C3H7N	PROPYLENEIMINE	75-55-8
m-METHYLBENZENAMINE	C7H9N	m-TOLUIDINE	108-44-1
N-METHYLBENZENAMINE	C7H9N	N-METHYLANILINE	100-61-8
o-METHYLBENZENAMINE	C7H9N	o-TOLUIDINE	95-53-4
p-METHYLBENZENAMINE	C7H9N	p-TOLUIDINE	106-49-0
2-METHYLBENZENAMINE	C7H9N	o-TOLUIDINE	95-53-4
3-METHYLBENZENAMINE	C7H9N	m-TOLUIDINE	108-44-1
4-METHYLBENZENAMINE	C7H9N	p-TOLUIDINE	106-49-0
METHYLBENZENE	C7H8	TOLUENE	108-88-3
4-METHYL-1,3-BENZENEDIAMINE	C7H10N2	TOLUENEDIAMINE	95-80-7
METHYL BENZOATE	C8H8O2	METHYL BENZOATE	93-58-3
2-METHYLBENZOFURAN	C9H8O	2-METHYLBENZOFURAN	4265-25-2
o-METHYLBENZOIC ACID	C8H8O2	o-TOLUIC ACID	118-90-1
2-METHYLBENZOIC ACID	C8H8O2	o-TOLUIC ACID	118-90-1
METHYLBENZOL	C7H8	TOLUENE	108-88-3
METHYLBIS(2-HYDROXYETHYL)AMINE	C5H13NO2	METHYL DIETHANOLAMINE	105-59-9
b-METHYLBIVINYL	C5H8	2-METHYL-1,3-BUTADIENE	78-79-5
METHYL BROMIDE	CH3Br	METHYL BROMIDE	74-83-9
2-METHYLBUTADIENE	C5H8	2-METHYL-1,3-BUTADIENE	78-79-5
2-METHYL-1,3-BUTADIENE	C5H8	2-METHYL-1,3-BUTADIENE	78-79-5
3-METHYL-1,2-BUTADIENE	C5H8	3-METHYL-1,2-BUTADIENE	598-25-4
2-METHYLBUTANE	C5H12	ISOPENTANE	78-78-4
2-METHYL-2-BUTANETHIOL	C5H12S	2-METHYL-2-BUTANETHIOL	1679-09-0
METHYL n-BUTANOATE	C5H10O2	METHYL n-BUTYRATE	623-42-7
3-METHYLBUTANOIC ACID	C5H10O2	ISOVALERIC ACID	503-74-2
3-METHYLBUTANOIC ACID, ETHYL ESTER	C7H14O2	ETHYL ISOVALERATE	108-64-5
3-METHYLBUTAN-1-OL	C5H12O	3-METHYL-1-BUTANOL	123-51-3
2-METHYL BUTANOL-2	C5H12O	tert-PENTYL-ALCOHOL	75-85-4
2-METHYL-2-BUTANOL	C5H12O	tert-PENTYL-ALCOHOL	75-85-4
2-METHYL-1-BUTANOL	C5H12O	2-METHYL-1-BUTANOL	137-32-6
2-METHYLBUTANOL	C5H12O	2-METHYL-1-BUTANOL	137-32-6
3-METHYLBUTAN-3-OL	C5H12O	2-METHYL-2-BUTANOL	75-85-4
3-METHYL BUTANOL	C5H12O	3-METHYL-1-BUTANOL	123-51-3
2-METHYL-4-BUTANOL	C5H12O	3-METHYL-1-BUTANOL	123-51-3
3-METHYL-1-BUTANOL	C5H12O	3-METHYL-1-BUTANOL	123-51-3
3-METHYL-2-BUTANOL	C5H12O	3-METHYL-2-BUTANOL	598-75-4
3-METHYL-2-BUTANONE	C5H10O	METHYL ISOPROPYL KETONE	563-80-4
3-METHYL BUTAN-2-ONE	C5H10O	METHYL ISOPROPYL KETONE	563-80-4
2-METHYL-1-BUTENE	C5H10	2-METHYL-1-BUTENE	563-46-2
3-METHYL-1-BUTENE	C5H10	3-METHYL-1-BUTENE	563-45-1
2-METHYL-2-BUTENE	C5H10	2-METHYL-2-BUTENE	513-35-9
2-METHYL-2-BUTENEDIOIC ACID	C5H6O4	CITRACONIC ACID	498-23-7
cis-METHYLBUTENEDIOIC ACID	C5H6O4	CITRACONIC ACID	498-23-7
(Z)-2-METHYL-2-BUTENEDIOIC ACID	C5H6O4	CITRACONIC ACID	498-23-7
2-METHYL-1-BUTENE-3-YNE	C5H6	2-METHYL-1-BUTENE-3-YNE	78-80-8
3-METHYL-3-BUTEN-2-ON	C5H8O	METHYL ISOPROPENYL KETONE	814-78-8
2-METHYL-1-BUTEN-3-ONE	C5H8O	METHYL ISOPROPENYL KETONE	814-78-8
3-METHYL-1-BUTYL ACETATE	C7H14O2	ISOPENTYL ACETATE	123-92-2
3-METHYLBUTYL ACETATE	C7H14O2	ISOPENTYL ACETATE	123-92-2
2-METHYL-BUTYLACRYLAAT	C8H14O2	n-BUTYL METHACRYLATE	97-88-1
2-METHYL-BUTYLACRYLAT	C8H14O2	n-BUTYL METHACRYLATE	97-88-1
3-METHYLBUTYL ETHANOATE	C7H14O2	ISOPENTYL ACETATE	123-92-2
METHYL sec-BUTYL ETHER	C5H12O	METHYL sec-BUTYL ETHER	6795-87-5
METHYL tert-BUTYL ETHER	C5H12O	METHYL tert-BUTYL ETHER	1634-04-4
METHYL n-BUTYL KETONE	C6H12O	2-HEXANONE	591-78-6
3-METHYL-1-BUTYNE	C5H8	3-METHYL-1-BUTYNE	598-23-2
METHYL n-BUTYRATE	C5H10O2	METHYL n-BUTYRATE	623-42-7
METHYL BUTYRATE	C5H10O2	METHYL n-BUTYRATE	623-42-7
2-METHYLBUTYRIC ACID	C5H10O2	2-METHYLBUTYRIC ACID	600-07-7
3-METHYLBUTYRIC ACID	C5H10O2	ISOVALERIC ACID	503-74-2
b-METHYLBUTYRIC ACID	C5H10O2	ISOVALERIC ACID	503-74-2
3-METHYLBUTYRIC ACID, ETHYL ESTER	C7H14O2	ETHYL ISOVALERATE	108-64-5
METHYL CARBITOL	C5H12O3	2-(2-METHOXYETHOXY)ETHANOL	111-77-3
METHYL-4-CARBOMETHOXY BENZOATE	C10H10O4	DIMETHYL TEREPHTHALATE	120-61-6
METHYLCATECHOL	C7H8O2	GUAIACOL	90-05-1
METHYLCHLORID	CH3Cl	METHYL CHLORIDE	74-87-3
METHYL CHLORIDE	CH3Cl	METHYL CHLORIDE	74-87-3

Name (Synonym)	Formula	Name in Tables	CAS Reg. No
METHYL CHLOROACETATE	C3H5ClO2	METHYL CHLOROACETATE	96-34-4
1-METHYL-2-CHLOROBENZENE	C7H7Cl	o-CHLOROTOLUENE	95-49-8
2-METHYLCHLOROBENZENE	C7H7Cl	o-CHLOROTOLUENE	95-49-8
METHYL CHLOROFORMATE	C2H3ClO2	METHYL CHLOROFORMATE	79-22-1
METHYL CHLOROSILANE	CH5ClSi	METHYL CHLOROSILANE	993-00-0
6-METHYL-m-CRESOL	C8H10O	2,5-XYLENOL	95-87-4
p-METHYL-CUMENE	C10H14	p-CYMENE	99-87-6
METHYL CYANIDE	C2H3N	ACETONITRILE	75-05-8
METHYL 2-CYANOACETATE	C4H5NO2	METHYL CYANOACETATE	105-34-0
METHYL CYANOACETATE	C4H5NO2	METHYL CYANOACETATE	105-34-0
METHYL CYANOETHANOATE	C4H5NO2	METHYL CYANOACETATE	105-34-0
METHYL b-CYANOETHYL ETHER	C4H7NO	3-METHOXYPROPIONITRILE	110-67-8
METHYLCYCLOHEXANE	C7H14	METHYLCYCLOHEXANE	108-87-2
1-METHYLCYCLOHEXANOL	C7H14O	1-METHYLCYCLOHEXANOL	590-67-0
cis-2-METHYLCYCLOHEXANOL	C7H14O	cis-2-METHYLCYCLOHEXANOL	7443-70-1
trans-2-METHYLCYCLOHEXANOL	C7H14O	trans-2-METHYLCYCLOHEXANOL	7443-52-9
cis-3-METHYLCYCLOHEXANOL	C7H14O	cis-3-METHYLCYCLOHEXANOL	5454-79-5
trans-3-METHYLCYCLOHEXANOL	C7H14O	trans-3-METHYLCYCLOHEXANOL	7443-55-2
cis-4-METHYLCYCLOHEXANOL	C7H14O	cis-4-METHYLCYCLOHEXANOL	7731-28-4
trans-4-METHYLCYCLOHEXANOL	C7H14O	trans-4-METHYLCYCLOHEXANOL	7731-29-5
N-METHYLCYCLOHEXYLAMINE	C7H15N	N-METHYLCYCLOHEXYLAMINE	100-60-7
METHYLCYCLOHEXYLAMINE	C7H15N	N-METHYLCYCLOHEXYLAMINE	100-60-7
METHYLCYCLOPENTADIENE	C6H8	METHYLCYCLOPENTADIENE	26519-91-5
METHYLCYCLOPENTANE	C6H12	METHYLCYCLOPENTANE	96-37-7
1-METHYLCYCLOPENTENE	C6H10	1-METHYLCYCLOPENTENE	693-89-0
3-METHYLCYCLOPENTENE	C6H10	3-METHYLCYCLOPENTENE	1120-62-3
4-METHYLCYCLOPENTENE	C6H10	4-METHYLCYCLOPENTENE	1759-81-5
METHYL DICHLOROSILANE	CH4Cl2Si	METHYL DICHLOROSILANE	75-54-7
METHYL-DICHLORSILAN	CH4Cl2Si	METHYL DICHLOROSILANE	75-54-7
N-METHYLDIETHANOLAMINE	C5H13NO2	METHYL DIETHANOLAMINE	105-59-9
METHYL DIETHANOLAMINE	C5H13NO2	METHYL DIETHANOLAMINE	105-59-9
N-METHYLDIETHANOLIMINE	C5H13NO2	METHYL DIETHANOLAMINE	105-59-9
METHYL 1,1-DIMETHYLETHYL ETHER	C5H12O	METHYL tert-BUTYL ETHER	1634-04-4
4-METHYL-1,2-DINITROBENZENE	C7H6N2O4	3,4-DINITROTOLUENE	610-39-9
2-METHYL-1,3-DINITROBENZENE	C7H6N2O4	2,6-DINITROTOLUENE	606-20-2
2-METHYL-1,4-DINITROBENZENE	C7H6N2O4	2,5-DINITROTOLUENE	619-15-8
1-METHYL-2,4-DINITROBENZENE	C7H6N2O4	2,4-DINITROTOLUENE	121-14-2
1-METHYL-3,5-DINITRO-BENZENE	C7H6N2O4	3,5-DINITROTOLUENE	618-85-9
METHYL DODECANOATE	C13H26O2	METHYL DODECANOATE	111-82-0
METHYL DODECYLATE	C13H26O2	METHYL DODECANOATE	111-82-0
METHYLE	C2H4O2	METHYL FORMATE	107-31-3
METHYLENE BICHLORIDE	CH2Cl2	DICHLOROMETHANE	75-09-2
2,2'-METHYLENEBIPHENYL	C13H10	FLUORENE	86-73-7
1,1-METHYLENEBIS(4-ISOCYANATOBENZENE)	C15H10N2O2	DIPHENYLMETHANE-4,4'-DIISOCYANATE	101-68-8
METHYLENEBIS(4-ISOCYANATOBENZENE)	C15H10N2O2	DIPHENYLMETHANE-4,4'-DIISOCYANATE	101-68-8
METHYLENE BROMIDE	CH2Br2	DIBROMOMETHANE	74-95-3
METHYLENE CHLOROBROMIDE	CH2BrCl	BROMOCHLOROMETHANE	74-97-5
METHYLENE CYANIDE	C3H2N2	MALONONITRILE	109-77-3
METHYLENE DIBROMIDE	CH2Br2	DIBROMOMETHANE	74-95-3
METHYLENE DICHLORIDE	CH2Cl2	DICHLOROMETHANE	75-09-2
METHYLENE DIIODIDE	CH2I2	DIIODOMETHANE	75-11-6
METHYLENE DIMETHYL ETHER	C3H8O2	METHYLAL	109-87-5
METHYLENE GLYCOL	CH2O	FORMALDEHYDE	50-00-0
METHYLENE IODIDE	CH2I2	DIIODOMETHANE	75-11-6
4-METHYLENE-2-OXETANONE	C4H4O2	DIKETENE	674-82-8
METHYLENE OXIDE	CH2O	FORMALDEHYDE	50-00-0
METHYLESTER KISELINY OCTOVE	C3H6O2	METHYL ACETATE	79-20-9
METHYLESTER KYSELINY BENZOOVE	C8H8O2	METHYL BENZOATE	93-58-3
METHYLESTER KYSELINY CHLOROCTOVE	C3H5ClO2	METHYL CHLOROACETATE	96-34-4
METHYLESTER KYSELINY METHAKRYLOVE	C5H8O2	METHYL METHACRYLATE	80-62-6
1-METHYLETHANETHIOL	C3H8S	ISOPROPYL MERCAPTAN	75-33-2
METHYL ETHANOATE	C3H6O2	METHYL ACETATE	79-20-9
N-METHYLETHANOLAMINE	C3H9NO	METHYLETHANOLAMINE	109-83-1
METHYLETHANOLAMINE	C3H9NO	METHYLETHANOLAMINE	109-83-1
METHYLETHENE	C3H6	PROPYLENE	115-07-1
1-METHYLETHYLAMINE	C3H9N	ISOPROPYLAMINE	75-31-0
METHYLETHYLBROMOMETHANE	C4H9Br	2-BROMOBUTANE	78-76-2
METHYLETHYLCARBINOL	C4H10O	sec-BUTANOL	78-92-2
1-METHYL-1-ETHYLCYCLOPENTANE	C8H16	1-METHYL-1-ETHYLCYCLOPENTANE	16747-50-5
METHYLETHYLENE	C3H6	PROPYLENE	115-07-1
METHYL ETHYLENE OXIDE	C3H6O	1,2-PROPYLENE OXIDE	75-56-9
METHYLETHYLENE GLYCOL	C3H8O2	1,2-PROPYLENE GLYCOL	57-55-6
2-METHYLETHYLENIMINE	C3H7N	PROPYLENEIMINE	75-55-8
METHYLETHYLENIMINE	C3H7N	PROPYLENEIMINE	75-55-8
METHYL ETHYL ETHER	C3H8O	METHYL ETHYL ETHER	540-67-0
METHYL ETHYL KETONE	C4H8O	METHYL ETHYL KETONE	78-93-3

Name (Synonym)	Formula	Name in Tables	CAS Reg. No
METHYLETHYLMETHANE	C4H10	n-BUTANE	106-97-8
METHYLETHYLOLAMINE	C3H9NO	METHYLETHANOLAMINE	109-83-1
3-METHYL-3-ETHYLPENTANE	C8H18	3-METHYL-3-ETHYLPENTANE	1067-08-9
N-(1-METHYLETHYL)-2-PROPANAMINE	C6H15N	DIISOPROPYLAMINE	108-18-9
METHYL FLUORIDE	CH3F	METHYL FLUORIDE	593-53-3
METHYLFLUOROFORM	C2H3F3	1,1,1-TRIFLUOROETHANE	420-46-2
N-METHYLFORMAMIDE	C2H5NO	N-METHYLFORMAMIDE	123-39-7
METHYLFORMAMIDE	C2H5NO	N-METHYLFORMAMIDE	123-39-7
METHYL FORMATE	C2H4O2	METHYL FORMATE	107-31-3
METHYLFORMIAAT	C2H4O2	METHYL FORMATE	107-31-3
METHYLFORMIAT	C2H4O2	METHYL FORMATE	107-31-3
METHYLGLUTARONITRILE	C6H8N2	METHYLGLUTARONITRILE	4553-62-2
METHYL GLYCOL	C3H8O2	1,2-PROPYLENE GLYCOL	57-55-6
2-METHYLHEPTANE	C8H18	2-METHYLHEPTANE	592-27-8
3-METHYLHEPTANE	C8H18	3-METHYLHEPTANE	589-81-1
4-METHYLHEPTANE	C8H18	4-METHYLHEPTANE	589-53-7
5-METHYLHEXAN-2-ONE	C7H14O	5-METHYL-2-HEXANONE	110-12-3
2-METHYLHEXANE	C7H16	2-METHYLHEXANE	591-76-4
3-METHYLHEXANE	C7H16	3-METHYLHEXANE	589-34-4
5-METHYL-1-HEXANOL	C7H16O	5-METHYL-1-HEXANOL	627-98-5
5-METHYL-2-HEXANONE	C7H14O	5-METHYL-2-HEXANONE	110-12-3
2-METHYL-5-HEXANONE	C7H14O	5-METHYL-2-HEXANONE	110-12-3
2-METHYL-1-HEXENE	C7H14	2-METHYL-1-HEXENE	6094-02-6
3-METHYL-1-HEXENE	C7H14	3-METHYL-1-HEXENE	3404-61-3
4-METHYL-1-HEXENE	C7H14	4-METHYL-1-HEXENE	3769-23-1
METHYL HEXYL KETONE	C8H16O	2-OCTANONE	111-13-7
METHYL HYDRIDE	CH4	METHANE	74-82-8
METHYL HYDROXIDE	CH4O	METHANOL	67-56-1
1-METHYL-4-HYDROXYBENZENE	C7H8O	p-CRESOL	106-44-5
METHYL-o-HYDROXYBENZOATE	C8H8O3	METHYL SALICYLATE	119-36-8
METHYL(b-HYDROXYETHYL)AMINE	C3H9NO	METHYLETHANOLAMINE	109-83-1
2-METHYL-2-p-HYDROXYPHENYLBUTANE	C11H16O	p-tert-AMYLPHENOL	80-46-6
N-METHYL-2,2'-IMINODIETHANOL	C5H13NO2	METHYL DIETHANOLAMINE	105-59-9
2,2'-(METHYLIMINO)DIETHANOL	C5H13NO2	METHYL DIETHANOLAMINE	105-59-9
N-METHYLIMINODIETHANOL	C5H13NO2	METHYL DIETHANOLAMINE	105-59-9
METHYLIMINODIETHANOL	C5H13NO2	METHYL DIETHANOLAMINE	105-59-9
1-METHYLINDENE	C10H10	1-METHYLINDENE	767-59-9
2-METHYLINDENE	C10H10	2-METHYLINDENE	2177-47-1
METHYL IODIDE	CH3I	METHYL IODIDE	74-88-4
METHYL ISOBUTENYL KETONE	C6H10O	MESITYL OXIDE	141-79-7
METHYLISOBUTYL CARBINOL	C6H14O	4-METHYL-2-PENTANOL	108-11-2
METHYL ISOBUTYL CARBINOL	C6H14O	2-METHYL-1-PENTANOL	105-30-6
METHYL-ISOBUTYL-CETONE	C6H12O	METHYL ISOBUTYL KETONE	108-10-1
METHYL ISOBUTYL ETHER	C5H120	METHYL ISOBUTYL ETHER	625-44-5
METHYLISOBUTYLKETON	C6H12O	METHYL ISOBUTYL KETONE	108-10-1
METHYL ISOBUTYL KETONE	C6H12O	METHYL ISOBUTYL KETONE	108-10-1
METHYL ISOCYANATE	C2H3NO	METHYL ISOCYANATE	624-83-9
METHYL ISOPROPENYL KETONE	C5H8O	METHYL ISOPROPENYL KETONE	814-78-8
METHYL ISOPROPENYL KETONE INHIBITED	C5H8O	METHYL ISOPROPENYL KETONE	814-78-8
1-METHYL-2-ISOPROPYLBENZENE	C10H14	o-CYMENE	527-84-4
1-METHYL-3-ISOPROPYLBENZENE	C10H14	m-CYMENE	535-77-3
1-METHYL-4-ISOPROPYLBENZENE	C10H14	p-CYMENE	99-87-6
m-METHYLISOPROPYLBENZENE	C10H14	m-CYMENE	535-77-3
p-METHYLISOPROPYL BENZENE	C10H14	p-CYMENE	99-87-6
1-METHYL-2-ISOPROPYLBENZOL	C10H14	o-CYMENE	527-84-4
1-METHYL-4-ISOPROPYL-1,3-CYCLOHEXADIENE	C10H16	alpha-TERPINENE	99-86-5
2-METHYL-5-ISOPROPYL-1,3-CYCLOHEXADIENE	C10H16	alpha-PHELLANDRENE	99-83-2
1-METHYL-4-ISOPROPYLCYCLOHEXADIENE-1,4	C1OH16	gamma-TERPINENE	99-85-4
METHYL ISOPROPYL ETHER	C4H10O	METHYL ISOPROPYL ETHER	598-53-8
METHYL ISOPROPYL KETONE	C5H10O	METHYL ISOPROPYL KETONE	563-80-4
METHYLJODID	CH3I	METHYL IODIDE	74-88-4
METHYLJODIDE	CH3I	METHYL IODIDE	74-88-4
METHYL KETONE	C3H6O	ACETONE	67-64-1
METHYLKYANID	C2H3N	ACETONITRILE	75-05-8
METHYL LAURINATE	C13H26O2	METHYL DODECANOATE	111-82-0
METHYLMALEIC ACID	C5H6O4	CITRACONIC ACID	498-23-7
METHYLMERCAPTAAN	CH4S	METHYL MERCAPTAN	74-93-1
METHYL MERCAPTAN	CH4S	METHYL MERCAPTAN	74-93-1
METHYLMETHACRYLAAT	C5H8O2	METHYL METHACRYLATE	80-62-6
METHYL-METHACRYLAT	C5H8O2	METHYL METHACRYLATE	80-62-6
METHYL METHACRYLATE	C5H8O2	METHYL METHACRYLATE	80-62-6
METHYL METHACRYLATE MONOMER, INHIBITED	C5H8O2	METHYL METHACRYLATE	80-62-6
N-METHYLMETHANAMINE	C2H7N	DIMETHYLAMINE	124-40-3
METHYLMETHANE	C2H6	ETHANE	74-84-0
METHYL METHANOATE	C2H4O2	METHYL FORMATE	107-31-3
METHYL-a-METHYLACRYLATE	C5H8O2	METHYL METHACRYLATE	80-62-6

Name (Synonym)	Formula	Name in Tables	CAS Reg. No
(R)-1-METHYL-4-(1-METHYLETHENYL)-CYCLOHEXENE	C10H16	D-LIMONENE	5989-27-5
1-METHYL-2-(1-METHYLETHYL)BENZENE	C10H14	o-CYMENE	527-84-4
1-METHYL-3-(1-METHYLETHYL)BENZENE	C10H14	m-CYMENE	535-77-3
1-METHYL-4-(1-METHYLETHYLIDENE)CYCLOHEXENE	C10H16	TERPINOLENE	586-62-9
METHYL-2-METHYL-2-PROPENOATE	C5H8O2	METHYL METHACRYLATE	80-62-6
2-METHYL-N-(2-METHYLPROPYL)-1-PROPANAMINE	C8H19N	DIISOBUTYLAMINE	110-96-3
METHYL MONOCHLORACETATE	C3H5ClO2	METHYL CHLOROACETATE	96-34-4
METHYL MONOCHLOROACETATE	C3H5ClO2	METHYL CHLOROACETATE	96-34-4
1-METHYLNAPHTHALENE	C11H10	1-METHYLNAPHTHALENE	90-12-0
2-METHYLNAPHTHALENE	C11H10	2-METHYLNAPHTHALENE	91-57-6
a-METHYLNAPHTHALENE	C11H10	1-METHYLNAPHTHALENE	90-12-0
b-METHYLNAPHTHALENE	C11H10	2-METHYLNAPHTHALENE	91-57-6
METHYL-NITRATE	CH3NO3	METHYL-NITRATE	598-58-3
METHYL-NITRITE	CH3NO2	METHYL-NITRITE	624-91-9
2-METHYLNITROBENZENE	C7H7NO2	o-NITROTOLUENE	88-72-2
3-METHYLNITROBENZENE	C7H7NO2	m-NITROTOLUENE	99-08-1
4-METHYLNITROBENZENE	C7H7NO2	p-NITROTOLUENE	99-99-0
m-METHYLNITROBENZENE	C7H7NO2	m-NITROTOLUENE	99-08-1
o-METHYLNITROBENZENE	C7H7NO2	o-NITROTOLUENE	88-72-2
p-METHYL NITROBENZENE	C7H7NO2	p-NITROTOLUENE	99-99-0
METHYL-NONADECYL-SULFIDE	C20H42S	METHYL-NONADECYL-SULFIDE	----------
2-METHYLNONANE	C10H22	2-METHYLNONANE	871-83-0
3-METHYLNONANE	C10H22	3-METHYLNONANE	5911-04-6
4-METHYLNONANE	C10H22	4-METHYLNONANE	17301-94-9
5-METHYLNONANE	C10H22	5-METHYLNONANE	15869-85-9
METHYL-NONYL-SULFIDE	C10H22S	METHYL-NONYL-SULFIDE	----------
METHYL-9-OCTADECENOATE	C19H36O2	METHYL OLEATE	112-62-9
METHYL (Z)-9-OCTADECENOATE	C19H36O2	METHYL OLEATE	112-62-9
METHYL cis-9-OCTADECENOATE	C19H36O2	METHYL OLEATE	112-62-9
METHYL-OCTADECYL-SULFIDE	C19H40S	METHYL-OCTADECYL-SULFIDE	----------
2-METHYLOCTANE	C9H20	2-METHYLOCTANE	3221-61-2
3-METHYLOCTANE	C9H20	3-METHYLOCTANE	2216-33-3
4-METHYLOCTANE	C9H20	4-METHYLOCTANE	2216-34-4
METHYL-OCTYL-SULFIDE	C9H20S	METHYL-OCTYL-SULFIDE	----------
METHYLOL	CH4O	METHANOL	67-56-1
METHYL OLEATE	C19H36O2	METHYL OLEATE	112-62-9
3-METHYLOLPENTANE	C6H14O	2-ETHYL-1-BUTANOL	97-95-0
METHYLOLPROPANE	C4H10O	n-BUTANOL	71-36-3
METHYL OXIRANE	C3H6O	1,2-PROPYLENE OXIDE	75-56-9
METHYL OXITOL	C3H8O2	2-METHOXYETHANOL	109-86-4
METHYL-3-OXOBUTYRATE	C5H8O3	METHYL ACETOACETATE	105-45-3
METHYL-PENTADECYL-SULFIDE	C16H34S	METHYL-PENTADECYL-SULFIDE	----------
4-METHYL-PENTAN-2-ON	C6H12O	METHYL ISOBUTYL KETONE	108-10-1
2-METHYLPENTANE	C6H14	2-METHYLPENTANE	107-83-5
3-METHYLPENTANE	C6H14	3-METHYLPENTANE	96-14-0
2-METHYL PENTANE-2,4-DIOL	C6H14O2	HEXYLENE GLYCOL	107-41-5
2-METHYL-2,4-PENTANEDIOL	C6H14O2	HEXYLENE GLYCOL	107-41-5
2-METHYLPENTANOL-1	C6H14O	2-METHYL-1-PENTANOL	105-30-6
4-METHYLPENTANOL-2	C6H14O	4-METHYL-2-PENTANOL	108-11-2
2-METHYL-1-PENTANOL	C6H14O	2-METHYL-1-PENTANOL	105-30-6
2-METHYL-4-PENTANOL	C6H14O	4-METHYL-2-PENTANOL	108-11-2
4-METHYL-2-PENTANOL	C6H14O	4-METHYL-2-PENTANOL	108-11-2
2-METHYL-2-PENTANOL-4-ONE	C6H12O2	DIACETONE ALCOHOL	123-42-2
4-METHYL-2-PENTANON	C6H12O	METHYL ISOBUTYL KETONE	108-10-1
2-METHYL-4-PENTANONE	C6H12O	METHYL ISOBUTYL KETONE	108-10-1
4-METHYL-2-PENTANONE	C6H12O	METHYL ISOBUTYL KETONE	108-10-1
2-METHYL-PENTENE-1	C6H12	2-METHYL-1-PENTENE	763-29-1
2-METHYL-1-PENTENE	C6H12	2-METHYL-1-PENTENE	763-29-1
2-METHYL-PENTENE-2	C6H12	2-METHYL-2-PENTENE	625-27-4
2-METHYL-2-PENTENE	C6H12	2-METHYL-2-PENTENE	625-27-4
3-METHYL-1-PENTENE	C6H12	3-METHYL-1-PENTENE	760-20-3
trans-3-METHYL-2-PENTENE	C6H12	trans-3-METHYL-2-PENTENE	616-12-6
3-METHYL-cis-2-PENTENE	C6H12	3-METHYL-cis-2-PENTENE	922-62-3
4-METHYL-1-PENTENE	C6H12	4-METHYL-1-PENTENE	691-37-2
4-METHYL-cis-2-PENTENE	C6H12	4-METHYL-cis-2-PENTENE	691-38-3
4-METHYL-trans-2-PENTENE	C6H12	4-METHYL-trans-2-PENTENE	674-76-0
4-METHYL-3-PENTENE-2-ONE	C6H10O	MESITYL OXIDE	141-79-7
4-METHYL-3-PENTEN-2-ON	C6H10O	MESITYL OXIDE	141-79-7
2-METHYL-2-PENTEN-4-ONE	C6H10O	MESITYL OXIDE	141-79-7
4-METHYL-3-PENTEN-2-ONE	C6H10O	MESITYL OXIDE	141-79-7
METHYL tert-PENTYL ETHER	C6H14O	METHYL tert-PENTYL ETHER	994-05-8
METHYL PENTYL KETONE	C7H14O	2-HEPTANONE	110-43-0
METHYL-PENTYL-SULFIDE	C6H14S	METHYL-PENTYL-SULFIDE	1741-83-9
4-(1-METHYL-1-PHENETHYL)PHENOL	C15H16O	p-CUMYLPHENOL	599-64-4
2-METHYLPHENOL	C7H8O	o-CRESOL	95-48-7
3-METHYLPHENOL	C7H8O	m-CRESOL	108-39-4

Name (Synonym)	Formula	Name in Tables	CAS Reg. No
4-METHYLPHENOL	C7H8O	p-CRESOL	106-44-5
m-METHYLPHENOL	C7H8O	m-CRESOL	108-39-4
o-METHYLPHENOL	C7H8O	o-CRESOL	95-48-7
p-METHYLPHENOL	C7H8O	p-CRESOL	106-44-5
N-METHYLPHENYLAMINE	C7H9N	N-METHYLANILINE	100-61-8
METHYLPHENYLAMINE	C7H9N	N-METHYLANILINE	100-61-8
4-METHYL-m-PHENYLENEDIAMINE	C7H10N2	TOLUENEDIAMINE	95-80-7
METHYL-m-PHENYLENE DIISOCYANATE	C9H6N2O2	TOLUENE DIISOCYANATE	26471-62-5
METHYLPHENYLENE ISOCYANATE	C9H6N2O2	TOLUENE DIISOCYANATE	26471-62-5
METHYL PHENYL ETHER	C7H8O	ANISOLE	100-66-3
as-METHYLPHENYLETHYLENE	C9H1O	alpha-METHYLSTYRENE	98-83-9
METHYL PHENYL KETONE	C8H8O	ACETOPHENONE	98-86-2
2-METHYL-1-PHENYLPROPANE	C10H14	ISOBUTYLBENZENE	538-93-2
2-METHYL-2-PHENYLPROPANE	C10H14	tert-BUTYLBENZENE	98-06-6
METHYL PHOSPHATE	C3H9O4P	TRIMETHYL PHOSPHATE	512-56-1
METHYL PHTHALATE	C10H10O4	DIMETHYL PHTHALATE	131-11-3
2-METHYL-1-PROPANAL	C4H8O	ISOBUTYRALDEHYDE	78-84-2
2-METHYLPROPANAL	C4H8O	ISOBUTYRALDEHYDE	78-84-2
2-METHYLPROPANENITRILE	C4H7N	ISOBUTYRONITRILE	78-82-0
a-METHYLPROPANENITRILE	C4H7N	ISOBUTYRONITRILE	78-82-0
2-METHYLPROPANETHIOL	C4H10S	ISOBUTYL MERCAPTAN	513-44-0
METHYL PROPANOATE	C4H8O2	METHYL PROPIONATE	554-12-1
2-METHYLPROPAN-1-OL	C4H10O	ISOBUTANOL	78-83-1
2-METHYL-1-PROPANOL	C4H10O	ISOBUTANOL	78-83-1
2-METHYL-2-PROPANOL	C4H10O	tert-BUTANOL	75-65-0
2-METHYL PROPANOL	C4H10O	ISOBUTANOL	78-83-1
2-METHYLPROPENAMIDE	C4H7NO	2-METHACRYLAMIDE	79-39-0
METHYL PROPENATE	C4H6O2	METHYL ACRYLATE	96-33-3
2-METHYLPROPENE	C4H8	ISOBUTENE	115-11-7
2-METHYLPROPENENITRILE	C4H5N	METHACRYLONITRILE	126-98-7
METHYL-2-PROPENOATE	C4H6O2	METHYL ACRYLATE	96-33-3
METHYL PROPENOATE	C4H6O2	METHYL ACRYLATE	96-33-3
2-METHYLPROPENOIC ACID	C4H6O2	METHACRYLIC ACID	79-41-4
2-METHYL-2-PROPENOIC ACID, ETHYL ESTER	C6H10O2	ETHYL METHACRYLATE	97-63-2
2-METHYL-2-PROPENOIC ACID METHYL ESTER	C5H8O2	METHYL METHACRYLATE	80-62-6
2-METHYLPROPIONALDEHYDE	C4H8O	ISOBUTYRALDEHYDE	78-84-2
METHYL PROPIONATE	C4H8O2	METHYL PROPIONATE	554-12-1
2-METHYLPROPIONIC ACID	C4H8O2	ISOBUTYRIC ACID	79-31-2
a-METHYLPROPIONIC ACID	C4H8O2	ISOBUTYRIC ACID	79-31-2
2-METHYLPROPIONIC ACID, ETHYL ESTER	C6H12O2	ETHYL ISOBUTYRATE	97-62-1
2-METHYLPROPIONITRILE	C4H7N	ISOBUTYRONITRILE	78-82-0
2-METHYL-1-PROPYL ACETATE	C6H12O2	ISOBUTYL ACETATE	110-19-0
2-METHYLPROPYL ACETATE	C6H12O2	ISOBUTYL ACETATE	110-19-0
2-METHYLPROPYL ACRYLATE	C7H12O2	ISOBUTYL ACRYLATE	106-63-8
2-METHYLPROPYL ALCOHOL	C4H10O	ISOBUTANOL	78-83-1
1-METHYLPROPYLAMINE	C4H11N	sec-BUTYLAMINE	13952-84-6
METHYL PROPYLATE	C4H8O2	METHYL PROPIONATE	554-12-1
METHYL PROPYL CARBINOL	C5H12O	2-PENTANOL	6032-29-7
METHYL-PROPYL-CETONE	C5H10O	2-PENTANONE	107-87-9
b-METHYLPROPYL ETHANOATE	C6H12O2	ISOBUTYL ACETATE	110-19-0
2-METHYL-2-PROPYLETHANOL	C6H14O	2-METHYL-1-PENTANOL	105-30-6
METHYL-PROPYL-ETHER	C4H10O	METHYL-PROPYL-ETHER	557-17-5
METHYL n-PROPYL ETHER	C4H10O	METHYL-PROPYL-ETHER	557-17-5
2-METHYLPROPYL ISOBUTYRATE	C8H16O2	ISOBUTYL ISOBUTYRATE	97-85-8
METHYL PROPYL KETONE	C5H10O	2-PENTANONE	107-87-9
METHYL-n-PROPYL KETONE	C5H10O	2-PENTANONE	107-87-9
2-METHYLPROPYLPROPANOIC ACID-2-METHYLPROPYL ESTER	C8H16O2	ISOBUTYL ISOBUTYRATE	97-85-8
METHYL-PROPYL-SULFIDE	C4H10S	METHYL-PROPYL-SULFIDE	3877-15-4
METHYLPROTOCATECHUALDEHYDE	C8H8O3	VANILLIN	121-33-5
2-METHYLPYRIDINE	C6H7N	2-METHYLPYRIDINE	109-06-8
3-METHYLPYRIDINE	C6H7N	3-METHYLPYRIDINE	108-99-6
4-METHYLPYRIDINE	C6H7N	4-METHYLPYRIDINE	108-89-4
a-METHYLPYRIDINE	C6H7N	2-METHYLPYRIDINE	109-06-8
1-METHYLPYRROLE	C5H7N	N-METHYLPYRROLE	96-54-8
N-METHYLPYRROLE	C5H7N	N-METHYLPYRROLE	96-54-8
N-METHYLPYRROLIDINE	C5H11N	N-METHYLPYRROLIDINE	120-94-5
1-METHYL-2-PYRROLIDINONE	C5H9NO	N-METHYL-2-PYRROLIDONE	872-50-4
N-METHYL-2-PYRROLIDINONE	C5H9NO	N-METHYL-2-PYRROLIDONE	872-50-4
1-METHYL-5-PYRROLIDINONE	C5H9NO	N-METHYL-2-PYRROLIDONE	872-50-4
N-METHYLPYRROLIDINONE	C5H9NO	N-METHYL-2-PYRROLIDONE	872-50-4
1-METHYL-2-PYRROLIDONE	C5H9NO	N-METHYL-2-PYRROLIDONE	872-50-4
N-METHYL-2-PYRROLIDONE	C5H9NO	N-METHYL-2-PYRROLIDONE	872-50-4
METHYLPYRROLIDONE	C5H9NO	N-METHYL-2-PYRROLIDONE	872-50-4
2-METHYLQUINOLINE	C10H9N	QUINALDINE	91-63-4
METHYL SALICYLATE	C8H8O3	METHYL SALICYLATE	119-36-8
METHYL SILANE	CH6SI	METHYL SILANE	992-94-9

Name (Synonym)	Formula	Name in Tables	CAS Reg. No
a-METHYLSTYREEN	C9H10	alpha-METHYLSTYRENE	98-83-9
alpha-METHYLSTYRENE	C9H10	alpha-METHYLSTYRENE	98-83-9
3-METHYLSTYRENE	C9H10	m-METHYLSTYRENE	100-80-1
m-METHYLSTYRENE	C9H10	m-METHYLSTYRENE	100-80-1
o-METHYLSTYRENE	C9H10	o-METHYLSTYRENE	611-15-4
p-METHYLSTYRENE	C9H10	p-METHYLSTYRENE	622-97-9
a-METHYL-STYROL	C9H10	alpha-METHYLSTYRENE	98-83-9
METHYL SULFATE	C2H6O4S	DIMETHYL SULFATE	77-78-1
METHYL SULFIDE	C2H6S	DIMETHYL SULFIDE	75-18-3
3-METHYL SULFOLANE	C5H10O2S	3-METHYL SULFOLANE	872-93-5
METHYL SULPHIDE	C2H6S	DIMETHYL SULFIDE	75-18-3
METHYL-TETRADECYL-SULFIDE	C15H32S	METHYL-TETRADECYL-SULFIDE	----------
N-METHYLTETRAHYDROPYRROLE	C5H11N	N-METHYLPYRROLIDINE	120-94-5
METHYLTHIOMETHANE	C2H6S	DIMETHYL SULFIDE	75-18-3
2-METHYLTHIOPHENE	C5H6S	2-METHYLTHIOPHENE	554-14-3
3-METHYLTHIOPHENE	C5H6S	3-METHYLTHIOPHENE	616-44-4
o-METHYLTOLUENE	C8H10	o-XYLENE	95-47-6
p-METHYLTOLUENE	C8H10	p-XYLENE	106-42-3
METHYL TRICHLORIDE	CHCl3	CHLOROFORM	67-66-3
METHYLTRICHLOROMETHANE	C2H3Cl3	1,1,1-TRICHLOROETHANE	71-55-6
METHYL TRICHLOROSILANE	CH3Cl3Si	METHYL TRICHLOROSILANE	75-79-6
METHYL-TRICHLORSILAN	CH3Cl3Si	METHYL TRICHLOROSILANE	75-79-6
METHYL-TRIDECYL-SULFIDE	C14H30S	METHYL-TRIDECYL-SULFIDE	----------
METHYL TRIFLUORIDE	CHF3	TRIFLUOROMETHANE	75-46-7
METHYLTRIMETHYLENE GLYCOL	C4H10O2	1,3-BUTANEDIOL	107-88-0
METHYL-UNDECYL-SULFIDE	C12H26S	METHYL-UNDECYL-SULFIDE	10059-13-9
1-METHYL-3-VINYLBENZENE	C9H10	m-METHYLSTYRENE	100-80-1
METHYL VINYL ETHER	C3H6O	METHYL VINYL ETHER	107-25-5
METIL	C2H4O2	METHYL FORMATE	107-31-3
METILAMIL ALCOHOL	C6H14O	4-METHYL-2-PENTANOL	108-11-2
METILAMINE	CH5N	METHYLAMINE	74-89-5
3-METIL-BUTANOLO	C5H12O	3-METHYL-1-BUTANOL	123-51-3
METILE	C3H6O2	METHYL ACETATE	79-20-9
METILETILCHETONE	C4H8O	METHYL ETHYL KETONE	78-93-3
METILISOBUTILCHETONE	C6H12O	METHYL ISOBUTYL KETONE	108-10-1
METILMERCAPTANO	CH4S	METHYL MERCAPTAN	74-93-1
METIL METACRILATO	C5H8O2	METHYL METHACRYLATE	80-62-6
4-METILPENTAN-2-OLO	C6H14O	4-METHYL-2-PENTANOL	108-11-2
4-METILPENTAN-2-ONE	C6H12O	METHYL ISOBUTYL KETONE	108-10-1
4-METIL-3-PENTEN-2-ONE	C6H10O	MESITYL OXIDE	141-79-7
a-METIL-STIROLO	C9H10	alpha-METHYLSTYRENE	98-83-9
METOPRYL	C4H10O	METHYL-PROPYL-ETHER	557-17-5
METRANIL	C5H8N4O12	PENTAERYTHRITOL TETRANITRATE	78-11-5
METYLAL	C3H8O2	METHYLAL	109-87-5
METYLENU CHLOREK	CH2Cl2	DICHLOROMETHANE	75-09-2
METYLESTER KYSELINY SALICYLOVE	C8H8O3	METHYL SALICYLATE	119-36-8
METYLOAMINA	CH5N	METHYLAMINE	74-89-5
METYLOCYKLOHEKSAN	C7H14	METHYLCYCLOHEXANE	108-87-2
METYLOETYLOKETON	C4H8O	METHYL ETHYL KETONE	78-93-3
METYLOIZOBUTYLOKETON	C6H12O	METHYL ISOBUTYL KETONE	108-10-1
METYLOPROPYLOKETON	C5H10O	2-PENTANONE	107-87-9
METYLOWY ALKOHOL	CH4O	METHANOL	67-56-1
METYLU CHLOREK	CH3Cl	METHYL CHLORIDE	74-87-3
METYLU JODEK	CH3I	METHYL IODIDE	74-88-4
MIAK	C7H14O	5-METHYL-2-HEXANONE	110-12-3
MIBC	C6H14O	4-METHYL-2-PENTANOL	108-11-2
MIBK	C6H12O	METHYL ISOBUTYL KETONE	108-10-1
3-MIC	C6H14O	4-METHYL-2-PENTANOL	108-11-2
MIC	C2H3NO	METHYL ISOCYANATE	624-83-9
MIERENZUUR	CH2O2	FORMIC ACID	64-18-6
MIGHTY 150	C10H8	NAPHTHALENE	91-20-3
MIK	C6H12O	METHYL ISOBUTYL KETONE	108-10-1
MILLER'S FUMIGRAIN	C3H3N	ACRYLONITRILE	107-13-1
MINERAL NAPHTHA	C6H6	BENZENE	71-43-2
MIPAX	C10H10O4	DIMETHYL PHTHALATE	131-11-3
MIPK	C5H10O	METHYL ISOPROPYL KETONE	563-80-4
MIRBANE OIL	C6H5NO2	NITROBENZENE	98-95-3
MME	C5H8O2	METHYL METHACRYLATE	80-62-6
MNA	C6H6N2O2	m-NITROANILINE	99-09-2
MNBK	C6H12O	2-HEXANONE	591-78-6
MNT	C7H7NO2	m-NITROTOLUENE	99-08-1
MOLTEN ADIPIC ACID	C6H10O4	ADIPIC ACID	124-04-9
MONASIRUP	C10H12O	ANETHOLE	104-46-1
MONDUR P	C7H5NO	PHENYL ISOCYANATE	103-71-9
MONDUR-TD	C9H6N2O2	TOLUENE DIISOCYANATE	26471-62-5
MONDUR-TD-80	C9H6N2O2	TOLUENE DIISOCYANATE	26471-62-5

Name (Synonym)	Formula	Name in Tables	CAS Reg. No
MONOAETHANOLAMIN	C2H7NO	MONOETHANOLAMINE	141-43-5
MONOALLYLAMINE	C3H7N	ALLYLAMINE	107-11-9
MONOAMYLAMINE MONOAMYLAMINE	C5H13N	n-PENTYLAMINE	110-58-7
MONOBENZYLAMINE	C7H9N	BENZYLAMINE	100-46-9
MONOBROMOBENZENE	C6H5Br	BROMOBENZENE	108-86-1
MONOBROMOETHANE	C2H5Br	BROMOETHANE	74-96-4
MONOBROMOMETHANE	CH3Br	METHYL BROMIDE	74-83-9
MONOBUTILAMINA	C4H11N	n-BUTILAMINE	109-73-9
MONOBUTYLAMINE	C4H11N	n-BUTYLAMINE	109-73-9
MONOCHLOORBENZEEN	C6H5Cl	MONOCHLOROBENZENE	108-90-7
MONOCHLORACETIC ACID	C2H3ClO2	CHLOROACETIC ACID	79-11-8
MONOCHLORBENZENE	C6H5Cl	MONOCHLOROBENZENE	108-90-7
MONOCHLORBENZOL	C6H5Cl	MONOCHLOROBENZENE	108-90-7
MONOCHLORETHANE	C2H5Cl	ETHYL CHLORIDE	75-00-3
MONOCHLORHYDRINE du GLYCOL	C2H5ClO	2-CHLOROETHANOL	107-07-3
MONOCHLOROACETALDEHYDE	C2H3ClO	CHLOROACETALDEHYDE	107-20-0
MONOCHLOROACETIC ACID	C2H3ClO2	CHLOROACETIC ACID	79-11-8
MONOCHLOROACETIC ACID METHYL ESTER	C3H5ClO2	METHYL CHLOROACETATE	96-34-4
MONOCHLOROACETYL CHLORIDE	C2H2Cl2O	CHLOROACETYL CHLORIDE	79-04-9
MONOCHLOROBENZENE	C6H5Cl	MONOCHLOROBENZENE	108-90-7
MONOCHLORODIFLUOROMETHANE	CHClF2	CHLORODIFLUOROMETHANE	75-45-6
MONOCHLOROETHANOIC ACID	C2H3ClO2	CHLOROACETIC ACID	79-11-8
2-MONOCHLOROETHANOL	C2H5ClO	2-CHLOROETHANOL	107-07-3
MONOCHLOROETHENE	C2H3Cl	VINYL CHLORIDE	75-01-4
MONOCHLOROETHYLENE	C2H3Cl	VINYL CHLORIDE	75-01-4
MONOCHLOROMETHANE	CH3Cl	METHYL CHLORIDE	74-87-3
MONO-CHLORO-MONO-BROMO-METHANE	CH2BrCl	BROMOCHLOROMETHANE	74-97-5
MONOCHLOROMONOFLUOROMETHANE	CH2ClF	CHLOROFLUOROMETHANE	593-70-4
MONOCHLOROPENTAFLUOROETHANE	C2ClF5	CHLOROPENTAFLUOROETHANE	76-15-3
MONOCHLOROTRIFLUOROETHYLENE	C2ClF3	CHLOROTRIFLUOROETHYLENE	79-38-9
MONOCHLOROTRIFLUOROMETHANE	CClF3	CHLOROTRIFLUOROMETHANE	75-72-9
"MONOCITE" METHACRYLATE MONOMER	C5H8O2	METHYL METHACRYLATE	80-62-6
MONOCLOROBENZENE	C6H5Cl	MONOCHLOROBENZENE	108-90-7
MONODODECYLAMINE	C12H27N	DODECYLAMINE	124-22-1
MONOETHANOLAMINE	C2H7NO	MONOETHANOLAMINE	141-43-5
MONOETHANOLETHYLENEDIAMINE	C4H12N2O	N-AMINOETHYL ETHANOLAMINE	111-41-1
MONOETHYLAMINE	C2H7N	ETHYLAMINE	75-04-7
MONOETHYLENE GLYCOL	C2H6O2	ETHYLENE GLYCOL	107-21-1
MONOETHYLENE GLYCOL DIMETHYL ETHER	C4H10O2	1,2-DIMETHOXYETHANE	110-71-4
MONOETHYL ETHER of DIETHYLENE GLYCOL	C6H14O3	2-(2-ETHOXYETHOXY)ETHANOL	111-90-0
MONOFLUOROETHANE	C2H5F	ETHYL FLUORIDE	353-36-6
MONOFLUOROETHYLENE	C2H3F	VINYL FLUORIDE	75-02-5
MONOFLUOROTRICHLOROMETHANE	CCl3F	TRICHLOROFLUOROMETHANE	75-69-4
MONOGLYME	C4H10O2	1,2-DIMETHOXYETHANE	110-71-4
MONOHYDROXYBENZENE	C6H6O	PHENOL	108-95-2
MONOHYDROXYMETHANE	CH4O	METHANOL	67-56-1
MONOIODURO di METILE	CH3I	METHYL IODIDE	74-88-4
MONOISOBUTYLAMINE	C4H11N	ISOBUTYLAMINE	78-81-9
MONO-ISO-PROPANOLAMINE	C3H9NO	1-AMINO-2-PROPANOL	78-96-6
MONOISOPROPYLAMINE	C3H9N	ISOPROPYLAMINE	75-31-0
MONOMETHYLACETAMIDE	C3H7NO	N-METHYLACETAMIDE	79-16-3
MONOMETHYLAMINE	CH5N	METHYLAMINE	74-89-5
N-MONOMETHYLAMINOETHANOL	C3H9NO	METHYLETHANOLAMINE	109-83-1
MONOMETHYLAMINOETHANOL	C3H9NO	METHYLETHANOLAMINE	109-83-1
N-MONOMETHYLANILINE	C7H9N	N-METHYLANILINE	100-61-8
MONOMETHYL ANILINE	C7H9N	N-METHYLANILINE	100-61-8
MONOMETHYL ETHER of ETHYLENE GLYCOL	C3H8O2	2-METHOXYETHANOL	109-86-4
MONOMETHYL ETHER HYDROQUINONE	C7H8O2	p-METHOXYPHENOL	150-76-5
MONOMETHYLFORMAMIDE	C2H5NO	N-METHYLFORMAMIDE	123-39-7
MONO-n-BUTYLAMINE	C4H11N	n-BUTYLAMINE	109-73-9
MONO-N-HEXYLAMINE	C6H15N	n-HEXYLAMINE	111-26-2
MONO-N-PROPYLAMINE	C3H9N	n-PROPYLAMINE	107-10-8
MONOPENTEK	C5H12O4	PENTAERYTHRITOL	115-77-5
MONOPHENOL	C6H6O	PHENOL	108-95-2
MONOPLEX DBS	C18H34O4	DIBUTYL SEBACATE	109-43-3
MONOPROPYLENE GLYCOL	C3H8O2	1,2-PROPYLENE GLYCOL	57-55-6
MONOPROPYL ETHER of ETHYLENE GLYCOL	C5H12O2	ETHYLENE GLYCOL MONOPROPYL ETHER	2807-30-9
MONOPYRROLE	C4H5N	PYRROLE	109-97-7
MOON	C3H8O3	GLYCEROL	56-81-5
MORPHOLINE	C4H9NO	MORPHOLINE	110-91-8
MORPHOLINE, AQUEOUS MIXTURE	C4H9NO	MORPHOLINE	110-91-8
MOTH BALLS	C10H8	NAPHTHALENE	91-20-3
MOTH FLAKES	C10H8	NAPHTHALENE	91-20-3
MOTOR BENZOL	C6H6	BENZENE	71-43-2
MOTTENHEXE	C2Cl6	HEXACHLOROETHANE	67-72-1
3MPA	C3H6O2S	3-MERCAPTOPROPIONIC ACID	107-96-0

Name (Synonym)	Formula	Name in Tables	CAS Reg. No
MPK	C5H10O	2-PENTANONE	107-87-9
MROWCZAN ETYLU	C3H6O2	ETHYL FORMATE	109-94-4
MTBE	C5H12O	METHYL tert-BUTYL ETHER	1634-04-4
MURIATIC ETHER	C2H5Cl	ETHYL CHLORIDE	75-00-3
MUTHMANNS LIQUID	C2H2Br4	1,1,2,2-TETRABROMOETHANE	79-27-6
MYRISIIC ALCOHOL	C14H30O	1-TETRADECANOL	112-72-1
NACCONATE-100	C9H6N2O2	TOLUENE DIISOCYANATE	26471-62-5
NADONE	C6H10O	CYCLOHEXANONE	108-94-1
NAFTALEN	C10H8	NAPHTHALENE	91-20-3
NAPHTHALENE	C10H8	NAPHTHALENE	91-20-3
1,2-(1,8-NAPHTHALENEDIYL)BENZENE	C16H10	FLUORANTHENE	206-44-0
NAPHTHALENE, 1-ETHYL-	C12H12	1-ETHYLNAPHTHALENE	1127-76-0
NAPHTHALENE-1,2,3,4-TETRAHYDRIDE	C10H12	1,2,3,4-TETRAHYDRONAPHTHALENE	119-64-2
NAPHTHALIN	C10H8	NAPHTHALENE	91-20-3
NAPHTHALINE	C10H8	NAPHTHALENE	91-20-3
D5,7,9-NAPHTHANTRIENE	C10H12	1,2,3,4-TETRAHYDRONAPHTHALENE	119-64-2
NAPHTHENE	C10H8	NAPHTHALENE	91-20-3
NAPHTHOELAN NAVY BLUE	C12H12N2	p-AMINODIPHENYLAMINE	101-54-2
1,2-(1,8-NAPHTHYLENE)BENZENE	C16H10	FLUORANTHENE	206-44-0
NAPHTHYLENEETHYLENE	C12H10	ACENAPHTHENE	83-32-9
NAPHTOELAN ORANGE R BASE	C6H6N2O2	m-NITROANILINE	99-09-2
NAPHTOELAN RED GG BASE	C6H6N2O2	p-NITROANILINE	100-01-6
NAPHTOL AS-KG	C7H9N	p-TOLUIDINE	106-49-0
NAPHTOL AS-KGLL	C7H9N	p-TOLUIDINE	106-49-0
NARCOTANE	C2HBrClF3	HALOTHANE	151-67-7
NARCOTANN NE-SPOFA	C2HBrClF3	HALOTHANE	151-67-7
NARCOTILE	C2H5Cl	ETHYL CHLORIDE	75-00-3
NARCYLEN	C2H2	ACETYLENE	74-86-2
NATRASCORB INJECTABLE	C6H8O6	ASCORBIC ACID	50-81-7
NATURAL WINTERGREEN OIL	C8H8O3	METHYL SALICYLATE	119-36-8
NAULI "GUM"	C10H12O	ANETHOLE	104-46-1
NAXOL	C6H12O	CYCLOHEXANOL	108-93-0
NC5	C6H11N	HEXANENITRILE	628-73-9
NCI-C00511	C2H4Cl2	1,2-DICHLOROETHANE	107-06-2
NCI-C00920	C2H6O2	ETHYLENE GLYCOL	107-21-1
NCI-C01854	C12H12N2	HYDRAZOBENZENE	122-66-7
NCI-C01865	C7H6N2O4	2,4-DINITROTOLUENE	121-14-2
NCI-C02039	C6H6ClN	p-CHLOROANILINE	106-47-8
NCI-C02108	C2H5NO	ACETAMIDE	60-35-5
NCI-C02200	C8H8	STYRENE	100-42-5
NCI-C03554	C2H2Cl4	1,1,2,2-TETRACHLOROETHANE	79-34-5
NCI-C03601	C8H4O3	PHTHALIC ANHYDRIDE	85-44-9
NCI-C03689	C4H8O2	1,4-DIOXANE	123-91-1
NCI-C03736	C6H7N	ANILINE	62-53-3
NCI-C03781	C3H9O4P	TRIMETHYL PHOSPHATE	512-56-1
NCI-C04535	C2H4Cl2	1,1-DICHLOROETHANE	75-34-3
NCI-C04579	C2H3Cl3	1,1,2-TRICHLOROETHANE	79-00-5
NCI-C04580	C2Cl4	TETRACHLOROETHYLENE	127-18-4
NCI-C04591	CS2	CARBON DISULFIDE	75-15-0
NCI-C04604	C2Cl6	HEXACHLOROETHANE	67-72-1
NCI-C04615	C3H5Cl	3-CHLOROPROPENE	107-05-1
NCI-C04626	C2H3Cl3	1,1,1-TRICHLOROETHANE	71-55-6
NCI-C04637	CCl3F	TRICHLOROFLUOROMETHANE	75-69-4
NCI-C05970	C6H6O2	1,3-BENZENEDIOL	108-46-3
NCI-C06111	C7H8O	BENZYL ALCOHOL	100-51-6
NCI-C06155	C4H9Cl	n-BUTYL CHLORIDE	109-69-3
NCI-C06224	C2H5Cl	ETHYL CHLORIDE	75-00-3
NCI-C06360	C7H7Cl	BENZYL CHLORIDE	100-44-7
NCI-C07272	C7H8	TOLUENE	108-88-3
NCI-C50055	C10H10O4	DIMETHYL TEREPHTHALATE	120-61-6
NCI-C50077	C3H6	PROPYLENE	115-07-1
NCI-C50088	C2H4O	ETHYLENE OXIDE	75-21-8
NCI-C50099	C3H6O	1,2-PROPYLENE OXIDE	75-56-9
NCI-C50124	C6H6O	PHENOL	108-95-2
NCI-C50135	C2H5ClO	2-CHLOROETHANOL	107-07-3
NCI-C50384	C5H8O2	ETHYL ACRYLATE	140-88-5
NCI-C50602	C4H6	1,3-BUTADIENE	106-99-0
NCI-C50646	C6H11NO	epsilon-CAPROLACTAM	105-60-2
NCI-C50680	C5H8O2	METHYL METHACRYLATE	80-62-6
NCI-C52459	C2H2Cl4	1,1,1,2-TETRACHLOROETHANE	630-20-6
NCI-C52904	C10H8	NAPHTHALENE	91-20-3
NCI-C53894	C2HCl5	PENTACHLOROETHANE	76-01-7
NCI-C54262	C2H2Cl2	1,1-DICHLOROETHYLENE	75-35-4
NCI-C54808	C6H8O6	ASCORBIC ACID	50-81-7
NCI-C54853	C4H10O2	2-ETHOXYETHANOL	110-80-5
NCI-C54886	C6H5Cl	MONOCHLOROBENZENE	108-90-7

733

Name (Synonym)	Formula	Name in Tables	CAS Reg. No
NCI-C54944	C6H4Cl2	o-DICHLOROBENZENE	95-50-1
NCI-C54955	C6H4Cl2	p-DICHLOROBENZENE	106-46-7
NCI-C54999	C8H12	VINYLCYCLOHEXENE	100-40-3
NCI-C55005	C6H10O	CYCLOHEXANONE	108-94-1
NCI-C55141	C3H6Cl2	1,2-DICHLOROPROPANE	78-87-5
NCI-C55174	C4H11NO2	DIETHANOLAMINE	111-42-2
NCI-C55209	C2H2O4	OXALIC ACID	144-62-7
NCI-C55276	C6H6	BENZENE	71-43-2
NCI-C55298	C9H7NO	8-HYDROXYQUINOLINE	148-24-3
NCI-C55301	C5H5N	PYRIDINE	110-86-1
NCI-C55367	C4H10O	tert-BUTANOL	75-65-0
NCI-C55481	C2H5Br	BROMOETHANE	74-96-4
NCI-C55492	C6H5Br	BROMOBENZENE	108-86-1
NCI-C55527	C4H8O	1,2-EPOXYBUTANE	106-88-7
NCI-C55572	C10H16	D-LIMONENE	5989-27-5
NCI-C55607	C5Cl6	HEXACHLOROCYCLOPENTADIENE	77-47-4
NCI-C55618	C9H14O	ISOPHORONE	78-59-1
NCI-C55696	C4H4O3	SUCCINIC ANHYDRIDE	108-30-5
NCI-C55834	C6H6O2	p-HYDROQUINONE	123-31-9
NCI-C55856	C6H6O2	1,2-BENZENEDIOL	120-80-9
NCI-C55878	C4H6O2	gamma-BUTYROLACTONE	96-48-0
NCI-C56133	C7H6O	BENZALDEHYDE	100-52-7
NCI-C56155	C7H5N3O6	2,4,6-TRINITROTOLUENE	118-96-7
NCI-C56177	C5H4O2	FURFURAL	98-01-1
NCI-C56202	C4H4O	FURAN	110-00-9
NCI-C56224	C5H6O2	FURFURYL ALCOHOL	98-00-0
NCI-C56279	C4H6O	trans-CROTONALDEHYDE	123-73-9
NCI-C56291	C4H8O	n-BUTYRALDEHYDE	123-72-8
NCI-C56326	C2H4O	ACETALDEHYDE	75-07-0
NCI-C56393	C8H10	ETHYLBENZENE	100-41-4
NCI-C56428	C8H11N	N,N-DIMETHYLANILINE	121-69-7
NCI-C56440	C3F6O	HEXAFLUOROACETONE	684-16-2
NCI-C56633	C9H4O5	TRIMELLITIC ANHYDRIDE	552-30-7
NCI-C60048	C12H14O4	DIETHYL PHTHALATE	84-66-2
NCI-C60082	C6H5NO2	NITROBENZENE	98-95-3
NCI-C60208	C2H2F2	1,1-DIFLUOROETHYLENE	75-38-7
NCI-C60219	CCl2O	PHOSGENE	75-44-5
NCI-C60220	C3H5Cl3	1,2,3-TRICHLOROPROPANE	96-18-4
NCI-C60231	C2H3ClO2	CHLOROACETIC ACID	79-11-8
NCI-C60388	C7H7NO3	o-NITROANISOLE	91-23-6
NCI-C60402	C2H8N2	ETHYLENEDIAMINE	107-15-3
NCI-C60537	C7H7NO2	p-NITROTOLUENE	99-99-0
NCI-C60560	C4H8O	TETRAHYDROFURAN	109-99-9
NCI-C60571	C6H14	n-HEXANE	110-54-3
NCI-C60822	C2H3N	ACETONITRILE	75-05-8
NCI-C60866	C4H10S	n-BUTYL MERCAPTAN	109-79-5
NCI-C60899	C5H10O3	DIETHYL CARBONATE	105-58-8
NCI-C60913	C3H7NO	N,N-DIMETHYLFORMAMIDE	68-12-2
NCI-C60935	C12H26S	n-DODECYL MERCAPTAN	112-55-0
NCI-C60968	C4H8O	ISOBUTYRALDEHYDE	78-84-2
NCI-C61018	C9H18O	1-NONANAL	124-19-6
NCI-C61029	C3H6O	n-PROPIONALDEHYDE	123-38-6
NCI-C61405	C6H16N2	HEXAMETHYLENEDIAMINE	124-09-4
NEANTINE	C12H14O4	DIETHYL PHTHALATE	84-66-2
NECATORINE	CCl4	CARBON TETRACHLORIDE	56-23-5
NEMA	C2Cl4	TETRACHLOROETHYLENE	127-18-4
NEMAFENE	C3H6Cl2	2,2-DICHLOROPROPANE	594-20-7
NEO-FAT 10	C10H20O2	n-DECANOIC ACID	334-48-5
NEO-FAT 12	C12H24O2	n-DODECANOIC ACID	143-07-7
NEO-FAT 18-61	C18H36O2	STEARIC ACID	57-11-4
NEO-FAT 18-S	C18H36O2	STEARIC ACID	57-11-4
NEO-FAT 8	C8H16O2	n-OCTANOIC ACID	124-07-2
NEOHEXANE	C6H14	2,2-DIMETHYLBUTANE	75-83-2
NEOL	C5H12O2	NEOPENTYL GLYCOL	126-30-7
NEOPENTANE	C5H12	NEOPENTANE	463-82-1
NEOPENTYLENE GLYCOL	C5H12O2	NEOPENTYL GLYCOL	126-30-7
NEOPENTYL GLYCOL	C5H12O2	NEOPENTYL GLYCOL	126-30-7
NEOTHYL	C4H10O	METHYL-PROPYL-ETHER	557-17-5
NEPHIS	C2H4Br2	1,2-DIBROMOETHANE	106-93-4
NIAX ISOCYANATE TDI	C9H6N2O2	TOLUENE DIISOCYANATE	26471-62-5
NINOL AA-62 EXTRA	C12H24O2	n-DODECANOIC ACID	143-07-7
NIOBE OIL	C8H8O2	METHYL BENZOATE	93-58-3
NIPAR S-20	C3H7NO2	2-NITROPROPANE	79-46-9
NIPAR S-20 SOLVENT	C3H7NO2	2-NITROPROPANE	79-46-9
NIPAR S-30 SOLVENT	C3H7NO2	2-NITROPROPANE	79-46-9
NISSAN AMINE BB	C12H27N	DODECYLAMINE	124-22-1

Name (Synonym)	Formula	Name in Tables	CAS Reg. No
m-NITRANILINE	C6H6N2O2	m-NITROANILINE	99-09-2
NITRATE de PROPYLE NORMAL	C3H7NO3	PROPYL-NITRATE	627-13-4
NITRATION BENZENE	C6H6	BENZENE	71-43-2
NITRIC ACID, ETHYL ESTER	C2H5NO3	ETHYL-NITRATE	625-58-1
NITRIC ACID, METHYL ESTER	CH3NO3	METHYL-NITRATE	598-58-3
NITRIC ACID, PROPYL ESTER	C3H7NO3	PROPYL-NITRATE	627-13-4
NITRIC ETHER	C2H5NO3	ETHYL-NITRATE	625-58-1
NITRILE ADIPICO	C6H8N2	ADIPONITRILE	111-69-3
NITRIL KYSELINY MALONOVE	C3H2N2	MALONONITRILE	109-77-3
NITRILOACETONITRILE	C2N2	CYANOGEN	460-19-5
NITRILO-2,2',2"-TRIETHANOL	C6H15NO3	TRIETHANOLAMINE	102-71-6
2,2',2"-NITRILOTRIETHANOL	C6H15NO3	TRIETHANOLAMINE	102-71-6
m-NITROAMINOBENZENE	C6H6N2O2	m-NITROANILINE	99-09-2
2-NITROANILINE	C6H6N2O2	o-NITROANILINE	88-74-4
3-NITROANILINE	C6H6N2O2	m-NITROANILINE	99-09-2
m-NITROANILINE	C6H6N2O2	m-NITROANILINE	99-09-2
o-NITROANILINE	C6H6N2O2	o-NITROANILINE	88-74-4
p-NITROANILINE	C6H6N2O2	p-NITROANILINE	100-01-6
2-NITROANISOLE	C7H7NO3	o-NITROANISOLE	91-23-6
o-NITROANISOLE	C7H7NO3	o-NITROANISOLE	91-23-6
NITROBENZEEN	C6H5NO2	NITROBENZENE	98-95-3
NITROBENZEN	C6H5NO2	NITROBENZENE	98-95-3
3-NITROBENZENAMINE	C6H6N2O2	m-NITROANILINE	99-09-2
NITROBENZENE	C6H5NO2	NITROBENZENE	98-95-3
NITROBENZOL	C6H5NO2	NITROBENZENE	98-95-3
3-NITROBENZOTRIFLUORIDE	C7H4F3NO2	3-NITROBENZOTRIFLUORIDE	98-46-4
m-NITROBENZOTRIFLUORIDE	C7H4F3NO2	3-NITROBENZOTRIFLUORIDE	98-46-4
1-NITROBUTANE	C4H9NO2	1-NITROBUTANE	627-05-4
2-NITROBUTANE	C4H9NO2	2-NITROBUTANE	600-24-8
NITROCARBOL	CH3NO2	NITROMETHANE	75-52-5
p-NITROCHLOORBENZEEN	C6H4ClNO2	p-CHLORONITROBENZENE	100-00-5
m-NITROCHLOROBENZENE	C6H4ClNO2	m-CHLORONITROBENZENE	121-73-3
o-NITROCHLOROBENZENE	C6H4ClNO2	o-CHLORONITROBENZENE	88-73-3
p-NITROCHLOROBENZENE	C6H4ClNO2	p-CHLORONITROBENZENE	100-00-5
p-NITROCHLOROBENZOL	C6H4ClNO2	p-CHLORONITROBENZENE	100-00-5
p-NITROCLOROBENZENE	C6H4ClNO2	p-CHLORONITROBENZENE	100-00-5
NITROETAN	C2H5NO2	NITROETHANE	79-24-3
NITROETHANE	C2H5NO2	NITROETHANE	79-24-3
NITROGLYCERINE	C3H5N3O9	NITROGLYCERINE	55-63-0
NITROISOPROPANE	C3H7NO2	2-NITROPROPANE	79-46-9
NITROMETAN	CH3NO2	NITROMETHANE	75-52-5
NITROMETHANE	CH3NO2	NITROMETHANE	75-52-5
o-NITROPHENYL	C7H7NO3	o-NITROANISOLE	91-23-6
m-NITROPHENYLAMINE	C6H6N2O2	m-NITROANILINE	99-09-2
1-NITROPROPANE	C3H7NO2	1-NITROPROPANE	103-03-2
2-NITROPROPANE	C3H7NO2	2-NITROPROPANE	79-46-9
b-NITROPROPANE	C3H7NO2	2-NITROPROPANE	79-46-9
NITROPROPANE	C3H7NO2	2-NITROPROPANE	79-46-9
2-NITROTOLUENE	C7H7NO2	o-NITROTOLUENE	88-72-2
3-NITROTOLUENE	C7H7NO2	m-NITROTOLUENE	99-08-1
4-NITROTOLUENE	C7H7NO2	p-NITROTOLUENE	99-99-0
m-NITROTOLUENE	C7H7NO2	m-NITROTOLUENE	99-08-1
o-NITROTOLUENE	C7H7NO2	o-NITROTOLUENE	88-72-2
p-NITROTOLUENE	C7H7NO2	p-NITROTOLUENE	99-99-0
3-NITROTOLUOL	C7H7NO2	m-NITROTOLUENE	99-08-1
4-NITROTOLUOL	C7H7NO2	p-NITROTOLUENE	99-99-0
m-NITROTRIFLUOROTOLUENE	C7H4F3NO2	3-NITROBENZOTRIFLUORIDE	98-46-4
m-NITROTRIFLUORTOLUOL	C7H4F3NO2	3-NITROBENZOTRIFLUORIDE	98-46-4
NITROUS ACID, METHYL ESTER	CH3NO2	METHYL-NITRITE	624-91-9
NIVITIN	C6H14O6	SORBITOL	50-70-4
NMP	C5H9NO	N-METHYL-2-PYRROLIDONE	872-50-4
NO BUNT	C6Cl6	HEXACHLOROBENZENE	118-74-1
NO BUNT 40	C6Cl6	HEXACHLOROBENZENE	118-74-1
NO BUNT 80	C6Cl6	HEXACHLOROBENZENE	118-74-1
NO BUNT LIQUID	C6Cl6	HEXACHLOROBENZENE	118-74-1
n-NONADECANE	C19H40	n-NONADECANE	629-92-5
1-NONADECANETHIOL	C19H40S	1-NONADECANETHIOL	----------
NONADECANOIC ACID	C19H38O2	NONADECANOIC ACID	646-30-0
1-NONADECANOL	C19H40O	1-NONADECANOL	1454-84-8
1-NONADECENE	C19H38	1-NONADECENE	18435-45-5
1-NONADECYNE	C19H36	1-NONADECYNE	26186-01-6
1-NONALDEHYDE	C9H18O	1-NONANAL	124-19-6
NONALOL	C9H20O	1-NONANOL	143-08-8
1-NONANAL	C9H18O	1-NONANAL	124-19-6
n-NONANE	C9H20	n-NONANE	111-84-2
1-NONANECARBOXYLIC ACID	C10H20O2	n-DECANOIC ACID	334-48-5

Name (Synonym)	Formula	Name in Tables	CAS Reg. No
NONANEDIOIC ACID	C9H16O4	AZELAIC ACID	123-99-9
n-NONANOIC ACID	C9H18O2	n-NONANOIC ACID	112-05-0
1-NONANOL	C9H20O	1-NONANOL	143-08-8
NONAN-1-OL	C9H20O	1-NONANOL	143-08-8
2-NONANOL	C9H20O	2-NONANOL	70419-06-6
1-NONENE	C9H18	1-NONENE	124-11-8
n-NONOIC ACID	C9H18O2	n-NONANOIC ACID	112-05-0
NONOX DPPD	C18H16N2	N,N'-DIPHENYL-p-PHENYLENEDIAMINE	74-31-7
NONYL ALCOHOL	C9H20O	1-NONANOL	143-08-8
sec-NONYL ALCOHOL	C9H20O	2,6-DIMETHYL-4-HEPTANOL	108-82-7
1-NONYL ALDEHYDE	C9H18O	1-NONANAL	124-19-6
n-NONYLAMINE	C9H21N	n-NONYLAMINE	112-20-9
n-NONYLBENZENE	C15H24	n-NONYLBENZENE	1081-77-2
NONYLCARBINOL	C10H22O	1-DECANOL	112-30-1
NONYL-DISULFIDE	C18H38S2	NONYL-DISULFIDE	----------
n-NONYLIC ACID	C9H18O2	n-NONANOIC ACID	112-05-0
n-NONYL MERCAPTAN	C9H20S	n-NONYL MERCAPTAN	1455-21-6
1-n-NONYLNAPHTHALENE	C19H26	1-n-NONYLNAPHTHALENE	26438-26-6
NONYLPHENOL	C15H24O	NONYLPHENOL	25154-52-3
NONYL-PROPYL-SULFIDE	C12H26S	NONYL-PROPYL-SULFIDE	----------
NONYL-SULFIDE	C18H38S	NONYL-SULFIDE	----------
1-NONYNE	C9H16	1-NONYNE	3452-09-3
NOPINEN	C10H16	beta-PINENE	127-91-3
NOPINENE	C10H16	beta-PINENE	127-91-3
2-NORBORNENE	C7H10	2-NORBORNENE	498-66-8
NORBORNYLENE	C7H10	2-NORBORNENE	498-66-8
NORCAMPHENE	C7H10	2-NORBORNENE	498-66-8
NORKOOL	C2H6O2	ETHYLENE GLYCOL	107-21-1
NORLEUCAMINE	C5H13N	n-PENTYLAMINE	110-58-7
NORVALAMINE	C4H11N	n-BUTYLAMINE	109-73-9
NO SCALD	C12H11N	DIPHENYLAMINE	122-39-4
NOVOSCABIN	C14H12O2	BENZYL BENZOATE	120-51-4
2-NP	C3H7NO2	2-NITROPROPANE	79-46-9
NPG	C5H12O2	NEOPENTYL GLYCOL	126-30-7
N-PHENYLACETAMIDE	C8H9NO	ACETANILIDE	103-84-4
NSC-2752	C4H4O4	FUMARIC ACID	110-17-8
NSC-3051	C2H5NO	N-METHYLFORMAMIDE	123-39-7
NSC-3138	C4H9NO	N,N-DIMETHYLACETAMIDE	127-19-5
NSC-5354	C6H8N2	o-PHENYLENEDIAMINE	95-54-5
NSC-5356	C3H7NO	N,N-DIMETHYLFORMAMIDE	68-12-2
NSC-60520	C3H4Cl2	2,3-DICHLOROPROPENE	78-88-6
NSC-65426	C3H8O2	1,3-PROPYLENE GLYCOL	504-63-2
NSC-744	C3H3Cl	PROPARGYL CHLORIDE	624-65-7
NSC-7764	C3H5NO	LACTONITRILE	78-97-7
NSC-8028	C4H11N	ISOBUTYLAMINE	78-81-9
NSC-8260	C4H6O	METHACROLEIN	78-85-3
NSC-87419	C7H11NO	CYCLOHEXYL ISOCYANATE	3173-53-3
NSC-8819	C3H4O	ACROLEIN	107-02-8
NTM	C10H10O4	DIMETHYL PHTHALATE	131-11-3
NUC SILICONE VS 7207	C8H24O4Si4	OCTAMETHYLCYCLOTETRASILOXANE	556-67-2
OAP	C6H19NSi2	HEXAMETHYLDISILAZANE	999-97-3
9,12-OCTADECADIENOIC ACID	C18H32O2	LINOLEIC ACID	60-33-3
cis-9,cis-12-OCTADECADIENOIC ACID	C18H32O2	LINOLEIC ACID	60-33-3
cis,cis-9,12-OCTADECADIENOIC ACID	C18H32O2	LINOLEIC ACID	60-33-3
n-OCTADECANE	C18H38	n-OCTADECANE	593-45-3
1-OCTADECANETHIOL	C18H38S	1-OCTADECANETHIOL	2885-00-9
OCTADECANOIC ACID	C18H36O2	STEARIC ACID	57-11-4
OCTADECANOIC ACID, BUTYL ESTER	C22H44O2	n-BUTYL STEARATE	123-95-5
1-OCTADECANOL	C18H38O	1-OCTADECANOL	112-92-5
n-OCTADECANOL	C18H38O	1-OCTADECANOL	112-92-5
OCTADECANOL	C18H38O	1-OCTADECANOL	112-92-5
1-OCTADECENE	C18H36	1-OCTADECENE	112-88-9
(Z)-9-OCTADECENOIC ACID METHYL ESTER	C19H36O2	METHYL OLEATE	112-62-9
n-OCTADECYL ALCOHOL	C18H38O	1-OCTADECANOL	112-92-5
OCTA DECYL ALCOHOL	C18H38O	1-OCTADECANOL	112-92-5
1-OCTADECYNE	C18H34	1-OCTADECYNE	629-89-0
OCTAFLUOROBUTENE-2	C4F8	OCTAFLUORO-2-BUTENE	360-89-4
OCTAFLUOROBUT-2-ENE	C4F8	OCTAFLUORO-2-BUTENE	360-89-4
OCTAFLUORO-2-BUTENE	C4F8	OCTAFLUORO-2-BUTENE	360-89-4
OCTAFLUOROCYCLOBUTANE	C4F8	OCTAFLUOROCYCLOBUTANE	115-25-3
OCTAFLUOROPROPANE	C3F8	OCTAFLUOROPROPANE	76-19-7
OCTAMETHYLCYCLOTETRASILOXANE	C8H24O4Si4	OCTAMETHYLCYCLOTETRASILOXANE	556-67-2
1-OCTANAL	C8H16O	1-OCTANAL	124-13-0
OCTANALDEHYDE	C8H16O	1-OCTANAL	124-13-0
1-OCTANAMINE	C8H19N	n-OCTYLAMINE	111-86-4
OCTAN AMYLU	C7H14O2	n-PENTYL ACETATE	628-63-7

736

Name (Synonym)	Formula	Name in Tables	CAS Reg. No
OCTAN n-BUTYLU	C6H12O2	n-BUTYL ACETATE	123-86-4
n-OCTANE	C8H18	n-OCTANE	111-65-9
1-OCTANECARBOXYLIC ACID	C9H18O2	n-NONANOIC ACID	112-05-0
1,8-OCTANEDICARBOXYLIC ACID	C10H18O4	SEBACIC ACID	111-20-6
tert-OCTANETHIOL	C8H18S	tert-OCTYL MERCAPTAN	141-59-3
OCTAN ETOKSYETYLU	C6H12O3	2-ETHOXYETHYL ACETATE	111-15-9
OCTAN ETYLU	C4H8O2	ETHYL ACETATE	141-78-6
OCTAN METYLU	C3H6O2	METHYL ACETATE	79-20-9
n-OCTANOIC ACID	C8H16O2	n-OCTANOIC ACID	124-07-2
1-OCTANOL	C8H18O	1-OCTANOL	111-87-5
2-OCTANOL	C8H18O	2-OCTANOL	123-96-6
n-OCTANOL	C8H18O	1-OCTANOL	111-87-5
OCTANOL	C8H18O	1-OCTANOL	111-87-5
2-OCTANONE	C8H16O	2-OCTANONE	111-13-7
OCTAN PROPYLU	C5H10O2	n-PROPYL ACETATE	109-60-4
1-OCTENE	C8H16	1-OCTENE	111-66-0
trans-2-OCTENE	C8H16	trans-2-OCTENE	13389-42-9
trans-3-OCTENE	C8H16	trans-3-OCTENE	14919-01-8
trans-4-OCTENE	C8H16	trans-4-OCTENE	14850-23-8
OCTIC ACID	C8H16O2	n-OCTANOIC ACID	124-07-2
OCTILIN	C8H18O	1-OCTANOL	111-87-5
n-OCTOIC ACID	C8H16O2	n-OCTANOIC ACID	124-07-2
OCTOWY ALDEHYD	C2H4O	ACETALDEHYDE	75-07-0
OCTOWY KWAS	C2H4O2	ACETIC ACID	64-19-7
OCTYL ACRYLATE	C11H20O2	2-ETHYLHEXYL ACRYLATE	103-11-7
OCTYL ALCOHOL, NORMAL-PRIMARY	C8H18O	1-OCTANOL	111-87-5
n-OCTYL ALDEHYDE	C8H16O	1-OCTANAL	124-13-0
n-OCTYLAMINE	C8H19N	n-OCTYLAMINE	111-86-4
n-OCTYLBENZENE	C14H22	n-OCTYLBENZENE	2189-60-8
OCTYL CARBINOL	C9H20O	1-NONANOL	143-08-8
OCTYL-DISULFIDE	C16H34S2	OCTYL-DISULFIDE	----------
n-OCTYL FORMATE	C9H18O2	n-OCTYL FORMATE	112-32-3
n-OCTYLIC ACID	C8H16O2	n-OCTANOIC ACID	124-07-2
n-OCTYL MERCAPTAN	C8H18S	n-OCTYL MERCAPTAN	111-88-6
tert-OCTYL MERCAPTAN	C8H18S	tert-OCTYL MERCAPTAN	141-59-3
T-OCTYL MERCAPTAN	C8H18S	tert-OCTYL MERCAPTAN	141-59-3
tert-OCTYLMERCAPTAN	C8H18S	tert-OCTYL MERCAPTAN	141-59-3
p-tert-OCTYLPHENOL	C14H22O	p-tert-OCTYLPHENOL	140-66-9
OCTYL-PROPYL-SULFIDE	C11H24S	OCTYL-PROPYL-SULFIDE	----------
OCTYL-SULFIDE	C16H34S	OCTYL-SULFIDE	----------
ODB	C6H4Cl2	o-DICHLOROBENZENE	95-50-1
ODCB	C6H4Cl2	o-DICHLOROBENZENE	95-50-1
OENANTHAL	C7H14O	1-HEPTANAL	111-71-7
OENANTHALDEHYDE	C7H14O	1-HEPTANAL	111-71-7
OENANTHIC ACID	C7H14O2	n-HEPTANOIC ACID	111-14-8
OENANTHIC ALDEHYDE	C7H14O	1-HEPTANAL	111-71-7
OENANTHOL	C7H14O	1-HEPTANAL	111-71-7
OENANTHYLIC ACID	C7H14O2	n-HEPTANOIC ACID	111-14-8
OIL of ANISEED	C10H12O	ANETHOLE	104-46-1
OIL of MIRBANE	C6H5NO2	NITROBENZENE	98-95-3
OIL of MYRBANE	C6H5NO2	NITROBENZENE	98-95-3
OIL OF NIOBE	C8H8O2	METHYL BENZOATE	93-58-3
OIL OF WINTERGREEN	C8H8O3	METHYL SALICYLATE	119-36-8
OIL SOLUBLE ANILINE YELLOW	C12H11N3	p-AMINOAZOBENZENE	60-09-3
OIL YELLOW AAB	C12H11N3	p-AMINOAZOBENZENE	60-09-3
OKTAMETHYLCYKLOTETRASILOXAN	C8H24O4Si4	OCTAMETHYLCYCLOTETRASILOXANE	556-67-2
OKTAN	C8H18	n-OCTANE	111-65-9
OKTANEN	C8H18	n-OCTANE	111-65-9
terc. OKTANTHIOL	C8H18S	tert-OCTYL MERCAPTAN	141-59-3
p-terc.OKTYLFENOL	C14H22O	p-tert-OCTYLPHENOL	140-66-9
OLAMINE	C2H7NO	MONOETHANOLAMINE	141-43-5
OLEFIANT GAS	C2H4	ETHYLENE	74-85-1
OLEIC ACID	C18H34O2	OLEIC ACID	112-80-1
ONCB	C6H4ClNO2	o-CHLORONITROBENZENE	88-73-3
ONT	C7H7NO2	o-NITROTOLUENE	88-72-2
O,O,O-TRIMETHYL PHOSPHATE	C3H9O4P	TRIMETHYL PHOSPHATE	512-56-1
OPTAL	C3H8O	n-PROPANOL	71-23-8
8-OQ	C9H7NO	8-HYDROXYQUINOLINE	148-24-3
ORANGE BASE CIBA II	C6H6N2O2	o-NITROANILINE	88-74-4
ORANGE BASE IRGA I	C6H6N2O2	m-NITROANILINE	99-09-2
ORANGE BASE IRGA II	C6H6N2O2	o-NITROANILINE	88-74-4
ORANGE GC BASE	C6H6ClN	m-CHLOROANILINE	108-42-9
ORANGE OIL	C8H10O	2-PHENYLETHANOL	60-12-8
ORANGE SALT CIBA II	C6H6N2O2	o-NITROANILINE	88-74-4
ORGANOL YELLOW	C12H11N3	p-AMINOAZOBENZENE	60-09-3
ORSIN	C6H8N2	p-PHENYLENEDIAMINE	106-50-3

737

Name (Synonym)	Formula	Name in Tables	CAS Reg. No
ORTHAMINE	C6H8N2	o-PHENYLENEDIAMINE	95-54-5
ORTHOCRESOL	C7H8O	o-CRESOL	95-48-7
ORTHODICHLOROBENZENE	C6H4Cl2	o-DICHLOROBENZENE	95-50-1
ORTHODICHLOROBENZOL	C6H4Cl2	o-DICHLOROBENZENE	95-50-1
ORTHOHYDROXYBENZOIC ACID	C7H6O3	SALICYLIC ACID	69-72-7
ORTHOTOLUIC ACID	C8H8O2	o-TOLUIC ACID	118-90-1
ORVINYLCARBINOL	C3H6O	ALLYL ALCOHOL	107-18-6
OSMOSOL EXTRA	C3H8O	n-PROPANOL	71-23-8
OSSIDO di MESITILE	C6H10O	MESITYL OXIDE	141-79-7
OTTANE	C8H18	n-OCTANE	111-65-9
1-OXA-4-AZACYCLOHEXANE	C4H9NO	MORPHOLINE	110-91-8
OXACYCLOBUTANE	C3H6O	1,3-PROPYLENE OXIDE	503-30-0
OXACYCLOPENTADIENE	C4H4O	FURAN	110-00-9
OXACYCLOPENTANE	C4H8O	TETRAHYDROFURAN	109-99-9
OXACYCLOPROPANE	C2H4O	ETHYLENE OXIDE	75-21-8
OXALIC ACID	C2H2O4	OXALIC ACID	144-62-7
OXALIC ACID, DIETHYL ESTER	C6H10O4	DIETHYL OXALATE	95-92-1
OXALIC ACID DINITRILE	C2N2	CYANOGEN	460-19-5
OXALONITRILE	C2N2	CYANOGEN	460-19-5
OXALYL CYANIDE	C2N2	CYANOGEN	460-19-5
OXANE	C2H4O	ETHYLENE OXIDE	75-21-8
3-OXAPENTANEDIOIC ACID	C4H6O5	DIGLYCOLIC ACID	110-99-6
3-OXAPENTANE-1,5-DIOL	C4H10O3	DIETHYLENE GLYCOL	111-46-6
3-OXA-1,5-PENTANEDIOL	C4H10O3	DIETHYLENE GLYCOL	111-46-6
OXAZOLE	C3H3NO	OXAZOLE	288-42-6
2-OXEPANONE	C6H10O2	epsilon-CAPROLACTONE	502-44-3
OXETAN	C3H6O	1,3-PROPYLENE OXIDE	503-30-0
OXIDATE LE	C8H8O2	METHYL BENZOATE	93-58-3
a,b-OXIDOETHANE	C2H4O	ETHYLENE OXIDE	75-21-8
OXIDOETHANE	C2H4O	ETHYLENE OXIDE	75-21-8
OXINE	C9H7NO	8-HYDROXYQUINOLINE	148-24-3
OXIRANE	C2H4O	ETHYLENE OXIDE	75-21-8
OXITOL	C4H10O2	2-ETHOXYETHANOL	110-80-5
2-OXOBORNANE	C10H16O	CAMPHOR	76-22-2
3-OXOBUTANOIC ACID ETHYL ESTER	C6H10O2	n-PROPYL ACRYLATE	925-60-0
3-OXOBUTANOIC ACID METHYL ESTER	C5H8O3	METHYL ACETOACETATE	105-45-3
a-OXODIPHENYLMETHANE	C13H10O	BENZOPHENONE	119-61-9
2-OXOHEXAMETHYLENIMINE	C6H11NO	epsilon-CAPROLACTAM	105-60-2
OXOLANE	C4H8O	TETRAHYDROFURAN	109-99-9
OXOLE	C4H4O	FURAN	110-00-9
3-OXO-I-GULOFURANOLACTONE	C6H8O6	ASCORBIC ACID	50-81-7
4-OXOPENTANOIC ACID	C5H8O3	LEVULINIC ACID	123-76-2
2-OXOPROPANOIC ACID	C3H4O3	PYRUVIC ACID	127-17-3
2-OXOPROPIONIC ACID	C3H4O3	PYRUVIC ACID	127-17-3
2-OXOPYRROLIDINE	C4H7NO	2-PYRROLIDONE	616-45-5
4-OXOVALERIC ACID	C5H8O3	LEVULINIC ACID	123-76-2
p-OXYBENZALDEHYDE	C7H6O2	p-HYDROXYBENZALDEHYDE	123-08-0
OXYBENZENE	C6H6O	PHENOL	108-95-2
OXYBENZOPYRIDINE	C9H7NO	8-HYDROXYQUINOLINE	148-24-3
OXYBIS	C2H4Cl2O	BIS(CHLOROMETHYL)ETHER	542-88-1
OXYBISACETIC ACID	C4H6O5	DIGLYCOLIC ACID	110-99-6
2,2'-OXYBISACETIC ACID	C4H6O5	DIGLYCOLIC ACID	110-99-6
1,1'-OXYBIS(BUTANE)	C8H18O	DI-n-BUTYL ETHER	142-96-1
2,2'-OXYBISBUTANE	C8H18O	DI-sec-BUTYL ETHER	6863-58-7
1,1'-OXYBISETHANE	C4H10O	DIETHYL ETHER	60-29-7
1,1'-(OXYBIS(2,1-ETHANEDIYLOXY))BISBUTANE	C12H26O3	DIETHYLENE GLYCOL DI-n-BUTYL ETHER	112-73-2
2,2'-OXYBISETHANOL	C4H10O3	DIETHYLENE GLYCOL	111-46-6
1,1'-OXYBISETHENE	C4H6O	DIVINYL ETHER	109-93-3
2,2'-(OXYBIS(ETHYLENEOXY))DIETHANOL	C8H18O5	TETRAETHYLENE GLYCOL	112-60-7
1,1'-OXYBISHEXANE	C12H26O	DI-n-HEXYL ETHER	112-58-3
OXYBISMETHANE	C2H6O	DIMETHYL ETHER	115-10-6
1,1'-OXYBISPENTANE	C10H22O	DI-n-PENTYL ETHER	693-65-2
1,1'-OXYBISPROPANE	C6H14O	DI-n-PROPYL ETHER	111-43-3
OXYBIS(TRIMETHYLSILANE)	C6H18OSi2	HEXAMETHYLDISILOXANE	107-46-0
OXYCARBON SULFIDE	COS	CARBONYL SULFIDE	463-58-1
o-OXYCHINOLIN	C9H7NO	8-HYDROXYQUINOLINE	148-24-3
OXYCHINOLIN	C9H7NO	8-HYDROXYQUINOLINE	148-24-3
OXYDE de CARBONE	CO	CARBON MONOXIDE	630-08-0
OXYDE de MESITYLE	C6H10O	MESITYL OXIDE	141-79-7
OXYDE de PROPYLENE	C3H6O	1,2-PROPYLENE OXIDE	75-56-9
OXYDE dETHYLE	C4H10O	DIETHYL ETHER	60-29-7
2,2'-OXYDIACETIC ACID	C4H6O5	DIGLYCOLIC ACID	110-99-6
OXYDIACETIC ACID	C4H6O5	DIGLYCOLIC ACID	110-99-6
2,2'-OXYDIETHANOL	C4H10O3	DIETHYLENE GLYCOL	111-46-6
OXYDIETHANOLIC ACID	C4H6O5	DIGLYCOLIC ACID	110-99-6
b,b'-OXYDIPROPIONITRILE	C6H8N2O	BIS(CYANOETHYL)ETHER	1656-48-0

Name (Synonym)	Formula	Name in Tables	CAS Reg. No
OXYFUME	C2H4O	ETHYLENE OXIDE	75-21-8
OXYFUME 12	C2H4O	ETHYLENE OXIDE	75-21-8
OXYMETHYLENE	CH2O	FORMALDEHYDE	50-00-0
OXYPHENIC ACID	C6H6O2	1,2-BENZENEDIOL	120-80-9
8-OXYQUINOLINE	C9H7NO	8-HYDROXYQUINOLINE	148-24-3
OXYQUINOLINE	C9H7NO	8-HYDROXYQUINOLINE	148-24-3
OXYTOL ACETATE	C6H12O3	2-ETHOXYETHYL ACETATE	111-15-9
m-OXYTOLUENE	C7H8O	m-CRESOL	108-39-4
o-OXYTOLUENE	C7H8O	o-CRESOL	95-48-7
p-OXYTOLUENE	C7H8O	p-CRESOL	106-44-5
PALATINOL A	C12H14O4	DIETHYL PHTHALATE	84-66-2
PALATINOL C	C16H22O4	DIBUTYL PHTHALATE	84-74-2
PALATINOL M	C10H10O4	DIMETHYL PHTHALATE	131-11-3
PARAAMINODIPHENYL	C12H11N	p-AMINODIPHENYL	92-67-1
PARABROMOTOLUENE	C7H7Br	p-BROMOTOLUENE	106-38-7
PARACETALDEHYDE	C6H12O3	PARALDEHYDE	123-63-7
PARACHLOROPHENOL	C6H5ClO	p-CHLOROPHENOL	106-48-9
PARACIDE	C6H4Cl2	p-DICHLOROBENZENE	106-46-7
PARA CRYSTALS	C6H4Cl2	p-DICHLOROBENZENE	106-46-7
PARACYMENE	C10H14	p-CYMENE	99-87-6
PARACYMOL	C10H14	p-CYMENE	99-87-6
PARADI	C6H4Cl2	p-DICHLOROBENZENE	106-46-7
PARADICHLORBENZOL	C6H4Cl2	p-DICHLOROBENZENE	106-46-7
PARADICHLOROBENZENE	C6H4Cl2	p-DICHLOROBENZENE	106-46-7
PARADICHLOROBENZOL	C6H4Cl2	p-DICHLOROBENZENE	106-46-7
PARAHYDROXYBENZALDEHYDE	C7H6O2	p-HYDROXYBENZALDEHYDE	123-08-0
PARAL	C6H12O3	PARALDEHYDE	123-63-7
PARALDEHYD	C6H12O3	PARALDEHYDE	123-63-7
PARALDEHYDE	C6H12O3	PARALDEHYDE	123-63-7
PARALDEIDE	C6H12O3	PARALDEHYDE	123-63-7
PARAMETHYL PHENOL	C7H8O	p-CRESOL	106-44-5
PARANAPHTHALENE	C14H10	ANTHRACENE	120-12-7
PARAPHENOLAZO ANILINE	C12H11N3	p-AMINOAZOBENZENE	60-09-3
PCHO	C6H12O3	PARALDEHYDE	123-63-7
PCL	C5Cl6	HEXACHLOROCYCLOPENTADIENE	77-47-4
PE	C5H12O4	PENTAERYTHRITOL	115-77-5
b-PEA	C8H10O	2-PHENYLETHANOL	60-12-8
PEA	C8H10O	2-PHENYLETHANOL	60-12-8
PEARL STEARIC	C18H36O2	STEARIC ACID	57-11-4
PEAR OIL	C7H14O2	ISOPENTYL ACETATE	123-92-2
PELAGOL GREY C	C6H6O2	1,2-BENZENEDIOL	120-80-9
PELAGOL GREY RS	C6H6O2	1,3-BENZENEDIOL	108-46-3
PELARGIC ACID	C9H18O2	n-NONANOIC ACID	112-05-0
PELARGON	C9H18O2	n-NONANOIC ACID	112-05-0
PELARGONIC ACID	C9H18O2	n-NONANOIC ACID	112-05-0
PELARGONIC ALCOHOL	C9H20O	1-NONANOL	143-08-8
PELARGONIC ALDEHYDE	C9H18O	1-NONANAL	124-19-6
PENNFLOAT M	C12H26S	n-DODECYL MERCAPTAN	112-55-0
PENNFLOAT S	C12H26S	n-DODECYL MERCAPTAN	112-55-0
PENN SALT TD-183	C4Cl4S	TETRACHLOROTHIOPHENE	6012-97-1
PENPHENE	C4Cl4S	TETRACHLOROTHIOPHENE	6012-97-1
1,4,7,10,13-PENTAAZATRIDECANE	C8H23N5	TETRAETHYLENEPENTAMINE	112-57-2
PENT-ACETATE	C7H14O2	n-PENTYL ACETATE	628-63-7
PENTACHLOORETHAAN	C2HCl5	PENTACHLOROETHANE	76-01-7
PENTACHLORAETHAN	C2HCl5	PENTACHLOROETHANE	76-01-7
PENTACHLORETHANE	C2HCl5	PENTACHLOROETHANE	76-01-7
PENTACHLOROETHANE	C2HCl5	PENTACHLOROETHANE	76-01-7
PENTACHLOROPHENYL CHLORIDE	C6Cl6	HEXACHLOROBENZENE	118-74-1
PENTACLOROETANO	C2HCl5	PENTACHLOROETHANE	76-01-7
n-PENTADECANE	C15H32	n-PENTADECANE	629-62-9
1-PENTADECANECARBOXYLIC ACID	C16H32O2	n-HEXADECANOIC ACID	57-10-3
1-PENTADECANETHIOL	C15H32S	1-PENTADECANETHIOL	----------
PENTADECANOIC ACID	C15H30O2	PENTADECANOIC ACID	1002-84-2
1-PENTADECANOL	C15H32O	1-PENTADECANOL	629-76-5
1-PENTADECENE	C15H30	1-PENTADECENE	13360-61-7
PENTADECYLIC ACID	C15H30O2	PENTADECANOIC ACID	1002-84-2
PENTADECYL-PROPYL-SULFIDE	C18H38S	PENTADECYL-PROPYL-SULFIDE	----------
1-PENTADECYNE	C15H28	1-PENTADECYNE	765-13-9
1,2-PENTADIENE	C5H8	1,2-PENTADIENE	591-95-7
cis-1,3-PENTADIENE	C5H8	cis-1,3-PENTADIENE	1574-41-0
trans-1,3-PENTADIENE	C5H8	trans-1,3-PENTADIENE	2004-70-8
1,4-PENTADIENE	C5H8	1,4-PENTADIENE	591-93-5
2,3-PENTADIENE	C5H8	2,3-PENTADIENE	591-96-8
PENTAERYTHRITE	C5H12O4	PENTAERYTHRITOL	115-77-5
PENTAERYTHRITOL	C5H12O4	PENTAERYTHRITOL	115-77-5
PENTAERYTHRITOL TETRANITRATE	C5H8N4O12	PENTAERYTHRITOL TETRANITRATE	78-11-5

Name (Synonym)	Formula	Name in Tables	CAS Reg. No
PENTAETHYLBENZENE	C16H26	PENTAETHYLBENZENE	605-01-6
PENTAFLUOROETHANE	C2HF5	PENTAFLUOROETHANE	354-33-6
PENTALIN	C2HCl5	PENTACHLOROETHANE	76-01-7
PENTAMETHYLENE	C5H10	CYCLOPENTANE	287-92-3
PENTAMETHYLENE GLYCOL	C5H12O2	1,5-PENTANEDIOL	111-29-5
PENTAMETHYLENEIMINE	C5H11N	PIPERIDINE	110-89-4
2,2,3,3,4-PENTAMETHYLPENTANE	C10H22	2,2,3,3,4-PENTAMETHYLPENTANE	16747-44-7
2,2,3,4,4-PENTAMETHYLPENTANE	C10H22	2,2,3,4,4-PENTAMETHYLPENTANE	16747-45-8
PENTAN	C5H12	n-PENTANE	109-66-0
n-PENTANAL	C5H10O	VALERALDEHYDE	110-62-3
PENTANAL	C5H10O	VALERALDEHYDE	110-62-3
1-PENTANAMINE	C5H13N	n-PENTYLAMINE	110-58-7
PENTANDIOIC ACID	C5H8O4	GLUTARIC ACID	110-94-1
n-PENTANE	C5H12	n-PENTANE	109-66-0
tert-PENTANE	C5H12	NEOPENTANE	463-82-1
PENTANE	C5H12	n-PENTANE	109-66-0
PENTANE, 1-BROMO-	C5H11Br	1-BROMOPENTANE	110-53-2
PENTANE-1,5-DIOL	C5H12O2	1,5-PENTANEDIOL	111-29-5
PENTANE, 1,1'-OXYBIS-	C10H22O	DI-n-PENTYL ETHER	693-65-2
3-PENTANECARBOXYLIC ACID	C6H12O2	2-ETHYL BUTYRIC ACID	88-09-5
PENTANEDINITRILE	C5H6N2	GLUTARONITRILE	544-13-8
1,5-PENTANEDIOIC ACID	C5H8O4	GLUTARIC ACID	110-94-1
PENTANEDIOIC ACID	C5H8O4	GLUTARIC ACID	110-94-1
1,5-PENTANEDIOL	C5H12O2	1,5-PENTANEDIOL	111-29-5
2,4-PENTANEDIONE	C5H8O2	ACETYLACETONE	123-54-6
PENTANEDIONE	C5H8O2	ACETYLACETONE	123-54-6
PENTANEN	C5H12	n-PENTANE	109-66-0
PENTANENITRILE	C5H9N	VALERONITRILE	110-59-8
2-PENTANETHIOL, 2,4,4-TRIMETHYL-	C8H18S	tert-OCTYL MERCAPTAN	141-59-3
n-PENTANOIC ACID	C5H10O2	VALERIC ACID	109-52-4
PENTANOIC ACID	C5H10O2	VALERIC ACID	109-52-4
PENTAN-1-OL	C5H12O	1-PENTANOL	71-41-0
1-PENTANOL	C5H12O	1-PENTANOL	71-41-0
PENTANOL-1	C5H12O	1-PENTANOL	71-41-0
2-PENTANOL	C5H12O	2-PENTANOL	6032-29-7
PENTANOL-2	C5H12O	2-PENTANOL	6032-29-7
PENTAN-3-OL	C5H12O	3-PENTANOL	584-02-1
3-PENTANOL	C5H12O	3-PENTANOL	584-02-1
PENTANOL-3	C5H12O	3-PENTANOL	584-02-1
N-PENTANOL	C5H12O	1-PENTANOL	71-41-0
tert-PENTANOL	C5H12O	2-METHYL-2-BUTANOL	75-85-4
1-PENTANOL ACETATE	C7H14O2	n-PENTYL ACETATE	628-63-7
2-PENTANONE	C5H10O	2-PENTANONE	107-87-9
3-PENTANONE	C5H10O	DIETHYL KETONE	96-22-0
PENTANONE-3	C5H10O	DIETHYL KETONE	96-22-0
PENTAPHEN	C11H16O	p-tert-AMYLPHENOL	80-46-6
PENTASOL	C5H12O	1-PENTANOL	71-41-0
PENTEK	C5H12O4	PENTAERYTHRITOL	115-77-5
1-PENTENE	C5H10	1-PENTENE	109-67-1
cis-2-PENTENE	C5H10	cis-2-PENTENE	627-20-3
trans-2-PENTENE	C5H10	trans-2-PENTENE	646-04-8
1-PENTENE-3-YNE	C5H6	1-PENTENE-3-YNE	646-05-9
1-PENTENE-4-YNE	C5H6	1-PENTENE-4-YNE	871-28-3
PENTIFORMIC ACID	C6H12O2	n-HEXANOIC ACID	142-62-1
R-PENTINE	C5H6	CYCLOPENTADIENE	542-92-7
PENTOLE	C5H6	CYCLOPENTADIENE	542-92-7
1-PENTYL ACETATE	C7H14O2	n-PENTYL ACETATE	628-63-7
n-PENTYL ACETATE	C7H14O2	n-PENTYL ACETATE	628-63-7
PENTYL ACETATE	C7H14O2	n-PENTYL ACETATE	628-63-7
sec-PENTYL ALCOHOL	C5H12O	2-PENTANOL	6032-29-7
PENTYL ALCOHOL	C5H12O	1-PENTANOL	71-41-0
tert-PENTYL-ALCOHOL	C5H12O	tert-PENTYL-ALCOHOL	75-85-4
n-PENTYLAMINE	C5H13N	n-PENTYLAMINE	110-58-7
PENTYLAMINE	C5H13N	n-PENTYLAMINE	110-58-7
n-PENTYLBENZENE	C11H16	n-PENTYLBENZENE	538-68-1
1-PENTYL BROMIDE	C5H11Br	1-BROMOPENTANE	110-53-2
n-PENTYL BROMIDE	C5H11Br	1-BROMOPENTANE	110-53-2
PENTYL BROMIDE	C5H11Br	1-BROMOPENTANE	110-53-2
3-PENTYLCARBINOL	C6H14O	2-ETHYL-1-BUTANOL	97-95-0
sec-PENTYLCARBINOL	C6H14O	2-ETHYL-1-BUTANOL	97-95-0
PENTYLCARBINOL	C6H14O	1-HEXANOL	111-27-3
PENTYLCYCLOHEXANE	C11H22	PENTYLCYCLOHEXANE	4292-92-6
1,5-PENTYLENE GLYCOL	C5H12O2	1,5-PENTANEDIOL	111-29-5
n-PENTYL FORMATE	C6H12O2	n-PENTYL FORMATE	638-49-3
PENTYL FORMATE	C6H12O2	n-PENTYL FORMATE	638-49-3
PENTYLFORMIC ACID	C6H12O2	n-HEXANOIC ACID	142-62-1

Name (Synonym)	Formula	Name in Tables	CAS Reg. No
n-PENTYL MERCAPTAN	C5H12S	n-PENTYL MERCAPTAN	110-66-7
PENTYL MERCAPTAN	C5H12S	n-PENTYL MERCAPTAN	110-66-7
1-PENTYLNAPHTHALENE	C15H18	1-PENTYLNAPHTHALENE	86-89-5
2-PENTYLNAPHTHALENE	C15H18	2-PENTYLNAPHTHALENE	93-22-1
p-tert-PENTYLPHENOL	C11H16O	p-tert-AMYLPHENOL	80-46-6
PENTYL-PROPYL-SULFIDE	C8H18S	PENTYL-PROPYL-SULFIDE	----------
PENTYL-SULFIDE	C10H22S	PENTYL-SULFIDE	----------
1-PENTYNE	C5H8	1-PENTYNE	627-19-0
2-PENTYNE	C5H8	2-PENTYNE	627-21-4
PERAWIN	C2Cl4	TETRACHLOROETHYLENE	127-18-4
PERBUTYL H	C4H10O2	t-BUTYL HYDROPEROXIDE	75-91-2
PERCHLORETHYLENE	C2Cl4	TETRACHLOROETHYLENE	127-18-4
PERCHLOROBENZENE	C6Cl6	HEXACHLOROBENZENE	118-74-1
PERCHLOROBUTADIENE	C4Cl6	HEXACHLORO-1,3-BUTADIENE	87-68-3
PERCHLOROCYCLOPENTADIENE	C5Cl6	HEXACHLOROCYCLOPENTADIENE	77-47-4
PERCHLOROETHANE	C2Cl6	HEXACHLOROETHANE	67-72-1
PERCHLOROMETHANE	CCl4	CARBON TETRACHLORIDE	56-23-5
PERCHLOROTHIOPHENE	C4Cl4S	TETRACHLOROTHIOPHENE	6012-97-1
PERFLUORO-2-BUTENE	C4F8	OCTAFLUORO-2-BUTENE	360-89-4
PERFLUOROACETIC ACID	C2HF3O2	TRIFLUOROACETIC ACID	76-05-1
PERFLUOROBUT-2-ENE	C4F8	OCTAFLUORO-2-BUTENE	360-89-4
PERFLUOROCYCLOBUTANE	C4F8	OCTAFLUOROCYCLOBUTANE	115-25-3
PERFLUOROETHANE	C2F6	HEXAFLUOROETHANE	76-16-4
PERFLUOROETHENE	C2F4	TETRAFLUOROETHYLENE	116-14-3
PERFLUOROETHYLENE	C2F4	TETRAFLUOROETHYLENE	116-14-3
PERFLUOROMETHANE	CF4	CARBON TETRAFLUORIDE	75-73-0
PERFLUOROPROPENE	C3F6	HEXAFLUOROPROPYLENE	116-15-4
PERFLUOROPROPYLENE	C3F6	HEXAFLUOROPROPYLENE	116-15-4
PERHYDROAZEPINE	C6H13N	HEXAMETHYLENEIMINE	111-49-9
2-PERHYDROAZEPINONE	C6H11NO	epsilon-CAPROLACTAM	105-60-2
PERIETHYLENENAPHTHALENE	C12H10	ACENAPHTHENE	83-32-9
PERK	C2Cl4	TETRACHLOROETHYLENE	127-18-4
PEROSSIDO di BUTILE TERZIARIO	C8H18O2	DI-t-BUTYL PEROXIDE	110-05-4
PEROXYDE de BUTYLE TERTIAIRE	C8H18O2	DI-t-BUTYL PEROXIDE	110-05-4
PERUSCABIN	C14H12O2	BENZYL BENZOATE	120-51-4
PESTMASTER	C2H4Br2	1,2-DIBROMOETHANE	106-93-4
PESTMASTER EDB-85	C2H4Br2	1,2-DIBROMOETHANE	106-93-4
PETROHOL	C3H8O	ISOPROPANOL	67-63-0
PETZINOL	C2HCl3	TRICHLOROETHYLENE	79-01-6
PG	C3H8O2	1,3-PROPYLENE GLYCOL	504-63-2
PG 12	C3H8O2	1,2-PROPYLENE GLYCOL	57-55-6
alpha-PHELLANDRENE	C10H16	alpha-PHELLANDRENE	99-83-2
beta-PHELLANDRENE	C1OH16	beta-PHELLANDRENE	555-10-2
PHENALGENE	C8H9NO	ACETANILIDE	103-84-4
PHENALGIN	C8H9NO	ACETANILIDE	103-84-4
PHENANTHREN	C14H10	PHENANTHRENE	85-01-8
PHENANTHRENE	C14H10	PHENANTHRENE	85-01-8
PHENANTRIN	C14H10	PHENANTHRENE	85-01-8
PHENE	C6H6	BENZENE	71-43-2
PHENETHANOL	C8H10O	2-PHENYLETHANOL	60-12-8
2-PHENETHYL ALCOHOL	C8H10O	2-PHENYLETHANOL	60-12-8
b-PHENETHYL ALCOHOL	C8H10O	2-PHENYLETHANOL	60-12-8
PHENETHYLENE	C8H8	STYRENE	100-42-5
p-PHENETIDINE	C8H11NO	p-PHENETIDINE	156-43-3
PHENETOLE	C8H10O	PHENETOLE	103-73-1
PHENIC ACID	C6H6O	PHENOL	108-95-2
PHENOHEP	C2Cl6	HEXACHLOROETHANE	67-72-1
PHENOL	C6H6O	PHENOL	108-95-2
PHENOL ALCOHOL	C6H6O	PHENOL	108-95-2
PHENOLCARBINOL	C7H8O	BENZYL ALCOHOL	100-51-6
PHENOL, 4-CHLORO-	C6H5ClO	p-CHLOROPHENOL	106-48-9
PHENOL, 2,3-DIMETHYL-	C8H10O	2,3-XYLENOL	526-75-0
PHENOL, p-(a-a-DIMETHYLBENZYL)-	C15H16O	p-CUMYLPHENOL	599-64-4
PHENOLE	C6H6O	PHENOL	108-95-2
PHENOL, p-ETHYL-	C8H10O	p-ETHYLPHENOL	123-07-9
PHENOL, 4-(1-METHYL-1-PHENETHYL)-	C15H16O	p-CUMYLPHENOL	599-64-4
PHENOL, THIO-	C6H6S	PHENYL MERCAPTAN	108-98-5
PHENOPYRIDINE	C9H7NO	8-HYDROXYQUINOLINE	148-24-3
PHENOXYBENZENE	C12H10O	DIPHENYL ETHER	101-84-8
PHENYLAMINE	C6H7N	ANILINE	62-53-3
4-PHENYLAMINODIPHENYLAMINE	C18H16N2	N,N'-DIPHENYL-p-PHENYLENEDIAMINE	74-31-7
p-PHENYLAMINODIPHENYLAMINE	C18H16N2	N,N'-DIPHENYL-p-PHENYLENEDIAMINE	74-31-7
p-PHENYLANILINE	C12H11N	p-AMINODIPHENYL	92-67-1
N-PHENYLANILINE	C12H11N	DIPHENYLAMINE	122-39-4
N-(PHENYLAZO)ANILINE	C12H11N3	1,3-DIPHENYLTRIAZENE	136-35-6
N-PHENYLBENEZENAMINE	C12H11N	DIPHENYLAMINE	122-39-4

Name (Synonym)	Formula	Name in Tables	CAS Reg. No
4-PHENYLBIPHENYL	C18H14	p-TERPHENYL	92-94-4
1-PHENYLBUTANE	C10H14	n-BUTYLBENZENE	104-51-8
2-PHENYLBUTANE	C10H14	sec-BUTYLBENZENE	135-98-8
PHENYLCARBIMIDE	C7H5NO	PHENYL ISOCYANATE	103-71-9
PHENYLCARBINOL	C7H8O	BENZYL ALCOHOL	100-51-6
PHENYL CARBONIMIDE	C7H5NO	PHENYL ISOCYANATE	103-71-9
PHENYL CARBOXYLIC ACID	C7H6O2	BENZOIC ACID	65-85-0
PHENYL CHLORIDE	C6H5Cl	MONOCHLOROBENZENE	108-90-7
PHENYL CHLOROFORM	C7H5Cl3	BENZOTRICHLORIDE	98-07-7
PHENYL CYANIDE	C7H5N	BENZONITRILE	100-47-0
PHENYLDIMETHYLCARBINOL	C9H12O	2-PHENYL-2-PROPANOL	617-94-7
4-PHENYLDIPHENYL	C18H14	p-TERPHENYL	92-94-4
PHENYLDODECAN	C18H30	n-DODECYLBENZENE	123-01-3
1,2-PHENYLENEDIAMINE	C6H8N2	o-PHENYLENEDIAMINE	95-54-5
1,3-PHENYLENEDIAMINE	C6H8N2	m-PHENYLENEDIAMINE	108-45-2
m-PHENYLENEDIAMINE	C6H8N2	m-PHENYLENEDIAMINE	108-45-2
o-PHENYLENEDIAMINE	C6H8N2	o-PHENYLENEDIAMINE	95-54-5
p-PHENYLENEDIAMINE	C6H8N2	p-PHENYLENEDIAMINE	106-50-3
PHENYLENEDIAMINE, PARA	C6H8N2	p-PHENYLENEDIAMINE	106-50-3
o-PHENYLENEDIOL	C6H6O2	1,2-BENZENEDIOL	120-80-9
PHENYLETHANE	C8H10	ETHYLBENZENE	100-41-4
2-PHENYLETHANOL	C8H10O	2-PHENYLETHANOL	60-12-8
b-PHENYLETHANOL	C8H10O	2-PHENYLETHANOL	60-12-8
1-PHENYLETHANONE	C8H8O	ACETOPHENONE	98-86-2
PHENYLETHENE	C8H8	STYRENE	100-42-5
2-PHENYLETHYL ALCOHOL	C8H10O	2-PHENYLETHANOL	60-12-8
b-PHENYLETHYL ALCOHOL	C8H10O	2-PHENYLETHANOL	60-12-8
PHENYLETHYL ALCOHOL	C8H10O	2-PHENYLETHANOL	60-12-8
PHENYLETHYLENE	C8H8	STYRENE	100-42-5
PHENYL ETHYL ETHER	C8H10O	PHENETOLE	103-73-1
PHENYL FLUORIDE	C6H5F	FLUOROBENZENE	462-06-6
PHENYLFLUOROFORM	C7H5F3	BENZOTRIFLUORIDE	98-08-8
N-PHENYLFORMAMIDE	C7H7NO	FORMANILIDE	103-70-8
PHENYL FORMAMIDE	C7H7NO	FORMANILIDE	103-70-8
PHENYLFORMIC ACID	C7H6O2	BENZOIC ACID	65-85-0
1-PHENYLHEXADECANE	C22H38	1-PHENYLHEXADECANE	1459-09-2
PHENYL HYDRATE	C6H6O	PHENOL	108-95-2
PHENYLHYDRAZIN	C6H8N2	PHENYLHYDRAZINE	100-63-0
PHENYLHYDRAZINE	C6H8N2	PHENYLHYDRAZINE	100-63-0
PHENYL HYDROXIDE	C6H6O	PHENOL	108-95-2
PHENYLIC ACID	C6H6O	PHENOL	108-95-2
PHENYL IODIDE	C6H5I	IODOBENZENE	591-50-4
PHENYL ISOCYANATE	C7H5NO	PHENYL ISOCYANATE	103-71-9
2-PHENYLISOPROPANOL	C9H12O	2-PHENYL-2-PROPANOL	617-94-7
PHENYL KETONE	C13H10O	BENZOPHENONE	119-61-9
PHENYL MERCAPTAN	C6H6S	PHENYL MERCAPTAN	108-98-5
PHENYLMETHANE	C7H8	TOLUENE	108-88-3
PHENYLMETHANOL	C7H8O	BENZYL ALCOHOL	100-51-6
PHENYLMETHYL ALCOHOL	C7H8O	BENZYL ALCOHOL	100-51-6
N-PHENYLMETHYLAMINE	C7H9N	N-METHYLANILINE	100-61-8
(PHENYLMETHYL)AMINE	C7H9N	BENZYLAMINE	100-46-9
PHENYL METHYL ETHER	C7H8O	ANISOLE	100-66-3
PHENYL METHYL KETONE	C8H8O	ACETOPHENONE	98-86-2
1-PHENYLNAPHTHALENE	C16H12	1-PHENYLNAPHTHALENE	605-02-7
1-PHENYLPENTADECANE	C21H36	1-PHENYLPENTADECANE	2131-18-2
PHENYL PERCHLORYL	C6Cl6	HEXACHLOROBENZENE	118-74-1
1-PHENYLPROPANE	C9H12	n-PROPYLBENZENE	103-65-1
2-PHENYLPROPANE	C9H12	CUMENE	98-82-8
2-PHENYL-2-PROPANOL	C9H12O	2-PHENYL-2-PROPANOL	617-94-7
2-PHENYLPROPENE	C9H10	alpha-METHYLSTYRENE	98-83-9
b-PHENYLPROPENE	C9H10	alpha-METHYLSTYRENE	98-83-9
2-PHENYLPROPYLENE	C9H10	alpha-METHYLSTYRENE	98-83-9
b-PHENYLPROPYLENE	C9H10	alpha-METHYLSTYRENE	98-83-9
1-PHENYLTETRADECANE	C20H34	1-PHENYLTETRADECANE	----------
PHENYLTRICHLOROMETHANE	C7H5Cl3	BENZOTRICHLORIDE	98-07-7
PHOSFLEX 179-C	C21H21O4P	TRI-o-CRESYL PHOSPHATE	78-30-8
PHOSGEN	CCl2O	PHOSGENE	75-44-5
PHOSGENE	CCl2O	PHOSGENE	75-44-5
PHOSPHORIC ACID, TRI-o-CRESYL ESTER	C21H21O4P	TRI-o-CRESYL PHOSPHATE	78-30-8
PHOSPHORIC ACID, TRIMETHYL ESTER	C3H9O4P	TRIMETHYL PHOSPHATE	512-56-1
PHOSPHORIC ACID, TRIPHENYL ESTER	C18H15O4P	TRIPHENYL PHOSPHATE	115-86-6
PHOSPHORIC ACID, TRIS(2-METHYLPHENYL) ESTER	C21H21O4P	TRI-o-CRESYL PHOSPHATE	78-30-8
PHOSPHORIC TRIS(DIMETHYLAMIDE)	C6H18N3OP	HEXAMETHYL PHOSPHORAMIDE	680-31-9
PHOSPHORYL HEXAMETHYLTRIAMIDE	C6H18N3OP	HEXAMETHYL PHOSPHORAMIDE	680-31-9
1,3-PHTHALANDIONE	C8H4O3	PHTHALIC ANHYDRIDE	85-44-9
m-PHTHALIC ACID	C8H6O4	ISOPHTHALIC ACID	121-91-5

Name (Synonym)	Formula	Name in Tables	CAS Reg. No
PHTHALIC ACID	C8H6O4	PHTHALIC ACID	88-99-3
PHTHALIC ACID ANHYDRIDE	C8H4O3	PHTHALIC ANHYDRIDE	85-44-9
PHTHALIC ACID, DIETHYL ESTER	C12H14O4	DIETHYL PHTHALATE	84-66-2
PHTHALIC ACID METHYL ESTER	C10H10O4	DIMETHYL PHTHALATE	131-11-3
PHTHALIC ANHYDRIDE	C8H4O3	PHTHALIC ANHYDRIDE	85-44-9
m-PHTHALIC DICHLORIDE	C8H4Cl2O2	ISOPHTHALOYL CHLORIDE	99-63-8
PHTHALOL	C12H14O4	DIETHYL PHTHALATE	84-66-2
m-PHTHALOYL CHLORIDE	C8H4Cl2O2	ISOPHTHALOYL CHLORIDE	99-63-8
PHTHALSAEUREANHYDRID	C8H4O3	PHTHALIC ANHYDRIDE	85-44-9
PHTHALSAEUREDIAETHYLESTER	C12H14O4	DIETHYL PHTHALATE	84-66-2
PHTHALSAEUREDIMETHYLESTER	C10H10O4	DIMETHYL PHTHALATE	131-11-3
2-PICOLINE	C6H7N	2-METHYLPYRIDINE	109-06-8
4-PICOLINE	C6H7N	4-METHYLPYRIDINE	108-89-4
a-PICOLINE	C6H7N	2-METHYLPYRIDINE	109-06-8
b-PICOLINE	C6H7N	3-METHYLPYRIDINE	108-99-6
g-PICOLINE	C6H7N	4-METHYLPYRIDINE	108-89-4
m-PICOLINE	C6H7N	3-METHYLPYRIDINE	108-99-6
o-PICOLINE	C6H7N	2-METHYLPYRIDINE	109-06-8
p-PICOLINE	C6H7N	4-METHYLPYRIDINE	108-89-4
PIMELIC KETONE	C6H10O	CYCLOHEXANONE	108-94-1
PINAKON	C6H14O2	HEXYLENE GLYCOL	107-41-5
2(10)-PINENE	C10H16	beta-PINENE	127-91-3
alpha-PINENE	C10H16	alpha-PINENE	80-56-8
beta-PINENE	C10H16	beta-PINENE	127-91-3
PIPERAZIDINE	C4H10N2	PIPERAZINE	110-85-0
PIPERAZIN	C4H10N2	PIPERAZINE	110-85-0
PIPERAZINE	C4H10N2	PIPERAZINE	110-85-0
PIPERAZINE, anhydrous	C4H10N2	PIPERAZINE	110-85-0
PIPERIDIN	C5H11N	PIPERIDINE	110-89-4
PIPERIDINE	C5H11N	PIPERIDINE	110-89-4
trans-PIPERYLENE	C5H8	trans-1,3-PENTADIENE	2004-70-8
PIRIDINA	C5H5N	PYRIDINE	110-86-1
PIRYDYNA	C5H5N	PYRIDINE	110-86-1
PLACIDOL E	C12H14O4	DIETHYL PHTHALATE	84-66-2
PLASTICIZER DIHEXYL ADIPATE	C18H34O4	DIHEXYL ADIPATE	110-33-8
PM 2763	C7H14O	DIISOPROPYL KETONE	565-80-0
PNCB	C6H4ClNO2	p-CHLORONITROBENZENE	100-00-5
PNT	C7H7NO2	p-NITROTOLUENE	99-99-0
POLAAX	C18H38O	1-OCTADECANOL	112-92-5
POLYCIZER 332	C22H44O2	n-BUTYL STEARATE	123-95-5
POLYCIZER DBP	C16H22O4	DIBUTYL PHTHALATE	84-74-2
POLYCIZER DBS	C18H34O4	DIBUTYL SEBACATE	109-43-3
POLYOXYMETHYLENE	C3H6O3	TRIOXANE	110-88-3
POLY-SOLV	C6H14O3	2-(2-ETHOXYETHOXY)ETHANOL	111-90-0
POLY-SOLV DB	C8H18O3	DIETHYLENE GLYCOL MONOBUTYL ETHER	112-34-5
POLY-SOLV DM	C5H12O3	2-(2-METHOXYETHOXY)ETHANOL	111-77-3
POLY-SOLV EE	C4H10O2	2-ETHOXYETHANOL	110-80-5
POLY-SOLV EE ACETATE	C6H12O3	2-ETHOXYETHYL ACETATE	111-15-9
POLY-SOLV EM	C3H8O2	2-METHOXYETHANOL	109-86-4
PONTAMINE DEVELOPER TN	C7H10N2	TOLUENEDIAMINE	95-80-7
PPD	C6H8N2	p-PHENYLENEDIAMINE	106-50-3
PRIMARY AMYL ACETATE	C7H14O2	n-PENTYL ACETATE	628-63-7
PRIMARY AMYL ALCOHOL	C5H12O	1-PENTANOL	71-41-0
PRIMARY DECYL ALCOHOL	C10H22O	1-DECANOL	112-30-1
PRIMARY OCTYL ALCOHOL	C8H18O	1-OCTANOL	111-87-5
PRIST	C3H8O2	2-METHOXYETHANOL	109-86-4
PROFUME	CH3Br	METHYL BROMIDE	74-83-9
PROPADIENE	C3H4	PROPADIENE	463-49-0
PROPALDEHYDE	C3H6O	n-PROPIONALDEHYDE	123-38-6
PROPANAL	C3H6O	n-PROPIONALDEHYDE	123-38-6
2-PROPANAMINE	C3H9N	ISOPROPYLAMINE	75-31-0
PROPANAMINE	C3H9N	n-PROPYLAMINE	107-10-8
1-PROPANAMINE, 2-METHYL-	C4H11N	ISOBUTYLAMINE	78-81-9
2-PROPANAMINE, N-(1-METHYLETHYL)-	C6H15N	DIISOPROPYLAMINE	108-18-9
PROPANE	C3H8	PROPANE	74-98-6
1-PROPANECARBOXYLIC ACID	C4H8O2	n-BUTYRIC ACID	107-92-6
1,2-PROPANEDIAMINE	C3H10N2	1,2-PROPANEDIAMINE	78-90-0
1,3-PROPANEDICARBOXYLIC ACID	C5H8O4	GLUTARIC ACID	110-94-1
PROPANEDINITRILE	C3H2N2	MALONONITRILE	109-77-3
PROPANEDIOIC ACID, DIETHYL ESTER	C7H12O4	DIETHYL MALONATE	105-53-3
PROPANE-1,2-DIOL	C3H8O2	1,2-PROPYLENE GLYCOL	57-55-6
PROPANE-1,3-DIOL	C3H8O2	1,3-PROPYLENE GLYCOL	504-63-2
1,3-PROPANEDIOL, 2-ETHYL-2-(HYDROXYMETHYL)-	C6H14O3	TRIMETHYLOLPROPANE	77-99-6
PROPANE, 1-METHOXY-	C4H10O	METHYL-PROPYL-ETHER	557-17-5
PROPANE, 2-METHOXY-2-METHYL-	C5H12O	METHYL tert-BUTYL ETHER	1634-04-4
PROPANENITRILE	C3H5N	PROPIONITRILE	107-12-0

Name (Synonym)	Formula	Name in Tables	CAS Reg. No
a-g-PROPANE OXIDE	C3H6O	1,3-PROPYLENE OXIDE	503-30-0
PROPANE, 1,1'-OXYBIS-	C6H14O	DI-n-PROPYL ETHER	111-43-3
2-PROPANETHIOL	C3H8S	ISOPROPYL MERCAPTAN	75-33-2
1,2,3-PROPANETRIOL	C3H8O3	GLYCEROL	56-81-5
1,2,3-PROPANETRIOL TRIACETATE	C9H14O6	GLYCERYL TRIACETATE	102-76-1
1,2,3-PROPANETRIOL, TRINITRATE	C3H5N3O9	NITROGLYCERINE	55-63-0
1,2,3-PROPANETRIYL NITRATE	C3H5N3O9	NITROGLYCERINE	55-63-0
PROPANOIC ACID	C3H6O2	PROPIONIC ACID	79-09-4
PROPANOIC ACID BUTYLESTER	C7H14O2	n-BUTYL PROPIONATE	590-01-2
PROPANOIC ACID, ETHENYL ESTER	C5H8O2	VINYL PROPIONATE	105-38-4
PROPANOIC ACID, 3-MERCAPTO-	C3H6O2S	3-MERCAPTOPROPIONIC ACID	107-96-0
PROPANOIC ACID, METHYL ESTER	C4H8O2	METHYL PROPIONATE	554-12-1
PROPANOIC ACID, 2-OXO-(9CI)	C3H4O3	PYRUVIC ACID	127-17-3
PROPANOIC ANHYDRIDE	C6H10O3	PROPIONIC ANHYDRIDE	123-62-6
PROPANOL-1	C3H8O	n-PROPANOL	71-23-8
1-PROPANOL	C3H8O	n-PROPANOL	71-23-8
PROPAN-2-OL	C3H8O	ISOPROPANOL	67-63-0
2-PROPANOL	C3H8O	ISOPROPANOL	67-63-0
i-PROPANOL	C3H8O	ISOPROPANOL	67-63-0
n-PROPANOL	C3H8O	n-PROPANOL	71-23-8
1,3-PROPANOLAMINE	C3H9NO	3-AMINO-1-PROPANOL	156-87-6
PROPANOLAMINE	C3H9NO	3-AMINO-1-PROPANOL	156-87-6
PROPANOLE	C3H8O	n-PROPANOL	71-23-8
PROPANOLEN	C3H8O	n-PROPANOL	71-23-8
PROPANOLI	C3H8O	n-PROPANOL	71-23-8
PROPANOLIDE	C3H4O2	beta-PROPIOLACTONE	57-57-8
PROPANOL, OXYBIS-	C6H14O3	DIPROPYLENE GLYCOL	25265-71-8
2-PROPANONE	C3H6O	ACETONE	67-64-1
PROPANONE	C3H6O	ACETONE	67-64-1
PROPARGYL ALCOHOL	C3H4O	PROPARGYL ALCOHOL	107-19-7
PROPARGYL CHLORIDE	C3H3Cl	PROPARGYL CHLORIDE	624-65-7
PROPELLANT 114	C2Cl2F4	1,2-DICHLOROTETRAFLUOROETHANE	76-14-2
PROPELLANT 12	CCl2F2	DICHLORODIFLUOROMETHANE	75-71-8
PROPELLANT 22	CHClF2	CHLORODIFLUOROMETHANE	75-45-6
PROPELLANT C318	C4F8	OCTAFLUOROCYCLOBUTANE	115-25-3
2-PROPENAL	C3H4O	ACROLEIN	107-02-8
PROP-2-EN-1-AL	C3H4O	ACROLEIN	107-02-8
2-PROPENAL, 2-METHYL-	C4H6O	METHACROLEIN	78-85-3
2-PROPENAMIDE	C3H5NO	ACRYLAMIDE	79-06-1
PROPENAMIDE	C3H5NO	ACRYLAMIDE	79-06-1
2-PROPEN-1-AMINE	C3H7N	ALLYLAMINE	107-11-9
2-PROPENAMINE	C3H7N	ALLYLAMINE	107-11-9
1-PROPENE	C3H6	PROPYLENE	115-07-1
PROPENE ACID	C3H4O2	ACRYLIC ACID	79-10-7
1-PROPENE, 3-IODO-	C3H5I	3-IODO-1-PROPENE	556-56-9
2-PROPENENITRILE	C3H3N	ACRYLONITRILE	107-13-1
PROPENENITRILE	C3H3N	ACRYLONITRILE	107-13-1
PROPENE OXIDE	C3H6O	1,2-PROPYLENE OXIDE	75-56-9
2-PROPENOIC ACID	C3H4O2	ACRYLIC ACID	79-10-7
PROPENOIC ACID	C3H4O2	ACRYLIC ACID	79-10-7
2-PROPENOIC ACID, ETHYL ESTER	C5H8O2	ETHYL ACRYLATE	140-88-5
2-PROPENOIC ACID-2-ETHYLHEXYL ESTER	C11H20O2	2-ETHYLHEXYL ACRYLATE	103-11-7
2-PROPENOIC ACID-2-HYDROXYETHYL ESTER	C5H8O3	2-HYDROXYETHYL ACRYLATE	818-61-1
2-PROPENOIC ACID, 2-METHYL-	C4H6O2	METHACRYLIC ACID	79-41-4
2-PROPENOIC ACID METHYL ESTER	C4H6O2	METHYL ACRYLATE	96-33-3
PROPENOIC ACID METHYL ESTER	C4H6O2	METHYL ACRYLATE	96-33-3
2-PROPENOIC ACID, 2-METHYL-, METHYL ESTER	C5H8O2	METHYL METHACRYLATE	80-62-6
2-PROPENOIC ACID-2-METHYLPROPYL ESTER	C7H12O2	ISOBUTYL ACRYLATE	106-63-8
2-PROPEN-1-OL	C3H6O	ALLYL ALCOHOL	107-18-6
PROPEN-1-OL-3	C3H6O	ALLYL ALCOHOL	107-18-6
1-PROPEN-3-OL	C3H6O	ALLYL ALCOHOL	107-18-6
PROPENOL	C3H6O	ALLYL ALCOHOL	107-18-6
2-PROPEN-1-ONE	C3H4O	ACROLEIN	107-02-8
2-PROPENYL ALCOHOL	C3H6O	ALLYL ALCOHOL	107-18-6
PROPENYL ALCOHOL	C3H6O	ALLYL ALCOHOL	107-18-6
4-PROPENYLANISOLE	C10H12O	ANETHOLE	104-46-1
p-1-PROPENYLANISOLE	C10H12O	ANETHOLE	104-46-1
cis-PROPENYLBENZENE	C9H10	cis-PROPENYLBENZENE	766-90-5
trans-PROPENYLBENZENE	C9H10	trans-PROPENYLBENZENE	873-66-5
2-PROPENYL CHLORIDE	C3H5Cl	3-CHLOROPROPENE	107-05-1
p-PROPENYLPHENYL METHYL ETHER	C10H12O	ANETHOLE	104-46-1
PROPINE	C3H4	METHYLACETYLENE	74-99-7
PROPIOLACTONE	C3H4O2	beta-PROPIOLACTONE	57-57-8
1,3-PROPIOLACTONE	C3H4O2	beta-PROPIOLACTONE	57-57-8
3-PROPIOLACTONE	C3H4O2	beta-PROPIOLACTONE	57-57-8
beta-PROPIOLACTONE	C3H4O2	beta-PROPIOLACTONE	57-57-8

Name (Synonym)	Formula	Name in Tables	CAS Reg. No
n-PROPIONALDEHYDE	C3H6O	n-PROPIONALDEHYDE	123-38-6
PROPIONATE de METHYLE	C4H8O2	METHYL PROPIONATE	554-12-1
PROPIONE	C5H10O	DIETHYL KETONE	96-22-0
PROPIONIC ACID	C3H6O2	PROPIONIC ACID	79-09-4
PROPIONIC ACID ANHYDRIDE	C6H10O3	PROPIONIC ANHYDRIDE	123-62-6
PROPIONIC ACID GRAIN PRESERVER	C3H6O2	PROPIONIC ACID	79-09-4
PROPIONIC ACID, 2-METHYL-	C4H8O2	ISOBUTYRIC ACID	79-31-2
PROPIONIC ACID, 2-METHYLENE-	C4H6O2	METHACRYLIC ACID	79-41-4
PROPIONIC ACID, PROPYL ESTER	C6H12O2	n-PROPYL PROPIONATE	106-36-5
PROPIONIC ALDEHYDE	C3H6O	n-PROPIONALDEHYDE	123-38-6
PROPIONIC ANHYDRIDE	C6H10O3	PROPIONIC ANHYDRIDE	123-62-6
PROPIONIC NITRILE	C3H5N	PROPIONITRILE	107-12-0
PROPIONITRILE	C3H5N	PROPIONITRILE	107-12-0
PROPIONITRILE, 2-HYDROXY-	C3H5NO	LACTONITRILE	78-97-7
PROPIONITRILE, 3-METHOXY-	C4H7NO	3-METHOXYPROPIONITRILE	110-67-8
b-PROPIONOLACTONE	C3H4O2	beta-PROPIOLACTONE	57-57-8
PROPIONYL OXIDE	C6H10O3	PROPIONIC ANHYDRIDE	123-62-6
2-PROPOXYETHANOL	C5H12O2	ETHYLENE GLYCOL MONOPROPYL ETHER	2807-30-9
b-PROPRIOLACTONE	C3H4O2	beta-PROPIOLACTONE	57-57-8
b-PROPROLACTONE	C3H4O2	beta-PROPIOLACTONE	57-57-8
1-PROPYL ACETATE	C5H10O2	n-PROPYL ACETATE	109-60-4
2-PROPYL ACETATE	C5H10O2	ISOPROPYL ACETATE	108-21-4
n-PROPYL ACETATE	C5H10O2	n-PROPYL ACETATE	109-60-4
PROPYL ACETATE	C5H10O2	n-PROPYL ACETATE	109-60-4
PROPYLACETIC ACID	C5H10O2	VALERIC ACID	109-52-4
n-PROPYL ACRYLATE	C6H10O2	n-PROPYL ACRYLATE	925-60-0
1-PROPYL ALCOHOL	C3H8O	n-PROPANOL	71-23-8
sec-PROPYL ALCOHOL	C3H8O	ISOPROPANOL	67-63-0
PROPYL ALCOHOL	C3H8O	n-PROPANOL	71-23-8
PROPYL ALDEHYDE	C3H6O	n-PROPIONALDEHYDE	123-38-6
n-PROPYL ALKOHOL	C3H8O	n-PROPANOL	71-23-8
i-PROPYLALKOHOL	C3H8O	ISOPROPANOL	67-63-0
2-PROPYLAMINE	C3H9N	ISOPROPYLAMINE	75-31-0
n-PROPYLAMINE	C3H9N	n-PROPYLAMINE	107-10-8
sec-PROPYLAMINE	C3H9N	ISOPROPYLAMINE	75-31-0
n-PROPYLBENZENE	C9H12	n-PROPYLBENZENE	103-65-1
PROPYL BENZENE	C9H12	n-PROPYLBENZENE	103-65-1
PROPYL BROMIDE	C3H7Br	1-BROMOPROPANE	106-94-5
n-PROPYL n-BUTYRATE	C7H14O2	n-PROPYL n-BUTYRATE	105-66-8
PROPYL BUTYRATE	C7H14O2	n-PROPYL n-BUTYRATE	105-66-8
PROPYLCARBINOL	C4H10O	n-BUTANOL	71-36-3
N-PROPYLCARBINYL CHLORIDE	C4H9Cl	n-BUTYL CHLORIDE	109-69-3
n-PROPYL CHLORIDE	C3H7Cl	n-PROPYL CHLORIDE	540-54-5
PROPYL CYANIDE	C4H7N	n-BUTYRONITRILE	109-74-0
n-PROPYLCYCLOHEXANE	C9H18	n-PROPYLCYCLOHEXANE	1678-92-8
n-PROPYLCYCLOPENTANE	C8H16	n-PROPYLCYCLOPENTANE	2040-96-2
PROPYLENE	C3H6	PROPYLENE	115-07-1
PROPYLENE ALDEHYDE	C4H6O	trans-CROTONALDEHYDE	123-73-9
PROPYLENE CHLORIDE	C3H6Cl2	1,2-DICHLOROPROPANE	78-87-5
PROPYLENE DIAMINE	C3H10N2	1,2-PROPANEDIAMINE	78-90-0
PROPYLENE DIBROMIDE	C3H6Br2	1,2-DIBROMOPROPANE	78-75-1
a,b-PROPYLENE DICHLORIDE	C3H6Cl2	1,2-DICHLOROPROPANE	78-87-5
PROPYLENE EPOXIDE	C3H6O	1,2-PROPYLENE OXIDE	75-56-9
1,2-PROPYLENE GLYCOL	C3H8O2	1,2-PROPYLENE GLYCOL	57-55-6
1,3-PROPYLENE GLYCOL	C3H8O2	1,3-PROPYLENE GLYCOL	504-63-2
a-PROPYLENEGLYCOL	C3H8O2	1,2-PROPYLENE GLYCOL	57-55-6
b-PROPYLENE GLYCOL	C3H8O2	1,3-PROPYLENE GLYCOL	504-63-2
PROPYLENE GLYCOL	C3H8O2	1,2-PROPYLENE GLYCOL	57-55-6
PROPYLENE GLYCOL USP	C3H8O2	1,2-PROPYLENE GLYCOL	57-55-6
1,2-PROPYLENEIMINE	C3H7N	PROPYLENEIMINE	75-55-8
PROPYLENE IMINE	C3H7N	PROPYLENEIMINE	75-55-8
1,2-PROPYLENE OXIDE	C3H6O	1,2-PROPYLENE OXIDE	75-56-9
1,3-PROPYLENE OXIDE	C3H6O	1,3-PROPYLENE OXIDE	503-30-0
PROPYLESTER KYSELINY DUSICNE	C3H7NO3	PROPYL-NITRATE	627-13-4
PROPYLESTER KYSELINY MASELNE	C7H14O2	n-PROPYL n-BUTYRATE	105-66-8
PROPYLESTER KYSELINY MRAVENCI	C4H8O2	n-PROPYL FORMATE	110-74-7
PROPYLESTER KYSELINY OCTOVE	C5H10O2	n-PROPYL ACETATE	109-60-4
b-PROPYL-a-ETHYLACROLEIN	C8H16O	2-ETHYLHEXANAL	123-05-7
n-PROPYL FORMATE	C4H8O2	n-PROPYL FORMATE	110-74-7
PROPYL FORMATE	C4H8O2	n-PROPYL FORMATE	110-74-7
PROPYLFORMIC ACID	C4H8O2	n-BUTYRIC ACID	107-92-6
4-PROPYLHEPTANE	C10H22	4-PROPYLHEPTANE	3178-29-8
PROPYL HYDRIDE	C3H8	PROPANE	74-98-6
PROPYLIC ALCOHOL	C3H8O	n-PROPANOL	71-23-8
PROPYLIC ALDEHYDE	C3H6O	n-PROPIONALDEHYDE	123-38-6
PROPYLIDENE CHLORIDE	C3H6Cl2	1,1-DICHLOROPROPANE	78-99-9

Name (Synonym)	Formula	Name in Tables	CAS Reg. No
i-PROPYL IODIDE	C3H7I	ISOPROPYL IODIDE	75-30-9
n-PROPYL IODIDE	C3H7I	n-PROPYL IODIDE	107-08-4
2-PROPYL MERCAPTAN	C3H8S	ISOPROPYL MERCAPTAN	75-33-2
n-PROPYLMERCAPTAN	C3H8S	n-PROPYLMERCAPTAN	107-03-9
n-PROPYL METHACRYLATE	C7H12O2	n-PROPYL METHACRYLATE	2210-28-8
PROPYL METHACRYLATE	C7H12O2	n-PROPYL METHACRYLATE	2210-28-8
PROPYL METHANOATE	C4H8O2	n-PROPYL FORMATE	110-74-7
PROPYLMETHANOL	C4H10O	n-BUTANOL	71-36-3
1-PROPYLNAPHTHALENE	C13H14	1-PROPYLNAPHTHALENE	2765-18-6
2-PROPYLNAPHTHALENE	C13H14	2-PROPYLNAPHTHALENE	2027-19-2
PROPYL NITRATE	C3H7NO3	PROPYL-NITRATE	627-13-4
PROPYLOWY ALKOHOL	C3H8O	n-PROPANOL	71-23-8
N-PROPYL-1-PROPANAMINE	C6H15N	DI-n-PROPYLAMINE	142-84-7
PROPYL PROPANOATE	C6H12O2	n-PROPYL PROPIONATE	106-36-5
n-PROPYL PROPIONATE	C6H12O2	n-PROPYL PROPIONATE	106-36-5
PROPYL-TETRADECYL-SULFIDE	C17H36S	PROPYL-TETRADECYL-SULFIDE	----------
n-PROPYLTHIOL	C3H8S	n-PROPYLMERCAPTAN	107-03-9
PROPYL-TRIDECYL-SULFIDE	C16H34S	PROPYL-TRIDECYL-SULFIDE	----------
PROPYL-UNDECYL-SULFIDE	C14H30S	PROPYL-UNDECYL-SULFIDE	----------
PROPYNE	C3H4	METHYLACETYLENE	74-99-7
PROZOIN	C3H6O2	PROPIONIC ACID	79-09-4
PRUSSIC ACID	CHN	HYDROGEN CYANIDE	74-90-8
PRUSSITE	C2N2	CYANOGEN	460-19-5
PSEUDOACETIC ACID	C3H6O2	PROPIONIC ACID	79-09-4
PSEUDOBUTYLBENZENE	C10H14	tert-BUTYLBENZENE	98-06-6
PSEUDOCUMENE	C9H12	1,2,4-TRIMETHYLBENZENE	95-63-6
PSEUDOCUMOL	C9H12	1,2,4-TRIMETHYLBENZENE	95-63-6
PSEUDOHEXYL ALCOHOL	C6H14O	2-ETHYL-1-BUTANOL	97-95-0
PSEUDOPINEN	C10H16	beta-PINENE	127-91-3
PTAP	C11H16O	p-tert-AMYLPHENOL	80-46-6
PX 104	C16H22O4	DIBUTYL PHTHALATE	84-74-2
PX 404	C18H34O4	DIBUTYL SEBACATE	109-43-3
PYRANTON	C6H12O2	DIACETONE ALCOHOL	123-42-2
PYRAZINE HEXAHYDRIDE	C4H10N2	PIPERAZINE	110-85-0
PYREN	C16H10	PYRENE	129-00-0
PYRENE	C16H10	PYRENE	129-00-0
PYRIDIN	C5H5N	PYRIDINE	110-86-1
PYRIDINE	C5H5N	PYRIDINE	110-86-1
PYRIDINE, 2,6-DIMETHYL-(9CI)	C7H9N	2,6-DIMETHYLPYRIDINE	108-48-5
b-PYRINE	C16H10	PYRENE	129-00-0
PYROACETIC ACID	C3H6O	ACETONE	67-64-1
PYROACETIC ETHER	C3H6O	ACETONE	67-64-1
PYROCATECHINIC ACID	C6H6O2	1,2-BENZENEDIOL	120-80-9
PYROCATECHUIC ACID	C6H6O2	1,2-BENZENEDIOL	120-80-9
PYROGALLIC ACID	C6H6O3	1,2,3-BENZENETRIOL	87-66-1
PYROGUAIAC ACID	C7H8O2	GUAIACOL	90-05-1
M-PYROL	C5H9NO	N-METHYL-2-PYRROLIDONE	872-50-4
2-PYROL	C4H7NO	2-PYRROLIDONE	616-45-5
PYROMELLITIC ACID	C10H6O8	PYROMELLITIC ACID	89-05-4
PYROMUCIC ALDEHYDE	C5H4O2	FURFURAL	98-01-1
PYROPENTYLENE	C5H6	CYCLOPENTADIENE	542-92-7
PYRORACEMIC ACID	C3H4O3	PYRUVIC ACID	127-17-3
PYROTARTARIC ACID NITRILE	C5H6N2	GLUTARONITRILE	544-13-8
PYRROLE	C4H5N	PYRROLE	109-97-7
PYRROLIDINE	C4H9N	PYRROLIDINE	123-75-1
alpha-PYRROLIDINONE	C4H7NO	2-PYRROLIDONE	616-45-5
alpha-PYRROLIDONE	C4H7NO	2-PYRROLIDONE	616-45-5
PYRROLIDONE	C4H7NO	2-PYRROLIDONE	616-45-5
2-PYRROLIDONE	C4H7NO	2-PYRROLIDONE	616-45-5
PYRROLYLENE	C4H6	1,3-BUTADIENE	106-99-0
PYRUVIC ACID	C3H4O3	PYRUVIC ACID	127-17-3
QO THFA	C5H10O2	TETRAHYDROFURFURYL ALCOHOL	97-99-4
QUINALDINE	C10H9N	QUINALDINE	91-63-4
QUINOL	C6H6O2	p-HYDROQUINONE	123-31-9
8-QUINOL	C9H7NO	8-HYDROXYQUINOLINE	148-24-3
QUINOLINE	C9H7N	QUINOLINE	91-22-5
QUINOPHENOL	C9H7NO	8-HYDROXYQUINOLINE	148-24-3
R 10	CCl4	CARBON TETRACHLORIDE	56-23-5
R 12	CCl2F2	DICHLORODIFLUOROMETHANE	75-71-8
R 12B1	CBrClF2	BROMOCHLORODIFLUOROMETHANE	353-59-3
R 12B2	CBr2F2	DIBROMODIFLUOROMETHANE	75-61-6
R 13	CClF3	CHLOROTRIFLUOROMETHANE	75-72-9
R 13B1	CBrF3	BROMOTRIFLUOROMETHANE	75-63-8
R 14	CF4	CARBON TETRAFLUORIDE	75-73-0
R 23	CHF3	TRIFLUOROMETHANE	75-46-7
R 30	CH2Cl2	DICHLOROMETHANE	75-09-2

Name (Synonym)	Formula	Name in Tables	CAS Reg. No
R 31	CH2ClF	CHLOROFLUOROMETHANE	593-70-4
R 40	CH3Cl	METHYL CHLORIDE	74-87-3
R 40B1	CH3Br	METHYL BROMIDE	74-83-9
R 113	C2Cl3F3	1,1,2-TRICHLOROTRIFLUOROETHANE	76-13-1
R 114	C2Cl2F4	1,2-DICHLOROTETRAFLUOROETHANE	76-14-2
R 114B2	C2Br2F4	1,2-DIBROMOTETRAFLUOROETHANE	124-73-2
R 116	C2F6	HEXAFLUOROETHANE	76-16-4
R 142B	C2H3ClF2	1-CHLORO-1,1-DIFLUOROETHANE	75-68-3
R 143a	C2H3F3	1,1,1-TRIFLUOROETHANE	420-46-2
R 161	C2H5F	ETHYL FLUORIDE	353-36-6
R 744	CO2	CARBON DIOXIDE	124-38-9
R 1132a	C2H2F2	1,1-DIFLUOROETHYLENE	75-38-7
R-C 318	C4F8	OCTAFLUOROCYCLOBUTANE	115-25-3
RC COMONOMER DBM	C12H18	HEXAMETHYLBENZENE	87-85-4
RC PLASTICIZER B-17	C22H44O2	n-BUTYL STEARATE	123-95-5
RCRA WASTE NUMBER P003	C3H4O	ACROLEIN	107-02-8
RCRA WASTE NUMBER P005	C3H6O	ALLYL ALCOHOL	107-18-6
RCRA WASTE NUMBER P014	C6H6S	PHENYL MERCAPTAN	108-98-5
RCRA WASTE NUMBER P016	C2H4Cl2O	BIS(CHLOROMETHYL)ETHER	542-88-1
RCRA WASTE NUMBER P022	CS2	CARBON DISULFIDE	75-15-0
RCRA WASTE NUMBER P023	C2H3ClO	CHLOROACETALDEHYDE	107-20-0
RCRA WASTE NUMBER P024	C6H6ClN	p-CHLOROANILINE	106-47-8
RCRA WASTE NUMBER P028	C7H7Cl	BENZYL CHLORIDE	100-44-7
RCRA WASTE NUMBER P031	C2N2	CYANOGEN	460-19-5
RCRA WASTE NUMBER P033	CClN	CYANOGEN CHLORIDE	506-77-4
RCRA WASTE NUMBER P054	C2H5N	ETHYLENEIMINE	151-56-4
RCRA WASTE NUMBER P063	CHN	HYDROGEN CYANIDE	74-90-8
RCRA WASTE NUMBER P064	C2H3NO	METHYL ISOCYANATE	624-83-9
RCRA WASTE NUMBER P067	C3H7N	PROPYLENEIMINE	75-55-8
RCRA WASTE NUMBER P069	C4H7NO	ACETONE CYANOHYDRIN	75-86-5
RCRA WASTE NUMBER P081	C3H5N3O9	NITROGLYCERINE	55-63-0
RCRA WASTE NUMBER P095	CCl2O	PHOSGENE	75-44-5
RCRA WASTE NUMBER P101	C3H5N	PROPIONITRILE	107-12-0
RCRA WASTE NUMBER P112	CN4O8	TETRANITROMETHANE	509-14-8
RCRA WASTE NUMBER U001	C2H4O	ACETALDEHYDE	75-07-0
RCRA WASTE NUMBER U002	C3H6O	ACETONE	67-64-1
RCRA WASTE NUMBER U003	C2H3N	ACETONITRILE	75-05-8
RCRA WASTE NUMBER U006	C2H3ClO	ACETYL CHLORIDE	75-36-5
RCRA WASTE NUMBER U007	C3H5NO	ACRYLAMIDE	79-06-1
RCRA WASTE NUMBER U008	C3H4O2	ACRYLIC ACID	79-10-7
RCRA WASTE NUMBER U009	C3H3N	ACRYLONITRILE	107-13-1
RCRA WASTE NUMBER U017	C7H6Cl2	BENZYL DICHLORIDE	98-87-3
RCRA WASTE NUMBER U023	C7H5Cl3	BENZOTRICHLORIDE	98-07-7
RCRA WASTE NUMBER U029	CH3Br	METHYL BROMIDE	74-83-9
RCRA WASTE NUMBER U031	C4H10O	n-BUTANOL	71-36-3
RCRA WASTE NUMBER U033	CF2O	CARBONYL FLUORIDE	353-50-4
RCRA WASTE NUMBER U034	C2HCl3O	TRICHLOROACETALDEHYDE	75-87-6
RCRA WASTE NUMBER U037	C6H5Cl	MONOCHLOROBENZENE	108-90-7
RCRA WASTE NUMBER U041	C3H5ClO	alpha-EPICHLOROHYDRIN	106-89-8
RCRA WASTE NUMBER U043	C2H3Cl	VINYL CHLORIDE	75-01-4
RCRA WASTE NUMBER U044	CHCl3	CHLOROFORM	67-66-3
RCRA WASTE NUMBER U045	CH3Cl	METHYL CHLORIDE	74-87-3
RCRA WASTE NUMBER U048	C6H5ClO	o-CHLOROPHENOL	95-57-8
RCRA WASTE NUMBER U050	C18H12	CHRYSENE	218-01-9
RCRA WASTE NUMBER U052	C7H8O	m-CRESOL	108-39-4
RCRA WASTE NUMBER U053	C4H6O	trans-CROTONALDEHYDE	123-73-9
RCRA WASTE NUMBER U056	C6H12	CYCLOHEXANE	110-82-7
RCRA WASTE NUMBER U057	C6H10O	CYCLOHEXANONE	108-94-1
RCRA WASTE NUMBER U067	C2H4Br2	1,2-DIBROMOETHANE	106-93-4
RCRA WASTE NUMBER U068	CH2Br2	DIBROMOMETHANE	74-95-3
RCRA WASTE NUMBER U069	C16H22O4	DIBUTYL PHTHALATE	84-74-2
RCRA WASTE NUMBER U075	CCl2F2	DICHLORODIFLUOROMETHANE	75-71-8
RCRA WASTE NUMBER U076	C2H4Cl2	1,1-DICHLOROETHANE	75-34-3
RCRA WASTE NUMBER U077	C2H4Cl2	1,2-DICHLOROETHANE	107-06-2
RCRA WASTE NUMBER U078	C2H2Cl2	1,1-DICHLOROETHYLENE	75-35-4
RCRA WASTE NUMBER U079	C2H2Cl2	trans-1,2-DICHLOROETHYLENE	156-60-5
RCRA WASTE NUMBER U080	CH2Cl2	DICHLOROMETHANE	75-09-2
RCRA WASTE NUMBER U083	C3H6Cl2	1,2-DICHLOROPROPANE	78-87-5
RCRA WASTE NUMBER U088	C12H14O4	DIETHYL PHTHALATE	84-66-2
RCRA WASTE NUMBER U092	C2H6AlCl	DIMETHYLALUMINUM CHLORIDE	1184-58-3
RCRA WASTE NUMBER U096	C9H12O2	CUMENE HYDROPEROXIDE	80-15-9
RCRA WASTE NUMBER U101	C8H10O	2,4-XYLENOL	105-67-9
RCRA WASTE NUMBER U102	C10H10O4	DIMETHYL PHTHALATE	131-11-3
RCRA WASTE NUMBER U103	C2H6O4S	DIMETHYL SULFATE	77-78-1
RCRA WASTE NUMBER U105	C7H6N2O4	2,4-DINITROTOLUENE	121-14-2
RCRA WASTE NUMBER U106	C7H6N2O4	2,6-DINITROTOLUENE	606-20-2

Name (Synonym)	Formula	Name in Tables	CAS Reg. No
RCRA WASTE NUMBER U108	C4H8O2	1,4-DIOXANE	123-91-1
RCRA WASTE NUMBER U109	C12H12N2	HYDRAZOBENZENE	122-66-7
RCRA WASTE NUMBER U110	C6H15N	DI-n-PROPYLAMINE	142-84-7
RCRA WASTE NUMBER U112	C4H8O2	ETHYL ACETATE	141-78-6
RCRA WASTE NUMBER U113	C5H8O2	ETHYL ACRYLATE	140-88-5
RCRA WASTE NUMBER U115	C2H4O	ETHYLENE OXIDE	75-21-8
RCRA WASTE NUMBER U117	C4H10O	DIETHYL ETHER	60-29-7
RCRA WASTE NUMBER U118	C6H10O2	ETHYL METHACRYLATE	97-63-2
RCRA WASTE NUMBER U120	C16H10	FLUORANTHENE	206-44-0
RCRA WASTE NUMBER U121	CCl3F	TRICHLOROFLUOROMETHANE	75-69-4
RCRA WASTE NUMBER U122	CH2O	FORMALDEHYDE	50-00-0
RCRA WASTE NUMBER U123	CH2O2	FORMIC ACID	64-18-6
RCRA WASTE NUMBER U124	C4H4O	FURAN	110-00-9
RCRA WASTE NUMBER U125	C5H4O2	FURFURAL	98-01-1
RCRA WASTE NUMBER U128	C4Cl6	HEXACHLORO-1,3-BUTADIENE	87-68-3
RCRA WASTE NUMBER U130	C5Cl6	HEXACHLOROCYCLOPENTADIENE	77-47-4
RCRA WASTE NUMBER U131	C2Cl6	HEXACHLOROETHANE	67-72-1
RCRA WASTE NUMBER U138	CH3I	METHYL IODIDE	74-88-4
RCRA WASTE NUMBER U140	C4H10O	ISOBUTANOL	78-83-1
RCRA WASTE NUMBER U147	C4H2O3	MALEIC ANHYDRIDE	108-31-6
RCRA WASTE NUMBER U149	C3H2N2	MALONONITRILE	109-77-3
RCRA WASTE NUMBER U152	C4H5N	METHACRYLONITRILE	126-98-7
RCRA WASTE NUMBER U153	CH4S	METHYL MERCAPTAN	74-93-1
RCRA WASTE NUMBER U154	CH4O	METHANOL	67-56-1
RCRA WASTE NUMBER U156	C2H3ClO2	METHYL CHLOROFORMATE	79-22-1
RCRA WASTE NUMBER U159	C4H8O	METHYL ETHYL KETONE	78-93-3
RCRA WASTE NUMBER U161	C6H12O	METHYL ISOBUTYL KETONE	108-10-1
RCRA WASTE NUMBER U162	C5H8O2	METHYL METHACRYLATE	80-62-6
RCRA WASTE NUMBER U165	C10H8	NAPHTHALENE	91-20-3
RCRA WASTE NUMBER U169	C6H5NO2	NITROBENZENE	98-95-3
RCRA WASTE NUMBER U171	C3H7NO2	2-NITROPROPANE	79-46-9
RCRA WASTE NUMBER U182	C6H12O3	PARALDEHYDE	123-63-7
RCRA WASTE NUMBER U184	C2HCl5	PENTACHLOROETHANE	76-01-7
RCRA WASTE NUMBER U190	C8H4O3	PHTHALIC ANHYDRIDE	85-44-9
RCRA WASTE NUMBER U191	C6H7N	2-METHYLPYRIDINE	109-06-8
RCRA WASTE NUMBER U194	C3H9N	n-PROPYLAMINE	107-10-8
RCRA WASTE NUMBER U196	C5H5N	PYRIDINE	110-86-1
RCRA WASTE NUMBER U201	C6H6O2	1,3-BENZENEDIOL	108-46-3
RCRA WASTE NUMBER U208	C2H2Cl4	1,1,1,2-TETRACHLOROETHANE	630-20-6
RCRA WASTE NUMBER U209	C2H2Cl4	1,1,2,2-TETRACHLOROETHANE	79-34-5
RCRA WASTE NUMBER U210	C2Cl4	TETRACHLOROETHYLENE	127-18-4
RCRA WASTE NUMBER U211	CCl4	CARBON TETRACHLORIDE	56-23-5
RCRA WASTE NUMBER U213	C4H8O	TETRAHYDROFURAN	109-99-9
RCRA WASTE NUMBER U220	C7H8	TOLUENE	108-88-3
RCRA WASTE NUMBER U221	C7H10N2	TOLUENEDIAMINE	95-80-7
RCRA WASTE NUMBER U223	C9H6N2O2	TOLUENE DIISOCYANATE	26471-62-5
RCRA WASTE NUMBER U225	CHBr3	TRIBROMOMETHANE	75-25-2
RCRA WASTE NUMBER U226	C2H3Cl3	1,1,1-TRICHLOROETHANE	71-55-6
RCRA WASTE NUMBER U227	C2H3Cl3	1,1,2-TRICHLOROETHANE	79-00-5
RCRA WASTE NUMBER U228	C2HCl3	TRICHLOROETHYLENE	79-01-6
RCRA WASTE NUMBER U234	C6H3N3O6	1,3,5-TRINITROBENZENE	99-35-4
REFRIGERANT 112	C2Cl4F2	1,1,2,2-TETRACHLORODIFLUOROETHANE	76-12-0
REFRIGERANT 113	C2Cl3F3	1,1,2-TRICHLOROTRIFLUOROETHANE	76-13-1
REFRIGERANT 12	CCl2F2	DICHLORODIFLUOROMETHANE	75-71-8
REFRIGERANT 14	CF4	CARBON TETRAFLUORIDE	75-73-0
REFRIGERANT 22	CHClF2	CHLORODIFLUOROMETHANE	75-45-6
RENAL PF	C6H8N2	p-PHENYLENEDIAMINE	106-50-3
RESORCIN	C6H6O2	1,3-BENZENEDIOL	108-46-3
RESORCINE	C6H6O2	1,3-BENZENEDIOL	108-46-3
RETARDER AK	C8H4O3	PHTHALIC ANHYDRIDE	85-44-9
RETARDER BA	C7H6O2	BENZOIC ACID	65-85-0
RETARDER ESEN	C8H4O3	PHTHALIC ANHYDRIDE	85-44-9
RETARDER PD	C8H4O3	PHTHALIC ANHYDRIDE	85-44-9
RETARDER W	C7H6O3	SALICYLIC ACID	69-72-7
RETARDEX	C7H6O2	BENZOIC ACID	65-85-0
RHOPLEX AC-33	C6H10O2	ETHYL METHACRYLATE	97-63-2
1,1,2-TRIPHENYLETHYLENE	C20H16	TRIPHENYLETHYLENE	58-72-0
ROSE OIL	C8H10O	2-PHENYLETHANOL	60-12-8
RUBINATE TDI	C9H6N2O2	TOLUENE DIISOCYANATE	26471-62-5
RUBINATE TDI 80/20	C9H6N2O2	TOLUENE DIISOCYANATE	26471-62-5
S.B.A.	C4H10O	sec-BUTANOL	78-92-2
SA	C7H6O3	SALICYLIC ACID	69-72-7
SAH	C7H6O2	SALICYLALDEHYDE	90-02-8
SALICYLAL	C7H6O2	SALICYLALDEHYDE	90-02-8
SALICYLALDEHYDE	C7H6O2	SALICYLALDEHYDE	90-02-8
SALICYLIC ACID	C7H6O3	SALICYLIC ACID	69-72-7

Name (Synonym)	Formula	Name in Tables	CAS Reg. No
SALICYLIC ACID, METHYL ESTER	C8H8O3	METHYL SALICYLATE	119-36-8
SALICYLIC ALDEHYDE	C7H6O2	SALICYLALDEHYDE	90-02-8
SALVO LIQUID	C7H6O2	BENZOIC ACID	65-85-0
SALVO POWDER	C7H6O2	BENZOIC ACID	65-85-0
SANTOFLEX IC	C6H8N2	p-PHENYLENEDIAMINE	106-50-3
SANTOWAX	C18H14	p-TERPHENYL	92-94-4
SANTOWAX M	C18H14	m-TERPHENYL	92-06-8
SAX	C7H6O3	SALICYLIC ACID	69-72-7
SCABANCA	C14H12O2	BENZYL BENZOATE	120-51-4
SCALDIP	C12H11N	DIPHENYLAMINE	122-39-4
SCHWEFELKOHLENSTOFF	CS2	CARBON DISULFIDE	75-15-0
SCINTILLAR	C8H10	p-XYLENE	106-42-3
SCONATEX	C2H2Cl2	1,1-DICHLOROETHYLENE	75-35-4
SEBACIC ACID	C10H18O4	SEBACIC ACID	111-20-6
SEBACIC ACID, DIBUTYL ESTER	C18H34O4	DIBUTYL SEBACATE	109-43-3
SENTRY GRAIN PRESERVER	C3H6O2	PROPIONIC ACID	79-09-4
SESQUIETHYLALUMINUM CHLORIDE	C6H15Al2Cl3	ETHYL ALUMINUM SESQUICHLORIDE	12075-68-2
SEXTONE	C6H10O	CYCLOHEXANONE	108-94-1
SEXTONE B	C7H14	METHYLCYCLOHEXANE	108-87-2
SF 1173	C8H24O4Si4	OCTAMETHYLCYCLOTETRASILOXANE	556-67-2
SHELL MIBK	C6H12O	METHYL ISOBUTYL KETONE	108-10-1
SHELLSOL 140	C9H20	n-NONANE	111-84-2
SHELL UNDRAUTTED A	C3H6O	ALLYL ALCOHOL	107-18-6
SILICONE SF 1173	C8H24O4Si4	OCTAMETHYLCYCLOTETRASILOXANE	556-67-2
SILVIC ACID	C20H30O2	ABIETIC ACID	514-10-3
SIONIT	C6H14O6	SORBITOL	50-70-4
SIONON	C6H14O6	SORBITOL	50-70-4
SIPOL L10	C10H22O	1-DECANOL	112-30-1
SIPOL L8	C8H18O	1-OCTANOL	111-87-5
SIPOL S	C18H38O	1-OCTADECANOL	112-92-5
SIPOMER DAM	C10H12O4	DIALLYL MALEATE	999-21-3
SIPONOL S	C18H38O	1-OCTADECANOL	112-92-5
SIRLENE	C3H8O2	1,2-PROPYLENE GLYCOL	57-55-6
SK 6048	C3H3Cl	PROPARGYL CHLORIDE	624-65-7
SOLACTOL	C5H10O3	ETHYL LACTATE	97-64-3
SOLAESTHIN	CH2Cl2	DICHLOROMETHANE	75-09-2
SOLAR WINTER BAN	C3H8O2	1,2-PROPYLENE GLYCOL	57-55-6
SOLFURO di CARBONIO	CS2	CARBON DISULFIDE	75-15-0
SOLMETHINE	CH2Cl2	DICHLOROMETHANE	75-09-2
SOLVANOL	C12H14O4	DIETHYL PHTHALATE	84-66-2
SOLVANOM	C10H10O4	DIMETHYL PHTHALATE	131-11-3
SOLVARONE	C10H10O4	DIMETHYL PHTHALATE	131-11-3
SOLVENT 111	C2H3Cl3	1,1,1-TRICHLOROETHANE	71-55-6
SOLVENT ETHER	C4H10O	DIETHYL ETHER	60-29-7
SOLVOSOL	C6H14O3	2-(2-ETHOXYETHOXY)ETHANOL	111-90-0
SORBICOLAN	C6H14O6	SORBITOL	50-70-4
SORBITE	C6H14O6	SORBITOL	50-70-4
d-SORBITOL	C6H14O6	SORBITOL	50-70-4
SORBITOL	C6H14O6	SORBITOL	50-70-4
SORBO	C6H14O6	SORBITOL	50-70-4
SORBOL	C6H14O6	SORBITOL	50-70-4
SORBOSTYL	C6H14O6	SORBITOL	50-70-4
SORVILANDE	C6H14O6	SORBITOL	50-70-4
SPECTRAR SPECTRAR	C3H8O	ISOPROPANOL	67-63-0
SPIROPENTANE	C5H8	SPIROPENTANE	157-40-4
STAFLEX DBM	C12H18	HEXAMETHYLBENZENE	87-85-4
STAFLEX DBP	C16H22O4	DIBUTYL PHTHALATE	84-74-2
STAFLEX DBS	C18H34O4	DIBUTYL SEBACATE	109-43-3
STAR	C3H8O3	GLYCEROL	56-81-5
STARFOL BS-100	C22H44O2	n-BUTYL STEARATE	123-95-5
STEARIC ACID	C18H36O2	STEARIC ACID	57-11-4
STEAROL	C18H38O	1-OCTADECANOL	112-92-5
STEARYL ALCOHOL	C18H38O	1-OCTADECANOL	112-92-5
STEPAN C40	C13H26O2	METHYL DODECANOATE	111-82-0
STERAFFINE	C18H38O	1-OCTADECANOL	112-92-5
STERILIZING GAS ETHYLENE OXIDE 100%	C2H4O	ETHYLENE OXIDE	75-21-8
STEROLAMIDE	C6H15NO3	TRIETHANOLAMINE	102-71-6
cis-STILBENE	C14H12	cis-STILBENE	645-49-8
(E)-STILBENE	C14H12	trans-STILBENE	103-30-0
trans-STILBENE	C14H12	trans-STILBENE	103-30-0
STIROLO	C8H8	STYRENE	100-42-5
STROBANE	C2H3Cl3	1,1,1-TRICHLOROETHANE	71-55-6
STYRENE	C8H8	STYRENE	100-42-5
STYRENE, m-METHYL-	C9H10	m-METHYLSTYRENE	100-80-1
STYRENE, p-METHYL-	C9H10	p-METHYLSTYRENE	622-97-9
STYRENE MONOMER	C8H8	STYRENE	100-42-5

Name (Synonym)	Formula	Name in Tables	CAS Reg. No
STYROL	C8H8	STYRENE	100-42-5
STYROLE	C8H8	STYRENE	100-42-5
STYROLENE	C8H8	STYRENE	100-42-5
STYRON	C8H8	STYRENE	100-42-5
STYROPOR	C8H8	STYRENE	100-42-5
SUBERANE	C7H14	CYCLOHEPTANE	291-64-5
SUCCINIC ACID	C4H6O4	SUCCINIC ACID	110-15-6
SUCCINIC ACID ANHYDRIDE	C4H4O3	SUCCINIC ANHYDRIDE	108-30-5
SUCCINIC ACID DINITRILE	C4H4N2	SUCCINONITRILE	110-61-2
SUCCINIC ANHYDRIDE	C4H4O3	SUCCINIC ANHYDRIDE	108-30-5
SUCCINIC DINITRILE	C4H4N2	SUCCINONITRILE	110-61-2
SUCCINODINITRILE	C4H4N2	SUCCINONITRILE	110-61-2
SUCCINONITRILE	C4H4N2	SUCCINONITRILE	110-61-2
SUCCINYL OXIDE	C4H4O3	SUCCINIC ANHYDRIDE	108-30-5
SULFATE de METHYLE	C2H6O4S	DIMETHYL SULFATE	77-78-1
SULFATE DIMETHYLIQUE	C2H6O4S	DIMETHYL SULFATE	77-78-1
SULFOLAN	C4H8O2S	SULFOLANE	126-33-0
SULFOLANE	C4H8O2S	SULFOLANE	126-33-0
SULFURE de METHYLE	C2H6S	DIMETHYL SULFIDE	75-18-3
SULFURIC ACID, DIMETHYL ESTER	C2H6O4S	DIMETHYL SULFATE	77-78-1
SULPHOCARBONIC ANHYDRIDE	CS2	CARBON DISULFIDE	75-15-0
SULPHOLANE	C4H8O2S	SULFOLANE	126-33-0
SULPHOXALINE	C4H8O2S	SULFOLANE	126-33-0
SUMINE 2005	C7H9N	BENZYLAMINE	100-46-9
SUMINE 2006	C7H9N	BENZYLAMINE	100-46-9
SUPERLYSOFORM	CH2O	FORMALDEHYDE	50-00-0
SUPEROL	C3H8O3	GLYCEROL	56-81-5
SUXIL	C4H4N2	SUCCINONITRILE	110-61-2
SWEET BIRCH OIL	C8H8O3	METHYL SALICYLATE	119-36-8
SYNOX TBC	C10H14O2	p-tert-BUTYLCATECHOL	98-29-3
SYNTHETIC GLYCERIN	C3H8O3	GLYCEROL	56-81-5
SYNTHETIC WINTERGREEN OIL	C8H8O3	METHYL SALICYLATE	119-36-8
T 100	C9H6N2O2	TOLUENE DIISOCYANATE	26471-62-5
TA 12	C8H6O4	TEREPHTHALIC ACID	100-21-0
TA-33MP	C8H6O4	TEREPHTHALIC ACID	100-21-0
TAR CAMPHOR	C10H8	NAPHTHALENE	91-20-3
TARTARIC ACID	C4H6O6	TARTARIC ACID	133-37-9
TBE	C2H2Br4	1,1,2,2-TETRABROMOETHANE	79-27-6
TBHP-70	C4H10O2	t-BUTYL HYDROPEROXIDE	75-91-2
TCE	C2H2Cl4	1,1,2,2-TETRACHLOROETHANE	79-34-5
1,1,1-TCE	C2H3Cl3	1,1,1-TRICHLOROETHANE	71-55-6
TCTP	C4Cl4S	TETRACHLOROTHIOPHENE	6012-97-1
TD-183	C4Cl4S	TETRACHLOROTHIOPHENE	6012-97-1
TDI	C9H6N2O2	TOLUENE DIISOCYANATE	26471-62-5
TDI 80-20	C9H6N2O2	TOLUENE DIISOCYANATE	26471-62-5
TDI-80	C9H6N2O2	TOLUENE DIISOCYANATE	26471-62-5
TEA	C6H15Al	TRIETHYL ALUMINUN	97-93-8
TEABERRY OIL	C8H8O3	METHYL SALICYLATE	119-36-8
90 TECHNICAL GLYCERINE	C3H8O3	GLYCEROL	56-81-5
TECZA	C6H18N4	TRIETHYLENE TETRAMINE	112-24-3
TEG	C6H14O4	TRIETHYLENE GLYCOL	112-27-6
TEGESTER BUTYL STEARATE	C22H44O2	n-BUTYL STEARATE	123-95-5
TELONE	C3H6Cl2	2,2-DICHLOROPROPANE	594-20-7
TEN	C6H15N	TRIETHYLAMINE	121-44-8
TENN-PLAS	C7H6O2	BENZOIC ACID	65-85-0
TENOX P GRAIN PRESERVATIVE	C3H6O2	PROPIONIC ACID	79-09-4
TEP	C6H15O4P	TRIETHYL PHOSPHATE	78-40-0
TEQUINOL	C6H6O2	p-HYDROQUINONE	123-31-9
TEREPHTHALIC ACID	C8H6O4	TEREPHTHALIC ACID	100-21-0
TEREPHTHALIC ACID METHYL ESTER	C10H10O4	DIMETHYL TEREPHTHALATE	120-61-6
TERETON	C3H6O2	METHYL ACETATE	79-20-9
m-TERPHENYL	C18H14	m-TERPHENYL	92-06-8
o-TERPHENYL	C18H14	o-TERPHENYL	84-15-1
p-TERPHENYL	C18H14	p-TERPHENYL	92-94-4
1,3-TERPHENYL	C18H14	m-TERPHENYL	92-06-8
alpha-TERPINENE	C10H16	alpha-TERPINENE	99-86-5
gamma-TERPINENE	C1OH16	gamma-TERPINENE	99-85-4
TERPINOLENE	C10H16	TERPINOLENE	586-62-9
TERTRAL D	C6H8N2	p-PHENYLENEDIAMINE	106-50-3
TESCOL	C2H6O2	ETHYLENE GLYCOL	107-21-1
TETA	C6H18N4	TRIETHYLENE TETRAMINE	112-24-3
1,4,7,10-TETRAAZADECANE	C6H18N4	TRIETHYLENE TETRAMINE	112-24-3
1,1,2,2-TETRABROMAETHAN	C2H2Br4	1,1,2,2-TETRABROMOETHANE	79-27-6
TETRABROMOACETYLENE	C2H2Br4	1,1,2,2-TETRABROMOETHANE	79-27-6
1,1,2,2-TETRABROMOETANO	C2H2Br4	1,1,2,2-TETRABROMOETHANE	79-27-6
S-TETRABROMOETHANE	C2H2Br4	1,1,2,2-TETRABROMOETHANE	79-27-6

Name (Synonym)	Formula	Name in Tables	CAS Reg. No
1,1,2,2-TETRABROMOETHANE	C2H2Br4	1,1,2,2-TETRABROMOETHANE	79-27-6
1,1,2,2-TETRABROOMETHAAN	C2H2Br4	1,1,2,2-TETRABROMOETHANE	79-27-6
TETRACAP	C2Cl4	TETRACHLOROETHYLENE	127-18-4
1,2,4,5-TETRACARBOXYBENZENE	C10H6O8	PYROMELLITIC ACID	89-05-4
1,1,2,2-TETRACHLOORETHAAN	C2H2Cl4	1,1,2,2-TETRACHLOROETHANE	79-34-5
1,1,2,2-TETRACHLORAETHAN	C2H2Cl4	1,1,2,2-TETRACHLOROETHANE	79-34-5
TETRACHLORETHANE	C2H2Cl4	1,1,2,2-TETRACHLOROETHANE	79-34-5
1,1,2,2-TETRACHLORETHANE	C2H2Cl4	1,1,2,2-TETRACHLOROETHANE	79-34-5
TETRACHLOROCARBON	CCl4	CARBON TETRACHLORIDE	56-23-5
1,1,2,2-TETRACHLORODIFLUOROETHANE	C2Cl4F2	1,1,2,2-TETRACHLORODIFLUOROETHANE	76-12-0
sym-TETRACHLOROETHANE	C2H2Cl4	1,1,2,2-TETRACHLOROETHANE	79-34-5
TETRACHLOROETHENE	C2Cl4	TETRACHLOROETHYLENE	127-18-4
1,1,1,2-TETRACHLOROETHANE	C2H2Cl4	1,1,1,2-TETRACHLOROETHANE	630-20-6
1,1,2,2-TETRACHLOROETHANE	C2H2Cl4	1,1,2,2-TETRACHLOROETHANE	79-34-5
TETRACHLOROETHYLENE	C2Cl4	TETRACHLOROETHYLENE	127-18-4
1,1,2,2-TETRACHLOROETHYLENE	C2Cl4	TETRACHLOROETHYLENE	127-18-4
TETRACHLOROMETHANE	CCl4	CARBON TETRACHLORIDE	56-23-5
TETRACHLOROTHIOFENE	C4Cl4S	TETRACHLOROTHIOPHENE	6012-97-1
TETRACHLOROTHIOPHENE	C4Cl4S	TETRACHLOROTHIOPHENE	6012-97-1
TETRACHLORURE D'ACETYLENE	C2H2Cl4	1,1,2,2-TETRACHLOROETHANE	79-34-5
1,1,2,2-TETRACLOROETANO	C2H2Cl4	1,1,2,2-TETRACHLOROETHANE	79-34-5
n-TETRADECANE	C14H30	n-TETRADECANE	629-59-4
1-TETRADECANETHIOL	C14H30S	1-TETRADECANETHIOL	2079-95-0
n-TETRADECANOIC ACID	C14H28O2	n-TETRADECANOIC ACID	544-63-8
TETRADECANOIC ACID	C14H28O2	n-TETRADECANOIC ACID	544-63-8
1-TETRADECANOL	C14H30O	1-TETRADECANOL	112-72-1
N-TETRADECANOL-1	C14H30O	1-TETRADECANOL	112-72-1
1-TETRADECENE	C14H28	1-TETRADECENE	1120-36-1
n-TETRADECOIC ACID	C14H28O2	n-TETRADECANOIC ACID	544-63-8
TETRADECYL ALCOHOL	C14H30O	1-TETRADECANOL	112-72-1
N-TETRADECYL ALCOHOL	C14H30O	1-TETRADECANOL	112-72-1
TETRADECYLAMINE	C14H31N	TETRADECYLAMINE	2016-42-4
1,2,3,4-TETRAETHYLBENZENE	C14H22	1,2,3,4-TETRAETHYLBENZENE	642-32-0
1,2,3,5-TETRAETHYLBENZENE	C14H22	1,2,3,5-TETRAETHYLBENZENE	----------
1,2,4,5-TETRAETHYLBENZENE	C14H22	1,2,4,5-TETRAETHYLBENZENE	635-81-4
TETRAETHYLENE GLYCOL	C8H18O5	TETRAETHYLENE GLYCOL	112-60-7
TETRAETHYLENE GLYCOL DIMETHYL ETHER	C10H22O5	TETRAETHYLENE GLYCOL DIMETHYL ETHER	143-24-8
TETRAETHYLENEPENTAMINE	C8H23N5	TETRAETHYLENEPENTAMINE	112-57-2
TETRAFINOL	CCl4	CARBON TETRACHLORIDE	56-23-5
TETRAFLUORETHYLENE	C2F4	TETRAFLUOROETHYLENE	116-14-3
TETRAFLUOROCARBON	CF4	CARBON TETRAFLUORIDE	75-73-0
1,1,2,2-TETRAFLUORO-1,2-DICHLOROETHANE	C2Cl2F4	1,2-DICHLOROTETRAFLUOROETHANE	76-14-2
TETRAFLUOROETHENE	C2F4	TETRAFLUOROETHYLENE	116-14-3
1,1,1,2-TETRAFLUOROETHANE	C2H2F4	1,1,1,2-TETRAFLUOROETHANE	811-97-2
TETRAFLUOROETHYLENE	C2F4	TETRAFLUOROETHYLENE	116-14-3
TETRAFLUOROMETHANE	CF4	CARBON TETRAFLUORIDE	75-73-0
TETRAFORM	CCl4	CARBON TETRACHLORIDE	56-23-5
TETRAGLYME	C10H22O5	TETRAETHYLENE GLYCOL DIMETHYL ETHER	143-24-8
1,2,3,4-TETRAHYDROBENZENE	C6H10	CYCLOHEXENE	110-83-8
TETRAHYDRO-p-DIOXIN	C4H8O2	1,4-DIOXANE	123-91-1
TETRAHYDRO-1,4-DIOXIN	C4H8O2	1,4-DIOXANE	123-91-1
TETRAHYDRO-2,5-DIOXOFURAN	C4H4O3	SUCCINIC ANHYDRIDE	108-30-5
TETRAHYDROFURAAN	C4H8O	TETRAHYDROFURAN	109-99-9
TETRAHYDROFURAN	C4H8O	TETRAHYDROFURAN	109-99-9
TETRAHYDRO-2-FURANCARBINOL	C5H10O2	TETRAHYDROFURFURYL ALCOHOL	97-99-4
TETRAHYDRO-2-FURANMETHANOL	C5H10O2	TETRAHYDROFURFURYL ALCOHOL	97-99-4
TETRAHYDROFURANNE	C4H8O	TETRAHYDROFURAN	109-99-9
TETRAHYDRO-2-FURANONE	C4H6O2	gamma-BUTYROLACTONE	96-48-0
TETRAHYDROFURFURYL ALCOHOL	C5H10O2	TETRAHYDROFURFURYL ALCOHOL	97-99-4
TETRAHYDROFURYLALKOHOL	C5H10O2	TETRAHYDROFURFURYL ALCOHOL	97-99-4
TETRAHYDRO-p-ISOXAZINE	C4H9NO	MORPHOLINE	110-91-8
TETRAHYDRO-1,4-ISOXAZINE	C4H9NO	MORPHOLINE	110-91-8
3a,4,7,7a-TETRAHYDRO-4,7-METHANOINDENE	C10H12	DICYCLOPENTADIENE	77-73-6
3-THIOPROPIONIC ACID	C3H6O2S	3-MERCAPTOPROPIONIC ACID	107-96-0
TETRAHYDRONAPHTHALENE	C10H12	1,2,3,4-TETRAHYDRONAPHTHALENE	119-64-2
1,2,3,4-TETRAHYDRONAPHTHALENE	C10H12	1,2,3,4-TETRAHYDRONAPHTHALENE	119-64-2
TETRAHYDRO-1,4-OXAZINE	C4H9NO	MORPHOLINE	110-91-8
TETRAHYDRO-2H-1,4-OXAZINE	C4H9NO	MORPHOLINE	110-91-8
TETRAHYDROPYRROLE	C4H9N	PYRROLIDINE	123-75-1
1,2,3,4-TETRAHYDROSTYRENE	C8H12	VINYLCYCLOHEXENE	100-40-3
TETRAHYDROTHIOFEN	C4H8S	TETRAHYDROTHIOPHENE	110-01-0
TETRAHYDROTHIOFEN-1,1-DIOXID	C4H8O2S	SULFOLANE	126-33-0
TETRAHYDROTHIOPHENE	C4H8S	TETRAHYDROTHIOPHENE	110-01-0
TETRAHYDROTHIOPHENE 1,1-DIOXIDE	C4H8O2S	SULFOLANE	126-33-0
2,3,4,5-TETRAHYDROTHIOPHENE-1,1-DIOXIDE	C4H8O2S	SULFOLANE	126-33-0
TETRAHYDROXYMETHYLMETHANE	C5H12O4	PENTAERYTHRITOL	115-77-5

Name (Synonym)	Formula	Name in Tables	CAS Reg. No
TETRAIDROFURANO	C4H8O	TETRAHYDROFURAN	109-99-9
TETRAKIS(HYDROXYMETHYL)METHANE	C5H12O4	PENTAERYTHRITOL	115-77-5
TETRALIN	C10H12	1,2,3,4-TETRAHYDRONAPHTHALENE	119-64-2
TETRALINA	C10H12	1,2,3,4-TETRAHYDRONAPHTHALENE	119-64-2
TETRALINE	C10H12	1,2,3,4-TETRAHYDRONAPHTHALENE	119-64-2
1,2,3,4-TETRAMETHYLBENZENE	C10H14	1,2,3,4-TETRAMETHYLBENZENE	488-23-3
1,2,3,5-TETRAMETHYLBENZENE	C10H14	1,2,3,5-TETRAMETHYLBENZENE	527-53-7
1,2,4,5-TETRAMETHYLBENZENE	C10H14	1,2,4,5-TETRAMETHYLBENZENE	95-93-2
2,2,3,3-TETRAMETHYLBUTANE	C8H18	2,2,3,3-TETRAMETHYLBUTANE	594-82-1
p-(1',1',3',3'-TETRAMETHYLBUTYL)FENOL	C14H22O	p-tert-OCTYLPHENOL	140-66-9
TETRAMETHYLENE	C4H8	CYCLOBUTANE	287-23-0
TETRAMETHYLENE CYANIDE	C6H8N2	ADIPONITRILE	111-69-3
1,4-TETRAMETHYLENE GLYCOL	C4H10O2	1,4-BUTANEDIOL	110-63-4
TETRAMETHYLENE IODIDE	C4H8I2	1,2-DIIODOBUTANE	628-21-7
TETRAMETHYLENE OXIDE	C4H8O	TETRAHYDROFURAN	109-99-9
TETRAMETHYLENESULFIDE	C4H8S	TETRAHYDROTHIOPHENE	110-01-0
TETRAMETHYLENE SULFONE	C4H8O2S	SULFOLANE	126-33-0
TETRAMETHYLENIMINE	C4H9N	PYRROLIDINE	123-75-1
2,2,3,3-TETRAMETHYLHEXANE	C10H22	2,2,3,3-TETRAMETHYLHEXANE	13475-81-5
2,2,3,4-TETRAMETHYLHEXANE	C10H22	2,2,3,4-TETRAMETHYLHEXANE	52897-08-2
2,2,3,5-TETRAMETHYLHEXANE	C10H22	2,2,3,5-TETRAMETHYLHEXANE	52897-09-3
2,2,4,4-TETRAMETHYLHEXANE	C10H22	2,2,4,4-TETRAMETHYLHEXANE	51750-65-3
2,2,4,5-TETRAMETHYLHEXANE	C10H22	2,2,4,5-TETRAMETHYLHEXANE	16747-42-5
2,2,5,5-TETRAMETHYLHEXANE	C10H22	2,2,5,5-TETRAMETHYLHEXANE	1071-81-4
2,3,3,4-TETRAMETHYLHEXANE	C10H22	2,3,3,4-TETRAMETHYLHEXANE	52897-10-6
2,3,3,5-TETRAMETHYLHEXANE	C10H22	2,3,3,5-TETRAMETHYLHEXANE	52897-11-7
2,3,4,4-TETRAMETHYLHEXANE	C10H22	2,3,4,4-TETRAMETHYLHEXANE	52897-12-8
2,3,4,5-TETRAMETHYLHEXANE	C10H22	2,3,4,5-TETRAMETHYLHEXANE	52897-15-1
3,3,4,4-TETRAMETHYLHEXANE	C10H22	3,3,4,4-TETRAMETHYLHEXANE	5171-84-6
TETRAMETHYLOLMETHANE	C5H12O4	PENTAERYTHRITOL	115-77-5
2,2,3,3-TETRAMETHYLPENTANE	C9H20	2,2,3,3-TETRAMETHYLPENTANE	7154-79-2
2,2,3,4-TETRAMETHYLPENTANE	C9H20	2,2,3,4-TETRAMETHYLPENTANE	1186-53-4
2,2,4,4-TETRAMETHYLPENTANE	C9H20	2,2,4,4-TETRAMETHYLPENTANE	1070-87-7
2,3,3,4-TETRAMETHYLPENTANE	C9H20	2,3,3,4-TETRAMETHYLPENTANE	16747-38-9
TETRAMETHYLSILANE	C4H12Si	TETRAMETHYLSILANE	75-76-3
TETRANAP	C10H12	1,2,3,4-TETRAHYDRONAPHTHALENE	119-64-2
TETRANITROMETHANE	CN4O8	TETRANITROMETHANE	509-14-8
TETRA OLIVE N2G	C14H10	ANTHRACENE	120-12-7
2,5,8,11-TETRAOXADODECANE	C8H18O4	TRIETHYLENE GLYCOL DIMETHYL ETHER	112-49-2
TETRAPHENYLETHYLENE	C26H20	TETRAPHENYLETHYLENE	632-51-9
TETRASOL	CCl4	CARBON TETRACHLORIDE	56-23-5
TETROLE	C4H4O	FURAN	110-00-9
TETROPIL	C2Cl4	TETRACHLOROETHYLENE	127-18-4
TETRYL FORMATE	C5H10O2	ISOBUTYL FORMATE	542-55-2
TEXACO LEAD APPRECIATOR	C6H12O2	tert-BUTYL ACETATE	540-88-5
THF	C4H8O	TETRAHYDROFURAN	109-99-9
THFA	C5H10O2	TETRAHYDROFURFURYL ALCOHOL	97-99-4
THIACETIC ACID	C2H4OS	THIOACETIC-ACID	507-09-5
THIACYCLOBUTANE	C3H6S	THIACYCLOBUTANE	287-27-4
THIACYCLOHEPTANE	C6H12S	THIACYCLOHEPTANE	----------
THIACYCLOHEXANE	C5H10S	THIACYCLOHEXANE	1613-51-0
THIACYCLOPENTADIENE	C4H4S	THIOPHENE	110-02-1
THIACYCLOPENTANE	C4H8S	TETRAHYDROTHIOPHENE	110-01-0
THIACYCLOPENTANE DIOXIDE	C4H8O2S	SULFOLANE	126-33-0
THIACYCLOPROPANE	C2H4S	THIACYCLOPROPANE	420-12-2
5-THIANONANE	C8H18S	DIBUTYL-SULFIDE	544-40-1
THIANONANE-5	C8H18S	DIBUTYL-SULFIDE	544-40-1
THIAPHENE	C4H4S	THIOPHENE	110-02-1
2-THIAPROPANE	C2H6S	DIMETHYL SULFIDE	75-18-3
THIIRANE	C2H4S	THIACYCLOPROPANE	420-12-2
THILANE	C4H8S	TETRAHYDROTHIOPHENE	110-01-0
THIOACETIC-ACID	C2H4OS	THIOACETIC-ACID	507-09-5
THIOCYCLOPENTANE-1,1-DIOXIDE	C4H8O2S	SULFOLANE	126-33-0
THIOETHANOL	C2H6S	ETHYL MERCAPTAN	75-08-1
THIOETHYL ALCOHOL	C2H6S	ETHYL MERCAPTAN	75-08-1
THIOFACO M-50	C2H7NO	MONOETHANOLAMINE	141-43-5
THIOFACO T-35	C6H15NO3	TRIETHANOLAMINE	102-71-6
THIOFAN	C4H8S	TETRAHYDROTHIOPHENE	110-01-0
THIOFENOL	C6H6S	PHENYL MERCAPTAN	108-98-5
THIOFURAM	C4H4S	THIOPHENE	110-02-1
THIOFURAN	C4H4S	THIOPHENE	110-02-1
THIOFURFURAN	C4H4S	THIOPHENE	110-02-1
THIOLACETIC ACID	C2H4OS	THIOACETIC-ACID	507-09-5
THIOLANE	C4H8S	TETRAHYDROTHIOPHENE	110-01-0
THIOLANE-1,1-DIOXIDE	C4H8O2S	SULFOLANE	126-33-0
THIOLE	C4H4S	THIOPHENE	110-02-1

Name (Synonym)	Formula	Name in Tables	CAS Reg. No
THIOMETHANOL	CH4S	METHYL MERCAPTAN	74-93-1
THIONOACETIC ACID	C2H4OS	THIOACETIC-ACID	507-09-5
THIOPHANE	C4H8S	TETRAHYDROTHIOPHENE	110-01-0
THIOPHANE DIOXIDE	C4H8O2S	SULFOLANE	126-33-0
THIOPHEN	C4H4S	THIOPHENE	110-02-1
THIOPHENE	C4H4S	THIOPHENE	110-02-1
THIOPHENOL	C6H6S	PHENYL MERCAPTAN	108-98-5
3-THIOPROPANOIC ACID	C3H6O2S	3-MERCAPTOPROPIONIC ACID	107-96-0
b-THIOPROPIONIC ACID	C3H6O2S	3-MERCAPTOPROPIONIC ACID	107-96-0
THIOTETROLE	C4H4S	THIOPHENE	110-02-1
3-THIOTOLENE	C5H6S	3-METHYLTHIOPHENE	616-44-4
THREAMINE	C3H9NO	1-AMINO-2-PROPANOL	78-96-6
THRETHYLENE	C2HCl3	TRICHLOROETHYLENE	79-01-6
TL 1450	C2H3NO	METHYL ISOCYANATE	624-83-9
TL 314	C3H3N	ACRYLONITRILE	107-13-1
TL 337	C2H5N	ETHYLENEIMINE	151-56-4
TL 350	C4H5N	VINYLACETONITRILE	109-75-1
TL 423	C3H5ClO2	ETHYL CHLOROFORMATE	541-41-3
TLA	C6H12O2	tert-BUTYL ACETATE	540-88-5
TMA	C3H9N	TRIMETHYLAMINE	75-50-3
TMAN	C9H4O5	TRIMELLITIC ANHYDRIDE	552-30-7
TMB	C9H12	MESITYLENE	108-67-8
TMP	C6H14O3	TRIMETHYLOLPROPANE	77-99-6
TNB	C6H3N3O6	1,3,5-TRINITROBENZENE	99-35-4
TNG	C3H5N3O9	NITROGLYCERINE	55-63-0
a-TNT	C7H5N3O6	2,4,6-TRINITROTOLUENE	118-96-7
TNT	C7H5N3O6	2,4,6-TRINITROTOLUENE	118-96-7
TNT-TOLITE	C7H5N3O6	2,4,6-TRINITROTOLUENE	118-96-7
TOCP	C21H21O4P	TRI-o-CRESYL PHOSPHATE	78-30-8
TOFK	C21H21O4P	TRI-o-CRESYL PHOSPHATE	78-30-8
TOLIT	C7H5N3O6	2,4,6-TRINITROTOLUENE	118-96-7
TOLITE	C7H5N3O6	2,4,6-TRINITROTOLUENE	118-96-7
p-TOLUALDEHYDE	C8H8O	p-TOLUALDEHYDE	104-87-0
TOLUEEN	C7H8	TOLUENE	108-88-3
TOLUEN	C7H8	TOLUENE	108-88-3
TOLUENE	C7H8	TOLUENE	108-88-3
TOLUENE, p-BROMO-	C7H7Br	p-BROMOTOLUENE	106-38-7
TOLUENEDIAMINE	C7H10N2	TOLUENEDIAMINE	95-80-7
TOLUENE, a-a-DICHLORO-	C7H6Cl2	BENZYL DICHLORIDE	98-87-3
TOLUENE DIISOCYANATE	C9H6N2O2	TOLUENE DIISOCYANATE	26471-62-5
TOLUENE HEXAHYDRIDE	C7H14	METHYLCYCLOHEXANE	108-87-2
TOLUENE TRICHLORIDE	C7H5Cl3	BENZOTRICHLORIDE	98-07-7
a-TOLUENOL	C7H8O	BENZYL ALCOHOL	100-51-6
o-TOLUIC ACID	C8H8O2	o-TOLUIC ACID	118-90-1
p-TOLUIC ACID	C8H8O2	p-TOLUIC ACID	99-94-5
m-TOLUIDIN	C7H9N	m-TOLUIDINE	108-44-1
o-TOLUIDIN	C7H9N	o-TOLUIDINE	95-53-4
p-TOLUIDIN	C7H9N	p-TOLUIDINE	106-49-0
m-TOLUIDINE	C7H9N	m-TOLUIDINE	108-44-1
o-TOLUIDINE	C7H9N	o-TOLUIDINE	95-53-4
p-TOLUIDINE	C7H9N	p-TOLUIDINE	106-49-0
2-TOLUIDINE	C7H9N	o-TOLUIDINE	95-53-4
3-TOLUIDINE	C7H9N	m-TOLUIDINE	108-44-1
4-TOLUIDINE	C7H9N	p-TOLUIDINE	106-49-0
o-TOLUIDYNA	C7H9N	o-TOLUIDINE	95-53-4
m-TOLUOL	C7H8O	m-CRESOL	108-39-4
o-TOLUOL	C7H8O	o-CRESOL	95-48-7
p-TOLUOL	C7H8O	p-CRESOL	106-44-5
TOLUOL	C7H8	TOLUENE	108-88-3
TOLUOLO	C7H8	TOLUENE	108-88-3
TOLU-SOL	C7H8	TOLUENE	108-88-3
o-TOLUYLIC ACID	C8H8O2	o-TOLUIC ACID	118-90-1
p-TOLUYLIC ACID	C8H8O2	p-TOLUIC ACID	99-94-5
p-TOLYL ALCOHOL	C7H8O	p-CRESOL	106-44-5
m-TOLYLAMINE	C7H9N	m-TOLUIDINE	108-44-1
o-TOLYLAMINE	C7H9N	o-TOLUIDINE	95-53-4
p-TOLYLAMINE	C7H9N	p-TOLUIDINE	106-49-0
TOLYLAMINE	C7H9N	p-TOLUIDINE	106-49-0
o-TOLYL CHLORIDE	C7H7Cl	o-CHLOROTOLUENE	95-49-8
TOLYL CHLORIDE	C7H7Cl	BENZYL CHLORIDE	100-44-7
1-p-TOLYLETHENE	C9H10	p-METHYLSTYRENE	622-97-9
o-TOLYL PHOSPHATE	C21H21O4P	TRI-o-CRESYL PHOSPHATE	78-30-8
TOPANEL	C4H6O	trans-CROTONALDEHYDE	123-73-9
TOTP	C21H21O4P	TRI-o-CRESYL PHOSPHATE	78-30-8
TOXILIC ACID	C4H4O4	MALEIC ACID	110-16-7
TOXILIC ANHYDRIDE	C4H2O3	MALEIC ANHYDRIDE	108-31-6

Name (Synonym)	Formula	Name in Tables	CAS Reg. No
TPP	C18H15O4P	TRIPHENYL PHOSPHATE	115-86-6
TRIACETALDEHYDE	C6H12O3	PARALDEHYDE	123-63-7
TRIACETIN	C9H14O6	GLYCERYL TRIACETATE	102-76-1
TRIAD	C2HCl3	TRICHLOROETHYLENE	79-01-6
TRIAETHANOLAMIN-NG	C6H15NO3	TRIETHANOLAMINE	102-71-6
TRIAETHYLAMIN	C6H15N	TRIETHYLAMINE	121-44-8
TRIASOL	C2HCl3	TRICHLOROETHYLENE	79-01-6
TRIBROMMETHAAN	CHBr3	TRIBROMOMETHANE	75-25-2
TRIBROMMETHAN	CHBr3	TRIBROMOMETHANE	75-25-2
TRIBROMOMETAN	CHBr3	TRIBROMOMETHANE	75-25-2
TRIBROMOMETHANE	CHBr3	TRIBROMOMETHANE	75-25-2
TRIBUTOXYBORANE	C12H27BO3	TRI-n-BUTYL BORATE	688-74-4
TRI-n-BUTOXYBORANE	C12H27BO3	TRI-n-BUTYL BORATE	688-74-4
TRI-n-BUTYLAMINE	C12H27N	TRI-n-BUTYLAMINE	102-82-9
TRIBUTYL BORATE	C12H27BO3	TRI-n-BUTYL BORATE	688-74-4
TRI-n-BUTYL BORATE	C12H27BO3	TRI-n-BUTYL BORATE	688-74-4
(TRIBUTYL)PEROXIDE	C8H18O2	DI-t-BUTYL PEROXIDE	110-05-4
TRICHLOORMETHAAN	CHCl3	CHLOROFORM	67-66-3
TRICHLOORMETHYLBENZEEN	C7H5Cl3	BENZOTRICHLORIDE	98-07-7
1,1,2-TRICHLORETHANE	C2H3Cl3	1,1,2-TRICHLOROETHANE	79-00-5
TRICHLORMETHAN	CHCl3	CHLOROFORM	67-66-3
TRICHLORMETHYLBENZOL	C7H5Cl3	BENZOTRICHLORIDE	98-07-7
TRICHLOR-METHYLSILAN	CH3Cl3Si	METHYL TRICHLOROSILANE	75-79-6
1,1,2-TRICHLORO-1,2,2-TRIFLUOROETHANE	C2Cl3F3	1,1,2-TRICHLOROTRIFLUOROETHANE	76-13-1
TRICHLOROACETALDEHYDE	C2HCl3O	TRICHLOROACETALDEHYDE	75-87-6
2,2,2-TRICHLOROACETALDEHYDE	C2HCl3O	TRICHLOROACETALDEHYDE	75-87-6
TRICHLOROACETIC ACID CHLORIDE	C2Cl4O	TRICHLOROACETYL CHLORIDE	76-02-8
TRICHLOROACETOCHLORIDE	C2Cl4O	TRICHLOROACETYL CHLORIDE	76-02-8
TRICHLOROACETYL CHLORIDE	C2Cl4O	TRICHLOROACETYL CHLORIDE	76-02-8
unsym-TRICHLOROBENZENE	C6H3Cl3	1,2,4-TRICHLOROBENZENE	120-82-1
1,2,4-TRICHLOROBENZENE	C6H3Cl3	1,2,4-TRICHLOROBENZENE	120-82-1
1,2,5-TRICHLOROBENZENE	C6H3Cl3	1,2,4-TRICHLOROBENZENE	120-82-1
1,3,4-TRICHLOROBENZENE	C6H3Cl3	1,2,4-TRICHLOROBENZENE	120-82-1
1,2,4-TRICHLOROBENZOL	C6H3Cl3	1,2,4-TRICHLOROBENZENE	120-82-1
TRICHLOROBROMOMETHANE	CBrCl3	BROMOTRICHLOROMETHANE	75-62-7
TRICHLOROETHANAL	C2HCl3O	TRICHLOROACETALDEHYDE	75-87-6
a-TRICHLOROETHANE	C2H3Cl3	1,1,1-TRICHLOROETHANE	71-55-6
b-TRICHLOROETHANE	C2H3Cl3	1,1,2-TRICHLOROETHANE	79-00-5
1,1,1-TRICHLOROETHANE	C2H3Cl3	1,1,1-TRICHLOROETHANE	71-55-6
1,1,2-TRICHLOROETHANE	C2H3Cl3	1,1,2-TRICHLOROETHANE	79-00-5
1,2,2-TRICHLOROETHANE	C2H3Cl3	1,1,2-TRICHLOROETHANE	79-00-5
TRICHLOROETHYLENE	C2HCl3	TRICHLOROETHYLENE	79-01-6
1,1,1-TRICHLOROFLUOROETHANE	C2H2Cl3F	1,1,1-TRICHLOROFLUOROETHANE	27154-33-2
TRICHLOROFLUOROMETHANE	CCl3F	TRICHLOROFLUOROMETHANE	75-69-4
TRICHLOROFORM	CHCl3	CHLOROFORM	67-66-3
TRICHLOROHYDRIN	C3H5Cl3	1,2,3-TRICHLOROPROPANE	96-18-4
TRICHLOROMETHANE	CHCl3	CHLOROFORM	67-66-3
TRICHLOROMETHYLBENZENE	C7H5Cl3	BENZOTRICHLORIDE	98-07-7
1-(TRICHLOROMETHYL)BENZENE	C7H5Cl3	BENZOTRICHLORIDE	98-07-7
TRICHLOROMETHYLSILANE	CH3Cl3Si	METHYL TRICHLOROSILANE	75-79-6
TRICHLOROMONOFLUOROMETHANE	CCl3F	TRICHLOROFLUOROMETHANE	75-69-4
TRICHLOROPHENYLMETHANE	C7H5Cl3	BENZOTRICHLORIDE	98-07-7
1,2,3-TRICHLOROPROPANE	C3H5Cl3	1,2,3-TRICHLOROPROPANE	96-18-4
a,a,a-TRICHLOROTOLUENE	C7H5Cl3	BENZOTRICHLORIDE	98-07-7
w,w,w-TRICHLOROTOLUENE	C7H5Cl3	BENZOTRICHLORIDE	98-07-7
TRICHLOROTRIETHYLDIALUMINIUM	C6H15Al2Cl3	ETHYL ALUMINUM SESQUICHLORIDE	12075-68-2
TRICHLOROTRIETHYLDIALUMINUM	C6H15Al2Cl3	ETHYL ALUMINUM SESQUICHLORIDE	12075-68-2
TRICHLOROTRIFLUOROETHANE	C2Cl3F3	1,1,2-TRICHLOROTRIFLUOROETHANE	76-13-1
1,1,2-TRICHLOROTRIFLUOROETHANE	C2Cl3F3	1,1,2-TRICHLOROTRIFLUOROETHANE	76-13-1
TRICLOROMETANO	CHCl3	CHLOROFORM	67-66-3
TRICLOROMETILBENZENE	C7H5Cl3	BENZOTRICHLORIDE	98-07-7
TRICLOROTOLUENE	C7H5Cl3	BENZOTRICHLORIDE	98-07-7
TRICRESYL PHOSPHATE	C21H21O4P	TRI-o-CRESYL PHOSPHATE	78-30-8
TRI-o-CRESYL PHOSPHATE	C21H21O4P	TRI-o-CRESYL PHOSPHATE	78-30-8
1-TRIDECANAL	C13H26O	1-TRIDECANAL	10486-19-8
n-TRIDECANE	C13H28	n-TRIDECANE	629-50-5
1-TRIDECANECARBOXYLIC ACID	C14H28O2	n-TETRADECANOIC ACID	544-63-8
1-TRIDECANETHIOL	C13H28S	1-TRIDECANETHIOL	----------
n-TRIDECANOL	C13H28O	1-TRIDECANOL	112-70-9
TRIDECANOL	C13H28O	1-TRIDECANOL	112-70-9
1-TRIDECANOL	C13H28O	1-TRIDECANOL	112-70-9
1-TRIDECENE	C13H26	1-TRIDECENE	2437-56-1
n-TRIDECYL ALCOHOL	C13H28O	1-TRIDECANOL	112-70-9
TRIDECYL ALCOHOL	C13H28O	1-TRIDECANOL	112-70-9
n-TRIDECYLBENZENE	C19H32	n-TRIDECYLBENZENE	123-02-4
1-TRIDECYNE	C13H24	1-TRIDECYNE	26186-02-7

Name (Synonym)	Formula	Name in Tables	CAS Reg. No
TRI(DIMETHYLAMINO)PHOSPHINE OXIDE	C6H18N3OP	HEXAMETHYL PHOSPHORAMIDE	680-31-9
TRIEN	C6H18N4	TRIETHYLENE TETRAMINE	112-24-3
TRIENTINE	C6H18N4	TRIETHYLENE TETRAMINE	112-24-3
TRI-ETHANE	C2H3Cl3	1,1,1-TRICHLOROETHANE	71-55-6
TRIETHANOLAMIN	C6H15NO3	TRIETHANOLAMINE	102-71-6
TRIETHANOLAMINE	C6H15NO3	TRIETHANOLAMINE	102-71-6
TRIETHYLALUMINUM	C6H15Al	TRIETHYL ALUMINUN	97-93-8
TRIETHYLALUMINUM SESQUICHLORIDE	C6H15Al2Cl3	ETHYL ALUMINUM SESQUICHLORIDE	12075-68-2
TRIETHYL ALUMINUN	C6H15Al	TRIETHYL ALUMINUN	97-93-8
TRIETHYLAMINE	C6H15N	TRIETHYLAMINE	121-44-8
1,2,3-TRIETHYLBENZENE	C12H18	1,2,3-TRIETHYLBENZENE	----------
1,2,4-TRIETHYLBENZENE	C12H18	1,2,4-TRIETHYLBENZENE	877-44-1
1,3,5-TRIETHYLBENZENE	C12H18	1,3,5-TRIETHYLBENZENE	102-25-0
TRIETHYLENEDIAMINE	C6H12N2	TRIETHYLENEDIAMINE	280-57-9
TRIETHYLENE GLYCOL	C6H14O4	TRIETHYLENE GLYCOL	112-27-6
TRIETHYLENE GLYCOL DIMETHYL ETHER	C8H18O4	TRIETHYLENE GLYCOL DIMETHYL ETHER	112-49-2
TRIETHYLENE TETRAMINE	C6H18N4	TRIETHYLENE TETRAMINE	112-24-3
TRIETHYLOLAMINE	C6H15NO3	TRIETHANOLAMINE	102-71-6
TRIETHYL PHOSPHATE	C6H15O4P	TRIETHYL PHOSPHATE	78-40-0
TRIETHYLTRICHLORODIALUMINUM	C6H15Al2Cl3	ETHYL ALUMINUM SESQUICHLORIDE	12075-68-2
TRIETILAMINA	C6H15N	TRIETHYLAMINE	121-44-8
TRIFLUOROACETIC ACID	C2HF3O2	TRIFLUOROACETIC ACID	76-05-1
1,1,1-TRIFLUORO-2-BROMO-2-CHLOROETHANE	C2HBrClF3	HALOTHANE	151-67-7
TRIFLUOROBROMOETHYLENE	C2BrF3	BROMOTRIFLUOROETHYLENE	598-73-2
1,1,1-TRIFLUORO-2-CHLORO-2-BROMOETHANE	C2HBrClF3	HALOTHANE	151-67-7
2,2,2-TRIFLUORO-1-CHLORO-1-BROMOETHANE	C2HBrClF3	HALOTHANE	151-67-7
TRIFLUOROCHLOROETHYLENE	C2ClF3	CHLOROTRIFLUOROETHYLENE	79-38-9
1,1,2-TRIFLUORO-2-CHLOROETHYLENE	C2ClF3	CHLOROTRIFLUOROETHYLENE	79-38-9
TRIFLUOROCHLOROMETHANE	CClF3	CHLOROTRIFLUOROMETHANE	75-72-9
a,a,a-TRIFLUORO-4-CHLOROTOLUENE	C7H4ClF3	p-CHLOROBENZOTRIFLUORIDE	98-56-6
1,1,1-TRIFLUOROETHANE	C2H3F3	1,1,1-TRIFLUOROETHANE	420-46-2
TRIFLUOROETHENE	C2HF3	TRIFLUOROETHENE	359-11-5
1,1,1-TRIFLUOROFORM	C2H3F3	1,1,1-TRIFLUOROETHANE	420-46-2
TRIFLUOROMETHANE	CHF3	TRIFLUOROMETHANE	75-46-7
(TRIFLUOROMETHYL)BENZENE	C7H5F3	BENZOTRIFLUORIDE	98-08-8
TRIFLUOROMETHYL CHLORIDE	CClF3	CHLOROTRIFLUOROMETHANE	75-72-9
p-(TRIFLUOROMETHYL)CHLOROBENZENE	C7H4ClF3	p-CHLOROBENZOTRIFLUORIDE	98-56-6
m-(TRIFLUOROMETHYL)NITROBENZENE	C7H4F3NO2	3-NITROBENZOTRIFLUORIDE	98-46-4
3-TRIFLUOROMETHYLNITROBENZENE	C7H4F3NO2	3-NITROBENZOTRIFLUORIDE	98-46-4
p-TRIFLUOROMETHYLPHENYL CHLORIDE	C7H4ClF3	p-CHLOROBENZOTRIFLUORIDE	98-56-6
a,a,a-TRIFLUORO-m-NITROTOLUENE	C7H4F3NO2	3-NITROBENZOTRIFLUORIDE	98-46-4
TRIFLUOROMONOBROMOMETHANE	CBrF3	BROMOTRIFLUOROMETHANE	75-63-8
TRIFLUOROMONOCHLOROCARBON	CClF3	CHLOROTRIFLUOROMETHANE	75-72-9
TRIFLUOROMONOCHLOROETHYLENE	C2ClF3	CHLOROTRIFLUOROETHYLENE	79-38-9
a,a,a-TRIFLUOROTOLUENE	C7H5F3	BENZOTRIFLUORIDE	98-08-8
w-TRIFLUOROTOLUENE	C7H5F3	BENZOTRIFLUORIDE	98-08-8
TRIFLUOROVINYLBROMIDE	C2BrF3	BROMOTRIFLUOROETHYLENE	598-73-2
TRIFLUOROVINYL CHLORIDE	C2ClF3	CHLOROTRIFLUOROETHYLENE	79-38-9
TRIGEN	C6H14O4	TRIETHYLENE GLYCOL	112-27-6
TRIGLYCOL	C6H14O4	TRIETHYLENE GLYCOL	112-27-6
TRIGONOX A-75	C4H10O2	t-BUTYL HYDROPEROXIDE	75-91-2
1,2,3-TRIHYDROXYBENZEN	C6H6O3	1,2,3-BENZENETRIOL	87-66-1
1,2,3-TRIHYDROXYBENZENE	C6H6O3	1,2,3-BENZENETRIOL	87-66-1
TRI(HYDROXYETHYL)AMINE	C6H15NO3	TRIETHANOLAMINE	102-71-6
1,1,1-TRI(HYDROXYMETHYL)PROPANE	C6H14O3	TRIMETHYLOLPROPANE	77-99-6
TRIHYDROXYPROPANE	C3H8O3	GLYCEROL	56-81-5
1,2,3-TRIHYDROXYPROPANE	C3H8O3	GLYCEROL	56-81-5
2,2',2"-TRIHYDROXYTRIETHYLAMINE	C6H15NO3	TRIETHANOLAMINE	102-71-6
TRIIODOMETHANE	CHI3	TRIIODOMETHANE	75-47-8
o-TRIKRESYLPHOSPHATE	C21H21O4P	TRI-o-CRESYL PHOSPHATE	78-30-8
TRIMELLIC ACID ANHYDRIDE	C9H4O5	TRIMELLITIC ANHYDRIDE	552-30-7
TRIMELLIC ACID-1,2-ANHYDRIDE	C9H4O5	TRIMELLITIC ANHYDRIDE	552-30-7
TRIMELLITIC ACID CYCLIC-1,2-ANHYDRIDE	C9H4O5	TRIMELLITIC ANHYDRIDE	552-30-7
TRIMELLITIC ANHYDRIDE	C9H4O5	TRIMELLITIC ANHYDRIDE	552-30-7
TRIMETHYLAMINE	C3H9N	TRIMETHYLAMINE	75-50-3
TRIMETHYLAMINOMETHANE	C4H11N	tert-BUTYLAMINE	75-64-9
as-TRIMETHYL BENZENE	C9H12	1,2,4-TRIMETHYLBENZENE	95-63-6
sym-TRIMETHYLBENZENE	C9H12	MESITYLENE	108-67-8
TRIMETHYL BENZENE	C9H12	MESITYLENE	108-67-8
1,2,3-TRIMETHYLBENZENE	C9H12	1,2,3-TRIMETHYLBENZENE	526-73-8
1,2,4-TRIMETHYLBENZENE	C9H12	1,2,4-TRIMETHYLBENZENE	95-63-6
1,2,5-TRIMETHYL BENZENE	C9H12	1,2,4-TRIMETHYLBENZENE	95-63-6
TRIMETHYL BENZOL	C9H12	MESITYLENE	108-67-8
2,6,6-TRIMETHYLBICYCLO(3.1.1)-2-HEPT-2-ENE	C10H16	alpha-PINENE	80-56-8
1,7,7-TRIMETHYLBICYCLO(2.2.1)-2-HEPTANONE	C10H16O	CAMPHOR	76-22-2
4,6,6-TRIMETHYLBICYKLO(3,1,1)HEPT-3-EN	C10H16	alpha-PINENE	80-56-8

755

Name (Synonym)	Formula	Name in Tables	CAS Reg. No
2,2,3-TRIMETHYLBUTANE	C7H16	2,2,3-TRIMETHYLBUTANE	464-06-2
2,3,3-TRIMETHYL-1-BUTENE	C7H14	2,3,3-TRIMETHYL-1-BUTENE	594-56-9
TRIMETHYLCARBINOL	C4H10O	tert-BUTANOL	75-65-0
TRIMETHYLCHLOROMETHANE	C4H9Cl	tert-BUTYL CHLORIDE	507-20-0
cis,cis-1,3,5-TRIMETHYLCYCLOHEXANE	C9H18	cis,cis-1,3,5-TRIMETHYLCYCLOHEXANE	1795-27-3
cis,trans-1,3,5-TRIMETHYLCYCLOHEXANE	C9H18	cis,trans-1,3,5-TRIMETHYLCYCLOHEXANE	1795-26-2
3,5,5-TRIMETHYL-2-CYCLOHEXEN-1-ON	C9H14O	ISOPHORONE	78-59-1
1,1,3-TRIMETHYL-3-CYCLOHEXENE-5-ONE	C9H14O	ISOPHORONE	78-59-1
3,5,5-TRIMETHYL-2-CYCLOHEXENE-1-ONE	C9H14O	ISOPHORONE	78-59-1
TRIMETHYLENE	C3H6	CYCLOPROPANE	75-19-4
TRIMETHYLENE DICHLORIDE	C3H6Cl2	1,3-DICHLOROPROPANE	142-28-9
TRIMETHYLENE GLYCOL	C3H8O2	1,3-PROPYLENE GLYCOL	504-63-2
TRIMETHYLENE OXIDE	C3H6O	1,3-PROPYLENE OXIDE	503-30-0
TRIMETHYLENOXID	C3H6O	1,3-PROPYLENE OXIDE	503-30-0
TRIMETHYL GLYCOL	C3H8O2	1,2-PROPYLENE GLYCOL	57-55-6
2,2,3-TRIMETHYLHEPTANE	C10H22	2,2,3-TRIMETHYLHEPTANE	52896-92-1
2,2,4-TRIMETHYLHEPTANE	C10H22	2,2,4-TRIMETHYLHEPTANE	14720-74-2
2,2,5-TRIMETHYLHEPTANE	C10H22	2,2,5-TRIMETHYLHEPTANE	20291-95-6
2,2,6-TRIMETHYLHEPTANE	C10H22	2,2,6-TRIMETHYLHEPTANE	1190-83-6
2,3,3-TRIMETHYLHEPTANE	C10H22	2,3,3-TRIMETHYLHEPTANE	52896-93-2
2,3,4-TRIMETHYLHEPTANE	C10H22	2,3,4-TRIMETHYLHEPTANE	52896-95-4
2,3,5-TRIMETHYLHEPTANE	C10H22	2,3,5-TRIMETHYLHEPTANE	20278-85-7
2,3,6-TRIMETHYLHEPTANE	C10H22	2,3,6-TRIMETHYLHEPTANE	4032-93-3
2,4,4-TRIMETHYLHEPTANE	C10H22	2,4,4-TRIMETHYLHEPTANE	4032-92-2
2,4,5-TRIMETHYLHEPTANE	C10H22	2,4,5-TRIMETHYLHEPTANE	20278-84-6
2,4,6-TRIMETHYLHEPTANE	C10H22	2,4,6-TRIMETHYLHEPTANE	2613-61-8
2,5,5-TRIMETHYLHEPTANE	C10H22	2,5,5-TRIMETHYLHEPTANE	1189-99-7
3,3,4-TRIMETHYLHEPTANE	C10H22	3,3,4-TRIMETHYLHEPTANE	2278-87-9
3,3,5-TRIMETHYLHEPTANE	C10H22	3,3,5-TRIMETHYLHEPTANE	7154-80-5
3,4,4-TRIMETHYLHEPTANE	C10H22	3,4,4-TRIMETHYLHEPTANE	2278-88-0
3,4,5-TRIMETHYLHEPTANE	C10H22	3,4,5-TRIMETHYLHEPTANE	2278-89-1
2,2,3-TRIMETHYLHEXANE	C9H20	2,2,3-TRIMETHYLHEXANE	16747-25-4
2,2,4-TRIMETHYLHEXANE	C9H20	2,2,4-TRIMETHYLHEXANE	16747-26-5
2,2,5-TRIMETHYLHEXANE	C9H20	2,2,5-TRIMETHYLHEXANE	3522-94-9
2,3,3-TRIMETHYLHEXANE	C9H20	2,3,3-TRIMETHYLHEXANE	16747-28-7
2,3,4-TRIMETHYLHEXANE	C9H20	2,3,4-TRIMETHYLHEXANE	921-47-1
2,3,5-TRIMETHYLHEXANE	C9H20	2,3,5-TRIMETHYLHEXANE	1069-53-0
2,4,4-TRIMETHYLHEXANE	C9H20	2,4,4-TRIMETHYLHEXANE	16747-30-1
3,3,4-TRIMETHYLHEXANE	C9H20	3,3,4-TRIMETHYLHEXANE	16747-31-2
1,2,3-TRIMETHYLINDENE	C12H14	1,2,3-TRIMETHYLINDENE	4773-83-5
1,7,7-TRIMETHYLNORCAMPHOR	C10H16O	CAMPHOR	76-22-2
TRIMETHYLOLPROPANE	C6H14O3	TRIMETHYLOLPROPANE	77-99-6
1,1,1-TRIMETHYLOLPROPANE	C6H14O3	TRIMETHYLOLPROPANE	77-99-6
2,2,3-TRIMETHYLPENTANE	C8H18	2,2,3-TRIMETHYLPENTANE	564-02-3
2,2,4-TRIMETHYLPENTANE	C8H18	2,2,4-TRIMETHYLPENTANE	540-84-1
2,3,3-TRIMETHYLPENTANE	C8H18	2,3,3-TRIMETHYLPENTANE	560-21-4
2,3,4-TRIMETHYLPENTANE	C8H18	2,3,4-TRIMETHYLPENTANE	565-75-3
2,4,4-TRIMETHYL-2-PENTANETHIOL	C8H18S	tert-OCTYL MERCAPTAN	141-59-3
2,4,4-TRIMETHYL-1-PENTENE	C8H16	2,4,4-TRIMETHYL-1-PENTENE	107-39-1
2,4,4-TRIMETHYL-2-PENTENE	C8H16	2,4,4-TRIMETHYL-2-PENTENE	107-40-4
TRIMETHYLPHENYLMETHANE	C10H14	tert-BUTYLBENZENE	98-06-6
TRI 2-METHYLPHENYL PHOSPHATE	C21H21O4P	TRI-o-CRESYL PHOSPHATE	78-30-8
TRIMETHYL PHOSPHATE	C3H9O4P	TRIMETHYL PHOSPHATE	512-56-1
2,4,6-TRIMETHYLPYRIDINE	C8H11N	2,4,6-TRIMETHYLPYRIDINE	108-75-8
TRIMETHYL SILANE	C3H10Si	TRIMETHYL SILANE	993-07-7
a,a,a'-TRIMETHYLTRIMETHYLENE GLYCOL	C6H14O2	HEXYLENE GLYCOL	107-41-5
1,1,1-TRIMETHYL-N-(TRIMETHYLSILYL)SILANAMINE	C6H19NSi2	HEXAMETHYLDISILAZANE	999-97-3
2,4,6-TRIMETHYL-1,3,5-TRIOXAAN	C6H12O3	PARALDEHYDE	123-63-7
2,4,6-TRIMETHYL-s-TRIOXANE	C6H12O3	PARALDEHYDE	123-63-7
2,4,6-TRIMETHYL-1,3,5-TRIOXANE	C6H12O3	PARALDEHYDE	123-63-7
s-TRIMETHYLTRIOXYMETHYLENE	C6H12O3	PARALDEHYDE	123-63-7
3,5,5-TRIMETIL-2-CICLOESEN-1-ONE	C9H14O	ISOPHORONE	78-59-1
2,4,6-TRIMETIL-1,3,5-TRIOSSANO	C6H12O3	PARALDEHYDE	123-63-7
TRINITRIN	C3H5N3O9	NITROGLYCERINE	55-63-0
TRINITROBENZEEN	C6H3N3O6	1,3,5-TRINITROBENZENE	99-35-4
TRINITROBENZENE	C6H3N3O6	1,3,5-TRINITROBENZENE	99-35-4
1,3,5-TRINITROBENZENE	C6H3N3O6	1,3,5-TRINITROBENZENE	99-35-4
TRINITROBENZOL	C6H3N3O6	1,3,5-TRINITROBENZENE	99-35-4
TRINITROGLYCERIN	C3H5N3O9	NITROGLYCERINE	55-63-0
TRINITROGLYCEROL	C3H5N3O9	NITROGLYCERINE	55-63-0
2,4,6-TRINITROTOLUEEN	C7H5N3O6	2,4,6-TRINITROTOLUENE	118-96-7
s-TRINITROTOLUENE	C7H5N3O6	2,4,6-TRINITROTOLUENE	118-96-7
sym-TRINITROTOLUENE	C7H5N3O6	2,4,6-TRINITROTOLUENE	118-96-7
TRINITROTOLUENE	C7H5N3O6	2,4,6-TRINITROTOLUENE	118-96-7
2,4,6-TRINITROTOLUENE	C7H5N3O6	2,4,6-TRINITROTOLUENE	118-96-7
s-TRINITROTOLUOL	C7H5N3O6	2,4,6-TRINITROTOLUENE	118-96-7

Name (Synonym)	Formula	Name in Tables	CAS Reg. No
sym-TRINITROTOLUOL	C7H5N3O6	2,4,6-TRINITROTOLUENE	118-96-7
TRIOSSIMETELENE	C3H6O3	TRIOXANE	110-88-3
sym-TRIOXANE	C3H6O3	TRIOXANE	110-88-3
TRIOXANE	C3H6O3	TRIOXANE	110-88-3
1,3,5-TRIOXANE	C3H6O3	TRIOXANE	110-88-3
5,8,11-TRIOXAPENTADECANE	C12H26O3	DIETHYLENE GLYCOL DI-n-BUTYL ETHER	112-73-2
3,6,9-TRIOXAUNDECANE	C8H18O3	DIETHYLENE GLYCOL DIETHYL ETHER	112-36-7
TRIOXYMETHYLEEN	C3H6O3	TRIOXANE	110-88-3
TRIOXYMETHYLEN	C3H6O3	TRIOXANE	110-88-3
m-TRIPHENYL	C18H14	m-TERPHENYL	92-06-8
p-TRIPHENYL	C18H14	p-TERPHENYL	92-94-4
TRIPHENYL PHOSPHATE	C18H15O4P	TRIPHENYL PHOSPHATE	115-86-6
TRIPHENYLETHYLENE	C20H16	TRIPHENYLETHYLENE	58-72-0
TRIPHENYLPHOSPHINE	C18H15P	TRIPHENYLPHOSPHINE	603-35-0
TRIPROPYLAMINE	C9H21N	TRIPROPYLAMINE	102-69-2
TRIS-N-BUTYLAMINE	C12H27N	TRI-n-BUTYLAMINE	102-82-9
TRIS(o-CRESYL)-PHOSPHATE	C21H21O4P	TRI-o-CRESYL PHOSPHATE	78-30-8
TRIS(DIMETHYLAMINO)PHOSPHINE OXIDE	C6H18N3OP	HEXAMETHYL PHOSPHORAMIDE	680-31-9
TRIS(DIMETHYLAMINO)PHOSPHORUS OXIDE	C6H18N3OP	HEXAMETHYL PHOSPHORAMIDE	680-31-9
TRIS(2-HYDROXYETHYL)AMINE	C6H15NO3	TRIETHANOLAMINE	102-71-6
TRIS(HYDROXYMETHYL)PROPANE	C6H14O3	TRIMETHYLOLPROPANE	77-99-6
1,1,1-TRIS(HYDROXYMETHYL)PROPANE	C6H14O3	TRIMETHYLOLPROPANE	77-99-6
TRIS(o-METHYLPHENYL)PHOSPHATE	C21H21O4P	TRI-o-CRESYL PHOSPHATE	78-30-8
TRIS(o-TOLYL)-PHOSPHATE	C21H21O4P	TRI-o-CRESYL PHOSPHATE	78-30-8
TRITHENE	C2ClF3	CHLOROTRIFLUOROETHYLENE	79-38-9
TRI-o-TOLYL PHOSPHATE	C21H21O4P	TRI-o-CRESYL PHOSPHATE	78-30-8
TRI-2-TOLYL PHOSPHATE	C21H21O4P	TRI-o-CRESYL PHOSPHATE	78-30-8
TROJCHLOROBENZEN	C6H3Cl3	1,2,4-TRICHLOROBENZENE	120-82-1
TROJKREZYLU FOSFORAN	C21H21O4P	TRI-o-CRESYL PHOSPHATE	78-30-8
TROLAMINE	C6H15NO3	TRIETHANOLAMINE	102-71-6
TROVIDUR	C2H3Cl	VINYL CHLORIDE	75-01-4
TUMEX	C9H7NO	8-HYDROXYQUINOLINE	148-24-3
TUTANE	C4H11N	sec-BUTYLAMINE	13952-84-6
TYRANTON	C6H12O2	DIACETONE ALCOHOL	123-42-2
U-1149	C4H4O4	FUMARIC ACID	110-17-8
U-5954	C4H9NO	N,N-DIMETHYLACETAMIDE	127-19-5
UC 7207	C8H24O4Si4	OCTAMETHYLCYCLOTETRASILOXANE	556-67-2
UCAR 17	C2H6O2	ETHYLENE GLYCOL	107-21-1
UCAR AMYL PHENOL 4T	C11H16O	p-tert-AMYLPHENOL	80-46-6
UCAR BUTYLPHENOL 4-T	C10H14O	p-tert-BUTYLPHENOL	98-54-4
UCON 112	C2Cl4F2	1,1,2,2-TETRACHLORODIFLUOROETHANE	76-12-0
UCON 113	C2Cl3F3	1,1,2-TRICHLOROTRIFLUOROETHANE	76-13-1
UCON 114	C2Cl2F4	1,2-DICHLOROTETRAFLUOROETHANE	76-14-2
UCON 12	CCl2F2	DICHLORODIFLUOROMETHANE	75-71-8
UCON 12/HALOCARBON 12	CCl2F2	DICHLORODIFLUOROMETHANE	75-71-8
UCON 22	CHClF2	CHLORODIFLUOROMETHANE	75-45-6
UCON FLUOROCARBON 113	C2Cl3F3	1,1,2-TRICHLOROTRIFLUOROETHANE	76-13-1
UCON REFRIGERANT 11	CCl3F	TRICHLOROFLUOROMETHANE	75-69-4
n-UNDECANAL	C11H22O	1-UNDECANAL	112-44-7
UNDECANAL	C11H22O	1-UNDECANAL	112-44-7
1-UNDECANAL	C11H22O	1-UNDECANAL	112-44-7
UNDECANALDEHYDE	C11H22O	1-UNDECANAL	112-44-7
n-UNDECANE	C11H24	n-UNDECANE	1120-21-4
1-UNDECANECARBOXYLIC ACID	C12H24O2	n-DODECANOIC ACID	143-07-7
n-UNDECANOL	C11H24O	1-UNDECANOL	112-42-5
1-UNDECANOL	C11H24O	1-UNDECANOL	112-42-5
1-UNDECENE	C11H22	1-UNDECENE	821-95-4
n-UNDECYL ALDEHYDE	C11H22O	1-UNDECANAL	112-44-7
UNDECYL ALDEHYDE	C11H22O	1-UNDECANAL	112-44-7
n-UNDECYLBENZENE	C17H28	n-UNDECYLBENZENE	6742-54-7
UNDECYLIC ALDEHYDE	C11H22O	1-UNDECANAL	112-44-7
UNDECYL MERCAPTAN	C11H24S	UNDECYL MERCAPTAN	5332-52-5
1-UNDECYNE	C11H20	1-UNDECYNE	2243-98-3
UNIFLEX BYS	C22H44O2	n-BUTYL STEARATE	123-95-5
UNION CARBIDE 7207	C8H24O4Si4	OCTAMETHYLCYCLOTETRASILOXANE	556-67-2
UNIPHAT A40	C13H26O2	METHYL DODECANOATE	111-82-0
UNIVERM	CCl4	CARBON TETRACHLORIDE	56-23-5
UNIVOL U 316S	C14H28O2	n-TETRADECANOIC ACID	544-63-8
URNERS LIQUID	C2H2Cl2O2	DICHLOROACETIC ACID	79-43-6
URSOL D	C6H8N2	p-PHENYLENEDIAMINE	106-50-3
USAF A-4600	C3H2N2	MALONONITRILE	109-77-3
USAF A-9442	C4H4N2	SUCCINONITRILE	110-61-2
USAF AN-7	C7H8O2	p-METHOXYPHENOL	150-76-5
USAF CZ-1	C5H8O3	LEVULINIC ACID	123-76-2
USAF DO-21	C8H16	2-ETHYL-1-HEXENE	1632-16-2
USAF DO-45	C6H14O2	ACETAL	105-57-7

Name (Synonym)	Formula	Name in Tables	CAS Reg. No
USAF DO-46	C6H15N3	N-AMINOETHYL PIPERAZINE	140-31-8
USAF DO-50	C3H9NO	METHYLETHANOLAMINE	109-83-1
USAF DO-52	C5H13NO2	METHYL DIETHANOLAMINE	105-59-9
USAF EK-1860	C4H4S	THIOPHENE	110-02-1
USAF EK-2122	C7H16S	n-HEPTYL MERCAPTAN	1639-09-4
USAF EK-218	C9H7N	QUINOLINE	91-22-5
USAF EK-3	C8H9NO	ACETANILIDE	103-84-4
USAF EK-356	C6H6O2	p-HYDROQUINONE	123-31-9
USAF EK-394	C6H8N2	p-PHENYLENEDIAMINE	106-50-3
USAF EK-4628	C6H14S	n-HEXYLMERCAPTAN	111-31-9
USAF EK-488	C2H3N	ACETONITRILE	75-05-8
USAF EK-496	C8H8O	ACETOPHENONE	98-86-2
USAF EK-600	C12H9N	DIBENZOPYRROLE	86-74-8
USAF EK-794	C9H7NO	8-HYDROXYQUINOLINE	148-24-3
USAF EK-P-583	C4H4O4	FUMARIC ACID	110-17-8
USAF EK-P-737	C2H4OS	THIOACETIC-ACID	507-09-5
USAF GY-2	C18H16N2	N,N'-DIPHENYL-p-PHENYLENEDIAMINE	74-31-7
USAF HC-1	C10H18O4	SEBACIC ACID	111-20-6
USAF KF-22	C4H5NO2	METHYL CYANOACETATE	105-34-0
USAF KF-25	C5H7NO2	ETHYL CYANOACETATE	105-56-6
USAF M-6	C7H6O2	p-HYDROXYBENZALDEHYDE	123-08-0
USAF MA-16	C7H5F3	BENZOTRIFLUORIDE	98-08-8
USAF MA-5	C7H4F3NO2	3-NITROBENZOTRIFLUORIDE	98-46-4
USAF RH-1	C4H7NO	2-METHACRYLAMIDE	79-39-0
USAF RH-7	C3H5NO	HYDRACRYLONITRILE	109-78-4
USAF RH-8	C4H7NO	ACETONE CYANOHYDRIN	75-86-5
USAF ST-40	C4H5N	METHACRYLONITRILE	126-98-7
USAF XR-19	C6H6S	PHENYL MERCAPTAN	108-98-5
USAF XR-20	C10H6O8	PYROMELLITIC ACID	89-05-4
VALAMINE	C4H11N	ISOBUTYLAMINE	78-81-9
VALERAL	C5H10O	VALERALDEHYDE	110-62-3
VALERALDEHYDE	C5H10O	VALERALDEHYDE	110-62-3
VALERIANIC ACID	C5H10O2	VALERIC ACID	109-52-4
VALERIANIC ALDEHYDE	C5H10O	VALERALDEHYDE	110-62-3
n-VALERIC ACID	C5H10O2	VALERIC ACID	109-52-4
VALERIC ACID	C5H10O2	VALERIC ACID	109-52-4
VALERIC ACID ALDEHYDE	C5H10O	VALERALDEHYDE	110-62-3
VALERIC ALDEHYDE	C5H10O	VALERALDEHYDE	110-62-3
VALERONE	C9H18O	DIISOBUTYL KETONE	108-83-8
n-VALERONITRILE	C5H9N	VALERONITRILE	110-59-8
VALERONITRILE	C5H9N	VALERONITRILE	110-59-8
VALERYLALDEHYDE	C5H10O	VALERALDEHYDE	110-62-3
VALINE ALDEHDYE	C4H8O	ISOBUTYRALDEHYDE	78-84-2
VANAY	C9H14O6	GLYCERYL TRIACETATE	102-76-1
VANILLA	C8H8O3	VANILLIN	121-33-5
VANILLALDEHYDE	C8H8O3	VANILLIN	121-33-5
VANILLIC ALDEHYDE	C8H8O3	VANILLIN	121-33-5
p-VANILLIN	C8H8O3	VANILLIN	121-33-5
VANILLIN	C8H8O3	VANILLIN	121-33-5
VANZOATE	C14H12O2	BENZYL BENZOATE	120-51-4
VAROX DCP-R	C18H22O2	DICUMYL PEROXIDE	80-43-3
VAROX DCP-T	C18H22O2	DICUMYL PEROXIDE	80-43-3
VCM	C2H3Cl	VINYL CHLORIDE	75-01-4
VCN	C3H3N	ACRYLONITRILE	107-13-1
VDC	C2H2Cl2	1,1-DICHLOROETHYLENE	75-35-4
VDF	C2H2F2	1,1-DIFLUOROETHYLENE	75-38-7
VENTOX	C3H3N	ACRYLONITRILE	107-13-1
VENZONATE	C14H12O2	BENZYL BENZOATE	120-51-4
VERMOESTRICID	CCl4	CARBON TETRACHLORIDE	56-23-5
VERSNELLER NL 63/10	C8H11N	N,N-DIMETHYLANILINE	121-69-7
VIDDEN D	C3H6Cl2	2,2-DICHLOROPROPANE	594-20-7
VINEGAR ACID	C2H4O2	ACETIC ACID	64-19-7
VINEGAR NAPHTHA	C4H8O2	ETHYL ACETATE	141-78-6
VINESTHENE	C4H6O	DIVINYL ETHER	109-93-3
VINESTHESIN	C4H6O	DIVINYL ETHER	109-93-3
VINETHEN	C4H6O	DIVINYL ETHER	109-93-3
VINETHENE	C4H6O	DIVINYL ETHER	109-93-3
VINETHER	C4H6O	DIVINYL ETHER	109-93-3
VINIDYL	C4H6O	DIVINYL ETHER	109-93-3
VINYDAN	C4H6O	DIVINYL ETHER	109-93-3
VINYL ACETATE	C4H6O2	VINYL ACETATE	108-05-4
VINYL ACETATE H.Q.	C4H6O2	VINYL ACETATE	108-05-4
VINYLACETONITRILE	C4H5N	VINYLACETONITRILE	109-75-1
VINYLACETYLENE	C4H4	VINYLACETYLENE	689-97-4
VINYL AMIDE	C3H5NO	ACRYLAMIDE	79-06-1
VINYL A MONOMER	C4H6O2	VINYL ACETATE	108-05-4

758

Name (Synonym)	Formula	Name in Tables	CAS Reg. No
VINYLBENZEN	C8H8	STYRENE	100-42-5
VINYLBENZENE	C8H8	STYRENE	100-42-5
VINYLBENZOL	C8H8	STYRENE	100-42-5
VINYL BROMIDE	C2H3Br	VINYL BROMIDE	593-60-2
VINYLCARBINOL	C3H6O	ALLYL ALCOHOL	107-18-6
VINYL CHLORIDE	C2H3Cl	VINYL CHLORIDE	75-01-4
VINYL CHLORIDE MONOMER	C2H3Cl	VINYL CHLORIDE	75-01-4
VINYL C MONOMER	C2H3Cl	VINYL CHLORIDE	75-01-4
VINYL CYANIDE	C3H3N	ACRYLONITRILE	107-13-1
VINYLCYCLOHEXENE	C8H12	VINYLCYCLOHEXENE	100-40-3
1-VINYLCYCLOHEX-3-ENE	C8H12	VINYLCYCLOHEXENE	100-40-3
4-VINYLCYCLOHEXENE	C8H12	VINYLCYCLOHEXENE	100-40-3
4-VINYLCYCLOHEXENE-1	C8H12	VINYLCYCLOHEXENE	100-40-3
1-VINYLCYCLOHEXENE-3	C8H12	VINYLCYCLOHEXENE	100-40-3
VINYLESTER KYSELINY OCTOVE	C4H6O2	VINYL ACETATE	108-05-4
VINYL ETHANOATE	C4H6O2	VINYL ACETATE	108-05-4
VINYLETHYLENE	C4H6	1,3-BUTADIENE	106-99-0
VINYL FLUORIDE	C2H3F	VINYL FLUORIDE	75-02-5
VINYL FORMATE	C3H4O2	VINYL FORMATE	692-45-5
VINYLFORMIC ACID	C3H4O2	ACRYLIC ACID	79-10-7
VINYLIDENE CHLORIDE	C2H2Cl2	1,1-DICHLOROETHYLENE	75-35-4
VINYLIDENE DIFLUORIDE	C2H2F2	1,1-DIFLUOROETHYLENE	75-38-7
VINYLIDINE CHLORIDE	C2H2Cl2	1,1-DICHLOROETHYLENE	75-35-4
VINYLKYANID	C3H3N	ACRYLONITRILE	107-13-1
VINYL METHYL ETHER	C3H6O	METHYL VINYL ETHER	107-25-5
VINYL-n-BUTYL ETHER	C6H12O	BUTYL VINYL ETHER	111-34-2
VINYL PROPIONATE	C5H8O2	VINYL PROPIONATE	105-38-4
m-VINYLSTYRENE	C10H10	m-DIVINYLBENZENE	108-57-6
m-VINYLTOLUENE	C9H10	m-METHYLSTYRENE	100-80-1
p-VINYLTOLUENE	C9H10	p-METHYLSTYRENE	622-97-9
VINYL TRICHLORIDE	C2H3Cl3	1,1,2-TRICHLOROETHANE	79-00-5
VITACIN	C6H8O6	ASCORBIC ACID	50-81-7
VITAMIN C	C6H8O6	ASCORBIC ACID	50-81-7
VITAMISIN	C6H8O6	ASCORBIC ACID	50-81-7
VITASCORBOL	C6H8O6	ASCORBIC ACID	50-81-7
VS 7207	C8H24O4Si4	OCTAMETHYLCYCLOTETRASILOXANE	556-67-2
VULKANOX 4020	C6H8N2	p-PHENYLENEDIAMINE	106-50-3
VYAC	C4H6O2	VINYL ACETATE	108-05-4
WECOLINE 1295	C12H24O2	n-DODECANOIC ACID	143-07-7
WEED DRENCH	C3H6O	ALLYL ALCOHOL	107-18-6
WEEVILTOX	CS2	CARBON DISULFIDE	75-15-0
WEGLA DWUSIARCZEK	CS2	CARBON DISULFIDE	75-15-0
WEGLA TLENEK	CO	CARBON MONOXIDE	630-08-0
WESTRON	C2H2Cl4	1,1,2,2-TETRACHLOROETHANE	79-34-5
WHITE TAR	C10H8	NAPHTHALENE	91-20-3
WICKENOL 122	C22H44O2	n-BUTYL STEARATE	123-95-5
WINTERGREEN OIL, SYNTHETIC	C8H8O3	METHYL SALICYLATE	119-36-8
WITCIZER 200	C22H44O2	n-BUTYL STEARATE	123-95-5
WITCIZER 201	C22H44O2	n-BUTYL STEARATE	123-95-5
WITCIZER 300	C16H22O4	DIBUTYL PHTHALATE	84-74-2
WOOD ALCOHOL	CH4O	METHANOL	67-56-1
WOOD ETHER	C2H6O	DIMETHYL ETHER	115-10-6
WORMWOOD	C4H6O4	SUCCINIC ACID	110-15-6
WORMWOOD ACID	C4H6O4	SUCCINIC ACID	110-15-6
XENYLAMIN	C12H11N	p-AMINODIPHENYL	92-67-1
XENYLAMINE	C12H11N	p-AMINODIPHENYL	92-67-1
XITIX	C6H8O6	ASCORBIC ACID	50-81-7
m-XYLENE	C8H10	m-XYLENE	108-38-3
o-XYLENE	C8H10	o-XYLENE	95-47-6
p-XYLENE	C8H10	p-XYLENE	106-42-3
1,2-XYLENE	C8H10	o-XYLENE	95-47-6
1,3-XYLENE	C8H10	m-XYLENE	108-38-3
1,4-XYLENE	C8H10	p-XYLENE	106-42-3
m-XYLENOL	C8H10O	2,4-XYLENOL	105-67-9
o-XYLENOL	C8H10O	2,3-XYLENOL	526-75-0
p-XYLENOL	C8H10O	2,5-XYLENOL	95-87-4
1,3,4-XYLENOL	C8H10O	3,4-XYLENOL	95-65-8
1,3,5-XYLENOL	C8H10O	3,5-XYLENOL	108-68-9
2,3-XYLENOL	C8H10O	2,3-XYLENOL	526-75-0
2,4-XYLENOL	C8H10O	2,4-XYLENOL	105-67-9
2,5-XYLENOL	C8H10O	2,5-XYLENOL	95-87-4
2,6-XYLENOL	C8H10O	2,6-XYLENOL	576-26-1
3,4-XYLENOL	C8H10O	3,4-XYLENOL	95-65-8
3,5-XYLENOL	C8H10O	3,5-XYLENOL	108-68-9
l-XYLOASCORBIC ACID	C6H8O6	ASCORBIC ACID	50-81-7
o-XYLOL	C8H10	o-XYLENE	95-47-6

Name (Synonym)	Formula	Name in Tables	CAS Reg. No
p-XYLOL	C8H10	p-XYLENE	106-42-3
ZACLONDISCOIDS	CHN	HYDROGEN CYANIDE	74-90-8
ZESET T	C4H6O2	VINYL ACETATE	108-05-4
ZIMCO	C8H8O3	VANILLIN	121-33-5
ZOBA BLACK D	C6H8N2	p-PHENYLENEDIAMINE	106-50-3
ZYTOX	CH3Br	METHYL BROMIDE	74-83-9

INORGANICS The following compilation for inorganics provides the compound list by name (synonym), formula in table, name in table, and CAS registry number. To locate property data for a specific compound, the user should use the name (synonym) to identify the formula and name. This formula and name are then used in the data tabulations to locate the property data of interest.

Name (Synonym)	Formula	Name in Tables	CAS Reg. No
ACIDE BROMHYDRIQUE	HBr	HYDROGEN BROMIDE	10035-10-6
ACIDE CHLORHYDRIQUE	HCl	HYDROGEN CHLORIDE	7647-01-0
ACIDE CYANHYDRIQUE	HCN	HYDROGEN CYANIDE	74-90-8
ACIDE FLUORHYDRIQUE	HF	HYDROGEN FLUORIDE	7664-39-3
ACIDE NITRIQUE	HNO3	NITRIC ACID	7697-37-2
ACIDE SULFHYDRIQUE	H2S	HYDROGEN SULFIDE	2148878.00
ACIDE SULFURIQUE	H2SO4	SULFURIC ACID	7664-93-9
ACIDO BROMIDRICO	HBr	HYDROGEN BROMIDE	10035-10-6
ACIDO CIANIDRICO	HCN	HYDROGEN CYANIDE	74-90-8
ACIDO CLORIDRICO	HCl	HYDROGEN CHLORIDE	7647-01-0
ACIDO FLUORIDRICO	HF	HYDROGEN FLUORIDE	7664-39-3
ACIDO NITRICO	HNO3	NITRIC ACID	7697-37-2
ACIDO SOLFORICO	H2SO4	SULFURIC ACID	7664-93-9
ACID-SPAR	CaF2	CALCIUM FLUORIDE	7789-75-5
ACTICARBONE	C	CARBON	7440-44-0
AERO	HCN	HYDROGEN CYANIDE	74-90-8
AGATE	SiO2	SILICON DIOXIDE	14808-60-7
AGENE	NCl3	NITROGEN TRICHLORIDE	10025-85-1
AKRO-MAG	MgO	MAGNESIUM OXIDE	1309-48-4
ALBONE	H2O2	HYDROGEN PEROXIDE	7722-84-1
ALCIDE	ClO2	CHLORINE DIOXIDE	10049-04-4
ALCOA SODIUM FLUORIDE	NaF	SODIUM FLUORIDE	7681-49-4
ALLBRI NATURAL COPPER	Cu	COPPER	7440-50-8
ALLUMINIO(CLORURO DI)	AlCl3	ALUMINUM CHLORIDE	7446-70-0
ALMITE	Al2O3	ALUMINUM OXIDE	1344-28-1
ALUM	Al2S3O12	ALUMINUM SULFATE	10043-01-3
g-ALUMINA	Al2O3	ALUMINUM OXIDE	1344-28-1
ALUMINUM	Al	ALUMINUM	7429-90-5
ALUMINUM 27	Al	ALUMINUM	7429-90-5
ALUMINUM A00	Al	ALUMINUM	7429-90-5
ALUMINUM BOROHYDRIDE	AlB3H12	ALUMINUM BOROHYDRIDE	16962-07-5
ALUMINUM CHLORIDE	AlCl3	ALUMINUM CHLORIDE	7446-70-0
ALUMINUM FLUORIDE	AlF3	ALUMINUM FLUORIDE	7784-18-1
ALUMINUM FLUORURE	AlF3	ALUMINUM FLUORIDE	7784-18-1
ALUMINUM HYDROBORATE	AlB3H12	ALUMINUM BOROHYDRIDE	16962-07-5
ALUMINUM IODIDE	AlI3	ALUMINUM IODIDE	7784-23-8
ALUMINUM OXIDE	Al2O3	ALUMINUM OXIDE	1344-28-1
a-ALUMINUM OXIDE	Al2O3	ALUMINUM OXIDE	1344-28-1
b-ALUMINUM OXIDE	Al2O3	ALUMINUM OXIDE	1344-28-1
g-ALUMINUM OXIDE	Al2O3	ALUMINUM OXIDE	1344-28-1
ALUMINUM SESQUIOXID	Al2O3	ALUMINUM OXIDE	1344-28-1
ALUMINUM SULFATE	Al2S3O12	ALUMINUM SULFATE	10043-01-3
ALUMINUM TRIBROMIDE	AlBr3	ALUMINUN BROMIDE	7727-15-3
ALUMINUM TRICHLORIDE	AlCl3	ALUMINUM CHLORIDE	7446-70-0
ALUMINUM TRIFLUORIDE	AlF3	ALUMINUM FLUORIDE	7784-18-1
ALUMINUM TRIIODIDE	AlI3	ALUMINUM IODIDE	7784-23-8
ALUMINUM TRISULFATE	Al2S3O12	ALUMINUM SULFATE	10043-01-3
ALUMINUMCHLORID	AlCl3	ALUMINUM CHLORIDE	7446-70-0
ALUMINUN BROMIDE	AlBr3	ALUMINUN BROMIDE	7727-15-3
AMALOX	ZnO	ZINC OXIDE	1314-13-2
AMCHLOR	NH4Cl	AMMONIUM CHLORIDE	12125-02-9
AMETHYST	SiO2	SILICON DIOXIDE	14808-60-7
AM-FOL	NH3	AMMONIA	7664-41-7
AMIDOSULFONIC ACID	H3NO3S	SULFAMIC ACID	5329-14-6
AMIDOSULFURIC ACID	H3NO3S	SULFAMIC ACID	5329-14-6
AMINOSULFONIC ACID	H3NO3S	SULFAMIC ACID	5329-14-6
AMMONERIC	NH4Cl	AMMONIUM CHLORIDE	12125-02-9
AMMONIA	NH3	AMMONIA	7664-41-7

Name (Synonym)	Formula	Name in Tables	CAS Reg. No
AMMONIA ANHYDROUS	NH3	AMMONIA	7664-41-7
AMMONIA GAS	NH3	AMMONIA	7664-41-7
AMMONIA SOLUTIONS	NH3	AMMONIA	7664-41-7
AMMONIAC	NH3	AMMONIA	7664-41-7
AMMONIACA	NH3	AMMONIA	7664-41-7
AMMONIAK	NH3	AMMONIA	7664-41-7
AMMONIUM BROMIDE	NH4Br	AMMONIUM BROMIDE	12124-97-9
AMMONIUM CARBAMATE	N2H6CO2	AMMONIUM CARBAMATE	----------
AMMONIUM CHLORIDE	NH4Cl	AMMONIUM CHLORIDE	12125-02-9
AMMONIUM CYANIDE	N2H4C	AMMONIUM CYANIDE	----------
AMMONIUM HYDROGENSULFIDE	NH5S	AMMONIUM HYDROGENSULFIDE	----------
AMMONIUM HYDROXIDE	NH5O	AMMONIUM HYDROXIDE	1336-21-6
AMMONIUM IODIDE	NH4I	AMMONIUM IODIDE	12027-06-4
AMMONIUM MURIATE	NH4Cl	AMMONIUM CHLORIDE	12125-02-9
AMMONIUMCHLORID	NH4Cl	AMMONIUM CHLORIDE	12125-02-9
AMONIAK	NH3	AMMONIA	7664-41-7
ANAC 110	Cu	COPPER	7440-50-8
ANAYODIN	NaI	SODIUM IODIDE	7681-82-5
ANHYDRIDE CARBONIQUE	CO2	CARBON DIOXIDE	124-38-9
ANHYDROUS AMMONIA	NH3	AMMONIA	7664-41-7
ANHYDROUS HYDRIODIC ACID	HI	HYDROGEN IODIDE	10034-85-2
ANHYDROUS HYDROBROMIC ACID	HBr	HYDROGEN BROMIDE	10035-10-6
ANHYDROUS HYDROCHLORIC ACID	HCl	HYDROGEN CHLORIDE	7647-01-0
ANIMAG	MgO	MAGNESIUM OXIDE	1309-48-4
ANTHIUM DIOXIDE	ClO2	CHLORINE DIOXIDE	10049-04-4
ANTIBULIT	NaF	SODIUM FLUORIDE	7681-49-4
ANTIMOINE (TRICHLORURE)	SbCl3	ANTIMONY TRICHLORIDE	10025-91-9
ANTIMONIO (TRICLORURO di)	SbCl3	ANTIMONY TRICHLORIDE	10025-91-9
ANTIMONOUS CHLORIDE	SbCl3	ANTIMONY TRICHLORIDE	10025-91-9
ANTIMONTRICHLORID	SbCl3	ANTIMONY TRICHLORIDE	10025-91-9
ANTIMONWASSERSTOFFES	SbH3	STIBINE	7803-52-3
ANTIMONY	Sb	ANTIMONY	7440-36-0
ANTIMONY BLACK	Sb	ANTIMONY	7440-36-0
ANTIMONY BUTTER	SbCl3	ANTIMONY TRICHLORIDE	10025-91-9
ANTIMONY CHLORIDE	SbCl3	ANTIMONY TRICHLORIDE	10025-91-9
ANTIMONY HYDRIDE	SbH3	STIBINE	7803-52-3
ANTIMONY PENTACHLORIDE	SbCl5	ANTIMONY PENTACHLORIDE	----------
ANTIMONY POWDER	Sb	ANTIMONY	7440-36-0
ANTIMONY REGULUS	Sb	ANTIMONY	7440-36-0
ANTIMONY TRIBROMIDE	SbBr3	ANTIMONY TRIBROMIDE	7789-61-9
ANTIMONY TRICHLORIDE	SbCl3	ANTIMONY TRICHLORIDE	10025-91-9
ANTIMONY TRIHYDRIDE	SbH3	STIBINE	7803-52-3
ANTIMONY TRIIODIDE	SbI3	ANTIMONY TRIIODIDE	7790-44-5
ANTIMONY TRIOXIDE	Sb2O3	ANTIMONY TRIOXIDE	----------
ANTIMOONTRICHLORIDE	SbCl3	ANTIMONY TRICHLORIDE	10025-91-9
ANTYMON	Sb	ANTIMONY	7440-36-0
ANTYMONOWODOR	SbH3	STIBINE	7803-52-3
AQUA FORTIS	HNO3	NITRIC ACID	7697-37-2
AQUACAT	Co	COBALT	7440-48-4
ARGENTUM	Ag	SILVER	7440-22-4
ARGON	Ar	ARGON	7440-37-1
ARSENIC	As	ARSENIC	7440-37-1
ARSENIC BUTTER	AsCl3	ARSENIC TRICHLORIDE	7784-34-1
ARSENIC FLUORIDE	AsF3	ARSENIC TRIFLUORIDE	7784-35-2
ARSENIC HYDRID	AsH3	ARSINE	7784-42-1
ARSENIC HYDRIDE	AsH3	ARSINE	7784-42-1
ARSENIC PENTAFLUORIDE	AsF5	ARSENIC PENTAFLUORIDE	7784-36-3
ARSENIC TRIBROMIDE	AsBr3	ARSENIC TRIBROMIDE	7784-33-0
ARSENIC TRICHLORIDE	AsCl3	ARSENIC TRICHLORIDE	7784-34-1
ARSENIC TRIFLUORIDE	AsF3	ARSENIC TRIFLUORIDE	7784-35-2
ARSENIC TRIHYDRIDE	AsH3	ARSINE	7784-42-1
ARSENIC TRIIODIDE	AsI3	ARSENIC TRIIODIDE	7784-45-4
ARSENIC TRIOXIDE	As2O3	ARSENIC TRIOXIDE	----------
ARSENIC(III) CHLORIDE	AsCl3	ARSENIC TRICHLORIDE	7784-34-1
ARSENIOUS CHLORIDE	AsCl3	ARSENIC TRICHLORIDE	7784-34-1
ARSENIURETTED HYDROGEN	AsH3	ARSINE	7784-42-1
ARSENOUS BROMIDE	AsBr3	ARSENIC TRIBROMIDE	7784-33-0
ARSENOUS CHLORIDE	AsCl3	ARSENIC TRICHLORIDE	7784-34-1
ARSENOUS FLUORIDE	AsF3	ARSENIC TRIFLUORIDE	7784-35-2
ARSENOUS HYDRIDE	AsH3	ARSINE	7784-42-1
ARSENOUS IODIDE	AsI3	ARSENIC TRIIODIDE	7784-45-4
ARSENOUS TRIBROMIDE	AsBr3	ARSENIC TRIBROMIDE	7784-33-0
ARSENOUS TRICHLORIDE	AsCl3	ARSENIC TRICHLORIDE	7784-34-1
ARSENOUS TRIIODIDE	AsI3	ARSENIC TRIIODIDE	7784-45-4
ARSENOWODOR	AsH3	ARSINE	7784-42-1
ARSENWASSERSTOFF	AsH3	ARSINE	7784-42-1

Name (Synonym)	Formula	Name in Tables	CAS Reg. No
ARSINE	AsH3	ARSINE	7784-42-1
ARWOOD COPPER	Cu	COPPER	7440-50-8
ASTATINE	At	ASTATINE	7440-68-8
AZOTE	NO2	NITROGEN DIOXIDE	10102-44-0
AZOTIC ACID	HNO3	NITRIC ACID	7697-37-2
AZOTO	NO2	NITROGEN DIOXIDE	10102-44-0
AZOTOWY KWAS	HNO3	NITRIC ACID	7697-37-2
BARIUM	Ba	BARIUM	7440-39-3
BENSULFOID	S	SULFUR	7704-34-9
BERTHOLITE	Cl2	CHLORINE	7782-50-5
BERYLLIUM	Be	BERYLLIUM	7440-41-7
BERYLLIUM BOROHYDRIDE	BeB2H8	BERYLLIUM BOROHYDRIDE	----------
BERYLLIUM BROMIDE	BeBr2	BERYLLIUM BROMIDE	7787-46-4
BERYLLIUM CHLORIDE	BeCl2	BERYLLIUM CHLORIDE	7787-47-5
BERYLLIUM DIBROMIDE	BeBr2	BERYLLIUM BROMIDE	7787-46-4
BERYLLIUM DICHLORIDE	BeCl2	BERYLLIUM CHLORIDE	7787-47-5
BERYLLIUM DIFLUORIDE	BeF2	BERYLLIUM FLUORIDE	7787-49-7
BERYLLIUM DIIODIDE	BeI2	BERYLLIUM IODIDE	7787-53-3
BERYLLIUM FLUORIDE	BeF2	BERYLLIUM FLUORIDE	7787-49-7
BERYLLIUM IODIDE	BeI2	BERYLLIUM IODIDE	7787-53-3
BERYLLIUM-9	Be	BERYLLIUM	7440-41-7
BICHLORURE de MERCURE	HgCl2	MERCURIC CHLORIDE	7487-94-7
BIFLUORIDEN	F2	FLUORINE	7782-41-4
BIOXYDE dAZOTE	NO	NITRIC OXIDE	10102-43-9
BISMUTH	Bi	BISMUTH	7440-69-9
BISMUTH TRIBROMIDE	BiBr3	BISMUTH TRIBROMIDE	7787-58-8
BISMUTH TRICHLORIDE	BiCl3	BISMUTH TRICHLORIDE	7787-60-2
BISMUTH-209	Bi	BISMUTH	7440-69-9
BISULFITE	SO2	SULFUR DIOXIDE	7446-09-5
BLAUSAEURE	HCN	HYDROGEN CYANIDE	74-90-8
BLAUWZUUR	HCN	HYDROGEN CYANIDE	74-90-8
BONAZEN	ZnSO4	ZINC SULFATE	7733-02-0
BORACIC ACID	BH3O3	BORIC ACID	10043-35-3
BORIC ACID	BH3O3	BORIC ACID	10043-35-3
BORINE CARBONYL	BH2CO	BORINE CARBONYL	----------
BORINE TRIAMINE	B3N3H6	BORINE TRIAMINE	----------
BOROETHANE	B2H6	DIBORANE	19287-45-7
BOROFAX	BH3O3	BORIC ACID	10043-35-3
BORON	B	BORON	7440-42-8
BORON BROMIDE	BBr3	BORON TRIBROMIDE	10294-33-4
BORON CHLORIDE	BCl3	BORON TRICHLORIDE	10294-34-5
BORON FLUORIDE	BF3	BORON TRIFLUORIDE	7637-07-2
BORON HYDRIDE	B2H6	DIBORANE	19287-45-7
BORON TRIBROMIDE	BBr3	BORON TRIBROMIDE	10294-33-4
BORON TRICHLORIDE	BCl3	BORON TRICHLORIDE	10294-34-5
BORON TRIFLUORIDE	BF3	BORON TRIFLUORIDE	7637-07-2
BORSAEURE	BH3O3	BORIC ACID	10043-35-3
BOV	H2SO4	SULFURIC ACID	7664-93-9
BRIMSTONE	S	SULFUR	7704-34-9
BROM	Br2	BROMINE	7726-95-6
BROME	Br2	BROMINE	7726-95-6
BROMIDE SALT OF POTASSIUM	KBr	POTASSIUM BROMIDE	7758-02-3
BROMIDE SALT OF SODIUM	NaBr	SODIUM BROMIDE	7647-15-6
BROMINE	Br2	BROMINE	7726-95-6
BROMINE CYANIDE	CNBr	CYANOGEN BROMIDE	506-68-3
BROMINE PENTAFLUORIDE	BrF5	BROMINE PENTAFLUORIDE	----------
BROMNATRIUM	NaBr	SODIUM BROMIDE	7647-15-6
BROMO	Br2	BROMINE	7726-95-6
BROMOCYAN	CNBr	CYANOGEN BROMIDE	506-68-3
BROMOCYANOGEN	CNBr	CYANOGEN BROMIDE	506-68-3
BROMODICHLOROFLUOROSILANE	SiBrCl2F	BROMODICHLOROFLUOROSILANE	----------
BROMOFLUORODICHLOROSILANE	SiBrCl2F	BROMODICHLOROFLUOROSILANE	----------
BROMOSILANE	SiH3Br	MONOBROMOSILANE	13465-73-1
BROMOTRIFLUOROSILANE	SiBrF3	TRIFLUOROBROMOSILANE	----------
BROMOWODOR	HBr	HYDROGEN BROMIDE	10035-10-6
BROMURE de CYANOGEN	CNBr	CYANOGEN BROMIDE	506-68-3
BROMWASSERSTOFF	HBr	HYDROGEN BROMIDE	10035-10-6
BRONZE POWDER	Cu	COPPER	7440-50-8
BROOM	Br2	BROMINE	7726-95-6
BROOMWATERSTOF	HBr	HYDROGEN BROMIDE	10035-10-6
BUFOPTO ZINC SULFATE	ZnSO4	ZINC SULFATE	7733-02-0
BURNISH GOLD	Au	GOLD	7440-57-5
BUTTER of ANTIMONY	SbCl3	ANTIMONY TRICHLORIDE	10025-91-9
BUTTER of ZINC	ZnCl2	ZINC CHLORIDE	7646-85-7
C.I. 77050	Sb	ANTIMONY	7440-36-0
C.I. 77056	SbCl3	ANTIMONY TRICHLORIDE	10025-91-9

Name (Synonym)	Formula	Name in Tables	CAS Reg. No
C.I. 77180	Cd	CADMIUM	7440-43-9
C.I. 77320	Co	COBALT	7440-48-4
C.I. 77400	Cu	COPPER	7440-50-8
C.I. 77480	Au	GOLD	7440-57-5
C.I. 77575	Pb	LEAD	7439-92-1
C.I. 77577	PbO	LEAD OXIDE	1317-36-8
C.I. 77640	PbS	LEAD SULFIDE	1314-87-0
C.I. 77775	Ni	NICKEL	7440-02-0
C.I. 77795	Pt	PLATINUM	7440-06-4
C.I. 77805	Se	SELENIUM	7782-49-2
C.I. 77820	Ag	SILVER	7440-22-4
C.I. PIGMENT BLACK 10	C	CARBON	7440-44-0
C.I. PIGMENT METAL 2	Cu	COPPER	7440-50-8
C.I. PIGMENT METAL 3	Au	GOLD	7440-57-5
C.I. PIGMENT METAL 4	Pb	LEAD	7439-92-1
C.I. PIGMENT YELLOW 46	PbO	LEAD OXIDE	1317-36-8
CAB-O-GRIP	Al2O3	ALUMINUM OXIDE	1344-28-1
CADDY	CdCl2	CADMIUM CHLORIDE	10108-64-2
CADMIUM	Cd	CADMIUM	7440-43-9
CADMIUM CHLORIDE	CdCl2	CADMIUM CHLORIDE	10108-64-2
CADMIUM DICHLORIDE	CdCl2	CADMIUM CHLORIDE	10108-64-2
CADMIUM DIFLUORIDE	CdF2	CADMIUM FLUORIDE	7790-79-6
CADMIUM DIIODIDE	CdI2	CADMIUM IODIDE	7790-80-9
CADMIUM FLUORIDE	CdF2	CADMIUM FLUORIDE	7790-79-6
CADMIUM FLUORURE	CdF2	CADMIUM FLUORIDE	7790-79-6
CADMIUM IODIDE	CdI2	CADMIUM IODIDE	7790-80-9
CADMIUM MONOXIDE	CdO	CADMIUM OXIDE	1306-19-0
CADMIUM OXIDE	CdO	CADMIUM OXIDE	1306-19-0
CAKE ALUM	Al2S3O12	ALUMINUM SULFATE	10043-01-3
CALAMINE	ZnO	ZINC OXIDE	1314-13-2
CALCICAT	Ca	CALCIUM	7440-70-2
CALCINED BRUCITE	MgO	MAGNESIUM OXIDE	1309-48-4
CALCINED MAGNESIA	MgO	MAGNESIUM OXIDE	1309-48-4
CALCINED MAGNESITE	MgO	MAGNESIUM OXIDE	1309-48-4
CALCIUM	Ca	CALCIUM	7440-70-2
CALCIUM DIFLUORIDE	CaF2	CALCIUM FLUORIDE	7789-75-5
CALCIUM FLUORIDE	CaF2	CALCIUM FLUORIDE	7789-75-5
CALOCHLOR	HgCl2	MERCURIC CHLORIDE	7487-94-7
CAMPILIT	CNBr	CYANOGEN BROMIDE	506-68-3
CARBAMIDE	CH4N2O	UREA	57-13-6
CARBAMIDE RESIN	CH4N2O	UREA	57-13-6
CARBAMIMIDIC ACID	CH4N2O	UREA	57-13-6
CARBON	C	CARBON	7440-44-0
CARBON BISULFIDE	CS2	CARBON DISULFIDE	75-15-0
CARBON BISULPHIDE	CS2	CARBON DISULFIDE	75-15-0
CARBON DIFLUORIDE OXIDE	CF2O	CARBONYL FLUORIDE	353-50-4
CARBON DIOXIDE	CO2	CARBON DIOXIDE	124-38-9
CARBON DISULFIDE	CS2	CARBON DISULFIDE	75-15-0
CARBON DISULPHIDE	CS2	CARBON DISULFIDE	75-15-0
CARBON FLUORIDE OXIDE	CF2O	CARBONYL FLUORIDE	353-50-4
CARBON HYDRIDE NITRIDE	HCN	HYDROGEN CYANIDE	74-90-8
CARBON MONOXIDE	CO	CARBON MONOXIDE	630-08-0
CARBON NITRIDE	C2N2	CYANOGEN	460-19-5
CARBON OXIDE	CO	CARBON MONOXIDE	630-08-0
CARBON OXIDE SULFIDE	COS	CARBONYL SULFIDE	463-58-1
CARBON OXYCHLORIDE	CCl2O	PHOSGENE	75-44-5
CARBON OXYFLUORIDE	CF2O	CARBONYL FLUORIDE	353-50-4
CARBON OXYSELENIDE	COSe	CARBON OXYSELENIDE	----------
CARBON OXYSULFIDE	COS	CARBONYL SULFIDE	463-58-1
CARBON SELENOSULFIDE	CSeS	CARBON SELENOSULFIDE	----------
CARBON SUBSULFIDE	C3S2	CARBON SUBSULFIDE	----------
CARBON SULFIDE	CS2	CARBON DISULFIDE	75-15-0
CARBON SULPHIDE	CS2	CARBON DISULFIDE	75-15-0
CARBONE (OXYCHLORURE de)	CCl2O	PHOSGENE	75-44-5
CARBONE (OXYDE de)	CO	CARBON MONOXIDE	630-08-0
CARBONE (SUFURE de)	CS2	CARBON DISULFIDE	75-15-0
CARBONIC ACID ANHYDRIDE	CO2	CARBON DIOXIDE	124-38-9
CARBONIC ACID GAS	CO2	CARBON DIOXIDE	124-38-9
CARBONIC ANHYDRIDE	CO2	CARBON DIOXIDE	124-38-9
CARBONIC DIFLUORIDE	CF2O	CARBONYL FLUORIDE	353-50-4
CARBONIC OXIDE	CO	CARBON MONOXIDE	630-08-0
CARBONIO (OSSICLORURO di)	CCl2O	PHOSGENE	75-44-5
CARBONIO (OSSIDO di)	CO	CARBON MONOXIDE	630-08-0
CARBONIO (SOLFURO di)	CS2	CARBON DISULFIDE	75-15-0
CARBONYL CHLORIDE	CCl2O	PHOSGENE	75-44-5
CARBONYL DIAMIDE	CH4N2O	UREA	57-13-6

Name (Synonym)	Formula	Name in Tables	CAS Reg. No
CARBONYL DIFLUORIDE	CF2O	CARBONYL FLUORIDE	353-50-4
CARBONYL FLUORIDE	CF2O	CARBONYL FLUORIDE	353-50-4
CARBONYL SULFIDE	COS	CARBONYL SULFIDE	463-58-1
CARBONYL SULFIDE-32S	COS	CARBONYL SULFIDE	463-58-1
CARBONYLCHLORID	CCl2O	PHOSGENE	75-44-5
CARBONYLDIAMINE	CH4N2O	UREA	57-13-6
CAUSTIC POTASH	KOH	POTASSIUM HYDROXIDE	1310-58-3
CAUSTIC SODA	NaOH	SODIUM HYDROXIDE	1310-73-2
CAVI-TROL	NaF	SODIUM FLUORIDE	7681-49-4
CDA 101	Cu	COPPER	7440-50-8
CDA 102	Cu	COPPER	7440-50-8
CDA 110	Cu	COPPER	7440-50-8
CDA 122	Cu	COPPER	7440-50-8
CELPHOS	PH3	PHOSPHINE	7803-51-2
CESIUM	Cs	CESIUM	7440-46-2
CESIUM BROMIDE	CsBr	CESIUM BROMIDE	7787-69-1
CESIUM CHLORIDE	CsCl	CESIUM CHLORIDE	7647-17-8
CESIUM FLUORIDE	CsF	CESIUM FLUORIDE	13400-13-0
CESIUM IODIDE	CsI	CESIUM IODIDE	7789-17-5
CESIUM MONOBROMIDE	CsBr	CESIUM BROMIDE	7787-69-1
CESIUM MONOCHLORIDE	CsCl	CESIUM CHLORIDE	7647-17-8
CESIUM MONOFLUORIDE	CsF	CESIUM FLUORIDE	13400-13-0
CESIUM MONOIODIDE	CsI	CESIUM IODIDE	7789-17-5
CHALCEDONY	SiO2	SILICON DIOXIDE	14808-60-7
CHEMIFLUOR	NaF	SODIUM FLUORIDE	7681-49-4
CHERTS	SiO2	SILICON DIOXIDE	14808-60-7
CHLOOR	Cl2	CHLORINE	7782-50-5
CHLOORWATERSTOF	HCl	HYDROGEN CHLORIDE	7647-01-0
CHLOR	Cl2	CHLORINE	7782-50-5
CHLORCYAN	CNCl	CYANOGEN CHLORIDE	506-77-4
CHLORE	Cl2	CHLORINE	7782-50-5
CHLORID AMONNY	NH4Cl	AMMONIUM CHLORIDE	12125-02-9
CHLORID ANTIMONITY	SbCl3	ANTIMONY TRICHLORIDE	10025-91-9
CHLORID DRASELNY	KCl	POTASSIUM CHLORIDE	7447-40-7
CHLORID KREMICITY	SiCl4	SILICON TETRACHLORIDE	10026-04-7
CHLORID MEDNY	CuCl	CUPROUS CHLORIDE	7758-89-6
CHLORID RTUTNATY	HgCl2	MERCURIC CHLORIDE	7487-94-7
CHLORIDE OF PHOSPHORUS	PCl3	PHOSPHORUS TRICHLORIDE	7719-12-2
CHLORINE	Cl2	CHLORINE	7782-50-5
CHLORINE CYANIDE	CNCl	CYANOGEN CHLORIDE	506-77-4
CHLORINE DIOXIDE	ClO2	CHLORINE DIOXIDE	10049-04-4
CHLORINE FLUORIDE	ClF3	CHLORINE TRIFLUORIDE	7790-91-2
CHLORINE FLUORIDE OXIDE	ClFO3	PERCHLORYL FLUORIDE	7616-94-6
CHLORINE HEPTOXIDE	Cl2O7	CHLORINE HEPTOXIDE	----------
CHLORINE MOL.	Cl2	CHLORINE	7782-50-5
CHLORINE MONOFLUORIDE	ClF	CHLORINE MONOFLUORIDE	7790-89-8
CHLORINE MONOXIDE	Cl2O	CHLORINE MONOXIDE	7791-21-1
CHLORINE NITRIDE	NCl3	NITROGEN TRICHLORIDE	10025-85-1
CHLORINE OXIDE	ClO2	CHLORINE DIOXIDE	10049-04-4
CHLORINE OXYFLUORIDE	ClFO3	PERCHLORYL FLUORIDE	7616-94-6
CHLORINE PENTAFLUORIDE	ClF5	CHLORINE PENTAFLUORIDE	13637-63-3
CHLORINE PEROXIDE	ClO2	CHLORINE DIOXIDE	10049-04-4
CHLORINE TRIFLUORIDE	ClF3	CHLORINE TRIFLUORIDE	7790-91-2
CHLORINE(IV) OXIDE	ClO2	CHLORINE DIOXIDE	10049-04-4
CHLORKU LITU	LiCl	LITHIUM CHLORIDE	7447-41-8
CHLOROCYAN	CNCl	CYANOGEN CHLORIDE	506-77-4
CHLOROCYANIDE	CNCl	CYANOGEN CHLORIDE	506-77-4
CHLOROCYANOGEN	CNCl	CYANOGEN CHLORIDE	506-77-4
CHLORODIBROMOFLUOROSILANE	SiBr2ClF	DIBROMOCHLOROFLUOROSILANE	----------
CHLOROFLUORODIBROMOSILANE	SiBr2ClF	DIBROMOCHLOROFLUOROSILANE	----------
CHLOROFORMYL CHLORIDE	CCl2O	PHOSGENE	75-44-5
CHLOROHYDRIC ACID	HCl	HYDROGEN CHLORIDE	7647-01-0
CHLOROPEROXYL	ClO2	CHLORINE DIOXIDE	10049-04-4
CHLOROPOTASSURIL	KCl	POTASSIUM CHLORIDE	7447-40-7
CHLOROSILANE	SiH3Cl	MONOCHLOROSILANE	13465-78-6
CHLOROSULFONIC ACID	ClHO3S	CHLOROSULFONIC ACID	7790-94-5
CHLOROTRIFLUORIDE	ClF3	CHLORINE TRIFLUORIDE	7790-91-2
CHLOROTRIFLUOROSILANE	SiClF3	TRIFLUOROCHLOROSILANE	----------
CHLOROWODOR	HCl	HYDROGEN CHLORIDE	7647-01-0
CHLORURE ANTIMONIEUX	SbCl3	ANTIMONY TRICHLORIDE	10025-91-9
CHLORURE ARSENIEUX	AsCl3	ARSENIC TRICHLORIDE	7784-34-1
CHLORURE d'ALUMINUM	AlCl3	ALUMINUM CHLORIDE	7446-70-0
CHLORURE d'ARSENIC	AsCl3	ARSENIC TRICHLORIDE	7784-34-1
CHLORURE de BORE	BCl3	BORON TRICHLORIDE	10294-34-5
CHLORURE DE CYANOGENE	CNCl	CYANOGEN CHLORIDE	506-77-4
CHLORURE de LITHIUM	LiCl	LITHIUM CHLORIDE	7447-41-8

764

Name (Synonym)	Formula	Name in Tables	CAS Reg. No
CHLORURE de ZINC	ZnCl2	ZINC CHLORIDE	7646-85-7
CHLORURE MERCURIQUE	HgCl2	MERCURIC CHLORIDE	7487-94-7
CHLORURE PERRIQUE	FeCl3	FERRIC CHLORIDE	7705-08-0
CHLORWASSERSTOFF	HCl	HYDROGEN CHLORIDE	7647-01-0
CHLORYL RADICAL	ClO2	CHLORINE DIOXIDE	10049-04-4
CHROME	Cr	CHROMIUM	7440-47-3
CHROMIUM	Cr	CHROMIUM	7440-47-3
CHROMIUM CARBONYL	CrC6O6	CHROMIUM CARBONYL	----------
CHROMIUM HEXACARBONYL	CrC6O6	CHROMIUM CARBONYL	----------
CHROMIUM OXYCHLORIDE	CrO2Cl2	CHROMIUM OXYCHLORIDE	----------
CHROMOSULFURIC ACID	ClHO3S	CHLOROSULFONIC ACID	7790-94-5
CIANURO di SODIO	NaCN	SODIUM CYANIDE	143-33-9
CLF II	C	CARBON	7440-44-0
CLORO	Cl2	CHLORINE	7782-50-5
CLORURO di MERCURIO	HgCl2	MERCURIC CHLORIDE	7487-94-7
CMB 200	C	CARBON	7440-44-0
CMB 50	C	CARBON	7440-44-0
COBALT	Co	COBALT	7440-48-4
COBALT CHLORIDE	CoCl2	COBALT CHLORIDE	7646-79-9
COBALT DICHLORIDE	CoCl2	COBALT CHLORIDE	7646-79-9
COBALT MURIATE	CoCl2	COBALT CHLORIDE	7646-79-9
COBALT NITROSYL TRICARBONYL	CoNC3O4	COBALT NITROSYL TRICARBONYL	----------
COBALT-59	Co	COBALT	7440-48-4
COBALTOUS CHLORIDE	CoCl2	COBALT CHLORIDE	7646-79-9
COBALTOUS DICHLORIDE	CoCl2	COBALT CHLORIDE	7646-79-9
COKE POWDER	C	CARBON	7440-44-0
COLLOIDAL CADMIUM	Cd	CADMIUM	7440-43-9
COLLOIDAL MANGANESE	Mn	MANGANESE	7439-96-5
COLLOIDAL MERCURY	Hg	MERCURY	7439-97-6
COLLOIDAL SULFUR	S	SULFUR	7704-34-9
COLLOKIT	S	SULFUR	7704-34-9
COLSUL	S	SULFUR	7704-34-9
COLUMBIA LCK	C	CARBON	7440-44-0
COLUMBIUM	Nb	NIOBIUM	7440-03-1
COLUMBIUM FLUORIDE	CbF5	COLUMBIUM FLUORIDE	----------
COLUMBIUM PENTAFLUORIDE	CbF5	COLUMBIUM FLUORIDE	----------
COMMON SALT	NaCl	SODIUM CHLORIDE	7647-14-5
COMPALOX	Al2O3	ALUMINUM OXIDE	1344-28-1
CONDUCTEX	C	CARBON	7440-44-0
COPPER	Cu	COPPER	7440-50-8
COPPER BICHLORIDE	CuCl2	CUPRIC CHLORIDE	7447-39-4
COPPER BRONZE	Cu	COPPER	7440-50-8
COPPER IODIDE	CuI	COPPER IODIDE	7681-65-4
COPPER MONOCHLORIDE	CuCl	CUPROUS CHLORIDE	7758-89-6
COPPER MONOIODIDE	CuI	COPPER IODIDE	7681-65-4
COPPER SLAG-AIRBORNE	Cu	COPPER	7440-50-8
COPPER SLAG-MILLED	Cu	COPPER	7440-50-8
COPPER(2+) CHLORIDE	CuCl2	CUPRIC CHLORIDE	7447-39-4
COPPER(II) CHLORIDE	CuCl2	CUPRIC CHLORIDE	7447-39-4
COPPER-AIRBORNE	Cu	COPPER	7440-50-8
COPPER-MILLED	Cu	COPPER	7440-50-8
COROSUL D AND S	S	SULFUR	7704-34-9
CORROSIVE MERCURY CHLORIDE	HgCl2	MERCURIC CHLORIDE	7487-94-7
CORROSIVE SUBLIMATE	HgCl2	MERCURIC CHLORIDE	7487-94-7
COSAN	S	SULFUR	7704-34-9
CREDO	NaF	SODIUM FLUORIDE	7681-49-4
CRYSTEX	S	SULFUR	7704-34-9
CUPRIC CHLORIDE	CuCl2	CUPRIC CHLORIDE	7447-39-4
CUPRIC DICHLORIDE	CuCl2	CUPRIC CHLORIDE	7447-39-4
CUPROUS BROMIDE	CuBr	CUPROUS BROMIDE	7787-70-4
CUPROUS CHLORIDE	CuCl	CUPROUS CHLORIDE	7758-89-6
CUPROUS DICHLORIDE	CuCl	CUPROUS CHLORIDE	7758-89-6
CUZ 3	C	CARBON	7440-44-0
CWN 2	C	CARBON	7440-44-0
CYAANWATERSTOF	HCN	HYDROGEN CYANIDE	74-90-8
CYANIDE of SODIUM	NaCN	SODIUM CYANIDE	143-33-9
CYANOBRIK	NaCN	SODIUM CYANIDE	143-33-9
CYANOBROMIDE	CNBr	CYANOGEN BROMIDE	506-68-3
CYANOGEN	C2N2	CYANOGEN	460-19-5
CYANOGEN BROMIDE	CNBr	CYANOGEN BROMIDE	506-68-3
CYANOGEN CHLORIDE	CNCl	CYANOGEN CHLORIDE	506-77-4
CYANOGEN FLUORIDE	CNF	CYANOGEN FLUORIDE	1495-50-7
CYANOGEN GAS	C2N2	CYANOGEN	460-19-5
CYANOGEN MONOBROMIDE	CNBr	CYANOGEN BROMIDE	506-68-3
CYANOGENE	C2N2	CYANOGEN	460-19-5
CYANOGRAN	NaCN	SODIUM CYANIDE	143-33-9

Name (Synonym)	Formula	Name in Tables	CAS Reg. No
CYANURE de SODIUM	NaCN	SODIUM CYANIDE	143-33-9
CYANWASSERSTOFF	HCN	HYDROGEN CYANIDE	74-90-8
CYCLON	HCN	HYDROGEN CYANIDE	74-90-8
CYCLONE B	HCN	HYDROGEN CYANIDE	74-90-8
CYJANOWODOR	HCN	HYDROGEN CYANIDE	74-90-8
CYMAG	NaCN	SODIUM CYANIDE	143-33-9
CYNKU TLENEK	ZnO	ZINC OXIDE	1314-13-2
DARAMMON	NH4Cl	AMMONIUM CHLORIDE	12125-02-9
DECABORANE	B10H14	DECABORANE	----------
DELICIA DELICIA	PH3	PHOSPHINE	7803-51-2
DENDRITIS	NaCl	SODIUM CHLORIDE	7647-14-5
DETIA GAS EX-B	PH3	PHOSPHINE	7803-51-2
DEUTERIUM	D2	DEUTERIUM	7782-39-0
DEUTERIUM CYANIDE	DCN	DEUTERIUM CYANIDE	----------
DEUTERIUM OXIDE	D2O	DEUTERIUM OXIDE	7789-20-0
DEUTERODIBORANE	B2D6	DEUTERODIBORANE	----------
DIALUMINUM SULPHATE	Al2S3O12	ALUMINUM SULFATE	10043-01-3
DIALUMINUM TRIOXIDE	Al2O3	ALUMINUM OXIDE	1344-28-1
DIALUMINUM TRISULFATE	Al2S3O12	ALUMINUM SULFATE	10043-01-3
DIAMIDE	N2H4	HYDRAZINE	302-01-2
DIBORANE	B2H6	DIBORANE	19287-45-7
DIBORANE HYDROBROMIDE	B2H5Br	DIBORANE HYDROBROMIDE	----------
DIBORANE MIXTURES	B2H6	DIBORANE	19287-45-7
DIBORON HEXAHYDRIDE	B2H6	DIBORANE	19287-45-7
DIBROMOCHLOROFLUOROSILANE	SiBr2ClF	DIBROMOCHLOROFLUOROSILANE	----------
DIBROMOFLUOROCHLOROSILANE	SiBr2ClF	DIBROMOCHLOROFLUOROSILANE	----------
DIBROMOSILANE	SiH2Br2	DIBROMOSILANE	13768-94-0
DICESIUM DICHLORIDE	CsCl	CESIUM CHLORIDE	7647-17-8
DICESIUM DIFLUORIDE	CsF	CESIUM FLUORIDE	13400-13-0
DICESIUM DIIODIDE	CsI	CESIUM IODIDE	7789-17-5
DICHLORO HEPTOXIDE	Cl2O7	CHLORINE HEPTOXIDE	----------
DICHLORO OXIDE	Cl2O	CHLORINE MONOXIDE	7791-21-1
DICHLOROBROMOFLUOROSILANE	SiBrCl2F	BROMODICHLOROFLUOROSILANE	----------
DICHLORODIFLUOROSILANE	SiCl2F2	DICHLORODIFLUOROSILANE	18356-71-3
DICHLOROFLUOROBROMOSILANE	SiBrCl2F	BROMODICHLOROFLUOROSILANE	----------
DICHLOROSILANE	SiH2Cl2	DICHLOROSILANE	4109-96-0
DICOPPER	CuCl	CUPROUS CHLORIDE	7758-89-6
DICYANOGEN	C2N2	CYANOGEN	460-19-5
DIDEUTERIUM OXIDE	D2O	DEUTERIUM OXIDE	7789-20-0
DIFLUORODICHLOROSILANE	SiCl2F2	DICHLORODIFLUOROSILANE	18356-71-3
DIFLUOROFORMALDEHYDE	CF2O	CARBONYL FLUORIDE	353-50-4
DIFLUOROSILANE	SiH2F2	DIFLUOROSILANE	13824-36-7
DIGERMANE	Ge2H6	DIGERMANE	13818-89-8
DIHYDROGEN DIOXIDE	H2O2	HYDROGEN PEROXIDE	7722-84-1
DIHYDROGEN OXIDE	H2O	WATER	7732-18-5
DIIODOSILANE	SiH2I2	DIIODOSILANE	----------
DINITROGEN MONOXIDE	N2O	NITROUS OXIDE	10024-97-2
DINITROGEN TETRAFLUORIDE	N2F4	TETRAFLUOROHYDRAZINE	10036-47-2
DINITROGEN TETROXIDE	N2O4	NITROGEN TETRAOXIDE	10544-72-6
DIPHOSGENE	CCl2O	PHOSGENE	75-44-5
DIPOTASSIUM DICHLORIDE	KCl	POTASSIUM CHLORIDE	7447-40-7
DIPPING ACID	H2SO4	SULFURIC ACID	7664-93-9
DISILANE	Si2H6	DISILANE	1590-87-0
DISILANYL CHLORIDE	Si2H5Cl	DISILANYL CHLORIDE	----------
DISILOXANE	Si2OH6	DISILOXANE	13597-73-4
DISODIUM DIFLUORIDE	NaF	SODIUM FLUORIDE	7681-49-4
DISODIUM SULFATE	Na2SO4	SODIUM SULFATE	7757-82-6
DISPAL	Al2O3	ALUMINUM OXIDE	1344-28-1
DITHIOCARBONIC ANHYDRIDE	CS2	CARBON DISULFIDE	75-15-0
DOTMENT 324	Al2O3	ALUMINUM OXIDE	1344-28-1
DOXCIDE 50	ClO2	CHLORINE DIOXIDE	10049-04-4
DRY ICE	CO2	CARBON DIOXIDE	124-38-9
DUS-TOP	MgCl2	MAGNESIUM CHLORIDE	7786-30-3
ELECTRONIC E-2	H2Se	HYDROGEN SELENIDE	7783-07-5
EMANAY ZINC OXIDE	ZnO	ZINC OXIDE	1314-13-2
EMAR	ZnO	ZINC OXIDE	1314-13-2
EMPLETS POTASSIUM CHLORIDE	KCl	POTASSIUM CHLORIDE	7447-40-7
ENSEAL	KCl	POTASSIUM CHLORIDE	7447-40-7
ETHANEDINITRILE	C2N2	CYANOGEN	460-19-5
EUROPIUM	Eu	EUROPIUM	7440-53-1
EVERCYN	HCN	HYDROGEN CYANIDE	74-90-8
EXHAUST GAS	CO	CARBON MONOXIDE	630-08-0
F1-TABS	NaF	SODIUM FLUORIDE	7681-49-4
FACTITIOUS AIR	N2O	NITROUS OXIDE	10024-97-2
FASERTON	Al2O3	ALUMINUM OXIDE	1344-28-1
FDA 0101	NaF	SODIUM FLUORIDE	7681-49-4

766

Name (Synonym)	Formula	Name in Tables	CAS Reg. No
FERMENICIDE LIQUID	SO2	SULFUR DIOXIDE	7746-09-5
FERMENICIDE POWDER	SO2	SULFUR DIOXIDE	7746-09-5
FERRIC CHLORIDE	FeCl3	FERRIC CHLORIDE	7705-08-0
FERROUS CHLORIDE	FeCl2	FERROUS CHLORIDE	7758-94-3
FLINT	SiO2	SILICON DIOXIDE	14808-60-7
FLORES MARTIS	FeCl3	FERRIC CHLORIDE	7705-08-0
FLORIDINE	NaF	SODIUM FLUORIDE	7681-49-4
FLOROCID	NaF	SODIUM FLUORIDE	7681-49-4
FLOWERS of SULPHUR	S	SULFUR	7704-34-9
FLOZENGES	NaF	SODIUM FLUORIDE	7681-49-4
FLUE GAS	CO	CARBON MONOXIDE	630-08-0
FLUOPHOSGENE	CF2O	CARBONYL FLUORIDE	353-50-4
FLUOR	F2	FLUORINE	7782-41-4
FLUORAL	NaF	SODIUM FLUORIDE	7681-49-4
FLUORID HLINITY	AlF3	ALUMINUM FLUORIDE	7784-18-1
FLUORID SODNY	NaF	SODIUM FLUORIDE	7681-49-4
FLUORIDENT	NaF	SODIUM FLUORIDE	7681-49-4
FLUORIGARD	NaF	SODIUM FLUORIDE	7681-49-4
FLUORINE	F2	FLUORINE	7782-41-4
FLUORINE CYANIDE	CNF	CYANOGEN FLUORIDE	1495-50-7
FLUORINE MONOXIDE	F2O	FLUORINE OXIDE	7783-41-7
FLUORINE OXIDE	F2O	FLUORINE OXIDE	7783-41-7
FLUORINEED	NaF	SODIUM FLUORIDE	7681-49-4
FLUORINSE	NaF	SODIUM FLUORIDE	7681-49-4
FLUORITAB	NaF	SODIUM FLUORIDE	7681-49-4
FLUORITE	CaF2	CALCIUM FLUORIDE	7789-75-5
FLUORO	F2	FLUORINE	7782-41-4
FLUOROBROMODICHLOROSILANE	SiBrCl2F	BROMODICHLOROFLUOROSILANE	----------
FLUOROCHLORODIBROMOSILANE	SiBrCl2F	BROMODICHLOROFLUOROSILANE	----------
FLUOROCYAN	CNF	CYANOGEN FLUORIDE	1495-50-7
FLUOROCYANIDE	CNF	CYANOGEN FLUORIDE	1495-50-7
FLUOROCYANOGEN	CNF	CYANOGEN FLUORIDE	1495-50-7
FLUORODIBROMOCHLOROSILANE	SiBr2ClF	DIBROMOCHLOROFLUOROSILANE	----------
FLUOROFORMYL FLUORIDE	CF2O	CARBONYL FLUORIDE	353-50-4
FLUOR-O-KOTE	NaF	SODIUM FLUORIDE	7681-49-4
FLUOROPHOSGENE	CF2O	CARBONYL FLUORIDE	353-50-4
FLUOROSILANE	SiH3F	MONOFLUOROSILANE	13537-33-2
FLUOROTRICHLOROSILANE	SiCl3F	TRICHLOROFLUOROSILANE	14965-52-7
FLUOROWODOR	HF	HYDROGEN FLUORIDE	7664-39-3
FLUORSPAR	CaF2	CALCIUM FLUORIDE	7789-75-5
FLUORURE de BORE	BF3	BORON TRIFLUORIDE	7637-07-2
FLUORURE de POTASSIUM	KF	POTASSIUM FLUORIDE	7789-23-3
FLUORURES ACIDE	F2	FLUORINE	7782-41-4
FLUORURI ACIDI	F2	FLUORINE	7782-41-4
FLUORWASSERSTOFF	HF	HYDROGEN FLUORIDE	7664-39-3
FLUORWATERSTOF	HF	HYDROGEN FLUORIDE	7664-39-3
FORMIC ANAMMONIDE	HCN	HYDROGEN CYANIDE	74-90-8
FORMONITRILE	HCN	HYDROGEN CYANIDE	74-90-8
FOSFORO(PENTACHLORURO di)	PCl5	PHOSPHORUS PENTACHLORIDE	10026-13-8
FOSFORO(TRICLORURO di)	PCl3	PHOSPHORUS TRICHLORIDE	7719-12-2
FOSFOROWODOR	PH3	PHOSPHINE	7803-51-2
FOSFOROXYCHLORID	POCl3	PHOSPHORUS OXYCHLORIDE	10025-87-3
FOSFORPENTACHLORIDE	PCl5	PHOSPHORUS PENTACHLORIDE	10026-13-8
FOSFORTHIOCHLORID	PSCl3	PHOSPHORUS THIOCHLORIDE	3982-91-0
FOSFORTRICHLORIDE	PCl3	PHOSPHORUS TRICHLORIDE	7719-12-2
FOSGEEN	CCl2O	PHOSGENE	75-44-5
FOSGEN	CCl2O	PHOSGENE	75-44-5
FOSGENE	CCl2O	PHOSGENE	75-44-5
FRANCIUM	Fr	FRANCIUM	7440-73-5
FUMING LIQUID ARSENIC	AsCl3	ARSENIC TRICHLORIDE	7784-34-1
G 2	Al2O3	ALUMINUM OXIDE	1344-28-1
GADOLINIUM	Gd	GADOLINIUM	7440-54-2
GALENA	PbS	LEAD SULFIDE	1314-87-0
GALLIUM	Ga	GALLIUM	7440-55-3
GALLIUM CHLORIDE	GaCl3	GALLIUM TRICHLORIDE	13450-90-3
GALLIUM TRICHLORIDE	GaCl3	GALLIUM TRICHLORIDE	13450-90-3
GERMANE	GeH4	GERMANIUM TETRAHYDRIDE	7782-65-2
GERMANIUM	Ge	GERMANIUM	7440-56-4
GERMANIUM BROMIDE	GeBr4	GERMANIUM BROMIDE	----------
GERMANIUM CHLORIDE	GeCl4	GERMANIUM CHLORIDE	----------
GERMANIUM HYDRIDE	GeH4	GERMANIUM TETRAHYDRIDE	7782-65-2
GERMANIUM TETRABROMIDE	GeBr4	GERMANIUM BROMIDE	----------
GERMANIUM TETRACHLORIDE	GeCl4	GERMANIUM CHLORIDE	----------
GERMANIUM TETRAHYDRIDE	GeH4	GERMANIUM TETRAHYDRIDE	7782-65-2
GLOVER	Pb	LEAD	7439-92-1
GLUCINIUM	Be	BERYLLIUM	7440-41-7

767

Name (Synonym)	Formula	Name in Tables	CAS Reg. No
GLUCINUM	Be	BERYLLIUM	7440-41-7
GOLD	Au	GOLD	7440-57-5
1721 GOLD	Cu	COPPER	7440-50-8
GOLD BRONZE	Cu	COPPER	7440-50-8
GRANMAG	MgO	MAGNESIUM OXIDE	1309-48-4
GROUND VOCLE SULPHUR	S	SULFUR	7704-34-9
HAFNIUM	Hf	HAFNIUM	7440-58-6
HAFNIUM POWDER	Hf	HAFNIUM	7440-58-6
HALITE	NaCl	SODIUM CHLORIDE	7647-14-5
HEAVY AMMONIA	ND3	HEAVY AMMONIA	----------
HEAVY WATER-D2	D2O	DEUTERIUM OXIDE	7789-20-0
HELIUM	He	HELIUM-4	7440-59-7
HELIUM-3	He	HELIUM-3	14762-55-1
HELIUM-4	He	HELIUM-4	7440-59-7
HEXACHLORODISILANE	Si2Cl6	HEXACHLORODISILANE	----------
HEXACHLORODISILOXANE	Si2OCl6	HEXACHLORODISILOXANE	----------
HEXAFLUORODISILANE	Si2F6	HEXAFLUORODISILANE	----------
HEXAFLUORURE de SOUFRE	SF6	SULFUR HEXAFLUORIDE	2551-62-4
HIOXYL	H2O2	HYDROGEN PEROXIDE	7722-84-1
HYDRARGYRUM BIJODATUM	HgI2	MERCURIC IODIDE	7774-29-0
HYDRAZINE	N2H4	HYDRAZINE	302-01-2
HYDRAZINE AQUEOUS SOLUTIONS	N2H4	HYDRAZINE	302-01-2
HYDRAZYNA	N2H4	HYDRAZINE	302-01-2
HYDRIODIC ACID	HI	HYDROGEN IODIDE	10034-85-2
HYDROBROMIC ACID MONOAMMONIATE	NH4Br	AMMONIUM BROMIDE	12124-97-9
HYDROCHLORIC ACID	HCl	HYDROGEN CHLORIDE	7647-01-0
HYDROCHLORIDE	HCl	HYDROGEN CHLORIDE	7647-01-0
HYDROCYANIC ACID	HCN	HYDROGEN CYANIDE	74-90-8
HYDROCYANIC ACID, SODIUM SALT	NaCN	SODIUM CYANIDE	143-33-9
HYDROFLUORIC ACID	HF	HYDROGEN FLUORIDE	7664-39-3
HYDROFLUORIDE	HF	HYDROGEN FLUORIDE	7664-39-3
HYDROGEN	H2	HYDROGEN	1333-74-0
HYDROGEN ANTIMONIDE	SbH3	STIBINE	7803-52-3
HYDROGEN ARSENIDE	AsH3	ARSINE	7784-42-1
HYDROGEN BROMIDE	HBr	HYDROGEN BROMIDE	10035-10-6
HYDROGEN CHLORIDE	HCl	HYDROGEN CHLORIDE	7647-01-0
HYDROGEN CYANIDE	HCN	HYDROGEN CYANIDE	74-90-8
HYDROGEN DIOXIDE	H2O2	HYDROGEN PEROXIDE	7722-84-1
HYDROGEN DISULFIDE	H2S2	HYDROGEN DISULFIDE	----------
HYDROGEN FLUORIDE	HF	HYDROGEN FLUORIDE	7664-39-3
HYDROGEN IODIDE	HI	HYDROGEN IODIDE	10034-85-2
HYDROGEN NITRATE	HNO3	NITRIC ACID	7697-37-2
HYDROGEN PEROXIDE	H2O2	HYDROGEN PEROXIDE	7722-84-1
HYDROGEN PHOSPHIDE	PH3	PHOSPHINE	7803-51-2
HYDROGEN SELENIDE	H2Se	HYDROGEN SELENIDE	7783-07-5
HYDROGEN SULFIDE	H2S	HYDROGEN SULFIDE	7783-06-4
HYDROGEN SULFURIC ACID	H2S	HYDROGEN SULFIDE	7783-06-4
HYDROGEN TELLURIDE	H2Te	HYDROGEN TELLURIDE	7783-09-7
HYDROGENE SULFURE	H2S	HYDROGEN SULFIDE	7783-06-4
HYDROOT	H2SO4	SULFURIC ACID	7664-93-9
HYDROPEROXIDE	H2O2	HYDROGEN PEROXIDE	7722-84-1
HYDROXYDE de POTASSIUM	KOH	POTASSIUM HYDROXIDE	1310-58-3
HYDROXYDE de SODIUM	NaOH	SODIUM HYDROXIDE	1310-73-2
HYDROXYLAMINE	NH3O	HYDROXYLAMINE	7803-49-8
HYPONITROUS ACID ANHYDRIDE	N2O	NITROUS OXIDE	10024-97-2
IDROGENO SOLFORATO	H2S	HYDROGEN SULFIDE	7783-06-4
INDIUM	In	INDIUM	7440-74-6
INHIBINE	H2O2	HYDROGEN PEROXIDE	7722-84-1
IODE	I2	IODINE	7553-56-2
IODINE	I2	IODINE	7553-56-2
IODINE HEPTAFLUORIDE	IF7	IODINE HEPTAFLUORIDE	----------
IODIO	I2	IODINE	7553-56-2
IODOSILANE	SiH3I	IODOSILANE	13598-42-0
IODURIL	NaI	SODIUM IODIDE	7681-82-5
IRIDIUM	Ir	IRIDIUM	7439-88-5
IRON	Fe	IRON	7439-89-6
IRON CHLORIDE	FeCl3	FERRIC CHLORIDE	7705-08-0
IRON DICHLORIDE	FeCl2	FERROUS CHLORIDE	7758-94-3
IRON PENTACARBONYL	FeC5O5	IRON PENTACARBONYL	----------
IRON PROTOCHLORIDE	FeCl2	FERROUS CHLORIDE	7758-94-3
IRON TRICHLORIDE	FeCl3	FERRIC CHLORIDE	7705-08-0
IRON(II) CHLORIDE	FeCl2	FERROUS CHLORIDE	7758-94-3
IRON(III) CHLORIDE	FeCl3	FERRIC CHLORIDE	7705-08-0
IRTRAN 3	CaF2	CALCIUM FLUORIDE	7789-75-5
ISOUREA	CH4N2O	UREA	57-13-6
JASAD	Zn	ZINC	7440-66-6

Name (Synonym)	Formula	Name in Tables	CAS Reg. No
JISC 3110	Al	ALUMINUM	7429-90-5
JOD	I2	IODINE	7553-56-2
JODID SODNY	NaI	SODIUM IODIDE	7681-82-5
JOOD	I2	IODINE	7553-56-2
K1-N	KI	POTASSIUM IODIDE	7681-11-0
KADMIUM	Cd	CADMIUM	7440-43-9
KADMIUMCHLORID	CdCl2	CADMIUM CHLORIDE	10108-64-2
KADMU TLENEK	CdO	CADMIUM OXIDE	1306-19-0
KAFAR COPPER	Cu	COPPER	7440-50-8
KALITABS	KCl	POTASSIUM CHLORIDE	7447-40-7
KALIUMHYDROXID	KOH	POTASSIUM HYDROXIDE	1310-58-3
KALIUMHYDROXYDE	KOH	POTASSIUM HYDROXIDE	1310-58-3
KAOCHLOR	KCl	POTASSIUM CHLORIDE	7447-40-7
KAON-CI	KCl	POTASSIUM CHLORIDE	7447-40-7
KAY CIEL	KCl	POTASSIUM CHLORIDE	7447-40-7
KHLADON 744	CO2	CARBON DIOXIDE	124-38-9
KHP 2	Al2O3	ALUMINUM OXIDE	1344-28-1
K-LOR	KCl	POTASSIUM CHLORIDE	7447-40-7
KLOTRIX	KCl	POTASSIUM CHLORIDE	7447-40-7
KNOLLIDE	KI	POTASSIUM IODIDE	7681-11-0
KOBALT	Co	COBALT	7440-48-4
KOBALT CHLORID	CoCl2	COBALT CHLORIDE	7646-79-9
KOHLENDIOXYD	CO2	CARBON DIOXIDE	124-38-9
KOHLENDISULFID (SCHWEFELKOHLENSTOFF)	CS2	CARBON DISULFIDE	75-15-0
KOHLENMONOXID	CO	CARBON MONOXIDE	630-08-0
KOHLENOXYD	CO	CARBON MONOXIDE	630-08-0
KOHLENSAEURE	CO2	CARBON DIOXIDE	124-38-9
KOOLMONOXYDE	CO	CARBON MONOXIDE	630-08-0
KOOLSTOFDISULFIDE (ZWAVELKOOLSTOF)	CS2	CARBON DISULFIDE	75-15-0
KOOLSTOFOXYCHLORIDE	CCl2O	PHOSGENE	75-44-5
K-PRENDE-DOME	KCl	POTASSIUM CHLORIDE	7447-40-7
KRYPTON	Kr	KRYPTON	7439-90-9
KWIK	Hg	MERCURY	7439-97-6
KYANID SODNY	NaCN	SODIUM CYANIDE	143-33-9
KYSELINA AMIDOSULFONOVA	H3NO3S	SULFAMIC ACID	5329-14-6
KYSELINA DUSICNE	HNO3	NITRIC ACID	7697-37-2
KYSELINA SULFAMINOVA	H3NO3S	SULFAMIC ACID	5329-14-6
L16	Al	ALUMINUM	7429-90-5
LANTHANUM	La	LANTHANUM	7439-91-0
LAUGHING GAS	N2O	NITROUS OXIDE	10024-97-2
LEAD	Pb	LEAD	7439-92-1
LEAD BROMIDE	PbBr2	LEAD BROMIDE	10031-22-8
LEAD CHLORIDE	PbCl2	LEAD CHLORIDE	7758-95-4
LEAD DIBROMIDE	PbBr2	LEAD BROMIDE	10031-22-8
LEAD DICHLORIDE	PbCl2	LEAD CHLORIDE	7758-95-4
LEAD DIFLUORIDE	PbF2	LEAD FLUORIDE	7783-46-2
LEAD DIIODIDE	PbI2	LEAD IODIDE	10101-63-0
LEAD FLAKE	Pb	LEAD	7439-92-1
LEAD FLUORIDE	PbF2	LEAD FLUORIDE	7783-46-2
LEAD IODIDE	PbI2	LEAD IODIDE	10101-63-0
LEAD OXIDE	PbO	LEAD OXIDE	1317-36-8
LEAD OXIDE YELLOW	PbO	LEAD OXIDE	1317-36-8
LEAD PROTOXIDE	PbO	LEAD OXIDE	1317-36-8
LEAD SULFIDE	PbS	LEAD SULFIDE	1314-87-0
LEAD(2+) CHLORIDE	PbCl2	LEAD CHLORIDE	7758-95-4
LEAD(II) CHLORIDE	PbCl2	LEAD CHLORIDE	7758-95-4
LEAD(II) OXIDE	PbO	LEAD OXIDE	1317-36-8
LEWIS-RED DEVIL LYE	NaOH	SODIUM HYDROXIDE	1310-73-2
LIPARITE	CaF2	CALCIUM FLUORIDE	7789-75-5
LIQUID BRIGHT PLATINUM	Pt	PLATINUM	7740-06-4
LITHARGE	PbO	LEAD OXIDE	1317-36-8
LITHARGE YELLOW L-28	PbO	LEAD OXIDE	1317-36-8
LITHIUM	Li	LITHIUM	7439-93-2
LITHIUM BROMIDE	LiBr	LITHIUM BROMIDE	7550-35-8
LITHIUM CHLORIDE	LiCl	LITHIUM CHLORIDE	7447-41-8
LITHIUM FLUORIDE	LiF	LITHIUM FLUORIDE	7789-24-4
LITHIUM FLUORURE	LiF	LITHIUM FLUORIDE	7789-24-4
LITHIUM IODIDE	LiI	LITHIUM IODIDE	10377-51-2
LITHIUM METAL	Li	LITHIUM	7439-93-2
LITHIUM MONOBROMIDE	LiBr	LITHIUM BROMIDE	7550-35-8
LITHIUM MONOCHLORIDE	LiCl	LITHIUM CHLORIDE	7447-41-8
LITHIUM MONOFLUORIDE	LiF	LITHIUM FLUORIDE	7789-24-4
LITHIUM MONOIODIDE	LiI	LITHIUM IODIDE	10377-51-2
LUCALOX	Al2O3	ALUMINUM OXIDE	1344-28-1
LUTECIUM	Lu	LUTECIUM	----------
LYE	NaOH	SODIUM HYDROXIDE	1310-73-2

Name (Synonym)	Formula	Name in Tables	CAS Reg. No
LYE	KOH	POTASSIUM HYDROXIDE	1310-58-3
M1	Cu	COPPER	7440-50-8
MA 100	C	CARBON	7440-44-0
MAGCAL	MgO	MAGNESIUM OXIDE	1309-48-4
MAGCHEM 100	MgO	MAGNESIUM OXIDE	1309-48-4
MAGLITE	MgO	MAGNESIUM OXIDE	1309-48-4
MAGNESIA	MgO	MAGNESIUM OXIDE	1309-48-4
MAGNESIA USTA	MgO	MAGNESIUM OXIDE	1309-48-4
MAGNESIO	Mg	MAGNESIUM	7439-95-4
MAGNESIUM	Mg	MAGNESIUM	7439-95-4
MAGNESIUM CHLORIDE	MgCl2	MAGNESIUM CHLORIDE	7786-30-3
MAGNESIUM DICHLORIDE	MgCl2	MAGNESIUM CHLORIDE	7786-30-3
MAGNESIUM OXIDE	MgO	MAGNESIUM OXIDE	1309-48-4
MAGNESIUM OXIDE FUME	MgO	MAGNESIUM OXIDE	1309-48-4
MAGOX	MgO	MAGNESIUM OXIDE	1309-48-4
MANGACAT	Mn	MANGANESE	7439-96-5
MANGAN	Mn	MANGANESE	7439-96-5
MANGAN NITRIDOVANY	Mn	MANGANESE	7439-96-5
MANGANESE	Mn	MANGANESE	7439-96-5
MANGANESE CHLORIDE	MnCl2	MANGANESE CHLORIDE	7773-01-5
MANGANESE DICHLORIDE	MnCl2	MANGANESE CHLORIDE	7773-01-5
MARMAG	MgO	MAGNESIUM OXIDE	1309-48-4
MASSICOT	PbO	LEAD OXIDE	1317-36-8
MASSICOTITE	PbO	LEAD OXIDE	1317-36-8
MATTING ACID	H2SO4	SULFURIC ACID	7664-93-9
MERCURE	Hg	MERCURY	7439-97-6
MERCURIC BROMIDE	HgBr2	MERCURIC BROMIDE	7789-47-1
MERCURIC CHLORIDE	HgCl2	MERCURIC CHLORIDE	7487-94-7
MERCURIC IODIDE	HgI2	MERCURIC IODIDE	7774-29-0
MERCURIO	Hg	MERCURY	7439-97-6
MERCURY	Hg	MERCURY	7439-97-6
MERCURY BICHLORIDE	HgCl2	MERCURIC CHLORIDE	7487-94-7
MERCURY DIBROMIDE	HgBr2	MERCURIC BROMIDE	7789-47-1
MERCURY DICHLORIDE	HgCl2	MERCURIC CHLORIDE	7487-94-7
MERCURY DIIODIDE	HgI2	MERCURIC IODIDE	7774-29-0
MERCURY PERCHLORIDE	HgCl2	MERCURIC CHLORIDE	7487-94-7
MERCURY, METALLIC	Hg	MERCURY	7439-97-6
MERRILLITE	Zn	ZINC	7440-66-6
METANA ALUMINUM PASTE	Al	ALUMINUM	7429-90-5
MET-SPAR	CaF2	CALCIUM FLUORIDE	7789-75-5
MICROGRIT WCA	Al2O3	ALUMINUM OXIDE	1344-28-1
MOLECULAR CHLORINE	Cl2	CHLORINE	7782-50-5
MOLYBDATE	Mo	MOLYBDENUM	7439-98-7
MOLYBDENUM	Mo	MOLYBDENUM	7439-98-7
MOLYBDENUM FLUORIDE	MoF6	MOLYBDENUM FLUORIDE	7783-77-9
MOLYBDENUM HEXAFLUORIDE	MoF6	MOLYBDENUM FLUORIDE	7783-77-9
MOLYBDENUM OXIDE	MoO3	MOLYBDENUM OXIDE	1313-27-5
MOLYBDENUM(VI) OXIDE	MoO3	MOLYBDENUM OXIDE	1313-27-5
MOLYBDIC ANHYDRIDE	MoO3	MOLYBDENUM OXIDE	1313-27-5
MOLYBDIC TRIOXIDE	MoO3	MOLYBDENUM OXIDE	1313-27-5
MONOBROMOSILANE	SiH3Br	MONOBROMOSILANE	13465-73-1
MONOCHLOROSILANE	SiH3Cl	MONOCHLOROSILANE	13465-78-6
MONOCHLOROSULFURIC ACID	ClHO3S	CHLOROSULFONIC ACID	7790-94-5
MONOFLUOROSILANE	SiH3F	MONOFLUOROSILANE	13537-33-2
MONOGERMANE	GeH4	GERMANIUM TETRAHYDRIDE	7782-65-2
MONOIODOSILANE	SiH3I	IODOSILANE	13598-42-0
MONOSILANE	SiH4	SILANE	7803-62-5
NATRIUM	Na	SODIUM	7440-23-5
NATRIUMCHLORID	NaCl	SODIUM CHLORIDE	7647-14-5
NATRIUMHYDROXID	NaOH	SODIUM HYDROXIDE	1310-73-2
NATRIUMHYDROXYDE	NaOH	SODIUM HYDROXIDE	1310-73-2
NATRIUMJODID	NaI	SODIUM IODIDE	7681-82-5
NATRIUMSULFAT	Na2SO4	SODIUM SULFATE	7757-82-6
NATURAL LEAD SULFIDE	PbS	LEAD SULFIDE	1314-87-0
NCI-C02119	CH4N2O	UREA	57-13-6
NCI-C02551	CdO	CADMIUM OXIDE	1306-19-0
NCI-C04591	CS2	CARBON DISULFIDE	75-15-0
NCI-C56417	BH3O3	BORIC ACID	10043-35-3
NCI-C60173	HgCl2	MERCURIC CHLORIDE	7487-94-7
NCI-C60219	CCl2O	PHOSGENE	75-44-5
NCI-C60311	Co	COBALT	7440-48-4
NCI-C60399	Hg	MERCURY	7439-97-6
NEODYMIUM	Nd	NEODYMIUM	7440-00-8
NEON	Ne	NEON	7740-01-9
NEPTUNIUM	Np	NEPTUNIUM	7439-99-8
NICHEL	Ni	NICKEL	7440-02-0

Name (Synonym)	Formula	Name in Tables	CAS Reg. No
NICKEL	Ni	NICKEL	7440-02-0
NICKEL CARBONYL	NiC4O4	NICKEL CARBONYL	----------
NICKEL DIFLUORIDE	NiF2	NICKEL FLUORIDE	10028-18-9
NICKEL FLUORIDE	NiF2	NICKEL FLUORIDE	10028-18-9
NICKEL TETRACARBONYL	NiC4O4	NICKEL CARBONYL	----------
NIOBIUM	Nb	NIOBIUM	7440-03-1
NIOBIUM-93	Nb	NIOBIUM	7440-03-1
NITRIC ACID	HNO3	NITRIC ACID	7697-37-2
NITRIC OXIDE	NO	NITRIC OXIDE	10102-43-9
NITRILOACETONITRILE	C2N2	CYANOGEN	460-19-5
NITRITO	NO2	NITROGEN DIOXIDE	10102-44-0
NITROGEN	N2	NITROGEN	7727-37-9
NITROGEN DIOXIDE	NO2	NITROGEN DIOXIDE	10102-44-0
NITROGEN DIOXIDE, DI-	N2O4	NITROGEN TETRAOXIDE	10544-72-6
NITROGEN FLUORIDE	NF3	NITROGEN TRIFLUORIDE	7783-54-2
NITROGEN MONOXIDE	NO	NITRIC OXIDE	10102-43-9
NITROGEN OXYCHLORIDE	NOCl	NITROSYL CHLORIDE	2696-92-6
NITROGEN OXYFLUORIDE	NOF	NITROSYL FLUORIDE	7789-25-5
NITROGEN PENTOXIDE	N2O5	NITROGEN PENTOXIDE	10102-03-1
NITROGEN PEROXIDE	NO2	NITROGEN DIOXIDE	10102-44-0
NITROGEN TETRAOXIDE	N2O4	NITROGEN TETRAOXIDE	10544-72-6
NITROGEN TRICHLORIDE	NCl3	NITROGEN TRICHLORIDE	10025-85-1
NITROGEN TRIFLUORIDE	NF3	NITROGEN TRIFLUORIDE	7783-54-2
NITROGEN TRIOXIDE	N2O3	NITROGEN TRIOXIDE	10544-73-7
NITRO-SIL	NH3	AMMONIA	7664-41-7
NITROSYL CHLORIDE	NOCl	NITROSYL CHLORIDE	2696-92-6
NITROSYL FLUORIDE	NOF	NITROSYL FLUORIDE	7789-25-5
NITROUS OXIDE	N2O	NITROUS OXIDE	10024-97-2
NORAL ALUMINUM	Al	ALUMINUM	7429-90-5
NORAL EXTRA FINE LINING GRADE	Al	ALUMINUM	7429-90-5
NORAL INK GRADE ALUMINUM	Al	ALUMINUM	7429-90-5
NORAL NON-LEAFING GRADE	Al	ALUMINUM	7429-90-5
NORDHAUSEN ACID	H2SO4	SULFURIC ACID	7664-93-9
NORIT	C	CARBON	7440-44-0
NP 2	Ni	NICKEL	7440-02-0
NUCHAR	C	CARBON	7440-44-0
OCTACHLOROTRISILANE	Si3Cl8	OCTACHLOROTRISILANE	----------
OIL OF VITRIOL	H2SO4	SULFURIC ACID	7664-93-9
OLOW	Pb	LEAD	7439-92-1
OMAHA	Pb	LEAD	7439-92-1
OMAHA & GRANT	Pb	LEAD	7439-92-1
ONYX	SiO2	SILICON DIOXIDE	14808-60-7
ORTHOBORIC ACID	BH3O3	BORIC ACID	10043-35-3
OSMIUM	Os	OSMIUM	7440-04-2
OSMIUM OXIDE PENTAFLUORIDE	OsOF5	OSMIUM OXIDE PENTAFLUORIDE	----------
OSMIUM TETROXIDE - WHITE	OsO4	OSMIUM TETROXIDE - WHITE	----------
OSMIUM TETROXIDE - YELLOW	OsO4	OSMIUM TETROXIDE - YELLOW	----------
OU-B	C	CARBON	7440-44-0
OXALIC ACID DINITRILE	C2N2	CYANOGEN	460-19-5
OXALONITRILE	C2N2	CYANOGEN	460-19-5
OXALYL CYANIDE	C2N2	CYANOGEN	460-19-5
OXAMMONIUM	NH3O	HYDROXYLAMINE	7803-49-8
OXYCARBON SULFIDE	COS	CARBONYL SULFIDE	463-58-1
OXYCHLORID FOSFORECNY	POCl3	PHOSPHORUS OXYCHLORIDE	10025-87-3
OXYDE de CARBONE	CO	CARBON MONOXIDE	630-08-0
OXYDE NITRIQUE	NO	NITRIC OXIDE	10102-43-9
OXYDOL	H2O2	HYDROGEN PEROXIDE	7722-84-1
OXYGEN	O2	OXYGEN	7782-44-7
OXYGEN FLUORIDE	F2O	FLUORINE OXIDE	7783-41-7
OXYMAG	MgO	MAGNESIUM OXIDE	1309-48-4
OZIDE	ZnO	ZINC OXIDE	1314-13-2
OZLO	ZnO	ZINC OXIDE	1314-13-2
OZON	O3	OZONE	10028-15-6
OZONE	O3	OZONE	10028-15-6
PALLADIUM	Pd	PALLADIUM	7440-05-3
PASCO	Zn	ZINC	7440-66-6
PEARSALL	AlCl3	ALUMINUM CHLORIDE	7446-70-0
PELIKAN C 11/1431a	C	CARBON	7440-44-0
PENTABORANE	B5H9	PENTABORANE	18433-84-6
PENTASULFURE de PHOSPHORE	P4S10	PHOSPHORUS PENTASULFIDE	1314-80-3
PERCHLORIC ACID	ClHO4	PERCHLORIC ACID	7601-90-3
PERCHLORIDE of MERCURY	HgCl2	MERCURIC CHLORIDE	7487-94-7
PERCHLORURE de FER	FeCl3	FERRIC CHLORIDE	7705-08-0
PERCHLORYL FLUORIDE	ClFO3	PERCHLORYL FLUORIDE	7616-94-6
PERFLUORO HYDRAZINE	N2F4	TETRAFLUOROHYDRAZINE	10036-47-2
PERHYDROL	H2O2	HYDROGEN PEROXIDE	7722-84-1

Name (Synonym)	Formula	Name in Tables	CAS Reg. No
PERICLASE	MgO	MAGNESIUM OXIDE	1309-48-4
PERONE	H2O2	HYDROGEN PEROXIDE	7722-84-1
PEROSSIDO di IDROGENO	H2O2	HYDROGEN PEROXIDE	7722-84-1
PEROXAN	H2O2	HYDROGEN PEROXIDE	7722-84-1
PEROXIDE	H2O2	HYDROGEN PEROXIDE	7722-84-1
PEROXYDE dHYDROGENE	H2O2	HYDROGEN PEROXIDE	7722-84-1
PFIKLOR	KCl	POTASSIUM CHLORIDE	7447-40-7
PHILOSOPHERS WOOL	ZnO	ZINC OXIDE	1314-13-2
PHOSGEN	CCl2O	PHOSGENE	75-44-5
PHOSGENE	CCl2O	PHOSGENE	75-44-5
PHOSPHINE	PH3	PHOSPHINE	7803-51-2
PHOSPHONIUM BROMIDE	PH4Br	PHOSPHONIUM BROMIDE	----------
PHOSPHONIUM CHLORIDE	PH4Cl	PHOSPHONIUM CHLORIDE	24567-53-1
PHOSPHONIUM IODIDE	PH4I	PHOSPHONIUM IODIDE	----------
PHOSPHORE(PENTACHLORURE de)	PCl5	PHOSPHORUS PENTACHLORIDE	10026-13-8
PHOSPHORE(TRICHLORURE de)	PCl3	PHOSPHORUS TRICHLORIDE	7719-12-2
PHOSPHORIC CHLORIDE	PCl5	PHOSPHORUS PENTACHLORIDE	10026-13-8
PHOSPHORIC SULFIDE	P4S10	PHOSPHORUS PENTASULFIDE	1314-80-3
PHOSPHOROTHIOIC TRICHLORIDE	PSCl3	PHOSPHORUS THIOCHLORIDE	3982-91-0
PHOSPHOROTHIONIC TRICHLORIDE	PSCl3	PHOSPHORUS THIOCHLORIDE	3982-91-0
PHOSPHOROUS SULFOCHLORIDE	PSCl3	PHOSPHORUS THIOCHLORIDE	3982-91-0
PHOSPHOROUS THIOCHLORIDE	PSCl3	PHOSPHORUS THIOCHLORIDE	3982-91-0
PHOSPHOROUS TRICHLORIDE SULFIDE	PSCl3	PHOSPHORUS THIOCHLORIDE	3982-91-0
PHOSPHORPENTACHLORID	PCl5	PHOSPHORUS PENTACHLORIDE	10026-13-8
PHOSPHORTRICHLORID	PCl3	PHOSPHORUS TRICHLORIDE	7719-12-2
PHOSPHORUS	P	PHOSPHORUS - WHITE	7723-14-0
PHOSPHORUS - WHITE	P	PHOSPHORUS - WHITE	7723-14-0
PHOSPHORUS BROMIDE	PBr3	PHOSPHORUS TRIBROMIDE	7789-60-8
PHOSPHORUS CHLORIDE	PCl3	PHOSPHORUS TRICHLORIDE	7719-12-2
PHOSPHORUS DICHLORIDE TRIFLUORIDE	PCl2F3	PHOSPHORUS DICHLORIDE TRIFLUORIDE	----------
PHOSPHORUS OXYCHLORIDE	POCl3	PHOSPHORUS OXYCHLORIDE	10025-87-3
PHOSPHORUS OXYTRICHLORIDE	POCl3	PHOSPHORUS OXYCHLORIDE	10025-87-3
PHOSPHORUS PENTACHLORIDE	PCl5	PHOSPHORUS PENTACHLORIDE	10026-13-8
PHOSPHORUS PENTASULFIDE	P4S10	PHOSPHORUS PENTASULFIDE	1314-80-3
PHOSPHORUS PENTOXIDE	P4O10	PHOSPHORUS PENTOXIDE	16752-60-6
PHOSPHORUS PERCHLORIDE	PCl5	PHOSPHORUS PENTACHLORIDE	10026-13-8
PHOSPHORUS PERSULFIDE	P4S10	PHOSPHORUS PENTASULFIDE	1314-80-3
PHOSPHORUS THIOBROMIDE	PSBr3	PHOSPHORUS THIOBROMIDE	----------
PHOSPHORUS THIOCHLORIDE	PSCl3	PHOSPHORUS THIOCHLORIDE	3982-91-0
PHOSPHORUS TRIBROMIDE	PBr3	PHOSPHORUS TRIBROMIDE	7789-60-8
PHOSPHORUS TRICHLORIDE	PCl3	PHOSPHORUS TRICHLORIDE	7719-12-2
PHOSPHORUS TRIHYDRIDE	PH3	PHOSPHINE	7803-51-2
PHOSPHORUS TRIOXIDE	P4O6	PHOSPHORUS TRIOXIDE	----------
PHOSPHORWASSERSTOFF	PH3	PHOSPHINE	7803-51-2
PHOSPHORYL CHLORIDE	POCl3	PHOSPHORUS OXYCHLORIDE	10025-87-3
PIECIOCHLOREK FOSFORU	PCl5	PHOSPHORUS PENTACHLORIDE	10026-13-8
PLATIN	Pt	PLATINUM	7440-06-4
PLATINUM	Pt	PLATINUM	7440-06-4
PLATINUM BLACK	Pt	PLATINUM	7440-06-4
PLATINUM SPONGE	Pt	PLATINUM	7440-06-4
PLOMB FLUORURE	PbF2	LEAD FLUORIDE	7783-46-2
PLUMBAGO	C	CARBON	7440-44-0
PLUMBOUS CHLORIDE	PbCl2	LEAD CHLORIDE	7758-95-4
PLUMBOUS FLUORIDE	PbF2	LEAD FLUORIDE	7783-46-2
PLUMBOUS OXIDE	PbO	LEAD OXIDE	1317-36-8
PLUMBOUS SULFIDE	PbS	LEAD SULFIDE	1314-87-0
POLONIUM	Po	POLONIUM	7440-08-6
POTASSA	KOH	POTASSIUM HYDROXIDE	1310-58-3
POTASSE CAUSTIQUE	KOH	POTASSIUM HYDROXIDE	1310-58-3
POTASSIO (IDROSSIDO di)	KOH	POTASSIUM HYDROXIDE	1310-58-3
POTASSIUM	K	POTASSIUM	7440-09-7
POTASSIUM (HYDROXYDE de)	KOH	POTASSIUM HYDROXIDE	1310-58-3
POTASSIUM BROMIDE	KBr	POTASSIUM BROMIDE	7758-02-3
POTASSIUM CHLORIDE	KCl	POTASSIUM CHLORIDE	7447-40-7
POTASSIUM FLUORIDE	KF	POTASSIUM FLUORIDE	7789-23-3
POTASSIUM FLUORURE	KF	POTASSIUM FLUORIDE	7789-23-3
POTASSIUM HYDRATE	KOH	POTASSIUM HYDROXIDE	1310-58-3
POTASSIUM HYDROXIDE	KOH	POTASSIUM HYDROXIDE	1310-58-3
POTASSIUM IODIDE	KI	POTASSIUM IODIDE	7681-11-0
POTASSIUM MONOCHLORIDE	KCl	POTASSIUM CHLORIDE	7447-40-7
POTAVESCENT	KCl	POTASSIUM CHLORIDE	7447-40-7
POTIDE	KI	POTASSIUM IODIDE	7681-11-0
PRESPERSION, 75 UREA	CH4N2O	UREA	57-13-6
PRUSSITE	C2N2	CYANOGEN	460-19-5
PS 1	Al2O3	ALUMINUM OXIDE	1344-28-1
PSEUDOTHIOUREA	CH4N2S	THIOUREA	62-56-6

Name (Synonym)	Formula	Name in Tables	CAS Reg. No
PSEUDOUREA	CH4N2O	UREA	57-13-6
PURE QUARTZ	SiO2	SILICON DIOXIDE	14808-60-7
QUARTZ	SiO2	SILICON DIOXIDE	14808-60-7
QUAZO PURO	SiO2	SILICON DIOXIDE	14808-60-7
QUECKSILBER	Hg	MERCURY	7439-97-6
QUECKSILBER CHLORID	HgCl2	MERCURIC CHLORIDE	7487-94-7
QUICK SILVER	Hg	MERCURY	7439-97-6
R 717	NH3	AMMONIA	7664-41-7
R 744	CO2	CARBON DIOXIDE	124-38-9
RADIUM	Ra	RADIUM	7440-14-4
RADON	Rn	RADON	10043-92-2
RAMOR	Tl	THALLIUM	7440-28-0
RANEY ALLOY	Ni	NICKEL	7440-02-0
RANEY NICKEL	Ni	NICKEL	7440-02-0
RC 172DBM	Al2O3	ALUMINUM OXIDE	1344-28-1
RCRA WASTE NUMBER P015	Be	BERYLLIUM	7440-41-7
RCRA WASTE NUMBER P022	CS2	CARBON DISULFIDE	75-15-0
RCRA WASTE NUMBER P031	C2N2	CYANOGEN	460-19-5
RCRA WASTE NUMBER P033	CNCl	CYANOGEN CHLORIDE	506-77-4
RCRA WASTE NUMBER P056	F2	FLUORINE	7782-41-4
RCRA WASTE NUMBER P076	NO	NITRIC OXIDE	10102-43-9
RCRA WASTE NUMBER P078	NO2	NITROGEN DIOXIDE	10102-44-0
RCRA WASTE NUMBER P095	CCl2O	PHOSGENE	75-44-5
RCRA WASTE NUMBER P096	PH3	PHOSPHINE	7803-51-2
RCRA WASTE NUMBER P106	NaCN	SODIUM CYANIDE	143-33-9
RCRA WASTE NUMBER U033	CF2O	CARBONYL FLUORIDE	353-50-4
RCRA WASTE NUMBER U133	N2H4	HYDRAZINE	302-01-2
RCRA WASTE NUMBER U134	HF	HYDROGEN FLUORIDE	7664-39-3
RCRA WASTE NUMBER U135	H2S	HYDROGEN SULFIDE	7783-06-4
RCRA WASTE NUMBER U151	Hg	MERCURY	7439-97-6
RCRA WASTE NUMBER U189	P4S10	PHOSPHORUS PENTASULFIDE	1314-80-3
RCRA WASTE NUMBER U204	SeO2	SELENIUM DIOXIDE	7446-08-4
RCRA WASTE NUMBER U219	CH4N2S	THIOUREA	62-56-6
RCRA WASTE NUMBER U246	CNBr	CYANOGEN BROMIDE	506-68-3
RED MERCURIC IODIDE	HgI2	MERCURIC IODIDE	7774-29-0
REKAWAN	KCl	POTASSIUM CHLORIDE	7447-40-7
RHENIUM	Re	RHENIUM	7440-15-5
RHENIUM HEPTOXIDE	Re2O7	RHENIUM HEPTOXIDE	----------
RHODIUM	Rh	RHODIUM	7440-16-6
RMC	Mg	MAGNESIUM	7439-95-4
ROSE QUARTZ	SiO2	SILICON DIOXIDE	14808-60-7
RTEC	Hg	MERCURY	7439-97-6
RUBIDIUM	Rb	RUBIDIUM	7440-17-7
RUBIDIUM BROMIDE	RbBr	RUBIDIUM BROMIDE	7789-39-1
RUBIDIUM CHLORIDE	RbCl	RUBIDIUM CHLORIDE	7791-11-9
RUBIDIUM FLUORIDE	RbF	RUBIDIUM FLUORIDE	13446-74-7
RUBIDIUM IODIDE	RbI	RUBIDIUM IODIDE	7790-29-6
RUBIDIUM MONOBROMIDE	RbBr	RUBIDIUM BROMIDE	7789-39-1
RUBIDIUM MONOCHLORIDE	RbCl	RUBIDIUM CHLORIDE	7791-11-9
RUBIDIUM MONOFLUORIDE	RbF	RUBIDIUM FLUORIDE	13446-74-7
RUBIDIUM MONOIODIDE	RbI	RUBIDIUM IODIDE	7790-29-6
RUBIGINE	HF	HYDROGEN FLUORIDE	7664-39-3
RUTHENIUM	Ru	RUTHENIUM	7440-18-8
RUTHENIUM PENTAFLUORIDE	RuF5	RUTHENIUM PENTAFLUORIDE	----------
SAEURE FLUORIDE	F2	FLUORINE	7782-41-4
SAL AMMONIA	NH4Cl	AMMONIUM CHLORIDE	12125-02-9
SAL AMMONIAC	NH4Cl	AMMONIUM CHLORIDE	12125-02-9
SALAMMONITE	NH4Cl	AMMONIUM CHLORIDE	12125-02-9
SALINE	NaCl	SODIUM CHLORIDE	7647-14-5
SALMIAC	NH4Cl	AMMONIUM CHLORIDE	12125-02-9
SALPETERSAEURE	HNO3	NITRIC ACID	7697-37-2
SALPETERZUUROPLOSSINGEN	HNO3	NITRIC ACID	7697-37-2
SALT	NaCl	SODIUM CHLORIDE	7647-14-5
SALT CAKE	Na2SO4	SODIUM SULFATE	7757-82-6
SAMARIUM	Sm	SAMARIUM	7440-19-9
SAND	SiO2	SILICON DIOXIDE	14808-60-7
SCANDIUM	Sc	SCANDIUM	7440-20-2
SCHWEFELDIOXYD	SO2	SULFUR DIOXIDE	7446-09-5
SCHWEFELKOHLENSTOFF	CS2	CARBON DISULFIDE	75-15-0
SCHWEFELSAEURELOESUNGEN	H2SO4	SULFURIC ACID	7664-93-9
SCHWEFELWASSERSTOFF	H2S	HYDROGEN SULFIDE	7783-06-4
SEA SALT	NaCl	SODIUM CHLORIDE	7647-14-5
SEDONEURAL	NaBr	SODIUM BROMIDE	7647-15-6
SELEN	Se	SELENIUM	7782-49-2
SELENINYL CHLORIDE	SeOCl2	SELENIUM OXYCHLORIDE	7791-23-3
SELENIOUS ANHYDRIDE	SeO2	SELENIUM DIOXIDE	7446-08-4

Name (Synonym)	Formula	Name in Tables	CAS Reg. No
SELENIUM	Se	SELENIUM	7782-49-2
SELENIUM CHLORIDE	SeCl4	SELENIUM TETRACHLORIDE	----------
SELENIUM CHLORIDE OXIDE	SeOCl2	SELENIUM OXYCHLORIDE	7791-23-3
SELENIUM DIOXIDE	SeO2	SELENIUM DIOXIDE	7446-08-4
SELENIUM FLUORIDE	SeF6	SELENIUM HEXAFLUORIDE	7783-79-1
SELENIUM HEXAFLUORIDE	SeF6	SELENIUM HEXAFLUORIDE	7783-79-1
SELENIUM HOMOPOLYMER	Se	SELENIUM	7782-49-2
SELENIUM HYDRIDE	H2Se	HYDROGEN SELENIDE	7783-07-5
SELENIUM OXIDE	SeO2	SELENIUM DIOXIDE	7446-08-4
SELENIUM OXYCHLORIDE	SeOCl2	SELENIUM OXYCHLORIDE	7791-23-3
SELENIUM TETRACHLORIDE	SeCl4	SELENIUM TETRACHLORIDE	----------
SHELL SILVER	Ag	SILVER	7440-22-4
SIARKI DWUTLENEK	SO2	SULFUR DIOXIDE	7446-09-5
SIARKOWODOR	H2S	HYDROGEN SULFIDE	7783-06-4
SILANE	SiH4	SILANE	7803-62-5
SILBER	Ag	SILVER	7440-22-4
SILICA FLOUR	SiO2	SILICON DIOXIDE	14808-60-7
SILICANE	SiH4	SILANE	7803-62-5
SILICIC ANHYDRIDE	SiO2	SILICON DIOXIDE	14808-60-7
SILICI-CHLOROFORME	SiHCl3	TRICHLOROSILANE	10025-78-2
SILICIO	SiCl4	SILICON TETRACHLORIDE	10026-04-7
SILICIUM	SiCl4	SILICON TETRACHLORIDE	10026-04-7
SILICIUMCHLOROFORM	SiHCl3	TRICHLOROSILANE	10025-78-2
SILICIUMTETRACHLORID	SiCl4	SILICON TETRACHLORIDE	10026-04-7
SILICIUMTETRACHLORIDE	SiCl4	SILICON TETRACHLORIDE	10026-04-7
SILICOBROMOFORM	SiHBr3	TRIBROMOSILANE	7789-57-3
SILICOCHLOROFORM	SiHCl3	TRICHLOROSILANE	10025-78-2
SILICON	Si	SILICON	7440-21-3
SILICON DIOXIDE	SiO2	SILICON DIOXIDE	14808-60-7
SILICON TETRACHLORIDE	SiCl4	SILICON TETRACHLORIDE	10026-04-7
SILICON TETRAFLUORIDE	SiF4	SILICON TETRAFLUORIDE	7783-61-1
SILICON TETRAHYDRIDE	SiH4	SILANE	7803-62-5
SILVER	Ag	SILVER	7440-22-4
SILVER CHLORIDE	AgCl	SILVER CHLORIDE	7783-90-6
SILVER IODIDE	AgI	SILVER IODIDE	7783-96-2
SILVER MONOCHLORIDE	AgCl	SILVER CHLORIDE	7783-90-6
SILVER MONOIODIDE	AgI	SILVER IODIDE	7783-96-2
SILYL BROMIDE	SiH3Br	MONOBROMOSILANE	13465-73-1
SIRNIK FOSFORECNY	P4S10	PHOSPHORUS PENTASULFIDE	1314-80-3
SLOW-K	KCl	POTASSIUM CHLORIDE	7447-40-7
SODA LYE	NaOH	SODIUM HYDROXIDE	1310-73-2
SODIO(IDROSSIDO di)	NaOH	SODIUM HYDROXIDE	1310-73-2
SODIUM	Na	SODIUM	7440-23-5
SODIUM BROMIDE	NaBr	SODIUM BROMIDE	7647-15-6
SODIUM CHLORIDE	NaCl	SODIUM CHLORIDE	7647-14-5
SODIUM CYANIDE	NaCN	SODIUM CYANIDE	143-33-9
SODIUM FLUORIDE	NaF	SODIUM FLUORIDE	7681-49-4
SODIUM HYDRATE	NaOH	SODIUM HYDROXIDE	1310-73-2
SODIUM HYDROXIDE	NaOH	SODIUM HYDROXIDE	1310-73-2
SODIUM IODIDE	NaI	SODIUM IODIDE	7681-82-5
SODIUM IODINE	NaI	SODIUM IODIDE	7681-82-5
SODIUM MONOBROMIDE	NaBr	SODIUM BROMIDE	7647-15-6
SODIUM MONOCHLORIDE	NaCl	SODIUM CHLORIDE	7647-14-5
SODIUM MONOIODIDE	NaI	SODIUM IODIDE	7681-82-5
SODIUM SULFATE	Na2SO4	SODIUM SULFATE	7757-82-6
SODIUM SULPHATE	Na2SO4	SODIUM SULFATE	7757-82-6
SOLFURO di CARBONIO	CS2	CARBON DISULFIDE	75-15-0
SPENT SULFURIC ACID	H2SO4	SULFURIC ACID	7664-93-9
SPIRIT of HARTSHORN	NH3	AMMONIA	7664-41-7
STANNIC BROMIDE	SnBr4	STANNIC BROMIDE	7789-67-5
STANNIC CHLORIDE	SnCl4	STANNIC CHLORIDE	7646-78-8
STANNIC HYDRIDE	SnH4	STANNIC HYDRIDE	----------
STANNIC IODIDE	SnI4	STANNIC IODIDE	7790-47-8
STANNOUS CHLORIDE	SnCl2	STANNOUS CHLORIDE	7772-99-8
STERLING	NaCl	SODIUM CHLORIDE	7647-14-5
STIBINE	SbH3	STIBINE	7803-52-3
STIBINE, TRICHLORO-	SbCl3	ANTIMONY TRICHLORIDE	10025-91-9
STIBIUM	Sb	ANTIMONY	7440-36-0
STICKMONOXYD	NO	NITRIC OXIDE	10102-43-9
STICKSTOFFDIOXID	NO2	NITROGEN DIOXIDE	10102-44-0
STIKSTOFDIOXYDE	NO2	NITROGEN DIOXIDE	10102-44-0
STINK DAMP	H2S	HYDROGEN SULFIDE	7783-06-4
STRONTIUM	Sr	STRONTIUM	7440-24-6
STRONTIUM OXIDE	SrO	STRONTIUM OXIDE	1314-11-0
SUBLIMAT	HgCl2	MERCURIC CHLORIDE	7487-94-7
SULEMA	HgCl2	MERCURIC CHLORIDE	7487-94-7

Name (Synonym)	Formula	Name in Tables	CAS Reg. No
SULFAMIC ACID	H3NO3S	SULFAMIC ACID	5329-14-6
SULFAMIDIC ACID	H3NO3S	SULFAMIC ACID	5329-14-6
SULFAN	SO3	SULFUR TRIOXIDE	7446-11-9
SULFATE de ZINC	ZnSO4	ZINC SULFATE	7733-02-0
SULFIDOTRICHLORID FOSFORECNY	PSCl3	PHOSPHORUS THIOCHLORIDE	3982-91-0
SULFINYL BROMIDE	SOBr2	THIONYL BROMIDE	----------
SULFINYL CHLORIDE	SOCl2	THIONYL CHLORIDE	7791-25-5
SULFINYL FLUORIDE	SOF2	SULFUROUS OXYFLUORIDE	----------
SULFONIC ACID, MONOCHLORIDE	ClHO3S	CHLOROSULFONIC ACID	7790-94-5
SULFONYL CHLORIDE	SO2Cl2	SULFURYL CHLORIDE	7791-25-5
SULFUR	S	SULFUR	7704-34-9
SULFUR BROMIDE OXIDE	SOBr2	THIONYL BROMIDE	----------
SULFUR CHLORIDE OXIDE	SOCl2	THIONYL CHLORIDE	7791-25-5
SULFUR DIOXIDE	SO2	SULFUR DIOXIDE	7446-09-5
SULFUR FLUORIDE	SF6	SULFUR HEXAFLUORIDE	2551-62-4
SULFUR FLUORIDE OXIDE	SOF2	SULFUROUS OXYFLUORIDE	----------
SULFUR HEXAFLUORIDE	SF6	SULFUR HEXAFLUORIDE	2551-62-4
SULFUR HYDRIDE	H2S	HYDROGEN SULFIDE	7783-06-4
SULFUR MONOCHLORIDE	S2Cl2	SULFUR MONOCHLORIDE	----------
SULFUR OXIDE	SO2	SULFUR DIOXIDE	7446-09-5
SULFUR PHOSPHIDE	P4S10	PHOSPHORUS PENTASULFIDE	1314-80-3
SULFUR TETRAFLUORIDE	SF4	SULFUR TETRAFLUORIDE	7783-60-0
SULFUR TRIOXIDE	SO3	SULFUR TRIOXIDE	7446-11-9
SULFURETED HYDROGEN	H2S	HYDROGEN SULFIDE	7783-06-4
SULFURIC ACID	H2SO4	SULFURIC ACID	7664-93-9
SULFURIC ACID, DISODIUM SALT	Na2SO4	SODIUM SULFATE	7757-82-6
SULFURIC ANHYDRIDE	SO3	SULFUR TRIOXIDE	7446-11-9
SULFURIC CHLOROHYDRIN	ClHO3S	CHLOROSULFONIC ACID	7790-94-5
SULFURIC OXIDE	SO3	SULFUR TRIOXIDE	7446-11-9
SULFURIC OXYCHLORIDE	SO2Cl2	SULFURYL CHLORIDE	7791-25-5
SULFUROUS ACID ANHYDRIDE	SO2	SULFUR DIOXIDE	7446-09-5
SULFUROUS ANHYDRIDE	SO2	SULFUR DIOXIDE	7446-09-5
SULFUROUS DIBROMIDE	SOBr2	THIONYL BROMIDE	----------
SULFUROUS DICHLORIDE	SOCl2	THIONYL CHLORIDE	7791-25-5
SULFUROUS DIFLUORIDE	SOF2	SULFUROUS OXYFLUORIDE	----------
SULFUROUS OXIDE	SO2	SULFUR DIOXIDE	7446-09-5
SULFUROUS OXYBROMIDE	SOBr2	THIONYL BROMIDE	----------
SULFUROUS OXYCHLORIDE	SOCl2	THIONYL CHLORIDE	7791-25-5
SULFUROUS OXYFLUORIDE	SOF2	SULFUROUS OXYFLUORIDE	----------
SULFURYL CHLORIDE	SO2Cl2	SULFURYL CHLORIDE	7791-25-5
SULOUREA	CH4N2S	THIOUREA	62-56-6
SULPHAMIC ACID	H3NO3S	SULFAMIC ACID	5329-14-6
SULPHOCARBONIC ANHYDRIDE	CS2	CARBON DISULFIDE	75-15-0
SULPHUR	S	SULFUR	7704-34-9
SULPHUR DIOXIDE, LIQUEFIED	SO2	SULFUR DIOXIDE	7446-09-5
SULPHURIC ACID	H2SO4	SULFURIC ACID	7664-93-9
SULSOL	S	SULFUR	7704-34-9
SUPER COBALT	Co	COBALT	7440-48-4
SUPER COSAN	S	SULFUR	7704-34-9
SUPERCEL 3000	CH4N2O	UREA	57-13-6
SUPEROXOL	H2O2	HYDROGEN PEROXIDE	7722-84-1
TABLE SALT	NaCl	SODIUM CHLORIDE	7647-14-5
TANTALUM	Ta	TANTALUM	7440-25-7
TECHNETIUM TC 99M SULFUR COLLOID	S	SULFUR	7704-34-9
TECNNETIUM	Tc	TECNNETIUM	7440-26-8
TELLOY	Te	TELLURIUM	13494-80-9
TELLUR	Te	TELLURIUM	13494-80-9
TELLURIC CHLORIDE	TeCl4	TELLURIUM TETRACHLORIDE	10026-07-0
TELLURIUM	Te	TELLURIUM	13494-80-9
TELLURIUM CHLORIDE	TeCl4	TELLURIUM TETRACHLORIDE	10026-07-0
TELLURIUM HEXAFLUORIDE	TeF6	TELLURIUM HEXAFLUORIDE	7783-80-4
TELLURIUM HYDRIDE	H2Te	HYDROGEN TELLURIDE	7783-09-7
TELLURIUM TETRACHLORIDE	TeCl4	TELLURIUM TETRACHLORIDE	10026-07-0
TESULOID	S	SULFUR	7704-34-9
TETRABORANE	B4H10	TETRABORANE	18283-93-7
TETRACHLOROSILANE	SiCl4	SILICON TETRACHLORIDE	10026-04-7
TETRACHLOROTELLURIUM	TeCl4	TELLURIUM TETRACHLORIDE	10026-07-0
TETRACHLORURE de SILICIUM	SiCl4	SILICON TETRACHLORIDE	10026-04-7
TETRACHLORURE de TITANE	TiCl4	TITANIUM TETRACHLORIDE	7550-45-0
TETRAFLUOROHYDRAZINE	N2F4	TETRAFLUOROHYDRAZINE	10036-47-2
TETRAFLUOROSILANE	SiF4	SILICON TETRAFLUORIDE	7783-61-1
TETRAFLUOROSULFURANE	SF4	SULFUR TETRAFLUORIDE	7783-60-0
TETRAHYDROPENTABORANE	B5H11	TETRAHYDROPENTABORANE	----------
TETRASILANE	Si4H10	TETRASILANE	----------
THALLIUM	Tl	THALLIUM	7440-28-0
THALLIUM MONOBROMIDE	TlBr	THALLOUS BROMIDE	----------

Name (Synonym)	Formula	Name in Tables	CAS Reg. No
THALLIUM MONOIODIDE	TlI	THALLOUS IODIDE	7790-30-9
THALLOUS BROMIDE	TlBr	THALLOUS BROMIDE	----------
THALLOUS IODIDE	TlI	THALLOUS IODIDE	7790-30-9
THENARDITE	Na2SO4	SODIUM SULFATE	7757-82-6
THIOCARBAMATE	CH4N2S	THIOUREA	62-56-6
THIOCARBAMIDE	CH4N2S	THIOUREA	62-56-6
THIOCHLORID FOSFORECNY	PSCl3	PHOSPHORUS THIOCHLORIDE	3982-91-0
THIOLUX	S	SULFUR	7704-34-9
THIONYL BROMIDE	SOBr2	THIONYL BROMIDE	----------
THIONYL CHLORIDE	SOÇl2	THIONYL CHLORIDE	7791-25-5
THIONYL DIBROMIDE	SOBr2	THIONYL BROMIDE	----------
THIONYL DICHLORIDE	SOCl2	THIONYL CHLORIDE	7791-25-5
THIONYL DIFLUORIDE	SOF2	SULFUROUS OXYFLUORIDE	----------
THIOPHOSPHORIC ANHYDRIDE	P4S10	PHOSPHORUS PENTASULFIDE	1314-80-3
THIOPHOSPHORIC TRICHLORIDE	PSCl3	PHOSPHORUS THIOCHLORIDE	3982-91-0
THIOPHOSPHORYL TRICHLORIDE	PSCl3	PHOSPHORUS THIOCHLORIDE	3982-91-0
b-THIOPSEUDOUREA	CH4N2S	THIOUREA	62-56-6
THIOUREA	CH4N2S	THIOUREA	62-56-6
2-THIOUREA	CH4N2S	THIOUREA	62-56-6
THIOVIT	S	SULFUR	7704-34-9
THREE ELEPHANT	BH3O3	BORIC ACID	10043-35-3
THU	CH4N2S	THIOUREA	62-56-6
THULIUM	Tm	THULIUM	7440-30-4
TIN	Sn	TIN	7440-31-5
TIN CHLORIDE	SnCl4	STANNIC CHLORIDE	7646-78-8
TIN DICHLORIDE	SnCl2	STANNOUS CHLORIDE	7772-99-8
TIN PERBROMIDE	SnBr4	STANNIC BROMIDE	7789-67-5
TIN PERCHLORIDE	SnCl4	STANNIC CHLORIDE	7646-78-8
TIN PROTOCHLORIDE	SnCl2	STANNOUS CHLORIDE	7772-99-8
TIN TETRABROMIDE	SnBr4	STANNIC BROMIDE	7789-67-5
TIN TETRACHLORIDE	SnCl4	STANNIC CHLORIDE	7646-78-8
TIN TETRAHYDRIDE	SnH4	STANNIC HYDRIDE	----------
TIN TETRAIODIDE	SnI4	STANNIC IODIDE	7790-47-8
TINTETRACHLORIDE	SnCl4	STANNIC CHLORIDE	7646-78-8
TITAANTETRACHLORID	TiCl4	TITANIUM TETRACHLORIDE	7550-45-0
TITANE (TETRACHLORURE de)	TiCl4	TITANIUM TETRACHLORIDE	7550-45-0
TITANIO TETRACHLORURO di	TiCl4	TITANIUM TETRACHLORIDE	7550-45-0
TITANIUM	Ti	TITANIUM	7440-32-6
TITANIUM CHLORIDE	TiCl4	TITANIUM TETRACHLORIDE	7550-45-0
TITANIUM TETRACHLORIDE	TiCl4	TITANIUM TETRACHLORIDE	7550-45-0
TITANTETRACHLORID	TiCl4	TITANIUM TETRACHLORIDE	7550-45-0
TL 262	PSCl3	PHOSPHORUS THIOCHLORIDE	3982-91-0
TL 822	CNBr	CYANOGEN BROMIDE	506-68-3
TL 898	HgCl2	MERCURIC CHLORIDE	7487-94-7
TLD 100	LiF	LITHIUM FLUORIDE	7789-24-4
TOP FLAKE	NaCl	SODIUM CHLORIDE	7647-14-5
TRIATOMIC OXYGEN	O3	OZONE	10028-15-6
TRIBROMOALUMINUM	AlBr3	ALUMINUN BROMIDE	7727-15-3
TRIBROMOARSINE	AsBr3	ARSENIC TRIBROMIDE	7784-33-0
TRIBROMOPHOSPHINE	PBr3	PHOSPHORUS TRIBROMIDE	7789-60-8
TRIBROMOSILANE	SiHBr3	TRIBROMOSILANE	7789-57-3
TRIBROMOSTIBINE	SbBr3	ANTIMONY TRIBROMIDE	7789-61-9
TRICESIUM	CsF	CESIUM FLUORIDE	13400-13-0
TRICESIUM TRICHLORIDE	CsCl	CESIUM CHLORIDE	7647-17-8
TRICESIUM TRIIODIDE	CsI	CESIUM IODIDE	7789-17-5
TRICHLOORSILAAN	SiHCl3	TRICHLOROSILANE	10025-78-2
TRICHLORAMINE	NCl3	NITROGEN TRICHLORIDE	10025-85-1
TRICHLORINE NITRIDE	NCl3	NITROGEN TRICHLORIDE	10025-85-1
TRICHLORO GERMANE	GeHCl3	TRICHLORO GERMANE	----------
TRICHLOROALUMINUM	AlCl3	ALUMINUM CHLORIDE	7446-70-0
TRICHLOROARSINE	AsCl3	ARSENIC TRICHLORIDE	7784-34-1
TRICHLOROFLUOROSILANE	SiCl3F	TRICHLOROFLUOROSILANE	14965-52-7
TRICHLOROMONOSILANE	SiHCl3	TRICHLOROSILANE	10025-78-2
TRICHLOROOXOVANADIUM	VOCl3	VANADIUM OXYTRICHLORIDE	7727-18-6
TRICHLOROPHOSPHINE SULFIDE	PSCl3	PHOSPHORUS THIOCHLORIDE	3982-91-0
TRICHLOROSILANE	SiHCl3	TRICHLOROSILANE	10025-78-2
TRICHLOROSTIBINE	SbCl3	ANTIMONY TRICHLORIDE	10025-91-9
TRICHLOROTRIFLUORODISILOXANE	Si2OCl3F3	TRICHLOROTRIFLUORODISILOXANE	----------
TRICHLORSILAN	SiHCl3	TRICHLOROSILANE	10025-78-2
TRICHLORURE dARSENIC	AsCl3	ARSENIC TRICHLORIDE	7784-34-1
TRICLOROSILANO	SiHCl3	TRICHLOROSILANE	10025-78-2
TRIFLUOROARSINE	AsF3	ARSENIC TRIFLUORIDE	7784-35-2
TRIFLUOROBROMOSILANE	SiBrF3	TRIFLUOROBROMOSILANE	----------
TRIFLUOROCHLOROSILANE	SiClF3	TRIFLUOROCHLOROSILANE	----------
TRIFLUOROSILANE	SiHF3	TRIFLUOROSILANE	13465-71-9
TRIFLUOROTRICHLORODISILOXANE	Si2OCl3F3	TRICHLOROTRIFLUORODISILOXANE	----------

Name (Synonym)	Formula	Name in Tables	CAS Reg. No
TRIFLUORURE de CHLORE	ClF3	CHLORINE TRIFLUORIDE	7790-91-2
TRIGERMANE	Ge3H8	TRIGERMANE	14691-44-2
TRIIODOARSINE	AsI3	ARSENIC TRIIODIDE	7784-45-4
TRIIODOSTIBINE	SbI3	ANTIMONY TRIIODIDE	7790-44-5
TRIPOTASSIUM TRICHLORIDE	KCl	POTASSIUM CHLORIDE	7447-40-7
TRISILANE	Si3H8	TRISILANE	7783-26-8
TRISILAZANE	Si3H9N	TRISILAZANE	----------
TROJCHLOREK FOSFORU	PCl3	PHOSPHORUS TRICHLORIDE	7719-12-2
TRONA	BBr3	BORON TRIBROMIDE	10294-33-4
TRONAMANG	Mn	MANGANESE	7439-96-5
TSIZP 34	CH4N2S	THIOUREA	62-56-6
T-STUFF	H2O2	HYDROGEN PEROXIDE	7722-84-1
TUNGSTEN	W	TUNGSTEN	7440-33-7
TUNGSTEN FLUORIDE	WF6	TUNGSTEN FLUORIDE	7783-82-6
TUNGSTEN HEXAFLUORIDE	WF6	TUNGSTEN FLUORIDE	7783-82-6
URANIUM	U	URANIUM	7440-61-1
URANIUM FLUORIDE	UF6	URANIUM FLUORIDE	7783-81-5
URANIUM HEXAFLUORIDE	UF6	URANIUM FLUORIDE	7783-81-5
UREA	CH4N2O	UREA	57-13-6
UREAPHIL	CH4N2O	UREA	57-13-6
UREOPHIL	CH4N2O	UREA	57-13-6
UREVERT	CH4N2O	UREA	57-13-6
USAF EK-497	CH4N2S	THIOUREA	62-56-6
USP SODIUM CHLORIDE	NaCl	SODIUM CHLORIDE	7647-14-5
VANADIUM	V	VANADIUM	7440-62-2
VANADIUM CHLORIDE	VCl4	VANADIUM TETRACHLORIDE	7632-51-1
VANADIUM OXYTRICHLORIDE	VOCl3	VANADIUM OXYTRICHLORIDE	7727-18-6
VANADIUM TETRACHLORIDE	VCl4	VANADIUM TETRACHLORIDE	7632-51-1
VANADIUM TRICHLORIDE OXIDE	VOCl3	VANADIUM OXYTRICHLORIDE	7727-18-6
VANADYL TRICHLORIDE	VOCl3	VANADIUM OXYTRICHLORIDE	7727-18-6
VANDEX	Se	SELENIUM	7782-49-2
VARIOFORM II	CH4N2O	UREA	57-13-6
VERAZINC	ZnSO4	ZINC SULFATE	7733-02-0
VI-CAD	CdCl2	CADMIUM CHLORIDE	10108-64-2
VITRIOL BROWN OIL	H2SO4	SULFURIC ACID	7664-93-9
VITRIOL, OIL OF	H2SO4	SULFURIC ACID	7664-93-9
VN 1	Nb	NIOBIUM	7440-03-1
WASSERSTOFFPEROXID	H2O2	HYDROGEN PEROXIDE	7722-84-1
WATER	H2O	WATER	7732-18-5
WATER2-H2	D2O	DEUTERIUM OXIDE	7789-20-0
WATERCARB	C	CARBON	7440-44-0
WATER-D2 (9CI)	D2O	DEUTERIUM OXIDE	7789-20-0
WATERSTOFPEROXYDE	H2O2	HYDROGEN PEROXIDE	7722-84-1
WEEVILTOX	CS2	CARBON DISULFIDE	75-15-0
WEGLA DWUSIARCZEK	CS2	CARBON DISULFIDE	75-15-0
WEGLA TLENEK	CO	CARBON MONOXIDE	630-08-0
WHITE COPPERAS	ZnSO4	ZINC SULFATE	7733-02-0
WHITE CRYSTAL	NaCl	SODIUM CHLORIDE	7647-14-5
WHITE VITRIOL	ZnSO4	ZINC SULFATE	7733-02-0
WITCARB 940	C	CARBON	7440-44-0
WOLFRAM	W	TUNGSTEN	7440-33-7
XE 340	C	CARBON	7440-44-0
XENON	Xe	XENON	7440-63-3
YELLOW LEAD OCHER	PbO	LEAD OXIDE	1317-36-8
YTTERBIUM	Yb	YTTERBIUM	7440-64-4
YTTRIUM	Yt	YTTRIUM	7440-65-5
ZINC	Zn	ZINC	7440-66-6
ZINC (CHLORURE de)	ZnCl2	ZINC CHLORIDE	7646-85-7
ZINC BUTTER	ZnCl2	ZINC CHLORIDE	7646-85-7
ZINC CHLORIDE	ZnCl2	ZINC CHLORIDE	7646-85-7
ZINC DICHLORIDE	ZnCl2	ZINC CHLORIDE	7646-85-7
ZINC DIFLUORURE	ZnF2	ZINC FLUORIDE	7783-49-5
ZINC FLUORIDE	ZnF2	ZINC FLUORIDE	7783-49-5
ZINC OXIDE	ZnO	ZINC OXIDE	1314-13-2
ZINC SULFATE	ZnSO4	ZINC SULFATE	7733-02-0
ZINC VITRIOL	ZnSO4	ZINC SULFATE	7733-02-0
ZINC WHITE	ZnO	ZINC OXIDE	1314-13-2
ZINCITE	ZnO	ZINC OXIDE	1314-13-2
ZINCO (CLORURO di)	ZnCl2	ZINC CHLORIDE	7646-85-7
ZINCOID	ZnO	ZINC OXIDE	1314-13-2
ZINKCHLORID	ZnCl2	ZINC CHLORIDE	7646-85-7
ZINKCHLORIDE	ZnCl2	ZINC CHLORIDE	7646-85-7
ZINKOSITE	ZnSO4	ZINC SULFATE	7733-02-0
ZINN	Sn	TIN	7440-31-5
ZINNTETRACHLORID	SnCl4	STANNIC CHLORIDE	7646-78-8
ZIRCAT	Zr	ZIRCONIUM	7440-67-7

Name (Synonym)	Formula	Name in Tables	CAS Reg. No
ZIRCONIUM	Zr	ZIRCONIUM	7440-67-7
ZIRCONIUM BROMIDE	ZrBr4	ZIRCONIUM BROMIDE	13777-25-8
ZIRCONIUM CHLORIDE	ZrCl4	ZIRCONIUM CHLORIDE	10026-11-6
ZIRCONIUM IODIDE	ZrI4	ZIRCONIUM IODIDE	13986-26-0
ZIRCONIUM TETRABROMIDE	ZrBr4	ZIRCONIUM BROMIDE	13777-25-8
ZIRCONIUM TETRACHLORIDE	ZrCl4	ZIRCONIUM CHLORIDE	10026-11-6
ZIRCONIUM TETRAIODIDE	ZrI4	ZIRCONIUM IODIDE	13986-26-0
ZWAVELWATERSTOF	H2S	HYDROGEN SULFIDE	7783-06-4
ZWAVELZUUROPLOSSINGEN	H2SO4	SULFURIC ACID	7664-93-9

Index

ABOUT THE AUTHOR

Carl Yaws, Ph.D, is a professor of chemical engineering at Lamar University. He has extensive industrial experience in process evaluation research, development, and design at laboratory, pilot plant, and production levels. He has worked for Exxon, Ethyl, and Texas Instruments. The author of 24 books, he holds several patents and has published more than 320 technical articles in process engineering, proper data and pollution prevention.